REFERENCE ☑ P9-BTO-538

**BFO750**  CAS:7440-41-7  **HR: 3**
**BERYLLIUM**
**DOT:** UN 1966/UN 1567
mf: Be  mw: 9.01

**PROP:** A silvery-white, relatively soft, lustrous metal, ductile at red heat. Unreactive to $H_2O$ and air; dissolves vigorously in dil acids. Be Reacts with aq alkalis or $H_2$. Mp: 1287–1292°, bp: 2970°, d: 1.85.

**SYNS:** BERYLLIUM-9 □ BERYLLIUM COMPOUNDS, n.o.s. (UN 1566) (DOT) □ BERYLLIUM, powder (UN 1567) (DOT) □ GLUCINIUM □ GLUCINUM □ RCRA WASTE NUMBER P015

**TOXICITY DATA** WITH **REFERENCE**
dnd-esc 30 μmol/L  MUREAV 89,95,81
dni-nml-ivn 30 μmol/kg  PHMCAA 12,298,70
dnd-hmn:hla 30 μmol/L  MUREAV 89,95,81
dnd-mus:ast 30 μmol/L  MUREAV 89,95,81
itr-rat TDLo:13 mg/kg:NEO  ENVRAL 21,63,80
ivn-rbt TDLo:20 mg/kg:ETA  LANCAO 1,463,50
ihl-hmn TCLo:300 mg/m³:PUL  AEHLAU 9,473,64
ivn-rat LD50:496 μg/kg  LAINAW 15,176,66

**CONSENSUS REPORTS:** NTP 7th Annual Report On Carcinogens. IARC Cancer Review: Group 1 IMEMDT 58,41,93; Human Sufficient Evidence IMEMDT 58,41,93; Animal Sufficient Evidence IMEMDT 1,17,72; Animal Sufficient Evidence IMEMDT 23,143,80; Animal Sufficient Evidence IMEMDT 58,41,93. Beryllium and its compounds are on the Community Right-To-Know List. Reported in EPA TSCA Inventory.

**OSHA PEL:** TWA 0.002 mg(Be)/m³; STEL 0.005 mg(Be)/m³/30M; CL 0.025 mg(Be)/m³
**ACGIH TLV:** TWA 0.002 mg/m³, Suspected Human Carcinogen.
**DFG TRK:** Animal Carcinogen, Suspected Human Carcinogen. Grinding of beryllium metal and alloys: 0.005 mg/m³ calculated as beryllium in that portion of dust that can possibly be inhaled; other beryllium compounds: 0.002 mg/m³ calculated as beryllium in that portion of dust that can possibly be inhaled
**NIOSH REL:** CL not to exceed 0.0005 mg(Be)/m³
**DOT CLASSIFICATION:** 6.1; *Label:* Poison (UN 1566); DOT Class: 6.1; *Label:* Poison, Flammable Solid (UN 1567)

**SAFETY PROFILE:** Confirmed carcinogen with experimental carcinogenic, neoplastigenic, and tumorigenic data. A deadly poison by intravenous route. Human systemic effects by inhalation: lung fibrosis, dyspnea, and weight loss. Human mutation data reported. See also BERYLLIUM COMPOUNDS. A moderate fire hazard in the form of dust or powder, or when exposed to flame or by spontaneous chemical reaction. Slight explosion hazard in the form of powder or dust. Incompatible with halocarbons. Reacts incandescently with fluorine or chlorine. Mixtures of the powder with $CCl_4$ or trichloroethylene will flash or spark on impact. When heated to decomposition in air it emits very toxic fumes of BeO. Reacts with Li and P.

For occupational chemical analysis use NIOSH: Beryllium, 7102; Elements, 7300.

---

**HR**: – Hazard Rating indicating relative hazard for toxicity, fire, and reactivity with 3 denoting the worst hazard level.
*See Introduction: paragraph 3, p. xiii*

**CAS**: – The American Chemical Society's Chemical Abstracts Service number. A complete CAS number cross-index is located in Section 2.
*See Introduction: paragraph 4, p. xiii.*

**Cited References**: – A code which represents the cited reference for the toxicity data. Complete bibliographic citations in order are listed in Section 4.
*See Introduction: paragraph 16, p. xxiii*

**Consensus Reports**: – Supply additional information to enable the reader to make knowledgeable evaluations of potential chemical hazards.
*See Introduction: paragraph 17, p. xxiii.*

**Safety Profiles**: – These are text summaries of the toxicity, flammability, reactivity, incompatibility, and other dangerous properties of the entry.
*See Introduction: paragraph 19, p.xxvi.*

**Analytical Methods**: – References to OSHA and NIOSH occupational analytical methods.
*See Introduction: paragraph 19, p.xxvii.*

# Sax's
# Dangerous
# Properties of
# Industrial Materials

## Ninth Edition

# Sax's Dangerous Properties of Industrial Materials

## Ninth Edition

## RICHARD J. LEWIS, SR.

## VAN NOSTRAND REINHOLD

I⟨T⟩P™  A Division of International Thomson Publishing Inc.

New York • Albany • Bonn • Boston • Detroit • London • Madrid • Melbourne
Mexico City • Paris • San Francisco • Singapore • Tokyo • Toronto

## DISCLAIMER

Extreme care has been taken in preparation of this work. However, neither the publisher nor the authors shall be held responsible or liable for any damages resulting in connection with or arising from the use of any of the information in this book.

Copyright © 1996 by Van Nostrand Reinhold

I(T)P® A division of International Thomson Publishing, Inc.
   The ITP logo is a trademark under license

Printed in the United States of America

For more information, contact:

Van Nostrand Reinhold
115 Fifth Avenue
New York, NY 10003

Chapman & Hall GmbH
Pappelallee 3
69469 Weinheim
Germany

Chapman & Hall
2-6 Boundary Row
London SEI 8HN
United Kingdom

International Thomson Publishing Asia
221 Henderson Road #05-10
Henderson Building
Singapore 0315

Thomas Nelson Australia
102 Dodds Street
South Melbourne, 3205
Victoria, Australia

International Thomson Publishing Japan
Hirakawacho Kyowa Building, 3F
2-2-1 Hirakawacho
Chiyoda-ku, 102 Tokyo
Japan

Nelson Canada
1120 Birchmount Road
Scarborough, Ontario
Canada M1K 5G4

International Thomson Editores
Campos Eliseos 385, Piso 7
Col. Polanco
11560 Mexico D.F. Mexico

1 2 3 4 5 6 7 8 9 10 TPP 01 00 99 98 97 96

**Library of Congress Cataloging-in-Publication Data**
Lewis, Richard J., Sr.
        Sax's dangerous properties of industrial materials / Richard J.
   Lewis, Sr. — 9th ed.
        p.      cm.
        Includes bibliographical references and index.
        ISBN SET            0–442–02025–2
        ISBN VOLUME I     0–442–02255–7
        ISBN VOLUME II    0–442–02256–5
        ISBN VOLUME III 0–442–02257–3
        1. Hazardous substances—Handbooks, manuals, etc.      I. Sax, N.
   Irving (Newton Irving).    Dangerous properties of industrial
   materials.    9th ed.      II. Title.
   T55.3.H3L53    1995
   604.7—dc20                                                  95-46514
                                                                CIP

*Dedicated to Grace Ross Lewis
for her invaluable participation
in every aspect of this revision.*

# Acknowledgments

I extend thanks to Marianne J. Russell, Nancy Olsen, and Bob Esposito for their encouragement. My thanks and best wishes to Renee Guilmette and especially Geraldine Albert for their expert professional advice and assistance in converting the manuscript to this volume.

# Contents

*Sax's Dangerous Properties of Industrial Materials,
Ninth Edition* is also available as a CD-ROM
subscription (with three quarterly updates).
For ordering information, please call:

1-800-575-5886

# Preface

This ninth edition of *Dangerous Properties of Industrial Materials,* a three-volume set, represents a major revision and updating of the eighth edition. The objective of the book, however, remains the same: to promote safety by providing the most up-to-date hazard information available. The growth in the availability of toxicological and hazard control reports continues unabated. This book cannot contain all the published data and continue to provide the accessibility for which it is known. To continue to provide complete hazard assessments for the maximum number of entries, data for each entry has been selectively reduced. In particular, carcinogenic and reproductive data lines above those required to establish the hazard of the entry have been excluded. Complete data for these entries are available in the books *Carcinogenically Active Chemicals* and *Reproductively Active Chemicals,* both available from the publisher, and in the CD-ROM version of this work.

Over two-thirds of the entries have been revised for this edition, and 2,000 new entries have been added. Some less useful entries will appear only in the CD-ROM editions of *Dangerous Properties of Industrial Materials.* Most of the new entries were selected because they are on the EPA TSCA Inventory. These are reported to be used in commerce in the United States. Emphasis was placed on adding and updating physical properties, updating all DOT Classifications, and the addition of references to both OSHA and NIOSH occupational analytical methods.

Numerous synonyms have been added to assist in locating the many materials that are known under a variety of systematic and common names. The synonym cross-index contains the entry name as well as each synonym. This index should be consulted first to locate a material by name. Synonyms are given in English as well as other major languages such as French, German, Dutch, Polish, Japanese, and Italian.

Many additional physical and chemical properties have been added. Whenever available, physical descriptions, formulas, molecular weights, melting points, boiling points, explosion limits, flash points, densities, autoignition temperatures, and the like have been supplied.

A court order has vacated the OSHA Air Standards set in 1989 and contained in 29CFR 1910.1000. OSHA has decided to enforce only pre-1989 air standards. I have elected to include both the Transitional Limits scheduled to be effective on December 31, 1992, and the Final Rule limits, that were scheduled to be effective on September 1, 1989. These represent the current best judgment as to appropriate workplace air levels. While they may not be enforceable by OSHA, they are better guides than the OSHA Air Standards adopted in 1969.

The following classes of data are new or have been updated for all entries for which they apply:

1. For the first time, OSHA and NIOSH occupational analytical methods are referenced by method name or number for over 700 entries.
2. Also new to this addition are over 100 biological entities or their toxic products.
3. ACGIH TLVs and BEIs reflect the latest recommendations and now include "intended changes."
4. German MAK and BAT reflect the latest recommendations.
5. NTP 7th Annual Report On Carcinogens entries are identified.
6. DOT classifications were updated reflecting the HM-181 rulemaking.
7. CAS numbers are provided for additional entries.

Each entry concludes with a safety profile, a textual summary of the hazards presented by the entry. The discussion of human exposures includes target organs and specific effects reported. Carcinogenic and reproductive assessments have been completely revised for this edition.

Fire and explosion hazards are briefly summarized in terms of conditions of flammable or reactive hazard. Where feasible, fire-fighting materials and methods are discussed. Materials that are known to be incompatible with an entry are listed here.

Also included in the safety profile are comments on disaster hazards that serve to alert users of materials to the dangers that may be encountered on entering storage premises during a fire or other emergency. Although the presence of water, steam, acid fumes, or powerful vibrations can cause the decomposition of many materials into dangerous compounds, of particular concern are high temperatures (such as those resulting from a fire) because these can cause many otherwise mild chemicals to emit highly toxic gases or vapors such as $NO_x$, $SO_x$, acids, and so forth, or evolve vapors of antimony, arsenic, mercury, and the like.

The book, which consists of three volumes, is divided as follows:

The first volume contains a CAS number cross-index, a synonym cross-index, and the complete citations for bibliographic references given in the data section.

Section 1 contains the DOT Guide number cross-index for the listed materials.

Section 2 contains the CAS number cross-index for CAS numbers for the listed materials.

Section 3 contains the prime name and synonym cross-index for the listed materials.

Section 4 contains the complete bibliographic references.

The main section of the book is contained in Volumes II and III. It lists and describes approximately 20,000 materials in alphabetical order by entry name.

Please refer to the Introduction for an explanation of the sources of data and codes used.

Every effort has been made to include the most current and complete information. The author welcomes comments or corrections to the data presented.

Richard J. Lewis, Sr.

# Key to Abbreviations

abs—absolute
ACGIH—American Conference of Governmental Industrial Hygienists
af—atomic formula
alc—alcohol
alk—alkaline
amorph—amorphous
anhyd—anhydrous
approx—approximately
aq—aqueous
atm—atmosphere
autoign—autoignition
aw—atomic weight
BEI—ACGIH Biological Exposure Indexes
bp—boiling point
b range—boiling range
CAS—Chemical Abstracts Service
cc—cubic centimeter
CC—closed cup
CL—ceiling concentration
COC—Cleveland open cup
compd(s)—compound(s)
conc—concentration, concentrated
contg—containing
cryst—crystal(s), crystalline
d—density
D—day(s)
decomp—decomposition
deliq—deliquescent
dil—dilute
DOT—U.S. Department of Transportation
EPA—U.S. Environmental Protection Agency
eth—ether
(F)—Fahrenheit
FCC—Food Chemical Codex
FDA—U.S. Food and Drug Administration
flam—flammable
flash p—flash point
fp—freezing point
g—gram
glac—glacial
gran—granular, granules
H—hour(s)
HR:—hazard rating
htd—heated
htg—heating
hygr—hygroscopic
IARC—International Agency for Research on Cancer
immisc—immiscible

incomp—incompatible
insol—insoluble
IU—International Unit
kg—kilogram (one thousand grams)
L—liter
lel—lower explosive limit
liq—liquid
M—minute(s)
$m^3$—cubic meter
mf—molecular formula
mg—milligram
misc—miscible
mL—milliliter
mm—millimeter
mod—moderately
mp—melting point
mppcf—million particles per cubic foot
mw—molecular weight
μ—micro
μg—microgram
ng—nanogram
NIOSH—National Institute for Occupational Safety and Health
nonflam—nonflammable
NTP—National Toxicology Program
OBS—obsolete
OC—open cup
org—organic
ORM—other regulated material (DOT)
OSHA—Occupational Safety and Health Administration
Pa—Pascals
PEL—permissible exposure level
pet, petr—petroleum
pg—picogram (one trillionth of a gram)
Pk—peak concentration
pmole—picomole
powd—powder
ppb—parts per billion (v/v)
pph—parts per hundred (v/v)(percent)
ppm—parts per million (v/v)
ppt—parts per trillion (v/v)
prac—practically
prep—preparation
PROP—properties
refr—refractive
rhomb—rhombic
S, sec—second(s)
sl, slt—slight
sltly—slightly

sol—soluble
soln—solution
solv(s)—solvent(s)
spar—sparingly
spont—spontaneous(ly)
STEL—short-term exposure limit
subl—sublimes
TCC—Tag closed cup
tech—technical
temp—temperature
TLV—Threshold Limit Value
TOC—Tag open cup
TWA—time weighted average
uel—upper explosive limit
unk—unknown, unreported
ULC, ulc—Underwriters Laboratory Classification

USDA—U.S. Department of Agriculture
vac—vacuum
vap—vapor
vap d—vapor density
vap press—vapor pressure
visc—viscosity
vol—volume
W—week(s)
Y—year(s)
%—percent(age)
>—greater than
<—less than
<=—less than or equal to
=>—greater than or equal to
°—degrees of temperature in Celsius (centigrade)
°F—temperature in Fahrenheit

# Introduction

The list of potentially hazardous materials includes drugs, food additives, preservatives, ores, pesticides, dyes, detergents, lubricants, soaps, plastics, extracts from plant and animal sources, plants and animals that are toxic by contract or consumption, and industrial intermediates and waste products from production processes. Some of the information refers to materials of undefined composition. The chemicals included are assumed to exhibit the reported toxic effect in their pure state unless otherwise noted. However, even in the case of a supposedly "pure" chemical, there is usually some degree of uncertainty as to its exact composition and the impurities that may be present. This possibility must be considered in attempting to interpret the data presented because the toxic effects observed could in some cases be caused by a contaminant. Some radioactive materials are included but the effect reported is the chemically produced effect rather than the radiation effect.

For each entry the following data are provided when available: the DPIM code, hazard rating, entry name, CAS number, DOT number, molecular formula, molecular weight, line structural formula, a description of the material and physical properties, and synonyms. Following this are listed: Toxicity Data with References for reports of primary skin and eye irritation, mutation, reproductive, carcinogenic, and acute toxic dose data. The Consensus Reports section contains, where available, NTP 7th Annual Report On Carcinogens notation, IARC reviews, NTP Carcinogenesis Testing Program results, EPA Extremely Hazardous Substances List, the EPA Genetic Toxicology Program, and the Community Right-to-Know List. We also indicate the presence of the material on the update of the EPA TSCA inventory of chemicals in use in the United States. The next grouping consists of the U.S. Occupational Safety and Health Administration's (OSHA) permissible exposure levels, the American Conference of Governmental Industrial Hygienists' (ACGIH) Threshold Limit Values (TLVs), German Research Society's (MAK) values, National Institute for Occupational Safety and Health (NIOSH) recommended exposure levels, and U.S. Department of Transportation (DOT) classifications. Each entry concludes with a Safety Profile that discusses the toxic and other hazards of the entry. The Safety Profile concludes with the OSHA and NIOSH occupational analytical method, referenced by method name or number.

1. *DPIM Entry Code* identifies each entry by a unique code consisting of three letters and three numbers, for example, AAA123. The first letter of the entry code indicates the alphabetical position of the entry. Codes beginning with "A" are assigned to entries indexed with the A's. Each listing in the cross-indexes is referenced to its appropriate entry by the DPIM entry code.

2. *Entry Name* is the name of each material, selected, where possible, to be a commonly used designation.

3. *Hazard Rating (HR):* is assigned to each material in the form of a number (1, 2, or 3) that briefly identifies the level of the toxicity or hazard. The letter "D" is used where the data available are insufficient to indicate a relative rating. In most cases a "D" rating is assigned when only in-vitro mutagenic or experimental reproductive data are available. Ratings are assigned on the basis of low (1), medium (2), or high (3) toxic, fire, explosive, or reactivity hazard.

The number "3" indicates an LD50 below 400 mg/kg or an LC50 below 100 ppm; or that the material is explosive, highly flammable, or highly reactive.

The number "2" indicates an LD50 of 400-4,000 mg/kg or an LC50 of 100-500 ppm; or that the material is flammable or reactive.

The number "1" indicates an LD50 of 4000-40,000 mg/kg or an LC50 of 500-5000 ppm; or that the material is combustible or has some reactivity hazard.

4. *Chemical Abstracts Service Registry Number (CAS)* is a numeric designation assigned by the American Chemical Society's Chemical Abstracts Service and uniquely identifies a specific chemical compound. This entry allows one to conclusively identify a material regardless of the name or naming system used.

5. *DOT:* indicates a four-digit hazard code assigned by the U.S. Department of Transportation. This code is recognized internationally and is in agreement with the United Nations coding system. The code is used on transport documents, labels, and placards. It is also used to determine the regulations for shipping the material.

6. *Molecular Formula (mf:)* or atomic formula (*af:*) designates the elemental composition of the

material and is structured according to the Hill System (see *Journal of the American Chemical Society,* 22(8): 478–494, 1900), in which carbon and hydrogen (if present) are listed first, followed by the other elemental symbols in alphabetical order. The formulas for compounds that do not contain carbon are ordered strictly alphabetically by element symbol. Compounds such as salts or those containing waters of hydration have molecular formulas incorporating the CAS dot-disconnect convention. In this convention, the components are listed individually and separated by a period. The individual components of the formula are given in order of decreasing carbon atom count, and the component ratios given. A lowercase "*x*" indicates that the ratio is unknown. A lowercase "*n*" indicates a repeating, polymer-like structure. The formula is obtained from one of the cited references or a chemical reference text, or derived from the name of the material.

7. *Molecular Weight* (*mw:*) or atomic weight (*aw:*) is calculated from the molecular formula, using standard elemental molecular weights (carbon = 12.01).

8. *Structural Formula* is a line formula indicating the structure of a given material.

9. *Properties* (**PROP:**) are selected to be useful in evaluating the hazard of a material and designing its proper storage and use procedures. A definition of the material is included where necessary. The physical description of the material may refer to the form, color, and odor to aid in positive identification. When available, the boiling point, melting point, density, vapor pressure, vapor density, and refractive index are given. The flash point, autoignition temperature, and lower and upper explosive limits are included to aid in fire protection and control. An indication is given of the solubility or miscibility of the material in water and common solvents. Unless otherwise indicated, temperature is given in Celsius, pressure in millimeters of mercury.

10. *Synonyms* for the entry name are listed alphabetically. Synonyms include other chemical names, common or generic names, foreign names (with the language in parentheses), or codes. Some synonyms consist in whole or in part of registered trademarks. These trademarks are not identified as such. The reader is cautioned that some synonyms, particularly common names, may be ambiguous and refer to more than one material. In the Synonym Cross-Index, entry names appear in boldface type; synonyms in lightface.

11. *Skin and Eye Irritation Data* lines include, in sequence, the tissue tested (skin or eye); the species of animal tested; the total dose, and, where applicable, the duration of exposure; for skin tests only, whether open or occlusive; an interpretation of the irritation response severity when noted by the author; and the reference from which the information was extracted. Only positive irritation test results are included.

Materials that are applied topically to the skin or to the mucous membranes can elicit either (a) systemic effects of an acute or chronic nature or (b) local effects, more properly termed "primary irritation." A primary irritant is a material that, if present in sufficient quantity for a sufficient period of time, will produce a nonallergic, inflammatory reaction of the skin or of the mucous membrane at the site of contract. Primary irritants are further limited to those materials that are not corrosive. Hence, concentrated sulfuric acid is not classified as a primary irritant.

*a. Primary Skin Irritation.* In experimental animals, a primary skin irritant is defined as a chemical that produces an irritant response on first exposure in a majority of the test subjects. However, in some instances compounds act more subtly and require either repeated contact or special environmental conditions (humidity, temperature, occlusion, etc.) to produce a response.

The most standard animal irritation test is the Draize procedure (*Journal of Pharmacology and Experimental Therapeutics,* 82: 377-419, 1944). This procedure has been modified and adopted as a regulatory test by the Consumer Product Safety Commission (CPSC) in 16 CFR 1500.41 (formerly 21 CFR 191.11). In this test a known amount (0.5 mL of a liquid, or 0.5 g of a solid or semisolid) of the test material is introduced under a one square-inch gauze patch. The patch is applied to the skin (clipped free of hair) of twelve albino rabbits. Six rabbits are tested with intact skin and six with abraded skin. The abrasions are minor incisions made through the stratum corneum but are not sufficiently deep to disturb the dermis or to produce bleeding. The patch is secured in place with adhesive tape, and the entire trunk of the animal is wrapped with an impervious material, such as rubberized cloth, for a 24-hour period. The animal is immobilized during exposure. After 24 hours the patches are removed and the resulting reaction evaluated for erythema, eschar, and edema formation. The reaction is again scored at the end of 72 hours (48 hours after the initial reading), and the two readings are averaged. A material producing any degree of positive reaction is cited as an irritant.

As the modified Draize procedure described above has become the standard test specified by the U.S. government, nearly all of the primary skin irritation data either strictly adheres to the test protocol or involves only simple modifications to it. When test procedures other than those described previously are reported in the literature, appropriate codes are included in the data line to indicate those deviations.

The most common modification is the lack of occlusion of the test patch, so that the treated area is left open to the atmosphere. In such cases the notation "open" appears in the irritation data line. Another frequent modification involves immersion of the whole arm or whole body in the test material or, more commonly, in a dilute aqueous solution of the test material. This type of test is often conducted on soap and detergent solutions. Immersion data are identified by the abbreviation "imm" in the data line.

The dose reported is based first on the lowest dose producing an irritant effect and second on the latest study published. The dose is expressed as follows:

(1) Single application by the modified Draize procedure is indicated by only a dose amount. If no exposure time is given, then the data are for the standard 72-hour test. For test times other than 72 hours, the dose data are given in milligrams (or another appropriate unit)/duration of exposure, for example, 10 mg/24 H.

(2) Multiple applications involve administration of the dose in divided portions applied periodically. The total dose of test material is expressed in milligrams (or another appropriate unit)/duration of exposure, with the symbol "I" indicating intermittent exposure, for example, 5 mg/6D-I.

The method of testing materials for primary skin irritation given in the Code of Federal Regulations does not include an interpretation of the response. However, some authors do include a subjective rating of the irritation observed. If such a severity rating is given, it is included in the data line as mild ("MLD"), moderate ("MOD"), or severe ("SEV"). The Draize procedure employs a rating scheme that is included here for informational purposes only, because other researchers may not categorize irritation response in this manner.

| Category | Code | Skin Reaction (Draize) |
|----------|------|------------------------|
| Slight (Mild) | MLD | Well-defined erythema slight edema (edges of area well defined by definite raising) |
| Moderate | MOD | Moderate-to-severe erythema and moderate edema (area raised approximately 1 mm) |
| Severe | SEV | Severe erythema (beet redness) to slight eschar formation (injuries in depth) and severe edema (raised more than 1 mm and extending beyond area of exposure) |

*b. Primary Eye Irritation.* In experimental animals, a primary eye irritant is defined as a chemical that produces an irritant response in the test subject on first exposure. Eye irritation study procedures that Draize developed have been modified and adopted as a regulatory test by CPSC in 16 CFR 1500.42. In this procedure, a known amount of the test material (0.1 mL of a liquid or 100 mg of a solid or paste) is placed in one eye of each of six albino rabbits; the other eye remains untreated, serving as a control. The eyes are not washed after instillation and are examined at 24, 48, and 72 hours for ocular reaction. After the recording of ocular reaction at 24 hours, the eyes may be further examined following the application of fluorescein. The eyes may also be washed with a sodium chloride solution (U.S.P. or equivalent) after the 24-hour reaction has been recorded.

A test is scored positive if any of the following effects are observed: (1) ulceration (besides fine stippling); (2) opacity of the cornea (other than slight dulling of normal luster); (3) inflammation of the iris (other than a slight deepening of the rugae or circumcorneal injection of the blood vessel); (4) swelling of the conjunctiva (excluding the cornea and iris) with eversion of the eyelid; or (5) a diffuse crimson-red color with individual vessels not clearly identifiable. A material is an eye irritant if four of six rabbits score positive. It is considered a nonirritant if none or only one of six animals exhibits irritation. If intermediate results are obtained, the test is performed again. Materials producing any degree of irritation in the eye are identified as irritants. When an author has designated a substance as either a mild, moderate, or severe eye irritant, this designation is also reported.

The dose reported is based first on the lowest dose producing an irritant effect and second on the latest study published. Single and multiple applications are indicated as described previously under "Primary Skin Irritation." Test times other than 72 hours are noted in the dose. All eye irritant test exposures are assumed to be continuous, unless the reference states that the eyes were washed after instillation. In this case, the notation "rns" (rinsed) is included in the data line.

Because Draize procedures for determining both skin and eye irritation specify rabbits as the test species, most of the animal irritation data are for rabbits, although any of the species listed in Table 2 may be used. We have endeavored to include as much human data as possible, since this information is directly applicable to occupational exposure, much of which comes from studies conducted on volunteers (for example, for cosmetic or soap ingredients) or from persons accidentally exposed. When accidental exposure, such as a spill, is cited, the line includes the abbreviation "nse" (nonstandard exposure). In these cases it is often very difficult to determine the precise amount of the material to which the individual was exposed. Therefore, for accidental exposures an estimate of the concentration or strength of the material, rather than a total dose amount, is generally provided.

12. *Mutation Data* lines include, in sequence, the mutation test system utilized, the species of the tested organism (and, where applicable, the route of administration or cell type), the exposure concentration or dose, and the reference from which the information was extracted.

A mutation is defined as any heritable change in genetic material. Unlike irritation, reproductive, tumorigenic, and toxic dose data, which report the results of whole-animal studies, mutation data also include studies on lower organisms such as bacteria, molds, yeasts, and insects, as well as in-vitro mammalian cell cultures. Studies of plant mutagenesis are not included. No attempt is made to evaluate the significance of the data or to rate the relative potency of the compound as a mutagenic risk to humans.

Each element of the mutation line is discussed as follows:

*a. Mutation Test System.* Several test systems are used to detect genetic alterations caused by chemicals. Additional test systems may be added as they are reported in the literature. Each test system is identified by the three-letter code shown in parentheses. For additional information about mutation tests, the reader may wish to consult the *Handbook of Mutagenicity Test Procedures,* edited by B. J. Kilbey, M. Legator, W. Nichols, and C. Ramel (Amsterdam: Elsevier Scientific Publishing Company/North-Holland Biomedical Press, 1977).

(1) The Mutation in Microorganisms (mmo) System utilizes the detection of heritable genetic alterations in microorganisms that have been exposed directly to the chemical.

(2) The Microsomal Mutagenicity Assay (mma) System utilizes an in-vitro technique that allows enzymatic activation of promutagens in the presence of an indicator organism in which induced mutation frequencies are determined.

(3) The Micronucleus Test (mnt) System utilizes the fact that chromosomes or chromosome fragments may not be incorporated into one or the other of the daughter nuclei during cell division.

(4) The Specific Locus Test (slt) System utilizes a method for detecting and measuring rates of mutation at any or all of several recessive loci.

(5) The DNA Damage (dnd) System detects the damage to DNA strands, including strand breaks, crosslinks, and other abnormalities.

(6) The DNA Repair (dnr) System utilizes methods of monitoring DNA repair as a function of induced genetic damage.

(7) The Unscheduled DNA synthesis (dns) System detects the synthesis of DNA during usually nonsynthetic phases.

(8) The DNA Inhibition (dni) System detects damage that inhibits the synthesis of DNA.

(9) The Gene Conversion and Mitotic Recombination (mrc) System utilizes unequal recovery of genetic markers in the region of the exchange during genetic recombination.

(10) The Cytogenetic Analysis (cyt) System utilizes cultured cells or cell lines to assay for chromosomal aberrations following the administration of the chemical.

(11) The Sister Chromatid Exchange (sce) System detects the interchange of DNA in cytological preparations of metaphase chromosomes between replication products at apparently homologous loci.

(12) The Sex Chromosome Loss and Nondisjunction (sln) System measures the nonseparation of homologous chromosomes at meiosis and mitosis.

(13) The Dominant Lethal Test (dlt). A dominant lethal is a genetic change in a gamete that kills the zygote produced by that gamete. In mammals, the dominant lethal test measures the reduction of litter size by examining the uterus and noting the number of surviving and dead implants.

(14) The Mutation in Mammalian Somatic Cells (msc) System utilizes the induction and isolation of mutants in cultured mammalian cells by identification of the gene change.

(15) The Host-Mediated Assay (hma) System uses two separate species, generally mammalian and bacterial, to detect heritable genetic alteration caused by metabolic conversion of chemical substances administered to host mammalian species in the bacterial indicator species.

(16) The Sperm Morphology (spm) System measures the departure from normal in the appearance of sperm.

(17) The Heritable Translocation Test (trn) measures the transmissibility of induced translocations to subsequent generations. In mammals, the test uses sterility and reduced fertility in the progeny of the treated parent. In addition, cytological analysis of the F1 progeny or subsequent progeny of the treated parent is carried out to prove the existence of the induced translocation. In *Drosophila*, heritable translocations are detected genetically using easily distinguishable phenotypic markers, and these translocations can be verified with cytogenetic techniques.

(18) The Oncogenic Transformation (otr) System utilizes morphological criteria to detect cytological differences between normal and transformed tumorigenic cells.

(19) The Phage Inhibition Capacity (pic) System utilizes a lysogenic virus to detect a change in the genetic characteristics by the transformation of the virus from noninfectious to infectious.

(20) The Body Fluid Assay (bfa) System uses two separate species, usually mammalian and bacterial. The test substance is first administered to the host, from whom body fluid (for example, urine, blood) is subsequently taken. This body fluid is then tested in-vitro, and mutations are measured in the bacterial species.

*b. Species.* Those test species that are peculiar to mutation data are designated by the three-letter codes shown as follows.

|  | Code | Species |
|---|---|---|
| Bacteria | bcs | *Bacillus subtilis* |
|  | esc | *Escherichia coli* |
|  | hmi | *Haemophilus influenzae* |
|  | klp | *Klebsiella pneumoniae* |
|  | sat | *Salmonella typhimurium* |
|  | srm | *Serratia marcescens* |
| Molds | asn | *Aspergillus nidulans* |
|  | nsc | *Neurospora crassa* |
| Yeasts | smc | *Saccharomyces cerevisiae* |
|  | ssp | *Schizosaccharomyces pombe* |
| Protozoa | clr | *Chlamydomonas reinhardi* |
|  | eug | *Euglena gracilis* |
|  | omi | Other microorganisms |

(*Table continued on facing page*)

|         | Code | Species |
|---------|------|---------|
| Insects | dmg  | *Drosophila melanogaster* |
|         | dpo  | *Drosophila pseudo-obscura* |
|         | grh  | Grasshopper |
|         | slw  | Silkworm |
|         | oin  | Other insects |
| Fish    | sal  | Salmon |
|         | ofs  | Other fish |

If the test organism is a cell type from a mammalian species, the parent mammalian species is reported, followed by a colon and the cell type designation. For example, human leukocytes are coded "hmn:leu." The various cell types currently cited in this edition are as follows:

| Designation | Cell Type |
|-------------|-----------|
| ast | Ascites tumor |
| bmr | Bone marrow |
| emb | Embryo |
| fbr | Fibroblast |
| hla | HeLa cell |
| kdy | Kidney |
| leu | Leukocyte |
| lng | Lung |
| lvr | Liver |
| lym | Lymphocyte |
| mmr | Mammary gland |
| ovr | Ovary |
| spr | Sperm |
| tes | Testis |
| oth | Other cell types not listed above |

In the case of host-mediated and body-fluid assays, both the host organism and the indicator organism are given as follows: host organism/indicator organism, for example, "ham/sat" for a test in which hamsters were exposed to the test chemical and *S. typhimurium* was used as the indicator organism.

For in-vivo mutagenic studies, the route of administration is specified following the species designation, for example, "mus-orl" for oral administration to mice. See Table 1 for a complete list of routes cited. The route of administration is not specified for in-vitro data.

*c. Units of Exposure.* The lowest dose producing a positive effect is cited. The author's calculations are used to determine the lowest dose at which a positive effect was observed. If the author fails to state the lowest effective dose, two times the control dose will be used. Ideally, the dose should be reported in universally accepted toxicological units such as milligrams of test chemical per kilogram of test animal body weight. Although this is possible in cases where the actual intake of the chemical by an organism of known weight is reported, it is not possible in many systems using insect and bacterial species. In cases where a dose is reported or where the amount can be converted to a dose unit, it is normally listed as milligrams per kilogram (mg/kg). However, micro-

grams (µg), nanograms (ng), or picograms (pg) per kilogram may also be used for convenience of presentation. Concentrations of gaseous materials in air are listed as parts per hundred (pph), per million (ppm), per billion (ppb), or per trillion (ppt).

Test systems using microbial organisms traditionally report exposure data as an amount of chemical per liter (L) or amount per plate, well, disc, or tube. The amount may be on a weight (g, mg, µg, ng, or pg) or molar (millimole (mmol), micromole (µmol), nanomole (nmol), or picomole (pmol)). These units describe the exposure concentration rather than the dose actually taken up by the test species. Insufficient data currently exist to permit the development of dose amounts from this information. In such cases, therefore, the material concentration units that the author used are reported.

Because the exposure values reported in host-mediated and body-fluid assays are doses delivered to the host organism, no attempt is made to estimate the exposure concentration to the indicator organism. The exposure values cited for host-mediated assay data are in units of milligrams (or other appropriate units of weight) of material administered per kilogram of host body weight, or in parts of vapor or gas per million (ppm) parts of air (or other appropriate concentrations) by volume.

13. *Toxicity Dose Data* lines include, in sequence, the route of exposure; the species of animal studied; the toxicity measure; the amount of material per body weight or concentration per unit of air volume and, where applicable, the duration of exposure; a descriptive notation of the type of effect reported; and the reference from which the information was extracted. Only positive toxicity test results are cited in this section.

All toxic-dose data appearing in the book are derived from reports of the toxic effects produced by individual materials. For human data, a toxic effect is defined as any reversible or irreversible noxious effect on the body, any benign or malignant tumor, any teratogenic effect, or any death that has been reported to have resulted from exposure to a material via any route. For humans, a toxic effect is any effect that was reported in the source reference. There is no qualifying limitation on the duration of exposure or for the quantity or concentration of the material, or is there a qualifying limitation on the circumstances that resulted from the exposure. Regardless of the absurdity of the circumstances that were involved in a toxic exposure, it is assumed that the same circumstances could recur. For animal data, toxic effects are limited to the production of tumors, benign (neoplastigenesis) or malignant (carcinogenesis); the production of changes in the offspring resulting from action on the fetus directly (teratogenesis); and death. There is no limitation on either the duration of exposure or on the quantity or concentration of the dose of the material reported to have caused these effects.

The report of the lowest total dose administered over the shortest time to produce the toxic effect was given preference, although some editorial liberty was taken so that additional references might be cited. No restrictions were placed on the amount of a material producing death in an experimental animal nor on the time period over which the dose was given.

Each element of the toxic dose line is discussed below.

*a. Route of Exposure or Administration.* Although many exposures to materials in the industrial community occur via the respiratory tract or skin, most studies in the published literature report exposures of experimental animals in which the test materials were introduced primarily through the mouth by pills, in food, in drinking water, or by intubation directly into the stomach. The abbreviations and definitions of the various routes of exposure reported are given in Table 1.

*b. Species Exposed.* Because the effects of exposure of humans are of primary concern, we have indicated, when available, whether the results were observed in man, woman, child, or infant. If no such distinction was made in the reference, the abbreviation "hmn" (human) is used. However, the results of studies on rats or mice are the most frequently reported and hence provide the most useful data for comparative purposes. The species and abbreviations used in reporting toxic dose data are listed alphabetically in Table 2.

*c. Description of Exposure.* In order to describe the administered dose reported in the literature, six

## TABLE 1. Routes of Administration to, or Exposure of, Animal Species to Toxic Substances

| Route | Abbreviation | Definition |
|---|---|---|
| Eyes | eye | Administration directly onto the surface of the eye. Used exclusively for primary irritation data. See *Ocular.* |
| Intraaural | ial | Administration into the ear |
| Intraarterial | iat | Administration into the artery |
| Intracerebral | ice | Administration into the cerebrum |
| Intracervical | icv | Administration into the cervix |
| Intradermal | idr | Administration within the dermis by hypodermic needle. |
| Intraduodenal | idu | Administration into the duodenum |
| Inhalation | ihl | Inhalation in chamber, by cannulation, or through mask. |
| Implant | imp | Placed surgically within the body location described in reference. |
| Intramuscular | ims | Administration into the muscle by hypodermic needle. |
| Intraplacental | ipc | Administration into the placenta |
| Intrapleural | ipl | Administration into the pleural cavity by hypodermic needle. |
| Intraperitoneal | ipr | Administration into the peritoneal cavity |
| Intrarenal | irn | Administration into the kidney |
| Intraspinal | isp | Administration into the spinal canal |
| Intratracheal | itr | Administration into the trachea |
| Intratesticular | itt | Administration into the testes |
| Intrauterine | iut | Administration into the uterus |
| Intravaginal | ivg | Administration into the vagina |
| Intravenous | ivn | Administration directly into the vein by hypodermic needle. |
| Multiple | mul | Administration into a single animal by more than one route. |
| Ocular | ocu | Administration directly onto the surface of the eye or into the conjunctival sac. Used exclusively for systemic toxicity data. |
| Oral | orl | Per os, intragastric, feeding, or introduction with drinking water. |
| Parenteral | par | Administration into the body through the skin. Reference cited is not specific about the route used. Could be ipr, scu, ivn, ipl, ims, irn, or ice. |
| Rectal | rec | Administration into the rectum or colon in the form of enema or suppository. |
| Subcutaneous | scu | Administration under the skin. |
| Skin | skn | Application directly onto the skin, either intact or abraded. Used for both systemic toxicity and primary irritant effects. |
| Unreported | unr | Dose, but not route, is specified in the reference. |

abbreviations are used. These terms indicate whether the dose caused death (LD) or other toxic effects (TD) and whether it was administered as a lethal concentration (LC) or toxic concentration (TC) in the inhaled air. In general, the term "Lo" is used where the number of subjects studied was not a significant number from the population or the calculated percentage of subjects showing an effect was listed as 100. The definition of terms is as follows:

*TDLo—Toxic Dose Low—*the lowest dose of a material introduced by any route, other than inhalation, over any given period of time and reported to produce any toxic effect in humans or to produce carcinogenic, neoplastigenic, or teratogenic effects in animals or humans.

*TCLo—Toxic Concentration Low—*the lowest concentration of a material in air to which humans or animals have been exposed for any given period of time that has produced any toxic effect in humans or produced a carcinogenic, neoplastigenic, or teratogenic effect in animals or humans.

*LDLo—Lethal Dose Low—*the lowest dose (other than LD50) of a material introduced by any route, other than inhalation, over any given period of time in one or more divided portions and reported to have caused death in humans or animals.

*LD50—Lethal Dose Fifty—*a calculated dose of a material that is expected to cause the death of 50% of an entire defined experimental animal population. It is determined from the exposure to the material, by any route other than inhalation, of a significant number from that population. Other lethal dose percentages, such as LD1, LD10, LD30, and LD99, may be published in the scientific literature for the specific purposes of the author. Such data would be published if these figures, in the absence of a calculated lethal dose (LD50), were the lowest found in the literature.

*LCLo—Lethal Concentration Low—*the lowest concentration of a material in air, other than LC50, that has been reported to have caused death in humans or animals. The reported concentrations may

## TABLE 2. Species
### (with assumptions for toxic dose calculation from nonspecific data*)

| Species | Abbreviation | Age | Weight | | Consumption Food (g/day) | (Approx.) Water (mL/day) | 1 ppm in Food Equals, in (mg/kg/day) | Approximate Gestation Period (days) |
|---|---|---|---|---|---|---|---|---|
| Bird, type not specified | brd | | 1 | kg | | | | |
| Bird, wild bird species | bwd | | 40 | g | | | | |
| Cat, adult | cat | | 2 | kg | 100 | 100 | 0.05 | 64 (59–68) |
| Child | chd | 1–13 Y | 20 | kg | | | | |
| Chicken, adult | ckn | 8 W | 800 | g | 140 | 200 | 0.175 | |
| Cattle | ctl | | 500 | kg | 10,000 | | 0.02 | 284 (279–290) |
| Duck, adult (domestic) | dck | 8 W | 2.5 | kg | 250 | 500 | 0.1 | |
| Dog, adult | dog | 52 W | 10 | kg | 250 | 500 | 0.025 | 62 (56–68) |
| Domestic animals (goat, sheep) | dom | | 60 | kg | 2,400 | | 0.04 | G: 152 (148–156) S: 146 (144–147) |
| Frog, adult | frg | | 33 | g | | | | |
| Guinea Pig, adult | gpg | | 500 | g | 30 | 85 | 0.06 | 68 |
| Gerbil | grb | | 100 | g | 5 | 5 | 0.05 | 25 (24–26) |
| Hamster | ham | 14 W | 125 | g | 15 | 10 | 0.12 | 16 (16–17) |
| Human | hmn | Adult | 70 | kg | | | | |
| Horse, donkey | hor | | 500 | kg | 10,000 | | 0.02 | H: 339 (333–345) D: 365 |
| Infant | inf | 0–1 Y | 5 | kg | | | | |
| Mammal (species unspecified in reference) | mam | | 200 | g | | | | |
| Man | man | Adult | 70 | kg | | | | |
| Monkey | mky | 2.5 Y | 5 | kg | 250 | 500 | 0.05 | 165 |
| Mouse | mus | 8 W | 25 | g | 3 | 5 | 0.12 | 21 |
| Non-mammalian species | nml | | | | | | | |
| Pigeon | pgn | 8 W | 500 | gm | | | | |
| Pig | pig | | 60 | kg | 2,400 | | 0.041 | 114 (112–115) |
| Quail (laboratory) | qal | | 100 | g | | | | |
| Rat, adult female | rat | 14 W | 200 | g | 10 | 20 | 0.05 | 22 |
| Rat, adult male | rat | 14 W | 250 | g | 15 | 25 | 0.06 | |
| Rat, adult | rat | 14 W | 200 | g | 15 | 25 | | |
| Rat, weanling | rat | 3 W | 50 | g | 15 | 25 | 0.3 | |
| Rabbit, adult | rbt | 12 W | 2 | kg | 60 | 330 | 0.03 | 31 |
| Squirrel | sql | | 500 | g | | | | 44 |
| Toad | tod | | 100 | g | | | | |
| Turkey | trk | 18 W | 5 | kg | | | | |
| Woman | wmn | Adult | 50 | kg | 270 | | | |

* Values given in Table 2 are within reasonable limits usually found in the published literature and are selected to facilitate calculations for data from publications in which toxic dose information has not been presented for an individual animal of the study. See, for example, *Association of Food and Drug Officials, Quarterly Bulletin,* volume 18, page 66, 1954; Guyton, *American Journal of Physiology,* volume 150, page 75, 1947; *The Merck Veterinary Manual,* 5th Edition, Merck & Co., Inc., Rahway, NJ, 1979; and The UFAW *Handbook on the Care and Management of Laboratory Animals,* 4th Edition, Churchill Livingston, London, 1972. Data for lifetime exposure are calculated from the assumptions for adult animals for the entire period of exposure. For definitive dose data, the reader must review the referenced publication.

be entered for periods of exposure that are less than 24 hours (acute) or greater than 24 hours (subacute and chronic).

*LC50—Lethal Concentration Fifty*—a calculated concentration of a material in air, exposure to which for a specified length of time is expected to cause the death of 50% of an entire defined experimental animal population. It is determined from the exposure to the material of a significant number from that population.

The following table summarizes the above information.

| Category | Exposure Time | Route of Exposure | TOXIC EFFECTS Human | TOXIC EFFECTS Animal |
|---|---|---|---|---|
| TDLo | Acute or chronic | All except inhalation | Any Non-lethal | CAR, NEO, NEO, ETA, TER, REP |
| TCLo | Acute or chronic | Inhalation | Any Non-lethal | CAR, NEO, ETA, ETA, TER, REP |
| LDLo | Acute or chronic | All except inhalation | Death | Death |
| LD50 | Acute | All except inhalation | Not applicable | Death (statistically determined) |
| LCLo | Acute or Chronic | Inhalation | Death | Death |
| LC50 | Acute | Inhalation | Not applicable | Death (statistically determined) |

*d. Units of Dose Measurement.* As in almost all experimental toxicology, the doses given are expressed in terms of the quantity administered per unit body weight, or quantity per skin surface area, or quantity per unit volume of the respired air. In addition, the duration of time over which the dose was administered is also listed, as needed. Dose amounts are generally expressed as milligrams (thousandths of a gram) per kilogram (mg/kg). In some cases, because of dose size and its practical presentation in the file, grams per kilogram (g/kg), micrograms (millionths of a gram) per kilogram ($\mu$g/kg), or nanograms (billionths of a gram) per kilogram (ng/kg) are used. Volume measurements of dose were converted to weight units by appropriate calculations. Densities were obtained from standard reference texts. Where densities were not readily available, all liquids were assumed to have a density of 1 g/mL. Twenty drops of liquid are assumed to be equal in volume to 1 mL.

All body weights have been converted to kilograms (kg) for uniformity. For those references in which the dose was reported to have been administered to an animal of unspecified weight or a given number of animals in a group (for example, feeding studies) without weight data, the weights of the respective animal species were assumed to be those listed in Table 2 and the dose is listed on a per-kilogram body-weight basis. Assumptions for daily food and water intake are found in Table 2 to allow approximation doses for humans and species of experimental animals in cases in which the dose was originally reported as a concentration in food or water. The values presented are selections that are reasonable for the species and convenient for dose calculations.

Concentrations of a gaseous material in air are generally listed as parts of vapor or gas per million parts of air by volume (ppm). However, parts per hundred (pph or percent), parts per billion (ppb), or parts per trillion (ppt) may be used for convenience of presentation. If the material is a solid or a liquid, the concentrations are listed preferably as milligrams per cubic meter ($mg/m^3$) but may, as applicable, be listed as micrograms per cubic meter ($\mu g/m^3$), nanograms per cubic meter ($ng/m^3$), or picograms (trillionths of a gram) per cubic meter ($pg/m^3$) of air. For those cases in which other measurements of contaminants are used, such as the number of fibers or particles, the measurement is spelled out.

Where the duration of exposure is available, time is presented as minutes (M), hours (H), days (D), weeks (W), or years (Y). Additionally, continuous (C) indicates that the exposure was continuous over the time administered, such as ad-libitum feeding studies or 24-hour, 7-day-per-week inhalation exposures. Intermittent (I) indicates that the dose was administered during discrete periods, such as daily twice weekly. In all cases, the total duration of exposure appears first after the kilogram body weight and a slash, and is followed by descriptive data; for example, 10 mg/kg/3W-I indicates ten milligrams per kilogram body weight administered over a period of three weeks, intermittently in a number of separate, discrete doses. This description is intended to provide the reader with enough information for an approximation of the experimental conditions, which can be further clarified by studying the reference cited.

*e. Frequency of Exposure.* Frequency of exposure to the test material depends on the nature of the experiment. Frequency of exposure is given in the case of an inhalation experiment, for human exposures (where applicable), or where CAR, NEO, ETA, REP, or TER is specified as the toxic effect.

*f. Duration of Exposure.* For assessment of tumorigenic effect, the testing period should be the life span of the animal, or should extend until statistically valid calculations can be obtained regarding tumor incidence. In the toxic dose line, the total dose causing the tumorigenic effect is given. The duration of exposure is included to give an indication of the testing period during which the animal was exposed to this total dose. For multigenerational studies, the time during gestation when the material was administered to the mother is also provided.

*g. Notations Descriptive of the Toxicology.* The toxic dose line thus far has indicated the route of entry, the species involved, the description of the dose, and the amount of the dose. The next entry found on this line when a toxic exposure (TD or TC) has been listed is the toxic effect. Following a colon will be one of the notations found in Table 3. These notations indicate the organ system affected or special effects that the material produced, for exam-

### Table 3. Notations Descriptive of the Toxicology

| Notation | Effects (not limited to effects listed) |
|---|---|
| ALR | Allergic systemic reaction such as might be experienced by individuals sensitized to penicillin. |
| BAH | Behavioral—includes wakefulness, euphoria, hallucinations, coma, etc. |
| BCM | Blood clotting mechanism effects—any effect that increases or decreases clotting time. |
| BLD | Blood effects—effect on all blood elements, electrolytes, pH, proteins, oxygen carrying or releasing capacity. |
| BPR | Blood pressure effects—any effect that increases or decreases any aspect of blood pressure. |
| CAR | Carcinogenic effects—see paragraph 9(g) in text. |
| CNS | Central nervous system effects—includes effects such as headaches, tremor, drowsiness, convulsions, hypnosis, anesthesia. |
| COR | Corrosive effects—burns, desquamation. |
| CUM | Cumulative effects—where material is retained by the body in greater quantities than is excreted, or the effect is increased in severity by repeated body insult. |
| CVS | Cardiovascular effects—such as an increase or decrease in the heart activity through effect on ventricle or auricle; fibrillation; constriction or dilation of the arterial or venous system. |
| DDP | Drug dependence effects—any indication of addiction or dependence. |
| ETA | Equivocal tumorigenic agent—see text. |
| EYE | Eye effects—irritation, diplopia, cataracts, eye ground, blindness by effects to the eye or the optic nerve. |
| GIT | Gastrointestinal tract effects—diarrhea, constipation, ulceration. |
| GLN | Glandular effects—any effect on the endocrine glandular system. |
| IRR | Irritant effects—any irritant effect on the skin, eye, or mucous membrane. |
| MLD | Mild irritation effects—used exclusively for primary irritation data. |
| MMI | Mucous membrane effects—irritation, hyperplasia, changes in ciliary activity. |
| MOD | Moderate irritation effects—used exclusively for primary irritation data. |
| MSK | Musculo-skeletal effects—such as osteoporosis, muscular degeneration. |
| NEO | Neoplastic effects—see text. |
| PNS | Peripheral nervous system effects. |
| PSY | Psychotropic effects—exerting an effect upon the mind. |
| PUL | Pulmonary system effects—effects on respiration and respiratory pathology. |
| RBC | Red blood cell effects—includes the several anemias. |
| REP | Reproductive effects—see text. |
| SEV | Severe irritation effects—used exclusively for primary irritation data. |
| SKN | Skin effects—such as erythema, rash, sensitization of skin, petechial hemorrhage. |
| SYS | Systemic effects—effects on the metabolic and excretory function of the liver or kidneys. |
| TER | Teratogenic effects—nontransmissible changes produced in the offspring. |
| UNS | Unspecified effects—the toxic effects were unspecific in the reference. |
| WBC | White blood cell effects—effects on any of the cellular units other than erythrocytes, including any change in number or form. |

ple, TER = teratogenic effect. No attempt was made to be definitive in reporting these effects because such definition requires detailed qualification that is beyond the scope of this book. The selection of the dose was based first on the lowest dose producing an effect and second on the latest study published.

14. *Reproductive Effects Data* lines include, in sequence, the reproductive effect reported, the route of exposure, the species of animal tested, the type of dose, the total dose amount administered, the time and duration of administration, and the reference from which the information was extracted. Only positive reproductive effects data for mammalian species are cited. Because of differences in the reproductive systems among species and the systems' varying responses to chemical exposures, no attempt is made to extrapolate animal data or to evaluate the significance of a substance as a reproductive risk to humans.

Each element of the reproductive effects data line is discussed below.

*a. Reproductive Effect.* For human exposure, the effects are included in the safety profile. The effects include those reported to affect the male or female reproductive systems, mating and conception success, fetal effects (including abortion), transplacental carcinogenesis, and post-birth effects on parents and offspring.

*b. Route of Exposure or Administration.* See Table 1 for a complete list of abbreviations and definitions of the various routes of exposure reported. For reproductive effects data, the specific route is listed either when the substance was administered to only one of the parents or when the substance was administered to both parents by the same route. However, if the substance was administered to each parent by a different route, the route is indicated as "mul" (multiple).

*c. Species Exposed.* Reproductive effects data are cited for mammalian species only. Species abbreviations are shown in Table 2. Also shown in Table 2 are approximate gestation periods.

*d. Type of Exposure.* Only two types of exposure, TDLo and TCLo, are used to describe the dose amounts reported for reproductive effects data.

*e. Dose Amounts and Units.* The total dose amount that was administered to the exposed parent is given. If the substance was administered to both parents, the individual amounts to each parent have been added together and the total amount shown. Where necessary, appropriate conversion of dose units has been made. The dose amounts listed are those for which the reported effects are statistically significant. However, human case reports are cited even when no statistical tests can be performed. The statistical test is that used by the author. If no statistic is reported, a Fisher's Exact Test is applied with significance at the 0.05 level, unless the author makes a strong case for significance at some other level.

Dose units are usually given as an amount administered per unit body weight or as parts of vapor or gas per million parts of air by volume. There is no limitation on either the quantity or concentration of the dose or the duration of exposure reported to have caused the reproductive effect(s).

*f. Time and Duration of Treatment.* The time when a substance is administered to either or both parents may significantly affect the results of a reproductive study, because there are differing critical periods during the reproductive cycles of each species. Therefore, to provide some indication of when the substance was administered, which should facilitate selection of specific data for analysis by the user, a series of up to four terms follows the dose amount. These terms indicate to which parent(s) and at what time the substance was administered. The terms take the general form:

($u$D male/$v$D pre/w-$x$D preg/$y$D post)

where $u$ = total number of days of administration to male prior to mating

$v$ = total number of days of administration to female prior to mating

$w$ = first day of administration to pregnant female during gestation

$x$ = last day of administration to pregnant female during gestation

$y$ = total number of days of administration to lactating mother after birth of offspring

If administration is to the male only, then only the first of the above four terms is shown following the total dose to the male, for example, 10 mg/kg (5D male). If administration is to the female only, then only the second, third, or fourth term, or any combination thereof, is shown following the total dose to the female, for example:

10 mg/kg (3D pre)

10 mg/kg (3D pre/4-7D preg)

10 mg/kg (3D pre/4-7D preg/5D post)

10 mg/kg (3D pre/5D post)

10 mg/kg (4-7D preg)

10 mg/kg (4-7D preg/5D post)

10 mg/kg (5D post) (NOTE: This example indicates administration was only to the lactating mother, and only after birth of the offspring.)

If administration is to both parents, then the first term and any combination of the last three terms are listed, for example, 10 mg/kg (5D male/3D pre/4–7D preg). If administration is continuous through two or more of the above periods, the above format is abbreviated by replacing the slash (/) with a dash (-). For example, 10 mg/kg (3D pre-5D post) indicates a total of 10 mg/kg administered to the female for three days prior to mating, on each day during gestation, and for five days following birth. Approximate gestation periods for various species are shown in Table 2.

*g. Multigeneration Studies.* Some reproductive studies entail administration of a substance to several consecutive generations, with the reproductive effects measured in the final generation. The protocols for such studies vary widely. Therefore, because of the inherent complexity and variability of these studies, they are cited in a simplified format as follows. The specific route of administration is reported if it was the same for all parents of all generations; otherwise the abbreviation "mul" is used. The total dose amount shown is that administered to the F0 generation only; doses to the F$n$ (where $n$ = 1, 2, 3, etc.) generations are not reported. The time and duration of treatment for multigeneration studies are not included in the data line. Instead, the dose amount is followed by the abbreviation "(MGN)", for example, 10 mg/kg (MGN). This code indicates a multigeneration study, and the reader must consult the cited reference for complete details of the study protocol.

15. *Carcinogenic Study Result.* Tumorigenic citations are classified according to the reported results of the study to aid the reader in selecting appropriate references for in-depth review and evaluation. The classification ETA (equivocal tumorigenic agent) denotes those studies reporting uncertain, but seemingly positive, results. The criteria for the three classifications are listed below. These criteria are used to abstract the data in individual reports on a consistent basis and do not represent a comprehensive evaluation of a material's tumorigenic potential to humans.

The following nine technical criteria are used to abstract the toxicological literature and classify studies that report positive tumorigenic responses. No attempts are made either to evaluate the various test procedures or to correlate results from different experiments.

(1) A citation is coded "CAR" (carcinogenic) when review of an article reveals that all the following criteria are satisfied:

(a) There is a statistically significant increase in the incidence of tumors in the test animals. The statistical test is that used by the author. If no statistic is reported, a Fisher's Exact Test is applied with significance at the 0.05 level, unless the author makes a strong case for significance at some other level.

(b) A control group of animals is used and the treated and control animals are maintained under identical conditions.

(c) The sole experimental variable between the groups is the administration or nonadministration of the test material (see (10) below).

(d) The tumors consist of autonomous populations of cells of abnormal cytology capable of invading and destroying normal tissues, or the tumors metastasize as confirmed by histopathology.

(2) A citation is coded "NEO" (neoplastic) when review of an article reveals that all the following criteria are satisfied:

(a) There is a statistically significant increase in the incidence of tumors in the test animals. The statistical test is that used by the author. If no statistic is reported, a Fisher's Exact Test is applied with significance at the 0.05 level, unless the author makes a strong case for significance at some other level.

(b) A control group of animals is used and the treated and control animals are maintained under identical conditions.

(c) The sole experimental variable between the groups is the administration or nonadministration of the test material.

(d) The tumors consist of autonomous populations of cells of abnormal cytology capable of invading and destroying normal tissues, or the tumors metastasize as confirmed by histopathology.

(3) A citation is coded "ETA" (equivocal tumorigenic agent) when some evidence of tumorigenic activity is presented, but one or more of the criteria listed in (1) or (2) above are lacking. Thus, a report with positive pathological findings, but with no mention of control animals, is coded "ETA."

(4) Because an author may make statements or draw conclusions based on a larger context than that of the particular data reported, papers in which the author's conclusions differ substantially from the evidence presented in the paper are subject to review.

(5) All doses except those for transplacental carcinogenesis are reported in one of the following formats.

(a) For all routes of administration other than inhalation: cumulative dose is reported in mg (or another appropriate unit)/kg/duration of administration.

Whenever the dose reported in the reference is not in the units discussed herein, conversion to this format is made. The total cumulative dose is derived from the lowest dose level that produces tumors in the test group.

(b) For inhalation experiments: concentration is reported in ppm (or $mg/m^3$)/total duration of exposure.

The concentration refers to the lowest concentration that produces tumors.

(6) Transplacental carcinogenic doses are reported in one of the following formats:

(a) For all routes of administration other than inhalation, cumulative dose is reported in mg/kg/(time of administration during pregnancy).

The cumulative dose is derived from the lowest single dose that produces tumors in the offspring. The test chemical is administered to the mother.

(b) For inhalation experiments: concentration is reported in ppm (or $mg/m^3$)/(time of exposure during pregnancy).

The concentration refers to the lowest concentration that produces tumors in the offspring. The mother is exposed to the test chemical.

(7) For the purposes of this listing, all test chemicals are reported as pure, unless stated to be otherwise by the author. This does not rule out the possibility that unknown impurities may have been present.

(8) A mixture of compounds whose test results satisfy the criteria in (1), (2), or (3) above is included if the composition of the mixture can be clearly defined.

(9) For tests involving promoters or initiators, a study is included if the following conditions are satisfied (in addition to the criteria previously mentioned in (1), (2), or (3)):

(a) The test chemical is applied first, followed by an application of a standard promoter. A positive control group in which the test animals are subjected to the same standard promoter under identical conditions is maintained throughout the duration of the experiment. The data are only used if positive and negative control groups are mentioned in the reference.

(b) A known carcinogen is first applied as an initiator, followed by application of the test chemical as a promoter. A positive control group in which the test animals are subjected to the same initiator under identical conditions is maintained throughout the duration of the experiment. The data are used only if positive and negative control groups are mentioned in the reference.

16. *Cited Reference* is the final entry of the irritation, mutation, reproductive, tumorigenic, and toxic dose data lines. This is the source from which the information was extracted. All references cited are publicly available. No governmental classified documents have been used for source information. All references have been given a unique six-letter CODEN character code (derived from the American Society for Testing and Materials *CODEN for Periodical Titles* and the CAS *Source Index*), which identifies periodicals, serial publications, and individual published works. For those references for which no CODEN was found, the corresponding six-letter code includes asterisks (*) in the last one or two positions following the first four or five letters of an acronym for the publication title. Following the CODEN designation (for most entries) are: the number of the volume, followed by a comma; the page number of the first page of the article, followed by a comma; and a two-digit number, indicating the year of publication in the twentieth century. When the cited reference is a report, the report number is listed. Where contributors have provided information on their unpublished studies, the CODEN consists of the first three letters of the last name, the initials of the first and middle names, and a number sign (#). The date of the letter supplying the information is listed. All CODEN acronyms are listed in alphabetical order and defined in the CODEN Section.

17. *Consensus Reports* lines supply additional information to enable the reader to make knowledgeable evaluations of potential chemical hazards. Two types of reviews are listed: (a) International Agency for Research on Cancer (IARC) monograph reviews, which are published by the United Nations World Health Organization (WHO) and (b) the National Toxicology Program (NTP).

*a. Cancer Reviews.* In the U.N. International Agency for Research on Cancer (IARC) monographs, information on suspected environmental carcinogens is examined, and summaries of available data with appropriate references are presented. Included in these reviews are synonyms, physical and chemical properties, uses and occurrence, and biological data relevant to the evaluation of carcinogenic risk to humans. The monographs in the series contain an evaluation of approximately 1200 materials. Single copies of the individual monographs (specify volume number) can be ordered from WHO Publications Centre USA, 49 Sheridan Avenue, Albany, NY 12210, telephone (518) 436-9686.

The format of the IARC data line is as follows. The entry "IARC Cancer Review:" indicates that the carcinogenicity data pertaining to a compound has been reviewed by the IARC committee. The committee's conclusions are summarized in three words. The first word indicates whether the data pertains to humans or to animals. The next two words indicate the degree of carcinogenic risk as defined by IARC.

For experimental animals the evidence of carcinogenicity is assessed by IARC and judged to fall into one of four groups defined as follows:

(1) Sufficient Evidence of carcinogenicity is provided when there is an increased incidence of malignant tumors: (a) in multiple species or strains; (b) in multiple experiments (preferably with different routes of administration or using different dose levels); or (c) to an unusual degree with regard to the incidence, site, or type of tumor, or age at onset. Additional evidence may be provided by data on dose-response effects.

(2) Limited Evidence of carcinogenicity is available when the data suggest a carcinogenic effect but are limited because: (a) the studies involve a single species, strain, or experiment; (b) the experiments are restricted by inadequate dosage levels, inadequate duration of exposure to the agent, inadequate period of follow-up, poor survival, the use of too few animals, or inadequate reporting; or (c) the neoplasms produced often occur spontaneously and, in the past, have been difficult to classify as malignant by histological criteria alone (for example, lung adenomas and adenocarcinomas, and liver tumors in certain strains of mice).

(3) Inadequate Evidence is available when, because of major qualitative or quantitative limitations, the studies cannot be interpreted as showing either the presence or absence of a carcinogenic effect.

(4) No Evidence applies when several adequate studies are available that show that within the limitations of the tests used, the chemical is not carcinogenic.

It should be noted that the categories *Sufficient Evidence* and *Limited Evidence* refer only to the strength of the experimental evidence that these chemicals are carcinogenic and not to the extent of their carcinogenic activity nor to the mechanism involved. The classification of any chemical may change as new information becomes available.

The evidence for carcinogenicity from studies in humans is assessed by the IARC committees and judged to fall into one of four groups defined as follows:

(1) Sufficient Evidence of carcinogenicity indicates that there is a causal relationship between the exposure and human cancer.

(2) Limited Evidence of carcinogenicity indicates that a causal relationship is credible, but that alternative explanations, such as chance, bias, or confounding, could not adequately be excluded.

(3) Inadequate Evidence, which applies to both positive and negative evidence, indicates that one of two conditions prevailed: (a) there are few pertinent data; or (b) the available studies, while showing evidence of association, do not exclude chance, bias, or confounding.

(4) No Evidence applies when several adequate studies are available that do not show evidence of carcinogenicity.

This cancer review reflects only the conclusion of the IARC committee based on the data available for the committee's evaluation. Hence, for some substances there may be a disparity between the IARC determination and the information on the tumorigenic data lines (see paragraph 15). Also, some substances previously reviewed by IARC may be reexamined as additional data become available. These substances will contain multiple IARC review lines, each of which is referenced to the applicable IARC monograph volume.

An IARC entry indicates that some carcinogenicity data pertaining to a compound have been reviewed by the IARC committee. It indicates whether the data pertain to humans or to animals and whether the results of the determination are positive, suspected, indefinite, or negative, or whether there are no data.

This cancer review reflects only the conclusion of the IARC committee, based on the data available at the time of the committee's evaluation. Hence, for some materials there may be disagreement between the IARC determination and the tumorigenicity information in the toxicity data lines.

*b. NTP Status.* The notation "NTP 7th Annual Report On Carcinogens" indicated that the entry is listed on the seventh report made to the U.S. Congress by the National Toxicology Program (NTP) as required by law. This listing implies that the entry is assumed to be a human carcinogen.

Another NTP notation indicates that the material has been tested by the NTP under its Carcinogenesis Testing Program. These entries are also identified as National Cancer Institute (NCI), which reported the studies before the NCI Carcinogenesis Testing Program was absorbed by NTP. To obtain additional information about NTP, the Carcinogenesis Testing Program, or the status of a particular material under test, contact the Toxicology Information and Scientific Evaluation Group, NTP/TRTP/NIEHS, Mail Drop 18-01, P.O. Box 12233, Research Triangle Park, NC 27709.

*c. EPA Extremely Hazardous Substances List.* This list was developed by the U.S. Environmental Protection Agency (EPA) as required by the Superfund Amendments and Reauthorization Act of 1986 (SARA). Title III Section 304 requires notification by facilities of a release of certain extremely hazardous substances. These 402 substances were listed by EPA in the *Federal Register* of November 17, 1986.

*d. Community Right-to-Know List.* This list was developed by the EPA as required by the Superfund Amendments and Reauthorization Act of 1986 (SARA). Title III, Sections 311-312 require manufacturing facilities to prepare Material Safety Data Sheets and notify local authorities of the presence of listed chemicals. Both specific chemicals and classes of chemicals are covered by these sections.

*e. EPA Genetic Toxicology Program (GENE-TOX).* This status line indicates that the material has had genetic effects reported in the literature during the period 1969-1979. The test protocol in the literature is evaluated by an EPA expert panel on mutations, and the positive or negative genetic effect of the substance is reported. To obtain additional information about this program, contact GENE-TOX program, USEPA, 401 M Street, SW, TS796, Washington, DC 20460, Telephone (202) 260-1513.

*f. EPA TSCA Status Line.* This line indicates that the material appears on the chemical inventory prepared by the Environmental Protection Agency in accordance with provisions of the Toxic Substances Control Act (TSCA). Materials reported in the inventory include those that are produced commercially in or are imported into this country. The reader should note, however, that materials already regulated by EPA under FIFRA and by the Food and Drug Administration under the Food, Drug, and Cosmetic Act, as amended, are not included in the TSCA inventory. Similarly, alcohol, tobacco, and explosive materials are not regulated under TSCA. TSCA regulations should be consulted for an exact definition of reporting requirements. For additional information about TSCA, contact EPA, Office of Toxic Substances, Washington, DC 20402. Specific questions about the inventory can be directed to the EPA Office of Industry Assistance, telephone (800) 424-9065.

18. *Standards and Recommendations* section contains regulations by agencies of the U.S. government or recommendations by expert groups. "OSHA" refers to standards promulgated under Section 6 of the Occupational Safety and Health Act of 1970. "DOT" refers to materials regulated for shipment by the Department of Transportation. Because of frequent changes to and litigation of federal regulations, it is recommended that the reader contact the applicable agency for information about the current standards for a particular material. Omission of a material or regulatory notation from this edition does not imply any relief from regulatory responsibility.

*a. OSHA Air Contaminant Standards.* The values given are for the revised standards that were published in January 13, 1989 and were scheduled to take effect from September 1, 1989 through December 31, 1992. These are noted with the entry "OSHA PEL:" followed by "TWA" or "CL," meaning either time-weighted average or ceiling value, respectively, to which workers can be exposed for a normal 8-hour day, 40-hour work week without ill effects. For some materials, TWA, CL, and Pk (peak) values are given in the standard. In those cases, all three are listed. Finally, some entries may be followed by the designation "(skin)." This designation indicates that the compound may be absorbed by the skin and that, even though the air concentration may be below the standard, significant additional exposure through the skin may be possible.

*b. ACGIH Threshold Limit Values.* The American Conference of Governmental Industrial Hygienists (ACGIH) Threshold Limit Values are noted with the entry "ACGIH TLV:" followed by "TWA" or "CL," meaning either time-weighted average or ceiling value, respectively, to which workers can be exposed for a normal 8-hour day, 40-hour work week without ill effects. The notation "CL" indicates a ceiling limit that must not be exceeded. The notation "skin" indicates that the material penetrates intact skin, and skin contact should be avoided even though the TLV concentration is not exceeded. STEL indicates a short-term exposure limit, usually a 15-minute time-weighted average, which should not be exceeded. Biological Exposure Indices (*BEI:*) are, according to the ACGIH, set to provide a warning level ". . .of biological response to the chemical, or warning levels of that chemical or its metabolic product(s) in tissues, fluids, or exhaled air of exposed workers...."

The latest annual TLV list is contained in the publication *Threshold Limit Values and Biological Exposure Indices.* This publication should be consulted for future trends in recommendations. The ACGIH TLVs are adopted in whole or in part by many countries and local administrative agencies throughout the world. As a result, these recommendations have a major effect on the control of workplace contaminant concentrations. The ACGIH may be contacted for additional information at Kemper Woods Center, 1330 Kemper Meadow Drive, Cincinnati, OH 45240.

*c. DFG MAK.* These lines contain the German Research Society's Maximum Allowable Concentration values. Those materials that are classified as to workplace hazard potential by the German Research Society are noted on this line. The MAK values are also revised annually and discussions of materials under consideration for MAK assignment are included in the annual publication together with the current values. *BAT:* indicates Biological Tolerance Value for a Working Material, which is defined as, ". . .the maximum permissible quantity of a chemical compound, its metabolites, or any deviation from the norm of biological parameters induced by these substances in exposed humans." *TRK:* values are Technical Guiding Concentrations for workplace control of carcinogens. For additional information, write to Deutsche Forschungsgemeinschaft (German Research Society), Kennedyallee 40, D-5300 Bonn 2, Federal Republic of Germany. The publication *Maximum Concentrations at the Workplace and Biological Tolerance Values for Working Materials Report No. 29* can be obtained from VCH Publishers, Inc., 303 N.W. 12th Ave, Deerfield Beach, FL 33442-1788 or Verlag Chemie GmbH, Buchauslieferung, P.O. Box 1260/1280, D-6940 Weinheim, Federal Republic of Germany.

*d. NIOSH REL.* This line indicates that a NIOSH criteria document recommending a certain occupational exposure has been published for this compound or for a class of compounds to which this material belongs. These documents contain extensive data, analysis, and references. The more recent publications can be obtained from the National Institute for Occupational Safety and Health, U.S. Department of Health and Human Services, 4676 Columbia Pkwy., Cincinnati, OK 45226.

*e. DOT Classification.* This is the hazard classification according to the U.S. Department of Transportation (DOT) or the International Maritime Organization (IMO). This classification gives an indication of the hazards expected in transportation, and serves as a guide to the development of proper labels, placards, and shipping instructions. The basic hazard classes include compressed gases, flammables, oxidizers, corrosives, explosives, radioactive materials, and poisons. Although a material may be designated by only one hazard class, additional hazards may be indicated by adding labels or by using other means as directed by DOT. Many materials are regulated under general headings such as "pesticides" or "combustible liquids" as defined in the regulations. These are not noted here, as their specific concentration or properties must be known for proper classification. Special regulations may govern shipment by air. This information should serve *only as a guide,* because the regulation of transported materials is carefully controlled in most countries by federal and local agencies. Because there are frequent changes to regulations, it is recommended that the reader contact the applicable agency for information about the current standards for a particular material. United States transportation regulations are found in 40 CFR, Parts 100 to 189. Contact the U.S. Department of Transportation, Materials Transportation Bureau, Washington, DC 20590.

19. *Safety Profiles* are text summaries of the reported hazards of the entry. The word "experimental" indicates that the reported effects resulted from a controlled exposure of laboratory animals to the substance. Toxic effects reported include carcinogenic, reproductive, acute lethal, and human nonlethal effects, skin and eye irritation, and positive mutation study results.

Human effects are identified either by *human* or more specifically by *man, woman, child,* or *infant.* Specific symptoms or organ systems effects are reported when available.

Carcinogenicity potential is denoted by the words "confirmed," "suspected," or "questionable." The substance entries are grouped into three classes based on experimental evidence and the opinion of expert review groups. The OSHA, IARC, ACGIH, and DFG MAK decision schedules are not related or synchronized. Thus, an entry may have had a recent review by only one group. The most stringent classification of any regulation or expert group is taken as governing.

Class I—Confirmed Carcinogens
These substances are capable of causing cancer in exposed humans. An entry was assigned to this class if it had one or more of the following data items present:
a. an OSHA regulated carcinogen
b. an ACGIH assignment as a human or animal carcinogen
c. a DFG MAK assignment as a confirmed human or animal carcinogen
d. an IARC assignment of human or animal sufficient evidence of carcinogenicity, or higher
e. NTP 7th Annual Report On Carcinogens

Class II—Suspected Carcinogens
These substances may be capable of causing cancer in exposed humans. The evidence is suggestive, but not sufficient to convince expert review committees. Some entries have not yet had expert review, but contain experimental reports of carcinogenic activity. In particular, an entry is included if it has positive reports of carcinogenic endpoint in two species. As more studies are published, many Class II carcinogens will have their carcinogenicity confirmed. On the other hand, some will be judged noncarcinogenic in the future. An entry was assigned to this class if it had one or more of the following data items present:
a. an ACGIH assignment of suspected carcinogen
b. a DFG MAK assignment of suspected carcinogen
c. an IARC assignment of human or animal limited evidence
d. two animal studies reporting positive carcinogenic endpoint in different species

Class III—Questionable Carcinogens
For these entries there is minimal published evidence of possible carcinogenic activity. The reported endpoint is often neoplastic growth with no spread or invasion characteristic of carcinogenic pathology. An even weaker endpoint is that of *equivocal tumorigenic agent* (ETA). Reports are assigned this designation when the study was defective. The study may have lacked control animals, may have used a very small sample size, often lack complete pathology reporting, or suffer many other study design defects. Many of these studies were designed for other than carcinogenic evaluation, and the reported carcinogenic effect is a byproduct of the study, not the goal. The data are presented because some of the substances studied may be carcinogens.

There is insufficient data to affirm or deny the possibility. An entry was assigned to this class if it had one or more of the following data items present:
a. an IARC assignment of inadequate or no evidence
b. a single human report of carcinogenicity
c. a single experimental carcinogenic report, or duplicate reports in the same species
d. one or more experimental neoplastic or equivocal tumorigenic agent report

Fire and explosion hazards are briefly summarized in terms of conditions of flammable or reactive hazard. Materials that are incompatible with the entry are listed here. Fire and explosion hazards are briefly summarized in terms of conditions of flammable or reactive hazard. Fire-fighting materials and methods are discussed where feasible. A material with a flash point of 100°F or less is considered dangerous; if the flash point is from 100 to 200°F, the flammability is considered moderate; if it is above 200°F, the flammability is considered low (the material is considered combustible).

Also included in the safety profile are disaster hazards comments, which serve to alert users of materials, safety professionals, researchers, supervisors, and firefighters to the dangers that may be encountered on entering storage premises during a fire or other emergency. Although the presence of water, steam, acid fumes, or powerful vibrations can cause many materials to decompose into dangerous compounds, we are particularly concerned with high temperatures (such as those resulting from a fire) because these can cause many otherwise inert chemicals to emit highly toxic gases or vapors such as $NO_x$, $SO_x$, acids, and so forth, or evolve vapors of antimony, arsenic, mercury, and the like.

The Safety Profile concludes with the OSHA and NIOSH occupational analytical methods, referenced by method name or number. The OSHA Manual of Analytical Methods can be ordered form the ACGIH, Kemper Woods Center, 1330 Kemper Meadow Drive, Cincinnati, OH 45240. The NIOSH Manual of Analytical Methods is available from NIOSH Publications Office, 4676 Columbia Parkway, Cincinnati, OH 45226.

# Sax's
# Dangerous
# Properties of
# Industrial Materials

## Ninth Edition

## Volume I
## Indexes

# DOT Guide Number Cross-Index

| | | | | |
|---|---|---|---|---|
| NA 0349 see GJU600 | NA 9206 see MOB399 | UN 0490 see NMP620 | UN 1086 see VNP000 | UN 1176 see BMC250 |
| NA 0473 see BAO900 | NA 9260 see AGX000 | UN 1001 see ACI750 | UN 1087 see MQL750 | UN 1178 see DHI000 |
| NA 0473 see LDP000 | NA 9269 see TLB750 | UN 1002 see AFG250 | UN 1088 see AAG000 | UN 1179 see EHA500 |
| NA 0473 see MMP000 | UN 0004 see ANS500 | UN 1003 see AFG250 | UN 1089 see AAG250 | UN 1180 see EHE000 |
| NA 1051 see HHS000 | UN 0027 see ERF500 | UN 1005 see AMY500 | UN 1090 see ABC750 | UN 1181 see EHG500 |
| NA 1247 see MLH750 | UN 0027 see PLL750 | UN 1006 see AQW250 | UN 1091 see ABC750 | UN 1182 see EHK500 |
| NA 1361 see CDI000 | UN 0028 see ERF500 | UN 1008 see BMG700 | UN 1092 see ADR000 | UN 1183 see DFK000 |
| NA 1361 see CDI250 | UN 0028 see PLL750 | UN 1009 see TJY100 | UN 1093 see ADX500 | UN 1184 see EIY600 |
| NA 1361 see CDJ000 | UN 0072 see CPR800 | UN 1010 see BOP100 | UN 1098 see AFV500 | UN 1185 see EJM900 |
| NA 1463 see CMK000 | UN 0074 see DUR800 | UN 1011 see BOR500 | UN 1099 see AFY000 | UN 1188 see EJH500 |
| NA 1463 see DUR800 | UN 0075 see DJE400 | UN 1013 see CBU250 | UN 1100 see AGB250 | UN 1189 see EJJ500 |
| NA 1499 see SIT750 | UN 0076 see DUY600 | UN 1014 see CBV250 | UN 1104 see AOD725 | UN 1190 see EKL000 |
| NA 1549 see AQE000 | UN 0079 see HET500 | UN 1015 see CBV000 | UN 1104 see AOD735 | UN 1192 see LAJ000 |
| NA 1549 see AQK000 | UN 0081 see DYG000 | UN 1016 see CBW750 | UN 1106 see AOJ000 | UN 1193 see MKA400 |
| NA 1556 see DFP200 | UN 0110 see GJU600 | UN 1017 see CDV750 | UN 1106 see PBV500 | UN 1194 see ENN000 |
| NA 1557 see ARI000 | UN 0114 see TEF500 | UN 1018 see CFX500 | UN 1108 see AOI800 | UN 1195 see EPB500 |
| NA 1557 see ARJ100 | UN 0118 see CPR800 | UN 1020 see CJI500 | UN 1110 see MGN500 | UN 1196 see EPY500 |
| NA 1574 see CAM300 | UN 0129 see LCM000 | UN 1022 see CLR250 | UN 1112 see AOL250 | UN 1198 see FMV000 |
| NA 1574 see CAM500 | UN 0130 see LEE000 | UN 1023 see HHJ500 | UN 1114 see BBL250 | UN 1199 see FPQ875 |
| NA 1649 see TCF000 | UN 0133 see MAW250 | UN 1026 see COO000 | UN 1123 see BPU750 | UN 1201 see FQT000 |
| NA 1665 see NMS000 | UN 0135 see MDC000 | UN 1027 see CQD750 | UN 1123 see BPV000 | UN 1203 see GBY000 |
| NA 1707 see TEL750 | UN 0143 see NGY000 | UN 1028 see DFA600 | UN 1123 see BPV100 | UN 1204 see NGY000 |
| NA 1760 see TGH250 | UN 0144 see NGY000 | UN 1029 see DFL000 | UN 1125 see BPX750 | UN 1206 see HBC500 |
| NA 1778 see SCO500 | UN 0146 see NMB000 | UN 1032 see DOQ800 | UN 1126 see BMX500 | UN 1207 see HEM000 |
| NA 1807 see PHS250 | UN 0147 see NMQ500 | UN 1033 see MJW500 | UN 1128 see BRK000 | UN 1208 see HEN000 |
| NA 1811 see PKU250 | UN 0151 see PBT050 | UN 1035 see EDZ000 | UN 1129 see BSU250 | UN 1212 see IIL000 |
| NA 1829 see SOR500 | UN 0153 see PIC800 | UN 1036 see EFU400 | UN 1130 see CBB500 | UN 1213 see IIJ000 |
| NA 1831 see SOI520 | UN 0154 see PID000 | UN 1037 see EHH000 | UN 1131 see CBV500 | UN 1214 see IIM000 |
| NA 1911 see DDI450 | UN 0155 see TML325 | UN 1038 see EIO000 | UN 1134 see CEJ125 | UN 1218 see IMS000 |
| NA 1967 see PAK230 | UN 0208 see TEG250 | UN 1039 see EMT000 | UN 1135 see EIU800 | UN 1219 see INJ000 |
| NA 1986 see AFJ000 | UN 0209 see TMN490 | UN 1040 see EJN500 | UN 1136 see CMY900 | UN 1220 see INE100 |
| NA 1987 see AFJ000 | UN 0214 see TMK500 | UN 1041 see EJO000 | UN 1143 see COB250 | UN 1221 see INK000 |
| NA 1993 see DHE800 | UN 0215 see TML000 | UN 1045 see FEZ000 | UN 1144 see COC500 | UN 1222 see IQP000 |
| NA 1993 see DHE900 | UN 0219 see SMP500 | UN 1046 see HAM500 | UN 1145 see CPB000 | UN 1223 see KEK000 |
| NA 1993 see FOP000 | UN 0220 see UTJ000 | UN 1048 see HHJ000 | UN 1146 see CPV750 | UN 1228 see CPW300 |
| NA 1999 see ARO500 | UN 0222 see ANN000 | UN 1049 see HHW500 | UN 1147 see DAE800 | UN 1228 see FPM000 |
| NA 2212 see ARM250 | UN 0224 see BAI000 | UN 1050 see HHL000 | UN 1148 see DBF750 | UN 1228 see HBD500 |
| NA 2212 see ARM260 | UN 0226 see CQH250 | UN 1052 see HHU500 | UN 1149 see BRH750 | UN 1228 see HES000 |
| NA 2212 see ARM268 | UN 0234 see SGP550 | UN 1053 see HIC500 | UN 1150 see DFH800 | UN 1228 see IMU000 |
| NA 2212 see ARM275 | UN 0235 see PIC500 | UN 1055 see IIC000 | UN 1152 see DFX000 | UN 1228 see LBX000 |
| NA 2212 see ARM280 | UN 0236 see PIC750 | UN 1060 see MFX600 | UN 1153 see EJE500 | UN 1228 see PBM000 |
| NA 2215 see MAK900 | UN 0282 see NHA500 | UN 1061 see MGC250 | UN 1154 see DHJ200 | UN 1228 see PML500 |
| NA 2672 see ANK250 | UN 0284 see GJU600 | UN 1063 see MIF765 | UN 1155 see EJU000 | UN 1228 see TGO750 |
| NA 2762 see AFK250 | UN 0285 see GJU600 | UN 1064 see MLE650 | UN 1156 see DJN750 | UN 1228 see TGP000 |
| NA 2783 see DXH325 | UN 0292 see GJU600 | UN 1065 see NCG500 | UN 1157 see DNI800 | UN 1228 see TGP250 |
| NA 2783 see MNH000 | UN 0293 see GJU600 | UN 1066 see NGP500 | UN 1158 see DNM200 | UN 1229 see MDJ750 |
| NA 2783 see PAK000 | UN 0318 see NGU500 | UN 1067 see NGU500 | UN 1159 see IOZ750 | UN 1230 see MGB150 |
| NA 2783 see TCF250 | UN 0340 see CCU250 | UN 1069 see NMH000 | UN 1160 see DOQ800 | UN 1231 see MFW100 |
| NA 2783 see TIQ250 | UN 0341 see CCU250 | UN 1070 see NGU000 | UN 1161 see MIF000 | UN 1233 see HFJ000 |
| NA 2809 see MCW250 | UN 0342 see CCU250 | UN 1072 see OQW000 | UN 1162 see DFE259 | UN 1234 see MGA850 |
| NA 2821 see PDN750 | UN 0343 see CCU250 | UN 1073 see OQW000 | UN 1163 see DSF400 | UN 1235 see MGC250 |
| NA 2845 see MOC000 | UN 0372 see GJU600 | UN 1075 see LGM000 | UN 1164 see TFP000 | UN 1237 see MHY000 |
| NA 2920 see DEV200 | UN 0385 see NFJ000 | UN 1076 see PGX000 | UN 1165 see DVQ000 | UN 1238 see MIG000 |
| NA 2922 see SHR000 | UN 0391 see CPR800 | UN 1077 see PMO500 | UN 1166 see DVR800 | UN 1239 see CIO250 |
| NA 2927 see EOP600 | UN 0394 see SMP500 | UN 1079 see SOH500 | UN 1167 see VOP000 | UN 1242 see DFS000 |
| NA 2927 see EOR000 | UN 0402 see PCD500 | UN 1080 see SOI000 | UN 1170 see EFU000 | UN 1243 see MKG750 |
| NA 2949 see SHR000 | UN 0411 see PBC250 | UN 1081 see TCH500 | UN 1171 see EES350 | UN 1244 see MKN000 |
| NA 3018 see MNH000 | UN 0452 see GJU600 | UN 1082 see CLQ750 | UN 1172 see EES400 | UN 1245 see HFG500 |
| NA 3018 see TCF250 | UN 0483 see CPR800 | UN 1083 see TLD500 | UN 1173 see EFR000 | UN 1246 see MKY500 |
| NA 9163 see ZTJ000 | UN 0484 see CQH250 | UN 1085 see VMP000 | UN 1175 see EGP500 | UN 1248 see MOT000 |
| NA 9202 see CBW750 | | | | |

| | | | | |
|---|---|---|---|---|
| UN 1249 see PBN250 | UN 1366 see DKE600 | UN 1498 see SIO900 | UN 1617 see LCK000 | UN 1702 see TBP750 |
| UN 1250 see MQC500 | UN 1369 see DSY600 | UN 1500 see SIQ500 | UN 1617 see LCK100 | UN 1704 see SOD100 |
| UN 1251 see BOY500 | UN 1376 see IHG100 | UN 1502 see PCE750 | UN 1618 see LCL000 | UN 1705 see TCF260 |
| UN 1255 see NAI500 | UN 1380 see PAT750 | UN 1503 see SJC000 | UN 1620 see LCU000 | UN 1708 see TGQ500 |
| UN 1256 see NAI500 | UN 1381 see PHP010 | UN 1504 see SJC500 | UN 1621 see LIC000 | UN 1708 see TGQ750 |
| UN 1257 see GBY000 | UN 1382 see PLT250 | UN 1505 see SJE000 | UN 1622 see ARD000 | UN 1708 see TGR000 |
| UN 1259 see NCZ000 | UN 1384 see SHR500 | UN 1506 see SMF500 | UN 1623 see MDF350 | UN 1709 see TGL750 |
| UN 1261 see NHM500 | UN 1385 see SJY500 | UN 1507 see SMK000 | UN 1624 see MCY475 | UN 1710 see TIO750 |
| UN 1262 see OCU000 | UN 1396 see AGX000 | UN 1509 see SMK500 | UN 1625 see MDF000 | UN 1712 see ZDS000 |
| UN 1264 see PAI250 | UN 1397 see AHE750 | UN 1510 see TDY250 | UN 1626 see PLU500 | UN 1713 see ZGA000 |
| UN 1265 see EIK000 | UN 1400 see BAH250 | UN 1511 see HIB500 | UN 1627 see MDE750 | UN 1714 see ZLS000 |
| UN 1265 see PBK250 | UN 1401 see CAL250 | UN 1512 see ZDA000 | UN 1629 see MCS750 | UN 1715 see AAX500 |
| UN 1267 see PCR250 | UN 1402 see CAN750 | UN 1513 see ZES000 | UN 1629 see MDE250 | UN 1716 see ACD750 |
| UN 1268 see PCS250 | UN 1403 see CAQ250 | UN 1514 see ZJJ000 | UN 1630 see MCW500 | UN 1717 see ACF750 |
| UN 1270 see NAI500 | UN 1407 see CDC000 | UN 1515 see ZLA000 | UN 1631 see MCX500 | UN 1718 see ADF250 |
| UN 1271 see PCT250 | UN 1408 see FBG000 | UN 1516 see ZLJ000 | UN 1636 see MDA250 | UN 1722 see AGB500 |
| UN 1272 see PIH750 | UN 1410 see LHS000 | UN 1517 see PIC750 | UN 1637 see MDC500 | UN 1723 see AGI250 |
| UN 1274 see PND000 | UN 1411 see LHS000 | UN 1541 see MLC750 | UN 1638 see MDC750 | UN 1724 see AGU250 |
| UN 1275 see PMT750 | UN 1413 see LHT000 | UN 1545 see AGJ250 | UN 1639 see MCV250 | UN 1725 see AGX750 |
| UN 1276 see PNC250 | UN 1414 see LHH000 | UN 1546 see DCG800 | UN 1640 see MDF250 | UN 1726 see AGY750 |
| UN 1277 see PND250 | UN 1415 see LGO000 | UN 1547 see AOQ000 | UN 1641 see MCT500 | UN 1727 see ANJ000 |
| UN 1278 see CKP750 | UN 1417 see LHP000 | UN 1548 see BBL000 | UN 1642 see MDA500 | UN 1728 see PBY750 |
| UN 1279 see PNJ400 | UN 1418 see MAC750 | UN 1550 see AQE250 | UN 1643 see NCP500 | UN 1729 see AOY250 |
| UN 1280 see PNL600 | UN 1419 see AHD250 | UN 1551 see AQG250 | UN 1644 see MCU000 | UN 1730 see AQD000 |
| UN 1281 see PNM500 | UN 1420 see PKT250 | UN 1553 see ARB250 | UN 1646 see MCU250 | UN 1731 see AQD000 |
| UN 1282 see POP250 | UN 1422 see PLS500 | UN 1554 see ARB250 | UN 1647 see BNM750 | UN 1732 see AQF250 |
| UN 1288 see COD750 | UN 1423 see RPA000 | UN 1555 see ARF250 | UN 1648 see ABE500 | UN 1733 see AQC500 |
| UN 1289 see SIK450 | UN 1426 see SFF500 | UN 1558 see ARA750 | UN 1650 see NBE500 | UN 1736 see BDM500 |
| UN 1292 see EPF550 | UN 1427 see SHO500 | UN 1559 see ARH500 | UN 1651 see AQN635 | UN 1737 see BEC000 |
| UN 1294 see TGK750 | UN 1428 see see500 | UN 1560 see ARF500 | UN 1653 see NDB500 | UN 1738 see BEE375 |
| UN 1295 see TJD500 | UN 1431 see SIK450 | UN 1561 see ARI750 | UN 1654 see NDN000 | UN 1739 see BEF500 |
| UN 1296 see TJO000 | UN 1432 see SJI500 | UN 1562 see ARE500 | UN 1656 see NDP400 | UN 1741 see BMG500 |
| UN 1297 see TLD500 | UN 1433 see TGE500 | UN 1562 see ARE750 | UN 1657 see NDR000 | UN 1742 see BMG750 |
| UN 1298 see TLN250 | UN 1435 see ZBJ000 | UN 1565 see BAK750 | UN 1658 see NDR500 | UN 1744 see BMP000 |
| UN 1299 see TOD750 | UN 1436 see ZBJ000 | UN 1567 see BFO750 | UN 1659 see NDS500 | UN 1745 see BMQ000 |
| UN 1300 see TOD750 | UN 1437 see ZRA000 | UN 1569 see BNZ000 | UN 1660 see NEG100 | UN 1746 see BMQ325 |
| UN 1301 see VLU250 | UN 1438 see AHD750 | UN 1570 see BOL750 | UN 1661 see NEN500 | UN 1747 see BSR000 |
| UN 1302 see EQF500 | UN 1439 see ANB500 | UN 1571 see BAI000 | UN 1661 see NEO000 | UN 1748 see HOV500 |
| UN 1303 see VPK000 | UN 1442 see PCD500 | UN 1572 see HKC000 | UN 1661 see NEO500 | UN 1749 see CDX750 |
| UN 1304 see IJQ000 | UN 1444 see ANR000 | UN 1573 see ARB750 | UN 1662 see NEX000 | UN 1750 see CEA500 |
| UN 1305 see TIN750 | UN 1445 see BAJ500 | UN 1575 see CAQ500 | UN 1663 see NIE500 | UN 1751 see CEA000 |
| UN 1307 see XGS000 | UN 1446 see BAN250 | UN 1577 see CGL750 | UN 1663 see NIF000 | UN 1752 see CEC250 |
| UN 1309 see AGX000 | UN 1447 see PCD750 | UN 1578 see CJA950 | UN 1664 see NMO500 | UN 1753 see CKM250 |
| UN 1310 see ANS500 | UN 1448 see PCK000 | UN 1578 see CJB250 | UN 1664 see NMO525 | UN 1754 see CLG500 |
| UN 1312 see BMD000 | UN 1449 see BAO250 | UN 1578 see CJB750 | UN 1664 see NMO550 | UN 1755 see CMK000 |
| UN 1313 see CAW500 | UN 1451 see CDE250 | UN 1578 see NFS525 | UN 1669 see PAW500 | UN 1756 see CMJ530 |
| UN 1314 see CAW500 | UN 1452 see CAO500 | UN 1579 see CLK235 | UN 1670 see PCF300 | UN 1756 see CMJ560 |
| UN 1318 see CNE000 | UN 1453 see CAP000 | UN 1580 see CKN500 | UN 1671 see PDN750 | UN 1757 see CMJ530 |
| UN 1320 see DUY600 | UN 1454 see CAU000 | UN 1583 see CKN500 | UN 1672 see PFJ400 | UN 1757 see CMJ560 |
| UN 1326 see HAC000 | UN 1456 see CAV250 | UN 1584 see PIE500 | UN 1673 see PEY000 | UN 1758 see CML125 |
| UN 1328 see HEI500 | UN 1457 see CAV500 | UN 1586 see CNN500 | UN 1673 see PEY250 | UN 1759 see FBI000 |
| UN 1332 see TDW500 | UN 1463 see CMK000 | UN 1587 see CNL000 | UN 1673 see PEY500 | UN 1760 see FBI000 |
| UN 1334 see NAJ500 | UN 1466 see FAY200 | UN 1589 see COO750 | UN 1674 see ABU500 | UN 1761 see DBU800 |
| UN 1336 see NHA500 | UN 1466 see IHB900 | UN 1591 see DEP600 | UN 1677 see ARD250 | UN 1762 see CPE500 |
| UN 1337 see NMB000 | UN 1467 see GLA000 | UN 1592 see DEP800 | UN 1678 see PKV000 | UN 1763 see CPR250 |
| UN 1338 see PHO500 | UN 1469 see LDO000 | UN 1593 see MJP450 | UN 1679 see PLC175 | UN 1764 see DEL000 |
| UN 1339 see PHQ750 | UN 1470 see LDS499 | UN 1594 see DKB110 | UN 1680 see PLC500 | UN 1765 see DEN400 |
| UN 1340 see PHS000 | UN 1471 see LHJ000 | UN 1595 see DUD100 | UN 1683 see SDM100 | UN 1766 see DGF200 |
| UN 1341 see PHS500 | UN 1472 see LHO000 | UN 1597 see DUQ180 | UN 1684 see SDP000 | UN 1767 see DEY800 |
| UN 1343 see PHT750 | UN 1474 see MAH000 | UN 1597 see DUQ200 | UN 1685 see ARD750 | UN 1768 see PHF250 |
| UN 1344 see PID000 | UN 1475 see PCE000 | UN 1597 see DUQ400 | UN 1685 see SEY100 | UN 1769 see DFF000 |
| UN 1346 see SCP000 | UN 1476 see MAH750 | UN 1597 see DUQ600 | UN 1686 see SEY200 | UN 1770 see BNG750 |
| UN 1347 see PID200 | UN 1484 see PKY300 | UN 1599 see DUY600 | UN 1686 see SEY500 | UN 1770 see BNH000 |
| UN 1348 see SGP550 | UN 1485 see PLA250 | UN 1600 see PGN100 | UN 1687 see SFA000 | UN 1771 see DYA800 |
| UN 1349 see PIC500 | UN 1486 see PLL500 | UN 1603 see EGV000 | UN 1688 see HKC500 | UN 1773 see FAU000 |
| UN 1350 see SOD500 | UN 1487 see PLM000 | UN 1604 see EEA500 | UN 1689 see SGA500 | UN 1775 see FDD125 |
| UN 1354 see TMK500 | UN 1488 see PLM500 | UN 1605 see EIY500 | UN 1690 see SHF500 | UN 1775 see HHS600 |
| UN 1355 see TML000 | UN 1489 see PLO500 | UN 1606 see IGN000 | UN 1691 see SME500 | UN 1776 see PHJ250 |
| UN 1356 see TMN490 | UN 1490 see PLP000 | UN 1607 see IGO000 | UN 1692 see SMN500 | UN 1777 see FLZ000 |
| UN 1357 see UTJ000 | UN 1491 see PLP250 | UN 1608 see IGM000 | UN 1692 see SMO500 | UN 1779 see FNA000 |
| UN 1358 see ZOA000 | UN 1492 see DWQ000 | UN 1611 see HCY000 | UN 1695 see CDN200 | UN 1780 see FOY000 |
| UN 1360 see CAW250 | UN 1493 see SDS000 | UN 1613 see HHS000 | UN 1697 see CEA750 | UN 1781 see HCQ000 |
| UN 1361 see CBT500 | UN 1494 see SFG000 | UN 1614 see HHS000 | UN 1698 see PDB000 | UN 1782 see HDE000 |
| UN 1362 see CBT500 | UN 1495 see SFS000 | UN 1616 see LCV000 | UN 1699 see CGN000 | UN 1783 see HEO000 |
| UN 1363 see CNR000 | UN 1496 see SFT500 | UN 1617 see ARC750 | UN 1701 see XRS000 | UN 1784 see HFX500 |

| | | | | |
|---|---|---|---|---|
| UN **2451** see NGW000 | UN **2525** see DJT200 | UN **2657** see SBR000 | UN **2751** see DJW600 | UN **2948** see AID500 |
| UN **2453** see FIB000 | UN **2526** see FPW000 | UN **2658** see SBO500 | UN **2752** see EKM200 | UN **2950** see MAC750 |
| UN **2456** see CKS000 | UN **2527** see IIK000 | UN **2659** see SFU500 | UN **2761** see AFK250 | UN **2951** see OPE000 |
| UN **2457** see DQT400 | UN **2528** see IIW000 | UN **2661** see HCL500 | UN **2761** see DHB400 | UN **2952** see ASL750 |
| UN **2458** see HCQ600 | UN **2529** see IJU000 | UN **2662** see HIH000 | UN **2785** see MPV400 | UN **2956** see TML750 |
| UN **2463** see AHB500 | UN **2530** see IJW000 | UN **2664** see DDP800 | UN **2785** see TET900 | UN **2965** see BMH000 |
| UN **2464** see BFT000 | UN **2531** see MDN250 | UN **2666** see EHP500 | UN **2789** see AAT250 | UN **2966** see MCN250 |
| UN **2465** see DGN200 | UN **2535** see MMA250 | UN **2668** see CDN500 | UN **2790** see AAT250 | UN **2967** see SNK500 |
| UN **2466** see PLE260 | UN **2538** see NHP990 | UN **2670** see TJD750 | UN **2798** see DGE400 | UN **2968** see MAS500 |
| UN **2467** see SJB400 | UN **2541** see TBE000 | UN **2671** see AMI000 | UN **2799** see PFW200 | UN **2969** see CCP000 |
| UN **2468** see TIQ750 | UN **2542** see THX250 | UN **2671** see AMI250 | UN **2799** see PFW210 | UN **2970** see BBS300 |
| UN **2470** see PEA750 | UN **2545** see HAC000 | UN **2671** see AMI500 | UN **2802** see CNK500 | UN **2972** see DVF400 |
| UN **2471** see OKK000 | UN **2547** see SJZ100 | UN **2673** see CEH250 | UN **2803** see GBG000 | UN **2975** see TFS750 |
| UN **2473** see ARA500 | UN **2548** see CDX250 | UN **2674** see DXE000 | UN **2805** see LHH000 | UN **2976** see TFT500 |
| UN **2474** see TFN500 | UN **2552** see HDA000 | UN **2676** see SLQ000 | UN **2806** see LHM000 | UN **2977** see UOJ000 |
| UN **2475** see VEP000 | UN **2553** see NAI500 | UN **2677** see RPZ000 | UN **2812** see AHG000 | UN **2978** see UOJ000 |
| UN **2477** see ISE000 | UN **2554** see CIU750 | UN **2678** see RPZ000 | UN **2815** see AKB000 | UN **2979** see UNS000 |
| UN **2478** see AGJ000 | UN **2555** see CCU250 | UN **2679** see LHI100 | UN **2817** see ANJ000 | UN **2980** see URS000 |
| UN **2478** see BMT150 | UN **2556** see CCU250 | UN **2680** see LHI100 | UN **2818** see ANT000 | UN **2981** see URA200 |
| UN **2478** see CHL250 | UN **2557** see CCU250 | UN **2681** see CDD750 | UN **2819** see PBW750 | UN **2984** see HIB005 |
| UN **2478** see CKA750 | UN **2558** see BNI000 | UN **2682** see CDD750 | UN **2820** see BSW000 | UN **2984** see HIB010 |
| UN **2478** see CKB000 | UN **2564** see TII250 | UN **2684** see DIY800 | UN **2822** see CKW000 | UN **2989** see LCV100 |
| UN **2478** see COI250 | UN **2565** see DGT600 | UN **2685** see DJI400 | UN **2823** see COB500 | UN **3022** see BOX750 |
| UN **2478** see FHC200 | UN **2567** see SJA000 | UN **2686** see DHO500 | UN **2826** see CLJ750 | UN **3023** see MKJ250 |
| UN **2478** see FLR100 | UN **2572** see PFI000 | UN **2687** see DGU200 | UN **2829** see HEU000 | UN **3023** see OFE030 |
| UN **2478** see IKG800 | UN **2573** see TEJ100 | UN **2688** see BNA825 | UN **2830** see LHK000 | UN **3054** see CPB625 |
| UN **2478** see IKH000 | UN **2574** see TNP500 | UN **2689** see CDT750 | UN **2831** see MIH275 | UN **3055** see AJU250 |
| UN **2478** see IKH099 | UN **2576** see PHU000 | UN **2692** see BMG400 | UN **2835** see SEM500 | UN **3056** see HBB500 |
| UN **2478** see OBG000 | UN **2579** see PIJ000 | UN **2698** see TDB000 | UN **2837** see SEG800 | UN **3057** see TJX500 |
| UN **2478** see TGM750 | UN **2580** see AGX750 | UN **2699** see TKA250 | UN **2838** see VNF000 | UN **3064** see NGY000 |
| UN **2478** see THY750 | UN **2581** see AGY750 | UN **2708** see MHV750 | UN **2839** see AAH750 | UN **3071** see CPW300 |
| UN **2478** see TKJ250 | UN **2582** see FAU000 | UN **2710** see DWT600 | UN **2840** see BSU500 | UN **3071** see FPM000 |
| UN **2478** see XSS260 | UN **2587** see QQS200 | UN **2713** see ADJ500 | UN **2841** see DCH200 | UN **3071** see HBD500 |
| UN **2480** see MKX250 | UN **2591** see XDS000 | UN **2716** see BST500 | UN **2842** see NFY500 | UN **3071** see HES000 |
| UN **2481** see ELS500 | UN **2599** see FOO515 | UN **2717** see CBA750 | UN **2845** see EOQ000 | UN **3071** see IMU000 |
| UN **2482** see PNP000 | UN **2603** see COY000 | UN **2719** see BAI750 | UN **2849** see CKP725 | UN **3071** see LBX000 |
| UN **2485** see BRQ500 | UN **2606** see MPI750 | UN **2720** see CMJ600 | UN **2851** see BMG800 | UN **3071** see PBM000 |
| UN **2487** see PFK250 | UN **2607** see ADR500 | UN **2721** see CNJ900 | UN **2853** see MAF600 | UN **3071** see PML500 |
| UN **2488** see CPN500 | UN **2608** see NIY000 | UN **2723** see MAE000 | UN **2854** see COE000 | UN **3071** see TGO750 |
| UN **2489** see MJP400 | UN **2610** see THN000 | UN **2724** see MAS900 | UN **2855** see ZIA000 | UN **3071** see TGP000 |
| UN **2490** see BII250 | UN **2611** see CKR500 | UN **2725** see NDG000 | UN **2858** see ZOA000 | UN **3071** see TGP250 |
| UN **2491** see EEC600 | UN **2612** see MOU830 | UN **2726** see NDG550 | UN **2859** see ANY250 | UN **3080** see AGJ000 |
| UN **2493** see HDG000 | UN **2614** see IMW000 | UN **2727** see TEK750 | UN **2862** see VDU000 | UN **3080** see BMT150 |
| UN **2495** see IDT000 | UN **2615** see EPC125 | UN **2728** see ZSA000 | UN **2864** see PLK900 | UN **3080** see CHL250 |
| UN **2496** see PMV500 | UN **2616** see IOI000 | UN **2729** see HCC500 | UN **2865** see OLS000 | UN **3080** see CKA750 |
| UN **2497** see SJF000 | UN **2617** see MIQ745 | UN **2733** see AOE200 | UN **2869** see TGG250 | UN **3080** see CKB000 |
| UN **2498** see FNK025 | UN **2618** see VQK650 | UN **2733** see BPY000 | UN **2870** see AHG875 | UN **3080** see COI250 |
| UN **2501** see TND250 | UN **2619** see DQP800 | UN **2733** see BPY250 | UN **2871** see AQB250 | UN **3080** see FHC200 |
| UN **2502** see VBA000 | UN **2621** see ABB500 | UN **2733** see DAG600 | UN **2872** see DDL800 | UN **3080** see FLR100 |
| UN **2503** see ZPA000 | UN **2622** see GGW000 | UN **2733** see HBL600 | UN **2873** see DDU600 | UN **3080** see IKG800 |
| UN **2505** see ANH250 | UN **2626** see CDU000 | UN **2733** see HFK000 | UN **2874** see FPU000 | UN **3080** see IKH000 |
| UN **2506** see ANJ500 | UN **2628** see PLG000 | UN **2733** see OEK010 | UN **2875** see HCL000 | UN **3080** see IKH099 |
| UN **2507** see CKO750 | UN **2629** see SHG500 | UN **2733** see PBV505 | UN **2876** see REA000 | UN **3080** see OBG000 |
| UN **2508** see MRD500 | UN **2630** see SJT500 | UN **2734** see AOE200 | UN **2877** see ISR000 | UN **3080** see TGM750 |
| UN **2509** see PKX750 | UN **2642** see FIC000 | UN **2734** see BPY000 | UN **2879** see SBT500 | UN **3080** see THY750 |
| UN **2511** see CKS750 | UN **2643** see MHR250 | UN **2734** see BPY250 | UN **2906** see IMG000 | UN **3080** see TKJ250 |
| UN **2512** see ALT000 | UN **2644** see MKW200 | UN **2734** see DAG600 | UN **2907** see CCK125 | UN **3080** see XIJ000 |
| UN **2512** see ALT250 | UN **2646** see HCE500 | UN **2734** see HBL600 | UN **2931** see VEZ000 | UN **3080** see XSS260 |
| UN **2514** see PEO500 | UN **2647** see MAO250 | UN **2734** see HFK000 | UN **2933** see CKT000 | UN **3083** see PCF750 |
| UN **2515** see BNL000 | UN **2648** see NIE600 | UN **2734** see PBV505 | UN **2936** see TFK250 | UN **3136** see CBY750 |
| UN **2516** see CBX750 | UN **2649** see BIK250 | UN **2738** see BQH850 | UN **2937** see PDE000 | UN **3149** see PCL500 |
| UN **2517** see CFX250 | UN **2650** see DFU000 | UN **2739** see BSW550 | UN **2938** see MHA750 | |
| UN **2521** see KFA000 | UN **2651** see MJQ000 | UN **2740** see PNH000 | UN **2943** see TCS500 | |
| UN **2522** see DPG600 | UN **2655** see PLH750 | UN **2746** see CBX109 | UN **2945** see MHV000 | |
| UN **2524** see ENY500 | UN **2656** see QMJ000 | UN **2750** see DGG400 | UN **2946** see ALT500 | |

# SECTION 2
# CAS Number
# Cross-Index

| | | | | |
|---|---|---|---|---|
| 50-00-0 see FMV000 | 51-03-6 see PIX250 | 52-62-0 see PBT000 | 54-96-6 see DCD000 | 56-95-1 see CDT125 |
| 50-01-1 see GKY000 | 51-05-8 see AIT250 | 52-66-4 see PAP500 | 55-03-8 see LFG050 | 56-97-3 see TLQ500 |
| 50-02-2 see SOW000 | 51-06-9 see AJN500 | 52-67-5 see MCR750 | 55-18-5 see NJW500 | 56-99-5 see CCK650 |
| 50-03-3 see HHQ800 | 51-12-7 see BET000 | 52-68-6 see TIQ250 | 55-21-0 see BBB000 | 57-06-7 see AGJ250 |
| 50-04-4 see CNS825 | 51-15-0 see FNZ000 | 52-76-6 see NNV000 | 55-22-1 see ILC000 | 57-09-0 see HCQ500 |
| 50-06-6 see EOK000 | 51-17-2 see BCB750 | 52-85-7 see FAB600 | 55-27-6 see NNP050 | 57-10-3 see PAE250 |
| 50-07-7 see AHK500 | 51-18-3 see TND500 | 52-86-8 see CLY500 | 55-31-2 see AES500 | 57-11-4 see SLK000 |
| 50-09-9 see ERE000 | 51-20-7 see BOL000 | 52-88-0 see MGR500 | 55-37-8 see DST200 | 57-12-5 see COI500 |
| 50-10-2 see ORQ000 | 51-21-8 see FMM000 | 52-89-1 see CQK250 | 55-38-9 see FAQ900 | 57-13-6 see USS000 |
| 50-11-3 see DJO800 | 51-28-5 see DUZ000 | 52-90-4 see CQK000 | 55-43-6 see DCR200 | 57-14-7 see DSF400 |
| 50-12-4 see MKB250 | 51-30-9 see IMR000 | 53-03-2 see PLZ000 | 55-48-1 see ARR500 | 57-15-8 see ABD000 |
| 50-13-5 see DAM700 | 51-34-3 see SBG000 | 53-06-5 see CNS800 | 55-51-6 see BIA750 | 57-22-7 see LEY000 |
| 50-14-6 see VSZ100 | 51-40-1 see NNO699 | 53-10-1 see VJZ000 | 55-52-7 see PDN000 | 57-24-9 see SMN500 |
| 50-18-0 see CQC650 | 51-41-2 see NNO500 | 53-16-7 see EDV000 | 55-55-0 see MGJ750 | 57-27-2 see MRO500 |
| 50-19-1 see HNJ000 | 51-42-3 see AES000 | 53-19-0 see CDN000 | 55-56-1 see BIM250 | 57-29-4 see NAG500 |
| 50-21-5 see LAG000 | 51-43-4 see VGP000 | 53-21-4 see CNF000 | 55-57-2 see PEO000 | 57-30-7 see SID000 |
| 50-23-7 see CNS750 | 51-44-5 see DER600 | 53-36-1 see DAZ117 | 55-63-0 see NGY000 | 57-33-0 see NBU000 |
| 50-24-8 see PMA000 | 51-45-6 see HGD000 | 53-39-4 see AOO125 | 55-65-2 see GKQ000 | 57-37-4 see BCA000 |
| 50-27-1 see EDU500 | 51-46-7 see TBJ000 | 53-43-0 see AOO450 | 55-68-5 see MCU750 | 57-39-6 see TNK250 |
| 50-28-2 see EDO000 | 51-48-9 see TFZ275 | 53-46-3 see DJM800 | 55-80-1 see DUH600 | 57-41-0 see DKQ000 |
| 50-29-3 see DAD200 | 51-50-3 see DCT050 | 53-46-3 see XCJ000 | 55-81-2 see MFC500 | 57-42-1 see DAM600 |
| 50-31-7 see TIK500 | 51-52-5 see PNX000 | 53-59-8 see CNF400 | 55-86-7 see BIE500 | 57-43-2 see AMX750 |
| 50-32-8 see BCS750 | 51-55-8 see ARR000 | 53-60-1 see PMI500 | 55-91-4 see IRF000 | 57-44-3 see BAG000 |
| 50-33-9 see BRF500 | 51-56-9 see HGH150 | 53-69-0 see DQI600 | 55-93-6 see DSU000 | 57-47-6 see PIA500 |
| 50-34-0 see HKR500 | 51-57-0 see MDT600 | 53-70-3 see DCT400 | 55-97-0 see HEA000 | 57-50-1 see SNH000 |
| 50-35-1 see TEH500 | 51-58-1 see NIJ400 | 53-79-2 see AEI000 | 55-98-1 see BOT250 | 57-52-3 see BLN500 |
| 50-36-2 see CNE750 | 51-60-5 see DQY909 | 53-84-9 see CNF390 | 56-04-2 see MPW500 | 57-53-4 see MQU750 |
| 50-37-3 see DJO000 | 51-61-6 see DYC400 | 53-86-1 see IDA000 | 56-10-0 see AJY250 | 57-55-6 see PML000 |
| 50-39-5 see HNY500 | 51-62-7 see BBK750 | 53-89-4 see BCP650 | 56-12-2 see PIM500 | 57-56-7 see HGU000 |
| 50-41-9 see CMX700 | 51-63-8 see BBK500 | 53-94-1 see HIU500 | 56-17-7 see CQJ750 | 57-57-8 see PMT100 |
| 50-44-2 see POK000 | 51-64-5 see AOA500 | 53-95-2 see HIP000 | 56-18-8 see AIX250 | 57-62-5 see CMA750 |
| 50-47-5 see DSI709 | 51-67-2 see TOG250 | 53-96-3 see FDR000 | 56-23-5 see CBY000 | 57-63-6 see EEH500 |
| 50-48-6 see EAH500 | 51-68-3 see DPE000 | 54-04-6 see MDI500 | 56-24-6 see TMI250 | 57-64-7 see PIA750 |
| 50-49-7 see DLH600 | 51-71-8 see PFC500 | 54-05-7 see CLD000 | 56-25-7 see CBE750 | 57-66-9 see DWW000 |
| 50-50-0 see EDP000 | 51-73-0 see DWX600 | 54-06-8 see AES639 | 56-29-1 see ERD500 | 57-67-0 see AHO250 |
| 50-52-2 see MOO250 | 51-74-1 see HGE000 | 54-11-5 see NDN000 | 56-34-8 see TCC250 | 57-68-1 see SNJ000 |
| 50-53-3 see CKP250 | 51-75-2 see BIE250 | 54-12-6 see TNW500 | 56-35-9 see BLL750 | 57-71-6 see OMY910 |
| 50-54-4 see QHA000 | 51-77-4 see GDG200 | 54-16-0 see HLJ000 | 56-36-0 see TIC000 | 57-74-9 see CDR750 |
| 50-55-5 see RDK000 | 51-78-5 see ALU500 | 54-21-7 see SJO000 | 56-37-1 see BFL300 | 57-83-0 see PMH500 |
| 50-56-6 see ORU500 | 51-79-6 see UVA000 | 54-25-1 see RJA000 | 56-38-2 see PAK000 | 57-85-2 see TBG000 |
| 50-59-9 see TEY000 | 51-80-9 see TDR750 | 54-30-8 see NOC000 | 56-40-6 see GHA000 | 57-88-5 see CMD750 |
| 50-60-2 see PDW400 | 51-82-1 see DRQ650 | 54-31-9 see CHJ750 | 56-45-1 see SCA355 | 57-91-0 see EDO500 |
| 50-62-4 see HFF500 | 51-83-2 see CBH250 | 54-35-3 see PAQ200 | 56-47-3 see DAQ800 | 57-92-1 see SLW500 |
| 50-63-5 see CLD250 | 51-84-3 see CMF250 | 54-36-4 see MCJ370 | 56-49-5 see MIJ750 | 57-94-3 see TOA000 |
| 50-65-7 see DFV400 | 51-85-4 see MCN500 | 54-42-2 see DAS000 | 56-53-1 see DKA600 | 57-95-4 see TNY750 |
| 50-67-9 see AJX500 | 51-93-4 see EQC600 | 54-47-7 see PII100 | 56-54-2 see QFS000 | 57-96-5 see DWM000 |
| 50-70-4 see SKV200 | 51-98-9 see ABU000 | 54-49-9 see HNB875 | 56-55-3 see BBC250 | 57-97-6 see DQJ200 |
| 50-71-5 see AFT750 | 52-01-7 see AFJ500 | 54-62-6 see AMG750 | 56-57-5 see NJF000 | 58-00-4 see AQP250 |
| 50-76-0 see AEB000 | 52-21-1 see SOV100 | 54-64-8 see MDI000 | 56-65-5 see ARQ500 | 58-08-2 see CAK500 |
| 50-78-2 see ADA725 | 52-24-4 see TFQ750 | 54-71-7 see PIF250 | 56-69-9 see HOO100 | 58-13-9 see PDC875 |
| 50-79-3 see DER400 | 52-26-6 see MRO750 | 54-77-3 see DTO000 | 56-72-4 see CNU750 | 58-14-0 see TGD000 |
| 50-81-7 see ARN000 | 52-28-8 see CNG500 | 54-80-8 see INS000 | 56-75-7 see CDP250 | 58-15-1 see DOT000 |
| 50-84-0 see DER100 | 52-31-3 see TDA500 | 54-84-2 see CMP900 | 56-81-5 see GGA000 | 58-18-4 see MPN500 |
| 50-89-5 see TFX790 | 52-43-7 see AFS500 | 54-85-3 see ILD000 | 56-82-6 see GFY200 | 58-20-8 see TBF600 |
| 50-90-8 see CFJ750 | 52-46-0 see AQO000 | 54-86-4 see NDW500 | 56-84-8 see ARN850 | 58-22-0 see TBF500 |
| 50-91-9 see DAR400 | 52-49-3 see BBV000 | 54-87-5 see SEE250 | 56-85-9 see GFO050 | 58-25-3 see LFK000 |
| 50-98-6 see EAW500 | 52-51-7 see BNT250 | 54-88-6 see TLE750 | 56-86-0 see GFO000 | 58-27-5 see MMD500 |
| 50-98-6 see EAX000 | 52-52-8 see AJK250 | 54-91-1 see BHJ250 | 56-89-3 see CQK325 | 58-28-6 see DLS600 |
| 50-99-7 see GFG000 | 52-53-9 see IRV000 | 54-92-2 see ILE000 | 56-92-8 see HGD500 | 58-32-2 see PCP250 |
| 51-02-5 see INT000 | 52-60-8 see DJR800 | 54-95-5 see PBI500 | 56-93-9 see BFM250 | 58-33-3 see PMI750 |

| | | | | |
|---|---|---|---|---|
| 84-80-0 see VTA000 | 87-44-5 see CCN000 | 89-65-6 see SAA025 | 91-62-3 see MPF800 | 93-56-1 see SMQ100 |
| 84-86-6 see ALI000 | 87-47-8 see PPQ625 | 89-68-9 see CLJ800 | 91-63-4 see QEJ000 | 93-58-3 see MHA750 |
| 84-89-9 see ALI240 | 87-48-9 see BNL750 | 89-69-0 see TIT750 | 91-64-5 see CNV000 | 93-59-4 see PCM000 |
| 84-96-8 see AFL500 | 87-51-4 see ICN000 | 89-72-5 see BSE000 | 91-66-7 see DIS700 | 93-60-7 see NDV000 |
| 84-97-9 see PCK500 | 87-52-5 see DYC000 | 89-73-6 see SAL500 | 91-71-4 see TFD750 | 93-62-9 see HKM500 |
| 85-01-8 see PCW250 | 87-56-9 see MRU900 | 89-78-1 see MCF750 | 91-75-8 see PDC000 | 93-65-2 see CIR500 |
| 85-22-3 see PAT850 | 87-59-2 see XMJ000 | 89-80-5 see MCG275 | 91-79-2 see DPJ200 | 93-68-5 see ABA000 |
| 85-31-4 see TFJ500 | 87-60-5 see CLK200 | 89-81-6 see MCF250 | 91-80-5 see TEO250 | 93-69-6 see TGX550 |
| 85-32-5 see GLS750 | 87-61-6 see TIK100 | 89-82-7 see POI615 | 91-81-6 see TMP750 | 93-70-9 see AAY600 |
| 85-34-7 see TIY500 | 87-62-7 see XNJ000 | 89-83-8 see TFX810 | 91-84-9 see WAK000 | 93-71-0 see CFK000 |
| 85-36-9 see AAM875 | 87-63-8 see CLK227 | 89-84-9 see DMG400 | 91-85-0 see NCD500 | 93-72-1 see TIX500 |
| 85-40-5 see TDB100 | 87-65-0 see DFY000 | 89-93-0 see MHM510 | 91-88-3 see HKS100 | 93-76-5 see TAA100 |
| 85-41-6 see PHX000 | 87-66-1 see PPQ500 | 89-94-1 see MPO000 | 91-93-0 see DCJ400 | 93-78-7 see TGF210 |
| 85-43-8 see TDB000 | 87-68-3 see HCD250 | 89-98-5 see CEI500 | 91-94-1 see DEQ600 | 93-79-8 see BSQ750 |
| 85-44-9 see PHW750 | 87-69-4 see TAF750 | 90-00-6 see PGR250 | 91-95-2 see BGK500 | 93-80-1 see TIW750 |
| 85-52-9 see BDL850 | 87-76-3 see TLN150 | 90-02-8 see SAG000 | 91-97-4 see DQS000 | 93-89-0 see EGR000 |
| 85-60-9 see BRP750 | 87-85-4 see HEC000 | 90-03-9 see CHW675 | 91-99-6 see DHF400 | 93-90-3 see MKQ250 |
| 85-68-7 see BEC500 | 87-86-5 see PAX250 | 90-04-0 see AOV900 | 92-00-2 see CJU200 | 93-91-4 see BDJ800 |
| 85-70-1 see BQP750 | 87-87-6 see TBQ500 | 90-05-1 see GKI000 | 92-04-6 see CHN500 | 93-92-5 see SMP600 |
| 85-71-2 see MOD000 | 87-90-1 see TIQ750 | 90-11-9 see BNS200 | 92-06-8 see TBC620 | 93-99-2 see PEL500 |
| 85-73-4 see PHY750 | 87-99-0 see XPJ000 | 90-12-0 see MMB750 | 92-09-1 see AJY750 | 94-01-9 see BHB100 |
| 85-79-0 see DDT200 | 88-04-0 see CLW000 | 90-13-1 see CIZ000 | 92-13-7 see PIF000 | 94-02-0 see EGR600 |
| 85-82-5 see FAG080 | 88-05-1 see TLG500 | 90-15-3 see NAW500 | 92-15-9 see ABA500 | 94-04-2 see VOU000 |
| 85-83-6 see SBC500 | 88-06-2 see TIW000 | 90-16-4 see BDH000 | 92-23-9 see LET000 | 94-07-5 see HLV500 |
| 85-84-7 see FAG130 | 88-09-5 see DHI400 | 90-17-5 see TIT000 | 92-24-0 see NAI000 | 94-09-7 see EFX000 |
| 85-86-9 see OHI200 | 88-10-8 see DIW400 | 90-20-0 see AKH000 | 92-26-2 see DBT200 | 94-11-1 see IOY000 |
| 85-91-6 see MGQ250 | 88-12-0 see EEG000 | 90-22-2 see VBK000 | 92-30-8 see TKE775 | 94-13-3 see HNU500 |
| 85-98-3 see DJC400 | 88-14-2 see FQF000 | 90-27-7 see PEP250 | 92-31-9 see AJP250 | 94-14-4 see MOT750 |
| 86-00-0 see NFP500 | 88-15-3 see ABI500 | 90-30-2 see PFT250 | 92-36-4 see MHJ300 | 94-15-5 see DNY000 |
| 86-13-5 see BDI000 | 88-18-6 see BSE460 | 90-33-5 see MKP500 | 92-39-7 see CJL100 | 94-17-7 see BHM750 |
| 86-21-5 see TMJ750 | 88-19-7 see TGN200 | 90-34-6 see PMC300 | 92-43-3 see PDM500 | 94-20-2 see CKK000 |
| 86-26-0 see PEG000 | 88-21-1 see SNO100 | 90-39-1 see SKX500 | 92-44-4 see NAN510 | 94-24-6 see BQA010 |
| 86-29-3 see DVX200 | 88-24-4 see MJN250 | 90-41-5 see BGE250 | 92-48-8 see MIP750 | 94-25-7 see BPZ000 |
| 86-30-6 see DWI000 | 88-26-6 see IFX200 | 90-42-6 see LBX100 | 92-49-9 see EHJ600 | 94-26-8 see BSC000 |
| 86-34-0 see MNZ000 | 88-27-7 see DEA100 | 90-43-7 see BGJ250 | 92-52-4 see BGE000 | 94-28-0 see FCD560 |
| 86-35-1 see EOL100 | 88-27-7 see FAB000 | 90-45-9 see AHS500 | 92-53-5 see PFS750 | 94-30-4 see AOV000 |
| 86-40-8 see XAK000 | 88-29-9 see ACL750 | 90-46-0 see XBJ000 | 92-54-6 see PFX000 | 94-35-9 see PFJ000 |
| 86-50-0 see ASH500 | 88-32-4 see BQI000 | 90-47-1 see XBS000 | 92-62-6 see BDS000 | 94-36-0 see BDS600 |
| 86-52-2 see CIP750 | 88-35-7 see AOS500 | 90-49-3 see PFB350 | 92-64-8 see CPJ500 | 94-41-7 see CDH000 |
| 86-54-4 see HGP495 | 88-41-5 see BQW490 | 90-50-6 see CMQ100 | 92-67-1 see AJS100 | 94-44-0 see NCR040 |
| 86-55-5 see NAV490 | 88-44-8 see AKQ000 | 90-64-2 see MAP000 | 92-69-3 see BGJ500 | 94-46-2 see IHP100 |
| 86-56-6 see DSU400 | 88-45-9 see PEY800 | 90-65-3 see PAP750 | 92-70-6 see HMX520 | 94-47-3 see PFB750 |
| 86-57-7 see NHQ000 | 88-51-7 see AJJ250 | 90-69-7 see LHY000 | 92-71-7 see DWI200 | 94-51-9 see DWS800 |
| 86-60-2 see ALI300 | 88-58-4 see DEC800 | 90-72-2 see TNH000 | 92-81-9 see ADI775 | 94-52-0 see NFD500 |
| 86-65-7 see NBE850 | 88-60-8 see BQV600 | 90-81-3 see EAW100 | 92-82-0 see PDB500 | 94-58-6 see DMD600 |
| 86-72-6 see CBN100 | 88-61-9 see XJJ000 | 90-82-4 see POH000 | 92-83-1 see XAT000 | 94-59-7 see SAD000 |
| 86-73-7 see FDI100 | 88-62-0 see AMT000 | 90-87-9 see HII600 | 92-84-2 see PDP250 | 94-62-2 see PIV600 |
| 86-74-8 see CBN000 | 88-63-1 see PFA250 | 90-89-1 see DIW000 | 92-87-5 see BBX000 | 94-63-3 see POS750 |
| 86-85-1 see MLH000 | 88-64-2 see AHQ300 | 90-94-8 see MQS500 | 92-88-6 see BGG500 | 94-67-7 see SAG500 |
| 86-86-2 see NAK000 | 88-67-5 see IEE000 | 90-98-2 see DES000 | 92-91-1 see MHP500 | 94-70-2 see PDK819 |
| 86-87-3 see NAK500 | 88-68-6 see AID620 | 91-01-0 see HKF300 | 92-92-2 see PEL600 | 94-74-6 see CIR250 |
| 86-88-4 see AQN635 | 88-69-7 see IQX100 | 91-02-1 see PGE760 | 92-93-3 see NFQ000 | 94-75-7 see DAA800 |
| 86-92-0 see THA300 | 88-72-2 see NMO525 | 91-04-3 see HLV100 | 92-94-4 see TBC750 | 94-76-8 see CLO600 |
| 86-93-1 see PGJ750 | 88-73-3 see CJB750 | 91-08-7 see TGM800 | 93-01-6 see HMU500 | 94-78-0 see PEK250 |
| 86-95-3 see QNA000 | 88-74-4 see NEO000 | 91-10-1 see DOJ200 | 93-05-0 see DJV200 | 94-80-4 see BQZ000 |
| 86-96-4 see QEJ800 | 88-75-5 see NIE500 | 91-15-6 see PHY000 | 93-07-2 see VHP600 | 94-81-5 see CLN750 |
| 86-97-5 see ALJ500 | 88-82-4 see TKQ250 | 91-16-7 see DOA200 | 93-08-3 see ABC500 | 94-82-6 see DGA000 |
| 86-98-6 see DGJ250 | 88-84-6 see GKO000 | 91-17-8 see DAE800 | 93-09-4 see NAV500 | 94-86-0 see IRY000 |
| 87-01-4 see DPJ800 | 88-85-7 see BRE500 | 91-19-0 see QRJ000 | 93-10-7 see QEA000 | 94-87-1 see CHO125 |
| 87-02-5 see AKI000 | 88-89-1 see PID000 | 91-20-3 see NAJ500 | 93-14-1 see RLU000 | 94-91-7 see DWS400 |
| 87-08-1 see PDT500 | 88-96-0 see BBO500 | 91-21-4 see TCU500 | 93-15-2 see AGE250 | 94-93-9 see DWY200 |
| 87-09-2 see AGK250 | 88-99-3 see PHW250 | 91-22-5 see QMJ000 | 93-16-3 see IKR000 | 94-96-2 see EKV000 |
| 87-10-5 see THW750 | 89-02-1 see DUR400 | 91-23-6 see NER000 | 93-17-4 see VIK100 | 95-01-2 see REF100 |
| 87-11-6 see ABI250 | 89-05-4 see PPQ630 | 91-33-8 see BDE250 | 93-18-5 see EEY500 | 95-04-5 see EHP000 |
| 87-12-7 see BOD600 | 89-19-0 see BQX250 | 91-38-3 see CIB700 | 93-19-6 see IJM000 | 95-06-7 see CDO250 |
| 87-13-8 see EEV200 | 89-25-8 see NNT000 | 91-40-7 see PEG500 | 93-23-2 see LBW000 | 95-08-9 see TJQ250 |
| 87-17-2 see SAH500 | 89-28-1 see SAU480 | 91-44-1 see DIL400 | 93-28-7 see EQS000 | 95-13-6 see IBX000 |
| 87-18-3 see BSH100 | 89-32-7 see PPQ635 | 91-49-6 see BPU500 | 93-29-8 see AAX750 | 95-14-7 see BDH250 |
| 87-19-4 see IJN000 | 89-37-2 see DVB850 | 91-51-0 see LFT100 | 93-40-3 see HGK600 | 95-16-9 see BDE500 |
| 87-25-2 see EGM000 | 89-52-1 see AAJ150 | 91-53-2 see SAV000 | 93-45-8 see NBF500 | 95-19-2 see HAT500 |
| 87-29-6 see API750 | 89-55-4 see BOE500 | 91-56-5 see ICR000 | 93-46-9 see NBL000 | 95-21-6 see MHK500 |
| 87-31-0 see DUR800 | 89-57-6 see AMM500 | 91-57-6 see MMC000 | 93-51-6 see MEK250 | 95-24-9 see AJE750 |
| 87-33-2 see CCK125 | 89-58-7 see NMS520 | 91-58-7 see CJA000 | 93-53-8 see COF000 | 95-25-0 see CDQ750 |
| 87-39-8 see AFU000 | 89-61-2 see DFT400 | 91-59-8 see NBE500 | 93-54-9 see EGQ000 | 95-26-1 see DQO800 |
| 87-42-3 see CKV500 | 89-63-4 see KDA050 | 91-60-1 see NAP500 | 93-55-0 see EOL500 | 95-29-4 see DNN900 |

95-30-7 see BDF250
95-31-8 see BQK750
95-32-9 see BDF750
95-33-0 see CPI250
95-35-2 see BHA500
95-38-5 see AHP500
95-39-6 see BFY250
95-41-0 see HFO700
95-45-4 see DBH000
95-46-5 see BOG260
95-47-6 see XHJ000
95-48-7 see CNX000
95-49-8 see CLK100
95-50-1 see DEP600
95-51-2 see CEH670
95-52-3 see FLZ100
95-53-4 see TGQ750
95-54-5 see PEY250
95-55-6 see ALT000
95-56-7 see BNV000
95-57-8 see CJK250
95-63-6 see TLL750
95-64-7 see XNS000
95-65-8 see XLJ000
95-68-1 see XMS000
95-69-2 see CLK220
95-70-5 see TGM000
95-71-6 see MKO250
95-73-8 see DGM700
95-74-9 see CLK215
95-76-1 see DEO300
95-77-2 see DFY425
95-78-3 see XNA000
95-79-4 see CLK225
95-80-7 see TGL750
95-82-9 see DEO295
95-83-0 see CFK125
95-85-2 see CEH250
95-86-3 see DCA200
95-87-4 see XKS000
95-88-5 see CLD750
95-92-1 see DJT200
95-93-2 see TDM750
95-94-3 see TBN750
95-95-4 see TIV750
96-05-9 see AGK500
96-08-2 see LFV000
96-09-3 see EBR000
96-10-6 see DHI885
96-11-7 see GGG000
96-12-8 see DDL800
96-13-9 see DDS000
96-14-0 see MNI500
96-17-3 see MJX500
96-18-4 see TJB600
96-19-5 see TJC000
96-20-8 see AJA250
96-21-9 see DDR800
96-22-0 see DJN750
96-23-1 see DGG400
96-24-2 see CDT750
96-27-5 see MRM750
96-29-7 see EMU500
96-31-1 see DUM200
96-32-2 see MHR250
96-33-3 see MGA500
96-34-4 see MIF775
96-37-7 see MIU500
96-40-2 see CFI625
96-45-7 see IAQ000
96-47-9 see MPO500
96-48-0 see BOV000
96-49-1 see GHM000
96-50-4 see AMS250
96-53-7 see TFS250

96-54-8 see MPB000
96-64-0 see SKS500
96-66-2 see TFD000
96-67-3 see HMY500
96-69-5 see TFC600
96-73-1 see CJB825
96-75-3 see NEP500
96-76-4 see DEG000
96-80-0 see DNP000
96-83-3 see IFY100
96-84-4 see IFZ800
96-88-8 see SBB000
96-91-3 see DUP400
96-93-5 see AKI250
96-96-8 see MFB000
96-99-1 see CJC500
97-00-7 see CGM000
97-02-9 see DUP600
97-05-2 see SOC500
97-06-3 see MMH400
97-11-0 see POO000
97-16-5 see DFY400
97-17-6 see DFK600
97-18-7 see TFD250
97-23-4 see MJM500
97-36-9 see OOI100
97-39-2 see DXP200
97-41-6 see EHM100
97-42-7 see CCM750
97-44-9 see ABX500
97-45-0 see MCD000
97-52-9 see NEQ000
97-53-0 see EQR500
97-54-1 see IKQ000
97-56-3 see AIC250
97-61-0 see MQJ750
97-62-1 see ELS000
97-63-2 see EMF000
97-64-3 see LAJ000
97-72-3 see IJW000
97-74-5 see BJL600
97-77-8 see DXH250
97-81-4 see ENZ000
97-84-7 see TDN000
97-85-8 see IIW000
97-86-9 see IIY000
97-88-1 see MHU750
97-90-5 see BKM250
97-93-8 see TJN750
97-94-9 see TJP250
97-95-0 see EGW000
97-96-1 see DHI000
97-99-4 see TCT000
98-00-0 see FPU000
98-01-1 see FPQ875
98-02-2 see FPM000
98-03-3 see TFM500
98-04-4 see TMB750
98-05-5 see BBL750
98-06-6 see BQJ750
98-07-7 see BFL250
98-08-8 see BDH500
98-09-9 see BBS750
98-10-2 see BBR500
98-11-3 see BBS250
98-12-4 see CPR250
98-13-5 see TJA750
98-14-6 see PDO250
98-16-8 see AID500
98-17-9 see TKE750
98-18-0 see MDM760
98-19-1 see DQU800
98-27-1 see BRU800
98-29-3 see BSK000
98-44-2 see AIE000

98-46-4 see NFJ500
98-47-5 see NFB500
98-50-0 see ARA250
98-51-1 see BSP500
98-52-2 see BQW000
98-53-3 see BQW250
98-54-4 see BSE500
98-55-5 see TBD750
98-56-6 see CEM825
98-57-7 see CKG750
98-59-9 see TGO250
98-60-2 see CEK375
98-64-6 see CEK000
98-67-9 see HNL600
98-71-5 see PFI500
98-72-6 see NIJ500
98-73-7 see BQK500
98-77-1 see PIY500
98-80-6 see BBM000
98-82-8 see COE750
98-83-9 see MPK250
98-84-0 see ALW250
98-85-1 see PDE000
98-86-2 see ABH000
98-87-3 see BAY300
98-88-4 see BDM500
98-89-5 see HDH100
98-91-9 see TFC550
98-92-0 see NCR000
98-94-2 see DRF709
98-95-3 see NEX000
98-96-4 see POL500
98-98-6 see PIB930
99-03-6 see AHR500
99-04-7 see TGP750
99-05-8 see AIH500
99-06-9 see HJI100
99-08-1 see NMO500
99-09-2 see NEN500
99-10-5 see REF200
99-11-6 see DMV400
99-24-1 see MKI100
99-28-5 see DDQ500
99-30-9 see RDP300
99-35-4 see TMK500
99-42-3 see HMY075
99-45-6 see MGC350
99-48-9 see MKY250
99-49-0 see MCD250
99-50-3 see POE200
99-53-6 see NFV010
99-54-7 see DFT600
99-55-8 see NMP500
99-56-9 see ALL500
99-57-0 see NEM500
99-59-2 see NEQ500
99-60-5 see CJC250
99-61-6 see NEV000
99-62-7 see DNN829
99-63-8 see IMO000
99-65-0 see DUQ200
99-66-1 see PNR750
99-69-4 see NHI600
99-71-8 see BSE250
99-72-9 see THD750
99-73-0 see DDJ600
99-75-2 see MPX850
99-76-3 see HJL500
99-77-4 see ENO000
99-79-6 see ELQ500
99-80-9 see MJG750
99-83-2 see MCC000
99-85-4 see MCB750
99-86-5 see MLA250
99-87-6 see CQI000

99-89-8 see IQZ000
99-91-2 see CEB250
99-92-3 see AHR240
99-93-4 see HIO000
99-94-5 see TGQ250
99-96-7 see SAI500
99-97-8 see TLG150
99-98-9 see DTL600
99-99-0 see NMO550
100-00-5 see NFS525
100-01-6 see NEO500
100-02-7 see NIF000
100-03-8 see CEJ600
100-06-1 see MDW750
100-07-2 see AOY250
100-09-4 see AOU600
100-10-7 see DOT400
100-14-1 see NFN400
100-15-2 see MMF800
100-16-3 see NIR000
100-17-4 see NER500
100-18-5 see DNN830
100-19-6 see NEL600
100-20-9 see TAV250
100-21-0 see TAN750
100-22-1 see BJF500
100-25-4 see DUQ600
100-29-8 see NID000
100-33-4 see DBM000
100-34-5 see BBN500
100-35-6 see CGV500
100-36-7 see DJI400
100-37-8 see DHO500
100-38-9 see DIY600
100-39-0 see BEC000
100-40-3 see CPD750
100-41-4 see EGP500
100-42-5 see SMQ000
100-43-6 see VQK590
100-44-7 see BEE375
100-45-8 see CPC625
100-47-0 see BCQ250
100-49-2 see HDH200
100-50-5 see FNK025
100-51-6 see BDX500
100-52-7 see BAY500
100-53-8 see TGO750
100-54-9 see NDW515
100-55-0 see NDW510
100-56-1 see PFM500
100-57-2 see PFN100
100-58-3 see PFL600
100-60-7 see MIT000
100-61-8 see MGN750
100-63-0 see PFI000
100-64-1 see HLI500
100-65-2 see PFJ250
100-66-3 see AOX750
100-68-5 see TFC250
100-69-6 see VQK560
100-72-5 see MDS500
100-73-2 see ADR500
100-74-3 see ENL000
100-75-4 see NLJ500
100-79-8 see DVR600
100-85-6 see BFM500
100-86-7 see DQQ200
100-88-9 see CPQ625
100-89-0 see TKM250
100-97-0 see HEI500
100-99-2 see TKR500
101-00-8 see TKT200
101-01-9 see TMS500
101-02-0 see TMU250
101-05-3 see DEV800

101-07-5 see DJK200
101-08-6 see DVO819
101-10-0 see CJQ300
101-14-4 see MJM200
101-20-2 see TIL500
101-21-3 see CKC000
101-25-7 see DVF400
101-26-8 see MDL600
101-27-9 see CEW500
101-31-5 see HOU000
101-33-7 see TJM000
101-37-1 see THN500
101-38-2 see CHR000
101-39-3 see MIO000
101-40-6 see PNN400
101-41-7 see MHA500
101-42-8 see DTP400
101-48-4 see PDX000
101-49-5 see BEN250
101-50-8 see AJS500
101-54-2 see PFU500
101-59-7 see AOS750
101-61-1 see MJN000
101-67-7 see DVK400
101-68-8 see MJP400
101-70-2 see BKO600
101-71-3 see PFR325
101-72-4 see PFL000
101-73-5 see HKF000
101-75-7 see PEI800
101-76-8 see BIM800
101-77-9 see MJQ000
101-79-1 see CEH125
101-80-4 see OPM000
101-82-6 see BFG600
101-83-7 see DGT600
101-84-8 see PFA850
101-85-9 see AOH000
101-86-0 see HFO500
101-87-1 see PET000
101-90-6 see REF000
101-92-8 see AAY250
101-93-9 see BJO500
101-96-2 see DEG200
101-97-3 see EOH000
101-99-5 see CBL750
102-01-2 see AAY000
102-06-7 see DWC600
102-07-8 see CBM250
102-08-9 see DWN800
102-09-0 see DVZ000
102-17-0 see APE000
102-19-2 see IHV000
102-20-5 see PDI000
102-22-7 see GDM400
102-24-9 see TKZ100
102-28-3 see AHQ000
102-29-4 see RDZ900
102-36-3 see IKH099
102-50-1 see MGO500
102-54-5 see FBC000
102-56-7 see AKD925
102-60-3 see QAT000
102-62-5 see DBF600
102-67-0 see TMY100
102-69-2 see TMY250
102-70-5 see THN000
102-71-6 see TKP500
102-76-1 see THM500
102-77-2 see BDG000
102-79-4 see BQM000
102-81-8 see DDU600
102-82-9 see THX250
102-83-0 see DDV200
102-85-2 see TIA750

| | | | | |
|---|---|---|---|---|
| 102-96-5 see NMC100 | 104-53-0 see HHP000 | 105-91-9 see NCP000 | 107-11-9 see AFW000 | 108-42-9 see CEH675 |
| 102-98-7 see PFP250 | 104-54-1 see CMQ740 | 105-95-3 see EJQ500 | 107-12-0 see PMV750 | 108-43-0 see CJK500 |
| 103-00-4 see CPG700 | 104-55-2 see CMP969 | 105-99-7 see AEO750 | 107-13-1 see ADX500 | 108-44-1 see TGQ500 |
| 103-03-7 see CBL000 | 104-57-4 see CBL000 | 106-11-6 see HKJ000 | 107-14-2 see CDN500 | 108-45-2 see PEY000 |
| 103-05-9 see BEC250 | 104-60-9 see PFP100 | 106-14-9 see HOG000 | 107-15-3 see EEA500 | 108-46-3 see REA000 |
| 103-07-1 see MNT000 | 104-61-0 see CNF250 | 106-19-4 see DWQ875 | 107-16-4 see HIM500 | 108-47-4 see LIY990 |
| 103-08-2 see ENW500 | 104-62-1 see PFC250 | 106-20-7 see DJA800 | 107-18-6 see AFV500 | 108-48-5 see LJA010 |
| 103-09-3 see OEE000 | 104-64-3 see HHQ000 | 106-21-8 see DTE600 | 107-19-7 see PMN450 | 108-50-9 see DTU800 |
| 103-11-7 see ADU250 | 104-65-4 see CMR500 | 106-22-9 see CMT250 | 107-20-0 see CDY500 | 108-55-4 see GFU000 |
| 103-14-0 see BDY750 | 104-67-6 see HBN200 | 106-22-9 see DTF410 | 107-21-1 see EJC500 | 108-57-6 see DXQ745 |
| 103-16-2 see AEY000 | 104-68-7 see PEQ750 | 106-23-0 see CMS845 | 107-22-2 see GIK000 | 108-58-7 see REA100 |
| 103-17-3 see CEP000 | 104-74-5 see DXY725 | 106-24-1 see DTD000 | 107-25-5 see MQL750 | 108-59-8 see DSM200 |
| 103-18-4 see AKE500 | 104-74-5 see LBX050 | 106-25-2 see DTD200 | 107-27-7 see CHC500 | 108-60-1 see BII250 |
| 103-23-1 see AEO000 | 104-75-6 see EKS500 | 106-27-4 see IHP400 | 107-29-9 see AAH250 | 108-62-3 see TDW500 |
| 103-24-2 see BJQ500 | 104-76-7 see EKQ000 | 106-29-6 see GDE810 | 107-30-2 see CIO250 | 108-64-5 see ISY000 |
| 103-26-4 see MIO500 | 104-78-9 see DIY800 | 106-30-9 see EKN050 | 107-31-3 see MKG750 | 108-65-6 see PNL265 |
| 103-27-5 see PFO000 | 104-82-5 see MHN300 | 106-31-0 see BSW550 | 107-35-7 see TAG750 | 108-67-8 see TLM050 |
| 103-28-6 see IJV000 | 104-83-6 see CFB600 | 106-32-1 see ENY000 | 107-36-8 see HKI500 | 108-68-9 see XLS000 |
| 103-29-7 see BFX500 | 104-85-8 see TGT750 | 106-33-2 see ELY700 | 107-37-9 see AGU250 | 108-69-0 see XOA000 |
| 103-30-0 see SLR100 | 104-88-1 see CEI600 | 106-34-3 see QFJ000 | 107-41-5 see HFP875 | 108-70-3 see TIK300 |
| 103-33-3 see ASL250 | 104-89-2 see END000 | 106-35-4 see EHA600 | 107-43-7 see GHA050 | 108-72-5 see THN775 |
| 103-34-4 see BKU500 | 104-90-5 see EOS000 | 106-36-5 see PNU000 | 107-44-8 see IPX000 | 108-73-6 see PGR000 |
| 103-36-6 see EHN000 | 104-91-6 see NLF200 | 106-37-6 see BNV775 | 107-45-9 see TDN250 | 108-74-7 see HDW100 |
| 103-37-7 see BED000 | 104-92-7 see AOY450 | 106-38-7 see BOG255 | 107-46-0 see HEE000 | 108-75-8 see TME272 |
| 103-38-8 see ISW000 | 104-93-8 see MGP000 | 106-40-1 see BMT325 | 107-49-3 see TCF250 | 108-77-0 see TJD750 |
| 103-41-3 see BEG750 | 104-94-9 see AOW000 | 106-41-2 see BNU750 | 107-58-4 see BPW050 | 108-78-1 see MCB000 |
| 103-44-6 see ELB500 | 104-96-1 see AMS675 | 106-42-3 see XHS000 | 107-61-9 see DVR000 | 108-80-5 see THS000 |
| 103-45-7 see PFB250 | 104-98-3 see UVJ440 | 106-43-4 see TGY075 | 107-64-2 see DXG625 | 108-82-7 see DNH800 |
| 103-48-0 see PDF750 | 105-01-1 see IIR100 | 106-44-5 see CNX250 | 107-66-4 see DEG700 | 108-83-8 see DNI800 |
| 103-50-4 see BEO250 | 105-07-7 see COK250 | 106-45-6 see TGP250 | 107-69-7 see TIO500 | 108-84-9 see HFJ000 |
| 103-52-6 see PFB800 | 105-08-8 see BKH325 | 106-46-7 see DEP800 | 107-70-0 see MEX250 | 108-86-1 see PEO500 |
| 103-53-7 see BEE250 | 105-10-2 see DTL800 | 106-47-8 see CEH680 | 107-71-1 see BSC250 | 108-87-2 see MIQ740 |
| 103-54-8 see CMQ730 | 105-11-3 see DVR200 | 106-48-9 see CJK750 | 107-72-2 see PBY750 | 108-88-3 see TGK750 |
| 103-56-0 see CMR850 | 105-12-4 see DVE260 | 106-49-0 see TGR000 | 107-74-4 see DTE400 | 108-89-4 see MOY250 |
| 103-58-2 see HHQ500 | 105-13-5 see MED500 | 106-50-3 see PEY500 | 107-75-5 see CMS850 | 108-90-7 see CEJ125 |
| 103-59-3 see CMR750 | 105-14-6 see NOB800 | 106-51-4 see QQS200 | 107-81-3 see BNU500 | 108-91-8 see CPF500 |
| 103-60-6 see PDS900 | 105-16-8 see DIB300 | 106-54-7 see CEK425 | 107-82-4 see BNP250 | 108-93-0 see CPB750 |
| 103-61-7 see CMQ800 | 105-18-0 see BJA250 | 106-55-8 see DTR400 | 107-83-5 see IKS600 | 108-94-1 see CPC000 |
| 103-64-0 see BOF000 | 105-21-5 see HBA550 | 106-58-1 see LIQ500 | 107-84-6 see CII000 | 108-95-2 see PDN750 |
| 103-65-1 see IKG000 | 105-28-2 see GIO000 | 106-61-6 see GGO000 | 107-87-9 see PBN250 | 108-98-5 see PFL850 |
| 103-69-5 see EGK000 | 105-29-3 see MNL775 | 106-63-8 see IIK000 | 107-88-0 see BOS500 | 108-99-6 see PIB920 |
| 103-70-8 see FNJ000 | 105-30-6 see AOK750 | 106-65-0 see SNB100 | 107-89-1 see AAH750 | 109-00-2 see PPH025 |
| 103-71-9 see PFK250 | 105-31-7 see HFZ000 | 106-67-2 see ELT500 | 107-91-5 see COJ250 | 109-01-3 see MOD250 |
| 103-72-0 see ISQ000 | 105-34-0 see MIQ000 | 106-67-2 see EMZ000 | 107-92-6 see BSW000 | 109-02-4 see MMA250 |
| 103-73-1 see PDM000 | 105-36-2 see EGV000 | 106-68-3 see ODI000 | 107-94-8 see CKS500 | 109-04-6 see BOB600 |
| 103-75-3 see EER500 | 105-37-3 see EPB500 | 106-69-4 see HES500 | 107-96-0 see MCQ000 | 109-06-8 see MOY000 |
| 103-76-4 see HKY500 | 105-38-4 see VQK000 | 106-70-7 see MHY700 | 107-98-2 see UBA000 | 109-08-0 see MOW750 |
| 103-79-7 see MHO100 | 105-39-5 see EHG500 | 106-71-8 see ADT111 | 107-99-3 see CGW000 | 109-09-1 see CKW000 |
| 103-81-1 see PDX750 | 105-40-8 see EMQ500 | 106-72-9 see DSD775 | 108-00-9 see DPC000 | 109-12-6 see PPH300 |
| 103-82-2 see PDY850 | 105-44-2 see MKW750 | 106-73-0 see MKK100 | 108-01-0 see DOY800 | 109-13-7 see BSC600 |
| 103-83-3 see DQP800 | 105-45-3 see MFX250 | 106-74-1 see ADT500 | 108-03-2 see NIX500 | 109-16-0 see MDN510 |
| 103-84-4 see AAQ500 | 105-46-4 see BPV000 | 106-75-2 see OPO000 | 108-05-4 see VLU250 | 109-17-1 see TCE400 |
| 103-85-5 see PGN250 | 105-52-2 see DKP400 | 106-83-2 see BRH250 | 108-09-8 see DQU600 | 109-19-3 see ISX000 |
| 103-88-8 see BMR100 | 105-53-3 see EMA500 | 106-86-5 see VNZ000 | 108-10-1 see HFG500 | 109-20-6 see GDK000 |
| 103-89-9 see ABJ250 | 105-54-4 see EHE000 | 106-87-6 see VOA000 | 108-11-2 see MKW600 | 109-21-7 see BQM500 |
| 103-90-2 see HIM000 | 105-55-5 see DKC400 | 106-88-7 see BOX750 | 108-13-4 see MAO000 | 109-27-3 see TEF500 |
| 103-93-5 see THA250 | 105-56-6 see EHP500 | 106-89-8 see EAZ500 | 108-16-7 see DPT800 | 109-29-5 see OKU000 |
| 103-94-6 see NHX100 | 105-57-7 see AAG000 | 106-90-1 see ECH500 | 108-18-9 see DNM200 | 109-31-9 see ASC000 |
| 103-95-7 see COU500 | 105-58-8 see DIX200 | 106-91-2 see ECI000 | 108-20-3 see IOZ750 | 109-42-2 see BSS100 |
| 104-01-8 see MFE250 | 105-59-9 see MKU250 | 106-92-3 see AGH150 | 108-21-4 see INE100 | 109-43-3 see DEH600 |
| 104-03-0 see NII510 | 105-60-2 see CBF700 | 106-93-4 see EIY500 | 108-22-5 see MQK750 | 109-44-4 see BJO225 |
| 104-04-1 see NEK000 | 105-64-6 see DNR400 | 106-94-5 see BNX750 | 108-23-6 see IOL000 | 109-46-6 see DEI000 |
| 104-06-3 see FNF000 | 105-66-8 see PNF100 | 106-95-6 see AFY000 | 108-24-7 see AAX500 | 109-52-4 see VAQ000 |
| 104-12-1 see CKB000 | 105-67-9 see XKJ500 | 106-96-7 see PMN500 | 108-26-9 see MOX100 | 109-53-5 see IJQ000 |
| 104-13-2 see AJA550 | 105-74-8 see LBR000 | 106-97-8 see BOR500 | 108-29-2 see VAV000 | 109-55-7 see AJQ100 |
| 104-14-3 see AKT250 | 105-75-9 see DEC600 | 106-99-0 see BOP500 | 108-30-5 see SNC000 | 109-56-8 see INN400 |
| 104-15-4 see TGO000 | 105-76-0 see DED600 | 107-00-6 see EFS500 | 108-31-6 see MAM000 | 109-57-9 see AGT500 |
| 104-19-8 see DPH400 | 105-77-1 see BSS550 | 107-01-7 see BOW255 | 108-32-7 see CBW500 | 109-59-1 see INA500 |
| 104-20-1 see MFF580 | 105-79-3 see IIT000 | 107-02-8 see ADR000 | 108-34-9 see MOX250 | 109-60-4 see PNC250 |
| 104-40-5 see NNC510 | 105-82-8 see AAG850 | 107-04-0 see CES500 | 108-36-1 see DDK050 | 109-61-5 see PNH000 |
| 104-45-0 see PNE250 | 105-83-9 see BGU750 | 107-05-1 see AGB250 | 108-38-3 see XHA000 | 109-62-6 see EMD000 |
| 104-46-1 see PMQ750 | 105-85-1 see CMT750 | 107-06-2 see EIY600 | 108-39-4 see CNW750 | 109-63-7 see BMH250 |
| 104-47-2 see MFF000 | 105-86-2 see GCY000 | 107-07-3 see EIU800 | 108-40-7 see TGO800 | 109-64-8 see TLR000 |
| 104-50-7 see OCE000 | 105-87-3 see DTD800 | 107-08-4 see PNO750 | | 109-65-9 see BMX500 |
| 104-51-8 see BQI750 | 105-90-8 see GDM450 | 107-10-8 see PND250 | | 109-66-0 see PBK250 |

| | | | | |
|---|---|---|---|---|
| 109-69-3 see BQQ750 | 110-77-0 see EPP500 | 111-97-7 see DGS600 | 114-49-8 see HOT500 | 117-34-0 see DVW800 |
| 109-70-6 see BNA825 | 110-78-1 see PNP000 | 112-03-8 see TLW500 | 114-63-6 see HJM000 | 117-39-5 see QCA000 |
| 109-71-7 see CHR400 | 110-80-5 see EES350 | 112-04-9 see OBI000 | 114-70-5 see SFA200 | 117-51-1 see HGK500 |
| 109-72-8 see BRR739 | 110-81-6 see DJC600 | 112-05-0 see NMY000 | 114-80-7 see POD000 | 117-52-2 see ABF500 |
| 109-73-9 see BPX750 | 110-82-7 see CPB000 | 112-06-1 see HBL000 | 114-83-0 see ACX750 | 117-55-5 see AMM125 |
| 109-74-0 see BSX250 | 110-83-8 see CPC579 | 112-07-2 see BPM000 | 114-85-2 see BFW250 | 117-61-3 see BBX500 |
| 109-75-1 see BOX500 | 110-85-0 see PIJ000 | 112-10-7 see IRL100 | 114-86-3 see PDF000 | 117-62-4 see ALH000 |
| 109-76-2 see PMK500 | 110-86-1 see POP250 | 112-12-9 see UKS000 | 114-90-9 see BGS250 | 117-79-3 see AIB000 |
| 109-77-3 see MAO250 | 110-87-2 see DMC200 | 112-14-1 see OEG000 | 115-02-6 see ASA500 | 117-80-6 see DFT000 |
| 109-78-4 see HGP000 | 110-88-3 see TMP000 | 112-15-2 see CBQ750 | 115-07-1 see PMO500 | 117-81-7 see DVL700 |
| 109-79-5 see BRR900 | 110-89-4 see PIL500 | 112-18-5 see DRR800 | 115-09-3 see MDD750 | 117-82-8 see DOF400 |
| 109-80-8 see PML350 | 110-91-8 see MRP750 | 112-19-6 see UMS000 | 115-10-6 see MJW500 | 117-83-9 see BHK000 |
| 109-81-9 see MJW100 | 110-93-0 see MKK000 | 112-23-2 see HBO500 | 115-11-7 see IIC000 | 117-84-0 see DVL600 |
| 109-82-0 see HHW000 | 110-94-1 see GFS000 | 112-24-3 see TJR000 | 115-18-4 see MHU100 | 117-89-5 see TKE500 |
| 109-83-1 see MGG000 | 110-95-2 see TDU500 | 112-25-4 see HFT500 | 115-19-5 see MHX250 | 117-96-4 see DCK000 |
| 109-84-2 see HHC000 | 110-96-3 see DNH400 | 112-26-5 see TKL500 | 115-20-8 see TIN500 | 117-97-5 see BLC500 |
| 109-85-3 see MEM500 | 110-97-4 see DNL600 | 112-27-6 see TJQ000 | 115-21-9 see EPY500 | 117-98-6 see AAW750 |
| 109-86-4 see EJH500 | 110-98-5 see OQM000 | 112-29-8 see BNB800 | 115-24-2 see ABD500 | 118-00-3 see GLS000 |
| 109-87-5 see MGA850 | 110-99-6 see ONQ100 | 112-30-1 see DAI600 | 115-25-3 see CPS000 | 118-02-5 see DVF300 |
| 109-89-7 see DHJ200 | 111-01-3 see SLG700 | 112-31-2 see DAG000 | 115-26-4 see BJE750 | 118-03-6 see ALI750 |
| 109-90-0 see ELS500 | 111-02-4 see SLG800 | 112-31-2 see DAG200 | 115-27-5 see CDS050 | 118-10-5 see CMP925 |
| 109-92-2 see EQF500 | 111-10-4 see MEP750 | 112-32-3 see OEY100 | 115-28-6 see CDS000 | 118-12-7 see TLU200 |
| 109-93-3 see VOP000 | 111-11-5 see MHY800 | 112-33-4 see AMC250 | 115-29-7 see EAQ750 | 118-28-5 see AJU500 |
| 109-94-4 see EKL000 | 111-12-6 see MND275 | 112-34-5 see DJF200 | 115-31-1 see IHZ000 | 118-33-2 see ALH500 |
| 109-95-5 see ENN000 | 111-13-7 see ODG000 | 112-35-6 see TJQ750 | 115-32-2 see BIO750 | 118-42-3 see PJB750 |
| 109-97-7 see PPS250 | 111-14-8 see HBE000 | 112-36-7 see DIW800 | 115-37-7 see TEN000 | 118-46-7 see ALJ750 |
| 109-99-9 see TCR750 | 111-15-9 see EES400 | 112-37-8 see UKA000 | 115-38-8 see ENB500 | 118-52-5 see DFE200 |
| 110-00-9 see FPK000 | 111-16-0 see PIG000 | 112-38-9 see ULS000 | 115-43-5 see AGQ875 | 118-55-8 see PGG750 |
| 110-01-0 see TDC730 | 111-17-1 see BHM000 | 112-40-3 see DXT200 | 115-44-6 see AFY500 | 118-58-1 see BFJ750 |
| 110-02-1 see TFM250 | 111-18-2 see TDR100 | 112-42-5 see UNA000 | 115-58-2 see PBS250 | 118-60-5 see ELB000 |
| 110-05-4 see BSC750 | 111-20-6 see SBJ500 | 112-43-6 see UMA000 | 115-63-9 see HFG400 | 118-61-6 see SAL000 |
| 110-12-3 see MKW450 | 111-21-7 see EJB500 | 112-44-7 see UJJ000 | 115-67-3 see PAH500 | 118-68-3 see AJB250 |
| 110-13-4 see HEQ500 | 111-22-8 see TJQ500 | 112-45-8 see ULJ000 | 115-69-5 see ALB000 | 118-71-8 see MAO350 |
| 110-15-6 see SMY000 | 111-25-1 see HFM500 | 112-48-1 see DDW400 | 115-76-4 see PMB250 | 118-74-1 see HCC500 |
| 110-16-7 see MAK900 | 111-26-2 see HFK000 | 112-49-2 see TKL875 | 115-77-5 see PBB750 | 118-75-2 see TBO500 |
| 110-17-8 see FOU000 | 111-27-3 see HFJ500 | 112-50-5 see EFL000 | 115-78-6 see THY500 | 118-79-6 see THV750 |
| 110-18-9 see TDQ750 | 111-28-4 see HCS500 | 112-53-8 see DXV600 | 115-79-7 see MSC100 | 118-90-1 see TGQ000 |
| 110-19-0 see IIJ000 | 111-29-5 see PBK750 | 112-54-9 see DXT000 | 115-82-2 see EKQ500 | 118-91-2 see CEL250 |
| 110-20-3 see ABE250 | 111-30-8 see GFQ000 | 112-55-0 see LBX000 | 115-84-4 see BKH625 | 118-92-3 see API500 |
| 110-22-5 see ACV500 | 111-31-9 see HES000 | 112-56-1 see BPL250 | 115-86-6 see TMT750 | 118-93-4 see HIN500 |
| 110-26-9 see MJL500 | 111-34-2 see VMZ000 | 112-57-2 see TCE500 | 115-90-2 see FAQ800 | 118-96-7 see TMN490 |
| 110-27-0 see IQN000 | 111-35-3 see EFG000 | 112-58-3 see DKO800 | 115-93-5 see CQL250 | 119-04-0 see NCF000 |
| 110-32-7 see AEQ000 | 111-36-4 see BRQ500 | 112-59-4 see HFN000 | 115-95-7 see LFY100 | 119-06-2 see DXQ200 |
| 110-36-1 see MSA300 | 111-40-0 see DJG600 | 112-60-7 see TCE250 | 115-96-8 see CGO500 | 119-07-3 see OEU000 |
| 110-38-3 see EHE500 | 111-41-1 see AJW000 | 112-61-8 see MJW000 | 115-98-0 see VQF000 | 119-12-0 see POP000 |
| 110-40-7 see DJY600 | 111-42-2 see DHF000 | 112-62-9 see OHW000 | 116-01-8 see DNX600 | 119-15-3 see DUW500 |
| 110-41-8 see MQI550 | 111-43-3 see PNM000 | 112-66-3 see DXV400 | 116-02-9 see TLO500 | 119-21-1 see CGC100 |
| 110-42-9 see MHY650 | 111-44-4 see DFJ050 | 112-69-6 see HCP525 | 116-06-3 see CBM500 | 119-26-6 see DVC400 |
| 110-43-0 see MGN500 | 111-45-5 see AGO000 | 112-70-9 see TJI750 | 116-09-6 see ABC000 | 119-27-7 see DUP800 |
| 110-44-1 see SKU000 | 111-46-6 see DJD600 | 112-72-1 see TBY250 | 116-11-0 see MFL300 | 119-32-4 see NMP000 |
| 110-45-2 see IHS000 | 111-48-8 see TFI500 | 112-73-2 see DDW200 | 116-14-3 see TCH500 | 119-33-5 see NFU500 |
| 110-46-3 see IMB000 | 111-49-9 see HDG000 | 112-80-1 see OHU000 | 116-15-4 see HDF000 | 119-34-6 see NEM480 |
| 110-49-6 see EJJ500 | 111-55-7 see EJD759 | 112-90-3 see OHM700 | 116-16-5 see HCL500 | 119-36-8 see MPI000 |
| 110-51-0 see POQ250 | 111-60-4 see EJM500 | 112-92-5 see OAX000 | 116-17-6 see TKT500 | 119-38-0 see DSK200 |
| 110-52-1 see DDL000 | 111-61-5 see EPF700 | 112-96-9 see OBG000 | 116-29-0 see CKM000 | 119-39-1 see PHV750 |
| 110-53-2 see AOF800 | 111-65-9 see OCU000 | 112-98-1 see TCE350 | 116-31-4 see VSK985 | 119-41-5 see ELH600 |
| 110-54-3 see HEN000 | 111-68-2 see HBL600 | 113-00-8 see GKW000 | 116-38-1 see EAE600 | 119-47-1 see MJO500 |
| 110-57-6 see BRG000 | 111-69-3 see AER250 | 113-15-5 see EDC000 | 116-52-9 see DGQ200 | 119-48-2 see DUO400 |
| 110-58-7 see PBV505 | 111-70-6 see HBL500 | 113-18-8 see CHG000 | 116-54-1 see DEM800 | 119-51-7 see ILI000 |
| 110-59-8 see VAV300 | 111-71-7 see HBB500 | 113-38-2 see EDR000 | 116-66-5 see MRU300 | 119-53-9 see BCP250 |
| 110-60-1 see BOS000 | 111-75-1 see BQC000 | 113-42-8 see PAM000 | 116-71-2 see DCU800 | 119-58-4 see TDO750 |
| 110-61-2 see SNE000 | 111-76-2 see BPJ850 | 113-45-1 see MNQ000 | 116-76-7 see APL750 | 119-60-8 see DGV600 |
| 110-62-3 see VAG000 | 111-77-3 see DJG000 | 113-48-4 see OES000 | 116-82-5 see AIY500 | 119-61-9 see BCS250 |
| 110-63-4 see BOS750 | 111-79-5 see MNB750 | 113-52-0 see DLH630 | 116-84-7 see AJH250 | 119-64-2 see TCX500 |
| 110-64-5 see BOX300 | 111-80-8 see MNC000 | 113-53-1 see DYC875 | 116-85-8 see AKE250 | 119-65-3 see IRX000 |
| 110-65-6 see BST500 | 111-82-0 see MLC800 | 113-59-7 see TAF675 | 116-90-5 see DCV000 | 119-67-5 see FNK010 |
| 110-66-7 see PBM000 | 111-83-1 see BNU000 | 113-69-9 see BDW000 | 117-03-3 see APM000 | 119-68-6 see MGQ000 |
| 110-67-8 see MFL750 | 111-84-2 see NMX000 | 113-92-8 see TAI500 | 117-05-5 see BDK750 | 119-72-2 see ALP750 |
| 110-68-9 see MHV000 | 111-86-4 see OEK000 | 113-98-4 see BFD000 | 117-08-8 see TBT150 | 119-79-9 see ALI250 |
| 110-69-0 see BSU500 | 111-87-5 see OEI000 | 114-03-4 see HOA575 | 117-10-2 see DMH400 | 119-80-2 see BHM300 |
| 110-70-3 see DRI600 | 111-90-0 see EOK000 | 114-03-4 see HON800 | 117-11-3 see AJE325 | 119-81-3 see NEK100 |
| 110-71-4 see DOE600 | 111-91-1 see BID750 | 114-07-8 see EDH500 | 117-12-4 see DMH200 | 119-84-6 see HHR500 |
| 110-73-6 see EGA500 | 111-92-2 see DDT800 | 114-26-1 see PMY300 | 117-14-6 see APK625 | 119-90-4 see DCJ200 |
| 110-74-7 see PNM500 | 111-94-4 see BIQ500 | 114-42-1 see FBY000 | 117-18-0 see TBR750 | 119-93-7 see TGJ750 |
| 110-75-8 see CHI250 | 111-96-6 see BKN750 | 114-45-4 see IRU000 | 117-27-1 see BIN500 | 120-02-5 see CBI250 |

| | | | | |
|---|---|---|---|---|
| 344-07-0 see PBE100 | 367-12-4 see FKT100 | 407-98-7 see FFW000 | 460-19-5 see COO000 | 476-70-0 see DNZ100 |
| 344-72-9 see AMU550 | 367-51-1 see SKH500 | 407-99-8 see FLR100 | 460-35-5 see TJY200 | 477-27-0 see ADE000 |
| 345-78-8 see POH250 | 368-39-8 see TJL600 | 408-35-5 see SIZ025 | 460-40-2 see TKH020 | 477-29-2 see DAN375 |
| 346-18-9 see PKL250 | 368-43-4 see BBT250 | 409-21-2 see SCQ000 | 461-05-2 see CCK655 | 477-30-5 see MIW500 |
| 348-67-4 see MDT730 | 368-47-8 see PGO500 | 420-04-2 see COH500 | 461-56-3 see FLR000 | 477-73-6 see GJI400 |
| 349-37-1 see DKH000 | 368-53-6 see TKB775 | 420-12-2 see EJP500 | 461-58-5 see COP125 | 478-15-9 see TDI750 |
| 350-03-8 see ABI000 | 368-68-3 see DKG100 | 420-23-5 see DRY289 | 461-78-9 see CLY250 | 478-84-2 see BNM250 |
| 350-30-1 see CHI950 | 368-97-8 see DKI400 | 420-46-2 see TJY900 | 461-89-2 see THR750 | 478-99-9 see LJI000 |
| 350-46-9 see FKL000 | 369-57-3 see BBO325 | 420-52-0 see TKF775 | 462-06-6 see FGA000 | 479-18-5 see DNC000 |
| 350-87-8 see DKH100 | 370-14-9 see FMS875 | 421-17-0 see TKB300 | 462-08-8 see AMI250 | 479-45-8 see TEG250 |
| 351-63-3 see DKG980 | 371-28-8 see CGZ000 | 421-20-5 see MKG250 | 462-18-0 see TJJ300 | 479-50-5 see DHU000 |
| 351-65-5 see DRL000 | 371-29-9 see FIN000 | 421-53-4 see TJZ000 | 462-27-1 see FIH100 | 479-92-5 see INY000 |
| 352-21-6 see AKF375 | 371-40-4 see FFY000 | 422-05-9 see PBE750 | 462-72-6 see FHA000 | 480-16-0 see MRN500 |
| 352-32-9 see FMC000 | 371-41-5 see FKV000 | 422-61-7 see PBF300 | 462-73-7 see FHB000 | 480-18-2 see DMD000 |
| 352-93-2 see EPH000 | 371-47-1 see SIE000 | 422-63-9 see PBE500 | 462-94-2 see PBK500 | 480-22-8 see APH250 |
| 353-03-7 see EKI000 | 371-62-0 see FIE000 | 422-64-0 see PBF000 | 462-95-3 see EFT500 | 480-30-8 see SKS700 |
| 353-13-9 see FMO000 | 371-67-5 see TJY275 | 423-62-1 see IEU075 | 463-04-7 see AOL500 | 480-54-6 see RFP000 |
| 353-16-2 see FHC200 | 371-69-7 see EKK550 | 424-40-8 see DJK100 | 463-40-1 see OAX100 | 480-63-7 see IKN300 |
| 353-17-3 see FHD000 | 371-78-8 see BLQ325 | 425-87-6 see CLR000 | 463-51-4 see KEU000 | 480-68-2 see NET550 |
| 353-18-4 see CON500 | 371-86-8 see PHF750 | 427-00-9 see DKX600 | 463-58-1 see CCC000 | 480-79-5 see IDG000 |
| 353-21-9 see FIZ000 | 372-09-8 see COJ500 | 427-01-0 see GCG300 | 463-71-8 see TFN500 | 480-81-9 see SBX500 |
| 353-36-6 see FIB000 | 372-18-9 see DKF800 | 427-45-2 see TNG050 | 463-82-1 see NCH000 | 481-06-1 see SAU500 |
| 353-42-4 see BMH000 | 372-48-5 see FLT100 | 427-51-0 see CQJ500 | 463-88-7 see VQR300 | 481-39-0 see WAT000 |
| 353-50-4 see CCA500 | 372-64-5 see BLO325 | 429-30-1 see TJX250 | 464-07-3 see BRU300 | 481-42-5 see PJH610 |
| 353-59-3 see BNA250 | 372-91-8 see FHC000 | 431-03-8 see BOT500 | 464-10-8 see NMQ000 | 481-49-2 see CCX550 |
| 354-06-3 see TJY000 | 373-02-4 see NCX000 | 431-97-0 see BLO300 | 464-45-9 see NCQ820 | 481-85-6 see MMC250 |
| 354-13-2 see TIJ175 | 373-14-8 see FJA000 | 432-04-2 see TNN775 | 464-48-2 see CBB000 | 482-41-7 see MKD250 |
| 354-21-2 see TIM000 | 373-88-6 see TKA750 | 434-03-7 see GEK500 | 464-49-3 see CBB250 | 482-44-0 see IHR300 |
| 354-23-4 see TJY750 | 373-91-1 see TKD375 | 434-05-9 see PMC700 | 464-72-2 see TEA600 | 482-49-5 see DYB000 |
| 354-32-5 see TJX500 | 375-01-9 see HAW100 | 434-07-1 see PAN100 | 465-16-7 see OHQ000 | 482-54-2 see CPB120 |
| 354-58-5 see TJE100 | 375-22-4 see HAX500 | 434-13-9 see LHW000 | 465-19-0 see BOM655 | 483-04-5 see AFG750 |
| 354-93-8 see PCH300 | 376-18-1 see HCO000 | 434-16-2 see DAK600 | 465-39-4 see BOM650 | 483-18-1 see EAL500 |
| 355-02-2 see PCH290 | 376-50-1 see DJT100 | 434-64-0 see PCH500 | 465-65-6 see NAG550 | 483-55-6 see HMI000 |
| 355-43-1 see PCH100 | 376-53-4 see PCG600 | 436-30-6 see MKD750 | 465-73-6 see IKO000 | 483-57-8 see FBP300 |
| 355-66-8 see OBK100 | 376-89-6 see HDC300 | 437-38-7 see PDW500 | 466-06-8 see POB500 | 483-63-6 see EHO500 |
| 356-12-7 see FDD150 | 377-38-8 see TCJ000 | 437-74-1 see XCS000 | 466-11-5 see DBC550 | 484-20-8 see MFN275 |
| 356-18-3 see DFM025 | 378-44-9 see BFV750 | 438-41-5 see MDQ250 | 466-24-0 see PIC250 | 484-23-1 see OJD300 |
| 356-27-4 see HAY000 | 379-52-2 see TMV850 | 438-60-8 see DDA600 | 466-40-0 see IKZ000 | 484-47-9 see TMS750 |
| 356-69-4 see PCH275 | 379-79-3 see EDC500 | 438-67-5 see EDV600 | 466-81-9 see EDG500 | 484-78-6 see HJD000 |
| 356-69-4 see PCH350 | 381-73-7 see DKF200 | 439-14-5 see DCK759 | 466-99-9 see DLW600 | 485-19-8 see SLX500 |
| 357-07-3 see ORG100 | 382-10-5 see HDC450 | 439-25-8 see MKD500 | 467-22-1 see CBQ625 | 485-31-4 see BGB500 |
| 357-08-4 see NAH000 | 382-21-8 see OBM000 | 440-17-5 see TKK250 | 467-36-7 see TES500 | 485-35-8 see CQL500 |
| 357-56-2 see AFJ400 | 383-73-3 see BLO000 | 440-58-4 see AAI750 | 467-60-7 see DWK400 | 485-47-2 see DMV200 |
| 357-57-3 see BOL750 | 388-72-7 see FFZ000 | 441-38-3 see BCP500 | 467-63-0 see TJK000 | 485-71-2 see CMP910 |
| 358-21-4 see PCG760 | 389-08-2 see EID000 | 442-51-3 see HAI500 | 468-28-0 see LIU000 | 485-89-2 see OPK300 |
| 358-52-1 see POA250 | 390-64-7 see PEV750 | 443-30-1 see DOT600 | 468-61-1 see DHQ200 | 486-17-9 see BRS000 |
| 358-74-7 see DJJ400 | 391-57-1 see FEE000 | 443-48-1 see MMN250 | 469-21-6 see DYE500 | 486-25-9 see FDO000 |
| 359-06-8 see FFR000 | 391-70-8 see TNV625 | 444-27-9 see TEV000 | 469-59-0 see JCS000 | 486-84-0 see MPA050 |
| 359-40-0 see OLK000 | 392-56-3 see HDB000 | 445-29-4 see FGH000 | 469-61-4 see CCR500 | 487-10-5 see ASN500 |
| 359-46-6 see TJX750 | 392-83-6 see BOJ750 | 446-35-5 see DKH900 | 469-62-5 see DAB879 | 487-19-4 see NDX300 |
| 359-48-8 see PCO250 | 393-52-2 see FGP000 | 446-86-6 see ASB250 | 469-65-8 see HJO500 | 487-26-3 see FBW150 |
| 359-83-1 see DOQ400 | 393-75-9 see CGM225 | 447-05-2 see PPJ900 | 469-79-4 see KFK000 | 487-53-6 see DHO600 |
| 360-53-2 see HAY500 | 395-28-8 see VGF000 | 447-14-3 see TKH310 | 469-81-8 see MRN675 | 487-54-7 see SAN200 |
| 360-54-3 see MKK750 | 395-47-1 see TKF525 | 447-25-6 see DAR150 | 470-67-7 see IKC100 | 487-89-8 see FNO100 |
| 360-68-9 see DKW000 | 396-01-0 see UVJ450 | 447-31-4 see CFJ100 | 470-82-6 see CAL000 | 487-93-4 see DPG109 |
| 360-89-4 see OBO000 | 398-32-3 see FKZ000 | 451-40-1 see PEB000 | 470-90-6 see CDS750 | 488-10-8 see JCA100 |
| 360-97-4 see AKK250 | 399-24-6 see FKQ100 | 453-13-4 see DKI800 | 471-03-4 see BIH500 | 488-17-5 see DNE000 |
| 361-09-1 see SFW000 | 400-44-2 see CHK750 | 453-18-9 see MKD000 | 471-25-0 see PMT275 | 488-23-3 see TDM250 |
| 361-37-5 see MLD250 | 400-99-7 see NMQ100 | 453-20-3 see HOI200 | 471-29-4 see MKI750 | 488-41-5 see DDP600 |
| 363-03-1 see PEL750 | 401-78-5 see BOJ500 | 454-41-1 see ALF600 | 471-34-1 see CAT775 | 488-81-3 see RIF000 |
| 363-13-3 see BEH000 | 402-26-6 see TKF530 | 455-14-1 see TKB750 | 471-35-2 see TDP250 | 489-84-9 see DSJ800 |
| 363-17-7 see FER000 | 402-31-3 see BLO270 | 455-16-3 see TGO500 | 471-46-5 see OLO000 | 489-98-5 see PIC800 |
| 363-20-2 see TIH800 | 402-51-7 see TKF535 | 455-80-1 see HHM500 | 471-53-4 see GIE000 | 490-02-8 see ARO000 |
| 363-24-6 see DVJ200 | 402-71-1 see THH450 | 456-59-7 see DNU100 | 471-77-2 see NBU800 | 490-31-3 see RLP000 |
| 363-42-8 see TLQ000 | 404-42-2 see DKF170 | 457-60-3 see NCJ500 | 471-95-4 see BON000 | 490-78-8 see DMG600 |
| 363-49-5 see HIK500 | 404-72-8 see ISL000 | 457-87-4 see EGI500 | 472-54-8 see NNT500 | 490-79-9 see GCU000 |
| 364-62-5 see AJH000 | 404-82-0 see PDM250 | 458-24-2 see ENJ000 | 474-07-7 see LFT800 | 490-91-5 see IQF000 |
| 364-71-6 see FFT000 | 404-86-4 see CBF750 | 458-88-8 see PNT000 | 474-25-9 see CDL325 | 491-07-6 see IKY000 |
| 364-76-1 see FKK100 | 405-22-1 see FPE100 | 459-22-3 see FLC000 | 474-86-2 see ECW000 | 491-36-1 see QFA000 |
| 364-98-7 see DCQ700 | 405-86-7 see FHV000 | 459-44-9 see TGM450 | 475-08-1 see CBK125 | 491-92-9 see RHZ000 |
| 365-26-4 see HKH500 | 406-20-2 see MKE000 | 459-72-3 see EKG500 | 475-20-7 see LID100 | 492-08-0 see PAB250 |
| 366-18-7 see BGO500 | 406-23-5 see AGG750 | 459-80-3 see GCW000 | 475-81-0 see TDI475 | 492-17-1 see BGF109 |
| 366-70-1 see PME500 | 406-90-6 see TKB250 | 459-99-4 see FIM000 | 475-83-2 see NOE500 | 492-18-2 see SIH500 |
| 366-71-2 see MKN750 | 407-25-0 see TJX000 | 460-07-1 see ACB250 | 476-32-4 see CDL000 | 492-39-7 see NNM510 |
| 366-93-8 see BHN000 | 407-83-0 see FHF000 | 460-12-8 see BOQ625 | 476-60-8 see LEX200 | 492-41-1 see NNM000 |

| | | | | |
|---|---|---|---|---|
| 541-28-6 see IHU200 | 547-91-1 see IEP200 | 556-56-9 see AGI250 | 573-56-8 see DVA200 | 587-85-9 see DWD800 |
| 541-33-3 see BRQ050 | 547-95-5 see MIG850 | 556-61-6 see ISE000 | 573-58-0 see SGQ500 | 587-98-4 see MDM775 |
| 541-41-3 see EHK500 | 548-00-5 see BKA000 | 556-64-9 see MPT000 | 573-83-1 see PLQ775 | 588-07-8 see CDZ050 |
| 541-42-4 see IQQ000 | 548-26-5 see BNK700 | 556-65-0 see TFF500 | 574-25-4 see TFJ825 | 588-42-1 see TJL250 |
| 541-47-9 see MHT500 | 548-42-5 see AEY375 | 556-67-2 see OCE100 | 574-66-3 see BCS400 | 588-59-0 see SLR000 |
| 541-53-7 see DXL800 | 548-43-6 see EAJ000 | 556-72-9 see PBJ750 | 576-24-9 see DFX500 | 589-16-2 see EGL000 |
| 541-58-2 see DUG200 | 548-57-2 see SBE500 | 556-82-1 see MHU110 | 576-26-1 see XLA000 | 589-18-4 see MHB250 |
| 541-59-3 see MAM750 | 548-61-8 see THP000 | 556-88-7 see NHA500 | 576-55-6 see TBJ500 | 589-38-8 see HEV500 |
| 541-64-0 see FPY000 | 548-62-9 see AOR500 | 556-89-8 see NMQ500 | 576-68-1 see MAW500 | 589-41-3 see HKQ025 |
| 541-66-2 see FMX000 | 548-68-5 see DHY400 | 557-04-0 see MAJ030 | 577-11-7 see DJL000 | 589-43-5 see DSE600 |
| 541-69-5 see PEY750 | 548-73-2 see DYF200 | 557-05-1 see ZMS000 | 577-55-9 see DNN800 | 589-59-3 see ITA000 |
| 541-73-1 see DEP599 | 548-93-6 see AKE750 | 557-07-3 see ZJS000 | 577-59-3 see NEL450 | 589-82-2 see HBF000 |
| 541-85-5 see EGI750 | 549-18-8 see EAI000 | 557-09-5 see ZEJ000 | 577-66-2 see EEO000 | 589-90-2 see DRG200 |
| 541-91-3 see MIT625 | 549-49-5 see QJJ100 | 557-11-9 see AGV000 | 577-71-9 see DVA400 | 589-92-4 see MIR625 |
| 541-95-7 see MOU500 | 550-24-3 see EAJ600 | 557-17-5 see MOU830 | 577-85-5 see HLC600 | 589-98-0 see OCY100 |
| 542-18-7 see CPI400 | 550-28-7 see AKL625 | 557-18-6 see DJO100 | 577-91-3 see PDM750 | 590-00-1 see PLS750 |
| 542-46-1 see CMU850 | 550-33-4 see RJF000 | 557-19-7 see NDB500 | 578-32-5 see DRK800 | 590-01-2 see BSJ500 |
| 542-54-1 see MNJ000 | 550-34-5 see NAH800 | 557-20-0 see DKE600 | 578-54-1 see EGK500 | 590-19-2 see BOP250 |
| 542-55-2 see IIR000 | 550-70-9 see TMX775 | 557-21-1 see ZGA000 | 578-57-4 see BMT400 | 590-21-6 see PMR750 |
| 542-56-3 see IJD000 | 550-74-3 see PIE000 | 557-30-2 see EEA000 | 578-94-9 see PDB000 | 590-28-3 see PLC250 |
| 542-58-5 see CGO600 | 550-82-3 see HNG500 | 557-34-6 see ZBS000 | 579-07-7 see PGA500 | 590-29-4 see PLG750 |
| 542-59-6 see EJI000 | 550-90-3 see LIQ000 | 557-40-4 see DBK000 | 579-10-2 see MFW000 | 590-46-5 see CCH850 |
| 542-62-1 see BAK750 | 550-99-2 see NCW000 | 557-48-2 see NMV760 | 579-56-6 see VGA300 | 590-63-6 see HOA500 |
| 542-63-2 see DIV000 | 551-06-4 see ISN000 | 557-66-4 see EFW000 | 579-66-8 see DIS650 | 590-86-3 see MHX500 |
| 542-69-8 see BRQ250 | 551-08-6 see BRQ100 | 557-91-5 see DDN800 | 580-17-6 see AMK725 | 590-88-5 see BOR750 |
| 542-75-6 see DGG950 | 551-11-1 see POC500 | 557-98-2 see CKS000 | 580-48-3 see CDQ325 | 590-92-1 see BOB250 |
| 542-76-7 see CKT250 | 551-16-6 see PCU500 | 557-99-3 see ACM000 | 580-74-5 see DVU000 | 590-96-5 see HMG000 |
| 542-78-9 see PMK000 | 551-36-0 see AMM750 | 558-13-4 see CBX750 | 581-08-8 see ABG350 | 591-08-2 see ADD250 |
| 542-88-1 see BIK000 | 551-58-6 see SOW500 | 558-17-8 see TLU000 | 581-28-2 see AHS000 | 591-09-3 see ACS750 |
| 542-90-5 see EPP000 | 551-74-6 see MAW750 | 558-25-8 see MDR750 | 581-29-3 see ADJ375 | 591-10-6 see BJJ500 |
| 542-92-7 see CPU500 | 551-92-8 see DSV800 | 560-53-2 see SMN002 | 581-64-6 see AKK750 | 591-11-7 see MKH500 |
| 543-20-4 see SNG000 | 552-30-7 see TKV000 | 561-27-3 see HBT500 | 581-88-4 see IKB000 | 591-12-8 see AOO750 |
| 543-21-5 see ACJ250 | 552-41-0 see PAC250 | 561-43-3 see IBP200 | 581-89-5 see NHQ500 | 591-12-8 see MKH250 |
| 543-38-4 see AKD500 | 552-46-5 see NBF000 | 561-78-4 see NEB000 | 582-08-1 see ASN750 | 591-17-3 see BOG300 |
| 543-39-5 see MLO250 | 552-86-3 see FQI000 | 562-09-4 see CIS000 | 582-17-2 see NAO500 | 591-21-9 see DRG000 |
| 543-49-7 see HBE500 | 552-89-6 see NEU500 | 562-10-7 see PGE775 | 582-25-2 see PKW760 | 591-24-2 see MIR600 |
| 543-53-3 see PPB550 | 552-94-3 see SAN000 | 562-74-3 see TBD825 | 582-52-5 see DVO100 | 591-27-5 see ALT500 |
| 543-59-9 see PBW500 | 553-24-2 see AJQ250 | 562-95-8 see TJS500 | 582-60-5 see DQM100 | 591-33-3 see ABG250 |
| 543-63-5 see BRS750 | 553-26-4 see BGO600 | 563-12-2 see EEH600 | 582-61-6 see BDL750 | 591-35-5 see DFY450 |
| 543-67-9 see PNQ750 | 553-27-5 see AOQ875 | 563-25-7 see DDY800 | 583-03-9 see BQJ500 | 591-50-4 see IEC500 |
| 543-80-6 see BAH500 | 553-53-7 see NDU500 | 563-41-7 see SBW500 | 583-15-3 see MCX500 | 591-51-5 see PFL500 |
| 543-81-7 see BFP000 | 553-68-4 see IAC000 | 563-43-9 see EFU050 | 583-39-1 see BCC500 | 591-60-6 see BPV250 |
| 543-82-8 see ILM000 | 553-69-5 see PGG350 | 563-45-1 see MHT250 | 583-52-8 see PLN300 | 591-62-8 see EHA100 |
| 543-87-3 see ILW100 | 553-84-4 see PCI750 | 563-46-2 see MHT000 | 583-57-3 see DRF800 | 591-76-4 see MKL250 |
| 543-90-8 see CAD250 | 553-97-9 see MHI250 | 563-47-3 see CIU750 | 583-58-4 see LJB000 | 591-78-6 see HEV000 |
| 543-94-2 see SME000 | 554-00-7 see DEO290 | 563-52-0 see CEV250 | 583-60-8 see MIR500 | 591-80-0 see PBQ750 |
| 544-13-8 see TLR500 | 554-12-1 see MOT000 | 563-54-2 see DGG800 | 583-63-1 see BDC250 | 591-87-7 see AFU750 |
| 544-16-1 see BRV500 | 554-13-2 see LGZ000 | 563-63-3 see SDI800 | 583-69-7 see BNL260 | 591-89-9 see PLU500 |
| 544-17-2 see CAS250 | 554-14-3 see MPV000 | 563-68-8 see TEI250 | 583-78-8 see DFX850 | 591-97-9 see CEU825 |
| 544-25-2 see COY000 | 554-18-7 see AOO800 | 563-71-3 see FBH100 | 583-80-2 see MQI000 | 592-01-8 see CAQ500 |
| 544-40-1 see BSM125 | 554-35-8 see GFC100 | 563-80-4 see MLA750 | 583-91-5 see HMR600 | 592-04-1 see MDA250 |
| 544-47-8 see CEQ800 | 554-68-7 see TJO050 | 564-00-1 see DHB800 | 584-02-1 see IHP010 | 592-05-2 see LCU000 |
| 544-62-7 see GGA915 | 554-70-1 see TJT775 | 564-25-0 see DYE425 | 584-03-2 see BOS250 | 592-31-4 see BSS250 |
| 544-63-8 see MSA250 | 554-76-7 see SLP600 | 564-36-3 see EDA600 | 584-08-7 see PLA000 | 592-35-8 see BQP250 |
| 544-76-3 see HCO600 | 554-84-7 see NIE600 | 564-94-3 see FNK150 | 584-09-8 see RPB200 | 592-41-6 see HFB000 |
| 544-85-4 see DYC900 | 554-92-7 see TKW750 | 565-74-2 see BNM100 | 584-26-9 see ADC750 | 592-42-7 see HCR500 |
| 544-92-3 see CNL000 | 554-99-4 see MJV000 | 565-80-0 see DTI600 | 584-42-9 see SIT850 | 592-45-0 see HCR000 |
| 544-97-8 see DUO200 | 555-06-6 see SEO500 | 566-09-6 see SKR500 | 584-79-2 see AFR250 | 592-57-4 see CPA500 |
| 545-06-2 see TII750 | 555-15-7 see NGC000 | 566-28-9 see ONO000 | 584-84-9 see TGM750 | 592-62-1 see MGS750 |
| 545-55-1 see TND250 | 555-16-8 see NEV500 | 567-47-5 see HMX000 | 584-93-0 see BOL250 | 592-65-4 see IJO000 |
| 545-91-5 see DWH600 | 555-21-5 see NIJ000 | 568-69-4 see TNJ750 | 584-94-1 see DSE509 | 592-76-7 see HBJ000 |
| 545-93-7 see QCS000 | 555-30-6 see DNA800 | 568-70-7 see HMF500 | 585-08-0 see PCX000 | 592-79-0 see FFV000 |
| 546-46-3 see ZFJ250 | 555-31-7 see AHC600 | 568-75-2 see HMF000 | 585-54-6 see AQZ900 | 592-84-7 see BRK000 |
| 546-48-5 see PAP110 | 555-37-3 see BRA250 | 568-81-0 see DQK200 | 586-06-1 see DMV800 | 592-85-8 see MCU250 |
| 546-68-9 see IRN200 | 555-57-7 see MOS250 | 569-57-3 see CLO750 | 586-11-8 see DVA600 | 592-87-0 see LEC500 |
| 546-71-4 see ENQ000 | 555-60-2 see CKA550 | 569-58-4 see AGW750 | 586-38-9 see AOU500 | 592-88-1 see AGS250 |
| 546-80-5 see TFW000 | 555-77-1 see TNF250 | 569-59-5 see PDD000 | 586-62-9 see TBE000 | 593-12-4 see FKS000 |
| 546-88-3 see ABB250 | 555-84-0 see NDY000 | 569-61-9 see RMK020 | 586-76-5 see BMU100 | 593-14-6 see FKT000 |
| 546-89-4 see LGO100 | 555-89-5 see NCM700 | 569-64-2 see AFG500 | 586-77-6 see BNF250 | 593-53-3 see FJK000 |
| 546-93-0 see MAC650 | 555-96-4 see BEQ000 | 569-65-3 see HGC500 | 586-92-5 see POQ500 | 593-54-4 see MOB000 |
| 547-32-0 see MRM250 | 556-12-7 see FPF000 | 569-77-7 see TDD500 | 586-95-8 see POR810 | 593-57-7 see DQG600 |
| 547-44-4 see SNQ550 | 556-18-3 see AIC825 | 570-22-9 see IAM000 | 586-98-1 see POR800 | 593-60-2 see VMP000 |
| 547-58-0 see MND600 | 556-22-9 see GII000 | 571-22-2 see HJB100 | 587-15-5 see DGV200 | 593-63-5 see CEC500 |
| 547-63-7 see MKX000 | 556-24-1 see ITC000 | 571-60-8 see NAN500 | 587-63-3 see DLR000 | 593-68-0 see EON000 |
| 547-64-8 see MLC600 | 556-52-5 see GGW500 | 572-48-5 see DXO000 | 587-84-8 see DVY000 | 593-70-4 see CHI900 |

| | | | | |
|---|---|---|---|---|
| 628-52-4 see CMJ000 | 636-09-9 see DKB160 | 654-42-2 see DSG700 | 688-74-4 see THX750 | 719-59-5 see AJE400 |
| 628-63-7 see AOD725 | 636-21-5 see TGS500 | 655-35-6 see CBR500 | 689-11-2 see BSS300 | 720-69-4 see DBY800 |
| 628-68-2 see DJD750 | 636-23-7 see DCE000 | 657-24-9 see DQR600 | 689-13-4 see FNO000 | 722-25-8 see TGV750 |
| 628-73-9 see HER500 | 636-26-0 see TFQ650 | 657-27-2 see LJO000 | 689-67-8 see GDE400 | 723-46-6 see SNK000 |
| 628-76-2 see DFX200 | 636-47-5 see SLI300 | 657-84-1 see SKK000 | 689-93-0 see EMK600 | 723-62-6 see APG550 |
| 628-77-3 see PBH125 | 636-53-3 see IMK100 | 659-70-1 see ITB000 | 689-97-4 see BPE109 | 724-31-2 see THV250 |
| 628-81-9 see EHA500 | 636-97-5 see NFH000 | 659-86-9 see IEZ800 | 689-98-5 see CGP125 | 724-34-5 see BDG325 |
| 628-83-1 see BSN500 | 637-01-4 see TDT250 | 660-27-5 see DNM400 | 690-02-8 see DTJ000 | 725-04-2 see FKX000 |
| 628-85-3 see DWU000 | 637-03-6 see PEG750 | 660-68-4 see DIS500 | 690-49-3 see TDV000 | 725-06-4 see FKY000 |
| 628-86-4 see MDC000 | 637-07-0 see ARQ750 | 661-11-0 see FIX000 | 690-94-8 see MKM300 | 728-40-5 see DEE800 |
| 628-87-5 see IBB100 | 637-12-7 see AHH825 | 661-18-7 see FFX000 | 691-35-0 see AGE625 | 728-84-7 see BNH100 |
| 628-89-7 see DKN000 | 637-23-0 see TNS200 | 661-54-1 see TKH030 | 691-60-1 see IRR100 | 728-88-1 see TGK200 |
| 628-91-1 see DWW500 | 637-32-1 see CKB500 | 661-69-8 see HEE500 | 691-88-3 see DEE200 | 729-46-4 see MRR090 |
| 628-92-2 see COY250 | 637-53-6 see TFA250 | 661-95-0 see DDO450 | 692-13-7 see BRA625 | 729-99-7 see AIE750 |
| 628-94-4 see AEN000 | 637-56-9 see PDL750 | 662-50-0 see HAX000 | 692-42-2 see DIS800 | 731-27-1 see DFL400 |
| 628-96-6 see EJG000 | 637-60-5 see MNU500 | 663-25-2 see HAY200 | 692-45-5 see VPF000 | 732-11-6 see PHX250 |
| 629-03-8 see DDO800 | 637-61-6 see CHQ500 | 664-95-9 see CPR000 | 692-56-8 see BLR125 | 734-88-3 see OOI200 |
| 629-04-9 see HBN100 | 637-92-3 see EHA550 | 665-66-7 see AED250 | 692-86-4 see EQD200 | 737-22-4 see DRY400 |
| 629-11-8 see HEP500 | 638-03-9 see TGS250 | 667-29-8 see TJX780 | 692-95-5 see AOI200 | 737-31-5 see SEN500 |
| 629-13-0 see DCL600 | 638-07-3 see EHH100 | 667-49-2 see TJX825 | 693-07-2 see CGY750 | 738-70-5 see TKZ000 |
| 629-14-1 see EJE500 | 638-10-8 see MIP800 | 668-37-1 see DHY200 | 693-13-0 see DNO400 | 738-99-8 see BJC250 |
| 629-15-2 see EJF000 | 638-16-4 see THS250 | 669-49-8 see XSS900 | 693-16-3 see OEK010 | 739-71-9 see DLH200 |
| 629-20-9 see CPS500 | 638-21-1 see PFV250 | 670-40-6 see POH500 | 693-21-0 see DJE400 | 741-58-2 see DNO800 |
| 629-25-4 see LBN000 | 638-23-3 see CBR675 | 670-54-2 see EEE500 | 693-30-1 see CHC000 | 742-20-1 see CPR750 |
| 629-27-6 see OFA100 | 638-29-9 see VBA000 | 671-04-5 see CGI500 | 693-33-4 see CDF450 | 743-45-3 see AGF250 |
| 629-33-4 see HFQ100 | 638-38-0 see MAQ000 | 671-16-9 see PME250 | 693-36-7 see DXG700 | 744-80-9 see BDS300 |
| 629-35-6 see DEE000 | 638-49-3 see AOJ500 | 671-35-2 see FMO100 | 693-54-9 see OFE050 | 745-65-3 see POC350 |
| 629-38-9 see MIE750 | 638-53-9 see TJI250 | 671-51-2 see ASH750 | 693-65-2 see PBX000 | 747-45-5 see QFS100 |
| 629-40-3 see OCW050 | 638-65-3 see SLL500 | 672-06-0 see MIA500 | 693-85-6 see DKG700 | 749-02-0 see SLE500 |
| 629-50-5 see TJH500 | 639-14-5 see GMG000 | 672-66-2 see DTN896 | 693-98-1 see MKT750 | 749-13-3 see TKK500 |
| 629-59-4 see TBX750 | 639-58-7 see CLU000 | 672-76-4 see TFV750 | 694-53-1 see SDX250 | 751-01-9 see CMQ625 |
| 629-60-7 see TJH750 | 640-15-3 see PHI500 | 673-04-1 see BJP250 | 694-59-7 see POS000 | 751-84-8 see PAQ100 |
| 629-62-9 see PAY750 | 640-19-7 see FFF000 | 673-06-3 see PEC500 | 694-83-7 see CPB100 | 751-94-0 see SHK000 |
| 629-64-1 see HBO000 | 640-68-6 see VBU000 | 673-19-8 see BSG250 | 694-85-9 see MPA075 | 751-97-3 see PPY250 |
| 629-70-9 see HCP100 | 641-16-7 see TDY600 | 673-31-4 see PGA750 | 695-06-7 see HDY600 | 752-58-9 see TMP250 |
| 629-78-7 see HAS100 | 641-38-3 see AGW476 | 674-81-7 see NKH000 | 695-34-1 see ALC250 | 753-73-1 see DUG825 |
| 629-82-3 see OEY000 | 641-81-6 see AQO500 | 674-82-8 see KFA000 | 695-53-4 see PMO250 | 753-90-2 see TKA500 |
| 630-08-0 see CBW750 | 642-15-9 see DUO350 | 675-09-2 see DRF400 | 695-64-7 see CJF825 | 756-79-6 see DSR400 |
| 630-10-4 see SBV000 | 642-44-4 see AFW500 | 675-10-5 see THL800 | 696-07-1 see IFP000 | 756-80-9 see PHH500 |
| 630-18-2 see PJA750 | 642-65-9 see DBF200 | 675-14-9 see TKK000 | 696-24-2 see DDR200 | 757-44-8 see DKD500 |
| 630-20-6 see TBQ000 | 642-72-8 see BCD750 | 675-20-7 see PIU000 | 696-28-6 see DGB600 | 757-58-4 see HCY000 |
| 630-56-8 see HNT500 | 642-78-4 see SLJ050 | 675-62-7 see DFS700 | 696-33-3 see IFE000 | 758-17-8 see FNW000 |
| 630-60-4 see OKS000 | 643-22-1 see EDJ500 | 676-46-0 see MAN250 | 696-59-3 see DON800 | 758-24-7 see BNA000 |
| 630-64-8 see HAN800 | 643-28-7 see INW100 | 676-54-0 see EPF600 | 697-91-6 see DES400 | 758-96-3 see DTS600 |
| 630-72-8 see TMK250 | 643-79-8 see PHV500 | 676-59-5 see DTQ089 | 698-49-7 see AMV790 | 759-73-9 see ENV000 |
| 630-76-2 see TEA750 | 643-84-5 see MAO600 | 676-75-5 see IEI000 | 698-76-0 see ODE300 | 759-94-4 see EIN500 |
| 630-93-3 see DNU000 | 644-06-4 see AEX850 | 676-83-5 see MOC250 | 698-87-3 see PGA600 | 760-19-0 see DHI880 |
| 631-06-1 see DVP400 | 644-08-6 see MJH905 | 676-97-1 see MOB399 | 699-10-5 see MHN350 | 760-21-4 see EGW500 |
| 631-07-2 see EOL000 | 644-26-8 see AOM000 | 676-98-2 see MOC000 | 699-17-2 see FPT000 | 760-23-6 see DEV100 |
| 631-27-6 see GHR609 | 644-31-5 see ACC250 | 676-99-3 see MJD275 | 700-02-7 see AEH250 | 760-56-5 see NJK000 |
| 631-40-3 see TED250 | 644-35-9 see PNS250 | 677-21-4 see TKH015 | 700-06-1 see ICP100 | 760-60-1 see IJF000 |
| 631-41-4 see TCB500 | 644-62-2 see DGM875 | 679-84-5 see BOF750 | 700-07-2 see MGD200 | 760-67-8 see EKO600 |
| 631-60-7 see MDE250 | 644-64-4 see DQZ000 | 680-31-9 see HEK000 | 701-73-5 see MJK500 | 760-79-2 see DQV300 |
| 631-61-8 see ANA000 | 644-97-3 see DGE400 | 681-71-0 see TMD699 | 702-54-5 see DJT400 | 760-93-0 see MDN699 |
| 631-67-4 see DUG450 | 645-05-6 see HEJ500 | 681-84-5 see MPI750 | 702-62-5 see DVO600 | 761-65-9 see DEC400 |
| 632-14-4 see TMJ250 | 645-08-9 see HJC000 | 681-99-2 see THY750 | 703-80-0 see ICY100 | 762-04-9 see DJW400 |
| 632-21-3 see TBN300 | 645-15-8 see BLA600 | 682-09-7 see TLX000 | 703-95-7 see TCQ500 | 762-13-0 see NNA100 |
| 632-22-4 see TDX250 | 645-43-2 see GKU000 | 682-11-1 see TLX110 | 704-01-8 see TDT000 | 762-16-3 see CBF705 |
| 632-58-6 see TBT100 | 645-48-7 see PGM750 | 683-10-3 see LBU200 | 705-86-2 see DAF200 | 762-21-0 see DHI850 |
| 632-69-9 see RMP175 | 645-55-6 see NEO510 | 683-18-1 see DDY200 | 706-14-9 see HKA500 | 762-42-5 see DOP400 |
| 632-73-5 see TDE750 | 645-56-7 see PNS500 | 683-45-4 see COL750 | 706-79-6 see DFM050 | 762-51-6 see FIQ000 |
| 632-99-5 see MAC250 | 645-59-0 see HHP100 | 683-60-3 see SHY300 | 707-61-9 see MNV800 | 763-29-1 see MNK000 |
| 633-03-4 see BAY750 | 645-62-5 see EKR000 | 683-72-7 see DEM300 | 708-06-5 see HMU200 | 763-69-9 see EJV500 |
| 633-59-0 see CLX250 | 645-88-5 see ALQ650 | 684-16-2 see HCZ000 | 709-55-7 see EGE500 | 764-05-6 see COO250 |
| 633-65-8 see BFN600 | 646-04-8 see PBQ250 | 684-82-2 see BQY300 | 709-98-8 see DGI000 | 764-39-6 see PBP500 |
| 633-96-5 see CMM220 | 646-06-0 see DVR800 | 684-93-5 see MNA750 | 710-04-3 see UKJ000 | 764-41-0 see DEV000 |
| 634-60-6 see AML600 | 646-07-1 see IKT000 | 685-09-6 see MNN000 | 710-25-8 see NGL500 | 764-42-1 see FOX000 |
| 634-66-2 see TBN740 | 646-13-9 see IJN100 | 685-91-6 see DHI200 | 712-48-1 see CGN000 | 764-48-7 see EJL500 |
| 634-90-2 see TBN745 | 646-20-8 see HBD000 | 687-46-7 see EHH200 | 712-50-5 see PES750 | 764-84-9 see AGS750 |
| 634-93-5 see TIJ750 | 646-25-3 see DAG650 | 687-48-9 see EII500 | 712-68-5 see NGI500 | 764-99-8 see DJE600 |
| 634-95-7 see DKD650 | 650-42-0 see PMD350 | 687-78-5 see TJW000 | 713-68-8 see PDV250 | 765-05-9 see EEE200 |
| 635-22-3 see CJA185 | 650-51-1 see TII500 | 687-80-9 see TJW250 | 713-95-1 see HBP450 | 765-09-3 see BOI500 |
| 635-41-6 see TKX250 | 650-52-2 see BLO280 | 687-81-0 see TJV785 | 715-48-0 see MMR000 | 765-12-8 see TDY800 |
| 635-85-8 see DOI400 | 652-04-0 see MHF250 | 688-71-1 see TMY750 | 716-79-0 see PEL250 | 765-15-1 see DYA200 |
| 635-93-8 see CLD800 | 652-67-5 see HID350 | 688-73-3 see TIB500 | 719-22-2 see DDV500 | 765-30-0 see CQE250 |

| | | | | |
|---|---|---|---|---|
| **982-57-0** see CDP500 | **1034-01-1** see OFA000 | **1080-12-2** see MLI800 | **1122-58-3** see DQB600 | **1184-78-7** see TLE100 |
| **985-13-7** see PAH260 | **1038-95-5** see TNN750 | **1080-32-6** see DIU600 | **1122-60-7** see NFV500 | **1185-55-3** see MQF500 |
| **985-16-0** see SGS500 | **1045-21-2** see DHQ500 | **1082-88-8** see MLH250 | **1122-82-3** see ISJ000 | **1185-57-5** see FAS700 |
| **987-65-5** see AEM100 | **1053-74-3** see ACR500 | **1083-48-3** see NFO750 | **1122-90-3** see ARJ755 | **1186-09-0** see TJU800 |
| **987-78-0** see CMF350 | **1057-81-4** see HBT000 | **1083-57-4** see HJS850 | **1122-91-4** see BMT700 | **1187-00-4** see BKM500 |
| **989-38-8** see RGW000 | **1059-86-5** see DAK800 | **1084-65-7** see MQQ050 | **1123-54-2** see AIB340 | **1187-33-3** see DEH300 |
| **990-73-8** see PDW750 | **1063-55-4** see BSZ000 | **1085-12-7** see HBO650 | **1123-61-1** see DFP800 | **1187-58-2** see MOS900 |
| **991-42-4** see NNF000 | **1064-48-8** see FAB830 | **1085-98-9** see DFL200 | **1123-85-9** see HGR600 | **1187-59-3** see MGA300 |
| **992-21-2** see MRV250 | **1066-17-7** see PKD250 | **1087-21-4** see IMK000 | **1123-91-7** see MHI300 | **1187-93-5** see TKG800 |
| **992-94-9** see MPI625 | **1066-26-8** see MRK609 | **1088-11-5** see CGA500 | **1124-11-4** see TDV725 | **1189-71-5** see CLG250 |
| **993-11-3** see TDT500 | **1066-30-4** see CMH000 | **1088-80-8** see SAX210 | **1124-31-8** see PLM700 | **1189-85-1** see BQV000 |
| **993-16-8** see MQC750 | **1066-33-7** see ANB250 | **1090-53-5** see BFK325 | **1124-33-0** see NJA500 | **1190-16-5** see COR325 |
| **993-43-1** see EOP600 | **1066-45-1** see CLT000 | **1095-03-0** see TMR500 | **1125-27-5** see DFJ800 | **1190-53-0** see BOM750 |
| **993-74-8** see BJG125 | **1066-48-4** see TBW025 | **1095-90-5** see MDP000 | **1125-78-6** see TCY000 | **1190-53-0** see BQL000 |
| **993-86-2** see EPY600 | **1066-57-5** see EPS000 | **1098-97-1** see BMB000 | **1125-88-8** see DOG700 | **1190-93-8** see ACQ250 |
| **994-31-0** see TJV000 | **1067-14-7** see TJS250 | **1099-87-2** see SJK400 | **1126-34-7** see AIF250 | **1191-15-7** see DNI600 |
| **994-43-4** see TJU250 | **1067-29-4** see BLT300 | **1099-87-2** see SJK410 | **1126-78-9** see BQH850 | **1191-16-8** see DOQ350 |
| **994-50-3** see HCX100 | **1067-33-0** see DBF800 | **1103-05-5** see DVY800 | **1126-79-0** see BSF750 | **1191-50-0** see SIO000 |
| **994-65-0** see TED650 | **1067-47-6** see COS800 | **1104-22-9** see MBX250 | **1128-05-8** see MPQ600 | **1191-79-3** see BAI800 |
| **996-05-4** see DWU200 | **1067-53-4** see TNJ500 | **1109-85-9** see MRN700 | **1128-08-1** see DLQ600 | **1191-80-6** see MDF250 |
| **996-08-7** see DDM400 | **1067-97-6** see TID500 | **1111-39-3** see ACH500 | **1128-67-2** see MHJ250 | **1191-96-4** see EHU500 |
| **996-31-6** see PLK650 | **1067-99-8** see COR500 | **1111-44-0** see DIH000 | **1129-41-5** see MIB750 | **1192-22-9** see ECC600 |
| **997-95-5** see DRQ200 | **1068-22-0** see DJW875 | **1111-78-0** see AND750 | **1129-50-6** see BSV500 | **1192-28-5** see CPW750 |
| **998-30-1** see TJM750 | **1068-27-5** see DRJ825 | **1112-63-6** see HCX050 | **1130-69-4** see NLQ500 | **1192-30-9** see TCS550 |
| **998-40-3** see TIA300 | **1068-57-1** see ACM750 | **1113-02-6** see DNX800 | **1131-18-6** see PFQ350 | **1192-75-2** see BJP450 |
| **999-21-3** see DBK200 | **1068-90-2** see AAK750 | **1113-14-0** see BLG500 | **1132-20-3** see CEC100 | **1192-89-8** see PFM250 |
| **999-29-1** see DCO600 | **1069-23-4** see DMU000 | **1113-21-9** see GDG300 | **1132-39-4** see PGH250 | **1193-02-8** see AIF750 |
| **999-61-1** see HNT600 | **1069-66-5** see PNX750 | **1113-38-8** see ANO750 | **1133-64-8** see NJK150 | **1193-54-0** see DFN800 |
| **999-81-5** see CMF400 | **1070-01-5** see TJI000 | **1113-59-3** see BOD550 | **1134-23-2** see EHT500 | **1194-02-1** see FGH100 |
| **999-97-3** see HED500 | **1070-11-7** see EDW875 | **1114-51-8** see DJX250 | **1134-35-6** see DQS100 | **1194-65-6** see DER800 |
| **1000-40-4** see BHL500 | **1070-19-5** see BQI250 | **1114-71-2** see PNF500 | **1134-47-0** see BAC275 | **1195-16-0** see TCZ000 |
| **1000-70-0** see BLQ950 | **1070-42-4** see BIG500 | **1115-12-4** see MDL250 | **1135-99-5** see DWO400 | **1195-79-5** see TLW250 |
| **1001-55-4** see COP750 | **1070-70-8** see TDQ100 | **1115-47-5** see ACQ270 | **1136-45-4** see MNV000 | **1196-57-2** see QSJ100 |
| **1001-58-7** see COM750 | **1070-75-3** see LGP875 | **1115-70-4** see DQR800 | **1136-84-1** see TCY275 | **1197-16-6** see CCI550 |
| **1001-62-3** see DNX300 | **1070-78-6** see TBT250 | **1115-99-7** see TJR250 | **1137-41-3** see AIR250 | **1197-18-8** see AJV500 |
| **1002-16-0** see AOL250 | **1071-22-3** see CON000 | **1116-01-4** see PHQ250 | **1137-79-7** see BGF899 | **1197-55-3** see AID700 |
| **1002-36-4** see HBS500 | **1071-23-4** see EED000 | **1116-54-7** see NKM000 | **1138-80-3** see CBR125 | **1198-27-2** see ALK000 |
| **1002-62-6** see SGB600 | **1071-27-8** see COR750 | **1116-76-3** see DVL000 | **1139-30-6** see CCN100 | **1198-55-6** see TBU500 |
| **1002-67-1** see DJE800 | **1071-39-2** see DNQ800 | **1117-41-5** see TMJ100 | **1141-37-3** see BHP125 | **1198-77-2** see SKO575 |
| **1002-84-2** see PAZ000 | **1071-73-4** see ABH750 | **1117-55-1** see HFS600 | **1141-88-4** see DXJ800 | **1198-97-6** see PGF800 |
| **1002-89-7** see ANU200 | **1071-83-6** see PHA500 | **1117-71-1** see MHR790 | **1142-70-7** see BOR000 | **1199-03-7** see QSJ000 |
| **1003-03-8** see CQA000 | **1071-93-8** see AEQ250 | **1117-94-8** see DGQ600 | **1145-73-9** see DUB800 | **1199-85-5** see CIF250 |
| **1003-10-7** see TDC800 | **1071-98-3** see DGS000 | **1117-99-3** see TIH825 | **1146-95-8** see DWF200 | **1200-89-1** see TCU650 |
| **1003-40-3** see AMJ000 | **1072-15-7** see SIX500 | **1118-12-3** see BSS310 | **1149-99-1** see LIO600 | **1204-06-4** see ICO000 |
| **1003-67-4** see MOY790 | **1072-52-2** see ASI000 | **1118-14-5** see TMI000 | **1154-59-2** see TBV000 | **1204-59-7** see AIW250 |
| **1003-73-2** see MOY550 | **1072-53-3** see EJP000 | **1118-27-0** see LGC000 | **1155-03-9** see ZUA000 | **1204-78-0** see MGF750 |
| **1003-78-7** see DUD400 | **1072-66-8** see ASI250 | **1118-39-4** see AAW500 | **1155-38-0** see MGZ000 | **1204-79-1** see AIW000 |
| **1004-22-4** see SJF500 | **1072-71-5** see TES250 | **1118-42-9** see DVV200 | **1155-49-3** see PNB250 | **1208-52-2** see MJP750 |
| **1004-40-6** see AMS750 | **1072-85-1** see FGX000 | **1118-46-3** see BSR250 | **1156-19-0** see TGJ500 | **1210-05-5** see DVV800 |
| **1005-30-7** see IEE100 | **1072-93-1** see VQA100 | **1118-58-7** see MNH750 | **1160-36-7** see TKP850 | **1210-35-1** see DCX300 |
| **1005-67-0** see BRV100 | **1073-05-8** see TLR750 | **1119-16-0** see MQJ500 | **1161-13-3** see CBR220 | **1210-56-6** see MLJ100 |
| **1005-93-2** see ELJ500 | **1073-06-9** see FGY000 | **1119-22-8** see DXR500 | **1162-06-7** see TMT000 | **1211-28-5** see PNS000 |
| **1006-59-3** see DJU700 | **1073-23-0** see DTV089 | **1119-34-2** see AQW000 | **1162-65-8** see AEU250 | **1211-40-1** see ALM000 |
| **1006-99-1** see CIH100 | **1073-26-3** see PMX320 | **1119-68-2** see PBX250 | **1163-19-5** see PAU500 | **1214-39-7** see BDX090 |
| **1007-22-3** see BJG150 | **1073-62-7** see BEQ500 | **1119-69-3** see DGS200 | **1163-36-6** see CMV400 | **1215-16-3** see BHO500 |
| **1007-33-6** see PNN300 | **1073-72-9** see MPV300 | **1119-85-3** see HFC000 | **1164-33-6** see HEL500 | **1215-83-4** see CMW500 |
| **1008-79-3** see PGE250 | **1073-79-6** see TCQ350 | **1119-94-4** see DYA810 | **1164-38-1** see ELF500 | **1218-34-4** see ADE075 |
| **1010-61-3** see HIZ000 | **1073-91-2** see TDW275 | **1119-97-7** see TCB200 | **1165-39-5** see AEV000 | **1218-35-5** see OKO500 |
| **1011-50-3** see QNJ000 | **1074-52-8** see ANR250 | **1120-01-0** see HCP900 | **1165-48-6** see DNV000 | **1219-20-1** see OOE100 |
| **1011-73-0** see DVA800 | **1074-95-9** see MCE250 | **1120-04-3** see OBG100 | **1166-52-5** see DXX200 | **1220-83-3** see SNL800 |
| **1011-95-6** see DWM800 | **1074-98-2** see MMP750 | **1120-10-1** see CJH500 | **1169-26-2** see HEI000 | **1220-94-6** see AKP250 |
| **1012-72-2** see DDV450 | **1075-30-5** see BFZ120 | **1120-21-4** see UJS000 | **1170-02-1** see EIV100 | **1221-56-3** see SKM000 |
| **1012-82-4** see HOJ000 | **1075-61-2** see AMF500 | **1120-46-3** see LDQ000 | **1172-18-5** see DAB800 | **1222-05-5** see GBU000 |
| **1014-69-3** see INR000 | **1075-76-9** see AOT100 | **1120-48-5** see DVJ600 | **1173-88-2** see MNV250 | **1222-57-7** see ZUA200 |
| **1014-98-8** see XSS260 | **1076-22-8** see MQN500 | **1120-71-4** see PML400 | **1174-83-0** see PHN250 | **1222-98-6** see NFS505 |
| **1016-75-7** see DKV200 | **1076-38-6** see HJY000 | **1121-30-8** see HOB500 | **1176-08-5** see DTO600 | **1223-31-0** see BLA750 |
| **1017-56-7** see TLW750 | **1076-44-4** see PBE250 | **1121-31-9** see MCQ700 | **1178-29-6** see MQR200 | **1223-36-5** see CJN750 |
| **1018-71-9** see CFC000 | **1076-56-8** see MPW650 | **1121-47-7** see FPR000 | **1178-99-0** see MFF625 | **1224-64-2** see NIM000 |
| **1021-11-0** see GLS700 | **1076-98-8** see DXO600 | **1121-53-5** see BFJ850 | **1179-69-7** see TEZ000 | **1225-55-4** see POF250 |
| **1024-57-3** see EBW500 | **1078-21-3** see PEE500 | **1121-70-6** see SIM100 | **1181-54-0** see MND500 | **1225-60-1** see AEG625 |
| **1024-65-3** see IHX200 | **1078-38-2** see ACO750 | **1121-92-2** see SMV500 | **1182-87-2** see EAQ050 | **1225-65-6** see DYB800 |
| **1025-15-6** see THS100 | **1078-79-1** see BGY000 | **1122-10-7** see DDP400 | **1184-53-8** see MIP250 | **1227-61-8** see DIB600 |
| **1027-14-1** see DHL800 | **1079-01-2** see MSC050 | **1122-17-4** see DFN700 | **1184-57-2** see MLG000 | **1227-61-8** see MCH250 |
| **1030-06-4** see CDO625 | **1079-21-6** see BGG250 | **1122-54-9** see ADA365 | **1184-58-3** see DOQ700 | **1229-29-4** see AEG750 |
| **1031-47-6** see AIX000 | **1079-33-0** see BDG250 | **1122-56-1** see CPB050 | **1184-64-1** see CBW200 | **1229-35-2** see MDT500 |

| | | | | |
|---|---|---|---|---|
| **1505-95-9** see IOU000 | **1622-61-3** see CMW000 | **1705-85-7** see MIN750 | **1817-68-1** see MHR050 | **1916-07-0** see MQF300 |
| **1508-65-2** see OPK000 | **1622-62-4** see FDD100 | **1706-01-0** see MKC750 | **1817-73-8** see DUS200 | **1916-59-2** see QCJ275 |
| **1510-31-2** see DKJ225 | **1622-79-3** see FKI000 | **1707-14-8** see MNV750 | **1820-50-4** see CPD650 | **1918-00-9** see MEL500 |
| **1515-76-0** see ABM250 | **1623-19-4** see THN750 | **1707-95-5** see ONY000 | **1824-81-3** see ALC500 | **1918-02-1** see PIB900 |
| **1516-17-2** see AAU750 | **1623-24-1** see PHE500 | **1708-39-0** see BBA000 | **1825-21-4** see MNH250 | **1918-13-4** see DGM600 |
| **1516-27-4** see HGI600 | **1623-99-0** see PGI100 | **1709-50-8** see DIU400 | **1825-58-7** see DRU200 | **1918-16-7** see CHS500 |
| **1516-32-1** see BSO500 | **1624-02-8** see BLS750 | **1709-70-2** see TMJ000 | **1825-62-3** see EFL500 | **1918-18-9** see DEV600 |
| **1518-15-6** see BIY600 | **1628-58-6** see DQC400 | **1711-42-8** see DVV550 | **1829-00-1** see CMP050 | **1919-43-3** see DGK000 |
| **1518-16-7** see TBW750 | **1629-58-9** see PBR250 | **1712-64-7** see IQP000 | **1833-31-4** see BEE800 | **1920-05-4** see DRS000 |
| **1518-58-7** see DJD000 | **1631-29-4** see CKC500 | **1713-07-1** see AAJ125 | **1835-04-7** see PMX600 | **1920-21-4** see DLI600 |
| **1519-47-7** see XSS000 | **1631-82-9** see CIS625 | **1715-33-9** see PMA100 | **1836-75-5** see DFT800 | **1921-70-6** see PMD500 |
| **1520-71-4** see TNA750 | **1632-16-2** see EKR500 | **1715-81-7** see APM750 | **1836-77-7** see NIW500 | **1923-76-8** see EQC000 |
| **1520-78-1** see TLU175 | **1633-83-6** see BOU250 | **1716-09-2** see LIN400 | **1837-57-6** see EDW500 | **1928-45-6** see DFZ000 |
| **1527-12-4** see PGL270 | **1634-02-2** see TBM750 | **1717-00-6** see FOO550 | **1838-08-0** see OBE000 | **1929-73-3** see DFY709 |
| **1528-74-1** see DUS000 | **1634-04-4** see MHV859 | **1718-34-9** see SIU000 | **1838-56-8** see HIO875 | **1929-77-7** see PNI750 |
| **1529-30-2** see TJV250 | **1634-73-7** see AMV375 | **1719-53-5** see DEY800 | **1838-59-1** see AGH000 | **1929-82-4** see CLP750 |
| **1530-48-9** see AGT750 | **1634-78-2** see OPK250 | **1722-62-9** see CBR250 | **1840-42-2** see FMI000 | **1929-86-8** see CLO200 |
| **1537-62-8** see CGY825 | **1637-24-7** see HFQ000 | **1725-01-5** see OLE100 | **1841-19-6** see PFU250 | **1929-88-0** see MHM500 |
| **1538-09-6** see BFC750 | **1638-22-8** see BSE450 | **1732-14-5** see PON500 | **1843-05-6** see HND100 | **1933-50-2** see DUI000 |
| **1541-60-2** see MHJ750 | **1639-09-4** see HBD500 | **1733-25-1** see IRN100 | **1845-11-0** see NAD500 | **1934-21-0** see FAG140 |
| **1544-46-3** see FFE000 | **1639-60-7** see PNA500 | **1733-63-7** see TMS100 | **1845-38-1** see TLO250 | **1936-40-9** see NOB700 |
| **1544-68-9** see ISM000 | **1639-66-3** see SOD300 | **1734-79-8** see NFT425 | **1847-24-1** see FDA100 | **1937-19-5** see AKC800 |
| **1551-44-6** see CPI300 | **1640-89-7** see EHU000 | **1734-91-4** see BGO000 | **1847-58-1** see SIB700 | **1937-37-7** see AQP000 |
| **1552-12-1** see CPR825 | **1641-74-3** see EQQ100 | **1738-25-6** see DPU000 | **1847-63-8** see NAD750 | **1940-18-7** see EHR500 |
| **1553-34-0** see THL500 | **1642-54-2** see DIW200 | **1739-53-3** see DFH300 | **1851-71-4** see BJU250 | **1940-42-7** see LEN050 |
| **1553-36-2** see BGX500 | **1643-19-2** see TBK500 | **1740-19-8** see DAK400 | **1851-77-0** see DKE400 | **1940-57-4** see BNK500 |
| **1553-60-2** see IJG000 | **1643-20-5** see DRS200 | **1745-81-9** see AGQ500 | **1852-14-8** see EJC100 | **1940-89-2** see FAC195 |
| **1555-58-4** see MHQ750 | **1646-26-0** see ACC100 | **1746-01-6** see TAI000 | **1852-16-0** see BPM660 | **1941-24-8** see MGN150 |
| **1557-57-9** see DGS300 | **1646-75-9** see MLX800 | **1746-09-4** see BLR000 | **1854-26-8** see DTG000 | **1941-27-1** see TBL250 |
| **1558-25-4** see CIY325 | **1646-87-3** see MLW750 | **1746-13-0** see AGR000 | **1861-32-1** see TBV250 | **1941-30-6** see TEC750 |
| **1559-35-9** see EKX500 | **1646-88-4** see AFK000 | **1746-77-6** see IOJ000 | **1863-63-4** see ANB100 | **1942-52-5** see DHO400 |
| **1561-49-5** see DGV650 | **1649-08-7** see DFA000 | **1746-81-2** see CKD500 | **1866-15-5** see ADC300 | **1943-16-4** see CLT250 |
| **1563-01-5** see DIJ300 | **1649-18-9** see FLU000 | **1750-46-5** see AJH750 | **1866-31-5** see AGC000 | **1943-79-9** see PFS350 |
| **1563-66-2** see CBS275 | **1653-19-6** see DEU400 | **1752-24-5** see IBJ100 | **1866-43-9** see RLZ000 | **1943-83-5** see IKH000 |
| **1563-67-3** see DLS800 | **1653-64-1** see MJT000 | **1752-30-3** see TFQ250 | **1867-66-9** see CKD750 | **1944-12-3** see FAQ100 |
| **1569-02-4** see EFF500 | **1656-16-2** see BLI250 | **1754-47-8** see PFV750 | **1867-72-7** see BQM309 | **1944-83-8** see MNX000 |
| **1569-69-3** see CPB625 | **1656-48-0** see OQQ000 | **1754-58-1** see PEV500 | **1871-96-1** see SBK500 | **1945-32-0** see AMD500 |
| **1570-45-2** see ELU000 | **1660-94-2** see MJT100 | **1755-52-8** see CBO625 | **1874-58-4** see BFL125 | **1945-91-1** see PMO800 |
| **1570-64-5** see CFE000 | **1661-03-6** see MAI250 | **1755-67-5** see CHD675 | **1874-62-0** see EFF000 | **1948-33-0** see BRM500 |
| **1571-30-8** see HOE000 | **1663-35-0** see VPZ000 | **1757-18-2** see DIX600 | **1875-92-9** see DQQ000 | **1949-07-1** see MQH250 |
| **1571-33-1** see PFV500 | **1665-48-1** see XVS000 | **1758-68-5** see DBO800 | **1877-24-3** see DHM309 | **1949-20-8** see OOE000 |
| **1576-35-8** see THE250 | **1665-59-4** see DJV300 | **1759-28-0** see MQM800 | **1877-52-7** see FAC165 | **1949-45-7** see MQR350 |
| **1582-09-8** see DUV600 | **1666-13-3** see BLF500 | **1760-24-3** see TLC500 | **1877-53-8** see FAC160 | **1951-12-8** see CEW000 |
| **1585-74-6** see CGC200 | **1668-19-5** see DYE409 | **1761-71-3** see MJQ260 | **1879-09-0** see BST000 | **1951-25-3** see AJK750 |
| **1586-92-1** see EER000 | **1668-54-8** see MGH800 | **1762-95-4** see ANW750 | **1881-37-4** see FIA500 | **1952-11-0** see GAC000 |
| **1587-41-3** see DFE550 | **1670-14-0** see BBM750 | **1764-39-2** see FHP000 | **1881-75-0** see FJQ000 | **1953-02-2** see MCI375 |
| **1589-49-7** see MFL000 | **1671-82-5** see MML750 | **1765-48-6** see EAE875 | **1881-76-1** see FJR000 | **1953-56-6** see DNX000 |
| **1591-30-6** see BGF250 | **1672-46-4** see DKN300 | **1768-31-6** see PAV225 | **1882-26-4** see PPH050 | **1953-99-7** see TBT200 |
| **1592-23-0** see CAX350 | **1674-62-0** see MHP400 | **1769-24-0** see MPF500 | **1884-64-6** see COP775 | **1954-28-5** see TJQ333 |
| **1594-56-5** see DVF800 | **1674-96-0** see ARR875 | **1769-99-9** see COL250 | **1885-14-9** see CBX109 | **1955-21-1** see DNF200 |
| **1596-52-7** see DVE000 | **1675-54-3** see BLD750 | **1770-80-5** see CDS025 | **1885-29-6** see APJ750 | **1955-45-9** see DTH000 |
| **1596-84-5** see DQD400 | **1677-46-9** see HMR500 | **1772-03-8** see GAT000 | **1886-45-9** see HII000 | **1961-77-9** see CDY275 |
| **1597-82-6** see PAL600 | **1678-25-7** see BBR750 | **1774-47-6** see TMG800 | **1886-81-3** see DXW400 | **1962-75-0** see DEH700 |
| **1599-49-1** see MNM450 | **1679-07-8** see CPW300 | **1777-84-0** see NEL000 | **1888-71-7** see HCM000 | **1965-09-9** see OQI000 |
| **1600-27-7** see MCS750 | **1680-21-3** see TJQ100 | **1778-08-1** see DUA800 | **1888-89-7** see DHC800 | **1965-29-3** see HKR000 |
| **1600-31-3** see PIE525 | **1682-39-9** see AKC250 | **1779-25-5** see CGB500 | **1888-91-1** see ACF100 | **1966-58-1** see DET400 |
| **1600-37-9** see PAY200 | **1688-71-7** see AJC750 | **1779-51-7** see BSR900 | **1888-94-4** see CHC675 | **1967-16-4** see CEX250 |
| **1600-44-8** see TDQ500 | **1689-82-3** see HJF000 | **1779-81-3** see TEV600 | **1892-29-1** see DXM600 | **1972-08-3** see TCM250 |
| **1603-40-3** see ALC000 | **1689-83-4** see HKB500 | **1780-40-1** see TBU250 | **1892-43-9** see CJO500 | **1972-28-7** see DIT300 |
| **1603-41-4** see AMA010 | **1689-84-5** see DDP000 | **1783-81-9** see ALX100 | **1892-54-2** see PDA500 | **1977-10-2** see DCS200 |
| **1604-01-9** see CNH730 | **1689-89-0** see HLJ500 | **1785-74-6** see DDD000 | **1892-80-4** see CBF825 | **1977-11-3** see HOU059 |
| **1605-51-2** see EIL000 | **1689-99-2** see DDM200 | **1786-81-8** see CMS250 | **1893-33-0** see FHG000 | **1981-58-4** see SJW500 |
| **1605-58-9** see DJQ200 | **1693-71-6** see THN250 | **1788-93-8** see MII250 | **1896-62-4** see BAY275 | **1982-36-1** see CJR809 |
| **1606-67-3** see PON000 | **1694-09-3** see FAG120 | **1789-58-8** see DFK000 | **1897-45-6** see TBQ750 | **1982-37-2** see MPE250 |
| **1607-17-6** see PCR150 | **1696-17-9** see BCM250 | **1797-74-6** see PMS500 | **1897-96-7** see LID000 | **1982-47-4** see CJQ000 |
| **1607-30-3** see DQW600 | **1696-20-4** see ACR750 | **1798-11-4** see NII000 | **1899-02-1** see TMB500 | **1982-67-8** see MDU100 |
| **1609-47-8** see DIZ100 | **1698-53-9** see DGE800 | **1798-49-8** see DVW000 | **1901-26-4** see MNS600 | **1983-10-4** see FME000 |
| **1610-17-9** see EGD000 | **1698-60-8** see PEE750 | **1798-50-1** see PIY750 | **1904-98-9** see POJ500 | **1984-06-1** see SIX600 |
| **1610-18-0** see MFL250 | **1701-69-5** see PMX300 | **1806-29-7** see BGG000 | **1907-13-7** see ABW750 | **1984-77-6** see LBM000 |
| **1615-80-1** see DJL400 | **1701-71-9** see PNV755 | **1807-55-2** see MJO000 | **1910-36-7** see HLV000 | **1984-87-8** see DVW900 |
| **1616-88-2** see MEO000 | **1701-73-1** see VBA100 | **1808-12-4** see BNW500 | **1910-42-5** see PAJ000 | **1985-12-2** see BNW825 |
| **1617-90-9** see VLF000 | **1702-17-6** see DGJ100 | **1809-19-4** see DEG800 | **1910-68-5** see MKW250 | **1985-84-8** see AEN750 |
| **1618-08-2** see DCP775 | **1703-58-8** see BOU500 | **1809-20-7** see DNQ600 | **1912-24-9** see ARQ725 | **1988-89-2** see PFD400 |
| **1619-34-7** see QUJ990 | **1704-04-7** see AJY000 | **1811-28-5** see PMH100 | **1912-26-1** see TJL500 | **1990-07-4** see CAT175 |
| **1622-32-8** see CGO125 | **1704-62-7** see DPA600 | **1812-30-2** see BMN750 | **1912-84-1** see OAZ000 | **1990-90-5** see AMA000 |

| | | | | |
|---|---|---|---|---|
| 2430-27-5 see PNX600 | 2497-18-9 see HFE100 | 2592-85-0 see DNU325 | 2666-17-3 see CMM070 | 2785-87-7 see MFM750 |
| 2431-50-7 see TIL360 | 2497-21-4 see HFE200 | 2592-95-2 see HJN650 | 2668-92-0 see NAQ500 | 2787-93-1 see HEW100 |
| 2432-90-8 see PHW550 | 2497-34-9 see ELD500 | 2593-15-9 see EFK000 | 2669-32-1 see DRU400 | 2797-51-5 see AJI250 |
| 2432-99-7 see AMW000 | 2498-27-3 see HNN500 | 2594-20-9 see IPA000 | 2669-35-4 see TID100 | 2799-07-7 see TNR475 |
| 2435-16-7 see HBP425 | 2498-75-1 see MGV250 | 2595-54-2 see DJI000 | 2674-91-1 see DSK600 | 2801-68-5 see DOK600 |
| 2435-64-5 see SNW800 | 2498-76-2 see MGV000 | 2597-03-7 see DRR400 | 2675-77-6 see CJA100 | 2802-70-2 see FJI510 |
| 2435-76-9 see DCQ600 | 2498-77-3 see MGU750 | 2597-09-3 see AGR125 | 2679-01-8 see MJU250 | 2804-00-4 see FLJ000 |
| 2436-66-0 see MPF750 | 2499-58-3 see HBL100 | 2597-54-8 see ACL000 | 2680-03-7 see DOP800 | 2807-30-9 see PNG750 |
| 2436-90-0 see CMT050 | 2499-95-8 see ADV000 | 2597-93-5 see EME050 | 2686-99-9 see TMD000 | 2808-86-8 see TBT750 |
| 2437-25-4 see DXT400 | 2507-91-7 see KFA100 | 2597-95-7 see MLG750 | 2687-25-4 see TGY800 | 2809-21-4 see HKS780 |
| 2437-29-8 see MAK600 | 2508-18-1 see FIT200 | 2598-25-6 see CJK000 | 2687-91-4 see EPC700 | 2812-73-9 see CLJ750 |
| 2438-32-6 see PJJ325 | 2508-20-5 see NKF500 | 2598-31-4 see HOE200 | 2688-84-8 see PDR490 | 2813-95-8 see ACE500 |
| 2438-49-5 see DTK800 | 2508-23-8 see NBD000 | 2598-70-1 see MNU100 | 2691-41-0 see CQH250 | 2814-20-2 see IQL000 |
| 2438-51-9 see PCZ000 | 2510-95-4 see DVX600 | 2598-71-2 see MFG200 | 2696-84-6 see PNE000 | 2818-88-4 see MHI500 |
| 2438-72-4 see BPP750 | 2511-10-6 see DJR700 | 2598-72-3 see MNR100 | 2696-92-6 see NMH000 | 2818-89-5 see DQO600 |
| 2438-88-2 see TBR250 | 2517-98-8 see ABR250 | 2598-73-4 see CJP500 | 2697-60-1 see AJS950 | 2820-51-1 see NDP400 |
| 2439-01-2 see ORU000 | 2519-30-4 see BMA000 | 2598-74-5 see MNP500 | 2697-65-6 see AJS900 | 2823-90-7 see FFO000 |
| 2439-10-3 see DXX400 | 2521-01-9 see EGT000 | 2598-75-6 see CJK100 | 2698-38-6 see EMR000 | 2823-91-8 see FFN000 |
| 2439-35-2 see DPB300 | 2522-81-8 see PCV350 | 2598-76-7 see MNP400 | 2698-41-1 see CEQ600 | 2823-93-0 see FFM000 |
| 2439-77-2 see AOT750 | 2524-03-0 see DTQ600 | 2600-55-7 see DVC200 | 2699-79-8 see SOU500 | 2823-94-1 see FFP000 |
| 2439-99-8 see BLG250 | 2524-04-1 see DJW600 | 2602-46-2 see CMO000 | 2700-89-2 see DFS200 | 2823-95-2 see FFQ000 |
| 2440-22-4 see HML500 | 2528-36-1 see DEG600 | 2605-44-9 see CAG775 | 2702-72-9 see SGH500 | 2824-10-4 see FFL000 |
| 2440-29-1 see BSV750 | 2530-10-1 see DUG425 | 2606-85-1 see MHL250 | 2703-13-1 see MOB599 | 2825-00-5 see DTU200 |
| 2440-45-1 see BJT250 | 2530-46-3 see DXY300 | 2606-87-3 see FJO000 | 2705-87-5 see AGC500 | 2825-15-2 see NHE100 |
| 2441-88-5 see FAR200 | 2530-83-8 see ECH000 | 2610-11-9 see CMO885 | 2706-28-7 see CMM758 | 2825-60-7 see FDB000 |
| 2442-10-6 see ODW028 | 2530-85-0 see TLC250 | 2610-86-8 see WAT209 | 2706-47-0 see TNC725 | 2825-82-3 see TLR675 |
| 2443-39-2 see ECD500 | 2530-99-6 see AGS500 | 2611-61-2 see UVA150 | 2706-47-0 see TNC750 | 2827-46-5 see TNJ825 |
| 2445-07-0 see USJ075 | 2532-17-4 see IEE050 | 2611-82-7 see FMU080 | 2706-50-5 see AOA750 | 2828-42-4 see PEQ500 |
| 2445-76-3 see HFV100 | 2532-50-5 see BLM750 | 2612-33-1 see CKU625 | 2709-56-0 see FMO129 | 2832-40-8 see AAQ250 |
| 2445-83-2 see MIP775 | 2533-82-6 see MGQ750 | 2614-06-4 see TEH510 | 2713-09-9 see FFS000 | 2834-84-6 see AKD250 |
| 2446-84-6 see DQH509 | 2536-18-7 see DUW503 | 2614-76-8 see BJY825 | 2719-13-3 see NIQ550 | 2835-39-4 see ISV000 |
| 2447-57-6 see AIE500 | 2536-31-4 see CDT000 | 2619-97-8 see GHE000 | 2719-23-5 see TEW100 | 2835-95-2 see AKJ750 |
| 2448-39-7 see AJJ500 | 2536-91-6 see MGD500 | 2621-62-7 see DGB480 | 2720-73-2 see PKV100 | 2835-99-6 see AKZ000 |
| 2448-68-2 see PII200 | 2537-36-2 see TCD000 | 2622-08-4 see TJF750 | 2724-69-8 see MOA500 | 2836-32-0 see SHT000 |
| 2449-49-2 see DSO800 | 2538-85-4 see HLI100 | 2622-21-1 see VNZ990 | 2726-03-6 see CMW250 | 2840-26-8 see AIA250 |
| 2450-71-7 see POA000 | 2539-17-5 see TBQ290 | 2622-26-6 see PIW000 | 2728-04-3 see DKD000 | 2842-37-7 see PFB500 |
| 2454-11-7 see FNK040 | 2539-53-9 see IKO100 | 2623-23-6 see MCG750 | 2731-16-0 see DBA200 | 2842-38-8 see CPG125 |
| 2454-37-7 see AKR000 | 2540-82-1 see DRR200 | 2624-03-5 see EDB200 | 2732-09-4 see FNR000 | 2845-82-1 see PPN500 |
| 2457-76-3 see CEG750 | 2541-68-6 see FJP000 | 2624-17-1 see SGB550 | 2735-04-8 see DNY500 | 2847-65-6 see TMV800 |
| 2461-15-6 see GGY100 | 2541-69-7 see MGW750 | 2624-31-9 see PLO000 | 2736-21-2 see MFS500 | 2850-61-5 see DDV250 |
| 2463-45-8 see CBW400 | 2545-59-7 see TAH900 | 2624-43-3 see FBP100 | 2736-23-4 see DGK900 | 2852-07-5 see TIL350 |
| 2463-53-8 see NNA300 | 2549-90-8 see EKY000 | 2624-44-4 see DIS600 | 2736-80-3 see DVD000 | 2854-16-2 see ALB250 |
| 2463-84-5 see NFT000 | 2549-93-1 see BGT750 | 2626-34-8 see DQO650 | 2738-18-3 see DSD800 | 2855-19-8 see DXU400 |
| 2465-27-2 see IBA000 | 2550-26-7 see PDF800 | 2628-16-2 see ABW550 | 2740-04-7 see DNV200 | 2856-75-9 see MOM750 |
| 2465-29-4 see ADK000 | 2551-62-4 see SOI000 | 2628-17-3 see VQA200 | 2743-38-6 see DDE300 | 2857-03-6 see TIB750 |
| 2466-76-4 see ACO250 | 2552-89-8 see AFH275 | 2629-78-9 see BLM000 | 2746-81-8 see PMI250 | 2858-66-4 see PAO500 |
| 2467-12-1 see TMO250 | 2556-10-7 see EES380 | 2631-37-0 see CQI500 | 2747-31-1 see DTK600 | 2862-16-0 see DAE200 |
| 2467-13-2 see TJR500 | 2556-73-2 see MHY750 | 2631-40-5 see MIA250 | 2748-88-1 see MSB500 | 2865-19-2 see DEA000 |
| 2467-15-4 see TJK250 | 2564-65-0 see BBF500 | 2631-68-7 see TJE200 | 2750-76-7 see RKA000 | 2865-70-5 see CJM750 |
| 2469-34-3 see SBY000 | 2564-83-2 see TDT800 | 2633-54-7 see TIR250 | 2752-65-0 see CBA125 | 2867-47-2 see DPG600 |
| 2469-55-8 see OPC100 | 2567-83-1 see TCD002 | 2634-33-5 see BCE475 | 2757-18-8 see TEM399 | 2869-09-2 see MMD000 |
| 2470-73-7 see MKQ000 | 2570-26-5 see PBA000 | 2636-26-2 see COQ399 | 2757-28-0 see TKS500 | 2869-10-5 see MMD250 |
| 2472-17-5 see ARR250 | 2571-22-4 see TOE175 | 2637-34-5 see TFP300 | 2758-06-7 see BNK325 | 2869-12-7 see MJA750 |
| 2475-33-4 see CMU770 | 2571-86-0 see CNR725 | 2637-37-8 see QOJ100 | 2759-71-9 see CQJ250 | 2869-59-2 see DCY800 |
| 2475-44-7 see BKP500 | 2578-75-8 see FQJ000 | 2638-94-0 see ASL500 | 2761-24-2 see PBZ000 | 2869-60-5 see MJB000 |
| 2475-45-8 see TBG700 | 2579-20-6 see BGT500 | 2639-63-6 see HFM700 | 2763-96-4 see AKT750 | 2869-83-2 see DHM500 |
| 2475-46-9 see MGG250 | 2580-78-1 see BMM500 | 2641-56-7 see DIV600 | 2767-41-1 see DFC200 | 2870-71-5 see MGR250 |
| 2479-46-1 see REF070 | 2581-34-2 see NFV000 | 2642-37-7 see ICD000 | 2767-47-7 see DUG800 | 2871-01-4 see ALO750 |
| 2482-80-6 see EQI600 | 2581-69-3 see NIK000 | 2642-50-4 see MFR000 | 2767-54-6 see BOI750 | 2872-48-2 see DBX000 |
| 2485-10-1 see FEN000 | 2583-80-4 see DXG025 | 2642-71-9 see EKN000 | 2767-55-7 see DJB000 | 2873-97-4 see DTH200 |
| 2487-01-6 see BRU750 | 2586-57-4 see CMO600 | 2642-98-0 see CML800 | 2767-61-5 see BOK750 | 2881-62-1 see SEG700 |
| 2487-40-3 see TFL500 | 2586-58-5 see CMO810 | 2644-70-4 see HGV000 | 2768-02-7 see TLD000 | 2883-98-9 see IHX400 |
| 2487-90-3 see TLB750 | 2586-60-9 see CMP000 | 2646-17-5 see TGW000 | 2771-13-3 see MFC600 | 2884-67-5 see EHO700 |
| 2488-01-9 see PEW725 | 2587-75-9 see ABX175 | 2646-38-0 see CDL375 | 2773-92-4 see DNX400 | 2884-67-5 see EQD875 |
| 2489-52-3 see FLY200 | 2587-81-7 see TJU150 | 2648-61-5 see DEN200 | 2778-04-3 see EAS000 | 2885-39-4 see ABN000 |
| 2489-77-2 see TMH750 | 2587-82-8 see THY850 | 2650-18-2 see FMU059 | 2778-42-9 see TDX300 | 2886-89-7 see BFY750 |
| 2490-89-3 see CBI500 | 2587-84-0 see DED400 | 2651-85-6 see DJH500 | 2781-10-4 see BJQ250 | 2892-51-5 see DMJ600 |
| 2491-06-7 see MPI100 | 2587-90-8 see MIW250 | 2652-77-9 see DWA500 | 2782-57-2 see DGN200 | 2893-78-9 see SGG500 |
| 2491-76-1 see CGD250 | 2589-01-7 see TKL250 | 2655-14-3 see DTN200 | 2782-70-9 see BES250 | 2896-87-9 see FBM000 |
| 2492-26-4 see SIG500 | 2589-02-8 see BIY000 | 2655-19-8 see DEG400 | 2782-91-4 see TDX000 | 2897-21-4 see SBU200 |
| 2494-55-5 see PMW750 | 2589-15-3 see SCD500 | 2656-72-6 see HBO600 | 2783-17-7 see DXW800 | 2897-46-3 see ABX000 |
| 2494-56-6 see BSY300 | 2589-47-1 see DNB000 | 2664-63-3 see TFJ000 | 2783-94-0 see FAG150 | 2898-11-5 see MBY000 |
| 2495-54-7 see HIQ500 | 2591-57-3 see ENI175 | 2665-12-5 see TJF500 | 2783-96-2 see SGQ000 | 2898-12-6 see CGA000 |
| 2496-92-6 see ISD000 | 2591-86-8 see FOH000 | 2665-30-7 see MOB699 | 2784-86-3 see HIJ400 | 2899-02-7 see DEU259 |
| 2497-07-6 see OQS000 | 2592-62-3 see HCK000 | 2666-14-0 see TNL750 | 2784-94-3 see BKF250 | 2901-66-8 see MPH300 |

| | | | | |
|---|---|---|---|---|
| 3564-09-8 see FAG018 | 3669-32-7 see CMQ475 | 3736-26-3 see HIF575 | 3883-43-0 see DFE600 | 4114-31-2 see EHG000 |
| 3564-14-5 see CMP880 | 3670-09-5 see BFH100 | 3736-86-5 see THL750 | 3886-91-7 see DTH700 | 4116-10-3 see CEA100 |
| 3565-15-9 see SHX500 | 3671-00-9 see MPD250 | 3736-92-3 see DWJ400 | 3891-30-3 see SLM000 | 4120-77-8 see AAM250 |
| 3565-26-2 see NLR000 | 3671-71-4 see FDZ000 | 3737-09-5 see DNN600 | 3898-08-6 see DWO000 | 4120-78-9 see PCY500 |
| 3566-00-5 see MIB250 | 3671-78-1 see FDX000 | 3737-35-7 see DHW400 | 3898-45-1 see PDK300 | 4124-30-5 see DEM825 |
| 3566-10-7 see ANZ000 | 3676-91-3 see TDP525 | 3737-66-4 see UAG075 | 3900-31-0 see FDB100 | 4128-71-6 see PEH750 |
| 3567-38-2 see PGE000 | 3680-02-2 see MQM750 | 3737-72-2 see UAG000 | 3913-02-8 see BSA500 | 4135-11-9 see PKC550 |
| 3567-65-5 see CMM320 | 3681-71-8 see HFE150 | 3741-14-8 see NGA500 | 3913-71-1 see DAI350 | 4147-51-7 see EPN500 |
| 3567-66-6 see CMS231 | 3683-12-3 see PFP600 | 3741-38-6 see COV750 | 3915-83-1 see NCO500 | 4148-16-7 see LJD500 |
| 3567-69-9 see HJF500 | 3684-97-7 see MMT750 | 3750-18-3 see HDY100 | 3917-15-5 see AGV250 | 4150-34-9 see TMW000 |
| 3567-76-8 see DKR200 | 3685-84-5 see AAE500 | 3750-26-3 see PMC750 | 3926-62-3 see SFU500 | 4154-69-2 see BOV625 |
| 3568-23-8 see IDA500 | 3686-43-9 see DLY700 | 3758-54-1 see CCP675 | 3930-19-6 see SMA000 | 4163-15-9 see CQG750 |
| 3568-56-7 see ENI500 | 3686-69-9 see PNA200 | 3759-07-7 see DRM000 | 3931-89-3 see TFN750 | 4164-06-1 see ARJ800 |
| 3570-54-5 see TCJ500 | 3687-13-6 see DGD400 | 3759-61-3 see NBH200 | 3938-45-2 see DQX300 | 4164-07-2 see ARJ760 |
| 3570-58-9 see CHC750 | 3687-31-8 see LCK100 | 3759-92-0 see FPI100 | 3941-06-8 see PDV700 | 4164-28-7 see DSV200 |
| 3570-61-4 see SKL000 | 3687-61-4 see CPN750 | 3761-42-0 see DSS800 | 3942-54-9 see CKF000 | 4166-00-1 see MRN000 |
| 3570-75-0 see NDY500 | 3687-67-0 see CMU320 | 3761-53-3 see FMU070 | 3947-65-7 see NCE500 | 4168-79-0 see TDU000 |
| 3570-80-7 see FEV100 | 3688-08-2 see ISK000 | 3765-65-9 see TBI600 | 3949-14-2 see MFO250 | 4169-04-4 see PNL300 |
| 3570-93-2 see CPB650 | 3688-11-7 see PFO750 | 3766-55-0 see AGT000 | 3952-78-1 see AFM400 | 4170-30-3 see COB250 |
| 3571-74-2 see AGT250 | 3688-35-5 see AJE000 | 3766-60-7 see CKF750 | 3953-10-4 see EGZ000 | 4171-13-5 see ENJ500 |
| 3572-06-3 see AAR500 | 3688-53-7 see FQN000 | 3766-81-2 see MOV000 | 3958-19-8 see TMQ600 | 4176-53-8 see ALS000 |
| 3572-35-8 see BJA825 | 3688-66-2 see CNG250 | 3768-60-3 see BJJ125 | 3958-60-9 see NFM700 | 4180-23-8 see PMR250 |
| 3572-47-2 see DVM000 | 3688-79-7 see MEB750 | 3771-19-5 see MCB500 | 3963-79-9 see AIC500 | 4185-47-1 see NFW000 |
| 3572-74-5 see DSQ600 | 3688-85-5 see CIP500 | 3772-23-4 see BRQ000 | 3963-95-9 see MDO250 | 4189-47-3 see BMS250 |
| 3572-80-3 see COV500 | 3689-24-5 see SOD100 | 3772-26-7 see IRG100 | 3965-55-7 see BKN000 | 4193-55-9 see BGW000 |
| 3575-31-3 see OEQ000 | 3689-50-7 see AFL750 | 3772-76-7 see MDU300 | 3967-55-3 see DFI800 | 4194-69-8 see ICH000 |
| 3577-01-3 see CCR890 | 3689-76-7 see CDY325 | 3775-55-1 see ALM250 | 3978-86-7 see DLV800 | 4196-86-5 see PBC000 |
| 3581-11-1 see DTQ800 | 3690-04-8 see AMF375 | 3775-85-7 see EJD000 | 3979-76-8 see MHO000 | 4203-77-4 see CMU825 |
| 3582-17-0 see DKH200 | 3690-12-8 see AKO750 | 3775-90-4 see BQD250 | 3982-20-5 see PIA000 | 4205-90-7 see DGB500 |
| 3583-47-9 see DFF600 | 3690-50-4 see MOW500 | 3778-73-2 see IMH000 | 3982-91-0 see TFO000 | 4205-91-8 see CMX760 |
| 3585-32-8 see DUI709 | 3691-16-5 see VMA000 | 3778-76-5 see TGJ150 | 3983-39-9 see HNO500 | 4212-43-5 see PCO100 |
| 3586-14-9 see MNV770 | 3691-35-8 see CJJ000 | 3781-28-0 see FLN000 | 3983-40-2 see HNP000 | 4212-94-6 see ING400 |
| 3586-69-4 see NES000 | 3691-74-5 see GBB500 | 3785-34-0 see BHD250 | 3999-10-8 see PGJ500 | 4213-32-5 see BHU750 |
| 3589-21-7 see TAL000 | 3691-78-9 see BBU625 | 3787-28-8 see TIJ250 | 4000-16-2 see THN800 | 4213-34-7 see BHU500 |
| 3590-07-6 see EGU000 | 3692-90-8 see PMN250 | 3789-77-3 see DRC800 | 4003-94-5 see NMC000 | 4213-40-5 see BIA000 |
| 3594-15-8 see DVN909 | 3693-22-9 see DDB600 | 3792-59-4 see SCC000 | 4005-51-0 see AMR250 | 4213-41-6 see BIC600 |
| 3598-37-6 see AAF750 | 3693-53-6 see MJT750 | 3794-64-7 see HAW000 | 4008-48-4 see NHF500 | 4213-44-9 see CLD500 |
| 3599-32-4 see CCK000 | 3694-45-9 see CEQ750 | 3794-83-0 see TEE250 | 4013-92-7 see IGG300 | 4213-45-0 see QDS000 |
| 3605-01-4 see TNR485 | 3695-77-0 see TMT500 | 3804-89-5 see SIJ000 | 4016-11-9 see EKM200 | 4219-24-3 see HFD000 |
| 3607-48-5 see IDR000 | 3697-24-3 see MIN500 | 3808-42-2 see RAG300 | 4016-14-2 see IPD000 | 4222-21-3 see CIK825 |
| 3608-75-1 see PIB925 | 3697-25-4 see AAE000 | 3810-74-0 see SLY500 | 4019-40-3 see TIP750 | 4222-26-8 see CIB725 |
| 3610-27-3 see AAV500 | 3697-27-6 see DRE800 | 3810-80-8 see LIB000 | 4024-34-4 see TGJ475 | 4222-27-9 see CIB625 |
| 3613-30-7 see DSM800 | 3697-30-1 see EGO500 | 3810-81-9 see BKS810 | 4027-14-9 see TIF500 | 4223-11-4 see VMU000 |
| 3613-73-8 see TCQ260 | 3698-54-2 see TDY000 | 3811-04-9 see PLA250 | 4027-17-2 see COI000 | 4224-87-7 see MIF762 |
| 3613-82-9 see TLQ750 | 3701-40-4 see NAO600 | 3811-06-1 see CBQ125 | 4028-32-4 see BHB950 | 4230-21-1 see PDS750 |
| 3613-89-6 see HEF300 | 3703-76-2 see CMX800 | 3811-10-7 see BCP685 | 4032-26-2 see DWY000 | 4230-97-1 see AGM500 |
| 3614-69-5 see FMU409 | 3703-79-5 see BQF250 | 3811-49-2 see MEC250 | 4033-46-9 see CCF250 | 4232-84-2 see THO550 |
| 3615-21-2 see DGO400 | 3704-09-4 see MQS225 | 3811-73-2 see MCQ750 | 4035-89-6 see TNJ300 | 4235-09-0 see PCY750 |
| 3615-24-5 see INM000 | 3704-42-5 see NIW400 | 3813-05-6 see BAV000 | 4039-32-1 see LGX000 | 4238-84-0 see LJH000 |
| 3622-76-2 see VOF000 | 3706-77-2 see DXN850 | 3817-11-6 see HJQ350 | 4044-65-9 see PFA500 | 4239-06-9 see TDG500 |
| 3622-84-2 see BQJ650 | 3710-30-3 see OBK000 | 3818-69-7 see HEF200 | 4051-27-8 see DHE000 | 4242-33-5 see FIP999 |
| 3624-87-1 see AJA500 | 3710-84-7 see DJN000 | 3818-88-0 see EAI875 | 4055-39-4 see MQX500 | 4245-77-6 see ENU000 |
| 3624-96-2 see BFX125 | 3714-62-3 see TBR500 | 3818-90-4 see CKI000 | 4055-40-7 see MQX750 | 4246-51-9 see DJD800 |
| 3625-18-1 see DQU200 | 3715-67-1 see EJA100 | 3819-00-9 see PII500 | 4065-45-6 see HLR700 | 4247-19-2 see DHC600 |
| 3626-28-6 see CMO830 | 3715-90-0 see THJ825 | 3820-53-9 see DTH800 | 4067-16-7 see PBD000 | 4247-30-7 see ECA000 |
| 3632-91-5 see MAG000 | 3715-92-2 see NKK500 | 3823-94-7 see MQF200 | 4074-88-8 see ADT250 | 4251-89-2 see EHJ000 |
| 3634-83-1 see XIJ000 | 3717-88-2 see FCB100 | 3825-26-1 see ANP625 | 4075-79-0 see PDY500 | 4253-22-9 see DEI200 |
| 3635-74-3 see DOZ000 | 3719-37-7 see BBE250 | 3836-23-5 see NNQ000 | 4075-81-4 see CAW400 | 4253-34-3 see MQB500 |
| 3639-66-5 see DOQ600 | 3721-28-6 see PET750 | 3837-54-5 see DRU000 | 4076-02-2 see DNU860 | 4254-22-2 see ACN875 |
| 3644-11-9 see MEX300 | 3724-65-0 see COB500 | 3837-55-6 see DSW600 | 4076-40-8 see MHL750 | 4255-24-7 see AGQ250 |
| 3644-32-4 see NII200 | 3724-89-8 see AKB500 | 3844-45-9 see FAE000 | 4079-68-9 see DJX800 | 4268-36-4 see MOV500 |
| 3644-37-9 see BGK000 | 3731-39-3 see DUH800 | 3844-60-8 see DRN400 | 4080-31-3 see CEG550 | 4269-88-9 see MJC775 |
| 3644-61-9 see MRW125 | 3731-51-9 see ALB750 | 3844-63-1 see NKL000 | 4091-50-3 see CNH375 | 4274-06-0 see MIH500 |
| 3647-17-4 see MFF250 | 3732-90-9 see MJF000 | 3848-24-6 see HEQ200 | 4093-35-0 see VCK100 | 4279-76-9 see PDR250 |
| 3647-19-6 see CJR250 | 3733-45-7 see ACD250 | 3851-22-7 see ZGW100 | 4095-45-8 see PDB250 | 4298-16-2 see DAR100 |
| 3647-69-6 see CHE000 | 3733-63-9 see BBV750 | 3852-09-3 see MFL400 | 4096-33-7 see HIE000 | 4301-50-2 see FDB200 |
| 3648-18-8 see DVJ800 | 3734-17-6 see DTO200 | 3855-45-6 see TKG525 | 4097-22-7 see DHA325 | 4302-87-8 see AJR000 |
| 3648-20-2 see DXQ400 | 3734-33-6 see DAP812 | 3861-47-0 see DNG200 | 4097-47-6 see IOX000 | 4309-66-4 see AMO000 |
| 3648-36-0 see CMM768 | 3734-48-3 see HCN000 | 3861-73-2 see PMC750 | 4098-71-9 see IMG000 | 4310-69-4 see EHI500 |
| 3653-48-3 see SIL500 | 3734-60-9 see APV750 | 3861-99-2 see CGM500 | 4100-38-3 see DRM600 | 4312-87-2 see FMZ000 |
| 3658-48-8 see BJQ709 | 3734-67-6 see CMM300 | 3862-73-5 see TJX900 | 4104-14-7 see BIM000 | 4312-97-4 see SFV250 |
| 3658-77-3 see FPK200 | 3734-95-0 see PHK250 | 3867-15-0 see PIN225 | 4104-85-2 see THO250 | 4314-63-0 see LJJ000 |
| 3658-77-3 see HKC575 | 3734-97-2 see AMX825 | 3871-82-7 see MNN250 | 4106-66-5 see DDB800 | 4317-14-0 see AMY000 |
| 3665-51-8 see HMW500 | 3735-23-7 see MNO750 | 3877-86-9 see EAH100 | 4109-96-0 see DGK300 | 4323-43-7 see HFG600 |
| 3666-67-9 see DDH600 | 3735-90-8 see PDC850 | 3878-19-1 see FQK000 | 4113-57-9 see CLH625 | 4328-17-0 see TMM775 |

5377-20-8 see MQQ500
5385-75-1 see DCR300
5388-42-1 see PFW750
5388-62-5 see CGL500
5390-07-8 see PCF500
5392-40-5 see DTC800
5397-31-9 see ELA000
5405-53-8 see DUW200
5408-74-2 see EQG500
5410-29-7 see NEX500
5410-78-6 see AOR640
5411-08-5 see PMQ000
5412-64-6 see BRU250
5413-60-5 see DLY400
5415-07-6 see OOK200
5416-74-0 see DYA400
5417-82-3 see EET100
5418-32-6 see HNJ500
5419-55-6 see IOI000
5421-46-5 see ANM500
5421-48-7 see AAS250
5423-12-1 see AKJ250
5424-19-1 see PGE765
5424-26-0 see TDF000
5427-20-3 see EMC500
5428-37-5 see FPH000
5428-90-0 see CKW500
5428-95-5 see MCL500
5430-13-7 see FGA100
5431-31-2 see HFR000
5431-33-4 see ECJ000
5432-28-0 see NKT500
5433-44-3 see THQ900
5434-57-1 see HFP700
5435-64-3 see ILJ100
5438-85-7 see ALK625
5440-19-7 see TJG750
5444-80-4 see TJM500
5450-96-4 see SMX000
5453-66-7 see HNG800
5455-98-1 see ECK000
5456-28-0 see SBP900
5459-93-8 see EHT000
5462-06-6 see MLJ050
5464-28-8 see DVR909
5464-79-9 see AKM750
5466-22-8 see AHR000
5468-75-7 see CMS212
5469-26-1 see PJB100
5470-11-1 see HLN000
5470-66-6 see MMP500
5471-51-2 see RBU000
5486-77-1 see AFS625
5488-45-9 see TIB000
5493-45-8 see DKM500
5503-08-2 see TEQ700
5503-41-3 see RHF150
5504-68-7 see MHC000
5510-99-6 see DEF800
5511-98-8 see ACI000
5518-62-7 see DWP250
5522-43-0 see NJA000
5534-09-8 see AFJ625
5536-17-4 see AQQ900
5544-25-2 see API125
5550-12-9 see GLS800
5556-57-0 see EOM700
5560-72-5 see DPX200
5566-34-7 see CDR575
5571-97-1 see DEM000
5576-62-5 see CDQ500
5579-85-1 see BMZ000
5581-40-8 see DRX400
5581-52-2 see AKY250
5585-39-7 see GDM000

5585-67-1 see DQB800
5585-71-7 see BCE250
5585-73-9 see BPT750
5586-87-8 see PKS500
5588-10-3 see OJY000
5588-33-0 see MON750
5591-33-3 see OCI000
5591-45-7 see NBP500
5593-20-4 see BFV765
5593-70-4 see BSP250
5598-13-0 see CMA250
5598-52-7 see PHE250
5598-53-8 see SEZ350
5610-40-2 see SBN350
5610-59-3 see SDR000
5613-68-3 see TNR490
5625-90-1 see MJQ750
5626-16-4 see DBA600
5628-99-9 see VLZ000
5632-47-3 see MRJ750
5633-14-7 see BBU800
5633-16-9 see LEF400
5637-83-2 see THR250
5638-76-6 see HGE820
5653-21-4 see NEJ600
5653-80-5 see DBE100
5665-94-1 see CEX275
5667-20-9 see HIT000
5675-31-0 see FLQ000
5684-13-9 see BIT000
5688-80-2 see TKX500
5694-72-4 see PDX250
5696-06-0 see ENC500
5697-56-3 see BGD000
5707-69-7 see MLC250
5709-67-1 see NFC700
5709-68-2 see MMH740
5711-19-3 see ABX125
5714-00-1 see ABG000
5714-22-7 see SOQ450
5716-15-4 see DHF200
5716-20-1 see BOV825
5728-52-9 see BGE125
5730-85-8 see CEH000
5736-15-2 see MRM000
5736-85-6 see PMY500
5743-04-4 see CAD275
5743-18-0 see CAK750
5746-86-1 see NNS500
5756-69-4 see HMP000
5760-50-9 see ULS400
5760-73-6 see LBV000
5763-61-1 see VIK050
5766-67-6 see EJB000
5776-49-8 see PHA750
5779-79-3 see AAF000
5786-21-0 see CMY650
5786-68-5 see QWJ500
5786-77-6 see TLQ250
5787-73-5 see NCL300
5789-17-3 see IAT100
5796-14-5 see DYC200
5796-89-4 see PCO150
5797-06-8 see PCM550
5798-79-8 see BMW250
5800-19-1 see MQQ000
5806-84-8 see BKX500
5809-59-6 see HJQ000
5810-88-8 see DJK400
5814-20-0 see TDL250
5826-73-3 see DRB400
5826-91-5 see DJX400
5827-03-2 see DKB600
5831-08-3 see MEU750
5831-09-4 see DRH400

5831-10-7 see DRH200
5831-11-8 see TLP000
5831-12-9 see MEV000
5831-16-3 see DLI200
5831-17-4 see DMF600
5834-17-3 see MDZ000
5834-25-3 see DDC000
5834-81-1 see PFN500
5836-10-2 see PNH750
5836-28-2 see TDF750
5836-29-3 see EAT600
5836-73-7 see DEQ000
5836-85-1 see DLR200
5837-17-2 see DLU600
5837-78-5 see TGA800
5840-95-9 see EMC000
5843-82-3 see OEM000
5847-48-3 see GHQ000
5847-52-9 see TIC750
5853-29-2 see CCX125
5854-93-3 see AFH750
5857-37-4 see CHK000
5857-94-3 see AJH500
5858-81-1 see CMS155
5863-35-4 see NHP500
5868-05-3 see NCW300
5870-29-1 see CPZ125
5870-93-9 see HBN150
5874-97-5 see MDM800
5878-19-3 see MDW300
5878-19-3 see MFL100
5882-48-4 see DPN800
5888-61-9 see THU000
5890-18-6 see LBO100
5892-15-9 see AIT000
5892-41-1 see AAD875
5892-48-8 see SEG000
5894-60-0 see HCQ000
5902-51-2 see BQT750
5902-52-3 see BQU500
5902-76-1 see MLG250
5902-79-4 see MLF500
5902-95-4 see CAM000
5903-13-9 see MME809
5905-52-2 see LAL000
5910-75-8 see DXQ000
5910-85-0 see HAV450
5910-89-4 see DTU400
5913-76-8 see CNG675
5913-82-6 see DOX000
5915-41-3 see BQB000
5921-54-0 see EMT600
5928-69-8 see LFW300
5929-01-1 see BDV750
5929-09-9 see BBU750
5934-20-3 see DPI000
5934-69-0 see DAB807
5936-28-7 see HGQ500
5949-18-8 see SJH000
5950-69-6 see HHI000
5954-50-7 see DSA800
5954-90-5 see MOB750
5956-63-8 see BFO100
5957-75-5 see TCM000
5959-42-2 see DWB000
5959-52-4 see ALJ000
5959-56-8 see ALK500
5959-98-8 see CHD700
5960-88-3 see DFG700
5965-13-9 see DKX000
5967-09-9 see BGQ000
5967-42-0 see MDJ000
5967-73-7 see MDP250
5968-79-6 see APE625
5970-32-1 see MCU000

5970-45-6 see ZCA000
5974-19-6 see MQN000
5975-73-5 see BGS750
5976-47-6 see CEF250
5976-61-4 see HKH850
5976-95-4 see TAF775
5977-35-5 see AFQ575
5978-92-7 see HOR500
5980-33-6 see HMB000
5980-86-9 see CFD250
5985-35-3 see HMH500
5985-35-3 see MDV250
5986-38-9 see DTD400
5987-82-6 see OPI300
5988-31-8 see AIT500
5988-91-0 see TCY300
5989-27-5 see LFU000
5989-54-8 see MCC500
5989-77-5 see DLL000
5991-71-9 see DKV700
6000-43-7 see GHK000
6000-44-8 see GHG000
6002-77-3 see PCC475
6004-24-6 see CDF750
6009-67-2 see DIF200
6011-62-7 see IGG600
6012-97-1 see TBV750
6018-32-2 see ARA000
6018-89-9 see NCX500
6023-26-3 see HIS000
6027-28-7 see BQA750
6028-07-5 see HAI300
6029-87-4 see FOT000
6030-03-1 see EMS000
6030-80-4 see ECW500
6032-29-7 see PBM750
6033-07-4 see HNQ000
6035-40-1 see NBP275
6036-95-9 see MCJ250
6044-68-4 see DOM600
6046-93-1 see CNI325
6047-17-2 see TIX750
6051-87-2 see NAU525
6052-82-0 see DKB100
6055-19-2 see CQC500
6055-52-3 see HEO100
6055-69-2 see ADH875
6062-26-6 see CLO000
6064-83-1 see PHA575
6065-01-6 see NHV500
6065-04-9 see NHU000
6065-09-4 see NHX000
6065-10-7 see NHW500
6065-11-8 see NHW000
6065-13-0 see NHB500
6065-14-1 see NHB000
6065-17-4 see NHD000
6065-18-5 see NIA000
6065-19-6 see NHZ000
6066-49-5 see BSH500
6071-81-4 see COP400
6080-56-4 see LCJ000
6087-56-5 see DCP300
6088-91-1 see IEG000
6091-11-8 see AQQ250
6091-44-7 see HET000
6098-44-8 see ABL000
6098-46-0 see BDP000
6104-30-9 see IIV000
6106-46-3 see SBH000
6109-70-2 see QUS000
6109-97-3 see AJV250
6111-78-0 see MGX250
6112-76-1 see MCQ100
6117-91-5 see BOY000

6120-10-1 see DQF000
6130-87-6 see TDW300
6130-92-3 see AMQ750
6130-93-4 see TDS000
6131-99-3 see HKC550
6147-53-1 see CNA500
6151-25-3 see QCA175
6152-43-8 see DPD400
6152-43-8 see DTO800
6152-95-0 see PAP100
6153-33-9 see MBW100
6155-81-3 see DSL000
6156-78-1 see MAQ250
6159-44-0 see UQT700
6159-55-3 see VGA000
6163-73-1 see TLA600
6164-98-3 see CJJ250
6165-01-1 see BRS500
6168-86-1 see ODY000
6169-12-6 see AQE500
6180-21-8 see LHC000
6186-91-0 see CHJ250
6190-33-6 see BFN750
6191-22-6 see ALR250
6192-13-8 see NBX500
6192-29-6 see BRS250
6198-57-8 see VGU000
6202-23-9 see DPX800
6211-32-1 see YCA000
6217-24-9 see DVY100
6219-71-2 see CEG625
6240-55-7 see DFU400
6247-46-7 see CMU800
6249-65-6 see HNK500
6258-06-6 see DKR000
6264-93-3 see AJJ800
6268-32-2 see NLG500
6269-50-7 see HKE500
6272-74-8 see LBD100
6273-75-2 see PDL500
6275-69-0 see DBM400
6279-54-5 see BRC750
6280-15-5 see MKI250
6280-75-7 see DQQ600
6280-99-5 see DED500
6283-24-5 see ABQ000
6283-25-6 see CJA180
6283-63-2 see DJV250
6285-05-8 see CKT500
6285-34-3 see DAJ450
6288-93-3 see PNW800
6292-55-3 see FDS000
6292-91-7 see MJS750
6295-12-1 see HJI000
6296-45-3 see CHF500
6300-37-4 see CMP090
6303-21-5 see PGY250
6304-07-0 see DHK200
6304-33-2 see TMQ250
6305-18-6 see MFH900
6305-43-7 see DDK600
6307-82-0 see CJC515
6310-09-4 see CDN525
6317-18-6 see MJT500
6318-57-6 see HJE400
6320-14-5 see CMM765
6325-54-8 see CIG250
6325-93-5 see NFB000
6332-68-9 see BAL275
6334-11-8 see TLH000
6334-30-1 see THO500
6334-96-9 see OPE040
6336-12-5 see TEP750
6341-85-1 see DHD200
6358-29-8 see CMO875

| | | | |
|---|---|---|---|
| 7440-17-7 see RPA000 | 7491-74-9 see NNE400 | 7641-77-2 see HEC500 | 7722-86-3 see PCN750 | 7780-06-5 see IOO000 |
| 7440-18-8 see RRU000 | 7492-37-7 see BEN800 | 7644-67-9 see ASO501 | 7722-88-5 see TEE500 | 7782-39-0 see DBB800 |
| 7440-21-3 see SCP000 | 7492-44-6 see BQV250 | 7645-25-2 see ARC750 | 7723-14-0 see PHO500 | 7782-41-4 see FEZ000 |
| 7440-22-4 see SDI500 | 7492-66-2 see CMS323 | 7646-69-7 see SHO500 | 7723-14-0 see PHP010 | 7782-44-7 see OQW000 |
| 7440-22-4 see SDI750 | 7492-67-3 see CMT300 | 7646-78-8 see TGC250 | 7726-95-6 see BMP000 | 7782-49-2 see SBO500 |
| 7440-23-5 see SEE500 | 7492-70-8 see BQP000 | 7646-79-9 see CNB599 | 7727-15-3 see AGX750 | 7782-49-2 see SBP000 |
| 7440-23-5 see SEF500 | 7493-57-4 see PDD400 | 7646-85-7 see ZFA000 | 7727-18-6 see VDP000 | 7782-50-5 see CDV750 |
| 7440-23-5 see SEF600 | 7493-74-5 see AGQ750 | 7646-93-7 see PKX750 | 7727-21-1 see DWQ000 | 7782-61-8 see IHC000 |
| 7440-25-7 see TAE750 | 7493-78-9 see AOG750 | 7647-01-0 see HHL000 | 7727-37-9 see NGP500 | 7782-63-0 see FBO000 |
| 7440-28-0 see TEI000 | 7495-45-6 see EES300 | 7647-01-0 see HHX000 | 7727-43-7 see BAP000 | 7782-64-1 see MAS750 |
| 7440-29-1 see TFS750 | 7495-93-4 see CAD500 | 7647-10-1 see PAD500 | 7727-54-0 see ANR000 | 7782-65-2 see GEI100 |
| 7440-31-5 see TGB250 | 7496-02-8 see NFT400 | 7647-14-5 see SFT000 | 7732-18-5 see WAT259 | 7782-68-5 see IDK000 |
| 7440-32-6 see TGF250 | 7499-32-3 see DLT000 | 7647-15-6 see SFG500 | 7733-02-0 see ZNA000 | 7782-75-4 see MAH775 |
| 7440-32-6 see TGF500 | 7505-62-6 see BBC750 | 7647-17-8 see CDD000 | 7738-94-5 see CMH250 | 7782-77-6 see NMR000 |
| 7440-33-7 see TOA750 | 7506-80-1 see HKS550 | 7647-18-9 see AQD000 | 7739-33-5 see TKU000 | 7782-78-7 see NMJ000 |
| 7440-36-0 see AQB750 | 7511-54-8 see EHM000 | 7647-19-0 see PHR750 | 7745-89-3 see CLK230 | 7782-79-8 see HHG500 |
| 7440-37-1 see AQW250 | 7518-35-6 see MAW800 | 7648-01-3 see EPF500 | 7751-31-7 see BHQ750 | 7782-87-8 see PLQ750 |
| 7440-38-2 see ARA750 | 7519-36-0 see NLL500 | 7651-40-3 see AGL875 | 7757-79-1 see PLL500 | 7782-89-0 see LGT000 |
| 7440-39-3 see BAH250 | 7521-80-4 see BSR000 | 7651-91-4 see DFO900 | 7757-81-5 see SJV000 | 7782-92-5 see SEN000 |
| 7440-41-7 see BFO750 | 7525-62-4 see EPG000 | 7652-64-4 see BLG400 | 7757-82-6 see SJY000 | 7782-94-7 see NEG000 |
| 7440-42-8 see BMD500 | 7529-27-3 see EJE000 | 7654-03-7 see NCQ100 | 7757-83-7 see SJZ000 | 7782-99-2 see SOO500 |
| 7440-43-9 see CAD000 | 7530-07-6 see OFI000 | 7659-86-1 see EKW300 | 7757-93-9 see CAW100 | 7783-00-8 see SBO000 |
| 7440-44-0 see CBT500 | 7531-39-7 see DJW200 | 7660-25-5 see LFI000 | 7758-01-2 see PKY300 | 7783-06-4 see HIC500 |
| 7440-45-1 see CCY250 | 7532-52-7 see ALM750 | 7660-71-1 see MDM000 | 7758-02-3 see PKY500 | 7783-07-5 see HIC000 |
| 7440-46-2 see CDC000 | 7532-60-7 see BQH800 | 7664-38-2 see PHB250 | 7758-05-6 see PLK250 | 7783-08-6 see SBN500 |
| 7440-47-3 see CMI750 | 7535-34-4 see MKK250 | 7664-39-3 see HHU500 | 7758-09-0 see PLM500 | 7783-18-8 see ANK600 |
| 7440-48-4 see CNA250 | 7538-45-6 see MCO250 | 7664-41-7 see AMY500 | 7758-16-9 see DXF800 | 7783-20-2 see ANU750 |
| 7440-50-8 see CNI000 | 7539-12-0 see AGS000 | 7664-93-9 see SOI500 | 7758-19-2 see SFT500 | 7783-28-0 see ANH500 |
| 7440-54-2 see GAF000 | 7542-37-2 see NCF500 | 7664-93-9 see SOI530 | 7758-23-8 see CAW110 | 7783-30-4 see MDC750 |
| 7440-55-3 see GBG000 | 7546-30-7 see MCW000 | 7664-98-4 see DWV000 | 7758-88-5 see CDA750 | 7783-33-7 see NCP500 |
| 7440-56-4 see GDU000 | 7549-37-3 see DOE000 | 7673-09-8 see TNE775 | 7758-89-6 see CNK250 | 7783-35-9 see MDG500 |
| 7440-57-5 see GIS000 | 7550-35-8 see LGY000 | 7680-73-1 see MOY875 | 7758-94-3 see FBI000 | 7783-36-0 see MDG250 |
| 7440-58-6 see HAC000 | 7550-45-0 see TGH350 | 7681-11-0 see PLK500 | 7758-95-4 see LCQ000 | 7783-40-6 see MAF500 |
| 7440-59-7 see HAM500 | 7553-56-2 see IDM000 | 7681-28-9 see GAX000 | 7758-97-6 see LCR000 | 7783-41-7 see ORA000 |
| 7440-61-1 see UNS000 | 7554-65-6 see MOX000 | 7681-34-7 see SAQ000 | 7758-98-7 see CNP250 | 7783-42-8 see TFL250 |
| 7440-62-2 see VCP000 | 7558-63-6 see MRF000 | 7681-38-1 see SEG800 | 7758-99-8 see CNP500 | 7783-46-2 see LDF000 |
| 7440-63-3 see XDS000 | 7558-79-4 see SJH090 | 7681-49-4 see SHF500 | 7761-45-7 see MQR100 | 7783-47-3 see TGD100 |
| 7440-64-4 see YDA000 | 7558-80-7 see SJH100 | 7681-52-9 see SHU500 | 7761-88-8 see SDS000 | 7783-48-4 see SMI500 |
| 7440-65-5 see YEJ000 | 7560-83-0 see MJC750 | 7681-53-0 see SHV000 | 7763-77-1 see CGX000 | 7783-49-5 see ZHS000 |
| 7440-66-6 see ZBJ000 | 7562-61-0 see UWJ100 | 7681-55-2 see SHV500 | 7764-50-3 see DKV175 | 7783-50-8 see FAX000 |
| 7440-67-7 see ZOA000 | 7563-42-0 see CAV000 | 7681-57-4 see SII000 | 7765-88-0 see MGD210 | 7783-53-1 see MAW000 |
| 7440-69-9 see BKU750 | 7568-37-8 see MHY550 | 7681-76-7 see MMN750 | 7770-47-0 see MKW000 | 7783-54-2 see NGW000 |
| 7440-70-2 see CAL250 | 7568-93-6 see HNF000 | 7681-82-5 see SHW000 | 7772-76-1 see ANR750 | 7783-55-3 see PHQ500 |
| 7440-74-6 see ICF000 | 7570-25-4 see VLY300 | 7681-93-8 see PIF750 | 7772-98-7 see SKI000 | 7783-56-4 see AQE000 |
| 7442-07-1 see AIS550 | 7570-26-5 see DUV720 | 7682-90-8 see BAS000 | 7772-99-8 see TGC000 | 7783-60-0 see SOR000 |
| 7446-07-3 see TAJ750 | 7572-29-4 see DEN600 | 7683-59-2 see DMV600 | 7773-01-5 see MAR000 | 7783-61-1 see SDF650 |
| 7446-08-4 see SBQ500 | 7575-48-6 see TMG750 | 7689-03-4 see CBB870 | 7773-06-0 see ANU650 | 7783-64-4 see ZQS000 |
| 7446-09-5 see SOH500 | 7578-36-1 see BMG750 | 7693-26-7 see PLJ250 | 7773-34-4 see DJB200 | 7783-66-6 see IDT000 |
| 7446-11-9 see SOR500 | 7580-37-2 see TDH250 | 7695-91-2 see TGJ055 | 7774-29-0 see MDD000 | 7783-70-2 see AQF250 |
| 7446-14-2 see LDY000 | 7580-67-8 see LHH000 | 7697-37-2 see NED500 | 7774-29-0 see MDD250 | 7783-71-3 see TAF250 |
| 7446-18-6 see TEM000 | 7585-39-9 see COW925 | 7697-37-2 see NEE500 | 7774-41-6 see ARC500 | 7783-79-1 see SBS000 |
| 7446-20-0 see ZNJ000 | 7585-41-3 see CMS148 | 7697-46-3 see PGF900 | 7774-65-4 see MCF515 | 7783-80-4 see TAK250 |
| 7446-27-7 see LDU000 | 7601-54-9 see SJH200 | 7698-91-1 see MGL600 | 7774-82-5 see TJJ400 | 7783-81-5 see UOJ000 |
| 7446-32-4 see AQJ250 | 7601-55-0 see DUM000 | 7698-97-7 see FAO200 | 7775-09-9 see SFS000 | 7783-82-6 see TOC550 |
| 7446-34-6 see SBT000 | 7601-87-8 see DRN600 | 7699-31-2 see DJL600 | 7775-11-3 see DXC200 | 7783-91-7 see SDN500 |
| 7446-70-0 see AGY750 | 7601-89-0 see PCE750 | 7699-41-4 see SCL000 | 7775-14-6 see SHR500 | 7783-92-8 see SDN399 |
| 7447-39-4 see CNK500 | 7601-90-3 see PCD250 | 7699-43-6 see ZSJ000 | 7775-19-1 see SII100 | 7783-93-9 see SDV000 |
| 7447-40-7 see PLA500 | 7613-16-3 see OCW000 | 7700-17-6 see COD000 | 7775-27-1 see SJE000 | 7783-95-1 see SDQ500 |
| 7447-41-8 see LHB000 | 7616-83-3 see MCY755 | 7704-34-9 see SOD500 | 7775-41-9 see SDQ000 | 7784-08-9 see SDM100 |
| 7447-44-1 see ALB500 | 7616-83-3 see MDG200 | 7704-71-4 see MAF750 | 7776-33-2 see SBU710 | 7784-13-6 see AGZ000 |
| 7448-86-4 see ONW100 | 7616-94-6 see PCF750 | 7704-99-6 see ZRA000 | 7778-18-9 see CAX500 | 7784-18-1 see AHB000 |
| 7450-97-7 see OKW000 | 7621-86-5 see ALV050 | 7705-07-9 see TGG250 | 7778-39-4 see ARB250 | 7784-19-2 see THQ500 |
| 7452-79-1 see EMP600 | 7631-86-9 see SCI000 | 7705-08-0 see FAU000 | 7778-43-0 see ARC000 | 7784-21-6 see AHB500 |
| 7456-24-8 see DUC400 | 7631-89-2 see ARD750 | 7705-12-6 see IHB675 | 7778-44-1 see ANE500 | 7784-27-2 see AHD900 |
| 7458-65-3 see HDS200 | 7631-90-5 see SFE000 | 7718-54-9 see NDH000 | 7778-50-9 see PKX250 | 7784-30-7 see PHB500 |
| 7460-84-6 see SLK500 | 7631-95-0 see DXE800 | 7718-98-1 see VEP000 | 7778-54-3 see HOV500 | 7784-33-0 see ARF250 |
| 7466-54-8 see AOV250 | 7631-98-3 see DXZ000 | 7719-09-7 see TFL000 | 7778-66-7 see PLK000 | 7784-34-1 see ARF500 |
| 7476-91-7 see PDF500 | 7631-99-4 see SIO900 | 7719-12-2 see PHT275 | 7778-74-7 see PLO500 | 7784-35-2 see ARI250 |
| 7483-25-2 see DDY400 | 7632-00-0 see SIQ500 | 7720-78-7 see FBN100 | 7778-80-5 see PLT000 | 7784-37-4 see MDF350 |
| 7487-28-7 see BJN500 | 7632-10-2 see DBB000 | 7721-01-9 see TAF000 | 7779-27-3 see HDW000 | 7784-40-9 see LCK000 |
| 7487-88-9 see MAJ250 | 7632-50-0 see ANF800 | 7722-06-7 see AMN300 | 7779-41-1 see AFJ700 | 7784-41-0 see ARK250 |
| 7487-94-7 see MCY475 | 7632-51-1 see VEF000 | 7722-64-7 see PLP000 | 7779-80-8 see IIS000 | 7784-42-1 see ARK500 |
| 7488-55-3 see TGF010 | 7633-57-0 see NKN000 | 7722-73-8 see HGN000 | 7779-86-4 see ZGJ100 | 7784-44-3 see DCG800 |
| 7488-56-4 see SBR000 | 7635-51-0 see DTN800 | 7722-84-1 see HIB000 | 7779-86-4 see ZIJ100 | 7784-45-4 see ARG750 |
| 7488-70-2 see TFZ300 | 7637-07-2 see BMG700 | 7722-84-1 see HIB010 | 7779-88-6 see ZJJ000 | 7784-46-5 see SEY500 |

| | | | | |
|---|---|---|---|---|
| 9000-30-0 see GLU000 | 9004-54-0 see DBC800 | 9014-02-2 see NBV500 | 10025-73-7 see CMJ250 | 10061-02-6 see DGH225 |
| 9000-36-6 see KBK000 | 9004-54-0 see DBD000 | 9014-92-0 see TAX500 | 10025-74-8 see DYG600 | 10070-95-8 see BHR500 |
| 9000-36-6 see SLO500 | 9004-54-0 see DBD200 | 9015-68-3 see ARN800 | 10025-76-0 see ERA500 | 10072-24-9 see QCS875 |
| 9000-40-2 see LIA000 | 9004-54-0 see DBD400 | 9015-73-0 see DHW600 | 10025-77-1 see FAW000 | 10072-25-0 see DFH000 |
| 9000-55-9 see PJJ000 | 9004-54-0 see DBD600 | 9016-00-6 see PJR000 | 10025-78-2 see TJD500 | 10072-50-1 see PKD050 |
| 9000-64-0 see BAF000 | 9004-54-0 see DBD700 | 9016-01-7 see OJM400 | 10025-82-8 see ICK000 | 10075-36-2 see IHO200 |
| 9000-65-1 see THJ250 | 9004-57-3 see EHG100 | 9016-45-9 see NND500 | 10025-85-1 see NGQ500 | 10083-53-1 see MGL500 |
| 9000-69-5 see PAO150 | 9004-62-0 see HKQ100 | 9016-45-9 see PKF000 | 10025-87-3 see PHQ800 | 10085-81-1 see BCH750 |
| 9000-70-8 see PCU360 | 9004-64-2 see HNV000 | 9016-45-9 see TAW250 | 10025-91-9 see AQC500 | 10086-50-7 see IJA000 |
| 9001-00-7 see BMO000 | 9004-65-3 see HNX000 | 9016-45-9 see TAW500 | 10025-97-5 see IGJ499 | 10087-89-5 see DWL400 |
| 9001-05-2 see CCP525 | 9004-66-4 see IGS000 | 9016-45-9 see TAX000 | 10025-98-6 see PLN750 | 10094-34-5 see BEL850 |
| 9001-13-2 see CMY725 | 9004-67-5 see MIF760 | 9016-45-9 see TAX250 | 10025-99-7 see PJD250 | 10097-26-4 see BSA750 |
| 9001-33-6 see FBS000 | 9004-70-0 see CCU250 | 9016-87-9 see PKB100 | 10026-03-6 see SBU000 | 10097-28-6 see SDH000 |
| 9001-37-0 see GFG100 | 9004-74-4 see MFJ750 | 9031-11-2 see GAV100 | 10026-04-7 see SCQ500 | 10099-58-8 see LAX000 |
| 9001-62-1 see GGA800 | 9004-74-4 see MFK000 | 9034-34-8 see HHK000 | 10026-06-9 see TGC282 | 10099-59-9 see LBA000 |
| 9001-67-6 see NCQ200 | 9004-74-4 see MFK250 | 9034-40-6 see LIU360 | 10026-07-0 see TAJ250 | 10099-60-2 see LBB000 |
| 9001-73-4 see PAG500 | 9004-81-3 see PJY000 | 9036-06-0 see PMJ100 | 10026-08-1 see TFT000 | 10099-66-8 see LIW000 |
| 9001-92-7 see BAC020 | 9004-82-4 see SIB500 | 9036-19-5 see GHS000 | 10026-10-5 see UQJ000 | 10099-67-9 see LIY000 |
| 9002-01-1 see SLW450 | 9004-86-8 see PKE750 | 9036-19-5 see IAH000 | 10026-11-6 see ZPA000 | 10099-70-4 see BOX250 |
| 9002-04-4 see TFU800 | 9004-95-9 see PJT300 | 9036-19-5 see OFM000 | 10026-12-7 see NEA000 | 10099-71-5 see MAL000 |
| 9002-07-7 see TNW000 | 9004-96-0 see PJY100 | 9036-19-5 see OFO000 | 10026-13-8 see PHR500 | 10099-72-6 see MAL500 |
| 9002-13-5 see UTU550 | 9004-96-0 see PJY250 | 9036-19-5 see OFQ000 | 10026-17-2 see CNC100 | 10099-73-7 see MAL750 |
| 9002-18-0 see AEX250 | 9004-98-2 see OIG000 | 9036-19-5 see OFS000 | 10026-18-3 see CNE250 | 10099-74-8 see LDO000 |
| 9002-60-2 see AES650 | 9004-98-2 see OIG040 | 9038-95-3 see GHY000 | 10028-15-6 see ORW000 | 10099-76-0 see LDW000 |
| 9002-64-6 see PAK250 | 9004-98-2 see OIK000 | 9038-95-3 see UBS000 | 10028-18-9 see NDC500 | 10101-41-4 see CAX750 |
| 9002-70-4 see SCA750 | 9004-98-2 see PJW500 | 9038-95-3 see UCA000 | 10029-04-6 see ELI500 | 10101-50-5 see SJC000 |
| 9002-72-6 see PJA250 | 9004-99-3 see PJV250 | 9038-95-3 see UCJ000 | 10031-13-7 see LCL000 | 10101-53-8 see CMK415 |
| 9002-83-9 see KDK000 | 9004-99-3 see PJW750 | 9038-95-3 see UDA000 | 10031-18-2 see MCX750 | 10101-68-5 see MAU750 |
| 9002-84-0 see TAI250 | 9005-00-9 see SLM500 | 9038-95-3 see UDJ000 | 10031-26-2 see IGQ000 | 10101-83-4 see SKC000 |
| 9002-86-2 see PKQ059 | 9005-00-9 see SLN000 | 9038-95-3 see UDS000 | 10031-27-3 see TAK600 | 10101-88-9 see TNM750 |
| 9002-88-4 see PJS750 | 9005-08-7 see PJU500 | 9038-95-3 see UEA000 | 10031-37-5 see CMK425 | 10101-97-0 see NDL000 |
| 9002-89-5 see PKP750 | 9005-25-8 see SLJ500 | 9038-95-3 see UEJ000 | 10031-43-3 see CNN000 | 10102-06-4 see URA200 |
| 9002-89-5 see PKP800 | 9005-27-0 see HLB400 | 9038-95-3 see UFA000 | 10031-53-5 see ERC000 | 10102-17-7 see SKI500 |
| 9002-92-0 see DXY000 | 9005-32-7 see AFL000 | 9039-53-6 see UVS500 | 10031-58-0 see TDS500 | 10102-18-8 see SJT500 |
| 9002-92-0 see LBS000 | 9005-35-0 see CAM200 | 9039-61-6 see RDA350 | 10031-59-1 see TEL750 | 10102-20-2 see SKC500 |
| 9002-92-0 see LBT000 | 9005-37-2 see PNJ750 | 9041-08-1 see HAQ550 | 10031-82-0 see EEL500 | 10102-23-5 see SBN505 |
| 9002-92-0 see LBU000 | 9005-38-3 see SEH000 | 9041-93-4 see BLY780 | 10031-96-6 see EQS100 | 10102-40-6 see DXE875 |
| 9002-93-1 see PKF500 | 9005-46-3 see SFQ000 | 9045-81-2 see PKQ100 | 10032-02-7 see GDG000 | 10102-43-9 see NEG100 |
| 9002-98-6 see PJX800 | 9005-49-6 see HAQ500 | 9046-56-4 see VGU700 | 10032-13-0 see HFQ575 | 10102-44-0 see NGR500 |
| 9002-98-6 see PJX825 | 9005-64-5 see PKG000 | 9047-13-6 see AOM150 | 10032-15-2 see HFR200 | 10102-45-1 see TEK750 |
| 9002-98-6 see PJX835 | 9005-64-5 see PKL000 | 9049-05-2 see CAO250 | 10034-81-8 see PCE000 | 10102-49-5 see IGN000 |
| 9002-98-6 see PJX845 | 9005-65-6 see PKL100 | 9049-76-7 see HNY000 | 10034-85-2 see HHI500 | 10102-50-8 see IGM000 |
| 9003-00-3 see ADY250 | 9005-66-7 see PKG500 | 9056-38-6 see NMB000 | 10034-93-2 see HGW500 | 10102-53-1 see ARB000 |
| 9003-01-4 see ADW200 | 9005-67-8 see PKL030 | 9060-10-0 see BLY760 | 10035-10-6 see HHJ000 | 10103-50-1 see ARD000 |
| 9003-04-7 see SJK000 | 9005-70-3 see TOE250 | 9061-82-9 see SFP000 | 10036-47-2 see TCI000 | 10103-60-3 see ARD600 |
| 9003-05-8 see PJK350 | 9005-71-4 see SKV195 | 9067-32-7 see HGN600 | 10038-98-9 see GDY000 | 10108-56-2 see BQW750 |
| 9003-07-0 see PMP500 | 9005-79-2 see GHK300 | 9074-07-1 see SBI860 | 10039-33-5 see DVM600 | 10108-64-2 see CAE250 |
| 9003-08-1 see MCB050 | 9005-81-6 see CCT250 | 9076-25-9 see PCC000 | 10039-54-0 see OLS000 | 10108-73-3 see CDB000 |
| 9003-11-6 see PJH630 | 9006-00-2 see PJH500 | 9082-00-2 see PJX900 | 10040-45-6 see SJJ175 | 10112-91-1 see MCY300 |
| 9003-13-8 see BRP250 | 9006-04-6 see ROH900 | 9082-07-9 see SFW300 | 10042-76-9 see SMK000 | 10118-76-0 see CAV250 |
| 9003-17-2 see PJL350 | 9006-42-2 see MQQ250 | 9084-06-4 see BLX000 | 10042-84-9 see SEP500 | 10118-90-8 see MQW250 |
| 9003-20-7 see AAX250 | 9007-12-9 see TFZ000 | 9086-60-6 see CCH000 | 10042-88-3 see TAM000 | 10119-31-0 see ZPJ000 |
| 9003-22-9 see AAX175 | 9007-13-0 see CAW500 | 9087-70-1 see PAF550 | 10043-01-3 see AHG750 | 10121-94-5 see MNS000 |
| 9003-34-3 see PJR500 | 9007-16-3 see ADW300 | 10004-44-1 see HLM000 | 10043-09-1 see BJN750 | 10124-36-4 see CAJ000 |
| 9003-39-8 see PKQ250 | 9007-40-3 see COC875 | 10008-90-9 see CLU500 | 10043-18-2 see BPL750 | 10124-37-5 see CAU000 |
| 9003-39-8 see PKQ500 | 9007-73-2 see FBB000 | 10010-36-3 see FLY000 | 10043-35-3 see BMC000 | 10124-43-3 see CNE125 |
| 9003-39-8 see PKQ750 | 9007-81-2 see FOO600 | 10017-37-5 see MGJ800 | 10043-52-4 see CAO750 | 10124-48-8 see MCW500 |
| 9003-39-8 see PKR000 | 9007-92-5 see GEW875 | 10022-31-8 see BAN250 | 10043-67-1 see AHF100 | 10124-50-2 see PKV500 |
| 9003-39-8 see PKR250 | 9008-54-2 see CMS225 | 10022-50-1 see NMT500 | 10045-34-8 see AGM060 | 10124-53-5 see MAK000 |
| 9003-39-8 see PKR500 | 9008-99-5 see PJA200 | 10022-60-3 see AEP500 | 10045-94-0 see MDF000 | 10124-56-8 see SHM500 |
| 9003-39-8 see PKR750 | 9009-54-5 see PKL500 | 10022-68-1 see CAH250 | 10045-95-1 see NCB000 | 10125-85-6 see DTO100 |
| 9003-39-8 see PKS000 | 9009-65-8 see POD750 | 10023-25-3 see DKS800 | 10048-13-2 see SLP000 | 10127-36-3 see BMM550 |
| 9003-53-6 see SMQ500 | 9009-86-3 see RJK000 | 10024-58-5 see DAH450 | 10048-32-5 see PAJ500 | 10137-69-6 see CPE500 |
| 9003-54-7 see ADY500 | 9010-06-4 see SEH450 | 10024-70-1 see MEF500 | 10048-95-0 see ARC250 | 10137-73-2 see CQB275 |
| 9003-55-8 see SMR000 | 9010-53-1 see UVJ475 | 10024-74-5 see BKQ500 | 10049-03-3 see FFD000 | 10137-74-3 see CAO500 |
| 9003-98-9 see EDK650 | 9010-98-4 see PJQ050 | 10024-78-9 see EMY000 | 10049-04-4 see CDW450 | 10137-90-3 see ENX000 |
| 9003-98-9 see PAF575 | 9011-04-5 see HCV500 | 10024-89-2 see MRQ600 | 10049-05-5 see CMJ300 | 10137-96-9 see EJK000 |
| 9004-06-2 see EAG875 | 9011-05-6 see UTU500 | 10024-93-8 see NBY000 | 10049-06-6 see TGG750 | 10137-98-1 see EJL000 |
| 9004-07-3 see CML850 | 9011-06-7 see CGW300 | 10024-97-2 see NGU000 | 10049-07-7 see RHK000 | 10138-01-9 see ERC550 |
| 9004-10-8 see IDF300 | 9011-13-6 see SEA500 | 10025-64-6 see ZKS100 | 10049-08-8 see RRZ000 | 10138-21-3 see DFK400 |
| 9004-17-5 see IDF325 | 9011-14-7 see PKB500 | 10025-65-7 see PJE000 | 10051-06-6 see PBD500 | 10138-34-8 see EBK000 |
| 9004-32-4 see SFO500 | 9011-18-1 see DBD750 | 10025-67-9 see SON510 | 10058-20-5 see SMZ000 | 10138-39-3 see AGF500 |
| 9004-34-6 see CCU150 | 9011-93-2 see LJQ000 | 10025-68-0 see SBS500 | 10060-12-5 see CMK450 | 10138-41-7 see ECX500 |
| 9004-51-7 see IGT000 | 9012-59-3 see CBA000 | 10025-69-1 see TGC275 | 10061-01-5 see DGH200 | 10138-47-3 see EGJ000 |
| 9004-53-9 see DBD800 | 9014-01-1 see BAC000 | 10025-70-4 see SMH525 | | 10138-52-0 see GAH000 |

| | | | | |
|---|---|---|---|---|
| 10138-60-0 see HFF000 | 10294-70-9 see TGD500 | 10453-89-1 see CML650 | 11032-12-5 see TOE750 | 12010-67-2 see BKY250 |
| 10138-62-2 see HGG000 | 10308-82-4 see AKD375 | 10457-58-6 see BQY000 | 11041-12-6 see CME400 | 12011-67-5 see IGQ750 |
| 10138-63-3 see PFC100 | 10308-83-5 see DBW100 | 10457-59-7 see IOR000 | 11043-98-4 see MQX000 | 12011-76-6 see DAC450 |
| 10138-79-1 see TNF000 | 10308-84-6 see GLC000 | 10457-90-6 see BNU725 | 11043-99-5 see MQX250 | 12012-50-9 see PLW200 |
| 10138-87-1 see EGJ500 | 10308-90-4 see DUX600 | 10465-27-7 see SAR000 | 11048-13-8 see NBR500 | 12012-95-2 see AGQ000 |
| 10138-89-3 see TLA000 | 10309-37-2 see BAD625 | 10466-65-6 see PLQ000 | 11048-92-3 see GME300 | 12013-56-8 see CAR750 |
| 10139-98-7 see DAZ140 | 10309-79-2 see MHN750 | 10467-10-4 see EMA000 | 11050-62-7 see IKV000 | 12013-82-0 see TIH250 |
| 10140-75-7 see DBF875 | 10309-97-4 see HNK900 | 10473-64-0 see DWL500 | 11051-88-0 see CNF159 | 12014-28-7 see CAI125 |
| 10140-87-1 see DFG159 | 10310-38-0 see EIU000 | 10473-98-0 see DWL525 | 11052-01-0 see EDC600 | 12014-93-6 see DEK600 |
| 10140-89-3 see DGH800 | 10311-84-9 see DBI099 | 10476-81-0 see SMF000 | 11054-63-0 see TNX650 | 12016-80-7 see CNC232 |
| 10140-91-7 see CGH675 | 10312-83-1 see MDW250 | 10476-85-4 see SMG500 | 11054-70-9 see LBF500 | 12018-18-7 see NDA100 |
| 10140-94-0 see CHH125 | 10318-23-7 see DPP400 | 10476-86-5 see SMJ500 | 11055-06-4 see FPD000 | 12018-19-8 see ZFA100 |
| 10140-97-3 see CJG750 | 10318-26-0 see DDJ000 | 10476-95-6 see AAW250 | 11056-06-7 see BLY000 | 12020-65-4 see ERC500 |
| 10141-05-6 see CNC500 | 10319-70-7 see AHN000 | 10477-72-2 see PCU425 | 11056-12-5 see CMS232 | 12024-21-4 see GBS050 |
| 10141-07-8 see COD100 | 10325-39-0 see DEU300 | 10482-16-3 see MKR000 | 11069-19-5 see DEV200 | 12025-32-0 see GEI000 |
| 10141-15-8 see COM250 | 10325-94-7 see CAH000 | 10484-36-3 see AOK000 | 11069-34-4 see MGS500 | 12029-98-0 see IDS300 |
| 10141-19-2 see VQU000 | 10326-21-3 see MAE000 | 10486-00-7 see SJB350 | 11071-15-1 see AQH000 | 12030-85-2 see PLL250 |
| 10143-20-1 see DSJ200 | 10326-24-6 see ZDS000 | 10488-36-5 see TGJ250 | 11072-93-8 see EDL500 | 12030-88-5 see PLE260 |
| 10143-22-3 see DTG200 | 10328-51-5 see CPL100 | 10497-05-9 see PHE750 | 11077-03-5 see PAF000 | 12031-80-0 see LHO000 |
| 10143-23-4 see DTI400 | 10329-95-0 see MRU080 | 10504-99-1 see DWM600 | 11078-23-2 see CNH800 | 12032-88-1 see MAV250 |
| 10143-38-1 see DYH100 | 10331-57-4 see DFD000 | 10519-11-6 see DAF100 | 11081-39-3 see PKB775 | 12033-59-9 see SBT100 |
| 10143-53-0 see DJF100 | 10339-31-8 see NGY700 | 10519-12-7 see DAF150 | 11082-38-5 see TNM850 | 12034-12-7 see SJZ100 |
| 10143-54-1 see DJF600 | 10339-55-6 see ELZ000 | 10534-86-8 see HBX000 | 11094-61-4 see ECX000 | 12035-39-1 see NDL500 |
| 10143-56-3 see DJG200 | 10347-38-3 see BLK750 | 10535-87-2 see PEF500 | 11096-42-7 see NND000 | 12035-51-7 see NDB875 |
| 10143-60-9 see DJK600 | 10347-81-6 see MAW850 | 10537-47-0 see DED000 | 11096-82-5 see PJN250 | 12035-52-8 see NCY100 |
| 10143-66-5 see DOB200 | 10350-81-9 see PMY000 | 10540-29-1 see NOA600 | 11097-69-1 see PJN000 | 12035-72-2 see NDJ500 |
| 10158-43-7 see DTM800 | 10356-76-0 see FHO000 | 10543-95-0 see HDA000 | 11097-82-8 see GCO200 | 12036-02-1 see OKI000 |
| 10161-84-9 see SBU900 | 10356-92-0 see DAS400 | 10544-63-5 see EHO200 | 11100-14-4 see EHO200 | 12036-10-1 see RSF875 |
| 10161-85-0 see DJA325 | 10361-03-2 see SII500 | 10544-72-6 see NGU500 | 11103-86-9 see PLW500 | 12037-82-0 see PHQ750 |
| 10163-15-2 see DXD600 | 10361-37-2 see BAK000 | 10545-99-0 see SOG500 | 11104-28-2 see PJM000 | 12038-67-4 see RGK000 |
| 10165-33-0 see AKM250 | 10361-44-1 see BKW250 | 10546-24-4 see MME500 | 11105-11-6 see TOD000 | 12039-52-0 see TEL500 |
| 10168-80-6 see ECY500 | 10361-76-9 see PLP750 | 10552-94-0 see NLQ000 | 11107-01-0 see TOB750 | 12041-76-8 see DET000 |
| 10168-81-7 see GAL000 | 10361-79-2 see PLX750 | 10557-85-4 see IEI600 | 11111-23-2 see LHX350 | 12041-87-1 see TDX860 |
| 10168-82-8 see HGH100 | 10361-80-5 see PLY250 | 10563-70-9 see TDL000 | 11111-49-2 see HDB500 | 12042-91-0 see AHA000 |
| 10169-00-3 see ANL500 | 10361-82-7 see SAR500 | 10578-16-2 see DKG600 | 11114-18-4 see LFN000 | 12044-52-9 see PJE750 |
| 10169-02-5 see CMM325 | 10361-83-8 see SAT200 | 10580-52-6 see VDA000 | 11114-20-8 see CCL350 | 12045-01-1 see DGQ300 |
| 10171-76-3 see BJM750 | 10361-84-9 see SBC000 | 10580-77-5 see BJN000 | 11114-46-8 see FBD000 | 12047-79-9 see BAN500 |
| 10171-78-5 see HNX800 | 10361-91-8 see YDJ000 | 10584-98-2 see DDY600 | 11114-92-4 see CNA750 | 12054-48-7 see NDE000 |
| 10176-39-3 see FND100 | 10361-92-9 see YES000 | 10588-01-9 see SGI000 | 11115-82-5 see EAT800 | 12056-53-0 see PLW150 |
| 10187-79-8 see MMK000 | 10361-93-0 see YFJ000 | 10589-74-9 see PBX500 | 11116-31-7 see BLY250 | 12057-74-8 see MAI000 |
| 10187-86-7 see AKY000 | 10361-95-2 see ZES000 | 10592-13-9 see HGP550 | 11116-32-8 see BLY500 | 12057-92-0 see MAT750 |
| 10192-29-7 see ANE250 | 10369-17-2 see HEJ375 | 10595-95-6 see MKB000 | 11118-72-2 see AQM000 | 12056-74-1 see SKG500 |
| 10192-30-0 see ANB600 | 10371-86-5 see MEL775 | 10598-82-0 see NFO700 | 11120-29-9 see PJO250 | 12058-85-4 see SJI500 |
| 10193-95-0 see BKM000 | 10377-48-7 see LHR000 | 10599-90-3 see CDO750 | 11121-08-7 see KDA025 | 12060-00-3 see LED000 |
| 10196-18-6 see NEF500 | 10377-60-3 see MAH000 | 10605-21-7 see MHC750 | 11121-57-6 see PII150 | 12063-27-3 see IHN050 |
| 10210-36-3 see GJG000 | 10377-66-9 see MAS900 | 11001-74-4 see QQS075 | 11133-98-5 see CNI600 | 12064-62-9 see GAP000 |
| 10210-68-1 see CNB500 | 10379-14-3 see CFG750 | 11002-90-7 see ART500 | 11135-81-2 see PLS500 | 12067-99-1 see PHU750 |
| 10212-25-6 see COW900 | 10380-28-6 see BLC250 | 11003-24-0 see ARY750 | 11138-49-1 see AHG000 | 12068-50-7 see HAF375 |
| 10213-74-8 see EGY000 | 10380-77-5 see COE125 | 11004-30-1 see SKQ500 | 11138-87-7 see LAZ000 | 12068-85-8 see IGV000 |
| 10213-75-9 see EKZ000 | 10387-13-0 see BIJ750 | 11005-02-0 see ORI300 | 11141-16-5 see PJM250 | 12069-69-1 see CNJ750 |
| 10213-77-1 see TNA000 | 10389-72-7 see DRC600 | 11005-63-3 see SMN000 | 12001-26-2 see MQS250 | 12070-08-5 see TGG000 |
| 10215-25-5 see TEU000 | 10397-75-8 see TDQ230 | 11005-70-2 see CCX625 | 12001-28-4 see ARM275 | 12070-09-6 see UOB100 |
| 10215-33-5 see BPS500 | 10402-33-2 see AGL000 | 11005-92-8 see CCX175 | 12001-29-5 see ARM268 | 12070-12-1 see TOB500 |
| 10217-52-4 see HGU500 | 10402-52-5 see PGB750 | 11005-94-0 see CNW100 | 12001-47-7 see BKV250 | 12070-14-3 see ZQC200 |
| 10218-83-4 see DDQ100 | 10402-53-6 see ECU550 | 11006-22-7 see FCA000 | 12001-65-9 see HEI650 | 12070-27-8 see BAH750 |
| 10222-01-2 see DDM000 | 10402-90-1 see PFB000 | 11006-31-8 see TAB300 | 12001-79-5 see VSZ500 | 12071-29-3 see SME100 |
| 10232-90-3 see DFJ500 | 10405-02-4 see KEA300 | 11006-33-0 see PGQ500 | 12001-85-3 see NAT000 | 12071-33-9 see UOC200 |
| 10232-91-4 see EGY500 | 10405-27-3 see DKF400 | 11006-33-0 see PGQ750 | 12001-89-7 see DGR200 | 12071-83-9 see ZMA000 |
| 10232-92-5 see DMM600 | 10415-75-5 see MDE750 | 11006-64-7 see ITG000 | 12002-03-8 see COF500 | 12075-68-2 see TJP775 |
| 10232-93-6 see DOF000 | 10415-87-9 see PFR200 | 11006-70-5 see OIS000 | 12002-19-6 see MCV250 | 12079-65-1 see CPV000 |
| 10233-88-2 see GJE000 | 10416-59-8 see TMF000 | 11006-76-1 see VRF000 | 12002-26-5 see TDJ100 | 12079-66-2 see CDD625 |
| 10235-09-3 see PHN500 | 10419-79-1 see CHG375 | 11011-73-7 see BML750 | 12002-43-6 see GEO000 | 12081-88-8 see PLG825 |
| 10238-21-8 see CEH700 | 10420-90-3 see HCS100 | 11013-97-1 see MKK600 | 12002-53-8 see BQT600 | 12089-29-1 see BGY720 |
| 10241-05-1 see MRD500 | 10421-48-4 see FAY200 | 11014-59-8 see LAU400 | 12003-96-2 see AHH125 | 12108-13-3 see MAV750 |
| 10256-92-5 see TCJ025 | 10421-48-4 see IHB900 | 11015-37-5 see MRA250 | 12005-86-6 see SHM000 | 12111-24-9 see CAY500 |
| 10262-69-8 see LIN800 | 10427-00-6 see FPV000 | 11016-29-8 see PJJ350 | 12007-25-9 see MAD250 | 12116-66-4 see HAE500 |
| 10265-92-6 see DTQ400 | 10428-19-0 see TBL500 | 11016-71-0 see RQF350 | 12007-33-9 see BMH659 | 12122-67-7 see EIR000 |
| 10290-12-7 see CNN500 | 10429-82-0 see EAK000 | 11016-72-1 see RQK000 | 12007-46-4 see SIX550 | 12124-97-9 see ANC250 |
| 10294-33-4 see BMG400 | 10431-47-7 see SBO100 | 11018-93-2 see TER250 | 12007-56-6 see CAN250 | 12124-99-1 see ANJ750 |
| 10294-34-5 see BMG500 | 10431-86-4 see BSY000 | 11024-24-1 see DKL400 | 12007-97-5 see MRC650 | 12125-01-8 see ANH250 |
| 10294-40-3 see BAK250 | 10433-59-7 see EAK500 | 11028-39-0 see MCE275 | 12008-41-2 see DXF200 | 12125-02-9 see ANE500 |
| 10294-41-4 see CDB250 | 10436-39-2 see TBT500 | 11028-71-0 see CNH625 | 12008-61-6 see ANG125 | 12125-03-0 see PLS760 |
| 10294-46-9 see CNO350 | 10443-70-6 see EHL500 | 11029-61-1 see GJO025 | 12009-21-1 see BAP750 | 12125-09-6 see PHA000 |
| 10294-48-1 see PCF775 | 10447-38-8 see DWM400 | 11031-48-4 see SAX500 | 12010-12-7 see BFQ750 | 12125-56-3 see NDE010 |
| 10294-54-9 see CDE500 | 10453-86-8 see BEP500 | 11032-05-6 see TAK750 | 12010-53-6 see PJH775 | 12125-77-8 see COY100 |

| | | | | |
|---|---|---|---|---|
| 12126-59-9 see ECU750 | 12622-79-6 see CMK750 | 13092-75-6 see SDJ000 | 13345-25-0 see BCT750 | 13463-41-7 see ZMJ000 |
| 12126-59-9 see PMB000 | 12623-78-8 see AEC000 | 13092-75-6 see SDR759 | 13345-50-1 see MCA025 | 13463-67-7 see TGG760 |
| 12134-29-1 see TIH750 | 12626-36-7 see CAI600 | 13093-88-4 see LEO000 | 13345-58-9 see HKU000 | 13464-37-4 see SEY200 |
| 12136-15-1 see TKW000 | 12626-81-2 see LED100 | 13098-39-0 see HDE500 | 13345-60-3 see MEW000 | 13464-42-1 see SEY100 |
| 12136-83-3 see SIP000 | 12627-35-9 see PAR250 | 13104-70-6 see HGL600 | 13345-61-4 see FNT000 | 13464-82-9 see ICJ000 |
| 12136-85-5 see RQF000 | 12629-02-6 see VSK000 | 13106-47-3 see BFP750 | 13345-62-5 see CIN750 | 13464-97-6 see HGZ000 |
| 12137-13-2 see NDJ475 | 12634-34-3 see MAB750 | 13106-76-8 see ANM750 | 13345-64-7 see TLK750 | 13465-07-1 see HHZ000 |
| 12139-70-7 see CNC750 | 12640-89-0 see SBT200 | 13115-28-1 see NHR500 | 13347-42-7 see CFI750 | 13465-33-3 see MCX700 |
| 12142-88-0 see NDL425 | 12642-23-8 see PJP750 | 13115-40-7 see FMU039 | 13354-35-3 see PGL000 | 13465-73-1 see BOE750 |
| 12161-82-9 see BFO250 | 12645-45-3 see IGJ300 | 13118-10-0 see AEY400 | 13355-00-5 see PLK800 | 13465-77-5 see HCH500 |
| 12164-01-1 see TNO250 | 12645-49-7 see IHB677 | 13121-70-5 see CQH650 | 13356-08-6 see BLU000 | 13465-95-7 see PCD750 |
| 12164-12-4 see SJU500 | 12656-85-8 see MRC000 | 13121-71-6 see ABW600 | 13360-45-7 see CES750 | 13466-78-9 see CCK500 |
| 12164-94-2 see ANA750 | 12663-46-6 see COW750 | 13127-50-9 see MHD250 | 13361-31-4 see IIQ100 | 13470-26-3 see VEK000 |
| 12165-69-4 see PHT750 | 12672-29-6 see PJM750 | 13138-21-1 see DCN875 | 13361-32-5 see AGC150 | 13472-33-8 see SJD500 |
| 12167-74-7 see CAW120 | 12674-11-2 see PJL800 | 13138-45-9 see NDG000 | 13365-38-3 see DOS800 | 13472-45-2 see SKN500 |
| 12172-73-5 see ARM262 | 12674-40-7 see TFT100 | 13147-25-6 see ELG500 | 13367-92-5 see DQG700 | 13473-90-0 see AHD750 |
| 12174-11-7 see PAE750 | 12679-83-3 see SPE500 | 13150-00-0 see SIC000 | 13380-94-4 see OPC000 | 13474-03-8 see AEM750 |
| 12192-57-3 see ART250 | 12684-33-2 see SCF500 | 13154-66-0 see PNV775 | 13382-33-7 see HAH000 | 13476-05-6 see ECZ000 |
| 12194-11-5 see BIR500 | 12688-25-4 see JDS000 | 13168-78-0 see NLF300 | 13394-86-0 see BLV250 | 13477-00-4 see BAJ500 |
| 12198-93-5 see ORU900 | 12706-94-4 see APF000 | 13171-21-6 see FAB400 | 13395-16-9 see BGR000 | 13477-09-3 see BAM250 |
| 12206-14-3 see SKF000 | 12709-98-7 see LDM000 | 13171-22-7 see DTP600 | 13400-13-0 see CDD500 | 13477-17-3 see CAI000 |
| 12208-54-7 see ANO900 | 12710-02-0 see LFP000 | 13171-25-0 see YCJ200 | 13402-08-9 see ACX500 | 13477-21-9 see CAJ500 |
| 12217-79-7 see CMP070 | 12712-28-6 see BIK750 | 13172-31-1 see DXH350 | 13402-51-2 see BFK750 | 13477-34-4 see CAU250 |
| 12218-77-8 see TFU000 | 12718-69-3 see TOC000 | 13183-79-4 see MPQ250 | 13403-01-5 see DVV109 | 13478-00-7 see NDG500 |
| 12219-87-3 see CMM200 | 12737-87-0 see PJO750 | 13194-48-4 see EIN000 | 13406-60-5 see DHZ050 | 13478-10-9 see FBJ000 |
| 12224-02-1 see BHC750 | 12738-76-0 see TAN000 | 13195-76-1 see TKR750 | 13410-01-0 see DXG000 | 13478-33-6 see CND900 |
| 12225-26-2 see CMS227 | 12758-40-6 see CCF125 | 13219-97-1 see MNY800 | 13411-16-0 see NDY400 | 13479-29-3 see HOP000 |
| 12228-13-6 see DVT800 | 12765-82-1 see SNQ700 | 13225-10-0 see MKI125 | 13412-64-1 see DGE200 | 13482-49-0 see TAM500 |
| 12232-67-6 see BFR000 | 12770-50-2 see BFP250 | 13232-74-1 see HET350 | 13413-18-8 see HEW150 | 13483-18-6 see BIJ250 |
| 12232-97-2 see BKW500 | 12772-68-8 see LJP500 | 13235-16-0 see CAW000 | 13422-55-4 see VSZ050 | 13484-13-4 see EBR500 |
| 12232-99-4 see SFD000 | 12774-81-1 see TBN100 | 13240-06-7 see ERB000 | 13422-81-6 see TDJ300 | 13492-01-8 see PET500 |
| 12244-57-4 see GJC000 | 12788-93-1 see ADF250 | 13242-44-9 see DOY600 | 13424-46-9 see LCM000 | 13494-80-9 see TAJ000 |
| 12244-59-6 see PLM575 | 12789-46-7 see PBW750 | 13244-33-2 see AIA500 | 13425-22-4 see ALX250 | 13494-80-9 see TAJ010 |
| 12245-39-5 see CPR840 | 12794-92-2 see AGY000 | 13244-35-4 see CIX200 | 13425-94-0 see ARM000 | 13494-90-1 see GBS000 |
| 12255-10-6 see NCY125 | 12798-63-9 see OKK500 | 13254-34-7 see FON200 | 13426-91-0 see DBU800 | 13494-91-2 see GBS100 |
| 12255-80-0 see NDJ399 | 13001-46-2 see PLK600 | 13256-06-9 see DCH600 | 13435-12-6 see TME750 | 13494-98-9 see YFS000 |
| 12256-33-6 see NDJ400 | 13004-56-3 see QWJ000 | 13256-07-0 see AOL000 | 13441-36-6 see PFY200 | 13495-01-7 see HAO000 |
| 12259-92-6 see ANT000 | 13007-92-6 see HCB000 | 13256-11-6 see MNU250 | 13442-07-4 see HIV000 | 13497-05-7 see SJN000 |
| 12263-85-3 see MGC225 | 13009-91-1 see TNK000 | 13256-12-7 see DVE400 | 13442-08-5 see HIV500 | 13497-91-1 see TBP500 |
| 12264-18-5 see CAR800 | 13009-99-9 see MAC000 | 13256-13-8 see NKF000 | 13442-09-6 see HIW000 | 13508-53-7 see DRC400 |
| 12265-93-9 see SHK500 | 13010-08-7 see NLC000 | 13256-15-0 see NJQ000 | 13442-10-9 see HIW500 | 13510-49-1 see BFU250 |
| 12266-58-9 see BGR500 | 13010-10-1 see NLC500 | 13256-19-4 see NKM500 | 13442-11-0 see CHM500 | 13517-10-7 see BMH500 |
| 12271-71-5 see ZDJ100 | 13010-47-4 see CGV250 | 13256-21-8 see NKQ500 | 13442-12-1 see CHN000 | 13517-17-4 see SFW500 |
| 12273-50-6 see PLJ000 | 13013-17-7 see PNB790 | 13256-22-9 see NLR500 | 13442-13-2 see DFM200 | 13517-26-5 see SJM500 |
| 12275-13-7 see EIR500 | 13018-50-3 see BKS825 | 13256-23-0 see NLH000 | 13442-14-3 see CCF750 | 13517-49-2 see SPA650 |
| 12275-58-0 see SGM100 | 13019-22-2 see DAI400 | 13256-32-1 see DTB200 | 13442-15-4 see HIX500 | 13520-61-1 see NDJ000 |
| 12291-11-1 see BLL250 | 13021-50-6 see EET600 | 13257-44-8 see NNE100 | 13442-16-5 see HIY000 | 13520-74-6 see SCR100 |
| 12328-03-9 see TJU600 | 13029-44-2 see DHB550 | 13261-62-6 see DPJ600 | 13442-17-6 see DVE200 | 13520-83-7 see URS000 |
| 12331-76-9 see PLU750 | 13037-20-2 see EOK550 | 13265-01-5 see DUM100 | 13444-71-8 see PCJ250 | 13520-90-6 see VEK100 |
| 12380-95-9 see TIH000 | 13037-86-0 see HBP285 | 13265-60-6 see DOP200 | 13444-85-4 see NGW500 | 13520-92-8 see ZPS000 |
| 12397-35-2 see CCB500 | 13045-94-8 see SAX200 | 13266-07-4 see ABX325 | 13444-89-8 see NMJ400 | 13520-96-2 see POS500 |
| 12400-16-7 see BFR250 | 13046-06-5 see THS500 | 13271-93-7 see DUL500 | 13444-90-1 see NMT000 | 13523-86-9 see VSA000 |
| 12401-86-4 see SIN500 | 13048-33-4 see HEQ100 | 13275-68-8 see EGI000 | 13445-50-6 see DWP229 | 13529-51-6 see HHH000 |
| 12408-07-0 see TJC800 | 13055-82-8 see DNA600 | 13277-59-3 see NEE000 | 13446-10-1 see PCJ750 | 13529-75-4 see LHE525 |
| 12427-38-2 see MAS500 | 13056-98-9 see PEU500 | 13279-22-6 see PEH250 | 13446-30-5 see MAK250 | 13530-65-9 see ZFJ100 |
| 12430-27-2 see CDE125 | 13058-67-8 see LIN000 | 13279-24-8 see EJN400 | 13446-34-9 see MAR250 | 13536-84-0 see UQA000 |
| 12436-28-1 see SHN000 | 13061-80-8 see HFF300 | 13280-07-4 see CEV800 | 13446-48-5 see ANO250 | 13536-85-1 see FFB000 |
| 12439-96-2 see VEZ100 | 13065-64-0 see DQA710 | 13284-86-1 see SIC500 | 13446-73-6 see RPK000 | 13537-18-3 see TFW500 |
| 12442-63-6 see CMB500 | 13067-93-1 see CON300 | 13286-32-3 see DID200 | 13446-74-7 see RPP000 | 13537-21-8 see CMI300 |
| 12504-13-1 see SML000 | 13071-27-7 see DUO600 | 13289-18-4 see HAN600 | 13446-75-8 see RPU000 | 13537-32-1 see PHJ250 |
| 12505-77-0 see DDI500 | 13071-79-9 see BSO000 | 13292-46-1 see RKP000 | 13448-22-1 see ODY100 | 13537-45-6 see HGU100 |
| 12529-66-7 see DNW200 | 13073-35-3 see AKB250 | 13292-87-0 see MPL250 | 13449-22-4 see BPX500 | 13539-59-8 see AQN750 |
| 12540-13-5 see CNI500 | 13073-86-4 see TCN250 | 13302-06-2 see TIF000 | 13450-90-3 see GBM000 | 13548-38-4 see CMJ600 |
| 12542-36-8 see GJM259 | 13074-65-2 see HFO600 | 13302-08-4 see TMW500 | 13450-92-5 see GDW000 | 13551-87-6 see NHH500 |
| 12542-85-7 see MGC230 | 13074-85-6 see SAS000 | 13311-84-7 see FMR000 | 13453-07-1 see GIW176 | 13551-92-3 see NHI000 |
| 12550-17-3 see AQI000 | 13074-91-4 see DYG800 | 13312-42-0 see BCT250 | 13453-30-0 see TEJ100 | 13552-44-8 see MJQ100 |
| 12558-71-3 see YFA100 | 13078-75-6 see DOK000 | 13316-70-6 see CDF500 | 13453-57-1 see LCQ300 | 13553-79-2 see RKU000 |
| 12558-92-8 see MDB250 | 13080-06-3 see BRJ125 | 13319-75-0 see BMG800 | 13453-78-6 see LHN000 | 13556-50-8 see DCL100 |
| 12602-23-2 see CNB495 | 13080-89-2 see SNZ000 | 13323-62-1 see DEJ000 | 13454-96-1 see PJE250 | 13560-89-9 see DAI460 |
| 12604-53-4 see FBE000 | 13082-24-1 see DQA700 | 13324-20-4 see CMS228 | 13455-36-2 see CND920 | 13561-08-5 see BJO000 |
| 12604-58-9 see FBP000 | 13084-45-2 see BJD750 | 13327-32-7 see BFS250 | 13455-50-0 see HGG500 | 13569-63-6 see RGP000 |
| 12607-70-4 see NCY600 | 13084-46-3 see DDI000 | 13329-71-0 see IQT000 | 13461-01-3 see AAF800 | 13569-65-8 see RHP000 |
| 12607-92-0 see AGX125 | 13084-47-4 see BJD500 | 13331-27-6 see NEY500 | 13463-30-4 see LEC000 | 13573-18-7 see SKN000 |
| 12607-93-1 see TAH750 | 13086-63-0 see SDW500 | 13344-50-8 see NLS000 | 13463-39-3 see NCZ000 | 13589-15-6 see MIH250 |
| 12609-89-1 see MRW800 | 13087-53-1 see IGC000 | 13345-21-6 see BCX250 | 13463-40-6 see IHG500 | 13590-71-1 see PGZ950 |

| | | | | |
|---|---|---|---|---|
| 17095-24-8 see RCU000 | 17523-77-2 see PLN500 | 17918-11-5 see PEH500 | 18591-81-6 see ALB625 | 19115-30-1 see HKA700 |
| 17097-76-6 see CAT125 | 17524-18-4 see THP750 | 17924-92-4 see ZAT000 | 18598-63-5 see MBX800 | 19120-62-8 see VDK000 |
| 17099-81-9 see HIA000 | 17526-24-8 see BBH500 | 17927-57-0 see CFK325 | 18649-64-4 see EQN000 | 19137-90-7 see EPC999 |
| 17109-49-8 see EIM000 | 17526-74-8 see GGY000 | 17959-11-4 see MDX250 | 18656-21-8 see IDJ600 | 19139-31-2 see DKP000 |
| 17124-74-2 see DAZ135 | 17548-36-6 see HCY500 | 17959-12-5 see MLX000 | 18656-25-2 see DWE200 | 19142-68-8 see CJL409 |
| 17125-80-3 see BAO750 | 17549-30-3 see DJD200 | 17960-21-3 see MMJ955 | 18662-53-8 see NEI000 | 19142-70-2 see POT000 |
| 17140-68-0 see DIH600 | 17557-23-2 see NCI300 | 17969-20-9 see CKK250 | 18683-91-5 see AHJ250 | 19142-71-3 see PPE500 |
| 17140-78-2 see DYB400 | 17560-51-9 see ZAK300 | 17977-68-3 see FOE000 | 18694-40-1 see MCH550 | 19143-00-1 see CPD500 |
| 17156-85-3 see TBJ475 | 17563-48-3 see BRA550 | 18046-21-4 see CKI750 | 18705-22-1 see PNV750 | 19146-99-7 see PPH000 |
| 17156-88-6 see DRJ400 | 17573-21-6 see BCY250 | 18067-13-5 see DAB750 | 18713-58-1 see IOL100 | 19155-52-3 see MFQ250 |
| 17160-71-3 see FDA880 | 17573-23-8 see DKU000 | 18109-81-4 see BOR350 | 18714-34-6 see NJB500 | 19168-23-1 see ANF000 |
| 17167-73-6 see AAH100 | 17575-20-1 see LAT000 | 18127-01-0 see BMK300 | 18727-07-6 see UTU400 | 19202-92-7 see TAL550 |
| 17168-82-0 see AQQ100 | 17575-21-2 see LAT500 | 18139-02-1 see FHX000 | 18760-80-0 see MNN350 | 19210-06-1 see ZGS000 |
| 17168-83-1 see DBY300 | 17575-22-3 see LAU000 | 18139-03-2 see BJU500 | 18774-85-1 see HFS500 | 19216-56-9 see AJP000 |
| 17168-85-3 see THP250 | 17576-63-5 see FKO000 | 18156-74-6 see TMF250 | 18787-40-1 see PIY000 | 19218-16-7 see BJP899 |
| 17169-60-7 see FBD500 | 17576-88-4 see BNE600 | 18172-33-3 see CQD000 | 18791-02-1 see DDS100 | 19237-84-4 see FPP100 |
| 17174-98-0 see COK300 | 17596-45-1 see BMB150 | 18179-67-4 see MFO000 | 18791-21-4 see PPI775 | 19245-07-9 see FAC155 |
| 17176-77-1 see DDH400 | 17597-95-4 see HFG700 | 18181-70-9 see IEN000 | 18810-58-7 see BAI000 | 19246-24-3 see TAK800 |
| 17185-68-1 see AQQ125 | 17598-65-1 see DBH200 | 18181-80-1 see IOS000 | 18815-73-1 see MQP250 | 19247-68-8 see DKK200 |
| 17210-48-9 see DNY400 | 17599-02-9 see DPC200 | 18186-71-5 see DXU200 | 18820-29-6 see MAV000 | 19267-68-6 see BIU260 |
| 17224-08-7 see DKF600 | 17599-08-5 see AIM000 | 18204-79-0 see TMF625 | 18829-55-5 see HBI770 | 19287-45-7 see DJY400 |
| 17224-09-8 see TCH250 | 17599-09-6 see AIU000 | 18207-29-3 see NLD800 | 18854-01-8 see DJV600 | 19311-91-2 see DJY400 |
| 17230-87-4 see CLS250 | 17601-12-6 see BLE250 | 18230-75-6 see TMF125 | 18857-59-5 see HMI500 | 19315-64-1 see HIN000 |
| 17230-88-5 see DAB830 | 17605-71-9 see MJU750 | 18237-15-5 see AME750 | 18868-43-4 see MRD250 | 19356-22-0 see PCN000 |
| 17236-22-5 see IFG000 | 17607-20-4 see BGW710 | 18237-16-6 see AJB500 | 18868-66-1 see EGP000 | 19361-41-2 see AAK400 |
| 17242-52-3 see PKV000 | 17608-59-2 see NKC000 | 18244-91-2 see TJU500 | 18869-73-3 see ACD500 | 19367-79-4 see MEP000 |
| 17243-64-0 see EOA500 | 17615-73-5 see AOT125 | 18252-65-8 see DEU115 | 18883-66-4 see SMD000 | 19379-90-9 see BDJ600 |
| 17243-64-0 see EOB000 | 17617-13-9 see DID800 | 18264-75-0 see ALN750 | 18886-42-5 see ENC600 | 19381-50-1 see NAX500 |
| 17247-77-7 see DLX800 | 17617-23-1 see FMQ000 | 18264-88-5 see DMA400 | 18897-36-4 see CAI350 | 19383-97-2 see PEK750 |
| 17264-01-6 see MIS750 | 17617-45-7 see PIE510 | 18268-70-7 see FCD512 | 18902-42-6 see NMI500 | 19387-91-8 see TGD250 |
| 17278-93-2 see MJA000 | 17639-93-9 see CKT000 | 18268-70-7 see PJC250 | 18905-29-8 see GLK100 | 19395-58-5 see MRN250 |
| 17284-75-2 see BQH500 | 17650-86-1 see AHL000 | 18273-30-8 see BJE250 | 18912-80-6 see DJF800 | 19395-78-9 see PCJ350 |
| 17300-62-8 see MAJ775 | 17654-88-5 see MCP500 | 18279-20-4 see DVU600 | 18917-82-3 see LDL000 | 19396-06-6 see PKE100 |
| 17309-87-4 see DSI800 | 17661-50-6 see TCB100 | 18279-21-5 see DKP200 | 18917-93-6 see LAL100 | 19408-46-9 see KCK000 |
| 17311-31-8 see DVQ800 | 17667-23-1 see DVO809 | 18283-93-7 see TBJ300 | 18921-70-5 see DBY500 | 19416-93-4 see PLK825 |
| 17321-77-6 see CDV000 | 17672-21-8 see HJC500 | 18309-32-5 see VIP000 | 18924-91-9 see TNI000 | 19423-89-3 see LDI000 |
| 17333-74-3 see TGM550 | 17673-25-5 see PGS250 | 18312-12-4 see IDZ100 | 18928-76-2 see TDH100 | 19441-09-9 see ANT300 |
| 17333-83-4 see CEJ500 | 17689-16-6 see TEY750 | 18318-83-7 see HFA620 | 18936-66-8 see BND750 | 19456-73-6 see DQF200 |
| 17333-84-5 see CEJ250 | 17692-39-6 see PDU250 | 18323-44-9 see CMV675 | 18936-75-9 see AIC750 | 19456-74-7 see DQE000 |
| 17333-86-7 see BBN250 | 17697-55-1 see ASP500 | 18355-50-5 see BGM000 | 18936-78-2 see FNS000 | 19456-75-8 see DQF400 |
| 17339-60-5 see BJG100 | 17702-41-9 see DAE400 | 18355-54-9 see EKD000 | 18943-30-1 see SMJ000 | 19456-77-0 see DPP709 |
| 17369-59-4 see PNO500 | 17702-57-7 see FNE600 | 18356-02-0 see MPJ000 | 18968-99-5 see PAP000 | 19464-55-2 see TKT750 |
| 17372-87-1 see BNH500 | 17710-62-2 see CJU125 | 18361-48-3 see LEV025 | 18972-56-0 see MAG250 | 19471-27-3 see DPP600 |
| 17380-19-7 see COP550 | 17719-22-1 see MPT300 | 18365-12-3 see PNH500 | 18976-74-4 see RCZ000 | 19471-28-4 see DQE200 |
| 17380-21-1 see COP525 | 17721-94-7 see BRZ200 | 18378-89-7 see MQW750 | 18979-94-7 see CHL500 | 19473-49-5 see MRK500 |
| 17381-88-3 see DBJ400 | 17721-95-8 see DTA400 | 18380-68-2 see DQG400 | 18984-80-0 see EQQ500 | 19477-24-8 see CCJ500 |
| 17397-89-6 see ECE500 | 17730-82-4 see BTA000 | 18413-14-4 see ELC000 | 18987-38-7 see TME600 | 19481-39-1 see DJQ800 |
| 17400-65-6 see DPQ400 | 17737-65-4 see CMX770 | 18417-89-5 see SAU000 | 18996-35-5 see MRL000 | 19481-40-4 see EIL500 |
| 17400-68-9 see DQE400 | 17750-93-5 see TCJ775 | 18429-70-4 see DQJ600 | 18997-62-1 see DUN800 | 19482-31-6 see PAS829 |
| 17400-69-0 see MJF500 | 17751-20-1 see DEX600 | 18429-71-5 see TLJ250 | 18999-28-5 see HBK500 | 19485-03-1 see BRG500 |
| 17400-70-3 see MJF750 | 17766-62-0 see BQK830 | 18431-36-2 see HFR100 | 19005-95-9 see ADE050 | 19493-75-5 see TJU750 |
| 17401-48-8 see DQH800 | 17766-63-1 see PFX600 | 18433-84-6 see PAT799 | 19009-56-4 see MIW000 | 19521-84-7 see BBO400 |
| 17406-45-0 see THG250 | 17766-66-4 see CKJ100 | 18444-66-1 see COE250 | 19010-66-3 see LCW000 | 19525-20-3 see TER500 |
| 17416-17-0 see DPQ800 | 17766-68-6 see MFH760 | 18454-12-1 see LCS000 | 19010-79-8 see CAI400 | 19526-81-9 see EAJ500 |
| 17416-18-1 see DPQ600 | 17766-70-0 see MFH770 | 18461-55-7 see AAU250 | 19025-95-7 see BKY500 | 19546-20-4 see BLB500 |
| 17416-20-5 see MJG000 | 17766-74-4 see THF300 | 18463-85-9 see DPO400 | 19039-44-2 see GLO000 | 19561-70-7 see HKW450 |
| 17416-21-6 see DQE600 | 17766-75-5 see THF310 | 18463-86-0 see DQM200 | 19044-88-3 see OJY100 | 19562-30-2 see PAF250 |
| 17418-58-5 see AKI750 | 17766-77-7 see TKY300 | 18466-11-0 see MOB250 | 19049-40-2 see BFT500 | 19590-85-3 see DAE695 |
| 17427-00-8 see ABV250 | 17766-79-9 see TEX220 | 18472-51-0 see CDT250 | 19056-01-0 see MLW250 | 19595-66-5 see DQF600 |
| 17433-31-7 see ADA000 | 17780-75-5 see CMY000 | 18472-87-2 see ADG250 | 19056-03-2 see PNV250 | 19597-69-4 see LGV000 |
| 17449-96-6 see CMW750 | 17784-47-3 see ADJ750 | 18479-58-8 see DLX000 | 19060-37-8 see PPA000 | 19598-90-4 see GAN000 |
| 17455-13-9 see COD500 | 17794-13-7 see TAO750 | 18497-13-7 see DLO400 | 19060-39-0 see PPA250 | 19622-19-6 see EIH500 |
| 17455-23-1 see DGT300 | 17796-82-6 see CPQ700 | 18501-44-5 see HDX500 | 19060-40-3 see PPD250 | 19624-22-7 see PAT750 |
| 17463-44-4 see AKP500 | 17804-35-2 see BAV575 | 18520-54-2 see TAR000 | 19060-43-6 see POV250 | 19689-86-2 see NFN000 |
| 17466-45-4 see PCU350 | 17804-49-8 see PMF540 | 18520-57-5 see TAR250 | 19060-44-7 see PPD000 | 19704-60-0 see BJX750 |
| 17471-59-9 see TDU750 | 17814-73-2 see MOR250 | 18523-48-3 see ASD375 | 19060-45-8 see POT750 | 19706-58-2 see DEI600 |
| 17471-82-8 see GKG300 | 17822-71-8 see DHP550 | 18523-69-8 see NGN000 | 19060-74-3 see POW250 | 19708-47-5 see THR100 |
| 17476-04-9 see LGS000 | 17822-72-9 see DHP450 | 18530-56-8 see HDP500 | 19060-75-4 see POW000 | 19716-21-3 see DQC600 |
| 17496-59-2 see DEL800 | 17822-73-0 see DHP500 | 18539-34-9 see DRR500 | 19060-76-5 see POW500 | 19721-74-5 see BIY250 |
| 17498-10-1 see DJN875 | 17822-74-1 see DHQ800 | 18559-59-6 see IDD100 | 19072-57-2 see DRK600 | 19750-95-9 see CJJ500 |
| 17501-44-9 see PBL750 | 17831-71-9 see ADT050 | 18559-60-9 see TMJ800 | 19083-81-9 see PPF250 | 19763-77-0 see DGY000 |
| 17508-17-7 see DVC700 | 17861-62-0 see NDG550 | 18559-63-2 see TKX125 | 19083-82-0 see PPG250 | 19767-45-4 see MDK875 |
| 17511-60-3 see TJG600 | 17869-27-1 see AMG500 | 18559-92-7 see DPO600 | 19089-24-8 see PQB750 | 19780-25-7 see EHO000 |
| 17513-40-5 see MGY500 | 17874-34-9 see DQV000 | 18559-94-9 see BQF500 | 19089-92-0 see HFM600 | 19780-35-9 see EJS000 |
| 17518-47-7 see NBN000 | 17902-23-7 see FLZ050 | 18559-95-0 see SMT000 | 19109-66-1 see TMA500 | 19789-69-6 see FKP000 |

| | | | | |
|---|---|---|---|---|
| 19792-18-8 see TBO765 | 20562-02-1 see SKS000 | 21075-41-2 see MNS250 | 21621-75-0 see MLW600 | 22248-79-9 see RAF100 |
| 19794-93-5 see THK875 | 20562-03-2 see CDG500 | 21082-50-8 see EME000 | 21621-78-3 see IRX100 | 22251-01-0 see FDT000 |
| 19816-89-8 see BMT500 | 20570-96-1 see BEQ250 | 21083-47-6 see DWN600 | 21626-24-4 see IGX875 | 22254-24-6 see IGG000 |
| 19836-78-3 see MND750 | 20600-96-8 see TDY100 | 21085-56-3 see MNO775 | 21638-36-8 see MMJ000 | 22260-42-0 see COF825 |
| 19841-73-7 see DWF865 | 20611-21-6 see BBT000 | 21087-64-9 see MQR275 | 21642-82-0 see PBS750 | 22260-51-1 see BNB325 |
| 19855-39-1 see DAD650 | 20624-25-3 see SGJ500 | 21107-27-7 see TLP275 | 21642-83-1 see PBS500 | 22262-18-6 see MPI205 |
| 19864-71-2 see IBP000 | 20627-28-5 see DQJ800 | 21109-95-5 see BAP250 | 21644-95-1 see AKK000 | 22262-19-7 see MPN275 |
| 19875-60-6 see LJE500 | 20627-31-0 see DQK800 | 21124-13-0 see MQB250 | 21645-51-2 see AHC000 | 22295-11-0 see TDU250 |
| 19879-06-2 see GJS000 | 20627-32-1 see TLK000 | 21140-85-2 see MFG600 | 21649-57-0 see CBO000 | 22295-99-4 see HCA000 |
| 19882-03-2 see HGI700 | 20627-33-2 see TLJ750 | 21172-28-1 see ALG500 | 21650-02-2 see ARX150 | 22298-04-0 see DLO000 |
| 19895-66-0 see DCI800 | 20627-34-3 see TLK500 | 21187-98-4 see DBL700 | 21658-26-4 see NIQ500 | 22298-29-9 see BFV760 |
| 19899-80-0 see HJP575 | 20645-04-9 see ZTS000 | 21208-99-1 see CQC250 | 21662-09-9 see DAI360 | 22304-30-9 see ASA000 |
| 19910-65-7 see BSD000 | 20652-39-5 see DNP600 | 21209-02-9 see MIT250 | 21667-01-6 see BHW250 | 22316-47-8 see CIR750 |
| 19926-22-8 see BBD000 | 20661-60-3 see NJH500 | 21224-57-7 see EOK500 | 21679-31-2 see TNN250 | 22323-45-1 see ZJA000 |
| 19932-64-0 see DCO509 | 20667-12-3 see SDU500 | 21224-77-1 see EFC000 | 21704-44-9 see ELY575 | 22326-31-4 see SNU500 |
| 19932-85-5 see BMV000 | 20675-51-8 see PBW400 | 21224-81-7 see THB250 | 21704-46-1 see CBI675 | 22345-47-7 see GJS200 |
| 19935-86-5 see PNR500 | 20680-07-3 see MHS375 | 21230-20-6 see BBR380 | 21711-65-9 see NLA000 | 22346-43-6 see HLX000 |
| 19937-59-8 see MQR225 | 20684-29-1 see FAC130 | 21232-47-3 see TBN550 | 21715-46-8 see CGQ500 | 22349-59-3 see DTJ200 |
| 19952-47-7 see AJE500 | 20685-78-3 see SPE000 | 21239-57-6 see TBN200 | 21722-83-8 see EHS000 | 22373-78-0 see MRE230 |
| 19982-87-7 see FMH000 | 20689-96-7 see ENS500 | 21243-26-5 see DLO200 | 21725-46-2 see BLW750 | 22392-07-0 see DKB175 |
| 19986-35-7 see CGF000 | 20691-83-2 see DSN200 | 21247-98-3 see BCV250 | 21736-83-4 see SLI325 | 22432-68-4 see TBQ255 |
| 19992-69-9 see PPL500 | 20691-84-3 see CCE750 | 21248-00-0 see BMU500 | 21738-42-1 see OLT000 | 22439-58-3 see ACH090 |
| 19996-03-3 see CFB500 | 20706-25-6 see PNA225 | 21248-01-1 see CEK500 | 21739-91-3 see CQK600 | 22445-73-4 see TDX835 |
| 20004-62-0 see HAL000 | 20719-22-6 see POX750 | 21254-73-9 see CPU750 | 21794-01-4 see RQP000 | 22451-06-5 see TAI800 |
| 20018-09-1 see DNF850 | 20719-23-7 see EKB000 | 21255-83-4 see BMP500 | 21813-99-0 see FQR100 | 22457-23-4 see EMP550 |
| 20056-92-2 see DXU800 | 20719-34-0 see TAP000 | 21256-18-8 see OLW600 | 21816-42-2 see DSY000 | 22471-42-7 see HFA000 |
| 20057-09-4 see PCW500 | 20719-36-2 see TAP250 | 21259-20-1 see FQS000 | 21820-82-6 see DMB000 | 22480-64-4 see BNX000 |
| 20064-00-0 see NFW470 | 20731-44-6 see NIA700 | 21259-76-7 see TFK270 | 21829-25-4 see AEC750 | 22494-42-4 see DKI600 |
| 20064-38-4 see CAB125 | 20737-02-4 see SDS500 | 21260-46-8 see BKW000 | 21842-58-0 see DLQ800 | 22506-53-2 see DUW120 |
| 20064-40-8 see CKJ825 | 20738-78-7 see DWX200 | 21267-72-1 see CDS275 | 21848-62-4 see NNR200 | 22514-23-4 see FMU225 |
| 20064-41-9 see MQB100 | 20740-05-0 see HFN500 | 21280-29-5 see LAQ100 | 21884-44-6 see LIV000 | 22545-60-4 see AEG250 |
| 20120-33-6 see MND550 | 20743-57-1 see BGF000 | 21282-96-2 see AAY750 | 21888-96-0 see DBE000 | 22571-95-5 see SPB500 |
| 20123-68-6 see ALV750 | 20762-60-1 see PKW000 | 21284-11-7 see COB000 | 21892-31-9 see PBD300 | 22578-17-2 see SJA500 |
| 20123-80-2 see DMI300 | 20762-98-5 see FQJ025 | 21299-86-5 see INU200 | 21898-19-1 see VHA350 | 22585-64-4 see BOL310 |
| 20139-55-3 see AAY300 | 20777-39-3 see LCA100 | 21308-79-2 see MNF250 | 21905-27-1 see AGQ775 | 22591-21-5 see DGF300 |
| 20153-98-4 see CNR750 | 20777-49-5 see DKV160 | 21340-68-1 see MIO975 | 21905-32-8 see HNX600 | 22604-10-0 see TAI050 |
| 20168-99-4 see CMP950 | 20780-49-8 see DTE800 | 21351-39-3 see UTU600 | 21905-40-8 see CKA575 | 22608-53-3 see TMD625 |
| 20170-20-1 see PAM500 | 20816-12-0 see OKK000 | 21351-79-1 see CDD750 | 21908-53-2 see MCT500 | 22609-73-0 see NDY600 |
| 20182-56-3 see BKD250 | 20819-54-9 see TMD400 | 21361-93-3 see NDP500 | 21916-66-5 see POB250 | 22653-19-6 see TDP500 |
| 20187-55-7 see BAV325 | 20820-44-4 see NHK650 | 21362-69-6 see MCH600 | 21917-91-9 see MNO250 | 22670-79-7 see PEJ250 |
| 20198-77-0 see DFJ400 | 20820-80-8 see EIF500 | 21368-68-3 see CBA800 | 21923-23-9 see CLJ875 | 22691-91-4 see DSV289 |
| 20228-27-7 see RSU450 | 20829-66-7 see EIV700 | 21372-60-1 see TKH025 | 21928-82-5 see MMT250 | 22692-30-4 see FHZ200 |
| 20228-97-1 see CIJ250 | 20830-75-5 see DKN400 | 21380-82-5 see ACI375 | 21961-08-0 see HFY000 | 22713-35-5 see DQP125 |
| 20231-45-2 see VSF400 | 20830-81-3 see DAC000 | 21413-28-5 see DWJ300 | 21970-53-6 see DQD200 | 22722-03-8 see ELY550 |
| 20236-55-9 see BAO900 | 20839-15-0 see ACZ000 | 21416-67-1 see RCA375 | 21985-87-5 see PBM100 | 22722-98-1 see SGK800 |
| 20240-62-4 see MHW000 | 20839-16-1 see PGS500 | 21416-87-5 see PIK250 | 21992-92-7 see MPI225 | 22750-53-4 see CAD325 |
| 20240-98-6 see MNT500 | 20854-03-9 see TGY250 | 21436-96-4 see XOJ000 | 22029-76-1 see IFT500 | 22750-56-7 see CDC125 |
| 20241-03-6 see DSR200 | 20856-57-9 see CDP750 | 21436-97-5 see TLG750 | 22042-96-2 see DJG700 | 22750-65-8 see BGR250 |
| 20248-45-7 see TNH500 | 20859-73-8 see AHE750 | 21447-39-2 see BHR400 | 22047-25-2 see ADA350 | 22750-69-2 see DCM600 |
| 20265-96-7 see CJR200 | 20866-13-1 see HGL630 | 21447-86-9 see BHQ760 | 22047-49-0 see OFU300 | 22750-85-2 see DPR200 |
| 20265-97-8 see AOX500 | 20917-34-4 see DSL400 | 21447-87-0 see BIA100 | 22056-53-7 see MPO800 | 22750-86-3 see DPR400 |
| 20268-52-4 see CEJ000 | 20917-49-1 see OBY000 | 21450-81-7 see AAS500 | 22059-60-5 see RSZ600 | 22750-93-2 see EOD000 |
| 20275-19-8 see BGH000 | 20917-50-4 see OCA000 | 21452-14-2 see AKR500 | 22071-15-4 see BDU500 | 22751-18-4 see NFX500 |
| 20281-00-9 see CDE325 | 20919-99-7 see HBD650 | 21457-22-7 see DME700 | 22086-53-9 see CGN325 | 22751-23-1 see NIJ200 |
| 20300-26-9 see GJM025 | 20921-41-9 see MNX850 | 21476-57-3 see BJI125 | 22089-22-1 see TNT500 | 22751-24-2 see NFA500 |
| 20302-25-4 see POV500 | 20921-50-0 see PGQ000 | 21482-59-7 see TDP275 | 22098-38-0 see NIG500 | 22754-97-8 see RPB100 |
| 20311-78-8 see EAO500 | 20929-99-1 see BJW250 | 21498-08-8 see LIA400 | 22103-31-7 see TAG875 | 22755-01-7 see OOS000 |
| 20325-40-0 see DOA800 | 20930-00-1 see BJW500 | 21535-47-7 see BMA625 | 22131-79-9 see AGN000 | 22755-25-5 see SIW625 |
| 20333-40-8 see BRF550 | 20930-10-3 see DVY900 | 21548-32-3 see DHH200 | 22137-01-5 see AJZ000 | 22755-34-6 see TNE275 |
| 20344-15-4 see MCH535 | 20941-65-5 see EPJ000 | 21554-20-1 see AJQ500 | 22144-77-0 see ZUS000 | 22755-36-8 see TLE500 |
| 20354-26-1 see BGD250 | 20977-05-3 see EDM000 | 21561-99-9 see MMY500 | 22151-75-3 see HOO875 | 22760-18-5 see POB300 |
| 20369-63-5 see TID150 | 20977-50-8 see FGV000 | 21564-17-0 see BOO635 | 22188-15-4 see NMG000 | 22771-17-1 see DGV900 |
| 20373-56-2 see DEW400 | 20982-74-5 see NMV740 | 21572-61-2 see TBO778 | 22189-32-8 see SKY500 | 22771-18-2 see MRH212 |
| 20389-01-9 see DNG800 | 20986-33-8 see IMM000 | 21586-21-0 see MNI525 | 22199-08-2 see SNI425 | 22781-23-3 see DQM600 |
| 20398-06-5 see EEE000 | 20991-79-1 see TEK300 | 21587-39-3 see MEC330 | 22199-30-0 see CMX820 | 22797-20-2 see DLG000 |
| 20404-94-8 see DDO400 | 21000-42-0 see DWX000 | 21590-92-1 see AHP125 | 22204-24-6 see POK575 | 22826-61-5 see TBJ250 |
| 20405-19-0 see MIH750 | 21001-46-7 see TCM400 | 21593-23-7 see CCX500 | 22204-53-1 see MFA500 | 22830-45-1 see FBL000 |
| 20408-97-3 see GFK000 | 21019-39-6 see HKB600 | 21595-62-0 see PPC250 | 22205-30-7 see DVM400 | 22832-87-7 see MQS560 |
| 20427-56-9 see RSK000 | 21035-25-6 see BCD325 | 21600-42-0 see DTV200 | 22223-55-8 see MQH100 | 22839-47-0 see ARN825 |
| 20427-59-2 see CNM500 | 21041-93-0 see CNC231 | 21600-43-1 see DJY000 | 22224-92-6 see FAK000 | 22862-76-6 see AOY000 |
| 20436-27-5 see OGI100 | 21059-46-1 see CAM675 | 21600-45-3 see HKV000 | 22225-32-7 see FDU000 | 22868-13-9 see SGR500 |
| 20519-92-0 see TLT100 | 21062-28-2 see WCJ750 | 21600-51-1 see CBP250 | 22232-71-9 see MBV250 | 22885-98-9 see MPQ500 |
| 20537-88-6 see AMD000 | 21064-50-6 see MHD000 | 21609-90-5 see LEN000 | 22235-85-4 see AGM250 | 22904-40-1 see LDB000 |
| 20539-85-9 see BJO125 | 21070-32-6 see BRW750 | 21615-29-2 see BKB000 | 22236-53-9 see CJC750 | 22916-47-8 see MQS550 |
| 20548-54-3 see CAY000 | 21070-33-7 see BRM750 | 21621-73-8 see MLW630 | 22238-17-1 see THY000 | 22936-86-3 see CQI750 |

| | | | | |
|---|---|---|---|---|
| 22952-87-0 see ASG675 | 23505-41-1 see DIN600 | 24219-97-4 see MQS220 | 24829-11-6 see DHD400 | 25265-77-4 see TEG500 |
| 22953-53-3 see BHV750 | 23509-16-2 see BAR750 | 24220-18-6 see DPS200 | 24848-81-5 see TMO500 | 25267-15-6 see PJQ250 |
| 22953-54-4 see BHW000 | 23510-39-6 see CHG400 | 24221-86-1 see EAW995 | 24851-98-7 see HAK100 | 25267-55-4 see CNQ375 |
| 22958-08-3 see ASE875 | 23518-13-0 see HIS140 | 24234-06-8 see ELA600 | 24853-80-3 see ASC250 | 25284-83-7 see MDR750 |
| 22960-71-0 see EEX500 | 23521-13-3 see DUA200 | 24237-00-1 see GMG100 | 24854-67-9 see DHC000 | 25287-52-9 see PFF750 |
| 22966-79-6 see EDR500 | 23521-14-4 see DUA400 | 24245-27-0 see DWC625 | 24869-88-3 see DOS300 | 25295-51-6 see DXB400 |
| 22967-92-6 see MLF550 | 23526-02-5 see BAC125 | 24264-08-2 see DJN700 | 24884-69-3 see MMP200 | 25301-02-4 see TDN750 |
| 23031-25-6 see TAN100 | 23535-89-9 see DEH800 | 24268-87-9 see NFW430 | 24891-41-6 see TLK600 | 25311-71-1 see IMF300 |
| 23031-32-5 see TAN250 | 23537-16-8 see RRA000 | 24268-89-1 see NFW460 | 24909-09-9 see DLC000 | 25312-34-9 see IFT400 |
| 23031-38-1 see POD875 | 23541-50-6 see DAC200 | 24279-91-2 see BGX750 | 24928-15-2 see PGS750 | 25316-40-9 see HKA300 |
| 23067-13-2 see EDI500 | 23560-59-0 see HBK700 | 24280-93-1 see MRX000 | 24928-17-4 see PGT250 | 25321-09-9 see DNN709 |
| 23093-74-5 see BON400 | 23564-05-8 see PEX500 | 24282-51-7 see HOI400 | 24934-91-6 see CDY299 | 25321-14-6 see DVG600 |
| 23103-98-2 see DOX600 | 23564-06-9 see DJV000 | 24283-57-6 see CFC500 | 24938-91-8 see TJJ250 | 25321-41-9 see XJA000 |
| 23107-11-1 see DAL100 | 23581-62-6 see EAV100 | 24301-86-8 see PEF750 | 24939-03-5 see PJY750 | 25322-20-7 see TBP750 |
| 23107-12-2 see DAL400 | 23593-75-1 see MRX500 | 24301-89-1 see DWB300 | 24951-05-1 see PIX750 | 25322-68-3 see PJT000 |
| 23107-96-2 see BDS500 | 23595-00-8 see TKF699 | 24305-27-9 see TNX400 | 24961-39-5 see BNO750 | 25322-68-3 see PJT200 |
| 23109-05-9 see AHI625 | 23605-05-2 see APN000 | 24342-55-0 see HGI200 | 24961-49-7 see DMK400 | 25322-68-3 see PJT225 |
| 23110-15-8 see FOZ000 | 23605-74-5 see HEQ600 | 24342-56-1 see MIQ725 | 24973-25-9 see DOJ800 | 25322-68-3 see PJT230 |
| 23115-33-5 see BLR500 | 23606-32-8 see SDL500 | 24345-16-2 see AQN650 | 25013-15-4 see VQK650 | 25322-68-3 see PJT240 |
| 23129-50-2 see HEY000 | 23657-27-4 see TNO300 | 24346-78-9 see HBP250 | 25013-16-5 see BQI000 | 25322-68-3 see PJT250 |
| 23135-22-0 see DSP600 | 23668-11-3 see PAB500 | 24356-60-3 see HMK000 | 25036-33-3 see PKP250 | 25322-68-3 see PJT500 |
| 23139-00-6 see MNA650 | 23668-76-0 see DNR200 | 24356-66-9 see AEH100 | 25038-54-4 see PJY500 | 25322-68-3 see PJT750 |
| 23139-02-8 see BNX125 | 23674-86-4 see DKJ300 | 24358-29-0 see CGJ250 | 25038-59-9 see PKF750 | 25322-68-3 see PJU000 |
| 23155-02-4 see PHA550 | 23679-20-1 see DBL000 | 24365-47-7 see LEX400 | 25046-79-1 see PMG000 | 25322-69-4 see PKI500 |
| 23180-57-6 see PAC000 | 23705-25-1 see PLN100 | 24365-61-5 see PGF125 | 25047-48-7 see CMV375 | 25322-69-4 see PKI550 |
| 23182-46-9 see PIX800 | 23712-05-2 see DGA800 | 24378-32-3 see PAY600 | 25056-70-6 see TIX250 | 25322-69-4 see PKI750 |
| 23184-66-9 see CFW750 | 23734-06-7 see CCL109 | 24382-04-5 see MAN700 | 25057-89-0 see MJY500 | 25323-30-2 see DFH800 |
| 23191-75-5 see FOO875 | 23734-88-5 see AKC625 | 24397-89-5 see AEA109 | 25061-59-0 see MBV710 | 25324-56-5 see TGE500 |
| 23209-59-8 see CAX260 | 23745-86-0 see PLG000 | 24407-55-4 see BKP200 | 25068-38-6 see EBF500 | 25331-92-4 see AFH550 |
| 23210-56-2 see IAG600 | 23746-34-1 see PKX500 | 24423-68-5 see DNL200 | 25068-38-6 see EBG000 | 25332-09-6 see PMS900 |
| 23210-58-4 see IAG625 | 23753-67-5 see BNU125 | 24425-13-6 see BQJ750 | 25068-38-6 see EBG500 | 25332-39-2 see CKJ000 |
| 23214-92-8 see AES750 | 23757-42-8 see AMQ500 | 24426-36-6 see IPG000 | 25068-38-6 see ECL000 | 25332-39-2 see THK880 |
| 23217-86-9 see AMV750 | 23777-55-1 see MJB300 | 24458-48-8 see NHK800 | 25068-38-6 see IPO000 | 25333-83-9 see PMX500 |
| 23233-88-7 see BOL325 | 23777-63-1 see DCF750 | 24477-37-0 see DBE885 | 25081-01-0 see CLF500 | 25339-09-7 see IKC050 |
| 23239-41-0 see SGB500 | 23777-80-2 see HCA275 | 24518-45-4 see DEN820 | 25086-29-7 see VQK595 | 25339-17-7 see IKK000 |
| 23239-51-2 see RLK700 | 23779-99-9 see TKG000 | 24519-85-5 see DLK700 | 25090-71-5 see DAT000 | 25339-56-4 see HBK350 |
| 23246-96-0 see RJZ000 | 23784-96-5 see CFB825 | 24526-64-5 see NMV700 | 25090-72-6 see DAT600 | 25339-57-5 see BOP100 |
| 23255-69-8 see FQR000 | 23795-03-1 see DWW200 | 24536-75-2 see TLA525 | 25090-73-7 see DAT200 | 25340-17-4 see DIU000 |
| 23255-93-8 see HGO500 | 23826-71-3 see MED100 | 24544-04-5 see DNN630 | 25103-09-7 see ILR000 | 25340-18-5 see TJO750 |
| 23256-30-6 see NGG000 | 23828-92-4 see AHJ500 | 24549-06-2 see MJY000 | 25103-12-2 see TKT000 | 25351-18-2 see TLY250 |
| 23256-50-0 see GKO750 | 23834-96-0 see BND325 | 24554-26-5 see NGM500 | 25103-58-6 see DXT800 | 25354-97-6 see HFP500 |
| 23257-56-9 see LFO000 | 23840-95-1 see ZTK300 | 24570-11-4 see LHX600 | 25122-46-7 see CMW300 | 25355-59-3 see MFG400 |
| 23257-58-1 see LFG100 | 23844-24-8 see BCL250 | 24570-12-5 see LHX510 | 25122-57-0 see CMW400 | 25355-61-7 see MMW775 |
| 23261-20-3 see DCI600 | 23902-87-6 see CKI180 | 24579-91-7 see CHN750 | 25134-21-8 see NAC000 | 25376-45-8 see TGL500 |
| 23273-02-1 see DOX200 | 23902-88-7 see BQK800 | 24589-78-4 see MQG750 | 25136-55-4 see DRO800 | 25377-72-4 see AOI800 |
| 23282-20-4 see NMV000 | 23902-89-8 see CKI040 | 24596-38-1 see IOT875 | 25136-85-0 see PJL600 | 25377-73-5 see DXV000 |
| 23291-96-5 see DAL350 | 23902-91-2 see CKI185 | 24596-39-2 see BRB450 | 25147-05-1 see CBT125 | 25382-52-9 see MJU350 |
| 23292-52-6 see TJX375 | 23903-11-9 see DWK900 | 24596-41-6 see BRB460 | 25151-00-2 see BSO700 | 25384-17-2 see AGV890 |
| 23292-85-5 see TMW250 | 23904-72-5 see CKI190 | 24600-36-0 see FMU000 | 25152-84-5 see DAE450 | 25387-67-1 see CBB875 |
| 23315-05-1 see EAG000 | 23904-74-7 see CKI050 | 24602-86-6 see DUJ400 | 25152-85-6 see HFE500 | 25389-94-0 see KAM000 |
| 23319-66-6 see TNI500 | 23904-87-2 see CKI090 | 24613-89-6 see CMI250 | 25154-52-3 see NNC500 | 25389-94-0 see KAV000 |
| 23324-72-3 see MKR500 | 23904-88-3 see CKI080 | 24621-17-8 see ZQB100 | 25154-54-5 see DUQ180 | 25395-31-7 see GGA100 |
| 23327-57-3 see NBS500 | 23917-55-7 see CKI030 | 24623-77-6 see AHC250 | 25155-18-4 see MHB500 | 25395-41-9 see HLT300 |
| 23339-04-0 see DRX600 | 23920-57-2 see CKI070 | 24632-47-1 see NDY350 | 25155-25-3 see BHL100 | 25402-50-0 see IDP000 |
| 23344-17-4 see SMB850 | 23930-19-0 see AFK875 | 24632-48-2 see NDY370 | 25155-30-0 see DXW200 | 25405-85-0 see PGT000 |
| 23362-09-6 see MPJ100 | 23947-60-6 see BRI750 | 24656-22-2 see PIC100 | 25167-31-1 see CEI325 | 25410-64-4 see LHX515 |
| 23389-74-4 see HEW050 | 23950-58-5 see DTT600 | 24671-27-4 see CKA600 | 25167-67-3 see BOW250 | 25410-69-9 see LHX498 |
| 23389-74-4 see PBX800 | 23978-09-8 see LFQ000 | 24684-41-1 see MJD750 | 25167-70-8 see TMA250 | 25413-64-3 see PNF750 |
| 23395-20-2 see HMS500 | 24017-47-8 see THT750 | 24684-42-2 see DLU200 | 25167-83-3 see TBS250 | 25417-20-3 see NBS700 |
| 23399-90-8 see TIY250 | 24027-84-7 see BCX500 | 24684-49-9 see MEV250 | 25167-93-5 see CJA950 | 25425-12-1 see CMS500 |
| 23412-26-2 see THD300 | 24040-34-4 see CEC700 | 24684-56-8 see DLP600 | 25168-04-1 see NMS000 | 25429-29-2 see PAV600 |
| 23414-72-4 see ZLA000 | 24049-18-1 see AGN250 | 24684-58-0 see ABN250 | 25168-05-2 see CLK130 | 25430-97-1 see DJL200 |
| 23422-53-9 see DSO200 | 24050-16-6 see MPR000 | 24690-46-8 see DTT400 | 25168-15-4 see TGF200 | 25439-20-7 see PEB775 |
| 23435-31-6 see DON400 | 24050-58-6 see DWK700 | 24699-40-9 see CNE375 | 25168-24-5 see BKK250 | 25450-02-6 see LHX500 |
| 23471-13-8 see EED600 | 24066-82-8 see ELX530 | 24704-64-1 see AHE875 | 25168-26-7 see ILO000 | 25455-73-6 see SDV700 |
| 23471-23-0 see MHS250 | 24089-00-7 see MND100 | 24729-96-2 see CMV690 | 25174-65-6 see BFE770 | 25474-92-4 see DXH400 |
| 23476-83-7 see POC000 | 24094-93-7 see CMJ850 | 24735-35-1 see ZTS100 | 25174-66-7 see CKI020 | 25480-76-6 see CIQ625 |
| 23483-74-1 see BNL275 | 24095-80-5 see BLP325 | 24748-25-2 see TDH775 | 25201-35-8 see PAV250 | 25481-21-4 see EJA500 |
| 23488-38-2 see TBK250 | 24096-53-5 see DGF000 | 24771-52-6 see PBH150 | 25214-70-4 see FMW330 | 25481-10-7 see TCG450 |
| 23489-00-1 see EKN100 | 24124-25-2 see LGJ300 | 24781-13-3 see SAK500 | 25230-72-2 see TNV575 | 25483-93-8 see DUF000 |
| 23489-01-2 see ENY100 | 24143-08-6 see AAL500 | 24800-44-0 see TMZ000 | 25238-02-2 see CLS125 | 25486-91-3 see DUF000 |
| 23489-02-3 see ENX100 | 24143-17-7 see OMK300 | 24813-03-4 see BHP150 | 25251-03-0 see PCC750 | 25486-92-4 see DLT400 |
| 23489-03-4 see DAJ300 | 24166-13-0 see CMY525 | 24815-24-5 see TLN500 | 25265-19-4 see PAN250 | 25487-36-9 see HGL650 |
| 23491-45-4 see MOD500 | 24167-76-8 see SGM000 | 24817-51-4 see PDF780 | 25265-19-4 see PAN500 | 25496-72-4 see GGR200 |
| 23495-12-7 see EJK500 | 24168-96-5 see IKN200 | 24818-79-9 see AHA150 | 25265-71-8 see DWS500 | 25498-49-1 see MEV500 |
| | | | | 25523-79-9 see XES000 |

| | | | | |
|---|---|---|---|---|
| 51787-42-9 see TLI500 | 52479-15-9 see DHU600 | 53222-15-4 see MPM000 | 54099-13-7 see DPB400 | 55294-15-0 see EAE675 |
| 51787-43-0 see TLI750 | 52479-18-2 see AKA500 | 53222-25-6 see AHT850 | 54099-14-8 see DHT200 | 55297-95-5 see TET800 |
| 51787-44-1 see TDL500 | 52485-79-7 see TAL325 | 53222-52-9 see ALC750 | 54099-23-9 see DHP000 | 55297-96-6 see DAR000 |
| 51799-29-2 see NDY360 | 52486-78-9 see MCV500 | 53230-00-5 see DWU800 | 54156-67-1 see IKW000 | 55299-24-6 see CGB000 |
| 51821-32-0 see HKT200 | 52549-17-4 see PLX400 | 53231-79-1 see PJF500 | 54240-36-7 see KGK300 | 55308-57-1 see PEU650 |
| 51877-12-4 see DDN100 | 52557-97-8 see MHY600 | 53296-30-3 see ERD000 | 54301-15-4 see ADL500 | 55335-06-3 see TJE890 |
| 51898-39-6 see AET250 | 52578-56-0 see BJE500 | 53306-53-9 see DSF200 | 54301-19-8 see HJE600 | 55365-87-2 see EPC150 |
| 51909-61-6 see GAA100 | 52581-71-2 see UHA000 | 53317-25-2 see CQK500 | 54323-85-2 see POE100 | 55380-34-2 see DRO000 |
| 51922-16-8 see GAA120 | 52583-02-5 see BNF000 | 53330-94-2 see ACO300 | 54328-07-3 see PLR500 | 55398-24-8 see BDO199 |
| 51938-12-6 see NLN000 | 52622-27-2 see CBF710 | 53365-77-8 see MPB175 | 54340-62-4 see BQD000 | 55398-25-9 see BDQ000 |
| 51938-14-8 see BRO000 | 52623-84-4 see SFS500 | 53370-90-4 see HFS759 | 54350-48-0 see EMJ500 | 55398-26-0 see BDO500 |
| 51938-15-9 see BRY000 | 52623-88-8 see MAP300 | 53378-71-5 see TCQ600 | 54363-49-4 see MNH500 | 55398-27-1 see EOJ000 |
| 51938-16-0 see MKP000 | 52645-53-1 see AHJ750 | 53378-72-6 see PBJ600 | 54405-61-7 see MGT000 | 55398-86-2 see MRG000 |
| 51938-42-2 see SKR875 | 52663-81-7 see DXS375 | 53384-39-7 see MJK750 | 54448-39-4 see PLL000 | 55448-20-9 see MAS250 |
| 51940-44-4 see PIZ000 | 52663-84-0 see FEO000 | 53398-80-4 see HFE700 | 54453-03-1 see CNL750 | 55467-31-7 see CCX725 |
| 52049-26-0 see CHA500 | 52670-52-7 see BSL450 | 53404-82-3 see TIE600 | 54460-46-7 see HCP500 | 55477-20-8 see CEV750 |
| 52061-60-6 see IQE000 | 52670-78-7 see MNW100 | 53421-36-6 see CNN750 | 54472-62-7 see INN500 | 55477-27-5 see CHO250 |
| 52093-21-7 see MQS579 | 52673-65-1 see BJG750 | 53421-38-8 see DWD200 | 54481-45-7 see ABQ250 | 55489-49-1 see CFL750 |
| 52096-16-9 see NEN300 | 52673-66-2 see BJI750 | 53422-49-4 see ASF800 | 54484-73-0 see EKS120 | 55489-49-1 see CFR000 |
| 52096-22-7 see BCD125 | 52684-23-8 see OIW000 | 53459-38-4 see MAG050 | 54484-91-2 see CFQ000 | 55509-78-9 see SBN510 |
| 52098-13-2 see HIS100 | 52694-54-9 see CPK625 | 53460-80-3 see CIL850 | 54504-70-0 see TEQ500 | 55520-67-7 see VOA550 |
| 52098-14-3 see HIS120 | 52712-76-2 see AJA375 | 53460-81-4 see CIQ400 | 54514-12-4 see TGM425 | 55541-30-5 see DBC500 |
| 52098-15-4 see HIS130 | 52716-12-8 see CEX255 | 53469-21-9 see PJM500 | 54524-31-1 see MCK300 | 55556-85-9 see NLK000 |
| 52098-16-5 see BES300 | 52740-16-6 see ARK780 | 53477-43-3 see MFG510 | 54527-84-3 see PCG550 | 55556-86-0 see DTA800 |
| 52098-17-6 see MDX300 | 52740-56-4 see MEA500 | 53499-68-6 see MLK750 | 54531-52-1 see BCJ150 | 55556-88-2 see DRN800 |
| 52098-18-7 see MDX310 | 52775-76-5 see MJU500 | 53516-81-7 see AMU125 | 54546-26-8 see BSR600 | 55556-91-7 see NLL000 |
| 52098-56-3 see EPM000 | 52777-39-6 see MGK750 | 53534-20-6 see DRI700 | 54567-24-7 see ASE500 | 55556-92-8 see NLU500 |
| 52112-09-1 see BLN750 | 52783-44-5 see HAT000 | 53555-01-4 see TIL750 | 54573-23-8 see CPW325 | 55556-93-9 see NLK500 |
| 52112-66-0 see AIX750 | 52794-97-5 see KCA000 | 53558-25-1 see PPP750 | 54593-27-0 see BJG500 | 55556-94-0 see MJH000 |
| 52112-67-1 see AJL750 | 52814-92-3 see MEC300 | 53569-62-3 see CBV000 | 54597-56-7 see MRJ600 | 55557-00-1 see DVE600 |
| 52112-68-2 see AJF750 | 52831-39-7 see FGJ000 | 53569-64-5 see NGY500 | 54605-45-7 see IDJ500 | 55557-02-3 see NKH500 |
| 52125-53-8 see EJV000 | 52831-41-1 see DLN000 | 53578-07-7 see IPI350 | 54622-43-4 see DYE550 | 55566-30-8 see TDI000 |
| 52129-71-2 see DUV400 | 52831-45-5 see FGB000 | 53581-53-6 see BNE250 | 54634-49-0 see NLP600 | 55636-92-5 see HKS600 |
| 52137-03-8 see PJU250 | 52831-55-7 see DLN200 | 53583-79-2 see EPD100 | 54643-52-6 see HOB000 | 55644-07-0 see DGS700 |
| 52152-93-9 see CCS550 | 52831-56-8 see FGD000 | 53597-25-4 see SAO200 | 54693-46-8 see HMK050 | 55644-07-0 see DGS800 |
| 52157-57-0 see CIO275 | 52831-58-0 see FGO000 | 53597-27-6 see FAO100 | 54708-51-9 see CJW500 | 55651-31-5 see DLP400 |
| 52171-37-6 see DOD400 | 52831-60-4 see FJT000 | 53597-29-8 see MQF750 | 54708-68-8 see CKD250 | 55651-36-0 see DLP200 |
| 52171-41-2 see DOD200 | 52831-62-6 see FGL000 | 53607-04-8 see SNL830 | 54746-50-8 see BPD000 | 55661-38-6 see ALF500 |
| 52171-42-3 see DOF800 | 52831-65-9 see FGG000 | 53609-64-6 see DNB200 | 54749-90-5 see CLX000 | 55688-38-5 see HLK500 |
| 52171-92-3 see DRE400 | 52831-67-1 see DLN400 | 53639-82-0 see PMC275 | 54767-75-8 see SOU600 | 55689-65-1 see OMU000 |
| 52171-93-4 see DLI000 | 52831-68-2 see FGF000 | 53663-14-2 see NHY500 | 54818-88-1 see DDD400 | 55719-85-2 see PFD250 |
| 52171-94-5 see DUF200 | 52833-75-7 see TKC000 | 53681-76-8 see MRA300 | 54824-17-8 see MQX775 | 55726-47-1 see EAU075 |
| 52175-10-7 see POJ100 | 52851-26-0 see THQ600 | 53716-49-7 see CCK800 | 54824-20-3 see MAB055 | 55738-54-0 see DPL000 |
| 52195-07-0 see PIJ600 | 52913-14-1 see TGO575 | 53734-79-5 see IBQ300 | 54827-17-7 see TDM800 | 55764-18-6 see OOK000 |
| 52205-73-9 see EDT100 | 52917-86-9 see CEE800 | 53746-45-5 see FAP100 | 54856-23-4 see BFV350 | 55769-64-7 see CNW125 |
| 52207-83-7 see AGH500 | 52917-87-0 see EEE025 | 53757-28-1 see NGM400 | 54889-82-6 see PNE750 | 55779-06-1 see FOK000 |
| 52207-87-1 see TGH690 | 52918-63-5 see DAF300 | 53757-29-2 see MMJ975 | 54897-62-0 see ELE500 | 55792-21-7 see PIO750 |
| 52210-18-1 see IFX300 | 52934-83-5 see RMK200 | 53757-31-6 see NGL000 | 54897-63-1 see NKE000 | 55818-96-7 see TDH500 |
| 52212-02-9 see PII250 | 52936-25-1 see EJC000 | 53760-19-3 see PAM775 | 54925-45-0 see AEL500 | 55837-18-8 see IJH000 |
| 52217-47-7 see ACL500 | 52955-41-6 see MRY100 | 53762-93-9 see VRA000 | 54927-63-8 see SCA000 | 55837-27-9 see PDW250 |
| 52222-87-4 see BDN200 | 53004-03-8 see DOU000 | 53778-51-1 see DXF400 | 54965-21-8 see VAD000 | 55837-29-1 see TGF175 |
| 52237-03-3 see BLG100 | 53011-73-7 see IGF200 | 53780-34-0 see DUK000 | 54965-24-1 see TAD175 | 55843-86-2 see IAY000 |
| 52279-59-1 see MRU750 | 53025-21-1 see PAG225 | 53797-35-6 see RIP000 | 54976-93-1 see SFV300 | 55852-84-1 see BAC260 |
| 52292-20-3 see PKN000 | 53043-14-4 see AOH250 | 53847-48-6 see TEL150 | 55011-46-6 see NMK000 | 55864-39-6 see ALG375 |
| 52315-07-8 see RLF350 | 53067-74-6 see DWC050 | 53858-86-9 see FBP050 | 55028-70-1 see AQT575 | 55870-36-5 see BIT350 |
| 52329-60-9 see AIH000 | 53112-33-7 see MPO750 | 53866-33-4 see DHA425 | 55042-15-4 see UOA000 | 55898-33-4 see LJP000 |
| 52338-90-6 see TJK500 | 53123-88-9 see RBK000 | 53885-35-1 see TGA525 | 55044-04-7 see ELN600 | 55902-04-0 see AGL750 |
| 52340-46-2 see CHL875 | 53125-86-3 see SKO000 | 53902-12-8 see RLK800 | 55049-48-4 see SFX725 | 55921-66-9 see ALA750 |
| 52351-96-9 see MEC500 | 53127-17-6 see EEI060 | 53908-27-3 see DYE400 | 55077-30-0 see AAC875 | 55939-60-1 see SHT500 |
| 52400-55-2 see EGC000 | 53130-67-9 see HIX200 | 53910-25-1 see PBT100 | 55079-83-9 see REP400 | 55981-23-2 see CEK875 |
| 52400-58-5 see DJU800 | 53142-01-1 see AHN625 | 53912-89-3 see PIT650 | 55080-20-1 see ABO500 | 55984-51-5 see NKV000 |
| 52400-60-9 see DJI200 | 53152-21-9 see BOO630 | 53935-32-3 see POE050 | 55090-44-3 see NKU000 | 56001-43-5 see NCN800 |
| 52400-61-0 see DPC400 | 53153-66-5 see MNB500 | 53939-28-9 see HCP050 | 55102-43-7 see ROF200 | 56011-02-0 see IHV050 |
| 52400-65-4 see MLL500 | 53164-05-9 see AAE625 | 53955-81-0 see MLL600 | 55102-44-8 see RFU600 | 56047-14-4 see RHK250 |
| 52400-66-5 see MHD750 | 53179-09-2 see APY500 | 53956-04-0 see GIE100 | 55112-89-5 see BIB250 | 56073-10-0 see TAC800 |
| 52400-76-7 see BDY500 | 53184-19-3 see SBU950 | 53962-20-2 see PAM789 | 55123-66-5 see AAL300 | 56090-02-9 see HEK550 |
| 52400-77-8 see CFP250 | 53198-41-7 see ABT750 | 53988-42-4 see AFK900 | 55124-14-6 see BJE325 | 56092-91-2 see COQ375 |
| 52400-80-3 see DDM800 | 53198-87-1 see EQZ000 | 53994-73-3 see PAG075 | 55158-44-6 see CNK700 | 56105-46-5 see IIW100 |
| 52400-82-5 see DHT800 | 53213-78-8 see BJP300 | 54025-36-4 see NNX600 | 55216-04-1 see BIO500 | 56124-62-0 see TJX350 |
| 52400-99-4 see DHT600 | 53214-97-4 see DNE875 | 54063-28-4 see CBA375 | 55217-61-3 see AKL750 | 56139-33-4 see BQU000 |
| 52401-02-2 see DPI400 | 53216-90-3 see GKC000 | 54083-22-6 see ROU800 | 55256-53-6 see ARU250 | 56172-46-4 see DTE000 |
| 52401-04-4 see DHT400 | 53221-83-3 see ADO750 | 54086-41-8 see DKN250 | 55257-88-0 see DXY750 | 56179-80-7 see ECR259 |
| 52463-83-9 see PIH100 | 53221-85-5 see ADO250 | 54096-45-6 see EBD600 | 55268-74-1 see BGB400 | 56179-83-0 see DLZ000 |
| 52470-25-4 see GLB300 | 53221-86-6 see ADO500 | 54099-11-5 see AED750 | 55268-75-2 see CCS600 | 56183-20-1 see MEB000 |
| 52479-14-8 see DIC600 | 53221-88-8 see ADQ000 | 54099-12-6 see AEE000 | 55283-68-6 see ENE500 | 56187-09-8 see HCN100 |

| | | | | |
|---|---|---|---|---|
| 64741-44-2 see GBW000 | 65041-92-1 see ACS500 | 66232-28-8 see BHT000 | 67050-64-0 see SFJ875 | 67633-94-7 see PFR400 |
| 64741-46-4 see NAQ560 | 65043-22-3 see IBW500 | 66232-30-2 see BHS750 | 67050-97-9 see EPC000 | 67634-15-5 see EIJ600 |
| 64741-49-7 see MQV755 | 65057-90-1 see TAC500 | 66267-18-3 see DMK200 | 67051-27-8 see ILU000 | 67639-45-6 see FHH025 |
| 64741-50-0 see MQV815 | 65057-91-2 see TAC750 | 66267-19-4 see DLD400 | 67055-59-8 see CCX600 | 67658-42-8 see EJS100 |
| 64741-51-1 see MQV785 | 65072-04-0 see VIZ000 | 66267-67-2 see HJG100 | 67057-34-5 see DEB600 | 67658-46-2 see PNL800 |
| 64741-52-2 see MQV810 | 65089-17-0 see CLW500 | 66276-87-7 see BHS500 | 67110-84-3 see CNH300 | 67664-94-2 see TJB750 |
| 64741-53-3 see MQV780 | 65094-73-7 see NFW435 | 66289-74-5 see DLJ500 | 67114-26-5 see SHY000 | 67694-88-6 see ECQ200 |
| 64741-54-4 see NAQ520 | 65098-93-3 see TEO800 | 66309-69-1 see CCS375 | 67176-33-4 see FEG000 | 67722-96-7 see ECG100 |
| 64741-55-5 see NAQ540 | 65141-46-0 see NDL800 | 66327-54-6 see VIZ150 | 67195-50-0 see BQU750 | 67730-10-3 see DWW700 |
| 64741-56-6 see RDK200 | 65146-47-6 see AIW500 | 66332-77-2 see IJJ000 | 67195-51-1 see ECA500 | 67730-11-4 see AKS250 |
| 64741-57-7 see GBW005 | 65184-10-3 see TAL560 | 66357-35-5 see RBF400 | 67196-02-5 see PIU800 | 67746-30-9 see HFA600 |
| 64741-58-8 see GBW025 | 65210-28-8 see CPG250 | 66408-78-4 see CMS324 | 67210-66-6 see DIL600 | 67749-11-5 see HCP600 |
| 64741-59-9 see DXG840 | 65210-29-9 see MNP250 | 66409-97-0 see BPW000 | 67227-20-7 see AOS000 | 67762-92-9 see SDF000 |
| 64741-61-3 see DXG810 | 65210-30-2 see DPH200 | 66409-98-1 see FNP000 | 67227-30-9 see FPX000 | 67774-31-6 see NKA850 |
| 64741-62-4 see CMU890 | 65210-31-3 see DIC800 | 66427-01-8 see HKF600 | 67239-27-4 see CQA100 | 67774-32-7 see FBU509 |
| 64741-63-5 see NAQ550 | 65210-32-4 see MLP750 | 66471-17-8 see ALW900 | 67239-28-5 see PMM300 | 67785-74-4 see UNA100 |
| 64741-68-0 see NAQ530 | 65210-33-5 see MOI500 | 66472-85-3 see SBU100 | 67242-54-0 see TCP600 | 67785-77-7 see DQQ400 |
| 64741-80-6 see RDK100 | 65210-37-9 see BKQ250 | 66486-68-8 see HBI725 | 67255-31-6 see HHH100 | 67801-38-1 see IGJ600 |
| 64741-83-9 see NAQ580 | 65216-94-6 see COM500 | 66499-61-4 see MCS600 | 67262-60-6 see MEM750 | 67814-76-0 see SBE800 |
| 64741-87-3 see NAQ570 | 65229-18-7 see HMQ500 | 66535-86-2 see CKL325 | 67262-61-7 see MEN000 | 67849-02-9 see LHR650 |
| 64741-88-4 see MQV850 | 65232-69-1 see TJL775 | 66547-10-2 see APE529 | 67262-62-8 see EES500 | 67856-65-9 see BKO000 |
| 64741-89-5 see MQV855 | 65235-63-4 see BND250 | 66552-77-0 see TDK885 | 67262-64-0 see EMI000 | 67856-66-0 see BJO250 |
| 64741-96-4 see MQV845 | 65235-79-2 see SDJ500 | 66634-53-5 see DSH700 | 67262-69-5 see EMG500 | 67856-68-2 see BIF500 |
| 64741-97-5 see MQV852 | 65268-91-9 see DHU400 | 66671-82-7 see MEB820 | 67262-71-9 see EMH500 | 67874-80-0 see DTF000 |
| 64742-03-6 see MQV860 | 65271-80-9 see MQY090 | 66686-30-4 see FDN000 | 67262-72-0 see EMH000 | 67874-81-1 see CCR525 |
| 64742-04-7 see MQV859 | 65272-47-1 see MKJ000 | 66731-42-8 see DMB200 | 67262-74-2 see IMX000 | 67880-14-2 see PKW250 |
| 64742-05-8 see MQV862 | 65277-42-1 see KFK100 | 66733-21-9 see EDC650 | 67262-75-3 see IPZ000 | 67880-17-5 see DCL200 |
| 64742-10-5 see MQV863 | 65296-81-3 see CCJ350 | 66734-13-2 see AFI980 | 67262-78-6 see DRT400 | 67880-20-0 see DCL400 |
| 64742-11-6 see MQV857 | 65313-33-9 see CIH900 | 66788-01-0 see DKU400 | 67262-79-7 see DSN800 | 67880-21-1 see DCM000 |
| 64742-17-2 see MQV872 | 65313-34-0 see DIV400 | 66788-03-2 see DNC400 | 67262-80-0 see MEN250 | 67880-26-6 see MHR750 |
| 64742-18-3 see MQV760 | 65313-35-1 see TNR500 | 66788-06-5 see DMK600 | 67262-82-2 see MEN500 | 67880-27-7 see LHE450 |
| 64742-19-4 see MQV770 | 65313-36-2 see EIY550 | 66788-11-2 see ECQ150 | 67292-61-9 see HIF000 | 67883-79-8 see HFE710 |
| 64742-20-7 see MQV765 | 65313-37-3 see TDP300 | 66788-41-8 see DXI800 | 67292-62-0 see HKA200 | 67924-63-4 see EDK700 |
| 64742-21-8 see MQV775 | 65400-79-5 see MPR250 | 66789-14-8 see ADG425 | 67292-63-1 see HIE550 | 68006-83-7 see ALD750 |
| 64742-44-5 see PCS260 | 65405-73-4 see GDM100 | 66793-67-7 see PCG650 | 67292-68-6 see MLI750 | 68037-57-0 see EHC900 |
| 64742-45-6 see PCS270 | 65405-76-7 see HFE300 | 66795-86-6 see EDL000 | 67293-64-5 see CIO500 | 68038-71-1 see BAC040 |
| 64742-46-7 see DXG830 | 65405-77-8 see SAJ000 | 66813-55-6 see TAI725 | 67293-75-8 see SHQ000 | 68039-38-3 see CMT500 |
| 64742-47-8 see KEK100 | 65405-84-7 see TDO255 | 66826-72-0 see DAC975 | 67293-86-1 see MEA750 | 68041-18-9 see HLY000 |
| 64742-52-5 see MQV790 | 65445-59-2 see DTA600 | 66826-73-1 see DGH500 | 67293-88-3 see DBB600 | 68070-90-6 see SKS299 |
| 64742-53-6 see MQV800 | 65454-27-5 see MCB100 | 66827-45-0 see ABN700 | 67298-49-1 see NCM275 | 68071-23-8 see EBU100 |
| 64742-54-7 see MQV795 | 65496-97-1 see SDX630 | 66827-50-7 see AGC750 | 67312-43-0 see SIW600 | 68085-85-8 see GJU600 |
| 64742-55-8 see MQV805 | 65521-60-0 see PLW275 | 66827-74-5 see NEP000 | 67330-25-0 see BRJ325 | 68107-26-6 see NKX500 |
| 64742-56-9 see MQV840 | 65546-74-9 see ROA400 | 66839-97-2 see PPT400 | 67335-42-6 see BBG000 | 68130-43-8 see SOJ000 |
| 64742-63-8 see MQV820 | 65567-32-0 see EOL050 | 66839-98-3 see MEF400 | 67335-43-7 see BBD750 | 68131-01-4 see HGM100 |
| 64742-64-9 see MQV835 | 65573-02-6 see IBQ075 | 66859-63-0 see EPC135 | 67360-94-5 see ARK750 | 68131-40-8 see TAY250 |
| 64742-65-0 see MQV825 | 65597-24-2 see CDX800 | 66862-11-1 see HEJ350 | 67360-95-6 see BCB000 | 68131-40-8 see TAZ000 |
| 64742-68-3 see MQV865 | 65597-25-3 see TJX650 | 66903-23-9 see BKR250 | 67360-95-6 see BCB250 | 68131-40-8 see TAZ100 |
| 64742-69-4 see MQV867 | 65654-13-9 see TEN725 | 66922-67-6 see EOD500 | 67401-56-3 see MQS565 | 68131-40-8 see TBA500 |
| 64742-70-7 see MQV868 | 65664-23-5 see PLU575 | 66922-79-0 see AQI500 | 67410-20-2 see TMW600 | 68132-21-8 see PCJ100 |
| 64742-71-8 see MQV870 | 65666-07-1 see SDX625 | 66941-43-3 see DSI200 | 67411-81-8 see DLF400 | 68133-72-2 see HFE600 |
| 64742-80-9 see DXG820 | 65700-59-6 see DAU200 | 66941-48-8 see AFV750 | 67418-30-8 see TGA275 | 68133-73-3 see IHS100 |
| 64742-81-0 see KEK110 | 65700-60-9 see DAU000 | 66941-49-9 see AFJ300 | 67445-50-5 see TMY850 | 68141-17-3 see MNB600 |
| 64742-86-5 see GBW010 | 65734-38-5 see ACN500 | 66941-53-5 see AGA250 | 67465-04-7 see BEA000 | 68141-57-1 see FHS000 |
| 64755-14-2 see BMK325 | 65763-32-8 see DLD800 | 66941-60-4 see AGD000 | 67465-26-3 see PNV760 | 68162-13-0 see DLD600 |
| 64771-59-1 see MCX600 | 65792-56-5 see NLN500 | 66941-77-3 see AGL250 | 67465-27-4 see EII600 | 68162-93-6 see MLM500 |
| 64781-77-7 see DCQ500 | 65793-50-2 see AIS250 | 66941-81-9 see AGM000 | 67465-28-5 see EPI400 | 68171-33-5 see IPS450 |
| 64808-48-6 see LHZ600 | 65860-38-0 see TCU000 | 66955-43-9 see PCL775 | 67465-39-8 see CHW000 | 68188-03-4 see CAB800 |
| 64817-78-3 see MSA750 | 65899-73-2 see TGF050 | 66955-44-0 see PCO175 | 67465-41-2 see BII500 | 68247-85-8 see BLY770 |
| 64819-51-8 see BQI300 | 65928-58-7 see SMP400 | 66964-37-2 see MHA000 | 67465-42-3 see EET500 | 68278-23-9 see EAE775 |
| 64838-75-1 see DMP000 | 65979-81-9 see ZKS000 | 66967-60-0 see MFR250 | 67465-43-4 see INE000 | 68291-97-4 see BCE750 |
| 64854-98-4 see CBD500 | 65986-80-3 see ENR500 | 66967-65-5 see DRW000 | 67465-44-5 see MES250 | 68307-81-3 see TMP175 |
| 64854-99-5 see CBD250 | 65996-92-1 see CMY900 | 66967-84-8 see HNG000 | 67466-28-8 see EJT000 | 68308-34-9 see COD750 |
| 64910-63-0 see BSM250 | 65996-93-2 see CMZ100 | 66969-02-6 see BNE500 | 67466-58-4 see HLO300 | 68334-30-5 see DHE900 |
| 64920-31-6 see DLD200 | 66007-89-4 see SFP500 | 66988-15-6 see DAI800 | 67479-03-2 see MCR250 | 68348-85-6 see DEJ849 |
| 64925-80-0 see PGR775 | 66009-08-3 see BIW500 | 66997-69-1 see ECR500 | 67479-04-3 see PPP550 | 68377-91-3 see BQE000 |
| 64953-12-4 see LBH200 | 66017-91-2 see PNR250 | 67011-19-6 see AIU250 | 67523-22-2 see DLF600 | 68379-32-8 see DEJ880 |
| 64977-44-2 see FJY000 | 66064-11-7 see AIZ000 | 67031-48-5 see AIU250 | 67526-05-0 see ENA000 | 68401-82-1 see EBE100 |
| 64977-46-4 see FKA000 | 66085-59-4 see NDY700 | 67032-45-5 see BPH750 | 67527-71-3 see MQU000 | 68411-30-3 see LGF825 |
| 64977-47-5 see FKB000 | 66104-23-2 see MPU500 | 67037-37-0 see DKH875 | 67536-44-1 see BKH125 | 68424-85-1 see QAT520 |
| 64977-48-6 see FKC000 | 66104-24-3 see BFP500 | 67049-51-8 see HFV000 | 67557-56-6 see BSX500 | 68426-46-0 see NKV500 |
| 64977-49-7 see FKD000 | 66147-68-0 see ADP750 | 67049-95-0 see OMY700 | 67557-57-7 see MMS000 | 68436-99-7 see TBJ400 |
| 64988-06-3 see EMF600 | 66147-69-1 see ADP000 | 67050-00-4 see BPA750 | 67590-46-9 see BRO750 | 68442-69-3 see UAK000 |
| 65002-17-7 see MCO775 | 66217-76-3 see MHB000 | 67050-04-8 see BPB500 | 67590-56-1 see EQD100 | 68448-47-5 see MDE500 |
| 65036-47-7 see BNG125 | 66231-56-9 see CNH650 | 67050-11-7 see BPC500 | 67590-57-2 see IRQ100 | 68475-76-3 see PKS750 |
| 65039-20-5 see CIS325 | 66232-25-5 see BHS250 | 67050-26-4 see BRJ250 | 67632-66-0 see TIS750 | 68476-30-2 see DHE800 |

| | | | | |
|---|---|---|---|---|
| 68476-33-5 see FOP050 | 68991-29-7 see OHJ100 | 69782-11-2 see NBK000 | 70983-41-4 see ZQS100 | 72704-51-9 see IGD075 |
| 68476-85-7 see LGM000 | 69011-63-8 see MAF000 | 69782-16-7 see PDS250 | 70992-03-9 see TBG750 | 72732-50-4 see DAD500 |
| 68480-17-1 see TLO530 | 69012-64-2 see SCH001 | 69782-18-9 see DOJ600 | 71002-67-0 see BRW500 | 72738-90-0 see MPL750 |
| 68527-78-6 see AOH100 | 69029-52-3 see LDC000 | 69782-24-7 see CJP250 | 71016-15-4 see MHW350 | 72812-40-9 see ISR100 |
| 68527-79-7 see ICS100 | 69091-15-2 see NKA700 | 69782-26-9 see DOJ400 | 71073-91-1 see ADW750 | 72820-33-8 see DIL200 |
| 68533-38-0 see BMD825 | 69091-16-3 see NKA695 | 69782-38-3 see NDQ000 | 71250-00-5 see CDD325 | 72850-64-7 see BEG300 |
| 68541-88-8 see TMX350 | 69095-83-6 see EOL600 | 69782-42-9 see NDW000 | 71251-30-4 see MRE750 | 72869-73-9 see TAD250 |
| 68555-34-0 see BKR100 | 69102-79-0 see PJF750 | 69782-43-0 see NDU000 | 71277-79-7 see DXD875 | 72957-64-3 see DDN150 |
| 68555-58-8 see PMB600 | 69103-91-9 see MFH000 | 69782-45-2 see ALU750 | 71292-84-7 see TCH325 | 72977-18-5 see AIQ875 |
| 68574-13-0 see TDK300 | 69103-95-3 see PFX250 | 69782-62-3 see ANY500 | 71359-62-1 see HAY059 | 72985-54-7 see BJV625 |
| 68574-15-2 see TDJ325 | 69103-96-4 see HJU000 | 69804-02-0 see AEG129 | 71359-64-3 see HDC425 | 72985-56-9 see CER825 |
| 68574-17-4 see TDJ350 | 69103-97-5 see PFX500 | 69806-50-4 see FDA885 | 71392-29-5 see CDM500 | 73074-20-1 see AMI750 |
| 68594-19-4 see BLI000 | 69112-96-5 see DFW000 | 69853-15-2 see DUO500 | 71435-43-3 see DKW400 | 73080-51-0 see IHR200 |
| 68594-24-1 see DXG050 | 69112-98-7 see NKG000 | 69853-71-0 see HJV500 | 71447-49-9 see LIU380 | 73118-23-7 see BKG500 |
| 68596-88-3 see DQL899 | 69112-99-8 see NKT000 | 69866-21-3 see APT375 | 71463-34-8 see AOC275 | 73118-24-8 see BKG750 |
| 68596-89-4 see HJH000 | 69113-00-4 see NJR000 | 69883-99-4 see DAT800 | 71500-21-5 see AGU400 | 73239-98-2 see NKO600 |
| 68596-94-1 see TGM590 | 69113-01-5 see NLZ000 | 69884-15-7 see DPF200 | 71522-58-2 see FOJ100 | 73297-70-8 see GJY100 |
| 68596-99-6 see TKQ000 | 69136-21-6 see EBV100 | 69928-30-9 see DSH400 | 71598-10-2 see NLD000 | 73309-75-8 see AGW625 |
| 68597-10-4 see COQ325 | 69180-59-2 see IFE875 | 69928-47-8 see PIR100 | 71609-22-8 see HKB000 | 73341-70-5 see GMC000 |
| 68602-57-3 see TJX800 | 69226-06-8 see DKC200 | 69929-16-4 see EES100 | 71628-96-1 see MCB600 | 73341-73-8 see MAB400 |
| 68647-73-4 see TAI150 | 69226-39-7 see AKG000 | 69946-37-8 see MID800 | 71631-15-7 see CMS135 | 73343-67-6 see AOY300 |
| 68674-44-2 see CFX125 | 69226-43-3 see DVN000 | 69975-86-6 see TEQ175 | 71653-63-9 see RSZ375 | 73343-69-8 see HNK800 |
| 68683-20-5 see IHX450 | 69226-44-4 see DVN400 | 70052-12-9 see DBE835 | 71677-48-0 see MMZ800 | 73343-70-1 see HNK700 |
| 68688-87-9 see MJD610 | 69226-45-5 see DVL800 | 70084-70-7 see TAC200 | 71700-95-3 see VIZ500 | 73343-71-2 see DHR900 |
| 68690-89-1 see NKW000 | 69226-46-6 see DVN200 | 70134-26-8 see DEU100 | 71731-58-3 see TGF075 | 73343-72-3 see DNF550 |
| 68743-79-3 see SLF500 | 69226-47-7 see TIG500 | 70142-16-4 see BMQ500 | 71752-66-4 see NJO500 | 73343-74-5 see DDP300 |
| 68771-70-0 see PPB500 | 69239-37-8 see DEI800 | 70145-54-9 see HLH000 | 71752-67-5 see NMA000 | 73361-47-4 see IHH300 |
| 68771-76-6 see POY250 | 69242-96-2 see MFN000 | 70145-55-0 see DWZ100 | 71752-68-6 see NLX000 | 73379-85-8 see KHK100 |
| 68772-00-9 see DBR000 | 69260-83-9 see DMX000 | 70145-56-1 see HLU000 | 71752-69-7 see NKO400 | 73419-42-8 see CAK250 |
| 68772-09-8 see MLU000 | 69260-85-1 see DML775 | 70145-60-7 see PEO750 | 71752-70-0 see NKK000 | 73452-31-0 see BLQ850 |
| 68772-10-1 see EKA500 | 69267-51-2 see NHQ950 | 70145-80-1 see AMJ625 | 71767-91-4 see HOS500 | 73452-32-1 see BSR825 |
| 68772-13-4 see AMG000 | 69321-16-0 see HLX550 | 70145-82-3 see MFJ000 | 71771-90-9 see DAP850 | 73454-79-2 see MHW250 |
| 68772-17-8 see AKO000 | 69343-45-9 see NCJ000 | 70145-83-4 see MFD250 | 71785-87-0 see NKC300 | 73487-24-8 see NJO200 |
| 68772-18-9 see MLU250 | 69352-67-6 see BKP000 | 70213-45-5 see ART125 | 71840-26-1 see MOB275 | 73506-32-8 see HGW100 |
| 68772-19-0 see MLT500 | 69352-90-5 see TDB200 | 70224-81-6 see MDH000 | 71856-48-9 see MGS700 | 73506-39-5 see SGS000 |
| 68772-21-4 see EKC500 | 69353-21-5 see GBA000 | 70247-32-4 see COS825 | 71901-54-7 see DTI709 | 73526-98-4 see BJU350 |
| 68772-29-2 see EKA000 | 69357-10-4 see PFG750 | 70247-50-6 see SIL600 | 71939-10-1 see PLR175 | 73529-25-6 see EEF500 |
| 68772-43-0 see AKO250 | 69357-11-5 see PFG000 | 70247-51-7 see SDR400 | 71963-77-4 see ARL425 | 73561-96-3 see MBW780 |
| 68772-49-6 see PEX750 | 69357-13-7 see PFG500 | 70277-99-5 see BNE325 | 71964-72-2 see DQK900 | 73599-90-3 see CEE850 |
| 68780-95-0 see TKO750 | 69357-14-8 see PFG250 | 70278-00-1 see DEO500 | 72007-81-9 see PFR300 | 73599-91-4 see BMT130 |
| 68786-66-3 see CFL200 | 69365-68-0 see QOJ000 | 70288-86-7 see ITD875 | 72017-28-8 see NAY500 | 73599-92-5 see HME100 |
| 68795-10-8 see FHV800 | 69365-73-7 see CCD750 | 70299-48-8 see ENA500 | 72017-60-8 see AQC000 | 73599-93-6 see HMK150 |
| 68797-80-7 see DBR200 | 69382-20-3 see BGC250 | 70301-54-1 see DLV400 | 72040-09-6 see CGL125 | 73599-94-7 see CEE825 |
| 68797-91-1 see POX500 | 69388-84-7 see PAP600 | 70301-64-3 see DLS000 | 72044-13-4 see MDA100 | 73599-95-8 see BMT100 |
| 68808-54-8 see AJR500 | 69402-04-6 see BJJ750 | 70301-64-3 see MOG250 | 72050-78-3 see DBE625 | 73622-67-0 see DBH800 |
| 68822-50-4 see PJL000 | 69462-47-1 see DSP650 | 70301-68-7 see DLV600 | 72064-79-0 see AAF625 | 73637-16-8 see ACA750 |
| 68825-98-9 see MRC600 | 69462-48-2 see TCV400 | 70303-46-7 see EPR200 | 72066-32-1 see MLF750 | 73639-62-0 see CIL775 |
| 68833-55-6 see MCW349 | 69462-51-7 see MLL650 | 70303-47-8 see BSO200 | 72074-66-9 see DMS400 | 73651-49-7 see DKX800 |
| 68844-77-9 see ARP675 | 69462-52-8 see MLL655 | 70324-20-8 see SDP025 | 72074-67-0 see DMS200 | 73665-15-3 see MRW500 |
| 68848-64-6 see LHP000 | 69462-53-9 see MLL660 | 70324-23-1 see DTY600 | 72074-69-2 see ECS000 | 73666-84-9 see TEF725 |
| 68880-55-7 see SLC300 | 69462-56-2 see TCV375 | 70324-35-5 see PLD730 | 72117-72-7 see COW780 | 73680-58-7 see AHB250 |
| 68908-87-2 see DQL899 | 69579-13-1 see RFU800 | 70343-15-6 see DUX560 | 72122-60-2 see MRR775 | 73684-69-2 see MRW750 |
| 68916-04-1 see BLV750 | 69598-87-4 see VIZ250 | 70348-66-2 see PNH550 | 72150-17-5 see SLI200 | 73688-63-8 see APM250 |
| 68916-09-6 see FDA000 | 69654-93-9 see DJC875 | 70356-03-5 see CCR850 | 72207-94-4 see EQF100 | 73688-85-4 see DPO275 |
| 68916-26-7 see OGM100 | 69657-51-8 see AEC725 | 70384-29-1 see PCB000 | 72209-26-8 see AEJ250 | 73693-97-7 see PIM750 |
| 68916-39-2 see WCB000 | 69712-56-7 see CCS373 | 70441-82-6 see MBZ000 | 72209-27-9 see AEK500 | 73696-62-5 see ALG250 |
| 68916-73-4 see CBA200 | 69745-66-0 see BRB500 | 70441-63-7 see MBY500 | 72214-01-8 see ELB400 | 73696-64-7 see BKU000 |
| 68916-94-9 see JEJ200 | 69766-15-0 see PIS000 | 70443-38-8 see DNC800 | 72254-58-1 see ALE750 | 73696-65-8 see ENH000 |
| 68916-95-0 see KCA050 | 69766-22-9 see MOK500 | 70458-96-7 see BAB625 | 72275-67-3 see FOL000 | 73698-75-6 see AIK500 |
| 68917-09-9 see OAF000 | 69766-47-8 see ARX770 | 70476-82-3 see MQY100 | 72299-02-6 see MIS500 | 73698-76-7 see AIK750 |
| 68917-15-7 see BFN990 | 69766-48-9 see ARX750 | 70501-82-5 see NLU480 | 72320-60-6 see EOS100 | 73698-77-8 see AIL000 |
| 68917-43-1 see HGA100 | 69766-49-0 see ARX500 | 70536-17-3 see AFI850 | 72378-89-3 see TGF000 | 73698-78-9 see AIL250 |
| 68917-50-0 see SOY100 | 69766-62-7 see TBF000 | 70548-53-7 see DBW200 | 72385-44-5 see TIT275 | 73713-75-4 see EAN500 |
| 68917-52-2 see SBE000 | 69780-81-0 see SHI000 | 70561-82-9 see DBW600 | 72443-10-8 see FOS300 | 73728-78-6 see TDJ250 |
| 68917-73-7 see WBJ700 | 69780-82-1 see AIQ880 | 70664-49-2 see ASF625 | 72497-31-5 see PIA400 | 73728-79-7 see TLF000 |
| 68937-41-7 see DYF500 | 69780-83-2 see AIQ885 | 70711-40-9 see BKB300 | 72571-82-5 see PII350 | 73728-82-2 see AKF250 |
| 68955-06-6 see TKS000 | 69780-84-3 see AIQ890 | 70711-41-0 see DBE875 | 72586-67-5 see MFX725 | 73747-22-5 see BEL550 |
| 68955-35-1 see NAQ510 | 69781-71-1 see EGG000 | 70715-92-3 see ABR125 | 72586-68-6 see DTN875 | 73747-29-2 see AJO750 |
| 68955-54-4 see PMC400 | 69781-89-1 see XVA000 | 70729-68-9 see TCE375 | 72589-96-9 see CAE375 | 73747-51-0 see BFZ170 |
| 68956-68-3 see VGU200 | 69781-93-7 see THC250 | 70786-64-0 see DSY800 | 72590-77-3 see HHQ850 | 73747-53-2 see ICW100 |
| 68956-82-1 see CNE000 | 69781-96-0 see PDV750 | 70816-59-0 see TEF600 | 72674-05-6 see AFN500 | 73747-54-3 see MQE100 |
| 68979-48-6 see BGW650 | 69782-00-9 see PGM250 | 70851-61-5 see MIW060 | 72676-73-4 see THH490 | 73758-18-6 see TJJ750 |
| 68988-02-3 see CNF340 | 69782-01-0 see PDU500 | 70858-14-9 see VLU210 | 72676-74-5 see THH480 | 73758-56-2 see DEK400 |
| 68990-15-8 see FAR000 | 69782-05-4 see PGL500 | 70907-61-8 see CPP750 | 72676-77-8 see THH460 | 73771-13-8 see SJJ190 |
| 68991-20-8 see AQY385 | 69782-09-8 see PDV000 | 70950-00-4 see GIT000 | 72676-78-9 see THH470 | 73771-52-5 see BJA000 |

| | | | | |
|---|---|---|---|---|
| 73771-72-9 see FJU000 | 73973-02-1 see PDB300 | 75881-20-8 see NKW800 | 77855-81-3 see MQT600 | 78109-87-2 see DII400 |
| 73771-73-0 see FJV000 | 73986-95-5 see DGA400 | 75881-22-0 see MMX200 | 77879-90-4 see GEO200 | 78109-88-3 see EEY000 |
| 73771-74-1 see FJW000 | 73987-00-5 see ACH375 | 75884-37-6 see FQQ100 | 77922-38-4 see TME260 | 78109-90-7 see PGE500 |
| 73771-79-6 see DNC600 | 73987-16-3 see TLO600 | 75888-03-8 see FPO100 | 77944-89-9 see CIK500 | 78110-10-8 see CLB250 |
| 73771-81-0 see QIS000 | 73987-51-6 see TGK500 | 75889-62-2 see DIU500 | 77945-03-0 see DIA000 | 78110-23-3 see MDK000 |
| 73785-34-9 see DBG200 | 73987-52-7 see EEP000 | 75965-74-1 see MFB400 | 77945-09-6 see DHX600 | 78110-37-9 see CEY250 |
| 73785-40-7 see NJS300 | 73990-29-1 see AHI500 | 75993-65-6 see MPI800 | 77966-20-2 see BPJ250 | 78110-38-0 see ARX875 |
| 73790-27-9 see MOK000 | 74007-80-0 see OPE100 | 76050-49-2 see FMR700 | 77966-25-7 see BPY500 | 78128-69-5 see BKD000 |
| 73791-29-4 see AIY750 | 74011-58-8 see EAU100 | 76059-11-5 see DVS100 | 77966-26-8 see DHJ800 | 78128-80-0 see BRO250 |
| 73791-32-9 see TMA750 | 74037-18-6 see DEU125 | 76059-13-7 see DVS300 | 77966-27-9 see DHK000 | 78128-81-1 see CPM250 |
| 73791-39-6 see ALV100 | 74037-31-3 see ASA250 | 76059-14-8 see DVS400 | 77966-28-0 see PPU750 | 78128-83-3 see MJF250 |
| 73791-40-9 see BRQ800 | 74037-60-8 see BHB500 | 76069-32-4 see TAB785 | 77966-30-4 see AFW250 | 78128-84-4 see HND000 |
| 73791-41-0 see DGA425 | 74038-45-2 see DHH600 | 76095-16-4 see EAO100 | 77966-31-5 see BDY250 | 78128-85-5 see PGC750 |
| 73791-42-1 see CKD800 | 74039-01-3 see AIW750 | 76145-76-1 see THG700 | 77966-32-6 see BEE750 | 78173-90-7 see SON520 |
| 73791-43-2 see IPS100 | 74039-02-4 see AMO750 | 76175-45-6 see TEY600 | 77966-34-8 see BEJ250 | 78173-91-8 see ARS135 |
| 73791-44-3 see MNU050 | 74039-79-4 see BLT500 | 76180-96-6 see AKT600 | 77966-38-2 see CEX500 | 78173-92-9 see ARS130 |
| 73791-45-4 see MOT800 | 74039-79-5 see BLT750 | 76206-37-6 see NBM500 | 77966-40-6 see CFH000 | 78186-37-5 see MQO250 |
| 73803-48-2 see CHK825 | 74039-80-8 see BLU250 | 76263-73-5 see POM700 | 77966-41-7 see CFI000 | 78186-61-5 see MLJ750 |
| 73806-23-2 see OMA000 | 74039-81-9 see BLU500 | 76298-68-5 see ACY700 | 77966-42-8 see CFL000 | 78193-30-3 see SON525 |
| 73806-49-2 see THJ750 | 74050-97-8 see HAG300 | 76319-15-8 see PAM800 | 77966-43-9 see CFR750 | 78218-16-3 see CGY500 |
| 73815-11-9 see MEW800 | 74093-43-9 see SDM550 | 76379-66-3 see DXU250 | 77966-44-0 see CFS000 | 78218-37-8 see CGF500 |
| 73816-43-0 see BKJ500 | 74195-73-6 see RGP450 | 76379-67-4 see OAX050 | 77966-45-1 see CFS250 | 78218-38-9 see CIE500 |
| 73816-74-7 see CHG250 | 74203-42-2 see POV750 | 76429-97-5 see DWP950 | 77966-46-2 see CFM250 | 78218-40-3 see CKO000 |
| 73816-75-8 see DGO000 | 74203-45-5 see NDF000 | 76429-98-6 see DCM875 | 77966-47-3 see CFM500 | 78218-42-5 see CIU250 |
| 73825-59-9 see DAQ000 | 74203-58-0 see PGA250 | 76487-32-6 see ALU875 | 77966-48-4 see CFM750 | 78218-43-6 see BQN500 |
| 73825-85-1 see SNS100 | 74203-59-1 see MOD750 | 76541-72-5 see CMU875 | 77966-49-5 see CFN000 | 78218-49-2 see TMU750 |
| 73826-58-1 see DGN600 | 74203-61-5 see SJK475 | 76556-13-3 see ANG625 | 77966-51-9 see CFU500 | 78219-33-7 see EOY100 |
| 73840-42-3 see MOA250 | 74220-04-5 see SBV500 | 76631-42-0 see NMV480 | 77966-52-0 see CFW250 | 78219-35-9 see MOI250 |
| 73870-33-4 see NLV500 | 74222-97-2 see SNW550 | 76648-01-6 see TAA400 | 77966-53-1 see CFW000 | 78219-38-2 see PFD500 |
| 73873-83-3 see OPY000 | 74273-75-9 see DCM700 | 76714-88-0 see DGC100 | 77966-54-2 see CGC000 | 78219-45-1 see PFD750 |
| 73926-79-1 see TGE750 | 74278-22-1 see KHU000 | 76738-28-8 see DFQ100 | 77966-55-3 see CGD000 | 78219-52-0 see PNT500 |
| 73926-80-4 see DDF000 | 74339-98-3 see DKX875 | 76822-96-3 see VRU000 | 77966-56-4 see CGE750 | 78219-53-1 see PNT750 |
| 73926-81-5 see DDG600 | 74340-04-8 see DKX900 | 76824-35-6 see FAB500 | 77966-58-6 see CGN250 | 78219-57-5 see EOV000 |
| 73926-83-7 see THU250 | 74356-00-6 see CCS371 | 76828-34-7 see ECJ100 | 77966-59-7 see CGP380 | 78219-58-6 see EOW000 |
| 73926-85-9 see BIV900 | 74469-00-4 see ARS125 | 77094-11-2 see AJQ600 | 77966-61-1 see CHS750 | 78219-61-1 see MON000 |
| 73926-87-1 see CET000 | 74512-62-2 see PNR800 | 77162-70-0 see NFW350 | 77966-62-2 see CID250 | 78219-62-2 see PDJ000 |
| 73926-88-2 see CHL000 | 74578-38-4 see UNJ810 | 77174-66-4 see CMW550 | 77966-63-3 see CIT750 | 78219-63-3 see PDJ250 |
| 73926-89-3 see CIC000 | 74749-73-8 see DUO800 | 77234-90-3 see DAP880 | 77966-67-7 see CKN750 | 78246-54-5 see HLX925 |
| 73926-90-6 see TJU850 | 74758-13-7 see DPI600 | 77251-47-9 see CPD625 | 77966-68-8 see CKT750 | 78265-89-1 see CAB500 |
| 73926-91-7 see DGK400 | 74758-19-3 see BJH500 | 77287-90-2 see COW675 | 77966-70-2 see DDM600 | 78265-90-4 see CAB750 |
| 73926-92-8 see DGK600 | 74764-93-5 see MNU750 | 77327-05-0 see DHA300 | 77966-71-3 see DHS800 | 78265-91-5 see CAB250 |
| 73926-94-0 see CLG000 | 74782-23-3 see OKS200 | 77337-54-3 see PNM650 | 77966-72-4 see DIC200 | 78265-95-9 see HLR000 |
| 73927-60-3 see EHX000 | 74790-08-2 see SLE875 | 77430-23-0 see MPS300 | 77966-73-5 see DIK600 | 78265-97-1 see PFF500 |
| 73927-86-3 see DEI400 | 74816-28-7 see CFM000 | 77469-44-4 see BIF625 | 77966-75-7 see BEJ500 | 78280-29-2 see DII600 |
| 73927-87-4 see DWW400 | 74816-32-3 see CFU250 | 77491-30-6 see SAD100 | 77966-77-9 see BRI500 | 78280-31-6 see AMN000 |
| 73927-88-5 see BSP000 | 74847-35-1 see PPQ650 | 77500-04-0 see AJQ675 | 77966-79-1 see DDU200 | 78281-06-8 see HLE650 |
| 73927-89-6 see TMV825 | 74920-78-8 see EKL250 | 77503-17-4 see FQQ400 | 77966-80-4 see DHT000 | 78308-37-9 see BQH750 |
| 73927-90-9 see QSJ800 | 74926-97-9 see BRR500 | 77523-56-9 see ACQ790 | 77966-81-5 see DIC400 | 78308-53-9 see MKS750 |
| 73927-91-0 see TID750 | 74926-98-0 see IQD000 | 77536-66-4 see ARM260 | 77966-82-6 see DHK800 | 78313-59-4 see BKS800 |
| 73927-92-1 see TNB500 | 74927-02-9 see BKQ750 | 77536-67-5 see ARM264 | 77966-83-7 see DIK800 | 78329-75-6 see SCA550 |
| 73927-93-2 see TIE000 | 74940-23-1 see HIE600 | 77536-68-6 see ARM280 | 77966-84-8 see DNM600 | 78329-87-0 see BQN250 |
| 73927-94-3 see IEF000 | 74940-26-4 see HIE700 | 77650-95-4 see DJX300 | 77966-85-9 see DTR800 | 78329-88-1 see BSI750 |
| 73927-95-4 see TIE500 | 74955-23-0 see HIE570 | 77680-87-6 see SED700 | 77966-90-6 see PIN000 | 78329-97-2 see EIA000 |
| 73927-96-5 see QTJ000 | 75016-34-1 see NKT100 | 77698-19-2 see NKL300 | 77966-93-9 see CGI750 | 78330-02-6 see ELK000 |
| 73927-97-6 see TDO000 | 75016-36-3 see NKT105 | 77698-20-5 see NLE400 | 77967-05-6 see BGH500 | 78338-31-5 see DTA690 |
| 73927-98-7 see TIG000 | 75034-93-4 see DKQ650 | 77791-20-9 see DHN800 | 77967-24-9 see DHL200 | 78338-32-6 see DTA700 |
| 73927-99-8 see TNC000 | 75038-71-0 see PLU550 | 77791-27-6 see DIN400 | 77967-25-0 see DHL400 | 78343-32-5 see FJI500 |
| 73928-00-4 see TKT850 | 75198-31-1 see NGI800 | 77791-37-8 see END500 | 77984-94-2 see CFR500 | 78350-94-4 see LHM750 |
| 73928-02-6 see MFP500 | 75236-19-0 see EPC950 | 77791-38-9 see EOU500 | 77985-16-1 see CFN500 | 78354-52-6 see MBU775 |
| 73928-03-7 see MPJ250 | 75321-19-6 see TMN000 | 77791-40-3 see MDK250 | 77985-17-2 see CKU000 | 78371-75-2 see XWS000 |
| 73928-04-8 see MPJ500 | 75321-20-9 see DVD400 | 77791-41-4 see MRT250 | 77985-21-8 see DHM400 | 78371-84-3 see BIP500 |
| 73928-11-7 see DCJ000 | 75348-40-2 see DHQ600 | 77791-42-5 see MLS250 | 77985-23-0 see DHV200 | 78371-85-4 see BIP750 |
| 73928-18-4 see TJS750 | 75348-49-1 see PPV750 | 77791-43-6 see MLS750 | 77985-24-1 see DHV400 | 78371-90-1 see CLK500 |
| 73928-21-9 see TJT250 | 75410-87-6 see ICZ100 | 77791-53-8 see BPR000 | 77985-25-2 see DHV600 | 78371-91-2 see CLK750 |
| 73940-79-1 see SJP500 | 75410-89-8 see BMK634 | 77791-55-0 see BQA500 | 77985-27-4 see PIO000 | 78371-92-3 see CLL250 |
| 73940-85-9 see DKC600 | 75411-83-5 see NKU500 | 77791-57-2 see CFR250 | 77985-28-5 see PIO250 | 78371-93-4 see CLL000 |
| 73940-86-0 see TMI100 | 75464-10-7 see PMW760 | 77791-58-3 see CFL500 | 77985-29-6 see PIO500 | 78371-94-5 see CLL750 |
| 73940-87-1 see BLS900 | 75524-40-2 see CHJ625 | 77791-63-0 see DEY000 | 77985-30-9 see PPW750 | 78371-95-6 see CLL500 |
| 73940-88-2 see TIE250 | 75530-68-6 see NDY650 | 77791-64-1 see DFF200 | 77985-31-0 see PPX000 | 78371-96-7 see CLM000 |
| 73940-89-3 see TIF750 | 75662-22-5 see DGW450 | 77791-66-3 see DIS000 | 77985-32-1 see PPX250 | 78371-97-8 see CLM500 |
| 73940-90-6 see CHU750 | 75738-58-8 see CCS300 | 77791-67-4 see DIC000 | 78003-71-1 see SKS800 | 78371-98-9 see CLM250 |
| 73941-35-2 see THV500 | 75775-83-6 see ADP500 | 77824-42-1 see AQE300 | 78100-57-9 see VFP200 | 78371-99-0 see CLM750 |
| 73953-53-4 see HHA100 | 75841-84-8 see MBV735 | 77824-43-2 see AQE320 | 78109-79-2 see BKC500 | 78372-00-6 see CLN000 |
| 73954-17-3 see BCA375 | 75881-16-2 see NKU550 | 77824-44-3 see AQE305 | 78109-80-5 see BPJ500 | 78372-01-7 see CLN500 |
| 73963-72-1 see CMP825 | 75881-19-5 see NKU580 | 77846-96-9 see CKU250 | 78109-81-6 see BPJ750 | 78372-02-8 see CLO500 |

| | | | | |
|---|---|---|---|---|
| **78372-03-9** see DIB000 | **83665-55-8** see IEL700 | **91480-89-6** see AKC550 | **100836-66-6** see MQO750 | **102128-91-6** see DIE600 |
| **78372-04-0** see DIJ200 | **83768-87-0** see VLU200 | **91480-90-9** see AJO800 | **100836-67-7** see MQP000 | **102128-92-7** see DIN000 |
| **78372-05-1** see DPE800 | **83876-50-0** see DLE500 | **91480-92-1** see AJM600 | **100836-70-2** see PEQ250 | **102129-02-2** see CIL000 |
| **78372-06-2** see DPG400 | **83876-56-6** see MET875 | **91480-97-6** see HMN100 | **101018-97-7** see DKT500 | **102129-03-3** see CIX000 |
| **78393-39-2** see CLN250 | **83876-62-4** see ABQ600 | **91480-98-7** see MEX275 | **101052-67-9** see DAB820 | **102129-21-5** see DIA400 |
| **78431-47-7** see POM710 | **83968-18-7** see LGF875 | **91481-02-6** see AJI300 | **101491-82-1** see EOF500 | **102129-33-9** see MKH750 |
| **78491-02-8** see IAS100 | **83997-16-4** see PEE100 | **91481-03-7** see AKO450 | **101491-83-2** see EOG000 | **102207-73-8** see AJL875 |
| **78499-27-1** see DLI650 | **84002-64-2** see MKI300 | **91481-04-8** see DTO300 | **101491-84-3** see EPV000 | **102207-75-0** see AJN375 |
| **78600-25-6** see ABX800 | **84012-64-6** see MJW300 | **91503-79-6** see FJT100 | **101491-85-4** see EQI000 | **102207-76-1** see AJO625 |
| **78776-28-0** see DEC775 | **84082-68-8** see NOG000 | **91682-96-1** see SLQ650 | **101491-86-5** see EQI500 | **102207-77-2** see AJP125 |
| **78822-08-9** see KDA000 | **84504-69-8** see MQY300 | **91724-16-2** see SME500 | **101516-88-5** see CNW105 | **102207-84-1** see DHO800 |
| **78831-88-6** see DWP900 | **84604-12-6** see RMP000 | **91845-41-9** see PCR000 | **101563-89-7** see DBE600 | **102207-85-2** see DHS600 |
| **78907-15-0** see BQG850 | **84604-12-6** see RNK000 | **92145-26-1** see MGU550 | **101563-93-3** see GKA000 | **102207-86-3** see DRT600 |
| **78907-16-1** see BPY625 | **84775-95-1** see RGA000 | **92202-07-8** see PJH615 | **101564-02-7** see FCN000 | **102259-67-6** see DIL000 |
| **78937-12-9** see CDE400 | **84777-85-5** see VRP775 | **92760-57-1** see TEW450 | **101564-67-4** see DWF875 | **102262-34-0** see SDM000 |
| **78937-14-1** see PLO100 | **84837-04-7** see SLB500 | **93023-34-8** see EIF450 | **101564-69-6** see DWK500 | **102280-81-9** see DWH500 |
| **79127-36-9** see HIS300 | **84928-98-3** see BQI125 | **93088-18-7** see VIK200 | **101607-48-1** see MPO400 | **102280-93-3** see MLH100 |
| **79127-47-2** see MSB100 | **84928-99-4** see TNC800 | **93164-88-6** see AAD250 | **101607-49-2** see MPO390 | **102338-56-7** see PJC000 |
| **79201-85-7** see PIB700 | **84929-34-0** see WBA000 | **93165-23-2** see VKA650 | **101651-36-9** see MDK750 | **102338-57-8** see PJC750 |
| **79307-93-0** see ASC130 | **85086-83-5** see AEA625 | **93334-51-1** see SLO100 | **101651-55-2** see BEI250 | **102338-88-5** see DMD200 |
| **79458-80-3** see EAJ100 | **85287-61-2** see CCS525 | **93405-68-6** see DEY400 | **101651-60-9** see CFQ250 | **102366-69-8** see PAN800 |
| **79543-29-6** see MIY200 | **85559-57-5** see DAQ110 | **93407-11-5** see DPO100 | **101651-61-0** see CFT500 | **102366-79-0** see DKV125 |
| **79796-14-8** see PLT750 | **85616-56-4** see DLI300 | **93431-23-3** see BLD325 | **101651-62-1** see CGD750 | **102367-57-7** see DNH500 |
| **79796-40-0** see TFR000 | **85622-95-3** see MQY110 | **93763-70-3** see PCJ400 | **101651-64-3** see CHG750 | **102395-05-1** see NCQ500 |
| **79818-59-0** see EAI800 | **85625-90-7** see MAQ600 | **93793-83-0** see HNT100 | **101651-65-4** see CIL750 | **102395-10-8** see NCV500 |
| **79867-78-0** see MRR850 | **85721-24-0** see CAP250 | **94110-08-4** see AJE350 | **101651-66-5** see CIU000 | **102395-72-2** see PAW600 |
| **79869-58-2** see PML500 | **85723-21-3** see AAI125 | **94362-44-4** see DCQ650 | **101651-69-8** see DEY200 | **102395-95-9** see TGV100 |
| **80039-73-2** see VGA025 | **85886-25-5** see SAO475 | **94857-19-9** see TDX320 | **101651-73-4** see EPL000 | **102418-04-2** see MCB250 |
| **80266-48-4** see DUI200 | **85923-37-1** see DRD850 | **95004-22-1** see UAG025 | **101651-76-7** see AAN500 | **102418-06-4** see NCK500 |
| **80382-23-6** see LII300 | **86045-52-5** see BLQ900 | **95524-59-7** see EMY100 | **101651-77-8** see AAO500 | **102418-13-3** see TAH500 |
| **80449-58-7** see ACF250 | **86166-58-7** see BQF750 | **95619-40-2** see DTS625 | **101651-78-9** see AAO750 | **102418-16-6** see TEG750 |
| **80471-63-2** see EBH400 | **86255-25-6** see SDT300 | **95770-03-9** see CGE250 | **101651-79-0** see EOG500 | **102418-17-7** see TEH000 |
| **80539-34-0** see VQU500 | **86341-95-9** see PLJ775 | **96081-07-1** see DYB300 | **101651-86-9** see KCU000 | **102419-73-8** see CML822 |
| **80611-44-5** see PGQ275 | **86425-12-9** see DGQ625 | **96231-64-0** see OJC000 | **101651-88-1** see BKC750 | **102420-56-4** see DLE400 |
| **80660-68-0** see BLN100 | **86451-37-8** see NMV450 | **96806-34-7** see NKJ050 | **101651-90-5** see CFU000 | **102433-83-0** see BNN125 |
| **80702-47-2** see RJF400 | **86539-71-1** see MEW775 | **96806-35-8** see NKJ100 | **101651-94-9** see CGE000 | **102488-99-3** see AEB750 |
| **80722-69-6** see PPS600 | **86641-76-1** see SLD600 | **96811-96-0** see DPI750 | **101651-96-1** see CGF250 | **102489-36-1** see THH560 |
| **80734-02-7** see LEJ500 | **86674-51-3** see NLP375 | **96860-89-8** see DJX350 | **101651-97-2** see CGG750 | **102489-45-2** see BEI750 |
| **80748-58-9** see IKA000 | **86803-90-9** see OBW200 | **97194-20-2** see DBV200 | **101651-98-3** see CHB250 | **102489-46-3** see BEI500 |
| **80790-68-7** see MRT100 | **86886-16-0** see DNQ700 | **97196-24-2** see XVJ000 | **101652-00-0** see CHQ000 | **102489-47-4** see BRA500 |
| **80830-42-8** see FAQ950 | **87050-94-0** see TJX600 | **97702-94-8** see DIB200 | **101652-01-1** see DEY600 | **102489-48-5** see CFO000 |
| **81098-60-4** see CMS237 | **87050-95-1** see HAY600 | **97702-95-9** see DIB800 | **101652-02-2** see DFD600 | **102489-49-6** see CFP000 |
| **81265-54-5** see IBW100 | **87209-80-1** see MQY350 | **97702-97-1** see DIE400 | **101652-05-5** see DFM600 | **102489-50-9** see CFQ500 |
| **81295-38-7** see SAF500 | **87425-02-3** see MGD100 | **97702-98-2** see DII800 | **101652-07-7** see DFM800 | **102489-51-0** see CFQ750 |
| **81296-95-9** see NCG715 | **87625-62-5** see POI100 | **97702-99-3** see DIJ000 | **101652-10-2** see DUA600 | **102489-52-1** see CFS500 |
| **81412-43-3** see TJJ500 | **88026-65-7** see SPD100 | **97703-02-1** see DIK200 | **101652-11-3** see DPG200 | **102489-53-2** see CFS750 |
| **81424-67-1** see CBG075 | **88208-15-5** see NLY750 | **97703-04-3** see DIK400 | **101652-13-5** see BMT150 | **102489-54-3** see CFT000 |
| **81486-22-8** see NEA100 | **88208-16-6** see NJY500 | **97703-08-7** see DIM800 | **101670-43-3** see HGM500 | **102489-55-4** see CFT250 |
| **81781-28-4** see TMO775 | **88321-09-9** see OMY925 | **97703-11-2** see DIN200 | **101670-51-3** see DIH800 | **102489-56-5** see CFT750 |
| **81789-85-7** see IBY600 | **88338-63-0** see CDM575 | **97805-00-0** see VIA875 | **101670-78-4** see BFC200 | **102489-58-7** see CHK125 |
| **81840-15-5** see DLF700 | **88746-71-8** see ARP625 | **97864-38-5** see DFE000 | **101688-07-7** see NHR100 | **102489-59-8** see CHR750 |
| **81852-50-8** see EGN100 | **88969-41-9** see DLX100 | **97919-22-7** see CMA600 | **101692-44-8** see DHR800 | **102489-61-2** see DII000 |
| **81866-63-9** see THC500 | **89022-11-7** see CBT175 | **98271-51-3** see MOL300 | **101809-53-4** see MRE500 | **102489-70-3** see TCI100 |
| **81994-68-5** see BQB825 | **89022-12-8** see MDY300 | **98459-16-6** see CCW800 | **101809-55-6** see ITF000 | **102492-24-0** see SDL000 |
| **82018-90-4** see NLX700 | **89194-77-4** see YGA700 | **99071-30-4** see HNK575 | **101809-59-0** see MBW250 | **102504-44-9** see EMG000 |
| **82177-75-1** see PIE550 | **89367-92-0** see IGE100 | **99422-01-2** see EPR100 | **101809-59-0** see MBW500 | **102504-45-0** see EMI500 |
| **82177-80-8** see SDM525 | **89591-51-5** see DUO300 | **99520-58-8** see IBZ100 | **101831-59-8** see NJI850 | **102504-64-3** see CGJ000 |
| **82198-80-9** see PMW770 | **89911-79-5** see NOC400 | **99814-12-7** see SDY675 | **101831-65-6** see MOG600 | **102504-65-4** see CIE250 |
| **82219-81-6** see CCS635 | **89947-76-2** see HLE750 | **99999-42-5** see CIC500 | **101831-88-3** see DPG000 | **102504-71-2** see DPS700 |
| **82419-36-1** see OGI300 | **89957-52-8** see BFW010 | **100215-34-7** see NCT500 | **101833-83-4** see DDF200 | **102505-08-8** see EDM500 |
| **82423-05-0** see COL125 | **89958-12-3** see SAF300 | **100447-46-9** see HDO500 | **101834-51-9** see BEB750 | **102516-61-0** see AKI900 |
| **82464-70-8** see INY100 | **89997-47-7** see SAF000 | **100466-04-4** see DKT400 | **101913-67-1** see TKH300 | **102516-65-4** see THH550 |
| **82508-31-4** see POH600 | **90028-00-5** see DKL300 | **100482-23-3** see CCK790 | **101931-68-4** see TKB285 | **102517-11-3** see PIE750 |
| **82508-32-5** see POH550 | **90043-86-0** see AKD775 | **100482-34-6** see PFI600 | **101952-86-7** see SHN150 | **102534-95-2** see EDW000 |
| **82547-81-7** see TAA420 | **90466-79-8** see BJA200 | **100620-36-8** see CGE500 | **101952-95-8** see SAY000 | **102571-36-8** see DPU600 |
| **82636-28-0** see CCW375 | **90566-09-9** see BGS825 | **100700-23-0** see AAN250 | **101975-70-6** see NNE500 | **102577-46-8** see TGF025 |
| **82643-25-2** see MQS600 | **90584-32-0** see DFJ100 | **100700-24-1** see AAN500 | **101976-64-1** see NEC500 | **102583-64-2** see TDB750 |
| **83053-57-0** see EER400 | **90729-15-0** see DWQ850 | **100700-25-2** see AAP250 | **101997-27-7** see PGQ350 | **102583-65-3** see SLN509 |
| **83053-59-2** see INA400 | **90742-91-9** see DFM875 | **100700-29-6** see MEI000 | **101997-51-7** see BEA275 | **102583-71-1** see ZVS000 |
| **83053-62-7** see DLU700 | **91216-69-2** see BHJ625 | **100700-34-3** see MFU250 | **102070-98-4** see PAN750 | **102584-01-0** see EJZ000 |
| **83053-63-8** see DLO950 | **91297-11-9** see DBL300 | **100733-36-6** see MPZ000 | **102071-30-7** see CKI175 | **102584-11-2** see POX250 |
| **83195-98-6** see TDI100 | **91315-15-0** see AFJ850 | **100836-53-1** see MDK500 | **102071-76-1** see BQF825 | **102584-14-5** see POY750 |
| **83335-32-4** see NJN300 | **91336-54-8** see MKK500 | **100836-62-2** see MMV750 | **102071-88-5** see MRR125 | **102584-42-9** see CKE250 |
| **83435-67-0** see DAM315 | **91480-86-3** see AKO400 | **100836-63-3** see BRV000 | **102107-61-9** see NHE600 | **102584-86-1** see XSJ000 |
| **83463-62-1** see BMY800 | **91480-88-5** see AKO100 | **100836-65-5** see MQO500 | **102128-78-9** see MHD300 | **102584-97-4** see CGT000 |

| | | | | |
|---|---|---|---|---|
| 102585-37-5 see BQG500 | 102585-60-4 see MFU000 | 102648-39-5 see LEW000 | 105735-71-5 see DUW100 | 108419-32-5 see AAT550 |
| 102585-38-6 see BQG750 | 102585-62-6 see SLQ500 | 102648-46-4 see POG000 | 106440-54-4 see CDB770 | 109509-25-3 see CHR850 |
| 102585-42-2 see CJQ500 | 102612-93-1 see WBS860 | 102916-22-3 see ZNS000 | 106440-55-5 see CDB772 | 109651-74-3 see CHR700 |
| 102585-43-3 see CJQ750 | 102612-94-2 see WBS855 | 103331-86-8 see TBA800 | 106742-36-3 see MOL400 | 110147-48-3 see MMA600 |
| 102585-52-4 see EFN500 | 102629-86-7 see MAC600 | 103416-59-7 see AAI100 | 108171-26-2 see PAH800 | 110335-28-9 see CGP375 |
| 102585-53-5 see EFM500 | 102646-51-5 see UBJ000 | 104639-49-8 see SIS000 | 108171-27-3 see PAH810 | 112945-52-5 see SCH000 |
| 102585-59-1 see MFU500 | 102647-16-5 see OIU499 | 104931-87-5 see HAI600 | 108278-70-2 see NKO900 | 730771-71-0 see HKI075 |

1080 see SHG500
A 00 see AGX000
3A see SMQ500
11A see NHG000
A-19 see BGY140
A 21 see DMV600
A-36 see DAE600
A 42 see TED500
A65 see PJA140
A 66 see PMA750
A 71 see CHG000
A 95 see AGX000
A 99 see AGX000
A-101 see CGA500
A-139 see BDC750
A 162 see MQD750
A 171 see TLD000
A 172 see TNJ500
A-310 see ALF250
A 348 see PAH500
A 350 see AJH129, DKR200
A 361 see ARQ725
A 363 see DOR400
A 3-80 see SMQ500
A 435 see SNQ550
A 446 see LFW300
A 468 see DLS800
A-502 see SNJ000
A 585 see PIK625
593-A see PIK075
688A see DDG800
A 688 see PDT250
A 820 see BQN600
883A see PHA550
A 884 see DBA800
A-980 see CEW500
A 995 see AGX000
A 999 see AGX000
A 1100 see TJN000
A 1141 see GGS000
1212A see GLU000
A-1348 see PEC250
A 1390 see MNW150
A 1530 see AQF000
A 1582 see AQF000
A 1803 see BHJ250
A-1981 see DTO200
A 2275 see DBJ100
A 2297 see TLP750
A-2371 see MQW750
2814-A see NCP875
A 3322 see TEO250
72-A34 see BQN600
A 4077 see FMP000
A-4700 see EDK875
A-4760 see EDM000
A 4766 see DSM500
A-4828 see TNT500
A 4942 see IMH000
A 5089 see DSM500
A-8103 see BHJ250
A-8506 see TFQ275
A 10846 see DUD800
A-11025 see RDF000
A 11032 see UVA000
A-12223 see PHK000
A-12253 see TAL350
A 19120 see BEX500
A-27053 see CBR500
A-32686 see POB500
A-45975 see TEF700
A-48257 see BOM600

A-91033 see DQA400
734571A see DGB500
A7301153 see HLF500
A 100 (pharmaceutical) see IGS000
A 15 (polymer) see AAX175
A 1 (sorbent) see AHE250
9AA see AHS500
AA-9 see DXW200
AA-497 see TCU000
AA 1099 see AGX000
AA1199 see AGX000
A 3733A see HAL000
A 3823A see MRE225
AAB see PEI000
AAoC see AAY600
AACAPTAN see CBG000
AACIFEMINE see EDU500
A. AESTIVALIS see PCU375
2-AAF see FDR000
AAF see FDR000
AAFERTIS see FAS000
AA223 LEDERLE see AFI625
AALINDAN see BBQ500
AAMANGAN see MAS500
A. AMOREUXI VENOM see AOO250
A. AMURENSIS see PCU375
AAN see AAY000
A. ANNUA see PCU375
AAOT YELLOW see CMS212
9AAP see AHT000
AAP see AIB300
AAPROTECT see BJK500
AARANE see CNX825
AARARRE see CNX825
o-AAT see AIC250
AAT see AIC250, PAK000
AATACK see TFS350
AATP see PAK000
AATREX see ARQ725
AATREX 4L see ARQ725
AATREX NINE-O see ARQ725
AATREX 80W see ARQ725
A. AUSTRALIS HECTOR VENOM see AOO265
AAVOLEX see BJK500
AAZIRA see BJK500
2-AB see BPY000
AB-15 see ALV750
AB-42 see COH250
AB 109 see DVU300
AB 35616 see CDQ250
ABACIL see CDT250
ABACIN see TKX000
ABACTRIM see TKX000
ABADOL see AMS250
ABADOLE see AMS250
ABAR see LEN000
ABASIN see ACE000
ABATE see TAL250
ABATHION see TAL250
ABAVIT see PFP500
ABBOCILLIN see BDY669
ABBOCORT see HHQ800
ABBOLEXIN see PGG350
ABBOMEEN E-2 see DMT400
ABBOMEEN E-2 AEROSOL see DMT400
ABBOTT-468 11 see CCS550
ABBOTT-22370 see TKX250
ABBOTT-28440 see DEW400
ABBOTT-30360 see FIW000
ABBOTT 30400 see PAP000
ABBOTT-35616 see CDQ250
ABBOTT 36581 see BOR350

ABBOTT 38414 see VGU700
ABBOTT 40566 see TCM250
ABBOTT-43326 see MQT550
ABBOTT 44090 see PNR750
ABBOTT-44747 see FOL000
ABBOTT-45975 see FPP100
ABBOTT'S A.P. 43 see DIF200
ABC 8/3 see OLM300
ABC 12/3 see TEQ175
ABCID see SNN300
ABC LANATOSIDE COMPLEX see LAU400
A. BELLADONNA see AHI635
ABELMOSCHUS MANIHOT (Linn.) Medik., extract see HGA550
ABENSANIL see HIM000
ABEREL see VSK950
ABESON NAM see DXW200
ABESTA see RDK000
ABG 3034 see BDX090
AB 50912 HEMIHYDROCHLORIDE see CCS300
ABICEL see CCU150
ABICOL see RDK000
ABIES ALBA OIL see AAC250
ABIES OIL see CBB900
ABIETIC ACID see AAC500
ABIETIC ACID, METHYL ESTER see MFT500
ABIGUANIL see AHO250
ABILIT see EPD500
ABIOL see HJL500
ABIROL see DAL300
"A" BLASTING POWDER see ERF500
ABMINTHIC see DJT800
ABOL see DOX600
ABOVIS see AAC875
ABRACOL S.L.G see OAV000
ABRAMYCIN see TBX000
ABRAREX see AHE250
ABRIAL LAVANDIN OIL see LCA000
ABRICYCLINE see TBX000
ABRIN see AAD000
ABRINS see AAD000
ABRODEN see SHX000
ABRODIL see SHX000
ABROMA AUGUSTA Linn., root extract see UIS300
ABROMEEN E-25 see CPG125
ABROMINE see GHA050
ABROVAL see BNP750
ABRUS PRECATORIUS L., seed kernel extract see AAD100
ABRUS PRECATORIUS L. (SEED) see RMK250
ABS (pyrolysis products) see ADX750
ABS see AFO500
ABSENTOL see TLP750
ABSETIL see TLP750
ABSIN see ACE000
ABSINTHIUM see ARL250
ABSOLUTE ETHANOL see EFU000
ABSOLUTE FRENCH ROSE see RMP000
ABSOLUTE MIMOSA see MQV000
ABSONAL see BDJ600
ABSONAL V see BDJ600
ABSORBABLE GELATIN SPONGE see PCU360
ABSTENSIL see DXH250
ABSTINYL see DXH250
ABURAMYCIN see OIS000
ABURAMYCIN B see CMK650
5-AC see ARY000
AC 8 see PJS750
AC 220 see MPP000
AC 394 see PJS750
AC 680 see PJS750
AC-1075 see AQQ750
AC 1198 see PMO250
AC 1220 see PJS750
AC 1370 see CCS525
AC 2770 see FND100
AC 3422 see EEH600
AC 5223 see DXX400
AC 5230 see ADA725
AC-12682 see DSP400
AC 18133 see EPC500
AC 18682 see IOT000
AC-18,737 see EAS000
AC 24055 see DUI000
AC 26,691 see CQL250
AC 38023 see FAB600
AC-43064 see DXN600

AC 47031 see PGW750
AC 47470 see DHH400
AC 52160 see TAL250
AC 64475 see DHH200
AC 84777 see ARW000
AC 85258 see DFE469
AC 92553 see DRN200
AC 921000 see BSO000
AC-R-11 see BHJ500
ACABEL see CDG250
ACACIA see AQQ500
ACACIA DEALBATA GUM see AQQ500
ACACIA (EXTRACT) see AAD250
ACACIA FARNESIANA (Linn.) Willd., extract excluding roots see AAD500
ACACIA GUM see AQQ500
ACACIA MOLLISSIMA TANNIN see MQV250
ACACIA PALIDA (PUERTO RICO) see LED500
ACACIA SENEGAL see AQQ500
ACACIA SYRUP see AQQ500
ACACIA VILLOSA see AAD750
ACADYL see BJZ000
ACAJOU (HAITI) see MAK300
A. CALIFORNICA see HGL575
ACALMID see CKE750
ACALO see CKE750
ACAMOL see HIM000
ACAMYLOPHENINE see NOC000
ACAMYLOPHENINE DIHYDROCHLORIDE see AAD875
A. CANADENSIS see PAM780
ACANTHOPHIA ANTARCTICUS VENOM see ARU875
ACAPRIN see PJA120
ACAR see DER000
ACARABEN 4E see DER000
ACARAC see MJL250
ACARACIDE see SOP500
ACARALATE see PNH750
ACARICYDOL E 20 see CJT750
ACARIN see BIO750
ACARITHION see TNP250
ACAROL see IOS000
ACARON see CJJ250
ACAVYL see BJZ000
ACCELERATE see DXD000
ACCELERATOR CZ see CPI250
ACCELERATOR EFK see ZHA000
ACCELERATOR L see BJK500
ACCELERATOR OTOS see OPQ100
ACCELERATOR THIURAM see TFS350
ACCELERINE see DSY600
ACCEL R see BKU500
ACCENT see MRL500
ACCICURE HBS see CPI250
ACCOBOND 3524 see MCB050
ACCOBOND 3900 see MCB050
ACCOBOND 3903 see MCB050
ACCO FAST RED KB BASE see CLK225
ACCONEM see DHH200
ACCOSPERSE TOLUIDINE RED XL see MMP100
ACCOTHION see DSQ000
ACCOTHION O-ANALOG see PHD750
ACCUCOL see PPN750
ACCUSAND see SCK600
ACCUZOLE see SNN500
ACD 7029 see MIB000
AC-DI-SOL NF see SFO500
ACDRILE see MBX800
4,10-ACE-1,2-BENZANTHRACENE see AAE000
8:9-ACE-1:2-BENZANTHRACENE see BAW000
dl-ACEBUTOLOL see AAE100
(±)-ACEBUTOLOL see AAE100
ACEBUTOLOL see AAE100
ACEBUTOLOL HYDROCHLORIDE see AAE125
ACECARBROMAL see ACE000
ACECLIDINE see AAE250
ACECLIDIN-HCL see QUS000
ACECOLINE see ABO000, CMF250
ACEDAPSONE see SNY500
ACEDOXIN see DKL800
ACEDRON see BBK500
ACEF see CCS250
ACEFEN see AAE500
ACEGLUTAMIDE ALUMINUM see AGX125
ACEITE CHINO (CUBA) see TOA275
ACELAN, combustion products see ADX750
ACEMETACIN see AAE625

ACENAPHTHALENE see AAE750
ACENAPHTHANTHRACENE see AAF000
5-ACENAPHTHENAMINE see AAF250
ACENAPHTHENE see AAF275
ACENAPHTHENEDIONE see AAF300
13H-ACENAPHTHO(1,8-ab)PHENANTHRENE see DCR600
ACENAPHTHYLENE see AAF500
ACENAPHTHYLENE, 1,2-DIHYDRO- see AAF275
1,2-ACENAPHTHYLENEDIONE see AAF300
ACENOCOUMARIN see ABF750
ACENOCOUMAROL see ABF750
ACENOCUMAROL see ABF750
ACENOKUMARIN see ABF750
ACENTERINE see ADA725
ACEOTHION see DSQ000
ACEPHAT (GERMAN) see DOP600
ACEPHATE see DOP600
ACEPHATE-MET see DTQ400
ACEPHENE see DPE000
ACEPRAMINE see AJD000
ACEPREVAL see AAF625
ACEPROMAZINA see ABH500
ACEPROMAZINE see ABH500
ACEPROMAZINE MALEATE see AAF750
ACEPROMETAZINE see AAF800
ACEPROMETHAZINE see AAF800
ACEPROMIZINA see ABH500
ACEPROSOL see AOO800
ACERDOL see CAV250
ACESAL see ADA725
ACETAAL (DUTCH) see AAG000
ACETACID RED B see HJF500
ACETACID RED 2BR see FAG020
ACETACID RED J see FMU070
3-ACETAET 4^B-PROPANOATE LEUCOMYCIN V see LEV025
ACETAGESIC see HIM000
ACETAL see AAG000, ADA725
ACETALDEHYD (GERMAN) see AAG250
ACETALDEHYDE see AAG250
ACETALDEHYDE, AMINE SALT see AAG500
ACETALDEHYDE AMMONIA see AAG500
ACETALDEHYDE DIBUTYL ACETAL see DDT400
ACETALDEHYDE DIMETHYL ACETAL see DOO600
ACETALDEHYDE, ((3,7-DIMETHYL-2,6-OCTADIENYL)OXY)-, (E)- see GDM100
ACETALDEHYDE, (3,7-DIMETHYL-6-OCTENYL)OXY)- see CMT300
ACETALDEHYDE-DI-n-PROPYL ACETAL see AAG850
ACETALDEHYDE, DIPROPYL ACETAL see AAG850
ACETALDEHYDE ETHYL cis-3-HEXENYL ACETAL see EKS100
ACETALDEHYDE ETHYL HEXYL ACETAL see EKS120
ACETALDEHYDE ETHYL LINALYL ACETAL see ELZ050
ACETALDEHYDE ETHYL PHENETHYL ACETAL see EES380
ACETALDEHYDE ETHYL 2-PHENYLETHYL ACETAL see EES380
ACETALDEHYDE-N-FORMYL-N-METHYLHYDRAZONE see AAH000
ACETALDEHYDE, (HEXYLOXY)-, DIMETHYL ACETAL see HFG700
ACETALDEHYDE-N-METHYL-N-FORMYLHYDRAZONE see AAH000
ACETALDEHYDE, N-METHYLHYDRAZONE see AAH100
ACETALDEHYDE METHYLHYDRAZONE see AAH100
ACETALDEHYDE OXIME see AAH250
ACETALDEHYDE, PHENETHYL PROPYL ACETAL see PDD400
ACETALDEHYDE SODIUM BISULFITE see AAH500
ACETALDEHYDE SODIUM SULFITE see AAH500
ACETALDEHYDE, TETRAMER see TDW500
ACETALDEHYDE, TRICHLORO-(9CI) see CDN550
ACETALDEHYDE, TRIMER see PAI250
ACETAL DIETHYLIQUE (FRENCH) see AAG000
ACETALDOL see AAH750
ACETALDOXIME see AAH250
ACETALE (ITALIAN) see AAG000
ACETALGIN see HIM000
ACETAL R see PDD400
ACETALIN see AAI000
ACETAMIDE, N-(4'-(ACETYLAMINO)(1,1'-BIPHENYL)-4-YL)-N-HYDROXY-(9CI)
   see HKB000
ACETAMIDE, N-(ACETYLOXY)-N-(1,1'-BIPHENYL)-4-YL-(9CI) see ABJ750
ACETAMIDE, N-(1-ADAMANTYL)- see AEE100
ACETAMIDE, 2-AMINO-N-(2-(2,5-DIMETHOXYPHENYL)-2-HYDROXYETHYL)-,
   MONOHYDROCHLORIDE, (±)- (9CI) see MQT530
ACETAMIDE, 2-AMINO-N-(β-HYDROXY-2,5-DIMETHOXYPHENETHYL)-, MONO-
   HYDROCHLORIDE, (±)- see MQT530
ACETAMIDE, N-((4-AMINOPHENYL)SULFONYL)-, MONOSODIUM SALT (9CI)
   see SNQ000
ACETAMIDE, N-(4-BROMOPHENYL)- see BMR100
ACETAMIDE, 2-CHLORO-N-(2,6-DIETHYLPHENYL)-N-(2-PROPOXYETHYL)- see
   PMB850

ACETAMIDE, 2-CHLORO-N-(2,6-DIMETHYLPHENYL)-N-(2-METHOXYETHYL)- see
   DSM500
ACETAMIDE, 2-CYANO-N-((ETHYLAMINO)CARBONYL)-2-(METHOXYIMINO)- see
   COM300
ACETAMIDE, DICHLORO- see DEM300
ACETAMIDE, 2,2-DICHLORO-(8CI,9CI) see DEM300
ACETAMIDE, N-(2,5-DICHLOROPHENYL)-(9CI) see DGB480
ACETAMIDE, 2-(DIETHYLAMINO)-N-(1,3-DIMETHYL-4-(o-FLUOROBENZOYL)-5-
   PYRAZOLYL)-, MONOHYDROCHLORIDE see AAI100
ACETAMIDE, N,N-DIETHYL-N'-(1,2,3,4-TETRAHYDRO-1-NAPHTHYL)- see
   TGI725
ACETAMIDE, N,N-DIMETHYLTHIO- see DUG450
ACETAMIDE, N-(2-ETHOXYPHENYL)-(9CI) see ABG350
ACETAMIDE, N-(4-(2-FLUOROBENZOYL)-1,3-DIMETHYL-1H-PYRAZOL-5-YL)-2-
   ((3-(2-METHYL-1-PIPERIDINYL)PROPYL)AM see AAI125
ACETAMIDE, N-(4-((2-HYDROXY-5-METHYLPHENYL)AZO)PHENYL)- see
   AAQ250
ACETAMIDE, N-(3-ISOTHIOCYANATOPHENYL)-(9CI) see ISG000
2-ACETAMIDE-4-MERCAPTOBUTYRIC ACID γ-THIOLACTONE see TCZ000
ACETAMIDE, N,N'-(1,4-CYCLOHEXYLENEDIMETHYLENE)BIS(2-(1-AZIRIDINYL)-
   see CPL100
ACETAMIDE, N-(5-NITRO-2-THIAZOLYL)- see ABY900
ACETAMIDE, N-PHENYL- see AAQ500
ACETAMIDE, N-SULFANILYL-, N-SODIUM deriv see SNQ000
ACETAMIDE, N-SULFANILYL-, SODIUM deriv see SNQ000
5-ACETAMIDE-1,3,4-THIADIAZOLE-2-SULFONAMIDE see AAI250
ACETAMIDE, N-(2-THIAZOLYL)- see TEW100
ACETAMIDE, α-TRICHLORO- see TII000
ACETAMIDINE, N,N'-DIPHENYL- see DVW750
3-ACETAMIDO-5-(ACETAMIDOMETHYL)-2,4,6-TRIIODOBENZOIC ACID see
   AAI750
3-ACETAMIDO-5-AMINO-2,4,6-TRIIODOBENZOIC ACID see AAJ125
4-ACETAMIDOANILINE see AHQ250
p-ACETAMIDOANILINE see AHQ250
p-ACETAMIDOAZOBENZENE see PEH750
p-ACETAMIDOBENZALDEHYDE THIOSEMICARBAZONE see FNF000
ACETAMIDOBENZENE see AAQ500
p-ACETAMIDOBENZENEARSONIC ACID, SODIUM SALT, TETRAHYDRATE see
   ARA000
p-ACETAMIDOBENZENESTIBONIC ACID SODIUM SALT see SLP500
4'-ACETAMIDOBENZIDINE see ACC000
2-ACETAMIDOBENZOIC ACID see AAJ150
o-ACETAMIDOBENZOIC ACID see AAJ150
((p-ACETAMIDOBENZOYL)OXY)TRIBUTYLSTANNANE see TIB750
4-ACETAMIDOBIPHENYL see PDY500
ACETAMIDOBIPHENYL see PDY000
N-(4'-ACETAMIDOBIPHENYLYL)ACETOHYDROXAMIC ACID see HKB000
2-ACETAMIDO-4,5-BIS(ACETOXYMERCURI)THIAZOLE see AAJ250
4-ACETAMIDOBUTYRIC ACID see AAJ350
3-ACETAMIDODIBENZFURANE see DDC000
3-ACETAMIDODIBENZOFURAN see DDC000
3-ACETAMIDODIBENZTHIOPHENE see DDD600
3-ACETAMIDODIBENZTHIOPHENE OXIDE see DDD800
7-ACETAMIDO-6,7-DIHYDRO-10-HYDROXY-1,2,3-TRIMETHOXY-BEN-
   ZO(a)HEPTALEN-9(5H)-ONE see ADE000
7-ACETAMIDO-6,7-DIHYDRO-1,2,3,10-TETRAMETHOXY-BENZO(a)HEPTALEN-
   9(5H)-ONE see CNG830
ACETAMIDOETHANE see EFM000
2-ACETAMIDOETHANOL see HKM000
1-ACETAMIDO-4-ETHOXYBENZENE see ABG750
2,2'-((5-ACETAMIDO-2-ETHOXYPHENYL)IMINO)DIETHANOL see BKB000
3-ACETAMIDOFLUORANTHENE see AAK400
2-ACETAMIDOFLUORENE see FDR000
1-(N-ACETAMIDOFLUOROMETHYL)-NAPHTHALENE see MME809
4-ACETAMIDO-2'-HYDROXY-5'-METHYLAZOBENZENE see AAQ250
3-ACETAMIDO-4-HYDROXY-PHENYLARSONIC ACID see ABX500
7-ACETAMIDO-10-HYDROXY-1,2,3-TRIMETHOXY-6,7-DIHYDROBEN-
   ZO(a)HEPTALEN-9(5H)-ONE see ADE000
ACETAMIDOMALONIC ACID DIETHYL ESTER see AAK750
2-ACETAMIDO-4-MERCAPTOBUTYRIC ACID THIOLACTONE see TCZ000
l-α-ACETAMIDO-β-MERCAPTOPROPIONIC ACID see ACH000
2-(3-ACETAMIDO-5-N-METHYL-ACETAMIDO-2,4,6-TRIIODOBENZAMIDO)2-DE-
   OXY-d-GLUCOSE see MQR300
6-ACETAMIDO-4-METHYL-1,2-DITHIOLO(4,3-B)PYRROL-5(4H)-ONE see ABI250
7-ACETAMIDO-1-METHYL-4-(p-(p-((1- METHYLPYRIDINIUM-4-
   YL)AMINO)BENZAMIDO) ANILINO)QUINOLINIUM DI-p-TOLUENESULFON see
   AAL000
5-ACETAMIDO-1-METHYL-3-(5-NITRO-2-FURYL)-s-TRIAZOLE see MMK000
3-ACETAMIDO-5-METHYLPYRROLIN-4-ONE(4,3-D)-1,2-DITHIOLE see ABI250
2-(2-ACETAMIDO-4-METHYLVALERAMIDO)-N-(1-FORMYL-4-GUANIDINOBUTYL)-
   4-METHYLVALERAMIDE see LEX400
(S)-2-(2-ACETAMIDO-4-METHYLVALERAMIDO)-N-(1-FORMYL-4-GUANIDINOBU-
   TYL)-4-METHYL-VALERAMIDE see AAL300
4-ACETAMIDO-3-NITROANISOLE see NEK100
p-ACETAMIDONITROBENZENE see NEK000

5-ACETAMIDO-3-(5-NITRO-2-FURYL)-6H-1,2,4-OXADIAZINE see AAL500
2-ACETAMIDO-4-(5-NITRO-2-FURYL)THIAZOLE see AAL750
4-ACETAMIDO-4-PENTEN-3-ONE-1-CARBOXAMIDE see PMC750
p-ACETAMIDOPHENACYL CHLORIDE see CEC000
3-ACETAMIDOPHENANTHRENE see PCY500
9-ACETAMIDOPHENANTHRENE see PCY750
2-ACETAMIDOPHENATHRENE see AAM250
2-ACETAMIDOPHENOL see HIL000
3-ACETAMIDOPHENOL see HIL500
4-ACETAMIDOPHENOL see HIM000
m-ACETAMIDOPHENOL see HIL500
o-ACETAMIDOPHENOL see HIL000
p-ACETAMIDOPHENOL see HIM000
1-(4-ACETAMIDOPHENOXY)-3-ISOPROPYLAMINO-2-PROPANOL see ECX100
4'-(ACETAMIDO)PHENYL-2-ACETOXYBENZOATE see SAN600
p-ACETAMIDOPHENYL ACETYLSALICYLATE see SAN600
1-(p-ACETAMIDOPHENYL)-3,3-DIMETHYLTRIAZENE see DUI000
3-ACETAMIDOPHENYL ISOTHIOCYANATE see ISG000
1-ACETAMIDO-2-PYRROLIDINONE see NNE400
4-ACETAMIDOSTILBENE see SMR500
trans-4-ACETAMIDOSTILBENE see SMR500
2-ACETAMIDO-5-SULFONAMIDO-1,3,4-THIADIAZOLE see AAI250
α-ACETAMIDO-γ-THIOBUTYROLACTONE see TCZ000
3-ACETAMIDO-2,4,6-TRIIODOBENZOIC ACID see AAM875
3-ACETAMIDO-2,4,6-TRIIODOBENZOIC ACID SODIUM SALT see AAN000
2-(3-ACETAMIDO-2,4,6-TRIIODO-5-(N-METHYLACETAMIDO)BENAZMIDO)-2-DE-
    OXY-d-GLUCOPYRANOSE see MQR300
2-(3-ACETAMIDO-2,4,6-TRIIODO-5-(N-METHYLACETAMIDO)BENZAMIDO)-2-DE-
    OXY-d-GLUCOSE see MQR300
2-((4-(3-ACETAMIDO-2,4,6-TRIIODOPHENOXY)BUTOXY)METHYL)BUTYRIC ACID
    SODIUM SALT see AAN250
2-(2-(3-ACETAMIDO-2,4,6-TRIIODOPHENOXY)ETHOXY)ACETIC ACID SODIUM
    SALT see AAN500
2-(2-(3-ACETAMIDO-2,4,6-TRIIODOPHENOXY)ETHOXY)BUTYRIC ACID SODIUM
    SALT see AAN750
2-(2-(3-ACETAMIDO-2,4,6-TRIIODOPHENOXY)ETHOXY)-2-(o-TOLYL)ACETIC
    ACID SODIUM SALT see AAO500
2-(2-(3-ACETAMIDO-2,4,6-TRIIODOPHENOXY)ETHOXY)-2-(p-TOLYL)ACETIC
    ACID SODIUM SALT see AAO750
2-((3-(3-ACETAMIDO-2,4,6-TRIIODOPHENOXY)PROPOXY)METHYL)BUTYRIC
    ACID SODIUM SALT see AAP250
ACETAMIDOTHIADIAZOLESULFONAMIDE see AAI250
2-ACETAMIDOTHIAZOLE see TEW100
ACETAMIN see SNY500
ACETAMINE DIAZO BLACK RD see DCJ200
ACETAMINE RUBINE B see CMP080
ACETAMINE YELLOW CG see AAQ250
ACETAMINE YELLOW 2R see DUW500
m-ACETAMINOANILINE see AHQ000
p-ACETAMINOBENZYLIDENETHIOSEMICARBAZONE see FNF000
3-ACETAMINODIBENZOTHIOPHENE see DDD600
ACET-o-AMINOFENOL (CZECH) see HIL000
p-ACETAMINOFENYL-2-HYDROXYETHYLSULFON see HLB500
p-ACETAMINOFENYL-β-HYDROXYETHYLSULFON see HLB500
2-ACETAMINOFLUORENE see FDR000
2-ACETAMINO-4-(5-NITRO-2-FURYL)THIAZOLE see AAL750
4-ACETAMINO-2-NITROPHENETOLE see NEL000
ACETAMINOPHEN see HIM000
2-ACETAMINOPHENANTHRENE see AAM250
3-ACETAMINOPHENANTHRENE see PCY500
9-ACETAMINOPHENANTHRENE see PCY750
2-ACETAMINOPHENOL see HIL000
p-ACETAMINOPHENOL see HIM000
1-(4-ACETAMINOPHENYL)-3,3-DIMETHYLTRIAZENE see DUI000
trans-4-ACETAMINOSTILBENE see SMR500
3-ACETAMIDOTOLUENE see ABI750
p-ACETAMIDOTOLUENE see ABJ250
ACETAMOX see AAI250
ACETANHYDRIDE see AAX500
ACETANIL see AAQ500
ACETANILID see AAQ500
ACETANILIDE see AAQ500
ACETANILIDE, 2-ACETYL- see AAY000
ACETANILIDE, 4'-BROMO- see BMR100
ACETANILIDE, p-BROMO- see BMR100
ACETANILIDE, 2',5'-DICHLORO- see DGB480
ACETANILIDE, 2'-ETHOXY- see ABG350
ACETANILIDE, 4'-(2-HYDROXYETHYLSULFONYL)- see HLB500
ACETANILIDE, 4-PHENYL-, N-ACETATE (ester) see ABJ750
ACETANILIDE-4-SULFONIC ACID, 3-AMINO- see AHQ300
p-ACETANISIDIDE, 2-NITRO- see NEK100
ACETANISOLE (FCC) see MDW750
ACETARSIN see ACN250
ACETARSOL see ABX500
ACETARSONE see ABX500

ACETARSONE DIETHYLAMINE SALT see ACN250
ACETATE d'AMYLE (FRENCH) see AOD725
ACETATE-AS see HHQ800
ACETATE BLUE G see TBG700
ACETATE BRILLIANT BLUE 4B see MGG250
ACETATE de BUTYLE (FRENCH) see BPU750
ACETATE de BUTYLE SECONDAIRE (FRENCH) see BPV000
ACETATE C-7 see HBL000
ACETATE C-8 see OEG000
ACETATE C-11 see UMS000
ACETATE C-12 see DXV400
ACETATE de CELLOSOLVE (FRENCH) see EES400
ACETATE CORTISONE see CNS825
ACETATE de CUIVRE (FRENCH) see CNI250
ACETATE DE VINYLE see VLU250
16-ACETATE DIGITOXIN see ACH375
(Z)-ACETATE-9-DODECEN-1-OL see GJU050
ACETATE de l'ETHER MONOETHYLIQUE DE L'ETHYLENE-GLYCOL
    (FRENCH) see EES400
ACETATE d'ETHYLGLYCOL (FRENCH) see EES400
ACETATE FAST ORANGE R see AKP750
ACETATE FAST PINK 3B see DBX000
ACETATE FAST RED 2B see AKE250
ACETATE FAST RUBINE B see CMP080
ACETATE FAST YELLOW G see AAQ250
ACETATE FAST YELLOW 5RL see CMP090
ACETATE of 4-(HYDROXYPHENYL)-2-BUTANONE see AAR500
ACETATE d'ISOBUTYLE (FRENCH) see IIJ000
ACETATE d'ISOPROPYLE (FRENCH) see INE100
ACETATE de L'ETHER MONOMETHYLIQUE de L'ETHYLENE-GLYCOL
    (FRENCH) see EJJ500
ACETATE of LIME see CAL750
ACETATE de METHYLE (FRENCH) see MFW100
ACETATE de METHYLE GLYCOL (FRENCH) see EJJ500
ACETATE P.A. see AGQ750
ACETATE PHENYLMERCURIQUE (FRENCH) see ABU500
ACETATE de PLOMB (FRENCH) see LCV000
1-ACETATE-1,2,3-PROPANETRIOL see GGO000
ACETATE de PROPYLE NORMAL (FRENCH) see PNC250
ACETATE RED VIOLET R see DBP000
ACETATE-REPLACING FACTOR see DXN800
ACETATE de TRIMETHYLPLOMB (FRENCH) see ABX125
ACETATE de TRIPHENYL-ETAIN (FRENCH) see ABX250
ACETATE de TRIPROPYLPLOMB (FRENCH) see ABX325
ACETATE TURQUOISE BLUE B see DMM400
ACETATE YELLOW 6G see MEB750
ACETATO(2-AMINO-5-NITROPHENYL)MERCURY see ABQ250
(ACETATO)(p-AMINOPHENYL)MERCURY see ABQ000
(ACETATO)BIS(HEPTYLOXY)PHOSPHINYLMERCURY see AAR750
(ACETATO)BIS(HEXYLOXY)PHOSPHINYLMERCURY see AAS000
ACETATO di CELLOSOLVE (ITALIAN) see EES400
(ACETATO)(DIETHOXYPHOSPHINYL)MERCURY see AAS250
ACETATO((5-HYDROXYMERCURI)-2-THIENYL)MERCURY see HLQ000
(ACETATO-o)HYDROXY-mu-2,5-THIOPHENEDIYLDIMERCURY see HLQ000
ACETATO(2-METHOXYETHYL)MERCURY see MEO750
ACETATO di METIL CELLOSOLVE (ITALIAN) see EJJ500
(ACETATO-O)ETHYLMERCURY see EMD000
ACETATOPHENYLMERCURATE(1-) AMMONIUM SALT see PFO550
(ACETATO)PHENYLMERCURY see ABU500
(ACETATO)PHENYL-MERCURY mixed with CHLOROETHYL-MERCURY (19:1)
    see PFO500
ACETATO di STAGNO TRIFENILE (ITALIAN) see ABX250
(ACETATO)(2,3,5,6-TETRAMETHYLPHENYL)MERCURY see AAS500
(ACETATO)(8-THEOPHYLLINYL)MERCURY see TEP750
(ACETATO)(TRIMETAARSENITO)DICOPPER see COF500
ACETATOTRIPHENYLSTANNANE see ABX250
ACETAZINE see ABH500
ACETAZOLAMID see AAI250
ACETAZOLAMIDE see AAI250
ACETAZOLAMIDE SODIUM see AAS750
ACETAZOLAMIDE SODIUM SALT see AAS750
ACETAZOLEAMIDE see AAI250
ACETCARBROMAL see ACE000
ACETDIMETHYLAMIDE see DOO800
ACETDRON see AOB250
ACETEIN see ACH000
ACETENE see EIO000
ACETETHYLANILIDE see EFQ500
ACETEUGENOL see EQS000
ACETHION see DIX000
ACETHIONE see DIX000
ACETHROPAN see AES650
ACETHYDRAZIDE see ACM750
ACETHYDROXAMSAEURE (GERMAN) see ABB250
ACETHYLPROMAZIN see ABH500

ACETIC ACID (aqueous solution) (DOT) see AAT250
ACETIC ACID see AAT250
ACETIC ACID, (N-ACETYL-N-(4-BIPHENYL)AMINO) ESTER see ABJ750
ACETIC ACID, (N-ACETYL-N-(2-FLUORENYL)AMINO) ESTER see ABL000
ACETIC ACID(N-ACETYL-N-(4-FLUORENYL)AMINO)ESTER see ABO500
ACETIC ACID (N-ACETYL-N-(2-PHENANTHRYL)AMINO)ESTER see ABK250
ACETIC ACID-(N-ACETYL-N-(p-STYRYLPHENYL)AMINO) ESTER see ABW500
ACETIC ACID, (ACETYLTHIO)-(9CI) see ACQ250
ACETIC ACID ALLYL ESTER see AFU750
ACETIC ACID, 4-ALLYLOXY-3-CHLOROPHENYL-, compounded with 2-AMI-NOETHANOL see MDI225
ACETIC ACID AMIDE see AAI000
ACETIC ACID, AMINO-sec-BUTYL- see IKX000
ACETIC ACID, 2,2′-(((4-(AMINOCARBONYL)PHENYL)ARSINIDENE)BIS(THIO))BIS- (9CI) see TFA350
ACETIC ACID, (p-AMINOPHENYL)- see AID700
ACETIC ACID, AMMONIUM SALT see ANA000
ACETIC ACID, AMYL ESTER see AOD725, AOD750
ACETIC ACID, ANHYDRIDE (9CI) see AAX500
ACETIC ACID, ANHYDRIDE with NITRIC ACID (1:1) see ACS750
ACETIC ACID ANILIDE see AAQ500
ACETIC ACID, BARIUM SALT see BAH500
ACETIC ACID, BENZ(a)ANTHRACENE-7,12-DIMETHANOL DIESTER see BBF750
ACETIC ACID, BENZ(a)ANTHRACENE-7-METHANOL ESTER see BBH500
ACETIC ACID, BENZOYL-, ETHYL ESTER see EGR600
ACETIC ACID BENZYL ESTER see BDX000
ACETIC ACID, 2,2′-((BIS(PHENYLMETHYL)STANNYLENE)BIS(THIO))BIS-, DIISO-OCTYL ESTER (9CI) see EDC560
ACETIC ACID, compd. with BORON FLUORIDE (BF3) (8CI) see BMG750
ACETIC ACID-1,3-BUTADIENYL ESTER see ABM250
ACETIC ACID-2-BUTOXY ESTER see BPV000
ACETIC ACID-2-(tert-BUTYL)-4,6-DINITRO-m-TOLYL ESTER see BRU750
ACETIC ACID n-BUTYL ESTER see BPU750
ACETIC ACID-tert-BUTYL ESTER see BPV100
ACETIC ACID, 3-BUTYL-5-METHYL-TETRAHYDRO-2H-PYRAN-4-YL ESTER see MHX000
ACETIC ACID-1-(BUTYLNITROSOAMINO)BUTYL ESTER see BPV325
ACETIC ACID, CADMIUM SALT see CAD250
ACETIC ACID, CADMIUM SALT, DIHYDRATE see CAD275
ACETIC ACID, C$_{7-9}$-BRANCHED ALKYL ESTERS, C$_8$-rich see AAT550
ACETIC ACID, CEDROL ESTER see CCR250
ACETIC ACID CHLORIDE see ACF750
ACETIC ACID, ((4-CHLORO-2-METHYL)PHENYL)THIO- see CLO600
ACETIC ACID, 3-CHLOROPROPYLENE ESTER see CIL900
ACETIC ACID, CHROMIUM (2+) SALT (8CI, 9CI) see CMJ000
ACETIC ACID, CINNAMYL ESTER see CMQ730
ACETIC ACID, CITRONELLYL ESTER see AAU000
ACETIC ACID, COBALT(2+) SALT see CNC000
ACETIC ACID, COBALT(3+) SALT see CNA300
ACETIC ACID, COBALT(2+) SALT, TETRAHYDRATE see CNA500
ACETIC ACID, CUPRIC SALT see CNI250
ACETIC ACID, CYANO-, ALLYL ESTER see AGC150
ACETIC ACID, CYANO-, ISOBUTYL ESTER see IIQ100
ACETIC ACID, CYANO-, 2-METHYLPROPYL ESTER see IIQ100
ACETIC ACID, CYANO-, 2-PROPENYL ESTER see AGC150
ACETIC ACID, CYANO-, TRIPHENYLSTANNYL ESTER see TMV825
ACETIC ACID, (1,2-CYCLOHEXYLENEDINITRILO)TETRA- see CPB120
ACETIC ACID, CYCLOHEXYLETHYL ESTER see EHS000
ACETIC ACID, 9-DECENYL ESTER see DAI450
ACETIC ACID, DIAMINOPROPANOLTETRA- see DCB100
ACETIC ACID, ((DIBENZYLSTANNYLENE)DITHIO)DI-, DIISOOCTYL ESTER see EDC560
ACETIC ACID, 2,2′-((DIBUTYLSTANNYLENE)BIS(THIO))BIS-, DINONYL ESTER see DEH650
ACETIC ACID, ((DIBUTYLSTANNYLENE)DITHIO)DI-, DINONYL ESTER (8CI) see DEH650
ACETIC ACID, (DIETHOXYPHOSPHINYL)-, ETHYL ESTER (9CI) see EIC000
ACETIC ACID, ((2,3-DIHYDRO-6,7-DICHLORO-2-METHYL-1-OXO-2-PHENYL-1H-INDEN-5-YL)OXY)-, (±)- see IBQ400
ACETIC ACID DIHYDROTERPINYL ESTER see DME400
ACETIC ACID, (((3,4-DIHYDROXY-2-ANTHRAQUINONYL)METHYL)IMINO)DI- see AFM400
ACETIC ACID, (3,4-DIMETHOXYPHENYL)- see HGK600
ACETIC ACID, DIMETHYL- see IJU000
ACETIC ACID DIMETHYLAMIDE see DOO800
ACETIC ACID, 2-(DIMETHYLAMINO)ETHYL ESTER see DOZ100
ACETIC ACID-α-(DIMETHYLAMINOMETHYL)BENZYL ESTER see ABN700
ACETIC ACID-1,3-DIMETHYLBUTYL ESTER see HFJ000
ACETIC ACID-2,6-DIMETHYL-m-DIOXAN-4-YL ESTER see ABC250
ACETIC ACID-1,1-DIMETHYLETHYL ESTER see BPV100
ACETIC ACID-3,7-DIMETHYL-6-OCTEN-1-YL ESTER see AAU000
ACETIC ACID, (4,6-DINITRO-2-s-BUTYLPHENYL) ESTER see ACE500
ACETIC ACID, (2,4-DINITRO-6-s-BUTYLPHENYL) ESTER see ACE500
ACETIC ACID-4,6-DINITRO-o-CRESOL ESTER see AAU250

ACETIC ACID, DIPHENYL-, 2-(BIS(2-HYDROXYETHYL)AMINO)ETHYL ESTER, HYDROCHLORIDE see DVW900
ACETIC ACID, DODECYL ESTER see DXV400
ACETIC ACID, ESTER with N-4-BIPHENYLYLACETOHYDROXAMIC ACID see ABJ750
ACETIC ACID ESTER with N-(FLUOREN-3-YL)ACETOHYDROXAMIC ACID see ABO250
ACETIC ACID, ESTER with N-(FLUOREN-4-YL)ACETOXYHYDROXAMIC ACID see ABO500
ACETIC ACID, ESTER with N-NITROSO-2,2′-IMINODIETHANOL see NKM500
ACETIC ACID ESTER with N-(2-PHENANTHRYL)ACETOHYDROXAMIC ACID see ABK250
ACETIC ACID-ESTER with N-(p-STYRYLPHENYL)ACETOHYDROXAMIC ACID see ABW500
ACETIC ACID, ESTER with N-(p-STYRYLPHENYL)ACETOHYDROXAMIC ACID see SMT000
ACETIC ACID, ETHENYL ESTER see VLU250
ACETIC ACID ETHENYL ESTER HOMOPOLYMER see AAX250
ACETIC ACID ETHENYL ESTER POLYMER with CHLORETHENE (9CI) see AAX175
ACETIC ACID-2-ETHOXYETHYL ESTER see EES400
ACETIC ACID, (ETHYLENEDINITRILO)TETRA-, CALCIUM (II) COMPLEX see CAR800
ACETIC ACID, (ETHYLENEDINITRILO)TETRA-, CALCIUM DISODIUM SALT see CAR780
ACETIC ACID, (ETHYLENEDINITRILO)TETRA-, DIPOTASSIUM SALT see EEB100
ACETIC ACID, (ETHYLENEDINITRILO)TETRA-, YTTRIUM complex see YFA100
ACETIC ACID, (ETHYLENEDINITRILO)TETRA-, ZINC(II) COMPLEX see ZGS100
ACETIC ACID, ETHYLENE ETHER see VLU250
ACETIC ACID α-ETHYLEXYL ESTER see OEE000
ACETIC ACID GERANIOL ESTER see DTD800
ACETIC ACID, GLACIAL see AAT250
ACETIC ACID-2,4-HEXADIEN-1-OL ESTER see AAU750
ACETIC ACID, HEXYLENE GLYCOL see HFQ000
ACETIC ACID HEXYL ESTER see HFI500
ACETIC ACID, (o-((2-HYDROXY-3-HYDROXYMERCU-RI)PROPYL)CARBAMOYL)PHENOXY- see NCM800
ACETIC ACID, (p-HYDROXYPHENYL)- see HNG600
ACETIC ACID, ((2-HYDROXY-1,3-TRIMETHYLENE)DINITRILO)TETRA- see DCB100
ACETIC ACID, ((2-HYDROXYTRIMETHYLENE)DINITRILO)TETRA-(8CI) see DCB100
ACETIC ACID, 3-INDOLYL ESTER see IDA600
ACETIC ACID, IRON(2+) SALT see FBH000
ACETIC ACID, ISOBUTYL ESTER see IIJ000
ACETIC ACID, ISOPENTYL ESTER see IHO850
ACETIC ACID, ISOPROPENYL ESTER see MQK750
ACETIC ACID ISOPROPYL ESTER see INE100
ACETIC ACID, ISOTHIOCYANATO-, ETHYL ESTER see ELX530
ACETIC ACID LEAD(2+) SALT see LCV000
ACETIC ACID, LEAD SALT see LCG000
ACETIC ACID, LEAD(2+) SALT TRIHYDRATE see LCJ000
ACETIC ACID LINALOOL ESTER see LFY100
ACETIC ACID, LITHIUM SALT see LGO100
ACETIC ACID, MAGNESIUM SALT see MAD000
ACETIC ACID MANGANESE(II) SALT (2:1) see MAQ000
ACETIC ACID-p-MENTHAN-8-OL ESTER see DME400
ACETIC ACID, MERCAPTO-, ACETATE (8CI) see ACQ250
ACETIC ACID, MERCAPTO-, 1,2-ETHANEDIYL ESTER see MCN000
ACETIC ACID, MERCURY(2+) SALT see MCS750
ACETIC ACID-3-METHOXYBUTYL ESTER see MHV750
ACETIC ACID, 2-(2-(2-METHOXYETHOXY)ETHOXY)ETHYL ESTER see AAV500
ACETIC ACID, 2-(2-METHOXYETHOXY) ETHYL ESTER see MIE750
ACETIC ACID, 2-METHOXY-1-METHYLETHYL ESTER see PNL265
ACETIC ACID, 2-METHYLAMYL ESTER see MGN300
ACETIC ACID METHYL ESTER see MFW100
ACETIC ACID-1-METHYLETHYL ESTER (9CI) see INE100
ACETIC ACID-2-METHYL-6-METHYLENE-7-OCTEN-2-YL ESTER see AAW500
ACETIC ACID METHYLNITROSAMINOMETHYL ESTER see AAW000
ACETIC ACID, (METHYL-ONN-AZOXY)METHYL ESTER see MGS750
ACETIC ACID, 2-METHYL-2,4-PENTANEDIOL DIESTER see HFQ000
ACETIC ACID, 2-METHYLPENTYL ESTER see MGN300
ACETIC ACID, α-METHYL-PHENETHYL ESTER see ABU800
ACETIC ACID-4-METHYLPHENYL ESTER see MNR250
ACETIC ACID-2-METHYL-2-PROPENE-1,1-DIOL DIESTER see AAW250
ACETIC ACID-2-METHYLPROPYL ESTER see IIJ000
ACETIC ACID-1-METHYLPROPYL ESTER (9CI) see BPV000
ACETIC ACID, MONOGLYCERIDE see GGO000
ACETIC ACID MYRCENYL ESTER see AAW500
ACETIC ACID, (2-NAPHTHYLOXY)- see NBJ100
ACETIC ACID, NICKEL(2+) SALT see NCX000
ACETIC ACID, NICKEL(+2) SALT, TETRAHYDRATE see NCX500
ACETIC ACID, NITRILOTRI-, IRON(III) chelate see IHC100
ACETIC ACID, (p-NITROBENZOYL)-, ETHYL ESTER see ENO100

ACETIC ACID, NITRO-, ETHYL ESTER see ENN100
ACETIC ACID, (p-NITROPHENYL)- see NII510
ACETIC ACID, ((4-NITROPHENYL)AMINO)OXO-(9CI) see NHX100
ACETIC ACID, OCTYL ESTER see OEG000
ACETIC ACID, ((OCTYLSTANNYLIDYNE)TRITHIO)TRI-, TRIS(2-ETHYLHEXYL) ESTER see OGI000
ACETIC ACID, OXIME see ABB250
ACETIC ACID, OXO-, METHYL ESTER see MKI550
ACETIC ACID, (4-OXO-2-THIAZOLIDINYLIDENE)-, BUTYL ESTER (9CI) see BPI300
ACETIC ACID, 2,2'-OXYBIS-(9CI) see ONQ100
ACETIC ACID, OXYDI- see ONQ100
ACETIC ACID, OXYDIETHYLENE ESTER see DJD750
ACETIC ACID, PALLADIUM(2+) SALT see PAD300
ACETIC ACID PALLADIUM SALT see PAD300
ACETIC ACID, PHENYL-, BUTYL ESTER see BQJ350
ACETIC ACID, PHENYL-, 3,7-DIMETHYL-2,6-OCTADIENYL ESTER, (E)-(8CI) see GDM400
ACETIC ACID, PHENYL-, 3,7-DIMETHYL-6-OCTENYL ESTER see CMU050
ACETIC ACID, PHENYL-, 1,5-DIMETHYL-1-VINYL-4-HEXENYL ESTER (8C1) see LGC050
ACETIC ACID-2-PHENYLETHYL ESTER see PFB250
ACETIC ACID PHENYLHYDRAZONE see ACX750
ACETIC ACID PHENYLMERCURY DERIV. see ABU500
ACETIC ACID PHENYLMETHYL ESTER see BDX000
ACETIC ACID, PHENYL-, SODIUM SALT see SFA200
ACETIC ACID, PHOSPHONO-, TRIETHYL ESTER (9CI) see EIC000
ACETIC ACID-2-PROPENYL ESTER see AFU750
ACETIC ACID, n-PROPYL ESTER see PNC250
ACETIC ACID-1-(PROPYLNITROSAMINO)PROPYL ESTER see ABT750
ACETIC ACID, SAMARIUM SALT see SAR000
ACETIC ACID, SILVER(1+) SALT see SDI800
ACETIC ACID, SODIUM SALT see SEG500
ACETIC ACID, SULFO-, 1-DODECYL ESTER, SODIUM SALT see SIB700
ACETIC ACID, SULFO-, DODECYL ESTER, S-SODIUM SALT (7CI) see SIB700
ACETIC ACID, TRIANHYDRIDE with ANTIMONIC ACID see AQJ750
ACETIC ACID, TRIETHYLENE GLYCOL DIESTER see EJB500
ACETIC ACID solution, >10% but not >80% acid, by weight (UN 2790) (DOT) see AAT250
ACETIC ACID, glacial or acetic acid solution, >80% acid, by weight (UN 2790) (DOT) see AAT250
ACETIC ACID-VETIVEROL ESTER see AAW750
ACETIC ACID VINYL ESTER see VLU250
ACETIC ACID, VINYL ESTER, CHLOROETHYLENE COPOLYMER see PKP500
ACETIC ACID, VINYL ESTER, POLYMER with CHLOROETHYLENE see AAX175
ACETIC ACID VINYL ESTER POLYMERS see AAX250
ACETIC ACID, ZINC SALT see ZBS000
ACETIC ACID, ZINC SALT, DIHYDRATE see ZCA000
ACETIC ALDEHYDE see AAG250
ACETIC ANHYDRIDE see AAX500
ACETIC CHLORIDE see ACF750
ACETIC-4-CHLOROANILIDE see CDZ100
ACETIC ETHER see EFR000
ACETIC OXIDE see AAX500
ACETIC PEROXIDE see PCL500
ACETICYL see ADA725
ACETIN, DI- see GGA100
ACETIDIN see EFR000
ACETILARSANO see ACN250
3-((3-ACETIL-4-(3-tert-BUTILAMINO)-2-HIDROXIPROPOXI)FENIL)-1,1-DIETILUREA HCl (SPANISH) see SBN475
ACETILE DIAZO BLACK N see DPO200
ACETILSALICILICO see ADA725
ACETILUM ACIDULATUM see ADA725
ACETIMIDIC ACID see AAI000
ACETIN see GGO000
ACETIODONE see AAN000
ACETIROMATE see ADD875
ACETISAL see ADA725
ACETISOEUGENOL see AAX750
ACETKARBROMAL see ACE000
ACETMIN see AOB500
ACETOACETAMIDE, 2-CHLORO-N-ETHYL- see CEA100
ACETOACETAMIDOBENZENE see AAY000
ACETOACETANILID see AAY000
ACETOACETANILIDE see AAY000
ACETOACETANILIDE, 2'-CHLORO- see AAY600
ACETOACETANILIDE, 4-CHLORO- see CEA050
ACETOACETANILIDE, o-CHLORO- see AAY600
ACETOACETANILIDE, p-CHLORO- see AAY250
ACETOACETANILIDE, 2',4'-DIMETHYL- see OOI100
o-ACETOACETANISIDE see ABA500
o-ACETOACETANISIDIDE, 4'-CHLORO- see AAY300

ACETOACET-o-ANISIDIN (CZECH) see ABA500
ACETOACET-o-CHLORANILIDE see AAY600
ACETOACET-p-CHLORANILIDE see AAY250
ACETOACET-o-CHLOROANILIDE see AAY600
ACETOACET-p-CHLOROANILIDE see AAY250
ACETOACET-4-CHLORO-2-METHYLANILIDE see AAY300
ACETOACET-2,4-DIMETHYLPHENYL see OOI100
ACETOACETIC ACID ANILIDE see AAY000
ACETOACETIC ACID-o-ANISIDIDE see ABA500
ACETOACETIC ACID BUTYL ESTER see BPV250
ACETOACETIC ACID, 4-CHLORO-, ETHYL ESTER see EHH100
ACETOACETIC ACID, 4-CHLORO-, METHYL ESTER see MIH600
ACETOACETIC ACID, ETHYL ESTER see EFS000
ACETOACETIC ACID, 2-HYDROXYETHYL ESTER, ACRYLATE (8CI) see AAY750
ACETOACETIC ANILIDE see AAY000
ACETOACETIC ESTER see EFS000
ACETOACETIC METHYL ESTER see MFX250
o-ACETOACETOCHLORANILIDE see AAY600
ACETOACETONE see ABX750
ACETOACETONITRILE, 2-PHENYL-(8CI) see PEA500
p-ACETOACETOPHENETIDIDE see AAZ000
ACETOACETOPHENONE see BDJ800
2-ACETOACETOXYETHYL ACRYLATE see AAY750
2',4'-ACETOACETOXYLIDIDE see OOI100
2,4-ACETOACETOXYLIDIDE see OOI100
ACETOACETO-m-XYLIDIDE see OOI100
ACETOACET-p-PHENETIDIDE see AAZ000
ACETOACET-o-TOLUIDIDE see ABA000
ACETOACET-m-XYLIDIDE see OOI100
2-ACETOACETYLAMINOANISOLE see ABA500
((ACETOACETYL)AMINO)BENZENE see AAY000
2-ACETOACETYLAMINOTOLUENE see ABA000
ACETOACETYLANILINE see AAY000
ACETOACETYL-o-ANISIDE see ABA500
ACETOACETYL-o-ANISIDINE see ABA500
ACETOACETYL-o-ANISINE see ABA500
ACETOACETYL-2-CHLOROANILIDE see AAY600
ACETOACETYL-2-METHYLANILIDE see ABA000
ACETOACETYL-m-XYLIDIDE see OOI100
p-ACETOAMINOANILINE see AHQ250
o-ACETOAMINOBENZOIC ACID see AAJ150
ACETOAMINOFLUORENE see FDR000
ACETO-m-AMINOTOLUENE see ABI750
ACETOANILIDE see AAQ500
ACETOARSENITE de CUIVRE (FRENCH) see COF500
ACETO AZIB see ASL750
ACETOBUTOLOL HYDROCHLORIDE see AAE125
ACETO-CAUSTIN see TII250
ACETOCID see SNP500
ACETO-CORT see HHQ800
ACETO DIPP see NBL000
ACETO DNPT 40 see DVF400
ACETO DNPT 80 see DVF400
ACETO DNPT 100 see DVF400
ACETOFERROCENE see ABA750
2-ACETOFLUORENE see FEI200
ACETOGUAIACONE see HLQ500
ACETOHEXAMIDE see ABB000
ACETO HMT see HEI500
ACETOHYDRAZIDE see ACM750
ACETOHYDROXAMIC ACID see ABB250
ACETOHYDROXAMIC ACID, N-(4'-ACETAMIDOBIPHENYL-4-YL)- see HKB000
ACETOHYDROXAMIC ACID, N-(4-BIPHENYLYL)-, ACETATE see ABJ750
ACETOHYDROXAMIC ACID, FLUOREN-2-YL-o-GLUCURONIDE see HIQ500
ACETOHYDROXAMIC ACID, N-(7-IODOFLUOREN-2-YL)-O-MYRISTOYL- see MSB100
ACETOHYDROXIMIC ACID see ABB250
ACETOHYDROXIMIC ACID, THIO-, METHYL ESTER see MPS270
ACETOIN see ABB500
ACETOL (1) see ABC000
ACETOL see ADA725
ACETOMESIDIDE see TLD250
ACETOMETHOXAN see ABC250
ACETOMETHOXANE see ABC250
ACETOMETHYLANILIDE see MFW000
ACETOMORFINE see HBT500
ACETOMORPHINE see HBT500
ACETONANIL see PJQ750, TLP500
ACETONANYL see TLP500
1-ACETONAPHTHALENE see ABC475
β-ACETONAPHTHALENE see ABC500
1'-ACETONAPHTHONE see ABC475
1-ACETONAPHTHONE see ABC475
2'-ACETONAPHTHONE see ABC500

2-ACETONAPHTHONE see ABC500
α-ACETONAPHTHONE see ABC475
β-ACETONAPHTHONE see ABC500
ACETONAPHTHONE see ABC500
ACETONCIANHIDRINEI (ROUMANIAN) see MLC750
ACETONCIANIDRINA (ITALIAN) see MLC750
ACETONCYAANHYDRINE (DUTCH) see MLC750
ACETONCYANHYDRIN (GERMAN) see MLC750
ACETONE see ABC750
ACETONE ANIL see TLP500
ACETONE BIS(ETHYL SULFONE) see ABD500
ACETONE-o-CARBANILOYLOXIME see PEQ500
ACETONE CHLOROFORM see ABD000
ACETONECYANHYDRINE (FRENCH) see MLC750
ACETONE CYANOHYDRIN (ACGIH,DOT) see MLC750
ACETONE-1,3-DICHLORO-1,1,3,3-TETRAFLUOROACETONE see DGL400
ACETONE DIETHYLSULFONE see ABD500
ACETONE, HEXACHLORO- see HCL500
ACETONE OILS (DOT) see ABC750
ACETONE-OXIME-N-PHENYLCARBAMATE see PEQ500
ACETONE OXIME PHENYLURETHANE see PEQ500
ACETONE PEROXIDE see ABE000
ACETONE SEMICARBAZONE see ABE250
ACETON (GERMAN, DUTCH, POLISH) see ABC750
ACETONIC ACID see LAG000
ACETONITRIL (GERMAN, DUTCH) see ABE500
ACETONITRILE see ABE500
ACETONITRILE, BENZOYLPHENYL- see OOK200
ACETONITRILE, (p-CHLOROPHENYL)- see CEP300
ACETONITRILE, 2,2'-IMINOBIS-(9CI) see IBB100
ACETONITRILE, IMINODI- see IBB100
ACETONITRILE, PHENYLACETO see PEA500
ACETONITRILETHIOL see MCK300
ACETONKYANHYDRIN (CZECH) see MLC750
ACETONOXIME see ABF000
ACETONYL see ADA725
ACETONYL ACETONE see HEQ500
p-ACETONYLANISOLE see AOV875
3-(α-ACETONYLBENZYL)-4-HYDROXYCOUMARIN see WAT200
3-(α-ACETONYLBENZYL)-4-HYDROXY-COUMARIN SODIUM SALT see WAT220
ACETONYL BROMIDE see BNZ000
ACETONYL CHLORIDE see CDN200
3-(α-ACETONYLFURFURYL)-4-HYDROXYCOUMARIN see ABF500
3-(α-ACETONYL-p-NITROBENZYL)-4-HYDROXY-COUMARIN see ABF750,
   ABF750
p-ACETONYLOXYBENZENEARSONIC ACID SODIUM SALT see SJK475
2-ACETONYLPIPERIDINE see PAO500
ACETO PAN see PFT250
ACETO PBN see PFT500
ACETOPHEN see ADA725
ACETO-p-PHENALIDE see ABG750
ACETOPHENAZINE see ABG000
ACETOPHENAZINE MALEATE see ABG000
p-ACETOPHENETIDE see ABG750
m-ACETOPHENETIDIDE see ABG250
o-ACETOPHENETIDIDE see ABG350
p-ACETOPHENETIDIDE see ABG750
ACETO-p-PHENETIDIDE see ABG750
ACETOPHENETIDIN see ABG750
ACETO-4-PHENETIDINE see ABG750
ACETOPHENETIDINE see ABG750
ACETOPHENETIN see ABG750
ACETOPHENONE see ABH000
ACETOPHENONE, 2-ANILINO-4'-(BENZYLOXY)-2-PHENYL- see AOR635
ACETOPHENONE, 2-ANILINO-4'-(2-(DIETHYLAMINO)ETHOXY)-2-PHENYL-, HY-
   DROCHLORIDE see AOR700
ACETOPHENONE, 2-BROMO-4'-CHLORO- see CJJ100
ACETOPHENONE, α-CHLORO-α-PHENYL- see CFJ100
ACETOPHENONE, 2-CHLORO-2-PHENYL-(8CI) see CFJ100
ACETOPHENONE, 2-DIAZO- see PET800
ACETOPHENONE, 2,2-DIPHENYL-(6CI,7CI,8CI) see TMS100
ACETOPHENONE, 2'-HYDROXY-(8CI) see HIN500
ACETOPHENONE, 2-HYDROXY-2-PHENYL- see BCP250
ACETOPHENONE, 4'-NITRO- see NEL600
ACETOPHENONE, OXIME see ABH150
ACETOPHENONE, 3'-PHENOXY- see PDR200
ACETOPHENONE, 4'-PHENYL- see MHP500
ACETOPHOS see DIW600
ACETOPROMAZINE see ABH500
ACETOPROPIONIC ACID see LFH000
ACETOPROPYL ALCOHOL see ABH750
ACETOPURPURINE 8B see CMO880
3-ACETOPYRIDINE see ABI000
ACETOPYRROTHINE see ABI250
ACETOQUAT CPB see HCP800

ACETOQUAT CPC see CCX000
ACETOQUAT CTAB see HCQ500
ACETOQUINONE BLUE L see TBG700
ACETOQUINONE BLUE R see TBG700
ACETOQUINONE LIGHT GOOSEBERRY RL see AKE250
ACETOQUINONE LIGHT GREEN BLUE JL see DMM400
ACETOQUINONE LIGHT HELIOTROPE NL see DBP000
ACETOQUINONE LIGHT ORANGE JL see AKP750
ACETOQUINONE LIGHT PINK RLZ see AKO350
ACETOQUINONE LIGHT PURE BLUE R see MGG250
ACETOQUINONE LIGHT RUBINE BLZ see CMP080
ACETOQUINONE LIGHT VIOLET N see AKP250
ACETOQUINONE LIGHT YELLOW see AAQ250
ACETOQUINONE LIGHT YELLOW 4JLZ see AAQ250
ACETOQUINONE LIGHT YELLOW 2RZ see DUW500
ACETOSAL see ADA725
ACETOSALIC ACID see ADA725
ACETOSALIN see ADA725
ACETO SDD 40 see SGM500
ACETOSPAN see AQX500
ACETOSULFAMIN see SNP500
ACETO TETD see TFS350
2-ACETOTHIENONE see ABI500
ACETOTHIOAMIDE see TFA000
2-ACETOTHIOPHENE see ABI500
ACETO TMTM see BJL600
4-ACETOTOLUIDE see ABJ250
o-ACETOTOLUIDE see ABJ000
p-ACETOTOLUIDE see ABJ250
ACETOTOLUIDE see ABI750
m-ACETOTOLUIDIDE see ABI750
o-ACETOTOLUIDIDE see ABJ000
p-ACETOTOLUIDIDE see ABJ250
ACETOVANILLONE see HLQ500
ACETOVANILONE see HLQ500
ACETOVANYLLON see HLQ500
ACETOXIME see ABF000
ACETOXON see DIW600
N-ACETOXY-4-ACETAMIDOBIPHENYL see ABJ750
N-ACETOXY-2-ACETAMIDOFLUORENE see ABL000
ACETOXY(2-ACETAMIDO-5-NITROPHENYL)MERCURY see ABK000
N-ACETOXY-2-ACETAMIDOPHENANTHRENE see ABK250
N-ACETOXY-4-ACETAMIDOSTILBENE see ABW500, SMT000
2-ACETOXY-4'-ACETAMINO)PHENYLBENZOATE see SAN600
N-ACETOXY-4-ACETYLAMINOBIPHENYL see ABJ750
N-ACETOXY-N-ACETYL-2-AMINOFLUORENE see ABL000
N-ACETOXY-2-ACETYLAMINOFLUORENE see ABL000
N-ACETOXY-2-ACETYLAMINOPHENANTHRENE see ABK250
trans-N-ACETOXY-4-ACETYL-AMINOSTILBENE see ABL250
2-ACETOXYACRYLONITRILE see ABL500
α-ACETOXYACRYLONITRILE see ABL500
2-ACETOXYBENZOIC ACID see ADA725
o-ACETOXYBENZOIC ACID see ADA725
N-ACETOXY-4-BIPHENYLACETAMIDE see ABJ750
3-β-ACETOXY-BIS NOR-Δ⁵-CHOLENIC ACID see ABM000
1-ACETOXY-1,3-BUTADIENE see ABM250
N-(α-ACETOXY)BUTYL-N-BUTYLNITROSAMINE see BPV325
1-ACETOXY-2-tert-BUTYLCYCLOHEXANE see BQW490
17-α-ACETOXY-6-CHLORO-6-DEHYDROPROGESTERONE see CBF250
17-α-ACETOXY-6-CHLORO-6-DEHYDROPROGESTERONE see CBF250
17-α-ACETOXY-6-CHLORO-6,7-DEHYDROPROGESTERONE see CBF250
ACETOXY-4'-CHLORO-3,5-DIIODOBENZANILIDE see CGB250
2-ACETOXY-2'-CHLORO-N-METHYL-DIETHYLAMINE see MFW750
17-α-ACETOXY-6-CHLORO-1-α,2-α-METHYLENEPREGNA-4,6-DIENE-3,20-DIONE
   see CQJ500
17-α-ACETOXY-6-CHLOROPREGNA-4,6-DIENE-3,20-DIONE see CBF250
17-α-ACETOXY-6-CHLORO-4,6-PREGNADIENE-3,20-DIONE see CBF250
ACETOXYCYCLOHEXIMIDE see ABN000
1'-ACETOXY-2',3'-DEHYDROESTRAGOLE see EQN225
17-α-ACETOXY-6-DEHYDRO-6-METHYLPROGESTERONE see VTF000
2-ACETOXY-3-DIETHYLCARBAMYL-9,10-DIMETHOXY-1,2,3,4,6,7-HEXAHYDRO-
   11B-BENZO(a)QUINOLIZINE see BCL250
ACETOXYDIETHYLPHENYLSTANNANE see DJV800
11-ACETOXY-15-DIHYDROCYCLOPENTA(a)PHENANTHRACEN-17-ONE see
   ABN250
1-ACETOXY-1,4-DIHYDRO-4-(HYDROXYAMINO)QUINOLINE ACETATE (ESTER)
   see ABN500
21-ACETOXY-7,18-DIHYDROXY-16,18-DIMETHYL-10-PHENYL-
   (11)CYTOCHALASA-6(12),13,19-TRIENE-1-ONE see PAM775
(11-β)-21-(ACETOXY)-11,17-DIHYDROXY-PREGN-4-ENE-3,20-DIONE see
   HHQ800
21-ACETOXY-11-β,17-α-DIHYDROXYPREGN-4-ENE-3,20-DIONE see HHQ800
21-ACETOXY-3,20-DIKETOPREGN-4-ENE see DAQ800
1-ACETOXYDIMERCURIO-1-PERCHLORATODIMERCURIOPROPEN-2-ONE see
   ABN625

6-ACETOXY-2,4-DIMETHYL-m-DIOXANE see ABC250
α-ACETOXY DIMETHYLNITROSAMINE see AAW000
3-(2-(5-ACETOXY-3,5-DIMETHYL-2-OXOCYCLOHEXYL)-2-HYDROXYETH-YL)GLUTARIMIDE see ABN000
β-ACETOXY-N,N-DIMETHYLPHENETHYLAMINE see ABN700
1′-ACETOXYELEMICIN see TLC850
1′-ACETOXYESTRAGOLE see ABN725
ACETOXYETHANE see EFR000
7-α-(ACETOXYETHANE)OXYFLAVONE see ELH600
4-ACETOXY-3-ETHOXYBENZALDEHYDE see EQF100
ACETOXYETHYL CHLORIDE see CGO600
N-ACETOXYETHYL-N-CHLOROETHYLMETHYLAMINE see MFW750
1-(2-ACETOXYETHYL)-4-(3-(2-CHLORO-10-PHENOTHIAZI-NYL)PROPYL)PIPERAZIME DIHYDROCHLORIDE see TFP250
1-ACETOXYETHYLENE see VLU250
17-β-ACETOXY-16-β-ETHYLESTR-4-EN-3-ONE see ELF110
1-ACETOXYETHYL 2-(2-FLUORO-4-BIPHENYLYL)PROPIONATE see FJT100
ACETOXYETHYL-TRIMETHYLAMMONIUM BROMIDE see CMF260
2-ACETOXYETHYLTRIMETHYLAMMONIUM CHLORIDE see ABO000
N-ACETOXY-2-FLUORENYLACETAMIDE see ABL000
N-ACETOXY-3-FLUORENYLACETAMIDE see ABO250
N-ACETOXY-4-FLUORENYLACETAMIDE see ABO500
N-ACETOXYFLUORENYLACETAMIDE see ABO250
N-ACETOXY-2-FLUORENYLBENZAMIDE see ABO750
1-ACETOXYHEXADECANE see HCP100
21-ACETOXY-17,α-HYDROXYPREGN-4-ENE-3,11,20-TRIONE see CNS825
21-ACETOXY-17,α-HYDROXY-3,11,20-TRIKETOPREGNENE-4 see CNS825
3-ACETOXYINDOLE see IDA600
(2-(4-ACETOXY-2-ISOPROPYL-5-METHYLPHENOXY)ETHYL)DIMETHYLAMINE HYDROCHLORIDE see TFY000
2-ACETOXYISOSUCCINODINITRILE see MFX000
ACETOXYL see BDS000
2′,4′-ACETOXYLIDIDE see ABO740
2′,6′-ACETOXYLIDIDE see ABP760
3′,4′-ACETOXYLIDIDE see ABP770
2,6-ACETOXYLIDIDE, 2-CHLORO-N-(2-METHOXYETHYL)- see DSM500
p-(ACETOXYMERCURI)ANILINE see ABQ000
(ACETOXYMERCURI)BENZENE see ABU500
2-(ACETOXYMERCURI)-5-(HYDROXYMERCURI)THIOPHENE see HLQ000
2′-(ACETOXYMERCURI)-4′-NITROACETANILIDE see ABK000
2-(ACETOXYMERCURI)-4-NITROANILINE see ABQ250
6-ACETOXYMERCURI-5-NITROGUAIACOL and 4,6-DIACETOXYMERCURI-5-NI-TROGUAIACOL see GKO500
1-ACETOXYMERCURIO-1-PERCHLORATOMERCURIOPROPEN-2-ONE see ABQ375
8-(ACETOXYMERCURIO)THEOPHYLLINE see TEP750
1-ACETOXY-2-METHOXY-4-ALLYLBENZENE see EQS000
4-ACETOXY-3-METHOXY-1-PROPENYLBENZENE see AAX750
4-ACETOXY-7-METHYLBENZ(c)ACRIDINE see ABQ600
10-ACETOXYMETHYL-1,2-BENZANTHRACENE see BBH500
6-ACETOXY METHYL BENZO(a)PYRENE see ACU500
N-(ACETOXY)METHYL-N,N-BUTYLNITROSAMINE see BRX500
ACETOXYMETHYLBUTYLNITROSAMINE see BRX500
N-(ACETOXY)METHYL-N-ETHYLNITROSAMINE see ENR500
ACETOXYMETHYLETHYLNITROSAMINE see ENR500
N-(ACETOXYMETHYL)-N-ISOBUTYLNITROSAMINE see ABR125
7-ACETOXYMETHYL-12-METHYLBENZ(a)ANTHRACENE see ABR250
ACETOXYMETHYL-METHYL-NITROSAMIN (GERMAN) see AAW000
N-α-ACETOXYMETHYL-N-METHYLNITROSAMINE see AAW000
ACETOXYMETHYL METHYLNITROSAMINE see AAW000
N-(1-ACETOXYMETHYL)-N-NITROSOETHYL AMINE see ENR500
N-ACETOXYMETHYL-N-NITROSOETHYLAMINE see ENR500
ACETOXYMETHYLPHENYLNITROSAMINE see ABR625
17-α-ACETOXY-6-METHYLPREGNA-4,6-DIENE-3,20-DIONE see VTF000
17-α-ACETOXY-6-METHYL-4,6-PREGNADIENE-3,20-DIONE see VTF000
17-α-ACETOXY-6-METHYLPREGNA-4,6-DIENE-3,20-DIONE see VTF000
17-α-ACETOXY-6-α-METHYLPREGN-4-ENE-3,20-DIONE see MCA000
17-α-ACETOXY-6-α-METHYLPROGESTERONE see MCA000
N-(ACETOXY)METHYL-N-n-PROPYLNITROSAMINE see PNR250
ACETOXYMETHYLPROPYLNITROSAMINE see PNR250
p-ACETOXYMETHYLTOLUENE see XQJ700
ACETOXYMETHYLTRIDEUTEROMETHYLNITROSAMINE see MMS000
N-ACETOXY-2-MYRISTOYL-AMINOFLUORENE see FEM000
p-ACETOXYNITROBENZENE see ABS750
1-ACETOXY-N-NITROSODIBUTYLAMINE see BPV325
1-ACETOXY-N-NITROSODIMETHYLAMINE see AAW000
1-ACETOXY-N-NITROSODIPROPYLAMINE see ABT750
1-ACETOXY-N-NITROSO-N-TRIDEUTEROMETHYLMETHYLAMINE see MMS000
17-β-ACETOXY-19-NOR-17-α-PREGN-4-EN-20-YN-3-ONE see ABU000
17-α-ACETOXY-19-NOR-17-α-PREGN-4-EN-20-YN-3-ONE see ABU000
2-ACETOXYPENTANE see AOD735
N-ACETOXY-4-PHENANTHRYLACETAMIDE see ABK250
3-ACETOXYPHENOL see RDZ900
4-(p-ACETOXYPHENYL)-2-BUTANONE see AAR500
ACETOXYPHENYLMERCURY see ABU500

2-ACETOXY-1-PHENYLPROPANE see ABU800
3-ACETOXYPHENYLTRIMETHYLAMMONIUM IODIDE see ABV250
2-ACETOXY-N-(3-(m-(1-PIPERIDINYLMETHYL)PHENOXY)PROPYL)ACETAMIDE HYDROCHLORIDE see HNT100
3-β-ACETOXYPREGN-6-ENE-20-CARBOXYLIC ACID see ABM000
1-ACETOXYPROPANE see PNC250
2-ACETOXYPROPANE see INE100
3-ACETOXYPROPENE see AFU750
N-(α-ACETOXY)PROPYL-N-N-PROPYLNITROSAMINE see ABT750
3-ACETOXYQUINUCLIDINE GLAUCOSTAT see AAE250
1′-ACETOXYSAFROLE see ACV000
1′-ACETOXY-N-(4-STILBENYL) ACETAMIDE see ABW500
4-ACETOXYSTYRENE see ABW550
p-ACETOXYSTYRENE see ABW550
4-ACETOXYTOLUENE see MNR250
α-ACETOXYTOLUENE see BDX000
p-ACETOXYTOLUENE see MNR250
ACETOXYTRIBUTYLPLUMBANE see THY850
ACETOXYTRIBUTYLSTANNANE see TIC000
ACETOXYTRICYCLOHEXYLSTANNANE see ABW600
ACETOXYTRIETHYLPLUMBANE see TJU150
ACETOXYTRIETHYLSTANNANE see ABW750
ACETOXYTRIETHYLTIN see ABW750
ACETOXYTRIHEXYLSTANNANE see ABX000
ACETOXYTRIHEXYLTIN see ABX000
ACETOXYTRIISOPROPYLSTANNANE see TKT750
ACETOXYTRIMETHYLPLUMBANE see ABX125
ACETOXYTRIMETHYLSTANNANE see TMI000
ACETOXYTRIOCTYLSTANNANE see ABX150
3-ACETOXY-1,2,4-TRIOXOLANE see VLU310
ACETOXYTRIPENTYLSTANNANE see ABX175
ACETOXYTRIPHENYLLEAD see TMT000
ACETOXY-TRIPHENYL-STANNAN (GERMAN) see ABX250
ACETOXY-TRIPHENYLSTANNANE see ABX250
ACETOXYTRIPHENYLSTANNANE see ABX250
ACETOXYTRIPHENYLTIN see ABX250
ACETOXYTRIPROPYLPLUMBANE see ABX325
ACETOXYTRIPROPYLSTANNANE see TNB000
ACETOZALAMIDE see AAI250
ACETO ZDBD see BIX000
ACETO ZDED see BJK500
ACETO ZDMD see BJK500
ACET-p-PHENALIDE see ABG750
ACETPHENARSINE see ABX500
p-ACETPHENETIDIN see ABG750
ACET-p-PHENETIDIN see ABG750
ACETPHENETIDIN see ABG750
ACETRIZOATE SODIUM see AAN000
ACETRIZOIC ACID see AAM875
ACETRIZOIC ACID SODIUM SALT see AAN000
ACET-THEOCIN see TEP000
α-ACETYLACETANILIDE see AAY000
ACETYLACETANILIDE see AAY000
ACETYLACETONATE-1,5-CYCLOOCTADIENE RHODIUM see CPR840
ACETYL ACETONE see ABX750
ACETYL ACETONE PEROXIDE with >9% by weight active oxygen (DOT) see PBL600
2-ACETYLACETOPHENONE see BDJ800
α-ACETYLACETOPHENONE see BDJ800
N-(ACETYLACETYL)ANILINE see AAY000
O-ACETYL-1-ACETYLBENZOCYCLOBUTENE OXIME see BFZ130
3-ACETYLACONITINE HYDROBROMIDE see ABX800
ACETYL ADALIN see ACE000
ACETYLADRIAMYCIN see DAC000
3-(ACETYLAMINO)-5-((ACETYLAMINO)METHYL)-2,4,6-TRIIODOBENZOIC ACID see AAI750
3-(ACETYLAMINO)-5-(ACETYLMETHYLAMINO)-2,4,6-TRIIODO-BENZOIC ACID see MQR350
2-((3-(ACETYLAMINO)-5-(ACETYLMETHYLAMINO)-2,4,6-TRIIODOBEN-ZOYL)AMINO)-2-DEOXY-d-GLUCOSE see MQR300
3-ACETYLAMINOANILINE see AHQ000
4-(ACETYLAMINO)ANILINE see AHQ250
m-(ACETYLAMINO)ANILINE see AHQ000
p-(ACETYLAMINO)ANILINE see AHQ250
4-ACETYLAMINOAZOBENZENE see PEH750
p-ACETYLAMINOBENZALDEHYDE THIOSEMICARBAZONE see FNF000
ACETYLAMINOBENZENE see AAQ500
N-ACETYL-p-AMINOBENZENEARSONIC ACID, SODIUM SALT, TETRAHYD-RATE see ARA000
N-ACETYL-4-AMINOBENZENESULFONAMIDE see SNP500
2-(ACETYLAMINO)BENZOIC ACID see AAJ150
N-ACETYLAMINOBENZOIC ACID see AAJ150
4-(ACETYLAMINO)BENZOIC ACID with 2-(DIMETHYLAMINO)ETHANOL (1:1) see DOZ000
2-ACETYLAMINOBIPHENYL see PDY000

α-ACETYLDIGITOXIN see ACH500
ACETYLDIGITOXIN-β see ACH750
ACETYLDIGOXIN-α see ACI000
α-ACETYLDIGOXIN see ACI000
β-ACETYLDIGOXIN see ACH750
ACETYLDIGOXIN-β see ACI250
β-ACETYLDIGOXIN see ACI250
ε-ACETYLDIGOXIN (GERMAN) see DKN600
3-ACETYL-1,5-DIHYDRO-4-HYDROXY-5-(1-METHYLPROPYL)-2H-PYRROL-2-ONE see VTA750
3-ACETYL-10-(3-DIMETHYLAMINOPROPYL)PHENOTHIAZINE see ABH500
2-ACETYL-10-(3-(DIMETHYLAMINO)PROPYL)PHENOTHIAZINE, MALEATE see AAF750
ACETYLDIMETHYLARSINE see ACI375
2-ACETYL-3,5-DIMETHYL-4-ETHYL-PYRROLE see EIL200
3-ACETYL-2,4-DIMETHYL-PYRROLE see ACI500
3-ACETYL-2,5-DIMETHYL-PYRROLE see ACI550
N-ACETYLDIPHENYLAMINE see PDX500
ACETYLEN see ACI750
ACETYLENE, dissolved (DOT) see ACI750
ACETYLENE see ACI750
ACETYLENE BLACK see CBT750
(ACETYLENECARBONYLOXY)TRIPHENYLTIN see TMW600
ACETYLENECARBOXYLIC ACID see PMT275
ACETYLENECARBOXYLIC ACID METHYL ESTER see MOS875
ACETYLENE CHLORIDE see ACJ000
ACETYLENE COMPOUNDS and ALKYNES see ACJ125
ACETYLENEDICARBOXAMIDE see ACJ250
ACETYLENEDICARBOXYLIC ACID DIAMIDE see ACJ250
ACETYLENEDICARBOXYLIC ACID, DIMETHYL ESTER see DOP400
ACETYLENEDICARBOXYLIC ACID MONOPOTASSIUM SALT see ACJ500
ACETYLENE DICHLORIDE see DFI210
trans-ACETYLENE DICHLORIDE see ACK000
ACETYLENE, METHYL- see MFX590
ACETYLENE, PHENYL- see PEB750
ACETYLENE TETRABROMIDE see ACK250
ACETYLENE TETRACHLORIDE see TBQ100
ACETYLENE TRICHLORIDE see TIO750
ACETYL ENHEPTIN see ABY900
ACETYLENOGEN see CAN750
N-ACETYL ETHANOLAMINE see HKM000
ACETYL ETHER see AAX500
N-ACETYL ETHYL CARBAMATE see ACL000
ACETYL ETHYLENE see BOY500
ACETYLETHYLENEIMINE see ACB250
N′-ACETYL ETHYLNITROSOUREA see ACL500
ACETYL ETHYL TETRAMETHYL TETRALIN see ACL750
ACETYLETHYL TETRAMETHYLTETRALIN see ACL750
ACETYLEUGENOL see EQS000
N-ACETYL-p-FENYLENDIAMIN (CZECH) see AHQ250
N-ACETYL-m-FENYLENEDIAMIN (CZECH) see AHQ000
1-ACETYLFERROCENE see ABA750
ACETYLFERROCENE see ABA750
2-ACETYLFLUORENE see FEI200
N-ACETYL-N-9H-FLUOREN-2-YL-ACETAMIDE see DBF200
ACETYL FLUORIDE see ACM000
21-ACETYL-6-α-FLUORO-16-α-METHYLPREDNISOLONE see PAL600
ACETYLFORMALDEHYDE see PQC000
ACETYLFORMIC ACID see PQC100
ACETYLFORMYL see PQC000
16-ACETYLGITOXIN see ACM250
N-ACETYL-l-GLUTAMINE ALUMINUM SALT see AGX125
5-ACETYL-1,1,2,3,3,6-HEXAMETHYLINDAN see HEJ400
N-ACETYLHOMOCYSTEINE THIOLACTONE see TCZ250
N-ACETYLHOMOCYSTEINTHIOLAKTON (GERMAN) see TCZ000
ACETYL HYDRAZIDE see ACM750
N-ACETYLHYDRAZINE see ACM750
ACETYL HYDROPEROXIDE see PCL500
2-ACETYLHYDROQUINONE see DMG600
ACETYLHYDROQUINONE see DMG600
ACETYLHYDROXAMIC ACID see ABB250
N-ACETYL-4-HYDROXY-m-ARSANILIC ACID see ABX500
N-ACETYL-4-HYDROXY-m-ARSANILIC ACID, CALCIUM SALT see CAL500
N-ACETYL-4-HYDROXYARSANILIC ACID compounded with DIETHYLAMINE (1:1) see ACN250
N-ACETYL-4-HYDROXY-m-ARSANILIC ACID DIETHYLAMINE SALT see ACN250
N-ACETYL-4-HYDROXY-m-ARSANILIC ACID SODIUM SALT see SEG000
2-ACETYL-7-(2-HYDROXY-3-sec-BUTYLAMINOPROPOXY)BENZOFURAN see ACN310
2-ACETYL-7-(2-HYDROXY-3-tert-BUTYLAMINOPROPOXY)BENZOFURAN see ACN320
2-ACETYL-4-(2-HYDROXY-3-tert-BUTYLAMINOPROPOXY)BENZOFURAN see ACN300
17-β-ACETYL-17-HYDROXYESTR-4-ENE-3-ONE HEXANOATE see GEK510

2-ACETYL-10-(3-(4-(β-HYDROXYETHYL)PIPERAZINYL)PROPYL)PHENOTHIAZINE see ABG000
2-ACETYL-10-(3-(4-(β-HYDROXYETHYL)PIPERIDINO)PROPYL)PHENOTHIAZINE see PII500
2-ACETYL-7-((2-HYDROXY-3-ISOPROPYLAMINO)PROPOXY)BENZOFURAN HYDROCHLORIDE see BAU255
3′-ACETYL-4′-(2-HYDROXY-3-(ISOPROPYLAMINO)PROPOXY)BUTYRANILIDE see AAE100
3′-ACETYL-4′-(2-HYDROXY-3-(ISOPROPYLAMINO)PROPOXY)BUTYRANILIDE HYDROCHLORIDE see AAE125
(±)-N-(3-ACETYL-4-(2-HYDROXY-3-((1-METHYLETHYL)AMINO)PROPOXY)PHENYL)BUTANAMIDE see AAE100
2-ACETYL-5-HYDROXY-3-OXO-4-HEXENOIC ACID Δ-LACTONE see MFW500
5-ACETYL-8-HYDROXYQUINOLINE see HOE200
ACETYL HYPOBROMITE see ACN875
2-ACETYL-7-(2-HYROXY-3-ISOPROPYLAMINOPROPOXY)BENZOFURAN see HLK600
ACETYLIDES see ACO000
1-ACETYLIMIDAZOLE see ACO250
N-ACETYLIMIDAZOLE see ACO250
ACETYLIN see ADA725
3-ACETYLINDOLE see ICY100
ACETYL-3-INDOLE see ICY100
ACETYL-5-INDOLE see ACO300
5-ACETYLINDOLE see ACO300
3-ACETYLINDOLE-5-CARBONITRILE see COP550
5-ACETYLINDOLINE see ACO320
ACETYL IODIDE see ACO500
ACETYLISOEUGENOL see AAX750
N-ACETYLISONIAZID see ACO750
ACETYL ISONIAZID see ACO750
1-ACETYL-2-ISONICOTINOYLHYDRAZINE see ACO750
N-ACETYLISONICOTINYLHYDRAZIDE see ACO750
ACETYLISOPENTANOYL see MKL300
ACETYL ISOVALERYL see MKL300
ACETYLKAPROLAKTAM see ACF100
ACETYLKIDAMYCIN see ACP000
ACETYL KITASAMYCIN see MRE750
(2-ACETYLLACTOYLOXYETHYL)TRIMETHYLAMMONIUM HEMI-1,5-NAPHTHALENEDISULFONATE see AAC875
(2-ACETYLLACTOYLOXYETHYL)TRIMETHYLAMMONIUM 1,5-NAPHTHALENEDISULFONATE see AAC875
ACETYLLANDROMEDOL see AOO375
N-ACETYL-l-LEUCYL-N-(4-((AMINOIMINOMETHYL)AMINO)-1-FORMYLBUTYL)-l-LEUCINAMIDE (9CI) see LEX450
N-ACETYL-l-LEUCYL-l-LEUCYL-l-ARGINAL see LEX400
1-ACETYLLYSERGIC ACID DIETHYLAMIDE BITARTRATE see ACP500
1-ACETYLLYSERGIC ACID ETHYLAMIDE see ACP750
d-1-ACETYL LYSERGIC ACID MONOETHYLAMIDE see ACP750
ACETYL MERCAPTAN see TFA500
ACETYLMERCAPTOACETIC ACID see ACQ250
N-ACETYL-3-MERCAPTOALANINE see ACH000
α-1-ACETYLMETHADOL see ACQ666
l-α-ACETYLMETHADOL see ACQ666
levo-α-ACETYLMETHADOL see ACQ666
l-α-ACETYLMETHADOL HYDROCHLORIDE see ACQ690
N-ACETYLMETHIONINE see ACQ275
dl-N-ACETYLMETHIONINE see ACQ270
N-ACETYL-dl-METHIONINE see ACQ270
ACETYL-dl-METHIONINE see ACQ270
N-ACETYL-l-METHIONINE see ACQ275
ACETYLMETHIONINE see ACQ275
3-ACETYL-6-METHOXY-2H-1-BENZOPYRAN see ACQ700
3-ACETYL-7-METHOXY-2H-1-BENZOPYRAN see ACQ730
3-ACETYL-8-METHOXY-2H-1-BENZOPYRAN see ACQ760
2-ACETYL-7-METHOXYNAPHTHO(2,1-b)FURAN see ACQ790
N-ACETYL-5-METHOXYTRYPTAMINE see MCB350
4-(N-ACETYL-N-METHYL)AMINO-4′-N-METHYLAMINOAZOBENZENE see MLK750
4-(N-ACETYL-N-METHYL)AMINO-4′-(N′,N′-DIMETHYLAMINO)AZOBENZENE see DPQ200
N-ACETYL-METHYLANILINE see MFW000
ACETYL METHYL BROMIDE see BNZ000
N-ACETYL-N-(METHYLCARBAMOYLOXY)-N′-METHYLUREA see CBG075
ACETYL METHYL CARBINOL see ABB500
o-ACETYL-β-METHYLCHOLINE CHLORIDE see ACR000
2-ACETYL-2-METHYL-1,3-DITHIOLANE see ACR050
α-ACETYL-6-METHYLERGOLINE-8-β-PROPIONAMIDE see ACR100
N-ACETYL-N-(2-METHYL-4-((2-METHYLPHENYL)AZO)PHENYL)ACETAMIDE see ACR300
ACETYL-METHYL-NITROSO-HARNSTOFF (GERMAN) see ACR400
N′-ACETYL-METHYLNITROSOUREA see ACR400
ACETYLMETHYLNITROSOUREA see ACR400
3-ACETYL-10-(3′-N-METHYL-PIPERAZINO-N′-PROPYL)PHENOTHIAZIN see ACR500

3-ACETYL-6-METHYL-2H-PYRAN-2,4(3H)-DIONE see MFW500
3-ACETYL-6-METHYLPYRANDIONE-2,4 see MFW500
3-ACETYL-6-METHYL-2,4-PYRANDIONE see MFW500
4-ACETYLMORPHOLINE see ACR750
N-ACETYLMORPHOLINE see ACR750
3-ACETYL-10-(3'-MORPHOLINOPROPYL)PHENOTHIAZIN see MMA600
N-ACETYL-N'-MYRISTOYLOXY-2-AMINOFLUORENE see ACS000
1-ACETYLNAPHTHALENE see ABC475
2-ACETYLNAPHTHALENE see ABC500
β-ACETYLNAPHTHALENE see ABC500
N-ACETYL-1-NAPHTHYLAMINE see NAK000
N-ACETYL-2-NAPHTHYLHYDROXYLAMINE see NBD000
N-ACETYL-N'-α-BROMO-α-ETHYLBUTYRYLCARBAMIDE see ACE000
N-ACETYLNEOMYCIN see ACS375
8-ACETYLNEOSOLANIOL see ACS500
N-ACETYL-N'-(p-HYDROXYMETHYL)PHENYLHYDRAZINE see ACN500
N-ACETYL-N'-ISONICOTINYL HYDRAZIDE see ADA000
ACETYL NITRATE see ACS750
p-ACETYLNITROBENZENE see NEL600
2-ACETYL-5-NITROFURAN see ACT250
O-ACETYL NITROHYDROXYAMINE see NEC500
O-ACETYL-N-NITROHYDROXYLAMINE see NEC500
ACETYLNITROPEROXIDE see PCL750
2-ACETYL-4-NITROPYRROLE see ACT300
2-ACETYL-5-NITROPYRROLE see ACT330
2-ACETYL-5-NITROTHIOPHENE see NML100
N'-ACETYL-N'-METHYL-4'-AMINO-N,N-DIMETHYL-4-AMINOAZOBENZENE see DPQ200
N'-ACETYL-N'-MONOMETHYL-4'-AMINO-N-ACETYL-N-MONOMETHYL-4-AMINOAZOBENZENE see DQH200
N'-ACETYL-N'-MONOMETHYL-4'-AMINO-N-MONOMETHYL-4-AMINOAZOBENZENE see MLK750
N-ACETYL-N'-PHENYLHYDRAZINE see ACX750
ACETYLON FAST BLUE G see TBG700
ACETYLON FAST PINK B see AKE250
ACETYLON FAST RED VIOLET R see DBP000
ACETYL OXIDE see AAX500
5-ACETYLOXINE see HOE200
7-ACETYL-5-OXO-5H-(1)BENZOPYRANO(2,3-b)PYRIDINE see ACU125
2-(ACETYLOXY)BENZOIC ACID see ADA725
2-(ACETYLOXY)BENZOIC ACID 4-(ACETYLAMINO)PHENYL ESTER see SAN600
2-(ACETYLOXY)BENZOIC ACID, mixed with 3,7-DIHYDRO-1,3,7-TRIMETHYL-1H-PURINE-2,6-DIONE and N-(4-ETHOXYPHENYL)ACETAMIDE see ARP250
N-(ACETYLOXY)-N-(1,1'-BIPHENYL)-4-YLACETAMIDE see ABJ750
2-(ACETYLOXY)-3-BROMO-N-(4-BROMOPHENYL)-5-CHLORO-BENZENECARBOTHIOAMIDE see BOL325
2-(ACETYLOXY)-N-(4-CHLOROPHENYL)-3,5-DIIODOBENZAMIDE see CGB250
17-(ACETYLOXY)-6-CHLOROPREGNA-4,6-DIENE-3,20-DIONE see CBF250
ACETYLOXYCYCLOHEXIMIDE see ABN000
(11-β,16-α)-21-(ACETYLOXY)-16,17-(CYCLOPENTYLIDENEBIS(OXY))-9-FLUORO-11-HYDROXYPREGNA-1,4-DIENE-3,20-DIONE see COW825
(11-β)-21-(ACETYLOXY)-11,17-DIHYDROXY-PREGN-4-ENE-3,20-DIONE (9CI) see HHQ800
4-(ACETYLOXY)-12,13-EPOXY-3,7,15-TRIHYDROXY-(3-α,4-β,7-β)-TRICHOTHEC-9-EN-8-ONE see FQR000
(16-β,17-β)-17-(ACETYLOXY)-16-ETHYL-ESTR-4-EN-3-ONE (9CI) see ELF110
N-(ACETYLOXY)-N-9H-FLUOREN-2-YL-TETRADECANAMIDE see FEM000
6-β-(ACETYLOXY)-3-β-(β-d-GLUCOPYRANOSYLOXY)-8,14-DIHYDROXYBUFA-4,20,22-TRIENOLIDE see SBF500
3-β,6-β-6-ACETYLOXY-3-(β-d-GLUCOPYRANOSYLOXY)-8,14-DIHYDROXYBUFA-4,20,22-TRIENOLIDE see SBF500
21-(ACETYLOXY)-11-β-HYDROXY-17-((1-OXOPENTYL)OXY)PREGNA-1,4-DIENE-3,20-DIONE see AAF625
21-(ACETYLOXY)-17-HYDROXY-PREGN-4-ENE-3,11,20-TRIONE (9CI) see CNS825
4-(4-(ACETYLOXY)-3-IODOPHENOXY)-3,5-DIIODO-BENZOIC ACID see ADD875
6-ACETYLOXYMETHYLBENZO(a)PYRENE see ACU500
(6-α)-17-(ACETYLOXY)-6-METHYLPREG-4-ENE-3,20-DIONE see MCA000
17-(ACETYLOXY)-6-METHYLPREGN-4-ENE-3,20-DIONE, (6-α)- mixed with (17-α)-19-NORPREGNA-1,3,5(10)-TRIEN-20-YNE-3,17-DIOL see POF275
(17-α)-17-(ACETYLOXY)-19-NORPREGN-4-EN-20-YN-3-ONE see ABU000
17-ACETYLOXY(17-α)-19-NORPREGN-4-ESTREN-17-β-OL-ACETATE-3-ONE see ABU000
2-(2-(ACETYLOXY)-1-OXOPROPOXY)-N,N,N-TRIMETHYLETHANAMINIUM 1,5-NAPHTHALENEDISULFONATE (2:1) see AAC875
(ACETYLOXY)PHENYL-CARBAMIC ACID see ING400
4-((4-(ACETYLOXY)PHENYL)CYCLOHEXYLIDENEMETHYL)PHENOL ACETATE see FBP100
β-ACETYLOXY-β-PHENYLETHYL DIMETHYLAMINE see ABN700
5-(1-ACETYLOXY-2-PROPENYL)-1,3-BENZODIOXOLE see ACV000
2-(ACETYLOXY)-N,N,N-TRIMETHYLETHANAMINIUM see CMF250
2-(ACETYLOXY)-N,N,N-TRIMETHYLETHANAMINIUM BROMIDE see CMF260
2-(ACETYLOXY)-N,N,N-TRIMETHYLETHANAMINIUM CHLORIDE see ABO000
(ACETYLOXY)TRIOCTYLSTANNANE see ABX150

(ACETYLOXY)TRIPHENYLPLUMBANE see TMT000
(ACETYLOXY)TRIPHENYL-STANNANE (9CI) see ABX250
ACETYL PEROXIDE, solid, or >25% in solution (DOT) see ACV500
ACETYL PEROXIDE see ACV500
ACETYL PEROXIDE, not >25% in solution (UN 2084) (DO) see ACV500
ACETYLPHENETIDIN see ABG750
N-ACETYL-p-PHENETIDINE see ABG750
ACETYLPHENETURIDE see ACX500
2-ACETYLPHENOL see HIN500
4-ACETYLPHENOL see HIO000
o-ACETYLPHENOL see HIN500
p-ACETYLPHENOL see HIO000
ACETYL PHENOL see PDY750
4-(N-ACETYL)PHENYLAMINO-1-ETHYL-2,2,6,6-TETRAMETHYLPIPERIDINE see EPL000
N-ACETYL-m-PHENYLENEDIAMINE see AHQ000
ACETYL-p-PHENYLENEDIAMINE see AHQ250
1-ACETYL-3-PHENYLETHYLACETYLUREA see ACX500
1-ACETYL-2-PHENYLHYDRAZINE see ACX750
β-ACETYLPHENYLHYDRAZINE see ACX750
ACETYLPHENYLHYDRAZINE see ACX750
1-(p-ACETYLPHENYL)-3-METHYL-3-NITROSOUREA see MFX725
cis-2-ACETYL-3-PHENYL-5-TOSYL-3,3a,4,5-TETRAHYDROPYRAZOLO(4,3-c)QUINOLINE see ACY700
1-ACETYL-2-PHENYL-1,5,5-TRIMETHYL-SEMIOXAMAZIDE see DVU100
12-O-ACETYL-PHORBOL-13-DECA-(Δ-2)-ENOATE see ACY750
12-O-ACETYL-PHORBOL-13-DECANOATE see ACZ000
ACETYLPHOSPHORAMIDOTHIOIC ACID-O,S-DIMETHYL ESTER see DOP600
1-ACETYL-2-PICOLINOLHYDRAZINE see ADA000
1-ACETYL-2-PICOLINOYLHYDRAZINE see ADA000
N-ACETYLPIPERIDIN (GERMAN) see ADA250
1-ACETYLPIPERIDINE see ADA250
ACETYLPROMAZINE see ABH500
ACETYLPROMAZINE MALEATE (1:1) see AAF750
3-ACETYLPROPANOL see ABH750
β-ACETYLPROPIONIC ACID see LFH000
ACETYLPROPIONYL see PBL350
2-ACETYL PYRAZINE see ADA350
2-ACETYLPYRIDINE see PPH200
3-ACETYLPYRIDINE see ABI000
4-ACETYLPYRIDINE see ADA365
β-ACETYLPYRIDINE see ABI000
ACETYLPYRIDINE see PPH200
ACETYL RED B see CMS231
ACETYL RED G see CMM300
ACETYL RED J see CMM300
4-ACETYLRESORCINOL see DMG400
ACETYLRESORCINOL see RDZ900
ACETYL ROSE 2GL see CMM300
ACETYLSAL see ADA725
ACETYLSALICYLIC ACID see ADA725
o-ACETYLSALICYLIC ACID, SODIUM SALT see ADA750
ACETYLSALICYLIC ACID SODIUM SALT see ADA750
ACETYLSALICYLSAEURE (GERMAN) see ADA725
ACETYLSALICYLSAEURE NATRIUMSALZ (GERMAN) see ADA750
N-ACETYL-SARCOLYSIL VALINE ETHYL ETHER see ARM000
ACETYLSELENO-2 BENZOIC ACID see SBU100
2-(ACETYLSELENO)BENZOIC ACID see SBU100
5-ACETYLSPIRO(BENZOFURAN-2(3H),1'-CYCLOPROPAN)-3-ONE see SLE880
o-ACETYLSTERIGMATOCYSTIN see ADB250
N¹-ACETYLSULFANILAMIDE see SNP500
N'-ACETYLSULFANILAMIDE see SNP500
N¹-ACETYLSULFANILAMIDE SODIUM see SNQ000
N¹-ACETYLSULFANILAMIDE SODIUM SALT see SNQ000
N-ACETYLSULFANILAMINE see SNP500
N-ACETYL-N-TETRADECANOYLOXY-2-AMINOFLUORENE see ACS000
6-ACETYL-1,1,4,4-TETRAMETHYL-7-ETHYL-1,2,3,4,-TETRALIN see ACL750
7-ACETYL-1,1,4,4-TETRAMETHYL-1,2,3,4-TETRAHYDRONAPHTHALENE see ACL750
(ACETYLTHIO)ACETIC ACID see ACQ250
ACETYLTHIOCHOLINE DIIODIDE see ADC300
S-ACETYLTHIOCHOLINE IODIDE see ADC300
ACETYLTHIOCHOLINE IODIDE see ADC300
S-ACETYLTHIOGLYCOLIC ACID see ACQ250
1-ACETYL-2-THIOHYDANTOIN see ADC750
7-α-ACETYLTHIO-3-OXO-17-α-PREGN-4-ENE-21,17-β-CARBOLACTONE see AFJ500
7-α-ACETYLTHIO-3-OXO-17-β-PREGN-4-ENE-21,17-β-CARBOLACTONE see AFJ500
2-ACETYLTHIOPHENE see ABI500
2-(ACETYLTHIO)-N,N,N-TRIMETHYLETHANAMINIUM IODIDE see ADC300
1-ACETYL-2-THIOUREA see ADD250
ACETYL THIOUREA see ADD250
p-ACETYLTOLUENE see MFW250
N-ACETYL-p-TOLUIDIDE see ABJ250

N-ACETYL-m-TOLUIDINE see ABI750
ACETYL-o-TOLUIDINE see ABJ000
ACETYL-p-TOLUIDINE see ABJ250
ACETYL TRIBUTYL CITRATE see THX100
ACETYL TRIETHYL CITRATE see ADD750
ACETYLTRIIODOTHYRONINE FORMIC ACID see ADD875
N-ACETYL TRIMETHYLCOLCHICINIC ACID see ADE000
N-ACETYL TRIMETHYLCOLCHICINIC ACID METHYL ETHER see CNG830
3-ACETYL-2,4,5-TRIMETHYL-PYRROLE see ADE050
ACETYL-l-TRP see ADE075
N-ACETYLTRYPTOPHAN see ADE075
(S)-N-ACETYLTRYPTOPHAN see ADE075
N-ACETYL-l-TRYPTOPHAN see ADE075
ACETYL-l-TRYPTOPHAN see ADE075
ACETYLTRYPTOPHAN see ADE075
ACETYLURETHANE see ACL000
ACETYLUREUM see PEC250
ACETYL YELLOW G see CMM758
ACETYLZIRCONIUM, ACETONATE see PBL750
AC GA see PJS750
ACH see CMF250
ACH CHLORIDE see ABO000
ACHILLEIC ACID see ADH000
ACHIOTE see APE100
ACHLESS see TKH750
ACHLETIN see HII500
ACHROCIDIN see ABG750
ACHROMYCIN see TBX000, TBX250
ACHROMYCIN HYDROCHLORIDE see TBX250
ACHROMYCIN (PURINE DERIVATIVE) see AEI000
ACHTL see TCZ000
AC 2197 HYDROCHLORIDE see DBA475
ACIBILIN see TAB250
ACICLOVIR see AEC700
ACID see DJO000
ACIDAL BLACK 10B see FAB830
ACIDAL BRIGHT PONCEAU 3R see FMU080
ACIDAL BRILLIANT RED 2G see CMM300
ACIDAL CARMINE V see CMM062
ACIDAL GREEN G see FAE950
ACID ALIZARINE PURE BLUE B see CMM090
ACID ALIZARINE PURE BLUE R see CMM080
ACID ALIZARINE SKY BLUE B see CMM090
ACIDAL LIGHT GREEN SF see FAF000
ACIDAL NAVY BLUE 3BR see FAB830
ACIDAL PONCEAU G see FMU070
ACID AMARANTH see FAG020
ACID AMIDE see NCR000
ACID AMMONIUM CARBONATE see ANB250
ACID AMMONIUM FLUORIDE see ANJ000
ACID AMMONIUM SULFATE see ANJ150
ACID ANTHRACENE RED G see CMM325
ACID ANTHRACENE RED GA-CF see CMM325
ACID ANTHRACENE YELLOW GR see CMM759
ACID ANTHRAQUINONE PURE BLUE see CMM090
ACID BLACK 1 see FAB830
ACID BLACK 10A see FAB830
ACID BLACK 12B see FAB830
ACID BLACK 10B see FAB830
ACID BLACK 10BA see FAB830
ACID BLACK BASE M see FAB830
ACID BLACK 10BN see FAB830
ACID BLACK 4BN see FAB830
ACID BLACK 4BNU see FAB830
ACID BLACK BRX see FAB830
ACID BLACK BX see FAB830
ACID BLACK H see FAB830
ACID BLACK JVS see FAB830
ACID BLUE 1 see ADE500
ACID BLUE 3 see CMM062
ACID BLUE 9 see FMU059
ACID BLUE 41 see CMM070
ACID BLUE 62 see CMM080
ACID BLUE 78 see CMM090
ACID BLUE 92 see ADE750
ACID BLUE 129 see CMM100
ACID BLUE A see ADE750
ACID BLUE BLACK 10B see FAB830
ACID BLUE BLACK B see FAB830
ACID BLUE BLACK BG see FAB830
ACID BLUE BLACK DOUBLE 600 see FAB830
ACID BLUE O see ERG100
ACID BLUE V see ADE500
ACID BLUE W see FAE100
ACID BRIGHT AZURE Z see ADE500

ACID BRIGHT RED see CMM300
ACID BRILLIANT BLUE VF see ADE500
ACID BRILLIANT BLUE Z see ADE500
ACID BRILLIANT GREEN BS see ADF000
ACID BRILLIANT GREEN SF see FAF000
ACID BRILLIANT PINK B see FAG070
ACID BRILLIANT RED see CMM300
ACID BRILLIANT RUBINE 2G see HJF500
ACID BRILLIANT SCARLET 3R see FMU080
ACID BRILLIANT SKY BLUE Z see ADE500
ACID BUTYL PHOSPHATE see ADF250
ACID CALCIUM PHOSPHATE see CAW110
ACID CARBOYS, EMPTY see ADF500
ACID CHROME BLUE BA see HJF500
ACID CHROME YELLOW GG see SIT850
ACID CHROME YELLOW 2GW see SIT850
ACID COPPER ARSENITE see CNN500
ACIDE ACETIQUE (FRENCH) see AAT250
ACIDE ACETYLSALICYLIQUE (FRENCH) see ADA725
ACIDE ACETYL SELENO-2 BENZOIQUE see SBU100
ACIDE ANISIQUE (FRENCH) see MPI000
ACIDE ARSENIEUX see ARI750
ACIDE ARSENIQUE LIQUIDE (FRENCH) see ARB250
ACIDE BENZOIQUE (FRENCH) see BCL750
ACIDE BENZOYL-2-PHENYLACETIQUE (FRENCH) see BDS500
ACIDE BROMACETIQUE (FRENCH) see BMR750
ACIDE BROMHYDRIQUE (FRENCH) see HHJ000
l'ACIDE BUCLOXIQUE (FRENCH) see CPJ000
ACIDE BUCLOXIQUE CALCIUM (FRENCH) see CPJ250
ACIDE CACODYLIQUE (FRENCH) see HKC000
ACIDE CARBOLIQUE (FRENCH) see PDN750
ACIDE CHLORACETIQUE (FRENCH) see CEA000
ACIDE CHLORHYDRIQUE (FRENCH) see HHL000
ACIDE 2-(4-CHLORO-2-METHYL-PHENOXY)PROPIONIQUE (FRENCH) see CIR500
ACIDE p-CHLOROPHENYL-2-THIAZOLE-ACETIQUE-4 (FRENCH) see CKK250
ACIDE CHROMIQUE (FRENCH) see CMH250
ACIDE CRESYLIQUE (FRENCH) see CNW500
ACIDE CYANACETIQUE (FRENCH) see COJ500
ACIDE CYANHYDRIQUE (FRENCH) see HHS000
l'ACIDE (CYCLOHEXYL-4, CHLORO-3, PHENYL)-4,OXO-4, BUTYRIQUE CALCIUM (FRENCH) see CPJ250
ACIDE DEHYDROCHOLIQUE (FRENCH) see DAL000
ACIDE-2,4-DICHLORO PHENOXYACETIQUE (FRENCH) see DAA800
ACIDE-2-(2,4-DICHLORO-PHENOXY) PROPIONIQUE (FRENCH) see DGB000
ACIDE DIMETHYLARSINIQUE (FRENCH) see HKC000
ACIDE DIMETHYL-ETHYL-ALLENOLIQUE ETHER METHYLIQUE (FRENCH) see DRU600
ACIDE DIPHENYLETHOXYACETIQUE (FRENCH) see EES300
ACIDE DIPHENYLHYDROXYACETIQUE see BBY990
ACIDE DISELINO SALICYLIQUE see SBU150
ACIDE ETHYLENEDIAMINETETRACETIQUE (FRENCH) see EIX000
ACIDE ETHYL-8 OXO-5 PIPERAZINYL-2 DIHYDRO-5,8 PYRIDO(2,3-d)PYRIMIDINE-6 CARBOXYLIQUE see PIZ000
ACIDE 1-ETIL-7-METIL-1,8-NAFTIRIDIN-4-ONE-3-CARBOSSILICO (ITALIAN) see EID000
ACIDE FLUORHYDRIQUE (FRENCH) see HHU500
ACIDE FLUOROSILICIQUE (FRENCH) see SCO500
ACIDE FLUOSILICIQUE (FRENCH) see SCO500
ACIDE FORMIQUE (FRENCH) see FNA000
ACIDE (ISOBUTYL-4 PHENYL)-2 PROPIONIQUE (FRENCH) see IIU000
ACIDE ISO-NICOTINIQUE (FRENCH) see ILC000
ACIDE ISONIPECOTIQUE see ILG100
ACIDE ISOPHTALIQUE (FRENCH) see IMJ000
ACIDE ISOVANILLIQUE (FRENCH) see HJC000
ACIDE o-METHOXYCINNAMIQUE (FRENCH) see MEJ775
ACIDE β-(1-METHOXY-4-NAPHTHOYL)PROPIONIQUE (FRENCH) see MEZ300
ACIDE METHYL-o-BENZOIQUE (FRENCH) see MPI000
ACIDE METHYL SELENO-2-BENZOIQUE see MPI200
ACIDE METIAZINIQUE (FRENCH) see MNQ500
ACIDE MONOCHLORACETIQUE (FRENCH) see CEA000
ACIDE-MONOFLUORACETIQUE (FRENCH) see FIC000
ACIDE NALIDIXICO (ITALIAN) see EID000
ACIDE NALIDIXIQUE (FRENCH) see EID000
ACIDE NAPHTHYLOXYACETIQUE see NBJ100
ACIDE NICOTINIQUE (FRENCH) see NCQ900
ACIDE NIFLUMIQUE (FRENCH) see NDX500
ACIDE NITRIQUE (FRENCH) see NED500
lACIDE OLEIQUE (FRENCH) see OHU000
ACIDE ORTHOVANILLIQUE see HJB500
ACIDE OXALIQUE (FRENCH) see OLA000
ACIDE PERACETIQUE (FRENCH) see PCL500
ACIDE PHENOXY-2-PROPIONIQUE see PDV725
ACIDE PHENYLBORIQUE (FRENCH) see BBM000
ACIDE-α-PHENYL-o-METHOXYCINNAMIQUE (FRENCH) see MFG600

ACIDE PHOSPHORIQUE (FRENCH) see PHB250
ACIDE PHTHALIQUE (FRENCH) see PHW250
ACIDE PICOLIQUE (FRENCH) see PIB930
ACIDE PICRAMIQUE (FRENCH) see DUP400
ACIDE PICRIQUE (FRENCH) see PID000
ACIDE PIPECOLIQUE see HDS300
ACIDE PIPERIDINE-CARBOXYLIQUE-2 see HDS300
ACIDE PIPERIDINE-CARBOXYLIQUE-4 see ILG100
ACIDE PROPIONIQUE (FRENCH) see PMU750
l'ACIDE RICINOLEIQUE (FRENCH) see RJP000
ACIDE SULFHYDRIQUE (FRENCH) see HIC500
ACIDE SULFURIQUE (FRENCH) see SOI500
d'ACIDE TANNIQUE (FRENCH) see TAD750
ACIDE TEREPHTHALIQUE (FRENCH) see TAN750
ACIDE THIOGLYCOLIQUE (FRENCH) see TFJ100
ACIDE TRICHLORACETIQUE (FRENCH) see TII250
ACIDE 2,4,5-TRICHLORO PHENOXYACETIQUE (FRENCH) see TAA100
ACIDE 2-(2,4,5-TRICHLORO-PHENOXY) PROPIONIQUE (FRENCH) see TIX500
ACIDE VANILLIQUE see VFF000
ACID FAST BLUE BS see CMM100
ACID FAST RED EGG see CMM300
ACID FAST RED FB see HJF500
ACID FAST RED 3G see CMM300
ACID FAST YELLOW AG see SGP500
ACID FAST YELLOW E5R see SGP500
ACID FAST YELLOW MR see CMM759
ACID FUCHSINE D see CMS231
ACID FUCHSIN FAST B see CMS231
ACID GREEN 3 see FAE950
ACID GREEN 40 see CMM200
ACID GREEN 50 see ADF000
ACID GREEN see FAE950
ACID GREEN A see FAF000
ACID GREEN B see FAE950
ACID GREEN 2G see FAE950
ACID GREEN G see FAE950
ACID GREEN L see FAE950
ACID GREEN N see FAF000
ACID GREEN S see FAE950
ACIDIC METANIL YELLOW see MDM775
ACIDINE RED G see CMM300
ACIDINE RED RD see NAO600
ACIDINE SCARLET GD see CMM320
ACID IV see ALH250
ACID LEAD ARSENATE see LCK000
ACID LEAD ORTHOARSENATE see LCK000
ACID LEATHER BLUE IC see FAE100
ACID LEATHER BLUE IGW see FAB830
ACID LEATHER BLUE R see ADE750
ACID LEATHER BLUE V see ADE500
ACID LEATHER BROWN 2G see XMA000
ACID LEATHER DARK BLUE G see FAB830
ACID LEATHER FAST BLUE BLACK G see FAB830
ACID LEATHER GREEN F see FAE950
ACID LEATHER GREEN 3G see FAE950
ACID LEATHER GREEN S see ADF000
ACID LEATHER LIGHT BROWN G see SGP500
ACID LEATHER ORANGE EXTRA see CMM220
ACID LEATHER ORANGE EXTRA G see CMM220
ACID LEATHER ORANGE EXTRA PRW see CMM220
ACID LEATHER ORANGE I see FAG010
ACID LEATHER RED 2BG see NAO600
ACID LEATHER RED BG see CMM330
ACID LEATHER RED GR see CMM320
ACID LEATHER RED KG see CMM300
ACID LEATHER RED KPR see FMU070
ACID LEATHER SCARLET G see CMM320
ACID LEATHER YELLOW CRS see CMM759
ACID LEATHER YELLOW PRW see MDM775
ACID LEATHER YELLOW R see MDM775
ACID LEATHER YELLOW T see FAG140
ACID METANIL YELLOW see MDM775
ACID NAFTOL RED G see CMM300
ACIDO ACETICO (ITALIAN) see AAT250
ACIDO 3-ACETILAMINO-2,4,6-TRIIODOBENZOICO (ITALIAN) see AAM875
ACIDO o-ACETIL-BENZOICO (ITALIAN) see ADA725
ACIDO ACETILSALICILICO (ITALIAN) see ADA725
ACIDO p-AMINOBENZOICO see AIH600
ACIDO-3-AMINO-2,4,6-TRIIODOBENZOICO (ITALIAN) see AMU625
ACIDO BROMIDRICO (ITALIAN) see HHJ000
ACIDO CIANIDRICO (ITALIAN) see HHS000
ACIDO CLORIDRICO (ITALIAN) see HHL000
ACIDO m-CLOROBENZOICO see CEL290
ACIDO p-CLOROBENZOICO see CEL300

ACIDO 1-(p-CLOROBENZOIL)-5-METOSSI-2-METIL-3-INDOLILACETOIDROSSAMI-CO (ITALIAN) see OLM300
ACIDO 2-(4-CLORO-2-METIL-FENOSSI)-PROPIONICO (ITALIAN) see CIR500
ACIDO (2,4-DICLORO-FENOSSI)-ACETICO (ITALIAN) see DAA800
ACIDO-2-(2,4-DICLORO-FENOSSI)-PROPIONICO (ITALIAN) see DGB000
ACIDO (3,6-DICLORO-2-METOSSI)-BENZOICO (ITALIAN) see MEL500
ACIDO DOISYNOLICO (SPANISH) see DYB000
ACIDO FENCLOZICO (ITALIAN) see CKK250
ACIDO-5-FENIL-5-ALLILBARBITURICO (ITALIAN) see AGQ875
ACIDO-5-FENIL-5-ETILBARBITURICO (ITALIAN) see EOK000
ACIDO FENOFIBRICO see CJP750
ACIDO FLUFENAMICO (ITALIAN) see TKH750
ACIDO FLUORIDRICO (ITALIAN) see HHU500
ACIDO FLUOSILICICO (ITALIAN) see SCO500
ACIDO FORMICO (ITALIAN) see FNA000
ACIDO FOSFORICO (ITALIAN) see PHB250
ACIDOGEN NITRATE see UTJ000
ACIDO-m-IDROSSIBENZOICO (ITALIAN) see HJI100
ACIDO INDOXAMICO (ITALIAN) see OLM300
ACIDO IPPURICO see HGB300
ACIDO-4-(ISONICOTINIL-IDRAZONE)PIMELICO (ITALIAN) see ILG000
ACIDO MANDELICO see MAP000
ACIDO MERCAPTOBENZOICO see TFC550
ACIDO-1-METIL-5-(p-TOLNIL)-PIRROL-2-ACETICO (SPANISH) see TGJ850
ACIDOMONOCLOROACETICO (ITALIAN) see CEA000
ACIDO MONOFLUOROACETIO (ITALIAN) see FIC000
ACIDOMYCIN see CCI500, CMP885
ACIDO NIFLUMICO (ITALIAN) see NDX500
ACIDO NITRICO (ITALIAN) see NED500
ACIDO ORTOCRESOTINICO (ITALIAN) see CNX625
ACIDO OSSALICO (ITALIAN) see OLA000
ACIDO 3-OSSI-5-METIL-BENZOICO (ITALIAN) see CNX625
ACIDO PICRICO (ITALIAN) see PID000
ACIDO PIPEMIDICO see PIZ000
ACID ORANGE 24 see XMA000
ACID ORANGE 7 see CMM220
ACID ORANGE see CMM220
ACID ORANGE A see CMM220
ACID ORANGE II see CMM220
ACID ORANGE NO. 3 see SGP500
ACIDO SALICILICO (ITALIAN) see SAI000
ACIDO SOLFORICO (ITALIAN) see SOI500
ACIDO TRICLOROACETICO (ITALIAN) see TII250
ACIDO (2,4,5-TRICLORO-FENOSSI)-ACETICO (ITALIAN) see TAA100
ACIDO 2-(2,4,5-TRICLORO-FENOSSI)-PROPIONICO (ITALIAN) see TIX500
ACIDO 1-(m-TRIFLUOROMETILFENIL)-N-NITROSO ANTRANILICO (ITALIAN) see TKF699
ACID OXALATE see AMX825
ACID PHLOXINE GA see CMM300
ACID PHOSPHINE CL see FAG010
ACID PONCEAU 4R see FMU080
ACID PONCEAU R see FMU070
ACID POTASSIUM SULFATE see PKX750
ACID POTASSIUM TARTRATE see PKU600
ACID PURE SKY BLUE ANTHRAQUINONE see CMM090
ACID QUININE HYDROCHLORIDE see QIJ000
ACID RED 1 see CMM300
ACID RED 18 see FMU080
ACID RED 26 see FMU070
ACID RED 33 see CMS231
ACID RED 85 see CMM320
ACID RED 92 see ADG250
ACID RED 97 see CMM325
ACID RED 99 see NAO600
ACID RED 114 see CMM330
ACID RED 388 see RGZ100
ACID RED 2A see CMS231
ACID RED ALIZARINE see SEH475
ACID RED B see CMS231
ACID RED 2G see CMM300
ACID RED GA see CMM300
ACID RED PG see CMM320
ACID ROSE 2GL see CMM300
ACID RUBINE see HJF500
ACID SCARLET see FMU070
ACID SCARLET GNA see CMP620
ACID SCARLET JN EXTRA PURE A see CMP620
ACID SCARLET 3R see FMU080
ACID SKY BLUE A see FAE000
ACID SKY BLUE ANTHRAQUINONE see CMM090
ACID SKY BLUE O see ERG100
ACID-SPAR see CAS000
ACID-TREATED HEAVY NAPHTHENIC DISTILLATE see MQV760
ACID-TREATED LIGHT NAPHTHENIC DISTILLATE see MQV770
ACID-TREATED LIGHT PARAFFINIC DISTILLATE see MQV775

ACID-TREATED RESIDUAL OIL see MQV872
ACID TURQUOISE BLUE A see ERG100
ACIDUM ACETYLSALICYLICUM see ADA725
ACIDUM FENCLOZICUM see CKK250
ACIDUM NICOTINICUM see NCQ900
ACID VIOLET see FAG120
ACID WOOL BLUE RL see ADE750
ACID YELLOW 23 see FAG140
ACID YELLOW 36 see MDM775
ACID YELLOW 3 see CMM510
ACID YELLOW 42 see CMM759
ACID YELLOW see CMM758
ACID YELLOW AT see CMM758
ACID YELLOW E see SGP500
ACID YELLOW G see CMM758
ACID YELLOW GEIGY see CMM758
ACID YELLOW G KOND see CMM758
ACID YELLOW K see CMM759
ACID YELLOW T see FAG140
ACID YELLOW TRA see FAG150
ACIFLOCTIN see AEN250
ACIFLUORFEN see CLS075
ACIFLUORFENE see CLS075
ACIGENA see HCL000
ACILAN FAST NAVY BLUE R see ADE750
ACILAN GREEN B see FAE950
ACILAN GREEN BS see ADF000
ACILAN GREEN SFG see FAF000
ACILAN NAPHTHOL RED G see CMM300
ACILAN NAPHTOL RED G see CMM300
ACILAN ORANGE II see CMM220
ACILAN PONCEAU RRL see FMU070
ACILAN RED SE see FAG020
ACILAN SCARLET V3R see FMU080
ACILAN TURQUOISE BLUE AE see FMU059
ACILAN YELLOW EXTRA see CMM758
ACILAN YELLOW GG see FAG140
ACILETTEN see CMS750
ACILLIN see AIV500
ACIMETION see MDT740
ACIMETTEN see ADA725
ACINETTEN see AEN250
ACINITRAZOL see ABY900
ACINITRAZOLE see ABY900
ACINOL see BEL900
ACINTENE A see PIH250
ACINTENE DP see MCC250
ACINTENE DP DIPENTENE see MCC250
ACINTENE O see PMQ750
ACIPEN V see PDT500
ACIPHENOCHINOLINE see PGG000
ACIPHENOCHINOLINIUM see PGG000
ACISAL see ADA725
AC 220J see MPP000
ACKEE see ADG400
ACKET see SAH000
ACL-59 see PLD000
ACL 60 see SGG500
ACL 70 see DGN200
ACL 85 see TIQ750
ACLACINOMYCIN A see APU500
ACLACINOMYCIN T see DAY835
ACLACINOMYCIN Y1 see ADG425
ACLACINOMYCIN Y see ADG425
ACLAR 22A see KDK000
ACLAR 33C see KDK000
ACLATONIUM NAPADISILATE see AAC875
ACM see AAE625
ACNA BLACK DF BASE see PFU500
ACNEGEL see BDS000
ACNESTROL see DKA600
ACNU see ALF500
ACOCANTHERIN see OKS000
ACODEEN see BOR350
ACOINE see PDN500
ACOKANTHERA (VARIOUS SPECIES) see BOO700
ACOLEN see DAL000
A. COLUMBIANUM see MRE275
ACON see VSK600
ACONINE see ADG500
ACONITANE see ADH750
ACONITE see MRE275
ACONITIC ACID see ADH000
ACONITIN CRISTALLISAT (GERMAN) see ADH750
ACONITINE (crystalline) see ADH750

ACONITINE HYDROCHLORIDE see ADH875
ACONITUM CARMICHAELI see ADI250
ACONITUM (Various Species) see MRE275
A. CONTORTRIX CONTORTRIX VENOM see SKW775
A. CONTORTRIX MOKASEN VENOM see NNX700
A. CORDATA see TOA275
ACORN TANNIN see ADI625
ACORTAN see AES650
ACORTO see AES650
ACORUS CALAMUS Linn., oil extract see OGL020
ACORUS CALAMUS OIL see OGL020
ACOVENOSIDE B see SGB550
ACP 6 see PJS750
ACP 322 see SIO500
ACPC see AJK250
ACP GRASS KILLER see TII500
ACP-M-728 see AJM000
AC 8 (POLYMER) see PJS750
ACQUINITE see ADR000, CKN500
ACRALDEHYDE see ADR000
ACRAMINE RED see DBN000
ACRAMINE YELLOW see AHS750
ACREX see CBW000
ACRIBEL, combustion products see ADX750
ACRICHINE see ARQ250, CFU750
ACRICID see BGB500
ACRIDAN see ADI775
ACRIDANE see ADI775
2-ACRIDINAMINE see AHS000
9-ACRIDINAMINE see AHS500
3-ACRIDINAMINE (9CI) see ADJ375
9-ACRIDINAMINE MONOHYDROCHLORIDE see AHS750
9-ACRIDINAMINE, 1,2,3,4-TETRAHYDRO-(9CI) see TCJ075
ACRIDINE see ADJ500
ACRIDINE-9-CARBOXAMIDE, N,N-DIETHYL-1,2,3,4-TETRAHYDRO- see ADJ550
4-ACRIDINECARBOXAMIDE, 9-((2-METHOXY-4-((PROPYLSULFO-
    NYL)AMINO)PHENYL)AMINO)-, HYDROCHLORIDE see MFN000
ACRIDINE-9-CARBOXAMIDE, 1,2,3,4-TETRAHYDRO-N,N-DIETHYL- see ADJ550
ACRIDINE, 9-(2-((2-CHLOROETHYL)AMINO)ETHYLAMINO)-6-CHLORO-2-ME-
    THOXY-, DIHYDROCHLORIDE, HYDRATE see QCS875
ACRIDINE, 9-((2-((2-CHLOROPROPYL)AMINO)ETHYL)AMINO)-2-METHOXY-, DI-
    HYDROCHLORIDE, HEMIHYDRATE see CKU250
2,6-ACRIDINEDIAMINE see DBN000
3,6-ACRIDINEDIAMINE see DBN600
3,9-ACRIDINEDIAMINE (9CI) see ADJ625
3,6-ACRIDINEDIAMINE, MONOHYDROCHLORIDE (9CI) see PMH250
ACRIDINE, 9,10-DIHYDRO-(9CI) see ADI775
ACRIDINE HYDROCHLORIDE see ADJ750
ACRIDINE MONOHYDROCHLORIDE see ADJ750
ACRIDINE, 9-(MORPHOLINOCARBONYL)-1,2,3,4-TETRAHYDRO- see MRR760
ACRIDINE MUSTARD see ADJ875
ACRIDINE ORANGE see BAQ250, BJF000
ACRIDINE ORANGE FREE BASE see BJF000
ACRIDINE ORANGE NO see BAQ250
ACRIDINE ORANGE R see BAQ250
ACRIDINE RED see ADK000
ACRIDINE RED 3B see ADK000
ACRIDINE RED, HYDROCHLORIDE see ADK000
ACRIDINE, 1,2,3,4-TETRAHYDRO-9-(MORPHOLINOCARBONYL)- see MRR760
ACRIDINE, 1,2,3,4-TETRAHYDRO-9-(PIPERIDINOCARBONYL)- see PIU100
ACRIDINE YELLOW BASE see DBT200
ACRIDINIUM CHLORIDE see ADJ750
ACRIDINO(2,1,9,8-klmna)ACRIDINE see ADK250
ACRIDINO(2,1,9,8-klmna)ACRIDINE SULFATE see DCK200
4′-(9-ACRIDINYLAMINO)METHANESULFON-m-ANISIDE MONOHYDROCHLORIDE
    see ADL500
4′-(9-ACRIDINYLAMINO)METHANESULPHON-m-ANISIDIDE see ADL750
4′-(9-ACRIDINYLAMINO)-2′-METHOXYMETHANESULFONANILIDE see ADM000
4′-(9-ACRIDINYLAMINO)-3′-METHOXYMETHANESULFONANILIDE see ADL750
N-(4-(ACRIDINYL-9-AMINO)-3-METHOXYPHENYL)ETHANESULFONAMIDE METH-
    ANESULFONATE see ADM500
N-(4-(9-ACRIDINYLAMINO)-3-METHOXYPHENYL)METHANESULFONAMIDE see
    ADM000
4′-(9 ACRIDINYLAMINO)-3′-METHYLMETHANESULFONANILIDE see ADN250
4′-(9-ACRIDINYLAMINO)METHYLSULFONYL-m-ANISIDINE see ADL750
N-(p-(9-ACRIDINYLAMINO)PHENYL)BUTANESULFONAMIDE, HYDROCHLORIDE
    see ADO250
N-(p-(9-ACRIDINYLAMINO)PHENYL)-1-ETHANESULFONAMIDE see ADO500
N-(p-(ACRIDIN-9-YLAMINO)PHENYL)-ETHANESULFONAMIDE, HYDROCHLORIDE
    see ADO750
N-(p-(ACRIDIN-9-YLAMINO)PHENYLHEXANESULFONAMIDE) HYDROCHLORIDE)
    see ADP000
N-(p-(ACRIDIN-9-YLAMINO)PHENYL)METHANESULFONAMIDE HYDROCHLO-
    RIDE see ADP500

N-(p-(ACRIDIN-9-YLAMINO)PHENYL)PENTANESULFONAMIDE HYDROCHLORIDE see ADP750
N-(p-(9-ACRIDINYLAMINO)PHENYL)-1-PROPANESULFONAMIDE see ADQ000
N-(p-(ACRIDIN-9-YLAMINO)PHENYL)PROPANESULFONAMIDE HYDROCHLO-RIDE see ADQ250
N'-9-ACRIDINYL-N-(2-CHLOROETHYL)-N-ETHYL-1,3-PROPANEDIAMINE DIHY-DROCHLORIDE see CGX250
N'-9-ACRIDINYL-N,N-DIMETHYL-1,4-BENZENEDIAMINE see DOS800
ACRIFLAVIN see DBX400
ACRIFLAVINE see XAK000
ACRIFLAVINE NEUTRAL see XAK000
ACRIFLAVINE mixture with PROFLAVINE see DBX400
ACRIFLAVINIUM CHLORIDE see DBX400
ACRIFLAVINIUM CHLORIDUM see DBX400
ACRIFLAVON see DBX400, XAK000
ACRILAFIL see ADY500
ACRINAMINE see ARQ250
ACRINOL see EDW500
ACRIQUINE see ARQ250
ACRISIN FS 017 see UTU500
ACRITET see ADX500
ACROART see PJS750
ACROFOL see SFV250
ACROL see AHP000
ACROLACTINE see EDW500
ACROLEIC ACID see ADS750
ACROLEIN, 2-METHYL- see MGA250
ACROLEIN see ADR000
ACROLEINA (ITALIAN) see ADR000
ACROLEIN ACETAL see DHH800
ACROLEIN CYANOHYDRIN see HJQ000
ACROLEIN DIACETATE see ADR250
ACROLEIN DIMER, stabilized (DOT) see ADR500
ACROLEIN DIMER see ADR500
ACROLEINE (DUTCH, FRENCH) see ADR000
ACROMONA see MMN250
ACROMYCINE see ADR750
ACRONINE see ADR750
ACRONIZE see CMA750
ACRONOL PHLOXINE FF see CMM765
ACRONYCINE see ADR750
ACROPOR see ADY250
ACROPOR AN 200 see ADY250
ACROPOR AN 450 see ADY250
ACROPOR AN see ADY250
ACROPORE see ADY250
ACROSTICHUM AUREUM Linn., extract see ADS150
ACRYL, combustion products see ADX750
ACRYLALDEHYD (GERMAN) see ADR000
ACRYLALDEHYDE see ADR000
ACRYLALDEHYDE DIETHYL ACETAL see DHH800
ACRYLAMIDE see ADS250
ACRYLAMIDE, N-BUTOXYMETHYL- see BPM660
ACRYLAMIDE, N-tert-BUTYL- see BPW050
ACRYLAMIDE, N-(ISOBUTOXYMETHYL)- see IIE100
ACRYLAMIDE, N-(METHOXYMETHYL)- see MEX300
ACRYLAMIDE, N-METHYL- see MGA300
ACRYLATE d'ETHYLE (FRENCH) see EFT000
ACRYLATE de METHYLE (FRENCH) see MGA500
ACRYL BRILLIANT GREEN B see AFG500
ACRYLIC ACID, inhibited (DOT) see ADS750
ACRYLIC ACID see ADS750
ACRYLIC ACID, 3-p-ANISOYL-3-BROMO-, SODIUM SALT, (E)- see CQK600
ACRYLIC ACID n-BUTYL ESTER (MAK) see BPW100
ACRYLIC ACID BUTYL ESTER see BPW100
ACRYLIC ACID CHLORIDE see ADZ000
ACRYLIC ACID, 2-CHLORO-, ETHYL ESTER see EHH200
ACRYLIC ACID-β-CHLOROETHYL ESTER see ADT000
ACRYLIC ACID-2-CYANOETHYL ESTER see ADT111
ACRYLIC ACID, DIESTER with TETRAETHYLENE GLYCOL see ADT050
ACRYLIC ACID, DIESTER with TRIETHYLENE GLYCOL see TJQ100
ACRYLIC ACID, 2-(DIMETHYLAMINO)ETHYL ESTER see DPB300
ACRYLIC ACID ESTER with HYDRACRYLONITRILE see ADT111
ACRYLIC ACID, 2-ETHOXYETHANOL DIESTER see ADT250
ACRYLIC ACID-2-ETHOXYETHANOL ESTER see ADT500
ACRYLIC ACID, 2-ETHOXYETHYL ESTER see ADT500
ACRYLIC ACID, ETHYLENE ESTER see EIP000
ACRYLIC ACID, ETHYLENE GLYCOL DIESTER see EIP000
ACRYLIC ACID ETHYL ESTER see EFT000
ACRYLIC ACID-2-ETHYLHEXYL ESTER see ADU250
ACRYLIC ACID-2-(5'-ETHYL-2-PYRIDYL)ETHYL ESTER see ADU750
ACRYLIC ACID, GLACIAL see ADS750
ACRYLIC ACID, HEPTYL ESTER see HBL100
ACRYLIC ACID, HEXAMETHYLENE ESTER see HEQ100
ACRYLIC ACID HEXYL ESTER see ADV000

ACRYLIC ACID-2-HYDROXYETHYL ESTER see ADV250
ACRYLIC ACID-2-HYDROXYPROPYL ESTER see HNT600
ACRYLIC ACID ISOBUTYL ESTER see IIK000
ACRYLIC ACID, ISODECYL ESTER see IKL000
ACRYLIC ACID, 2-METHOXYETHYL ESTER see MEM250
ACRYLIC ACID-2-METHOXYETHYL ESTER see MIF750
ACRYLIC ACID, 2-METHYL- see MDN250
ACRYLIC ACID METHYL ESTER (MAK) see MGA500
ACRYLIC ACID, 2-METHYL-, METHYL ESTER see MLH750
ACRYLIC ACID-2-(METHYLTHIO)ETHYL ESTER see MPT250
ACRYLIC ACID METHYLTHIOETHYL ESTER see MPT250
ACRYLIC ACID-1-METHYLTRIMETHYLENE ESTER see BRG500
ACRYLIC ACID-5-NORBORNEN-2-METHYL ESTER see BFY250
ACRYLIC ACID-5-NORBORNEN-2-YLMETHYL ESTER see BFY250
ACRYLIC ACID, OXYBIS(ETHYLENEOXYETHYLENE) ESTER see ADT050
ACRYLIC ACID, OXYDIETHYLENE ESTER (8CI) see ADT250
ACRYLIC ACID, PENTAERITHRITOL TRIESTER see PBC750
ACRYLIC ACID, POLYMERS see ADW200
ACRYLIC ACID, POLYMER with SUCROSEPOLYALLYL ETHER see ADW300
ACRYLIC ACID, POLYMER, ZINC SALT see ADW100
ACRYLIC ACID, PROPYLENEBIS(OXYPROPYLENE) ESTER see TMZ100
ACRYLIC ACID RESIN see ADW200
ACRYLIC ACID, TELOMER with TRICHLOROACETIC ACID see ADW750
ACRYLIC ACID ((3a,4,7,7a-TETRAHYDRO)-4,7-METHANOINDENYL) ESTER see DGW400
ACRYLIC ACID, TETRAMETHYLENE ESTER (6CI,7CI,8CI) see TDQ100
ACRYLIC ACID TRIDECYL ESTER see ADX000
ACRYLIC ACID-N,N-DIETHYLAMINOETHYL ESTER see DHT125
ACRYLIC ALDEHYDE see ADR000
ACRYLIC AMIDE see ADS250
ACRYLIC POLYMER see ADW200
ACRYLIC RESIN see ADW200
ACRYLITE see PKB500
ACRYLNITRIL (GERMAN, DUTCH) see ADX500
ACRYLOAMIDE see ADX250
ACRYLON see ADX500
ACRYLONITRILE, inhibited (DOT) see ADX500
ACRYLONITRILE see ADX500
ACRYLONITRILE-BUTADIENE-STYRENE (pyrolysis products) see ADX750
ACRYLONITRILE MONOMER see ADX500
ACRYLONITRILE POLYMER with 1,3-BUTADIENE, and STYRENE, COMBUS-TION PRODUCTS see ADX750
ACRYLONITRILE POLYMER with CHLOROETHYLENE see ADY250
ACRYLONITRILE POLYMER with STYRENE see ADY500
ACRYLONITRILE-STYRENE COPOLYMER see ADY500
ACRYLONITRILE-STYRENE POLYMER see ADY500
ACRYLONITRILE-STYRENE RESIN see ADY500
ACRYLOPHENONE see PMQ250
ACRYLOPHENONE, 4'-METHOXY- see ONW100
2-ACRYLOXYETHYLDIMETHYLSULFONIUM METHYL SULFATE see ADY750
ACRYLOYL CHLORIDE see ADZ000
2-(ACRYLOYLOXY)ETHANOL see ADV250
ACRYLSAEUREAETHYLESTER (GERMAN) see EFT000
ACRYLSAEUREMETHYLESTER (GERMAN) see MGA500
ACRYLYL CHLORIDE see ADZ000
ACRYPET see PKB500
ACRYSOL A 1 see ADW200
ACRYSOL A 3 see ADW200
ACRYSOL A 5 see ADW200
ACRYSOL AC 5 see ADW200
ACRYSOL ASE-75 see ADW200
ACRYSOL WS-24 see ADW200
ACS see ADY500, AJD000
AC 1370 SODIUM see CCS525
ACT see AEB000
ACTAEA (VARIOUS SPECIES) see BAF325
ACTAMER see TFD250
ACTASAL see CMG000
ACTEDRON see BBK000
ACTELIC see DIN800
ACTELLIC see DIN800
ACTELLIFOG see DIN800
ACTEMIN see AOB500
ACTEROL see NHH000
ACTH see AES650
ACTHAR see AES650
ACTI-AID see CPE750
ACTICARBONE see CBT500
ACTICEL see SCH000
ACTI-CHLORE see CDP000
ACTIDILAT see TMX775
ACTIDIONE see CPE750
ACTIDIONE TGF see CPE750
ACTIDOL see TMX775
ACTIDONE see CPE750

ACTILIN see NCF000
ACTINE see EHP000
ACTINIC RADIATION see AEA000
ACTINOBOLIN see AEA109
ACTINOCHRYSIN see AEA750
ACTINOGAN see AEA250
ACTINOLITE ASBESTOS see ARM260
ACTINOMYCIN 23-21 see AEA625
ACTINOMYCIN see AEA500
ACTINOMYCIN 1048A see AEC000
ACTINOMYCIN BV see AEC200
ACTINOMYCIN C see AEA750
ACTINOMYCIN D see AEB000
ACTINOMYCINDIOIC D ACID, DILACTONE see AEB000
ACTINOMYCIN DV see AEC200
ACTINOMYCIN I see AEB000
ACTINOMYCIN J1 see AEC200
ACTINOMYCIN K see AEB500
ACTINOMYCIN 2104L see AEB750
ACTINOMYCIN L see AEB750
ACTINOMYCIN S3 see AEC200
ACTINOMYCIN S see AEC000
ACTINOMYCIN-V see AEC200
ACTINOMYCIN X2 see AEC200
ACTINOSPECTACIN DIHYDROCHLORIDE PENTAHYDRATE see SKY500
ACTINOXANTHIN see AEC250
ACTINOXANTHINE see AEC250
ACTIOL see MBX800
ACTISPRAY see CPE750
ACTITHIAZIC ACID see CCI500, CMP885
ACTIVATED ALUMINUM OXIDE see AHE250
ACTIVATED ATTAPULGITE see PAE750
ACTIVATED CARBON see CBT500, CDI000
ACTIVE ACETYL ACETATE see EFS000
ACTIVE DICUMYL PEROXIDE see DGR600
ACTIVIN see DYF450
ACTIVOL see ALT250
ACTOL see NDX500
ACTON see AES650
ACTONAR see AES650
ACTOR Q see DVR200
ACTOZINE see BCA000, PGA750
ACTRAPID see IDF300
AC-17 TRIHYDRATE see AER666
ACTRIL see HKB500
AC-TRY see ADE075
ACTYBARYTE see BAP000
ACTYLOL see LAJ000
ACULEACIN A see AEC625
ACUPAN see NBS500
ACYCLOGUANOSINE see AEC700
ACYCLOGUANOSINE SODIUM (OBS.) see AEC725
ACYCLOVIR see AEC700
ACYCLOVIR SODIUM SALT see AEC725
ACYLANID see ACH500
ACYLATE-1 see ING400
ACYLATE see ING400
ACYLPYRIN see ADA725
ACYTOL see LAJ000
5-ACZ see ARY000
AD 1 see AGX000
A-20D see GLU000
AD 32 see TJX350
AD 122 see OEM000
AD-205 see DSH000
AD-810 see BCE750
AD-1590 see DLI650
AD see AEB000
ADAB see DPO200
ADAKANE 12 see DXT200
ADALAT see AEC750
ADALIN see BNK000
ADAM AND EVE see ITD050
1-ADAMANTAMINE see TJG250
1-ADAMANTANAMINE see TJG250
1-ADAMANTANAMINE HYDROCHLORIDE see AED250
ADAMANTANAMINE HYDROCHLORIDE see AED250
1-ADAMANTANEACETIC ACID-2-(DIETHYLAMINO)ETHYL ESTER, ETHYL IO-
    DIDE see AED750
1-ADAMANTANEACETIC ACID-3-(DIMETHYLAMINO)PROPYL ESTER, ETHYL
    IODIDE see AEE000
ADAMANTINE HYDROCHLORIDE see AED250
N-(1-ADAMANTYL)ACETAMIDE see AEE100
1-ADAMANTYLAMINE HYDROCHLORIDE see AED250
ADAMANTYLAMINE HYDROCHLORIDE see AED250

5-(1-ADAMANTYL)-2,4-DIAMINO-6-METHYLPYRIMIDINE ETHYLSULFONATE see
    AEF250
N-1-ADAMANTYL-N-(2-(DIMETHYLAMINO)ETHOXY)ACETAMIDE HYDROCHLO-
    RIDE see AEF500
2,2′-(1,3-ADAMANTYLENE)N,N,N′,N′-TETRAMETHYL-ETHYLAMINE DIHYDRO-
    CHLORIDE see BJG750
N-(2-ADAMANTYL)-2-MERCAPTOACETAMIDINE HYDROCHLORIDE see
    AEG000
S-(N-(1-ADAMANTYLMETHYLAMIDINO)METHYL)PHOSPHOROTHIOATE MONO-
    SODIUM SALT see AEG129
N-(1-ADAMANTYLMETHYL)-2-MERCAPTOACETAMIDINE HYDROCHLORIDE see
    AEG250
(2-(2-ADAMANTYLOXY)ETHYL)TRIETHYL-AMMONIUMIODIDE see DHP000
(2-(1-ADAMANTYLOXY)PROPYL)DIMETHYLETHYLAMMONIUM IODIDE see
    DPU600
3-(1-ADAMANTYL)PROPYLAMINE HYDROCHLORIDE see AMC500
N-(3-(1-ADAMANTYL)PROPYL)-2-MERCAPTOACETAMIDINE HYDROCHLORIDE
    HYDRATE (10:10:3) see AEG500
ADAME see DPB300
ADAMSITE see PDB000
ADAMYCIN see HOH500
ADANON see MDO750
ADANON HYDROCHLORIDE see MDP000, MDP750
ADANTON HYDROCHLORIDE see AEG625
ADAPIN see AEG750
ADAPTOL see AEG875
ADC AURAMINE O see IBA000
ADC BRILLIANT GREEN CRYSTALS see BAY750
ADCHEM GMO see GGR200
ADC MALACHITE GREEN CRYSTALS see AFG500
ADC RHODAMINE B see FAG070
ADC TOLUIDINE RED B see MMP100
ADDEX-THAM see TEM500
ADDISOMNOL see BNK000
ADDITIN 30 see PFT250
ADDUKT HEXACHLORCYKLOPENTADIENU S CYKLOPENTADIENEM (CZECH)
    see HCN000
2-ADE see PDR490
ADELFAN see RDK000
ADELFA (PUERTO RICO) see OHM875
ADELPHANE see RDK000
ADELPHIN see RDK000
ADELPHIN-ESIDREX-K see RDK000
ADEMINE see UVJ450
ADEMOL see TKG750
ADENINE see AEH000
ADENINE ARABINOSIDE see AEH100, AQQ900
ADENINE ARABINOSIDE 5′-MONOPHOSPHATE see AQQ905
ADENINE ARABINOSIDE MONOPHOSPHATE see AQQ905
ADENINE, N-BENZYL- see BDX090
ADENINE-FLAVIN DINUCLEOTIDE see RIF100
ADENINE-FLAVINE DINUCLEOTIDE see RIF100
ADENINE, N-FURFURYL- see FPT250
ADENINE-NICOTINAMIDE DINUCLEOTIDE see CNF390
ADENINE-1-N-OXIDE see AEH250
ADENINE-RIBOFLAVIN DINUCLEOTIDE see RIF100
ADENINE-RIBOFLAVINE DINUCLEOTIDE see RIF100
ADENINE RIBOSIDE see AEH750
ADENINE SULFATE see AEH500
ADENINIMINE see AEH000
ADENINSULFAT see AEH500
ADENIUM (VARIOUS SPECIES) see DBA450
ADENOCK see ZVJ000
ADENOHYPOPHYSEAL GROWTH HORMONE see PJA250
ADENOSIN (GERMAN) see AEH750
β-d-ADENOSINE see AEH750
β-ADENOSINE see AEH750
ADENOSINE see AEH750
ADENOSINE-3′-(α-AMINO-p-METHOXYHYDROCINNAMAMIDO)-3′-DEOXY-N,N-DI-
    METHYL see AEI000
ADENOSINE-5′-CARBOXAMIDE see AEI250
ADENOSINE, 2-CHLORO- see CEF100
ADENOSINE-5′-(N-CYCLOBUTYL)CARBOXAMIDE see AEI500
ADENOSINE-5′-(N-CYCLOPENTYL)CARBOXAMIDE see AEI750
ADENOSINE-5′-(N-CYCLOPROPYL)CARBOXAMIDE see AEJ000
ADENOSINE-5′-(N-CYCLOPROPYL)CARBOXAMIDE-N′-OXIDE see AEJ250
ADENOSINE-5′-(N-CYCLOPROPYLMETHYL)CARBOXAMIDE see AEJ500
ADENOSINE 5′-(DIHYDROGEN PHOSPHATE), SODIUM SALT see AEM750
ADENOSINE-5′-(N-(2-(DIMETHYLAMINO)ETHYL))CARBOXAMIDE see AEJ750
ADENOSINE-5′-(N,N-DIMETHYL)CARBOXAMIDE HYDRATE see AEK000
ADENOSINE 5′-DIPHOSPHATE see AEK100
ADENOSINE DIPHOSPHATE see AEK100
ADENOSINE 5′-DIPHOSPHORIC ACID see AEK100
ADENOSINE DIPHOSPHORIC ACID see AEK100
ADENOSINE-5′-(N-ETHYL)CARBOXAMIDE HEMIHYDRATE see AEK250

ADENOSINE-5′-(N-ETHYL)CARBOXAMIDE-N′-OXIDE see AEK500
ADENOSINE-5′-(N-HEXYL)CARBOXAMIDE HEMIHYDRATE see AEK750
ADENOSINE-5′-(N-(2-HYDROXYETHYL))CARBOXAMIDE see AEL000
ADENOSINE-5′-(N-ISOPROPYL)CARBOXAMIDE see AEL250
ADENOSINE-5′-(N-METHOXY)CARBOXAMIDE HYDRATE see AEL500
ADENOSINE-5′-(N-METHYL)CARBOXAMIDE HEMIHYDRATE see AEL750
ADENOSINE, N-METHYL-, mixed with SODIUM NITRITE (1:4) see SIS650
ADENOSINE-5′-MONOPHOSPHATE see AOA125
ADENOSINE-5′-MONOPHOSPHATE POTASSIUM SALT see AEM500
ADENOSINE 5′-MONOPHOSPHATE SODIUM SALT see AEM750
ADENOSINE-5′-MONOPHOSPHORIC ACID see AOA125
ADENOSINE-5-MONOPHOSPHORIC ACID see AOA125
ADENOSINE-5′-MONOPHOSPHORIC ACID POTASSIUM SALT see AEM500
ADENOSINE-5′-PHOSPHATE see AOA125
ADENOSINE PHOSPHATE see AOA125
ADENOSINE-5′-PHOSPHATE POTASSIUM SALT see AEM500
ADENOSINE-5′-PHOSPHORIC ACID see AOA125
ADENOSINE-5′-PHOSPHORIC ACID POTASSIUM SALT see AEM500
ADENOSINE-5′-(N-PROPYL)CARBOXAMIDE see AEM000
ADENOSINE 5′-PYROPHOSPHATE see AEK100
ADENOSINE PYROPHOSPHATE see AEK100
ADENOSINE 5′-PYROPHOSPHORIC ACID see AEK100
ADENOSINE 5′-(TETRAHYDROGENTRIPHOSPHATE), DISODIUM SALT see AEM100
ADENOSINE-5′-(TETRAHYDROGENTRIPHOSPHATE) SODIUM SALT see AEM250
ADENOSINE, 5′-(TRIHYDROGEN DIPHOSPHATE) (9CI) see AEK100
ADENOSINE, 5′-(TRIHYDROGEN PYROPHOSPHATE) see AEK100
ADENOSINE 5′-(TRIHYDROGEN PYROPHOSPHATE), 5′-5′-ESTER with RIBO-FLAVINE see RIF100
ADENOSINE-5′-TRIPHOSPHATE see ARQ500
ADENOSINE TRIPHOSPHATE see ARQ500
ADENOSINE TRIPHOSPHATE DISODIUM see AEM100
ADENOSINE-5′-TRIPHOSPHORIC ACID see ARQ500
ADENOVITE see AOA125
ADENYL see AOA125
ADENYLIC ACID see AOA125
tert-ADENYLIC ACID see AOA125
5′-ADENYLIC ACID POTASSIUM SALT see AEM500
5′-ADENYLIC ACID, SODIUM SALT see AEM750
5′-ADENYLPHOSPHORIC ACID see AEK100
ADENYLPYROPHOSPHORIC ACID see ARQ500
ADEPHOS see ARQ500
ADEPSINE OIL see MQV750
ADERGON see AOR500
ADERMINE see PPK250
ADERMINE HYDROCHLORIDE see PPK500
ADETOL see ARQ500
ADETPHOS see AEM100
ADHERE see MIQ075
ADIABEN see CKK000
ADIAZINE see PPP500
ADILACTETTEN see AEN250
ADINOL T see SIY000
ADIPAMIDE see AEN000
ADIPAN see AOB250, BBK000
ADIPAN HEXAMETHYLENDIAMINU see HEG120
ADIPARTHROL see AOB250
ADIPEX see MDQ500, MDT600
ADIPHENIN see DHX800, THK000
ADIPHENINE see DHX800
ADIPIC ACID see AEN250
ADIPIC ACID, BIS(2-(2-BUTOXYETHOXY)ETHYL) ESTER see DDT500
ADIPIC ACID BIS(3,4-EPOXY-6-METHYLCYCLOHEXYLMETHYL) ESTER see AEN750
ADIPIC ACID, BIS(2-ETHOXYETHYL) ESTER see BJO225
ADIPIC ACID BIS(2-ETHYLHEXYL) ESTER see AEO000
ADIPIC ACID-3-CYCLOHEXENYLMETHANOL DIESTER see AEO250
ADIPIC ACID DIALLYL ESTER see AEO500
ADIPIC ACID DIAMIDE see AEN000
ADIPIC ACID DIBUTYL ESTER see AEO750
ADIPIC ACID DI-(3-CARBOXY-2,4,6-TRIIODOANILIDE) DISODIUM see BGB325
ADIPIC ACID, DIESTER with 6-METHYL-3-CYCLOHEXENE-1-METHANOL see BKR100
ADIPIC ACID-(DI-2-(2-ETHYLBUTOXY)ETHYL) ESTER see AEP250
ADIPIC ACID DI(2-ETHYLBUTYL) ESTER see AEP500
ADIPIC ACID DIETHYL ESTER see AEP750
ADIPIC ACID, DI(2-HEXYLOXYETHYL) ESTER see AEQ000
ADIPIC ACID DIHYDRAZIDE see AEQ250
ADIPIC ACID DIISOPENTYL ESTER see AEQ500
ADIPIC ACID DIISOPROPYL ESTER see DNL800
ADIPIC ACID DINITRILE see AER250
ADIPIC ACID DI-2-PROPYNYL ESTER see AEQ750
ADIPIC ACID, compd. with 1,6-HEXANEDIAMINE (1:1) see HEG120

ADIPIC ACID, 6-METHYL-3-CYCLOHEXENYL-METHANOL DIESTER see BKR100
ADIPIC ACID, METHYL VINYL ESTER see MQL000
ADIPIC ACID, MONOETHYL ESTER see ELC600
ADIPIC ACID NITRILE see AER250
ADIPIC ACID, POLYMER with 1,4-BUTANEDIOL and METHYLENEDI-p-PHE-NYLENE ISOCYANATE see PKM250
ADIPIC ACID, POLYMER with 1,4-BUTANEDIOL, METHYLENEDI-p-PHENYLENE ISOCYANATE and 2,2′-(p-PHENYLENEDIOXY)DIETHANOL see PKM500
ADIPIC ACID, POLYMER with ETHYLENE GLYCOL and METHYLENEDI-p-PHENYLENE ISOCYANATE see PKL750
ADIPIC ACID, UREA mixed with CARBOXYMETHYLCELLULOSE ACIDS see AER000
ADIPIC DIAMIDE see AEN000
ADIPIC DIHYDRAZIDE see AEQ250
ADIPIC KETONE see CPW500
ADIPINIC ACID see AEN250
ADIPINSAEURE-DI-(3-CARBOXY-2,4,6-TRIJOD-ANILID) DINATRIUM (GERMAN) see BGB325
ADIPIODONE MEGLUMINE see BGB315
ADIPLON see PPO000
ADIPODINITRILE see AER250
ADIPOL 2EH see AEO000
5,5′-(ADIPOLYDIMINO)BIS(2,4,6-TRIIODO-N-METHYLISOPHTHALAMIC ACID) see IDJ500
ADIPONITRILE see AER250
ADIPOSETTIN see NNW500
ADIPRAZINE see HEP000
ADITYL see ACE000
ADJUDETS see BBK500
ADK (CZECH) see AGL500
AD1M see AGX000
ADM see AES750
ADMA 2 see DRR800
ADMER PB 02 see PMP500
ADMEX 741 see FAB920
ADMEX 746 see FAB920
ADM HYDROCHLORIDE see HKA300
ADMUL see OAV000
ADNEPHRINE see VGP000
2-ADO see DDB600
ADO see AGX000
ADOBACILLIN see AIV500
ADOBIOL see AER500
ADOGEN 442 see QAT550
ADOGEN 448 see QAT550
ADOGENEN 142 see OBC000
ADOGEN 442-100 P see QAT550
ADOL 34 see OBA000
ADOL 68 see OAX000
ADOL 80 see OBA000
ADOL 85 see OBA000
ADOL 90 see OBA000
ADOL 320 see OBA000
ADOL 330 see OBA000
ADOL 340 see OBA000
ADOL see HCP000, OAX000, OBA000
ADONAL see EOK000
ADONA TRIHYDRATE see AER666
ADONIDIN see AER750
ADONIS (VARIOUS SPECIES) see PCU375
ADONITOL see RIF000
ADOPON see NOC000
ADORM see TDA500
5′-ADP see AEK100
ADP see AEK100
ADPHEN see DKE800
ADP (NUCLEOTIDE) see AEK100
ADR see HKA300
ADRAN see IIU000
ADRAXONE see AES639
ADRENAL see VGP000
ADRENAL CORTEX HORMONE see AES650
ADRENALEX see CNS800
1-ADRENALIN see VGP000
ADRENALIN BITARTRATE see AES000
ADRENALIN CHLORIDE see AES500
dl-ADRENALINE see EBB500
d-ADRENALINE see AES250
l-(+)-ADRENALINE see AES250
ADRENALINE ACID TARTRATE see AES000
(−)-ADRENALINE ACID TARTRATE see AES000
1-ADRENALINE BITARTRATE see AES000
1-ADRENALINE-d-BITARTRATE see AES000
ADRENALINE BITARTRATE see AES000

(— )-ADRENALINE BITARTRATE see AES000
1-ADRENALINE CHLORIDE see AES500
1-ADRENALINE HYDROCHLORIDE see AES500
dl-ADRENALINE HYDROCHLORIDE see AES625
(±)-ADRENALINE HYDROCHLORIDE see AES625
(— )-ADRENALINE HYDROCHLORIDE see AES500
1-ADRENALINE HYDROGEN TARTRATE see AES000
ADRENALINE HYDROGEN TARTRATE see AES000
(— )-ADRENALINE HYDROGEN TARTRATE see AES000
l-ADRENALINE TARTRATE see AES000
ADRENALINE TARTRATE see AES000
(— )-ADRENALINE TARTRATE see AES000
ADRENALIN HYDROCHLORIDE see AES500
ADRENALIN-MEDIHALER see VGP000
ADRENALONE see MGC350
ADRENAMINE see VGP000
ADRENAN see VGP000
ADRENAPAX see VGP000
ADRENASOL see VGP000
ADRENATRATE see VGP000
ADRENOCHROME see AES639
ADRENOCHROME SULFONATE AC 17 TRIHYDRATE see AER666
ADRENOCORTICOTROPHIC HORMONE see AES650
ADRENOCORTICOTROPHIN see AES650
ADRENOCORTICOTROPIC HORMONE see AES650
ADRENOCORTICOTROPIN see AES650
ADRENODIS see VGP000
ADRENOHORMA see VGP000
ADRENOMONE see AES650
ADRENON see MGC350
ADRENONE see MGC350
ADRENOR see NNO500
ADRENOTROPHIN see AES650
ADRENUTOL see VGP000
ADRESON see CNS800, CNS825
ADREVIL see PEU000
ADRIACIN see HKA300
ADRIAMICINA see MDO250
ADRIAMYCIN, HYDROCHLORIDE see HKA300
ADRIAMYCIN-HCl see AES750
ADRIAMYCIN see AES750, HKA300
ADRIAMYCIN-14-OCTANOATEHYDROCHLORIDE see AET250
ADRIAMYCIN SEMIQUINONE see AES750
ADRIANOL see SPC500
ADRIBLASTIN see HKA300
ADRIBLASTINA see AES750
ADRIBLASTINE see HKA300
ADRINE see VGP000
ADRIXINE see BBK500
ADROIDIN see PAN100
ADRONAL see CPB750
ADROYD see PAN100
ADRUCIL see FMM000
ADSORBONAC see CAR780
ADULSIN see MIF760
ADUMBRAN see CFZ000
ADUVEX 248 see HND100
ADVASTAB 46 see HND100
ADVASTAB 401 see BFW750
ADVASTAB 405 see MJO500
ADVASTAB 800 see TFD500
ADVASTAB 802 see DXG700
ADVASTAB DBTM see DEJ100
ADVASTAB 17 MO see BKK750
ADVASTAB PS 802 see DXG700
ADVASTAB T290 see DEJ100
ADVASTAB T340 see DEJ100
ADVAWAX 140 see OAV000
ADVENTAN see FQJ100
ADYNOL see ARQ500
AE see AGX000
AED see CQJ750
AENH (GERMAN) see ENV000
A 38414 (ENZYME) see VGU700
AEORLIN see BQF500
AERBRON see POF500
AERO see MCB000
AERO-CYANAMID see CAQ250
AERO CYANAMID GRANULAR see CAQ250
AERO CYANAMID SPECIAL GRADE see CAQ250
AERO CYANATE see PLC250
AERO liquid HCN see HHS000
AEROL 1 (pesticide) see TIQ250
AEROLITE 300 see UTU500
AEROLITE A 300 see UTU500

AEROLITE FFD see UTU500
AEROLITE MF 15 see MCB050
AEROMATT see CAT775
AEROMONAS HYDROPHILA A₃ ENDOTOXIN see AET500
AEROSEB-DEX see SOW000
AEROSEB-HC see CNS750
AEROSIL see SCH000
AEROSOL GPG see DJL000
AEROSOL MA80 see DKP800
AEROSOL of THERMOVACUUM CADMIUM see CAK000
AEROSPORIN see PKC500
AEROTEX 92 see MCB050
AEROTEX 3700 see MCB050
AEROTEX GLYOXAL 40 see GIK000
AEROTEX M 3 see MCB050
AEROTEX MW see MCB050
AEROTEX RESIN MW see MCB050
AEROTEX UM see MCB050
AEROTHENE MM see MJP450
AEROTHENE TT see MIH275
AEROXANTHATE 343 see SIA000
AEROXANTHATE 350 see PKV100
AEROXANTHATE see PKV100
α-AESCIN see EDL000
β-AESCIN see EDL500
AESCIN (GERMAN) see EDK875
AESCIN SODIUM SALT see EDM000
AESCIN TRIETHANOLAMINE SALT see EDM500
AESCULETIN DIMETHYL ETHER see DRS800
AESCULUS (VARIOUS SPECIES) see HGL575
α-AESCUSAN see EDL000
β-AESCUSAN see EDL500
AESCUSAN see EDK875
AESCUSAN SODIUM SALT see EDM000
AESTOCIN see DPE200
AET see AJY250
AET-2HBR see AJY250
AET BROMIDE see AJY250
AET DICHLORIDE see AJY500
AET DIHYDROBROMIDE see AJY250
AETHALDIAMIN (GERMAN) see EEA500
AETHANETHIOL (GERMAN) see EMB100
AETHANOL (GERMAN) see EFU000
AETHANOLAMIN (GERMAN) see EEC600
AETHER see EJU000
2-AETHINYLBUTANOL see EQL000
AETHIONIN see EEI000
AETHISTERON see GEK500
AETHOBROMID DES α,α-DICYCLOPENTYLESSIGSAEURE-β′-DIAETHYLAMINO
    AETHYLESTER (GERMAN) see DGW600
AETHON see ENY500
AETHOPHYLLINUM see HLC000
AETHOPROPAZIN see DIR000
AETHOSUXIMIDE (GERMAN) see ENG500
AETHOXEN see VMA000
2-AETHOXY-AETHYLACETAT (GERMAN) see EES400
1-AETHOXY-4-(1-CAETO-2-HYDROXYAETHYL)-NAPHTALAENE SUCCINATE see
    SNB500
3-(AETHOXYCARBONYLAMINOPHENYL)-N-PHENYL-CARBAMAT (GERMAN)
    see EEO500
S-AETHOXY-CARBONYLTHIAMIN HYDROCHLORID (GERMAN) see EEQ500
2-AETHOXY-6,9-DIAMINOACRIDINLACTAT (GERMAN) see EDW500
1-p-AETHOXYPHENYL-3-DIAETHYLAMINO-INDAN CITRAT (GERMAN) see
    DHR800
p-AETHOXYPHYLHARNSTOFF (GERMAN) see EFE000
AETHOXYSILATRAN see EFJ600
5-AETHOXY-3-TRICHLORMETHYL-1,2,4-THIADIAZOL (GERMAN) see EFK000
AETHUSA CYNAPIUM see FMU200
AETHYLACETAT (GERMAN) see EFR000
AETHYL ACETOXYMETHYLNITROSAMIN (GERMAN) see ENR500
AETHYLACRYLAT (GERMAN) see EFT000
AETHYL-AETHANOL-NITROSOAMIN (GERMAN) see ELG500
AETHYLALKOHOL (GERMAN) see EFU000
AETHYLAMINE (GERMAN) see EFU400
2-AETHYLAMINO-4-sek.BUTYLAMINO-6-CHLOR-1,3,5-TRIAZIN see THR600
4-AETHYLAMINO-2-tert-BUTYLAMINO-6-METHYLTHIO-s-TRIAZIN (GERMAN) see
    BQC750
2-AETHYLAMINO-5-BUTYL-4-YL-DIMETHYLSULFAMAT (GERMAN) see BRJ000
2-AETHYLAMINO-4-CHLOR-6-ISOPROPYLAMINO-1,3,5-TRIAZIN (GERMAN) see
    ARQ725
2-AETHYLAMINO-6-CHLOR-4-METHYL-4-PHENYL-4H-3,1-BENZOXAZIN (GER-
    MAN) see CGQ500
2-AETHYLAMINO-4-ISOPROPYLAMINO-6-CHLOR-1,3,5-TRIAZIN (GERMAN) see
    ARQ725
2-AETHYLAMINO-3-PHENYL-NOR-CAMPHAN (GERMAN) see EOM000

AETHYLANILIN (GERMAN) see EGK000
AETHYLBENZOL (GERMAN) see EGP500
AETHYLBUTYLKETON (GERMAN) see EHA600
AETHYL-N-BUTYL-NITROSOAMIN (GERMAN) see EHC000
AETHYL-tert-BUTYL-NITROSOAMIN (GERMAN) see NKD500
AETHYLCARBAMAT (GERMAN) see UVA000
AETHYLCHLORID (GERMAN) see EHH000
AETHYL-CHLORVYNOL see CHG000
1-AETHYL-CYCLOHEXANOL-(1) (GERMAN) see EHR500
AETHYL-2-(3',5'-DIJOD-4'-OXYBENZOYL)-3 CUMARON see EID200
O-AETHYL-S,S-DIPHENYL-DITHIOPHOSPHAT (GERMAN) see EIM000
S-AETHYL-N,N-DIPROPYLTHIOLCARBAMAT (GERMAN) see EIN500
AETHYLEN-BIS-THIURAMMONOSULFID (GERMAN) see ISK000
AETHYLENBROMID (GERMAN) see EIY500
AETHYLENCHLORID (GERMAN) see EIY600
AETHYLENECHLORHYDRIN (GERMAN) see EIU800
AETHYLENEDIAMIN (GERMAN) see EEA500
AETHYLENGLYKOLAETHERACETAT (GERMAN) see EES400
AETHYLENGLYKOLMETHYLAETHERACETAT (GERMAN) see EJJ500
AETHYLENGLYKOL-MONOMETHYLAETHER (GERMAN) see EJH500
AETHYLENIMIN (GERMAN) see EJM900
AETHYLENIMINO-2-OXYBUTEN (GERMAN) see VMA000
AETHYLENOXID (GERMAN) see EJN500
AETHYLENSULFID (GERMAN) see EJP500
N-AETHYLFORMAMID see EKK600
AETHYLFORMIAT (GERMAN) see EKL000
AETHYLHARNSTOFF und NATRIUMNITRIT (GERMAN) see EQE000
AETHYLHARNSTOFF und NITRIT (GERMAN) see EQE000
S-AETHYL-N-HEXAHYDRO-1H-AZEPINTHIOLCARBAMAT (GERMAN) see
    EKO500
1-AETHYLHEXANOL (GERMAN) see EKQ000
AETHYLIDENCHLORID (GERMAN) see DFF809
AETHYLIS see EHH000
AETHYLIS CHLORIDUM see EHH000
2-(O-AETHYL-N-ISOPROPYLAMINDOTHIOPHOSPHORYLOXY)-BENZOSAEURE-
    ISOPROPYLESTER (GERMAN) see IMF300
AETHYL-ISOPROPYL-NITROSOAMIN (GERMAN) see ELX500
AETHYLMERCAPTAN (GERMAN) see EMB100
AETHYLMETHYLKETON (GERMAN) see MKA400
2-AETHYL-6-METHYL-N-(1-METHYL-2-METHOXYAETHYL)-CHLORACETANILID
    (GERMAN) see MQQ450
O-AETHYL-O-(3-METHYL-4-METHYLTHIOPHENYL)-ISOPROPYLAMIDO-PHOS-
    PHORSAEURE ESTER (GERMAN) see FAK000
3-β-AETHYL-1-METHYL-4-PHENYL-4-α-PIPERIDYLPROPIONAT HYDROCHLORID
    (GERMAN) see NOE550
3-β-AETHYL-1-METHYL-4-PHENYL-4-α-PROPIONYLOXYPIPERIDIN HYDROCHLO-
    RID (GERMAN) see NOE550
O-AETHYL-O-n(4-NITROPHENYL)-PHENYL-MONOTHIOPHOSPHONAT (GER-
    MAN) see EBD700
AETHYLNITROSO-HARNSTOFF (GERMAN) see ENV000
AETHYLNITROSOURETHAN (GERMAN) see NKE500
N-AETHYL-N'-NITRO-N-NITROSOGUANIDIN (GERMAN) see ENU000
Z-(3-AETHYL-4-OXO-5-PIPERIDINO-THIAZOLIDIN-2-YLIDEN)-ESSIGSAURE-AETHY-
    LESTER (GERMAN) see EOA500
3-AETHYL-PENTANOL-(3) (GERMAN) see TJP550
5-AETHYL-5-PENTYL-(2')-BARBITURSAEURE (GERMAN) see PBS250
S-AETHYL-N-PHENYL-DITHIOCARBAMAT see EOK550
5-AETHYL-5-PHENYL-HEXAHYDROPYRIMIDIN-4,6-DION (GERMAN) see
    DBB200
4-AETHYL-1-PHOSPHA-2,6,7-TRIOXABICYCLO(2.2.2)OCTAN (GERMAN) see
    TNI750
4-AETHYL-1-PHOSPHA-2,6,7-TRIOXABICYCLO(2.2.2)OCTAN-1-OXID (GERMAN)
    see ELJ500
AETHYL-4-PICOLYLNITROSAMIN (GERMAN) see NLH000
N-AETHYLPIPERIDIN (GERMAN) see EOS500
N-(1-AETHYLPROPYL)-3,4-DIMETHYL-2,6-DINITROANILIN (GERMAN) see
    DRN200
N-(1-AETHYLPROPYL)-2,6-DINITRO-3,4-XYLIDIN (GERMAN) see DRN200
AETHYLPROPYLKETON (GERMAN) see HEV500
AETHYLRHODANID (GERMAN) see EPP000
O-AETHYL-S-(2-DIMETHYLAMINOAETHYL)-METHYLPHOSPHONOTHIOATE
    (GERMAN) see EIF500
AETHYLSENFOEL (GERMAN) see ISK000
1-N-AETHYLSISOMICIN see SBD000
O-AETHYL-S-PHENYL-AETHYL-DITHIOPHOSPHONAT (GERMAN) see FMU045
S-2-AETHYLSULFINYL-1-METHYL AETHYL-O,O DIMETHYL-MONOTHIOPHOSP-
    HAT see DSK600
O-AETHYL-O-(2,4,5-TRICHLORPHENYL)-AETHYLTHIONOPHOSPHONAT (GER-
    MAN) see EPY000
AETHYLTRICHLORPHON (GERMAN) see EPY600
N-AETHYL-N-(2,4,6-TRIJOD-3-AMINOPHENYL)-SUCCINAMIDSAEURE (GERMAN)
    see AMV375
AETHYLURETHAN (GERMAN) see UVA000
AETHYL-VINYL-NITROSOAMIN (GERMAN) see NKF000
AETHYLZINNTRICHLORID (GERMAN) see EPS000

AETINA see EPQ000
AETIVA see EPQ000
AETM (GERMAN) see ISK000
AETT see ACL750
AF 10 see FMW330
AF 101 see DXQ500
AF 260 see AHC000
AF 425 see CKI000
AF 594 see CPN750
AF 864 see BBW500
AF 983 see BAV325
AF 1161 see CKJ000, THK880
AF 1890 see DEL200
AF 2259 see IJJ000
AF-2 (preservative) see FQN000
AF see XAK000
AFASTOGEN BLUE 5040 see DNE400
AFATIN see BBK500
AFAXIN see VSK600
AFBI see AEU250
A.F. BLUE No. 1 see FMU059
A.F. BLUE No. 2 see FAE100
AFCOLAC B 101 see SMR000
AFCOLENE 666 see SMQ500
AFCOLENE see SMQ500
AFCOLENE S 100 see SMQ500
AFESIN see CKD500
A.F. GREEN NO. 1 see FAE950
A.F. GREEN No. 2 see FAF000
AF 438 HYDROCHLORIDE see OOE100
AFIBRIN see AJD000
AFICIDA see DOX600
AFICIDE see BBQ500
A-FIL CREAM see TGG760
AFILINE see MPU250
AFI-PHYLLIN see DNC000
AFI-TIAZIN see PDP250
AFKO-HIST see WAK000
AFKO-SAL see SAH000
AFL 1081 see FFF000
AFL 1082 see FFH000
AFL see AEW500
AFLATOXICOL see AEW500
AFLATOXICOL NATURAL EPIMER see AEW500
AFLATOXIN see AET750
AFLATOXIN B1 see AEU250
AFLATOXIN B2 see AEU750
AFLATOXIN B see AEU250
AFLATOXIN B1-2,3-DICHLORIDE see AEU500
AFLATOXIN B1 DICHLORIDE see AEU500
AFLATOXIN G1 see AEV000
AFLATOXIN G2 see AEV500
AFLATOXIN G1 mixed with AFLATOXIN B1 see AEV250
AFLATOXIN M1 see AEW000
AFLATOXIN Ro see AEW500
A. FLAVA see HGL575
AFLIX see DRR200
AFLON see TAI250
AFLOQUALONE see AEW625
A. FLOS-AQUAE TOXIN see AON825
AFLOXAN see POF550
AFLUON see MAF500
AFNOR see CJJ000
A.F. ORANGE No. 1 see FAG010
A.F.ORANGE No. 2 see TGW000
A. FORDII see TOA275
AFOS see DJI000
AFRAZINE see AEX000
A.F. RED No. 1 see FAG018
A.F. RED No. 5 see XRA000
AFRICAN COFFEE TREE see CCP000
AFRICAN LILAC TREE see CDM325
AFRIDOL BLUE see AEW750
AFRIN see AEX000
AFRIN HYDROCHLORIDE see AEX000
AFTATE see TGB475
AF 1312/TS see CEQ625
A.F. VIOLET No 1 see FAG120
A.F YELLOW No. 2 see FAG130
A.F. YELLOW No. 3 see FAG135
A.F. YELLOW NO. 4 see FAG140
A.F. YELLOW NO. 5 see FAG150
AG 3 see CBR500, CBT500
AG 5 see CBT500
8 AG see AJO500

AG-629 see SLE880
AG. 5895 see VGA100
AG 58107 see TAI600
AG 3 (ADSORBENT) see CBT500
AG 5 (ADSORBENT) see CBT500
AGALITE see TAB750
AGALLO FORTE see MEP250
AGALLOL see MEP250
AGALLOLAT see MEP250
AGALOL see MEP250
AGAR see AEX250
AGAR-AGAR see AEX250
AGAR AGAR FLAKE see AEX250
AGAR-AGAR GUM see AEX250
AGARIN see AKT750
AGASTEN see FOS100
AGATE see SCI500, SCJ500
AGC see GFA000
AGE see AGH150
AGEDAL see DPH600
AGEFLEX AMA see AGK500
AGEFLEX BGE see BRK750
AGEFLEX CGE see GGS000
AGEFLEX EGDM see BKM250
AGEFLEX FA-2 see DHT125
AGEFLEX FA-10 see IKL000
AGEFLEX FM-1 see DPG600
AGEFLEX FM-4 see BQD250
AGEFLEX FM 246 see DXY200
AGEFLEX n-HA see ADV000
AGEFLOC WT 20 see DTS500
AGELFLEX FM-10 see IKM000
AGENAP see NAR000
AGENE see NGQ500
AGENT 504 see DAI600
AGENT AT 717 see PKQ250
AGENT BLUE see HKC000
AGENT ORANGE see AEX750
AGERATOCHROMENE see AEX850
AGERITE 150 see HKF000
AGERITE see AEY000, BLE500
AGERITE ALBA see AEY000
AGERITEDPPD see BLE500
AGERITE ISO see HKF000
AGERITE MA see PJQ750
AGERITE POWDER see PFT500
AGERITE RESIN D see TLP500
AGERITE WHITE see NBL000
AGEROPLAS see BKB250
AGGLUTININ see AAD000
AGIDOL 3 see DEA100, FAB000
AGIDOL 7 see MJN250
AGIDOL see BFW750
AGILENE see DRK600, PJS750
AGIOLAN see VSK600
AGI TALC, BC 1615 see TAB750
AGKISTRODON ACUTUS VENOM see HGM600
AGKISTRODON CONTORTRIX CONTROTRIX VENOM see SKW775
AGKISTRODON CONTORTRIX MOKASEN VENOM see NNX700
AGKISTRODON CONTORTRIX VENOM see AEY125
AGKISTRODON PISCIVORUS PISCIVORUS VENOM see EAB200
AGKISTRODON PISCIVORUS VENOM see AEY130
AGKISTRODON RHODOSTOMA VENOM see AEY135
A. GLABRA see HGL575
AGLICID see BSQ000
AGLUMIN see DIS600
AGM-9 see TJN000
AGOFOLLIN see EDR000
AGORAL see PDO750
AGOSTILBEN see DKA600
AGOTAN see PGG000
AGOVIRIN see TBG000
AGP see AKC625
AGRAZINE see PDP250
AGREFLAN see DUV600
AGRIA 1050 see DSQ000
AGRIBON see SNN300
AGRICIDE MAGGOT KILLER (F) see CDV100
AGRICULTURAL LIMESTONE see CAO000
AGRIDIP see CNU750
AGRIFLAN 24 see DUV600
AGRIMYCIN 17 see SLW500
AGRI-MYCIN see SLY500
AGRION see SGH500
AGRISIL see EPY000

AGRISOL G-20 see BBQ500
AGRISTREP see SLY500
AGRISYNTH B2D see BOX300
AGRITAN see DAD200
AGRITOL see BAC040
AGRITOX see CIR250, EPY000
AGRIYA 1050 see DSQ000
AGRIZAN see CNK559
A-GRO see MNH000
AGROCERES see HAR000
AGROCIDE see BBQ500
AGROCLAVINE see AEY375
AGROFORM see UTU500
AGROFOROTOX see TIQ250
AGROMICINA see TBX000
AGRONAA see NAK500
AGRONEXIT see BBQ500
AGROSAN see ABU500
AGROSAND see ABU500
AGROSAN GN 5 see ABU500
AGROSOL see MLF250
AGROSOL S see CBG000
AGROTECT see DAA800
AGROTHION see DSQ000
AGROXONE 3 see SIL500
AGROXONE see CIR250
AGROX 2-WAY and 3-WAY see CBG000
AGRYPNAL see EOK000
AGSTONE see CAO000
AGUATHOL see DXD000
A″2-GUTTIFERIN see GME300
AH-42 see TEO250
AH 289 see CDR000
AH 501 see PAJ000
AH 1932 see PFY105
AH 3232 see CDQ250
AH 3365 see BQF500
AH 19065 see RBF400
AH see DBM800
AH 5158A see HMM500
AHA see ABB250
MAKAHALA (HAWAII) see DAC500
AHCOCID CARMINE 2G see CMM300
AHCOCID FAST SCARLET R see FMU070
AHCO DIRECT BLACK GX see AQP000
AHCOQUINONE BLUE IR BASE see HOK000
AHCOQUINONE RED S see SEH475
AHCOQUINONE SKY BLUE B see CMM090
AHCOVAT BLUE BCF see DFN425
AHCOVAT BRILLIANT VIOLET 2R see DFN450
AHCOVAT BRILLIANT VIOLET 4R see DFN450
AHCOVAT BROWN BR see CMU770
AHCOVAT NAVY BLUE BR see VGP100
AHCOVAT OLIVE ARN see DUP100
AHCOVAT OLIVE R see DUP100
AHCOVAT PRINTING GOLDEN YELLOW see DCZ000
AHCOVAT PRINTING NAVY BLUE XSA see VGP100
AHCOVAT PRINTING ORANGE R see CMU815
AHCOVAT RUBINE R see CMU825
AHCTL see TCZ000
AHE POI (HAWAII) see EAI600
AH-289 HYDROCHLORIDE see CDR250
A. HIPPOCASTANUM see HGL575
AHOUAI des ANTILLES see YAK350
AHR 85 see GKK000
AHR 233 see MFD500
AHR 376 see AEY400
AHR-619 see ENL100, SLU000
AHR-1680 see DMB000
AHR-1767 see ENC600
AHR-3053 see CCH125
AHR 3219 see EQN600
AHR 5850D MONOHYDRATE see AHK625
AHYDOL (RUSSIAN) see TMJ000
A 446 HYDROCHLORIDE see LFW300
A 66 HYDROCHLORIDE see MNV750
A-HYDROCORT see HHR000
AHYGROSCOPIN-B see VGZ000
AHYPNON see MKA250
AI3-09071 see NAQ000
AI3-14631 see GKW100
AI 318284 see MRQ750
AI3-18285 see CBF725
AI3-18581 see TNI300
AI 3-22542 see DKC800

AI3-25449 see MFL750
AI 3-29024 see AFR750
AI3-29128 see PHK000
AI3-29158 see AHJ750
AI 3-29183 see DEG150
AI3-32960 see HCS600
AI3-34872 see CNG760
AI3-34886 see ECB200
AI3-35349 see TJF400
AI3-35937 see HCP050
AI3-35966 see ISZ000
AI3-36401 see HJG100
AI3-36420 see ODO000
AI3-36537 see CPI350
AI3-36543 see MLM500
AI3-36558 see MLM600
AI3-36559 see MLM700
AI3-36561 see DSP650
AI3-36563 see TCV400
AI3-36564 see MLL650
AI3-36565 see MLL655
AI3-36566 see MLL660
AI3-36570 see TCV375
AI3-37220 see CPD625
AI3-50172 see HEF500
AI 3-51254 see MGE100
AI3-70087 see HKQ500
AI3-70736 see BJK650
AI3-36329-A see MIS500
AIA see PNR800
AIB see MGB000
AIBN see ASL750
AIC see AKK250
AICA see AKK250
AI3-36175-Ga see EMY100
AIGLONYL see EPD500
A-250-II see CBF680
A 11725 II see MRW750
AIL du CANADA (CANADA) see WBS850
AIMAX see MLJ500
AIMSAN see DRR400, MEG250
AIP see AHE750
AIPTASIA PALLIDA VENOM see SBI800
AIPYSURUS LAEVIS VENOM see SBI880
AIPYSURUS LAEVIS VENOM (AUSTRALIA) see AFG000
AIR, refrigerated liquid see AFG250
AIRBRON see ACH000
AIRDALE BLUE IN see FAE100
AIREDALE BLACK 2BG see FAB830
AIREDALE BLACK BHD see CMN800
AIREDALE BLACK ED see AQP000
AIREDALE BLUE 2BD see CMO000
AIREDALE BLUE D see CMO500
AIREDALE BLUE RL see ADE750
AIREDALE BLUE RWD see CMO600
AIREDALE BROWN BSD see CMO820
AIREDALE CARMOISINE see HJF500
AIREDALE GREEN BD see CMO840
AIREDALE GREEN BWD see CMO830
AIREDALE ORANGE II see CMM220
AIREDALE RED FD see CMO870
AIREDALE RED KD see CMO885
AIREDALE RED PGM see CMM320
AIREDALE RED RM see NAO600
AIREDALE SCARLET 3BD see CMO875
AIREDALE SCARLET GM see CMM325
AIREDALE VIOLET ND see CMP000
AIREDALE YELLOW E see SGP500
AIREDALE YELLOW 3GM see CMM759
AIREDALE YELLOW T see FAG140
AIR-FLO GREEN see CNN500
AIRLOCK see CAU500
AIRONE see ZMA000
AIR, compressed (UN 1002) (DOT) see AFG250
AIR, refrigerated liquid (cryogenic liquid) (UN 1003) (DOT) see AFG250
AIR, refrigerated liquid, (cryogenic liquid) non-pressurized (UN 1003) (DOT)
    see AFG250
AISELAZINE see HGP500
AISEMIDE see CHJ750
A. ITALICUM see ITD050
AITC see AGJ250
AI3-26730-X see TJK800
AIZEN ACID PHLOXINE PB see ADG250
AIZEN AMARANTH see FAG020
AIZEN ASTRA PHLOXINE FF see CMM765

AIZEN AURAMINE see IBA000
AIZEN BRILLIANT ACID BLUE AFH see ERG100
AIZEN BRILLIANT ACID PURE BLUE VH see ADE500
AIZEN BRILLIANT BLUE FCF see FMU059
AIZEN BRILLIANT SCARLET 3RH see FMU080
AIZEN CATHILON ORANGE GL see CMM764
AIZEN CATHILON ORANGE GLH see CMM764
AIZEN CATHILON PINK FG see CMM768
AIZEN CATHILON PINK FGH see CMM768
AIZEN CRYSTAL VIOLET see AOR500
AIZEN CRYSTAL VIOLET EXTRA PURE see AOR500
AIZEN DIAMOND GREEN GH see BAY750
AIZEN DIRECT BLACK BH see CMN800
AIZEN DIRECT BLUE 2BH see CMO000
AIZEN DIRECT DARK GREEN BH see CMO830
AIZEN DIRECT DEEP BLACK EH see AQP000
AIZEN DIRECT DEEP BLACK GH see AQP000
AIZEN DIRECT DEEP BLACK RH see AQP000
AIZEN DIRECT FAST RED FH see CMO870
AIZEN DIRECT GREEN BH see CMO840
AIZEN DIRECT SKY BLUE 5BH see CMO500
AIZEN EOSINE GH see BNH500, BNK700
AIZEN ERYTHROSINE see FAG040
AIZEN FOOD BLUE No. 2 see FAE000
AIZEN FOOD GREEN No. 3 see FAG000
AIZEN FOOD ORANGE No. 1 see FAG010
AIZEN FOOD ORANGE No. 2 see TGW000
AIZEN FOOD RED No. 5 see XRA000
AIZEN FOOD VIOLET No 1 see FAG120
AIZEN FOOD YELLOW NO. 5 see FAG150
AIZEN MAGENTA see MAC250
AIZEN MALACHITE GREEN see AFG500
AIZEN METANIL YELLOW see MDM775
AIZEN METHYLENE BLUE BH see BJI250
AIZEN NAPHTHOL ORANGE I see FAG010
AIZEN ORANGE I see FAG010
AIZEN PONCEAU RH see FMU070
AIZEN PRIMULA BROWN BRLH see CMO750
AIZEN PRIMULA RED 4BH see CMO885
AIZEN RHODAMINE BH see FAG070
AIZEN RHODAMINE 6GCP see RGW000
AIZEN TARTRAZINE see FAG140
AIZEN URANINE see FEW000
AIZEN VICTORIA BLUE BOH see VKA600
AJAN see NBS500
AJAX GMO see GGR200
AJAX, LEMON (scouring powder) see AFG625
AJI CABALLERO (PUERTO RICO) see PCB275
AJI de GALLINA (PUERTO RICO) see PCB275
AJI GUAGUAO (CUBA) see PCB275
AJINOMOTO see MRL500
AJI PICANTE (PUERTO RICO) see PCB275
AJMALICINE see AFG750
AJMALICINE HYDROCHLORIDE see AFH000
AJMALICINE MONOHYDROCHLORIDE see AFH000
AJMALINE see AFH250
AJMALINE BIS(CHLOROACETATE) (ester) HYDROCHLORIDE see AFH275
AJMALINE HYDROCHLORIDE see AFH280
AJO see WBS850
AK, flower extract see CAZ075
'AKA'AKAI (HAWAII) see WBS850
'AKA'AKAI-PILAU (HAWAII) see WBS850
AKADAMA see CAT775
AK (ADSORBENT) see CBT500
AKAR see DER000
AKARITHION see TNP250
AKARITOX see CKM000
AKEE see ADG400
AKETDRIN see AOB250
AKHNOT see WAT000
AKI see ADG400
AKINETON see BGD500
AKINETON HYDROCHLORIDE see BGD750
AKINOPHYL see BGD500, BGD750
AKIRIKU RHODAMINE B see FAG070
AKLAVIN see DAY835
AKLOMIX-3 see HMY000
AKLONIN (GERMAN) see PCV750
AKLONINE see PCV750
AKNADI EXTRACT see SLO100
AKOIN HYDROCHLORID (GERMAN) see PDN500
AKOTIN see NCQ900
AK PS see AFH500
AKRA, flower extract see CAZ075
AKRICHIN see ARQ250

AKROCHEM ETU-22 see IAQ000
AKROFOL see SFV250
AKROLEIN (CZECH) see ADR000
AKROLEINA (POLISH) see ADR000
AKRO-MAG see MAH500
AKRO-ZINC BAR 85 see ZKA000
AKRYLAMID (CZECH) see ADS250
AKRYLONITRYL (POLISH) see ADX500
AKRYLOYLOXYETHYLESTER KYSELINY ACETOCTOVE see AAY750
AKSA, combustion products see ADX750
AKTAMIN see NNO500
AKTAMIN HYDROCHLORIDE see NNP000
AKTEDRIN see AOB250
AKTEDRON see AOB500
AKTIKON see ARQ725
AKTIKON PK see ARQ725
AKTINIT A see ARQ725
AKTINIT PK see ARQ725
AKTINIT S see BJP000
AKTIVAN see FNF000
AKTIVEX see CCX000
AKTIVIN see CDP000
AKTON see DIX600
AKULON see PJY500
AK-33X see MAV750
AKYL DIMETHYL BENZYL AMMONIUM SACCHARINATE see BBA625
AKYPOROX O 50 see PJY100
AKYPOSAL TLS see SON000
AKZO CHEMIE MANEB see MAS500
AL-50 see RDP300
AL-100 see ZVJ000
AL-1021 see FGV000
AL 1076 see PFJ000
AL-1612 see AFH550
'ALA-AUMOE (HAWAII) see DAC500
ALABASTER see CAX750
ALABASTER NO. 3 see FAG150
ALACHLOR (USDA) see CFX000
ALACIL see AFW500
ALACINE see PFC750
ALAIXOL II see OAT000
AL-ALCHILI (ITALIAN) see DNI600
ALAMANDA MORADA FALSA (PUERTO RICO) see ROU450
ALAMINE 4 see DXW000
ALAMINE 6 see HCO500
ALAMINE 7 see OBC000
ALAMINE 11 see OHM700
ALAMINE 308 see DVL000
ALAMINE 336 see DVL000
ALAMINE 7D see OBC000
ALAMON see HOR470
ALANAP-3 see SIO500
ALANE see AHB500
ALANEX see CFX000
(R,S)-ALANINE see AFH600
dl-α-ALANINE see AFH600
dl-ALANINE see AFH600
(+-)-ALANINE see AFH600
ALANINE, 3-(((3-AMINO-4-HYDROXYPHENYL)PHENYLARSINO)THIO)- see
  AKI900
ALANINE, N-BENZYLOXYCARBONYL-3-PHENYL-, VINYL ESTER, l- see
  CBR235
ALANINE, N-CARBOXY-3-PHENYL-, N-BENZYL 1-VINYL ESTER, l- see
  CBR235
DL-ALANINE, N-(2,6-DIMETHYLPHENYL)-N-(METHOXYACETYL)-, METHYL ES-
  TER (9CI) see MDM100
ALANINE, 3,3'-DISELENOBIS-(9CI) see SBP600
ALANINE, 3,3'-DISELENODI-, dl- see SBU200
ALANINE, N-l-γ-GLUTAMYL-3-(METHYLENECYCLOPROPYL)- see HOW100
ALANINE, 3-(4-(4-HYDROXY-3-IODOPHENOXY)-3,5-DIIODOPHENYL)-, l- see
  LGK050
ALANINE MUSTARD see BHN500
ALANINE NITROGEN MUSTARD see BHV250, PED750
ALANINE, 3-PHENYL- see PEC750
l-ALANINE, PHENYL- see PEC750
ALANINE, PHENYL- see PEC750
β-ALANINOL see PMM250
l-ALANOSINE see AFH750
ALANOSINE see AFH750
4-N-d-ALANYL-2,4-DIAMINO-2,4-DIDEOXY-l-ARABINOSE see AFI500
β-ALANYL-l-HISTIDINE see CCK665
N-l-ALANYL-3-(5-OXO-7-OXABICYCLO(4.1.0)HEPT-2-YL)-l-ALANINE see BAC175
ALANYL-2,6-XYLIDIDE see TGI699
ALA de PICO (MEXICO) see RBZ400
ALAR-85 see DQD400

ALAR see DQD400
ALATHON 14 see PJS750
ALATHON 15 see PJS750
ALATHON 1560 see PJS750
ALATHON 6600 see PJS750
ALATHON 7026 see PJS750
ALATHON 7040 see PJS750
ALATHON 7050 see PJS750
ALATHON 7140 see PJS750
ALATHON 7511 see PJS750
ALATHON see PJS750, PKF750
ALATHON 5B see PJS750
ALATHON 71XHN see PJS750
ALAUN (GERMAN) see AGX000
ALAZIN see PFC750
ALAZINE see PFC750
ALAZOPEPTIN see AFI625
ALBACAR 5970 see CAT775
ALBACAR see CAT775
ALBA-DOME see AEY000
ALBAFIL see CAT775
ALBAGEL PREMIUM USP 4444 see BAV750
ALBAGLOS see CAT775
ALBAGLOS SF see CAT775
ALBALON LIQUIFILM see NCW000
ALBAMINE see SNP500
ALBAMIX see SMB000
ALBAMYCIN see NOB000, SMB000
ALBAMYCIN SODIUM see NOB000
ALBARICOQUE (SPANISH) see AQP890
ALBEGO see CFY250
ALBEMAP see BBK500
ALBENDAZOLE (USDA) see VAD000
ALBEXAN see SNM500
ALBIGEN A see PKQ250
ALBIOTIC see LGD000
ALBITOCIN see AFI750
ALBOLINE see MQV750
ALBOMYCIN A1 see GJU800
ALBON see SNN300
ALBONE 35 see HIB010
ALBONE 50 see HIB010
ALBONE 70 see HIB010
ALBONE see HIB000
ALBONE 35CG see HIB010
ALBONE 50CG see HIB010
ALBONE 70CG see HIB010
ALBORAL see DCK759
ALBOSAL see SNM500
AL BROMOHYDRATE see AGY000
ALBSAPOGENIN see GMG000
ALBUCID see SNP500
ALBUCIDE see SNQ000
ALBUCID SOLUBLE see SNQ000
ALBUMIN see AFI850
ALBUMIN MACRO AGGREGATES see AFI850
ALBUTEROL see BQF500
ALCA see AHA135
ALCALASE see BAC000
ALCANFOR see CBB250
ALCHLOQUIN see CHR500
ALCIDE see CDW450
ALCIDE LD see SFT500
ALCLOFENAC SODIUM SALT see AGN250
ALCLOMETASONE DIPROPIONATE see AFI980
ALCLOPHENAC see AGN000
ALCLOXA see AHA135
ALCOA 331 see AHC000
ALCOA F 1 see AHE250
ALCOA SODIUM FLUORIDE see SHF500
ALCOBAM NM see SGM500
ALCOBAM ZM see BJK500
ALCOBON see FHI000
ALCOGUM see ADW200
ALCOHOL, anhydrous see EFU000
ALCOHOL, dehydrated see EFU000
ALCOHOL see EFU000
ALCOHOL C-8 see OEI000
ALCOHOL C-9 see NNB500
ALCOHOL C-10 see DAI600
ALCOHOL C-11 see UMA000, UNA000
ALCOHOL C-12 see DXV600
ALCOHOL C-16 see HCP000
ALCOHOL, DENATURED see AFJ000
ALCOHOLS, C7-9 see LGF875

ALCOHOLS, C₁₁.₁₅-SECONDARY, ETHOXYLATED see TAZ100
ALCOHOLS, C12-14-SECONDARY, ETHOXYLATED PROPOXYLATED see TBA800
ALCOHOLS, N.O.S. see AFJ250
ALCOHOLS, toxic, n.o.s. (UN 1986) (DOT) see EFU000
ALCOHOLS, n.o.s. (UN 1987) (DOT) see EFU000
ALCOID see AFJ400
ALCOOL ALLILCO (ITALIAN) see AFV500
ALCOOL ALLYLIQUE (FRENCH) see AFV500
ALCOOL AMILICO (ITALIAN) see IHP000
ALCOOL AMYLIQUE (FRENCH) see AOE000
ALCOOL BUTYLIQUE (FRENCH) see BPW500
ALCOOL BUTYLIQUE SECONDAIRE (FRENCH) see BPW750
ALCOOL BUTYLIQUE TERTIAIRE (FRENCH) see BPX000
ALCOOL ETHYLIQUE (FRENCH) see EFU000
ALCOOL ETILICO (ITALIAN) see EFU000
l'ALCOOL n-HEPTYLIQUE PRIMAIRE (FRENCH) see HBL500
ALCOOL ISOAMYLIQUE (FRENCH) see IHP000
ALCOOL ISOBUTYLIQUE (FRENCH) see IIL000
ALCOOL ISOPROPILICO (ITALIAN) see INJ000
ALCOOL ISOPROPYLIQUE (FRENCH) see INJ000
ALCOOL METHYL AMYLIQUE (FRENCH) see MKW600
ALCOOL METHYLIQUE (FRENCH) see MGB150
ALCOOL METILICO (ITALIAN) see MGB150
ALCOOL PROPILICO (ITALIAN) see PND000
ALCOOL PROPYLIQUE (FRENCH) see PND000
ALCOPAN-250 see PAG200
ALCOPHOBIN see DXH250
ALCOPOL O see DJL000
ALCOTEX 88/05 see PKP750
ALCOTEX 88/10 see PKP750
ALCOWAX 6 see PJS750
ALCUONIUM DICHLORIDE see DBK400
ALCURONIUM CHLORIDE see DBK400
ALDABAN see TAI450
ALDACOL Q see PKQ250
ALDACTAZIDE see AFJ500
ALDACTIDE see AFJ500
ALDACTONE see AFJ500
ALDACTONE A see AFJ500
ALDADIENE-KALIUM see PKZ000
ALDADIENE-POTASSIUM see PKZ000
ALDANIL see FMW000
ALDECARB see CBM500
ALDECIN see AFJ625
ALDEHYDE-14 see UJJ000
ALDEHYDE ACETIQUE (FRENCH) see AAG250
ALDEHYDE ACRYLIQUE (FRENCH) see ADR000
ALDEHYDE AMMONIA see AAG500
ALDEHYDE B see COU500
ALDEHYDE BUTYRIQUE (FRENCH) see BSU250
ALDEHYDE C-6 see HEM000
ALDEHYDE C-8 see OCO000
ALDEHYDE C-9 see NMW500
ALDEHYDE C-14 see HBN200
ALDEHYDE C-18 see CNF250
ALDEHYDE C10 see DAG000
ALDEHYDE C-10 DIMETHYLACETAL see AFJ700
ALDEHYDE C-12, MNA see MQI550
ALDEHYDE C-12 MNA DIMETHYL ACETAL see MNB600
ALDEHYDE C-14, MYRISTIC see TBX500
ALDEHYDECOLLIDINE see EOS000
ALDEHYDE C-14 PEACH see HBN200
ALDEHYDE C-14 PURE see UJA800
ALDEHYDE CROTONIQUE (FRENCH) see COB260
ALDEHYDE C-11, UNDECYLENIC see ULJ000
ALDEHYDE-2-ETHYLBUTYRIQUE (FRENCH) see DHI000
ALDEHYDE FORMIQUE (FRENCH) see FMV000
ALDEHYDE ISOVALERIANIQUE see MHX500
ALDEHYDE M.N.A. see MQI550
ALDEHYDE PROPIONIQUE (FRENCH) see PMT750
ALDEHYDES see AFJ800
ALDEHYDINE see EOS000
ALDEHYDODICHLOROMALEIC ACID see MRU900
ALDEIDE ACETICA (ITALIAN) see AAG250
ALDEIDE ACRILICA (ITALIAN) see ADR000
ALDEIDE BUTIRRICA (ITALIAN) see BSU250
ALDEIDE FORMICA (ITALIAN) see FMV000
ALDER BUCKTHORN see MBU825
ALDERLIN see INS000
ALDERLIN HYDROCHLORIDE see INT000
ALDICARB (USDA) see CBM500
ALDICARBE (FRENCH) see CBM500
ALDICARB OXIME see MLX800
ALDICARB SULFOXIDE see MLW750

ALDIFEN see DUZ000
AL-DIISOBUTYL see DNI600
ALDIMORPH see AFJ850
ALDINAMID see POL500
ALDO-28 see OAV000
ALDO 40 see GGR200
ALDO-72 see OAV000
ALDO see CLY500
ALDO HMS see OAV000
ALDOL see AAH750
ALDOMET see DNA800, MJE780
ALDOMETIL see DNA800, MJE780
ALDOMIN see DNA800, MJE780
ALDO MO-FG see GGR200
ALDO MS see OAV000
ALDO MSA see OAV000
ALDO MSLG see OAV000
ALDOMYCIN see NGE500
ALDOSPERSE L 9 see DXY000
ALDOXIME see AAH250
2-ALDOXIME PYRIDINIUM-N-METHYL METHANESULPHONATE see PLX250
ALDOXYCARB see AFK000
ALDREX 30 see AFK250
ALDREX see AFK250
ALDRIN, cast solid (DOT) see AFK250
ALDRICH see DOJ200
ALDRIN see AFK250
ALDRINE (FRENCH) see AFK250
ALDRITE see AFK250
ALDROSOL see AFK250
ALDYL A see PJS750
ALECOR see ECU600
ALELAILA (PUERTO RICO) see CDM325
ALENTIN see BSM000
ALENTOL see AOB250
ALEPSIN see DNU000
ALERMINE see CLD250
ALERYL see BBV500
ALESTEN see SNP500
ALETAMINE HYDROCHLORIDE see AGQ250
ALEUDRIN see DMV600
ALEURITES (VARIOUS SPECIES) see TOA275
ALEURITIC ACID, tech see TKP000
ALEVAIRE see TDN750
ALEVIATIN see DKQ000
ALEXAN see AQQ750, COW900
ALEXANDRIAN LAUREL see MBU780
ALF see AKR500
ALFACALCIDOL see HJV000
ALFACILLIN see PDD350
ALFACRON see IEN000
ALFADIONE see AFK500
ALFALFA MEAL see AFK750
ALFAMAT see GFA000
ALFANAFTILAMINA (ITALIAN) see NBE700
ALFA-NAFTYLOAMINA (POLISH) see NBE700
ALFATESINE (FRENCH) see AFK500
ALFATIL see CCR850
ALFA-TOX see DCM750, MEI500
ALFATROFIN see AES650
ALFAXALONE see AFK875
ALFENAMIN see AHA875
ALFENOL 3 see PKF500
ALFENOL 9 see PKF500
ALFEPROL (RUSSIAN) see AGW250
ALFIBRATE see AHA150
ALFICETYN see CDP250
ALFIDE see SOX875
ALFIMID see DYC800
ALFLORONE see FHH100
ALFOCILLIN see PDD350
ALFOL 8 see OEI000
ALFOL-10 see ODE000
ALFOL 12 see DXV600
ALFONAL K see AFK900
ALFUCIN see NGE500
ALGAFAN see PNA500
ALGAMON see SAH000
ALGAROBA see LIA000
ALGEON 22 see CFX500
ALGERIAN IVY see AFK950
ALGERIL see PMX250
ALGIAMIDA see SAH000
ALGIDON see MDP750
ALGIL see DAM700

ALGIMYCIN see ABU500
ALGIN (polysaccharide) see SEH000
ALGIN see CAM200, SEH000
ALGINATE KMF see SEH000
ALGIN GUM see CAO250
ALGINIC ACID see AFL000
ALGINIC ACID HYDROGEN SULFATE SODIUM SALT (9CI) see SEH450
ALGIPON L-1168 see SEH000
ALGISTAT see DFT000
ALGLOFLON see TAI250
ALGOCOR see EID200
ALGODON de SEDA (CUBA, PUERTO RICO) see COD675
ALGOFLON SV see TAI250
ALGOFRENE 22 see CFX500
ALGOFRENE TYPE 1 see TIP500
ALGOFRENE TYPE 2 see DFA600
ALGOFRENE TYPE 5 see DFL000
ALGOFRENE TYPE 67 see ELN500
ALGOFRENE TYPE 6 see CFX500
ALGOL ORANGE RF see CMU815
ALGOLYSIN see MDP750
ALGONEURINA see TES800
ALGOSEDIV see TEH500
ALGOTROPYL see HIM000
ALGRAIN see EFU000
ALGUSI, extract see AHI630
ALGYLEN see TIO750
ALHELI EXTRANJERO (PUERTO RICO) see OHM875
ALIANT see TGX550
ALICYANATE see PLC250
ALIDOCHLOR see CFK000
ALIMEMAZINE see AFL500
ALIMEMAZINE-S,S-DIOXIDE see AFL750
ALIMET see HMR600
ALIMEZINE see AFL500
ALINAMIN see DXO300
ALINAMIN F see FQJ100
ALINDOR see BRF500
ALIOMYCIN see AFM000
ALIPHATIC and AROMATIC EPOXIDES see AFM250
ALIPHATIC CHLORINATED HYDROCARBONS see CDV250
ALIPORINA see TEY000
ALIPUR see AFM375
ALIPUR-O see CPT000
ALIQUAT 7 see TLW500
ALIQUAT 203 see DGX200
ALIQUAT 207 see DXG625
ALIQUAT 264 see QAT550
ALIQUAT 336 see MQH000
ALIQUAT 336N see MQH000
ALIQUAT 336-PTC see MQH000
ALISOBUMAL see AGI750
ALITHON 7050 see PJS750
ALITIA S see TES800
ALITON see DXO300
ALIVAL see NMV725
ALIVAL (ANTIDEPRESSANT) see NMV725
ALIZANTHRENE BLUE RC see DFN425
ALIZANTHRENE NAVY BLUE R see VGP100
ALIZANTHRENE NAVY BLUE RT see VGP100
ALIZARIN see DMG800
ALIZARINA see DMG800
ALIZARIN B see DMG800
ALIZARINBORDEAUX see TDD000
ALIZARIN CARMINE (BIOLOGICAL STAIN) see SEH475
ALIZARIN COMPLEXON see AFM400
ALIZARIN COMPLEXONE see AFM400
ALIZARINE see DMG800
ALIZARINE ACID BLUE B see CMM090
ALIZARINE 3B see DMG800
ALIZARINE B see DMG800
ALIZARINE BLUE A see CMM070
ALIZARINE BLUE AR see CMM070
ALIZARINE BLUE BLACK OCBN see CMP880
ALIZARINE BLUE BLACK OCGN see CMP880
ALIZARINE BLUE BLACK OCGP see CMP880
ALIZARINE BLUE GRL see CMM090
ALIZARINE BLUE OCB see CMP880
ALIZARINE BORDEAUX see TDD000
ALIZARINE BORDEAUX B see TDD000
ALIZARINE BRILLIANT SAPPHIRE R see CMM080
ALIZARINE BRILLIANT SKY BLUE R see CMM080
ALIZARINE BROWN HD see TKN500
ALIZARINE BROWN R see TKN500
ALIZARINE CARMINE INDICATOR see SEH475

ALIZARINE CHROME RED G see CMM325
ALIZARINE COMPLEXON see AFM400
ALIZARINE COMPLEXONE see AFM400
ALIZARINE CYANINE GREEN BASE see BLK000
ALIZARINE DIRECT BLUE AR see CMM070
ALIZARINE DIRECT BLUE ARA see CMM070
ALIZARINE DIRECT PURE BLUE R see CMM080
ALIZARINE FAST BLUE 2B see CMM090
ALIZARINE FAST BLUE BE see CMM100
ALIZARINE FAST BLUE RFE see CMM080
ALIZARINE FAST LIGHT BLUE C see CMM090
ALIZARINE FLUORINE BLUE see AFM400
ALIZARINE GR see SIT850
ALIZARINE INDICATOR see DMG800
ALIZARINE LAKE RED IPX see DMG800
ALIZARINE LAKE RED 2P see DMG800
ALIZARINE LAKE RED 3P see DMG800
ALIZARINE LIGHT BLUE AR see CMM090
ALIZARINE L PASTE see DMG800
ALIZARINE NAC see DMG800
ALIZARINE PASTE 20% BLUISH see DMG800
ALIZARINE PURE BLUE B see CMM090
ALIZARINE RED see DMG800
ALIZARINE RED A see SEH475
ALIZARINE RED AS see SEH475
ALIZARINE RED B2 see DMG800
ALIZARINE RED B see DMG800
ALIZARINE RED INDICATOR see SEH475
ALIZARINE RED IP see DMG800
ALIZARINE RED IPP see DMG800
ALIZARINE RED L see DMG800
ALIZARINE RED S (BIOLOGICAL STAIN) see SEH475
ALIZARINE RED S SODIUM SALT see SEH475
ALIZARINE RED SW see SEH475
ALIZARINE RED SZ see SEH475
ALIZARINE RED W see SEH475
ALIZARINE RED WA see SEH475
ALIZARINE RED for WOOL see SEH475
ALIZARINE RED WS see SEH475
ALIZARINE S see SEH475
ALIZARINE SAPPHIRE AR see CMM070
ALIZARINE S EXTRA CONC. A EXPORT see SEH475
ALIZARINE S EXTRA PURE A see SEH475
ALIZARINE SKY BLUE B see CMM090
ALIZARINE SKY BLUE BS-CF see CMM090
ALIZARINE SKY BLUE R see CMM080
ALIZARINE SUPRA BLUE R see CMM080
ALIZARINE SUPRA SKY RA see CMM080
ALIZARINE VIOLET 3B BASE see HOK000
ALIZARINE YELLOW AGP see SIT850
ALIZARINE YELLOW 2G see SIT850
ALIZARINE YELLOW G see SIT850
ALIZARINE YELLOW GG see SIT850
ALIZARINE YELLOW GGW see SIT850
ALIZARINE YELLOW GM see SIT850
ALIZARIN FLUORINE BLUE see AFM400
ALIZARINKOMPLEXON see AFM400
ALIZARINPRIMEVEROSIDE see DMG800
ALIZARIN RED see DMG800
ALIZARIN RED S see SEH475
ALIZARINROT-S see SEH475
ALIZARIN S see SEH475
ALIZARINSULFONATE see SEH475
ALIZARIN YELLOW see SIT850
ALIZARIN YELLOW G see SIT850
ALIZARIN YELLOW GG see SIT850
ALIZARIN YELLOW G SODIUM SALT see SIT850
ALIZAROL YELLOW GW see SIT850
ALKALIES see AFM500
ALKALI GRASS see DAE100
ALKALINE DEOXYRIBONUCLEASE see EDK650, PAF575
ALKALINE DNASE see EDK650, PAF575
ALKALI RESISTANT RED DARK see NAY000
ALKALOID C see AFG750
ALKALOID H 3, from COLCHICUM ANTUMNALE see MIW500
ALKALOID II see AFG750
ALKALOIDS see AFM750
ALKALOID SALTS see AFM750
ALKALOIDS, VERATRUM see VIZ000
ALKALOID V see CPT750
ALKAMID see PJY500
ALKANES see AFN250
ALKAPOL PEG-200 see PJT000
ALKAPOL PEG-300 see PJT000
ALKAPOL PEG-400 see MFD500

ALKAPOL PEG-600 see PJT000
ALKAPOL PEG-6000 see PJT000
ALKAPOL PEG-8000 see PJT000
ALKAPOL PPG-1200 see PKI500
ALKARAU see RDK000
ALKARSODYL see HKC500
ALKASERP see RDK000
ALKATHENE 200 see PJS750
ALKATHENE 22 300 see PJS750
ALKATHENE 17/04/00 see PJS750
ALKATHENE see PJS750
ALKATHENE ARN 60 see PJS750
ALKATHENE RXDG33 see TAI250
ALKATHENE WJG 11 see PJS750
ALKATHENE WNG 14 see PJS750
ALKATHENE XDG 33 see PJS750
ALKATHENE XJK 25 see PJS750
ALK-AUBS see DXH250
ALKAVERVIR see VIZ000
ALKAZENE 42 see EHY000
α-ALKENESULFONIC ACID see AFN500
ALKENYL DIMETHYLETHYL AMMONIUM BROMIDE see AFN750
ALK-ENZYME see BAC000
ALKERAN (RUSSIAN) see BHV000
ALKERAN see PED750
ALKIRON see MPW500
ALKOFEN MBP see MHR050
ALKOHOL (GERMAN) see EFU000
ALKOHOLU ETYLOWEGO (POLISH) see EFU000
ALKOTEX see PKP750
ALKOVERT see PIB250
3-(ALKYLAMINO)PROPIONITRILE see AFO200
ALKYL ARYL POLYETHER ALCOHOLS see ARL875
ALKYLARYLPOLYGLYKOLAETHER (GERMAN) see ARL875
ALKYL ARYL SULFONATE see AFO250
ALKYLBENZENESULFONATE see AFO500
p-n-ALKYLBENZENESULFONIC ACID DERIVATIVE, SODIUM SALT see
    AFO750
ALKYLBENZENESULFONIC ACID SODIUM SALT see AFP000
p-N-ALKYLBENZENSULFONAN SODNY (CZECH) see AFO750
ALKYL(C$_{14-16}$)DIMETHYLBENZYL AMMONIUM CHLORIDES see AFP100
ALKYL(C$_8$H$_{17}$ to C$_{18}$H$_{37}$) DIMETHYL-3,4-DICHLOROBENZYL AMMONIUM CHLO-
    RIDE see AFP750
ALKYL(C$_6$H$_{18}$)DIMETHYL-3,4-DICHLOROBENZYLAMMONIUM CHLORIDE see
    AFP750
ALKYL(C-13) POLYETHOXYLATES(ETHOXY-6) see TJJ250
ALKYL(C-14) POLYETHOXYLATES (ETHOXY-7) see TBY750
ALKYL(C$_{9-15}$)TOLYL METHYLTRIMETHYL AMMONIUM CHLORIDE see DYA600
ALKYL DIMETHYL BENZALKONIUM CHLORIDE see AFP075
ALKYL DIMETHYL BENZALKONIUM SACCHARINATE see BBA625
ALKYL DIMETHYLBENZYL AMMONIUM CHLORIDE see AFP250
ALKYLDIMETHYLETHYLBENZYL AMMONIUM CHLORIDE see BBA500
ALKYLDIMETHYL(PHENYLMETHYL)QUATERNARY AMMONIUM CHLORIDES
    see AFP250
ALKYL((ETHYLPHENYL)METHYL)DIMETHYL QUATERNARY AMMONIUM
    CHLORIDES see BBA500
ALKYLNITRILE see AFQ000
ALKYL PHENOL POLYGLYCOL ETHERS see ARL875
ALKYL PHENOXY POLYETHOXY ETHANOLS see ARL875
ALKYL PHENYL POLYETHYLENE GLYCOL ETHER see AFQ250
ALKYL PYRIDINES R see AFQ500
ALKYROM see AFQ575
ALLACYL see AFW500
ALLAMANDA see AFQ625
ALLAMANDA CATHARTICA see AFQ625
ALLANTOXANIC ACID, POTASSIUM SALT see AFQ750
ALLBRI ALUMINUM PASTE and POWDER see AGX000
ALLBRI NATURAL COPPER see CNI000
ALLEDRYL see BBV500
(+)-ALLELRETHONYL (+)-cis,trans-CHRYSANTHEMATE see AFR250
ALLENE see AFR000
ALLEOSIDE A DIHYDRATE see HAO000
ALLERCLOR see TAI500
ALLERCUR see CMV400
ALLERCURE HYDROCHLORIDE see CMV400
ALLERGAN 211 see DAS000
ALLERGAN see CKV625
ALLERGAN B see BBV500
ALLERGEFON MALEATE see CGD500
ALLERGEN see LJR000
ALLERGEVAL see BBV500
ALLERGICAL see BBV500
ALLERGICAN see CLX300
ALLERGIN see BBV500, TAI500
ALLERGINA see BBV500

ALLERGISAN see CLX300, TAI500
ALLERGIVAL see BBV500
ALLERON see PAK000
(+)-cis-ALLETHRIN see AFR500
d-ALLETHRIN see AFR250
d-trans ALLETHRIN see AFR250, BGC750
ALLETHRIN see AFR250
trans-(+)-ALLETHRIN see AFR750
ALLETHRIN I see AFR250
ALLETHRIN RACEMIC MIXTURE see AFS000
d-ALLETHROLONE CHRYSANTHEMUMATE see AFR750
(+)-ALLETHRONYL (+)-trans-CHRYSANTHEMUMATE see AFR750
ALLETONE see HKC575
ALLEVIATE see AFR250
ALLICIN see AFS250
ALLIDOCHLOR see CFK000
ALLIED PE 617 see PJS750
ALLIED WHITING see CAT775
ALLIGATOR LILY see BAR325
ALLIGATOR PEAR OIL see ARW800
ALLILE (CLORURO DI) (ITALIAN) see AGB250
ALLIL-GLICIDIL-ETERE (ITALIAN) see AGH150
1-ALLILOSSI-2,3 EPOSSIPROPANO (ITALIAN) see AGH150
ALLILOWY ALKOHOL (POLISH) see AFV500
ALLIONAL see AFT000
ALLISAN see RDP300
ALLIUM SATIVUM Linn., powder see GBU850
ALLIUM (Various Species) see WBS850
ALLOBARBITAL see AFS500
ALLOBARBITONE see AFS500
ALLOCAINE see AIL750, AIT250
ALLOCLAMIDE see AFS625
ALLOCLAMIDE HYDROCHLORIDE see AFS640
ALLODAN see AFS750
ALLODENE see BBK000
ALLOFENYL see AGQ875
ALLOFERIN see DBK400
ALLOMALEIC ACID see FOU000
ALLOMALEIC ACID DIMETHYL ESTER see DSB600
ALLOMYCIN see AHL000
ALLONAL see AFT000
ALLO-OCIMENOL see LFX000
1-(p-ALLOPHANOYLBENZYL)-2-METHYLHYDRAZINE HYDROBROMIDE see
    MKN750
ALLOPHENYLUM see AGQ875
ALLOPSEUDOCODEINE HYDROCHLORIDE see AFT250
ALLOPURINOL see ZVJ000
ALLOPYDIN see AGN000
ALLORPHINE see AFT500
ALLOTROPAL see EQL000
ALLOXAN see AFT750
ALLOXAN MONOHYDRATE see MDL500
ALLOXAN-5-OXIME see AFU000
ALLOXANTIN see AFU250
2-ALLOXYETHANOL (CZECH) see AGO000
ALLOZYM see ZVJ000
ALLPHASEM see AGQ875
ALLTEX see CDV100
ALLTOX see CDV100
ALLUMINIO(CLORURO DI) (ITALIAN) see AGY750
ALLUMINIO DIISOBUTIL-MONOCLORURO (ITALIAN) see CGB500
ALLURAL see ZVJ000
ALLUVAL see BNP750
ALLYL ACETATE see AFU750
ALLYLACETIC ACID see PBQ750
ALLYL ADIPATE see AEO500
ALLYL AL see AFV500
ALLYL ALCOHOL see AFV500
ALLYL ALDEHYDE see ADR000
ALLYLALKOHOL (GERMAN) see AFV500
5-ALLYL-5-(1-(ALLYTHIO)ETHYL)BARBITURIC ACID SODIUM SALT see
    AFV750
ALLYLAMINE see AFW000
2-(ALLYLAMINO)-6'-CHLORO-o-ACETOTOLUIDIDE HYDROCHLORIDE see
    AFW250
2-(ALLYLAMINO)-2'-CHLORO-6'-METHYLACETANILIDE HYDROCHLORIDE see
    AFW250
1-ALLYL-6-AMINO-3-ETHYL-2,4(1H,3H)-PYRIMIDINEDIONE see AFW500
1-ALLYL-6-AMINO-3-ETHYLURACIL see AFW500
p-ALLYLANISOLE see AFW750
ALLYLBARBITAL see AGI750
ALLYLBARBITONE see AGI750
ALLYLBARBITURAL see AFS500
ALLYLBARBITURIC ACID see AGI750
ALLYLBENZENE see AFX000

ALLYL BENZENE SULFONATE see AFX250
5-ALLYL-1,3-BENZODIOXOLE see SAD000
5-ALLYL-5-BENZYL-2-THIOBARBITURIC ACID SODIUM SALT see AFX500
ALLYL-BIS(β-CHLOROETHYL)AMINE HYDROCHLORIDE see AFX750
ALLYL BROMIDE see AFY000
ALLYL BUTANOATE see AFY250
5-ALLYL-5-(2-BUTENYL)-2-THIOBARBITURIC ACID SODIUM SALT see AFY300
5-ALLYL-5-sec-BUTYLBARBITURIC ACID see AFY500
5-ALLYL-5-sec-BUTYL-2-THIOBARBITURIC ACID see AFY750
ALLYL-sec-BUTYL THIOBARBITURIC ACID see AFY750
5-ALLYL-5-sec-BUTYL-2-THIOBARBITURIC ACID SODIUM SALT see AGA000
5-ALLYL-5-(1-BUTYLTHIO)ETHYL)BARBITURIC ACID SODIUM SALT see AGA250
ALLYL BUTYRATE see AFY250
ALLYL CAPROATE see AGA500
ALLYL CAPRYLATE see AGM500
ALLYL CARBAMATE see AGA750
N-ALLYL CARBAMIC ACID-3-DIMETHYLAMINOPHENYL ESTER HYDROCHLORIDE see AGB000
ALLYLCARBAMIC ESTER of m-OXYPHENYLDIMETHYLAMINE HYDROCHLORIDE see AGB000
ALLYLCARBAMIDE see AGV000
ALLYLCATECHOL METHYLENE ETHER see SAD000
ALLYLCHLORID (GERMAN) see AGB250
ALLYL CHLORIDE see AGB250
ALLYL CHLOROCARBONATE see AGB500
ALLYL CHLOROFORMATE (DOT) see AGB500
ALLYL CHLOROFORMATE see AGB500
ALLYLCHLOROHYDRIN ETHER see AGB750
ALLYL (3-CHLORO-2-HYDROXYPROPYL) ETHER see AGB750
ALLYL CINERIN see AFR250
ALLYL CINNAMATE see AGC000
ALLYL COMPOUNDS see AGC125
ALLYL CYANIDE see BOX500
ALLYL CYANOACETATE see AGC150
6-ALLYL-α-CYANOERGOLINE-8-PROPIONAMIDE see AGC200
ALLYL CYCLOHEXANEACETATE see AGC250
ALLYL CYCLOHEXANEPROPIONATE see AGC500
5-ALLYL-5-(2-CYCLOHEXEN-1-YL)-2-THIOBARBITURIC ACID see TES500
ALLYL CYCLOHEXYLACETATE see AGC250
2-ALLYL-2-CYCLOHEXYLACETIC ACID-3-(DIETHYLAMINO)-2,2-DIMETHYLPROPYLESTER HYDROCHLORIDE see AGC750
3-ALLYLCYCLOHEXYL PROPIONATE see AGC500
5-ALLYL-5-(2-CYCLOPENTENYL)-2-THIOBARBITURIC ACID see AGD000
N-ALLYL-7,8-DEHYDRO-4,5-EPOXY-3,6-DIHYDROXYMORPHINAN see AFT500
N-ALLYL-N-DESMETHYLMORPHINE see AFT500
I-ALLYL-(6-DIAZO-5-OXO)-I-NORLEUCYL-(6-DIAZO-5-OXO)-I-NORLEUCINE see AFI625
ALLYL DIGLYCOL CARBONATE see AGD250
6-ALLYL-6,7-DIHYDRO-5H-DIBENZ(c,e)AZEPINE see ARZ000
6-ALLYL-6,7-DIHYDRO-5H-DIBENZ(c,e)AZEPINE PHOSPHATE see AGD500
6-ALLYL-6,7-DIHYDRO-3,9-DICHLORO-5H-DIBENZ(c,e)AZEPINE see AGD750
I-N-ALLYL-7,8-DIHYDRO-14-HYDROXYNORMORPHINONE see NAG550
N-ALLYL-7,8-DIHYDRO-14-HYDROXYNORMORPHINONE, HYDROCHLORIDE see NAH000
6-ALLYL-6,7-DIHYDRO-6-METHYL-5H-DIBENZ(c,e)AZEPINIUM IODIDE see AGE000
N-ALLYL-2,12-DIHYDROXY-1,11-EPOXYMORPHINENE-13 HYDROCHLORIDE see NAG500
4-ALLYL-1,2-DIMETHOXYBENZENE see AGE250
1-ALLYL-3,4-DIMETHOXYBENZENE see AGE250
1-ALLYL-2,5-DIMETHOXY-3,4-METHYLENEDIOXYBENZENE see AGE500
ALLYLDIMETHYLARSINE see AGE625
1-ALLYL-1-(3,7-DIMETHYLOCTYL)PIPERIDINIUM BROMIDE see AGE750
1-ALLYL-1-(3,7-DIMETHYLOCTYL)-PIPERIDIUMBROMID (GERMAN) see AGE750
β-ALLYL-N,N-DIMETHYLPHENETHYLAMINE see AGF000
ALLYLDIOXYBENZENE METHYLENE ETHER see SAD000
5-ALLYL-1,3-DIPHENYLBARBITURIC ACID see AGF250
5-ALLYL-1,3-DIPHENYL-2,4,6(1H,3H,5H)-PYRIMIDINETERIONE see AGF250
ALLYL DISULFIDE see AGF300
ALLYL DISULPHIDE see AGF300
ALLYLE (CHLORURE D') (FRENCH) see AGB250
ALLYL ENANTHATE see AGH250
ALLYLENE see MFX590
17-ALLYL-4,5-α-EPOXY-3,14-DIHYDROXYMORPHINAN-6-ONE see NAG550
17-ALLYL-4,5-α-EPOXY-3,14-DIHYDROXYMORPHINAN-6-ONE HYDROCHLORIDE see NAH000
ALLYL-3,4-EPOXY-6-METHYLCYCLOHEXANECARBOXYLATE see AGF500
ALLYL-2,3-EPOXYPROPYL ETHER see AGH150
ALLYL-9,10-EPOXYSTEARATE see ECO500
ALLYLESTER KYSELINY CHLORMRAVENCI see AGB500
ALLYLESTER KYSELINY METHAKRYLOVE see AGK500
ALLYLESTER KYSELINY MRAVENCI see AGH000
ALLYLETHER see DBK000
ALLYL ETHER of PROPYLENE GLYCOL see PNK000

1-ALLYL-3-ETHYL-6-AMINOTETRAHYDROPYRIMIDINEDIONE see AFW500
ALLYL ETHYL ETHER see AGG000
2-ALLYL-5-ETHYL-2'-HYDROXY-9-METHYL-6,7-BENZOMORPHAN see AGG250
ALLYL FLUORIDE see AGG500
ALLYL FLUOROACETATE see AGG750
ALLYL FORMATE see AGH000
α-ALLYL GLYCEROL ETHER see AGP500
ALLYLGLYCIDAETHER (GERMAN) see AGH150
ALLYL GLYCIDYL ETHER see AGH150
4-ALLYLGUAIACOL see EQR500
ALLYL HEPTANOATE see AGH250
ALLYL HEPTOATE see AGH250
ALLYL HEPTYLATE see AGH250
ALLYL HEXAHYDROPHENYLPROPIONATE see AGC500
ALLYL HEXANOATE (FCC) see AGA500
ALLYL HOMOLOG of CINERIN I see AFR250
ALLYL HONOLOG of CINERIN I see BGC750
ALLYLHYDRAZINE HYDROCHLORIDE see AGH500
ALLYL HYDROPEROXIDE see AGH750
4-ALLYL-1-HYDROXY-2-METHOXYBENZENE see EQR500
2-ALLYL-4-HYDROXY-3-METHYL-2-CYCLOPENTEN-1-ONE see AFR750
d,I-2-ALLYL-4-HYDROXY-3-METHYL-2-CYCLOPENTEN-1-ONE-d,I-CHRYSANTHEMUM MONOCARBOXYLATE see AFR250
N-ALLYL-3-HYDROXYMORPHINAN see AGI000
I-N-ALLYL-14-HYDROXYNORDIHYDROMORPHINONE see NAG550
I-N-ALLYL-14-HYDROXYNORDIHYDROMORPHINONE HYDROCHLORIDE see NAH000
ALLYLHYDROXYPHENYLARSINE OXIDE see AGQ775
ALLYLIC ALCOHOL see AFV500
ALLYLIDENE DIACETATE see ADR250
ALLYL IODIDE see AGI250
ALLYL α-IONONE see AGI500
ALLYLISOBUTYLBARBITAL see AGI750
ALLYLISOBUTYLBARBITURATE see AGI750
5-ALLYL-5-ISOBUTYLBARBITURIC ACID see AGI750
ALLYL ISOCYANATE see AGJ000
ALLYLISOPROPYLACETYLCARBAMIDE see IQX000
ALLYLISOPROPYLACETYLUREA see IQX000
5-ALLYL-5-ISOPROPYLBARBITURATE see AFT000
5-ALLYL-5-ISOPROPYLBARBITURIC ACID see AFT000
ALLYLISOPROPYLBARBITURIC ACID see AFT000
ALLYLISOPROPYLMALONYLUREA see AFT000
ALLYL ISORHODANIDE see AGJ250
ALLYL ISOSULFOCYANATE see AGJ250
ALLYL ISOTHIOCYANATE, stabilized (DOT) see AGJ250
ALLYL ISOTHIOCYANATE see AGJ250
ALLYL ISOVALERATE see ISV000
ALLYL ISOVALERIANATE see ISV000
3-ALLYL-4-KETO-2-METHYLCYCLOPENTENYL CHRYSANTHEMUMMONOCARBOXYLATE see AFR250
ALLYLLITHIUM see AGJ375
ALLYL MERCAPTAN see AGJ500
α-ALLYLMERCAPTO-α, α-DIETHYLACETAMIDE see AGJ750
2-ALLYLMERCAPTO-2-ETHYLBUTYRAMIDE see AGJ750
2-ALLYLMERCAPTOISOBUTYRAMIDE see AGK000
α-ALLYLMERCAPTOISOBUTYRAMIDE see AGK000
ALLYLMERCAPTOMETHYLPENICILLIN see AGK250
ALLYLMERCAPTOMETHYLPENICILLINIC ACID see AGK250
ALLYL MESYLATE see AGK750
ALLYL METHACRYLATE see AGK500
ALLYL METHANESULFONATE see AGK750
4-ALLYL-1-METHOXYBENZENE see AFW750
5-ALLYL-1-METHOXY-2,3-(METHYLENEDIOXY)BENZENE see MSA500
4-ALLYL-2-METHOXYPHENOL see EQR500
4-ALLYL-2-METHOXYPHENOL ACETATE see EQS000
4-ALLYL-2-METHOXYPHENOL FORMATE see EQS100
4-ALLYL-2-METHOXYPHENYLPHENYLACETATE see AGL000
5-ALLYL-5-(1-METHYLALLYL)-2-THIOBARBITURIC ACID SODIUM SALT see AGL250
5-ALLYL-5-(1-METHYLBUTYL)BARBITURIC ACID see SBM500
5-ALLYL-5-(1-METHYLBUTYL)BARBITURIC ACID SODIUM DERIVATIVE see SBN000
5-ALLYL-5-(1-METHYLBUTYL)BARBITURIC ACID SODIUM SALT see SBN000
(R+)-5-ALLYL-5-(1-METHYLBUTYL)BARBITURIC ACID SODIUM SALT see SBL500
5-ALLYL-5-(1-METHYLBUTYL)MALONYLUREA see SBM500
5-ALLYL-5-(1-METHYLBUTYL)MALONYLUREA SODIUM SALT see SBN000
5-ALLYL-5-(1-METHYLBUTYL)-2-THIOBARBITURATE SODIUM see SOX500
5-ALLYL-5-(1-METHYLBUTYL)-2-THIOBARBITURIC ACID see AGL375
5-ALLYL-5-(1-METHYLBUTYL)-2-THIO-BARBITURIC ACID SODIUM SALT see SOX500
ALLYL 3-METHYLBUTYRATE see ISV000
2-ALLYL-3-METHYL-2-CYCLOPENTEN-1-ON-4-YL-N,N'-DIMETHYL-KARBAMAT (CZECH) see AGL500
4-ALLYL-1,2-METHYLENEDIOXYBENZENE see SAD000

1-ALLYL-3,4-METHYLENEDIOXYBENZENE see SAD000
2-ALLYL-3-METHYL-4-HYDROXY-2-CYCLOPENTEN-1-ONE DIMETHYLCARBAMATE see AGL500
5-ALLYL-1-METHYL-5-(1-METHYL-2-PENTYNYL)BARBITURIC ACID SODIUM SALT see MDU500
N-ALLYL-3-METHYL-N-α-METHYLPHENETHYL-6-OXO-1(6H)-PYRIDAZINE ACETAMIDE see AGL750
3-ALLYL-2-METHYL-4-OXO-2-CYCLOPENTEN-1-YL CHRYSANTHEMATE see AFR250
dl-3-ALLYL-2-METHYL-4-OXOCYCLOPENT-2-ENYL dl-cis trans CHRYSANTHEMATE see AFR250
(±)-3-ALLYL-2-METHYL-4-OXO-2-CYCLOPENTENYL 2,2,3,3-TETRAMETHYLCYCLOPROPANE CARBOXYLATE see TAL575
(±)-5-ALLYL-5-(1-METHYL-2-PENTYNYL)-2-THIOBARBITURIC ACID see AGL875
dl-1-ALLYL-1-METHYL-4-PHENYL-4-PIPERIDINOL PROPIONATE HYDROCHLORIDE see AGV890
5-ALLYL-5-(1-METHYLPROPENYL)BARBITURIC ACID see AGM000
5-ALLYL-5-(1-METHYLPROPYL) BARBITURIC ACID see AFY500
5-ALLYL-5-(2′-METHYL-N-PROPYL) BARBITURIC ACID see AGI750
ALLYL MONOSULFIDE see AGS250
ALLYLMORPHINE HYDROCHLORIDE see NAG500
ALLYL MUSTARD OIL see AGJ250
ALLYLNITRILE see BOX500
1-ALLYL-2-NITROIMIDAZOLE see AGM060
4-(ALLYLNITROSAMINO)-1-BUTANOL see HJS400
3-(ALLYLNITROSAMINO)-1,2-PROPANEDIOL see NJY500
1-(ALLYLNITROSAMINO)-2-PROPANONE see AGM125
4-(ALLYLNITROSOAMINO)BUTRIC ACID see CCJ375
N-ALLYL-N-NITROSO-3-BUTENYLAMINE see BPD000
1-ALLYL-1-NITROSOUREA see NJK000
N-ALLYLNORMORPHINE see AFT500
N-ALLYLNORMORPHINE HYDROCHLORIDE see NAG500
8-ALLYL-(±)-1-α-H,5-α-H-NORTHROPAN-3-α-OL see AGM250
ALLYL OCTANOATE see AGM500
2-ALLYLOXYBENZAMIDE see AGM750
o-(ALLYLOXY)BENZAMIDE see AGM750
ALLYLOXYCARB see DBI800
2-(ALLYLOXY)-4-CHLORO-N-(2-DIETHYLAMINO)ETHYL)BENZAMIDE see AFS625
2-ALLYLOXY-4-CHLORO-N-(2-(DIETHYLAMINO)ETHYL)BENZAMIDE HYDROCHLORIDE see AFS640
(4-ALLYLOXY-3-CHLOROPHENYL)ACETIC ACID see AGN000
(4-(ALLYLOXY)-3-CHLOROPHENYL)ACETIC ACID SODIUM SALT see AGN250
1-ALLYLOXY-3-CHLORO-2-PROPANOL see AGB750
(±)-1-(β-(ALLYLOXY)-2,4-DICHLOROPHENETHYL)IMIDAZOLE see FPB875
2-(ALLYLOXY)-N-(2-(DIETHYLAMINO)ETHYL)-α,α,α-TRIFLUORO-p-TOLUAMIDE see FDA875
o-ALLYLOXY-N,N-DIETHYLBENZAMIDE see AGN500
o-ALLYLOXY-N,N-DIMETHYLBENZAMIDE see AGN750
1-ALLYLOXY-2,3-EPOXY-PROPAAN (DUTCH) see AGH150
1-ALLYLOXY-2,3-EPOXYPROPAN (GERMAN) see AGH150
1-(ALLYLOXY)-2,3-EPOXYPROPANE see AGH150
2-ALLYLOXYETHANOL see AGO000
o-ALLYLOXY-N-(β-HYDROXYETHYL)BENZAMIDE see AGO250
6-ALLYLOXY-2-METHYLAMINO-4-(N-METHYLPIPERAZINO)-5-METHYLTHIOPYRIMIDINE see AGO500
2-(o-ALLYLOXYPHENOXY)-2-HYDROXY-N-ISOPROPYL-1-PROPYLAMINE HYDROCHLORIDE see THK750
1-(o-ALLYLOXY)PHENOXY)-3-(ISOPROPYLAMINO)-2-PROPANOL see CNR500
(−)-1-(o-ALLYLOXYPHENOXY)-3-ISOPROPYLAMINO-2-PROPANOLHYDROCHLORIDE see AGP000
1-(o-ALLYLOXYPHENOXY)-3-ISOPROPYLAMINOPROPAN-2-OL HYDROCHLORIDE see THK750
(+)-1-(o-ALLYLOXYPHENOXY)-3-ISOPROPYLAMINO-2-PROPANOLHYDROCHLORIDE see AGO750
3-(ALLYLOXYPHENOXY)-1,2-PROPANEDIOL see AGP250
3-ALLYLOXY-1,2-PROPANEDIOL see AGP500
1-ALLYLOXY-2,3-PROPANEDIOL see AGP500
3-ALLYLOXYPROPIONITRILE see AGP750
β-ALLYLOXY-PROPIONITRILE see AGP750
ALLYL PALLADIUM CHLORIDE DIMER see AGQ000
α-ALLYL PHENETHYLAMINEHYDROCHLORIDE see AGQ250
2-ALLYL PHENOL see AGQ500
o-ALLYL PHENOL see AGQ500
ALLYL PHENOXYACETATE see AGQ750
1-(o-ALLYLPHENOXY)-3-(ISOPROPYLAMINO)-2-PROPANOL see AGW250
1-(o-ALLYLPHENOXY)-3-(ISOPROPYLAMINO)-2-PROPANOL HYDROCHLORIDE see AGW000
ALLYL PHENYLACETATE see PMS500
ALLYL-3-PHENYLACRYLATE see AGC000
ALLYL PHENYL ARSINIC ACID see AGQ775
5-ALLYL-5-PHENYLBARBITURIC ACID see AGQ875
ALLYL PHENYL ETHER see AGR000
1-ALLYL-2-PHENYL-ETHYLAMINE HYDROCHLORIDE see AGQ250

2-ALLYL-2-PHENYL-4-PENTENOIC ACID 2-(DIETHYLAMINO)ETHYL ESTER HYDROCHLORIDE see AGR125
1-ALLYL-3-PHENYL-2-THIOUREA see AGR250
ALLYL PHENYL THIOUREA see AGR250
ALLYL PHOSPHATE see THN750
ALLYLPRODINE HYDROCHLORIDE see AGV890
ALLYL PROPYL DISULFIDE see AGR500
ALLYLPROPYMAL see AFT000
m-ALLYLPYROCATECHIN METHYLENE ETHER see SAD000
4-ALLYLPYROCATECHOL FORMALDEHYDE ACETAL see SAD000
ALLYLPYROCATECHOL METHYLENE ETHER see SAD000
1-ALLYLQUINALDINUM BROMIDE see AGR750
ALLYLRETHRONYL dl-cis-trans-CHRYSANTHEMATE see AFR250
ALLYLRHODANID (GERMAN) see AGS750
ALLYLSENFOEL (GERMAN) see AGJ250
ALLYL SEVENOLUM see AGJ250
ALLYLSUCCINIC ANHYDRIDE see AGS000
ALLYL SULFIDE see AGS250
ALLYL SULFOCYANIDE see AGS750
12-ALLYL-7,7a,8,9-TETRAHYDRO-3,7a-DIHYDROXY-4aH-8,9c-IMINOETHANOPHENANTHRO(4,5-bcd)FURANONE see NAG550
ALLYL-TETRA-HYDROGERANYL-PIPERIDINIUMBROMID (GERMAN) see AGE750
3-(ALLYL-(TETRAHYDRONAPHTHYL)AMINO)-N,N-DIETHYLPROPIONAMIDE HYDROCHLORIDE see AGS375
3-(ALLYL-(TETRAHYDRONAPHTHYL)AMINO)-N,N-DIETHYL-PROPIONAMIDE see AGS375
1-ALLYLTHEOBROMINE see AGS500
ALLYLTHEOBROMINE see AGS500
ALLYLTHIOCARBAMIDE see AGT500
ALLYL THIOCARBONIMIDE see AGJ250
ALLYL THIOCYANATE see AGS750
2-ALLYLTHIO-2-ETHYLBUTYRAMIDE see AGJ750
ALLYLTHIOMETHYLPENICILLIN see AGK250
4-ALLYLTHIOSEMICARBAZIDE see AGT000
2-(ALLYLTHIO)-2-THIAZOLINE see AGT250
1-ALLYLTHIOUREA see AGT500
1-ALLYL-2-THIOUREA see AGT500
N-ALLYLTHIOUREA see AGT500
ALLYL TRI-N-BUTYLPHOSPHONIUMCHLORIDE see AGT750
ALLYL TRICHLORIDE see TJB600
ALLYLTRICHLOROSILANE, stabilized (DOT) see AGU250
ALLYL TRICHLOROSILANE see AGU250
ALLYLTRIMETHYLAMMONIUM CHLORIDE see HGI600
ALLYL 3,5,5-TRIMETHYLHEXANOATE see AGU400
ALLYL TRIMETHYLHEXANOATE see AGU400
ALLYLTRIPHENYL STANNANE see AGU500
ALLYLTRIPHENYLTIN see AGU500
9-ALLYL-2-(4-(2-TRITYLETHYL)-1-PIPERAZINYL)-9H-PURINEDIMETHANESULFONATE see AGU750
1-ALLYLUREA see AGV000
N-ALLYLUREA see AGV000
ALLYLUREA see AGV000
4-ALLYLVERATROLE see AGE250
ALLYL VINYL ETHER see AGV250
N-ALLYNORATROPINE see AGM250
ALLYXYCARB see DBI800
(ALL-Z)-5,8,11,14-EICOSATETRAENOIC ACID see AQS750
ALMAZINE see CFC250
ALMECILLIN see AGK250
ALMEDERM see HCL000
ALMEFRIN see SPC500
ALMEFROL see VSZ450
(±)-ALMINOPROFEN see MGC200
ALMITE see AHE250
ALMITRINA (SPANISH) see BGS500
ALMOCARPINE see PIF000, PIF250
ALMOCETAMIDE see SNQ000
ALMOND ARTIFICIAL ESSENTIAL OIL see BAY500
ALMOND OIL BITTER, FFPA (FCC) see BLV500
ALMORA see MAG000
ALNOVIN see FAO100
ALNOX see AFS500
ALNOXIN see PQC500
ALOBARBITAL see AFS500
ALOCASIA (VARIOUS SPECIES) see EAI600
ALOCHLOR see CFX000
ALODAN (GEROT) see DAM700
ALOE see AGV875
ALOGINAN see FOS100
ALON see AHE250
A. LONGIFLORA see BOO700
ALOPERIDIN see CLY500
ALOPERIDOLO see CLY500
ALOSITOL see ZVJ000

ALOTEC see MDM800
A 1588LP see AQF000
ALPEN see AIV500
ALPEN-N see SEQ000
ALPERIDINE HYDROCHLORIDE see AGV890
ALPEROX C see LBR000
ALPHA CHYMAR see CML850
ALPHA-CHYMAR OPHTH see CML850
ALPHACILINA see AOD000
ALPHACILLIN see AOD000
ALPHADIONE see AFK500
ALPHALIN see VSK600
ALPHA MEDOPA see DNA800
ALPHAMIN see FOS100
ALPHANAPHTHYL THIOUREA see AQN635
ALPHANAPHTYL THIOUREE (FRENCH) see AQN635
ALPHAPRODINE see NEA500
ALPHAPRODINE HYDROCHLORIDE see NEB000
ALPHASOL OT see DJL000
ALPHASPRA see NAK500
ALPHASTEROL see VSK600
ALPHAXALONE see AFK875
ALPHAZURINE see FMU059
ALPHAZURINE A (6Cl) see ERG100
ALPHAZURINE 2G see ADE500
ALPHEBA see AGQ875
ALPHENAL see AGQ875
ALPHENATE see AGQ875
ALPHEX FIT 221 see PJS750
AL-PHOS see AHE750
ALPINE TALC USP, BC 127 see TAB750
ALPINE TALC USP, BC 141 see TAB750
ALPINE TALC USP, BC 662 see TAB750
ALPINYL see HIM000
ALPRAZOLAM see XAJ000
ALPRENOL HYDROCHLORIDE see AGW000
ALPRENOLOL see AGW250
ALPROSTADIL see POC500
ALPROSTADIL-α-CYCLODEXTRIN CLATHRATE see AGW275
ALQOVERIN see BRF500
ALQUEQUENJE (PUERTO RICO) see JBS100
ALRATO see AQN635
ALRHEUMAT see BDU500
ALRHEUMUM see BDU500
ALSERIN see RDK000
ALSEROXYLON see AGW300
ALSTONINE HYDROCHLORIDE see AGW375
ALTABACTINA see FPI000
ALTADIOL see EDS100
ALTAFUR see FPI000
ALTAN see DMH400
ALTAX see BDE750
ALTAZINE BLACK BH see CMN800
ALTCO SPERSE FAST YELLOW GFN NEW see AAQ250
ALTERNARIOL see AGW476
ALTERNARIOL and ALTERNARIOL MONOMETHYL ETHER (1:1) see AGW550
ALTERNARIOL MONOMETHYL ETHER and ALTERNARIOL (1:1) see AGW550
ALTERTON see CKE750
ALTERUNGSSCHUTZMITTEL ZMB see ZIS000
ALTEZOL see AKO500
ALTHESIN see AFK500
ALTHOSE HYDROCHLORIDE see MDP000
ALTICINA see PDD350
ALTOCHROME MILLING SCARLET G see CMM320
ALTOCHROME SCARLET G see CMM325
ALTOCYL BRILLIANT BLUE B see MGG250
ALTODEL see POB000
ALTODOR see DIS600
ALTOSID (GERMAN) see AGW625
ALTOSID see KAJ000
ALTOSIDE see AGW625
ALTOSID IGR see KAJ000
ALTOSID SR 10 see KAJ000
ALTOTAL see PQC500
ALTOWHITES see KBB600
ALTOX see AFK250
ALTOZAR see EQD000
ALTRAD see EDO000
ALTRETAMINE see HEJ500
ALUDRINE see DMV600
ALUFENAMINE see AHA875
ALUFIBRATE see AHA150
ALUGEL 34TN see AHH825
ALULINE see ZVJ000
ALUM see AHG750

ALUMIGEL see AHC000
β-ALUMINA see AHE250, AHG000
β''-ALUMINA see AHG000
γ-ALUMINA see AHE250
α-ALUMINA (OSHA) see AHE250
ALUMINA see AHE250
ALUMINA FIBRE see AGX000
ALUMINA HYDRATE see AHC000
ALUMINA HYDRATED see AHC000
ALUMINATE(1−), TETRAHYDRO-, SODIUM, (T-4)-(9Cl) see SEM500
α-ALUMINA TRIHYDRATE see AHC000
ALUMINA TRIHYDRATE see AHC000
ALUMINIC ACID see AHC000
ALUMINIUM BRONZE see AGX000
ALUMINIUM CHLOROHYDROXYALLANTOINATE see AHA135
ALUMINIUM STEARATE see AHH825
ALUMINON see AGW750
ALUMINOPHOSPHORIC ACID see PHB500
ALUMINUM 27 see AGX000
ALUMINUM see AGX000
ALUMINUM A00 see AGX000
ALUMINUM ACEGLUTAMIDE see AGX125
ALUMINUM(III) ACETYLACETONATE see TNN000
ALUMINUM ACETYLACETONATE see TNN000
ALUMINUM ACID PHOSPHATE see PHB500
ALUMINUM ALLOY, Al,Be see BFP250
ALUMINUM AMMONIUM SULFATE see AGX250
ALUMINUM AZIDE see AGX300
ALUMINUM BERYLLIUM ALLOY see BFP250
ALUMINUM BOROHYDRIDE in devices (DOT) see AHG875
ALUMINUM BOROHYDRIDE (DOT) see AHG875
ALUMINUM BOROHYDRIDE see AGX500
ALUMINUM BROMHYDROXIDE see AGY000
ALUMINUM BROMIDE see AGX750
ALUMINUM BROMIDE HYDROXIDE see AGY000
ALUMINUM BROMIDE, anhydrous (UN 1725) (DOT) see AGX750
ALUMINUM BROMIDE, solution (UN 2580) (DOT) see AGX750
ALUMINUM BROMOHYDROL see AGY000
ALUMINUM CALCIUM SILICATE see AGY100
ALUMINUM CARBIDE see AGY250
ALUMINUM CHLORATE see AGY500
ALUMINUM CHLORHYDRATE see AHA000
ALUMINUM CHLORHYDROL see AHA000
ALUMINUM CHLORHYDROXIDE see AHA000
ALUMINUMCHLORID (GERMAN) see AGY750
ALUMINUM CHLORIDE (1:3) see AGY750
ALUMINUM CHLORIDE, anhydrous (DOT) see AGY750
ALUMINUM CHLORIDE, solution (DOT) see AGY750
ALUMINUM CHLORIDE see AGY750
ALUMINUM(III) CHLORIDE, HEXAHYDRATE see AGZ000
ALUMINUM CHLORIDE HEXAHYDRATE see AGZ000
ALUMINUM CHLORIDE HYDROXIDE see AHA000
ALUMINUM CHLORIDE NITROMETHANE see AHA125
ALUMINUM CHLORIDE OXIDE see AHA130
ALUMINUM, CHLORO((2,5-DIOXO-4-IMIDAZOLIDI-NYL)UREATO)TETRAHYDROXYDI- see AHA135
ALUMINUM CHLOROHYDROXIDE see AHA000
ALUMINUM CHLOROHYDROXYALLANTOINATE see AHA135
ALUMINUM, CHLOROTETRAHYDROXY((2-HYDROXY-5-OXO-2-IMIDAZOLIN-4-YL)UREATO)DI- see AHA135
ALUMINUM CLOFIBRATE see AHA150
ALUMINUM COMPOUNDS see AHA175
ALUMINUM DEHYDRATED see AGX000
ALUMINUM DEXTRAN see AHA250
ALUMINUM, DICHLOROETHYL- see EFU050
ALUMINUM DISTEARATE (ACGIH) see AHA275
ALUMINUM DISTEARATE see AHA275
ALUMINUM ETHYLATE see AHA750
ALUMINUM FLAKE see AGX000
ALUMINUM FLUFENAMATE see AHA875
ALUMINUM FLUORIDE see AHB000
ALUMINUM FLUOROSULFATE, HYDRATE see AHB250
ALUMINUM FLUORURE (FRENCH) see AHB000
ALUMINUM FORMATE see AHB375
ALUMINUM FOSFIDE (DUTCH) see AHE750
ALUMINUM HEXAFLUOROSILICATE see THH000
ALUMINUM HYDRATE see AHC000
ALUMINUM HYDRIDE see AHB500
ALUMINUM HYDRIDE-DIETHYL ETHER see AHB625
ALUMINUM HYDRIDE-TRIMETHYL AMINE see AHB750
ALUMINUM HYDROBORATE see AHG875
ALUMINUM(III) HYDROXIDE see AHC000
ALUMINUM HYDROXIDE see AHC000
ALUMINUM HYDROXIDE CHLORIDE see AHA000
ALUMINUM HYDROXIDE DISTEARATE see AHA275

ALUMINUM HYDROXIDE GEL see AHC000
ALUMINUM HYDROXIDE OXIDE see AHC250
ALUMINUM, HYDROXYBIS(OCTADECANOATO-O)-(9CI) see AHA275
ALUMINUM, HYDROXYBIS(STEARATO)- see AHA275
ALUMINUM HYDROXYBROMIDE see AGY000
ALUMINUM HYDROXYCHLORIDE see AHA000
ALUMINUM HYDROXYDISTEARATE see AHA275
ALUMINUM IODIDE see AHC500
ALUMINUM ISOPROPOXIDE see AHC600
ALUMINUM(III) ISOPROPYLATE see AHC600
ALUMINUM LITHIUM HYDRIDE see LHS000
ALUMINUM MAGNESIUM PHOSPHIDE see AHD250
ALUMINUM METAHYDROXIDE see AHC250
ALUMINUM METAL (OSHA) see AGX000
ALUMINUM METHYL see AHD500
ALUMINUM MONOPHOSPHATE see PHB500
ALUMINUM MONOPHOSPHIDE see AHE750
ALUMINUM MONOSTEARATE see AHA250
ALUMINUM, molten (NA 9260) (DOT) see AGX000
ALUMINUMNATRIUMLACTAT (GERMAN) see AHF750
ALUMINUM(III) NITRATE (1:3) see AHD750
ALUMINUM NITRATE (DOT) see AHD750
ALUMINUM(III) NITRATE, NONAHYDRATE (1:3:9) see AHD900
ALUMINUM NITRATE NONAHYDRATE see AHD900
ALUMINUM NITRIDE see AHE000
ALUMINUM OXIDE (2:3) see AHE250
α-ALUMINUM OXIDE see AHE250
β-ALUMINUM OXIDE see AHE250
γ-ALUMINUM OXIDE see AHE250
ALUMINUM OXIDE see AHE250, EAL100
ALUMINUM OXIDE CHLORIDE see AHA130
ALUMINUM OXIDE HYDRATE see AHC000
ALUMINUM OXIDE SILICATE see AHF500
ALUMINUM OXIDE TRIHYDRATE see AHC000
ALUMINUM OXYCHLORIDE see AHA130
ALUMINUM PHOSPHATE (1:1) see PHB500
ALUMINUM PHOSPHATE, solution (DOT) see PHB500
ALUMINUM PHOSPHATE see PHB500
ALUMINUM PHOSPHIDE see AHE750
ALUMINUM PHOSPHINATE see AHE875
ALUMINUM PICRATE see AHF000
ALUMINUM POTASSIUM ALUM see AHF100
ALUMINUM POTASSIUM DISULFATE see AHF100
ALUMINUM POTASSIUM SULFATE see AHF100
ALUMINUM POTASSIUM SULFATE, ALUM see AHF100
ALUMINUM POTASSIUM SULFATE, ANHYDROUS see AHF100
ALUMINUM POWDER see AGX000
ALUMINUM POWDER, uncoated (UN 1396) (DOT) see AGX000
ALUMINUM POWDER, coated (UN 1309) (DOT) see AGX000
ALUMINUM PYRO POWDERS (OSHA) see AGX000
ALUMINUM SESQUIOXIDE see AHE250
ALUMINUM(III) SILICATE (2:1) see AHF500
ALUMINUM SODIUM FLUORIDE see SHF000
ALUMINUM SODIUM HYDRIDE see SEM500
ALUMINUM SODIUM LACTATE see AHF750
ALUMINUM SODIUM OXIDE see AHG000
ALUMINUM SODIUM SULFATE see AHG500
ALUMINUM STEARATE (ACGIH) see AHA250
ALUMINUM STEARATE see AHH825
ALUMINUM SULFATE (2:3) see AHG750
ALUMINUM TETRAHYDROBORATE see AGX500, AHG875
ALUMINUM THALLIUM SULFATE see AHH000
ALUMINUM-TITANIUM ALLOY (1:1) see AHH125
ALUMINUM TRIACETYLACETONATE see TNN000
ALUMINUM TRIBROMIDE see AGX750
ALUMINUM, TRIBROMOTRIMETHYLDI- see MGC225
ALUMINUM TRICHLORIDE see AGY750
ALUMINUM TRICHLORIDEHEXAHYDRATE see AGZ000
ALUMINUM, TRICHLOROTRIMETHYLDI- see MGC230
ALUMINUM, TRIETHYL- see TJN750
ALUMINUM TRIFLUORIDE see AHB000
ALUMINUM TRIHYDRAT see AHC000
α-ALUMINUM TRIHYDRIDE see AHB500
ALUMINUM TRIHYDRIDE see AHB500
ALUMINUM TRIHYDROXIDE see AHC000
ALUMINUM TRINITRATE see AHD750
ALUMINUM TRINITRATE NONAHYDRATE see AHD900
ALUMINUM TRIPROPYL see AHH750
ALUMINUM, TRIPROPYL- see TMY100
ALUMINUM TRIS(ACETYLACETONATE) see TNN000
ALUMINUM, TRIS(2-METHYLPROPYL)-(9CI) see TKR500
ALUMINUM TRISTEARATE see AHH825
ALUMINUM (TRIS(2-((3-TRIFLUOROMETHYL)PHENYL)AMINO)BENZOATO-N,o)-o-
 PYRIN see AHA875
ALUMINUM TRISULFATE see AHG750

ALUMINUM WELDING FUMES (OSHA) see AGX000
ALUMITE see AHE250
ALUM POTASSIUM see AHF100
A 21 LUNDBECK see KFK000
ALUNDUM see AHE250
ALUNEX see TAI500
ALUPENT see MDM800
ALUPHOS see PHB500
ALURAL see BNP750
ALURATE see AFT000
ALURENE see CLH750
ALUSAL see AHC000
ALUTOR M 70 see PKB500
ALUTYL see PGG000
ALUZINE see CHJ750
ALVEDON see HIM000
ALVINOL see CHG000
ALVIT see DHB400
ALVO see OLW600
ALVYL see PKP750
ALYPIN see AHI250
ALYPINE see AHI250
ALYSINE see SJO000
ALZODEF see CAQ250
AM-715 see BAB625
AM-2604 A see VRP775
1-A-2-MA (RUSSIAN) see AKM250
AMABEVAN see CBJ000
AMACEL BLUE BNN see MGG250
AMACEL BLUE GG see TBG700
AMACEL BRILLIANT BLUE B see MGG250
AMACEL CERISE B see DBX000
AMACEL DEVELOPED NAVY SD see DCJ200
AMACEL GREEN BLUE B see DMM400
AMACEL GREEN BLUE G see DMM400
AMACEL HELIOTROPE R see DBP000
AMACEL PINK B see AKE250
AMACEL PURE BLUE B see TBG700
AMACEL RUBINE B see CMP080
AMACEL VIOLET 6B see AKP250
AMACEL YELLOW G see AAQ250
AMACEL YELLOW RR see DUW500
AMACID AMARANTH see FAG020
AMACID BLACK 10BR see FAB830
AMACID BLUE FG CONC see FMU059
AMACID BLUE V see ADE500
AMACID BRILLIANT BLUE see FAE100
AMACID CHROME BLUE R see HJF500
AMACID FAST BLUE R see ADE750
AMACID FAST YELLOW RS see CMM759
AMACID FUCHSINE 4B see CMS231
AMACID GREEN B see FAE950
AMACID GREEN G see FAF000
AMACID LAKE SCARLET 2R see FMU070
AMACID MILLING RED PGS see CMM320
AMACID MILLING RED PRS see CMM330
AMACID MILLING SCARLET G see CMM325
AMACID ORANGE Y see CMM220
AMACID PHLOXINE see CMM300
AMACID WOOL GREEN S see ADF000
AMACID YELLOW M see MDM775
AMACID YELLOW RG see CMM758
AMACID YELLOW T see FAG140
A. MACULATUM see ITD050
AMADIL see HIM000
AMAIZO W 13 see SLJ500
AMAL see AMX750
AMALOX see ZKA000
AMANIL BLACK GL see AQP000
AMANIL BLACK WD see AQP000
AMANIL BLUE 2BX see CMO000
AMANIL BLUE RW see CMO600
AMANIL CHLORAMINE RED 8BS see CMO880
AMANIL DEVELOPED BLACK BHSW see CMN800
AMANIL FAST BROWN HP see CMO820
AMANIL FAST RED 8BL see CMO885
AMANIL FAST RED 8BLW see CMO885
AMANIL FAST RED FS see CMO870
AMANIL FAST SCARLET 3B see CMO875
AMANIL FAST VIOLET N see CMP000
AMANIL GREEN B see CMO840
AMANIL GREEN LT see CMO830
AMANIL RAYON BROWN B see CMO820
AMANIL SKY BLUE see CMO250, CMO500
AMANIL SUPRA BLUE 9GL see CMO650

AMANIL SUPRA BROWN LBL see CMO750
AMANITA RUBESCENS TOXIN see AHI500
α-AMANITIN (8CI, 9CI) see AHI625
α-AMANITINE see AHI625
AMANO N-AP see GGA800
AMANTADINE see TJG250
AMANTADINE HYDROCHLORIDE see AED250
AMANTHRENE BLUE BCL see DFN425
AMANTHRENE BRILLIANT VIOLET RR see DFN450
AMANTHRENE BROWN BR see CMU770
AMANTHRENE GOLDEN YELLOW see DCZ000
AMANTHRENE NAVY BLUE BN see VGP100
AMANTHRENE OLIVE R see DUP100
AMANTHRENE ORANGE R see CMU815
AMANTHRENE SUPRA NAVY BLUE BN see VGP100
AMANTHRENE SUPRA NAVY BLUE BNR see VGP100
AMAPALO AMRILLO (PUERTO RICO) see PLW800
AMAPLAST GREEN OZ see BLK000
AMAPLAST RED VIOLET P 2R see DBP000
AMARANT see FAG020
AMARANTH see FAG020
AMARANTHE USP (biological stain) see FAG020
AMARBEL EXTRACT see AHI630
AMAREX see MQQ250
AMARSAN see ABX500
AMARTHOL FAST RED B BASE see NEQ000
AMARTHOL FAST RED TR BASE see CLK220, CLK235
AMARTHOL FAST RED TR SALT see CLK235
AMARTHOL FAST SCARLET G BASE see NMP500
AMARTHOL FAST SCARLET G SALT see NMP500
AMARTHOL FAST SCARLETT GG BASE see DEO295
AMARYLLIS see AHI635
AMASUST see AMX750
AMATIN see HCC500
AMATOL see AHI750
AMAX see BDG000
AMAZE see IMF300
AMAZOLON see AED250
AMAZON YELLOW X2485 see DEU000
AMB see AOC500
AMBACAMP see BAB250
AMBAM see ANZ000
AMBATHIZON see FNF000
AMBAZON see AHI875
AMBAZONE see AHI875
AMBEN see AIH600
AMBENONIUM CHLORIDE see MSC100
AMBENONIUM DICHLORIDE see MSC100
AMBENYL see BAU750
AMBER see AHJ000
AMBER ACID see SMY000
AMBERGRIS TINCTURE see AHJ000
AMBEROL ST 140F see AHC000
AMBESIDE see SNM500
AMBESTIGMIN CHLORIDE see MSC100
AMBIBEN see AJM000
AMBILHAR see NML000
AMBISTRIN see SLY200
AMBITERIC D 40 see LBU200
AMBLOSIN see AIV500
AMBOCHLORIN see CDO500
AMBOCLORIN see CDO500
AMBOFEN see CDP250
AMBOMYCIN see AFI625
AMBOX see BGB500
AMBRA see AHJ000
AMBRACYN see TBX250
AMBRAMICINA see TBX000
AMBRAMYCIN see TBX000
AMBRATE see EQN300
AMBRETTOLID see OKU100
AMBRETTOLIDE see OKU100
AMBROXOL see AHJ250
AMBROXOL HYDROCHLORIDE see AHJ500
AMBRUNATE see MNQ500
AMBUCETAMID see DDT300
AMBUCETAMIDE see DDT300
AMBUSH see AHJ750, CBM500
AMBUTYROSIN A see BSX325
AMBUYROSIN A see BSX325
AMBYTHENE see PJS750
6-AMC see CML800
AMCAP see AOD125
AMCHA see AJV500
trans-AMCHA see AJV500

AMCHEM 70-25 see BQN600
AMCHEM 68-250 see CDS125
AMCHEM A-280 see BQN600
AMCHEM GRASS KILLER see TII250
AMCHEM R 14 see PKL750
AMCHEM 2,4,5-TP see TIX500
AMCHLOR see ANE500
AMCIDE see ANU650
AMCILL see AIV500, AOD125
AMCILL-S see SEQ000
AMCINONIDE see COW825
AMCO see PMP500
AMD see DNA800, MJE780
AMDEX see BBK500
AMDON GRAZON see PIB900
AMDRAM see DBA800
AME and AOH (1:1) see AGW550
AMEBACILIN see FOZ000
AMEBAN see CBJ000
AMEBARSONE see CBJ000
AMEBICIDE see EAN000
AMEBIL see CHR500
AMECHOL see ACR000
AMEDEL see BHJ250
AMEDRINE see DBA800
AMEISENATOD see BBQ500
AMEISENMITTEL MERCK see BBQ500
AMEISENSAEURE (GERMAN) see FNA000
AMEPROMAT see MQU750
AMERCIAN CYANAMID 18133 see EPC500
AMERCIDE see CBG000
AMERFIL see PMP500
AMERICAINE see EFX000
AMERICAN ALLSPICE see CCK675
AMERICAN BITTERSWEET see AHJ875
AMERICAN CL-26691 see CQL250
AMERICAN CYANAMID 12,503 see POP000
AMERICAN CYANAMID 12880 see DSP400
AMERICAN CYANAMID 12,008 see DJN600
AMERICAN CYANAMID 18682 see IOT000
AMERICAN CYANAMID 18706 see DNX600
AMERICAN CYANAMID-38023 see FAB600
AMERICAN CYANAMID-43073 see MJG500
AMERICAN CYANAMID 4,049 see MAK700
AMERICAN CYANAMID 47031 see PGW750
AMERICAN CYANAMID 5223 see DXX400
AMERICAN CYANAMID AC 43,064 see DXN600
AMERICAN CYANAMID AC 43,913 see PHN250
AMERICAN CYANAMID AC 52,160 see TAL250
AMERICAN CYANAMID CL-24055 see DUI000
AMERICAN CYANAMID CL-26,691 see CQL250
AMERICAN CYANAMID CL-38,023 see FAB600
AMERICAN CYANAMID CL-43913 see PHN250
AMERICAN CYANAMID CL-47,300 see DSQ000
AMERICAN CYANAMID CL-47470 see DHH400
AMERICAN CYANAMID E.I. 43,913 see PHN250
AMERICAN CYANIMID 24,055 see DUI000
AMERICAN ELDER see EAI100
AMERICAN HELLEBORE see VIZ000
AMERICAN HOLLY see HGF100
AMERICAN LAUREL see MRU359
AMERICAN MANDRAKE see MBU800
AMERICAN MEZEREON see LEF100
AMERICAN NIGHTSHADE see PJJ315
AMERICAN PENICILLIN see BFD250
AMERICAN PENNYROYAL OIL see PAR500
AMERICAN VERATRUM see VIZ000
AMERICAN WHITE HELLEBORE see FAB100
AMERICIUM see AHK000
AMERICIUM TRICHLORIDE see AHK250
AMERIZOL see TOA000
AMERLATE P see IPS500
AMERLATE W see IPS500
AMEROL see AMY050
AMEROX OE-20 see PJW500
AMET (GERMAN) see AAI750
AMETANTRONE ACETATE see BKB300
AMETHOCAINE see BQA010
AMETHOCAINE HYDROCHLORIDE see TBN000
AMETHONE HYDROCHLORIDE see DIF200
AMETHOPTERIN see MDV500, MDV750
AMETHOPTERIN SODIUM see MDV600
AMETHYST see SCI500, SCJ500
AMETOTERINA see ABY900
AMETOX see SKI500

AMETRIODINIC ACID see AAI750
AMETYCIN see AHK500
AMEX 820 see BQN600
AMEX see BQN600
AMEZINIUMMETILSULFAT (GERMAN) see MQS100
AMEZINIUM METILSULFATE see MQS100
AMFENACO (SPANISH) see AIU500
AMFENAC SODIUM MONOHYDRATE see AHK625
AMFETAMINA see AOB250
AMFETAMINE see AOB250
d-AMFETASUL see BBK500
AMFETYLINE HYDROCHLORIDE see CBF825
AMFH see AAH100
AMFIPEN see AIV500
AM-FOL see AMY500
AMFOMYCIN see AOB875
AMGABA see AJC375
AMIANTHUS see ARM250
AMIBIARSON see CBJ000
AMICAL 48 see DNF850
AMICAR see AJD000
AMICARDINE see AHK750
AMICETIN see AHL000
AMICETIN CITRATE see AHL250
AMICHLOPHENE see CJN750
AMICIDE see ANU650
AMICIN see NHG000
AMIDANTEL see DPF200
AMIDATE see HOU100
AMIDAZIN see EPQ000
AMIDAZOPHEN see DOT000
AMIDEFRINE MESYLATE see AHL500
AMIDE of GABA see AJC375
AMIDEPHRINE MESYLATE see AHL500
AMIDEPHRINE MONOMETHANESULFONATE see AHL500
AMIDE PP see NCR000
AMIDES see AHL750
AMIDINE BLUE 4B see CMO250
p-AMIDINOBENZOIC ACID BUTYL ESTER see AHL875
p-AMIDINOBENZOIC ACID HEXYL ESTER see AHL880
p-AMIDINOBENZOIC ACID PENTYL ESTER see AHL885
p-AMIDINOBENZOIC ACID PROPYL ESTER see AHL890
N-AMIDINO-2-(2,6-DICHLOROPHENYL)ACETAMIDE HYDROCHLORIDE see
    GKU300
N-(2-AMIDINOETHYL)-3-AMINOCYCLOPENTANOCARBOXAMIDE see AHN625
N''-(2-AMIDINOETHYL)-4-FORMAMIDO-1,1',1''-TRIMETHYL-(N,4':N',4''-TERPYR-
    ROLE)-2-CARBOXAMIDE HYDROCHLORIDE see DXG600
N''-(2-AMIDINOETHYL)-4-FORMAMIDO-1,1',1''-TRIMETHYL-(N,4':N',4''-TERPYR-
    ROLE)-2-CARBOXAMIDE see SLI300
1-AMIDINOHYDRAZONO-4-THIOSEMICARBAZONO-2,5-CYCLOHEXADIENE see
    AHI875
S-(AMIDINOMETHYL) HYDROGEN THIOSULFATE see AHN000
AMIDINOMYCIN see AHN625
3-AMIDINO-1-p-NITROPHENYLUREA HYDROCHLORIDE see NIJ400
4-AMIDINO-1-(NITROSAMINOAMIDINO)-1-TETRAZENE see TEF500
4-AMIDINO-1-(NITRSOAMINOAMIDINO)-1-TETRAZENE see TEF500
N'-AMIDINOSULFANILAMIDE see AHO250
AMIDITHION see AHO750
AMID KYSELINY AKRYLOVE see ADS250
AMID KYSELINY OCTOVE (POLISH) see AAI000
AMID KYSELINY PROPIONOVE see PMU250
AMID KYSELINY STAVELOVE (CZECH) see OLO000
AMID KYSELINY TRICHLOROCTOVE see TII000
o-AMIDOAZOTOLUOL (GERMAN) see AIC250
o-AMIDOBENZOIC ACID see API500
AMIDOCYANOGEN see COH500
AMIDOFEBRIN see DOT000
AMIDOFOS see COD850
AMIDO-G-ACID see NBE850
AMIDOL see AHP000
AMIDOLINE see AHP125
AMIDON see ILD000
AMIDO NAPHTHOL RED 2G see CMM300
AMIDO NAPHTHOL RED G see CMM300
AMIDO NAPHTHOL RED GA see CMM300
AMIDONE see MDO750
AMIDONE HYDROCHLORIDE see MDP000
AMIDON HYDROCHLORIDE see MDP750
AMIDOPEROXYMONOSULFURIC ACID see HLN100
AMIDOPHEN see DOT000
AMIDOPHENAZONE see DOT000
AMIDOPHOS see COD850
AMIDOPROCAIN see PME000
AMIDOPYRAZOLINE see DOT000
AMIDOPYRIN see DOT000

AMIDO RED 2G see CMM300
AMIDOSAL see SAH000
AMIDOSULFONIC ACID see SNK500
AMIDOSULFONIC PERACID see HLN100
AMIDOSULFURIC ACID see SNK500
AMIDO SULFURYL AZIDE see AHP250
AMIDOUREA HYDROCHLORIDE see SBW500
AMIDOX see DAA800
AMIDOXAL see SNN500
AMIDO YELLOW E see SGP500
AMIDO YELLOW EA see SGP500
AMIDO YELLOW EA-CF see SGP500
AMIDRINE see ILM000
AMIDRYL see BBV500
AMID-SAL see SAH000
AMID-THIN see NAK000
AMIFENATSOL HYDROCHLORIDE see DCA600
AMIFUR see NGE500
AMIKACIN see APS750
AMIKACIN SULFATE see APT000
AMIKAPRON see AJV500
AMIKHELLIN HYDROCHLORIDE see AHP375
AMIKIN see APT000
AMIKLIN see APT000
AMIKOL 65 see UTU500
AMILAC 3 see MCB050
AMILAN see NOH000
AMILAN CM 1001 see PJY500
AMILAR see PKF750
AMILPHENOL see AON000
AMINACRINE see AHS500
AMINACRINE HYDROCHLORIDE see AHS750
AMINARSON see CBJ000
AMINARSONE see CBJ000
AMINASINE see CKP250
AMINATE BASE see AKC750
AMINAZIN see CKP250
AMINAZINE see CKP250
AMINAZIN MONOHYDROCHLORIDE see CKP500
AMINE 220 see AHP500
AMINE AB see OBC000
AMINE BB see DXW000
AMINES, FATTY see AHP760
AMINE 2,4,5-T FOR RICE see TAA100
AMINIC ACID see FNA000
AMINICOTIN see NCR000
AMINITROZOL see ABY900
AMINITROZOLE see ABY900
5-AMINOACENAPHTHENE see AAF250
2-AMINO-4-ACETAMINIFENETOL (CZECH) see AJT750
3-AMINOACETANILID (CZECH) see AHQ000
4'-AMINOACETANILID (CZECH) see AHQ250
3'-AMINOACETANILIDE see AHQ000
4'-AMINOACETANILIDE see AHQ250
4-AMINOACETANILIDE see AHQ250
m-AMINOACETANILIDE see AHQ000
p-AMINOACETANILIDE see AHQ250
3-AMINOACETANILIDE-4-SULFONIC ACID see AHQ300
AMINOACETIC ACID see GHA000
AMINOACETONITRILE SULFATE see AHR000
2-AMINOACETOPHENONE see AHR250
3'-AMINOACETOPHENONE see AHR500
4'-AMINOACETOPHENONE see AHR240
β-AMINOACETOPHENONE see AHR500
ω-AMINOACETOPHENONE see AHR250
m-AMINOACETOPHENONE see AHR500
p-AMINOACETOPHENONE see AHR240
m-AMINOACETYLBENZENE see AHR500
p-AMINOACETYLBENZENE see AHR240
2-AMINOACRIDINE see AHS000
3-AMINOACRIDINE see ADJ375
5-AMINOACRIDINE see AHS500
9-AMINOACRIDINE see AHS500
2-AMINOACRIDINE (EUROPEAN) see ADJ375
3-AMINOACRIDINE (EUROPEAN) see AHS000
5-AMINOACRIDINE HYDROCHLORIDE see AHS750
AMINOACRIDINE HYDROCHLORIDE see AHS750
9-AMINOACRIDINE MONOHYDROCHLORIDE see AHS750
9-AMINOACRIDINE PENICILLIN see AHT000
4'-(2-AMINO-9-ACRIDINYLAMINO)METHANESULFONANILIDE see AHT825
1-AMINOADAMANTANE see TJG250
AMINOADAMANTANE HYDROCHLORIDE see AED250
1-AMINOADAMANTENE HYDROCHLORIDE see AED250
1-AMINOADAMATANE see TJG250
2-AMINOADENINE see POJ500

2-AMINOAETHANOL (GERMAN) see EEC600
2-AMINOAETHYLISOSELENOURONIUMBROMID-HYDROBROMID (GERMAN)
    see AJY000
β-AMINOAETHYLISOTHIURONIUM-CHLORID-HYDROCHLORID (GERMAN) see
    AJY500
β-AMINOAETHYL-ISOTHIURONIUM DIHYDROBROMID(GERMAN) see AJY250
β-AMINOAETHYL-MORPHOLIN (GERMAN) see AKA750
4-AMINO-6-((2-AMINO-1,6-DIMETHYLPYRIMIDINIUM-4-YL)AMINO)-1-METHYL-
    QUINALDINIUM BIS(METHYL SULFATE) see AQN625
6-AMINO-2-(3′-AMINOFENYL)BENZIMIDAZOL HYDROCHLORID (CZECH) see
    AHT900
6-AMINO-2-(4′-AMINOFENYL)BENZIMIDAZOL HYDROCHLORID (CZECH) see
    AHT950
4-AMINO-N-(AMINOIMINOMETHYL)BENZENESULFONAMIDE see AHO250
3-AMINO-N-(3-AMINO-3-IMINOPROPYL)CYCLOPENTANECARBOXAMIDE see
    AHN625
6-AMINO-4-((3-AMINO-4-(((4-((1-METHYLPYRIDINIUM-4-
    YL)AMINO)PHENYL)AMINO)CARBONYL)PHENYL)AMINO)-1-METH see
    AHT850
4-AMINO-N-(2′-AMINOPHENYL)BENZAMIDE see DBQ125
6-AMINO-2-(3′-AMINOPHENYL)BENZIMIDAZOLE, DIHYDROCHLORIDE see
    AHT900
6-AMINO-2-(4′-AMINOPHENYL)BENZIMIDAZOLE DIHYDROCHLORIDE see
    AHT950
2-AMINOANILINE see PEY250
3-AMINOANILINE see PEY000
4-AMINOANILINE see PEY500
m-AMINOANILINE see PEY000
p-AMINOANILINE see PEY500
3-AMINOANILINE DIHYDROCHLORIDE see PEY750
4-AMINOANILINE DIHYDROCHLORIDE see PEY650
m-AMINOANILINE DIHYDROCHLORIDE see PEY750
p-AMINOANILINE DIHYDROCHLORIDE see PEY650
3-AMINO-p-ANISIC ACID see AIA250
2-AMINOANISOLE see AOV900
3-AMINOANISOLE see AOV890
4-AMINOANISOLE see AOW000
m-AMINOANISOLE see AOV890
o-AMINOANISOLE see AOV900
p-AMINOANISOLE see AOW000
2-AMINOANISOLE HYDROCHLORIDE see AOX250
o-AMINOANISOLE HYDROCHLORIDE see AOX250
4-AMINOANISOLE-3-SULFONIC ACID see AIA500
1-AMINOANTHRACENE see APG050
2-AMINOANTHRACENE see APG100
α-AMINOANTHRACENE see APG050
β-AMINOANTHRACENE see APG100
1-AMINO-9,10-ANTHRACENEDIONE see AIA750
2-AMINO-9,10-ANTHRACENEDIONE see AIB000
1-AMINOANTHRACHINON (CZECH) see AIA750
1-AMINOANTHRAQUINONE see AIA750
2-AMINOANTHRAQUINONE see AIB000
1-AMINO-9,10-ANTHRAQUINONE see AIA750
2-AMINO-9,10-ANTHRAQUINONE see AIB000
α-AMINOANTHRAQUINONE see AIA750
β-AMINOANTHRAQUINONE see AIB000
N-(4-AMINOANTHRAQUINONYL)BENZAMIDE see AIB250
4-AMINOANTIPYRENE see AIB300
AMINOANTIPYRIN see AIB300
4-AMINOANTIPYRINE see AIB300
AMINOANTIPYRINE see AIB300
4-AMINO-1-ARABINOFURANOSYL-2-OXO-1,2-DIHYDROPYRIMIDIN see AQQ750
4-AMINO-1-ARABINOFURANOSYL-2-OXO-1,2-DIHYDROPYRIMIDINE see
    AQQ750
4-AMINO-1-β-D-ARABINOFURANOSYL-2(1H)-PYRIMIDINON see AQQ750
4-AMINO-1-β-D-ARABINOFURANOSYL-2(1H)-PYRIMIDINONE see AQQ750
2-AMINO-4-ARSENOSOPHENOL see OOK100
2-AMINO-4-ARSENOSOPHENOL HYDROCHLORIDE see ARL000
AMINOARSON see CBJ000
6-AMINO-8-AZAPURINE see AIB340
4-AMINO-1,1′-AZOBENZENE see PEI000
4-AMINOAZOBENZENE see PEI000
p-AMINOAZOBENZENE see PEI000
AMINOAZOBENZENE see PEI000
4-AMINOAZOBENZENE-3,4′-DISULFONIC ACID see AJS500
4-AMINOAZOBENZENE-3,4′-DISULFONIC ACID DISODIUM SALT see CMM758
4-AMINOAZOBENZENE HYDROCHLORIDE see PEI250
p-AMINOAZOBENZENE HYDROCHLORIDE see PEI250
4-AMINOAZOBENZOL see PEI000
p-AMINOAZOBENZOL see PEI000
6-AMINO-3,4′-AZODI-BENZENESULFONIC ACID see AJS500
AMINOAZOPHENAZONE see AIB300
4′-AMINO-2,3′-AZOTOLUENE see AIC250
4′-AMINO-2:3′-AZOTOLUENE see AIC250
p-AMINO-2′:3-AZOTOLUENE see AIC000

2′-AMINO-2:5′-AZOTOLUENE see TGW500
4′-AMINO-3,2′-AZOTOLUENE see AIC000
4′-AMINO-4,2′-AZOTOLUENE see AIC500
4′-AMINO-4-3′-AZOTOLUENE see TGW750
2-AMINO-5-AZOTOLUENE see AIC250
o-AMINOAZOTOLUENE (MAK) see AIC250
AMINOAZOTOLUENE (indicator) see AIC250
o-AMINOAZOTOLUENO (SPANISH) see AIC250
o-AMINOAZOTOLUOL see AIC250
AMINOBENZ see AKF000
10-AMINOBENZ(a)ACRIDINE see AIC750
4-AMINOBENZALDEHYDE see AIC825
p-AMINOBENZALDEHYDE see AIC825
4-AMINOBENZALDEHYDE THIOSEMICARBAZONE see AID250
p-AMINOBENZALDEHYDETHIOSEMICARBAZONE see AID250
m-AMINOBENZAL FLUORIDE see AID500
2-AMINOBENZAMIDE see AID620
3-AMINOBENZAMIDE see AID625
m-AMINOBENZAMIDE see AID625
o-AMINOBENZAMIDE see AID620
3-AMINO-BENZAMIDE (9CI) see AID625
1-AMINO-4-BENZAMIDOANTHRAQUINONE see AIB250
10-AMINO-1,2-BENZANTHRACENE see BBB750
5-AMINO-1:2-BENZANTHRACENE see BBC000
AMINOBENZENE see AOQ000
4-AMINOBENZENEACETIC ACID see AID700
(±)-α-AMINO-BENZENEACETIC ACID, DECYL ESTER, HYDROCHLORIDE see
    PFF500
(±)-α-AMINO-BENZENEACETIC ACID HEPTYL ESTER HYDROCHLORIDE see
    PFF750
(±)-α-AMINOBENZENEACETIC ACID,3-METHYLBUTYL ESTER HYDROCHLO-
    RIDE (9CI) see PCV750
(±)-α-AMINO-BENZENEACETIC ACID NONYL ESTER HYDROCHLORIDE see
    PFG250
(±)-α-AMINO-BENZENEACETIC ACID OCTYL ESTER HYDROCHLORIDE see
    PFG500
(±)-α-AMINO-BENZENEACETIC ACID PENTYL ESTER HYDROCHLORIDE see
    PFG750
4-AMINOBENZENEARSONIC ACID see ARA250
p-AMINOBENZENEARSONIC ACID see ARA250
p-AMINOBENZENEAZODIMETHYLANILINE see DPO200
p-AMINO BENZENE DIAZONIUMPERCHLORATE see AID750
2-AMINO-BENZENE-1,4-DISULFONIC ACID see AIE000
2-AMINO-1,4-BENZENEDISULFONIC ACID see AIE000
1-AMINO-2,5-BENZENEDISULFONIC ACID see AIE000
2-AMINO-p-BENZENEDISULFONIC ACID see AIE000
(S)-α-AMINOBENZENEPROPANOIC ACID see PEC750
4-AMINOBENZENESTIBONIC ACID see SLP600
p-AMINOBENZENESTIBONIC ACID see SLP600
p-AMINOBENZENESULFAMIDE see SNM500
3-(p-AMINOBENZENESULFAMIDO)-6-METHOXYPYRIDAZINE see AKO500
2-(p-AMINOBENZENESULFANAMIDE)-3-METHOXYPYRAZINE see MFN500
p-AMINOBENZENESULFONACETAMIDE see SNP500
3-AMINOBENZENESULFONAMIDE see MDM760
4-AMINOBENZENESULFONAMIDE see SNM500
m-AMINOBENZENESULFONAMIDE see MDM760
p-AMINOBENZENESULFONAMIDE see SNM500
6-(4-AMINOBENZENESULFONAMIDO)-4,5-DIMETHOXYPYRIMIDINE see AIE500
5-(p-AMINOBENZENESULFONAMIDO)-3,4-DIMETHYLISOOXALE see SNN500
5-(p-AMINOBENZENESULFONAMIDO)-3,4-DIMETHYLISOXAZOLE see SNN500
2-(p-AMINOBENZENESULFONAMIDO)-4,5-DIMETHYLOXAZOLE see AIE750
6-(4-AMINOBENZENESULFONAMIDO)-2,4-DIMETHYLPYRIMIDINE see SNJ350
6-(p-AMINOBENZENESULFONAMIDO)-2,4-DIMETHYLPYRIMIDINE see SNJ350
2-(p-AMINOBENZENESULFONAMIDO)-4,6-DIMETHYLPYRIMIDINE see SNJ000
2-(p-AMINOBENZENESULFONAMIDO)-5-METHOXYPYRIMIDINE SODIUM SALT
    see MFO000
2-(p-AMINOBENZENESULFONAMIDO)-5-METHYLTHIADIAZOLE see MPQ750
3-(p-AMINOBENZENESULFONAMIDO)-2-PHENYLPYRAZOLE see AIF000
2-p-AMINOBENZENESULFONAMIDOQUINOXALINE see QTS000
2-(p-AMINOBENZENESULFONAMIDO)THIAZOLE see TEX250
2-AMINOBENZENESULFONIC ACID see SNO100
1-AMINOBENZENE-3-SULFONIC ACID see SNO000
3-AMINO-BENZENESULFONIC ACID see SNO000
4-AMINOBENZENESULFONIC ACID see SNN600
m-AMINOBENZENESULFONIC ACID see SNO000
o-AMINOBENZENESULFONIC ACID see SNO100
p-AMINOBENZENESULFONIC ACID see SNN600
m-AMINOBENZENESULFONIC ACID SODIUM SALT see AIF250
p-AMINOBENZENESULFONYL-2-AMINO-4,5-DIMETHYLOXAZOLE see AIE750
6-(p-AMINOBENZENESULFONYL)AMINO-2,4-DIMETHYLPYRIMIDINE see SNJ350
p-AMINOBENZENESULFONYLGUANIDINE see AHO250
N-(4-AMINOBENZENESULFONYL)-N′-BUTYLUREA see BSM000
1-(4-AMINOBENZENESULFONYL)UREA see SNQ550
p-AMINOBENZENESULFONYLUREA see SNQ550
m-AMINOBENZENESULPHONAMIDE see MDM760

5-(p-AMINOBENZENESULPHONAMIDE)-3,4-DIMETHYLISOXAZOLE see SNN500
2-p-AMINOBENZENESULPHONAMIDO-4,6-DIMETHOXYPYRIMIDINE see SNI500
5-(p-AMINOBENZENESULPHONAMIDO)-3,4-DIMETHYLISOXAZOLE see SNN500
3-p-AMINOBENZENESULPHONAMIDO-7-METHOXYPYRIDAZINE see AKO500
2-p-AMINOBENZENESULPHONAMIDOQUINOXALINE see QTS000
2-(p-AMINOBENZENESULPHONAMIDO)THIAZOLE see TEX250
N-p-AMINOBENZENESULPHONYLGUANIDINE MONOHYDRATE see AHO250
2-AMINOBENZENETHIOL see AIF500
4-AMINOBENZENETHIOL see AIF750
p-AMINOBENZENETHIOL see AIF750
2-AMINOBENZIMIDAZOLE see AIG000
2-AMINO-6-BENZIMIDAZOLYL PHENYLKETONE see AIH000
2-AMINOBENZOIC ACID see API500
3-AMINOBENZOIC ACID see AIH500
4-AMINOBENZOIC ACID see AIH600
γ-AMINOBENZOIC ACID see AIH600
m-AMINOBENZOIC ACID see AIH500
o-AMINOBENZOIC ACID see API500
p-AMINOBENZOIC ACID see AIH600
AMINOBENZOIC ACID see AIH600
p-AMINOBENZOIC ACID-2-N-AMYLAMINOETHYL ESTER see PBV750
p-AMINOBENZOIC ACID BUTYL ESTER see BPZ000
p-AMINOBENZOIC ACID-3-(DIBUTYLAMINO)PROPYL ESTER, HYDROCHLORIDE see AIT000
p-AMINOBENZOIC ACID-2-(2-(2-(2-(2-(DIETHYLAMINO)ETHOXY)ETHOXY)ETHOXY)ETHYL ESTER, HYDROCHLORIDE see AIK500
p-AMINOBENZOIC ACID-2-(2-(2-(2-(DIETHYLAMINO)ETHOXY)ETHOXY)ETHYL ESTER, HYDROCHLORIDE see AIK750
p-AMINOBENZOIC ACID-2-(2-(2-(DIETHYLAMINO)ETHOXY)ETHOXY)ETHYL ESTER, HYDROCHLORIDE see AIL000
p-AMINOBENZOIC ACID-2-(2-(DIETHYLAMINO)ETHOXY)ETHYL ESTER, HYDROCHLORIDE see AIL250
(p-AMINOBENZOIC ACID-3-(β-DIETHYLAMINO)ETHOXY)PROPYL ESTER see AIL500
p-AMINOBENZOIC ACID-2-DIETHYLAMINOETHYL ESTER see AIL750
4-AMINOBENZOIC ACID DIETHYLAMINOETHYL ESTER see AIL750
4-AMINOBENZOIC ACID 2-(DIETHYLAMINO)ETHYL ESTER, HYDROCHLORIDE see AIT250
p-AMINOBENZOIC ACID-2-DIETHYLAMINOETHYL ESTER, HYDROCHLORIDE see AIT250
p-AMINOBENZOIC ACID-N-1-DIETHYLAMINO-1-ISOBUTYLETHANOL METHANESULFONATE see LEU000
p-AMINOBENZOIC ACID-β-DIETHYLAMINOISOHEXYL ESTER METHANESULFONATE see LEU000
p-AMINOBENZOIC ACID-3-(DIETHYLAMINO)PROPYL ESTER HYDROCHLORIDE see AIM000
p-AMINOBENZOIC ACID-2-(DIISOPROPYLAMINO)ETHYL ESTER, HYDROCHLORIDE see AIT500
p-AMINOBENZOIC ACID-3-(DIISOPROPYLAMINO)PROPYL ESTER HYDROCHLORIDE see AIM250
p-AMINOBENZOIC ACID 3-(DIMETHYLAMINO)-1,2-DIMETHYLPROPYL ESTER, HYDROCHLORIDE see AIT750
p-AMINOBENZOIC ACID-(2-(DIMETHYLAMINO)-1-PHENYL)ETHYL ESTER see AIU250
p-AMINOBENZOIC ACID-2-(DIPROPYLAMINO)ETHYL ESTER HYDROCHLORIDE see AIN000
p-AMINOBENZOIC ACID 3-(DIPROPYLAMINO)PROPYL ESTER, HYDROCHLORIDE see AIU000
4-AMINOBENZOIC ACID ETHYL ESTER see EFX000
o-AMINOBENZOIC ACID, ETHYL ESTER see EGM000
p-AMINOBENZOIC ACID ETHYL ESTER see EFX000
3-AMINOBENZOIC ACID ETHYL ESTER METHANESULFONATE see EFX500
p-AMINO-BENZOIC ACID 1-ETHYL-4-PIPERIDYL ESTER, HYDROCHLORIDE see EOV000
p-AMINOBENZOIC ACID ISOBUTYL ESTER see MOT750
2-AMINOBENZOIC ACID METHYL ESTER see APJ250
o-AMINOBENZOIC ACID METHYL ESTER see APJ250
p-AMINOBENZOIC ACID METHYL ESTER see AIN150
p-AMINOBENZOIC ACID-(2-METHYL-2-(1-METHYL-HEPTYLAMINO))PROPYL ESTER HYDROCHLORIDE see MOK500
p-AMINOBENZOIC ACID-3-(3-METHYLPIPERIDINO) PROPYL ESTER HYDROCHLORIDE see MOK500
p-AMINO-BENZOIC ACID-1-METHYL-4-PIPERIDYL ESTER, HYDROCHLORIDE see MON000
4-AMINOBENZOIC ACID, MODOSODIUM SALT see SEO500
p-AMINOBENZOIC ACID MONOGLYCERYL ESTER see GGQ000
p-AMINO-BENZOIC ACID-1-PHENETHYL-4-PIPERIDYL ESTER HYDROCHLORIDE see PDJ000
2-AMINOBENZOIC ACID-3-PHENYL-2-PROPENYL ESTER see API750
p-AMINOBENZOIC ACID PHOSPHATE see AIQ875
p-AMINOBENZOIC ACID-3-PIPERIDINOPROPYL ESTER HYDROCHLORIDE see PIS000

m-AMINOBENZOIC ACID-2-(2-PIPERIDYL)ETHYL ESTER HYDROCHLORIDE see AIQ880
o-AMINOBENZOIC ACID 2-(2-PIPERIDYL)ETHYL ESTER HYDROCHLORIDE see AIQ885
p-AMINOBENZOIC ACID-2-(2-PIPERIDYL)ETHYL ESTER HYDROCHLORIDE see AIQ890
p-AMINOBENZOIC ACID SODIUM SALT see SEO500
p-AMINOBENZOIC ACID-N,N-DIETHYLLEUCINOL ESTER METHANESULFONATE see LEU000
p-AMINOBENZOIC DIETHYLAMINOETHYLAMIDE see AJN500
2-(p-AMINOBENZOLSULFONAMIDO)-4,5-DIMETHYLOXAZOL (GERMAN) see AIE750
6-(4′-AMINOBENZOL-SULFONAMIDO)-2,4-DIMETHYLPYRIMIDIN (GERMAN) see SNJ000
(p-AMINOBENZOLSULFONYL)-2-AMINO-4,6-DIMETHYLPYRIMIDIN (GERMAN) see SNJ000
(p-AMINOBENZOLSULFONYL)-4-AMINO-2,6-DIMETHYLPYRIMIDIN (GERMAN) see SNJ350
(p-AMINOBENZOLSULFONYL)-4-AMINO-2-METHYL-6-METHOXY-PYRIMIDIN (GERMAN) see MDU300
(p-AMINOBENZOLSULFONYL)-2-AMINO-4-METHYLPYRIMIDIN (GERMAN) see ALF250
4-AMINO-BENZOLSULFONYL-METHYLCARBAMAT (GERMAN) see SNQ500
2-AMINOBENZONITRILE see APJ750
3-AMINOBENZONITRILE see AIR125
m-AMINOBENZONITRILE see AIR125
o-AMINOBENZONITRILE see APJ750
p-AMINOBENZONITRILE (8CI) see COK125
4-AMINOBENZONITRILE (9CI) see COK125
p-AMINOBENZOPHENONE see AIR250
3-AMINOBENZO-6,7-QUINAZOLINE-4-ONE see AIS250
3-AMINOBENZO(g)QUINAZOLIN-4(3H)-ONE see AIS250
2-AMINOBENZOTHIAZOLE see AIS500
6-AMINO-2-BENZOTHIAZOLETHIOL see AIS550
3-AMINOBENZOTRIFLUORIDE see AID500
m-AMINOBENZOTRIFLUORIDE see AID500
p-AMINOBENZOTRIFLUORIDE see TKB750
2-AMINOBENZOXAZOLE see AIS600
3-(p-AMINOBENZOXY)-1-DI-n-BUTYLAMINOPROPANE see BOO750
3-(p-AMINOBENZOXY)-1-DI-n-BUTYLAMINOPROPANE SULFATE see BOP000
1-AMINO-4-BENZOYLAMINOANTHRACHINON (CZECH) see AIB250
4-AMINO-1-BENZOYLAMINOANTHRAQUINONE see AIB250
1-AMINO-4-(BENZOYLAMINO)ANTHRAQUINONE see AIB250
p-AMINOBENZOYLAMINOMETHYLHYDROCOTARNINE see AIS625
2-AMINO-3-BENZOYL BENZENEACETIC ACID see AIU500
2-AMINO-3-BENZOYLBENZENEACETIC ACID SODIUM SALT HYDRATE see AHK625
2-AMINO-5-BENZOYLBENZIMIDAZOLE see AIH000
p-AMINOBENZOYLDIBUTYLAMINOPROPANOL see BOO750
AMINOBENZOYLDIBUTYLAMINOPROPANOL HYDROCHLORIDE see AIT000
p-AMINOBENZOYLDIBUTYLAMINOPROPANOL SULFATE see BOP000
p-AMINOBENZOYLDIETHYLAMINOETHANOL see AIL750
p-AMINOBENZOYLDIETHYLAMINOETHANOL HYDROCHLORIDE see AIT250
p-AMINO BENZOYL DIETHYL AMINO PROPANOL HYDROCHLORIDE see AIM000
o-AMINOBENZOYL DI(ISOPROPYLAMINO)ETHANOL HYDROCHLORIDE see IJZ000
p-AMINO BENZOYL DI-ISO-PROPYL AMINO ETHANOL HYDROCHLORIDE see AIT500
p-AMINO BENZOYL DIISOPROPYL AMINO PROPANOL HYDROCHLORIDE see AIM250
p-AMINOBENZOYLDIMETHYLAMINO-1,2-DIMETHYLPROPANOL HYDROCHLORIDE see AIT750
p-AMINO BENZOYL DI-N-PROPYL AMINO ETHANOL HYDROCHLORIDE see AIN000
p-AMINO BENZOYL DI-N-PROPYL AMINOPROPANOL HYDROCHLORIDE see AIU000
o-AMINOBENZOYLFORMIC ANHYDRIDE see ICR000
β-4-AMINOBENZOYLOXY-β-PHENYLETHYL DIMETHYLAMINE see AIU250
2-AMINO-3-BENZOYLPHENYLACETIC ACID see AIU500
2-AMINOBENZTHIAZOLE see AIS500
4-(4-AMINOBENZYL)ANILINE see MJQ000
α-(α-AMINOBENZYL)BENZYL ALCOHOL HYDROCHLORIDE see DWB000
2-AMINO-6-BENZYLMERCAPTOPURINE see BFL125
2-AMINO-6-BENZYL-MP see BFL125
d-(− )-α-AMINOBENZYLPENICILLIN see AIV500
AMINOBENZYLPENICILLIN see AIV500
d-(−)-α-AMINOBENZYLPENICILLIN SODIUM SALT see SEQ000
α-AMINOBENZYLPENICILLIN TRIHYDRATE see AOD125
AMINOBENZYLPENICILLIN TRIHYDRATE see AOD125
4-AMINO-N-(1-BENZYL-4-PIPERIDYL)-5-CHLORO-o-ANISAMIDE HYDROXYSUCCINATE see AIV625
2-AMINO-6-(BENZYLTHIO)PURINE see BFL125
2-AMINOBIPHENYL see BGE250
4-AMINOBIPHENYL see AJS100

o-AMINOBIPHENYL see BGE250
p-AMINOBIPHENYL see AJS100
4-AMINOBIPHENYL DIHYDROCHLORIDE see BGE300
4-AMINOBIPHENYL ETHER see PDR500
4-AMINO-3-BIPHENYLOL see AIV750
4′-AMINO-4-BIPHENYLOL see AIW000
3-AMINO-3-BIPHENYLOL HYDROCHLORIDE see AIW250
4-AMINO-3-BIPHENYLOL HYDROGEN SULFATE see AKF250
N-(4′-AMINO(1,1′-BIPHENYL)-4-YL)-ACETAMIDE see ACC000
2-AMINO-5-BIPHENYLYLIMIDAZOLEHYDROCHLORIDE see AIW500
2-(4′-AMINO-1,1′-BIPHENYL-4-YL)-2H-NAPHTHO(1,2-d)TRIAZOLE-6,8-DISULFON-
    IC ACID, DIPOTASSIUM SALT see AIW750
5-AMINO(3,4′-BIPYRIDIN)-6-(1H)-ONE see AOD375
2-AMINO-4-(p-BIS(2-CHLOROETHYL)AMINO)PHENYL)BUTYRIC ACID see
    AJE000
5-AMINO-1-BIS(DIMETHYLAMIDE)PHOSPHORYL-3-PHENYL-1,2,4-TRIAZOLE see
    AIX000
5-AMINO-1-BIS(DIMETHYLAMIDO)PHOSPHORYL-3-PHENYL-1,2,4-TRIAZOLE see
    AIX000
5-AMINO-1-(BIS(DIMETHYLAMINO)PHOSPHINYL)-3-PHENYL-1,2,4-TRIAZOLE see
    AIX000
AMINOBIS(PROPYLAMINE) see AIX250
1-AMINO-4-BROMANTHRACHINON-2-SULFONAN SODNY (CZECH) see
    DKR000
1-AMINO-2-BROM-4-HYDROXYANTHRACHINON (CZECH) see AIY500
2-AMINO-5-BROMOBENZOXAZOLE see AIX500
2-AMINO-6-BROMOBENZOXAZOLE see AIX750
2-AMINO-6-BROMO-5-CHLOROBENZOXAZOLE see AIY250
2-AMINO-5-BROMO-6-CHLOROBENZOXAZOLE see AIY000
4-AMINO-5-BROMO-N-(2-(DIETHYLAMINO)ETHYL)-o-ANISAMIDE see VCK100
1-AMINO-2-BROMO-4-HYDROXYANTHRAQUINONE see AIY500
1-AMINO-2-BROMO-4-(2-(2-HYDROXYETHYL)SULFONYL-4-METHYLPHENYLAMI-
    NO)ANTHRAQUINONE see AIY750
2-AMINO-5-BROMO-6-PHENYL-4(1H)-PYRIMIDINONE see AIY850
N-AMINO-2-(m-BROMOPHENYL)SUCCINIMIDE see AIZ000
1-AMINO-3-BROMOPROPANE HYDROBROMIDE see AJA000
1-AMINO-BUTAAN (DUTCH) see BPX750
1-AMINOBUTAN (GERMAN) see BPX750
1-AMINOBUTANE see BPX750
2-AMINOBUTANE see BPY000
(S)-AMINOBUTANEDIOIC ACID see ARN850
4-AMINOBUTANOIC ACID see PIM500
2-AMINO-1-BUTANOL see AJA250
2-AMINOBUTAN-1-OL see AJA250
4-AMINO-2-(4-BUTANOYLHEXAHYDRO-1H-1,4-DIAZEPIN-1-YL)-6,7-DIMETHOXY-
    QUINAZOLINE HYDROCHLORIDE see AJA375
3-AMINO-2-BUTOXYBENZOIC ACID-2-DIETHYLAMINOETHYL ESTER HYDRO-
    CHLORIDE see AJA500
4-AMINO-2-BUTOXY-BENZOIC ACID 2-(DIETHYLAMINO)ETHYL ESTER HY-
    DROCHLORIDE see SPB800
4-AMINO-2-BUTOXY-BENZOIC ACID 2-(DIETHYLAMINO)ETHYL ESTER, MONO-
    HYDROCHLORIDE see SPB800
2-AMINO-n-BUTYL ALCOHOL see AJA250
4-AMINO-N-((BUTYLAMINO)CARBONYL)BENZENESULFONAMIDE see BSM000
dl-4-AMINO-α-(tert-BUTYLAMINO)-3-CHLORO-5-(TRIFLUOROMETH-
    YL)PHENETHYL ALCOHOL HCl see MAB300
4-AMINO-α-((tert-BUTYLAMINO)METHYL)-3-CHLORO-5-(TRIFLUOROMETH-
    YL)BENZYL ALCOHOL HCl see KGK300
4-AMINO-α-((tert-BUTYLAMINO)METHYL)-3,5-DICHLOROBENZYL ALCOHOL HY-
    DROCHLORIDE see VHA350
4-AMINO-α-((tert-BUTYLAMINO)METHYL)-3,5-DICHLOROBENZYLALKOHOL-HY-
    DROCHLORID (GERMAN) see VHA350
N¹-3-(((4-AMINOBUTYL)AMINO)PROPYL)BLEOMYCINAMIDE see BLY500
1-AMINO-4-BUTYLBENZENE see AJA550
p-AMINOBUTYLBENZENE see AJA550
(4-AMINOBUTYL)DIETHOXYMETHYLSILANE see AJA750
p-AMINO-β-sec-BUTYL-N,N-DIMETHYLPHENETHYLAMINE see AJB000
3-(2-AMINOBUTYL)INDOLE ACETATE see AJB250
3-(4-AMINOBUTYL)INDOLE HYDROCHLORIDE see AJB500
Δ-AMINOBUTYLMETHYLDIETHOXYSILANE see AJA750
4-AMINO-6-tert-BUTYL-3-(METHYLTHIO)-1,2,4-TRIAZIN-5-ONE see MQR275
4-AMINO-6-tert-BUTYL-3-METHYLTHIO-as-TRIAZIN-5-ONE see MQR275
5-AMINO-N-BUTYL-2-PROPARGYLOXYBENZAMIDE see AJC000
5-AMINO-N-BUTYL-2-(2-PROPYNYLOXY)BENZAMIDE see AJC000
(4-AMINOBUTYL)TRIETHOXYSILANE see AJC250
4-AMINOBUTYRAMIDE see AJC375
4-AMINOBUTYRIC ACID see PIM500
γ-AMINO-N-BUTYRIC ACID see PIM500
γ-AMINOBUTYRIC ACID CETYL ESTER see AJC500
4-AMINOBUTYRIC ACID LACTAM see PPT500
γ-AMINOBUTYRIC ACID LACTAM see PPT500
4-AMINOBUTYRIC ACID METHYL ESTER see AJC625
γ-AMINOBUTYRIC LACTAM see PPT500
γ-AMINOBUTYROLACTAM see PPT500
4′-AMINOBUTYROPHENONE see AJC750

p-AMINOBUTYROPHENONE see AJC750
4-AMINO-2-(4-BUTYRYLHEXAHYDRO-1H-1,4-DIAZEPIN-1-YL)-6,7-DIMETHOXY-
    QUINAZOLINE HYDROCHLORIDE see BON350
AMINOCAINE see AIT250
6-AMINOCAPROIC ACID see AJD000
ε-AMINOCAPROIC ACID see AJD000
ω-AMINOCAPROIC ACID see AJD000
AMINOCAPROIC ACID see AJD000
AMINOCAPROIC LACTAM see CBF700
p-AMINO CAPROPHENONE see AJD250
AMINOCARB see DOR400
4-AMINO-5-CARBAMYL-3-BENZYLTHIAZOLE-2(3H)-THIONE see AJD500
AMINOCARBE (FRENCH) see DOR400
3-AMINO-9-p-CARBETHOXYAMINOPHENYL-10-METHYLPHENANTHRIDINIUM
    ETHANESULPHONATE see CBQ125
2-AMINO-α-CARBOLINE see AJD750
AMINO-α-CARBOLINE see AJD750
(3-((AMINOCARBONYL)AMINO)-2-METHOXYPROPYL)CHLOROMERCURY see
    CHX250
2,2′-((4-((AMINOCARBONYL)AMINO)PHENYL)ARSINIDENE)BIS(THIO)BISACETIC
    ACID see CBI250
2,2′-(((4-((AMINOCARBONYL)AMINO)PHENYL)ARSINIDENEBIS(THIO))BIS) BEN-
    ZOIC ACID see TFD750
(4-((AMINOCARBONYL)AMINO)PHENYL)ARSONIC ACID see CBJ000
N-(AMINOCARBONYL)BENZENEACETAMIDE see PEC250
N-(AMINOCARBONYL)-2-BROMO-2-ETHYLBUTANAMIDE see BNK000
N-(AMINOCARBONYL)-2-BROMO-3-METHYLBUTANAMIDE see BNP750
(6R-(6-α,7-β(R*)))-4-(AMINOCARBONYL)-1-((2-CARBOXY-8-OXO-7-((PHENYLSUL-
    FOACETYL)AMINO)-5-THIO-1-AZABICYCLO(4.2.0)OCT-2-EN-3-YL)METHYL)-
    PYRIDINIUM HYDROXIDE, inner salt, MONOSODIUM SALT see CCS550
(Z)-N-(AMINOCARBONYL)-2-ETHYL-2-BUTENAMIDE see EHP000
N-(AMINOCARBONYL)HYDROXYLAMINE see HOO500
N-(AMINOCARBONYL)-2-(1-METHYLETHYL)-4-PENTENAMIDE see IQX000
N-(AMINOCARBONYL)-4-((2-METHYLHYDRAZINO)METHYL)BENZAMIDE MO-
    NOHYDROBROMIDE see MKN750
N-(AMINOCARBONYL)-N-NITROGLYCINE see NKI500
N-((AMINOCARBONYL)OXY)ETHANIMIDOTHIOIC ACID, METHYL ESTER see
    MPS250
(6R-(6-α,7-β(Z)))-3-(((AMINOCARBONYL)OXY)METHYL)-7-((2-FURA-
    NYL(METHYOXYIMINO)ACETYL)AMINO)-8-OXO-5-THIA-1-AZABICY-
    CLO(4.2.0)OCT-2-ENE-2-CARBOXYLIC ACID MONOSODIUM SALT see
    SFQ300
(6R-(6-α,7-β(Z)))-3-(((AMINOCARBONYL)OXY)METHYL)-7-((2-FURA-
    NYL(METHYOXYIMINO)ACETYL)AMINO)-8-OXO-5-THIA-1-AZABICY-
    CLO(4.2.0)OCT-2-ENE-2-CARBOXYLIC ACID see CCS600
(6R-cis)-3-(((AMINOCARBONYL)OXY)METHYL)-7-METHYOXY-8-OXO-7-((2-THIE-
    NYLACETYL)AMINO)-5-THIA-1-AZABICYCLO(4.2.0)OCT-2-ENE-2-CARBOXYLIC
    ACID MONOSODIUM SALT see CCS510
2-(((AMINOCARBONYL)OXY)METHYL)-2-METHYLPENTYL ESTER BUTYL CAR-
    BAMIC ACID see MOV500
2-((AMINOCARBONYL)OXY)-N,N,N-TRIMETHYLETHANAMINIUM CHLORIDE see
    CBH250
2-((AMINOCARBONYL)OXY)-N,N,N-TRIMETHYL-1-PROPANAMINIUM CHLORIDE
    see HOA500
(((4-(AMINOCARBONYL)PHENYL)ARSINIDINE)BIS(THIO))BIACETIC ACID see
    TFA350
1-(((4-(AMINOCARBONYL)PYRIDINIO)METHOXY)METHYL)-2-((HYDROXYIMI-
    NO)METHYL) PYRIDINIUM 2Cl see HGE900
(((4-(AMINOCARBONYL)PYRIDINO)METHOXY)-2-((HYDROXYIMI-
    NO)METHYL)PYRIDINIUM DICHLORIDE see HGE900
2-AMINO-6-p-CARBOXYAMINOPHENYL)-5-METHYLPHENANTHRIDINIUM ETH-
    ANESULFONATE ETHYL ESTER see CBQ125
1-AMINO-2-CARBOXYBENZENE see API500
1-AMINO-4-CARBOXYBENZENE see AIH600
(6R-(6-α,7-α))-7-((((2-AMINO-2-CARBOXYETHYL)THIO)ACETYL)AMINO)-7-ME-
    THOXY-3-(((1-METHYL-1H-TETRAZOL-5-YL)THIO)METHYL)-8-OXO-5-THIA-1-
    AZABICYCLO(4.2.0)OCT-2-ENE-2-CARBOXYLIC ACID MONOSODIUM SALT
    see CCS365
(6R-cis)-7-((((4-(2-AMINO-1-CARBOXY-2-OXOETHYL)-1,3-DITHIETAN-2-
    YL)CARBONYL)AMINO)-7-METHOXY-3-(((1-METHYL-1H-TETRAZOL-5-
    YL)THIO)METHYL)-8-OXO-5-THIA-1-AZABICYCLO(4.2.0)OCT-2-ENE-2-CAR-
    BOXYLIC ACID MONOSODIUM SALT see CCS371
N²-(((+)-5-AMINO-5-CARBOXYPENTYLAMINO)METHYL)TETRACYCLINE see
    MRV250
3-AMINO-N-(α-CARBOXYPHENETHYL)SUCCINAMIC ACID N-METHYL ESTER,
    stereoisomer see ARN825
AMINOCARDOL see TEP500
AMINOCHLORAMBUCIL see AJE000
2-AMINO-6′-CHLORO-o-ACETOTOLUIDIDE, HYDROCHLORIDE see AJE250
4-AMINO-6-p-CHLOROANILINO-1,2-DWUHYDRO-2,2-DWUMETHYLO-1,3,5-TRO-
    JAZYNA see COX400
1-AMINO-5-CHLOROANTHRAQUINONE see AJE325
1-AMINO-2-CHLOROBENZENE see CEH670
1-AMINO-3-CHLOROBENZENE see CEH675
1-AMINO-4-CHLOROBENZENE see CEH680

m-AMINOCHLOROBENZENE see CEH675
1-AMINO-4-CHLOROBENZENE HYDROCHLORIDE see CJR200
4-AMINO-2-CHLOROBENZOIC ACID see CEG750
3-AMINO-4-CHLOROBENZOIC ACID 2-((DIMETHYLAMINO)ETHYL) ESTER HY-
    DROCHLORIDE see AJE350
2-AMINO-5-CHLOROBENZOPHENONE see AJE400
2-AMINO-4-CHLOROBENZOTHIAZOLE see AJE500
2-AMINO-6-CHLOROBENZOTHIAZOLE see AJE750
3-AMINO-4-CHLOROBENZOTRIFLUORIDE see CEG800
2-AMINO-4-CHLOROBENZOXAZOLE see AJF250
2-AMINO-5-CHLOROBENZOXAZOLE see AJF500
2-AMINO-6-CHLOROBENZOXAZOLE see AJF750
2-AMINO-7-CHLOROBENZOXAZOLE see AJG750
4-AMINO-5-CHLORO-N-(2-(DIETHYLAMINO)ETHYL)-N-ANISAMIDE see AJH000
4-AMINO-5-CHLORO-N-(2-(DIETHYLAMINO)ETHYL)-o-ANISAMIDE DIHYDRO-
    CHLORIDE MONOHYDRATE see MQQ300
4-AMINO-5-CHLORO-N-(2-(DIETHYLAMINO)ETHYL)-2-METHOXYBENZAMIDE see
    AJH000
5-AMINO-4-CHLORO-2,3-DIHYDRO-3-OXO-2-PHENYLPYRIDAZINE see PEE750
4-AMINO-5-CHLORO-N-(2-(ETHYLAMINOETHYL)-o-ANISAMIDE) see AJH125
4-AMINO-5-CHLORO-N-(2-ETHYLAMINOETHYL)-2-METHOXYBENZAMIDE see
    AJH125
6-AMINO-2-(2-CHLOROETHYL)-2,3-DIHYDRO-4H-1,3-BENZOXAZIN-4-ONE see
    DKR200
6-AMINO-2-(2-CHLOROETHYL)-2,3-DIHYDRO-4H-1,3-BENZOXAZIN-4-ONEHY-
    DROCHLORIDE see AJH129
4-AMINO-5-CHLORO-N-(2-ETILAMINOETIL)-2-METOSSIBENZAMIDE (ITALIAN)
    see AJH125
4-AMINO-3-CHLOROFENYLOMETHYLOSULFON see CIX200
1-AMINO-5-CHLORO-4-HYDROXYANTHRAQUINONE see AJH250
3-AMINO-5-CHLORO-4-HYDROXYBENZENESULFONIC ACID see AJH500
2-AMINO-5-CHLORO-6-HYDROXYBENZOXAZOLE see AJH750
2-AMINO-5-CHLORO-6-METHOXYBENZOXAZOLE see AJI000
4-AMINO-5-CHLORO-2-METHOXY-N-(1-BENZYL-4-PIPERIDYL)BENZAMIDE MAL-
    ATE see CMV325
4-AMINO-5-CHLORO-2-METHOXY-N-(4-PIPERIDYL)BENZAMIDE see DBA250
2-AMINO-2'-CHLORO-6'-METHYLACETANILIDE, HYDROCHLORIDE see AJE250
1-AMINO-3-CHLORO-2-METHYLBENZENE see CLK200
1-AMINO-5-CHLORO-4-METHYLBENZENE see CLK215
1-AMINO-2-CHLORO-6-METHYLBENZENE see CLK200
1-AMINO-3-CHLORO-6-METHYLBENZENE see CLK225
2-AMINO-3-CHLORO-1,4-NAPHTHOQUINONE see AJI250
1-AMINO-2-CHLORO-4-NITROBENZENE see CJA175
2-AMINO-4-CHLOROPHENOL (DOT) see CEH250
γ-AMINO-β-(p-CHLOROPHENYL)BUTYRIC ACID see BAC275
1-(4-AMINO-5-(2-CHLOROPHENYL)-2-METHYL-1H-PYRROL-3-YL)ETHANONE see
    AJI300
1-(4-AMINO-5-(p-CHLOROPHENYL)-2-METHYL-1H-PYRROL-3-YL)ETHANONE see
    AJI330
4-AMINO-3-CHLOROPHENYLMETHYLSULFONE see CIX200
4-AMINO-3-CHLOROPHENYLMETHYLSULPHONE see CIX200
(2-AMINO-5-CHLOROPHENYL)PHENYLMETHANONE see AJE400
5-AMINO-4-CHLORO-2-PHENYL-3(2H)-PYRIDAZINONE see PEE750
2-AMINO-5-((p-CHLOROPHENYL)THIOMETHYL)-2-OXAZOLINE see AJI500
(±)-1-AMINO-3-CHLORO-2-PROPANOL HYDROCHLORIDE see AJI600
dl-1-AMINO-3-CHLORO-2-PROPANOL HYDROCHLORIDE see AJI600
4-AMINO-N-(5-CHLORO-2-QUINOXALINYL)BENZENESULFONAMIDE see
    CMA600
2-AMINO-5-CHLOROTHIAZOLE see AJI650
4-AMINO-2-CHLOROTOLUENE see CLK215
2-AMINO-3-CHLOROTOLUENE see CLK227
2-AMINO-4-CHLOROTOLUENE see CLK225
2-AMINO-5-CHLOROTOLUENE see CLK220
2-AMINO-6-CHLOROTOLUENE see CLK200
2-AMINO-5-CHLOROTOLUENE HYDROCHLORIDE see CLK235
6-AMINO-4-CHLORO-m-TOLUENESULFONIC ACID see AJJ250
4-AMINO-3-CHLOROTRIFLUOROMETHYL-α-((tert-BUTYLAMI-
    NO)METHYL)BENZYLALCOHOL HYDROCHLORIDE see KGK300
dl-1-(4-AMINO-3-CHLORO-5-TRIFLUOROMETHYLPHENYL)-2-tert-BUTYLAMINOE-
    THANOL HYDROCHLORIDE see MAB300
3-AMINO-4-CHLORO-α-α-α-TRIFLUOROTOLUENE see CEG800
AMINOCHLORTHENOXAZINE see DKR200
AMINOCHLORTHENOXAZIN HYDROCHLORIDE see AJH129
6-AMINOCHRYSENE see CML800
3-AMINO-4-CLORO-BENZOATO di DIMETILAMINOETILE CLORIDRATO (ITAL-
    IAN) see AJE350
6-AMINOCOUMARIN COUMARIN-3-CARBOXYLIC ACID SALT see AJJ500
6-AMINOCOUMARIN HYDROCHLORIDE see AJJ750
4-AMINO-m-CRESOL see AKZ000
5-AMINO-o-CRESOL see AKJ750
3-AMINO-p-CRESOL METHYL ESTER see MGO750
m-AMINO-p-CRESOL, METHYL ESTER see MGO750
4-AMINO-5-CYANO-7-(d-RIBOFURANOSYL)-7H-PYRROLO(2,3-d)PYRIMIDINE see
    VGZ000
2-AMINO-2,4,6-CYCLOHEPTATRIEN-1-ONE see AJJ800

d-7-(2-AMINO-2-(1,4-CYCLOHEXADIEN-1-YL)ACETAMIDO)-3-METHYL-8-OXO-5-
    THIA-1-AZABICYCLO(4.2.0)OCT-2-ENE-2-CARBOXYLIC ACID HYDRATE see
    SBN450
(6R-(6-α,7-β(R*)))-7-((AMINO-1,4-CYCLOHEXADIEN-1-YLACETAL)AMINO)-3-
    METHYL-8-OXO-5-THIA-1-AZABICYCLO(4.2.0)OCT-2-ENE-2-CARBOXYLIC
    ACID DIHYDRATE see CCS535
(6R-(6-α,7-β(R*)))-7-((AMINO-1,4-CYCLOHEXADIEN-1-YLACETAL)AMINO)-3-
    METHYL-8-OXO-5-THIA-1-AZABICYCLO(4.2.0)OCT-2-ENE-2-CARBOXYLIC
    ACID see CCS530
AMINOCYCLOHEXANE see CPF500
6-(1-AMINOCYCLOHEXANECARBOXAMIDO)PENICILLANIC ACID see AJJ875
AMINOCYCLOHEXANE HYDROCHLORIDE see CPA775
(1-AMINOCYCLOHEXYL)PENICILLIN see AJJ875
AMINOCYCLOHEXYLPENICILLIN see AJJ875
1-AMINO-2-(o-CYCLOHEXYLPHENOXY)PROPIONALDOXIME see AJK000
AMINOCYCLOPENTANE see CQA000
1-AMINO-1-CYCLOPENTANECARBOXYLIC ACID see AJK250
1-AMINOCYCLOPENTANE-1-CARBOXYLIC ACID see AJK250
4-AMINO-N-CYCLOPROPYL-3,5-DICHLOROBENZAMIDE see AJK500
4-AMINO-DAB see DPO200
AMINODARONE see AJK750
1-AMINODECANE see DAG600
2-AMINO-2-DEOXY-d-GALACTOSE HYDROCHLORIDE see GAT000
o-3-AMINO-3-DEOXY-α-d-GLUCOPYRANOSYL-(1-6)-o-(2,6-DIAMINO-2,6-DIDE-
    OXY-α-d-GLUCOPYRANOSYL-(1-4))-2-DEOXY-d-STREPTAMINE SULFATE
    (1:1) see KBA100
o-3-AMINO-3-DEOXY-α-d-GLUCOPYRANOSYL-(1-4)-o-(2,6-DIAMINO-2,6-DIDE-
    OXY)-α-d-GLUCOPYRANOSYL-(1-6)-2-DEOXY-d-STREPTAMINE see BAU270
o-3-AMINO-3-DEOXY-α-d-GLUCOPYRANOSYL-(1-6)-o-(2,6-DIAMINO-2,3,4,6-TET-
    RADEOXY-α-d-erythro-HEXOPYRANOSYL-(1-4))-2-DEOXY-d-STREPTAMINE
    SULFATE (SALT) see PAG050
2'-AMINO-2'-DEOXYKANAMYCIN see BAU270
AMINODEOXYKANAMYCIN see BAU270
AMINODEOXYKANAMYCIN SULFATE see KBA100
4-AMINO-4-DEOXY-N¹⁰-METHYLPTEROYLGLUTAMATE see MDV500
4-AMINO-4-DEOXY-N¹⁰-METHYLPTEROYLGLUTAMIC ACID see MDV500
4-AMINO-1-(2-DEOXY-β-d-erythro-PENTOFURANOSYL)-s-TRIAZIN-2(1H)-ONE see
    ARY125
4-AMINO-4-DEOXYPTEROYLGLUTAMATE see AMG750
2-AMINO-9-(2-DEOXY-β-d-RIBOFURANOSYL)-9H-PURINE-6-THIOL HYDRATE see
    TFJ250
β-d-5-(2-AMINO-2-DEOXY-l-XYLONAMIDO)-1,5-DIDEOXY-1-(3,4-DIHYDRO-5-HY-
    DROXYMETHYL)-2,4-DIOXO-1(2H)-PYRIMIDINYL ALLOFURANURONIC ACID,
    MONOCARBAMATE (ester) see PKE100
4-AMINO-N-(DIAMINOMETHYLENE)BENZENESULFONAMIDE see AHO250
9-AMINO-1,2,5,6-DIBENZANTHRACENE see AJL250
7-AMINODIBENZ(a,h)ANTHRACENE see AJL250
3-AMINODIBENZOFURAN see DDB600
1-AMINO-2,4-DIBROMANTHRACHINON (CZECH) see AJL500
1-AMINO-2,4-DIBROMOANTHRAQUINONE see AJL500
trans-4-((2-AMINO-3,5-DIBROMOBENCIL)AMINO) CICLOHEXANOL (SPANISH)
    see AHJ250
2-AMINO-5,7-DIBROMOBENZOXAZOLE see AJL750
trans-4-((2-AMINO-3,5-DIBROMOBENZYL)AMINO)CYCLOHEXANOL HYDRO-
    CHLORIDE see AHJ500
N-(2-AMINO-3,4-DIBROMOCICLOHEXIL)-trans-4-AMINOCICLOHEXANOL (SPAN-
    ISH) see AHJ250
N-(2-AMINO-3,4-DIBROMOCYCLOHEXYL)-trans-4-AMINOCYCLOHEXANOL see
    AHJ250
2-AMINO-3,5-DIBROMO-N-CYCLOHEXYL-N-METHYL-BENZENEMETHANAMINE
    MONOHYDROCHLORIDE (9CI) see BMO325
2-AMINO-4-(2-DIBUTYLAMINOETHOXY)PYRIMIDINE see AJL875
2-AMINO-4-DIBUTYLAMINOETHOXYPYRIMIDINE see AJL875
2-AMINO-4-DICHLOROARSINOPHENOL HYDROCHLORIDE see DFX400
1-AMINO-3,4-DICHLOROBENZENE see DEO300
3-AMINO-2,5-DICHLOROBENZOIC ACID see AJM000
2-AMINO-5,6-DICHLOROBENZOXAZOLE see AJM500
α-AMINO-γ-(p-DICHLOROETHYLAMINO)-PHENYLBUTYRIC ACID see AJE000
3-AMINO-1-(3,4-DICHLORO-α-METHYLBENZYL)-2-PYRAZOLIN-5-ONE see
    EAE675
5-AMINO-2-(1-(3,4-DICHLOROPHENYL)-ETHYL)-2,4-DIHYDRO-3H-PYRAZOL-3-
    ONE see EAE675
1-(4-AMINO-5-(3,4-DICHLOROPHENYL)-2-METHYL-1H-PYRROL-3-YL)ETHANONE
    see AJM600
2-AMINO-5-((3,4-DICHLOROPHENYL)THIOMETHYL)-2-OXAZOLINE see AJM750
5-AMINO-9-(DIETHYLAMINO)BENZO(a)PHENOXAZIN-7-IUM SULFATE (2:1) see
    AJN250
2-AMINO-4-(2-DIETHYLAMINOETHOXY)PYRIMIDINE see AJN375
2-AMINO-4-DIETHYLAMINOETHOXYPYRIMIDINE see AJN375
p-AMINO-N-(2-DIETHYLAMINOETHYL)BENZAMIDE see AJN500
4-AMINO-N-(2-DIETHYLAMINOETHYL)-BENZAMIDE (9CI) see AJN500
p-AMINO-N-(2-(DIETHYLAMINO)ETHYL)BENZAMIDE HYDROCHLORIDE see
    PME000
4-AMINO-N-(2-(DIETHYLAMINO)ETHYL)BENZAMIDE MONOHYDROCHLORIDE
    see PME000

p-AMINO-N-(2-DIETHYLAMINOETHYL)BENZAMIDE SULFATE see AJN750
N-(2-AMINO-5-DIETHYLAMINOPHENETHYL)METHANE SULFONAMIDEHYDRO-
  CHLORIDE see AJO000
1-AMINO-3-(DIETHYLAMINO)PROPANE see DIY800
2-AMINO-4-γ-DIETHYLAMINOPROPYLAMINO-5,6-DIMETHYLPYRIMIDINE see
  AJQ000
p-AMINODIETHYLANILINE see DJV200
p-AMINO DIETHYLANILINE HYDROCHLORIDE see AJO250
4-AMINODIFENIL (SPANISH) see AJS100
p-AMINODIFENYLAMIN (CZECH) see PFU500
4-AMINODIFENYLETHER see PDR500
N-(4-AMINO-9,10-DIHYDRO-9,10-DIOXO-1-ANTHRACENTY)-BENZAMIDE see
  AIB250
2-AMINO-1,9-DIHYDRO-9-((2-HYDROXYETHOXY)METHYL)-6H-PURIN-6-ONE see
  AEC700
2-AMINO-1,9-DIHYDRO-9-((2-HYDROXYETHOXY)METHYL)-6H-PURIN-6-ONE
  MONOSODIUM SALT see AEC725
6-AMINO-1,2-DIHYDRO-1-HYDROXY-2-IMINO-4-PIPERIDINOPYRIMIDINE see
  DCB000
(s)-7-AMINO-6,7-DIHYDRO-10-HYDROXY-1,2,3-TRIMETHOXYBEN-
  ZO(a)HEPTALEN-9(5H)-ONE see TLN750
4-o-(3-AMINO-3,4-DIHYDRO-6-((METHYLAMINO)METHYL)-2H-PYRAN-2-YL)-2-DE-
  OXY-6-o-(3-DEOXY-4-C-METHYL-3-(METHYLAMINO)-β-I-ARABINOPYRANO-
  SYL)-d-STREPTAMINE (2S-cis)-, SULFATE (salt) see GAA120
3-AMINO-1,5-DIHYDRO-5-METHYL-1-β-d-RIBOFURANOSYL-1,4,5,6,8-PENTAA-
  ZAACENAPHTHYLENE see TJE870
α-AMINO-2,3-DIHYDRO-3-OXO-5-ISOXAZOLEACETIC ACID see AKG250
(s)-7-AMINO-6,7-DIHYDRO-1,2,3,10-TETRAMETHOXYBENZO(a)HEPTALEN-9(5H)-
  ONE see TLO000
5-AMINO-1,6-DIHYDRO-7H-v-TRIAZOLO(4,5-d)PYRIMIDIN-7-ONE see AJO500
5-AMINO-1,4-DIHYDRO-7H-1,2,3-TRIAZOLO(4,5-d)PYRIMIDIN-7-ONE (9CI) see
  AJO500
l-2-AMINO-1-(3,4-DIHYDROXYPHENYL)ETHANOL see NNO500
2-AMINO-3-(3,4-DIHYDROXYPHENYL)PROPANOIC ACID see DNA200
2-AMINO-3,5-DIIODOBENZOIC ACID see API800
2-AMINO-4-(2-DIISOBUTYLAMINOETHOXY)PYRIMIDINE see AJO625
2-AMINO-4-DI-ISOBUTYLAMINOETHOXYPYRIMIDINE see AJO625
3-AMINO-2-(2-(DIISOPROPYLAMI-
  NO)ETHOXY)BUTYROPHENONEDIHYDROCHLORIDE see AJO750
1-(4-AMINO-5-(3,4-DIMETHOXYPHENYL)-2-METHYL-1H-PYRROL-3-
  YL)ETHANONE see AJO800
2-AMINO-1-(2,5-DIMETHOXYPHENYL)-1-PROPANOL HYDROCHLORIDE see
  MDW000
4-AMINO-N-(4,6-DIMETHOXY-2-PYRIMIDINYL)BENZENESUL-FONAMIDE see
  SNI500
4-AMINO-N-(2,6-DIMETHOXY-4-PYRIMIDINYL)BENZENESULFONAMIDE see
  SNN300
4-AMINO-N-(5,6-DIMETHOXY-4-PYRIMIDINYL)BENZENESULFONAMIDE see
  AIE500
1-(4-AMINO-6,7-DIMETHOXY-2-QUINAZOLINYL)-4-(2-FURANYLCARBO-
  NYL)PIPERAZINE HYDROCHLORIDE see FPP100
1-(4-AMINO-6,7-DIMETHOXY-2-QUINAZOLINYL)-4-(2-FURANYLCARBONYL)) PI-
  PERAZINE see AJP000
1-(4-AMINO-6,7-DIMETHOXY-2-QUINAZOLINYL)-4-(2-FUROYL)PIPERAZINE
  MONOHYDROCHLORIDE see FPP100
1-(4-AMINO-6,7-DIMETHOXY-2-QUINAZOLINYL)-4-((TETRAHYDRO-2-FURA-
  NYL)CARBON YL)PIPERAZINE HCl 2H₂O see TEF700
4-AMINO-4'-DIMETHYLAMINOAZOBENZENE see DPO200
4'-AMINO-N,N-DIMETHYL-4-AMINOAZOBENZENE see DPO200
2-AMINO-4-(2-DIMETHYLAMINOETHOXY)PYRIMIDINE see AJP125
2-AMINO-4-DIMETHYLAMINOETHOXYPYRIMIDINE see AJP125
3-AMINO-7-DIMETHYLAMINO-2-METHYLPHENAZATHIONIUM CHLORIDE see
  AJP250
3-AMINO-7-DIMETHYLAMINO-2-METHYLPHENAZINE HYDROCHLORIDE see
  AJQ250
3-AMINO-7-(DIMETHYLAMINO)-2-METHYL-PHENAZINE MONOHYDROCHLORIDE
  see AJQ250
3-AMINO-7-(DIMETHYLAMINO)-2-METHYL-PHENOSELENAZIN-5-IUM CHLORIDE
  see SBU950
3-AMINO-7-(DIMETHYLAMINO)PHENOTHIAZIN-5-IUM CHLORIDE see DUG700
1-AMINO-3-DIMETHYLAMINOPROPANE see AJQ100
AMINODIMETHYLAMINOTOLUAMINOZINE HYDROCHLORIDE see AJQ250
p-AMINODIMETHYLANILINE see DTL800
4-AMINO-2',3-DIMETHYLAZOBENZENE see AIC250
4'-AMINO-2,3'-DIMETHYLAZOBENZENE see AIC250
4-AMINO-1,3-DIMETHYLBENZENE see XMS000
3-AMINO-1,4-DIMETHYLBENZENE see XNA000
1-AMINO-2,4-DIMETHYLBENZENE see XMS000
1-AMINO-2,5-DIMETHYLBENZENE see XNA000
AMINODIMETHYLBENZENE see XMA000
4-AMINO-1,3-DIMETHYLBENZENE HYDROCHLORIDE see XOJ000
3-AMINO-1,4-DIMETHYLBENZENE HYDROCHLORIDE see XOS000
5-AMINO-1,4-DIMETHYLBENZENE HYDROCHLORIDE see XOS000
1-AMINO-2,4-DIMETHYLBENZENE HYDROCHLORIDE see XOJ000
1-AMINO-2,5-DIMETHYLBENZENE HYDROCHLORIDE see XOS000

3-AMINO-1,4-DIMETHYL-γ-CARBOLINE see TNX275
2-AMINODIMETHYLETHANOL see IIA000
4-AMINO-6-(1,1-DIMETHYLETHYL)-3-(METHYLTHIO)-1,2,4-TRIAZIN-5(4H)-ONE
  see MQR275
4-AMINO-3',5'-DIMETHYL-4'-HYDROXYAZOBENZENE see AJQ500
2-AMINO-3,4-DIMETHYLIMIDAZO(4,5-f)QUINOLINE see AJQ600
2-AMINO-3,8-DIMETHYL-3H-IMIDAZO(4,5-f)QUINOXALINE see AJQ675
2-AMINO-3,8-DIMETHYLIMIDAZO(4,5-f)QUINOXALINE see AJQ675
4-AMINO-N-(3,4-DIMETHYL-5-ISOXAZOLYL)BENZENESULPHONAMIDE see
  SNN500
2-AMINO-6-DIMETHYL-4-(p-(p-((p-((1-METHYLPYRIDINIUM-3-
  YL)CARBAMOYL)PHENYL)CARBABENZAMIDO)ANILINO)PYRIMIDIMI see
  AJQ750
4-AMINO-N-(4,5-DIMETHYL-2-OXAZOLYL)BENZENESULFONAMIDE see AIE750
p-AMINO-N,α-DIMETHYLPHENETHYLAMINE see AJR000
2-AMINO-4,5-DIMETHYLPHENOL see AMW750
1-(4-AMINO-1,2-DIMETHYL-5-PHENYL-1H-PYRROL-3-YL)ETHANONE see
  AJR100
4-AMINO-1,2-DIMETHYL-5-PHENYLPYRROL-3-YLETHANONE see AJR100
5-AMINO-1,3-DIMETHYL-4-PYRAZOLYL o-FLUOROPHENYL KETONE see
  AJR400
(5-AMINO-1,3-DIMETHYL-1H-PYRAZOL-4-YL)(2-FLUOROPHENYL)METHANONE
  see AJR400
3-AMINO-1,4-DIMETHYL-5H-PYRIDO(4,3-b)INDOLE see TNX275
3-AMINO-1,4-DIMETHYL-5H-PYRIDO(4,3-b)INDOLE ACETATE see AJR500
4-AMINO-N-(2,6-DIMETHYL-4-PYRIMIDINYL)BENZENESULFONAMIDE (9CI) see
  SNJ350
4-AMINO-N-(4,6-DIMETHYL-2-PYRIMIDINYL)BENZENESULFONAMIDE, MONOSO-
  DIUM SALT see SJW500
2-AMINO-4,6-DINITROPHENOL see DUP400
4-AMINO-3,5-DINITROTOLUENE see DVI100
2-AMINO-4,6-DINITROTOLUENE see AJR750
2-AMINODIPHENYL see BGE250
4-AMINODIPHENYL see AJS100
o-AMINODIPHENYL see BGE250
p-AMINODIPHENYL see AJS100
4-AMINODIPHENYLAMINE see PFU500
p-AMINODIPHENYLAMINE see PFU500
2-AMINODIPHENYLENE OXIDE see DDB600
2-AMINODIPHENYLENOXYD (GERMAN) see DDB800
2-AMINO-1,2-DIPHENYLETHANOL HYDROCHLORIDE see DWB000
2-AMINODIPHENYL ETHER see PDR490
4-AMINODIPHENYL ETHER see PDR500
2-AMINO-6-((1,2-DIPHENYLETHYL)AMINO)-3-PYRIDINECARBAMIC ACID ETHYL
  ESTER, MONOHYDROCHLORIDE see DAB200
p-AMINODIPHENYLIMIDE see PEI000
2-AMINODIPYRIDO(1,2-a:3',2'-d)IMIDAZOLE see DWW700
2-AMINODIPYRIDO(1,2-a:3',2'-d)IMIDAZOLE HYDROCHLORIDE see AJS225
2-AMINO-4,6-DIPYRROLIDINOTRIAZINE see AJS250
4-AMINO-3,4'-DISULFOAZOBENZENE see AJS500
2-AMINO-1,4-DISULFOBENZENE see AIE000
1-AMINO-3,6-DISULFO-8-NAFTYLESTER KYSELINA p-TOLUENSULFONOVE
  (CZECH) see AKH500
5-AMINO-1,2,4-DITHIAZOLE-3-THIONE see IBL000
1-AMINODODECANE see DXW000
4-AMINO-1-DODECYLQUINALDINIUM ACETATE see AJS750
AMINODUR see TEP500
2-AMINOETANOLO (ITALIAN) see EEC600
1-AMINOETHANE see EFU400
AMINOETHANE see EFU400
2-AMINO-ETHANESELENOL HYDROCHLORIDE see AJS900
2-AMINOETHANESELENOSULFURIC ACID see AJS950
2-AMINOETHANESULFONIC ACID see TAG750
2-AMINOETHANESULFONO-p-PHENETIDINE HYDROCHLORIDE see TAG875
2-AMINOETHANETHIOL see AJT250
2-AMINO-ETHANETHIOL DIHYDROGEN PHOSPHATE(ester), MONOSODIUM
  SALT see AKB500
2-AMINOETHANETHIOSULFURIC ACID see AJT500
1-AMINOETHANOL see AAG500
2-AMINOETHANOL (MAK) see EEC600
2-AMINOETHANOL compounded with 6-CYCLOHEXYL-1-HYDROXY-4-METHYL-
  2(1H)-PYRIDINONE (1:1) see BAR800
2-AMINOETHANOL HYDROCHLORIDE see EEC700
β-AMINOETHANOL HYDROCHLORIDE see EEC700
3-AMINO-4-ETHOXYACETANILIDE see AJT750
(L)-3-(2-AMINOETHOXY)ALANINE see OLK200
4-AMINOETHOXYBENZENE see PDK890
2-AMINO-3-ETHOXYCARBONYL-6-BENZYL-4,5,6,7-TETRAHYDROTHIENO(2,3-
  c)PYRIDINE HYDROCHLORIDE see TGE165
2-AMINO-3-ETHOXYCARBONYL-5-BENZYL-4,5,6,7-TETRAHYDROTHIENO (2,3-
  c)PYRIDINE HYDROCHLORIDE see AJU000
2-(2-AMINOETHOXY)ETHANOL see AJU250
2-AMINOETHOXYETHANOL see AJU250
5-AMINO-6-ETHOXY-2-NAPHTHALENESULFONIC ACID see AJU500
1-(1-AMINOETHYL)ADAMANTANE HYDROCHLORIDE see AJU625

α-AMINOETHYL ALCOHOL see AAG500
β-AMINOETHYL ALCOHOL see EEC600
3-((2-((2-AMINOETHYL)AMINO)ETHYL)AMINO)PROPIONITRILE see LBV000
(3-(2-AMINOETHYL)AMINOPROPYL)TRIMETHOXYSILANE see TLC500
2-AMINOETHYLAMMONIUM PERCHLORATE see AJU875
1-AMINO-4-ETHYLBENZENE see EGL000
β-AMINOETHYLBENZENE see PDE250
o-AMINOETHYLBENZENE see EGK500
4-(2-AMINOETHYL)-1,2-BENZENEDIOL HYDROCHLORIDE see DYC600
α-(1-AMINOETHYL)BENZENEMETHANOL HYDROCHLORIDE see PMJ500
4-(2-AMINOETHYL)-1,2,3-BENZENETRIOL HYDROCHLORIDE see HKG000
α-(1-AMINOETHYL)-BENZYL ALCOHOL see NNM000
dl-α-(1-AMINOETHYL)BENZYL ALCOHOL see NNM500
α-(1-AMINOETHYL)BENZYL ALCOHOL HYDROCHLORIDE see NNN000,
  PMJ500
3-(2-AMINOETHYL)-1-BENZYL-5-METHOXY-2-METHYLINDOLE HYDROCHLO-
  RIDE see BEM750
3-AMINO-9-ETHYLCARBAZOLE see AJV000
3-AMINO-N-ETHYLCARBAZOLE see AJV000
3-AMINO-9-ETHYLCARBAZOLEHYDROCHLORIDE see AJV250
trans-4-AMINOETHYLCYCLOHEXANE-1-CARBOXYLIC ACID see AJV500
4-(2-AMINOETHYL)-6-DIAZO-2,4-CYCLOHEXADIENONE HYDROCHLORIDE see
  DCQ575
α-(1-AMINOETHYL)-2,5-DIMETHOXYBENZYL ALCOHOL HYDROCHLORIDE see
  MDW000
α-(1-AMINOETHYL)-2,4-DIMETHOXYBENZYL ALCOHOL HYDROCHLORIDE see
  AJV850
2-AMINOETHYL DISULFIDE DIHYDROCHLORIDE see CQJ750
AMINOETHYLENE see EJM900
2-AMINOETHYL ESTER CARBAMIMIDOTHIOIC ACID DIHYDROBROMIDE see
  AJY250
AMINOETHYLETHANEDIAMINE see DJG600
N-AMINOETHYLETHANOLAMINE see AJW000
AMINOETHYL ETHANOLAMINE see AJW000
N-(2-AMINOETHYL)ETHYLENEDIAMINE see DJG600
6-AMINO-1-ETHYL-4-p-((((1-ETHYLPYRIDINIUM-4-YL)AMINO)2-AMINOPHE-
  NYL)CARBAMOYL)ANILINO)QUINOLINIUM DIIODIDE see AJW250
6-AMINO-1-ETHYL-4-(p-(p-((1-ETHYLPYRIDINIUM-4-
  YL)AMINO)BENZAMIDO)ANILINO)QUINOLINIUM DIIODIDE see AJW500
6-AMINO-1-ETHYL-4-(p-((p-((1-ETHYLPYRIDINIUM-4-
  YL)AMINO)PHENYL)CARBAMOYL)ANILINOQUINOLINIUM) DIBROMIDE see
  AJW750
β-AMINOETHYLGLYOXALINE see HGD000
1-AMINO-2-ETHYLHEXAN (CZECH) see EKS500
1-α-(1-AMINOETHYL)-m-HYDROXYBENZYL ALCOHOL see HNB875
1-α-(1-AMINOETHYL)-m-HYDROXYBENZYL ALCOHOL BITARTRATE see
  HNC000
(−)-α-(1-AMINOETHYL)-m-HYDROXYBENZYL ALCOHOL BITARTRATE see
  HNC000
1-α-(1-AMINOETHYL)-m-HYDROXYBENZYL ALCOHOL HYDROGEN-d-TAR-
  TRATE see HNC000
3-(β-AMINOETHYL)-5-HYDROXYINDOLE see AJX500
4-(2-AMINOETHYL)IMIDAZOLE see HGD000
β-AMINOETHYLIMIDAZOLE see HGD000
4-(2-AMINOETHYL)IMIDAZOLE BIS(DIHYDROGEN PHOSPHATE) see HGE000
4-(2-AMINOETHYL)IMIDAZOLE DI-ACID PHOSPHATE see HGE000
4-AMINOETHYLIMIDAZOLE HYDROCHLORIDE see HGE500
3-(2-AMINOETHYL)INDOLE see AJX000
(AMINO-2 ETHYL)-3-INDOLE (FRENCH) see AJX000
3-(1-AMINOETHYL)INDOLE HYDROCHLORIDE see LIU100
3-(2-AMINOETHYL)INDOLE HYDROCHLORIDE see AJX250
3-(2-AMINOETHYL)INDOL-5-OL see AJX500
3-(2-AMINOETHYL)INDOL-5-OL CREATININE SULFATE see AJX750
AMINOETHYLISOSELENOURONIUM BROMIDE HYDROBROMIDE see AJY000
2-β-AMINOETHYLISOTHIOUREA see AJY250
S-β-AMINOETHYLISOTHIOURONIC DIHYDROCHLORIDE see AJY500
2-(β-AMINOETHYL)ISOTHIOURONIUM BROMIDE HYDROBROMIDE see AJY250
2-AMINOETHYLISOTHIOURONIUM DIBROMIDE see AJY250
2-AMINOETHYLISOTHIOURONIUMDICHLORIDE see AJY500
2-AMINOETHYLISOTHIURONIUM BROMIDE HYDROBROMIDE see AJY250
β-AMINOETHYLISOTHIURONIUM BROMIDE HYDROBROMIDE see AJY250
S-(2-AMINOETHYL)ISOTHIURONIUM BROMIDE HYDROBROMIDE see AJY250
S-(β-AMINOETHYL)ISOTHIURONIUM BROMIDE HYDROBROMIDE see AJY250
2-AMINOETHYLISOTHIURONIUM DIHYDROBROMIDE see AJY250
2-AMINOETHYL MERCAPTAN see AJT250
4-AMINO-N-ETHYL-m-(β-METHANESULFONAMIDOETHYL)-m-TOLUIDINE see
  AJY750
3-(2-AMINOETHYL)-5-METHOXYBENZOFURAN HYDROCHLORIDE see AJZ000
6-(2-AMINOETHYL)-5-METHOXYBENZOFURAN HYDROCHLORIDE see AKA000
6-(β-AMINOETHYL)-5-METHOXYBENZOFURANHYDROCHLORIDE see AKA000
α-(1-AMINOETHYL)-4-METHOXYBENZYL ALCOHOL HYDROCHLORIDE see
  AKA250
α-(1-AMINOETHYL)-4-METHOXYBENZYLALCOHOL HYDROCHLORIDE see
  AKA250
3-(2-AMINOETHYL)-5-METHOXYINDOLE see MFS400

3-(2-AMINOETHYL)-5-METHOXYINDOLE HYDROCHLORIDE see MFT000
3-(2-AMINOETHYL)-6-METHOXYINDOLE HYDROCHLORIDE see MFS500
3-(2-AMINOETHYL)-5-METHOXY-2-METHYLBENZO-FURAN, HYDROCHLORIDE
  see MGH750
2-AMINOETHYLMETHYLAMINE see MJW100
2-AMINOETHYL-2-METHYL-1,3-BENZODIOXOLE HYDROCHLORIDE see
  AKA500
2-(AMINOETHYL)-2-METHYL-1,3-BENZODIOXOLE HYDROCHLORIDE see
  AKA500
1-(4-AMINO-1-ETHYL-2-METHYL-5-PHENYL-1H-PYRROL-3-YL)ETHANONE see
  AKA600
N-AMINOETHYLMORPHOLINE see AKA750
2-AMINO-5-ETHYLNONANE see EMY000
p-(2-AMINOETHYL)PHENOL see TOG250
p-β-AMINOETHYLPHENOL see TOG250
4-(2-AMINOETHYL)PHENOL HYDROCHLORIDE see TOF750
p-(2-AMINOETHYL)PHENOL MONOCHLORIDE see TOF750
1-(2-AMINOETHYL)PIPERAZINE see AKB000
N-(2-AMINOETHYL)PIPERAZINE see AKB000
N-(β-AMINOETHYL)PIPERAZINE see AKB000
N-AMINOETHYLPIPERAZINE see AKB000
AMINOETHYLPIPERAZINE see AKB000
6-AMINO-3-ETHYL-1-(2-PROPENYL)-2,4(1H,3H-)-PYRIMIDINEDIONE see AFW500
4-(2-AMINOETHYL)PYROCATECHOL see DYC400
4-(2-AMINOETHYL)PYROCATECHOL HYDROCHLORIDE see DYC600
2-AMINOETHYLSULFONIC ACID see TAG750
5-AMINO-2-ETHYLTETRAZOL see AKB125
2-AMINO-4-(ETHYLTHIO)BUTYRIC ACID see AKB250, EEI000
dl-2-AMINO-4-(ETHYLTHIO)BUTYRIC ACID see EEI000
l-2-AMINO-4-(ETHYLTHIO)BUTYRIC ACID see AKB250
S-(2-AMINOETHYL)THIOPHOSPHATEMONOSODIUM SALT see AKB500
2-AMINOETHYL-2-THIOPSEUDOUREA DICHLORIDE see AJY500
2-(2-AMINOETHYL)-2-THIOPSEUDOUREA DIHYDROCHLORIDE see AJY500
2-(2-AMINOETHYL)-2-THIOPSEUDOUREA HYDROBROMIDE see AJY250
3-AMINO-α-ETHYL-2,4,6-TRIIODOHYDROCINNAMIC ACID see IFY100
3′-AMINO-N-ETHYL-2′,4′,6′-TRIIODOSUCCINANILIC ACID see AMV375
AMINOFENAZONE (ITALIAN) see DOT000
p-AMINOFENETOL see PDK890
5-AMINO-3-FENIL-1-BIS(-DIMETILAMINO)-FOSFORIL-1,2,4-TRIAZOLO (ITALIAN)
  see AIX000
m-AMINOFENOL (CZECH) see ALT500
p-AMINOFENOL (CZECH) see ALT250
5-AMINO-3-FENYL-1-BIS(DIMETHYL-AMINO)-FOSFORYL-1,2,4-TRIAZOOL
  (DUTCH) see AIX000
m-AMINOFENYLMOCOVINA HYDROCHLORID (CZECH) see ALY750
AMINOFILINA (SPANISH) see TEP500
AMINOFLUOREN (GERMAN) see FDI000
2-AMINOFLUORENE see FDI000
2-AMINO-N-FLUOREN-2-YLACETAMIDE see AKC000
2-AMINO-5-FLUOROBENZOXAZOLE see AKC250
4-AMINO-4′-FLUORODIPHENYL see AKC500
6-AMINO-2-(FLUOROMETHYL)-3-(2-METHYLPHENYL)-4(3H)-QUINAZOLINONE
  (9CI) see AEW625
6-AMINO-2-FLUOROMETHYL-3-(o-TOLYL)-4(3H)-QUINAZOLINONE see AEW625
2-AMINO-1-(3-FLUOROPHENYL)ETHANOL HYDROBROMIDE see AKS500
(2-AMINO-6-(((4-FLUOROPHENYL)METHYL)AMINO)-3-PYRIDINYL)CARBAMIC
  ACID ETHYL ESTER MALEATE see FMP100
1-(4-AMINO-5-(3-FLUOROPHENYL)-2-METHYL-1H-PYRROL-3-YL)ETHANONE see
  AKC550
1-(4-AMINO-5-(4-FLUOROPHENYL)-2-METHYL-1H-PYRROL-3-YL)ETHANONE see
  AKC560
3-AMINO-2-FLUORO-PROPANOIC ACID HYDROCHLORIDE see FFT100
4-AMINO-5-FLUORO-2(1H)-PYRIMIDINONE see FHI000
AMINOFORM see HEI500
AMINOFORMAMIDINE see GKW000
AMINOFORMAMIDINE HYDROCHLORIDE see GKY000
AMINOFOSTINE see AMD000
2-AMINOGLUTARAMIC ACID see GFO050
l-2-AMINOGLUTARAMIDIC ACID see GFO050
α-AMINOGLUTARIC ACID see GFO000
l-2-AMINOGLUTARIC ACID see GFO000
p-AMINOGLUTETHIMIDE see AKC600
AMINOGLUTETHIMIDE see AKC600
AMINOGLUTETHIMIDE PHOSPHATE see AKC625
1-AMINOGLYCEROL see AMA250
AMINOGLYCOL see ALB000
AMINOGUANIDINE see AKC750
AMINOGUANIDINE HYDROCHLORIDE see AKC800
AMINOGUANIDINE SULFATE see AKD250
AMINOGUANIDINE SULPHATE see AKD250
AMINO GUANIDINIUM NITRATE see AKD375
l,2-AMINO-4-(GUANIDINOOXY)BUTYRIC ACID see AKD500
2-AMINO-4-(GUANIDINOOXY)-l-BUTYRIC ACID see AKD500
1-AMINOHEPTANE see HBL600
3-AMINOHEPTANE see HBM500

5-AMINO-2-MERCAPTO-1,3,4-THIADIAZOLE see AKM000
2-AMINO-5-MERCAPTO-1,3,4-THIADIAZOLE see AKM000
AMINOMERCURIC CHLORIDE see MCW500
2-AMINOMESITYLENE see TLG500
AMINOMESITYLENE see TLG500
2-AMINOMESITYLENE HYDROCHLORIDE see TLH000
AMINOMESITYLENE HYDROCHLORIDE see TLH000
6-AMINO-1-METALLYL-3-METHYLPYRIMIDINE-2,4-DIONE see AKL625
AMINOMETHANAMIDINE see GKW000
AMINOMETHANAMIDINE HYDROCHLORIDE see GKY000
AMINOMETHANE see MGC250
6-AMINOMETHAQUALONE see AKM125
1-AMINO-2-METHOXYANTHRAQUINONE see AKM250
4-AMINO-3-METHOXYAZOBENZENE see MFB000, MFF500
3-AMINO-4-METHOXY BENZANILIDE see AKM500
1-AMINO-2-METHOXYBENZENE see AOV900
1-AMINO-4-METHOXYBENZENE see AOW000
2-AMINO-5-METHOXY BENZENESULFONIC ACID see AIA500
3-AMINO-4-METHOXYBENZOIC ACID see AIA250
2-AMINO-4-METHOXYBENZOTHIAZOLE see AKM750
2-AMINO-5-METHOXYBENZOXAZOLE see AKN000
4-AMINO-4′-METHOXY-3-BIPHENYLOL see HLR500
4-AMINO-4′-METHOXY-3-BIPHENYLOLHYDROCHLORIDE see AKN250
2-AMINO-3-METHOXYDIPHENYLENE OXIDE see AKN500
2-AMINO-3-METHOXYDIPHENYLENOXYD (GERMAN) see MDZ000
4-AMINO-N-(2-METHOXYETHYL)-7-((2-METHOXYETHYL)AMINO-2-PHENYL)-6-PTERIDINECARBOXAMIDE see AKN750
3′-(l-α-AMINO-p-METHOXYHYDROCINNAMAMIDO)-3′-DEOXY-N,N-DIMETHYLA-DENOSINE see AEI000
3′-(α-AMINO-p-METHOXYHYDROCINNAMAMIDO)-3′-DEOXY-N,N-DIMETHYLADE-NOSINE MONOHYDROCHLORIDE see POK250
1-AMINO-2-METHOXY-5-METHYLBENZENE see MGO750
6-AMINO-8-METHOXY-1-METHYL-4-(p-(p-((1-METHYLPYRIDINIUM-4-YL)AMINO)BENZAMIDO)ANILINOQUINOLINIUM) DI-p-TOL see AKO000
1-(4-AMINO-5-(4-METHOXY-3-METHYLPHENYL)-2-METHYL-1H-PYRROL-3-YL)ETHANONE see AKO100
7-AMINO-4-(2-METHOXY-p-(p-((1-METHYLPYRIDINIUM-4-YL)AMINO)BENZAMIDO)ANILINO)-1-METHYLQUINOLINIUM)) DIBROMID see AKO250
4-AMINO-N-(6-METHOXY-2-METHYL-4-PYRIMIDINYL)BENZENESULFONAMIDE see MDU300
7-AMINO-9-α-METHOXYMITOSANE see AHK500
2-AMINO-1-METHOXYNAPHTHALENE see MFA000
2-AMINO-1-METHOXY-4-NITROBENZENE see NEQ500
3-AMINO-4-METHOXYNITROBENZENE see NEQ500
1-AMINO-2-METHOXY-4-OXYANTHRAQUINONE see AKO350
1-(4-AMINO-5-(4-METHOXYPHENYL)-2-METHYL-1H-PYRROL-3-YL)ETHANONE see AKO430
1-(4-AMINO-5-(o-METHOXYPHENYL)-2-METHYL-1H-PYRROL-3-YL)ETHANONE see AKO450
1-(4-AMINO-5-(3-METHOXYPHENYL)-2-METHYL-1H-PYRROL-3-YL)ETHANONE see AKO400
(S)-3′-((2-AMINO-3-(4-METHOXYPHENYL)-1-OXOPROPYL)AMINO)-3′-DEOXY-N,N-DIMETHYLADENOSINE see AEI000
4-AMINO-6-METHOXY-1-PHENYLPYRIDAZINIUM-METHYLSULFAT (GERMAN) see MQS100
4-AMINO-6-METHOXY-1-PHENYLPYRIDAZINIUM METHYL SULFATE see MQS100
4-AMINO-N-(6-METHOXY-3-PYRIDAZINYL)-BENZENESULFONAMIDE see AKO500
4-AMINO-2-METHOXY-5-PYRIMIDINEMETHANOL see AKO750
4-AMINO-N-(6-METHOXY-4-PYRIMIDINYL)-BENZENESULFONAMIDE (9CI) see SNL800
4-AMINO-N-(5-METHOXY-2-PYRIMIDINYL)BENZENESULFONAMIDE SODIUM SALT see MFO000
3-AMINO-4-METHOXYTOLUEN (CZECH) see MFQ500
3-AMINO-4-METHOXYTOLUENE see MGO750
3-AMINOMETHYLALIZARIN-N,N-DIACETIC ACID see AFM400
1-AMINO-4-(METHYLAMINO)-9,10-ANTHRACENEDIONE see AKP250
4-AMINO-1-METHYLAMINOANTHRAQUINONE see AKP250
dl-α-AMINO-β-METHYLAMINOPROPIONIC ACID see AKP500
4-AMINO-2-METHYLANILINE see TGM000
2-AMINO-4-METHYLANISOLE see MGO750
1-AMINO-2-METHYL-9,10-ANTHRACENEDIONE see AKP750
1-AMINO-2-METHYLANTHRAQUINONE see AKP750
2-AMINO-1-METHYLBENZENE see TGQ750
3-AMINO-1-METHYLBENZENE see TGQ500
4-AMINO-1-METHYLBENZENE see TGR000
1-AMINO-2-METHYLBENZENE see TGQ750
2-AMINO-1-METHYLBENZENE HYDROCHLORIDE see TGS500
1-AMINO-2-METHYLBENZENE HYDROCHLORIDE see TGS500
β-(AMINOMETHYL)BENZENEPROPANOIC ACID see PEE500
β-(AMINOMETHYL)-BENZENEPROPANOIC ACID HYDROCHLORIDE see GAD000
2-AMINO-5-METHYLBENZENESULFONIC ACID see AKQ000

2-AMINO-4-METHYLBENZOTHIAZOLE see AKQ500
2-AMINO-6-METHYLBENZOTHIAZOLE see MGD500
2-AMINO-5-METHYLBENZOXAZOLE see AKQ750
3-AMINO-α-METHYLBENZYL ALCOHOL see AKR000
m-AMINO-α-METHYLBENZYL ALCOHOL see AKR000
8-(4-AMINO-1-METHYLBUTYLAMINO)-6-METHOXYQUINOLINE see PMC300
8-((4-AMINO-1-METHYLBUTYL)AMINO)-6-METHOXYQUINOLINE DIPHOSPHATE see AKR250
2-AMINO-3-METHYL-α-CARBOLINE see ALD750
3-AMINO-1-METHYL-γ-CARBOLINE see ALD500
2-AMINO-4-METHYL-5-CARBOXANILIDOTHIAZOLE see AKR500
β-(AMINOMETHYL)-4-CHLOROBENZENEPROPANOIC ACID see BAC275
β-(AMINOMETHYL)-p-CHLOROHYDROCINNAMIC ACID see BAC275
trans-4-AMINOMETHYL-1-CYCLOHEXANECARBOXYLIC ACID see AJV500
trans-1-AMINOMETHYLCYCLOHEXANE-4-CARBOXYLIC ACID see AJV500
trans-p-(AMINOMETHYL)CYCLOHEXANECARBOXYLICACID see AJV500
trans-4-(((4-(AMINOMETHYL)CYCLOHEXYL)CARBONYL)OXY)-BENZENEPROPA-NOIC ACID HYDROCHLORIDE see CDF380
trans-4-(((4-(AMINOMETH-YL)CYCLOHEXYL)CARBONYL)OXY)BENZENEPROPANOIC ACID see CDF375
4-AMINO-3-METHYL-N,N-DIETHYLANILINEHYDROCHLORIDE see AKR750
2-AMINOMETHYL-3,4-DIHYDRO-2H-PYRAN see AKS000
2-AMINOMETHYL-2,3-DIHYDRO-4H-PYRAN see AKS000
α-(AMINOMETHYL)-3,4-DIHYDROXYBENZYL ALCOHOL see ARL500, ARL750
l-α-(AMINOMETHYL)-3,4-DIHYDROXYBENZYL ALCOHOL see NNO500
2-AMINO-6-METHYLDIPYRIDO(1,2-a:3′,2′-d)IMIDAZOLE see AKS250
2-AMINO-6-METHYLDIPYRIDO(1,2-a:3′,2′-d)IMIDAZOLE HYDROCHLORIDE see AKS275
N-AMINOMETHYLDOPA see CBQ500
α-AMINO-2-METHYLENE-CYCLOPROPANEPROPANOIC ACID (9CI) see MJP500
α-AMINOMETHYLENECYCLOPROPANEPROPIONIC ACID see MJP500
l-α-AMINO-β-METHYLENECYCLOPROPANEPROPIONIC ACID see MJP500
α-AMINO-β-(2-METHYLENECYCLOPROPYL)PROPIONIC ACID see MJP500
2-AMINO-4,5-METHYLENEHEX-5-ENOIC ACID see MJP500
α-AMINOMETHYL-3-FLUOROBENZYLALCOHOL HYDROBROMIDE see AKS500
4-AMINO-10-METHYLFOLIC ACID see MDV500
6-AMINO-3-METHYLHEPTANE see ILM000
2-AMINO-6-METHYLHEPTANE see ILM000
4-AMINO-2-METHYL-3-HEXANOL see AKS750
β-(AMINOMETHYL)HYDROCINNAMIC ACID see PEE500
β-(AMINOMETHYL)-HYDROCINNAMIC ACID HYDROCHLORIDE see GAD000
α-(AMINOMETHYL)-m-HYDROXYBENZYL ALCOHOL see AKT000
α-(AMINOMETHYL)-p-HYDROXYBENZYL ALCOHOL see AKT250
α-(AMINOMETHYL)-3-HYDROXYBENZYLALCOHOL HYDROCHLORIDE see AKT500
(±)-α-(AMINOMETHYL)-m-HYDROXYBENZYL ALCOHOL HYDROCHLORIDE see NNT100
5-AMINOMETHYL-3-HYDROXYISOXAZOLE see AKT750
2-AMINO-3-METHYLIMIDAZO(4,5-f)QUINOLINE see AKT600
2-AMINO-3-METHYLIMIDAZO(4,5-f)QUINOLINE DIHYDROCHLORIDE see AKT620
5-(AMINOMETHYL)-3-ISOXAZOLOL see AKT750
5-(AMINOMETHYL)-3(2H)-ISOXAZOLONE see AKT750
4-AMINO-N-(5-METHYL-3-ISOXAZOLYL)BENZENESULFONAMIDE see SNK000
5-AMINOMETHYL-3-ISOXYZOLE see AKT750
l-α-AMINO-γ-METHYLMERCAPTOBUTYRIC ACID see MDT750
6-AMINO-3-METHYL-1-(2-METHYLALLYL)-2,4(1H,3H)-PYRIMIDINEDIONE see AKL625
6-AMINO-3-METHYL-1-(2-METHYLALLYL)URACIL see AKL625
1-(4-AMINO-2-METHYL-5-(2-METHYLPHENYL)-1H-PYRROL-3-YL)ETHANONE see AKT800
1-(4-AMINO-2-METHYL-5-(3-METHYLPHENYL)-1H-PYRROL-3-YL)ETHANONE see AKT830
1-(4-AMINO-2-METHYL-5-(4-METHYLPHENYL)-1H-PYRROL-3-YL)ETHANONE see AKT850
6-AMINO-3-METHYL-1-(2-METHYL-2-PROPENYL)-2,4(1H,3H)-PYRIMIDINEDIONE see AKL625
4-AMINO-2-METHYL-1-NAPHTHALENOL see AKX500
4-AMINO-2-METHYL-1-NAPHTHOL see AKX500
1-AMINO-2-METHYL-5-NITROBENZENE see NMP500
3-AMINO-4-METHYL-5-(5-NITRO-2-FURYL)-s-TRIAZOLE see AKY000
2-AMINO-4-((E)-2-(1-METHYL-5-NITRO-1H-IMIDAZOL-2-YL)ETHENYL)PYRIMIDINE see TJF000
2-AMINO-6-(1-METHYL-4-NITRO-5-IMIDAZOLYL)MERCAPTOPURINE see AKY250
2-AMINO-6-(1′-METHYL-4′-NITRO-5′-IMIDAZOLYL)MERCAPTOPURINE see AKY250
(E)-2-AMINO-4-(2-(1-METHYL-5-NITROIMIDAZOL-2-YL)VINYL)PYRIMIDINE see TJF000
2-(AMINOMETHYL)NORBORNANE see AKY750
2-AMINO-2-METHYLOL-1,3-PROPANEDIOL see TEM500
2-AMINO-4-METHYLOXAZOLE see AKY875
2-AMINO-3-METHYLPENTANOIC ACID see IKX000
2-AMINO-4-METHYLPENTANOIC ACID see LES000

5-AMINO-2-METHYLPHENOL see AKJ750
4-AMINO-3-METHYLPHENOL see AKZ000
cis-2-AMINO-5-METHYL-4-PHENYL-1-PYRROLINE see ALA000
trans-2-AMINO-5-METHYL-4-PHENYL-1-PYRROLINE see ALA250
1-(4-AMINO-2-METHYL-5-PHENYL-1H-PYRROL-3-YL)ETHANONE HYDROCHLORIDE see ALA300
8-AMINO-2-METHYL-4-PHENYL-1,2,3,4-TETRAHYDROISOQUINOLINE see NMV700
8-AMINO-2-METHYL-4-PHENYL-1,2,3,4-TETRAHYDROISOQUINOLINE MALEATE see NMV725
4-AMINO-3-METHYL-6-PHENYL-1,2,4-TRIAZIN-5(4H)-ONE see ALA500
2-AMINO-4-(N-METHYLPIPERAZINO)-5-METHYLTHIO-6-CHLOROPYRIMIDINE see ALA750
1-AMINO-2-METHYLPROPANE see IIM000
2-AMINO-2-METHYLPROPANE see BPY250
2-AMINO-2-METHYL-1,3-PROPANEDIOL see ALB000
2-AMINO-2-METHYLPROPANOIC ACID see MGB000
2-AMINO-2-METHYL-1-PROPANOL see IIA000
2-AMINO-2-METHYLPROPAN-1-OL see IIA000
1-AMINO-2-METHYL-2-PROPANOL see ALB250
2-AMINO-2-METHYLPROPANOL see IIA000
S-2-AMINO-2-METHYLPROPYL DIHYDROGEN PHOSPHOROTHIOATE see ALB500
S-(2-AMINO-2-METHYLPROPYL)PHOSPHOROTHIOATE see ALB500
(−)-α-(AMINOMETHYL)PROTOCATECHUYL ALCOHOL see NNO500
3-AMINO-6-METHYL-4-PYRIDAZINETHIOL see ALB625
2-AMINOMETHYLPYRIDINE see ALB750
2-AMINO-3-METHYLPYRIDINE see ALC000
2-AMINO-4-METHYLPYRIDINE see ALC250
2-AMINO-5-METHYLPYRIDINE see AMA010
2-AMINO-6-METHYLPYRIDINE see ALC500
4-((3-AMINO-4-((4-((1-METHYLPYRIDINIUM-4-YL)AMINO)BENZOYL)AMINO)PHENYL)AMINO)-1-METHYLQUINOLINIUM)DIBROMIDE see ALC750
2-AMINO-3-METHYL-9H-PYRIDO(2,3-b)INDOLE see ALD750
3-AMINO-1-METHYL-5H-PYRIDO(4,3-b)INDOLE see ALD500
3-AMINO-1-METHYL-5H-PYRIDO(4,3-b)INDOLE ACETATE see ALE750
4-AMINO-N-(4-METHYL-2-PYRIMIDINYL)-BENZENESULFONAMIDE see ALF250
4-AMINO-N-(4-METHYL-2-PYRIMIDINYL)-BENZENESULFONAMIDE MONOSODIUM SALT see SJW475
N′-((4-AMINO-2-METHYL-5-PYRIMIDINYL)METHYL)-N-(2-CHLOROETHYL)-N-NITROSOUREA HCl see ALF500
3-(4-AMINO-2-METHYL-5-PYRIMIDINYL)METHYL-1-(2-CHLOROETHYL)-1-NITROSOUREA see NDY800
1-(4-AMINO-2-METHYLPYRIMIDIN-5-YL)METHYL-3-(2-CHLOROETHYL)-3-NITROSOUREA see ALF500
3-((4-AMINO-2-METHYL-5-PYRIMIDINYL)METHYL)-1-(2-CHLOROETHYL)-1-NITROSOUREA HYDROCHLORIDE see ALF500
3-((4-AMINO-2-METHYL-5-PYRIMIDINYL)METHYL)-5-(2-HYDROXYETHYL)-4-METHYLTHIAZOLIUM CHLORIDE see TES750
3-(4-AMINO-2-METHYLPYRIMIDYL-5-METHYL)-4-METHYL-5,β-HYDROXYETHYLTHIAZOLIUM NITRATE see TET500
4-AMINO-N-METHYL-1,2,5-SELENADIAZOLE-3-CARBOXAMIDE see MGL600
2-AMINO-4-(METHYLSELENYL)BUTYRIC ACID see SBU725
2-AMINO-4-(METHYLSULFINYL)BUTYRIC ACID see ALF600
2-AMINOMETHYLTETRAHYDROPYRAN see ALF750
2-AMINO-N-(3-METHYL-2-THIAZOLIDINYLIDENE)ACETAMIDE see ALG250
2-AMINO-4-(METHYLTHIO)BUTYRIC ACID see MDT750
l(−)-AMINO-γ-METHYLTHIOBUTYRIC ACID see MDT750
5-AMINO-3-METHYLTHIO-1,2,4-OXADIAZOLE see ALG375
4-AMINO-3-METHYLTOLUENE see XMS000
4-AMINO-3-METHYLTOLUENE HYDROCHLORIDE see XOJ000
2-AMINO-3-METHYLTOLUENE HYDROCHLORIDE see XOS000
6-AMINO-2-METHYL-3-(o-TOLYL)-1(3H)-QUINAZOLINONE see AKM125
α-AMINOMETHYL-m-TRIFLUOROMETHYLBENZYL ALCOHOL see ALG500
2-AMINO-4-METHYLVALERIC ACID see LES000
dl-2-AMINO-4-METHYLVALERIC ACID see LER000
l,2-AMINO-4-METHYLVALERIC ACID see LES000
α-AMINO-β-METHYLVALERIC ACID see IKX000
α-AMINO-γ-METHYLVALERIC ACID see LES000
AMINOMETRADINE see AFW500
AMINOMETRAMIDE see AFW500
AMINOMONOPERSULFURIC ACID see HLN100
2-AMINO-6-MP see AMH250
1-AMINONAFTALEN (CZECH) see NBE700
2-AMINONAFTALEN (CZECH) see NBE500
1-AMINONAPHTHALENE see NBE700
2-AMINONAPHTHALENE see NBE500
3-AMINO-2-NAPHTHALENECARBOXYLIC ACID see ALJ000
6-AMINO-NAPHTHALENE-1,3-DISULFONIC ACID see ALH500
7-AMINO-1,3-NAPHTHALENEDISULFONIC ACID see NBE850
2-AMINO-1,5-NAPHTHALENEDISULFONIC ACID see ALH000
3-AMINO-1,5-NAPHTHALENEDISULFONIC ACID see ALH250
7-AMINO-1,5-NAPHTHALENEDISULFONIC ACID see ALH250
2-AMINO-4,8-NAPHTHALENEDISULFONIC ACID see ALH250

2-AMINO-1-NAPHTHALENESULFONIC ACID see ALH750
4-AMINO-1-NAPHTHALENESULFONIC ACID see ALI000
5-AMINO-1-NAPHTHALENESULFONIC ACID see ALI240
7-AMINO-1-NAPHTHALENESULFONIC ACID see ALI300
5-AMINO-2-NAPHTHALENESULFONIC ACID see ALI250
1-AMINONAPHTHALENE-4-SULFONIC ACID see ALI000
1-AMINO-6-NAPHTHALENESULFONIC ACID see ALI250
7-AMINO-1,3,6-NAPHTHALENETRISULFONIC ACID see ALI750
AMINONAPHTHALENOL see ALJ250
3-AMINO-2-NAPHTHOIC ACID see ALJ000
4-AMINO-1-NAPHTHOIC ACID 2-(DIETHYLAMINO)ETHYL ESTER HYDROCHLORIDE see NAH800
2-AMINO-1-NAPHTHOL see ALJ250
5-AMINO-2-NAPHTHOL see ALJ500
8-AMINO-2-NAPHTHOL see ALJ750
2-AMINO-1-NAPHTHOL HYDROCHLORIDE see ALK250
4-AMINO-1-NAPHTHOL HYDROCHLORIDE see ALK500
1-AMINO-2-NAPHTHOL HYDROCHLORIDE see ALK000
1-AMINO-4-NAPHTHOL HYDROCHLORIDE see ALK500
2-AMINO-1-NAPHTHOL PHOSPHATE (ESTER) SODIUM SALT see BGU000
AMINONAPHTHOL SULFONIC ACID J see AKI000
AMINONAPHTHOL SULFONIC ACID S see AKH750
2-AMINO-1,4-NAPHTHOQUINONE IMINE HYDROCHLORIDE see ALK625
2-AMINO-1-NAPHTHYL ESTER SULFURIC ACID see ALK750
2-AMINO-1-NAPHTHYLGLUCOSIDURONIC ACID see ALL000
2-AMINO-1-NAPHTHYL HYDROGEN SULFATE see ALK750
2-AMINO-1-NAPHTHYL HYDROGEN SULPHATE see ALK750
6-AMINONICOTINAMIDE see ALL250
AMINONICOTINAMIDE see ALL250
6-AMINONICOTINIC ACID AMIDE see ALL250
6-AMINO-NICOTINSAEUREAMID (GERMAN) see ALL250
6-AMINONIKOTINSAEUREAMID (GERMAN) see ALL250
4-AMINO-2-NITROANILINE see ALL750
2-AMINO-4-NITROANILINE see ALL500
2-AMINO-5-NITROANISOL (CZECH) see NEQ000
2-AMINO-4-NITROANISOLE see NEQ500
2-AMINO-5-NITROANISOLE see NEQ000
1-AMINO-2-NITROBENZENE see NEO000
1-AMINO-3-NITROBENZENE see NEN500
3-AMINONITROBENZENE see IEB000
1-AMINO-4-NITROBENZENE see NEO500
m-AMINONITROBENZENE see NEN500
p-AMINONITROBENZENE see NEO500
2-AMINO-5-NITRO BENZENESULFONIC ACID see NEP500
2-AMINO-5-NITROBENZENESULFONIC ACID AMMONIUM SALT see ALL800
2-AMINO-4-NITRO-BENZOIC ACID see NES500
4-AMINO-4′-NITROBIPHENYL see ALM000
4-AMINO-4′-NITRO-3-BIPHENYLOL HYDROCHLORIDE see AKI500
4-AMINO-4′-NITRODIPHENYL SULFIDE see AOS750
2-AMINO-5-(5-NITRO-2-FURYL)-1,3,4-OXADIAZOLE see ALM250
5-AMINO-2-(5-NITRO-2-FURYL)-1,3,4-THIADIAZOLE see NGI500
2-AMINO-5-(5-NITRO-2-FURYL)-1,3,4-THIADIAZOLE see NGI500
2-AMINO-4-(5-NITRO-2-FURYL)THIAZOLE see ALM500
5-AMINO-3-(5-NITRO-2-FURYL)-s-TRIAZOLE see ALM750
3-AMINO-6-(2-(5-NITRO-2-FURYL)VINYL)-1,2,4-TRIAZINE see FPF000
3-AMINO-6-(2-(5-NITRO-2-FURYL)VINYL)-as-TRIAZINE see FPF000
1-AMINO-3-NITRO GUANIDINE see ALN750
4-AMINO-2-NITROPHENOL see NEM480
2-AMINO-4-NITROPHENOL see NEM500
2-AMINO-5-NITROPHENOL see NEM500
2-AMINO-4-NITROPHENOL SODIUM SALT see ALO500
2-((4-AMINO-2-NITROPHENYL)AMINO)ETHANOL see ALO750
2-AMINO-4-(p-NITROPHENYL)THIAZOLE see ALP000
l-2-AMINO-3-((N-NITROSO)HYDROXYLAMINO)PROPIONIC ACID see AFH750
4-AMINO-4′-NITRO-2,2′-STILBENEDISULFONIC ACID see ALP750
2-AMINO-5-NITROTHIAZOLE see ALQ000
AMINONITROTHIAZOLE see ALQ000
AMINONITROTHIAZOLUM see ALQ000
4-AMINO-2-NITROTOLUENE see NMP000
2-AMINO-4-NITROTOLUENE see NMP500
4-AMINO-N[10]-METHYLPTEROYLGLUTAMIC ACID see MDV500
4-AMINO-N[10]-METHYLPTEROYLGLUTAMIC ACID DISODIUM SALT see MDV600
1-AMINOOCTADECANE see OBC000
2-AMINOOCTANE see OEK010
23-AMINO-O[4]-DEACETYL-23-DEMETHOXYVINCALEUKOBLASTINE SULFATE see VGU750
1-(4-AMINO-4-OXO-3,3-DIPHENYLBUTYL)-1-METHYLPIPERIDINIUM BROMIDE (9CI) see RDA375
(4-((2-AMINO-2-OXOETHYL)AMINO)PHENYL)ARSONIC ACID see CBJ750
N-(2-AMINO-2-OXOETHYL)-2-DIAZOACETAMIDE see CBK000
α-((2-AMINO-1-OXOPROPYL)AMINO)-5-OXO-7-OXABICYCLO(4.1.0)HEPTANE-2-PROPANOIC ACID see BAC175
N-(p-(((2-AMINO-4-OXO-6-PTERIDINYL)METHYL)-N-NITROSOAMINO)BENZOYL)-l-GLUTAMIC ACID see NLP000

N-((4-AMINOPHENYL)SULFONYL)ACETAMIDE see SNP500
2-(p-AMINOPHENYLSULFONYLAMINO)-4,5-DIMETHYL-OXAZOLE see AIE750
5-(4-AMINOPHENYLSULPHONAMIDO)-3,4-DIMETHYLISOXAZOLE see SNN500
3-(p-AMINOPHENYLSULPHONAMIDO)-5-METHYLISOXAZOLE see SNK000
2-AMINO-5-PHENYLTHIOMETHYL-2-OXAZOLINE see ALY500
3-AMINO-3-PHENYL-1,2,4-TRIAZOLE see ALY675
5-AMINO-3-PHENYL-1,2,4-TRIAZOLE-1-YL-N,N,N',N'-TETRAMETHYLPHOSPHO-DIAMIDE see AIX000
5-AMINO-3-PHENYL-1,2,4-TRIAZOLYL-1-BIS(DIMETHYLAMIDO)PHOSPHATE see AIX000
5-AMINO-3-PHENYL-1,2,4-TRIAZOLYL-N,N,N'N'-TETRAMETHYL-PHOSPHONAM-IDE see AIX000
p-(5-AMINO-3-PHENYL-1H-1,2,4-TRIAZOL-1-YL)-N,N,N'-TETRAMETHYL PHOS-PHONIC DIAMIDE see AIX000
1-(m-AMINOPHENYL)UREA HYDROCHLORIDE see ALZ750
AMINOPHON see ALZ000
AMINOPHYLLINE see TEP000, TEP500
4-AMINO-3-PICOLINE see AMA000
6-AMINO-3-PICOLINE see AMA010
2-AMINO-4-PICOLINE see ALC250
α-AMINO-γ-PICOLINE see ALC250
(4-AMINOPIPERIDINO)METHYL INDOL-3-YL KETONE see AMA100
2-AMINO-PROPAAN (DUTCH) see INK000
2-AMINOPROPAN (GERMAN) see INK000
1-AMINOPROPANE see PND250
2-AMINOPROPANE see INK000
1-AMINOPROPANE-1,3-DICARBOXYLIC ACID see GFO000
3-AMINO-1,3-PROPANEDIOL see AMA250
2-AMINO-PROPANO (ITALIAN) see INK000
3-AMINO-1-PROPANOL see PMM250
1-AMINO-2-PROPANOL see AMA500
1-AMINOPROPAN-2-OL see AMA500
3-AMINOPROPANOL see PMM250
γ-AMINOPROPANOL see PMM250
AMINOPROPANOL PYROCATECHOLHYDROCHLORIDE see AMB000
3-AMINOPROPENE see AFW000
4-(2-AMINOPROPIONAMIDO)-3,4,4a.5,6,7-HEXAHYDRO-5,6,8-TRIHYDROXY-3-METHYLISOCOUMARIN see AEA109
α-(2-AMINOPROPIONAMIDO)-5-OXO-7-OXABICYCLO(4.1.0)HEPTANE-2-PROPI-ONIC ACID see BAC175
(+-)-2-AMINOPROPIONIC ACID see AFH600
dl-2-AMINOPROPIONIC ACID see AFH600
dl-α-AMINOPROPIONIC ACID see AFH600
2-AMINO PROPIONITRILE see AMB250
3-AMINOPROPIONITRILE see AMB500
β-AMINOPROPIONITRILE see AMB500
β-AMINOPROPIONITRILE FUMARATE see AMB750
2-AMINO-2',6'-PROPIONOXYLIDIDE see TGI699
p-AMINOPROPIOPHENONE see AMC000
3-AMINOPROPOXY-2-ETHOXY ETHANOL see AMC250
1-(3-AMINOPROPYL)ADAMANTANEHYDROCHLORIDE see AMC500
3-AMINOPROPYL ALCOHOL see PMM250
2-((3-AMINOPROPYL)AMINO)-ETHANETHIOL, DIHYDROGEN PHOSPHATE ES-TER (9CI) see AMD000
2-((3-AMINOPROPYL)AMINO) ETHANETHIOL, DIHYDROGEN PHOSPHATE (es-ter-HYDRATE) see AMC750
S-ω-(3-AMINOPROPYLAMINO)ETHYL DIHYDROGEN PHOSPHOROTHIOATE see AMD000
S-(2-(3-AMINOPROPYLAMINO)ETHYL) PHOSPHOROTHIOATE see AMD000
S,2-(3-AMINOPROPYLAMINO)ETHYL-PHOSPHOROTHIOIC ACID see AMD000
2-(3-AMINOPROPYLAMINO)ETHYL THIOPHOSPHATE see AMD000
AMINOPROPYL AMINOETHYLTHIOPHOSPHATE see AMD000
S-3-(ω-AMINOPROPYLAMINO)-2-HYDROXYPROPYL DIHYDROGENPHOSPHO-ROTHIOATE see AMD250
1-AMINO-4-PROPYLBENZENE see PNE000
β-AMINOPROPYLBENZENE see AOA250
N-(3-AMINOPROPYL)-1,4-BUTANEDIAMINE see SLA000
N-(3-AMINOPROPYL)-1,4-BUTANEDIAMINE, PHOSPHATE see AMD500
N-(3-AMINOPROPYL)-1,4-BUTANEDIAMINE TRIHYDROCHLORIDE see SLG600
N-(3-AMINOPROPYL)-1,4-DIAMINOBUTANE see SLA000
(3-AMINOPROPYL)DIETHOXYMETHYLSILANE see AME000
3-AMINOPROPYLENE see AFW000
N-(1-(3-AMINOPROPYL)-10-GUANIDINO-2-HYDROXYDECYL)-9-GUANIDINONO-NANAMIDE see EQS500
3-(2-AMINOPROPYL)INDOLE see AME500
3-(γ-AMINOPROPYL)-INDOLEHYDROCHLORIDE see AME750
N-(3-AMINOPROPYL)MORFOLIN see AMF250
4-AMINOPROPYLMORPHOLINE see AMF250
N-(3-AMINOPROPYL)MORPHOLINE see AMF250
AMINOPROPYLON see AMF375
AMINOPROPYLONE see AMF375
m-(2-AMINOPROPYL)PHENOL see AMF500
p-AMINO-1-PROPYL-4-PIPERIDYL ESTER HYDROCHLORIDE BENZOIC ACID see PNT500

6-AMINO-1-PROPYL-4-(p-((p-((1-PROPYLPYRIDINIUM-4-YL)AMINO)-2-AMINOPHE-NYL)CARBAMOYL)ANILINO)QUINOLINIUM) DII see AMF750
6-AMINO-1-PROPYL-4-(p-((p-((1-PROPYLPYRIDINIUM-4-YL)AMINO)PHENYL)CARBAMOYL)ANILINO)QUINOLINIUM)DIBROMIDE) see AMG000
α-(1-AMINOPROPYL)PROTOCATECHUYL ALCOHOL HYDROCHLORIDE see ENX500
3-AMINOPROPYLSILATRAN (CZECH) see AMG500
(3-AMINOPROPYL)TRIETHOXYSILANE see TJN000
(γ-AMINOPROPYL)TRIETHOXYSILANE see TJN000
1-(3-AMINOPROPYL)-2,8,9-TRIOXA-5-AZA-1-SILABICYCLO(3.3.3) UNDECANE see AMG500
AMINOPTERIDINE see AMG750
AMINOPTERIN see AMG750
4-AMINOPTEROYLGLUTAMIC ACID see AMG750
6-AMINO-1H-PURINE see AEH000
6-AMINO-3H-PURINE see AEH000
6-AMINOPURINE see AEH000
6-AMINO-9H-PURINE see AEH000
2-AMINOPURINE-6-THIOL see AMH250
2-AMINO-6-PURINETHIOL see AMH250
2-AMINOPURINE-6(1H)-THIONE see AMH250
1-(6-AMINO-9H-PURIN-9-YL)-N-CYCLOBUTYL-1-DEOXYRIBOFURANURONAMIDE see AEI500
1-(6-AMINO-9H-PURIN-9-YL)-N-CYCLOPROPYL-1-DEOXY-2,3-DIHYDROXYRIBO-FURANURONAMIDE DIACETATE see AMH500
1-(6-AMINO-9H-PURIN-9-YL)-N-CYCLOPROPYL-1-DEOXYRIBOFURANURONAM-IDE-N-OXIDE see AEJ250
1-(6-AMINO-9H-PURIN-9-YL)-N-CYCLOPROPYL-1-DEOXYRIBOFURANURONAM-IDE see AEJ000
1-(6-AMINO-9H-PURIN-9-YL)-N-CYCLOPROPYLMETHYL-1-DEOXYRIBOFURANU-RONAMIDE see AEJ500
1-(6-AMINO-9H-PURIN-9-YL)-1-DEOXY-2,3-DIHYDROXY-N-ETHYLRIBOFURANU-RONAMIDE DIACETATE see AMH750
1-(6-AMINO-9H-PURIN-9-YL)-1-DEOXY-N,N-DIMETHYLRIBOFURANURONAMIDE HYDRATE see AEK000
1-(6-AMINO-9H-PURIN-9-YL)-1-DEOXY-N-ETHYLRIBOFURANURONAMIDE HEMIH-YDRATE see AEK250
1-(6-AMINO-9H-PURIN-9-YL)-1-DEOXY-N-ETHYLRIBOFURANURONAMIDE- N-OX-IDE see AEK500
1-(6-AMINO-9H-PURIN-9-YL)-1-DEOXY-N-HEXYLRIBOFURANURONAMIDE HEMIH-YDRATE see AEK750
1-(6-AMINO-9H-PURIN-9-YL)-1-DEOXY-N-(2-HYDROXYETH-YL)RIBOFURANURONAMIDE see AEL000
1-(6-AMINO-9H-PURIN-9-YL)-1-DEOXY-N-ISOPROPYLRIBOFURANURONAMIDE see AEL250
1-(6-AMINO-9H-PURIN-9-YL)-1-DEOXY-N-METHOXYRIBOFURANURONAMIDE HY-DRATE see AEL500
1-(6-AMINO-9H-PURIN-9-YL)-1-DEOXY-N-METHYLRIBOFURANURONAMIDE HEM-IHYDRATE see AEL750
1-(6-AMINO-9H-PURIN-9-YL)-1-DEOXY-N-PROPYLRIBOFURANURONAMIDE see AEM000
β-d-1-(6-AMINO-9H-PURIN-9-YL)-1-DEOXYRIBOFURANURONAMIDE see AEI250
1-(6-AMINO-9H-PURIN-9-YL)-N-(2-(DIMETHYLAMINO)-ETHYL-1-DEOXYRIBOFU-RANURONAMIDE) see AEJ750
4-AMINOPYRAZOLO(3,4-d)PYRIMIDINE see POM600
4-AMINOPYRAZOLOPYRIMIDINE see POM600
1-AMINOPYRENE see PON000
3-AMINOPYRENE see PON000
AMINO-2-PYRIDINE see AMI000
2-AMINOPYRIDINE see AMI000
3-AMINOPYRIDINE see AMI250
AMINO-3-PYRIDINE see AMI250
AMINO-4-PYRIDINE see AMI500
4-AMINOPYRIDINE see AMI500
α-AMINOPYRIDINE see AMI000
m-AMINOPYRIDINE (DOT) see AMI250
γ-AMINOPYRIDINE see AMI500
o-AMINOPYRIDINE see AMI000
p-AMINOPYRIDINE see AMI500
AMINOPYRIDINE see POP100
3-AMINOPYRIDINE HYDROCHLORIDE see AMI750
4-AMINOPYRIDINE HYDROCHLORIDE see AMJ000
4-AMINO-PYRIDINEN-OXIDE see AMJ250
4-AMINOPYRIDINE-1-OXIDE see AMJ250
4-AMINO-N-(2-(4-(2-PYRIDINYL)-1-PIPERAZINYL)ETHYL)BENZAMIDE see AMJ500
5-AMINO-5-(4-PYRIDINYL)-2(1H)-PYRIDINONE see AOD375
2-AMINO-9H-PYRIDO(2,3-B)INDOLE see AJD750
2-AMINO-5-(4-PYRIDYL)-1,3,4-THIADIAZOLEHYDROCHLORIDE see AMJ625
2-AMINOPYRIMIDINE see PPH300
4-AMINO-2(1H)-PYRIMIDINONE see CQM600
4-AMINO-N-2-PYRIMIDINYLBENZENESULFONAMIDE see PPP500
4-AMINO-N-2-PYRIMIDINYL-BENZENESULFONAMIDE MONOSILVER(1+) SALT see SNI425

4-AMINO-N-(2-PYRIMIDINYL)BENZENESULFONAMIDE SILVER SALT see SNI425

2-(2-AMINO-4-PYRIMIDINYLVINYL)QUINOXALINE-N,N′-DIOXIDE see AMJ750

AMINOPYRINE see DOT000, POP100

AMINOPYRINE-BARBITAL see AMK250

AMINOPYRINE mixed with SODIUM NITRITE (1:1) see DOT200

AMINOPYRINE SODIUM SULFONATE see AMK500

AMINOQUIN see RHZ000

3-AMINOQUINOLINE see AMK725

2-AMINO-4-((2-QUINOXALINYL-N,N′-DIOXIDE)VINYL)PYRIMIDINES see AMJ750

2-AMINORESORCINOL HYDROCHLORIDE see AML600

AMINOREXFUMARATE see ALX250

6-AMINO-9-β-d-RIBOFURANOSYL-9H-PURINE see AEH750

2-AMINO-9-β-d-RIBOFURANOSYL-9H-PURINE-6-THIOL see TFJ500

2-AMINO-9-(β-d-RIBOFURANOSYL)PURINE-6-THIOL see TFJ500

4-AMINO-1-β-d-RIBOFURANOSYL-2(1H)-PYRIMIDINONE see CQM500

4-AMINO-7-β-d-RIBOFURANOSYL-7H-PYRROLO(2,3-D)PYRIMIDINE see TNY500

4-AMINO-7-(β-d-RIBOFURANOSYL)-PYRROLO(2,3-D)PYRIMIDINE see TNY500

4-AMINO-7-β-d-RIBOFURANOSYL-7H-PYRROLO(2,3-d)PYRIMIDINE-5-CARBONITRILE see VGZ000

4-AMINO-7-β-d-RIBOFURANOSYL-7H-PYRROLO(2,3-d)PYRIMIDINE-5-CARBOXAMIDE see SAU000

2-AMINO-9-β-d-RIBOFURANOSYL-6-SELENO-OH-PURIN-6(1H)-ONE see SBU700

4-AMINO-1-β-d-RIBOFURANOSYL-1,3,5-TRIAZIN-2(1H)-ONE see ARY000

4-AMINO-1-β-d-RIBOFURANOSYL-d-TRIAZIN-2(1H)-ONE see ARY000

5-AMINO-2-β-d-RIBOFURANOSYL-as-TRIAZIN-3(2H)-ONE see AMM000

AMINO-S ACID see AMM125

p-AMINOSALICYLATE SODIUM see SEP000

4-AMINOSALICYLIC ACID see AMM250

5-AMINOSALICYLIC ACID see AMM500

m-AMINOSALICYLIC ACID see AMM500

p-AMINOSALICYLIC ACID see AMM250

AMINOSALICYLIC ACID see AMM250

4-AMINOSALICYLIC ACID-2-(DIETHYLAMINO)ETHYL ESTER HYDROCHLORIDE see AMM750

p-AMINOSALICYLIC ACID, 2-(DIETHYLAMINO)ETHYL ESTER, HYDROCHLORIDE see AMM750

p-AMINOSALICYLIC ACID, 2-(DIMETHYLAMINO)ETHYL ESTER HYDROCHLORIDE see AMN000

p-AMINOSALICYLIC ACID HYDRAZIDE see AMN250

p-AMINOSALICYLIC ACID SODIUM SALT see SEP000

p-AMINOSALICYLSAEURE (GERMAN) see AMM500

p-AMINOSALICYLSAEUREDIAETHYLAMINOAETHYLESTER-CHLORHYDRAT (GERMAN) see AMM750

p-AMINOSALICYLSAEURES SALZ (GERMAN) see TMJ750

4-AMINO-1,2,5-SELENADIAZOLE-3-CARBOXAMIDE see AMN300

2-AMINOSELENOAZOLIN (GERMAN) see AMN500

2-AMINOSELENOAZOLINE see AMN500

4-AMINOSEMICARBAZIDE see CBS500

AMINOSIDIN see NCF500

AMINOSIDINE SULFATE see APP500

AMINOSIDINE SULPHATE see APP500

AMINOSIDIN SULFATE see APP500

AMINOSIN see AGT500

2-AMINO-4-(S-METHYLSULFONIMIDOYL)-BUTANOIC ACID (9CI) see MDU100

4-AMINOSTILBENE see SLQ900

p-AMINOSTILBENE see SLQ900

trans-4-AMINOSTILBENE see AMO000

2-(p-AMINOSTYRYL)-6-(p-ACETYLAMINOBENZOYLAMINO)QUINOLINE METHOACETATE see AMO250

dl-AMINOSUCCINIC ACID see ARN830

l-AMINOSUCCINIC ACID see ARN850

o-AMINOSULFANILIC ACID see PFA250

1-AMINO-2-SULFO-4-(4′-AMINO-3′-SULFOANILINO)ANTHRAQUINONE see SOB600

2,(4′-AMINO-3′-SULFO-1,1′-BIPHENYL-4-YL)-2H-NAPHTHO(1,2-4)TRIAZOLE-6,8-DISULFONIC ACID, TRIPOTASS see AMO750

6-AMINO-5-SULFOMETHYL-2-NAPHTHALENESULFONIC ACID see AMP000

1-AMINO-4-SULFONAPHTHALENE see ALI000

1-AMINO-6-SULFONAPHTHALENE see ALI250

AMINOSULFONIC ACID see SNK500

4-(AMINOSULFONYL)BENZOIC ACID see SNL840

3-(AMINOSULFONYL)-5-(BUTYLAMINO)-4-PHENOXY-3-(AMINOSULFONYL)-5-(BUTYLAMINO)-4-PHENOXYBENZOIC ACID see BON325

3-(AMINOSULFONYL)-4-CHLORO-N-(2,3-DIHYDRO-2-METHYL-1H-INDOL-1-YL)-BENZAMIDE (9CI) see IBV100

5-(AMINOSULFONYL)-4-CHLORO-N-(2,6-DIMETHYLPHENYL)-2-HYDROXY BENZAMIDE (9CI) see CLF325

5-(AMINOSULFONYL)-4-CHLORO-2-((2-FURANYLMETHYL)AMINO)BENZOIC ACID see CHJ750

5-(AMINOSULFONYL)-N-((1-ETHYL-2-PYRROLIDINYL)METHYL)-2-METHOXYBENZAMIDE see EPD500

3-(AMINOSULFONYL)-4-PHENOXY-5-(1-PYRROLIDINYL)-BENZOIC ACID see PDW250

O-(4-(AMINOSULFONYL)PHENYL) O,O-DIMETHYL PHOSPHOROTHIOATE see CQL250

N-(5-(AMINOSULFONYL)-1,3,4-THIADIAZOL-2-YL)ACETAMIDE see AAI250

4-(4-AMINO-3-SULFOPHENYLAZO)BENZENESULFONIC ACID see AJS500

2-AMINO-5-((4-SULFOPHENYL)AZO)-BENZENESULFONIC ACID see AJS500

3,3′-(2-AMINOTEREPHTHALOYBIS(IMINO-p-PHENYLENECARBONYLIMINO))BIS(1-ETHYLPYRIDINIUM), DI-p-TOLUENESULFONATE see AMQ000

3,3′-(2-AMINOTEREPHTHALOYLBIS(IMINO(3-AMINO-p-PHENYLENE)CARBONYLIMINO))BIS(1-ETHYLPYRIDINIUM), DI-p-TOLUE see AMP500

3,3′-(2-AMINOTEREPHTHALOYLBIS(IMINO(3-AMINO-p-PHENYLENE)CARBONYLIMINO))BIS(1-PROPYLPYRIDINIUM), DI-p-TOLU see AMP750

3,3′-(2-AMINOTEREPHTHALOYLBIS(IMINO-p-PHENYLENECARBONYLIMINO))BIS(1-METHYLPYRIDINIUM), DI-p-TOLUENESULFONATE see AMQ250

o-2-AMINO-2,3,4,6-TETRADEOXY-6-(METHYLAMINO)-α-d-glycero-HEX-4-ENOPYRANOSYL-(1-4)-o-(3-DEOXY-4-C-METHYL-3-(METHYLAMINO)-β-I-ARABINOPYRANOSYL-(1-6))-2-DEOXY-d-STREPTAMINE see GAA100

d-o-2-AMINO-2,3,4,6-TETRADEOXY-6-(METHYLAMINO)-α-d-erythro-HEXOPYRANOSYL-(1-4)-o-(3-DEOXY-4-C-METHYL-3-(METHYLAMINO)-β-I-ARABINOPYRANOSYL-(1-6))-2-DEOXY-STREPTAMINE SULFATE see MQS600

9-AMINO-1,2,3,4-TETRAHYDROACRIDINE see TCJ075

5-AMINO-6,7,8,9-TETRAHYDROACRIDINE (European) see TCJ075

8-AMINO-1,2,3,4-TETRAHYDRO-2-METHYL-4-PHENYLISOQUINOLINE MALEATE see NMV725

2-AMINO-1,2,3,4-TETRAHYDRONAFTALEN (CZECH) see TCY250

5-AMINO-2,2,4,4-TETRAKIS(TRIFLUOROMETHYL)IMIDAZOLIDINE see AMQ500

4-AMINO-2,2,5,5-TETRAKIS(TRIFLUOROMETHYL)-3-IMIDAZOLINE see AMQ500

2-AMINOTETRALIN see TCY250

AMINOTETRALIN (CZECH) see TCY250

1-AMINO-2,2,6,6-TETRAMETHYLPIPERIDINE see AMQ750

5-AMINO-1H-TETRAZOLE see AMR000

5-AMINOTETRAZOLE see AMR000

AMINOTETRAZOLE see AMR000

2-AMINO-1,3,4-THIADIAZOLE see AMR250

AMINOTHIADIAZOLE see AMR250

2-AMINO-1,3,4-THIADIAZOLEHYDROCHLORIDE see AMR500

2-AMINO-1,3,4-THIADIAZOLE, MONOHYDROCHLORIDE see AMR500

5-AMINO-1,3,4-THIADIAZOLE-2-THIOL see AKM000

2-AMINO-1,3,4-THIADIAZOLE-5-THIOL see AKM000

5-AMINO-1,3,4-THIADIAZOLINE-2-THIONE see AKM000

2-AMINO-Δ²-1,3,4-THIADIAZOLINE-5-THIONE see AKM000

5-AMINO-1,2,3,4-THIATRIAZOLE see AMS000

2-AMINOTHIAZOLE see AMS250

AMINOTHIAZOLE see AMS250

2-AMINO-2-THIAZOLINE see TEV600

4-AMINO-N-2-THIAZOLYLBENZENESULFONAMIDE see TEX250

2-(((1-(2-AMINO-4-THIAZOLYL)-2-((2-METHYL-4-OXO-1-SULFO-3-AZETIDINYL)AMINO)-2-OXOETHYLIDENE)AMINO)OXY)-2-METHYLPROPANOIC ACID, (2S-(2-α,3β(Z))- see ARX875

3-AMINOTHIOANISOLE see ALX100

4-AMINOTHIOANISOLE see AMS675

m-AMINOTHIOANISOLE see ALX100

p-AMINOTHIOANISOLE see AMS675

o-AMINOTHIOFENOLAT ZINECNATY (CZECH) see BGV500

α-AMINO-2-THIOPHENEPROPANOIC ACID see TEY250

2-AMINOTHIOPHENOL see AIF500

4-AMINOTHIOPHENOL see AIF750

o-AMINOTHIOPHENOL see AIF500

p-AMINOTHIOPHENOL see AIF750

6-AMINO-2-THIOURACIL see AMS750

N-AMINOTHIOUREA see TFQ000

N-(4-(((AMINOTHIOXOMETHYL)HYDRAZONO)METHYL)PHENYL)ACETAMIDE see FNF000

3-AMINOTOLUEN (CZECH) see TGQ500

4-AMINOTOLUEN (CZECH) see TGR000

2-AMINOTOLUENE see TGQ750

3-AMINOTOLUENE see TGQ500

4-AMINOTOLUENE see TGR000

m-AMINOTOLUENE see TGQ500

o-AMINOTOLUENE see TGQ750

p-AMINOTOLUENE see TGR000

2-AMINOTOLUENE HYDROCHLORIDE see TGS500

4-AMINOTOLUENE HYDROCHLORIDE see TGS750

o-AMINOTOLUENE HYDROCHLORIDE see TGS500

α-AMINO-p-TOLUENESULFONAMIDE, MONOACETATE see MAC000

4-AMINOTOLUENE-3-SULFONIC ACID see AKQ000

6-AMINO-m-TOLUENESULFONIC ACID see AKQ000

4-AMINO-o-TOLUENESULFONIC ACID see AMT250

2-AMINO-p-TOLUENESULFONIC ACID see AMT000

p-AMINO-α-TOLUIC ACID see AID700

5-AMINO-o-TOLUIDINE see TGL750

3-AMINO-p-TOLUIDINE see TGL750

p-((4-AMINO-m-TOLYL)AZO)BENZENESULFONAMIDE see SNX000

AMINOTRATE PHOSPHATE see TJL250

AMINOTRIACETIC ACID see AMT500

3-AMINO-1H-1,2,4-TRIAZOLE see AMY050

3-AMINO-1,2,4-TRIAZOLE see AMY050
2-AMINO-1,3,4-TRIAZOLE see AMY050
2-AMINOTRIAZOLE see AMY050
3-AMINOTRIAZOLE see AMY050
3-AMINO-1,2,4-TRIAZOLE (ACGIH) see AMY050
3-AMINO-s-TRIAZOLE see AMY050
AMINOTRIAZOLE see AMY050
AMINOTRIAZOLE (PLANT REGULATOR) see AMY050
AMINO TRIAZOLE WEEDKILLER 90 see AMY050
7-AMINO-1h-v-TRIAZOLO(4,5-d)PYRIMIDINE see AIB340
5-AMINO-v-TRIAZOLO(4,5-d)PYRIMIDIN-7-OL see AJO500
5-AMINO-1H-v-TRIAZOLO(d)PYRIMIDIN-7-OL see AJO500
AMINOTRIAZOL-SPRITZPULVER see AMY050
3-AMINO-1-TRICHLORO-2-PENTANOL see AMU000
((3-AMINO-2,4,6-TRICHLOROPHENYL)METHYLENE) HYDRAZIDE BENZENESUL-
    FONIC ACID see AMU125
4-AMINO-3,5,6-TRICHLORO-2-PICOLINIC ACID see PIB900
4-AMINO-3,5,6-TRICHLOROPICOLINIC ACID see PIB900
3-AMINO-1-TRICHLORO-2-PROPANOL see AMU500
4-AMINO-3,5,6-TRICHLORPICOLINSAEURE (GERMAN) see PIB900
1-AMINOTRICYCLO(3.3.1.1³⁷)DECANE see TJG250
2-AMINO-4-(TRIFLUOROMETHYL)-5-THIAZOLECARBOXYLIC ACID ETHYL ES-
    TER see AMU550
30-AMINO-3,14,25-TRIHYDROXY-3,9,14,20,25-PENTAAZATRIACONTANE-
    2,10,13,21,24-PENTAONE see DAK200
3-AMINO-2,4,6-TRIIODO-BENZOIC ACID see AMU625
N-(3-AMINO-2,4,6-TRIIODOBENZOYL)-N-(2-CARBOXYETHYL)ANILINE see
    AMU750
3-((3-AMINO-2,4,6-TRIIODOBENZOYL)PHENYLAMINO)PROPIONIC ACID see
    AMU750
2-(3-AMINO-2,4,6-TRIIODOBENZYL)BUTYRIC ACID see IFY100
4-((3-AMINO-2,4,6-TRIIODOPHENYL)ETHYLAMINO)-4-OXO-BUTANOIC ACID see
    AMV375
3-(3-AMINO-2,4,6-TRIIODOPHENYL)-2-ETHYLPROPANOIC ACID see IFY100
β-(3-AMINO-2,4,6-TRIIODOPHENYL)-α-ETHYLPROPIONIC ACID see IFY100
2-(3-AMINO-2,4,6-TRIIODOPHENYL)VALERIC ACID see AMV750
N-(3-AMINO-2,4,6-TRIJODBENZOYL)-N-PHENYL-β-AMINOPROPIONSAEURE
    (GERMAN) see AMU750
1-AMINO-2,4,6-TRIMETHYLBENZEN (CZECH) see TLG500
2-AMINO-1,3,5-TRIMETHYLBENZENE see TLG500
1-AMINO-2,4,5-TRIMETHYLBENZENE see TLG250
2-AMINO-1,3-5-TRIMETHYLBENZENE HYDROCHLORIDE see TLH000
1-AMINO-2,4,5-TRIMETHYLBENZENE HYDROCHLORIDE see TLG750
4-AMINO-a,a,4-TRIMETHYLCYCLOHEXANEMETHAMINE see MCD750
AMINOTRI(METHYLENEPHOSPHONIC ACID) see NEI100
AMINOTRI(METHYLENEPHOSPHONIC ACID) PENTASODIUM SALT see
    PBP200
AMINOTRIMETHYLOMETHANE see TEM500
AMINOTRI(METHYLPHOSPHONIC ACID) see NEI100
AMINOTRIS(HYDROXYMETHYL)METHANE see TEM500
AMINOTRIS(METHANEPHOSPHONIC ACID) see NEI100
AMINOTRIS(METHYLPHOSPHONIC ACID) see NEI100
4-AMINOTROPOLONE see AMV790
5-AMINOTROPOLONE see AMV800
2-AMINOTROPONE see AJJ800
AMINO-TS-ACID see AMV875
11-AMINOUNDECANOIC ACID see AMW000
AMINOUNDECANOIC ACID see AMW000
11-AMINOUNDECYLIC ACID see AMW000
6-AMINOURACIL see AMW250
AMINOURACIL MUSTARD see BIA250
AMINOUREA see HGU000
AMINOUREA HYDROCHLORIDE see SBW500
p-AMINO VALEROPHENONE see AMW500
AMINOX see AMM250
2-(4′-AMINOXENYL)NAFTO-α,β-TRIAZOL-6,8-DISULFONAN DRASELNY
    (CZECH) see AIW750
2-(4′-AMINOXENYL)NAFTO-α,β-TRIAZOL-6,8,3′TRISULFONAN DRASELNY
    (CZECH) see AMO750
4-AMINO-1,3-XYLENE see XMS000
2-AMINO-1,4-XYLENE see XNA000
4-AMINO-1,3-XYLENE HYDROCHLORIDE see XOJ000
2-AMINO-1,4-XYLENE HYDROCHLORIDE see XOS000
2-AMINO-4,5-XYLENOL see AMW750
3-AMINO-4-(2-(2,6-XYLYLOXY)ETHYL)-4H-1,2,4-TRIAZOLE see AMX000
AMINOZIDE see DQD400
1,2-AMINOZOPHENYLENE see BDH250
AMINZOL SOLUBLE see ALQ000
AMIODARONE see AJK750
AMIODOXYL BENZOATE see AMX250
AMIOYL see BCA000
AMIP 15m see TAI250
AMIPAN T see DUO400
AMIPAQUE see MQR300
AMIPENIX S see AIV500

AMIPHENAZOLE HYDROCHLORIDE see DCA600
AMIPHOS see DOP200
AMI-PILO see PIF250
AMIPOLNE see AAE500
AMIPOL 6S see LBU200
AMIPRESS see HMM500
AMIPROL see DCK759
AMIPTAN see AHK750
AMIPURIMYCIN HYDRATE see AMX500
AMIPYLO see AMF375
AMIRAL see CJO250
AMIDOTRIZOATE MEGLUMINE see AOO875
AMIDOTRIZOIC ACID see DCK000
AMISEPAN see SBG500
AMISOMETRADIN see AKL625
AMISOMETRADINE see AKL625
AMISTURA P see PIF250
AMISYL see BCA000
AMITAKON see BCA000
AMITAL see AMX750
AMITHIOZONE see FNF000
AMITID see EAI000
AMITIOZON see FNF000
AMITOL see AMY050
AMITON see DJA400
AMITON OXALATE see AMX825
AMITRAZ see MJL250
AMITRAZE see MJL250
AMITRAZ ESTRELLA see MJL250
AMITRENE see BBK250, BBK500
AMITRIL see AMY050, EAI000
AMITRIL T.L. see AMY050
AMITRIPTILINE see EAH500
AMITRIPTYLIN (GERMAN) see EAH500
AMITRIPTYLINE see EAH500
AMITRIPTYLINE CHLORIDE see EAI000
AMITRIPTYLINE-N-OXIDE see AMY000
AMITRIPTYLINOXIDE see AMY000
AMITROL 90 see AMY050
AMITROL see AMY050
AMITROLE see AMY050
AMITROL-T see AMY050
AMITRYPTYLINE HYDROCHLORIDE see EAI000
AMIXICOTYN see NCR000
AMIZIL HYDROCHLORIDE see BCA000
AMIZOL see AMY050
AMIZOL D see AMY050
AMIZOL DP NAU see AMY050
AMIZOL F see AMY050
AMMAT see ANU650
AMMATE see ANU650
AMMELIDE see MCB000
AMMICARDINE see AHK750
AMMIDIN see IHR300
AMMI-KHELLIN see AHK750
AMMINE PENTAHYDROXO PLATINUM see AMY250
4-AMMINOANTIPIRINA see AIB300
AMMIPURAN see AHK750
AMMISPASMIN see AHK750
AMMIVIN see AHK750
AMMIVISNAGEN see AHK750
AMMN see AAW000
AMMO see RLF350
AMMOFORM see HEI500
AMMOIDIN see XDJ000
AMMONERIC see ANE500
AMMONIA, anhydrous, liquefied (DOT) see AMY500
AMMONIA see AMY500
AMMONIA ANHYDROUS see AMY500
AMMONIA AQUEOUS see ANK250
AMMONIAC (FRENCH) see AMY500
AMMONIACA (ITALIAN) see AMY500
AMMONIA GAS see AMY500
AMMONIAK (GERMAN) see AMY500
AMMONIA SOLUTIONS, relative density <0.880 at 15 degrees C in water,
    with >50% ammonia (DOT) see AMY500
AMMONIA SOLUTIONS, with >10% but not >35% ammonia (UN 2672)
    (DOT) see ANK250
AMMONIA SOLUTIONS, with >35% but not >50% ammonia (UN 2073)
    (DOT) see ANK250
AMMONIATED GLYCYRRHIZIN see GIE100
AMMONIATED MERCURY see MCW500
AMMONIA WATER 29% see ANK250

AMMONIIUM, BENZYLDIMETHYL(2-(2-(4-(1,1,3,3-TETRAMETHYLBU-TYL)TOLYLOXY)ETHOXY)ETHYL)-, CHLORIDE, MONOHYDRATE see MHB500
AMMONIO (DICROMATO DI) (ITALIAN) see ANB500
AMMONIOFORMALDEHYDE see HEI500
2-AMMONIOTHIAZOLE NITRATE see AMZ125
AMMONIUM ACETATE see ANA000
AMMONIUM ACID ARSENATE see DCG800
AMMONIUM ACID SULFATE see ANJ500
AMMONIUM-AETHYL-CARBAMOYL-PHOSPHONAT (GERMAN) see ANG750
AMMONIUM, ALKYL(C₁₂-C₁₆)DIMETHYLBENZYL-, CHLORIDES see QAT520
AMMONIUM, ALLYLTRIMETHYL-, CHLORIDE see HGI600
AMMONIUM ALUMINUM FLUORIDE see THQ500
AMMONIUM AMIDOSULFONATE see ANU650
AMMONIUM AMIDOSULPHATE see ANU650
AMMONIUM AMINOFORMATE see AND750
AMMONIUM (AMINYLENIUM BIS [TRIHYDROBORATE]) see ANA500
AMMONIUM ARSENATE, solid (DOT) see DCG800
AMMONIUM AURINTRICARBOXYLATE see AGW750
AMMONIUM AZIDE see ANA750
AMMONIUM-2-(BENZAMIDOOXY)ACETATE see ANB000
AMMONIUM BENZAMIDOOXYACETATE see ANB000
AMMONIUM BENZOATE see ANB100
AMMONIUM, BENZYLBIS(2-HYDROXYETHYL)DODECYL-, CHLORIDE see BDJ600
AMMONIUM ((3-N-BENZYLCARBAMOYLOXY)PHENYL)TRIMETHYL METHYL-SULFATE see BED500
AMMONIUM, BENZYLDIETHYL((2,6-XYLYLCARBAMOYL)METHYL)-, BENZOATE see DAP812
AMMONIUM, BENZYLHEXADECYLDIMETHYL-, CHLORIDE see BEL900
AMMONIUM, BENZYLTRIETHYL-, CHLORIDE see BFL300
AMMONIUM, BENZYLTRIMETHYL-, METHOXIDE see TLM100
AMMONIUM BICARBONATE (1:1) see ANB250
AMMONIUMBICHROMAAT (DUTCH) see ANB500
AMMONIUM BICHROMATE see ANB500
AMMONIUM BIFLUORIDE see ANJ000
AMMONIUM BIS(2,3-DIBROMOPROPYL)PHOSPHATE see MRH214
AMMONIUM BISULFATE see ANJ500
AMMONIUM BISULFIDE see ANJ750
AMMONIUM BISULFITE see ANB600
AMMONIUM BITHIOLICUM see IAD000
AMMONIUM BOROFLUORIDE see ANH000
AMMONIUM BROMATE (DOT) see ANC000
AMMONIUM BROMATE see ANC000
AMMONIUM BROMIDE see ANC250
AMMONIUM BROMO SELENATE see ANC750
AMMONIUM, (2-BUTYRYLOXYETHYL)TRIMETHYL-, IODIDE see BSY300
AMMONIUM CADMIUM CHLORIDE see AND250
AMMONIUM CALCIUM ARSENATE see AND500
AMMONIUM CARBAMATE see AND750
AMMONIUM CARBAZOATE see ANS500
AMMONIUMCARBONAT (GERMAN) see ANE000
AMMONIUM CARBONATE see ANB250, ANE000
AMMONIUM, (3-CARBOXY-2-HYDROXYPROPYL)TRIMETHYL-, CHLORIDE, (±)- see CCK655
AMMONIUM, (3-CARBOXY-2-HYDROXYPROPYL)TRIMETHYL-, CHLORIDE, (−)- see CCK660
AMMONIUM CARBOXYMETHYL CELLULOSE see CCH000
AMMONIUM, (CARBOXYMETHYL)DODECYLDIMETHYL-, HYDROXIDE, inner salt (7CI,8CI) see LBU200
AMMONIUM, (CARBOXYMETHYL)HEXADECYLDIMETHYL-, HYDROXIDE, inner salt (8CI) see CDF450
AMMONIUM, (CARBOXYMETHYL)TRIMETHYL-, CHLORIDE see CCH850
AMMONIUM, (CARBOXYMETHYL)TRIMETHYL-, CHLORIDE, HYDRAZIDE see GEQ500
AMMONIUM CHLORATE see ANE250
AMMONIUMCHLORID (GERMAN) see ANE500
AMMONIUM CHLORIDE see ANE500
AMMONIUM CHLOROHEPTENE ARSONATE see CEI250
AMMONIUM CHLOROPALLADATE(II) see ANE750
AMMONIUM CHLOROPALLADATE(IV) see ANF000
AMMONIUM CHLOROPLATINATE see ANF250
AMMONIUM CHROMATE(VI) see ANF500, NCQ550
AMMONIUM CHROMATE see ANF500, NCQ550
AMMONIUM CHROME ALUMS see ANF625
AMMONIUM CHROMIC SULFATE see ANF750
AMMONIUM CITRATE see ANF800
AMMONIUM CITRATE, DIBASIC (DOT) see ANF800
AMMONIUM CRYOLITE see THQ500
AMMONIUM CYANIDE see ANG000
AMMONIUM DECAHYDRODECABORATE (2−) see ANG125
AMMONIUM, DECYLTRIMETHYL-, BROMIDE see DAJ500
AMMONIUM, DIALLYLDIMETHYL-, CHLORIDE, POLYMERS see DTS500
AMMONIUM, (4-((4,6-DIAMINO-m-TOLYL)IMINO)-2,5-CYCLOHEXADIEN-1-YLI-DENE)DIMETHYL-, CHLORIDE, MONOHYDRATE see TGU500

AMMONIUMDICHROMAAT (DUTCH) see ANB500
AMMONIUMDICHROMAT (GERMAN) see ANB500
AMMONIUM DICHROMATE(VI) see ANB500
AMMONIUM DICHROMATE see ANB500
AMMONIUM DIFLUORIDE see ANJ000
AMMONIUM DIFLUORIDE mixed with HYDROCHLORIC ACID see ANG250
AMMONIUM, DIMETHYLDIOCTADECYL-, CHLORIDE see DXG625
AMMONIUM DIMETHYL DITHIOCARBAMATE see ANG500
AMMONIUM-3,5-DINITRO-1,2,4-TRIAZOLIDE see ANG625
AMMONIUM DISULFATONICKELATE(II) see NCY050
AMMONIUM DNOC see DUT800
AMMONIUM-N-DODECYL SULFATE see SOM500
AMMONIUM DODECYL SULFATE see SOM500
AMMONIUM, DODECYLTRIMETHYL-, BROMIDE see DYA810
AMMONIUM ETHYL CARBAMOYLPHOSPHONATE see ANG750
AMMONIUM ETHYL CARBAMOYLPHOSPHONATE solution see ANG750
AMMONIUM, ETHYLTRIMETHYL-, IODIDE see EQC600
AMMONIUM FERRIC OXALATE see ANG925
AMMONIUM FERRIOXALATE see ANG925
AMMONIUM FLUOALUMINATE see THQ500
AMMONIUM FLUOBORATE see ANH000
AMMONIUM FLUORIDE see ANH250
AMMONIUM FLUORIDE comp. with HYDROGEN FLUORIDE (1:1) see ANJ000
AMMONIUM FLUOROBORATE see ANH000
AMMONIUM FLUOROSILICATE (DOT) see COE000
AMMONIUM FLUORURE (FRENCH) see ANH250
AMMONIUM FLUOSILICATE see COE000
AMMONIUM FORMATE see ANH500
AMMONIUMGLUTAMINAT (GERMAN) see MRF000
AMMONIUM GLYCYRRHIZINATE see GIE100
AMMONIUM HEXACHLOROPALLADATE see ANF000
AMMONIUM HEXACHLOROPLATINATE(IV) see ANF250
AMMONIUM HEXACYANOFERRATE(II) see ANH875
AMMONIUM HEXAFLUOROALUMINATE see THQ500
AMMONIUM HEXAFLUOROFERRATE see ANI000
AMMONIUM HEXAFLUOROSILICATE see COE000
AMMONIUM HEXAFLUOROTITANATE see ANI250
AMMONIUM HEXAFLUOROVANADATE see ANI500
AMMONIUM, HEXAMETHYLENEBIS((CARBOXYMETHYL)DIMETHYL)-, DICHLO-RIDE, DIDODECYL ESTER see HEF200
AMMONIUM, HEXAMETHYLENEBIS(TRIMETHYL-, DIBENZENESULFONATE see HEG100
AMMONIUM HEXANITRO COBALTATE see ANI750
AMMONIUM HYDROFLUORIDE see ANJ000
AMMONIUM HYDROGEN BIFLUORIDE see ANJ000
AMMONIUM HYDROGEN CARBONATE see ANB250
AMMONIUM HYDROGEN DIFLUORIDE see ANJ000
AMMONIUM HYDROGEN FLUORIDE see ANJ000
AMMONIUM HYDROGEN FLUORIDE, solid (UN 1727) (DOT) see ANJ000
AMMONIUM HYDROGEN FLUORIDE, solution (UN 2817) (DOT) see ANJ000
AMMONIUM HYDROGEN SULFATE see ANJ500
AMMONIUM HYDROGEN SULFIDE see ANJ750
AMMONIUM HYDROGEN SULFITE see ANB600
AMMONIUM HYDROSULFIDE, solution (DOT) see ANJ750
AMMONIUM HYDROSULFIDE see ANJ750
AMMONIUM HYDROXIDE see ANK250
AMMONIUM, (2-HYDROXYPROPYL)TRIMETHYL-, CHLORIDE see MIM300
AMMONIUM HYPOPHOSPHITE see ANK500
AMMONIUM HYPOSULFITE see ANK600
AMMONIUM ICHTHOSULFONATE see IAD000
AMMONIUM IMIDOBISSULFATE see ANK650
AMMONIUM IMIDODISULFONATE see ANK650
AMMONIUM IMIDOSULFONATE see ANK650
AMMONIUM IODATE see ANK750
AMMONIUM IODIDE see ANL000
AMMONIUM ISETHIONATE see ANL100
AMMONIUM LANTHANUM NITRATE see ANL500
AMMONIUM LAURYL SULFATE see SOM500
AMMONIUM MAGNESIUM ARSENATE see ANL750
AMMONIUM MAGNESIUM CHROMATE see ANM000
AMMONIUM MANDELATE see ANM250
AMMONIUM MERCAPTAN see ANJ750
AMMONIUM MERCAPTOACETATE see ANM500
AMMONIUM, (2-MERCAPTOETHYL)TRIMETHYL-, IODIDE ACETATE see ADC300
AMMONIUM METAVANADATE (DOT) see ANY250
AMMONIUM-3-METHYL-2,4,6-TRINITROPHENOXIDE see ANM625
AMMONIUM MOLYBDATE see ANM750
AMMONIUM MONOHYDROGEN SULFATE see ANJ500
AMMONIUM MONOSULFITE see ANB600
AMMONIUM MURIATE see ANE500
AMMONIUM NICKEL SULFATE see NCY050
AMMONIUM(I) NITRATE(1:1) see ANN000
AMMONIUM NITRATE see ANN000

AMMONIUM NITRATE, with not >0.2% of combustible substances (UN 1942) (DOT) see ANN000
AMMONIUM NITRATE, with >0.2% combustible substances (UN 0222) (DOT) see ANN000
AMMONIUM NITRATE, liquid (hot concentrated solution) (UN 2426) (DOT) see ANN000
AMMONIUM NITRITE see ANO250
AMMONIUM aci-NITROMETHANE see ANO400
AMMONIUM-N-NITROSOPHENYLHYDROXYLAMINE see ANO500
AMMONIUM ORTHOPHOSPHITE see ANS250
AMMONIUM OXALATE see ANO750
AMMONIUM OXOFLUOROMOLYBDATE see ANO875
AMMONIUM PARAMOLYBDATE see ANM750
AMMONIUM PARATUNGSTATE HEXAHYDRATE see ANO900
AMMONIUM PENTADECAFLUOROOCTANATE see ANP625
AMMONIUM PENTA PEROXODICHROMATE see ANP000
AMMONIUM PERCHLORATE (DOT) see PCD500
AMMONIUM PERCHLORATE see ANP250
AMMONIUM PERCHLORYL AMIDE see ANP500
AMMONIUM PERFLUOROCAPRILATE see ANP625
AMMONIUM PERFLUOROCAPRYLATE see ANP625
AMMONIUM PERFLUOROOCTANOATE see ANP625
AMMONIUM-m-PERIODATE see ANP750
AMMONIUM PERMANGANATE see PCJ750
AMMONIUM PEROXO BORATE see ANQ250
AMMONIUM PEROXO DISULFATE see ANQ500
AMMONIUM PEROXYCHROMATE see ANQ750
AMMONIUM PEROXYDISULFATE see ANR000
AMMONIUM PERSULFATE see ANR000
AMMONIUM PHENYLDITHIOCARBAMATE see ANR250
AMMONIUM, PHENYLTRIMETHYL-, BROMIDE see TMB000
AMMONIUM PHOSPHATE see ANR500
AMMONIUM PHOSPHATE, DIBASIC see ANR500
AMMONIUM PHOSPHATE, MONOBASIC see ANR750
AMMONIUM PHOSPHIDE see ANS000
AMMONIUM PHOSPHITE see ANS250
AMMONIUM PICRATE see ANS500
AMMONIUM PICRATE, wetted with not <10% water, by weight (UN 1310) (DOT) see ANS500
AMMONIUM PICRATE, dry or wetted with <10% water, by weight (UN 0004) (DOT) see ANS500
AMMONIUM PICRONITRATE see ANS500
AMMONIUM PLATINIC CHLORIDE see ANF250
AMMONIUM POLYSULFIDE, solution (DOT) see ANT000
AMMONIUM POLYSULFIDE (solution) see ANT000
AMMONIUM POTASSIUM SELENIDE mixed with AMMONIUM POTASSIUM SULFIDE see ANT250
AMMONIUM POTASSIUM SULFIDE mixed with AMMONIUM POTASSIUM SEL-ENIDE see ANT250
AMMONIUM REINECKATE HYDRATE see ANT300
AMMONIUM RHODANATE see ANW750
AMMONIUM RHODANIDE see ANW750
AMMONIUM SACCHARIN see ANT500
AMMONIUM SALICYLATE see ANT600
AMMONIUM SALTPETER see ANN000
AMMONIUM SALTS of PHOSPHATIDIC ACIDS see ANU000
AMMONIUMSALZ der AMIDOSULFONSAEURE (GERMAN) see ANU650
AMMONIUMSALZ des α-METHYL-β-HYDROXY-Δᵅˑᵝ-BUTYCOLACTAM (GER-MAN) see MKS750
AMMONIUM SILICOFLUORIDE see COE000
AMMONIUM SILICON FLUORIDE see COE000
AMMONIUM STEARATE see ANU200
AMMONIUM SULFAMATE see ANU650
AMMONIUM SULFATE (2:1) see ANU750
AMMONIUM SULFATE, and CHROMIC SULFATE, TETRACOSAHYDRATE see ANF625
AMMONIUM SULFHYDRATE see ANJ750
AMMONIUM SULFIDE, solution, red see ANT000
AMMONIUM SULFIDE (POLY-) see ANT000
AMMONIUM SULFOCYANATE see ANW750
AMMONIUM SULFOCYANIDE see ANW750
AMMONIUM SULFOICHTHYOLATE see IAD000
AMMONIUM SULPHAMATE see ANU650
AMMONIUM SULPHATE see ANU750
AMMONIUM TARTRATE (DOT) see DCH000
AMMONIUM-d-TARTRATE see DCH000
AMMONIUM TETRACHLOROPALLADATE see ANE750
AMMONIUM TETRACHLOROPLATINATE see ANV800
AMMONIUM, TETRADECYLTRIMETHYL-, BROMIDE see TCB200
AMMONIUM, TETRAETHYL-, PERCHLORATE see TCD002
AMMONIUM TETRAFLUOROBORATE(1-) see ANH000
AMMONIUM TETRAFLUOROBORATE see ANH000
AMMONIUM, TETRAMETHYL-, HYDROXIDE see TDK500
AMMONIUM TETRANITROPLATINATE(II) see ANW250
AMMONIUM TETRAPEROXO CHROMATE see ANW500

AMMONIUM THIOCYANATE see ANW750
AMMONIUM THIOGLYCOLATE see ANM500
AMMONIUM THIOGLYCOLLATE see ANM500
AMMONIUM THIOSULFATE see ANK600
AMMONIUM TRICHLOROACETATE see ANX750
AMMONIUM TRIFLUOROSTANNITE see ANX800
AMMONIUM, TRIMETHYLTETRADECYL-, BROMIDE see TCB200
AMMONIUM-2,4,5-TRINITROIMIDAZOLIDE see ANX875
AMMONIUM TRIOXALATOFERRATE(III) see ANG925
AMMONIUM TRISULFIDE see ANT000
AMMONIUM VANADATE see ANY250
AMMONIUM VANADI-ARSENATE see ANY500
AMMONIUM VANADO-ARSENATE see ANY750
AMMONIUMYL, DIBUTYL-, HEXACHLOROSTANNATE(2-) (2:1) see BIV900
AMMONYL BR 1244 see BEO000
AMMONYX 4 see DTC600
AMMONYX 2200 see QAT550
AMMONYX see AFP250
AMMONYX CA SPECIAL see DTC600
AMMONYX CPC see CCX000
AMMONYX DME see EKN500
AMMONYX G see BEL900
AMMONYX LO see DRS200
AMMONYX T see BEL900
AMMOPHYLLIN see TEP500
AMN see AOL000
AMNESTROGEN see ECU750
AMNICOTIN see NCR000
AMNOSED see TDA500
AMNUCOL see SEH000
AMOBAM see ANZ000
AMOBARBITAL see AMX750
AMOBARBITONE see AMX750
AMOBEN see AJM000
AMOCO 1010 see PMP500
AMOCO 610A4 see PJS750
AMOCO NT-45 PROCESS OIL see DXG830
AMOEBAL see ABX500
AMOENOL see CHR500
AMOGLANDIN see POC500
AMOKIN see CLD000
AMOLANONE HYDROCHLORIDE see DIF200
AMONIAK (POLISH) see AMY500
A. MONTANA see TOA275
AMONYX AO see DRS200
1A-2MO-4OA see AKO350
AMOPYROQUIN DIHYDROCHLORIDE see PMY000
A1-MORIN see MRN500
AMORPHAN see NNW500
AMORPHOUS AESCIN see EDK875
AMORPHOUS CROCIDOLITE ASBESTOS see ARM275
AMORPHOUS QUARTZ see SCK600
AMORPHOUS SILICA see DCJ800, SCK600
AMORPHOUS SILICA DUST see SCH000
AMORPHOUS SILICA FUME see SCH001
AMOSENE see MQU750
AMOSITE ASBESTOS see ARM262
AMOSITE (OBS.) see ARM250
AMOSPAN see AMX750
AMOSULALOL HYDROCHLORIDE see AOA075
AMOSYT see DYE600
AMOTRIL see ARQ750
AMOTRIPHENE HYDROCHLORIDE see TNJ750
AMOX see NGS500
AMOXAPINE see AOA095
AMOXEPINE see AOA095
AMOXICILLIN mixed with POTASSIUM CLAVULANATE (2:1) see ARS125
AMOXICILLIN TRIHYDRATE see AOA100
AMOXONE see DAA800
5'-AMP see AOA125
5-AMP see AOA125
A5MP see AOA125
AMP (nucleotide) see AOA125
AMP see AOA125
AMPACET E/C see EHG100
AMPAZINE see DQA600
AMPD see ALB000
AMPERIL see AIV500, AOD125
AMPHAETEX see BBK500
AMPHATE see AOB500
AMPHEDRINE see BBK500
AMPHEDROXY see DBA800
AMPHEDROXYN see DBA800
AMPHENAZOLE HYDROCHLORIDE see DCA600
AMPHENICOL see CDP250

AMPHENIDONE see ALY250
AMPHEPRAMONUM HYDROCHLORIDE see DIP600
AMPHEREX see BBK500
dl-AMPHETAMINE see BBK000
d-AMPHETAMINE see AOA500
AMPHETAMINE see AOA250
(+)-AMPHETAMINE see AOA500
AMPHETAMINE HYDROCHLORIDE see AOA750
dl-AMPHETAMINE PHOSPHATE see AOB500
AMPHETAMINE PHOSPHATE see AOB500
dl-AMPHETAMINE SALT with FINE RESIN see AOB000
dl-AMPHETAMINE SULFATE see AOB250
d-AMPHETAMINE SULFATE see BBK500
l-AMPHETAMINE SULFATE see BBK750
(±)-AMPHETAMINE SULFATE see AOB250
(+)-AMPHETAMINE SULFATE see BBK500
(−)-AMPHETAMINE SULFATE see BBK750
AMPHETANE PHOSPHATE see AOB500
AMPHIBOLE see ARM250
AMPHICOL see CDP250
AMPHISOL HYDROCHLORIDE see DCA600
AMPHITOL 24B see LBU200
AMPHITOL 20BS see LBU200
AMPHOIDS S see BBK250
AMPHOJEL see AHC000
AMPHOMORONAL see AOC500
AMPHOMYCIN see AOB875
AMPHORDS S see BBK250
AMPHOS see AOB500
AMPHOTERGE K-2 see AOC250
AMPHOTERIC-17 see AOC275
AMPHOTERIC-2 see AOC250
AMPHOTERICIN beta see AOC500
AMPHOTERICIN B see AOC500
AMPHOTERICIN B, METHYL ESTER HYDROCHLORIDE see AOC750
AMPHOTERICINE B see AOC500
AMPHOZONE see AOC500
AMPI-BOL see AIV500
AMPICHEL see AOD125
AMPICILLIN, (5-METHYL-2-OXO-1,3-DIOXOLEN-4-YL)METHYL ESTER, HYDRO-
    CHLORIDE see LEJ500
AMPICILLIN (USDA) see AIV500
d-AMPICILLIN see AIV500
d-(−)-AMPICILLIN see AIV500
AMPICILLIN A see AIV500
AMPICILLIN ACID see AIV500
AMPICILLIN ANHYDRATE see AIV500
AMPICILLIN-OXACILLIN MIXTURE see AOC875
AMPICILLIN PIVALOYLOXYMETHYL ESTER HYDROCHLORIDE see AOD000
AMPICILLIN SODIUM see SEQ000
AMPICILLIN SODIUM SALT see SEQ000
AMPICILLIN TRIHYDRATE see AOD125
AMPICIN see AIV500
AMPIKEL see AIV500, AOD125
AMPIMED see AIV500
AMPINOVA see AOD125
AMPIPENIN see AIV500
AMPLIACTIL see CKP250
AMPLIACTIL MONOHYDROCHLORIDE see CKP500
AMPLICITIL see CKP250
AMPLIGRAM see TEY000
AMPLIN see AOD125
AMPLISOM see AIV500
AMPLIT see IFZ900
AMPLITAL see AIV500
AMPLIVIX see EID200
5′-AMP POTASSIUM SALT see AEM500
AMPROLENE see EJN500
AMPROTROPINE PHOSPHATE see AOD250
5′-AMP SODIUM SALT see AEM750
AMP SODIUM SALT see AEM750
AMPY-PENYL see AIV500
AMPYRONE see AIB300
AMPYROX see SBH500
AMRINONE see AOD375
AMRITAMYCIN see MQX250
AMROOD, extract see POH800
AMS see ANU650
m-AMSA see ADL750, ADM000
AMSA see ADL750
AMSACRINE see ADL750
m-AMSA HYDROCHLORIDE see ADL500
m-AMSA METHANESULFONATE see ADL750
AMSCO H-J see NAI500
AMSCO H-SB see NAI500

AMSECLOR see CDP250
AMSIDINE see ADL750
AMSINCKIA INTERMEDIA see TAG250
AMSINE see ADL750
AMSONIC ACID see FCA100
AMSONIC ACID DISODIUM SALT see FCA200
AMSPEC-KR see AQF000
AMSTAT see AJV500
AMSUBIT see IAD000
AMSUSTAIN see AOA500, BBK500
AMTHIO see ANW750
AMUDANE see GKE000
AMUNO see IDA000
AMYBAL see AMX750
AMYDRICAINE see AHI250
AMYGDALIC ACID see MAP000
AMYGDALIN see AOD500
AMYGDALINIC ACID see MAP000
AMYGDALONITRILE see MAP250
AMYL ACETATE (DOT) see AOD725
AMYL ACETATE (mixed isomers) see AOD750
n-AMYL ACETATE see AOD725
sec-AMYL ACETATE see AOD735
AMYL ACETIC ESTER see AOD725
AMYL ACID PHOSPHATE (DOT) see PBW750
tert-AMYL ALCOHOL (DOT) see PBV000
N-AMYL ALCOHOL see AOE000
sec-AMYL ALCOHOL see PBM750
AMYL ALCOHOL see AOE000
AMYL ALCOHOL, NORMAL see AOE000
AMYL ALDEHYDE see VAG000
N-AMYLALKOHOL (CZECH) see AOE000
AMYLAMINE (DOT) see PBV505
AMYLAMINE (mixed isomers) see PBV500
iso-AMYLAMINE see AOE200
n-AMYLAMINE see PBV505
AMYLAMINE see PBV505
AMYLAMINES (DOT) see PBV500
2-N-AMYLAMINOETHYL-p-AMINOBENZOATE see PBV750
AMYLAZETAT (GERMAN) see AOD725
AMYL AZIDE see AOE500
AMYLBARBITONE see AMX750
5-n-AMYL-1:2-BENZANTHRACENE see AOE750
tert-AMYLBENZENE see AOF000
AMYL BENZOATE see IHP100, PBW800
4-AMYL-N-BENZOHYDRYLPYRIDINIUM BROMIDE see AOF250
p-AMYLBENZOIC ACID see PBW000
AMYL BIPHENYL see AOF500
d-AMYL BROMIDE see AOF750
n-AMYL BROMIDE see AOF800
AMYL BROMIDE see AOF800
n-AMYL BUTYRATE see AOG000
AMYL BUTYRATE see AOG000
γ-N-AMYLBUTYROLACTONE see CNF250
AMYLCAINE see PBV750
AMYL CAPROATE see PBW450
AMYL CAPRONATE see PBW450
AMYLCARBINOL see HFJ500
AMYL CHLORIDE (DOT) see PBW500
n-AMYL CHLORIDE see PBW500
α-AMYL CINNAMALDEHYDE see AOG500
AMYL CINNAMATE see AOG600
AMYL CINNAMIC ACETATE see AOG750
α-AMYLCINNAMIC ALCOHOL see AOH000
α-AMYL CINNAMIC ALDEHYDE see AOG500
α-AMYLCINNAMYL ALCOHOL see AOH000
AMYL CINNAMYLIDENE METHYL ANTHRANILATE see AOH100
6-n-AMYL-m-CRESOL see AOH250
4-tert-AMYLCYCLOHEXANONE see AOH750
AMYLCYCLOHEXYL ACETATE (mixed isomers) see AOI000
N-AMYLDICHLORARSINE see AOI200
AMYLDICHLORARSINE see AOI200
AMYL-p-DIMETHYLAMINOBENZOATE see AOI250
AMYLDIMETHYL-p-AMINO BENZOIC ACID see AOI500
AMYL DIMETHYL PABA see AOI250
AMYLEINE see AOM000
AMYLENE see AOI800
tert-AMYLENE see IHP150
2,4-AMYLENEGLYCOL see PBL000
AMYLENE HYDRATE see PBV000
n-AMYLENE PENTENE see AOI800
AMYLENES, MIXED see AOJ000
AMYLESTER KYSELINY DUSICNE see AOL250
2-AMYLESTER KYSELINY OCTOVE see AOD735
AMYLESTER KYSELINY OCTOVE see AOD725

sek.AMYLESTER KYSELINY OCTOVE see AOD735
n-AMYL ETHER see PBX000
AMYL ETHER see PBX000
AMYLETHYLCARBINOL see OCY100
AMYL ETHYL KETONE see ODI000
AMYLETHYLMETHYLCARBINOL see MND050
n-AMYL FORMATE see AOJ500
AMYL FORMATE see AOJ500
o-n-AMYL HARMOL HYDROCHLORIDE see AOJ750
AMYL HARMOL HYDROCHLORIDE see AOJ750
AMYL HEXANOATE see IHU100, PBW450
n-AMYLHYDRAZINE HYDROCHLORIDE see PBX250
AMYL HYDRIDE (DOT) see PBK250
tert-AMYL HYDROPEROXIDE see PBX325
AMYL HYDROSULFIDE see PBM000
3-AMYL-1-HYDROXY-6,6,9-TRIMETHYL-6H-DIBENZO(b,d)PYRAN see CBD625
n-AMYL IODIDE see IET500
AMYL IODIDE see IET500
AMYLISOEUGENOL see AOK000
AMYL KETONE see ULA000
AMYL LACTATE see AOK250
AMYL LAURATE see AOK500
AMYL MERCAPTAN (DOT) see PBM000
n-AMYL MERCAPTAN see PBM000
AMYL METHYL ALCOHOL see AOK750
AMYL METHYL CARBINOL see HBE500
AMYL-METHYL-CETONE (FRENCH) see MGN500
AMYL METHYL KETONE (DOT) see MGN500
n-AMYL METHYL KETONE see MGN500
n-AMYL-N-METHYLNITROSAMINE see AOL000
AMYL NITRATE see AOL250
AMYL NITRITE (DOT) see AOL500
n-AMYL NITRITE see AOL500
1-AMYL-1-NITROSOUREA see PBX500
n-AMYLNITROSOUREA see PBX500
n-AMYL-N-NITROSOURETHANE see AOL750
N-(n-AMYL)-N'-NITRO-N-NITROSOGUANIDINE see NLC500
AMYLOBARBITAL see AMX750
AMYLOBARBITONE see AMX750
AMYLOCAINE see AOM000
AMYLOFENE see EOK000
AMYLOMAIZE VII see SLJ500
AMYLOPECTIN, HYDROGEN SULFATE see AOM150
AMYLOPECTINE SULPHATE see AOM150
AMYLOPECTIN SULFATE see AOM150
AMYLOPECTIN SULFATE (SN-263) see AOM150
β-AMYLOSE see CCU150
AMYLOWY ALKOHOL (POLISH) see IHP000
AMYLOXYISOEUGENOL see AOK000
2-sec-AMYLPHENOL see AOM500
4-n-AMYLPHENOL see AOM250
4-sec-AMYLPHENOL see AOM750
4-tert-AMYLPHENOL see AON000
o-AMYLPHENOL see AOM325
p-tert-AMYLPHENOL see AON000
AMYL PHENOL 4T see AON000
α-AMYL-β-PHENYLACROLEIN see AOG500
α-N-AMYL-β-PHENYLACRYL ACETATE see AOG750
3-sec-AMYLPHENYL-N-METHYLCARBAMATE see AON250
2-AMYL-3-PHENYL-2-PROPEN-1-OL see AOH000
AMYL POTASSIUM XANTHATE see PKV100
AMYL PROPIONATE see AON350
AMYL SALICYLATE see SAK000
AMYLSINE see PBV750
AMYL SULFHYDRATE see PBM000
AMYL THIOALCOHOL see PBM000
n-AMYL THIOCYANATE see AON500
AMYL TRICHLOROSILANE see PBY750
AMYLTRIETHOXYSILANE see PBZ000
AMYLTRIMETHYLAMMONIUM IODIDE see TMA500
AMYLUM see SLJ500
AMYL-Δ-VALEROLACTONE see DAF200
AMYLVINYLCARBINOL see ODW000
AMYL VINYL CARBINOL ACETATE see ODW028
AMYL VINYL CARBINYL ACETATE see ODW028
AMYL ZIMATE see BJK500
AMYRIS OIL see WBJ650
AMYTAL see AMX750
AMYTAL SODIUM see AON750
6-AN see ALL250
AN 23 see DJO800
AN-148 see MDP750
AN 1041 see LJR000
AN 1087 see EOY000
AN 1324 see GFM200

AN see AAQ500
6-ANA see ALL250
ANA see NAK500
ANABACTYL see CBO250
ANABAENA FLOS-AQUAE TOXIN see AON825
ANABASIDE HYDROCHLORIDE see PIT650
ANABASIN see AON875
(−)-ANABASIN see AON875
ANABASIN CHLORIDE see PIT650
ANABASINE see AON875
ANABASINE MONOHYDROCHLORIDE see PIT650
ANABASIN HYDROCHLORIDE see PIT650
ANABAZIN see AON875
ANABET see CNR675
ANABOLIN see DAL300
ANAC 110 see CNI000
ANACARDONE see DJS200
ANACEL see TBN000
ANACETIN see CDP250
ANACOBIN see VSZ000
ANACORDONE see DJS200
ANADOLOR see AIT250
ANADOMIS GREEN see CMJ900
ANADREX see DBB000
ANADROL see PAN100
ANADROYD see PAN100
ANAESTHETIC ETHER see EJU000
ANAFEBRINA see DOT000
ANAFLEX see UTU500
ANAFLON see HIM000
ANAFRANIL see CDU750, CDV000
ANAGESTONE ACETATE mixed with MESTRANOL (10:1) see AOO000
ANAGIARDIL see MMN250
ANAHIST see RDU000
ANALEPTIN see HLV500
ANALETIL see GCE600
ANALEXIN see PGG350, PGG355
ANALGIZER see DFA400
O-ANALOG of DIMETHOATE see DNX800
ANALUD see PEW000
ANALUX see DPE000
ANAMENTH see TIO750
ANAMID see SAH000
ANANASE see BMO000
ANANSIOL see MNM500
ANAPAC see ABG750
A. NAPELLUS see MRE275
ANAPHRANIL see CDV000
ANAPOLON see PAN100
ANAPRAL see TLN500
ANAPREL see TLN500
ANAPRILIN see ICC000
ANARCON see AFT500
ANASCLEROL see VLF000
ANASTERON see PAN100
ANASTERONAL see PAN100
ANASTERONE see PAN100
ANASTRESS see MQU750
ANATASE see OBU100
ANATENSIN see GGS000
ANATENSOL see TJW500
ANATHYLMON see MQU750
ANATOLA see VSK600
ANATOXIN-a see AOO120
ANATOXIN I see AOO120
ANATRAN see ABH500, HII500
ANATROPIN mixed with MESTRANOL (10:1) see AOO000
ANAUTINE see DYE600
ANAVAR see AOO125
ANAYODIN see IEP200, SHW000
ANAZOLENE, SODIUM see ADE750
ANC 113 see OLW400
ANCAMINE TL see MJQ000
ANCEF see CCS250
ANCHOIC ACID see ASB750
ANCHRED STANDARD see IHD000
ANCILLIN see AOD125
ANCISTRODON PISCIVORUS VENOM see AOO135
ANCITABINE see COW875
ANCITABINE HYDROCHLORIDE see COW900
ANCOBON see FHI000
ANCOLAN see HGC500
ANCOLAN DIHYDROCHLORIDE see MBX250
ANCOR EN 80/150 see IGK800
ANCORTONE see PLZ000

ANCOTIL see FHI000
ANCROD see VGU700
ANCYLOL see DNG000
ANCYTABINE see COW875
ANDAKSIN see MQU750
ANDANTOL see AEG625
ANDAXIN see MQU750
ANDERE see BQL000
ANDHIST see RDU000
ANDIAMINE see HFF500
ANDRAMINE see DYE600
ANDRANE see CCR510
ANDREZ see SMR000
ANDROCTONUS AMOREUXI VENOM see AOO250
ANDROCTONUS AUSTRALIS HECTOR VENOM see AOO265
ANDRODIOL see AOO475
ANDROFLUORENE see AOO275
ANDROFLUORONE see AOO275
ANDROFURAZANOL see AOO300
ANDROGEN see TBG000
ANDROLIN see TBF500
ANDROMEDOTOXIN see AOO375
ANDROMETH see MPN500
ANDRONAQ see TBF500
ANDROSAN see MPN500, TBG000
ANDROSAN (tablets) see MPN500
ANDROSTALONE see MJE760
ANDROSTANAZOL see AOO400
ANDROSTANAZOLE see AOO400
5α-ANDROSTAN-3α,17β-DIOL, 2β,16β-DIPIPECOLINIO-, DIBROMIDE, DIACE-
   TATE see PAF625
ANDROSTANE-2-CARBONITRILE, 4,5-EPOXY-17-HYDROXY-3-OXO-, (2-α-4-α-5-
   α-17-β)- see EBY600
5-α-ANDROSTANE-2-α-CARBONITRILE, 4-α-5-EPOXY-17-β-HYDROXY-3-OXO-
   see EBY600
5-α-ANDROSTANE-17-α-METHYL-17-β-OL-3-ONE see MJE760
ANDROSTAN-3-ONE, 17-HYDROXY-17-METHYL-, (5-α-17-β)-(9CI) see MJE760
ANDROSTEN see MPN500
Δ⁴-ANDROSTEN-3,17-DIONE see AOO425
ANDROST-2-ENE-2-CARBONITRILE, 3,17-DIHYDROXY-4,17-DIMETHYL-4,5-EP-
   OXY-, (4-α-5-α- 17-β)- see EBH400
ANDROST-5-ENE-3,17-DIOL, DIPROPANOATE, (3-β,17-β)- (9CI) see AOO410
ANDROST-5-ENE-3-β,17-β-DIOL, DIPROPIONATE see AOO410
ANDROSTENEDIOL DIPROPIONATE see AOO410
4-ANDROSTENE-3,17-DIONE see AOO425
Δ⁴-ANDROSTENE-3,17-DIONE see AOO425
ANDROSTENEDIONE see AOO425
Δ-4-ANDROSTENEDIONE see AOO425
4-ANDROSTENE-17-α-METHYL-17-β-OL-3-ONE see MPN500
Δ⁴-ANDROSTENE-17-β-PROPIONATE-3-ONE see TBG000
ANDROST-4-EN-17β-OL-3-ONE see TBF500
Δ⁴-ANDROSTEN-17(β)-OL-3-ONE see TBF500
ANDROSTENOLONE see AOO450
ANDROST-4-EN-3-ONE, 17-(3-CYCLOPENTYL-1-OXOPROPOXY)-, (17-β)- (9CI)
   see TBF600
ANDROST-4-EN-3-ONE, 17-β-HYDROXY-1-α-7-α-DIMERCAPTO-17-METHYL-, 1,7-
   DIACETATE see TFK300
ANDROSTEROLO see AOO275
ANDROSTESTONE-M see AOO475
ANDROSTESTON-M see AOO475
ANDROTARDYL see TBF750
ANDROTESTON see TBG000
ANDROTEST P see TBG000
ANDROTEX see AOO425
ANDRUSOL see TBF500
ANDRUSOL-P see TBG000
ANDUR see PKL500
ANECOTAN see DFA400
ANECTINE see CMG250, HLC500
ANECTINE CHLORIDE see HLC500
ANELIX see HIM000
ANELMID see DJT800
ANEMONE see PAM780
ANEMONE (VARIOUS SPECIES) see PAM780
ANERGAN see ABH500
ANERTAN see MPN500, TBG000
ANERTAN (tablets) see MPN500
ANERVAL see BRF500
ANESTACON see DHK400
ANESTACON HYDROCHLORIDE see DHK600
ANESTHENYL see MGA850
ANESTHESIA ETHER see EJU000
ANESTHESIN see EFX000
ANESTHESOL see AIT250
ANESTHETIC COMPOUND No. 347 see EAT900

ANESTHETIC ETHER see EJU000
ANESTHONE see EFX000
ANESTIL see AIT250
ANETAIN see BQA010
ANETHAINE see TBN000
trans-ANETHOL see PMR250
ANETHOLE (FCC) see PMQ750
cis-ANETHOLE see PMR000
trans-ANETHOLE see PMR250
ANETHOLE TRITHIONE see AOO490
ANETHOLTRITHION see AOO490
ANEURAL see MQU750
ANEURIMEC see DXO300
ANEURINE see TES750
ANEURINE DISULFIDE see TES800
ANEUXRAL see MQU750
ANEXOL see CDP000
ANFETAMINA see AOB250
ANFLAGEN see IIU000
ANFOTERICO LB see LBU200
ANFRAM 3PB see TNC500
ANFT see ALM500
ANG 66 see AOP250
ANGECIN see FQC000
ANGEL DUST see AOO500
α-ANGELICA LACTONE see AOO750
β,γ-ANGELICA LACTONE see MKH250
β-ANGELICA LACTONE see MKH500
ANGELICA LACTONE see MKH250
Δ¹-ANGELICA LACTONE see MKH500
Δ²-ANGELICA LACTONE see MKH250
ANGELICA OIL, root see AOO760
ANGELICA ROOT OIL see AOO760
ANGELICIN (coumarin derivative) see FQC000
ANGELIKA OEL see AOO760
ANGELI'S SULFONE see AOO800
ANGELI SULFONE see AOO800
ANGEL'S TRUMPET see AOO825
ANGEL TULIP see SLV500
ANGEL WINGS see CAL125
ANGIBID see NGY000
ANGICAP see PBC250
ANGIFLAN see DBX400
ANGIGRAFIN see AOO875
ANGINAL see PCP250
ANGININ see PPH050
ANGININE see NGY000, PPH050
ANGINON see AHI875
ANGINYL see DNU600
ANGIOCICLAN see POD750
ANGIOGRAFIN see AOO875
ANGIOKAPSUL see ARQ750
ANGIOLINGUAL see NGY000
ANGIOMIN see XCS000
ANGIOPAC see VLF000
ANGIOTENSIN see AOO900
ANGIOTONIN see AOO900
ANGIOXINE see PPH050
ANGITET see PBC250
ANGITRIT see TJL250
ANGLISLITE see LDY000
ANGOLAMYCIN see AOO925
ANGORIN see NGY000
ANGORLISIN see ELH600
ANG.-STERANTHREN (GERMAN) see DCR800
ANG-STERANTHRENE see DCR800
ANGUIDIN see AOP250
ANGUIDINE see AOP250
ANGUIFUGAN see DJT800
ANHIBA see HIM000
ANHISTABS see WAK000
ANHISTAN see FOS100
ANHISTOL see WAK000
ANHITOL 24B see LBU200
ANHYDRIDE ACETIQUE (FRENCH) see AAX500
ANHYDRIDE ARSENIEUX see ARI750
ANHYDRIDE ARSENIQUE (FRENCH) see ARH500
ANHYDRIDE CARBONIQUE (FRENCH) see CBU250
ANHYDRIDE CARBONIQUE et OXYDE d'ETHYLENE MELANGES (FRENCH)
   see EJO000
ANHYDRIDE CHROMIQUE (FRENCH) see CMK000
ANHYDRIDE KYSELINY 4-CHLOR-1,2,3,6-TETRAHYDROFTA-LOVE (CZECH)
   see CFG500
ANHYDRIDE PHTHALIQUE (FRENCH) see PHW750
ANHYDRIDES see AOP500

ANHYDRIDE VANADIQUE (FRENCH) see VDU000
ANHYDRID KYSELINY 3,6-ENDOMETHYLEN-Δ(SUP 4)-TETRAHYDROFTALOVE (CZECH) see BFY000
ANHYDRID KYSELINY KROTONOVE see COB900
ANHYDRID KYSELINY MASELNE see BSW550
ANHYDRID KYSELINY OCTOVE see AAX500
ANHYDRID KYSELINY TETRAHYDROFTALOVE (CZECH) see TDB000
ANHYDRID KYSELINY TRIFLUOROCTOVE (CZECH) see TJX000
2,2'-ANHYDRO-1-β-d-ARABINOFURANOSYLCYTOSINE HYDROCHLORIDE see COW900
2,2'-ANHYDROARABINOSYLCYTOSINE see COW875
2,2'-ANHYDROARABINOSYLCYTOSINE HYDROCHLORIDE see COW900
ANHYDROARA C see COW875
ANHYDRO-4,4'-BIS(DIETHYLAMINO)TRIPHENYLMETHANOL-2',4''-DISULPHONIC ACID, MONOSODIUM SALT see ADE500
ANHYDRO-4-CARBAMOYL-5-HYDROXY-1-β-d-RIBOFURANOSYL-IMIDAZOLIUM-HYDROXIDE see BMM000
2,2'-ANHYDROCYTARABINE HYDROCHLORIDE see COW900
2,2'-ANHYDROCYTIDINE see COW875
ANHYDROCYTIDINE see COW875
2,2'-ANHYDROCYTIDINE HYDROCHLORIDE see COW900
3,6-ANHYDRO-d-GALACTAN see CCL250
ANHYDROGITALIN see GEU000
ANHYDROGLUCOCHLORAL see GFA000
10-(1',5'-ANHYDROGLUCOSYL)ALOE-EMODIN-9-ANTHRONE see BAF825
ANHYDROHEXITOL SESQUIOLEATE see SKV170
ANHYDROHYDROXYPROGESTERONE see GEK500
ANHYDROL see EFU000
ANHYDROMYRIOCIN see AOP750
ANHYDRONE see PCE000
ANHYDROSORBITOL STEARATE see SKV150
ANHYDRO-o-SULFAMINEBENZOIC ACID see BCE500
ANHYDROTRIMELLIC ACID see TKV000
ANHYDROUS AMMONIA see AMY500
ANHYDROUS BORAX see DXG035
ANHYDROUS CALCIUM SULFATE see CAX500
ANHYDROUS CHLORAL see CDN550
ANHYDROUS CHLOROBUTANOL see ABD000
ANHYDROUS HYDRIODIC ACID see HHI500
ANHYDROUS HYDROBROMIC ACID see HHJ000
ANHYDROUS HYDROCHLORIC ACID see HHL000
ANHYDROUS IRON OXIDE see IHD000
ANHYDROUS OXIDE of IRON see IHD000
ANHYDROUS SODIUM ARSANILATE see ARA500
ANHYDROXYPROGESTERONE see GEK500
ANI see ISN000
ANICON KOMBI see CIR250
ANICON M see CIR250
ANIDRIDE ACETICA (ITALIAN) see AAX500
ANIDRIDE CROMICA (ITALIAN) see CMK000
ANIDRIDE CROMIQUE (FRENCH) see CMJ900
ANIDRIDE FTALICA (ITALIAN) see PHW750
ANILANA, combustion products see ADX750
ANILAZIN see DEV800
ANILAZINE see DEV800
ANILID KYSELINY ACETOCTOVE see AAY000
ANILID KYSELINY SALICYLOVE (POLISH) see SAH500
ANILIN (CZECH) see AOQ000
ANILINA (ITALIAN, POLISH) see AOQ000
ANILINE see AOQ000
ANILINE, N-ACETYL- see AAQ500
ANILINE ANTIMONYL TARTRATE see AOQ250
ANILINE, p-ARSENOSO- see ARJ755
ANILINE, p-ARSENOSO-N,N-BIS(2-CHLOROETHYL)- see ARJ760
ANILINE, p-ARSENOSO-N,N-BIS(2-HYDROXYETHYL)- see ARJ770
ANILINE, p-ARSENOSO-N,N-DIETHYL- see ARJ800
ANILINE, p-ARSENOSO-, DIHYDRATE see ALV100
p-ANILINEARSONIC ACID see ARA250
ANILINE, 4,4'-AZODI- see ASK925
ANILINE, p-(7-BENZOFURYLAZO)-N,N-DIMETHYL- see BCL100
ANILINE, N,N-BIS(2-CHLOROETHYL)-2,3-DIMETHOXY- see BIC600
ANILINE, N,N-BIS(2-(2,3-EPOXYPROPOXY)ETHOXY)- see BJN850
ANILINE, N,N-BIS(2-(2,3-EPOXYPROPOXY)ETHYL)- see BJN875
ANILINE, p-((4-BROMO-3-ETHYLPHENYL)AZO)-N,N-DIMETHYL- see BNK100
ANILINE, p-((3-BROMO-4-ETHYLPHENYL)AZO)-N,N-DIMETHYL- see BNK275
ANILINE, p-(m-BROMOPHENYLAZO)-N,N-DIMETHYL- see BNE600
ANILINE, p-((4-BROMO-m-TOLYL)AZO)-N,N-DIMETHYL- see BNQ110
ANILINE, p-((3-BROMO-p-TOLYL)AZO)-N,N-DIMETHYL- see BNQ100
ANILINE, 4-BUTYL- see AJA550
ANILINE, N-sec-BUTYL-4-tert-BUTYL-2,6-DINITRO- see BQN600
ANILINE, p-((p-BUTYLPHENYL)AZO)-N,N-DIMETHYL- see BRB450
ANILINE, p-((p-tert-BUTYL)PHENYL)AZO)-N,N-DIMETHYL- see BRB460
ANILINE-3-CARBOXYLIC ACID see AIH500
ANILINE CARMINE POWDER see FAE100

ANILINE CHLORIDE see BBL000
ANILINE, N-(2-CHLOROETHYL)-N-ETHYL- see EHJ600
ANILINE, p-CHLORO-, HYDROCHLORIDE see CJR200
ANILINE, 5-CHLORO-2-MERCAPTO-, HYDROCHLORIDE see CHU100
ANILINE, 2-CHLORO-4-(METHYLSULFONYL)- see CIX200
ANILINE, 4-CHLORO-2-NITRO- see KDA050
ANILINE, 4-CHLORO-3-NITRO- see CJA185
ANILINE, 2-CHLORO-5-NITRO- see CJA180
ANILINE, N-(2-CYANOETHYL)-N-ETHYL- see EHQ500
ANILINE, 2,4-DICHLORO- see DEO290
ANILINE, p-DICHLOROARSINO-, HYDROCHLORIDE see AOR640
ANILINE, 2,6-DIETHYL- see DIS650
ANILINE, 2,6-DIISOPROPYL- see DNN630
ANILINE, 2,5-DIMETHOXY- see AKD925
ANILINE, N,N-DIMETHYL-p-(3,5-DIFLUOROPHENYLAZO)- see DKH100
ANILINE-2,5-DISULFONIC ACID see AIE000
ANILINE DYES see AOQ500
ANILINE, p-ETHOXY- see PDK890
ANILINE, o-ETHYL-(8CI) see EGK500
ANILINE, p-((o-ETHYLPHENYL)AZO)-N,N-DIMETHYL- see EIF450
ANILINE, 4-FLUORO-3-NITRO- see FKK100
ANILINE-FORMALDEHYDE CONDENSATE see FMW330
ANILINE-FORMALDEHYDE POLYMER see FMW330
ANILINE GREEN see AFG500, BAY750
ANILINE HYDROCHLORIDE (DOT) see BBL000
ANILINE, o-ISOPROPYL- see INW100
ANILINE, 2-METHOXY-4-NITRO- see NEQ000
ANILINE, m-(METHYLTHIO)- see ALX100
ANILINE, p-(METHYLTHIO)- see AMS675
ANILINE MUSTARD see AOQ875
ANILINE, N,N'-DIMETHYL-4,4'-METHYLENEDI- see MJO000
ANILINE, 4-NITRO- see NEO500
ANILINE, N-NITRO- see NEO510
ANILINE OIL see AOQ000
ANILINE OIL DRUMS, EMPTY see AOR000
ANILINE, 2,3,4,5,6-PENTANITRO- see PBM100
ANILINE, 2-PHENOXY- see PDR490
ANILINE, o-PHENOXY-(8CI) see PDR490
ANILINE, p,p'-(m-PHENYLENEDIOXY)DI- see REF070
ANILINE, POLYMER with FORMALDEHYDE (8CI) see FMW330
''ANILINE SALT'' see BBL000
p-ANILINESULFONAMIDE see SNM500
ANILINE-4-SULFONIC ACID see SNN600
m-ANILINESULFONIC ACID see SNO000
ANILINE-p-SULFONIC ACID see SNN600
ANILINE-p-SULFONIC AMIDE see SNM500
ANILINE, 4,4'-(SULFONYLBIS(p-PHENYLENEOXY))DI- see SNZ000
ANILINE-p-SULPHONIC ACID see SNN600
ANILINE, 2,3,4-TRIFLUORO- see TJX900
ANILINE, 2,4,6-TRINITRO- see PIC800
ANILINE VANADATE, DIHYDRATE see AOR250
ANILINE VIOLET see AOR500
ANILINE VIOLET PYOKTANINE see AOR500
ANILINE YELLOW see PEI000
ANILINIUM CHLORIDE see BBL000
ANILINIUM NITRATE see AOR625
ANILINIUM PERCHLORATE see AOR630
p-ANILINOANILINE see PFU500
ANILINOBENZENE see DVX800
2-ANILINOBENZOIC ACID see PEG500
o-ANILINOBENZOIC ACID see PEG500
2-ANILINO-4'-(BENZYLOXY)-2-PHENYLACETOPHENONE see AOR635
4'-(α-ANILINOBENZYLOXY)-2-PHENYLACETOPHENONE see AOR635
α-ANILINO-p-BENZYLOXYPHENYL BENZYL KETONE see AOR635
4-ANILINODICHLOROARSINE, HYDROCHLORIDE see AOR640
2-ANILINO-4'-(2-(DIETHYLAMINO)ETHOXY)-2-PHENYLACETOPHENONE HYDRO-CHLORIDE see AOR700
α-ANILINO-p-DIETHYLAMINOETHOXYPHENYL BENZYL KETONE HYDROCHLO-RIDE see AOR700
2-ANILINO-5-(2,4-DINITROANILINO)BENZENESULFONIC ACID SODIUM SALT see SGP500
1-ANILINO-2,5-DISULFONIC ACID see AIE000
ANILINOETHANE see EGK000
2-ANILINOETHANOL see AOR750
(2-ANILINOETHYL)HYDRAZONE DIHYDROCHLORIDE see AOS000
ANILINOMETHANE see MGN750
1-ANILINONAPHTHALENE see PFT250
2-ANILINONAPHTHALENE see PFT500
ANILINONAPHTHALENE see PFT500
2-ANILINO-5-NITROBENZENESULFONIC ACID see AOS500
ANILINO (p-NITROPHENYL) SULFIDE see AOS750
4-ANILINOPHENOL see AOT000
p-ANILINOPHENOL see AOT000
3-ANILINOPROPIONITRILE see AOT100
β-ANILINOPROPIONITRILE see AOT100

4-((4-ANILINO-5-SULFO-1-NAPHTHYL)AZO)-5-HYDROXY-2,7-NAPHTHALENEDI-FULFONIC ACID TRISODIUM see ADE750
ANILINO-2-SULFONIC ACID see SNO100
ANILINO-o-SULFONIC ACID see SNO100
6-(p-ANILINOSULFONYL)METANILAMIDE see AOT125
ANILINO-o-SULPHONIC ACID see SNO100
ANILITE see AOT250
ANILIX see CKL500
ANIMAG see MAH500
ANIMAL CONIINE see PBK500
ANIMAL GALACTOSE FACTOR see OJV500
ANIMAL OIL see BMA750
ANIMAL STARCH see GHK300
ANIMERT see CKL750
ANIMERT V-10 see CKL750
ANIMERT V-101 see CKL750
ANIMERT V-10K see CKL750
ANISALACETONE see MLI400
2-ANISALDEHYDE see AOT525
o-ANISALDEHYDE see AOT525
p-ANISALDEHYDE see AOT530
o-ANISAMIDE see AOT750
o-ANISAMIDE, N-((1-ETHYL-2-PYRROLIDINYL)METHYL)-5-(ETHYLSULFONYL)- see EPD100
ANISE ALCOHOL see MED500
ANISE CAMPHOR see PMQ750
ANISEED OIL see AOU250
ANISENE see CLO750
ANISE OIL see AOU250
4-ANISIC ACID see AOU600
m-ANISIC ACID see AOU500
o-ANISIC ACID see MPI000
p-ANISIC ACID see AOU600
p-ANISIC ACID, ETHYL ESTER see AOV000
o-ANISIC ACID, HYDRAZIDE see AOV250
p-ANISIC ACID, HYDRAZIDE see AOV500
ANISIC ACID HYDRAZIDE see AOV500
p-ANISIC ACID, METHYL ESTER see AOV750
o-ANISIC ACID, METHYL ESTER (7CI,8CI) see MLH800
ANISIC ALCOHOL see MED500
ANISIC ALDEHYDE see AOT530
ANISIC HYDRAZIDE see AOV500
ANISIC KETONE see AOV875
2-ANISIDINE see AOV900
4-ANISIDINE see AOW000
m-ANISIDINE see AOV890
o-ANISIDINE see AOV900
p-ANISIDINE see AOW000
m-ANISIDINE ANTIMONYL TARTRATE see AOW500
o-ANISIDINE ANTIMONYL TARTRATE see AOW750
p-ANISIDINE ANTIMONYL TARTRATE see AOX000
o-ANISIDINE HYDROCHLORIDE see AOX250
p-ANISIDINE HYDROCHLORIDE see AOX500
2-ANISIDINE NITRATE see MEA600
o-ANISIDINE NITRATE see NEQ500
ANISKETONE see AOV875
ANIS OEL (GERMAN) see AOU250
p-ANISOL ALCOHOL see MED500
ANISOLE see AOX750
ANISOLE, 2-AMINO-5-NITRO- see NEQ000
ANISOLE, o-BROMO- see BMT400
ANISOLE, p-BROMO- see AOY450
ANISOLE, p-(3-BROMOPROPENYL)-, (E)- see BMT300
ANISOLE, 2,4-DIAMINO-, HYDROGEN SULFATE see DBO400
ANISOLE, 2,4-DIAMINO-, SULFATE see DBO400
ANISOLE, 2-ISOPROPYL-5-METHYL- see MPW650
ANISOLE, 2,4,6-TRINITRO- see TMK300
ANISOMYCIN see AOY000
ANISOPIROL see HAH000
ANISOPYRADAMINE see DBM800
ANISOTROPINE METHOBROMIDE see LJS000
(E)-3-p-ANISOYL-3-BROMOACRYLIC ACID SODIUM SALT see CQK600
3-p-ANISOYL-3-BROMOACRYLIC ACID, SODIUM SALT see SIK000
ANISOYL CHLORIDE see AOY250
p-ANISOYLHYDRAZINE see AOV500
ANISOYLHYDRAZINE see AOV500
3-ANISOYL-2-MESITYLBENZOFURAN see AOY300
ANISTADIN see HII500
ANISYL ACETATE see AOY400
2-(p-ANISYL)ACETIC ACID see MFE250
ANISYLACETONE see MFF580
ANISYLACETONITRILE see MFF000
ANISYL ALCOHOL (FCC) see MED500
m-ANISYLAMINE see AOV890
o-ANISYLAMINE see AOV900

p-ANISYLAMINE see AOW000
o-ANISYLAMINE HYDROCHLORIDE see AOX250
ANISYL BROMIDE see AOY450, BMT400
N-(o-ANISYL)-2-(p-BUTOXYPHENOXY)-N-(2(DIETHYLAMINO)ETHYL)ACETAMIDE HYDROCHLORIDE see APA000
ANISYL-N-BUTYRATE see MED750
ANISYL FORMATE see MFE250
ANISYLIDENE ACETONE see MLI400
ANISYL METHYL KETONE see AOV875
ANISYL PHENYLACETATE see APE000
p-ANISYL 3-PYRIDYL KETONE see MED100
p-ANISYL 4-PYRIDYL KETONE see MFH930
p-ANISYOL CHLORIDE see AOY250
ANIT see ISN000
ANITSOTROPINE METHYLBROMIDE see LJS000
ANKERBIN see SLJ050
ANKILOSTIN see PCF275
ANN (GERMAN) see AAW000
ANNALINE see CAX750
ANNANOX CK see SCQ000
ANNATTO EXTRACT see APE100
ANNONA MURICATA see SKV500
ANNUAL POINSETTIA see EQX000
(6)ANNULENE see BBL250
ANODYNON see EHH000
ANOFEX see DAD200
ANOL see CPB750
ANON BL see LBU200
ANOPROLIN see ZVJ000
ANOREXIDE see BBK000
ANOVIGAM see EEM000
ANOVLAR 21 see EEH520
β-ANOXIN see DNM400
ANOZOL see DJX000
ANP 235 see DPE000
ANP 246 see CJN750
ANP 3548 see BPM750
ANP 3624 see TGA600
ANP 4364 see DFO600
ANPARTON see ARQ750
235 ANP HYDROCHLORIDE see AAE500
ANPROLENE see EJN500
ANPROLINE see EJN500
ANQI see ALK625
ANQUIL see FLK100, RDK000
ANSADOL see SAH500
ANSAL see AQN250
ANSAMITOCIN P-4 see APE529
ANSAR 160 see HKC500
ANSAR 170 see MRL750
ANSAR 184 see DXE600
ANSAR 560 see HKC500
ANSAR see HKC000
ANSAR DSMA LIQUID see DXE600
ANSATIN see TKH750
ANSEPRON see PGA750
ANSIACAL see MDQ250
ANSIATAN see MQU750
ANSIBASE RED KB see CLK225
ANSIL see MQU750
ANSILAN see CGA000
ANSIOLISINA see CFZ000, DCK759
ANSIOWAS see MQU750
ANSIOXACEPAM see CFZ000
ANSOLYSEN see PBT000
ANSOLYSEN BITARTRATE see PBT000
ANSOLYSEN TARTRATE see PBT000
ANSUL ETHER 181AT see PBO500
ANT-1 see TKH750
ANTABUS see DXH250
ANTABUSE see DXH250
ANTADIX see DXH250
ANTADOL see BRF500
ANTAENYL see DXH250
ANTAETHAN see DXH250
ANTAETHYL see DXH250
ANTAETIL see DXH250
ANTAGE W 400 see MJO500
ANTAGE W 500 see MJN250
ANTAGONATE see TAI500
ANTAGOSAN see PAF550
ANTAGOTHYROID see TFR250
ANTAGOTHYROIL see TFR250
ANTAK see DAI600, ODE000
ANTALCOL see DXH250

ANTALERGAN see WAK000
ANTALLIN see CAR780
ANTALVIC see PNA500
ANTAMINE see WAK000
ANTAN see NAH500
ANTAROX A-200 see PKF500
ANTARSIN see DNV600
ANTASTEN see PDC000
ANTAZOLINE see PDC000
ANTEBOR see SNQ000
ANTEMOQUA see AJX500
ANTEMOVIS see AJX500
ANTENE see BJK500
ANTEPAR see PIJ500
ANTERGAN see BEM500
ANTERGAN HYDROCHLORIDE see PEN000
ANTERGYL see SEO500
ANTERIOR PITUITARY GROWTH HORMONE see PJA250
ANTERON see SCA750
ANTETAN see DXH250
ANTETHYL see DXH250
ANTETIL see DXH250
ANTEX-490 see SCA750
ANTEYL see DXH250
ANTHALLAN HYDROCHLORIDE see APE625
ANTHANTHREN (GERMAN) see APE750
ANTHANTHRENE see APE750
ANTHECOLE see PIJ500
ANTHELMYCIN see APF000
ANTHELONE U see UVJ475
ANTHELVET see TDX750
ANTHER see IHV050
ANTHGLUTIN see GFO100
ANTHIO see DRR200
ANTHIOLIMINE see LGU000
ANTHIOMALINE see LGU000
ANTHIOMALINE NONAHYDRATE see AQE500
ANTHION see DWQ000
ANTHIPHEN see MJM500
ANTHISAN see WAK000
ANTHISAN MALEATE see DBM800
ANTHIUM DIOXIDE see CDW450
ANTHON see TIQ250
ANTHOPHYLITE see ARM264
ANTHRA(9,1,2-cde)BENZO(h)CINNOLINE see APF750
ANTHRA(2,1-d:6,5-d')BISTHIAZOLE-6,12-DIONE, 2,8-DIPHENYL- see VGP200
ANTHRACEN (GERMAN) see APG500
1-ANTHRACENAMINE see APG050
2-ANTHRACENAMINE see APG100
ANTHRACENE see APG500
ANTHRACENE BROWN FD see TKN500
ANTHRACENE BROWN FF see TKN500
ANTHRACENE BROWN G see TKN500
ANTHRACENE BROWN N see TKN500
ANTHRACENE BROWN S see TKN500
ANTHRACENE BROWN WH see TKN500
ANTHRACENE BROWN WL see TKN500
ANTHRACENE-9-CARBOXYLIC ACID see APG550
9,10-ANTHRACENEDIONE see APK250
9,10-ANTHRACENEDIONE, 1-AMINO-4-HYDROXY-(9CI) see AKE250
9,10-ANTHRACENEDIONE, 1-AMINO-4-HYDROXY-2-METHOXY-(9CI) see
  AKO350
9,10-ANTHRACENEDIONE, 1-AMINO-4-(METHYLAMINO)-(9CI) see AKP250
9,10-ANTHRACENEDIONE, 1-BROMO-4-(METHYLAMINO)- see BNN550
9,10-ANTHRACENEDIONE, 2-CHLORO- see CEI100
9,10-ANTHRACENEDIONE, 2,6-DIAMINO- see APK850
9,10-ANTHRACENEDIONE, 1,5-DIAMINOCHLORO-4,8-DIHYDROXY- see
  CMP070
9,10-ANTHRACENEDIONE, 1,8-DIAMINO-4,5-DIHYDROXY- see DBQ250
9,10-ANTHRACENEDIONE, 1,4-DIAMINO-2-METHOXY-(9CI) see DBX000
9,10-ANTHRACENEDIONE, 1,4-DIAMINO-5-NITRO-(9CI) see DBY700
9,10-ANTHRACENEDIONE, 1,5-DICHLORO- see DEO700
9,10-ANTHRACENEDIONE, 1,8-DICHLORO- see DEO750
9,10-ANTHRACENEDIONE, 1,2-DIHYDROXY- see DMG800
9,10-ANTHRACENEDIONE, 1,8-DIHYDROXY-4,5-DINITRO-(9CI) see DMN400
9,10-ANTHRACENEDIONE, 1,8-DIHYDROXY-4-((4-(2-HYDROXYETH-
  YL)PHENYL)AMINO)-5-NITRO- see CMP060
9,10-ANTHRACENEDIONE, 1,8-DIHYDROXY-2,4,5,7-TETRANITRO- see CML600
9,10-ANTHRACENEDIONE, 1,5-DIPHENOXY- see DVW100
9,10-ANTHRACENEDIONE, 1-(PHENYLTHIO)-(9CI) see PGL000
9,10-ANTHRACENEDIONE, 1,2,5,8-TETRAHYDROXY-(9CI) see TDD000
9,10-ANTHRACENEDIONE, 1,2,3-TRIHYDROXY-(9CI) see TKN500
ANTHRACENE PRINTING BROWN see TKN500
2-ANTHRACENESULFONIC ACID, 1-AMINO-4-(4-AMINO-3-SULFOANILINO)-9,10-
  DIHYDRO-9,10-DIOXO- see SOB600

2-ANTHRACENESULFONIC ACID, 1-AMINO-4-((4-AMINO-3-SULFOPHE-
  NYL)AMINO)-9,10-DIHYDRO-9,10-DIOXO- see SOB600
2-ANTHRACENESULFONIC ACID, 1-AMINO-4-(3-((4,6-DICHLORO-s-TRIAZIN-2-
  YL)AMINO)-4-SULFOANILINO)-9,10-DIHYDRO-9,10-DIOXO- see CMS228
2-ANTHRACENESULFONIC ACID, 1-AMINO-9,10-DIHYDRO-9,10-DIOXO-4-(2,4,6-
  TRIMETHYLANILINO)-, MONOSODIUM SALT see CMM100
2-ANTHRACENESULFONIC ACID, 1-AMINO-9,10-DIHYDRO-4-(p-(N-METHYLACE-
  TAMIDO)ANILINO)-9,10-DIOXO-, MONOSODIUM SALT see CMM070
2-ANTHRACENESULFONIC ACID, 9,10-DIHYDRO-1-AMINO-4-(CYCLOHEXYLAM-
  INO)-9,10-DIOXO-, MONOSODIUM SALT see CMM080
2-ANTHRACENESULFONIC ACID, 9,10-DIHYDRO-3,4-DIHYDROXY-9,10-DIOXO-,
  MONOSODIUM SALT see SEH475
1,4,9,10-ANTHRACENETETRAOL see LEX200
1,4,9,10-ANTHRACENETETROL (9CI) see LEX200
1,8,9-ANTHRACENETRIOL see APH250
1,8,9-ANTHRACENETRIOL TRIACETATE see APH500
9(10H)-ANTHRACENONE, 1,8-DIHYDROXY-10-(1-OXOPROPYL)- see PMW760
ANTHRACHINON-1,8-DISULFONAN DRASELNY (CZECH) see DLJ600
ANTHRACHINON-1,5-DISULFONAN SODNY (CZECH) see DLJ700
ANTHRACHINON-1-SULFONAN SODNY (CZECH) see DLJ800, SER000
ANTHRACIN see APG500
ANTHRACITE PARTICLES see CMY760
1-ANTHRACYLAMINE see APG050
2-ANTHRACYLAMINE see APG100
ANTHRADIONE see APK250
ANTHRAFLAVIC ACID see DMH600
ANTHRAFLAVIN see DMH600
ANTHRAGALLIC ACID see TKN500
ANTHRAGALLOL see TKN500
ANTHRALAN BLUE B see CMM070
ANTHRALAN YELLOW RRT see SGP500
ANTHRALIN see APH250
1-ANTHRAMINE see APG050
2-ANTHRAMINE see APG100
ANTHRAMYCIN see API000
ANTHRAMYCIN-11-METHYL ETHER see API125
ANTHRAMYCIN METHYL ETHER see API125
ANTHRA(2,1,9-mna)NAPHTH(2,3-h)ACRIDINE-5,10,15-TRIONE see CMU810
ANTHRANILAMIDE see AID620
ANTHRANILIC ACID see API500
ANTHRANILIC ACID, N-ACETYL- see AAJ150
ANTHRANILIC ACID, N-(2-BENZYLIDENEHEPTYLIDENE)-, METHYL ESTER see
  AOH100
ANTHRANILIC ACID, N-(3-(p-tert-BUTYLPHENYL)-2-METHYLPROPYLIDENE)-,
  METHYL ESTER see LFT100
ANTHRANILIC ACID, CINNAMYL ESTER see API750
ANTHRANILIC ACID, 3,5-DIIODO- see API800
ANTHRANILIC ACID, LINALYL ESTER see APJ000
ANTHRANILIC ACID, METHYL ESTER see APJ250
ANTHRANILIC ACID, PHENETHYL ESTER see APJ500
ANTHRANILIMIDIC ACID see AID620
m-ANTHRANILONITRILE see AIR125
ANTHRANILONITRILE see APJ750
ANTHRANOL CHROME YELLOW 2G see SIT850
ANTHRANOL CHROME YELLOW 5GS see SIT850
ANTHRANTHRENE see APE750
ANTHRAPOLE AZ see BQK250
ANTHRA(1,9-cd)PYRAZOL-6(2H)-ONE see APK000
β-ANTHRAQUINOLINE see NAZ000
9,10-ANTHRAQUINONE see APK250
ANTHRAQUINONE see APK250
ANTHRAQUINONE, 1-AMINO-4-HYDROXY- see AKE250
ANTHRAQUINONE, 1-AMINO-4-HYDROXY-2-METHOXY- see AKO350
ANTHRAQUINONE, 1,4-BIS((2-HYDROXYETHYL)AMINO)-5,8-DIHYDROXY- see
  DMM400
ANTHRAQUINONE, 1,4-BIS(METHYLAMINO)- see BKP500
ANTHRAQUINONE BLUE see IBV050
ANTHRAQUINONE BLUE SKY see CMM090
ANTHRAQUINONE BRILLIANT GREEN CONCENTRATE ZH see APK500
ANTHRAQUINONE, 2-BROMO-1,5-DIAMINO-4,8-DIHYDROXY- see BNC800
ANTHRAQUINONE, 1-BROMO-4-(METHYLAMINO)- see BNN550
ANTHRAQUINONE, 2-CHLORO- see CEI100
ANTHRAQUINONE DEEP BLUE see IBV050
ANTHRAQUINONE, 2,6-DIAMINO- see APK850
ANTHRAQUINONE, 1,8-DIAMINO-4,5-DIHYDROXY- see DBQ250
ANTHRAQUINONE, 1,4-DIHYDROXY-5,8-BIS((2-HYDROXYETHYL)AMINO)- see
  DMM400
ANTHRAQUINONE, 1,8-DIHYDROXY-4-(p-(2-HYDROXYETHYL)ANILINO)-5-NITRO-
  see CMP060
ANTHRAQUINONE, 1,8-DIHYDROXY-2,4,5,7-TETRANITRO- see CML600
1,2-ANTHRAQUINONEDIOL see DMG800
1,8-ANTHRAQUINONEDISULFINIC ACID see APK635
1,5-ANTHRAQUINONEDISULFONIC ACID see APK625
ANTHRAQUINONEDISULFONIC ACID, DIPOTASSIUM SALT see DLJ600
9,10-ANTHRAQUINONE-2-SODIUM SULFONATE see SER000

2-ANTHRAQUINONESULFONATE SODIUM see SER000
ANTHRAQUINONE-2-SULFONATE SODIUM SALT see SER000
2-ANTHRAQUINONESULFONIC ACID SODIUM SALT see SER000
ANTHRAQUINONE, 1-THIOPHENYL see PGL000
α-ANTHRAQUINONYLAMINE see AIA750
β-ANTHRAQUINONYLAMINE see AIB000
ANTHRAQUINONYLAMINOANTHRAQUINONE see IBI000
((N-ANTHRAQUINON-2-YL)AMINOMETHYLENE)DIMETHYLAMMONIUM CHLO-
   RIDE see APK750
1,4-ANTHRAQUINONYLDIAMINE see DBP000
1,5-ANTHRAQUINONYLDIAMINE see DBP200, DBP400
1,8-ANTHRAQUINONYLDIAMINE see DBP400
2,6-ANTHRAQUINONYLDIAMINE see APK850
N,N''''-(2,6-ANTHRAQUINONYLENE)BIS(N,N-DIETHYLACETAMIDE) see APL250
2,2'-(1,4-ANTHRAQUINONYLENEDIIMINO)BIS(5-METHYLBENZENESULFONIC
   ACID) DISODIUM SALT see APL500
1,1'-(ANTHRAQUINON-1,4-YLENEDIIMINO)DIANTHRAQUINONE see APL750
1,1'-(ANTHRAQUINON-1,5-YLENEDIIMINO)DIANTHRAQUINONE see APM000
4,4'-(1,4-ANTHRAQUINONYLENEDIIMINODIPHENYL-1,4-ENEDIOX-
   O)BENZENESULFONIC ACID see APM250
ANTHRA RED G see CMM325
ANTHRARUFIN see DMH200
ANTHRASORB see CBT500
1,8,9-ANTHRATRIOL see APH250
ANTHRAVAT GOLDEN YELLOW see DCZ000
ANTHRAVAT NAVY BLUE BR see VGP100
5,9,14,18-ANTHRAZINETETRONE, 7,16-DICHLORO-6,15-DIHYDRO- see
   DFN425
5,9,14,18-ANTHRAZINETETRONE, 6,15-DIHYDRO- see IBV050
ANTHRIMIDE see IBI000
9-ANTHROIC ACID see APG550
9-ANTHRONOL see APM750
ANTHROPODEOXYCHOLIC ACID see CDL325
ANTHROPODESOXYCHOLIC ACID see CDL325
ANTHROPODODESOXYCHOLIC ACID see CDL325
2-ANTHRYLAMINE see APG100
ANTHURIUM see APM875
ANTIAETHAN see DXH250
ANTIANGOR see CBR500
α-ANTIARBIN see APN000
ANTIB see FNF000
ANTIBASON see MPW500
ANTIBIOCIN see PDT750
ANTIBIOTIC 60-6 see CCX725
ANTIBIOTIC 899 see VRA700
ANTIBIOTIC 1037 see VGZ000
ANTIBIOTIC 1142 see NCP875
ANTIBIOTIC 1600 see APP500
ANTIBIOTIC 1719 see ASO501
ANTIBIOTIC 66-40 see SDY750
ANTIBIOTIC 29275 see CBF680
ANTIBIOTIC 33876 see APF000
ANTIBIOTIC 67-694 see RMF000
ANTIBIOTIC 281471 see MRW800
ANTIBIOTIC 6761-31 see TFQ275
ANTIBIOTIC D-45 see SLW475
ANTIBIOTIC A-246 see FPC000
ANTIBIOTIC A-649 see OIU499
ANTIBIOTIC 833A see PHA550
ANTIBIOTIC A-5283 see PIF750
ANTIBIOTIC A 8506 see TFQ275
ANTIBIOTIC A-64922 see OIU499
ANTIBIOTIC A see API125
ANTIBIOTIC A 130A see LEJ700
ANTIBIOTIC A 3733A see HAL000
ANTIBIOTIC A 28695 A see SCA000
ANTIBIOTIC AD 32 see TJX350
ANTIBIOTIC A-250-II see CBF680
ANTIBIOTIC AK PS see AFH500
ANTIBIOTIC AM-2604 A see VRP775
ANTIBIOTIC APM see XFS600
ANTIBIOTIC A-399-Y4 see VGZ000
ANTIBIOTIC AY 22989 see RBK000
ANTIBIOTIC B 599 see CMK650
ANTIBIOTIC B-14437 see SAU000
ANTIBIOTIC B 21085 see GEO200
ANTIBIOTIC B-98891 see MQU000
ANTIBIOTIC N-329 B see VBZ000
ANTIBIOTIC BAY-f 1353 see MQS200
ANTIBIOTIC BB-K 8 see APS750
ANTIBIOTIC BB-K 8 SULFATE see APT000
ANTIBIOTIC B 41D see MQT600
ANTIBIOTIC BL-640 see APT250
ANTIBIOTIC BL-S 640 see APT250
ANTIBIOTIC BU 2231A see TAC500

ANTIBIOTIC BU 2231B see TAC750
ANTIBIOTIC CC 1065 see APT375
ANTIBIOTIC CGP 9000 see CCS530
ANTIBIOTIC DC 11 see TEF725
ANTIBIOTIC DE 3936 see LIF000
ANTIBIOTIC E212 see VGZ000
ANTIBIOTIC 1163 F.I. see LIN000
ANTIBIOTIC FN 1636 see PEC750
ANTIBIOTIC FR 1923 see APT750
ANTIBIOTIC G-52 see GAA100
ANTIBIOTIC G-52 SULFATE see APU000, GAA120
ANTIBIOTIC HA-9 see TFC500
ANTIBIOTIC KA 66061 see SLF500
ANTIBIOTIC KM 208 see BAC175
ANTIBIOTIC KW 1062 see MQS579
ANTIBIOTIC KW-1070 see FOK000
ANTIBIOTIC LA 7017 see MQW750
ANTIBIOTIC M 4365A2 see RMF000
ANTIBIOTIC MA 144A1 see APU500
ANTIBIOTIC MA 144A see APU500
ANTIBIOTIC MA 144A2 see TAH675
ANTIBIOTIC MA 144B2 see TAH650
ANTIBIOTIC MA 144M1 see MAB250
ANTIBIOTIC MA 144S2 see APV000
ANTIBIOTIC MA 144T1 see DAY835
ANTIBIOTIC MA 144U2 see MAX000
ANTIBIOTIC MM 14151 see CMV250
ANTIBIOTIC No. 899 see VRF000
ANTIBIOTIC NSC-71936 see CMQ725
ANTIBIOTIC NSC 70845 see NMV500
ANTIBIOTIC OS 3966A see RMK200
ANTIBIOTIC PA147 see APV750
ANTIBIOTIC PA-106 see AOY000
ANTIBIOTIC PA-93 see SMB000
ANTIBIOTIC PA 11481 see VRA700
ANTIBIOTIC PA 114 B1 see VRA700
ANTIBIOTIC Ro 21 6150 see LEJ700
ANTIBIOTIC 20-798RP see DAC300
ANTIBIOTIC 6059-S see LBH200
ANTIBIOTIC S 15-1A see RAG300
ANTIBIOTIC SF 733 see XQJ650
ANTIBIOTIC SF 837 see MBY150
ANTIBIOTIC S 7481F1 see CQH100
ANTIBIOTIC SF 837 A1 see MBY150
ANTIBIOTIC SF 837 A₁ see MBY150
ANTIBIOTIC SF 767B see NCF500
ANTIBIOTIC 1719 SODIUM SALT see ASO510
ANTIBIOTIC 66-40 SULFATE see APY500
ANTIBIOTIC T-1384 see NCP875
ANTIBIOTIC 205T3 see NMV500
ANTIBIOTIC T see COB000
ANTIBIOTIC TM 25 see HOH500
ANTIBIOTIC TM 481 see LIF000
ANTIBIOTIC U-12,241 see CMS232
ANTIBIOTIC U 18496 see ARY000
ANTIBIOTIC U 48160 see QQS075
ANTIBIOTICUM PA147 (GERMAN) see APV750
ANTIBIOTIC 44 VI see MRW800
ANTIBIOTIC W-847-A see MCA250
ANTIBIOTIC WR 141 see GJS000
ANTIBIOTIC X 146 see TFQ275
ANTIBIOTIC X 537 see LBF500
ANTIBIOTIC X-465A see CDK250
ANTIBIOTIC X465A SODIUM SALT see CDK500
ANTIBIOTIC XK 62-2 see MQS579
ANTIBIOTIC XK 41C see MCA250
ANTIBIOTIC XS-89 see SMM500
ANTIBIOTIC YL-704 A3 see JDS200
ANTIBIOTIC YL 704 B₁ see MBY150
ANTIBIOTIC YL-704 B3 see LEV025
ANTIBIOTIQUE see NCF000
ANTIBULIT see SHF500
ANTICANITIC VITAMIN see AIH600
ANTICARIE see HCC500
ANTICHLOR see SKI500
ANTI-CHROMOTRICHIA FACTOR see AIH600
ANTIDEPRIN see DLH600
ANTIDEPRIN HYDROCHLORIDE see DLH630
ANTIDUROL see DAM700
ANTIEGENE MB see BCC500
ANTIETANOL see DXH250
ANTI-ETHYL see DXH250
ANTIETIL see DXH250
ANTIFEBRIN see AAQ500
ANTIFEEDANT 24005 see DUI000

ANTIFEEDING COMPOUND 24,055 see DUI000
ANTIFOAM FD 62 see SCR400
ANTIFOLAN see MDV500
ANTIFORMIN see SHU500
ANTIGENE RDF see PJQ750
ANTI-GERM 77 see BEN000
ANTIGESTIL see DKA600
ANTIHELMYCIN see AQB000
ANTIHEMORRHAGIC VITAMIN see VTA000
ANTIHIST see DBM800
ANTIHISTAL see PDC000
ANTI-INFECTIVE VITAMIN see VSK600
ANTI-INFLAMMATORY HORMONE see CNS750
ANTIKNOCK-33 see MAV750
ANTIKOL see DXH250
ANTIKREIN see PAF550
ANTILEPSIN see DNU000
ANTILIPID see ARQ750
ANTILYSIN see PAF550
ANTILYSINE see PAF550
ANTIMALARINA see ARQ250
ANTIMICINA see ILD000
ANTIMIGRANT C 45 see SEH000
ANTIMILACE see TDW500
ANTIMIT see BIE500
ANTIMOINE FLUORURE (FRENCH) see AQE000
ANTIMOINE (TRICHLORURE d') see AQC500
ANTIMOL see SFB000
ANTIMONIAL SAFFRON see AQF500
ANTIMONIC "ACID" see AQF750
ANTIMONIC CHLORIDE see AQD000
ANTIMONIC OXIDE see AQF750
ANTIMONIC SULFIDE see AQF500
ANTIMONIO (PENTACLORURO DI) (ITALIAN) see AQD000
ANTIMONIO (TRICLORURO di) see AQC500
ANTIMONIOUS OXIDE see AQF000
ANTIMONOUS CHLORIDE (DOT) see AQC500
ANTIMONOUS CHLORIDE see AQC500
ANTIMONOUS FLUORIDE see AQE000
ANTIMONOUS SULFATE see AQJ250
ANTIMONOUS SULFIDE see AQL500
ANTIMONPENTACHLORID (GERMAN) see AQD000
ANTIMONTRICHLORID see AQC500
ANTIMONWASSERSTOFFES (GERMAN) see SLQ000
ANTIMONY see AQB750
ANTIMONY(III) ACETATE see AQJ750
ANTIMONY AMMONIA TRIACETIC ACID see AQC000
ANTIMONY, BIS(TRICHLORO) compounded with 1 mole of OCTAMETHYL PYROPHOSPHORAMIDE see AQC250
ANTIMONY BLACK see AQB750
ANTIMONY BUTTER see AQC500
ANTIMONY CHLORIDE (DOT) see AQC500
ANTIMONY(III) CHLORIDE see AQC500
ANTIMONY(V) CHLORIDE see AQD000
ANTIMONY CHLORIDE see AQC500
ANTIMONY COMPOUNDS see AQD500
ANTIMONY DIMERCAPTOSUCCINATE(IV) see AQD750
ANTIMONY DIMERCAPTOSUCCINATE see AQD750
ANTIMONY EMETINE IODIDE see EAM000
ANTIMONY(III) FLUORIDE (1:3) see AQE000
ANTIMONY(V) FLUORIDE see AQF250
ANTIMONY FLUORIDE see AQF250
ANTIMONY GLANCE see AQL500
ANTIMONY HYDRIDE see SLQ000
ANTIMONY LACTATE, solid (DOT) see AQE250
ANTIMONY LACTATE see AQE250
ANTIMONYL ANILINE TARTRATE see AOQ250
ANTIMONYLBRENZEATECHINDISULFOSAURES NATRIUM (GERMAN) see AQH500
ANTIMONYL-2,4-DIHYDROXY-5-HYDROXYMETHYL PYRIMIDINE see AQE300
ANTIMONYL-2,4-DIHYDROXY PYRIMIDINE see AQE305
ANTIMONYL-7-FORMYL-8-HYDROXYQUINOLINE-5-SULPHONATE see AQE320
ANTIMONY LITHIUM THIOMALATENONAHYDRATE see AQE500
ANTIMONYL POTASSIUM TARTRATE see AQG250
ANTIMONY, compounded with NICKEL (1:1) see NCY100
ANTIMONY NITRIDE see AQE750
ANTIMONY ORANGE see AQL500
ANTIMONY(3+) OXIDE see AQF000
ANTIMONY OXIDE see AQF000
ANTIMONY PENTACHLORIDE (DOT) see AQD000
ANTIMONY PENTACHLORIDE see AQD000
ANTIMONY PENTAFLUORIDE (DOT) see AQF250
ANTIMONY(V) PENTAFLUORIDE see AQF250
ANTIMONY PENTAOXIDE see AQF750
ANTIMONY PENTASULFIDE see AQF500

ANTIMONY PENTOXIDE see AQF750
ANTIMONY PERCHLORIDE see AQD000
ANTIMONY PEROXIDE see AQF000
ANTIMONY POTASSIUM DIMETHYLCYSTEINOTARTRATE see AQG000
dl-ANTIMONY POTASSIUM TARTRATE see AQG750
d-ANTIMONY POTASSIUM TARTRATE see AQG500
l-ANTIMONY POTASSIUM TARTRATE see AQH000
meso-ANTIMONY POTASSIUM TARTRATE see AQH250
ANTIMONY POTASSIUM TARTRATE see AQG250
ANTIMONY POWDER (DOT) see AQB750
ANTIMONY PYROCATECHOL SODIUM DISULFONATE see AQH500
ANTIMONY RED see AQF500
ANTIMONY REGULUS see AQB750
ANTIMONY SESQUIOXIDE see AQF000
ANTIMONY SESQUISULFIDE see AQL500
ANTIMONY SODIUM DIMETHYL CYSTEINO TARTRATE see AQH750
ANTIMONY(III) SODIUM GLUCONATE see AQI000
ANTIMONY(V) SODIUM GLUCONATE see AQI250
ANTIMONY SODIUM GLUCONATE see AQH800
ANTIMONY SODIUM OXIDE-l-(+)-TARTRATE see AQI750
ANTIMONY SODIUM PROPYLENE DIAMINE TETRAACETIC ACID DIHYDRATE see AQI500
ANTIMONY SODIUM TARTRATE see AQI750
ANTIMONY(III) SULFATE (2:3) see AQJ250
ANTIMONY SULFIDE see AQF500, AQL500
ANTIMONY TARTRATE see AQJ500
ANTIMONY TELLURIDE see AQL750
ANTIMONY TRIACETATE see AQJ750
ANTIMONY TRIBROMIDE, solid or solution (DOT) see AQK000
ANTIMONY TRIBROMIDE see AQK000
ANTIMONY TRICHLORIDE, liquid (DOT) see AQC500
ANTIMONY TRICHLORIDE, solid (DOT) see AQC500
ANTIMONY TRICHLORIDE, solution (DOT) see AQC500
ANTIMONY TRICHLORIDE see AQC500
ANTIMONY TRICHLORIDE OXIDE see AQK250
ANTIMONY TRIETHYL see AQK500
ANTIMONY TRIFLUORIDE, solid or solution (DOT) see AQE000
ANTIMONY TRIFLUORIDE see AQE000
ANTIMONY TRIHYDRIDE see SLQ000
ANTIMONY TRIIODIDE see AQK750
ANTIMONY TRIMETHYL see AQL000
ANTIMONY TRIOXIDE see AQF000
ANTIMONY TRIPHENYL see AQL250
ANTIMONY TRIPHENYLDICHLORIDE see DGO800
ANTIMONY TRISULFATE see AQJ250
ANTIMONY TRISULFIDE see AQL500
ANTIMONY TRISULFIDE COLLOID see AQL500
ANTIMONY TRITELLURIDE see AQL750
ANTIMONY VERMILION see AQL500
ANTIMONY WHITE see AQF000
ANTIMOONPENTACHLORIDE (DUTCH) see AQD000
ANTIMOONTRICHLORIDE see AQC500
ANTIMOSAN see AQH500
ANTIMUCIN WDR see ABU500
ANTIMYCIN see AQM000, CMS775
ANTIMYCIN A1 see DUO350
ANTIMYCIN A3 see BLX750
ANTIMYCIN A4 see AQM260
ANTIMYCIN A see AQM250
ANTIMYCOIN see AQM500
ANTINOLO RED B see DNT300
ANTINONIN see DUS700
ANTINOSIN see TDE750
ANTIO see DRR200
ANTIOK S see DXG700
ANTI OX see MJO500
ANTIOXIDANT 116 see PFT500
ANTIOXIDANT 1 see MJO500
ANTIOXIDANT 29 see BFW750
ANTIOXIDANT 330 see TMJ000
ANTIOXIDANT 425 see MJN250
ANTIOXIDANT 736 see TFD000
ANTIOXIDANT 754 see IFX200
ANTIOXIDANT AS see TFD500
ANTIOXIDANT D see HLI500
ANTIOXIDANT DBPC see BFW750
ANTIOXIDANT HS see PJQ750
ANTIOXIDANT HSL see PJQ750
ANTIOXIDANT LTDP see TFD500
ANTIOXIDANT MB (CZECH) see BCC500
ANTIOXIDANT No. 33 see DEG000
ANTIOXIDANT PBN see PFT500
ANTIOXIDANT TOD (CZECH) see BLB500
ANTIOXIDANT ZMB see ZIS000
ANTIPAR see DHF600, DII200

ANTI-PELLAGRA VITAMIN see NCQ900
ANTIPERZ see TII500
ANTIPHEN see MJM500
ANTI-PICA see FDA880, HAF400
ANTIPIRICULLIN see AQM250
ANTIPRESSINE DIHYDROCHLORIDE see DQB800
ANTIPREX 461 see ADW200
ANTIPREX A see ADW200
ANTIPYONIN see SFF000
ANTIPYRINE see AQN000
ANTIPYRINE SALICYLATE see AQN250
N-ANTIPYRINYL-2-(DIMETHYLAMINO)PROPIONAMIDE see AMF375
N-((ANTIPYRINYLISOPROPYLAMINO)METHYL)NICOTINAMIDE see AQN500
(ANTIPYRINYLMETHYLAMINO)METHANESULFONIC ACID SODIUM SALT see AMK500
ANTIRAD see AJY250
ANTIRADON see AJY250
ANTIREN see PIJ000
ANTIREX see EAE600
ANTI-RUST see SIQ500
ANTISACER see DKQ000, DNU000
ANTISAL 1a see TGK750
ANTISEPSIN see BMR100
ANTISEPTOL see BEN000
ANTISOL 1 see PCF275
ANTISTERILITY VITAMIN see VSZ450
ANTISTINE see PDC000
ANTISTOMINUM see BBV500
ANTISTREPT see SNM500
ANTI-STRESS see EQL000
ANTITANIL see DME300
ANTI-TETANY SUBSTANCE 10 see DME300
ANTITROMBOSIN see BJZ000
ANTITUBERKULOSUM see ILD000
ANTIULCERA MASTER see CDQ500
ANTI-UV P see HND100
ANTIVERM see PDP250
ANTIVITIUM see DXH250
ANTIXEROPHTHALMIC VITAMIN see VSK600
ANTLERMICIN A see TEF725
ANTOBAN see PIJ500
ANTOFIN see AFT500
ANTOL see EGV000
ANTOMIN see BBV500
ANTORA see PBC250
ANTORPHINE see AFT500
ANTOSTAB see SCA750
ANTOX see AQF000
ANTOXYLIC ACID see ARA250
ANTOZITE 67E see DQV250
ANTOZITE 67 see DQV250
ANTRACOL see ZMA000
ANTRACROMO BROWN D see TKN500
ANTRAMYCIN see API000
ANTRANCINE 12 see BQI000
ANTRAPUROL see DMH400
ANTRENIL see ORQ000
ANTRENYL see ORQ000
ANTRENYL BROMIDE see ORQ000
ANTROMBIN K see WAT209
ANTRYCIDE see AQN625
ANTRYCIDE METHYL SULFATE see AQN625
ANTRYPOL see BAT000
ANTU see AQN635
ANTURAT see AQN635
ANTURIO see APM875
ANTUSSAN see DBE200
ANTX-a see AOO120
ANTYMON (POLISH) see AQB750
ANTYMONOWODOR (POLISH) see SLQ000
ANTYWYLEGACZ see CMF400
ANU see PBX500
ANURAL see MQU750
ANUSPIRAMIN see BRF500
A. NUTTALLIANA see PAM780
ANVITOFF see AJV500
ANXIETIL see MQU750
ANXINE see GGS000
ANXIOLIT see CFZ000
ANZIEF see ZVJ000
ANZON-TMS see AQF000
AO 29 see BFW750
AO-40 see TMJ000
AO 425 see MJN250
AO 754 see IFX200

AO A1 see AGX000
1A-4OA see AKE250
AOAA see ALQ650
A. OBLONGIFOLIA see BOO700
AOH see AGW476
AOH and AME (1:1) see AGW550
AO 4K see BFW750
AOM see ASP250
AOMB see BCC500
A. OPPOSITIFOLIA see BOO700
AORAL see VSK600
AOS see AFN500
4-AP see AMI500
AP-14 see PAM500
AP 43 see DIF200
AP 50 see AQF000
AP-237 see BTA000
AP 407 see AOD250
A1-0109 P see AHE250
AP see PEK250
6-APA see PCU500
APACHLOR see CDS750
A. PACHYPODA see BAF325
APACIL see AMM250
APADODINE see DXX400
APADON see HIM000
APADRIN see MRH209
APAETP see AMD000
A. PALAESTINUM see ITD050
APAMIDE see HIM000
APAMIDON see FAB400
APAMIN see AQN650
APAMINE see AQN650, DBA800
APAP see HIM000
APARKAN see BBV000
APARKAZIN see DHF600
APARSIN see BBQ500
APAS see AMM250
APASCIL see MQU750
APATATE DRAPE see TES750
A. PATENS see PAM780
APAURIN see DCK759
APAVAP see DGP900
APAVINPHOS see MQR750
APAZONE see AQN750
APAZONE DIHYDRATE see ASA000
APC (pharmaceutical) see ARP250
APC see ABG750, DBI800
APCO 2330 see PEY000
APD see DBB500
'APE (HAWAII) see EAI600
APELAGRIN see NCQ900
APESAN see IPU000
APETAIN see BBK500
APEX 4 see BSL600
APEX 462-5 see TNC500
APEXOL see VSK600
APFO see ANP625
APGA see AMG750
APH see ACX750
APHAMITE see PAK000
APHENYLBARBIT see EOK000
APHOLATE see AQO000
APHOSAL see GFA000
APHOX see DOX600
APHOXIDE see TND250
APHRODINE see YBJ000
APHRODINE HYDROCHLORIDE see YBS000
APHROSOL see YBJ000
APHTIRIA see BBQ500
APIGENIN see CDH250
APIGENINE see CDH250
APIGENOL see CDH250
API No.2 FUEL OIL see DHE800
APIOL see AGE500
APIRACHOL see TGJ150
APIRACOHL see TGJ150
APIROLIO 1476 C see PAV600
APLAKIL see CFZ000
APLIDAL see BBQ500
APL-LUSTER see TEX000
APM see XFS600
β-APN see AMB750
APN see AQO000
APO see TND250

APOATROPIN see AQO250
APOATROPINE see AQO250
APOCHOLIC ACID see AQO500
APOCID MILLING RED G see CMM320
APOCODEINE see AQO750
APOCYNAMARIN see SMM500
APOCYNINE see HLQ500
APOLAN see ARQ750
APOLON B₆ see PII100
APOMINE BLACK GX see AQP000
APOMINE GREEN B see CMO840
APOMORFIN see AQP250
APOMORPHINE see AQP250
APONORIN see HII500
APOPEN see PDT500
APOPLON see RDK000
APORMORPHINE see AQP250
APORMORPHINE CHLORIDE see AQP500
6A-β-APORMPHINE-10,11-DIOL HYDROCHLORIDE see AQP500
6A-β-APORPHINE-10,11-DIOL see AQP250
APOSULFATRIM see TKX000
APOTHESINE see AQP750
3-α,16-α-APOVINCAMINIC ACID ETHYL ESTER see EGM100
A-POXIDE see MDQ250
APOZEPAM see DCK759
4-APP see POM600
APPA see PHX250
APPALACHIAN TEA see HGF100
APPEX see RAF100
APPLE SEEDS see AQP875
APPLE of SODOM (extract) see AQP800
APPL-SET see NAK500
APPRESINUM see HGP500
APPRESSIN see HGP495
APRELAZINE see HGP500
APREN S see TES800
APRESAZIDE see HGP500
APRESINE see HGP500
APRESOLIN see HGP495, HGP500
APRESOLINE-ESIDRIX see HGP500
APRESOLINE HYDROCHLORIDE see HGP500
APREZOLIN see HGP495, HGP500
APRICOT PITS see AQP890
APRIDOL see EQL000
APRIL FOOLS see PAM780
APRINDINE HYDROCHLORIDE see FBP850
APRINOX see BEQ625
APROBARBITAL see AFT000
APROBARBITAL SODIUM see BOQ750
APROBARBITONE see AFT000
APROBARBITONE SODIUM see BOQ750
APROBIT see DQA400
APROCARB see PMY300
APROL 160 see MND100
APROL 161 see MND050
APRON 2E see MDM100
APRON see MDM100
APRONAL see IQX000
APRONALIDE see IQX000
APRON FL see MDM100
APROTININ see PAF550
APROZAL see AFT000
6-APS see PCU500
AP-S see ANT000
APSICAL see RDK000
APSIN VK see PDT750
APTAL see CFE250
APTIN see AGW000
APTROL SULFATE see AQQ000
A. PULSATILLA see PAM780
APURIN see ZVJ000
APURINA see DWW000
APUROL see ZVJ000
APV see CBR000
APYONINE AURAMINE BASE see IBB000
AQ 110 see TKX125
AQD see TMJ800
4HAQO see HIY500
AQUA AMMONIA see ANK250
AQUACAL see CAX350
AQUACAT see CNA250
AQUA CERA see HKJ000
AQUACHLORAL see CDO000
AQUACIDE see DWX800
AQUACRINE see EDV000

AQUA-1,2-DIAMINOETHANE DIPEROXO CHROMIUM(IV) see AQQ100
AQUA-1,2-DIAMINOPROPANEDIPEROXOCHROMIUM(IV) DIHYDRATE see AQQ125
AQUAFIL see SCH000
AQUA FORTIS see NED500
AQUAKAY see MMD500
AQUA-KLEEN see DAA800
AQUALINE see ADR000
AQUALOSE see PKE700
AQUA MEPHYTON see VTA000
AQUAMOLLIN see EIV000
AQUAMYCETIN see CDP250
AQUAMYCIN see ACJ250
AQUAPEL (POLYSACCHARIDE) see SLJ500
AQUAPHOR see CLF325
AQUAPLAST see SFO500
AQUA REGIA see HHM000
AQUAREX METHYL see SIB600
AQUARILLS see CFY000
AQUARIUS see CFY000
AQUASOL see VSP000
AQUASYNTH see VSK600
AQUATAG see BDE250
AQUATENSEN see MIV500
AQUATHOL see EAR000
AQUATIN see CLU000
AQUA-VEX see TIX500
AQUAVIRON see TBG000
AQUAZINE see BJP000
AQUILIDE A see POI100
AQUINONE see MMD500
AR2 see AGX000
AR 3 see CBT500
AR-11 see HNP500
AR-12 see MID000
AR-13 see HNR000
AR-16 see TGH665
AR-17 see HNP000
AR-19 see AGB000
AR-21 see HNQ500
AR-22 see BED250
AR-23 see BED500
AR-25 see HNS500
AR-32 see DQY909
AR-33 see HNQ000
AR-35 see HNS000
AR-41 see DQX800
AR-45 see AQQ250
A 60-20R see PJS750
A 60-70R see PJS750
AR 12008 see DIO200
ARA-A see AEH100
ARA-ATP see ARQ500
ARABIC GUM see AQQ500
ARABINOCYTIDINE see AQQ750
9-β-d-ARABINO FURANOSYL ADENINE see AQQ900
9-(β-d-ARABINOFURANOSYL)ADENINE-5′-(DIHYDROGEN PHOSPHATE) see AQQ905
9-β-d-ARABINOFURANOSYLADENINEMONOHYDRATE see AEH100
9-β-d-ARABINOFURANOSYLADENINE 5′-TRIPHOSPHATE see ARQ500
1-β-D-ARABINOFURANOSYL-4-AMINO-2(1H)PYRIMIDINONE see AQQ750
1-β-d-ARABINOFURANOSYL-2,2′-ANHYDRO-CYTOSINE HYDROCHLORIDE see COW900
1-(β-D-ARABINOFURANOSYL)CYTOSINE see AQQ750
1-β-ARABINOFURANOSYLCYTOSINE see AQQ750
1-ARABINOFURANOSYLCYTOSINE see AQQ750
1-β-d-ARABINOFURANOSYLCYTOSINE HYDROCHLORIDE see AQR000
1-β-d-ARABINOFURANOSYLCYTOSINE-5′-PALMITATE see AQS875
1-β-d-ARABINOFURANOSYLCYTOSINE-5′-PALMITOYL ESTER see AQS875
N-(1-β-d-ARABINOFURANOSYL-1,2-DIHYDRO-2-OXO-4-PYRIMIDIN-YL)DOCOSANAMIDE see EAU075
9-β-d-ARABINOFURANOSYL-9H-PURINE-6-AMINE MONOHYDRATE see AEH100
1-β-d-ARABINOFURANOSYL-2′,3′,5′-TRIACETATE see AQR500
9-ARABINOSYLADENINE see AQQ900
β-d-ARABINOSYLADENINE see AQQ900
ARABINOSYLADENINE see AQQ900
5′-ARABINOSYLADENINE MONOPHOSPHATE see AQQ905
ARABINOSYLADENINE MONOPHOSPHATE see AQQ905
β-D-ARABINOSYLCYTOSINE see AQQ750
ARABINOSYLCYTOSINE HYDROCHLORIDE see AQR000
ARABINOSYLCYTOSINE PALMITATE see AQS875
ARABITIN see AQQ750
d-ARABOASCORBIC ACID see SAA025
ARABOASCORBIC ACID see SAA025

ARAB RAT DETH see WAT200
ARA-C see AQQ750
ARACET APV see PKP750
ARACHIC ACID see EAF000
ARACHIDIC ACID see EAF000
ARACHIDONIC ACID see AQS750
ARACHIS OIL see PAO000
ARACID see CJR500
ARACIDE see SOP500
ARA-CP see AQS875
ARA-C PALMITATE see AQS875
ARACTIDINE see AQQ750
ARA-CYTIDINE see AQQ750
ARACYTIDINE-5′-PALMITATE see AQS875
ARACYTIN see AQQ750
ARAGONITE see CAO000
ARALDIT DY 026 see BOS100
ARALDITE ACCELERATOR 062 see DQP800
ARALDITE ERE 1359 see REF000
ARALDITE 6010 mixed with ERR 4205 (1:1) see OPI200
ARALDITE HARDENER 972 see MJQ000
ARALDITE HARDENER HY 951 see TJR000
ARALDITE HY 951 see TJR000
ARALEN see CLD000
ARALEN DIPHOSPHATE see CLD250
ARALEN PHOSPHATE see CLD250
ARALO see PAK000
ARAMINE see HNB875, HNC000
ARAMITE see SOP500
ARAMITEARARAMITE-15W see SOP500
ARANCIO CROMO (ITALIAN) see LCS000
ARASAN see TFS350
ARATAN see MEP250
ARATEN PHOSPHATE see AQT250
ARATHANE see AQT500
ARATRON see SOP500
ARBAPROSTIL see AQT575
ARBESTAB DSTDP see DXG700
ARBITEX see BBQ500
ARBOCEL see CCU150
ARBOCEL BC 200 see CCU150
ARBOCELL B 600/30 see CCU150
ARBOGAL see DSQ000
ARBOL DEL QUITASOL (CUBA) see CDM325
ARBOL de PERU (MEXICO) see PCB300
ARBORICID see BSQ750
ARBOROL see DUS700
ARBOTECT see TEX000
ARBRE FRICASSE (HAITI) see ADG400
ARBUZ see PAG500
ARCACIL see PDT750
ARCADINE see BCA000
ARCASIN see PDT750
ARCHIDONATE see AQS750
ARCHIDYN see RKP000
ARCOBAN see MQU750
ARCOMONOL TABLETS see MQY400
ARCOSOLV see DWT200
ARCOTRATE see PBC250
ARCTON 0 see CBY250
ARCTON 3 see CLR250
ARCTON 4 see CFX500
ARCTON 6 see DFA600
ARCTON 7 see DFL000
ARCTON 9 see TIP500
ARCTON 22 see CFX500
ARCTON 33 see FOO509
ARCTON 63 see FOO000
ARCTON 114 see FOO509
ARCTON see CBY750
ARCTUVIN see HIH000
ARCUM R-S see RDK000
ARDALL see SJO000
ARDAP see RLF350
ARDEX see BBK500
ARDUAN see PII250
ARECA CATECHU Linn., fruit extract see BFW000
ARECA CATECHU Linn., nut extract see BFW000
ARECA CATECHU see BFW000
ARECAIDINE METHYL ESTER see AQT750
ARECA NUT see AQT650
ARECHIN see CLD250
ARECOLINE see AQT750
ARECOLINE BASE see AQT750
ARECOLINE BROMIDE see AQU000

ARECOLINE HYDROBROMIDE see AQU000
ARECOLINE HYDROCHLORIDE see AQU250
AREDION see CKM000
AREGINAL see EKL000
ARELIZ see PDW250
ARESIN see CKD500
ARESKAP 100 see AQU500
ARESKET 300 see AQU750
ARESKLENE 400 see AQV000
ARESOL see RLU000
ARETAN 6 see MEP250
ARETAN see MEP250
ARETAN-NIEUW see BKS810
ARETIT (the phenol) see ACE500
ARETIT see ACE500, BRE500
AREZIN see CKD500
AREZINE see CKD500
ARFICIN see RKP000
ARFONAD see TKW500
ARFONAD CAMPHORSULFONATE see TKW500
ARFONAD ROCHE see TKW500
ARGAMINE see AQW000
ARGEMONE OIL mixed with MUSTARD OIL see OGS000
ARGENT FLUORURE (FRENCH) see SDQ500
ARGENTIC FLUORIDE see SDQ500
ARGENTIUM CREDE see SDI750
ARGENTOUS OXIDE see SDU500
ARGENTUM see SDI500
ARGEZIN see ARQ725
d-ARGININE HYDROCHLORIDE see AQV500
l-ARGININE HYDROCHLORIDE see AQW000
ARGININE HYDROCHLORIDE see AQW000
ARGININE, MONOHYDROCHLORIDE, d- see AQV500
l-ARGININE MONOHYDROCHLORIDE see AQW000
ARGININE MONOHYDROCHLORIDE see AQW000
d-ARGININE, MONOHYDROCHLORIDE (9CI) see AQV500
ARGIVENE see AQW000
ARGO BRAND CORN STARCH see SLJ500
ARGOBYL see TKP100
ARGON see AQW250
ARGONAL see PFM250
ARGUN see AGN000
ARHEOL see OHG000
ARIAVIT RED 2G see CMM300
ARIBINE see MPA050
ARICHIN see CFU750
ARIGAL C see MCB050
ARILAT see CBM750
ARILATE see BAV575, CBM750
ARINAMINE see DXO300
ARIOTOX see TDW500
ARISAN see CKF750
ARISTAMID see SNJ350
ARISTAMIDE see SNJ350
ARISTOCORT see AQX250
ARISTOCORT ACETONIDE see AQX500
ARISTODERM see AQX500
ARISTOGEL see AQX500
ARISTOGYN see SNJ350
ARISTOLICHIA INDICA L., ALCOHOLIC EXTRACT see AQY000
ARISTOLOCHIC ACID see AQY250
ARISTOLOCHIC ACID SODIUM SALT see AQY125
ARISTOLOCHINE see AQY250
ARISTOPHYLLIN see DNC000
ARIZOLE see PIH750, PMQ750
ARKITROPIN see MDL000
ARKLONE P see FOO000
ARKOFIX NG see DTG000
ARKOFIX NM see MCB050
ARKOPAL N-090 see PKF000
ARKOTINE see DAD200
ARKOZAL see BSQ000
ARLACEL 40 see MRJ800
ARLACEL 60 see SKV150
ARLACEL 80 see SKV100
ARLACEL 83 see SKV170
ARLACEL 161 see OAV000
ARLACEL 169 see OAV000
ARLACEL C see SKV170
ARLACIDE G see CDT250
ARLANTHRENE GOLDEN YELLOW see DCZ000
ARLANTHRENE VIOLET 4R see DFN450
ARLEF see TKH750
ARLIDIN HYDROCHLORIDE see DNU200
ARLOSOL GREEN B see BLK000

ARLOSOL YELLOW S see CMS245
ARMAC 18D see OAP300
ARMAC OD see OAP300
ARMAZAL see COH250
ARMCO IRON see IGK800
ARMEEN 18 see OBC000
ARMEEN L-7 see HBM490
ARMEEN 12D see DXW000
ARMEEN 16D see HCO500
ARMEEN 18D see OBC000
ARMEEN DM-12D see DRR800
ARMEEN DM16D see HCP525
ARMEEN DM 18D see DTC400
ARMEEN O see OHM700
ARMENIAN BOLE see IHD000
ARMINE see ENQ000
ARMODOUR see PKQ059
ARMOFILM see OBC000
ARMOFOS see SKN000
ARMOISE OIL see AQY385
ARMOSTAT 801 see OAV000
ARMOTAN MO see SKV100
ARMOTAN MS see SKV150
ARMOTAN PML-20 see PKG000
ARMOTAN PMO-20 see PKL100
ARMSTRONG'S ACID see AQY400
ARMSTRONG'S S ACID see AQY400
ARMYL see MRV250
ARNAUDON'S GREEN see CMK300
ARNAUDON'S GREEN (HEMIHEPTAHYDRATE) see CMK300
ARNEEL 8 see OCW100
ARNICA see AQY500
ARNITE A see PKF750
ARNOSULFAN see SNN300
ARO see CBT750
AROALL see SJO000
AROCHLOR 1221 see PJM000
AROCHLOR 1242 see PJM500
AROCHLOR 1254 see PJN000
AROCHLOR 1260 see PJN250
AROCLOR 54 see CLD250
AROCLOR 1016 see PJL750, PJL800
AROCLOR 1221 see PJL750
AROCLOR 1232 see PJL750, PJM250
AROCLOR 1242 see PJL750, PJM500
AROCLOR 1248 see PJL750, PJM750
AROCLOR 1254 see PJL750, PJN000
AROCLOR 1260 see PJL750, PJN250
AROCLOR 1262 see PJL750, PJN500
AROCLOR 1268 see PJL750, PJN750
AROCLOR 2565 see PJL750, PJO000
AROCLOR 4465 see PJL750, PJO250
AROCLOR 5442 see PJL750, PJP750
AROCLOR see PJL750
AROFLOW see CBT750
AROFT see AEW625
AROFUTO see AEW625
AROGEN see CBT750
MAKAROL see DKA600
AROLON see ADW200
AROMA BLANCA (CUBA, HAWAII) see LED500
AROMATIC AMINES see AQY750
AROMATIC CASTOR OIL see CCP250
AROMATIC SPIRITS of AMMONIA see AQZ000
AROMEX see CBT750
AROMOX DMMC-W see DRS200
ARON see ADW200
ARON A 10H see ADW200
ARON COMPOUND HW see PKQ059
AROSOL see PER000
AROSURF TA 100 see DXG625
AROTONE see CBT750
AROVEL see CBT750
AROVIT see VSP000
ARPEZINE see PIJ500
ARPHONAD see TKW500
ARQUAD 18 see TLW500
ARQUAD 18-50 see TLW500
ARQUAD DM14B-90 see TCA500
ARQUAD DM18B-90 see DTC600
ARQUAD DMMCB-75 see AFP250
ARQUAD R 40 see DXG625
ARQUAD 2HT see QAT550
ARQUAD 2HT75 see QAT550
ARQUEL see DGM875

ARRESIN see CKD500
ARRESTEN see MQQ050
ARRET see CMG000, LIH000
ARRHENAL see DXE600
ARROW see CBT750
ARROWROOT STARCH see SLJ500
ARROW WOOD see MBU825
ARSACETIN see AQZ900
ARSACETIN SODIUM SALT see AQZ900
ARSACETIN SODIUM SALT, TETRAHYDRATE see ARA000
ARSACETIN TETRAHYDRATE see ARA000
ARSAMBIDE see CBJ000
ARSAMIN see ARA500
ARSAMINOL see SAP500
ARSAN see HKC000
4-ARSANILIC ACID see ARA250
p-ARSANILIC ACID see ARA250
ARSANILIC ACID see ARA250
ARSANILIC ACID, N-ACETYL-, SODIUM SALT see AQZ900
p-ARSANILIC ACID, N,N-BIS(2-CHLOROETHYL)- see BIA300
p-ARSANILIC ACID, N,N-BIS(2-HYDROXYETHYL)- see BKD600
p-ARSANILIC ACID, BISMUTH, SODIUM SALT see BKX500
p-ARSANILIC ACID, N,N-DIETHYL- see DIS775
ARSANILIC ACID, MONOSODIUM SALT see ARA500
ARSANILIC ACID SODIUM SALT see ARA500
ARSAPHENAN see ACN250
ARSECLOR see DFX400
ARSECODILE see HKC500
ARSENAMIDE see TFA350
ARSENATE see ARB250
ARSENATE of IRON, FERRIC see IGN000
ARSENATE of IRON, FERROUS see IGM000
ARSENATE of LEAD see LCK000
ARSENENOUS ACID, CALCIUM SALT (2:1) see CAM300
ARSENENOUS ACID, POTASSIUM SALT see PKV500
ARSEN (GERMAN, POLISH) see ARA750
ARSENIATE de MAGNESIUM (FRENCH) see ARD000
ARSENIATE de PLOMB (FRENCH) see ARC750
ARSENIC-75 see ARA750
ARSENIC, metallic (DOT) see ARA750
ARSENIC see ARA750
ARSENIC ACID, liquid (DOT) see ARB250
ARSENIC ACID, solid (DOT) see ARB250, ARC500
m-ARSENIC ACID see ARB000
o-ARSENIC ACID see ARB250
ARSENIC ACID see ARH500
ARSENIC ACID ANHYDRIDE see ARH500
ARSENIC ACID, CALCIUM SALT (2:3) see ARB750
ARSENIC ACID, DISODIUM SALT see ARC000
ARSENIC ACID, DISODIUM SALT, HEPTAHYDRATE see ARC250
ARSENIC ACID, HEMIHYDRATE see ARC500
ARSENIC ACID, LEAD(2+) SALT (2:3) see LCK100
ARSENIC ACID, LEAD SALT see ARC750
ARSENIC ACID, MAGNESIUM SALT see ARD000
ARSENIC ACID, METHYLPHENYL-(9CI) see HMK200
ARSENIC ACID, MONOPOTASSIUM SALT see ARD250
ARSENIC ACID, MONOSODIUM SALT see ARD500, ARD600
ARSENIC ACID, SODIUM SALT see ARD750
ARSENIC ACID, SODIUM SALT (9CI) see ARD500
ARSENIC ACID, TRICESIUM SALT see CDC375
ARSENIC(V) ACID, TRISODIUM SALT, HEPTAHYDRATE (1:3:7) see ARE000
ARSENIC ACID, ZINC SALT see ZDJ000
ARSENICAL solution see FOM050
ARSENICAL DIP, liquid (DOT) see ARE250
ARSENICAL DIP see ARE250
ARSENICAL DUST see ARE500
ARSENICAL FLUE DUST see ARE500, ARE750
ARSENICALS see ARA750, ARF750
ARSENIC ANHYDRIDE see ARH500
ARSENIC BISULFIDE see ARF000
ARSENIC BLACK see ARA750
ARSENIC BLANC see ARI750
ARSENIC(III) BROMIDE see ARF250
ARSENIC BUTTER see ARF500
ARSENIC(III) CHLORIDE see ARF500
ARSENIC CHLORIDE see ARF500
ARSENIC COMPOUNDS see ARF750
ARSENIC DICHLOROETHANE see DFH200
ARSENIC DIETHYL see ARG000
ARSENIC DIMETHYL see ARG250
ARSENIC DISULFIDE see ARJ100
ARSENIC FLUORIDE see ARI250
ARSENIC HEMISELENIDE see ARG500
ARSENIC HYDRID see ARK250
ARSENIC HYDRIDE see ARK250

ARSENIC IODIDE see ARG750
ARSENIC(III) OXIDE see ARI750
ARSENIC(V) OXIDE see ARH500
ARSENIC OXIDE see ARH500, ARI750
ARSENIC PENTASULFIDE see ARH250
ARSENIC PENTOXIDE see ARH500
ARSENIC PHOSPHIDE see ARH750
ARSENIC SESQUIOXIDE see ARI750
ARSENIC SESQUISULFIDE see ARI000
ARSENIC SULFIDE (DOT) see ARJ100
ARSENIC SULFIDE see ARI000
ARSENIC SULFIDE YELLOW see ARI000
ARSENIC SULPHIDE see ARI000
ARSENIC TERSULPHIDE see ARI000
ARSENIC TRIBROMIDE see ARF250
ARSENIC TRIFLUORIDE see ARI250
ARSENIC TRIHYDRIDE see ARK250
ARSENIC TRIIODIDE see ARG750
ARSENIC TRIIODIDE mixed with MERCURIC IODIDE see ARI500
ARSENIC TRIOXIDE see ARI750
ARSENIC TRIOXIDE mixed with SELENIUM DIOXIDE (1:1) see ARJ000
ARSENIC TRISULFIDE (DOT) see ARI000
ARSENIC TRISULFIDE see ARI000, ARJ100
ARSENICUM ALBUM see ARI750
ARSENIC YELLOW see ARI000
ARSENIDES see ARJ250
ARSENIGEN SAURE see ARI750
ARSENIOUS ACID see ARI750
ARSENIOUS ACID, CALCIUM SALT see CAM500
ARSENIOUS ACID ($H_3AsO_3$), TRISODIUM SALT (8CI) see SEY200
ARSENIOUS ACID, SODIUM SALT see ARJ500, SEY500
ARSENIOUS ACID, SODIUM SALT POLYMERS see ARJ500
ARSENIOUS ACID, STRONTIUM SALT see SME500
ARSENIOUS ACID, TRISILVER(1+) SALT see SDM100
ARSENIOUS ACID, ZINC SALT (9CI) see ZDS000
ARSENIOUS CHLORIDE see ARF500
ARSENIOUS and MERCURIC IODIDE, solution (DOT) see ARI500
ARSENIOUS OXIDE see ARI750
ARSENIOUS SULPHIDE see ARI000
ARSENIOUS TRIOXIDE see ARI750
ARSENITE see ARI750
ARSENITE de POTASSIUM (FRENCH) see PKV500
ARSENITE de SODIUM (FRENCH) see SEY500
ARSENIURETTED HYDROGEN see ARK250
ARSENO 39 see ARL000
ARSENO-BISMULAK see BKX500
ARSENOLITE see ARI750
1-(p-ARSENOPHENYL)UREA see CBI500
ARSENOPYRITE see ARJ750
ARSENOSAN see ARL000
p-ARSENOSOANILINE see ARJ755
4-ARSENOSOANILINE, DIHYDRATE see ALV100
ARSENOSOBENZENE see PEG750
p-ARSENOSO-N,N-BIS(2-CHLOROETHYL)ANILINE see ARJ760
p-ARSENOSO-N,N-BIS(2-HYDROXYETHYL)ANILINE see ARJ770
p-ARSENOSO-N,N-DIETHYLANILINE see ARJ800
4-ARSENOSO-2-NITROPHENOL see NHE600
p-ARSENOSOPHENOL see HNG800
p-ARSENOSOTOLUENE see TGV100
ARSENOUS ACID see ARI750
ARSENOUS ACID ANHYDRIDE see ARI750
ARSENOUS ACID, SODIUM SALT (9CI) see SEY500
ARSENOUS ACID, TRISILVER(1+) SALT (9CI) see SDM100
ARSENOUS ACID, TRISODIUM SALT see SEY200
ARSENOUS ANHYDRIDE see ARI750
ARSENOUS BROMIDE see ARF250
ARSENOUS CHLORIDE see ARF500
ARSENOUS FLUORIDE see ARI250
ARSENOUS HYDRIDE see ARK250
ARSENOUS IODIDE see ARG750
ARSENOUS OXIDE see ARI750
ARSENOUS OXIDE ANHYDRIDE see ARI750
ARSENOUS SULFIDE see ARI000
ARSENOUS TRIBROMIDE see ARF250
ARSENOUS TRICHLORIDE (9CI) see ARF500
ARSENOUS TRIIODIDE (9CI) see ARG750
ARSENOWODOR (POLISH) see ARK250
ARSENOXIDE see ARL000
ARSENOXIDE SODIUM see ARJ900
ARSENPHENOLAMINE HYDROCHLORIDE see SAP500
ARSENTRIOXIDE see ARI750
ARSENWASSERSTOFF (GERMAN) see ARK250
ARSEVAN see NCJ500
ARSINE see ARK250

ARSINE, (p-AMINOPHENYL)DICHLORO-, HYDROCHLORIDE see AOR640
ARSINE, (p-AMINOPHENYL)OXO-, DIHYDRATE see ALV100
ARSINE, AMYLDICHLORO- see AOI200
ARSINE BORON TRIBROMIDE see ARK500
ARSINE, sec-BUTYLDICHLORO- see BQY300
ARSINE, CHLORO(2-CHLOROVINYL)PHENYL- see PER600
ARSINE, (2-CHLOROETHYL)DICHLORO- see CGV275
ARSINE, DICHLOROHEPTYL- see HBN600
ARSINE, DICHLOROHEXYL- see HFP600
ARSINE, DICHLOROISOPENTYL- see IHQ100
ARSINE, DICHLOROPENTYL- see AOI200
ARSINE, DICHLOROPROPYL- see PNH650
ARSINE, DIIODOMETHYL- see MGQ775
ARSINE, DIPHENYLHYDROXY- see DVY100
ARSINE, ETHYLENEBIS(DIPHENYL)- see EIQ200
ARSINE, HYDROXYDIPHENYL- see DVY100
ARSINE OXIDE, ALLYLHYDROXYPHENYL- see AGQ775
ARSINE OXIDE, BUTYLHYDROXYISOPROPYL- see BRQ800
ARSINE OXIDE, (o-CHLOROPHENYL)(3-(2,4-DICHLOROPHENOXY)-2-HYDROXY-PROPYL)HYDROXY- see DGA425
ARSINE OXIDE; (m-CHLOROPHENYL)HYDROXY(β-HYDROXYPHENETHYL)- see CKA575
ARSINE OXIDE, (p-CHLOROPHENYL)HYDROXYMETHYL- see CKD800
ARSINE OXIDE, DIBUTYLHYDROXY- see DDV250
ARSINE OXIDE, DIETHYLHYDROXY- see DIS850
ARSINE OXIDE, HYDROXYDIMETHYL-, SODIUM SALT, TRIHYDRATE see HKC550
ARSINE OXIDE, HYDROXY(2-HYDROXYPROPYL)PHENYL- see HNX600
ARSINE OXIDE, HYDROXYISOBUTYLISOPROPYL- see IPS100
ARSINE OXIDE, HYDROXYMETHYLPHENETHYL- see MNU050
ARSINE OXIDE, HYDROXYMETHYLPHENYL- see HMK200
ARSINE OXIDE, HYDROXYMETHYLPROPYL- see MOT800
ARSINE, OXO(4-CARBOXY)PHENYL- see CCI550
ARSINE SELENIDE, TRIMETHYL- see TLH250
ARSINE SULFIDE, DIMETHYLDI- see DQG700
ARSINE-TRI-1-PIPERIDINIUM CHLORIDE see ARK750
ARSINETTE see LCK000, LCK100
ARSINIC ACID, DIBUTYL-(9CI) see DDV250
ARSINIC ACID, DIMETHYL-(9CI) see HKC000
ARSINIC ACID, DIMETHYL-, SODIUM SALT (9CI) see HKC500
ARSINOSOLVIN see ARA500
ARSINOTRIS PIPERIDINIUM TRICHLORIDE see ARK750
ARSINYL see DXE600
ARSION see CEI250
ARSODENT see ARI750
ARSONATE liquid see MRL750
ARSONIC ACID see ABX500
ARSONIC ACID, (4-(ACETYLAMINO)PHENYL)-, MONOSODIUM SALT (9CI) see AQZ900
ARSONIC ACID, (4-AMINOPHENYL)-, MONOSODIUM SALT (9CI) see ARA500
ARSONIC ACID, CALCIUM SALT (1:1) see ARK780
ARSONIC ACID, COPPER(2+) SALT (1:1) (9CI) see CNN500
ARSONIC ACID, (4-HYDROXYPHENYL)-, polymer with FORMALDEHYDE see BCJ150
ARSONIC ACID, POTASSIUM SALT see PKV500
ARSONIC ACID, SODIUM SALT (9CI) see ARJ500
ARSONIUM, (3-HYDROXYPHENYL)DIETHYLMETHYL-, IODIDE, METHYLCARBA-MATE see MID900
ARSONIUM, TETRAPHENYL-, CHLORIDE see TEA300
((p-ARSONOPHENYL)CARBAMOYL)DITHIOCARBAMIC ACID see DXM100
4-ARSONOPHENYLGLYCINAMIDE see CBJ750
p-ARSONOPHENYLUREA see CBJ000
(3-(p-ARSONOPHENYL)UREIDO)DITHIOBENZOIC ACID see ARK800
ARSONOUS DICHLORIDE, ETHYL-(9CI) see DFH200
ARSONOUS DICHLORIDE, METHYL-(9CI) see DFP200
ARSONOUS DICHLORIDE, (1-METHYLPROPYL)-(9CI) see BQY300
ARSONOUS DICHLORIDE, PHENYL-(9CI) see DGB600
ARSONOUS DIIODIDE, METHYL-(9CI) see MGQ775
N-((p-ARSONPHENYL)CARBAMOYL)DITHIOGLYCINE see DXM100
ARSPHEN see ABX500
ARSPHENAMINE see SAP500
ARSPHENAMINE METHYLENESULFOXYLIC ACID SODIUM SALT see NCJ500
ARSPHENOXIDE see ARL000
ARSYCODILE see HKC500
ARSYNAL see DXE600
ART 2 see CBT500
ARTAM see PGG000
ARTANE see BBV000
ARTANE HYDROCHLORIDE see BBV000
ARTANE TRIHEXYPHENIDYL see BBV000
ARTEANNUIN see ARL375
ARTEGODAN see PAH250
ARTEMETHER see ARL425
ARTEMISIA OIL see ARL250
ARTEMISIA OIL (WORMWOOD) see ARL250

ARTEMISINE see ARL375
ARTEMISININ see ARL375
ARTEMISININELACTOL METHYL ETHER see ARL425
dl-ARTERENOL see ARL750
d-ARTERENOL see ARL500
l-ARTERENOL see NNO500
ARTERENOL see NNO500
l-ARTERENOL BITARTRATE see NNO699
dl-ARTERENOL HYDROCHLORIDE see NNP050
ARTERIOFLEXIN see ARQ750
ARTERIOVINCA see VLF000
ARTEROCOLINE see ABO000, CMF250
ARTEROCYN see PLV750
ARTERODY see BBJ750
ARTEROSOL see ARQ750
ARTES see ARQ750
ARTEVIL see ARQ750
d-ARTHIN see VSZ100
ARTHODIBROM see NAG400
ARTHO LM see MLH000
ARTHROBID see SOU550
ARTHROCHIN see CLD000
ARTHROCINE see SOU550
ARTHROPAN see CMG000
ARTHRYTIN OXOATE see AMX250
ARTIC see MIF765
ARTIFICIAL ALMOND OIL see BAY500
ARTIFICIAL ANT OIL see FPQ875
ARTIFICIAL BARITE see BAP000
ARTIFICIAL CINNAMON OIL see CCO750
ARTIFICIAL GUM see DBD800
ARTIFICIAL HEAVY SPAR see BAP000
ARTIFICIAL MUSTARD OIL see AGJ250
ARTIFICIAL SILK BLACK G see CMN240
ARTIFICIAL SILK BLACK GN see CMN240
ARTIFICIAL SILK BLACK GR see CMN240
ARTIFICIAL SWEETENING SUBSTANZ GENDORF 450 see SJN700
ARTISIL BLUE BSG see MGG250
ARTISIL BLUE GREEN GP see DMM400
ARTISIL BLUE SAP see TBG700
ARTISIL BRILLIANT PINK RFS see AKO350
ARTISIL BRILLIANT ROSE 5BP see DBX000
ARTISIL DIRECT RED 3BP see AKE250
ARTISIL DIRECT YELLOW G see AAQ250
ARTISIL ORANGE 3RP see AKP750
ARTISIL RED 3BP see AKE250
ARTISIL VIOLET 2RP see DBP000
ARTISIL YELLOW G see AAQ250
ARTISIL YELLOW 2GN see AAQ250
ARTIZIN see BRF500
ARTO-ESPASMOL see DNU100
ARTOLON see MQU750
ARTOMEY see VCK100
ARTOMYCIN see TBX250
ARTONIL see BBW750
ARTOSIN see BSQ000
ARTOZIN see BSQ000
ARTRACIN see IDA000
ARTRIBID see SOU550
ARTRIL 300 see IIU000
ARTRINOVO see IDA000
ARTRIONA see CNS825
ARTRIVIA see IDA000
ARTRIZONE see BRF500
ARTROBIONE see CMG000
ARTROFLOG see HNI500
ARTROPAN see BRF500
A. RUBRA see BAF325
ARUMEL see FMM000
ARUM (Various Species) see ITD050
ARUSAL see IPU000
ARVIN see VGU700
ARVYNOL see CHG000
ARWIN see VGU700
ARWOOD COPPER see CNI000
ARYL ALKYL POLYETHER ALCOHOL see ARL875
ARYLAM see CBM750
ARZENE see PEG750
AS-15 see SLY500
A.S. 1.398 see BAC020
AS-17665 see NDY500
2-ASe see AMN500
AS see CCU250
ASA 158-5 see PJA130
ASA see ADA725

A.S.A. see ADA725
ASABAINE see DJM800, XCJ000
ASA COMPOUND see ABG750
A.S.A. EMPIRIN see ADA725
ASAGRAEA OFFICINALIS see VHZ000
ASAGRAN see ADA725
ASAHISOL 1527 see AAX250
ASALIN see ARM000
ASALINE see ARM000
ASAMEDOL see DHS200
ASAMID see ENG500
ASARON see IHX400
α-ASARONE see IHX400
ASARONE see IHX400
trans-ASARONE see IHX400
ASARONE, trans- see IHX400
ASARUM CAMPHOR see IHX400
ASATARD see ADA725
ASAZOL see MRL750
ASB 516 see AAX250
ASBEST (GERMAN) see ARM250
ASBESTINE see TAB750
7-45 ASBESTOS see ARM268
ASBESTOS (ACGIH) see ARM260, ARM262, ARM264, ARM268, ARM275, ARM280
ASBESTOS see ARM250
ASBESTOS, ACTINOLITE see ARM260
ASBESTOS, AMOSITE see ARM262
ASBESTOS, ANTHOPHYLITE see ARM264
ASBESTOS, ANTHOPHYLLITE see ARM266
ASBESTOS, CHRYSOTILE see ARM268
ASBESTOS, CROCIDOLITE see ARM275
ASBESTOS FIBER see ARM250
ASBESTOS, TREMOLITE see ARM280
ASC-4 see THW750
ASCABIN see BCM000
ASCABIOL see BCM000
ASCAREX SYRUP see PIJ500
ASCARIDOL see ARM500
ASCARIDOLE (organic peroxide) (DOT) see ARM500
ASCARIDOLE see ARM500
ASCARISIN see ARM500
ASCARYL see HFV500
ASCENSIL see ALC250
ASCEPTICHROME see MCV000
AS 61CL see ADY500
ASCLEPIN see ACF000
ASCOFURANONE see ARM750
ASCOPHEN see ARP250
l(+)-ASCORBIC ACID see ARN000
l-ASCORBIC ACID see ARN000
ASCORBIC ACID see ARN000
ASCORBUTINA see ARN000
ASCORPHYLLINE see HLC000
ASCOSERP see RDK000
ASCOSERPINA see RDK000
ASCOSIN see ARN250
ASCUMAR see ABF750
ASCURON see BJI000
ASEBOTOXIN see AOO375
ASECRYL see GIC000
ASENDIN see AOA095
ASEPSIN see BMR100
ASEPTA HERBAN see HDP500
ASEPTICHROME see MCV000
ASEPTOFORM see HJL500
ASEPTOFORM E see HJL000
ASEPTOFORM P see HNU500
ASEPTOLAN see SAH500
ASEX see SFS000
ASHLENE see NOH000
ASIATICOSIDE see ARN500
ASIDON 3 see TEH500
ASILAN see SNQ500
A 171 (SILANE DERIVATIVE) see TLD000
ASIPRENOL see DMV600
ASL-603 see BMV750
ASMACORIL see PER700
ASMADION see TEH500
ASMALAR see DMV600
ASMATANE MIST see AES000, VGP000
ASMAVAL see TEH500
ASM MB see BCC500
ASOZIN see MGQ750
ASP 47 see SOD100

ASP 51 see TED500
ASPALON see ADA725
ASPAMINOL HYDROCHLORIDE see ARN700
ASPARA see MAI600
l-ASPARAGIC ACID see ARN850
ASPARAGIC ACID see ARN850
l-ASPARAGINASE see ARN800
ASPARAGINASE see ARN800
1-ASPARAGINASE X see ARN800
l-ASPARAGINASI (ITALIAN) see ARN800
ASPARAGINATE CALCIUM see CAM675
l-ASPARAGINE AMIDOHYDROLASE see ARN800
l-ASPARAGINIC ACID see ARN850
ASPARAGINIC ACID see ARN850
ASPARA K see PKV600
ASPARTAME see ARN825
ASPARTAT see MAI600
(S)-ASPARTIC ACID see ARN850
dl-ASPARTIC ACID see ARN830
l-ASPARTIC ACID see ARN850
l-(+)-ASPARTIC ACID see ARN850
(l)-ASPARTIC ACID see ARN850
ASPARTIC ACID see ARN850
ASPARTIC ACID DISODIUM SALT see SEZ350
l-ASPARTIC ACID, DISODIUM SALT (9CI) see SEZ350
l-ASPARTIC ACID, POTASSIUM SALT (9CI) see PKV600
l-ASPARTIC ACID, SODIUM SALT (9CI) see SEZ355
ASPARTYLPHENYLALANINE METHYL ESTER see ARN825
N-l-α-ASPARTYL-l-PHENYLALANINE 1-METHYL ESTER (9CI) see ARN825
ASPERASE see ARN875
ASPERFILLUS ALKALINE PROTEINASE see SBI860
ASPERGILLIC ACID see ARO000
ASPERGILLIN see ARO250
ASPERGILLOPEPTIDASE B see SBI860
ASPERGUM see ADA725
ASPHALT see ARO500
ASPHALT (CUT BACK) see ARO750
ASPHALT, at or above its Fp (DOT) see ARO500
ASPHALT FUMES (ACGIH) see ARO500
ASPHALT, PETROLEUM see ARO500, PCR500
ASPHALTUM see ARO500
A. SPICATA see BAF325
ASPICULAMYCIN see ARP000
ASPIRDROPS see ADA725
ASPIRIN see ADA725
ASPIRIN ACETAMINOPHEN ESTER see SAN600
ASPIRINE see ADA725
ASPIRIN-ISOPROPYLANTIPYRINE see PNR800
ASPIRIN-dl-LYSINE see ARP125
ASPIRIN-NATRIUM (GERMAN) see ADA750
ASPIRIN, PHENACETIN, and CAFFEINE see ARP250
ASPON see TED500
ASPON-CHLORDANE see CDR750
ASPOR see EIR000
ASPORUM see EIR000
ASPRO see ADA725
ASPRON see HBT500
ASSAM TEA see ARP500
ASSIFLAVINE see DBX400
ASSIMIL see MJE760
ASSIPRENOL see DMV600
ASSUGRIN see SGC000
ASSUGRIN FEINUSS see SGC000
ASSUGRIN VOLLSUSS see SGC000
ASTA 5122 see SOD000
ASTA see CQC650
ASTA B518 see CQC650
ASTA C see CNR750
ASTA CD 072 see CNR825
ASTARIL see SLP600
ASTA Z 4828 see TNT500
ASTA Z 4942 see IMH000
ASTA Z 7557 see ARP625
ASTEMIZOL (GERMAN) see ARP675
ASTEMIZOLE see ARP675
ASTERIC see ADA725
ASTEROMYCIN see ARP000
ASTHENTHILO see DKL800
ASTHMA METER MIST see VGP000
ASTHMA WEED see CCJ825
ASTMAHALIN see VGP000
ASTMAMASIT see DNC000
A-STOFF see CDN200
ASTOMIN see MLP250
ASTRA CHRYSOIDINE R see PEK000

ASTRACILLIN see PDD350
ASTRA DIAMOND GREEN GX see BAY750
ASTRAFER see IGT000
ASTRA FUCHSINE B see MAC250
ASTRALON see PKQ059
ASTRANTIAGENIN D see GMG000
ASTRAPHLOXIN see CMM765
ASTRA PHLOXINE see CMM765
ASTRA PHLOXINE G see CMM765
ASTRA PHLOXINE G EXTRA see CMM765
ASTRAPHLOXIN G see CMM765
ASTRATONE see EJQ500
ASTRAZOLO see SNN500
ASTRAZON ORANGE see CMM764
ASTRAZON ORANGE G see CMM764
ASTRAZON PINK FG see CMM768
ASTRESS see CFZ000
ASTRIDINE see CCK125
ASTRINGEN see AHA000
ASTROBAIN see OKS000
ASTROBOT see DGP900
ASTROCAR see DJS200
ASTROMEL NW 6A see MCB050
ASTROPHYLLIN see DNC000
ASTROTIA STOKESII VENOM see SBI890
ASTRUMAL see PLO500
ASTYN see EHP000
ASUCCIN see SMY000
ASUCROL see CKK000
ASUGRYN see SGC000
ASULAM see SNQ500
ASULFIDINE see PPN750
ASULFOX F see SNQ500
ASULOX 40 see SNQ500
ASULOX see SNQ500
ASUNTHOL see CNU750
A. SUPERBA (AUSTRALIA) VENOM see ARU750
A. SUPERBA VENOM see ARV625
ASURO see DBD750
ASVERIN-C see ARP875
ASVERIN CITRATE see ARP875
ASVERINE CITRATE see ARP875
AS XVII see KEA300
ASYMMETRICAL TRIMETHYL BENZENE see TLL750
ASYMMETRIN see FNO000
3,A-T see AMY050
AT 7 see HCL000
A.T. 10 see DME300
AT-90 see AMY050
AT 101 see HID350
AT-290 see PED750
AT 327 see BLV000
AT-581 see ARQ000
AT 717 see PKQ250
AT-2266 see EAU100
o-AT see AIC250
AT see AGK250, AMY050
ATA see AMY050
ATABRINE see ARQ250
ATABRINE DIHYDROCHLORIDE see CFU750
ATABRINE HYDROCHLORIDE see CFU750
ATACTIC POLY(ACRYLIC ACID) see ADW200
ATACTIC POLYPROPYLENE see PMP500
ATACTIC POLYSTYRENE see SMQ500
ATACTIC POLY(VINYL CHLORIDE) see PKQ059
ATADIOL see CKE750
ATALCO C see HCP000
ATALCO O see OBA000
ATALCO S see OAX000
ATAMASCO LILY see RBA500
ATARA see CJR909
ATARAX see CJR909, HOR470
ATARAX DIHYDROCHLORIDE see HOR470
ATARAX HYDROCHLORIDE see VSF000
ATARAXOID see CJR909
ATARAXOID DIHYDROCHLORIDE see HOR470
ATARAZOID see CJR909
ATA-Sb see AQC000
ATAZINA see CJR909
ATAZINAX see ARQ725
ATC see TEV000
AT 327 CITRATE see ARP875
ATCOTIBINE see ILD000
ATCP see PIB900
ATDA see AMR250

ATDA HYDROCHLORIDE see AMR500
ATEC see ADD750
ATECULON see ARQ750
ATEM see IGG000
ATEMPOL see EQL000
ATENOLOL see TAL475
ATENSIN see GGS000
ATENSINE see DCK759
ATERAX see CJR909
ATERAX DIHYDROCHLORIDE see HOR470
ATERIAN see AHO250
ATERIOSAN see ARQ750
ATEROCYN see PLV750
ATEROID see SEH450
ATEROSAN see PPH050
ATGARD see DGP900
ATHAPROPAZINE see DIR000
ATHEBRATE see ARQ750
ATHERILINE see ARQ325
ATHEROLINE see ARQ325
ATHEROLIP see AHA150
ATHEROMIDE see ARQ750
ATHEROPRONT see ARQ750
ATHOPROPAZIN see DIR000
ATHRANID-WIRKSTOFF see ARQ750
ATHROMBIN see WAT220
ATHROMBINE-K see WAT200
ATHROMBON see PFJ750
ATHYLEN (GERMAN) see EIO000
ATHYLENGLYKOL (GERMAN) see EJC500
ATHYLENGLYKOL-MONOATHYLATHER (GERMAN) see EES350
ATHYL-GUSATHION see EKN000
ATHYMIL see BMA625
ATIC VAT BLUE BC see DFN425
ATIC VAT BLUE XRN see IBV050
ATIC VAT BRILLIANT PURPLE 4R see DFN450
ATIC VAT OLIVE R see DUP100
ATILEN see DCK759
ATILON see SNY500
ATIPI see ARQ500
ATIRAN see MEP250
ATIRIN see CCS250
ATIVAN see CFC250
AT Liquid see AMY050
ATLACIDE see SFS000
ATLANTIC see CBT750
ATLANTIC ACID FAST BLUE B see CMM090
ATLANTIC ARTIFICIAL SILK BLACK G see CMN240
ATLANTIC BLACK BD see AQP000
ATLANTIC BLACK C see AQP000
ATLANTIC BLACK E see AQP000
ATLANTIC BLACK EA see AQP000
ATLANTIC BLACK GAC see AQP000
ATLANTIC BLACK GG see AQP000
ATLANTIC BLACK GXCW see AQP000
ATLANTIC BLACK GXOO see AQP000
ATLANTIC BLACK SD see AQP000
ATLANTIC BLUE 2B see CMO000
ATLANTIC BLUE RW see CMO600
ATLANTIC BRILLIANT YELLOW MN see CMP050
ATLANTIC BROWN BCW see CMO820
ATLANTIC BROWN BP see CMO820
ATLANTIC BROWN D 3Y see CMO810
ATLANTIC CONGO RED see SGQ500
ATLANTIC DARK GREEN see CMO830
ATLANTIC FAST RED F see CMO870
ATLANTIC GREEN 2B see CMO840
ATLANTIC GREEN WT see CMO830
ATLANTICHROME YELLOW 2G see SIT850
ATLANTIC RESIN FAST BROWN BRL see CMO750
ATLANTIC SCARLET 3B see CMO875
ATLANTIC SKY BLUE A see CMO500
ATLANTIC VIOLET N see CMP000
ATLAS "A" see SEY500
ATLAS G 924 see SLL000
ATLAS G-2133 see DXY000
ATLAS G-2142 see PJY100
ATLAS G-2144 see PJY100
ATLAS G 2146 see HKJ000
ATLAS G 3802 see PJT300
ATLAS G 3816 see PJT300
ATLASOL SPIRIT RED3 see MMP100
ATLAS WHITE TITANIUM DIOXIDE see TGG760
ATLATEST see TBF750
ATLOX 1087 see PKL100

ATLOX 4862 see BLX000
AuTM see GJC000
ATMONIL see AOR500
ATMOS 150 see OAV000
ATMUL 67 see OAV000
ATMUL 84 see OAV000
ATMUL 124 see OAV000
ATOCIN see PGG000
A″-TOMATIDINE see THG250
ATOMIT see CAO000, CAT775
ATOMITE see CAT775
ATONIN O see ORU500
ATOPHAN see PGG000
ATOPHAN-NATRIUM (GERMAN) see SJH000
ATOPHAN SODIUM see SJH000
ATOREL see IDE000
ATOSIL see DQA400
ATOVER see PPH050
ATOXAN see CBM750
ATOXICOCAINE see AIT250
ATOXYL see ARA500
ATOXYLIC ACID see ARA250
5′-ATP see ARQ500
ATP (nucleotide) see ARQ500
ATP see ARQ500
ATP DISODIUM see AEM100
ATP DISODIUM SALT see AEM100
AT-17 PHOSPHATE see MLP250
ATP Na SALT see AEM250
ATRAL see PJA120
ATRALYMON see ECB200
ATRANEX see ARQ725
ATRASINE see ARQ725
ATRATOL see SFS000
ATRATOL A see ARQ725
ATRATOL 80W see ARQ700
ATRATON see EGD000
ATRATONE see EGD000
ATRAVET see AAF750, ABH500
ATRAXINE see MQU750
ATRAZIN see ARQ725
ATRAZINE see ARQ725
ATRED see ARQ725
ATREX see ARQ725
ATRIPHOS see ARQ500
A. TRISPERMA see TOA275
ATRIVYL see MMN250
ATROCHIN see SBG000
ATROL see DPA000
ATROLEN see ARQ750
ATROMID see ARQ750
ATROMIDIN see ARQ750
ATROMID S see ARQ750
ATROPA BELLADONNA see DAD880
ATROPAMIN see AQO250
ATROPAMINE see AQO250
ATROPIN (GERMAN) see ARR000
ATROPINE see ARR000
(−)-ATROPINE see HOU000
ATROPINE METHOBROMIDE see MGR250
ATROPINE METHONITRATE see MGR500
ATROPINE METHYLBROMIDE see MGR250
ATROPINE METHYL NITRATE see MGR500
ATROPINE OCTABROMIDE see OEM000
ATROPINE-N-OCTYL BROMIDE see OEM000
ATROPINE SULFATE (1:1) see ARR250
ATROPINE SULFATE (2:1) see ARR500
ATROPIN-N-OCTYLBROMID (GERMAN) see OEM000
ATROPIN SIRAN (CZECH) see ARR500
ATROPINSULFAT (GERMAN) see ARR500
ATROPYLTROPEINE see AQO250
ATROQUIN see SBG000
ATROTON see EGD000
ATROVENT see IGG000
ATROVIS see ARQ750
AT SEFEN see DPE000
ATSETOZIN see ABH500
ATTAC 6 see CDV100
ATTAC 6-3 see CDV100
ATTACLAY see PAE750
ATTACLAY X 250 see PAE750
ATTACOTE see PAE750
ATTAGEL 40 see PAE750
ATTAGEL 50 see PAE750
ATTAGEL 150 see PAE750

ATTAGEL see PAE750
ATTAPULGITE see PAE750
ATTAR of ROSE see RNA000
ATTAR ROSE see RNA000
ATTASORB see PAE750
ATUL ACID BLACK 10BX see FAB830
ATUL ACID BLACK BX see FAB830
ATUL ACID GERANINE G see CMM300
ATUL ACID ORANGE II see CMM220
ATUL ACID SCARLET 3R see FMU080
ATUL CONGO RED see SGQ500
ATUL CRYSTAL RED F see HJF500
ATUL DEVELOPED BLACK BT see CMN800
ATUL DIRECT BLACK E see AQP000
ATUL DIRECT BLUE 2B see CMO000
ATUL DIRECT BROWN CN see CMO810
ATUL DIRECT DARK GREEN P see CMO830
ATUL DIRECT GREEN B see CMO840
ATUL DIRECT SKY BLUE see CMO500
ATUL DIRECT VIOLET N see CMP000
ATUL FAST YELLOW R see DOT300
ATUL INDIGO CARMINE see FAE100
ATUL OIL ORANGE T see TGW000
ATUL OIL RED G see OHA000
ATUL ORANGE R see PEJ500
ATUL SUNSET YELLOW FCF see FAG150
ATUL TARTRAZINE see FAG140
ATUL VULCAN FAST PIGMENT ORANGE G see CMS145
ATUMIN see ARR750
ATURBANE HYDROCHLORIDE see ARR875
ATX II see ARS000
ATYSMAL see ENG500
AU 3 see CBT500
AU-95722 see AMD000
l'AUBIER de TILIA SYLVESTRIS (FRENCH) see SAW300
l'AUBIER de TILLEUL (FRENCH) see SAW300
AUBYGEL GS see CCL250
AUBYGUM DM see CCL250
AUCUBA JAPONICA see JBS050
AUGMENTIN (antibiotic) see ARS125
AUGMENTIN see ARS125
'AUKO'I (HAWAII) see CNG825
AULES see TFS350
AULIGEN see BJU000
AURAMINE (MAK) see IBA000, IBB000
AURAMINE BASE see IBB000
AURAMINE HYDROCHLORIDE see IBA000
AURAMINE O (BIOLOGICAL STAIN) see IBA000
AURAMINE YELLOW see IBA000
AURAMYCIN A see ARS130
AURAMYCIN B see ARS135
AURANILE see DKQ000, DNU000
AURANOFIN see ARS150
AURANTIA see HET500
AURANTICA see MRN500
AURANTIN see AEA500
AURATE(1-), TETRACHLORO-, SODIUM see GIZ100
AURATE(1-), TETRACHLORO-, SODIUM, (SP-4-1)-(9CI) see GIZ100
AUREINE see ARS500
AUREMETINE see ARS750
AUREOCICLINA see CMB000
AUREOCINA see CMA750
AUREOCYCLINE see CMB000
AUREOFUSCIN see ART000
AUREOLIC ACID see MQW750
AUREOLIN see PLI750
AUREOMYCIN see CMA750
AUREOMYCIN A-377 see CMA750
AUREOMYCIN HYDROCHLORIDE see CMB000
AUREOMYKOIN see CMA750
AUREOTAN see ART250
AUREOTHIN see DTU200
AURIC CHLORIDE see GIW176
AURICIDINE see GJG000
AURINE-TRICARBOXYLATE d'AMMONIUM (FRENCH) see AGW750
AURINTRICARBOXYLIC ACID AMMONIUM SALT see AGW750
AURIPIGMENT see ARI000
AURLELIC ACID see MQW750
AUROCIDIN see GJG000
AUROLIN see GJG000
AUROMOMYCIN see ART125
AUROMYOSE see ART250
AUROPAN see NBU000
AUROPEX see GJG000
AUROPIN see GJG000

AURORA YELLOW see CAJ750
AUROSAN see GJG000
AUROTAN see ART250
1-AUROTHIO-d-GLUCOPYRANOSE see ART250
AUROTHIOGLUCOSE see ART250
AUROTHION see GJG000
AUROVERTIN see ART500
AURUMINE see ART250
AUSOVIT see DXO300
AUSTIOX see TGG760
AUSTOCYSTIN D see ARU250
AUSTRACIL see CDP250
AUSTRACOL see CDP250
AUSTRALIAN BROWN SNAKE VENOM see TEG650
AUSTRALIAN COPPERHEAD SNAKE VENOM see ARU750
AUSTRALIAN DEATH ADDER SNAKE VENOM see ARU875
AUSTRALIAN GUM see AQQ500
AUSTRALIAN KING BROWN SNAKE VENOM see ARV000
AUSTRALIAN KING COBRA SNAKE VENOM see ARV125
AUSTRALIAN RED-BELLIED BLACK SNAKE VENOM see ARV250
AUSTRALIAN ROUGH SCALED SNAKE VENOM see ARV375
AUSTRALIAN TAIPAN SNAKE VENOM see ARV500
AUSTRALIAN TIGER SNAKE VENOM see ARV550
AUSTRALIS see ARV000
AUSTRALOL see IQZ000
AUSTRANAL see MCK500
AUSTRAPEN see AIV500
AUSTRAPINE see RDK000
AUSTRASTAPH see SLJ050
AUSTRELAPS SUPERBA (AUSTRALIA) VENOM see ARU750
AUSTRELAPS SUPERBA VENOM see ARV625
AUSTRELAPS SUPERBUS VENOM see ARV625
AUSTRIAN CINNABAR see LCS000
AUSTROMINAL see EOK000
AUSTROVIT PP see NCR000
AUTAN see DKC800
AUTHRON see ART250
AUTOMIN see BBV500
AUTOMOBILE EXHAUST CONDENSATE see GCE000
AUTOMOTIVE DIESEL OIL see DHE900, FOP000
AUTUMN CROCUS see CNX800
AUXEOMYCIN see CMB000
AUXILSON see DBC510
AUXINUTRIL see PBB750
AUXOBIL see DYE700
AV00 see AGX000
AV000 see AGX000
AVACAN see AAD875
AVACAN HYDROCHLORIDE see ARV750
AVADEX see DBI200
AVADEX BW see DNS600
AVADYL see NOC000
AVAGAL see DJM800, XCJ000
AVAPENA see CKV625
AVAZYME see CML850
AVELLANO (DOMINICAN REPUBLIC) see TOA275
AVENGE see ARW000
AVENTOX see TJL500
AVERMIN see AOR500
A. VERNALIS see PCU375
AVERSAN see DXH250
AVERTIN see ARW250, THV000
AVERZAN see DXH250
AVESYL see GGS000
A/VI see EBY500
AVIBEST C see ARM268
AVIBON see VSK600
AVICADE see RLF350
AVICEL 101 see CCU150
AVICEL 102 see CCU150
AVICEL see CCU150
AVICEL PH 101 see CCU150
AVICEL PH 105 see CCU150
AVICOL see ARW750, PAX000
AVIL-RETARD see TMK000
AVIOMARIN see DYE600
AVIROL 118 CONC see SIB600
AVISUN see PMP500
AVITA see VSK600
AVITEX C see HCP900
AVITEX SF see HCP900
AVITOL see VSK600
AVITROL see AMI500
AVLOCLOR see CLD000, CLD250
AVLON see DBX400, XAK000

AVLOSULPHONE see SOA500
AVLOTANE see HCI000
AVOCADO OIL see ARW800
AVOCAN see NOC000
AVOGODRITE see PKY000
AVOLIN see DTR200
AVOMINE see DQA400
AVON GREEN A-4379 see BAY750
AVOXYL see GGS000
AW-14′2333 see HOU059
AW-14-2446 see CMW600
AWD 19-166 see BMB125
AWELYSIN see SLW450
AWPA #1 see CMY825
AW 15 (POLYSACCHARIDE) see HKQ100
AX 363 see CAT775
AXEROPHTHAL see VSK985
AXIOM see DIX600
AXM see ABN000
AXURIS see AOR500
AY-5406 see BCA000
AY-6108 see AIV500
AY 8682 see CQH625
AY 9944 see BHN000
AY 21011 see ECX100
AY 22989 see RBK000
AY 24034 see LIU360
AY-25,329 see NMV750
AY-57,062 see ABH500
AY-61122 see MLJ500
AY 61123 see ARQ750
AY-62014 see BPT750
AY 62021 see CLX250
AY-62022 see MBZ100
AY 64043 see ICB000, ICC000
AY 136155 see MBZ100
AYAA see AAX250
AYAF see AAX250
AYAPANIN see MEK300
AYERMATE see MQU750
AYERST 62013 see TJQ333
AYFIVIN see BAC250
AYUSH-47 see ARX125
8-AZAADENINE see AIB340
10-AZAANTHRACENE see ADJ500
9-AZAANTHRACENE see ADJ500
5-AZA-10-ARSENAANTHRACENE CHLORIDE see PDB000
12-AZABENZ(a)ANTHRACENE see BAW750
AZABENZENE see POP250
3-AZABENZONITRILE see NDW515
AZABICYCLANE CITRATE see ARX150
3-AZABICYCLO(3.2.2)NONANE see ARX300
3-AZABICYCLO(3.2.2)NONANE, 3-(CHLOROACETYL)- see CEC100
3-AZABICYCLO(3.2.2)NONANE, 3-(IODOACETYL)- see IDZ100
1-AZABICYCLO(2.2.2)OCTAN-3-OL, BENZILATE (9CI) see QVA000
1-AZABICYCLO(2.2.2)OCTAN-3-OL, BENZYLATE (ESTER), HYDROCHLORIDE
    see QWJ000
1-AZABICYCLO(2.2.2)OCTAN-3-OL (9CI) see QUJ990
1-AZABICYCLO(3.2.1)OCTAN-6-OL DIPHENYLACETATE HYDROCHLORIDE see
    ARX500
AZABICYCLO(2.2.1)OCTAN-3-OL, DIPHENYLACETATE (ester), HYDROGEN
    SULFATE (2:1) DIHYDRATE see QVJ000
1-AZABICYCLO(3.2.1)OCTAN-6-OL-9-FLUORENECARBOXYLATE HYDROCHLO-
    RIDE see ARX750
AZABICYCLOOCTANOL METHYL BROMIDE DIPHENYLACETATE see ARX770
1-(3-AZABICYCLO(3.3.0)OCT-3-YL)-3-(p-TOLYLSULFONYL)UREA see DBL700
AZACITIDINE see ARY000
AZACOSTEROL DIHYDROCHLORIDE see ARX800
AZACOSTEROL HYDROCHLORIDE see ARX800
AZACTAM see ARX875
1-AZACYCLOHEPTANE see HDG000
AZACYCLOHEPTANE see HDG000
2-AZACYCLOHEPTANONE see CBF700
AZACYCLOHEXANE see PIL500
AZACYCLONOL HYDROCHLORIDE see PIY750
β-1-AZACYCLOOCTYLETHYLGUANIDINE SULFATE see GKS000
1-AZACYCLOOCT-2-YL METHYL GUANIDINE see GKO800
1-AZA-2,4-CYCLOPENTADIENE see PPS250
AZACYCLOPENTANE see PPS500
AZACYCLOPROPANE see EJM900
AZACYCLOTRIDECAN-2-ONE see COW930
2-AZACYCLOTRIDECANONE see COW930
5′-AZACYTIDINE see ARY000
5-AZACYTIDINE see ARY000
6-AZACYTIDINE see AMM000

AZACYTIDINE see ARY000
5-AZA-2′-DEOXYCYTIDINE see ARY125
5-AZADEOXYCYTIDINE see ARY125
7-AZADIBENZ(a,h)ANTHRACENE see DCS400
14-AZADIBENZ(a,j)ANTHRACENE see DCS800
7-AZADIBENZ(a,j)ANTHRACENE see DCS600
7-AZA-7H-DIBENZO(c,g)FLUORENE see DCY000
AZADIENO see MJL250
9-AZAFLUORENE see CBN000
AZAFORM see MJL250
8-AZAGUANINE see AJO500
AZAGUANINE-8 see AJO500
AZAGUANINE see AJO500
2-AZAHYPOXANTHINE see ARY500
1-AZAINDENE see ICM000
3-AZAINDOLE see BCB750
AZALEA see RHU500
AZALINE see ARM000
AZALOMYCIN F see ARY750
AZALONE see PDC000
AZAMETHONE see MKU750
AZAMETHONIUM BROMIDE see MKU750
AZAMETON see MKU750
AZAN see AJO500
1-AZANAPHTHALENE see QMJ000
2-AZANAPHTHALENE see IRX000
AZANIDAZOLE see TJF000
AZANIL RED SALT TRD see CLK235
AZANIN see ASB250
AZANOL FAST ACID BLACK 10B see FAB830
3-AZAPENTANE-1,5-DIAMINE see DJG600
AZAPERONE (USDA) see FLU000
AZAPETINE see ARZ000
AZAPETINE PHOSPHATE see AGD500
AZAPICYL see ADA000
AZAPLANT see AMY050
AZAPLANT KOMBI see AMY050
AZAPROPAZON (GERMAN) see AQN750
AZAPROPAZON DIHYDRAT (GERMAN) see ASA000
AZAPROPAZONE (anhydrous) see AQN750
AZAPROPAZONE see ASA000
AZAPROPAZONE SODIUM see ASA250
AZAPROPAZON NATRIUMSALZ (GERMAN) see ASA250
8-AZAPURINE, 6-AMINO- see AIB340
8-AZAPURINE, 6-HYDROXY-2-THIO- see HOJ100
AZARIBINE see THM750
AZASERIN see ASA500
l-AZASERINE see ASA500
AZASERINE see ASA500
6-AZASPIRO(3,4)OCTANE-5,7-DIONE see ASA750
AZA-6-SPIRO(3,4)OCTANE-DIONE-5,7 (FRENCH) see ASA750
4-(3-AZASPIRO(5.5)UNDEC-3-YL)-4′-FLUORO-BUTYROPHENONE HYDROCHLO-
    RIDE see ASA875
AZASTEROL see ARX800
AZATADINE DIMALEATE see DLV800
AZATADINE MELEATE see DLV800
AZATHIOPRINE see ASB250
8-AZA-2-THIOXANTHINE see HOJ100
8-AZATHIOXANTHINE see MLY000
AZATIOPRIN see ASB250
N-(4-AZA-endo-TRICYCLO(5.2.1.5$^{2,6}$)-DECAN-4-YL)-4-CHLORO-3-SULFAMOYL-
    BENZAMIDE see CHK825
6-AZAURACIL see THR750
4(6)-AZAURACIL see THR750
6-AZAURACILRIBOSIDE see RJA000
6-AZAURACIL-β-d-RIBOSIDE see RJA000
6-AZAURIDINE see RJA000
AZAURIDINE see RJA000
8-AZAXANTHINE see THT350
AZBOLEN ASBESTOS see ARM264, ARM266
AZDEL see PMP500
AZDID see BRF500
AZELAIC ACID see ASB750
AZELAIC ACID DI(2-ETHYLHEXYL)ESTER see BJQ500
AZELAIC ACID DIHEXYL ESTER see ASC000
AZELASTIN see ASC130
AZELASTINE see ASC125
AZELASTINE HYDROCHLORIDE see ASC130
AZEPERONE see FLU000
AZEPHEN see ASC250
AZEPINE PHOSPHATE see AGD500
2H-AZEPIN-2-ONE, 1-ACETYLHEXAHYDRO- see ACF100
AZEPROMAZINE see ABH500
AZETALDEHYDSCHWEFLIGSAUREN NATRIUMS (GERMAN) see AAH500
l-2-AZETIDINECARBOXYLIC ACID see ASC500

AZETYLAMINOFLUOREN (GERMAN) see FDR000
AZG see AJO500
AZIDANIL see TBJ250
AZIDE see SFA000
AZIDES see ASC750
AZIDINE BLUE 3B see CMO250
AZIDITHION see ASD000
AZIDOACETIC ACID see ASD375
AZIDOACETONE see ASD500
AZIDOACETO NITRILE see ASE000
AZIDOBENZENE see PEH000
N-AZIDO CARBONYL AZEPINE see ASE250
AZIDOCARBONYL GUANIDINE see ASE500
AZIDOCODEINE see ASE875
AZIDODIMETHYL BORANE see ASF500
2-AZIDO-3,5-DINITROFURAN see ASF625
AZIDODITHIOCARBONIC ACID (DOT) see ASF750
AZIDODITHIOFORMIC ACID see ASF750
(5-α,6-β)-6-AZIDO-4,5-EPOXY-17-METHYL-MORPHINAN-3,14-DIOL see HJE600
2-AZIDOETHANOL NITRATE see ASF800
2-AZIDOETHYL NITRATE see ASF800
AZIDOETHYL NITRATE (DOT) see ASF800
AZIDO FLUORINE see FFA000
2-AZIDO-4-ISOPROPYLAMINO-6-METHYLTHIO-1,3,5-TRIAZINE see ASG250
2-AZIDO-4-ISOPROPYLAMINO-6-METHYLTHIO-s-TRIAZINE see ASG250
N-AZIDO METHYL AMINE see ASG500
2-AZIDOMETHYLBENZENEDIAZONIUM TETRAFLUOROBORATE see ASG625
4-AZIDO-N-(1-METHYLETHYL)-6-(METHYLTHIO)-1,3,5-TRIAZIN-2-AMINE see ASG250
4-AZIDO-N-(1-METHYLETHYL)-6-(METHYLTHIO)-1,3,5-TRIAZIN-2-AMINI see ASG250
AZIDOMORPHINE see ASG675
3-AZIDO-1,2,4-TRIAZOLE see ASH250
5-AZIDOTETRAZOLE see ASH000
AZIDOTHIOCARBONIC ACID see ASF750
AZIJNZUUR (DUTCH) see AAT250
AZIJNZUURANHYDRIDE (DUTCH) see AAX500
AZIMETHYLENE see DCP800
AZIMIDOBENZENE see BDH250
AZIMINOBENZENE see BDH250
AZINDOLE see BCB750
AZINE see POP250
AZINE BRILLIANT BLUE RW see CMO600
AZINE DARK GREEN BH/C see CMO830
AZINE DEEP BLACK EW see AQP000
AZINE DIAZO BLACK BHK see CMN800
AZINE FAST BLACK D see CMN230
AZINE FAST RED FC see CMO870
AZINE GREEN BX see CMO840
AZINE SKY BLUE 5B see CMO500
AZINFOS-ETHYL (DUTCH) see EKN000
AZINFOS-METHYL (DUTCH) see ASH500
AZINOS see EKN000
AZINOTHRICIN see ASH425
AZINPHOS-AETHYL (GERMAN) see EKN000
AZINPHOS ETHYL see EKN000
AZINPHOS-ETILE (ITALIAN) see EKN000
AZINPHOS METHYL, liquid (DOT) see ASH500
AZINPHOS METHYL see ASH500
AZINPHOS-METILE (ITALIAN) see ASH500
AZIONYL see ARQ750
AZIPROTRYN see ASG250
AZIPROTRYNE see ASG250
AZIRANE see EJM900
AZIRIDIN (GERMAN) see EJM900
AZIRIDINE see EJM900
1-AZIRIDINEACETAMIDE, N,N'-(1,4-CYCLOHEXYLENEDIMETHYLENE)BIS- see CPL100
AZIRIDINE, 1-(3-(BIS(2-CHLOROETHYL)AMINO-p-TOLUOYL))- see BHQ760
AZIRIDINE, 1,1'-CARBONYLBIS- see BJP450
1-AZIRIDINECARBOXAMIDE, N,N-DIMETHYL- see EJA100
1-AZIRIDINECARBOXAMIDE, N-METHYL- see EJN400
1-AZIRIDINECARBOXAMIDE, N,N'-HEXAMETHYENEBIS- see HEF500
1-AZIRIDINECARBOXAMIDE, N,N'-1,6-HEXANEDIYLBIS-(9CI) see HEF500
1-AZIRIDINECARBOXANILIDE see PEH250
AZIRIDINE CARBOXYLIC ACID ETHYL ESTER see ASH750
1-AZIRIDINE ETHANOL see ASI000
AZIRIDINE, 1,1'-ISOPHTHALOYLBIS(2-METHYL)- see BLG400
AZIRIDINE, 1,1'-(1,3-PHENYLENEDICARBONYL)BIS(2-METHYL)-(9CI) see BLG400
1-AZIRIDINEPROPANENITRILE see ASI250
1-AZIRIDINE PROPIONITRILE see ASI250
AZIRIDINE-1,3,5,2,4,6-TRIAZATRIPHOSPHORINE DERIVATIVE see AQO000
1-AZIRIDINYL m-(BIS(2-CHLOROETHYL)AMINO)PHENYL KETONE see ASI300
1-(1-AZIRIDINYL)-N-(m-CHLOROPHENYL)FORMAMIDE see CJR250

6-(1-AZIRIDINYL)-4-CHLORO-2-PHENYLPYRIMIDINE see CGW750
2-(1-AZIRIDINYL)ETHANOL see ASI000
4-AZIRIDINYL-3-HYDROXY-1,2-BUTENE see VMA000
1-(1-AZIRIDINYL)-N-(p-METHOXYPHENYL)FORMAMIDE see MFF250
α-(1-AZIRIDINYLMETHYL)BENZYL ALCOHOL see PEH500
2-(1-AZIRIDINYL)-1-PHENYLETHANOL see PEH500
1-AZIRIDINYL PHOSPHINE OXIDE (TRIS) (DOT) see TND250
1-AZIRIDINYLPHOSPHONITRILE TRIMER see AQO000
AZIRIDINYLQUINONE see ASK875
2-(1-AZIRIDINYL)-1-VINYLETHANOL see VMA000
AZIRIDYL BENZOQUINONE see BDC750
AZIDOTRIMETHYLSILANE see TMF100
AZIRPOTRYNE see ASG250
AZIUM see SFA000, SOW000
AZO-33 see ZKA000
AZOAETHAN (GERMAN) see ASN250
AZOAMINE PINK O see NEQ000
AZOAMINE RED ZH see NEO500
AZOAMINE SCARLET see NEQ500
p-AZOANILINE see ASK925
AZOBASE DCA see DEO295
AZOBASE MNA see NEN500
AZOBENZEEN (DUTCH) see ASL250
AZOBENZENE see ASL250
AZOBENZENE, 4-ANILINO- see PEI800
AZOBENZENE OXIDE see ASO750
AZOBENZIDE see ASL250
AZOBENZOL see ASL250
4,4'-AZOBISBENZENAMINE see ASK925
AZOBISBENZENE see ASL250
4,4'-AZOBIS(4-CYANOPENTANOIC ACID) see ASL500
4,4'-AZOBIS(4-CYANOVALERIC ACID) see ASL500
AZOBIS(CYANOVALERIC ACID) see ASL500
α,α'-AZOBISISOBUTYLONITRILE see ASL750
AZOBISISOBUTYLONITRILE see ASL750
AZOBIS ISOBUTYRAMIDE HYDROCHLORIDE see ASM000
2,2'-AZOBIS(ISOBUTYRONITRILE) see ASL750
AZOBISISOBUTYRONITRILE see ASL750
2,2'-AZOBIS(2-METHYLPROPANIMIDAMIDE) DIHYDROCHLORIDE see ASM050
2,2'-AZOBIS(2-METHYLPROPIONAMIDINE) DIHYDROCHLORIDE see ASM050
2,2'-AZOBIS(2-METHYLPROPIONITRILE) see ASL750
AZOCARBONITRILE see DGS300
AZOCARD BLACK EW see AQP000
AZOCARD BLUE 2B see CMO000
AZOCARD BLUE BH see CMN800
AZOCARD DARK GREEN B see CMO830
AZOCARD FAST RED F see CMO870
AZOCARD GREEN B see CMO840
AZOCARD RED CONGO see SGQ500
AZOCARD VIOLET N see CMP000
AZOCHLORAMIDE see ASM250
AZOCHROMOL BLUE BLACK EB see CMP880
AZOCHROMOL YELLOW 5G see SIT850
AZOCYCLOTIN see THT500
AZO DARK BLUE C 2B see FAB830
AZO DARK BLUE HR see FAB830
AZO DARK BLUE S see FAB830
AZO DARK BLUE SH see FAB830
4,4'-AZODIANILINE see ASK925
AZODIBENZENE see ASL250
AZODIBENZENEAZOFUME see ASL250
AZODICARBONAMIDE see ASM300
2,2'-AZODIISOBUTYRONITRILE see ASL750
α,α'-AZODIISOBUTYRONITRILE see ASL750
AZODIISOBUTYRONITRILE (DOT) see ASL750
AZODIISOBUTYRONITRILE see ASL750
AZODINE see PDC250
N,N'-(AZODI-4,1-PHENYLENE)BIS(N-METHYLACETAMIDE) see DQH200
AZODIUM see PDC250
AZODOX-55 see ZKA000
AZODRIN-71 see MRH209
AZODRIN (OSHA) see MRH209
"AZODRIN" see ASN000
AZODRIN PESTICIDE see MRH209
AZO DYE No. 6945 see MIH500
AZODYNE see PDC250
AZOENE FAST BLUE BASE see DCJ200
AZOENE FAST ORANGE GR BASE see NEO000
AZOENE FAST ORANGE RD SALT see CEG800
AZOENE FAST RED B BASE see NEQ000
AZOENE FAST RED 3GL BASE see KDA050
AZOENE FAST RED KB BASE see CLK225
AZOENE FAST RED TR BASE see CLK220
AZOENE FAST RED TR SALT see CLK235
AZOENE FAST SCARLET GC BASE see NMP500

AZOENE FAST SCARLET GC SALT see NMP500
AZOEN FAST SCARLET 2G BASE see DEO295
AZO ETHANE see ASN250
AZOFENE see BDJ250
AZOFENOL 4K see CMP090
AZOFIX BLUE B SALT see DCJ200
AZOFIX SCARLET GG SALT see DEO295
AZOFIX SCARLET G SALT see NMP500
AZOFORMALDOXIME see ASN375
AZOFOS see MNH000
AZO FUCHSINE see CMS231
AZO-GANTANOL see SNK000
AZO GANTRISIN see PDC250, SNN500
AZO GASTANOL see PDC250
AZOGEN BLACK D see CMN230
AZOGEN DEVELOPER A see NAX000
AZOGEN DEVELOPER H see TGL750
AZOGENE ECARLATE R see NEQ500
AZOGENE FAST ORANGE GR see NEO000
AZOGENE FAST RED TR see CLK220, CLK235
AZOGENE FAST SCARLET G see NMP500
AZO GERANINE 2G see CMM300
AZO GERANINE 2GA see CMM300
AZOGNE FAST BLUE B see DCJ200
AZO GRENADINE see CMS231
AZOIC DIAZO COMPONENT 12 see NMP500
AZOIC DIAZO COMPONENT 32 see CLK225
AZOIC DIAZO COMPONENT 37 see NEO500
AZOIC DIAZO COMPONENT 46 see CLK200
AZOIC DIAZO COMPONENT 6 see NEO000
AZOIC DIAZO COMPONENT 9 see KDA050
AZOIC DIAZO COMPONENT 11 BASE see CLK220, CLK235
AZOIC DIAZO COMPONENT 13 BASE see NEQ500
AZOIC RED 36 see MGO750
AZOIMIDE see HHG500
AZOLAN see AMY050
AZOLASTONE see MQY110
AZOLE see AMY050, PPS250
AZOLID see BRF500
AZOLMETAZIN see SNJ000
AZO MAGENTA G see CMS231
AZO-MANDELAMINE see PDC250
AZOMETHANE see ASN400
AZO MILLING RED G see CMM325
AZOMINE see PDC250
AZOMINE BLACK BH see CMN800
AZOMINE BLACK EWO see AQP000
AZOMINE BLUE 2B see CMO000
AZOMINE GREEN B see CMO840
AZOMYCIN see NHG000
1,1'-AZONAPHTHALENE see ASN500
2,2'-AZONAPHTHALENE see ASN750
AZONAPHTHOL RED J see CMM300
8-AZONIABICYCLO(3.2.1)OCTANE, 8-(2-FLUOROETHYL)-3-((HYDROXYDIPHE-NYLACETYL)OXY)-8-METHYL-, BROMIDE, (endo,syn)-, MONOHYDRATE see FMR300
AZONIASPIRO(3-α-BENZILOYLOXY-NORTROPAN-8,1'-PYRROLIDINE)-CHLORIDE see KEA300
AZONIASPIRO COMPOUND XVII see KEA300
AZOPHENYLENE see PDB500
AZOPHLOXIN see CMM300
AZOPHLOXINE see CMM300
AZO PHLOXINE GA see CMM300
AZO PHLOXINE GA-CF see CMM300
AZOPHOS see MNH000
AZOPYRIN see PPN750
AZO RED R see FAG020
AZORESORCIN see HNG500
AZO RHODINE 2G see CMM300
AZORUBIN see HJF500
AZOSALT R see PFU500
AZOSEMIDE see ASO375
AZOSEPTALE see TEX250
AZOSSIBENZENE (ITALIAN) see ASO750
AZO-STANDARD see PDC250
AZO-STAT see PDC250
AZOSULFIZIN see SNN500
AZOTA CABALLO see GIW200
AZOTE (FRENCH) see NGR500
AZOTHIOPRINE see ASB250
AZOTIC ACID see NED500
AZOTO (ITALIAN) see NGR500
2:3'-AZOTOLUENE see DQH000
AZOTOMYCIN see ASO510, ASO501
AZOTOMYCIN SODIUM see ASO510

AZOTOMYCIN SODIUM SALT see ASO510
AZOTOWY KWAS (POLISH) see NED500
AZOTOX see DAD200
AZOTOYPERITE see BIE500
AZOTREX see PDC250
AZOTURE de SODIUM (FRENCH) see SFA000
AZOTU TLENKI (POLISH) see NGT500
AZOXYAETHAN (GERMAN) see ASP000
AZOXYBENZEEN (DUTCH) see ASO750
AZOXYBENZENE see ASO750
AZOXYBENZIDE see ASO750
AZOXYBENZOL (GERMAN) see ASO750
AZOXYDIBENZENE see ASO750
AZOXYETHANE see ASP000
AZOXYISOPROPANE see ASP510
AZOXYMETHANE see ASP250
1,1'-AZOXYPROPANE see ASP500
1-AZOXYPROPANE see ASP500
2-AZOXYPROPANE see ASP510
AZQ see ASK875
AZS see ASA500
AZTEC BPO see BDS000
AZTHREONAM see ARX875
AZTREONAM see ARX875
AZUCENA de MEJICO (MEXICO) see AHI635
AZULENE see ASP600
AZULENE, 1,2,3,4,5,6,7,8-OCTAHYDRO-1,4-DIMETHYL-7-(1-METHYLETHYLI-DENE)-, MONOEPOXIDE see EBU100
AZULENO(5,6,7-cd)PHENALENE see ASP750
AZULFIDINE see PPN750
AZULON see DSJ800
6-AZUR see RJA000
AZUR see RJA000
AZURE A see DUG700
AZUREN see ILD000
AZURENE see BNU725
6-AZURIDINE see RJA000
AZUROL ALIZARINOVY SW (CZECH) see DKX800
AZZURRO DIRETTO 3B see CMO250


B-9 see SNA500
B10 see SFO500
B-12 see VSZ000
23B see DHY400
B-28 see AJO500
B32 see HCL000
B-45 see AOF250
B-71 see PKC000
B 75 see IGS000
B/77 see DNX600
B-325 see BEX750
B-343 see IHQ100
B 377 see DVC200
B-436 see PEV750
B-500 see QMJ000
B 518 see CQC650
B 577 see HKK000
B586 see CKE750
B-622 see DEV800
693B see SLP600
B 995 see DQD400
B 1500 see MCA100
B-1,776 see BSH250
B-1843 see BLG500
B2740 see THD275
B-3015 see SAZ000
Bi 3411 see CDO000
B-4130 see AAI750
B 4992 see BKS810
B 5333 see CFY250
B-5477 see EAT810
B-9002 see HMV000
B-10094 see CON300
B 11163 see DTQ800
B-11420 see IGA000
B-14437 see SAU000
B-15000 see IFY000
B 17476 see RQF350
B-21085 see GEO200
B-66256 see USJ000
B 66347 see CLW625
B 67347 see CLW625
70-314B see BQN600
B 77488 see BAT750

BALSAMS, CANADA see FBS200
BALSAMS, COPAIBA see CNH792
BALSAMS, TOLU see BAF000
BALSAM TOLU see BAF000
BAMBERMYCIN see MRA250
BAMBICORT see HHQ800
BAMCH see BGT750
BAMD 400 see MQU750
BAMETAN SULFATE see BOV825
BAMETHANE see BQF250
BAMETHAN SULFATE see BOV825
BAMIFYLLINE HYDROCHLORIDE see THL750
B-AMIN see TES750
BAMIPHYLLINE HYDROCHLORIDE see THL750
BAMN see BRX500
BANABIN see CKF500
BANABIN-SINTYAL see CKF500
BANANA OIL see IHO850
BANANOTE see MDW750
BANASIL see RDK000
BANCEMINE 115-60 see MCB050
BANCEMINE 125-60 see MCB050
BANCEMINE SM 947 see MCB050
BANCEMINE SM 975 see MCB050
BANCEMINE SM 970 see MCB050
BANDANE see BAF250
BANDOL see CBQ625
BANEBERRY see BAF325
BANEX see MEL500
BANGTON see CBG000
BAN-HOE see CBM000
BANICOL see BEL900
BANISIL see RDK000
BANISTERINE see HAI500
BANLEN see MEL500
BANMINTH see TCW750
BANOCIDE see DIW200
BANOL see CGI500
BANOL TUCO SOK see CGI500
BANOL TURF FUNGICIDE see PNI250
BANROT see PEW750
BANTENOL see MHL000
BANTHIN see DJM800, XCJ000
BANTHINE see XCJ000
BANTHINE BROMIDE see DJM800, XCJ000
BANTHIONINE see MDT740
BANTROL see HKB500
BANUCALAD (HAWAII) see TOA275
BANVEL see MEL500
BANVEL HERBICIDE see MEL500
BANVEL T see TIK000
6-BAP see BDX090
BAP see BDX090, NJM500
BAPC see BAB250
BAP (GROWTH STIMULANT) see BDX090
2-BAP HYDROCHLORIDE see BEA000
BAPN see AMB500
BAPN FUMARATE see AMB750
BAPTITOXIN see CQL500
BAPTITOXINE see CQL500
BARACOUMIN see BJZ000
BARAMINE see BBV500
BARATOL see TLG000
BARAZAE see SNN500
BARAZAN see BAB625
BARBADOS GOOSEBERRY see JBS100
BARBADOS LILY see AHI635
BARBADOS NUT see CNR135
BARBADOS PRIDE see CAK325
BARBALLYL see AFS500
BARBALOIN see BAF825
BARBAMATE see CEW500
BARBAMIL see AMX750
BARBAMYL see AMX750
BARBAMYL ACID see AMX750
BARBAN see CEW500
BARBANE see CEW500
BARBAPIL see EOK000
BARBASCO see RNZ000
BARBENYL see EOK000
BARBEXACLONE see CPO500
BARBEXACLONUM see CPO500
BARBIDAL see AFS500
BARBIDORM see ERD500
BARBILEHAE (BARBILETTAE) see EOK000

BARBIMON see AMK250
BARBIPHENYL see EOK000
BARBITA see EOK000
BARBITAL see BAG000, BAG500
BARBITAL Na see BAG250
BARBITAL SODIUM see BAG250, BAG500
BARBITAL SOLUBLE see BAG250
BARBITONE see BAG000, BAG500
BARBITONE SODIUM see BAG250
BARBITURATES see BAG500
BARBITURIC ACID, 1-(1-(1-CYCLOHEXYL-N-METHYL-2-PROPANAMINE)-5-ETH-YL-5-PHENYL see CPO500
BARBITURIC ACID, 5-NITRO- see NET550
BARBONAL see EOK000
BARBONIN HYDROCHLORIDE see PAH260
BARBOPHEN see EOK000
BARBOSEC see SBM500, SBN000
BARDAC 22 see DGX200
BAR-DEX see AOB500
BARDIOL see EDO000
BARECO POLYWAX 2000 see PJS750
BARECO WAX C 7500 see PJS750
m-BARENE see NBV100
BARHIST see DPJ400
BARIDIUM see PDC250
BARIDOL see BAP000
B. ARIETANS VENOM see BLV075
BARIO (PEROSSIDO di) (ITALIAN) see BAO250
BARITE see BAP000
BARITOP see BAP000
BARITRATE see PBC250
BARIUM see BAH250
BARIUM ACETATE see BAH500
BARIUM ACETYLIDE see BAH750
BARIUM AZIDE see BAI000
BARIUM AZIDE, wetted with not <50% water, by weight (UN 1571) (DOT) see BAI000
BARIUM AZIDE, dry or wetted with <50% water, by weight (UN 0224) (DOT) see BAI000
BARIUM BENZOATE see BAI500
BARIUM BICHROMATE see BAL500
BARIUM BINOXIDE see BAO250
BARIUM BIS(TETRAFLUOROBORATE) see BAL750
BARIUM BROMATE see BAI750
BARIUM CADMIUM LAURATE see BAI770
BARIUM CADMIUM STEARATE see BAI800
BARIUM CAPRYLATE see BAI825
BARIUM CARBIDE see BAJ000
BARIUM CARBONATE (1:1) see BAJ250
BARIUM CARBONATE see BAJ250
BARIUM CHLORATE see BAJ500
BARIUM CHLORIDE see BAK000
BARIUM CHLORITE see BAK125
BARIUM CHROMATE (1:1) see BAK250
BARIUM CHROMATE(VI) see BAK250
BARIUM CHROMATE OXIDE see BAK250
BARIUM COMPOUNDS (soluble) see BAK500
BARIUM CYANIDE, solid (DOT) see BAK750
BARIUM CYANIDE see BAK750
BARIUM CYANOPLATINITE see BAL000
BARIUM DIACETATE see BAH500
BARIUM DIAZIDE see BAL250
BARIUM DIBENZYLPHOSPHATE see BAL275
BARIUM DICHLORIDE see BAK000
BARIUM DICHROMATE see BAL500
BARIUM DICYANIDE see BAK750
BARIUM DINITRATE see BAN250
BARIUM DIOXIDE see BAO250
BARIUM DISTEARATE see BAO825
BARIUM FLUOBORATE see BAL750
BARIUM FLUORIDE see BAM000
BARIUM FLUOROSILICATE see BAO750
BARIUM FLUOSILICATE see BAO750
BARIUM HEXAFLUOROSILICATE(2-) see BAO750
BARIUM HEXAFLUOROSILICATE see BAO750
BARIUM HYDRIDE see BAM250
BARIUM HYDROXIDE see BAM500
BARIUM HYPOPHOSPHITE see BAM750
BARIUM IODATE see BAN000
BARIUM MONOXIDE see BAO000
BARIUM(II) NITRATE (1:2) see BAN250
BARIUM NITRATE (DOT) see BAN250
BARIUM NITRIDE see BAN500
BARIUM NITROSYLPENTACYANOFERRATE see PAY600
BARIUM OCTANOATE see BAI825

BARIUM OCTOATE see BAI825
BARIUM OXIDE see BAO000
BARIUM PERCHLORATE (DOT) see PCD750
BARIUM PERMANGANATE see PCK000
BARIUMPEROXID (GERMAN) see BAO250
BARIUM PEROXIDE see BAO250
BARIUMPEROXYDE (DUTCH) see BAO250
BARIUMPOLYSULFID see BAO300
BARIUM POLYSULFIDE see BAO300
BARIUM PROTOXIDE see BAO000
BARIUM RHODANIDE see BAO500
BARIUMSILICOFLUORID see BAO750
BARIUM SILICOFLUORIDE see BAO750
BARIUM SILICON FLUORIDE see BAO750
BARIUM STEARATE see BAO825
BARIUM STYPHNATE see BAO900
BARIUM SULFATE see BAP000
BARIUM SULFIDE see BAO300, BAP250
BARIUM SUPEROXIDE see BAO250
BARIUM TETRAFLUOROBORATE see BAL750
BARIUM THIOCYANATE see BAP500
BARIUM ZIRCONATE see BAP750
BARIUM ZIRCONIUM(IV) OXIDE see BAP750
BARIUM ZIRCONIUM OXIDE see BAP750
BARIUM ZIRCONIUM TRIOXIDE see BAP750
BARIZON see MOV000
BARLENE 125 see DRR800
BARNETIL see EPD100
BAROLUB FTO see HOG000
BAROS CAMPHOR see BMD000
BAROSPERSE see BAP000
BAROTRAST see BAP000
BARPENTAL see NBU000
BARQUAT MB-50 see AFP250
BARQUAT MB 80 see QAT520
BARQUAT SB-25 see DTC600
BARQUINOL see CHR500
BARRAGE see DAA800
BARRA SUPER see BLX000
BARRICADE see RLF350
BARSEB HC see CNS750
BAR-TIME see BBK250
BARTOL see EOK000
BARYTA see BAO000
BARYTA WHITE see BAP000
BARYTA YELLOW see BAK250
BARYTES see BAP000
BARYUM FLUORURE (FRENCH) see BAM000
BAS 2903H see CDS275
BAS-3050 see MNS500
BAS 3220 see DJV000
BAS-3460 see MHC750
BAS 4239 see CEQ500
BAS 67054 see MHC750
BAS see BEM750
BASACRYL RED GL see CMM770
BASAGRAN see MJY500
BASALIN see FDA900
BASAMAIZE see CDS275
BASAMID see DSB200
BASAMID-FLUID see VFU000
BASAMID G see DSB200
BASAMID-GRANULAR see DSB200
BASAMID P see DSB200
BASAMID-PUDER see DSB200
BASANITE see BRE500
BASCHEM 12 see MAG750
BASCOREZ see AAX250
BASCURAT see BOV825
BASE 661 see SMR000
BASECIL see MPW500
BASEDOL see AMS250
BASE LP 12 see DXY000
BASE OIL see PCR250
BASERGIN see LJL000
BASETHYRIN see MPW500
BAS 352 F see RMA000
BAS 2203F see TJJ500
BAS 2205-F see DUJ400
BAS 32500F see PEX500
BAS 36801F see MAP300
BASF see UTU500
BASFAPON see DGI400
BASFAPON B see DGI400, DGI600
BASFAPON/BASFAPON N see DGI400

BASF-GRUNKUPFER see CNK559
BASF III see SMQ500
BASF-MANEB SPRITZPULVER see MAS500
BASFUNGIN see MLX850
BASFUNGINE see MLX850
BASF URSOL D see PEY500
BASF URSOL EG see ALT500
BASF URSOL ERN see NAW500
BASF URSOL 3GA see ALT000
BASF URSOL P BASE see ALT250
BAS-290-H see CDS275
BAS 351-H see MJY500
BAS 392-H see FDA900
BAS 2900H see CDS275
BASIC ALUMINUM CHLORATE see AHA000
BASIC BISMUTH NITRATE see BKW100
BASIC BLUE 9 see BJI250
BASIC BLUEK see VKA600
BASIC BRIGHT GREEN see BAY750
BASIC CHROMIC SULFATE see NBW000
BASIC CHROMIC SULPHATE see NBW000
BASIC CHROMIUM CARBONATE see CMJ100
BASIC CHROMIUM SULFATE see NBW000
BASIC CHROMIUM SULPHATE see NBW000
BASIC COPPER CARBONATE see CNJ750
BASIC COPPER CHLORIDE see CNK559
BASIC CUPRIC CARBONATE see CNJ750
BASIC FUCHSIN see MAC250
BASIC FUCHSINE see MAC250
BASIC GREEN 4 see AFG500
BASIC LEAD ACETATE see LCH000
BASIC LEAD CHROMATE see LCS000
BASIC MAGENTA see MAC250
BASIC MAGENTA E-200 see MAC250
BASIC MERCURIC SULFATE see MDG000
BASIC NICKEL(II) CARBONATE see NCY600
BASIC NICKEL CARBONATE see NCY500
BASIC ORANGE 21 see CMM764
BASIC ORANGE 3RN see BAQ250, BJF000
BASIC PANCREATIC TRYPSIN INHIBITOR see PAF550
BASIC PARAFUCHSINE see RMK020
BASIC RED 1 see RGW000
BASIC RED 2 see GJI400
BASIC RED 13 see CMM768
BASIC RED 29 see CMM770
BASIC RHODAMINE YELLOW see RGW000
BASIC RHODAMINIC YELLOW see RGW000
BASIC ROSE 2S see CMM768
BASIC VIOLET 14 see MAC250
BASIC VIOLET 10 see FAG070
BASIC VIOLET 3 see AOR500
BASIC VIOLET BN see AOR500
BASIC VIOLET K see MQN025
BASIC ZINC CHROMATE see ZFJ100, ZFJ130
BASIC ZIRCONIUM CHLORIDE see ZSJ000
BASIL OIL see BAR250
BASIL OIL, EUROPEAN TYPE (FCC) see BAR250
BASIL OIL, SWEET see BAR250
BASINEX see DGI400
BASKET FLOWER see BAR325
BASLE GREEN see COF500
BASOFORTINA see MJV750, PAM000
BASOLAN see MCO500
BASON see BNK350
BASORA CORRA see BAR500
BASOTECT see MCB050
B. ASPER VENOM see BMI000
BASSA see MOV000
BASSINET (CANADA) see FBS100
BASTARD ACACIA see GJU475
BASUDIN see DCM750
BASUDIN 10 G see DCM750
BATASAN see ABX250
BATAZINA see BJP000
BATEL see BFW250
BATHYRAN see MCK500
BATILOL see GGA915
BATRACHOTOXIN see BAR750
BATRAFEN see BAR800
BATRILEX see PAX000
BATROXOBIN see RDA350
B. ATROX VENOM see BMI125
BATTRE AUTOUR (HAITI) see JBS100
BATYL ALCOHOL see GGA915
BAU see CBT500

BAUMYCIN A1 see BAR825
BAUMYCIN A2 see BAR830
BAUXITE RESIDUE see IHD000
BAVISTIN see MHC750
BAX see BAU750
BAXACOR see DHS200
BAXARYTMON see PMJ525
BAX 1400Z see PMO250
BAX 2793Z see THL750
BAY-E-393 see SOD100
BAY 1040 see AEC750
BAY 1470 see DMW000
BAY 1521 see DPH600
BAY 2353 see DFV400
BAY 3517 see AJV500
BAY 4059 see BOL325
BAY 4503 see PMX250
BAY 4934 see MGQ750
BAY e 5009 see EMR600
BAY 5097 see MRX500
BAY 5621 see BAT750
BAY 5821 see DJY200
BAY-a 7168 see NDY600
BAY 9010 see PMY300
BAY 9015 see DFD000
BAY 9017 see DJR800
BAY 9026 see DST000
BAY 9027 see ASH500
BAY 10756 see DAO600
BAY 11405 see MNH000
BAY 15203 see DAO800, MIW100
BAY 15922 see TIQ250
BAY 16225 see EKN000
BAY 18436 see DAP400
BAY 19149 see DGP900
BAY 21097 see DAP000
BAY 23129 see PHI500
BAY 23323 see OQS000
BAY 23655 see DSK600
BAY 25141 see FAQ800
BAY 25634 see EAT600
BAY 29492 see LIN400
BAY 29493 see FAQ900
BAY 30130 see DGI000
BAY 32394 see TNH750
BAY 32651 see MIB250
BAY 33051 see DRR400
BAY 33819 see BIM000
BAY 34042 see ENI500
BAY 34727 see COQ399
BAY 37341 see DJR800
BAY 37342 see DST200
BAY 38156 see PHM750
BAY 39731 see MIA250
BAY 41637 see MOV000
BAY 41831 see DSQ000
BAY 42247 see PHD750
BAY 42696 see DQE800
BAY 42903 see EOO000
BAY 44646 see DOR400
BAY 45432 see DNX800
BAY 46131 see ZMA000
BAY 47531 see DFL200
BAY 48130 see DLS800
BAY 50282 see DBI800
BAY 50519 see EOP500
BAY 61597 see MQR275
BAY 62863 see DLS800
BAY 68138 see FAK000
BAY 70143 see CBS275
BAY 70533 see CFC750
BAY 71628 see DTQ400
BAY 75546 see BAS000
BAY 77049 see DJY200
BAY 77488 see BAT750
BAY 79770 see CDP750
BAY-92114 see IMF300
BAY 105807 see MIA250
BAY d8815 see DPF200
BAY A 1040 see AEC750
BAY BUE 1452 see THT500
BAY BUSH (BAHAMAS) see CNH789
BAYCAIN see BMA125
BAYCAINE see BMA125
BAYCALNE see BMA125

BAYCARB see MOV000
BAYCARON see MCA100
BAYCHROM A see CMK415
BAYCHROM F see CMK415
BAYCID see FAQ900
BAYCLEAN see AFP250
BAYCOVIN see DIZ100
BAY DIC 1468 see MQR275
BAY-DRW 1139 see ALA500
BAY E-601 see MNH000
BAY E-605 see PAK000
BAY-E 9736 see NDY700
BAY ENE 11183 B see EAT600
BAYER 73 see DFV400, DFV600
BAYER-186 see CMW700
BAYER 205 see BAT000
BAYER 693 see SLP600
BAYER 1219 see AFL500
BAYER 1355 see PDC850
BAYER 1362 see BSZ000
BAYER 1420 see PMM000
BAYER 2353 see DFV400
BAYER 2502 see NGG000
BAYER 3231 see TND000
BAYER 4245 see DET000
BAYER 4964 see ORU000
BAYER 5072 see DOU600
BAYER 5080 see DOR400
BAYER 5081 see EPY000
BAYER 5312 see EPQ000
BAYER 5360 see MMN250
BAYER 6159H see MQR275
BAYER 6443H see MQR275
BAYER 8169 see DAO500, DAO600
BAYER 9007 see FAQ900
BAYER 9013 see DST200
BAYER 9015 see DFD000
BAYER 9051 see TDX750
BAYER 15080 see BDD000
BAYER 15922 see TIQ250
BAYER 16259 see EKN000
BAYER 17147 see ASH500
BAYER 18510 see DRR400
BAYER 19639 see DXH325
BAYER 20315 see DAP600
BAYER 21/116 see MIW100
BAYER 21/199 see CNU750
BAYER 23655 see DSK600
BAYER 25/154 see DAP400
BAYER 25648 see DFV600
BAYER 25820 see HMV000
BAYER 29492 see LIN400
BAYER 29952 see MOB599
BAYER 32394 see TNH750
BAYER 33172 see FQK000
BAYER 34727 see COQ399
BAYER 36205 see ORU000
BAYER 37289 see EPY000
BAYER 37341 see DJR800
BAYER 37342 see DST200
BAYER 37344 see DST000
BAYER 38819 see BIM000
BAYER 38920 see HCK000
BAYER 39007 see PMY300
BAYER 39731 see MIA250
BAYER 41637 see MOV000
BAYER 41831 see DSQ000
BAYER 42903 see EOO000
BAYER 44646 see DOR400
BAYER 45,432 see DNX800
BAYER 46131 see ZMA000
BAYER 47531 see DFL200
BAYER 62863 see DLS800
BAYER 25 634 see EAT600
BAYER 70533 see CFC750
BAYER 71628 see DTQ400
BAYER 78418 see EIM000
BAYER 94337 see MQR275
BAYER A-128 see PAF550
BAYER A 139 see BDC750
BAYER B-186 see CMW700
BAYER 41367C see MOV000
BAYER-E 393 see SOD100
BAYER E-605 see PAK000
BAYER G4073 see BGW750

BAYERITIAN see TGG760
BAYER L 13/59 see TIQ250
BAYER 1440 L see DNA800, MJE780
BAYER R39 SOLUBLE see BDC750
BAYER S767 see FAQ800
BAYER S 4400 see EPY000
BAYER S 5660 see DSQ000
BAYERTITAN see TGG760
BAY 6681 F see CJO250
BAY G 2821 see EAE675
BAYGON see PMY300
BAYGON, NITROSO derivative see PMY310
BAY H 4502 see BGA825
BAY HOL 0574 see EON500
BAY-HOX-1901 see EPR000
BAY LEAF OIL see BAT500, LBK000
BAYLETON see CJO250
BAYLUSCID see DFV400, DFV600
BAYLUSCIDE see DFV600
BAY-MEB-6447 see CJO250
BAYMIX 50 see CNU750
BAY NTN 8629 see DGC800
BAY-NTN-9306 see SOU625
BAY OIL see BAT500, LBK000
BAYOL F see MQV750
BAYPRESOL see DNA800, MJE780
BAYRE 77488 see BAT750
BAYRITES see BAP000
BAYROGEL see HKK000
BAYRUSIL see DJY200
BAY S 2758 see MIB250
BAY-SRA-12869 see IMF300
BAYTAN see MEP250
BAYTEX see FAQ900
BAYTHION see BAT750
BAYTITAN see TGG760
BAY VA 1470 see DMW000
BAY-VA 4059 see BOL325
BAZUDEN see DCM750
BB-8 see OAH000
BBAL see NJO200
BBC 12 see DDL800
BBC see BAV575, BMW250
BBCE see BIQ500
BBH see BBQ500
BB-K8 see APT000
''B'' BLASTING POWDER see ERF500
BBN see BMW250, HJQ350
BBNOH see HJQ350
BBP see BEC500
BC 7 see PJT300
BC 10 see PJT300
BC 20 see PJT300, UTU500
BC 27 see MCB050
BC 40 see UTU500
BC 71 see MCB050
BC 77 see UTU500
BC 336 see MCB050
BCF-BUSHKILLER see TAA100
B-CHLORO-N,N-DIMETHYLAMINODIBORANE see CGD399
BCM see MHC750
BCME see BIK000
BCM (NH) see BCD325
BiCNU see BIF750
BCNU see BIF750
BCP see BQW825
BCPE see BIN000
BCPE mixed with SPAS see CKL500
BCPN see BQQ250
BC 20 (POLYMER) see UTU500
BCS COPPER FUNGICIDE see CNP250
BC 20 TX see PJT300
BC 30 TX see PJT300
B 41D see MQT600
BD 40A see FNE100
BDCM see BND500
BDH 312 see GGS000
BDH 1298 see VTF000
BDH 6146 see TKH050
BDH 29-790 see SMQ500
BDH 29-801 see TAI250
(BDH) see EQL000
B.D.H. 200 HYDROCHLORIDE see EEQ000
6-Bz-1-DIBROMBENZANTHRON (CZECH) see DDK875

B-DIETHYLAMINOETHYLAMINOPHENYLACETIC ACID ISOAMYL ESTER see NOC000
BDMA see DQP800
BDM-CHLORIDE (RUSSIAN) see AFP075
BD(a,h)P see DCY200
5-BDU see BNC750
BDU see BNC750
BE 100 see IAB000
BE 1293 see CLF325
BE see BNI500
Be 724-A see BEQ625
BEACH APPLE see MAO875
BEACILLIN see BFC750
BEAD TREE see CDM325
BEAM see MQC000
BEAMETTE see PMP500
BEAN SEED PROTECTANT see CBG000
BEAN TREE see GIW195
BEAUTYLEAF see MBU780
BEAUVERIA BASSIANA see BAT830
BEAUVERIN see BAT830
BEAVER POISON see WAT325
BEBUXINE see CQH325
BEC 001 see AQP800
BECAMPICILLIN see BAB250
BECANTAL see DEE600
BECANTEX see DEE600
BECANTYL see DEE600
BECAPTAN see AJT250
BECAPTAN DISULFURE (FRENCH) see MCN500
BECILAN see PPK500
BECKAMINE 21-511 see UTU500
BECKAMINE APH see MCB050
BECKAMINE APM see MCB050
BECKAMINE G 82 see MCB050
BECKAMINE J 101 see MCB050
BECKAMINE J 820 see MCB050
BECKAMINE J 1012 see MCB050
BECKAMINE J 820-60 see MCB050
BECKAMINE L 105-60 see MCB050
BECKAMINE MA-S see MCB050
BECKAMINE NF 5 see UTU500
BECKAMINE P 136 see UTU500
BECKAMINE P 138 see UTU500
BECKAMINE P 138-60 see UTU500
BECKAMINE P 196M see UTU500
BECKAMINE PM see MCB050
BECKAMINE PM-N see MCB050
BECLACIN see AFJ625
BECLAMID see BEG000
BECLAMIDE see BEG000
BECLOFORTE see AFJ625
BECLOMETASONE-17,21-DIPROPIONATE see AFJ625
BECLOMETASONE DIPROPIONATE see AFJ625
BECLOMETHASONE DIPROPIONATE see AFJ625
BECLOVAL see AFJ625
BECLOVENT see AFJ625
BECOREL see PAN100
BECOTIDE see AFJ625
BEECHWOOD CRESOATE see BAT850
BEESIX see PPK250
BEESWAX see BAU000
BEESWAX, WHITE see BAU000
BEESWAX, YELLOW see BAU000
BEET-KLEEN see CBM000, CKC000, DTP400
BEETLE 55 see UTU500
BEETLE 60 see UTU500
BEETLE 65 see UTU500
BEETLE 80 see UTU500
BEETLE 336 see MCB050
BEETLE 338 see MCB050
BEETLE 212-9 see UTU500
BEETLE 3735 see MCB050
BEETLE BC 27 see MCB050
BEETLE BC 71 see MCB050
BEETLE BC 309 see MCB050
BEETLE BC 371 see MCB050
BEETLE BE 336 see MCB050
BEETLE BE 645 see MCB050
BEETLE BE 669 see MCB050
BEETLE BE 670 see MCB050
BEETLE BE 681 see MCB050
BEETLE BE 683 see MCB050
BEETLE BE 685 see UTU500
BEETLE BE 687 see MCB050

BEETLE BE 3021 see MCB050
BEETLE BE 3735 see MCB050
BEETLE BE 3747 see MCB050
BEETLE BT 309 see MCB050
BEETLE BT 323 see MCB050
BEETLE BT 336 see MCB050
BEETLE BT 370 see MCB050
BEETLE BT 670 see MCB050
BEETLE BU 700 see UTU500
BEETLE RESIN 323 see MCB050
BEETLE XB 1050 see UTU500
BEET SUGAR see SNH000
BEFLAVINE see RIK000
BEFUNOLOL see HLK600
BEFUNOLOL HYDROCHLORIDE see BAU255
BEHA see AEO000
BEHEN see MBU800
N⁴-BEHENOYL-1-β-d-ARABINOFURANOSYLCYTOSINE see EAU075
N⁴-BEHENOYLCYTOSINE ARABINOSIDE see EAU075
BEHENOYLCYTOSINE ARABINOSIDE see EAU075
BEHP see DVL700
BEI-1293 see CLF325
BEIENO (MEXICO) see HAQ100
BEIVON see TES750
BEJUCO AHOJA VACA (DOMINICAN REPUBLIC) see YAK300
BEJUCO DO PEO (PUERTO RICO) see CDH125
BEJUCO de LOMBRIZ (CUBA) see PGQ285
BEK see BJU000
BEKADID see OOI000
BEKANAMYCIN see BAU270
BEKANAMYCIN SULFATE see KBA100
BEKAPTAN see MCN750
BEKLAMID see BEG000
BELACID MILLING RED G see CMM325
BELACID MILLING YELLOW R see CMM759
BELACID PHLOXINE G see CMM300
BELAMINE BLACK GX see AQP000
BELAMINE BLUE 2B see CMO000
BELAMINE DIAZO BLACK BH see CMN800
BELAMINE FAST BROWN BP see CMO820
BELAMINE FAST RED 8 BL see CMO885
BELAMINE FAST RED FC see CMO870
BELAMINE GREEN BX see CMO840
BELAMINE SKY BLUE A see CMO500
BELDAVRIN see HOT500
BELFENE see LJR000
BELGANYL see BAT000
BELGENINE see BAU325
BELLADONA (HAWAII) see AOO825
BELLADONNA see BAU500, DAD880
BELLADONNA LILY see AHI635
BELLASTHMAN see DMV600
BELL CML(E) see CAU500
BELL MINE see CAT225
BELL MINE PULVERIZED LIMESTONE see CAO000
BELLUAINE (CANADA) see SED550
BELLYACHE BUSH see CNR135
BELMARK see FAR100
BELOC see SBV500
BELOPHAR KLA see OOI200
BELOPHOR OD see DXB450
BELORAN see BDJ600
BELOSIN see NOC000
BELSEREN see CDQ250
BELT see CDR750
BELUSTINE see CGV250
BEMACO see CLD000
BEMAPHATE see CLD000, CLD250
BEMASULPH see CLD000
BEMEGRIDE see MKA250
BEN-30 see BAV000
BENA see BAU750, BBV500
BENACHLOR see BBV500
BENACTIZINA (ITALIAN) see DHU900
BENACTIZINE HYDROCHLORIDE see BCA000
BENACTYZIN (CZECH) see BCA000
BENACTYZIN see DHU900
BENACTYZINE see DHU900
BENACTYZINE CHLORIDE see BCA000
BENACTYZINE HYDROCHLORIDE see BCA000
BENADON see BBV500, PPK500
BENADRIN see BBV500
BENADRYL see BAU750, BBV500
BENADRYL HYDROCHLORIDE see BAU750
BEN-A-HIST see PDC000

BENAKTIN see BCA000
BENALGIN see BBW500
BEN-ALLERGIN see BBV500
BENANSERIN HYDROCHLORIDE see BEM750
BENAPON see BBV500
BENASPIR see ADA725
BENAZALOX see BAV000
BENAZIDE see IKC000
BENAZOLIN see BAV000
BENAZOL P see HML500
BENAZYL see RDK000
BENCARBATE see DQM600
BENCHINOX see BDD000
BENCICLANE see BAV250
BENCIDAL BLACK E see AQP000
BENCIDAL BLUE 2B see CMO000
BENCIDAL BLUE 3B see CMO250
BENCIDAL DARK GREEN B see CMO830
BENCIDAL FAST BLACK G see CMN240
BENCIDAL FAST RED F see CMO870
BENCIDAL FAST VIOLET N see CMP000
BENCIDAL GREEN B see CMO840
BENCIDAL NAVY BLUE BH see CMN800
BENCONASE see AFJ625
BEN-CORNOX see BAV000
BENCYCLANE see BAV250
BENCYCLANE FUMARATE see BAV250
BENDAZAC see BAV325
BENDAZOL see BEA825
BENDAZOLE see BEA825
BENDAZOLIC ACID see BAV325
BENDECTIN see BAV350
BENDEX see BLU500
BENDIGON see RDK000
BENDIOCARB see DQM600, MHZ000
BENDIOXIDE see MJY500
BENDOPA see DNA200
BENDRALAN see PDD350
BENDROFLUAZIDE see BEQ625
BENDROFLUMETHIAZIDE see BEQ625
BENDYLATE see BAU750
BENECARDIN see AHK750
BENECID see DWW000
BENEMID see DWW000
BENEPEN see PAQ100
BENESAL see SAH000
BENETACIL see PAQ100
BENETHAMINE PENICILLIN see PAQ100
BENETHAMINE PENICILLIN G see PAQ100
BENETOLIN see PAQ100
BENFOS see DGP900
BENGAL GELATIN see AEX250
BENGAL ISINGLASS see AEX250
BENGUINOX see BDD000
BEN-HEX see BBQ500
BENHEXAL see DPH000
BENICOT see NCR000
BENIHINAL see FNK150
BENIROL see BBA500
BENKFURAN see NGE000
BENLATE 50 see BAV575
BENLATE and SODIUM NITRITE see BAV500
BENMOXINE see NCQ100
BENNIE see AOB250
BENOCTEN see BAU750
BENODAINE HYDROCHLORIDE see BCI500
BENODIN see BBV500
BENODINE see BBV500
BENOMYL see BAV575
BENOMYL 50W see BAV575
BENOPAN see BAV000
BENOQUIN see AEY000
BENORAL see SAN600
BENORILATE see SAN600
BENORTAN see SAN600
BENORYLATE see SAN600
BENOVOCYLIN see EDP000
BENOXAPROFEN see OJI750
BENOXIL see OPI300
BENOXINATE HYDROCHLORIDE see OPI300
BENOXYL see BDS000
BENOZIL see DAB800
BEN-P see BFC750
BENPERIDOL see FGU000, FLK100
BENPROPERINE PHOSPHATE see PJA130

BENQUINOX see BDD000
BENSECAL see BAV000
BENSERAZIDE HYDROCHLORIDE see SCA400
BENSULFOID see SOD500
BENSULIDE see DNO800
BENSYLYT see DDG800
BENSYLYTE see PDT250
BENT see MDQ250
BENTANEX see EEO500
BENTANIDOL see BFW250
BENTAZEPAM see BAV625
BENTAZON see MJY500
BENTHIOCARB see SAZ000
BENTHIOZONE see FNF000
BENTIROMIDE see CML835
BENTONE see KBB600
BENTONITE 2073 see BAV750
BENTONITE see BAV750
BENTONITE MAGMA see BAV750
BENTONYL see TJL250
BENTOX 10 see BBQ500
BENTOX see BAU255
BENTRIDE see BEQ625
BENTROL see DNF400, HKB500
BENURIDE see PFB350
BENURON see BEQ625
BEN-U-RON see HIM000
BENURYL see DWW000
BENVIL see MOV500
BENYLAN see BBV500
BENYLATE see BCM000
(5R,6R)-BENXYLPENICILLIN see BDY669
BENZABAR see TIK500
BENZAC 1281 see DOR800
BENZAC see BDS000, PJQ000, TIK500
BENZ(1)ACEANTHRENE see BAW000
BENZ(j)ACEANTHRYLENE, 1,2-DIHYDRO-3,6-DIMETHYL-(9CI) see DRD850
BENZ(j)ACEANTHRYLENE, 1,2-DIHYDRO-5-METHYL- see MIK000
BENZ(j)ACEANTHRYLENE, 1,2,6,7,8,9,10,12b-OCTAHYDRO-3-METHYL- see HDR500
1,2-BENZACENAPHTHENE see FDF000
BENZ(k)ACEPHENANTHRENE see AAF000
3,4-BENZ(e)ACEPHENANTHRYLENE see BAW250
BENZ(e)ACEPHENANTHRYLENE see BAW250
BENZACILLIN see BFC750
BENZACIN see BAW500
BENZACINE see BAW500
BENZACINE HYDROCHLORIDE see BAW500
BENZACIN HYDROCHLORIDE see BAW500
BENZACONINE see PIC250
BENZ(a)ACRIDIN-10-AMINE see AIC750
3,4-BENZACRIDINE see BAW750
7,8-BENZACRIDINE (FRENCH) see BAW750
BENZ(c)ACRIDINE see BAW750
3,4-BENZACRIDINE-9-ALDEHYDE see BAX250
BENZ(c)ACRIDINE-7-CARBONITRILE see BAX000
BENZ(c)ACRIDINE-7-CARBOXALDEHYDE see BAX250
BENZ(a)ACRIDINE-5,6-DIOL, 5,6-DIHYDRO-12-METHYL-, (Z)- see DLE500
BENZ(c)ACRIDINE 3,4-DIOL-1,2-EPOXIDE-1 see EBH875
BENZ(c)ACRIDINE 3,4-DIOL-1,2-EPOXIDE-2 see EBH850
BENZ(c)ACRIDINE, 3-METHOXY-7-METHYL- see MET875
BENZ(c)ACRIDIN-4-OL, 7-METHYL-, ACETATE (ESTER) see ABQ600
N'-BENZ(c)ACRIDIN-7-YL-N-(2-CHLOROETHYL)-N-ETHYL-1,2-ETHANEDIAMINE DIHYDROCHLORIDE see EHI500
α-(BENZ(c)ACRIDIN-7-YL)-N-(p-(DIMETHYLAMINO)PHENYL)NITRONE see BAY250
α-(9-(3,4-BENZACRIDYL))-N-(p-DIMETHYLAMINO-PHENYL)-NITRONE see BAY250
BENZADONE BLUE RC see DFN425
BENZADONE BLUE RS see IBV050
BENZADONE BRILLIANT PURPLE 2R see DFN450
BENZADONE BRILLIANT PURPLE 4R see DFN450
BENZADONE BROWN BR see CMU770
BENZADONE GOLDEN YELLOW see DCZ000
BENZADONE GREY M see CMU475
BENZADONE OLIVE R see DUP100
BENZADOX see ANB000
BENZAHEX see BBQ750
BENZAIDIN see BFW250
BENZAKNEW see BDS000
BENZALACETON (GERMAN) see SMS500
BENZALACETONE see SMS500
trans-BENZALACETONE see BAY275
2-BENZALACETOPHENONE see CDH000
BENZAL ALCOHOL see BDX500

BENZAL-(BENZYL-CYANID) (GERMAN) see DVX600
BENZAL CHLORIDE see BAY300
BENZALDEHYDE see BAY500
BENZALDEHYDE, 4-(ACETYLOXY)-3-ETHOXY- see EQF100
BENZALDEHYDE, 4-AMINO- see AIC825
BENZALDEHYDE, p-BROMO- see BMT700
BENZALDEHYDE, 4-BROMO-(9CI) see BMT700
BENZALDEHYDE, p-CHLORO- see CEI600
BENZALDEHYDE CYANOHYDRIN see MAP250
BENZALDEHYDE, 2,4-DIHYDROXY- see REF100
BENZALDEHYDE, DIMETHYL ACETAL see DOG700
BENZALDEHYDE, 3-ETHOXY-2-HYDROXY- see NOC100
BENZALDEHYDE, 4-ETHOXY-3-HYDROXY- see IKO100
BENZALDEHYDE FFC see BBM500
BENZALDEHYDE GLYCERYL ACETAL (FCC) see BBA000
BENZALDEHYDE GREEN see AFG500, BAY750
BENZALDEHYDE, 2-METHOXY-(9CI) see AOT525
BENZALDEHYDE, 2,3,4-TRIHYDROXY- see TKN800
BENZALDEHYDKYANHYDRIN (CZECH) see MAP250
BENZALETAS see BEL900
BENZAL GLYCERYL ACETAL see BBA000
BENZALIN see DLY000
BENZALKONIUM BROMIDE see BEO000
BENZALKONIUM CHLORIDE see AFP250, BBA500
BENZALKONIUM SACCHARINATE see BBA625
BENZALMALONONITRILE see BBA750
BENZAMIDE see BBB000
BENZAMIDE, o-AMINO- see AID620
BENZAMIDE, 4-AMINO-5-BROMO-N-(2-(DIETHYLAMINO)ETHYL)-2-METHOXY- (9CI) see VCK100
BENZAMIDE, 4-AMINO-5-CHLORO-N-(1-(3-(4-FLUOROPHENOXY)PROPYL)-3-ME-THOXY-4-PIPERIDINYL)-2-METHOXY-, MONOHYDRATE, cis- see CMS237
BENZAMIDE, 2-AMINO-(9CI) see AID620
BENZAMIDE, 5-BROMO-N-(4-BROMOPHENYL)-2-HYDROXY- see BOD600
BENZAMIDE, N-(5-CHLORO-4-((4-CHLOROPHENYL)CYANOMETHYL)-2-ME-THYLPHENYL)-2-HYDROXY-3,5- DIIODO- see CFC100
BENZAMIDE, 2-CYANO- see COK300
BENZAMIDE, o-CYANO-(8CI) see COK300
BENZAMIDE, 3,5-DIBROMO-N-(4-BROMOPHENYL)-2-HYDROXY- see THW750
BENZAMIDE, N-((1-ETHYL-2-PYRROLIDINYL)METHYL)-5-(ETHYLSULFONYL)-2-METHOXY- see EPD100
BENZAMIDE, 2-HYDROXY-N-PHENYL- see SAH500
BENZAMIDE, N,N'-(DITHIODI-2,1-PHENYLENE)BIS- see BDK800
BENZAMIDEPHENYLHYDRAZONE HYDROCHLORIDE see PEK675
BENZAMIDINE, HYDROCHLORIDE see BBM750
BENZAMIDOACETIC ACID see HGB300
1-BENZAMIDO-5-CHLORO-ANTHRAQUINONE see BDK750
dl-α-BENZAMIDO-p-(2-(DIETHYLAMINO)ETHOXY)-N,N-DIPROPYLHYDROCINNA-MAMIDE see TGF175
dl-4-BENZAMIDO-N,N-DIPROPYLGLUTARAMIC ACID see BGC625
dl-4-BENZAMIDO-N,N-DIPROPYLGLUTARAMIC ACID SODIUM SALT see PMH575
(S)-p-(α-BENZAMIDO-p-HYDROXYHYDROCINNAMAMIDO)BENZOIC ACID see CML835
BENZAMIDOOXY ACETIC ACID, AMMONIUM SALT see ANB000
1-BENZAMIDO-1-PHENYL-3-PIPERIDINOPROPANE HYDROCHLORIDE see DKK800
N-(3-BENZAMIDO-3-PHENYL)PROPYL PIPERIDINE HYDROCHLORIDE see DKK800
BENZAMIL BLACK E see AQP000
BENZAMIL SUPRA BROWN BRLL see CMO750
BENZAMIN BLACK DS see CMN230
BENZAMINE BLUE see CMO250
BENZAMINE BLUE see CMO250
BENZAMPHETAMINE see AOB250
BENZANIDINE see BFW250
BENZANIL BLACK BH see CMN800
BENZANIL BLUE 2B see CMO000
BENZANIL BLUE RW see CMO600
BENZANIL BROWN BS see CMO820
BENZANIL DARK GREEN BW see CMO830
BENZANIL FAST BLACK D see CMN230
BENZANIL FAST BLACK G see CMN240
BENZANIL FAST RED F see CMO870
BENZANIL FAST RED K see CMO885
BENZANIL GREEN B see CMO840
BENZANIL GREEN BN see CMO840
BENZANILIDE, 3-AMINO-4-METHOXY- see AKM500
BENZANILIDE, 2'-CHLORO-2-(2-(DIETHYLAMINO)ETHOXY)- see DHP450
BENZANILIDE, 4'-CHLORO-2-(2-(DIETHYLAMINO)ETHOXY)- see DHP550
BENZANILIDE, 2',2''''-DITHIOBIS- see BDK800
BENZANIL SCARLET 3B see CMO875
BENZANIL SKY BLUE see CMO500
BENZANIL VIOLET N see CMP000
BENZANOL BLUE RW see CMO600
BENZANOL BRILLIANT SCARLET 3B see CMO875

BENZANOL FAST RED F see CMO870
BENZ(a)ANTHRACEN-7-ACETIC ACID, METHYL ESTER see BBC500
BENZ(a)ANTHRACEN-7-ACETONITRILE see BBB500
BENZ(a)ANTHRACEN-7-AMINE see BBB750
BENZ(a)ANTHRACEN-8-AMINE see BBC000
1,2:5,6-BENZANTHRACENE see DCT400
1,2-BENZANTHRACENE see BBC250
2,3-BENZANTHRACENE see NAI000
1,2-BENZ(a)ANTHRACENE see BBC250
BENZ(a)ANTHRACENE see BBC250
BENZ(b)ANTHRACENE see NAI000
BENZANTHRACENE see BBC250
1,2-BENZANTHRACENE-10-ACETIC ACID, METHYL ESTER see BBC500
1,2-BENZANTHRACENE-10-ALDEHYDE see BBC750
BENZ(a)ANTHRACENE, 8-BROMO-7,12-DIMETHYL- see BNF315
BENZ(a)ANTHRACENE-7-CARBOXALDEHYDE see BBC750
BENZ(a)ANTHRACENE-7,12-DICARBOXALDEHYDE see BBD000
BENZ(a)ANTHRACENE-10,11-DIHYDRODIOL see BBF000
BENZ(a)ANTHRACENE-1,2-DIHYDRODIOL see BBD250
BENZ(a)ANTHRACENE-3,4-DIHYDRODIOL see BBD500
(+)-(3S,4S)trans-BENZ(a)ANTHRACENE-3,4-DIHYDRODIOL see BBD750
BENZ(a)ANTHRACENE-5,6-DIHYDRODIOL see BBE250
BENZ(a)ANTHRACENE-5,6-trans-DIHYDRODIOL see BBE250
trans-BENZ(a)ANTHRACENE-8,9-DIHYDRODIOL see BBE750
BENZ(a)ANTHRACENE, 3,4-DIHYDROXY-1,2-EPOXY-1,2,3,4-TETRAHYDRO-, (Z),
  (+)- see DMO500
BENZ(a)ANTHRACENE-7,12-DIMETHANOL see BBF500
BENZ(a)ANTHRACENE-7,12-DIMETHANOLDIACETATE see BBF750
(−)(3R,4R)-trans-BENZ(a)ANTHRACENE-3,4-DIOL see BBG000
BENZ(a)ANTHRACENE 3,4-DIOL-1,2-EPOXIDE-2 see DLE000
BENZ(a)ANTHRACENE-7-ETHANOL see BBG500
BENZ(a)ANTHRACENE-7-METHANEDIOLDIACETATE (ester) see BBG750
BENZ(a)ANTHRACENE-7-METHANETHIOL see BBH000
BENZ(a)ANTHRACENE-7-METHANOL see BBH250
BENZ(a)ANTHRACENE-7-METHANOL ACETATE see BBH500
7H-BENZ(de)ANTHRACENE-7-ONE see BBI250
BENZ(a)ANTHRACENE-5,6-OXIDE see EBP000
BENZ(a)ANTHRACENE, 8-PHENYL see PEK750
BENZ(a)ANTHRACENE-7-THIOL see BBH750
BENZ(a)ANTHRACEN-5-OL see BBI000
7H-BENZ(de)ANTHRACEN-7-ONE see BBI250
N-(BENZ(a)ANTHRACEN-5-YLCARBAMOYL)GLYCINE see BBI750
N-(BENZ(a)ANTHRACEN-7-YLCARBAMOYL)GLYCINE see BBJ000
BENZ(a)ANTHRACEN-7-YL-OXIRANE see ONC000
1-(BENZ(a)ANTHRACEN-7-YL)-2,2,2-TRICHLOROETHANONE see TIJ000
BENZ(a)ANTHRACEN-7-YL TRICHLOROMETHYL KETONE see TIJ000
BENZ(a)ANTHRA-5,6-OXIDE see EBP000
1,2-BENZANTHRAZEN (GERMAN) see BBC250
1,2-BENZANTHRENE see BBC250
2,3-BENZANTHRENE see NAI000
BENZANTHRENE see BBC250
BENZANTHRENONE see BBI250
BENZANTHRONE see BBI250
1,2-BENZANTHRYL-10-CARBAMIDOACETIC ACID see BBJ000
1,2-BENZANTHRYL-3-CARBAMIDOACETIC ACID see BBI750
1,2-BENZANTHRYL-10-ISOCYANATE see BBJ250
1,2-BENZANTHRYL-10-MERCAPTAN see BBH750
1,2-BENZANTHRYL-10-METHYLMERCAPTAN see BBH000
7-BENZANTHRYLOXIRANE see ONC000
BENZANTINE see BBV500
BENZAR see BAV000
BENZARONE see BBJ500
BENZATHINE see BHB300
BENZATHINE BENZYLPENICILLIN see BFC750
BENZATHINE PENICILLIN see BFC750
BENZATHINE PENICILLIN G see BFC750
BENZATIN see BHB300
BENZATROPINE METHANESULFONATE see TNU000
BENZAZIDE see BDL750
BENZAZIMIDE see BDH000
BENZAZIMIDOL HYDRATE see HJN650
BENZAZIMIDONE see BDH000
1-BENZAZINE see QMJ000
2-BENZAZINE see IRX000
1-BENZAZOLE see ICM000
BENZAZOLINE see BBW750
BENZAZOLINE HYDROCHLORIDE see BBJ750
BENZAZON VII see NGC400
BENZBROMARON see DDP200
BENZBROMARONE see DDP200
1,2-BENZCARBAZOLE see BCG250
BENZ-o-CHLOR see DER000
BENZCHLOROPROPAMIDE see BEG000
BENZCHLORPROPAMID see BEG000
3,4-BENZCHRYSENE see PIB750

BENZCURINE IODIDE see PDD300
15,16-BENZDEHYDROCHOLANTHRENE see DCR400
BENZEDREX (SKF) see PNN300
dl-BENZEDRINE see BBK000
(±)-BENZEDRINE see BBK000
BENZEDRINE see BBK000
d-BENZEDRINE SULFATE see BBK500
l-BENZEDRINE SULFATE see BBK750
BENZEDRINE SULFATE see BBK250
BENZEDRYNA see AOB250
BENZEEN (DUTCH) see BBL250
BENZEHIST see BAU750
BENZEN (POLISH) see BBL250
BENZENEACETIC ACID see PDY850
BENZENAMINE see AOQ000
BENZENAMINE, 4-((4-AMINOPHENYL)(4-IMINO-2,5-CYCLOHEXADIEN-1-YLI-
  DENE)METHYL)-2-METHYL- see MAC500
BENZENAMINE, 4,4′-AZOBIS-(9CI) see ASK925
BENZENAMINE, 2,6-BIS(1-METHYLETHYL)- see DNN630
BENZENAMINE, 4-((3-BROMOPHENYL)AZO)-N,N-DIMETHYL-(9CI) see BNE600
BENZENAMINE, 4-BUTYL-(9CI) see AJA550
BENZENAMINE, N-BUTYL-(9CI) see BQH850
BENZENAMINE, 2-CHLORO-4,6-DINITRO- see CGL325
BENZENAMINE, N-(2-CHLOROETHYL)-N-ETHYL-(9CI) see EHJ600
BENZENAMINE, 4-CHLORO-, HYDROCHLORIDE see CJR200
BENZENAMINE, 2-CHLORO-4-METHYL- see CLK210
BENZENAMINE, 4-CHLORO-4-(METHYLSULFONYL)- see CIX200
BENZENAMINE, 4-CHLORO-2-NITRO-(9CI) see KDA050
BENZENAMINE, 2,4-DICHLORO-(9CI) see DEO290
BENZENAMINE, 2,6-DICHLORO-N-2-IMIDAZOLIDINYLIDENE- (9CI) see DGB500
BENZENAMINE, 4-((3,4-DICHLOROPHENYL)AZO)-N,N-DIMETHYL-(9CI) see
  DFD400
BENZENAMINE, 2,6-DIETHYL-(9CI) see DIS650
BENZENAMINE, N,N-DIETHYL-(9CI) see DIS700
BENZENAMINE, 4-((3,4-DIETHYLPHENYL)AZO)-N,N-DIMETHYL-(9CI) see
  DJB400
BENZENAMINE, 2,5-DIMETHOXY-(9CI) see AKD925
BENZENAMINE, N,N-DIMETHYL-4′-BROMO-3′-ETHYL-4-(PHENYLAZO)- see
  BNK100
BENZENAMINE, N,N-DIMETHYL-3′-BROMO-4′-ETHYL-4-(PHENYLAZO)- see
  BNK275
BENZENAMINE, N,N-DIMETHYL-4′-BROMO-3′-METHYL-4-(PHENYLAZO)- see
  BNQ110
BENZENAMINE, N,N-DIMETHYL-3′-BROMO-4′-METHYL-4-(PHENYLAZO)- see
  BNQ100
BENZENAMINE, N,N-DIMETHYL-4′-CHLORO-3′-ETHYL-4-(PHENYLAZO)- see
  CGW105
BENZENAMINE, N,N-DIMETHYL-3′-CHLORO-4′-ETHYL-4-(PHENYLAZO)- see
  CGW100
BENZENAMINE, N,N,-DIMETHYL-(9CI) see DQF800
BENZENAMINE, 4-(1,1-DIMETHYLETHYL)-N-(1-METHYLPROPYL)-2,6-DINITRO-
  (9CI) see BQN600
BENZENAMINE, N,N-DIMETHYL-2′-ETHYL-4-(PHENYLAZO)- see EIF450
BENZENAMINE, 3-ETHOXY-(9CI) see PDK800
BENZENAMINE, 4-ETHOXY-(9CI) see PDK890
BENZENAMINE, 4-ETHOXY-, HYDROCHLORIDE (9CI) see PDL750
BENZENAMINE, 2-ETHYL-(9CI) see EGK500
BENZENAMINE, 2-ETHYL-6-METHYL- see MJY000
BENZENAMINE, 4-((2-ETHYLPHENYL)AZO)-N,N-DIMETHYL- see EIF450
BENZENAMINE, 4-FLUORO-(9CI) see FFY000
BENZENAMINE, 4-FLUORO-3-NITRO- see FKK100
BENZENAMINE HYDROCHLORIDE see BBL000
BENZENAMINE, 2-METHOXY-(9CI) see AOV900
BENZENAMINE, 3-METHOXY-(9CI) see AOV890
BENZENAMINE, 2-METHOXY-, HYDROCHLORIDE (9CI) see AOX250
BENZENAMINE, 4-METHOXY-N-(4-METHOXYPHENYL)- see BKO600
BENZENAMINE, 2-METHOXY-4-NITRO-(9CI) see NEQ000
BENZENAMINE, 4-(6-METHYL-2-BENZOTHIAZOLYL)-(9CI) see MHJ300
BENZENAMINE, N-METHYL-(9CI) see MGN750
BENZENAMINE, 4,4′-METHYLENEBIS- see MJQ000
BENZENAMINE, 4,4′-METHYLENEBIS-, DIHYDROCHLORIDE see MJQ100
BENZENAMINE, 4,4′-METHYLENEBIS(N-METHYL)-(9CI) see MJO000
BENZENAMINE, 2-(1-METHYLETHYL)-(9CI) see INW100
BENZENAMINE, 2-METHYL-5-NITRO- see NMP500
BENZENAMINE, 2-METHYL-5-NITRO-, MONOHYDROCHLORIDE see NMP600
BENZENAMINE, N-(2-METHYL-2-NITROPROPYL)-p-NITROSO-(9CI) see NHK800
BENZENAMINE, N-METHYL-N,2,4,6-TETRANITRO-(9CI) see TEG250
BENZENAMINE, 3-(METHYLTHIO)-(9CI) see ALX100
BENZENAMINE, 4-(METHYLTHIO)-(9CI) see AMS675
BENZENAMINE, N-NITRO-(9CI) see NEO510
BENZENAMINE, 4-NITRO-(9CI) see NEO500
BENZENAMINE, 2,3,4,5,6-PENTANITRO- see PBM100
BENZENAMINE, 2-PHENOXY-(9CI) see PDR490
BENZENAMINE, 4-PHENOXY-(9CI) see PDR500
BENZENAMINE, 4,4′-(1,3-PHENYLENEBIS(OXY))BIS- see REF070

1,2-BENZENEDICARBOXYLIC ACID see PHW250
BENZENE-1,3-DICARBOXYLIC ACID see IMJ000
1,4-BENZENEDICARBOXYLIC ACID see TAN750
m-BENZENEDICARBOXYLIC ACID see IMJ000
o-BENZENEDICARBOXYLIC ACID see PHW250
p-BENZENEDICARBOXYLIC ACID see TAN750
1,2-BENZENEDICARBOXYLIC ACID ANHYDRIDE see PHW750
1,2-BENZENEDICARBOXYLIC ACID BI(2-METHOXYETHYL)ESTER (9CI) see DOF400
1,4-BENZENEDICARBOXYLIC ACID, BIS(2-ETHYLHEXYL)ESTER (9CI) see BJS500
1,2-BENZENEDICARBOXYLIC ACID, BIS(1-METHYLETHYL) ESTER see PHW600
1,2-BENZENEDICARBOXYLIC ACID, BUTYL PHENYLMETHYL ESTER see BEC500
1,2-BENZENEDICARBOXYLIC ACID, DECYL OCTYL ESTER see OEU000
1,4-BENZENEDICARBOXYLIC ACID, DIBUTYL ESTER see DEH700
BENZENE-o-DICARBOXYLIC ACID DI-n-BUTYL ESTER see DEH200
o-BENZENEDICARBOXYLIC ACID, DIBUTYL ESTER see DEH200
1,2-BENZENEDICARBOXYLIC ACID, DICYCLOHEXYL ESTER see DGV700
1,2-BENZENEDICARBOXYLIC ACID, DIDODECYL ESTER (9CI) see PHW550
1,2-BENZENEDICARBOXYLIC ACID, DIETHYL ESTER see DJX000
1,4-BENZENEDICARBOXYLIC ACID, DIETHYL ESTER see DKB160
1,2-BENZENEDICARBOXYLIC ACID DIHEXYL ESTER see DKP600
1,2-BENZENEDICARBOXYLIC ACID, DIISOOCTYL ESTER see ILR100
1,2-BENZENEDICARBOXYLIC ACID DIMETHYL ESTER see DTR200
1,3-BENZENEDICARBOXYLIC ACID, DIMETHYL ESTER see IML000
1,4-BENZENE DICARBOXYLIC ACID DIMETHYL ESTER (9CI) see DUE000
1,2-BENZENEDICARBOXYLIC ACID DIOCTYL ESTER see DVL600
o-BENZENEDICARBOXYLIC ACID DIOCTYL ESTER see DVL600
1,2-BENZENEDICARBOXYLIC ACID, DIPROPYL ESTER see DWV500
1,2-BENZENEDICARBOXYLIC ACID, DITRIDECYL ESTER see DXQ200
1,2-BENZENEDICARBOXYLIC ACID, 3,4,5,6-TETRACHLORO-(9CI) see TBT100
BENZENE, 1,2-DICHLORO- see DEP600
BENZENE, (DICHLOROFLUOROMETHYL)- see DFL100
BENZENE, 2,4-DICHLORO-1-METHYL-(9CI) see DGM700
BENZENE, o-DIETHOXY- see CCP900
BENZENE, 1,2-DIETHOXY-(9CI) see CCP900
BENZENE, 2,4-DIFLUORO-1-NITRO- see DKH900
BENZENE, p-DIHYDROXY- see HIH000
BENZENE, 1-((DIIODOMETHYL)SULFONYL)-4-METHYL- see DNF850
BENZENE-1,3-DIISOCYANATE see BBP000
BENZENE, 1,3-DIISOCYANATO- see BBP000
BENZENE-, 1,3-DIISOCYANATOMETHYL- see TGM740
BENZENE, p-DIISOPROPYL- see DNN830
1,3-BENZENEDIMETHANOL, 2-HYDROXY-5-METHYL- see HLV100
BENZENE, 1,3-DIMETHOXY- see REF025
BENZENE, m-DIMETHOXY- see REF025
BENZENE, 1,3-DIMETHYL-, BENZYLATED see DQL820
BENZENE, 1-(1,1-DIMETHYLETHYL)-2,6-DINITRO-3,4,5-TRIMETHYL- see MRW272
BENZENE, 1,4-DIMETHYL-2-NITRO- see NMS520
p-BENZENEDINITRILE see BBP250
BENZENE, p-DINITROSO- see DVE260
BENZENE, 1,4-DINITROSO-(9CI) see DVE260
1,2-BENZENEDIOL see CCP850
1,3-BENZENEDIOL see REA000
1,4-BENZENEDIOL see HIH000
m-BENZENEDIOL see REA000
o-BENZENEDIOL see CCP850
p-BENZENEDIOL see HIH000
(1,2-BENZENEDIOLATO-O)PHENYLMERCURY see PFO750
1,4-BENZENEDIOL, 2-BROMO- see BNL260
1,3-BENZENEDIOL, DIACETATE see REA100
1,3-BENZENEDIOL, DIBENZOATE see BHB100
1,3-BENZENEDIOL, 5-(2-((1,1-DIMETHYLETHYL)AMINO)-1-HYDROXYETHYL)-(9CI) see TAN100
BENZENE-1,3-DIOL, 2,4-DINITROSO- see DVF300
1,2-BENZENEDIOL, 4-(1-HYDROXY-2-(METHYLAMINO)ETHYL)-, HYDROCHLORIDE, (R)- (9CI) see AES500
1,3-BENZENEDIOL, 2-METHYL-(9CI) see MPH400
1,3-BENZENEDIOL, MONOACETATE see RDZ900
1,2-BENZENEDIOL, MONO(METHYLCARBAMATE) (9CI) see HNK900
1,3-BENZENEDIOL, 4-(PHENYLAZO)- see CMP600
1,3-BENZENEDIOL, 2,4,6-TRINITRO-, BARIUM SALT, HYDRATE (2:1:1) see BAO900
BENZENE-1,3-DISULFONYL CHLORIDE FLUORIDE see FLY200
BENZENE, DIVINYL- see DXQ740
BENZENEETHANOL, α-METHYL-(9CI) see PGA600
BENZENEETHANOL, β-METHYL-(9CI) see HGR600
BENZENE, 1,1′-(1,2-ETHENEDIYL)BIS-(9CI) see SLR000
BENZENE, 1,1′-(1,2-ETHENEDIYL)BIS-, (E)-(9CI) see SLR100
BENZENE, ETHENYL-, HOMOPOLYMER (9CI) see SMQ500
BENZENE, 1-ETHENYL-4-METHYL-(9CI) see VQK700
BENZENE, ETHOXY- see PDM000

BENZENE, (2-(1-ETHOXYETHOXY)ETHYL)- see EES380
BENZENE, 1-(ETHOXYMETHYL)-2-METHOXY- see EMF600
BENZENE, 1-FLUORO-2-METHYL-(9CI) see FLZ100
BENZENEFORMIC ACID see BCL750
BENZENE HEXACHLORIDE-α-isomer see BBQ000
α-BENZENEHEXACHLORIDE see BBQ000
β-BENZENEHEXACHLORIDE see BBR000
BENZENE HEXACHLORIDE-γ-isomer see BBQ500
γ-BENZENE HEXACHLORIDE see BBQ500
BENZENEHEXACHLORIDE (mixed isomers) see BBQ750
BENZENE HEXACHLORIDE see BBP750
trans-α-BENZENEHEXACHLORIDE see BBR000
Δ-BENZENEHEXACHLORIDE see BFW500
BENZENEIODIDE see IEC500
BENZENE, 1-(1-ISOCYANATO-1-METHYLETHYL)-3-(1-METHYLETHENYL)- see IKG800
BENZENE ISOPROPYL see COE750
BENZENE-1-ISOTHIOCYANATE see ISQ000
BENZENEMETHANAMINE-N-(2-CHLOROETHYL)-2-ETHOXY-5-NITRO see CGX325
BENZENEMETHANAMINE, 3,4-DIMETHOXY- see VIK050
BENZENEMETHANAMINIUM, N-(2-((2,6-DIMETHYLPHENYL)AMINO)-2-OXOE-THYL)-N,N-DIETHYL-, BENZOATE see DAP812
BENZENEMETHANAMINIUM, N-(4-((2,4-DISULFOPHENYL)(4-(ETH-YL(PHENYLMETHYL)AMINO)PHENYL)METHYLENE)-2, 5-CYCLOHEXADIEN-1-YLIDENE)-N-ETHYL-, HYDROXIDE, inner salt, SODIUM SALT see ERG100
BENZENEMETHANAMINIUM, N-DODECYL-N,N-BIS(2-HYDROXYETHYL)-, CHLO-RIDE (9CI) see BDJ600
BENZENEMETHANAMINIUM, N-DODECYL-N,N-DIMETHYL-, BROMIDE (9CI) see BEO000
BENZENEMETHANAMINIUM, N-HEXADECYL-N,N-DIMETHYL-, CHLORIDE see BEL900
BENZENEMETHANAMINIUM, N,N,N-TRIETHYL-, CHLORIDE (9CI) see BFL300
BENZENEMETHANAMINIUM, N,N,N-TRIMETHYL-, CHLORIDE (9CI) see BFM250
BENZENEMETHANAMINIUM, N,N,N-TRIMETHYL-, METHOXIDE (9CI) see TLM100
BENZENEMETHANESULFONYL FLUORIDE see TGO300
BENZENEMETHANOIC ACID see BCL750
BENZENEMETHANOL see BDX500
BENZENEMETHANOL, α-(1-AMINOETHYL)-3-HYDROXY-, (R-(R*,S*))-( 9CI) see HNB875
BENZENEMETHANOL, α-(((2-(3,4-DIMETHOXYPHENYL)ETHYL)AMINO)METHYL)-4-HYDROXY-, (R)- see DAP850
BENZENEMETHANOL, 2,4-DIMETHYL-, ACETATE see DQP500
BENZENEMETHANOL, 2-((4-(DIMETHYLAMINO)PHENYL)AZO)-(9CI) see HMB595
BENZENEMETHANOL, α-ETHYNYL-, CARBAMATE (9CI) see PGE000
BENZENEMETHANOL, α-ETHYNYL-4-METHOXY- see HKA700
BENZENEMETHANOL, α-METHYL- see PDE000
BENZENEMETHANOL, 4-METHYL-, ACETATE see XQJ700
BENZENEMETHANOL, α-METHYL-, ACETATE (9CI) see SMP600
BENZENEMETHANOL, α-(1-(METHYLAMINO)ETHYL)-, (R*,S*)-(±)-(9CI) see EAW100
BENZENEMETHANOL, α-(1-(METHYLAMINO)ETHYL)-, HYDROCHLORIDE, (R-(R*,S*))- see EAW500
BENZENEMETHANOL, α-(1-(METHYLAMINO)ETHYL)-, (S-(R*,S*))-(9CI) see EAW200
BENZENEMETHANOL, 4-(1-METHYLETHYL)-, ACETATE see IOF050
BENZENEMETHANOL, 2-NITRO-(9CI) see NFM550
BENZENEMETHANOL, 3-PHENOXY-(9CI) see HMB625
BENZENEMETHANOL, α-(TRICHLOROMETHYL)-(9CI) see TIR800
BENZENE, METHOXY- see AOX750
BENZENE, 2-METHOXY-1,3,5-TRINITRO-(9CI) see TMK300
BENZENE, METHYL- see TGK750
BENZENE, (2-(3-METHYLBUTOXY)ETHYL)- see IHV050
BENZENE, METHYLDINITRO- see DVG600
BENZENE, 1-METHYL-4-(1-METHYLETHYL)-2-(1-PROPENYL)- see VIP100
BENZENE, 1-METHYL-2-NITROSO-(9CI) see NLW500
BENZENE, 1-METHYL-3-PHENOXY- see MNV770
BENZENE, 2-METHYL-1,3,5-TRINITRO- see TMN490
BENZENENITRILE see BCQ250
BENZENE, 1,1′-OXYBIS-, HEXACHLORO derivatives (9CI) see CDV175
BENZENE, PENTABROMOETHYL- see PAT850
BENZENEPHOSPHONIC ACID see PFV500
BENZENEPHOSPHONIC ACID, DIOCTYL ESTER see PFV750
BENZENEPROPANAL see HHP000
BENZENEPROPANAL, 4-(1,1-DIMETHYLETHYL)- see BMK300
BENZENEPROPANAL, α-α-DIMETHYL-4-ETHYL- see EIJ600
BENZENEPROPANAL, 4-METHOXY-α-METHYL- see MLJ050
BENZENEPROPANAMINE, N-(1,1-DIMETHYLETHYL)-α-METHYL-γ-PHENYL-, HY-DROCHLORIDE (9CI) see TBC200
BENZENEPROPANENITRILE (9CI) see HHP100
BENZENEPROPANENITRILE, β-OXO-α-PHENYL-(9CI) see OOK200

BENZENEPROPANOIC ACID, 4-NITRO-β-OXO-, ETHYL ESTER (9CI) see ENO100

BENZENEPROPANOIC ACID, β-OXO-, ETHYL ESTER (9CI) see EGR600

3-BENZENEPROPANOL see HHP050

BENZENEPROPANOL CARBAMATE see PGA750

BENZENEPROPANOL, PROPANOATE (9CI) see HHQ550

BENZENEPROPIONITRILE see HHP100

BENZENE, (2-(1-PROPOXYETHOXY)ETHYL)- see PDD400

BENZENESELENIC ACID see BBR325

BENZENESELENONIC ACID see PGH500

BENZENESULFANILIDE see BBR750

BENZENESULFINIC ACID, p-CHLORO- see CEJ600

BENZENESULFINYL AZIDE see BBR380

BENZENE SULFINYL CHLORIDE see BBR390

BENZENESULFOHYDRAZIDE see BBS300

BENZENESULFONAMIDE see BBR500

BENZENESULFONAMIDE, 3-AMINO- see MDM760

BENZENESULFONAMIDE, 4-AMINO-N-(AMINOCARBONYL)- (9CI) see SNQ550

BENZENESULFONAMIDE, N-(5-AMINO-1-(CHLOROACETYL)PENTYL)-4-METHYL-, (S)-(9CI) see THH500

BENZENESULFONAMIDE, 4-AMINO-N-(5-CHLORO-2-QUINOXALINYL)- see CMA600

BENZENESULFONAMIDE, 4-AMINO-N-(DIAMINOMETHYLENE)- see AHO250

BENZENESULFONAMIDE, 4-AMINO-N-(5-ETHYL-1,3,4-THIADIAZOL-2-YL)-, MONOSODIUM SALT (9CI) see SJW300

BENZENESULFONAMIDE, 4-AMINO-N-(5-METHYL-1,3,4-THIADIAZOL-2-YL)- see MPQ750

BENZENESULFONAMIDE, 4-AMINO-N-2-PYRIDINYL-, MONOSODIUM SALT (9CI) see PPO250

BENZENESULFONAMIDE, N-BUTYL- see BQJ650

BENZENESULFONAMIDE, 4-CHLORO-N-((CYCLOHEXYLAMINO)CARBONYL)- see CDR550

BENZENESULFONAMIDE, 2-CHLORO-5-(2,3-DIHYDRO-1-HYDROXY-3-OXO-1H-ISOINDOL-1-YL)- (9CI) see CLY600

BENZENESULFONAMIDE, 2-CHLORO-5-(1-HYDROXY-3-OXO-1-ISOINDOLINYL)- see CLY600

BENZENESULFONAMIDE, 4-CHLORO-N-(((1-METHYLE-THYL)AMINO)CARBONYL)- (9CI) see CDY100

BENZENESULFONAMIDE, N-CHLORO-4-METHYL-, SODIUM SALT (9CI) see CDP000

BENZENESULFONAMIDE, N-(3-CHLORO-2-OXO-1-(PHENYLMETHYL)PROPYL)-4-METHYL-,(S)- see THH450

BENZENESULFONAMIDE, p-((p-(DIMETHYLAMINO)PHENYL)AZO)- see SNW800

BENZENESULFONAMIDE, p-HYDROXY-, O-ESTER with O,O-DIMETHYL PHOSPHOROTHIOATE see CQL250

BENZENESULFONAMIDE, 4-METHYL-N-PHENYL-(9CI) see TGN600

2-BENZENESULFONAMIDO-5-tert-BUTYL-1,3,4-THIADIAZOLE see GFM200

2-BENZENESULFONAMIDO-5-(β-METHOXYETHOXY)PYRIMIDINE SODIUM SALT see GHK200

2-BENZENESULFONAMIDO-5-TERTIOBUTYL-1-THIA-3,4-DIAZOLE see GFM200

BENZENESULFONANILIDE see BBR750

BENZENESULFONATE de 4-CHLOROPHENYLE (FRENCH) see CJR500

BENZENE SULFONCHLORIDE see BBS750

BENZENESULFONIC ACID see BBS250

BENZENESULFONIC ACID, 4-ACETAMIDO-2-AMINO- see AHQ300

BENZENESULFONIC ACID, ALKYL DERIVATIVES see AFO500

BENZENESULFONIC ACID, o-AMINO- see SNO100

BENZENESULFONIC ACID, 2-AMINO-4-CHLORO-4-CHLORO-5-METHYL- see AJJ250

BENZENESULFONIC ACID, 4-AMINO-(9CI) see SNN600

BENZENESULFONIC ACID, 2-AMINO-5-NITRO-, AMMONIUM SALT see ALL800

BENZENESULFONIC ACID, 2-AMINO-5-((4-SULFOPHENYL)AZO)-, DISODIUM SALT see CMM758

BENZENESULFONIC ACID, 5-BENZOYL-4-HYDROXY-2-METHOXY- see HLR700

BENZENESULFONIC ACID BUTYL AMIDE see BQJ650

BENZENESULFONIC (ACID) CHLORIDE see BBS750

BENZENESULFONIC ACID, 5-CHLORO-2-((2-HYDROXY-1-NAPHTHALE-NYL)AZO)-4-METHYL-, MONOSODIUM SALT see CMS150

BENZENESULFONIC ACID, 4-CHLOROPHENYL ESTER see CJR500

BENZENESULFONIC ACID, 2,5-DIAMINO- see PEY800

BENZENESULFONIC ACID, 3-(4-DIETHYLAMINO-2-HYDROXYPHENYLAZO)-4-HYDROXY- see DIJ300

BENZENESULFONIC ACID, 5-((2,4-DINITROPHENYL)AMINO)-2-(PHENYLAMINO)-, MONOSODIUM SALT see SGP500

BENZENESULFONIC ACID, DODECYL- see LBU100

BENZENESULFONIC ACID, DODECYL ESTER see DXW400

BENZENESULFONIC ACID, DODECYL-, compd. with 2,2',2''-NITRILO-TRIS(ETHANOL) (1:1) see TJK800

BENZENESULFONIC ACID, 2,2'-(1,2-ETHANEDIYL)BIS(5-AMINO-, DISODIUM SALT see FCA200

BENZENESULFONIC ACID, 2,2'-(1,2-ETHYLENEDIYL)BIS(5-AMINO-(9CI) see FCA100

BENZENESULFONIC ACID, HYDRAZIDE see BBS300

BENZENESULFONIC ACID, p-HYDROXY- see HNL600

BENZENESULFONIC ACID, 4-((2-HYDROXY-1-NAPHTHALENYL)AZO)-, MONO-SODIUM SALT see CMM220

BENZENESULFONIC ACID, 4-HYDROXY-, ZINC SALT (2:1) see ZIJ300

BENZENESULFONIC ACID, p-HYDROXY-, ZINC SALT (2:1) see ZIJ300

BENZENESULFONIC ACID, METHYL ESTER see MHB300

BENZENESULFONIC ACID, OXYBIS-, DIHYDRAZIDE (9CI) see OPE000

BENZENESULFONIC ACID, SODIUM SALT see SJH050

BENZENESULFONIC HYDRAZIDE see BBS300

BENZENESULFONOHYDRAZIDE see BBS300

BENZENE, 1,1'-SULFONYLBIS(4-FLUORO-3-NITRO)- see DKH250

BENZENESULFONYL CHLORIDE see BBS750

BENZENESULFONYL CHLORIDE, m-(FLUOROSULFONYL)- see FLY200

2-(BENZENESULFONYL)ETHANOL see BBT000

BENZENESULFONYL FLUORIDE, 4-METHYL-(9CI) see TGO500

BENZENESULFONYL HYDRAZIDE see BBS300

BENZENESULFONYL HYDRAZINE see BBS300

BENZENESULPHONAMIDE see BBR500

BENZENE SULPHONOHYDRAZIDE see BBS300

BENZENE SULPHONYL CHLORIDE (DOT) see BBS750

BENZENESULPHONYL FLUORIDE see BBT250

1,2,4,5-BENZENETETRACARBOXYLIC ACID see PPQ630

1,2,4,5 BENZENETETRACARBOXYLIC 1,2:4,5 DIANHYDRIDE see PPQ635

BENZENE, 1,2,3,5-TETRACHLORO- see TBN745

BENZENETETRAHYDRIDE see CPC579

BENZENETHIOL, o-METHOXY- see MFQ300

BENZENETHIOL, 2-METHOXY-(9CI) see MFQ300

BENZENETHIOL, 3-METHYL-(9CI) see TGO800

1,3,5-BENZENETRIAMINE see THN775

1,2,4-BENZENETRICARBOXYLIC ACID see TKU700

1,2,4-BENZENETRICARBOXYLIC ACID ANHYDRIDE see TKV000

1,2,4-BENZENETRICARBOXYLIC ACID, CYCLIC 1,2-ANHYDRIDE see TKV000

1,2,4-BENZENETRICARBOXYLIC ACID, TRIS(2-ETHYLHEXYL)ESTER see TJR600

1,2,4-BENZENETRICARBOXYLIC ANHYDRIDE see TKV000

BENZENE, 1,2,3-TRICHLORO- see TIK100

BENZENE, 1,3,5-TRICHLORO- see TIK300

BENZENE, 1,2,4-TRICHLORO-3-METHYL- see TJD600

BENZENE, 1,3,5-TRICHLORO-2,4,6-TRINITRO- see TJE200

BENZENE, (TRIFLUOROETHENYL)-(9CI) see TKH310

BENZENETRIFUROXAN see BBU125

cis-BENZENE TRIIMINE see THQ600

BENZENE, 1,2,4-TRIMETHOXY-5-PROPENYL-, (E)- see IHX400

BENZENE, 1,2,4-TRIMETHOXY-5-PROPENYL-, trans- see IHX400

BENZENE, 1,3,5-TRIMETHYL- see TLM050

BENZENE, 1,3,5-TRIMETHYL-2,4-DINITRO- see DUW505

BENZENE, 1,3,5-TRIMETHYL-2,4,6-TRINITRO- see TMI800

1,2,3-BENZENETRIOL see PPQ500

1,2,4-BENZENETRIOL see BBU250

1,3,5-BENZENETRIOL see PGR000

BENZENE-1,3,5-TRIOL see PGR000

BENZENE-s-TRIOL see PGR000

1,2,4-BENZENETRIOL, TRIACETATE see HLF600

BENZENE TRIOZONIDE see BBU500

BENZENOL see PDN750

BENZENOSULFOCHLOREK (POLISH) see BBS750

BENZENOSULPHOCHLORIDE see BBS750

BENZENYL CHLORIDE see BFL250

BENZENYL FLUORIDE see BDH500

BENZENYL TRICHLORIDE see BFL250

BENZETAMOPHYLLINE HYDROCHLORIDE see THL750

BENZETHACIL see BFC750

BENZETHIDIN see BBU625

BENZETHIDINE see BBU625

BENZETHONIUM CHLORIDE see BEN000

BENZETHONIUM CHLORIDE MONOHYDRATE see BBU750

BENZETIMIDE see BBU800

BENZETONIUM CHLORIDE see BEN000

BENZEX see BBQ750

10,11-BENZFLUORANTHENE see BCJ500

2,3-BENZFLUORANTHENE see BAW250

3,4-BENZFLUORANTHENE see BAW250

BENZ(j)FLUOROANTHRENE see BCJ500

BENZHEXOL see PAL500

BENZHEXOL CHLORIDE see BBV000

BENZHEXOL HYDROCHLORIDE see BBV000

BENZHORMOVARINE see EDP000

BENZHYDRAMINE see BBV500

BENZHYDRAMINE HYDROCHLORIDE see BAU750

BENZHYDRAMINUM see BBV500

BENZHYDRAZIDE see BBV250

BENZHYDRIL see BBV500

BENZHYDROL see HKF300

BENZHYDRYL see BBV500

BENZHYDRYL ALCOHOL see HKF300

o-BENZHYDRYLDIMETHYLAMINOETHANOL see BBV500

o-BENZHYDRYLDIMETHYLAMINOETHANOL-8-CHLOROTHEOPHYLLINATE see DYE600

1-BENZHYDRYL-4-(2-(2-HYDROXYETHOXY)ETHYL)PIPERAZINE see BBV750

BENZHYDRYL METHYL KETONE see MJH910

N-BENZHYDRYL-N-METHYL PIPERAZINE see EAN600

(N-BENZHYDRYL)(N'-METHYL)DIETHYLENEDIAMINE see EAN600

N-BENZHYDRYL-N'-METHYLPIPERAZINE HYDROCHLORIDE see MAX275

N-BENZHYDRYL-N'-METHYLPIPERAZINE MONOHYDROCHLORIDE see MAX275

2-(BENZHYDRYLOXY)-N,N-DIMETHYLETHYLAMINE see BBV500

2-(BENZHYDRYLOXY)-N,N-DIMETHYLETHYLAMINE with 8-CHLOROTHEOPHYL-LINE see DYE600

2-(BENZHYDRYLOXY)-N,N-DIMETHYLETHYLAMINEHYDROCHLORIDE see BAU750

4-(BENZHYDRYLOXY)-1-METHYLPIPERIDINE see LJR000

BENZHYDRYL PHENYL KETONE see TMS100

BENZIDAMINE HYDROCHLORIDE see BBW500

BENZIDAZOL see BBW750

BENZIDIN (CZECH) see BBX000

BENZIDINA (ITALIAN) see BBX000

BENZIDINE see BBX000

3,3'-BENZIDINEDICARBOXYLIC ACID see BFX250

3,3'-BENZIDINE DICARBOXYLIC ACID, DISODIUM SALT see BBX250

γ,γ'-3,3'-BENZIDINE DIOXYDIBUTYRIC ACID see DEK400

3,3'-BENZIDINE-γ,γ'-DIOXYDIBUTYRIC ACID see DEK400

2,2'-BENZIDINEDISULFONIC ACID see BBX500

BENZIDINE HYDROCHLORIDE see BBX750

BENZIDINE LACQUER YELLOW G see DEU000

BENZIDINE ORANGE 45-2850 see CMS145

BENZIDINE ORANGE 45-2880 see CMS145

BENZIDINE ORANGE see CMS145

BENZIDINE ORANGE TONER see CMS145

BENZIDINE ORANGE WD 265 see CMS145

BENZIDINE SULFATE see BBY000

BENZIDINE-3-SULFURIC ACID see BBY250

BENZIDINE SULPHATE and HYDRAZINE-BENZENE see BBY300

BENZIDINE-3-SULPHURIC ACID see BBY250

BENZIDINE YELLOW 20544 see CMS208

BENZIDINE YELLOW see DEU000

BENZIDINE YELLOW AAOT see CMS212

BENZIDINE YELLOW ABZ 249 see CMS212

BENZIDINE YELLOW G see CMS212

BENZIDINE YELLOW GE see CMS208

BENZIDINE YELLOW GGT see CMS212

BENZIDINE YELLOW GR see CMS208

BENZIDINE YELLOW L see CMS212

BENZIDINE YELLOW LEMON 12221 see CMS208

BENZIDINE YELLOW OT (6CI) see CMS212

BENZIDINE YELLOW OTYA 8055 see CMS212

BENZIDINE YELLOW TONER YT-378 see DEU000

BENZIDIN-3-YL ESTER SULFURIC ACID see BBY500

BENZIDIN-3-YL HYDROGEN SULFATE see BBY500

BENZIES see AOB250

BENZIL see BBY750

BENZILAN see DER000

BENZILATE DU DIETHYLAMINO-ETHANOL CHLORHYDRATE (FRENCH) see BCA000

1-BENZILBIGUANIDE CLORIDRATO (ITALIAN) see BEA850

BENZILE (CLORURO di) (ITALIAN) see BEE375

N-β-(BENZILFENILAMINO)ETILPIPERIDINA CLORIDRATO (ITALIAN) see BEA275

1-(2-(2-BENZILFENOSSI)-1-METILETIL)-PIPERIDINA see MOA600

1-(2-(2-BENZILFENOSSI)-1-METILETIL)-PIPERIDINA FOSFATO (ITALIAN) see PJA130

BENZILIC ACID see BBY990

BENZILIC ACID-β-DIETHYLAMINOETHYL ESTER see DHU900

BENZILIC ACID-β-DIETHYLAMINOETHYL ESTER HYDROCHLORIDE see BCA000

BENZILIC ACID ESTER with 1-ETHYL-3-HYDROXY-1-METHYLPIPERIDINIUM BROMIDE see PJA000

BENZILIC ACID, ester with ETHYL (2-HYDROXYETHYL)DIMETHYLAMMONIUM CHLORIDE see ELF500

BENZILIC ACID ester with 3-HYDROXY-1,1-DIMETHYLPIPERIDINIUM BROMIDE see CBF000

BENZILIC ACID ester with 2-(HYDROXYMETHYL)-1,1-DIMETHYLPIPERIDINIUM METHYL SULFATE see CDG250

BENZILIC ACID-1-METHYL-3-PIPERIDYL ESTER see MON250

BENZILIC ACID, 3-QUINUCLIDINYL ESTER, HYDROCHLORIDE see QWJ000

(2-BENZILOXYETHYL)DIMETHYLOCTYLAMMONIUM BROMIDE see DSH000

8-α-BENZILOYLOXY-6,10-ETHANO-5-AZONIASPIRO(4.5)DECANE CHLORIDE see KEA300

8-BENZILOYLOXY-6,10-ETHANO-5-AZONIASPIRO(4.5)DECANE CHLORIDE see BCA375

4-BENZILOYLOXY-1,1,2,2,6-PENTAMETHYLPIPERIDINIUM CHLORIDE (β FORM) see BCB000

4-BENZILOYLOXY-1,1,2,6,6-PENTAMETHYLPIPERIDINIUM CHLORIDE (α FORM) see BCB250

BENZILSAEURE-(N,N-DIMETHYL-2-HYDROXYMETHYL-PIPERIDINIUM)-ESTER-ME-THYLSULFAT (GERMAN) see CDG250

BENZILSAEURE-DIMETHYL-OCTYL-AMMONIUM-AETHYLESTER BROMIDE (GERMAN) see DSH000

BENZILSAEURE-DIMETHYL-PENTYL-AMMONIUM-AETHYLESTER BROMIDE (GERMAN) see DSH200

BENZILYLOXYETHYLDIMETHYLETHYLAMMONIUM CHLORIDE see ELF500

4-BENZILYLOXY-1,2,2,6-TETRAMETHYLPIPERIDINE METHOCHLORIDE (α FORM) see BCB250

4-BENZILYLOXY-1,2,2,6-TETRAMETHYLPIPERIDINE METHOCHLORIDE (β FORM) see BCB000

1H-BENZIMIDAZOL-5-AMINE, 2-(4-AMINOPHENYL)- see ALV050

o-BENZIMIDAZOLE see BCB750

BENZIMIDAZOLE see BCB750

2-BENZIMIDAZOLEACETONITRILE see BCC000

BENZIMIDAZOLE, 5-AMINO-2-(p-AMINOPHENYL)- see ALV050

BENZIMIDAZOLE, 2-AMINO-5-BENZOYL- see AIH000

2-BENZIMIDAZOLECARBAMIC ACID, 5-BENZOYL-, METHYL ESTER see MHL000

2-BENZIMIDAZOLECARBAMIC ACID, 5-(p-FLUOROBENZOYL)-, METHYL ES-TER see FDA887

BENZIMIDAZOLE-2-CARBAMIC ACID, METHYL ESTER see MHC750

1H-BENZIMIDAZOLE, 5-CHLORO-6-(2,3-DICHLOROPHENOXY)-2-(METHYLTHIO)-see CFL200

1H-BENZIMIDAZOLE (9CI) see BCB750

BENZIMIDAZOLE, 1-(2-DIETHYLAMINOETHYL)-2-(p-ETHOXYBENZYL)-5-NITRO-, HYDROCHLORIDE see EQN750

BENZIMIDAZOLE, 5,6-DIMETHYL- see DQM100

BENZIMIDAZOLE METHYLENE MUSTARD see BCC250

BENZIMIDAZOLE MUSTARD see BCC250

1H-BENZIMIDAZOLE, 2-PHENYL-(9CI) see PEL250

2-BENZIMIDAZOLETHIOL see BCC500

2-BENZIMIDAZOLINONE see ONI100

1H-BENZIMIDAZOLIUM HEXAKIS-1-DODECYL-3-METHYL-2-PHENYL-(CYANO-C)FERRATE(1−) see TNH750

BENZIMIDAZOLIUM-1-NITROIMIDATE see BCD125

2-BENZIMIDAZOLYLACETONITRILE see BCC000

N-2-(BENZIMIDAZOLYL) CARBAMATE see MHC750

1H-BENZIMIDAZOL-2-YLCARBAMIC ACID METHYL ESTER see MHC750

1-(2-BENZIMIDAZOLYL)-3-METHYLUREA see BCD325

4-(2-BENZIMIDAZOLYL)THIAZOLE see TEX000

BENZIMINAZOLE see BCB750

BENZIN B70 see NAI500

BENZINDAMINE see BCD750

BENZINDAMINE HYDROCHLORIDE see BBW500

1H-BENZ(6,7)INDAZOLO(2,3,4-fgh)NAPHTH(2'',3'':6',7')INDOLO(3',2':5,6)ANTHR A(2,1,9-mna) ACRIDINE-5,8,13,25-TETRAONE see CMU475

BENZ(e)INDENO(1,2-b)INDOLE see BCE000

BENZ(b)INDENO(1,2-d)PYRAN-3,6a,9,10(6H)-TETROL, 7,11b-DIHYDRO- see LFT800

BENZ(cd)INDOL-2(1H)-ONE see NAX100

BENZINDOPYRINE HYDROCHLORIDE see BCE250

1-BENZINE see QMJ000

BENZINE (LIGHT PETROLEUM DISTILLATE) see PCT250

BENZINE (OBS.) see BBL250

BENZIN (OBS.) see BBL250

BENZINOFORM see CBY000

BENZINOL see TIO750

BENZIODARON see EID200

BENZIODARONE see EID200

1H-BENZ(de)ISOQUINOLINE-1,3(2H)-DIONE, 2-(2-(DIMETHYLAMINO)ETHYL)-5-NI-TRO- see MQX775

1,2-BENZISOTHIAZOLIN-3-ONE see BCE475

3-BENZISOTHIAZOLINONE-1,1-DIOXIDE see BCE500

1,2-BENZISOTHIAZOLIN-3-ONE 1,1-DIOXIDE AMMONIUM SALT see ANT500

1,2-BENZISOTHIAZOL-3(2H)-ONE see BCE475

1,2-BENZISOTHIAZOL-3(2H)-ONE-1,1-DIOXIDE see BCE500

1,2-BENZISOTHIAZOL-3(2H)-ONE-1,1-DIOXIDE, CALCIUM SALT see CAM750

BENZISOTRIAZOLE see BDH250

1,2-BENZISOXAZOLE-3-METHANESULFONAMIDE see BCE750

BENZITRAMIDE see BCE825

3,4-BENZOACRIDINE see BAW750

1,2-BENZOANTHRACENE see BBC250

BENZO(a)ANTHRACENE see BBC250

BENZOANTHRACENE see BBC250

BENZO(a)ANTHRACENE-5,6-OXIDE see EBP000

7H-BENZO(de)ANTHRACEN-7-ONE see BBI250

BENZOANTHRONE see BBI250

BENZO(A)PYRAZINE see QRJ000

BENZOATE see BCL750

17-BENZOATE-3-n-BUTYRATE d'OESTRADIOL (FRENCH) see EDP500

BENZOATE of MONOETHANOLAMINE see MRH300

BENZOATE d'OESTRADIOL (FRENCH) see EDP000

BENZOATE d'OESTRONE (FRENCH) see EDV500
BENZOATE of SODA see SFB000
BENZOATE SODIUM see SFB000
1-BENZOAZO-2-NAPHTHOL see PEJ500
BENZOBARBITAL see BDS300
BENZO(g)(1,3)BENZODIOXOLO(5,6-a)QUINOLIZINIUM,5,6-DIHYDRO-9,10-DIME-THOXY-, CHLORIDE (9CI) see BFN600
BENZO(f)(1)BENZOTHIENO(3,2-b)QUINOLINE see BCF500
BENZO(h)(1)BENZOTHIENO(3,2-b)QUINOLINE see BCF750
BENZO(e)(1)BENZOTHIOPYRANO(4,3-b)INDOLE see BCG000
BENZO BLACK BLUE BH see CMN800
BENZO BLACK BLUE FBH see CMN800
BENZO BLUE see CMO250
BENZO BLUE GS see CMO000
BENZO BLUE RWA see CMO600
BENZO BLUE RWS see CMO600
BENZO(B)NAPHTHACENE see PAV000
BENZO BRILLIANT RED 8BS see CMO880
BENZO BROWN D 3GA-CF see CMO810
BENZOCAINE see EFX000
11H-BENZO(a)CARBAZOLE see BCG250
8,9-BENZO-γ-CARBOLINE see BDB500
BENZOCHINAMIDE see BCL250
BENZO-CHINON (GERMAN) see QQS200
BENZOCHLORPROPAMID see BEG000
2,3-BENZOCHRYSENE see BCG500
BENZO(a)CHRYSENE see PIB750
BENZO(b)CHRYSENE see BCG500
BENZO(c)CHRYSENE see BCG750
BENZO(d,e,f)CHRYSENE see BCS750
BENZO(g)CHRYSENE see BCH000
BENZO(10,11)CHRYSENO(1,2-b)OXIRENE-6-β,7-α-DIHYDRO see BCV750
BENZO CONGO RED see SGQ500
5,6-BENZOCOUMARIN-3-CARBONIC ACID ETHYL ETHER see OOI200
5,6-BENZOCOUMARIN-3-CARBOXYLIC ACID ETHYL ESTER see OOI200
N-6-(3,4-BENZOCOUMARINYL)ACETAMIDE see BCH250
BENZOCTAMINE HYDROCHLORIDE see BCH750
1-BENZOCYCLOBUTENYL n-BUTYL KETONE see BCH800
BENZO(de)CYCLOPENT(a)ANTHRACENE see BCI000
1H-BENZO(a)CYCLOPENT(b)ANTHRACENE see BCI250
BENZO DARK GREEN B see CMO830
BENZO DARK GREEN BA-CF see CMO830
BENZO DEEP BLACK E see AQP000
BENZO DEEP BROWN NZ see CMO820
BENZODIAPIN see MDQ250
1,4-BENZODIAZINE see QRJ000
1,3-BENZODIAZOLE see BCB750
1H,3H-BENZO(1,2-c:4,5-c')DIFURAN-1,3,5,7-TETRONE see PPQ635
1,2-BENZODIHYDROPYRONE (FCC) see HHR500
BENZODIOXANE HYDROCHLORIDE see BCI500
2-((1,4-BENZODIOXAN-2-YLMETHYL)AMINO)ETHANOL HYDROCHLORIDE see HKN500
3-(((1,4-BENZODIOXAN-2-YL)METHYL)AMINO)-N-METHYLPROPIONAMIDE see MHD300
3-(((1,4-BENZODIOXAN-2-YL)METHYL)AMINO)-1-MORPHOLINO-1-PROPANONE see MRR125
1-(1,4-BENZODIOXAN-2-YLMETHYL)PIPERIDINEHYDROCHLORIDE see BCI500
2,3-BENZODIOXIN-1,4-DIONE see PHY500
1,4-BENZODIOXINE, 2,3-DIHYDRO-, 6-(4-FLUOROPHENYL)- see DKT500
3,4-BENZODIOXOLE-5-CARBOXALDEHYDE see PIW250
1,3-BENZODIOXOLE-5-(2-PROPEN-1-OL) see BCJ000
4-(1,3-BENZODIOXOL-5-YL)-3-BUTEN-2-ONE see MJR250
2-(4-(1,3-BENZODIOXOL-5-YLMETHYL)-1-PIPERAZINYL)PYRIMIDINE see TNR485
1-(5-(1,3-BENZODIOXOL-5-YL)-1-OXO-2,4-PENTADIENYL)PIPERIDINE (E,E)- (9CI) see PIV600
1,3-BENZODIOXOL-5-YL-OXO-2,4-PENTADIENYL-PIPERINE see PIV600
1,3-BENZODITHIOLIUM PERCHLORATE see BCJ125
BENZODODECINIUM BROMIDE see BEO000
BENZODOL see BCJ150
BENZOE-DIAETHYL (GERMAN) see BQA000
BENZOEPIN see EAQ750
BENZOESAEURE (GERMAN) see BCL750
BENZOESAEURE (NA-SALZ) (GERMAN) see SFB000
BENZOESTROFOL see EDP000
BENZO FAST BLACK G see CMN240
BENZO FAST RED 8BL see CMO885
BENZO FAST RED F see CMO870
BENZO(h)FLAVONE see NBI100
7,8-BENZOFLAVONE (7CI) see NBI100
BENZOFLEX 9-88 see DWS800
BENZOFLEX 9-98 see DWS800
BENZOFLEX S-552 see PBC000
BENZOFLEX P 200 see PKE750
BENZOFLEX P-600 see PKE750

BENZOFLEX 9-88 SG see DWS800
11,12-BENZOFLUORANTHENE see BCJ750
BENZO(1)FLUORANTHENE see BCJ500
2,3-BENZOFLUORANTHENE see BAW250
3,4-BENZOFLUORANTHENE see BAW250
7,8-BENZOFLUORANTHENE see BCJ500
8,9-BENZOFLUORANTHENE see BCJ750
BENZO(b)FLUORANTHENE see BAW250
BENZO(j)FLUORANTHENE see BCJ500
11,12-BENZO(k)FLUORANTHENE see BCJ750
BENZO(k)FLUORANTHENE see BCJ750
BENZO(e)FLUORANTHENE see BAW250
2,3-BENZOFLUORANTHRENE see BAW250
BENZO(jk)FLUORENE see FDF000
BENZOFOLINE see EDP000
BENZOFORM BLACK BCN-CF see AQP000
BENZOFORM BLACK RRA-CF see CMN240
BENZOFURAN, 3-(3,5-DIBROMO-4-HYDROXYBENZOYL)-2-MESITYL- see DDP300
BENZOFURAN, 3-(p-(2-(DIETHYLAMINO)ETHOXY)BENZOYL)-2-ETHYL- see EDV700
BENZOFURAN, 3-(p-(2-(DIETHYLAMINO)ETHOXY)BENZOYL)-2-MESITYL-, HY-DROCHLORIDE see DHR900
BENZOFURAN, 3-(3,5-DIIODO-4-HYDROXYBENZOYL)-2-ETHYL- see EID200
BENZOFURAN, 3-(3,5-DIIODO-4-HYDROXYBENZOYL)-2-MESITYL- see DNF550
BENZOFURAN, (2-ETHYL-3-(4'-HYDROXYBENZOYL)) see BBJ500
BENZOFURAN, 3-(p-HYDROXYBENZOYL)-2-MESITYL- see HNK700
BENZOFURAN, 4-(p-HYDROXYBENZOYL)-2-MESITYL- see HNK800
BENZOFURAN, 2-MESITYL-3-(p-ANISOYL)- see AOY300
BENZOFURAN, 2-MESITYL-3-(p-METHOXYBENZOYL)- see AOY300
2,3-BENZOFURAN see BCK250
BENZO(b)FURAN see BCK250
BENZOFURAN see BCK250
2-BENZOFURANCARBOXYLIC ACID see CNU875
1-(2-BENZOFURANYL)ETHANONE see ACC100
BENZO(b)FURAN-2-YL METHYL KETONE see ACC100
2-BENZOFURANYL METHYL KETONE see ACC100
BENZOFUR D see PEY500
BENZOFURFURAN see BCK250
BENZOFUR GG see ALT000
BENZOFUR MT see TGL750
22H-BENZOFURO(3A,3-H)(1,5,10)TRIAZACYCLOEICOSINE-3,14,22-TRIONE,4,5,6,7,8,9,10,11,12,13,20A,21,23,24-TETRADECAHYDRO-17,19-ETH-ENO-, HYDROCHLORIDE see LIP000
BENZOFUROLINE see BEP500
BENZOFUR P see ALT250
p-(7-BENZOFURYLAZO)-N,N-DIMETHYLANILINE see BCL100
BENZO GREEN B see CMO840
BENZO GREEN BG-CF see CMO840
BENZO GREEN CA-CF see CMO840
BENZO GREEN GA-CF see CMO840
BENZO GREY LBGV see CMN230
BENZOGUANAMINE see BCL250
BENZO-GYNOESTRYL see EDP000
BENZO(a)HEPTALEN-9(5H)-ONE, 7-AMINO-6,7-DIHYDRO-1,2,3-TRIMETHOXY-10-(METHYLTHIO)-, (S)- see DBA200
BENZO(a)HEPTALEN-9(5H)-ONE, 6,7-DIHYDRO-1,2,3,10-TETRAMETHOXY-7-(METHYLAMINO)-, (S)- see MIW500
BENZOHEXONIUM see HEG100
BENZOHYDRAZIDE see BBV250
BENZOHYDRAZINE see BBV250
BENZOHYDROL see HKF300
BENZOHYDROQUINONE see HIH000
BENZOHYDROXAMATE see BCL500
BENZOHYDROXAMIC ACID see BCL500
2-(BENZOHYDRYLOXY)-N,N-DIMETHYLETHYLAMINE see BBV500
BENZOIC ACID (DOT) see BCL750
BENZOIC ACID see BCL750
BENZOIC ACID, 2-(ACETYLAMINO)-(9CI) see AAJ150
BENZOIC ACID, 2-(ACETYLSELENO)- see SBU100
BENZOIC ACID AMIDE see BBB000
BENZOIC ACID, 4-AMINO- see AIH600
BENZOIC ACID, 2-AMINO-3,5-DIIODO-(9CI) see API800
BENZOIC ACID, compd. with 2-AMINOETHANOL (1:1) see MRH300
BENZOIC ACID, 5-((4'-((2-AMINO-8-HYDROXY-6-SULFO-1-NAPHTHALE-NYL)AZO)(1,1'-BIPHENYL)-4-YL )AZO)-2-HYDROXY-, DISODIUM SALT see CMO870
BENZOIC ACID, 4-((AMINOIMINOMETHYL)AMINO)-, 2-(METHOXYCARBO-NYL)PHENYL ESTER see CBT175
BENZOIC ACID, 4-((AMINOIMINOMETHYL)AMINO)-, 2-METHOXY-4-(2-PROPE-NYL)PHENYL ESTER see MDY300
BENZOIC ACID, 2-(6-AMINO-3-IMINO-3H-XANTHEN-9-YL)-, METHYL ESTER, MONOHYDROCHLORIDE see RGP600
BENZOIC ACID, p-AMINO-, METHYL ESTER see AIN150
BENZOIC ACID, 2-AMINO-, 2-PHENYLETHYL ESTER see APJ500

BENZOIC ACID, 4-(AMINOSULFONYL)-(9CI) see SNL840
BENZOIC ACID, 5-(AMINOSULFONYL)-2,4-DICHLORO- see DGK900
BENZOIC ACID, AMMONIUM SALT see ANB100
BENZOIC ACID, 4-ARSENOSO- see CCI550
BENZOIC ACID AZIDE see BDL750
BENZOIC ACID, 2-BENZOYL- see BDL850
BENZOIC ACID, o-BENZOYL- see BDL850
BENZOIC ACID, BENZYL ESTER see BCM000
BENZOIC ACID, p-BROMO- see BMU100
BENZOIC ACID, 4-BROMO-(9CI) see BMU100
BENZOIC ACID-n-BUTYL ESTER see BQK250
BENZOIC ACID, 2-((2-CARBOXYPHENYL)AMINO)-4-CHLORO-, DISODIUM SALT see LHZ600
BENZOIC ACID, CHLORIDE see BDM500
BENZOIC ACID, 3-CHLORO- see CEL290
BENZOIC ACID, m-CHLORO- see CEL290
BENZOIC ACID, p-CHLORO- see CEL300
BENZOIC ACID, 4-CHLORO-(9CI) see CEL300
BENZOIC ACID, 5-CHLORO-2-HYDROXY- see CLD825
BENZOIC ACID, 3-CHLORO-2-HYDROXYPROPYL ESTER see CKQ500
BENZOIC ACID, 2-CHLORO-5-NITRO, METHYL ESTER see CJC515
BENZOIC ACID, 5-(2-CHLORO-4-(TRIFLUOROMETHYL)PHENOXY)-2-NITRO- see CLS075
BENZOIC ACID, CINNAMYL ESTER see CMQ750
BENZOIC ACID, 3,5-DIACETAMIDO-2,4,6-TRIIODO-, compd. with 1-DEOXY-1-(METHYLAMINO)-d-GLUCITOL see AOO875
BENZOIC ACID, 3,5-DIAMINO-, DIHYDROCHLORIDE see DBQ200
BENZOIC ACID, 5-((4'-((2,6-DIAMINO-3-(8-HYDROXY-3,6-DISULFO-7-((4-SULFO-1-NAPHTHALENYL) AZO)-2-NAPHTHALENYL)AZO)-5-METHYLPHE-NYL)AZO)(1,1'-BIPHENYL)-4-YL)AZO)-2-HYDROXY-, TETRASODIUM SALT see CMO820
BENZOIC ACID, 5-((4'-((2,6-DIAMINO-3-METHYL-5-((4-SULFOPHE-NYL)AZO)PHENYL)AZO)(1,1'-BIPHE NYL)-4-YL)AZO)-2-HYDROXY-, DISODI-UM SALT see CMO810
BENZOIC ACID, 5-((4'-((2,6-DIAMINO-3-METHYL-5-((4-SULFOPHE-NYL)AZO)PHENYL)AZO)(1,1'-BIPHE NYL)-4-YL)AZO)-2-HYDROXY-3-METHYL-, DISODIUM SALT see CMO825
BENZOIC ACID, 2,4-DICHLORO- see DER100
BENZOIC ACID, 3,6-DICHLORO-2-HYDROXY-(9CI) see DGK250
BENZOIC ACID, 2,4-DICHLORO-5-SULFAMOYL- see DGK900
BENZOIC ACID, DIESTER with DIETHYLENE GLYCOL see DJE000
BENZOIC ACID DIESTER with DIPROPYLENE GLYCOL see DWS800
BENZOIC ACID DIESTER with POLYETHYLENE GLYCOL 600 see PKE750
BENZOIC ACID DIETHYLAMIDE see BCM250
BENZOIC ACID, 3,4-DIHYDROXY- see POE200
BENZOIC ACID, 3,5-DIHYDROXY-(9CI) see REF200
BENZOIC ACID, 2,5-DIHYDROXY-, MONOSODIUM SALT (9CI) see GCU050
BENZOIC ACID, 3,5-DIIODO-4-HYDROXY- see DNF300
BENZOIC ACID, 3,4-DIMETHOXY- see VHP600
BENZOIC ACID, 3,4-DIMETHYL- see DQM850
BENZOIC ACID, 3,5-DIMETHYL- see MDJ748
BENZOIC ACID, p-(DIMETHYLAMINO)- see DOU650
BENZOIC ACID,2-((3-(4-(1,1-DIMETHYLETHYL)PHENYL)-2-METHYLPROPYLI-DENE)AMINO)-, METHYLESTER see LFT100
BENZOIC ACID, 2,6-DIMETHYL-4-PROPOXY-, 2-METHYL-2-(1-PYRROLIDI-NYL)PROPYL ESTER, HYDROCHLORIDE see DTS625
BENZOIC ACID, o-((3-(4,6-DIMETHYL-2-PYRIMIDINYL)UREIDO)SULFONYL)-, METHYL ESTER see SNW550
BENZOIC ACID-n-DIPROPYLENE GLYCOL DIESTER see DWS800
BENZOIC ACID, 2,2'-DISELENOBIS- see SBU150
BENZOIC ACID, 2,2'-DITHIOBIS-(9CI) see BHM300
BENZOIC ACID, 3,3'-DITHIOBIS(6-NITRO- see DUV700
BENZOIC ACID, 2,2'-DITHIODI- see BHM300
BENZOIC ACID ESTRADIOL see EDP000
BENZOIC ACID-1-ETHYL-4-PIPERIDYL ESTER, HYDROCHLORIDE see EOW000
BENZOIC ACID-2-(1-ETHYL-2-PIPERIDYL)ETHYL ESTER HYDROCHLORIDE see EOY100
BENZOIC ACID, 2-FORMYL-(9CI) see FNK010
BENZOIC ACID, p-GUANIDINO-, 4-ALLYL-2-METHOXYPHENYL ESTER see MDY300
BENZOIC ACID, p-GUANIDINO-, 4-METHYL-2-OXO-2H-1-BENZOPYRAN-7-YL ESTER see GLC100
BENZOIC ACID, HEXYL ESTER see HFL500
BENZOIC ACID HYDRAZIDE, 3-HYDRAZONE with DAUNORUBICIN, MONO-HYDROCHLORIDE see ROZ000
BENZOIC ACID, p-HYDRAZINO- see HHB100
BENZOIC ACID, 4-HYDRAZINO-(9CI) see HHB100
BENZOIC ACID, 4-HYDROXY-3,5-DIMETHOXY- see SPE700
BENZOIC ACID, 2-HYDROXY-, 4-(1,1-DIMETHYLETHYL)PHENYL ESTER see BSH100
BENZOIC ACID, 2-HYDROXY-3,5-DINITRO-(9CI) see HKE600
BENZOIC ACID, p-HYDROXY-, HEPTYL ESTER see HBO650
BENZOIC ACID, 4-HYDROXY-, HEPTYL ESTER (9CI) see HBO650
BENZOIC ACID, 2-HYDROXY-, HYDRAZIDE see HJL100

BENZOIC ACID(4-(HYDROXYIMINO)-2,5-CYCLOHEXADIEN-1-YLIDENE) HYDRA-ZIDE see BDD000
BENZOIC ACID, 2-HYDROXY-, 3-METHYL-2-BUTENYL ESTER see PMB600
BENZOIC ACID, 2-HYDROXY-, 4-METHYLPHENYL ESTER (9CI) see THD850
BENZOIC ACID, 2-HYDROXY-, compounded with MORPHOLINE (1:1) see SAI100
BENZOIC ACID, 2-HYDROXY-4-NITRO- see NJF100
BENZOIC ACID, 4-HYDROXY-3-NITRO-, METHYL ESTER see HMY075
BENZOIC ACID, 2-HYDROXY-5-((3-NITROPHENYL)AZO)-, MONOSODIUM SALT see SIT850
BENZOIC ACID-m-HYDROXYPHENYL ESTER see HNH500
BENZOIC ACID, p-IODO- see IEE025
BENZOIC ACID, ISOPROPYL ESTER see IOD000
BENZOIC ACID, 2-METHOXY-, METHYL ESTER see MLH800
BENZOIC ACID, 2-(6-(METHYLAMINO)-3-(METHYLIMINO)-3H-XANTHEN-9-YL)-, MONOPERCHLORATE see RGW100
BENZOIC ACID, 3-METHYL-2-BUTENYL ESTER see MHU150
BENZOIC ACID, 1-(3-METHYL)BUTYL ESTER see IHP100
BENZOIC ACID, 4-METHYLPHENYL ESTER see TGX100
BENZOIC ACID-2-(4-METHYLPIPERIDINO)ETHYL ESTER HYDROCHLORIDE see MOI250
BENZOIC ACID-1-METHYL-4-PIPERIDYL ESTER HYDROCHLORIDE see MON500
BENZOIC ACID, 2-(METHYLSELENO)- see MPI200
BENZOIC ACID, o-(METHYLSELENO)- see MPI200
BENZOIC ACID, o-(METHYLSELENO)-, see MPI205
BENZOIC ACID, o-(METHYLTELLURO)-, SODIUM SALT see MPN275
BENZOIC ACID NITRILE see BCQ250
BENZOIC ACID, 3-NITRO-(9CI) see NFG000
BENZOIC ACID, PENTAFLUORO- see PBD275
BENZOIC ACID, PENTYL ESTER see PBV800
BENZOIC ACID, PEROXIDE see BDS000
BENZOIC ACID-1-PHENETHYL-4-PIPERIDYL ESTER HYDROCHLORIDE see PDJ250
BENZOIC ACID, 2-(2-PHENYLETHYLPIPERIDINO) ETHYL ESTER, HYDRO-CHLORIDE see PFD500
BENZOIC ACID-3-(2-PHENYLETHYLPIPERIDINO) PROPYL ESTER, HYDRO-CHLORIDE see PFD750
BENZOIC ACID, o-(PHENYLHYDROXYARSINO)- see PFI600
BENZOIC ACID, PHENYLMETHYL ESTER see BCM000
BENZOIC ACID, o-(PHENYLTHIO)- see PGL270
BENZOIC ACID, 2-(PHENYLTHIO)-(9CI) see PGL270
BENZOIC ACID, 2-(PHOSPHONOOXY)- (9CI) see PHA575
BENZOIC ACID, 1-PROPYL-1-PIPERIDYL ESTER HYDROCHLORIDE see PNT750
BENZOIC ACID, SODIUM SALT see SFB000
BENZOIC ACID p-SULFAMIDE see SNL840
BENZOIC ACID, p-SULFAMOYL- see SNL840
BENZOIC ACID, TETRAESTER with PENTAERYTHRITOL see PBC000
BENZOIC ACID, THIO- see TFC550
BENZOIC ACID, p-TOLYL ESTER see TGX100
BENZOIC ACID, 3,4,5-TRIHYDROXY-, ETHYL ESTER (9CI) see EKM100
BENZOIC ACID, 3,4,5-TRIMETHOXY-, METHYL ESTER see MQF300
BENZOIC ACID, 2,4,6-TRIMETHYL- see IKN300
BENZOIC ACID, VINYL ESTER see VMK000
BENZOIC ACID-N,N-DIETHYLAMIDE see BCM250
BENZOIC ALDEHYDE see BAY500
BENZOIC-3-CHLORO-N-ETHOXY-2,6-DIMETHOXYBENZIMIDIC ANHYDRIDE see BCP000
BENZOIC ETHER see EGR000
BENZOIC HYDRAZIDE see BBV250
o-BENZOIC SULPHIMIDE see BCE500
BENZOIC TRICHLORIDE see BFL250
BENZOIMIDAZOLE see BCB750
BENZOIN see BCP250
4,5-BENZO-2,3-1',2'-INDENOINDOLE (FRENCH) see BCE000
BENZOINOXIM (CZECH) see BCP500
α-BENZOIN OXIME see BCP500
BENZOKETCTRIAZINE see BDH000
BENZOL (DOT) see BBL250
BENZOLE see BBL250
BENZO LEATHER BLACK E see AQP000
BENZOLENE see BBL250
BENZOLIN see NOF500
BENZOLINE see PCT250
BENZOLIN HYDROCHLORIDE see NOF500
BENZOLO (ITALIAN) see BBL250
BENZOMATE see BCP000
BENZOMETAN see BCP650
BENZOMETH-AMINE BROMIDE see BCP685
BENZON 00 see HND100
BENZONAL see BDS300
BENZO(a)NAPHTHO(8,1,2-cde)NAPHTHACENE see BCP750
BENZO(h)NAPHTHO(1,2-f,s-3)QUINOLINE see BCQ000
BENZONE see BRF500

BENZONITRILE (DOT) see BCQ250
BENZONITRILE see BCQ250
BENZONITRILE, 3-AMINO-(9CI) see AIR125
BENZONITRILE, p-FLUORO- see FGH100
BENZONITRILE, p-ISOPROPYL- see IOD050
BENZONITRILE, 3-METHYL-(9CI) see TGT250
BENZONITRILE, 4-(1-METHYLETHYL)- see IOD050
BENZONITRILE, p-NITRO- see NFI010
BENZONITRILE, 4-NITRO-(9CI) see NFI010
BENZOPARADIAZINE see QRJ000
BENZOPENICILLIN see BDY669
BENZO(rst)PENTAPHENE see BCQ500
BENZO(rst)PENTAPHENE-5-CARBOXALDEHYDE see BCQ750
BENZOPERIDOL see FGU000, FLK100
BENZOPEROXIDE see BDS000
1,12-BENZOPERYLENE see BCR000
BENZO(ghi)PERYLENE see BCR000
BENZO(h)PHENALENO(1,9-bc)ACRIDINE see BCR500
BENZO(a)PHENALENO(1,9-hi)ACRIDINE see BCR250
BENZO(a)PHENALENO(1,9-i,j)ACRIDINE see BCR500
BENZO(c)PHENALENO(1,9-i,j)ACRIDINE see BCR250
1,2-BENZOPHENANTHRENE see CML810
BENZO(1)PHENANTHRENE see TMS000
2,3-BENZOPHENANTHRENE see BBC250
3,4-BENZOPHENANTHRENE see BCR750
9,10-BENZOPHENANTHRENE see TMS000
BENZO(a)PHENANTHRENE see BBC250, CML810
BENZO(b)PHENANTHRENE see BBC250
BENZO(c)PHENANTHRENE see BCR750
BENZO(def)PHENANTHRENE see PON250
(±)-BENZO(c)PHENANTHRENE-3,4-DIHYDRODIOL see BCS100
(+)-BENZO(c)PHENANTHRENE-3,4-DIOL-1,2-EPOXIDE-1 see BCS103
(±)-BENZO(c)PHENANTHRENE-3,4-DIOL-1,2-EPOXIDE-1 see BMK634
(+)-BENZO(c)PHENANTHRENE-3,4-DIOL-1,2-EPOXIDE-2 see BCS105
(−)-BENZO(c)PHENANTHRENE-3,4-DIOL-1,2-EPOXIDE-2 see BCS110
BENZO(c)PHENANTHRENE-3-α-4-β-DIOL, 1,2,3,4-TETRAHYDRO-1-β,2-β-EPOXY-, (±)- see BMK634
BENZO(c)PHENANTHRENE-3,4-DIOL, 1,2,3,4-TETRAHYDRO-1,2-EPOXY-, (Z)-(+)-(1R,2S,3R,4S)- see BCS103
BENZO(c)PHENANTHRENE-3,4-DIOL, 1,2,3,4-TETRAHYDRO-1,2-EPOXY-, (E)-(−)-(1R,2S,3S,4R)- see BCS110
BENZO(c)PHENANTHRENE-3,4-DIOL, 1,2,3,4-TETRAHYDRO-1,2-EPOXY-1, (E)-(+)-(1S,2R,3R,4S)- see BCS105
BENZO(rst)PHENANTHRO(10,1,2-cde)PENTAPHENE-9,18-DIONE, DICHLORO- see DFN450
BENZO(c)PHENATHRENE-8-CARBOXALDEHYDE see BCS000
BENZOPHENONE 12 see HND100
BENZOPHENONE-2 see BCS325
BENZOPHENONE-3 see MES000
BENZOPHENONE 4 see HLR700
BENZOPHENONE see BCS250
BENZOPHENONE-2-CARBOXYLIC ACID see BDL850
BENZOPHENONE, 2-HYDROXY-4-(OCTYLOXY)- see HND100
p-BENZOPHENONE, METHYL- see MHF750
BENZOPHENONE, OXIME see BCS400
BENZOPHENONE PINACOL see TEA600
BENZOPHENOXIME see BCS400
BENZOPHOSPHATE see BDJ250
BENZOPINACOL see TEA600
BENZOPINACONE see TEA600
BENZOPIPERILONE (ITALIAN) see BCP650
3,4-BENZOPIRENE (ITALIAN) see BCS750
BENZOPROPYL see AHI250
2H-1-BENZOPYRAN, 3-ACETYL- see BCS500
2H-1-BENZOPYRAN, 3-ACETYL-5-METHOXY- see MEC340
2H-1-BENZOPYRAN, 6,7-DIMETHOXY-2,2-DIMETHYL- see AEX850
3H-2-BENZOPYRAN-7-CARBOXYLIC ACID, 4,6-DIHYDRO-8-HYDROXY-3,4,5-TRIMETHYL-6-OXO-, (3R-trans)- see CMS775
2H-1-BENZOPYRAN-6-OL, 3,4-DIHYDRO-2,5,7,8-TETRAMETHYL-2-(4,8,12-TRIMETHYLTRIDECYL)-, ACETATE see TGJ055
2H-1-BENZOPYRAN-5-OL, 2-METHYL-2-(4-METHYL-3-PENTENYL)-7-PENTYL- see PBW400
2H-1-BENZOPYRAN-2-ONE see CNV000
4H-1-BENZOPYRAN-4-ONE, 7-((6-O-(6-DEOXY-α-I-MANNOPYRANOSYL)-β-D-GLUCOPYRANOSYL)OXY)-2,3-DIHYDRO-5-HYDROXY-2-(3-HYDROXY-4-METHOXYPHENYL)-, MONOMETHYL ETHER see MKK600
4H-1-BENZOPYRAN-4-ONE, 2,3-DIHYDRO-2-PHENYL-(9CI) see FBW150
2H-1-BENZOPYRAN-2-ONE, 6,7-DIHYDROXY-4-METHYL-(9CI) see MJV800
2H-1-BENZOPYRAN-2-ONE, 7-(DIMETHYLAMINO)-4-METHYL-(9CI) see DPJ800
4H-1-BENZOPYRAN-4-ONE, 8-((DIMETHYLAMINO)METHYL)-7-METHOXY-2,3-DIMETHYL-, HYDROCHLORIDE see DRL600
2H-1-BENZOPYRAN-2-ONE, 4-HYDROXY-3-(1-(4-NITROPHENYL)-3-OXOBUTYL)- see ABF750
2H-1-BENZOPYRAN-2-ONE, 7-METHOXY- see MEK300

4H-1-BENZOPYRAN-4-ONE, 7-METHOXY-2,3-DIMETHYL-8-(4-MORPHOLINYLMETHYL)-, HYDROCHLORIDE see DSM600
4H-1-BENZOPYRAN-4-ONE, 2-PHENYL- see PER700
2-(5H-(1)BENZOPYRANO(2,3-b)PYRIDIN-7-YL)PROPIONIC ACID see PLX400
1-(2H-1-BENZOPYRAN-3-YL)ETHANONE see BCS500
2H-1-BENZOPYRAN-3-YL METHYL KETONE see BCS500
1-(2H-1-BENZOPYRAN-8-YLOXY)-3-ISOPROPYLAMINO-2-PROPANOL see HLK800
1,2-BENZOPYRENE see BCT000
3,4-BENZOPYRENE see BCS750
4,5-BENZOPYRENE see BCT000
6,7-BENZOPYRENE see BCS750
BENZO(a)PYRENE see BCS750
BENZO(e)PYRENE see BCT000
BENZO(a)PYRENE-6-CARBOXALDEHYDE THIOSEMICARBAZONE see BCT500
BENZO(a)PYRENE-6-CARBOXYALDEHYDE see BCT250
BENZO(e)PYRENE-4,5-DIHYDRODIOL see DMK400
(E)-BENZO(a)PYRENE-4,5-DIHYDRODIOL see DLC600
BENZO(a)PYRENE-7,8-DIHYDRODIOL see BCT750
BENZO(a)PYRENE-7,8-DIHYDRODIOL-9,10-EPOXIDE (anti) see BCU000
anti-BENZO(a)PYRENE-7,8-DIHYDRODIOL-9,10-OXIDE see BCU000
BENZO(a)PYRENE-7,8-DIHYDRO-7,8-EPOXY see BCV750
BENZO(a)PYRENE, 4,5-DIHYDROXY-4,5-DIHYDRO- see DLB800
anti(±)BENZO(a)PYRENE-DIOL-EPOXIDE see BCU250
anti-BENZO(a)PYRENE-DIOLEPOXIDE see DMQ000
BENZO(e)PYRENE, 9,10-DIOL-11,12-EPOXIDE 1 (cis) see DMR150
BENZO(a)PYRENE DIOL EPOXIDE ANTI see BCU250
BENZO(a)PYRENE-1,6-DIONE see BCU500
BENZO(a)PYRENE-3,6-DIONE see BCU750
BENZO(a)PYRENE-6,12-DIONE see BCV000
1,6-BENZO(a)PYRENEDIONE see BCU500
3,6-BENZO(a)PYRENEDIONE see BCU750
6,12-BENZO(a)PYRENEDIONE see BCV000
BENZO(a)PYRENE-4,5-EPOXIDE see BCV500
BENZO(a)PYRENE-7,8-EPOXIDE see BCV750
BENZO(a)PYRENE-6-METHANOL see BCV250
BENZO(a)PYRENE MONOPICRATE see BDV750
BENZO(a)PYRENE-11,12-OXIDE see BCW250
BENZO(a)PYRENE-4,5-OXIDE see BCV500
BENZO(a)PYRENE-7,8-OXIDE see BCV750
BENZO(a)PYRENE-9,10-OXIDE see BCW000
BENZO(a)PYRENE-1,6-QUINONE see BCU500
BENZO(a)PYRENE-3,6-QUINONE see BCU750
BENZO(a)PYRENE-6,12-QUINONE see BCV000
6,12-BENZOPYRENE QUINONE see BCV000
BENZO(a)PYRENE, 7,8,9,10-TETRAHYDRO-7-β,8-α-9-α-10-α-TETRAHYDROXY- see TDD750
BENZO(a)PYRENE-7-β,8-α-9-α-10-α-TETRAOL see TDD750
BENZO(a)PYREN-11-OL see BCY750
BENZO(a)PYREN-12-OL see BCZ000
BENZO(a)PYREN-10-OL see BCY500
BENZO(a)PYREN-2-OL see BCX000
BENZO(a)PYREN-3-OL see BCX250
BENZO(a)PYREN-5-OL see BCX500
BENZO(a)PYREN-6-OL see BCX750
BENZO(a)PYREN-7-OL see BCY000
BENZO(a)PYREN-9-OL see BCY250
BENZO(1,2)PYRENO(4,5-b)OXIRENE-3b,4b-DIHYDRO see BCV500
BENZO(b)PYRIDINE see QMJ000
BENZO(c)PYRIDINE see IRX000
5,6-BENZOPYRIDO(2′,3′:1,2)CARBAZOLE see BDB000
7,8-BENZOPYRIDO(2′,3′:1,2)CARBAZOLE see BDA750
13H-BENZO(g)PYRIDO(2,3-a)CARBAZOLE see BDB000
7H-BENZO(c)PYRIDO(2,3-g)CARBAZOLE see BDA250
5,6-BENZOPYRIDO(3′,2′:1,2)CARBAZOLE see BDB250
5,6-BENZOPYRIDO(3′,2′:3,4)CARBAZOLE see BDA250
1,2-BENZOPYRIDO(3′,2′:5,6)CARBAZOLE see BDA000
3,4-BENZOPYRIDO(3′,2′:5,6)CARBAZOLE see BDA500
13H-BENZO(g)PYRIDO(3,2-a)CARBAZOLE see BDB250
7H-BENZO(a)PYRIDO(3,2-g)CARBAZOLE see BDA000
7H-BENZO(c)PYRIDO(3,2-g)CARBAZOLE see BDA500
13H-BENZO(g)PYRIDO(3,2-i)CARBAZOLE see BDA750
11H-BENZO(g)PYRIDO(4,3-b)INDOLE see BDB500
1,2-BENZOPYRONE see CNV000
2,3-BENZOPYRROLE see ICM000
BENZOPYRROLE see ICM000
1-BENZOPYRYLIUM, 2-(3,4-DIHYDROXYPHENYL)-3,5,7-TRIHYDROXY-, CHLORIDE see COI750
BENZOQUIN see AEY000
BENZOQUINAMIDE see BCL250
1,4-BENZOQUINE see QQS200
BENZOQUINOL see HIH000
2,3-BENZOQUINOLINE see ADJ500
BENZO(b)QUINOLINE see ADJ500

2H-BENZO(a)QUINOLIZIN-2-ONE, 1,3,4,6,7,11b-HEXAHYDRO-3-ISOBUTYL-9,10-DIMETHOXY- see TBJ275
1,2-BENZOQUINONE see BDC250
1,4-BENZOQUINONE see QQS200
BENZOQUINONE (DOT) see BDC250, QQS200
o-BENZOQUINONE see BDC250
p-BENZOQUINONE see QQS200
p-BENZOQUINONE AMIDINOHYDRAZONE THIOSEMICARBAZONE see AHI875
BENZOQUINONE AZIRIDINE see BDC750
BENZOQUINONE-1,4-BIS(CHLOROIMINE)(1,4-BIS(CHLORIMIDO)-2,5-CYCLOHEXADIENE) see BDD125
p-BENZOQUINONE, 2,6-DI-tert-BUTYL- see DDV500
1,4-BENZOQUINONE DIIMINE see BDD200
1,4-BENZOQUINONE DIOXINE see DVR200
BENZOQUINONE GUANYLHYDRAZONE THIOSEMICARBAZONE see AHI875
p-BENZOQUINONE, compounded with HYDROQUINONE see QFJ000
p-BENZOQUINONE IMINE see BDD500
p-BENZOQUINONE MONOIMINE see BDD500
1,2-BENZOQUINONE MONOXIME see NLF300
1,4-BENZOQUINONE-N'-BENZOYLHYDRAZONE OXIME see BDD000
p-BENZOQUINONE OXIME BENZOYLHYDRAZONE see BDD000
p-BENZOQUINONE, 2-PHENYL- see PEL750
BENZOQUINONE, TRIMETHYL-(6CI) see POG400
p-BENZOQUINONE, 2,3,5-TRIMETHYL-(8CI) see POG400
p-BENZOQUINONIMINE see BDD500
BENZO RED 3B see CMO875
2,1,3-BENZOSELENADIAZOLE, 5,6-DIMETHYL- see DQO650
2,1,3-BENZOSELENADIAZOLE, 5-METHYL- see MHI300
BENZOSELENAZOLIUM, 3-ETHYL-2-(3-(3-ETHYL-2-BENZOSELENAZOLINYLIDENE)-2-METHYLPROPENYL)-, IODIDE see DJQ300
BENZO SKY BLUE A-CF see CMO500
BENZO SKY BLUE S see CMO500
o-BENZOSULFIMIDE see BCE500
BENZOSULFONAMIDE see BBR500
BENZOSULFONAZOLE see BDE500
BENZO-2-SULPHIMIDE see BCE500
BENZOSULPHIMIDE see BCE500
3,4-BENZOTETRACENE see BCG500
3,4-BENZOTETRAPHENE see BCG500
BENZO(c)TETRAPHENE see BCG500
BENZO-1,2,3-THIADIAZOLE-1,1-DIOXIDE see BDE000
BENZOTHIAMIDE see BBM250
BENZOTHIAZIDE see BDE250
BENZOTHIAZOLE see BDE500
BENZOTHIAZOLE, 6-AMINO-2-MERCAPTO- see AIS550
BENZOTHIAZOLE, 2-(p-AMINOPHENYL)-6-METHYL- see MHJ300
BENZOTHIAZOLE, 5-CHLORO-2-METHYL- see CIH100
BENZOTHIAZOLE DISULFIDE see BDE750
BENZOTHIAZOLE, 2-MERCAPTO-6-AMINO- see AIS550
2-BENZOTHIAZOLESULFENAMIDE, N,N-DIISOPROPYL- see DNN900
7-BENZOTHIAZOLESULFONIC ACID see ALX000
7-BENZOTHIAZOLESULFONIC ACID, 2,2'-(1-TRIAZENE-1,3-DIYLDI-4,1-PHENYLENE)BIS(6-METHYL-), DISODIUM SALT see CMP050
2-BENZOTHIAZOLETHIOL see BDF000
2-BENZOTHIAZOLETHIOL, ZINC SALT (2:1) see BHA750
2(3H)-BENZOTHIAZOLETHIONE see BDF000
BENZOTHIAZOLE-2-THIONE see BDF000
(p-(2-BENZOTHIAZOLYL)BENZYL)PHOSPHONIC ACID DIETHYL ESTER see DIU500
2-BENZOTHIAZOLYL-N,N-DIETHYLTHIOCARBAMYL SULFIDE see BDF250
2-BENZOTHIAZOLYL DISULFIDE see BDE750
BENZOTHIAZOLYL DISULFIDE see BDE750
2-BENZOTHIAZOLYL MERCAPTAN see BDF000
2-BENZOTHIAZOLYL MORPHOLINODISULFIDE see BDF750
2-BENZOTHIAZOLYL-N-MORPHOLINOSULFIDE see BDG000
N-(2-BENZOTHIAZOLYL)-N'-METHYL-N'NITROSOUREA see NKR000
N-(2-BENZOTHIAZOLYL)-N'-METHYLUREA see MHM500
2-BENZOTHIAZOLYLSULFENYL MORPHOLINE see BDG000
4-(2-BENZOTHIAZOLYLTHIO)MORPHOLINE see BDG000
BENZOTHIAZYL-2-CYCLOHEXYLSULFENAMIDE see CPI250
BENZO(b)THIEN-4-YL METHYLCARBAMATE see BDG250
4-BENZOTHIENYL METHYLCARBAMATE see BDG250
BENZOTHIOAMIDE see BBM250
BENZO(b)THIOPHENE-4-OL METHYLCARBAMATE see BDG250
6-BENZOTHIOPURINE see BDG325
BENZOTHIOZANE see FNF000
BENZOTHIOZON see FNF000
BENZOTRIAZINEDITHIOPHOSPHORIC ACID DIMETHOXY ESTER see ASH500
BENZOTRIAZINE derivative of an ETHYL DITHIOPHOSPHATE see EKN000
BENZOTRIAZINE derivative of a METHYL DITHIOPHOSPHATE see ASH500
1,2,3-BENZOTRIAZIN-4(1H)-ONE see BDH000
3H-1,2,3-BENZOTRIAZIN-4-ONE see BDH000
1,2,3-BENZOTRIAZOLE see BDH250
1H-BENZOTRIAZOLE see BDH250
1H-BENZOTRIAZOLE, 1-HYDROXY-, AMMONIUM SALT see HJN600

1H-BENZOTRIAZOLE, 6-NITRO- see NFJ000
2-(2H-BENZOTRIAZOL-2-YL)-4-METHYLPHENOL see HML500
BENZOTRICHLORIDE (DOT, MAK) see BFL250
BENZOTRIFLUORIDE see BDH500
BENZOTRIFLUORIDE, 4-CHLORO-3,5-DINITRO- see CGM225
3,5-BENZOTRIFLUORODIAMINE see TKB775
BENZOTRIFUROXAN see BBU125
2,5,8-BENZOTRIOXACYCLOUNDECIN-1,9-DIONE, 3,4,6,7-TETRAHYDRO-(9CI) see DJD700
BENZO(b)TRIPHENYLENE see BDH750
BENZOTRIS(c)FURAZAN-2-OXIDE see BBU125
BENZOTROPINE see BDI000
BENZOTROPINE MESYLATE see TNU000
BENZOTROPINE METHANESULFONATE see TNU000
BENZOURACIL see QEJ800
BENZO VIOLET N see CMP000
BENZOXALE see TMP750
BENZOXAMATE see BCP000
2-BENZOXAXOLOL see BDJ000
2H-3,1-BENZOXAZINE-2,4(1H)-DIONE, 6-NITRO- see NHK600
BENZOXAZOLE see BDI500
BENZOXAZOLE, 2-CHLORO- see CEM850
3(2H)-BENZOXAZOLEPROPANESULFONIC ACID, 2-(4-(1,3-DIBUTYLTETRAHYDRO-4,6-DIOXO-2-THIOXO-5(2H)-PYRIMIDINYLIDENE)-2-BUTENYLIDENE)-, SODIUM SALT see EDD500
2-BENZOXAZOLETHIOL see MCK900
2-BENZOXAZOLINONE see BDJ000
BENZOXAZOLINONE see BDJ000
S-((3-BENZOXAZOLINYL-6-CHLORO-2-OXO)METHYL) O,O-DIETHYLPHOSPHORODITHIOATE see BDJ250
2(3H)-BENZOXAZOLONE see BDJ000
BENZOXAZOLONE see BDJ000
BENZOXINE see BFW250
BENZOXONIUM CHLORIDE see BDJ600
3-BENZOXY-1-(2-METHYLPIPERIDINO)PROPANE see PIV750
3-BENZOXY-1-(2-METHYLPIPERIDINO)PROPANE HYDROCHLORIDE see IJZ000
dl-3-BENZOXY-1-(2-METHYLPIPERIDINO)PROPANE HYDROCHLORIDE see IJZ000
BENZO YELLOW TZ see CMP050
BENZOYL see BDS000
BENZOYLACETIC ACID ETHYL ESTER see EGR600
BENZOYL-ACETON see BDJ800
BENZOYLACETONE see BDJ800
BENZOYLACONINE see PIC250
BENZOYL ALCOHOL see BDX500
BENZOYLAMIDE see BBB000
2-BENZOYLAMIDOFLUORENE-2'-CARBOXYLATE see FEN000
5-BENZOYLAMINO-1-CHLOROANTHRAQUINONE see BDK750
(±)-α-(BENZOYLAMINO)-4-(2-(DIETHYLAMINO)ETHOXY)-N,N-DIPROPYLBENZENEPROPANAMIDE see TGF175
(±)-4-(BENZOYLAMINO)-5-(DIPROPYLAMINO)-5-OXO-PENTANOIC ACID see BGC625
2-BENZOYLAMINOFLUORENE see FDX000
2-BENZOYLAMINOFLUORENE-2'-CARBOXYLATE see FEN000
4-((2-(BENZOYLAMINO)-3-(4-HYDROXYPHENYL)-1-OXOPROPYL)AMINO)BENZOIC ACID see CML835
(S)-4-(((2-BENZOYLAMINO)-3-(4-HYDROXYPHENYL)-1-OXOPROPYL)AMINO)BENZOIC ACID see CML835
o-(BENZOYLAMINO)PHENYL DISULFIDE see BDK800
BENZOYL AZIDE see BDL750
BENZOYLBENZENE see BCS250
N-2 (5-BENZOYL-BENZIMIDAZOLE) CARBAMATE de METHYLE see MHL000
5-BENZOYL-2-BENZIMIDAZOLECARBAMIC ACID METHYL ESTER see MHL000
N-(BENZOYL-5, BENZIMIDAZOLYL)-2, CARBAMATE de METHYLE see MHL000
(5-BENZOYL-1H-BENZIMIDAZOL-2-YL)-CARBAMIC ACID METHYL ESTER see MHL000
2-BENZOYLBENZOIC ACID see BDL850
α-BENZOYL-ω-(BENZOYLOXY)POLY(OXY-1,2-ETHANEDIYL) see PKE750
α-BENZOYLBENZYL CYANIDE see OOK200
BENZOYLCARBINOL TRIMETHYLACETATE see PCV350
BENZOYLCARBINYL TRIMETHYLACETATE see PCV350
BENZOYL CHLORIDE (DOT) see BDM500
BENZOYL CHLORIDE see BDM500
BENZOYL CHLORIDE, METHOXY-(9CI) see AOY250
BENZOYL CHLORIDE, 2-NITRO-(9CI) see NFL000
BENZOYL CYANIDE-o-(DIETHOXYPHOSPHINOTHIOYL)OXIME see BAT750
N-BENZOYL-N-DEACETYL COLCHICINE see BDV250
N-BENZOYL-N-(3,4-DICHLOROPHENYL)-I-ALANINE ETHYL ESTER see EGS000
BENZOYLDIETHYLAMINE see BCM250
4-BENZOYL-3,5-DIMETHYL-N-NITROSOPIPERAZINE see NJL850
BENZOYLENEUREA see QEJ800
1-BENZOYL-5-ETHYL-5-PHENYLBARBITURIC ACID see BDS300

1-BENZYLBIGUANIDE HYDROCHLORIDE see BEA850
N-BENZYLBIGUANIDE HYDROCHLORIDE see BEA850
BENZYLBIGUANIDE HYDROCHLORIDE see BEA850
BENZYLBIS(β-CHLOROETHYL)AMINE see BIA750
BENZYL-BIS(2-CHLOROETHYL)AMINE HYDROCHLORIDE see EAK000
5-BENZYL-2,2-BIS(TRIFLUOROMETHYL)-4-METHYLOXAZOLIDINE HYDRATE
  see BEB750
BENZYL BLUE R see ADE750
BENZYL BROMIDE see BEC000
BENZYL n-BUTANOATE see BED000
BENZYL-tert-BUTANOL see BEC250
BENZYL BUTYL PHTHALATE see BEC500
1-(1-BENZYLBUTYL)PYRROLIDINE HYDROCHLORIDE see PNS000
BENZYL n-BUTYRATE see BED000
N-BENZYLCARBAMIC ACID-3-DIMETHYLAMINOPHENYL ESTER METHIODIDE
  see HNO000
N-BENZYL-CARBAMIC ACID-3-(TRIMETHYLAMMONIO)PHENYL ESTER, ME-
  THYLSULFATE see BED500
BENZYL CARBAMIC ACID-m-(TRIMETHYLAMMONIO)PHENYL ESTER IODIDE
  see HNO000
BENZYLCARBAMIC ESTER of 3-OXYPHENYLDIMETHYLAMINE HYDROCHLO-
  RIDE see BED250
BENZYLCARBAMIC ESTER of 3-OXYPHENYLTRIMETHYLAMMONIUM ME-
  THYLSULFATE see BED500
BENZYLCARBAMIDE see BFN125
1-(2-(BENZYLCARBAMOYL)ETHYL)-2-ISONICOTINOYLHYDRAZINE see BET000
N'-β-BENZYLCARBAMOYLETHYL-N²-ISONICOTINOYLHYDRAZINE see BET000
3-BENZYL-4-CARBAMOYLMETHYLSYDNONE see BED750
3-(N'-BENZYLCARBAMOYLOXY)-N,N-DIMETHYL-ANILINE HYDROCHLORIDE
  see BED250
(3-(BENZYLCARBAMOYLOXY)PHENYL)TRIMETHYLAMMONIUM IODIDE see
  HNO000
(2-(2-BENZYLCARBAMYL)ETHYL)-HYDRAZIDE ISONICOTINIC ACID see
  BET000
BENZYL CARBINOL see PDD750
BENZYLCARBINOL ISOBUTYRATE see PDF750
BENZYLCARBINYL ACETATE see PFB250
BENZYLCARBINYL ANTHRANILATE see APJ500
BENZYL CARBINYL BENZOATE see PFB750
BENZYLCARBINYL BUTYRATE see PFB800
BENZYLCARBINYL CINNAMATE see BEE250
BENZYLCARBINYL FORMATE see PFC250
BENZYLCARBINYL ISOBUTYRATE see PDF750
BENZYLCARBINYL 2-METHYLBUTYRATE see PDF780
BENZYLCARBINYL PROPIONATE see PDK000
BENZYLCARBINYL-α-TOLUATE see PDI000
BENZYLCARBONYL CHLORIDE see BEF500
1-BENZYL-3-CARBOXYPYRIDINIUM CHLORIDE BENZYL ESTER see SAA000
BENZYL-C₁₂-C₁₆-ALKYLDIMETHYL AMMONIUM CHLORIDES see QAT520
BENZYL "CELLOSOLVE" see EJI500
BENZYLCELOSOLV see EJI500
BENZYLCHLORID (GERMAN) see BEE375
BENZYL CHLORIDE see BEE375
BENZYL-α-CHLOROACETATE see BEE500
BENZYL CHLOROACETATE see BEE500
BENZYL CHLOROCARBONATE (DOT) see BEF500
N-BENZYL-6'-CHLORO-2-(DIETHYLAMINO)-o-ACETOTOLUIDIDE HYDROCHLO-
  RIDE see BEE750
BENZYLCHLORODIMETHYLSILANE see BEE800
2-(N-BENZYL-2-CHLOROETHYLAMINO)-1-PHENOXYPROPANE see PDT250
2-(N-BENZYL-2-CHLOROETHYLAMINO)-1-PHENOXYPROPANE HYDROCHLO-
  RIDE see DDG800
BENZYL(2-CHLOROETHYL)-(1-METHYL-2-PHENOXYETHYL)AMINE see PDT250
BENZYL(2-CHLOROETHYL)(1-METHYL-2-PHENOXYETHYL)AMINE HYDRO-
  CHLORIDE see DDG800
BENZYL CHLOROFORMATE (DOT) see BEF500
BENZYL CHLOROFORMATE see BEF500
2-BENZYL-4-CHLOROPHENOL see BEF750, CJU250
o-BENZYL-p-CHLOROPHENOL see CJU250
BENZYLCHLOROPHENOL see BEF750
2-BENZYL-4-CHLOROPHENOL, SODIUM SALT see SFB200
1-BENZYL-4-(3-(p-CHLOROPHENYL)-3-PHENYLPROPIONYL)PIPERAZINE see
  BFE770
N-BENZYL-β-CHLOROPROPANAMIDE see BEG000
N-BENZYL-3-CHLOROPROPIONAMIDE see BEG000
N-BENZYL-β-CHLOROPROPIONAMIDE see BEG000
BENZYL 2-CHLORO-4-(TRIFLUOROMETHYL)-5-THIAZOLECARBOXYLATE see
  BEG300
BENZYL CINNAMATE see BEG750
BENZYL CYANIDE see PEA750
1-BENZYL-2(1H)-CYCLOHEPTIMIDAZOLONE see BEH000
BENZYLCYCLOHEPTIMIDAZOL-2(1H)-ONE see BEH000
3-((1-BENZYLCYCLOHEPTYL)OXY)-N,N-DIMETHYLPROPYLAMINE FUMARATE
  see BAV250

N-(3-(1-BENZYL-CYCLOHEPTYLOXY)-PROPYL)-N,N-DIMETHYL-AMMONIUM-HY-
  DROGENFUMARAT (GERMAN) see BAV250
3-BENZYL-3,3-α,4,5,6,6-α,9,10,12,15-DECAHYDRO-6,12,15-TRIHYDROXY-
  4,10,12-TRIMETHYL-5-METHYLENE-1H-CYCLOUNDEC(d)ISOINDOLE-1,11(2H)-
  DIONE, 15-ACETATE see ZUS000
BENZYL DICHLORIDE see BAY300
2-(BENZYL(2-(DIETHYLAMINO)ETHYL)AMINO)ACETANILIDE DIHYDROCHLO-
  RIDE see BEI250
2-(BENZYL(2-(DIETHYLAMINO)ETHYL)AMINO)-o-ACETOTOLUIDIDE DIHYDRO-
  CHLORIDE see BEI500
2-(BENZYL(2-(DIETHYLAMINO)ETHYL)AMINO)-6'-CHLORO-o-ACETOTOLUIDIDE
  DIHYDROCHLORIDE see BEI750
(2-(BENZYL(3-DIETHYLAMINO)PROPYL)AMINO)-o-ACETOTOLUIDIDE DIHYDRO-
  CHLORIDE see BEJ250
2-(BENZYL(3-(DIETHYLAMINO)PROPYL)AMINO)-2',6'-ACETOXYLIDIDE DIHY-
  DROCHLORIDE see BEJ500
3-BENZYL-3,4-DIHYDRO-6-(TRIFLUOROMETHYL)-2H-1,2,4-BENZOTHIADIAZINE-
  7-SULFONAMIDE 1,1-DIOXIDE see BEQ625
BENZYL-N,N-DIMETHYLAMINE see DQP800
N-BENZYLDIMETHYLAMINE see DQP800
BENZYLDIMETHYLAMINE see DQP800
BENZYLDIMETHYLAMINE METHIODIDE see BFM750
2-(BENZYL(2-DIMETHYL AMINOETHYL)AMINO)PYRIDINE see TMP750
2-(BENZYL(2-(DIMETHYLAMINO)ETHYL)AMINO)PYRIDINE HYDROCHLORIDE
  see POO750
2-(BENZYL(2-(DIMETHYLAMINO)ETHYL)AMINO)PYRIDINE MONOHYDROCHLO-
  RIDE see POO750
N-BENZYL-N-DIMETHYLAMINOETHYL α-AMINOPYRIDINE MONOHYDROCHLO-
  RIDE see POO750
N-BENZYL-N-DIMETHYLAMINOETHYL, ANILINE, HYDROCHLORIDE see
  PEN000
1-BENZYL-3-(3-(DIMETHYLAMINO)PROPOXY)-1H-INDAZOLE see BCD750
1-BENZYL-3-(3-(DIMETHYLAMINO)PROPOXY)-1H-INDAZOLE HYDROCHLORIDE
  see BBW500
1-BENZYL-3-γ-DIMETHYLAMINOPROPOXY-1H-INDAZOLE HYDROCHLORIDE
  see BBW500
BENZYL 1-(2-(DIMETHYLAMINO)PROPYL)PYRROL-2-YL, CITRATE KETONE
  see BEL525
BENZYLDIMETHYLAMMONIUM HEXAFLUOROARSENATE see BEL550
BENZYL DIMETHYL CARBINOL see DQQ200
BENZYLDIMETHYL CARBINYL ACETATE see BEL750
BENZYL DIMETHYLCARBINYL n-BUTYRATE see BEL850
BENZYL DIMETHYLCARBINYL BUTYRATE see BEL850
BENZYLDIMETHYLCETYLAMMONIUM CHLORIDE see BEL900
BENZYLDIMETHYLDODECYLAMMONIUM BROMIDE see BEO000
BENZYLDIMETHYLDODECYLAMMONIUM CHLORIDE see BEM000
BENZYLDIMETHYLEICOSANYLAMMONIUM CHLORIDE see BEM250
1-BENZYL-2,3-DIMETHYL-GUANIDINE SULFATE (1:1/2) see BFW250
BENZYLDIMETHYLHEXADECYLAMMONIUM CHLORIDE see BEL900
BENZYLDIMETHYLOCTYLAMMONIUM CHLORIDE see OEW000
1-BENZYL-2,5-DIMETHYL SEROTONIN HYDROCHLORIDE see BEM750
BENZYLDIMETHYLSTEARYLAMMONIUM CHLORIDE see DTC600
BENZYLDIMETHYL-p-(1,1,3,3-TETRAMETHYLBUTYL)PHENOXYETHOXY-ETHY-
  LAMMONIUM CHLORIDE see BEN000
BENZYLDIMETHYL(2-(2-(p-(1,1,3,3-TETRAMETHYLBU-
  TYL)PHENOXY)ETHOXY)ETHYL) AMMONIUM CHLORIDE see BEN000
2-BENZYLDIOXOLAN see BEN250
dl-1-BENZYL-4-(2,6-DIOXO-3-PHENYL-3-PIPERIDYL)PIPERIDINE HYDROCHLO-
  RIDE see BBU800
(+)-1-BENZYL-4-(2,6-DIOXO-3-PHENYL-3-PIPERIDYL)PIPERIDINE HYDROCHLO-
  RIDE see DBE000
1-BENZYL DIPROPYL KETONE see BEN800
BENZYL DODECANOATE see BEU750
BENZYLDODECYLBIS(2-HYDROXYETHYL)AMMONIUM CHLORIDE (6CI,7CI)
  see BDJ600
BENZYLDODECYLDIMETHYL AMMONIUM BROMIDE see BEO000
BENZYLE (CHLORURE de) (FRENCH) see BEE375
BENZYLENE CHLORIDE see BAY300
BENZYL ETHANOATE see BDX000
BENZYL ETHER see BEO250
5-BENZYL-5-ETHYLBARBITURIC ACID see BEA500
8-BENZYL-7-(2-(ETHYL(2-HYDROXYETHYL)AMINO)ETHYL)THEOPHYLLINE, HY-
  DROCHLORIDE see THL750
BENZYLETS see BCM000
BENZYL FAST BLUE R see ADE750
BENZYL FAST RED 2BG see NAO600
BENZYL FAST RED BG see CMM330
BENZYL FAST RED GRG see CMM320
BENZYL FAST YELLOW RS see CMM759
BENZYL FORMATE see BEP250
BENZYLFUROLINE see BEP500
5-BENZYL-3-FURYLMETHYL(+)-trans-CHRYSANTHEMATE see BEP750
5-BENZYL-3-FURYLMETHYL (+)-cis-CHRYSANTHEMATE see RDZ875
5-BENZYL-3-FURYL METHYL(±)-cis,trans-CHRYSANTHEMATE see BEP500

(5-BENZYL-3-FURYL) METHYL-2,2-DIMETHYL-3-(2-METHYLPROPENYL)-CYCLO-PROPANECARBOXYLATE see BEP500
3-BENZYL-4-HEPTANONE see BEN800
BENZYLHYDRAZINE see BEQ000
BENZYLHYDRAZINE DIHYDROCHLORIDE see BEQ250
BENZYLHYDRAZINE HYDROCHLORIDE see BEQ500
BENZYLHYDROFLUMETHIAZIDE see BEQ625
BENZYL HYDROQUINONE see AEY000
BENZYLHYDROSULFIDE see TGO750
1-BENZYL-2-(HYDROXYACETYL)INDOLE see BES300
BENZYL-o-HYDROXYBENZOATE see BFJ750
2-BENZYL-4-HYDROXYMETHYL-1,3-DIOXANE see PDX250
2-BENZYL-4-HYDROXYMETHYL-1,3-DIOXOLANE see PDX250
4-BENZYL-α-(p-HYDROXYPHENOL)-β-METHYL-1-PIPERIDINEETHANOL see IAG600
4-BENZYL-α-(p-HYDROXYPHENYL)-β-METHYL-1-PIPERIDINE-ETHANOL-(L)-(+)-TARTRATE see IAG625
4-BENZYL-α-(p-HYDROXYPHENYL)-β-METHYL-1-PIPERIDINEETHANOL TAR-TRATE see IAG625
trans-BENZYLIDENACETONE see BAY275
BENZYLIDENEACETALDEHYDE see CMP969
2-BENZYLIDENEACETAMIDE see CMP970
BENZYLIDENE ACETONE see SMS500
trans-BENZYLIDENEACETONE see BAY275
2-BENZYLIDENEACETOPHENONE see CDH000
S,S'-BENZYLIDENE BIS(O,O-DIMETHYL PHOSPHORODITHIOATE) see BES250
1,1'-(BENZYLIDENEBIS((2-METHOXY-p-PHENYLENE))(AZO))DI-2-NAPHTHOL see DOO400
3-BENZYLIDENE-2-BUTANONE see MNS600
BENZYLIDENE CHLORIDE (DOT) see BAY300
BENZYLIDENE CHLORIDE see BAY300
4,6-o-BENZYLIDENE-β-d-GLUCOPYRANOSIDE PODOPHYLLOTOXIN see BER500
BENZYLIDENE GLYCEROL see BBA000
2-BENZYLIDENE-1-HEPTANOL see AOH000
BENZYLIDENEMETHYLPHOSPHORODITHIOATE see BES250
BENZYLIDENEPHENYLACETONITRILE see DVX600
BENZYLIDYNE CHLORIDE see BFL250
BENZYLIDYNE FLUORIDE see BDH500
2-BENZYL-2-IMIDAZOLINE see BBW750
2-BENZYL-4,5-IMIDAZOLINE see BBW750
2-BENZYL-4,5-IMIDAZOLINE HYDROCHLORIDE see BBW750
BENZYLIMIDAZOLINE HYDROCHLORIDE see BBJ750
2-BENZYL-2-IMIDAZOLINE MONOHYDROCHLORIDE see BBJ750
((1-BENZYL-1H-INDAZOL-3-YL)OXY)ACETIC ACID see BAV325
4-(1-BENZYL-3-INDOLETHYL)PYRIDINE HYDROCHLORIDE see BCE250
1-BENZYL-2-INDOLYL HYDROXYMETHYL KETONE see BES300
BENZYL ISOAMYL ETHER see BES500
BENZYL ISOBUTYRATE (FCC) see IJV000
BENZYL ISOEUGENOL see BES750
BENZYL ISOEUGENOL ETHER see BES750
N-BENZYL-β-(ISONICOTINOYLHYDRAZINE)PROPIONAMIDE see BET000
N-BENZYL-β-(ISONICOTINYLHYDRAZINO)PROPIONAMIDE see BET000
BENZYL ISOPENTYL ETHER see BES500
BENZYLISOPROPYL PROPIONATE see DQQ400
BENZYL-ISOTHIOCYANATE see BEU250
BENZYLISOTHIOUREA HYDROCHLORIDE see BEU500
2-BENZYLISOTHIOURONIUM CHLORIDE see BEU500
BENZYLISOTHIOURONIUM CHLORIDE see BEU500
BENZYL ISOVALERATE (FCC) see ISW000
BENZYLKYANID see PEA750
BENZYL LAURATE see BEU750
BENZYL MERCAPTAN see TGO750
6-BENZYLMERCAPTOPURINE see BDG325
BENZYL METHANOATE see BEP250
2-BENZYL-4-METHANOL-1,3-DIOXANE see PDX250
4-BENZYL-α-(4-METHOXYPHENYL)-β-METHYL-1-PIPERIDINEETHANOL see BEU800
BENZYL-2-METHOXY-4-PROPENYLPHENYL ETHER see BES750
1-BENZYL-2-METHYL-3-(2-AMINOETHYL)-5-METHOXYINDOLE HYDROCHLO-RIDE see BEM750
BENZYL-3-METHYLBUTANOATE see ISW000
BENZYL-3-METHYL BUTYRATE see ISW000
BENZYL METHYL CARBINOL see PGA600
BENZYLMETHYLCARBINYL ACETATE see ABU800
1-BENZYL-2-METHYLHYDRAZINE see MHN750
1-BENZYL-1-(5-METHYL-3-ISOXAZOIYLCARBONYL)HYDRAZINE see IKC000
1-BENZYL-2-(5-METHYL-3-ISOXAZOIYL-CARBONYL)HYDRAZINE see IKC000
N'-BENZYL N-METHYL-5-ISOXAZOLECARBOXYLHYDRAZIDE-3 see IKC000
BENZYL METHYL KETONE see MHO100
1-BENZYL-2-METHYL-5-METHOXYTRYPTAMINE HYDROCHLORIDE see BEM750
4-BENZYL-1-(1-METHYL-4-PIPERIDYL)-3-PHENYL-3-PYRAZOLIN-5-ONE see BCP650
BENZYL-2-METHYL PROPIONATE see IJV000

N-BENZYL-N-METHYL-2-PROPYNYLAMINE see MOS250
N-BENZYL-N-METHYL-2-PROPYNYLAMINE HYDROCHLORIDE see BEX500
BENZYLMETHYLPROPYNYLAMINE HYDROCHLORIDE see BEX500
(1-BENZYL-3-METHYL-5-PYRAZOLYLOXYETHYL)TRIMETHYLAMMONIUM IO-DIDE see BEX750
BENZYL MONOCHLORACETATE see BEE500
6-BENZYL-MP see BDG325
BENZYL MUSTARD OIL see BEU250, BFL000
N-BENZYL-N',N'-DIETHYL-N-1-NAPHTHYLETHYLENEDIAMINE see BEJ825
N-BENZYL-N',N'-DIETHYL-N-2-NAPHTHYLETHYLENEDIAMINE see BEJ830
N-BENZYL-N',N''-DIMETHYLGUANIDINE SULFATE see BFW250
N-BENZYL-N',N'-DIMETHYL-N-1-NAPHTHYLETHYLENEDIAMINE see BEM325
N-BENZYL-N',N'-DIMETHYL-N-2-NAPHTHYLETHYLENEDIAMINE see BEM330
N-BENZYL-N',N'-DIMETHYL-N-PHENYLETHYLENEDIAMINE see BEM500
N-BENZYL-N',N'-DIMETHYL-N-2-PYRIDYLETHYLENE DIAMINE see TMP750
N-BENZYL-N',N'-DIMETHYL-N-2-PYRIDYL-ETHYLENEDIAMINE HYDROCHLO-RIDE see POO750
BENZYL NICOTINATE see NCR040
BENZYL NITRATE see BFA250
BENZYL NITRILE see PEA750
1-BENZYL-1-NITROSOUREA see NJM000
BENZYL NORMECHLORETHAMINE see BIA750
N-BENZYL-N'-PHENYLACETYLHYDRAZIDE see PDY870
BENZYL OXIDE (CZECH) see BEO250
BENZYLOXYCARBONYL CHLORIDE see BEF500
N-BENZYLOXYCARBONYLGLYCINE see CBR125
BENZYLOXYCARBONYLGLYCINE see CBR125
(BENZYLOXYCARBONYL)PHENYLALANINE see CBR220
I-N-BENZYLOXYCARBONYL-3-PHENYLALANINE-1,2-DIBROMOETHYL ESTER see CBR225
N-BENZYLOXYCARBONYL-I-PHENYLALANINE VINYL ESTER see CBR235
2-BENZYLOXYETHANOL see EJI500
5-BENZYLOXY-3-ISONIPECOTOYLINDOLE see BFC200
p-BENZYLOXYPHENOL see AEY000
BENZYLPENCILLINDIBENZYLETHYLENEDIAMINE SALT see BFC750
BENZYLPENICILLIN see BDY669
BENZYLPENICILLIN BENZATHINE see BFC750
BENZYLPENICILLIN G see BDY669
BENZYLPENICILLINIC ACID see BDY669
BENZYLPENICILLINIC ACID POTASSIUM SALT see BFD000
BENZYL PENICILLINIC ACID SODIUM SALT see BFD250
BENZYLPENICILLIN NOVOCAINE SALT see PAQ200
BENZYLPENICILLIN POTASSIUM see BFD000
BENZYLPENICILLIN POTASSIUM SALT see BFD000
BENZYLPENICILLIN PROCAINE see PAQ200
BENZYLPENICILLIN SODIUM see BFD250
N-BENZYL-N-PHENOXYISOPROPYL-β-CHLORETHYLAMINE HYDROCHLORIDE see DDG800
1-(2-BENZYLPHENOXY)-2-PIPERIDINOPROPANE PHOSPHATE see PJA130
BENZYL PHENYLACETATE see BFD400
BENZYL γ-PHENYLACRYLATE see BEG750
N-β-(BENZYL-PHENYLAMINO)ETHYLPIPERIDINE HYDROCHLORIDE see BEA275
p-BENZYLPHENYL CARBAMATE see PFR325
BENZYL PHENYLFORMATE see BCM000
BENZYL PHENYL KETONE see PEB000
4-BENZYLPIPERAZINYL β-(p-CHLOROPHENYL)PHENETHYL KETONE see BFE770
2-(4-BENZYL-PIPERIDINO)-1-(4-HYDROXYPHENYL)-1-PROPANOL TARTRATE (2:1) see IAG625
(+)-2-(1-BENZYL-4-PIPERIDYL)-2-PHENYLGLUTARIMIDE HYDROCHLORIDE see DBE000
(+)-3-(1-BENZYL-4-PIPERIDYL)-3-PHENYLPIPERIDINE-2,6-DIONE HYDROCHLO-RIDE see DBE000
2-BENZYLPYRIDINE see BFG600
4-BENZYLPYRIDINE see BFG750
BENZYL PYRIDINE-3-CARBOXYLATE see NCR040
BENZYL-(α-PYRIDYL)-DIMETHYLAETHYLENDIAMIN (GERMAN) see TMP750
1-BENZYL-3-(2-(4-PYRIDYL)ETHYL)INDOLE HYDROCHLORIDE see BCE250
BENZYL 4-PYRIDYL KETONE THIOSEMICARBAZONE see BFH100
N-BENZYL-N-α-PYRIDYL-N',N'-DIMETHYL-AETHYLENDIAMIN-HYDROCHLORID (GERMAN) see POO750
BENZYL RED BR see CMM330
BENZYL RED GR see CMM320
BENZYL RED GS see CMM325
BENZYL RED MG see CMM325
4-BENZYL RESORCINOL see BFI400
BENZYLRODIURAN see BEQ625
BENZYL SALICYLATE see BFJ750
BENZYL SCARLET 3BS see CMO875
BENZYLSENFOEL (GERMAN) see BEU250
BENZYL SILANE see BFJ825
BENZYL SODIUM see BFJ850
BENZYLSTEARYLDIMETHYLAMMONIUM CHLORIDE see DTC600
BENZYL SULFITE see BFK000

BENZYLSULFONYL FLUORIDE see TGO300
BENZYL SULFOXIDE see DDH800
BENZYLSULPHONYL FLUORIDE see TGO300
3-BENZYLSYDNONE-4-ACETAMIDE see BED750
BENZYLT see PDT250
1-BENZYL-2-(3-(4,5,6,7-TETRAHYDROBENZISOXAZOY-LYL)CARBONYL)HYDRAZINE HYDROCHLORIDE see BFK325
2-(4-BENZYL-1,2,3,6-TETRAHYDROPYRIDINO)-1-(4'-METHOXYPHENYL)-1-PROPANOL see RCA450
D-BENZYL TG see EDC560
S-BENZYL THIOBENZOATE see BFK750
BENZYL THIOCYANATE see BFL000
6-BENZYLTHIOGUANINE see BFL125
BENZYLTHIOGUANINE see BFL125
BENZYLTHIOL see TGO750
3-((BENZYLTHIO)METHYL)-6-CHLORO-1,2,4-BENZOTHIADIAZINE-7-SULFON-AMIDE-1,1-DIOXIDE see BDE250
3-BENZYLTHIOMETHYL-6-CHLORO-2H-1,2,4-BENZOTHIADIAZINE-7-SULFON-AMIDE-1,1-DIOXIDE see BDE250
3-BENZYLTHIOMETHYL-6-CHLORO-7-SULFAMOYL-1,2,4-BENZOTHIADIAZINE-1,1-DIOXIDE see BDE250
3-BENZYLTHIOMETHYL-6-CHLORO-7-SULFAMYL-1,2,4-BENZOTHIADIAZINE-1,1-DIOXIDE see BDE250
3-BENZYLTHIOMETHYL-6-CHLORO-7-SULFAMYL-2H-1,2,4-BENZOTHIADIAZINE-1,1-DIOXIDE see BDE250
2-BENZYL-2-THIO-PSEUDOUREA HYDROCHLORIDE see BEU500
BENZYL THIOPSEUDOUREA HYDROCHLORIDE see BEU500
6-(BENZYLTHIO)PURINE see BDG325
S-BENZYLTHIURONIUM CHLORIDE see BEU500
BENZYLTHIURONIUM CHLORIDE see BEU500
BENZYL TRICHLORIDE see BFL250
BENZYLTRIETHYLAMMONIUM CHLORIDE see BFL300
3-BENZYL-6-TRIFLUOROMETHYL-7-SULFAMOYL-3,4-DIHYDRO-1,2,4-BENZOTHI-ADIAZINE-1,1-DIOXIDE see BEQ625
BENZYLTRIMETHYLAMMONIUM CHLORIDE see BFM250
BENZYLTRIMETHYLAMMONIUM HYDROXIDE see BFM500
BENZYL TRIMETHYL AMMONIUM IODIDE see BFM750
BENZYLTRIMETHYLAMMONIUM METHOXIDE see TLM100
1-BENZYLUREA see BFN125
N-BENZYLUREA see BFN125
BENZYLUREA see BFN125
BENZYL VIOLET see FAG120
BENZYL VIOLET 3B see FAG120
BENZYLYT see DDG800
3,4-BENZYPYRENE see BCS750
BENZYRIN see BBW500
BENZYTOL see CLW000
BEOCID-ISOPTAL see SNQ000
BEOSIT see EAQ750
BEP see BJP899, BKH625
BEPANTHEN see PAG200
BEPANTHENE see PAG200
BEPANTOL see PAG200
BEPARON see TCC000
BEPERIDEN see BGD500
BEPROCHINE see RHZ000
BERBERIN see BFN500
BERBERINE see BFN500
BERBERINE CHLORIDE see BFN600
BERBERINE CHLORIDE DIHYDRATE see BFN550
BERBERINE HYDROCHLORIDE see BFN600
BERBERINE HYDROCHLORIDE BIHYDRATE see BFN550
BERBERINE SULFATE (2:1) see BFN625
BERBERINE SULFATE see BFN625
BERBERINE SULFATE TRIHYDRATE see BFN750
BERBERINIUM CHLORIDE see BFN600
BERBERIN SULFATE see BFN625
BERBINIUM, 7,8,13,13a-TETRADEHYDRO-9,10-DIMETHOXY-2,3-(METHYLENEDI-OXY)-, CHLORIDE see BFN600
BERCEMA see EIR000
BERCEMA FERTAM 50 see FAS000
BERCEMA NMC50 see CBM750
BERCULON A see FNF000
BERELEX see GEM000
BERGAMIOL see LFY100
BERGAMOT MINT OIL see BFN990
BERGAMOT OIL rectified see BFO000
BERGAMOTTE OEL (GERMAN) see BFO000
BERGAPTEN see MFN275
BERGENIN HYDRATE see BFO100
BERKAZON see FNF000
BERKENDYL see CDP000
BERKFLAM B 10E see PAU500
BERKFURIN see NGE000
BERKMYCEN see HOH500

BERKOLOL see ICC000
BERKOMINE see DLH600, DLH630
BERKOZIDE see BEQ625
BERLINER see BAC040
BERLISON F see HHQ800
BERMAT see CJJ250
BERNARENIN see VGP000
BERNICE see CNE750
BERNIES see CNE750
BERNOCAINE see AIT250
BERNSTEINSAEURE (GERMAN) see SMY000
BERNSTEINSAEURE-ANHYDRID (GERMAN) see SNC000
BERNSTEINSAEURE-2,2-DIMETHYLHYDRAZID (GERMAN) see DQD400
BEROCILLIN see AOD000
BEROL 28 see PJT300
BEROL 478 see DJL000
BEROMYCIN 400 see PDT750
BEROMYCIN (penicillin) see PDT750
BEROMYCIN see PDT500, PDT750
BERONALD see CHJ750
BEROTEC see FAQ100
BEROTEC HYDROBROMIDE see FAQ100
BERRY ALDER see MBU825
BERSAMA ABYSSINICA Fres. ssp. ABYSSINICA, leaf extract see BFO125
BERSEN see DDT300
BERTHOLITE see CDV750
BERTHOLLET SALT see PLA250
BERTRANDITE see BFO250
BERUBIGEN see VSZ000
BERYL see BFO500
BERYLLIA see BFT250
BERYLLIUM-9 see BFO750
BERYLLIUM see BFO750
BERYLLIUM ACETATE see BFP000
BERYLLIUM ACETATE, BASIC see BFT500
BERYLLIUM ACETATE, NORMAL see BFP000
BERYLLIUM ALUMINOSILICATE see BFO500
BERYLLIUM ALUMINUM ALLOY see BFP250
BERYLLIUM ALUMINUM SILICATE see BFO500
BERYLLIUM CARBONATE (1:1) see BFP750
BERYLLIUM CARBONATE see BFP500
BERYLLIUM CARBONATE, BASIC see BFP500
BERYLLIUM CHLORIDE see BFQ000
BERYLLIUM COMPOUND with NIOBIUM (12:1) see BFQ750
BERYLLIUM COMPOUNDS see BFQ500
BERYLLIUM COMPOUNDS, n.o.s. (UN 1566) (DOT) see BFO750
BERYLLIUM COMPOUND with TITANIUM (12:1) see BFR000
BERYLLIUM COMPOUND with VANADIUM (12:1) see BFR250
BERYLLIUM-COPPER-COBALT ALLOY see CNK700
BERYLLIUM DICHLORIDE see BFQ000
BERYLLIUM DIFLUORIDE see BFR500
BERYLLIUM DIHYDROXIDE see BFS250
BERYLLIUM DINITRATE see BFT000
BERYLLIUM FLUORIDE see BFR500
BERYLLIUM HYDRATE see BFS250
BERYLLIUM HYDRIDE see BFR750
BERYLLIUM HYDROGEN PHOSPHATE (1:1) see BFS000
BERYLLIUM HYDROXIDE see BFS250
BERYLLIUM LACTATE see LAH000
BERYLLIUM MANGANESE ZINC SILICATE see BFS750
BERYLLIUM MONOXIDE see BFT250
BERYLLIUM-NICKEL ALLOY see NCY000
BERYLLIUM NITRATE see BFT000
BERYLLIUM ORTHOSILICATE see SCN500
BERYLLIUM OXIDE see BFT250
BERYLLIUM OXIDE ACETATE see BFT500
BERYLLIUMOXIDE CARBONATE see BFP500
BERYLLIUM OXYACETATE see BFT500
BERYLLIUM OXYFLUORIDE see BFT750
BERYLLIUM PERCHLORATE see BFU000
BERYLLIUM PHOSPHATE see BFS000
BERYLLIUM SILICATE see SCN500
BERYLLIUM SILICATE HYDRATE see BFO250
BERYLLIUM SILICIC ACID see SCN500
BERYLLIUM SULFATE (1:1) see BFU250
BERYLLIUM SULFATE TETRAHYDRATE (1:1:4) see BFU500
BERYLLIUM SULPHATE TETRAHYDRATE see BFU500
BERYLLIUM TETRAHYDROBORATE see BFU750
BERYLLIUM TETRAHYDROBORATETRIMETHYLAMINE see BFV000
BERYLLIUM, powder (UN 1567) (DOT) see BFO750
BERYLLIUM ZINC SILICATE see BFV250
BERYL ORE see BFO500
BESAN see POH000
BESANTIN see MHL000
BESTATIN see BFV300

BE-STILL TREE see YAK350
BETABION see TES750
BETACIDE P see HNU500
BETADID see HJS850
BETADINE see PKE250
BETADRENOL see BQB250
BETADRENOL HYDROCHLORIDE see BQB250
BETAFEDRINA see BBK500
BETAFEDRINE see BBK500
BETAFEN see AOB250
BETAHISTINE see HGE820
BETAHISTINE MESILATE see BFV350
BETAHISTINE MESYLATE see BFV350
BETAINE see GHA050
BETAINE CEPHALORIDINE see TEY000
BETAINE HYDRAZIDE HYDROCHLORIDE see GEQ500
BETAINE LAURYLDIMETHYLAMINOACETATE see LBU200
BETAISODONA see PKE250
BETALGIL see DTL200
BETALIN 12 CRYSTALLINE see VSZ000
BETALING see BFW250
BETALIN S see TES750
BETALOC see SBV500
BETAMEC see DNO800
BETAMETHASONE see BFV750
BETAMETHASONE ACETATE mixed with BETAMETHASONE SODIUM PHOS-
     PHATE see CCS675
BETAMETHASONE 17-BENZOATE see BFV760
BETAMETHASONE BENZOATE see BFV760
BETAMETHASONE 17,21-DIPROPIONATE see BFV765
BETAMETHASONE DIPROPIONATE see BFV765
BETAMETHASONE-21-DISODIUM PHOSPHATE see BFV770
BETAMETHASONE DISODIUM PHOSPHATE see BFV770
BETAMETHASONE SODIUM PHOSPHATE see BFV770
BETAMETHASONE SODIUM PHOSPHATE mixed with BETAMETHASONE AC-
     ETATE see CCS675
BETAMETHASONE 17-VALERATE see VCA000
BETAMETHASONE VALERATE see VCA000
BETA METHYL DIGOXIN see MJD300
BETA-NAFTYLOAMINA (POLISH) see NBE500
BETANAL see MEG250
BETANAL AM see EEO500
BETANAPHTHOL ORANGE see CMM220
BETA-NEG see ICC000
BETANEX see EEO500
BETANIDINE SULFATE see BFV900
BETANIDIN SULFATE see BFV900
BETANIDOLE see BFW250
BETAPAL see NBJ100
BETAPEN see PAQ100
BETAPEN-VK see PDT750
d-BETAPHEDRINE see BBK500
BETAPRONE see PMT100
BETAPTIN see AGW000
BETAPYRIMIDUM see DJS200
BETARUNDUM see SCQ000
BETARUNDUM ST-S see SCQ000
BETARUNDUM UF see SCQ000
BETARUNDUM ULTRAFINE see SCQ000
BETASAN see DNO800
BETATRON see DXO300
BETAXIN see TES750
BETAXINA see EID000
BETAXOLOL HYDROCHLORIDE see KEA350
BETAZED see BRF500
BETEL LEAVES see BFV975
BETEL NUT, polyphenol fraction see BFW010
BETEL NUT see AQT650, BFW000
BETEL NUT TANNIN see BFW050
BETEL QUID see BFW120
BETEL QUID EXTRACT see BFW125
BETEL TOBACCO EXTRACT see BFW135
BETHABARRA WOOD see HLY500
BETHAINE CHOLINE CHLORIDE see HOA500
BETHAMETHASONE 17-BENZOATE see BFV760
BETHANECHOL CHLORIDE see HOA500
BETHANID see BFW250
BETHANIDINE, HEMISULFATE see BFW250
BETHANIDINE SULFATE see BFW250
BETHIAMIN see TES750
BETNELAN see BFV750
BETNESOL see BFV770
BETNOVATE see VCA000
BETNOVATEAT see VCA000
BETOXON see NBJ100

BETRAMIN see BBV500
BETRILOL see BON400
BETSOLAN see BFV750
BETULA OIL see MPI000
BEVALOID 35 see BLX000
BEVATINE-12 see VSZ000
BEVIDOX see VSZ000
BEVONIUM METHYL SULFATE see CDG250
BEVONIUM METILSULFATE see CDG250
BEWON see TES750
BEXANE see CLN750
BEXIDE see BJU000
BEXOL see BBQ500
BEXON see MMN250
BEXONE see CLN750
BEXT see BJU000, DKE400
BEXTON see CHS500
BEXTRENE XL 750 see SMQ500
BEZITRAMIDE see BCE825
BF 121 see FOJ100
BF 200 see CAT775
BF 5930 see OJW000
B(b)F see BAW250
B(j)F see BCJ500
BFE 60 see BAU255
B 169-FERRICYANIDE see TNH750
BFH see BRK100
BFP see BJE750
BFPO see BJE750
BFV see FMV000
BG 5930 see OJW000
BG 6080 see CBT500
B. GABONICA VENOM see BLV080
BGE (OSHA) see BRK750
BGE see BRK750
B''-GUTTIFERIN see CBA125
BH 6 see BGS250
3-tert-BHA see BQI010
BHA (FCC) see BQI000
BHANG see CBD750
BHB see BPI125
BHBN see HJQ350
α-BHC see BBQ000
β-BHC see BBR000
γ-BHC see BBQ500
BHC (USDA) see BBP750
BHC see BBQ500, BJZ000
Δ-BHC see BFW500
BH 2,4-D see DAA800
BHD see BDN125
BH DALAPON see DGI400
BHEN see BRO000
B-HERBATOX see SFS000
BHFT see BEQ625
BHIMSAIM CAMPHOR see BMD000
BH MCPA see CIR250
BH MECOPROP see CIR500
BHP see DNB200
BHPBN see HIE600
BHT (food grade) see BFW750
BI-58 see DSP400
4',4'''-BIACETANILIDE see BFX000
BIACETYL see BOT500
BIACETYLMONOXIME see OMY910
BIALAMICOL HYDROCHLORIDE see BFX125
BIALCOL see BDJ600
BIALFLAVINA see DBX400
BIALLYL see HCR500
BIALLYLAMICOL DIHYDROCHLORIDE see BFX125
BIALLYLAMICOL HYDROCHLORIDE see BFX125
BIALMINAL see EOK000
BIALPIRINIA see ADA725
BIALZEPAM see DCK759
4,4'-BIANILINE see BBX000
o,p'-BIANILINE see BGF109
p,p-BIANILINE see BBX000
N,N'-BIANILINE see HHG000
BIANISIDINE see TGJ750
1,1'-BIANTHRACENE-9,9',10,10'-TETRAONE, 2,2'-DIMETHYL- see DQR350
5,5'-BIANTHRANILIC ACID see BFX250
(3,3'-BIANTHRA(1,9-cd)PYRAZOLE)-6,6'(1H,1'H)-DIONE, 1,1'-DIETHYL- see
     CMU825
BIARISON see POB300
BIARSAN see POB300
1,1'-BIAZIRIDINYL see BFX325

BIAZOLINA see CCS250
BIBENZAL see SLR000
BIBENZENE see BGE000
α-α'-BIBENZHYDROL see TEA600
(Δ²˒²'(3H,3'H)-BIBENZO(b)THIOPHENE)-3,3'-DIONE see DNT300
((Δ²˒²'(3H,3'H))-BIBENZO(b)THIOPHENE)-3,3'-DIONE, /S CALCOLOID PRINTING
   ORANGE RE see CMU815
BIBENZYL see BFX500
BIBENZYLIDENE see SLR000
BIBENZYLIDINE see SLR000
BIBESOL see DGP900
BIC see BRQ500, IAN000
BICAM ULV see DQM600
BICA-PENICILLIN see BFC750
BICARBONATE of SODA see SFC500
BICARBURET of HYDROGEN see BBL250
BICARBURETTED HYDROGEN see EIO000
BICARNESINE see CCK655
BICEP see MQQ450
BICETONIUM see BEL900
BICHE PRIETO (MEXICO) see CNG825
BICHLORACETIC ACID see DEL000
BICHLORENDO see MQW500
BICHLORIDE of MERCURY see MCY475
BICHLORURE d'ETHYLENE (FRENCH) see EIY600
BICHLORURE de MERCURE (FRENCH) see MCY475
BICHLORURE de PROPYLENE (FRENCH) see PNJ400
BICHOL see DYE700
BICHROMATE d'AMMONIUM (FRENCH) see ANB500
BICHROMATE OF POTASH see PKX250
BICHROMATE of SODA see SGI000
BICHROMATE de SODIUM (FRENCH) see SGI000
BICILLIN see BFC750
BICIRON see RGP450
BICKIE-MOL see HIM000
BICOL see PPN100
BICOLASTIC A 75 see SMQ500
BICOLENE C see PJS750
BICOLENE H see SMQ500
BICOLENE P see PMP500
BICORTONE see PLZ000
BICP see CEX250
BI-CURVAL see ZJS300
BICYCLO(4.4.0)DECANE see DAE800
BICYCLO(0.3.5)DECA-1,3,5,7,9-PENTAENE see ASP600
BICYCLO(5.3.0)-DECA-2,4,6,8,10-PENTAENE see ASP600
BICYCLO(5.3.0)DECAPENTAENE see ASP600
BICYCLO(2,2,2)-1,4-DIAZAOCTANE see DCK400
BICYCLO(2.2.1)HEPTADIENE see NNG000
BICYCLO(2.2.1)HEPTANE, 2,2-DIMETHYL-3-METHYLENE-(9CI) see CBA500
BICYCLO(2.2.1)HEPTANE-1-METHANESULFONIC ACID, 7,7-DIMETHYL-2-OXO-,
   (1S)-(9CI) see RFU100
BICYCLO(2.2.1)HEPTANE, 2-METHOXY-1,7,7-TRIMETHYL-, exo- see IHX500
BICYCLO(2.2.1)HEPTANE, 2,2,5,6-TETRACHLORO-1,7,7-
   TRIS(CHLOROMETHYL)-, (5-endo,6-exo)- see THH575
BICYCLO(3.1.1)HEPTAN-2-OL, 2,6,6-TRIMETHYL-, (1-α-2-α-5α– see PIH050
BICYCLO(2.2.1)HEPTAN-2-OL, 1,7,7-TRIMETHYL-, endo-(9CI) see BMD000
BICYCLO(2.2.1)HEPTAN-2-ONE, 1,3,3-TRIMETHYL-, (1R)-(9CI) see FAM300
BICYCLO(3.1.1)HEPT-2-ENE-2-CARBOXALDEHYDE, 6,6-DIMETHYL- see
   FNK150
BICYCLO(2.2.1)HEPT-5-ENE-2-CARBOXALDEHYDE, 3-PROPYL- see CML620
cis-BICYCLO(2.2.1)HEPT-5-ENE-2,3-DICARBOXYLIC ACID, DIMETHYL ESTER
   see DRB400
(endo,endo)-BICYCLO(2.2.1)HEPT-5-ENE-2,3-DICARBOXYLIC ACID DIMETHYL
   ESTER see DRB400
BICYCLO(2.2.1)HEPTENE-2-DICARBOXYLIC ACID, 2-ETHYLHEXYLIMIDE see
   OES000
BICYCLO(2.2.1)HEPT-5-ENE-2,3-DICARBOXYLIC ACID, 1,4,5,6,7,7-HEXACHLO-
   RO-, DIBUTYL ESTER see CDS025
cis-BICYCLO(2,2,1-HEPTENE-2,3-DICARBOXYLIC ACID) METHYL ESTER see
   DRB400
BICYCLO(2.2.1)-HEPT-5-ENE-2,3-DICARBOXYLIC ANHYDRIDE see BFY000
BICYCLO(3.1.1)HEPT-2-ENE-2-METHANOL, 6,6-DIMETHYL-, ACETATE, (1S)-
   see MSC050
BICYCLO(2.2.1)HEPT-5-ENE-2-METHYLOL ACRYLATE see BFY250
trans-BICYCLO(2.2.1)HEPT-5-ENYL-2,3-DICARBONYLCHLORIDE see NNI000
α-BICYCLO(2.2.1)HEPT-5-EN-1-YL-α-PHENYL-PIPERIDINEPROPANOL HYDRO-
   CHLORIDE see BGD750
α-(BICYCLO(2.2.1)HEPT-5-EN-2-YL)-α-PHENYL-1-PIPERIDINEPROPANOL HY-
   DROCHLORIDE see BGD750
1-BICYCLOHEPTENYL-1-PHENYL-3-PIPERIDINO-PROPANOL-1 see BGD500
α-(BICYCLO(2.2.1)HEPT-5-EN-2-YL)-α-PHENYL-1-PIPERIDINO PROPANOL see
   BGD500
1-BICYCLOHEPTENYL-1-PHENYL-3-PIPERIDINOPROPANOL-1 HYDROCHLORIDE
   see BGD750

N'-BICYCLO(2.2.1)HEPT-2-YL-N-(2-CHLOROETHYL)-N-NITROSOUREA see
   CHE500
(1,1'-BICYCLOHEXYL)-2-ONE see LBX100
BICYCLO(4,3,0)NONA-3,7-DIENE see TCU250
BICYCLONONADIENE DIEPOXIDE see BFY750
BICYCLONONALACTONE see OBV200
BICYCLO(4.2.0)OCTA-1,3,5-TRIENE, 7-ACETYL- see BFZ120
BICYCLO(4.2.0)OCTA-1,3,5-TRIENE, 7-BENZOYL- see BFZ180
BICYCLO(4.2.0)OCTA-1,3,5-TRIENE, 7-VALERYL- see BCH800
BICYCLO(4.2.0)OCTA-1,3,5-TRIEN-7-YL BENZYL KETONE see BFZ100
BICYCLO(4.2.0)OCTA-1,3,5-TRIEN-7-YL BENZYL KETONE OXIME see BFZ110
BICYCLO(4.2.0)OCTA-1,3,5-TRIEN-7-YL METHYL KETONE see BFZ120
BICYCLO(4.2.0)OCTA-1,3,5-TRIEN-7-YL METHYL KETONE O-ACETYLOXIME
   see BFZ130
BICYCLO(4.2.0)OCTA-1,3,5-TRIEN-7-YL METHYL KETONE O-ALLYLOXIME see
   BFZ140
BICYCLO(4.2.0)OCTA-1,3,5-TRIEN-7-YL METHYL KETONE O-BUTYLOXIME see
   BFZ150
BICYCLO(4.2.0)OCTA-1,3,5-TRIEN-7-YL METHYL KETONE O-(2-HYDROXYPRO-
   PYL)OXIME see HNT550
BICYCLO(4.2.0)OCTA-1,3,5-TRIEN-7-YL METHYL KETONE O-METHYLOXIME
   see MFX550
BICYCLO(4.2.0)OCTA-1,3,5-TRIEN-7-YL METHYL KETONE OXIME see BFZ160
1-BICYCLO(4.2.0)OCTA-1,3,5-TRIEN-7-YL-1-PENTANONE see BCH800
1-BICYCLO(4.2.0)OCTA-1,3,5-TRIEN-7-YL-1-PENTANONE OXIME see BFZ170
BICYCLO(4.2.0)OCTA-1,3,5-TRIEN-7-YL PENTYL KETONE OXIME see BFZ170
BICYCLO(4.2.0)OCTA-1,3,5-TRIEN-7-YL PHENYL KETONE see BFZ180
BICYCLOPENTADIENE see DGW000
BICYCLOPENTADIENE DIOXIDE see BGA250
BICYCLO(2.1.0)PENT-2-ENE see BGA650
BIDERON see DGC800
BIDIPHEN see TFD250
BIDIRL see DGQ875
BIDISIN see CFC750
BIDRIN see DGQ875
BIEBRICH SCARLET BPC see SBC500
BIEBRICH SCARLET RED see SBC500
BIEBRICH SCARLET R MEDICINAL see SBC500
BIETASERPINE BITARTRATE see DIH000
BIETHYLENE see BOP500
1,1'-BI(ETHYLENE OXIDE) see BGA750
BIETHYLXANTHOGENTRISULFIDE see BJU000
BIFEX see PMY300
BIFLORINE see FOW000
BIFLUORIDEN (DUTCH) see FEZ000
BIFLUORURE de POTASSIUM (FRENCH) see PKU250
BIFONAZOL see BGA825
BIFONAZOLE see BGA825
BIFORMYCHLORAZIN see TKL100
BIFORMYLCHLORAZIN see TKL100
BIFORON see BQL000
BIFURON see NGG500
BIG DIPPER see DVX800
BIGITALIN see GEU000
BIG LEAF IVY see MRU359
BIGUANIDE, 1,1'-HEXAMETHYLENEBIS(5-(p-CHLOROPHENYL))-, DIGLUCO-
   NATE see CDT250
BIGUANIDE, 1-o-TOLYL- see TGX550
BIGUANIDINE, CHROMATE see BJW825
BIGUMAL see CKB250
BIGUNAL see BQL000
BIHEXYL see DXT200
BIHOROMYCIN (crystalline) see BGB250
10204-BII see FBP300
BI-KELLINA see AHK750
BIKLIN see APT000
BILAGEN see TGZ000
BILARCIL see TIQ250
BILCOLIC see MKP500
BILENE see DYE700
BILETAN see DXN800
BILEVON see HCL000
BILEVON M see DFD000
BILHARCID see PIJ600
BILICANTE see MKP500
BILIDREN see DAL000
BILIMIN see SKM000
BILIMIRO see IGA000
BILIMIRON see IGA000
BILINEURINE see CMF000
BILIOGNOST see PDM750
BILI-ORAL see EQC000
BILIGRAFIN FORTE see BGB315
BILIGRAFIN NATRIUM (GERMAN) see BGB325
BILIGRAFIN SODIUM see BGB325

BILIRON see SGD500
BILISCOPIN see IGD100
BILISELECTAN see PDM750
BILITRAST see TDE750
BILIVISTAN NATRIUM (GERMAN) see BGB350
BILIVISTAN SODIUM see BGB350
BILOBORN see MRH209
BILOBRAN see MRH209
BILOCOL see DYE700
BI-LOFT, combustion products see ADX750
BILOPAC see SKO500
BILOPAQUE see SKO500
BILOPTIN see SKM000
BILOPTINON see SKM000
BILOSTAT see DAL000
BILTRICIDE see BGB400
BIM see MQC000
Δ²,²'-BIMALONONITRILE see EEE500
6,6'-BIMETANILIC ACID see BBX500
BIMETHYL see EDZ000
BINAPACRYL see BGB500
(1,2'-BINAPHTHALENE)-1,2'-DIAMINE see BGC000
(1,1'-BINAPHTHALENE)-2,2'-DIAMINE see BGB750
(1,1'-BINAPHTHALENE)-2,2'-DIOL see BGC100
1,1'-BI-2-NAPHTHOL see BGC100
(±)-1,1-BI-2-NAPHTHOL see CDM625
β-BINAPHTHOL see BGC100
(8,8'-BI-1H-NAPHTHO(2,3-c)PYRAN)-3,3'-DIACETIC ACID, 3,3',4,4'-TETRAHY-
    DRO-9,9',10,10'-TETRAHYDRO-7,7'-DIMETHOXY-1,1'-DIOXO-, DIMETHYL ES-
    TER see VRP200
2,3,1',8'-BINAPHTHYLENE see BCJ750
β,β-BINAPHTHYLENEETHENE see PIB750
BINAZIN see TGJ150
BINAZINE see TGJ150
BINDAN see PFJ750
BINDAZAC see BAV325
BINDON see ONY000
BINDON ATHYLATHER see BGC250
BINDON ETHYL ETHER see BGC250
BINITROBENZENE see DUQ200
BINOCTAL see AMX750
BINODALINE HYDROCHLORIDE see BGC500
BINODALIN HYDROCHLORID (GERMAN) see BGC500
BINOSIDE see BGC625
BINOTAL see AIV500
BINOTAL SODIUM see SEQ000
BINOVA see DXO300
BIO 1,137 see BIT250
BIO 5,462 see EAQ750
BIOACRIDIN see DBX400
BIOALETRINA (PORTUGUESE) see BGC750
S-BIOALLETHRIN see AFR750
S-trans-BIOALLETHRIN see AFR750
BIOALLETHRIN see AFR250, BGC750
BIOALTRINA see AFR250
BIOBAMAT see MQU750
BIOBAN-C see CAW400
BIO-BEADS S-S 2 see SMQ500
BIOCALC see CAT225
BIOCAMYCIN see HGP550
BIOCETIN see CDP250
N-1386 BIOCIDE see BLM500
BIOCIDE see ADR000
BIO-CLAVE see CJU250
BIOCOLINA see CMF750
BIOCORT ACETATE see CNS825
BIOCORTAR see HHQ800
BIO-DAC 50-22 see DGX200
BIO-DES see DKA600
BIODIASTASE 1000 see BGC825
BIODOPA see DNA200
BIOEPIDERM see VSU100
BIOFANAL see NOH500
BIOFERMIN see DWX600
BIOFLAVONOID see RSU000
BIOFUREA see NGE500
BIOGASTRONE see BGD000, CBO500
BIOGRISIN-FP see GKE000
BIO (JAPANESE) see SDY600
BIOMET 204 see TMV850
BIOMET TBTO see BLL750
BIOMINE 1651 see MCB050
BIOMIORAN see CDQ750
BIOMITSIN see CMA750
BIOMYCIN see CMA750

BIOMYDRIN see SPC500
BIONIC see NCQ900
BIOPAL CVL-10 see NND000
BIOPAL NR-20 see NND000
BIOPAL VRO 10 see NND000
BIOPAL VRO 20 see NND000
BIOPHEDRIN see EAW000
BIOPHENICOL see CDP250
BIO-PHYLLINE see HLC000
BIOPRASE see BAC000
BIOQUAT 80 see QAT520
BIOQUAT 501 see QAT520
BIO-QUAT 50-24 see AFP250
BIOQUIN 1 see BLC250
BIOQUIN see BLC250, QPA000
BIORAL see BGD000, CBO500
BIORENINE see VGP000
BIORESMETHRIN see BEP750
BIORESMETHRINE see BEP750
BIORESMETRINA (PORTUGUESE) see BEP750
BIORPHEN see MJH900
BIOSCLERAN see ARQ750
BIOSECHS see PII100
BIOSEDAN see NBU000
BIOSEPT see CCX000
BIOSERPINE see RDK000
BIOSHIK see TGD250
BIOS II see VSU100
BIO-SOFT D-40 see DXW200
BIO-SOFT S 100 see LBU100
BIOSOL VETERINARY see NCG000
BIOSONE see GIE000
BIOSORB 130 see HND100
BIOSTAT see HOH500
BIOSTAT PA see HOH500
BIOSTEROL see VSK600
BIOSUPRESSIN see HOO500
BIOTERTUSSIN see CMW500
BIO-TESTICULINA see TBG000
BIO-TETRA see TBX000
BIOTHION see TAL250
d-BIOTIN see VSU100
d-(+)-BIOTIN see VSU100
BIOTIN see VSU100
(+)-BIOTIN see VSU100
BIOTIRMONE see SKJ300
BIOTROL see BAC040
2,2'-BIOXIRANE see BGA750
(S-(R*,R*))-2,2'-BIOXIRANE see BOP750
(R*,S*)-2,2'-BIOXIRANE see DHB800
BIOXIRANE see BGA750
BIOXONE see BGD250
BIOXYDE d'AZOTE (FRENCH) see NEG100
BIOXYDE de PLOMB (FRENCH) see LCX000
BIPANAL see SBN000
BIPC (the herbicide) see CEX250
BIPERIDEN see BGD500
BIPERIDEN HYDROCHLORIDE see BGD750
BIPERIDINE HYDROCHLORIDE see BGD750
2,2'-BIPHENOL see BGG000
o,o'-BIPHENOL see BGG000
p,p'-BIPHENOL see BGG500
1,1'-BIPHENYL see BGE000
BIPHENYL see BGE000
4-BIPHENYLACETAMIDE see PDY500
N-4-BIPHENYLACETAMIDE see PDY500
4-BIPHENYLACETHYDROXAMIC ACID see ACD000
(1,1'-BIPHENYL)-4-ACETIC ACID see BGE125
(4-BIPHENYL) ACETIC ACID see BGE125
4-BIPHENYLACETIC ACID see BGE125
p-BIPHENYLACETIC ACID see BGE125
(1,1'-BIPHENYL)-4-ACETIC ACID, 2-FLUOROETHYL ESTER see FDB200
4-BIPHENYLACETIC ACID, 2-FLUOROETHYL ESTER see FDB200
4-BIPHENYLACETIC ACID, 2-FLUORO-α-METHYL- see FLG100
4-BIPHENYLACETIC ACID, 2-FLUORO-α-METHYL-, 1-ACETOXYETHYL ESTER
    see FJT100
(1,1'-BIPHENYL)-4-ACETIC ACID, 2-FLUORO-α-METHYL-, 1-(ACETYLOX-
    Y)ETHYL ESTER see FJT100
(1,1'-BIPHENYL)-4-ACETIC ACID, 2-FLUORO-α-METHYL- (9CI) see FLG100
2-BIPHENYLAMINE see BGE250
(1,1'-BIPHENYL)-4-AMINE see AJS100
4-BIPHENYLAMINE see AJS100
o-BIPHENYLAMINE see BGE250
p-BIPHENYLAMINE see AJS100
BIPHENYLAMINE see AJS100

(1,1'-BIPHENYL)-2-AMINE (9CI) see BGE250
4-BIPHENYLAMINE, DIHYDROCHLORIDE see BGE300
4-BIPHENYLAMINE, 4,4'-DIMETHOXY- see BKO600
2-BIPHENYLAMINE, HYDROCHLORIDE see BGE325
(1,1'-BIPHENYL)-4-AMINE, N-HYDROXY- see BGI250
N-4-BIPHENYLBENZAMIDE see BGF000
BIPHENYL, mixed with BIPHENYL OXIDE (3:7) see PFA860
BIPHENYL, 3-BROMO- see BMW290
4-BIPHENYLCARBOXYLIC ACID see PEL600
(1,1'-BIPHENYL)-4-CARBOXYLIC ACID (9CI) see PEL600
(1,1'-BIPHENYL)-2,4'-DIAMINE see BGF109
2,4'-BIPHENYLDIAMINE see BGF109
4,4'-BIPHENYLDIAMINE see BBX000
(1,1'-BIPHENYL)-4,4'-DIAMINE (9CI) see BBX000
(1,1'-BIPHENYL)-4,4'-DIAMINE, DIHYDROCHLORIDE see BBX750
(1,1'-BIPHENYL)-4,4'-DIAMINE SULFATE (1:1) see BBY000
4,4'-BIPHENYLDICARBONITRILE see BGF250
2,2'-BIPHENYLDICARBOXALDEHYDE see DVV800
4-BIPHENYLDIMETHYLAMINE see BGF899
2,2'-BIPHENYLDIOL see BGG000
(1,1'-BIPHENYL)-2,5-DIOL see BGG250
2,5-BIPHENYLDIOL see BGG250
4,4'-BIPHENYLDIOL see BGG500
BIPHENYL-DIPHENYL ETHER mixture see PFA860
(1,1'-BIPHENYL)-2,2'-DISULFONIC ACID, 4,4'-BIS((4,5-DIHYDRO-3-METHYL-5-
   OXO-1-PHENYL-1H-PYRAZOL-4-YL)AZO)-, DISODIUM SALT see CMM759
(1,1'-BIPHENYL)-2,2'-DISULFONIC ACID, 4,4'-BIS((2-HYDROXY-1-NAPHTHALE-
   NYL)AZO)-, DISODIUM SALT see CMM325
N,N'-(1,1'-BIPHENYL)-4,4'-DIYLBIS-ACETAMIDE 4',4'''-BIACETANILIDE see
   BFX000
2,2'-(1,1'-BIPHENYL-4,4'-DIYLBIS(2-HYDROXY-4,4-DIMETHYL)-MORPHOLINIUM
   DIBROMIDE see HAQ000
2,2'-((1,1'-BIPHENYL)-4,4'-DIYLDI-2,1-ETHENEDIYL)BIS-BENZENESULFONIC
   ACID DISODIUM SALT see TGE150
1,1'-(p'p'-BIPHENYLENEBIS(CARBONYLMETHYL))DI-2-PICOLINIUM DIBROMIDE
   see BGH000
4,4'-BIPHENYLENEBIS(2-OXOETHYLENE)BIS(DIMETHYL(2-HYDROXYETH-
   YL)AMMONIUM) DIBROMIDE see BGH250
4,4'-BIPHENYLENEBIS(3-OXOPROPYLENE)BIS(DIMETHYL(2-HYDROXYETH-
   YL)AMMONIUM)DIBROMIDE see BGH500
4,4'-BIPHENYLENEDIAMINE see BBX000
o-BIPHENYLENEMETHANE see FDI100
4-BIPHENYLHYDROXYLAMINE see BGI250
o-BIPHENYLMETHANE see FDI100
4,4'-BIPHENYLMETHANEBISMALEIMIDE see BKL800
BIPHENYL, 4-METHYL- see MJH905
N-(p-BIPHENYLMETHYL)-ATROPINIUM BROMIDE see PEM750
N,4-BIPHENYL-METHYL-dl-TROPEYL-α-TROPINIUMBROMID (GERMAN) see
   PEM750
p-BIPHENYLMETHYL-(dl-TROPYL-α-TROPINIUM)BROMIDE see PEM750
1,1-BIPHENYL, NONABROMO- see NMV735
2-BIPHENYLOL see BGJ250
4-BIPHENYLOL see BGJ500
(1,1'-BIPHENYL)-2-OL see BGJ250
o-BIPHENYLOL see BGJ250
(1,1'-BIPHENYL)-4-OL, 3-(1-PYRROLIDINYLMETHYL)- see PPT400
(1,1'-BIPHENYL)-2-OL, SODIUM SALT see BGJ750
2-BIPHENYLOL, SODIUM SALT see BGJ750
BIPHENYL OXIDE see PFA850
1,1'-BIPHENYL, mixed with 1,1'-OXYBIS(BENZENE) see PFA860
(2-BIPHENYLOXY)TRIBUTYLTIN see BGK000
BIPHENYL, PENTACHLORO- see PAV600
1,1'-BIPHENYL, PENTACHLORO-(9CI) see PAV600
BIPHENYL, POLYCHLORO- see PJL750
BIPHENYL SELENIUM see PGH250
1,1'-BIPHENYL, 2,2',6,6'-TETRACHLORO- see TBO600
BIPHENYL, 2,2',6,6'-TETRACHLORO- see TBO600
3,3',4,4'-BIPHENYLTETRAMINE see BGK500
3,3',4,4'-BIPHENYLTETRAMINE TETRAHYDROCHLORIDE see BGK750
N-(2-BIPHENYLYL)ACETAMIDE see PDY000
N-(3-BIPHENYLYL)ACETAMIDE see PDY250
N-(4-BIPHENYLYL)ACETAMIDE see PDY500
N-(4-BIPHENYLYL)ACETOHYDROXAMIC ACETATE see ABJ750
N-(4-BIPHENYLYL)ACETOHYDROXAMIC ACID ACETATE see ABJ750
N-4-BIPHENYLYLBENZAMIDE see BGF000
N-4-BIPHENYLYL BENZENESULFONAMIDE see BGL000
N-4-BIPHENYLYLBENZENESULFONAMIDE see BGL000
N-4-BIPHENYLYLBENZOHYDROXAMIC ACID see HJP500
1-(α-(4-BIPHENYLYL)BENZYL)IMIDAZOLE see BGA825
3-(4-BIPHENYLYLCARBONYL)PROPIONIC ACID see BGL250
1-(1,1'-BIPHENYL)-4-YL-2-((4-(DICHLOROACETYL)PHENYL)AMINO)-2-HYDROX-
   YETHANONE see BGL400
1-BIPHENYLYL-3,3-DIMETHYLTRIAZENE see BGL500
N,N'-4,4'-BIPHENYLYLENEBISACETAMIDE see BFX000

7,7'-(4,4'-BIPHENYLYLENEBIS(CARBONYLIMINO)) BIS(1-ETHYLQUINOLINIUM)DI-
   p-TOLUENESULFONATE see BGM000
7,7'-(p,p'-BIPHENYLENEBIS(CARBONYLIMINO))BIS(2-ETHYLQUINOLINIUM) DI-
   TOSYLATE see BGM000
(4,4'-BIPHENYLYLENEBIS(2-OXOETHYLENE))-2-PICOLINIUM DIBROMIDE see
   BGH000
(4,4'-BIPHENYLYLENEBIS(2-OXOETHYLENE))-3-PICOLINIUM DIBROMIDE see
   BGP750
2,2'-BIPHENYLYLENE SULFIDE see TES300
1-(1,1'-BIPHENYL)-4-YLETHANONE see MHP500
4-BIPHENYLYL ETHYLKETONE see BGM100
N-(4-BIPHENYLYL)FORMOHYDROXAMIC ACID see HLE650
N-4-BIPHENYLYL-N-HYDROXYBENZENESULFONAMIDE see BGN000
N-4-BIPHENYLYLHYDROXYLAMINE see BGI250
4-BIPHENYLYL METHYL KETONE see MHP500
4-(4-BIPHENYLYL)-4-OXOBUTYRIC ACID see BGL250
((1,1'-BIPHENYL)-2-YLOXY)TRIBUTYL-(9CI) STANNANE see BGK000
((2-BIPHENYLYLOXY)TRIBUTYL)STANNANE see BGK000
2-(2-BIPHENYLYLOXY)TRIETHYLAMINE HYDROCHLORIDE see BGO000
1-((4-BIPHENYLYL)PHENYLMETHYL)-1H-IMIDAZOLE see BGA825
BIPINAL SODIUM see SBN000
BIPINDOGENIN-I-RHAMNOSID (GERMAN) see RFZ000
BIPOTASSIUM CHLORAZEPATE see CDQ250
BIPOTASSIUM CHROMATE see PLB250
2,2'-BIPYRIDINE see BGO500
4,4'-BIPYRIDINE see BGO600
α,α'-BIPYRIDINE see BGO500
BIPYRIDINE see BGO500
2,2'-BIPYRIDINE, 4,4'-DIMETHYL- see DQS100
BIPYRIDINIUM, 1,1'-DIMETHYL-4,4'-, DICHLORIDE see PAJ000
2,2'-BIPYRIDYL see BGO500
4,4-BIPYRIDYL see BGO600
4,4'-BIPYRIDYL see BGO600
α,α'-BIPYRIDYL see BGO500
γ,γ'-BIPYRIDYL see BGO600
BIPYROMUCYL see FPZ000
BIQUIN DURULES see QFS100
BIRCH TAR OIL see BGO750
BIRCH TAR OIL, RECTIFIED (FCC) see BGO750
BIRD of PARADISE see CAK325
BIRD PEPPER see PCB275
BIRLANE see CDS750
BIRMI, LEAF EXTRACT see KCA100
BIRNENOEL see AOD725
BIRTHWORT see AQY250
BIRUTAN see RSU000
(–)-α-BISABOLOL see BGO775
BISABOLOL see BGO775
2,7-BIS(ACETAMIDO)FLUORENE see BGP250
BIS(4-ACETAMIDOPHENYL)SULFONE see SNY500
BIS(p-ACETAMIDOPHENYL) SULFONE see SNY500
BIS-4-ACETAMINO PHENYL SELENIUMDIHYDROXIDE see BGP500
BIS(ACETATO-O)(ω-(2-(ACETYLAMINO)-5-NITRO-1,3-PHENYLENE)DI)-MERCURY
   see BGQ250
BIS(ACETATO-O,O')OXOZIRCONIUM see ZTS000
BIS(ACETATO)PALLADIUM see PAD300
BIS(ACETATO)TETRAHYDROXYTRILEAD see LCH000
BIS(ACETATO)TRIHYDROXYTRILEAD see LCJ000
BIS(ACETO)DIHYDROXYTRILEAD see LCH000
BIS(ACETO)DIOXOURANIUM DIHYDRATE see UQT700
BIS(ACETO-O)DIOXOURANIUM DIHYDRATE see UQT700
4,4'-BISACETOPHENONE-α,α'-DI(3-METHYLPYRIDINIUM) DIBROMIDE see
   BGP750
BIS-(ACETOXYAETHYL)NITROSAMIN (GERMAN) see NKM500
BIS(ACETOXY)CADMIUM see CAD250
2,6-BIS(ACETOXYMERCURI)-4-NITROACETANILIDE see BGQ250
9,10-BISACETOXYMETHYL-1,2-BENZANTHRACENE see BBF750
BIS-(p-ACETOXYPHENYL)-CYCLOHEXYLIDENEMETHANE see FBP100
BIS(p-ACETOXYPHENYL)-2-PYRIDYLMETHANE see PPN100
BIS(ACETYLACETONATO) TITANIUM OXIDE see BGQ750
BIS(ACETYL ACETONE)COPPER see BGR000
2,5-BIS(ACETYLAMINO)FLUORENE see BGR250
3,5-BIS(ACETYLAMINO)-2,4,6-TRIIODOBENZOIC ACID see DCK000
4,4'-BIS(N-ACETYL-N-METHYLAMINO)AZOBENZENE see DQH200
1,1'-((2-β,3-α,5-α,16-β,17-β)-3,17-BIS(ACETYLOXY)ANDROS see BGR325
1,1'-(3,17-BIS(ACETYLOXY)ANDROSTANE-2,16-DIYL)BIS(1-METHYLPIPERIDINI-
   UM) DIBROMIDE see PAF625
17,21-BIS(ACETYLOXY)-2-BROMO-6-β,9-DIFLUORO-11-β-HYROXYPREGNA-1,4-
   DIEN-3,20-DIONE see HAG325
BIS(ACETYLOXY)DIBUTYLSTANNANE see DBF800
BIS(ACETYLOXY)MERCURY see MCS750
2,2-BIS((ACETYLOXY)METHYL)-1,3-PROPANEDIOL DIACETATE see NNR400
BISACETYLPALLADIUM see PAD300

BIS((4-(BIS(2-CHLOROETHYL)AMINO)BENZENE)ACETATE)OESTRA-1,3,5(10)-TRIENE-3,17-DIOL(17-β) see EDR500
2,5-BIS(BIS(2-CHLOROETHYL)AMINOMETHYL)HYDROQUINONE see BHB750
BIS((p-(BIS(2-CHLOROETHYL)AMINO)PHENYL)ACETATE)ESTRADIOL see EDR500
BIS((p-(BIS(2-CHLOROETHYL)AMINO)PHENYL)ACETATE)ESTRA-1,3,5(10)-TRIENE-3,17-β-DIOL see EDR500
BIS((p-(BIS(2-CHLOROETHYL)AMINO)PHENYL)ACETATE)OESTRADIOL see EDR500
BIS((p-BIS(2-CHLOROETHYL)AMINOPHENYL)ACETATE)OESTRA-1,3,5(10)-TRIENE-3,17-β-DIOL see EDR500
BIS(BISDIMETHYLAMINOPHOSPHONOUS)ANHYDRIDE see OCM000
4,4'-BIS((4-BIS((2-HYDROXYETHYL)AMINO)-6-CHLORO-s-TRIAZIN-2-YL)AMINO)-2,2'-STILBENEDISULFONIC ACID, D see BHB950
2,6-BIS(BIS(2-HYDROXYETHYL)AMINO)-4,8-DIPIPERIDINOPYRIMIDO(5,4-d)PYRIMIDINE see PCP250
4,4'-BIS((4-BIS(2-HYDROXYETHYL)AMINO-6-METHOXY-s-TRIAZIN-2-YL)AMINO)-2,2'-STILBENEDISULFONIC ACID DIS see BHC500
4,4'-BIS((4-BIS((2-HYDROXYETHYL)AMINO)-6-(m-SULFOANILINO)-s-TRIAZIN-2-YL)AMINO)-2,2'-STILBENEDISULF see BHC750
BIS(BIS(β-HYDROXYETHYL)SULFONIUMMETHYL)SULFIDE DICHLORIDE see BHD000
BIS-(B-o-METHOXYPHENYL-ISOPROPYL)-AMINE LACTATE see BKP200
1,2-BIS(BROMOACETOXY)ETHANE see BHD250
2,2-BIS(BROMOMETHYL)-1,3-PROPANEDIOL see DDQ400
3,12-BIS(3-BROMO-1-OXOPROPYL)-3,12-DIAZA-6,9-DIAZONIADISPI-RO(5.2.5.2)HEXADECANE DICHLORIDE see SLD600
BIS(p-BROMOPHENYL) ETHER see BHJ000
1,4-BIS(3-BROMOPROPIONYL)-PIPERAZINE see BHJ250
2,3,4,5-BIS(2-BUTENYLENE)TETRAHYDROFURFURAL see BHJ500
BISBUTENYLENETETRAHYDROFURFURAL see BHJ500
2,3,4,5-BIS(Δ²-BUTENYLENE)TETRAHYDROFURFURAL see BHJ500
4,5-BIS(2-BUTENYLOXY)-2-IMIDAZOLIDINONE see BHJ625
BIS(2-BUTOXYETHYL) ETHER see DDW200
BIS(BUTOXYETHYL) ETHER see DDW200
BIS(2-BUTOXYETHYL)PHTHALATE see BHK000
BIS(BUTOXYMALEOYLOXY)DIBUTYLSTANNANE see BHK250
BIS(BUTOXYMALEOYLOXY)DIOCTYLSTANNANE see BHK500
BIS(BUTYLCARBITOL)FORMAL see BHK750
2,3,4,5-BIS(2-BUTYLENE)TETRAHYDRO-2-FURALDEHYDE see BHJ500
2,3:4,5-BIS(2-BUTYLENE)TETRAHYDRO-2-FURFURAL see BHJ500
2,3,4,5-BIS(Δ²-BUTYLENE)TETRAHYDROFURFURAL see BHJ500
BIS-Δ²-BUTYLENETETRAHYDROFURFURAL see BHJ500
BIS(2-BUTYL)ETHER see BRH760, OPE030
BIS(3-tert-BUTYL-4-HYDROXY-6-METHYLPHENYL) SULFIDE see TFC600
α-α'-BIS(tert-BUTYLPEROXY)DIISOPROPYLBENZENE see BHL100
2,6-BIS(tert-BUTYL)PHENOL see DEG100
BIS-(4-tert-BUTYLPYRIDINE)-1-METHYLETHER DICHLORIDE see SAB800
BIS(n-BUTYL)SEBACATE see DEH600
BIS(BUTYLTHIO)DIMETHYL STANNANE see BHL500
BIS(BUTYLTHIO)DIMETHYLTIN see BHL500
BIS-BUTYLXANTHOGEN see BSS550
BIS(10-CAMPHORSULFONATO)DIETHYLSTANNANE see DKC600
1,3-BIS(CARBAMOYLTHIO)-2-(N,N-DIMETHYLAMINO)PROPANE HYDROCHLORIDE see BHL750
S-(1,2-BIS(CARBETHOXY)ETHYL)-O,O-DIMETHYL DITHIOPHOSPHATE see MAK700
BIS(1-(CARBO-β-DIETHYLAMINOETHOXY)-1-PHENYLCYCLOPENTANE)ETHANE DISULFONATE see CBG250
BIS(CARBONATO(2-))DIHYDROXYTRIBERYLLIUM see BFP500
(Z,Z)-BIS((3-CARBOXYACRYLOYL)OXY)DIOCTYL-STANNANE DIISOOCTYL ESTER (8CI) see BKL000
BIS-β-CARBOXYETHYLGERMANIUM SESQUIOXIDE see CCF125
BIS(2-CARBOXYETHYL) SULFIDE see BHM000
3,6-BIS(CARBOXYMETHYL)-3,5-DIAZOOCTANEDIOIC ACID see EIX000
BIS(CARBOXYMETHYL)ETHER see ONQ100
N,N-BIS(CARBOXYMETHYL)GLYCINE see AMT500
N,N-BIS(CARBOXYMETHYL)GLYCINE DISODIUM SALT see DXF000
N,N-BIS(CARBOXYMETHYL)GLYCINE SODIUM SALT see SEP500
N,N-BIS(CARBOXYMETHYL)GLYCINE TRISODIUM SALT MONOHYDRATE see NEI000
p-(BIS(CARBOXYMETHYLMERCAPTO)ARSINO)BENZAMIDE see TFA350
BIS(CARBOXYMETHYLMERCAPTO)(p-CARBAMYLPHENYL)-ARSINE see TFA350
BIS(CARBOXYMETHYLMERCAPTO)(p-UREIDOPHENYL)ARSINE see CBI250
BIS(CARBOXYMETHYLTHIO)(p-UREIDOPHENYL)ARSINE see CBI250
BIS(2-CARBOXYPHENYL) DISULFIDE see BHM300
BIS(o-CARBOXYPHENYL) DISULFIDE see BHM300
p-(BIS(o-CARBOXYPHENYLMERCAPTO)-ARSINO)-PHENYLUREA see TFD750
BIS(3-CARBOXYPROPIONYL)PEROXIDE see BHM500
BIS(3-CARBOXYPROPIONYL) PEROXIDE see SNC500
N,N-BIS-(β-CHLORAETHYL)-AMIN (GERMAN) see BHN750
3-(BIS(2-CHLORAETHYL)AMINOMETHYL)BENZOXAZOLON-(2) (GERMAN) see BHQ750

4-(BIS-(2-CHLORAETHYL)AMINOMETHYL)-1-PHENYL-2,3-DIMETHYLPYRAZOLON HYDROCHLORID (GERMAN) see BHR500
N,N-BIS-(β-CHLORAETHYL)-N',O-PROPYLEN-PHOSPHORSAEURE-ESTER-DIAMID (GERMAN) see CQC650
BIS-(4-CHLORBUT-1-YL)-ETHER see OPE040
BIS-(2-CHLORETHYL)VINYLFOSFONAT (CZECH) see VQF000
BIS-(5-CHLOR-2-HYDROXYPHENYL)-METHAN see MJM500
BIS(p-CHLOROBENZOYL) PEROXIDE see BHM750
trans-N,N'-BIS(2-CHLOROBENZYL)-1,4-CYCLOHEXANEBIS(METHYLAMINE) DIHYDROCHLORIDE see BHN000
1,3-BIS(p-CHLOROBENZYLIDENE)AMINO)GUANIDINE see RLK890
BIS(4-CHLOROBUTYL) ETHER see OPE040
1,2-BIS((CHLOROCARBONYL)OXY)ETHANE see EIQ000
1,2-BIS(2-CHLOROETHOXY)ETHANE see TKL500
BIS(2-CHLOROETHOXY)METHANE see BID750
N,N-BIS(β-CHLOROETHYL)-dl-ALANINE HYDROCHLORIDE see BHN500
BIS-β-CHLOROETHYLAMINE see BHN750
N,N-BIS(2-CHLOROETHYL)AMINE HYDROCHLORIDE see BHO250
BIS(2-CHLOROETHYL)AMINE HYDROCHLORIDE see BHO250
BIS(β-CHLOROETHYL)AMINE HYDROCHLORIDE see BHO250
4'-(BIS(2-CHLOROETHYL)AMINO)ACETANILIDE see BHO500
4-(BIS(2-CHLOROETHYL)AMINO)-BENZENEACETIC ACID (9CI) see PCU425
4-(BIS(2-CHLOROETHYL)AMINO)BENZENEBUTANOIC ACID see CDO500
4-(BIS(2-CHLOROETHYL)AMINO)-BENZENEPENTANOIC ACID (9CI) see BHY625
4-(BIS(2-CHLOROETHYL)AMINO)BENZOIC ACID see BHP125
p-(BIS(2-CHLOROETHYL)AMINO)BENZOIC ACID see BHP125
1-((BIS(2-CHLOROETHYL)AMINO)BENZOYL)PIPERIDINE see BHP150
1,6-BIS(CHLOROETHYLAMINO)-1,6-BIS-DEOXY-d-MANNITOL see MAW500
2-(BIS(2-CHLOROETHYL)AMINO)-3-(2-CHLOROETHYL)TETRAHYDRO-2H-1,3,2-OXAPHOSPHORINE-2-OXIDE see TNT500
1,6-BIS(CHLOROETHYLAMINO)-1,6-DESOXY-d-MANNITOLDIHYDROCHLORIDE see MAW750
1,6-BIS(CHLOROETHYLAMINO)-1,6-DIDEOXY-d-MANNITE see MAW500
1,6-BIS((2-CHLOROETHYL)AMINO)-1,6-DIDEOXY-d-MANNITOL see MAW500
1,6-BIS(β-CHLOROETHYLAMINO)-1,6-DIDEOXY-d-MANNITOL see MAW500
1,6-BIS-(CHLOROETHYLAMINO)-1,6-DIDEOXY-d-MANNITOLDIHYDROCHLORIDE see MAW750
2-(p-BIS(2-CHLOROETHYL)AMINO)-N-ETHYLACETAMIDE see PCV775
9-(2-(BIS(2-CHLOROETHYL)AMINO)ETHYLAMINO)-6-CHLORO-2-METHOXYACRIDINE DIHYDROCHLORIDE see DFH000
4-(BIS(2-CHLOROETHYL)AMINO)-N-ETHYL-BENZENEACETAMIDE see PCV775
3-(2-(BIS(2-(CHLOROETHYL)AMINO)ETHYL))-1,3-DIAZASPIRO(4.5)DECANE-2,4-DIONE see SLD900
3-(2-(BIS(2-(CHLOROETHYL)AMINO)ETHYL))-5,5-PENTAMETHYLENEHYDANTOIN see SLD900
2-(BIS-2-CHLOROETHYL)AMINO)ETHYL VINYL SULFONE see BHP500
4'-(BIS(2-CHLOROETHYL)AMINO)-2-FLUORO ACETANILIDE see BHP750
2-(BIS(2-CHLOROETHYL)AMINO)-4-HYDROPEROXYTETRAHYDRO-2H-1,3,2-OXAZAPHOSPHORINE see HIF000
2-(BIS(2-CHLOROETHYL)AMINO)-4-HYDROXYTETRAHYDRO-2H-1,3,2-OXAZAPHOSPHORINE see HKA200
3-BIS(2-CHLOROETHYL)AMINO-4-METHYLBENZOIC ACID see AFQ575
3-(BIS(2-CHLOROETHYL)AMINOMETHYL)-2-BENZOXAZOLINONE see BHQ750
1-(3-(BIS(2-CHLOROETHYL)AMINO-4-METHYLBENZOYL)AZIRIDINE) see BHQ760
1-(3-(BIS(2-CHLOROETHYL)AMINO)-4-METHYLBENZOYL)MORPHOLINE see BHR400
9-(4-(BIS-β-CHLOROETHYLAMINO)-1-METHYLBUTYLAMINO)-6-CHLORO-2-METHOXYACRIDINE see QDJ000
9-(4-BIS(2-CHLOROETHYL)AMINO-1-METHYLBUTYLAMINO)-6-CHLORO-2-METHOXYACRIDINE DIHYDROCHLORIDE see QDS000
4-((4-(BIS(2-CHLOROETHYL)AMINO)-1-METHYLBUTYL)AMINO)-7-CHLOROQUINOLINE, DIHYDROCHLORIDE see CLD500
2-(BIS(2-CHLOROETHYL)AMINOMETHYL)-5,5-DIMETHYLBENZIMIDAZOLE HYDROCHLORIDE see BCC250
4-(BIS(2-CHLOROETHYL)AMINOMETHYL)-2,3-DIMETHYL-1-PHENYL-3-PYRAZOLIN-5-ONE HYDROCHLORIDE see BHR500
3-(o-((BIS(2-CHLOROETHYL)AMINO)METHYL)PHENYL)ALANINE DIHYDROCHLORIDE see ARQ000
2-((BIS(2-CHLOROETHYL)AMINO)METHYL)-PHENYLALANINE DIHYDROCHLORIDE (9CI) see ARQ000
o-BIS(2-CHLOROETHYL)AMINOMETHYLPHENYLALANINE HYDROCHLORIDE see ARQ000
5-(BIS(2-CHLOROETHYL)AMINO)-6-METHYLURACIL see DYC700
2-BIS(2-CHLOROETHYL)AMINONAPHTHALENE see BIF250
1-BIS(2-CHLOROETHYL)AMINO-1-OXA-2-AZA-5-OXAPHOSPHORIDINE MONOHYDRATE see CQC500
2-(BIS(2-CHLOROETHYL)AMINO)-1-OXA-3-AZA-2-PHOSPHOCYCLOHEXANE 2-OXIDE MONOHYDRATE see CQC500
2-(BIS(2-CHLOROETHYL)AMINO)-2H-1,3,2-OXAAZAPHOSPHORINE 2-OXIDE see CQC650
4-(BIS(2-CHLOROETHYL)AMINO)PHENOL see BHR750
p-(BIS(2-CHLOROETHYL)AMINO)PHENOL BENZOATE see BHV500
p-(BIS(2-CHLOROETHYL)AMINO)PHENOL-p-BROMOBENZOATE see BHV750

BIS(2,2-DIETHOXYETHYL)DISELENIDE see BJA200
BIS-DIETHYLAMID KYSELINY FTALOVE see GCE600
1,4-BIS(DIETHYLAMINO)-2-BUTYNE see BJA250
α,α'-BIS(DIETHYLAMINO)-5,5'-DIALLYL-m,m'-BITOLYL-4,4'-DIOL DIHYDRO-
   CHLORIDE see BFX125
2,6-BIS(2-(DIETHYLAMINO)ETHOXY)-9,10-ANTHRACENEDIONE DIHYDROCHLO-
   RIDE see BJA500
BIS(1-(2-DIETHYLAMINOETHOXYCARBONYL)-1-PHENYLCYCLOPEN-
   TANE)ETHANE DISULFONATE see CBG250
2,7-BIS(2-(DIETHYLAMINO)ETHOXY)-FLUOREN-9-ONE DIHYDROCHLORIDE see
   TGB000
2,7-BIS(2-(DIETHYLAMINO)ETHOXY)-9H-FLUOREN-9-ONE DIHYDROCHLORIDE
   see TGB000
1-(BIS(2-DIETHYLAMINO)ETHYL)AMINO)-5-CHLORO-3-(p-CHLOROPHE-
   NYL)INDOLE DIHYDROCHLORIDE HEMIHYDRATE see BJA750
1-(BIS(2-DIETHYLAMINO)ETHYL)AMINO)-3-PHENYLINDOLE DIHYDROCHLO-
   RIDE see BJA809
N,N'-BIS(2-DIETHYLAMINOETHYL)OXAMIDE BIS(2-CHLOROBENZYL CHLO-
   RIDE) see MSC100
1,3-BIS(DIETHYLAMINO)-2-(α-PHENYL-α-CYCLOHEXYLMETHYL)PROPANE HY-
   DROCHLORIDE see LFK200
3,6-BIS(3-DIETHYLAMINOPROPOXY)PYRIDAZINE BISMETHIODIDE see BJA825
(BIS(DIETHYLAMINO)THIOXOMETHYL) DISULPHIDE see DXH250
BIS(DIETHYLDITHIOCARBAMATO)CADMIUM see BJB500
BIS(DIETHYLDITHIOCARBAMATO)MERCURY see BJB750
BIS(DIETHYLDITHIOCARBAMATO)ZINC see BJC000
BISDIETHYLENE TRIAMINE COBALT(III) PERCHLORATE see BJC500
BIS-O,O-DIETHYLPHOSPHORIC ANHYDRIDE see TCF250
BIS-O,O-DIETHYLPHOSPHOROTHIONIC ANHYDRIDE see SOD100
BIS(N,N-DIETHYLTHIOCARBAMOYL) DISULFIDE see DXH250
BIS(DIETHYLTHIOCARBAMOYL) DISULFIDE see DXH250
BIS(DIETHYLTHIOCARBAMOYL)DISULFIDE mixed with SODIUM NITRITE see
   SIS200
BIS(N,N-DIETHYLTHIOCARBAMOYL) DISULPHIDE see DXH250
BIS(DIETHYLTHIO)CHLORO METHYL PHOSPHONATE see BJD000
BIS(2,2-DIFLUOROAMINO)-1,3-BIS(DINITRO-FLUOROETHOXY)PROPANE see
   SPA500
BIS(DIFLUOROAMINO)DIFLUOROMETHANE see BJD250
1,1-BIS(DIFLUOROAMINO)-2,2-DIFLUORO-2-NITRO-ETHYL METHYL ETHER see
   BJD375
1,2-BIS(DIFLUOROAMINO)ETHANOL see BJD500
1,2-BIS(DIFLUOROAMINO)ETHYL VINYL ETHER see BJD750
4,4-BIS(DIFLUOROAMINO)-3-FLUOROIMINO-1-PENTENE see BJE000
1,2-BIS(DIFLUOROAMINO)-N-NITROETHYLAMINE see BJE250
BIS(DIFLUOROBORYL)METHANE see BJE325
(T-4-(R)(R)-BIS(N-(2-DIHYDROXY-3,3-DIMETHYL-1-OXOBUTYL-β-ALANIN-
   ATO)ZINC see ZKS000
2,2-BISDIHYDROXYMETHYL-1,3-PROPANEDIOL TETRANITRATE see PBC250
1,4-BIS(3,4-DIHYDROXPHENYL)-2,3-DIMETHYLBUTANE see NBR000
N,N'-BIS(2-(3',4'-DIHYDROXYPHENYL)-2-HYDROXYETH-
   YL)HEXAMETHYLENEDIAMINE SULFATE see HFG650
N,N'-BIS(2-(3',4'-DIHYDROXYPHENYL)-2-HYDROXYETH-
   YL)HEXAMETHYLENEDIAMINE DIHYDROCHLORIDE see HFG600
BIS(DIHYDROXYPHENYL)SULFIDE see BJE500
BIS(DIMETHYLAMIDO)FLUORO PHOSPHATE see BJE750
BIS(DIMETHYLAMIDO)PHOSPHORYL FLUORIDE see BJE750
2,8-BISDIMETHYLAMINOACRIDINE see BJF000
3,6-BIS(DIMETHYLAMINO)ACRIDINE see BJF000
BIS(DIMETHYLAMINO)-3-AMINO-5-PHENYLTRIAZOLYL PHOSPHINE OXIDE see
   AIX000
1,4-BIS(DIMETHYLAMINO)BENZENE see BJF500
p-BIS(DIMETHYLAMINO)BENZENE see BJF500
4,4'-BIS(DIMETHYLAMINO)BENZHYDRYLIDENIMINE HYDROCHLORIDE see
   IBA000
4,4'-BIS(DIMETHYLAMINO)BENZOHYDROL see TDO750
4,4'-BIS(DIMETHYLAMINO)BENZOPHENONE see MQS500
p,p'-BIS(N,N-DIMETHYLAMINO)BENZOPHENONE see MQS500
4,4'-BIS(DIMETHYLAMINO)BENZOPHENONE-IMINE HYDROCHLORIDE see
   IBA000
BIS(DIMETHYLAMINOBORANE)ALUMINUM TETRAHYDROBORATE see
   BJG000
BIS((DIMETHYLAMINO)CARBONOTHIOYL) DISULPHIDE see TFS350
2,2'-BIS(DIMETHYLAMINO) DIETHYLSULPHIDE DIHYDROCHLORIDE see
   BJG100
BIS(DIMETHYLAMINO)DIMETHYLSTANNANE see BJG125
4,4'-BIS(DIMETHYLAMINO)DIPHENYLMETHANE see MJN000
p,p'-BIS(DIMETHYLAMINO)DIPHENYLMETHANE see MJN000
3,5-BIS-DIMETHYLAMINO-1,2,4-DITHIAZOLIUM CHLORIDE see BJG150
1,2-BIS(DIMETHYLAMINO)-ETHANE (DOT) see TDQ750
BIS(2-DIMETHYLAMINOETHOXY)ETHANE see BJG250
3,6-BIS(2-(DIMETHYLAMINO)ETHOXY)-9H-XANTHEN-9-ONE DIHYDROCHLORIDE
   see BJG500
1,3-BIS(2-DIMETHYLAMINOETHYL)ADAMANTANE DIHYDROCHLORIDE see
   BJG750

1-(BIS(2-(DIMETHYLAMINO)ETHYL)AMINO)-5-METHYL-3-PHENYLINDOLE DIHY-
   DROCHLORIDE see BJH000
1-(BIS(2-(DIMETHYLAMINO)ETHYL)AMINO)-3-PHENYLINDOLE DIHYDROCHLORI-
   DEHYDRATE see BJH500
BIS(2-DIMETHYLAMINOETHYL) ETHER see BJH750
BIS(2-DIMETHYLAMINOETHYL)SUCCINATE BIS(METHOCHLORIDE) see
   HLC500
BIS(β-DIMETHYLAMINOETHYL)SUCCINATE BIS(METHYLIODIDE) see BJI000
BISDIMETHYLAMINOFLUOROPHOSPHINE OXIDE see BJE750
BIS(DIMETHYLAMINO)FLUOROPHOSPHATE see BJE750
BIS(DIMETHYLAMINO)ISOPROPYLMETHACRYLATE see BJI125
3,7-BIS(DIMETHYLAMINO)-4-NITRO-PHENOTHIAZIN-5-IUM, CHLORIDE see
   MJI250
3,7-BIS(DIMETHYL AMINO)PHENAZA THIONIUM CHLORIDE see BJI250
3,7-BIS(DIMETHYLAMINO)PHENOTHIAZIN-5-IUM CHLORIDE see BJI250
BIS(p-(N,N-DIMETHYLAMINO)PHENYL)KETONE see MQS500
p,p'-BIS(N,N-DIMETHYLAMINOPHENYL)METHANE see MJN000
BIS(p-(N,N-DIMETHYLAMINOPHENYL)METHANE see MJN000
BIS(p-DIMETHYLAMINOPHENYL)METHANE see MJN000
α,α-BIS(p-DIMETHYLAMINOPHENYL)METHANOL see TDO750
BIS(4-(DIMETHYLAMINO)PHENYL)METHANONE see MQS500
BIS(p-DIMETHYLAMINOPHENYL)METHYLENEIMINE see IBB000
1,1-BIS(p-DIMETHYLAMINOPHENYL)METHYLENIMINEHYDROCHLORIDE see
   IBA000
BIS(DIMETHYLAMINO)PHOSPHONOUS ANYHDRIDE see OCM000
BIS(DIMETHYLAMINO)PHOSPHORIC ANHYDRIDE see OCM000
1,3-BIS(2-DIMETHYLAMINOPROPYL)ADAMANTANE DIHYDROCHLORIDE see
   BJI750
N,N'-BIS(3-DIMETHYLAMINOPROPYL)DITHIOOXAMIDE see BJJ000
BIS(DIMETHYLAMINO)SULFOXIDE see BJJ125
BIS(DIMETHYLARSINYLDIAZOMETHYL)MERCURY see BJJ200
BIS-DIMETHYL ARSINYL OXIDE see BJJ250
BIS-DIMETHYL ARSINYL SULFIDE see BJJ500
BIS(α,α-DIMETHYLBENZYL)PEROXIDE see DGR600
1,2-BIS(3,7-DIMETHYL-5-n-BUTOXY-1-AZA-5-BORA-4,6-DIOXOCYCLOOC-
   TYL)ETHANE see BJJ750
BIS(1,3-DIMETHYL)BUTYLAMINE see TDN250
BIS(DIMETHYLCARBAMODITHIOATO-S,S')LEAD see LCW000
BIS(DIMETHYLCARBAMODITHIOATO-S,S')ZINC see BJK500
BIS(DIMETHYLDITHIOCARBAMATE de ZINC) (FRENCH) see BJK500
BIS(DIMETHYLDITHIOCARBAMATO)NICKEL see BJK250
BIS(DIMETHYLDITHIOCARBAMATO)ZINC see BJK500
BIS(DIMETHYLDITHIOCARBAMIATO)LEAD see LCW000
3,5-BIS(1,1-DIMETHYLETHYL)-4-HYDROXY-BENZENEMETHANOL see IFX200
3,5-BIS(1,1-DIMETHYLETHYL)-4-HYDROXY-, 1,2-ETHANEDIYLBIS (OXY-2,1-ETH-
   ANEDIYL) ESTER BENZENEPROPANOIC ACID see BJK550
((3,5-BIS(1,1-DIMETHYLETHYL)-4-HYDROXYPHEN-
   YL)METHYLENE)PROPANEDINITRILE see DED000
2,6-BIS(1,1-DIMETHYLETHYL)-4-METHYLPHENOL see BFW750
2,4-BIS(1,1-DIMETHYLETHYL)-6-(1-PHENYLETHYL)PHENOL see BJK650
2-((3,5-BIS(1,1-DIMETHYL)-4-HYDROXYPHEN-
   YL)METHYLENE)PROPANEDINITRILE see DED000
1,1'-BIS(3,5-DIMETHYLMORPHOLINOCARBONYLMETHYL)-4,4'-BIPYRIDINIUM-DI-
   CHLORID (GERMAN) see BJK750
1,1'-BIS(3,5-DIMETHYLMORPHOLINOCARBONYLMETHYL)-4,4'-BIPYRIDYNIUM
   DICHLORIDE see BJK750
1,1'-BIS(2-(3,5-DIMETHYL-4-MORPHOLINYL)-2-OXOETHYL)-4,4'-BIPYRIDINIUM
   DICHLORIDE see BJK750
11,11-BIS((3,7-DIMETHYL-2,6-OCTADIENYL)OXY)-1-UNDECENE see UNA100
N,N'-BIS(1,4-DIMETHYLPENTYL)-p-PHENYLENEDIAMINE see BJL000
2,5-BIS(1,1-DIMETHYLPROPYL)HYDROQUINONE see DCH400
1,3-BIS(N,N-DIMETHYL-2-PYRROLIDINIUM)PROPANE DIBENZENESULFONATE
   see COG500
1,4-BIS(DIMETHYLSILYL)BENZENE see PEW725
BIS(DIMETHYLSILYL) ETHER see TDP775
BIS(DIMETHYLSILYL) OXIDE see TDP775
BISDIMETHYL STIBINYL OXIDE see BJL250
BIS(DIMETHYL THALLIUM)ACETYLIDE see BJL500
BIS(DIMETHYL-THIOCARBAMOYL)-DISULFID (GERMAN) see TFS350
BIS(DIMETHYLTHIOCARBAMOYL) DISULFIDE see TFS350
BIS(DIMETHYLTHIOCARBAMOYL)DISULFIDE and NITROSO-
   TRIS(DIMETHYLDITHIOCARBAMATO)IRON see FAZ000
BIS(DIMETHYLTHIOCARBAMOYL)SULFIDE see BJL600
BIS(DIMETHYLTHIOCARBAMOYLTHIO)METHYL-ARSINE see USJ075
BIS(DIMETHYLTHIOCARBAMYL) MONOSULFIDE see BJL600
BIS(N,N-DIMETIL-DITIOCARBAMMATO) DI ZINCO (ITALIAN) see BJK500
BIS(β-(N,N-DIMORPHOLINO)ETHYL)SELENIDE DIHYDROCHLORIDE see
   MRU255
1,3-BIS(2,2,-DINITRO-2-FLUOROETHOXY)-N,N,N,N'- TETRAFLUORO-2,2-PRO-
   PANEDIAMINE see SPA500
(±)-1,2-BIS(3,5-DIOXOPIPERAZINE-1-YL)PROPANE see PIK250
(±)-1,2-BIS(3,5-DIOXOPIPERAZINYL)PROPANE see PIK250
2,6-BIS(DIPHENYLHYDROXYMETHYL)PIPERIDINE see BJM250
2,6-BIS(DIPHENYLHYDROXYMETHYL)PYRIDINE see BJM500
3-(BIS(3,3-DIPHENYLPROPYL)AMINO)PROPANE-1-OL see BJM625

3-(BIS(3,3-DIPHENYLPROPYL)AMINO)-1-PROPANOL see BJM625

BIS-O,O-DI-n-PROPYLPHOSPHOROTHIONIC ANHYDRIDE see TED500

BIS(4-(DITHIOCARBOXY)-1-PIPERAZINEACETATO(2−))-MERCURY(2−), DISODI-UM see SJR500

BIS(1,3-DITHIOCYANATO-1,1,3,3-TETRABUTYLDISTANNOXANE) see BJM700

BIS(DITHIOPHOSPHATE de O,O-DIETHYLE) de S,S′-(1,4-DIOXANNE-2,3-DIYLE) (FRENCH) see DVQ709

BIS(DODECANOLOXY)DIOCTYLSTANNANE see DVJ800

BIS(DODECANOYLOXY)DI-n-BUTYLSTANNANE see DDV600

BIS(DODECYLOXYCARBONYLETHYL) SULFIDE see TFD500

2,6-BIS-(DWUBENZYLOHYDROKSYMETYLO)-PIPERYDYNA (POLISH) see BIV000

BIS(2,5-ENDOMETHYLENECYCLOHEXYLMETHYL)AMINE see BJM750

BIS(3,4-EPOXYBUTYL) ETHER see BJN000

BIS(2,3-EPOXYCYCLOPENTYL) ETHER see BJN250

BIS(2,3-EPOXYCYCLOPENTYL) ETHER mixed with DIGLYCIDYL ETHER of BISPHENOL A (1:1) see OPI200

BIS(3,4-EPOXY-6-METHYLCYCLOHEXYLMETHYL)ADIPATE see AEN750

BIS(3,4-EPOXY-2-METHYLPROPYL)ETHER see BJN500

1,3-BIS(2,3-EPOXYPROPOXY)BENZENE see REF000

m-BIS(2,3-EPOXYPROPOXY)BENZENE see REF000

1,4-BIS(2,3-EPOXYPROPOXY)BUTANE see BOS100

2,2′-BIS(2,3-EPOXYPROPOXY)-N-tert-BUTYLDIPROPYLAMINE see DKM400

1,3-BIS(2,3-EPOXYPROPOXY)-2,2-DIMETHYLPROPANE see NCI300

2,3-BIS(2,3-EPOXYPROPOXY)-1,4-DIOXANE see BJN750

N,N-BIS(2-(2,3-EPOXYPROPOXY)ETHOXY)ANILINE see BJN850

1,2-BIS(2,3-EPOXYPROPOXY)ETHOXY)ETHANE see TJQ333

N,N-BIS(2-(2,3-EPOXYPROPOXY)ETHYL)ANILINE see BJN875

2,2-BIS(p-(2,3-EPOXYPROPOXY)PHENYL)PROPANE mixed with 2,2′-OXYBIS(6-OXABICYCLO(3.1.0)HEXANE) see OPI200

N,N-BIS(2,3-EPOXYPROPYL)ANILINE see DKM120

BIS(2,3-EPOXYPROPYL)ANILINE see DKM120

1,3-BIS(2,3-EPOXYPROPYL)-5,5-DIMETHYLHYDANTOIN see DKM130

BIS(2,3-EPOXYPROPYL)ETHER see DKM200

BIS(2,3-EPOXYPROPYL)-5-ETHYL-5-METHYLHYDANTOIN see ECI200

2,2-BIS(4-(2,3-EPOXYPROPYLOXY)PHENYL)PROPANE see BLD750

BIS(EPOXYPROPYL)PHENYLAMINE see DKM120

2,6-BIS(2,3-EPOXYPROPYL)PHENYL-2,3-EPOXYPROPYLETHER see BJO000

BIS(2,6-(2,3-EPOXYPROPYL)PHENYL GLYCIDYL ETHER see BJO000

N,N′-BIS(2,3-EPOXYPROPYL)PIPERAZINE see DHE100

N,N-BIS(2,3-EPOXYPROPYL)-p-TOLUENESULFONAMIDE see DKM800

BISEPTOL see TKX000

BIS(ETHOXYCARBONYLDIAZOMETHYL)MERCURY see BJO125

S-(1,2-BIS(ETHOXY-CARBONYL)-ETHYL)-O,O-DIMETHYL-DITHIOFOSFAAT (DUTCH) see MAK700

S-(1,2-BIS(ETHOXYCARBONYL)-ETHYL)-O,O-DIMETHYL PHOSPHORODITHIOATE see MAK700

S-1,2-BIS(ETHOXYCARBONYL)ETHYL-O,O-DIMETHYL THIOPHOSPHATE see MAK700

1,2-BIS(3-ETHOXYCARBONYL-2-THIOUREIDO) BENZENE see DJV000

1,2-BIS-(3-ETHOXYCARBONYLTHIOUREIDO)BENZENE see DJV000

BIS(2-ETHOXYETHYL) ADIPATE see BJO225

BIS(2-ETHOXYETHYL)ETHER see DIW800

BIS(2-ETHOXYETHYL)NITROSOAMINE see BJO250

N′,N(sup 2)-BIS(p-ETHOXYPHENYL)ACETAMIDINE see BJO500

N,N′-BIS(p-ETHOXYPHENYL)ACETAMIDINE see BJO500

N,N′-BIS(4-ETHOXYPHENYL)ETHANIMIDAMIDE see BJO500

1,4-BIS-(2,3-ETHOXYPROPYL)PIPERAZINE see DHE100

BIS(ETHOXYTHIOCARBONYL)TRISULFIDE see DKE400

2,4-BIS(ETHYLAMINO)-6-CHLORO-s-TRIAZINE see BJP000

4,6-BIS(ETHYLAMINO)-2-METHOXY-s-TRIAZINE see BJP250

2,4-BIS(ETHYLAMINO)-6-METHOXY-s-TRIAZINE see BJP250

1,2-BIS(ETHYLAMMONIO)ETHANE PERCHLORATE see BJP300

BIS(N-ETHYLDITHIOCARBANILATO)ZINC see ZHA000

BIS(ETHYLENEDIAMINE)(MERCURICTETRATHIOCYANATO)COPPER see BJP425

ω,ω′-BIS-(ETHYLENEIMINOSULPHONYL)PROPANE see BJP899

BISETHYLENEUREA see BJP450

BIS(ETHYLENIMIDO)PHOSPHORYLURETHAN see EHV500

2,6-BIS(ETHYLEN-IMINO)-4-AMINO-s-TRIAZINE see BJP500

2,5-BIS-ETHYLENIMINOBENZOQUINONE see BGW750

1,3-BIS(ETHYLENIMINOSULFONYL)PROPANE see BJP899

BIS(ETHYLHEXANOATE)TRIETHYLENE GLYCOL see FCD560

BIS(2-ETHYLHEXANOYLOXY)DIBUTYL STANNANE see BJQ250

BIS(2-ETHYLHEXYL) ADIPATE see AEO000

BIS-2-ETHYLHEXYLAMIN see DJA800

2-(BIS(2-ETHYLHEXYL)AMINO)ETHANOL see DJK200

BIS(2-ETHYLHEXYL) AZELATE see BJQ500

BIS(2-ETHYLHEXYL)-1,2-BENZENEDICARBOXYLATE see DVL700

BIS(2-ETHYLHEXYL) ESTER PHOSPHOROUS ACID CADMIUM SALT see CAD500

BIS(ETHYLHEXYL) ESTER of SODIUM SULFOSUCCINIC ACID see DJL000

BIS(2-ETHYLHEXYL)ETHER see DJK600

BIS(2-ETHYLHEXYL) FUMARATE see DVK600

BIS(2-ETHYLHEXYL)HYDROGEN PHOSPHATE see BJR750

BIS(2-ETHYLHEXYL) HYDROGEN PHOSPHITE see BJQ709

BIS(2-ETHYLHEXYL) ISOPHTHALATE see BJQ750

BIS(2-ETHYLHEXYL) MALEATE see BJR000

BIS(2-ETHYLHEXYL)ORTHOPHOSPHORIC ACID see BJR750

BIS(2-ETHYLHEXYLOXYCARBONYLMETHYLTHIO)DIBUTYLSTANNANE see BKK250, DDY600

BIS(2-ETHYLHEXYLOXYCARBONYLMETHYLTHIO)DIMETHYLSTANNANE see BKK500

BIS((2-(ETHYL)HEXYLOXY)MALEOYLOXY) DI(n-BUTYL)STANNANE see BJR250

BIS(2-ETHYLHEXYL) PHENYL PHOSPHATE see BJR625

BIS(2-ETHYLHEXYL) PHOSPHATE see BJR750

BIS(2-ETHYLHEXYL)PHOSPHORIC ACID see BJR750

BIS(2-ETHYLHEXYL)PHTHALATE see DVL700

BIS(2-ETHYLHEXYL) SEBACATE see BJS250

BIS(2-ETHYLHEXYL)SODIUM PHOSPHATE see TBA750

1,4-BIS(2-ETHYLHEXYL) SODIUM SULFOSUCCINATE see DJL000

BIS(2-ETHYLHEXYL)SODIUM SULFOSUCCINATE see DJL000

BIS(2-ETHYLHEXYL)-S-SODIUM SULFOSUCCINATE see DJL000

1,4-BIS(2-ETHYLHEXYL)SULFOBUTANEDIOIC ACID ESTER, SODIUM SALT see DJL000

BIS(2-ETHYLHEXYL) TEREPHTHALATE see BJS500

BIS(2-ETHYLHEXYLTHIOGLYCOLATE)DIBUTYLTIN see DDY600

BIS(2-ETHYLHEXYLTHIOGLYCOLATE)DIOCTYLTIN see DVM800

BIS(ETHYLMERCURI) PHOSPHATE see BJT250

N,N′-BIS(1-ETHYL-3-METHYLPENTYL)-p-PHENYLENEDIAMINE see BJT500

BIS(ETHYLPHENYLCARBAMODITHIOATO-S,S′)-(T-4)-ZINC see ZHA000

2,2-BIS(p-ETHYLPHENYL)-1,1-DICHLOROETHANE see DJC000

1,1-BIS(p-ETHYLPHENYL)-2,2-DICHLOROETHANE see DJC000

BIS(N-ETHYL-N-PHENYL)UREA see DJC400

2,2-BIS(ETHYLSULFONYL)BUTANE see BJT750

2,2-BIS(ETHYLSULFONYL)PROPANE see ABD500

BIS(ETHYLXANTHIC)DISULFIDE see BJU000

BIS(ETHYLXANTHOGEN) DISULFIDE see BJU000

BIS(ETHYLXANTHOGEN) TETRASULFIDE see BJU250

BIS(ETHYLXANTHOGEN) TRISULFIDE see DKE400

S-(1,2-BIS(ETOSSI-CARBONIL)-ETIL)-O,O-DIMETIL-DITIOFOSFATO (ITALIAN) see MAK700

BISEXOVIS see AOO410

BISEXOVISTER see AOO410

BISFEROL A (GERMAN) see BLD500

1,6-BIS(9 FLUORENYLDIMETHYL-AMMONIUM)HEXANE BROMIDE see HEG000

BIS(2-FLUORO-2,2-DINITROETHOXY)DIMETHYLSILANE see BJU350

BIS(2-FLUORO-2,2-DINITROETHYL)AMINE see BJU500

BIS(4-FLUORO-3-NITROPHENYL)SULFONE see DKH250

1,1-BIS(FLUOROOXY)HEXAFLUOROPROPANE see BJV625

2,2-BIS(FLUOROOXY)HEXAFLUOROPROPANE see BJV630

1,1-BIS(FLUOROOXY)TETRAFLUOROETHANE see BJV635

1-(4,4-BIS(p-FLUOROPHENYL)BUTYL)-4-(2-OXO-1-BENZIMIDAZOLI-NYL)PIPERIDINE see PIH000

trans-4-(4,4-BIS(p-FLUOROPHENYL)BUTYL)-1-(2-(4′-PHENYLCYCLOHEXYLAMI-NO)ETHYL)PIPERAZI

8-(4,4-BIS(p-FLUOROPHENYL)BUTYL)-1-PHENYL-1,3,8-TRIAZASPI-RO(4,5)DECAN-4-ONE see PFU250

trans-2-(4-(4,4-BIS(p-FLUOROPHENYL)BUTYL)PIPERAZINYL)-N-(4′-PHENYLCY-CLOHEXYL)ACETAMIDE DIHYDROCHLORIDE see BJW000

1-(1-(4,4-BIS(p-FLUOROPHENYL)BUTYL)-4-PIPERIDYL)-2-BENZIMIDAZOLINONE see PIH000

(3)-1-(BIS(p-FLUOROPHENYL)METHYL)-4-CINNAMYLPIPERAZINE DIHYDRO-CHLORIDE see FDD080

(E)-1-(BIS(4-FLUOROPHENYL)METHYL)-4-(3-PHENYL-2-PROPENYL)PIPERAZINE DIHYDROCHLORIDE see FDD080

1,1-BIS(4-FLUOROPHENYL)-2-PROPYNYL-N-CYCLOHEPTYLCARBAMATE see BJW250

1,1-BIS(4-FLUOROPHENYL)-2-PROPYNYL-N-CYCLOOCTYL CARBAMATE see BJW500

BIS(3-FLUOROSALICYLALDEHYDE)-ETHYLENEDIIMINE-COBALT see EIS000

1,4-BIS(1-FORMAMIDO-2,2,2-TRICHLOROETHYL)PIPERAZINE see TKL100

N,N′-BIS(1-FORMAMIDO-2,2,2-TRICHLOROETHYL)PIPERAZINE see TKL100

BIS(N-FORMYL-p-AMINOPHENYL)SULFONE see BJW750

BIS(FORMYLMETHYL) MERCURY see BJW800

1,3-BIS(4-FORMYLPYRIDINIUM)-PROPANE BISOXIDE DIBROMIDE see TLQ500

1,3-BIS(4-FORMYLPYRIDINIUM)-PROPANE BISOXIME DICHLORIDE see TLQ750

2-(BIS(FURFURYLIDENAMINO))METHYLFURAN see FPS000

3,5-BIS(2-FURYL)-1H-1,2,4-TRIAZOLE see AMX000

3,5-BIS-d-GLUCONAMIDO-2,4,6-TRIIODO-N-METHYLBENZAMIDE see IFS400

m-BIS(GLYCIDYLOXY)BENZENE see REF000

1,4-BIS(GLYCIDYLOXY)BUTANE see BOS100

2,3-BIS(GLYCIDYLOXY)-1,4-DIOXANE see BJN750

1,2-BIS(GLYCIDYLOXY)ETHANE see EEA600

BIS(4-GLYCIDYLOXYPHENYL)DIMETHYLMETHANE see BLD750

2,2-BIS(p-GLYCIDYLOXYPHENYL)PROPANE see BLD750

BISGUANIDINIUM CARBONATE see GKW100

BIS(GUANIDINIUM) CHROMATE see BJW825

(BIS-(HEPTYLOXY)PHOSPHINYL)MERCURY ACETATE see AAR750

BIS(HEXANOYLOXY)DI-n-BUTYLSTANNANE see BJX750

N-BIS(2-HYDROXYPROPYL)NITROSAMINE see DNB200
BIS(3-HYDROXY-1-PROPYNYL)MERCURY see BKJ250
BIS(1-HYDROXY-2(1H)-PYRIDINETHIONATO)ZINC see ZMJ000
BIS(8-HYDROXYQUINOLINE-5-SULFONIC ACID) MANGANESE(II) see BKJ275
1,3-BIS(1-HYDROXY-2,2,2-TRICHLOROETHYL)UREA see DGQ200
BIS(2-HYDROXY-3,5,6-TRICHLOROPHENYL)METHANE see HCL000
BISIBUTIAMINE see BKJ325
BISINA see CKF500
BIS(3-INDOLEMETHYLENEMORPHOLINIUM)HEXACHLOROSTANNATE see BKJ500
BIS(ISOBUTYL)ALUMINUM CHLORIDE see CGB500
BIS(ISOBUTYL)HYDROALUMINUM see DNI600
BIS(4-ISOCYANATOCYCLOHEXYL)METHANE see MJM600
1,3-BIS(ISOCYANATOMETHYL)BENZENE see XIJ000
BIS(1-ISOCYANATO-1-METHYLETHYL)BENZENE see TDX300
BIS(1,4-ISOCYANATOPHENYL)METHANE see MJP400
BIS(4-ISOCYANATOPHENYL)METHANE see MJP400
BIS(p-ISOCYANATOPHENYL)METHANE see MJP400
1,3-BIS-(ISOKYANATOMETHYL)BENZEN see XIJ000
BIS(ISONICOTINALDOXIME 1-METHYL) ETHER DICHLORIDE see BGS250
BIS(ISOOCTYLOXYCARBONYLMETHYLTHIO)DIBUTYL STANNANE see BKK250
BIS(ISOOCTYLOXYCARBONYLMETHYLTHIO)DIMETHYLSTANNANE see BKK500
BIS(ISOOCTYLOXYCARBONYLMETHYLTHIO)DIOCTYL STANNANE see BKK750
BIS(ISOOCTYLOXYMALEOYLOXY)DIOCTYLSTANNANE see BKL000
BIS(ISOPROPYLAMIDO) FLUOROPHOSPHATE see PHF750
2,4-BIS(ISOPROPYLAMINO)-6-CHLORO-s-TRIAZINE see PMN850
2,4-BIS(ISOPROPYLAMINO)-6-ETHYLTHIO-s-TRIAZINE see EPN500
2,4-BIS(ISOPROPYLAMINO)-6-METHOXY-s-TRIAZINE see MFL250
4,6-BIS(ISOPROPYLAMINO)-2-METHYLMERCAPTO-s-TRIAZINE see BKL250
2,4-BIS(ISOPROPYLAMINO)-6-METHYLMERCAPTO-s-TRIAZINE see BKL250
2,4-BIS(ISOPROPYLAMINO)-6-METHYLTHIO-1,3,5-TRIAZINE see BKL250
2,4-BIS(ISOPROPYLAMINO)-6-METHYLTHIO-s-TRIAZINE see BKL250
2,4-BIS(ISOPROPYLAMINO)-6-(METHYLTHIO)-s-TRIAZINE mixed with METHA-
  NEARSONIC ACID MONOSODIUM SALT (1:4) see BKL500
BIS(ISOPROPYLBENZENE)CHROMIUM see DGR200
BIS(LACTATO)MAGNESIUM see LAL100
BIS(LAUROYLOXYCARBONYLMETHYLTHIO)DIOCTYLSTANNANE see DVN000
BIS(LAUROYLOXY)DI(n-BUTYL)STANNANE see DDV600
BIS(LAUROYLOXY)DIBUTYLSTANNANE see DDV600
BIS(LAUROYLOXY)DIOCTYLSTANNANE see DVJ800
1,3-BISMALEIMIDE-4-METHYLBENZENE see TGY770
BISMALEIMIDE S see BKL800
1,3-BISMALEIMIDO BENZENE see BKL750
4,4-BIS(MALEIMIDO)DIPHENYLMETHANE see BKL800
4,4'-BIS(MALEIMIDOPHENYL)METHANE see BKL800
BIS(4-MALEIMIDOPHENYL)METHANE see BKL800
BIS(p-MALEIMIDOPHENYL)METHANE see BKL800
2,4-BISMALEIMIDOTOLUENE see TGY770
BISMARSEN see BKV250
BISMATE see BKW000
BIS(MERCAPTOACETATE)-1,4-BUTANEDIOL see BKM000
BIS(MERCAPTOACETATE)DIOCTYLTIN BIS(BUTYL) ESTER see DVM200
BIS(MERCAPTOACETATE)DIOCTYLTIN BIS(2-ETHYLHEXYL) ESTER see DVM800
BIS(MERCAPTOACETATE)DIOCTYL-TIN BIS(ISOOCTYL) ESTER see BKK750
BIS(MERCAPTOBENZIMIDAZOLATO)ZINC see ZIS000
BIS(MERCAPTOBENZOTHIAZOLATO)ZINC see BHA750
BIS(MERCAPTO)DIOCTYLTIN BIS(DODECYL) ESTER see DVM400
1,2-BIS(MESYLOXY)ETHANE see BKM125
1,4-BIS(2'-MESYLOXYETHYLAMINO)-1,4-DIDEOXYMESOERYTHRITOL see LJD500
1,2-BIS(METHACRYLOYLOXY)ETHANE see BKM250
1,4-BIS(METHANESULFONOXY)BUTANE see BOT250
BIS(METHANE SULFONYL)-d-MANNITOL see BKM500
(1,4-BIS(METHANESULFONYLOXY)BUTANE) see BOT250
BISMETHIN see PEV750
3,4-BIS(METHOXY)BENZYL CHLORIDE see BKM750
3,5-BIS(METHOXYCARBONYL)BENZENESULFONIC ACID, SODIUM SALT see BKN000
1,2-BIS(METHOXYCARBONYL)ETHYNE see DOP400
N-(2-(2,3-BIS-(METHOXYCARBONYL)-GUANIDINO)-5-(PHENYLTHIO)-PHENYL)-2-
  METHOXYACETAMIDE see BKN250
1,2-BIS(METHOXYCARBONYLTHIOUREIDO)BENZENE see PEX500
1,2-BIS(3-(METHOXYCARBONYL)-2-THIOUREIDO)BENZENE see PEX500
o-BIS(3-METHOXYCARBONYL-2-THIOUREIDO)BENZENE see PEX500
3,6-BIS(β-METHOXYETHOXY)-2,5-BIS(ETHYLENEIMINO)-p-BENZOQUINONE see BDC750
2,5-BISMETHOXYETHOXY-3,6-BISETHYLENEIMINO-1,4-BENZOQUINONE see BDC750
3,6-BIS(β-METHOXYETHOXY)-2,5-BIS(ETHYLENIMINO)-p-BENZOQUINONE see BDC750
BIS(2-(2-METHOXYETHOXY)ETHYL) ETHER see PBO500

4,4'-BIS((4-(2-METHOXYETHOXY)-6-(N-METHYL-N-2-SULFOETHYL)AMINO-s-
  TRIAZIN-2-YL)AMINO)-2,2'-STILB see BKN500
BIS(2-METHOXYETHYL)ESTER MALEIC ACID see DOF000
BIS(2-METHOXY ETHYL)ETHER see BKN750
BIS(2-METHOXYETHYL)ETHER see PBO500
BIS(2-METHOXYETHYL)NITROSOAMINE see BKO000
BIS(2-METHOXYETHYL) PHTHALATE see DOF400
BIS(METHOXYETHYL) PHTHALATE see DOF400
BIS(METHOXYMALEOYLOXY)DIBUTYLSTANNANE see BKO250
BIS(METHOXYMALEOYLOXY)DIOCTYLSTANNANE see BKO500
2,6-BIS(p-METHOXYPHENETHYL)-1-METHYLPIPERIDINE ETHANESULFONATE
  see MJE800
BIS(4-METHOXYPHENYL)AMINE see BKO600
BIS(p-METHOXYPHENYL)AMINE see BKO600
4,4'-BIS(4-METHOXY-6-PHENYLAMINO-2-s-TRIAZINYLAMINO)-2,2'-STILBENEDI-
  SULFONIC ACID see BKO750
1,5-BIS(o-METHOXYPHENYL)-3,7-DIAZAADMANTAN-9-ONE see BKP000
3,4-BIS(p-METHOXYPHENYL)-3-HEXENE see DJB200
N,N'-BIS(4-METHOXYPHENYL)-N''-(4-ETHOXYPHENYL)GUANIDINE HYDRO-
  CHLORIDE see PDN500
2,2-BIS(p-METHOXYPHENYL)-1,1,1-TRICHLOROETHANE see MEI450
1,1-BIS(p-METHOXYPHENYL)-2,2,2-TRICHLOROETHANE see MEI450
1,10-BIS(N-METHYL-N-(1'-ADAMANTYL)AMINO)DECANE DIIODOMETHYLATE
  see DAE500
1,4-BIS(METHYLAMINO)-9,10-ANTHRACENEDIONE see BKP500
1,2-BIS(METHYLAMINO)ETHANE see DRI600
BIS-(2-METHYL-4-AMINO-6-QUINOLYLOXY)ETHANE DIHYDROCHLORIDE see
  EJB100
BIS(N-METHYLANILINE)METHANE see MJO000
BIS(N-METHYLANILINO)METHAN (GERMAN) see MJO000
N,N'-(BIS(2-(2-METHYL-1,3-BENZODIOXOL-2-YL)ETHYL))ETHYLENEDIAMINE DI-
  HYDROCHLORIDE see BKQ250
BIS(α-METHYLBENZYL)AMINE see BKQ500
N,N'-BIS(α-METHYLBENZYL)ETHYLENEDIAMINE see DQQ600
3,6-BIS(5-(3-METHYL-2-BUTENYL)INDOL-3-YL)-2,5-DIHYDROXY-p-BENZOQUI-
  NONE see CNF159
BIS(3-METHYLBUTYL) ADIPATE see AEQ500
2,6-BIS(1-METHYL.BUTYL)PHENOL see BKQ750
1,4-BIS(METHYLCARBAMYLOXY)-2-ISOPROPYL-5-METHYLBENZENE see
  BKR000
BIS((6-METHYL-3-CYCLOHEXEN-1-YL)METHYL) ESTER HEXANEDIOIC ACID
  see BKR100
BIS(3-METHYLCYCLOHEXYL PEROXIDE) see BKR250
1,1-BIS((3,4-METHYLENEDIOXYPHENOXY)METHYL-N,N-DIMETHYL-1-BUTANOL
  CITRATE see BKR500
α,α-BIS((3,4-(METHYLENEDIOXY)PHENOXY)METHYL)-1-PIPERIDINEBUTANO-
  LACETATE CITRATE see BKR750
1-(1,3-BIS(3,4-(METHYLENEDIOXY)PHENOXY)-2-PROPYL)PYRROLIDINE CI-
  TRATE see BKS500
α-((2-BIS(1-METHYLETHYL)AMINO)ETHYL)-α-PHENYL-2-PYRIDINEACETAMIDE
  PHOSPHATE (9CI) see RSZ600
1,4-BIS(1-METHYLETHYL)BENZENE see DNN830
BIS(1-METHYLETHYL)DIAZENE 1-OXIDE see ASP510
N,N'-BIS(1-METHYLETHYL)-6-METHYL-THIO-1,3,5-TRIAZINE-2,4-DIAMINE see
  BKL250
2,6-BIS(1-METHYLETHYL)PHENOL see DNR800
3,5-BIS(1-METHYLETHYL)PHENOL METHYLCARBAMATE see DNS200
3,5-BIS(1-METHYLETHYL)PHENYL ESTER METHYL CARBAMIC ACID see
  DNS200
3,5-BIS(1-METHYLETHYL)PHENYL METHYLCARBAMATE see DNS200
O,O-BIS(1-METHYLETHYL)-S-(PHENYLMETHYL)PHOSPHOROTHIOATE see
  BKS750
O,O-BIS(1-METHYLETHYL)-S-(2-((PHENYLSULFONYL) AMI-
  NO)ETHYL)PHEOSPHORODITHIOATE see DNO800
N,N'-BIS(1-METHYLETHYL)THIOUREA see DNS800
BIS(2-METHYLGLYCIDYL) ETHER see BJN500
BIS(6-METHYLHEPTYL)ESTER of PHTHALIC ACID see ILR100
N,N'-BIS(5-METHYL-3-HEPTYL)-p-PHENYLENEDIAMINE see BJT500
BIS(3-METHYLHEXYL)PHTHALIC ACID ESTER see DSF200
1,1-BIS(2-METHYL-4-HYDROXY-5-tert-BUTYLPHENYL)BUTANE see BRP750
BIS(2-METHYL-3-HYDROXY-4-METHOXYMETHYL-5-METHYLPYRIDYL)DISULFIDE
  DIHYDROCHLORIDE see BKS800
3,5-BIS-METHYLKARBOXY-BENZENSULFONAN SODNY (CZECH) see BKN000
BIS(METHYLMERCURIC)SULFATE see BKS810
BIS-(METHYLMERCURY)-SULFATE see BKS810
BIS-(METHYLMERKURI)SULFAT see BKS810
N,N-BIS(N-METHYL-N-PHENYL-tert-BUTYLACETAMIDO)-β-HYDROXYETHYLA-
  MINE see DTL200
N,N'-BIS((1-METHYL-4-PHENYL-4-PIPERIDINYL)METHYL)-DECANEDIAMIDE (9CI)
  see BKS825
N,N'-BIS(1-METHYL-4-PHENYL-4-PIPERIDYLMETHYL)SEBACAMIDE see BKS825
N,N'-BIS(2-METHYLPHENYL)THIOUREA see DXP600
BIS(METHYLPROPYL)CARBAMOTHIOIC ACID-S-ETHYL ESTER see EID500
N-BISMETHYLPTEROYLGLUTAMIC ACID see MDV500
BIS(2-METHYL PYRIDINE)SODIUM see BKT250

N,N-BIS(METHYLQUECKSILBER)-p-TOLUOL-SULFAMID see MLH100
N,N-BIS(METHYLSULFONEPROPOXY)AMINE HYDROCHLORIDE see YCJ000
1,6-BIS-o-METHYLSULFONYL-d-MANNITOL see BKM500
BIS((METHYLSULFONYL)OXY)DIBUTYLSTANNANE see DEI400
BIS((METHYLSULFONYL)OXY)DIPROPYLSTANNANE see DWW400
BIS(3-METHYLSULFONYLOXYPROPYL)AMINE p-TOLUENESULFONATE see IBQ100
N,N'-BIS(3-METHYL-2-THIAZOLIDINYLIDENE)UREA see BKU000
BIS(METHYLXANTHOGEN) DISULFIDE see DUN600
1,3-BIS(3,4-METILENDIOSSIFENOSSI)-2-AMINOPROPANO (ITALIAN) see MJS250
1,3-BIS-(3,4-METILENDIOSSIFENOSSI)-2-(3-DIMETILAMINOPROPIL)PROPAN-2-OLO CITRATO (ITALIAN) see BKR500
1,3-BIS-(3,4-METILENDIOSSIFENOSSI)-2-PIRROLIDINOPROPANO CITRATO (ITALIAN) see BKS500
BIS(MONOISOPROPYLAMINO)FLUOROPHOSPHATE see PHF750
BIS(MONOISOPROPYLAMINO)FLUOROPHOSPHINE OXIDE see PHF750
BIS(4-MORPHOLINECARBODITHIOATO)MERCURY see BKU250
N,N'-BISMORPHOLINE DISULFIDE see BKU500
BISMORPHOLINO DISULFIDE see BKU500
BIS(MORPHOLINO-)METHAN (GERMAN) see MJQ750
BISMORPHOLINO METHANE see MJQ750
5,7-BIS(MORPHOLINOMETHYL)-2-HYDROXY-3-ISOPROPYL-2,4,6-CYCLOHEPTA-TRIEN-1-ONE DIHYDROCHLORIDE see IOW500
BIS(MORPHOLINOTHIOCARBONYL)DISULFIDE see MRR090
BISMUTH-209 see BKU750
BISMUTH see BKU750
BISMUTH AMIDE OXIDE see BKV000
BISMUTH ARSPHENAMINE SULFONATE see BKV250
BISMUTH CHROMATE see DDT250
BISMUTH COMPOUNDS see BKV750
BISMUTH DIMETHYL DITHIOCARBAMATE see BKW000
BISMUTH EMETINE IODIDE see EAM500
BISMUTH HYDROXIDE NITRATE OXIDE see BKW100
BISMUTH MAGISTERY see BKW100
BISMUTH NITRATE see BKW250
BISMUTH NITRIDE see BKW500
BISMUTHOUS OXIDE see BKW600
BISMUTH(3+) OXIDE see BKW600
BISMUTH OXIDE see BKW600
BISMUTH PENTAFLUORIDE see BKW750
BISMUTH PERCHLORATE see BKW850
BISMUTH PLUTONIDE see BKX000
BISMUTH POTASSIUM SODIUM TARTRATE (SOLUBLE) see BKX250
BISMUTH SESQUIOXIDE see BKW600
BISMUTH SESQUISULFIDE see DDI200
BISMUTH SESQUITELLURIDE see BKY000
BISMUTH SODIUM-p-AMINOPHENYLARSONATE see BKX500
BISMUTH SODIUM THIOGLYCOLLATE see BKX750
BISMUTH STANNATE PENTAHYDRATE see BKY250
BISMUTH SUBNITRATE see BKW100
BISMUTH SUBNITRICUM see BKW100
BISMUTH(3+) SULFIDE see DDI200
BISMUTH TELLURIDE see BKY000
BISMUTH TELLURIDE, UNDOPED see BKY000
BISMUTH TIN OXIDE see BKY250
BISMUTH TRIOXIDE see BKW600
BISMUTH TRISODIUM THIOGLYCOLLATE see BKY500
BISMUTH VIOLET see AOR500
BISMUTH WHITE see BKW100
BISMUTH YELLOW see BKW600
BISMUTHYL NITRATE see BKW100
BIS-β-NAPHTHOL see BGC100
1,4-BIS(N,N'-DIETHYLENE PHOSPHAMIDE)PIPER-AZINE see BJC250
BIS(NITRATO)DIOXOURANIUM HEXAHYDRATE see URS000
2,5-BIS-(NITRATOMERCURIMETHYL)-1,4-DIOXANE see BKZ000
BIS(NITRATO-O,O')DIOXO URANIUM (solid) see URA200
BIS(NITRATO-O)OXOZIRCONIUM see BLA000
BIS-p-NITRO BENZENE DIAZO SULFIDE see BLA250
1,5-BIS(5-NITRO-2-FURANYL)-1,4-PENTADIEN-3-ONE, (AMINOIMINOME-THYL)HYDRAZONE see PAF500
sym-BIS(5-NITRO-2-FURFURYLIDENE) ACETONE GUANYLHYDRAZONE see PAF500
BIS(5-NITROFURFURYLIDENE)ACETONE GUANYLHYDRAZONE see PAF500
1,5-BIS(5-NITRO-2-FURYL)-3-PENTADIENONE AMIDINONHYDRAZONE see PAF500
1,5-BIS(5-NITRO-2-FURYL)-3-PENTADIENONE GUANYLHYDRAZONE see PAF500
BIS(4-NITROPHENYL) PHOSPHATE see BLA600
BIS(p-NITROPHENYL) PHOSPHATE see BLA600
BIS(p-NITROPHENYL)SULFIDE see BLA750
N,N'-BIS(4-NITROPHENYL)UREA compd. with 4,6-DIMETHYL-2-PYRIMIDINOL (1:1) see NCW100
BIS(2-NITRO-4-TRIFLUOROMETHYLPHENYL) DISULFIDE see BLA800
BIS(NONYLOXYMALEOYLOXY)DIOCTYLSTANNANE see DEI800

BIS-N,N'-(3-PHENYLPROPYL-2)-PIPERAZINE DIHYDROCHLORIDE see BLF250
BIS-N,N,N',N'-TETRAMETHYLPHOSPHORODIAMIDIC ANHYDRIDE see OCM000
BIS(OCTANOYLOXY)DI-n-BUTYL STANNANE see BLB250
BIS(OCTANOYLOXY)DI-n-BUTYLTIN see BLB250
2,2-BIS(3'-tert-OCTYL)-4'-HYDROXYPHENYLPROPANE see BLB500
BIS(1-OCTYL) MALEATE see DVK800
BIS(2-OCTYL)PHTHALATE see BLB750
BISODIUM TARTRATE see BLC000
BISOFLEX 81 see DVL700
BISOFLEX 91 see DVJ000
BISOFLEX DOA see AEO000
BISOFLEX DOP see DVL700
BISOFLEX DOS see BJS250
BIS-(2-OKTYL)ESTER KYSELINY FTALOVE see BLB750
BIS(OLEOYLOXY)DIBUTYLSTANNANE see DEJ000
BISOLVOMYCIN see HOI000
BISOLVON see BMO325
BISOLVON HYDROCHLORIDE see BMO325
BISOMEL see IQN000
BISOMER 2HEA see ADV250
BIS(2-OXO-9-BORNANESULFONIC ACID) DIETHYLSTANNYL ESTER see DKC600
BIS(1-OXODODECYL)PEROXIDE see LBR000
BIS-(2-OXOPROPYL)-N-NITROSAMINE see NJN000
BIS(1-OXOPROPYL)PEROXIDE see DWQ800
BIS(8-OXYQUINOLINE)COPPER see BLC250
BIS(PANTOTHENAMIDOETHYL) DISULFIDE see PAG150
BIS(PENTACHLOR-2,4-CYCLOPENTADIEN-1-YL) see DAE425
BIS(PENTACHLORO-2,4-CYCLOPENTADIEN-1-YL) see DAE425
BIS(PENTACHLOROCYCLOPENTADIENYL) see DAE425
BIS(PENTACHLOROPHENOL), ZINC SALT see BLC500
BIS(PENTAFLUOROETHYL)ETHER see PCG760
BIS(PENTA FLUORO PHENYL)ALUMINUM BROMIDE see BLC750
BISPENTAFLUOROSULFUR OXIDE see BLD000
BIS(PENTAMETHYLENETHIURAM)-TETRASULFIDE see TEF750
BIS(2,4-PENTANEDIONATO)CHROMIUM see BLD250
BIS(2,4-PENTANEDIONATO)COPPER see BGR000
BIS(2,4-PENTANEDIONATO-O,O')ZINC see ZCJ000
BIS(2,4-PENTANEDIONATO)TITANIUM OXIDE see BGQ750
4,5-BIS(4-PENTENYLOXY)-2-IMIDAZOLIDINONE see BLD325
BISPHENOL A see BLD500
BISPHENOL A DIGLYCIDYL ETHER see BLD750
BISPHENOL C see IPK000
BISPHENOL DIGLYCIDYL ETHER, MODIFIED see BLE000
BIS(PHENOXARSIN-10-YL) ETHER see OMY850
BIS(10-PHENOXARSYL) OXIDE see OMY850
10,10'-BIS(PHENOXYARSINYL) OXIDE see OMY850
BIS(10-PHENOXYARSINYL) OXIDE see OMY850
BIS(p-PHENOXYPHENYL)DIPHENYLSTANNANE see BLE250
BIS(p-PHENOXYPHENYL)DIPHENYLTIN see BLE250
1,4-BIS(PHENYL AMINO)BENZENE see BLE500
BISPHENYL-(2-CHLORPHENYL)-1-IMIDAZOLYL-METHAN (GERMAN) see MRX500
2,6-BIS(1-PHENYLETHYL)-4-METHYLPHENOL see MHR050
N,N'-BIS(PHENYLISOPROPYL)PIPERAZINE DIHYDROCHLORIDE see BLF250
BIS(PHENYLMERCURI)METHYLENEDINAPHTHALENESULFONATE see PFN000
BIS(PHENYLMERCURYLAURYL)SULFIDE see PFN750
BIS(PHENYLSELENIDE) see BLF500
BIS(PHENYLTHIO)DIMETHYLTIN see BLF750
1,3-BIS((PHENYL)TRIAZENO)BENZENE see BLG000
4,4'-BIS(4-PHENYL-2H-1,2,3-TRIAZOL-2-YL)-2,2'-STILBENEDISULFONIC ACID DI-POTASSIUM SALT see BLG100
N,N-BIS(PHOSPHONOMETHYL)GLYCINE see BLG250
2,6-BIS(PICRYLAMINO)-3,5-DINITROPYRIDINE see PQC525
BIS(PIPERIDINOTHIOCARBONYL) TETRASULFIDE see TEF750
3,8-BIS(1-PIPERIDINYLMETHYL)-2,7-DIOXASPIRO(4.4)NONANE-1,6-DIONE see BLG350
2,6-BIS(1-PIPERIDYLMETHYL)-4-(α,α-DIMETHYLBENZYL)PHENOL DIHYDRO-BROMIDE see BHR750
BIS(2-PROPANOL)AMINE see DNL600
BIS(2-PROPOXYETHYL)-1,4-DIHYDRO-2,6-DIMETHYL-4-(3-NITROPHENYL)-3,5-PY-RIDINEDICARBOXYLATE see NDY600
2,4-BIS(PROPYLAMINO)-6-CHLOR-1,3,5-TRIAZIN (GERMAN) see PMN850
N,N'-BISPROPYLENEISOPHTHALAMIDE see BLG400
BIS(PROPYLOXY)DIAZENE see DNQ700
trans-1,2-BIS(n-PROPYLSULFONYL)ETHYLENE see BLG500
2,2-BIS(((3-PYRIDINYLCARBONYL)OXY)METHYL)-1,3-PROPANEDIYL ESTER of 3-PYRIDINECARBOXYLIC ACID see NCW300
BIS(2-PYRIDYLTHIO)ZINC, 1,1'-DIOXIDE see ZMJ000
BIS(8-QUINOLINATO)COPPER see BLC250
BIS(8-QUINOLINOLATO)COPPER see BLC250
BIS(8-QUINOLINOLATO-N',O⁸)-COPPER see BLC250
BIS(SALICYLALDEHYDE)ETHYLENEDIIMINE COBALT(II) see BLH250
BIS(S-(DIETHOXYPHOSPHINOTHIOYL)MERCAPTO)METHANE see EEH600
BIS(SUCCINYLDICHLOROCHOLINE) see HLC500

BIS(5-SULFO-8-QUINOLINOLATO-N',O⁸) MANGANESE(II) see BKJ275
2,2'-BIS-6-TERC.BUTYL-p-KRESYLMETHAN (CZECH) see MJO500
2,2-BIS-3'-TERC. OKTYL-4'-HYDROXYFENYLPROPAN (CZECH) see BLB500
BISTERIL see PDC250
BISTER ML see LBU200
BIS(2,3,3,3-TETRACHLOROPROPYL) ETHER see OAL000
BIS(TETRADECANOYLOXY)DIBUTYLSTANNANE see BLH309
1,3-BIS(TETRAHYDRO-2-FURANYL)-5-FLUORO-2,4-PYRIMIDINEDIONE see BLH325
1,3-BIS(TETRAHYDRO-2-FURYL)-5-FLUOROURACIL see BLH325
BIS(TETRAKIS(HYDROXYMETHYL)PHOSPHONIUM)SULFATE (salt) see TDI000
1,6-BIS(5-TETRAZOLYL)HEXAAZ-1,5-DIENE see BLI000
3,4-BIS(1,2,3,4-THIATRIAZOL-5-YL THIO) MALEIMIDE see BLI250
BIS(1,2,3,4-THIATRIAZOL-5-YL THIO)METHANE see BLI500
I-3,3-BIS(3'-THIENYL)-2-PROPENYL-(3-HYDROXY-3-PHENYLPROPYL-2)AMINE see DXI800
BIS(THIOCARBAMOYL)DISULFIDE see TFS500
BISTHIOCARBAMYL HYDRAZINE see BLJ250
BIS(THIOCYANATO)-MERCURY see MCU250
p,p'-BIS(α-THIOL CARBAMYLACETAMIDO)BIPHENYL see MCL500
BIS(THIOUREA) see BLJ250
BISTOLUENE DIAZO OXIDE see BLJ500
1,4-BIS(p-TOLYLAMINO)ANTHRAQUINONE see BLK000
BIS-1,4-p-TOLYLAMINOANTHRCHINON (CZECH) see BLK000
N-BIS(p-TOLYLSULFONYL)AMIDOMETHYL MERCURY see BLK250
1,3-BIS(o-TOLYL)-2-THIOUREA see DXP600
BISTON see DCV200
BIS(TRIBENZYLSTANNYL)SULFIDE see BLK750
BIS(TRIBENZYLTIN) SULFIDE see BLK750
BIS-(TRI-N-BUTYLCIN)OXID (CZECH) see BLL750
BIS(TRIBUTYLOXIDE) of TIN see BLL750
BIS(TRI-N-BUTYLPHOSPHINE) DICHLORONICKEL see BLS250
BIS(TRIBUTYL(SEBACOYLDIOXY))TIN see BLL000
BIS((TRI-n-BUTYLSTANNYL)CYCLOPENTADIENYL)IRON see BLL250
1,1'-BIS(TRIBUTYLSTANNYL)FERROCENE see BLL250
BIS(TRIBUTYLSTANNYL)OXIDE see BLL750
BIS(TRIBUTYLTIN) ITACONATE see BLL500
BIS(TRIBUTYL TIN)OXIDE see BLL750
BIS(TRIBUTYLTIN)SULFIDE see HCA700
BIS(TRI-N-BUTYLZINN)-OXYD (GERMAN) see BLL750
BIS-2,3,5-TRICHLOR-6-HYDROXYFENYLMETHAN (CZECH) see HCL000
m-BIS(TRICHLORMETHYL)BENZENE see BLL825
BIS(TRICHLOROACETYL)PEROXIDE see BLM000
BIS-2,4,5-TRICHLORO BENZENE DIAZO OXIDE see BLM250
1,3-BIS(2,2,2-TRICHLORO-1-HYDROXYETHYL)UREA see DGQ200
BIS(3,5,6-TRICHLORO-2-HYDROXYPHENYL)METHANE see HCL000
1,3-BIS(TRICHLOROMETHYL)BENZENE see BLL825
1:4-BIS(TRICHLOROMETHYL) BENZENE see HCM500
m-BIS(TRICHLOROMETHYL)BENZENE see BLL825
BIS(TRICHLOROMETHYL)SULFONE see BLM500
BISTRICHLOROMETHYLTRISULFID (CZECH) see BLM750
BIS(TRICHLORO METHYL)TRISULFIDE see BLM750
1,3-BIS(2,4,5-TRICHLOROPHENOXY)-1,1,3,3-TETRABUTYLDISTANNOXANE see OPE100
BIS(2,3,5-TRICHLOROPHENYLTHIO)ZINC see BLN000
BIS(TRIETHYLENETETRAMINE)TUNGSTATONICKEL see BLN100
BIS(TRIETHYL TIN)ACETYLENE see BLN250
BIS(TRIETHYLTIN) SULFATE see BLN500
BIS(TRIFLUOROACETIC) ANHYDRIDE see TJX000
BIS(TRIFLUOROACETOXY)DIBUTYLTIN see BLN750
BIS(TRIFLUOROACETYL)PEROXIDE see BLO000
BIS(2,2,2-TRIFLUOROETHYL)ETHER see HDC000
BIS(TRIFLUOROETHYL)ETHER see HDC000
3,5-BIS(TRIFLUOROMETHYL)ANILINE see BLO250
1,3-BIS(TRIFLUOROMETHYL)BENZENE see BLO270
BIS(TRIFLUOROMETHYL)CHLOROPHOSPHINE see BLO280
BIS(TRIFLUOROMETHYL)CYANOPHOSPHINE see BLO300
BIS(TRIFLUOROMETHYL)DISULFIDE see BLO325
1,1-BIS(TRIFLUOROMETHYL)ETHENE see HDC450
2,2-BIS(TRIFLUOROMETHYL)-4-METHYL-5-PHENYL-OXAZOLIDINE HYDRATE see BLP250
BIS(TRIFLUOROMETHYL)NITROXIDE see BLP300
2-(3,5-BIS(TRIFLUOROMETHYL)PHENYL)-N-METHYL-HYDRAZINECARBOTH-IOAMIDE (9CI) see BLP325
1-((3,5-BIS-TRIFLUOROMETHYL)PHENYL)-4-METHYL-THIOSEMICARBAZIDE see BLP325
BIS(TRIFLUOROMETHYL)PHOSPHORUS(III) AZIDE see BLP500
α,α-BIS(TRIFLUOROMETHYL)-1-PIPERIDINEMETHANOL HYDRATE see BLQ250
BIS(TRIFLUOROMETHYL)SULFIDE see BLQ325
2,2-BIS(TRIFLUOROMETHYL)THIAZOLIDINE HYDRATE see BLQ500
2,2'-BIS(1,6,7-TRIHYDROXY-3-METHYL-5-ISOPROPYL-8-ALDEHYDONAPHTHAL-ENE see GJM000
BIS(TRIISOBUTYLSTANNANE) see HDY100
BISTRIMATE see BKX750

N,N'-BIS(3-(3,4,5-TRIMETHOXYBENZOYLOXY)PROPYL)HOMOPIPERAZINE DIHY-DROCHLORIDE see CNR750
1,4-BIS(3-(3,4,5-TRIMETHOXYBENZOYLOXY)-PROPYL)PERHYDRO-1,4-DIAZE-PINE DIHYDROCHLORIDE see CNR750
2,3-BISTRIMETHYLACETOXYMETHYL-1-METHYLPYRROLE see BLQ600
α,ω-BIS(TRIMETHYL AMMONIUM)HEXANE DIBROMIDE see HEA000
α,ω-BIS(TRIMETHYLAMMONIUM)HEXANE DICHLORIDE see HEA500
BIS(TRIMETHYLHEXYL)TIN DICHLORIDE see BLQ750
N,o-BIS(TRIMETHYLSILYL)ACETAMIDE see TMF000
BIS(TRIMETHYLSILYL)ACETAMIDE see TMF000
BIS(TRIMETHYLSILYL)AMINE see HED500
N,N'-BIS(TRIMETHYLSILYL)AMINOBORANE see BLQ850
cis-BIS(TRIMETHYLSILYLAMINO)TELLURIUM TETRAFLUORIDE see BLQ900
BIS(TRIMETHYLSILYL)CARBODIIMIDE see BLQ950
BIS(TRIMETHYLSILYL)CHROMATE see BLR000
1,2-BIS(TRIMETHYLSILYL)HYDRAZINE see BLR125
BIS(TRIMETHYLSILYL)MERCURY see BLR140
BISTRIMETHYL SILYL OXIDE see BLR250
BIS(TRIMETHYLSILYL)PEROXOMONOSULFATE see BLR500
BIS(1,3,7-TRIMETHYL-8-XANTHINYL)MERCURY see MCT000
N,N'-BIS(2,2,2-TRINITROETHYL)UREA see BLR625
BIS(2,4,6-TRINITRO-PHENYL)-AMIN (GERMAN) see HET500
BIS(TRINITROPHENYL)SULFIDE see BLR750
BISTRIPERCHLORATO SILICON OXIDE see BLS000
BIS(TRIPHENYLPHOSPHINE)DICHLORONICKEL see BLS250
BIS(TRIPHENYL PHOSPHINE) NICKEL DITHIOCYANATE see BLS500
BIS(TRIPHENYL SILYL)CHROMATE see BLS750
BIS(TRIPHENYLTIN)ACETYLENEDICARBOXYLATE see BLS900
BIS(TRIPHENYLTIN)SULFATE see BLT000
BIS(TRIPHENYLTIN)SULFIDE see BLT250
BIS(TRIPROPYLTIN)OXIDE see BLT300
BIS(TRIS(p-CHLOROPHENYL)PHOSPHINE)MERCURIC CHLORIDE COMPLEX see BLT500
BIS(TRIS(p-DIMETHYLAMINOPHENYL)PHOSPHINE)MERCURIC CHLORIDE COMPLEX see BLT750
BIS(TRIS(p-DIMETHYLAMINOPHENYL)PHOSPHINE OXIDE)STANNIC CHLORIDE COMPLEX see BLT775
BIS(TRIS(β,β-DIMETHYLPHENETHYL)TIN)OXIDE see BLU000
BIS(TRIS(p-METHOXYPHENYL)PHOSPHINE)MERCURIC CHLORIDE COMPLEX see BLU250
BIS(TRIS(2-METHYL-2-PHENYLPROPYL)TIN)OXIDE see BLU000
BIS(TRIS(p-METHYLTHIOPHENYL)PHOSPHINE)MERCURIC CHLORIDE COM-PLEX see BLU500
BISTRIUM CHLORIDE see HEA500
BISULFAN see BOT250
BISULFITE see HIC600, SOH500
BISULFITE de SODIUM (FRENCH) see SFE000
BISULPHANE see BOT250
BISULPHITE see HIC600
BIS(2-VINYLOXYETHYL)ETHER see DJE600
N,N-BIS(2,4-XYLYLIMINOMETHYL)METHYLAMINE see MJL250
BITEMOL see BJP000
BITEMOL S 50 see BJP000
BITHALLIUM TRISULFATE see TEM100
BITHIODINE see ARP875
BITHION see TAL250
BITHIONOL see TFD250
BITHIONOL SULFIDE see TFD250
BITIN see TFD250
BITIODIN see BLV000
BITIRAZINE see DIW000
BITIS ARIETANS VENOM see BLV075
BITIS GABONICA VENOM see BLV080
BITOKSYBACILLIN see BAC040
BITOLTEROL MESILATE see BLV125
BITOLTEROL MESYLATE see BLV125
4,4'-BI-o-TOLUIDINE see TGJ750
(m,o'-BITOLYL)-4-AMINE see BLV250
BITOSCANATE see PFA500
BITREX see DAP812
β-BITTER ACID see LIU000
BITTER ALMOND OIL see BLV500
BITTER ALMOND OIL CAMPHOR see BCP250
BITTER CUCUMBER see FPD100
BITTER FENNEL OIL see FAP000
BITTER GOURD see FPD100
BITTER ORANGE OIL see BLV750
BITTER SALTS see MAJ500
BITTERSWEET see AHJ875
BITUMEN (MAK) see ARO500
BIVERM see PDP250
BIVINYL see BOP500
BIXA ORELLANA see APE100
BIZMUTHIOL II (CZECH) see MCP500
BK8 see TOC000

BK 15 see TOB750
BKF see MJO500
B-K LIQUID see SHU500
B-K POWDER see HOV500
BL 9 see DXY000
BL 15 see HKQ100
BL 25 see MCB050
BL 35 see MCB050
BL 139 see DOY400
BL 191 see PBU100
BL 434 see MCB050
γ-BL see BOV000
BLA see LCH000
BLACAR 1716 see PKQ059
1743 BLACK see BMA000
11557 BLACK see IHC550
BLACK ACACIA see GJU475
BLACK ALG AETRINE see QAT520
BLACK AND WHITE BLEACHING CREAM see HIH000
BLACK ANTIMONY see AQL500
BLACK BIRCH OIL see SOY100
BLACK BLASTING POWDER see ERF500
BLACK CALLA see ITD050
BLACK COPPER OXIDE see CNO250
BLACK DOGWOOD see MBU825
BLACK EYED SUSAN see RMK250
BLACK GOLD F 89 see IHC550
BLACK IRON BM see IHC550
BLACK LEAD see CBT500
BLACK LEAF see NDN000
BLACK LOCUST see GJU475
BLACK MANGANESE OXIDE see MAS000
BLACK 4EMBL see AQP000
BLACK 2EMBL see AQP000
BLACK NIGHTSHADE see DAD880
BLACK OXIDE of IRON see IHD000
BLACK PEARLS see CBT750
BLACK PEPPER OIL see BLW250
BLACK PN see BMA000
BLACK POWDER, compressed (DOT) see ERF500, PLL750
BLACK POWDER, granular or as a meal (UN 0027) (DOT) see ERF500, PLL750
BLACK POWDER, in pellets (UN 0028) (DOT) see ERF500, PLL750
BLACK WIDOW SPIDER VENOM see BLW500
BLACOSOLV see TIO750
BLADAFUM see SOD100
BLADAFUME see SOD100
BLADAFUN see SOD100
BLADAN see EEH600, HCY000, PAK000, TCF250
BLADAN BASE see HCY000
BLADAN-M see MNH000
BLADDERON see FCB100
BLADDERPOD LOBELIA see CCJ825
BLADEX see BLW750
BLADEX G see DFY800
BLADEX H see TAH900
BLADEX 80WP see BLW750
BLAETTERALKOHOL see HFE000
BLANC DE FARD see BKW100
BLANC FIXE see BAP000
BLANCOL see BLX000
BLANCOL DISPERSANT see BLX000
BLANDLUBE see MQV750
BLANKOPHOR BBH see CMP200
BLANKOPHOR HZPA see DXB450
BLANKOPHOR MBBH see CMP200
BLANOSE BWM see SFO500
BLA-S see BLX500
BLASCORID see MOA600, PJA130
BLASTICIDEN-S-LAURYLSULFONATE see BLX250
BLASTICIDIN see BLX500
BLASTICIDIN S see BLX500
BLASTING GELATIN (DOT) see NGY000
BLASTING OIL see NGY000
BLASTING POWDER see ERF500, PLL750
BLASTMYCIN see BLX750
BLASTOESTIMULINA see ARN500
BLASTOMYCIN see BLX750
BLATTANEX see PMY300
L-BLAU 1 see IBV050
L-BLAU 3 see ADE500, CMM062
L-BLAU 2 (GERMAN) see FAE100
BLAUES PYOKTANIN see AOR500
BLAUSAEURE (GERMAN) see HHS000
BLAUWZUUR (DUTCH) see HHS000

BLEACHING POWDER, containing 39% or less chlorine (DOT) see HOV500
BLEACHING POWDER see HOV500
BLEDO CARBONERO (CUBA) see PJJ315
BLEIACETAT (GERMAN) see LCV000
BLEIAZETAT (GERMAN) see LCJ000
BLEIPHOSPHAT (GERMAN) see LDU000
BLEISTEARAT (GERMAN) see LDX000
BLEISTIFTBAUMS (GERMAN) see EQY000
BLEISULFAT (GERMAN) see LDY000
BLEKIT EVANSA (POLISH) see BGT250
BLEKIT TURKUSOWY A see ERG100
BLEMINOL see ZVJ000
BLENDED RED OXIDES of IRON see IHD000
BLENOXANE see BLY000, BLY780
BLEO see BLY000
BLEOCIN see BLY000
BLEOMYCETIN see BLY500
BLEOMYCIN see BLY000
BLEOMYCIN A2 see BLY250
BLEOMYCIN A5 see BLY500
BLEOMYCIN A COMPLEX see BLY750
BLEOMYCIN B2 see BLY760
BLEOMYCIN PEP see BLY770
BLEOMYCIN SULFATE see BLY780
BLEOMYCIN, SULFATE (salt) (9CI) see BLY780
BLEPH-10 see SNP500
BLEPH 10 see SNQ000
BLEU BRILLIANT FCF see FMU059
BLEU DIAMINE see CMO250
BLEU PATENTE V see ADE500, CMM062
BLEU SOLANTHRENE see IBV050
BLEX see DIN800
BLEXANE see BLY780
BL H368 see BEQ625
BLIGHIA SAPIDA see ADG400
BLIGHTOX see EIR000
BLISTER FLOWER see FBS100
BLISTERING BEETLES see CBE250
BLISTERING FLIES see CBE250
BLISTER WORT see FBS100
BLITEX see EIR000
BLITOX 50 see CNK559
BLITOX see CNK559
BLIZENE see EIR000
BLM see BLY000
BLM-PEP see BLY770
BLO see BOV000
BLOC see FAK100
BLOCADREN see DDG800
BLOCAN see SBH500
BLOCKADE see DWS200
BLON see BOV000
BLOODBERRY see ROA300
BLOODSTONE see HAO875
BLOTIC see MKA000
BLO-TROL see THX100
BLOX see LIH000
BLOXANTH see ZVJ000
BL P 152 see PDD350
BLP-1011 see DGE200
BLP 1322 see HMK000
BLS 640 see APT250
BLUE 1084 see ADE500
1085 BLUE see ADE500
1206 BLUE see FAE000
1311 BLUE see FAE100
11388 BLUE see FMU059
12070 BLUE see FAE100
BLUE ANTHRAQUINONE PIGMENT see IBV050
BLUE 'APE (HAWAII) see XCS800
BLUE ASBESTOS (DOT) see ARM275
BLUE 2B see CMO000
BLUEBELL see CMV390
BLUEBERRY ROOT see BMA150
BLUE BH see CMN800
BLUE BLACK 12B see FAB830
BLUE BLACK BN see BMA000
BLUE BLACK SX see FAB830
BLUE BN BALSE see DCJ200
BLUECAIN see BMA125
BLUE CARDINAL FLOWER see CCJ825
BLUE CHAMOMILE OIL see CDH500
BLUE COHOSH see BMA150
BLUE COPPER-50 see CNK559
BLUE COPPER see CNK559, CNP250

BLUE COPPERRAS see CNP500
BLUE CROSS see CGN000
BLUE DEVIL WEED see VQZ675
BLUE EMB see CMO250
BLUE GINSENG see BMA150
BLUE JESSAMINE see CMV390
BLUE K see DFN425
BLUE O see IBV050
BLUE OIL see AOQ000, COD750
BLUE-OX see ZLS000
BLUE POWDER see ZBJ000
BLUE STAR see AQF000
BLUE STONE see CNP250
BLUESTONE see CNP500
BLUE TARO (HAWAII) see XCS800
BLUE URS see ADE500
BLUE VITRIOL see CNP250, CNP500
BLUE VRS see ADE500
BLUE ZN 3 see CMM062
BLUTENE see AJP250
BLUTENE CHLORIDE see AJP250
BLUTON see IIU000
BM 1 see HNI500
BM 3055 see IBP200
7-BMBA see BNO750
BMC see BRS750, MHC750
BMD see BAC260
BMF 1 see MCB050
BMF 1 (AMINOPLAST) see MCB050
BMIH see IKC000
BMOO see BRT000
BN 30 see MCB050
BN see BFW000
B-NINE see DQD400
BNM see BAV575
B. N. MEXICANUS VENOM see BMJ500
BNP 30 see BRE500
BNP see NIY500
BNPP see BLA600
BNS see NMC100
BNU see BSA250
BO 714 see TCZ000
BO-ANA see FAB600
BOB see NKL300
BOCEP VITI see GJU050
BOEA see BKA000
B.O.E.A. see BKA000
BOG MANGANESE see MAS000
BOH see HHC000
BOISAMBRENE see FMV200
BOISAMBRENE FORTE see FMV100
BOIS BLEUDE HONQRIE see FBW000
BOIS D'ARC (FRENCH) see MRN500
BOIS D'INDE see LBK000
BOIS GENTIL (CANADA) see LAR500
BOIS d'INDE see BAT500
BOIS JAMBETTE (HAITI) see GIW200
BOIS JOLI (CANADA) see LAR500
BOIS de PLOMB (CANADA) see LEF100
BOL-148 see BNM250
BOL see BNM250
BOLATRON see PKQ059
BOLDIN see DNZ100
(+)-(S)-BOLDINE see DNZ100
(S)-BOLDINE see DNZ100
(+)-BOLDINE see DNZ100
BOLDINE see DNZ100
BOLDINE DIMETHYL ETHER see TDI475
BOLDO LEAF OIL see BMA600
BOLERO see SAZ000
BOLETIC ACID see FOU000
BOLETIC ACID DIMETHYL ESTER see DSB600
BOLINAN see PKQ250
BOLLS-EYE see HKC000, HKC500
BOLSTAR see SOU625
BOLVIDON see BMA625
BOMBITA see DBA800
BOMYL see SOY000
BON see HMX520
BONA see HMX520
BONABOL see XQS000
BON ACID see HMX520
BONADETTES see HGC500
BONADOXIN see HGC500
BONAMID see PJY500

BONAMINE see HGC500
BONAPAR see PGG350
BONAPHTHON see BNS750
BONAPICILLIN see AIV500
BONARE see CFZ000
BONAZEN see ZNA000
BONBONNIER (HAITI) see LAU600
BONBRAIN see TEH500
BOND CH 18 see AAX250
BONDELANE A see SNW500
BONDOLANE A see SNW500
BONE OIL see BMA750
BONGAY see HGL575
BONIBAL see DXH250
BONICOR see HNY500
BONIDE BLUE DEATH RAT KILLER see PHP010
BONIDE KRAB CRABGRASS KILLER see PLC250
BONIDE RYATOX see RSZ000
BONIDE TOPZOL RAT BAITS and KILLING SYRUP see RCF000
BONIFEN see BMB000
BONINE see MBX500
BONITON see AEH750
BONJELA see BEL900
BONLOID see PKQ059
BONNECOR see BMB125
BONOFORM see TBQ100
BONOMOLD OE see HJL000
BONOMOLD OP see HNU500
BON RED YELLOW SHADE see CMS148
BOOKSAVER see AAX250
BOOMER-RID see SMN500
BOOTS BTS 27419 see MJL250
BOP see NJN000
BORACIC ACID see BMC000
BORACSU see SFF000
BORANE-AMMONIA see BMB150
BORANE, COMPOUND with N,N-DIMETHYLMETHANAMINE (1:1) see BMB250
BORANE, COMPOUND with TRIMETHYLAMINE (1:1) see BMB250
BORANE with DIMETHYLAMINE (1:1) see DOR200
BORANE, compound with DIMETHYLSULFIDE see MPL250
BORANE-HYDRAZINE see BMB260
BORANE, compounded with MORPHOLINE see MRQ250
BORANE-PHOSPHORUS TRIFLUORIDE see BMB270
BORANE-PYRIDINE see POQ250
BORANES see BMB280
BORANE-TETRAHYDROFURAN see BMB300
BORANE, TRIFLUORO-, DIHYDRATE see BMG800
BORASSUS FLABELLIFER Linn., extract see BMB325
BORATES, TETRA, SODIUM SALT, anhydrous (OSHA, ACGIH) see SFF000, SFE500
BORATES, TETRA, SODIUM SALT, anhydrous (OSHA) see DXG035
BORATE(1-), TETRAFLUORO-, CADMIUM (2:1) (9CI) see CAG000
BORATE(1-), TETRAFLUORO-, COBALT(2+) (8CI,9CI) see CNC050
BORATE(1-), TETRAFLUORO-, HYDROGEN see FDD125, HHS600
BORATE(1-), TETRAHYDRO-, ALUMINUM (3:1) (9CI) see AHG875
BORAX (8CI) see SFF000
BORAX DECAHYDRATE see SFF000
BORAX GLASS see DXG035
BORAZINE see BMB500
BORAZOLE see BMB500
BORDEAUX see FAG020
BORDEAUX ARSENITE see BMB750
BORDEAUX EMBL see CMO885
BORDEAUX RRN see CMU820
BORDEN 2123 see AAX250
BORDERMASTER see CIR250
BOREA see BMM650
BORER SOL see EIY600
BORESTER 2 see THX750
BORESTER O see TLN000
BORIC ACID see BMC000
BORIC ACID, DISODIUM SALT see DXG035
BORIC ACID, ETHYL ESTER see BMC250
BORIC ACID (H2-B4-O7), CALCIUM SALT (1:1) (8CI) see CAN250
BORIC ACID, MONOSODIUM SALT see SII100
BORIC ACID, PHENYLMERCURY SILVER derivative see PFP250
BORIC ACID, SODIUM SALT see SJD000
BORIC ACID, TRI-n-AMYL ESTER see TMQ000
BORIC ACID, TRI-sec-BUTYL ESTER see THX750, THY000
BORIC ACID, TRIBUTYL ESTER see THX500
BORIC ACID, TRI-o-CHLOROPHENYL ESTER see TIY750
BORIC ACID, TRI-o-CRESYL ESTER see TJF500
BORIC ACID, TRIETHYL ESTER see TJK250, TJP500
BORIC ACID, TRIHEXYL ESTER see TKM000
BORIC ACID, TRIISOBUTYL ESTER see TKR750

BORIC ACID, TRIISOPROPYL ESTER see IOI000
BORIC ACID, TRIOCTADECYL ESTER see TMN750
BORIC ACID, TRI-n-OCTYL ESTER see TMO250
BORIC ACID, TRIOLEYL ESTER see BMC500
BORIC ACID, TRI-n-PENTYL ESTER see TMQ000
BORIC ACID, TRIS(2-AMINOETHYL) ESTER see TJK750
BORIC ACID, TRIS(1-AMINO-2-PROPYL) ESTER see TKT200
BORIC ACID, TRIS(2-ETHYLHEXYL) ESTER see TJR500
BORIC ACID, TRIS(1-METHYLHEPTYL) ESTER see TMO500
BORIC ACID, TRIS(4-METHYL-2-PENTYL) ESTER see BMC750
BORIC ACID, TRIS(PHENYLCYCLOHEXYL) ESTER see TMR750
BORIC ACID, TRISTEARYL ESTER see TMN750
BORIC ANHYDRIDE see BMG000
BORICIN see SFF000
BOR-IND see IDA400
BORNANE, 2,2,5-endo,6-exo,8,9,10-HEPTACHLORO- see THH575
BORNANE, 2-METHOXY-, exo-(8Cl) see IHX500
10-BORNANESULFONIC ACID, 2-OXO-, (1S,4R)-(+)- see RFU100
1-2-BORNANOL see NCQ820
2-BORNANOL, endo- see BMD000
2-BORNANONE see CBA750
(+)-2-BORNANONE see CBB250
d-2-BORNANONE see CBB250
BORNATE see IHZ000
BORNEO CAMPHOR see BMD000
BORNEOL (DOT) see BMD000
BORNEOL see BMD000
trans-BORNEOL see BMD000
(1S,2R,4S)-(−)-1-BORNEOL see NCQ820
(−)-BORNEOL see NCQ820
l-BORNYL ACETATE see BMD100
BORNYL ACETATE see BMD100
1-BORNYL ALCOHOL see NCQ820
BORNYL ALCOHOL see BMD000
S-((N-BORNYLAMIDIN)METHYL) HYDROGEN THIOSULFATE see BMD250
BORNYL ISOVALERATE see HOX100
BORNYVAL see HOX100
BOROETHANE see DDI450
BOROFAX see BMC000
BOROFLUORIC ACID see FDD125, HHS600
BOROHYDRURE de POTASSIUM (FRENCH) see PKY250
BOROHYDRURE de SODIUM (FRENCH) see SFF500
BOROLIN see PIB900
BORON see BMD500
BORON AZIDE DICHLORIDE see BMD750
BORON AZIDE DIIODIDE see BMD825
BORON BROMIDE see BMG400
BORON BROMIDE DIIODIDE see BME250
BORON CALCIUM OXIDE see CAN250
BORON CHLORIDE see BMG500
BORON COMPOUNDS see BME500
BORON DIBROMIDE IODIDE see BME750
BORON FLUORIDE see BMG700
BORON FLUORIDE, compd. with ACETIC ACID see BMG750
BORON FLUORIDE DIHYDRATE see BMG800
BORON HYDRIDE see DDI450
BORON OXIDE see BMG000
BORON PHOSPHIDE see BMG250
BORON SESQUIOXIDE see BMG000
BORON TRIAZIDE see BMG325
BORON TRIBROMIDE see BMG400
BORON TRICHLORIDE see BMG500
BORON TRIFLUORIDE see BMG700
BORON TRIFLUORIDE–ACETIC ACID COMPLEX see BMG750
BORON TRIFLUORIDE DIETHYL ETHERATE see BMH250
BORON TRIFLUORIDE DIHYDRATE (DOT) see BMG800
BORON TRIFLUORIDE DIHYDRATE see BMG800
BORON TRIFLUORIDE-DIMETHYL ETHER see BMH000
BORON TRIFLUORIDE DIMETHYL ETHERATE (DOT) see BMH000
BORON TRIFLUORIDE ETHERATE see BMH250
BORONTRIFLUORIDE MONOETHYLAMINE see EFU500
BORON TRIIODIDE see BMH500
BORON TRIOXIDE see BMG000
BORON TRISULFIDE see BMH659
BOROPHENYLIC ACID see BBM000
BOROXIN, TRIMETHOXY- see TKZ100
BORRELIDIN see BMH750
BORSAEURE (GERMAN) see BMC000
BORSIL P see SCK600
BORTRAN see RDP300
BORTRYSAN see DEV800
BORUTA BLACK A see FAB830
BOSAN SUPRA see DAD200
BOSMIN see VGP000
BOTHROPS ASPER VENOM see BMI000

BOTHROPS ATROX VENOM see BMI125
BOTHROPS COLOMIBIENSIS VENOM see BMI250
BOTHROPS GODMANI VENOM see BMI500
BOTHROPS LATERALIS VENOM see BMI750
BOTHROPS NASUTUS VENOM see BMJ000
BOTHROPS NIGROVIRIDIS NEGROVIRIDIS VENOM see BMJ250
BOTHROPS NUMMIFER MEXICANUS VENOM see BMJ500
BOTHROPS OPHYOMEGA VENOM see BMJ750
BOTHROPS PICADOI VENOM see BMK000
BOTHROPS SCHLEGLII VENOM see BMK250
BOTHROPS VENOM PROTEINASE see RDA350
BOTRAN see RDP300
BOTROPASE see RDA350
BOTRYODIPLODIN see BMK290
(−)-BOTRYODIPLODIN see BMK290
BOTULINUM NEUROTOXIN see BMM292
BOTULINUSTOXIN see CMY030
BOURBONAL see EQF000
BOURGEONAL see BMK300
BOURREAU DES ARBRES (CANADA) see AHJ875
BOUTON d'OR (CANADA) see FBS100
BOUVARDIN see BMK325
BOV see SOI500
BOVERIN see BAT830
BOVERINE see BAT830
BOVIDERMOL see DAD200
BOVINOCIDIN see NIY500
BOVINOX see TIQ250
BOVIZOLE see TEX000
BOVOFLAVIN see DBX400
BOVOLIDE see BMK500
BOY see HBT500
BOYGON see PMY300
B P 1 see MCB050
BP2 see AFJ625
BP 400 see MOO750
BP-400 see MOP000
B(a)P see BCS750
B(e)P see BCT000
anti-BPDE see BCU250
BPDE see BCU250
BPDE-syn see DMR000
BP-4,5-DIHYDRODIOL see DLB800
B(e)P-4,5-DIHYDRODIOL see DMK400
BP-7,8-DIHYDRODIOL see BCT750, DML000, DML200
BP-9,10-DIHYDRODIOL see DLC400
B(E)P 9,10-DIHYDRODIOL see DMK600
trans-BP-7,8-DIHYDRODIOL DIACETATE see DBG200
BP-7,8-DIHYDRODIOL-9,10-EPOXIDE (anti) see BCU000
anti-BP-7,8-DIHYDRODIOL-9,10-OXIDE see BCU000
(+)-BP-7,α,8-β-DIOL-9,α,10,α-EPOXIDE 1 see DMP600
(−)BP-7,β,8,α-DIOL-9,β,10-β-EPOXIDE 1 see DMP800
B(e)P DIOL EPOXIDE-1 see DMR150
(+)-BP-7-β,8-α-DIOL-9-α,10-α-EPOXIDE 2 see BMK620
(−)-BP 7-α,8-β-DIOL-9-β,10-β-EPOXIDE 2 see DVO175
B(E)P DIOL EPOXIDE-2 see DMR200
BP 7,8-DIOL-9,10-EPOXIDE 2 see DMP900
B(e)P 9,10-DIOL-11,12-EPOXIDE-1 see DMR150
anti-BP-DIOLEPOXIDE see DMQ000
BP DIOL EPOXIDE ANTI see BCU250
BPE-I see PJS750
BP-4,5-EPOXIDE see BCV500
BP 7,8-EPOXIDE see BCV750
B(a)P EPOXIDE I see DMR000
B(a)P EPOXIDE II see BMK630
BPG 400 see PKK500
BPG 800 see PKK750
B(E)P H4-9,10-DIOL see DNC400
B(c)PH DIOL EPOXIDE-1 see BMK634
B(c)PH DIOL EPOXIDE-2 see BMK635
B(e)P H4-9,10-EPOXIDE see ECQ150
BP-3-HYDROXY see BCX250
BP-KLP see SMQ500
BPL see PMT100
BPMC see MOV000
BP 4,5-OXIDE see BCV500
BP 7,8-OXIDE see BCV750
BP-9,10-OXIDE see BCW000
BP-11,12-OXIDE see BCW250
BPPS see SOP000
BP-3,6-QUINONE see BCU750
BP-6,12-QUINONE see BCV000
BPZ see FLL000
BR 700 see CKI750
BR 750 see GKO750

BR-931 see CLW500
BRACE see PHK000
BRACKEN FERN, CHLOROFORM FRACTION see BMK750
BRACKEN FERN, DRIED see BML000
BRACKEN FERN TANNIN see BML250
BRACKEN FERN TOXIC COMPONENT see SCE000
BRACKEN FERN, TANNIN-FREE see TAE250
BRADILAN see TDX860
BRADOPHEN see BDJ600
BRADYL see NAC500
BRALEN KB 2-11 see PJS750
BRALEN RB 03-23 see PJS750
BRAMYCIN see BML750
BRAN ABSOLUTE see WBJ700
BRASILAMINA BLACK GN see AQP000
BRASILAMINA BLUE 2B see CMO000
BRASILAMINA BLUE 3B see CMO250
BRASILAMINA BLUE RW see CMO600
BRASILAMINA CONGO 4B see SGQ500
BRASILAMINA FAST RED F see CMO870
BRASILAMINA GREEN B see CMO830
BRASILAMINA GREEN G see CMO840
BRASILAMINA VIOLET 3R see CMP000
BRASILAN AZO RUBINE 2NS see HJF500
BRASILAN BLACK BS see FAB830
BRASILAN FUCHSINE D see CMS231
BRASILAN METANIL YELLOW see MDM775
BRASILAN ORANGE A see CMM220
BRASILAZET BLUE GR see TBG700
BRASILAZINA OIL RED B see SBC500
BRASILAZINA OIL SCARLET see OHA000
BRASILAZINA OIL SCARLET 6G see XRA000
BRASILAZINA OIL YELLOW G see PEI000
BRASILAZINA OIL YELLOW R see AIC250
BRASILAZINA ORANGE Y see PEK000
BRASILAZOL BLACK BH see CMN800
BRASIL (CUBA) see CAK325
BRASILETTO (BAHAMAS) see CAK325
BRASILIN see LFT800
BRASORAN see ASG250
BRASSICOL see PAX000
BRAUNOSAN H see PKE250
BRAUNSTEIN (GERMAN) see MAS000
BRAVO see TBQ750
BRAVO 6F see TBQ750
BRAVO-W-75 see TBQ750
BRAXIN C see POI100
BRAXORONE see BML825
BRAZILETTO see LFT800
BRAZILIAN PEPPER TREE see PCB300
BRAZILIN see LFT800
BREADFRUIT VINE see SLE890
BRECHWEINSTEIN see AQJ500
BRECOLANE NDG see DJD600
BREDININ see BMM000
BREDININE see BMM000
BREK see LIH000
BRELLIN see GEM000
BREMIL see CFY000
BRENAL see AAE500
BRENDIL see VGK000
BRENOL see ART250
BRENTAMINE FAST BLUE B BASE see DCJ200
BRENTAMINE FAST ORANGE GR BASE see NEO000
BRENTAMINE FAST RED B BASE see NEQ000
BRENTAMINE FAST RED TR BASE see CLK220
BRENTAMINE FAST RED TR SALT see CLK235
BREON 202 see CGW300
BREON 351 see AAX175
BREON see PKQ059
BREON CS 100/30 see CGW300
BRESIT see ARQ750
BRESTAN see ABX250
BRESTANOL see CLU000
BRETHINE see TAN250
BRETOL see EKN500
BRETYLAN see BMV750
BRETYLATE see BMV750
BRETYLIUM-p-TOLUENESULFONATE see BMV750
BRETYLIUM TOSYLATE see BMV750
BRETYLOL see BMV750
BREVIMYTAL see MDU500
BREVINYL see DGP900
BREVIRENIN see VGP000
BREVITAL SODIUM see MDU500

BRIANIL see IPU000
BRICAN see TAN100
BRICANYL see TAN100, TAN250
BRICAR see TAN100
BRICARIL see TAN100
BRICK OIL see CMY825
BRICYN see TAN100
BRIDAL see BEM500
BRIER (BAHAMAS) see CAK325
BRIETAL SODIUM see MDU500
BRIGHT RED see CHP500, CMS150
BRIGHT RED G TONER see CMS148
BRIJ 30 see DXY000
BRIJ 38 see PJT300
BRIJ 52 see PJT300
BRIJ 56 see PJT300
BRIJ 58 see PJT300
BRIJ 92 see OIG000
BRIJ 98 see PJW500
BRIJ 92((2)-OLEYL) see OIG000
BRIJ 96((10) OLEYL) see OIG040
BRIJ W1 see PJT300
BRILLIANT 15 see CAT775
BRILLIANT ACID BLACK BNA EXPORT see BMA000
BRILLIANT ACID BLACK BN EXTRA PURE A see BMA000
BRILLIANT ACID BLUE A EXPORT see ADE500
BRILLIANT ACID BLUE AS see ERG100
BRILLIANT ACID BLUE N EXTRA see ERG100
BRILLIANT ACID BLUE V EXTRA see ADE500
BRILLIANT ACID BLUE VS see ADE500
BRILLIANT ACID RED G see CMM300
BRILLIANT ACID ROSAMINE 2G see CMM300
BRILLIANT ACRIDINE ORANGE E see BJF000
BRILLIANT ALIZARINE CYANINE R see CMM080
BRILLIANT ALIZARINE LIGHT BLUE 3FR see CMM080
BRILLIANT ALIZARINE SKY BLUE BS see CMM100
BRILLIANT BLACK see BMA000
BRILLIANT BLACK A see BMA000
BRILLIANT BLACK BN see BMA000
BRILLIANT BLACK NAF see BMA000
BRILLIANT BLACK N.FQ see BMA000
BRILLIANT BLUE see FMU059
BRILLIANT BLUE FCD No. 1 see FAE000
BRILLIANT BLUE FCF see FAE000
BRILLIANT BLUE GS see ADE500
BRILLIANT BLUE R see BMM500
BRILLIANT CHROME LEATHER BLACK H see AQP000
BRILLIANT COLACID RED G see CMM300
BRILLIANT CRESYL BLUE see BMM550
BRILLIANT CRESYL BLUE BB see BMM550
BRILLIANT CRIMSON RED see HJF500
BRILLIANT FAST YELLOW see DOT300
BRILLIANT FAT SCARLET R see CMS242
BRILLIANT GREEN 3EMBL see FAE950
BRILLIANT GREEN SULFATE see BAY750
BRILLIANT LAKE M see CMS160
BRILLIANT LAKERED R see CMS160
BRILLIANT MILLING RED see NAO600
BRILLIANT OIL ORANGE R see PEJ500
BRILLIANT OIL ORANGE R BASE see CMM760
BRILLIANT OIL ORANGE Y BASE see PEK000
BRILLIANT OIL SCARLET B see XRA000
BRILLIANT OIL YELLOW see IBB000
BRILLIANT ORANGE GR see CMU820
BRILLIANT PINK AS see CMM765
BRILLIANT PINK B see FAG070
BRILLIANT PONCEAU G see FMU070
BRILLIANT PONCEAU 3R see FMU080
BRILLIANT RED see CHP500
BRILLIANT RED 5SKH see PMF540
BRILLIANT RED TONER RA see CMS160
BRILLIANTSAEURE GRUEN BS see ADF000
BRILLIANT SAFRANINE BR see GJI400
BRILLIANT SAFRANINE G see GJI400
BRILLIANT SAFRANINE GR see GJI400
BRILLIANT SCARLET see CHP500, FMU080
BRILLIANT SCARLET G see CMS160
BRILLIANTSCHWARZ BN (GERMAN) see BMA000
BRILLIANT TONER Z see CHP500
BRILLIANT TONING RED AMINE see AJJ250
BRILLIANT VIOLET 5B see AOR500
BRILLIANT VIOLET K see DFN450
BRILLIANT YELLOW SLURRY see DEU000
BRIMSTONE see SOD500
BRINDERDIN see RDK000

BRIPADON see FLG000
BRIQUEST 543-33S see DJG700
BRISERINE see RDK000
BRISPEN see DGE200
BRISTACICLIN α see TBX000
BRISTACIN see PPY250
BRISTACYCLINE see TBX000, TBX250
BRISTAMIN HYDROCHLORIDE see DTO800
BRISTAMYCIN see EDJ500
BRISTOL A-649 see OIU499
BRISTOL LABORATORIES BC 2605 see CQF079
BRISTOPHEN see MNV250
BRISTURIC see BEQ625
BRISTURON see BEQ625
BRITACIL see AIV500
BRITAI see CFH825, CMV500
BRITISH ALUMINUM AF 260 see AHC000
BRITISH ANTILEWISITE see BAD750
BRITISH EAST INDIAN LEMONGRASS OIL see LEG000
BRITOMYA M see CAT775
BRITON see TIQ250
BRITTEN see TIQ250
BRITTOX see DDP000
BRL 152 see PDD350
BRL 556 see IBP200
BRL 1341 see AIV500
BRL 1383 see SGS500
BRL 1400 see DSQ800, MNV250
BRL-1621 see SLJ000
BRL-1702 see DGE200
BRL-2064 see CBO250
BRL 3475 see CBO000
BRL 25000 see ARS125
BRL 147777 see MFA300
BRL see AIV500
BRL 14151K see PLB775
BRL-1621 SODIUM SALT see SLJ050
BRL 2333 TRIHYDRATE see AOA100
BR 55N see PAU500
BROBAMATE see MQU750
BROCADISIPAL see MJH900, OJW000
BROCADOPA see DNA200
BROCASIPAL see MJH900, OJW000
BROCIDE see EIY600
BROCKMANN, ALUMINUM OXIDE see AHE250
BROCSIL see PDD350
BRODAN see CMA100
BRODIAR see DDS600
BRODIFACOUM see TAC800
BROFAREMINE HYDROCHLORIDE see BMM625
BROFENE see BNL250
BROGDEX 555 see SGM500
BROM (GERMAN) see BMP000
BROMACETOCARBAMIDE see BNK000
BROMACETYLENE see BMS500
BROMACIL see BMM650
BROMADAL see BNK000
BROMADEL see BNK000
BROMADIALONE see BMN000
BROMADIOLONE see BMN000
BROMADRYL see BMN250
BROMAL HYDRATE see THU500
BROMALLYLENE see AFY000
5-(2'-BROMALLYL)-5-ISOPROPYLBARBITURIC ACID see QCS000
BROMAMID see BMN350
BROMAMIDE (pharmaceutical) see BMN350
BROMAMIDE see BMN350, BMT250
p-BROMANILID KYSELINY 5-BROMSALICYLOVE see BOD600
4-BROMANILINU (CZECH) see BMT325
p-BROMANISOLE see AOY450
BROMANMINAN SODNY (CZECH) see DKR000
BROMANYLPROMIDE see BMN350
BROMARAL see BNP750
BROMAT see HCQ500
BROMATES see BMN500
BROMATE de SODIUM (FRENCH) see SFG000
BROMAZEPAM see BMN750
BROMAZIL see BMM650
10-BROM-1,2-BENZANTHRACEN (GERMAN) see BMT750
3-BROMBENZANTHRONE see BMU000
2-BROMBENZOTRIFLUORID (CZECH) see BOJ750
3-BROMBENZOTRIFLUORID (CZECH) see BOJ500
BROMBENZYL CYANIDE see BMW250
N-p-BROMBENZYL-N-α-PYRIDYL-N',N'-DIMETHYL-AETHYLENDIAMIN-HYDRO-
  CHLORIDE (GERMAN) see HGA500

N-p-BROMBENZYL-N-α-PYRIDYL-N'-METHYL-N'-AETHYL-AETHYLENDIAMIN-MA-
  LEINAT (GERMAN) see BMW000
BROMCARBAMIDE see BNP750
BROMCHLOPHOS see NAG400
BROMCHLORENONE see BMZ000
BROMCHOLITIN see TDI475
BROMDEFENURON see MHS375
O-(4-BROM-2,5-DICHLOR-PHENYL)-O,O-DIMETHYL-MONOTHIOPHOSPHAT
  (GERMAN) see BNL250
d-2-BROM-DIETHYLAMIDE of LYSERGIC ACID see BNM250
BROME (FRENCH) see BMP000
BROMEK DWUMETYLOLAURYLOBENZYLOAMONIOWY see BEO000
BROMELAIN see BMO000
BROMELAINS see BMO000
BROMELIA see EEY500
BROMELIN see BMO000
BROMEOSIN see BMO250
BROMETHOL see ARW250, THV000
BROMEX see CES750, DFK600, NAG400
1-p-BROMFENYL-3,3-DIMETHYLTRIAZEN (CZECH) see BNW250
2-BROMFLUORBENZEN (CZECH) see FGX000
3-BROMFLUORBENZEN (CZECH) see FGY000
BROMHEXINE CHLORIDE see BMO325
BROMHEXINE HYDROCHLORIDE see BMO325
BROMIC ACID, AMMONIUM SALT see ANC000
BROMIC ACID, POTASSIUM SALT see PKY300
BROMIC ACID, SODIUM SALT see SFG000
BROMIC ETHER see EGV400
BROMIDES see BMO750
BROMIDE SALT OF POTASSIUM see PKY500
BROMIDE SALT of SODIUM see SFG500
BROMID UHLICITY see CBX750
BROMINAL see DDP000
BROMINAL M & PLUS see CIR250
BROMINE, solution (DOT) see BMP000
BROMINE see BMP000
BROMINE AZIDE see BMP250
BROMINE CYANIDE see COO500
BROMINE DIOXIDE see BMP500
BROMINE FLUORIDE see BMP750
BROMINE NITRIDE see BMP250
BROMINE PENTAFLUORIDE see BMQ000
BROMINE PERCHLORATE see BMQ250
BROMINE TRIFLUORIDE see BMQ325
BROMINE(1) TRIFLUOROMETHANESULFONATE see BMQ500
BROMINE TRIOXIDE see BMQ750
BROMINEX see DDP000
BROMINIL see DDP000
5-BROMISATIN (CZECH) see BNL750
5-BROM-3-ISOPROPYL-6-METHYL-URACIL (GERMAN) see BNM000
BROMISOVAL see BNP750
α-BROMISOVALERYLUREA see BNP750
BROMISOVALERYLUREA see BNP750
BROMISOVALUM see BNP750
BROMIZOVAL see BNP750
BROMKAL 80 see OAH000
BROMKAL 80-9D see NMV735
BROMKAL 83-10DE see PAU500
BROMKAL 82-ODE see PAU500
BROMKAL P 67-6HP see TNC500
BROM LSD see BNM250
BROMLYSERGAMIDE see BNM250
2-BROM-d-LYSERGIC ACID DIETHYLAMINE see BNM250
BROM-METHAN (GERMAN) see MHR200
BROMNATRIUM (GERMAN) see SFG500
5-BROM-5-NITRO-1,3-DIOXAN (GERMAN) see BNT000
BROMO (ITALIAN) see BMP000
4'-BROMOACETANILIDE see BMR100
4-BROMOACETANILIDE see BMR100
p-BROMO-N-ACETANILIDE see BMR100
p-BROMOACETANILIDE see BMR100
α-BROMOACETIC ACID see BMR750
BROMOACETIC ACID, solid or solution (DOT) see BMR750
BROMOACETIC ACID see BMR750
BROMOACETIC ACID ETHYLENE ESTER see BHD250
BROMOACETIC ACID, ETHYL ESTER see EGV000
BROMOACETIC ACID METHYL ESTER see MHR250
BROMOACETONE, liquid (DOT) see BNZ000
BROMOACETONE (DOT) see BNZ000
BROMOACETONE see BNZ000
BROMOACETONE OXIME see BMS000
1-BROMOACETOXY-2-PROPANOL see BMS250
1-BROMOACETYL-α-α-DIPHENYL-4-PIPERIDINEMETHANOL see BMS300
BROMOACETYLENE see BMS500
BROMOACETYLENYLETHYLMETHYLCARBINOL see BNK350

BROMO ACID see BNH500, BNK700
2-BROMOACROLEIN see BMT000
3-BROMOADAMANTYL DIAZOMETHYL KETONE see EEE025
1-BROMO-3-ADAMANTYL ETHOXYMETHYL KETONE see BMT100
1-BROMO-3-ADAMANTYL HYDROXYMETHYL KETONE see BMT130
1-(2-BROMO-1-ADAMANTYL)-N-METHYL-2-PROPYLAMINE HYDROCHLORIDE see BNO000
1-(3-BROMO-1-ADAMANTYL)-N-METHYL-2-PROPYLAMINE HYDROCHLORIDE see BNO250
5-(2-BROMOALLYL)-5-sec-BUTYLBARBITURIC ACID see BOR000
γ-BROMOALLYLENE see PMN500
3-BROMOALLYL ISOCYANATE see BMT150
5-(2'-BROMOALLYL)-5-(1'-METHYL-N-PROPYL)BARBITURIC ACID see BOR000
BROMOAMINE see BMT250
3'-BROMO-trans-ANETHOLE see BMT300
BROMOANILIDE see BMR100
4-BROMOANILINE see BMT325
p-BROMOANILINE see BMT325
2-BROMOANISOLE see BMT400
4-BROMOANISOLE see AOY450
o-BROMOANISOLE see BMT400
p-BROMOANISOLE see AOY450
BROMOANTIFEBRIN see BMR100
BROMOAPROBARBITAL see QCS000
1-BROMOAZIRIDINE see BMT500
BROMO B see BNK700
4-BROMOBENZALDEHYDE see BMT700
p-BROMOBENZALDEHYDE see BMT700
10-BROMO-1,2-BENZANTHRACENE see BMT750
3-BROMOBENZ(d,e)ANTHRONE see BMU000
3-BROMO-7H-BENZ(DE)ANTHRACEN-7-ONE see BMU000
4-BROMO-BENZENAMINE (9CI) see BMT325
BROMOBENZENE (DOT) see PEO500
4-BROMOBENZENEACETONITRILE see BNV750
2-BROMO-1,4-BENZENEDIOL see BNL260
β-(p-BROMOBENZHYDRYLOXY)ETHYLDIMETHYLAMINE HYDROCHLORIDE see BNW500
2-(4-BROMOBENZOHYDRYLOXY)ETHYLDIMETHYLAMINE HYDROCHLORIDE see BNW500
4-BROMOBENZOIC ACID see BMU100
p-BROMOBENZOIC ACID see BMU100
6-BROMOBENZO(a)PYRENE see BMU500
2-BROMOBENZOTRIFLUORIDE see BOJ750
3-BROMOBENZOTRIFLUORIDE see BOJ500
m-BROMOBENZOTRIFLUORIDE see BOJ500
o-BROMOBENZOTRIFLUORIDE see BOJ750
5-BROMO-2-BENZOXAZOLINONE see BMU750
6-BROMO-2-BENZOXAZOLINONE see BMV000
p-BROMOBENZOYL AZIDE see BMV250
p-BROMOBENZOYLTHIOHYDROXIMIC ACID-5-DIETHYLAMINOETHYL ESTER HYDROCHLORIDE see DHZ000
4-BROMOBENZYLCYANIDE see BNV750
α-BROMOBENZYL CYANIDE see BMW250
p-BROMOBENZYL CYANIDE see BNV750
2-((p-BROMOBENZYL)(2-(DIMETHYLAMINO)ETHYL)AMINO)PYRIDINE HYDROCHLORIDE see HGA500
(o-BROMOBENZYL)ETHYLDIMETHYLAMMONIUM-p-TOLUENESULFONATE see BMV750
N-p-BROMOBENZYL-N',N'-DIMETHYL-N-2-PYRIDYLETHYLENE-DIAMINE HYDROCHLORIDE see HGA500
N-p-BROMOBENZYL-N'-ETHYL-N'-METHYL-N-2-PYRIDYLETHYLENEDIAMINE MALEATE see BMW000
α-BROMOBENZYLNITRILE see BMW250
BROMOBENZYLNITRILE see BMW250
3-BROMOBENZYLTRIFLUORIDE see BOJ500
o-BROMOBENZYLTRIFLUORIDE see BOJ750
3-BROMOBIPHENYL see BMW290
3-(3-(4'-BROMO(1,1'-BIPHENYL)-4-YL)3-HYDROXY-1-PHENYLPROPYL)-4-HYDROXY-2H-1-BENZOPYRAN-2-ONE see BMN000
3-(3-(4'-BROMO-1,1'-BIPHENYL-4-YL)-1,2,3,4-TETRAHYDRO-1-NAPHTHYL)-4-HYDROXYCOUMARIN see TAC800
3-(3-(4'-BROMOBIPHENYL-4-YL)-1,2,3,4-TETRAHYDRONAPHTH-1-YL)-4-HYDROXYCOUMARIN see TAC800
α-BROMO-β,β-BIS(p-ETHOXYPHENYL)STYRENE see BMX000
4-BROMO-7-BROMOMETHYLBENZ(a)ANTHRACENE see BMX250
4-BROMO-α-(4-BROMOPHENYL)-α-HYDROXYBENZENEACETIC ACID-1-METHYLETHYL ESTER see IOS000
2-BROMO-6-(N-(p-BROMOPHENYL)THIOCARBAMOYL)-4-CHLORO-BENZOIC ACID see BOL325
1-BROMOBUTANE see BMX500
2-BROMOBUTANE see BMX750
4-BROMO-1-BUTENE see BMX825
3-BROMO-3-BUTEN-2-ONE see MHS400
5-BROMO-3-sec-BUTYL-6-METHYLURACIL see BMM650
2-BROMOBUTYRIC ACID see BMY250

α-BROMOBUTYRIC ACID see BMY250
4-BROMOBUTYRONITRILE see BMY500
BROMOCARBAMIDE see BNP750
BROMOCET see HCP800
BROMOCHLOROACETONITRILE see BMY800
BROMOCHLOROACETYLENE see BMY825
6-BROMO-5-CHLORO-2-BENZOXAZOLINONE see BMZ000
6-BROMO-5-CHLOROBENZOXAZOLONE see BMZ000
1-BROMO-1-CHLORO-2,2-DIFLUOROETHENE see BNA000
2-BROMO-2-CHLORO-1,1-DIFLUOROETHYLENE see BNA000
1-BROMO-1-CHLORO-2,2-DIFLUOROETHYLENE see BNA000
BROMOCHLORODIFLUOROMETHANE see BNA250
3-BROMO-1-CHLORO-5,5-DIMETHYLHYDANTOIN see BNA325
3-BROMO-1-CHLORO-5,5-DIMETHYL-2,4-IMIDAZOLIDINEDIONE see BNA325
3-BROMO-N-(2-CHLOROMERCURICYCLOHEXYL)PROPIONAMIDE see CET000
BROMOCHLOROMETHANE see CES650
BROMOCHLOROMETHYL CYANIDE see BMY800
O-(4-BROMO-2-CHLOROPHENYL)-O-ETHYL-S-PROPYL PHOSPHOROTHIOATE see BNA750
3-(4-BROMO-3-CHLOROPHENYL)-1-METHOXY-1-METHYLUREA see CES750
N'-(4-BROMO-3-CHLOROPHENYL)-N-METHOXY-N-METHYLUREA see CES750
2-BROMO-4-(2-CHLOROPHENYL)-9-METHYL-6H-THIENO(3,2-f)(1,2,4)TRIAZOLO(4,3-a)(1,4)DIAZEPINE see LEJ600
N-(4-BROMO-3-CHLOROPHENYL)-N'-METHOXY-N'-METHYLUREA see CES750
1-BROMO-3-CHLOROPROPANE see BNA825
2-BROMO-2-CHLORO-1,1,1-TRIFLUOROETHANE see HAG500
BROMOCHLOROTRIFLUOROETHANE see HAG500
BROMOCRIPTIN see BNB250
BROMOCRIPTINE see BNB250
BROMOCRIPTINE MESILATE see BNB325
BROMOCYAN see COO500
BROMOCYANOGEN see COO500
1-BROMO-12-CYCLOTRIDECADIEN-4,8,10-TRIYNE see BNB750
1-BROMODECANE see BNB800
BROMODEOXYGLYCEROL see MRF275
5-BROMO-2-DEOXYURIDINE see BNC750
5-BROMO-2'-DEOXYURIDINE see BNC750
5-BROMODEOXYURIDINE see BNC750
BROMODEOXYURIDINE see BNC750
5-BROMODESOXYURIDINE see BNC750
2-BROMO-1,8-DIAMINO-4,5-DIHYDROXYANTHRAQUINONE see BND250
2-BROMO-1,5-DIAMINO-4,8-DIHYDROXYANTHRAQUINONE see BNC800
BROMODIBORANE see BND325
BROMODICHLOROMETHANE see BND500
4-BROMO-2,5-DICHLOROPHENOL see LEN050
4-BROMO-2,5-DICHLOROPHENOL-o-ESTER with O,O-DIETHYL PHOSPHOROTHIOATE see EGV500
O-(4-BROMO-2,5-DICHLOROPHENYL)-O,O-DIETHYL PHOSPHOROTHIOATE see EGV500
O-(4-BROMO-2,5-DICHLOROPHENYL)-O,O-DIETHYLPHOSPHOROTHIONATE see EGV500
4-BROMO-2,5-DICHLOROPHENYL DIMETHYL PHOSPHOROTHIONATE see BNL250
o-(4-BROMO-2,5-DICHLOROPHENYL)-o-ETHYL PHENYLPHOSPHONOTHIOATE see BND750
O-(4-BROMO-2,5-DICHLOROPHENYL)-O-METHYL PHENYLPHOSPHONOTHIOATE see LEN050
O-(4-BROMO-2,5-DICLORO-FENIL)-O,O-DIMETIL-MONOTIOFOSFATO (ITALIAN) see BNL250
2-BROMO-9,10-DIDEHYDRO-N,N-DIETHYL-6-METHYLERGOLINE-8-β-CARBOXAMIDE see BNM250
BROMODIETHYLACETYLCARBAMIDE see BNK000
BROMODIETHYLACETYLUREA see BNK000
5-BROMO-2-(2-(DIETHYLAMINO)ETHOXY)BENZANILIDE see DHP400
BROMODIETHYLGOLD see DJJ850
7-BROMO-1,3-DIHYDRO-5-(2-PYRIDYL)-2H-1,4-BENZDIAZEPIN-2-ONE see BMN750
dl-4-BROMO-2,5-DIMETHOXYAMPHETAMINE HYDROBROMIDE see BNE250
2-BROMO-3,5-DIMETHOXYANILINE see BNE325
dl-4-BROMO-2,5-DIMETHOXY-α-METHYLPHENETHYLAMINE HYDROBROMIDE see BNE250
2-BROMO-N,N-DIMETHYL-1-ADAMANATANEMETHANAMINE HYDROCHLORIDE-HEMIHYDRATE see BNE500
2-BROMO-N,N-DIMETHYL-1-ADAMANTANEPROPANAMINE HYDROCHLORIDE see BNF000
3'-BROMO-4-DIMETHYLAMINOAZOBENZENE see BNE600
2-((p-BROMO-α-(2-DIMETHYLAMINO)ETHYL)BENZYL)PYRIDINE BIMALEATE see BNE750
(±)-2-(p-BROMO-α-(2-(DIMETHYLAMINO)ETHYL)BENZYL)PYRIDINE MALEATE see DNW759
2-(p-BROMO-α-(2-(DIMETHYLAMINO)ETHYL)BENZYL)PYRIDINE MALEATE (1:1) see BNE750
(+)-2-(p-BROMO-α-(2-(DIMETHYLAMINO)ETHYL)BENZYL)PYRIDINE MALEATE see DXG100

2-BROMO-1-(N,N-DIMETHYLAMINOMETHYL)ADAMANTANE HYDROCHLORIDE-HEMIHYDRATE see BNE500
2-BROMO-1-(3-DIMETHYLAMINOPROPYL)ADAMANTANE HYDROCHLORIDE see BNF000
4-BROMODIMETHYLANILINE see BNF250
4-BROMO-N,N-DIMETHYL ANILINE see BNF250
5-BROMO-9,10-DIMETHYL-1,2-BENZANTHRACENE see BNF315
3-BROMO-7,12-DIMETHYLBENZ(a)ANTHRACENE see BNF300
4-BROMO-7,12-DIMETHYLBENZ(a)ANTHRACENE see BNF310
1-BROMO-3,3-DIMETHYL-2-BUTANONE see PJB100
p-BROMO-α,α-DIMETHYLPHENETHYLAMINE HYDROCHLORIDE see BNF750
4-BROMO-2,6-DIMETHYLPHENOL see BOL303
α-BROMO-β-DIMETHYLPROPANOYLUREA see BNP750
3-BROMO-5,7-DIMETHYL PYRAZOLYL-2-PYRIMIDINEPHOSPHOROTHIOIC ACID-O,O-DIETHYL ESTER see BAS000
6-BROMO-2,4-DINITROANILINE see DUS200
2-BROMO-4,6-DINITROANILINE see DUS200
6-BROMO-2,4-DINITROBENZENEDIAZONIUM HYDROGEN SULFATE see BNG125
3-BROMO-2,7-DINITRO-5-BENZO(b)-THIOPHENEDIAZONIUM-4-OLATE see BNG250
2-(1-(4-BROMODIPHENYL)ETHOXY)-N,N-DIMETHYLETHYLAMINE HYDROCHLORIDE see BMN250
BROMODIPHENYLMETHANE see BNG750
BROMODIPHENYLMETHANE (solution) see BNH000
2-BROMO-1,3-DIPHENYL-1,3-PROPANEDIONE see BNH100
3-BROMO-DMBA see BNF300
4-BROMO-DMBA see BNF310
BROMO DNA see DUS200
BROMOEOSIN see BMO250
BROMOEOSINE see BNH500, BNK700
3-BROMO-1,2-EPOXYPROPANE see BNI000
α-BROMOERGOCRIPTINE see BNB250
2-BROMOERGOCRYPTINE see BNB250
BROMOERGOCRYPTINE see BNB250
2-BROMO-α-ERGOCRYPTINE METHANESULFONATE see BNB325
2-BROMO-α-ERGOKRYPTIN see BNB250
2-BROMO-α-ERGOKRYPTINE-MESILATE (GERMAN) see BNB325
BROMOETHANE see EGV400
2-BROMO-ETHANEPHOSPHORIC ACID BIS(2-BROMOETHYL) ESTER see TNE600
α-BROMOETHANOIC ACID see BMR750
BROMOETHANOIC ACID see BMR750
2-BROMO ETHANOL see BNI500
BROMOETHANOL see BNI500
BROMOETHENE see VMP000
2-BROMO-2-ETHYLBUTYRLUREA see BNK000
(α-BROMO-α-ETHYLBUTYRYL)CARBAMIDE see BNK000
1-BROMO-ETHYL-BUTYRYL-UREA see BNK000
2-BROMO-2-ETHYLBUTYRYLUREA see BNK000
(α-BROMO-α-ETHYLBUTYRYL)UREA see BNK000
4'-BROMO-3'-ETHYL-4-DIMETHYLAMINOAZOBENZENE see BNK100
3'-BROMO-4'-ETHYL-4-DIMETHYLAMINOAZOBENZENE see BNK275
2-BROMO-N-ETHYL-N,N-DIMETHYLBENZENEMETHANAMINIUM 4-METHYLBENZENESULFONATE see BMV750
BROMOETHYLENE see VMP000
BROMOETHYLENE POLYMER see PKQ000
2-BROMO ETHYL ETHYL ETHER see BNK250
2-BROMOETHYL-3-NITROANISOLE see NFN000
p-((4-BROMO-3-ETHYLPHENYL)AZO)-N,N-DIMETHYLANILINE see BNK100
p-((3-BROMO-4-ETHYLPHENYL)AZO)-N,N-DIMETHYLANILINE see BNK275
(2-BROMOETHYL)TRIMETHYLAMMONIUM BROMIDE see BNK325
BROMOETHYNE see BMS500
2-BROMOETHYNYL-2-BUTANOL see BNK350
BROMOETHYNYLETHYLMETHYLCARBINOL see BNK350
4-BROMO-2-FENILINDAN-1,3-DIONE (ITALIAN) see BNW750
BROMOFLOR see CDS125
9-BROMOFLUORENE see BNK500
BROMOFLUORESCEIC ACID see BMO250, BNH500, BNK700
BROMO FLUORESCEIN see BNH500, BNK700
BROMOFLUOROFORM see TJY100
1-BROMO-8-FLUOROOCTANE see FKS000
10-BROMO-11b-(2-FLUOROPHENYL)2,3,7,11b-TETRAHYDROOXAZOLO(3,2-d)(1,4)BENZODIAZEPIN-6(5H)-ONE see HAG800
BROMOFORM see BNL000
BROMOFORME (FRENCH) see BNL000
BROMOFORMIO (ITALIAN) see BNL000
BROMOFOS see BNL250
BROMOFOS-ETHYL see EGV500
BROMOFOSMETHYL see BNL250
BROMOFUME see EIY500
BROMO-O-GAS see MHR200
1-BROMOHEPTANE see HBN100
2-BROMO-2,3,3,4,4,4-HEXAFLUOROBUTYRIC ACID METHYL ESTER see EKO000

BROMOHEXANE see HFM500
BROMOHEXYLMERCURY see HFR100
α-BROMOHYDRIN see MRF275
2-BROMOHYDROQUINONE see BNL260
BROMOHYDROQUINONE see BNL260
3-BROMO-6-HYDROXYBENZ-p-BROMANILIDE see BOD600
BROMO(2-HYDROXYETHYL)MERCURY see EED600
BROMO(2-HYDROXYETHYL)MERCURY AMMONIA SALT see BNL275
2-BROMO-12'-HYDROXY-2'-(1-METHYLETHYL)-5'-α-(2-METHYLPROPYL)ERGOTAMIN-3',6',18-TRIONE see BNB250
5-BROMOINDOLE-2,3-DIONE see BNL750
2-BROMOISOBUTANE see BQM250
5-BROMO-3-ISOPROPYL-6-METHYL, 2,4-PYRIMIDINEDIONE (FRENCH) see BNM000
5-BROMO-3-ISOPROPYL-6-METHYLURACIL see BNM000
5-BROMO-3-ISOPROPYL-6-METIL-URACIL (ITALIAN) see BNM000
3'-BROMOISOSAFROLE see BOA750
2-BROMOISOVALERIC ACID see BNM100
α-BROMOISOVALERIC ACID see BNM100
α-BROMOISOVALERIC ACID UREIDE see BNP750
α-BROMOISOVALEROYLUREA see BNP750
(α-BROMOISOVALERYL)UREA see BNP750
9-α-BROMO-11-KETOPROGESTERONE see BML825
BROMOL see THV750
2-BROMO-d-LYSERGIC ACID DIETHYLAMIDE see BNM250
BROMOLYSERGIDE see BNM250
2-(BROMOMERCURI)ETHANOL see EED600
2-(BROMOMERCURI) ETHANOL-AMMONIA (1:0.8 moles) compound see BNL275
7-BROMOMESOBENZANTHRONE see BMU000
BROMOMETANO (ITALIAN) see MHR200
BROMOMETHANE see MHR200
BROMOMETHANE mixed with DIBROMOETHANE see BNM750
1-BROMO-2-METHOXYBENZENE see BMT400
1-BROMO-4-METHOXYBENZENE see AOY450
4-(7-BROMO-5-METHOXY-2-BENZOFURANYL)PIPERIDINE HYDROCHLORIDE see BMM625
3-BROMO-3-(4-METHOXYBENZOYL)ACRYLIC ACID SODIUM SALT see SIK000
2-(5-BROMO-2-METHOXYBENZYLOXY)TRIETHYLAMINE see BNN125
BROMO(METHOXYCARBONYL) MERCURY see MHS250
(Z)-3-BROMO-4-(4-METHOXYPHENYL)-4-OXO-2-BUTENOIC ACID, SODIUM SALT see SIK000
2-BROMO-N-METHYL-1-ADAMANTANEETHYLAMINE MALEATE see BNN250
2-BROMO-N-METHYL-1-ADAMANTANEMETHANAMINE HYDROCHLORIDE see BNN500
1-BROMO-4-(METHYLAMINO)ANTHRAQUINONE see BNN550
3-BROMO-1-(2-METHYLAMINOPROPYL)ADAMANTANE HYDROCHLORIDE see BNO250
2-BROMO-1-(2-METHYLAMINOPROPYL)ADAMANTANE HYDROCHLORIDE see BNO000
9-BROMOMETHYLANTHRACENE see BNO500
7-BROMOMETHYLBENZ(a)ANTHRACENE see BNO750
1-BROMO-2-METHYLBENZENE see BOG260
(BROMOMETHYL)BENZENE see BEC000
p-BROMO-α-METHYLBENZHYDRYL-2-DIMETHYLAMINOETHYL ETHER HYDROCHLORIDE see BMN250
6-BROMOMETHYLBENZO(a)PYRENE see BNP000
1-BROMO-3-METHYL BUTANE see BNP250
2-BROMO-3-METHYLBUTANOIC ACID see BNM100
BROMOMETHYL tert-BUTYL KETONE see PJB100
2-BROMO-3-METHYLBUTYRIC ACID see BNM100
2-BROMO-3-METHYLBUTYRYLUREA see BNP750
9-(BROMOMETHYL)-10-CHLOROANTHRACENE see BNP850
10-BROMOMETHYL-9-CHLOROANTHRACENE see BNP850
7-BROMOMETHYL-4-CHLOROBENZ(a)ANTHRACENE see BNQ000
BROMOMETHYL p-CHLOROPHENYL KETONE see CJJ100
4'-BROMO-3'-METHYL-4-DIMETHYLAMINOAZOBENZENE see BNQ110
3'-BROMO-4'-METHYL-4-DIMETHYLAMINOAZOBENZENE see BNQ100
7-BROMOMETHYL-6-FLUOROBENZ(a)ANTHRACENE see BNQ250
2-BROMO METHYL FURAN see BNQ500
7-BROMO METHYL-12-METHYLBENZ(a)ANTHRACENE see BNR000
7-BROMOMETHYL-1-METHYLBENZ(a)ANTHRACENE see BNQ750
12-BROMOMETHYL-7-METHYLBENZ(a)ANTHRACENE see BNR250
2-BROMOMETHYL-5-METHYLFURAN see BNR325
BROMOMETHYL METHYL KETONE see BNZ000
5-BROMO-6-METHYL-3-(1-METHYLPROPYL)-2,4(1H,3H)-PYRIMIDINEDIONE see BMM650
5-BROMO-6-METHYL-3-(1-METHYLPROPYL)URACIL see BMM650
1-(BROMOMETHYL)-4-NITROBENZENE see NFN000
p-(BROMOMETHYL)NITROBENZENE see BEC000
1-BROMO-3-METHYLPENTIN-3-OL see BNK350
1-BROMO-3-METHYL-1-PENTYN-3-OL see BNK350
2-((p-BROMO-α-METHYL-α-PHENYLBENZYL)OXY)-N,N-DIMETHYLETHYLAMINE HYDROCHLORIDE see BMN250
1-BROMO-2-METHYLPROPANE see BNR750

2-BROMO-2-METHYL PROPANE see BNS000
2-BROMO-2-METHYLPROPANE (DOT) see BQM250
2-(BROMOMETHYL)-TETRAHYDROFURAN see TCS550
1-BROMONAPHTHALENE see BNS200
α-BROMONAPHTHALENE see BNS200
6-BROMO-1,2-NAPHTHOQUINONE see BNS750
BROMONE see BMN000
8-β-((5-BROMONICOTINOYLOXY)METHYL)-1,6-DIMETHYL-10-α-METHOXYERGO-
  LINE see NDM000
5-BROMO-5-NITRO-1,3-DIOXANE see BNT000
5-BROMO-5-NITRO-m-DIOXANE see BNT000
2-BROMO-2-NITROPANE-1,3-DIOL see BNT250
2-BROMO-2-NITROPROPAN-1,3-DIOL see BNT250
2-BROMO-2-NITRO-1,3-PROPANEDIOL see BNT250
3-BROMO-4-NITROQUINOLINE-1-OXIDE see BNT500
α-BROMO-p-NITROTOLUENE see NFN000
β-BROMO-β-NITROTRIMETHYLENEGLYCOL see BNT250
1-BROMOOCTANE see BNU000
1-BROMO-2-OXIMINOPROPANE see BMS000
3-BROMO-2-OXOPROPANOIC ACID see BOD550
α-BROMOPARANITROTOLUENE see NFN000
4-BROMO-PDMT see BNW250
1-BROMOPENTABORANE (9) see BNU125
4-BROMO-1,2,2,6,6-PENTAMETHYLPIPERIDINE see BNU250
1-BROMOPENTANE see AOF800
2-BROMOPENTANE see BNU500
4-(2-(5-BROMO-2-PENTYLOXYBENZYLOXY)ETHYL)MORPHOLINE see BNU660
2-(5-BROMO-2-PENTYLOXYBENZYLOXY)TRIETHYLAMINE see BNU700
BROMOPERIDOL see BNU725
4-BROMOPHENACYL BROMIDE see DDJ600
p-BROMOPHENACYL BROMIDE see DDJ600
dl-BROMOPHENIRAMINE MALEATE see DNW759
BROMOPHENIRAMINE MALEATE see BNE750
4-BROMOPHENOL see BNU750
o-BROMOPHENOL see BNV000
p-BROMOPHENOL see BNU750
BROMOPHENOL see BNU800
BROMO PHENOLS see BNV250
2-(4-BROMOPHENYL)ACETONITRILE see BNV750
4-BROMOPHENYLACETONITRILE see BNV750
α-BROMOPHENYLACETONITRILE see BMW250
p-BROMOPHENYLACETONITRILE see BNV750
p-BROMOPHENYLAMINE see BMT325
3-((4-BROMOPHENYL)AMINO)-N,N-DIMETHYL-PROPANAMIDE (9CI) see
  BMN350
p-(m-BROMOPHENYLAZO)-N,N-DIMETHYLANILINE see BNE600
p-BROMOPHENYL BROMIDE see BNV775
(±)-(Z)-γ-(4-BROMOPHENYL)-N,N-DIMETHYL-2-PYRIDINEPROPANAMINE 2-BU-
  TENEDIOATE (1:1) see DNW759
(S)-γ-(4-BROMOPHENYL)-N,N-DIMETHYL-2-PYRIDINEPROPANAMINE (Z)-2-BU-
  TENEDIOATE (1:1) see DXG100
3-(4-BROMOPHENYL)-N,N-DIMETHYL-3-(3-PYRIDINYL)-2-PROPEN-1-AMINE see
  ZBA500
(Z)-3-(4-BROMOPHENYL)-N,N-DIMETHYL-3-(3-PYRIDINYL)-2-PROPEN-1-AMINE
  DIHYDROCHLORIDE see ZBA525
3-(p-BROMOPHENYL)-N,N-DIMETHYL-3-(3-PYRIDYL)ALLYLAMINE see ZBA500
1-(4-BROMOPHENYL)-3,3-DIMETHYLTRIAZENE see BNW250
p-BROMOPHENYL ESTER ISOTHIOCYANIC ACID see BNW825
α-BROMO-β-PHENYLETHYLENE see BOF000
BROMOPHENYL HYDRAMINE HYDROCHLORIDE see BNW500
4-BROMOPHENYL HYDRAZINE HYDROCHLORIDE see BNW550
3-(α-(p-(p-BROMOPHENYL)-β-HYDROXYPHENETHYL)BENZYL)-4-HYDROXYCOU-
  MARIN see BMN000
4-(4-(4-BROMOPHENYL)-4-HYDROXYPIPERIDINO)-4'-FLUOROBUTYROPHENONE
  see BNU725
4-(4-(4-BROMOPHENYL)-4-HYDROXYPIPERIDINO)-4'-FLUOROBUTYROPHENONE
  see BNU725
4-(4-(p-BROMOPHENYL)-4-HYDROXYPIPERIDINOL)-4'-FLUOROBUTYROPHE-
  NONE see BNU725
4-(4-(4-BROMOPHENYL)-4-HYDROXY-1-PIPERIDINYL)-1-(4-FLUOROPHENYL)-1-
  BUTANONE see BNU725
4-BROMO-2-PHENYL-1,3-INDANDIONE see BNW750
5-BROMO-2-PHENYL-1,3-INDANDIONE see UVJ400
5-BROMO-2-PHENYLINDAN-1,3-DIONE see UVJ400
5-BROMO-2-PHENYL-1H-INDENE-1,3(2H)-DIONE see UVJ400
p-BROMOPHENYL ISOTHIOCYANATE see BNW825
p-BROMO PHENYL LITHIUM see BNX000
BROMOPHENYLMETHANE see BEC000
3-(p-BROMOPHENYL)-1-METHOXY-1-METHYLUREA see PAM785
N'-(4-BROMOPHENYL)-N-METHOXY-N-METHYLUREA see PAM785
o-BROMOPHENYL METHYL ETHER see BMT400
p-BROMOPHENYL METHYL ETHER see AOY450
3-(p-BROMOPHENYL)-1-METHYL-1-METHOXYUREA see PAM785
3-(p-BROMOPHENYL)-1-METHYL-1-NITROSOUREA see BNX125
1-(p-BROMOPHENYL)-3-METHYLUREA see MHS375

1-(p-BROMOPHENYL)-3-METHYLUREA mixed with SODIUM NITRITE see
  SIQ675
2-(m-BROMOPHENYL)-N-(4-MORPHOLINOMETHYL)SUCCINIMIDE see BNX250
1-(p-BROMOPHENYL)-1-PHENYL-1-(2-DIMETHYLAMINOETHOXY)ETHANE HY-
  DROCHLORIDE see BMN250
2-(1-(4-BROMOPHENYL)-1-PHENYLETHOXY)-N,N-DIMETHYLETHANAMINE HY-
  DROCHLORIDE see BMN250
(2-(1-p-BROMOPHENYL-1-PHENYLETHOXY)ETHYL)DIMETHYLETHYLAMINE HY-
  DROCHLORIDE see BMN250
(Z)-3-(4'-BROMOPHENYL)-3-(3''-PYRIDYL)DIMETHYLALLYLAMINE see ZBA500
BROMOPHOS see BNL250
BROMOPHOSETHYL see EGV500
1-BROMOPINACOLIN see PJB100
BROMOPINACOLIN see PJB100
1-BROMOPINACOLONE see PJB100
α-BROMOPINACOLONE see PJB100
BROMOPINACOLONE see PJB100
9-α-BROMOPREGN-4-ENE-3,11,20-TRIONE see BML825
9-BROMOPREGN-4-ENE-3,11,20-TRIONE see BML825
BROMOPRIDA see VCK100
BROMOPRIDE see VCK100
3-BROMO-1-PROPANAMINE HYDROBROMIDE see AJA000
1-BROMOPROPANE see BNX750
2-BROMOPROPANE see BNY000
1-BROMOPROPANE (DOT) see BNX750
3-BROMO-1,2-PROPANEDIOL see MRF275
1-BROMO-2-PROPANONE see BNZ000
BROMO-2-PROPANONE see BNZ000
2-BROMOPROPENALDEHYDE see BMT000
1-BROMO-2-PROPENE see AFY000
3-BROMOPROPENE see AFY000
(E)-p-(3-BROMOPROPENYL)ANISOLE see BMT300
5-(3-BROMO-1-PROPENYL)-1,3-BENZODIOXOLE see BOA750
5-(2-BROMO-2-PROPENYL)-5-(1-METHYLETHYL)-2,4,6(1H,3H,5H)-PYRIMIDINE-
  TRIONE see QCS000
3-BROMOPROPIONIC ACID see BOB250
α-BROMOPROPIONIC ACID see BOB000
β-BROMOPROPIONIC ACID see BOB250
3-BROMOPROPIONITRILE see BOB500
N,N³-DI(β-BROMOPROPIONYL)-N'¹,N²-DISPIROTRIPIPERAZINIUM DICHLORIDE
  see SLD600
2-BROMOPROPIOPHENONE see BOB550
α-BROMOPROPIOPHENONE see BOB550
3-BROMOPROPYLAMINE HYDROBROMIDE see AJA000
BROMOPROPYLATE see IOS000
3-BROMOPROPYL CHLORIDE see BNA825
3-BROMOPROPYLENE see AFY000
3-BROMO-1-PROPYNE see PMN500
3-BROMOPROPYNE (DOT) see PMN500
2-BROMOPYRIDINE see BOB500
3-BROMOPYRIDINE see BOC510
7-BROMO-5-(2-PYRIDYL)-3H-1,4-BENZODIAZEPIN-2(1H)-ONE see BMN750
3-(2-(5-BROMO-2-PYRIDYLOXY)ETHYL)THIAZOLIDINE HYDROCHLORIDE see
  BOD000
2-(6-(5-BROMO-2-PYRIDYL OXY)HEXYL)AMINOETHANE THIOL HYDROCHLO-
  RIDE see BOD500
1-BROMO-2,5-PYRROLIDINEDIONE see BOF500
3-BROMOPYRUVATE see BOD550
3-BROMOPYRUVIC ACID see BOD550
β-BROMOPYRUVIC ACID see BOD550
BROMOPYRUVIC ACID see BOD550
2-BROMOQUIN see QJJ100
2-BROMOQUINOL see BNL260
5-BROMOSALICYL-4-BROMOANILIDE see BOD600
5-BROMOSALICYLIC ACID see BOE500
BROMO SELTZER see ABG750
BROMOSILANE see BOE750
β-BROMOSTYRENE see BOF000
ω-BROMOSTYRENE see BOF000
BROMOSTYROL see BOF000
BROMOSTYROLENE see BOF000
N-BROMOSUCCIMIDE see BOF500
N-BROMOSUCCINIMIDE see BOF500
BROMOSULFALEIN see HAQ600
BROMOSULFOPHTHALEIN see HAQ600
BROMOSULPHALEIN see HAQ600
BROMOSULPHTHALEIN see HAQ600
BROMOTALEINA see HAQ600
3-BROMO-1,1,2,2-TETRAFLUOROPROPANE see BOF750
3-BROMOTETRAHYDROTHIOPHENE-1,1-DIOXIDE see BOG000
N-BROMOTETRAMETHYL GUANIDINE see BOG250
5-BROMO-2-(2-(3-THIAZOLIDINYL)ETHOXY)PYRIDINE HYDROCHLORIDE see
  BOD000
p-BROMOTHIOBENZOHYDROXIMIC ACID-S-DIETHYLAMINOETHYL ESTER HY-
  DROCHLORIDE see DHZ000

2-BROMOTOLUENE see BOG260
3-BROMOTOLUENE see BOG300
5-BROMOTOLUENE see BOG300
α-BROMOTOLUENE (DOT) see BEC000
ω-BROMOTOLUENE see BEC000
m-BROMOTOLUENE see BOG300
o-BROMOTOLUENE see BOG260
p-BROMOTOLUENE see BOG255
α-BROMO-α-TOLUNITRILE see BMW250
p-((4-BROMO-m-TOLYL)AZO)-N,N-DIMETHYLANILINE see BNQ110
p-((3-BROMO-p-TOLYL)AZO)-N,N-DIMETHYLANILINE see BNQ100
BROMOTRIBUTYLSTANNANE see TIC250
BROMOTRICHLOROMETHANE see BOH750
3-BROMO-1,1,1-TRICHLORO PROPANE see BOI000
3-BROMOTRICYCLOQUINAZOLINE see BOI250
1-BROMOTRIDECANE see BOI500
BROMOTRIETHYLSTANNANE see BOI750
BROMOTRIETHYLSTANNANE compounded with 2-PIPECOLINE (1:1) see
    TJU850
BROMOTRIFLUOROETHENE see BOJ000
BROMO TRIFLUOROETHYLENE see BOJ000
BROMOTRIFLUOROMETHANE see TJY100
3-BROMOTRIFLUOROMETHYLBENZENE see BOJ500
m-BROMO(TRIFLUOROMETHYL)BENZENE see BOJ500
m-BROMO-α,α,α-TRIFLUOROTOLUENE see BOJ500
o-BROMO-α,α,α-TRIFLUOROTOLUENE see BOJ750
BROMOTRIPENTYLSTANNANE see BOK250
BROMOTRIPHENYLMETHANE see TNP600
BROMOTRIPROPYLSTANNANE see BOK750
5-BROMOURACIL see BOL000
5-BROMOURACIL-2-DEOXYRIBOSIDE see BNC750
5-BROMOURACIL DEOXYRIBOSIDE see BNC750
BROMOURACIL DEOXYRIBOSIDE see BNC750
BROMOVAL see BNP750
α-BROMOVALERIC ACID see BOL250
BROMOVALEROCARBAMIDE see BNP750
BROMOVALERYLUREA see BNP750
BROMOWODOR (POLISH) see HHJ000
BROMOXIL see BNP750
4-BROMO-2,6-XYLENOL see BOL303
BROMOXYNIL see DDP000
BROMOXYNIL OCTANOATE see DDM200
BROMPERIDOL see BNU725
p-BROMPHENACYL-8 see DDJ600
dl-BROMPHENIRAMINE MALEATE see DNW759
d-BROMPHENIRAMINE MALEATE see DXG100
(±)-BROMPHENIRAMINE MALEATE see DNW759
3-(4-BROMPHENYL)-1-METHOXYHARNSTOFF (GERMAN) see PAM785
omega-BROMPINAKOLIN see PJB100
β-BROMSTYROL see BOF000
BROMSULFALEIN see HAQ600
BROMSULFAN see HAQ600
BROMSULFOPHTHALEIN see HAQ600
BROMSULFTHALEIN see HAQ600
BROMSULPHALEIN see HAQ600
BROMSULPHTHALEIN see HAQ600
BROM-TETRAGNOST see HAQ600
BROMTHALEIN see HAQ600
BROMURAL see BNP750
BROMURE de CYANOGEN (FRENCH) see COO500
BROMURE d'ETHYLE see EGV400
BROMURE de METHYLE (FRENCH) see MHR200
BROMURE de VINYLE (FRENCH) see VMP000
BROMURE de XYLYLE (FRENCH) see XRS000
BROMURO di ETILE (ITALIAN) see EIY500
BROMURO di METILE (ITALIAN) see MHR200
BROMURO de OXITROPIO(SPANISH) see ONI000
BROMUVAN see BNP750
BROMVALERYLUREA see BNP750
BROMVALETONE see BNP750
BROMVALETONUM see BNP750
BROMVALUREA see BNP750
BROMWASSERSTOFF (GERMAN) see HHJ000
BROMYL see BNP750
BROMYL FLUORIDE see BOL310
BRONCHIOCAIN see BQH250
BRONCHOCAIN see BQH250
BRONCHOCAINE see BQH250
BRONCHODIL see DNA600
BRONCHOLYSIN see ACH000
BRONCHOSELECTAN see AAN000
BRONCHOSPASMIN see DNA600
BRONCOVALEAS see BQF500
BRONKAID MIST see VGP000
BRONKEPHRINE see DMV600

BRONKEPHRINE HYDROCHLORIDE see ENX500
BRONOCOT see BNT250
BRONOPOL see BNT250
BRONOSOL see BNT250
BRONOX see TJL500
BRONTYL see HOA000
BRONZE BROMO see BNH500, BNK700
BRONZE GREEN TONER A-8002 see AFG500
BRONZE ORANGE see CMS150
BRONZE ORANGE TONER see CMS150
BRONZE POWDER see CNI000
BRONZE RED RO see CHP500
BRONZE SCARLET see CHP500
BROOM (DUTCH) see BMP000
BROOM ABSOLUTE see GCM000
O-(4-BROOM-2,5-DICHLOOR-FENYL)-O,O-DIMETHYL-MONOTHIOFOSFAAT
    (DUTCH) see BNL250
5-BROOM-3-ISOPROPYL-6-METHYL-URACIL DUTCH) see BNM000
BROOMMETHAAN (DUTCH) see MHR200
BROOMWATERSTOF (DUTCH) see HHJ000
BROPIRAMINE see AIY850
BROPIRIMINE see AIY850
BROSERPINE see RDK000
BROTIANIDE see BOL325
BROTIZOLAM see LEJ600
BROTOPON see CLY500
BROVALIN see BNP750
BROVALUREA see BNP750
BROVARIN see BNP750
BROVEL see ECU600
1545 BROWN see CMP250
11460 BROWN see XMA000
BROWN ACETATE see CAL750
BROWN ASBESTOS (DOT) see ARM275
BROWN COPPER OXIDE see CNO000
BROWN FK see CMP250
BROWN HEMATITE see LFW000
BROWN IRON ORE see LFW000
BROWN IRONSTONE CLAY see LFW000
BROWN SALT NV see MIH500
BROWN SK see CMU770
BROXALAX see PPN100
BROXIL see PDD350
BROXURIDINE see BNC750
BROXYKINOLIN see DDS600
BROXYNIL see DDP000
BROXYQUINOLINE see DDS600
BRUCEANTIN see BOL500
BRUCELLA MELITENSIS ENDOTOXIN see BOL600
BRUCINA (ITALIAN) see BOL750
BRUCINE (DOT) see BOL750
BRUCINE see BOL750
(−)-BRUCINE see BOL750
BRUCINE IODOMETHYLATE see BOM000
BRUCINE IODOMETHYLE (FRENCH) see BOM000
BRUCINE METHIODIDE see BOM000
BRUCITE see NBT000
BRUDR see BNC750
BRUFANEUXOL see DOT000
BRUFANIC see IIU000
BRUFEN see IIU000
BRUGMANSIA ARBOREA see AOO825
BRUGMANSIA SANGUINEA see AOO825
BRUGMANSIA SUAVEOLENS see AOO825
BRUGMANSIA X CANDIDA see AOO825
BRUINSTEEN (DUTCH) see MAS000
BRULAN see BSN000
BRUMIN see WAT200
BRUNEOMYCIN see SMA000
BRUOMOPHOS (RUSSIAN) see BNL250
BRUSH BUSTER see MEL500
BRUSH-OFF 445 MLD VOLATILE BRUSH KILLER see TAA100
BRUSH-RHAP see DAA800
BRUSH RHAP see TAA100
BRUSHTOX see TAA100
BRYAMYCIN see TFQ275
BS 572 see DNU100
BS 4231 see PMG000
BS 5930 see OJW000
BS 5933 see DRR500
BS 6825 see TGX500
BS 7020a see DAZ140
BS 7029 see CQH625
BS 7051 see HBT000
BS 7331 see TGJ250

BS100-141 see GKU300
BS see BSL600
BSA see BBR500, TMF000
BSB-S 40 see SMQ500
BSB-S-E see SMQ500
BSC-REFINE D see BBS750
B-SELEKTONON see DAA800
B-SELEKTONON M see CIR250
BSF see HAQ600
BSF SIMES see HAQ600
BSP see HAQ600
BSP SODIUM see HAQ600
BT 31 see DEJ100
BT 621 see CBS000
BT see BPG325, BPU000
BTB 202 see BAC040
BTB see BAC040
BTC 471 see BBA500
BTC 835 see QAT520
BTC 1010 see DGX200
BTC see AFP250
BTF see BBU125
BTFMEA see EFU500
621-BT HYDROCHLORIDE HYDRATE see TGJ150
BTKH see BEU500
BTM see BFM250
BTO see BLL750
BTPABA see CML835
B-1,3,5-TRICHLOROBORAZINE see TIL300
B-TRIMETHYLBORAZINE see TLN100
BTS 18322 see FLG100
BTS 27,419 see MJL250
BTS see PQC100
BTT see BSO750
BTZ see DKU875
BU-6 see BGS250
BU 533 see PIN100
BU 700 see UTU500
BU 2231A see TAC500
BU2AE see DDU600
BUBAN 37 see DUV400
BUBBIE BLOSSOMS see CCK675
BUBBY BUSH see CCK675
BUBURONE see IIU000
BUCACID AZURE BLUE see FMU059
BUCACID BLUE BLACK see FAB830
BUCACID BRILLIANT SCARLET 3R see FMU080
BUCACID FAST CRIMSON see CMM300
BUCACID FAST WOOL BLUE R see ADE750
BUCACID GUINEA GREEN BA see FAE950
BUCACID INDIGOTINE B see FAE100
BUCACID METANIL YELLOW see MDM775
BUCACID ORANGE A see CMM220
BUCACID PATENT BLUE AF see ERG100
BUCACID PATENT BLUE VF see ADE500
BUCACID TARTRAZINE see FAG140
BUCACID WOOL GREEN see ADF000
BUCARBAN see BSM000
BUCB see DJF200
BUCCALSONE see HHR000
BUCETIN see HJS850
BUCK BRUSH see SED550
BUCKEYE see HGL575
BUCKTHORN see BOM125, MBU825
BUCLIZINE DIHYDROCHLORIDE see BOM250
BUCLODIN see BOM250
BUCLOSINSAEURE (GERMAN) see CPJ000
BUCLOXIC ACID see CPJ000
BUCLOXIC ACID CALCIUM see CPJ250
BUCLOXIC ACID CALCIUM SALT see CPJ250
BUCLOXINSAEURE KALZIUM (FERMAN) see CPJ250
BUCLOXONIC ACID see CPJ000
BUCLOXONIC ACID CALCIUM SALT see CPJ250
BUCOLOM see BQW825
BUCOLOME see BQW825
BUCROL see BSM000
BUCS see BPJ850
BUCTRIL see DDP000
BUCTRIL INDUSTRIAL see DDP000
dl-BUCUMOLOL HYDROCHLORIDE see BOM510
BUCUMOLOL HYDROCHLORIDE see BOM510
BUDAMIN MF 55I see MCB050
BUDAMIN MF 60I see MCB050
BUDESONIDE see BOM520
BUDIPIN (GERMAN) see BOM530

BUDIPINE see BOM530
BUD-NIP see CKC000
BUDOFORM see CHR500
BUDORM see BPF500
5-BUDR see BNC750
BUDR see BNC750
BUDRALAZINE see DQU400
BUENO see MRL750
BUFAPTO METHALOSE see MIF760
BUFEDIL see BOM600
BUFEMID see BGL250
BUFEN see ABU500
BUFENCARB see BTA250
BUFETOLOL HYDROCHLORIDE see AER500
BUFEXAMIC ACID see BPP750
BUFF-A-COMP see ABG750
BUFLOMEDIL see BOM600
BUFLOMEDIL HYDROCHLORIDE see BOM600
BUFOGENIN see BOM650
BUFOGENIN B see BOM655
BUFON see DKA600
BUFONAMIN see BOM750, BQL000
BUFOPTO ZINC SULFATE see ZNA000
BUFORMIN see BRA625
BUFORMINE see BRA625
BUFORMIN HYDROCHLORIDE see BOM750, BQL000
BUFOTALIN see BON000
BUFOTALINE see BON000
BUFOTENIN see DPG109
BUFURALOL see BQD000
BUG MASTER see CBM750
BUHACH see POO250
BUIS de SAPIA (CANADA) see YAK500
BUKARBAN see BSM000
BUKS see BFW750
BUKSAMIN see AKF375
BULANA, combustion products see ADX750
BULAN and PROLAN MIXTURE (2:1) see BON250
d-BULBOCAPNINE see HLT000
BULBOCAPNINE see HLT000
BULBONIN see BQL000
BULBOSAN see TJE200
BULEN A 30 see PJS750
BULEN A see PJS750
BULGARIAN see RNF000
BULKALOID see MIF760
BULKOSOL see BON300
BULL FLOWER see MBU550
BULPUR see PLC250
BUMADIZON CALCIUM SALT HEMIHYDRATE see CCD750
BUMETANIDE see BON325
BUMEX see BON325
BUNAIOD see EQC000
BUNAMIODYL see EQC000
BUNAZOCINE HYDROCHLORIDE see BON350
BUNAZOSIN HYDROCHLORIDE see BON350
BUNGARUS CAERULEUS VENOM see BON365
BUNGARUS FASCIATUS VENOM see BON367
BUNGARUS MULTICINCTUS VENOM see BON370
BUNIODYL see EQC000
BUNITROLOL HYDROCHLORIDE see BON400
BUNK see PJJ300
BUNSENITE see NDF500
BUNT-CURE see HCC500
BUNT-NO-MORE see HCC500
BUPATOL see BOV825
BUPHENINE HYDROCHLORIDE see DNU200
BUPICAINE HYDROCHLORIDE (+) see BON750
BUPICAINE HYDROCHLORIDE (±) see BOO000
dl-BUPIVACAINE see BSI250
d(+)-BUPIVACAINE see BOO250
l(−)-BUPIVACAINE see BOO500
BUPIVACAINE see BSI250
BUPIVACAINE HYDROCHLORIDE see BOO000
BUPLEURUM FALCATUM L see SAF300
BUPLEURUM MARGINATUM WALL. EX. DC., EXTRACT see BOO625
BUPRANOLOL HYDROCHLORIDE see BQB250
BUPRENORPHINE see TAL325
BUPRENORPHINE HYDROCHLORIDE see BOO630
BUPROPION HYDROCHLORIDE see WBJ500
BUR see EHA100
BURCOL see BSM000
BURESE see CNE750
BUREX (CZECH) see PEE750
BURFOR see NCW300

BURGODIN see BCE825
BURINE see BON325
BURINEX see BON325
BURITAL SODIUM see SOX500
BURLEY TOBACCO see TGH725, TGH750
BURMA GREEN B see AFG500
BURN BEAN see NBR800
BURNISH GOLD see GIS000
BURNOL see DBX400, XAK000
BURNT ALUM see AHF100
BURNTISLAND RED see IHD000
BURNT LIME see CAU500
BURNT SIENNA see IHD000
BURNT UMBER see IHD000
BUROFLAVIN see DBX400
BURONIL see FKI000
BURPLEURUM FALCATUM LINN. VAR. MARGINATUM (WALL. EX. DC.) CL.,
    EXTRACT see BOO625
BURSINE see CMF800
BURTOLIN see DMC600
BURTONITE 44 see FPQ000
BURTONITE-V-40-E see CCL250
BURTONITE V-7-E see GLU000
BUSAN 72A see BOO635
BUSCAPINA see SBG500
BUSCAPINE see SBG500
BUSCOL see SBG500
BUSCOLAMIN see SBG500
BUSCOLYSINE see SBG500
BUSCOPAN see SBG500
BUSCOPAN COMPOSITUM see BOO650
BUSERELIN see LIU420
BUSHMAN'S POISON see BOO700
BUSONE see BRF500
BUSPIRONE see PPP250
BUSTREN see SMQ500
BUSTREN K 500 see SMQ500
BUSTREN K 525-19 see SMQ500
BUSTREN U 825 see SMQ500
BUSTREN U 825E11 see SMQ500
BUSTREN Y 825 see SMQ500
BUSTREN Y 3532 see SMQ500
BUTABARB see BPF000
BUTABARBITAL see BPF000
BUTABARBITAL SODIUM see BPF250
BUTABARBITONE see BPF000
BUTABITAL see AFY500
BUTACAINE see BOO750
BUTACAINE SULFATE see BOP000
BUTACARB see DEG400
BUTACARBE (FRENCH) see DEG400
BUTACHLOR see CFW750
BUTACIDE see PIX250
BUTACOMPREN see BRF500
BUTACOTE see BRF500
BUTA-1,3-DIEEN (DUTCH) see BOP500
BUTADIEEN (DUTCH) see BOP500
BUTA-1,3-DIEN (GERMAN) see BOP500
BUTADIEN (POLISH) see BOP500
BUTADIENDIOXYD (GERMAN) see BGA750
1,2-BUTADIENE see BOP250
BUTA-1,3-DIENE see BOP500
1,3-BUTADIENE see BOP500
α-γ-BUTADIENE see BOP500
BUTADIENE see BOP100
1,3-BUTADIENE, 2-CHLORO-, POLYMERS see PJQ050
1,3-BUTADIENE DIEPOXIDE see BGA750
l-BUTADIENE DIEPOXIDE see BOP750
BUTADIENE DIEPOXIDE see BGA750
BUTADIENE DIMER see CPD750
dl-BUTADIENE DIOXIDE see DHB600
BUTADIENE DIOXIDE see BGA750
1,1′,1′′,1′′′-(1,3-BUTADIENE-1,4-DIYLIDENE)TETRABENZENE see TEA400
BUTADIENE MONOEPOXIDE see EBJ500
BUTADIENE MONOXIDE see EBJ500
BUTADIENE PEROXIDE see BOQ250
BUTADIENES, inhibited (DOT) see BOP100
1,3-BUTADIENE-STYRENE COPOLYMER see SMR000
1,3-BUTADIENE-STYRENE POLYMER see SMR000
BUTADIENE-STYRENE POLYMER see SMR000
BUTADIENE-STYRENE RESIN see SMR000
BUTADIENE-STYRENE RUBBER (FCC) see SMR000
BUTADIENE SULFONE see DMF000
1,3-BUTADIENE, 1,1,4,4-TETRAPHENYL- see TEA400
BUTADIEN-FURFURAL COPOLYMER see BHJ500

N-2,3-BUTADIENYL-N-METHYLBENZYLAMINE HYDROCHLORIDE see BOQ500
BUTADIONE see BOT500
1,3-BUTADIYNE see BOQ625
BUTAFLOGIN see HNI500
BUTAFUME see BPY000
BUTAKON 85-71 see SMR000
BUTAL see BSU250
BUTALAMINE HYDROCHLORIDE see PEU000
BUTALAN see BRF500
BUTALBARBITAL see AGI750
BUTALBITAL see AGI750
BUTALBITAL SODIUM see BOQ750
BUTALDEHYDE see BSU250
BUTALGIN see MDP750
BUTALGINA see BRF500
BUTALIDON see BRF500
BUTALIN see BQN600
BUTALLYLONAL see BOR000
BUTALLYLONAL SODIUM see BOR250
BUTALYDE see BSU250
BUTAMBEN see BPZ000
BUTAMID see BSQ000
BUTAMIN see AIT750
BUTAMIRATE CITRATE see BOR350
BUTAMYRATE CITRATE see BOR350
n-BUTANAL (CZECH) see BSU250
BUTANAL see BSU250
1-BUTANAL, 3-METHYL- see MHX500
BUTANAL, 4-(OCTAHYDRO-4,7-METHANO-5H-INDEN-5-YLIDENE)- see
    OBW100
BUTANAL OXIME see BSU500
BUTANAMIDE, 2-CHLORO-N-METHYL-3-OXO-(9CI) see CEA100
BUTANAMIDE, N-(4-CHLORO-2-METHYLPHENYL)-3-OXO- see AAY300
BUTANAMIDE, 4-CHLORO-3-OXO-N-PHENYL- see CEA050
BUTANAMIDE, N-(4-CHLOROPHENYL)-3-OXO-(9CI) see AAY250
BUTANAMIDE, 2,2′-((3,3′-DICHLORO(1,1′-BIPHENYL)-4,4′-DIYL)BIS(AZO))BIS(N-
    (2,4-DIMETHYLPHENYL)-3-OXO)- see CMS208
BUTANAMIDE, 2,2′-((3,3′-DICHLORO(1,1′-BIPHENYL)-4,4′-DIYL)BIS(AZO))BIS(N-
    (2-METHYLPHEN YL)-3-OXO)- see CMS212
BUTANAMIDE, N-(2,4-DIMETHYLPHENYL)-3-OXO-(9CI) see OOI100
BUTANAMIDE, N-(4-ETHOXYPHENYL)-3-HYDROXY- see HJS850
BUTANAMIDE, 3-OXO-N-PHENYL-(9CI) see AAY000
1-BUTANAMINE see BPX750
2-BUTANAMINE see BPY000
2-BUTANAMINE, 3-METHYL-(9CI) see AOE200
1-BUTANAMINE, 1,1,2,2,3,3,4,4,4-NONAFLUORO-N,N-
    BIS(NONAFLUOROBUTYL)-(9CI) see HAS000
1-BUTANAMINIUM, N-N-N-TRIBUTYL-, HYDROXIDE (9CI) see TBK750
1-BUTANAMINIUM, N-N-N-TRIBUTYL-, NITRATE (9CI) see TBL250
1,3-BUTANDIOL (GERMAN) see BOS500
n-BUTANE (DOT) see BOR500
BUTANE see BOR500
BUTANEAMIDE, N-(2-CHLOROPHENYL)-3-OXO- see AAY600
BUTANE, 1,4-BIS(2,3-EPOXYPROPOXY)- see BOS100
1-BUTANECARBOXYLIC ACID see VAQ050
BUTANECARBOXYLIC ACID see VAQ000
1,3-BUTANEDIAMINE see BOR750
1,4-BUTANEDIAMINE see BOS000
1,4-BUTANEDIAMINE, N-(3-AMINOPROPYL)-, TRIHYDROCHLORIDE see
    SLG600
1,4-BUTANEDIAMINE, 2-METHYL-, POLYMER with α-HYDRO-ω-HYDROXYPO-
    LY(OXY-1,4-BUTANEDIYL) and 1,1′-METHYLENEBIS(4-ISOCYANATOCYCLO-
    HEXANE) see PKN500
1,4-BUTANEDICARBOXAMIDE see AEN000
1,4-BUTANEDICARBOXYLIC ACID see AEN250
BUTANE, 1,1-DICHLORO- see BRQ050
BUTANE DIEPOXIDE see BGA750
1,4-BUTANE DIGLYCIDYL ETHER see BOS100
BUTANE, 1,4-DIIODO- see TDQ400
1,4-BUTANEDINITRILE see SNE000
BUTANEDIOIC ACID see SMY000
BUTANEDIOIC ACID, 2,3-BIS(BENZOYLOXY)-, (R-(R*,R*))- see DDE300
BUTANEDIOIC ACID DIBUTYL ESTER see SNA500
BUTANEDIOIC ACID, DIETHYL ESTER see SNB000
BUTANEDIOIC ACID, 2,3-DIHYDROXY- see TAF750
BUTANEDIOIC ACID, 2,3-DIHYDROXY-, (R-(R*,R*))-, MONOPOTASSIUM SALT
    (9CI) see PKU600
BUTANEDIOIC ACID, 2,3-DIMERCAPTO-(9CI) see DNV610
BUTANEDIOIC ACID, DIMETHYL ESTER see SNB100
BUTANEDIOIC ACID DIPROPYL ESTER see DWV800
BUTANEDIOIC ACID, DISODIUM SALT (9CI) see SJW100
BUTANEDIOIC ACID, HYDROXY-(9CI) see MAN000
BUTANEDIOIC ACID, HYDROXY-, DIBUTYL ESTER, (+-)- see DED500
BUTANEDIOIC ACID MONO(7-CHLORO-2,3-DIHYDRO-2-OXO-5-PHENYL-1H-1,4-
    BENZODIAZEPIN-3-YL) ESTER see CFY500

BUTANEDIOIC ACID MONO(2,2-DIMETHYLHYDRAZIDE) see DQD400
BUTANEDIOIC ACID MONO(3-((2-ETHYLHEXYL)AMINO)-1-METHYL-3-OXOPRO-PYL) ESTER (9CI) see BPF825
BUTANEDIOIC ACID, SULFO-, 1,4-DIHEXYL ESTER, SODIUM SALT (9CI) see DKP800
BUTANEDIOIC ACID, SULFO-, 1,4-DIOCTYL ESTER, SODIUM SALT (9CI) see SOD300
BUTANEDIOIC ANHYDRIDE see SNC000
1,2-BUTANEDIOL see BOS250
1,3-BUTANEDIOL see BOS500
BUTANE-1,3-DIOL see BOS500
BUTANE-1,4-DIOL see BOS750
1,4-BUTANEDIOL see BOS750
2,3-BUTANEDIOL see BOT000
1,2-BUTANEDIOL, CYCLIC CARBONATE see BOT200
1,3-BUTANEDIOL, CYCLIC SULFITE see MQG500
1,3-BUTANEDIOL DIACRYLATE see BRG500
1,4-BUTANEDIOL DIACRYLATE see TDQ100
1,4-BUTANEDIOL DIGLYCIDYL ETHER see BOS100
BUTANE-1:4-DIOL DIGLYCIDYL ETHER see BOS100
BUTANEDIOL DIGLYCIDYL ETHER see BOS100
2,3-BUTANEDIOL, 1,4-DIMERCAPTO-, (R*,S*)-(9CI) see DXN350
1,4-BUTANEDIOL DIMETHANESULPHONATE see BOT250
1,4-BUTANEDIOL DIMETHYL SULFONATE see BOT250
1,4-BUTANEDIOL, POLYMER with 1,6-DIISOCYANATOHEXANE see PKO750
1,4-BUTANEDIOL POLYMER with 1,1'-METHYLENEBIS(4-ISOCYANATOBEN-ZENE) see PKP000
2,3-BUTANEDIONE see BOT500
BUTANEDIONE (DOT) see BOT500
2,3-BUTANEDIONE, MONOOXIME see OMY910
2,3-BUTANEDIONE 2-OXIME see OMY910
1,3-BUTANEDIONE, 1-PHENYL- see BDJ800
N,N'-(1,4-BUTANEDIYL)BIS-1-AZIRIDIENCARBOXAMIDE see TDQ225
2,2'-(1,4-BUTANEDIYL)BISOXIRANE see DHD800
2,2'-(1,4-BUTANEDIYLBIS(OXYMETHYLENE))BISOXIRANE see BOS100
BUTANEFRINE HYDROCHLORIDE see ENX500
BUTANE, 1-IODO-3-METHYL- see IHU200
BUTANE, 1-METHOXY-(9CI) see BRU780
BUTANE MIXTURES (DOT) see BOR500
BUTANEN (DUTCH) see BOR500
n-BUTANENITRILE see BSX250
BUTANENITRILE see BSX250
BUTANENITRILE, 4-CHLORO-(9CI) see CEU300
BUTANE, 1,1'-OXYBIS(4-CHLORO-(9CI) see OPE040
BUTANE, 2,2'-OXYBIS-(9CI) see BRH760, OPE030
BUTANESULFONE see BOU250
1,4-BUTANESULTONE (MAK) see BOU250
BUTANE SULTONE see BOU250
Δ-BUTANE SULTONE see BOU250
1,2,3,4-BUTANETETRACARBOXYLIC ACID see BOU500
BUTANETETRACARBOXYLIC ACID see BOU500
5H,6H-6,5A,13A,14-(1,2,3,4)BUTANETETRACYCLOOCTA(1,2-B:5,6-B')DINAPHTHALENE see LIV000
BUTANETHIOL (OSHA) see BRR900
tert-BUTANETHIOL see MOS000
1-BUTANETHIOL, TIN(2+) SALT see BOU550
1,2,3-BUTANETRIOL, TRINITRATE see MKI300
1,2,4-BUTANETRIOL, TRINITRATE see BOU700
1,2,4-BUTANETRIOL TRINITRATE see BOU700
BUTANEX see CFW750
BUTANI (ITALIAN) see BOR500
BUTANILICAINE HYDROCHLORIDE see BQA750
BUTANIMIDE see SND000
BUTANOIC ACID see BSW000
BUTANOIC ACID, 2-AMINO-4-(METHYLSULFINYL)-(9CI) see ALF600
BUTANOIC ACID, ANHYDRIDE (9CI) see BSW550
BUTANOIC ACID, 2-BROMO-3-METHYL-(9CI) see BNM100
BUTANOIC ACID-2-BUTOXY-1-METHYL-2-OXOETHYL ESTER (9CI) see BQP000
BUTANOIC ACID, 4-CHLORO-3-OXO-, ETHYL ESTER (9CI) see EHH100
BUTANOIC ACID, 4-CHLORO-3-OXO-, METHYL ESTER (9CI) see MIH600
BUTANOIC ACID, CYCLOHEXYL ESTER (9CI) see CPI300
BUTANOIC ACID, 3,7-DIMETHYL-2,6-OCTADIENYL ESTER, (E)-(9CI) see GDE810
BUTANOIC ACID, 4,4'-DISELENOBIS(2-AMINO)-(9CI) see SBU710
BUTANOIC ACID ETHYL ESTER see EHE000
BUTANOIC ACID, HEPTYL ESTER see HBN150
BUTANOIC ACID, 2,4-HEXADIENYL ESTER see HCS600
BUTANOIC ACID, 2-HYDROXY-4-(METHYLTHIO)-(9CI) see HMR600
BUTANOIC ACID, 4-HYDROXY-, MONOLITHIUM SALT see LHM800
BUTANOIC ACID, 3-METHYLBUTYL ESTER (9CI) see IHP400
BUTANOIC ACID, 3-METHYL-, HEXYL ESTER see HFQ575
BUTANOIC ACID, 2-METHYL-, 2-PHENYLETHYL ESTER see PDF780
BUTANOIC ACID, 3-METHYL-, 3-PHENYL-2-PROPENYL ESTER see PEE200

BUTANOIC ACID, 3-METHYL-, 1,7,7-TRIMETHYLBICYCLO(2.2.1)HEPT-2-YL ES-TER, endo- acid- see HOX100
BUTANOIC ACID, 3-OXO-, 5-METHYL-2-(1-METHYLETHYL)CYCLOHEXYL ES-TER, (1R-(1-α-2-β, 5α-) see MCG850
BUTANOIC ACID, 3-OXO-, 2-((1-OXO-2-PROPENYL)OXY)ETHYL ESTER (9CI) see AAY750
BUTANOIC ACID, 2-OXO-, SODIUM SALT (9CI) see SIY600
BUTANOIC ACID PENTYL ESTER see AOG000
BUTANOIC ACID, 1,2,3-PROPANETRIYL ESTER see TIG750
BUTANOIC ACID, PROPYL ESTER (9CI) see PNF100
BUTANOIC ACID 2,2,2-TRICHLORO-1-(DIMETHOXYPHOSPHINYL)ETHYL ES-TER see BPG000
BUTANOIC ANHYDRIDE see BSW550
BUTAN-1-OL see BPW500
1-BUTANOL see BPW500
BUTAN-2-OL see BPW750
2-BUTANOL see BPW750
sec-BUTANOL (DOT) see BPW750
BUTANOL (DOT) see BPW500
BUTANOL (FRENCH) see BPW500
n-BUTANOL see BPW500
tert-BUTANOL see BPX000
2-BUTANOL ACETATE see BPV000
3-BUTANOLAL see AAH750
BUTANOL-2-AMINE see AJA250
BUTANOL (4)-BUTYL-NITROSAMINE see HJQ350
2-BUTANOL, 3,3-DIMETHYL- see BRU300
BUTANOLEN (DUTCH) see BPW500
1-BUTANOL, 2,2,3,3,4,4,4-HEPTAFLUORO- see HAW100
4-BUTANOLIDE see BOV000
1-BUTANOL, 3-METHYL-, NITRATE (9CI) see ILW100
1-BUTANOL, 4-(NITROSO-2-PROPENYLAMINO)- see HJS400
BUTANOLO (ITALIAN) see BPW500
2,3-BUTANOLONE see ABB500
2-BUTANOL-3-ONE see ABB500
BUTANOL SECONDAIRE (FRENCH) see BPW750
BUTANOL TERTIAIRE (FRENCH) see BPX000
BUTANONE 2 (FRENCH) see MKA400
2-BUTANONE (OSHA) see MKA400
1-BUTANONE, 1-(4-AMINOPHENYL)-(9CI) see AJC750
2-BUTANONE, AZINE see EMT600
2-BUTANONE, 1-BROMO-3,3-DIMETHYL- see PJB100
2-BUTANONE, 1,1-DICHLORO-3,3-DIMETHYL- see DGF300
2-BUTANONE, 3,3-DIMETHYL-1-(METHYLTHIO)-, OXIME see DSS900
2-BUTANONE, 4-(4-HYDROXY-3-METHOXYPHENYL)- see VFP100
2-BUTANONE, 4-(4-HYDROXYPHENYL)- see RBU000
2-BUTANONE, 4-(6-METHOXY-2-NAPHTHALENYL)- see MFA300
2-BUTANONE, 4-(p-METHOXYPHENYL)-(6CI,7CI,8CI) see MFF580
2-BUTANONE, (1-METHYLPROPYLIDENE)HYDRAZONE see EMT600
2-BUTANONE, OXIME see EMU500
2-BUTANONE OXIME HYDROCHLORIDE see BOV625
1-BUTANONE, 1-(4-PYRIDYL)- see PNV755
2-BUTANONE, SEMICARBAZONE see MKA750
BUTANOVA see HNI500
BUTANOX LPT see MKA500
BUTANOX M 50 see MKA500
BUTANOX M 105 see MKA500
BUTAPERAZINE DIMALEATE see BSZ000
BUTAPHENE see BRE500
BUTAPIRAZOL see BRF500
BUTAPIRONE see HNI500
BUTAPYRAZOLE see BRF500
BUTARECBON see BRF500
n-BUTARSAMIDE see CBK750
BUTARTRINA see BRF500
BUTATAB see BPF000
BUTATAL see BPF000
BUTATENSIN see MBW750
BUTAZATE see BIX000
BUTAZATE 50-D see BIX000
BUTAZINA see BRF500
BUTAZOLIDINE SODIUM see BOV750
BUTAZONA see BRF500
BUTAZONE see BRF500
BUTEA FRONDOSA, seed extract see BOV800
BUTEDRIN see BOV825
BUTEDRINE see BQF250
BUTELLINE see BOP000
2-BUTENAL see COB250
(E)-2-BUTENAL see COB260
trans-2-BUTENAL see COB260
1-BUTENE see BOW250
2-BUTENE see BOW255
cis-2-BUTENE see BOW500
trans-2-BUTENE see BOW750

2-BUTENE, 1,3-DICHLORO- see DEU650
1-BUTENE, 3,4-DICHLORO- see DEV100
BUTENE, DICHLORO- see DEV200
2-BUTENE, 1,4-DIHYDROXY- see BOX300
2-BUTENEDINITRILE, (E)- see FOX000
(Z)-2-BUTENEDIOATE (1:1) 1,2,3,4-TETRAHYDRO-2-METHYL-4-PHENYL-8-ISO-
    QUINOLINAMINE see NMV725
cis-BUTENEDIOIC ACID see MAK900
(E)-BUTENEDIOIC ACID see FOU000
(Z)-BUTENEDIOIC ACID see MAK900
trans-BUTENEDIOIC ACID see FOU000
2-BUTENEDIOIC ACID BIS(1,3-DIMETHYLBUTYL) ESTER see DKP400
2-BUTENEDIOIC ACID BIS(2-ETHYLHEXYL) ESTER see DVK600
2-BUTENEDIOIC ACID BIS(1-METHYLETHYL) ESTER see BOX250
2-BUTENEDIOIC ACID, DIBUTYL ESTER see DED600
(Z)-2-BUTENEDIOIC ACID DIETHYL ESTER see DJO200
trans-BUTENEDIOIC ACID DIMETHYL ESTER see DSB600
2-BUTENEDIOIC ACID (Z)-, DIOCTYL ESTER (9CI) see DVK800
BUTENEDIOIC ACID, METHYL-, (E)- see MDI250
2-BUTENEDIOIC ACID, MONO(2-(2-HYDROXYETHOXY)ETHYL) ESTER see
    MAL500
2-BUTENEDIOIC ACID, MONO(2-HYDROXYPROPYL) ESTER see MAL750
cis-BUTENEDIOIC ANHYDRIDE see MAM000
2-BUTENE-1,4-DIOL see BOX300
2-BUTENE-1,4-DIOL, DIMETHANESULFONATE, (E)- see DNX000
2-BUTENENITRILE see COQ750
3-BUTENE NITRILE see BOX500
1-BUTENE-4-NITRILE see BOX500
3-BUTENE-2-ONE see BOY500
1-BUTENE OXIDE see BOX750
trans-2-BUTENE OZONIDE see BOX825
1-BUTENE, 2,3,4-TRICHLORO- see TIL360
2-BUTENOIC ACID see COB500
α-BUTENOIC ACID see COB500
trans-2-BUTENOIC ACID see ERE000
2-BUTENOIC ACID, ANHYDRIDE (9CI) see COB900
2-BUTENOIC ACID, 3-BROMO-4-(4-METHOXYPHENYL)-4-OXO-, SODIUM SALT,
    (E)- (9CI) see CQK600
2-BUTENOIC ACID, 4-BROMO-, METHYL ESTER (9CI) see MHR790
2-BUTENOIC ACID, 2-BUTENYLIDENE ESTER, (E,E,E)-(8CI,9CI) see COD100
2-BUTENOIC ACID, 2,3-DICHLOR-4-OXO-, (Z)-(9CI) see MRU900
2-BUTENOIC ACID-3,7-DIMETHYL-6-OCTENYL ESTER see CMT500
2-BUTENOIC ACID, ETHENYL ESTER see VNU000
2-BUTENOIC ACID, 3-(((ETHYLAMINO)METHOXYPHOSPHINOTHIOYL)OXY)-, 1-
    METHYLETHYL ESTER, (E)-, mixt. with 2,2-DICHLOROETHENYL DIMETHYL
    PHOSPHATE see SAD100
2-BUTENOIC ACID, ETHYL ESTER see EHO200
trans-2-BUTENOIC ACID ETHYL ESTER see COB750
2-BUTENOIC ACID, ETHYL ESTER, (E)-(9CI) see COB750
2-BUTENOIC ACID, HEXYL ESTER see HFM600
2-BUTENOIC ACID, 2-METHYL-, (E)-(9CI) see TGA700
trans-2-BUTENOIC ACID METHYL ESTER see COB825
2-BUTENOIC ACID, 2-METHYL-, ETHYL ESTER, (E)- see TGA800
2-BUTENOIC ACID, 2-METHYL-, 3-HEXENYL ESTER, (E,Z)- see HFE710
2-BUTENOIC ACID, 2-METHYL, 1-ISOPROPYL ESTER (E)- see IRN100
2-BUTENOIC ACID, 2-METHYL-, METHYL ESTER, (E)- see MPW700
2-BUTENOL see BOY000
2-BUTEN-1-OL see BOY000
3-BUTEN-2-OL see MQL250
3-BUTENO-β-LACTONE see KFA000
2-BUTEN-1-OL, 3-METHYL-, ACETATE see DOQ350
2-BUTEN-1-OL, 3-METHYL-, BENZOATE see MHU150
2-BUTEN-1-OL, 3-METHYL-, SALICYLATE see PMB600
3-BUTEN-2-OL, 4-(2,6,6-TRIMETHYL-1-CYCLOHEXEN-1-YL)- see IFT500
3-BUTEN-2-OL, 4-(2,6,6-TRIMETHYL-2-CYCLOHEXEN-1-YL)- see IFT400
3-BUTEN-2-OL, 4-(2,6,6-TRIMETHYL-2-CYCLOHEXEN-1-YL)-, ACETATE see
    IFX300
3-BUTEN-2-ONE see BOY500
1-BUTEN-3-ONE see BOY250, MQM100
3-BUTEN-2-ONE, 3-BROMO- see MHS400
3-BUTEN-2-ONE, 4-(4-HYDROXY-3-METHOXYPHENYL)- see MLI800
3-BUTEN-2-ONE, 4-(4-METHOXYPHENYL)- see MLI400
3-BUTEN-2-ONE, 4-(3,4-(METHYLENEDIOXY)PHENYL)-(6CI,7CI,8CI) see
    MJR250
3-BUTEN-2-ONE, 3-METHYL-4-PHENYL- see MNS600
3-BUTEN-2-ONE, 4-PHENYL-, (E)- see BAY275
3-BUTEN-2-ONE, 4-(2,5,6,6-TETRAMETHYL-2-CYCLOHEXEN-1-YL)-(9CI) see
    IGW500
3-BUTEN-2-ONE, 4-(2,4,6-TRIMETHYL-3-CYCLOHEXEN-1-YL)- see IGJ600
β-BUTENONITRILE see BOX500
2-BUTENYL ALCOHOL see BOY000
2-BUTENYL CHLORIDE see CEU825
cis-1-BUTENYL CYANIDE see PBQ275
2-BUTEN-1-YL DIAZOACETATE see BPA250
5-(1-BUTENYL)-5-ETHYLBARBITURIC ACID see BPA500

5-(2-BUTENYL)-5-ETHYL-2-THIOBARBITURIC ACID SODIUM SALT see BPA750
BUTENYL(3-HYDROXYPROPYL)NITROSAMINE see BPC600
(2-BUTENYLIDENE)ACETIC ACID see SKU000
2-BUTENYLIDENE CROTONATE see COD100
5-(1-BUTENYL)-5-ISOPROPYLBARBITURIC ACID see BPB500
5-(2-BUTENYL)-5-(1-METHYLBUTYL)-2-THIOBARBITURIC ACID SODIUM SALT
    see BPC500
3-(3-BUTENYLNITROSAMINO)-1-PROPANOL see BPC600
2-BUTENYLPHENOL (mixed isomers) see BPC750
3-BUTENYL-(2-PROPENYL)-N-NITROSAMINE see BPD000
1-(2′-BUTENYL)THEOBROMINE see BSM825
BUTEN-3-YNE see BPE109
3-BUTEN-1-YNYL DIETHYL ALUMINUM see BPE250
3-BUTEN-1-YNYL DIISOBUTYL ALUMINUM see BPE500
2-BUTEN-1-YNYL TRIETHYL LEAD see BPE750
BUTERAZINE see DQU400
BUTETHAL see BPF500
BUTETHAL SODIUM see BPF750
BUTETHAMINE HYDROCHLORIDE see IAC000
BUTETHANOL see TBN000
BUTFORMIN see BRA625
BUTIBATOL see BOV825
BUTIBUFEN see IJH000
BUTICAPS see BPF000
BUTIDIONA see BRF500
BUTIFOS see BSH250
n-BUTILAMINA (ITALIAN) see BPX750
BUTILATE see EID500
BUTILCHLOROFOS see BPG000, DDP000
o-(4-terz.-BUTIL-2-CLORO-FENIL)-o-METIL-FOSFORAMMIDE (ITALIAN) see
    COD850
BUTILE (ACETATI di) (ITALIAN) see BPU750
BUTILENE see HNI500
BUTIL METACRILATO (ITALIAN) see MHU750
BUTILOPAN see IJH000
2-BUTIN HEXAMETHYL-DEWAR-BENZOL (GERMAN) see HEC500
BUTINOX see BLL750
BUTIPHOS see BSH250
BUTIROSIN A see BSX325
BUTISAN see CDS275
BUTISANE see CDS275
BUTISERPAZIDE-25 see RDK000
BUTISERPAZIDE-50 see RDK000
BUTISERPINE see RDK000
BUTISOL see BPF000
BUTISOL SODIUM see BPF250
BUTISULFINA see BSM000
BUTOBARBITAL see BPF500
BUTOBARBITAL SODIUM see BPF750
BUTOBARBITONE SODIUM see BPF750
BUTOBARBITONE see BPF500
BUTOBARBITURAL see BPF500
BUTOBEN see BSC000, DTC800
BUTOBENDINE DIHYDROCHLORIDE see CNW125
BUTOCARBOXIM (GERMAN) see MPU250
BUTOCIDE see PIX250
BUTOCTAMIDE HYDROGEN SUCCINATE see BPF825
BUTOCTAMIDE SEMISUCCINATE see BPF825
BUTOFLIN see DAF300
BUTOKSYETYLOWY ALKOHOL (POLISH) see BPJ850
BUTOLAN see PFR325
BUTOLEN see PFR325
BUTONATE see BPG000
BUTONE see BRF500
BUTOPHEN see BPG250
BUTOPYRONOXYL see BRT000
BUTORPHANOL TARTRATE see BPG325
2-BUTOSSI-ETANOLO (ITALIAN) see BPJ850
BUTOX see DAF300
BUTOXICARBOXIM (GERMAN) see SOB500
BUTOXIDE see PIX250
BUTOXON see DGA000
BUTOXONE see DGA000
BUTOXONE AMINE see DGA000
BUTOXONE ESTER see DGA000
BUTOXONE SB see EAK500
BUTOXY ACETYLENE see BPG500
2-BUTOXY-AETHANOL (GERMAN) see BPJ850
2-BUTOXY-3-AMINOBENZOIC ACID β-DIETHYLAMINOETHYL ESTER HYDRO-
    CHLORIDE see AJA500
2-N-BUTOXYBENZAMIDE see BPG750
o-BUTOXYBENZAMIDE see BPG750
2-BUTOXYBENZOESAEURE-4′-DIAETHYLAMINO-L′-METHYL-BUTYLAMID (1′)
    HYDROCHLORID (GERMAN) see BPJ500

2-BUTOXYBENZOESAEURE-3'-DIAETHYLAMINOPROPYLAMID-(1') HYDRO-
CHLORID (GERMAN) see BPJ750
p-BUTOXYBENZOIC ACID-3-(2-METHYLPIPERIDINO)PROPYL ESTER HYDRO-
CHLORIDE see BPH750
1-(2-(4-BUTOXYBENZOYL)ETHYL)PIPERIDINE HYDROCHLORIDE see BPR500
(−)-8-(p-BUTOXYBENZYL)-3-α-HYDROXY-1-α-H,5-α-H-TROPANIUM BROMIDE
TROPATE (ester) see BPI125
BUTOXYBENZYL HYOSCYAMINE BROMIDE see BPI125
1-(1-(p-n-BUTOXYBENZYL)HYOSCYAMINIUM) BROMIDE see BPI125
p-BUTOXYBENZYL HYOSCYAMINIUM BROMIDE see BPI125
1-BUTOXYBUTANE see BRH750
tert-BUTOXYCARBONYL AZIDE (DOT) see BQI250
t-BUTOXYCARBONYL AZIDE see BQI250
2-BUTOXYCARBONYLMETHYLENE-4-OXOTHIAZOLIDONE see BPI300
1-(4-BUTOXY-3-CHLORO-5-METHYLPHENYL)-3-(1-PIPERIDINYL)1-PROPANONE
HYDROCHLORIDE (9CI) see BPI625
4'-BUTOXY-3'-CHLORO-5'-METHYL-3-PIPERIDINO-PROPIOPHENONE HYDRO-
CHLORIDE see BPI625
4'-BUTOXY-2'-CHLORO-2-PYRROLIDINYL ACETANILIDE HYDROCHLORIDE see
BPI750
BUTOXYCINCHONINIC ACID DIETHYLETHYLENEDIAMIDE HYDROCHLORIDE
see NOF500
2-BUTOXY-CYCLOPROPANECARBOXYLIC ACID BUTYL ESTER see BQM750
4'-BUTOXY-2-(DIETHYLAMINO)ACETANILIDE HYDROCHLORIDE see BPJ000
2-BUTOXY-N-(2-(DIETHYLAMINO)ETHYL)CINCHONINAMIDE see DDT200
2-BUTOXY-N-(β-DIETHYLAMINOETHYL)CINCHONINAMIDE see DDT200
2-BUTOXY-N-(2-DIETHYLAMINOETHYL)CINCHONINAMIDE HYDROCHLORIDE
see NOF500
2-N-BUTOXY-N-(2-DIETHYLAMINOETHYL)CINCHONINAMIDE HYDROCHLORIDE
see NOF500
2-BUTOXY-N-(2-DIETHYLAMINOETHYL)CINCHONINIC ACID AMIDE HYDRO-
CHLORIDE see NOF500
2-BUTOXY-N-(2-(DIETHYLAMINO)ETHYL)-N-(2,6-XYLYL)CINCHONINAMIDE HY-
DROCHLORIDE see BPJ250
2-BUTOXY-N-((2-(DIETHYLAMINO)ETHYL)-N-(2,6-XYLYL)-4-QUINOLINECARBOX-
AMIDE HYDROCHLORIDE see BPJ250
o-BUTOXY-N-(5-(DIETHYLAMINO)-2-PENTYL)BENZAMIDE HYDROCHLORIDE
see BPJ500
o-BUTOXY-N-(3-(DIETHYLAMINO)PROPYL)BENZAMIDE HYDROCHLORIDE see
BPJ750
BUTOXYDIETHYLENE GLYCOL see DJF200
BUTOXYDIGLYCOL see DJF200
2-BUTOXY-1-ETHANOL see BPJ850
2-sec-BUTOXYETHANOL see EJJ000
2-BUTOXYETHANOL see BPJ850
n-BUTOXYETHANOL see BPJ850
BUTOXYETHANOL see BPJ850
2-BUTOXYETHANOL ACETATE see BPM000
2,4,5-T BUTOXYETHANOL ESTER see TAH900
2-BUTOXYETHANOL PHOSPHATE see BPK250
2-BUTOXYETHANOL PHTHALATE (2:1) see BHK000
BUTOXYETHENE see VMZ000
2-BUTOXYETHOXY ACRYLATE see BPK500
1-BUTOXY-2-ETHOXYETHANE see BPK750
2-(2-BUTOXYETHOXY)ETHANOL see DJF200
2-(2-BUTOXYETHOXY)ETHANOL ACETATE see BQP500
2-(2-(2-BUTOXYETHOXY)ETHOXY)ETHANOL see TKL750
α-(2-(2-BUTOXYETHOXY)ETHOXY)-4,5-METHYLENEDIOXY-2-PROPYLTOLUENE
see PIX250
α-(2-(2-n-BUTOXYETHOXY)-ETHOXY)-4,5-METHYLENEDIOXY-2-PROPYLTO-
LUENE see PIX250
5-((2-(2-BUTOXYETHOXY)ETHOXY)METHYL)-6-PROPYL-1,3-BENZODIOXOLE
see PIX250
2-(2-BUTOXYETHOXY)ETHYL ACETATE see BQP500
2-(2-BUTOXY ETHOXY)ETHYL THIOCYANATE see BPL250
2-(2-(BUTOXY)ETHOXY)ETHYL THIOCYANIC ACID ESTER see BPL250
1-(2-BUTOXYETHOXY)-2-PROPANOL see BPL500
1-BUTOXY ETHOXY-2-PROPANOL see BPL500
3-(2-BUTOXYETHOXY)PROPANOL see BPL750
2-BUTOXYETHYL ACETATE see BPM000
BUTOXYETHYL-2,4-DICHLOROPHENOXYACETATE see DFY709
2,4,5-T BUTOXYETHYL ESTER see TAH900
2-BUTOXYETHYL ESTER ACETIC ACID see BPM000
β-BUTOXYETHYL PHTHALATE see BHK000
BUTOXYETHYL 2,4,5-T see TAH900
2-BUTOXYETHYLVINYL ETHER see VMU000
(3-n-BUTOXY-2-HYDROXYPROPYL)PHENYL ETHER see BPP250
BUTOXYL see MHV750
N-(BUTOXYMETHYL)ACRYLAMIDE see BPM660
N-BUTOXYMETHYLAKRYLAMID see BPM660
N-BUTOXYMETHYL-2-CHLORO-2',6'-DIETHYLACETANILIDE see CFW750
N-(BUTOXYMETHYL)-2-CHLORO-N-(2,6-DIETHYLPHENYL)ACETAMIDE see
CFW750
N-(BUTOXYMETHYL)-2-PROPENAMIDE see BPM660

2-(p-BUTOXYPHENOXY)-N-(2-(DIETHYLAMINO)ETHYL)-2,5'-DIETHOXYACETANI-
LIDE MONOHYDROCHLORIDE see BPM750
2-(p-BUTOXYPHENOXY)-N-(2-(DIETHYLAMINO)ETHYL)-N-(2,4-DIMETHOXYPHE-
NYL)ACETAMIDE HYDROCHLORIDE see BPN000
2-(p-BUTOXYPHENOXY)-N-(2-(DIETHYLAMINO)ETHYL)-N-(2,5-DIMETHOXYPHE-
NYL)ACETAMIDE HYDROCHLORIDE see BPN250
2-(p-BUTOXYPHENOXY)-N-(2-(DIMETHYLAMINO)ETHYL)-N-(2,6-DIMETHYLPHE-
NYL)ACETAMIDE HYDROCHLORIDE see BPO250
3-n-BUTOXY-1-PHENOXY-2-PROPANOL see BPP250
1-BUTOXY-3-PHENOXY-2-PROPANOL see BPP250
BUTOXYPHENYL see BSF750
4-BUTOXYPHENYLACETOHYDROXAMIC ACID see BPP750
p-BUTOXYPHENYLACETOHYDROXAMIC ACID see BPP750
4'-BUTOXY-2-PIPERIDINOACETANILIDE HYDROCHLORIDE see BPR000
4-BUTOXY-3-(PIPERIDINO)PROPIOPHENONE HYDROCHLORIDE see BPR250
4'-BUTOXY-3-PIPERIDINO PROPIOPHENONE HYDROCHLORIDE see BPR500
4-n-BUTOXY-β-(1-PIPERIDYL)PROPIOPHENONE HYDROCHLORIDE see BPR500
BUTOXYPOLYPROPYLENE GLYCOL see BRP250
BUTOXYPROPANEDIOL POLYMER see BRP250
3-BUTOXYPROPANENITRILE see BPT000
3-BUTOXY PROPANOIC ACID see BPS000
3-BUTOXY-1-PROPANOL see BPS500
1-BUTOXY-2-PROPANOL see BPS250
n-BUTOXYPROPANOL (mixed isomers) see BPS750
BUTOXYPROPANOL (mixed isomers) see BPS750
3-BUTOXYPROPIONIC ACID see BPS000
3-BUTOXYPROPIONITRILE see BPT000
4'-BUTOXY-2-PYRROLIDINYLACETANILIDE HYDROCHLORIDE see BPT250
2-BUTOXYQUINOLINE-4-CARBOXYLIC ACID DIETHYLAMINOETHYLAMIDE see
DDT200
BUTOXYRHODANODIETHYL ETHER see BPL250
1-BUTOXY-2-(2-THIOCYANATOETHYXY)ETHANE see BPL250
2-BUTOXY-2'-THIOCYANODIETHYL ETHER see BPL250
β-BUTOXY-β'-THIOCYANODIETHYL ETHER see BPL250
1-BUTOXY-2-(2-THIOCYANOETHOXY)ETHANE see BPL250
BUTOXYTRIETHYLENE GLYCOL see TKL750
BUTOXYTRIGLYCOL see TKL750
1-BUTOXY-2-(VINYLOXY)ETHANE see VMU000
BUTOZ see BRF500
BUTRALIN see BQN600
BUTRALINE see BQN600
BUTRATE see BPF000
BUTRIPTYLINE HYDROCHLORIDE see BPT750
BUTRIZOL see BPU000
BUTROPIPAZON see FLL000
BUTROPIPAZONE see FLL000
BUTROPIUM BROMIDE see BPI125
BUTTER of ANTIMONY see AQC500, AQD000
BUTTER CRESS see FBS100
BUTTERCUP see FBS100
BUTTERCUP YELLOW see CMK500, PLW500, ZFJ100
BUTTER DAISY see FBS100
BUTTERSAEURE (GERMAN) see BSW000
BUTTER YELLOW see AIC250, DOT300
BUTTER of ZINC see ZFA000
BUTURON see CKF750
BUTVAR see PKI000
N-BUTYLACETANILIDE see BPU500
BUTYLACETAT (GERMAN) see BPU750
1-BUTYL ACETATE see BPU750
2-BUTYL ACETATE see BPV000
n-BUTYL ACETATE see BPU750
sec-BUTYL ACETATE see BPV000
BUTYL ACETATE see BPU750
tert-BUTYL ACETATE see BPV100
BUTYLACETATEN (DUTCH) see BPU750
BUTYLACETIC ACID see HEU000
BUTYL ACETOACETATE see BPV250
N-BUTYL-N-(1-ACETOXYBUTYL)NITROSAMINE see BPV325
N-BUTYL-N-(ACETOXYMETHYL)NITROSAMINE see BRX500
BUTYL ACETOXYMETHYLNITROSAMINE see BRX500
n-BUTYL-3,o-ACETYL-12-β-13-α-DIHYDROJERVINE see BPW000
n-BUTYL ACID PHOSPHATE see ADF250
3-BUTYLACROLEIN see HBI770
β-BUTYLACROLEIN see HBI770
N-tert-BUTYLACRYLAMIDE see BPW050
n-BUTYL ACRYLATE see BPW100
BUTYL ACRYLATE see BPW100
BUTYLACRYLATE, INHIBITED (DOT) see BPW100
tert-BUTYL-1-ADAMANTANE PEROXYCARBOXYLATE see BPW250
BUTYL ADIPATE see AEO750
BUTYLAETHYLMALONSAEURE-AETHYL-DIAETHYLAMINOAETHYL-DI-ESTER
(GERMAN) see BRJ125
2-BUTYL ALCOHOL see BPW750
BUTYL ALCOHOL (DOT) see BPW500

n-BUTYL ALCOHOL see BPW500
sec-BUTYL ALCOHOL see BPW750
tert-BUTYL ALCOHOL see BPX000
sec-BUTYL ALCOHOL ACETATE see BPV000
BUTYL ALCOHOL HYDROGEN PHOSPHITE see DEG800
n-BUTYL ALDEHYDE see BSU250
sec-BUTYL ALLYL BARBITURIC ACID see AFY500
BUTYLALYLONAL see BOR000
n-BUTYL AMIDO SULFURYL AZIDE see BPX500
n-BUTYLAMIN (GERMAN) see BPX750
(+)-2-BUTYLAMINE see BPY100
BUTYLAMINE (OSHA) see BPX750
S-2-BUTYLAMINE see BPY100
sec-BUTYLAMINE, (S)- see BPY100
n-BUTYLAMINE see BPX750
sec-BUTYLAMINE see BPY000
BUTYLAMINE, tertiary see BPY250
tert-BUTYLAMINE see BPY250
tert-BUTYLAMINE, HYDROCHLORIDE see MGJ800
BUTYLAMINE OLEATE see OIA200
2-(BUTYLAMINO)-p-ACETOPHENETIDIDE HYDROCHLORIDE see BPY500
3-(tert-BUTYLAMINO)ACETYLINDOLE HYDROCHLORIDE HYDRATE see BPY625
3-((tert-BUTYLAMINO)ACETYL)INDOLE HYDROCHLORIDE HYDRATE see BPY625
2-tert-BUTYLAMINO-4-AETHYLAMINO-6-CHLOR-1,3,5-TRIAZIN (GERMAN) see BQB000
BUTYL-p-AMINOBENZOATE see BPZ000
p-BUTYLAMINOBENZOIC ACID-2-(DIETHYLAMINO)ETHYL ESTER MONOHYDROCHLORIDE see BQA000
p-(BUTYLAMINO)BENZOIC ACID, 2-(DIMETHYLAMINO)ETHYL ESTER see BQA010
p-(BUTYLAMINO)BENZOIC ACID, 2-(DIMETHYLAMINO)ETHYL ESTER, HYDROCHLORIDE see TBN000
p-BUTYLAMINOBENZOYL-2-DIMETHYLAMINOETHANOL see BQA010
p-BUTYLAMINOBENZOYL-2-DIMETHYLAMINOETHANOL HYDROCHLORIDE see TBN000
N-((BUTYLAMINO)CARBONYL)-4-METHYLBENZENESULFONAMIDE see BSQ000
2-(BUTYLAMINO)-2'-CHLOROACETANILIDE HYDROCHLORIDE see BQA500
2-(BUTYLAMINO)-6'-CHLORO-o-ACETOTOLUIDIDE MONOHYDROCHLORIDE see BQA750
2-tert-BUTYLAMINO-4-CHLORO-6-ETHYLAMINO-s-TRIAZINE see BQB000
2-(BUTYLAMINO)-6'-CHLORO-o-HEXANOTOLUIDIDE HYDROCHLORIDE see CAB500
1-(tert-BUTYLAMINO)-3-(2-CHLORO-5-METHYLPHENOXY)-2-PROPANOL HYDROCHLORIDE see BQB250
2-(BUTYLAMINO)-N-(2-CHLORO-6-METHYLPHENYL)ACETAMIDE HYDROCHLORIDE see BQA750
(±)-α-tert-BUTYLAMINO-3-CHLOROPROPIOPHENONE HYDROCHLORIDE see WBJ500
1-(BUTYLAMINO)CYCLOHEXYLPHOSPHONIC ACID DIBUTYL ESTER see ALZ000
(−)-1-(tert-BUTYLAMINO)-3-(o-CYCLOPENTYLPHENOXY)-2-PROPANOL SULFATE see PAP230
5-BUTYLAMINO-2-(2-DIETHYLAMINOETHYL)-1H-ISOINDOLE-1,3(2H)-DIONE HYDROCHLORIDE see BQB825
4-BUTYLAMINO-N-(2-(DIETHYLAMINO)ETHYL)PHTHALIMIDE HYDROCHLORIDE see BQB825
2-BUTYLAMINOETHANOL see BQC000
2-sec-BUTYLAMINO-4-ETHYLAMINO-6-METHOXY-1,3,5-TRIAZINE see BQC250
2-tert-BUTYLAMINO-4-ETHYLAMINO-6-METHOXY-1,3,5-TRIAZINE see BQC500
2-sec-BUTYLAMINO-4-ETHYLAMINO-6-METHOXY-s-TRIAZINE see BQC250
2-tert-BUTYLAMINO-4-ETHYLAMINO-6-METHOXY-s-TRIAZINE see BQC500
2-tert-BUTYLAMINO-4-ETHYLAMINO-6-METHYLMERCAPTO-s-TRIAZINE see BQC750
2-tert-BUTYLAMINO-4-ETHYLAMINO-6-METHYLTHIO-s-TRIAZINE see BQC750
2-tert-BUTYLAMINO-1-(7-ETHYL-2-BENZOFURANYL)ETHANOL HYDROCHLORIDE see BQD000·
3-(2-(tert-BUTYLAMINO)ETHYL)-6-HYDROXYBENZYL ALCOHOL SULFATE (2:1) see BQD125
2-(tert-BUTYLAMINO)ETHYL METHACRYLATE see BQD250
tert-BUTYL AMINO ETHYL METHACRYLATE see BQD250
(5-(2-(tert-BUTYLAMINO)-1-HYDROXYETHYL)-2-HYDROXYPHENYL)UREA HYDROCHLORIDE see BQD500
6-(2-(tert-BUTYLAMINO)-1-HYDROXYETHYL)-3-HYDROXY-2-PYRIDINEMETHANOL DIHYDROCHLORIDE see POM800
4-(2-(tert-BUTYLAMINO)-1-HYDROXYETHYL)-o-PHENYLENE DI-p-TOLUATE MESILATE see BLV125
2-(tert-BUTYLAMINO)-1-(4-HYDROXY-3-HYDROXYMETHYLPHENYL)ETHANOL see BQF500
2-BUTYLAMINO-1-p-HYDROXYPHENYLETHANOL see BQF250
o-(3-tert-BUTYLAMINO-2-HYDROXYPROPOXY)BENZONITRILE HYDROCHLORIDE see BON400

5-(3-tert-BUTYLAMINO-2-HYDROXY)PROPOXY-3,4-DIHYDROCARBOSTYRIL HYDROCHLORIDE see MQT550
5-(3-tert-BUTYLAMINO-2-HYDROXY)-PROPOXY)-3,4-DIHYDRO-(2(1H)-CHINOLINON-HYDROCHLORID (GERMAN) see MQT550
8-(3-tert-BUTYLAMINO-2-HYDROXY)PROPOXY-5-METHYLCOUMARIN HYDROCHLORIDE see BOM510
(±)-2-(3'-tert-BUTYLAMINO-2'-HYDROXYPROPYLTHIO)-4-(5'-CARBAMOYL-2'-THIENYL)THIAZOLE HYDROCHLORIDE see BQE000
2-(tert-BUTYLAMINO)-1-(3-INDOLYL)-1-PROPANONE HYDROCHLORIDE HYDRATE see BQG850
2-(tert-BUTYLAMINO)-1-(3-INDOLYL)-1-PROPANONE MONOHYDROCHLORIDE, MONOHYDRATE see BQG850
α-((tert-BUTYLAMINO)METHYL)-o-CHLOROBENZYL ALCOHOL HYDROCHLORIDE see BQE250
α-(tert-BUTYLAMINO)METHYL-2-CHLOROBENZYL ALCOHOL HYDROCHLORIDE see BQE250
α-((BUTYLAMINO)METHYL)-3,5-DIHYDROXYBENZYL ALCOHOL, SULFATE (2:1) see TAN250
α-((BUTYLAMINO)METHYL)-4-HYDROXYBENZENEMETHANOL see BQF250
α-((BUTYLAMINO)METHYL)-p-HYDROXYBENZYL ALCOHOL see BQF250
α-((BUTYLAMINO)METHYL)-p-HYDROXYBENZYL ALCOHOL SULFATE see BOV825
α-1-((tert-BUTYLAMINO)METHYL)-4-HYDROXY-m-XYLENE-α,α-DIOL see BQF500
α'-((tert-BUTYL AMINO)METHYL)-4-HYDROXY-m-XYLENE-α,α'-DIOL see BQF500
1-(tert-BUTYLAMINO)3-(3-METHYL-2-NITROPHENOXY)-2-PROPANOL see BQF750
1-(BUTYLAMINO)-3-((4-METHYLPHENYL)AMINO)-2-PROPANOL (9CI) see BQH800
2-(BUTYLAMINO)-2-METHYL-1-PROPANOL BENZOATE (ester) HYDROCHLORIDE see BQF825
2-(BUTYLAMINO)-2-METHYL-1-PROPANOL BENZOATE HYDROCHLORIDE see BQF825
2-(BUTYLAMINO)-N-METHYL-N-(1-(2,6-XYLYLOXY)-2-PROPYL) ACETAMIDE HYDROCHLORIDE see BQG250
2-(sec-BUTYLAMINO)-N-METHYL-N-(1-(2,4-XYLYLOXY)-2-PROPYL)ACETAMIDE HYDROCHLORIDE see BQG500
(−)-1-(tert-BUTYLAMINO)-3-((4-MORPHOLINO-1,2,5-THIADIAZOL-3-YL)OXY)-2-PROPANOL MALEATE see TGB185
2-(BUTYLAMINO)-N-(1-PHENOXY-2-PROPYL)ACETAMIDE HYDROCHLORIDE see BQG750
3-(BUTYLAMINO)-4-PHENOXY-5-SULFAMOYLBENZOIC ACID see BON325
3-(tert-BUTYLAMINO)PROPIONYLINDOLE HYDROCHLORIDE HYDRATE see BQG850
p-BUTYLAMINO SALICYLIC ACID-2-(DIETHYLAMINO)ETHYL ESTER HYDROCHLORIDE see BQH250
4-(BUTYLAMINO)SALICYLIC ACID 2-(DIETHYLAMINO)ETHYL ESTER HYDROCHLORIDE see BQH250
4-(BUTYLAMINO)-SALICYLIC ACID 2-(DIETHYLAMINO)ETHYL ESTER MONOHYDROCHLORIDE see BQH250
p-BUTYLAMINOSALICYLIC ACID-2-(DIMETHYLAMINO)ETHYL ESTER HYDROCHLORIDE see BQH500
p-BUTYLAMINOSALICYLIC ACID-1-ETHYL-4-PIPERIDYL ESTER HYDROCHLORIDE see BQH750
1-(tert-BUTYLAMINO)-3-((5,6,7,8-TETRAHYDRO-cis-6,7-DIHYDROXY-1-NAPHTHYL)OXY)-2-PROPANOL see CNR675
1-(tert-BUTYLAMINO)-3-(o-((TETRAHYDROFURFURYL)OXY)PHENOXY)-2-PROPANOL HYDROCHLORIDE see AER500
1-(BUTYLAMINO)-3-p-TOLUIDINO-2-PROPANOL see BQH800
tert-BUTYLAMMONIUM CHLORIDE see MGJ800
BUTYLAMMONIUM OLEATE see OIA200
BUTYLAMYLNITROSAMIN (GERMAN) see BRY250
N-n-BUTYLANILINE (DOT) see BQH850
N-(n-BUTYL)ANILINE see BQH850
N-BUTYLANILINE see BQH850
p-n-BUTYLANILINE see AJA550
BUTYLATE see EID500
3-tert-BUTYLATED HYDROXYANISOLE see BQI010
BUTYLATED HYDROXYANISOLE see BQI000
BUTYLATED HYDROXYTOLUENE see BFW750
BUTYLATE-2,4,5-T see BSQ750
N-BUTYL-N-2-AZIDOETHYLNITRAMINE see BQI125
tert-BUTYL AZIDOFORMATE see BQI250
2-t-BUTYLAZO-2-HYDROXY-5-METHYLHEXANE see BQI300
5-n-BUTYL-1,2-BENZANTHRACENE see BQI500
8-BUTYLBENZ(a)ANTHRACENE see BQI500
4-BUTYLBENZENAMINE see AJA550
N-BUTYLBENZENAMINE (9CI) see BQH850
n-BUTYLBENZENE see BQI750
sec-BUTYLBENZENE see BQJ000
tert-BUTYLBENZENE see BQJ250
BUTYLBENZENEACETATE see BQJ350
α-BUTYLBENZENEMETHANOL see BQJ500
N-BUTYLBENZENESULFONAMIDE see BQJ650
2-tert-BUTYLBENZIMIDAZOLE see BQJ750

5-BUTYL-2-BENZIMIDAZOLECARBAMIC ACID METHYL ESTER see BQK000
N-(BUTYL-5-BENZIMIDAZOLYL)-2-CARBAMATE de METHYLE (FRENCH) see BQK000
(4-BUTYL-1H-BENZIMIDAZOL-2-YL)-CARBAMIC ACID METHYL ESTER see BQK000
n-BUTYL BENZOATE see BQK250
BUTYL BENZOATE see BQK250
2-BUTYL-3-BENZOFURANYL p-((2-DIETHYLAMINO)ETHOXY)-m,m-DIIODOPHENYL KETONE see AJK750
p-tert-BUTYL BENZOIC ACID see BQK500
N-tert-BUTYL-2-BENZOTHIAZOLESULFENAMIDE see BQK750
α-BUTYLBENZYL ALCOHOL see BQJ500
1-(p-tert-BUTYLBENZYL)-4-(p-CHLORODIPHENYLMETHYL)PIPERAZINE DIHYDROCHLORIDE see BOM250
1-(p-tert-BUTYLBENZYL-4-p-CHLORO-α-PHENYLBENZYL)PIPERAZINE DIHYDROCHLORIDE see BOM250
1-(p-tert-BUTYLBENZYL)-4-(3-(p-CHLOROPHENYL)-3-PHENYLPROPIONYL)PIPERAZINE see BQK800
n-BUTYL BENZYL PHTHALATE see BEC500
BUTYL BENZYL PHTHALATE see BEC500
4-(p-tert-BUTYLBENZYL)PIPERAZINYL β-(p-CHLOROPHENYL) KETONE see BQK800
4-(p-tert-BUTYLBENZYL)PIPERAZINYL 3,4,5-TRIMETHOXYPHENYL KETONE see BQK830
1-(p-tert-BUTYLBENZYL)-4-(3,4,5-TRIMETHOXYBENZOYL)PIPERAZINE see BQK830
tert-BUTYL-BICYCLOPHOSPHATE see BQK850
BUTYLBIGUANIDE see BRA625
1-BUTYLBIGUANIDE HYDROCHLORIDE see BQL000
N-BUTYLBIGUANIDE HYDROCHLORIDE see BQL000
N-BUTYL-BIS(2-CHLOROETHYLAMINE) HYDROCHLORIDE see BIB250
sec-BUTYLBIS(2-CHLOROETHYL)AMINE HYDROCHLORIDE see BQL500
tert-BUTYLBIS(2-CHLOROETHYL)AMINE HYDROCHLORIDE see BQL750
sec-BUTYL-BIS(β-CHLOROETHYL)AMINE HYDROCHLORIDE see BQL500
BUTYLBIS(β-CHLOROETHYL)AMINE HYDROCHLORIDE see BIB250
tert-BUTYLBIS(β-CHLOROETHYL)AMINE HYDROCHLORIDE see BQL750
N-BUTYL-N,N-BIS(HYDROXY ETHYL)AMINE see BQM000
n-BUTYL BORATE see THX750
BUTYL BORATE see THX750
sec-BUTYL-BROM-ALLYL BARBITURIC ACID SODIUM SALT see BOR250
1-BUTYL BROMIDE see BNR750
n-BUTYL BROMIDE (DOT) see BMX500
BUTYL BROMIDE (DOT) see BMX500
iso-BUTYL BROMIDE see BNR750
sec-BUTYL BROMIDE see BMX750
tert-BUTYL BROMIDE see BQM250
3-sek.BUTYL-5-BROM-6-METHYLURACIL (GERMAN) see BMM650
5-sec-BUTYL-5-(β-BROMOALLYL)BARBITURIC ACID see BOR000
tert-BUTYL BROMOMETHYL KETONE see PJB100
N-BUTYL-1-BUTANAMINE see DDT800
N-tert-BUTYL-1,4-BUTANEDIAMINE DIHYDROCHLORIDE see BQM309
BUTYL BUTANEDIOATE see SNA500
n-BUTYL n-BUTANOATE see BQM500
BUTYL-BUTANOL(4)-NITROSAMIN see HJQ350
BUTYL-BUTANOL-NITROSAMINE see HJQ350
BUTYL-2-BUTOXYCYCLOPROPANE-1-CARBOXYLATE see BQM750
p-(N-BUTYL-2-(BUTYLAMINO)ACETAMIDO)BENZOIC ACID BUTYL ESTER HYDROCHLORIDE see BQN250
N-BUTYL-2-(BUTYLAMINO)-2′,6′-PROPIONOXYLIDIDE HYDROCHLORIDE see BQN500
N-sec-BUTYL-4-tert-BUTYL-2,6-DINITROANILINE see BQN600
BUTYL BUTYRATE (FCC) see BQM500
n-BUTYL n-BUTYRATE see BQM500
n-BUTYL BUTYRATE see BQM500
BUTYL BUTYROLACTATE see BQP000
γ-n-BUTYL-γ-BUTYROLACTONE see OCE000
N-n-BUTYL-N-4-(1,4-BUTYROLACTONE)NITROSAMINE see NJO200
BUTYL BUTYRYL LACTATE see BQP000
BUTYL CAPROATE see BRK900
BUTYLCAPTAX see BSN325
BUTYL CARBAMATE see BQP250
tert-BUTYLCARBAMIC ACID ESTER with 3-(m-HYDROXYPHENYL)-1,1-DIMETHYLUREA see DUM800
1-(BUTYLCARBAMOYL)-2-BENZIMIDAZOLECARBAMIC ACID METHYL ESTER and SODIUM NITRITE (1:6) see BAV500
1-(BUTYLCARBAMOYL)-2-BENZIMIDAZOLECARBAMIC ACID, METHYL ESTER see BAV575
1-(BUTYLCARBAMOYL)-2-BENZIMIDAZOL-METHYLCARBAMAT (GERMAN) see BAV575
1-(N-BUTYLCARBAMOYL)-2-(METHOXY-CARBOXAMIDO)-BENZIMIDAZOL (GERMAN) see BAV575
N′-(BUTYLCARBAMOYL)SULFANILAMIDE see BSM000
N′-(BUTYLCARBAMOYL)SULFANILAMIDE see BSM000
N-BUTYLCARBINOL see AOE000
dl-sec-BUTYLCARBINOL see MHS750

BUTYL CARBITOL see DJF200
BUTYL CARBITOL ACETATE see BQP500
BUTYLCARBITOL FORMAL see BHK750
BUTYL CARBITOL 6-PROPYLPIPERONYL ETHER see PIX250
BUTYL CARBITOL RHODANATE see BPL250
BUTYL CARBITOL THIOCYANATE see BPL250
BUTYL-CARBITYL (6-PROPYLPIPERONYL) ETHER see PIX250
BUTYL CARBOBUTOXYMETHYL PHTHALATE see BQP750
5-BUTYL-2-(CARBOMETHOXYAMINO)BENZIMIDAZOLE see BQK000
1-((tert-BUTYLCARBONYL-4-CHLOROPHENOXY)METHYL)-1H-1,2,4-TRIAZOLE see CJO250
4-BUTYL-4-(β-CARBOXYPROPIONYL-OXYMETHYL)-1,2-DIPHENYL-3,5-PYRAZOLIDINEDIONE see SOX875
N-BUTYL-(3-CARBOXY PROPYL)NITROSAMINE see BQQ250
4-tert-BUTYLCATECHOL see BSK000
BUTYL CELLOSOLVE see BPJ850
BUTYL CELLOSOLVE ACETATE see BPM000
BUTYL CELLOSOLVE ACRYLATE see BPK500
BUTYL "CELLOSOLVE" PHTHALATE see BHK000
BUTYL CHEMOSEPT see BSC000
o-(4-tert BUTYL-2-CHLOOR-FENYL)-o-METHYL-FOSFORZUUR-N-METHYL-AMIDE (DUTCH) see COD850
BUTYL-CHLORHYDRINETHER (CZECH) see BQT500
BUTYL CHLORIDE (DOT) see BQQ750
n-BUTYL CHLORIDE see BQQ750
sec-BUTYL CHLORIDE see CEU250
tert-BUTYL CHLORIDE see BQR000
3-tert-BUTYL-5-CHLOR-6-METHYLURACIL (GERMAN) see BQT750
α-BUTYL-p-CHLOROBENZYL ESTER of SUCCINIC ACID see CKI000
β-sec-BUTYL-3-CHLORO-N,N-DIMETHYL-4-ETHOXYPHENETHYLAMINE see BQR250
β-sec-BUTYL-5-CHLORO-N,N-DIMETHYL-2-METHOXYPHENETHYLAMINE see BQS000
β-sec-BUTYL-3-CHLORO-N,N-DIMETHYL-4-METHOXYPHENETHYLAMINE see BQR750
β-sec-BUTYL-p-CHLORO-N,N-DIMETHYLPHENETHYLAMINE see BQS250
β-sec-BUTYL-5-CHLORO-2-ETHOXY-N,N-DIISOPROPYLPHENETHYLAMINE see BQT000
1-(β-sec-BUTYL-5-CHLORO-2-ETHOXYPHENETHYL)PIPERIDINE see BQT250
BUTYL (3-CHLORO-2-HYDROXYPROPYL) ETHER see BQT500
t-BUTYL-CHLORO-2-METHYL-CYCLOHEXANECARBOXYLATE see BQT600
3-tert-BUTYL-5-CHLORO-6-METHYLURACIL see BQT750
tert-BUTYL CHLOROPEROXYFORMATE see BQU000
4-tert-BUTYL-2-CHLORO PHENYL METHYL METHYL PHOSPHORAMIDATE see COD850
4-tert.-BUTYL 2-CHLOROPHENYL METHYLPHOSPHORAMIDATE de METHYLE (FRENCH) see COD850
o-(4-tert-BUTYL-2-CHLOROPHENYL)-o-METHYL PHOSPHORAMIDOTHIONATE see BQU500
o-(4-tert-BUTYL-2-CHLOR-PHENYL)-o-METHYL-PHOSPHORSAEURE-N-METHYL AMID (GERMAN) see COD850
3-tert-BUTYLCHOLANTHRENE see BQU750
tert-20-BUTYLCHOLANTHRENE see BQU750
tert-BUTYL CHROMATE see BQV000
α-BUTYLCINNAMALDEHYDE see BQV250
n-BUTYL CINNAMATE see BQV500
α-BUTYLCINNAMIC ALDEHYDE see BQV250
BUTYL CINNAMIC ALDEHYDE see BQV250
BUTYL CITRATE see THY100
6′-tert-BUTYL-m-CRESOL see BQV600
2-tert-BUTYL-p-CRESOL see BQV750
4-tert-BUTYLCYCLOHEXANOL see BQW000
2-tert-BUTYLCYCLOHEXANOL ACETATE see BQW490
2-sec-BUTYLCYCLOHEXANONE see MOU800
p-tert-BUTYLCYCLOHEXANONE see BQW250
2-tert-BUTYLCYCLOHEXYL ACETATE see BQW490
4-tert-BUTYLCYCLOHEXYL ACETATE see BQW500
p-tert-BUTYLCYCLOHEXYL ACETATE see BQW500
N-BUTYL CYCLOHEXYL AMINE see BQW750
5-BUTYL-1-CYCLOHEXYLBARBITURIC ACID see BQW825
N-(4-tert-BUTYL CYCLOHEXYL)-3,3-DIPHENYL PROPYLAMINE HYDROCHLORIDE see BQX000
5-BUTYL-1-CYCLOHEXYL-2,4,6(1H,3H,5H)-PYRIMIDINETRIONE see BQW825
5-n-BUTYL-1-CYCLOHEXYL-2,4,6-TRIOXOPERHYDROPYRIMIDINE see BQW825
BUTYL 2,4-D see BQZ000
BUTYL DECYL PHTHALATE see BQX250
N-tert-BUTYL-1,4-DIAMINOBUTANE DIHYDROCHLORIDE see BQM309
tert-BUTYL DIAZOACETATE see BQX750
10-n-BUTYL-1,2,5,6-DIBENZACRIDINE (FRENCH) see BQY000
14-n-BUTYL DIBENZ(a,h)ACRIDINE see BQY000
n-BUTYL-2-DIBUTYLTHIOUREA see BQY250
sec-BUTYLDICHLORARSINE see BQY300
sec-BUTYLDICHLOROARSINE see BQY300
BUTYLDICHLOROBORANE see BQY500
N-BUTYL-2,2′-DICHLORODIETHYLAMINE see DEV300

N-sec-BUTYL-2,2'-DICHLORODIETHYLAMINE, HYDROCHLORIDE see BQL500
N-tert-BUTYL-2,2'-DICHLORO-DIETHYLAMINE HYDROCHLORIDE see BQL750
BUTYL (2,4-DICHLOROPHENOXY)ACETATE see BQZ000
BUTYL DICHLOROPHENOXYACETATE see BQZ000
1-BUTYL-3-(3,4-DICHLOROPHENYL)-1-METHYLUREA see BRA250
N-BUTYLDIETHANOLAMINE see BQM000
2-(BUTYL(2-(DIETHYLAMINO)ETHYL)AMINO)-6'-CHLORO-o-ACETOTOLUIDIDE HYDROCHLORIDE see BRA500
o-BUTYL DIETHYLENE GLYCOL see DJF200
n-BUTYLDIETHYLTIN IODIDE see BRA550
tert-BUTYLDIFLUOROPHOSPHINE see BRA600
BUTYL DIGLYME see DDW200
1-BUTYLDIGUANIDE see BRA625
BUTYLDIGUANIDE see BRA625
1-BUTYLDIGUANIDE HYDROCHLORIDE see BQL000
p-tert-BUTYLDIHYDROCINNAMALDEHYDE see BMK300
BUTYL-3,4-DIHYDRO-2,2-DIMETHYL-4-OXO-2H-PYRAN-6-CARBOXYLATE see BRT000
n-BUTYL-12-β-13-α-DIHYDROJERVINE-3-ACETATE see BPW000
9-BUTYL-1,9-DIHYDRO-6H-PURINE-6-THIONE see BRS500
2-N-BUTYL-3',5'-DIIODO-4'-N-DIETHYLAMINOETHOXY-3-BENZOYLBENZOFU-RAN see AJK750
2-BUTYL-3-(3,5-DIIODO-4-(2-DIETHYLAMINOETHOXY)BENZOYL)BENZOFURAN see AJK750
4'-n-BUTYL-4-DIMETHYLAMINOAZOBENZENE see BRB450
4'-tert-BUTYL-4-DIMETHYLAMINOAZOBENZENE see BRB460
3-BUTYL-1-(2-(DIMETHYLAMINO)ETHOXY)ISOQUINOLINE HYDROCHLORIDE see DNX400
4-(1-sec-BUTYL-2-(DIMETHYLAMINO)ETHYL)PHENOL see BRB500
2-(1-sec-BUTYL-2-(DIMETHYLAMINO)ETHYL)QUINOLINE see BRB750
2-(1-sec-BUTYL-2-(DIMETHYLAMINO)ETHYL)QUINOXALINE see BRC000
2-(1-sec-BUTYL-2-(DIMETHYLAMINO)ETHYL)THIOPHENE see BRC250
5-n-BUTYL-2-DIMETHYLAMINO-4-HYDROXY-6-METHYLPYRIMIDINE see BRD000
6-BUTYL-5-DIMETHYLAMINO-5H-INDENO(5,6-d)-1,3-DIOXOLE HYDROCHLORIDE see BRC500
2-n-BUTYL-3-DIMETHYLAMINO-5,6-METHYLENEDIOXYINDENE HYDROCHLO-RIDE see BRC500
BUTYL-3-((DIMETHYLAMINO)METHYL)-4-HYDROXYBENZOATE see BRC750
5-BUTYL-2-(DIMETHYLAMINO)-6-METHYL-4-PYRIMIDINOL see BRD000
5-BUTYL-2-(DIMETHYLAMINO)-6-METHYL-4(1H)-PYRIMIDINONE see BRD000
2-(4-tert-BUTYL-2,6-DIMETHYLBENZYL)-2-IMIDAZOLINE HYDROCHLORIDE see OKO500
2-(4-tert-BUTYL-2,6-DIMETHYLBENZYL)-2-IMIDAZOLINE MONOHYDROCHLO-RIDE see OKO500
8-tert-BUTYL-4,6-DIMETHYLCOUMARIN see DQV000
6-BUTYL-2,4-DIMETHYLDIHYDROPYRANE see GMG100
N-(7-BUTYL-4,9-DIMETHYL-2,6-DIOXO-8-HYDROXY-1,5-DIOXONAN-3-YL)-3-FOR-MAMIDOSALICYLAMIDE see DAM000
β-sec-BUTYL-N,N-DIMETHYL-2-ETHOXY-5-FLUOROPHENETHYLAMINE see BRD500
β-sec-BUTYL-N,N-DIMETHYL-5-FLUORO-2-METHOXYPHENETHYLAMINE see BRD750
2-(4-tert-BUTYL-2,6-DIMETHYL-3-HYDROXYBENZYL)-2-IMIDAZOLINIUM CHLO-RIDE see AEX000
1-BUTYL-3,3-DIMETHYL-1-NITROSOUREA see BRE000
β-sec-BUTYL-N,N-DIMETHYLPHENETHYLAMINE see BRE250
β-sec-BUTYL-N,N-DIMETHYLPHENETHYLAMINE HYDROCHLORIDE see BRE255
6-tert-BUTYL-2,4-DIMETHYLPHENOL see BST000
2-sek.BUTYL-4,6-DINITROFENYLESTER KYSELINY OCTOVE (CZECH) see ACE500
2-sec-BUTYL-4,6-DINITROPHENOL see BRE500
o-tert-BUTYL-4,6-DINITROPHENOL see DRV200
2-sec-BUTYL-4,6-DINITROPHENOL ACETATE (ester) see ACE500
2-sec-BUTYL-4,6-DINITROPHENOL AMMONIUM SALT see BPG250
2-sec-BUTYL-4,5-DINITROPHENOL ISOPROPYL CARBONATE see CBW000
2-sec-BUTYL-4,6-DINITROPHENOL- 2,2',2''-NITRILOTRIETHANOL SALT see BRE750
o-sec-BUTYL-4,6-DINITROPHENOLTRIETHANOLAMINE SALT see BRE750
6-sec-BUTYL-2,4-DINITROPHENYLACETATE see ACE500
2-sec-BUTYL-4,6-DINITROPHENYLACETATE see ACE500
2-tert-BUTYL-4,6-DINITROPHENYL ACETATE see DVJ400
2-sec-BUTYL-4,6-DINITROPHENYL-3,5-DIMETHYLACRYLATE see BGB500
2-sec-BUTYL-4,6-DINITROPHENYL ISOPROPYL CARBONATE see CBW000
2-sec-BUTYL-4,6-DINITROPHENYL-3-METHYL-2-BUTENOATE see BGB500
2-sec-BUTYL-4,6-DINITROPHENYL-3-METHYLCROTONATE see BGB500
2-sec-BUTYL-4,5-DINITROPHENYL SENECIOATE see BGB500
BUTYL DIOXITOL see DJF200
4-BUTYL-1,2-DIPHENYL-3,5-DIOXO PYRAZOLIDINE see BRF500
4-BUTYL-1,2-DIPHENYLPYRAZOLIDINE-3,5-DIONE see BRF500
4-BUTYL-1-1,2-DIPHENYL-3,5-PYRAZOLIDINEDIONE with 4-(DIMETHYLAMINO)-1,2-DIHYDRO-1,5-DIMETHYL-2-PHENYL-3H-PYRAZOL-3-ONE see IGI000
4-BUTYL-1,2-DIPHENYL-3,5-PYRAZOLIDINEDIONE SODIUM SALT see BOV750
BUTYL DISELENIDE see BRF550

6-BUTYLDODECAHYDRO-7,14-METHANO-2H,6H-DIPYRIDO(1,2-a: 1',2'-e)(1,5)DIAZOCINE see BSL450
BUTYLE (ACETATE de) (FRENCH) see BPU750
α-BUTYLENE see BOW250
β-BUTYLENE see BOW255
γ-BUTYLENE see IIC000
BUTYLENE see BOW250
1,2-BUTYLENE CARBONATE see BOT200
1,3-BUTYLENE DIACRYLATE see BRG500
1,4-BUTYLENE DIACRYLATE see TDQ100
BUTYLENE DIACRYLATE see TDQ100
1,4-BUTYLENEDIAMINE see BOS000
BUTYLENEDIAMINE see BOS000
2-BUTYLENE DICHLORIDE see BRG000
1,2-BUTYLENE GLYCOL see BOS250
1,4-BUTYLENE GLYCOL see BOS750
2,3-BUTYLENE GLYCOL see BOT000
β-BUTYLENE GLYCOL see BOS000
1,3-BUTYLENE GLYCOL (FCC) see BOS500
BUTYLENE GLYCOL BIS(MERCAPTOACETATE) see BKM000
1,3-BUTYLENE GLYCOL DIACRYLATE see BRG500
1,4-BUTYLENE GLYCOL DIACRYLATE see TDQ100
BUTYLENE HYDRATE see BPW750
1,2-BUTYLENE OXIDE see BOX750
1,2-BUTYLENE OXIDE, stabilized (DOT) see BOX750
BUTYLENE OXIDE see BOX750, TCR750
1,4-BUTYLENE SULFONE see BOU250
BUTYLENIN see IIU000
BUTYL-9,10-EPOXYSTEARATE see BRH250
2,4,5-T-N-BUTYL ESTER see BSQ750
2,4-d,n-BUTYL ESTER mixed with 2,4,5-T,n-BUTYL ESTER (1:1) see AEX750
2,4,5-T,n-BUTYL ESTER mixed with 2,4-d,n-BUTYL ESTER see AEX750
BUTYL ESTER 2,4-D see BQZ000
n-BUTYL ESTER of 3,4-DIHYDRO-2,2-DIMETHYL-4-OXO-2H-PYRAN-6-CARBOX-YLIC ACID see BRT000
N-BUTYLESTER KYSELINI-2,4,5-TRICHLORFENOXYOCTOVE (CZECH) see BSQ750
BUTYLESTER KYSELINY MRAVENCI see BRK000
terc.BUTYLESTER KYSELINY PEROXYBENZOOVE (CZECH) see BSC500
BUTYL ETHANOATE see BPU750
BUTYL ETHER (DOT) see BRH750
n-BUTYL ETHER see BRH750
sec-BUTYL ETHER see BRH760, OPE030
BUTYL ETHYL ACETALDEHYDE see BRI000
BUTYL ETHYL ACETIC ACID see BRI250
n-BUTYL-2-(ETHYLAMINO)-2',6'-ACETOXYLIDIDE HYDROCHLORIDE see BRI500
5-n-BUTYL-2-ETHYLAMINO-4-HYDROXY-6-METHYL-PYRIMIDINE see BRI750
5-BUTYL-2-(ETHYLAMINO)-6-METHYL-4(1H)-PYRIMIDINONE see BRI750
5-BUTYL-2-ETHYLAMINO-6-METHYLPYRIMIDIN-4-YL DIMETHYLSULPHAMATE see BRJ000
5-sec-BUTYL-5-ETHYLBARBITURIC ACID see BPF000
5-BUTYL-5-ETHYLBARBITURIC ACID see BPF500
5-sec-BUTYL-5-ETHYLBARBITURIC ACID SODIUM SALT see BPF250
BUTYL ETHYLENE see HFB000
o-BUTYL ETHYLENE GLYCOL see BPJ850
tert-BUTYL ETHYL ETHER see EHA550
n-BUTYL ETHYL KETONE see EHA600
BUTYLETHYLMALONIC ACID-2-(DIETHYLAMINO)ETHYL ETHYL ESTER see BRJ125
5-sec-BUTYL-5-ETHYLMALONYL UREA see BPF000
5-sec-BUTYL-5-ETHYL-1-METHYLBARBITURIC ACID see BRJ250
BUTYLETHYL-PROPANEDIOIC ACID-2-(DIETHYLAMINO)ETHYL ETHYL ESTER (9CI) see BRJ125
2-BUTYL-2-ETHYL-1,3-PROPANEDIOL see BKH625
5-BUTYL-5-ETHYL-2,4,6(1H,3H,5H)-PYRIMIDINETRIONE (9CI) see BPF500
5-BUTYL-5-ETHYL-2,4,6(1H,3H,5H)-PYRIMIDINETRIONE MONOSODIUM SALT (9CI) see BPF750
BUTYLETHYLTHIOCARBAMIC ACID S-PROPYL ESTER see PNF500
2-sec.-BUTYLFENOL (CZECH) see BSE000
p-tert-BUTYLFENOL (CZECH) see BSE500
2-sek.BUTYLFENYLESTER KYSELINY METHYLKARBAMINOVE (CZECH) see MOV000
p-terc.BUTYLFENYLESTER KYSELINY SALICYLOVE see BSH100
BUTYL FLUFENAMATE see BRJ325
BUTYL FORMAL see VAG000
tert-BUTYL FORMAMIDE see BRJ750
BUTYL FORMATE (DOT) see BRK000
n-BUTYL FORMATE see BRK000
N-n-BUTYL-N-FORMYLHYDRAZINE see BRK100
1-BUTYL-3-(2-FUROYL)UREA see BRK250
n-BUTYL GLYCIDYL ETHER see BRK750
BUTYL GLYCIDYL ETHER see BRK750
BUTYL GLYCOL see BPJ850
BUTYLGLYCOL (FRENCH, GERMAN) see BPJ850

BUTYL GLYCOL PHTHALATE see BHK000
4-tert-BUTYLHEXAHYDROPHENYL ACETATE see BQW500
n-BUTYL HEXANOATE see BRK900
BUTYL HEXANOATE see BRK900
n-BUTYLHYDRAZINE HYDROCHLORIDE see BRL500
O,O-tert-BUTYL HYDROGEN MONOPEROXY MALEATE see BRM000
terc. BUTYLHYDROPEROXID (CZECH) see BRM250
tert-BUTYLHYDROPEROXIDE see BRM250
N-BUTYL-N-(1-HYDROPEROXYBUTYL)NITROSAMINE see HIE600
tert-BUTYLHYDROQUINONE see BRM500
BUTYL HYDROXIDE see BPW500
tert-BUTYL HYDROXIDE see BPX000
6-HYDROXY-4-BUTYLAMINOQUINOLINE-1-OXIDE see BRM750
2(3)-tert-BUTYL-4-HYDROXYANISOLE see BQI000
3-tert-BUTYL-4-HYDROXYANISOLE see BRN000
tert-BUTYL-4-HYDROXYANISOLE see BQI000
BUTYLHYDROXYANISOLE see BQI000
tert-BUTYLHYDROXYANISOLE see BQI000
n-BUTYL-o-HYDROXYBENZOATE see BSL250
BUTYL-o-HYDROXYBENZOATE see BSL250
BUTYL-p-HYDROXYBENZOATE see BSC000
BUTYL p-HYDROXYBENZOATE see DTC800
α-n-BUTYL-β-HYDROXY-Δ^{α,β}-BUTENOLID (GERMAN) see BRO250
n-BUTYL-(4-HYDROXYBUTYL)NITROSAMINE see HJQ350
N-BUTYL-N-(4-HYDROXYBUTYL)NITROSAMINE see HJQ350
N-(7-BUTYL-8-HYDROXY-4,9-DIMETHYL-2,6-DIOXO-1,5-DIOXONAN-3-YL)-3-FOR-MAMIDOSALICYLAMIDE see DAM000
BUTYL(2-HYDROXYETHYL)NITROSOAMINE see BRO000
3-BUTYL-4-HYDROXY-2(5H)FURANONE see BRO250
BUTYLHYDROXYISOPROPYLARSINE OXIDE see BRQ800
4-BUTYL-4-HYDROXYMETHYL-1,2-DIPHENYL-3,5-PYRAZOLIDINEDIONE HYDROGEN SUCCINATE see SOX875
2-(tert-BUTYL)-2-(HYDROXYMETHYL)-1,3-PROPANEDIOL, CYCLIC PHOSPHITE (1:1) see BRO750
BUTYLHYDROXYOXOSTANNANE see BSL500
4-BUTYL-2-(4-HYDROXYPHENYL)-1-PHENYL-3,5-DIOXOPYRAZOLIDINE see HNI500
4-BUTYL-2-(p-HYDROXYPHENYL)-1-PHENYL-3,5-PYRAZOLIDINEDIONE see HNI500
4-BUTYL-1-(4-HYDROXYPHENYL)-2-PHENYL-3,5-PYRAZOLIDINEDIONE see HNI500
4-BUTYL-1-(p-HYDROXYPHENYL)-2-PHENYL-3,5-PYRAZOLIDINEDIONE see HNI500
α-BUTYL-ω-HYDROXYPOLY(OXY(METHYL-1,2-ETHANEDIYL)) see BRP250
BUTYL α-HYDROXYPROPIONATE see BRR600
BUTYLHYDROXYTOLUENE see BFW750
2-(tert-BUTYL)-2-(HYEROXYMETHYL)-1,3-PROPANEDIOL, CYCLIC PHOSPHATE (1:1) see BQK850
BUTYLHYOSCINE see SBG500
N-BUTYLHYOSCINE BROMIDE see SBG500
N-BUTYLHYOSCINIUM BROMIDE see SBG500
tert-BUTYL HYPOCHLORITE see BRP500
4,4'-BUTYLIDENEBIS(6-tert-BUTYL-m-CRESOL) see BRP750
4,4'-BUTYLIDENEBIS(6-tert-BUTYL-3-METHYLPHENYL) see BRP750
4,4'-BUTYLIDENEBIS(3-METHYL-6-tert-BUTYLPHENOL) see BRP750
(11-β,16-α)-16,17-(BUTYLIDENEBIS(OXY))-11,21-DIHYDROXYPREGNA-1,4-DIENE-3,20-DIONE see BOM520
6,6'-BUTYLIDENEBIS(2,4-XYLENOL) see BRQ000
BUTYLIDENE CHLORIDE see BRQ050
16-α,17-α-BUTYLIDENEDIOXY-11-β,21-DIHYDROXY-1,4-PREGNADIENE-3,20-DI-ONE see BOM520
3-BUTYLIDENE PHTHALIDE see BRQ100
n-BUTYLIDENE PHTHALIDE see BRQ100
BUTYLIDENE PHTHALIDE see BRQ100
6-t-BUTYL-3-(2-IMIDAZOLIN-2-YLMETHYL)-2,4-DIMETHYLPHENOL see ORA100
6-tert-BUTYL-3-(2-IMIDAZOLIN-2-YLMETHYL)-2,4-DIMETHYLPHENOL HYDROCHLORIDE see AEX000
N-BUTYLIMIDODICARBONIMIDIC DIAMIDE MONOHYDROCHLORIDE (9CI) see BQL000
N-BUTYL-2,2'-IMINODIETHANOL see BQM000
n-BUTYL IODIDE see BRQ250
sec-BUTYL IODIDE see IEH000
tert-BUTYL IODIDE see TLU000
BUTYL ISOBUTYRATE see BRQ350
n-BUTYL ISOCYANATE see BRQ500
tert-BUTYL ISOCYANIDE see BRQ750
tert-BUTYLISONITRILE see BRQ750
tert-BUTYLISOPENTANAL see ILJ100
n-BUTYL ISOPENTANOATE see ISX000
BUTYL(ISOPROPYL)ARSINIC ACID see BRQ800
tert-BUTYL ISOPROPYL BENZENE HYDROPEROXIDE (DOT) see BRR250
tert-BUTYL ISOPROPYL BENZENE HYDROPEROXIDE see BRR250
2-sec-BUTYL-6-ISOPROPYLPHENOL see BRR500
2-((3-BUTYL-1-ISOQUINOLINYL)OXY)-N,N-DIMETHYLETHANAMINE MONOHYDROCHLORIDE see DNX400

1-BUTYL ISOVALERATE see ISX000
n-BUTYL ISOVALERATE see ISX000
BUTYL ISOVALERATE see ISX000
BUTYL ISOVALERIANATE see ISX000
BUTYL KETONE see NMZ000
2-tert-BUTYL-p-KRESOL (CZECH) see BQV750
n-BUTYL LACTATE see BRR600
BUTYL LACTATE see BRR600
n-BUTYL LAEVULINATE see BRR700
BUTYL LAEVULINATE see BRR700
n-BUTYL LEVULINATE see BRR700
BUTYL LEVULINATE see BRR700
BUTYL LITHIUM see BRR739
tert-BUTYL LITHIUM see BRR750
BUTYLMALONIC ACID MONO(1,2-DIPHENYLHYDRAZIDE) CALCIUM SALT HEMIHYDRATE see CCD750
BUTYL-MALONSAEURE-MONO-(1,2-DIPHENYL-HYDRAZID)-CALCIUM-SEMIHYDRAT (German) see CCD750
n-BUTYL MERCAPTAN (ACGIH,DOT) see BRR900
n-BUTYL MERCAPTAN see BRR900
BUTYL MERCAPTAN see BRR900
tert-BUTYL MERCAPTAN see MOS000
p-BUTYLMERCAPTOBENZHYDRYL-β-DIMETHYLAMINOETHYLSULPHIDE see BRS000
2-(BUTYLMERCAPTO)ETHYL VINYL ETHER see VNA000
BUTYLMERCAPTOMETHYLPENICILLIN see BRS250
9-BUTYL-6-MERCAPTOPURINE see BRS500
n-BUTYLMERCURIC CHLORIDE see BRS750
S-(BUTYLMERCURIC)-THIOGLYCOLIC ACID, SODIUM SALT see SFJ500
n-BUTYL MESITYL OXIDE OXALATE see BRT000
n-BUTYLMESITYLOXID OXALATE see BRT000
BUTYLMETHACRYLAAT (DUTCH) see MHU750
BUTYL-2-METHACRYLATE see MHU750
N-BUTYL METHACRYLATE see MHU750
2-tert-BUTYL-4-METHOXYPHENOL see BRN000
3-tert-BUTYL-4-METHOXYPHENOL see BQI010
BUTYL-p-METHYLBENZENESULFONATE see BSP750
n-BUTYL-α-METHYLBENZYLAMINE see BRU250
BUTYL 3-METHYLBUTYRATE see ISX000
tert-BUTYL METHYL CARBINOL see BRU300
6-tert-BUTYL-3-METHYL-2,4-DINITRO ANISOLE see BRU500
2-tert-BUTYL-5-METHYL-4,6-DINITROPHENYL ACETATE see BRU750
BUTYL METHYL ETHER (DOT) see BRU780
1-BUTYL-2-METHYL-HYDRAZINE DIHYDROCHLORIDE see MHW250
p-tert-BUTYL-α-METHYLHYDROCINNAMALDEHYDE see LFT000
p-tert-BUTYL-α-METHYLHYDROCINNAMIC ALDEHYDE see LFT000
n-BUTYL METHYL KETONE see HEV000
BUTYL METHYL KETONE see HEV000
tert-BUTYL METHYL KETONE see DQU000
4-tert-BUTYL-2-METHYLPHENOL see BRU800
2-tert-BUTYL-4-METHYLPHENOL see BQV750
2-tert-BUTYL-6-METHYLPHENOL see BRU790
1-BUTYL-3-(p-METHYLPHENYLSULFONYL)UREA see BSQ000
2-sec-BUTYL-2-METHYL-1,3-PROPANEDIOL DICARBAMATE see MBW750
BUTYL-2-METHYL-2-PROPENOATE see MHU750
N-BUTYL-2-METHYL-2-PROPYL-1,3-PROPANEDIOL DICARBAMATE see MOV500
N-N-BUTYL-2-METHYL-2-PROPYL-1,3-PROPANEDIOL DICARBAMATE see MOV500
3-BUTYL-5-METHYL-TETRAHYDRO-2H-PYRAN-4-YL ACETATE see MHX000
tert-BUTYL-N-(3-METHYL-2-THIAZOLIDINYLIDENE)CARBAMATE see BRV000
2-sec-BUTYL-2-METHYLTRIMETHYLENE DICARBAMATE see MBW750
BUTYLMIN see SBG500
BUTYL MONOSULFIDE see BSM125
4-BUTYLMORPHOLINE see BRV100
N-(n-BUTYL)MORPHOLINE see BRV100
N-BUTYLMORPHOLINE see BRV100
9-BUTYL-6-MP see BRS500
BUTYL MYRISTATE see MSA300
BUTYL NAMATE see SGF500
N-BUTYL-N'-(3,4-DICHLOROPHENYL)-N-METHYLUREA see BRA250
n-BUTYL-N'-(2-FUROYL) see BRK250
BUTYL NITRATE see BRV325
BUTYL NITRITE (DOT) see BRV500
n-BUTYL NITRITE see BRV500
sec-BUTYL NITRITE see BRV750
tert-BUTYL NITRITE see BRV760
tert-BUTYLNITROACETATE see DSV289
tert-BUTYL NITROACETYLENE see BRW000
tert-BUTYL-p-NITRO PEROXY BENZOATE see BRW250
BUTYL-p-NITROPHENYL ESTER of ETHYLPHOSPHONIC ACID see BRW500
6-BUTYL-4-NITROQUINOLINE-1-OXIDE see BRW750
4-(n-BUTYLNITROSAMINO)-1-BUTANOL see HJQ350
4-(BUTYLNITROSAMINO)-1-BUTANOL see HJQ350
2-(BUTYLNITROSAMINO)ETHANOL see BRO000
4-(N-BUTYLNITROSAMINO)-4-HYDROXYBUTYRIC ACID LACTONE see NJO200

4-(BUTYLNITROSOAMINO)BUTANOIC ACID see BQQ250
1-(BUTYLNITROSOAMINO)BUTYL ACETATE see BPV325
BUTYLNITROSOAMINOMETHYL ACETATE see BRX500
1-(BUTYLNITROSOAMINO)-2-PROPANONE see BRY000
N-BUTYL-N-NITROSO AMYL AMINE see BRY250
n-BUTYL-N-NITROSO-1-BUTAMINE see BRY500
N-BUTYL-N-NITROSOBUTYRAMIDE see NJO150
N-BUTYL-N-NITROSO ETHYL CARBAMATE see BRZ000
BUTYLNITROSOHARNSTOFF (GERMAN) see BSA250
N-BUTYL-N-NITROSOPENTYLAMINE see BRY250
4-tert-BUTYL-1-NITROSOPIPERIDINE see BRZ200, NJO300
N-BUTYL-N-NITROSOSUCCINAMIC ACID ETHYL ESTER see EHC800
1-sec-BUTYL-1-NITROSOUREA see NJO500
1-BUTYL-1-NITROSOUREA see BSA250
N-n-BUTYL-N-NITROSOUREA see BSA250
n-BUTYLNITROSOUREA see BSA250
1-BUTYL-1-NITROSOURETHAN see BRZ000
N-BUTYL-N-NITROSOURETHAN see BRZ000
N-BUTYL-N'-NITRO-N-NITROSOGUANIDINE see NLC000
n-BUTYLNORSYMPATHOL see BQF250
BUTYLNORSYMPATOL see BOV825
BUTYL-NOR-SYMPATOL see BQF250
n-BUTYLNORSYNEPHRINE see BQF250
n-BUTYL-N'-p-TOLUENESULFONYLUREA see BSQ000
N-n-BUTYL-N'-TOSYLUREA see BSQ000
BUTYLOCAINE see TBN000
n-BUTYL OCTADECANOATE see BSL600
BUTYL OCTADECANOATE see BSL600
2-BUTYL-1-OCTANOL see BSA500
2-BUTYLOCTYL ALCOHOL see BSA500
2-BUTYLOCTYL ESTER METHACRYLIC ACID see BSA750
BUTYLOHYDROKSYANIZOL (POLISH) see BQI000
BUTYL OLEATE see BSB000
BUTYLONE see NBU000
BUTYLOWY ALKOHOL (POLISH) see BPW500
BUTYL OXITOL see BPJ850
N-BUTYL-N-(1-OXOBUTYL)NITROSAMINE see NJO150
N-BUTYL-N-(2-OXOBUTYL)NITROSAMINE see BSB500
N-BUTYL-N-(3-OXOBUTYL)NITROSAMINE see BSB750
BUTYL 4-OXOPENTANOATE see BRR700
4-tert-BUTYL-1-OXO-1-PHOSPHA-2,6,7-TRIOXABICYCLO(2.2.2)OCTANE see BQK850
BUTYL(2-OXOPROPYL)NITROSOAMINE see BRY000
tert-BUTYLOXYCARBONYL AZIDE see BQI250
2-(n-BUTYLOXYCARBONYLMETHYLENE)THIAZOLID-4-ONE see BPI300
α-BUTYLOXYCINCHONINIC ACID DIETHYLETHYLENEDIAMIDE see DDT200
BUTYL PARABEN see BSC000
n-BUTYL PARAHYDROXYBENZOATE see BSC000
BUTYL PARASEPT see BSC000
N-BUTYL-N-PENTYLNITROSAMINE see BRY250
tert-BUTYL PERACETATE see BSC250
t-BUTYL PERACETATE see BSC250
terc.BUTYLPERBENZOAN (CZECH) see BSC500
tert-BUTYL PERBENZOATE see BSC500
t-BUTYL PERBENZOATE see BSC500
tert-BUTYL PERISOBUTYRATE see BSC600
tert-BUTYL PEROXIDE see BSC750
tert-BUTYL PEROXYACETATE, >76% in solution (DOT) see BSC250
t-BUTYL PEROXYACETATE see BSC250
t-BUTYL PEROXY BENZOATE see BSC500
n-BUTYL PEROXYDICARBONATE, >52% in solution (DOT) see BSC800
sec-BUTYL PEROXYDICARBONATE see BSD000
BUTYL PEROXYDICARBONATE see BSC800
tert-BUTYL PEROXYISOBUTYRATE, >77% in solution (DOT) see BSC600
tert-BUTYL PEROXYISOBUTYRATE see BSC600
tert-BUTYL PEROXYPIVALATE see BSD250
t-BUTYL PEROXYPIVALATE see BSD250
tert-BUTYL PERPIVALATE see BSD250
BUTYLPHEN see BSE500
2-n-BUTYLPHENOL see BSE440
2-t-BUTYLPHENOL see BSE460
4-n-BUTYLPHENOL see BSE450
4-sec BUTYL PHENOL see BSE250
4-t-BUTYLPHENOL see BSE250
p-tert-BUTYLPHENOL (MAK) see BSE500
o-sec-BUTYLPHENOL see BSE000
p-sec-BUTYLPHENOL see BSE250
2-(p-tert-BUTYLPHENOXY)CYCLOHEXYL PROPARGYL SULFITE see SOP000
2-(p-tert-BUTYLPHENOXY)CYCLOHEXYL 2-PROPYNYL SULFITE see SOP000
4'-(3-(4'-tert-BUTYLPHENOXY)-2-HYDROXYPROPOXY)BENZOIC ACID see BSE750
2-(4-tert-BUTYLPHENOXY)ISOPROPYL 2-CHLOROETHYL SULFITE see SOP500
2-(p-BUTYLPHENOXY)ISOPROPYL 2-CHLOROETHYL SULFITE see SOP500
BUTYLPHENOXYISOPROPYL CHLOROETHYL SULFITE see SOP500

2-(p-tert-BUTYLPHENOXY)ISOPROPYL 2'-CHLOROETHYL SULPHITE see SOP500
2-(p-tert-BUTYLPHENOXY)-1-METHYLETHYL 2-CHLOROETHYL ESTER of SULPHUROUS ACID see SOP500
2-(p-BUTYLPHENOXY)-1-METHYLETHYL 2-CHLOROETHYL SULFITE see SOP500
2-(p-tert-BUTYLPHENOXY)-1-METHYLETHYL-2-CHLOROETHYL SULFITE ESTER see SOP500
2-(p-tert-BUTYLPHENOXY)-1-METHYLETHYL 2'-CHLOROETHYL SULPHITE see SOP500
2-(p-tert-BUTYLPHENOXY)-1-METHYLETHYL SULPHITE of 2-CHLOROETHANOL see SOP500
1-(p-tert-BUTYLPHENOXY)-2-PROPANOL-2-CHLOROETHYL SULFITE see SOP500
n-BUTYL PHENYLACETATE see BQJ350
BUTYL PHENYL ACETATE see BBA000
BUTYL PHENYLACETATE see BQJ350
α-n-BUTYL-β-PHENYLACROLEIN see BQV250
n-BUTYL PHENYLACRYLATE see BQV500
p-((p-BUTYLPHENYL)AZO)-N,N-DIMETHYLANILINE see BRB450
p-((p-tert-BUTYLPHENYL)AZO)-N,N-DIMETHYLANILINE see BRB460
o-sec-BUTYLPHENYL CARBAMATE see BSF250
4-BUTYL-1-PHENYL-3,5-DIOXOPYRAZOLIDINE see MQY400
BUTYL PHENYL ETHER see BSF750
S-p-tert-BUTYLPHENYL-o-ETHYL ETHYLPHOSPHONODITHIOATE see BSG000
2-sec-BUTYLPHENYL N-METHYLCARBAMATE see MOV000
3-sec-BUTYLPHENYL-N-METHYLCARBAMATE see BSG250
3-tert-BUTYLPHENYL N-METHYLCARBAMATE see BSG300
m-sec-BUTYLPHENYL-N-METHYLCARBAMATE see BSG250
o-sec-BUTYLPHENYL METHYLCARBAMATE see MOV000
β-(4-tert-BUTYLPHENYL)-α-METHYLPROPIONALDEHYDE see LFT000
2-tert-BUTYL-3-PHENYL OXAZIRANE see BSH000
α-sec-BUTYL-α-PHENYL-I-PIPERIDINEBUTYRONITRILE HYDROCHLORIDE see EQZ000
4-BUTYL-1-PHENYL-3,5-PYRAZOLIDINEDIONE see MQY400
p-tert-BUTYLPHENYL SALICYLATE see BSH100
BUTYL PHOSPHITE see MRF500
BUTYL PHOSPHORIC ACID see ADF250
BUTYL PHOSPHOROTRITHIOATE see BSH250
n-BUTYL PHTHALATE (DOT) see DEH200
BUTYL PHTHALATE BUTYL GLYCOLATE see BQP750
3-n-BUTYLPHTHALIDE see BSH500
3-BUTYLPHTHALIDE see BSH500
BUTYLPHTHALIDE see BSH500
BUTYL PHTHALYL BUTYL GLYCOLATE see BQP750
5-BUTYL PICOLINIC ACID see BSI000
5-BUTYLPICOLINIC ACID CALCIUM SALT HYDRATE see FQR100
1-BUTYL-2',6'-PIPECOLOXYLIDIDE (±) see BOO000
1-BUTYL-2',6'-PIPECOLOXYLIDIDE see BSI250
d-(+)-1-BUTYL-2',6'-PIPECOLOXYLIDIDE see BOO250
l-(−)-1-BUTYL-2',6'-PIPECOLOXYLIDIDE see BOO500
1-BUTYL-2',6'-PIPECOLOXYLIDIDE HYDROCHLORIDE (+) see BON750
(±)-1-BUTYL-2',6'-PIPECOLOXYLIDIDE MONOHYDROCHLORIDE, MONOHYDRATE see BOO000
p-(N-BUTYL-2-(PIPERIDINO)ACETAMIDO)BENZOIC ACID BUTYL ESTER HYDROCHLORIDE see BSI750
BUTYL POTASSIUM XANTHATE see PKY850
BUTYLPROPANEDIOIC ACID MONO(1,2-DIPHENYLHYDRAZIDE) CALCIUM SALT HEMIHYDRATE see CCD750
BUTYL PROPANOATE see BSJ500
BUTYL-2-PROPENOATE see BPW100
n-BUTYL PROPIONATE see BSJ500
BUTYL PROPIONATE see BSJ500
9-BUTYL-9H-PURINE-6-THIOL see BRS500
3-n-BUTYLPYRIDINE see BSJ550
3-BUTYLPYRIDINE see BSJ550
5-BUTYL-2-PYRIDINECARBOXYLIC ACID see BSI000
5-BUTYL-2-PYRIDINECARBOXYLIC ACID CALCIUM SALT HYDRATE see FQR100
BUTYL 4-PYRIDYL KETONE see VBA100
BUTYLPYRIN see BRF500
4-tert-BUTYLPYROCATECHOL see BSK000
p-tert-BUTYLPYROCATECHOL see BSK000
4-tert-BUTYLPYROKATECHIN (CZECH) see BSK000
n-BUTYLPYRROLIDINE see BSK250
N-BUTYL-α-PYRROLIDINE-CARBOXY-MESIDIDE HYDROCHLORIDE see PQB750
n-BUTYL RHODANATE see BSN500
n-BUTYL SALICYLATE see BSL250
BUTYL SALICYLATE see BSL250
N-BUTYLSCOPOLAMINE BROMIDE see SBG500
BUTYLSCOPOLAMINE BROMIDE see SBG500
n-BUTYLSCOPOLAMINE TANNATE see BSL325
N-BUTYLSCOPOLAMINIUM BROMIDE see SBG500
N-BUTYLSCOPOLAMMONIUM BROMIDE see SBG500

BUTYLSCOPOLAMMONIUM BROMIDE see SBG500
N-BUTYLSCOPOLAMMONIUM BROMIDE combined with SODIUM SULPYRINE
  (1:25) see BOO650
17-BUTYLSPARTEIN see BSL450
BUTYL STANNOIC ACID see BSL500
n-BUTYL STEARATE see BSL600
BUTYL STEARATE see BSL600
n-BUTYL-k-STROPHANTHIDIN see BSL750
1-BUTYL-3-SULFANILYL UREA see BSM000
N-BUTYLSULFANILYLUREA see BSM000
n-BUTYL-SULFIDE see BSM125
BUTYL SULFIDE see BSM125
1-BUTYLSULFONIMIDOCYCLOHEXAMETHYLENE see BSM250
BUTYLSYMPATHOL see BQF250
BUTYL-2,4,5-T see BSQ750
BUTYL TEGOSEPT see BSC000
BUTYL n-TETRADECANOATE see MSA300
BUTYL TETRADECANOATE see MSA300
1-BUTYL THEOBROMINE see BSM825
N-(5-tert-BUTYL-1,3,4-THIADIAZOL-2-YL)BENZENESULFONAMIDE see GFM200
1-(5-tert-BUTYL-1,3,4-THIADIAZOL-2-YL)-3-DIMETHYLHARNSTOFF (GERMAN)
  see BSN000
1-(5-(tert-BUTYL)-1,3,4-THIADIAZOL-2-YL)-1,3-DIMETHYLUREA see BSN000
2-BUTYLTHIOBENZOTHIAZOLE see BSN325
BUTYLTHIOBUTANE see BSM125
n-BUTYL THIOCYANATE see BSN500
5-((1-(BUTYLTHIO)ETHYL)-5-ETHYLBARBITURIC ACID SODIUM SALT see
  SFJ875
2-(BUTYLTHIO)ETHYL VINYL ETHER see VNA000
S-((tert-BUTYLTHIO)METHYL)-O,O-DIETHYLPHOSPHORODITHIOATE see
  BSO000
n-BUTYLTHIOMETHYLPENICILLIN see BRS250
2-((p-(BUTYLTHIO)-α-PHENYLBENZYL)THIO)-N,N-DIMETHYLETHYLAMINE see
  BRS000
(BUTYLTHIO)TRIOCTYLSTANNANE see BSO200
(BUTYLTHIO)TRIPROPYLSTANNANE see TMY850
n-BUTYL THIOUREA see BSO500
n-BUTYLTIN TRICHLORIDE see BSO750
BUTYLTIN TRI(DODECANOATE) see BSO750
BUTYLTIN TRILAURATE see BSO750
n-BUTYLTIN TRIS(DIBUTYLDITHIOCARBAMATE) see BSP000
BUTYL TITANATE see BSP250
p-tert-BUTYLTOLUENE see BSP500
n-BUTYL-p-TOLUENESULFONATE see BSP750
BUTYL-p-TOLUENESULFONATE see BSP750
1-BUTYL-3-(p-TOLYL SULFONYL)UREA see BSQ000
BUTYL TOSYLATE see BSP750
1-BUTYL-3-TOSYLUREA see BSQ000
4-N-BUTYL-4H-1,2,4-TRIAZOLE see BPU000
4-BUTYL-s-TRIAZOLE see BPU000
BUTYLTRICHLOROGERMANE see BSQ500
N-BUTYL (2,4,5-TRICHLOROPHENOXY)ACETATE see BSQ750
BUTYL-2,4,5-TRICHLOROPHENOXYACETATE see BSQ750
BUTYLTRICHLOROSILANE see BSR000
BUTYL TRICHLORO STANNANE see BSR250
3-tert-BUTYLTRICYCLOQUINAZOLINE see BSR500
BUTYL-2-((3-(TRIFLUOROMETHYL)PHENYL)AMINO)BENZOATE see BRJ325
BUTYL-o-((m-(TRIFLUOROMETHYL)PHENYL)AMINO)BENZOATE see BRJ325
BUTYL 2-(4-(5-TRIFLUOROMETHYL-2-PYRIDINYLOXY)PHENOXY)PROPANOATE
  see FDA885
BUTYLTRI(LAUROYLOXY)STANNANE see BSO750
5-tert-BUTYL-1,2,3-TRIMETHYL-4,6-DINITROBENZENE see MRW272
2-BUTYL-4,4,6-TRIMETHYL-1,3-DIOXANE see BSR600
tert-BUTYL TRIMETHYLPEROXYACETATE see BSD250
1-BUTYL-N-(2,4,6-TRIMETHYLPHENYL)-2-PYRROLIDINECARBOXAMIDE MONO-
  HYDROCHLORIDE see PQB750
N-tert-BUTYL-N-TRIMETHYLSILYLAMINOBORANE see BSR825
5-tert-BUTYL-2,4,6-TRINITROXYLENE see TML750
5-tert-BUTYL-2,4,6-TRINITRO-m-XYLENE (DOT) see TML750
4-(tert-BUTYL)-2,6,7-TRIOXA-1-PHOSPHABICYCLO(2.2.2)OCTANE see BRO750
4-(tert-BUTYL)-2,6,7-TRIOXA-1-PHOSPHABICYCLO(2.2.2)OCTAN-1-ONE see
  BQK850
n-BUTYLTRIPHENYLPHOSPHONIUM BROMIDE see BSR900
BUTYLTRIS(DIBUTYLDITHIOCARBAMATO)STANNANE see BSP000
BUTYLTRIS(2-ETHYLHEXYLOXYCARBONYLMETHYLTHIO)STANNANE see
  BSS000
BUTYLTRIS(ISOOCTYLOXYCARBONYLMETHYLTHIO)STANNANE see BSS000
BUTYL 10-UNDECENOATE see BSS100
BUTYL UNDECYLENATE see BSS100
N-BUTYLUREA see BSS250
sec-BUTYLUREA see BSS300
tert-BUTYLUREA see BSS310
1-BUTYLUREA and SODIUM NITRITE (2:1) see BSS500
1-BUTYLURETHAN see EHA100
1-BUTYLURETHANE see EHA100

N-BUTYLURETHANE see EHA100
BUTYLURETHANE see EHA100
BUTYL VINYL ETHER (inhibited) see VMZ000
BUTYL VINYL ETHER see VMZ000
BUTYL-XANTHIC ACID POTASSIUM SALT see PKY850
sec-BUTYLXANTHIC ACID SODIUM SALT see DXM000
BUTYLXANTHIC DISULFIDE see BSS550
5-tert-BUTYL-m-XYLENE see DQU800
6-tert-BUTYL-2,4-XYLENOL see BST000
BUTYL ZIMATE see BIX000
BUTYL ZIRAM see BIX000
BUTYN see BOO750
1-BUTYNE see EFS500
2-BUTYNE see COC500
2-BUTYNEDIAMIDE see ACJ250
2-BUTYNEDINITRILE see DGS000
2-BUTYNE-1,4-DIOL see BST500
1,4-BUTYNEDIOL (DOT) see BST500
2-BUTYNE-1-THIOL see BST750
BUTYNOIC ACID, 3-PHENYL-2-PROPENYL ESTER see CMQ800
1-BUTYN-3-OL see EQM600
BUTYN-1-OL-3-ESTER of m-CHLOROPHENYLCARBAMIC ACID see CEX250
1-BUTYN-3-OL, 3-METHYL- see MHX250
3-BUTYN-2-OL, 2-PHENYL- see EQN230
BUTYNORATE see DDV600
BUTYN SULFATE see BOP000
3-BUTYNYL-m-CHLOROCARBANILATE see CEX250
2-BUTYNYL-4-CHLORO-m-CHLOROCARBANILATE see CEW500
1-BUTYN-3-YL-m-CHLOROPHENYLCARBAMATE see CEX250
2-BUTYNYLENEDIAMINE, N,N,N',N'-TETRAETHYL- see BJA250
1,1'-(2-BUTYNYLENE)DIPYRROLIDINE see DWX600
BUTYNYL-3N-3-CHLOROPHENYLCARBAMATE mixed with 3-CYCLOOCTYL-
  1,1-DIMETHYL UREA see AFM375
3-BUTYN-1-YL-p-TOLUENE SULFONATE see BSU000
2-BUTYOXY-N-(2-(DIETHYLAMINO)ETHYL)-4-QUINOLINECARBOXAMIDE see
  DDT200
BUTYRAC see DGA000, EAK500
BUTYRAC ESTER see DGA000
BUTYRAL see BSU250
BUTYRALDEHYD (GERMAN) see BSU250
BUTYRALDEHYDE (CZECH) see BSU250
n-BUTYRALDEHYDE see BSU250
n-BUTYRALDEHYDE OXIME see BSU500
BUTYRALDOXIME (DOT) see BSU500
N-BUTYRALDOXIME see BSU500
BUTYRAMIDE, N,N-DIMETHYL- see DQV300
3-BUTYRAMIDO-α-ETHYL-2,4,6-TRIIODOCINNAMIC ACID SODIUM SALT see
  EQC000
3-BUTYRAMIDO-α-ETHYL-2,4,6-TRIIODOHYDROCINNAMIC ACID SODIUM SALT
  see SKO500
5'-BUTYRAMIDO-2'-(2-HYDROXY-3-ISOPROPYLAMINOPROPOX-
  Y)ACETOPHENONE see AAE100
2-(3-BUTYRAMIDO-2,4,6-TRIIODOPHENYLMETHYLENE)BUTYRIC ACID SODIUM
  SALT see EQC000
2-(3-BUTYRAMIDO-2,4,6-TRIIODOPHENYL)PROPIONIC ACID see BSV250
BUTYRANHYDRID see BSW550
n-BUTYRANILIDE see BSV500
BUTYRANILIDE, 4'-CHLORO-2'-METHYL-3-OXO- see AAY300
BUTYRANILIDE, 4'-ETHOXY-3-HYDROXY- see HJS850
BUTYRATE SODIUM see SFN600
(BUTYRATO)PHENYLMERCURY see BSV750
BUTYRHODANID (GERMAN) see BSN500
n-BUTYRIC ACID see BSW000
BUTYRIC ACID, 4-ACETAMIDO- see AAJ350
BUTYRIC ACID, 2-AMINO-4-(METHYLSULFINYL)- see ALF600
n-BUTYRIC ACID ANHYDRIDE see BSW550
BUTYRIC ACID ANHYDRIDE see BSW550
BUTYRIC ACID, 2-BROMO-3-METHYL- see BNM100
BUTYRIC ACID, 4-CHLORO-, TRIBUTYLSTANNYL ESTER see TID000
BUTYRIC ACID, CINNAMYL ESTER see CMQ800
BUTYRIC ACID, CYCLOHEXYL ESTER see CPI300
BUTYRIC ACID, 3,7-DIMETHYL-2,6-OCTADIENYL ESTER, (E)- see GDE810
BUTYRIC ACID-3,7-DIMETHYL-6-OCTENYL ESTER see DTF800
BUTYRIC ACID, α-α-DIMETHYLPHENETHYL ESTER see BEL850
BUTYRIC ACID, 4,4'-DISELENOBIS(2-AMINO)- see SBU710
BUTYRIC ACID ESTER with BUTYL LACTATE see BQP000
BUTYRIC ACID, HEPTYL ESTER see HBN150
BUTYRIC ACID, HEXYL ESTER see HFM700
BUTYRIC ACID, 2-HYDROXY-4-(METHYLTHIO)- see HMR600
BUTYRIC ACID ISOBUTYL ESTER see BSW500
BUTYRIC ACID LACTONE see BOV000
BUTYRIC ACID, 2-METHYL-, PHENETHYL ESTER (8CI) see PDF780
BUTYRIC ACID NITRILE see BSX250
BUTYRIC ACID, 2-OXO-, SODIUM SALT see SIY600
BUTYRIC ACID, PROPYL ESTER see PNF100

BUTYRIC ACID TRIESTER with GLYCERIN see TIG750
BUTYRIC ACID, VINYL ESTER see VNF000
BUTYRIC ALDEHYDE see BSU250
n-BUTYRIC ANHYDRIDE see BSW550
BUTYRIC ANHYDRIDE see BSW550
BUTYRIC ETHER see EHE000
BUTYRIC or NORMAL PRIMARY BUTYL ALCOHOL see BPW500
BUTYRINASE see GGA800
17-BUTYRLOXY-11-β-HYDROXY-21-PROPIONYLOXY-4-PREGNENE-3,20-DIONE
    see HHQ850
γ-BUTYROLACTAM see PPT500
BUTYROLACTAM see PPT500
α-BUTYROLACTONE see BOV000
β-BUTYROLACTONE see BSX000
γ-BUTYROLACTONE (FCC) see BOV000
BUTYRON see CKF750
BUTYRONE (DOT) see DWT600
BUTYRONITRILE (DOT) see BSX250
BUTYRONITRILE see BSX250
BUTYRONITRILE, 4-CHLORO- see CEU300
BUTYRONITRILE, 4-(DIETHOXYMETHYLSILYL)- see COR500
BUTYRONITRILE, 4-(TRICHLOROSILYL)- see COR750
BUTYRONITRILE, 4-(TRIETHOXYSILYL)- see COS800
BUTYROPHENONE, 4'-AMINO- see AJC750
BUTYROPHENONE, 4-(4-(p-CHLOROPHENYL)-4-HYDROXYPIPERIDINO)-4'-(DI-
    METHYLAMINO)- see CKA600
BUTYROSIN A see BSX325
4-BUTYROTHIOLACTONE see TDC800
N-(1-BUTYROXYMETHYL)METHYLNITROSAMINE see BSX500
N-(1-BUTYROXYMETHYL)-N-NITROSOMETHYLAMINE see BSX500
12-o-BUTYROYL-PHORBOLDODECANOATE see BSX750
3-(3-BUTYRYLAMINO-2,4,6-TRIIODOPHENYL)-2-ETHYLACRYLIC ACID SODIUM
    SALT see EQC000
1-n-BUTYRYLAZIRIDINE see BSY000
1-BUTYRYLAZIRIDINE see BSY000
BUTYRYL CHLORIDE see BSY250
BUTYRYLCHOLINE IODIDE see BSY300
1-N-BUTYRYL-4-CINNAMYL PIPERAZINE HYDROCHLORIDE see BTA000
1-BUTYRYL-4-CINNAMYLPIPERAZINE HYDROCHLORIDE see BTA000
2-BUTYRYL-β-(N,N-DIISOPROPYL)PHENOXYETHYLAMINE HYDROCHLORIDE
    see DNN000
2-BUTYRYL-10-(3-DIMETHYLAMINOPROPYL)PHENOTHIAZINE MALEATE see
    BTA125
BUTYRYLETHYLENEIMINE see BSY000
BUTYRYLETHYLENIMINE see BSY000
BUTYRYL LACTONE see BOV000
BUTYRYL NITRATE see BSY750
BUTYRYL OXIDE see BSW550
BUTYRYLPERAZINE DIMALEATE see BSZ000
1-BUTYRYL-4-(PHENYLALLYL)PIPERAZINE HYDROCHLORIDE see BTA000
BUTYRYLPROMAZINE MALEATE see BTA125
4-BUTYRYLPYRIDINE see PNV755
BUTYRYL TRIGLYCERIDE see TIG750
BUVETZONE see BRF500
BUX see BTA250
BUX-TEN see BTA250
2-n-BUYTLAMINOETHANOL see BQC000
BUZEPIDE METHIODIDE see BTA325
BUZON see BRF500
BUZULFAN see BOT250
BVU see BNP750
BW 47-83 see EAN600
BW 5071 see AMH250
BW 56-72 see TKZ000
BW 50-197 see MQR100
BW 56-158 see ZVJ000
B.W. 57-233 see SBE500
BW 57-322 see ASB250
BW 57-323H see AKY250
BW 57-323 see AKY250
BW 58-271 see RLZ000
BW 58-283b see DBX875
BW 58-283 see DBX875
B-W see SJU000
BW 467-C-60 see BFW250
B.W. 356-C-61 see KFA100
BW 33-T-57 see MKW250
BW-197U see MQR100
BW 248U see AEC700
BW 283U see DBX875
BW-21-Z see AHJ750
B-2847-Y see THD300
BYKOMYCIN see NCD550
BYLADOCE see VSZ000
2,2'-BYPYRIDIN see BGO500

BZ 55 see BSQ000
B-3-Zh see CMS215
BZ see BBU800, QVA000
BZCF see BEF500
BZF-60 see BDS000
BZI see BCB750
BZL see BTA500
BZQ see BCL250
BZT see BEN000, CBF825

C 2 see DXY725, LBX050
C 6 see HEA000
C-10 see BPH750
C 45 see CGG500
C-56 see HCE500
C 78 see BQE250
C 172 see PAQ060
C-272 see BLG500
C 283 see LEF300, NFW500
4-C-32 see TGA525
C-410 see NFW200
C-492 see NFW460
C-516 see NFW100
C-541 see NFW430
C 570 see FAB400
C 609 see NFW450
C 661 see MNX260
C-666 see CAB125
C-684 see NFW470
C-702 see NFW400
C 702 see NFW425
C 709 see DGQ875
C-776 see DTP600
C-829 see NFW435
C-835 see NFW350
C-847 see CEW500
C-854 see CJT750
C-908 see ABW550
C 1,006 see CJT750
C 1120 see DLS800
C-1228 see OIU499
C 1414 see MRH209
C 1686 see MDK000
C 1739 see DOK600
C 1863 see DIN000
C 1983 see CJQ000
C 2018 see CCU250
C 2039 see DII800
C 2046 see DIB200
C 2047 see DIK200
C 2048 see MDK750
C 2052 see DIE600
C 2053 see DIE400
C 2054 see EOG500
C 2057 see DIJ000
C 2059 see DUK800
C 2060 see DIM000
C 2061 see MDK250
C 2085 see MRT250
C 2094 see MNX250
C 2095 see CJQ750
C 2096 see CJQ500
C 2097 see DIK400
C 2098 see DIB800
C 2102 see EOU500
C 2103 see END500
C 2126 see EOG000
C 2127 see EQI500
C 2136 see MDK500
C 2137 see EOF500
C 2138 see EPV000
C 2140 see MQO500
C 2141 see MQO750
C 2142 see EQI000
C 2242 see CIS250
C 2446 see AHO750
C 3037 see CFM250
C 3039 see DDM600
C 3049 see CFU250
C 3053 see DEX600
C 3054 see DHO800
C 3057 see DFF200
C 3058 see CKT750
C 3059 see CHS750

C 3061 see CFM000
C 3062 see PIN000
C 3063 see CGP380
C 3065 see DHK800
C 3067 see CIK500
C 3068 see CFO000
C 3069 see CIJ250
C 3070 see CFL500
C 3071 see CGN250
C 3072 see CFL000
C 3073 see CLC125
C 3074 see CFW250
C 3078 see CFW000, CLC100
C 3080 see DHK200
C 3085 see PIM750
C 3087 see PPU750
C 3089 see DHL400
C 3094 see DHK000
C 3095 see DHJ800
C 3101 see CFM500
C 3102 see DHL200
C 3103 see DDU200
C 3104 see AJE250
C 3115 see CFH000
C 3117 see BDY250
C 3120 see CFI000
C 3121 see BPJ000
C 3124 see AFW250
C 3125 see BPR000
C-3126 see PAM785
C 3127 see DIP800
C 3130 see BPT250
C 3133 see CGE750
C 3134 see MQO250
C 3135 see DHN800
C 3136 see BEE750
C 3137 see DHT000
C 3138 see CKO000
C 3139 see CIU250
C 3140 see CGJ000
C 3141 see DHX600
C 3144 see DII600
C 3145 see DIJ200
C 3150 see DHS600
C 3152 see CFM750
C 3155 see DWC000
C 3156 see CHR750
C 3158 see CIE250
C 3160 see BQN500
C 3162 see CIE500
C 3164 see BRI500
C 3167 see CID250
C 3172 see TNK250
C 3173 see CFP000
C 3181 see BSI750
C 3182 see CLL250
C 3183 see BIP500
C 3184 see CLL500
C 3186 see CLL750
C 3187 see BPI750
C 3189 see CKU000
C 3191 see CFL750
C 3192 see BQN250
C 3193 see CLO500
C 3199 see CLB250
C 3201 see CFN000
C 3205 see CGT000
C 3206 see CAB250
C 3207 see CAB500
C 3208 see CAB750
C 3209 see DHO000
C 3211 see CHK125
C 3213 see CLM500
C 3214 see CLM000
C 3215 see DPG400
C 3218 see BIP750
C 3219 see XWS000
C 3221 see DHV600
C 3222 see DIA000
C 3223 see DHS400
C 3229 see CLM750
C 3230 see DIA800
C 3234 see DIB000
C 3235 see DII400
C 3246 see CLM250

C 3247 see CLL000
C 3249 see CFS500
C 3253 see CFQ500
C 4200 see AMN000
C 4201 see AMM750
C 4207 see BQH500
C 4208 see BQH250
C 4211 see BQH750
74C48 see CBQ125
C 4910 see DEY200
C 4920 see CFT500
C 4924 see DII000
C 4926 see CFT750
C 4928 see DIH800
C-5068 see HGP495
C 5123 see DHS800
C 5124 see EPC875
C 5125 see DHM400
C 5126 see CFN500
C 5290 see CFT000
C 5296 see BEI750
C 5307 see DHV200
C 5308 see DHV400
C 5309 see PIO000
C 5310 see PIO500
53-11 C see EQQ100
C 5311 see PIO250
C 5312 see PPW750
C 5318 see PPX250
C 5319 see PPX000
C 5320 see CLN000
C 5324 see CLN250
C 5326 see CLN500
53-32C see TGA525
C 5334 see DIN400
C 5342 see DIC400
C 5346 see DIC200
C 5347 see CFS750
C 5348 see BEI250
C 5351 see BEI500
C 5353 see BEJ250
C 5354 see BEJ500
C 5364 see CEX500
C 5365 see DEY000
C 5366 see CFR000
C 5384 see CFQ750
C 5385 see CFT250
C 5388 see BRA500
C 5397 see CFR500
C 5398 see CFS250
C 5400 see CFR250
C 5401 see CFR750
C 5402 see CFS000
C 5405 see CIL000
C 5406 see CIU000
C 5407 see CIX000
C 5410 see CHG750
C 5412 see CFQ250
C 5413 see BQA500
C 5414 see BPY500
C 5415 see CIL750
C 5416 see DIC000
C 5417 see CGD750
C 5420 see CIW250
C 5422 see BPR250
C 5458 see CGE500
C 5501 see CGE250
C 5968 see HGP495
C 6005 see MQP000
C 6257 see BQG750
C 6259 see BQG250
C 6260 see BQG500
C 6304 see PEQ250
C-6313 see CES750
C 6379 see DBA800
C 6575 see MLS250
C 6583 see MLS750
C 6606 see DXY400
C 6608 see DIM800
C 6610 see DIN200
C 6866 see BIE500
C-6989 see NIX000
C 7019 see ASG250
C-7441 see DKQ200
C 7441 see OJD300

75-20 C see NCQ100
8057HC see DSQ000
C 8514 see CJJ250
C 9295 see MKU750
C-9491 see IEN000
C-10015 see CDS750, DRP600
11925 C see PCY300
C-12669 see MIW500
C 20684 see EQN750
295 C 51 see TMX775
467-C-60 see BFW250
356C61 see KFA100
C 45 (pharmaceutical) see CGG500
CA 16 see PJT300
CA 33 see CAM200
CA 105 see MCB050
CA 80-15 see CCU250
CA 70203 see TKL100
CA see BMW250
CAA see COJ500
CABADON M see VSZ000
CABBLEMONE see DXU830
CABELLOS de ANGEL (CUBA, PUERTO RICO) see CMV390
CABEZA de BURRO see EAI600
CABEZA de VIEJO (MEXICO) see CMV390
CAB-O-GRIP see AHE250
CAB-O-GRIP II see SCH000
CAB-O-LITE 100 see WCJ000
CAB-O-LITE 130 see WCJ000
CAB-O-LITE 160 see WCJ000
CAB-O-LITE F 1 see WCJ000
CAB-O-LITE P 4 see WCJ000
CABLONGA see YAK350
CABRAL see PGG350
CABREUVA OIL see CAB800
CABRONAL see EOK000
CAB-O-SIL see SCH000
CAB-O-SPERSE see SCH000
CACHALOT L-50 see DXV600
CACHALOT O-1 see OBA000
CACHALOT O-3 see OBA000
CACHALOT O-8 see OBA000
CACHALOT O-15 see OBA000
CACHALOT C-50 see HCP000
C-8 ACID see OCY000
C ACID see ALH250
CACODYLATE de SODIUM (FRENCH) see HKC500
CACODYL HYDRIDE see DQG600
CACODYLIC ACID (DOT) see HKC000
CACODYLIC ACID SODIUM SALT see HKC500
CACODYL NEW see DXE600
CACODYL SULFIDE see CAC250
CACP see PJD000
CACTINOMYCIN see AEA750
C. ADAMANTEUS VENOM see EAB225
CADAVERIN see PBK500
CADAVERINE see PBK500
CADCO 0115 see SMQ500
CADDY see CAE250
CADE OIL RECTIFIED see JEJ000
CADET see BDS000
CADIA DEL PERRO see CAC500
CADIZEM see DNU600
CADMINATE see CAI750
CADMIUM see CAD000
CADMIUM ACETATE (DOT) see CAD250
CADMIUM(II) ACETATE see CAD250
CADMIUM ACETATE DIHYDRATE see CAD275
CADMIUM AMIDE see CAD325
CADMIUM AZIDE see CAD350
CADMIUM BARIUM STEARATE see BAI800
CADMIUM BIS(2-ETHYLHEXYL) PHOSPHITE see CAD500
CADMIUM, BIS(1-HYDROXY-2(1H)-PYRIDINETHIONATO)- see CAI350
CADMIUM, BIS(SALICYLATO)- see CAI400
CADMIUM BROMIDE see CAD600
CADMIUM CAPRYLATE see CAD750
CADMIUM CARBONATE see CAD800
CADMIUM CHLORATE see CAE000
CADMIUM CHLORIDE see CAE250
CADMIUM CHLORIDE, DIHYDRATE see CAE375
CADMIUM CHLORIDE, HYDRATE (2:5) see CAE425
CADMIUM CHLORIDE, MONOHYDRATE see CAE500
CADMIUM COMPOUNDS see CAE750
CADMIUM DIACETATE see CAD250
CADMIUM DIACETATE DIHYDRATE see CAD275

CADMIUM DIAMIDE see CAD325
CADMIUM DIAZIDE see CAD350
CADMIUM DIBROMIDE see CAD600
CADMIUM DICHLORIDE see CAE250
CADMIUM DICYANIDE see CAF500
CADMIUM DIETHYL DITHIOCARBAMATE see BJB500
CADMIUM DILAURATE see CAG775
CADMIUM DINITRATE see CAH000
CADMIUM DODECANOATE see CAG775
CADMIUM(II) EDTA COMPLEX see CAF750
CADMIUM FLUOBORATE see CAG000
CADMIUM FLUORIDE see CAG250
CADMIUM FLUOROBORATE see CAG000
CADMIUM FLUOROSILICATE see CAG500
CADMIUM FLUORURE (FRENCH) see CAG250
CADMIUM FLUOSILICATE see CAG500
CADMIUM FUME see CAH750
CADMIUM GOLDEN 366 see CAJ750
CADMIUM GOLDEN see CMS215
CADMIUM HEXAFLUOROSILICATE (7CI) see CAG500
CADMIUM LACTATE see CAG750
CADMIUM LAURATE see CAG775
CADMIUM LEMON see CMS215
CADMIUM LEMON YELLOW 527 see CAJ750
CADMIUM MONOCARBONATE see CAD800
CADMIUM MONOSULFIDE see CAJ750
CADMIUM MONOTELLURIDE see CAJ800
CADMIUM MONOXIDE see CAH500
CADMIUM(II) NITRATE see CAH000
CADMIUM NITRATE see CAH000
CADMIUM(II) NITRATE TETRAHYDRATE (1:2:4) see CAH250
CADMIUM NITRIDE see TIH000
CADMIUM OCTADECANOATE see OAT000
CADMIUM ORANGE see CAJ750
CADMIUM OXIDE see CAH500
CADMIUM OXIDE FUME see CAH750
CADMIUM PHOSPHATE see CAI000
CADMIUM PHOSPHIDE see CAI125
CADMIUM PRIMROSE 819 see CAJ750
CADMIUM PRIMROSE see CMS215
CADMIUM PROPIONATE see CAI250
CADMIUM PT see CAI350
CADMIUM 2-PYRIDINETHIONE see CAI350
CADMIUM SALICYLATE see CAI400
CADMIUM SELENIDE see CAI500
CADMIUM SELENIDE SULFIDE see CAI600
CADMIUM SILICON FLUORIDE see CAG500
CADMIUM(II) STEARATE see OAT000
CADMIUM STEARATE see OAT000
CADMIUM SUCCINATE see CAI750
CADMIUM SULFATE (1:1) see CAJ000
CADMIUM SULFATE see CAJ000
CADMIUM SULFATE (1:1) HYDRATE (3:8) see CAJ250
CADMIUM SULFATE OCTAHYDRATE see CAJ250
CADMIUM SULFATE TETRAHYDRATE see CAJ500
CADMIUM SULFIDE see CAJ750
CADMIUM SULFIDE SELENIDE see CAI600
CADMIUM SULFIDE mixed with ZINC SULFIDE (1:1) see CMS215
CADMIUM SULFOSELENIDE see CAI600
CADMIUM SULPHATE see CAJ000
CADMIUM SULPHIDE see CAJ750
CADMIUM SULPHOSELENIDE see CAI600
CADMIUM TELLURIDE see CAJ800
CADMIUM TETRAFLUOROBORATE (7CI) see CAG000
CADMIUM THERMOVACUUM AEROSOL see CAK000
CADMIUM-THIONEINE see CAK250
CADMIUM salt of 2,4,5-TRIBROMOIMIDAZOLE see THV500
CADMIUM YELLOW 892 see CAJ750
CADMIUM YELLOW 000 see CAJ750
CADMIUM YELLOW see CAJ750
CADMIUM YELLOW CONC. DEEP see CAJ750
CADMIUM YELLOW CONC. GOLDEN see CAJ750
CADMIUM YELLOW CONC. LEMON see CAJ750
CADMIUM YELLOW CONC. PRIMROSE see CAJ750
CADMIUM YELLOW 10G CONC. see CAJ750
CADMIUM YELLOW OZ DARK see CAJ750
CADMIUM YELLOW PRIMROSE 47-4100 see CAJ750
CADMOPUR GOLDEN YELLOW N see CAJ750
CADMOPUR YELLOW see CAJ750
CI-ADO see CEF100
CADOX see BDS000, BSC750, MKA500
CADOX TBH see BRM250
CADOX TS 40,50 see BIX750
CADOX TS see BIX750
CADPX PS see BHM750

CALCIUM HYPOCHLORITE see HOV500
CALCIUM HYPOPHOSPHITE see CAT250
CALCIUM IODATE see CAT500
CALCIUM LACTATE see CAT600
CALCIUM METHANEARSONATE see CAM000
CALCIUM METHIONATE see CAT700
CALCIUM MOLYBDATE see CAT750
CALCIUM MOLYBDENUM OXIDE (CaMoO₄) see CAT750
CALCIUM MONOCARBONATE see CAT775
CALCIUM MONOCHROMATE see CAP500
CALCIUM MONOSILICATE see CAW850
CALCIUM NEMBUTAL see CAV000
CALCIUM(II) NITRATE (1:2) see CAU000
CALCIUM NITRATE (DOT) see CAU000
CALCIUM(II) NITRATE TETRAHYDRATE (1:2:4) see CAU250
CALCIUM NITRIDE see TIH250
CALCIUM ORTHOARSENATE see ARB750
CALCIUM OXIDE see CAU500
CALCIUM OXYCHLORIDE see HOV500
CALCIUM PANTHOTHENATE (FCC) see CAU750
d-CALCIUM PANTOTHENATE see CAU750
CALCIUM-d-PANTOTHENATE see CAU750
CALCIUM PANTOTHENATE see CAU750
CALCIUM PANTOTHENATE, CALCIUM CHLORIDE DOUBLE SALT see CAU780
CALCIUM PENTOBARBITAL see CAV000
CALCIUM PERMANGANATE see CAV250
CALCIUM PEROXIDE see CAV500
CALCIUM PEROXODISULPHATE see CAW000
CALCIUM-2-(m-PHENOXYPHENYL)PROPIONATE DIHYDRATE see FAP100
CALCIUM-2-PHENYL-4-(p-CHLOROPHENYL)-5-THIAZOLEACETATE see CAP250
CALCIUM PHOSPHATE, DIBASIC see CAW100
CALCIUM PHOSPHATE, MONOBASIC see CAW110
CALCIUM PHOSPHATE, TRIBASIC see CAW120
CALCIUM PHOSPHIDE see CAW250
CALCIUM PHOSPHINATE see CAT250
CALCIUM PHOSPHONOMYCIN HYDRATE see CAW376
CALCIUM PHOTOPHOR see CAW250
CALCIUM POLYSILICATE see CAW850
CALCIUM PROPIONATE see CAW400
CALCIUM PYROPHOSPHATE see CAW450
CALCIUM RESINATE see CAW500
CALCIUM RESINATE (UN 1313) (DOT) see CAW500
CALCIUM RESINATE, fused (UN 1314) (DOT) see CAW500
CALCIUM RHODANID (GERMAN) see CAY250
CALCIUMRHODANID see CAY250
CALCIUM (−)-(1R,2S)-(1,2-EPOXYPROPYL)PHOSPHONATE HYDRATE see CAW376
CALCIUM SACCHARIN see CAM750
CALCIUM SACCHARINA see CAM750
CALCIUM SACCHARINATE see CAM750
CALCIUM SALTPETER see CAU000
CALCIUM SILICATE, synthetic nonfibrous (ACGIH) see CAW850
CALCIUM SILICATE see CAW850
CALCIUM SILICOFLUORIDE see CAX250
CALCIUM SODIUM METAPHOSPHATE see CAX260
CALCIUM STEARATE see CAX350
CALCIUM SULFATE see CAX500
CALCIUM(II) SULFATE DIHYDRATE (1:1:2) see CAX750
CALCIUM SULFIDE see CAY000
CALCIUM SULFOCYANATE see CAY250
CALCIUM SUPEROXIDE see CAV500
CALCIUM TETRABORATE see CAN250
CALCIUM THIOCYANATE see CAY250
CALCIUM TITRIPLEX see CAR780
CALCIUM TRISODIUM CHEL 330 see CAY500
CALCIUM TRISODIUM DIETHYLENE TRIAMINE PENTAACETATE see CAY500
CALCIUM TRISODIUM DTPA see CAY500
CALCIUM TRISODIUM PENTETATE see CAY500
CALCIUM TRISODIUM SALT of DIETHYLENETRIAMINEPENTAACETIC ACID see CAY500
CALCIUM VALPROATE see CAY675
CALCO 2246 see MJO500
CALCOCHROME ALIZARINE RED SC see SEH475
CALCOCHROME BLUE BLACK BC see CMP880
CALCOCHROME YELLOW 2G see SIT850
CALCOCID ALIZARINE BLUE SKY see CMM090
CALCOCID AMARANTH see FAG020
CALCOCID BLUE AX see ERG100
CALCOCID BLUE BLACK see FAB830
CALCOCID BLUE BLACK 2R see FAB830
CALCOCID BLUE EG see FMU059
CALCOCID BRILLIANT SCARLET 3RN see FMU080
CALCOCID ERYTHROSINE N see FAG040
CALCOCID FAST BLUE SR see ADE750

CALCOCID GREEN G see FAE950
CALCOCID MILLING RED G see CMM325
CALCOCID MILLING RED RC see NAO600
CALCOCID MILLING YELLOW R see CMM759
CALCOCID ORANGE Y see CMM220
CALCOCID PHLOXINE 2G see CMM300
CALCOCID 2RIL see FMU070
CALCOCID URANINE B4315 see FEW000
CALCOCID VIOLET 4BNS see FAG120
CALCOCID YELLOW MCG see FAG140
CALCOCID YELLOW MXXX see MDM775
CALCOCID YELLOW XX see FAG140
CALCODUR BROWN BRL see CMO750
CALCODUR RED 8BL see CMO885
CALCOFLUOR WHITE MR see DXB450
CALCOGAS ORANGE NC see PEJ500
C 10 ALCOHOL see DAI600
CALCOLAKE SCARLET 2R see FMU070
CALCOLOID BLUE BLC see DFN425
CALCOLOID BLUE BLD see DFN425
CALCOLOID BLUE BLFD see DFN425
CALCOLOID BLUE BLR see DFN425
CALCOLOID BLUE RS see IBV050
CALCOLOID BROWN BR see CMU770
CALCOLOID DIAZO BLACK BHL see CMN800
CALCOLOID GOLDEN YELLOW see DCZ000
CALCOLOID NAVY BLUE see VGP100
CALCOLOID NAVY BLUE 2GC see CMU500
CALCOLOID NAVY BLUE NTC see VGP100
CALCOLOID OLIVE R see DUP100
CALCOLOID OLIVE RC see DUP100
CALCOLOID OLIVE RL see DUP100
CALCOLOID VIOLET 4RD see DFN450
CALCOLOID VIOLET 4RP see DFN450
CALCOMINE BLACK see AQP000
CALCOMINE BLACK EXL see AQP000
CALCOMINE BLUE 2B see CMO000
CALCOMINE BROWN B see CMO820
CALCOMINE CATECHU 2B see CMO820
CALCOMINE DARK GREEN BG see CMO830
CALCOMINE DIAZO BLACK BHD see CMN800
CALCOMINE DIAZO BLACK BTCW see CMN800
CALCOMINE GREEN BY see CMO840
CALCOMINE RED FC see CMO870
CALCOMINE SCARLET 3B see CMO875
CALCOMINE VIOLET N see CMP000
CALCON see HLI100
CALCO OIL ORANGE 7078 see PEJ500
CALCO OIL RED D see SBC500
CALCO OIL SCARLET BL see XRA000
CALCOSYN PINK B see AKE250
CALCOSYN SAPPHIRE BLUE R see MGG250
CALCOSYN YELLOW GC see AAQ250
CALCOSYN YELLOW GCN see AAQ250
CALCOTONE ORANGE R see CMS145
CALCOTONE RED see IHD000
CALCOTONE RED 3B see NAY000
CALCOTONE TOLUIDINE RED YP see MMP100
CALCOTONE WHITE T see TGG760
CALCOTONE YELLOW GP see CMS212
CALCOZINE BLUE ZF see BJI250
CALCOZINE BRILLIANT GREEN G see BAY750
CALCOZINE CHRYSOIDINE Y see PEK000
CALCOZINE FUCHSINE HO see MAC250
CALCOZINE MAGENTA N see RMK020
CALCOZINE MAGENTA RTN see MAC250
CALCOZINE MAGENTA XX see MAC250
CALCOZINE ORANGE YS see PEK000
CALCOZINE RED BG Liquid see CMM765
CALCOZINE RED BX see FAG070
CALCOZINE RED 6G see RGW000
CALCOZINE RED Y see GJI400
CALCOZINE RHODAMINE BX see FAG070
CALCOZINE RHODAMINE 6GX see RGW000
CALCOZINE VIOLET 6BN see AOR500
CALCOZINE VIOLET C see AOR500
CALCOZINE YELLOW OX see IBA000
CALCYAN see CAQ500
CALCYANIDE see CAQ500
CALDAN see BHL750
CALDEDON NAVY BLUE AR see VGP100
C-8 ALDEHYDE see OCO000
C-9 ALDEHYDE see NMW500
C-16 ALDEHYDE see ENC000
C-10 ALDEHYDE see DAG000

C-12 ALDEHYDE, LAURIC see DXT000
C-14 ALDEHYDE, MYRISTIC see TBX500
CALDON see BRE500
CALEDON BLUE RN see IBV050
CALEDON BLUE XRC see DFN425
CALEDON BLUE XRN see IBV050
CALEDON BRILLIANT BLUE RN see IBV050
CALEDON BRILLIANT PURPLE 4R see DFN450
CALEDON BRILLIANT PURPLE 4RP see DFN450
CALEDON DARK BLUE G see CMU500
CALEDON DARK BROWN 3R see CMU770
CALEDONE OLIVE RP see DUP100
CALEDON GOLDEN YELLOW see DCZ000
CALEDON GREY M see CMU475
CALEDON JADE GREEN 2G see APK500
CALEDON NAVY BLUE ART see VGP100
CALEDON NAVY BLUE 2R see VGP100
CALEDON OLIVE R see DUP100
CALEDON PAPER BLUE RN see IBV050
CALEDON PRINTING BLUE RN see IBV050
CALEDON PRINTING BLUE XRN see IBV050
CALEDON PRINTING NAVY G see CMU500
CALEDON PRINTING PURPLE 4R see DFN450
CALEDON PRINTING YELLOW see DCZ000
CALF KILL see MRU359
CALFLO E see CAW850
CALGINATE see CAM200
CALGLUCOL see CAS750
CALGLUCON see CAS750
CALGON 261 see DTS500
CALGON see SHM500
CALGON 261LV see DTS500
CALGON POLYMER 261 see DTS500
CALIBENE see SOX875
CALIBRITE see CAT775
CALICO BUSH see MRU359
CALICO YELLOW see MRN500
CALIDRIA RG 144 see ARM268
CALIDRIA RG 100 see ARM268
CALIDRIA RG 600 see ARM268
CALIFORNIA CHEMICAL COMPANY RE5305 see BSG250
CALIFORNIA FERN see PJJ300
CALIFORNIA PEPPER TREE see PCB300
CALIGRAN M see MAP300
CALIXIN see TJJ500
CALLA see CAY800
CALLA LILY see CAY800
CALLA PALUSTRIS see WAT300
CAL-LIGHT SA see CAT775
CALMADIN see MQU750
CALMATHION see MAK700
CALMAX see MQU750
CALMDAY see CGA500
CALMINAL see EOK000
CALMINOL see MNM500
CALMIREN see MQU750
CALMIXENE see MOO750
CALMOCITENE see DCK759
CALMODEN see MDQ250
CALMONAL see HGC500
CALMORE see TEH500
CALMOREX see TEH500
CALMOS see CAT775
CALMOSINE see EJA379
CALMOTE see CAT775
CALMOTIN see BNP750
CALNEGYT see CAY875
CALOCAIN see MNQ000
CALOCHLOR see MCY475
CALOCID GREEN S see ADF000
CALOCID GREEN SB see ADF000
CALO-CLOR see CAY950
CALOFIL A 4 see CAT775
CALOFORT S see CAT775
CALOFORT U see CAT775
CALOFOR U 50 see CAT775
CALOGREEN see MCW000, MCY300
CALOMEL see MCW000, MCY300
CALOMELANO (ITALIAN) see MCW000
CALOMEL and MAGNESIUM SULFATE (5:8) see CAZ000
CALOPAKE F see CAT775
CALOPAKE HIGH OPACITY see CAT775
CALOPHYLLUM INOPHYLLUM see MBU780
CALOSAN see MCW000
CALOTAB see MCY300

CALOTROPIS PROCERA (Ait.) R.Br., flower extract see CAZ075
CALOTROPIS (VARIOUS SPECIES) see COD675
CALOXOL CP2 see CAU500
CALOXOL W3 see CAU500
CALPANATE see CAU750
CALPHOSAN see CAT600
CALPLUS see CAO750
CALPOL see HIM000
CALPURNINE see CAZ125
CALSEEDS see CAT775
CALSIL see CAW850
CALSMIN see DLY000
CALSOFT F-90 see DXW200
CALSOFT LAS 99 see LBU100
CALSOL see EIV000
CALSTAR see CAX350
CALTAC see CAO750
CALTEC see CAT775
CALTHA (VARIOUS SPECIES) see MBU550
CALTHOR see AJJ875
CALVACIN see CBA000
CALVISKEN see VSA000
CALVIT see CAT225
CALX see CAU500
CALXYL see CAU500
CALYCANTH see CCK675
CALYCANTHINE, HYDROCHLORIDE see CBA075
CALYCANTHUS (VARIOUS SPECIES) see CCK675
CALYSTIGINE see PAE100
CAM see CDP250
CAMA see CAM000
CAMATROPINE see MDL000
CAMAZEPAM see CFY250
CAMBAXIN see BAB250
CAMBOGIC ACID see CBA125
CAMCOLIT see LGZ000
CAMEL-CARB see CAT775
CAMELIA OIL see CBA200
CAMELLIA SINENSIS see ARP500
CAMEL-TEX see CAT775
CAMEL-WITE see CAT775
CAMFOSULFONATO del d-3-4-(1'DIBENZIL-2-CHETO-IMIDAZOLIDO)-1,2-TRIME-TILTHIOPHANIUM (ITALIAN) see TKW500
CAMILAN see SNY500
CAMILICHIGUI (MEXICO) see DHB309
CAMITE see BMW250
CAMIVERINE see CBA375
CAMOFORM HYDROCHLORIDE see BFX125
CAMOMILE OIL, ENGLISH TYPE (FCC) see CDH750
CAMOMILE OIL GERMAN see CDH500
CAMPANA (CUBA, PUERTO RICO) see AOO825
CAMPAPRIM A 1544 see AMY050
CAMPBELLINE OIL ORANGE see PEJ500
CAMPECHE (PUERTO RICO) see LED500
CAMPHANE, 2-HYDROXY- see BMD000
2-CAMPHANOL see BMD000
1-2-CAMPHANOL see NCQ820
2-CAMPHANONE see CBA750
d-2-CAMPHANONE see CBB250
CAMPHECHLOR see CDV100
CAMPHENE see CBA500
CAMPHERSULFOSAEURE see RFU100
CAMPHIDONIUM see TLG000
CAMPHOCHLOR see CDV100
CAMPHOCLOR see CDV100
CAMPHOFENE HUILEUX see CDV100
CAMPHOGEN see CQI000
CAMPHOL see BMD000
CAMPHOPHYLINE see CNR125
CAMPHOR, synthetic (ACGIH, DOT) see CBA750
dl-CAMPHOR see CBA800
d-(+)-CAMPHOR see CBB250
d-CAMPHOR see CBB250
l-CAMPHOR see CBB000
l-(—)-CAMPHOR see CBB000
CAMPHOR-natural see CBA750
CAMPHOR see CBA750
(±)-CAMPHOR see CBA800
(+)-CAMPHOR see CBB250
(1R,4R)-(+)-CAMPHOR see CBB250
CAMPHORATED OIL see CBB375
CAMPHOR LINIMENT see CBB375
CAMPHOR OIL see CBB500
CAMPHOR OIL, RECTIFIED see CBB500
CAMPHOR OIL WHITE see CBB500

CAMPHOR OIL YELLOW see CBB500
(+)-β-CAMPHORSULFONIC ACID see RFU100
d-10-CAMPHORSULFONIC ACID see RFU100
d-CAMPHORSULFONIC ACID see RFU100
CAMPHORSULFONIC ACID see RFU100
(+)-CAMPHORSULFONIC ACID see RFU100
CAMPHOR TAR see NAJ500
CAMPHOR USP see CBB250
CAMPHOZONE see DJS200
CAMPILIT see COO500
CAMPOSAN see CDS125
CAMPOVITON 6 see PPK500
CAMPTOTHECIN see CBB870
20(S)-CAMPTOTHECINE see CBB870
CAMPTOTHECINE see CBB870
CAMPTOTHECIN, SODIUM SALT see CBB875
CAMUZULENE see DRV000
CAMYLOFINE see NOC000
CAMYLOFINE DIHYDROCHLORIDE see AAD875
CAMYLOFINE HYDROCHLORIDE see AAD875
CAMYLOFIN HYDROCHLORIDE see AAD875
CAMYLOPIN see NOC000
CANACERT AMARANTH see FAG020
CANACERT BRILLIANT BLUE FCF see FAE000
CANACERT ERYTHROSINE BS see FAG040
CANACERT INDIGO CARMINE see FAE100
CANACERT SUNSET YELLOW FCF see FAG150
CANACERT TARTRAZINE see FAG140
CANADA MOONSEED see MRN100
CANADIAN BALSAM see FBS200
CANADIAN FIR NEEDLE OIL see CBB900
CANADIEN 2000 see BMN000
α-CANADINE see TCJ800
CANADINE see TCJ800
(−)-CANADINE see TCJ800
CANADOL see PCT250
CANAFISTOLA (CUBA) see GIW300
CANANGA see CAL125
CANARIO (PUERTO RICO) see AFQ625
CANARY CHROME YELLOW 40-2250 see LCR000
CANARY IVY see AFK950
CANAVANIN see AKD500
l-CANAVANINE see AKD500
CANCARB see CBT750
CANCER JALAP see PJJ315
CANDAMIDE see LGZ000
CANDASEPTIC see CFE250
CANDELILLA (MEXICO) see SDZ475
CANDEPTIN see LFF000
CANDEREL see ARN825
CANDEX see ARQ725, NOH500
CANDIDA ALBICANS GLYCOPROTEINS see CBC375
CANDIDIN see CBC500
CANDIDINE see CBC500
CANDIMON see LFF000
CANDIO-HERMAL see NOH500
CANDLEBERRY see TOA275
CANDLENUT see TOA275
CANDLE SCARLET 2B see SBC500
CANDLE SCARLET B see SBC500
CANDLE SCARLET G see SBC500
CANDLETOXIN A see CBD250
CANDLETOXIN B see CBD500
CANESCINE see RDF000
CANESCINE 10-METHOXYDERIVATIVE see MEK700
CANESTEN see MRX500
CANE SUGAR see SNH000
CANNABICHROME see PBW400
CANNABICHROMENE see PBW400
CANNABIDIOL see CBD599
(−)-CANNABIDIOL see CBD599
(−)-trans-CANNABIDIOL see CBD599
CANNABINOL see CBD625
CANNABIS see CBD750
CANNABIS RESIN see CBD750
CANNABIS SMOKE RESIDUE see CBD760
CANNANBICHROMENE see PBW400
CANNE-A-GRATTER (HAITI) see DHB309
CANNE-MADERE (HAITI) see DHB309
CANNOGENIN-α-l-THEVETOSIDE see EAQ050
C. ANNUUM see PCB275
CANOCENTA see CKL325
CANOGARD see DGP900
CANQUIL-400 see MQU750
CANRENOATE-K see PKZ000

CANRENOATE POTASSIUM see PKZ000
CANTABILINE see MKP500
CANTABILINE SODIUM see HMB000
CANTHARIDES see CBE250
CANTHARIDES CAMPHOR see CBE750
CANTHARIDIN see CBE750
CANTHARIDINE see CBE750
CANTHARONE see CBE750
CANTHOXAL see MLJ050
3C ANTIBIOTIC see HLT100
CANTIL see CBF000
CANTREX see KAL000, KAM000, KAV000
CANTRIL see CBF000
CAN-TROL see CLN750
CANTROL see CLO000
CAO 1 see BFW750
CAO 3 see BFW750
CAOBA (CUBA, DOMINICAN REPUBLIC, PUERTO RICO) see MAK300
CAOUTCHOUC see ROH900
CAOUTCHOUC (HAITI) see ROU450
CAP see CBF250, CDP250, CEA750
CAPARISIDE see TFA350
CAPAROL see BKL250
CAPARSOLATE see TFA350
CAPATHYN see DPJ400
CAPAURIDINE see TDI750
CAPAURINE, dl- see TDI750
CAPAURINE, (±)- see TDI750
CAPE BELLADONNA see AHI635
CAPE GOOSEBERRY see JBS100
CAPEN see MCI375
CAPERASE see CCP525
CAPER SPURGE see EQW000
CAPILAN see DNU100
CAPISTEN see BDU500
CAPITUS see ENG500
CAPLENAL see ZVJ000
CAPMUL see PKG000, PKL030
CAPMUL POE-O see PKL100
CAPOBENATE see CBF625
CAPOBENATE SODIUM see CBF625
CAPORIT see HOV500
CAPOTEN see MCO750
CAPOTILLO (MEXICO) see CAL125
CAP-P see CDP700
CAP-PALMITATE see CDP700
CAPRALDEHYDE see DAG000
CAPRALENSE see AJD000
CAPRAMOL see AJD000
CAPRAN 80 see PJY500
CAPREOMYCIN DISULFATE see CBF675
CAPREOMYCIN IA see CBF680
n-CAPRIC ACID see DAH400
CAPRIC ACID see DAH400
CAPRIC ACID ETHYL ESTER see EHE500
CAPRIC ACID METHYL ESTER see MHY650
CAPRIC ALCOHOL see DAI600
CAPRIC ALDEHYDE see DAG000
CAPRIN see ADA725
CAPRINALDEHYDE see DAG000
CAPRINIC ACID see DAH400
CAPRINIC ACID, SODIUM SALT see SGB600
CAPRINIC ALCOHOL see DAI600
CAPRINIC ALDEHYDE see DAG000
CAPROALDEHYDE see HEM000
CAPROAMIDE see HEM500
CAPROAMIDE POLYMER see PJY500
CAPROCID see AJD000
CAPROCIN see CBF675
CAPRODAT see IPU000
n-CAPROIC ACID see HEU000
CAPROIC ACID see HEU000
CAPROIC ALDEHYDE see HEM000
CAPROIC TRIGLYCERIDE see GGK000
CAPROKOL see HFV500
6-CAPROLACTAM see CBF700
ω-CAPROLACTAM (MAK) see CBF700
CAPROLACTAM see CBF700
CAPROLACTAM OLIGOMER see PJY500
ε-CAPROLACTAM POLYMERE (GERMAN) see PJY500
6-CAPROLACTONE see HDY600
γ-CAPROLACTONE see HDY600
epsilon-CAPROLACTONE see LAP000
CAPROLACTONE see LAP000
CAPROLATTAME (FRENCH) see CBF700

CAPROLIN see CBM750
CAPROLISIN see AJD000
CAPROLON see NOH000
CAPROLYL PEROXIDE see CBF705
CAPROMYCIN see CBF680
CAPRON see HNT500, PJY500
CAPRONALDEHYDE see HEM000
CAPRONAMIDE see HEM500
CAPRONIC ACID see HEU000
CAPRONITRILE see HER500
CAPROYL ALCOHOL see HFJ500
n-CAPROYLALDEHYDE see HEM000
1-CAPROYLAZIRIDINE see HEW000
CAPROYLETHYLENEIMINE see HEW000
CAPRYL ALCOHOL see OCY090, OEI000
CAPRYLAMINE see OEK000, OEK010
CAPRYLATE d'OESTRADIOL (FRENCH) see EDQ500
2-CAPRYL-4,6-DINITROPHENYL CROTONATE see AQT500
CAPRYLDINITROPHENYL CROTONATE see AQT500
n-CAPRYLIC ACID see OCY000
CAPRYLIC ACID see OCY000
CAPRYLIC ACID, METHYL ESTER see MHY800
CAPRYLIC ACID SODIUM SALT see SIX600
CAPRYLIC ACID TRIGLYCERIDE see TMO000
CAPRYLIC ALCOHOL see OEI000
CAPRYLIC/CAPRIC TRIGLYCERIDE see CBF710
CAPRYLIC ETHER see OEY000
4-CAPRYLMORPHOLINE see CBF725
CAPRYLNITRILE see OCW100
CAPRYLONE see HBO790
CAPRYLONITRILE see OCW100
CAPRYL PEROXIDE see CBF705
CAPRYL o-PHTHALATE see BLB750
CAPRYLYL ACETATE see OEG000
CAPRYLYLAMINE see OEK000
CAPRYLYL PEROXIDE (DOT) see CBF705
CAPRYLYL PEROXIDE see CBF705
CAPRYLYL PEROXIDE, solution see OFI000
CAPRYLYL PEROXIDE SOLUTION (DOT) see CBF705
CAPRYNIC ACID see DAH400
CAPSAICIN see CBF750
CAPSAICINE see CBF750
CAPSEBON see CAJ750
CAPSICUM (VARIOUS SPECIES) see PCB275
CAPSINE see DUS700
CAPSTAT see CBF680
CAPTAF see CBG000
CAPTAFOL see CBF800
CAPTAGON HYDROCHLORIDE see CBF825
CAPTAMINE HYDROCHLORIDE see DOY600
CAPTAN see CBG000
CAPTANCAPTENEET 26,538 see CBG000
CAPTANE see CBG000
CAPTAN-STREPTOMYCIN 7.5-0.1 POTATO SEED PIECE PROTECTANT see CBG000
CAPTAX see BDF000
CAPTEX 300 see CBF710
CAPTEX see CBG000
CAPTODIAME see BRS000
CAPTODIAMIN see BRS000
CAPTODIAMINE see BRS000
CAPTOFOL see CBF800
CAP-O-TRAN see MQU750
CAPUT MORTUUM see IHD000
CAPVAL see NOA000
CAPVAL HYDROCHLORIDE see NOA500
CARACEMIDE see CBG075
CARACHOL see SGD500
CARADATE 30 see MJP400
CARADRIN see POB500
CARAGARD see BQC500
CARAIBE (HAITI) see EAI600, XCS800
CARAMIFENE (ITALIAN) see CBG375
CARAMIPHEN see PET250
CARAMIPHENE HYDROCHLORIDE see PET250
CARAMIPHEN ETHANE DISULFONATE see CBG250
CARAMIPHEN HYDROCHLORIDE see CBG375
CARASTAY see CCL250
CARASTAY G see CCL250
CARAWAY OIL see CBG500
CARBACHOL see CBH250
CARBACHOL CHLORIDE see CBH250
CARBACHOLIN see CBH250
CARBACHOLINE CHLORIDE see CBH250
CARBACOLINA see CBH250

CARBACRYL see ADX500
CARBADINE see EIR000
CARBADIPIMIDINE MALEATE see CCK790
CARBADOX (USDA) see FOI000
CARBAETHOXYDIGOXIN (GERMAN) see EEP000
CARBAICA see AKK625
CARBAM see VFU000
CARBAMALDEHYDE see FMY000
CARBAMAMIDINE see GKW000
CARBAMATE see FAS000
CARBAMATE de l'ETHINYLCYCLOHEXANOL (FRENCH) see EEH000
CARBAMATE de METHYLPENTINOL (FRENCH) see MNM500
CARBAMATE du PROPINYLCYCLOHEXANOL (FRENCH) see POA250
CARBAMATES see CBH750
CARBAMAZEPEN see DCV200
CARBAMAZEPINE see DCV200
CARBAMAZINE see DIW000
CARBAMEZEPINE see DCV200
CARBAMIC ACID, ALLYL ESTER see AGA750
CARBAMIC ACID, BUTYL ESTER see BQP250
CARBAMIC ACID-2-sec-BUTYL-2-METHYLTRIMETHYLENE ESTER see MBW750
CARBAMIC ACID-3-DIMETHYLAMINOPHENYL ESTER, METHOSULFATE see HNP500
CARBAMIC ACID, DIMETHYL-(9CI) DQY950
CARBAMIC ACID, DIMETHYLDITHIO-, ANHYDROSULFIDE see BJL600
CARBAMIC ACID, DIMETHYLDITHIO-, 2,4-DINITROPHENYL ESTER see DVB850
CARBAMIC ACID, DIMETHYLDITHIO-, ZINC SALT (2:1) see BJK500
CARBAMIC ACID, DIMETHYL-, ester with (m-HYDROXYPHEN-YL)TRIMETHYLAMMONIUM BROMIDE see POD000
CARBAMIC ACID, N,N-DIMETHYL-, m-ISOPROPYLPHENYL ESTER see DQX300
CARBAMIC ACID, DITHIO-, N,N-DIMETHYL-, DIMETHYLAMINOMETHYL ESTER see DRQ650
CARBAMIC ACID, ESTER with CHOLINE CHLORIDE see CBH250
CARBAMIC ACID, ESTER with 2-(HDYROXYMETHYL)-1-METHYLPENTYLISOPROPYLCARBAMATE see IPU000
CARBAMIC ACID, ESTER with 2-(HYDROXYMETHYL)-2-METHYLPENTYL BU-TYLCARBAMATE see MOV500
CARBAMIC ACID, ESTER with 2-METHYL-2-PROPYL-1,3-PROPANEDIOL BU-TYLCARBAMATE see MOV500
CARBAMIC ACID, ESTER with 2-METHYL-2-PROPYL-1,3-PROPANEDIOL ISO-PROPYLCARBAMATE see IPU000
CARBAMIC ACID, ETHYLENEBIS(DITHIO)-, MANGANESE SALT see MAS500
CARBAMIC ACID, ETHYL ESTER see UVA000
CARBAMIC ACID-1-ETHYL-1-METHYL-2-PROPYNYL ESTER see MNM500
CARBAMIC ACID-2-ETHYNYL-2-BUTYL ESTER see MNM500
CARBAMIC ACID, (5-(4-FLUOROBENZOYL)-1H-BENZIMIDAZOL-2-YL)-, METHYL ESTER (9CI) see FDA887
CARBAMIC ACID, HEXAMETHYLENEBIS(DITHIO-, DISODIUM SALT see HEF400
CARBAMIC ACID HYDRAZIDE see HGU000
CARBAMIC ACID, (2-HYDROXYETHYL)-, 2-HYDROXYETHYL ESTER see HKS550
CARBAMIC ACID, 2-HYDROXY-3-(o-METHOXYPHENOXY)PROPYL ESTER see GKK000
CARBAMIC ACID-β-HYDROXYPHENETHYL ESTER see PFJ000
CARBAMIC ACID, (3-HYDROXYPHENYL)-, ETHYL ESTER (9CI) see ELE600
CARBAMIC ACID, ISOPROPYL ESTER see IOJ000
CARBAMIC ACID, METHYL-, 3-tert-BUTYLPHENYL ESTER see BSG300
CARBAMIC ACID, N-METHYL-, 3-DIETHYLARSINOPHENYL ESTER, METH-IODIDE see MID900
CARBAMIC ACID-N-METHYL-3-DIMETHYLAMINOPHENYL ESTER METHIODIDE see HNO500
CARBAMIC ACID-1-METHYLETHYL ESTER see IOJ000
CARBAMIC ACID, METHYL-, o-HYDROXYPHENYL ESTER see HNK900
CARBAMIC ACID, METHYL-, o-ISOPROPYLPHENYL ESTER see MIA250
CARBAMIC ACID, METHYL-, 2-(1-METHYLETHYL)PHENYL ESTER see MIA250
CARBAMIC ACID, METHYL-, 3-METHYLPHENYL ESTER (9CI) see MIB750
CARBAMIC ACID, METHYL-, NAPHTHALENYL ESTER see MME800
CARBAMIC ACID, METHYLNITROSO-, o-ISOPROPOXYPHENYL ESTER see PMY310
CARBAMIC ACID, METHYLNITROSO-, 2-(1-METHYLETHOXY)PHENYL ESTER (9CI) see PMY310
CARBAMIC ACID, (3-METHYLPHENYL)-, 3-((METHOXYCARBO-NYL)AMINO)PHENYL ESTER (9CI) see MEG250
CARBAMIC ACID, METHYL-, 3-TOLYL ESTER see MIB750
CARBAMIC ACID-3-PHENYLPROPYL ESTER see PGA750
CARBAMIC ACID, 1-PHENYL-2-PROPYNYL ESTER see PGE000
CARBAMIC ACID, PROPYL ESTER see PNG250
CARBAMIC ACID, THIO, S-ESTER with 2-MERCAPTOACETANILIDE see MCL500
CARBAMIC ACID, 2,2,2-TRICHLOROETHYL ESTER see TIO500

CARBAMIC ACID, (m-TRIMETHYLAMMONIO)PHENYL ESTER, METHYLSULFATE see HNP500
CARBAMIC ACID, VINYL ESTER see VNK000
CARBAMIC ESTER of 3-OXYPHENYLTRIMETHYLAMMONIUM METHYLSULFATE see HNP500
CARBAMIDAL see DJS200
CARBAMIDE see USS000
CARBAMIDE PEROXIDE see HIB500
CARBAMIDE PHENYLACETATE see PEC250
CARBAMIDE RESIN see USS000
CARBAMIDINE see GKW000
CARBAMIDINE HYDROCHLORIDE see GKY000
p-CARBAMIDOBENZENEARSONIC ACID see CBJ000
p-CARBAMIDOPHENYL ARSENOUS ACID see CBI500
p-CARBAMIDOPHENYL ARSENOUS OXIDE see CBI500
4-CARBAMIDOPHENYL BIS(CARBOXYMETHYLTHIO)ARSENITE see CBI250
p-CARBAMIDOPHENYL-BIS(2-CARBOXYPHENYLMERCAPTO)ARSINE see TFD750
4-CARBAMIDOPHENYL BIS(o-CARBOXYPHENYLTHIO)ARSENITE see TFD750
p-CARBAMIDOPHENYL-DI(1'-CARBOXYPHENYL-2')THIOARSENITE see TFD750
4-CARBAMIDOPHENYLOXOARSINE see CBI500
CARBAMIDSAEURE-AETHYLESTER (GERMAN) see UVA000
CARBAMIDIC ACID see USS000
CARBAMIMIDOTHIOIC ACID-2-(DIMETHYLAMINO)ETHYL ESTER DIHYDROCHLORIDE see NNL400
CARBAMIMIDOTHIOIC ACID, ETHYL ESTER with METAPHOSPHORIC ACID (1:1) see ELY575
CARBAMIMIDOTHIOIC ACID, ETHYL ESTER, MONO(DIETHYL PHOSPHATE) see CBI675
CARBAMIMIDOTHIOIC ACID, METHYL ESTER, SULFATE (9CI) see MPV790
CARBAMINE see CBM750
CARBAMINOCHOLINE CHLORIDE see CBH250
p-CARBAMINO PHENYL ARSONIC ACID see CBJ000
CARBAMINOPHENYL-p-ARSONIC ACID see CBJ000
CARBAMINOTHIOGLYCOLIC ACID ANILIDE see MCL500
CARBAMINOYLCHOLINE CHLORIDE see CBH250
CARBAMIOTIN see CBH250
CARBAMMATO di FENILETINILCARBINOLO see PGE000
CARBAMODITHIOIC ACID, DIETHYL-(9CI) see DJC800
CARBAMODITHIOIC ACID, DIETHYL-, compd. with N-ETHYLETHANAMINE (1:1) (9CI) see DJD000
CARBAMODITHIOIC ACID, METHYL-, MONOSODIUM SALT, DIHYDRATE (9CI) see SIL550
CARBAMODITHIOIC ACID, PHENYL-, ETHYL ESTER see EOK550
CARBAMOHYDROXAMIC ACID see HOO500
CARBAMOHYDROXIMIC ACID see HOO500
CARBAMOHYDROXYAMIC ACID see HOO500
CARBAMOL see UTU500
CARBAMONITRILE see COH500
CARBAMOTHIOIC ACID-S,S'-(2-(DIMETHYLAMINO)-1,3-PROPANEDIYL) ESTER, MONOHYDROCHLORIDE (9CI) see BHL750
4'-(4-CARBAMOYL-9-ACRIDINYLAMINO)-3'-METHOXY-1-PROPANESULFONANILIDE HYDROCHLORIDE see MFN000
(p-CARBAMOYLAMINO)PHENYLARSINOBIS(2-THIO-ACETIC ACID) see CBI250
2-CARBAMOYLANILINE see AID620
N-CARBAMOYLARSANILIC ACID see CBJ000
p-CARBAMOYLBENZAMIDE see TAN600
8-CARBAMOYL-3-(2-CHLOROETHYL)IMIDAZO(5,1-d)-1,2,3,5-TETRAZIN-4(3H)-ONE see MQY110
γ-CARBAMOYL CHOLINE CHLORIDE see CBH250
CARBAMOYLCHOLINE CHLORIDE see CBH250
5-CARBAMOYL-5H-DIBENZ(b,f)AZEPINE see DCV200
5-CARBAMOYL-5H-DIBENZO(b,f)AZEPINE see DCV200
5-CARBAMOYLDIBENZO(b,f)AZEPINE see DCV200
N-CARBAMOYL-2-(2,6-DICHLOROPHENYL)ACETAMIDINE HYDROCHLORIDE see PFV000
N-(2-CARBAMOYLETHYL)ARSANILIC ACID SODIUM SALT see SJG000
1-CARBAMOYLFORMIMIDIC ACID see OLO000
4'-CARBAMOYL-2-FORMYL-1,1'-(OXYDIMETHYLENE)DI-PYRIDINIUM-DICHLORIDE-2-OXIME see HGE900
CARBAMOYLHYDRAZINE see HGU000
N-CARBAMOYLHYDROXYLAMINE see HOO500
p-((CARBAMOYLMETHYL)AMINO)-BENZENEARSONIC ACID see CBJ750
N-(CARBAMOYLMETHYL)ARSANILIC ACID see CBJ750
N-(CARBAMOYLMETHYL)-2-DIAZOACETAMIDE see CBK000
1-p-CARBAMOYLMETHYLPHENOXY-3-ISOPROPYLAMINO-2-PROPANOL see TAL475
2-CARBAMOYL-2-NITROACETONITRILE see CBK125
N-CARBAMOYL-N-NITROSOGLYCINE see NKI500
CARBAMOYL OXIME see HOO500
1-CARBAMOYLOXY-2-HYDROXY-3-(o-METHYLPHENOXY)PROPANE see CBK500
3-CARBAMOYLOXY-3-METHYL-4-PENTYNE see MNM500
(3-(N-CARBAMOYLOXY)PHENYL)DIETHYLMETHYL-AMMONIUM METHOSULFATE see HNK550

1-CARBAMOYLOXY-3-PHENYLPROPANE see PGA750
((m-CARBAMOYLOXY)PHENYL)TRIMETHYLAMMONIUM METHYLSULFATE see HNP500
2-CARBAMOYLOXYPROPYLTRIMETHYLAMMONIUM CHLORIDE see HOA500
1-CARBAMOYLOXY-1-(2-PROPYNYL)CYCLOHEXANE see POA250
10-(3-(4-CARBAMOYLPIPERIDINE)PROPYL)-2-(METHANESULFONYL)PHENOTHIAZINE see MQR000
N-(1-CARBAMOYLPROPYL)ARSANILIC ACID see CBK750
4-CARBAMOYL-1-β-d-RIBOFURANOSYL-IMIDAZOLIUM-5-OLATE see BMM000
CARBAMULT see CQI500
4-CARBAMYLAMINOPHENYLARSONIC ACID see CBJ000
N-CARBAMYL ARSANILIC ACID see CBJ000
CARBAMYL CHLORIDE, N,N-DIMETHYL- see DQY950
CARBAMYLCHOLINE CHLORIDE see CBH250
5-CARBAMYL-5H-DIBENZO(b,f)AZEPINE see DCV200
5-CARBAMYLDIBENZO(b,f)AZEPINE see DCV200
CARBAMYLHYDRAZINE see HGU000
CARBAMYLHYDRAZINE HYDROCHLORIDE see SBW500
CARBAMYL HYDROXAMATE see HOO500
CARBAMYLMETHYLCHOLINE CHLORIDE see HOA500
4-CARBAMYLPHENYL BIS(CARBOXYMETHYLTHIO)ARSENITE see TFA350
1-CARBAMYL-2-PHENYLHYDRAZINE see CBL000
2-CARBAMYL PYRAZINE see POL500
CARBANIL see PFK250
CARBANILALDEHYDE see FNJ000
CARBANILIC ACID, DITHIO-, ETHYL ESTER see EOK550
d-(—)-CARBANILIC ACID (1-ETHYLCARBAMOYL)ETHYL ESTER see CBL500
CARBANILIC ACID ETHYL ESTER see CBL750
CARBANILIC ACID, m-HYDROXY-, ETHYL ESTER see ELE600
CARBANILIC ACID ISOPROPYL ESTER see CBM000
CARBANILIDE see CBM250
CARBANILIDE, 3,3'-BIS((5-((4,6,8-TRISULFO-1-NAPHTHYL)CARBAMOYL)-o-TOLYL)CARBAMOYL)- see BAT000
m-CARBANILOYLOXYCARBANILIC ACID ETHYL ESTER see EEO500
CARBANOCHLORIDIC ACID, NAPHTHYL ESTER see NBH200
CARBANOLATE see CBM500, CGI500
CARBANTHRENE BLUE BCF see DFN425
CARBANTHRENE BLUE BCS see DFN425
CARBANTHRENE BLUE 2R see IBV050
CARBANTHRENE BLUE RBCF see DFN425
CARBANTHRENE BLUE RCS see DFN425
CARBANTHRENE BLUE RS see IBV050
CARBANTHRENE BLUE RSP see IBV050
CARBANTHRENE BRILLIANT VIOLET 4R see DFN450
CARBANTHRENE BROWN BR see CMU770
CARBANTHRENE GOLDEN YELLOW see DCZ000
CARBANTHRENE NAVY BLUE G see CMU500
CARBANTHRENE NAVY BLUE RA see VGP100
CARBANTHRENE OLIVE R see DUP100
CARBANTHRENE RED BROWN 5R see CMU800
CARBANTHRENE RED G 2B see CMU825
CARBANTHRENE RED G 2BP see CMU825
CARBANTHRENE VIOLET 2R see DFN450
CARBANTHRENE VIOLET 2RP see DFN450
CARBARSONE (USDA) see CBJ000
CARBARSONE OXIDE see CBI500
CARBARYL (ACGIH,DOT,OSHA) see CBM750
CARBARYL see CBM750
CARBASED see ACE000
CARBASONE see CBJ000
CARBATENE see MQQ250
CARBATHIONE see VFU000
CARBATOX-60 see CBM750
CARBATOX-75 see CBM750
CARBATOX see CBM750
CARBAURINE see PFR325
CARBAVINE see CBM875
CARBAVUR see CBM750
CARBAX see BIO750
CARBAZALDEHYDE see FNN000
CARBAZAMIDE see HGU000
CARBAZEPINE see DCV200
CARBAZIC ACID, ETHYL ESTER see EHG000
CARBAZIC ACID HYDRAZIDE see CBS500
CARBAZIDE see CBS500
CARBAZILQUINONE see BGX750
CARBAZINC see BJK500
CARBAZINE see ADI775
CARBAZOCHROME SODIUM SULFONATE TRIHYDRATE see AER666
9H-CARBAZOLE see CBN000
CARBAZOLE see CBN000
9H-CARBAZOLE, 9-ETHENYL-(9CI) see VNK100
CARBAZOLE, 3-(p-HYDROXYANILINO)- see CBN100
9H-CARBAZOLE, 1,3,6,8-TETRANITRO- see TDY120
CARBAZOLE, 9-VINYL- see VNK100

4-(3-CARBAZOLYLAMINO)PHENOL see CBN100
CARBAZONE, DIPHENYLTHIO- see DWN200
CARBAZOTIC ACID see PID000
CARBECIN see CBO250
CARBENDAZIM see MHC750
CARBENDAZIME see MHC750
CARBENDAZIME and SODIUM NITRITE (1:1) see SIQ700
CARBENDAZIM and SODIUM NITRITE (5:1) see CBN375
CARBENDAZOL see MHC750
CARBENDAZOLE see MHC750
CARBENDAZYM see MHC750
CARBENICILLIN DISODIUM SALT see CBO250
CARBENICILLIN PHENYL see CBN750
CARBENICILLIN PHENYL ESTER see CBN750
CARBENICILLIN PHENYL SODIUM see CBO000
CARBENICILLIN SODIUM see CBO250
CARBENOXALONE, DISODIUM SALT see CBO500
CARBENOXOLONE see BGD000
CARBENOXOLONE, DISODIUM SALT see CBO500
CARBENOXOLONE SODIUM see CBO500
CARBESTROL see CBO625
CARBETAMEX see CBL500
CARBETAMID (GERMAN) see CBL500
CARBETAMIDE see CBL500
CARBETHOXYACETIC ESTER see EMA500
3-CARBETHOXYAMINO-5-DIMETHYLAMINOACETYL-10,11-DIHYDRODI-
BENZ(b,f)AZEPINE HYDROCHLORIDE see BMB125
3-CARBETHOXYAMINOPHENOL see ELE600
α-CARBETHOXY-β,β-BISCYCLOPROPYL ACRYLONITRILE see DGX000
N-CARBETHOXYETHYLENIMINE see ASH750
1-CARBETHOXY HYDRAZINE see EHG000
N-(CARBETHOXY)HYDRAZINE see EHG000
CARBETHOXYHYDRAZINE see EHG000
CARBETHOXY MALAOXON see OPK250
CARBETHOXY MALATHION see MAK700
2-CARBETHOXYMETHYLENE-3-METHYL-5-PIPERIDINO-4-THIAZOLIDONE see
MNG000
CARBETHOXYMETHYL ISOTHIOCYANATE see ELX530
p-CARBETHOXYPHENOL see HJL000
1(4-CARBETHOXYPHENYL)-3,3-DIMETHYLTRIAZENE see CBP250
2-(N-(4-CARBETHOXY-4-PHENYL)PIPERIDINO)PROPIOPHENONE HYDROCHLO-
RIDE see CBP325
CARBETHOXYSYRINGOYL METHYLRESERPATE see RCA200
S-CARBETHOXYTHIAMINE HYDROCHLORIDE see EEQ500
CARBETOVUR see MAK700
CARBETOX see MAK700
CARBICRON see DGQ875
CARBIDE 6-12 see EKV000
CARBIDE BLACK D see CMN230
CARBIDE BLACK DU see CMN230
CARBIDE BLACK E see AQP000
CARBIDIUM ETHANESULFONATE see CBQ125
CARBIDIUM ETHANESULPHONATE see CBQ125
CARBIDOPA see CBQ500
CARBIDOPA MONOHYDRATE see CBQ529
CARBILAZINE see DIW000
CARBIMIDE see COH500
CARBIN see CEW500
CARBINAMINE see MGC250
CARBINOL see MGB150
CARBINOLBASE DES KRISTALLVIOLETT (GERMAN) see TJK000
CARBINOLBASE des METHYLVIOLETT (GERMAN) see MQN250
CARBINOXAMIDE MALEATE see TAI500
CARBINOXAMINE DIPHENYLDISULFONATE see CBQ575
p-CARBINOXAMINE MALEATE see CGD500
CARBINOXAMINE MALEATE see CGD500
CARBIPHENE HYDROCHLORIDE see CBQ625
CARBITAL 90 see CAT775
CARBITOL see CBR000, DJD600
CARBITOL ACETATE see CBQ750
CARBITOL CELLOSOLVE see CBR000
CARBITOL SOLVENT see CBR000
CARBIUM see CAT775
CARBIUM MM see CAT775
CARBON, activated (DOT) see CBT500
CARBON, animal or vegetable origin (DOT) see CBT500
CARBON, ACTIVATED see CDI000
CARBOBENZOXY CHLORIDE see BEF500
N-CARBOBENZOXYGLYCINE-1,2-DIBROMOETHYL ESTER see CBR175
N-CARBOBENZOXYGLYCINE VINYL ESTER see CBR200
N-CARBOBENZOXY-I-LEUCINE-1,2-DIBROMOETHYL ESTER see CBR210
N-CARBOBENZOXY-I-LEUCINE VINYL ESTER see CBR215
CARBOBENZOXYLGLYCINE see CBR125
N-CARBOBENZOXY-I-PHENYLALANINE see CBR220
CARBOBENZOXY-I-PHENYLALANINE see CBR220

CARBOBENZOXYPHENYLALANINE see CBR220
N-CARBOBENZOXY-I-PHENYLALANINE-1,2-DIBROMOETHYL ESTER see
CBR225
N-CARBOBENZOXY-I-PHENYLALANINE VINYL ESTER see CBR235
N-CARBOBENZOXY-I-PROLINE-1,2-DIBROMOETHYL ESTER see CBR245
N-CARBOBENZOXY-I-PROLINE VINYL ESTER see CBR247
N-CARBOBENZOYLGLYCINE see CBR125
CARBOBENZOYL GLYCINE see CBR125
CARBOBENZYLOXY CHLORIDE see BEF500
N-CARBOBENZYLOXYGLYCINE see CBR125
CARBOBENZYLOXYGLYCINE see CBR125
2-CARBO-n-BUTOXY-6,6-DIMETHYL-5,6-DIHYDRO-1,4-PYRONE see BRT000
CARBOCAINE see SBB000
CARBOCAINE HYDROCHLORIDE see CBR250
CARBOCHOL see CBH250
CARBOCHOLIN see CBH250
CARBOCHROMENE HYDROCHLORIDE see CBR500
CARBOCISTEINE see CBR675
CARBOCIT see CBR675
CARBO-CORT see CMY800
CARBOCROMENE see CBR500
CARBOCYSTEINE see CBR675
CARBODIHYDRAZIDE see CBS500
CARBODIIMIDE, BIS(TRIMETHYLSILYL)-(7CI,8CI) see BLQ950
CARBODIS see CBT750
CARBOETHOXYHYDRAZINE see EHG000
CARBOETHOXYMETHYL ISOTHIOCYANATE see ELX530
N¹-CARBOETHOXY-N²-PHTHALAZINO HYDRAZINE see CBS000
CARBOETHOXYPHTHALAZINO HYDRAZINE see CBS000
CARBOFENOTHION (DUTCH) see TNP250
CARBOFLUORENE AMINO ESTER see CBS250
CARBOFOS see MAK700
CARBOFRAX M see SCQ000
CARBOFURAN see CBS275
CARBOGEN (8CI) see CBV250
CARBOHYDRAZIDE see CBS500
N-CARBOISOPROPOXY-o-ACETYL-N-PHENYL CARBAMATE see ING400
CARBOLAC 1 see CBT750
CARBOLAC see CBT750
CARBOLIC ACID see PDN750
CARBOLITH see LGZ000
CARBOLON see SCQ000
CARBOLSAEURE (GERMAN) see PDN750
CARBOMAL see BNK000
CARBOMATE see CBM750
CARBOMER 934 see ADW300
CARBOMER 940 see ADW200
CARBOMER 934P see ADW200
CARBOMET see CBT750
CARBOMETHENE see KEU000
2-CARBOMETHOXYANILINE see APJ250
o-CARBOMETHOXYANILINE see APJ250
2-β-CARBOMETHOXY-3-β-BENZOXYTROPANE see CNE750
4'-CARBOMETHOXY-2,3'-DIMETHYLAZOBENZENE see CBS750
4'-CARBOMETHOXY-2,3'-DIMETHYLAZOBENZOL see CBS750
N-CARBOMETHOXYMETHYLIMINOPHOSPHORYL CHLORIDE see CBT125
α-2-CARBOMETHOXY-1-METHYLVINYL DIMETHYL PHOSPHATE see MQR750
2'-CARBOMETHOXYPHENYL 4-GUANIDINOBENZOATE see CBT175
2-CARBOMETHOXY-1-PROPEN-2-YL DIMETHYL PHOSPHATE see MQR750
N-(3-CARBOMETHOXYPROPYL)-N-(1-ACETOXYBUTYL)NITROSAMINE see
MEI000
p-CARBOMETHOXYTOLUENE see MPX850
CARBOMYCIN see CBT250
CARBOMYCIN A see CBT250
CARBON-12 see CBT500
CARBON see CBT500
CARBONA see CBY000
CARBONATE MAGNESIUM see MAC650
(CARBONATO)DIHYDROXYDICOPPER see CNJ750
CARBONAZIDIC ACID, 1,1-DIMETHYLETHYL ESTER see BQI250
CARBONAZIDODITHIOIC ACID see ASF750
CARBON BICHLORIDE see PCF275
CARBON BISULFIDE (DOT) see CBV500
CARBON BISULPHIDE see CBV500
CARBON BLACK see CBT750
CARBON BLACK, ACETYLENE see CBT750
CARBON BLACK BV and V see CBT750
CARBON BLACK, CHANNEL see CBT750
CARBON BLACK, FURNACE see CBT750
CARBON BLACK, LAMP see CBT750
CARBON BLACK, THERMAL see CBT750
CARBON BROMIDE see CBX750
CARBON CHLORIDE see CBY000
CARBON CHLOROSULFIDE see TFN500
CARBON D see DXD200

CARBON DICHLORIDE see PCF275
CARBON DIFLUORIDE OXIDE see CCA500
CARBON DIOXIDE see CBU250
CARBON DIOXIDE, mixture with NITROGEN OXIDE (N₂O) see CBV000
CARBON DIOXIDE–NITROUS OXIDE mixture (DOT) see CBV000
CARBON DIOXIDE mixed with NITROUS OXIDE see CBV000
CARBON DIOXIDE-OXYGEN mixture (DOT) see CBV250
CARBON DIOXIDE mixed with OXYGEN see CBV250
CARBON DIOXIDE, solid (UN 1845) (DOT) see CBU250
CARBON DIOXIDE, refrigerated liquid (UN 2187) (DOT) see CBU250
CARBON DISULFIDE see CBV500
CARBON DISULPHIDE see CBV500
CARBONE (OXYCHLORURE de) (FRENCH) see PGX000
CARBONE (OXYDE de) (FRENCH) see CBW750
CARBONE (SUFURE de) (FRENCH) see CBV500
CARBON FERROCHROMIUM see FBD000
CARBON FLUORIDE see CBY250
CARBON FLUORIDE OXIDE see CCA500
CARBON HEXACHLORIDE see HCI000
CARBON HYDRIDE NITRIDE (CHN) see HHS000
CARBONIC ACID, ALLYL ESTER, DIESTER with DIETHYLENE GLYCOL see AGD250
CARBONIC ACID, AMMONIUM SALT see ANE000
CARBONIC ACID ANHYDRIDE see CBU250
CARBONIC ACID, BARIUM SALT (1:1) see BAJ250
CARBONIC ACID BERYLLIUM SALT (1:1) see BFP750
CARBONIC ACID-2-sec-BUTYL-4,6-DINITROPHENYL-2,4-DINITROPHENYL ESTER (8CI) see DVC200
CARBONIC ACID-2-sec-BUTYL-4,6-DINITROPHENYL ISOPROPYL ESTER see CBW000
CARBONIC ACID, CADMIUM SALT see CAD800
CARBONIC ACID, CALCIUM SALT (1:1) see CAO000
CARBONIC ACID, CALCIUM SALT (1:1) see CAT775
CARBONIC ACID, CHROMIUM SALT see CMJ100
CARBONIC ACID, COBALT(2+) SALT (1:1) see CNB475
CARBONIC ACID, COPPER(2+) SALT (1:1) see CBW200
CARBONIC ACID, CYCLIC 3-CHLOROPROPYLENE ESTER see CBW400
CARBONIC ACID, CYCLIC ETHYLENE ESTER see GHM000
CARBONIC ACID, CYCLIC ETHYLETHYLENE ESTER see BOT200
CARBONIC ACID CYCLIC PROPYLENE ESTER see CBW500
CARBONIC ACID, DIAMMONIUM SALT see ANE000
CARBONIC ACID, DICESIUM SALT see CDC750
CARBONIC ACID, DIESTER with 1-(2,3-DIMETHYLPHENYL)-3-(2-HYDROXYETHYL)UREA see PLK580
CARBONIC ACID DIHYDRAZIDE see CBS500
CARBONIC ACID, DILITHIUM SALT see LGZ000
CARBONIC ACID, DIPHENYL ESTER see DVZ000
CARBONIC ACID, DIPOTASSIUM SALT see PLA000
CARBONIC ACID, DIRUBIDIUM SALT see RPB200
CARBONIC ACID, DISODIUM SALT see SFO000
CARBONIC ACID, DITHALLIUM(1+) SALT see TEJ000
CARBONIC ACID, DITHIO-, O-ISOPROPYL ESTER, POTASSIUM SALT see IRG050
CARBONIC ACID, DITHIO-, O-PENTYL ESTER, POTASSIUM SALT see PKV100
CARBONIC ACID GAS see CBU250
CARBONIC ACID, compd. with GUANIDINE (1:2) see GKW100
CARBONIC ACID, LEAD(2+) SALT (1:1) see LCP000
CARBONIC ACID LITHIUM SALT see LGZ000
CARBONIC ACID, MAGNESIUM SALT see MAC650
CARBONIC ACID METHYL-4-(o-TOLYLAZO)-o-TOLYL ESTER see CBS750
CARBONIC ACID, MONOAMMONIUM SALT see ANB250
CARBONIC ACID MONOSODIUM SALT see SFC500
CARBONIC ACID, NICKEL SALT (1:1) see NCY500
CARBONIC ACID, NICKEL SALT, BASIC see NCY600
CARBONIC ACID, TRITHIO-, CYCLIC ETHYLENE ESTER see EJQ100
CARBONIC ACID, cyclic VINYLENE ESTER see VOK000
CARBONIC ACID, ZINC SALT (1:1) see ZEJ050
CARBONIC ANHYDRASE INHIBITOR NO. 6063 see AAI250
CARBONIC ANHYDRIDE see CBU250
CARBONIC DIFLUORIDE see CCA500
CARBONIC DIHYDRAZIDE see CBS500
CARBONIC OXIDE see CBW750
CARBONIMIDIC DICHLORIDE, PHENYL-(9CI) see PFJ400
CARBONIMIDIC DIHYDRAZIDE, BIS((4-CHLOROPHENYL)METHYLENE)- see RLK890
4,4'-CARBONIMIDOYLBIS(N,N-DIMETHYLBENZENAMINE) see IBB000
4,4'-CARBONIMIDOYLBIS(N,N-DIMETHYLBENZENAMINE)MONOHYDROCHLORIDE see IBA000
CARBON IODIDE see CBY500
CARBONIO (OSSICLORURO di) (ITALIAN) see PGX000
CARBONIO (OSSIDO di) (ITALIAN) see CBW750
CARBONIO (SOLFURO di) (ITALIAN) see CBV500
CARBON MONOXIDE (ACGIH,OSHA) see CBW750

CARBON MONOXIDE see CBW750
CARBON MONOXIDE, refrigerated liquid (cryogenic liquid) (NA 9202) (DOT) see CBW750
CARBON MONOXIDE (UN 1016) (DOT) see CBW750
CARBON NITRIDE see COO000
CARBON NITRIDE ION (CN¹) see COI500
CARBONOCHLORIDE ACID, 1,2-ETHANEDIYL ESTER see EIQ000
CARBONOCHLORIDE ACID-1-METHYL ESTER see IOL000
CARBONOCHLORIDIC ACID, 9H-FLUOREN-9-YLMETHYL ESTER see FEI100
CARBONOCHLORIDIC ACID, 1-NAPHTHALENYL ESTER see NBH200
CARBONOCHLORIDIC ACID, OXYDI-2,1-ETHANEDIYL ESTER see OPO000
CARBONOCHLORIDIC ACID PHENYL ESTER see CBX109
CARBONOCHLORIDIC ACID, PROPYL ESTER see PNH000
CARBONOCHLORIDIC ACID TRICHLOROMETHYL ESTER see TIR920
CARBONOCHLORIDIC ACID, 2,4,6-TRICHLOROPHENYL ESTER (9CI) see TIY800
CARBONODITHIOIC ACID, O-ETHYL ESTER, POTASSIUM SALT see PLF000
CARBONODITHIOIC ACID, O-(1-METHYLETHYL) ESTER, POTASSIUM SALT (9CI) see IRG050
CARBONODITHIOIC ACID, O-(1-METHYLETHYL)ESTER, SODIUM SALT (9CI) see SIA000
CARBONODITHIOIC ACID, O-PENTYL ESTER, POTASSIUM SALT (9CI) see PKV100
CARBONOHYDRAZIDE see CBS500
CARBONOHYDRAZONIC DIHYDRAZIDE, MONONITRATE (9CI) see THN800
CARBON OIL see BBL250
CARBONOTHIOIC DICHLORIDE (9CI) see TFN500
CARBONOTHIOIC DIHYDRAZIDE see TFE250
CARBONOTRITHIOIC ACID, METHYL TRICHLOROMETHYL ESTER see TIS500
CARBONOTRITHIOIC ACID-2-PROPENYLTRICHLOROMETHYL ESTER see TIR750
CARBON OXIDE see CBU250
CARBON OXIDE (CO) see CBW750
CARBON OXIDE SULFIDE see CCC000
CARBON OXYCHLORIDE see PGX000
CARBON OXYFLUORIDE see CCA500
CARBON OXYSULFIDE see CCC000
CARBON S see SGM500
CARBON SILICIDE see SCQ000
CARBON SULFIDE see CBV500
CARBON SULPHIDE (DOT) see CBV500
CARBON TET see CBY000
CARBON TETRABROMIDE see CBX750
CARBON TETRACHLORIDE see CBY000
CARBON TETRAFLUORIDE see CBY250
CARBON TETRAIODIDE see CBY500
CARBON TRIFLUORIDE see CBY750
CARBONYL AZIDE see CCA000
CARBONYLBIS(1-AZIRIDINE) see BJP450
CARBONYLBIS(AZIRIDINE) see BJP450
CARBONYLCHLORID (GERMAN) see PGX000
CARBONYL CHLORIDE see PGX000
CARBONYL CHLORIDE, THIO- see TFN500
CARBONYL CYANIDE, 3-CHLOROPHENYLHYDRAZONE see CKA550
CARBONYL DIAMIDE see USS000
CARBONYLDIAMINE see USS000
CARBONYL DIAZIDE see CCA000
CARBONYL DICYANIDE see OOM400
CARBONYL DIFLUORIDE see CCA500
CARBONYLDIHYDRAZINE see CBS500
CARBONYL DIISOTHIOCYANATE see CCA125
CARBONYL FLUORIDE see CCA500
CARBONYL IRON see IGK800
CARBONYL POTASSIUM see CCB500
CARBONYLS see CCB609
CARBONYL SULFIDE see CCC000
CARBONYL SULFIDE-³²S see CCC000
CARBOPHOS see MAK700
CARBOPLATIN see CCC075
CARBOPOL 934 see ADW200, ADW300
CARBOPOL 940 see ADW200
CARBOPOL 941 see ADW200
CARBOPOL 960 see ADW200
CARBOPOL 961 see ADW200
CARBOPOL EXTRA see CBT500
CARBOPOL M see CBT500
CARBOPOL 934P see ADW200
CARBOPOL Z 4 see CBT500
CARBOPOL Z EXTRA see CBT500
CARBOPROST see CCC100
CARBOPROST TROMETHAMINE see CCC110
CARBOQUONE see BGX750
CARBORAFFIN see CDI000
CARBORAFINE see CDI000
*m*-CARBORANE see NBV100

CARBORANYLMETHYLETHYL SULFIDE see DEJ600
CARBORANYLMETHYLPROPYL SULFIDE see DEJ800
CARBOREX 2 see CAT775
CARBORUNDEUM see SCQ000
CARBORUNDUM see SCQ000
CARBOSET 515 see ADW200
CARBOSET see ADW200
CARBOSET RESIN NO. 515 see ADW200
CARBOSIEVE see CBT500
CARBOSORBIT R see CBT500
CARBOSPOL see AGJ250
CARBOSTESIN see BOO000
CARBOSTYRIL see CCC250
CARBOSTYRIL, THIO- see QOJ100
2-CARBOSYPYRIDINE see PIB930
CARBOTHIALDIN see DSB200
CARBOTHIALDINE see DSB200
CARBOWAX 1000 see PJT250
CARBOWAX 1500 see PJT500
CARBOWAX 4000 see PJT750
CARBOWAX 6000 see PJU000
CARBOWAX see PJT000, PJT200
CARBOXAMIDOACETAMIDE see MAO000
1-p-(CARBOXAMIDOPHENYL)-3,3-DIMETHYLTRIAZINE see CCC325
5-CARBOXANILIDO-2,3-DIHYDRO-6-METHYL-1,4-OXATHIIN see CCC500
CARBOXIDE see CAT225
CARBOXIN (USDA) see CCC500
CARBOXINE see CCC500
2-CARBOXYACETANILIDE see AAJ150
CARBOXYACETIC ACID see CCF750
CARBOXYACETYLENE see PMT275
1-(p-CARBOXYAETHYLPHENYL)-3,3-DIMETHYLTRIAZEN (GERMAN) see CBP250
2-CARBOXYANILINE see API500
3-CARBOXYANILINE see AIH500
4-CARBOXYANILINE see AIH600
o-CARBOXYANILINE see API500
p-CARBOXYANILINE see AIH600
CARBOXYANILINE see API500
(3-(α-CARBOXY-o-ANISAMIDO)-2-(2-HYDROXYETHOXY)PROPYL)HYDROXY-MERCURY MONOSODIUM SALT see HLO300
(3-(α-CARBOXY-p-ANISAMIDO)-2-HYDROXYPROPYL)HYDROXYMERCURY see HLP500
3-(α-CARBOXY-o-ANISAMIDO)-2-METHOXYPROPYL HYDROXYMERCURY, MONOSODIUM SALT see SIH500
9-CARBOXYANTHRACENE see APG550
2-CARBOXYBENZALDEHYDE see FNK010
o-CARBOXYBENZALDEHYDE see FNK010
CARBOXYBENZENE see BCL750
4-CARBOXYBENZENESULFONAMIDE see SNL840
p-CARBOXYBENZENESULFONAMIDE see SNL840
p-CARBOXYBENZENESULFONDICHLOROAMIDE see HAF000
CARBOXYBENZENESULFONYL AZIDE see CCD625
(o-CARBOXYBENZOYL)-p-AMINOPHENYLSULFONAMIDOTHIAZOLE see PHY750
N⁴-(o-CARBOXYBENZOYL)-N′-2-THIAZOLYL-SULFANILAMIDE SULFAPHTHALA-ZOLE see PHY750
CARBOXYBENZYLPENICILLIN PHENYL ESTER SODIUM SALT see CBO000
CARBOXYBENZYLPENICILLIN SODIUM see CBO250
4-CARBOXYBIPHENYL see PEL600
p-CARBOXYBROMOBENZENE see BMU100
N-(2-CARBOXYCAPROYL)HYDRAZOBENZENE CALCIUM SALT HEMIHYDRATE see CCD750
α-CARBOXYCAPROYL-N,N′-DIPHENYLHYDRAZINE CALCIUM SALT HEMIHYD-RATE see CCD750
(6R-(6-α,7-β(R*)))-1-((2-CARBOXY-7-(((((5-CARBOXY-1H-IMIDAZOL-4-YL)CARBONYL)AMINO)PHENYLACETYLE)AMINO)-8-OXO-5-THIA-1-AZABICY-CLO(4.2.0)OCT-2-EN-3-YL)METHYL)-4-(2-SULFOETHYL)-PYRIDINIUM HY-DROXODIE, inner salt, MONOSODIUM SALT see CCS525
p-CARBOXYCHLOROBENZENE see CEL300
CARBOXYCYCLOHEXANE see HDH100
N-(1-((1-CARBOXY-5-DIAZO-5-OXOPENTYL)CARBAMOYL)-5-DIAZO-4-OXOPEN-TYL)-GLUTAMINE SODIUM SALT see ASO510
2-CARBOXY-1-((2,6-DICARBOXY-2,3-DIHYDRO-4(1H)-PYRIDINYLI-DENE)ETHYLIDENE)5,6-DIHYDROXY-1H-INDOLIUM, HYDROXIDE, INNER SALT, SULFATE (SALT) see BFV900
2-CARBOXY-4′-(DIMETHYLAMINO)AZOBENZENE see CCE500
3′-CARBOXY-4-DIMETHYLAMINOAZOBENZENE see CCE750
N-(2-CARBOXY-3,3-DIMETHYL-7-OXO-4-THIA-1-AZABICYCLO(3.2.0)HEPT-6-YL)-2-PHENYL-MALONAMIC ACID DISODIUM SALT see CBO250
2-CARBOXYDIPHENYLAMINE see PEG500
2-CARBOXYDIPHENYLARSINOUS ACID see PFI600
CARBOXYETHANE see PMU750
S-2-CARBOXYETHYL-l-CYSTEINE see CCF250
CARBOXYETHYLGERMANIUM SESQUIOXIDE see CCF125

2-CARBOXYETHYLGERMASESQUIOXANE see CCF125
3-CARBOXY-1-ETHYL-7-METHYL-1,8-NAPHTHIDIN-4-ONE see EID000
N-(2-CARBOXYETHYL)-N-NITRO-β-ALANINE see NHI600
4′-(2-CARBOXYETHYL)PHENYL-trans-4-AMINOMETHYLCYCLOHEXANE CAR-BOXYLATE HYDROCHLORIDE see CDF380
2-(4-(1-CARBOXYETHYL)PHENYL)-1-ISOINDOLINONE see IDA400
3-((2-CARBOXYETHYL)THIO)ALANINE see CCF250
12-CARBOXYEUDESMA-3,11(13)-DIENE see EQR000
2-CARBOXYFURAN see FQF000
3-CARBOXY-4-HYDROXYBENZENESULFONIC ACID see SOC500
2′-CARBOXY-2-HYDROXY-4-METHOXYBENZOPHENONE(o-(2-HYDROXY-p-ANIS-OYL)BENZOIC ACID) see HLS500
3-(3-CARBOXY-4-HYDROXYPHENYL)-2-PHENYL-4,5-DIHYDRO-3H-BENZ(e)INDOLE see FAO100
(±)-(3-CARBOXY-2-HYDROXYPROPYL)TRIMETHYLAMMONIUM CHLORIDE see CCK655
(3-CARBOXY-2-HYDROXYPROPYL)TRIMETHYLAMMONIUM CHLORIDE see CCK650
(−)-(3-CARBOXY-2-HYDROXYPROPYL)TRIMETHYLAMMONIUM CHLORIDE see CCK660
l-(3-CARBOXY-2-HYDROXYPROPYL)TRIMETHYLAMMONIUM CHLORIDE see CCK660
3-CARBOXY-5-HYDROXY-1-p-SULFOPHENYL-4-p-SULFOPHENYLAZOPYRA-ZOLE TRISODIUM SALT see FAG140
(R)-3-CARBOXY-2-HYDROXY-N,N,N-TRIMETHYL-1-PROPANAMINIUM CHLORIDE see CCK660
3-CARBOXY-2-HYDROXY-N,N,N-TRIMETHYL-1-PROPANAMINIUM CHLORIDE (9CI) see CCK650
1-(4′-CARBOXYLAMIDOPHENYL)-3,3-DIMETHYLTRIAZINE see CCC325
(3-(4-(CARBOXYLATOMETHOXY)PHENYL)-2-HYDROXYPROPYL)HYDROXY-MERCURATE(1-), SODIUM see CCF500
l-N-CARBOXYLEUCINE-N-BENZYL-1-(1,2-DIBROMOETHYL) ESTER see CBR210
l-N-CARBOXYLEUCINE N-BENZYL 1-VINYL ESTER see CBR215
6-CARBOXYL-4-HYDROXYLAMINOQUINOLINE-1-OXIDE see CCF750
6-CARBOXYL-4-NITROQUINOLINE-1-OXIDE see CCG000
(CARBOXYMETHOXY)AMINE see ALQ650
(3-(o-(CARBOXYMETHOXY)BENZAMIDO)-2-METHOXYPROPYL)HYDROXY MER-CURY, MONOSODIUM SALT compounded with THEOPHYLLINE see SAQ000
(4-(CARBOXY METHOXY)-3-CHLOROPHENYL)(5,5-DIETHYL-2,4,6(1H,3H,5H)-PY-RIMIDINETRIONATO)-O²-MERCURY, MONO see CCG500
4-CARBOXYMETHYLANILINE see AID700
4-CARBOXYMETHYLBIPHENYL see BGE125
CARBOXYMETHYL CARBAMIMONIOTHIOATE CHLORIDE see CCH199
1-CARBOXYMETHYL-1-CARBOXYETHOXYETHYL-2-COCO-IMIDAZOLINIUM BE-TAINE see AOC250
CARBOXYMETHYL CELLULOSE see SFO500
CARBOXYMETHYL CELLULOSE, AMMONIUM SALT see CCH000
CARBOXYMETHYLCELLULOSE NORDIC see CCH000
CARBOXYMETHYL CELLULOSE, SODIUM see SFO500
CARBOXYMETHYL CELLULOSE, SODIUM SALT see SFO500
S-(CARBOXYMETHYL)CYSTEINE see CBR675
S-CARBOXYMETHYLCYSTEINE see CCH125
l-CARBOXYMETHYLCYSTEINE see CBR675
4-o-(CARBOXYMETHYL)-1-DEOXY-1,4-DIHYDRO-4-HYDROXY-1-OXO-RIFAMYCIN γ-LACTONE see RKP400
N-(CARBOXYMETHYL)-N,N-DIMETHYL-1-HEXADECANAMINIUM HYDROXIDE in-ner salt see CDF450
N-(CARBOXYMETHYL)-N-(2-(2,6-DIMETHYLPHENYL)AMINO)-2-OXOETHYL)-GLY-CINE (9CI) see LFO300
3,3′-(CARBOXYMETHYLENE)BIS(4-HYDROXYCOUMARIN) ETHYL ESTER see BKA000
N-(CARBOXYMETHYL)GLYCINE see IBH000
(CARBOXYMETHYL)HEXADECYLDIMETHYLAMMONIUM HYDROXIDE, inner salt (7CI) see CDF450
(o-CARBOXYMETHYL)HYDROXYLAMINE see ALQ650
((CARBOXYMETHYLIMINO)BIS(ETHYLENENITRILO))TETRAACETIC ACID see DJG800
2-CARBOXYMETHYLISOTHIOURONIUM CHLORIDE see CCH199
N-(γ-CARBOXYMETHYLMERCAPTOMERCURI-β-ME-THOXY)PROPYLCAMPHORAMIC ACID DISODIUM SALT see TFK270
2-(CARBOXYMETHYLMERCAPTO)PHENYLSTIBONIC ACID see CCH250
2-(CARBOXYMETHYLMERCAPTO)PHENYL-STIBONSAEURE (GERMAN) see CCH250
N-(CARBOXYMETHYL)-N′-(2-HYDROXYETHYL)-N,N′-ETHYLENEDIGLYCINE see HKS500
1-(CARBOXYMETHYL)-1-NITROSOUREA see NKI500
CARBOXYMETHYLNITROSOUREA see CCH500, NKI500
6-(5-CARBOXY-3-METHYL-2-PENTENYL)-7-HYDROXY-5-METHOXY-4-METHYL-PHTHALIDE see MRX000
4-o-(CARBOXYMETHYL)RIFAMYCIN see RKK000
(CARBOXYMETHYLTHIO)ACETIC ACID see MCM750
3-(CARBOXYMETHYLTHIO)ALANINE see CBR675
3-((CARBOXYMETHYL)THIO)ALANINE see CCH125
l-3-((CARBOXYMETHYL)THIO)ALANINE see CBR675

5-CARBOXYMETHYL-3-p-TOLYL-THIAZOLIDINE-2,4-DIONE-2-ACETOPHENONE HYDRAZONE see CCH800
(CARBOXYMETHYL)TRIMETHYLAMMONIUM CHLORIDE see CCH850
(CARBOXYMETHYL)TRIMETHYLAMMONIUM CHLORIDE HYDRAZIDE see GEQ500
(CARBOXYMETHYL)TRIMETHYLAMMONIUM HYDROXIDE, inner salt see GHA050
(CARBOXYMETHYL)TRIMETHYLAMMONIUM IODIDE-6-(DIMETHYLAMINO)-4-ISO-PROPYL-m-TOLYL ESTER see DQD500
(CARBOXYMETHYL)TRIMETHYLAMMONIUM IODIDE-5-(DIMETHYLAMINO)-4-ISO-PROPYL-o-TOLYL ESTER see DOW875
(CARBOXYMETHYL)TRIMETHYLAMMONIUM IODIDE-4-(DIMETHYLAMINO)-3-ISO-PROPYLPHENYL ESTER see IOU200
N-CARBOXY-3-MORPHOLINOSYDNONIMINE ETHYL ESTER see MRN275
1-CARBOXYNAPHTHALENE see NAV490
1-CARBOXY-4-NITROBENZENE see CCI250
6-CARBOXY-4-NITROQUINOLINE-1-OXIDE see CCG000
3-(3-CARBOXY-1-OXOPROPOXY)-11-OXOOLEAN-12-EN-29-OIC ACID, DISODI-UM SALT (3-β,20-β) see CBO500
21-(3-CARBOXY-1-OXOPROPOXY)-5-β-PREGNANE-3,20-DIONE SODIUM SALT see VJZ000
3-CARBOXY-2,4-PENTADIENALLACTOL see APV750
2-(5-CARBOXYPENTYL)-4-THIAZOLIDONE see CCI500
3-CARBOXYPHENOL see HJI100
4-CARBOXYPHENOL see SAI500
o-CARBOXYPHENYL ACETATE see ADA725
I-N-CARBOXY-3-PHENYLALANINE-N-BENZYL ESTER see CBR220
I-N-CARBOXY-3-PHENYLALANINE N-BENZYL 1-VINYL ESTER see CBR235
p-CARBOXYPHENYLAMINE see AIH600
2-((2-CARBOXYPHENYL)AMINO)-4-CHLOROBENZOIC ACID DISODIUM SALT see LHZ600
p-CARBOXY PHENYLARSENOXIDE see CCI550
9-(2-CARBOXYPHENYL)-3,6-BIS(METHYLAMINO)XANTHYLIUM PERCHLORATE see RGW100
(p-CARBOXYPHENYL)CHLOROMERCURY see CHU500
9-o-CARBOXYPHENYL-6-DIETHYLAMINO-3-ETHYLIMINO-3-ISOXANTHENE, 3-ETHOCHLORIDE see FAG070
(9-(o-CARBOXYPHENYL)-6-(DIETHYLAMINO)-3H-XANTHEN-3-YLIDENE) DIETHY-LAMMONIUM CHLORIDE see FAG070
(4-CARBOXYPHENYL)HYDRAZINE see HHB100
p-CARBOXYPHENYLHYDRAZINE see HHB100
9-(o-CARBOXYPHENYL)-6-HYDROXY-3-ISOXANTHENONE see FEV000
9-o-CARBOXYPHENYL-6-HYDROXY-3-ISOXANTHONE, DISODIUM SALT see FEW000
(o-CARBOXYPHENYL)HYDROXY-MERCURY SODIUM SALT see SHT500
9-(o-CARBOXYPHENYL)-6-HYDROXY-2,4,5,7-TETRAIODO-3-ISOXANTHONE see FAG040
9-(o-CARBOXYPHENYL)-6-HYDROXY-3H-XANTHEN-3-ONE see FEV000
4'-CARBOXYPHENYLMETHANESULFONANILIDE, SODIUM SALT see CCJ000
o-CARBOXYPHENYL PHOSPHATE see PHA575
((o-CARBOXYPHENYL)THIO)ETHYLMERCURY SODIUM SALT see MDI000
4-CARBOXYPHTHALATO(1,2-DIAMINOCYCLOHEXANE)PLATINUM(II) see CCJ350
4-CARBOXYPHTHALIC ANHYDRIDE see TKV000
3-o-(β-CARBOXYPROPIONYL)-11-OXO-18-β-OLEAN-12-EN-30-OIC ACID, DISODI-UM SALT see CBO500
3-β-(3-CARBOXYPROPIONYLOXY)-11-OXO-OLEAN-12-EN-30-OIC ACID see BGD000
3-CARBOXYPROPYL(2-PROPENYL)NITROSAMINE see CCJ375
3-CARBOXYPYRIDINE see NCQ900
4-CARBOXYPYRIDINE see ILC000
5-CARBOXYRESORCINOL see REF200
trans-β-CARBOXYSTYRENE see CMP980
4-CARBOXYTHIAZOLIDINE see TEV000
2-CARBOXYTHIOPHENE see TFM600
6-CARBOXYURACIL see OJV500
CARBOXY VINYL POLYMER see CCJ400
CARBRITAL see NBU000
CARBUTAMID see BSM000
CARBUTAMIDE see BSM000
CARBUTEN see MBW750
CARBUTEROL HYDROCHLORIDE see BQD500
CARBYL see CBH250
CARBYL SULFATE see DXI500
CARBYNE see CEW500
CARCHOLIN see CBH250
CARCINOCIDIN see CCN250
CARCINOLIPIN see CCJ500
CARDAMINE see DJS200
CARDAMIST see NGY000
CARDAMON see CCJ625
CARDAMON OIL see CCJ625
CARDELMYCIN see SMB000
CARDENAL de MACETA (MEXICO) see CCJ825
CARDIAGEN see DJS200

CARDIAMID see DJS200
CARDIAMINA see DJS200
neo-CARDIAMINE see GCE600
CARDIAMINE see DJS200
CARDIDIGIN see DKL800
CARDIEM see DNU600
CARDIGIN see DKL800
CARDILATE see VSA000
CARDIMON see DJS200
CARDINAL FLOWER see CCJ825
CARDINOPHILLIN see CCN500
CARDINOPHYLLIN see CCN500
CARDIO see CCK125
CARDIOFILINA see TEP500
CARDIOGRAFIN see AOO875
CARDIO-GREEN see CCK000
CARDIO-KHELLIN see AHK750
CARDIOLANATA see LAU400
CARDIOLIPOL see NCW300
CARDIOMIN see TEP500
CARDIOMONE see AOA125
CARDION see POB500
CARDIOQUIN see GAX000
CARDIORYTHMINE see AFH250
CARDIOSERPIN see RDK000
CARDIOTRAST see DNG400
CARDIOVITAL see GCE600
CARDIOVITE see POB500
CARDIS see CCK125
CARDITIN see PEV750
CARDITIVO see RDK000
CARDITOXIN see DKL800
CARDIVIX see EID200
CARDOGENEN-(20:22)-DIOL-(3-β,14) (GERMAN) see DMJ000
CARDOGENEN-(20:22)-TRIOL-(3-β,12,14) (GERMAN) see DKN300
β-CARDONE see CCK250
CARDOPHYLIN see TEP500
CARDOPHYLLIN see TEP500
CARDOVERINA see PAH250
CARDOXIN see PCP250
CARENA see TEP500
3-CARENE see CCK500
S-3-CARENE see CCK500
Δ³-CARENE see CCK500
CARFECILLIN see CBN750
CARFECILLIN SODIUM see CBO000
CARFENE see ASH500
CARFIMAT see PGE000
CARFIMATE see PGE000
CARFLOC D 1000 see CNH125
CARIAQUILLO (PUERTO RICO) see LAU600
CARICIDE see DIW000, DIW200
CARIDOROL see ICC000
CARINA see PKQ059
CARINEX GP see SMQ500
CARINEX HR see SMQ500
CARINEX HRM see SMQ500
CARINEX SB 59 see SMQ500
CARINEX SB 61 see SMQ500
CARINEX SL 273 see SMQ500
CARINEX TGX/MF see SMQ500
CARIOMIN see TEP500
CARIOTA (CUBA) see FBW100
CARISOL see IPU000
CARISOMA see IPU000
CARISOPRODATE see IPU000
CARISOPRODATUM see IPU000
CARISOPRODOL see IPU000
CARITROL see DIW200
CARLONA 900 see PJS750
CARLONA 58-030 see PJS750
CARLONA 18020 FA see PJS750
CARLONA P see PMP500
CARLONA PXB see PJS750
CARLSODAL see IPU000
CARLSOMA see IPU000
CARLYTENE see TFY000
CARMAZINE see DXI400
CARMAZON see POB500
CARMETHOSE see SFO500
CARMINAPH see PEJ500
CARMIN BLUE VS see ADE500
CARMINE BLUE AF see ERG100
CARMINE BLUE (BIOLOGICAL STAIN) see FAE100
CARMINE BLUE V see CMM062

CARMINE BLUE VF see ADE500
CARMINOMICIN I see CCK625
CARMINOMYCIN see KBU000
CARMINOMYCIN HYDROCHLORIDE see KCA000
CARMINOMYCIN I see CCK625
CARMIN (PUERTO RICO) see ROA300
CARMOFUR see CCK630
CARMOISIN (GERMAN) see HJF500
CARMOISINE ALUMINUM LAKE see HJF500
CARMOISINE SUPRA see HJF500
CARMOL HC see HHQ800
CARMUBRIS see BIF750
CARMUSTIN see BIF750
CARMUSTINE see BIF750
CARNACID-COR see GEW000
CARNATION RED TONER B see NAY000
CARNELIO HELIO RED see MMP100
CARNELIO ORANGE G see CMS145
CARNELIO RUBINE LAKE see SEH475
CARNELIO YELLOW GX see DEU000
dl-CARNITINE CHLORIDE see CCK655
l-CARNITINE CHLORIDE see CCK660
CARNITINE CHLORIDE see CCK650
(±)-CARNITINE CHLORIDE see CCK655
(R)-CARNITINE HYDROCHLORIDE see CCK660
d,l-CARNITINE HYDROCHLORIDE see CCK655
dl-CARNITINE HYDROCHLORIDE see CCK655
l-CARNITINE HYDROCHLORIDE see CCK660
(±)-CARNITINE HYDROCHLORIDE see CCK655
l-CARNOSINE see CCK665
CARNOSINE see CCK665
CAROB BEAN GUM see LIA000
CAROB FLOUR see LIA000
CAROFAM see EID200
CAROID see PAG500
CAROLINA ALLSPICE see CCK675
CAROLINA JASMINE see YAK100
CAROLINA PINK see PIH800
CAROLINA TEA see HGF100
CAROLINA WILD WOODBINE see YAK100
CAROLINA YELLOW JASMINE see YAK100
CAROLYSINE see BIE500
CARPENE see DXX400
CARPHENOL see PFR325
CARPIPRAMINE DIHYDROCHLORIDE see CCK775
CARPIPRAMINE DIHYDROCHLORIDE MONOHYDRATE see CCK780
CARPIPRAMINE HYDROCHLORIDE see CCK775
CARPIPRAMINE MALEATE see CCK790
CARPOLENE see ADW200
CARPOLIN see CBM750
CARPROFEN see CCK800
CARQUEJOL see CCL109
kappa-CARRAGEEN see CCL350
CARRAGEEN see CCL250
CARRAGEENAN, CALCIUM(II) SALT see CAO250
CARRAGEENAN, DEGRADED see CCL500
CARRAGEENAN (FCC) see CCL250
kappa-CARRAGEENAN see CCL350
CARRAGEENAN GUM see CCL250
CARRAGEENAN, SODIUM SALT see SFP000
kappa-CARRAGEENIN see CCL350
CARRAGHEANIN see CCL250
CARRAGHEEN see CCL250
CARRAGHEENAN see CCL250
CARREL-DAKIN SOLUTION see SHU500
CARRIOMYCIN, SODIUM SALT see SFP500
CARROT SEED OIL see CCL750
CARRSERP see RDK000
CARRTIME see BBK500
CARSIL see SDX625
CARSODOL see IPU000
CARSONOL SLS see SIB600
CARSONON N-9 see PKF000
CARSONON PEG-4000 see PJT750
CARSOQUAT SDQ-25 see DTC600
CARSORON see DER800
CARSTAB 700 see HND100
CARSTAB DLTDP see TFD500
CARTAGYL see ARQ750
CARTAP HYDROCHLORIDE see BHL750
CARTEOLOL see CCL800
CARTEOLOL HYDROCHLORIDE see MQT550
CARTOSE see GFG000
CARTWHEELS see AOB250
CARUBICIN see CCK625

CARUSIS P see CAT775
CARVACROL see CCM000
CARVACRON see HII500
CARVANIL see CCK125
CARVASEPT see CEX275
CARVASIN see CCK125
1-CARVEOL see MKY250
CARVIL see MOV000
4-CARVOMENTHENOL see TBD825
3-CARVOMENTHENONE see MCF250
1-CARVONE see CCM120
(R)-CARVONE see CCM120
(S)-(+)-CARVONE see CCM100
(S)-CARVONE see CCM100
d(+)-CARVONE see CCM100
d-CARVONE see CCM100
l(−)-CARVONE see CCM120
(+)-CARVONE see CCM100
CARVONE see MCD250
(−)-CARVONE see CCM120
1-CARVYL ACETATE see CCM750
l-CARVYL PROPIONATE see MCD000
CARYLDERM see CBM750
CARYNE see CEW500
CARYOLYSIN see BIE250
CARYOLYSINE see BIE500
CARYOLYSINE HYDROCHLORIDE see BIE500
β-CARYOPHYLLENE (FCC) see CCN000
CARYOPHYLLENE see CCN000
CARYOPHYLLENE ACETATE see CCN050
β-CARYOPHYLLENE EPOXIDE see CCN100
CARYOPHYLLENE EPOXIDE see CCN100
β-CARYOPHYLLENE OXIDE see CCN100
CARYOPHYLLENE OXIDE see CCN100
(−)-CARYOPHYLLENE OXIDE see CCN100
CARYOPHYLLIC ACID see EQR500
CARYOTA (VARIOUS SPECIES) see FBW100
CARZAZO (DOMINICAN REPUBLIC) see CAK325
CARZENID see SNL840
CARZENIDE see SNL840
CARZINOCIDIN see CCN250
CARZINOPHILIN see CCN500
CARZINOPHILIN A see CCN750
CARZOL see CJJ250
CARZOL SP see DSO200
CARZONAL see FLZ050, FMM000
CASALIS GREEN see CMJ900
CASANTIN see DHF600, DII200
CASATE see GCU050
CASATE SODIUM see GCU050
CASCABELILLO (PUERTO RICO) see RBZ400
CASCAMITE see UTU500
CASCARA see MBU825
CASCARITA see CMV390
CASCO 5H see UTU500
CASCO PR 335 see UTU500
CASCO RESIN see UTU500
CASCO UL 30 see UTU500
CASCO WS 114-79 see UTU500
CASCO WS 138-43 see UTU500
CASCO WS 138-44 see UTU500
CASEIN and CASEINATE SALTS (FCC) see SFQ000
CASEIN-SODIUM see SFQ000
CASEINS, SODIUM COMPLEXES see SFQ000
CASEIN, SODIUM COMPLEX see SFQ000
CASHES see PJJ300
CASHMILON, combustion products see ADX750
CASIFLUX VP 413-004 see WCJ000
CASIMAN (MEXICO, PUERTO RICO) see SLE890
CASORON 133 see DER800
CASPAN see MDD750
CASSAINE HYDROCHLORIDE see CCO675
CASSAPPRET SR see PKF750
CASSAVA see CCO680
CASSE (HAITI) see GIW300
CASSEL BROWN see MAT500
CASSEL GREEN see MAT250
CASSELLA 532 see OJD300
CASSELLA 4489 see CBR500
CASSENA see HGF100
CASSIA ALDEHYDE see CMP969
CASSIA FISTULA see GIW300
CASSIA OCCIDENTALIS see CNG825
CASSIA OIL see CCO750
CASSIAR AK see ARM268

CASSIA TORA Linn., leaf extract see CCO800
CASSURIT HML see MCB050
CASSURIT LR see DTG000
CASSURIT MLP see MCB050
CASSURIT MLS see MCB050
CASSURIT MT see MCB050
CASTANEA SATIVA MILL TANNIN see CDM250
CASTOR BEAN see CCP000
CASTOR BEANS (DOT) see CCP000
CASTOR FLAKE (DOT) see CCP000
CASTOR MEAL (DOT) see CCP000
CASTOR OIL see CCP250
CASTOR OIL AROMATIC see CCP250
CASTOR OIL PLANT see CCP000
CASTOR POMACE (DOT) see CCP000
CASTRIX see CCP500
CASTRON see PDN000
CASWELL NO. 481G see CDT250
CAT (herbicide) see BJP000
CATACIDE see DIW000
CATALASE from MICROCOCCUS LYSODEIKTICUS see CCP525
CATALIN CAO-3 see BFW750
CATALOID see SCH000
CATALYTIC CRACKED CLARIFIED OIL see CMU890
CATALYTIC-DEWAXED HEAVY NAPHTHENIC DISTILLATE see MQV865
CATALYTIC-DEWAXED HEAVY PARAFFINIC DISTILLATE see MQV868
CATALYTIC-DEWAXED LIGHT NAPHTHENIC DISTILLATE see MQV867
CATALYTIC-DEWAXED LIGHT PARAFFINIC DISTILLATE see MQV870
CATAMINE AB see AFP250
CATANAC SP see CCP675
CATANAC SP ANTISTATIC AGENT see CCP675
CATANIL see CKK000
CATAPRES see CMX760
CATAPRESAN see CMX760
CATAPYRIN see AFW500
CAT CRACKED CLARIFIED OIL-DECANTED OIL see CMU890
d-CATECHIN see CCP875
d-(+)-CATECHIN see CCP875
CATECHIN see CCP850, CCP875
(+)-CATECHIN see CCP875
CATECHIN (FLAVAN) see CCP875
CATECHIN HYDRATE see DMD000
CATECHINIC ACID see CCP875
d-CATECHOL see CCP875
CATECHOL see CCP850
(+)-CATECHOL see CCP875
CATECHOL see CCP875
CATECHOL DIETHYL ETHER see CCP900
CATECHOL (FLAVAN) see CCP875
CATECHUIC ACID see CCP875
CATENULIN see NCF500
CATERGEN see CCP875
CATESBY'S VINE (BAHAMAS) see YAK300
CATEUDYL see QAK000
CAT-FLOC see DTS500
CATHILON PINK FGH see CMM768
CATHINE see NNM510
CATHOCIN see SMB000
CATHOMYCIN see SMB000
CATHOMYCIN SODIUM see NOB000
CATHOMYCIN SODIUM LYOVAC see NOB000
CATIGENE T80 see QAT520
CATILAN see CDP250
CATION AB see TLW500
CATIONIC ORANGE ZH see CMM764
CATIONIC PINK 2S see CMM768
CATIONIC ROSE 2S see CMM768
CATIONIC SP see CCP675
CATOLIN 14 see MJO500
CATOVITAN see PNS000
CATRAL see PDN000
CATRAN see PDN000
CATRON HYDROCHLORIDE see PDN250
CATRONIACID see PDN250
CATRONIAZIDE see PDN000
C. ATROX VENOM see WBJ600
CAT'S BLOOD see ROA300
CATS EYES see PAM780
CAULOPHYLLUM THALICTROIDES, glycoside extract see CCQ125
CAULOPHYLLUM THALICTROIDES see BMA150
CAURITE see DTG700
CAUSOIN see DKQ000
CAUSTIC BARLEY see VHZ000
CAUSTIC POTASH, dry, solid, flake, bead, or granular (DOT) see PLJ500
CAUSTIC POTASH, liquid or solution (DOT) see PLJ500

CAUSTIC POTASH see PLJ500
CAUSTIC SODA, bead (DOT) see SHS000
CAUSTIC SODA, dry (DOT) see SHS000
CAUSTIC SODA, flake (DOT) see SHS000
CAUSTIC SODA, granular (DOT) see SHS000
CAUSTIC SODA, liquid (DOT) see SHS000
CAUSTIC SODA, solid (DOT) see SHS000
CAUSTIC SODA, solution (DOT) see SHS000
CAUSTIC SODA see SHS000
CAUSTIC SODA, solution see SHS500
CAUTIVA (PUERTO RICO) see AFQ625
CAVALITE BRILLIANT BLUE R see BMM500
CAV-ECOL see WBJ700
CAVINTON see EGM100
CAVI-TROL see SHF500
CAVODIL see PDN000
CAVONLY see TDA500
CAVUMBREN see BGB315
CAYENNE PEPPER see PCB275
4-CB see TBO700
CB-154 see BNB250, BNB325
CB 304 see THM750
CB-337 see HMC000
804 CB see CPJ000
CB 1331 see PCU425
CB 1348 see CDO500
CB 1356 see BHY625
CB-1385 see AJE000
CB 1506 see CHC750
1522 CB see ABH500
1613-CB see BTA125
CB 1639 see AJK250
1664 CB see AAF800
1678 CB see IDA500, PMX500
CB 1689 see CQA000
CB 1729 see BHT250
C.B. 2041 see BOT250
CB2058 see DNX200
CB 2095 see DNX000
CB 2511 see BKM500
CB 2562 see TFU500
CB 3008 see BHV000
CB 3025 see BHV250, PED750
3026 C.B. see SAX200
CB-3026 see SAX200
CB-3307 see BHT750
CB 4261 see CFG750
CB 4306 see CDQ250
4311 CB see DKV700
4361 CB see CFG750
CB-4564 see CQC500
CB 4564 see CQC650
CB-4835 see BIA250
CB 8000 see DEQ200
CB 8019 see IPU000
8065 C.B. see MPN000
8089 CB see FGU000
8089 C.B. see FLK100
CB 10286 see CCC325
C 34647Ba see BAC275
C-39089-Ba see THK750
C 49249Ba see EEH575
2-CBA see CEL250
CBBP see THY500
CBC 806495 see TFQ750
CBC 900139 see SBE500
CBC 906288 see TND250
CB 804 CALCIUM see CPJ250
CBD 90 see TIQ750
CBD see CBD599
CBDCA see CCC075
C12BET see LBU200
C16BET see CDF450
C. BICOLOR see CAL125
CBN see CBD625, CEW500
C. BONDUC see CAK325
CBPC see CBO250
CBS-1114 see PEK675
CBS see CPI250
CB 1348 SODIUM SALT see CDO625
CBSP see HAQ600
(CBZ)GLY see CBR125
CC 914 see CBI250
CC-1065 see APT375
CC 11511 see DYC800

CCA see LHZ600
"C" CARRIE see CNE750
CCC see CAQ250
CCC G-WHITE see CAT775
CCC No. AA OOLITIC see CAT775
CCCP see CKA550
CCC PLANT GROWTH REGULANT see CMF400
C. CERASTES VENOM see CCX620
CCH see HOV500
CCHO see CPD000
C 7337 CIBA see PDW400
CCK 179 see DLL400
CCL see PAG075
CCN52 see RLF350
C.C. No. 914 see CBI250
CCNU see CGV250
CCP see CKA550
CCR see CAT775
CCRG 81010 see MQY110
CCS 203 see BPW500
CCS 301 see BPW750
CCS see CJT750
CCUCOL see ASB250
CCW see CAT775
CD 2 see LFK000
CD 68 see CDR750
CD-3400 see MKR100
CD see TGD000
CDA 101 see CNI000
CDA 102 see CNI000
CDA 110 see CNI000
CDA 122 see CNI000
CD 15006 A see BOS100
CDAA see CFK000
CDAAT see CFK000
3',4'-Cl2-DAB see DFD400
CDA: CETYLCIDE see EKN500
CDB 60 see DGN200
CDB 63 see SGG500
CDBAC see BEL900
CDBM see CFK500
CDC see CDL325
CDCA see CDL325
CDDP see PJD000
CDEC see CDO250
2-CDF see PBT100
C. DIURNUM see DAC500
CDM see CJJ250
CDNA see RDP300
CDP see LFK000
CDP-CHOLIN see CMF350
CDP-CHOLINE see CMF350
CDP-COLINA see CMF350
67/20CDRI see CCW750
C. DRUMMONDII see CAK325
CDT see BJP000, COW935
CDTA see CPB120
114 C.E. see HNM000
264CE see AFS625
305CE see FDA875
746 CE see ECU550
CE 3624 see TGA600
C-14919 E-2 see MAB400
CE see TEG250
CEBESINE see OPI300
CEBETOX see MIW250
CEBOLLA see WBS850
CEBOLLEJA (MEXICO) see GJU460
CEBROGEN see GFO050
CEBRUM see MDQ250
CECALGINE TBV see SEH000
CECARBON see CBT500
CE CE CE see CMF400
CECENU see CGV250
CECIL see CNE750
CECLOR see CCR850
CECOLENE see TIO750
CEDAD see BCA000
CEDAR LEAF OIL see CCQ500
CEDARWOOD OIL ATLAS see CCQ750
CEDARWOOD OIL MOROCCAN see CCQ750
CEDARWOOD OIL (VIRGINIA) see CCR000
CEDILANID see LAU000
CEDOCARD see CCK125
CEDRAMBER see CCR525

CEDRANE, 8,9-EPOXIDE see CCR510
8-β-H-CEDRAN-8-OL ACETATE see CCR250
CEDRANYL ACETATE see CCR250
CEDR-8-ENE see CCR500
α-CEDRENE see CCR500
CEDR-8-ENE EPOXIDE see CCR510
CEDROL FORMATE see CCR524
CEDROL METHYL ETHER see CCR525
CEDRO OIL see LEI000
CEDRUS ATLANTICA OIL see CCQ750
CEDRYL ACETATE see CCR250
CEDRYL FORMATE see CCR524
CEE see PMB000
CEE DEE see HCQ500
CEENU see CGV250
CEEPRYN see CCX000, CDF750
CEEPRYN CHLORIDE see CCX000
CEFACETRILE SODIUM see SGB500
CEFACIDAL see CCS250
CEFACLOR see CCR850, PAG075
CEFACLOR HYDRATE see CCR850
CEFADOL see CCR875
CEFADYL see HMK000
CEFA-ISKIA see ALV000
CEFALOGLYCIN see CCR890
CEFALOJECT see HMK000
CEFALORIDIN see TEY000
CEFALORIDINE see TEY000
CEFALORIZIN see TEY000
CEFALOTHINE SODIUM see SFQ500
CEFALOTIN see CCX250
CEFALOTINA SODICA (SPANISH) see SFQ500
CEFALOTO see ALV000
CEFAMANDOLE NAFATE see FOD000
CEFAMANDOLE SODIUM see CCR925
CEFAMANDOL NAFATO see FOD000
CEFAMEDIN see CCS250
CEFAMEZIN see CCS250
CEFAPIRIN (GERMAN) see CCX500
CEFAPIRIN SODIUM see HMK000
CEFAPRIN SODIUM see HMK000
CEFATIN see OAV000
CEFATOXIME SODIUM see CCR950
CEFATREXYL see HMK000
CEFATRIZINE see APT250
CEFAZEDONE SODIUM SALT see RCK000
CEFAZIL see CCS250
CEFAZINA see CCS250
CEFAZOLIN see CCS250
CEFAZOLINE SODIUM see CCS250
CEFAZOLIN SODIUM SALT see CCS250
CEFBUPERAZONE SODIUM see TAA400
CEFEDRIN see CCX600
CEFLORIN see TEY000
CEFMENOXIME HEMIHYDROCHLORIDE see CCS300
CEFMETAZOLE see CCS350
CEFMETAZOLE SODIUM see CCS360
CEFMINOX see CCS365
CEFOPERAZONE SODIUM see CCS369
CEFOTAN see CCS371
CEFOTAXIME SODIUM see CCR950
CEFOTETAN see CCS373
CEFOTETAN DISODIUM SALT see CCS371
CEFOTIAM DIHYDROCHLORIDE see CCS375
CEFOTIAM HYDROCHLORIDE see CCS375
CEFOXITIN see CCS500
CEFOXITIN SODIUM SALT see CCS510
CEFOXOTIN SODIUM see CCS510
CEFPIMIZOLE SODIUM see CCS525
CEFRACYCLINE SUSPENSION see TBX000
CEFRACYCLINE TABLETS see TBX250
CEFRADINE see SBN440
CEFROXADIN see CCS530
CEFROXADIN DIHYDRATE see CCS535
CEFROXADINE see CCS530
CEFSULODIN SODIUM see CCS550
CEFTEZOLE SODIUM see CCS560
CEFTIZOXIME SODIUM see EBE100
CEFTIZOXIM-NATRIUM (GERMAN) see EBE100
CEFUROXIM see CCS600
CEFUROXIME see CCS600
CEFUROXIME AXETIL see CCS625
CEFUROXIME SODIUM see SFQ300
CEFUROXIME SODIUM SALT see SFQ300
CEFZONAME SODIUM see CCS635

CEGLUTION see LGZ000
CEKIURON see DXQ500
CEKUBARYL see CBM750
CEKU C.B. see HCC500
CEKUDAZIM see MHC750
CEKUDIFOL see BIO750
CEKUFON see TIQ250
CEKUGIB see GEM000
CEKUMETA see TDW500
CEKUMETHION see MNH000
CEKUQUAT see PAJ000
CEKUSAN see BJP000, DGP900
CEKUSIL see ABU500
CEKUSIL UNIVERSAL A see MEO750
CEKUSIL UNIVERSAL C see MEP250
CEKUTHOATE see DSP400
CEKUTROTHION see DSQ000
CEKUZINA-S see BJP000
CEKUZINA-T see ARQ725
CELA 50 see TKL100
CELA S-2225 see EGV500
CELA S-2957 see CLJ875
CELA A-36 see DAE600
CELACOL M20 see MIF760
CELACOL M450 see MIF760
CELACOL M see MIF760
CELACOL MM see MIF760
CELACOL MM 10P see MIF760
CELACOL M 20P see MIF760
CELAMERCK S-2957 see CLJ875
CELANAR see PKF750
CELANDINE see CCS650
CELANEX see BBQ500
CELANOL DOS 75 see DJL000
CELANTHRENE BRILLIANT BLUE see MGG250
CELANTHRENE FAST BLUE 2G see DMM400
CELANTHRENE FAST PINK 3B see DBX000
CELANTHRENE PURE BLUE BRS see TBG700
CELANTHRENE RED 3BN see AKE250
CELANTHRENE RED VIOLET R see DBP000
CELA S 1942 see BNL250
CELASTROL-METHYLETHER see PMD525
CELASTRUS SCANDENS see AHJ875
CELATHION see CLJ875
CELATOX see TNX375
CELA W 524 see TKL100
CELERY OIL see OGL100
CELERY SEED OIL see OGL100
CELESTAN-DEPOT see CCS675
CELESTODERM see VCA000
CELESTONE see BFV750
CELESTONE CHRONODOSE see CCS675
CELESTONE SOLOSPAN see CCS675
CELESTONE SOLUSPAN see CCS675
CELEX see CCU250
CELGARD 2500 see PMP500
CELINHOL -A see OAV000
CELIOMYCIN see VQZ000
CELIPROLOL HYDROCHLORID (GERMAN) see SBN475
CELIPROLOL HYDROCHLORIDE see SBN475
CELITE see DCJ800
CELLAPRET see MIF760
CELLATIVE see AAE500
CELLEX MX see CCU150
CELLIDRIN see ZVJ000
CELLITAZOL B see DCJ200
CELLITON BLUE FFR see MGG250
CELLITON BLUE G see TBG700
CELLITON BLUE GREEN B see DMM400
CELLITON BLUE RN see IBV050
CELLITON BRILLIANT YELLOW 8G see MEB750
CELLITON DISCHARGE YELLOW GL see AAQ250
CELLITON DISCHARGE YELLOW 5RL see CMP090
CELLITON DISCHARGING RUBINE BL see CMP080
CELLITON FAST BLUE GREEN B see DMM400
CELLITON FAST BLUE GREEN BA-CF see DMM400
CELLITON FAST PINK BA-CF see AKE250
CELLITON FAST PINK BN see AKE250
CELLITON FAST PINK FF3B see DBX000
CELLITON FAST PINK FF3BA-CF see DBX000
CELLITON FAST PINK RF see AKO350
CELLITON FAST PINK RFA-CF see AKO350
CELLITON FAST RED VIOLET see DBP000
CELLITON FAST RUBINE B see CMP080
CELLITON FAST RUBINE BA-CF see CMP080

CELLITON FAST VIOLET 6B see AKP250
CELLITON FAST VIOLET B see DBY700
CELLITON FAST VIOLET 6BA-CF see AKP250
CELLITON FAST VIOLET BA-CF see DBY700
CELLITON FAST YELLOW G see AAQ250
CELLITON FAST YELLOW GA see AAQ250
CELLITON FAST YELLOW GA-CF see AAQ250
CELLITON FAST YELLOW 5R see CMP090
CELLITON FAST YELLOW RR see DUW500
CELLITON ORANGE R see AKP750
CELLITON ROSE FF3B see DBX000
CELLITON RUBINE B see CMP080
CELLITON RUBY B see CMP080
CELLITON VIOLET B see DBY700
CELLITON YELLOW G see AAQ250
CELLITON YELLOW 5R see CMP090
CELLMIC S see OPE000
CELLOCIDIN see ACJ250
CELLOFAS see SFO500
CELLOFOR (CZECH) see DWO800
CELLOGEL C see SFO500
CELLOGRAN see MIF760
CELLOIDIN see CCU250
CELLON see TBQ100
CELLOPHANE see CCT250
CELLOSIZE 4400H16 see HKQ100
CELLOSIZE QP3 see HKQ100
CELLOSIZE QP 1500 see HKQ100
CELLOSIZE QP 4400 see HKQ100
CELLOSIZE QP 30000 see HKQ100
CELLOSIZE QP see HKQ100
CELLOSIZE UT 40 see HKQ100
CELLOSIZE WP 300 see HKQ100
CELLOSIZE WP 4400 see HKQ100
CELLOSIZE WP see HKQ100
CELLOSIZE WP 300H see HKQ100
CELLOSIZE WP 400H see HKQ100
CELLOSIZE WPO 9H17 see HKQ100
CELLOSOLVE (DOT) see EES350
CELLOSOLVE ACETATE (DOT) see EES400
CELLOSOLVE ACRYLATE see ADT500
CELLOSOLVE SOLVENT see EES350
CELLOTHYL see MIF760
CELLPRO see SFO500
CELLRYL see CCT825
CELLU-BRITE see DXB450
CELLUFIX FF 100 see SFO500
CELLUFLEX see CGO500
CELLUFLEX 179C see TNP500
CELLUFLEX DOP see DVL600
CELLUFLEX DPB see DEH200
CELLUFLEX FR-2 see TNG750
CELLUFLEX TPP see TMT750
CELLUGEL see SFO500
CELLULASE AP3 see CCT900
CELLULOSE 248 see CCU150
CELLULOSE (ACGIH,OSHA) see CCU150
α-CELLULOSE see CCU150
CELLULOSE CRYSTALLINE see CCU150
CELLULOSE ETHYL see EHG100
CELLULOSE ETHYLATE see EHG100
CELLULOSE GEL see CCU100
CELLULOSE GLYCOLIC ACID, SODIUM SALT see SFO500
CELLULOSE GUM see SFO500
CELLULOSE HYDROXYETHYLATE see HKQ100
CELLULOSE, 2-HYDROXYETHYL ETHER see HKQ100
CELLULOSE HYDROXYETHYL ETHER see HKQ100
CELLULOSE METHYL see MIF760
CELLULOSE METHYLATE see MIF760
CELLULOSE, MICROCRYSTALLINE see CCU100
CELLULOSE NITRATE see CCU250
CELLULOSE, NITRATE (9CI) see CCU250
CELLULOSE, POWDERED see CCU150
CELLULOSE SODIUM GLYCOLATE see SFO500
CELLULOSE TETRANITRATE see CCU250
CELLUMETH see MIF760
CELLUPHOS 4 see TIA250
CELLU-QUIN see BLC250
CELLUTATE RED VIOLET RH see DBP000
CELMER see ABU500, MEP250
CELMIDE see EIY500
CELMIDOL see CCR875
CELMONE see NAK500
CELOCURINE see BJI000
CELOGEN BSH see BBS300

CELOGEN OT see OPE000
CELON A see EIX000
CELON ATH see EIX000
CELON E see EIV000
CELON H see EIV000
CELON IS see EIV000
CELONTIN see MLP800
CELOSPOR see SGB500
CELPHIDE see AHE750
CELPHOS see AHE750, PGY000
CELTHIGN see MAK700
CELUFI see CCU150
CELUTATE BLUE BLT see MGG250
CELUTATE GREEN BLUE BGH see DMM400
CELUTATE PINK B see AKE250
CELUTATE PINK BN see AKE250
CELUTATE PINK BY see AKE250
CELUTATE YELLOW GH see AAQ250
CEMENT (rubber) see CCW250
CEMENT BLACK see MAS000
CEMENT, PORTLAND see PKS750
CEMENT, RUBBER see CCW250
CEMIDON see ILD000
CEMULSOL 1050 see PJY100
CEMULSOL D-8 see PJY100
CEMULSOL A see PJY100
CEMULSOL C 105 see PJY100
CENESTIL see PMS825
CENITRON OB see OPE000
CENOL GARDEN DUST see RNZ000
CENOMYCIN see CCS510
CENSTIM see DLH630
CENSTIN see DLH600, DLH630
CENTBUCRIDINE HYDROCHLORIDE see CCW375
CENTBUTINDOLE see CCW500
CENTCHROMAN HYDROCHLORIDE see CCW750
CENTEDEIN see MNQ000
CENTEDRIN see RLK000
CENTELASE see ARN500
CENTIMIDE see HCQ500
CENTPHENAQUIN see CCW800
CENTRALGIN see DAM700
CENTRALINE BLUE 3B see CMO250
CENTRALITE II see DRB200
CENTRAX see DAP700
CENTREDIN see MNQ000
CENTRINE see DOY400
CENTROFENOXINA see DPE000
CENTROPHENOXINE see AAE500
CENTRUROIDES SUFFUSUS SUFFUSUS VENOM see CCW925
CENTURINA see AOD000
CENTURY 1240 see SLK000
CENTURY CD FATTY ACID see OHU000
CENTYL see BEQ625
CEP see CDS125
2-CEPA see CDS125
CEPA CABALLERO (CUBA) see MQW525
CEPACILINA see BFC750
CEPACILLINA see BFC750
CEPACOL see CDF750
CEPACOL CHLORIDE see CCX000
CEPALORIDIN see TEY000
CEPALORIN see TEY000
CEPAVERIN see PAH250
CEPH 87/4 see TEY000
CEPHA see CDS125
CEPHACETRILE SODIUM see SGB500
(−)-CEPHAELINE DIHYDROCHLORIDE see CCX125
CEPHAELINE HYDROCHLORIDE see CCX125
CEPHAELINE METHYL ETHER see EAL500
CEPHALEXIN see ALV000
CEPHALOGLYCIN see CCR890
d-CEPHALOGLYCINE see CCR890
CEPHALOGLYCINE see CCR890
CEPHALOMYCIN see CCX175
CEPHALORIDIN see TEY000
CEPHALORIDINE see TEY000
(3(R))-CEPHALOTAXINE-4-METHYL-2-HYDROXY-2-(4-HYDROXY-4-METHYLPEN-TYL)BUTANEDIOATE (ESTER) see HGI575
CEPHALOTHIN see CCX250
CEPHALOTHIN SODIUM see HMK000, SFQ500
CEPHALOTIN see CCX250
CEPHA 10LS see CDS125
CEPHAMANDOLE NAFATE see FOD000
CEPHAMYCIN see CCS365

CEPHAOGLYCIN ACID see CCR890
CEPHAPIRIN see CCX500
CEPHARANTHIN see CCX550
CEPHARANTHINE see CCX550
CEPHATREXYL see HMK000
CEPHEDRINE see CCX600
CEPHOXITIN see CCS500
CEPHRADIN see SBN440
CEPHRADINE see SBN440
CEPHROL see CMT250, DTF410
CEPHUROXIME see CCS600
CEPO see CCU150
CEPO CFM see CCU150
CEPORAN see TEY000
CEPOREX see ALV000
CEPOREXIN see ALV000
CEPOREXINE see ALV000
CEPORINE see TEY000
CEPO S 20 see CCU150
CEPO S 40 see CCU150
CEPOVENIN see SFQ500
CEPRIM see CCX000
CEQUARTYL see BBA500
CERAMIC FIBRE see AHF500
CERAPHYL 230 see DNL800
CERAPHYL 368 see OFE100
CERAPHYL 375 see ISC550
CERASINE YELLOW GG see DOT300
CERASIN RED see OHA000
CERASINROT see OHA000
CERASTES CERASTES VENOM see CCX620
CERASYNT see HKJ000
CERASYNT 1000-D see OAV000
CERASYNT PA see SLL000
CERASYNT PN see SLL000
CERASYNT S see OAV000
CERASYNT SD see OAV000
CERASYNT SE see OAV000
CERASYNT WM see OAV000
CERAZOL (suspension) see TEX250
CERBERIGENIN see DMJ000
CERBEROSID (GERMAN) see CCX625
CERBEROSIDE see CCX625
CERBROSIDE see CCX625
CER-o-CILLIN see AGK250
CERCINE see DCK759
CERCOBIN see DJV000
CERCOBIN METHYL see PEX500
CEREB see CMF350
CEREBON see DPE000
CEREBROFORTE see NNE400
CEREDON see BDD000
CERELINE see BDD000
CERELOSE see GFG000
CERENOX see BDD000
CEREPAP see DNA200
CEREPAX see CFY750
CERESAN see ABU500, CHC500
CERESAN M see EME500
CERESAN UNIVERSAL see ABU500
CERESAN UNIVERSAL-FEUCHTBEIZE see BKS810
CERESAN-UNIVERSAL NASSBEIZE see MEP250
CERESAN UNIVERSAL NAZBEIZE see MEP250
CERESOL see ABU500
CERES ORANGE G see CMP600
CERES ORANGE GN see CMP600
CERES ORANGE R see PEJ500
CERES ORANGE RR see XRA000
CERESPAN see PAH250
CERES RED 7B see EOJ500
CERES RED BB see SBC500
CERES RED G 102 see CMS242
CERES RED G see CMS242
CERES YELLOW R see PEI000
CEREWET see BKS810
CEREXIN A see CCX725
CEREZA (SPANISH) see AQP890
CERFA 114 see HNM000
CERIA see CCY000
CERIC DIOXIDE see CCY000
CERIC OXIDE see CCY000
CERIMAN see SLE890
CERIMAN de MEJICO (CUBA) see SLE890
CERISE B see MAC250
CERISE TONER X1127 see FAG070

CERISOL SCARLET G see XRA000
CERISOL YELLOW AB see FAG130
CERISOL YELLOW GR see CMP600
CERISOL YELLOW TB see FAG135
CERIT FAC 3 see HOG000
CERIUM see CCY250
CERIUM ACETATE see CCY500
CERIUM AZIDE see CCY699
CERIUM(III) CHLORIDE see CCY750
CERIUM CHLORIDE see CCY750
CERIUM(III) CITRATE see CCZ000
CERIUM CITRATE see CCZ000
CERIUM COMPOUNDS see CDA250
CERIUM DIOXIDE see CCY000
CERIUM EDETATE see CDA500
CERIUM FLUORIDE see CDA750
CERIUM FLUORURE (FRENCH) see CDA750
CERIUM(3+) NITRATE see CDB000
CERIUM(III) NITRATE see CDB000
CERIUM NITRATE see CDB000
CERIUM(III) NITRATE, HEXAHYDRATE (1:3:6) see CDB250
CERIUM NITRATE, HEXAHYDRATE see CDB250
CERIUM NITRIDE see CDB325
CERIUM(4+) OXIDE see CCY000
CERIUM(III) TETRAHYDROALUMINATE see CDB500
CERIUM TRIACETATE see CCY500
CERIUM TRICHLORIDE see CCY750
CERIUM TRIFLUORIDE see CDA750
CERIUM TRINITRATE see CDB000
CERIUM TRINITRATE HEXAHYDRATE see CDB250
CERIUM TRISULFIDE see DEK600
CERM-1766 see IRP000
CERM-1841 see TKJ500
3024 CERM see MFG250
CERM 10,137 see TGJ885
CERNILTON see CDB760
CERNITIN GBX see CDB770
CERNITIN T-60 see CDB772
CERN KYPOVA 27 see DUP100
CERN KYPOVA 8 see CMU475
CERN KYSELA 1 see FAB830
CERN OSTAZINOVA H-N (CZECH) see DGN600
CERN PRIMA 17 see CMN230
CERN PRIMA 19 see CMN240
CERN PRIMA 38 see AQP000
CERN REAKTIVNI 8 see CMS227
CEROTINE PONCEAU 3B see SBC500
CEROTINORANGE G see PEJ500
CEROTINSCHARLACH G see XRA000
CEROTINSCHARLACH R see OHA000
CEROUS ACETATE see CCY500
CEROUS CHLORIDE see CCY750
CEROUS CITRATE see CCZ000
CEROUS FLUORIDE see CDA750
CEROUS NITRATE see CDB000
CEROUS NITRATE HEXAHYDRATE see CDB250
CEROXIN GL see HOG000
CEROXONE see BJK750
CERTICOL BLACK PNW see BMA000
CERTICOL CARMOISINE S see HJF500
CERTICOL PONCEAU MXS see FMU070
CERTICOL PONCEAU 4RS see FMU080
CERTICOL PONCEAU SXS see FAG050
CERTICOL RED B see CMS231
CERTICOL SUNSET YELLOW CFS see FAG150
CERTICOL TARTRAZOL YELLOW S see FAG140
CERTINAL see ALT250
CERTIQUAL ALIZARINE see DMG800
CERTIQUAL EOSINE see BNH500, BNK700
CERTIQUAL FLUORESCEINE see FEW000
CERTIQUAL OIL RED see OHA000
CERTIQUAL ORANGE I see FAG010
CERTIQUAL ORANGE II see CMM220
CERTIQUAL RHODAMIEN see FAG070
CERTOL see DNF400
CERTOLAKE SUNSET YELLOW see FAG150
CERTOMYCIN see NCP550
CERTOX see SMN500
CERTROL see HKB500
CERUBIDIN see DAC000
CERUBIDINE see DAC200
CERULENIN see ECE500
CERULIGNOL see MFM750
CERUSSETE see LCP000
CERUTIL see AAE500

CERVAGEM see CDB775
CERVEN BRILANTNI OSTACETOVA F-LB (CZECH) see AKI750
CERVEN BRILANTNI OSTAZINOVA S-5B (CZECH) see DGN800
CERVEN BRILANTNI OSTAZINOVA H-3B (CZECH) see CHG250
CERVEN DISPERZNI 11 see DBX000
CERVEN DISPERZNI 15 see AKE250
CERVEN DISPERZNI 4 see AKO350
CERVEN DISPERZNI 60 see AKI750
CERVEN 2G see CMM300
CERVEN KOSENILOVA A see FMU080
CERVEN KUMIDINOVA see FAG018
CERVEN KYPOVA 13 see CMU825
CERVEN KYSELA 114 see CMM330
CERVEN KYSELA 1 see CMM300
CERVEN KYSELA 26 see FMU070
CERVEN KYSELA 27 see FAG020
CERVEN PIGMENT 3 see MMP100
CERVEN PIGMENT 53 see CMS150
CERVEN PIGMENT 57 see CMS155
CERVEN POTRAVINARSKA 10 see CMM300
CERVEN POTRAVINARSKA 1 see FAG050
CERVEN POTRAVINARSKA 2 see CMP620
CERVEN POTRAVINARSKA 9 see FAG020
CERVEN ROZPOUSTEDLOVA 23 see OHA000
CERVEN ROZPOUSTEDLOVA 24 see SBC500
CERVEN ZASADITA 1 see RGW000
CERVEN ZASADITA 2 see GJI400
CERVICUNDIN see EQJ500
CERVOLIDE see OKW110
CERVOXAN see DOZ000
CES see ECU750, SOP500
CESALIN see CDB800
CESIUM-133 see CDC000
CESIUM see CDC000
CESIUM ACETYLIDE see CDC125
CESIUM ARSENATE see CDC375
CESIUM BROMIDE see CDC500
CESIUM BROMOXENATE see CDC699
CESIUM CARBONATE see CDC750
CESIUM CHLORIDE see CDD000
CESIUM CYANOTRIDECAHYDRODECABORATE (2-) see CDD325
CESIUM FLUORIDE see CDD500
CESIUM GRAPHITE see CDD625
CESIUM HYDRATE see CDD750
CESIUM HYDROXIDE (ACGIH, OSHA) see CDD750
CESIUM HYDROXIDE see CDD750
CESIUM HYDROXIDE DIMER see CDD750
CESIUM IODIDE see CDE000
CESIUM LITHIUM TRIDECAHYDRONONABORATE see CDE125
CESIUM MONOCHLORIDE see CDD000
CESIUM MONOFLUORIDE see CDD500
CESIUM MONOIODIDE see CDE000
CESIUM(I) NITRATE (1:1) see CDE250
CESIUM NITRATE (DOT) see CDE250
CESIUM NITRIDE see TIH750
CESIUM OXIDE see CDE325
CESIUM PENTACARBONYLVANADATE (3-) see CDE400
CESIUM SELENIDE see DEJ400
CESIUM SULFATE see CDE500
CESOL see BGB400
CESTRUM (VARIOUS SPECIES) see DAC500
CET see BJP000, CCX250
CETAB see HCQ500
CETACORT see CNS750
CETADOL see HIM000
CETAFFINE see HCP000
CETAIN see AIT250
CETAL see HCP000
CETALKONIUM CHLORIDE see BEL900
CETALOL CA see HCP000
CETAMIUM see CCX000
n-CETANE see HCO600
CETANE see HCO600
CETAPHARM see HCP800
CETARIN see MKR250
CETAROL see HCQ500
CETASOL see HCP800
CETAVLON see HCQ500
CETAZOL see HCP800
CETETH 1 see PJT300
CETETH 2 see PJT300
CETETH see PJT300
CETHYLOSE see MIF760
CETHYTIN see MIF760
CETIL LIGHT RED GG see CMM300

CETIN see HCP700
CETOBEMIDON see KFK000
CETOBEMIDONE see KFK000
CETOCIRE see PJT300
CETOCYLINE see CDF250
CETOL see BEL900
CETOMACROGOL 1000 see PJT300
CETONAL see TDO255
CETRAMIN see CNH125
CETRAXATE see CDF375
CETRAXATE HYDROCHLORIDE see CDF380
CETRIMIDE see HCQ500
CETRIMONIUM BROMIDE see HCQ500
CETYL ACETATE see HCP100
CETYL ALCOHOL see HCP000
CETYL ALCOHOL ETHOXYLATE see PJT300
CETYLAMIN (GERMAN) see HCO500
CETYLAMINE see HCO500, HCQ500
CETYLAMINE-HF see CDF400
CETYLAMINE HYDROFLUORIDE see CDF400
CETYLAMINHYDROFLUORID (GERMAN) see CDF400
CETYL-γ-AMINOBUTYRATE see AJC500
CETYL BETAINE see CDF450
CETYLDIETHYLETHYLAMMONIUM BROMIDE see CDF500
CETYLDIMETHYLAMINE see HCP525
CETYL DIMETHYL ETHYL AMMONIUM BROMIDE see EKN500
CETYL ETHYL DIMETHYLAMMONIUM BROMIDE see EKN500
CETYL 2-ETHYLHEXANOATE see HCP550
CETYL GABA see AJC500
CETYLIC ACID see PAE250
CETYLIC ALCOHOL see HCP000
CETYLOL see HCP000
CETYLON see BEL900
CETYL PALMITATE see HCP700
CETYL POLY(OXYETHYLENE) ETHER see PJT300
1-CETYLPYRIDINIUM BROMIDE see HCP800
N-CETYLPYRIDINIUM BROMIDE see HCP800
CETYLPYRIDINIUM BROMIDE see HCP800
1-CETYLPYRIDINIUM CHLORIDE see CCX000
N-CETYLPYRIDINIUM CHLORIDE see CCX000
CETYLPYRIDINIUM CHLORIDE see CCX000
CETYLPYRIDINIUM CHLORIDE MONOHYDRATE see CDF750
CETYL SODIUM SULFATE see HCP900
CETYL SULFATE SODIUM SALT see HCP900
CETYLTRIETHYLAMMONIUM BROMIDE see CDF500
N-CETYLTRIMETHYLAMMONIUM BROMIDE see HCQ500
CETYLTRIMETHYLAMMONIUM BROMIDE see HCQ500
CETYLUREUM see PEC250
CETYL ZEPHIRAN see BEL900
CEVADENE see CDG000
CEVADIC ACID see TGA700
CEVADILLA see VHZ000
CEVADIN see CDG000
CEVADINE see CDG000, VHZ000
CEVANE-3-β,4-β,7-α,14,15-α,16-β,20-HEPTOL,4,9-EPOXY-, 15-((+)-2-HYDROXY-2-METHYLBUTYRATE) 3-((−)-2-METHYLBUTYRATE) see VHF000
CEVANOL see BCA000
CEVIAN A 678 see AAX250
CEVIAN HL see ADY500
CEVIN see EBL000
CEVINE see EBL000
CEVITAMIC ACID see ARN000
CEVITAMIN see ARN000
CEX see ALV000
CEYLON ISINGLASS see AEX250
CEZ SODIUM see CCS250
CF 8 see CBT500
CF 125 see CDT000
CFC 22 see CFX500
CFC 31 see CHI900
CFC 112a see TBP000
CFC-112 see TBP050
CFC 133a see TJY175
CFC 142b see CFX250
CFC see PGE000
CF 8 (CARBON) see CBT500
C. FERTILIS see CCK675
C. FLORIDUS see CCK675
CFNU see CPL750
C. FRUTESCENS see PCB275
C.F.S. see TEM000
CFT 1201 see AGR125
CFV see CDS750
CFX see CCS500
CG 113 see PMB850

CG 117 see MDM100
CG-120 see SFX725
CG 201 see CDG250
CG 315 see THJ500
CG 601 see MRU080
CG-1283 see MQW500
CG 3117 see MFA500
CGA 10832 see CQG250
CGA-12223 see PHK000
CGA 15324 see BNA750
CGA-18762 see PMF600
CGA-24705 see MQQ450
CGA 26351 see CDS750
CGA 26423 see PMB850
CGA-43089 see COP700
CGA 48988 see MDM100
CGA 89317 see CFL200
CGA 92194 see OKS200
CG 10213 GO see SAY950
C. GIGANTEA see COD675
C. GILLIESII see CAK325
CGP 7174E see CCS550
CGP 2175 see MQR144
CGP 9000 see CCS530
CGP-11305A see BMM625
CGP-9000 DIHYDRATE see CCS535
C-GREEN 10 see BLK000
CGS 10787B see CDG300
CGT see COX400
CGTA see CPB120
CH 02 see HGM100
CH 800 see CKI750
CH 3565 see TIQ000
CH see SBW500
CHA 02 see HGM100
CH-777-A see NCP875
CHA see CPF500
α-CHACONINE see CDG500
CHAETOGLOBOSIN A see CDG750
CHALCEDONY see SCI500, SCJ500
CHALCONE see CDH000
CHALCONE, 2-CHLORO-(CI,7CI,8CI) see CLF150
CHALCONE, 4-CHLORO-(6CI,7CI,8CI) see CLF100
CHALCONE, 4-METHYL-(6CI,7CI,8CI) see MIF762
CHALCONE, 4-NITRO- see NFS505
CHALICE VINE see CDH125
CHALK see CAO000
CHALOTHANE see HAG500
CHALOXYD MEKP-HA 1 see MKA500
CHALOXYD MEKP-LA 1 see MKA500
CHAMAZULEN see DRV000
CHAMAZULENE see DRV000
CHAMBER CRYSTALS see NMJ000
CHAMELEON MINERAL see PLP000
CHAMICO BEJUCO (CUBA) see CDH125
CHAMOMILE see CDH250
CHAMOMILE-GERMAN OIL see CDH500
CHAMOMILE OIL see CDH500
CHAMOMILE OIL (ROMAN) see CDH750
CHANGALA see CNT350
CHANNEL BLACK see CBT750
CHANNING'S SOLUTION see NCP500
CHAPCO Cu-NAP see NAS000
CHARAS see CBD750
CHARCOAL see CDI250
CHARCOAL, ACTIVATED (DOT) see CDI000
CHARCOAL (BRIQUETTES) see CDI250
CHARCOAL SCREENINGS (DOT) see CDI250
CHARCOAL, SHELL (DOT) see CDJ000
CHARCOAL (SHELL) see CDJ000
CHARCOAL WOOD (DOT) see CDI250
CHARGER E see GHS000
CHARTREUSIN see CDK250
CHARTREUSIN, SODIUM SALT see CDK500
CHAULMOOGRIC ACID, SODIUM SALT see SFR000
CHAVICOL METHYL ETHER see AFW750
1,3-CHBP see BNA825
CHEBUTAN see KGK000
CHEELOX BF see EIV000
CHEELOX BF ACID see EIX000
CHEELOX BR-33 see EIV000
CHEL 138 see EIV100
CHEL 300 see AMT500
CHEL 330 see DJG800
CHEL 600 see CPB120

CHEL 330 ACID see DJG800
CHELADRATE see EIX500
CHELAFER see FBC100
CHELAFRIN see VGP000
CHELAPLEX III see EIX500
CHELATON III see EIX500
CHEL DTPA see DJG800
CHELEN see EHH000
CHELIDONINE see CDL000
CHELIDONIUM MAJUS L. see CCS650
CHEL-IRON see FBC100
CHELLIN see AHK750
CHELLINA (ITALIAN) see AHK750
β-CHELOCARDIN see CDF250
CHELON 100 see EIV000
CHEMAGRO 1,776 see BSH250
CHEMAGRO 2353 see DFV400
CHEMAGRO 5461 see BJD000
CHEMAGRO 25141 see FAQ800
CHEMAGRO 37289 see EPY000
CHEMAGRO D-113 see DFS200
CHEMAGRO B-1776 see BSH250, TIG250
CHEMAGRO B-1843 see BLG500
CHEMAGRO B-9002 see HMV000
CHEMAGRO R-5461 see BJD000
CHEMAID see HKC500
CHEMANOX 11 see BFW750
CHEMANOX 21 see MJO500
CHEMANOX 22 see MJN250
CHEMATHION see MAK700
CHEMAX NP SERIES see NND500
CHEM BAM see DXD200
CHEMCARB see CAD800, CAT775
CHEMCOCCIDE see RLK890
CHEMCOLOX 200 see EIV000
CHEMCOLOX 340 see EIX000
CHEMCOR see PJS750
CHEM DM ACID see HKS000
CHEMESTER 300-OC see PJY100
CHEMETRON FIRE SHIELD see AQF000
CHEM FISH see RNZ000
CHEMFORM see DKC800, DMC600, MEI450, SLW500
CHEM-FROST see SFS500
CHEM-HOE see CBM000
CHEMIAZID see ILD000
CHEMICAL 109 see AQN635
CHEMICAL MACE see CEA750
CHEMICETIN see CDP250
CHEMICETINA see CDP250
CHEMI-CHARL see SHM500
CHEMICTIVE BRILLIANT RED 5B see PMF540
CHEMIFLUOR see SHF500
CHEMIOCHIN see CFU750
CHEMIOFURAN see NGE000
CHEMIPEN see PDD350
CHEMIPEN-C see PDD350
CHEMITHRENE BROWN BR see CMU770
CHEMITRIM see TKX000
CHEMLINK 160 see PER250
CHEMLON see PJY500
CHEM-MITE see RNZ000
CHEM NEB see MAS500
CHEMOCCIDE see RLK890
CHEMOCHIN see CLD000
CHEMOCIN see CNK559
CHEMOFURAN see NGE500
CHEMOSAN see MGC350
CHEMOSEPT see TEX250
CHEMOTHERAPY CENTER No. 606 see CBI500
CHEMOUAG see SNN500
CHEMOX GENERAL see BRE500
CHEMOX P.E. see BRE500
CHEMOX PE see DUZ000
CHEMOX SELECTIVE see BPG250
CHEMPAR see CNK559
CHEM PELS C see SEY500
CHEM-PHENE see CDV100
CHEMPLEX 3006 see PJS750
CHEMRAT see PIH175
CHEM RICE see DGI000
CHEMSECT DNOC see DUS700
CHEM-SEN 56 see SEY500
CHEM-TOL see PAX250
CHEMYSONE see HHQ800
CHEM ZINEB see EIR000

CHENDAL see CDL325
CHENDOL see CDL325
CHENIC ACID see CDL325
CHENIX see CDL325
CHENOCEDON see CDL325
CHENODEOXYCHOLIC ACID see CDL325
CHENODEOXYCHOLIC ACID SODIUM SALT see CDL375
CHENODESOXYCHOLIC ACID see CDL325
CHENODESOXYCHOLIC ACID SODIUM SALT see CDL375
CHENODESOXYCHOLSAEURE (GERMAN) see CDL325
CHENODEX see CDL325
CHENODIOL see CDL325
CHENOFALK see CDL325
CHENOPODIUM AMBROSIOIDES see SAF000
CHENOPODIUM OIL see CDL500
CHENOSAURE see CDL325
CHENOSSIL see CDL325
CHEPIROL see KGK000
CHEQUE see MQS225
CHERRY see AQP890
CHERRY BARK OAK see CDL750
CHERRY LAUREL OIL see CDM000
CHERRY PEPPER see PCB275
CHERTS see SCI500, SCJ500
CHESTNUT COMPOUND see CNJ750
CHESTNUT TANNIN see CDM250
CHETAZOLIDIN see KGK000
CHETIL see KGK000
17-CHETOVIS see AOO450
CHEVRON 9006 see DTQ400
CHEVRON ACETONE see ABC750
CHEVRON ORTHO 9006 see DTQ400
CHEVRON RE5305 see BSG250
CHEVRON RE 5655 see MOU750
CHEVRON RE 12,420 see DOP600
CHEWING TOBACCO see SED400
CHEXMATE see HKC000
CHFB see CHK750
CHICAGO ACID S see AKH750
CHICAGO BLUE RW see CMO600
CHICLIDA see HGC500
1,4-CHIDM see BKH325
CHILDREN'S BANE see WAT325
CHILE PEPPER see PCB275
CHILE SALTPETER see SIO900
CHIMASSORB 81 see HND100
CHIMCOCCIDE see RLK890
CHIMIPAL AE 3 see DXY000
CHIMOREPTIN see DLH630
CHINABERRY see CDM325
CHINACRIN HYDROCHLORIDE see CFU750
CHINA GREEN (BIOLOGICAL STAIN) see AFG500
CHINALDINE see QEJ000
CHINALPHOS see DJY200
CHINA TREE see CDM325
CHINAWOOD OIL see TOA510
CHINAWOOD OIL TREE see TOA275
CHINE APE see EAI600
CHINESE INKBERRY see DAC500
CHINESE ISINGLASS see AEX250
CHINESE LANTERN PLANT see JBS100
CHINESE RED see LCS000
CHINESE SEASONING see MRL500
CHINESE WHITE see ZKA000
CHINGAMIN see CLD000, CLD250
CHINIDIN (GERMAN) see QFS000
CHINIDIN DURULES see QFS100
CHINIDINE SULFATE see QHA000
CHINIDIN VUFB see QFS100
CHININ (GERMAN) see QHJ000
CHININDIHYDROCHLORID (GERMAN) see QIJ000
CHININ HYDROBROMID (GERMAN) see QJJ100
CHINIOFON see IEP200
CHINOFER see IGS000
CHINOFORM see CHR500
CHINOFUNGIN see TGB475
CHINOGELB see CMM510
CHINOGELB EXTRA see CMM510
CHINOGELB WASSERLOESLICH see CMM510
CHINOIN 103 see CPG500
CHINOIN-127 see CDM500
CHINOIN-170 see CDM575
CHINOIN see EID000
CHINOLEINE see QMJ000
CHINOLIN see QMJ000

CHINOLINE see QMJ000
CHINOLINE YELLOW D SOL. IN SPIRITS see CMS245
CHINOLINE YELLOW ZSS see CMS245
CHINOMETHIONATE see ORU000
p-CHINON (GERMAN) see QQS200
CHINON (DUTCH, GERMAN) see QQS200
CHINONE see QQS200
CHINON I (GERMAN) see BGW750
CHINONOXIM-BENZOYLHYDRAZON (GERMAN) see BDD000
CHINONOXIME-BENZOYLHYDRAZONE see BDD000
CHINORTA see NIM500
CHINOTILIN see KGK400
CHINOXONE see OPK300
3-CHINUCLIDYLBENZILATE see QVA000
CHIP see IGG775
CHIP-CAL see ARB750
CHIP-CAL GRANULAR see ARB750
CHIPCO 26019 see GIA000
CHIPCO BUCTRIL see DDP000
CHIPCO CRAB-KLEEN see DDP000
CHIPCO CRAB KLEEN see DXE600
CHIPCO THIRAM 75 see TFS350
CHIPCO TURF HERBICIDE "D" see DAA800
CHIPCO TURF HERBICIDE MCPP see CIR500
CHIPMAN 3,142 see TBR750
CHIPMAN 6199 see AMX825
CHIPMAN 6200 see DJA400
CHIPMAN 11974 see BDJ250
CHIPMAN R-6, 199 see AMX825
CHIPTOX see CIR250
CHIRAL BINAPHTHOL see CDM625
CHIRONEX FLECKERI TOXIN see CDM700
CHISSO 507B see PMP500
CHISSONOX 201 see ECB000
CHISSONOX 206 see VOA000
CHITA ROOT EXTRACT see PJH615
CHITRAKA ROOT EXTRACT see PJH615
CHKHZ 18 see DVF400
CHLODITAN see CDN000
CHLODITHANE see CDN000
CHLOFENVINPHOS see CDS750
CHLOFEXAMIDE see CJN750
CHLOFUCID see DEL300
CHLOMAPHENE see CMX500
CHLOMIN see CDP250
CHLOMYCOL see CDP250
CHLONIXIN see CMX770
CHLOOR (DUTCH) see CDV750
3-CHLOORANILINEN (DUTCH) see CEH675
2-CHLOORBENZALDEHYDE (DUTCH) see CEI500
o-CHLOORBENZALDEHYDE (DUTCH) see CEI500
CHLOORBENZEEN (DUTCH) see CEJ125
CHLOORBENZIDE (DUTCH) see CEP000
(4-CHLOOR-BENZYL)-(4-CHLOOR-FENYL)-SULFIDE (DUTCH) see CEP000
2-CHLOOR-1,3-BUTADIEEN (DUTCH) see NCI500
(4-CHLOOR-BUT-2-YN-YL)-N-(3-CHLOOR-FENYL)-CARBAMAAT (DUTCH) see
   CEW500
CHLOORDAAN (DUTCH) see CDR750
O-2-CHLOOR-1-(2,4-DICHLOOR-FENYL)-VINYL-O,O-DIETHYLFOSFAAT (DUTCH)
   see CDS750
(2-CHLOOR-3-DIETHYLAMINO-1-METHYL-3-OXO-PROP-1-EN-YL)-DIMETHYL-
   FOSFAAT see FAB400
2-CHLOOR-4-DIMETHYLAMINO-6-METHYL-PYRIMIDINE (DUTCH) see CCP500
1-CHLOOR-2,4-DINITROBENZEEN (DUTCH) see CGM000
1-CHLOOR-2,3-EPOXY-PROPAAN (DUTCH) see EAZ500
CHLOORETHAAN (DUTCH) see EHH000
2-CHLOORETHANOL (DUTCH) see EIU800
CHLOORFACINON (DUTCH) see CJJ000
3-((4-(4-CHLOOR-FENOXY)-FENOXY)-FENYL)-1,1-DIMETHYLUREUM (DUTCH)
   see CJQ000
CHLOORFENSON (DUTCH) see CJT750
(4-CHLOOR-FENYL)-BENZEEN-SULFONAAT (DUTCH) see CJR500
(4-CHLOOR-FENYL)-4-CHLOOR-BENZEEN-SULFONAAT (DUTCH) see CJT750
3-(4-CHLOOR-FENYL)-1,1-DIMETHYLUREUM (DUTCH) see CJX750
2(2-(4-CHLOOR-FENYL-2-FENYL)-ACETYL)-INDAAN-1,3-DION (DUTCH) see
   CJJ000
N-(3-CHLOOR-FENYL)-ISOPROPYL CARBAMAAT (DUTCH) see CKC000
CHLOOR-HEXAVIET see HEA500
CHLOOR-METHAAN (DUTCH) see MIF765
2-(4-CHLOOR-2-METHYL-FENOXY)-PROPIONZUUR (DUTCH) see CIR500
1-CHLOOR-4-NITROBENZEEN (DUTCH) see NFS525
O-(4-CHLOOR-3-NITRO-FENYL)-O,O-DIMETHYLMONOTHIOFOSFAAT (DUTCH)
   see NFT000
O-(3-CHLOOR-4-NITRO-FENYL)-O,O-DIMETHYL-MONOTHIOFOSFAAT (DUTCH)
   see MIJ250

CHLOORPIKRINE (DUTCH) see CKN500
CHLOORTHION (DUTCH) see MIJ250
CHLOORWATERSTOF (DUTCH) see HHL000
CHLOPHEDIANOL HYDROCHLORIDE see CMW700
CHLOPHEN see PJL750
CHLOR (GERMAN) see CDV750
CHLORACETAMID (GERMAN) see CDY850
α-CHLORACETESSIGSAEUREAETHYLAMID see CEA100
CHLORACETIC ACID see CEA000
CHLORACETONE see CDN200
CHLORACETONITRILE see CDN500
CHLORACETOPHENONE see CEA750
CHLORACETYL CHLORIDE see CEC250
5-CHLOR-2-ACETYL THIOPHEN see CDN525
CHLORACON see BEG000
CHLORACTIL see CKP500
2-CHLORAETHANOL (GERMAN) see EIU800
α-CHLOR-6'-AETHYL-n-(2-METHOXY-1-METHYLAETHYL)-ACET-o-TOLUIDIN
   (GERMAN) see MQQ450
N-(2-CHLORAETHYL)-N'-(2 CHLOROETHYL)-N'-o-PROPYLEN-PHOSPHORSAEU-
   REESTER-DIAMID (GERMAN) see IMH000
2-CHLORAETHYL-PHOSPHONSAEURE (GERMAN) see CDS125
2-CHLORAETHYL-TRIMETHYLAMMONIUMCHLORID see CMF400
CHLORAK see TIQ250
CHLORAKON see BEG000
3-CHLORAKRYLAN SODNY see SFV250
CHLORAL, anhydrous, inhibited (DOT) see CDN550
CHLORAL see CDN550
CHLORAL ALCOHOLATE see TIO000
CHLORALDEHYDE see DEM200
CHLORALDURAT see CDO000
CHLORAL ETHYLALCOHOLATE see TIO000
CHLORAL, ETHYL HEMIACETAL see TIO000
CHLORAL HYDRATE see CDO000
2-CHLORALLYL DIETHYLDITHIOCARBAMATE see CDO250
CHLORALLYL DIETHYLDITHIOCARBAMATE see CDO250
CHLORALLYLENE see AGB250
CHLORALONE see CDP000
CHLORALOSANE see GFA000
α-CHLORALOSE see GFA000
CHLORAMBEN see AJM000
CHLORAMBUCIL see CDO500
CHLORAMBUCIL SODIUM SALT see CDO625
CHLORAMEISENSAEUREAETHYLESTER (GERMAN) see EHK500
CHLORAMEISENSAEURE METHYLESTER (GERMAN) see MIG000
CHLORAMEX see CDP250
CHLORAMFICIN see CDP250
CHLORAMFILIN see CDP250
CHLORAMIDE see CDO750
CHLORAMIFENE see CMX500
CHLORAMIN see BIE500
CHLORAMINE (inorganic compound) see CDO750
CHLORAMINE see BIE500, CDO750
CHLORAMINE B see SFV275
CHLORAMINE BLACK BH see CMN800
CHLORAMINE BLACK C see AQP000
CHLORAMINE BLACK EC see AQP000
CHLORAMINE BLACK ERT see AQP000
CHLORAMINE BLACK EX see AQP000
CHLORAMINE BLACK EXR see AQP000
CHLORAMINE BLACK SD see CMN230
CHLORAMINE BLACK XO see AQP000
CHLORAMINE BLUE see CMO250
CHLORAMINE BLUE 2B see CMO000
CHLORAMINE BRILLIANT RED 8B see CMO880
CHLORAMINE CARBON BLACK S see AQP000
CHLORAMINE CARBON BLACK SJ see AQP000
CHLORAMINE CARBON BLACK SN see AQP000
CHLORAMINE FAST BROWN BRL see CMO750
CHLORAMINE FAST RED 5BL see CMO885
CHLORAMINE FAST RED F see CMO870
CHLORAMINE FAST RED FS see CMO870
CHLORAMINE FAST RED K see CMO885
CHLORAMINE GREEN 2B see CMO840
CHLORAMINE GREEN B see CMO840
CHLORAMINE GREEN BC see CMO840
CHLORAMINE GREEN 3G see CMO840
CHLORAMINE RED 3B see CMO875
CHLORAMINE RED 8B see CMO880
CHLORAMINE SKY BLUE A see CMO500
CHLORAMINE SKY BLUE 4B see CMO500
CHLORAMINE T see CDP000
CHLORAMINE-T see SFV550
CHLORAMIN HYDROCHLORIDE see BIE500
1-CHLOR-5-AMINOANTHRACHINON (CZECH) see AJE325

CHLORAMINOPHEN see CDO500
CHLORAMINOPHENE see CDO500
CHLORAMIPHENE see CMX500, CMX700
CHLORAMIPHENE CITRATE see CMX700
CHLORAMP (RUSSIAN) see PIB900
d-CHLORAMPHENICOL see CDP250
d-threo-CHLORAMPHENICOL see CDP250
CHLORAMPHENICOL see CDP250
CHLORAMPHENICOL ACID SUCCINATE see CDP725
CHLORAMPHENICOL HEMISUCCINATE see CDP725
CHLORAMPHENICOL HYDROGEN SUCCINATE see CDP725
CHLORAMPHENICOL MONOPALMITATE see CDP700
CHLORAMPHENICOL MONOSUCCINATE see CDP725
CHLORAMPHENICOL MONOSUCCINATE SODIUM SALT see CDP500
CHLORAMPHENICOL PALMITATE see CDP700
CHLORAMPHENICOL SODIUM MONOSUCCINATE see CDP500
CHLORAMPHENICOL SODIUM SUCCINATE see CDP500
CHLORAMPHENICOL SUCCINATE see CDP725
CHLORAMPHENICOL SUCCINATE SODIUM see CDP500
CHLORAMPHENICOL-SUKZINAT-NATRIUM (GERMAN) see CDP500
CHLORAMSAAR see CDP250
CHLORANAUTINE see DYE600
CHLORANIFORMETHAN see CDP750
CHLORANIFORMETHANE see CDP750
CHLORANIL see TBO500
4-CHLORANILIN (CZECH) see CEH680
2-(2-CHLORANILIN)-4,6-DICHLOR-1,3,5-TRIAZIN (GERMAN) see DEV800
m-CHLORANILINE see CEH675
o-CHLORANILINE see CEH670
p-CHLORANILINE see CEH680
CHLORANOCRYL see DFO800
1-CHLORANTHRACHINON (CZECH) see CEI000
CHLORANTINE FAST RED see CMO885
CHLORANTINE FAST RED 5B (6CI) see CMO885
CHLORAQUINE see CLD000
CHLORARSENOL see CEI250
CHLORARSOL see DFX400
CHLORASAN see CDP000
CHLORASEN see DFX400
CHLORASEPTINE see CDP000
CHLORASOL see CDP250
CHLORA-TABS see CDP250
CHLORATE de CALCIUM (FRENCH) see CAO500
CHLORATE of POTASH (DOT) see PLA250
CHLORATE de POTASSIUM (FRENCH) see PLA250
CHLORATES see CDQ000
CHLORATE SALT of MAGNESIUM see MAE000
CHLORATE SALT of SODIUM see SFS000
CHLORATE of SODA (DOT) see SFS000
2-(3-(2-CHLORATHYL)-3-NITROSOUREIDO)ATHYLMETHANSULFONAT (GER-
     MAN) see CHF250
CHLORAX see SFS000
CHLORAZAN see CDP000
CHLORAZENE see CDP000
CHLORAZEPAM see CDQ250
CHLORAZEPATE DIPOTASSIUM see CDQ250
CHLORAZIN see CKP500
CHLORAZINE see CDQ325
CHLORAZOL BLACK BH see CMN800
CHLORAZOL BLACK E see AQP000
CHLORAZOL BLACK EA see AQP000
CHLORAZOL BLACK E (BIOLOGICAL STAIN) see AQP000
CHLORAZOL BLACK EN see AQP000
CHLORAZOL BLUE 3B see CMO250
CHLORAZOL BLUE B see CMO000
CHLORAZOL BLUE RW see CMO600
CHLORAZOL BRILLIANT PURPURINE 8B see CMO880
CHLORAZOL BROWN LF see CMO820
CHLORAZOL BURL BLACK E see AQP000
CHLORAZOL DARK GREEN PL see CMO830
CHLORAZOL DIAZO BLACK SD see CMN230
CHLORAZOL FAST RED FP see CMO870
CHLORAZOL FAST RED FS see CMO870
CHLORAZOL GREEN BN see CMO840
CHLORAZOL GREEN BNP see CMO840
CHLORAZOL LEATHER BLACK BH see CMN800
CHLORAZOL LEATHER BLACK ENP see AQP000
CHLORAZOL ORANGE BROWN X see CMO810
CHLORAZOL PAPER GREEN BN see CMO840
CHLORAZOL SILK BLACK G see AQP000
CHLORAZOL SKY BLUE FF see BGT250
CHLORAZOL VIOLET N see CMP000
CHLORAZOL VISCOSE BLACK B see CMN240
CHLORAZOL YELLOW DP see CMP050
CHLORAZOL YELLOW 2G see CMP050

CHLORAZONE see CDP000
CHLORBENSID (GERMAN) see CEP000
CHLORBENSIDE see CEP000
CHLORBENXIDE see CEP000
2-CHLORBENZALDEHYD (GERMAN) see CEI500
CHLORBENZENE see CEJ125
p-CHLORBENZENESULFOCHLORID (CZECH) see CEK375
p-CHLORBENZENSULFONAN SODNY (CZECH) see CEK250
1-p-CHLORBENZHYDRYL-m-METHYLBENZYLPIPERAZINE DIHYDROCHLORIDE
     see MBX250
CHLORBENZIDE see CEP000
CHLORBENZILATE see DER000
p-CHLORBENZOIC ACID see CEL300
CHLORBENZOL see CEJ125
o-CHLORBENZONITRIL (CZECH) see CEM000
CHLORBENZOSAMINE DIHYDROCHLORIDE see CDQ500
CHLORBENZOXAMINE DIHYDROCHLORIDE see CDQ500
5-CHLORBENZOXAZOLIN-2-ON see CDQ750
CHLORBENZOXYETHAMINE DIHYDROCHLORIDE see CDQ500
1-CHLOR-5-BENZOYLAMINOANTHRACHINON (CZECH) see BDK750
N-(p-CHLORBENZOYL)-γ-(2,6-DIMETHYLANILINO)-BUTTERSAEURE (GERMAN)
     see CLW625
(1-(p-CHLORBENZOYL)-5-METHOXY-2-METHYLINDOL-3-ACE-
     TOXY)ESSIGSAEURE (GERMAN) see AAE625
N-p-CHLORBENZOYL-5-METHOXY-2-METHYLINDOLE-3-ACETIC ACID see
     IDA000
5-CHLORBENZOZAZOLIN-2-ON see CDQ750
(4-CHLOR-BENZYL)-(4-CHLOR-PHENYL)-SULFID (GERMAN) see CEP000
4-CHLOR-BENZYL-CYANID see CEP300
1-p-CHLORBENZYL-2-METHYL-BENZIMIDAZOL (GERMAN) see CDY325
p-CHLORBENZYL-α-PYRIDYL-DIMETHYL-AETHYLENDIAMIN (GERMAN) see
     CKV625
CHLORBICYCLENE (FRENCH) see DAM700
CHLORBROMURON see CES750
CHLORBUFAM see CEX250
CHLORBUFAN mixed with CYCEURON see AFM375
CHLORBUPHAM see CEX250
2-CHLOR-1,3-BUTADIEN (GERMAN) see NCI500
4-CHLORBUTAN-1-OL (GERMAN) see CEU500
CHLORBUTANOL see ABD000
(4-CHLOR-BUT-2-IN-YL)-N-(3-CHLOR-PHENYL)-CARBAMAT (GERMAN) see
     CEW500
CHLORBUTOL see ABD000
CHLORCARVACROL see CEX275
CHLORCHOLINCHLORID see CMF400
CHLORCHOLINE CHLORIDE see CMF400
p-CHLOR-m-CRESOL see CFE250
CHLORCYAN see COO750
CHLORCYCLINE see CFF500
CHLORCYCLIZINE see CFF500
CHLORCYCLIZINE DIHYDROCHLORIDE see CDR000
CHLORCYCLIZINE HYDROCHLORIDE see CDR250
CHLORCYCLIZINE HYDROCHLORIDE A see CDR500
CHLORCYCLIZINIUM CHLORIDE see CDR250
CHLORCYCLOHEXAMIDE see CDR550
4-(3-CHLOR-4-CYCLOHEXYL-PHENYL)-4-OXO-BUTTERSAEURE KALZIUM
     (GERMAN) see CPJ250
6-CHLOR-N-CYCLOPROPYL-N'-(1-METHYLETHYL)-1,3,5-TRIZAINE-2,4-DIAMINE
     see CQI750
α-CHLORDAN see CDR675
γ-CHLORDAN see CDR575, CDR750
cis-CHLORDAN see CDR675
CHLORDAN see CDR750
trans-CHLORDAN see CDR575
α(cis)-CHLORDANE see CDR675
α-CHLORDANE see CDR675
CHLORDANE, liquid (DOT) see CDR750
γ(trans)-CHLORDANE see CDR575
cis-CHLORDANE see CDR675
CHLORDANE see CDR750
CHLORDECONE see KEA000
CHLORDENE see HCN000
7-CHLOR-4-(4-(DIAETHYLAMINO)-1-METHYLBUTYLAMINO)-CHINOLINDIPHOSP-
     HAT (GERMAN) see CLD250
(2-CHLOR-3-DIAETHYLAMINO-1-METHYL-3-OXO-PROP-1-EN-YL)-DIMETHYL-
     PHOSPHAT see FAB400
CHLORDIAZACHEL see MDQ250
CHLORDIAZEPOXIDE see LFK000
CHLORDIAZEPOXIDE HYDROCHLORIDE see MDQ250
CHLORDIAZEPOXIDE MONOHYDROCHLORIDE see MDQ250
CHLORDIAZEPOXIDE, NITROSATED see SIS000
CHLORDIAZEPOXIDE mixed with SODIUM NITRITE (1:1) see SIS000
O-2-CHLOR-1-(2,4-DICHLOR-PHENYL)-VINYL-O,O-DIAETHYLPHOSPHAT (GER-
     MAN) see CDS750

7-CHLOR-2,3-DIHYDRO-1-METHYL-5-PHENYL-1H-1,4-BENZODIAZEPIN HYDRO-CHLORID (GERMAN) see MBY000
CHLORDIMEFORM see CJJ250
CHLORDIMEFORM HYDROCHLORIDE see CJJ500
2-CHLOR-11-(2-DIMETHYAMINOAETHOXY)-DIBENZO(b,f)-THIEPIN (GERMAN) see ZUJ000
2-CHLOR-4-DIMETHYLAMINO-6-METHYLPYRIMIDIN (GERMAN) see CCP500
CHLORDIMETHYLETHER (CZECH) see CIO250
1-CHLOR-2,4-DINITROBENZENE see CGM000
CHLORE (FRENCH) see CDV750
CHLOREFENIZON (FRENCH) see CJT750
CHLORENDIC ACID see CDS000
CHLORENDIC ACID DIBUTYL ESTER see CDS025
CHLORENDIC ANHYDRIDE see CDS050
CHLORENDIC IMIDE see CDS100
CHLOREPIN see CIR750
1-CHLOR-2,3-EPOXY-PROPAN (GERMAN) see EAZ500
CHLORESENE see BBQ500
CHLORESSIGSAEURE-N-ISOBUTINYLANILID (GERMAN) see CDS275
CHLORESSIGSAEURE-N-ISOPROPYLANILID (GERMAN) see CHS500
CHLORESSIGSAEURE-N-(METHOXYMETHYL)-2,6-DIAETHYLANILID (GERMAN) see CFX000
CHLORESTROLO see CLO750
CHLORETHAMINACIL see BIA250
CHLORETHAMINE see BIE500
CHLOR-ETHAMINE see EIW000
2-CHLORETHANOL (GERMAN) see EIU800
CHLORETHAZINE see BIE500
CHLORETHENE see VNP000
CHLORETHEPHON see CDS125
CHLORETHIAZOL see CHD750
2-(2-CHLORETHOXY)ETHYL 2'-CHLORETHYL ETHER see TKL500
CHLORETHYL see EHH000
2-CHLORETHYLACETAT see CGO600
CHLORETHYLBENZMETHOXAZONE see CDS250
CHLORETHYLENE see VNP000
2-CHLORETHYLESTER KYSELINY CHLORMRAVENCI see CGU199
2-CHLORETHYLISOKYANAT see IKH000
2-CHLORETHYLPHOSPHONIC ACID see CDS125
2-CHLORETHYL VINYL ETHER see CHI250
CHLORETIN see CDS275
CHLORETONE see ABD000
CHLOREX see DFJ050
CHLOREXTOL see PJL750
CHLORFACINON (GERMAN) see CJJ000
CHLORFENAC see TIY500
CHLORFENAMIDINE see CJJ250
CHLORFENETHOL see BIN000
CHLORFENIDIM see CJX750
p-CHLORFENOL (CZECH) see CJK750
2-(4'-CHLORFENOXY)ETHANOL (CZECH) see CJO500
CHLORFENPROP-METHYL see CFC750
CHLORFENSON see CJT750
CHLORFENSONE see CJT750
CHLORFENSULFID (GERMAN) see CDS500
CHLORFENSULFIDE see CDS500
CHLORFENVINFOS see CDS750
CHLORFENVINPHOS see CDS750
1-p-CHLORFENYL-3,3-DIMETHYLTRIAZEN (CZECH) see CJI100
3-CHLOR-p-FENYLENDIAMIN (CZECH) see CEG600
p-CHLORFENYLISOKYANAT (CZECH) see CKB000
m-CHLORFENYLISOKYANAT (CZECH) see CKA750
p-CHLORFENYLMERKAPTOMETHYLCHLORID (CZECH) see CFB750
p-CHLORFENYLMONOGLYKOLETHER (CZECH) see CJO500
p-CHLORFENYLSILATRAN (CZECH) see CKM750
CHLORFLURAZOLE see DGO400
CHLORFLURECOL see CDT000
CHLORFLURECOL-METHYL see CDT000
CHLORFLURECOL-METHYL ESTER see CDT000
CHLORFLURENOL see CDT000
CHLORFLURENOL METHYL ESTER see CDT000
CHLORFONIUM see THY500
CHLORFOS see TIQ250
CHLOR-N-(2-FURYLMETHYL)-5-SULFAMYLANTHRANILSAEURE (GERMAN) see CHJ750
CHLORGUANIDE see CKB250
CHLORGUANIDE HYDROCHLORIDE see CKB500
CHLORGUANIDE TRIAZINE see COX400
CHLORHEXAMIDE see CDR550
CHLORHEXIDIN (CZECH) see BIM250
CHLORHEXIDINE see BIM250
CHLORHEXIDINE ACETATE see CDT125
CHLORHEXIDINE DIACETATE see CDT125
CHLORHEXIDINE DIGLUCONATE see CDT250
CHLORHEXIDINE GLUCONATE see CDT250

CHLORHEXIDIN GLUKONATU see CDT250
6-CHLORHEXYLISOKYANAT see CHL250
CHLORHYDRATE de ACETOXY-THYMOXY-ETHYL-DIMETHYLAMINE (FRENCH) see TFY000
CHLORHYDRATE d'AMIKHELLINE (FRENCH) see AHP375
CHLORHYDRATE d'ANILINE (FRENCH) see BBL000
CHLORHYDRATE de 4-CHLOROORTHOTOLUIDINE (FRENCH) see CLK235
CHLORHYDRATE de CONESSINE (FRENCH) see CNH660
CHLORHYDRATE de N-(DIETHOXY-2,5-PHENYL)-N-DIETHYLAMINO-2-ETHYL BUTOXY-4-PHENOXYACETAMIDE (FRENCH) see BPM750
CHLORHYDRATE de DIETHYLAMINOETHYLTHEOPHYLLINE (FRENCH) see DIH600
CHLORHYDRATE de (N-ETHYL,N,β-CHLORETHYL)AMINO-METHYLBENZODI-OXANE (FRENCH) see CGX625
CHLORHYDRATE d'HISTAMINE (FRENCH) see HGE500
CHLORHYDRATE de (NAPHTHYLOXY-1)-4 HYDROXY-3 BUTYRAMIDOXIME (FRENCH) see NAC500
CHLORHYDRATE de NICOTINE (FRENCH) see NDP400
CHLORHYDRATE de PAPAVERINE (FRENCH) see PAH250
CHLORHYDRATE de PHENETHYL-8-OXA-1-DIAZA-3,8-SPIRO(4,5)DECANONE-2 (FRENCH) see DAI200
CHLORHYDRATE de α-PHENYL-α-(β'-DIETHYLAMINOETHYL)GLUTARIMIDE (FRENCH) see ARR875
CHLORHYDRATE de PIPERIDINOMETHYLCYCLOHEXANE (FRENCH) see PIR000
CHLORHYDRATE de (PIPERONYL-4-PIPERAZINO)-1)-(PHENYL-1-PYRROLI-DONE-2-CARBOXAMIDE-4) (FRENCH) see PGA250
CHLORHYDRATE de RAUGALLINE (FRENCH) see AFH280
CHLORHYDRATE de TETRACYCLINE (FRENCH) see TBX250
CHLORHYDRATE de (TRIMETHOXY-2-4-6) PHENYL-(PYRROLIDINE-3) PROPY-LACETONE (FRENCH) see BOM600
α-CHLORHYDRIN see CDT750
CHLORHYDRIN see CDT750
CHLORHYDROL see AHA000
CHLORHYDROXYALUMINUM ALLANTOINATE see AHA135
3-CHLOR-4-HYDROXYBIFENYL (CZECH) see CHN500
2-CHLOR-9-HYDROXYFLUOREN-CARBONSAEURE-(9)-METHYLESTER (GER-MAN) see CDT000
CHLORIAZID see CLH750
CHLORIC ACID, solution, containing not more than 10% acid (DOT) see CDU000
CHLORIC ACID see CDU000
CHLORIC ACID, AMMONIUM SALT see ANE250
CHLORIC ACID, BARIUM SALT see BAJ500
CHLORIC ACID, COPPER SALT see CNJ900
CHLORIC ACID, STRONTIUM SALT see SMF500
CHLORIC ACID, THALLIUM(1+) SALT see TEJ100
CHLORICOL see CDP250
CHLORID AMONNY (CZECH) see ANE500
CHLORID ANILINU (CZECH) see BBL000
CHLORID ANTIMONITY see AQC500
CHLORIDAZON see PEE750
CHLORID-N-BUTYLCINICITY (CZECH) see BSR250
CHLORID CHROMITY HEXAHYDRAT see CMK450
CHLORID DI-n-BUTYLCINICITY (CZECH) see DDY200
CHLORID DRASELNY (CZECH) see PLA500
CHLORIDEAZEPOXIDE HYDROCHLORIDE see MDQ250
CHLORIDE de CHOLINE (FRENCH) see CMF750
CHLORIDE of DIAMINOMETHYLPHENYLDIMETHYL-p-BENZOQUINONE-DIIMINE see DCE800
CHLORIDE of LIME (DOT) see HOV500
CHLORIDE of PHOSPHORUS see PHT275
CHLORIDES see CDU250
CHLORIDE of SULFUR (DOT) see SON510
CHLORIDE of SULFUR see SOG500
CHLORID FENYLRTUTNATY (CZECH) see PFM500
CHLORIDIAZEPIDE see LFK000
CHLORIDIAZEPOXIDE see LFK000
CHLORIDIN see TGD000
CHLORIDINE see TGD000
CHLORID KREMICITY (CZECH) see SCQ500
CHLORID KYSELINY-p-CHLORBENSULFONOVE (CZECH) see CEK375
CHLORID KYSELINY CHLORMETHANSULFONOVE (CZECH) see CHY000
CHLORID KYSELINY CHLOROCTOVE see CEC250
CHLORID KYSELINY DICHLOROCTOVE see DEN400
CHLORID KYSELINY DIMETHYLKARBAMINOVE see DQY950
CHLORID MEDNY (CZECH) see CNK250
CHLORID RTUTNATY (CZECH) see MCY475
CHLORID TRIBENZYLCINICITY (CZECH) see CLP000
CHLORID TRI-n-BUTYLCINICITY (CZECH) see CLP500
CHLORIDUM see EHH000
CHLORIERTE BIPHENYLE, CHLORGEHALT 42% (GERMAN) see PJM500
CHLORIERTE BIPHENYLE, CHLORGEHALT 54% (GERMAN) see PJN000
CHLORIERTES CAMPHEN see CDU325
CHLOR-IFC see CKC000

CHLORIMIPRAMINE see CDU750
CHLORIMIPRAMINE HYDROCHLORIDE see CDV000
CHLORINAT see CEW500
CHLORINATED BIPHENYL see PJL750
CHLORINATED CAMPHENE see CDV100
CHLORINATED DIBENZO DIOXINS see CDV125
CHLORINATED DIPHENYL see PJL750
CHLORINATED DIPHENYLENE see PJL750
CHLORINATED DIPHENYL OXIDE see CDV175
CHLORINATED HC, ALIPHATIC see CDV250
CHLORINATED HC AROMATIC see CDV500
CHLORINATED HYDROCARBONS, ALIPHATIC see CDV250
CHLORINATED HYDROCARBONS, AROMATIC see CDV500
CHLORINATED HYDROCHLORIC ETHER see DFF809
CHLORINATED LIME (DOT) see HOV500
CHLORINATED NAPHTHALENES see CDV575
CHLORINATED PARAFFINS (C23, 43% CHLORINE) see PAH810
CHLORINATED PARAFFINS (C12, 60% CHLORINE) see PAH800
CHLORINATED POLYETHER POLYURETHAN see CDV625
CHLORINDAN see CDR750
CHLORINDANOL see CDV700
CHLORINE see CDV750
CHLORINE AZIDE see CDW000
CHLORINE CYANIDE see COO750
CHLORINE DIOXIDE, not hydrated (DOT) see CDW450
CHLORINE DIOXIDE see CDW450
CHLORINE FLUORIDE see CDX750
CHLORINE FLUORIDE (ClF$_5$) see CDX250
CHLORINE FLUORIDE OXIDE see PCF750
CHLORINE MOL. see CDV750
CHLORINE NITRIDE see NGQ500
CHLORINE(IV) OXIDE see CDW450
CHLORINE OXIDE see CDW450
CHLORINE OXYFLUORIDE see PCF750
CHLORINE PENTAFLUORIDE (DOT) see CDX250
CHLORINE PENTAFLUORIDE see CDX250
CHLORINE PERCHLORATE see CDX500
CHLORINE PEROXIDE see CDW450
CHLORINE SULFIDE see SOG500
CHLORINE TETROXYFLUORIDE see FFD000
CHLORINE TRIFLUORIDE see CDX750
CHLORINE(1)TRIFLUOROMETHANESULFONATE see CDX800
CHLOR-IPC see CKC000
CHLORISONDAMINE see CDY000
CHLORISONDAMINE CHLORIDE see CDY000
CHLORISONDAMINE DIMETHOCHLORIDE see CDY000
CHLORISOPROPAMIDE see CDY100
CHLORITES see CDY250
5-CHLOR-7-JOD-8-8HYDROXY-CHINOLIN (GERMAN) see CHR500
CHLOR KIL see CDR750
CHLORKU LITU (POLISH) see LHB000
CHLORMADINON see CDY275
CHLORMADINON ACETATE see CBF250
CHLORMADINONE see CDY275
CHLORMADINONE ACETATE see CBF250
CHLORMADINONU (POLISH) see CBF250
CHLORMENE see TAI500
CHLORMEPHOS see CDY299
CHLORMEQUAT see CMF400
CHLORMEQUAT CHLORIDE see CMF400
CHLORMEROPRIN see CHX250
CHLOR-METHAN (GERMAN) see MIF765
CHLORMETHANSULFOCHLORID (CZECH) see CHY000
CHLORMETHAZANONE see CKF500
CHLORMETHAZONE see CKF500
CHLORMETHIAZOLE see CHD750
CHLORMETHINE see BIE250
CHLORMETHINE HYDROCHLORIDE see BIE500
CHLORMETHINE-N-OXIDE HYDROCHLORIDE see CFA750
CHLORMETHINUM see BIE500
3-(3-CHLOR-4-METHOXYPHENYL)-1,1-DIMETHYLHARNSTOFF (GERMAN) see
    MQR225
N'-(3-CHLOR-4-METHOXY-PHENYL)-N,N-DIMETHYLHARNSTOFF (GERMAN) see
    MQR225
CHLORMETHYL-METHYL-DIETHOXYSILAN (CZECH) see ClO000
α-(CHLORMETHYL)-2-METHYL-5-NITRO-IMIDAZOL-1-AETHANOL (GERMAN) see
    OJS000
4-(4-CHLOR-2-METHYLPHENOXY)-BUETTERSAEURE (GERMAN) see CLN750
4-(4-CHLOR-2-METHYLPHENOXY)-BUTTERSAEURE (GERMAN) see CLN750
4-(4-CHLOR-2-METHYL-PHENOXY)-BUTTERSAEURE NATRIUMSALZ (GERMAN)
    see CLO000
2-(4-CHLOR-2-METHYL-PHENOXY)-PROPIONSAEURE (GERMAN) see CIR500
3-(3-CHLOR-4-METHYLPHENYL)-1,1-DIMETHYLHARNSTOFF (GERMAN) see
    CIS250

N-(3-CHLOR-METHYLPHENYL)-2-METHYLPENTANAMID (GERMAN) see
    SKQ400
3-CHLOR-2-METHYL-PROP-1-EN (GERMAN) see CIU750
CHLORMETHYL-TRIETHOXYSILAN (CZECH) see CIY500
CHLORMEZANONE see CKF500
CHLORMIDAZOLE see CDY325
CHLORMITE see PNH750
CHLORNAFTINA see BIF250
CHLORNAPHAZIN see BIF250
α-CHLORNAPHTHALENE see CIZ000
CHLORNAPHTHIN see BIF250
1-CHLOR-5-NITROANTHRACHINON (CZECH) see CJA250
1-CHLOR-4-NITROBENZOL (GERMAN) see NFS525
CHLORNITROFEN see NIW500
CHLORNITROMYCIN see CDP250
O-(3-CHLOR-4-NITRO-PHENYL)-O,O-DIMETHYL-MONOTHIOPHOSPHAT (GER-
    MAN) see MIJ250
O-(4-CHLOR-3-NITRO-PHENYL)-O,O-DIMETHYL-MONOTHIOPHOSPHAT (GER-
    MAN) see NFT000
2-CHLOROACETALDEHYDE see CDY500
CHLOROACETALDEHYDE see CDY500
CHLOROACETALDEHYDE MONOMER see CDY500
2-CHLOROACETAMIDE see CDY850
α-CHLOROACETAMIDE see CDY850
N-CHLOROACETAMIDE see CDY825
CHLOROACETAMIDE see CDY850
CHLOROACETAMIDE OXIME see CDZ000
3'-CHLOROACETANILIDE see CDZ050
4'-CHLOROACETANILIDE see CDZ100
m-CHLOROACETANILIDE see CDZ050
α-CHLOROACETIC ACID see CEA000
CHLOROACETIC ACID see CEA000
CHLOROACETIC ACID BENZYL ESTER see BEE500
CHLOROACETIC ACID CHLORIDE see CEC250
CHLOROACETIC ACID, ETHYL ESTER see EHG500
CHLOROACETIC ACID METHYL ESTER see MIF775
CHLORO-ACETIC ACID, PHENETHYL ESTER see PDF500
CHLOROACETIC ACID SODIUM SALT see SFU500
CHLOROACETIC ACID, solid (UN 1751) (DOT) see CEA000
CHLOROACETIC ACID, liquid (UN 1750) (DOT) see CEA000
CHLOROACETIC CHLORIDE see CEC250
2'-CHLOROACETOACETANILIDE see AAY600
4'-CHLOROACETOACETANILIDE see AAY250
4-CHLOROACETOACETANILIDE see CEA050
γ-CHLOROACETOACETANILIDE see CEA050
o-CHLOROACETOACETANILIDE see AAY600
p-CHLOROACETOACETANILIDE see AAY250
γ-CHLOROACETO ACETIC ACID ANILIDE see CEA050
α-CHLOROACETOACETIC ACID MONOETHYLAMIDE see CEA100
CHLOROACETONE, stabilized (DOT) see CDN200
CHLOROACETONE see CDN200
2-CHLOROACETONITRILE see CDN500
α-CHLOROACETONITRILE see CDN500
CHLOROACETONITRILE (DOT) see CDN500
1-CHLOROACETOPHENONE see CEA750
2'-CHLOROACETOPHENONE see CEB000
4-CHLOROACETOPHENONE see CEB250
4'-CHLOROACETOPHENONE see CEB250
α-CHLOROACETOPHENONE see CEA750
CHLOROACETOPHENONE, liquid or solid (DOT) see CEA750
ω-CHLOROACETOPHENONE see CEA750
p-CHLOROACETOPHENONE see CEB250
3'-CHLORO-p-ACETOTOLUIDIDE see CEB750
2-CHLORO-10-(3-(4-(2-ACETOXYETHYL)PIPERAZINYL)PROPYL)PHENOTHIAZINE
    see TFP250
6-CHLORO-17-α-ACETOXY-4,6-PREGNADIENE-3,20-DIONE see CBF250
Δ⁶-6-CHLORO-17-α-ACETOXYPROGESTERONE see CBF250
6-CHLORO-Δ⁶-17-ACETOXYPROGESTERONE see CBF250
6-CHLORO-Δ⁶-(17-α)ACETOXYPROGESTERONE see CBF250
(CHLOROACETOXY)TRIBUTYLSTANNANE see TIC750
4'-(CHLOROACETYL)ACETANILIDE see CEC000
4'-CHLOROACETYL ACETANILIDE see CEC000
N-(CHLOROACETYL)-3-AZABICYCLO(3.2.1)NONANE see CEC100
CHLOROACETYL CHLORIDE see CEC250
N-CHLOROACETYLDIETHYLAMINE see DIX400
1-CHLOROACETYL-α-α-DIPHENYL-4-PIPERIDINEMETHANOL see CEC300
CHLOROACETYLENE see CEC500
2-CHLOROACETYLFLUORENE see CEC700
I-N-(α-(CHLOROACETYL)PHENETHYL)-p-TOLUENESULFONAMIDE see THH450
N-(CHLOROACETYL)-3-PHENYL-N-(p-TOLYLSULFONYL)ALANINE see THH550
5-CHLORO-2-ACETYL THIOPHEN see CDN525
2-CHLOROACRYLIC ACID see CEE500
α-CHLOROACRYLIC ACID see CEE500
CHLOROACRYLIC ACID see CEE500
2-CHLOROACRYLIC ACID ETHYL ESTER see EHH200

2-CHLOROACRYLIC ACID, METHYL ESTER see MIF800
cis-β-CHLOROACRYLIC ACID SODIUM SALT see SFV250
2-CHLOROACRYLONITRILE see CEE750
α-CHLOROACRYLONITRILE see CEE750
CHLOROACRYLONITRILE see CEE750
(3-CHLORO-1-ADAMANTANOYL)DIAZOMETHANE see CEE800
3-CHLOROADAMANTYL DIAZOMETHYL KETONE see CEE800
1-CHLORO-3-ADAMANTYL ETHOXYMETHYL KETONE see CEE825
1-CHLORO-3-ADAMANTYL HYDROXYMETHYL KETONE see CEE850
2-CHLOROADENOSINE see CEF100
2-CHLOROADENOSINE-5'-SULFAMATE see CEF125
CHLOROAETHAN (GERMAN) see EHH000
β-CHLORO ALLYL ALCOHOL see CEF250
pi-CHLORO ALLYL ALCOHOL see CEF500
α-CHLOROALLYL CHLORIDE see DGG950
γ-CHLOROALLYL CHLORIDE see DGG950
2-CHLOROALLYL DIETHYLDITHIOCARBAMATE see CDO250
2-CHLOROALLYL-N,N-DIETHYLDITHIOCARBAMATE see CDO250
CHLOROALLYLENE see AGB250
1-(3-CHLOROALLYL)-3,5,7-TRIAZA-1-AZONIAADAMANTANE CHLORIDE see
   CEG550
CHLOROALONIL see TBQ750
CHLOROALOSANE see GFA000
CHLOROAMBUCIL see CDO500
CHLOROAMINE see CDO750
3-CHLORO-4-AMINOANILINE see CEG600
3-CHLORO-4-AMINOANILINE SULFATE see CEG625
5-CHLORO-1-AMINOANTHRAQUINONE see AJE325
2-CHLORO-4-AMINOBENZOIC ACID see CEG750
4-CHLORO-3-AMINOBENZOTRIFLUORIDE see CEG800
4'-CHLORO-4-AMINOBIPHENYL ETHER see CEH125
3-CHLORO-4-AMINODIPHENYL see CEH000
4-CHLORO-4'-AMINODIPHENYL ETHER see CEH125
2-CHLORO-3-AMINO-1,4-NAPHTHOQUINONE see AJI250
p-CHLORO-o-AMINOPHENOL see CEH250
3-CHLORO-2-AMINOTOLUENE see CLK227
4-CHLORO-2-AMINOTOLUENE see CLK225
5-CHLORO-2-AMINOTOLUENE see CLK220
2-CHLORO-4-AMINOTOLUENE see CLK215
5-CHLORO-2-AMINOTOLUENE HYDROCHLORIDE see CLK235
2-CHLORO-4-AMINOTOLUENE-5-SULFONIC ACID see AJJ250
CHLOROAMITRIPTYLINE HYDROCHLORIDE see CEH500
2-CHLOROANILINE see CEH670
3-CHLOROANILINE see CEH675
4-CHLOROANILINE see CEH680
3-CHLOROANILINE (ITALIAN) see CEH675
m-CHLOROANILINE, liquid see CEH675
o-CHLOROANILINE, liquid see CEH670
p-CHLOROANILINE, liquid see CEH680
m-CHLOROANILINE see CEH675
o-CHLOROANILINE see CEH670
p-CHLOROANILINE see CEH680
m-CHLOROANILINE, solid see CEH675
o-CHLOROANILINE, solid see CEH670
p-CHLOROANILINE, solid see CEH680
4-CHLOROANILINE HYDROCHLORIDE see CJR200
p-CHLOROANILINE HYDROCHLORIDE see CJR200
p-CHLOROANILINIUM CHLORIDE see CJR200
(o-CHLOROANILINO)DICHLOROTRIAZINE see DEV800
1-((p-(2-(CHLORO-o-ANISAMIDO)ETHYL)PHENYL)SULFONYL)-3-CYCLOHEXYL
   UREA see CEH700
3-CHLOROANISIDINE see CEH750
1-CHLORO-9,10-ANTHRACENEDIONE see CEI000
2-CHLORO-9,10-ANTHRACENEDIONE see CEI100
1-CHLOROANTHRAQUINONE see CEI000
2-CHLOROANTHRAQUINONE see CEI100
1-CHLORO-9,10-ANTHRAQUINONE see CEI000
α-CHLOROANTHRAQUINONE see CEI000
CHLOROARSENOL see CEI250
1-CHLOROAZIRIDINE see CEI325
CHLOROBEN see DEP600
CHLOROBENZAL see BAY300
4-CHLOROBENZALCHLORIDE see TJD650
p-CHLOROBENZALCHLORIDE see TJD650
2-CHLOROBENZALDEHYDE see CEI500
4-CHLOROBENZALDEHYDE see CEI600
α-CHLOROBENZALDEHYDE see BDM500
o-CHLOROBENZALDEHYDE see CEI500
p-CHLOROBENZALDEHYDE see CEI600
2-CHLOROBENZAL MALONONITRILE see CEQ600
o-CHLOROBENZAL MALONONITRILE see CEQ600
1-CHLORO-5-BENZAMIDO-ANTHRAQUINONE see BDK750
10-CHLORO-1,2-BENZANTHRACENE see CEJ000
7-CHLOROBENZ(a)ANTHRACENE see CEJ000
CHLOROBENZEN (POLISH) see CEJ125

3-CHLOROBENZENAMINE see CEH675
4-CHLOROBENZENAMINE see CEH680
2-CHLORO-BENZENAMINE (9CI) see CEH670
4-CHLOROBENZENAMINE HYDROCHLORIDE see CJR200
CHLOROBENZENE see CEJ125
4-CHLOROBENZENEACETONITRILE see CEP300
4-CHLOROBENZENEAMINE see CEH680
o-CHLOROBENZENECARBOXALDEHYDE see CEI500
p-CHLOROBENZENECARBOXALDEHYDE see CEI600
4-CHLORO-1,3-BENZENEDIAMINE see CJY120
2-CHLORO-1,4-BENZENEDIAMINE see CEG600
2-CHLORO-1,4-BENZENEDIAMINE SULFATE see CEG625
2-CHLOROBENZENEDIAZONIUM SALTS see CEJ500
m-CHLOROBENZENEDIAZONIUM SALTS see CEJ250
o-CHLOROBENZENEDIAZONIUM SALTS see CEJ500
p-CHLOROBENZENESULFINIC ACID see CEJ600
p-CHLOROBENZENESULFONAMIDE see CEK000
4-CHLOROBENZENESULFONATE de 4-CHLOROPHENYLE (FRENCH) see
   CJT750
p-CHLOROBENZENESULFONIC ACID-p-CHLOROPHENYL ESTER see CJT750
p-CHLOROBENZENESULFONIC ACID, SODIUM SALT see CEK250
p-CHLOROBENZENESULFONYL CHLORIDE see CEK375
N-(p-CHLOROBENZENESULFONYL)-N'-PROPYLUREA see CKK000
1-(p-CHLOROBENZENESULFONYL)-3-PROPYLUREA see CKK000
4-CHLOROBENZENETHIOL see CEK425
6-CHLOROBENZENO(a)PYRENE see CEK500
1-(p-CHLOROBENZHYDRYL)-4-(p-tert-BUTYLBENZYL)DIETHYLENEDIAMINE DI-
   HYDROCHLORIDE see BOM250
1-p-CHLOROBENZHYDRYL-4-p-(tert)-BUTYLBENZYLPIPERAZINE DIHYDRO-
   CHLORIDE see BOM250
1-(p-CHLOROBENZHYDRYL)-4-(2-(2-HYDROXYETHOX-
   Y)ETHYL)DIETHYLENEDIAMINE see CJR909
1-(p-CHLOROBENZHYDRYL)-4-(2-(2-HYDROXYETHOX-
   Y)ETHYL)DIETHYLENEDIAMINE HYDROCHLORIDE see VSF000
1-(p-CHLOROBENZHYDRYL)-4-(2-(2-HYDROXYETHOXY)ETHYL)PIPERAZINE see
   CJR909
1-(p-CHLOROBENZHYDRYL)-4-(m-METHYLBENZYL)DIETHYLENEDIAMINE see
   HGC500
1-p-CHLOROBENZHYDRYL-4-m-METHYLBENZYLPIPERAZINE see HGC500
1-(4-CHLOROBENZHYDRYL)-4-METHYLPIPERAZINE see CFF500
1-(4-CHLOROBENZHYDRYL)-4-METHYLPIPERAZINE DIHYDROCHLORIDE see
   CDR000
1-(p-CHLOROBENZHYDRYL)-4-METHYLPIPERAZINE HYDROCHLORIDE see
   CDR250
N-(4-CHLOROBENZHYDRYL)-N'-(HYDROXYETHOXYETHYL)PIPERAZINE see
   CJR909
N¹-(4'-CHLOROBENZHYDRYL)-N⁴-SPIROMORPHOLINO-PIPERAZINIUM CHLO-
   RIDE HYDROCHLORIDE see CEK875
1-(p-CHLOROBENZIL)-2-PIRROLIDIL-METIL-BENZIMIDAZOLO CLORIDATO (ITAL-
   IAN) see CMV400
2-CHLOROBENZO(e)(1)BENZOTHIOPYRANO(4,3-b)INDOLE see CEL000
2-CHLOROBENZOIC ACID see CEL250
3-CHLOROBENZOIC ACID see CEL290
4-CHLOROBENZOIC ACID see CEL300
m-CHLOROBENZOIC ACID see CEL290
o-CHLOROBENZOIC ACID see CEL250
p-CHLOROBENZOIC ACID see CEL300
4-CHLOROBENZOIC ACID-3-ETHYL-7-METHYL-3,7-DIAZABICYCLO(3.3.1)NON-9-
   YL ESTER HYDROCHLORIDE see YGA700
o-CHLOROBENZOIC ACID NICKEL(II) SALT see CEL500
CHLOROBENZOL (DOT) see CEJ125
o-CHLOROBENZONITRILE see CEM000
p-CHLOROBENZONITRILE see CEM250
6-CHLORO-2H-1,2,4-BENZOTHIADIAZINE-7-SULFONAMIDE-1,1-DIOXIDE see
   CLH750
2-CHLOROBENZOTHIAZOLE see CEM500
1-CHLOROBENZOTRIAZOL see CEM625
4-CHLOROBENZOTRICHLORIDE see TIR900
p-CHLOROBENZOTRICHLORIDE see TIR900
p-CHLOROBENZOTRIFLUORIDE see CEM825
5-CHLORO-2-BENZOXAZOLAMINE see AJF500
2-CHLOROBENZOXAZOLE see CEM850
5-CHLOROBENZOXAZOLIDONE see CDQ750
5-CHLORO-2-BENZOXAZOLINONE see CDQ750
6-CHLORO-2-BENZOXAZOLINONE see CDQ750
5-CHLOROBENZOXAZOL-2-ONE see CDQ750
5-(4-CHLOROBENZOYL)-1,4-DIMETHYL-1H-PYRROLE-2-ACETIC ACID SODIUM
   SALT DIHYDRATE see ZUA300
1-(4-CHLOROBENZOYL)-N-HYDROXY-5-METHOXY-2-METHYL-1H-INDOLE-3-
   ACETAMIDE see OLM300
1-(p-CHLOROBENZOYL)-5-METHOXY-2-METHYLINDOLE-3-ACETIC ACID see
   IDA000
1-(4-CHLOROBENZOYL)-5-METHOXY-2-METHYL-1H-INDOLE-3-ACETIC ACID
   CARBOXYMETHYL ESTER see AAE625

1-(p-CHLOROBENZOYL)-5-METHOXY-2-METHYLINDOLE-3-ACETOHYDROXAMIC ACID see OLM300

((1-(4-CHLOROBENZOYL)-5-METHOXY-2-METHYLINDOLE-3-YL)ACETOXY)ACETIC ACID see AAE625

1-(p-CHLOROBENZOYL)-5-METHOXY-2-METHYL-3-INDOLYLACETOHYDROXAM-IC ACID see OLM300

1-(p-CHLOROBENZOYL)-2-METHYL-5-METHOXYINDOLE-3-ACETIC ACID see IDA000

1-(p-CHLOROBENZOYL)-2-METHYL-5-METHOXY-3-INDOLE-ACETIC ACID see IDA000

α-(1-(p-CHLOROBENZOYL)-2-METHYL-5-METHOXY-3-INDOLYL)ACETIC ACID see IDA000

2-(p-CHLOROBENZOYL)-1-(2-MORPHOLINOETHYL)PYRROLE MONOHYDRO-CHLORIDE see CEO100

1-(2-(2-CHLOROBENZOYL)-4-NITROPHENYL)-2-(DIETHYLAMINOME-THYL)IMIDAZOLE FUMARATE see NMV400

p-CHLOROBENZOYL PEROXIDE (DOT) see BHM750

CHLOROBENZYLATE see DER000

p-CHLOROBENZYL-p-CHLOROPHENYL SULFIDE see CEP000

4-CHLOROBENZYL-4-CHLOROPHENYL SULPHIDE see CEP000

p-CHLOROBENZYL-p-CHLOROPHENYL SULPHIDE see CEP000

4-CHLOROBENZYL CYANIDE see CEP300

p-CHLOROBENZYL CYANIDE see CEP300

α-(p-CHLOROBENZYL)-4-DIETHYLAMINOETHOXY-4'-METHYLBENZHYDROL see TMP500

S-(4-CHLOROBENZYL)-N,N-DIETHYLTHIOCARBAMATE see SAZ000

2-((p-CHLOROBENZYL)(2-(DIMETHYLAMINO)ETHYL)AMINO)PYRIDINE see CKV625

4-(p-CHLOROBENZYL)-2-((2-DIMETHYLAMINO)ETHYL)-1(2H)-PHTHALAZINONE HYDROCHLORIDE see CEP675

(2-(p-CHLOROBENZYL)-3-DIMETHYLAMINOMETHYL-2-BUTANOL HYDROCHLO-RIDE see CMW500

4-(p-CHLOROBENZYL)-2-(HEXAHYDRO-1-METHYL-1H-AZEPIN-4-YL)-1-(2H)-PHTHALAZINONE HCl see ASC130

4-(p-CHLOROBENZYL)-2-(HEXAHYDRO-1-METHYL-1H-AZEPIN-4-YL)-1-(2H)-PHTHALAZINONE see ASC125

p-CHLOROBENZYL-3-HYDROXYCROTONATE DIMETHYL PHOSPHATE see CEQ500

2-CHLOROBENZYLIDENEACETOPHENONE see CLF150

(4-CHLOROBENZYLIDENE)ACETOPHENONE see CLF100

4-CHLOROBENZYLIDENE CHLORIDE see TJD650

o-CHLOROBENZYLIDENE MALONITRILE see CEQ600

2-CHLOROBENZYLIDENE MALONONITRILE see CEQ600

o-CHLOROBENZYLIDENE MALONONITRILE see CEQ600

1-(4-CHLOROBENZYL)-1H-INDAZOLE-3-CARBOXYLIC ACID see CEQ625

1-(p-CHLOROBENZYL)-1H-INDAZOLE-3-CARBOXYLIC ACID see CEQ625

4-CHLOROBENZYL ISOTHIOCYANATE see CEQ750

N-(4-CHLOROBENZYL)-N',N'-DIMETHYL-N-(2-PYRIDYL)ETHYLENEDIAMINE see CKV625

N-3'-CHLOROBENZYL-N'-ETHYLUREA see LII400

3-(p-CHLOROBENZYL)OCTAHYDRO-QUINOLIZINE TARTRATE (1:1) see CMX920

1-(p-CHLOROBENZYLOXYCARBONYL)-1-PROPEN-2-YL-DIMETHYLPHOSPHATE see CEQ500

1-(1-(2-((3-CHLOROBENZYL)OXY)PHENYL)VINYL)-1H-IMIDAZOLE HYDROCHLO-RIDE see CMW550

α-CHLOROBENZYL PHENYL KETONE see CFJ100

p-CHLOROBENZYLPSEUDOTHIURONIUM CHLORIDE see CEQ800

1-(p-CHLOROBENZYL)-2-(1-PYRROLIDINYLMETHYL)BENZIMIDAZOLE HYDRO-CHLORIDE see CMV400

1-p-CHLOROBENZYL-PYRROLIDYL-METHYLENE-BENZIMIDAZOLE HYDRO-CHLORIDE see CMV400

3-(p-CHLOROBENZYL)QUINOLIZIDINE TARTRATE see CMX920

5-(o-CHLOROBENZYL)-4,5,6,7-TETRAHYDROTHIENO(3,2-c)PYRIDINE HYDRO-CHLORIDE see TGA525

2-(4-CHLOROBENZYL)-2-THIOPSEUDOUREA HYDROCHLORIDE see CEQ800

7-CHLOROBICYCLO(3.2.0)HEPTA-2,6-DIEN-6-YL DIMETHYL PHOSPHATE see HBK700

2-CHLORO-1,1'-BIPHENYL see CGM750

CHLORO 1,1-BIPHENYL see PJL750

CHLORO BIPHENYL see PJL750

3-CHLOROBIPHENYLAMINE see CEH000

2-CHLORO-N,N-BIS(2-CHLOROETHYL)-1-PROPYLAMINE HYDROCHLORIDE see NOB700

2-CHLORO-4,6-BIS(DIETHYLAMINO)-s-TRIAZINE see CDQ325

2-CHLORO-4,6-BIS(ETHYLAMINO)-1,3,5-TRIAZINE see BJP000

1-CHLORO-3,5-BISETHYLAMINO-2,4,6-TRIAZINE see BJP000

2-CHLORO-4,6-BIS(ETHYLAMINO)-s-TRIAZINE see BJP000

4-CHLORO-2,6-BIS-ETHYLENEIMINOPYRIMIDINE see EQI600

2-CHLORO-1,1-BIS(FLUOROOXY)TRIFLUOROETHANE see CER825

CHLOROBIS(2-METHYLPROPYL)ALUMINUM see CGB500

CHLOROBLE M see MAS500

2-CHLOROBMN see CEQ600

1-CHLORO-3-BROMO-BUTENE-1 see CES250

1-CHLORO-2-BROMOETHANE see CES500

sym-CHLOROBROMOETHANE see CES500

CHLOROBROMOMETHANE see CES650

4-CHLORO-7-BROMOMETHYLBENZ(a)ANTHRACENE see BNQ000

1-(3-CHLORO-4-BROMOPHENYL)-3-METHYL-3-METHOXYUREA see CES750

1-CHLORO-3-BROMOPROPANE (DOT) see BNA825

ω-CHLOROBROMOPROPANE see BNA825

trans-CHLORO(2-(3-BROMOPROPIONAMIDO)CYCLOHEXYL)MERCURY see CET000

CHLOROBRUMURON see CES750

CHLOROBUFAM see CEX250

1-CHLORO-1,3-BUTADIENE see CET250

2-CHLORO-1,3-BUTADIENE see NCI500

2-CHLOROBUTA-1,3-DIENE see NCI500

1-CHLOROBUTADIENE see CET250

2-CHLORO-1,3-BUTADIENE HOMOPOLYMER (9Cl) see PJQ050

2-CHLORO-1,3-BUTADIENE POLYMER see PJQ050

CHLOROBUTADIENE POLYMER see PJQ050

2-CHLOROBUTANE see CEU250

1-CHLOROBUTANE (DOT) see BQQ750

4-CHLOROBUTANENITRILE see CEU300

4-CHLORO-1-BUTANE-OL see CEU500

3-CHLOROBUTANOIC ACID see CEW000

4-CHLORO-1-BUTANOL see CEU500

4-CHLOROBUTANOL see CEU500

CHLOROBUTANOL see ABD000

1-CHLORO-2-BUTANONE see CEU750

3-CHLORO-1-BUTENE see CEV250

1-CHLORO-2-BUTENE see CEU825

2-CHLORO-2-BUTENE see CEV000

1-CHLORO-1-BUTEN-3-ONE see CEV500

CHLOROBUTIN see CDO500

CHLOROBUTINE see CDO500

N-(3-CHLORO-1-sec-BUTYLACETONYL)-p-TOLUENESULFONAMIDE see THH470

o-CHLORO-α-((tert-BUTYLAMINO)METHYL)BENZYLALCOHOL HYDROCHLO-RIDE see BQE250

4-CHLORO-2-(tert-BUTYLAMINO)-6-(4-METHYLPIPERAZINO)-5-METHYLTHIOPYRI-MIDINE see CEV750

5-CHLORO-3-tert-BUTYL-6-METHYLURACIL see BQT750

4-CHLORO-2-BUTYNOL see CEV800

CHLORO-2-BUTYNYL-m-CHLOROCARBAMATE see CEW500

4-CHLOROBUT-2-YNYL-m-CHLOROCARBANILATE see CEW500

4-CHLORO-2-BUTYNYL-m-CHLOROCARBANILATE see CEW500

4-CHLOROBUT-2-YNYL-3-CHLOROPHENYLCARBAMATE see CEW500

4-CHLORO-2-BUTYNYL-N-(3-CHLOROPHENYL)CARBAMATE see CEW500

3-CHLOROBUTYRIC ACID see CEW000

β-CHLOROBUTYRIC ACID see CEW000

4-CHLOROBUTYRIC ACID TRIBUTYLSTANNYL ESTER see TID000

4-CHLOROBUTYRONITRILE see CEU300

γ-CHLOROBUTYRONITRILE see CEU300

CHLOROCAIN see CBR250

CHLOROCAINE see AIT250

CHLOROCAMPHENE see CDV100

CHLOROCAPS see CDP250

m-CHLORO CARBANILIC ACID-4-CHLORO-2-BUTYNYL ESTER see CEW500

3-CHLOROCARBANILIC ACID, ISOPROPYL ESTER see CKC000

m-CHLOROCARBANILIC ACID, ISOPROPYL ESTER see CKC000

m-CHLOROCARBANILIC ACID-1-METHYL-2-PROPYNYL ESTER see CEX250

CHLOROCARBONATE D'ETHYLE (FRENCH) see EHK500

CHLOROCARBONATE de METHYLE (FRENCH) see MIG000

CHLOROCARBONIC ACID METHYL ESTER see MIG000

N-(CHLOROCARBONYLOXY)TRIMETHYLUREA see CEX255

3-CHLOROCARPIPRAMINE DIHYDROCHLORIDE see DKV309

5-CHLOROCARVACROL see CEX275

2-CHLOROCHALCONE see CLF150

4-CHLOROCHALCONE see CLF100

p-CHLOROCHALCONE see CLF100

CHLOROCHIN see CLD000

3-CHLOROCHLORDENE see HAR000

6'-CHLORO-2-(p-CHLOROBENZYL(2-(DIETHYLAMINO)ETHYL)AMINO)-o-ACETO-LUIIDIDE DIHYDROCHLORIDE see CEX500

1-CHLORO-2-(β-CHLOROETHOXY)ETHANE see DFJ050

6-CHLORO-9-((2-((2-CHLOROETHYL)AMINO)ETHYL)AMINO)-2-METHOXYACRI-DINE 2-HYDROCHLORIDE SESQUIHYDRATE see CEY250

2-CHLORO-N-(2-CHLOROETHYL)-N,N-DIMETHYLETHANAMINIUM CHLORIDE see DQS500

13-CHLORO-N-(2-CHLOROETHYL)-N,11-DINITROSO-10-OXO-5,6-DITHIA-2,9,11-TRIAZATRIDECANAMIDE see BIF625

2-CHLORO-N-(2-CHLOROETHYL)ETHANAMINE HYDROCHLORIDE see BHO250

7-CHLORO-10-(3-(N-(2-CHLOROETHYL)-N-ETHYL)AMINOPROPYLAMINO)-2-ME-THOXY-BENZO(B)(1,5)NAPHTHYRIDINE DIHYDROCHLORIDE see IAE000

6-CHLORO-9-(3-(2-CHLOROETHYL)MERCAPTOPROPYLAMINO)-2-METHOXYA-CRIDINE HYDROCHLORIDE see CFA250

2-CHLORO-N-(2-CHLOROETHYL)-N-METHYLETHANAMINE HYDROCHLORIDE see BIE500

2-CHLORO-N-(2-CHLOROETHYL)-N-METHYLETHANAMINE-N-OXIDE see CFA500

2-CHLORO-N-(2-CHLOROETHYL)-N-METHYLETHANAMINE-N-OXIDE HYDRO-CHLORIDE see CFA750

1-CHLORO-2-(β-CHLOROETHYLTHIO)ETHANE see BIH250

CHLORO(CHLOROMETHOXY)METHANE see BIK000

9-CHLORO-10-CHLOROMETHYL ANTHRACENE see CFB500

1-CHLORO-4-CHLOROMETHYLBENZENE see CFB600

cis-2-CHLORO-3-(CHLOROMETHYL)OXIRANE see DAC975

trans-2-CHLORO-3-(CHLOROMETHYL)OXIRANE see DGH500

1-CHLORO-4-(CHLOROMETHYLTHIO)BENZENE see CFB750

2-CHLORO-5-CHLOROMETHYLTHIOPHENE see CFB825

5-CHLORO-N-(2-CHLORO-4-NITROPHENYL)-2-HYDROXYBENZAMIDE see DFV400

5-CHLORO-N-(2-CHLORO-4-NITROPHENYL)-2-HYDROXYBENZAMIDE with 2-AMINOETHANOL (1:1) see DFV600

3-CHLORO-4-(3-CHLORO-2-NITROPHENYL)PYRROLE see CFC000

5-CHLORO-2'-CHLORO-4'-NITROSALICYLANILIDE see DFV400

N-(5-CHLORO-4-((4-CHLOROPHENYL)CYANOMETHYL)-2-METHYLPHENYL)-2-HYDROXY-3, 5-DIIODOBENZAMIDE see CFC100

7-CHLORO-5-(o-CHLOROPHENYL)-1,3-DIHYDRO-3-HYDROXY-2H-1,4-BENZO-DIAZEPIN-2-ONE see CFC250

7-CHLORO-5-(2-CHLOROPHENYL)-1,3-DIHYDRO-3-HYDROXY-2H-1,4-BENZO-DIAZEPIN-2-ONE see CFC250

7-CHLORO-5-(2-CHLOROPHENYL)-1,3-DIHYDRO-3-HYDROXY-1-METHYL-2H-1,4-BENZODIAZEPIN-2-ONE see MLD100

5-CHLORO-3-(4-CHLOROPHENYL)-4'-FLUORO-2'-METHYLSALICYLANILIDE see CFC500

7-CHLORO-5-(2-CHLOROPHENYL)-3-HYDROXY-1H-1,4-BENZODIAZEPIN-2(3H)-ONE see CFC250

7-CHLORO-5-(2-CHLOROPHENYL)-3-HYDROXY-1-METHYL-2,3-DIHYDRO-1H-1,4-BENZODIAZEPIN-2-ONE see MLD100

2-CHLORO-3-(4-CHLOROPHENYL)METHYLPROPIONATE see CFC750

1-CHLORO-4-(((4-CHLOROPHENYL)METHYL)THIO)BENZENE see CEP000

8-CHLORO-6-(o-CHLOROPHENYL)-1-METHYL-4H-s-TRIAZOLO(4,3-a)(1,4)BENZODIAZEPINE see THS800

8-CHLORO-6-(2-CHLOROPHENYL)-1-METHYL-4H-(1,2,4)TRIAZOLO(4,3-a)(1,4)BENZODIAZEPINE see THS800

2-CHLORO-3-(4-CHLOROPHENYL)PROPIONIC ACID METHYL ESTER see CFC750

4-CHLORO-α-(4-CHLOROPHENYL)-α-(TRICHLOROME-THYL)BENZENEMETHANOL see BIO750

CHLORO(2-CHLOROVINYL)MERCURY see CFD250

CHLOROCHOLINE CHLORIDE see CMF400

CHLOROCID see CDP250

CHLOROCIDE see CEP000

CHLOROCIDIN C TETRAN see CDP250

CHLOROCOL see CDP250

4-CHLORO-m-CRESOL see CFE250

6-CHLORO-m-CRESOL see CFE250, CFE500

p-CHLORO-m-CRESOL see CFE250

4-CHLORO-o-CRESOL see CFE000

p-CHLOROCRESOL see CFE250

CHLOROCRESOL see CFE250

4-CHLORO-o-CRESOXYACETIC ACID see CIR250

CHLOROCTAN SODNY (CZECH) see SFU500

CHLOROCYAN see COO750

CHLOROCYANIDE see COO750

CHLOROCYANOACETYLENE see CFE750

2-CHLORO-1-CYANOETHANOL see CHU000

CHLOROCYANOGEN see COO750

2-CHLORO-α-CYANO-6-METHYLERGOLINE-8-PROPIONAMIDE see CFF100

2-CHLORO-4-(1-CYANO-1-METHYLETHYLAMINO)-6-ETHYLAMINO-1,3,5-TRIA-ZINE see BLW750

endo-3-CHLORO-exo-6-CYANO-2-NORBORNANONE-o-(METHYLCARBA-MOYL)OXIME see CFF250

3-CHLORO-6-CYANO-2-NORBORNANONE-o-(METHYLCARBAMOYL)OXIME see CFF250

2-exo-CHLORO-6-endo-CYANO-2-NORBORNANONE-o-(METHYLCARBA-MOYL)OXIME2-CARBONITRILE see CFF250

3-CHLORO-6-CYANONORBORNANONE-2-OXIME-o,N-METHYLCARBAMATE see CFF250

CHLOROCYCLAMIDE-R see CDR550

CHLOROCYCLINE see CFF500

CHLOROCYCLIZINE see CFF500

CHLOROCYCLIZINE HYDROCHLORIDE see CDR500

2-CHLOROCYCLOHEPTANONE see CFF600

α-CHLOROCYCLOHEPTANONE see CFF600

CHLOROCYCLOHEXANE see CPI400

2-CHLOROCYCLOHEXANONE see CFG250

α-CHLOROCYCLOHEXANONE see CFG250

4-CHLORO-4-CYCLOHEXENE-1,2-DICARBOXYLIC ANHYDRIDE see CFG500

7-CHLORO-5-(CYCLOHEXEN-1-YL)-1,3-DIHYDRO-1-METHYL-2H-1,4-BENZO-DIAZEPIN-2-ONE see CFG750

7-CHLORO-5-(1-CYCLOHEXENYL)-1-METHYL-2-OXO-2,3-DIHYDRO-1H-(1,4)-BEN-ZO(f)DIAZEPINE see CFG750

6'-CHLORO-2-(CYCLOHEXYLAMINO)-o-ACETOTOLUIDIDE HYDROCHLORIDE see CFH000

5-CHLORO-N-(2-(4-((((CYCLOHEXYLAMINO) CARBON-YL)AMINO)SULFONYL)PHENYL)ETHYL)-2-METHOXYBENZAMIDE see CEH700

3-(3-CHLORO-4-CYCLOHEXYLBENZOYL)PROPIONIC ACID see CPJ000

3-(3-CHLORO-4-CYCLOHEXYLBENZOYL)PROPIONIC ACID CALCIUM SALT see CPJ250

(Z)-3-(2-CHLOROCYCLOHEXYL)-1-(2-CHLOROETHYL)-1-NITROSOUREA see CFH500

cis-3-(2-CHLOROCYCLOHEXYL)-1-(2-CHLOROETHYL)-1-NITROSOUREA see CFH500

cis-N'-(2-CHLOROCYCLOHEXYL)-N-(2-CHLOROETHYL)-N-NITROSOUREA see CFH500

trans-3-(2-CHLOROCYCLOHEXYL)-1-(2-CHLOROETHYL)-1-NITROSOUREA see CFH750

trans-N'-(2-CHLOROCYCLOHEXYL)-N-(2-CHLOROETHYL)-N-NITROSOUREA see CFH750

6-CHLORO-5-CYCLOHEXYL-2,3-DIHYDRO-1H-INDENE-1-CARBOXYLIC ACID (9CI) see CMV500

(±)-6-CHLORO-5-CYCLOHEXYL-2,3-DIHYDRO-1H-INDENE-1-CARBOXYLIC ACID (9CI) see CFH825

(±)-6-CHLORO-5-CYCLOHEXYLINDAN-1-CARBOXYLIC ACID see CFH825

(±)-6-CHLORO-5-CYCLOHEXYL-1-INDANCARBOXYLIC ACID see CFH825

6-CHLORO-5-CYCLOHEXYL-1-INDANCARBOXYLIC ACID see CFH825, CMV500

6'-CHLORO-2-(N-CYCLOHEXYL-N-METHYLAMINO)-o-ACETOTOLUIDIDE HY-DROCHLORIDE see CFI000

3-CHLORO-4-CYCLOHEXYL-α-OXO-BENZENEBUTANOIC ACID see CPJ250

3-CHLORO-4-CYCLOHEXYL-α-OXOBENZENEBUTANOIC ACID see CPJ000

3-CHLORO-4-CYCLOHEXYL-α-OXOBENZENEBUTANOIC ACID CALCIUM SALT see CPJ250

4-(3-CHLORO-4-CYCLOHEXYLPHENYL)-4-OXO-BUTYRIC ACID see CPJ000

4-(3-CHLORO-4-CYCLOHEXYLPHENYL)-4-OXOBUTYRIC ACID CALCIUM SALT see CPJ250

3-CHLOROCYCLOPENTENE see CFI625

4-CHLORO-2-CYCLOPENTYL PHENOL see CFI750

2-CHLORO-4-CYCLOPROPYLAMINO-6-ISOPROPYLAMINO-1,3,5-TRIAZINE see CQI750

2-CHLORO-4-CYCLOPROPYLAMINO-6-ISOPROPYLAMINO-sec-TRIAZINE see CQI750

2-((4-CHLORO-6-(CYCLOPROPYLAMINO)-1,3,5-TRIAZIN-2-YL)AMINO)-2-METHYL-PROPANENITRILE see PMF600

2-(4-CHLORO-6-(CYCLOPROPYLAMINO)-s-TRIAZIN-2-YL)AMINO-2-METHYLPRO-PIONITRILE see PMF600

2-(3-CHLORO-4-CYCLOPROPYLMETHOXYPHENYL)ACETIC ACID LYSINE SALT (d,l) see CFJ000

3-CHLORO-4-CYCLO-PROPYLMETHOXYPHENYLACETIC ACID LYSINE SALT (d,l) see CFJ000

7-CHLORO-1-CYCLOPROPYLMETHYL-1,3-DIHYDRO-5-(2-FLUOROPHENYL)-2H-1,4-BENZODIAZEPIN-2-ONE see FMR100

7-CHLORO-1-(CYCLOPROPYLMETHYL)-1,3-DIHYDRO-5-PHENYL-2H-1,4-BENZO-DIAZEPIN-2-ONE see DAP700

7-CHLORO-1-CYCLOPROPYLMETHYL-5-PHENYL-1H-1,4-BENZODIAZEPIN-2(3H)-ONE see DAP700

6-CHLORO-1-(CYCLOPROPYLMETHYL)-4-PHENYL-2(1H)-QUINAZOLINONE see CQF125

CHLORODANE see CDR750

CHLORODECANE see DAJ200

6-CHLORO-6-DEHYDRO-17-α-ACETOXYPROGESTERONE see CBF250

6-CHLORO-Δ⁶-DEHYDRO-17-ACETOXYPROGESTERONE see CBF250

6-CHLORO-6-DEHYDRO-17-α-HYDROXYPROGESTERONE ACETATE see CBF250

7-CHLORO-6-DEMETHYLTETRACYCLINE see MIJ500

7-CHLORO-6-DEMETHYLTETRACYCLINE HYDROCHLORIDE see DAI485

CHLORODEN see DEP600

α-CHLORODEOXYBENZOIN see CFJ100

CHLORODEOXYGLYCEROL see CDT750

CHLORODEOXYGLYCEROL DIACETATE see CIL900

7(S)-CHLORO-7-DEOXYLINCOMYCIN see CMV675

7(S)-CHLORO-7-DEOXYLINCOMYCIN-2-PHOSPHATE see CMV690

5-CHLORO-2'-DEOXYURIDINE see CFJ750

5-CHLORODEOXYURIDINE see CFJ750

CHLORODIABINA see CKK000

2-CHLORO-N,N-DIALLYLACETAMIDE see CFK000

α-CHLORO-N,N-DIALLYLACETAMIDE see CFK000

4-CHLORO-1,2-DIAMINOBENZENE see CFK125

1-CHLORO-2,4-DIAMINOBENZENE see CJY120

α-(2-CHLORO-4-(4,6-DIAMINO-2,2-DIMETHYL-S-TRIAZINE-1(2H)-YL)PHENOXY)-N,N-DIMETHYL-m-TOLUAMIDE ETHANESULFONATE see THR500

CHLORODIAZEPOXIDE see LFK000

1-CHLORODIBENZO-p-DIOXIN see CDV125

2-((8-CHLORODIBENZO(b,f)THIEPIN)-10-YL)OXY-N,N-DIMETHYLETHANAMINE see ZUJ000

CHLORODIBORANE see CFK325

CHLORODIBROMOMETHANE see CFK500

3-CHLORO-1,2-DIBROMOPROPANE see DDL800

1-CHLORO-2,3-DIBROMOPROPANE see DDL800

CHLORO(DIBUTOXYPHOSPHINYL)MERCURY see CFK750

6'-CHLORO-2-(DIBUTYLAMINO)-o-ACETOTOLUIDIDE, HYDROCHLORIDE see CFL000

1-CHLORO-2-(2,2-DICHLORO-1-(4-CHLOROPHENYL)ETHYL)BENZENE see CDN000

1-CHLORO-2,2-DICHLOROETHYLENE see TIO750

6-CHLORO-3-(DICHLOROMETHYL)-3,4-DIHYDRO-2H-1,2,4-BENZOTHIADIAZINE-7-SULFONAMIDE-1,1-DIOXIDE see HII500

6-CHLORO-3-(DICHLOROMETHYL)3,4-DIHYDRO-7-SULFAMYL-1,2,4-BENZOTHIA-DIAZINE-1,1-DIOXIDE see HII500

5-CHLORO-6-(2,3-DICHLOROPHENOXY)-2-(METHYLTHIO)-1H-BENZIMIDAZOLE see CFL200

6-CHLORO-5-(2,3-DICHLOROPHENOXY)-2-METHYLTHIO-BENZIMIDAZOLE see CFL200

5-CHLORO-2-(2,4-DICHLOROPHENOXYPHENOL) see TIQ000

O-(2-CHLORO-1-(2,5-DICHLOROPHENYL))-O,O-DIETHYL ESTER PHOSPHORO-THIOIC ACID see DIX600

β-2-CHLORO-1-(2',4'-DICHLOROPHENYL) VINYL DIETHYLPHOSPHATE see CDS750

2-CHLORO-1-(2,4-DICHLOROPHENYL)VINYL DIETHYL PHOSPHATE see CDS750

O-(2-CHLORO-1-(2,5-DICHLOROPHENYL)VINYL)-O,O-DIETHYL PHOSPHOROTH-IOATE see DIX600

CHLORODIETHOXYSILANE see DHF800

CHLORODIETHYLALUMINUM see DHI885

2'-CHLORO-2-(DIETHYLAMINO)ACETANILIDE HYDROCHLORIDE see CFL500

3'-CHLORO-2-(DIETHYLAMINO)ACETANILIDE HYDROCHLORIDE see CFL750

4'-CHLORO-2-(DIETHYLAMINO)ACETANILIDE HYDROCHLORIDE see CFM000

5'-CHLORO-2-(DIETHYLAMINO)-o-ACETOTOLUIDIDE HYDROCHLORIDE see CFM750

3'-CHLORO-2-(DIETHYLAMINO)-o-ACETOTOLUIDIDE HYDROCHLORIDE see CFM250

6'-CHLORO-2-(DIETHYLAMINO)-m-ACETOTOLUIDIDE HYDROCHLORIDE see CFN000

4'-CHLORO-2-(DIETHYLAMINO)-o-ACETOTOLUIDIDE HYDROCHLORIDE see CFM500

6'-CHLORO-3-(DIETHYLAMINO)-o-BUTYROTOLUIDIDE HYDROCHLORIDE see CFN500

6'-CHLORO-2-(2-(DIETHYLAMINO)ETHOXY)-o-ACETOTOLUIDIDE HYDROCHLO-RIDE see CFO000

5-CHLORO-2-(2-(DIETHYLAMINO)ETHOXY)BENZANILIDE see CFO250

5-CHLORO-2-(2-(2-(DIETHYLAMINO)ETHOXY)ETHYL)-2-METHYL-1,3-BENZODI-OXOLE see CFO750

2-CHLORO-1-(p-(β-DIETHYLAMINOETHOXY)PHENYL)-1,2-DIPHENYLETHYLENE see CMX700

6'-CHLORO-2-(2-(DIETHYLAMINO)ETHYL)AMINO-o-ACETOTOLUIDIDE HYDRO-CHLORIDE see CFP000

5-CHLORO-2-(2-(2-(DIETHYLAMINO)ETHYLAMINO)ETHYL)-2-METHYL-1,3-BEN-ZODIOXOLE DIHYDROCHLORIDE see CFP250

2-CHLORO-4-(DIETHYLAMINO)-6-(ETHYLAMINO)-s-TRIAZINE see TJL500

8-CHLORO-2-(2-(DIETHYLAMINO)ETHYL)-2H-(1)-BENZOTHIOPYRANO(4,3,2-cd)INDAZOLE-5-METHANOL MONOMETHANE SULFONATE see CFP750

8-CHLORO-2-(2-(DIETHYLAMINO)ETHYL)-2H-(1)BENZOTHIOPYRANO(4,3,2-cd)INDAZOLE-5-METHANOL-N-OXIDE see CFQ000

2'-CHLORO-2-((2-(DIETHYLAMINO)ETHYL)ETHYLAMINO)ACETANILIDE DIHY-DROCHLORIDE see CFQ250

6'-CHLORO-2-((2-(DIETHYLAMINO)ETHYL)ETHYLAMINO)-o-ACETOTOLUIDIDE HYDROCHLORIDE see CFQ500

7-CHLORO-1-(2-(DIETHYLAMINO)ETHYL)-5-(2-FLUOROPHENYL)-1H-1,4-BENZO-DIAZEPIN-2(3H)-ONE see FMQ000

6'-CHLORO-2-((2-(DIETHYLAMINO)ETHYL)ISOPROPYLAMINO)-o-ACETOTOLUI-DIDE HYDROCHLORIDE see CFQ750

2'-CHLORO-2-(2-(DIETHYLAMINO)ETHYL)METHYLAMINOACETANILIDE DIHY-DROCHLORIDE see CFR000

4'-CHLORO-2-(2-(DIETHYLAMINO)ETHYL)METHYLAMINOACETANILIDE DIHY-DROCHLORIDE see CFR250

5'-CHLORO-2-(2-(DIETHYLAMINO)ETHYL)METHYLAMINO-o-ACETOTOLUIDIDE DIHYDROCHLORIDE see CFS000

6'-CHLORO-2-(2-(DIETHYLAMINO)ETHYL)METHYLAMINO-m-ACETOTOLUIDIDE DIHYDROCHLORIDE see CFS250

3'-CHLORO-2-(2-(DIETHYLAMINO)ETHYL)METHYLAMINO-o-ACETOTOLUIDIDE DIHYDROCHLORIDE see CFR500

4'-CHLORO-2-(2-(DIETHYLAMINO)ETHYL)METHYLAMINO-o-ACETOTOLUIDIDE DIHYDROCHLORIDE see CFR750

6'-CHLORO-2-((2-(DIETHYLAMINO)ETHYL)METHYLAMINO)-o-ACETOTOLUIDIDE HYDROCHLORIDE see CFS500

6'-CHLORO-2-((2-(DIETHYLAMINO)ETHYL)OCTYLAMINO)-o-ACETOTOLUIDIDE HYDROCHLORIDE see CFS750

6'-CHLORO-2-((2-(DIETHYLAMINO)ETHYL)(2-PHENOXYETHYL)AMINO)-o-ACETO-TOLUIDIDE HYDROCHLORIDE see CFT000

(4-CHLORO-N-(2-DIETHYLAMINO)ETHYL)-2-(2-PROPENYLOXY)BENZAMIDE see AFS625

4-CHLORO-N-(2-DIETHYLAMINO)ETHYL)-2-(2-PROPENYLOXY)-BENZAMIDE (9CI) see AFS625

6'-CHLORO-2-((2-(DIETHYLAMINO)ETHYL)PROPYLAMINO)-o-ACETOTOLUIDIDE HYDROCHLORIDE see CFT250

2'-CHLORO-2-(2-(DIETHYLAMINO)ETHYLTHIO)ACETANILIDE HYDROCHLORIDE see CFT500

6'-CHLORO-2-(2-(DIETHYLAMINO)ETHYLTHIO)-o-ACETOTOLUIDIDE HYDRO-CHLORIDE see CFT750

7-CHLORO-10-(3-(DIETHYLAMINO)-2-HYDROXYPROPYL)ISOALLOXAZINE SUL-FATE see CFU000

2'-CHLORO-2-(DIETHYLAMINO)-5'-METHYLACETANILIDE HYDROCHLORIDE see CFN000

4'-CHLORO-2-(DIETHYLAMINO)-N-METHYLACETANILIDE HYDROCHLORIDE see CFU250

3'-CHLORO-2-(DIETHYLAMINO)-6'-METHYLACETANILIDE HYDROCHLORIDE see CFM750

2'-CHLORO-2-(DIETHYLAMINO)-2'-METHYLACETANILIDE HYDROCHLORIDE see CFM250

4'-CHLORO-2-(DIETHYLAMINO)-2'-METHYLACETANILIDE HYDROCHLORIDE see CFM500

6'-CHLORO-2-(DIETHYLAMINO)-N-METHYL-o-ACETOTOLUIDIDE HYDROCHLO-RIDE see CFU500

6-CHLORO-9-((4-(DIETHYLAMINO)-1-METHYLBUTYL)AMINO)-2-METHOXYACRI-DINE DIHYDROCHLORIDE see CFU750

2-CHLORO-5-(ω-DIETHYLAMINO-α-METHYLBUTYLAMINO)-7-METHOXYACRI-DINE DIHYDROCHLORIDE see CFU750

3-CHLORO-9-(4'-DIETHYLAMINO-1'-METHYLBUTYLAMINO)-7-METHOXYACRI-DINE DIHYDROCHLORIDE see CFU750

6-CHLORO-9-((4-DIETHYL AMINO)-1-METHYL BUTYL)AMINO)-2-METHOXYA-CRIDINE see ARQ250

7-CHLORO-4-(4-DIETHYLAMINO-1-METHYLBUTYLAMINO)QUINOLINE see CLD000

7-CHLORO-4-((4'-DIETHYLAMINO-1'-METHYLBUTYL)AMINO)QUINOLINE DIPHOS-PHATE see CLD250

7-CHLORO-4((4-(DIETHYLAMINO)-1-METHYLBUTYL)AMINO)-QUINOLINE PHOS-PHATE (1: 1) see AQT250

2-CHLORO-10-(3-DIETHYLAMINOPROPYL)PHENOTHIAZINE see CLY750

2-CHLORO-10-(3'-DIETHYLAMINOPROPYL)PHENOTHIAZINE HYDROCHLORIDE see CLZ000

2'-CHLORO-2-(DIETHYLAMINO)-5'-TRIFLUOROMETHYLACETANILIDE HYDRO-CHLORIDE see CFW000

4'-CHLORO-2-(DIETHYLAMINO)-3'-TRIFLUOROMETHYLACETANILIDE HYDRO-CHLORIDE see CFW250

CHLORODIETHYLBORANE see CFW625

2-CHLORO-2',6'-DIETHYL-N-(BUTOXYMETHYL)ACETANILIDE see CFW750

2-CHLORO-2-DIETHYLCARBAMOYL-1-METHYLVINYL DIMETHYLPHOSPHATE see FAB400

1-CHLORO-DIETHYLCARBAMOYL-1-PROPEN-2-YL DIMETHYL PHOSPHATE see FAB400

2-CHLORO-2',6'-DIETHYL-N-(METHOXYMETHYL)ACETANILIDE see CFX000

2-CHLORO-N-(2,6-DIETHYLPHENYL)-N-(METHOXYMETHYL)ACETAMIDE see CFX000

2-CHLORO-N-(2,6-DIETHYLPHENYL)-N-(2-PROPOXYETHYL)ACETAMIDE see PMB850

2-CHLORO-2',6'-DIETHYL-N-(2-PROPOXYETHYL)ACETANILIDE see PMB850

CHLORODIFLUOROACETYL HYPOCHLORITE see CFX125

CHLORODIFLUOROBROMOMETHANE (DOT) see BNA250

1-CHLORO-1,1-DIFLUOROETHANE see CFX250

CHLORODIFLUOROETHANES (DOT) see CFX250

CHLORODIFLUOROMETHANE (ACGIH,DOT,OSHA) see CFX500

CHLORODIFLUOROMETHANE see CFX500

CHLORODIFLUOROMETHANE and CHLOROPENTAFLUOROETHANE MIXTURE (DOT) see FOO510

1-CHLORO-3,3-DIFLUORO-2-METHOXYCYCLOPROPENE see CFX625

2-CHLORO-1-(DIFLUOROMETHOXY)-1,1,2-TRIFLUOROETHANE see EAT900

CHLORODIFLUOROMONOBROMOMETHANE see BNA250

10-CHLORO-5,10-DIHYDROARSACRIDINE see PDB000

6-CHLORO-3,4-DIHYDRO-2H-1,2,4-BENZOTHIADIAZINE-7-SULFONAMIDE- 1,1-DI-OXIDE see CFY000

4'-CHLORO-2-((3-(10,11-DIHYDRO-5H-DIBENZ(b,f)AZEPIN-5-YL)PROPYL)METHYLAM INO)ACETOPHENONE HCl see IFZ900

1-(8-CHLORO-10,11-DIHYDRODIBENZO(b,f)THIEPIN-10-YL)-4-METHYLPIPERA-ZINE see ODY100

12-CHLORO-7,12-DIHYDRO-8,11-DIMETHYLBENZO(a)PHENARSAZINE see CGH500

7-CHLORO-1,3-DIHYDRO-3-(N,N-DIMETHYLCARBAMOYL)-1-METHYL-5-PHENYL-2H-1,4-BENZODIAZEPIN-2-ONE see CFY250

3-CHLORO-10,11-DIHYDRO-N,N-DIMETHYL-5H-DIBENZ(b,f)AZEPINE-5-PROPA-NAMINE MONOHYDROCHLORIDE see CDV000

5-CHLORO-2,3-DIHYDRO-1,1-DIMETHYL-1H-INDOLIUM BROMIDE (9CI) see MIH300

4-CHLORO-2,6-DINITROANILINE see CGL500
2-CHLORO-4,6-DINITROANILINE see CGL325
4-CHLORO-1,3-DINITROBENZENE see CGM000
6-CHLORO-1,3-DINITROBENZENE see CGM000
1-CHLORO-2,4-DINITROBENZENE see CGM000
CHLORODINITROBENZENE (mixed isomers) (DOT) see CGL750
CHLORODINITROBENZENE see CGL750
4-CHLORO-2,5-DINITROBENZENE DIAZONIUM-6-OXIDE see CGM199
1-CHLORO-2,4-DINITROBENZOL (GERMAN) see CGM000
4-CHLORO-3,5-DINITROBENZOTRIFLUORIDE see CGM225
1-CHLORO-2,4-DINITRONAPHTHALENE see DUS600
4-CHLORO-3,5-DINITRO-α-α-α-TRIFLUOROTOLUENE see CGM225
p-CHLORO-2,4-DIOXA-5-ETHYL-p-THIONO-3-PHOSPHABICYCLO(4.4.0)DECANE see CGM375
p-CHLORO-2,4-DIOXA-5-METHYL-p-THIONO-3-PHOSPHABICY-CLO(4.4.0)DECANE see CGM400
CHLORO((3-(2,4-DIOXO-3-IMIDAZOLIDINYL)-2-METHOXY)PROPYL)MERCURY see CHV750
CHLORO((3-(2,4-DIOXO-5-IMIDAZOLIDINYL)-2-METHOXY)PROPYL) MERCURY see CGM450
6-CHLORO-1,3-DIOXO-5-ISOINDOLINESULFONAMIDE see CGM500, CLG825
CHLORO((3-(2,4-DIOXO-1-METHYL-3-IMIDAZOLIDINYL)-2-ME-THOXY)PROPYL)MERCURY see CHW250
CHLORO((3-(2,4-DIOXO-3-METHYL-1-IMIDAZOLIDINYL)-2-ME-THOXY)PROPYL)MERCURY see CHW000
CHLORO((3-(2,4-DIOXO-3-METHYL-5-IMIDAZOLIDINYL)-2-ME-THOXY)PROPYL)MERCURY see CHW500
2-CHLORODIPHENYL see CGM750
o-CHLORODIPHENYL see CGM750
2-CHLORO-2,2-DIPHENYLACETIC ACID-2-(DIETHYLAMINO)ETHYL ESTER HY-DROCHLORIDE see DHW200
CHLORODIPHENYLARSINE see CGN000
1-(o-CHLORO-α,α-DIPHENYLBENZYL)IMIDAZOLE see MRX500
CHLORODIPHENYL (42% Cl) (OSHA) see PJM500
CHLORODIPHENYL (54% Cl) (OSHA) see PJN000
CHLORODIPHENYL (21% Cl) see PJM000
CHLORODIPHENYL (32% Cl) see PJM250
CHLORODIPHENYL (41% Cl) see PJL800
CHLORODIPHENYL (48% Cl) see PJM750
CHLORODIPHENYL (62% Cl) see PJN500
CHLORODIPHENYL (68% Cl) see PJN750
CHLORODIPHENYL (60% Cl) see PJN250
2-CHLORO-1,2-DIPHENYLETHANONE see CFJ100
2-(4-(2-CHLORO-1,2-DIPHENYLETHENYL)PHENOXY)-N,N-DIETHYLETHANAMINE see CMX500
1-(p-CHLORODIPHENYLMETHYL)-4-(2-(2-HYDROXYETHOXY)ETHYL)PIPERAZINE see CJR909
4-CHLORODIPHENYL SULFONE see CKI625
R-CHLORODIPHENYL SULPHONE see CKI625
2-(p-(2-CHLORO-1,2-DIPHENYL VINYL)PHENOXY)TRIETHYLAMINE CITRATE (1:1) see CMX700
2-CHLORO-N,N-DI-2-PROPENYLACETAMIDE see CFK000
6′-CHLORO-2-(DIPROPYLAMINO)-o-ACETOTOLUIDIDE HYDROCHLORIDE see CGN250
CHLORODIPROPYLBORANE see CGN325
CHLORODRACYLIC ACID see CEL300
CHLOROEPOXYETHANE see CGX000
3-CHLORO-1,2-EPOXYPROPANE see EAZ500
1-CHLORO-2,3-EPOXYPROPANE see EAZ500
CHLOROETENE see MIH275
2-CHLORO-1-ETHANAL see CDY500
2-CHLOROETHANAL see CDY500
2-CHLOROETHANAMIDE see CDY850
CHLOROETHANE see EHH000
2-CHLOROETHANEPHOSPHONIC ACID see CDS125
2-CHLOROETHANE SULFOCHLORIDE see CGO125
2-CHLOROETHANESULFONYL CHLORIDE see CGO125
β-CHLOROETHANESULFONYL CHLORIDE see CGO125
CHLOROETHANOIC ACID see CEA000
2-CHLOROETHANOL (MAK) see EIU800
Δ-CHLOROETHANOL see EIU800
2-CHLOROETHANOL ACETATE see CGO600
2-CHLOROETHANOL ACRYLATE see ADT000
2-CHLOROETHANOL-2-(p-tert-BUTYLPHENOXY)-1-METHYLETHYL SULFITE see SOP500
2-CHLOROETHANOL ESTER with 2-(p-tert-BUTYLPHENOXY)-1-METHYLETHYL SULFITE see SOP500
2-CHLOROETHANOL HYDROGEN PHOSPHATE ESTER with 3-CHLORO-7-HY-DROXY-4-METHYLCOUMARIN see DFH600
2-CHLORO-ETHANOL-4-METHYLBENZENESULFONATE (9CI) see CHI125
2-CHLOROETHANOL PHOSPHATE see CGO500
2-CHLOROETHANOL PHOSPHATE DIESTER ESTER with 3-CHLORO-7-HY-DROXY-4-METHYLCOUMARIN see DFH600
2-CHLOROETHANOL PHOSPHITE (3:1) see PHO000
2-CHLORO-ETHANOL, PHOSPHOROTHIOATE (3:1) see PHN500

2-CHLORO-ETHANOL-p-TOLUENESULFONATE (8CI) see CHI125
CHLOROETHENE see MIH275, VNP000
CHLOROETHENE HOMOPOLYMER see PKQ059
(2-CHLOROETHENYL) ARSONOUS DICHLORIDE see CLV000
1,1′,1′′-(1-CHLORO-1-ETHENYL-2-YLIDENE)-TRIS(4-METHOXYBENZENE) see CLO750
17-α-CHLOROETHINYL-17-β-HYDROXYESTRA-4,9-DIEN-3-ONE see CHP750
2-(2-CHLOROETHOXY)ETHANOL see DKN000
(2-CHLOROETHOXY)ETHENE see CHI250
2-(2-CHLOROETHOXY)ETHYL 2′-CHLOROETHYL ETHER see TKL500
2-CHLOROETHYL ACETATE see CGO600
β-CHLOROETHYL ACETATE see CGO600
2-CHLOROETHYL ACRYLATE see ADT000
β-CHLOROETHYL ACRYLATE see ADT000
CHLOROETHYL ACRYLATE see ADT000
2-CHLOROETHYL ALCOHOL see EIU800
β-CHLOROETHYL ALCOHOL see EIU800
2-CHLOROETHYLAMINE see CGP125
2-CHLOROETHYLAMINE HYDROCHLORIDE see CGP250
β-CHLOROETHYLAMINE HYDROCHLORIDE see CGP250
2-CHLORO-4-ETHYLAMINEISOPROPYLAMINE-s-TRIAZINE see ARQ725
3′-CHLORO-2-ETHYLAMINO-o-ACETOTOLUIDIDE HYDROCHLORIDE see CGP375
6′-CHLORO-2-(ETHYLAMINO)-o-ACETOTOLUIDIDE HYDROCHLORIDE see CGP380
2-CHLORO-4-ETHYLAMINO-6-(1-CYANO-1-METHYL)ETHYLAMINO-s-TRIAZINE see BLW750
7-((2-((2-CHLOROETHYL)AMINO)ETHYL)AMINO)BENZ(c)ACRIDINE DIHYDRO-CHLORIDE HYDRATE see CGP500
9-(2-((2-CHLOROETHYL)AMINO)ETHYLAMINO)-6-CHLORO-2-METHOXYACRI-DINE, DIHYDROCHLORIDE see QCS875
2-CHLOROETHYLAMINOETHYL DEHYDROABIETATE HYDROCHLORIDE see CGQ250
N-(2-CHLOROETHYL)AMINOETHYL-4-ETHOXYNITROBENZENE see CGX325
6′-CHLORO-2-(ETHYLAMINO)-o-HEXANOTOLUIDIDE HYDROCHLORIDE see CAB750
2-CHLORO-4-ETHYLAMINO-6-ISOPROPYLAMINO-1,3,5-TRIAZINE see ARQ725
1-CHLORO-3-ETHYLAMINO-5-ISOPROPYLAMINO-2,4,6-TRIAZINE see ARQ725
1-CHLORO-3-ETHYLAMINO-5-ISOPROPYLAMINO-s-TRIAZINE see ARQ725
2-CHLORO-4-ETHYLAMINO-6-ISOPROPYLAMINO-s-TRIAZINE see ARQ725
2′-CHLORO-2-(ETHYLAMINO)-6′-METHYLACETANILIDE, HYDROCHLORIDE see CGP380
N-(2-CHLOROETHYL)AMINOMETHYL-4-HYDROXYNITROBENZENE see CGQ280
N-(2-CHLOROETHYL)AMINOMETHYL-4-METHOXYNITROBENZENE see CGQ400
2-(((2-CHLOROETHYL)AMINO)METHYL)-4-NITROPHENOL see CGQ280
2-CHLORO-3-(ETHYLAMINO)-1-METHYL-3-OXO-1-PROPENYL DIMETHYL ESTER PHOSPHORIC ACID see DTP600
2-CHLORO-3-(ETHYLAMINO)-1-METHYL-3-OXO-1-PROPENYL DIMETHYL PHOS-PHATE see DTP600
6-CHLORO-2-ETHYLAMINO-4-METHYL-4-PHENYL-4H-3,1-BENZOXAZINE see CGQ500
9-((3-((2-CHLOROETHYL)AMINO)PROPYL)AMINO)ACRIDINE DIHYDROCHLO-RIDE HYDRATE see CGR000
7-((3-((2-CHLOROETHYL)AMINO)PROPYL)AMINO)BENZ(c)ACRIDINE DIHYDRO-CHLORIDE SESQUIHYDRATE see CGR250
7-((3-((2-CHLOROE-THYL)AMINO)PROPYL)AMINO)BENZO(b)(1,10)PHENANTHROLINE DIHYDRO-CHLORIDE see CGR500
7-((3-((2-CHLOROE-THYL)AMINO)PROPYL)AMINO)BENZO(b)(1,8)PHENANTHROLINE DIHYDRO-CHLORIDE HYDRATE see CGR750
10-((2-CHLOROETHYLAMINO)PROPYLAMINO)-2-METHOXY-7-CHLOROBEN-ZO(b)-(1,5)-NAPHTHYRIDINE see CGS500
4-((3-((2-CHLOROETHYL)AMINO)PROPYL)AMINO)-6-METHOXYQUINOLINE HY-DROCHLORIDE see CGS750
2-(4-CHLORO-6-ETHYLAMINO-1,3,5-TRIAZINE-2-YLAMINO)-2-METHYLPROPIONI-TRILE see BLW750
2-(4-CHLORO-6-ETHYLAMINO-s-TRIAZINE-2-YLAMINO)-2-METHYL-PROPIONI-TRILE see BLW750
2-((4-CHLORO-6-(ETHYLAMINO)-1,3,5-TRIAZIN-2-YL)AMINO)-2-METHYL-PROPAN-ENITRILE see BLW750
2-((4-CHLORO-6-(ETHYLAMINO)-s-TRIAZIN-2-YL)AMINO)-2-METHYLPROPIONI-TRILE see BLW750
CHLOROETHYLAMINOURACIL see DYC700
6′-CHLORO-2-(ETHYLAMINO)-o-VALEROTOLUIDIDE HYDROCHLORIDE see CGT000
2-CHLOROETHYLAMMONIUM CHLORIDE see CGP250
2-(2-CHLOROETHYL)-3-AZA-4-CHROMANONE see CDS250
CHLOROETHYLBENZENE see EHH500
3-(2-CHLOROETHYL)-2-(BIS(2-CHLOROETHYL)AMINO)PERHYDRO-2H-1,3,2-OX-AZAPHOSPHORINE-2-OXIDE see TNT500
β-CHLOROETHYL-β-(BIS(β-HYDROXYETHYL)SULFONIUM)ETHYL SULFIDE CHLORIDE see BKD750

1-(2-CHLOROETHYL)-3-MORPHOLINO-1-NITROSOUREA see MRR775
N-(2-CHLOROETHYL)-N'-(2-CHLOROETHYL)-N',O-PROPYLENEPHOSPHORIC ACID DIAMIDE see IMH000
N-(2-CHLOROETHYL)-N'-(2-CHLOROETHYL)-N',O-PROPYLENEPHOSPHORIC ACID ESTER DIAMIDE see IMH000
N-(2-CHLOROETHYL)-N'-(7-CHLORO-2-METHOXYBENZO(b)-1,5-NAPHTHYRIDIN-10-YL)-1,3-PROPANEDIAMINE see CGS500
N-(2-CHLOROETHYL)-N'-CYCLODODECYL-N-NITROSOUREA see CGV000
N-(2-CHLOROETHYL)-N'-CYCLOHEXYL-N-NITROSOUREA see CGV250
N-(2-CHLOROETHYL)-N'-(2,6-DIOXO-3-PIPERIDINYL)-N-NITROSOUREA see CGW250
trans-N-(2-CHLOROETHYL)-N'-(2-HYDROXYCYCLOHEXYL)-N-NITROSOUREA see CHA750
cis-N-(2-CHLOROETHYL)-N'-(4-HYDROXYCYCLOHEXYL)-N-NITROSOUREA see CHA500
trans-N-(2-CHLOROETHYL)-N'-(4-HYDROXYCYCLOHEXYL)-N-NITROSOUREA see CHB000
N-(2-CHLOROETHYL)-N'-(trans-2-HYDROXYCYCLOHEXYL)-N-NITROSOUREA see CHA750
N-(2-CHLOROETHYL)-N'-(trans-4-HYDROXYCYCLOHEXYL)-N-NITROSOUREA see CHB000
N-(2-CHLOROETHYL)-N-NITROSOACETAMIDE see CHE250
N-(β-CHLOROETHYL)-N-NITROSOACETAMIDE see CHE250
4-((((2-CHLOROETHYL)NITROSOAMINO)CARBONYL)AMINO)CYCLOHEXANE CARBOXYLIC ACID, ETHYL ESTER see CHF000
2-((((2-CHLOROETHYL)NITROSOAMINO)CARBONYL)AMINO)-2-DEOXY-d-GLUCOPYRANOSE see CLX000
2-((((2-CHLOROETHYL)NITROSOAMINO)CARBONYL)AMINO)-2-DEOXY-d-GLUCOSE see CLX000
(17-β)-17-((((2-CHLOROETHYL)NITROSOAMINO)CARBONYL)OXY)ESTR-4-EN-3-ONE see NNX600
(2-CHLOROETHYL)NITROSOCARBAMIC ACID-2-FLUOROETHYL ESTER see FIH000
N-2-CHLOROETHYL-N-NITROSO-CARBAMIC ACID METHYLESTER see MIH250
N-2-CHLOROETHYL-N-NITROSOCARBOMOYL AZIDE see CHE325
N-(2-CHLOROETHYL)-N-NITROSOETHYLCARBAMATE see CHF500
1-(2-CHLOROETHYL)-1-NITROSO-3-(2-NORBORNYL)UREA see CHE500
1-(2-CHLOROETHYL)-1-NITROSO-3-RIBOPYRANOSYLUREA-2',3',4'-TRIACETATE see ROF200
1-(2-CHLOROETHYL)-1-NITROSOUREA see CHE750
N-(2-CHLOROETHYL)-N-NITROSOUREA see CHE750
trans-4-(3-(2-CHLOROETHYL))-3-NITROSOUREIDOCYCLOHEXANE CARBOXYLIC ACID ETHYL ESTER see CHF000
2-(3-(2-CHLOROETHYL)-3-NITROSOUREIDO)-2-DEOXY-d-GLUCOSOPYRANOSE see CLX000
2-(3-(2-CHLOROETHYL)3-NITROSOUREIDO)ETHYL METHANE SULFONATE see CHF250
2-(3-(2-CHLOROETHYL)-3-NITROSOUREIDO)-d-GLUCO-PYRANOSE see CLX000
N-(β-CHLOROETHYL)-N-NITROSOURETHAN see CHF500
2-CHLOROETHYL-N-NITROSOURETHANE see CHF500
trans-N-(2-CHLOROETHYL)-N'-(3-METHYLCYCLOHEXYL)-N-NITROSOUREA see CHD500
N-(2-CHLOROETHYL)-N'-(trans-4-METHYLCYCLOHEXYL)-N-NITROSOUREA see CHD250
6-CHLORO-N-ETHYL-N'-(1-METHYLETHYL)-1,3,5-TRIAZINE-2,4-DIAMINE (9CI) see ARQ725
1-(2-CHLOROETHYL)-3-(2-NORBORNYL)-1-NITROSOUREA see CHE500
CHLOROETHYLOWY ALKOHOL (POLISH) see EIU800
1-CHLORO-3-ETHYL-1-PENTEN-4-YN-3-OL see CHG000
CHLOROETHYLPHENAMIDE see BEG000
4-(4-CHLORO-6-ETHYLPHENYLAMINO)-2-s-TRIAZINYLAMINO-5-HYDROXY-6-(4-METHYL-2-SULFOPHENYLAZO)-2,7-NAPHTHALEN see CHG250
p-((4-CHLORO-3-ETHYLPHENYL)AZO)-N,N-DIMETHYLANILINE see CGW105
p-((3-CHLORO-4-ETHYLPHENYL)AZO)-N,N-DIMETHYLANILINE see CGW100
(2-CHLOROETHYL)PHOSPHONIC ACID DIETHYL ESTER see CHG375
(2-CHLOROETHYL)PHOSPHONIC ACID MONOETHYL ESTER see CHG400
3-(2-CHLOROETHYL)-2-PICOLINE HYDROCHLORIDE see CHD700
1-(2-CHLOROETHYL)PIPERIDINE HYDROCHLORIDE see CHG500
β-CHLOROETHYLPIPERIDINE HYDROCHLORIDE see CHG500
2'-CHLORO-2-(ETHYL(2-PIPERIDINOETHYL)AMINO)ACETANILIDE DIHYDROCHLORIDE see CHG750
7-(3-(2-CHLOROETHYL-n-PROPYLAMINO)PROPYLAMINO)BENZO(b)(1,10)-PHENATHROLINE HYDROCHLORIDE see CHH000
2-CHLORO-5-ETHYL-4-PROPYL-2-THIONO-1,3,2-DIOXAPHOSPHORINANE see CHH125
(CHLORO-2-ETHYL)-1-(RIBOFURANOSYLISOPROPYLIDENE-2'-3'-PARANITROBENZOATE-5')-3-NITROSOUREA see RFU600
(CHLORO-2-ETHYL)-1-(RIBOPYRANOSYLTRIACETATE-2',3',4')-3-NITROSOUREA see ROF200
2-CHLOROETHYLSULFONYL CHLORIDE see CGO125
2-CHLOROETHYL SULFUROUS ACID-2-(4-(1,1-DIMETHYLETHYL)PHENOXY)-1-METHYLETHYL ESTER see SOP500
2-CHLOROETHYL SULPHITE of 1-(p-tert-BUTYLPHENOXY)-2-PROPANOL see SOP500

2-((3-(2-CHLOROETHYL)TETRAHYDRO-2H-1,3,2-OXAZAPHOSPHORIN-2-YL)AMINO)-ETHANOL, METHANESULFONATE (ester), p-OXIDE see SOD000
1-CHLORO-2-(ETHYLTHIO)ETHANE see CGY750
2-((2-CHLOROETHYL)THIO)ETHANOL see CHC000
2-(2-CHLOROETHYL)THIOETHYLBIS(2-HYDROXYETHYL)-CHLORIDE see BKD750
2-CHLORO-N-(6-ETHYL-o-TOLYL)-N-(2-METHOXY-1-METHYLETHYL)-ACETAMIDE see MQQ450
2-CHLOROETHYL TOSYLATE see CHI125
N-(2-CHLOROETHYL)-α,α,α-TRIFLUORO-2,6-DINITRO-N-PROPYL-p-TOLUIDINE see FDA900
(2-CHLOROETHYL)TRIMETHYLAMMONIUM CHLORIDE see CMF400
(β-CHLOROETHYL)TRIMETHYLAMMONIUM CHLORIDE see CMF400
2-CHLOROETHYL VINYL ETHER see CHI250
CHLOROETHYNE see ACJ000
17-α-CHLOROETHYNLY-19-NOR-4,9-ANDROSTADIEN-17-β-OL-3-ONE see CHP750
17-α-CHLOROETHYNYL-17-β-HYDROXY-19-NOR-4,9-ANDROSTADIEN-3-ONE see CHP750
CHLOROETHYNYL NORGESTREL mixed with MESTRANOL (20:1) see CHI750
4-CHLORO-α-ETHYNYL-α-PHENYLBENZENEMETHANOL CARBAMATE see CKI250
(CHLORO-2-ETIL)-1-(RIBOFURANOSILISOPROPILIDENE-2',3'-PARANITROBENZOATO)-3-NITROSOUREA see RFU600
CHLOROFENIZON see CJT750
3-(4-(4-CHLORO-FENOSSIL))-1,1-DIMETIL-UREA (ITALIAN) see CJQ000
CHLOROFENSULPHIDE see CKL500
CHLOROFENVINPHOS see CDS750
p-CHLOROFENYLESTER KYSELINY BENZENSULFONOVE (CZECH) see CJR500
N-(7-CHLORO-2-FLUORENYL)ACETAMIDE see CHI825
N-2-(7-CHLORO)FLUORENYLACETAMIDE see CHI825
2-CHLORO-N-(9-FLUORENYL)DIETHYLAMINE HYDROCHLORIDE see FEE100
1-CHLORO-10-FLUORO-DECANE see FHN000
21-CHLORO-9-FLUORO-11-β,17-DIHYDROXY-16-β-METHYLPREGNA-1,4-DIENE-3,20-DIONE-17-PROPIONATE see CMW300
21-CHLORO-9-FLUORO-17-HYDROXY-16-β-METHYLPREGNA-1,4-DIENE-3,11,20-TRIONE BUTYRATE see CMW400
CHLOROFLUOROMETHANE see CHI900
2-CHLORO-1-FLUORO-4-NITROBENZENE see CHI950
3-CHLORO-4-FLUORONITROBENZENE see CHI950
1-CHLORO-8-FLUOROOCTANE see FKT000
1-CHLORO-5-FLUOROPENTANE see FFW000
7-CHLORO-5-(o-FLUOROPHENYL)-1,3-DIHDYRO-1-METHYL-2H-1,4-BENZO-DIAZEPIN-2-ONE see FDB100
7-CHLORO-5-(2-FLUOROPHENYL)-1-METHYL-1H-1,4-BENZODIAZEPIN-2(3H)-ONE see FDB100
8-CHLORO-6-(2-FLUOROPHENYL)-1-METHYL-4H-IMIDAZO(1,5-a)(1,4)BENZODIAZEPINE see MQT525
6-(3-(2-CHLORO-6-FLUOROPHENYL)-5-METHYL-4-ISOXAZOLECARBOXAMI-DO)PENICILLANIC ACID SODIUM SALT see CHJ000
3-(2-CHLORO-6-FLUOROPHENYL)-5-METHYL-4-ISOXAZOLYLPENICILLIN SODIUM MONOHYDRATE see CHJ000
3-CHLORO-2-FLUORO-1-PROPENE see CHJ250
3-CHLORO-2-FLUOROPROPENE see CHJ250
2-CHLORO-4'-FLUORO-α-(PYRIMIDIN-5-YL)BENZHYDRYL ALCOHOL see CHJ300
(+-)-2-CHLORO-4'-FLUORO-α-(PYRIMIDIN-5-YL)BENZHYDRYL ALCOHOL see CHJ300
2-CHLORO-4'-FLUORO-α-(PYRIMIDIN-5-YL)DIPHENYLMETHANOL see CHJ300
CHLORO FLUORO SULFONE see SOT500
21-CHLORO-9-FLUORO-11-β,16-α-17-TRIHYDROXYPREGN-4-ENE-3,20-DIONE cyclic 16,17-ACETAL with ACETONE see HAF300
CHLOROFLURAZOLE see DGO400
CHLOROFLURENOL-METHYL ESTER see CDT000
CHLOROFOLIN see CKN000
CHLOROFORM see CHJ500
CHLOROFORMAMIDINIUM CHLORIDE see CHJ599
CHLOROFORMAMIDINIUM NITRATE see CHJ625
CHLOROFORME (FRENCH) see CHJ500
CHLOROFORMIC ACID ALLYL ESTER see AGB500
CHLOROFORMIC ACID BENZYL ESTER see BEF500
CHLOROFORMIC ACID 2-CHLOROETHYL ESTER see CGU199
CHLOROFORMIC ACID DIMETHYLAMIDE see DQY950
CHLOROFORMIC ACID, ESTER with 1-NAPHTHOL see NBH200
CHLOROFORMIC ACID ETHYL ESTER see EHK500
CHLOROFORMIC ACID 2-FLUOROETHYL ESTER see FIH100
CHLOROFORMIC ACID ISOPROPYL ESTER see IOL000
CHLOROFORMIC ACID 1-MAPHTHYL ESTER see NBH200
CHLOROFORMIC ACID METHYL ESTER see MIG000
CHLOROFORMIC ACID PHENYL ESTER see CBX109
CHLOROFORMIC ACID PROPYL ESTER see PNH000
CHLOROFORMIC DIGITALIN see DKN400

CHLOROFORMYL CHLORIDE see PGX000
CHLOROFOS see TIQ250
CHLOROFTALM see TIQ250
CHLORO-2-FURANYL MERCURY see CHK000
4-CHLORO-N-FURFURYL-5-SULFAMOYLANTHRANILIC ACID see CHJ750
CHLORO(2-FURYL)MERCURY see CHK000
6'-CHLORO-2-(2-FURYLMETHYL)AMINO-o-ACETOTOLUIDIDE HYDROCHLORIDE
see CHK125
4-CHLORO-N-(2-FURYLMETHYL)-5-SULFAMOYLANTHRANILIC ACID see
CHJ750
CHLOROFYL see CKN000
CHLOROGENIC ACID see CHK175
CHLOROGERMANE see CHK250
CHLOROGUANIDE see CKB250
CHLOROGUANIDE HYDROCHLORIDE see CKB500
CHLOROGUANIDINE HYDROCHLORIDE see CKB500
5-CHLORO-3(H)-2-BENZOXAZOLONE see CDQ750
(2-CHLORO-1-HEPTENYL)ARSONIC ACID MONOAMMONIUM SALT see
CEI250
2-CHLORO-1,1,1,4,4,4-HEXAFLUOROBUTENE-2 see CHK750
endo-4-CHLORO-N-(HEXAHYDRO-4,7-METHANOISOINDOL-2-YL)-3-SULFAMOYL-
BENZAMIDE see CHK825
2-CHLORO-10-(3-(HEXAHYDROPYRROLO(1,2-a)PYRAZIN-2(1H)-YL)-PROPIO-
NYL)PHENOTHIAZINE 2HCl see NMV750
CHLORO(2-HEXANAMIDOCYCLOHEXYL)MERCURY, (E)- see CHL000
trans-CHLORO(2-HEXANAMIDOCYCLOHEXYL)MERCURY see CHL000
CHLOROHEXYL ISOCYANATE see CHL250
4-CHLORO-2-HEXYLPHENOL see CHL500
CHLOROHYDRIC ACID see HHL000
α-CHLOROHYDRIN see CDT750
dl-α-CHLOROHYDRIN see CHL875
epi-CHLOROHYDRIN see EAZ500
α-CHLOROHYDRIN DIACETATE see CIL900
CHLOROHYDROQUINONE see CHM000
5-CHLORO-4-(HYDROXYAMINO)QUINOLINE-1-OXIDE see CHM500
6-CHLORO-4-(HYDROXYAMINO)QUINOLINE-1-OXIDE see CHM750
7-CHLORO-4-(HYDROXYAMINO)QUINOLINE-1-OXIDE see CHN000
3-CHLORO-4-HYDROXYBIPHENYL see CHN500
CHLORO(2-HYDROXY-3,5-DINITROPHENYL)MERCURY see CHN750
3-CHLORO-4-HYDROXYDIPHENYL see CHN500
5-CHLORO-2-HYDROXYDIPHENYLMETHANE see CJU250
2-CHLORO-N-(2-HYDROXYETHYL)ANILINE see CHO125
4-CHLORO-6-(2-HYDROXYETHYLPIPERAZINO-2-METHYLAMINO-5)-METHYL-
THIOPYRIMIDINE see CHO250
5-CHLORO-3-(4-(2-HYDROXYETHYL)-1-PIPERAZINYL)CARBONYLMETHYL-2-
BENZOTHIAZOLINONE see CJH750
2-CHLORO-10-3-(1-(2-HYDROXYETHYL)-4-PIPERAZINYL)PROPYL PHENOTHI-
AZINE see CJM250
5-CHLORO-8-HYDROXY-7-IODOQUINOLINE see CHR500
2-CHLORO-4-(HYDROXY MERCURI)PHENOL see CHO750
5-CHLORO-4-HYDROXYMETANILIC ACID see AJH500
2-CHLORO-9-HYDROXY-9-METHYLCARBOXYLATEFLUORENE see CDT000
3-CHLORO-7-HYDROXY-4-METHYLCOUMARIN BIS(2-CHLOROE-
THYL)PHOSPHATE see DFH600
3-CHLORO-7-HYDROXY-4-METHYL-COUMARIN-O,O-DIETHYL PHOSPHOROTH-
IOATE see CNU750
3-CHLORO-7-HYDROXY-4-METHYL-COUMARIN-O-ESTER with O,O-DIETHYL
PHOSPHOROTHIOATE see CNU750
(−)-N-((5-CHLORO-8-HYDROXY-3-METHYL-1-OXO-7-ISOCHROMA-
NYL)CARBONYL)-3-PHENYLALANINE see CHP250
6-CHLORO-17-α-HYDROXY-16-α-METHYLPREGNA-4,6-DIENE-3,20-DIONE see
CHP375
5-CHLORO-2-((2-HYDROXY-1-NAPHTHALENYL)AZO)-4-METHYLBENZENE SUL-
FONIC ACID, BARIUM SALT (2:1) see CHP500
5-CHLORO-2-((2-HYDROXY-1-NAPHTHALENYL)AZO)-4-METHYLBENZENESUL-
FONIC ACID SODIUM SALT see CMS150
5-CHLORO-2-((2-HYDROXY-1-NAPHTHALENYL)AZO)-4-METHYLBENZENE SUL-
PHONIC ACID, BARIUM SALT see CHP500
5-CHLORO-2-((2-HYDROXY-1-NAPHTHYL)AZO)-p-TOLUENE SULFONIC ACID,
BARIUM SALT see CHP500
21-CHLORO-17-HYDROXY-19-NOR-17-α-PREGNA-4,9-DIEN-20-YN-3-ONE see
CHP750
7-CHLORO-3-HYDROXY-5-PHENYL-1,3-DIHYDRO-2H-1,4-BENZODIAZEPIN-2-ONE
see CFZ000
(3-CHLORO-4-HYDROXYPHENYL)HYDROXYMERCURY see CHO750
CHLORO(o-HYDROXYPHENYL)MERCURY see CHW675
CHLORO(p-HYDROXYPHENYL)MERCURY see CHW750
7-CHLORO-10-(2-HYDROXY-3-PIPERIDINOPROPYL)ISOALLOXAZINE SULFATE
see CHQ000
6-CHLORO-17-HYDROXYPREGNA-4,6-DIENE-3,20-DIONE see CDY275
6-CHLORO-17-α-HYDROXYPREGNA-4,6-DIENE-3,20-DIONE ACETATE see
CBF250
6-CHLORO-17-α-HYDROXY-Δ⁶-PROGESTERONE ACETATE see CBF250
1-(3-CHLORO-2-HYDROXYPROPYL)-5-IODO-2-METHYL-4-NITROIMIDAZOLE see
CIN000

1-(3-CHLORO-2-HYDROXYPROPYL)-2-METHYL-5-NITROIMIDAZOLE see OJS000
5-CHLORO-8-HYDROXYQUINOLINE see CLD600
2-CHLORO-HYDROXYTOLUENE see CFE250
6-CHLORO-3-HYDROXYTOLUENE see CFE250
5-CHLORO-4-(2-IMIDAZOLIN-2-YLAMINO)-2,1,3-BENZOTHIADIAZOLE HYDRO-
CHLORIDE see TGH600
4-CHLOROIMINO-2,5-CYCLOHEXADIENE-1-ONE see CHQ500
4-CHLOROIMINO-2,6-DIBROMO-2,5-CYCLOHEXADIENE-1-ONE see CHQ750
4-CHLOROIMINO-2,6-DICHLORO-2,5-CYCLOHEXADIENE-1-ONE see CHR000
3-CHLOROIMIPRAMINE see CDU750
3-CHLOROIMIPRAMINE HYDROCHLORIDE see CDV000
CHLOROIMIPRAMINE MONOHYDROCHLORIDE see CDV000
CHLOROIN see CLD250
7-CHLORO-4-INDANOL see CDV700
CHLOROIODOACETYLENE see CHR325
CHLOROIODOETHYNE see CHR325
5-CHLORO-7-IODO-8-HYDROXYQUINOLINE see CHR500
3-CHLORO-1-IODOPROPYNE see CHR400
CHLOROIODOQUINE see CHR500
5-CHLORO-7-IODO-8-QUINOLINOL see CHR500
2-CHLOROISOBUTANE see BQR000
3'-CHLORO-2-ISOBUTYLAMINO-p-ACETOTOLUIDIDE HYDROCHLORIDE see
CHR700
6'-CHLORO-2-(ISOBUTYLAMINO)-o-ACETOTOLUIDIDE HYDROCHLORIDE see
CHR750
3'-CHLORO-3-ISOBUTYLAMINO-o-PROPIONOTOLUIDIDE HYDROCHLORIDE see
CHR850
α-CHLOROISOBUTYLENE see IKE000
γ-CHLOROISOBUTYLENE see CIU750
4'-CHLOROISOBUTYROPHENONE see IOL100
p-CHLOROISOBUTYROPHENONE see IOL100
2-CHLORO-N-ISOPROPYLACETANILIDE see CHS500
α-CHLORO-N-ISOPROPYLACETANILIDE see CHS500
N-(3-CHLORO-1-ISOPROPYLACETONYL)-p-TOLUENESULFONAMIDE see
THH555
6'-CHLORO-2-(ISOPROPYLAMINO)-o-ACETOTOLUIDIDE HYDROCHLORIDE see
CHS750
2'-CHLORO-2-(ISOPROPYLAMINO)-6'-METHYLACETANILIDE HYDROCHLORIDE
see CHS750
2-CHLORO-N-ISOPROPYL-N-PHENYLACETAMIDE see CHS500
CHLOROJECT L see CDP250
3-CHLORO-LACTONITRILE see CHU000
CHLOROMADINONE ACETATE see CBF250
CHLOROMAX see CDP250
CHLOROMELAMINE see TNE775
5-CHLORO-2-MERCAPTOANILINE HYDROCHLORIDE see CHU100
S-(6-CHLORO-3-(MERCAPTOMETHYL)-2-BENZOXAZOLINONE)-O,O-DIETHYL
PHOSPHORODITHIOATE see BDJ250
(CHLOROMERCURI)BENZENE see PFM500
p-(CHLOROMERCURI)BENZOIC ACID see CHU500
p-CHLOROMERCURIC BENZOIC ACID see CHU500
N-(2-CHLOROMERCURICYCLOHEXYL) HEXANAMIDE, (E)- see CHL000
N-(2-CHLOROMERCURICYCLOHEXYL)PROPIONAMIDE see CIC000
2-(CHLOROMERCURI)-4,6-DINITROPHENOL see CHN750
N-(CHLOROMERCURI)FORMANILIDE see CHU750
2-CHLOROMERCURIFURAN see CHK000
3-(3-CHLOROMERCURI-2-METHOXY-1-PROPYL)-5,5-DIMETHYLHYDANTOIN see
CHV250
1-(3-CHLOROMERCURI-2-METHOXY-1-PROPYL)-HYDANTOIN see CHV500
3-(3-CHLOROMERCURI-2-METHOXY-1-PROPYL)HYDANTOIN see CHV750
1-(3-CHLOROMERCURI-2-METHOXY)PROPYLHYDANTOIN see CHV500
3-(3-CHLOROMERCURI-2-METHOXY-1-PROPYL)-1-METHYLHYDANTOIN see
CHW250
3-(3-(CHLOROMERCURI)-2-METHOXYPROPYL)-1-METHYLHYDANTOIN see
CHW250
1-(3-CHLOROMERCURI-2-METHOXY-1-PROPYL)-3-METHYLHYDANTOIN see
CHW000
5-(3-CHLOROMERCURI-2-METHOXY-1-PROPYL)-3-METHYLHYDANTOIN see
CHW500
1-(3-(CHLOROMERCURI)-2-METHOXYPROPYL)UREA see CHX250
(3-(CHLOROMERCURI)-2-METHOXYPROPYL)UREA see CHX250
o-CHLOROMERCURIPHENOL see CHW675
p-CHLOROMERCURIPHENOL see CHW750
p-(CHLOROMERCURI)PHENOL see CHW750
3-(CHLOROMERCURI)PYRIDINE see CKW500
1-(3-(CHLOROMERCURY)-2-METHOXYPROPYL)HYDANTOIN see CHV500
CHLOROMERIDIN see CHX250
CHLOROMERODRIN see CHX250
CHLOROMETHANE see MIF765
CHLOROMETHANE mixed with DICHLOROMETHANE see CHX750
CHLOROMETHANE SULFONATE d'ETHYLE (FRENCH) see CHC750
CHLOROMETHANE SULFONYL CHLORIDE see CHY000
CHLOROMETHAPYRILENE see CHY250
N-(4-(2-(5-CHLORO-2-METHOXYBENZAMIDO)ETHYL)PHENYLSULFONYL)-N'-CY-
CLOHEXYLUREA see CEH700

4-(CHLOROMETHYL)NITROBENZENE see NFN400
p-(CHLOROMETHYL)NITROBENZENE see NFN400
4-(CHLOROMETHYL)-2-(o-NITROPHENYL)-1,3-DIOXOLANE see CIQ400
2-CHLORO-2-METHYL-N-NITROSOETHANAMINE see CIQ500
2-CHLORO-N-METHYL-N-NITROSOETHYLAMINE see CIQ500
N-CHLORO-5-METHYL-2-OXAZOLIDINONE see CIQ625
2-(CHLOROMETHYL)OXIRANE see EAZ500
CHLOROMETHYLOXIRANE see EAZ500
2-CHLORO-N-METHYL-3-OXOBUTANAMIDE see CEA100
7-CHLORO-1-METHYL-2-OXO-5-PHENYL-3H-1,4-BENZODIAZEPINE see DCK759
4-CHLORO-2-METHYLPHENOL see CIR000
4-CHLORO-3-METHYLPHENOL see CFE250
2-CHLORO-5-METHYLPHENOL see CFE500
(4-CHLORO-2-METHYLPHENOXY)ACETIC ACID see CIR250
4-CHLORO-2-METHYLPHENOXYACETIC ACID, ETHYL ESTER see EMR000
4-CHLORO-2-METHYLPHENOXYACETIC ACID SODIUM SALT see SIL500
4-(4-CHLORO-2-METHYLPHENOXY)BUTANOIC ACID see CLN750
4-(4-CHLORO-2-METHYLPHENOXY)BUTANOIC ACID, SODIUM SALT see CLO000
4-(4-CHLORO-2-METHYLPHENOXY)BUTYRIC ACID see CLN750
γ-(4-CHLORO-2-METHYLPHENOXY)BUTYRIC ACID see CLN750
4-(4-CHLORO-2-METHYLPHENOXY)BUTYRIC ACID SODIUM SALT see CLO000
CHLOROMETHYLPHENOXYBUTYRIC ACID SODIUM SALT see CLO000
1-(2-CHLORO-5-METHYLPHENOXY)-3-((1,1-DIMETHYLETHYL)AMINO)-2-PROPANOL HYDROCHLORIDE see BQB250
2-(4-CHLORO-2-METHYLPHENOXY)PROPANOIC ACID (R) (9CI) see CIR325
2-(4-CHLORO-2-METHYLPHENOXY)PROPIONIC ACID see CIR500
(+)-α-(4-CHLORO-2-METHYLPHENOXY) PROPIONIC ACID see CIR500
4-CHLORO-2-METHYLPHENOXY-α-PROPIONIC ACID see CIR500
N-(3-CHLORO-2-METHYLPHENYL)ANTHRANILIC ACID see CLK325
7-CHLORO-N-METHYL-5-PHENYL-3H-1,4-BENZODIAZEPIN-2-AMINE-4-OXIDE see LFK000
7-CHLORO-1-METHYL-5-PHENYL-1H-1,5-BENZODIAZEPINE-2,4(3H,5H)-DIONE see CIR750
7-CHLORO-1-METHYL-5-PHENYL-2H-1,4-BENZODIAZEPIN-2-ONE see DCK759
2-((p-CHLORO-α-METHYL-α-PHENYLBENZYL)OXY)-N,N-DIMETHYLAMINE HYDROCHLORIDE see CIS000
(+)-2-(2-((p-CHLORO-α-METHYL-α-PHENYLBENZYL)OXY)ETHYL)-1-METHYL PYRROLIDINE FUMARATE see FOS100
7-CHLORO-1-METHYL-5-PHENYL-1,3-DIHYDRO-2H-1,4-BENZODIAZEPIN-2-ONE see DCK759
N'-(4-CHLORO-2-METHYLPHENYL)-N,N-DIMETHYLMETHANIMIDAMIDE see CJJ250
3-(3-CHLORO-4-METHYLPHENYL)-1,1-DIMETHYL-UREA see CIS250
7-CHLORO-N-METHYL-5-PHENYL-EH-1,4-BENZODIAZEPIN-2-AMINE-4-OXIDE, MONOHYDROCHLORIDE see MDQ250
2-CHLORO-5-METHYLPHENYLHYDROXYLAMINE see CIS325
5-CHLORO-1-METHYL-3-PHENYL-1H-IMIDAZO(4,5-b)PYRIDIN-2(3H)-ONE see MNT100
CHLOROMETHYL PHENYL KETONE see CEA750
N'-(4-CHLORO-2-METHYLPHENYL)-METHANIMIDAMIDE MONOHYDROCHLORIDE see CJJ500
N-(3-CHLORO-4-METHYLPHENYL)-2-METHYLPENTANAMIDE see SKQ400
N-(3-CHLORO-4-METHYLPHENYL)-N',N'-DIMETHYLUREA see CIS250
CHLOROMETHYLPHENYLSILANE see CIS625
4-CHLORO-2-METHYLPHENYLTHIOGLYCOLIC ACID see CLO600
8-CHLORO-1-METHYL-6-PHENYL-4H-s-TRIAZOLO(4,3-a)(1,4)BENZODIAZEPINE see XAJ000
2-CHLORO-11-(4-METHYLPIPERAZINO)DIBENZO(b,f)(1,4)THIAZEPINE see CIS750
2-CHLORO-11-(4'-METHYL)PIPERAZINO-DIBENZO(b,f)(1,4)THIAZEPINE HYDROCHLORIDE see CIT000
7-CHLORO-3-(4-METHYL-1-PIPERAZINYL)-4H-1,2,4-BENZOTHIADIAZINE-1,1-DIOXIDE see CIT625
8-CHLORO-11-(4-METHYL-1-PIPERAZINYL)-5H-DIBENZO(b,e)(1,4)DIAZEPINE see CMY650
2-CHLORO-11-(4-METHYL-1-PIPERAZINYL)-DIBENZO(b,f)(1,4)OXAZEPINE see DCS200
2-CHLORO-11-(4-METHYL-1-PIPERAZINYL)-DIBENZO(b,f)(1,4)OXOAZEPINE see DCS200
2-CHLORO-11-(4-METHYL-1-PIPERAZINYL)DIBENZO(b,f)(1,4)THIAZEPINE see CIS750
2-CHLORO-11-(4-METHYL-1-PIPERAZINYL)DIBENZO(b,f)(1,4)THIAZEPINE HYDROCHLORIDE see CIT000
2-CHLORO-10-(3-(4-METHYL-1-PIPERAZINYL)PROPYL)PHENOTHIAZINE see PMF500
3-CHLORO-10-(3-(1-METHYL-4-PIPERAZINYL)PROPYL)PHENOTHIAZINE see PMF500
2-CHLORO-10-(3-(1-METHYL-4-PIPERAZINYL)-PROPYL)-PHENOTHIAZINE see PMF500
CHLORO-3 (N-METHYLPIPERAZINYL-3 PROPYL)-10 PHENOTHIAZINE (FRENCH) see PMF500
2-CHLORO-10-(3-(4-METHYL-1-PIPERAZINYL)-PROPYL)-10H-PHENOTHIAZINE-(Z)-2-BUTENEDIOATE (1:2) see PMF250

2-CHLORO-10-(3-(1-METHYL-4-PIPERAZINYL)PROPYL)PHENOTHIAZINE, DIMALEATE see PMF250
2-CHLORO-10-(3-(4-METHYL-1-PIPERAZINYL)PROPYL)PHENOTHIAZINE DIMALEATE see PMF250
2-CHLORO-10-(3-(1-METHYL-4-PIPERAZINYL)PROPYL)PHENOTHIAZINE EDISYLATE see PME700
2-CHLORO-10-(3-(4-METHYL-1-PIPERAZINYL)PROPYL)PHENOTHIAZINE 1,2-ETHANEDISULFONATE (1:1) see PME700
2-CHLORO-10-(3-(4-METHYL-1-PIPERAZINYL)PROPYL)PHENOTHIAZINE MALEATE see PMF250
6'-CHLORO-2-(2-METHYLPIPERIDINO)-o-ACETOTOLUIDIDE HYDROCHLORIDE see CIT750
o-CHLORO-2-(METHYL(2-(PIPERIDINO)ETHYL)AMINO)ACETANILIDE DIHYDROCHLORIDE see CIU000
6'-CHLORO-3-(2-METHYLPIPERIDINO)-o-PROPIONOTOLUIDIDE HYDROCHLORIDE see CIU250
3'-CHLORO-5'-METHYL-3-PIPERIDINO-4'-PROPOXY-PROPIOPHENONE HYDROCHLORIDE see CIU325
1-CHLORO-2-METHYLPROPANE see CIU500
2-CHLORO-2-METHYLPROPANE see BQR000
1-CHLORO-2-METHYL-1-PROPENE see IKE000
3-CHLORO-2-METHYL-1-PROPENE see CIU750
1-CHLORO-2-METHYLPROPENE see IKE000
3-CHLORO-2-METHYLPROPENE see CIU750
1-(3-CHLORO-5-METHYL-4-PROPOXYPHENYL)-3-(1-PIPERIDINYL)1-PROPANONE HYDROCHLORIDE (9CI) see CIU325
2'-CHLORO-6'-METHYL-2-(PROPYLAMINO)ACETANILIDE HYDROCHLORIDE see CKT750
2-CHLORO-5-(1-METHYLPROPYL)PHENYL METHYLCARBAMATE see MOU750
2-CHLORO-N-(1-METHYL-2-PROPYNYL)ACETANILIDE see CDS275
2-CHLORO-N-(1-METHYL-2-PROPYNYL)-ACETANILIDE (8CI) see CDS275
2-CHLORO-N-(1-METHYL-2-PROPYNYL)-N-PHENYLACETAMIDE see CDS275
2-(CHLOROMETHYL) PYRIDINE HYDROCHLORIDE see PIC000
3-(CHLOROMETHYL) PYRIDINE HYDROCHLORIDE see CIV000
5'-CHLORO-2-(METHYL(2-(PYRROLIDINYL)ETHYL)AMINO)-O-ACETOTOLUIDIDE DIHYDROCHLORIDE see CIW250
o-CHLORO-2-(METHYL(2-(PYRROLIDINYL)ETHYL)AMINO)PROPIONANILIDE DIHYDROCHLORIDE see CIX000
6-CHLORO-2-METHYL-4-QUINAZOLINONE see MII750
2-CHLORO-4-(METHYLSULFONYL)ANILINE see CIX200
2-CHLOROMETHYLTHIOPHENE see CIY250
4-(CHLOROMETHYL)TOLUENE see MHN300
p-(CHLOROMETHYL)TOLUENE see MHN300
CHLOROMETHYL(TRICHLORO)SILANE see CIY325
(CHLOROMETHYL)TRICHLOROSILANE see CIY325
(CHLOROMETHYL)TRIETHOXYSILANE see CIY500
3-CHLORO-4-METHYL-UMBELLIFERONE BIS(2-CHLOROETHYL)PHOSPHATE see DFH600
3-CHLORO-4-METHYLUMBELLIFERONE-O-ESTER with O,O-DIETHYL PHOSPHOROTHIOATE see CNU750
3'-CHLORO-2-METHYL-p-VALEROTOLUIDIDE see SKQ400
N-CHLORO-3-MORPHOLINONE see CIY899
CHLOROMROWCZAN 1-NAFTYLU (CZECH) see NBH200
CHLOROMYCETIN see CDP250
CHLOROMYCETIN SUCCINATE see CDP725
CHLORONAFTINA see BIF250
1-CHLORONAPHTHALENE see CIZ000
2-CHLORONAPHTHALENE see CJA000
α-CHLORONAPHTHALENE see CIZ000
β-CHLORONAPHTHALENE see CJA000
CHLORONAPHTHINE see BIF250
CHLORONASE see CKK000
CHLORONEB see CJA100
CHLORONEBE (FRENCH) see CJA100
CHLORONEOANTERGAN see CKV625
CHLORONITRIN see CDP250
4-CHLORO-2-NITROANILINE see KDA050
4-CHLORO-3-NITROANILINE see CJA185
2-CHLORO-4-NITROANILINE see CJA175
N-CHLORO-4-NITROANILINE see CJA150
2-CHLORO-5-NITROANILINE see CJA180
p-CHLORO-o-NITROANILINE see KDA050
o-CHLORO-p-NITROANILINE see CJA175
2-CHLORO-4-NITRO-ANISOLE see CJA200
o-CHLORO-p-NITROANISOLE see CJA200
CHLORONITROANISOLE see CJA200
1-CHLORO-5-NITRO-9,10-ANTHRACENEDIONE see CJA250
5-CHLORO-1-NITROANTHRAQUINONE see CJA250
1-CHLORO-5-NITROANTHRAQUINONE see CJA250
4-CHLORO-2-NITROBENZENAMINE see KDA050
2-CHLORO-1-NITROBENZENE see CJB750
4-CHLORO-1-NITROBENZENE see NFS525
1-CHLORO-2-NITROBENZENE see CJB750
2-CHLORONITROBENZENE see CJB750
1-CHLORO-3-NITROBENZENE see CJB250

1-CHLORO-4-NITROBENZENE see NFS525
4-CHLORONITROBENZENE see NFS525
CHLORONITROBENZENE, ortho, liquid (DOT) see CJA950
o-CHLORONITROBENZENE, liquid (DOT) see CJB750
m-CHLORONITROBENZENE (DOT) see CJB250
m-CHLORONITROBENZENE see CJB250
CHLORO-m-NITROBENZENE see CJB250
o-CHLORONITROBENZENE see CJB750
CHLORO-o-NITROBENZENE see CJB750
p-CHLORONITROBENZENE see NFS525
CHLORONITROBENZENE see CJA950
2-CHLORO-4-NITROBENZENEAZO-2'-AMINO-4'-METHOXY-5'-METHYLBENZENE
    see MIH500
2-CHLORO-5-NITROBENZENESULFONIC ACID see CJB825
2-CHLORO-3,5-NITROBENZENESULFONIC ACID, SODIUM SALT see CJC000
4-CHLORO-3-NITROBENZOIC ACID see CJC500
2-CHLORO-4-NITROBENZOIC ACID see CJC250
2-CHLORO-5-NITROBENZOIC ACID METHYL ESTER see CJC515
2-CHLORO-5-NITROBENZYL ALCOHOL see CJC549
6-CHLORO-2-NITROBENZYL BROMIDE see CJC600
4-CHLORO-2-NITROBENZYL CHLORIDE see CJC625
2-CHLORO-6-NITROBENZYL CHLORIDE see CJC610
2-CHLORO-2-NITROBUTANE see CJC750
3-CHLORO-4-(2'-NITRO-3'-CHLOROPHENYL)PYRROLE see CFC000
1-CHLORO-1-NITROETHANE see CJC800
2-((o-CHLORO-α-(NITROMETHYL)BENZYL)THIO)ETHYLAMINE HYDROCHLO-
    RIDE see NEC000
2-CHLORO-4-NITROPHENOL see CJD250
2-CHLORO-4-NITROPHENYLAMIDE-6-CHLOROSALICYLIC ACID see DFV400
4-CHLORO-2-NITROPHENYL p-CHLOROPHENYL ETHER see CJD600
N-(2-CHLORO-4-NITROPHENYL)-5-CHLOROSALICYLAMIDE see DFV400
O-(2-CHLORO-4-NITROPHENYL) O,O-DIMETHYL PHOSPHOROTHIOATE see
    NFT000
O-(3-CHLORO-4-NITROPHENYL) O,O-DIMETHYL PHOSPHOROTHIOATE see
    MIJ250
2-CHLORO-5-NITROPHENYL ESTER ACETIC ACID see CJD625
1-(4-CHLORO-3-NITROPHENYL)-2-ETHOXY-2-((4-(METHYLTHI-
    O)PHENYL)AMINO)ETHAN ONE see CJD630
o-(2-CHLORO-4-NITROPHENYL)-o-ISOPROPYL ETHYLPHOSPHONOTHIOATE
    see CJD650
CHLORONITROPROPAN (POLISH) see CJD750
1-CHLORO-1-NITROPROPANE see CJE000
1-CHLORO-2-NITROPROPANE see CJD750
2-CHLORO-2-NITROPROPANE see CJE250
CHLORONITROPROPANE see CJD750, CJE000
3'-CHLORO-5-NITROSALICYLANILIDE see CJF500
1-CHLORO-1-NITROSOCYCLOHEXANE see CJF825
α-CHLORO-p-NITROSTYRENE see CJG750
6-CHLORO-2-NITROTOLUENE see CJG825
2-CHLORO-4-NITROTOLUENE see CJG800
2-CHLORO-6-NITROTOLUENE see CJG825
α-CHLORO-p-NITROTOLUENE see NFN400
4-CHLORO-3-NITRO-α,α,α-TRIFLUOROTOLUENE see NFS700
6-CHLORO-N(sup 2),N(sup 2),N(sup 4)-TRIETHYL-1,3,5-TRIAZINE-2,4-DIAMINE
    (IUPAC) see TJL500
9-CHLORONONANOIC ACID see CJH500
6-CHLORO-N,N,N',N'-TETRAETHYL-1,3,5-TRIAZINE-2,4-DIAMINE see CDQ325
6-CHLORO-N,N,N'-TRIETHYL-1,3,5-TRIAZINE-2,4-DIAMINE see TJL500
2-CHLORO-N,N,N'-TRIFLUOROPROPIONAMIDINE see CLS125
CHLOROOXIRANE see CGX000
5-CHLORO-1-(1-(3-(2-OXO-1-BENZIMIDAZOLINYL)PROPYL)-4-PIPERIDYL)-2-BEN-
    ZIMIDAZOLINONE see DYB875
4-CHLORO-2-OXO-3(2H)-BENZOTHIAZOLEACETIC ACID see BAV000
4-CHLORO-2-OXOBENZOTHIAZOLIN-3-YL ACETIC ACID see BAV000
4-((5-CHLORO-2-OXO-3(2H)-BENZOTHIAZOLYL)ACETYL)-1-PIPERAZINEETHA-
    NOL see CJH750
3-(6-CHLORO-2-OXOBENZOXAZOLIN-3-YL)METHYL-O,O-DIETHYL PHOSPHO-
    ROTHIOLOTHIONATE see BDJ250
N-(4-CHLORO-3-OXOBUTYL)-p-TOLUENESULFONAMIDE see THH370
exo-5-CHLORO-6-OXO-endo-2-NORBORNANECARBONITRILE-o-(METHYLCAR-
    BAMOYL)OXIME see CFF250
4-CHLORO-3-OXO-N-PHENYLBUTANAMIDE see CEA050
9-CHLORO-5-OXO-7-(1H-TETRAZOL-5-YL)-5H-1-BENZOPYRANO(2,3-b)PYRIDINE
    SODIUM SALT PENTAHYDRATE see THK850
CHLOROPARACIDE see CEP000
CHLORO-PDMT see CJI100
CHLOROPENTAFLUOROBENZENE see PBE100
CHLOROPENTAFLUOROETHANE see CJI500
CHLOROPENTAHYDROXYDIALUMINUM see AHA000
1-CHLOROPENTANE see PBW500
CHLOROPEPTIDE see CJI609
CHLOROPERALGONIC ACID see CJH500
CHLOROPERFLUOROBENZENE see PBE100
CHLOROPEROXYL see CDW450
CHLOROPEROXYTRIFLUOROMETHANE see CJI809

CHLOROPHACINONE see CJJ000
CHLOROPHEN see PAX250
4'-CHLOROPHENACYL BROMIDE see CJJ100
p-CHLOROPHENACYL BROMIDE see CJJ100
CHLOROPHENAMADIN see CJJ250
CHLOROPHENAMIDINE see CJJ250
CHLOROPHENAMIDINE HYDROCHLORIDE see CJJ500
CHLOROPHENE see CJU250
4-CHLOROPHENE-1,3-DIAMINE see CJY120
1-(p-CHLOROPHENETHYL)-6,7-DIMETHOXY-2-METHYL-1,2,3,4-TETRAHYDROI-
    SOQUINOLINE see MDV000
1-(p-CHLOROPHENETHYL)HYDRAZINE HYDROGEN SULFATE see CJK000
1-(p-CHLOROPHENETHYL)-2-METHYL-6,7-DIMETHOXY-1,2,3,4-TETRAHYDROI-
    SOQUINOLINE see MDV000
1-(p-CHLOROPHENETHYL)-1, 2,3,4-TETRAHYDRO-6,7-DIMETHOXY-2-METHYLI-
    SOQUINOLINE see MDV000
2-(o-CHLOROPHENETHYL)-3-THIOSEMICARBAZIDE see CJK100
4-CHLOROPHENIRAMINE see CLX300
2-CHLOROPHENOL see CJK250
3-CHLOROPHENOL see CJK500
4-CHLOROPHENOL see CJK750
o-CHLOROPHENOL, liquid see CJK250
m-CHLOROPHENOL see CJK500
o-CHLOROPHENOL see CJK250
p-CHLOROPHENOL see CJK750
o-CHLOROPHENOL, solid see CJK250
CHLOROPHENOLS see CJL000
CHLOROPHENOTHAN see DAD200
CHLOROPHENOTHANE see DAD200
2-CHLOROPHENOTHIAZINE see CJL100
2-(2-(4-(2-(2-CHLORO-10-PHENOTHIAZINYL)METHYL)PROPYL)-1-PIPERAZI-
    NYL)ETHOXY)ETHANOL see CJL409
1-(3-(3-CHLOROPHENOTHIAZIN-10-YL)PROPYL)-ISONIPECOTAMIDE see
    CJL500
4-(3-(2-CHLOROPHENOTHIAZIN-10-YL)PROPYL)-1-PIPERAZINEETHANOL see
    CJM250
4-(3-(2-CHLOROPHENOTHIAZIN-10-YL)PROPYL)-1-PIPERAZINEETHANOL DIHY-
    DROCHLORIDE see PCO500
4-(3-(2-CHLOROPHENOTHIAZIN-10-YL)PROPYL)-1-PIPERAZINEETHANOL MALE-
    ATE see PCO850
8-(3-(2-CHLOROPHENOTHIAZIN-10-YL))PROPYL-1-THIA-4,8-DIAZASPI-
    RO(4.5)DECAN-3-ONE HYDROCHLORIDE see SLB000
CHLOROPHENOTOXUM see DAD200
10-CHLORO-10H-PHENOXARSINE see CJM750
10-CHLOROPHENOXARSINE see CJM750
(4-CHLOROPHENOXY)ACETIC ACID see CJN000
p-CHLOROPHENOXYACETIC ACID see CJN000
(p-CHLOROPHENOXY)ACETIC ACID 2-(DIMETHYLAMINO)ETHYL ESTER HY-
    DROCHLORIDE see AAE500
p-CHLOROPHENOXYACETIC ACID-β-DIMETHYLAMINOETHYL ESTER see
    DPE000
p-CHLOROPHENOXYACETIC ACID-2-ISOPROPYLHYDRAZIDE see CJN250
1-(p-CHLOROPHENOXYACETYL)-2-ISOPROPYL HYDRAZINE see CJN250
4-(4-CHLOROPHENOXY)ANILINE see CEH125
p-(p-CHLOROPHENOXY)ANILINE see CEH125
10-CHLOROPHENOXYARSINE see CJM750
4-(4-CHLOROPHENOXY)-BENZENAMINE (9CI) see CEH125
2-(4-CHLOROPHENOXY)-N-(2-(DIETHYLAMINO)ETHYL)ACETAMIDE see CJN750
2-(p-CHLOROPHENOXY)-N-(2-(DIETHYLAMINO)ETHYL)ACETAMIDE see CJN750
2-(p-CHLOROPHENOXY)-N-(2-(DIETHYLAMINO)ETHYL) ACETAMIDE COM-
    POUND with 4-BUTYL-1,2-DIPHENYL-3,5-PYRAZOLIDINEDIONE (1:1) see
    CMW750
(p-CHLOROPHENOXY)DIMETHYL-ACETIC ACID see CMX000
1-(4-CHLOROPHENOXY)-3,3-DIMETHYL-1-(1H-1,2,4-TRIAZOL-1-YL)-2-BUTANONE
    see CJO250
1-(4-CHLOROPHENOXY)-3,3-DIMETHYL-1-(1,2,4-TRIAZOL-1-YL)-2-BUTAN-2-ONE
    see CJO250
2-(4-CHLOROPHENOXY)ETHANOL see CJO500
2-(p-CHLOROPHENOXY)ETHANOL see CJO500
(2-(p-CHLOROPHENOXY)ETHYL)HYDRAZINE HYDROCHLORIDE see CJP250
1-(2-(o-CHLOROPHENOXY)ETHYL)HYDRAZINE HYDROGEN SULFATE see
    CJP500
3-(CHLOROPHENOXY)-2-HYDROXYPROPYL CARBAMATE see CJQ250
CHLORO-4 PHENOXYISOBUTYRATE D'HYDROXY-4 N-DIMETHYLBUTYRAM-
    IDE (FRENCH) see LGK200
α-(p-CHLOROPHENOXY)ISOBUTYRIC ACID see CMX000
α-p-CHLOROPHENOXYISOBUTYRYL ETHYL ESTER see ARQ750
N-2(p-CHLOROPHENOXY)ISOBUTYRYL-N'-MORPHOLINOMETHYLUREA see
    PJB500
2-(4-CHLOROPHENOXY)-2-METHYLPROPANOIC ACID see CMX000
2-(4-CHLOROPHENOXY)-2-METHYLPROPANOIC ACID, 3,4-DIHYDRO-2,5,7,8-
    TETRAMETHYL-2-(4,8,12-TRIMETHYLTRIDECYL)-2H-1-BENZOPYRAN-6-YL-ES-
    TER, (2r(4r,8r)) see TGJ000
2-(4-CHLOROPHENOXY)-2-METHYLPROPANOIC ACID ETHYL ESTER see
    ARQ750

2-(4-CHLOROPHENOXY)-2-METHYLPROPANOIC ACID-1,3-PROPANEDIYL ESTER see SDY500

2-(4-CHLOROPHENOXY-2-METHYL)PROPIONIC ACID see CIR500

2-(p-CHLOROPHENOXY)-2-METHYLPROPIONIC ACID see CMX000

2-(4-CHLOROPHENOXY)-2-METHYL-PROPIONIC ACID 4-(DIMETHYLAMINO)-4-OXOBUTYL ESTER (9CI) see LGK200

2-(p-CHLOROPHENOXY)-2-METHYLPROPIONIC ACID ETHYL ESTER see ARQ750

2-(p-CHLOROPHENOXY)-2-METHYLPROPIONIC ACID TRIMETHYLENE ESTER see SDY500

N-2(p-CHLOROPHENOXY)-2-METHYLPROPIONYL-N'-MORPHOLINOMETHYLUREA see PJB500

2-(4-(4-CHLOROPHENOXY)PHENOXY)PROPIONIC ACID see CJP750

3-(p-CHLOROPHENOXY)PHENYL)-1,1-DIMETHYLUREA see CJQ000

N'-4-(4-CHLOROPHENOXY)PHENYL-N,N-DIMETHYLUREA see CJQ000

1-(4-(4-CHLORO-PHENOXY)PHENYL)-3,3-D'METHYLUREE (FRENCH) see CJQ000

3-(4-CHLOROPHENOXY)-1,2-PROPANEDIOL-1-CARBAMATE see CJQ250

3-(p-CHLOROPHENOXY)-1,2-PROPANEDIOL-1-CARBAMATE see CJQ250

2-(3-CHLOROPHENOXY)PROPANOIC ACID see CJQ300

2-(3-CHLOROPHENOXY)PROPIONIC ACID see CJQ300

2-(m-CHLOROPHENOXY)PROPIONIC ACID see CJQ300

N-(1-(o-CHLOROPHENOXY)-2-PROPYL)-2-(DIETHYLAMINO)-N-ETHYLACETAMIDE HYDROCHLORIDE see CJQ500

N-(1-(o-CHLOROPHENOXY)-2-PROPYL)-2-(DIETHYLAMINO)-N-METHYLACETAMIDE HYDROCHLORIDE see CJQ750

CHLOROPHENTERMINE see CLY250

CHLOROPHENTERMINE HYDROCHLORIDE see ARW750

N-(4-CHLOROPHENYL)ACETAMIDE see CDZ100

N-(2-CHLOROPHENYL)ACETOACETAMIDE see AAY600

2-(4-CHLOROPHENYL)ACETONITRILE see CEP300

(4-CHLOROPHENYL)ACETONITRILE see CEP300

p-CHLOROPHENYLACETONITRILE see CEP300

2-CHLORO-2-PHENYLACETOPHENONE see CFJ100

2-(α-p-CHLOROPHENYLACETYL)INDANE-1,3-DIONE see CJJ000

3-CHLOROPHENYLAMINE see CEH675

4-CHLOROPHENYLAMINE see CEH680

m-CHLOROPHENYLAMINE see CEH675

4-CHLOROPHENYLAMINE HYDROCHLORIDE see CJR200

p-CHLOROPHENYLAMINE HYDROCHLORIDE see CJR200

β-(p-CHLOROPHENYL)-γ-AMINOBUTYRIC ACID see BAC275

N-(((4-CHLOROPHENYL)AMINO)CARBONYL)-2,6-DIFLUOROBENZAMIDE see CJV250

N-(3-CHLOROPHENYL)-1-AZIRIDINECARBOXAMIDE see CJR250

4-CHLOROPHENYL BENZENESULFONATE see CJR500

p-CHLOROPHENYL BENZENESULFONATE see CJR500

4-CHLOROPHENYL BENZENESULPHONATE see CJR500

p-CHLOROPHENYL BENZENESULPHONATE see CJR500

1-(α-(2-CHLOROPHENYL)BENZHYDRYL)IMIDAZOLE see MRX500

1-(p-CHLORO-α-PHENYLBENZYL)HEXAHYDRO-4-METHYL-1H-1,4-DIAZEPINE DIHYDROCHLORIDE see CJR809

1-((p-CHLORO-α-PHENYLBENZYL)HEXAHYDRO-4-METHYL)-1H-1,4-DIAZEPINE HYDROCHLORIDE see HGI200

1-(p-CHLORO-α-PHENYLBENZYL)-4-(2-((2-HYDROXYETHOXY)ETHYL)PIPERAZINE) see CJR909

1-(p-CHLORO-α-PHENYLBENZYL)-4-(m-METHYLBENZYL)PIPERAZINE see HGC500

1-(p-CHLORO-α-PHENYLBENZYL)-4-(m-METHYLBENZYL)PIPERAZINE HYDROCHLORIDE see MBX500

1-(p-CHLORO-α-PHENYLBENZYL)-4-METHYLPIPERAZINE see CFF500

1-(p-CHLORO-α-PHENYLBENZYL)-4-METHYL-PIPERAZINE DIHYDROCHLORIDE see CDR000

1-(p-CHLORO-α-PHENYLBENZYL)-4-METHYLPIPERAZINE HYDROCHLORIDE see CDR500

2-(α-(p-CHLOROPHENYL)BENZYLOXY)-N,N-DIMETHYLETHYLAMINE HYDROCHLORIDE see CJR959

3-α-((p-CHLORO-α-PHENYLBENZYL)OXY)-1-α-H,5-α-H-TROPANE HYDROCHLORIDE see CMB125

2-(2-(4-(p-CHLORO-α-PHENYLBENZYL)-1-PIPERAZINYL)ETHOXY)ETHANOL see CJR909

2-(2-(4-(p-CHLORO-α-PHENYLBENZYL)-1-PIPERAZINYL)ETHOXY)ETHANOLDIHYDROCHLORIDE see HOR470

2-(2-(2-(4-(p-CHLORO-α-PHENYLBENZYL)-1-PIPERAZINYL)ETHOXY)ETHOXY)ETHANOL DIMALEATE see HHK100

1-(4-CHLOROPHENYL)BIGUANIDINIUM HYDROGEN DICHROMATE see CJT125

1-(o-CHLOROPHENYL)-2-tert-BUTYLAMINO ETHANOL HYDROCHLORIDE see BQE250

N-(3-CHLORO PHENYL) CARBAMATE de 4-CHLORO 2-BUTYNYLE (FRENCH) see CEW500

N-(3-CHLORO PHENYL) CARBAMATE D'ISOPROPYLE (FRENCH) see CKC000

(3-CHLOROPHENYL)CARBAMIC ACID 4-CHLORO-2-BUTYNYL ESTER see CEW500

N-(3-CHLOROPHENYL)CARBAMIC ACID, ISOPROPYL ESTER see CKC000

(3-CHLOROPHENYL)CARBAMIC ACID, 1-METHYLETHYL ESTER see CKC000

3-CHLOROPHENYLCARBAMIC ACID-1-METHYLPROPYNYL ESTER see CEX250

3-CHLOROPHENYL-N-CARBAMOYLAZIRIDINE see CJR250

p-CHLOROPHENYL CHLORIDE see DEP800

4-CHLOROPHENYL-4-CHLOROBENZENESULFONATE see CJT750

p-CHLOROPHENYL-p-CHLOROBENZENE SULFONATE see CJT750

4-CHLOROPHENYL-4-CHLOROBENZENESULPHONATE see CJT750

4-CHLOROPHENYL-4'-CHLOROBENZYL SULFIDE see CEP000

2-(o-CHLOROPHENYL)-2-(p-CHLOROPHENYL)-1,1-DICHLOROETHANE see CDN000

1-(o-CHLOROPHENYL)-4-(3-(p-CHLOROPHENYL)-3-PHENYLPROPIONYL)PIPERAZINE see CKI020

α-(2-CHLOROPHENYL)-α-(4-CHLOROPHENYL)-5-PYRIMIDINEMETHANOL see FAK100

(2-CHLOROPHENYL)-α-(4-CHLOROPHENYL)-5-PYRIMIDINEMETHANOL see FAK100

p-CHLOROPHENYL-N-(4'-CHLOROPHENYL)THIOCARBAMATE see CJU125

4-CHLORO-α-PHENYLCRESOL see BEF750

4-CHLORO-α-PHENYL-o-CRESOL see CJU250

1-(p-CHLOROPHENYL)-4,6-DIAMINO-2,2-DIMETHYL-1,2-DIHYDRO-s-TRIAZINE see COX400

5-(4'-CHLOROPHENYL)-2,4-DIAMINO-6-ETHYLPYRIMIDINE see TGD000

1-(3-CHLOROPHENYL)-2-((4-(DICHLOROACETYL)PHENYL)AMINO)-2-HYDROXYETHANONE see CJU300

(o-CHLOROPHENYL)(3-(2,4-DICHLOROPHENOXY)-2-HYDROXYPROPYL)HYDROXYARSINEOXIDE see DGA425

3-(4-CHLOROPHENYL)-4',5-DICHLOROSALICYLANILIDE see TIP750

N-(3-CHLOROPHENYL)DIETHANOLAMINE see CJU200

N-(m-CHLOROPHENYL)DIETHANOLAMINE see CJU200

2-(p-CHLOROPHENYL)-1-(p-(β-DIETHYLAMINOETHOXY)PHENYL)-1-(p-TOLYL)ETHANOL see TMP500

2-p-CHLOROPHENYL-1-(p-(2-DIETHYLAMINOETHOXY)PHENYL)-1-p-TOLYLETHANOL see TMP500

1-(4-CHLOROPHENYL)-3-(2,6-DIFLUOROBENZOYL)UREA see CJV250

1-p-CHLOROPHENYL-1,2-DIHYDRO-2,2-DIMETHYL-4,6-DIAMINO-s-TRIAZINE see COX400

1-(4-CHLOROPHENYL)-1,6-DIHYDRO-6,6-DIMETHYL-1,3,5-TRIAZINE-2,4-DIAMINE MONOHYDROCHLORIDE see COX325

5-p-CHLOROPHENYL-2,3-DIHYDRO-5H-IMIDAZO(2,1-A)ISOINDOL-5-OL see MBV250

5-(o-CHLOROPHENYL)-1,3-DIHYDRO-7-NITRO-2H-1,4-BENZODIAZEPIN-2-ONE see CMW000

(4-CHLOROPHENYL)(DIMETHOXYPHOSPHINYL)METHYL PHOSPHORIC ACID DIMETHYL ESTER see CMU875

1-(m-CHLOROPHENYL)-3-N,N-DIMETHYLCARBAMOYL-5-METHOXYPYRAZOLE see CJW500

1-(3-CHLOROPHENYL)-3,N,N-DIMETHYLCARBAMOYL-5-METHOXYPYRAZOLE see CJW500

1-(4-CHLOROPHENYL)-2,3-DIMETHYL-4-DIMETHYLAMINO-2-BUTANOL see CMW459

1-p-CHLOROPHENYL-2,3-DIMETHYL-4-DIMETHYLAMINO-2-BUTANOL HYDROCHLORIDE see CMW500

β-(p-CHLOROPHENYL)-α,α-DIMETHYLETHYLAMINE see CLY250

2-(4-CHLOROPHENYL)-1,1-DIMETHYLETHYL 2-AMINOPROPANOATE HYDROCHLORIDE see CJX000

(+)-1-(3-CHLOROPHENYL)-2-((1,1-DIMETHYLETHYL)AMINO)1-PROPANONE HYDROCHLORIDE (9CI) see WBJ500

N-(2-(4-CHLOROPHENYL)-1,1-DIMETHYLETHYL)-2-(DIETHYLAMINO)-PROPANAMIDE HYDROCHLORIDE see CGI125

N-(4-CHLOROPHENYL)-2,2-DIMETHYLPENTANAMIDE see CGL250

S-(p-CHLOROPHENYL)-O,O-DIMETHYL PHOSPHOTHIOATE see FOR000

1-(4-CHLOROPHENYL)-3,3-DIMETHYLTRIAZENE see CJI100

1-(p-CHLOROPHENYL)-3,3-DIMETHYL-TRIAZENE see CJI100

3-(4-CHLOROPHENYL)-1,1-DIMETHYLUREA see CJX750

3-(p-CHLOROPHENYL)-1,1-DIMETHYLUREA see CJX750

1-(p-CHLOROPHENYL)-3,3-DIMETHYLUREA see CJX750

N'-(4-CHLOROPHENYL)-N,N-DIMETHYLUREA see CJX750

3-(4-CHLOROPHENYL)-1,1-DIMETHYLUREA TRICHLOROACETATE see CJY000

3-(p-CHLOROPHENYL)-1,1-DIMETHYLUREA compounded with TRICHLOROACETIC ACID (1:1) see CJY000

1-(4-CHLORO PHENYL)-3,3-DIMETHYLUREE (FRENCH) see CJX750

N-(4-CHLOROPHENYL)-2,2-DIMETHYLVALEROAMIDE see CGL250

1-((2-CHLOROPHENYL)DIPHENYLMETHYL)-1H-IMIDAZOLE see MRX500

4-CHLORO-1,2-PHENYLENEDIAMINE see CFK125

4-CHLORO-1,3-PHENYLENEDIAMINE see CJY120

4-CHLOROPHENYLENE-1,3-DIAMINE see CJY120

4-CHLORO-m-PHENYLENEDIAMINE see CJY120

4-CHLORO-o-PHENYLENEDIAMINE see CFK125

p-CHLORO-o-PHENYLENEDIAMINE see CFK125

2-CHLORO-p-PHENYLENEDIAMINE see CEG600

o-CHLORO-p-PHENYLENEDIAMINE see CEG600

2-CHLORO-p-PHENYLENEDIAMINE SULFATE see CEG625

2-CHLORO-1-PHENYLETHANONE see CEA750

1-(4-CHLOROPHENYL)ETHANONE see CEB250

5-(o-CHLOROPHENYL)-7-ETHYL-1,3-DIHYDRO-1-METHYL-2H-THIENO(2,3-e)-1,4-DIAZEPIN-2-ONE see CKA000

p-CHLORO-β-PHENYLETHYLHYDRAZINE DIHYDROGEN SULFATE see CJK000

o-CHLORO-β-PHENYLETHYLHYDRAZINE DIHYDROGEN SULPHATE see CJZ000

5-(2-CHLOROPHENYL)-7-ETHYL-1-METHYL-1,3-DIHYDRO-2H-THIENO(2,3-e)(1,4)DIAZEPIN-2-ONE see CKA000

6-(o-CHLOROPHENYL)-8-ETHYL-1-METHYL-4H-sec-TRIAZOLO(3,4-c)THIENO(2,3-e)(1,4)-DIAZEPINE see EQN600

5-(4-CHLOROPHENYL)-6-ETHYL-2,4-PYRIMIDINEDIAMINE see TGD000

α-(2-CHLOROPHENYL)-α-(4-FLUOROPHENYL)-5-PYRIMIDINEMETHANOL see CHJ300

CHLORO(N-PHENYLFORMAMIDO)MERCURY see CHU750

β-(4-CHLOROPHENYL)GABA see BAC275

3-CHLORO-7-d-(2-PHENYLGLYCINAMIDO)-3-CEPHEM-4-CARBOXYLIC ACID see PAG075

1-(p-CHLOROPHENYL)-1,2,3,4,5,6-HEXAHYDRO-2,5-BENZODIAZOCINE, DIHYDROCHLORIDE see CKA500

1-(p-CHLOROPHENYL)-1,2,3,4,5,6-HEXAHYDRO-2,5-BENZODIAZOCINO DIHYDROCHLORIDE see CKA500

4-(2-CHLOROPHENYLHYDRAZONE)-3-METHYL-5-ISOXAZOLONE see MLC250

4-(2-CHLOROPHENYLHYDRAZONE)-3-METHYL-5(4H)-ISOXAZOLONE see MLC250

((3-CHLOROPHENYL)HYDRAZONO)PROPANEDINITRILE see CKA550

4-(p-CHLOROPHENYL)-4-HYDROXY-N,N-DIMETHYL-α,α-DIPHENYL-1-PIPERIDINE BUTYRAMIDE HCl see LIH000

4-(4-CHLOROPHENYL)-4-HYDROXY-N,N-DIMETHYL-α,α-DIPHENYL-1-PIPERIDINEBUTANAMIDE HYDROCHLORIDE see LIH000

(m-CHLOROPHENYL)HYDROXY(β-HYDROXYPHENETHYL)ARSINE OXIDE see CKA575

(p-CHLOROPHENYL)HYDROXYMETHYLARSINE OXIDE see CKD800

2-(3-(4-(p-CHLOROPHENYL)-4-HYDROXYPIPERIDINE)PROPYL)-7-FLUORO-3-METHYLCHROMONE see HJU000

4-(4-(p-CHLOROPHENYL)-4-HYDROXYPIPERIDINO)-4'-(DIMETHYLAMINO)BUTYROPHENO NE see CKA600

4-(4-(4-CHLOROPHENYL)-4-HYDROXY-1-PIPERIDINYL)-1-(4-FLUOROPHENYL)-1-BUTANONE see CLY500

4-(4-(p-CHLOROPHENYL)-4-HYDROXY-1-PIPERIDYL)-N,N-DIMETHYL-2,2-DIPHENYLBUTYRAMIDE HCl see LIH000

α-(p-CHLOROPHENYL)-α-2-IMIDAZOLIN-2-YL-2-PYRIDINEMETHANOL MALEATE see SBC800

N'-(4-CHLOROPHENYL)-N-ISOBUTINYL-N-METHYLUREA see CKF750

m-CHLOROPHENYL ISOCYANATE see CKA750

p-CHLOROPHENYL ISOCYANATE see CKB000

1-(p-CHLOROPHENYL)-5-ISOPROPYLBIGUANIDE see CKB250

1-(p-CHLOROPHENYL)-5-ISOPROPYLBIGUANIDE DIHYDROCHLORIDE see PAE600

1-(p-CHLOROPHENYL)-5-ISOPROPYLBIGUANIDE HYDROCHLORIDE see CKB500

N-3-CHLOROPHENYLISOPROPYLCARBAMATE see CKC000

4-CHLOROPHENYL ISOPROPYL KETONE see IOL100

p-CHLOROPHENYL ISOPROPYL KETONE see IOL100

4-CHLOROPHENYLLITHIUM see CKC325

N-(p-CHLOROPHENYL)MALEIMIDE see CKC500

p-CHLORO-PHENYL MERCAPTAN see CEK425

CHLOROPHENYLMETHANE see BEE375

1-(2-((4-CHLOROPHENYL)METHOXY)-2-(2,4-DICHLOROPHENYL)ETHYL)-1H-IMIDAZOLE NITRATE see EAE000

1-(3-(CHLOROPHENYL)-5-METHOXY-N-METHYL)-1H-PYRAZOLE-3-CARBOXAMIDE see CKD250

3-(4-CHLOROPHENYL)-1-METHOXY-1-METHYLUREA see CKD500

N'-(4-CHLOROPHENYL)-N-METHOXY-N-METHYLUREA see CKD500

1-(1-(2-((3-CHLOROPHENYL)METHOXY)PHENYL)ETHENYL)-1H-IMIDAZOLE HYDROCHLORIDE see CMW550

1-(1-(2-((3-CHLOROPHENYL)METHOXY)PHENYL)ETHENYL)-1H-IMIDAZOLE MONOHYDROCHLORIDE (9CI) see CMW550

2-(4-CHLOROPHENYL)-2-(METHYLAMINO)CYCLOHEXANONE see KEK200

2-(o-CHLOROPHENYL)-2-(METHYLAMINO)CYCLOHEXANONE HYDROCHLORIDE see CKD750

α-(4-CHLOROPHENYL)-2-((METHYLAMINO)METHYL)-BENZENEMETHANOL HYDROCHLORIDE (9CI) see CID825

1-(p-CHLOROPHENYL)-2-METHYL-2-AMINOPROPANE see CLY250

1-(p-CHLOROPHENYL)-2-METHYL-2-AMINOPROPANE HYDROCHLORIDE see ARW750

(4-CHLOROPHENYL)METHYLARSINIC ACID see CKD800

2-(4-CHLOROPHENYL)-α-METHYL-5-BENZOXAZOLEACETIC ACID see OJI750

2-(α-(p-CHLOROPHENYL)-α-METHYLBENZYLOXY)-N,N-DIETHYLETHYLAMINE see CKE000

2-(α-(p-CHLOROPHENYL)-α-METHYLBENZYLOXY)-N,N-DIMETHYLAMINE see CIS000

1-(2-(α-(p-CHLOROPHENYL)-α-METHYLBENZYLOXY))ETHYL PYRROLIDINE see CKE250

2-(4-CHLOROPHENYL)-3-METHYL-2,3-BUTANEDIOL see CKE750

2-p-CHLOROPHENYL-3-METHYL-2,3-BUTANEDIOL see CKE750

2-CHLOROPHENYL-N-METHYLCARBAMATE see CKF000

o-CHLOROPHENYL METHYLCARBAMATE see CKF000

S-((4-CHLOROPHENYL)METHYL)DIETHYLCARBAMOTHIOATE see SAZ000

1-((4-CHLOROPHENYL)METHYL)-1H-INDOLE-3-CARBOXYLIC ACID see CEQ625

6-(3-(o-CHLOROPHENYL)-5-METHYL-4-ISOXAZOLECARBOXAMIDEO)-3,3-DIMETHYL-7-OXO-4-THIA-1-AZABICYCLO(3.2.0)HEPTANE-2-CARBOXYLIC ACID, MONOSODIUM SALT see SLJ050

6-(3-(o-CHLOROPHENYL)-5-METHYL-4-ISOXAZOLECARBOXAMIDEO)-3,3-DIMETHYL-7-OXO-4-THIA-1-AZABICYCLO(3.2.0)HEPTANE-2-CARBOXYLIC ACID, SODIUM SALT, MONOHYDRATE see SLJ000

3-o-CHLOROPHENYL-5-METHYL-4-ISOXAZOLYLPENICILLIN SODIUM see SLJ050

2-(4-CHLOROPHENYL)-3-METHYL-4-METATHIAZANONE-1,1-DIOXIDE see CKF500

1-((4-CHLOROPHENYL)METHYL)-2-METHYL-1H-BENZIMIDAZOLE (9CI) see CDY325

3-(p-CHLOROPHENYL)-1-METHYL-1-(1-METHYL-2-PROPYNYL)UREA see CKF750

N'-(4-CHLOROPHENYL)-N-METHYL-N-(1-METHYL-2-PROPYNYL)-UREA see CKF750

N-((4-CHLOROPHENYL)METHYL)-N',N'-DIMETHYL-N-2-PYRIDINYL-1,2-ETHANEDIAMINE (9CI) see CKV625

3-(p-CHLOROPHENYL)-1-METHYL-1-NITROSOUREA see MMW775

1-(p-CHLOROPHENYL)-3-METHYL-3-NITROSOUREA see MMW775

2-(p-CHLOROPHENYL)-4-METHYL-2,4-PENTANEDIOL see CKG000

2-(p-CHLOROPHENYL)-4-METHYLPENTANE-2,4-DIOL see CKG000

4-CHLORO-2-(PHENYLMETHYL)PHENOL see CJU250

4-CHLORO-2-(PHENYLMETHYL)PHENOL SODIUM SALT see SFB200

3-CHLORO-N-(PHENYLMETHYL)PROPANAMIDE see BEG000

1-(4-CHLOROPHENYL)-2-METHYL-1-PROPANONE see IOL100

1-(3-CHLOROPHENYL)-4-(2-(5-METHYL-1H-PYRAZOL-3-YL)ETHYL)PIPERAZINE DIHYDROCHLORIDE see MCH535

p-CHLOROPHENYL METHYL SULFIDE see CKG500

4-CHLOROPHENYL METHYL SULFONE see CKG750

p-CHLOROPHENYL METHYL SULFONE see CKG750

4-CHLOROPHENYL METHYL SULFOXIDE see CKH000

p-CHLOROPHENYL METHYL SULFOXIDE see CKH000

(±)-1-(2-(((4-CHLOROPHENYL)METHYL)THIO)-2-(2,4-DICHLOROPHENYL)ETHYL)-1H-IMIDAZOLE MONONITRATE see SNH480

6-CHLORO-3-(((PHENYLMETHYL)THIO)METHYL)-2H-1,2,4-BENZOTHIADIAZINE-7-SULFONAMIDE DIOXIDE see BDE250

N-(4-CHLOROPHENYL)-N'-(3,4-DICHLOROPHENYL)UREA see TIL500

N-(p-CHLOROPHENYL)-N',N'-DIMETHYLUREA see CJX750

N-4-CHLOROPHENYL-N⁵-ISOPROPYLDIGUANIDE HYDROCHLORIDE see CKB500

5-(o-CHLOROPHENYL)-7-NITRO-1H-1,4-BENZODIAZEPIN-2(3H)-ONE see CMW000

N-(4-CHLOROPHENYL)-N'-METHOXY-N-METHYLUREA see CKD500

N-(4-CHLOROPHENYL)-3-OXOBUTANAMIDE see AAY250

1-p-CHLOROPHENYL PENTYL SUCCINATE see CKI000

β-(p-CHLOROPHENYL)PHENETHYL 4-(o-CHLOROPHENYL)PIPERAZINYL KETONE see CKI020

β-(p-CHLOROPHENYL)PHENETHYL 4-(2-HYDROXYPROPYL)PIPERAZINYL KETONE see CKI180

β-(p-CHLOROPHENYL)PHENETHYL 4-(o-METHOXYPHENYL)PIPERAZINYL KETONE see CKI185

β-(p-CHLOROPHENYL)PHENETHYL 4-(m-METHYLBENZYL)PIPERAZINYL KETONE see CKI030

β-(p-CHLOROPHENYL)PHENETHYL 4-PHENETHYLPIPERAZINYL KETONE see CKI040

β-(p-CHLOROPHENYL)PHENETHYL 4-(2-PYRIDYL)PIPERAZINYL KETONE see CKI050

β-(p-CHLOROPHENYL)PHENETHYL 4-(2-PYRIMIDYL)PIPERAZINYL KETONE see CKI060

β-(p-CHLOROPHENYL)PHENETHYL 4-(2-THIAZOLYL)PIPERAZINYL KETONE see CKI070

β-(p-CHLOROPHENYL)PHENETHYL 4-(p-TOLYL)PIPERAZINYL KETONE see CKI190

β-(p-CHLOROPHENYL)PHENETHYL 4-(o-TOLYL)PIPERAZINYL KETONE see CKI090

β-(p-CHLOROPHENYL)PHENETHYL 4-(m-TOLYL)PIPERAZINYL KETONE see CKI080

2-CHLORO-4-PHENYLPHENOL see CHN500

6-CHLORO-2-PHENYLPHENOL, SODIUM SALT see SFV500

2(2-(4-CHLOROPHENYL)-2-PHENYLACETYL)INDAN-1,3-DIONE see CJJ000

2-((p-CHLOROPHENYL)PHENYLACETYL)-1,3-INDANDIONE see CJJ000

2-((4-CHLOROPHENYL)PHENYLACETYL)-1H-INDENE-1,3(2H)-DIONE see CJJ000

1-o-CHLOROPHENYL-1-PHENYL-3-DIMETHYLAMINO-1-PROPANOL HYDROCHLORIDE see CMW700

2-(1-(4-CHLOROPHENYL)-1-PHENYLETHOXY)-N,N-DIMETHYLETHANAMINE HYDROCHLORIDE see CIS000

(1-(p-CHLOROPHENYL)-1-PHENYL)ETHYL (β-DIMETHYLAMINOETHYL) ETHER HYDROCHLORIDE see CIS000

2-CHLOROPROPANOL see CKR500
3-CHLOROPROPANOL see CKP725
1-CHLORO-2-PROPANOL with 2-CHLORO-1-PROPANOL see CKR750
1-CHLORO-2-PROPANONE see CDN200
CHLOROPROPANONE see CDN200
3-CHLOROPROPANONITRILE see CKT250
7-CHLORO-1-PROPARGYL-5-PHENYL-2H-1,4-BENZODIAZEPIN-2-ONE see PIH100
1-CHLORO-1-PROPENE see PMR750
2-CHLORO-1-PROPENE see CKS000
3-CHLORO-1-PROPENE see AGB250
1-CHLOROPROPENE see PMR750
1-CHLORO-2-PROPENE see AGB250
1-CHLORO PROPENE-2 see AGB250
3-CHLOROPROPENE see AGB250
2-CHLOROPROPENE (DOT) see CKS000
2-CHLORO-2-PROPENE-1-THIOL DIETHYLDITHIOCARBAMATE see CDO250
2-CHLORO-2-PROPENOIC ACID METHYL ESTER (9CI) see MIF800
cis-3-CHLOROPROPENOIC ACID, SODIUM SALT see SFV250
2-CHLORO-2-PROPEN-1-OL see CEF250
3-CHLORO-2-PROPEN-1-OL see CEF500
2-CHLORO-2-PROPENYL DIETHYLCARBAMODITHIOATE see CDO250
2-CHLORO-4-(2-PROPENYLOXY)BENZENEACETIC ACID see AGN000
2-CHLORO-2-PROPENYL TRIFLUOROMETHANE SULFONATE see CKS325
CHLOROPROPHAM see CKC000
CHLOROPROPHENPYRIDAMINE see CLX300
CHLOROPROPHENYLPYRIDAMINE MALEATE see TAI500
2-CHLOROPROPIONATE SODIUM SALT see CKT100
3-CHLOROPROPIONIC ACID see CKS500
α-CHLOROPROPIONIC ACID see CKS750
β-CHLOROPROPIONIC ACID see CKS500
2-CHLOROPROPIONIC ACID METHYL ESTER see CKT000
2-CHLOROPROPIONIC ACID SODIUM SALT see CKT100
α-CHLOROPROPIONIC ACID SODIUM SALT see CKT100
3-CHLOROPROPIONITRILE see CKT250
β-CHLOROPROPIONITRILE see CKT250
N-(3-CHLOROPROPIONYL)BENZYLAMINE see BEG000
p-CHLOROPROPIOPHENONE see CKT500
2-CHLOROPROPYL ALCOHOL see CKR500
6'-CHLORO-2-(PROPYLAMINO)-o-ACETOTOLUIDIDE HYDROCHLORIDE see CKT750
6'-CHLORO-2-(PROPYLAMINO)-o-BUTYROTOLUIDIDE HYDROCHLORIDE see CKU000
4-CHLORO-4-((PROPYLAMINO)CARBONYL)BENZENESULFONAMIDE see CKK000
9-((2-((2-CHLOROPROPYL)AMINO)ETHYL)AMINO)-2-METHOXYACRIDINE DIHYDROCHLORIDE HEMIHYDRATE see CKU250
2-CHLORO-4-(2-PROPYLAMINO)-6-ETHYLAMINO-s-TRIAZINE see ARQ725
CHLOROPROPYLATE see PNH750
3-CHLOROPROPYL BROMIDE see BNA825
3-CHLORO-1-PROPYLENE see AGB250
3-CHLOROPROPYLENE see AGB250
α-CHLOROPROPYLENE see AGB250
1-CHLORO-2,3-PROPYLENE DINITRATE see CKU625
3-CHLOROPROPYLENE GYLCOL see CDT750
3-CHLORO-1,2-PROPYLENE OXIDE see EAZ500
γ-CHLOROPROPYLENE OXIDE see EAZ500
CHLOROPROPYLENE OXIDE see EAZ500
N-(3-CHLOROPROPYL)-α-METHYLPHENETHYLAMINE HYDROCHLORIDE see PKS500
3-CHLOROPROPYNE see CKV275
CHLOROPROPYNENITRILE see CFE750
CHLOROPROTHIXENE see TAF675
CHLOROPTIC see CDP250
6-CHLOROPURINE see CKV500
6-CHLORO-9H-PURINE see CKV500
6-CHLORO-1H-PURINE (9CI) see CKV500
CHLOROPYRAMINE see CKV625
CHLOROPYRIBENZAMINE see CKV625
2-CHLOROPYRIDINE see CKW000
3-CHLOROPYRIDINE see CKW250
α-CHLOROPYRIDINE see CKW000
m-CHLOROPYRIDINE see CKW250
o-CHLOROPYRIDINE see CKW000
6-CHLORO-2-PYRIDINECARBOXYLIC ACID see CKN375
2-CHLOROPYRIDINE-N-OXIDE see CKW325
2-(4-(3-(3-CHLORO-10H-PYRIDO(3,2-b)-1,4-BENZOTHIAZINE-10-YL)PROPYL))-1-PIPERAZINYLETHANOL see CMY535
2-(4-(3-(3-CHLORO-10H-PYRIDO(3,2-b)(1,4)BENZOTHIAZIN-1-OYL)PROPYL)-1-PIPERAZINYL) ETHANOL see CMY535
4-(3-(3-CHLORO-10H-PYRIDO(3,2-b)(1,4)-BENZOTHIAZIN-10-YL)PROPYL)-1-PIPERAZINE ETHANOL see CMY535
CHLORO-3-PYRIDYLMERCURY see CKW500
4-(2-((6-CHLORO-2-PYRIDYL)THIO)ETHYL)MORPHOLINE MONOHYDROCHLORIDE see FMU225

CHLOROPYRILENE see CHY250
1-CHLORO-2,5-PYRROLIDINEDIONE see SND500
4-CHLORO-N-((1-PYRROLIDINYLAMINO)CARBONYL)BENZENESULFONAMIDE (9CI) see GHR609
6'-CHLORO-2-(PYRROLIDINYL)-o-DIACETOTOLUIDIDE HYDROCHLORIDE see CLB250
6'-CHLORO-2-PYRROLIDINYL-o-HEXANOTOLUIDIDE HYDROCHLORIDE see CAB250
4'-CHLORO-2-PYRROLIDINYL-α,α,α-TRIFLUORO-m-ACETOTOLUIDIDE, HYDROCHLORIDE see CLC125
6'-CHLORO-2-PYRROLIDINYL-α,α,α-TRIFLUORO-m-ACETOTOLUIDINE, HYDROCHLORIDE see CLC100
2'-CHLORO-2-PYRROLIDINYL-5'-TRIFLUOROMETHYLACETANILIDE HYDROCHLORIDE see CLC100
4'-CHLORO-2-PYRROLIDINYL-3'-TRIFLUOROMETHYLACETANILIDE HYDROCHLORIDE see CLC125
3-CHLORO-4-(3-PYRROLIN-1-YL)HYDRATROPIC ACID see PJA220
CHLOROQUINALDOL see CLC500
6-CHLORO-4(3H)-QUINAZOLINONE see CLC750
6-CHLORO-4-QUINAZOLINONE see CLC750
CHLOROQUINE see CLD000
CHLOROQUINE DIPHOSPHATE see CLD250
CHLOROQUINE MUSTARD see CLD500
CHLOROQUINE PHOSPHATE see AQT250, CLD250
CHLOROQUINIUM see CLD000
5-CHLORO-8-QUINOLINOL see CLD600
4-((7-CHLORO-4-QUINOLINYL)AMINO)-2-(1-PYRROLIDINYLMETHYL)PHENOL DIHYDROCHLORIDE see PMY000
N⁴-(7-CHLORO-4-QUINOLINYL)-N¹,N¹-DIETHYL-1,4-PENTANEDIAMINE see CLD000
2-((4-(7-CHLORO-4-QUINOLYL)AMINO)PENTYL)-ETHYLAMINO)ETHANOL see PJB750
4-((7-CHLORO-4-QUINOLYL)AMINO)-α-1-PYRROLIDINYL-o-CRESOL DIHYDROCHLORIDE see PMY000
4-CHLORORESORCINOL see CLD750
7-CHLORO-3-β-d-RIBOFURANOSYL-3H-IMIDAZO(4,5-b)PYRIDINE see AEB500
CHLOROS see SHU500
5-CHLOROSALICYLALDEHYDE see CLD800
5-CHLOROSALICYLIC ACID see CLD825
CHLORO-S.C.T.Z. see CHD750
CHLOROSILANE see DGK300
CHLOROSILANES see CLE250
2'-CHLORO-4-STILBENYL-N,N-DIMETHYLAMINE see CGK500
3'-CHLORO-4-STILBENYL-N,N-DIMETHYLAMINE see CGK750
4'-CHLORO-4-STILBENYL-N,N-DIMETHYLAMINE see CGL000
CHLOROSTOP see PKQ059
o-CHLOROSTYRENE see CLE750
CHLOROSTYRENE see CLE600
4-CHLOROSTYRYL PHENYL KETONE see CLF100
o-CHLOROSTYRYL PHENYL KETONE see CLF150
p-CHLOROSTYRYL PHENYL KETONE see CLF100
CHLOROSULFACIDE see CEP000
N-(4'-CHLORO-3'-SULFAMOYLBENZENESULFONYL)-N-METHYL-2-AMINOMETHYL-2-METHYLTETRAHYDROFURAN see MCA100
6-CHLORO-7-SULFAMOYL-2H-1,2,4-BENZOTHIADIAZINE-1,1-DIOXIDE see CLH750
6-CHLORO-7-SULFAMOYL-3,4-DIHYDRO-2H-1,2,4-BENZOTHIADIAZINE-1,1-DIOXIDE see CFY000
4-CHLORO-5-SULFAMOYL-2',6'-SALICYLOXYLIDIDE see CLF325
N-CHLOROSULFINYLIMIDE see CLF500
4-CHLORO-4'-(6-SULFO-2H-NAPHTHO(1,2-d)TRIAZOL-2-YL)-2,2'-STILBENEDISULFONIC ACID TRISODIUM SALT see CLG000
CHLOROSULFONIC ACID see CLG500
CHLOROSULFONIC ACID (with or without sulfur trioxide) (UN 1754) (DOT) see CLG500
CHLOROSULFONIC ANHYDRIDE see PPR500
5-(CHLOROSULFONYL)-2,4-DICHLOROBENZOIC ACID see CLG200
1-CHLOROSULFONYL-5-DIMETHYLAMINONAPHTHALENE see DPN200
CHLOROSULFONYL FLUORIDE see SOT500
CHLOROSULFONYLISOCYANATE see CLG250
1-(4-CHLORO-o-SULFO-5-TOLYLAZO)-2-NAPHTHOL,BARIUM SALT see CHP500
CHLOROSULFURIC ACID see CLG500
4-CHLORO-5-SULPHAMOYLPHTHALIMIDE see CGM500, CLG825
CHLOROSULTHIADIL see CFY000
7-CHLOROTETRACYCLINE see CMA750
CHLOROTETRACYCLINE HYDROCHLORIDE see CMB000
CHLOROTETRAFLUOROETHANE see CLH000
7-CHLORO-1,2,3,4-TETRAHYDRO-2-METHYL-3-(2-METHYLPHENYL)-4-OXO-6-QUINAZOLINESULFONAMIDE see ZAK300
CHLOROTETRAHYDROXY((2-HYDROXY-5-OXO-2-IMIDAZOLIN-4-YL)UREATO)DIALUMINUM see AHA135
9-CHLORO-7-(1H-TETRAZOL-5-YL)-5H-1-BENZOPYRANO(2,3-b)PYRIDIN-5-ONE SODIUM PENTAHYDRATE see THK850

2-CHLORO-5-(1H-TETRAZOL-5-YL)-N(sup 4)-2-THENYLSULFANILAMIDE see ASO375
CHLOROTHALIDONE see CLY600
CHLOROTHALONIL see TBQ750
CHLOROTHANE NU see MIH275
CHLOROTHEN see CHY250
CHLOROTHENE (inhibited) see MIH275
CHLOROTHENE see MIH275
CHLOROTHENE NU see MIH275
CHLOROTHENE VG see MIH275
2-((5-CHLORO-2-THENYL)(2-DIMETHYLAMINOETHYL)AMINO)PYRIDINE see CHY250
CHLOROTHENYLPYRAMINE see CHY250
8-CHLORO-THEOPHYLLINE compounded with 4-(DIPHENYLMETHOXY)-1-METHYLPIPERIDINE (1:1) see PIZ250
5-CHLORO-1,2,3-THIADIAZOLE see CLH625
CHLOROTHIAMIDE see DGM600
CHLOROTHIAZID see CLH750
CHLOROTHIAZIDE see CLH750
4-CHLOROTHIOANISOLE see CKG500
p-CHLOROTHIOANISOLE see CKG500
p-CHLOROTHIOCARBANILIC ACID-o-(p-CHLOROPHENYL) ESTER see CJU125
CHLOROTHIOFORMIC ACID ETHYL ESTER see CLJ750
2,3,4,5-CHLOROTHIOPHENE see TBV750
4-CHLOROTHIOPHENOL see CEK425
p-CHLOROTHIOPHENOL see CEK425
(Z)-3-(2-CHLORO-9H-THIOXANTHEN-9-YLIDENE)-N,N-DIMETHYL-1-PROPANAMINE see TAF675
4-(3-(2-CHLOROTHIOXANTHEN-9-YLIDENE)PROPYL)-1-PIPERAZINEETHANOL DIHYDROCHLORIDE see CLX250
6-CHLOROTHYMOL see CLJ800
CHLOROTHYMOL see CLJ800
4-CHLORO-o-TOLOXYACETIC ACID see CIR250
2-CHLOROTOLUENE see CLK100
4-CHLOROTOLUENE see TGY075
α-CHLOROTOLUENE see BEE375
p-CHLOROTOLUENE (DOT) see TGY075
ω-CHLOROTOLUENE see BEE375
ar-CHLOROTOLUENE see CLK130
o-CHLOROTOLUENE see CLK100
p-CHLOROTOLUENE see TGY075
CHLOROTOLUENE see CLK130
CHLOROTOLUENES see CLK130
4-CHLORO-2-TOLUIDINE see CLK220
6-CHLORO-2-TOLUIDINE see CLK227
3-CHLORO-o-TOLUIDINE see CLK200
4-CHLORO-o-TOLUIDINE see CLK220
5-CHLORO-o-TOLUIDINE see CLK225
6-CHLORO-o-TOLUIDINE see CLK227
2-CHLORO-p-TOLUIDINE see CLK210
3-CHLORO-p-TOLUIDINE see CLK215
4-CHLORO-2-TOLUIDINE HYDROCHLORIDE see CLK235
4-CHLORO-o-TOLUIDINE HYDROCHLORIDE (DOT) see CLK235
4-CHLORO-o-TOLUIDINE HYDROCHLORIDE see CLK235
3-CHLORO-p-TOLUIDINE HYDROCHLORIDE see CLK230
2-(2-CHLORO-p-TOLUIDINO)-2-IMIDAZOLINE NITRATE see TGJ885
CHLOROTOLURON see CIS250
N-(3-CHLORO-o-TOLYL)ANTHRANILIC ACID see CLK325
1-(6-CHLORO-o-TOLYL)-3-CYCLOHEXYL-3-(2-(DIETHYLAMINO)ETHYL)UREA HYDROCHLORIDE see CLK500
1-(6-CHLORO-o-TOLYL)-3-(3-(DIBUTYLAMINO)PROPYL)UREA HYDROCHLORIDE see CLK750
1-(6-CHLORO-o-TOLYL)-3-(2-(DIETHYLAMINO)ETHYL)-3-METHYLUREA see CLL000
1-(6-CHLORO-o-TOLYL)-3-(2-(DIETHYLAMINO)ETHYL)UREA HYDROCHLORIDE see CLL250
1-(6-CHLORO-o-TOLYL)-1-(2-(DIETHYLAMINO)ETHYL)-3-(2,6-XYLYL)UREA HYDROCHLORIDE see CLL750
1-(6-CHLORO-o-TOLYL)-3-(2-(DIETHYLAMINO)ETHYL)-3-(2,6-XYLYL)UREA HYDROCHLORIDE see CLL500
1-(6-CHLORO-o-TOLYL)-3-(3-(DIETHYLAMINO)PROPYL)UREA see CLM000
1-(6-CHLORO-o-TOLYL)-3-(2-(DIMETHYLAMINO)ETHYL)-3-ISOPROPYLUREA HYDROCHLORIDE see CLM250
1-(6-CHLORO-o-TOLYL)-3-(2-(DIMETHYLAMINO)ETHYL)UREA HYDROCHLORIDE see CLM500
1-(6-CHLORO-o-TOLYL)-3-(3-(DIMETHYLAMINO)PROPYL)UREA HYDROCHLORIDE see CLM750
N′-(4-CHLORO-o-TOLYL)-N,N-DIMETHYLFORMAMIDINE see CJJ250
N′-(4-CHLORO-o-TOLYL)-N,N-DIMETHYLFORMAMIDINE HYDROCHLORIDE see CJJ500
1-(6-CHLORO-o-TOLYL)-3-(4-METHOXYBENZYL)-3-(2-PIPERIDINOETHYL)UREA see CLN000
1-(6-CHLORO-o-TOLYL)-3-(4-METHOXYBENZYL)-3-(2-(PYRROLIDINYL)ETHYL)UREA HYDROCHLORIDE see CLN250

1-(4-CHLORO-o-TOLYL)-3-(p-METHYLBENZYL)-3-(2-PYRROLIDINYLETHYL)UREA HYDROCHLORIDE see CLN500
((4-CHLORO-o-TOLYL)OXY)ACETIC ACID see CIR250
((4-CHLORO-o-TOLYL)OXY)ACETIC ACID, ETHYL ESTER see EMR000
((4-CHLORO-O-TOLYL)OXY)-ACETIC ACID with 2,2′-IMINODIETHANOL (1:1) see MIH750
(p-CHLORO-o-TOLYLOXY)ACETIC ACID SODIUM SALT see SIL500
4-((4-CHLORO-o-TOLYL)OXY)BUTYRIC ACID see CLN750
4-(4-CHLORO-o-TOLYLOXY)BUTYRIC ACID see CLN750
(4-CHLORO-o-TOLYLOXY)BUTYRIC ACID SODIUM SALT see CLO000
2-(p-CHLORO-o-TOLYLOXY)PROPIONIC ACID see CIR500
2-((4-CHLORO-o-TOLYL)OXY)PROPIONIC ACID POTASSIUM SALT see CLO200
1-(6-CHLORO-o-TOLYL)-3-(2-PYRROLIDINYLETHYL)UREA HYDROCHLORIDE see CLO500
CHLOROTOLYLTHIOGLYCOLIC ACID see CLO600
CHLOROTRIANISENE see CLO750
CHLOROTRIANIZEN see CLO750
CHLOROTRIAZINE see TJD750
CHLOROTRIBENZYLSTANNANE see CLP000
CHLOROTRIBUTYLGERMANIUM see CLP250
CHLOROTRIBUTYLSTANNANE see CLP500
1-CHLORO-4-(TRICHLOROMETHYL)BENZENE see TIR900
3-CHLORO-3-TRICHLOROMETHYLDIAZIRINE see CLP625
2-CHLORO-6-(TRICHLOROMETHYL)PYRIDINE see CLP750
(Z)-2-CHLORO-1-(2,4,5-TRICHLOROPHENYL)VINYL DIMETHYL PHOSPHATE see RAF100
2-CHLORO-1-(2,4,5-TRICHLOROPHENYL)VINYL DIMETHYL PHOSPHATE see TBW100
2-CHLORO-1-(2,4,5-TRICHLOROPHENYL)VINYL PHOSPHORIC ACID DIMETHYL ESTER see TBW100
2-CHLOROTRIETHYLAMINE see CGV500
β-CHLOROTRIETHYLAMINE see CGV500
2-CHLOROTRIETHYLAMINE HYDROCHLORIDE see CLQ250
CHLORO(TRIETHYLPHOSPHINE)GOLD see CLQ500
CHLOROTRIETHYLSTANNANE see TJV000
CHLOROTRIETHYLTIN see TJV000
CHLOROTRIFLUORIDE see CDX750
2-CHLORO-1,1,1-TRIFLUOROETHANE see TJY175
1-CHLORO-2,2,2-TRIFLUOROETHANE see TJY175
CHLOROTRIFLUOROETHENE HOMOPOLYMER see KDK000
2-CHLORO-1,1,2-TRIFLUOROETHYL DIFLUOROMETHYL ETHER see EAT900
2-CHLORO-1,1,2-TRIFLUOROETHYLENE see CLQ750
1-CHLORO-1,2,2-TRIFLUOROETHYLENE see CLQ750
CHLOROTRIFLUOROETHYLENE see CLQ750
CHLOROTRIFLUOROETHYLENE POLYMER see KDK000
CHLOROTRIFLUOROETHYLENE POLYMERS see KDK000
2-CHLORO-1,1,2-TRIFLUOROETHYL METHYL ETHER see CLR000
CHLOROTRIFLUOROMETHANE see CLR250
CHLOROTRIFLUOROMETHANE mixed with TRIFLUOROMETHANE see FOO515
CHLOROTRIFLUOROMETHANE and TRIFLUOROMETHANE AZEOTROPIC MIXTURE (DOT) see FOO515
2-CHLORO-5-(TRIFLUOROMETHYL)ANILINE see CEG800
4-CHLOROTRIFLUOROMETHYLBENZENE see CEM825
p-CHLOROTRIFLUOROMETHYLBENZENE see CEM825
3-CHLORO-3-TRIFLUOROMETHYLDIAZIRINE see CLR825
5-(2-CHLORO-4-(TRIFLUOROMETHYL)PHENOXY)-2-NITROBENZOIC ACID see CLS075
3-CHLORO-1,1,1-TRIFLUOROPROPANE see TJY200
1-CHLORO-3,3,3-TRIFLUOROPROPANE see TJY200
6-CHLORO-α-α-α-TRIFLUORO-m-TOLUIDINE see CEG800
4-(4-CHLORO-α,α,α-TRIFLUORO-m-TOLYL)-1-(4,4-BIS(p-FLUOROPHENYL)BUTYL)-4-PIPERIDINOL see PAP250
4-(4-(4-CHLORO-α,α,α-TRIFLUORO-m-TOLYL)-4-HYDROXYPIPERIDINO)BUTYROPHENONE-4′-FL see CLS250
5-(2-CHLORO-α-α-α-TRIFLUORO-p-TOLYLOXY)-2-NITROBENZOIC ACID (IUPAC) see CLS075
CHLORO(TRIISOBUTYL)STANNANE see CLS500
7-CHLORO-4,6,2′-TRIMETHOXY-6′-METHYLGRIS-2′-EN-3,4′-DIONE see GKE000
1-CHLORO-4-(TRIMETHYL)-BENZENE (9CI) see CEM825
2-CHLORO-N,N,N-TRIMETHYLETHANAMINIUM CHLORIDE see CMF400
CHLOROTRIMETHYLPLUMBANE see TLU175
CHLOROTRIMETHYLSILANE see TLN250
CHLOROTRIMETHYLSTANNANE see CLT000
CHLOROTRIMETHYLTIN see CLT000
CHLOROTRINITROMETHANE see CLT250
CHLOROTRIPHENYLMETHANE see CLT500
CHLOROTRIPHENYLSILANE see TMU000
CHLOROTRIPHENYLSTANNANE see CLU000
CHLOROTRIPHENYLTIN see CLU000
CHLOROTRIPROPYLPLUMBANE see TNA750
CHLOROTRIPROPYLSTANNANE see CLU250
CHLOROTRISIN see CLO750
CHLOROTRIS(p-METHOXYPHENYL)ETHYLENE see CLO750

1-(o-CHLOROTRITYL)IMIDAZOLE see MRX500
(CHLOROTRITYL)IMIDAZOLE see MRX500
CHLORO(TRIVINYL)STANNANE see CLU500
CHLOROUS ACID, SILVER(1+) SALT see SDN500
CHLOROUS ACID, SODIUM SALT (8CI,9CI) see SFT500
CHLOROVINYLARSINE DICHLORIDE see CLV000
β-CHLOROVINYLBICHLOROARSINE see CLV000
2-CHLOROVINYLDICHLOROARSINE see CLV000
(2-CHLOROVINYL)DICHLOROARSINE see CLV250
(2-CHLOROVINYL)DIETHOXYARSINE see CLV250
2-CHLOROVINYL DIETHYL PHOSPHATE see CLV375
β-CHLOROVINYL ETHYLETHYNYL CARBINOL see CHG000
(2-CHLOROVINYL)MERCURIC CHLORIDE see CFD250
3-(β-CHLOROVINYL)-1-PENTYN-3-OL see CHG000
CHLOROVULES see CDP250
CHLOROWODOR (POLISH) see HHL000
CHLOROWODORKU 10-γ-DWUMETYLOAMINOPROPYLO-7-CHLOROBENZO(b)-(1,8)-NAFTYRYDONU-5 (POLISH) see CGG500
CHLOROX see SHU500
4(2-CHLORO-9H-XANTHEN-9-YLIDENE)-1-METHYLPIPERIDINE METHANESULFO-NATE see CMX860
CHLOROXAZONE see CDQ750
CHLOROXIFENIDIM see CJQ000
CHLOROXINE see DEL300
CHLOROXONE see DAA800
CHLOROXURON see CJQ000
CHLOROXYLAM see CGI500
α-CHLORO-p-XYLENE see MHN300
4-CHLORO-3,5-XYLENOL see CLW000
p-CHLORO-m-XYLENOL see CLW000
CHLORO-XYLENOL see CLW000
6-CHLORO-3,4-XYLENYL N-METHYLCARBAMATE see CGI500
(4-CHLORO-6-(2,3-XYLIDINO)-2-PYRIMIDINYLTHIO)ACETIC ACID see CLW250
2-((4-CHLORO-6-(2,3-XYLIDINO)-2-PYRIMIDINYL)THIO)-N-(2-HYDROXYETH-YL)ACETAMIDE see CLW500
4-(p-CHLORO-N-2,6-XYLYLBENZAMIDO)BUTYRIC ACID see CLW625
2-CHLORO-4,5-XYLYL ESTER, CARBAMIC ACID see CGI500
6-CHLORO-3,4-XYLYL N-METHYLCARBAMATE see CGI500
CHLOROXYPHOS see TIQ250
CHLOROXYQUINOLINE see CLD600, DEL300
CHLOROZIRCONYL see ZSJ000
CHLOROZODIN see ASM250
CHLOROZONE see CDP000
CHLOROZOTOCIN see CLX000
CHLORPARACIDE see CEP000
CHLORPENTHIXOL DIHYDROCHLORIDE see CLX250
CHLORPERALGONIC see CJH500
CHLORPERAZINE see PMF500
CHLORPERPHENTHIXENE DIHYDROCHLORIDE see CLX250
CHLORPHACINON (ITALIAN) see CJJ000
CHLORPHEDIANOL HYDROCHLORIDE see CMW700
CHLORPHENAMIDINE see CJJ250
CHLORPHENAMINE see CLX300
CHLORPHENESIN CARBAMATE see CJQ250
CHLORPHENIRAMINE see CLX300
S-(+)-CHLORPHENIRAMINE MALEATE see PJJ325
d-CHLORPHENIRAMINE MALEATE see PJJ325
(+)-CHLORPHENIRAMINE MALEATE see PJJ325
CHLORPHENIRAMINE MALEATE see TAI500
o-CHLORPHENOL (GERMAN) see CJK250
CHLORPHENOXAMINE HYDROCHLORIDE see CIS000
2-(4-(4′-CHLORPHENOXY)-PHENOXY)-PROPIONSAEURE (GERMAN) see CJP750
3-(4-(4-CHLOR-PHENOXY)-PHENYL)-1,1-DIMETHYLHARNSTOFF (GERMAN) see CJQ000
CHLORPHENPROP-METHYL see CFC750
CHLORPHENTERAMINE see CLY250
CHLORPHENTERMINE see CLY250
CHLORPHENTERMINE HYDROCHLORIDE see ARW750
CHLORPHENVINFOS see CDS750
CHLORPHENVINPHOS see CDS750
6-(o-CHLORPHENYL)-8-AETYL-1-METHYL-4H-sec-TRIAZOLO(3,4-c)THIENO(2,3-e)(1,4)DIAZEPIN (GERMAN) see EQN600
(4-CHLOR-PHENYL)-BENZOLSULFONAT (GERMAN) see CJR500
3-CHLORPHENYL-CARBAMIDSAURE-BUTIN-(1)-YL(3)-ESTER (GERMAN) see CEX250
4-CHLORPHENYL-4′-CHLORBENZOLSULFONAT (GERMAN) see CJT750
(4-CHLOR-PHENYL)-4-CHLOR-BENZOL-SULFONATE (GERMAN) see CJT750
3-(4-CHLORPHENYL)-2-CHLORPROPIONSAEUREMETHYLESTER (GERMAN) see CFC750
3-(4-CHLOR-PHENYL)-1,1-DIMETHYL-HARNSTOFF (GERMAN) see CJX750
N-(4-CHLORPHENYL)-2,2-DIMETHYLPENTAMID (GERMAN) see CGL250
1-(p-CHLOR-PHENYL)-3,3-DIMETHYL-TRIAZEN (GERMAN) see CJI100
N-(4-CHLOR-PHENYL)-2,2-DIMETHYL-VALERIANSAEUREAMID (GERMAN) see CGL250

4-(2-CHLORPHENYL)-2-ETHYL-9-METHYL-6H-THIENO(3,2-f)(1,2,4)TRIAZOLO(4,3-a)(1,4)DIAZEPINE see EQN600
γ-(4-(p-CHLORPHENYL)-4-HYDROXPIPERIDINO)-p-FLUORBUTYROPHENONE see CLY500
N-(3-CHLOR-PHENYL)-ISOPROPYL-CARBAMAT (GERMAN) see CKC000
4-CHLOR-PHENYL-ISOTHIOCYANAT (GERMAN) see ISH000
3-(4-CHLORPHENYL)-1-METHOXY-1-METHYLHARNSTOFF (GERMAN) see CKD500
3-(4-CHLORPHENYL)-1-METHYL-1-ISOBUTINYLHARNSTOFF (GERMAN) see CKF750
2-(p-CHLORPHENYL)-3-METHYL-1,3-PERHYDROTHIAZIN-4-ON-1,1-DIOXIDE see CKF500
N-(4-CHLORPHENYL)-N′-METHYL-N′-ISOBUTINYLHARNSTOFF (GERMAN) see CKF750
1-(4-CHLORPHENYL)-1-PHENYL-ACETYL-INDAN-1,3-DION (GERMAN) see CJJ000
((4-CHLORPHENYL)-1-PHENYL)-ACETYL-1,3-INDANDION (GERMAN) see CJJ000
2(2-(4-CHLOR-PHENYL-2-PHENYL)ACETYL)INDAN-1,3-DION (GERMAN) see CJJ000
4-(p-CHLORPHENYLTHIO)-BUTANOL (GERMAN) see CKK500
4-CHLORPHENYL-2′,4′,5′-TRICHLORPHENYLAZOSULFID (GERMAN) see CDS500
CHLORPHONIUM CHLORIDE see THY500
CHLORPHTHALIDOLONE see CLY600
CHLORPHTHALIDONE see CLY600
CHLOR-O-PIC see CKN500
CHLORPIKRIN (GERMAN) see CKN500
CHLORPROETHAZINE see CLY750
CHLORPROETHAZINE HYDROCHLORIDE see CLZ000
CHLORPROHEPTADIEN see CEH500
CHLORPROHEPTADIENE HYDROCHLORIDE see CEH500
CHLORPROMAZIN see CKP250
CHLORPROMAZINE see CKP250
CHLORPROPAMID see CKK000
CHLORPROPAMIDE see CKK000
3-CHLORPROPAN-1-OL see CKP725
3-CHLORPROPEN (GERMAN) see AGB250
CHLORPROPHAM see CKC000
CHLORPROPHAME (FRENCH) see CKC000
CHLORPROPHENPYRIDAMINE see CLX300
N-(3-CHLORPROPYL)-1-METHYL-2-PHENYL-AETHYLAMIN-HYDROCHLORID (GERMAN) see PKS500
CHLORPROTHIXEN see TAF675
α-CHLORPROTHIXENE see TAF675
cis-CHLORPROTHIXENE see TAF675
CHLORPROTHIXENE see TAF675
CHLORPROTIXEN see TAF675
CHLORPROTIXENE see TAF675
CHLORPROTIXINE see TAF675
CHLORPYRIFOS see CMA100
CHLORPYRIFOS-ETHYL see CMA100
CHLORPYRIFOS-METHYL see CMA250
CHLORPYRIPHOS see CMA100
CHLORPYRIPHOS-ETHYL see CMA100
CHLOR-PZ see CKP250
CHLORQUINALDOL see CLC500
CHLORQUINOL see DEL300
CHLORQUINOX see CMA500
CHLORSAEURE (GERMAN) see SFS000
CHLORSAL see CLH750
CHLORSEPTOL see CDP000
2-(4′′-CHLOR-4′-STILBYL)NAFTOTRIAZOL-6,2′,2′′-TRISULFONAN SODNY (CZECH) see CLG000
N-CHLORSUCCINIMIDE see SND500
CHLORSUCCINYLCHOLIN (GERMAN) see HLC500
4-CHLOR-5-SULFAMOYL-2′,6′-SALICYLOXYLIDID (GERMAN) see CLF325
CHLORSULFAQUINOXALINE see CMA600
CHLORSULFONAMIDO DIHYDROBENZOTHIADIAZINE DIOXIDE see CFY000
CHLORSULPHACIDE see CEP000
CHLORTALIDONE see CLY600
CHLORTEN see MIH275
CHLORTETRACYCLINE see CMA750
CHLORTETRACYCLINE, 6-DEMETHYL- see MIJ500
CHLORTETRACYCLINE HYDROCHLORIDE see CMB000
4-CHLORTETRAHYDROFTALANHYDRID (CZECH) see CFG500
CHLORTETRIN see DAI485
CHLORTHAL-DIMETHYL see TBV250
CHLORTHALIDON see CLY600
CHLORTHALIDONE see CLY600
CHLORTHAL-METHYL see TBV250
CHLORTHALONIL (GERMAN) see TBQ750
CHLORTHIAZIDE see CLH750
CHLORTHIEPIN see EAQ750
p-CHLORTHIOFENOL (CZECH) see CEK425

CHLORTHION METHYL see MIJ250
CHLORTHIOPHOS see CLJ875
CHLORTHYMOL see CLJ800
CHLORTION (CZECH) see MIJ250
3-CHLOR-2-TOLUIDIN (CZECH) see CLK200
2-CHLOR-4-TOLUIDIN (CZECH) see CLK210
α-CHLORTOLUOL (GERMAN) see BEE375
CHLORTOLURON see CIS250
N′-(4-CHLOR-o-TOLYL)-N,N-DIMETHYLFORMAMIDIN (GERMAN) see CJJ250
CHLORTOX see CDR750
CHLORTRIANISEN see CLO750
CHLORTRIFLUORAETHYLEN (GERMAN) see CLQ750
CHLOR-TRIMETON see CLD250, CLX300, TAI500
CHLOR-TRIMETON MALEATE see TAI500
CHLOR-TRIPOLON see CLX300, TAI500
CHLORTROPBENZYL see CMB125
CHLORURE d'ALUMINUM (FRENCH) see AGY750
CHLORURE ANTIMONIEUX see AQC500
CHLORURE d'ARSENIC (FRENCH) see ARF500
CHLORURE ARSENIEUX (FRENCH) see ARF500
CHLORURE de BENZENYLE (FRENCH) see BFL250
CHLORURE de 1-BENZYL-3-BENZYL-CARBOXY-PYRIDINIUM (FRENCH) see SAA000
CHLORURE de BENZYLE (FRENCH) see BEE375
CHLORURE de BORE (FRENCH) see BMG500
CHLORURE de BUTYLE (FRENCH) see BQQ750
CHLORURE de CHLORACETYLE (FRENCH) see CEC250
CHLORURE de CHROMYLE (FRENCH) see CML125
CHLORURE DE BENZYLIDENE see BAY300
CHLORURE DE CYANOGENE see COO750
CHLORURE de DICHLORACETYLE (FRENCH) see DEN400
CHLORURE de l'ETHYLAL TRIMETHYLAMMONIUM PROPANEDIOL (FRENCH) see MJH800
CHLORURE d'ETHYLE (FRENCH) see EHH000
CHLORURE d'ETHYLENE (FRENCH) see EIY600
CHLORURE d'ETHYLIDENE (FRENCH) see DFF809
CHLORURE de FUMARYLE (FRENCH) see FOY000
CHLORURE de LITHIUM (FRENCH) see LHB000
CHLORURE de MAGNESIUM HYDRATE (FRENCH) see MAE500
CHLORURE MERCUREUX (FRENCH) see MCW000
CHLORURE MERCUREUX see MCY300
CHLORURE MERCURIQUE (FRENCH) see MCY475
CHLORURE de METHALLYLE (FRENCH) see CIU750
CHLORURE de METHYLE (FRENCH) see MIF765
CHLORURE de METHYLENE (FRENCH) see MJP450
CHLORURE PERRIQUE see FAU000
CHLORURE de SUCCINILCOLINE (FRENCH) see HLC500
CHLORURE de VINYLE (FRENCH) see VNP000
CHLORURE de VINYLIDENE (FRENCH) see VPK000
CHLORURE de ZINC (FRENCH) see ZFA000
CHLORURIT see CLH750
CHLORVINPHOS see DGP900
CHLORWASSERSTOFF (GERMAN) see HHL000
CHLORXYLAM see DET600
CHLORYL see EHH000
CHLORYL ANESTHETIC see EHH000
CHLORYLEA see TIO750
CHLORYL PERCHLORATE see CMB500
CHLORYL RADICAL see CDW450
CHLORZIDE see CFY000
CHLORZOXAZONE see CDQ750
CHLOTAZOLE see CMB675
CHLOTHIXEN see TAF675
CHLOTRIDE see CLH750
CHLOTRIMAZOLE see MRX500
CHLZ see CLX000
CHNU-I see NKJ050
CHOB-i-QUT see CNT350
CHOCOLA A see VSK600
CHOCOLATE EMBL see CMO820
CHOKE CHERRY see AQP890
CHOKEGARD see HCP050
CHOLAGON see DAL000
CHOLAIC ACID see TAH250
CHOLALIN see CME750
CHOLAN DH see DAL000
CHOLANORM see CDL325
CHOLAXINE see SKV200
CHOLECALCIFEROL see CMC750
CHOLEDYL see CMF500
CHOLEGYL see CMF500
CHOLEIC ACID see DAQ400
CHOLEPULVIS see TDE750
CHOLERA ENDOTOXIN see VJZ100
CHOLERA ENTERO-EXOTOXIN see CMC800

CHOLERA ENTEROTOXIN see CMC800
CHOLERAGEN see CMC800
CHOLEREBIC see DAQ400
CHOLESOLVIN see SDY500
(3-β)CHOLESTA-5,7-DIEN-3-OL see DAK600
5,7-CHOLESTADIEN-3-β-OL see DAK600
CHOLESTA-5,7-DIEN-3-β-OL ACETATE see DAK800
(3-β,5-β)-CHOLESTAN-3-OL see DKW000
3-β-CHOLESTANOL see DKW000
epi-CHOLESTANOL see EBA100
CHOLESTAN-3-OL, (3-α-5α-)-(9CI) see EBA100
α-CHOLESTANOL (7CI) see EBA100
5-α-CHOLESTAN-3-α-OL (8CI) see EBA100
5:6-CHOLESTEN-3-β-OL see CMD750
CHOLEST-5-EN-3-β-OL see CMD750
5-CHOLESTEN-3-β-OL see CMD750
Δ⁵-CHOLESTEN-3-β-OL see CMD750
5-CHOLESTEN-3-β-OL 3-(p-(BIS(2-CHLOROETHYL)AMINO)PHENYL)ACETATE see CME250
CHOLESTERIN see CMD750
CHOLESTEROL see CMD750
Δ⁵·⁷-CHOLESTEROL see DAK600
Δ⁷-CHOLESTEROL see DAK600
CHOLESTEROL BASE H see CMD750
CHOLESTEROL-5-α,6-α-EPOXIDE see EBM000
CHOLESTEROL-α-EPOXIDE see EBM000
CHOLESTEROL mixed with OROTIC ACID mixed with CHOLIC ACID (2:2:1) see OJV525
CHOLESTEROL-α-OXIDE see EBM000
CHOLESTEROL OXIDE see EBM000
CHOLESTERYL ALCOHOL see CMD750
CHOLESTERYL-p-BIS(2-CHLOROETHYL)AMINO PHENYLACETATE see CME250
CHOLESTERYL-14-METHYLHEXADECANOATE see CCJ500
CHOLESTRIN see CMD750
CHOLESTROL see CMD750
CHOLESTYRAMINE see CME400
CHOLESTYRAMINE CHLORIDE see CME400
CHOLESTYRAMINE RESIN see CME400
CHOLETH 24 see PKE700
CHOLEXAMIN see CME675
CHOLEXAMINE see CME675
CHOLIBIL see CNG835
CHOLIC ACID see CME750
CHOLIC ACID mixed with CHOLESTEROL mixed with OROTIC ACID (1:2:2) see OJV525
CHOLIC ACID, MONOSODIUM SALT see SFW000
CHOLIFLAVIN see DBX400
CHOLIMED see DAL000
CHOLINE see CMF000
CHOLINE ACETATE see CMF250
CHOLINE ACETATE (ESTER) see CMF250
CHOLINE ACETATE (ESTER), BROMIDE see CMF260
CHOLINE, ACETYL-, BROMIDE see CMF260
CHOLINE, S-ACETYLTHIO-, IODIDE see ADC300
CHOLINE CARBAMATE CHLORIDE see CBH250
CHOLINE CHLORHYDRATE see CMF750
CHOLINE CHLORIDE (FCC) see CMF750
CHOLINE CHLORIDE ACETATE see ABO000
CHOLINE, CHLORIDE CARBAMATE(ESTER) see CBH250
CHOLINE CHLORINE CARBAMATE see CBH250
CHOLINE 5′-CYTIDINE DIPHOSPHATE see CMF350
CHOLINE CYTIDINE DIPHOSPHATE see CMF350
CHOLINE DICHLORIDE see CMF400
CHOLINE HYDROCHLORIDE see CMF750
CHOLINE HYDROXIDE see CMF800
CHOLINE, HYDROXIDE, 5′-ESTER with CYTIDINE 5′-(TRIHYDROGEN PYRO-PHOSPHATE), inner salt see CMF350
CHOLINE, INNER SALT, METHYLPHOSPHONOFLUORIDATE see MKF250
CHOLINE IODIDE SUCCINATE (2:1) see BJI000
CHOLINE ION see CMF000
CHOLINE 1,5-NAPHTHALENEDISULFONATE (2:1), DILACTATE, DIACETATE see AAC875
CHOLINE, PROPIONATE, IODIDE see PMW750
CHOLINE SALICYLATE see CMG000
CHOLINE SALICYLATE B see CMG000
CHOLINE, SALICYLATE (SALT) see CMG000
CHOLINE SALICYLIC ACID SALT see CMG000
CHOLINE SUCCINATE (ester) see CMG250
CHOLINE SUCCINATE DICHLORIDE see HLC500
CHOLINE SUCCINATE (2:1) (ESTER) see CMG250
CHOLINE THEOPHYLLINATE see CMF500
CHOLINE, with THEOPHYLLINE (1:1) see CMF500
CHOLINE THEOPHYLLINE SALT see CMF500
CHOLINE-2,6-XYLYL ETHER BROMIDE see XSS900

CHOLINIUM CHLORIDE see CMF750
CHOLINOPHYLLINE see CMF500
CHOLIT-URSAN see DMJ200
CHOLLY see CNE750
CHOLOGON see DAL000
CHOLOGRAF-N-METHYLGLUCAMINE see BGB315
CHOLOLIN see DAL000
CHOLOREBIC see DAQ400
CHOLOVUE see IFP800
CHOLOXIN see SKJ300
CHOLSAEURE (GERMAN) see CME750
CHOLUMBRIN see TDE750
CHOLYLTAURINE see TAH250
CHONDROITIN POLYSULFATE SODIUM see SFW300
CHONDRON see SFW300
CHONDRUS see CCL250
CHONDRUS EXTRACT see CCL250
CHONGRASS see PJJ315
CHOPSUI POTATO see YAG000
CHORAFURONE see AHK750
C. HORRIDUS HORRIDUS VENOM see TGB150
CHORYLEN see TIO750
CHOT see PBC250
CHOU PUANT (CANADA) see SDZ450
CHP see PNN300
CHP-PHENOBARBITALAT (GERMAN) see CPO500
CHQ see DEL300
CHRISTENSENITE see SCK000
CHRISTMAS BERRY TREE see PCB300
CHRISTMAS CANDLE see SDZ475
CHRISTMAS FLOWER see EQX000
CHRISTMAS ROSE see CMG700
CHROMALUM HEXAHYDRATE see CMG800
CHROMAL YELLOW M see SIT850
6-CHROMANOL, 2,5,7,8-TETRAMETHYL-2-(4,8,12-TRIMETHYLTRIDECYL)-, ACE-
  TATE see TGJ055
2-CHROMANONE see HHR500
CHROMAR see XHS000
CHROMARGYRE see MCV000
CHROMATE(1-), DIAMMINETETRAKIS(ISOTHIOCYANATO)-, AMMONIUM, HY-
  DRATE see ANT300
CHROMATE OF POTASSIUM see PLB250
CHROMATE de PLOMB (FRENCH) see LCR000
CHROMATE of SODA see DXC200
CHROMATEX RED J see MMP100
CHROMAZINE BLUE BLACK B see CMP880
CHROME see CMI750
CHROME ALUM see CMG850
CHROME ALUM (DODECAHYDRATE) see CMG850
CHROME BLUE BLACK BF see CMP880
CHROMEDIA CC 31 see CCU150
CHROMEDIA CF 11 see CCU150
CHROME FAST BLUE 2R see HJF500
CHROME FAST BROWN FC see TKN500
CHROME FAST CYANIDE G see CMP880
CHROME FAST CYANINE BP see CMP880
CHROME FAST CYANINE GN see CMP880
CHROME FAST CYANINE GNN see CMP880
CHROME FAST CYANINE GP see CMP880
CHROME FAST CYANINE GSS see CMP880
CHROME FAST RED F see CMO870
CHROME FAST RED FB see CMO870
CHROME FAST RED FW see CMO870
CHROME FERROALLOY see FBD000
CHROME FLUORURE see CMJ530, CMJ560
CHROME GREEN see CMJ900, LCR000
CHROME IRON NICKEL BLACK SPINEL see CMS135
CHROME LEATHER BLACK BH see CMN800
CHROME LEATHER BLACK CR see CMN800
CHROME LEATHER BLACK D see CMN230
CHROME LEATHER BLACK DS see CMN800
CHROME LEATHER BLACK E see AQP000
CHROME LEATHER BLACK EC see AQP000
CHROME LEATHER BLACK EM see AQP000
CHROME LEATHER BLACK G see AQP000
CHROME LEATHER BLACK GNA see CMN240
CHROME LEATHER BLUE 2B see CMO000
CHROME LEATHER BLUE 3B see CMO250
CHROME LEATHER BRILLIANT BLACK ER see AQP000
CHROME LEATHER BROWN BRLL see CMO750
CHROME LEATHER BROWN BS see CMO820
CHROME LEATHER DARK BLUE BHM see CMN800
CHROME LEATHER DARK GREEN N see CMO830
CHROME LEATHER DARK GREEN S see CMO830
CHROME LEATHER FAST RED N see CMO870

CHROME LEATHER GREEN B see CMO830
CHROME LEATHER PURE BLUE see CMO500
CHROME LEATHER RED 5B see CMO885
CHROME LEATHER RED F see CMO870
CHROME LEATHER RED F EXTRA see CMO870
CHROME LEATHER SCARLET 3BS see CMO875
CHROME LEMON see LCR000
CHROME OCHER see CMJ900
CHROME ORANGE see LCS000
CHROME ORE see CMI500
CHROME OXIDE see CMJ900
CHROME OXIDE GREEN see CMJ900
CHROME RED ALIZARINE see SEH475
CHROME (TRIOXYDE de) (FRENCH) see CMK000
CHROME VERMILION see MRC000
CHROME YELLOW see LCR000
CHROME YELLOW 2G see SIT850
CHROME YELLOW 2GR see SIT850
CHROMIA see CMJ900
CHROMIC ACETATE(III) see CMH000
CHROMIC ACETATE see CMH000
CHROMIC ACETYLACETONATE see TNN250
CHROMIC(III) ACID see CMH260
CHROMIC(VI) ACID see CMH250, CMK000
CHROMIC ACID see CMH250, CMJ900, CMK000
CHROMIC ACID, BARIUM SALT (1:1) see BAK250
CHROMIC ACID, BIS(TRIPHENYLSILYL) ESTER see BLS750
CHROMIC ACID, CALCIUM SALT (1:1) see CAP500
CHROMIC ACID, CALCIUM SALT (1:1), DIHYDRATE see CAP750
CHROMIC ACID, CHROMIUM(3+) SALT (3:2) see CMI250
CHROMIC ACID, COPPER-ZINC-COMPLEX see CNQ750
CHROMIC ACID, DIAMMONIUM SALT see ANF500, NCQ550
CHROMIC ACID, DI-tert-BUTYL ESTER see BQV000
CHROMIC ACID, DILITHIUM SALT see LHD000
CHROMIC ACID, DIPOTASSIUM SALT see PKX250
CHROMIC ACID, DISODIUM SALT see SGI000
CHROMIC ACID, DISODIUM SALT, DECAHYDRATE see SFW500
CHROMIC ACID GREEN see CMJ900
CHROMIC ACID, LEAD and MOLYBDENUM SALT see LDM000
CHROMIC ACID, LEAD(2+) SALT (1:1) see LCR000
CHROMIC ACID LEAD SALT with LEAD MOLYBDATE see LDM000
CHROMIC ACID, MERCURY ZINC COMPLEX see ZJA000
CHROMIC ACID, solid (NA 1463) (DOT) see CMK000
CHROMIC ACID, POTASSIUM ZINC SALT (2:2:1) see PLW500
CHROMIC ACID, STRONTIUM SALT (1:1) see SMH000
CHROMIC ACID, solution (UN 1755) (DOT) see CMK000
CHROMIC ACID, ZINC SALT (1:2) see CMK500
CHROMIC ACID, ZINC SALT see ZFJ100
CHROMIC ACID, ZINC SALT, BASIC see ZFJ130
CHROMIC AMMONIUM SULFATE see ANF625
CHROMIC ANHYDRIDE see CMK000
CHROMIC CHLORIDE see CMJ250
CHROMIC CHLORIDE HEXAHYDRATE see CMK450
CHROMIC CHLORIDE STEARATE see CMH300
CHROMIC CHROMATE see CMI250
CHROMIC FLUORIDE see CMJ530, CMJ560
CHROMIC FLUORIDE, solid (UN1756) (DOT) see CMJ530, CMJ560
CHROMIC FLUORIDE, solution (UN1757) (DOT) see CMJ530, CMJ560
CHROMIC (III) HYDROXIDE see CMH260
CHROMIC NITRATE see CMJ600
CHROMIC OXIDE see CMJ900
CHROMIC OXYCHLORIDE see CML125
CHROMIC PERCHLORATE see CMI300
CHROMIC PHOSPHATE see CMK300
CHROMIC SULFATE see CMK415
CHROMIC SULPHATE see CMK415
CHROMIC TRIFLUORIDE see CMJ530, CMJ560
CHROMIC TRIOXIDE see CMK000
CHROMIS ACID ($H_2Cr_2O_7$), DISODIUM SALT, DIHYDRATE (9CI) see SGI500
CHROMITAN B see CMK415
CHROMITAN MS see CMK415
CHROMITAN NA see CMK415
CHROMITE (mineral) see CMI500
CHROMITE see CMI500
CHROMITE ORE see CMI500
CHROMIUM see CMI750
CHROMIUM(2+) ACETATE see CMJ000
CHROMIUM(III) ACETATE see CMH000
CHROMIUM(II) ACETATE see CMJ000
CHROMIUM ACETATE see CMH000
CHROMIUM ACETATE HYDRATE see CMJ000
CHROMIUM(3+) ACETYLACETONATE see TNN250
CHROMIUM(III) ACETYLACETONATE see TNN250
CHROMIUM ACETYLACETONATE see TNN250
CHROMIUM ALLOY, BASE, Cr,C,Fe,N,Si (FERROCHROMIUM) see FBD000

CHROMIUM ALLOY, Cr,C,Fe,N,Si see FBD000
CHROMIUM, BIS(BENZENE)-(8CI) see BGY700
CHROMIUM, BIS(eta⁶)-BENZENE)-(9CI) see BGY700
CHROMIUM(1+), BIS(BENZENE)-, IODIDE (8CI) see BGY720
CHROMIUM(1+), BIS(eta⁶)-BENZENE)-, IODIDE (9CI) see BGY720
CHROMIUM, BIS(BENZENE)IODO- see BGY720
CHROMIUM CARBONATE see CMJ100
CHROMIUM CARBONYL (MAK) see HCB000
CHROMIUM CARBONYL (OC-6-11) (9CI) see HCB000
CHROMIUM(II) CHLORIDE (1:2) see CMJ300
CHROMIUM(III) CHLORIDE (1:3) see CMJ250
CHROMIUM(II) CHLORIDE see CMJ300
CHROMIUM CHLORIDE, anhydrous see CMJ250
CHROMIUM CHLORIDE see CMJ250, CMJ300
CHROMIUM(III) CHLORIDE, HEXAHYDRATE (1:3:6) see CMK450
CHROMIUM CHLORIDE, HEXAHYDRATE (8CI,9CI) see CMK450
CHROMIUM CHLORIDE, HEXAUREA see HEZ800
CHROMIUM CHLORIDE OXIDE see CML125
CHROMIUM CHROMATE (MAK) see CMI250
CHROMIUM-COBALT-MOLYBDENUM ALLOY see VSK000
CHROMIUM COMPOUNDS see CMJ500
CHROMIUM DIACETATE see CMJ000
CHROMIUM DICHLORIDE see CMJ300
CHROMIUM DICHLORIDE DIOXIDE see CML125
CHROMIUM DIOXIDE DICHLORIDE see CML125
CHROMIUM(VI) DIOXYCHLORIDE see CML125
CHROMIUM(II), DIPHENYL- see BGY700
CHROMIUM(III), DIPHENYL-, IODIDE see BGY720
CHROMIUM DISODIUM OXIDE see DXC200
CHROMIUM(III) FLUORIDE see CMJ530, CMJ560
CHROMIUM HEXACARBONYL see HCB000
CHROMIUM(3+), HEXAKIS(UREA-O)-, TRICHLORIDE, (OC-6-11)-(9CI) see HEZ800
CHROMIUM(3+), HEXAKIS(UREA)-, TRICHLORIDE (8CI) see HEZ800
CHROMIUM(III) HEXA-UREA CHLORIDE see HEZ800
CHROMIUM(III) HYDROXIDE see CMH260
CHROMIUM HYDROXIDE SULFATE see NBW000
CHROMIUM III SULFATE see CMK415
CHROMIUM LEAD OXIDE see LCS000
CHROMIUM LITHIUM OXIDE see LHD000
CHROMIUM METAL (OSHA) see CMI750
CHROMIUM MONOPHOSPHATE see CMK300
CHROMIUM NICKEL OXIDE see NDA100
CHROMIUM (3+) NITRATE see CMJ600
CHROMIUM NITRATE (DOT) see CMJ600
CHROMIUM(III) NITRATE see CMJ600
CHROMIUM NITRATE see CMJ600
CHROMIUM NITRIDE see CMJ850
CHROMIUM ORTHOPHOSPHATE see CMK300
CHROMIUM(VI) OXIDE (1:3) see CMK000
CHROMIUM(III) OXIDE (2:3) see CMJ900
CHROMIUM(3+) OXIDE see CMJ900
CHROMIUM(III) OXIDE see CMJ900
CHROMIUM(VI) OXIDE see CMK000
CHROMIUM OXIDE see CMJ900, CMK000
CHROMIUM OXIDE, NICKEL OXIDE, and IRON OXIDE FUME see IHE000
CHROMIUM OXYCHLORIDE see CML125
CHROMIUM PENTAFLUORIDE see CMK275
CHROMIUM PERCHLORATE see CMI300
CHROMIUM PHOSPHATE see CMK300
CHROMIUM POTASSIUM ZINC OXIDE see CMK400
CHROMIUM SESQUICHLORIDE see CMK450
CHROMIUM SESQUIOXIDE see CMJ900
CHROMIUM SODIUM OXIDE see DXC200, SGI000
CHROMIUM (III) SULFATE (2:3) see CMK415
CHROMIUM SULFATE (2:3) see CMK415
CHROMIUM SULFATE see NBW000
CHROMIUM SULFATE, BASIC see NBW000
CHROMIUM(III) SULFATE, HEXAHYDRATE (2:3:6) see CMG800
CHROMIUM SULFATE, PENTADECAHYDRATE see CMK425
CHROMIUM SULPHATE (2:3) see CMK415
CHROMIUM, SULPHATE see CMK415, NBW000
CHROMIUM, TETRACHLORO-μ-HYDROXY(μ-(OCTADECANOATO-O: O'))DI- see CMH300
CHROMIUM, TETRACHLORO-μ-HYDROXY(μ-STEARATO)DI- see CMH300
CHROMIUM TRIACETATE see CMH000
CHROMIUM TRIACETYLACETONATE see TNN250
CHROMIUM TRICHLORIDE see CMJ250
CHROMIUM TRICHLORIDE HEXAHYDRATE see CMK450
CHROMIUM TRIFLUORIDE see CMJ530, CMJ560
CHROMIUM TRIHYDROXIDE see CMH260
CHROMIUM TRINITRATE see CMJ600
CHROMIUM(3+) TRIOXIDE see CMJ900
CHROMIUM(6+) TRIOXIDE see CMK000
CHROMIUM TRIOXIDE, anhydrous (DOT) see CMK000

CHROMIUM TRIOXIDE see CMK000
CHROMIUM TRIOXIDE, anhydrous (UN 1463) (DOT) see CMK000
CHROMIUM TRIPERCHLORATE see CMI300
CHROMIUM TRIS(ACETYLACETONATE) see TNN250
CHROMIUM TRIS(BENZOYLACETONATE) see TNN500
CHROMIUM TRIS(2,4-PENTANEDIONATE) see TNN250
CHROMIUM YELLOW see LCR000
CHROMIUM ZINC OXIDE see ZFA100, ZFJ100
CHROMIUM(6+)ZINC OXIDE HYDRATE (1:2:6:1) see CMK500
CHROMOCARD BLUE BLACK B see CMP880
CHROMOCOR see PER700
CHROMOFLAVINE see DBX400, XAK000
CHROMOL YELLOW G see SIT850
CHROMOL YELLOW N see SIT850
CHROMOMYCIN see OIS000
CHROMOMYCIN A3 see CMK650
CHROMOMYCIN SODIUM see CMK750
CHROMOMYSIN A₃ see CMK650
CHROMONAR HYDROCHLORIDE see CBR500
CHROMOSMON see BJI250
CHROMOSORB T see TAI250
CHROMOSULFURIC ACID (UN 2240) (DOT) see CLG500
anti-CHROMOTRICHIA FACTOR see AIH600
CHROMOTRICHIA FACTOR see AIH600
CHROMO (TRIOSSIDO di) (ITALIAN) see CMK000
CHROMOUS ACETATE see CMJ000
CHROMOUS ACETATE MONOHYDRATE see CMJ000
CHROMOUS CHLORIDE see CMJ300
CHROMOXYCHLORID (GERMAN) see CML125
CHROMSAEUREANHYDRID (GERMAN) see CMK000
CHROMTRIOXID (GERMAN) see CMK000
CHROMYL AZIDE CHLORIDE see CML000
CHROMYLCHLORID (GERMAN) see CML125
CHROMYL CHLORIDE see CML125
[(CHROMYLDIOXY)IODO]BENZENE see PFJ775
CHROMYL NITRATE see CML325
CHROMYL PERCHLORATE see CML500
CHRONICIN FOAM see CDP725
CHRONIUM ZINCATE see ZFJ130
CHRONIUM ZINC OXIDE (9CI) see ZFJ130
CHRONOGYN see DAB830
CHROOMOXYLCHLORIDE (DUTCH) see CML125
CHROOMTRIOXYDE (DUTCH) see CMK000
CHROOMZUURANHYDRIDE (DUTCH) see CMK000
CHRYSAMMIC ACID see CML600
CHRYSAMMINIC ACID see CML600
CHRYSANTHAL see CML620
CHRYSANTHEMIC ACID see CML650
CHRYSANTHEMUM CINERAREAEFOLIUM see POO250
CHRYSANTHEMUMDICARBOXYLIC ACID MONOMETHYL ESTER PYRETHRO-LONE ESTER see POO100
CHRYSANTHEMUMIC ACID see CML650
(+)-trans-CHRYSANTHEMUMIC ACID ESTER of (+—)-ALLETHROLONE see BGC750
CHRYSANTHEMUMMONOCARBOXYLIC ACID see CML650
CHRYSANTHEMUM MONOCARBOXYLIC ACID PYRETHROLONE ESTER see POO050
CHRYSAZIN see DMH400
6-CHRYSENAMINE see CML800
CHRYSENE see CML810
CHRYSENEX see CML800
α-CHRYSIDINE see BAW750
CHRYSOIDIN see PEK000
CHRYSOIDIN A see DBP999
CHRYSOIDINE(II) see PEK000
CHRYSOIDINE see PEK000
CHRYSOIDINE A see PEK000
CHRYSOIDINE B see PEK000
CHRYSOIDINE C CRYSTALS see PEK000
CHRYSOIDINE G see PEK000
CHRYSOIDINE GN see PEK000
CHRYSOIDINE HR see PEK000
CHRYSOIDINE J see PEK000
CHRYSOIDINE M see PEK000
CHRYSOIDINE ORANGE see PEK000
CHRYSOIDINE PRL see PEK000
CHRYSOIDINE PRR see PEK000
CHRYSOIDINE R BASE see CMM760
CHRYSOIDINE 3RN BASE see CMM760
CHRYSOIDINE SL see PEK000
CHRYSOIDINE SPECIAL (biological stain and indicator) see PEK000
CHRYSOIDINE SS see PEK000
CHRYSOIDINE Y see PEK000
CHRYSOIDINE Y BASE NEW see PEK000
CHRYSOIDINE Y CRYSTALS see PEK000

CHRYSOIDINE Y EX see PEK000
CHRYSOIDINE YGH see PEK000
CHRYSOIDINE YL see PEK000
CHRYSOIDINE YN see PEK000
CHRYSOIDINE Y SPECIAL see PEK000
CHRYSOIDIN FB see PEK000
CHRYSOIDIN Y see PEK000
CHRYSOIDIN YN see PEK000
CHRYSOMYKINE see CMA750
CHRYSON see BEP500
CHRYSONEX see CML800
CHRYSOTILE ASBESTOS see ARM268
CHRYSRON see BEP500
CHRYTEMIN see DLH630
CHRYZOIDYNA F.B. (POLISH) see PEK000
CHUANGHSINMYCIN see CML820
CHUANGHSINMYCIN SODIUM see CML822
CHUANGXIMYCIN SODIUM see CML822
CHUANGXINMYCIN see CML820
CHUANLIANSU see CML825
CHWASTOKS see SIL500
CHWASTOX see CIR250, SIL500
191C49 HYDROCHLORIDE see DJP500
1C50 HYDROCHLORIDE see EIJ000
CHYMAR see CML850
CHYMEX see CML835
CHYMOTEST see CML850
α-CHYMOTRYPSIN see CML850
CHYMOTRYPSIN A see CML850
CHYMOTRYPSIN B see CML850
CI-2 see MAV750
C.I. 23 see OHA000
C.I. 79 see FMU070
C.I. 184 see FAG020
C.I. 185 see FMU080
C.I. 258 see SBC500
CI-337 see ASA500
CI 366 see ENG500
CI395 see AOO500
CI-406 see PAN100
C.I. 440 see TKH750
C.I. 456 see CIP500
CI-473 see XQS000
CI-505 see BQM309
C.I. 515 see GLS700
C.I. 556 see SNY500
CI 581 see CKD750
CI-588 see BGO500
CI 624 see MGD200
CI-628 see NHP500
C.I. 633 see CGB250
C.I. 640 see FAG140
C.I. 671 see FMU059
CI-683 see POL475
CI-705 see QAK000
C.I. 712 see ADE500, CMM062
CI-719 see GCK300
CI 720 see TDO260
C.I. 749 see FAG070
C.I. 766 see FEW000
C.I. 801 see CMM510
CI 881 see BKB300
CI-914 see CML890
C.I. 918 see CMM510
C.I. 925 see AJP250
C.I. 1106 see IBV050
C.I. 1956 see CKN000
C.I. 7581 see FAE100
C.I. 10000 see NLB000
C.I. 10020 see NAX500
C.I. 10305 see PID000
C.I. 10315 see DUX800
C.I. 10345 see DUW500
C.I. 10355 see DVX800
C.I. 10360 see HET500
C.I. 10385 see SGP500
C.I. 11000 see PEI000
C.I. 11020 see DOT300
C.I. 11025 see DPO200
C.I. 11050 see DHM500
C.I. 11115 see CMP080
C.I. 11160 see AIC250
C.I. 11270 see DBP999, PEK000
C.I. 11285 see NBG500
C.I. 11350 see PEJ600

C.I. 11380 see FAG130
C.I. 11390 see FAG135
C.I. 11855 see AAQ250
C.I. 11860 see OHK000
C.I. 11920 see CMP600
C.I. 12055 see PEJ500
C.I. 12100 see TGW000
C.I. 12120 see MMP100
C.I. 12140 see XRA000
C.I. 12150 see CMS242
C.I. 12156 see DOK200
C.I. 12355 see NAY000
C.I. 13015 see CMM758
C.I. 13020 see CCE500
C.I. 13025 see MND600
C.I. 13065 see MDM775
C.I. 13390 see ADE750
C.I. 14025 see SIT850
C.I. 14600 see FAG010
C.I. 14640 see CMP880
C.I. 14700 see FAG050
C.I. 14720 see HJF500
C.I. 14815 see CMP620
C.I. 15510 see CMM220
C.I. 15585 see CMS150
C.I. 15850 see CMS155
C.I. 15985 see FAG150
C.I. 16150 see FMU070
C.I. 16155 see FAG018
C.I. 16185 see FAG020
C.I. 16255 see FMU080
C.I. 18050 see CMM300
C.I. 19140 see FAG140
C.I. 19540 see CMP050
C.I. 20285 see CMP500
C.I. 20470 see FAB830
C.I. 21090 see DEU000
C.I. 21095 see CMS212
C.I. 21100 see CMS208
C.I. 21110 see CMS145
C.I. 22120 see SGQ500
C.I. 22245 see CMM320
C.I. 22310 see CMO870
C.I. 22570 see CMP000
C.I. 22590 see CMN800
C.I. 22610 see CMO000
C.I. 22890 see CMM325
C.I. 22910 see CMM759
C.I. 23050 see CMO880
C.I. 23060 see DEQ600
C.I. 23285 see NAO600
C.I. 23630 see CMO875
C.I. 23635 see CMM330
C.I. 23850 see CMO250
C.I. 23860 see BGT250
C.I. 24110 see DCJ200
C.I. 24280 see CMO600
C.I. 24400 see CMO500
C.I. 24401 see CMO650
C.I. 26050 see EOJ500
C.I. 26090 see CMP090
C.I. 26100 see OHA000
C.I. 26105 see SBC500
C.I. 27700 see CMN230
C.I. 28160 see CMO885
C.I. 28440 see BMA000
C.I. 30110 see CMO810
C.I. 30120 see CMO825
C.I. 30145 see CMO750
C.I. 30235 see AQP000
C.I. 30280 see CMO830
C.I. 30295 see CMO840
C.I. 35255 see CMN240
C.I. 35570 see AKH000
C.I. 35660 see CMO820
C.I. 35811 see AKD925
C.I. 37010 see DEO295
C.I. 37020 see DBR400
C.I. 37025 see NEO000
C.I. 37030 see NEN500
C.I. 37035 see NEO500
C.I. 37040 see KDA050
C.I. 37050 see CEG800
C.I. 37077 see TGQ750
C.I. 37085 see CLK235

| | | |
|---|---|---|
| C.I. 37105 see NMP500 | C.I. 62105 see CMM090 |
| C.I. 37107 see TGR000 | C.I. 62130 see CMM070 |
| C.I. 37115 see AOX250 | C.I. 62500 see DMM400 |
| C.I. 37125 see NEQ000 | C.I. 63285 see CMP070 |
| C.I. 37130 see NEQ500 | C.I. 63340 see TGS000 |
| C.I. 37200 see MIH500 | C.I. 64500 see TBG700 |
| C.I. 37225 see BBX000 | C.I. 65010 see IBJ000 |
| C.I. 37230 see TGJ750 | C.I. 67300 see VGP200 |
| C.I. 37240 see PFU500 | C.I. 69005 see DUP100 |
| C.I. 37270 see NBE500 | C.I. 69020 see CMU800 |
| C.I. 37275 see AIA750 | C.I. 69500 see CMU810 |
| C.I. 37500 see NAX000 | C.I. 69800 see IBV050 |
| C.I. 38480 see AJU500 | C.I. 69825 see DFN425 |
| C.I. 40621 see DXB450 | C.I. 70300 see APK000 |
| C.I. 40645 see TGE155 | C.I. 70320 see CMU825 |
| C.I. 41000 see IBA000 | C.I. 70800 see CMU770 |
| C.I. 42000 see AFG500 | C.I. 70802 see CMU770 |
| C.I. 42040 see BAY750 | C.I. 71000 see CMU475 |
| C.I. 42045 see ADE500 | C.I. 71105 see CMU820 |
| C.I. 42051 see CMM062 | C.I. 71200 see CMU500 |
| C.I. 42053 see FAG000 | C.I. 73015 see FAE100 |
| C.I. 42080 see ERG100 | C.I. 73300 see DNT300 |
| C.I. 42085 see FAE950 | C.I. 73335 see CMU815 |
| C.I. 42090 see FAE000, FMU059 | C.I. 73670 see CMU320 |
| C.I. 42095 see FAF000 | C.I. 74265 see CMS140 |
| C.I. 42500 see RMK020 | C.I. 75300 see COG000 |
| C.I. 42510 see MAC250 | C.I. 75410 see TKN750 |
| C.I. 42535 see MQN025 | C.I. 75440 see MQF250 |
| C.I. 42555 see AOR500 | C.I. 75470 see CNF050 |
| C.I. 42581 see VQU500 | C.I. 75480 see HMX600 |
| C.I. 42640 see FAG120 | C.I. 75490 see HLY500 |
| C.I. 44040 see VKA600 | C.I. 75500 see WAT000 |
| C.I. 44050 see PFT250 | C.I. 75620 see FBW000 |
| C.I. 44090 see ADF000 | C.I. 75660 see MRN500 |
| C.I. 45010 see DIS200 | C.I. 75670 see QCA000 |
| C.I. 45160 see RGW000 | C.I. 75720 see QCJ000 |
| C.I. 45330 see FEV000 | C.I. 75730 see RSU000 |
| C.I. 45380 see BNH500, BNK700 | C.I. 76000 see AOQ000 |
| C.I. 45405 see CMM000 | C.I. 76005 see AHQ250 |
| C.I. 45410 see ADG250 | C.I. 76010 see PEY250 |
| C.I. 45430 see FAG040 | C.I. 76020 see ALL500 |
| C.I. 46000 see XAK000 | C.I. 76025 see PEY000 |
| C.I. 46005 see BAQ250, BJF000 | C.I. 76027 see CJY120 |
| C.I. 47000 see CMS245 | C.I. 76030 see NIM550 |
| C.I. 47005 see CMM510 | C.I. 76035 see TGL750 |
| C.I. 47031 see PGW750 | C.I. 76042 see TGM000 |
| C.I. 48015 see CMM768 | C.I. 76043 see DCE600, TGM400 |
| C.I. 48035 see CMM764 | C.I. 76050 see DBO000 |
| C.I. 48070 see CMM765 | C.I. 76051 see DBO400 |
| C.I. 50040 see AJQ250 | C.I. 76060 see PEY500 |
| C.I. 50240 see GJI400 | C.I. 76061 see PEY650 |
| C.I. 50411 see DCE800 | C.I. 76065 see CEG600 |
| C.I. 50435 see DCE800 | C.I. 76066 see CEG625 |
| C.I. 51010 see BMM550 | C.I. 76070 see ALL750 |
| C.I. 51180 see AJN250 | C.I. 76075 see DTL800 |
| C.I. 52005 see DUG700 | C.I. 76085 see PFU500 |
| C.I. 52040 see AJP250 | C.I. 76500 see CCP850 |
| C.I. 57000 see TKN250 | C.I. 76505 see REA000 |
| C.I. 58000 see DMG800 | C.I. 76515 see PPQ500 |
| C.I. 58005 see SEH475 | C.I. 76520 see ALT000 |
| C.I. 58200 see TKN500 | C.I. 76535 see ALO000 |
| C.I. 58205 see TKN750 | C.I. 76545 see ALT500 |
| C.I. 58500 see TDD000 | C.I. 76555 see NEM480 |
| C.I. 58900 see MEB750 | C.I. 76605 see NAW500 |
| C.I. 59040 see TNM000 | C.I. 76645 see NAO500 |
| C.I. 59100 see DCZ000 | C.I. 77000 see AGX000 |
| C.I. 59815 see VGP100 | C.I. 77002 see AHC000 |
| C.I. 59820 see CMU750 | C.I. 77050 see AQB750 |
| C.I. 59825 see JAT000 | C.I. 77052 see AQF000 |
| C.I. 59830 see APK500 | C.I. 77056 see AQC500 |
| C.I. 60010 see DFN450 | C.I. 77060 see AQL500 |
| C.I. 60700 see AKP750 | C.I. 77061 see AQF500 |
| C.I. 60710 see AKE250 | C.I. 77086 see ARI000 |
| C.I. 60755 see AKO350 | C.I. 77099 see BAJ250 |
| C.I. 60767 see CMP060 | C.I. 77103 see BAK250 |
| C.I. 61100 see DBP000 | C.I. 77120 see BAP000 |
| C.I. 61105 see AKP250 | C.I. 77160 see BKW600 |
| C.I. 61200 see BMM500 | C.I. 77169 see BKW100 |
| C.I. 61205 see CMS228 | C.I. 77172 see DDI200 |
| C.I. 61505 see MGG250 | C.I. 77180 see CAD000 |
| C.I. 61565 see BLK000 | C.I. 77185 see CAD250 |
| C.I. 62015 see DBX000 | C.I. 77199 see CAJ750 |
| C.I. 62030 see DBY700 | C.I. 77205 see CMS215 |
| C.I. 62045 see CMM080 | C.I. 77223 see CAP500, CAP750 |
| C.I. 62059 see CMM100 | C.I. 77231 see CAX750 |

C.I. 77265 see CBT500
C.I. 77266 see CBT750
C.I. 77288 see CMJ900
C.I. 77295 see CMJ250
C.I. 77305 see CMK415
C.I. 77320 see CNA250
C.I. 77322 see CND125
C.I. 77323 see CND825
C.I. 77353 see CNB475
C.I. 77400 see CNI000
C.I. 77402 see CNO000
C.I. 77403 see CNO250
C.I. 77410 see COF500
C.I. 77450 see CNQ000
C.I. 77480 see GIS000
C.I. 77491 see IHD000
C.I. 77504 see CMS135
C.I. 77575 see LCF000
C.I. 77577 see LDN000
C.I. 77578 see LDS000
C.I. 77580 see LCX000
C.I. 77600 see LCR000
C.I. 77601 see LCS000
C.I. 77605 see MRC000
C.I. 77610 see LCU000
C.I. 77620 see LCV100
C.I. 77622 see LDU000
C.I. 77630 see LDY000
C.I. 77640 see LDZ000
C.I. 77713 see MAC650
C.I. 77718 see TAB750
C.I. 77726 see MAT250
C.I. 77727 see MAT500
C.I. 77728 see MAS000
C.I. 77755 see PLP000
C.I. 77760 see MCT500
C.I. 77764 see MCW000
C.I. 77775 see NCW500
C.I. 77777 see NDF500
C.I. 77779 see NCY500
C.I. 77795 see PJD500
C.I. 77805 see SBO500
C.I. 77820 see SDI500
C.I. 77847 see SMM000
C.I. 77864 see TGC000
C.I. 77891 see TGG760
C.I. 77901 see TOC750
C.I. 77938 see VDU000
C.I. 77940 see VEZ000
C.I. 77945 see ZBJ000
C.I. 77947 see ZKA000
C.I. 77955 see ZFJ100
C.I. 3/11855 see AAQ250
C.I. 45380:2 see BMO250
C.I. 52015 (CZECH) see BJI250
C.I. 61 570 (CZECH) see APL500
C.I. 59815 (CZECH) see CMU750
C.I. ACID BLACK 1 (7CI) see FAB830
C.I. ACID BLACK 1, DISODIUM SALT (8CI) see FAB830
C.I. ACID BLUE 1 see ADE500
C.I. ACID BLUE 3 see ADE500, CMM062
C.I. ACID BLUE 41 see CMM070
C.I. ACID BLUE 71 see CMM070
C.I. ACID BLUE 74 see FAE100
C.I. ACID BLUE 78 see CMM090
C.I. ACID BLUE 92 see ADE750
C.I. ACID BLUE 129 see CMM100
C.I. ACID BLUE 3, CALCIUM SALT (2:1) (8CI) see CMM062
C.I. ACID BLUE 7 (7CI) see ERG100
C.I. ACID BLUE 62 (8CI) see CMM080
C.I. ACID BLUE 9, DIAMMONIUM SALT see FMU059
C.I. ACID BLUE 9, DISODIUM SALT see FAE000
C.I. ACID BLUE 1, SODIUM SALT see ADE500
C.I. ACID BLUE 7, SODIUM SALT (8CI) see ERG100
C.I. ACID BLUE 92, TRISODIUM SALT see ADE750
C.I. ACID GREEN 1 see NAX500
C.I. ACID GREEN 3 see FAE950
C.I. ACID GREEN 40 see CMM200
C.I. ACID GREEN 5 see FAF000
C.I. ACID GREEN 5, DISODIUM SALT see FAF000
C.I. ACID GREEN 3, MONOSODIUM SALT see FAE950
C.I. ACID GREEN 50, MONOSODIUM SALT see ADF000
C.I. ACID GREEN 3, SODIUM SALT see FAE950
C.I. ACID ORANGE 20 see FAG010
C.I. ACID ORANGE 3 see SGP500

C.I. ACID ORANGE 52 see MND600
C.I. ACID ORANGE 7 see CMM220
C.I. ACID ORANGE 20, MONOSODIUM SALT see FAG010
C.I. ACID ORANGE 7, MONOSODIUM SALT see CMM220
C.I. ACID RED 1 see CMM300
C.I. ACID RED 2 see CCE500
C.I. ACID RED 18 see FMU080
C.I. ACID RED 26 see FMU070
C.I. ACID RED 27 see FAG020
C.I. ACID RED 33 see CMS231
C.I. ACID RED 51 see FAG040
C.I. ACID RED 85 see CMM320
C.I. ACID RED 87 see BNK700
C.I. ACID RED 92 see ADG250
C.I. ACID RED 97 see CMM325
C.I. ACID RED 98 see CMM000
C.I. ACID RED 99 see NAO600
C.I. ACID RED 114 see CMM330
C.I. ACID RED 114, DISODIUM SALT see CMM330
C.I. ACID RED 14, DISODIUM SALT see HJF500
C.I. ACID RED 1, DISODIUM SALT see CMM300
C.I. ACID RED 26, DISODIUM SALT see FMU070
C.I. ACID RED 85, DISODIUM SALT see CMM320
C.I. ACID RED 99, DISODIUM SALT see NAO600
C.I. ACID RED 97, DISODIUM SALT (8CI) see CMM325
C.I. ACID YELLOW 23 see FAG140
C.I. ACID YELLOW 36 see MDM775
C.I. ACID YELLOW 3 see CMM510
C.I. ACID YELLOW 42 see CMM759
C.I. ACID YELLOW 73 see FEW000
C.I. ACID YELLOW 9 see CMM758
C.I. ACID YELLOW 42 DISODIUM SALT see CMM759
C.I. ACID YELLOW 9, DISODIUM SALT (8CI) see CMM758
C.I. ACID YELLOW 36 MONOSODIUM SALT see MDM775
C.I. ACID YELLOW 23, TRISODIUM SALT see FAG140
CIAFOS see COQ399
CIANATIL MALEATE see COS899
CIANAZIL see COH250
CIANIDANOL see CCP875
CIANURO di SODIO (ITALIAN) see SGA500
CIANURO di VINILE (ITALIAN) see ADX500
CIATYL see CLX250
C.I. AZOIC COUPLING COMPONENT 107 see TGR000
C.I. AZOIC COUPLING COMPONENT 1 see NAX000
C.I. AZOIC DIAZO COMPONENT 112 see BBX000
C.I. AZOIC DIAZO COMPONENT 113 see TGJ750
C.I. AZOIC DIAZO COMPONENT 114 see NBE700
C.I. AZOIC DIAZO COMPONENT 11 see CLK235
C.I. AZOIC DIAZO COMPONENT 12 see NMP500
C.I. AZOIC DIAZO COMPONENT 13 see NEQ500
C.I. AZOIC DIAZO COMPONENT 21 see MIH500
C.I. AZOIC DIAZO COMPONENT 22 see PFU500
C.I. AZOIC DIAZO COMPONENT 37 see BIN500, NEO500
C.I. AZOIC DIAZO COMPONENT 3 see DEO295
C.I. AZOIC DIAZO COMPONENT 48 see DCJ200
C.I. AZOIC DIAZO COMPONENT 5 see NEQ000
C.I. AZOIC DIAZO COMPONENT 6 see NEO000
C.I. AZOIC DIAZO COMPONENT 7 see NEN500
C.I. AZOIC DIAZO COMPONENT 9 see KDA050
C.I. AZOIC RED 83 see MGO750
C.I. 11160B see AIC250
C.I. 11320B see CMM760
C.I. 41000B see IBB000
C.I. 42555B see TJK000
CIBA 34 see MAW850
CIBA 570 see FAB400
CIBA 709 see DGQ875
CIBA 1414 see MRH209
CIBA 1983 see CJQ000
CIBA 2059 see DUK800
CIBA 2446 see AHO750
CIBA-3126 see PAM785
CIBA 5968 see HGP495, HGP500
CIBA 6313 see CES750
CIBA 7115 see KFK000
CIBA 8353 see DVS000
CIBA 8514 see CJJ250, KEA000
CIBA 9295 see MKU750
CIBA 9491 see IEN000
CIBA 11925 see PCY300
CIBA 12223 see PHK000
CIBA 32644 see NML000
CIBA 12669A see MIW500
CIBA 34,647-Ba see BAC275
CIBA 39089-Ba see THK750

CIBA 17309 BA see DAL300
CIBA 32644-BA see NML000
CIBA 34276 BA see MAW850
CIBA 36278-BA see SGB500
CIBA 42155-BA see AGO750
CIBA 42244-BA see AGP000
CIBA C-768 see DTP800
CIBA C-776 see DTP600
CIBA C-2307 see DOL800
CIBA C 7019 see ASG500
CIBA C-7824 see MPG250
CIBA C-9491 see IEN000
CIBACET BLUE GREEN C see DMM400
CIBACET BLUE GREEN CB see DMM400
CIBACET BRILLIANT BLUE BG NEW see MGG250
CIBACET BRILLIANT PINK 4BN see DBX000
CIBACET BRILLIANT VIOLET 3B see DBY700
CIBACETE BRILLIANT PINK 4BN see DBX000
CIBACETE DIAZO NAVY BLUE 2B see DCJ200
CIBACETE RED 3B see AKE250
CIBACETE YELLOW GBA see AAQ250
CIBACET RED 3B see AKE250
CIBACET RED E3B see AKE250
CIBACET RUBINE BS see CMP080
CIBACET RUBINE R see CMP080
CIBACET SAPPHIRE BLUE G see TBG700
CIBACET TURQUOISE BLUE 2G see DMM400
CIBACET TURQUOISE BLUE 4G see DMM400
CIBACET TURQUOISE BLUE G see DMM400
CIBACET VIOLET 2R see DBP000
CIBACET YELLOW GBA see AAQ250
CIBACET YELLOW 2GC see AAQ250
CIBA CO. 2825 see PIT250
CIBACRON BLACK B-D see CMS227
CIBACTHEN see AES650
CIBA-GEIGY C-9491 see IEN000
CIBA-GEIGY C-10015 see DRP600
CIBA-GEIGY GS 13005 see DSO000
CIBA-GEIGY GS 19851 see IOS000
CIBA 2696GO see BLP325
CIBA GO.4350 see CMV375
CIBA 34276 HYDROCHLORIDE see MAW850
CIBAMIN M 84 see MCB050
CIBAMIN M 100 see MCB050
CIBAMIN ML 100GB see MCB050
CIBANONE BLUE FG see DFN425
CIBANONE BLUE FGF see DFN425
CIBANONE BLUE FGL see DFN425
CIBANONE BLUE FRS see IBV050
CIBANONE BLUE FRSN see IBV050
CIBANONE BLUE GF see DFN425
CIBANONE BLUE RS see IBV050
CIBANONE BRILLIANT BLUE FR see IBV050
CIBANONE BRILLIANT ORANGE GR see CMU820
CIBANONE BROWN BR see CMU770
CIBANONE BROWN FBR see CMU770
CIBANONE GOLDEN YELLOW see DCZ000
CIBANONE NAVY BLUE FRA see VGP100
CIBANONE NAVY BLUE RA see VGP100
CIBANONE OLIVE F2R see DUP100
CIBANONE OLIVE 2R see DUP100
CIBANONE RED 6B see CMU825
CIBANONE RED F 6B see CMU825
CIBANONE VIOLET F 4R see DFN450
CIBANONE VIOLET F 2RB see DFN450
CIBANONE VIOLET 2R see DFN450
CIBANONE VIOLET 4R see DFN450
CIBA ORANGE R see CMU815
CIBA ORANGE RDL see CMU815
CIBA ORANGE RP see CMU815
CIBA PINK B see DNT300
C.I. BASIC BLUE 12 see AJN250
C.I. BASIC BLUE 17 see AJP250
C.I. BASIC BLUE 9 see BJI250
C.I. BASIC GREEN 4 see AFG500
C.I. BASIC GREEN 1, SULFATE (1:1) see BAY750
C.I. BASIC ORANGE 14 see BAQ250, BJF000
C.I. BASIC ORANGE 1 see CMM760
C.I. BASIC ORANGE 21 see CMM764
C.I. BASIC ORANGE 2 see PEK000
C.I. BASIC ORANGE 3 see PEK000
C.I. BASIC ORANGE 2, MONOHYDROCHLORIDE see PEK000
C.I. BASIC RED 12 see CMM765
C.I. BASIC RED 13 see CMM768
C.I. BASIC RED 1 see RGW000

C.I. BASIC RED 29 see CMM770
C.I. BASIC RED 2 see GJI400
C.I. BASIC RED 5 see AJQ250
C.I. BASIC RED 1, MONOHYDROCHLORIDE see RGW000
C.I. BASIC RED 5, MONOHYDROCHLORIDE see AJQ250
C.I. BASIC RED 9, MONOHYDROCHLORIDE see RMK020
C.I. BASIC VIOLET 14 see MAC250
C.I. BASIC VIOLET 10 see FAG070
C.I. BASIC VIOLET 1 see MQN025
C.I. BASIC VIOLET 3 see AOR500
C.I. BASIC VIOLET 14, FREE BASE see MAC500
C.I. BASIC VIOLET 14, MONOHYDROCHLORIDE (8CI) see MAC250
C.I. BASIC YELLOW 2 see IBA000
C.I. BASIC YELLOW 2, FREE BASE see IBB000
C.I. BASIC YELLOW 2, MONOHYDROCHLORIDE see IBA000
CIBA THIOCRON see AHO750
CIC see BAR800, IKH000
C.I. 15800 CA SALT see CMS160
CI-628 CITRATE see NHP500
CICLACILLIN see AJJ875
CICLACILLUM see AJJ875
CICLIZINA see EAN600
CICLOBIOTIC see MDO250
CICLOESANO (ITALIAN) see CPB000
CICLOESANOLO (ITALIAN) see CPB750
CICLOESANONE (ITALIAN) see CPC000
6-CICLOESIL-2,4-DINITRO-FENOLO (ITALIAN) see CPK500
1-CICLOESIL-3-p-TOLILSOLFONILUREA (ITALIAN) see CPR000
CICLONIUM IODIDE see OLW400
CICLOPIROX ETHANOLAMINE SALT (1:1) see BAR800
CICLOPIROXOLAMIN see BAR800
CICLOPIROXOLAMINE see BAR800
CICLORAL see BSM000
CICLOSERINA (ITALIAN) see CQH000
CICLOSOM see TIQ250
CICLOSPASMOL see DNU100
CICLOSPORIN see CQH100
CICP see CKC000
CICUTA BULBIFERA L. see WAT325
CICUTA DOUGLASII see WAT325
CICUTAIRE (CANADA) see WAT325
CICUTA MACULATA see WAT325
CICUTIN see PNT000
CICUTINE see PNT000
CICUTOXIN see CMN000
CIDAL see SAH000
CIDALON see IHZ000
CIDAMEX see AAI250
CIDANCHIN see CLD000
CIDANDOPA see DNA200
CIDEFERRON see CMN125
CIDEMUL see DRR400
C.I. DEVELOPER 13 see PEY500
C.I. DEVELOPER 15 see PFU500
C.I. DEVELOPER 17 see NEO500
C.I. DEVELOPER 1 see NNT000
C.I. DEVELOPER 4 see REA000
C.I. DEVELOPER 5 see NAX000
C.I. DEVELOPER 8 see HMX520
C.I. DEVELOPER 20 (obs.) see HMX520
CIDEX see GFQ000
CIDIAL see DRR400
C.I. DIRECT BLACK 17 see CMN230
C.I. DIRECT BLACK 19 see CMN240
C.I. DIRECT BLACK 38 see AQP000
C.I. DIRECT BLACK 19, DISODIUM SALT see CMN240
C.I. DIRECT BLACK 38, DISODIUM SALT see AQP000
C.I. DIRECT BLACK 17, MONOSODIUM SALT see CMN230
C.I. DIRECT BLUE 14 see CMO250
C.I. DIRECT BLUE 15 see CMO500
C.I. DIRECT BLUE 218 see CMO650
C.I. DIRECT BLUE 2 see CMN800
C.I. DIRECT BLUE 53 see BGT250
C.I. DIRECT BLUE 22 (7CI) see CMO600
C.I. DIRECT BLUE 22, DISODIUM SALT (8CI) see CMO600
C.I. DIRECT BLUE 14, TETRASODIUM SALT see CMO250
C.I. DIRECT BLUE 15, TETRASODIUM SALT see CMO500
C.I. DIRECT BLUE 6, TETRASODIUM SALT see CMO000
C.I. DIRECT BLUE 2, TRISODIUM SALT see CMN800
C.I. DIRECT BROWN 1:2 CMO810
C.I. DIRECT BROWN 154 see CMO825
C.I. DIRECT BROWN 31 see CMO820
C.I. DIRECT BROWN see CMO750
C.I. DIRECT BROWN 1A, DISODIUM SALT see CMO810
C.I. DIRECT BROWN 78, DIAMMONIUM SALT see FMU059

C.I. DIRECT BROWN 154, DISODIUM SALT (8CI) see CMO825
C.I. DIRECT BROWN 31, TETRASODIUM SALT see CMO820
C.I. DIRECT GREEN 1 see CMO830
C.I. DIRECT GREEN 6 (7CI) see CMO840
C.I. DIRECT GREEN 6, DISODIUM SALT see CMO840
C.I. DIRECT RED 28 see SGQ500
C.I. DIRECT RED 39 see CMO875
C.I. DIRECT RED 46 see CMO880
C.I. DIRECT RED 81 see CMO885
C.I. DIRECT RED 1 (6CI,7CI) see CMO870
C.I. DIRECT RED 28, DISODIUM SALT see SGQ500
C.I. DIRECT RED 1, DISODIUM SALT (8CI) see CMO870
C.I. DIRECT RED 46, TETRASODIUM SALT see CMO880
C.I. DIRECT VIOLET 1, DISODIUM SALT see CMP000
C.I. DIRECT YELLOW 9 see CMP050
C.I. DIRECT YELLOW 9, DISODIUM SALT see CMP050
C.I. 45350 DISODIUM SALT see FEW000
C.I. DISPERSE BLACK 3 see DPO200
C.I. DISPERSE BLACK 6 see DCJ200
C.I. DISPERSE BLACK 6 DIHYDROCHLORIDE see DOA800
C.I. DISPERSE BLUE 1 see TBG700
C.I. DISPERSE BLUE 27 see CMP060
C.I. DISPERSE BLUE 3 see MGG250
C.I. DISPERSE BLUE 56 see CMP070
C.I. DISPERSE BLUE 59 see CMP070
C.I. DISPERSE BLUE 71 see CMP070
C.I. DISPERSE BLUE 78 see BKP500
C.I. DISPERSE BLUE 7 see DMM400
C.I. DISPERSE ORANGE 11 see AKP750
C.I. DISPERSE RED 11 see DBX000
C.I. DISPERSE RED 13 see CMP080
C.I. DISPERSE RED 15 see AKE250
C.I. DISPERSE RED 4 see AKO350
C.I. DISPERSE RED 71 see AKI750
C.I. DISPERSE RED 83 see AKI750
C.I. DISPERSE RED 60 (8CI) see AKI750
C.I. DISPERSE VIOLET 1 see DBP000
C.I. DISPERSE VIOLET 4 see AKP250
C.I. DISPERSE VIOLET 8 see DBY700
C.I. DISPERSE YELLOW 13 see MEB750
C.I. DISPERSE YELLOW 1 see DUW500
C.I. DISPERSE YELLOW 3 see AAQ250
C.I. DISPERSE YELLOW 7 see CMP090
CIDOCETINE see CDP250
CIDOXEPIN HYDROCHLORIDE see AEG750
CIDREX see CFY000
C.I. FLUORESCENT BRIGHTENER 260 see CMP200
C.I. FLUORESCENT BRIGHTENER 46 see TGE155
C.I. FLUORESCENT BRIGHTENER 9 see DXB450
C.I. FLUORESCENT BRIGHTENING AGENT 46, SODIUM SALT see TGE155
C.I. FOOD BLACK 1, TETRASODIUM SALT see BMA000
C.I. FOOD BLUE 1 see FAE100
C.I. FOOD BLUE 2 see FAE000, FMU059
C.I. FOOD BLUE 3 see ADE500
C.I. FOOD BLUE 5 see CMM062
C.I. FOOD BROWN 1 see CMP250
C.I. FOOD BROWN 3, DISODIUM SALT see CMP500
C.I. FOOD GREEN 1 see FAE950
C.I. FOOD GREEN 2 see FAF000
C.I. FOOD GREEN 3 see FAG000
C.I. FOOD GREEN 4 see ADF000
C.I. FOOD ORANGE 3 see CMP600
C.I. FOOD RED 1 see FAG050
C.I. FOOD RED 2 see CMP620
C.I. FOOD RED 3 see HJF500
C.I. FOOD RED 5 see FMU070
C.I. FOOD RED 6 see FAG018
C.I. FOOD RED 7 see FMU080
C.I. FOOD RED 9 see FAG020
C.I. FOOD RED 12 see CMS231
C.I. FOOD RED 15 see FAG070
C.I. FOOD RED 16 see CMS242
C.I. FOOD RED 10 see CMM300
C.I. FOOD RED 1, DISODIUM SALT see FAG050
C.I. FOOD RED 2, DISODIUM SALT see CMP620
C.I. FOOD RED 6, DISODIUM SALT see FAG018
C.I. FOOD VIOLET 2 see FAG120
C.I. FOOD VIOLET 3 see VQU500
C.I. FOOD YELLOW 11 see FAG135
C.I. FOOD YELLOW 13 see CMM510
C.I. FOOD YELLOW 10 see FAG130
C.I. FOOD YELLOW 2 see CMM758
C.I. FOOD YELLOW 3 see CMM510, FAG150
C.I. FOOD YELLOW 4 see FAG140
C.I. FOOD YELLOW 3, DISODIUM SALT see FAG150

C.I. 45350 (FREE ACID) see FEV000
CIGARETTE REFINED TAR see CMP800
CIGARETTE SMOKE CONDENSATE see SEC000
CIGARETTE TAR see CMP800
CIGUE (CANADA) see PJJ300
CI 624 HYDROCHLORIDE see MGD210
CI-IPC see CKC000
CILAG 61 see HDY000
CILEFA BLACK B see BMA000
CILEFA ORANGE S see FAG150
CILEFA PINK B see FAG040
CILEFA PONCEAU 4R see FMU080
CILEFA RED G see FAG020
CILEFA RUBINE 2B see FAG020
CILEFA YELLOW R see CMM758
CILEFA YELLOW T see FAG140
CILIFOR see TEY000
CILLA BLUE EXTRA see TBG700
CILLA FAST BLUE FFR see MGG250
CILLA FAST BLUE GREEN B see DMM400
CILLA FAST PINK BN see AKE250
CILLA FAST PINK FF3B see DBX000
CILLA FAST PINK RF see AKO350
CILLA FAST RED VIOLET RN see DBP000
CILLA FAST RUBINE B see CMP080
CILLA FAST VIOLET 6B see AKP250
CILLA FAST VIOLET B see DBY700
CILLA FAST YELLOW G see AAQ250
CILLA FAST YELLOW 5R see CMP090
CILLA FAST YELLOW RR see DUW500
CILLA ORANGE R see AKP750
CILLENTA see BFC750
CILLORAL see BDY669, BFD000
CILOPEN see BDY669
CILOSTAZOL see CMP825
CIM see CDU750
CIMAGEL see DXY000
CIMCOOL WAFERS see HMJ500
CIMETIDINE see TAB250
CIMEXAN see MAK700
C.I. MORDANT BLACK 3 see CMP880
C.I. MORDANT BLACK 3, MONOSODIUM SALT see CMP880
C.I. MORDANT BROWN 42 see TKN500
C.I. MORDANT ORANGE 1, MONOSODIUM SALT see SIU000
C.I. MORDANT RED 11 see DMG800
C.I. MORDANT RED 3 see SEH475
C.I. MORDANT RED 57 see CMO870
C.I. MORDANT VIOLET 26 see TDD000
C.I. MORDANT VIOLET 39, TRIAMMONIUM SALT (8CI) see AGW750
C.I. MORDANT YELLOW 1 see SIT850
C.I. MORDANT YELLOW 1, MONOSODIUM SALT see SIT850
CI-583 NA see SIF425
CINAMINE see TLN500
CINAMONIN see CMP885
CINANSERIN HYDROCHLORIDE see CMP900
CINATABS see TLN500
C.I. NATURAL BROWN 1 see FBW000
C.I. NATURAL BROWN 7 see WAT000
C.I. NATURAL BROWN 8 see MAT500
C.I. NATURAL ORANGE 6 see HMX600
C.I. NATURAL RED 1 see QCA000
C.I. NATURAL YELLOW 11 see MRN500
C.I. NATURAL YELLOW 14 see MQF250
C.I. NATURAL YELLOW 16 see HLY500
C.I. NATURAL YELLOW 1 see CDH250
C.I. NATURAL YELLOW 10 see QCA000
C.I. NATURAL YELLOW 8 see MRN500
CINCAINE HYDROCHLORIDE see NOF500
CINCHOCAINE see DDT200
CINCHOCAINE HYDROCHLORIDE see NOF500
CINCHOCAINIUM CHLORIDE see NOF500
CINCHONAN-9-OL, (8-α-9R)-(9CI) see CMP910
CINCHONAN-9-OL, 6′-METHOXY-, DIHYDROCHLORIDE, (8-α-9R)-, mixt. with
    2-(ACETYLOXY) BENZOIC ACID and 2-HYDROXY-1,2,3-PROPANETRICAR-
    BOXYLIC ACID, TRILITHIUM SALT see TGJ350
CINCHONAN-9-OL, 6′-METHOXY-, (9S)-, SULFATE (1:1) (SALT) (9CI) see
    QFS100
CINCHONAN-9-OL, 6′-METHOXY-, (9S)-, SULFATE (2:1) (SALT) (9CI) see
    QHA000
(8S,9R)-CINCHONIDINE see CMP910
(–)-CINCHONIDINE see CMP910
CINCHONIDINE see CMP910
d-CINCHONINE see CMP925
CINCHOPHENE see PGG000
CINCHOPHENIC ACID see PGG000

CINCHOPHEN SODIUM see SJH000
CINCHOPHEN, SODIUM SALT see SJH000
CINCHOVATINE see CMP910
CINCO NEGRITOS (MEXICO) see LAU600
CINCOPHEN see PGG000
CINDOMET see CMP950
CINEB see EIR000
CINENE see MCC250
1,4-CINEOL see IKC100
1,8-CINEOL see CAL000
1,4-CINEOLE see IKC100
1,8-CINEOLE see CAL000
CINEOLE see CAL000
CINEPAZIDE MALEATE see VGK000
CINERIN I ALLYL HOMOLOG see AFR250
CINERIN I or II see POO250
CINERUBIN A see TAH675
CINERUBIN B see TAH650
CINERUBINE A see TAH675
CINERUBINE B see TAH650
CINMETACIN see CMP950
CINMETHACIN see CMP950
CINNAMAL see CMP969
CINNAMALDEHYDE see CMP969
CINNAMAMIDE see CMP970
CINNAMEIN see BEG750
CINNAMENE see SMQ000
CINNAMENOL see SMQ000
CINNAMIC ACID see CMP975
(E)-CINNAMIC ACID see CMP980
CINNAMIC ACID, (E)- see CMP980
trans-CINNAMIC ACID see CMP980
trans-CINNAMIC ACID BENZYL ESTER see BEG750
CINNAMIC ACID-n-BUTYL ESTER see BQV500
CINNAMIC ACID, CINNAMYL ESTER see CMQ850
CINNAMIC ACID-3-(DIETHYLAMINO) PROPYL ESTER see AQP750
CINNAMIC ACID-1,5-DIMETHYL-1-VINYL-4-HEXEN-1-YL ESTER see LGA000
CINNAMIC ACID-1,5-DIMETHYL-1-VINYL-4-HEXENYL ESTER see LGA000
CINNAMIC ACID, o-HYDROXY-, (E)- see CNU850
CINNAMIC ACID, p-(1H-IMIDAZOL-1-YLMETHYL)-, SODIUM SALT, (E)- see SHV100
CINNAMIC ACID, ISOBUTYL ESTER see IIQ000
CINNAMIC ACID, LINALYL ESTER see LGA000
CINNAMIC ACID, NICKEL(II) SALT see CMQ000
CINNAMIC ACID, p-NITRO- see NFT440
CINNAMIC ACID, 3-PHENYLPROPYL ESTER (7CI,8CI) see PGB800
CINNAMIC ACID, 3,4,5-TRIMETHOXY- see CMQ100
CINNAMIC ALCOHOL see CMQ740
CINNAMIN see AQN750
CINNAMOHYDROXAMIC ACID see CMQ475
CINNAMON BARK OIL see CCO750
CINNAMON BARK OIL, CEYLON TYPE (FCC) see CCO750
CINNAMONIN see CCI500
CINNAMONITRILE see CMQ500
CINNAMON OIL see CCO750
CINNAMOPHENONE see CDH000
CINNAMOYLHYDROXAMIC ACID see CMQ475
1-CINNAMOYL-2-METHOXY-5-METHOXY-3-INDOLYLACETIC ACID see CMP950
1-CINNAMOYL-5-METHOXY-2-METHYLINDOLE-3-ACETIC ACID see CMP950
14-CINNAMOYLOXYCODEINONE see CMQ625
CINNAMYCIN see CMQ725
CINNAMYL ACETATE see CMQ730
CINNAMYL ALCOHOL see CMQ740
CINNAMYL ALCOHOL ANTHRANILATE see API750
CINNAMYL ALCOHOL, BENZOATE see CMQ750
CINNAMYL ALCOHOL, CINNAMATE see CMQ850
CINNAMYL ALCOHOL, FORMATE see CMR500
CINNAMYL ALCOHOL, SYNTHETIC see CMQ740
CINNAMYL ALDEHYDE see CMP969
CINNAMYL-2-AMINOBENZOATE see API750
CINNAMYL-o-AMINOBENZOATE see API750
CINNAMYL ANTHRANILATE (FCC) see API750
CINNAMYL BENZOATE see CMQ750
CINNAMYL BUTYRATE see CMQ800
CINNAMYL CINNAMATE see CMQ850
1-CINNAMYL-4-(DIPHENYLMETHYL)PIPERAZINE see CMR100
trans-1-CINNAMYL-(4-DIPHENYLMETHYL)PIPERAZINE see CMR100
d-CINNAMYLEPHEDRINE HYDROCHLORIDE see CMR250
CINNAMYLEPHEDRINE HYDROCHLORIDE, DEXTRO see CMR250
CINNAMYLESTER KYSELINY SKORICOVE see CMQ850
CINNAMYL FORMATE see CMR500
CINNAMYL ISOBUTYRATE see CMR750
CINNAMYL ISOVALERATE see CMR800, PEE200
CINNAMYL METHANOATE see CMR500
CINNAMYL 3-METHYL BUTYRATE see PEE200

CINNAMYL NITRILE see CMQ500
CINNAMYL PROPIONATE see CMR850
CINNARIZIN see ARQ750
CINNARIZINE see CMR100
CINNARIZINE CLOFIBRATE see CMS125
CINNIMIC ALDEHYDE see CMP969
CINNOPROPAZONE see AQN750
C.I. No. 77278 see CMJ900
C.I. No. 46005:1 see BJF000
CINOBAC see CMS130
CINOPAL see BGL250
CINOPOP see BGL250
CINOXACIN see CMS130
CIN-QUIN see QFS000, QHA000
CINU see CGV250
CINX see CMS130
CIODRIN see COD000
CIODRIN VINYL PHOSPHATE see COD000
CIOVAP see COD000
C.I. OXIDATION BASE 12 see DBO000
C.I. OXIDATION BASE 16 see PEY250
C.I. OXIDATION BASE 17 see ALT000
C.I. OXIDATION BASE 19 see AHQ250
C.I. OXIDATION BASE 10 see PEY500
C.I. OXIDATION BASE 21 see DUP400
C.I. OXIDATION BASE 22 see ALL750
C.I. OXIDATION BASE 26 see CCP850
C.I. OXIDATION BASE 2 see PFU500
C.I. OXIDATION BASE 200 see TGL750
C.I. OXIDATION BASE 20 see TGL750
C.I. OXIDATION BASE 31 see REA980
C.I. OXIDATION BASE 32 see PPQ500
C.I. OXIDATION BASE 33 see NAW500
C.I. OXIDATION BASE 35 see TGL750
C.I. OXIDATION BASE 4 see TGM400
C.I. OXIDATION BASE 7 see ALT500
C.I. OXIDATION BASE see TGL750
C.I. OXIDATION BASE 12A see DBO400
C.I. OXIDATION BASE 13A see CEG625
C.I. OXIDATION BASE 10A see PEY650
C.I. OXIDATION BASE 6A see ALT250
CIPC see CKC000
CIPE see PJS750
C.I. PIGMENT BLACK 13 see CND125
C.I. PIGMENT BLACK 14 see MAS000
C.I. PIGMENT BLACK 15 see CNO250
C.I. PIGMENT BLACK 16 see ZBJ000
C.I. PIGMENT BLACK 10 see CBT500
C.I. PIGMENT BLACK 30 see CMS135
C.I. PIGMENT BLACK 6 see CBT750
C.I. PIGMENT BLACK 7 see CBT750
C.I. PIGMENT BLUE 34 see CNQ000
C.I. PIGMENT BLUE 60 see IBV050
C.I. PIGMENT BROWN 8 see MAS000
C.I. PIGMENT GREEN 12 see NAX500
C.I. PIGMENT GREEN 17 see CMJ900
C.I. PIGMENT GREEN 36 see CMS140
C.I. PIGMENT GREEN 38 see CMS140
C.I. PIGMENT GREEN 41 see CMS140
C.I. PIGMENT GREEN 21 (9CI) see COF500
C.I. PIGMENT METAL 2 see CNI000
C.I. PIGMENT METAL 3 see GIS000
C.I. PIGMENT METAL 4 see LCF000
C.I. PIGMENT METAL 6 see ZBJ000
C.I. PIGMENT ORANGE 13 see CMS145
C.I. PIGMENT ORANGE 21 see LCS000
C.I. PIGMENT ORANGE 20 see CAJ750
C.I. PIGMENT ORANGE 43 see CMU820
C.I. PIGMENT RED 101 see IHD000
C.I. PIGMENT RED 104 see LDM000, MRC000
C.I. PIGMENT RED 105 see LDS000
C.I. PIGMENT RED 107 see AQL500
C.I. PIGMENT RED 195 see CMU825
C.I. PIGMENT RED 23 see NAY000
C.I. PIGMENT RED 3 see MMP100
C.I. PIGMENT RED 48:1 see CMS148
C.I. PIGMENT RED 53 see CMS150
C.I. PIGMENT RED 64:1 see CMS160
C.I. PIGMENT RED 83 see DMG800
C.I. PIGMENT RED see CHP500, LCS000
C.I. PIGMENT RED 48, BARIUM SALT (1:1) (8CI) see CMS148
C.I. PIGMENT RED 64, CALCIUM SALT (2:1) (8CI) see CMS160
C.I. PIGMENT RED 57 (7CI) see CMS155
C.I. PIGMENT RED 57, DISODIUM SALT (8CI) see CMS155
C.I. PIGMENT VIOLET 31 see DFN450

C.I. PIGMENT WHITE 11 see AQF000
C.I. PIGMENT WHITE 17 see BKW100
C.I. PIGMENT WHITE 18 see CAT775
C.I. PIGMENT WHITE 10 see BAJ250
C.I. PIGMENT WHITE 21 see BAP000
C.I. PIGMENT WHITE 25 see CAX750
C.I. PIGMENT WHITE 3 see LDY000
C.I. PIGMENT WHITE 4 see ZKA000
C.I. PIGMENT WHITE 6 see TGG760
C.I. PIGMENT YELLOW 12 see DEU000
C.I. PIGMENT YELLOW 13 see CMS208
C.I. PIGMENT YELLOW 14 see CMS212
C.I. PIGMENT YELLOW 31 see BAK250
C.I. PIGMENT YELLOW 32 see SMH000
C.I. PIGMENT YELLOW 33 see CAP500, CAP750
C.I. PIGMENT YELLOW 34 see LCR000
C.I. PIGMENT YELLOW 35 see CMS215
C.I. PIGMENT YELLOW 36 see ZFJ100
C.I. PIGMENT YELLOW 37 see CAJ750
C.I. PIGMENT YELLOW 46 see LDN000
C.I. PIGMENT YELLOW 48 see LCU000
C.I. PIGMENT YELLOW 40 see PLI750
C.I. PIGMENT YELLOW see ARI000
CIPLAMYCETIN see CDP250
CIPOVIOL W 72 see PKP750
CIPRIL (PUERTO RICO) see SLJ650
CIPROMID see CQJ250
CIRAM see BJK500
CIRANTIN see HBU000
CIRCAIN see DNC000
CIRCAIR see DNC000
CIRCANOL see DLL400
CIRCOSOLV see TIO750
CIRCULIN see CMS225
C.I. REACTIVE BLACK 8 see CMS227
C.I. REACTIVE BLUE 19 see BMM500
C.I. REACTIVE BLUE 4 see CMS228
C.I. REACTIVE BLUE 19, DISODIUM SALT see BMM500
C.I. REACTIVE RED 2 see PMF540
C.I. REACTIVE YELLOW 14 see RCZ000
C.I. REACTIVE YELLOW 73 see CMS230
C.I. RED 33, DISODIUM SALT see CMS231
C.I. REDUCING AGENT 6 see CMS231
CIRENE BRILLIANT BLUE R see ADE750
CIROLEMYCIN see CMS232
CIRPONYL see MQU750
CIRRASOL 185A see NMY000
CIRRASOL ALN-WF see PJT300
CIRRASOL-OD see HCQ500
CISAPRIDE see CMS237
CISCLOMIPHENE see CMX500
CISMETHRIN see RDZ875
C.I. SOLVENT BLUE 18 see TBG700
C.I. SOLVENT BLUE 69 see DMM400
C.I. SOLVENT BLUE 78 see BKP500
C.I. SOLVENT BLUE 7 see PEI000
C.I. SOLVENT BLUE 93 see BKP500
C.I. SOLVENT BROWN 1 see NBG500
C.I. SOLVENT BROWN PR see NBG500
C.I. SOLVENT GREEN 3 see BLK000
C.I. SOLVENT GREEN 7 see TNM000
C.I. SOLVENT ORANGE 15 see BJF000
C.I. SOLVENT ORANGE 2 see TGW000
C.I. SOLVENT ORANGE 3 see PEK000
C.I. SOLVENT ORANGE 7 see XRA000
C.I. SOLVENT ORANGE 1 (8CI) see CMP600
C.I. SOLVENT RED 19 see EOJ500
C.I. SOLVENT RED 23 see OHA000
C.I. SOLVENT RED 24 see SBC500
C.I. SOLVENT RED 41 see MAC500
C.I. SOLVENT RED 43 see BMO250
C.I. SOLVENT RED 53 see AKE250
C.I. SOLVENT RED 80 see DOK200
C.I. SOLVENT RED see CMS242
C.I. SOLVENT VIOLET 11 see DBP000
C.I. SOLVENT VIOLET 12 see AKP250
C.I. SOLVENT VIOLET 13 see HOK000
C.I. SOLVENT VIOLET 26 see DBX000
C.I. SOLVENT VIOLET 9 see TJK000
C.I. SOLVENT YELLOW 12 see OHK000
C.I. SOLVENT YELLOW 14 see PEJ500
C.I. SOLVENT YELLOW 1 see PEI000
C.I. SOLVENT YELLOW 2 see DOT300
C.I. SOLVENT YELLOW 33 see CMS245
C.I. SOLVENT YELLOW 34 see IBB000

C.I. SOLVENT YELLOW 3 see AIC250
C.I. SOLVENT YELLOW 4 see PEJ600
C.I. SOLVENT YELLOW 52 see DUW500
C.I. SOLVENT YELLOW 5 see FAG130
C.I. SOLVENT YELLOW 7 see HJF000
C.I. SOLVENT YELLOW 92 see AAQ250
C.I. SOLVENT YELLOW 94 see FEV000
C.I. SOLVENT YELLOW 99 see AAQ250
C.I. SOLVENT YELLOW 1, MONOHYDROCHLORIDE see PEI250
CISPLATINO (SPANISH) see PJD000
CISPLATYL see PJD000
CISTANCHE TUBULOSA Wight (extract) see CMS248
CISTAPHOS see AKB500
CISTEAMINA (ITALIAN) see AJT250
CITANEST see CMS250
CITANEST HYDROCHLORIDE see CMS250
CITARABINA see AQQ750
CITARIN see TDX750
CITARIN L see LFA020
CITEXAL see QAK000
CITEX BCL 462 see DDM300
CITGRENILE see MNB500
CITHROL PO see PJY100
CITICHOLINE see CMF350
CITICOLINE see CMF350
CITIDIN DIFOSFATO de COLINA see CMF350
CITIDOLINE see CMF350
CITIFLUS see ARQ750
CITILAT see AEC750
CITIOLASE see TCZ000
CITIOLONE see TCZ000
CITIREUMA see SOU550
CITOBARYUM see BAP000
CITOCOR see DJS200
CITODON see ERD500
CITOFUR see FLZ050
CITOL see ALT250
CITOMULGAN M see OAV000
CITOPAN see ERD500
CITOSARIN see AJJ875
CITOSULFAN see BOT250
CITOX see DAD200
CITRACETAL see CMS324
CITRACONIC ACID see CMS320
CITRACONIC ACID ANHYDRIDE see CMS322
CITRACONIC ANHYDRIDE see CMS322
CITRA-FORT see ABG750
CITRAL α see GCU100
α-CITRAL see GCU100
CITRAL (FCC) see DTC800
(E)-CITRAL see GCU100
trans-CITRAL see GCU100
CITRAL DIETHYL ACETAL see CMS323
CITRAL DIMETHYL ACETAL see DOE000
CITRAL ETHYLENE GLYCOL ACETAL see CMS324
CITRAL METHYLANTHRANILATE, SCHIFF'S BASE see CMS325
CITRAM see AMX825, DJA400
CITRAMON see ARP250
CITRATE de ACETOXY-THYMOXY-ETHYL-DIMETHYLAMINE (FRENCH) see MRN600
CITRAZINIC ACID see DMV400
CITRAZON see BCP000
CITREOVIRIDIN see CMS500
CITREOVIRIDINE see CMS500
CITRETTEN see CMS750
CITRIC ACID, anhydrous see CMS750
CITRIC ACID see CMS750
CITRIC ACID, ACETYL TRIETHYL ESTER see ADD750
CITRIC ACID, AMMONIUM IRON(3+) SALT see FAS700
CITRIC ACID, AMMONIUM SALT see ANF800
CITRIC ACID, COPPER(2+) SALT (8CI) see CNK625
CITRIC ACID, DYSPROSIUM(3+) salt (1:1) see DYG800
CITRIC ACID, GALLIUM SALT (1:1) see GBO000
CITRIC ACID, MONOSODIUM SALT see MRL000
CITRIC ACID, SAMARIUM SALT see SAS000
CITRIC ACID, SODIUM SALT see MRL000
CITRIC ACID, TRIBUTYL ESTER see THY100
CITRIC ACID, TRIBUTYL ESTER, ACETATE see THX100
CITRIC ACID, TRILITHIUM SALT see TKU500
CITRIC ACID, TRIPOTASSIUM SALT see PLB750
CITRIC ACID, YTTRIUM SALT (3:1) see YFA000
CITRIC ACID, ZINC SALT (2:3) see ZFJ250
CITRIDIC ACID see ADH000
CITRININ see CMS775
CITRO see CMS750

CITRODYLE see OHC100
CITROFLEX 2 see TJP750
CITROFLEX 4 see THY100
CITROFLEX A 2 see ADD750
CITROFLEX A 4 see THX100
CITROFLEX A see THX100
CITROFLUYL see MRL000
CITRONELLAL see CMS845
CITRONELLAL HYDRATE see CMS850
CITRONELLA OIL see CMT000
CITRONELLENE see CMT050
CITRONELLIC ACID see CMT125
α-CITRONELLOL see DTF400
CITRONELLOL see CMT250, DTF410
CITRONELLOXYACETALDEHYDE see CMT300
α-CITRONELLYL ACETATE see RHA000
CITRONELLYL ACETATE (FCC) see AAU000
CITRONELLYL-2-BUTENOATE see CMT500
CITRONELLYL BUTYRATE see CMT600
CITRONELLYL-α-CROTONATE see CMT500
CITRONELLYL ETHYL ETHER see EES100
CITRONELLYL FORMATE see CMT750
CITRONELLYL ISOBUTYRATE see CMT900
CITRONELLYL NITRILE see CMU000
CITRONELLYLOXYACETALDEHYDE see CMT300
CITRONELLYL PHENYLACETATE see CMU050
CITRONELLYL PROPIONATE see CMU100
CITRON YELLOW see PLW500, ZFJ100
CITROSODINE see TNL000
CITROVIOL see DTC000
CITRULLAMON see DKQ000, DNU000
l-CITRULLINE and SODIUM NITRITE (2:1) see SIS100
CITRUS AURANTIUM see LBE000
CITRUS FIX see DAA800
CITRUS HYSTRIX DC., fruit peel extract see CMU300
CITRUS RED No. 2 see DOK200
CITRYLIDENEACETONE see POH525
CITTERAL see SEQ000
C.I. VAT BLACK 1 see CMU320
C.I. VAT BLACK 27 see DUP100
C.I. VAT BLACK 28 see IBJ000
C.I. VAT BLACK 8 see CMU475
C.I. VAT BLUE 4 see IBV050
C.I. VAT BLUE 6 see DFN425
C.I. VAT BLUE 16 see CMU500
C.I. VAT BLUE 18 see VGP100
C.I. VAT BLUE 22 see CMU750
C.I. VAT BROWN 1 see CMU770
C.I. VAT BROWN 25 see CMU800
C.I. VAT BROWN 44 see CMU770
C.I. VAT GREEN 2 see APK500
C.I. VAT GREEN 3 see CMU810
C.I. VAT ORANGE 5 see CMU815
C.I. VAT ORANGE 7 see CMU820
C.I. VAT RED 13 see CMU825
C.I. VAT RED 41 see DNT300
C.I. VAT VIOLET 1 (8CI) see DFN450
C.I. VAT YELLOW 2 see VGP200
C.I. VAT YELLOW see DCZ000
CIVET see OGM100
CIVET ABSOLUTE see OGM100
cis-CIVETONE see CMU850
CIVETONE see CMU850
CIZARON see TMP750
CK3 see CBT750
ChKhZ 9 see BBS300
CL 337 see ASA500
CL 369 see CKD750
CL-395 see PDC890
CL-845 see PJA170
CL 10304 see AJD000
CL 12503 see POP000
CL 12,625 see PIF750
CL 12880 see DSP400
CL 13494 see AKO500
CL 13,900 see AEI000
CL-14377 see MDV500
CL 18133 see EPC500
CL 24055 see DUI000
CL 26691 see CQL250
CL 27,319 see MFD500
CL-34699 see COW825
CL-38023 see FAB600
CL 40881 see EDW875
CL-43,064 see DXN600

CL 47300 see DSQ000
CL-47,470 see DHH400
CL 52160 see TAL250
CL 59806 see MQW250
CL-62362 see DCS200
CL 64475 see DHH200
CL 65336 see AJV500
CL 67772 see AOA095
CL-71563 see DCS200
CL82204 see BGL250
CL 227193 see SJJ200
CL 232315 see MQY100
CL 1950675526 see TES000
CL 67310465 see PBT100
CLAFEN see CQC500, CQC650
CLAIRFORMIN see CMV000
CLAIRSIT see PCF300
CLAM POISON DIHYDROCHLORIDE see SBA500
CLANDILON see DNU100
CLANICLOR see CMU875
CLAODICAL see CCK125
CLAPHENE see CQC650
CLARIFIED OILS (PETROLEUM), CATALYTIC CRACKED see CMU890
CLARIFIED SLURRY OIL see CMU890
CLARIPEX see ARQ750
CLARK I see CGN000
CLARO 5591 see SLJ500
CLARY SAGE OIL see OGQ200
CLAUDELITE see ARI750
CLAUDETITE see ARI750
CLAVACIN see CMV000
CLAVELLINA (PUERTO RICO) see CAK325
CLAVULANIC ACID SODIUM SALT see CMV250
CLAYTON YELLOW see CMP050
CL-911C see DVP400
CL 912C see LFG100
CLEARASIL BENZOYL PEROXIDE LOTION see BDS000
CLEARASIL BP ACNE TREATMENT see BDS000
CLEAREL see SLJ500
CLEARJEL see SLJ500
CLEARTUF see PKF750
CLEARY 3336 see DJV000
CLEBOPRIDE HYDROGEN MALATE see CMV325
CLEBOPRIDE MALATE see CMV325
CLEFNON see CAT775
CLEISTANTHIN A see CMV375
CLELAND'S REAGENT see DXO775, DXO800
CLEMANIL see FOS100
CLEMASTINE FUMARATE see FOS100
CLEMASTINE HYDROGEN FUMARATE see FOS100
CLEMATIS see CMV390
CLEMATITE AUX GEAUX (CANADA) see CMV390
CLEMIZOLE HDYROCHLORIDE see CMV400
CLENBUTEROL HYDROCHLORIDE see VHA350
CLENIL-A see AFJ625
CLEOCIN see CMV675
CLEOCIN PHOSPHATE see CMV690
C. LEPTOSEPALA see MBU550
CLERA see NCW000
CLERIDIUM 150 see PCP250
CLESTOL see DJL000
CLEVE'S ACID-1,6 see ALI250
CLEVE'S BETA-ACID see ALI250
CLEXANE see HAQ500
CLF II see CBT500
CLIACIL see PDT750
CLIDANAC see CFH825, CMV500
CLIFT see MCI750
CLIMATERINE see DKA600
CLIMBING BITTERSWEET see AHJ875
CLIMBING LILY see GEW800
CLIMBING ORANGE ROOT see AHJ875
CLIMESTRONE see ECU750
CLIN see SJO000
CLINDAMYCIN see CMV675
CLINDAMYCINE (FRENCH) see CMV675
CLINDAMYCIN-2-PALMITATE MONOHYDROCHLORIDE see CMV680
CLINDAMYCIN-2-PHOSPHATE see CMV690
CLINDAMYCIN PHOSPHATE see CMV690
CLINDROL 101CG see BKE500
CLINDROL LT 15-73-1 see BKF500
CLINDROL SDG see HKJ000
CLINDROL SEG see EJM500
CLINDROL SUPERAMIDE 100L see BKE500
CLINESTROL see DKB000

CLINOFIBRATE see CMV700
CLINORIL see SOU550
CLINOXAN see CFG750
CLIOQUINOL see CHR500
CLIOXANIDE see CGB250
CLIQUINOL see CHR500
CLIRADON see KFK000
CLIRADONE see KFK000
CLISTIN see CGD500
CLISTINE MALEATE see CGD500
CLISTIN MALEATE see CGD500
CLIXODYNE see HIM000
CL 19217 4090L 7-5525 see EGI000
CLM see SFY500
CLOAZEPAM see CMW000
CLOBAZAM see CIR750
CLOBBER see CQJ250
CLOBENZEPAM HYDROCHLORIDE see CMW250
CLOBERAT see ARQ750
CLOBETASOL-17-PROPIONATE see CMW300
CLOBETASOL PROPIONATE see CMW300
CLOBETASONE-17-BUTYRATE see CMW400
CLOBETASONE BUTYRATE see CMW400
CLOBRAT see ARQ750
CLOBREN-SF see ARQ750
CLOBUTINOL see CMW459
CLOBUTINOL HYDROCHLORIDE see CMW500
CLOCAPRAMINE DIHYDROCHLORIDE see DKV309
CLOCARPRAMINE DIHYDROCHLORIDE see DKV309
CLOCETE see AAE500
CLOCONAZOLE HYDROCHLORIDE see CMW550
CLODAZONE see CMW600
CLOFAR see ARQ750
CLOFEDANOL HYDROCHLORIDE see CMW700
CLOFENOTANE see DAD200
CLOFENOXIN see DPE000
CLOFEXAMIDE see CJN750
CLOFEXAMIDE PHENYLBUTAZONE see CMW750
CLOFEXAMIDE-PHENYLBUTAZONE MIXTURE see CMW750
CLOFEZON see CMW750
CLOFEZONE see CMW750
CLOFIBRAM see ARQ750
CLOFIBRAT see ARQ750
CLOFIBRATO (SPANISH) see ARQ750
CLOFIBRATO de CINARIZINA (SPANISH) see CMS125
CLOFIBRIC ACID see CMX000
CLOFIBRINIC ACID see CMX000
CLOFIBRINSAEURE (GERMAN) see CMX000
CLOFINIT see ARQ750
CLOFIPRONT see ARQ750
CLOFLUPEROL HYDROCHLORIDE see CLS250
CLOFUZID see DEL300
CLOMEPHENE B see CMX500
CLOMETHIAZOLE see CHD750
CLOMETHIAZOLUM see CHD750
CLOMID see CMX700
CLOMIDAZOLE see CDY325
CLOMIFEN CITRATE see CMX700
CLOMIFENE see CMX500
CLOMIFENO see CMX700
CLOMIPHENE see CMX500
racemic-CLOMIPHENE CITRATE see CMX700
CLOMIPHENE CITRATE see CMX700
CLOMIPHENE DIHYDROGEN CITRATE see CMX700
CLOMIPHENE-R see CMX700
CLOMIPHINE see CMX700
CLOMIPRAMINE see CDU750
CLOMIPRAMINE HYDROCHLORIDE see CDV000
CLOMIVID see CMX700
CLOMPHID see CMX700
CLONAZEPAM see CMW000
CLONIDIN see DGB500
CLONIDINE see DGB500
CLONIDINE HYDROCHLORIDE see CMX760
CLONITARLID see DFV600
CLONITRALID see DFV400
CLONIXIC ACID see CMX770
CLONIXIN see CMX770
CLONIXINE see CMX770
CLONT see MMN250
CLOPANE HYDROCHLORIDE see CPV609
CLOPENTHIXOL DIHYDROCHLORIDE see CLX250
CLOPEPRAMINE HYDROCHLORIDE see IFZ900
CLOPERASTINA CLORIDRATO (ITALIAN) see CMX820
CLOPERASTINE see CMX800

CLOPERASTINE HYDROCHLORIDE see CMX820
CLOPERIDONE HYDROCHLORIDE see CMX840
CLOPHEDIANOL HYDROCHLORIDE see CMW700
CLOPHEN see PJL750
CLOPHEN A60 see PJN250
CLOPHENOXATE see DPE000
CLOPIDOL see CMX850
CLOPIPAZAN MESYLATE see CMX860
CLOPIXOL see CLX250
CLOPOXIDE see LFK000
CLOPRADONE see EQO000
CLOPROMAZINA (ITALIAN) see CKP250
CLOPROP see CJQ300
CLOPYRALID see DGJ100
CLOQUINOZINE TARTRATE see CMX920
CLORALIO see CDN550
CLORAMIDINA see CDP250
CLORAMIN see BIE250
CLORARSEN see DFX400
CLORAZEPATE DIPOTASSIUM see CDQ250
CLOR CHEM T-590 see CDV100
CLORDAN (ITALIAN) see CDR750
CLORDIAZEPOSSIDO (ITALIAN) see LFK000
CLORDION see CBF250
CLOREPIN see CIR750
CLORESTROLO see CLO750
CLOREX see DFJ050
CLORFENIRAMINA see CLX300
CLORGYLINE HYDROCHLORIDE see CMY000
CLORHIDRATO de CELIPROLOL (SPANISH) see SBN475
CLORIDRATO DI-2-BENZIL-4,5-IMIDAZOLINA (ITALIAN) see BBW750
CLORILAX see CKF500
CLORINA see CDP000
CLORINDANOL see CDV700
CLORMETAZANONE see CKF500
CLORMETHAZON see CKF500
CLORNAPHAZINE see BIF250
CLORO (ITALIAN) see CDV750
CLOROAMFENICOLO (ITALIAN) see CDP250
4-CLORO-3-AMINOBENZOATO di DIMETILAMINOETILE CLORIDRATO (ITALIAN) see AJE350
CLOROBEN see DEP600
2-CLOROBENZALDEIDE (ITALIAN) see CEI500
CLOROBENZENE (ITALIAN) see CEJ125
(4-CLORO-BENZIL)-(4-CLORO-FENIL)-SOLFURO (ITALIAN) see CEP000
1-p-CLORO-BENZOIL-5-METOXI-2-METILINDOL-3-ACIDO ACETICO (SPANISH) see IDA000
2-CLORO-1,3-BUTADIENE (ITALIAN) see NCI500
(4-CLORO-BUT-2-IN-IL)-N-(3-CLORO-FENIL)-CARBAMMATO (ITALIAN) see CEW500
CLOROCHINA see CLD000
CLORODANE see CDR750
O-2-CLORO-1-(2,4-DICLORO-FENIL)-VINYL-O,O-DIETILFOSFATO (ITALIAN) see CDS750
(2-CLORO-3-DIETILAMINO-1-METIL-3-OXO-PROP-1-EN-IL)-DIMETIL-FOSFATO see FAB400
CLORODIFENILI, CLORO 42% (ITALIAN) see PJM500
CLORODIFENILI, CLORO 54% (ITALIAN) see PJN000
p-CLORO-α-(2-DIMETILAMINO)-1-METILETIL)-α-METIL FENETIL ALCOOL (ITALIAN) see CMW459
p-CLORO-α-(2-DIMETILAMINO)-1-METILETIL)-α-METIL FENETIL ALCOOL CLORIDRATO (ITALIAN) see CMW500
2-CLORO-4-DIMETILAMINO-6-METIL-PIRIMIDINA (ITALIAN) see CCP500
2-CLORO-10-(3-DIMETILAMINOPROPIL)FENOTIAZINA (ITALIAN) see CKP250
1-CLORO-2,4-DINITROBENZENE (ITALIAN) see CGM000
1-CLORO-2,3-EPOSSIPROPANO (ITALIAN) see EAZ500
CLOROETANO (ITALIAN) see EHH000
2-CLOROETANOLO (ITALIAN) see EIU800
(CLORO-2-ETIL)-1-CICLOESIL-3-NITROSOUREA (ITALIAN) see CGV250
(CLORO-2-ETIL)-1-(RIBOPIRANOSILTRIACETATO-2′,3′,4′)-3-NITROSOUREA (ITALIAN) see ROF200
1-(2-(p-CLORO-α-FENILBENZILOSSI)ETIL)PIPERIDINA CLORIDRATO (ITALIAN) see CMX820
(4-CLORO-FENIL)-BENZOL-SOLFONATO (ITALIAN) see CJR500
(4-CLORO-FENIL)-4-CLORO-VENZOL-SOLFONATO (ITALIAN) see CJT750
3-(4-CLORO-FENIL)-1,1-DIMETIL-UREA (ITALIAN) see CJX750
2(2-(4-CLORO-FENIL-2-FENIL)-ACETIL)INDAN-1,3-DIONE (ITALIAN) see CJJ000
α-(p-CLOROFENIL)-α-FENIL-2-PIPERIDILMETANOLO CLORIDRATO (ITALIAN) see CKI175
N-(3-CLORO-FENIL)-ISOPROPIL-CARBAMMATO (ITALIAN) see CKC000
CLOROFORMIO (ITALIAN) see CHJ500
CLOROFOS (RUSSIAN) see TIQ250
CLOROMETANO (ITALIAN) see MIF765
7-CLORO-2-METILAMINO-5-FENIL-3H-1,4-BENZOIDIAZEPINA 4-OSSIDO (ITALIAN) see LFK000

7-CLORO-3-METIL-2H-1,2,4-BENZOTIODIAZINE-1,1-DIOSSIDO (ITALIAN) see DCQ700
3-CLORO-2-METIL-PROP-1-ENE (ITALIAN) see CIU750
CLOROMISAN see CDP250
1-CLORO-4-NITROBENZENE (ITALIAN) see NFS525
O-(4-CLORO-3-NITRO-FENIL)-O,O-DIMETIL-MONOIIOFOSFATO (ITALIAN) see NFT000
O-(3-CLORO-4-NITRO-FENIL)-O,O-DIMETIL-MONOTIOFOSFATO (ITALIAN) see MIJ250
CLOROPHENE see CJU250
CLOROPICRINA (ITALIAN) see CKN500
CLOROPIRIL see CLX300, TAI500
CLOROPRENE (ITALIAN) see NCI500
CLOROSAN see CDP000
CLOROSINTEX see CDP250
CLOROTEPINE see ODY100
CLOROTETRACICLINA CLORIDRATO (ITALIAN) see CMB000
CLOROTRISIN see CLO750
CLOROX see SHU500
CLORPROPAMIDE (ITALIAN) see CKK000
CLORTERMINE HYDROCHLORIDE see DRC600
CLORTOKEM see CIS250
CLORTRAN see ABD000
CLORURO DI ETILE (ITALIAN) see EHH000
CLORURO di ETHENE (ITALIAN) see EIY600
CLORURO di ETILIDENE (ITALIAN) see DFF809
CLORURO di MERCURIO (ITALIAN) see MCY475
CLORURO MERCUROSO (ITALIAN) see MCW000
CLORURO di METALLILE (ITALIAN) see CIU750
CLORURO di METILE (ITALIAN) see MIF765
CLORURO di SUCCINILCOLINA (ITALIAN) see HLC500
CLORURO di VINILE (ITALIAN) see VNP000
CLOSANTEL see CFC100
CLOSTRIDIUM BOTULINUM NEUROTOXIN see BMM292
CLOSTRIDIUM BOTULINUM TOXIN see CMY030
CLOSTRIDIUM BUTYRICUM MIYAIRI, powder see CMY050
CLOSTRIDIUM DIFFICILE TOXIN see CMY070
CLOSTRIDIUM DIFFICILE TOXIN B see CMY090
CLOSTRIDIUM NOVYI α-TOXIN see CMY130
CLOSTRIDIUM OEDEMATIENS TYPE A TOXIN see CMY150
CLOSTRIDIUM PERFRINGENS EXOTOXIN see CMY170
CLOSTRIDIUM PERFRINGENS β-TOXIN see CMY190
CLOSTRIDIUM PERFRINGENS (Welchii) TYPE A ENTEROTOXIN see CMY220
CLOSTRIDIUM SORDELLII TOXIN see CMY240
CLOSTRIDIUM TETANI TOXIN see CMY260
CLOSTRIDIUM TETANI, TYPE BE TOXIN see CMY280
CLOSTRIDIUM TETANI, TYPE S TOXIN see CMY300
CLOTAM see CLK325
CLOTEPIN see ODY100
CLOTHEPIN see ODY100
CLOTIAZEPAM see CKA000
CLOTRIDE see CLH750
CLOTRIMAZOL see MRX500
CLOUT see DXE600
CLOVE BUD OIL see CMY475
CLOVE LEAF OIL see CMY500
CLOVE LEAF OIL MADAGASCAR see CMY500
CLOVE OIL, stem see CMY510
CLOWN TREACLE see WBS850
CLOXACILLIN SODIUM MONOHYDRATE see SLJ000
CLOXACILLIN SODIUM SALT see SLJ050
CLOXAPEN see SLJ000, SLJ050
CLOXAZEPINE see DCS200
CLOXAZOLAM see CMY525
CLOXAZOLAZEPAM see CMY525
CLOXYPEN see SLJ000
CLOXYPENDYL see CMY535
CLOZAPIN see CMY650
CLOZAPINE see CMY650
CL 251931 SODIUM SALT see CCS635
CLUDR see CFJ750
CLUSIA ROSEA see BAE325
CLY-503 see SDY500
CLYSAR see PMP500
CM 6912 see EKF600
pCMA see CIF250
CMA see CBF250, CIF250
CMB 50 see CBT500
CMB 200 see CBT500
CMC 7H see SFO500
S-CMC see CCH125
CMC see SFO500
CM-CELLULOSE Na SALT see SFO500
CMC SODIUM SALT see SFO500
CMDP see MQR750

CME 74770 see TKL100
CME see CBD750
C-METON see TAI500
C. MEXICANA see CAK325
CMF see EHF500
CMH see MAE500
C. MITIS see FBW100
CML 21 see CAU500
CML 31 see CAU500
CMME see CIO250
CMMP see SKQ400
CMPABN see MEI000
CMPF see MIT600
CMPP see CIR500
CM S 2957 see CLJ875
CMU see CJX750
CMW BONE CEMENT see PKB500
CMZ SODIUM see CCS360
CN 009 see CDB760
CN 447 see DEJ000
CN 3123 see SNL850
CN 8676 see EEI000
CN-15,757 see ASA500
CN-27,554 see TKH750
CN-35355 see XQS000
CN-36337 see CIP500
CN 38703 see QAK000
CN 59,567 see CGB250
CN-11-2936 see DWS200
CN-25,253-2 see AOO500
CN-52,372-2 see CKD750
CN-55945-27 see NHP500
CN see CEA750
CNA see RDP300
CNC see NAS000
CNCC see BIF625
C. NOCTURNUM see DAC500
CNP 1032 see NIW500
CNP see NIW500
CNU see CHE750
CNUEMS see CHF250
CNU-ETHANOL see CHB750
CO 12 see DXV600
CO-1214 see DXV600
CO-1670 see HCP000
CO-1895 see OAX000
CO-1897 see OAX000
COAGULASE see CMY725
COAKUM see PJJ315
COAL CONVERSION MATERIALS, SRC-II HEAVY DISTILLATE see CMY750
COAL DUST see CMY760
COAL FACINGS see CMY760
COAL GAS (UN 1023) (DOT) see HHJ500
COAL, GROUND BITUMINOUS (DOT) see CMY760
COAL LIQUID see PCR250
COAL-MILLED see CMY760
COAL NAPHTHA see BBL250
COAL OIL see KEK000, PCR250
COAL SLAG-MILLED see CMY760
COAL TAR see CMY800
COAL TAR, AEROSOL see CMY800
COAL TAR CREOSOTE see CMY825
COAL TAR DISTILLATE see CMY900
COAL TAR DISTILLATES, flammable (DOT) see CMY900
COAL TAR DISTILLATES see CMY900
COAL TAR OIL (DOT) see CMY825
COAL TAR OIL see CMY825
COAL TAR PITCH VOLATILES see CMZ100
COAL TAR PITCH VOLATILES: PHENANTHRENE see PCW250
COAL TAR SOLUTION USP see CMY800
COAPT see MIQ075
COATHYLENE HA 1671 see PJS750
COATHYLENE PF 0548 see PMP500
COBADEX see CNS750
COBADOCE FORTE see VSZ000
COBALIN see VSZ000
COBALT-59 see CNA250
COBALT see CNA250
COBALT(2+) ACETATE see CNC000
COBALT(3+) ACETATE see CNA300
COBALT(III) ACETATE see CNA300
COBALT(II) ACETATE see CNC000
COBALT ACETATE see CNA300, CNC000
COBALT ACETATE TETRAHYDRATE see CNA500
COBALT ALLOY, Co,Cr see CNA750

COBALT(III) AMIDE see CNB000
COBALTATE(3), HEXAKIS(NITRITO-N)-, TRIPOTASSIUM, (OC-6-11)-(9CI) see PLI750
COBALTATE(3-), HEXAKIS(NITRITO-N)-, TRISODIUM (OC-6-11)- (9CI) see SFX750
COBALTATE(3-), HEXANITRO-, TRISODIUM see SFX750
COBALT(II) AZIDE see CNB099
COBALT, BIS(CARBONATO(2-))HEXAHYDROXYPENTA- see CNB495
COBALT, BIS(N-9H-FLUOREN-2-YL-N-HYDROXYACETAMIDATO-O,O')-(9CI) see FDU875
COBALT BLACK see CND125
COBALT BOROFLUORIDE see CNC050
COBALT BORON TETRAFLUORIDE see CNC050
COBALT(II) BROMIDE see CNB250
COBALT CARBONATE (1:1) see CNB475
COBALT(2+) CARBONATE see CNB475
COBALT CARBONATE see CNB475
COBALT CARBONATE, COBALT DIHYDROXIDE (2:3) see CNB495
COBALT CARBONATE HYDROXIDE see CNB495
COBALT CARBONYL see CNB500
COBALT(III) CHLORIDE see CNB750
COBALT(II) CHLORIDE see CNB599
COBALT(2+) CHLORIDE HEXAHYDRATE see CNB800
COBALT(II) CHLORIDE HEXAHYDRATE see CNB800
COBALT CHLORIDE, HEXAHYDRATE (8CI, 9CI) see CNB800
COBALT-CHROMIUM ALLOY see CNA750
COBALT-CHROMIUM-MOLYBDENUM ALLOY see VSK000
COBALT COMPOUNDS see CNB850
COBALT DIACETATE see CNC000
COBALT DIACETATE TETRAHYDRATE see CNA500
COBALT DIBROMIDE see CNB250
COBALT DICHLORIDE see CNB599
COBALT DICHLORIDE HEXAHYDRATE see CNB800
COBALT DIFLUORIDE see CNC100
COBALT DIHYDROXIDE see CNC231
COBALT DINITRATE see CNC500
COBALT DIPERCHLORATE HEXAHYDRATE see CND900
COBALT(2)-EDATHAMIL see DGQ400
COBALT(II)FLUOBORATE see CNC050
COBALT N-FLUOREN-2-YLACETOHYDROXAMATE see FDU875
COBALT(II) FLUORIDE see CNC100
COBALT-HISTIDINE see BJY000
COBALT HYDROCARBONYL see CNC230
COBALT(2+) HYDROXIDE see CNC231
COBALT(II) HYDROXIDE see CNC231
COBALT HYDROXIDE OXIDE see CNC232
COBALT HYDROXIDE OXIDE (CoO(OH)) see CNC232
COBALTIC ACETATE see CNA300
COBALTIC-COBALTOUS OXIDE see CND020
COBALTIC OXIDE see CND825
COBALTIC POTASSIUM NITRITE see PLI750
COBALT-METHYLCOBALAMIN see VSZ050
COBALT MONOCARBONATE see CNB475
COBALT MONOOXIDE see CND125
COBALT MONOSULFIDE see CNE200
COBALT MONOXIDE see CND125
COBALT MURIATE see CNB599
COBALT NAPHTHENATE, POWDER (DOT) see NAR500
COBALT(II) NITRATE see CNC500
COBALT(II) NITRIDE see CNC750
COBALT NITROPRUSSIDE see CND000
COBALT NITROSOPENTACYANOFERRATE(3) see CND000
COBALT NITROSYLPENTACYANOFERRATE see PAY610
COBALTOCENE see BIR529
COBALTO-COBALTIC OXIDE see CND020
COBALTO-COBALTIC TETROXIDE see CND020
COBALT OCTACARBONYL see CNB500
COBALTOSIC OXIDE see CND020
COBALTOUS ACETATE TETRAHYDRATE see CNA500
COBALTOUS BROMIDE see CNB250
COBALTOUS CARBONATE see CNB475
COBALTOUS CHLORIDE see CNB599
COBALTOUS CHLORIDE, HEXAHYDRATE see CNB800
COBALTOUS DIACETATE see CNC000
COBALTOUS DICHLORIDE see CNB599
COBALTOUS FLUORIDE see CNC100
COBALTOUS HYDROXIDE see CNC231
COBALTOUS NITRATE see CNC500
COBALTOUS OXIDE see CND125
COBALTOUS PERCHLORATE, HEXAHYDRATE see CND900
COBALTOUS PHOSPHATE see CND920
COBALTOUS SULFATE see CNE125
COBALTOUS SULFIDE see CNE200
COBALTOUS TETRAFLUOROBORATE see CNC050
COBALT(2+) OXIDE see CND125

COBALT(3+) OXIDE see CND825
COBALT(III) OXIDE see CND825
COBALT(II) OXIDE see CND125
COBALT OXIDE see CND125, CND020
COBALT OXIDE (8CI,9CI) see CND825
COBALT OXIDE HYDROXIDE (CoOOH) see CNC232
COBALT OXYHYDROXIDE see CNC232
COBALT(II) PERCHLORATE, HEXAHYDRATE see CND900
COBALT PERCHLORATE HEXAHYDRATE see CND900
COBALT PEROXIDE see CND825
COBALT(II) PHOSPHATE see CND920
COBALT PHOSPHATE see CND920
COBALT RESINATE, precipitated see CNE000
COBALT SALTS see FDU875
COBALT SESQIOXIDE see CND825
COBALT SESQUIOXIDE see CND825
COBALT(II) SULFATE (1:1) see CNE125
COBALT SULFATE (1:1) see CNE125
COBALT (2+) SULFATE see CNE125
COBALT SULFATE see CNE125
COBALT(II) SULFIDE see CNE200
COBALT SULFIDE (amorphous) see CNE200
COBALT SULFIDE see CNE200
COBALT(II) SULPHATE see CNE125
COBALT TETRACARBONYL see CNB500
COBALT TETRACARBONYL DIMER see CNB500
COBALT TETRAOXIDE see CND020
COBALT TRIACETATE see CNA300
COBALT TRIFLUORIDE see CNE250
COBALT TRIOXIDE see CND825
COBALT YELLOW see PLI750
COBAMIN see VSZ000
COBAN see MRE230
COBEN see CNE375
COBEN P see PIC100
COBEX (polymer) see PKQ059
COBEX see CNE500
COBEXO see CNE500
COBH see BDD000
COBINAMIDE, COBALT-METHYL derivative, HYDROXIDE, DIHYDROGEN PHOSPHATE (ester), inner salt, 3'-ESTER with 5,6-DIMETHYL-1-α-D-RIBOFU-RANOSYLBENZIMIDAZOLE see VSZ050
COBIONE see VSZ000
COBOX see CNK559
COBOX BLUE see CNK559
COBRATEC #99 see BDH250
COBRATEC TT 100 see MHK000
COCAFURIN see NGE500
COCAIN-CHLORHYDRAT (GERMAN) see CNF000
β-COCAINE see CNE750
l-COCAINE see CNE750
COCAINE see CNE750
(−)-COCAINE see CNE750
COCAINE CHLORIDE see CNF000
l-COCAINE HYDROCHLORIDE see CNF000
COCAINE HYDROCHLORIDE see CNF000
(−)-COCAINE HYDROCHLORIDE see CNF000
COCAINE MURIATE see CNF000
CO CAP IMIPRAMINE 25 see DLH630
COCARTRIT see CLD000
COCCIDOT see DUP300
C. OCCIDENTALIS see CCK675
COCCIDINE A see DUP300
COCCIDIOSTAT C see CMX850
COCCINE see FMU080
COCCOCLASE see PPO000
COCCULIN see PIE500
COCCULUS solid (DOT) see PIE500
COCCULUS see PIE500
COCHENILLE DYE see CNF050
COCHENILLEROT A see FMU080
COCHIN see LEG000
COCHINEAL (dye) see CNF050
COCHINEAL see CNF050
COCHINEAL RED A see FMU080
COCHINEAL TINCTURE see CNF050
COCHLIOBOLIN see CNF109
COCHLIOBOLIN A see CNF109
COCHLIODINOL see CNF159
COCO-DIAZINE see PPP500
COCO DIETHANOLAMIDE see BKE500
COCONUT ALDEHYDE see CNF250
COCONUT BUTTER see CNR000
COCONUT DIMETHYL AMINE OXIDE see CNF325

COCONUT MEAL PELLETS, containing 6–13% moisture and no more than 10% residual fat (DOT) see CNR000
COCONUT OIL (FCC) see CNR000
COCONUT OIL AMIDE of DIETHANOLAMINE see BKE500
COCONUT PALM OIL see CNR000
N-COCOPYRROLIDINONE see CNF340
COCUM see PJJ315
COD see CPR825
CODAL see MQQ450
CODECARBOXYLASE see PII100
CODECHINE see BBQ500
CODE H 133 see DER800
CODEHYDRASE I see CNF390
CODEHYDRASE II see CNF400
CODEHYDROGENASE I see CNF390
CODEHYDROGENASE II see CNF400
CODEINE see CNF500
CODEINE HYDROCHLORIDE see CNF750
CODEINE NICOTINATE (ESTER) see CNG250
CODEINE PHOSPHATE see CNG500
CODEINE PHOSPHATE SESQUIHYDRATE see CNG675
CODEINE SULFATE see CNG750
CODEINONE, DIHYDRO-, TARTRATE see DKX050
CODELCORTONE see PMA000
CODELSOL see PLY275
CODEMPIRAL see ABG750
CO-DERGOCRINE MESYLATE see DLL400
CODETHYLINE see ENK000
CODETHYLINE HYDROCHLORIDE see DVO700
CODHYDRINE see DKW800
CODIAZINE see PPP500
CODIBARBITA see EOK000
CODLELURE see CNG760
CODLEMONE see CNG760
CODYLIN see TCY750
COE 536 see HKC575
COENZYME I see CNF390
COENZYME II see CNF400
COENZYME Q$_{10}$ see UAH000
COENZYME R see VSU100
CO-ESTRO see ECU750
COFFEBERRY see MBU825
COFFEE SENNA see CNG825
COFFEIN (GERMAN) see CAK500
COFFEINE see CAK500
CO-FRAM see SNL850
COGESIC see DTO200
COGILOR BLUE 512.12 see FAE000
COGILOR RED 321.10 see FAG070
COGNAC OIL see EKN050
COGOMYCIN see FPC000
COHASAL-1H see SEH000
COHEDUR A see MCB050
COHORTAN see BDJ600
COHOSH see BAF325
CO-HYDELTRA see PMA000
COHYDRIN see DKW800
COIR DEEP BLACK C see AQP000
COKAN see PJJ315
COKE see CNE750
COKE POWDER see CBT500
S-COL see SCK600
COLACE see DJL000
COLACID BLACK 10A see FAB830
COLACID BLUE A see ADE750
COLACID ORANGE see CMM220
COLACID PONCEAU 4R see FMU080
COLACID PONCEAU SPECIAL see FMU070
COLACID RED 2A see CMS231
COLALIN see CME750
COLAMINE see EEC600
COLAMINE HYDROCHLORIDE see EEC700
COLCEMID see MIW500
COLCEMIDE see MIW500
COLCHAMIN see MIW500
COLCHAMINE see MIW500
COLCHICIN (GERMAN) see CNG830
COLCHICINA (ITALIAN) see CNG830
7-α-H-COLCHICINE see CNG830
COLCHICINE see CNG830
COLCHICINE, 7-DEACETAMIDO-7-(METHYLAMINO)- see MIW500
COLCHICINE, N-DEACETYL-10-DEMETHOXY-10-METHYLTHIO- see DBA200
COLCHICINE, DEACETYL-N-METHYL- see MIW500
COLCHICOSIDE see DAN375
COLCHICUM AUTUMNALE see CNX800

COLCHICUM SPECIOSUM see CNX800
COLCHICUM VERNUM see CNX800
COLCHINEOS see CNG830
COLCHINIC ACID TRIMETHYL see TLO000
COLCHISOL see CNG830
COLCIN see CNG830
COLCOTHAR see IHD000
COLDAN see NCW000
COLDRIN see CMW700
COLEBENZ see BCM000
COLECALCIFEROL see CMC750
COLEMANITE see CAN250
COLEMID see MIW500
COLEP see MOB699
COLEPAX see IFY100
COLEPUR see DDS600
COLESTERINEX see PPH050
COLESTYRAMIN see CME400
COL-EVAC see SFC500
COLEYTL see CBH250
COLFARIT see ADA725
COLIBIL see CNG835
COLIMYCIN see PKD250
COLIMYCIN M see SFY500
COLIOPAN see BPI125
COLIPAR see DDS600
COLISONE see PLZ000
COLISTICINA see PKD250
COLISTIMETHATE SODIUM see SFY500
COLISTIN see PKD250
COLISTINASE see BAC000
COLISTIN SODIUM METHANESULFONATE see SFY500
COLISTIN SULFOMETHAT see SFY500
COLISTIN SULFOMETHATE SODIUM see SFY500
COLISTRIMETHATE SODIUM see SFY500
COLITE see CMF350
COLLARGOL see SDI750
2,4,6-COLLIDINE see TME272
α-γ,α'-COLLIDINE see TME272
γ-COLLIDINE see TME272
s-COLLIDINE see TME272
sym-COLLIDINE see TME272
COLLIDINE, ALDEHYDECOLLIDINE see EOS000
COLLIRON I.V. see IHG000
COLLOCARB see CBT750
COLLODION see CCU250
COLLODION COTTON see CCU250
COLLODION WOOL see CCU250
COLLOID 775 see CCL250
COLLOIDAL ARSENIC see ARA750
COLLOIDAL CADMIUM see CAD000
COLLOIDAL FERRIC OXIDE see IHD000
COLLOIDAL GOLD see GIS000
COLLOIDAL MANGANESE see MAP750
COLLOIDAL MERCURY see MCW250
COLLOIDAL SELENIUM see SBO500
COLLOIDAL SILICA see SCH000
COLLOIDAL SILICON DIOXIDE see SCH000
COLLOIDAL SULFUR see SOD500
COLLOIDOX see CNK559
COLLOKIT see SOD500
COLLOMIDE see SNM500
COLLONE AC see PJT300
COLLOWELL see SFO500
COLLOXYLIN see CCU250
COLLUNOSOL see TIV750
COLLUNOVAR see NCJ500
COLLUNOVER see NCJ500
COLLUSUL-HC see HHQ800
COLOCASIA ESCULENTA see EAI600
COLOCASIA GIGANTEA see EAI600
COLOGNE EARTH see MAT500
COLOGNE SPIRIT see EFU000
COLOGNE UMBER see MAT500
COLOGNE YELLOW see LCR000
COLOMBIAN BLACK TOBACCO CIGARETTE REFINED TAR see CMP800
COLONATRAST see BAP000
COLONIAL SPIRIT see MGB150
COLORADO RIVER HEMP see SBC550
COLORFIX see CNH125
COLORINES (MEXICO) see NBR800
COLOR-SET see TIX500
COLPOVISTER see EDU500
COLPRO see MBZ100
COLPRONE see MBZ100

COLSALOID see CNG830
COLSUL see SOD500
COLSULANYDE see SNM500
COLTIROT see TOG500
COLTS FOOT see PCR000
COLUMBIA BLACK EP see AQP000
COLUMBIA CARBON see CBT750
COLUMBIA FAST BLACK G see CMN240
COLUMBIA FAST BLACK GB see CMN240
COLUMBIA FAST RED F see CMO870
COLUMBIA LCK see CBT500
COLUMBIAN SPIRITS (DOT) see MGB150
COLUMBIUM PENTACHLORIDE see NEA000
COLUMBIUM POTASSIUM FLUORIDE see PLN500
COLUTOID see GEK500
COLYER PECTIN see PAO150
COLY-MYCIN see PKD250
COLY-MYCIN INJECTABLE see SFY500
COLYMYSIN S see PKD250
COLYONAL see DBD750
COLYSTINMETHANSULFONAT (GERMAN) see SFY500
COMAC see CNM500
COMACID BLUE BLACK B see FAB830
COMBANTRIN see POK575
COMBETIN see SMN000
COMBINACE see CAM200
COMBINAL K1 see VTA000
COMBOT EQUINE see TIQ250
COMBRETODENDRON AFRICANUM (Welw), extract see CNH275
COMBUSTION IMPROVER-2 see MAV750
COMELIAN see CNR750
COMESA see EQL000, MNM500
COMESTROL see DKA600
COMESTROL ESTROBENE see DKA600
COMFREY, RUSSIAN see RRK000, RRP000
COMITAL see DKQ000
COMITE see SOP000
COMITIADONE see PEC250
COMMISTERONE see HKG500
COMMON GROUNDSEL see RBA400
COMMON SALT see SFT000
COMMON SENSE COCKROACH and RAT PREPARATIONS see PHP010
COMMOTIONAL see ABG750
COMPALOX see AHE250
COMPAZINE see PMF250, PMF500
COMPENDIUM see BMN750
COMPERLAN LD see BKE500
COMPITOX see CIR500
COMPLAMEX see XCS000
COMPLAMIN see XCS000
COMPLEMIX see DJL000
COMPLEXONE see EIV000
COMPLEXON I see AMT500
COMPLEXON II see EIX000
COMPLEXON III see EIX500
COMPLEXON IV see CPB120
COMPOCILLIN G see BDY669
COMPOCILLIN-VK see PDT750
COMPOSE 134 P (FRENCH) see MLK800
COMPOUND 42 see WAT200
COMPOUND 118 see AFK250
COMPOUND 269 see EAT500
COMPOUND 338 see DER000
COMPOUND 347 see EAT900
COMPOUND 497 see DHB400
COMPOUND 604 see DFT000
COMPOUND-666 see BBP750
COMPOUND 711 see IKO000
COMPOUND 864 see YCJ000
COMPOUND 889 see DVL700
COMPOUND 923 see DFY400
COMPOUND 1081 see FFF000
COMPOUND 1189 see KEA000
COMPOUND 1275 see PDV700
COMPOUND 13-61 see OOI200
COMPOUND 01748 see DJT800
COMPOUND 1836 see CLV375
COMPOUND 19-28 see DXB450
COMPOUND 2046 see MQR750
COMPOUND 3422 see PAK000
COMPOUND-3916 see DRB400
COMPOUND 3956 see CDV100
COMPOUND-4018 see CNL500
COMPOUND 4049 see MAK700
COMPOUND 4072 see CDS750

COMPOUND 47-83 see EAN600
COMPOUND 48/80 see CNH375
COMPOUND-4992 see BKS810
COMPOUND 6515 see DJA300
COMPOUND 6890 see TLQ000
COMPOUND 7215 see FBP300
COMPOUND 7744 see CBM750
COMPOUND 8958 see CQH500
COMPOUND 10854 see COF250
COMPOUND 17309 see DAL300
COMPOUND 20-438 see CNH300
COMPOUND 33355 see MKB750
COMPOUND 33,828 see MLJ500
COMPOUND 38,174 see INS000
COMPOUND 42339 see ADR750
COMPOUND 64716 see CMS130
COMPOUND 69/183 see CNH500
COMPOUND 74-637 see ALU875
COMPOUND 90459 see OJI750
COMPOUND M-81 see PHI500
COMPOUND S-6,999 see NNF000
COMPOUND 593A see PIK075
COMPOUND B DICAMBA see MEL500
COMPOUND C-9491 see IEN000
COMPOUND E see CNS800
COMPOUND E ACETATE see CNS825
COMPOUND F-2 see ZAT000
COMPOUND-1452-F see EME500
COMPOUND F see CNS750
COMPOUND F ACETATE see HHQ800
COMPOUND G-11 see HCL000
COMPOUND HP1275 see PDV700
COMPOUND 26539 HYDROCHLORIDE see EPL600
COMPOUND 6-12 INSECT REPELLENT see EKV000
COMPOUND 14045 METHIODIDE see CNH550
COMPOUND 14045 METHOCHLORIDE see EAI875
COMPOUND 14045 METHSULFATE see TJG225
COMPOUND No. 1080 see SHG500
COMPOUND 88R see SOP500
COMPOUND R-242 see CKI625
COMPOUND R-25788 see DBJ600
COMPOUND 2339 RP see PEN000
COMPOUND SN see SAY000
COMPOUND 33T57 see MKW250
COMPOUND UC-20047 A see CFF250
COMPOUND W see TKL100
COMPOUNE 732 see BQT750
COMPTIE see CNH789
COMYCETIN see CDP250
d-CON see WAT200
CON A see CNH625
CONAC A see CPI250
CONAC S see CPI250
CONCANAVALIN A see CNH625
CONCEP see COP700
CONCEP II see OKS200
CONCHININ see QFS000
CONCILIUM see FGU000
CONCLYTE CALCIUM see CAT600
CONCO AAS-35 see DXW200
CONCOGEL 2 CONCENTRATE see SIY000
CONCO NI-90 see PKF000
CONCO NIX-100 see PKF500
CONCO SULFATE C see HCP900
CONCO SULFATE WA see SIB600
CONCO XAL see DRS200
CONCTASE C see CNH650
CONDACAPS see VSZ100
CONDENSATE PL see BKE500
CONDENSATES (PETROLEUM), VACUUM TOWER (9CI) see MQV755
CONDITION see DCK759
CONDITIONER 1 see OBA000
CONDOCAPS see VSZ100
CONDOL see VSZ100
CONDUCTEX see CBT500, CBT750
CONDUCTIVE POLYMER 261 see DTS500
CONDYLON see CNG830
CONDY'S CRYSTALS see PLP000
CONESSINE DIHYDROBROMIDE see DOX000
CONESSINE HYDROCHLORIDE see CNH660
CONEST see ECU750
CONESTORAL see EDV600
CONESTRON see ECU750
CONFECTIONER'S SUGAR see SNH000
CONFORTID see IDA000

CONGOBLAU 3B see CMO250
CONGO BLUE see CMO250
CONGOCIDIN see NCP875
CONGOCIDINE see NCP875
CONGOCIDINE DIHYDROCHLORIDE see NCQ000
CONGO RED see SGQ500
CONGO RED R-138 see NAY000
γ-CONICEIN see CNH730
γ-CONICEINE see CNH730
d-CONICINE see PNT000
CONIGON BC see EIV000
CONIIN see PNT000
CONIINE see PNT000
(+)-CONIINE see PNT000
α-CONINE see PNT000
CONINE see PNT000
CONIUM MACULATUM see CNH750, PJJ300
CONJES see ECU750
CONJUGATED EQUINE ESTROGEN see PMB000
CONJUGATED ESTROGENS see ECU750
CONJUNCAIN see OPI300
CONJUTABS see ECU750
7-CON-o-METHYLNOGAROL see MCB600
CONOCO C-50 see DXW200
CONOCO DBCL see DXW600
CONOTRANE see PFN000
CONOVID see EAP000
CONOVID E see EAP000
CONQUERORS see HGL575
CONQUININE see QFS000
CONRAXIN H see IDJ600
CONRAY 30 see IGC000
CONRAY 60 see IGC000
CONRAY 280 see IGC000
CONRAY see IGC000
CONRAY MEGLUMIN see IGC000
CONRAY MEGLUMINE 282 see IGC000
CONSDRIN see SPC500
CONSDRIN HYDROCHLORIDE see SPC500
CONSTAPHYL see DGE200
CONSTONATE see DJL000
CONT see MMN250
CONTALAX see PPN100
CONTAVERM see PDP250
CONTEBEN see FNF000
CONTERGAN see TEH500
CONTIMET 30 see TGF250
CONTINAL see NBU000
CONTINENTAL see CBT750, KBB600
CONTINEX see CBT750
CONTIZELL see PKQ059
CONTRAC see BMN000
CONTRA CREME see ABU500
CONTRADOL see ABG750
CONTRALGIN see BQA010
CONTRALIN see DXH250
CONTRAMINE see DJD000
CONTRAPAR see KFA100
CONTRAPOT see DXH250
CONTRATHION see PLX250
CONTRHEUMA RETARD see ADA725
CONTRISTAMINE HYDROCHLORIDE see CIS000
CONTRIX 28 see IGC000
CONTROL see MFD500
CONTROVLAR see EEH520
CONTUREX see SHX000
CONVALLAOTOXIN see CNH780
CONVALLARIA MAJALIS see LFT700
CONVALLARIN see CNH775
CONVALLATON see CNH780
CONVALLATOXIGENIN see SMM500
CONVALLATOXIN see CNH780
CONVALLATOXOL see CNH785
CONVALLATOXOSIDE see CNH780
CONVAL LILY see LFT700
CONVALLOTOXIN see CNH780
CONVALLOTOXOL see CNH785
CONVALOTOXOL see CNH785
CONVENIXA see TLP750
CONVUL see DKQ000
CONVULEX see PNX750
COOLSPAN see EPD500
COOMASSIE BLUE see ADE750
COOMASSIE BLUE MEDICINAL see ADE750
COOMASSIE BLUE RL see ADE750

COOMASSIE MILLING SCARLET G see CMM325
COOMASSIE MILLING SCARLET GP see CMM325
COOMASSIE RED PG see CMM320
COOMASSIE RED PGP see CMM320
COOMASSIE RED R see NAO600
COOMASSIE VIOLET see FAG120
COOMASSIE YELLOW R see CMM759
COOMASSIE YELLOW RP see CMM759
COONTIE see CNH789
CO-OP HEXA see HCC500
COPAGEL PB 25 see SFO500
COPAIBA BALSAM see CNH792
COPAIBA OIL see CNH792
COPAIBA OLEORESIN see CNH792
COPAL (CUBA) see PCB300
COPAL Z see SMQ500
COPANOIC see IFY100
COPAROGIN see FLZ050
COPELLIDIN see END000
COPEY see BAE325
COPHARCILIN see AIV500
COPIAMYCIN see CNH800
COPIRENE see KGK000
COPOX see CNO000
COPPER-8 see BLC250
COPPER see CNI000
COPPER(2+) ACETATE see CNI250
COPPER(II) ACETATE see CNI250
COPPER ACETATE see CNI250
COPPER(2+) ACETATE, MONOHYDRATE see CNI325
COPPER(II) ACETATE MONOHYDRATE see CNI325
COPPER ACETOARSENITE (DOT) see COF500
COPPER ACETOARSENITE, solid (DOT) see COF500
COPPER(II) ACETYLACETONATE see BGR000
COPPER ACETYLIDE see CNI500
COPPER-AIRBORNE see CNI000
COPPER ALLOY, Cu, Be see CNI600
COPPER ALLOY, Cu, Be, Co see CNK700
COPPER ARSENATE (BASIC) see CNI900
COPPER ARSENATE HYDROXIDE see CNI900
COPPER ARSENITE, solid (DOT) see CNN500
COPPERAS see FBN100, FBO000
COPPER-BERYLLIUM ALLOY see CNI600
COPPER BICHLORIDE see CNK500
COPPER BIS(ACETYLACETONATE) see BGR000
COPPER BIS(ACETYLACETONE) see BGR000
COPPER, BIS(ETHYLENEDIAMINE)(MERCURICTETRATHIOCYANATO)- see BJP425
COPPER(2+), BIS(ETHYLENEDIAMINE)-, TETRAK-IS(THIOCYANATO)MERCURATE(2-), POLYMERS see BJP425
COPPER BIS(2,4-PENTANEDIONATE) see BGR000
COPPER BLUE see CNQ000
COPPER BRONZE see CNI000
COPPER BROWN see CNO250
COPPER CARBIDE see CNI500
COPPER CARBONATE (1:1) see CBW200
COPPER(II) CARBONATE see CBW200
COPPER CARBONATE see CBW200
COPPER(II) CARBONATE HYDROXIDE (2:1:2) see CNJ750
COPPER CARBONATE HYDROXIDE see CNJ750
COPPER CHELATE of N-HYDROXY-2-ACETYLAMINOFLUORENE see HIP500
COPPER CHLORATE (DOT) see CNJ900
COPPER CHLORATE see CNJ900
COPPER(II) CHLORIDE (1:2) see CNK500
COPPER(2+) CHLORIDE see CNK500
COPPER(II) CHLORIDE see CNK500
COPPER(I) CHLORIDE see CNK250
COPPER CHLORIDE, BASIC see CNK559
COPPER CHLORIDE, mixed with COPPER OXIDE, HYDRATE see CNK559
COPPER CHLORIDE OXIDE see CNK559
COPPER CHLORIDE OXIDE, HYDRATE (9CI) see CNK559
COPPER(I) CHLOROACETYLIDE see CNK599
COPPER CHLOROXIDE see CNK559
COPPER CHROMATE see CNK609
COPPER(I) CITRATE see CNK625
COPPER CITRATE see CNK625
COPPER-COBALT-BERYLLIUM see CNK700
COPPER COMPOUNDS see CNK750
COPPER CYANAMIDE see CNL250
COPPER CYANIDE (DOT) see CNL250
COPPER(II) CYANIDE see CNL250
COPPER(I) CYANIDE see CNL000
COPPER CYANIDE see CNL000
COPPER(2+) DIACETATE see CNI250
COPPER DIACETATE see CNI250

COPPER DIACETATE MONOHYDRATE see CNI325
COPPER DIACETYLACETONATE see BGR000
COPPER DIHYDROXIDE see CNM500
COPPER DIMETHYLDITHIOCARBAMATE see CNL500
COPPER DINITRATE see CNM750
COPPER DINITRATE TRIHYDRATE see CNN000
COPPER(II)-1,3-DI(5-TETRAZOLYL)TRIAZENIDE see CNL625
COPPER EDTA COMPLEX see CNL750
COPPER-ETHYLENEDIAMINE COMPLEX see DBU800
COPPERFINE-ZINC see CNP500
COPPER FUME see CNM000
COPPER, (1,3,8,16,18,24-HEXABROMO-2,4,9,10,11,15,17,22,23,25-DECACHLO-
    ROPHTHALOCYAN INATO(2-))- see CMS140
COPPER(I) HYDRIDE see CNM250
COPPER(2+) HYDROXIDE see CNM500
COPPER HYDROXIDE see CNM500
COPPER HYDROXIDE SULFATE see CNM600
COPPER-8-HYDROXYQUINOLATE see BLC250
COPPER HYDROXYQUINOLATE see BLC250
COPPER-8-HYDROXYQUINOLINATE see BLC250
COPPER-8-HYDROXYQUINOLINE see BLC250
COPPER LONACOL see ZJS300
COPPER-MILLED see CNI000
COPPER MONOCARBONATE see CBW200
COPPER MONOCHLORIDE see CNK250
COPPER MONOOXIDE see CNO250
COPPER MONOSULFATE see CNP250
COPPER MONOSULFIDE see CNQ000
COPPER MONOXIDE see CNO250
COPPER NAPHTHENATE see NAS000
COPPER(2+) NITRATE see CNM750
COPPER(II) NITRATE see CNM750
COPPER(II) NITRATE, TRIHYDRATE (1:2:3) see CNN000
COPPER(I) NITRIDE see CNN250
COPPER NORDOX see CNO000
COPPER OC FUNGICIDE see CNK559
COPPER 1,3,5-OCTATRIEN-7-YNIDE see CNN399
COPPER ORTHOARSENITE see CNN500
COPPER(II) OXALATE see CNN755
COPPER(I) OXALATE see CNN750
COPPER OXALATE see CNN755
COPPER(2+) OXIDE see CNO250
COPPER(II) OXIDE see CNO250
COPPER(I) OXIDE see CNO000
COPPER (2+) OXINATE see BLC250
COPPER OXINATE see BLC250
COPPER OXINE see BLC250
COPPER OXYCHLORIDE see CNK559
COPPER OXYCHLORIDE-ZINEB mixture see ZJS300
COPPER OXYQUINOLATE see BLC250
COPPER OXYQUINOLINE see BLC250
COPPER OXYSULFATE see CNM600
COPPER(II) PERCHLORATE see CNO350
COPPER(I) PERCHLORATE see CNO325
COPPER(II) PERCHLORATE, DIHYDRATE see CNO500
COPPER(II) PHOSPHINATE see CNO750
COPPER(I) POTASSIUM CYANIDE see PLC175
COPPER-8-QUINOLATE see BLC250
COPPER QUINOLATE see BLC250
COPPER-8-QUINOLINOL see BLC250
COPPER-8-QUINOLINOLATE see BLC250
COPPER QUINOLINOLATE see BLC250
COPPER SALT 2-HYDROXY-1,2,3-PROPANETRICARBOXYLIC ACID (1:2) see
    CNK625
COPPERSAN see CNK559
COPPER SARDEX see CNO000
COPPER SLAG-AIRBORNE see CNI000
COPPER SLAG-MILLED see CNI000
COPPER SODIUM CYANIDE see SFZ100
COPPER SORBATE see CNP000
COPPER complex with trans-N-(p-STYRYLPHENYL)ACETOHYDROXAMIC ACID
    see SMU000
COPPER(II) SULFATE (1:1) see CNP250
COPPER SULFATE see CNP250
COPPER(II) SULFATE PENTAHYDRATE (1:1:5) see CNP500
COPPER(2+) SULFIDE see CNQ000
COPPER (II) SULFIDE see CNQ000
COPPER SULFIDE see CNQ000
COPPER(I) TETRAHYDROALUMINATE see CNQ250
COPPER-2,4,5-TRICHLOROPHENOLATE see CNQ375
COPPER TRICHLOROPHENOLATE see CNQ375
COPPER UVERSOL see NAS000
COPPER-ZINC ALLOYS see CNQ500
COPPER-ZINC CHROMATE COMPLEX see CNQ750
COPPESAN see CNK559

COPPESAN BLUE see CNK559
COPRA (DOT) see CNR000
COPRAMAT see ZJS300
COPRANTOL see CNK559
COPRA (OIL) see CNR000
COPRA PELLETS (DOT) see CNR000
COPREN see GFW000
COPREX see CNK559
COPROL see DJL000
COPROSAN BLUE see CNK559
COPROSTAN-3-β-OL see DKW000
COPROSTANOL see DKW000
COPROSTEROL see DKW000
COPSAMINE see WAK000
COPTICIDE see SNM500
COP-TOX see CNK559
COQUE MOLLE (HAITI) see JBS100
COQUERET (CANADA) see JBS100
COQUES DU LEVANT (FRENCH) see PIE500
CORACON see DJS200
CORADON see WAK000
CORAETHAMIDE see DJS200
CORAETHAMIDUM see DJS200
CORAFIL see CNR125
CORAL BEAD PLANT see RMK250
CORAL BEAN see NBR800
CORAL BERRY see ROA300
CORALEPT see DJS200
CORALITOS (CUBA) see ROA300
CORAL PLANT see CNR135
CORAL SNAKE VENOM see CNR150
CORAL VEGETAL (CUBA) see CNR135
CORALYNE SULFOACETATE see CNR250
CORAMINE see DJS200
CORANIL BROWN H EPS see SGP500
CORANIL DIRECT BLACK B see CMN240
CORATOL see POB500
CORAVITA see DJS200
CORAX see CBT750, MDQ250
CO-RAX see WAT200
CORAX P see CBT750
CORAZON de CABRITO (CUBA) see CAL125
CORAZONE see DJS200
CORCAT see PJX000
CORCHORGENIN see SMM500
CORCHORIN see SMM500
CORCHOSIDE A AGLYCON see SMM500
CORCHSULARIN see SMM500
CORDABROMIN see HNY500
CORDALEROMIN see HNY500
CORDALIN see HLC000
CORDIAMID see DJS200
CORDIAMIN see DJS200
CORDIAMINE see DJS200
CORDILAN see LAU400
CORDILOX see VHA450
CORDIPIN see AEC750
CORDITON see DJS200
CORDOVAL see GEW000
CORDULAN see CMD750
CORDYNIL see DJS200
COREDIOL see DJS200
COREINE see CCL250
CORESPIN see DJS200
CORETAL see CNR500, THK750
CORETHAMIDE see DJS200
CORETONE see DJS200
CORETONIN see GCE600
CORFLEX 880 see ILR100
CORGARD see CNR675
CORGLYCON see CNH780
CORGLYCONE see CNH780
CORGLYKON see CNH780
CORIACID SCARLET R see CMM325
CORIAL EM FINISH F see CCU250
CORIAMYRTIN see CNR725
CORIAMYRTINE see CNR725
CORIAMYRTIONE see CNR725
CORIANDER OIL see CNR735
CORIARI MYRTIFOLIA see CNR740
CORICIDIN see ABG750
CORIFORTE see ABG750
CORIL see ELH600
CORINE see CNE750
CORINTH FLOUR see AIB250

CORISOL see VGP000
CORIZIUM see NGG500
CORLAN see HHR000
CORLIN see CNS800, ELH600
CORLUTIN see PMH500
CORLUTIN L.A. see HNT500
CORLUVITE see PMH500
CORMALONE see SOV100
CORMED see DJS200
CORMELIAN see CNR750
CORMELIAN-DIGOTAB see CNR825
CORMID see DJS200
CORMOGRIZIN see GJU800
CORMOTYL see DJS200
CORNE CABRITE (HAITI) see YAK300
CORN LILY see FAB100
CORNOCENTIN see EDB500, LJL000
CORN OIL see CNS000
CORNOTONE see DJS200
CORNOX CWK see BAV000
CORNOX-M see CIR250
CORNOX RD see DGB000
CORNOX RK see DGB000
CORN PRODUCTS see SLJ500
CORN SUGAR see GFG000
CORNUCOPIA see AOO825
CORODANE see OPC000
CORODIL see HNY500
CORODILAN see DHS200
CORODINOC see DUU600
CORONA COROZATE see BJK500
CORONAL see DNC000
CORONAL-CRINOS see EID200
CORONARIDINE HYDROCHLORIDE see CNS200
(−)-CORONARIDINE MONOHYDROCHLORIDE see CNS200
CORONARIN see DNC000
CORONARINE see PCP250
CORONATE EH see HEG300
CORONATE MR 200 see PKB100
CORONIN see AHK750
COROPHOS see MRH209
COROPHYLLIN-N see HLC000
COROSANIN see ELH600
COROSORBIDE see CCK125
COROSUL D AND S see SOD500
COROTHION see PAK000
COROTONIN see DJS200
COROTRAN see CJT750
COROVIT see DJS200
COROVLISS see CCK125
COROXON see CIK750
COROZATE see BJK500
CORPAX see PEV750
CORPHYLLIN see DNC000
CORPORIN see PMH500
CORPS PRALINE see MAO350
CORPS R. 261 see DBA200
CORPS 2339 R P (FRENCH) see PEN000
CORPUS LUTEUM HORMONE see PMH500
CORRIGAST see HKR500
CORRIGEN see OHQ000
CORRONAROBETIN see TEH500
CORROSIVE MERCURY CHLORIDE see MCY475
CORROSIVE SUBLIMATE see MCY475
CORRY'S SLUG DEATH see TDW500
CORSODYL see CDT250
CORSONE see SOW000
CORSTILINE see AES650
CORTACET see DAQ800
CORTACREAM see HHQ800
CORTADREN see CNS800, CNS825
CORTAID see HHQ800
CORTAN see PLZ000
CORTANCYL see PLZ000
CORTATE see DAQ800
CORT-DOME see CNS750
CORTEF ACETATE see HHQ800
CORTELAN see CNS825
Δ-CORTELAN see PLZ000
CORTELL see HHQ800
CORTENIL see DAQ800
CORTES see HHQ800
CORTESAN see DAQ800
CORTEXAL see IRA000
CORTEX ALDEHYDE see PDR000

CORTEXONE see DAQ600
CORTEXONE ACETATE see DAQ800
COR-THEOPHYLLINE see DNC000
CORTHION see PAK000
CORTHIONE see PAK000
CORTICOTROPHIN see AES650
CORTICOTROPIN see AES650
CORTICOTROPIN-LIKE SUBSTANCES see AES650
CORTIDELT see PLZ000
CORTIFAR see DAQ800
CORTIFOAM see HHQ800
CORTIGEN see DAQ800
CORTILAN-NEU see CDR750
CORTINAQ see DAQ800
CORTINAZINE see ILD000
CORTINELLUS SHIITAKE EXTRACT (JAPANESE) see JDJ100
CORTIPHYSON see AES650
CORTIPRED see SOV100
CORTIRON see DAQ800
CORTISAL see CNS800, CNS825
CORTISATE see CNS800, CNS825
CORTISOL see CNS750
Δ¹-CORTISOL see PMA000
CORTISOL ACETATE see HHQ800
CORTISOL ALCOHOL see CNS750
CORTISOL HEMISUCCINATE SODIUM SALT see HHR000
CORTISOL SODIUM HEMISUCCINATE see HHR000
CORTISOL-21-SODIUM SUCCINATE see HHR000
CORTISOL SODIUM SUCCINATE see HHR000
CORTISOL SUCCINATE, SODIUM SALT see HHR000
CORTISONE see CNS800
Δ¹-CORTISONE see PLZ000
Δ-CORTISONE see PLZ000
CORTISONE-21-ACETATE see CNS825
CORTISONE ACETATE see CNS825
CORTISONE MONOACETATE see CNS825
CORTISPRAY see CNS750
CORTISTAB see CNS825
CORTISTAL see CNS800
CORTISYL see CNS825
CORTIVIS see DAQ800
CORTIVITE see CNS800, CNS825
CORTIXYL see DAQ800
CORTOCIN-F see FDB000
CORTOGEN see CNS800, CNS825
CORTOGEN ACETATE see CNS825
CORTONE see CNS800, CNS825
Δ-CORTONE see PLZ000
CORTONE ACETATE see CNS825
CORTRIL ACETATE see HHQ800
CORTRIL ACETATE-AS see HHQ800
CORTROPHIN see AES650
CORTROPHYSON see AES650
CORTUSSIN see RLU000
CORUNDUM see EAL100
CORUNDUM FUME see CNT250
CORVASAL see MRN275
CORVASYMTON see SPD000
CORVATON see MRN275
CORVIC 55/9 see PKQ059
CORVIC 236581 see AAX175
CORVITAN see DJS200
CORVITIN see DBA800
CORVITOL see DJS200
CORVITONE see DJS200
CORWIN see XAH000
CORYBAN-D see ABG750
CORYDALOID see CNT325
CORYDININE see FOW000
CORYLON see HMB500
CORYLONE see HMB500
CORYLOPHYLINE see GFG100
CORYNINE see YBJ000
CORYSTIBIN see AQH500
CORYWAS see DJS200
CORYZOL see DPJ400
COSAN see SOD500
COSBIOL see SLG700
COSCOPIN see NBP275, NOA000
COSCOPIN HYDROCHLORIDE see NOA500
COSCOTABS see NOA000
COSCOTABS HYDROCHLORIDE see NOA500
COSDEN 550 see SMQ500
COSDEN 945E see SMQ500
COSLAN see XQS000

COSMEGEN see AEB000
COSMETIC BLUE LAKE see FAE000
COSMETIC BRILLIANT PINK BLUISH D CONC see FAG070
COSMETIC CORAL RED KO BLUISH see CHP500
COSMETIC GREEN BLUE R25396 see ADE500
COSMETIC WHITE see BKW100
COSMETIC WHITE C47-5175 see TGG760
COSMETOL see CCP250
COSMIC see TJJ500
COSMOPEN see BDY669, BFD000
COSMOPHLOXINE F see CMM765
COSPANON see TKP100
COSTUS OIL see CNT350
COSTUS ROOT see CNT350
COTALMON see AMK250
COTARNIN see CNT625
COTARNINE see CNT625
COTEL see VSZ000
COTINAZIN see ILD000
COTININ see FBW000
COTNION-ETHYL see EKN000
COTNION METHYL see ASH500
COTOFILM see HCL000
COTOFOR see EPN500
COTONE see PLZ000
COTORAN see DUK800
COTORAN MULTI see MQQ450
COTORAN MULTI 50WP see DUK800
CO-TRIFAMOLE see SNL850
CO-TRIMOXAZOLE see SNK000, TKX000
COTTON AIDE HC see HKC000
COTTON DUST see CNT750
COTTONEX see DUK800
COTTON GREEN B see CMO840
COTTON RED L see SGQ500
COTTONSEED OIL (unhydrogenated) see CNU000
COTTON VIOLET R see CMP000
COUER SAIGNANT (HAITI) see CAL125
COUMADIN see WAT200
COUMADIN SODIUM see WAT220
COUMAFENE see WAT200
COUMAFURYL see ABF500
COUMAMYCIN see CNV500
COUMAPHOS see CNU750
COUMAPHOS-O-ANALOG see CIK750
COUMAPHOS OXYGEN ANALOG (USDA) see CIK750
4-COUMARIC ACID see CNU825
p-COUMARIC ACID see CNU825
trans-o-COUMARIC ACID see CNU850
COUMARIN, 6,7-DIHYDROXY-4-METHYL- see MJV800
COUMARIN, 7-HYDROXY-3-(p-HYDROXYPHENYL)-4-PHENYL- see HNK600
COUMARILIC ACID see CNU875
COUMARIN, 7-METHOXY-(8CI) see MEK300
COUMARIN, 4-METHYL-7-HYDROXY-, p-GUANIDINOBENZOATE see GLC100
COUMARIN 1 see DIL400
COUMARIN 4 see MKP500
COUMARIN 311 see DPJ800
COUMARIN see CNV000
cis-o-COUMARINIC ACID LACTONE see CNV000
COUMARINIC ANHYDRIDE see CNV000
COUMARONE see BCK250
COUMATETRALYL see EAT600
COUMERMYCIN A1 see CNV500
COUNTER see BSO000
COUNTER 15G SOIL INSECTICIDE see BSO000
COUNTER 15G SOIL INSECTICIDE-NEMATICIDE see BSO000
COUNTRY WALNUT see TOA275
COURLENE-X3 see PJS750
COURLOSE A 590 see SFO500
COVALLATOXOL see CNH785
COVATIN see BRS000
COVATIX see BRS000
CO-VIDARABINE see PBT100
COVI-OX see VSZ450
COVIT see VSZ000
COVOL 971 see PKP750
COVOL see PKP750
COWBUSH (BAHAMAS) see LED500
COW GARLIC see WBS850
COWSLIP see MBU550
COXIGON see OJI750
COXISTAT see NGE500
COXYSAN see CNK559
COYDEN see CMX850
COYOTILLO see BOM125

COZYMASE see CNF390
COZYMASE I see CNF390
COZYMASE II see CNF400
COZYME see PAG200
3CP see CJQ300
4-CP see CJN000
CP 34 see TFM250
CP 105 see ECI000
CP 261 see DTS500
CP 3438 see TFD250
CP 4572 see CDO250
CP 6,343 see CFK000
CP 12574 see TGD250
CP 14,957 see OAN000
CP 15,336 see DBI200
CP 16171 see FAJ100
CP 19699 see CON300
CP 23426 see DNS600
CP 25017 see NHK800
CP 31393 see CHS500
CP 34089 see SOU650
CP 40294 see MOB699
CP-40507 see MOB750
CP 43858 see TIP750
CP 47114 see DSQ000
CP 48985 see CFC500
CP 49674 see DOP200
CP 50144 see CFX000
CP 53619 see CFW750
CP 53926 see DRR200
CP-12,252-1 see NBP500
CP-15-639-2 see CBO250
CP-16533-1 see IRV000
CP 10423-18 see TCW750
CP-15467-61 see LGZ000
CIP see CKV500
CP see CCJ400, CQC650
3-CPA see CJQ300
3-CPA see CJQ300
6-CPA see CKN375
CPA see CJN000, CQC650, CQJ500
C. PALUSTRIS see MBU550
CPAS see CDS500
CPAS mixed with BCPE see CKL500
CPB see BSS550, CJR500
CP BASIC SULFATE see CNP250
CPBS see CJR500
CPBU 7 see GHR609
CPC 3005 see SLJ500
CPC 6448 see SLJ500
CPCA see BIO750
CPCBS see CJT750
C.P. CHROME LIGHT 2010 see LCS000
C.P. CHROME ORANGE DARK 2030 see LCS000
C.P. CHROME ORANGE MEDIUM 2020 see LCS000
C.P. CHROME YELLOW LIGHT see LCR000
4-Cl-m-PD see CJY120
4-Cl-o-PD see CFK125
CPDC see PJD000
CPDD see PJD000
CPE 16 see PJS750
CPE 25 see PJS750
CPE see PJS750
CPH see CBL000, CDP250
CPIB see ARQ750
CPIRON see FBJ100
CP 1044 J3 see BPP750
CP 261LV see DTS500
CPMC see CKF000
CPNU-I see NKJ100
2-Cl-P-PD see CEG625
C15-PREDIOXIN see TIL750
C. PROCERA see COD675
CP 556S see SOU600
CPS see MIG850
CPSA see CDP725
CP 45899 SODIUM SALT see PAP600
CdPT see CAI350
CPT see CLK215, TAF675
C.P. TITANIUM see TGF250
CP 49952 p-TOLUENESULFONATE see SOU675
C.P.TOLUIDINE TONER A-2989 see MMP100
C.P.TOLUIDINE TONER A-2990 see MMP100
C.P.TOLUIDINE TONER DARK RS-3340 see MMP100
C.P.TOLUIDINE TONER DEEP X-1865 see MMP100
C.P.TOLUIDINE TONER LIGHT RS-3140 see MMP100

C.P.TOLUIDINE TONER RT-6101 see MMP100
C.P.TOLUIDINE TONER RT-6104 see MMP100
C. PULCHERRIMA see CAK325
CPX see TAF675
CPZ see CCS369, CKP250, CKP500
C.P. ZINC YELLOW X-883 see ZFJ100
CoQ₁₀ see UAH000
CQ see CLD250
C-QUENS see CBF250
CR 39 see AGD250
CR 200 see PKB100
CR 242 see BGC625
CR 409 see BJE750
CR-604 see POF550
CR-605 see TGF175
CR/662 see BLV000
CR 2024 see MCB050
CR 3029 see MAS500
CR see DDE200
CRAB-E-RAD see DXE600
CRAB'S EYES see AAD000, RMK250
CRADEX see BBK500
CRAG 341 see GII000
CRAG 974 see DSB200
CRAG DCU-73w see DGQ200
CRAG EXPERIMENTAL HERBICIDE 2 see DGQ200
CRAG FLY REPELLENT see BRP250
CRAG FRUIT FUNGICIDE 341 see GII000
CRAG FRUIT FUNGICIDE 34 see GIO000
CRAG FUNGICIDE 658 see CNQ750
CRAG FUNGICIDE 974 see DSB200
CRAG HERBICIDE 1 see CNW000
CRAG HERBICIDE 2 see DGQ200
CRAG HERBICIDE see CNW000
CRAG NEMACIDE see DSB200
CRAG SESONE see CNW000
CRAG SEVIN see CBM750
CRAG 85W see DSB200
CRAIN see FBS100
CRALO-E-RAD see DXE600
CRAMCILLIN-S see PDD350
CRAMPOL see ACX500
CRAMPOLE see ACX500
CRANOMYCIN see CNW100
CRANOMYCIN HYDROCHLORIDE see CNW105
CRASTIN S 330 see PKF750
CRATAEGUS, EXTRACT see EDK600
CRATECIL see EOK000
CRAVITEN see CNW125
CRAWHASPOL see TIO750
CRC 7001 see CEK875
CRD-401 see DNU600
CREAM of TARTAR see PKU600
CREATININE SULFATE compounded with 3-(2-AMINOETHYL)INDOLE-5-OL
  (1:1:1), MONOHYDRATE see AJX750
CREATININE SULFATE compounded with 3-(2-AMINOETHYL)INDOL-5-OL
  (1:1:1) see AJX750
C RED 2 see CMS242
CREDO see SHF500
CREIN see DAL300
CREMODIAZINE see PPP500
CREMOMERAZINE see ALF250
CREMOMETHAZINE see SNJ000
CREMOR TARTARI see PKU600
p-CRESOL see MEK250
CREOSOTE see CMY825
CREOSOTE, from COAL TAR see CMY825
CREOSOTE OIL see CMY825
CREOSOTE P1 see CMY825
CREOSOTUM see CMY825
CRESAN UNIVERSAL TROCKENBEIZE see MEP000
m-CRESIDINE see MGO500
p-CRESIDINE see MGO750
CRESIDINE see MGO750
CRESOATE, WOOD see BAT850
CRESODIOL see GGS000
2-CRESOL see CNX000
3-CRESOL see CNW750
4-CRESOL see CNX250
m-CRESOL see CNW750
o-CRESOL see CNX000
p-CRESOL see CNX250
CRESOL see CNW500
p-CRESOL ACETATE see MNR250
p-CRESOL, 2,6-BIS(α-METHYLBENZYL)- see MHR050

p-CRESOL, 2,6-DI-tert-BUTYL-α-(DIMETHYLAMINO)- see DEA100, FAB000
o-CRESOL, 4,6-DI-tert-BUTYL-α-PHENYL- see DEG150
o-CRESOL, DINITRO-, SODIUM SALT see SGP550
CRESOL DIPHENYL PHOSPHATE see TGY750
CRESOL FAST VIOLET see AJN250
CRESOL FLYCIDYL ETHER see TGZ100
o-CRESOL GLYCERYL ETHER see GGS000
CRESOLI (ITALIAN) see CNW500
p-CRESOL METHYL ETHER see MGP000
o-CRESOL, 4-NITRO- see NFV010
m-CRESOL, 4-NITRO-, DIMETHYL PHOSPHATE see PHD750
p-CRESOL, 2-NITRO-α-α-α-TRIFLUORO- see NMQ100
p-CRESOL, α-PHENYL-, CARBAMATE see PFR325
o-CRESOLPHTHALEIN see CNX400
CRESON see RLU000
CRESOPUR see BAV000
CRESORCINOL DIISOCYANATE see TGM750
CRESOSSIDIOLO see GGS000
CRESOSSIPROPANDIOLO see GGS000
2,3-CRESOTIC ACID see CNX625
o-CRESOTIC ACID see CNX625
CRESOTIC ACID see CNX625
CRESOTINE BLUE 2B see CMO000
CRESOTINE BLUE 3B see CMO250
CRESOTINE DARK GREEN B see CMO830
CRESOTINE FAST RED F see CMO870
CRESOTINE GREEN B see CMO840
CRESOTINE PURE BLUE see CMO500
2,3-CRESOTINIC ACID see CNX625
β-CRESOTINIC ACID see CNX625
o-CRESOTINIC ACID see CNX625
CRESOTINIC ACID see CNX625
CRESOTOL see DUU600
CRESOXYDIOL see GGS000
CRESOXYPROPANEDIOL see GGS000
CRESSA CRETICA Linn., extract see CNX700
CRESTABOLIC see AOO475
CRESTANIL see MQU750
CRESTOMYCIN see NCF500
CRESTOXO see CDV100
p-CRESYL ACETATE (FCC) see MNR250
p-CRESYL BENZOATE see TGX100
p-CRESYL CAPRYLATE see THB000
CRESYL DIPHENYL PHOSPHATE see TGY750
CRESYL FAST VIOLET see AJN250
o-CRESYL-α-GLYCERYL ETHER see GGS000
CRESYLGLYCIDE ETHER see TGZ100
CRESYL GLYCIDYL ETHER see TGZ100
m-CRESYLIC ACID see CNW750
o-CRESYLIC ACID see CNX000
p-CRESYLIC ACID see CNX250
CRESYLIC ACID see CNW500
CRESYLIC CREOSOTE see CMY825
p-CRESYL ISOBUTYRATE see THA250
CRESYLITE see TML500
m-CRESYL METHYLCARBAMATE see MIB750
p-CRESYL METHYL ETHER see MGP000
p-CRESYL OCTANOATE see THB000
o-CRESYL PHOSPHATE see TMO600
CRESYL PHOSPHATE see TNP500
p-CRESYL SALICYLATE see THD850
CRINovrL see EDV000
CRILL 2 see MRJ800
CRILL 3 see SKV150
CRILL 10 see PKL100
CRILL 16 see SKV170
CRILL 26 see SLL000
CRILL K 3 see SKV150
CRILL K 16 see SKV170
CRILLON L.D.E. see BKE500
CRIMIDIN (GERMAN) see CCP500
CRIMIDINA (ITALIAN) see CCP500
CRIMIDINE see CCP500
CRIMSON ANTIMONY see AQL500
CRIMSON EMBL see HJF500
CRIMSON SX see FMU080
CRINODORA see CFU750
CRINOTHENE see PKB500
CRINUM (VARIOUS SPECIES) see SLB250
CRINURYL see DFP600
CRISALBINE see GJG000
CRISALIN see DUV600
CRISAPON see DGI400
CRISATRINA see ARQ725
CRISAZINE see ARQ725

CRISEOCICLINE see TBX000
CRISEOCIL see MCH525
CRISODIN see MRH209
CRISODRIN see MRH209
CRISONAR see CFY750
CRISPATINE see FOT000
CRISPIN see THJ750
CRISPIN RED GM see CMM325
CRISQUAT see PAJ000
CRISTALLOSE see SJN700
CRISTALLOVAR see EDV000
CRISTALOMICINA see KAM000, KAV000
CRISTAPEN see BFD000
CRISTAPURAT see DKL800
CRISTERONE T see TBF500
CRISTOBALITE see SCI500, SCJ000
CRISTOXO 90 see CDV100
CRISULFAN see EAQ750
CRISURON see DXQ500
CRITTOX see EIR000
CROCEOMYCIN see HAL000
CROCIDOLITE (DOT) see ARM275
CROCIDOLITE ASBESTOS see ARM275
CROCOITE see LCR000
CROCUS see CNX800
CROCUS MARTIS ADSTRINGENS see IHD000
CRODACID see MSA250
CRODACOL A.10 see OBA000
CRODACOL-CAS see HCP000
CRODACOL-O see OBA000
CRODACOL-S see OAX000
CRODAMINE 1.18D see OBC000
CRODAMOL IPM see IQN000
CRODAMOL IPP see IQW000
CRODET O 6 see PJY100
CROFLEX see CBT750
CROLAC see CBT750
CROLEAN see ADR000
CROMARIL see PER700
CROMILE, CLORURO di (ITALIAN) see CML125
CROMOGLYCATE see CNX825
CROMOGLYCATE DISODIUM see CNX825
CROMOLYN SODIUM see CNX825
CROMOLYN SODIUM SALT see CNX825
CROMO, OSSICLORURO di (ITALIAN) see CML125
CROMOPHTAL BLUE A 3R see IBV050
CROMOSAN see MAD000
CRONETAL see DXH250
CRONETON see EPR000
CRONIL see EHP000
CROP RIDER see DAA800
CROTALARIA BERTEROANA see RBZ400
CROTALARIA INCANA see RBZ400
CROTALARIA JUNCEA see RBZ400
CROTALARIA RETUSA see RBZ400
CROTALARIA SPECTABILIS see RBZ400
CROTALINE see MRH000
CROTALUS ADAMANTEUS VENOM see EAB225
CROTALUS ATROX VENOM see WBJ600
CROTALUS CERASTES VENOM see CNX830
CROTALUS DURISSUS TERRIFICUS VENOM see CNY000
CROTALUS HORRIDUS HORRIDUS VENOM see TGB150
CROTALUS HORRIDUS VENOM see CNY300
CROTALUS RUBER RUBER VENOM see CNY325
CROTALUS SCUTULATUS SCUTULATUS VENOM see CNY350
CROTALUS SCUTULATUS VENOM see CNY375
CROTALUS VIRIDIS CERERUS VENOM see CNY390
CROTALUS VIRIDIS CONCOLOR VENOM see CNY750
CROTALUS VIRIDIS HELLERI VENOM see COA000
CROTALUS VIRIDIS VIRIDIS TOXIN see VRP200
CROTALUS VIRIDIS VIRIDIS VENOM see PLX100
CROTILIN see DAA800
CROTOCIN see COB000
CROTONAL see COB260
CROTONALDEHYDE, stabilized (DOT) see COB250
CROTONALDEHYDE see COB250
(E)-CROTONALDEHYDE see COB260
CROTONALDEHYDE see COB260
CROTONAMIDE, 2-CHLORO-N,N-DIETHYL-3-HYDROXY-, DIMETHYL PHOSPHATE see FAB400
CROTONATE de 2,4-DINITRO 6-(1-METHYL-HEPTYL)-PHENYLE (FRENCH) see AQT500
CROTONATE d'ETHYLE (FRENCH) see COB750
α-CROTONIC ACID see COB500
CROTONIC ACID see COB500

CROTONIC ACID, solid see COB500
CROTONIC ACID ANHYDRIDE see COB900
CROTONIC ACID, 4-BROMO-, METHYL ESTER see MHR790
CROTONIC ACID, 2-BUTENYLIDENE ESTER see COD100
α-CROTONIC ACID ETHYL ESTER see COB750
CROTONIC ACID, ETHYL ESTER see EHO200
CROTONIC ACID GERNAIOL ESTER see DTE000
CROTONIC ACID, 2-METHYL-, (E)- see TGA700
(E)-CROTONIC ACID METHYL ESTER see COB825
CROTONIC ACID, 2-METHYL-, ETHYL ESTER, (E)-(8CI) see TGA800
CROTONIC ACID, 2-METHYL-, METHYL ESTER, (E)-(8CI) see MPW700
CROTONIC ACID, VINYL ESTER see VNU000
CROTONIC ALDEHYDE see COB250, COB260
CROTONIC ANHYDRIDE see COB900
CROTONIC NITRILE see COC300
CROTONIQUE NITRILE see COC300
CROTONITRILE see COC300
CROTONOEL (GERMAN) see COC250
CROTON OIL see COC250
CROTONONITRILE see COC300
CROTON RESIN see COC250
CROTON TIGLIUM L. OIL see COC250
CROTONYL ALCOHOL see BOY000
CROTONYLENE see COC500
α-(N-CROTONYL-N-ETHYL)AMINO-BUTYRIC ACID mixed with BUTRIC ACID, α-(N-CROTONYL-N-PROPYL)AMINO see COC750
CROTONYL-N-ETHYL-o-TOLUIDINE see EHO500
CROTOXIN see COC875
CROTOXYPHOS see COD000
CROTURAL see EHP000
CROTYL ALCOHOL see BOY000
CROTYL CHLORIDE see CEU825
CROTYLIDENE ACETIC ACID see SKU000
CROTYLIDENE DICROTONATE see COD100
1-CROTYL THEOBROMINE see BSM825
CROVARIL see HNI500
CROW BERRY see PJJ315
CROWFOOD see FBS100
CROWN 18 see COD575
12-CROWN-4 see COD475
15-CROWN-5 see PBO000
18-CROWN-6 see COD500
CROWN BEAUTY see BAR325
CROWN FLOWER see COD675
CROYSULFONE see SOA500
CRP-401 see DNU600
CRTRON see VSZ100
CRUDE ARSENIC see ARI750
CRUDE COAL TAR see CMY800
CRUDE ERGOT see EDB500
CRUDE OIL see PCR250
CRUDE PETROLEUM see PCR250
CRUDE SAIKOSIDE see SAF300
CRUDE SHALE OILS see COD750
CRUFOMATE see COD850
CRUFOMATE A see COD850
CRUFORMATE see COD850
CRUMERON, combustion products see ADX750
CRUNCH see CBM750
CRUSTECDYSON see HKG500
CRYOFLEX see BHK750
CRYOFLUORAN see FOO509
CRYOFLUORANE see FOO509
CRYOGENINE see CBL000
CRYOLITE see SHF000
CRYOPOLYTHENE see PJS750
CRYPTOCILLIN see MNV250
CRYPTOCRYSTALLINE QUARTZ see SCK600
CRYPTOCYANINE IODIDE see KHU050
CRYPTOCYANINE O. A. 2 see KHU050
CRYPTOGIL OL see PAX250
CRYPTOGYL NA (ITALIAN) see PAX750
CRYPTOHALITE see COE000
CRYPTOSTEGIA GRANDIFLORA see ROU450
CRYPTOSTEGIA MADAGASCARIENSIS see ROU450
CRYSALBA see CAX500
CRYSTALETS see VSK900
CRYSTALLINA see VSZ100
CRYSTALLINE DEHYDROXY SODIUM ALUMINUM, CARBONATE see DAC450
CRYSTALLINE DIGITALIN see DKL800
CRYSTALLINE LIENOMYCIN see LFP000
CRYSTALLINIC ACID see SMB000
CRYSTALLIZED VERDIGRIS see CNI250
CRYSTALLOMYCIN see COE100
CRYSTALLOSE see SJN700

CRYSTAL O see CCP250
CRYSTAL PROPANIL-4 see DGI000
CRYSTALS of VENUS see CNI250
CRYSTAL VIOLET see AOR500
CRYSTAL VIOLET AO see AOR500
CRYSTAL VIOLET AON see AOR500
CRYSTAL VIOLET 10B see AOR500
CRYSTAL VIOLET 6B see AOR500
CRYSTAL VIOLET BASE see AOR500
CRYSTAL VIOLET 5BO see AOR500
CRYSTAL VIOLET 6BO see AOR500
CRYSTAL VIOLET BP see AOR500
CRYSTAL VIOLET BPC see AOR500
CRYSTAL VIOLET CHLORIDE see AOR500
CRYSTAL VIOLET EXTRA PURE see AOR500
CRYSTAL VIOLET EXTRA PURE APN see AOR500
CRYSTAL VIOLET EXTRA PURE APNX see AOR500
CRYSTAL VIOLET FN see AOR500
CRYSTAL VIOLET HL2 see AOR500
CRYSTAL VIOLET O see AOR500
CRYSTAL VIOLET PURE DSC see AOR500
CRYSTAL VIOLET PURE DSC BRILLIANT see AOR500
CRYSTAL VIOLET SS see AOR500
CRYSTAL VIOLET TECHNICAL see AOR500
CRYSTAL VIOLET USP see AOR500
CRYSTAMET see SJU000
CRYSTAMIN see VSZ000
CRYSTAPEN see BFD000, BFD250
CRYSTAR see ADA725, SCQ000
CRYSTEX see SOD500
CRYSTHION 2L see ASH500
CRYSTHYON see ASH500
CRYSTIC PREFIL S see CAT775
CRYSTODIGIN see DKL800
CRYSTOGEN see EDV000
CRYSTOIDS see HFV500
CRYSTOL CARBONATE see SFO000
CRYSTOLON 37 see SCQ000
CRYSTOLON 39 see SCQ000
CRYSTOSERPINE see RDK000
CRYSTOSOL see MQV750
CRYSTWEL see VSZ000
CRY-O-VAC L see PJS750
CS-61 see DON700
CS 359 see BOM510
CS 370 see CMY525
CS 386 see MQR760
CS-430 see HAG800
CS-439 see ALF500
CS-600 see LII300
CS 708 see BON250
CS-847 see CEW500
CS 1170 see CCS350
60-CS-16 see CMF400
CS 4030 see QVA000
50-CS-46 see EME050
CS 12602 see TCJ075
C-Sn-9 see BLL750
CS see CEQ600
CS 645A see BIN500
CSAC see EES400
CSC see SEC000
CSF-GIFTWEIZEN see TEM000
CSI PASTE see DTG700
CS LAFARGE see CAW850
C-3 SODIUM SALT see CBF625
CSP see CNP500
ChS-RR2 see BOS100
CS 1170 SODIUM see CCS360
C. SUFFUSUS SUFFUSUS VENOM see CCW925
CT-1341 see AFK500
CT 3318 see COE125
4365 CT see MLI750
CT 4436 see MFO250
CT see CCX250, CLH750
CTA see CLO750
CTAB see HCQ500
CTC see CMA750
CTCP see CNQ375
CTFE see CLQ750
CTH see CLK230
C-TOXIFERINE 1 see THI000
CTR 6669 see MHC750
CTT see CCS373
CTX see CCR950, CQC650

CU-56 see CNK559
CUBAN LILLY see SLH200
CUBE see RNZ000
CUBEB OIL see COE175
CUBE EXTRACT see RNZ000
CUBE-PULVER see RNZ000
CUBE ROOT see RNZ000
CUBES see DJO000
CUBIC NITER see SIO900
CUBISOL see DNM400
CUBOR see RNZ000
CUCKOOPINT see ITD050
CUCUMBER ALCOHOL see NMV780
CUCUMBER ALDEHYDE see NMV760
CUCUMBER DUST see ARB750
CUCURBITACIN E see COE250
CUCURBITACINE (D) see EAH100
CUCURBITACINE-E see COE250
CUEMID see CME400
CUENTAS de ORO (PUERTO RICO) see GIW200
CUIPU (MEXICO) see CNR135
CULLEN EARTH see MAT500
CULMINAL K 42 see MIF760
CULPEN see CHJ000
CULVERAM CDG see LBU200
CUM see COE750
CUMA see BJZ000
CUMAFOS (DUTCH) see CNU750
CUMAFURYL (GERMAN) see ABF500
CUMALDEHYDE see COE500
CUMAN see BJK500
CUMAN L see BJK500
p-CUMARIC ACID see CNU825
CUMATE see CNL500
CUMATETRALYL (GERMAN, DUTCH) see EAT600
CUMEEN (DUTCH) see COE750
CUMEENHYDROPEROXYDE (DUTCH) see IOB000
psi-CUMENE see TLL750
CUMENE see COE750
CUMENE ALDEHYDE see COF000
CUMENE HYDROPEROXIDE (DOT) see IOB000
CUMENE HYDROPEROXIDE, TECHNICALLY PURE (DOT) see IOB000
CUMENE PEROXIDE see DGR600
m-CUMENOL see IQX090
p-CUMENOL see IQZ000
m-CUMENOL METHYLCARBAMATE see COF250
CUMENT HYDROPEROXIDE see IOB000
p-(p-CUMENYLAZO)-N,N-DIMETHYLANILINE see IOT875
CUMENYL HYDROPEROXIDE see IOB000
m-CUMENYL METHYLCARBAMATE see COF250
o-CUMENYL METHYLCARBAMATE see MIA250
CUMERTILIN SODIUM see MCS000
CUMIC ALCOHOL see CQI250
p-CUMIC ALDEHYDE see COE500
CUMID see BJZ000
o-CUMIDINE see INW100
psi-CUMIDINE see TLG250
psi-CUMIDINE HYDROCHLORIDE see TLG750
CUMINALDEHYDE see COE500
CUMINIC ACETALDEHYDE see IRA000
CUMINIC ALCOHOL see CQI250
CUMINIC ALDEHYDE (FCC) see COE500
CUMIN OIL see COF325
CUMINOL see CQI250
CUMINYL ACETATE see IOF050
CUMINYL ALCOHOL see CQI250
CUMINYL ALDEHYDE see COE500
CUMINYL NITRILE see IOD050
CUMMIN see COF325
CUMOLHYDROPEROXID (GERMAN) see IOB000
psi-CUMOQUINONE see POG400
CUMOQUINONE see POG400
CUMYL ACETALDEHYDE see IRA000
α-CUMYL ALCOHOL see DTN100
CUMYL ALCOHOL see CQI250
α-CUMYL HYDROPEROXIDE see IOB000
CUMYL HYDROPEROXIDE see IOB000
CUMYL HYDROPEROXIDE, TECHNICAL PURE (DOT) see IOB000
CUMYL PEROXIDE see DGR600
p-(α-CUMYL)PHENOL see COF400
p-CUMYLPHENOL see COF400
CUNAPSOL see NAS000
CUNDEAMOR (CUBA, PUERTO RICO) see FPD100
CUNILATE 2472 see BLC250
CUNILATE see BLC250

CUPEY see BAE325
CUPFERRON see ANO500
CUP-OF-GOLD see CDH125
CUPPER OXIDE (RUSSIAN) see CNO000
CUPRAL 45 see CNK559
CUPRAL see SGJ000
CUPRAMAR see CNK559
CUPRAMER see CNK559
CUPRANIL BROWN BCW see CMO820
CUPRANIL BROWN BCWR see CMO820
CUPRANTOL see CNK559
CUPRATE(1-), DICYANO-, POTASSIUM see PLC175
CUPRATE(4-), (mu-((3,3'-((3,3'-DIHYDROXY(1,1'-BIPHENYL)-4,4'-
  DIYL)BIS(AZO))BIS(5-AMINO-4-HYDROXY-2,7-NAPHTHALENEDISULFONA-
  TO))(8-)))DI-, TETRASODIUM see CMO650
CUPRATE(2-), TRIS(CYANO-C)-, DISODIUM see SFZ100
CUPRAVET see CNK559
CUPRAVIT see CNK559
CUPRAVIT BLAU see CNM500
CUPRAVIT BLUE see CNM500
CUPRAVIT FORTE see CNK559
CUPRAVIT GREEN see CNK559
CUPRENIL see MCR750
CUPRIC ACETATE see CNI250
CUPRIC ACETATE MONOHYDRATE see CNI325
CUPRIC ACETOARSENITE see COF500
CUPRIC ACETYLACETONATE see BGR000
CUPRIC ARSENITE see CNN500
CUPRIC CARBONATE (1:1) see CBW200
CUPRIC CARBONATE see CBW200, CNJ750
CUPRIC CHELATE of 2-N-HYDROXYFLUORENYL ACETAMIDE see HIP500
CUPRIC CHLORIDE see CNK500
CUPRIC CITRATE see CNK625
CUPRIC CYANIDE (DOT) see CNL250
CUPRIC DIACETATE see CNI250
CUPRIC DICHLORIDE see CNK500
CUPRIC DINITRATE see CNM750
CUPRIC DIPERCHLORATE TETRAHYDRATE see CNO500
CUPRICELLULOSE see CCU150
CUPRIC GREEN see CNN500
CUPRIC HYDROXIDE see CNM500
CUPRIC-8-HYDROXYQUINOLATE see BLC250
CUPRICIN see CNL000
CUPRIC NITRATE (DOT) see CNM750
CUPRIC NITRATE TRIHYDRATE see CNN000
CUPRICOL see CNK559
CUPRIC OXALATE see CNN750, CNN755
CUPRIC OXIDE see CNO250
CUPRIC OXIDE CHLORIDE see CNK559
CUPRIC-8-QUINOLINOLATE see BLC250
CUPRIC SULFATE see CNP250
CUPRIC SULFATE PENTAHYDRATE see CNP500
CUPRIC SULFIDE see CNQ000
CUPRIETHYLENEDIAMINE, solution (DOT) see DBU800
CUPRIETHYLENE DIAMINE see DBU800
CUPRIMINE see MCR750
CUPRINOL see NAS000
CUPRITOX see CNK559
CUPROCIN see ZJS300
CUPROCITROL see CNK625
CUPROKYLT see CNK559
CUPROL see CNK559
CUPRON (CZECH) see BCP500
CUPRONE see BCP500
CUPROSAN see CNK559
CUPROSANA see CNK559
CUPROSAN BLUE see CNK559
CUPROUS ARSENATE, BASIC see CNI900
CUPROUS CHLORIDE see CNK250
CUPROUS CYANIDE see CNL000
CUPROUS DICHLORIDE see CNK250
CUPROUS OXIDE see CNO000
CUPROUS POTASSIUM CYANIDE see PLC175
CUPROVINOL see CNK559
CUPROX see CNK559
CUPROXAT see CNM600
CUPROXAT FLOWABLE see CNM600
CUPROXOL see CNK559
CUPROZAN see ZJS300
CURACIT see BJI000
CURACRON see BNA750
CURALIN M see MJM200
CURAMAGUEY (CUBA) see YAK300
CURANTYL see PCP250
CURARE see COF750

CURARIL see GGS000
CURARIN see COF825
(+)-CURARINE see COF825
CURARINE see COF825
CURARIN-HAF see TOA000
CURARYTHAN see GGS000
CURATERR see CBS275
CURATIN see AEG750
CURAX see CPI250
CURBISET see CDT000
CURCUMA LONGA Linn., rhizome extract see COF850
CURCUMA OIL see COG000
CURCUMIN see COG000
CURCUMINE see COG000
CURENE 442 see MJM200
CURENE see PKL500
C. URENS see FBW100
CURE-RITE 18 see OPQ100
CURESAN see MEP250
CURETAN see MEP250
CURETARD A see DWI000
CUREX FLEA DUSTER see RNZ000
CURITAN see DXX400
CURITHANE 103 see MGA500
CURITHANE see MJQ000
CURITHANE C126 see DEQ600
CURL FLOWER see CMV390
CURLY HEADS see CMV390
CUROL BRIGHT RED 4R see FMU080
CUROL ORANGE see CMM220
CUROL PURE BLUE B see CMM090
CURON FAST YELLOW 5G see FAG140
CUROSAJIN see HGI100
CURRAL see AFS500
CURTACAIN see TBN000
CURZATE see COM300
CURZATE M see MAS500
CUSCOHYGRINE BIS(METHYL BENZENESULFONATE) see COG500
CUSCUTA REFLEXA Roxb., extract excluding roots see AHI630
CUSITER see TIL500
CUSTOS see MHC750
CUTAMIN BLACK CG see CMN240
CUTAMIN BLUE CR see CMO600
CUTAMIN BRILLIANT RED CG see CMM320
CUTAMIN DARK BLUE CB see CMN800
CUTAMIN RED CF see CMO870
CUTICURA ACNE CREAM see BDS000
CUTISAN see TIL500
CUTISTEROL see FDB000
CUT LEAF PHILODENDRON see SLE890
CUTTING OILS see COH000
CUVALIT see LJE500
CUZ 3 see CBT500
CV 399 see MGH800
CV 1006 see CDF380
CV 3317 see DAM315
C. VESICARIA see CAK325
C. VIRIDIS VIRIDIS TOXIN see VRP200
C. VIRIDIS VIRIDIS VENOM see PLX100
CVK see PDD350
CVMP see RAF100
CVP see CDS750
CW 524 see TKL100
C-WEISS 7 (GERMAN) see TGG760
CWN 2 see CBT500
C-WR VIOLET 8 see MAC250
CX-59 see HKE000
CXA-DPS see CBQ575
CXD see CCS530
CXM see CCS600, SFQ300
CXM-AX see CCS625
CY-39 see PHU500
CY 116 see AJD000
CY 216 see HAQ500
CY 222 see HAQ500
CY see CQC650
CYAANWATERSTOF (DUTCH) see HHS000
CYACETACID see COH250
CYACETACIDE see COH250
CYACETAZID see COH250
CYACETAZIDE see COH250
CYALANE see DXN600
CYAMEMAZINE MALEATE see COS899
CYAMEPROMAZINE MALEATE see COS899
CYAMOPSIS GUM see GLU000

CYANACETAMIDE see COJ250
CYANACETATE ETHYLE (GERMAN) see EHP500
CYANACETHYDRAZIDE see COH250
CYANACETIC ACID HYDRAZIDE see COH250
CYANACETOHYDRAZIDE see COH250
CYANACETYLHYDRAZIDE see COH250
CYANAMID 24055 see DUI000
CYANAMIDE see CAQ250, COH500
CYANAMIDE CALCIQUE (FRENCH) see CAQ250
CYANAMIDE, CALCIUM SALT (1:1) see CAQ250
CYANAMID GRANULAR see CAQ250
CYANAMID SPECIAL GRADE see CAQ250
CYANASET see MJM200
CYANATES see COH750
CYANATOTRIBUTYLSTANNANE see COI000
CYANAZIDE see COH250
CYANAZINE see BLW750
CYANEA CAPILLATA TOXIN see COI125
CYANESSIGSAEURE (GERMAN) see COJ500
CYANHYDRINE d'ACETONE (FRENCH) see MLC750
CYANIC ACID, POTASSIUM SALT see PLC250
CYANIC ACID, SODIUM SALT see COI250
CYANIC ACID, TRIMETHYLSTANNYL ESTER see TMI100
CYANIDE see COI500
CYANIDE ANION see COI500
CYANIDE (CN¹) see COI500
CYANIDE(CN¹) see COI500
CYANIDE(CN¹) ION see COI500
CYANIDE ION see COI500
CYANIDELONON 1522 see QCA000
CYANIDE of POTASSIUM see PLC500
CYANIDES (OSHA) see PLC500
CYANIDE of SODIUM see SGA500
CYANIDE, dry (UN 1588) see COI500
CYANIDE SOLUTIONS (DOT) see COI500
CYANIDIN see COI750
CYANIDIN CHLORIDE see COI750
CYANIDINE see THR525
CYANIDOL see COI750
CYANIDOL CHLORIDE see COI750
CYANINE ACID BLUE R see ADE750
CYANINE ACID BLUE R NEW see ADE750
CYANINE FAST SCARLET G see CMM325
CYANINE FAST YELLOW M see CMM759
CYANINE GREEN G BASE see BLK000
CYANITE see AHF500
CYANIZIDE see COH250
2-CYANOACETAMIDE see COJ250
CYANOACETAMIDE see COJ250
CYANOACETHYDRAZIDE see COH250
CYANOACETIC ACID see COJ500
CYANOACETIC ACID, ALLYL ESTER see AGC150
CYANOACETIC ACID ETHYL ESTER see EHP500
CYANOACETIC ACID HYDRAZIDE see COH250
CYANOACETIC ACID, ISOBUTYL ESTER see IIQ100
CYANOACETIC ACID METHYL ESTER see MIQ000
CYANOACETIC ESTER see EHP500
α-CYANOACETOHYDRAZIDE see COH250
CYANOACETOHYDRAZIDE see COH250
CYANOACETONITRILE see MAO250
CYANOACETYL CHLORIDE see COJ625
CYANOACETYLHYDRAZIDE see COH250
1-(CYANOACETYL)MORPHOLINE see MRR750
4-CYANOACETYLMORPHOLINE see MRR750
2-CYANOACRYLIC ACID, METHYL ESTER see MIQ075
α-CYANOACRYLIC ACID METHYL ESTER see MIQ075
CYANOAMINE see COH500
2-CYANOANILINE see APJ750
3-CYANOANILINE see AIR125
4-CYANOANILINE see COK125
m-CYANOANILINE see AIR125
o-CYANOANILINE see APJ750
p-CYANOANILINE see COK125
CYANO-B12 see VSZ000
7-CYANOBENZ(c)ACRIDINE see BAX000
4-CYANOBENZALDEHYDE see COK250
p-CYANOBENZALDEHYDE see COK250
2-CYANOBENZAMIDE see COK300
o-CYANOBENZAMIDE see COK300
CYANOBENZENE see BCQ250
p-CYANOBENZENECARBOXALDEHYDE see COK250
7-CYANOBENZO(c)ACRIDINE see BAX000
4-CYANOBENZONITRILE see BBP250
α-CYANOBENZYL PHENYL KETONE see OOK200

5-o-CYANOBENZYL-4,5,6,7-TETRAHYDROTHIENO(3,2-c)PYRIDINE METHANE-
SULFONATE see TDC725
CYANOBORANE OLIGOMER see COK659
CYANOBRIK see SGA500
CYANOBROMIDE see COO500
N-CYANO-2-BROMOETHYLCYCLOHEXYLAMINE see COL000
1-CYANOBUTANE see VAV300
1-CYANO-1-BUTENE see PBQ275
CYANO-CARBAMIC ACID METHYL ESTER, DIMER see MIQ350
CYANOCOBALAMIN see VSZ000
p-CYANOCUMENE see IOD050
2-CYANO-N-(2-CYANOETHYL)ETHANAMINE see BIQ500
CYANOCYCLINE A see COL125
1-(1-CYANOCYCLOHEXYL)PIPERIDINE see PIN225
α-CYANODEOXYBENZOIN see OOK200
N-CYANODIALLYLAMINE see DBJ200
5-CYANO-10,11-DIHYDRO-5-(3-DIMETHYLAMINOPROPYL)-5H-DIBENZO
(a,d)CYCLOHEPTENE HYDROCHLORIDE see COL250
4-CYANO-2,6-DIIODOPHENOL see HKB500
4-CYANO-2,6-DIJODOPHENOL (GERMAN) see HKB500
4-CYANO-2,6-DIJODPHENOL CAPRYSAEUREESTER (GERMAN) see DNG200
4-CYANO-2,6-DIJODPHENOL LITHIUMSALZ (GERMAN) see DNF400
CYANO-3-(DIMETHYLAMINO-3-METHYL-2-PROPYL)-10-PHENOTHIAZINE MALE-
ATE see COS899
CYANODIMETHYLARSINE see COL750
α-CYANO-2,6-DIMETHYLERGOLINE-8-PROPIONAMIDE see COM075
2-CYANO-3,3-DIPHENYLACRYLIC ACID, ETHYL ESTER see DGX000
α-CYANODIPHENYLMETHANE see DVX200
2-CYANO-3,3-DIPHENYL-2-PROPENOIC ACID, ETHYL ESTER see DGX000
1'-(3-CYANO-3,3-DIPHENYLPROPYL)(1,4'-BIPIPERIDINE)-4'-CARBOXAMIDE see
PJA140
1-(3-CYANO-3,3-DIPHENYLPROPYL)-4-(2-OXO-3-PROPIONYL-1-BENZIMIDAZOLI-
NYL)PIPERIDINE see BCE825
1-(3-CYANO-3,3-DIPHENYLPROPYL)-4-PHENYLISONIPECOTIC ACID ETHYL ES-
TER HYDROCHLORIDE see LIB000
1-(1-(3-CYANO-3,3-DIPHENYLPROPYL)-4-PIPERIDYL)-3-PROPIONYL-2-BENZIMI-
DAZOLINONE see BCE825
1-CYANO-3,4-EPITHIOBUTANE see EBD600
CYANOETHANE see PMV750
2-CYANOETHANOL see HGP000
2-(2-CYANOETHOXY)ETHYL ESTER ACRYLIC ACID see COM125
4-CYANOETHOXY-2-METHYL-2-PENTANOL see COM250
CYANOETHYDRAZIDE see COH250
2-CYANOETHYL ACRYLATE see ADT111
CYANOETHYL ACRYLATE see ADT111
2-CYANOETHYL ALCOHOL see HGP000
β-CYANOETHYLAMINE see AMB500
2-CYANO-N-((ETHYLAMINO)CARBONYL)-2-(METHOXYIMINO)-ACETAMIDE see
COM300
N-(2-CYANOETHYL)ANILINE see AOT100
N-(β-CYANOETHYL)ANILINE see AOT100
N-(CYANOETHYL)ANILINE see AOT100
1-(2-CYANOETHYL)AZIRIDINE see ASI250
N-(2-CYANOETHYL)AZIRIDINE see ASI250
(2-CYANOETHYL)BENZENE see HHP100
2-(N-(2-CYANOETHYL)-N-CYCLOHEXYL)AMINO-ETHANOL see CPJ500
N-(CYANOETHYL)DIETHYLENETRIAMINE see COM500
CYANOETHYLENE see ADX500
1-(2-CYANOETHYL)ETHYLENIMINE see ASI250
N-(β-CYANOETHYL)ETHYLENIMINE see ASI250
2-CYANOETHYL-2'-FLUOROETHYLETHER see CON500
N-(β-CYANOETHYL)-N-(β-HYDROXYETHY)-ANILINE see CPJ500
β-CYANOETHYLMERCAPTAN see COM750
2-CYANOETHYL PROPENOATE see ADT111
2-CYANOETHYLTRICHLOROSILANE see CON000
β-CYANOETHYLTRICHLOROSILANE see CON000
(2-CYANOETHYL)TRIETHOXYSILANE see CON250
β-CYANOETHYLTRIETHOXYSILANE see CON250
CYANOFENPHOS see CON300
p-CYANOFLUOROBENZENE see FGH100
2-CYANO-2'-FLUORODIETHYL ETHER see CON500
CYANOFORMYL CHLORIDE see CON825
CYANOGAS see CAQ500
CYANOGEN see COO000
CYANOGENAMIDE see COH500
CYANOGEN AZIDE see COO250
CYANOGEN BROMIDE see COO500
CYANOGEN CHLORIDE (ACGIH,OSHA) see COO750
CYANOGEN CHLORIDE, inhibited (DOT) see COO750
CYANOGEN CHLORIDE see COO750
CYANOGENE (FRENCH) see COO000
CYANOGEN FLUORIDE see COO825
CYANOGEN GAS (DOT) see COO000
CYANOGEN IODIDE see COP000
CYANOGEN MONOBROMIDE see COO500

CYANOGEN NITRIDE see COH500
CYANOGRAN see SGA500
CYANOGUANIDINE see COP125
S-1-CYANO-2-HYDROXY-3-BUTENE see COP400
CYANOHYDROXYMERCURY see COP500
2-CYANO-10-(3-(4-HYDROXYPIPERIDINO)PROPYL)PHENOTHIAZINE see PIW000
2-CYANO-10-(3-(4-HYDROXY-1-PIPERIDYL)PROPYL)PHENOTHIAZINE see PIW000
CYANO-3-((HYDROXY-4-PIPERIDYL-1)-3 PROPYL)-10-PHENOTHIAZINE (FRENCH) see PIW000
CYANOIMINOACETIC ACID see COJ250
5-CYANO-3-INDOLYL ISOPROPYL KETONE see COP525
5-CYANO-3-INDOLYLMETHYL KETONE see COP550
α-CYANO-6-ISOBUTYLERGOLINE-8-PROPIONAMIDE see COP600
α-CYANOISOPROPYLAMIDE OF THE O,O-DIETHYLTHIOPHOSPHORYL ACE-TIC ACID see PHK250
CYANOLIT see MIQ075
CYANOMETHANE see ABE500
CYANOMETHANOL see HIM500
1-CYANO-2-METHOXYETHANE see MFL750
α-((CYANOMETHOXY)IMINO)-BENZACETONITRILE see COP700
α-((CYANOMETHOXY)IMINO)BENZENEACETONITRILE see COP700
O-(4-CYANO-2-METHOXYPHENYL)-O,O-DIMETHYL PHOSPHOROTHIOATE see DTQ800
CYANOMETHYL ACETATE see COP750
5-CYANO-10-METHYL-1,2-BENZANTHRACENE see MGY000
7-CYANO-10-METHYL-1,2-BENZANTHRACENE see MGY250
10-CYANOMETHYL-1,2-BENZANTHRACENE see BBB500
7-CYANO-12-METHYL-BENZ(a)ANTHRACENE see MIQ250
8-CYANO-7-METHYLBENZ(a)ANTHRACINE see MGY000
(CYANOMETHYL)BENZENE see PEA750
N-(CYANOMETHYL)DIMETHYLAMINE see DOS200
S-(((1-CYANO-1-METHYL-ETHYL)CARBAMOYL)METHYL) O,O-DIETHYL PHOS-PHOROTHIOATE see PHK250
S-N-(1-CYANO-1-METHYLETHYL)CARBAMOYLMETHYL DIETHYL PHOSPHORO-THIOLATE see PHK250
17-α-CYANOMETHYL-17-β-HYDROXY-ESTRA-4,9(10)-DIEN-3-ONE see SMP400
3-(CYANOMETHYL)INDOLE see ICW000
CYANO(METHYLMERCURI)GUANIDINE see MLF250
1-CYANO-2-METHYL-3-(2-(((5-METHYL-4-IMIDAZO-LYL)METHYL)THIO)ETHYL)GUANIDINE see TAB250
2-CYANO-1-METHYL-3-(2-(((5-METHYLIMIDAZOL-4-YL)METHYL)THIO)ETHYL)GUANIDINE see TAB250
α-CYANO-1-METHYL-β-OXO-PYRROLE-2-PROPIONANILIDE compounded with 2,2′,2″-NITRILOTRIETHANOL see CDG300
5-CYANO-5-METHYLTETRAZOLE see COP759
CYANONITRENE see COP775
3-CYANONITROBENZENE see NFH500
4-CYANONITROBENZENE see NFI010
m-CYANONITROBENZENE see NFH500
p-CYANONITROBENZENE see NFI010
2-CYANO-4-NITROBENZENEDIAZONIUM HYDROGEN SULFATE see COQ325
N-CYANO-N-NITROSOETHYLAMINE see ENT500
N-CYANO-N′-METHYL-N″-(2-(((5-METHYL-1H-IMIDAZOL-4-YL)METHYL)THIO)ETHYL)GUANIDINE see TAB250
1-CYANOOCTANE see OES300
3-(3-CYANO-1,2,4-OXADIAZOL-5-YL)-4-CYANO FURAZAN-2-(5-) OXIDE see COQ375
α-CYANO-3-PHENOXYBENZYL-2-(4-CHLOROPHENYL)ISOVALERATE PYDRIN see FAR100
α-CYANO-3-PHENOXYBENZYL-2-(4-CHLOROPHENYL)-3-METHYLBUTYRATE see FAR100
(±)-α-CYANO-3-PHENOXYBENZYL 2,2-DIMETHYL-3-(2,2-DICHLOROVI-NYL)CYCLOPROPANE CARBOXYLATE see RLF350
α-CYANO-3-PHENOXYBENZYL 2,2,3,3-TETRAMETHYL-1-CYCLOPROPANECAR-BOXYLATE see DAB825
CYANO(3-PHENOXYPHENYL)METHYL 4-CHLORO-α-(1-METHYLE-THYL)BENZENEACETATE see FAR100
CYANOPHENPHOS see CON300
O-(4-CYANOPHENYL) O,O-DIMETHYL PHOSPHOROTHIOATE see COQ399
O-p-CYANOPHENYL O,O-DIMETHYL PHOSPHOROTHIOATE see COQ399
o-(4-CYANOPHENYL)-o-ETHYL PHENYLPHOSPHONOTHIOATE see CON300
o-p-CYANOPHENYL-o-ETHYL PHENYLPHOSPHONOTHIOATE see CON300
CYANOPHENYLMETHYL-β-d-GLUCOPYRANOSIDURONIC ACID see LAS000
CYANOPHOS see COQ399, DGP900
1-CYANOPROPANE see BSX250
2-CYANOPROPANE see IJX000
2-CYANOPROPENE-1 see MGA750
1-CYANOPROPENE see COC300, COQ750
1-CYANO-2-PROPEN-1-OL see HJQ000
3-CYANOPROPYLDICHLOROMETHYLSILANE see COR325
(3-CYANOPROPYL)DIETHOXY(METHYL) SILANE see COR500
2-CYANO-2-PROPYL NITRATE see COR600
(3-CYANOPROPYL) TRICHLOROSILANE see COR750
(3-CYANOPROPYL) TRIETHOXYSILANE see COS800

3-CYANOPYRIDINE see NDW515
2-CYANO-3-(4-PYRIDYL)-1-(1,2,3,TRIMETHYLPROPYL)GUANIDINE see COS500
CYANOSIN see ADG250
CYANOSIN (ACID DYE) see ADG250
CYANOSINE see ADG250
α-CYANOSTILBENE see DVX600
2-(5-CYANOTETRAZOLE)PENTAMMINECOBALT(III) PERCHLORATE see COS825
2-CYANOTOLUENE see TGT500
3-CYANOTOLUENE see TGT250
4-CYANOTOLUENE see TGT750
α-CYANOTOLUENE see PEA750
ω-CYANOTOLUENE see PEA750
m-CYANOTOLUENE see TGT250
o-CYANOTOLUENE see TGT500
p-CYANOTOLUENE see TGT750
CYANOTOXIN see SMM500
CYANOTRICHLOROMETHANE see TII750
CYANOTRIMEPRAZINE MALEATE see COS899
2-CYANO-1,2,3-TRIS(DIFLUOROAMINO)PROPANE see COT000
CYANOTUBERICIDIN see VGZ000
α-CYANOVINYL ACETATE see ABL500
CYANOX 425 see MJN250
CYANOX see COQ399
CYANOX LTDP see TFD500
CYANOX-STDP see DXG700
CYANSAN see COI250
CYANTIN see NGE000
CYANURAMIDE see MCB000
CYANURCHLORIDE see TJD750
CYANURE see COI500
CYANURE d'ARGENT (FRENCH) see SDP000
CYANURE de CALCIUM (FRENCH) see CAQ500
CYANURE de CUIVRE (FRENCH) see CNL250
CYANURE DE SODIUM see SGA500
CYANURE de MERCURE (FRENCH) see MDA250
CYANURE de METHYL (FRENCH) see ABE500
CYANURE de PLOMB (FRENCH) see LCU000
CYANURE de POTASSIUM (FRENCH) see PLC500
CYANURE de SODIUM (FRENCH) see SGA500
CYANURE de VINYLE (FRENCH) see ADX500
CYANURE de ZINC (FRENCH) see ZGA000
CYANURIC ACID see THS000
CYANURIC ACID CHLORIDE see TJD750
CYANURIC ACID TRIGLYCIDYL ESTER see TKL250
CYANURIC CHLORIDE (DOT) see TJD750
CYANURIC FLUORIDE see TKK000
CYANURIC TRIAMIDE see MCB000
CYANURIC TRIAZIDE (DOT) see THR250
CYANURIC TRIAZIDE see THR250
CYANURIC TRICHLORIDE (DOT) see TJD750
CYANUROTRIAMIDE see MCB000
CYANUROTRIAMINE see MCB000
CYANURYL CHLORIDE see TJD750
CYANWASSERSTOFF (GERMAN) see HHS000
CYAP see COQ399
CYASORB UV 9 see MES000
CYASORB UV 531 see HND100
CYAZID see COH250
CYAZIDE see COH250
CYAZIN see ARQ725
CYBIS see EID000
CYCASIN see COU000
CYCASIN ACETATE see MGS750
CYCAS REVOLUTA GLUCOSIDE see COU000
CYCEURON plus CHLORBUFAN see AFM375
CYCHLORAL see CPR000
CYCLACILLIN see AJJ875
CYCLADIENE see DAL600
CYCLAINE see COU250
CYCLAINE HYDROCHLORIDE see COU250
CYCLAL CETYL ALCOHOL see HCP000
CYCLALIA see CMS850
CYCLAMAL see COU500
CYCLAMATE see CPQ625, SGC000
CYCLAMATE CALCIUM see CAR000
CYCLAMATE, CALCIUM SALT see CAR000
CYCLAMATE SODIUM see SGC000
CYCLAMEN ALDEHYDE see COU500
CYCLAMEN ALDEHYDE DIETHYL ACETAL see COU510
CYCLAMEN ALDEHYDE DIMETHYL ACETAL see COU525
CYCLAMIC ACID see CPQ625
CYCLAMIC ACID SODIUM SALT see SGC000
CYCLAMID see CPR000
CYCLAMIDE see ABB000, CPR000

CYCLAMIDOMYCIN see COV125
CYCLAN see CAR000
CYCLANDELATE see DNU100
CYCLAPEN see AJJ875
CYCLAPROP see TJG600
CYCLATE see BOV825
CYCLAZOCINE see COV500
CYCLE see PMF600
CYCLERGINE see DNU100
CYCLIC DIMETHYLSILOXANE PENTAMER see DAF350
CYCLIC DIMETHYLSILOXANE TETRAMER see OCE100
CYCLIC ETHYLENE ACETAL-2-BUTANONE see EIO500
CYCLIC ETHYLENE CARBONATE see GHM000
CYCLIC ETHYLENE (DIETHOXYPHOSPHINOTHIOYL)DITHIOIMIDOCARBONATE
  see DXN600
CYCLIC ETHYLENE(DIETHOXYPHOSPHINOTHIOYL)DITHIOIMIDOCARBONATE
  see PGW750
CYCLIC ETHYLENE ESTER of (DIETHOXYPHOSPHINO-
  THIOYL)DITHIOIMIDOCARBONIC ACID see DXN600
CYCLIC ETHYLENE P,P-DIETHYL PHOSPHONODITHIOIMIDOCARBONATE see
  PGW750
CYCLIC ETHYLENE SULFITE see COV750
CYCLIC ETHYLENE TRITHIOCARBONATE see EJQ100
CYCLIC (HYDROXYMETHYL)ETHYLENE ACETAL ACETONE see DVR600
CYCLIC METHYLETHYLENE CARBONATE see CBW500
CYCLIC-S,S-(6-METHYL-2,3-QUINOXALINEDIYL) DITHIOCARBONATE see
  ORU000
CYCLIC NEOPENTANETETRAYL BIS(2,4-DI-tert-BUTYLPHENYL)ESTER PHOS-
  PHOROUS ACID see COV800
CYCLIC N′,O-PROPYLENE ESTER of N,N-BIS(2-CHLOROE-
  THYL)PHOSPHORODIAMIDIC ACID MONOHYDRATE see CQC500
CYCLIC-1,2-PROPYLENE CARBONATE see CBW500
CYCLIC PROPYLENE CARBONATE see CBW500
CYCLIC PROPYLENE (DIETHOXYPHOSPHINYL)DITHIOIMIDOCARBONATE see
  DHH400
CYCLIC SODIUM TRIMETAPHOSPHATE see TKP750
CYCLIC TETRAMETHYLENE SULFONE see SNW500
CYCLISCHES TRINATRIUMMETAPHOSPHAT (GERMAN) see TKP750
CYCLIZINE see EAN600
CYCLIZINE CHLORIDE see MAX275
CYCLIZINE HYDROCHLORIDE see MAX275
α-CYCLOAMYLOSE see COW925
CYCLOATE see EHT500
CYCLOBARBITAL see TDA500
CYCLOBARBITAL-SALICYLAMIDE COMPLEX see COV825
CYCLOBARBITOL see TDA500
CYCLOBARBITONE see TDA500
CYCLOBENZAPRINE see PMH600
CYCLOBENZAPRINE HYDROCHLORIDE see DPX800
CYCLOBRAL see DNU100
CYCLOBUTANE see COW000
CYCLOBUTANE, 1-ACETYL-2,2-DIMETHYL-3-ETHYL-, (cis)- see EII700
cis-(1,1-CYCLOBUTANEDICARBOSYLATO)DIAMMINEPLATINUM(II) see CCC075
1,1-CYCLOBUTANEDICARBOXYLATE DIAMMINE PLATINUM(II) see CCC075
cis-(1,1-CYCLOBUTANEDICARBOXYLATO)DIAMMINEPLATINUM(II) see CCC075
CYCLOBUTANE, 1,2-DICHLOROHEXAFLUORO- see DFM025
CYCLOBUTANE, 1,2-DICHLORO-1,2,3,3,4,4-HEXAFLUORO-(9CI) see DFM025
CYCLOBUTENE see COW250
CYCLOBUTYLENE see COW250
17-CYCLOBUTYLMETHYL-3-HYDROXY-6-METHYLENE-8-β-METHYLMORPHINAN
  see COW675
(8-β)-17-(CYCLOBUTYLMETHYL)-6-METHYLENEMORPHINAN-3-OL METHANE-
  SULFONATE see COW675
CYCLOBUTYROL see COW700
CYCLOCAPRON see AJV500
CYCLOCEL see CMF400
CYCLOCHEM GMS see OAV000
CYCLOCHEM INEO see ISC550
CYCLOCHLOROTINE see COW750
α-CYCLOCITRYLIDENEACETONE see IFW000
β-CYCLOCITRYLIDENEACETONE see IFX000
α-CYCLOCITRYLIDENE-4-METHYLBUTAN-3-ONE see COW780
CYCLO-CMP HYDROCHLORIDE see COW900
CYCLOCORT see COW825
2,2′-o-CYCLOCYTIDINE see COW875
2,2′-CYCLOCYTIDINE see COW875
o-2,2′-CYCLOCYTIDINE see COW875
CYCLOCYTIDINE see COW875, COW900
2,2′-o-CYCLOCYTIDINE HYDROCHLORIDE see COW900
2,2′-CYCLOCYTIDINE HYDROCHLORIDE see COW900
CYCLOCYTIDINE HYDROCHLORIDE see COW900
o-2,2′-CYCLOCYTIDINE MONOHYDROCHLORIDE see COW900
CYCLODAN see EAQ750
β-CYCLODEXTRIN see COW925
CYCLODODECALACTAM see COW930

CYCLODODECANE, (ETHOXYMETHOXY)- see FMV100
CYCLODODECANE, (METHOXYMETHOXY)- see FMV200
CYCLODODECANE, 1-METHOXY-1-METHYL- see MEU300
cis,trans,trans-CYCLODODECA-1,5,9-TRIENE see COW935
1,5,9-CYCLODODECATRIENE (Z,E,E) see COW935
CYCLODODECYL-2,6-DIMETHYLMORPHOLINE ACETATE see COX000
N-CYCLODODECYL-2,6-DIMETHYLMORPHOLINIUM ACETATE see COX000
CYCLODOL see BBV000
CYCLODORM see TDA500
CYCLOESTROL see DLB400
CYCLOFENIL see FBP100
CYCLOFENYL see FBP100
CYCLOGEST see PMH500
CYCLOGUANIL see COX400
CYCLOGUANIL HYDROCHLORIDE see COX325
CYCLOGUANYL see COX400
CYCLOGYL see CPZ125
3-CYCLOHENENYL CYANIDE see CPC625
β-CYCLOHEPTAAMYLOSE see COW925
CYCLOHEPTAAMYLOSE see COW925
9-CYCLOHEPTADECEN-1-ONE see CMU850
9-CYCLOHEPTADECEN-1-ONE, (Z)-(8CI,9CI) see CMU850
CYCLOHEPTAGLUCOSAN see COW925
CYCLOHEPTANE see COX500
CYCLOHEPTANECARBAMIC ACID-1,1-BIS(p-FLUOROPHENYL)-2-PROPYNYL
  ESTER see BJW250
CYCLOHEPTANONE see SMV000
CYCLOHEPTANONE, 2-CHLORO- see CFF600
6H-CYCLOHEPTA(b)QUINOLINE-11-CARBOXAMIDE, N,N-DIETHYL-7,8,9,10-TET-
  RAHYDRO- see TCO100
6H-CYCLOHEPTA(b)QUINOLINE, 7,8,9,10-TETRAHYDRO-11-(MORPHOLINO-
  CARBONYL)- see MRU077
1,3,5-CYCLOHEPTATRIENE see COY000
CYCLOHEPTATRIENE (DOT) see COY000
CYCLOHEPTATRIENE MOLYBDENUM TRICARBONYL see COY100
2,4,6-CYCLOHEPTATRIEN-1-ONE, 2-HYDROXY-5-ISOPROPYL-(8CI) see TFV750
CYCLOHEPTENE see COY250
5-(1-CYCLOHEPTEN-1-YL)-5-ETHYLBARBITURIC ACID see COY500
CYCLOHEPTENYL ETHYLBARBITURIC ACID see COY500
CYCLOHEPTENYLETHYLMALONYLUREA see COY500
5-(1-CYCLOHEPTEN-1-YL)-5-ETHYL-2,4,6(1H,3H,5H)-PYRIMIDINETRIONE (9CI)
  see COY500
CYCLOHEXAAN (DUTCH) see CPB000
CYCLOHEXADECANOLIDE see OKU000
2,5-CYCLOHEXADIEN-1,4-DIONE, 2,3,5-TRIMETHYL- see POG400
1,3-CYCLOHEXADIENE see CPA500
3,5-CYCLOHEXADIENE-1,2-DIONE see BDC250
2,5-CYCLOHEXADIENE-1,4-DIONE see QQS200
1,4-CYCLOHEXADIENEDIONE see QQS200
CYCLOHEXADIENEDIONE see QQS200
2,5-CYCLOHEXADIENE-1,4-DIONE DIOXIME see DVR200
2,5-CYCLOHEXADIENE-1,4-DIONE, 2-PHENYL-(9CI) see PEL750
2,5-CYCLOHEXADIENE-1,4-DIONE, 2,3,5,6-TETRAAZIDO- see TBJ250
1,4-CYCLOHEXADIENE DIOXIDE see QQS200
CYCLOHEXADIENE-1-ETHANOL, 4-(1-METHYLETHYL)-, FORMATE see IHX450
CYCLOHEXADIENOL-4-ONE-1-SULFONATE de DIETHYLAMINE (FRENCH) see
  DIS600
2,5-CYCLOHEXADIEN-1-ONE, 2,6-DICHLORO-4-((p-HYDROXYPHENYL)IMINO)-,
  SODIUM SALT see SGG650
2,5-CYCLOHEXADIEN-1-ONE, 4-IMINO- see BDD500
CYCLOHEXAMETHYLENE CARBAMIDE see CPB050
CYCLOHEXAMETHYLENIMINE see HDG000
CYCLOHEXAN (GERMAN) see CPB000
CYCLOHEXANAMIDE see CPB050
CYCLOHEXANAMINE see CPF500
CYCLOHEXANAMINE HYDROCHLORIDE see CPA775
CYCLOHEXANAMINE, N-METHYL-(9CI) see MIT000
CYCLOHEXANAMINE, 4,4′-METHYLENEBIS-(9CI) see MJQ260
CYCLOHEXANE see CPB000
CYCLOHEXANEACETIC ACID, 1-(HYDROXYMETHYL)-, MONOSODIUM SALT
  (9CI) see SHL500
1,3-CYCLOHEXANEBIS(METHYLAMINE) (8CI) see BGT500
CYCLOHEXANECARBAMIC ACID, 1,1-DIPHENYL-2-BUTYNYL ESTER see
  DVY900
CYCLOHEXANECARBAMIC ACID, 1-METHYL-1-PHENYL-2-PROPYNYL ESTER
  see MNX850
CYCLOHEXANECARBINOL see HDH200
CYCLOHEXANECARBOXAMIDE see CPB050
CYCLOHEXANECARBOXYLIC ACID see HDH100
CYCLOHEXANECARBOXYLIC ACID, CHLORO-2-METHYL-, tert-BUTYL ESTER
  see BQT600
CYCLOHEXANECARBOXYLIC ACID, 4(or 5)-CHLORO-2-METHYL-, tert-BUTYL
  ESTER (8CI) see BQT600
CYCLOHEXANECARBOXYLIC ACID, 4(or 5)-CHLORO-2-METHYL-, 1,1-DIME-
  THYLETHYL ESTER (9CI) see BQT600

CYCLOHEXANECARBOXYLIC ACID-(2-HYDROXYETHYL) ESTER see HKQ500
CYCLOHEXANECARBOXYLIC ACID, LEAD SALT see NAS500
CYCLOHEXANECARBOXYLIC ACID, TRIBUTYLSTANNYL ESTER see TID100
CYCLOHEXANE, CHLORO- see CPI400
1,2-CYCLOHEXANEDIAMINE see CPB100
1,3-CYCLOHEXANEDIAMINE see DBQ800
1,2-CYCLOHEXANEDIAMINE-N,N,N′,N′-TETRAACETIC ACID see CPB120
1,2-CYCLOHEXANEDIAMINETETRAACETIC ACID see CPB120
(CYCLOHEXANE-1,2-DIAMMINE)(4-CARBOXYPHTHLATO)PLATINUM(II) see CCJ350
(Z)-(CYCLOHEXANE-1,2-DIAMMINE)ISOCITRATOPLATINUM(II) see PGQ275
CYCLOHEXANE, 1,2-DIBROMO-4-(1,2-DIBROMOETHYL)- see DDM300
1,2-CYCLOHEXANEDICARBOXYLIC ACID, BIS(2,3-EPOXYPROPYL) ESTER see DKM500
1,2-CYCLOHEXANEDICARBOXYLIC ACID, BIS(OXIRANYLMETHYL)ESTER (9CI) see DKM500
Δ$^{1 \cdot 4 \cdot α}$-CYCLOHEXANEDIMALONONITRILE see BIY600
CYCLOHEXANEDIMETHANAMINE (9CI) see BGT500
1,3-CYCLOHEXANEDIONE, 5-(2-(ETHYLTHIO)PROPYL)-2-(1-OXOPROPYL)- see EPR100
2,2′-CYCLOHEXANE-1,1-DIYLBIS(p-PHENYLENEOXY)BIS(2-METHYLBUTYRIC ACID) see CMV700
CYCLOHEXANEETHANOL, ACETATE see EHS000
CYCLOHEXANE ETHYL ACETATE see EHS000
CYCLOHEXANEFORMAMIDE see CPB050
CYCLOHEXANE, ISOCYANATO-(9CI) see CPN500
CYCLOHEXANE, 5-ISOCYANATO-1-(ISOCYANATOMETHYL)-1,3,3-TRIMETHYL- (9CI) see IMG000
CYCLOHEXANEMETHANOL see HDH200
CYCLOHEXANEMETHANOL, α-α-4-TRIMETHYL-(9CI) see MCE100
CYCLOHEXANE OXIDE see CPD000
CYCLOHEXANE, PIPERIDINOMETHYL-, CAMPHOSULFATE see PIQ750
CYCLOHEXANE, PIPERIDINOMETHYL-, HYDROCHLORIDE see PIR000
CYCLOHEXANESPIRO-5′-HYDANTOIN see DVO600
CYCLOHEXANESULFAMIC ACID, CALCIUM SALT see CAR000
CYCLOHEXANESULFAMIC ACID, MONOSODIUM SALT see SGC000
CYCLOHEXANESULPHAMIC ACID see CPQ625
CYCLOHEXANESULPHAMIC ACID, MONOSODIUM SALT see SGC000
CYCLOHEXANETHIOL see CPB625
CYCLOHEXANE, 1-(TRICHLOROSILYL)- see CPR250
1,2,3-CYCLOHEXANETRIONE TRIOXIME see CPB650
CYCLOHEXANOIC ACID see HDH100
CYCLOHEXANOL see CPB750
CYCLOHEXANOL ACETATE see CPF000
CYCLOHEXANOLAZETAT (GERMAN) see CPF000
1-CYCLOHEXANOL-α-BUTYRIC ACID see COW700
trans-(±)-CYCLOHEXANOL-2-((DIMETHYLAMINO)METHYL)-1-(3-METHOXYPHE-NYL) see THJ755
CYCLOHEXANOL, 2-(1,1-DIMETHYLETHYL)-, ACETATE see BQW490
CYCLOHEXANOL, 1-ETHYNYL-2-(1-METHYLPROPYL)-, ACETATE see EQN300
CYCLOHEXANOL, 1-((1-HYDROPEROXYCYCLOHEXYL)DIOXY)- see CPC300
CYCLOHEXANOL, p-ISOPROPYL- see IOO300
CYCLOHEXANOL, 2-ISOPROPYL-5-METHYL- see MCF750
CYCLOHEXANOL, 2-METHYL-5-(1-METHYLETHENYL)-, ACETATE,(1-α-2-β,5α- (9CI) see DKV160
CYCLOHEXANOL, 1-METHYL-4-(1-METHYLETHENYL)-(9CI) see TBD775
CYCLOHEXANOL, 3,3,5-TRIMETHYL-, cis- see TLO510
CYCLOHEXANON (DUTCH) see CPC000
CYCLOHEXANONE see CPC000
CYCLOHEXANONE-Δ see CPC250
CYCLOHEXANONE, 2-sec-BUTYL-(7CI,8CI) see MOU800
CYCLOHEXANONE, 4-(1-ETHOXYETHENYL)-3,3,5,5-TETRAMETHYL- see EES370
CYCLOHEXANONE ISO-OXIME see CBF700
CYCLOHEXANONE, 5-METHYL-2-(1-METHYLETHYLIDENE)-, (R)- see POI615
CYCLOHEXANONE, 2-(1-METHYLPROPYL)- see MOU800
CYCLOHEXANONE OXIME see HLI500
CYCLOHEXANONE PEROXIDE see CPC300
CYCLOHEXANYL ACETATE see CPF000
CYCLOHEXATRIENE see BBL250
CYCLOHEXENE see CPC579
CYCLOHEXENEBUTANAL, α-2,2,6-TETRAMETHYL- see TDO255
3-CYCLOHEXENE-1-CARBONITRILE see CPC625
3-CYCLOHEXENE-1-CARBOXALDEHYDE see FNK025
CYCLOHEXENECARBOXALDEHYDE see TCJ100
3-(CYCLOHEXENE-1-CARBOXALDEHYDE 4-(4-HYDROXY-4-METHYLPENTYL)-METHYLPHENYL)-3-CYCLOHEXEN-1-CARBOXALDEHYDE see LJE200
3-CYCLOHEXENE-1-CARBOXALDEHYDE, 1-METHYL-4-(4-METHYLPENTYL)- see VIZ150
3-CYCLOHEXENE-1-CARBOXYLIC ACID see CPC650
1-CYCLOHEXENE-1-CARBOXYLIC ACID, 3,4,5 see BML000
4-CYCLOHEXENE-1,2-DICARBOXIMIDE see TDB100
4-CYCLOHEXENE-1,2-DICARBOXIMIDE, N-(2,6-DIOXO-3-PIPERIDYL)- see TDB200
1-CYCLOHEXENE-1,2-DICARBOXYLIC ACID DIMETHYL ESTER see DUF400

CYCLOHEXENE EPOXIDE see CPD000
CYCLOHEX-1-ENE-1-METHANOL, 4-(1-METHYLETHENYL)- see PCI550
3-CYCLOHEXENE-1-METHANOL, 1,2,4(or 1,3,5)-TRIMETHYL- see ISR100
1,2-CYCLOHEXENE OXIDE see CPD000
CYCLOHEXENE-1-OXIDE see CPD000
CYCLOHEXENE OXIDE see CPD000
CYCLOHEXENE, 4-(TRICHLOROSILYL)- see CPE500
CYCLOHEXENE, 1-VINYL- see VNZ990
2-CYCLOHEXEN-1-ONE see CPD250
CYCLOHEXENONE see CPD250
5-Δ$^{2,3}$-CYCLOHEXENYL-5-ALLYL-2-THIOBARBITURIC ACID see TES500
2-((4-CYCLOHEXEN-3-YLBUTYL)AMINO)ETHANETHIOL HYDROGEN SULFATE (ESTER) see CPD500
S-2-((4-CYCLOHEXEN-3-YLBUTYL)AMINO)ETHYL THIOSULFATE see CPD500
1-(2-CYCLOHEXEN-1-YLCARBONYL)-2-METHYLPIPERIDINE see CPD625
5-(2-CYCLOHEXEN-1-YL)DIHYDRO-5-(2-PROPENYL)-2-THIOXO-4,6(1H,5H)-PYRI-MIDINEDIONE (9CI) see TES500
5-(1-CYCLOHEXEN-1-YL)-1,5-DIMETHYLBARBITURIC ACID see ERD500
5-(1-CYCLOHEXEN-1-YL)-1,5-DIMETHYLBARBITURIC ACID SODIUM SALT see ERE000
5-(1-CYCLOHEXEN-1-YL)-1,5-DIMETHYL-2,4,6(1H,3H,5H)-PYRIMIDINETRIONE see ERD500
5-(1-CYCLOHEXEN-1-YL)-1,5-DIMETHYL-2,4,6(1H,3H,5H,)-PYRIMIDINETRIONE MONOSODIUM SALT see ERE000
5-(1-CYCLOHEXEN-1-YL)-1,5-DIMETHYL-2,4,6(1H,3H,5H)-PYRIMIDINETRIONE SODIUM SALT (9CI) see ERE000
3-(3-CYCLOHEXENYL)-2,4-DIOXASPIRO(5.5)UNDEC-8-ENE see CPD650
5-(1-CYCLOHEXEN-1-YL)-5-ETHYLBARBITURIC ACID see TDA500
5-(1-CYCLOHEXENYL)-5-ETHYLBARBITURIC ACID see TDA500
CYCLOHEXENYL-ETHYL BARBITURIC ACID see TDA500
CYCLOHEXENYLETHYLENE see CPD750
5-(1-CYCLOHEXEN-1-YL)-5-ETHYL-2,4,6(1H,3H,5H)-PYRIMIDINETRIONE see TDA500
2-CYCLOHEXENYL HYDROPEROXIDE see CPE125
5-(1-CYCLOHEXENYL-1)-1-METHYL-5-METHYLBARBITURIC ACID see ERD500
5-(Δ-1,2-CYCLOHEXENYL)-5-METHYL-N-METHYL-BARBITURSAEURE (GERMAN) see ERD500
p-(2-CYCLOHEXEN-1-YLOXY)BENZOIC ACID, 3-(2-METHYL-1-PYRROLIDI-NYL)PROPYL ESTER see UAG025
1-(2-(1-CYCLOHEXEN-1-YL)PHENOXY)-3-((1-METHYLETHYL)AMINO)-2-PROPA-NOL HYDROCHLORIDE see ERE100
CYCLOHEXENYLTRICHLOROSILANE (DOT) see CPE500
CYCLOHEXENYL TRICHLOROSILANE see CPE500
CYCLOHEXIMIDE see CPE750
CYCLOHEXYL ACETATE see CPF000
CYCLOHEXYLACETIC ACID ALLYL ESTER see AGC250
CYCLOHEXYL ALCOHOL see CPB750
CYCLOHEXYLALLYL-ESSIGSAEUREESTER DES 3-DIAETHYLAMINO-2,2-DI-METHYL-1-PROPANOL (GERMAN) see AGC750
CYCLOHEXYLAMIDOSULPHURIC ACID see CPQ625
CYCLOHEXYLAMINE see CPF500
CYCLOHEXYLAMINESULPHONIC ACID see CPQ625
CYCLOHEXYLAMINO ACETIC ACID see CPG000
2-(CYCLOHEXYLAMINO)ETHANOL see CPG125
2-(2-(CYCLOHEXYLAMINO))ETHYL-2-METHYL-1,3-BENZODIOXOLE HYDRO-CHLORIDE see CPG250
3-(2-(CYCLOHEXYLAMINO)ETHYL)-2-(3,4-METHYLENEDIOXYPHENYL)-4-THIA-ZOLIDINONE HYDROCHLORIDE see WBS860
4-(CYCLOHEXYLAMINO)-1-(NAPHTHALENYLOXY)-2-BUTANOL see CPG500
dl-1-CYCLOHEXYL-2-AMINOPROPANE HYDROCHLORIDE see CPG625
1-CYCLOHEXYLAMINO-2-PROPANOL see CPG700
1-(CYCLOHEXYLAMINO)-2-PROPANOL BENZOATE (ESTER) HYDROCHLO-RIDE see COU250
CYCLOHEXYLAMMONIUM STEARATE see CPH500
3-(5-CYCLOHEXYL-o-ANISOYL)-PROPIONIC ACID SODIUM SALT see MEK350
CYCLOHEXYLBENZENE see PER750
N-CYCLOHEXYL-2-BENZOTHIAZOLESULFENAMIDE see CPI250
N-CYCLOHEXYL-2-BENZOTHIAZYLSULFENAMIDE see CPI250
2-(α-CYCLOHEXYLBENZYL)-N,N,N′,N′-TETRAETHYL-1,3-PROPANEDIAMINE HY-DROCHLORIDE see LFK200
N-CYCLOHEXYL-l-BUTANESULFONAMIDE see BSM250
CYCLOHEXYL BUTANOATE see CPI300
CYCLOHEXYL BUTYRATE see CPI300
N-CYCLOHEXYLCARBAMIC ACID 1-PHENYL-1-(3,4-XYLYL)-2-PROPYNYL ES-TER see PGQ000
CYCLOHEXYLCARBINOL see HDH200
2-CYCLOHEXYLCARBONYL-1,2,3,6,7,11b-HEXAHYDRO-4H-PYRAZINO(2,1-a)ISOQUINOLIN-4-ONE see BGB400
1-(CYCLOHEXYLCARBONYL)-3-METHYLPIPERIDINE see CPI350
((CYCLOHEXYLCARBONYL)OXY)TRIBUTYLSTANNANE see TID100
4-(CYCLOHEXYLCARBONYL)PYRIDINE see CPI375
CYCLOHEXYLCARBOXAMIDE see CPB050
CYCLOHEXYL CARBOXYAMIDE see CPB050
CYCLOHEXYLCARBOXYLIC ACID see HDH100
CYCLOHEXYL CHLORIDE see CPI400

4-(4-CYCLOHEXYL-3-CHLOROPHENYL)-4-OXOBUTYRIC ACID see CPJ000
4-(4-CYCLOHEXYL-3-CHLOROPHENYL)-4-OXOBUTYRIC ACID CALCIUM SALT see CPJ250
CYCLOHEXYLCYANOETHYLETHANOLAMINE see CPJ500
N-CYCLOHEXYLCYCLOHEXANAMINE see DGT600
2-CYCLOHEXYLCYCLOHEXANONE see LBX100
N-CYCLOHEXYL-N-DIETHYLTHIOCARBONYL SULFONAMIDE see CPK000
N-CYCLOHEXYLDIMETHYLAMINE see DRF709
CYCLOHEXYLDIMETHYLAMINE see DRF709
3-CYCLOHEXYL-6-(DIMETHYLAMINO)-1-METHYL-s-TRIAZINE-2,4(1H,3H)-DIONE see HFA300
3-CYCLOHEXYL-6-(DIMETHYLAMINO)-1-METHYL-1,3,5-TRIAZINE-2,4(1H,3H)-DIONE see HFA300
2-CYCLOHEXYL-4,6-DINITROFENOL (DUTCH) see CPK500
6-CYCLOHEXYL-2,4-DINITROPHENOL see CPK500
2-CYCLOHEXYL-4,6-DINITROPHENOL see CPK500
(+)-1-CYCLOHEXYL-4-(1,2-DIPHENYLETHYL)PIPERAZINE DIHYDROCHLORIDE see CPK625
(±)-1-CYCLOHEXYL-4-(1,2-DIPHENYLETHYL)PIPERAZINE DIHYDROCHLORIDE see MRU757
(S)-1-CYCLOHEXYL-4-(1,2-DIPHENYLETHYL)-PIPERAZINE DIHYDROCHLORIDE see CPK625
1,2-CYCLOHEXYLENEDIAMINETETRAACETIC ACID see CPB120
N,N'-(1,4-CYCLOHEXYLENEDIMETHYLENE)BIS(2-(1-AZIRIDINYL)ACETAMIDE) see CPL100
(1,2-CYCLOHEXYLENEDINITRILO)TETRAACETIC ACID see CPB120
CYCLOHEXYLENE OXIDE see CPD000
CYCLOHEXYLESTER KYSELINY OCTOVE see CPF000
2-CYCLOHEXYLETHANOL see CPL250
CYCLOHEXYLETHYL ACETATE see EHS000
CYCLOHEXYLETHYL ALCOHOL see CPL250
CYCLOHEXYLETHYLCARBAMOTHIOIC ACID-S-ETHYL ESTER see EHT500
1-CYCLOHEXYL-4-((ETHYL-p-METHOXY-α-METHYLPHENETHYL)AMINO)-1-BUTANONE HYDROCHLORIDE see SBN300
CYCLOHEXYLETHYLTHIOCARBAMIC ACID-S-ETHYL ESTER see EHT500
3-CYCLOHEXYL-1-(2-FLUOROETHYL)-1-NITROSOUREA see CPL750
N'-CYCLOHEXYL-N-(2-FLUOROETHYL)-N-NITROSOUREA see CPL750
CYCLOHEXYL FLUOROETHYL NITROSOUREA see CPL750
(+)-α-CYCLOHEXYL-α-HYDROXY-BENZENEACETIC ACID-1-METHYL-3-PIPERIDINYL ESTER, HCl see PMS800
α-CYCLOHEXYL-β-HYDROXY-Δ^{α,β}-BUTENOLID (GERMAN) see CPM250
3-CYCLOHEXYL-4-HYDROXY-2(5H)FURANONE see CPM250
6-CYCLOHEXYL-1-HYDROXY-4-METHYL-2(1H)-PYRIDINONE compounded with 2-AMINOETHANOL (1:1) see BAR800
6-CYCLOHEXYL-1-HYDROXY-4-METHYL-2(1H)-PYRIDON, 2-AMINOETHANOL-SALZ (GERMAN) see BAR800
6-CYCLOHEXYL-1-HYDROXY-4-METHYL-2(1H)-PYRIDONE, 2-AMINOETHANOL-SALT see BAR800
6-CYCLOHEXYL-1-HYDROXY-4-METHYL-2(1H)-PYRIDONE ETHANOLAMINE SALT see BAR800
4-(β-CYCLOHEXYL-β-HYDROXYPHENETHYL)-1,1-DIMETHYLPIPERAZINIUM METHYLSULFATE see HFG400
N-(β-CYCLOHEXYL-β-HYDROXY-β-PHENYLETHYL)-N'-METHYLPIPERAZINE DIMETHYLSULFATE see HFG400
1-(3-CYCLOHEXYL-3-HYDROXY-3-PHENYLPROPYL)-1-METHYL-PIPERIDINIUM IODIDE see CPM750
1-(3-CYCLOHEXYL-3-HYDROXY-3-PHENYLPROPYL)-1-METHYL-PYRROLIDINIUM METHYL SULFATE see TJG225
1-(3-CYCLOHEXYL-3-HYDROXY-3-PHENYLPROPYL)-1-METHYL-PYRROLIDINIUM CHLORIDE see EAI875
1-(3-CYCLOHEXYL-3-HYDROXY-3-PHENYLPROPYL)-1-METHYL-PYRROLIDINIUM IODIDE see CNH550
(±)-N-((3-CYCLOHEXYL-3-HYDROXY-3-PHENYL)PROPYL)-N-METHYLPYRROLIDINIUM CHLORIDE see EAI875
2,2'-(CYCLOHEXYLIDENEBIS(4,1-PHENYLENEOXY)BIS(2-METHYLBUTANOIC ACID) see CMV700
2,2'-(4,4'-CYCLOHEXYLIDENEDIPHENOXY)-2,2'-DIMETHYLDIBUTYRIC ACID see CMV700
2,2'-CYCLOHEXYLIMINODIETHANOL see DMT400
CYCLOHEXYL ISOCYANATE see CPN500
2-(N-CYCLOHEXYL-N-ISOPROPYLAMINOMETHYL)-1,3,4-OXADIAZOLE see CPN750
CYCLOHEXYLISOPROPYLMETHYLAMINE HYDROCHLORIDE see PNN300
CYCLOHEXYL-ISOTHIOCYANAT (GERMAN) see ISJ000
CYCLOHEXYL MERCAPTAN (DOT) see CPB625
CYCLOHEXYLMETHANE see MIQ740
CYCLOHEXYLMETHANOL see HDH200
CYCLOHEXYLMETHYLAMINE see MIT000
1-CYCLOHEXYL-2-METHYLAMINOPROPAN (GERMAN) see PNN400
1-CYCLOHEXYL-2-METHYLAMINOPROPANE HYDROCHLORIDE see PNN300
1,1-CYCLOHEXYL-2-METHYLAMINOPROPANE-5,5-PHENYLETHYLBARBITURATE see CPO500
N-(2-CYCLOHEXYL-1-METHYLETHYL)-3,3-DIPHENYLPROPYLAMINE HYDROCHLORIDE see CPP000

1-CYCLOHEXYL-3-((p-(2-(5-METHYL-3-ISOXAZOLECARBOXAMIDO)ETHYL)PHENYL)SULFONYL)UREA see DBE885
4-(CYCLOHEXYLMETHYL)-α-(4-METHOXYPHENYL)-β-METHYL-1-PIPERIDINEETHANOL see RCA435
CYCLOHEXYL METHYLPHOSPHONOFLUORIDATE see MIT600
1-(CYCLOHEXYLMETHYL)PIPERIDINE HYDROCHLORIDE see PIR000
2-(4-CYCLOHEXYLMETHYLPIPERIDINO)-1-(4'-METHOXYPHENYL)-1-PROPANOL see RCA435
1-CYCLOHEXYL-N-METHYL-2-PROPANAMINE see PNN400
1-CYCLOHEXYL-1-NITROSOUREA see NJV000
1-(2-CYCLOHEXYLPHENOXY)-1-(2-IMIDAZOLINYL)ETHANE HYDROCHLORIDE see CPP750
α-CYCLOHEXYL-α-PHENYL-1-PIPERIDINEPROPANOL see PAL500
α-CYCLOHEXYL-α-PHENYL-1-PIPERIDINEPROPANOL HYDROCHLORIDE see BBV000
2-CYCLOHEXYL-2-PHENYL-4-PIPERIDINOMETHYL-DIOXOLANE-1,3 METHIODIDE see OLW400
2-CYCLOHEXYL-2-PHENYL-1-PIPERIDINO-1-PROPANOL see PAL500
1-CYCLOHEXYL-1-PHENYL-3-PIPERIDINO-PROPANOL, METHYLIODIDE see CPM750
1-CYCLOHEXYL-1-PHENYL-3-PYRROLIDINO-1-PROPANOL see CPQ250
1-CYCLOHEXYL-1-PHENYL-3-PYRROLIDINO-1-PROPANOL METHSULFATE see TJG225
1-CYCLOHEXYL-1-PHENYL-3-PYRROLIDINO-1-PROPANOL METHYL CHLORIDE see EAI875
1-CYCLOHEXYL-1-PHENYL-3-(1-PYRROLIDINYL)-1-PROPANOL see CPQ250
α-CYCLOHEXYL-α-(2-(PIPERIDINO)ETHYL)-BENZYLALCOHOL METHYLIODIDE see CPM750
CYCLOHEXYL 4-PYRIDYL KETONE see CPI375
1-CYCLOHEXYL-2-PYRROLIDINONE see CPQ275
N-CYCLOHEXYLPYRROLIDINONE see CPQ275
N-CYCLOHEXYLPYRROLIDONE see CPQ275
CYCLOHEXYLSULFAMIC ACID (9CI) see CPQ625
N-CYCLOHEXYLSULFENYLPHTHALIMIDE see CPQ700
CYCLOHEXYL SULPHAMATE SODIUM see SGC000
N-CYCLOHEXYLSULPHAMIC ACID see CPQ625
CYCLOHEXYLSULPHAMIC ACID see CPQ625
CYCLOHEXYLSULPHAMIC ACID, CALCIUM SALT see CAR000
3-CYCLOHEXYLSYDNONE IMINE MONOHYDROCHLORIDE see CPQ650
6-(4-(1-CYCLOHEXYL-1H-TETRAZOL-5-YL)BUTOXY)-3,4-DIHYDRO-2(1H)-QUINOLINONE see CMP825
N-(CYCLOHEXYLTHIO)PHTHALIMIDE see CPQ700
1-CYCLOHEXYL-3-p-TOLUENESULFONYLUREA see CPR000
1-CYCLOHEXYL-3-p-TOLYSULFONYLUREA see CPR000
CYCLOHEXYLTRICHLOROSILANE see CPR250
1-CYCLOHEXYLTRIMETHYLAMINE see CPR500
CYCLOL ACRYLATE see BFY250
CYCLOLEUCINE see AJK250
CYCLOLYT see DNU100
CYCLOMALTOHEPTAOSE see COW925
CYCLOMANDOL see DNU100
CYCLOMEN see DAB830
CYCLOMETHIAZIDE see CPR750
CYCLOMIDE DIN 295/S see BKF500
CYCLOMORPH see COX000
CYCLOMYCIN see CQH000, TBX000
CYCLON see HHS000
CYCLONAL see ERD500
CYCLONAL SODIUM see ERE000
CYCLONAMINE see DIS600
CYCLONE B see HHS000
CYCLONITE see CPR800
CYCLONITE, desensitized (UN 0483) (DOT) see CPR800
CYCLONITE, wetted (UN 0072) (DOT) see CPR800
CYCLONIUM IODIDE see OLW400
CYCLONOL see TLO500
cis,cis-CYCLOOCTA-1,5-DIENE see CPR825
1,5-CYCLOOCTADIENE (Z,Z) see CPR825
(1,5-CYCLOOCTADIENE)(2,4-PENTANEDIONATO)RHODIUM see CPR840
CYCLOOCTAFLUOROBUTANE see CPS000
CYCLOOCTANECARBAMIC ACID-1,1-BIS(p-FLUOROPHENYL)-2-PROPYNYL ESTER see BJW500
1,3,5,7-CYCLOOCTATETRAENE see CPS500
3-CYCLOOCTYL-1,1-DIMETHYLHARNSTOFF (GERMAN) see CPT000
3-CYCLOOCTYL-1,1-DIMETHYLUREA see CPT000
3-CYCLOOCTYL-1,1-DIMETHYL UREA mixed with BUTYNYL-3N-3-CHLOROPHENYLCARBAMATE see AFM375
N-CYCLOOCTYL-N',N'-DIMETHYLUREA see CPT000
CYCLOOXABUTANE see OMW000
CYCLOPAMINE see CPT750
CYCLOPAN see ERD500
CYCLOPAR see TBX250
CYCLOPENIL see FBP100
CYCLOPENTACYCLOHEPTENE see ASP600
CYCLOPENTADECANONE see CPU250

CYCLOPENTADECANONE, 3-METHYL- see MIT625
1,3-CYCLOPENTADIENE see CPU500
CYCLOPENTADIENE see CPU500
1,3-CYCLOPENTADIENE, DIMER see DGW000
pi-CYCLOPENTADIENYL COMPOUND with NICKEL see NDA500
CYCLOPENTADIENYL GOLD(1) see CPU750
CYCLOPENTADIENYLMANGANESE TRICARBONYL see CPV000
CYCLOPENTADIENYL SODIUM see CPV500
CYCLOPENTA(c)FURO(3′,2′:4,5)FURO(2,3-h)(1)BENZOPYRAN-1,11-DIONE,
  2,3,6a,8,9,9a-HEXAHYDRO-8,9-DICHLORO-4-METHOXY-, (6aS-(6a-α-8-β,9-α-
  9aα-))- see AEU500
α,β-CYCLOPENTAMETHYLENETETRAZOLE see PBI500
CYCLOPENTAMINE HYDROCHLORIDE see CPV609
CYCLOPENTANAMINE see CQA000
CYCLOPENTA(de)NAPHTHALENE see AAF500
CYCLOPENTANE see CPV750
CYCLOPENTANEACETIC ACID, 3-OXO-2-PENTYL-, METHYL ESTER see
  HAK100
1,3-CYCLOPENTANEDISULFONYL DIFLUORIDE see CPW250
CYCLOPENTANETHIOL see CPW300
4,5-CYCLOPENTANOFURAZAN-N-OXIDE see CPW325
CYCLOPENTANONE see CPW500
CYCLOPENTANONE-2-α,3-α-EPITHIO-5-α-ANDROSTAN-17-β-YL METHYL ACE-
  TAL see MCH600
CYCLOPENTANONE, 2-(2-HEXENYL)- see HFE513
CYCLOPENTANONE, 2-HEXYL- see HFO600
CYCLOPENTANONE OXIME see CPW750
CYCLOPENTANONE, 3-(2-OXOPROPYL)-2-PENTYL- see OOO100
CYCLOPENTENE see CPX750
CYCLOPENTENE, 1,2-DICHLOROHEXAFLUORO- see DFM050
CYCLOPENTENE, 1,2-DICHLORO-3,3,4,4,5,5-HEXAFLUORO-(9CI) see DFM050
1-CYCLOPENTENE-1-PROPANOL, 5-(1-METHYLETHENYL)-β,β,2-TRIMETHYL-,
  PROPANOATE see MJW300
2-CYCLOPENTENE-1-TRIDECANOIC ACID, SODIUM SALT see SFR000
6,7-CYCLOPENTENO-1,2-BENZANTHRACENE see BCI250
2-CYCLOPENTEN-1-ONE, 2-HEXYL- see HFO700
2-CYCLOPENTEN-1-ONE, 3-METHYL-2-(2-PENTENYL)-, (Z)- see JCA100
2-CYCLOPENTEN-1-ONE, 2-PENTYL- see PBW600
2-CYCLOPENTENYL-4-HYDROXY-3-METHYL-2-CYCLOPENTEN-1-ONE CHRY-
  SANTHEMATE see POO000
3-(2-CYCLOPENTEN-1-YL)-2-METHYL-4-OXO-2-CYCLOPENTEN-1-YL CHRYSAN-
  THEMUMATE see POO000
3-(2-CYCLOPENTENYL)-2-METHYL-4-OXO-2-CYCLOPENTENYL CHRYSANTHE-
  MUMMONOCARBOXYLATE see POO000
N-(1-CYCLOPENTEN-1-YL)-MORPHOLINE see CPY800
CYCLOPENTENYL PROPIONATE MUSK see MJW300
CYCLOPENTENYLRETHONYL CHRYSANTHEMATE see POO000
CYCLOPENTHIAZIDE see CPR750
CYCLOPENTIMINE see PIL500
CYCLOPENTOLATE HYDROCHLORIDE see CPZ125
CYCLOPENTYLAMINE see CQA000
CYCLOPENTYL 3,4-DIHYDROXYPHENYL KETONE see CQA100
3-(α-CYCLOPENTYL-4,6-DIMETHOXY-m-TOLUOYL)-PROPIONIC ACID SODIUM
  SALT see DOB325
2-CYCLOPENTYL-4,6-DINITROPHENOL see CQB250
3-CYCLOPENTYL ENOL ETHER of NORETHINDRONE ACETATE see QFA275
CYCLOPENTYL ETHER see CQB275
α-CYCLOPENTYLMANDELIC ACID (1-ETHYL-2-PYRROLIDINYL)METHYL ESTER
  HYDROCHLORIDE see PJI575
α-CYCLOPENTYLMANDELIC ACID-1-METHYL-3-PYRROLIDINYL ESTER HY-
  DROCHLORIDE see AEY400
CYCLOPENTYL MERCAPTAN see CPW300
3-CYCLOPENTYLMETHYL HYDROCHLOROTHIAZIDE DERIV see CPR750
3-(CYCLOPENTYLOXY)-19-NOR-17-α-PREGNA-3,5-DIEN-20-YN-17-OL ACETATE
  (ester) see QFA275
3-(CYCLOPENTYLOXY)-19-NOR-17-α-PREGNA-1,3,5(10)-TRIEN-20-YN-17-OL see
  QFA250
S-2-((5-CYCLOPENTYLPENTYL)AMINO)ETHYL THIOSULFATE see CQC250
1-(2-CYCLOPENTYLPHENOXY)-3-((1,1-DIMETHYLETHYL)AMINO)-2-PROPANOL,
  (S)- see PAP225
α-CYCLOPENTYL-α-PHENYL-1-PIPERIDINEPROPANOL HYDROCHLORIDE see
  CQH500
α-CYCLOPENTYL-2-THIOPHENEGLYCOLATE DIETHYL(2-HYDROXYETH-
  YL)METHYLAMMONIUM BROMIDE see PBS000
α-CYCLOPENTYL-2-THIOPHENEGLYCOLIC ACID-2-(DIETHYLAMINO)ETHYL ES-
  TER HYDROCHLORIDE see DHW400
CYCLOPHENYL see FBP100
CYCLOHEPTAGLUCAN see COW925
CYCLOPHOSPHAMIDE see CQC650
CYCLOPHOSPHAMIDE HYDRATE see CQC500
CYCLOPHOSPHAMIDE and MNU (1:2) see CQC600
CYCLOPHOSPHAMIDE-N-MONOCHLOROETHYL derivative see TNT500
CYCLOPHOSPHAMIDE MONOHYDRATE see CQC500
CYCLOPHOSPHAMIDUM see CQC500, CQC650
CYCLOPHOSPHAN see CQC500, CQC650

CYCLOPHOSPHANE see CQC500
CYCLOPHOSPHANUM see CQC500
CYCLOPHOSPHORAMIDE see CQC650
α-CYCLOPIAZONIC ACID see CQD000
CYCLOPIAZONIC ACID see CQD000
CYCLOPRATE see HCP500
5H-CYCLOPROPA(3,4)BENZ(1,2-e)AZULEN-5-ONE, 1,1a,1b,4,4a,7a,7b,8,9,9a-
  DECAHYDRO-4a,7-β, 9,9a-TETRAHYDROXY-3-(HYDROXYMETHYL)-1,1,6,8-
  TETRAMETHYL-, 9-ACETATE 9a-LAURATE see PGS500
CYCLOPROPANAMINE, 2-PHENYL-, trans-(±)-, SULFATE (2:1) see PET500
CYCLOPROPANE, liquefied (DOT) see CQD750
CYCLOPROPANE see CQD750
CYCLOPROPANE, 1-(N-AMINO)CARBAMOYL-2-METHYL- see MIV300
CYCLOPROPANECARBOXYLIC ACID, 3-(2-CHLORO-3,3,3-TRIFLUORO-1-PRO-
  PENYL)-2,2-DIMETHYL-, CYANO(3-PHENOXYPHENYL)METHYL ESTER see
  GJU600
CYCLOPROPANECARBOXYLIC ACID, 2,2-DIMETHYL-3-(2-METHYL-1-PROPE-
  NYL)-, ETHYLESTER (9CI) see EHM100
CYCLOPROPANECARBOXYLIC ACID, 2,2-DIMETHYL-3-(2-METHYLPROPENYL)-
  , p-(METHOXYMETHYL) BENZYL ESTER see MNG525
CYCLOPROPANECARBOXYLIC ACID, 2,2-DIMETHYL-3-(2-METHYLPROPENYL)-
  see CML650
CYCLOPROPANECARBOXYLIC ACID, 2,2-DIMETHYL-3-(2-METHYLPROPENYL)-
  , ETHYL ESTER (8CI) see EHM100
CYCLOPROPANECARBOXYLIC ACID, 2,2-DIMETHYL-3-(2-METHYLPROPENYL)-
  , (2-METHYL-5-(2-PROPYNYL)-3-FURYL)METHYL ESTER see PMN700
CYCLOPROPANECARBOXYLIC ACID, 2,2-DIMETHYL-3-(2-METHYL-1-PROPE-
  NYL)-, ETHYLESTER see EHM100
CYCLOPROPANECARBOXYLIC ACID, HEXADECYL ESTER see HCP500
CYCLOPROPANE, METHOXY-(9CI) see CQE750
CYCLOPROPYLAMINE see CQE250
N-CYCLOPROPYL-4-AMINO-3,5-DICHLOROBENZAMIDE see AJK500
N-CYCLOPROPYL-3,5-DICHLORO-4-AMINOBENZAMIDE see AJK500
1-(4-(2-(CYCLOPROPYLMETHOXY)ETHYL)PHENOXY)-3-ISOPROPYLAMINOPRO-
  PAN-2-OL HYDROCHLORIDE see KEA350
1-N-CYCLOPROPYLMETHYL-3,14-DIHYDROXYMORPHINAN see CQF079
2-CYCLOPROPYLMETHYL-5,9-DIMETHYL-2′-HYDROXY-6,7-BENEOMORPHAN
  see COV500
3-CYCLOPROPYLMETHYL-6(eq),11(ax)-DIMETHYL-2,6-METHANO-3-BENZAZO-
  CIN-8-OL see COV500
(5-α)-17-(CYCLOPROPYLMETHYL-4,5-EPOXY-3,14-DIHYDROXY-MORPHINAN-6-
  ONE) (9CI) see CQF099
CYCLOPROPYL METHYL ETHER see CQE750
3-(CYCLOPROPYLMETHYL)1-1,2,3,4,5,6-HEXAHYDRO-6,11-DIMETHYL-2,6-
  METHANO-3-BENZAZOCIN-8-OL see COV500
N-CYCLOPROPYLMETHYL-14-HYDROXYDIHYDROMORPHINONE see CQF099
2-CYCLOPROPYLMETHYL-2′-HYDROXY-5,9-DIMETHYL-6,7-BENZOMORPHAN
  see COV500
(−)-17-CYCLOPROPYLMETHYLMORPHINAN-3,4-DIOL see CQF079
17-(CYCLOPROPYLMETHYL)MORPHINAN-3-OL see CQG750
N-CYCLOPROPYLMETHYLNOROXYMORPHONE see CQF099
1-CYCLOPROPYLMETHYL-4-PHENYL-6-CHLORO-2(1H)-QUINAZOLINONE see
  CQF125
N-(CYCLOPROPYLMETHYL)-α,α,α-TRIFLUORO-2,6-DINITRO-N-PROPYL-p-TOLUI-
  DINE see CQF250
1-(o-CYCLOPROPYLPHENOXY)-3-(ISOPROPYLAMINO)-2-PROPANOL HYDRO-
  CHLORIDE see PMF525
dl-1-(o-CYCLOPROPYLPHENOXY)-3-ISOPROPYLAMINO-2-PROPANOL HYDRO-
  CHLORIDE see PMF535
CYCLORPHAN see CQG750
CYCLORYL 21 see SIB600
CYCLORYL OS see OFU200
CYCLORYL TAWF see SON000
CYCLORYL WAT see SON000
CYCLOSAN see MCW000, MCY300
CYCLO-d-SERINE see CQH000
CYCLOSERINE see CQH000
CYCLOSIA see CMS850
CYCLOSPASMOL see DNU100
CYCLOSPORIN see CQH100
CYCLOSPORIN A see CQH100
CYCLOSPORINE see CQH100
CYCLOSPORINE A see CQH100
CYCLOSTIN see CQC650
CYCLOTEN see HMB500
CYCLOTETRAMETHYLENE OXIDE see TCR750
CYCLOTETRAMETHYLENE SULFONE see SNW500
CYCLOTETRAMETHYLENETETRANITRAMINE (dry or unphlegmatized) (DOT)
  see CQH250
CYCLOTETRAMETHYLENE TETRANITRAMINE see CQH250
CYCLOTETRAMETHYLENETETRANITRAMINE, wetted (UN 0226) (DOT) see
  CQH250
CYCLOTETRAMETHYLENETETRANITRAMINE, desensitized (UN 0483) (DOT)
  see CQH250
CYCLOTON V see HCQ500

CYCLOTRIMETHYLENENITRAMINE see CPR800
CYCLOTRIMETHYLENETRINITRAMINE see CPR800
CYCLOTRIMETHYLENETRINITRAMINE, desensitized (UN 0483) (DOT) see CPR800
CYCLOTRIMETHYLENETRINITRAMINE, wetted (UN 0072) (DOT) see CPR800
CYCLOURON see CPT000
CYCLOVIROBUXIN D see CQH325
CYCLOVIROBUXINE see CQH325
CYCLOVIROBUXINE D see CQH325
CYCLURON see CPT000
CYCOCEL see CMF400
CYCOCEL-EXTRA see CMF400
CYCOGAN see CMF400
CYCOGAN EXTRA see CMF400
CYCOLAMIN see VSZ000
CYCRIMINE HYDROCHLORIDE see CQH500
CYCTEINAMINE see AJT250
CYDRIN see CQK500
CYDTA see CPB120
CYFEN see DSQ000
CYFLEE see CQL250, FAB600
CYFOS see IMH000
CYGON see DSP400
CYGON INSECTICIDE see DSP400
CYHALOTHRIN see GJU600
CYHALOTHRINE see GJU600
CYHEPTAMIDE see CQH625
CYHEPTAMINE see CQH625
CYHEXATIN see CQH650
3-CYJANOPIRYDYNA see NDW515
CYJANOWODOR (POLISH) see HHS000
CYKAZINE see COU000
CYKLOHEKSAN (POLISH) see CPB000
CYKLOHEKSANOL (POLISH) see CPB750
CYKLOHEKSANON (POLISH) see CPC000
CYKLOHEKSEN (POLISH) see CPC579
CYKLOHEXANTHIOL see CPB625
CYKLOHEXYLAMINACETAT (CZECH) see CPG000
CYKLOHEXYLESTER KYSELINY THIOKYANOOCTOVE (CZECH) see TFF000
CYKLOHEXYLMERKAPTAN (CZECH) see CPB625
CYKLOHEXYLTHIOKYANOACETAT (CZECH) see TFF000
CYKLONIT see CPR800
CYKOBEMINET see VSZ000
CY-L 500 see CAQ250
CYLAN see CAR000, DXN600, PGW750
CYLERT see PAP000
CYLOCIDE see AQQ750, AQR000
CYLPHENICOL see CDP250
CYMAG see SGA500
CYMARIGENIN see SMM500
CYMARIN see CQH750
CYMARINE see CQH750
3-β-(β-d-CYMAROSYLOXY)-5,14-DIHYDROXY-19-OXO-5-β-CARD-20(22)-ENOLIDE see CQH750
CYMATE see BJK500
CYMBI see AIV500, AOD125
CYMBUSH see RLF350
CYMEL 200 see MCB050
CYMEL 202 see MCB050
CYMEL 235 see MCB050
CYMEL 245 see MCB050
CYMEL 255 see MCB050
CYMEL 285 see MCB050
CYMEL 300 see MCB050
CYMEL 301 see MCB050
CYMEL 303 see HDY500, MCB050
CYMEL 305 see MCB050
CYMEL 323 see MCB050
CYMEL 325 see MCB050
CYMEL 327 see MCB050
CYMEL 350 see MCB050
CYMEL 370 see MCB050
CYMEL 373 see MCB050
CYMEL 380 see MCB050
CYMEL 385 see MCB050
CYMEL 412 see MCB050
CYMEL 428 see MCB050
CYMEL 481 see MCB050
CYMEL 482 see MCB050
CYMEL 1080 see MCB050
CYMEL 1116 see MCB050
CYMEL 1130 see MCB050
CYMEL 1133 see MCB050
CYMEL 1135 see MCB050
CYMEL 1156 see MCB050

CYMEL 1158 see MCB050
CYMEL 1161 see MCB050
CYMEL 1168 see MCB050
CYMEL 1370 see MCB050
CYMEL 243-3 see MCB050
CYMEL 247-10 see MCB050
CYMEL 7273-7 see MCB050
CYMEL see MCB000, MCB050
CYMEL C 1156 see MCB050
CYMEL HM 6 see MCB050
CYMEL 265J see MCB050
CYMEL 266J see MCB050
CYMEL 1130-235J see MCB050
CYMEL 1130-254J see MCB050
CYMEL 1130-285J see MCB050
CYMEL 401 RESIN see MCB050
CYMEL XM 1116 see MCB050
p-CYMENE see CQI000
CYMENE see CQI000
p-CYMENE-7-CARBOXALDEHYDE see IRA000
2-p-CYMENOL see CCM000
3-p-CYMENOL see TFX810
p-CYMEN-3-OL see TFX810
m-CYMEN-4-OL see IQJ000
p-CYMEN-7-OL see CQI250
CYMETHION see MDT750
CYMETOX see MIW250
CYMIDON see KFK000
CYMOL see CQI000
CYMONIC ACID see FIC000
CYMOXANIL see COM300
m-CYM-5-YL METHYLCARBAMATE see CQI500
CYNARON see MDT740
CYNCAL 80 see QAT520
CYNEM see EPC500
CYNKOMIEDZIAN see ZJS300
CYNKOTOX see EIR000
CYNKU TLENEK (POLISH) see ZKA000
CYNOGAN see BMM650
CYNOTOXIN see SMM500
CYOCEL see CMF400
CYODRIN see COD000
CYOLAN see DXN600
CYOLANE see PGW750
CYOLANE INSECTICIDE see DXN600, PGW750
CYOMETRINIL see COP700
CYP see CON300
CYPENTIL see PIL500
CYPERKILL see RLF350
CYPERMETHRIN see RLF350
CYPERUS SCARIOSUS OIL see NAE505
CYPIP see DIW000
CYPONA see DGP900
CYPRAZINE see CQI750
CYPRESS OIL see CQJ000
CYPREX see DXX400
CYPREX 65W see DXX400
CYPROHEPTADIENE HYDROCHLORIDE see PCI250
CYPROHEPTADINE see PCI250, PCI500
CYPROHEPTADINE HYDROCHLORIDE see PCI250
CYPROME ETHER see CQE750
CYPROMID see CQJ250
CYPRON see MQU750
CYPROSTERONE ACETATE see CQJ500
CYPROTERONE ACETATE see CQJ500
CYPROTERON-R ACETATE see CQJ500
CYRAL see DBB200
CYREDIN see VSZ000
CYREN see DKA600
CYREN B see DKB000
CYREZ 933 see MCB050, UTU500
CYREZ 963 see MCB050
CYREZ 966 see MCB050
CYREZ see MCB050
CYREZ 963 P see MCB050
CYREZ 963P-A see MCB050
CYREZ 963 RESIN see HDY500
CYRSTHION see EKN000
CYSTAMIN see HEI500
CYSTAMINE see MCN500
CYSTAMINE DIHYDROCHLORIDE see CQJ750
CYSTAMINE "MCCLUNG" see PDC250
CYSTAPHOS see AKB500
CYSTAPHOS SODIUM SALT see AKB500
CYSTEAMIDE see AJT250

CYSTEAMINE see AJT250
CYSTEAMINE HYDROCHLORIDE see MCN750
CYSTEAMINHYDROCHLORID (GERMAN) see MCN750
CYSTEIN see CQK000
CYSTEINAMINE DISULFIDE see MCN500
l-(+)-CYSTEINE see CQK000
l-CYSTEINE see CQK000
CYSTEINE see CQK000
CYSTEINE CHLORHYDRATE see CQK250
CYSTEINE DISULFIDE see CQK325
CYSTEINE ETHYL ESTER HYDROCHLORIDE see EHU600
CYSTEINE-GERMANIC ACID see CQK100
CYSTEINE HYDRAZIDE see CQK125
l-CYSTEINE HYDROCHLORIDE see CQK250
CYSTEINE HYDROCHLORIDE see CQK250
l-CYSTEINE MONOHYDROCHLORIDE (FCC) see CQK250
(l-CYSTEINE)TETRAHYDROXYGERMANIUM see CQK100
CYSTIN see CQK325
CYSTINAMIN (GERMAN) see MCN500
l-CYSTINE see CQK325
(—)-CYSTINE see CQK325
CYSTINE ACID see CQK325
CYSTINEAMINE see MCN500
l-CYSTINE-BIS(N,N-β-CHLOROETHYL)HYDRAZIDEHYDROBROMIDE see
    CQK500
CYSTISINE see CQL500
CYSTO-CONRAY see IGC000
CYSTOGEN see HEI500
CYSTOGRAFIN see AOO875
CYSTOIDS ANTHELMINTIC see HFV500
CYSTOKON see AAN000
CYSTOPYRIN see PDC250
CYSTORELIN see LIU360
CYSTURAL see PDC250
CYTACON see VSZ000
CYTADREN see AKC600
CYTAMEN see VSZ000
CYTARABIN see AQQ750
CYTARABINA see AQQ750
CYTARABINE see AQQ750
CYTARABINE HYDROCHLORIDE see AQR000
CYTARABINOSIDE see AQQ750
CYTEL see DSQ000
CYTEMBENA see CQK600, SIK000
CYTEN see DSQ000
CYTHIOATE see CQL250
CYTHION see MAK700
CYTIDINDIPHOSPHOCHOLIN see CMF350
CYTIDINE see CQM500
CYTIDINE 5′-(CHOLINE DIPHOSPHATE) see CMF350
CYTIDINE CHOLINE DIPHOSPHATE see CMF350
CYTIDINE 5′-DIPHOSPHATE CHOLINE see CMF350
CYTIDINE DIPHOSPHATE CHOLINE see CMF350
CYTIDINE DIPHOSPHATE CHOLINE ESTER see CMF350
CYTIDINE DIPHOSPHATE CHOLIN ESTER see CMF350
CYTIDINE 5′-DIPHOSPHOCHOLINE see CMF350
CYTIDINE DIPHOSPHOCHOLINE see CMF350
CYTIDINE DIPHOSPHORYLCHOLINE see CMF350
CYTIDOLINE see CMF350
CYTISINE see CQL500
CYTITONE see CQL500
CYTOBION see VSZ000
CYTOCHALASIN B see CQM125
CYTOCHALASIN D see ZUS000
CYTOCHALASIN E see CQM250
CYTOCHALASIN-H see PAM775
CYTOPHOSPHAN see CQC500, CQC650
CYTOSAR see AQQ750
CYTOSAR HYDROCHLORIDE see AQR000
CYTOSAR-U see AQQ750
CYTOSINE β-D-ARABINOSIDE see AQQ750
CYTOSINE-β-ARABINOSIDE see AQQ750
CYTOSINEARABINOSIDE see AQQ750
CYTOSINE ARABINOSIDE HYDROCHLORIDE see AQR000
CYTOSINE ARABINOSIDE PALMITATE see AQS875
CYTOSINE, 1-β-D-ARABINOSYL- see AQQ750
CYTOSINE (8CI) see CQM600
CYTOSINE RIBOSIDE see CQM500
CYTOSINIMINE see CQM600
CYTOSTASAN see CQM750
CYTOTEC see MJE775
CYTOVIRIN see BLX500
CYTOXAL ALCOHOL see CQN000
CYTOXAN see CQC500, CQC650
CYTOXYL ALCOHOL CYCLOHEXYLAMMONIUM SALT see CQN000

CYTROL see AMY050
CYTROL AMITROLE-T see AMY050
CYTROLANE see DHH400
CYTROLE see AMY050
CYURAM DS see TFS350
CYZINE PREMIX see ABY900
CYZONE see PFL000
CZON see CCS635
CZT see CLX000
CZTEROCHLOREK WEGLA (POLISH) see CBY000
2,3,7,8-CZTEROCHLORODWUBENZO-p-DWUOKSYNY (POLISH) see TAI000
1,1,2,2-CZTEROCHLOROETAN (POLISH) see TBQ100
CZTEROCHLOROETYLEN (POLISH) see PCF275
CZTEROETHLEK OLOWIU (POLISH) see TCF000


$D_2$ see DBB800
D-13 see AHL000
2,4-D see DAA800
D-40 see AFO250
D 50 see AAX250, DAA800
D-50 see POL500
D 109 see COU250
D 206 see DYB600
D 268 see DRW000
D301 see BDJ600
D-365 see IRV000
D-638 see BPJ750
D-649 see BPJ500
D-695 see BKC500
D-701 see BKD000
D-703 see EEY000
D 735 see CCC500
dl-832 see TDI600
838-D see MHQ775
D 854 see CJT750
D 860 see BSQ000
D 100-2 see MCB050
D-1126 see FMU225
D 1221 see CBS275
D 1308 see EOS100
D-1410 see DSP600
D 1593 see CIP500
D-10,242 see DAB200
D-13,312 see TAL560
D-D see DGG000
DL-8280 see OGI300
D-90-A see CGL250
DA-241 see PEO750
DA 339 see DWA500
DA-398 see MCH550
DA-688 see GDG200
DA-1773 see SJJ175
DA 2370 see PEW000
DA see CGN000, DNA200
2,4-DAA see DBO000
1,2-DAA (RUSSIAN) see DBO800
DAAB see DWO800
DAAE see DCN800
2,4-DAA SULFATE see DBO400
DAB see DOT300
DABCO S-25 see DCK400
DABCO see DCK400
DABCO CRYSTAL see DCK400
DABCO EG see DCK400
DABCO 33LV see DCK400
DABCO R-8020 see DCK400
DABI see DOT600
DABICYCLINE see HOH500
DAB-O-LITE P 4 see WCJ000
DAB-N-OXIDE see DTK600
DABROSIN see ZVJ000
DABYLEN see BAU500, BBV500
DAC 2797 see TBQ750
DACAMINE see DAA800, TAA100
DACAMOX see DAB400
DACARBAZINE see DAB600
2,4-D ACETATE see DFY800
2,4-D ACID see DAA800
DACONATE 6 see MRL750
DACONIL see TBQ750
DACONIL 2787 FLOWABLE FUNGICIDE see TBQ750
DACORENE HYDROCHLORIDE see BGO000
DACORTIN see PLZ000
DACOSOIL see TBQ750

DACOTE see CAT775
DACOVIN see PKQ059
DAC PRO see DGL200
DACTHAL see TBV250
DACTIL see EOY000
DACTIL HYDROCHLORIDE see EOY000
DACTIN see DFE200
DACTINOL see RNZ000
DACTINOMYCIN see AEB000
DACTINOMYCIN (10%), ACTINOMYCIN C2 (45%), and ACTINOMYCIN C3 (45%) mixture see AEA750
DAD see DCI600
DADA see DNM400
DADDS see SNY500
DADEX see BBK500
DADIBUTOL see TGA500
DADOX d-CITRAMINE see BBK500
DADPE see OPM000
DADPM see MJQ000
DADPS see SOA500
DAEP see DOP200
DAF 68 see DVL700
DAFEN see LJR000
DAFF see BJR625
DAFFODIL see DAB700
DAFTAZOL HYDROCHLORIDE see DCA600
DAG see DCI600
DAGADIP see TNP250
DAGC see AGD250
DAGENAN see PPO000
DAGUTAN see SJN700
1,1-DAH see DBK100
1,2-DAH HYDROCHLORIDE see DBK120
DAI CARI XBN see BQK250
DAICEL 1150 see SFO500
DAI-EI ACID PURE BLUE VX see CMM062
DAIFLOIL 3 see KDK000
DAIFLOIL 10 see KDK000
DAIFLOIL 20 see KDK000
DAIFLOIL 50 see KDK000
DAIFLOIL 100 see KDK000
DAIFLON 22 see CFX500
DAIFLON see CLQ750
DAIFLON CTF3-D 55P see KDK000
DAIFLON CTFE see KDK000
DAIFLON D 45S see KDK000
DAIFLON M 300 see KDK000
DAIFLON M 300P see KDK000
DAIFLON S 3 see FOO000
DAILON see DXQ500
DAIMETON see SNL800
DAINICHI BENZIDINE YELLOW 2GR see CMS208
DAINICHI BENZIDINE YELLOW GRT see DEU000
DAINICHI BRILLIANT SCARLET G see CMS160
DAINICHI BRILLIANT SCARLET RG see CMS160
DAINICHI CHROME ORANGE R see LCS000
DAINICHI CHROME YELLOW G see LCR000
DAINICHI FAST ORANGE RR see CMS145
DAINICHI FAST RED B BASE see NEQ000
DAINICHI FAST SCARLET G BASE see NMP500
DAINICHI LAKE RED C see CHP500
DAINICHI PERMANENT RED 4 R see MMP100
DAIOMIN see TES800
DAIPIN see DAB750
DAIRYLIDE YELLOW AAA see DEU000
DAISAZIN see TES800
DAISEN see EIR000
DAISHIKI AMARANTH see FAG020
DAISHIKI BRILLIANT SCARLET 3R see FMU080
DAISOLAC see PJS750
DAITO ORANGE BASE R see NEN500
DAITO ORANGE SALT RD see CEG800
DAITO RED BASE B see NEQ000
DAITO RED BASE 3GL see KDA050
DAITO RED BASE TR see CLK220
DAITO RED SALT TR see CLK235
DAITO SCARLET BASE G see NMP500
DAIYA FOIL see PKF750
DAKINS SOLUTION see SHU500
DAKTARIN see MQS550
DAKTIN see DFE200
DAKTOSE B see DIB300
DAKURON see SKQ400
DALACIN C see CMV675
DALAPON 85 see DGI400

DALAPON (USDA) see DGI400
DALAPON see DGI600
DALAPON SODIUM see DGI600
DALAPON SODIUM SALT see DGI600
DALF see MNH000
DALGOL see EQL000
DALMADORM see DAB800
DALMADORM HYDROCHLORIDE see DAB800
DALMANE see DAB800
DALMATE see DAB800
DALMATIAN SAGE OIL see SAE500
DALMATION INSECT FLOWERS see POO250
DAL-E-RAD 100 see DXE600
DAL-E-RAD see MRL750
DALTOGEN see TKP500
DALTOLITE FAST ORANGE G see CMS145
DALTOLITE FAST YELLOW GT see DEU000
DALYSEP see MFN500
DALZIC see ECX100
DAM-57 see LJH000
DAM see OMY910
DAMA de DIA (PUERTO RICO) see DAC500
DAMA de NOCHE (PUERTO RICO) see DAC500
DAMANTOYLDIAZOMETHANE see DAB807
DAMC see DPJ800
DAMILAN see EAH500
DAMILEN HYDROCHLORIDE see EAI000
2,4-D AMINE SALT see DFY800
7-(D-2-AMINO-2-(1,4-CYCLOHEXADIENYL)ACETAMIDE)-3-METHOXY-3-CEPHEM-4-CARBOXYLIC ACID see CCS530
DAMINOZIDE (USDA) see DQD400
2,4-D AMMONIUM SALT see DAB815
1,4-DA-2-MOA see DBX000
DAMPA D see DAB820
DAMP-ES see AEF250
DA-2-N see DSU800
DAN see DSU600
DANA see NJW500
DANABOL see DAL300
DANAMID see PJY500
DANAMINE see DJS200
DANANTIZOL see MCO500
DANAZOL see DAB830
DANDELION (JAMAICA) see CNG825
DANERAL see TMK000
DANEX see TIQ250
DANFIRM see AAX250
DANIFOS see DIX800
DANILON see SOX875
DANILONE see PFJ750
DANINON see CKL500
DANITOL see DAB825
DANIZOL see MMN250
DANOCRINE see DAB830
DANOL see DAB830
DANSYL see DPN200
DANSYL CHLORIDE see DPN200
DANTAFUR see NGE000
DANTEN see DKQ000, DNU000
DANTHION see PAK000
DANTHRON see DMH400
DANTINAL see DKQ000
DANTOIN see DFE200, DNU000
DANTOINAL KLINOS see DKQ000
DANTOINE see DKQ000
DANTOROLENE SODIUM see DAB840
DANTRIUM see DAB840
DANTRIUM HEMIHEPTAHYDRATE see DAB840
DANTROLENE see DAB845
DANTROLENE SODIUM see DAB840
DANTRON see DMH400
DANU see DPN400
DANUVIL 70 see PKQ059
DAONIL see CEH700
DA79P see LGF875
DAP see DOT000, POJ500
DAPA see DNM400, DOU600
DAPACRYL see BGB500
DAPAZ see MQU750
DAPHENE see DSP400
DAPHNE MEZEREUM see LAR500
DAPHNETOXIN see DAB850
DAPLEN see PJS750
DAPLEN AD see PMP500
DAPLEN 1810 H see PJS750

DAPM see MJQ000
DAPOCEL see DNM400
DAPON 35 see DBL200
DAPON R see DBL200
DAPPU 100 see IMK000
DAPRISAL see ABG750
DAPSONE see SOA500
DAPTAZILE HYDROCHLORIDE see DCA600
DAPTAZOLE HYDROCHLORIDE see DCA600
DARACLOR see TGD000
DARAL see VSZ100
DARAMIN see ANT500, CAM750
DARAMMON see ANE500
DARAN see CGW300
DARAN CR 6795H see CGW300
DARANIDE see DEQ200
DARAPRAM see TGD000
DARAPRIM see TGD000
DARAPRIME see TGD000
DARATAK see AAX250
DARBID see DAB875
DAR-CHEM 14 see SLK000
DARCIL see PDD350
DARCO see CBT500
DAREBON see RDK000
DARENTHIN see BMV750
DARID QH see SEH000
DARILOID QH see SEH000
DARK GREEN EMBL see CMO830
DAROLON see ACE000
DAROPERVAMIN see DBA800
DARVAN 1 see BLX000
DARVAN No. 1 see BLX000
DARVIC 110 see PKQ059
DARVIS CLEAR 025 see PKQ059
DARVON see DAB879
DARVON COMPOUND see ABG750
DARVON HYDROCHLORIDE see PNA500
DARVON-N see DAB880, DYB400
DAS see AOP250, DOU600
DASANIDE see DEQ200
DASANIT see FAQ800
DASD see FCA100
DASEN see SCA625
DASERD see GGS000
DASEROL see GGS000
DASHEEN see EAI600
DASIKON see ABG750
DASKIL see NCQ900
DATC see DBI200
DATHROID see LFG050
DATRIL see HIM000
DATURA STRAMONIUM see SLV500
DATURINE see HOU000
DAUCUS OIL see CCL750
DAUNAMYCIN see DAC000
DAUNOBLASTIN see DAC200
DAUNOBLASTINA see DAC200
DAUNOMYCIN see DAC000
DAUNOMYCIN BENZOYLHYDRAZONE see ROU800
DAUNOMYCIN CHLOROHYDRATE see DAC200
DAUNOMYCIN HYDROCHLORIDE see DAC200
DAUNOMYCINOL see DAC300
DAUNORUBICIN, BENZOYLHYDRAZONE, MONOHYDROCHLORIDE see
  ROZ000
DAUNORUBICIN see DAC000
DAUNORUBICINE see DAC000
DAUNORUBICIN HYDROCHLORIDE see DAC200
DAUNORUBICINOL see DAC300
DAURAN see AFJ400
DAVA see VGU750
DAVANA OIL see DAC400
DAVISON SG-67 see SCH000
DAVITAMON D see VSZ100
DAVITAMON PP see NCQ900
DAVITIN see VSZ100
DAVOSIN see AKO500
DAWE'S DESTROL see DKA600
DAWSON 100 see MHR200
DAWSONITE see DAC450
DAXAD 11 see BLX000
DAXAD 15 see BLX000
DAXAD 18 see BLX000
DAXAD No. 11 see BLX000
DAY BLOOMING JESSAMINE see DAC500

DAYFEN see DOZ000, LJR000
DAZOMET see DSB200
DAZZEL see DCM750
DB 1 see MAD100
2,4-DB see DGA000
DB 133 see FQL200
DB 134 see DNF500
DB 135 see DSC100
DB 136 see DNF450
DB 138 see EID250
DB-905 see TBS000
DB 2041 see IDJ500
Df B see DAK200
DB see CQK100
1,2,5,6-DBA see DCT400
DB(a,c)A see BDH750
DB(a,h)A see DCT400
DBA see DCT400, DQJ200
DB(a,h)AC see DCS400
DB(a,j)AC see DCS600
DBA-1,2-DIHYDRODIOL see DMK200
trans-DBA-3,4-DIHYDRODIOL see DLD400
DBA-5,6-EPOXIDE see EBP500
DBB see DDL000
7H-DB(c,g)C see DCY000
DBCP see DDL800
DBD see ASH500, DDJ000
DBDPO see PAU500
DBE see EIY500
DBED see BHB300
DBED DIACETATE see DDF800
DBED DIHYDROCHLORIDE see DDG400
DBED DIPENCILLIN G see BFC750
DBED PENICILLIN see BFC750
DBF see DEC400, DJY100
1,1-DBH see DEC725
DBH see BBP750, BBQ500, DDO800
DBHMD see DEC699
DB 2182 HYDROCHLORIDE see IFZ900
DBI see PDF000
DBI-TD see PDF250
DBM see DDP600, DED600
DBMP see BFW750
2,6-DBN see DER800
DBN (the herbicide) see DER800
DBN see BRY500
DBNA see BRY500
DBNPA see DDM000
DBOT see DEF400
DB(a,e)P see NAT500
DB(a,i)P see BCQ500
DB(a,l)P see DCY400
DBP see DEH200, DES000
DBPC (technical grade) see BFW750
DBQ see DDV500
2,4-DB SODIUM SALT see EAK500
D.B.T.C. see DDY200
DBTL see DDV600
2,4-D BUTOXYETHANOL ESTER see DFY709
2,4-D 2-BUTOXYETHYL ESTER see DFY709
2,4-D BUTOXYETHYL ESTER see DFY709
2,4-D BUTYL ESTER see BQZ000
2,4-D BUTYRIC see DGA000
DBV see BRA625
DC-11 see TEF725
DC 360 see SCR400
DC 0572 see AHI875
3,4-DCA see DEO300
DC-38-A see GEO200
DCA 70 see AAX250
DCA see DAQ800, DEL000, DEO300, DFE200
DCAA see AFH275
1,4-DCB see DEV000
DC-45-B2 see TMO775
DCB see COC750, DEP600, DEQ600, DER800, DES000, DEV000
DCBA see BIA750
D and C BLUE No. 3 see ERG100
D&C BLUE No. 4 see FAE000, FMU059
D and C BLUE No. 9 see DFN425
D&C BLUE NUMBER 1 see BJI250
DCBN see DGM600
DCDD see DAC800
2,3-DCDT see DBI200
1-1-DCE see VPK000
1,2-DCE see EIY600

DCEE see DFJ050
D and C GREEN 1 see NAX500
D&C GREEN No. 4 see FAF000
D&C GREEN No. 6 see BLK000
D&C GREEN No. 8 see TNM000
DCH 21 see ERE200
DCHA see DGT600
DCHFB see DFM000
DCI see DFN400
DCI LIGHT MAGNESIUM CARBONATE see MAC650
DCIP (nematocide) see BII250
DCIP see BII250
DCL see DFN500
DCM see DFO000, DFO800, DGQ200, MJP450
DCMA-13-50-9 see CMS135
DCMA see DFO800
DCMC see DAI000
DCMO see CCC500
DCMOD see DLV200
DCMU see DXQ500
DCNA (fungicide) see RDP300
DCNA see RDP300
DCNB see DFT600
DCNU see CLX000
D&C ORANGE No. 2 see TGW000
D&C ORANGE No. 3 see FAG010
D and C ORANGE No. 4 see CMM220
D and C ORANGE NUMBER 15 see DMG800
2,4-DCP see DFX800
DCP see DFX800, DGG500
DCPA see DGI000, DGW400
DCPC see BIN000
DCPE see BIN000
DCPM see NCM700
cis-DCPO see DAC975
trans-DCPO see DGH500
D&C RED 2 see FAG020
D&C RED No. 3 see FAG040
D&C RED No. 5 see FMU070
D and C RED NO. 6 see CMS155
D and C RED No. 8 see CMS150
D&C RED No. 9 see CHP500
D&C RED No. 14 see TGX000
D & C RED NO. 17 see OHA000
D&C RED No. 19 see FAG070
D&C RED No. 21 see BMO250
D&C RED No. 22 see BNH500
D and C RED NO. 28 see ADG250
D and C RED NO.31 see CMS160
D and C RED No. 33 see CMS231
D and C RED NO. 35 see MMP100
D.C.S. see BGJ750
DCTA see CPB120
DCU see DGQ200
DC-38-V see GEO200
D+C VIOLET No. 2 see HOK000
D and C YELLOW NO. 11 see CMS245
D and C YELLOW No. 10 see CMM510
D and C YELLOW NO. 5 see FAG140
D&C YELLOW No. 7 see FEV000
D&C YELLOW No. 8 see FEW000
DD 234 see DAB750
DDA see DRR800
DDC see DQY950, SGJ000
cis-DDCP see DAD040
trans(+)-DDCP see DAD050
trans(−)-DDCP see DAD075
2,4′-DDD see CDN000
o,p′-DDD see CDN000
p,p′-DDD see BIM500
DDD see BIM500
DDDM see MJM500
p,p′-DDE see BIM750
DDE see BIM750
DDETA see HMQ500
2,4-D DIMETHYLAMINE SALT see DFY800
DDM see DSU000, MJM500, MJQ000
DD-METHYL ISOTHIOCYANATE MIXTURE see MLC000
DD MIXTURE see DGG000
DDMP see MQR100
DDN see LBU200
DDNO see DRS200
DDNP see DUR800
DDOA see ABC250
cis-DDP see PJD000

DDP see PJD000
DDS see DJC875, DXV000, SOA500
DDS A see DXV000
DD SOIL FUMIGANT see DGG000
o,p′-DDT see BIO625
p,p′-DDT see DAD200
DDT see DAD200
DDT DEHYDROCHLORIDE see BIM750
DDVF see DGP900
DDVP see DGP900
D.E. see DCJ800
DEA see DHF000, DIS700
N-DEACETYLCHOLCHICEINE see TLN750
DEACETYLCHOLCHICEINE see TLN750
N-DEACETYLCOLCHICINE see TLO000
DEACETYLCOLCHICINE see TLO000
DEACETYLCOLCHICINE I-TARTRATE see DBA175
N-DEACETYLCOLCHICINE I-TARTRATE(1:1), HYDRATE see DBA175
N-DEACETYL-10-DEMETHOXY-10-METHYLTHIOCOLCHICINE see DBA200
DEACETYLDEMETHYLTHYMOXAMINE see DAD500
DEACETYL-HT-2 TOXIN see DAD600
DEACETYLLANATOSIDE B see DAD650
DEACETYL-LANATOSIDE B (8Cl) see DAD650
DEACETYLLANATOSIDE C see DBH200
3-DE(2-(ACETYLMETHYLAMINO)PROPIONYLOXY)-3-HYDROXYMAYTANSINE ISOVALERATE (ESTER) see APE529
N-DEACETYL-N-METHYLCOLCHICINE see MIW500
DEACETYL-N-METHYLCOLCHICINE see MIW500
DEACETYLMETHYLCOLCHICINE see MIW500
N-DEACETYLMETHYLTHIOCOLCHICINE see DBA200
N-DEACETYLTHIOCOLCHICINE see DBA200
DEACETYLTHYMOXAMINE see DAD850
DEACTIVATOR E see DJD600
DEACTIVATOR H see DJD600
DEADLY NIGHTSHADE see BAU500, DAD880
DEAD MEN'S FINGERS see WAT315
DEADOPA see DNA200
DEAE see DHO500
DEAE-D see DHW600
DEALCA TP1 see GLU000
DEALCA TP2 see GLU000
DEALKYLPRAZEPAM see CGA500
DEAMELIN S see GHR609
DEAMINOHYDROXYTUBERCINDIN see DAE200
3′-DEAMINO-3′-MORPHOLINO-ADRIAMYCIN see MRT100
3′-DEAMINO-3′-(4-MORPHOLINYL)DAUNORUBICIN see MRT100
DEANER see DOZ000
DEANOL see DOY800
DEANOL-p-ACETAMIDOBENZOATE see DOZ000
DEANOL ACETAMIDOBENZOATE see DOZ000
DEANOL-p-CHLOROPHENOXYACETATE see DPE000
DEANOLESTERE see DPE000
DEANOX see IHD000
DEA OXO-5 see DBA800
DEAPASIL see AMM250
DEASERPYL see MEK700
DEATH CAMAS see DAE100
DEATH-OF-MAN see WAT325
7-DEAZAADENOSINE see TNY500
7-DEAZAADENOSINE-7-CARBOXAMIDE see SAU000
7-DEAZAINOSINE see DAE200
DEB see BGA750, DKA600
DEBA see BAG000
DEBANTIC see RAF100
DEBECACIN see DCQ800
DEBECACIN SULFATE see PAG050
DEBECILLIN see BFC750
DEBECYLINA see BFC750
DEBENAL see PPP500
DEBENAL-M see ALF250
DEBENDOX see BAV350
DEBENDRIN see BBV500
DEBETROL see SKJ300
DEBLASTON see PIZ000
DEBRICIN see FBS000
DEBRIDAT see TKU675
DEBRISOQUIN HYDROBROMIDE see DAI475
DEBRISOQUIN SULFATE see IKB000
DEBROUSSAILLANT 600 see DAA800
DEBROUSSAILLANT CONCENTRE see TAA100
DEBROXIDE see BDS000
DEC see DAE800, DIX200
DECABANE see DER800
DECABORANE(14) see DAE400
DECABORANE see DAE400

DECABROMOBIPHENYL ETHER see PAU500
DECABROMOBIPHENYL OXIDE see PAU500
DECABROMODIPHENYL OXIDE see PAU500
DECABROMOPHENYL ETHER see PAU500
n-DECACHLOR see DAE425
1,1′,2,2′,3,3′,4,4′,5,5′-DECACHLOROBI-2,4-CYCLOPENTADIEN-1-YL see DAE425
DECACHLOROBI-2,4-CYCLOPENTADIEN-1-YL see DAE425
1,2,3,5,6,7,8,9,10,10-DECACHLORO(5.2.1.0²·⁶.0³·⁹.0⁵·⁸)DECANO-4-ONE see KEA000
DECACHLOROKETONE see KEA000
DECACHLORO-1,3,4-METHENO-2H-CYCLOBUTA(cd)PENTALEN-2-ONE see KEA000
DECACHLOROOCTAHYDROKEPONE-2-ONE see KEA000
DECACHLOROOCTAHYDRO-1,3,4-METHENO-2H-CYCLOBUTA(cd)PENTALEN-2-ONE see KEA000
1,1a,3,3a,4,5,5,5a,5b,6-DECACHLOROOCTAHYDRO-1,3,4-METHENO-2H-CY-CLOBUTA(cd)PENTALEN-2-ONE see KEA000
DECACHLOROPENTACYCLO(5.3.0.0²·⁶.0⁴·¹⁰.0⁵·⁹)DECAN-3-ONE see KEA000
DECACHLOROPENTACYCLO(5.2.1.0²·⁶.0³·⁹.0⁵·⁸)DECAN-4-ONE see KEA000
DECACHLOROTETRACYCLODECANONE see KEA000
DECACHLOROTETRAHYDRO-4,7-METHANOINDENEONE see KEA000
DECACIL see LFK000
DECACURAN see DAF600
DECADERM see SOW000
(E,E)-2,4-DECADIENAL see DAE450
(2E,4E)-2,4-DECADIENAL see DAE450
(2E,4E)-DECADIENAL see DAE450
trans,trans-2,4-DECADIENAL see DAE450
2,4-DECADIENOIC ACID, ETHYL ESTER, (E,Z)- see EHV100
1,6-DECADIEN-3-OL, 3,7,9-TRIMETHYL- see IIW100
DECADONIUM DIIODIDE see DAE500
DECADRON see SOW000
DECADRON PHOSPHATE see DAE525
DECAETHOXY OLEYL ETHER see OIG040
DECAFENTIN see DAE600
DECAFLUOROBUTYRAMIDINE see DAE625
DECAHYDRO-4-α-HYDROXY-2,8,8-TRIMETHYL-2-NAPHTHOIC ACID, γ-LAC-TONE see LAQ100
trans-N-(DECAHYDRO-2-METHYL-5-ISOQUINOLYL)-3,4,5-TRIMETHOXYBENZAM-IDE see DAE700
cis-N-(DECAHYDRO-2-METHYL-5-ISOQUINOLYL)-3,4,5-TRIMETHOXYBENZAMIDE see DAE695
DECAHYDRONAPHTHALENE see DAE800
DECAHYDRONAPHTHALEN-2-OL see DAF000
DECAHYDRO-2-NAPHTHALENOL see DAF000
DECAHYDRO-β-NAPHTHOL see DAF000
trans-DECAHYDRO-β-NAPHTHOL see DAF000
DECAHYDRO-β-NAPHTHYL ACETATE see DAF100
DECAHYDRO-β-NAPHTHYL FORMATE see DAF150
DECAHYDRONAPTHOL-2 see DAF000
DECAHYDRO-4a,7,9-TRIHYDROXY-2-METHYL-6,8-BIS(METHYLAMINO)-4H-PYRA-NO(2,3-b)(1,4)BENZODIOXIN-4-ONE DIHYDROCHLORIDE, (2R-(2-α,4a-β,5a-β,6-β,7-β,8-β,9-α,9a-α,10a-β))- see SLI325
γ-N-DECALACTONE see HKA500
Δ-DECALACTONE see DAF200
n-DECALDEHYDE see DAG000
DECALDEHYDE see DAG000
DECALIN (DOT) see DAE800
DECALIN see DAE800
2-DECALINOL see DAF000
DECALIN SOLVENT see DAE800
DECALINYL FORMATE see DAF150
2-DECALOL see DAF000
DECAMETHIONIUM IODIDE see DAF800
DECAMETHONIUM see DAF600
DECAMETHONIUM BROMIDE see DAF600
DECAMETHONIUM DIBROMIDE see DAF600
DECAMETHONIUM DIIODIDE see DAF800
DECAMETHONIUM IODIDE see DAF800
DECAMETHRIN see DAF300
DECAMETHRINE see DAF300
DECAMETHYLCYCLOPENTASILOXANE see DAF350
N,N′-DECAMETHYLENEBIS((1-ADAMANTYL)DIMETHYLAMMONIUM, DIIODIDE see DAE500
1,1′-DECAMETHYLENEBIS(1-METHYLPIPERIDINIUM IODIDE) see DAF450
DECAMETHYLENEBIS(TRIMETHYLAMMONIUM BROMIDE) see DAF600
DECAMETHYLENE-1,10-BISTRIMETHYLAMMONIUM DIBROMIDE see DAF600
DECAMETHYLENEBIS(TRIMETHYLAMMONIUM DIIODIDE) see DAF800
DECAMETHYLENEBIS(TRIMETHYLAMMONIUM IODIDE) see DAF800
DECAMINE see DAA800
DECAMINE 4T see TAA100
1-DECANAL see DAG000
1-DECANAL (mixed isomers) see DAG200
n-DECANAL see DAG000

DECANAL see DAG000
DECANALDEHYDE see DAG000
DECANAL, DIMETHYLACETAL see AFJ700
DECANAL DIMETHYL ACETAL see DAI600
n-DECANE (DOT) see DAG400
DECANE see DAG400
1-DECANEAMINE see DAG600
DECANE, 1-BROMO- see BNB800
1-DECANECARBOXYLIC ACID see UKA000
1,10-DECANEDIAMINE see DAG650
DECANEDINITRILE see SBK500
DECANEDIOIC ACID see SBJ500
DECANEDIOIC ACID, BIS(2-ETHYLHEXYL) ESTER see BJS250
DECANEDIOIC ACID, DIBUTYL ESTER see DEH600
DECANEDIOIC ACID, compd. with 1,6-HEXANEDIAMINE (1:1) (9CI) see HEG130
1,10-DECANEDIOL, 2,2,9,9-TETRAMETHYL- see TDO260
DECANE, 1-(ETHENYLOXY)- see EEE200
DECANE, HENEICOSAFLUORO-1-IODO- see IEU075
DECANE, 1-IODO-1,1,2,2,3,3,4,4,5,5,6,6,7,7,8,8,9,9,10,10-HENEICOSAFLUO-RO- see IEU075
DECANE, 1-METHOXY- see MIW075
n-DECANOIC ACID see DAH400
DECANOIC ACID see DAH400
DECANOIC ACID-4-(4-CHLOROPHENYL)-1-(4-(4-FLUOROPHENYL)-4-OXYBU-TYL)-4-PIPERIDINYL ESTER see HAG300
DECANOIC ACID, DIESTER with TRIETHYLENE GLYCOL (mixed isomers) see DAH450
DECANOIC ACID, ETHYL ESTER see EHE500
DECANOIC ACID, 4-HYDROXY-4-METHYL-, γ-LACTONE see MIW050
DECANOIC ACID, METHYL ESTER see MHY650
DECANOIC ACID, SODIUM SALT see SGB600
1-DECANOL (FCC) see DAI600
DECANOL (mixed isomers) see DAI800
n-DECANOL see DAI600
DECANOL see DAI600
DECANOLIDE-1,4 see HKA500
DECANOLIDE-1,5 see DAF200
2-DECANONE see OFE050
DECANOPHENONE, 10-FLUORO- see FKQ100
4-DECANOYLMORPHOLINE see CBF725
9-DECAOCTENOIC ACID, TRIBUTYLSTANNYL ESTER see TIA000
DECAPRYN see PGE775
DECAPRYN SUCCINATE see PGE775
DECAPS see VSZ100
10-DECARBAMOYLMITOMYCIN C see DAI000
DECARBAMOYLMITOMYCIN C see DAI000
DECARBAMYLMITOMYCIN C see DAI000
DECARBOFURAN see DLS800
m-DECARBOROCARBORANE see NBV100
2-DECARBOXAMIDO-2-ACETYL-4-DESDIMETHYLAMINO-4-AMINO-9-METHYL-5A,6-ANHYDROTETRACYCLINE see CDF250
DECARBOXYCYSTEINE see AJT250
DECARBOXYCYSTINE see MCN500
DECARIS see LFA020
DECARPYN SUCCINATE (1:1) see PGE775
DECASERPIL see MEK700
DECASERPINE see MEK700
DECASERPYL see MEK700
DECASERPYL PLUS see MEK700
DECASONE see SOW000
DECASPIRIDE HYDROCHLORIDE see DAI200
DECASPRAY see SOW000
DECATOL see IOX400
n-DECATYL ALCOHOL see DAI600
DECCO SALT NO 5 see TNE775
DECCOTANE see BPY000
DECELITH H see PKQ059
DECEMTHION P-6 see PHX250
trans-2-DECEN-1-AL see DAI350
2-DECENAL see DAI350
cis-4-DECEN-1-AL (FCC) see DAI360
cis-4-DECENAL see DAI360
DECENALDEHYDE see DAI350
9-DECEN-1-OL see DAI400
1-DECEN-10-OL see DAI400
ω-DECENOL see DAI400
9-DECEN-1-OL, ACETATE see DAI450
DECENTAN see CJM250
9-DECENYL ACETATE see DAI450
DECENYL ACETATE see DAI450
DECHAN see DGU200
DECHLORANE 605 see DAI460
DECHLORANE 4070 see MQW500
DECHLORANE A-O see AQF000

DECHLORANE PLUS 2520 see DAI460
DECHLORANE PLUS 515 see DAI460
DECHLORANE PLUS see DAI460
DECHOLIN see DAL000
DECHOLIN SODIUM SALT see SGD500
DECICAINE see TBN000
DECIMEMIDE see DAJ400
DECINCAN see VLF000
DECIS see DAF300
DECLINAX see DAI475, IKB000
DECLOMYCIN see MIJ500
DECLOMYCIN HYDROCHLORIDE see DAI485
DECLOXIZINE see BBV750
DECOFOL see BIO750
n-DECOIC ACID see DAH400
DECONTRACTIL see GGS000
DECORPA see GLU000
DECORTANCYL see PLZ000
DECORTIN see DAQ800, PLZ000
DECORTIN H see PMA000
DECORTISYL see PLZ000
DECORTON see DAQ800
DECOSERPYL see MEK700
DECOSTERONE see DAQ800
DECOSTRATE see DAQ800
DECROTOX see COD000
DECTANCYL see SOW000
DECURVON see IDF300
DE-CUT see DMC600
n-DECYL ACRYLATE see DAI500
DECYL ACRYLATE see DAI500
DECYL ALCOHOL (mixed isomers) see DAI800
n-DECYL ALCOHOL see DAI600
DECYL ALCOHOL see DAI600
1-DECYL ALDEHYDE see DAG000, UJJ000
n-DECYL ALDEHYDE see DAG000
DECYL ALDEHYDE see DAG000
DECYLALDEHYDE DMA see AFJ700
DECYLAMINE see DAG600
DECYL BENZENE SODIUM SULFONATE see DAJ000
1-DECYL BROMIDE see BNB800
n-DECYL BROMIDE see BNB800
DECYL BROMIDE see BNB800
DECYL BUTYL PHTHALATE see BQX250
DECYL CHLORIDE (mixed isomers) see DAJ200
N-DECYL-N,N-DIMETHYL-1-DECANAMINIUM CHLORIDE (Cl) see DGX200
DECYLENIC ALCOHOL see DAI400
1-DECYL-1-ETHYLPIPERIDINIUM BROMIDE see DAJ300
n-DECYLIC ACID see DAH400
DECYLIC ACID see DAH400
DECYLIC ALCOHOL see DAI600
DECYLIC ALDEHYDE see DAG000
DECYL METHYL ETHER see MIW075
cis-2-DECYL-3-(5-METHYLHEXYL)OXIRANE see ECB200
DECYL OCTYL ALCOHOL see OAX000
n-DECYL n-OCTYL PHTHALATE see OEU000
DECYL OCTYL PHTHALATE see OEU000
4-(DECYLOXY)-3,5-DIMETHOXYBENZAMIDE see DAJ400
4-n-DECYLOXY-3,5-DIMETHOXYBENZOIC ACID AMIDE see DAJ400
4-DECYLOXY-2-HYDROXYPHENYL 4-DECYLOXYPHENYL KETONE see DAJ450
DECYLTRIMETHYLAMMONIUM BROMIDE see DAJ500
DECYLTRIPHENYLPHOSPHONIUM BROMOCHLOROTRIPHENYLSTANNATE see DAE600
(DECYL-TRIPHENYL-PHOSPHONIUM)-TRIPHENYL-BROM-CHLOR-STANNAT (GERMAN) see DAE600
DECYL VINYL ETHER see EEE200
DEDC see SGJ000
DEDELO see DAD200
DEDEVAP see DGP900
DEDK see SGJ000
DEDORAN see MCI500
DED-WEED see CIR250, DAA800, DGI400, TIX500
DED-WEED BRUSH KILLER see TAA100
DED-WEED CRABGRASS KILLER see PLC250
DED-WEED LV-69 see DAA800
DED-WEED LV-6 BRUSH KIL and T-5 BRUSH KIL see TAA100
DEDYL see DNM400
DEE-OSTEROL see VSZ100
DEEP CRIMSON MADDER 10821 see DMG800
DEEP FASTONA RED see MMP100
DEEP LEMON YELLOW see SMH000
DEER BERRY see HGF100
DEE-RON see VSZ100
DEE-RONAL see VSZ100

DEE-ROUAL see VSZ100
DEER'S TONGUE see DAJ800
DEERTONGUE INCOLORE see DAJ800
DEET see DKC800
16-DEETHYL-3-o-DEMETHYL-16-METHYL-3-o-(1-OXO-PROPYL)MONENSIN see DAK000
DEETILATO METOCLOPRAMIDE (ITALIAN) see AJH125
DEETILMETOCLOPRAMIDE (ITALIAN) see AJH125
DEF see BSH250
DEF DEFOLIANT see BSH250
DEFEKTON see CCK775, CCK780
DE-FEND see DSP400
DEFEROXAMINE see DAK200
DEFEROXAMINE MESILATE see DAK300
DEFEROXAMINE MESYLATE see DAK300
DEFEROXAMINE METHANESULFONATE see DAK300
DEFEROXAMINUM see DAK200
DEFERRIOXAMINE see DAK200
DEFERRIOXAMINE B see DAK200
DEFIBRASE see RDA350
DEFIBRASE R see RDA350
DEFICOL see PPN100
DEFILIN see DJL000
DEFILTRAN see AAI250
DEFLAMENE see FDB000
DEFLAMON-WIRKSTOFF see MMN250
DEFLEXOL see AJF500
DEFLOGIN see HNI500
DEFLORIN see TEY000
DEFOAMER S-10 see SCP000
DE-FOL-ATE see MAE000, SFS000
DEFOLIANT 713 see DKE400
DEFOLIANT 2929 RP see TFH500
DEFOLIT see TEX600
DEFONIN see ILD000
DEFRADIN HYDRATE see SBN450
DEFTOR see MQR225
DEFY see DFY800
DEG see DJD600
DEGALAN S 85 see PKB500
DEGALOL see DAQ400
DEGLARESIN see MCB050
DEGLARESIN N 12 see MCB050
DEGMVE (RUSSIAN) see DJG400
DEGRANOL see MAW500
DEGRASSAN see DUS700
DE-GREEN see BSH250
DEGUELIA ROOT see DBA000
DEGUSSA see CBT750
D.E.H. 20 see DJG600
DEH 24 see TJR000
D.E.H. 26 see TCE500
DEH see HHH000
DEHA see AEO000, DJN000
DEHACODIN see DKW800
DEHERBAN see DAA800
DEHIDROBENZPERIDOL see DYF200
DEHISTIN see TMP750
DEHISTIN HYDROCHLORIDE see POO750
DEHP see DVL700
DEHPA EXTRACTANT see BJR750
DEHYCHOL see DAL000
DEHYDOL LS 4 see DXY000
DEHYDRACETIC ACID see MFW500
DEHYDRATIN see AAI250
DEHYDRITE see PCE000
DEHYDROABIETIC ACID see DAK400
DEHYDRO-ABIETIC ACID-2-(2-(CHLOROETHYL)AMINO)ETHYL ESTER see CGQ250
DEHYDROACETIC ACID (FCC) see MFW500
DEHYDROACETIC ACID, SODIUM SALT see SGD000
trans-DEHYDROANDROSTERONE see AOO450
DEHYDROBENZPERIDOL see DYF200
6-DEHYDRO-6-CHLORO-17-α-ACETOXYPROGESTERONE see CBF250
DEHYDROCHOLATE SODIUM see SGD500
7-DEHYDROCHOLESTERIN see DAK600
DEHYDROCHOLESTERIN (GERMAN) see DAK600
7-DEHYDROCHOLESTEROL see DAK600
DEHYDROCHOLESTEROL see DAK600
7-DEHYDROCHOLESTEROL ACETATE see DAK800
7-DEHYDROCHOLESTERYL ACETATE see DAK800
7-DEHYDROCHOLESTROL, ACTIVATED see CMC750
DEHYDROCHOLIC ACID see DAL000
DEHYDROCHOLIC ACID, SODIUM SALT see SGD500
DEHYDROCHOLSAEURE (GERMAN) see DAL000

Δ¹-DEHYDROCORTISOL see PMA000
1-DEHYDROCORTISONE see PLZ000
Δ-1-DEHYDROCORTISONE see PLZ000
5-DEHYDROEPIANDROSTERONE see AOO450
DEHYDROEPIANDROSTERONE see AOO450
DEHYDROEPIANDROSTERONE SODIUM SULFATE DIHYDRATE see DAL030
DEHYDROEPIANDROSTERONE SULFATE SODIUM see DAL040
DEHYDROERGOTAMINE see DLK800
DEHYDROFOLLICULINIC ACID see BIT000
DEHYDROHELIOTRIDINE see DAL060
DEHYDROHELIOTRINE see DAL100
1-DEHYDROHYDROCORTISONE see PMA000
Δ¹-DEHYDROHYDROCORTISONE see PMA000
11-DEHYDRO-17-HYDROXYCORTICOSTERONE see CNS800
11-DEHYDRO-17-HYDROXYCORTICOSTERONE ACETATE see CNS825
11-DEHYDRO-17-HYDROXYCORTICOSTERONE-21-ACETATE see CNS825
5,6-DEHYDROISOANDROSTERONE see AOO450
DEHYDROISOANDROSTERONE see AOO450
6-DEHYDRO-6-METHYL-17-α-ACETOXYPROGESTERONE see VTF000
1,2-DEHYDRO-3-METHYLCHOLANTHRENE see DAL200
DEHYDRO-3-METHYLCHOLANTHRENE see DAL200
1-DEHYDRO-16-α-METHYL-9-α-FLUOROHYDROCORTISONE see SOW000
A1-DEHYDROMETHYLTESTERONE see DAL300
DEHYDROMETHYLTESTERONE see DAL300
1-DEHYDRO-17-α-METHYLTESTOSTERONE see DAL300
DEHYDROMONOCROTALINE see DAL350
DEHYDRONIVALENOL see VTF500
DEHYDRONIVALENOL MONOACETATE see ACH075
DEHYDROPHELOMYCIN D1 see BLY760
DEHYDRORETRONECINE see DAL400
DEHYDROSTILBESTROL see DAL600
DEHYDRO-p-TOLUIDINE see MHJ300
Δ³-DEHYDRO-3,4-TRIMETHYLENE-ISOBENZANTHRENE-2 see BCI000
DEHYQUART C see DXY725, LBX050
DEHYQUART CBB see BEL900
DEHYQUART CDB see BEL900
DEHYQUART STC-25 see DTC600
DEHYSTOLIN see DAL000
DEIDROBENZPERIDOLO see DYF200
DEIDROCOLICO VITA see DAL000
DEINAIT see CJR909
DEIQUAT see DWX800
DEISOVALERYL BLASTMYCIN see DAM000
DEJO see DJT800
DEK see DJN750
DE-KALIN see DAE800
DEKALINA (POLISH) see DAE800
DEKAMETHYLCYKLOPENTASILOXAN see DAF350
DEKAMETRIN (HUNGARIAN) see DAF300
DEKORTIN see PLZ000
DEKRYSIL see DUS700
DEKSONAL see DOU600
DEL see AJH000
DELAC J see DWI000
DELAC S see CPI250
DELACURARINE see TOA000
DELADIOL see EDS100
DELAGIL see CLD000, CLD250
DELAHORMONE UNIMATIC see EDS100
DELALUTIN see HNT500
DELAN see DLK200
DELAN-COL see DLK200
DELAPRIL HYDROCHLORIDE see DAM315
DELATESTRYL see MPN500, TBF750
DELATESTRYL and DEPO-MEDROXYPROGESTERONE ACETATE see DAM325
DELATESTRYL and DEPO-PROVERA see DAM325
DELAVAN see MHB500
DELCORTOL see PMA000
DELEAF DEFOLIANT see TIG250
DELESTROGEN see EDS100
DELESTROGEN 4X see EDS100
DELGESIC see ADA725
DELICIA see AHE750, PGY000
DELIVA see ARQ750
DELLIPSOIDS see BBK500
DELMOFULVINA see GKE000
DELMONEURINA see BET000
DELNAV see DVQ709
DELONIN AMIDE see NCR000
DELOWAS S see TBD000
DELOWAX OM see TBD000
DELPET 50M see PKB500
m-DELPHENE see DKC800
DELPHENE see DKC800

DELPHINIC ACID see ISU000
DELPHINIDOL see DAM400
DELSENE see MHC750
DELSENE M see MAS500
DELSTEROL see CMC750
DELTA see CJJ000
DELTA-CORTEF see PMA000
DELTACORTELAN see PLZ000
DELTACORTENOL see PMA000
DELTACORTISONE see PLZ000
DELTACORTONE see PLZ000
DELTACORTRIL see PMA000
DELTA-DOME see PLZ000
DELTA F see PMA000
DELTAFLUORENE see SOW000
DELTALIN see VSZ100
DELTAMETHRIN see DAF300
DELTA-MVE see MKB750
DELTAMYCIN A see CBT250
DELTAN see DUD800
DELTA-STAB see PMA000
DELTATHIONE see GFW000
DELTAZINA see PPP500
DELTILEN see SOV100
DELTISONE see PLZ000
DELTOIN see ENC500
DELTOSIDE see AHK750
DELTRATE-20 see PBC250
DELTYL see IQW000
DELTYLEXTRA see IQN000
DELTYL PRIME see IQW000
DELURSAN see DMJ200
DELUSSA BLACK FW see CBT750
DELVEX see DJT800 ·
DELVINAL SODIUM see VKP000
DELYSID see DJO000
DEM see AJH125
DEMA see BIE500
DEMAROL see DAM600
DEMASORB see DUD800
DEMAVET see DUD800
DEMECLOCYCLINE see MIJ500
DEMECLOCYCLINE HYDROCHLORIDE see DAI485
DEMECOLCIN see MIW500
DEMECOLCINE see MIW500
DEMEFLINE see DNV200
DEMEPHION see MIW250
DEMEROL see DAM600, DAM700
DEMEROL HYDROCHLORIDE see DAM700
DEMESO see DUD800
DEMETHON-METHYL (MAK) see MIW100
4-DEMETHOXYDAUNOMYCIN see DAN200
4-DEMETHOXYDAUNORUBICIN see DAN000
11-DEMETHOXYRESERPINE see RDF000
N-DEMETHYLACLACINOMYCIN A see DAN200
DEMETHYLAMITRIPTYLENE see NNY000
6-DEMETHYLCHLOROTETRACYCLINE see MIJ500
6-DEMETHYL-7-CHLOROTETRACYCLINE see MIJ500
DEMETHYLCHLOROTETRACYCLINE see MIJ500
DEMETHYLCHLOROTETRACYCLINE HYDROCHLORIDE see DAI485
DEMETHYLCHLORTETRACYCLIN see MIJ500
6-DEMETHYLCHLORTETRACYCLINE see MIJ500
6-DEMETHYL-7-CHLORTETRACYCLINE see MIJ500
DEMETHYLCHLORTETRACYCLINE see MIJ500
DEMETHYLCHLORTETRACYCLINE BASE see MIJ500
DEMETHYLCHLORTETRACYCLINE HYDROCHLORIDE see DAI485
2-DEMETHYLCOLCHICINE see DAN300
O²-DEMETHYLCOLCHICINE see DAN300
O¹⁰-DEMETHYLCOLCHICINE see ADE000
3-DEMETHYLCOLCHICINE GLUCOSIDE see DAN375
o-DEMETHYLDAUNOMYCIN see KBU000
(6R,25)-5-o-DEMETHYL-28-DEOXY-6,28-EPOXY-25-(1-METHYLE-
   THYL)MILBEMYCIN B see MQT600
(11R(2R,5S,6R),12R)-10-DEMETHYL-19-DE((TETRAHYDRO-5-METHOXY-6-METH-
   YL-2H-PYRAN-2-YL)OXY)-12-METHYL-11-o-(TETRAHYDRO-5-METHOXY-6-
   METHYL-2H-PYRAN-2-YL)-DIANEMYCIN see LEJ700
1-DEMETHYLDIAZEPAM see CGA500
N-DEMETHYLDIAZEPAM see CGA500
DEMETHYLDIAZEPAM see CGA500
DEMETHYLDOPAN see BIA250
4-DEMETHYLEPIODOPHYLLOTOXIN-β,d-ETHYLIDENEGLUCOSIDE see EAV500
DEMETHYL-EPIODOPHYLLOTOXIN ETHYLIDENE GLUCOSIDE see EAV500
4'-DEMETHYLEPIPODOPHYLLOTOXIN-9-(4,6-O-ETHYLIDENE-β-d-GLUCOPYRA-
   NOSIDE see EAV500

4'-DEMETHYLEPIPODOPHYLLOTOXIN ETHYLIDENE-β,d-GLUCOSIDE see EAV500
4-DEMETHYL-EPIPODOPHYLLOTOXIN-β,d-ETHYLIDEN-GLUCOSIDE see EAV500
4'-DEMETHYLEPIPODOPHYLLOTOXIN-9-(4,6-O-2-THENYLIDENE-β-d-GLUCOPY-RANOSIDE see EQP000
4'-DEMETHYL-EPIPODOPHYLLOTOXIN-β-d-THENYLIDENE-GLUCOSIDE see EQP000
4'-O-DEMETHYL-1-O-(4,6-O-ETHYLIDENE-β,d-GLUCOPYRANO-SYL)EPIPODOPHYLLOTOXIN see EAV500
DEMETHYLIMIPRAMINE see DSI709
DEMETHYLMISONIDAZOLE see NHI000
1'-DEMETHYL-(s)-NICOTINE see NNR500
N-DEMETHYLORPHENADRINE HYDROCHLORIDE see TGJ250
4'-DEMETHYL 1-O-(4,6-O,O-(2-THENYLIDENE)-β-d-GLUCOPYRANO-SYL)EPIPODOPHYLLOTOXIN see EQP000
6-DEMETIL-7-CLOROTETRACICLINA see MIJ500
DEMETON see DAO500, DAO600
DEMETON-O-METHYL see DAO800
DEMETON METHYL see MIW100
DEMETON-O-METHYL SULFOXIDE see DAP000
DEMETON-METHYL SULPHOXIDE see DAP000
DEMETON-O-METILE (ITALIAN) see DAO800
DEMETON-O see DAO500
DEMETON-O + DEMETON-S see DAO600
DEMETON-S see DAP200
DEMETON-S-METHYL see DAP400
DEMETON-S-METHYLSULFON (GERMAN) see DAP600
DEMETON-S-METHYLSULFONE see DAP600
DEMETON-S-METHYL-SULFOXID (GERMAN) see DAP000
DEMETON-S-METHYL SULFOXIDE see DAP000
DEMETON-S-METHYL-SULPHONE see DAP600
DEMETON-S-METILE (ITALIAN) see DAP400
DEMETRACICLINA see DAI485
DEMETRIN see DAP700
DEMISE see DFY800
DEMOCRACIN see TBX000
DEMOLOX see AOA095
DEMORPHAN see DBE200
DEMOSAN see CJA100
DEMOS-L40 see DSP400
DEMOTIL see DAP800
DEMOX see DAO600
DEMSODROX see DUD800
DEN see NJW500
DENA see NJW500
DENAMONE see VSZ450
DENAPON see CBM750
DENAPON, NITROSATED (JAPANESE) see NBJ500
DENATONIUM BENZOATE see DAP812
DENATURED ALCOHOL (DOT) see AFJ000
DENATURED SPIRITS see AFJ000
DENDREPAR see PEM750
DENDRID see DAS000
DENDRITIS see SFT000
DENDROASPIS ANGUSTICEPS VENOM see DAP815
DENDROASPIS JAMESONI VENOM see DAP820
DENDROASPIS VIRIDIS VENOM see GJU500
DENDROBAN-12-ONE HYDROCHLORIDE see DAP825
DENDROBINE HYDROCHLORIDE see DAP825
DENDROCALAMUS MEMBRANACEUS Munro, extract excluding roots see DAP840
DENEGYT see DAJ400
DENKA F 90 see SCK600
DENKA FB 44 see SCK600
DENKALAC 61 see AAX175
DENKA QP3 see SMQ500
DENKA VINYL SS 80 see PKQ059
DENOPAMINE see DAP850
DENSIC C 500 see SCQ000
DENSINFLUAT see TIO750
DENUDATINE see DAP875
DENVER RESEARCH CENTER No. DRC-4575 see AMU125
DENYL see DKQ000, DNU000
DENYLSODIUM see DNU000
DENZIMOL HYDROCHLORIDE see DAP880
DEOBASE see DAP900, KEK000
DEODOPHYLL see CKN000
DEODORIZED KEROSENE see DAP900
DEODORIZED KEROSINE see DAP900
DEODORIZED WINTERIZED COTTONSEED OIL see CNU000
DEOFED see DBA800, MDT600
DEORLENE GREEN JJO see BAY750
DEOSAN see SHU500
DEOVAL see DAD200

DEOXIN-1 see GFG100
11-DEOXO-12-β,13-α-DIHYDRO-11-α-HYDROXYJERVINE see DAQ000
11-DEOXO-12-β,13-α-DIHYDRO-11-β-HYDROXYJERVINE see DAQ002
9-DEOXO-16,16-DIMETHYL-9-METHYLENE-PGE2 see DAQ100
9-DEOXO-16,16-DIMETHYL-9-METHYLENEPROSTAGLANDIN E2 see DAQ100
11-DEOXOGLYCYRRHETINIC ACID HYDROGEN MALEATE SODIUM SALT see DAQ110
11-DEOXOJERVINE see CPT750
2-DEOXY-3-ARABINO-HEXOSE see DAR600
2-DEOXY-d-ARABINO-HEXOSE see DAR600
6-DEOXY-6-AZIDODIHYDROISOCODEINE see ASE875
6-DEOXY-6-AZIDODIHYDROISOMORPHINE see ASG675
DEOXYBENZOIN see PEB000
DEOXYCHOLATE SODIUM see SGE000
DEOXYCHOLATIC ACID see DAQ400
7-α-DEOXYCHOLIC ACID see DAQ400
DEOXYCHOLIC ACID (FCC) see DAQ400
DEOXYCHOLIC ACID SODIUM SALT see SGE000
4'-DEOXYCIRRAMYCIN A¹ see RMF000
o-3-DEOXY-4-C-METHYL-3-(METHYLAMINO)-β-l-ARABINOPYRANOSYL-(1-6)-o-(2,6-DIAMINO)-2,3,4,6-TETRADEOXY-α-d-glycero-HEX-4-ENDOPYRANOSYL-(1-4)-2-DEOXY-d-STREPTAMINE HYDROCHLORIDE see SDY755
o-3-DEOXY-4-C-METHYL-3-(METHYLAMINO)-β-l-ARABINOPYRANOSYL-(1-6)-o-(2,6-DIAMINO)-2,3,4,6-TETRADEOXY-α-d-glycero-HEX-4-ENOPYRANOSYL-(1-4)-2-DEOXY-d-STREPTAMINE see SDY750
o-3-DEOXY-4-C-METHYL-3-(METHYLAMINO)-β-l-ARABINOPYRANOSYL-(1-6)-o-(2,b-DIAMINO-2,3,4,6-TETRADEOXY-α-d-glycero-HEX-4-ENOPYRANOSYL-(1-4))-2-DEOXY-N¹-ETHYL-d-STREPTAMINE SULFATE (2:5) (salt) see NCP550
2'-DEOXYCOFORMYCIN see PBT100
DEOXYCOFORMYCIN see PBT100
11-DEOXYCORTICOSTERONE see DAQ600
11-DEOXYCORTICOSTERONE ACETATE see DAQ800
11-DEOXYCORTICOSTERONE-21-ACETATE see DAQ800
DEOXYCORTONE ACETATE see DAQ800
N-DEOXYDEMOXAPAM see CGA500
2-((2-DEOXY-2-(3,6-DIAMINOHEXANAMIDO)-α-d-GLUOPYRANOSYL)AMINO)-3,3a,5,6,7,7a-HEXAHYDRO-7-HYDROXY-4H-IMIDAZO(4,5-C)PYRIDIN-4-ONE-6'-CARBAMATE see RAG300
14-DEOXY-14-((2-DIETHYLAMINOETHYL)MERCAPTOACETOXY)-MUTILIN HY-DROGEN FUMARATE see DAR000
14-DEOXY-14-((2-DIETHYLAMINOETHYL-THIO)-ACETOXY)MUTILINE see TET800
6-DEOXY-7,8-DIHYDROMORPHINE see DKX600
DEOXYEPHEDRINE see DBA800, DBB000
dl-DEOXYEPHEDRINE HYDROCHLORIDE see DAR100
d-DEOXYEPHEDRINE HYDROCHLORIDE see MDT600
(12R,13S)-9-DEOXY-12,13-EPOXY-12,13-DIHYDRO-9-OXO-3-ACETATE-4B-(3-ME-THYLBUTANOATE) LEUCOMYCIN V HYDROCHLORIDE see MAC600
9-DEOXY-12,13-EPOXY-9-OXOLEUCOMYCIN V 3-ACETATE 4ᴮ-(3-METHYLBU-TANOATE) see CBT250
6-DEOXY-6-FLUOROGLUCOSE see DAR150
2'-DEOXY-5-FLUOROURIDINE see DAR400
5'-DEOXY-5-FLUOROURIDINE see DYE415
DEOXYFLUOROURIDINE see DAR400
2-DEOXYGLUCOSE see DAR600
2-DEOXY-d-GLUCOSE (FRENCH) see DAR600
d-2-DEOXYGLUCOSE see DAR600
2-DEOXY-d-GLUCOSE see DAR600
2-DEOXYGLYCEROL see PML250
2'-DEOXYGUANOSINE see DAR800
DEOXYGUANOSINE see DAR800
α-6-DEOXY-5-HYDROXYTETRACYCLINE see DYE425
7''-DEOXY-7''-HYDROXY-TRIOXACARCIN A see TMO775
2'-DEOXY-5-IODOURIDINE see DAS000
7-(6-o-(6-DEOXY-α-l-MANNOPYRANOSYL)-β-d-GLUCOPYRANO-SIDE)HESPERETIN see HBU000
(3-β)-2-((6-DEOXY-α-l-MANNOPYRANOSYL)OXY)-14-HYDROXYBUFA-4,20,22-TRIENOLIDE see POB500
3-β-((6-DEOXY-α-l-MANNOSYL)OXY)-14-HYDROXYBUFA-4,20,22-TRIENOLIDE see POB500
1-DEOXY-1-(METHYLAMINO)-d-GLUCITOL 5-ACETAMIDO-2,4,6-TRIIODO-N-ME-THYLISOPHTHALAMATE (SALT) see IGC000
3-((6-DEOXY-3-O-METHYL-α-l-GLUCOPYRANOSYL)OXY)-14-HYDROXY-19-OXO-CARD-20(22)-ENOLIDE see EAQ050
1-DEOXY-1-(METHYLNITROSAMINO)-d-GLUCITOL see DAS400
2-DEOXY-2-(((METHYLNITROSOAMINO)CARBONYL)AMINO)-d-GLUCOPYRA-NOSE see SMD000
2-DEOXY-2-(3-METHYL-3-NITROSOUREIDO)-α(and β)-d-GLUCOPYRANOSE see SMD000
2-DEOXY-2-(3-METHYL-3-NITROSOUREIDO)-d-GLUCOPYRANOSE see SMD000
1-DEOXY-1-(N-NITROSOMETHYLAMINO)-d-GLUCITOL see DAS400
4-DEOXYNIVALENOL see VTF500
DEOXYNIVALENOL see VTF500
DEOXYNIVALENOL MONOACETATE see ACH075
DEOXYNOREPHEDRINE see AOB250, BBK000

DES (synthetic estrogen) see DKA600
DESACCHROMIN DISPERSION see PJA200
DESACE see DBH200
DESACETYLBUFOTALIN see BOM655
DESACETYLCHOLCHICENE see TLN750
N-DESACETYLCOLCHICINE see TLO000
DESACETYLCOLCHICINE see TLO000
DESACETYLCOLCHICINE-d-TARTRATE see DBA175
DESACETYLLANATOSIDE B see DAD650
DESACETYLLANATOSIDE C see DBH200
DESACETYL-LANTOSID B (GERMAN) see DAD650
N-DESACETYL-N-METHYLCOLCHICINE see MIW500
N-DESACETYLMETHYLCOLCHICINE see MIW500
DESACETYLMETHYLCOLCHICINE see MIW500
N-DESACETYLTHIOCOLCHICINE see DBA200
DESACETYLTHYMOXAMINE see DAD850
DESACETYLVINBLASTINE AMIDE SULFATE see VGU750
DESACI see DBH200
DESADRENE see SOW000
DESAGLYBUZOLE see GFM200
DESALKYLPRAZEPAM see CGA500
DESAMETASONE see SOW000
DESAMINE see DBA800
DESBENZYL CLEBOPRIDE see DBA250
DESCETYLDIGILANIDE C see DBH200
DESCHLOROBIOMYCIN see TBX000
DESCORTERONE see DAQ800
DESCOTONE see DAQ800
N-DESCYCLOPROPYLMETHYLPRAZEPAM see CGA500
DESD see DKB000
DESDANINE see COV125
DESDEMIN see CHJ750
DES DISODIUM SALT see DKA400
DESENEX see ULS000
DESENTOL see BBV500
DESERIL see MLD250
DESERNYL see MLD250
DESERPIDINE, 10-METHOXY- see MEK700
DESERPINE see RDF000, RDK000
DESERTALC 57 see TAB750
DESERT RED see CHP500
DESERT ROSE see DBA450
DESERYL see MLD250
DESETHYLAPRINDINE HYDROCHLORIDE see DBA475
DESFEDRIN see DBA800
DESFERAL see DAK200
DESFERAL METHANESULFONATE see DAK300
DESFERRAL see DAK200
DESFERRIN see DAK200
DESFERRIOXAMINE see DAK200
DESFERRIOXAMINE B see DAK200
DESFERRIOXAMINE B MESYLATE see DAK300
DESFERRIOXAMINE B METHANESULFONATE see DAK300
DESFERRIOXAMINE METHANESULFONATE see DAK300
DESGLUCO-DIGITALINUM VERUM (GERMAN) see SMN275
DESGLUCODIGITONIN see DBA500
DESGLUCO-TRANSVAALIN see POB500
DESICAL P see CAU500
DES-I-CATE see DXD000
DESICCANT L-10 see ARB250
DESIMEX i see LBU200
DESIMIPRAMINE see DSI709
DESINFECT see CDP000
DESIPRAMIN see DSI709
DESIPRAMINE (D4) see DSI709
DESIPRAMINE HYDROCHLORIDE see DLS600
DESLANATOSIDE see DBH200
DESLANOSIDE see DBH200
DESMA see DKA600
DESMECOLCHINE see MIW500
DESMECOLCINE see MIW500
DESMEDIPHAM see EEO500
11-DESMETHOXYRESERPINE see RDF000
DESMETHOXYRESERPINE see RDF000
DESMETHYLAMITRIPTYLINE see NNY000
N-DESMETHYLDIAZEPAM see CGA500
DESMETHYLDIAZEPAM see CGA500
DESMETHYLDOPAN see BIA250
DESMETHYLDOXEPIN see DBA600
DESMETHYLIMIPRAMINE see DSI709
DESMETHYLIMIPRAMINE HYDROCHLORIDE see DLS600
DESMETHYLMISONIDAZOLE see DBA700, NHI000
DESMETRYN (GERMAN, DUTCH) see INR000
DESMETRYNE see INR000
DESMODUR 44 see MJP400

DESMODUR H see DNJ800
DESMODUR N see DNJ800
DESMODUR PU 1520A20 see PKB100
DESMODUR T80 see TGM750
DESMODUR T100 see TGM740
DESMODUR 44V20 see PKB100
2,4-DES-Na see CNW000
DES-N see TNJ300
2,4-DES-NATRIUM (GERMAN) see CNW000
DESO see EPI500
DESOLET see SFS000
DESOMORPHINE see DKX600
DESORMONE see DAA800, DGB000
14-DESOSSI-14-((2-DIETILAMINOETIL)MERCAPTO-ACETOSSI)MUTILIN IDROGE-
   NO FUMARATO (ITALIAN) see TET800
DESOSSIEFEDRINA see DBA800
DES-OXA-D see DBA800
DESOXEDRINE see DBA800
DES-3,4-OXIDE see DKB100
DES-α,β-OXIDE see DKB100
DESOXIN see DBA800
DESOXO-5 see DBA800, MDT600
DESOXON 1 see PCL500
DESOXYBENZOIN see PEB000
DESOXYCHOLIC ACID see DAQ400
DESOXYCHOLSAEURE (GERMAN) see DAQ400
DESOXYCORTICOSTERONE see DAQ600
DESOXYCORTICOSTERONE ACETATE see DAQ800
DESOXYCORTONE see DAQ600
DESOXYCORTONE ACETATE see DAQ800
14-DESOXY-14-((DIETHYLAMINOETHYL)-MERCAPTO ACETOXYL)-MUTILIN HY-
   DROGEN FUMARATE see TET800
DESOXYEPHEDRINE see DBB000
dl-DESOXYEPHEDRINE HYDROCHLORIDE see DAR100
d-DESOXYEPHEDRINE HYDROCHLORIDE see MDT600
l-DESOXYEPHEDRINE HYDROCHLORIDE see MDQ500
DESOXYEPHEDRINE HYDROCHLORIDE see DBA800
DESOXYFED see DBA800, MDT600
DESOXYN see BBK500, DBA800, DBB000, MDT600
DESOXYNE see MDT600
DESOXYNIVALENOL see VTF500
racemic-DESOXYNOREPHEDRINE see BBK000
DESOXYNOREPHEDRINE see AOA250, AOB250
(±)-DESOXYNOREPHEDRINE see BBK000
3-DESOXYNORLUTIN see NNV000
DESOXYPHED see DBA800
2-DESOXYPHENOBARBITAL see DBB200
DESOXYPHENOBARBITONE see DBB200
DESOXYPYRIDOXIME HYDROCHLORIDE see DAY825
DESOXYPYRIDOXINE see DAY800
5-DESOXYQUERCETIN see FBW000
DESOXYRIBONUCLEASE see EDK650, PAF575
DESPHEN see CDP250
DE-SPROUT see DMC600
DESSIN see CBW000
DESSON see CLW000
DESTENDO see BCA000
DESTIM see DBA800, MDT600
DESTOLIT see DMJ200
DESTOMYCIN A see DBB400
DESTONATE 20 see DBB400
DESTRIOL see EDU500
DESTROL see DKA600
DESTRONE see EDV000
DESTRUXOL APPLEX see DXE000
DESTRUXOL BORER-SOL see EIY600
DESTRUXOL ORCHID SPRAY see NDN000
DESTUN see TKF750
DESURIC see DDP200
DESYL CHLORIDE see CFJ100
DESYPHED see MDT600
DESYREL see CKJ000, THK875, THK880
m-DET see DKC800
DET see DKC800
m-DETA see DKC800
DETA see DJG600
DETAL see DUS700
DETALUP see VSZ100
DETAMIDE see DKC800
DETARIL see DWK400
DETERGENT 66 see SIB600
DETERGENT ALKYLATE see PEW500
DETERGENT HD-90 see DXW200
DETF see TIQ250
DETHMORE see WAT200

DETHYLANDIAMINE see TEO000
DETHYRONA see SKJ300
DETIA GAS EX-B see AHE750, PGY000
DETICENE see DAB600
DETIGON see CMW700
DETIGON-BAYER see CMW700
DETMOL-EXTRAKT see BBQ500
DETMOL MA 96% see MAK700
DETMOL MA see MAK700
DETMOL U.A. see CMA100
DETOX 25 see BBQ500
DETOX see DAD200
DETOXAN see DAD200
DETOXARGIN see AQW000
DETRALFATE see DBB500
DETRAVIS see DAI485
DETREOMYCINE see CDP250
DETREOPAL see CDP700
DETREX see DBA800
DETTOL see CLW000
DETYROXIN see SKJ300
DEURSIL see DMJ200
DEUSLON-A see EDU500
DEUTERIOMORPHINE see DBB600
DEUTERIUM see DBB800
DEUTERIUM FLUORIDE see DBC000
DEVAL RED K see CLK220
DEVAL RED TR see CLK220
DEVEGAN see ABX500
DEVELIN see PNA500
DEVELOPER 11 see PEY000
DEVELOPER 13 see PEY500
DEVELOPER 14 see TGL750
DEVELOPER A see NAX000
DEVELOPER AMS see NAX000
DEVELOPER B see TGL750
DEVELOPER BN see NAX000
DEVELOPER BON see HMX520
DEVELOPER DB see TGL750
DEVELOPER DBJ see TGL750
DEVELOPER H see TGL750
DEVELOPER MC see TGL750
DEVELOPER MT see TGL750
DEVELOPER MT-CF see TGL750
DEVELOPER MTD see TGL750
DEVELOPER P see NEO500
DEVELOPER PF see PEY500
DEVELOPER R see REA000
DEVELOPER SODIUM see NAX000
DEVELOPER T see TGL750
DEVELOPER Z see NNT000
DEVICARB see CBM750
DEVICOPPER see CNK559
DEVICORAN see HID350
DEVIGON see DSP400
DEVIKOL see DGP900
DEVIL'S APPLE see MBU800, SLV500
DEVIL'S BACKBONE see SDZ475
DEVIL'S CLAWS see FBS100
DEVILS HAIR see CMV390
DEVIL'S IVY see PLW800
DEVILS THREAD see CMV390
DEVINCAN see VLF000
DEVIPON see DGI400
DEVISULPHAN see EAQ750
DEVITHION see MNH000
DEVOL ORANGE B see NEO000
DEVOL ORANGE R see NEN500
DEVOL RED E see NEQ000
DEVOL RED F see KDA050
DEVOL RED GG see NEO500
DEVOL RED K see CLK235
DEVOL RED TA SALT see CLK235
DEVOL RED TR see CLK235
DEVOL SCARLET A (FREE BASE) see DEO295
DEVOL SCARLET B see NMP500
DEVOL SCARLET G SALT see NMP500
DEVONIUM see AOD000
DEVORAN see BBQ500
DEVOTON see MFW100
DEX see BJU000
DEXA see SOW000
DEXACORT see DAE525, SOW000
DEXA-CORTIDELT see SOW000
DEXA-CORTIDELT HOSTACORTIN H see PMA000

DEXADELTONE see SOW000
DEXADRESON see DAE525
DEXAGRO see DAE525
DEXAIME see BBK500
DEXALINE see BBK500
DEXALME see BBK500
DEXAMBUTOL see EDW875
DEXAMED see BBK500
DEXAMETH see SOW000
DEXAMETHASONE ALCOHOL see SOW000
DEXAMETHASONE 17,21-DIPROPIONATE see DBC500
DEXAMETHASONE DIPROPIONATE see DBC500
DEXAMETHASONE DISODIUM PHOSPHATE see DAE525
DEXAMETHASONE-21-ISONICOTINATE see DBC510
DEXAMETHASONE ISONICOTINATE see DBC510
DEXAMETHASONE SODIUM HEMISULFATE see DBC550
DEXAMETHASONE SODIUM PHOSPHATE see DAE525
DEXAMETHASONE SODIUM SULFATE see DBC550
DEXAMETHASONE-17-VALERATE see DBC575
DEXAMETHASONE VALERATE see DBC575
DEXAMETHAZONE SODIUM PHOSPHATE see DAE525
DEXAMINE see BBK500
DEXAMPHAMINE see BBK500
DEXAMPHETAMINE see AOA500, BBK500
DEXAMPHETAMINE SULFATE see BBK500
DEXAMYL see BBK500
DEXA-SCHEROSON (INJECTABLE) see DBC550
DEXBENZETIMIDE HYDROCHLORIDE see DBE000
DEXBROMPHENIRAMINE MALEATE see DXG100
DEXCHLOROPHENIRAMINE MALEATE see PJJ325
DEXCHLORPHENIRAMINE MALEATE see PJJ325
DEXEDRINA see BBK500
DEXEDRINE see AOA500
DEXEDRINE SULFATE see BBK500
DEXETIMIDE HYDROCHLORIDE see DBE000
DEXIES see BBK500
DEXIUM see DMI300
DEXON see DOU600
DEXON E 117 see PMP500
DEXONE see SOW000
DEXOPHRINE see DBA800
DEXOVAL see DBA800, MDT600
DEXOXADROL HYDROCHLORIDE see DVP400
DEXPANTHENOL (FCC) see PAG200
DEXTELAN see SOW000
DEXTIM see MDT600
DEXTRAN 1 see DBC800
DEXTRAN 2 see DBD000
DEXTRAN 5 see DBD200
DEXTRAN 10 see DBD400
DEXTRAN 11 see DBD600
DEXTRAN 70 see DBD700
DEXTRAN see DBD700
DEXTRAN ION COMPLEX see IGS000
DEXTRANS see DBD800
DEXTRAN SULFATE SODIUM see DBD750
DEXTRAN SULFATE SODIUM ALUMINIUM see DBB500
DEXTRARINE see DBD750
DEXTRAVEN see DBD700
DEXTRIFERRON see IGT000
DEXTRIFERRON INJECTION see IGT000
β-DEXTRIN see COW925
DEXTRINS see DBD800
DEXTROAMPHETAMINE SULFATE see BBK500
DEXTROBENZETIMIDE HYDROCHLORIDE see DBE000
DEXTRO CALCIUM PANTOTHENATE see CAU750
DEXTROCHLORPHENIRAMINE MALEATE see PJJ325
DEXTROFER 75 see IGS000
DEXTROID see SKJ300
DEXTROMETHADONE see DBE100
DEXTROMETHORPHAN see DBE150
DEXTROMETHORPHAN BROMIDE see DBE200
DEXTROMETHORPHAN HYDROBROMIDE see DBE200
DEXTRO-α-METHYLPHENETHYLAMINE SULFATE see BBK500
DEXTROMETORPHAN HYDROBROMIDE see DBE200
DEXTROMORAMIDE see AFJ400
DEXTROMYCETIN see CDP250
DEXTROMYCIN HYDROCHLORIDE see DBE600
DEXTRONE see DWX800, PAJ000
DEXTRONE-X see PAJ000
DEXTRO-1-PHENYL-2-AMINOPROPANE SULFATE see BBK500
DEXTRO-β-PHENYLISOPROPYLAMINE SULFATE see BBK500
DEXTROPROPOXYPHENE see DAB879
DEXTROPROPOXYPHENE HYDROCHLORIDE see PNA500
DEXTROPROPOXYPHENE NAPSYLATE see DBE625

DEXTROPROXYPHEN HYDROCHLORIDE see PNA500
DEXTROPUR see GFG000
DEXTRORPHAN see DBE800
DEXTRORPHAN TARTRATE see DBE825
DEXTROSE (FCC) see GFG000
DEXTROSE, anhydrous see GFG000
DEXTROSOL see GFG000
DEXTROSULPHENIDOL see MPN000
DEXTROTHYROXINE SODIUM see SKJ300
DEXTROTUBOCURARINE CHLORIDE see TOA000
DEXTROXIN see SKJ300
DEXULATE see DBD750
DEXURON see PAJ000
DEZIBARBITUR see EOK000
DEZONE see SOW000
DF 118 see DKW800, DKX000
DF 468 see PIM500
DF 469 see AAJ350
DF 493 see NOB800
DF-521 see RDA350
DFA see DVX800
DFD 0173 see PJS750
DFD 0188 see PJS750
DFD 2005 see PJS750
DFD 6005 see PJS750
DFD 6032 see PJS750
DFD 6040 see PJS750
DFDJ 5505 see PJS750
α-DFMO see DBE835, DKH875
DFO see DAK200
DFOA see DAK200
DFOM see DAK200
DFP see IRF000
DFT see DWN800
5'-DFUR see DYE415
DFV see DKF130
2-DG see DAR600
DGE see DKM200
DGNB 3825 see PJS750
DH 245 see AOO300
DH-524 see DGA800
DHA see AOO450, DAK400, MFW500
DHAQ see MQY090
DHAQ DIACETATE see DBE875
DHA-SODIUM see SGD000
DHA-S SODIUM see DAL040
2,5-DHBA see GCU000
DHBP see DYF200
β-DHC see DLO880
DHC see DAL000
DHK see DLR000
DHMS see DME000
DHNT see BKH500
DHPN see DNB200
DHPT see MHJ300
DHPTA see DCB100
DHS see MFW500
DHT₂ see DME300
5-β-DHT see HJB100
DHUTRA see SLV500
D 201 HYDROCHLORIDE see AEG625
DIABARIL see CKK000
DIABASE RED B see NEQ000
DIABASE SCARLET G see NMP500
DIABASIC MAGENTA see MAC250
DIABASIC MALACHITE GREEN see AFG500
DIABASIC RHODAMINE B see FAG070
DIABECHLOR see CKK000
DIABEFAGOS see DQR800
DIABEN see BSQ000
DIABENAL see CKK000
DIABENESE see CKK000
DIABENEZA see CKK000
DIABENOR see DBE885
DIABENYL see BBV500
DIABETA see CEH700
DIABETAMID see BSQ000
DIABETOL see BSQ000
DIABETORAL see CKK000
DIABET-PAGES see CKK000
DIABINESE see CKK000
DIABORAL see BSM000, CPR000
DIABRIN see BOM750, BQL000
DIABUTAL see NBU000
DIABUTON see BSQ000

DIABYLEN see BBV500
DIACARB see AAI250
DIACELLITON FAST BLUE BORDEAUZ B see CMP080
DIACELLITON FAST BLUE GREEN B see DMM400
DIACELLITON FAST BLUE R see TBG700
DIACELLITON FAST BORDEAUX B see CMP080
DIACELLITON FAST BRILLIANT BLUE B see MGG250
DIACELLITON FAST GREY G see DCJ200
DIACELLITON FAST PINK B see AKE250
DIACELLITON FAST VIOLET B see DBY700
DIACELLITON FAST VIOLET BF see AKP250
DIACELLITON FAST VIOLET 5R see DBP000
DIACELLITON FAST YELLOW G see AAQ250
DIACEL NAVY DC see DCJ200
DIACEPAN see DCK759
DIACEPHIN see HBT500
DIACESAL see SAN000
4,4'-DIACETAMIDODIPHENYL SULFONE see SNY500
2,7-DIACETAMIDOFLUORENE see BGP250
2-DIACETAMIDOFLUORENE see DBF200
N-1-DIACETAMIDOFLUORENE see DBF000
2,4-DIACETAMIDO-6-(5-NITRO-2-FURYL)-s-TRIAZINE see DBF400
3,5-DIACETAMIDO-2,4,6-TRIIODOBENZOIC ACID see DCK000
3,5-DIACETAMIDO-2,4,6-TRIIODOBENZOIC ACID, SODIUM SALT see SEN500
α-5-DIACETAMIDO-2,4,6-TRIIODO-m-TOLUIC ACID see AAI750
10,040 DIACETATE see CDT125
1,3-DIACETATE GLYCEROL see DBF600
1,2-DIACETATE 1,2,3-PROPANETRIOL see DBF600
DIACETATOPALLADIUM see PAD300
DIACETATOZIRCONIC ACID see ZTS000
DIACETAZOTOL see ACR300
DIACETIC ETHER see EFS000
1,2-DI-ACETIN see DBF600
1,3-DIACETIN see DBF600
2,3-DIACETIN see DBF600
DIACETIN see DBF600, GGA100
DIACETONALCOHOL (DUTCH) see DBF750
DIACETONALCOOL (ITALIAN) see DBF750
DIACETONALKOHOL (GERMAN) see DBF750
DIACETONE see DBF750
DIACETONE ACRYLAMIDE see DTH200
DIACETONE ALCOHOL see DBF750
DIACETONE ALCOHOL PEROXIDES, >57% in solution with >9% hydrogen peroxide (DOT) see HMK050
DIACETONE-ALCOOL (FRENCH) see DBF750
DIACETONE PEROXIDES, solid, or >25% in solution (DOT) see ACV500
DIACETOTOLUIDE see ACR300
o-DIACETOTOLUIDIDE, 4''-(o-TOLYLAZO)-(8CI) see ACR300
1,3-DIACETOXYBENZENE see REA100
DIACETOXYBUTYLTIN see DBF800
1,2-DIACETOXY-3-CHLOROPROPANE see CIL900
DIACETOXYDIBUTYLPLUMBANE see DED400
DIACETOXYDIBUTYL STANNANE see DBF800
DIACETOXYDIBUTYLTIN see DBF800
1,1-DIACETOXY-2,3-DICHLOROPROPANE see DBF875
trans-7,8-DIACETOXY-7,8-DIHYDROBENZO(a)PYRENE see DBG200
4-β,15-DIACETOXY-3-α,8-α-DIHYDROXY-12,13-EPOXYTRICHOTHEC-9-ENE see NCK000
DIACETOXYDIMETHYLPLUMBANE see DSL400
2-(4,4'-DIACETOXYDIPHENYLMETHYL)PYRIDINE see PPN100
(4,4'-DIACETOXYDIPHENYL)(2-PYRIDYL)METHANE see PPN100
4,4'-DIACETOXYDIPHENYLPYRID-2-YLMETHANE see PPN100
3-α,17-β-DIACETOXY-2-β,16-β-DIPIPERIDINO-5-α-ANDROSTANE DIMETHO-BROMIDE see PAF625
3-β,17-β-DIACETOXY-17-α-ETHYNYL-4-OESTRENE see EQJ500
4-β,15-DIACETOXY-3-α-HYDROXY-12,13-EPOXYTRICHOTHEC-9-ENE see AOP250
DIACETOXYMERCURIPHENOL see HNK575
DIACETOXYMERCURY see MCS750
7-DIACETOXYMETHYLBENZ(a)ANTHRACENE see BBG750
4,15-DIACETOXY-8-(3-METHYLBUTYRYLOXY)-12,13-EPOXY-Δ-9-TRICHOTHE-CEN-3-OL see FQS000
4-β,15-DIACETOXY-8-α-(3-METHYLBUTYRYLOXY)-3-α-HYDROXY-12,13-EPOXY-TRICHOTHEC-9-ENE see FQS000
3-β,17-β-DIACETOXY-19-NOR-17-α-PREGN-4-EN-20-YNE see EQJ500
DIACETOXYPALLADIUM see PAD300
DI-(4-ACETOXYPHENYL)-2-PYRIDYLMETHANE see PPN100
DI-(p-ACETOXYPHENYL)-2-PYRIDYLMETHANE see PPN100
1,3-DIACETOXYPROPENE see PMO800
1,1-DIACETOXYPROPENE-2 see ADR250
3,3-DIACETOXYPROPENE see ADR250
DIACETOXYPROPENE see ADR250
4,15-DIACETOXYSCIRPEN-3-OL see AOP250
DIACETOXYSCIRPENOL see AOP250
DIACETOXYTETRABUTYLDISTANNOXANE see BGQ000

DIACETYL (FCC) see BOT500
DIACETYLAMINOAZOTOLUENE see ACR300
4,4'-DIACETYLAMINOBIPHENYL see BFX000
2,7-DIACETYLAMINOFLUORENE see BGP250
N,N-DIACETYL-2-AMINOFLUORENE see DBF200
N-DIACETYL-2-AMINOFLUORENE see DBF200
2-DIACETYLAMINOFLUORENE see DBF200
DI-(p-ACETYLAMINOPHENYL)SULFONE see SNY500
3,5-DIACETYLAMINO-2,4,6-TRIJODBENZOSAEURE NATRIUM (GERMAN) see SEN500
4,4'-DIACETYLBENZIDINE see BFX000
N,N'-DIACETYL BENZIDINE see BFX000
2-((2,4-DIACETYL-5-BENZOFURANYL)OXY)TRIETHYLAMINE HYDROCHLORIDE see DBG499
N,O-DIACETYL-N-(4-BIPHENYLYL)HYDROXYLAMINE see ABJ750
DIACETYLCHOLINE see CMG250
DIACETYLCHOLINE CHLORIDE see HLC500
DIACETYLCHOLINE DICHLORIDE see HLC500
DIACETYLCHOLINE DIIODIDE see BJI000
N,N'-DIACETYLDAPSONE see SNY500
DIACETYLDAPSONE see SNY500
N,N'-DIACETYL-4,4'-DIAMINODIPHENYL SULFONE see SNY500
4,4'-DIACETYLDIAMINODIPHENYL SULFONE see SNY500
2,6-DIACETYL-7,9-DIHYDROXY-8,9b-DIMETHYL-1,3(2H,9bH)-DIBENZOFURAN-DIONE see UWJ000
N,N'-DIACETYL-3,3'-DIMETHYLBENZIDINE see DRI400
DIACETYL DIOXIME see DBH000
1,2-DIACETYLETHANE see HEQ500
α,β-DIACETYLETHANE see HEQ500
N,N-DIACETYL-2-FLUORENAMINE see DBF200
DIACETYL GLYCERINE see DBF600
DIACETYLGLYCEROL see GGA100
O,O'-DIACETYL 4-HYDROXYAMINOQUINOLINE-1-OXIDE see ABN500
DIACETYLLANATOSID C (GERMAN) see DBH200
DIACETYLLANATOSIDE see DBH200
DIACETYLMANGANESE see MAQ000
DIACETYLMETHANE see ABX750
DIACETYLMONOOXIME see OMY910
DIACETYLMONOXIME see OMY910
DIACETYLMORFIN see HBT500
DIACETYLMORPHINE see HBT500
DIACETYLMORPHINE HYDROCHLORIDE see DBH400
N,N'-DIACETYL-N,N'-DINITRO-1,2-DIAMINOETHANE see DBG900
1,3-DIACETYLOXYPROPENE see PMO800
DIACETYL PEROXIDE (MAK) see ACV500
N,O-DIACETYL-N-(p-STYRYLPHENYL)HYDROXYLAMINE see ABW500
trans-N,o-DIACETYL-N-(p-STYRYLPHENYL)HYDROXYLAMINE see ABL250
3,4-DI(ACETYLTHIOMETHYL)-5-HYDROXY-6-METHYLPYRIDINE HYDROBRO-MIDE see DBH800
N,N-DIACETYL-o-TOLYLAZO-o-TOLUIDINE see ACR300
DIACID see BNK000
DIACID ALIZARINE SKY BLUE B see CMM090
DIACID ALIZARIN SKY BLUE B see CMM090
DIACID BLUE BLACK 10B see FAB830
DIACID LIGHT BLUE BR see CMM070
DIACID METANIL YELLOW see MDM775
DIACID ORANGE II see CMM220
DIACOTTON BLACK BH see CMN800
DIACOTTON BLUE BB see CMO000
DIACOTTON BRILLIANT BLUE RW see CMO600
DIACOTTON CONGO RED see SGQ500
DIACOTTON DARK GREEN see CMO830
DIACOTTON DEEP BLACK see AQP000
DIACOTTON DEEP BLACK RX see AQP000
DIACOTTON FAST BLACK D see CMN230
DIACOTTON FAST RED F see CMO870
DIACOTTON GREEN B see CMO840
DIACOTTON SKY BLUE 5B see CMO500
DIACRID see DBX400
DIACROMO BLUE G see CMP880
DIACRYALTE DIETHYLENE GLYCOL see ADT250
DIACTOL see VSZ100
DIACYCINE see TBX250
DIADEM CHROME BLUE R see HJF500
DIADILAN see PFJ750
DIADOL see AFS500
DI-ADRESON F see PMA000
DI-ADRESON-F-AQUOSUM see PMA100
DIAETHANOLAMIN (GERMAN) see DHF000
DIAETHANOLAMIN-3,5-DIJODPYRIDON-(4)-ESSIGSAEURE (GERMAN) see DNG400
DIAETHANOLNITROSAMIN (GERMAN) see NKM000
1,1-DIAETHOXY-AETHAN (GERMAN) see AAG000
2-DIAETHOXYPHOSPHINYL-THIOAETHYL-TRIMETHYL-AMMONIUM-JODID (GER-MAN) see TLF500

DIAETHYLACETAL (GERMAN) see AAG000
DIAETHYLAETHER (GERMAN) see EJU000
DIAETHYLALLYLACETAMIDE (GERMAN) see DJU200
DIAETHYLAMIN (GERMAN) see DHJ200
DIAETHYLAMINOAETHANOL (GERMAN) see DHO500
o-DIAETHYLAMINOAETHOXY-BENZANILID (GERMAN) see DHP200
o-DIAETHYLAMINOAETHOXY-5-CHLOR-BENZANILID (GERMAN) see CFO250
1-DIAETHYLAMINO-AETHYLAMINO-4-METHYL-THIOXANTHONHYDROCHLORID (GERMAN) see SBE500
DIAETHYLAMINOAETHYL DIPHENYLCARBAMAT HYDROCHLORID (GERMAN) see DHY200
DIAETHYLAMINOAETHYL-THEOPHYLLIN (GERMAN) see CNR125
7-(β-DIAETHYLAMINO-AETHYL)-THEOPHYLLIN-HYDROCHLORID (GERMAN) see DIH600
DIAETHYLAMMONIUM-DIAETHYLDITHIOCARBAMAT see DJD000
DIAETHYLANILIN (GERMAN) see DIS700
O,O-DIAETHYL-O-(4-BROM-2,5-DICHLOR)-PHENYL-MONOTHIOPHOSPHAT (GERMAN) see EGV500
DIAETHYLCARBONAT (GERMAN) see DIX200
O,O-DIAETHYL-O-(CHINOXALYL-(2))-MONOTHIOPHOSPHAT (GERMAN) see DJV200
O,O-DIAETHYL-O-(3-CHLOR-4-METHYL-CUMARIN-7-YL)-MONOTHIOPHOSPHAT (GERMAN) see CNU750
O,O-DIAETHYL-o-(α-CYANBENZYLIDEN-AMINO)-THIONPHOSPHAT (GERMAN) see BAT750
O,O-DIAETHYL-o-(α-CYANO-BENZYLIDENAMINO)-MONOTHIOPHOSPHAT (GER-MAN) see BAT750
N-DIAETHYL CYSTEAMIN (GERMAN) see DIY600
O,O-DIAETHYL-O-(2,5-DICHLOR-4-BROMPHENYL)-THIONOPHOSPHAT (GER-MAN) see EGV500
O,O-DIAETHYL-O-1-(4,5-DICHLORPHENYL)-2-CHLOR-VINYL-PHOSPHAT (GER-MAN) see CDS750
O,O-DIAETHYL-O-2,4-DICHLOR-PHENYL-MONOTHIOPHOSPHAT (GERMAN) see DFK600
O,O-DIAETHYL-S((2,5-DICHLOR-PHENYL-THIO)-METHYL)-DITHIOPHOSPHAT (GERMAN) see PDC750
O,O-DIAETHYL-O-2,4-DICHLORPHENYL-THIONOPHOSPHAT (GERMAN) see DFK600
1,2-DIAETHYLHYDRAZINE (GERMAN) see DJL400
O,O-DIAETHYL-O-(2-ISOPROPYL-4-METHYL-PYRIMIDIN-6-YL)-MONOTHIOPHOSP-HAT (GERMAN) see DCM750
O,O-DIAETHYL-O-(2-ISOPROPYL-4-METHYL)-6-PYRIMIDYL-THIONOPHOSPHAT (GERMAN) see DCM750
O,O-DIAETHYL-O-(4-METHYL-COUMARIN-7-YL)-MONOTHIOPHOSPHAT (GER-MAN) see PKT000
O,O-DIAETHYL-O-(3-METHYL-1H-PYRAZOL-5-YL)-PHOSPHAT (GERMAN) see MOX250
O,O-DIAETHYL-O-4-METHYLSULFINYL-PHENYL-MONOTHIOPHOSPHAT (GER-MAN) see FAQ800
DIAETHYL-NICOTINAMID (GERMAN) see DJS200
O,O-DIAETHYL-O-(4-NITROPHENYL)-MONOTHIOPHOSPHAT (GERMAN) see PAK000
O,O'-DIAETHYL-p-NITROPHENYLPHOSPHAT (GERMAN) see NIM500
DIAETHYL-p-NITROPHENYLPHOSPHORSAEUREESTER (GERMAN) see NIM500
DIAETHYLNITROSAMIN (GERMAN) see NJW500
O,O-DIAETHYL-N-PHTALIMIDOTHIOPHOSPHAT (GERMAN) see DJX200
DI-AETHYL-PROPANEDIOL (GERMAN) see PMB250
O,O-DIAETHYL-O-(PYRAZIN-2YL)-MONOTHIOPHOSPHAT (GERMAN) see EPC500
O,O-DIAETHYL-O-(2-PYRAZINYL)-THIONOPHOSPHAT (GERMAN) see EPC500
O,O-DIAETHYL-S-(2-AETHYLTHIO-AETHYL)-DITHIOPHOSPHAT (GERMAN) see DXH325
O,O-DIAETHYL-S-(2-AETHYLTHIO-AETHYL)-MONOTHIOPHOSPHAT (GERMAN) see DAP200
O,O-DIAETHYL-S-(AETHYLTHIO-METHYL)-DITHIOPHOSPHAT (GERMAN) see PGS000
O,O-DIAETHYL-S-(6-CHLOR-2-OXO-BEN(b)-1,3-OXALIN-3-YL)-METHYL-DIT HIO-PHOSPHAT (GERMAN) see BDJ250
O,O-DIAETHYL-S-((4-CHLOR-PHENYL-THIO)-METHYL)DITHIOPHOSPHAT (GER-MAN) see TNP250
O,O-DIAETHYL-S-(1-METHYLAETHY)L-CARBAMOYL-METHYL-MONOTHIO-PHOSPHAT (GERMAN) see PHK250
O,O-DIAETHYL-S-(3-METHYL-2,4-DIOXO-5-OXA-3-AZA-HEPTYL)-DITHIOPHOSP-HAT (GERMAN) see DJI000
O,O-DIAETHYL-S-(4-OXOBENZOTRIAZIN-3-METHYL)-DITHIOPHOSPHAT (GER-MAN) see EKN000
O,O-DIAETHYL-S-((4-OXO-3H-1,2,3-BENZOTRIAZIN-3-YL)-METHYL)-DITHIO-PHOSPHAT (GERMAN) see EKN000
O,O-DIAETHYL-S-(3-THIA-PENTYL)-DITHIOPHOSPHAT (GERMAN) see DXH325
DIAETHYLSULFAT (GERMAN) see DKB110
DIAETHYLTHIOPHOSPHORSAEUREESTER des AETHYLTHIOGLYKOL (GER-MAN) see DAP200, DAO500
O,O-DIAETHYL-O-3,5,6-TRICHLOR-2-PYRIDYLMONOTHIOPHOSPHAT see CMA100
DIAETHYLZINNDICHLORID (GERMAN) see DEZ000

DIAFEN (antihistamine) see DVW700
DIAFEN see DVW700, LJR000
DIAFEN HYDROCHLORIDE see DVW700
DIAFORM UR see UTU500
DIAFURON see NGG500
DIAGINOL see AAN000
DIAGNORENOL see SHX000
DIAGRABROMYL see BNP750
DIAK 7 see THS100
DIAKARB see AAI250
DIAKARMON see SKV200
DIAKOL DM see UTU500
DIAKOL F see UTU500
DIAKOL M see UTU500
DIAKON see MLH750, PKB500
DIAL see AFS500, CFU750
DIAL-A-GESIC see HIM000
DIALFERIN see DBK400
DIALICOR see DHS200
DIALIFOR see DBI099
DIALKYL 79 PHTHALATE see LGF875
DIALKYLPHTHALATE C7-C9 see LGF875
DIALKYLZINCS see DBI159
DIALLAAT (DUTCH) see DBI200
DIALLAT (GERMAN) see DBI200
DIALLATE see DBI200
N,N-DIALLYDICHLOROACETAMIDE see DBJ600
DIALLYINITROSAMIN (GERMAN) see NJV500
DIALLYL see HCR500
DIALLYLAMINE see DBI600
4-DIALLYLAMINO-3,5-DIMETHYLPHENYL-N-METHYLCARBAMATE see DBI800
4-(DIALLYLAMINO)-3,5-XYLENOL METHYLCARBAMATE (ester) see DBI800
4-DIALLYL-AMINO-3,5-XYLYL N-METHYLCARBAMATE see DBI800
DIALLYLBARBITAL see AFS500
5,5-DIALLYLBARBITURIC ACID see AFS500
DIALLYLBARBITURIC ACID see AFS500
1,1-DIALLYL-3-(1,4-BENZODIOXAN-2-YLMETHYL)-3-METHYLUREA see DBJ100
2-(N,N-DIALLYLCARBAMYLMETHYL)AMINOMETHYL-1,4-BENZODIOXAN see DBJ100
N,N-DIALLYL-2-CHLOROACETAMIDE see CFK000
N,N-DIALLYL-α-CHLOROACETAMIDE see CFK000
N,N-DIALLYLCHLOROACETAMIDE see CFK000
DIALLYLCHLOROACETAMIDE see CFK000
DIALLYLCYANAMIDE see DBJ200
DIALLYLDIBROMO STANNANE see DBJ400
N,N-DIALLYL-2,2-DICHLOROACETAMIDE see DBJ600
DIALLYL DIGLYCOL CARBONATE see AGD250
DIALLYL DISULFIDE see AGF300
DIALLYL DISULPHIDE see AGF300
DIALLYL ETHER see DBK000
DIALLYLETHER ETHYLENGLYKOLU (CZECH) see EJE000
1,1-DIALLYLHYDRAZINE see DBK100
DIALLYLHYDRAZINE see DBK100
1,2-DIALLYLHYDRAZINE DIHYDROCHLORIDE see DBK120
DIALLYLMAL see AFS500
DIALLYL MALEATE see DBK200
DIALLYL MONOSULFIDE see AGS250
DIALLYLNITROSAMINE see NJV500
DIALLYLNORTOXIFERINE DICHLORIDE see DBK400
N,N′-DIALLYLNORTOXIFERINIUM DICHLORIDE see DBK400
2,6-DIALLYLPHENOL see DBK800
DIALLYL PHOSPHITE see DBL000
DIALLYL PHTHALATE see DBL200
DIALLYL SELENIDE see DBL300
DIALLYL SULFIDE see AGS250
DIALLYL THIOETHER see AGS250
DIALLYLTIN DIBROMIDE see DBJ400
DIALUMINOUS RED 4B see CMO885
DIALUMINUM DIPOTASSIUM SULFATE see AHF100
DIALUMINUM SULPHATE see AHG750
DIALUMINUM TRIOXIDE see AHE250
DIALUMINUM TRISULFATE see AHG750
DIALUX see ADY500
DIAMARIN see DYE600
DIAMAZO see ACR300
DIAMELKOL see MCB050
DIAMET see SIL500
DIAMIFEN HYDROCHLORIDE see DHY200
DIAMICRON see DBL700
DIAMIDAFOS see PEV500
DIAMIDE see DRP800, HGS000
DIAMIDFOS see PEV500
DIAMIDINE see DBL800
4,4′-DIAMIDINODIPHENOXYPENTANE see DBM000
4,4′-DIAMIDINO-α,ω-DIPHENOXYPENTANE see DBM000

4,4′-DIAMIDINODIPHENOXYPENTANE DI(β-HYDROXYETHANESULFONATE see DBL800
4,4′-DIAMIDINO-α,ω-DIPHENOXYPENTANE ISETHIONATE see DBL800
4,4′-DIAMIDINO-1,3-DIPHENOXYPROPANE DIHYDROCHLORIDE see DBM400
4,4′-DIAMIDINO-α,γ-DIPHENOXYPROPANE DIHYDROCHLORIDE see DBM400
4,4′-DIAMIDINOSTILBENE see SLS000
DIAMIDINO STILBENE see SLS000
4,4′-DIAMIDINOSTILBENE DIHYDROCHLORIDE see SLS500
DIAMINE see HGS000
2,4-DIAMINEANISOLE see DBO000
DIAMINE BLACK BH see CMN800
DIAMINE BLACK BHM see CMN800
DIAMINE BLUE 2B see CMO000
DIAMINE BLUE 3B see CMO250
DIAMINE BROWN 3GN-CF see CMO825
DIAMINE DARK GREEN B see CMO830
DIAMINE DARK GREEN N see CMO830
DIAMINE DEEP BLACK EC see AQP000
(4,4′-DIAMINE)-3,3′-DIMETHYL(1,1′-BIPHENYL) see TGJ750
DIAMINE DIPENICILLIN G see BFC750
DIAMINE DIRECT BLACK E see AQP000
DIAMINE FAST BLACK B see CMN240
DIAMINE FAST RED F see CMO870
DIAMINE FAST RED FA-CF see CMO870
DIAMINE FAST RED N see CMO870
DIAMINE FAST RED OJCD see CMO870
DIAMINE GREEN B see CMO840
DIAMINE SCARLET 3BA-CF see CMO875
DIAMINE SKY BLUE CI see CMO500
DIAMINE SKY BLUE FF see BGT250
DIAMINE VIOLET N see CMP000
DIAMINIDE MALEATE see DBM800
2,6-DIAMINOACRIDINE see DBN000
2,8-DIAMINOACRIDINE see DBN600
3,6-DIAMINOACRIDINE see DBN600
3,7-DIAMINOACRIDINE see DBN000
3,9-DIAMINOACRIDINE see ADJ625
3,6-DIAMINOACRIDINE mixture with 3,6-DIAMINO-10-METHYLACRIDINIUM CHLORIDE see DBX400
2,5-DIAMINOACRIDINE (EUROPEAN) see ADJ625
3,6-DIAMINOACRIDINE HEMISULFATE see PMH100
3,6-DIAMINOACRIDINE HYDROCHLORIDE HEMIHYDRATE see DBN200
3,6-DIAMINOACRIDINE MONOHYDROCHLORIDE see PMH250
2,8-DIAMINOACRIDINIUM see DBN600
3,6-DIAMINOACRIDINIUM see DBN600
3,6-DIAMINOACRIDINIUM CHLORIDE see PMH250
3,6-DIAMINOACRIDINIUM CHLORIDE HYDROCHLORIDE see PMH250
2,8-DIAMINOACRIDINIUM CHLORIDE MONOHYDROCHLORIDE see PMH250
1,2-DIAMINOAETHAN (GERMAN) see EEA500
2,6-DIAMINO-4-(((AMINO-CARBONYL)OXY)METHYL)-3a,4,8,9-TETRAHYDRO-1H,10H-PYRROLO(1,2-c)PURINE-10,10-DIOL (3aS-(3a-α-4-α,10aR*)) see SBA600
4,4′-DIAMINO-1,1′-ANIHRIMIDE see IBD000
2,4-DIAMINOANISOL see DBO000
2,4-DIAMINOANISOLE see DBO000
2,4-DIAMINOANISOLE BASE see DBO000
m-DIAMINOANISOLE 1,3-DIAMINO-4-METHOXYBENZENE see DBO000
2,5-DIAMINOANISOLE SULFATE see MEB820
2,4-DIAMINOANISOLE SULPHATE see DBO400
2,4-DIAMINO-ANISOL SULPHATE see DBO400
1,4-DIAMINO-9,10-ANTHRACENEDIOL see DBO600
1,4-DIAMINOANTHRACENE-9,10-DIOL see DBO600
1,4-DIAMINO-9,10-ANTHRACENEDIONE see DBP000
2,6-DIAMINOANTHRACHINON see APK850
1,4-DIAMINOANTHRACHINON (CZECH) see DBP000
1,5-DIAMINOANTHRACHINON (CZECH) see DBP200
1,8-DIAMINOANTHRACHINON (CZECH) see DBP400
1,2-DIAMINOANTHRAQUINONE see DBO800
1,4-DIAMINOANTHRAQUINONE see DBP000
1,5-DIAMINOANTHRAQUINONE see DBP200, DBP400
1,8-DIAMINOANTHRAQUINONE see DBP400
2,6-DIAMINOANTHRAQUINONE see APK850
1,5-DIAMINO-9,10-ANTHRAQUINONE see DBP200
2,6-DIAMINO-9,10-ANTHRAQUINONE see APK850
1,5-DIAMINOANTHRARUFIN see DBP909
4,8-DIAMINOANTHRARUFIN see DBP909
3,7-DIAMINO-5-AZAANTHRACENE see DBN600
2,4-DIAMINOAZOBENZEN (CZECH) see DBP999
4,4′-DIAMINOAZOBENZENE see ASK925
p-DIAMINOAZOBENZENE see ASK925
DIAMINOAZOBENZENE see DBP999
2,4-DIAMINOAZOBENZENE HYDROCHLORIDE see PEK000
2′,4-DIAMINOBENZANILIDE see DBQ125
1,2-DIAMINOBENZENE see PEY250
1,3-DIAMINOBENZENE see PEY000

1,4-DIAMINOBENZENE see PEY500
m-DIAMINOBENZENE see PEY000
o-DIAMINOBENZENE see PEY250
p-DIAMINOBENZENE see PEY500
1,3-DIAMINOBENZENE DIHYDROCHLORIDE see PEY750
1,4-DIAMINOBENZENE DIHYDROCHLORIDE see PEY650
m-DIAMINOBENZENE DIHYDROCHLORIDE see PEY750
p-DIAMINOBENZENE DIHYDROCHLORIDE see PEY650
1,3-DIAMINOBENZENESULFONIC ACID see PFA250
2,4-DIAMINOBENZENESULFONIC ACID see PFA250
2,5-DIAMINOBENZENE SULFONIC ACID see PEY800
1,3-DIAMINOBENZENE-4-SULFONIC ACID see PFA250
1,3-DIAMINOBENZENE-6-SULFONIC ACID see PFA250
3,3'-DIAMINOBENZIDENE see BGK500
3,3'-DIAMINOBENZIDINE TETRAHYDROCHLORIDE see BGK750
3,5-DIAMINOBENZOIC ACID DIHYDROCHLORIDE see DBQ200
3,5-DIAMINOBENZOTRIFLUORIDE see TKB775
6,6'-DIAMINO-m,m'-BIPHENOL see DMI400
4,4'-DIAMINO-1,1'-BIPHENYL see BBX000
4,4'-DIAMINOBIPHENYL see BBX000
o,p'-DIAMINOBIPHENYL see BGF109
p,p'-DIAMINOBIPHENYL see BBX000
4,4'-DIAMINOBIPHENYL-3,3'-DICARBOXYLIC ACID see BFX250
4,4'-DIAMINO-3,3'-BIPHENYLDICARBOXYLIC ACID see BFX250
4,4'-DIAMINO-3,3'-BIPHENYLDICARBOXYLIC ACID DISODIUM SALT see
    BBX250
4,4'-DIAMINO-3,3'-BIPHENYLDIOL see DMI400
4,4'-DIAMINO-2,2'-BIPHENYLDISULFONIC ACID see BBX500
4,4'-DIAMINOBIPHENYL-2,2'-DISULFONIC ACID see BBX500
4,4'-((4,4'-DIAMINO-(1,1'-BIPHENYL)-3,3'-DIYL)BIS(OXY))BISBUTANOIC ACID, DI-
    HYDROCHLORIDE see DEK000
4,4'-(3,3'-DIAMINO-p,p'-BIPHENYLENEDIOXY)DIBUTYRIC ACID see DEK400
4,4'-DIAMINOBIPHENYLOXIDE see OPM000
4,4'-DIAMINO-3-BIPHENYL-3-SULFONIC ACID see BBY250
1,5-DIAMINOBIURET see IBC000
DIAMINOBIURET see IBC000
1,5-DIAMINOBIURET DIHYDRAZIDE see IBC000
1,3-DIAMINOBUTANE see BOR750
1,4-DIAMINOBUTANE see BOS000
1,4-DIAMINOBUTANE DIHYDROCHLORIDE see POK325
3,4-DIAMINO-1-CHLOROBENZENE see CFK125
3,4-DIAMINOCHLOROBENZENE see CFK125
1,2-DIAMINO-4-CHLOROBENZENE see CFK125
4,6-DIAMINO-1-(p-CHLOROPHENYL)-1,2-DIHYDRO-2,2-DIMETHYL-s-TRIAZINE
    MONOHYDROCHLORIDE see COX325
2,4-DIAMINO-5-(4-CHLOROPHENYL)-6-ETHYLPYRIMIDINE see TGD000
2,4-DIAMINO-5-p-CHLOROPHENYL-6-ETHYLPYRIMIDINE see TGD000
DI-(4-AMINO-3-CHLOROPHENYL)METHANE see MJM200
4,5-DIAMINOCHRYSAZIN see DBQ250
1,8-DIAMINOCHRYSAZINE see DBQ250
DIAMINOCILLIAN see BFC750
DI-(4-AMINO-3-CLOROFENIL)METANO (ITALIAN) see MJM200
2,5-DIAMINO-2,4,6-CYCLOHEPTATRIEN-1-ONE HDYROCHLORIDE see DCF710
1,2-DIAMINOCYCLOHEXANE see CPB100
1,3-DIAMINOCYCLOHEXANE see DBQ800
1,2-DIAMINOCYCLOHEXANE-N,N'-TETRAACETIC ACID see CPB120
1,2-DIAMINOCYCLOHEXANEPLATINUM(II) CHLORIDE see DAD040
1,2-DIAMINOCYCLOHEXANETETRAACETIC ACID see CPB120
DI(p-AMINOCYCLOHEXYL)METHANE see MJQ260
2,5-DIAMINO-2,4,6-CYCLOPHETATRIEN-1-ONE see DCF700
1,12'-DIAMINODODECANE see DXW800
2,4-DIAMINO-5-(p-(p-((p-(2,4-DIAMINO-1-ETHYLPYRIMIDINIUM-5-
    YL)PHENYL)CARBAMOYLCINNAMAMIDO)PHENYL)-1-ETH see DBR000
2,4-DIAMINO-5-(p-(p-(((2,4-DIAMINO-1-METHYLPYRIMIDINIUM-5-
    YL)PHENYL)CARBAMOYL)CINNAMAMIDO)PHENYL-1-M see DBR200
4,4'-DIAMINO-1,1'-DIANTHRAQUINONYLAMINE see IBD000
4,4'-DIAMINO-1,1'-DIANTHRAQUINONYLIMINE see IBD000
4,4'-DIAMINO-1,1'-DIANTHRIMID (CZECH) see IBD000
4,4'-DIAMINO-1,1'-DIANTHRIMIDE see IBD000
1,4-DIAMINO-2,6-DICHLOROBENZENE see DBR400
4,4'-DIAMINO-3,3'-DICHLOROBIPHENYL see DEQ600
4,4'-DIAMINO-3,3'-DICHLORODIPHENYL see DEQ600
4,4'-DIAMINO-3,3'-DICHLORODIPHENYLMETHANE see MJM200
2,4-DIAMINO-5-(3,4-DICHLOROPHENYL)-6-METHYLPYRIMIDINE see MQR100
2,4-DIAMINO-5-(3',4'-DICHLOROPHENYL)-6-METHYLPYRIMIDINE see MQR100
2,4-DIAMINO-6-(2,5-DICHLOROPHENYL)-s-TRIAZINE MALEATE see MQY300
4:4'-DIAMINO-2:2'-DICHLOROSTILBENE see DGK400
4:4'-DIAMINO-3:3'-DICHLOROSTILBENE see DGK600
4,4'-DIAMINODICYCLOHEXYLMETHANE see MJQ260
p,p'-DIAMINODICYCLOHEXYLMETHANE see MJQ260
o-2,6-DIAMINO-2,6-DIDEOXY-α-d-GLUCOPYRANOSYL-(1-4)-O-(β-d-RIBOFURANO-
    SYL-(1-5))-2-DEOXY-d-STREPTAMINE see XQJ650
o-2,6-DIAMINO-2,6-DIDEOXY-α-d-GLUCOPYRANOSYL-(1-4)-o-(β-d-XYLOFURA-
    NOSYL-(1-5))-N'-(4-AMINO-2-HYDROXY-1-OXOBUTYL)-2-DEOXY-d-STREPTA-
    MINE see BSX325

o-2,6-DIAMINO-2,6-DIDEOXY-β-l-IDOPYRANOSYL-(1-3)-o-(β-d-RIBOFURANOSYL-
    (1-5)-o-(2,6-DIAMINO-2,6-DIDEOXY-α-d-GLUCOPYRANOSYL-(1-4))-2-DEOXYD-
    STREPTAMINE SULFATE (SALT) see NCD550
2,2'-DIAMINODIETHYLAMINE see DJG600
1,3-DIAMINO-7-(DIETHYLAMINO)-8-METHYLPHENOXAZIN-5-IUM CHLORIDE see
    BMM550
β,β'-DIAMINODIETHYL DISULFIDE see MCN500
3,8-DIAMINO-5-(3-(DIETHYLMETHYLAMMONIO)PROPYL)-6-PHENYLPHENAN-
    THRIDINIUM DIIODIDE see PMT000
DIAMINODIFENILSULFONA (SPANISH) see SOA500
p,p'-DIAMINODIFENYLMETHAN see MJQ000
3,3'-DIAMINODIFENYLSULFON (CZECH) see SOA000
1,4-DIAMINO-9,10-DIHYDROXYANTHRACEN (CZECH) see DBO600
1,5-DIAMINO-4,8-DIHYDROXY-9,10-ANTHRACENEDIONE see DBP909
1,8-DIAMINO-4,5-DIHYDROXYANTHRACHINON see DBQ250
4,8-DIAMINO-1,5-DIHYDROXYANTHRAQUINONE see DBP909
4,5-DIAMINO-1,8-DIHYDROXYANTHRAQUINONE see DBQ250
1,8-DIAMINO-4,5-DIHYDROXYANTHRAQUINONE see DBQ250
1,5-DIAMINO-4,8-DIHYDROXYANTHRAQUINONE see DBP909
leuco-1,5-DIAMINO-4,8-DIHYDROXYANTHRAQUINONE see DBP909
1,8-DIAMINO-4,5-DIHYDROXY-9,10-ANTHRAQUINONE see DBQ250
3,3'-DIAMINO-4,4'-DIHYDROXYARSENOBENZENE DIHYDROCHLORIDE see
    SAP500
3,3'-DIAMINO-4,4'-DIHYDROXY ARSENOBENZENE METHYLENESULFOXYLATE
    SODIUM see NCJ500
1,5-DIAMINO-4,8-DIHYDROXY-3-(p-METHOXYPHENYL)ANTHRAQUINONE see
    DBT000
2,4-DIAMINO-6-DIMETHOXYPHOSPHINOTHIONYLTHIOMETHYL-s-TRIAZINE see
    ASD000
3,6-DIAMINO-2,7-DIMETHYLACRIDINE see DBT200
2,8-DIAMINO-3,7-DIMETHYLACRIDINE see DBT200
4,4'-DIAMINO-3,3'-DIMETHYLBIPHENYL see TGJ750
4,4'-DIAMINO-3,3'-DIMETHYLBIPHENYL DIHYDROCHLORIDE see DQM000
4,4'-DIAMINO-3,3'-DIMETHYLDIPHENYL see TGJ750
2,2'-DIAMINO-1,1'-DINAPHTHYL see BGB750
1:2'-DIAMINO-1'：2-DINAPHTHYL see BGC000
2,4'-DIAMINODIPHENYL see BGF109
4,4'-DIAMINODIPHENYL see BBX000
p-DIAMINODIPHENYL see BBX000
O,O'-DIAMINO DIPHENYL DISULFIDE see DXJ800
4,4'-DIAMINODIPHENYL-2,2'-DISULFONIC ACID see BBX500
4,4-DIAMINODIPHENYL ETHER see OPM000
p,p'-DIAMINODIPHENYL ETHER see OPM000
DIAMINODIPHENYL ETHER see OPM000
4,4'-DIAMINODIPHENYLMETHAN see MJQ000
2,4'-DIAMINODIPHENYLMETHAN (GERMAN) see MJP750
2,4'-DIAMINODIPHENYLMETHANE see MJP750
4,4'-DIAMINODIPHENYLMETHANE see MJQ000
4,4'-DIAMINODIPHENYLMETHANE (DOT) see MJQ000
o,p'-DIAMINODIPHENYLMETHANE see MJP750
p,p'-DIAMINODIPHENYLMETHANE see MJQ000
DIAMINODIPHENYLMETHANE see MJQ000
4,4'-DIAMINODIPHENYL OXIDE see OPM000
4,4'-DIAMINODIPHENYL SULFIDE see TFI000
p,p'-DIAMINODIPHENYL SULFIDE see TFI000
3,3'-DIAMINODIPHENYL SULFONE see SOA000
4,4'-DIAMINODIPHENYL SULFONE see SOA500
DIAMINO-4,4'-DIPHENYL SULFONE see SOA500
p,p'-DIAMINODIPHENYL SULFONE see SOA500
p,p'-DIAMINODIPHENYLSULFONE-N,N'-DI(DEXTROSE SODIUM SULFONATE)
    see AOO800
p,p'-DIAMINODIPHENYL SULPHIDE see TFI000
DIAMINO-4,4'-DIPHENYL SULPHONE see SOA500
p,p-DIAMINODIPHENYL SULPHONE see SOA500
p,p'-DIAMINODIPHENYLTRICHLOROETHANE see BGU500
4:4'-DIAMINO-3-DIPHENYLYL HYDROGEN SULFATE see BBY250
4,4'-DIAMINO-3-DIPHENYLYL HYDROGEN SULFATE see BBY500
3,3-DIAMINODIPROPYLAMINE see AIX250
3,3'-DIAMINODIPROPYLAMINE see AIX250
DIAMINODITOLYL see TGJ750
1,2-DIAMINO-ETHAAN (DUTCH) see EEA500
1,2-DIAMINOETHANE see EEA500
1,2-DIAMINOETHANE COPPER COMPLEX see DBU800
1,2-DIAMINOETHANE DIHYDROCHLORIDE see EIW000
1,2-DIAMINO-ETHANO (ITALIAN) see EEA500
2,5-DIAMINO-7-ETHOXYACRIDINE LACTATE see EDW500
6,9-DIAMINO-2-ETHOXYACRIDINE LACTATE MONOHYDRATE see EDW500
4,6-DIAMINO-1-(2-ETHYLPHENYL)-2-METHYL-2-PROPYL-s-TRIAZINE HYDRO-
    CHLORIDE see DBV200
3,8-DIAMINO-5-ETHYL-6-PHENYLPHENANTHRIDINIUM BROMIDE see DBV400
2,7-DIAMINO-10-ETHYL-9-PHENYLPHENANTHRIDINIUM BROMIDE see DBV400
3,8-DIAMINO-5-ETHYL-6-PHENYLPHENANTHRIDINIUM CHLORIDE see HGI000
2,4-DIAMINO-6-ETHYL-5-PHENYLPYRIMIDINE see DCA300

4-(((4-(((4-(2,4-DIAMINO-1-ETHYLPYRIMIDINIUM-5-YL)PHENYL)AMINO)CARBONYL)PHENYL)AMINO)-1-ETHYLQUINOLINIU see DBW000

2,7-DIAMINOFLUORENE see FDM000

DIAMINOGENE VELOUR BLACK B see CMN800

DIAMINOGUANIDINIUM NITRATE see DBW100

1,6-DIAMINOHEXANE see HEO000

1,6-DIAMINOHEXANE DIHYDROCHLORIDE see HEO100

(I)-2,6-DIAMINO-1-HEXANETHIOL DIHYDROCHLORIDE see DBW200

2,6-DIAMINOHEXANOIC ACID HYDROCHLORIDE see LJO000

S-2,6-DIAMINOHEXYL DIHYDROGEN PHOSPHOROTHIOATE DIHYDRATE see DBW600

4,6-DIAMINO-2-HYDROXY-1,3-CYCLOHEXANE-3,6'-DIAMINO-3,6'-DIDEOXYDI-α-d-GLUCOSIDE see KAL000

4,6-DIAMINO-2-HYDROXY-1,3-CYCLOHEXYLENE 3,6'-DIAMINO-3,6'-DIDEOXYDI-d-GLUCOPYRANOSIDE see KAL000

α,2-DIAMINO-3-HYDROXY-γ-OXOBENZENEBUTANOIC ACID see HJD000

4,4'-DIAMINO-IMINO-1,1'BIANTHRAQUINONE see IBD000

4,4'-DIAMINO-1,1'-IMINOBISANTHRAQUINONE see IBD000

1,8-DIAMINO-p-MENTHANE see MCD750

1,4-DIAMINO-2-METHOXY-9,10-ANTHRACENEDIONE see DBX000

1,4-DIAMINO-2-METHOXYANTHRAQUINONE see DBX000

2,4-DIAMINO-1-METHOXYBENZENE see DBO000, DBO400

2,4-DIAMINO-1-METHOXYBENZENE SULPHATE see DBO400

1,3-DIAMINO-4-METHOXYBENZENE SULPHATE see DBO400

3,6-DIAMINO-9-(2-(METHOXYCARBONYL)PHENYL)XANTHYLIUM CHLORIDE see RGP600

2,8-DIAMINO-10-METHYLACRIDINIUM CHLORIDE see XAK000

3,6-DIAMINO-10-METHYLACRIDINIUM CHLORIDE see XAK000

3,6-DIAMINO-10-METHYLACRIDINIUM CHLORIDE with 3,6-ACRIDINEDIAMINE see DBX400

2,8-DIAMINO-10-METHYLACRIDINIUM CHLORIDE mixture with 2,8-DIAMINOA-CRIDINE see DBX400

2,4-DIAMINO-5-METHYLAZOBENZENE see CMM760

2,4-DIAMINO-1-METHYLBENZENE see TGL750

1,3-DIAMINO-4-METHYLBENZENE see TGL750

2,4-DIAMINO-5-METHYL-6-(BUT-2-YL)PYRIDO(2,3-d)PYRIMIDINE see DBX875

2,4-DIAMINO-5-METHYL-6-sec-BUTYLPYRIDO(2,3-d)PYRIMIDINE see DBX875

2,4-DIAMINO-1-METHYLCYCLOHEXANE see DBY750

1,3-DI(AMINOMETHYL)CYCLOHEXANE see BGT500

2,4-DIAMINOMETHYLCYCLOHEXANE see DBY000

3,7'-DIAMINO-N-METHYLDIPROPYLAMINE see BGU750

3-((2-((DIAMINOMETHYLENE)AMINO)-4-THIAZOLYL)METHYL)THIO)-N²-SULFA-MOYLPROPIONAMIDINE see FAB500

N,-(DIAMINOMETHYLENE)SULFANILAMIDE see AHO250

2,4-DIAMINO-2-METHYLPENTANE see MNI525

1,2-DIAMINO-2-METHYLPROPANE AQUADIPEROXO CHROMIUM(IV) see DBY300

N-(4-(((2,4-DIAMINO-5-METHYL-6-QUINAZOLINYL)METHYL)AMINO)BENZOYL)-l-ASPARTIC ACID see DBY500

N-(p-(((2,4-DIAMINO-5-METHYL-6-QUINAZOLINYL)METHYL)AMINO)BENZOYL)-l-ASPARTIC ACID see DBY500

DIAMINON see MDO750

1,5-DIAMINONAPHTHALENE see NAM000

1,7-DIAMINO-8-NAPHTHOL-3,6-DISULPHONIC ACID see DBY600

DIAMINON HYDROCHLORIDE see MDP000, MDP750

1,4-DIAMINO-5-NITRO-9,10-ANTHRACENEDIONE see DBY700

1,4-DIAMINO-5-NITRO ANTHRAQUINONE see DBY700

1,4-DIAMINO-2-NITROBENZENE see ALL750

2,4-DIAMINONITROBENZENE see NIM550

1,2-DIAMINO-4-NITROBENZENE see ALL500

4,6-DIAMINO-2-(5-NITRO-2-FURYL)-S-TRIAZINE see DBY800

1,5-DIAMINOPENTANE see PBK500

DIAMINOPHEN see DHW200

2,4-DIAMINOPHENOL see DCA200

2,4-DIAMINOPHENOL HYDROCHLORIDE see AHP000

3,7-DIAMINOPHENOTHIAZIN-5-IUM CHLORIDE see AKK750

2,6-DIAMINO-3-PHENYLAZOPYRIDINE see PEK250

2,6-DIAMINO-3-PHENYLAZOPYRIDINE HYDROCHLORIDE see PDC250

2,6-DIAMINO-3-(PHENYLAZO)PYRIDINE MONOHYDROCHLORIDE see PDC250

4,4'-((4,6-DIAMINO-m-PHENYLENE)BIS(AZO))DIBENZENESULFONIC ACID, DI-SODIUM SALT mixed with p-((4,6-DIAMINO-m-TOLYL)AZO)-BENZENESUL-FONIC ACID, SODIUM SALT see CMP250

4,4'-DIAMINOPHENYL ETHER see OPM000

2,7-DIAMINO-9-PHENYL-10-ETHYLPHENANTHRIDINIUM BROMIDE see DBV400

2,7-DIAMINO-9-PHENYL-10-ETHYLPHENANTHRIDINIUM CHLORIDE see HGI000

2,4-DIAMINO-5-PHENYL-6-ETHYLPYRIMIDINE see DCA300

DI-(4-AMINOPHENYL)METHANE see MJQ000

2,7-DIAMINO-9-PHENYLPHENANTHRIDINE ETHOBROMIDE see DBV400

DI(p-AMINOPHENYL) SULFIDE see TFI000

3,3'-DIAMINOPHENYL SULFONE see SOA000

DI(4-AMINOPHENYL)SULFONE see SOA500

DI(p-AMINOPHENYL) SULFONE see SOA500

DI(p-AMINOPHENYL)SULPHIDE see TFI000

DI(4-AMINOPHENYL)SULPHONE see SOA500

DI(p-AMINOPHENYL)SULPHONE see SOA500

2,4-DIAMINO 5-PHENYLTHIAZOL CHLORHYDRATE (FRENCH) see DCA600

2,4-DIAMINO-5-PHENYLTHIAZOLE HYDROCHLORIDE see DCA600

2,4-DIAMINO-5-PHENYLTHIAZOLE MONOHYDROCHLORIDE see DCA600

2,4-DIAMINO-6-PIPERIDINILPIRIMIDINA-3-OSSIDO (ITALIAN) see DCB000

2,4-DIAMINO-6-PIPERIDINOPYRIMIDINE-3-OXIDE see DCB000

1,2-DIAMINOPROPANE see PMK250

1,3-DIAMINOPROPANE see PMK500

DIAMINOPROPANOL TETRA ACETIC ACID see DCB100

DI-β-AMINOPROPIONITRILE FUMARATE see AMB750

DI(3-AMINOPROPOXY)ETHANE see DCB200

m-DI-(2-AMINOPROPYL)BENZENE DIHYDROCHLORIDE see DCC100

p-DI-(2-AMINOPROPYL)BENZENE DIHYDROCHLORIDE see DCC125

DI(3-AMINOPROPYL) ETHER of DIETHYLENE GLYCOL see DJD800

4-(((4-(((4-(2,4-DIAMINO-1-PROPYLPYRIMIDINIUM-5-YL)PHENYL)AMINO)CARBONYL)PHENYL)AMINO)-1-PROPYLQUINOLINIUM) see DCC200

DIAMINOPROPYLTETRAMETHYLENEDIAMINE see DCC400

l-(+)-N-(p-(((2,4-DIAMINO-6-PTERIDI-NYL)METHYL)METHYLAMINO)BENZOYL)GLUTAMIC ACID see MDV500

2,6-DIAMINOPURINE see POJ500

2,6-DIAMINOPYRIDINE see DCC800

3,4-DIAMINOPYRIDINE see DCD000

DIAMINO-3,4-PYRIDINE see DCD000

DIAMINOPYRITAMIN see TGD000

2,4-DIAMINOSOLE SULPHATE see DBO400

4: 4'-DIAMINOSTILBENE see SLR500

4,4'-DIAMINO-2,2'-STILBENEDISULFONIC ACID see FCA100

p,p'-DIAMINOSTILBENE-o,o'-DISULFONIC ACID DISODIUM SALT see FCA200

DIAMINOSTILBENE DISULPHONATE DISODIUM SALT see FCA200

2,4-DIAMINOTOLUEN (CZECH) see TGL750

2,4-DIAMINO-1-TOLUENE see TGY800

2,3-DIAMINOTOLUENE see TGY800

2,4-DIAMINOTOLUENE see TGL750

2,5-DIAMINOTOLUENE see TGM000

2,6-DIAMINOTOLUENE see TGM100

3,4-DIAMINOTOLUENE see TGM250

DIAMINOTOLUENE see TGL500, TGL750

2,4-DIAMINOTOLUENE DIHYDROCHLORIDE see DCE000

2,5-DIAMINOTOLUENE DIHYDROCHLORIDE see DCE200

2,6-DIAMINOTOLUENE DIHYDROCHLORIDE see DCE400

2,5-DIAMINOTOLUENE SULFATE see DCE600

p-DIAMINOTOLUENE SULFATE see DCE600

2,5-DIAMINOTOLUENE SULPHATE see DCE600

2,4-DIAMINOTOLUOL see TGL750

(4-((4,6-DIAMINO-m-TOLYL)IMINO)-2,5-CYCLOHEXADIEN-1-YLI-DENE)DIMETHYLAMMONIUM CHLORIDE H₂O see TGU500

(4-((4,6-DIAMINO-m-TOLYL)IMINO)-2,5-CYCLOHEXADIEN-1-YLIDENE) DIMETHY-LAMMONIUM CHLORIDE see DCE800

4,6-DIAMINO-s-TRIAZINE-2-METHANETHIOL S-ESTER with O,O-DIMETHYL-PHOSPHORODITHIOATE see ASD000

4,6-DIAMINO-1,3,5-TRIAZINE-2(1H)-THIONE see DCF000

4,6-DIAMINO-1,3,5-TRIAZINE-2-THIONE see DCF000

4,6-DIAMINO-s-TRIAZINE-2-THIONE see DCF000

S-((4,6-DIAMINO-1,3,5-TRIAZIN-2-YL)-METHYL)-O,O-DIMETHYL-DITHIOFOSFAAT (DUTCH) see ASD000

S-((4,6-DIAMINO-1,3,5-TRIAZIN-2-YL)-METHYL)-O,O-DIMETHYL-DITHIOPHOSPHAT (GERMAN) see ASD000

S-((4,6-DIAMINO-s-TRIAZIN-2-YL)METHYL)-O,O-DIMETHYL PHOSPHORODITH-IOATE see ASD000

4,6-DIAMINO-1,3,5-TRIAZIN-2-YLMETHYL-O,O-DIMETHYL PHOSPHORODITH-IOATE see ASD000

S-(4,6-DIAMINO-1,3,5-TRIAZIN-2-YLMETHYL)-O,O-DIMETHYL PHOSPHORODITH-IOATE see ASD000

S-(4,6-DIAMINO-1,3,5-TRIAZIN-2-YLMETHYL) DIMETHYL PHOSPHOROTHIOLO-THIONATE see ASD000

3,5-DIAMINO-s-TRIAZOLE see DCF200

3,10-DIAMINOTRICYCLO(5.2.1.0²·⁶)DECANE see DCF600

2,4-DIAMINO-5-(3,4,5-TRIMETHOXYBENZYL)PYRIMIDINE see TKZ000

2,5-DIAMINOTROPONE see DCF700

2,5-DIAMINOTROPONE HYDROCHLORIDE see DCF710

1,3-DIAMINOUREA see CBS500

DIAMMIDE SEBACICA della 4-FENIL-4-AMMINOMETIL-N-METILPIPERIDINA (ITALIAN) see BKS825

DIAMMINEBORONIUM HEPTAHYDROTETRABORATE see DCF725

DIAMMINEBORONIUM TETRAHYDROBORATE see DCF750

cis-DIAMMINE(1,1-CYCLOBUTANEDICARBOXYLATO)PLATINUM(II) see CCC075

DIAMMINE(1,1-CYCLOBUTANEDICARBOXYLATO)PLATINUM (II) see CCC075

trans-DIAMMINEDICHLOROPLATINUM(II) see DEX000

cis-DIAMMINEDICHLOROPLATINUM see PJD000

cis-DIAMMINEDINITRATO PLATINUM (II) see DCF800

DIAMMINEMALONATO PLATINUM (II) see DCG000

DIAMMINEPALLADIUM (II) NITRATE see DCG600

DIAMMINETETRACHLOROPLATINUM (OC-6-22) (9CI) see TBO776

cis-DIAMMINOTETRACHLOROPLATINUM see TBO776

DIASETYLMORFIIMI see HBT500
DIASONE HYDROCHLORIDE see MDP000
DIASTATIN see NOH500
DIASTYL see DKA600
DIASULFA see SNN300
DIASULFYL see SNN300
DIAT (GERMAN) see DCK000
DIATER see DXQ500
DIATERR-FOS see DCM750
DIATO BLUE BASE B see DCJ200
DIATOMACEOUS EARTH see DCJ800
DIATOMACEOUS EARTH, NATURAL see DCJ800
DIATOMACEOUS SILICA see DCJ800
DIATOMITE see DCJ800
DIATRAST see DNG400
DIA-TUSS see TCY750
DIATRIN HYDROCHLORIDE see DCJ850
DIATRIZOATE MEGLUMINE see AOO875
DIATRIZOATE METHYLGLUCAMINE see AOO875
DIATRIZOATE SODIUM SALT see SEN500
DIATRIZOESAURE (GERMAN) see DCK000
DIATRIZOIC ACID see DCK000
6,12-DIAZAANTHANTHRENE see ADK250
6,12-DIAZAANTHANTHRENE SULFATE see DCK200
6,12-DIAZAANTHANTHRENE SULPHATE see DCK200
1,2-DIAZABENZENE see OJW200
1,3-DIAZABENZENE see PPO750
1,4-DIAZABENZENE see POL490
10'-(β-(1,4-DIAZABICYCLO(4.3.0)NONANYL-4)-PROPIONYL)-2'-CHLOROPHEN-
    THIAZIN DICHLORIDE see NMV750
1,4-DIAZABICYCLO(2,2,2)OCTANE see DCK400
1,4-DIAZABICYCLO(2.2.2)OCTANE HYDROGEN PEROXIDATE see DCK500
DIAZACHEL (OBS.) see MDQ250
DIAZACOSTEROL HYDROCHLORIDE see ARX800
1,4-DIAZACYCLOHEPTANE see HGI900
1,3-DIAZA-2,4-CYCLOPENTADIENE see IAL000
7,14-DIAZADIBENZ(a,h)ANTHRACENE see DDC800
4,11-DIAZADIBENZO(a,h)PYRENE see NAU500
1,12-DIAZADIBENZO(a,i)PYRENE see NAU000
1,2-DIAZA-3,4:9,10-DIBENZPYRENE see APF750
2,5-DIAZAHEXANE see DRI600
1,6-DIAZA-3,4,8,9,12,13-HEXAOXABICYCLO(4.4.4)TETRADECANE see DCK700
1,3-DIAZAINDENE see BCB750
2,3-DIAZAINDOLE see BDH250
DIAZAJET see DCM750
DIAZALE see THE500
DIAZAMINE BLACK D see CMN230
DIAZAMINE GOLDEN YELLOW T see CMP050
DIAZAN see FNF000
1,4-DIAZANAPHTHALENE see QRJ000
3,6-DIAZAOCTANE-1,8-DIAMINE see TJR000
4,5-DIAZAPHENANTHRENE see PCY250
DIAZATOL see DCM750
DIAZENE 42 see EHY000
DIAZEPAM see DCK759
1H-1,4-DIAZEPINE, HEXAHYDRO- see HGI900
1H-1,4-DIAZEPINE, HEXAHYDRO-1-(p-CHLORO-α-PHENYLBENZYL)-4-METHYL-
    see HGI100
DIAZETARD see DCK759
DIAZETOXYSKIRPENOL (GERMAN) see AOP250
DIAZETYLMORPHINE see HBT500
DIAZIDE see DCM750
1,3-DIAZIDOBENZENE see DCL100
1,4-DIAZIDOBENZENE see DCL125
p-DIAZIDOBENZENE (DOT) see DCL125
1,2-DIAZIDOCARBONYL HYDRAZINE see DCL200
DIAZIDODIBENZALACETON (CZECH) see BJA000
DIAZIDODIBENZALACETON see BGW720
2,5-DIAZIDO-3,6-DICHLOROBENZOQUINONE see DCL300
DIAZIDODICYANOMETHANE see DCM000
DIAZIDODIMETHYLSILANE see DCL350
1,1-DIAZIDOETHANE see DCL400
1,2-DIAZIDOETHANE see DCL600
DIAZIDOMALONONITRILE see DCM000
1,3-DIAZIDO-2-NITROAZAPROPANE see DCM499
1,3-DIAZIDOPROPENE see DCM600
2,6-DIAZIDOPYRAZINE see DCM700
DIAZIL see DHU900
1,2-DIAZINE see OJW200
1,4-DIAZINE see POL490
m-DIAZINE see PPO750
p-DIAZINE see POL490
DIAZINE BLACK BHC see CMN800
DIAZINE BLACK DR see CMN230
DIAZINE BLACK E see AQP000

DIAZINE BLACK H see CMN800
DIAZINE BLACK HDW see CMN800
DIAZINE BLACK HNJ see CMN800
DIAZINE BLUE 2B see CMO000
DIAZINE BLUE 3B see CMO250
DIAZINE DARK GREEN BO see CMO830
DIAZINE DARK GREEN P see CMO830
DIAZINE DIRECT BLACK E see AQP000
DIAZINE DIRECT BLACK G see AQP000
DIAZINE FAST RED 8BK see CMO885
DIAZINE FAST RED F see CMO870
DIAZINE GREEN B see CMO840
DIAZINE GREEN DB see CMO840
DIAZINE VIOLET N see CMP000
DIAZINON see DCM750
DIAZINONE see DCM750
DIAZINON mixed with METHOXYCHLOR see MEI500
DIAZIQUONE see ASK875
3,6-DIAZIRIDINYL-2,5-BIS(CARBOETHOXYAMINO)-1,4-BENZOQUINONE see
    ASK875
DIAZIRINE see DCP800
DIAZIRINE-3,3-DICARBOXYLIC ACID see DCM875
DIAZITOL see DCM750
DIAZO see DUR800
DI-AZO see PDC250
DIAZO A see DCP300
DIAZOACETALDEHYDE see DCN200
2-(DIAZOACETAMINO)-N-ETHYLACETAMIDE see DCN600
l-DIAZOACETATE (ESTER) SERINE see ASA500
DIAZOACETATE (ESTER)-l-SERINE see ASA500
DIAZO-ACETIC ACID ESTER with SERINE see ASA500
DIAZOACETIC ACID, ETHYL ESTER see DCN800
DIAZOACETIC ESTER see DCN800
N-DIAZOACETILGLICINA-AMIDE (ITALIAN) see CBK000
N-DIAZOACETILGLICINA-IDRAZIDE (ITALIAN) see DCO800
DIAZOACETONITRILE see DCN875
2-DIAZOACETOPHENONE see PET800
α-DIAZOACETOPHENONE see PET800
omega-DIAZOACETOPHENONE see PET800
DIAZOACETOPHENONE see PET800
2-((DIAZOACETYL)AMINO)-N-ETHYLACETAMIDE see DCN600
2-((DIAZOACETYL)AMINO)-N-METHYLACETAMIDE see DCO200
DIAZOACETYL AZIDE see DCO509
N-(DIAZOACETYL)GLYCINAMIDE see CBK000
DIAZOACETYLGLYCINAMIDE see CBK000
N-DIAZOACETYLGLYCINE AMIDE see CBK000
DIAZOACETYLGLYCINE AMIDE see CBK000
N-DIAZOACETYLGLYCINEETHYLAMIDE see DCN600
N-DIAZOACETYLGLYCINE ETHYL ESTER see DCO600
DIAZOACETYLGLYCINE ETHYL ESTER see DCO600
DIAZOACETYLGLYCINE HYDRAZIDE see DCO800
N-(DIAZOACETYL)GLYCINE HYDRAZINE see DCO800
N-DIAZOACETYLGLYCINE METHYLAMIDE see DCO200
N-DIAZOACETYL GLYCYLHYDRAZIDE see DCO800
N-(1-(DIAZOACETYL)-2-METHYLBUTYL)-p-TOLUENESULFONAMIDE see
    THH480
o-DIAZOACETYL-l-SERINE see ASA500
DIAZOAMINOBENZEN (CZECH) see DWO800
p-DIAZOAMINOBENZENE see DWO800
DIAZOAMINOBENZENE see DWO800
DIAZOAMINOBENZOL (GERMAN) see DWO800
6-DIAZO-2-(2-(4-AMINO-4-CARBOXYBUTYRAMIDO)-6-DIAZO-5-OXOHEXANAMI-
    DO)-HEXANOIC ACID SODIUM see ASO510
DIAZOBEN see MND600
DIAZOBENZENE see ASL250
DIAZO BLACK BH see CMN800
DIAZO BLACK BHN-CF see CMN800
DIAZO BLACK BHSW see CMN800
DIAZO BLACK BHSWK see CMN800
DIAZO BLACK CR see CMN800
DIAZO BLACK D see CMN230
DIAZOBLEU see BGT250
DIAZOCARD CHRYSOIDINE G see PEK000
20,25-DIAZOCHOLESTEROL DIHYDROCHLORIDE see ARX800
DIAZODICYANOIMIDAZOLE see DCP880
DIAZODICYANOMETHANE see DCP775
4-DIAZO-N,N-DIMETHYLANILIN CHLOROZINCATE see DCP300
2-DIAZO-4,6-DINITROBENZENE-1-OXIDE see DUR800
DIAZODINITROPHENOL (dry) (DOT) see DUR800
DIAZODINITROPHENOL, wetted with not <40% H2O or mixture of alcohol &
    H2O (UN 0074) (DOT) see DUR800
DIAZODIPHENYLMETHANE (DOT) see DVZ100
DIAZO DIRECT BLACK N see CMN800
DIAZOESSIGSAEURE-AETHYLESTER (GERMAN) see DCN800
DIAZO FAST BLACK BH see CMN800

DIBENZ(c,f)INDENO(1,2,3-ij)(2,7)NAPHTHYRIDINE see DCX400
1,2,3,4-DIBENZNAPHTHALENE see TMS000
1,2,5,6-DIBENZOACRIDINE see DCS400
DIBENZO(a,j)ACRIDINE see DCS600
1,2:3,4-DIBENZOANTHRACENE see BDH750
1,2:5,6-DIBENZOANTHRACENE see DCT400
DIBENZO(a,c)ANTHRACENE see BDH750
DIBENZO(a,h)ANTHRACENE see DCT400
DIBENZOAZEPINE see DCS200
3,4,5,6-DIBENZOCARBAZOLE see DCY000
7H-DIBENZO(a,g)CARBAZOLE see DCX600
7H-DIBENZO(a,i)CARBAZOLE see DCX800
7H-DIBENZO(c,g)CARBAZOLE see DCY000
DIBENZO(b,def)CHRYSENE see DCY200
DIBENZO(def,p)CHRYSENE see DCY400
DIBENZO-(drf,mno)CHRYSENE see APE750
DIBENZO(def,mno)CHRYSENE-12-CARBOXALDEHYDE see FNK000
DIBENZO(def,p)CHRYSENE-10-CARBOXALDEHYDE see DCY800
DIBENZO(b,def)CHRYSENE-7-CARBOXALDEHYDE see DCY600
DIBENZO(b,def)CHRYSENE-7,14-DIONE see DCZ000
DIBENZO-18-CROWN-6 see COD575
DIBENZO(a,d)(1,4)-CYCLOHEPTADIENE-5-CARBOXAMIDE see CQH625
DIBENZO(a,d)CYCLOHEPTADIENE-5-CARBOXAMIDE see CQH625
DIBENZO(a,d)CYCLOHEPTADIEN-5-ONE see DCX300
4-(5-DIBENZO(a,e)CYCLOHEPTATRIENYLIDENE)PIPERIDINE HYDROCHLORIDE
  see PCI250
DIBENZOCYCLOHEPTENONE see DCX300
5H-DIBENZO(a,d)CYCLOHEPTEN-5-ONE, 10,11-DIHYDRO- see DCX300
(2-((5H-DIBENZO(a,d)CYCLOHEPTEN-5-YLIDENEAMINO)OXY)-1-METHYLE-
  THYL)ETHYLDIMETHYL AMMONIUM see LHX600
(2-((5H-DIBENZO(a,d)CYCLOHEPTEN-5-YLIDENEAMINO)OXY)-1-METHYLE-
  THYL)TRIMETHYLAMMONIUM see LHX510
3-(5H-DIBENZO(a,d)CYCLOHEPTEN-5-YLIDENE)-N,N-DIMETHYL-1-PROPANA-
  MINE HYDROCHLORIDE see DPX800
4-(5H-DIBENZO(a,d)CYCLOHEPTEN-5-YLIDENE)-1-METHYLPIPERIDINE see
  PCI500
4-(5-DIBENZO(a,d)CYCLOHEPTEN-5-YLIDINE)-1-METHYLPIPERIDINE see PCI500
N-3-(5H-DIBENZO(a,d)CYCLOHEPTEN-5-YL)PROPYL-N-METHYLAMINE see
  DDA600
1,3(2H,9bH)-DIBENZOFURANDIONE, 2,6-DIACETYL-7,9-DIHYDROXY-8,9b-DI-
  METHYL-, (9bR)- UWJ100
DIBENZO(1,4)DIOXIN see DDA800
DIBENZO(b.e)(1,4)DIOXIN see DDA800
DIBENZO-p-DIOXIN see CDV125, DDA800
DIBENZODIOXIN see DDA800
2,3,5,6-DIBENZOFLUORANTHENE see DCR300
DIBENZO(a,e)FLUORANTHENE see DCR300
1,2,5,6-DIBENZOFLUORENE see DDB000
13H-DIBENZO(a,g)FLUORENE see DDB000
13H-DIBENZO(a,i)FLUORENE see BCB200
DIBENZO(a,jk)FLUORENE see BCJ750
DIBENZO(b,jk)FLUORENE see BCJ750
1,1'-(2,8-DIBENZOFURADIYL)BIS(2-(DIMETHYLAMINOETHANONE)) DIHYDRO-
  CHLORIDE HYDRATE (2:5) see DDB400
2-DIBENZOFURANAMINE see DDB600
3-DIBENZOFURANAMINE see DDB800
3(9bH)-DIBENZOFURANONE, 2,6-DIACETYL-8,9b-DIMETHYL-1,7,9-TRIHY-
  DROXY-, D- see UWJ100
3(9bH)-DIBENZOFURANONE, 2,6-DIACETYL-1,7,9-TRIHYDROXY-8,9b-DIMETH-
  YL-, D- see UWJ100
3-DIBENZOFURANYLACETAMIDE see DDC000
N-3-DIBENZOFURANYLACETAMIDE see DDC000
DIBENZOFURANYLAMINE see DDB800
2-DIBENZOFURANYLPHENYL METHANONE see DDC100
DIBENZOFURAN, 2,3,4,7,8-PENTACHLORO- see PAW100
1,2,5,6-DIBENZONAPHTHALENE see CML810
DIBENZOPARADIAZINE see PDB500
DIBENZOPARATHIAZINE see PDP250
DIBENZO(h,rst)PENTAPHENE see DDC200
DIBENZO(cd,lm)PERYLENE see DDC400
1,2,3,4-DIBENZOPHENANTHRENE see BCH000
1,2:6,7-DIBENZOPHENANTHRENE see BCG500
1,2:7,8-DIBENZOPHENANTHRENE see PIB750
2,3:7,8-DIBENZOPHENANTHRENE see BCG500
DIBENZO-2,3,7,8-PHENANTHRENE see BCG500
DIBENZO(a,i)PHENANTHRENE see PIB750
DIBENZO(a,c)PHENAZINE see DDC600
DIBENZO(a,h)PHENAZINE see DDC800
DIBENZO PQD see DVR200
6H-DIBENZO(b,d)PYRAN-6-ONE, 1-METHYL-3,7,9-TRIHYDROXY-and 3,9-DIHY-
  DROXY-7-METHOXY-1-METHYL-DIBENZO(b,d)PYRAN-6-ONE (1:1) see
  AGW550
DIBENZOPYRAZINE see PDB500
DIBENZO(c,f)PYRAZINO(1,2-a)AZEPINE, 1,2,3,4,10,14b-HEXAHYDRO-2-METH-
  YL- see MQS220

1,2:3,4-DIBENZOPYRENE see DCY400
1,2,4,5-DIBENZOPYRENE see NAT500
1,2,6,7-DIBENZOPYRENE see DCY200
1,2,7,8-DIBENZOPYRENE see BCQ500
1,2,9,10-DIBENZOPYRENE see DCY400
2,3:4,5-DIBENZOPYRENE see DCY400
3,4,8,9-DIBENZOPYRENE see DCY200
3,4:9,10-DIBENZOPYRENE see BCQ500
DIBENZO(a,d)PYRENE see DCY400
DIBENZO(a,e)PYRENE see NAT500
DIBENZO(a,h)PYRENE see DCY200
DIBENZO(a,i)PYRENE see BCQ500
DIBENZO(a,l)PYRENE see DCY400
DIBENZO(b,h)PYRENE see BCQ500
DIBENZO(cd,mk)PYRENE see APE750
DIBENZO(a,b)PYRENE-7,14-DIONE see DCZ000
2,3,7,8-DIBENZOPYRENE-1,6-QUINONE see DCZ000
DIBENZO(b,e)PYRIDINE see ADJ500
DIBENZO(b,d)PYRROLE see CBN000
DIBENZOPYRROLE see CBN000
DIBENZO(a,g)QUINOLIZINIUM, 5,6-DIHYDRO-2,3,9,10-TETRAMETHOXY-, HY-
  DROXIDE see PAE100
DIBENZOSUBERAN-5-ONE see DCX300
2,3:6,7-DIBENZOSUBERONE see DCX300
DIBENZOSUBERONE see DCX300
DIBENZOSUBERONE OXIME see DDD000
DIBENZOTHIAZEPINE see CIS750
DIBENZO-1,4-THIAZINE see PDP250
DIBENZOTHIAZINE see PDP250
DI-2-BENZOTHIAZOLYLDISULFIDE see BDE750
2,2'-DIBENZOTHIAZYLDISULFIDE see BDE750
DIBENZOTHIAZYL DISULFIDE see BDE750
N-2-DIBENZOTHIENYLACETAMIDE see DDD400
N-3-DIBENZOTHIENYLACETAMIDE see DDD600
N-3-DIBENZOTHIENYLACETAMIDE-5-OXIDE see DDD800
DIBENZOTHIEN-2-YL METHYL KETONE see ACH090
DIBENZO(b,e)THIEPIN-μ¹¹⁽⁶ᴴ⁾,γ-PROPYLAMINE, N,N-DIMETHYL-, 5-OXIDE, MALE-
  ATE (1:1), (Z)- see POD800
3-DIBENZO(b,e)THIEPIN-11(6H)-YLIDENE-N,N-DIMETHYL-1-PROPAMINE see
  DYC875
DIBENZO(b,e)THIEPIN-11(6H)-YLIDENE-N,N-DIMETHYL-1-PROPANAMINE, HY-
  DROCHLORIDE see DPY200
DIBENZO(b,d)THIOPHENE see TES300
DIBENZOTHIOPHENE see TES300
1,1'-(2,8-DIBENZOTHIOPHENEDIYL)BIS(2-(DIMETHYLAMINO)ETHANONE) DIHY-
  DROCHLORIDE TRIHYDRATE see DDE000
DIBENZOTHIOXIN see PDQ750
1,4-DIBENZOTHIOXINE see PDQ750
DIBENZ(b,f)(1,4)OXAZEPINE see DCE200
DIBENZO(b,f)OXEPIN-2-ACETIC ACID, 10,11-DIHYDRO-α-8-DIMETHYL-11-OXO-
  see DLI650
DIBENZ(b,e)OXEPIN-Δ¹¹⁽⁶ᴴ⁾,γ-PROPYLAMINE see DBA600
DIBENZOYL see BBY750
2,2'-DIBENZOYLAMINODIPHENYL DISULFIDE see BDK800
DIBENZOYLDIETHYLENEGLYCOL ESTER see DJE000
DIBENZOYL DIPROPYLENE GLYCOL ESTER see DWS800
DIBENZOYLPEROXID (GERMAN) see BDS000
DIBENZOYL PEROXIDE (MAK) see BDS000
DIBENZOYLPEROXYDE (DUTCH) see BDS000
DIBENZOYLTARTARIC ACID see DDE300
DIBENZOYLTHIAZYL DISULFIDE see BDE750
1,2,3,4-DIBENZPHENANTHRENE see BCH000
1,2,5,6-DIBENZPHENANTHRENE see BCG750
1,2,3,4-DIBENZPHENAZINE see DDC600
1,2:5,6-DIBENZPHENAZINE see DDC800
DIBENZ(a,h)PHENAZINE see DDC800
1,2,3,4-DIBENZPYRENE see DCY400
1,2:7,8-DIBENZPYRENE see BCQ500
3,4,8,9-DIBENZPYRENE see DCY200
4,5,6,7-DIBENZPYRENE see DCY400
3,4:9,10-DIBENZPYRENE see BCQ500
DIBENZ(a,i)PYRENE see BCQ500
1',2',6',7'-DIBENZPYRENE-7,14-QUINONE see DCZ000
DIBENZTHIAZYL DISULFIDE see BDE750
DIBENZYL see BFX500
N,N-DIBENZYLAMINOETHYL CHLORIDE HYDROCHLORIDE see DCR200
DIBENZYLBUTYLSULFONIUM IODIDE MERCURIC IODIDE see DDF000
DIBENZYLBUTYLSULFONIUM IODIDE with MERCURY IODIDE (1:1) see
  DDF000
DIBENZYLCHLORETHAMINE HYDROCHLORIDE see DCR200
DIBENZYL CHLORETHYLAMINE see DCT050
DIBENZYLCHLOROETHYLAMINE HYDROCHLORIDE see DCR200
N,N-DIBENZYL-β-CHLOROETHYLAMINE see DCT050
N,N-DIBENZYL-2-CHLOROETHYLAMINE HYDROCHLORIDE see DCR200
N,N-DIBENZYL-β-CHLOROETHYLAMINE HYDROCHLORIDE see DCR200

1,3-DIBENZYLDECAHYDRO-2-OXOIMIDAZO(4,5-c)THIENO(1,2-a)THIOLIUM 10-CAMPHORSULFONATE see TKW500
1,3-DIBENZYLDECAHYDRO-2-OXO-IMIDAZO(4,5-c)THIENO(1,2-a)THIOLIUM-2-OXO-10-BORANESULFONATE see TKW500
1-1,3-DIBENZYLDECAHYDRO-2-OXOIMIDAZO(4,5-c)THIENO(1,2-a)THIOLIUM-2-OXO-10-BORANESULFONATE see DDF200
9,10-DIBENZYL-1,2,5,6-DIBENZANTHRACENE see DDF400
7,14-DIBENZYLDIBENZ(a,h)ANTHRACENE see DDF400
DIBENZYL(5-DIBENZYLAMINO-2,4-PENTADIENYLIDENE)AMMONIUM CHLORIDE SESQUIHYDRATE see DDF600
DIBENZYLDISULFID (CZECH) see DXH200
N,N'-DIBENZYLDITHIOOXAMIDE see DDF700
DIBENZYLENE see DDG800
DIBENZYLETHER (CZECH) see BEO250
N,N'-DIBENZYLETHYLENEDIAMINE see BHB300
N,N'-DIBENZYLETHYLENEDIAMINE BIS(BENZYL PENICILLIN) see BFC750
N,N'-DIBENZYLETHYLENEDIAMINE DIACETATE see DDF800
N,N'-DIBENZYLETHYLENEDIAMINE DIHYDROCHLORIDE see DDG400
DIBENZYLETHYLENEDIAMINE-DI-PENICILLIN G see BFC750
N,N'-DIBENZYLETHYLENEDIAMINE, compounded with PENICILLIN G (1:2) see BFC750
DIBENZYLETHYLSULFONIUM IODIDE MERCURIC IODIDE see DDG600
DIBENZYLETHYLSULFONIUM IODIDE with MERCURY IODIDE (1:1) see DDG600
DIBENZYLIN see DDG800
DIBENZYLINE see PDT250
DIBENZYLINE HYDROCHLORIDE see DDG800
1-3,4-(1',3'-DIBENZYL-2'-KETO-IMIDAZOLIDO)-1,2-TRIMETHYLENE THIOPHANI-UM CAMPHOR SULFONATE see DDF200
d-3,4-(1',3'-DIBENZYL-2'-KETOIMIDAZOLIDO)-1,2-TRIMETHYLENETHIOPHANIUM-d-CAMPHORSULFONATE see TKW500
DIBENZYLMERCURY see DDH000
DIBENZYL PEROXYDICARBONATE, >87% with water (DOT) see DDH200
DIBENZYL PEROXYDICARBONATE see DDH200
DIBENZYL PHOSPHITE see DDH400
2,2-DIBENZYL-4-(2-PIPERIDYL)-1,3-DIOXOLANE HYDROCHLORIDE see DDH600
DIBENZYLSULFOXIDE see DDH800
DIBENZYL SULPHOXIDE see DDH800
DIBENZYLTIN BIS(DIBUTYLDITHIOCARBAMATE) see BIW250
DIBENZYLTIN S,S'-BIS(ISOOCTYLMERCAPTOACETATE) see EDC560
DIBENZYRAN see DDG800
DIBERAL see DDH900
DIBESTIL see DKB000
DIBESTROL see DKA600
DIBETOS see BQL000
DI-1,2-BIS(DIFLUOROAMINO)ETHYL ETHER see DDI000
DIBISMUTH TRIOXIDE see BKW600
DIBISMUTH TRISULFIDE see DDI200
DIBONDRIN see BBV500
DIBORANE see DDI450
DIBORANE MIXTURES (NA 1911) see DDI450
DIBORON HEXAHYDRIDE see DDI450
DIBORON OXIDE see DDI500
DIBORON TETRACHLORIDE see DDI600
DIBORON TETRAFLUORIDE see DDI800
DIBOVAN see DAD200
DIBP see DNJ400
DIBROLUUR see BNP750
DIBROM see NAG400
1,2-DIBROMAETHAN (GERMAN) see EIY500
DIBROMANNIT see DDP600
d-DIBROMANNITOL see DDP600
DIBROMANNITOL see DDP600
DIBROMANTIN see DDI900
DIBROMANTINE see DDI900
3,9-DIBROMBENZANTHRONE see DDK875
1,4-DIBROMBUTAN (GERMAN) see DDL000
1,2-DIBROM-3-CHLOR-PROPAN (GERMAN) see DDL800
DIBROMCHLORPROPAN see DDL800
O-(1,2-DIBROM-2,2-DICHLORAETHYL)-O,O-DIMETHYL-PHOSPHAT (GERMAN) see NAG400
DIBROMDULCITOL see DDJ000
DIBROMOACETONITRILE see DDJ400
2,4'-DIBROMOACETOPHENONE see DDJ600
α,p-DIBROMOACETOPHENONE see DDJ600
DIBROMOACETYLENE see DDJ800
2-((4-(DIBROMOACETYL)PHENYL)AMINO)-2-ETHOXY-1-(4-NITROPHE-NYL)ETHANONE see DDJ850
2,4-DIBROMO-1-ANTHRAQUINONYLAMINE see AJL500
3,9-DIBROMO-7H-BENZ(de)ANTHRACEN-7-ONE see DDJ875
1,3-DIBROMOBENZENE see DDK050
1,4-DIBROMOBENZENE see BNV775
m-DIBROMOBENZENE see DDK050
p-DIBROMOBENZENE see BNV775

DIBROMOBENZENE see DDJ900
4,4'-DIBROMOBENZILIC ACID ISOPROPYL ESTER see IOS000
2,2'-DIBROMOBIACETYL see DDK600
α,α'-DIBROMOBIACETYL see DDK600
DIBROMOBICYCLOHEPTANE (mixed isomers) see DDK800
DIBROMOBICYCLOHEPTANE see DDK800
3,5-DIBROMO-N-(4-BROMOPHENYL)-2-HYDROXYBENZAMIDE see THW750
1,4-DIBROMO-1,3-BUTADIYNE see DDK875
1,4-DIBROMOBUTANE see DDL000
1,4-DIBROMO-2-BUTENE see DDL400
1,4-trans-DIBROMOBUTENE-2 see DDL600
trans-1,4-DIBROMOBUT-2-ENE see DDL600
DIBROMOBUTENE see DDL600
DIBROMOCHLOROMETHANE see CFK500
1,2-DIBROMO-3-CHLORO-2-METHYLPROPANE see MBV720
1,2-DIBROMO-3-CHLOROPROPANE see DDL800
DIBROMOCHLOROPROPANE see DDL800
3,4'-DIBROMO-5-CHLOROTHIOSALICYLANILIDE ACETATE (ESTER) see BOL325
1,2-DIBROMO-3-CLORO-PROPANO (ITALIAN) see DDL800
α,α-DIBROMO-α-CYANOACETAMIDE see DDM000
DIBROMOCYANOACETAMIDE see DDM000
2,6-DIBROMO-4-CYANOPHENOL see DDP000
2,6-DIBROMO-4-CYANOPHENYL OCTANOATE see DDM200
1,2-DIBROMO-4-(1,2-DIBROMOETHYL)CYCLOHEXANE see DDM300
DIBROMODIBUTYLSTANNANE see DDM400
DIBROMODIBUTYLTIN see DDM400
1,2-DIBROMO-2,2-DICHLOROETHYL DIMETHYL PHOSPHATE see NAG400
O-(1,2-DIBROMO-2,2-DICLORO-ETIL)-O,O-DIMETIL-FOSTATO (ITALIAN) see NAG400
1,6-DIBROMO-1,6-DIDEOXYDULCITOL see DDJ000
1,6-DIBROMODIDEOXYDULCITOL see DDJ000
1,6-DIBROMO-1,6-DIDEOXYGALACTITOL see DDJ000
1,6-DIBROMO-1,6-DIDEOXY-d-GALACTITOL see DDJ000
1,6-DIBROMO-1,6-DIDEOXY-d-MANNITOL see DDP600
1,6-DIBROMO-1,6-d-DIDESOXYMANNITOL see DDP600
2',6'-DIBROMO-2-(DIETHYLAMINO)-p-ACETOTOLUIDIDE HYDROCHLORIDE see DDM600
5,6-DIBROMO-2-(2-(2-(DIETHYLAMINO)ETHYLAMINO)ETHYL)-2-METHYL-1,3-BEN-ZODIOXOLE DIHYDROCHLORIDE see DDM800
2',6'-DIBROMO-2-(DIETHYLAMINO)-4'-METHYLACETANILIDE HYDROCHLORIDE see DDM600
DIBROMODIFLUOROMETHANE see DKG850
1,2-DIBROMO-1,2-DIISOCYANATOETHANE POLYMERS see DDN100
2,2-DIBROMO-1,3-DIMETHYLCYCLOPROPANOIC ACID see DDN150
N,N'-DIBROMODIMETHYLHYDANTOIN see DDI900
1,3-DIBROMO-5,5-DIMETHYL-2,4-IMIDAZOLIDINEDIONE see DDI900
DIBROMODIMETHYL STANNANE see DUG800
DIBROMODIPHENYLSTANNANE see DDN200
1,6-DIBROMODULCITOL see DDJ000
DIBROMODULCITOL see DDJ000
2,3-DIBROMO-5,6-EPOXY-7,8-DIOXABICYCLO(2.2.2)OCTANE see DDN700
1,2-DIBROMOETANO (ITALIAN) see EIY500
1,1-DIBROMOETHANE see DDN800
α,β-DIBROMOETHANE see EIY500
1,2-DIBROMOETHANE (MAK) see EIY500
sym-DIBROMOETHANE see EIY500
1-(1,2-DIBROMOETHYL)-3,4-DIBROMOCYCLOHEXANE see DDM300
DIBROMOFLUOROPHOSPHINE SULFIDE see TFO250
1,2-DIBROMOHEPTAFLUOROISOBUTYL METHYL ETHER see DDO400
1,2-DIBROMO-1,1,2,3,3,3-HEXAFLUOROPROPANE see DDO450
1,2-DIBROMOHEXAFLUOROPROPANE see DDO450
1,6-DIBROMOHEXAN (GERMAN) see DDO800
1,6-DIBROMOHEXANE see DDO800
3,5-DIBROMO-4-HYDROXYBENZONITRILE see DDP000
3-(3,5-DIBROMO-4-HYDROXYBENZOYL-2-ETHYLBENZOFURAN see DDP200
(DIBROMO-3,5 HYDROXY-4 BENZOYL)-3 MESITYL-2 BENZOFURANNE see DDP300
2,7-DIBROMO-4-HYDROXYMERCURIFLUORESCEINE DISODIUM SALT see MCV000
3,5-DIBROMO-4-HYDROXYPHENYLCYANIDE see DDP000
3,5-DIBROMO-4-HYDROXYPHENYL-2-ETHYL-3-BENZOFURANYL KETONE see DDP200
(3,5-DIBROMO-4-HYDROXYPHENYL)(2-ETHYL-3-BENZOFURANYL)METHANONE see DDP200
3,5-DIBROMO-4-HYDROXYPHENYL 2-MESITYL-3-BENZOFURANYL KETONE see DDP300
(3,5-DIBROMO-4-HYDROXYPHENYL)(2-(2,4,6-TRIMETHYLPHENYL)-3-BENZOFU-RANYL)M ETHANONE see DDP300
5,7-DIBROMO-8-HYDROXYQUINOLINE see DDS600
DIBROMOMALEINIMIDE see DDP400
1,6-DIBROMOMANNITOL see DDP600
2,7-DIBROMOMESOBENZANTHRONE see DDK875
DIBROMOMETHANE see DDP800
N,N-DIBROMOMETHYLAMINE see DDQ100

DIBROMOMETHYLBORANE see DDQ125
DIBROMONEOPENTYL GLYCOL see DDQ400
2,2-DIBROMO-3-NITRILOPROPIONAMIDE see DDM000
2,6-DIBROMO-4-NITROPHENOL see DDQ500
3,4-DIBROMONITROSOPIPERIDINE see DDQ800
DIBROMONORBORNANE see DDK800
3,5-DIBROMO-4-OCTANOYLOXYBENZONITRILE see DDM200
DIBROMOPENTAERYTHRITOL see DDQ400
2-(3,5-DIBROMO-2-PENTYLOXYBENZYLOXY)TRIETHYLAMINE see DDR100
1,2-DIBROMOPERFLUOROETHANE see FOO525
2,4-DIBROMOPHENOL see DDR150
DIBROMOPHENYLARSINE see DDR200
DIBROMOPROPANAL see DDS400
1,2-DIBROMOPROPANE see DDR400
1,3-DIBROMOPROPANE see TLR000
2,3-DIBROMOPROPANE see DDR600
α,γ-DIBROMOPROPANE see TLR000
ω,ω'-DIBROMOPROPANE see TLR000
1,3-DIBROMOPROPANE polymer with N,N,N',N'-TETRAMETHYL-1,6-HEXANEDI-
    AMINE see HCV500
2,3-DIBROMO-1-PROPANOL see DDS000
1,3-DIBROMO-2-PROPANOL see DDR800
2,3-DIBROMOPROPANOL see DDS000
2,3-DIBROMO-1-PROPANOL HYDROGEN PHOSPHATE AMMONIUM SALT see
    MRH214
2,3-DIBROMO-1-PROPANOL PHOSPHATE see TNC500
2,3-DIBROMO-1-PROPANOL, PHOSPHATE (3:1) see TNC500
2,3-DIBROMOPROPANOYL CHLORIDE see DDS100
2,3-DIBROMOPROPENE see DDS200
2,3-DIBROMOPROPIONALDEHYDE see DDS400
2,3-DIBROMOPROPIONYL CHLORIDE see DDS100
α,β-DIBROMOPROPIONYL CHLORIDE see DDS100
(2,3-DIBROMOPROPYL) PHOSPHATE see TNC500
5,7-DIBROMO-8-QUINOLINOL see DDS600
4',5-DIBROMOSALICYLANILIDE see BOD600
3,4-DIBROMOSULFOLANE see DDT000
1,2-DIBROMO-1,1,2,2-TETRAFLUOROETHANE see FOO525
1,2-DIBROMOTETRAFLUOROETHANE see FOO525
sym-DIBROMOTETRAFLUOROETHANE see FOO525
3,4-DIBROMOTETRAHYDROTHIOPHENE-1,1-DIOXIDE see DDT000
DIBROMOXYQUINOLINE see DDS600
DIBROMSALAN see BOD600
DIBROMURE d'ETHYLENE (FRENCH) see EIY500
1,2-DIBROOM-3-CHLOORPROPAAN (DUTCH) see DDL800
O-(1,2-DIBROOM-2,2-DICHLOOR-ETHYL)-O,O-DIMETHYL-FOSFAAT (DUTCH)
    see NAG400
1,2-DIBROOMETHAAN (DUTCH) see EIY500
DIBUCAIN see NOF500
DIBUCAINE see DDT200
DIBUCAINE HYDROCHLORIDE see NOF500
DIBULINESULFAT see DDW000
DIBULINE SULFATE see DDW000
DIBUSMUTH DICHROMIUM NONAOXIDE see DDT250
DIBUTADIAMIN DIHYDROCHLORIDE see BQM309
DIBUTALIN see BQN600
DIBUTAMID (GERMAN) see DDT300
DIBUTAMIDE see DDT300
2,3:4,5-DI(2-BUTENYL)TETRAHYDROFURFURAL see BHJ500
DIBUTIL see DIR000
DIBUTIN see ILD000
DIBUTOLINE see DDW000
DIBUTOLINE SULFATE see DDW000
1,1-DIBUTOXYETHANE see DDT400
1,2-DIBUTOXYETHANE see DDW400
DIBUTOXYETHOXYETHYL ADIPATE see DDT500
2,2'-DIBUTOXYETHYL ETHER see DDW200
DI(BUTOXYETHYL)PHTHALATE see BHK000
(DIBUTOXYPHOSPHINYL)MERCURY CHLORIDE see CFK750
DI(BUTOXYTHIOCARBONYL) DISULFIDE see BSS550
DIBUTYL ACETAL see DDT400
DIBUTYL ACID PHOSPHATE see DEG700
DI-N-BUTYL ADIPATE see AEO750
DIBUTYL ADIPATE see AEO750
DIBUTYL ADIPINATE see AEO750
DIBUTYLAMID KYSELINY MRAVENCI see DEC400
DI(n-BUTYL)AMINE (DOT) see DDT800
DI-n-BUTYLAMINE see DDT800
n-DIBUTYLAMINE see DDT800
DIBUTYLAMINE, HEXACHLOROSTANNATE (2:1) see BIV900
DIBUTYLAMINE, HEXAFLUOROARSENATE(1-) see DDV225
DIBUTYLAMINE, 4-HYDROXY-N-NITROSO- see HJQ350
2-(DIBUTYLAMINO)-2',6'-ACETOXYLIDIDE HYDROCHLORIDE see DDU200
1-(((DIBUTYLAMINO)CARBONYL)OXY)-N-ETHYL-N,N-DIMETHYLETHANAMINIUM
    SULFATE (2:1) see DDW000
2-N-DIBUTYLAMINOETHANOL see DDU600

2-DIBUTYLAMINOETHANOL see DDU600
N,N-DI-n-BUTYLAMINOETHANOL (DOT) see DDU600
2-DI-n-BUTYLAMINOETHANOL see DDU600
DIBUTYLAMINOETHANOL see DDU600
β-N-DIBUTYLAMINOETHYL ALCOHOL see DDU600
5-((2-(DIBUTYLAMINO)-ETHYL)AMINO)-3-PHENYL-1,2,4-OXADIAZOLE HYDRO-
    CHLORIDE see PEU000
α-DIBUTYL-AMINO-4-METHOXYBENZENEACETAMIDE (9CI) see DDT300
2-DIBUTYLAMINO-2-(p-METHOXYPHENYL)ACETAMIDE see DDT300
α-DIBUTYLAMINO-α-(p-METHOXYPHENYL)ACETAMIDE see DDT300
α-DIBUTYL-AMINO-p-METHOXYPHENYLACETAMIDE see DDT300
3-((DIBUTYLAMINO)METHYL)-4,5,6-TRIHYDROXYPHTHALIDE HYDROCHLORIDE
    see APE625
3-(DIBUTYLAMINO)-1-PROPANOL-p-AMINOBENZOATE see BOO750
3-DIBUTYLAMINO-1-PROPANOL-4-AMINOBENZOATE (ESTER) SULFATE
    (SALT) (2:1) see BOP000
3-(DIBUTYLAMINO)-1-PROPANOL-p-AMINOBENZOATE (ESTER) SULFATE (2:1)
    see BOP000
3-(DIBUTYLAMINO)PROPYLAMINE see DDV200
3-DIBUTYLAMINOPROPYL-p-AMINOBENZOATE see BOO750
3'-DIBUTYLAMINOPROPYL-4-AMINOBENZOATE SULFATE see BOP000
DIBUTYLAMINOPROPYL-p-AMINOBENZOATE SULFATE see BOP000
DI-N-BUTYLAMMONIUM HEXAFLUOROARSENATE see DDV225
DIBUTYLARSINIC ACID see DDV250
DIBUTYLATED HYDROXYTOLUENE see BFW750
5,5-DIBUTYLBARBITURIC ACID see DDV400
DIBUTYLBARBITURIC ACID see DDV400
p-DI-tert-BUTYLBENZENE see DDV450
DIBUTYL-1,2-BENZENEDICARBOXYLATE see DEH200
2,5-DI-tert-BUTYLBENZENE-1,4-DIOL see DEC800
2,6-DI-tert-BUTYL-p-BENZOQUINONE see DDV500
DIBUTYLBIS((3-CARBOXYACRYLOYL)OXY)-STANNANE DIMETHYL ESTER
    (Z,Z) (8CI) see BKO250
DIBUTYLBIS((2-ETHYLHEXANOYL)OXY)-STANNANE see BJQ250
DIBUTYLBIS((2-ETHYL-1-OXOHEXYL)OXY)-STANNANE (9CI) see BJQ250
DIBUTYLBIS(LAUROYLOXY)STANNANE see DDV600
DIBUTYLBIS(LAUROYLOXY)TIN see DDV600
DIBUTYLBIS(OCTANOYLOXY)STANNANE see BLB250
DIBUTYLBIS(OLEOYLOXY)STANNANE see DEJ000
DIBUTYLBIS((1-OXO-9-OCTADECENYL)OXY)STANNANE (Z,Z) see DEJ000
DIBUTYLBIS((1-OXOOCTYL)OXY)STANNANE see BLB250
DIBUTYLBIS(TRIFLUOROACETOXY)STANNANE see BLN750
DIBUTYL BUTANEPHOSPHONATE see DDV800
O,O-DIBUTYL-1-BUTYLAMINO-CYCLOHEXYLPHOSPHONATE see ALZ000
DIBUTYL BUTYLPHOSPHONATE see DDV800
DI-n-BUTYL-CARBAMYLCHOLINE SULPHATE see DDW000
(2-DIBUTYLCARBAMYLOXYETHYL)-DIMETHYLETHYLAMMONIUM SULFATE
    see DDW000
DIBUTYL CARBITOL see DDW200
DIBUTYLCARBITOLFORMAL see BHK750
DIBUTYL-o-(o-CARBOXYBENZOYL) GLYCOLATE see BQP750
DIBUTYL-o-CARBOXYBENZOYLOXYACETATE see BQP750
DIBUTYL CELLOSOLVE see DDW400
DIBUTYL CELLOSOLVE PHTHALATE see BHK000
DIBUTYL CHLORENDATE see CDS025
2,6-DI-tert-BUTYL-p-CRESOL (OSHA, ACGIH) see BFW750
9,10-DI-n-BUTYL-1,2,5,6-DIBENZANTHRACENE see DDX200
DI-n-BUTYL(DIBUTYRYLOXY)STANNANE see DDX600
DIBUTYLDICHLOROGERMANE see DDY000
DIBUTYLDICHLOROSTANNANE see DDY200
DIBUTYLDICHLOROTIN see DDY200
DIBUTYL (DIETHYLENE GLYCOL BISPHTHALATE) see DDY400
DIBUTYLDI(2-ETHYLHEXYLOXYCARBONYLMETHYLTHIO)STANNANE see
    DDY600
DIBUTYLDIFLUOROSTANNANE see DDY800
DIBUTYL(DIFORMYLOXY)STANNANE see DDZ000
2,2-DIBUTYLDIHYDRO-6H-1,3,2-OXATHIASTANNIN-6-ONE see DEJ200
DIBUTYLDIHYDROXYSTANNANE-3,3'-THIODIPROPIONATE see DEA400
DIBUTYLDIIODOSTANNANE see DEA000
2,6-DI-tert-BUTYL-α-(DIMETHYLAMINO)-p-CRESOL see DEA100, FAB000
2,2-DIBUTYL-1,3-DIOXA-7,9-DITHIA-2-STANNACYCLODODECAN see DEA200
2,2-DIBUTYL-1,3-DIOXA-2-STANNA-7,9-DITHIACYCLODODECAN-4,12-DIONE see
    DEA200
2,2-DIBUTYL-1,3-DIOXA-2-STANNA-7-THIACYCLODECAN-4,10-DIONE see
    DEA400
2,2-DIBUTYL-1,3,2-DIOXASTANNEPIN-4,7-DIONE see DEJ100
2,2-DIBUTYL-1,3,7,2-DIOXATHIASTANNECANE-4,10-DIONE see DEA400
DIBUTYLDIPENTANOYLOXYSTANNANE see DEA600
DIBUTYLDIPROPIONYLOXYSTANNANE see DEB400
DI-n-BUTYL-DISELENIDE see BRF550
DIBUTYL DISELENIDE see BRF550
DIBUTYLDISELENIUM see BRF550
DIBUTYLDITHIOCARBAMIC ACID, NICKEL SALT see BIW750
DIBUTYLDITHIOCARBAMIC ACID SODIUM SALT see SGF500
DIBUTYLDITHIOCARBAMIC ACID-S-TRIBUTYLSTANNYL ESTER see DEB600

DIBUTYLTIN OCTANOATE see BLB250
DI-n-BUTYLTIN OXIDE see DEF400
DIBUTYLTIN OXIDE see DEF400
DIBUTYLTIN SULFIDE see DEI200
DIBUTYLTIN TETRACHLOROPHTHALATE see DEH800
DIBUTYLTIN 3,3'-THIODIPROPIONATE see DEA400
6,6-DIBUTYL-4,8,11-TRIOXO-5,7,12-TRIOXA-6-STANNATRIDECA-2,9-DIENOIC ACID METHYL ESTER see BKO250
DIBUTYL XANTHOGEN DISULFIDE see BSS550
DIBUTYLZINN-S,S'-BIS(ISOOCTYLTHIOGLYCOLAT) (GERMAN) see BKK250
DI-n-BUTYL-ZINN DI-2-AETHYLHEXYL THIOGLYKOLAT (GERMAN) see DDY600
DI-n-BUTYL-ZINN-DICHLORID (GERMAN) see DDY200
DIBUTYL-ZINN-DILAURAT (GERMAN) see DDV600
DI-N-BUTYL-ZINN-DI(MONOBUTYL)MALEINAT (GERMAN) see BHK250
DI-n-BUTYLZINN-DIMONOMETHYLMALEINAT (GERMAN) see BKO250
DI-n-BUTYL-ZINN-DI(MONONONYL)MALEINAT (GERMAN) see DEI800
DI-n-BUTYL-ZINN-OXYD (GERMAN) see DEF400
DI-n-BUTYLZINN THIOGLYKOLAT (GERMAN) see DEF200
DIC 1468 see MQR275
DIC see DAB600
DICA see DEL200
DICACODYL SULFIDE see CAC250
DICAESIUM SELENIDE see DEJ400
DICAIN see BQA010
DICAINE see BQA010
DICALCIUM PHOSPHATE see CAW100
DICALITE see SCH000
DICAMBA (DOT) see MEL500
DICAMOYLMETHTANE see MBW750
DICANDIOL see MQU750
DICAPRYL 1,2-BENZENEDICARBOXYLATE see BLB750
DICAPRYL PHTHALATE see BLB750
DICAPRYLYL PEROXIDE see CBF705
DICAPTOL see BAD750
1,7-DICARBADODECABORANE (12) see NBV100
m-DICARBADODECABORANE (12) see NBV100
DICARBADODECABORANYLMETHYLETHYL SULFIDE see DEJ600
DICARBADODECABORANYLMETHYLPROPYL SULFIDE see DEJ800
DICARBAM see CBM750
2,2-DI(CARBAMOYLOXYMETHYL)PENTANE see MQU750
2,2-DICARBAMYLOXYMETHYL-3-METHYLPENTANE see MBW750
DICARBAZAMIDE see IBC000
S-(1,2-DICARBETHOXYETHYL)-O,O-DIMETHYLDITHIOPHOSPHATE see MAK700
DICARBETHOXYMETHANE see EMA500
DICARBOETHOXYETHYL-O,O-DIMETHYL PHOSPHORODITHIOATE see MAK700
DI(CARBOMETHOXY)ACETYLENE see DOP400
DICARBOMETHOXYZINC see ZBS000
DICARBONIC ACID DIETHYL ESTER see DIZ100
2,3-DICARBONITRILO-1,4-DIATHIAANTHRACHINON (GERMAN) see DLK200
DI-mu-CARBONYLHEXACARBONYLDICOBALT see CNB500
DICARBONYL MOLYBDENUM DIAZIDE see DEJ849
DICARBONYLPYRAZINE RHODIUM(1) PERCHLORATE see DEJ859
DICARBONYLTUNGSTEN DIAZIDE see DEJ880
DICARBOXIDINE HYDROCHLORIDE see DEK000
o-DICARBOXYBENZENE see PHW250
3,5-DICARBOXYBENZENESULFONIC ACID, SODIUM SALT see DEK200
3,3'-DICARBOXYBENZIDINE see BFX250
DICARBOXYDINE see DEK400
((1,2-DICARBOXYETHYL)THIO)GOLD DISODIUM SALT see GJC000
4,5-DICARBOXYIMIDAZOLE see IAM000
DICARBOXYMETHANE see CCC750
DICAROCIDE see DIW200
DICARZOL see DSO200
DICATRON see PDN000
DICERIUM TRISULFIDE see DEK600
DICESIUM CARBONATE see CDC750
DICESIUM DICHLORIDE see CDD000
DICESIUM DIFLUORIDE see CDD500
DICESIUM DIIODIDE see CDE000
DICESIUM SELENIDE see DEJ400
DICESIUM SULFATE see CDE500
DICESTAL see MJM500
DICHA see DGT600
DICHA (CUBA) see DHB309
DICHAN (CZECH) see DGU200
DICHAPETULUM CYMOSUM (HOOK) ENGL see PLG000
DICHINALEX see CLD000
DICHLOBENIL (DOT) see DER800
DICHLOFENAMIDE see DEQ200
DICHLOFENTHION see DFK600
DICHLOFENTION see DFK600
DICHLOFLUANID see DFL200
DICHLOFLUANIDE see DFL200

DICHLONE (DOT) see DFT000
1,4-DICHLOORBENZEEN (DUTCH) see DEP800
p-DICHLOORBENZEEN (DUTCH) see DEP800
1,1-DICHLOOR-2,2-BIS(4-CHLOOR FENYL)-ETHAAN (DUTCH) see BIM500
1,1-DICHLOORETHAAN (DUTCH) see DFF809
1,2-DICHLOORETHAAN (DUTCH) see EIY600
2,2'-DICHLOORETHYLETHER (DUTCH) see DFJ050
DICHLOORFEEN see MJM500
(2,4-DICHLOOR-FENOXY)-AZIJNZUUR (DUTCH) see DAA800
2-(2,4-DICHLOOR-FENOXY)-PROPIONZUUR (DUTCH) see DGB000
(3,4-DICHLOOR-FENYL-AZO)-THIOUREUM (DUTCH) see DEQ000
3-(3,4-DICHLOOR-FENYL)-1,1-DIMETHYLUREUM (DUTCH) see DXQ500
3-(3,4-DICHLOOR-FENYL)-1-METHOXY-1-METHYLUREUM (DUTCH) see DGD600
3,6-DICHLOOR-2-METHOXY-BENZOEIZUUR (DUTCH) see MEL500
1,1-DICHLOOR-1-NITROETHAAN (DUTCH) see DFU000
(2,2-DICHLOOR-VINYL)-DIMETHYL-FOSFAAT (DUTCH) see DGP900
DICHLOORVO (DUTCH) see DGP900
2,6-DICHLOQUINONE see DES400
DICHLOR see DGG800
2,5-DICHLORACETANILID see DGB480
DICHLORACETIC ACID see DEL000
DICHLORACETYL CHLORIDE see DEN400
1,1-DICHLORAETHAN (GERMAN) see DFF809
1,2-DICHLOR-AETHAN (GERMAN) see EIY600
1,2-DICHLOR-AETHEN (GERMAN) see DFI210
p-DI-(2-CHLORAETHYL)-AMINO-dl-PHENYL-ALANIN (GERMAN) see BHT750
o-(p-DI(2-CHLORAETHYL)-AMINOPHENYL)-dl-TYROSIN-DIHYDROCHLORID (GERMAN) see DFH100
DICHLORALANTIPYRIN see SKS700
DICHLORALANTIPYRINE see SKS700
S-(2,3-DICHLOR-ALLYL)-N,N-DIISOPROPYL-MONOTHIOCARBAMAAT (DUTCH) see DBI200
2,3-DICHLORALLYL-N,N-(DIISOPROPYL)-THIOCARBAMAT (GERMAN) see DBI200
DICHLORALPHENAZONE see SKS700
DICHLORAL UREA see DGQ200
DICHLORAMINE see BIE250
DICHLORAN (amine fungicide) see RDP300
DICHLORAN see RDP300
2,4-DICHLORANILIN see DEO290
3,4-DICHLORANILIN see DEO300
3,4-DICHLORANILINE see DEO300
1-(3,4-DICHLORANILINO)-1-FORMYLAMINO-2,2,2-TRICHLORAETHAN (GERMAN) see CDP750
1,5-DICHLORANTHRACHINON see DEO700
1,8-DICHLORANTHRACHINON see DEO750
DICHLORANTIN see DFE200
3,3'-DICHLORBENZIDIN (CZECH) see DEQ600
4,4'-DICHLORBENZILSAEUREAETHYLESTER (GERMAN) see DER000
1,4-DICHLOR-BENZOL (GERMAN) see DEP800
p-DICHLORBENZOL (GERMAN) see DEP800
o-DICHLOR BENZOL see DEP600
2,6-DICHLORBENZONITRIL (GERMAN) see DER800
1-(2,4-DICHLORBENZYL)INDAZOLE-3-CARBOXYLIC ACID see DEL200
1,1-DICHLOR-2,2-BIS(4-CHLOR-PHENYL)-AETHAN (GERMAN) see BIM500
2,3-DICHLOR-1,3-BUTADIEN (CZECH) see DEU400
2,2'-DICHLOR-DIAETHYLAETHER (GERMAN) see DFJ050
3,3'-DICHLOR-4,4'-DIAMINO-DIPHENYLAETHER (GERMAN) see BGT000
3,3'-DICHLOR-4,4'-DIAMINODIPHENYLMETHAN (GERMAN) see MJM200
DICHLOR-DIFENYLSILAN see DFF000
DICHLORDIMETHYLAETHER (GERMAN) see BIK000
DICHLOREMULSION see EIY600
DICHLOREN (GERMAN) see BIE250
DICHLOREN see BIE500
DICHLOREN HYDROCHLORIDE see BIE500
DICHLORETHANOIC ACID see DEL000
2,2'-DICHLORETHYL ETHER see DFJ050
β,β-DICHLOR-ETHYL-SULPHIDE see BIH250
DICHLORFENIDIM see DXQ500
2,6-DICHLORFENOL (CZECH) see DFY000
3,4-DICHLORFENYLAMID KYSELINY 3,5-DICHLORSALICYLOVE see TBV000
DICHLOR-FENYLARSIN see DGB600
3,4-DICHLORFENYLISOKYANAT see IKH099
N-DICHLORFLUORMETHYLTHIO-N',N'-DIMETHYLAMINOSULFONSAEUREANILID (GERMAN) see DFL200
N-(DICHLOR-FLUOR-METHYL-THIO)-N',N'-DIMETHYL-N-PHENYL-SCHWEFEL-SAEUREDIAMID (GERMAN) see DFL200
DICHLORFOS (POLISH) see DGP900
DICHLORHYDRATE de1-p.CHLORBENZHYDRYL-4-(2-(2-HYDROXYETHOX-Y)ETHYL)PIPERAZINE see HOR470
DICHLORHYDRATE de DIMETHOXY-3,4 BENZYL PIPERAZINE (FRENCH) see VIK200
5,7-DICHLOR-8-HYDROXYCHINOLIN see DEL300
DI-CHLORICIDE see DEP800

DICHLOROBIS(2-ETHOXYCYCLOHEXYL)SELENIUM see DEU125
1,1-DICHLORO-2,2-BIS(4-ETHYLPHENYL)ETHANE see DJC000
2,2-DICHLORO-1,1-BIS(p-ETHYLPHENYL)ETHANE see DJC000
1,1-DICHLORO-2,2-BIS(p-ETHYLPHENYL)ETHANE see DJC000
α,α-DICHLORO-2,2-BIS(p-ETHYLPHENYL)ETHANE see DJC000
1,1-DICHLORO-2,2-BIS(PARACHLOROPHENYL)ETHANE (DOT) see BIM500
cis-DICHLOROBIS(PYRROLIDINE)PLATINUM(II) see DEU200
N,N'-DICHLOROBIS(2,4,6-TRICHLOROPHENYL) UREA see DEU259
DICHLOROBORANE see DEU300
DICHLOROBROMOMETHANE see BND500
O-(2,5-DICHLORO-4-BROMOPHENYL)-O-METHYL PHENYLTHIOPHOSPHONATE
    see LEN000
2,3-DICHLORO-1,3-BUTADIENE see DEU400
1,4-DICHLORO-1,3-BUTADIYNE see DEU509
1,1-DICHLOROBUTANE see BRQ050
3,4-DICHLORO-1-BUTENE see DEV100
1,3-DICHLORO-2-BUTENE see DEU650
1,4-DICHLORO-2-BUTENE see BRG000, DEV000
1,4-DICHLOROBUTENE-2 (MAK) see DEV000
DICHLOROBUTENE see DEV200
1,4-DICHLOROBUTENE-2 (trans) see BRG000
2,2'-DICHLORO-N-BUTYLDIETHYLAMINE see DEV300
DICHLOROBUTYLENE see DEV200
1,4-DICHLOROBUTYNE see DEV400
1,4-DICHLORO-2-BUTYNE see DEV400
3,4-DICHLOROCARBANILIC ACID METHYL ESTER see DEV600
4,5-DICHLOROCATECHOL see DGJ200
2,3-DICHLOROCHINOXALIN-6-KARBONYLCHLORID (CZECH) see DGK000
2,4-DICHLORO-6-o-CHLORANILINO-s-TRIAZINE see DEV800
DICHLOROCHLORDENE see CDR750
2,4-DICHLORO-6-(2-CHLOROANILINO)-1,3,5-TRIAZINE see DEV800
2,4-DICHLORO-6-(o-CHLOROANILINO)-s-TRIAZINE see DEV800
1-(2,4-DICHLORO-β-((p-CHLOROBENZYL)OXY)-PHENETHYL)IMIDAZOLE NI-
    TRATE see EAE000
(±)-1-(2,4-DICHLORO-β-((4-CHLOROBENZYL)THIO)PHENETHYL)IMIDAZOLE NI-
    TRATE see SNH480
(±)-1-(2,4-DICHLORO-β-((2-CHLORO-3-ETHENYL)OXY)PHENETHYL)IMIDAZOLE
    see TGF050
1,1-DICHLORO-2-CHLOROETHYLENE see TIO750
2,4-DICHLORO-α-(CHLOROMETHYLENE)BENZYL ALCOHOL DIETHYL PHOS-
    PHATE see CDS750
1,1-DICHLORO-2-(o-CHLOROPHENYL)-2-(p-CHLOROPHENYL)ETHANE see
    CDN000
4',5-DICHLORO-N-(4-CHLOROPHENYL)-2-HYDROXY-(1,1'-BIPHENYL)-3-CAR-
    BOXAMIDE see TIP750
4,6-DICHLORO-N-(2-CHLOROPHENYL)-1,3,5-TRIAZIN-2-AMINE see DEV800
2,4-DICHLORO-5-CHLOROSULPHONYLBENZOIC ACID see CLG200
DICHLORO(2-CHLOROVINYL)ARSINE see CLV000
DICHLORO(2-CHLOROVINYL)ARSINE OXIDE see DEW000
5-(3,4-DICHLOROCINNAMOYL)-4,7-DIMETHOXY-6-(2-DIMETHYLAMINOETHOX-
    Y)BENZOFURAN MALEATE see DEW200
DICHLOROCTAN SODNY (CZECH) see SGG000
DICHLOROCYANURIC ACID see DGN200
2,6-DICHLORO-2,5-CYCLOHEXADIENE-1,4-DIONE see DES400
DICHLORO(1,2-CYCLOHEXANEDIAMINE)PLATINUM see DAD040
3,4'-DICHLOROCYCLOPROPANECARBOXANILIDE see CQJ250
(SP-4-2)-trans(+)-DICHLORO(1,2-CYCLOHEXANEDIAMINE-N,N')- (9CI) see
    DAD050
2,6-DICHLORO-N-CYCLOPROPYL-N-ETHYL ISONICOTINAMIDE see DEW400
3,3'-DICHLORO-4,4'-DIAMINO(1,1-BIPHENYL) see DEQ600
3,3'-DICHLORO-4,4'-DIAMINOBIPHENYL see DEQ600
DICHLORO(1,2-DIAMINOCYCLOHEXANE)PLATINUM see DAD040
cis-DICHLORO-1,2-DIAMINOCYCLOHEXANE PLATINUM(II) see DAD040
DICHLORO(1,2-DIAMINOCYCLOHEXANE)PLATINUM(II) see DAD040
trans(−)-DICHLORO-1,2-DIAMINOCYCLOHEXANEPLATINUM(II) see DAD075
3,3'-DICHLORO-4,4'-DIAMINODIPHENYL ETHER see BGT000
3,3'-DICHLORO-4,4'-DIAMINODIPHENYLMETHANE see MJM200
N-(3,5-DICHLORO-4-(2,4-DIAMINO-6-PTERIDINYL METH-
    YL)METHYLAMINO)BENZOYL)GLUTAMIC ACID see DFO000
cis-DICHLORODIAMMINE PLATINUM(II) see PJD000
trans-DICHLORODIAMMINEPLATINUM(II) see DEX000
2,7-DICHLORODIBENZODIOXIN see DAC800
2,7-DICHLORODIBENZO(b,e)(1,4)DIOXIN see DAC800
2,7-DICHLORODIBENZO-p-DIOXIN see DAC800
4,4'-DICHLORODIBUTYL ETHER see OPE040
DICHLORODIBUTYLGERMANE see DDY000
DICHLORODIBUTYLSTANNANE see DDY200
DICHLORODIBUTYLTIN see DDY200
1-(2,4-DICHLORO-β-((2,4-DICHLOROBENZYL)OXY)PHENETHYL)-IMIDAZOLE NI-
    TRATE see MQS560
1-(2,4-DICHLORO-β-(2,6-DICHLOROBENZYLOXY)PHENETHYL)IMIDAZOLE NI-
    TRATE see IKN200
1,1-DICHLORO-2,2-DICHLOROETHANE see TBQ100
1,1-DICHLORO-2,2-DI(4-CHLOROPHENYL)ETHANE see BIM500
2,3-DICHLORO-5,6-DICYANOBENZOQUINONE see DEX400

DICHLORODI-pi-CYCLOPENTADIENYLHAFNIUM see HAE500
DICHLORODICYCLOPENTADIENYLHAFNIUM see HAE500
DICHLORODI-pi-CYCLOPENTADIENYLTITANIUM see DGW200
DICHLORODICYCLOPENTADIENYLTITANIUM see DGW200
2,2'-DICHLORO DIETHYLAMINE HYDROCHLORIDE see BHO250
β,β'-DICHLORODIETHYLAMINE HYDROCHLORIDE see BHO250
2',6'-DICHLORO-2-(DIETHYLAMINO)ACETANILIDE HYDROCHLORIDE see
    DEX600
7,8-DICHLORO-10-(2-(DIETHYLAMINO)ETHYL)ISOALLOXAZINE HYDROCHLO-
    RIDE see DEX800
2',6'-DICHLORO-2-(2-(DIETHYLAMINO)ETHYL)METHYLAMINOACETANILIDE DI-
    HYDROCHLORIDE see DEY000
2',6'-DICHLORO-2-(2-(DIETHYLAMINO)ETHYL)THIOACETANILIDE HYDROCHLO-
    RIDE see DEY200
7,8-DICHLORO-10-(3-(DIETHYLAMINO)-2-HYDROXYPROPYL)ISOALLOXAZINE
    SULFATE see DEY400
7,8-DICHLORO-10-(4-(DIETHYLAMINO)-1-METHYLBUTYL)ISOALLOXAZINE HY-
    DROCHLORIDE see DEY600
β,β'-DICHLORODIETHYLANILINE see AOQ875
β,β'-DICHLORODIETHYL ETHER see DFJ050
DICHLORODI(2-ETHYLHEXYL)STANNANE see DJL200
β,β'-DICHLORODIETHYL-N-METHYLAMINE see BIE250
β,β'-DICHLORODIETHYL-N-METHYLAMINE HYDROCHLORIDE see BIE500
DICHLORODIETHYLSILANE see DEY800
DICHLORODIETHYLSTANNANE see DEZ000
DI-o-(CHLORODIETHYLSTANNYLOXO)BIS(CHLORO-DIETHYLTIN) see BIY500
2,2'-DICHLORODIETHYL SULFIDE see BIH250
DICHLORODIETHYLTIN see DEZ000
1,2-DICHLORO-1,1-DIFLUOROETHANE see DFA000
1,1-DICHLORO-2,2-DIFLUOROETHYLENE see DFA300
DICHLORODIFLUOROETHYLENE see DFA200
2,2-DICHLORO-1,1-DIFLUOROETHYL METHYL ETHER see DFA400
DICHLORODIFLUOROMETHANE see DFA600
DICHLORODIFLUOROMETHANE with 1,1-DIFLUOROETHANE see DFB400
DICHLORODIFLUOROMETHANE and DIFLUOROETHANE AZEOTROPIC MIX-
    TURE (DOT) see DFB400
DICHLORODIFLUOROMETHANE–TRICHLOROTRIFLUOROETHANE MIXTURE
    see DFC000
DICHLORODIFLUOROMETHANE with TRICHLOROTRIFLUOROETHANE see
    DFC000
2,2-DICHLORO-1,1-DIFLUORO-1-METHOXYETHANE see DFA400
DICHLORODIHEXYLSTANNANE see DFC200
2,5-DICHLORO-4-(4,5-DIHYDRO-3-METHYL-5-OXO-1H-PYRAZOL-1-
    YL)BENZENESULFONIC ACID see DFQ200
(2,5-DICHLORO-3,6-DIHYDROXY-p-BENZOQUINOLATO)MERCURY see DFC800
(2,5-DICHLORO-3,6-DIHYDROXY-p-BENZOQUINONE), MERCURY SALT see
    DFC800
2,5-DICHLORO-3,6-DIHYDROXY-p-BENZOQUINONE, MERCURY SALT see
    DFC800
cis-DICHLORO-trans-DIHYDROXYBISISOPROPYLAMINE PLATINUM (IV) see
    IGG775
5,5'-DICHLORO-2,2'-DIHYDROXY-3,3'-DINITROBIPHENYL see DFD000
5,5'-DICHLORO-2,2'-DIHYDROXYDIPHENYLMETHANE see MJM500
DICHLORODIISONONYL STANNANE see BLQ750
β,β'-DICHLORODIISOPROPYL ETHER see BII250
DICHLORODIISOPROPYL ETHER see BII250
DICHLORODIISOPROPYLSTANNANE see DNT000
3,4-DICHLORO-2,5-DILITHIOTHIOPHENE see DFD200
1,4-DICHLORO-2,5-DIMETHOXYBENZENE see CJA100
3',4'-DICHLORO-4-DIMETHYLAMINOAZOBENZENE see DFD400
7,8-DICHLORO-10-(2-(DIMETHYLAMINO)ETHYL)ISOALLOXAZINE SULFATE see
    DFD600
7,8-DICHLORO-10-(3-(DIMETHYLAMINO)PROPYL)ISOALLOXAZINE HYDRO-
    CHLORIDE see DFE000
1,1-DICHLORO-N-((DIMETHYLAMINO)SULFONYL)-1-FLUORO-N-PHENYLMETH-
    ANE SULFENAMIDE see DFL200
2,2-DICHLORO-3,3-DIMETHYLBUTANE see DFE100
1,1-DICHLORO-3,3-DIMETHYL-2-BUTANONE see DGF300
sym-DICHLORODIMETHYL ETHER (DOT) see BIK000
1,3-DICHLORO-5,5-DIMETHYL HYDANTOIN see DFE200
DICHLORODIMETHYLHYDANTOIN see DFE200
1,3-DICHLORO-5,5-DIMETHYL-2,4-IMIDAZOLIDINEDIONE see DFE200
cis-DICHLORO(4,5-DIMETHYL-O-PHENYLENEDIAMMINE)PLATINUM(II) see
    DFE229
DICHLORO(4,5-DIMETHYL-o-PHENYLENEDIAMMINE)PLATINUM(II) see DFE229
3,5-DICHLORO-N-(1,1-DIMETHYL-2-PROPYNYL)BENZAMIDE see DTT600
3,5-DICHLORO-2,6-DIMETHYL-4-PYRIDINOL see CMX850
DICHLORODIMETHYLSILANE see DFE259
DICHLORODIMETHYLSTANNANE see DUG825
3,6-DICHLORO-3,6-DIMETHYLTETRAOXANE see DFE300
DICHLORODIMETHYLTIN see DUG825
(trans-4)-DICHLORO(4,4-DIMETHYLZINC
    5((((METHYLAMINO)CARBONYL)OXY)IMINO)PENTANENITRILE) see DFE469
4,4'-DICHLORO-6,6'-DINITRO-O,O'-BIPHENOL see DFD000
3,3'-DICHLORO-5,5'-DINITRO-O,O'-BIPHENOL (FRENCH) see DFD000

5,5'-DICHLORO-3,3'-DINITRO(1,1'-BIPHENYL)-2,2'-DIOL see DFD000
DICHLORODINITROMETHANE see DFE550
DICHLORODIOCTYLSTANNANE see DVN300
trans-2,3-DICHLORO-1,4-DIOXANE see DFE600
trans-2,3-DICHLORO-p-DIOXANE see DFE600
DICHLORODIOXOCHROMIUM see CML125
4,5-DICHLORO-3,6-DIOXO-1,4-CYCLOHEXADIENE-1,2-DICARBONITRILE see DEX400
DICHLORODIPENTYLSTANNANE see DVV200
DI-p-CHLORODIPHENOXYMETHANE see NCM700
p,p'-DICHLORODIPHENYLACETIC ACID see BIL500
DICHLORODIPHENYLACETIC ACID see BIL500
o,p'-DICHLORODIPHENYLDICHLOROETHANE see CDN000
p,p'-DICHLORODIPHENYLDICHLOROETHANE see BIM500
DICHLORODIPHENYL DICHLOROETHANE see BIM500
p,p'-DICHLORODIPHENYLDICHLOROETHYLENE see BIM750
2,2'-((3,3'-DICHLORO(1,1'-DIPHENYL)-4,4'-DIYL)BIS(AZO)BIS(3-OXO-N-PHENYL-BUTANAMIDE see DEU000
DICHLORODIPHENYLETHANOL see BIN000
p,p'-DICHLORODIPHENYLMETHYLCARBINOL see BIN000
DICHLORODIPHENYL OXIDE see DFE800
DICHLORO DIPHENYLSILANE see DFF000
DICHLORODIPHENYLSTANNANE see DWO400
DICHLORODIPHENYLSTANNANE complex with PYRIDINE (1:2) see DWA400
p,p'-DICHLORODIPHENYL SULFIDE see CEP000
4,4'-DICHLORODIPHENYLTRICHLOROETHANE see DAD200
DICHLORODIPHENYLTRICHLOROETHANE (DOT) see DAD200
p,p'-DICHLORODIPHENYLTRICHLOROETHANE see DAD200
DICHLORODIPHENYLTRICHLOROETHANE see DAD200
2,2-DICHLORO-N,N-DI-2-PROPENYLACETAMIDE see DBJ600
2',6'-DICHLORO-2-(DIPROPYLAMINO)ACETANILIDE HYDROCHLORIDE see DFF200
DICHLORODIPROPYLSTANNANE see DFF400
DICHLORODIPROPYLTIN see DFF400
cis-DICHLORO(DIPYRIDINE)PLATINUM(II) see DFF500
DICHLORODIPYRIDINEPLATINUM(II) (Z) see DFF500
4,5-DICHLORO-1,3-DISULFAMOYLBENZENE see DEQ200
1,4-DICHLORO-2,3-EPOXYBUTANE see DFF600
cis-1,3-DICHLORO-1,2-EPOXYPROPANE see DAC975
trans-1,3-DICHLORO-1,2-EPOXYPROPANE see DGH500
1,1-DICHLOROETHANE see DFF809
1,2-DICHLOROETHANE see EIY600
α,β-DICHLOROETHANE see EIY600
DICHLORO-1,2-ETHANE (FRENCH) see EIY600
DICHLOROETHANE see DFF800
sym-DICHLOROETHANE see EIY600
DICHLOROETHANOIC ACID see DEL000
1,2-DICHLOROETHANOL ACETATE see DFG159
DICHLOROETHANOYL CHLORIDE see DEN400
1,1-DICHLOROETHENE see VPK000
1,1-DICHLOROETHENE POLYMER with CHLOROETHENE see CGW300
2,2-DICHLOROETHENOL DIMETHYL PHOSPHATE see DGP900
2,2-DICHLOROETHENYL DIETHYL PHOSPHATE see DFG200
2,2-DICHLOROETHENYL DIMETHYL PHOSPHATE see DGP900
1,1'-DICHLOROETHENYLIDENE)BIS(4-CHLOROBENZENE) see BIM750
2,2-DICHLOROETHENYL PHOSPHORIC ACID DIMETHYL ESTER see DGP900
DICHLOROETHER see DFJ050
1,2-DICHLOROETHYL ACETATE see DFG159
DICHLOROETHYLALUMINUM see EFU000
2,2-DICHLOROETHYLAMINE see DFG700
DI-2-CHLOROETHYLAMINE HYDROCHLORIDE see BHO250
p-(DI-2-CHLOROETHYLAMINE)PHENYL BUTYRIC ACID SODIUM SALT see CDO625
9-(2-(DI(2-CHLOROETHYL)AMINO)ETHYLAMINO)-6-CHLORO-2-METHOXYACRI-DINE see DFH000
2-(DI-2-CHLOROETHYL)AMINOMETHYL-5,6-DIMETHYLBENZIMIDAZOLE see BCC250
2-(DI-2-CHLOROETHYL)AMINO)-1-OXA-3-AZA-2-PHOSPHACYCLOHEXANE-2-OX-IDE MONOHYDRATE see CQC500
p-N,N-DI-(2-CHLOROETHYL)AMINOPHENYL ACETIC ACID see PCU425
p-DI-(2-CHLOROETHYL)-AMINO-d-PHENYLALANINE see SAX200
p-DI(2-CHLOROETHYL)AMINO-d-PHENYLALANINE see SAX200
o-DI-2-CHLOROETHYLAMINO-dl-PHENYLALANINE see BHT250
p-DI(2-CHLOROETHYL)AMINO-dl-PHENYLALANINE see BHT750
3-p-(DI(2-CHLOROETHYL)AMINO)-PHENYL-l-ALANINE see PED750
p-DI(2-CHLOROETHYL)AMINO-l-PHENYLALANINE see PED750
p-N,N-DI(2-CHLOROETHYL)AMINOPHENYLALANINE see PED750
γ-(p-DI(2-CHLOROETHYL)AMINOPHENYL)BUTYRIC ACID see CDO500
p-(N,N-DI-2-CHLOROETHYL)AMINOPHENYL BUTYRIC ACID see CDO500
p-N,N-DI-(β-CHLOROETHYL)AMINOPHENYL BUTYRIC ACID see CDO500
N,N-DICHLOROETHYL-γ-p-AMINOPHENYLBUTYRIC ACID see CDO500
o-(p-DI-(2-CHLOROETHYL)AMINOPHENYL)-dl-TYROSINE DIHYDROCHLORIDE see DFH100
p-N,N-DI-(2-CHLOROETHYL)AMINOPHENYLVALERIC ACID see BHY625

N,N-DI-(2-CHLOROETHYL)AMINO-N,O-PROPYLENE PHOSPHORIC ACID ESTER DIAMIDE MONOHYDRATE see CQC500
5-(DI-2-CHLOROETHYL)AMINOURACIL see BIA250
5-(DI-(β-CHLOROETHYL)AMINO)URACIL see BIA250
N,N-DI(2-CHLOROETHYL)ANILINE see AOQ875
DICHLOROETHYLARSINE see DFH200
DICHLOROETHYLBENZENE see EHY500
DI-(2-CHLOROETHYL)BENZYLAMINE see BIA750
DICHLOROETHYLBORANE see DFH300
O,O-DI(2-CHLOROETHYL)-O-(3-CHLORO-4-METHYLCOUMARIN-7-YL) PHOS-PHATE see DFH600
O,O-DI(2-CHLOROETHYL)-7-(3-CHLORO-4-METHYLCOUMARINYL)PHOSPHATE see CIK750
DI-(2-CHLOROETHYL)-3-CHLORO-4-METHYL-7-COUMARINYL PHOSPHATE see DFH600
DI-(2-CHLOROETHYL)-3-CHLORO-4-METHYLCOUMARIN-7-YL PHOSPHATE see DFH600
1,1-DICHLOROETHYLENE see VPK000
1,2-DICHLOROETHYLENE see DFI200, DFI210
DICHLORO-1,2-ETHYLENE (FRENCH) see DFI210
trans-1,2-DICHLOROETHYLENE (MAK) see ACK000
cis-DICHLOROETHYLENE see DFI200
DICHLOROETHYLENE see DFH800, EIY600
sym-DICHLOROETHYLENE see DFI210
trans-DICHLOROETHYLENE see ACK000
1,2-DICHLOROETHYLENE CARBONATE see DFI800
DICHLORO(ETHYLENEDIAMINE)PLATINUM(II) see DFJ000
1,1-DICHLOROETHYLENE-MONOCHLOROETHYLENE POLYMER see CGW300
1,1-DICHLOROETHYLENE POLYMER with CHLOROETHYLENE see CGW300
S-(1,2-DICHLOROETHYLENEYL)-I-CYSTEINE see DGP000
DI(2-CHLOROETHYL) ESTER, MALEIC ACID see DFJ200
β,β'-DICHLOROETHYL ETHER see DFJ050
DI(β-CHLOROETHYL)ETHER see DFJ050
2,2'-DICHLOROETHYL ETHER (MAK) see DFJ050
DICHLOROETHYL ETHER see DFJ050
sym-DICHLOROETHYL ETHER see DFJ050
DI-2-CHLOROETHYL FORMAL see BID750
DICHLOROETHYL FORMAL see BID750
1,2-DICHLOROETHYL HYDROPEROXIDE see DFJ100
DI-2-CHLOROETHYL MALEATE see DFJ200
2,3-DICHLORO-N-ETHYLMALEINIMIDE see DFJ400
DI(2-CHLOROETHYL)METHYLAMINE see BIE250
DI(2-CHLOROETHYL)METHYLAMINE HYDROCHLORIDE see BIE500
2-(1,2-DICHLOROETHYL)-4-METHYL-1,3-DIOXOLANE see DFJ500
2-N,N-DI(2-CHLOROETHYL)NAPHTHYLAMINE see BIF250
N,N-DI(2-CHLOROETHYL)-β-NAPHTHYLAMINE see BIF250
DI(2-CHLOROETHYL)-β-NAPHTHYLAMINE see BIF250
DICHLOROETHYL-β-NAPHTHYLAMINE see BIF250
DI((CHLORO-2-ETHYL)-2-N-NITROSO-N-CARBAMOYL)-N,N-CYSTAMINE see BIF625
DICHLOROETHYL OXIDE see DFJ050
DICHLOROETHYLPHENYLSILANE see DFJ800
DICHLOROETHYLPHOSPHINE see EOQ000
DICHLOROETHYLPHOSPHINE SULFIDE see EOP600
N,N-DI(2-CHLOROETHYL)-N,o-PROPYLENE-PHOSPHORIC ACID ESTER DIAM-IDE see CQC650
DICHLOROETHYLSILANE see DFK000
DI-2-CHLOROETHYL SULFIDE see BIH250
β,β'-DICHLOROETHYL SULFIDE see BIH250
2,2'-DICHLOROETHYL SULPHIDE (MAK) see BIH250
2-2'-DI(3-CHLOROETHYLTHIO)DIETHYL ETHER see DFK200
DICHLOROETHYLVINYLSILANE see DFK400
DICHLOROETHYNE see DEN600
DICHLOROFEN see MJM500
3-(3,4-DICHLORO-FENIL)-1-METOSSI-1-METIL-UREA (ITALIAN) see DGD600
DICHLOROFENTHION see DFK600
1,1-DICHLORO-1-FLUOROETHANE see FOO550
DICHLOROFLUOROMETHANE see DFL000
(DICHLOROFLUOROMETHYL)BENZENE see DFL100
4',5-DICHLORO-N-(4-FLUORO-2-METHYLPHENYL)-2-HYDROXY-(1,1'-BIPHENYL)-3-CARBOXAMIDE see CFC500
N-((DICHLOROFLUOROMETHYL)THIO)-N-((DIMETHYLAMI-NO)SULFONYL)ANILINE see DFL200
N'-DICHLOROFLUOROMETHYLTHIO-N,N-DIMETHYL-N'-(4-TOLYL)SULFAMIDE see DFL400
N-(DICHLOROFLUOROMETHYLTHIO)-N-(DIMETHYLSULFAMOYL)ANILINE see DFL200
N-(DICHLOROFLUOROMETHYLTHIO)-N',N'-DIMETHYL-N-PHENYLSULFAMIDE see DFL200
α-α-DICHLORO-α-FLUOROTOLUENE see DFL100
α-β-DICHLORO-β-FORMYL ACRYLIC ACID see MRU900
2,2'-DICHLORO-N-FURFURYLDIETHYLAMINE HYDROCHLORIDE see FPX000
DICHLOROGERMANE see DFL600
N,N-DICHLOROGLYCINE see DFL709
1,6-DICHLORO-2,4-HEXADIYNE see DFL800

2,3-DICHLORO-1,1,1,4,4,4-HEXAFLUOROBUTENE-2 see DFM000
2,3-DICHLOROHEXAFLUORO-2-BUTENE see DFM000
2,3-DICHLOROHEXAFLUOROBUTENE-2 see DFM000
1,2-DICHLORO-1,2,3,3,4,4-HEXAFLUOROCYCLOBUTANE see DFM025
1,2-DICHLOROHEXAFLUOROCYCLOBUTANE see DFM025
1,2-DICHLOROHEXAFLUOROCYCLOPENTENE see DFM050
1,2-DICHLORO-3,3,4,4,5,5-HEXAFLUOROCYCLOPENTENE see DFM050
4,5-DICHLORO-3,3,4,5,6,6-HEXAFLUORO-1,2-DIOXANE see DFM099
α-DICHLOROHYDRIN see DGG400
DICHLOROHYDRIN see DGG400
6,7-DICHLORO-4-(HYDROXYAMINO)QUINOLINE-1-OXIDE see DFM200
3,6-DICHLORO-2-HYDROXYBENZOIC ACID see DGK250
3,4-DICHLORO-2-HYDROXYCROTONOLACTONE see MRU900
3,4-DICHLORO-2-HYDROXYCROTONOLACTONIC ACID see MRU900
6,7-DICHLORO-10-(3-(N-(2-HYDROXYETHYL)ETHYLAMINO))ISOALLOXAZINE
  SULFATE see DFM600
6,7-DICHLORO-10-(3-(N-(2-HYDROXYETHYL)METHYLAMINO)PROPYL) ISOAL-
  LOXAZINE SULFATE see DFM800
d-(−)-threo-2,2-DICHLORO-N-(β-HYDROXY-α-(HYDROXYMETHYL))-p-NITRO-
  PHENETHYLACETAMIDE see CDP250
d-(−)-2,2-DICHLORO-N-(β-HYDROXY-α-(HYDROXYMETHYL)-p-NITROPHENYLE-
  THYL)ACETAMIDE see CDP250
2,6-DICHLORO-4-((p-HYDROXYPHENYL)IMINO)-2,5-CYCLOHEXADIEN-1-ONE
  SODIUM SALT see SGG650
DI-(5-CHLORO-2-HYDROXYPHENYL)METHANE see MJM500
5,7-DICHLORO-8-HYDROXYQUINALDINE see CLC500
5,7-DICHLORO-8-HYDROXYQUINOLINE see DEL300
DICHLOROHYDROXYQUINOLINE see DEL300
2,6-DICHLORO-N-2-IMIDAZOLIDINYLIDENE-BENZENAMINE HYDROCHLORIDE
  see CMX760
1-(2,5-DICHLORO-6-(1-(1H-IMIDAZOL-1-YL)VINYL)PHENOXY)-3-(ISOPROPYLAMI-
  NO)-2-PROPANOL HYDROCHLORIDE see DFM875
3,3′-DICHLOROINDANTHRONE see DFN425
7,16-DICHLOROINDANTHRONE see DFN425
DICHLOROINDANTHRONE see DFN425
2,6-DICHLOROINDOPHENOL, SODIUM SALT see SGG650
O-(2,5-DICHLORO-4-IODOPHENYL) O,O-DIMETHYL PHOSPHOROTHIOATE see
  IEN000
2,2′-DICHLORO-N-ISOBUTYL-DIETHYLAMINE HYDROCHLORIDE see BIE750
DICHLOROISOCYANURATE see DGN200
DICHLOROISOCYANURIC ACID, dry or dichloroisocyanuric acid salts (DOT)
  see DGN200
DICHLOROISOCYANURIC ACID see DGN200
DICHLOROISOCYANURIC ACID POTASSIUM SALT see PLD000
DICHLOROISOCYANURIC ACID SODIUM SALT (DOT) see SGG500
sym-DICHLOROISOPROPYL ALCOHOL see DGG400
3,4-DICHLORO-α-((ISOPROPYLAMINO)METHYL)BENZYL ALCOHOL see
  DFN400
2,2′-DICHLORO-N-ISOPROPYLDIETHYLAMINE HYDROCHLORIDE see IPG000
2,2′-DICHLOROISOPROPYL ETHER see BII250
DICHLOROISOPROPYL ETHER (DOT) see BII250
DICHLOROISOPROPYL ETHER see BII250
DICHLOROISOVIOLANTHRONE see DFN450
DICHLOROKELTHANE see BIO750
DICHLOROLAWSONE see DFN500
DICHLOROMALEALDEHYDIC ACID see MRU900
2,3-DICHLOROMALEIC ALDEHYDE ACID see MRU900
DICHLOROMALEIC ANHYDRIDE see DFN700
DICHLOROMALEIMIDE see DFN800
DICHLOROMALEINIMIDE see DFN800
DICHLOROMAPHARSEN see DFX400
3′,4′-DICHLORO-2-METHACRYLANILIDE see DFO800
DICHLOROMETHANE (MAK, DOT) see MJP450
DICHLOROMETHANETHIOSULFONIC ACID-S-TRICHLOROMETHYL ESTER see
  DFS600
DICHLOROMETHAZANONE see CKF500
3′5′-DICHLOROMETHOTREXATE see DFO000
DICHLOROMETHOTREXATE see DFO000
3,6-DICHLORO-2-METHOXYBENZOIC ACID see MEL500
2,5-DICHLORO-6-METHOXYBENZOIC ACID see MEL500
((2,3-DICHLORO-4-METHOXYPHENYL)-2-FURANYLMETHANONE)-O-(2-(DIETHYL-
  AMINO)ETHYL) OXIME,MONOMETHANE SULFONATE see DFO600
(DICHLORO-2,3-METHOXY-4) PHENYL FURYL-2-O-(DIETHYLAMINOETHYL)-CE-
  TONE-OXIME (FRENCH) see DFO600
3′,4′-DICHLORO-2-METHYLACRYLANILIDE see DFO800
N,N-DICHLOROMETHYLAMINE see DFO900
9,10-DI(CHLOROMETHYL)ANTHRACENE see BIJ750
DICHLOROMETHYLARSINE see DFP200
1,5-DICHLORO-3-METHYL-3-AZAPENTANE HYDROCHLORIDE see BIE500
2,4-DICHLORO-1-METHYLBENZENE see DGM700
(DICHLOROMETHYL)BENZENE see BAY300
4,4′-DICHLORO(METHYL BENZHYDROL) see BIN000
4,4′-DICHLORO-α-METHYLBENZHYDROL see BIN000
4,4′-DICHLORO-α-METHYLBENZOHYDROL see BIN000
p-(DICHLOROMETHYL)BENZYL CHLORIDE see TJD650

1-(2,4-DICHLORO-β-(p-METHYLBENZYLOXY)PHENETHYL)IMIDAZOLE NITRATE
  see DFP500
DICHLOROMETHYL tert-BUTYL KETONE see DGF300
3-DICHLOROMETHYL-6-CHLORO-7-SULFAMOYL-3,4-DIHYDRO-1,2,4-BENZOTHI-
  ADIAZINE-1,1-DIOXIDE see HII500
3-DICHLOROMETHYL-6-CHLORO-7-SULFAMYL-3,4-DIHYDRO-1,2,4-BENZOTHIA-
  DIAZINE-1,1-DIOXIDE see HII500
DICHLOROMETHYL CYANIDE see DEN000
2,2′-DICHLORO-N-METHYLDIETHYLAMINE see BIE250
2,2′-DICHLORO-N-METHYLDIETHYLAMINE HYDROCHLORIDE see BIE500
2,2′-DICHLORO-N-METHYLDIETHYLAMINE-N-OXIDE see CFA500
2,2′-DICHLORO-N-METHYLDIETHYLAMINE N-OXIDE HYDROCHLORIDE see
  CFA750
1-(DICHLOROMETHYLDIMETHYLSILYL)-1-HEXYN-3-OL see HGA000
N-(DICHLOROMETHYLENE)ANILINE see PFJ400
2,3-DICHLORO-4-(2-METHYLENEBUTYRL)PHENOXY ACETIC ACID see
  DFP600
(2,3-DICHLORO-4-(2-METHYLENEBUTYRYL)PHENOXY)ACETIC ACID see
  DFP600
4,4′-DICHLORO-2,2′-METHYLENEDIPHENOL see MJM500
(2,3-DICHLORO-4-(2-METHYLENE-1-OXOBUTYL)PHENOXY)ACETIC ACID see
  DFP600
α,α-DICHLOROMETHYL ETHER see DFQ000
sym-DICHLOROMETHYL ETHER see BIK000
3,4-DICHLORO-α-(((1-METHYLETHYL)AMINO)METHYL)BENZENEMETHANOL
  see DFN400
1,3-DICHLORO-5,5′-METHYLHYDANTOIN see DFE200
5,7-DICHLORO-2-METHYL-8-HYDROXYQUINOLINE see CLC500
2,3-DICHLORO-N-METHYLMALEIMIDE see DFP800
DICHLORO-N-METHYLMALEIMIDE see DFP800
α,α-DICHLOROMETHYL METHYL ETHER see DFQ000
DICHLOROMETHYL METHYL KETONE see DGG500
d-threo-2-(DICHLOROMETHYL)-α-(p-NITROPHENYL)-2-OXAZOLINE-4-METHANOL
  see DFQ100
6,7-DICHLORO-2-METHYL-1-OXO-2-PHENYL-5-INDANYLOXYACETIC ACID see
  IBQ400
2,5-DICHLORO-4-(3-METHYL-5-OXO-2-PYRAZOLIN-1-YL) BENZENESULFONIC
  ACID see DFQ200
3,3-DICHLOROMETHYLOXYCYCLOBUTANE see BIK325
2-((2,6-DICHLORO-3-METHYLPHENYL)AMINO)-BENZOIC ACID (9CI) see
  DGM875
2-((2,6-DICHLORO-3-METHYLPHENYL)AMINO)BENZOIC ACID ETHOXYMETHYL
  ESTER see DGN000
2-((2,6-DICHLORO)-3-METHYLPHENYL)AMINO-BENZOIC ACID MONOSODIUM
  SALT see SIF425
DICHLORO(4-METHYL-O-PHENYLENEDIAMMINE)PLATINUM(II) see DFQ400
DICHLOROMETHYLPHENYLSILANE see DFQ800
DICHLOROMETHYL PHOSPHINE see EOQ000
1,2-DICHLORO-2-METHYLPROPENE see IIQ200
DICHLORO(1-METHYLPROPYL)ARSINE see BQY300
5,7-DICHLORO-2-METHYL-8-QUINOLINOL see CLC500
DICHLOROMETHYLSILANE see DFS000
4-(DICHLOROMETHYLSILYL)BUTYRONITRILE see COR325
1,2-DICHLORO-1-(METHYLSULFONYL)ETHYLENE see DFS200
O-(DICHLORO(METHYLTHIO)PHENYL) O,O-DIETHYL PHOSPHOROTHIOATE (3
  isomers) see CLJ875
DICHLOROMETHYL TRICHLOROMETHYLTHIOSULFONE see DFS600
2,2′-DICHLORO-1″-METHYLTRIETHYLAMINE see IOF300
2,2′-DICHLORO-1″-METHYLTRIETHYLAMINE HYDROCHLORIDE see IPG000
DICHLOROMETHYL-3,3,3-TRIFLUOROPROPYLSILANE see DFS700
DICHLOROMETHYLVINYLSILANE see DFS800
DICHLOROMONOETHYLALUMINUM see EFU050
DICHLOROMONOFLUOROMETHANE (OSHA, DOT) see DFL000
2,3-DICHLORO-1,4-NAPHTHALENEDIONE see DFT000
2,3-DICHLORO-1,4-NAPHTHAQUINONE see DFT000
2,3-DICHLORO-1,4-NAPHTHOQUINONE see DFT000
2,3-DICHLORONAPHTHOQUINONE-1,4 see DFT000
2,3-DICHLORONAPHTHOQUINONE see DFT000
2,3-DICHLORO-α-NAPHTHOQUINONE see DFT000
DICHLORONAPHTHOQUINONE see DFT000
2,6-DICHLORO-4-NITROANILINE see RDP300
2,6-DICHLORO-4-NITROBENZENAMINE (9CI) see RDP300
1,4-DICHLORO-2-NITROBENZENE see DFT400
2,5-DICHLORONITROBENZENE see DFT400
3,4-DICHLORONITROBENZENE see DFT600
1,2-DICHLORO-4-NITROBENZENE see DFT600
2′,4′-DICHLORO-4-NITROBIPHENYL ETHER see DFT800
2,4-DICHLORO-4′-NITRODIPHENYL ETHER see DFT800
1,1-DICHLORO-1-NITROETHANE see DFU000
DICHLORONITROETHANE see DFU000
1,2-DICHLORO-3-NITRONAPHTHALENE see DFU400
2,4-DICHLORO-6-NITROPHENOL see DFU600
2,4-DICHLORO-6-NITROPHENOL ACETATE see DFU800
2,4-DICHLORO-1-(4-NITROPHENOXY)BENZENE see DFT800
6,7-DICHLORO-4-NITROQUINOLINE-1-OXIDE see DFV200

3-(3,5-DICHLOROPHENYL)-5-METHYL-5-VINYL-2,4-OXAZOLIDINEDIONE see RMA000

N-(3,4-DICHLOROPHENYL)-N'-(4-CHLOROPHENYL)UREA see TIL500

N-(3,4-DICHLOROPHENYL)-N'-(4-((1-ETHYL-3-PIPERIDYL)AMINO)-6-METHYL-2-PYRIMIDINYL)GUANIDINE see WCJ750

N-(3,4-DICHLOROPHENYL)-N'-HYDROXYUREA see DGD085

N'-3,4-DICHLOROPHENYL-N⁵-ISOPROPYLDIGUANIDE HYDROCHLORIDE see DGD100

2,4-DICHLOROPHENYL-4-NITROPHENYL ETHER see DFT800

2,4-DICHLOROPHENYL-p-NITROPHENYL ETHER see DFT800

N-(3,4-DICHLOROPHENYL)-N'-METHYL-N'-METHOXYUREA see DGD600

DICHLOROPHENYLPHOSPHINE see DGE400

DICHLOROPHENYLPHOSPHINE SULFIDE see PFW200, PFW210

N-(3,4-DICHLOROPHENYL)PROPANAMIDE see DGI000

1-(2-(2,4-DICHLOROPHENYL)-2-(2-PROPENYLOXY)ETHYL)-1H-IMIDAZOLE see FPB875

N-(3,4-DICHLOROPHENYL)PROPIONAMIDE see DGI000

4,5-DICHLORO-2-PHENYL-3(2H)-PYRIDAZINONE see DGE800

1-(3,5-DICHLOROPHENYL)-2,5-PYRROLIDINEDIONE see DGF000

N-(3,5-DICHLOROPHENYL)SUCCINIMIDE see DGF000

2-(3,4-DICHLOROPHENYL)TETRAHYDRO-3-METHYL-4H-1,3-THIAZIN-4-ONE-1,1-DIOXIDE see DEM000

((2,5-DICHLOROPHENYLTHIO)METHANETHIOL)-S-ESTER with O,O-DIMETHYL PHOSPHORODITHIOATE see MNO750

2,5-DICHLOROPHENYLTHIOMETHYL O,O-DIETHYL PHOSPHORODITHIOATE see PDC750

S-((3,4-DICHLOROPHENYLTHIO)METHYL)-O,O-DIETHYL PHOSPHORODITH-IOATE see DJA200

S-(2,5-DICHLOROPHENYLTHIOMETHYL) O,O-DIETHYL PHOSPHORODITHIOATE see PDC750

S-(2,5-DICHLOROPHENYLTHIOMETHYL) DIETHYL PHOSPHOROTHIOLOTHION-ATE see PDC750

S-(((2,5-DICHLOROPHENYL)THIO)METHYL) O,O-DIMETHYL PHOSPHORODITH-IOATE see MNO750

(RS)-2-(3,5-DICHLOROPHENYL)-2-(2,2,2-TRICHLOROETHYL)OXIRANE see TJK100

DI-(p-CHLOROPHENYL)TRICHLOROMETHYLCARBINOL see BIO750

DICHLOROPHENYLTRICHLOROSILANE (DOT) see DGF200

(DICHLOROPHENYL)TRICHLOROSILANE see DGF200

DICHLOROPHOS see DGP900

DICHLOROPHOSPHORIC ACID, ETHYL ESTER see EOR000

3,6-DICHLOROPICOLINIC ACID see DGJ100

α-α-DICHLOROPINACOLIN see DGF300

DICHLOROPINACOLIN see DGF300

DICHLOROPINAKOLIN see DGF300

DICHLOROPROP see DGB000

1,1-DICHLOROPROPANE see DGF400

1,2-DICHLOROPROPANE see PNJ400

1,3-DICHLOROPROPANE see DGF800

2,2-DICHLOROPROPANE see DGF900

α,β-DICHLOROPROPANE see PNJ400

1,2-DICHLOROPROPANE mixed with 1,3-DICHLOROPROPENE and ISOTHIO-CYANATOMETHANE see MLC000

DICHLOROPROPANE-DICHLOROPROPENE MIXTURE see DGG000

2,3-DICHLORO-1-PROPANOL see DGG600

1,3-DICHLORO-2-PROPANOL see DGG400

2,3-DICHLOROPROPANOL see DGG600

1,2-DICHLOROPROPANOL-3 see DGG600

1,2-DICHLORO-3-PROPANOL see DGG600

1,3-DICHLORO-2-PROPANOL-2 (DOT) see DGG400

1,3-DICHLORO-2-PROPANOL PHOSPHATE (3:1) see FQU875

2,3-DICHLOROPROPANOL PHOSPHATE (3:1) see TNG750

1,1-DICHLOROPROPANONE see DGG500

1,3-DICHLORO-2-PROPANONE see BIK250

2,3-DICHLORO-1-PROPENE see DGH400

1,2-DICHLOROPROPENE see DGG800

1,3-DICHLOROPROPENE see DGG950

(Z)-1,3-DICHLOROPROPENE see DGH200

(E)-1,3-DICHLOROPROPENE see DGH225

2,3-DICHLOROPROPENE see DGH400

DICHLOROPROPENE (DOT) see DGG950

cis-1,3-DICHLOROPROPENE see DGH200

DICHLOROPROPENE see DGG700

trans-1,3-DICHLOROPROPENE see DGH225

1,3-DICHLOROPROPENE and 1,2-DICHLOROPROPANE MIXTURE see DGG000

cis-1,3-DICHLOROPROPENE OXIDE see DAC975

trans-1,3-DICHLOROPROPENE OXIDE see DGH500

2,3-DICHLORO-2-PROPENE-1-THIOL DIISOPROPYLCARBAMATE see DBI200

S-(2,3-DICHLORO-2-PROPENYL)ESTER, BIS(1-METHYLETHYL) CARBAMO-THIOIC ACID see DBI200

1,2-DICHLORO-3-PROPIONAL see DGH800

2,3-DICHLORO PROPIONALDEHYDE see DGH800

α,β-DICHLOROPROPIONALDEHYDE see DGH800

3',4'-DICHLOROPROPIONANILIDE see DGI000

3,4-DICHLOROPROPIONANILIDE see DGI000

DICHLOROPROPIONANILIDE see DGI000

2,2-DICHLOROPROPIONIC ACID see DGI400

α,α-DICHLOROPROPIONIC ACID see DGI400

α-DICHLOROPROPIONIC ACID see DGI400

2,2-DICHLOROPROPIONIC ACID, SODIUM SALT see DGI600

α,α-DICHLOROPROPIONIC ACID SODIUM SALT see DGI600

2,2-DICHLOROPROPIONIC ACID, 2-(2,4,5-TRICHLOROPHENOXY)ETHYL ESTER see PBK000

1,3-DICHLOROPROPYENE-1 see DGG950

1,2-DICHLOROPROPYLENE see DGG800

1,3-DICHLOROPROPYLENE see DGG950

2,3-DICHLOROPROPYLENE see DGH400

α,γ-DICHLOROPROPYLENE see DGG950

cis-1,3-DICHLOROPROPYLENE see DGH200

DICHLOROPROPYLENE see DGG700, DGG950, DGI700

trans-1,3-DICHLOROPROPYLENE see DGH225

3,6-DICHLORO-2-PYRIDINECARBOXYLIC ACID see DGJ100

4,5-DICHLOROPYROCATECHOL see DGJ200

3,4-DICHLORO-2,5-PYRROLIDINEDIONE see DFN800

5,7-DICHLORO-8-QUINALDINOL see CLC500

4,7-DICHLOROQUINOLINE see DGJ250

DICHLOROQUINOLINOL see DEL300

2,6-DICHLORO-p-QUINONE see DES400

2,6-DICHLOROQUINONE CHLOROIMIDE see CHR000

2,3-DICHLOROQUINOXALINE see DGJ950

2,3-DICHLOROQUINOXALINE-6-CARBONYLCHLORIDE see DGK000

DICHLOROSAL see CFY000

3,5-DICHLOROSALICYLIC ACID see DGK200

3,6-DICHLOROSALICYLIC ACID see DGK250

DICHLOROSILANE see DGK300

2,2'-DICHLORO-4,4'-STILBENAMINE see DGK400

2,2'-DICHLORO-4,4'-STILBENEDIAMINE see DGK400

3,3'-DICHLORO-4,4'-STILBENEDIAMINE see DGK600

DICHLORO-S-TRIAZINE-2,4,6(1H,3H,5H)-TRIONE POTASSIUM DERIV see PLD000

2,4-DICHLORO-5-SULFAMOYLBENZOIC ACID see DGK900

p-DICHLOROSULFAMOYLBENZOIC ACID see HAF000

3,4-DICHLORO-5-SULFAMYLBENZENESULFONAMIDE see DEQ200

p-(N,N-DICHLOROSULFAMYL)BENZOIC ACID see HAF000

DICHLOROSULFANE see SOG500

3,4-DICHLOROSULFOLANE see DGL200

2,4-DICHLORO-5-SULPHAMOYLBENZOIC ACID see DGK900

4,6-DICHLORO-2',4',5',7'-TETRABROMOFLUORESCEIN DIPOTASSIUM SALT see CMM000

DICHLOROTETRAFLUOROACETONE see DGL400

sym-DICHLOROTETRAFLUOROACETONE see DGL400

DICHLOROTETRAFLUOROETHANE (OSHA, ACGIH) see FOO509

1,2-DICHLORO-1,1,2,2-TETRAFLUOROETHANE (MAK) see FOO509

DICHLOROTETRAFLUOROETHANE see DGL600

sym-DICHLOROTETRAFLUOROETHANE see FOO509

1,3-DICHLORO-1,1,3,3-TETRAFLUORO-2-PROPANONE see DGL400

2,3-DICHLOROTETRAHYDROFURAN see DGL800

3,4-DICHLOROTETRAHYDROTHIOPHENE-1,1-DIOXIDE see DGL200

(2,3-DICHLORO-4-(2-THENOYL)PHENOXY)ACETIC ACID see TGA600

endo-2,5-DICHLORO-7-THIABICYCLO(2.2.1) HEPTANE see DGL875

(2,3-DICHLORO-4-(2-THIENYLCARBONYL)PHENOXY)ACETIC ACID see TGA600

2,6-DICHLOROTHIOBENZAMIDE see DGM600

DICHLOROTHIOCARBONYL see TFN500

DICHLOROTHIOLANE DIOXIDE see DGL200

(2,3-DICHLORO-4-(2-THIOPHENECARBONYL)PHENOXY)ACETIC ACID see TGA600

DICHLOROTITANOCENE see DGW200

2,4-DICHLOROTOLUENE see DGM700

α-α-DICHLOROTOLUENE see BAY300

N-(2,6-DICHLORO-m-TOLYL)ANTHRANILIC ACID see DGM875

N-(2,6-DICHLORO-m-TOLYL)ANTHRANILIC ACID ETHOXYMETHYL ESTER see DGN000

1,3-DICHLORO-s-TRIAZINE-2,4,6(1H,3H,5H)-TRIONE see DGN200

1,3-DICHLORO-s-TRIAZINE-2,4,6(1H,3H,5H)TRIONE POTASSIUM SALT see PLD000

2-(4,6-DICHLORO-s-TRIAZIN-2-YLAMINO)-4-(4-AMINO-3-SULFO-1-ANTHRAQUINO-NYLAMINO)BENZENESULFONIC ACID, DISODIUM see DGN400

4-(4,6-DICHLORO-s-TRIAZIN-2-YLAMINO)-5-HYDROXY-6-(2-HYDROXY-5-NITRO-PHENYLAZO)-2,7-NAPHTHALENEDISULFONIC A see DGN600

5-(3,5-DICHLORO-s-TRIAZINYLAMINO)-4-HYDROXY-3-PHENYLAZO-2,7-NA-PHTHALENEDISULFONIC ACID see DGN800

2-(6-(4,6-DICHLORO-s-TRIAZINYL)METHYLAMINO-1-HYDROXY-3-SULFONA-PHTHYLAZO)-1,5-NAPHTHALENEDISULFONIC ACID see DGO000

4,4'-DICHLORO-α-(TRICHLOROMETHYL)BENZHYDROL see BIO750

2,2'-DICHLOROTRIETHYLAMINE see BID250

2,2-DICHLORO-1,1,1-TRIFLUOROETHANE see TJY500

4,5-DICHLORO-2-TRIFLUOROMETHYLBENZIMIDAZOLE see DGO400

5,6-DICHLORO-2-TRIFLUOROMETHYLBENZIMIDAZOLE-1-CARBOXYLATE see DGA200

5,6-DICHLORO-2-(TRIFLUOROMETHYL)-1H-BENZIMIDAZOLE-1-CARBOXYLIC ACID PHENYL ESTER see DGA200

1,3-DICHLORO-6-TRIFLUOROMETHYL-9-(3-(DIBUTYLAMINO)-1-HYDROXYPRO-PYL)PHENANTHRENE HCl see HAF500

1-(1,3-DICHLORO-6-TRIFLUOROMETHYL-9-PHENANTHRYL)-3-(DI-N-BUTYLAMI-NO)PROPANOL HYDROCHLORIDE see HAF500

DICHLORO(m-TRIFLUOROMETHYLPHENYL)ARSINE see DGO600

DICHLOROTRIPHENYLANTIMONY see DGO800

DICHLOROTRIPHENYLSTIBINE see DGO800

DICHLOROVAS see DGP900

2,2-DICHLOROVINYL ALCOHOL, DIMETHYL PHOSPHATE see DGP900

DICHLOROVINYLARSINE CHLORIDE see BIQ250

DICHLOROVINYLCHLOROARSINE (DOT) see BIQ250

S-DICHLOROVINYL-I-CYSTEINE see DGP000

2,2-DICHLOROVINYL DIETHYL PHOSPHATE see DFG200

2,2-DICHLOROVINYL DIMETHYL PHOSPHATE see DGP900

2,2-DICHLOROVINYL DIMETHYL PHOSPHORIC ACID ESTER see DGP900

I-3-((1,2-DICHLOROVINYL)THIO)ALANINE see DGP000

DICHLOROVOS see DGP900

DICHLOROXIN see DEL300

5,7-DICHLOROXINE see DEL300

α,α'-DICHLOROXYLENE see XSS250

α,α'-DICHLORO-m-XYLENE see DGP200

α,α'-DICHLORO-o-XYLENE see DGP400

α,α'-DICHLORO-p-XYLENE see DGP600

DICHLOROXYLENE see XSS250

DICHLORPHEN see MJM500

DICHLORPHENAMIDE see DEQ200

2,4-DICHLORPHENOXYACETIC ACID see DAA800

2-(1-(2,6-DICHLORPHENOXY)ATHYL)-2-IMIDAZOLIN-HYDROCHLORID (GER-MAN) see LIA400

(2,4-DICHLOR-PHENOXY)-ESSIGSAEURE (GERMAN) see DAA800

2-(2,4-DICHLOR-PHENOXY)-PROPIONSAEURE (GERMAN) see DGB000

(+)-2-(2,4-DICHLORPHENOXY)PROPIONSAFEURE (GERMAN) see DGB100

(3,4-DICHLOR-PHENYL-AZO)-THIOHARNSTOFF (GERMAN) see DEQ000

3-(3,4-DICHLORPHENYL)-1-N-BUTYL-HARNSTOFF (GERMAN) see BRA250

2,4-DICHLORPHENYL "CELLOSOLVE" see DGP800

3-(3,4-DICHLOR-PHENYL)-1,1-DIMETHYL-HARNSTOFF (GERMAN) see DXQ500

3-(3,4-DICHLOR-PHENYL)-1-METHOXY-1-METHYL-HARNSTOFF (GERMAN) see DGD600

3-(4,5-DICHLORPHENYL)-1-METHOXY-1-METHYLHARNSTOFF (GERMAN) see DGD600

2,4,-DICHLORPHENYL-4-NITROPHENYLAETHER (GERMAN) see DFT800

DICHLORPHENYLOXYUREA see DGD085

1-(2-(2,4-DICHLORPHENYL)-2-(PROPENYLOXY)AETHYL)-1H-IMIDAZOLE see FPB875

DICHLORPHOS see DGP900

ω,ω-DICHLORPINAKOLIN see DGF300

DICHLORPROP see DGB000

DICHLORPROPAN-DICHLORPROPENGEMISCH (GERMAN) see DGG000

DICHLORPROPEN-GEMISCH (GERMAN) see DGG800

DICHLOR STAPENOR see DGE200

DICHLORSULFOFENYL-METHYLPYRAZOLON (CZECH) see DFQ200

DICHLOR-s-TRIAZIN-2,4,6(1H,3H,5H)TRIONE POTASSIUM see PLD000

DICHLORURE de TRIMETHYLAMMONIUM-1-(β-N-METHYLINDOYL-3'') ETHYL-4'-PYRIDINIUM)-3 PROPANE see MKW100

O-(2,2-DICHLORVINYL)-O,O-DIETHYLPHOSPHAT (GERMAN) see DFG200

(2,2-DICHLOR-VINYL)-DIMETHYL-PHOSPHAT (GERMAN) see DGP900

O-(2,2-DICHLORVINYL)-O,O-DIMETHYLPHOSPHAT (GERMAN) see DGP900

DICHLORVOS see DGP900

DICHLORVOS-ETHYL see DFG200

DICHLOSALE see DFV400

DICHLOTIAZID see CFY000

DICHLOTRIDE see CFY000

DICHOLINE SUCCINATE see CMG250

DICHROMIC ACID, ZINC SALT (1:1) see ZFA102

DICHROMIUM SULFATE see CMK415

DICHROMIUM SULPHATE see CMK415

DICHROMIUM TRIOXIDE see CMJ900

DICHROMIUM TRISULFATE see CMK415

DICHROMIUM TRISULPHATE see CMK415

DICHRONIC see DEO600

DICHYSTROLUM see DME300

DI(2-CIANOETIL)AMMINA (ITALIAN) see BIQ500

DI(2-CIANOETIL)METILAMMINA see MHQ750

DI(CIANOMETIL)AMMINA see IBB100

DICK (GERMAN) see DFH200

DICLOCIL see DGE200

DICLOFENAC SODIUM see DEO600

DICLONDAZOLIC ACID see DEL200

DICLONIA see BPR500

DICLOPHENAC SODIUM see DEO600

DICLORALUREA see DGQ200

DICLORAN see RDP300

S-(2,3-DICLORO-ALLIL)-N,N-DIISOPROPIL-MONOTIOCARBAMMATO (ITALIAN) see DBI200

1,4-DICLOROBENZENE (ITALIAN) see DEP800

p-DICLOROBENZENE (ITALIAN) see DEP800

1,1-DICLORO-2,2-BIS(4-CLORO-FENIL)-ETANO (ITALIAN) see BIM500

3,3'-DICLORO-4,4'-DIAMINODIFENILMETANO (ITALIAN) see MJM200

1,1-DICLOROETANO (ITALIAN) see DFF809

1,2-DICLOROETANO (ITALIAN) see EIY600

2,2'-DICLOROETILETERE (ITALIAN) see DFJ050

(3,4-DICLORO-FENIL-AZO)-TIOUREA (ITALIAN) see DEQ000

3-(3,4-DICLORO-FENIL)-1,1-DIMETIL-UREA (ITALIAN) see DXQ500

d-threo-2-DICLOROMETIL-4-((4'-NITROFENIL)-OSSIMETIL)-2-OSSAZOLINA (ITAL-IAN) see DFQ100

1,1-DICLORO-1-NITROETANO (ITALIAN) see DFU000

(2,2-DICLORO-VINIL)DIMETILFOSFATO (ITALIAN) see DGP900

DICLOTRIDE see CFY000

DICLOXACILLIN SODIUM MONOHYDRATE see DGE200

DICLOXACILLIN SODIUM SALT see DGE200

DICO see OOI000

DICOBALT BORIDE see DGQ300

DICOBALT CARBONYL see CNB500

DICOBALT EDETATE see DGQ400

DICOBALT EDTA see DGQ400

DICOBALT OCTACARBONYL see CNB500

DICOBALT OXIDE see CND825

DICOBALT TRIOXIDE see CND825

DICODID see OOI000

DICOFERIN see DGQ500

DICOFOL see BIO750

DICOL see DJD600

DICONIRT D see SGH500

DICOPHANE see DAD200

DICOPPER(I) ACETYLIDE see DGQ600

DICOPPER DICHLORIDE see CNK250

DICOPPER DIHYDROXYCARBONATE see CNJ750

DICOPPER(I)-1,5-HEXADIYNIDE see DGQ625

DICOPPER(I) KETENIDE see DGQ650

DICOPPER MONOXIDE see CNO000

DICOPUR see DAA800

DICOPUR-M see CIR250

DICORANTIL see DNN600

DICORTOL see PMA000

DICORVIN see DKA600

DICOTEX 80 see SIL500

DICOTEX see CIR250

DICOTOX see DAA800

DICOUMARIN see BJZ000

DICOUMAROL see BJZ000

DICRESOL see DGQ700

DICRESYL see MIB750

DICRESYL N-METHYLCARBAMATE see MIB750

DICRESYLOLPROPANE see IPK000

DICRODEN see MQH250

DICROTALIC ACID see HMC000

DICROTOFOS (DUTCH) see DGQ875

DICROTONYL PEROXIDE see DGQ859

DICROTOPHOS see DGQ875

DICRYL see DFO800

DICTYCIDE see COH250

DICTYZIDE see COH250

DICUMACYL see BKA000

DICUMAN see BJZ000

DICUMARINE see BJZ000

DICUMENE CHROMIUM see DGR200

DICUMENYLCHROMIUM see DGR200

DICUMYLMETHANE see DGR400

DI-α-CUMYL PEROXIDE see DGR600

DICUMYL PEROXIDE (DOT) see DGR600

DI-CUP see DGR600

DI-CUP 40 KF see DGR600

DI-CUPR see DGR600

DICUPRAL see DXH250

DICURAN see CIS250

DICURONE see GFM000

DICYANDIAMIDE-FORMALDEHYDE ADDUCT see CNH125

DICYANDIAMIDE-FORMALDEHYDE POLYMER see CNH125

DICYANDIAMIDE-FORMALDEHYDE RESIN see CNH125

DICYANDIAMIN see COP125

DICYANOACETYLENE see DGS000

2,2'-DICYANO-2,2'-AZOPROPANE see ASL750

1,2-DICYANOBENZENE see PHY000

1,3-DICYANOBENZENE see PHX550

1,4-DICYANOBENZENE see BBP250

m-DICYANOBENZENE see PHX550

o-DICYANOBENZENE see PHY000
p-DICYANOBENZENE see BBP250
1,4-DICYANOBUTANE see AER250
1,4-DICYANO-2-BUTENE see DGS200
β,β-DICYANO-o-CHLOROSTYRENE see CEQ600
DICYANODIAMIDE see COP125
DICYANODIAZENE see DGS300
2,2′-DICYANODIETHYLAMINE see BIQ500
β,β′-DICYANODIETHYL ETHER see OQQ000
β,β′-DICYANODIETHYL SULFIDE see DGS600
2,3-DICYANO-1,4-DITHIA-ANTHRAQUINONE see DLK200
1,2-DICYANOETHANE see SNE000
s-DICYANOETHANE see SNE000
trans-1,2-DICYANOETHENE see FOX000
DI-(2-CYANOETHYL)AMINE see BIQ500
(E)-1,2-DICYANOETHYLENE see FOX000
DI(2-CYANOETHYL)SULFIDE see DGS600
DICYANOFURAZAN see DGS700
DICYANOFURAZAN-N-OXIDE see DGS800
DICYANOFUROXAN see DGS800
DICYANOGEN see COO000
DICYANOGEN-N,N-DIOXIDE see DGT000
1,6-DICYANOHEXANE see OCW050
DICYANOMETHANE see MAO250
1,8-DICYANOOCTANE see SBK500
1,5-DICYANOPENTANE see HBD000
1,3-DICYANOPROPANE see TLR500
1,3-DICYANOTETRACHLOROBENZENE see TBQ750
DICYANOTRISULFIDE see SOI200
cis-DICYCLOBUTYLAMMINEDICHLOROPLATINUM(II) see DGT200
DICYCLOHEXANO-18-CROWN-6 see DGV100
DICYCLOHEXANO-24-CROWN-8 see DGT300
DICYCLOHEXYL ADIPATE see DGT500
DICYCLOHEXYLAMINE (DOT) see DGT600
N,N-DICYCLOHEXYLAMINE see DGT600
DICYCLOHEXYLAMINE NITRITE see DGU200
DICYCLOHEXYLAMINE PENTANOATE see DGU400
1-(2-(DICYCLOHEXYLAMINO)ETHYL)-1-METHYL-PIPERIDINIUM BROMIDE see MJC775
1-(2-(DICYCLOHEXYLAMINO)ETHYL)-1-METHYL-PIPERIDINIUM CHLORIDE see CAL075
DICYCLOHEXYLAMINONITRITE see DGU200
cis-DICYCLOHEXYLAMMINEDICHLOROPLATINUM(II) see DGU709
DICYCLOHEXYLAMMONIUM NITRITE see DGU200
N,N-DICYCLOHEXYL-2-BENZOTHIAZOLESULFENAMIDE see DGU800
DICYCLOHEXYL-18-CROWN-6 see DGV100
2-(2,2-DICYCLOHEXYLETHYL)PIPERIDINE MALEATE see PCH800
DICYCLOHEXYLFLUOROPHOSPHATE see DGV200
DICYCLOHEXYL FLUOROPHOSPHONATE see DGV200
DICYCLOHEXYL KETONE see DGV600
DICYCLOHEXYLMETHANE-4,4′-DIISOCYANATE see MJM600
DICYCLOHEXYL PEROXIDE CARBONATE see DGV650
DICYCLOHEXYL PEROXYDICARBONATE, technically pure (UN 2152) (DOT) see DGV650
DICYCLOHEXYL PEROXYDICARBONATE, not >91% with water (UN 2153) (DOT) see DGV650
DICYCLOHEXYL PHTHALATE see DGV700
DICYCLOHEXYLTIN OXIDE see DGV900
DICYCLOMINE HYDROCHLORIDE see ARR750
DICYCLOPENTADIENE see DGW000
DICYCLOPENTADIENE DIEPOXIDE see BGA250
DICYCLOPENTADIENE DIOXIDE see BGA250
DICYCLOPENTADIENYLCOBALT see BIR529
DICYCLOPENTADIENYLDICHLOROTITANIUM see DGW200
DICYCLOPENTADIENYLHAFNIUM DICHLORIDE see HAE500
DI-2,4-CYCLOPENTADIEN-1-YL IRON see FBC000
DICYCLOPENTADIENYL IRON (OSHA, ACGIH) see FBC000
DI-pi-CYCLOPENTADIENYLNICKEL see NDA500
DI-pi-CYCLOPENTADIENYLTITANIUM see TGH500
DICYCLOPENTADIENYLTITANIUMDICHLORIDE see DGW200
DICYCLOPENTA(c,lmn)PHENANTHREN-1(9H)-ONE, 2,3-DIHYDRO- see DGW300
DICYCLOPENTENYL ACRYLATE see DGW400
DICYCLOPENTENYLOXYETHYL METHACRYLATE see DGW450
α,α-DICYCLOPENTYL-ACETIC ACID-DIETHYLAMINO-ETHYLESTER BROMOC-TYLATE see PAR600
DICYCLOPENTYLACETIC ACID-β-DIETHYLAMINOETHYL ESTER ETHOBROM-IDE see DGW600
(2-(DICYCLOPENTYLACETOXY)ETHYL)TRIETHYLAMMONIUM BROMIDE see DGW600
N-(2-((DICYCLOPENTYLACETYL)OXY)ETHYL)-N,N-DIETHYL-1-OCTANAMINIUM BROMIDE (9CI) see PAR600
2-((DICYCLOPENTYLACETYL)OXY)-N,N,N-TRIETHYL-ETHANAMINIUM BROMIDE see DGW600
cis-DICYCLOPENTYLAMMINEDICHLOROPLATINUM(II) see BIS250

α,α-DICYCLOPENTYLESSIGSAURE-DIAETHYLAMINO-AETHYLESTER-BROMOC-TYLAT (GERMAN) see PAR600
DICYCLOPROPYLDIAZOMETHANE see DGW875
3,3-DICYCLOPROPYL-2-(ETHOXYCARBONYL)ACRYLONITRILE see DGX000
DICYKLOHEXYLAMIN (CZECH) see DGT600
DICYKLOHEXYLAMINKAPRONAT (CZECH) see DGU400
DICYKLOHEXYLAMIN NITRIT (CZECH) see DGU200
N,N-DICYKLOHEXYLBENZTHIAZOLSULFENAMID (CZECH) see DGU800
DICYKLOPENTADIEN (CZECH) see DGW000
DICYNENE see DIS600
DICYNIT (CZECH) see DGU200
DICYNONE see DIS600
DICYSTEINE see CQK325
DID 47 see OMY850
DID 95 see CJM750
DIDAKENE see PCF275
DIDANDIN see DVV600
DIDAN-TDC-250 see DKQ000
DIDECANOYLTRIETHYLENE GLYCOL ESTER (mixed isomers) see DAH450
DIDECYL DIMETHYL AMMONIUM CHLORIDE see DGX200
DI-N-DECYL PHTHALATE see DGX600
DIDECYL PHTHALATE see DGX600
7,8-DIDEHYDROCHOLESTEROL see DAK600
7,8-DIDEHYDRO-3-(2-(DIETHYLAMINO)ETHOXY)-4,5-α-EPOXY-17-METHYLMOR-PHINAN-6-α-OL see DIE300
9,10-DIDEHYDRO-N,N-DIETHYL-2-BROMO-6-METHYLERGOLINE-8-β-CARBOX-AMIDE see BNM250
9,10-DIDEHYDRO-N,N-DIETHYL-6-METHYL-ERGOLINE-8-β-CARBOXAMIDE see DJO000
9,10-DIDEHYDRO-N,N-DIETHYL-6-METHYL-ERGOLINE-8-β-CARBOXAMIDE-d-TARTRATE with METHANOL (1:2) see LJG000
12,13-DIDEHYDRO-13,14-DIHYDRO-α-ERYTHROIDINE see EDG500
8,9-DIDEHYDRO-6,8-DIMETHYLERGOLINE see AEY375
(5-α,6-α)-7,8-DIDEHYDRO-4,5-EPOXY-3-ETHOXY-17-METHYLMORPHINAN-6-OL see ENK000
7,8-DIDEHYDRO-4,5-α-EPOXY-3-ETHOXY-17-METHYLMORPHINAN-6-α-OL HY-DROCHLORIDE DIHYDRATE see ENK500
7,8-DIDEHYDRO-4,5-α-EPOXY-14-HYDROXY-3-METHOXY-17-METHYLMORPHI-NAN-6-ONE-N-OXIDE see DGY000
7,8-DIDEHYDRO-4,5-α-EPOXY-14-HYDROXY-3-METHOXY-17-METHYLMORPHINAN-5-α-6-ONE see HJX500
7,8-DIDEHYDRO-4,5-α-EPOXY-14-HYDROXY-3-METHOXY-17-METHYLMORPHI-NAN-6-ONE see HJX500
7,8-DIDEHYDRO-4,5-α-EPOXY-3-METHOXY-17-METHYL-MORPHINAN-6-α-OL PHOSPHATE SESQUIHYDRATE (3:3:2) see CNG675
7,8-DIDEHYDRO-4,5-α-EPOXY-17-METHYLMORPHINAN-3,6-α-DIOL see MRP000
7,8-DIDEHYDRO-4,5-α-EPOXY-17-METHYLMORPHINAN-3,6-α-DIOL HYDRO-CHLORIDE see MRO750
7,8-DIDEHYDRO-4,5-α-EPOXY-17-METHYLMORPHINE HYDROCHLORIDE see MRO750
7,8-DIDEHYDRO-4,5-α-EPOXY-17-METHYL-3-(2-MORPHOLINOETHOX-Y)MORPHINAN-6-α-OL see TCY750
7,8-DIDEHYDRO-4,5-α-EPOXY-17-METHYL-3-(2-PIPERIDINOETHOX-Y)MORPHINAN-6-α-OL see PIT600
9,10-DIDEHYDRO-N-ETHYL-1,6-DIMETHYLERGOLINE-8-β-CARBOXAMIDE see MLD500
16,17-DIDEHYDRO-21-ETHYL-4-METHYL-7,20-CYCLOATIDANE-11-β,15-β-DIOL see DAP875
9,10-DIDEHYDRO-N-ETHYL-6-METHYLERGOLINE-8-β-CARBOXAMIDE, N-ETHYL-LYSERGAMIDE see LJI000
3,8-DIDEHYDRO-HELIOTRIDINE see DAL060
9,10-DIDEHYDRO-N-(α-(HYDROXYMETHYL)ETHYL)-6-METHYLERGOLINE-8-β-CARBOXAMIDE see LJL000
9,10-(DIDEHYDRO-N-(1-HYDROXYMETHYL)PROPYL)-1,6-DIMETHYLERGOLINE-8-β-CARBOXAMIDE see MLD250
9,10-DIDEHYDRO-N-(α-(HYDROXYMETHYL)PROPYL)-6-METHYL-ERGOLINE-8-β-CARBOXAMIDE see PAM000
13,19-DIDEHYDRO-12-HYDROXY-SENECIONAN-11,16-DIONE see SBX500
13,19-DIDEHYDRO-12-HYDROXY-SENECIONAN-11,16-DIONE HYDROCHLORIDE see SBX525
5,6-DIDEHYDROISOANDROSTERONE see AOO450
13,14-DIDEHYDROMATRIDIN-15-ONE HYDROBROMIDE see SKS800
13,14-DIDEHYDRO-MATRIDIN-15-ONE MONOHYDROBROMIDE see SKS800
N′-((8-α)-9,10-DIDEHYDRO-6-METHYLERGOLIN-8-YL)-N,N-DIETHYL-UREA (Z)-2-BETENEDIOATE see LJE500
3-(9,10-DIDEHYDRO-6-METHYLERGOLIN-8-YL)-1,1-DIETHYLUREA HYDROGEN MALEATE see LJE500
3-(9,10-DIDEHYDRO-6-METHYLERGOLIN-8-α-YL)-1,1-DIETHYLUREA MALEATE (1:1) see LJE500
16,17-DIDEHYDRO-19-METHYLOXAYOHIMBAN-16-CARBOXYLIC ACID METHYL ESTER see AFG750
3,8-DIDEHYDRORETRONECINE see DAL400
9,10-DIDEHYDRO-N,N,6-TRIMETHYLERGOLINE-8-β-CARBOXAMIDE see LJH000
4,4′-DIDEMETHYL-4,4′-DI-2-PROPENYLTOXIFERINE I DICHLORIDE see DBK400
DIDEMNIN B see DHA300

2-DIETHYLAMINOETHYL DIPHENYLTHIOLACETATE HYDROCHLORIDE see DHY400

2-DIETHYLAMINOETHYL-2,2-DIPHENYLVALERATE HYDROCHLORIDE see PBM500

S-DIETHYLAMINOETHYL ESTER-p-BROMOTHIOBENZO HYDROXIMIC ACID HYDROCHLORIDE see DHZ000

2-DIETHYLAMINOETHYLESTER KYSELINY DIFENYLOCTOVE see DHX800

2-DIETHYLAMINOETHYL ESTER KYSELINY DIFENYLOCTOVE (CZECH) see THK000

2-DIETHYLAMINOETHYLESTER KYSELINY METHAKRYLOVE see DIB300

2-(DIETHYLAMINO)ETHYL ESTER-2-PHENYL BENZENE ACETIC ACID see THK000

1-(2-(DIETHYLAMINO)ETHYL)-2-(p-ETHOXYBENZYL)-5-BENZIMIDAZOLYL METHYL KETONE see DHZ050

1-(2-(DIETHYLAMINO)ETHYL)-2-(p-ETHOXYBENZYL)-5-NITROBENZIMIDAZOLE HYDROCHLORIDE see EQN750

1-(2-(DIETHYLAMINO)ETHYL)-2-((p-ETHOXYPHENYL)THIO)BENZIMIDAZOLE HYDROCHLORIDE see DHZ100

N-(2-(DIETHYLAMINO)ETHYL)-2-ETHOXY-N-(2,6-XYLYL)CINCHONINAMIDE HYDROCHLORIDE see DIA000

N-(2-(DIETHYLAMINO)ETHYL)-2-ETHOXY-N-(2,6-XYLYL)-4-QUINOLINECARBOXAMIDE HYDROCHLORIDE see DIA000

5-(2-(DIETHYLAMINO)ETHYL)-3-(α-ETHYLBENZYL)-1,2,4-OXADIAZOLECITRATE see POF500

3-(2-(DIETHYLAMINO)ETHYL)-5-ETHYL-5-PHENYLBARBITURIC ACID HYDROCHLORIDE see HEL500

N-(2-(DIETHYLAMINO)ETHYL)-2-ETHYL-2-PHENYLMALONAMIC ACID ETHYL ESTER see FAJ150

3-(2-(DIETHYLAMINO)ETHYL)-5-(2-FURYL)-1-PHENYL-1H-PYRAZOLINE HYDROCHLORIDE see DIA400

N-(2-(DIETHYLAMINO)ETHYL)-2,2′,4,4′,6,6′-HEXAMETHYLBENZANILIDE HYDROCHLORIDE see DIA800

DIETHYLAMINOETHYL-3-HYDROXY-4-AMINOBENZOATE see DHO600

7-(β-DIETHYLAMINOETHYL)-8-(α-HYDROXYBENZYL) THEOPHYLLINE HYDROCHLORIDE see BTA500

N-(2-(DIETHYLAMINO)ETHYL)-β-ISODURYLAMIDE HYDROCHLORIDE see DII400

2-(DIETHYLAMINO)ETHYLMERCAPTAN see DIY600

β-DIETHYLAMINOETHYL MERCAPTAN see DIY600

N-(2-(DIETHYLAMINO)ETHYL)-N-MESITYL-β-ISODURYLAMIDE HYDROCHLORIDE see DIA800

1-(2-(DIETHYLAMINO)ETHYL)-3-MESITYL-1-METHYLUREA HYDROCHLORIDE see DIB000

2-(DIETHYLAMINO)-N-ETHYL-N-(1-MESITYLOXY-2-PROPYL)ACETAMIDE HYDROCHLORIDE see DIB200

2-(N,N-DIETHYLAMINO)ETHYL METHACRYLATE see DIB300

2-(DIETHYLAMINO)ETHYL METHACRYLATE see DIB300

β-(DIETHYLAMINO)ETHYL METHACRYLATE see DIB300

DIETHYLAMINOETHYL METHACRYLATE see DIB300

N-(DIETHYLAMINOETHYL)-2-METHOXY-4-AMINO-5-BROMOBENZAMIDE see VCK100

N-(DIETHYLAMINOETHYL)-2-METHOXY-4-AMINO-5-CHLOROBENZAMIDE see AJH000

N-(2-(DIETHYLAMINO)ETHYL)-2-METHOXY-5-(METHYLSULFONYL)BENZAMIDE HYDROCHLORIDE see TGA375

N-(2-(DIETHYLAMINO)ETHYL)-2-(p-METHOXYPHENOXY)ACETAMIDE see MCH250, DIB600

2-(DIETHYLAMINO)-N-ETHYL-N-(1-(p-METHOXYPHENOXY)-2-PROPYL)ACETAMIDE HYDROCHLORIDE see DIB800

5-(2-(DIETHYLAMINO)ETHYL)-3-(p-METHOXYPHENYL)-1,2,4-OXADIAZOLE see CPN750

2-(2-(DIETHYLAMINO)ETHYL)METHYLAMINO-o-ACETANISIDIDE DIHYDROCHLORIDE see DIC000

2-(2-(DIETHYLAMINO)ETHYL)METHYLAMINO-o-ACETOTOLUIDIDE DIHYDROCHLORIDE see DIC200

2-(2-(DIETHYLAMINO)ETHYL)METHYLAMINO-2′,6′-ACETOXYLIDIDE DIHYDROCHLORIDE see DIC400

2-((2-(DIETHYLAMINO)ETHYL)METHYLAMINO)ETHYL-2-METHYL-1,3-BENZODIOXOLE DIMALEATE see DIC600

2-(2-(DIETHYLAMINO)ETHYL)-2-METHYL-1,3-BENZODIOXOLE HYDROCHLORIDE see DIC800

3-(β-DIETHYLAMINOETHYL)-4-METHYL-7-(CARBETHOXYMETHOXY)-COUMARIN HYDROCHLORIDE see CBR500

1-DIETHYLAMINOETHYL-2-METHYL-3-PHENYL-1,2,3,4-TETRAHYDRO-4-QUINAZOLINONE see DID800

2-DIETHYLAMINOETHYL-3-METHYL-2-PHENYLVALERATE METHYLBROMIDE see VBK000

N-(2-(DIETHYLAMINO)ETHYL)-5-(METHYLSULFONYL)-o-ANISAMIDE HYDROCHLORIDE see TGA375

1-DIETHYLAMINOETHYL-2-METHYL-1,2,3,4-TETRAHYDRO-3-(o-TOLYL)-4-QUINAZOLINONE OXALATE see DIE200

DIETHYLAMINOETHYLMORPHINE see DIE300

N-DIETHYLAMINOETHYL-β-(1-NAPHTHYL)-β-TETRAHYDROFURYL ISOBUTYRATE see NAE000

2-DIETHYLAMINOETHYL-N′-PHENYLISOTHIURONIUM BROMIDE HYDROBROMIDE see DAZ135

o-(2-(DIETHYLAMINO)ETHYL)OXIME-5H-DIBENZO(a,d)CYCLOHEPTEN-5-ONE MONOHYDROCHLORIDE see LHX515

1-(2-(DIETHYLAMINO)ETHYL)-2-p-PHENETIDINO-5-BENZIMIDAZOLYL METHYL KETONE see DIE350

10-(2-DIETHYLAMINOETHYL)PHENOTHIAZINE see DII200

β-DIETHYLAMINOETHYL PHENOTHIAZINE-10-CARBOXYLATE HYDROCHLORIDE see THK600

β-DIETHYLAMINOETHYL PHENOTHIAZINE-N-CARBOXYLATE HYDROCHLORIDE see THK600

10-(2-DIETHYLAMINO)ETHYLPHENOTHIAZINE HYDROCHLORIDE see DHF600

DIETHYLAMINOETHYLPHENO-THIAZINYL-10-DITHIOCARBOXYLATE see PDP500

2-(DIETHYLAMINO)-N-ETHYL-N-(1-PHENOXY-2-PROPYL)ACETAMIDE HYDROCHLORIDE see DIE400

3-(DIETHYLAMINO)-N-ETHYL-N-(1-PHENOXY-2-PROPYL)PROPIONAMIDE HYDROCHLORIDE see DIE600

3-(β-DIETHYLAMINOETHYL)-3-PHENYL-2-BENZOFURANONE HYDROCHLORIDE see DIF200

DIETHYLAMINOETHYL-1-PHENYLCYCLOPENTANE-1-CARBOXYLATE ETHANE DISULFONATE see CBG250

DIETHYLAMINOETHYL-1-PHENYLCYCLOPENTANE-1-CARBOXYLATE HYDROCHLORIDE see PET250

N-(2-DIETHYLAMINOETHYL)-2-PHENYLGLYCINE ISOPENTYL ESTER see NOC000

β-DIETHYLAMINOETHYL-2-PHENYLHEXAHYDROBENZOATE HYDROCHLORIDE see KDU100

2-DIETHYLAMINOETHYL-2-PHENYL-3-METHYLVALERATE METHYL BROMIDE see VBK000

5-β-DIETHYLAMINOETHYL-3-PHENYL-1,2,4-OXADIAZOLE CITRATE see OOE000

5-(2-(DIETHYLAMINO)ETHYL)-3-PHENYL-1,2,4-OXADIAZOLE HYDROCHLORIDE see OOE100

2-(DIETHYLAMINO)ETHYL)-2-PHENYL-4-PENTENOIC ACID ETHYL ESTER see DIF600

5-(2-(DIETHYLAMINO)ETHYL)-3-(α-PHENYLPROPYL)-1,2,4-OXADIAZOLE CITRATE see POF500

2-(2-(DIETHYLAMINO)ETHYL)-1-PHENYL-2-THIOPSEUDOUREA DIHYDROBROMIDE see DAZ135

2-DIETHYLAMINOETHYL-α-PHENYLVALERATE HYDROCHLORIDE see PGP500

(2-DIETHYLAMINO)ETHYLPHOSPHOROTHIOIC ACID-O,O-DIETHYL ESTER see DJA400

S-(2-(DIETHYLAMINO)ETHYL)PHOSPHOROTHIOIC ACID-O,O-DIETHYL ESTER see DJA400

2-DIETHYLAMINOETHYL-α-PROPYLTOLUATE HYDROCHLORIDE see PGP500

1-(2-(DIETHYLAMINO)ETHYL)RESERPINE BITARTRATE see DIH000

2-(DIETHYLAMINO)ETHYL TETRAHYDRO-α-(1-NAPHTHYLMETHYL)-2-FURANPROPIONATE see NAE000

2-(DIETHYLAMINO)ETHYL TETRAHYDRO-α-(1-NAPHTHYLMETHYL)-2-FURANPROPIONATE OXALATE (1:1) see NAE100

7-(2-DIETHYLAMINO)ETHYLTHEOPHYLLINE see CNR125

7-(DIETHYLAMINOETHYL)THEOPHYLLINE see CNR125

DIETHYLAMINOETHYL THEOPHYLLINE see CNR125

7-(2-(N,N-DIETHYLAMINO)ETHYL)THEOPHYLLINE HYDROCHLORIDE see DIH600

7-(2-(DIETHYLAMINO)ETHYL)THEOPHYLLINE HYDROCHLORIDE see DIH600

2-((2-(DIETHYLAMINO)ETHYL)THIOACETANILIDE HYDROCHLORIDE see DIH800

2-(2-(DIETHYLAMINO)ETHYL)THIO-o-ACETOTOLUIDIDE HYDROCHLORIDE see DII000

N-(DIETHYLAMINOETHYL)THIODIPHENYLAMINE see DII200

N-(2-(DIETHYLAMINO)ETHYL)-2,4,6-TRIMETHYLBENZAMIDE HYDROCHLORIDE see DII400

β-DIETHYLAMINOETHYL XANTHENE-9-CARBOXYLATE METHOBROMIDE see DJM800

β-DIETHYLAMINOETHYL-9-XANTHENECARBOXYLATE METHOBROMIDE see DJM800

β-DIETHYLAMINOETHYL 9-XANTHENECARBOXYLATE METHOBROMIDE see XCJ000

β-DIETHYLAMINOETHYL XANTHENE-9-CARBOXYLATE METHOBROMIDE see XCJ000

N-(2-(DIETHYLAMINO)ETHYL)-2,6-XYLIDINE DIHYDROCHLORIDE see DII600

2-(DIETHYLAMINO)-N-ETHYL-N-(1-(2,4-XYLYLOXY)-2-PROPYL)ACETAMIDE HYDROCHLORIDE see DII800

2-(DIETHYLAMINO)-N-ETHYL-N-(1-(3,5-XYLYLOXY)-2-PROPYL) ACETAMIDE HYDROCHLORIDE see DIJ000

1-(2-(DIETHYLAMINO)ETHYL)-3-(2,6-XYLYL)UREA HYDROCHLORIDE see DIJ200

2-(DIETHYLAMINO)-N-(4-(2-FLUOROBENZOYL)-1,3-DIMETHYL-1H-PYRAZOL-5-YL)ACETAMIDE HYDROCHLORIDE see AAI100

3-(4-DIETHYLAMINO-2-HYDROXYPHENYLAZO)-4-HYDROXYBENZENESULFONIC ACID see DIJ300

4-(DIETHYLAMINO)-2-ISOPROPYL-2-PHENYLVALERONITRILE see DIK000

2-(DIETHYLAMINO)-N-(1-MESITYLOXY-2-PROPYL)-N-METHYLACETAMIDE HYDROCHLORIDE see DIK200

N,N-DIETHYLCARBANILIDE see DJC400
DIETHYL CARBINOL see IHP010
DIETHYLCARBINOL see IHP010
DIETHYL CARBITOL see DIW800
1,1'-DIETHYL-4,4'-CARBOCYANINE IODIDE see KHU050
DIETHYL CARBONATE (DOT) see DIX200
DIETHYL CARBONATE see DIX200
DIETHYLCETONE (FRENCH) see DJN750
O,O-DIETHYL-O-(2-CHINOXALYL)PHOSPHOROTHIOATE see DJY200
O,O-DIETHYL-O-(3-CHLOOR-4-METHYL-CUMARIN-7-YL)MONOTHIOFOSFAAT
    (DUTCH) see CNU750
N,N-DIETHYLCHLORACETAMIDE see DIX400
DIETHYLCHLOROALUMINUM see DHI885
O,O-DIETHYL-O-(2-CHLORO-1-(2',4'-DICHLOROPHENYL)VINYL) PHOSPHATE
    see CDS750
O,O-DIETHYL-O-(2-CHLORO-1,2,5-DICHLOROPHENYLVINYL) PHOSPHOROTH-
    IOATE see DIX600
DIETHYL(2-CHLOROETHYL)AMINE see CGV500
DIETHYL-β-CHLOROETHYLAMINEHYDROCHLORIDE see CLQ250
DIETHYL-3-CHLORO-4-METHYL-7-COUMARINYL PHOSPHATE see CIK750
O,O-DIETHYL-O-(3-CHLORO-4-METHYLCOUMARIN-7-YL)PHOSPHATE see
    CIK750
O,O-DIETHYL-O-(3-CHLORO-4-METHYL-7-COUMARINYL)PHOSPHOROTHIOATE
    see CNU750
O,O-DIETHYL-O-(3-CHLORO-4-METHYLCOUMARINYL-7) THIOPHOSPHATE see
    CNU750
O,O-DIETHYL-O-(3-CHLORO-4-METHYL-2-OXO-2H-BENZOPYRAN-7-
    YL)PHOSPHOROTHIOATE see CNU750
S,S-DIETHYL(CHLOROMETHYL)PHOSPHONODITHIOATE see BJD000
O,O-DIETHYL-3-CHLORO-4-METHYL-7-UMBELLIFERONE THIOPHOSPHATE see
    CNU750
O,O-DIETHYL-O-(3-CHLORO-4-METHYLUMBELLIFERYL)PHOSPHOROTHIOATE
    see CNU750
DIETHYL-3-CHLORO-4-METHYLUMBELLIFERYL THIONOPHOSPHATE see
    CNU750
DIETHYL CHLOROPHOSPHATE see DIY000
DIETHYLCHLOROTHIOPHOSPHATE see DJW600
DIETHYL-2-CHLOROVINYL PHOSPHATE see CLV375
O,O-DIETHYL-O-(2-CHLOROVINYL) PHOSPHATE see CLV375
DIETHYLCHLORTHIOFOSFAT (CZECH) see DJW600
DIETHYLCYANOPHOSPHATE see DJW800
DIETHYL CYANOPHOSPHONATE see DJW800
P,P-DIETHYL CYCLIC ETHYLENE ESTER OF PHOSPHONODITHIOIMIDOCAR-
    BONIC ACID see PGW750
p,p-DIETHYL CYCLIC PROPYLENE ESTER of PHOSPHONODITHIOIMIDOCAR-
    BONIC ACID see DHH400
DIETHYLCYCLOHEXANE (mixed isomers) see DIY200
DIETHYLCYSTEAMIN see DIY600
N,N-DIETHYL CYSTEAMINE see DIY600
N-DIETHYL CYSTEAMINE see DIY600
DIETHYLCYSTEAMINE see DIY600
DIETHYL-1,10-DECANEDIOATE see DJY600
DIETHYL DECANEDIOATE see DJY600
N,N-DIETHYL-1,3-DIAMINOPROPANE see DIY800
DIETHYLDIAZENE-1-OXIDE see ASP000
DIETHYL DICARBONATE see DIZ100
O,O-DIETHYL-O-(2,4-DICHLOOR-FENYL)-MONOTHIOFOSFAAT (DUTCH) see
    DFK600
O,O-DIETHYL O-2,5-DICHLORO-4-BROMOPHENYL-PHOSPHOROTHIOATE see
    EGV500
O,O-DIETHYL O-(2,5-DICHLORO-4-BROMOPHENYL) THIOPHOSPHATE see
    EGV500
O,O-DIETHYL-O-2,4,5-DICHLORO-(METHYLTHIO)PHENYL THIONOPHOSPHATE
    see CLJ875
O,O-DIETHYL-O-(2,4-DICHLOROPHENYL) PHOSPHOROTHIOATE see DFK600
DIETHYL 2,4-DICHLOROPHENYL PHOSPHOROTHIONATE see DFK600
O,O-DIETHYL-O-2,4-DICHLOROPHENYL THIOPHOSPHATE see DFK600
DIETHYLDICHLOROSILANE (DOT) see DEY800
DIETHYLDICHLOROSTANNANE see DEZ000
DIETHYL DI(DIMETHYLAMIDO)PYROPHOSPHATE (symmetrical) see DIV200
DIETHYL DI(DIMETHYLAMIDO)PYROPHOSPHATE (unsymmetrical) see DJA300
O,O-DIETHYL O-(2-DIETHYLAMINO-6-METHYL-4-PYRIMIDIN-
    YL)PHOSPHOROTHIOATE see DIN600
2,2'-DIETHYLDIHEXYLAMINE see DJA800
O,O-DIETHYL-(1,2-DIHYDRO-1,3-DIOXO-2H-ISOINDOL-2-
    YL)PHOSPHONOTHIOATE see DJX200
5,5-DIETHYLDIHYDRO-2H-1,3-OXAZINE-2,4(3H)-DIONE see DJT400
O,O-DIETHYL O-(2,3-DIHYDRO-3-OXO-2-PHENYL-6-PYRIDAZI-
    NYL)PHOSPHOROTHIOATE see POP000
DIETHYLDIIODOSTANNANE see DJB000
DIETHYL (DIMETHOXYPHOSPHINOTHIOYLTHIO) BUTANEDIOATE see MAK700
DIETHYL (DIMETHOXYPHOSPHINOTHIOYLTHIO)SUCCINATE see MAK700
α,α'-DIETHYL-4,4'-DIMETHOXYSTILBENE see DJB200
trans-α,α'-DIETHYL-4,4'-DIMETHOXYSTILBENE see DJB200

3',4'-DIETHYL-4-DIMETHYLAMINOAZOBENZENE see DJB400
8,8-DIETHYL-N,N-DIMETHYL-2-AZA-8-GERMASPIRO(4.5)DECANE-2-PROPANA-
    MINE DIHYDROCHLORIDE see SLD800
p'-DIETHYL-p-DIMETHYL THIOPYROPHOSPHATE see DJB600
p-DIETHYL-p'-DIMETHYLTHIOPYROPHOSPHATE see DJB600
$N^3,N^3$-DIETHYL-2,4-DINITRO-6-(TRIFLUOROMETHYL)-1,3-BENZENEDIAMINE see
    CNE500
3,3-DIETHYL-2,4-DIOXO-5-METHYLPIPERIDINE see DNW400
DIETHYLDIPHENYL DICHLOROETHANE see DJC000
1,3-DIETHYL-1,3-DIPHENYLUREA see DJC400
sym-DIETHYLDIPHENYLUREA see DJC400
DIETHYLDISULFID (CZECH) see DJC600
DIETHYLDISULFIDE see DJC600
N,N-DIETHYL-4,4-DI-2-THIENYL-3-BUTEN-2-AMINE HYDROCHLORIDE see
    DJP500
N,N-DIETHYL-3,3-DI-2-THIENYL-1-METHYLALLYLAMINE HYDROCHLORIDE see
    DJP500
DIETHYLDITHIO BIS(THIONOFORMATE) see BJU000
DIETHYL DITHIOCARBAMATE see DJC800
DIETHYLDITHIOCARBAMATE SODIUM see SGJ000
DIETHYLDITHIOCARBAMIC ACID see DJC800
DIETHYLDITHIOCARBAMIC ACID ANHYDROSULFIDE with DIMETHYLTHIO-
    CARBAMIC ACID see DJC875
DIETHYLDITHIOCARBAMIC ACID-2-CHLOROALLYL ESTER see CDO250
DIETHYLDITHIOCARBAMIC ACID DIETHYLAMINE SALT see DJD000
DIETHYLDITHIOCARBAMIC ACID LEAD(II) SALT see DJD200
DIETHYLDITHIOCARBAMIC ACID SELENIUM(II) SALT see DJD400
DIETHYLDITHIOCARBAMIC ACID SODIUM see SGJ000
DIETHYLDITHIOCARBAMIC ACID, SODIUM SALT see SGJ000
DIETHYLDITHIOCARBAMIC ACID SODIUM SALT TRIHYDRATE see SGJ500
DIETHYLDITHIOCARBAMIC ACID TELLURIUM SALT see EPJ000
DIETHYLDITHIOCARBAMIC ACID ZINC SALT see BJC000
DIETHYLDITHIOCARBAMIC ANHYDRIDE of O,O-DIISOPROPYL THIONOPHOS-
    PHORIC ACID see DKB600
DIETHYLDITHIOCARBAMIC ANHYDROSULFIDE see DKB600
DIETHYLDITHIOCARBAMIC SODIUM TRIHYDRATE see SGJ500
DIETHYLDITHIOCARBAMINIC ACID see DJC800
2-(N,N-DIETHYLDITHIOCARBAMYL)BENZOATHIAZOLE see BDF250
O,O-DIETHYLDITHIOFOSFORECNAN SODNY (CZECH) see PHG750
O,O-DIETHYL 1,3-DITHIOLAN-2-YLIDENEPHOSPHORAMIDOTHIOATE see
    DXN600
DIETHYL-N-1,3-DITHIOLANYL-2-IMINO PHOSPHATE see DXN600
DIETHYLDITHIONE see DJC800
O,O-DIETHYL-DITHIOPHOSPHORIC ACID, p-CHLOROPHENYLTHIOMETHYL ES-
    TER see TNP250
O,O-DIETHYLDITHIOPHOSPHORYLACETIC ACID-N-MONOISOPROPYLAMIDE
    see IOT000
3-DIETHYLDITHIOPHOSPHORYLMETHYL-6-CHLOROBENZOXAZOLONE-2 see
    BDJ250
DIETHYL DIXANTHOGEN see BJU000
DIETHYL EMME see EEV200
DIETHYLENDIAMINE see DPJ200
1,4-DIETHYLENEDIAMINE see PIJ000
N,N-DIETHYLENE DIAMINE (DOT) see PIJ000
1,4-DIETHYLENE DIOXIDE see DVQ000
DIETHYLENE DIOXIDE see DVQ000
DIETHYLENE ETHER see DVQ000
DIETHYLENE GLYCOL see DJD600
DIETHYLENE GLYCOL, BIS(ALLYL CARBONATE)- see AGD250
DIETHYLENE GLYCOL, BISCHLOROFORMATE see OPO000
DIETHYLENE GLYCOL BISPHTHALATE see DJD700
DIETHYLENE GLYCOL-n-BUTYL ETHER see DJF200
DIETHYLENE GLYCOL BUTYL ETHER ACETATE see BQP500
DIETHYLENE GLYCOL DIACETATE see DJD750
DIETHYLENE GLYCOL DIACRYLATE see ADT250
DIETHYLENE GLYCOL DI(3-AMINOPROPYL) ETHER see DJD800
DIETHYLENE GLYCOL DIBENZOATE see DJE000
DIETHYLENE GLYCOL DI-n-BUTYL ETHER see DDW200
DIETHYLENE GLYCOL DIBUTYL ETHER see DDW200
DIETHYLENE GLYCOL, DIESTER with BUTYLPHTHALATE see DDY400
DIETHYLENE GLYCOL DIETHYL ETHER see DIW800
DIETHYLENE GLYCOL DIMETHYL ETHER see BKN750
DIETHYLENE GLYCOL DINITRATE, containing at least 25% phlegmatizer
    (DOT) see DJE400
DIETHYLENE GLYCOL DINITRATE see DJE400
DIETHYLENE GLYCOL DIVINYL ETHER see DJE600
DIETHYLENE GLYCOL ETHYL ETHER see CBR000
DIETHYLENE GLYCOL ETHYL METHYL ETHER see DJE800
DIETHYLENE GLYCOL ETHYLVINYL ETHER see DJF000
DIETHYLENE GLYCOL-n-HEXYL ETHER see HFN000
DIETHYLENE GLYCOL METHYL ETHER see DJG000
DIETHYLENE GLYCOL MONOBUTYL ETHER see DJF200
DI(ETHYLENE GLYCOL MONOBUTYL ETHER)PHTHALATE see DJF400
DIETHYLENE GLYCOL MONO-2-CYANOETHYL ETHER see DJF600
DIETHYLENE GLYCOL, MONOESTER with STEARIC ACID see HKJ000

DIETHYLENE GLYCOL MONOETHYL ETHER see CBR000
DIETHYLENE GLYCOL MONOETHYL ETHER ACETATE see CBQ750
DIETHYLENE GLYCOL MONOHEPTYL ETHER see HBP275
DIETHYLENE GLYCOL MONOHEXYL ETHER see HFN000
DIETHYLENE GLYCOL, MONO(HYDROGEN MALEATE) see MAL500
DIETHYLENE GLYCOL MONOISOBUTYL ETHER see DJF800
DIETHYLENE GLYCOL MONOMETHYL ETHER see DJG000
DIETHYLENE GLYCOL MONOMETHYL ETHER ACETATE see MIE750
DIETHYLENE GLYCOL-MONO-2-METHYLPENTYL ETHER see DJG200
DIETHYLENE GLYCOL MONOMETHYLPENTYL ETHER see DJG200
DIETHYLENE GLYCOL MONOPHENYL ETHER see PEQ750
DIETHYLENE GLYCOL MONOSTEARATE see HKJ000
DIETHYLENE GLYCOL MONOVINYL ETHER see DJG400
DIETHYLENE GLYCOL PHENYL ETHER see PEQ750
DIETHYLENE GLYCOL STEARATE see HKJ000
DIETHYLENE GLYCOL VINYL ETHER see DJG400
DIETHYLENE IMIDE OXIDE see MRP750
DIETHYLENE IMIDOXIDE see MRP750
DIETHYLENE IMINEAMIDOTHIOPHOSPHORIC ACID see DNU850
2,6-DIETHYLENEIMINO-4-CHLOROPYRIMIDINE see EQI600
DI(ETHYLENE OXIDE) see DVQ000
DIETHYLENE OXIDE see TCR750
DIETHYLENE OXIMIDE see MRP750
1,3,-DI(ETHYLENESULPHAMOYL)PROPANE see BJP899
DIETHYLENETRIAMINE see DJG600
1,1,4,7,7-DIETHYLENETRIAMINEPENTAACETIC ACID see DJG800
DIETHYLENETRIAMINEPENTAACETIC ACID see DJG800
DIETHYLENETRIAMINE PENTAACETIC ACID, CALCIUM TRISODIUM SALT see CAY500
DIETHYLENETRIAMINEPENTA(METHYLENEPHOSPHONIC ACID), SODIUM SALT see DJG700
(DIETHYLENETRINITRILO)PENTAACETIC ACID see DJG800
DIETHYLENEUREA see BJP450
N,N′-DIETHYLENEUREA see BJP450
DIETHYLEN-GLYCOL MONOVINYL ESTER see DJG400
DIETHYLENGLYKOLDINITRATE (CZECH) see DJE400
DIETHYLENIMIDE OXIDE see MRP750
α,α′-DIETHYL-α,α′-EPOXYBIBENZYL-4,4′-DIOL see DKB100
DIETHYL-β,γ-EPOXYPROPYLPHOSPHONATE see DJH200
DIETHYLESTER KYSELINY ACETYLAMINOMALONOVE (CZECH) see AAK750
DIETHYLESTER KYSELINY SIROVE see DKB110
DIETHYL ESTER of PYROCARBONIC ACID see DIZ100
DIETHYL ESTER SULFURIC ACID see DKB110
N,N-DIETHYLETHANAMINE see TJO000
N,N-DIETHYL-1,2-ETHANEDIAMINE see DJI400
DIETHYL ETHANEDIOATE see DJT200
(R*,S*)-4,4′-(1,2-DIETHYL-1,2-ETHANEDIYL)BIS-BENZENESULFONIC ACID DIPOTASSIUM SALT see SPA650
DIETHYL ETHANE PHOSPHONITE see DJH500
N,N-DIETHYLETHANOLAMINE see DHO500
DIETHYLETHANOLAMINE see DHO500
(E)-1,1′-(1,2-DIETHYL-1,2-ETHENE-DIYL)BIS(4-METHOXYBENZENE) see DJB200
4,4′-(1,2-DIETHYL-1,2-ETHENEDIYL)BIS-PHENOL see DKA600
trans-4,4′-(1,2-DIETHYL-1,2-ETHENEDIYL)BISPHENOL see DKA600
4,4′-(1,2-DIETHYL-1,2-ETHENEDIYL)BISPHENOL-(E)-BIS(DIHYDROGEN PHOSPHATE) see DKA200
trans-4,4′-(1,2-DIETHYL-1,2-ETHENEDIYL)BISPHENOL DIPROPIONATE see DKB000
DIETHYL ETHER (DOT) see EJU000
DIETHYLETHEROXODIPEROXOCHROMIUM(VI) see DJH800
DIETHYLETHOXYALUMINUM see EER000
O,O-DIETHYL S-(N-ETHOXYCARBONYL-N-METHYLCARBAMOYLMETHYL) PHOSPHOROTHIOLOTHIONATE see DJI000
DIETHYL (ETHOXYMETHYLENE)MALONATE see EEV200
O,O-DIETHYL-O-(2-ETHTHIOETHYL)PHOSPHOROTHIOATE see DAO500
DIETHYL 2-ETHTHIOETHYL THIONOPHOSPHATE see DAO500
N,N-DIETHYLETHYLENEDIAMINE see DJI400
4,4′-(1,2-DIETHYLETHYLENE)DIPHENOL see DLB400
O,O-DIETHYL 2-ETHYLMERCAPTOETHYL THIOPHOSPHATE see DAO600
O,O-DIETHYL-2-ETHYLMERCAPTOETHYL THIOPHOSPHATE, THIONO ISOMER see SPF000
(T-4)-DIETHYL[2-(1-ETHYL-1-(2H-PYRROL-5-YL)PROPYL)-1H-PYRROLATO-N′,N²]-BORON see MRW275
O,O-DIETHYL S-(2-(ETHYLSULFINYL)ETHYL) PHOSPHORODITHIOATE see OQS000
O,O-DIETHYL-O-(2-ETHYLSULFONYLETHYL)PHOSPHOROTHIOATE see SPF000
DIETHYL-2-ETHYLSULFONYLETHYL THIONOPHOSPHATE see SPF000
O,O-DIETHYL-2-ETHYLTHIOETHYL PHOSPHORODITHIOATE see DXH325
O,O-DIETHYL-2-ETHYLTHIO ETHYL PHOSPHOROTHIOATE see DAO500
O,O-DIETHYL-O-2-(ETHYLTHIO)ETHYL PHOSPHOROTHIOATE see DAO500
DIETHYL-2-(ETHYLTHIO(ETHYL PHOSPHOROTHIONATE)) see DAO500
O,O-DIETHYL-ETHYLTHIOMETHYL PHOSPHORODITHIOATE see PGS000
N,N′-DIETHYL-p-FENYLENDIAMIN see DJV200
DIETHYL FLUOROPHOSPHATE see DJJ400
DIETHYL FORMAMIDE see DJJ600

DIETHYL FUMARATE see DJJ800
DIETHYL GALLIUM HYDRIDE see DJJ829
N,N-DIETHYLGLYCINE-2,6-DIMETHOXYPHENYL ESTER HYDROCHLORIDE see FAC166
N,N-DIETHYLGLYCINE MESITYL ESTER HYDROCHLORIDE see FAC170
N,N-DIETHYLGLYCINE-2,6-XYLYL ESTER HYDROCHLORIDE see FAC050
N,N-DIETHYLGLYCINONITRILE see DHJ600
DIETHYL GLYCOL DIMETHYL ETHER see BKN750
DIETHYLGOLD BROMIDE (DOT) see DJJ850
DIETHYL GOLD BROMIDE see DJJ850
DIETHYLGUANIDINE HYDROCHLORIDE DIHYDRATE see DJJ875
3,3′-DIETHYLHEPTAMETHINETHIACYANINE IODIDE see DJK000
DIETHYL HEXAFLUOROGLUTARATE see DJK100
DIETHYL HEXANEDIOATE see AEP750
DI-2-ETHYLHEXYL ADIPATE see AEO000
DI(2-ETHYLHEXYL)AMINE see DJA800
2-DI-(2-ETHYLHEXYL)AMINOETHANOL see DJK200
O,O′-DI(2-ETHYLHEXYL) DITHIOPHOSPHORIC ACID see DJK400
DI-(2-ETHYLHEXYL) ETHER see DJK600
DI(2-ETHYLHEXYL) FUMARATE see DVK600
DI-2-ETHYLHEXYL ISOPHTHALATE see BJQ750
DI-(2-ETHYLHEXYL)MALEATE see BJR000
DI(2-ETHYLHEXYL)ORTHOPHTHALATE see DVL700
DI(2-ETHYLHEXYL) PEROXYDICARBONATE see DJK800
DI(2-ETHYLHEXYL)PHENYL PHOSPHATE see BJR625
DI(2-ETHYLHEXYL)PHOSPHATE see BJR750
DI-2(ETHYLHEXYL)PHOSPHORIC ACID see BJR750
DI-(2-ETHYLHEXYL)PHOSPHORIC ACID (DOT) see BJR750
DI(2-ETHYLHEXYL)PHTHALATE see DVL700
DI-2-ETHYLHEXYL)SEBACATE see BJS250
DI-(2-ETHYLHEXYL) SODIUM SULFOSUCCINATE see DJL000
DI-2-ETHYLHEXYLTIN DICHLORIDE see DJL200
DI(2-ETHYLHEXYL)TIN DICHLORIDE see DJL200
1,2-DIETHYLHYDRAZINE see DJL400
sym-DIETHYLHYDRAZINE see DJL400
N-N′-DIETHYLHYDRAZINE see DJL400
1,2-DIETHYLHYDRAZINE DIHYDROCHLORIDE see DJL600
DIETHYL HYDROGEN PHOSPHITE see DJW400
DIETHYLHYDROXY ARSINE OXIDE see DIS850
N,N-DIETHYL-2-HYDROXYBENZAMIDE see DJY400
N,N-DIETHYL-2-((HYDROXYDIPHENYLACETYL)METHYLAMINO)-N-METHYL-ETHANAMINIUM BROMIDE (9CI) see BCP685
N,N-DIETHYL-N-(β-HYDROXYETHYL)AMINE see DHO500
DIETHYL(2-HYDROXYETHYL)AMMONIUM CHLORIDE DIPHENYLCARBAMATE see DHY200
DIETHYL(2-HYDROXYETHYL)METHYLAMMONIUM BROMIDE α-PHENYLCYCLOHEXANEGLYCOLATE see ORQ000
DIETHYL(2-HYDROXYETHYL)METHYLAMMONIUMBROMIDE XANTHENE-9-CARBOXYLATE see XCJ000
DIETHYL(2-HYDROXYETHYL)METHYLAMMONIUM BROMIDE XANTHENE-9-CARBOXYLATE see DJM800
DIETHYL(2-HYDROXYETHYL)METHYLAMMONIUM-3-METHYL-2-PHENYLVALERATE BROMIDE see VBK000
N,N-DIETHYLHYDROXYLAMINE see DJN000
DIETHYLHYDROXYLAMINE see DJN000
O,O-DIETHYL N-HYDROXYNAPHTHALIMIDE PHOSPHATE see HMV000
DIETHYL(m-HYDROXYPHENYL)ARSINE METHIODIDE see DJN400
DIETHYL(m-HYDROXYPHENYL)METHYLAMMONIUM BROMIDE see HNK000
DIETHYL(m-HYDROXYPHENYL)METHYLAMMONIUM IODIDE DIMETHYLCARBAMATE see HNK500
DIETHYL(m-HYDROXYPHENYL)METHYLARSONIUM IODIDE see DJN400
O,O-DIETHYL-7-HYDROXY-3,4-TETRAMETHYLENE COUMARINYL PHOSPHOROTHIOATE see DXO000
DIETHYL HYDROXYTIN HYDROPEROXIDE see DJN489
4,4′-(1,2-DIETHYLIDENE-1,2-ETHANEDIYL)BISPHENOL see DAL600
4,4′-(DIETHYLIDENEETHYLENE)DIPHENOL see DAL600
p,p′-(DIETHYLIDENEETHYLENE)DIPHENOL see DAL600
DIETHYL ISOPHTHALATE see IMK100
O,O-DIETHYL-O-(2-ISOPROPYL-4-METHYL-PYRIMIDIN-6-YL)MONOTHIOFOSFAAT (DUTCH) see DCM750
O,O-DIETHYL-O-(2-ISOPROPYL-4-METHYL-6-PYRIMIDIN-YL)PHOSPHOROTHIOATE see DCM750
O,O-DIETHYL-O-(2-ISOPROPYL-6-METHYL-4-PYRIMIDINYL) PHOSPHOROTHIOATE see DCM750
DIETHYL 4-(2-ISOPROPYL-6-METHYLPYRIMIDINYL)PHOSPHOROTHIONATE see DCM750
O,O-DIETHYL-O-(2-ISOPROPYL-4-METHYL-6-PYRIMIDYL)PHOSPHOROTHIOATE see DCM750
O,O-DIETHYL-O-(2-ISOPROPYL-4-METHYL-6-PYRIMIDYL) THIONOPHOSPHATE see DCM750
O,O-DIETHYL-2-ISOPROPYL-4-METHYLPYRIMIDYL-6-THIOPHOSPHATE see DCM750
DIETHYLKETENE see DJN700
O,O-DIETHYL-O-(2-KETO-4-METHYL-7-α′,β′-BENZO-α′-PYRANYL) THIOPHOSPHATE see PKT000

DIETHYL KETONE see DJN750
DIETHYL LEAD DIACETATE see DJN800
DIETHYL LEAD DINITRATE see DJN875
N,N-DIETHYLLEUCINON-p-AMINOBENZOIC ACID METHANESULFONATE see LEU000
N,N-DIETHYLLYSERGAMIDE see DJO000
DIETHYL MAGNESIUM see DJO100
DIETHYL MALEATE see DJO200
DIETHYL MALONATE (FCC) see EMA500
DIETHYLMALONYLUREA see BAG000
DIETHYLMALONYLUREA SODIUM see BAG250
DIETHYL(2-MERCAPTOETHYL)AMINE see DIY600
DIETHYL MERCAPTOSUCCINATE-O,O-DIMETHYL DITHIOPHOSPHATE, S-ESTER see MAK700
DIETHYL MERCAPTOSUCCINATE-O,O-DIMETHYL PHOSPHORODITHIOATE see MAK700
DIETHYL MERCAPTOSUCCINATE-O,O-DIMETHYL THIOPHOSPHATE see MAK700
DIETHYL MERCAPTOSUCCINATE-S-ESTER with O,O-DIMETHYLPHOSPHORODITHIOATE see MAK700
DIETHYL MERCAPTOSUCCINIC ACID O,O-DIMETHYL PHOSPHORODITHIOATE see MAK700
DIETHYL MERCURY see DJO400
N,N-DIETHYLMETANILAN SODNY (CZECH) see DHM000
N,N-DIETHYL-2-(4-(6-METHOXY-2-PHENYL-1H-INDEN-3-YL)PHENOXY)-ETHANAMINE HYDROCHLORIDE see MFG260
N,N-DIETHYL-3-(4-METHOXYPHENYL)-1,2,4-OXADIAZOLE-5-ETHANAMINE see CPN750
(((2-(DIETHYLMETHYLAMMONIO)-1-METHYL)ETHOXY)ETHYL)TRIMETHYLAMMONIUM DIIODIDE see MQF750
5,5-DIETHYL-1-METHYLBARBITURIC ACID see DJO800
N,N-DIETHYL-3-METHYLBENZAMIDE see DKC800
DIETHYL METHYL CARBINOLURETHAN see ENF000
O,O-DIETHYL S-(N-METHYL-N-CARBOETHOXYCARBAMOYLMETHYL) DITHIOPHOSPHATE see DJI000
O,O-DIETHYL-O-(4-METHYLCOUMARIN-7-YL)-MONOTHIOFOSFAAT (DUTCH) see PKT000
O,O-DIETHYL-O-(4-METHYL-7-COUMARINYL) PHOSPHOROTHIOATE see PKT000
O,O-DIETHYL-O-(4-METHYL-7-COUMARINYL) THIONOPHOSPHATE see PKT000
O,O-DIETHYL-O-(4-METHYLCOUMARINYL-7) THIOPHOSPHATE see PKT000
1,1-DIETHYL-2-METHYL-3-DIPHENYLMETHYLENEPYRROLIDINIUM BROMIDE see PAB750
N,N-DIETHYL-1-METHYL-3,3-DI-2-THIENYLALLYLAMINE HYDROCHLORIDE see DJP500
DIETHYL (4-METHYL-1,3-DITHIOLAN-2-YLIDENE)PHOSPHOROAMIDATE see DHH400
O,O-DIETHYL-O-6-METHYL-2-ISOPROPYL-4-PYRIMIDINYL PHOSPHOROTHIOATE see DCM750
O,O-DIETHYL-O-(4-METHYL-7-KUMARINYL) ESTER KYSELINY THIOFOSFORESCNE (CZECH) see PKT000
DIETHYLMETHYL METHANE see MNI500
DIETHYLMETHYL(2-(N-METHYLBENZILAMIDO)ETHYL)AMMONIUM BROMIDE see BCP685
O,O-DIETHYL-O-(3-METHYL-4-(METHYLTHIO)PHENYL)PHOSPHOROTHIOATE see LIN400
1,1-DIETHYL-3-METHYL-3-NITROSOUREA see DJP600
O,O-DIETHYL-O-(4-METHYL-2-OXO-2H-1-PHOSPHOROTHIOIC ACID BENZOPYRAN-7-YL)ESTER (9CI) see PKT000
N,N'-DI(1-ETHYL-3-METHYLPENTYL)-p-PHENYLENEDIAMINE see BJT500
N,N-DIETHYL-α-METHYL-10H-PHENOTHIAZINE-10-ETHANAMINE see DIR000
N,N-DIETHYL-2-(2-(2-METHYL-5-PHENYL-1H-PYRROL-1-YL)PHENOXY)-ETHANAMINE see LEF400
DIETHYLMETHYLPHOSPHINE see DJQ200
O,S-DIETHYL METHYLPHOSPHONOTHIOATE see DJR700
N,N-DIETHYL-4-METHYL-1-PIPERAZINECARBOXAMIDE see DIW000
N,N-DIETHYL-4-METHYL-1-PIPERAZINE CARBOXAMIDE CITRATE see DIW200
N,N-DIETHYL-4-METHYL-1-PIPERAZINECARBOXAMIDE DIHYDROGEN CITRATE see DIW200
N,N-DIETHYL-4-METHYL-1-PIPERAZINECARBOXAMIDE-2-HYDROXY-1,2,3-PROPANETRICARBOXYLATE see DIW200
3,3-DIETHYL-5-METHYLPIPERIDINE-2,4-DIONE see DNW400
3,3-DIETHYL-5-METHYL-2,4-PIPERIDINEDIONE see DNW400
O,O-DIETHYL-O-(3-METHYL-1H-PYRAZOL-5-YL)-FOSFAAT (DUTCH) see MOX250
DIETHYL-3-METHYL-5-PYRAZOLYL PHOSPHATE see MOX250
O,O-DIETHYL-O-(3-METHYL-5-PYRAZOLYL) PHOSPHATE see MOX250
5,5-DIETHYL-1-METHYL-2,4,6(1H,3H,5H)-PYRIMIDINETRIONE see DJO800
3,3'-DIETHYL-9-METHYLSELENOCARBOCYANINE IODIDE see DJQ300
O,O-DIETHYL-O-(p-(METHYLSULFINYL)PHENYL) PHOSPHOROTHIOATE see FAQ800
O,O-DIETHYL-O-p-(METHYLSULFINYL)PHENYL THIOPHOSPHATE see FAQ800
DIETHYLMETHYLSULFONIUM IODIDEMERCURIC IODIDE (ADDITION COMPOUND) see DJQ800

DIETHYLMETHYL SULFONIUM IODINE with MERCURY IODIDE (1:1) see DJQ800
N,N-DIETHYL-4-METHYLTETRAMETHYLENEDIAMINE see DJQ850
O,S-DIETHYL METHYLTHIOPHOSPHONATE see DJR700
O,O-DIETHYL-O-(4-(METHYLTHIO)-3,5-XYLYL)PHOSPHOROTHIOATE see DJR800
N,N-DIETHYL-5-METHYL-(1,2,4)TRIAZOLO(1,5-a)PYRIMIDINE-7-AMINE see DIO200
N,N-DIETHYL-N-METHYL-2-(2-(TRIMETHYLAMMONIO)ETHOXY)-1-PROPANAMINIUM DIIODIDE see MQF750
O,O-DIETHYL-O-(4-METHYLUMBELLIFERONE) ESTER OF THIOPHOSPHORIC ACID see PKT000
O,O-DIETHYL-O-(4-METHYLUMBELLIFERONE) PHOSPHOROTHIOATE see PKT000
DIETHYL (4-METHYLUMBELLIFERYL) THIONOPHOSPHATE see PKT000
N,N-DIETHYL-N-METHYL-2-((9H-XANTHEN-9-YLCARBONYL)OXY)ETHANAMINIUM BROMIDE see DJM800
α,α-DIETHYL-1-NAPHTHALENEACETIC ACID SODIUM SALT see DJS100
O,O-DIETHYL-o-NAPHTHALIMIDE PHOSPHOROTHIOATE see NAQ500
O,O-DIETHYL-o-NAPHTHALOXIMIDO PHOSPHOROTHIOATE see NAQ500
O,O-DIETHYL-o-NAPHTHALOXIMIDOPHOSPHOROTHIONATE see NAQ500
O,O-DIETHYL-o-NAPHTHYLAMIDOPHOSPHOROTHIOATE see NAQ500
N,N-DIETHYL-N'-(2,5-DIETHOXYPHENYL)-N'-(4-BUTOXYPHENOXYACETYL) ETHYLENEDIAMINE HCl see BPM750
N,N'-DIETHYL-N,N'-DINITROSOETHYLENEDIAMINE see DJB800
N,N'-DIETHYL-N,N'-DIPHENYLUREA see DJC400
N,N-DIETHYL-N'-(2-(2-ETHYL-1,3-BENZODIOXOL-2-YL)ETHYL) ETHYLENEDIAMINE DIMALEATE see DJI200
DIETHYL-NICOTAMIDE see DJS200
N,N-DIETHYLNICOTINAMIDE see DJS200
N,N-DIETHYL-N'-2-INDANYL-N'-PHENYL-1,3-PROPANEDIAMINE HYDROCHLORIDE see FBP850
DIETHYLNITRAMINE see DJS500
O,O-DIETHYL-O-(4-NITRO-FENIL)-MONOTHIOFOSFAAT (DUTCH) see PAK000
DIETHYL-p-NITROFENYL ESTER KYSELINY FOSFORECNE (CZECH) see NIM500
O,O-DIETHYL-O-p-NITROFENYLESTER KYSELINYTHIOFOSFORECNE (CZECH) see PAK000
O,O-DIETHYL-O-p-NITROFENYLFOSFAT (CZECH) see NIM500
O,O-DIETHYL-O-p-NITROFENYLTIOFOSFAT (CZECH) see PAK000
O,O-DIETHYL O-p-NITROPHENYL PHOSPHATE see NIM500
DIETHYL p-NITROPHENYL PHOSPHATE see NIM500
O,S-DIETHYL-O-(4-NITROPHENYL)PHOSPHOROTHIOATE see DJT000
O,O-DIETHYL-O-4-NITROPHENYLPHOSPHOROTHIOATE see PAK000
O,O-DIETHYL-O-(4-NITROPHENYL) PHOSPHOROTHIOATE see PAK000
O,S-DIETHYL-O-(p-NITROPHENYL) PHOSPHOROTHIOATE see DJT000
O,O-DIETHYL-O-(p-NITROPHENYL) PHOSPHOROTHIOATE see PAK000
O,S-DIETHYL-O-(4-NITROPHENYL)PHOSPHOROTHIOIC ACID ESTER see DJT000
O,S-DIETHYL-O-(p-NITROPHENYL)PHOSPHOROTHIOIC ACID ESTER see DJT000
DIETHYL-4-NITROPHENYL PHOSPHOROTHIONATE see PAK000
O,O-DIETHYL-O-(p-NITROPHENYL)THIONOPHOSPHATE see PAK000
DIETHYL-p-NITROPHENYLTHIONOPHOSPHATE see PAK000
O,S-DIETHYL-O-(4-NITROPHENYL)THIOPHOSPHATE see DJT000
O,O-DIETHYL-O-4-NITROPHENYL THIOPHOSPHATE see PAK000
O,O-DIETHYL-O-p-NITROPHENYL THIOPHOSPHATE see PAK000
DIETHYL-p-NITROPHENYLTHIOPHOSPHATE see PAK000
N,N-DIETHYLNITROSAMINE see NJW500
DIETHYLNITROSAMINE see NJW500
DIETHYLNITROSOAMINE see NJW500
O,N-DIETHYL-N-NITROSOHYDROXYLAMINE see NKC500
N,N-DIETHYL-N'-((8-α)-6-METHYLERGOLIN-8-YL)UREA (Z)-2-BUTENEDIOATE see DLR150
N,N-DIETHYL-N'-(1-NITRO-9-ACRIDINYL)-1,2-ETHANEDIAMINE DIHYDROCHLORIDE (9CI) see NFW100
N,N-DIETHYL-N'-(1-NITRO-9-ACRIDINYL)-1,3-PROPANEDIAMINE DIHYDROCHLORIDE (9CI) see NFW200
N,N-DIETHYL-N'-(2-(2-PHENYL-1,3-BENZODIOXOL-2-YL)ETHYL)ETHYLENEDIAMINE DIMALEATE see DJU800
N,N-DIETHYL-N'-PHENYLETHYLENEDIAMINE see DJV300
N,N-DIETHYL-N'-((8-α)-6-PROPYLERGOLIN-8-YL)UREA see DJX300
N,N-DIETHYL-N'-((8-α)-6-PROPYLERGOLIN-8-YL)UREA (Z)-2-BUTENEDIOATE see DJX350
N,N-DIETHYL-N'-(1,2,3,4-TETRAHYDRO-1-NAPHTHYL)ACETAMIDE see TGI725
N,N-DIETHYL-N'-2-(TETRAHYDRO-1,2,3,4-NAPHTHYL)-GLYCINAMIDE see TGI725
N,N-DIETHYL-N'-2-(TETRALYL)-GLYCINAMIDE see TGI725
N,N-DIETHYL-N,N',N'-TETRAMETHYL-N,N'-(2-METHYL-3-OXAPENTAMETHYLENE)BIS(AMMONIUM IODIDE) see MQF750
DIETHYL OCTAFLUOROADIPATE see DJT100
DIETHYL OCTAFLUOROHEXANEDIOATE see DJT100
DIETHYLOLAMINE see DHF000
O,O-DIETHYL-O,2-PYRAZINYL PHOSPHOROTHIOATE see EPC500

O,O-DIETHYL-S-ETHYLMERCAPTOMETHYL DITHIOPHOSPHONATE see PGS000

O,O-DIETHYL-S-ETHYL PHOSPHOROTHIOATE see TJU800

O,O-DIETHYL-S-((ETHYLSULFINYL)ETHYL)PHOSPHORODITHIOATE see OQS000

O,O-DIETHYL-S-(2-ETHYLTHIO-ETHYL)-DITHIOFOSFAAT (DUTCH) see DXH325

O,O-DIETHYL-S-(2-ETHYLTHIO-ETHYL)-MONOTHIOFOSFAAT (DUTCH) see DAP200

O,O-DIETHYL-S-2-(ETHYLTHIO)ETHYL PHOSPHORODITHIOATE see DXH325

O,O-DIETHYL-S-(2-ETHYLTHIOETHYL) PHOSPHORODITHIOATE see DXH325

O,O-DIETHYL-S-(2-ETHYLTHIO)ETHYL PHOSPHOROTHIOATE see DAP200

O,O-DIETHYL-S-(2-(ETHYLTHIO)ETHYL) PHOSPHOROTHIOLATE (USDA) see DAP200

O,O-DIETHYL-S-(2-ETHYLTHIOETHYL) THIOTHIONOPHOSPHATE see DXH325

O,O-DIETHYL-S-ETHYLTHIOMETHYL DITHIOPHOSPHONATE see PGS000

O,O-DIETHYL-S-ETHYLTHIOMETHYL THIOTHIONOPHOSPHATE see PGS000

O,O-DIETHYL-S-(N-ISOPROPYLCARBAMOYLMETHYL) DITHIOPHOSPHATE see IOT000

O,O-DIETHYL-S-ISOPROPYLCARBAMOYLMETHYL PHOSPHORODITHIOATE see IOT000

O,O-DIETHYL-S-(N-ISOPROPYLCARBAMOYLMETHYL) PHOSPHORODITHIOATE see IOT000

O,O-DIETHYL-S-2-ISOPROPYLMERCAPTOMETHYLDITHIOPHOSPHATE see DJN600

O,O-DIETHYL-S-(ISOPROPYLMERCAPTOMETHYL) PHOSPHORODITHIOATE see DJN600

O,O-DIETHYL-S-(ISOPROPYLTHIOMETHYL) PHOSPHORODITHIOATE see DJN600

O,O-DIETHYL-S-(3-METHYL-2,4-DIOXO-5-OXA-3-AZA-HEPTYL)-DITHIOFOSFAAT (DUTCH) see DJI000

O,O-DIETHYL-S-p-NITROFENYLESTER KYSELINY THIOFOSFORECNE (CZECH) see DJS800

O,O-DIETHYL-S-(4-NITROPHENYL) PHOSPHOROTHIOATE see DJS800

O,O-DIETHYL-S-(4-NITROPHENYL)PHOSPHOROTHIOIC ACID ESTER see DJS800

O,O-DIETHYL-S-(4-NITROPHENYL)THIOPHOSPHATE see DJS800

DIETHYL SODIUM DITHIOCARBAMATE see SGJ000

O,O-DIETHYL-S-(4-OXO-3H-1,2,3-BENZOTRIAZINE-3-YL)-METHYL-DITHIOPHOSPHATE see EKN000

O,O-DIETHYL-S-(4-OXOBENZOTRIAZINO-3-METHYL)PHOSPHORODITHIOATE see EKN000

O,O-DIETHYL-S-((4-OXO-3H-1,2,3-BENZOTRIAZIN-3-YL)-METHYL)-DITHIO FOSFAAT (DUTCH) see EKN000

DIETHYLSTANNIUM DIIODIDE see DJB000

DIETHYLSTANNIUMDIJODID (GERMAN) see DJB000

DIETHYLSTANNYL DICHLORIDE see DEZ000

N,N-DIETHYL-4-STILBENAMINE see DKA000

2,2'-DIETHYL-4,4'-STILBENEDIOL see DKA600

α,α'-DIETHYL-(E)-4,4'-STILBENEDIOL see DKA600

α,α'-DIETHYL-4,4'-STILBENEDIOL see DKA600

trans-α,α'-DIETHYL-4,4'-STILBENEDIOL see DKA600

α,α'-DIETHYLSTILBENEDIOL see DKA600

α,α'-DIETHYL-4,4'-STILBENEDIOL BIS(DIHYDROGEN PHOSPHATE) see TEE300

α,α'-DIETHYL-(E)-4,4'-STILBENEDIOL BIS(DIHYDROGEN PHOSPHATE) see DKA200

α,α'-DIETHYL-4,4'-STILBENEDIOL DIPALMITATE see DKA800

α,α'-DIETHYL-4,4'-STILBENEDIOL, DIPROPIONATE see DKB000

trans-α,α'-DIETHYL-4,4'-STILBENEDIOL DIPROPIONATE see DKB000

α,α'-DIETHYL-4,4'-STILBENEDIOL trans-DIPROPIONATE see DKB000

α,α'-DIETHYL-4,4'-STILBENEDIOL DIPROPIONYL ESTER see DKB000

α,α'-DIETHYL-4,4'-STILBENEDIOL DISODIUM SALT see DKA400

DIETHYLSTILBENE DIPROPIONATE see DKB000

DIETHYLSTILBESTEROL see DKA600

trans-DIETHYLSTILBESTEROL see DKA600

DIETHYLSTILBESTEROL DIPHOSPHATE see DKA200

DIETHYLSTILBESTEROL DIPROPIONATE see DKB000

DIETHYLSTILBESTEROL DISODIUM SALT see DKA400

DIETHYLSTILBESTROL see DKA600

trans-DIETHYLSTILBESTROL see DKA600

DIETHYLSTILBESTROL DIMETHYL ETHER see DJB200

DIETHYLSTILBESTROL DIPALMITATE see DKA800

DIETHYLSTILBESTROL DIPHOSPHATE see DKA200

DIETHYLSTILBESTROL DIPHOSPHATE TETRASODIUM see TEE300

DIETHYLSTILBESTROL DIPROPIONATE see DKB000

DIETHYLSTILBESTROL and ETHISTERONE see EEI050

DIETHYLSTILBESTROL, IODINE DERIVATIVE see TDE000

DIETHYLSTILBESTROL PHOSPHATE see DKA200

DIETHYLSTILBESTROL PHOSPHATE TETRASODIUM see TEE300

DIETHYLSTILBESTROL and PRANONE see EEI050

DIETHYLSTILBESTROL PROPIONATE see DKB000

DIETHYLSTILBESTRYL DIPHOSPHATE see DKA200

DIETHYLSTILBOESTEROL see DKA600

trans-DIETHYLSTILBOESTEROL see DKA600

DIETHYLSTILBOESTROL-3,4-OXIDE see DKB100

DIETHYLSTILBOESTROL-α,β-OXIDE see DKB100

O,O-DIETHYL-S-2-TRIMETHYLAMMONIUM ETHYLPHOSPHONOTHIOLATE IODIDE see TLF500

DIETHYL SUCCINATE (FCC) see SNB000

DIETHYL SULFATE see DKB110

DIETHYLSULFID (CZECH) see EPH000

DIETHYL SULFIDE (DOT) see EPH000

DIETHYL SULFIDE-2,2'-DICARBOXYLIC ACID see BHM000

DIETHYL SULFITE see DKB119

DIETHYLSULFONDIMETHYLMETHANE see ABD500

DIETHYLSULFONMETHYLETHYLMETHANE see BJT750

DIETHYL SULPHOXIDE see EPI500

DIETHYL TELLURIDE see DKB150

DIETHYLTELLURIUM see DKB150

DIETHYL TEREPHTHALATE see DKB160

α,α'-DIETHYL-3,3',5,5'-TETRAFLUORO-4,4'-STILBENEDIOL (E)- see TCH325

trans-α,α'-DIETHYL-3,3',5,5'-TETRAFLUORO-4,4'-STILBENEDIOL see TCH325

2,5-DIETHYLTETRAHYDROFURAN see DKB165

5,5-DIETHYLTETRAHYDRO-2H-1,3-OXAZINE-2,4(3H)-DIONE see DJT400

O,O-DIETHYL-O-(7,8,9,10-TETRAHYDRO-6-OXOBENZO(C)CHROMAN-3-YL)PHOSPHOROTHIOATE see DXO000

O,O-DIETHYL-O-(7,8,9,10-TETRAHYDRO-6-OXO-6H-DIBENZO(b,d)PYRAN-3-YL)PHOSPHOROTHIOATE see DXO000

O,O-DIETHYL-O-(3,4-TETRAMETHYLENECOUMARINYL-7) THIOPHOSPHATE see DXO000

DIETHYL TETRAOXOSULFATE see DKB110

DIETHYL THALLIUM PERCHLORATE see DKB175

3,3'-DIETHYLTHIADICARBOCYANINE IODIDE see DJT800

DIETHYLTHIADICARBOCYANINE IODIDE see DJT800

DIETHYLTHIAMBUTENE HYDROCHLORIDE see DJP500

N,N'-DIETHYLTHIOCARBAMIDE see DKC400

N,N-DIETHYLTHIOCARBAMYL-O,O-DIISOPROPYLDITHIOPHOSPHATE see DKB600

DIETHYLTHIOETHER see EPH000

2,2-DIETHYL-3-THIOMORPHOLINONE see DKC200

DIETHYL THIOPHOSPHORIC ACIDESTER of 3-CHLORO-4-METHYL-7-HYDROXYCOUMARIN see CNU750

DIETHYLTHIOPHOSPHORYL CHLORIDE (DOT) see DJW600

1,3-DIETHYLTHIOUREA see DKC400

1,3-DIETHYL-2-THIOUREA see DKC400

N,N'-DIETHYLTHIOUREA see DKC400

DIETHYLTIN CHLORIDE see DEZ000

DIETHYLTIN DI(10-CAMPHORSULFONATE) see DKC600

DIETHYLTIN DICAPRYLATE see DIV600

DIETHYLTIN DICHLORIDE see DEZ000

DIETHYLTIN DIIODIDE see DJB000

DIETHYLTIN DIOCTANOATE see DIV600

N,N-DIETHYL-m-TOLUAMIDE see DKC800

DIETHYL-m-TOLUAMIDE see DKC800

N,N-DIETHYL-o-TOLUAMIDE see DKD000

DIETHYLTOLUAMIDE see DKC800

1,3-DIETHYL-1-TRIAZENE see DKD200

1,3-DIETHYLTRIAZENE see DKD200

DIETHYL-TRIAZENE see DKD200

N,N'-DIETHYLTRIAZENE see DKD200

DIETHYL TRIAZENE see DKD200

1,3-DIETHYLTRIAZINE see DKD200

O,O-DIETHYL O-3,5,6-TRICHLORO-2-PYRIDYL PHOSPHOROTHIOATE see CMA100

DIETHYL (2-(TRIETHOXYSILYL)ETHYL)PHOSPHONIC ACID see DKD500

N⁴,N'-DIETHYL-α,α,α-TRIFLUORO-3,5-DINITRO-TOLUENE-2,4-DIAMINE see CNE500

N,N-DIETHYL-4-(α-(α,α,α-TRIFLUORO-o-TOLYL) BENZYLOXY)PENTYLAMINE CITRATE see DKD600

1,1-DIETHYLUREA see DKD650

N,N-DIETHYLUREA see DKD650

asym-DIETHYLUREA see DKD650

N,N-DIETHYLVANILLAMIDE see DKE200

DIETHYL XANTHOGENATE see BJU000

DIETHYLXANTHOGEN DISULFIDE see BJU000

DI(ETHYLXANTHOGEN)TRISULFIDE see DKE400

DIETHYLZINC see DKE600

2-α,17-α-DIETHYNYL-A-NOR-5-α-ANDROSTANE-2-β,17-β-DIOL DIHEMISUCCINATE see SDY675

DIETHYXIME see DHZ000

DIETIL see CAR000

DIETILAMIDE-CARBOPIRIDINA see DJS200

DIETILAMINA (ITALIAN) see DHJ200

α-DIETILAMINO-2,6-DIMETILACETANILIDE (ITALIAN) see DHK400

5-(2-DIETILAMINOETIL)-3-FENIL-1,2,4-OXADIEZOLO CLORIDRATO (ITALIAN) see OOE100

O,O-DIETIL-O-(4-BROMO-2,5-DICLORO-FENIL)-MONOTIOFOSFATO (ITALIAN) see EGV500

O,O-DIETIL-O-(3-CLORO-4-METIL-CUMARIN-7-IL-MONOTIOFOSFATO) (ITALIAN) see CNU750

O,O-DIETIL-O-(2,4-DICLORO-FENIL)-MONOTIOFOSFATO (ITALIAN) see DFK600

5,5-DIETILDIIDRO-1,3-OSSAZIN-2,4-DIONE (ITALIAN) see DJT400

DIETILESTILBESTROL (SPANISH) see DKA600

O,O-DIETIL-O-(2-ISOPROPIL-4-METIL-PIRIMIDIN-6-IL)-MONOTIOFOSFATO (ITAL-IAN) see DCM750

O,O-DIETIL-O-(4-METILCUMARIN-7-IL)-MONOTIOFOSFATO (ITALIAN) see PKT000

O,O-DIETIL-O-(3-METIL-1H-PIRAZOL-5-IL)-FOSFATO (ITALIAN) see MOX250

O,O-DIETIL-O-(4-NITRO-FENIL)-MONOTIOFOSFATO (ITALIAN) see PAK000

DIETILPROPANDIOLO see PMB250

O,O-DIETIL-S-((2-CIAN-2-METIL-ETIL)-CARBAMOIL)-METIL-MONOTIOFOSFATO (ITALIAN) see PHK250

O,O-DIETIL-S-((4-CLORO-FENIL-TIO)-METILE)-DITIOFOSFATO (ITALIAN) see TNP250

O,O-DIETIL-S-((6-CLORO-2-OXO-BENZOSSAZOLIN-3-IL)-METIL)-DITIOFOSFATO (ITALIAN) see BDJ250

O,O-DIETIL-S-(2-ETILTIO-ETIL)-DITIOFOSFATO (ITALIAN) see DXH325

O,O-DIETIL-S-(2-ETILTIO-ETIL)-MONOTIOFOSFATO (ITALIAN) see DAP200

O,O-DIETIL-S-(ETILTIO-METIL)-DITIOFOSFATO (ITALIAN) see PGS000

O,O-DIETIL-S-(N-ETOSSI-CARBONIL-N-METIL-CARBAMOIL-METIL)-DITIOFOSFA-TO (ITALIAN) see DJI000

O,O-DIETIL-S-((4-OXO-3H-1,2,3-BENZOTRIAZIN-3-IL)-METIL)-DITIOFOSFATO (ITALIAN) see EKN000

1,1-DIETOSSIETANO (ITALIAN) see AAG000

DIETREEN see RAF100

DIETROL see DKE800

DIETROXINE see DJT400

O,O-DIETYL-O-4-METHYLKUMARINYL(7)TIOFOSFAT (CZECH) see PKT000

O,O-DIETYL-S-2-ETYLMERKAPTOETYLTIOFOSFAT (CZECH) see DAP200

DIF 4 see DRP800

DIFACIL see DHX800, THK000

DIFEDRYL see BBV500

DIFENAMIZOLE see PAM500

DIFENHYDRAMIN see BBV500

DIFENHYDRAMINE HYDROCHLORIDE see BAU750

DIFENIDOL see DWK200

DIFENIDOL HYDROCHLORIDE see CCR875

DIFENIDOLIN see CCR875

DIFENIDRAMINA (ITALIAN) see BBV500

DIFENILDICHETOPIRAZOLIDINA (ITALIAN) see DWA500

DIFENILHIDANTOINA (SPANISH) see DKQ000

DIFENIL-METAN-DIISOCIANATO (ITALIAN) see MJP400

DIFENIN see DKQ000, DNU000

1,5-DIFENOXYANTHRACHINON see DVW100

DIFENSON see CJT750

DIFENTHOS see TAL250

DIFENYL-DIHYDROXYSILAN see DMN450

N,N'-DIFENYLETHYLENDIAMIN see DWB400

N,N'-DIFENYL-p-FENYLENDIAMIN (CZECH) see BLE500

2-(DIFENYL-HYDROXYACETOXY)ETHYL-DIETHYLAMMONIUMCHLORID (CZECH) see BCA000

DIFENYLIN see BGF109

DIFENYLMETHAAN-DIISSOCIANAAT (DUTCH) see MJP400

DIFENYLSULFON (CZECH) see PGI750

DIFENZOQUAT METHYL SULFATE see ARW000

DIFETOIN see DNU000

DIFEXAMIDE METHIODIDE see BTA325

DIFFLAM see BBW500

DIFFOLLISTEROL see EDP000

DIFHYDAN see DKQ000, DNU000

DIFLAVINE (ACRIDINE) see DBN000

DIFLUBENZURON see CJV250

DIFLUCORTOLONE 21-VALERATE see DKF130

DIFLUCORTOLONE VALERATE see DKF130

DIFLUCORTOLONVALERIANAT (GERMAN) see DKF130

DIFLUNISAL see DKI600

DIFLUORO see MJD275

2,4'-DIFLUOROACETANILIDE see DKF170

DIFLUOROACETIC ACID see DKF200

DIFLUOROAMINE see DKF400

3-DIFLUOROAMINO-1,2,3-TRIFLUORODIAZIRIDINE see DKF600

DIFLUOROAMMONIUM HEXAFLUOROARSENATE see DKF620

m-DIFLUOROBENZENE see DKF800

p-DIFLUOROBENZENE see DKG000

3,4-DIFLUOROBENZENEARSONIC ACID see DKG100

2,10-DIFLUOROBENZO(rst)PENTAPHENE see DKG400

1,1-DIFLUORO-1-CHLOROETHANE see CFX250

DIFLUOROCHLOROETHANES (DOT) see CFX250

DIFLUOROCHLOROMETHANE see CFX500

DIFLUORODIAZENE see DKG600

DIFLUORODIAZIRINE see DKG700

2,10-DIFLUORODIBENZO(a,i)PYRENE see DKG400

DIFLUORODIBROMOMETHANE see DKG850

1,1-DIFLUORO-2,2-DICHLOROETHYLENE see DFA300

DIFLUORODICHLOROMETHANE see DFA600

2',4'-DIFLUORO-4-DIMETHYLAMINOAZOBENZENE see DKG980

2',5'-DIFLUORO-4-DIMETHYLAMINOAZOBENZENE see DKH000

3',4'-DIFLUORO-4-DIMETHYLAMINOAZOBENZENE see DRL000

3',5'-DIFLUORO-4-DIMETHYLAMINOAZOBENZENE see DKH100

DIFLUORODIMETHYLSTANNANE see DKH200

4,4'-DIFLUORO-3,3-DINITRODIPHENYL SULFONE see DKH250

p,p'-DIFLUORO-m,m'-DINITRODIPHENYL SULFONE see DKH250

1,1-DIFLUOROETHANE see ELN500

DIFLUOROETHANE see ELN500

1,1-DIFLUOROETHENE see VPP000

1,1-DIFLUOROETHYLENE (DOT, MAK) see VPP000

DIFLUORO-N-FLUOROMETHANIMINE see DKH825

DIFLUOROFORMALDEHYDE see CCA500

2',4'-DIFLUORO-4-HYDROXY-(1,1'-BIPHENYL)-3-CARBOXYLIC ACID see DKI600

2',4'-DIFLUORO-4-HYDROXY-3-BIPHENYLCARBOXYLIC ACID see DKI600

2',4'-DIFLUORO-4-HYDROXY-(1',1-DIPHENYL)-3-CARBOXYLIC ACID see DKI600

6-α,9-DIFLUORO-11-β-HYDROXY-16-α-METHYL-21-VALERYLOXY-1,4-PREGNADI-ENE-3,20-DIONE see DKF130

6-α,9-α-DIFLUORO-16-α-HYDROXYPREDNISOLONE-16,17-ACETONIDE see SPD500

6-α,2-DIFLUORO-11-β-HYDROXY-21-VALERYLOXY-16-α-METHYL-1,4-PREGNADI-ENE-3,20-DIONE see DKF130

DIFLUOROMETHYLENE DIHYPOFLUORITE see DKH830

2-(DIFLUOROMETHYL)ORNITHINE see DBE835

α-DIFLUOROMETHYLORNITHINE see DBE835

dl-α-DIFLUOROMETHYLORNITHINE see DKH875

α-DIFLUOROMETHYLORNITHINE HYDROCHLORIDE see EAE775

2-(DIFLUOROMETHYL)-dl-ORNITHINE HYDROCHLORIDE see EAE775

DIFLUOROMETHYLPHOSPHINE OXIDE see MJD275

DIFLUOROMONOCHLOROMETHANE see CFX500

2,4-DIFLUORONITROBENZENE see DKH900

DIFLUOROPHENYLARSINE see DKI400

p-((3,5-DIFLUOROPHENYL)AZO)-N,N-DIMETHYLANILINE see DKH100

5-(2,4-DIFLUOROPHENYL)SALICYLIC ACID see DKI600

6-α,9-α-DIFLUOROPREDNISOLONE 17-BUTYRATE 21-ACETATE see DKJ300

1,3-DIFLUORO-2-PROPANOL see DKI800

1,2-DIFLUORO-1,1,2,2-TETRACHLOROETHANE see TBP050

1,1-DIFLUORO-1,2,2,2-TETRACHLOROETHANE see TBP000

1,1-DIFLUORO-1,2,2-TRICHLOROETHANE see TIM000

3,8-DIFLUOROTRICYCLOQUINAZOLINE see DKJ200

3,3-DIFLUORO-2-(TRIFLUOROMETHYL)ACRYLIC ACID, METHYL ESTER see MNN000

3,3-DIFLUORO-2-(TRIFLUOROMETHYL)-2-PROPENOIC ACID, METHYL ESTER see MNN000

6-α,9-DIFLUORO-11-β,17,21-TRIHYDROXYPREGNA-1,4-DIENE-3,20-DIONE-21-ACETATE-17-BUTYRATE see DKJ300

1,1-DIFLUOROUREA see DKJ225

DIFLUPREDNATE see DKJ300

DIFLUPYL see IRF000

DIFLUREX see TGA600

DIFLURON see CJV250

DIFLUROPHATE see IRF000

DIFO see BJE750

DIFOLATAN see CBF800

DIFOLLICULINE see EDP000

DIFONATE see FMU045

DIFORENE see DOZ000

7,12-DIFORMYLBENZ(a)ANTHRACENE see BBD000

1,4-DIFORMYLBENZENE see TAN500

2,2'-DIFORMYLBIPHENYL see DVV800

N,N'-DIFORMYL-p,p'-DIAMINODIPHENYLSULFONE see BJW750

1,2-DIFORMYLHYDRAZIN (GERMAN) see DKJ600

1,2-DIFORMYLHYDRAZINE see DKJ600

2,4'-DIFORMYL-1,1'-(OXYDIMETHYLENE) DIPYRIDINIUM DICHLORIDE, DIOXIME see HGL650

DIFOSAN see CBF800

DIFOSGEN see TIR920

DIFURAN see PAF500

DI-2-FURANYLETHANEDIONE see FPZ000

DIFURAZONE see PAF500

N,N-DIFURFURAL-n-PHENYLENEDIAMINE see DKK200

2,2'-DIFURFURYL ETHER see FPX025

DIFURFURYL ETHER (7CI) see FPX025

DI-2-FURYLGLYOXAL see FPZ000

DI-2-FURYLGLYOXAL MONOXIME see FQB000

DIGACIN see DKN400

DIGADOLINIUM TRIOXIDE see GAP000

DIGALLIUM TRIOXIDE see GBS050

DIGALLIUM TRISULFATE see GBS100

DIGAMMACAINE see DKK800

DIGENEA SIMPLEX MUCILAGE see AEX250

11,11-DIGERANYLOXY-1-UNDECENE see UNA100

DIGERMIN see DUV600

DIGIBUTINA see BRF500
DIGILANID A see LAT000
DIGILANID B see LAT500
DIGILANID C see LAU000
DIGILANIDE B see LAT500
DIGILANIDES see LAU400
DIGILONG see DKL800
DIGIMED see DKL800, LAU400
DIGIMERCK see DKL800
DIGISIDIN see DKL800
DIGITALIN see DKL800
DIGITALINE (FRENCH) see DKL800
DIGITALINE CRISTALLISEE see DKL800
DIGITALINE NATIVELLE see DKL800
DIGITALINUM VERUM see DKL800
DIGITALIS see DKL200, FOM100
DIGITALIS GLYCOSIDE see DKN400
DIGITALIS LANATA STANDARD see DKL300
DIGITALIS PURPUREA see FOM100
DIGITALIS PURPUREA, LEAF see DKL200
DIGITANNOID see DKL200
DIGITIN see DKL400
DIGITONIN see DKL400
DIGITOPHYLLIN see DKL800
DIGITOXIGENIN see DMJ000
DIGITOXIGENIN + 2 DIGITOXOSE + ACETYL-DIGILANIDOBOSE (GERMAN) see LAT000
DIGITOXIGENIN + 2-DIGITOXOSE + 1-ACETYL-(4)-DIGITOSE (GERMAN) see ACH750
DIGITOXIGENIN + 2-DIGITOXOSE + ACETYL-(3)-DIGITOXOSE (GERMAN) see ACH750
DIGITOXIGENINE see DMJ000
DIGITOXIGENIN-TRIDIGITOXOSID (GERMAN) see DKL800
DIGITOXIGENIN TRIDIGITOXOSIDE see DKL800
DIGITOXIN see DKL800
DIGITOXOSIDE see DKL875
DIGLYCERIDE ACETIC ACID see DBF600
DIGLYCEROL TETRANITRATE see TDY100
1,4-DIGLYCIDLOXYBUTANE see BOS100
N,N-DIGLYCIDYLANILIN (CZECH) see DKM120
N-N-DIGLYCIDYLANILINE see DKM120
DIGLYCIDYL BISPHENOL A ETHER see BLD750
N,N'-DIGLYCIDYL-5,5-DIMETHYLHYDANTOIN see DKM130
DIGLYCIDYLESTER KYSELINY HEXAHYDROFTALOVE (CZECH) see DKM500
DIGLYCIDYL ETHER see DKM200
DIGLYCIDYL ETHER of N,N-BIS(2-HYDROXYETHOXYETHYL)ANILINE see BJN850
DIGLYCIDYL ETHER of 2,2-BIS(4-HYDROXYPHENYL)PROPANE see BLD750
DIGLYCIDYL ETHER of 2,2-BIS(p-HYDROXYPHENYL)PROPANE see BLD750
DIGLYCIDYL ETHER of N,N-BIS(2-HYDROXYPROPYL)-tert-BUTYLAMINE see DKM400
DIGLYCIDYL ETHER of BISPHENOL A see BLD750
DIGLYCIDYL ETHER of BISPHENOL A mixed with BIS(2,3-EPOXYCYCLOPENTYL) ETHER (1:1) see OPI200
DIGLYCIDYL ETHER of 4,4'-ISOPROPYLIDENEDIPHENOL see BLD750
DIGLYCIDYL ETHER of NEOPENTYL GLYCOL see NCI300
DIGLYCIDYL ETHER of PHENYLDIETHANOLAMINE see BJN875
DIGLYCIDYLETHYLENE GLYCOL see EEA600
DIGLYCIDYL HEXAHYDROPHTHALATE see DKM500
1,3-DIGLYCIDYLOXYBENZENE see REF000
1,2-DIGLYCIDYLOXYETHANE see EEA600
N-N-DIGLYCIDYLPHENYLAMINE see DKM120
DIGLYCIDYL PIPERAZINE see DHE100
DIGLYCIDYL RESORCINOL ETHER see REF000
N,N-DIGLYCIDYL-p-TOLUENESULFONAMIDE see DKM800
N,N-DIGLYCIDYL-p-TOLUENESULFONAMIDE see DKM800
DIGLYCIDYLTRIETHYLENE GLYCOL see TJQ333
DIGLYCIN see IBH000
DIGLYCINE see IBH000
DIGLYCOL see DJD600
DIGLYCOLAMINE see AJU250
DIGLYCOL CHLORHYDRIN see DKN000
DIGLYCOL DIMETHACRYLATE see BKM250
DIGLYCOLDINITRAAT (DUTCH) see DJE400
DIGLYCOL (DINITRATE de) (FRENCH) see DJE400
DIGLYCOLIC ACID (6CI) see ONQ100
DIGLYCOLIC ACID DI-(3-CARBOXY-2,4,6-TRIIODOANILIDE) DISODIUM see BGB350
DIGLYCOL MONOBUTYL ETHER see DJF200
DIGLYCOL MONOBUTYL ETHER ACETATE see BQP500
DIGLYCOL MONOETHYL ETHER see CBR000
DIGLYCOL MONOETHYL ETHER ACETATE see CBQ750
DIGLYCOL MONOMETHYL ETHER see DJG000
DIGLYCOL MONOSTEARATE see HKJ000

DIGLYCOLSAEURE-DI-(3-CARBOXY-2,4,6-TRIJOD-ANILID) DINATRIUM (GERMAN) see BGB350
DIGLYCOL STEARATE see HKJ000
DIGLYKOKOLL see IBH000
DIGLYKOLDINITRAT (GERMAN) see DJE400
DIGLYME see BKN750
DIGOLD(I) KETENIDE see DKN250
DIGORID A see ACI000
DIGORID B see ACI250
DIGOXIGENIN see DKN300
DIGOXIGENINE see DKN300
DIGOXIGENIN-TRIDIGITOXOSID (GERMAN) see DKN400
DIGOXIGENIN + ZUCKERKETTE WIE BEI ACETYL-DIGITOXIN-α (GERMAN) see ACI250
DIGOXIGENIN + ZUCKERKETTE WIE BEI ACETYL-DIGITOXIN A (GERMAN) see ACI000
DIGOXIN see DKN400
ε-DIGOXIN ACETATE see DKN600
DIGOXINE see DKN400
DIGOXIN PENTAFORMATE see DKN875
DIGUANIDINIUM CARBONATE see GKW100
DIGUANYL see CKB500
DIHYDROPYRONE see BRT000
((DIHDYROXYPROPYL)THIO)METHYLMERCURY see MLG750
DIHEPTYL ETHER see HBO000
DIHEPTYL KETONE see HBO790
DIHEPTYLMERCURY see DKO000
DIHEXYL see DXT200
DI-N-HEXYLAMINE see DKO600
DIHEXYLAMINE see DKO600
DI-N-HEXYL AZELATE see ASC000
DIHEXYL trans-BUTENEDIOATE see DKP000
DIHEXYL ETHER see DKO800
DIHEXYL FUMARATE see DKP000
DI-n-HEXYL KETONE see TJJ300
DIHEXYL KETONE see TJJ300
DIHEXYL LEAD DIACETATE see DKP200
DIHEXYL MALEATE see DKP400
DIHEXYLOXYETHYL ADIPATE see AEQ000
DI-(2-(2-HEXYLOXY)ETHYL)ESTER KYSELINY ADIPOVE see AEQ000
DI-(2-HEXYLOXYETHYL)ESTER KYSELINY JANTAROVE see SMZ000
DI-2-HEXYLOXYETHYL SUCCINATE see SMZ000
DI-n-HEXYL PHTHALATE see DKP600
DIHEXYL PHTHALATE see DKP600
DIHEXYL SODIOSULFOSUCCINATE see DKP800
DIHEXYL SODIUM SULFOSUCCINATE see DKP800
DIHEXYL SULFOSUCCINATE SODIUM SALT see DKP800
DIHEXYLTIN DICHLORIDE see DFC200
DIHIDRAL see BBV500
DIHIDROBENZPERIDOL see DYF200
DIHIDROCLORURO de BENZIDINA (SPANISH) see BBX750
DIHYCON see DKQ000
DI-HYDAN see DKQ000, DNU000
DIHYDANTOIN see DKQ000, DNU000
DIHYDRALAZIN see OJD300
DIHYDRALAZINE see OJD300
DIHYDRALAZINE HYDROCHLORIDE see DKQ200
DIHYDRALAZINE SULFATE see DKQ600
DIHYDRALLAZIN see OJD300
DIHYDRAZINECOBALT(II) CHLORATE see DKQ400
1,4-DIHYDRAZINONAPHTHALAZINE see OJD300
1,4-DIHYDRAZINOPHTHALAZINE see OJD300
DIHYDRAZINOPHTHALAZINE see OJD300
1,4-DIHYDRAZINOPHTHALAZINE HYDROCHLORIDE see DKQ200
1,4-DIHYDRAZINOPHTHALAZINE SULFATE see DKQ600
DIHYDREL see DKQ650
DIHYDRIN see DKW800
1,2-DIHYDRO-5-ACENAPHTHYLENAMINE see AAF250
9,10-DIHYDROACRIDINE see ADI775
DIHYDROAFLATOXIN B1 see AEU750
DIHYDROAFLATOXIN G1 see AEV500
DIHYDROAMBRETTOLIDE see OKU000
9,10-DIHYDRO-1-AMINO-4-BROMO-9,10-DIOXO-2-ANTHRACENE SULFONIC ACID SODIUM SALT see DKR000
2,3-DIHYDRO-6-AMINO-2-(2-CHLOROETHYL)-4H-1,3-BENZOXAZIN-4-ONE see DKR200
9,10-DIHYDRO-1-AMINO-4-(CYCLOHEXYLAMINO)-9,10-DIOXO-2-ANTHRACENE-SULFONIC ACID SODIUM SALT see CMM080
2,4-DIHYDRO-5-AMINO-2-(1-(3,4-DICHLOROPHENYL)ETHYL)-3H-PYRAZOL-3-ONE see EAE675
1,9-DIHYDRO-2-AMINO-9-((2-HYDROXYETHOXY)METHYL)-6H-PURIN-6-ONE SODIUM SALT see AEC725
9,10-DIHYDRO-1-AMINO-4-(3-(2-HYDROXYETHYL)AMINOSULFONYL-4-METHYL-PHENYLAMINO)-9,10-DIOXO-2-ANTHRACENE SULFON see DKR400
DIHYDROANETHOLE see PNE250

N,N-DIHYDRO-1,1,1′,2′-ANTHRAQUINONE-AZINE see IBV050
22,23-DIHYDROAVERMECTIN B1 see ITD875
DIHYDROAZIRENE see EJM900
DIHYDRO-1H-AZIRINE see EJM900
DIHYDROBAIKIANE see HDS300
1,2-DIHYDROBENZ(1)ACEANTHRYLENE see BAW000
1,2-DIHYDROBENZ(e)ACEANTHRYLENE see AAE000
4,5-DIHYDROBENZ(k)ACEPHENANTHRYLENE see AAF000
5,6-DIHYDROBENZ(c)ACRIDINE-7-CARBOXYLIC ACID see TEF775
1,2-DIHYDROBENZ(a)ANTHRACENE see DKS400
3,4-DIHYDROBENZ(a)ANTHRACENE see DKS600
1a,11b-DIHYDROBENZ(3,4)ANTHRA(1,2-b)OXIRENE see EBP000
5,6-DIHYDROBENZENE(e)ACEANTHRYLENE see AAE000
7,11b-DIHYDROBENZ(b)INDENO(1,2-d)PYRAN-3,6a,9,10(6H)-TETROL see LFT800
1,2-DIHYDROBENZO(a)ANTHRACENE see DKS400
3,4-DIHYDROBENZO(a)ANTHRACENE see DKS600
6,13-DIHYDROBENZO(e)(1)BENZOTHIOPYRANO(4,3-b)INDOLE see DKS800
2,3-DIHYDRO-1H-BENZO(h,i)CHRYSENE see DKT400
6-β,7-α-DIHYDROBENZO(10,11)CHRYSENO(1,2-b)OXIRENE see BCV750
(2,3-DIHYDRO-1,4-BENZODIOXIN-6-YL)(4-FLUOROPHENYL)METHANONE see DKT500
N-((2,3-DIHYDRO-1,4-BENZODIOXIN-2-YL)METHYL)-N-METHYL-N′,N′-DI-2-PRO-PENYL-UREA (9CI) see DBJ100
9,10-DIHYDROBENZO(e)PYRENE see DKU400
7,8-DIHYDROBENZO(a)PYRENE see DKU000
trans-4,5-DIHYDROBENZO(a)PYRENE-4,5-DIOL see DLC600
9,10-DIHYDROBENZO(a)PYRENE-9,10-DIOL see DLC000
3,4-DIHYDRO-2H-1,4-BENZOTHIAZINE HYDROCHLORIDE see DKU875
7,8-DIHYDRO-N-BENZYLADENINE see DKV125
1,3-DIHYDRO-7-BROMO-5-(2-PYRIDYL)-2H-1,4-BENZODIAZEPIN-2-ONE see BMN750
DIHYDROBUTADIENE SULPHONE see SNW500
2,3-DIHYDRO-5-CARBOXANILIDO-6-METHYL-1,4-OXATHIIN see CCC500
2,3-DIHYDRO-5-CARBOXANILIDO-6-METHYL-1,4-OXATHIIN-4,4-DIOXIDE see DLV200
1,6-DIHYDROCARVEOL see DKV150
DIHYDROCARVEOL see DKV150
DIHYDROCARVEOL ACETATE see DKV160
DIHYDROCARVEYL ACETATE see DKV160
d-DIHYDROCARVONE see DKV175
DIHYDROCARVYL ACETATE see DKV160
5,10-DIHYDRO-7-CHLOR-10-(2-(DIMETHYLAMINO)ETHYL)-11H-DIBEN-ZO(b,e)(1,4)DIAZEPIN-11-ONE HCl see CMW250
DIHYDROCHLORIDE-1-NITRO-9-((2-DIMETHYLAMINO)-1-METHYLETHYLAMINO)-ACRIDINE see NFW435
DIHYDROCHLORIDE SALT OF DIETHYLENEDIAMINE see PIK000
2,3-DIHYDRO-6-CHLORO-2-(2-CHLOROETHYL)-4H-1,3-BENZOXAZIN-4-ONE see DKV200
(±)-2,3-DIHYDRO-6-CHLORO-5-CYCLOHEXYL-1H-INDENE-1-CARBOXYLIC ACID see CFH825
1′-(3-(10,11-DIHYDRO-3-CHLORO-5H-DIBENZ(b,f)AZEPIN-5-YL)PROPYL)-(1,4′-BI-PIPERIDINE)-4′-CARBOX see DKC309
1,3-DIHYDRO-7-CHLORO-5-(o-FLUOROPHENYL)-3-HYDROXY-1-(2-HYDROXY-ETHYL)-2H-1,4-BENZODIAZEPIN-2-ONE see DKV400
1,3-DIHYDRO-7-CHLORO-3-HYDROXY-1-METHYL-5-PHENYL-2H-1,4-BENZO-DIAZEPIN-2-ONE see CFY750
2,3-DIHYDRO-7-CHLORO-1H-INDEN-4-OL (9CI) see CDV700
3,3a-DIHYDRO-7-CHLORO-2-METHYL-2H,9H-ISOXAZOLO(3,2-b)(1,3)BENZOXAZIN-9-ONE see CIL500
2,3-DIHYDRO-7-CHLORO-1-METHYL-5-PHENYL-1H-1,4-BENZODIAZEPINE see CGA000
4,5-DIHYDRO-N-(2-CHLORO-4-METHYLPHENYL)-1H-IMIDAZOL-2-AMINE MONO-NITRATE see TGJ885
2,3-DIHYDRO-7-CHLORO-2-OXO-5-PHENYL-1H-1,4-BENZODIAZEPINE-3-CAR-BOXYLIC ACID MONOPOTASSIUM SALT see DKV700
4,5-DIHYDRO-5-((p-CHLOROPHENYL)THIOMETHYL)OXAZOLAMINE see AJI500
3,4-DIHYDRO-6-CHLORO-7-SULFAMYL-1,2,4-BENZOTHIADIAZINE-1,1-DIOXIDE see CFY000
DIHYDROCHLOROTHIAZID see CFY000
3,4-DIHYDROCHLOROTHIAZIDE see CFY000
DIHYDROCHLOROTHIAZIDE see CFY000
6,12,b-DIHYDROCHOLANTHRENE see DKV800
meso-DIHYDROCHOLANTHRENE see DKV800
DIHYDROCHOLESTEROL see DKW000
1,2-DIHYDROCHRYSENE see DKW200
3,4-DIHYDROCHRYSENE see DKW400
DIHYDROCHRYSENE see DKW200
(E)-1,2-DIHYDRO-1,2-CHRYSENEDIOL see DLD200
trans-1,2-DIHYDROCHRYSENE-1,2-DIOL see DLD200
DIHYDROCINNAMALDEHYDE see HHP000
6,7-DIHYDROCITRONELLAL see TCY300
DIHYDROCITRONELLAL see TCY300
DIHYDROCITRONELLOL see DTE600
DIHYDROCITRONELLYL ACETATE see DTE800

7,8-DIHYDROCODEINE see DKW800
DIHYDROCODEINE see DKW800
DIHYDROCODEINE ACID TARTRATE see DKX000
DIHYDROCODEINE BITARTRATE see DKX000
DIHYDROCODEINE TARTRATE (1:1) see DKX000
DIHYDROCODEINE TARTRATE see DKX000
DIHYDROCODEINONE see OOI000
DIHYDROCODEINONE BITARTRATE see DKX050
3,4-DIHYDROCOUMARIN see HHR500
DIHYDROCOUMARIN see HHR500
DIHYDROCUMINYL ALCOHOL see PCI550
DIHYDROCUMINYL ALDEHYDE see DKX100
10,11-DIHYDRO-5-CYANO-N,N-DIMETHYL-5H-DIBENZO(a,d)CYCLOHEPTENE-5-PROPYLAMINE HCl see COL250
3,4-DIHYDRO-6-(4-(1-CYCLOHEXYL-1H-TETRAZOL-5-YL)BUTOXY)-2(1H)-QUI-NOLINONE see CMP825
13-DIHYDRODAUNOMYCIN see DAC300
DIHYDRODAUNOMYCIN see DAC300
13-DIHYDRODAUNORUBICIN see DAC300
12-β,13-α-DIHYDRO-11-DEOXO-11-β-HYDROXYJERVINE see DAQ002
DIHYDRODEOXYMORPHINE see DKX600
DIHYDRODESOXYMORPHINE-D see DKX600
9,10-DIHYDRO-4,5-DIAMINO-1-HYDROXY-2,7-ANTHRACENE DISULFONIC ACID DISODIUM SALT see DKX800
9,10-DIHYDRO-8a,10,-DIAZONIAPHENANTHRENE DIBROMIDE see DWX800
9,10-DIHYDRO-8a,10a-DIAZONIAPHENANTHRENE(1,1′-ETHYLENE-2,2′-BIPYRI-DYLIUM)DIBROMIDE see DWX800
trans-10,11-DIHYDRODIBENZ(a,e)ACEANTHRYLENE-10,11-DIOL see DKX900
trans-1,2-DIHYDRODIBENZ(a,e)ACEANTHRYLENE-1,2-DIOL see DKX875
9,10-DIHYDRO-1,2,5,6-DIBENZANTHRACENE see DKY400
5,6-DIHYDRODIBENZ(a,h)ANTHRACENE see DKY000
7,14-DIHYDRODIBENZ(a,h)ANTHRACENE see DKY400
5,6-DIHYDRODIBENZ(a,j)ANTHRACENE see DKY200
10,11-DIHYDRO-5-DIBENZ(b,f)AZEPINE see DKY800
3,4-DIHYDRO-1,2,5,6-DIBENZCARBAZOLE see DLA000
12,13-DIHYDRO-7H-DIBENZO(a,g)CARBAZOLE see DLA000
5,8-DIHYDRODIBENZO(a,def)CHRYSENE see DLA100
7,14-DIHYDRODIBENZO(b,def)CHRYSENE see DLA120
10,11-DIHYDRO-5H-DIBENZO(a,d)CYCLOHEPTENE-5-CARBOXAMIDE see CQH625
10,11-DIHYDRO-5H-DIBENZO(a,d)CYCLOHEPTEN-5-ONE see DCX300
10,11-DIHYDRO-5H-DIBENZO(a,d)CYCLOHEPTEN-5-ONE OXIME see DDD000
3,10-DIHYDRO-5H-DIBENZO(a,d)CYCLOHEPTEN-5-YLIDENE-N,N-DIMETHYL-1-PROPANAMINE see EAH500
3-(10,11-DIHYDRO-5H-DIBENZO(a,d)CYCLOHEPTEN-5-YLIDENE)-1-ETHYL-2-ME-THYLPYRROLIDINE see PJA190
3-(10,11-DIHYDRO-5H-DIBENZO(a,d)CYCLOHEPTEN-5-YLIDENE)-1-ETHYL-2-METHYL-PYRROLIDINE HYDROCHLORIDE see TMK150
4-((10,11-DIHYDRO-5H-DIBENZO(a,d)CYCLOHEPTEN-5-YL)OXY)-1-METHYLPI-PERIDINE HYDROGEN MALEATE see HBT000
4-((10,11-DIHYDRO-5H-DIBENZO(a,d)CYCLOHEPTEN-5-YL)OXY)-1-METHYLPI-PERIDINE, MALEATE (1:1) see HBT000
3-α-((10,11-DIHYDRO-5H-DIBENZO(a,d)CYCLOHEPTEN-5-YL)OXY)-8-METHYL-TROPANIUM BROMIDE see DAZ140
5,10-DIHYDRO-3,4:8,9-DIBENZOPYRENE see DLA120
5,8-DIHYDRO-3,4:9,10-DIBENZOPYRENE see DLA100
2,3-DIHYDRODICYCLOPENTA(c,lmn)PHENANTHREN-1(9H)-ONE see DGW300
2,3-DIHYDRO-1-DIETHYLAMINOETHYL-2-METHYL-3-PHENYL-4(3H)-QUINAZOLI-NONE OXALATE see DID800
2,3-DIHYDRO-1-DIETHYLAMINOETHYL-2-METHYL-3-(o-TOLYL)-4(3H)-QUINAZOLI-NONE OXALATE see DIE200
DIHYDRO-5,5-DIETHYL-2H-1,3-OXAZINE-2,4(3H)-DIONE see DJT400
DIHYDRODIETHYLSTILBESTROL see DLB400
1,4-DIHYDRO-4-(2-(DIFLUOROMETHOXY)PHENYL)-2,6-DIMETHYL-3,5-PYRIDI-NEDICARBOXYLIC ACID DIMETHYL ESTER see RSZ375
(+)-(3S,4S)-trans-3,4-DIHYDRO-3,4-DIHYDROXYBENZ(a)ANTHRACENE see BBD750
(−)(3R,4R)-trans-3,4-DIHYDRO-3,4-DIHYDROXYBENZ(a)ANTHRACENE see BBG000
(−)(3R,4R)trans-3,4-DIHYDRO-3,4-DIHYDROXYBENZO(a)ANTHRACENE see BBG000
(+)-(3S,4S)-trans-3,4-DIHYDRO-3,4-DIHYDROXYBENZO(a)ANTHRACENE see BBD750
trans-3,4-DIHYDRO-3,4-DIHYDROXYBENZO(a)ANTHRACENE see BBD500
4,5-DIHYDRO-4,5-DIHYDROXYBENZO(e)PYRENE see DMK400
4,5-DIHYDRO-4,5-DIHYDROXYBENZO(a)PYRENE see DLB800
trans-4,5-DIHYDRO-4,5-DIHYDROXYBENZO(a)PYRENE see DLC600
9,10-DIHYDRO-9,10-DIHYDROXYBENZO(a)PYRENE see DLC000
(±)-9,10-DIHYDRO-9,10-DIHYDROXYBENZO(a)PYRENE see DLC400
trans-1,2-DIHYDRO-1,2-DIHYDROXYCHRYSENE see DLD200
(E)-1,2-DIHYDRO-1,2-DIHYDROXYDIBENZ(a,h)ANTHRACENE see DMK200
trans-3,4-DIHYDRO-3,4-DIHYDROXYDIBENZ(a,h)ANTHRACENE see DLD400
trans-3,4-DIHYDRO-3,4-DIHYDROXYDIBENZO(a,h)ANTHRACENE see DLD400
trans-12,13-DIHYDRO-12,13-DIHYDROXYDIBENZO(a,e)FLUORANTHENE see DKX875

1-(2-(4-(3,4-DIHYDRO-6-METHOXY-2-PHENYL-1-NAPHTHALE-NYL)PHENOXY)ETHYL)PYRROLIDENE see NAD500

1-(2-(3,4-DIHYDRO-6-METHOXY-2-PHENYL-1-NAPHTHALE-NYL)PHENOXY)ETHYL)-PYRROLIDINE HCl see NAD750

1-(2-(p-(3,4-DIHYDRO-6-METHOXY-2-PHENYL-1-NAPH-THYL)PHENOXY)ETHYL)PYRROLIDINE see NAD500

1-(2-(p-3,4-DIHYDRO-6-METHOXY-2-PHENYL-1-NAPH-THYL)PHENOXY)ETHYL)PYRROLIDINE HYDROCHLORIDE see NAD750

(R)-5,6-DIHYDRO-4-METHOXY-2-STYRYL-2H-PYRAN-2-ONE see GJI250

3,12-DIHYDRO-6-METHOXY-3,3,12-TRIMETHYL-7H-PYRANO(2,3-C)ACRIDIN-7-ONE see ADR750

10,11-DIHYDRO-5-(3-(METHYLAMINO)PROPYL)-5H-DIBENZ(b,f)AZEPINE HY-DROCHLORIDE see DLS500

1,2-DIHYDRO-3-METHYL-BENZ(j)ACEANTHRYLENE see MIJ750

1,2-DIHYDRO-3-METHYLBENZ(j)ACEANTHRYLENE COMPOUND with 2,4,6-TRINITROPHENOL (1:1) see MIL750

2,3-DIHYDRO-2-METHYLBENZOPYRANYL-7,N-METHYLCARBAMATE see DLS800

9,10-DIHYDRO-7-METHYLBENZO(a)PYRENE see DLT000

4,5-DIHYDRO-2-((2-METHYLBENZO(b)THIEN-3-YL)METHYL)-1H-IMIDAZOLE HY-DROCHLORIDE see MHJ500

1':2'-DIHYDRO-4'-METHYL-3:4-BENZPYRENE see DLT000

DIHYDRO-5-(1-METHYLBUTYL)-5-(2-PROPENYL)-2-THIOXO-4,6(1H,5H)-PYRIMIDI-NEDIONE (9CI) see AGL375

5,6-DIHYDRO-2-METHYL-3-CARBOXANILIDO-1,4-OXATHIIN (GERMAN) see CCC500

5,6-DIHYDRO-2-METHYL-3-CARBOXANILIDO-1,4-OXATHIIN-4,4-DIOXID (GER-MAN) see DLV200

11,12-DIHYDRO-3-METHYLCHOLANTHRENE see DLT400

6,12b-DIHYDRO-3-METHYLCHOLANTHRENE see DLT200

meso-DIHYDRO-3-METHYLCHOLANTHRENE see DLT200

(E)-11,12-DIHYDRO-3-METHYLCHOLANTHRENE-11,12-DIOL see DML800

9,10-DIHYDRO-3-METHYLCHOLANTHRENE-1,9,10-TRIOL see DLT600

9,10-DIHYDRO-3-METHYL-CHOLANTHRENE-1,9,10-TRIOL see TKO750

1,2-DIHYDRO-5-METHYL-1,2-CHRYSENEDIOL see DLF400

7,8-DIHYDRO-5-METHYL-7,8-CHRYSENEDIOL see DLF600

16,17-DIHYDRO-11-METHYL-15H-CYCLOPENTA(a)PHENANTHRENE see MJD750

16,17-DIHYDRO-7-METHYL-15H-CYCLOPENTA(a)PHENANTHRENE see MIV250

15,16-DIHYDRO-11-METHYL-17H-CYCLOPENTA(a)PHENANTHREN-17-OL see DLT800

16,17-DIHYDRO-11-METHYLCYCLOPENTA(a)PHENANTHREN-15-ONE see DLU200

15,16-DIHYDRO-11-METHYLCYCLOPENTA(a)PHENANTHREN-17-ONE see MJE500

15,16-DIHYDRO-11-METHYL-17H-CYCLOPENTA(a)PHENANTHREN-17-ONE see MJE500

15,16-DIHYDRO-7-METHYL-17H-CYCLOPENTA(a)PHENANTHREN-17-ONE see DLU400

15,16-DIHYDRO-7-METHYLCYCLOPENTA(a)PHENANTHREN-17-ONE see DLU400

10,11-DIHYDRO-N-METHYL-5H-DIBENZO(a,d)CYCLOHEPTANE-Δ,γ-PROPYLA-MINE see NNY000

16,17-DIHYDRO-17-METHYLENE-15H-CYCLOPENTA(a)PHENANTHRENE see DLU600

DIHYDROMETHYL-α-IONONE see TLO530

15,16-DIHYDRO-11-METHYL-15-METHOXYCYCLOPENTA(a)PHENANTHREN-17-ONE see DLU700

15,16-DIHYDRO-11-METHYL-6-METHOXY-17H-CYCLOPENTA(a)PHENANTHREN-17-ONE see MEV250

2,3-DIHYDRO-2-METHYL-1-(MORPHOLINOACETYL)-3-PHENYL-4(1H)-QUINAZOLI-NONE HYDROCHLORIDE see HGL630

1,4-DIHYDRO-1-METHYL-7-(2-(5-NITRO-2-FURYL)VINYL)-4-OXO-1,8-NAPHTHYRI-DINE-3-CARBOXYLIC ACID, POTASSIUM see DLU800

2,3-DIHYDRO-N-METHYL-7-NITRO-2-OXO-5-PHENYL-1H-1,4-BENZODIAZEPINE-1-CARBOXAMIDE see DLU900

1,3-DIHYDRO-1-METHYL-7-NITRO-5-PHENYL-2H-1,4-BENZODIAZEPIN-2-ONE see DLV000

5,6-DIHYDRO-2-METHYL-1,4-OXATHIIN-3-CARBOXANILIDE see CCC500

2,3-DIHYDRO-6-METHYL-1,4-OXATHIIN-5-CARBOXANILIDE see CCC500

5,6-DIHYDRO-2-METHYL-1,4-OXATHIIN-3-CARBOXANILIDE-4,4-DIOXIDE see DLV200

N-(4,5-DIHYDRO-4-METHYL-5-OXO-1,2-DITHIOLO(4,3-B)PYRROL-6-YL)ACETAMIDE see ABI250

2-(1,2-DIHYDRO-1-METHYL-2-OXO-3H-INDOLE-3-YLI-DENE)HYDRZAINECARBOTHIOAMIDE see MKW250

3,4-DIHYDRO-2-METHYL-4-OXO-3-o-TOLYLQUINAZOLINE see QAK000

(±)-DIHYDRO-5-(1-METHYL-2-PENTYNYL)-5-(2-PROPENYL)-2-THIOXO-4,6(1H,5H)-PYRIMIDINEDIONE see AGL875

3,6-DIHYDRO-α-((2-METHYLPHENOXY)METHYL)-1(2H)-PYRIDINEETHANOL HY-DROCHLORIDE see TGK225

2,3-DIHYDRO-2-METHYL-9-PHENYL-1H-INDENO(2,1-c)PYRIDINE HYDROBRO-MIDE see NOE525

5,6-DIHYDRO-2-METHYL-N-PHENYL-1,4-OXATHIIN-3-CARBOXAMIDE see CCC500

5,6-DIHYDRO-2-METHYL-N-PHENYL-1,4-OXATHIIN-3-CARBOXAMIDE-4,4-DIOXIDE see DLV200

5,11-DIHYDRO-11-((4-METHYL-1-PIPERAZINYL)ACETYL)-6H-PYRIDO(2,3-b)(1,4)BENZODIAZEPIN-6-ONE DIHYDROCHLORIDE see GCE500

10,11-DIHYDRO-2-(4-METHYL-1-PIPERAZINYL)-11-(2-ATHIAZOLYL)-PYRIDAZI-NO(3,4-b)(1,4)BENZOXAZEPINE see DLV400

10,11-DIHYDRO-2-(4-METHYL-1-PIPERAZINYL)-11-(3,4-XYLYL)PYRIDAZINO(3,4-b)(1,4)BENZOXAZEPINE MALEATE see DLV600

9,10-DIHYDRO-10-(1-METHYL-4-PIPERIDINYLIDENE)-9-ANTHRACENOL HYDRO-CHLORIDE see WAJ000

3,4-DIHYDRO-4-METHYL-2-PIPERIDINYL-1(2H)-NAPTHALENONE HYDROCHLO-RIDE see PIR100

1,2-DIHYDRO-2-(1-METHYL-4-PIPERIDINYL)-5-PHENYL-4-(PHENYLMETHYL)-3H-PYRAZOL-3-ONE (9CI) see BCP650

6,11-DIHYDRO-11-(1-METHYL-4-PIPERIDYLIDENE)-5H-BENZO(5,6)CYCLOHEPTA(1,2-b) PYRIDINE DIMALEATE see DLV800

3,4-DIHYDROMETHYL-2H-PYRAN see MJE750

(S)-(+)-5,6-DIHYDRO-6-METHYL-2H-PYRAN-2-ONE see PAJ500

4,9-DIHYDRO-1-METHYL-3H-PYRIDO(3,4-b)INDOL-7-OL MONOHYDROCHLO-RIDE see HAI300

1,5-DIHYDRO-5-METHYL-1-β-d-RIBOFURANOSYL-1,4,5,6,8-PENTAAZAACENA-PHTHYLEN-3-AMINE see TJE870

4,5-DIHYDRO-2-METHYLTHIAZOLE see DLV900

cis-(−)-3,5-DIHYDRO-3-METHYL-2H-THIOPYRANO(4,3,2-cd)INDOLE-2-CARBOX-YLIC ACID see CML820

2,3-DIHYDRO-5-METHYL-2-THIOXO-4(1H)-PYRIMIDINONE see TFQ650

2,3-DIHYDRO-6-METHYL-2-THIOXO-4(1H)-PYRIMIDINONE see MPW500

DIHYDROMORPHINE HYDROCHLORIDE see DNU310

DIHYDROMORPHINONE see DLW600

DIHYDROMORPHINONE HYDROCHLORIDE see DNU300

DIHYDROMORPHINON-N-OXYD-DITARTARAT (GERMAN) see EBY500

3,4-DIHYDROMORPHOL see PCW500

2,3-DIHYDRO-1-(MORPHOLINOACETYL)-3-PHENYL-4(1H)-QUINAZOLINONE see MRN250

2,3-DIHYDRO-1-(MORPHOLINOACETYL)-3-PHENYL-4(1H)-QUINAZOLINONE HY-DROCHLORIDE see PCJ350

DIHYDROMYRCENE see CMT050

DIHYDROMYRCENOL see DLX000

DIHYDROMYRCENYL ACETATE see DLX100

4,5-DIHYDRONAPHTH(1,2-k)ACEPHENANTHRYLENE see PCW000

3,4-DIHYDRO-1(2H)-NAPHTHALENONE see DLX200

4,5-DIHYDRO-2-(1-NAPHTHALENYLMETHYL)-1H-IMIDAZOLE MONOHYDRO-CHLORIDE see NCW000

4,5-DIHYDRO-2-(1-NAPHTHALENYLMETHYL)-1H-IMIDAZOLE MONONITRATE (9CI) see NAH550

2,3-DIHYDRO-2-(1-NAPHTHALENYL)-4(1H)-QUINAZOLINONE see DLX300

2,3-DIHYDRO-2-(1-NAPHTHYL)-4(1H)-QUINAZOLINONE see DLX300

DIHYDRONE HYDROCHLORIDE see DLX400

DIHYDRONEOPINE see DKW800

1,2-DIHYDRO-5-NITRO-ACENAPHTHYLENE see NEJ500

1,2-DIHYDRO-2-(5′-NITROFURYL)-4-HYDROXY-CHINAZOLIN-3-OXID (GERMAN) see DLX800

1,2-DIHYDRO-2-(5′-NITROFURYL)-4-HYDROXYQUINAZOLINE-3-OXIDE see DLX800

4,5-DIHYDRO-N-NITRO-1-NITROSO-1H-IMIDAZOL-2-AMINE see NLB700

1,3-DIHYDRO-7-NITRO-5-PHENYL-2H-1,4-BENZODIAZEPIN-2-ONE see DLY000

3,6-DIHYDRO-2-NITROSO-2H-1,2-OXAZINE see NJX000

DIHYDRO-1-NITROSO-2,4(1H,3H)-PYRIMIDINEDIONE see NJY000

2,5-DIHYDRO-1-NITROSO-1H-PYRROLE see NLQ000

5,6-DIHYDRO-1-NITROSOURACIL see NJY000

1,2-DIHYDRO-2-(5-NITRO-2-THIENYL)-4(3H)-QUINAZOLINONE see DLY200

1,2-DIHYDRO-2-(5-NITRO-2-THIENYL)QUINAZOLIN-4(3H)-ONE see DLY200

DIHYDRONORDICYCLOPENTADIENYL ACETATE see DLY400

DIHYDRONORGUAIARETIC ACID see NBR000

3,6-DIHYDRO-1,2,2H-OXAZINE see DLY700

DIHYDROOXIRENE see EJN500

2,3-DIHYDRO-3-OXOBENZISOSULFONAZOLE see BCE500

2,3-DIHYDRO-3-OXOBENZISOSULPHONAZOLE see BCE500

5,13-DIHYDRO-5-OXOBENZO(e)(2)BENZOPYRANO(4,3-b)INDOLE see DLY800

S-(3,4-DIHYDRO-4-OXO-1,2,3-BENZOTRIAZIN-3-YLMETHYL) O,O-DIETHYL PHOSPHORODITHIOATE see EKN000

3,4-DIHYDRO-4-OXO-3-BENZOTRIAZINYLMETHYL O,O-DIETHYL PHOSPHORO-DITHIOATE see EKN000

S-(3,4-DIHYDRO-4-OXO-1,2,3-BENZOTRIAZIN-3-YLMETHYL)-O,O-DIMETHYL PHOSPHORODITHIOATE see ASH500

S-(3,4-DIHYDRO-4-OXO-BENZO(α)(1,2,3)TRIAZIN-3-YLMETHYL)-O,O-DIMETHYL PHOSPHORODITHIOATE see ASH500

6,11-DIHYDRO-11-OXO-DIBENZ(b,e)OXEPIN-3-ACETIC ACID see OMU000

3-(1,3-DIHYDRO-1-OXO-2H-ISOINDOL-2-YL)-2,6-DIOXOPIPERIDINE see DVS600

4-(1,3-DIHYDRO-1-OXO-2H-ISOINDOL-2-YL)-α-METHYLBENZENEACETIC ACID see IDA400

3-(1,3-DIHYDRO-1-OXO-2H-ISOINDOL-2-YL)-2-OXOPIPERIDINE see EAJ100

O-(1,6)-(DIHYDRO-6-OXO-1-PHENYLPYRIDAZIN-3-YL), O,O-DIETHYL PHOSPHO-ROTHIOATE see POP000

2,5-DIHYDROXYBENZENESULFONIC ACID with N-ETHYLETHANAMINE see DIS600
3,3'-DIHYDROXYBENZIDINE see DMI400
2,4-DIHYDROXYBENZOFENON (CZECH) see DMI600
2,5-DIHYDROXYBENZOIC ACID see GCU000
3,4-DIHYDROXYBENZOIC ACID see POE200
3,5-DIHYDROXYBENZOIC ACID see REF200
1,4-DIHYDROXY-BENZOL (GERMAN) see HIH000
2,4-DIHYDROXYBENZOPHENONE see DMI600
2,2'-DIHYDROXYBINAPHTHALENE see BGC100
2,2'-DIHYDROXYBIPHENYL see BGG000
2,5-DIHYDROXYBIPHENYL see BGG250
2,6-DIHYDROXY-5-BIS(2-CHLOROETHYL)AMINOPYRAMIDINE see BIA250
1,4-DIHYDROXY-5,8-BIS((2-HYDROXYETHYL)AMINO)-9,10-ANTHRACENEDIONE see DMM400
1,4-DIHYDROXY-5,8-BIS(2-((2-HYDROXYETHYL)AMINO)ETHYLAMINO)-9,10-ANTHRACENEDIONE DIACETATE see DBE875
1,4-DIHYDROXY-5,8-BIS((2-((HYDROXYETHYL)AMINO)ETHYL)AMINO)-9,10-ANTHRACENEDIONE (9CI) see MQY090
(4,5-DIHYDROXY-1,3-BIS(HYDROXYMETHYL))-2-IMIDAZOLIDINONE see DTG000
2,5-DIHYDROXY-3,6-BIS(5-(3-METHYL-2-BUTENYL)-1H-INDOL-3-YL)-2,5-CYCLO-HEXADIENE-1,4-DIONE see CNF159
3-β,14-β-DIHYDROXYBUFA-4,20,22-TRIENOLIDE 3-RHAMNOSIDE see POB500
1,3-DIHYDROXYBUTANE see BOS500
1,4-DIHYDROXYBUTANE see BOS750
2,3-DIHYDROXYBUTANE see BOT000
2,3-DIHYDROXYBUTANEDIOIC ACID see TAF750
2,3-DIHYDROXYBUTANEDIOIC ACID, DIAMMONIUM SALT see DCH000
2,3-DIHYDROXY-(R-(R*,R*))-BUTANEDIOIC ACID DISODIUM SALT (9CI) see BLC000
(R*,S*)-2,3-DIHYDROXY-BUTANEDIOIC ACID ION(2−) (9CI) see TAF775
1,4-DIHYDROXY-1,4-BUTANEDISULFONIC ACID, DISODIUM SALT see SMX000
1,4-DIHYDROXY-2-BUTENE see BOX300
DIHYDROXYBUTENEDIOIC ACID see DMW200
2,6-DIHYDROXY-4-CARBOXYPYRIDINE see DMV400
(3-β,5-β)-3,14-DIHYDROXY-CARD-20(22)-ENOLIDE see DMJ000
3,β,14-DIHYDROXY-5,β-CARD-20(22)ENOLIDE see DMJ000
3-β,14-DIHYDROXY-5-β-CARD-20(22)-ENOLIDE-3-FORMATE see FNK050
(±)-2,3-DIHYDROXYCHLOROPROPANE see CHL875
DIHYDROXYCHLOROTHIAZIDUM see CFY000
DIHYDROXY 3-12 CHOLANATE de Na (FRENCH) see SGE000
3,12-DIHYDROXYCHOLANIC ACID see DAQ400
3-α,12-α-DIHYDROXYCHOLANIC ACID see DAQ400
3-α,7-α-DIHYDROXYCHOLANIC ACID see CDL325
3-α,7-β-DIHYDROXYCHOLANIC ACID see DMJ200
3,7-DIHYDROXYCHOLAN-24-OIC ACID see DMJ200
(3-α,5-β,7-β)-3,7-DIHYDROXYCHOLAN-24-OIC ACID see DMJ200
3-α,12-α-DIHYDROXY-5-β-CHOLAN-24-OIC ACID see DAQ400
3-α,7-α-DIHYDROXY-5-β-CHOLAN-24-OIC ACID see CDL325
3-α,7-β-DIHYDROXY-6-β-CHOLAN-24-OIC ACID see DMJ200
3-α,12-α-DIHYDROXY-5-β-CHOLANOIC ACID see DAQ400
3-α,7-β-DIHYDROXY-5-β-CHOLANOIC ACID see DMJ200
3-α-12-α-DIHYDROXY-5-β-CHOLAN-24-OIC ACID with DI-BENZ(a,h)ANTHRACENE see DCT800
(3-α,5-β,12-α)-3,12-DIHYDROXY-CHOLAN-24-OIC ACID MONOSODIUM SALT see SGE000
3-α,12-α-DIHYDROXY-5-β-CHOLAN-24-OIC ACID SODIUM SALT see SGE000
3-α,12-α-DIHYDROXYCHOLANSAEURE (GERMAN) see DAQ400
3-α,7-α-DIHYDROXYCHOLANSAEURE (GERMAN) see DMJ200
1,25-DIHYDROXYCHOLECALCIFEROL see DMJ400
1-α,25-DIHYDROXYCHOLECALCIFEROL see DMJ400
1a,25-DIHYDROXYCHOLECALCIFEROL see DMJ400
3-α,12-α-DIHYDROXY-5-β-CHOL-8(14)-EN-24-OIC ACID see AQO500
3,4-DIHYDROXYCINNAMIC ACID see CAK375
DIHYDROXYCODEINONE HYDROCHLORIDE see DLX400
DI-(4-HYDROXY-3-COUMARINYL)METHANE see BJZ000
3,4-DIHYDROXY-3-CYCLOBUTENE-1,2-DIONE see DMJ600
3,4-DIHYDROXYCYCLOBUTENE-1,2-DIONE see DMJ600
DIHYDROXYCYCLOBUTENEDIONE see DMJ600
(−)-3,14-DIHYDROXY-N-(CYCLOBUTYLMETHYL)MORPHINAN see CQF079
2,3-DIHYDROXY-2,4,6-CYCLOHEPTATRIEN-1-ONE see HON500
1,8-DIHYDROXY-4,5-DIAMINOANTHRAQUINONE see DBQ250
1,5-DIHYDROXY-4,8-DIAMINOANTHRACHINON (CZECH) see DBP909
1,5-DIHYDROXY-4,8-DIAMINOANTHRAQUINONE see DBP909
16,17-DIHYDROXYDIBENZANTHRONE see DMJ800
DIHYDROXYDIBENZANTHRONE see DMJ800
2,2'-DIHYDROXY-5,5'-DICHLORODIPHENYLMETHANE see MJM500
2,2'-DIHYDROXYDIETHYLAMINE see DHF000
4,4'-DIHYDROXY-α,β-DIETHYLDIPHENYLETHANE see DLB400
β,β'-DIHYDROXYDIETHYL ETHER see DJD600
DIHYDROXYDIETHYL ETHER see DJD600
4,4'-DIHYDROXY-α,α'-DIETHYL-STILBEN-DIPHOSPHATE TETRASODIUM see TEE300
4,4'-DIHYDROXYDIETHYLSTILBENE see DKA600
4,4'-DIHYDROXY-α,β-DIETHYLSTILBENE see DKA600

4,4'-DIHYDROXY-α,β-DIETHYLSTILBENE DIPROPIONATE see DKB000
DIHYDROXYDIETHYLSTILBENE DIPROPIONATE see DKB000
4,4'-DIHYDROXY-α,β-DIETHYLSTILBENE PALMITATE see DKA800
β,β'-DIHYDROXYDIETHYL SULFIDE see TFI500
trans-10,11-DIHYDROXY-10,11-DIHYDROBENZ(a)ANTHRACENE see BBF000
trans-1,2-DIHYDROXY-1,2-DIHYDROBENZ(a)ANTHRACENE see BBD250
trans-3,4-DIHYDROXY-3,4-DIHYDROBENZ(a)ANTHRACENE see BBD500
trans-5,6-DIHYDROXY-5,6-DIHYDROBENZ(a)ANTHRACENE see BBE250
trans-8,9-DIHYDROXY-8,9-DIHYDROBENZ(a)ANTHRACENE see BBE750
trans-1,2-DIHYDROXY-1,2-DIHYDROBENZ(a,h)ANTHRACENE see DMK200
(E)-3,4-DIHYDROXY-3,4-DIHYDROBENZ(a)ANTHRACENE-7,12-DIMETHANOL see DML775
trans-1,2-DIHYDROXY-1,2-DIHYDROBENZO(a,h)ANTHRACENE see DMK200
trans-4,5-DIHYDROXY-4,5-DIHYDROBENZO(e)PYRENE see DMK400
trans-9,10-DIHYDROXY-9,10-DIHYDROBENZO(e)PYRENE see DMK600
trans-4,5-DIHYDROXY-4,5-DIHYDROBENZO(a)PYRENE see DLC600
(+)-trans-7,8-DIHYDROXY-7,8-DIHYDROBENZO(a)PYRENE see DML400
(−)-trans-7,8-DIHYDROXY-7,8-DIHYDROBENZO(a)PYRENE see DML200
(+,−)-trans-7,8-DIHYDROXY-7,8-DIHYDROBENZO(a)PYRENE see DML000
trans-1,2-DIHYDROXY-1,2-DIHYDROCHRYSENE see DLD200
trans-3,4-DIHYDROXY-3,4-DIHYDRO-7,12-DIHYDROXYMETHYL-BENZ(a)ANTHRACENE see DML775
trans-3,4-DIHYDROXY-3,4-DIHYDRO-7-METHYLBENZ(c)ACRIDINE see MGU550
11,12-DIHYDROXY-11,12-DIHYDRO-3-METHYLCHOLANTHRENE (E) see DML800
trans-3,4-DIHYDROXY-3,4-DIHYDRO-7-METHYL-12-HYDROXYMETHYL-BENZ(a)ANTHRACENE see DMX000
5,8-DIHYDROXY-1,4-DIHYDROXYETHYLAMINOANTHRAQUINONE see DMM400
6-β,14-DIHYDROXY-3,4-DIMETHOXY-N-METHYLMORPHINAN see ORE000
(1'-α)-7',12'-DIHYDROXY-6,6'-DIMETHOXY-2,2,2',2'-TETRAMETHYLTUBOCU-RARANIUM (9CI) see COF825
d-7',12'-DIHYDROXY-6,6'-DIMETHOXY-2,2',2'-TRIMETHYLTUBOCURARANIUM CHLORIDE see TOA000
3,4-DIHYDROXY-α-(DIMETHYLAMINOMETHYL) BENZYL ALCOHOL see MJV000
2,5-DIHYDROXY-3-DIMETHYLAMINO-5-METHYL-2-CYCLOPENTEN-1-ONE see DXS200
2,4-DIHYDROXY-3,3-DIMETHYLBUTYRONITRILE see DMM600
N-(2,4-DIHYDROXY-3,3-DIMETHYLBUTYRYL)-β-ALANINE CALCIUM see CAU750
(R)-N-(2,4-DIHYDROXY-3,3-DIMETHYLBUTYRYL)-β-ALANINE ZINC SALT (2:1) see ZKS000
(R)-4-((2,4-DIHYDROXY-3,3-DIMETHYL-1-OXOBUTYL)AMINO)-BUTANOIC ACID CALCIUM SALT (2:1) see CAT125
2,2'-DIHYDROXYDINAPHTHYL see BGC100
1,8-DIHYDROXY-4,5-DINITROANTHRAQUINONE see DMN400
4,4'-DIHYDROXYDIPHENYLAMINE see IBJ100
4,4'-DIHYDROXYDIPHENYLDIMETHYLMETHANE see BLD500
p,p'-DIHYDROXYDIPHENYLDIMETHYLMETHANE see BLD500
4,4'-DIHYDROXYDIPHENYLDIMETHYLMETHANE DIGLYCIDYL ETHER see BLD750
p,p'-DIHYDROXYDIPHENYLDIMETHYLMETHANE DIGLYCIDYL ETHER see BLD750
4,4'-DIHYDROXY-γ,Δ-DIPHENYLHEXANE see DLB400
2,2'-DIHYDROXY-N-(4,5-DIPHENYLOXAZOLE-2-YL)DIETHYLAMINE MONOHY-DRATE see BKB250
2,2-(4,4'-DIHYDROXYDIPHENYL)PROPANE see BLD500
4,4'-DIHYDROXY-2,2-DIPHENYLPROPANE see BLD500
4,4'-DIHYDROXYDIPHENYL-2,2-PROPANE see BLD500
4,4'-DIHYDROXYDIPHENYLPROPANE see BLD500
p,p'-DIHYDROXYDIPHENYLPROPANE see BLD500
DIHYDROXYDIPHENYLSILANE see DMN450
4,4'-DIHYDROXYDIPHENYL SULFIDE see TFJ000
9,10-DIHYDROXY-9,10-DI-n-PROPYL-9,10-DIHYDRO-1,2:5,6-DIBENZANTHRA-CENE see DLK600
2,2'-DIHYDROXYDIPROPYL ETHER see OQM000
2,2'-DIHYDROXY-DI-n-PROPYLNITROSOAMINE see DNB200
2,4-DIHYDROXY-3,5-DI(4-SULPHO-1-NAPHTHYLAZO)BENZYL ALCOHOL, DISO-DIUM SALT see CMP500
(+)-trans-7-β,8-α-DIHYDROXY-9-α,10-α-EPOXY-7,8,9,10-TETRAHDYROBEN-ZO(a)PYRENE see BMK620
(±)-trans-8-β,9-α-DIHYDROXY-10-β,11-β-EPOXY-8,9,10,11-TETRAHYDROB see DMP200
(±)trans-8-β,9-α-DIHYDROXY-10-α,11-α-EPOXY-8,9,10,11-TETRAHYDROBE see DMO600
(±)-trans-1,β,2,α-DIHYDROXY-3,α,4,α-EPOXY-1,2,3,4-TETRAHYDROBENZ see DMP000
(+)-trans-3,4-DIHYDROXY-1,2-EPOXY-1,2,3,4-TETRAHYDROBENZ(a) ANTHRA-CENE see DMO800
(±)-3-α,4-β-DIHYDROXY-1-α,2-α-EPOXY-1,2,3,4-TETRAHYDRO-BENZ(a)ANTHRACENE see DLE000
(E)-(+)-3,4-DIHYDROXY-1,2-EPOXY-1,2,3,4-TETRAHYDROBENZ(a)ANTHRACENE see DMO800
(±)-cis-3,4-DIHYDROXY-1,2-EPOXY-1,2,3,4-TETRA-HYDRO-BENZ(a)ANTHRACENE see DMO500

(+)cis-7,α,8,β-DIHYDROXY-9,α,10,α-EPOXY-7,8,9,10-TETRAHYDROBENZO see DMP600

(−)-cis-7,β,8,α-DIHYDROXY-9,β,10,β-EPOXY-7,8,9,10-TETRAHYDROBENZO see DMP800

(±)-9,β,10,α-DIHYDROXY-11,α,12,α-EPOXY-9,10,11,12-TETRAHYDROBENZO see DMR200

(±)-9-α-10-β-DIHYDROXY-11-β,12-β-EPOXY-9,10,11,12-TETRAHYDROBENZO see DMR150

(±)-7,β,8,α-DIHYDROXY-9,β,10,β-EPOXY-7,8,9,10-TETRAHYDROBENZO(a) see DMR000

(±)-cis-3,4-DIHYDROXY-1,2-EPOXY-1,2,3,4-TETRAHYDROBEN-ZO(a)ANTHRACENE see DMO500

(+)-trans-3,4-DIHYDROXY-1,2-EPOXY-1,2,3,4-TETRAHYDROBEN-ZO(a)ANTHRACENE see DMO800

(±)-3-α-4-β-DIHYDROXY-1-β,2-β-EPOXY-1,2,3,4-TETRAHYDROBENZ O(c)PHENANTHRENE see BMK634

(±)-(E)-7,8-DIHYDROXY-9,10-EPOXY-7,8,9,10-TETRAHYDROBENZO(a)PYRENE see DMP900

(±)-trans-7,8-DIHYDROXY-9,10-EPOXY-7,8,9,10-TETRAHYDRO-BEN-ZO(a)PYRENE see DMQ000

anti-(±)-7-β,8-α-DIHYDROXY-9-α,10-α-EPOXY-7,8,9,10-TETRAHYDROBEN-ZO(a)PYRENE see BMK630

(±)-7-α,8-β-DIHYDROXY-9-α,10-α-EPOXY-7,8,9,10-TETRAHYDROBEN-ZO(a)PYRENE see DMR000

(−)-7-α,8-β-DIHYDROXY-9-β,10-β-EPOXY-7,8,9,10-TETRAHYDROBEN-ZO(a)PYRENE see DVO175

(±)-3-α,4-β-DIHYDROXY-1-α,2-α-EPOXY-1,2,3,4-TETRAHYDRO-BENZ(c)PHENANTHRENE see BMK635

(±)-1,β,2,α-DIHYDROXY-3-α,4,α-EPOXY-1,2,3,4-TETRAHYDROCHRYSENE see DMS200

(±)-1,β,2,α-DIHYDROXY-3-β,4,β-EPOXY-1,2,3,4-TETRAHYDROCHRYSENE see DMS400

trans-1,2-DIHYDROXY-anti-3,4-EPOXY-1,2,3,4-TETRAHYDROCHRYSENE see DMS000

( ±)-1-β,2-α-DIHYDROXY-3-α,4-α-EPOXY-1,2,3,4-TETRAHYDROPHENANTHRENE see ECS000

3,17-β-DIHYDROXY-1,3,5(10)-ESTRATRIENE see EDO000

3,17-β-DIHYDROXYESTRA-1,3,5(10)-TRIENE see EDO000

3,17-DIHYDROXYESTRATRIENE see EDO500

DIHYDROXYESTRIN see EDO000

1,2-DIHYDROXYETHANE see EJC500

DI-β-HYDROXYETHOXYETHANE see TJQ000

DI(2-HYDROXYETHYL)AMINE see DHF000

N,N-DI(2-HYDROXYETHYL)ANILINE see BKD500

N,N-DI(β-HYDROXYETHYL)ANILINE see BKD500

DIHYDROXYETHYLANILINE see BKD500

1,2-DIHYDROXYETHYLBENZENE see SMQ100

α-β-DIHYDROXYETHYLBENZENE see SMQ100

DI(2-HYDROXYETHYL)BENZYLDODECYLAMMONIUM CHLORIDE see BDJ600

N,N-DIHYDROXYETHYL-3-CHLOROANILINE see CJU200

N,N-DI(2-HYDROXYETHYL)CYCLOHEXYLAMINE see DMT400

N,N-DI(2-HYDROXYETHYL)CYCLOHEXYLAMINE see DMT400

2,2′-DIHYDROXYETHYL ETHER see DJD600

DI(HYDROXYETHYL) ETHER DINITRATE see DJE400

N,N-DIHYDROXYETHYL GLYCINE see DMT500

β,β′-DIHYDROXYETHYL SULFIDE see TFI500

N,N-DIHYDROXYETHYL-m-TOLUIDINE see DHF400

DI(HYDROXYETHYL)-o-TOLYLAMINE see DMT800

3,17-β-DIHYDROXY-17-α-ETHYNYL-1,3,5(10)-ESTRATRIENE see EEH500

3,17-β-DIHYDROXY-17-α-ETHYNYL-1,3,5(10)-OESTRATRIENE see EEH500

3′,6′-DIHYDROXYFLUORAN see FEV000

DIHYDROXYFLUORANE see FEV000

11-β,17-β-DIHYDROXY-9-α-FLUORO-17-α-METHYL-4-ANDROSTER-3-ONE see AOO275

2,2′-DIHYDROXY-3,3′,5,5′,6,6′-HEXACHLORODIPHENYLMETHANE see HCL000

2,2′-DIHYDROXY-3,5,6,3′,5′,6′-HEXACHLORODIPHENYLMETHANE see HCL000

3,4-DIHYDROXY-1,5-HEXADIENE see DMU000

1-6,10a-β-DIHYDROXY-1,2,3,9,10,10a-HEXAHYDRO-4H(10),4a-IMINOETHANO-PHENANTHRENE TARTRATE see NNJ600

1,6-DIHYDROXYHEXANE see HEP500

1,8-DIHYDROXY-4-(p-(2-HYDROXYETHYL)ANILINO)-5-NITROANTHRAQUINONE see CMP060

1,8-DIHYDROXY-4-(4′-β-HYDROXYETHYL)ANILINO-6-NITROANTHROQUINONE see CMP060

(2′S,3′R,6′R)-DIHYDROXY-2′-(HYDROXYMETHYL)-2′,4′,6′-TRIMETHYL-SPI-RO(CYCLOPROPANE-1,5′-(5H)INDEN)-7′(6′H)-ONE, 2′,3′-DIHYDRO-3′,6′- see LIO600

1,8-DIHYDROXY-3-HYDROXYMETHYL-10-(6-HYDROXYMETHYL-3,4,5-TRIHY-DROXY-2-PYRANYL)ANTHRONE see BAF825

7-(3,5-DIHYDROXY-2-(3-HYDROXY-1-OCTENYL)CYCLOPENTYL)-5-HEPTENOIC ACID, TRIMETHAMINE SALT see POC750

7-(3,5-DIHYDROXY-2-(3-HYDROXY-1-OCTENYL)CYCLOPENTYL)-5-HEPTENOIC ACID, THAM see POC750

7-(3,5-DIHYDROXY-2-(3-HYDROXY-1-OCTENYL)CYCLOPENTYL)-5-HEPTENOIC ACID see POC500

5,7-DIHYDROXY-2-(4-HYDROXYPHENYL)-4H-1-BENZOPYRAN-4-ONE see CDH250

2,3-DIHYDROXY-2-((4-HYDROXYPHENYL)METHYL)BUTANEDIOIC ACID see HJO500

dl-3,4-DIHYDROXY-N-3-(4-HYDROXYPHENYL)-1-METHYL-n-PROPYL PHENETH-YLAMINE HYDROCHLORIDE see DXS375

d-(+)-2,4-DIHYDROXY-N-(3-HYDROXYPROPYL)-3,3-DIMETHYLBUTYRAMIDE see PAG200

2,2-DIHYDROXY-1,3-INDANDIONE see DMV200

2,2-DIHYDROXY-1H-INDENE-1,3(2H)-DIONE see DMV200

2,6-DIHYDROXYISONICOTINIC ACID see DMV400

3,5-DIHYDROXY-α-((ISOPROPYLAMINO)METHYL)BENZYL ALCOHOL see DMV800

3,4-DIHYDROXY-α-((ISOPROPYLAMINO)METHYL)BENZYL ALCOHOL see DMV600

3,4-DIHYDROXY-α-((ISOPROPYLAMINO)METHYL)BENZYL ALCOHOL HYDRO-CHLORIDE see IMR000

(±)-3,4-DIHYDROXY-α-((ISOPROPYLAMINO)METHYL)BENZYL ALCOHOL HY-DROCHLORIDE see IQS500

3,5-DIHYDROXY-α-((ISOPROPYLAMINO)METHYL)BENZYL ALCOHOL SULFATE see MDM800

β,β′-DIHYDROXYISOPROPYL CHLORIDE see CDT750

2,2′-DIHYDROXYISOPROPYL ETHER see OQM000

3,5-DIHYDROXY-4-ISOVALERYL-2,6,6-TRIS(3-METHYL-2-BUTENYL)-2,4-CYCLO-HEXADIEN-1-ONE see LIU000

5,6-DIHYDRO-2-(2,6-XYLIDINO)-4H-1,3-THIAZINE see DMW000

DIHYDROXYMALEIC ACID see DMW200

1,2-DIHYDROXY-3-(2-METHOXYPHENOXY)PROPANE see RLU000

2-(3,4-DIHYDROXY-5-METHOXYPHENYL)-3,5,7-TRIHYDROXYBENZOPYRYLIUM, ACID ANION see PCU000

3,4′-DIHYDROXY-2-(METHYLAMINO)ACETOPHENONE see MGC350

3,4-DIHYDROXY-α-METHYLAMINOACETOPHENONE see MGC350

3,4-DIHYDROXY-α-((METHYLAMINO)METHYL)BENZYL ALCOHOL see VGP000

(±)-3,4-DIHYDROXY-α-((METHYLAMINO)METHYL)BENZYL ALCOHOL HYDRO-CHLORIDE see AES625

(− )-3,4-DIHYDROXY-α-(((METHYLAMINO)METHYL)BENZYL) ALCOHOL (+)-TARTRATE (1: 1) SALT see AES000

3-DI(HYDROXYMETHYL)AMINO-6-(5-NITRO-2-FURYLETHENYL)-1,2,4-TRIAZINE see BKH500

3-DI(HYDROXYMETHYL)AMINO-6-(2-(5-NITRO-2-FURYL)VINYL)-1,2,4-TRIAZINE see BKH500

7: 12-DIHYDROXYMETHYLBENZ(a)ANTHRACENE see BBF500

1,3-DIHYDROXY-5-METHYLBENZENE see MPH500

6,7-DIHYDROXY-4-METHYL-2H-1-BENZOPYRAN-2-ONE see MJV800

2,6-DIHYDROXY-4-METHYL-5-BIS(2-CHLOROETHYL)AMINOPYRIMIDINE see DYC700

2,2-(DIHYDROXYMETHYL)-1-BUTANOL, MONOALLYL ETHER see TLX110

cis-1,2-DIHYDROXY-3-METHYLCHOLANTHRENE see MIK750

6,7-DIHYDROXY-4-METHYLCOUMARIN see MJV800

(E)-3,4-DIHYDROXY-7-METHYL-3,4-DIHYDROBENZ(a)ANTHRACENE-12-METHA-NOL see DMX000

2,12-DIHYDROXY-4-METHYL-11,16-DIOXOSENECIONANIUM see DMX200

DI-4-HYDROXY-3,3′-METHYLENEDICOUMARIN see BJZ000

DIHYDROXYMETHYL FURATRIZINE see BKH500

(5Z,11-α,13E,15S,17Z)-11,15-DIHYDROXY-15-METHYL-9-OXO-PROSTA-5,13-DIEN-1-OIC ACID see MOV800

11-α,17-β-DIHYDROXY-17-METHYL-3-OXOANDROSTA-1,4-DIENE-2-CARBOXAL-DEHYDE see FNK040

14,16-DIHYDROXY-3-METHYL-7-OXO-trans-BENZOXACYCLOTETRADEC-11-EN-1-ONE see ZAT000

(11-α-13E)-(±)-11,16-DIHYDROXY-16-METHYL-9-OXOPROST-13-EN-1-OIC ACID METHYL ESTER see MJE775

2,4-DIHYDROXY-2-METHYLPENTANE see HFP875

1,2-DIHYDROXY-3-(2-METHYLPHENOXY)PROPANE see GGS000

α,β-DIHYDROXY-γ-(2-METHYLPHENOXY)PROPANE see GGS000

3,4-DIHYDROXY-3-METHYL-4-PHENYL-1-BUTYNE see DMX800

d-N,N′-DI(1-HYDROXYMETHYLPROPYL)ETHYLENEDIAMINE DIHYDROCHLO-RIDE mixed with SODIUM NITRITE see SIS500

N,N′-DIHYDROXYMETHYLUREA see DTG700

1,8-DIHYDROXY-10-MYRISTOYL-9-ANTHRONE see MSA750

2,7-DIHYDROXYNAPHTHALENE see NAO500

N,3-DIHYDROXY-N-(1-NAPHTHALENYLOXY)BUTANINIDAMIDE HYDROCHLO-RIDE see NAC500

d-threo-N-(1,1′-DIHYDROXY-1-p-NITROPHENYLISOPRO-PYL)DICHLOROACETAMIDE see CDP250

2,2′-DIHYDROXY-N-NITROSODIETHYLAMINE see NKM000

3,4-DIHYDROXYNOREPHEDRINE HYDROCHLORIDE see AMB000

3,17-α-DIHYDROXYOESTRA-1,3,5(10)-TRIENE see EDO500

3,17-β-DIHYDROXY-1,3,5(10)-OESTRATRIENE see EDO000

3,17-β-DIHYDROXYOESTRA-1,3,5-TRIENE see EDO000

DIHYDROXYOESTRIN see EDO000

(11-β)-11,21-DIHYDROXY-17-(1-OXOBUTOXY)-PREGN-4-ENE-3,20-DIONE see HHQ825

(5Z,11-α,13E,15S)-11,15-DIHYDROXY-9-OXOPROSTA-5,13-DIEN-1-OIC ACID see DVJ200

(11-α,13E,15S)-11,15-DIHYDROXY-9-OXO-PROST-13-EN-1-OIC ACID (9CI) see POC350

11,15-DIHYDROXY-9-OXO-PROST-13-EN-1-OIC ACID, (11-α,13E,15S)-, and α-CYCLODEXTRIN see AGW275

1,8-DIHYDROXY-10-(1-OXOTETRADECYL)-9(10H)-ANTHRACENONE see MSA750

1,2-DIHYDROXY-2-OXO-N-(2,6-XYLYL)-3-PYRIDINECARBOXAMIDE see XLS300

trans-7,8-DIHYDROXY-9,10-OXY-7,8,9,10-TETRAHYDROBENZO(a)PYRENE see BCU250

anti-r-7,trans-8-DIHYDROXY-trans-9,10-OXY-7,8,9,10-TETRAHYDROBENZO(a)PYRENE see BCU250

2,2′-DIHYDROXY-3,3′,5,5′,6-PENTACHLOROBENZANILIDE see DMZ000

1,5-DIHYDROXYPENTANE see PBK750

3,5-DIHYDROXYPHENOL see PGR000

3,4-DIHYDROXYPHENYLALANINE see DNA200

β-(3,4-DIHYDROXYPHENYL)-α-ALANINE see DNA200

l-3,4-DIHYDROXYPHENYL-α-ALANINE see DNA200

3-(3,4-DIHYDROXYPHENYL)-l-ALANINE see DNA200

l-3-(3,4-DIHYDROXYPHENYL)ALANINE see DYC200

l-3,4-DIHYDROXYPHENYLALANINE see DNA200

3,4-DIHYDROXY-l-PHENYLALANINE see DNA200

3,4-DIHYDROXYPHENYL-l-ALANINE see DNA200

l-α-DIHYDROXYPHENYLALANINE see DNA200

l-β-(3,4-DIHYDROXYPHENYL)ALANINE see DNA200

β-(3,4-DIHYDROXYPHENYL)-l-ALANINE see DNA200

l-DIHYDROXYPHENYL-l-ALANINE see DNA200

DIHYDROXY-l-PHENYLALANINE see DNA200

(−)-3-(3,4-DIHYDROXYPHENYL)-l-ALANINE see DNA200

(−)-3,4-DIHYDROXYPHENYLALANINE see DNA200

l-3-(3,4-DIHYDROXYPHENYL)ALANINE METHYL ESTER see DYC300

1-(3,4-DIHYDROXYPHENYL)-2-AMINO-1-BUTANOL HYDROCHLORIDE see ENX500

l-1-(3,4-DIHYDROXYPHENYL)-2-AMINOETHANOL see NNO500

3,4-DIHYDROXYPHENYL)-1-AMINO-2-ETHANOL-1-HYDROCHLORIDE see NNP050

3,4-DIHYDROXYPHENYLAMINOPROPANOL HYDROCHLORIDE see AMB000

1-(3,5-DIHYDROXYPHENYL)-2-tert-BUTYLAMINOETHANOL SULPHATE see TAN250

2-(3,4-DIHYDROXYPHENYL)-2,3-DIHYDRO-3,5,7-TRIHYDROXY-4H-1-BENZOPYRAN-4-ONE see DMD000

(2R-trans)-2-(3,4-DIHYDROXYPHENYL)-2,3-DIHYDRO-3,5,7-TRIHYDROXY-4H-1-BENZOPYRAN-4-ONE see DMD000

α-(3,4-DIHYDROXYPHENYL)-β-DIMETHYLAMINOETHANOL see MJV000

l-3,4-DIHYDROXYPHENYLETHANOLAMINE see NNO500

DIHYDROXYPHENYLETHANOLISOPROPYLAMINE see DMV600

1-(2,4-DIHYDROXYPHENYL)ETHANONE see DMG400

3,4-DIHYDROXYPHENYLETHYLMETHYLAMINE HYDROCHLORIDE see EAZ000

meso-3,4-DI(p-HYDROXYPHENYL)-n-HEXANE see DLB400

γ,Δ-DI(p-HYDROXYPHENYL)-HEXANE see DLB400

3,4′(4,4′-DIHYDROXYPHENYL)HEX-3-ENE see DKA600

1-(3,4-DIHYDROXYPHENYL)-1-HYDROXY-2-AMINOBUTANE HYDROCHLORIDE see ENX500

1-(3,5-DIHYDROXY-PHENYL-2-((1-(4-HYDROXYBENZYL)ETHYL)AMINO)-ETHANOL) HYDROBROMIDE see FAQ100

α-(3,4-DIHYDROXYPHENYL)-α-HYDROXY-β-DIMETHYLAMINOETHANE see MJV000

7-(3-(2-(3,5-DIHYDROXYPHENYL-2-HYDROXY-ETHYLAMINO)PROPYL))THEOPHYLLINE HYDROCHLORIDE see DNA600

1-(3,4-DIHYDROXYPHENYL)-2-ISOPROPYLAMINOETHANOL see DMV600

(±)1-(3,4-DIHYDROXYPHENYL)-2-ISOPROPYLAMINOETHANOL HYDROCHLORIDE see IQS500

1-(3,5-DIHYDROXYPHENYL)-2-(ISOPROPYLAMINO)ETHANOL SULFATE see MDM800

dl-α-3,4-DIHYDROXYPHENYL-β-ISOPROPYLAMINOETHANOL SULFATE see IRU000

l-(−)-3-(3,4-DIHYDROXYPHENYL)-2-METHYLALANINE see DNA800

l(−)-β-(3,4-DIHYDROXYPHENYL)-α-METHYLALANINE see DNA800

3,4-DIHYDROXYPHENYL-1-METHYLAMINO-2-ETHANE HYDROCHLORIDE see EAZ000

1-1-(3,4-DIHYDROXYPHENYL)-2-METHYLAMINOETHANOL see VGP000

1-1-(3,4-DIHYDROXYPHENYL)-2-METHYLAMINO-1-ETHANOL HYDROCHLORIDE see AES500

1-(3,4-DIHYDROXYPHENYL)-2-(METHYLAMINO)-ETHANONE (9CI) see MGC350

2-(3,4-DIHYDROXYPHENYL)-2,3,4,5,7-PENTAHYDROXY-1-BENZOPYRAN see HBA259

(DIHYDROXYPHENYL)PHENYL MERCURY see PFO250

α-DI(p-HYDROXYPHENYL)PHTHALIDE see PDO750

2,2-DI(4-HYDROXYPHENYL)PROPANE see BLD500

β-DI-p-HYDROXYPHENYLPROPANE see BLD500

3,4-DIHYDROXYPHENYLPROPANOLAMINE HYDROCHLORIDE see AMB000

3-(3,4-DIHYDROXYPHENYL)-2-PROPENOIC ACID (9CI) see CAK375

2-(3,4-DIHYDROXYPHENYL)-3,5,7-TRIHYDROXY-4H-1-BENZOPYRAN-4-ONE see QCA000

2-(3,4-DIHYDROXYPHENYL)-3,5,7-TRIHYDROXY-1-BENZOPYRYLIUM CHLORIDE see COI750

DIHYDROXYPHTHALOPHENONE see PDO750

17,21-DIHYDROXYPREGNA-1,4-DIENE-3,11,20-TRIONE see PLZ000

17α,21-DIHYDROXY-4-PREGNENE-3,11,20-TRIONE see CNS800

17,21-DIHYDROXY-PREGN-4-ENE-3,11,20-TRIONE 21-ACETATE see CNS825

17,21-DIHYDROXYPREGN-4-ENE-3,11,20-TRIONE ACETATE see CNS825

1,2-DIHYDROXYPROPANE see PML000

1,3-DIHYDROXYPROPANE see PML250

2,3-DIHYDROXYPROPYL ACETATE see GGO000

(17,21-α)-17,21-DIHYDROXY-4-PROPYLAJMALANIUM see PNC875

17R,21-α-DIHYDROXY-4-PROPYLAJMALANIUM HYDROGEN TARTRATE see DNB000

(17R,21-α)-17,21-DIHYDROXY-4-PROPYLAJMALINIUM BROMIDE see PNC925

2,3-DIHYDROXYPROPYLAMINE see AMA250

DI(2-HYDROXY-n-PROPYL)AMINE see DNB200

4-(2,3-DIHYDROXYPROPYLAMINO)-2-(5-NITRO-2-THIENYL)QUINAZOLINE see DNB600

2,3-DIHYDROXYPROPYL CHLORIDE see CDT750

7-(2,3-DIHYDROXYPROPYL)-3,7-DIHYDRO-1,3-DIMETHYL-1H-PURINE-2,5-DIONE see DNC000

N,N-DI-(2-HYDROXYPROPYL)NITROSAMINE see DNB200

1-((2,3-DIHYDROXYPROPYL)NITROSAMINO)-2-PROPANONE see NJY550

4-(o-(2′,3′-DIHYDROXYPROPYLOXYCARBONYL)PHENYL)-AMINO-8-TRIFLUORO-METHYLQUINOLINE see TKG000

7-(2,3-DIHYDROXYPROPYL)THEOPHYLLINE see DNC000

DIHYDROXYPROPYL THEOPHYLLINE see DNC000

DIHYDROXYPROPYL THEOPYLIN (GERMAN) see DNC000

(1,2-DIHYDROXY-3-PROPYL)THIOPHYLLIN see DNC000

2,3-DIHYDROXYPROPYL-N-(8-(TRIFLUOROMETHYL)-4-QUINO-LYL)ANTHRANILATE see TKG000

2,4-DIHYDROXY-2H-PYRAN-Δ-3(6H),α-ACETIC ACID-3,4-LACTONE see CMV000

(2,4-DIHYDROXY-2H-PYRAN-3(6H)-YLIDENE)ACETIC ACID-3,4-LACTONE see CMV000

5,7-DIHYDROXY-PYRIDOTETRAZOLE-6-CARBONITRILE see DND900

2,4-DIHYDROXYPYRIMIDINE see UNJ800

4,8-DIHYDROXYQUINALDIC ACID see DNC200

4,8-DIHYDROXYQUINALDINIC ACID see DNC200

2,4-DIHYDROXYQUINAZOLINE see QEJ800

4,8-DIHYDROXYQUINOLINE-2-CARBOXYLIC ACID see DNC200

2,3-DIHYDROXYQUINOXALINE see QRS000

8,8′-DIHYDROXY-RUGULOSIN see LIV000

12,18-DIHYDROXY-SENECIONAN-11,16-DIONE see RFP000

3′,6′-DIHYDROXYSPIRO(ISOBENZOFURAN-1(3H),9′(9H)-XANTHEN)-3-ONE see FEV000

2,2′-DIHYDROXY-3,3′,5,5′-TETRACHLORODIPHENYLSULFIDE see TFD250

trans-9,10-DIHYDROXY-9,10,11,12-TETRAHYDROBENZO(e)PYRENE see DNC400

trans-1,2-DIHYDROXY-1,2,3,4-TETRAHYDROCHRYSENE see DNC600

trans-3,4-DIHYDROXY-1,2,3,4-TETRAHYDRODIBENZ(a,h)ANTHRACENE see DNC800

trans-3,4-DIHYDROXY-1,2,3,4-TETRAHYDRODIBENZO(a,h)ANTHRACENE see DNC800

1,8-DIHYDROXY-2,4,5,7-TETRANITROANTHRAQUINONE (chrysamminic acid) (DOT) see CML600

5,7-DIHYDROXYTETRAZOLO(1,5-a)PYRIDINE-6-CARBONITRILE see DND900

2,3-DIHYDROXYTOLUENE see DNE000

2,5-DIHYDROXYTOLUENE see MKO250

2,6-DIHYDROXYTOLUENE see MPH400

3,5-DIHYDROXYTOLUENE see MPH500

2,4-DIHYDROXY-1,3,5-TRINITROBENZENE see SMP500

1,3-DIHYDROXY-2,4,6-TRINITROBENZENE see SMP500

3,5-DIHYDROXY-2,6,6-TRIS(3-METHYL-2-BUTENYL)-4-(3-METHYL-1-OXOBUTYL)-2,4-CYCLOHEXADIEN-1-ONE see LIU000

2,5-DIHYDROXY-3-UNDECYL-1,4-BENZOQUINONE see EAJ600

2,5-DIHYDROXY-3-UNDECYL-2,5-CYCLOHEXADIENE-1,4-DIONE (9CI) see EAJ600

6-(6,10-DIHYDROXYUNDECYL)-β-RESORCYLIC ACID-μ-LACTONE see RBF100

DIHYDROXYVIOLANTHRON (CZECH) see DMJ800

16,17-DIHYDROXYVIOLANTHRONE see DMJ800

1-α,25-DIHYDROXYVITAMIN D3 see DMJ400

1-α-DIHYDROXYVITAMIN D3 see HJV000

DIHYDROXYVITAMIN D3 see DMJ400

1,4-DIIDROBENZENE (ITALIAN) see HIH000

DIIIDROBENZO(1-4)TIAZINA CLORIDRATO (ITALIAN) see DKU875

DIIIDRO-5,5-DIETIL-2H-1,3-OSSAZIN-2,4(3H)-DIONE (ITALIAN) see DJT400

DIIIDROXI-1,4-BENZENESULFONATO-3-DI-ETILAMMONIUM (ITALIAN) see DIS600

1,4-DIIMIDO-2,5-CYCLOHEXADIENE see BDD200

1,3-DIIMINOISOINDOLIN (CZECH) see DNE400

1,3-DIIMINOISOINDOLINE see DNE400

DIINDIUM TRIOXIDE see ICI100

DIIODBENZOTEPH see DNE800

DIIODOACETYLENE see DNE500

3,5-DIIODOANTHRANILIC ACID see API800

1,2-DIIODOBENZENE see DNE700

2,6-DIIODO-1,4-BENZENEDIOL see DNF200
2,6-DIIODO-p-BENZOQUINONE see DNG800
DIIODOBENZOTEF see DNE800
N-2,5-DIIODOBENZOYL-N',N',N'',N''-DIETHYLENEPHOSPHORTRIAMIDE see
    DNE800
1,4-DIIODO-1,3-BUTADIYNE see DNE875
1,4-DIIODOBUTANE see TDQ400
cis-DIIODODIAMMINEPLATINUM (II) see DNF000
DIIODOETHYNE see DNE500
2,6-DIIODOHYDROQUINONE see DNF200
3,5-DIIODO-4-HYDROXYBENZOIC ACID see DNF300
3,5-DIIODO-4-HYDROXYBENZONITRILE see HKB500
3,5-DIIODO-4-HYDROXYBENZONITRILE, LITHIUM SALT see DNF400
3,5-DIIODO-4-HYDROXYBENZONITRILE OCTANOATE see DNG200
DIIODO-3,3 HYDROXY-4 BENZOYL 2 FURANNE see DNF500
(DIIODO-3,5 HYDROXY-4 BENZOYL)-3 MESITYL-2 BENZOFURANNE see
    DNF550
3,5-DIIODO-4-HYDROXYPHENYL 2,5-DIMETHYL-3-FURYL KETONE see
    DNF450
3,5-DIIODO-4-HYDROXYPHENYL 2-ETHYL-3-BENZOFURANYL KETONE see
    EID200
3,5-DIIODO-4-HYDROXYPHENYL 5-ETHYL-2-FURYL KETONE see EID250
3,5-DIIODO-4-HYDROXYPHENYL 2-FURYL KETONE see DNF500
3,5-DIIODO-4-HYDROXYPHENYL 2-MESITYL-3-BENZOFURANYL KETONE see
    DNF550
β-(3,5-DIIODO-4-HYDROXYPHENYL)-α-PHENYLPROPIONIC ACID see PDM750
DIIODOHYDROXYQUIN see DNF600
5,7-DIIODO-8-HYDROXYQUINOLINE see DNF600
DIIODOHYDROXYQUINOLINE see DNF600
3,5-DIIODO-4-(3'-IODO-4'-ACETOXYPHENOXY)BENZOIC ACID see TKP850
DIIODOMETHANE see DNF800
DIIODOMETHYLARSINE see MGQ775
DIIODOMETHYLATE de la BIS(PIPERIDINOMETHYL-COUMARANYL-5)CETONE
    (FRENCH) see COE125
1-((DIIODOMETHYL)SULFONYL)-4-METHYLBENZENE see DNF850
DIIODOMETHYL p-TOLYL SULFONE see DNF850
DIIODOMETILATO del BISPIPERIDINOMETILCUMARANIL-5-CHETONE (ITALIAN)
    see COE125
2,6-DIIODO-4-NITROPHENOL see DNG000
3,5-DIIODO-4-OCTANOYLOXYBENZONITRILE see DNG200
5,7-DIIODO-OXINE see DNF600
3,5-DIIODO-4-OXO-1(4H)PYRIDINEACETIC ACID-2,2'-IMINODIETHANOL SALT
    see DNG400
1,5-DIIODOPENTANE see PBH125
3,5-DIIODO-α-PHENYLPHLORETIC ACID see PDM750
1,3-DIIODOPROPANE see TLR050
3,5-DIIODO-4-PYRIDONE-N-ACETATE BIS(HYDROXYETHYL)AMMONIUM see
    DNG400
3,5-DIIODO-4-PYRIDONE-N-ACETIC ACID, DIETHANOLAMINE SALT see
    DNG400
DIIODOQUIN see DNF600
2,6-DIIODOQUINOL see DNF200
5,7-DIIODO-8-QUINOLINOL see DNF600
DIIODOQUINONE see DNG800
3,5-DIIODOSALICYLIC ACID see DNH000
DIISOAMYL ADIPATE see AEQ500
DIISOAMYLMERCURY see DNL200
DIISOBUTENE see TMA250
DIISOBUTILCHETONE (ITALIAN) see DNI800
DIISOBUTYL ADIPATE see DNH125
DIISOBUTYLALUMINIUM HYDRIDE see DNI600
DIISOBUTYLALUMINUM CHLORIDE see CGB500
DIISOBUTYLALUMINUM HYDRIDE see DNI600
DIISOBUTYLALUMINUM MONOCHLORIDE see CGB500
DIISOBUTYLAMINE see DNH400
DIISOBUTYLAMINOBENZOYLOXYPROPYL THEOPHYLLINE see DNH500
α-((DIISOBUTYLAMINO)METHYL)THEOPHYLLINE-8-ETHANOL BENZOATE (es-
    ter) see DNH500
DIISOBUTYL CARBINOL see DNH800
DI-ISOBUTYLCETONE (FRENCH) see DNI800
DIISOBUTYLCHLOROALUMINUM see CGB500
p-DIISOBUTYLCRESOXYETHYLDIMETHYLBENZYLAMMONIUM CHLORIDE
    MONOHYDRATE see MHB500
DIISOBUTYLENE see TMA250
DIISOBUTYLENE OXIDE see DNI200
DIISOBUTYLESTER KYSELINY FTALOVE see DNJ400
DIISOBUTYL FUMARATE see DNI400
DIISOBUTYLHYDROALUMINUM see DNI600
DIISOBUTYLKETON (DUTCH, GERMAN) see DNI800
DIISOBUTYL KETONE see DNI800
DI-ISO-BUTYLNITROSAMINE see DRQ200
DIISOBUTYLOXOSTANNANE see DNJ000
p-DIISOBUTYLPHENOXYETHOXYETHYLDIMETHYLBENZYLAMMONIUM CHLO-
    RIDE MONOHYDRATE see BBU750

DIISOBUTYLPHENOXYETHOXYETHYLDIMETHYL BENZYL AMMONIUM CHLO-
    RIDE see BEN000
DIISOBUTYL PHTHALATE see DNJ400
DIISOBUTYLSULFIDE HYDRATE see IJO000
DIISOBUTYLTHIOCARBAMIC ACID-S-ETHYL ESTER see EID500
DIISOBUTYLTIN OXIDE see DNJ000
DIISOBUTYRYL PEROXIDE see DNJ600
o,o'-DIISOBUTYRYLTHIAMINE DISULFIDE see BKJ325
DIISOCARB see EID500
4-4'-DIISOCYANATE de DIPHENYLMETHANE (FRENCH) see MJP400
DI-ISOCYANATE de TOLUYLENE see TGM750
1,3-DIISOCYANATOBENZENE see BBP000
4,4'-DIISOCYANATO-3,3'-DIMETHOXY-1,1'-BIPHENYL see DCJ400
4,4'-DIISOCYANATO-3,3'-DIMETHYL-1,1'-BIPHENYL see DQS000
4,4'-DIISOCYANATODIPHENYLMETHANE see MJP400
1,6-DIISOCYANATOHEXANE see DNJ800
1,6-DIISOCYANATOHEXANE HOMOPOLYMER see HEG300
DI-ISO-CYANATOLUENE see TGM750
DIISOCYANATOMETHANE see DNK100
2,6-DIISOCYANATO-1-METHYLBENZENE see TGM800
DIISOCYANATOMETHYLBENZENE see TGM740
2,4-DIISOCYANATO-1-METHYLBENZENE (9CI) see TGM750
1,5-DIISOCYANATONAPHTHALENE see NAM500
2,4-DIISOCYANATOTOLUENE see TGM750
2,6-DIISOCYANATOTOLUENE see TGM800
DIISOCYANATOTOLUENE see TGM740
DIISOCYANAT-TOLUOL see TGM750
4,4'-(2,3-DIISOCYANO-1,3-BUTADIENE-1,4-DIYL)BIS-1,2-BENZENEDIOL see
    XCS700
2,3-DIISONITROSOBUTANE see DBH000
DIISONONYLTIN DICHLORIDE see BLQ750
DIISOOCTYL ACID PHOSPHATE see DNK800
DIISOOCTYL ((DIOCTYLSTANNYLENE)DITHIO)DIACETATE see BKK750
DIISOOCTYL PHOSPHATE (DOT) see DNK800
DIISOOCTYL PHTHALATE see ILR100
DIISOPENTYLMERCURY see DNL200
DIISOPENTYLOXOSTANNANE see DNL400
DIISOPENTYLRTUT see DNL200
DIISOPENTYLTIN OXIDE see DNL400
DIISOPHENOL see DNG000
DIISOPROPANOLAMINE see DNL600
DIISOPROPANOLNITROSAMINE see DNB200
N,N'-DIISOPROPIL-FOSFORODIAMMIDO-FLUORURO (ITALIAN) see PHF750
DIISOPROPOXYPHOSPHORYL FLUORIDE see IRF000
((DIISOPROPROXYPHOSPHINOTHIOYL)THIO)TRICYCLOHEXYL STANNANE see
    DNT200
s-DIISOPROPYLACETONE see DNI800
DIISOPROPYL ADIPATE see DNL800
DI(ISOPROPYLAMIDO)PHOSPHORYLFLUORIDE see PHF750
DIISOPROPYLAMINE see DNM200
DIISOPROPYLAMINE DICHLORACETATE see DNM400
DIISOPROPYLAMINE, compd. with DICHLOROACETIC ACID (1:1) see
    DNM400
DIISOPROPYLAMINE DICHLOROETHANOATE see DNM400
2-(DIISOPROPYLAMINO)-2',6'-ACETOXYLIDIDE HYDROCHLORIDE see
    DNM600
2-DIISOPROPYLAMINOETHANOL see DNP000
2-(2-(DIISOPROPYLAMINO)ETHOXY)BUTYROPHENONE HYDROCHLORIDE see
    DNN000
S-(2-DIISOPROPYLAMINOETHYL)-O-ETHYL METHYL PHOSPHONOTHIOLATE
    see EIG000
α-(2-(DIISOPROPYLAMINO)ETHYL)-α-PHENYL-2-PYRIDINEACETAMIDE see
    DNN600
α-(2-DIISOPROPYLAMINOETHYL)-α-PHENYL-2-PYRIDINEACETAMIDE PHOS-
    PHATE see RSZ600
β-DIISOPROPYLAMINOETHYL-9-XANTHENECARBOXYLATE METHOBROMIDE
    see HKR500
2,6-DIISOPROPYLAMINO-4-METHOXYTRIAZINE see MFL250
γ-DIISOPROPYLAMINO-α-PHENYL-α-(2-PYRIDYL)BUTYRAMIDE see DNN600
DIISOPROPYLAMMINE-trans-DIHYDROXYMALONATOPLATINUM(IV) see IGG775
DIISOPROPYLAMMONIUM DICHLOROACETATE see DNM400
DIISOPROPYLAMMONIUM DICHLOROETHANOATE see DNM400
2,6-DIISOPROPYL ANILINE see DNN630
1,3-DIISOPROPYLBENZENE see DNN829
1,4-DIISOPROPYLBENZENE see DNN830
m-DIISOPROPYLBENZENE see DNN829
o-DIISOPROPYLBENZENE see DNN800
p-DIISOPROPYLBENZENE see DNN830
DIISOPROPYLBENZENE see DNN709
DIISOPROPYLBENZENE HYDROPEROXIDE, not more than 72% in solution
    (DOT) see DNS000
DIISOPROPYLBENZENE PEROXIDE see DGR600
1,3-DIISOPROPYLBENZENE SODIUM SALT, DIHYDROPEROXIDE see DNN840
1,4-DIISOPROPYLBENZENE SODIUM SALT, DIISOPEROXIDE see DNN850
p-DIISOPROPYLBENZOL see DNN830

N,N-DIISOPROPYL-2-BENZOTHIAZOLESULFENAMIDE see DNN900
DIISOPROPYLBERYLLIUM see DNO200
DIISOPROPYLCARBAMIC ACID, ETHYL ESTER see DNP600
DIISOPROPYLCARBODIIMIDE see DNO400
N,N'-DIISOPROPYL-DIAMIDO-FOSFORZUUR-FLUORIDE (DUTCH) see PHF750
N,N'-DIISOPROPYL-DIAMIDO-PHOSPHORSAEURE-FLUORID (GERMAN) see PHF750
N,N'-DIISOPROPYLDIAMIDOPHOSPHORYL FLUORIDE see PHF750
DIISOPROPYL-1,3-DITHIOL-2-YLIDENEMALONATE see MAO275
O,O-DIISOPROPYL DITHIOPHOSPHORIC ACID ESTER of-N,N-S-DIETHYLTHIO-CARBAMOYL-O,O-DIISOPROPYL PHOSPHOROTHIOATE see DKB600
N-(2-(O,O-DIISOPROPYLDITHIOPHOSPHORYL)ETHYL)BENZENESULFONAMIDE see DNO800
N-(β-O,O-DIISOPROPYLDITHIOPHOSPHORYLETHYL)BEZENESULFONAMIDE see DNO800
DIISOPROPYL ESTER of DITHIOCARBAMYL PHOSPHOROTHIOIC ACID see DKB600
DIISOPROPYL ESTER SULFURIC ACID see DNO900
N,N-DIISOPROPYL ETHANOLAMINE see DNP000
DIISOPROPYL ETHANOLAMINE see DNP000
DIISOPROPYL ETHER see IOZ750
N,N-DIISOPROPYL ETHYL CARBAMATE see DNP600
DIISOPROPYL ETHYL CARBAMATE see DNP600
N,N-DIISOPROPYL ETHYLENEDIAMINE see DNP700
DIISOPROPYL FLUOROPHOSPHATE see IRF000
O,O-DIISOPROPYL FLUOROPHOSPHATE see IRF000
DIISOPROPYL FLUOROPHOSPHONATE see IRF000
DIISOPROPYLFLUOROPHOSPHORIC ACID ESTER see IRF000
DIISOPROPYLFLUORPHOSPHORSAEUREESTER (GERMAN) see IRF000
DIISOPROPYL FUMARATE see DNQ200
DIISOPROPYL HYDROGEN PHOSPHITE see DNQ600
DIISOPROPYL(2-HYDROXYETHYL)METHYLAMMONIUMBROMIDE with XAN-THENE-9-CARBOXYLATE see HKR500
DIISOPROPYL HYPONITRITE see DNQ700
DIISOPROPYLIDENE ACETONE see PGW250
sym-DIISOPROPYLIDENE ACETONE see PGW250
1:2,5:6-DI-O-ISOPROPYLIDENE-α-D-GLUCOFURANOSE see DVO100
1:2,5:6-DI-O-ISOPROPYLIDEN-α-D-GLUCOFURANOSE see DVO100
DIISOPROPYL KETONE see DTI600
DIISOPROPYLMERCURY see DNQ800
DIISOPROPYL METHANEPHOSPHONATE see DNQ875
N,N'-DIISOPROPYL-6-METHOXY-1,3,5-TRIAZINE-2,4-DIYLDIAMINE see MFL250
DIISOPROPYL METHYLPHOSPHONATE see DNQ875
O,O-DIISOPROPYL-o,p-NITROPHENYL PHOSPHATE see DNR309
DIISOPROPYL-p-NITROPHENYL PHOSPHATE see DNR309
DIISOPROPYLNITROSAMIN (GERMAN) see NKA000
DIISOPROPYL OXIDE see IOZ750
DIISOPROPYLOXOSTANNANE see DNR200
DIISOPROPYL PARAOXON see DNR309
DIISOPROPYL PERDICARBONATE see DNR400
DIISOPROPYL PEROXYDICARBONATE see DNR400
2,6-DIISOPROPYLPHENOL see DNR800
3,5-DIISOPROPYLPHENOL METHYLCARBAMATE see DNS200
DIISOPROPYLPHENYLHYDROPEROXIDE (solution) see DNS000
3,5-DIISOPROPYLPHENYL METHYLCARBAMATE see DNS200
3,5-DIISOPROPYLPHENYL-N-METHYLCARBAMATE see DNS200
DIISOPROPYL PHOSPHITE see DNQ600
DIISOPROPYL PHOSPHOFLUORIDATE see IRF000
O,O-DIISOPROPYL PHOSPHONATE see DNQ600
DIISOPROPYLPHOSPHONATE see DNQ600
N,N'-DIISOPROPYLPHOSPHORODIAMIDIC FLUORIDE see PHF750
S-(O,O-DIISOPROPYL PHOSPHORODITHIOATE) ESTER of N-(2-MERCAPTOE-THYL)BENZENESULFONAMIDE see DNO800
DIISOPROPYL PHOSPHOROFLUORIDATE see IRF000
O,O'-DIISOPROPYL PHOSPHORYL FLUORIDE see IRF000
DIISOPROPYL PHTHALATE see PHW600
O,O-DIISOPROPYL-S-BENZYL PHOSPHOROTHIOLATE see BKS750
O,O-DIISOPROPYL-S-BENZYL THIOPHOSPHATE see BKS750
O,O-DIISOPROPYL-S-DIETHYLDITHIOCARBAMOYLPHOSPHORODITHIOATE see DKB600
O,O-DIISOPROPYL-S-TRICYCLOHEXYLTIN PHOSPHORODITHIOATE see DNT200
DI-ISOPROPYLSULFAT (GERMAN) see DNO900
DI-ISOPROPYLSULFATE see DNO900
N-DIISOPROPYLTHIOCARBAMIC ACID-S-2,3,3-TRICHLOROALLYL ESTER see DNS600
N-DIISOPROPYLTHIOCARBAMIC ACID S-2,3,3-TRICHLORO-2-PROPENYL ES-TER see DNS600
DI-ISOPROPYLTHIOLOCARBAMATE de S-(2,3-DICHLOROALLYLE) (FRENCH) see DBI200
1,3-DIISOPROPYLTHIOUREA see DNS800
DIISOPROPYL THIOUREA see DNS800
N,N'-DIISOPROPYLTHIOUREA see DNS800
DIISOPROPYLTIN DICHLORIDE see DNT000
DIISOPROPYLTIN OXIDE see DNR200

N,N-DIISOPROPYL-2,3,3-TRICHLORALLYL-THIOLCARBAMAT (GERMAN) see DNS600
DIISOPROPYLTRICHLOROALLYLTHIOCARBAMATE see DNS600
DIISOPYRAMIDE PHOSPHATE see RSZ600
1,4-DIISOTHIOCYANATOBENZENE see PFA500
1,2-DIISOTHIOCYANATOETHANE see ISK000
4,4'-DIISOTHIOINDIGO see DNT300
3,5-DIJOD-4-HYDROXY-BENZONITRIL (GERMAN) see HKB500
3,5-DIJOD-4-HYDROXY-BENZONITRIL CAPRYSAEUREESTER (GERMAN) see DNG200
3,5-DIJOD-4-HYDROXY-BENZONITRILE LITHIUMSALZ (GERMAN) see DNF400
DIKAIN see BQA010
DIKAIN HYDROCHLORIDE see TBN000
3,5-DIKARBOXYBENZENSULFONAN SODNY (CZECH) see DEK200
DIKETENE, inhibited (DOT) see KFA000
DIKETENE see KFA000
2,3-DIKETOBUTANE see BOT500
2,5-DIKETOHEXANE see HEQ500
1,3-DIKETOHYDRINDENE see IBS000
2,3-DIKETOINDOLINE see ICR000
DIKETONE ALCOHOL see DBF750
2,5-DIKETOPYRROLIDINE see SND000
2,5-DIKETOTETRAHYDROFURAN see SNC000
DIKOL see PDM750
DIKONIT see SGG500
DIKOTEKS see SIL500
DIKOTEX 30 see SIL500
DI-KU-SHUANG see MJL750
DILABIL see DAL000
DILABIL SODIUM see SGD500
DILACORAN see IRV000
DILACTONE ACTINOMYCINDIOIC D ACID see AEB000
DILAFURANE see EID200
DILAHIL see DAL000
DILAN see BON250
DILANGIL see MAW250
DILANGIO see POD750
DILANTHANUM OXIDE see LBA100
DILANTHANUM TRIOXIDE see LBA100
DILANTIN see DKQ000, DNU000
DILANTIN DB see DEP600
DILANTINE see DKQ000
DILANTIN SODIUM see DNU000
DILAPHYLLIN see HLC000
DILATAN KORE see ELH600
DILATIN see DNU100
DILATIN DB see DEP600
DILATOL HYDROCHLORIDE see DNU200
DILATYL see DNU200
DILAUDID see DNU300
DILAUDID HYDROCHLORIDE see DNU300, DNU310
DILAUROYL PEROXIDE see LBR000
DILAUROYL PEROXIDE, TECHNICAL PURE (DOT) see LBR000
DILAURYLESTER KYSELINY β',β'-THIODIPROPIONOVE see TFD500
DILAURYL PHTHALATE see PHW550
DILAURYL 3,3'-THIODIPROPIONATE see TFD500
DILAURYL β',β'-THIODIPROPIONATE see TFD500
DILAURYL β-THIODIPROPIONATE see TFD500
DILAURYL THIODIPROPIONATE see TFD500
DILA-VASAL see EID200
DILAVASE see VGA300, VGF000
DILAZEP/β-ACETYLDIGOXIN see CNR825
DILAZEP DIHYDROCHLORIDE see CNR750
DILCIT see HFG550
DILEAD(II) LEAD(IV) OXIDE see LDS000
DI-LEN see DNU000
DILENE see BIM500
DILEXPAL see HFG550
DILIC see HKC000
1,3-DILITHIOBENZENE see DNU325
DILITHIUM-1,1-BIS(TRIMETHYLSILYL)HYDRAZIDE see DNU350
DILITHIUM CARBONATE see LGZ000
DILITHIUM CHROMATE see LHD000
DILITURIC ACID see NET550
DILL FRUIT OIL see DNU400
DILL HERB OIL see DNU400
DILL OIL see DNU400
DILL SEED OIL see DNU400
DILL SEED OIL, EUROPEAN TYPE see DNU400
DILL WEED OIL see DNU400
DILOMBRIN see DJT800
DILOR see DLO880, DNC000
DILOSPAN S see PGR000
DILOSYN see MDT500, MPE250
DILOXOL see GGS000

DILTIAZEM HYDROCHLORIDE see DNU600
DILUEX see PAE750
DILURAN see AAI250
DILURGEN see SIG000
DILVASENE see FMX000
DILYN see RLU000
DILZEM see DNU600
DIMACIDE YELLOW N-5RL see SGP500
DIMAGNESIUM PHOSPHATE see MAH775
1,3-DIMALEIMIDOBENZENE see BKL750
4,4'-DIMALEIMIDOPHENYLMETHANE see BKL800
p,p'-DIMALEIMIDOPHENYLMETHANE see BKL800
2,4-DIMALEIMIDOTOLUENE see TGY770
DIMALONE see DRB400
DIMANGANESE TRIOXIDE see MAT500
DIMANIN C see SGG500
DIMANTINE see DTC400
DIMAPP see DQA400
DIMAPYRIN see DOT000
DIMAS see DQD400
DIMATE 267 see DSP400
DIMATIF see DNU850
DIMAVAL see DNU860
DIMAYAL see DNU860
DIMAZ see DXH325
DIMAZINE see DSF400
DIMAZON see ACR300
DIMEBOLIN see TCQ260
DIMEBOLINE see TCQ260
DIMEBON see TCQ260
DIMEBON DIHYDROCHLORIDE see DNU875
DIMEBONE see TCQ260
DIMECRON 100 see FAB400
DIMECRON see FAB400
DIMECROTIC ACID MAGNESIUM SALT see DOK400
DIMEDROL see BBV500
DIMEDRYL see BBV500
DIMEFADANE see DRX400
DIMEFLINE see DNV000
DIMEFLINE HYDROCHLORIDE see DNV200
DIMEFOX see BJE750
DIMEGLUMINE IOCARMATE see IDJ500
DIMELIN see ABB000
DIMELONE see DRB400
DIMELOR see ABB000
DIMEMORFAN PHOSPHATE see MLP250
DIMENFORMON see EDO000
DIMENFORMON BENZOATE see EDP000
DIMENFORMON DIPROPIONATE see EDR000
DIMENFORMONE see EDP000
DIMENFORMON PROLONGATUM see EDO000
DIMENHYDRINATE see DYE600
DIMENOXADOL HYDROCHLORIDE see DPE200
DIMEPHENTHIOATE see DRR400
DIMEPHENTHOATE see DRR400
DIMERAY see IDJ500
DIMERCAPROL PROPANOL see BAD750
2,3-DIMERCAPTOBUTANEDIOIC ACID see DNV610
(R*,S*)-2,3-DIMERCAPTOBUTANEDIOIC ACID see DNV800
(R*,S*)-1,4-DIMERCAPTO-2,3-BUTANEDIOL see DXN350
dl-threo-DIMERCAPTO-2,3-BUTANEDIOL see DXO775
d-threo-1,4-DIMERCAPTO-2,3-BUTANEDIOL see DXO800
(R*,R*)-(±)-1,4-DIMERCAPTO-2,3-BUTANEDIOL (9CI) see DXO775
1,2-DIMERCAPTOETHANE see EEB000
DIMERCAPTOL see BAD750
2,3-DIMERCAPTOL-1-PROPANOL see BAD750
1,2-DIMERCAPTO-4-METHYLBENZENE see TGN000
4,5-DI(MERCAPTOMETHYL)-2-METHYL-3-PYRIDINOL DITHIOACETATE HYDRO-
   BROMIDE see DBH800
1,2-DIMERCAPTOPROPANE see PML300
1,3-DIMERCAPTOPROPANE see PML350
2,3-DIMERCAPTOPROPANE see PML300
2,3-DIMERCAPTOPROPANE SODIUM SULPHONATE see DNU860
2,3-DIMERCAPTO-1-PROPANESULFONIC ACID SODIUM SALT see DNU860
2,3-DIMERCAPTOPROPANESULFONIC ACID SODIUM SALT see DNU860
2,3-DIMERCAPTOPROPAN-1-OL see BAD750
2,3-DIMERCAPTOPROPANOL see BAD750
DIMERCAPTOPROPANOL see BAD750
2,3-DIMERCAPTOPROPYL-p-TOLYSULFIDE see DNV600
4,5-DIMERCAPTOPYRIDOXINDI-THIOACETAT HYDROBROMID (GERMAN) see
   DBH800
2,3-DIMERCAPTOSUCCINIC ACID see DNV610
α-β-DIMERCAPTOSUCCINIC ACID see DNV610
meso-2,3-DIMERCAPTOSUCCINIC ACID see DNV800
meso-DIMERCAPTOSUCCINIC ACID see DNV800

DIMERCAPTOSUCCINIC ACID see DNV610
meso-DIMERCAPTOSUCCINIC ACID SODIUM SALT see DNU860
2,5-DIMERCAPTO-1,3,4-THIADIAZOLE see TES250
DIMERCUROUS METHANE ARSONATE see DNW000
DIMERCURY DICHLORIDE see MCY300
DIMERCURY IMIDE OXIDE see DNW200
DIMER CYKLOPENTADIENU (CZECH) see DGW000
DIMERIN see DNW400
DIMER X see IDJ500
DIMESTROL see DJB200
1,6-DIMESYL-d-MANNITOL see BKM500
1,4-DIMESYLOXYBUTANE see BOT250
1,4-DI(MESYLOXYETHYLAMINO)ERYTHRITOL see LJD500
DIMET see DXE600
DIMETACRINE see DNW700
DIMETACRINE BITARTRATE see DRM000
DIMETACRIN HYDROGENTARTRATE see DRM000
DIMETAN see DRL200
dl-DIMETANE MALEATE see DNW759
DIMETATE see DSP400
DIMETAZINA see SNN300
DIMETHACHLON see DGF000
DIMETHACHLOR see DSM500
DIMETHACHLORE see DSM500
DIMETHACIN see DNW700
DIMETHACINE see DNW700
DIMETHACRINE TARTRATE see DRM000
DIMETHADIONE see PMO250
DIMETHAEN see DPA000
2,5-DIMETHANESULFOMYLOXYHEXANE see DSU000
1,6-DIMETHANESULFONATE-d-MANNITOL see BKM500
1,4-DIMETHANESULFONATE THREITOL see TFU500
(2s,3s)-1,4-DIMETHANESULFONATE TREITOL see TFU500
1,4-DIMETHANESULFONOXYBUTANE see BOT250
cis-1,4-DIMETHANE SULFONOXY-2-BUTENE see DNW800
trans-1,4-DIMETHANE SULFONOXY-2-BUTENE see DNX000
1,4-DIMETHANESULFONOXY-2-BUTYNE see DNX200
1,4-DIMETHANESULFONOXY-1,4-DIMETHYLBUTANE see DSU000
1,6-DIMETHANE-SULFONOXY-d-MANNITOL see BKM500
1:3-DIMETHANESULFONOXYPROPANE see TLR250
1,4-DI(METHANESULFONYLOXY)BUTANE see BOT250
DIMETHANESULFONYL PEROXIDE see DNX300
1,6-DIMETHANESULPHONOXY-1,6-DIDEOXY-d-MANNITOL see BKM500
1,3-DIMETHANESULPHONOXYPROPANE see TLR250
1,4-DIMETHANESULPHONYLOXYBUTANE see BOT250
DIMETHESTERONE see DRT200
DIMETHICONE 350 see PJR000
DIMETHINDENE MALEATE see FMU409
DIMETHINDEN MALEATE see FMU409
DIMETHIOTAZINE see FMU039
DIMETHIRIMOL see BRD000
DIMETHISOQUIN HYDROCHLORIDE see DNX400
DIMETHISTERON see DRT200
DIMETHISTERONE see DRT200
DIMETHISTERONE and ETHINYL ESTRADIOL see DNX500
DIMETHOAT (DUTCH) see DSP400
DIMETHOAT (GERMAN) see DSP400
DIMETHOATE (USDA) see DSP400
DIMETHOATE O-ANALOG see DNX800
DIMETHOATE-ETHYL see DNX600
DIMETHOATE OXYGEN ANALOG see DNX800
DIMETHOATE PO ISOLOGUE see DNX800
DIMETHOAT TECHNISCH 95% see DSP400
DIMETHOCAINE see DNY000
DIMETHOGEN see DSP400
DIMETHOTHIAZINE see DUC400
DIMETHOTHIAZINE MESYLATE see FMU039
DIMETHOTHIAZINE METHANESULFONATE see FMU039
DIMETHOXANE see ABC250
DIMETHOXON see DNX800
1,2-DIMETHOXY-4-ALLYLBENZENE see AGE250
3,4'-DIMETHOXY-4-AMINOAZOBENZENE see DNY400
2,6-DIMETHOXY-4-(p-AMINOBENZENESULFONAMIDO)PYRIMIDINE see SNN300
(trans)-2,5-DIMETHOXY-4'-AMINOSTILBENE see DON400
2,5-DIMETHOXYAMPHETAMINE HYDROCHLORIDE see DOJ800
3,4-DIMETHOXYAMPHETAMINE HYDROCHLORIDE see DOK000
2,4-DIMETHOXYANILINE see DNY500
2,5-DIMETHOXYANILINE see AKD925
2,3-DIMETHOXYANILINE MUSTARD see BIC600
1,5-DIMETHOXY-9,10-ANTHRACENEDIONE see DNY800
1,5-DIMETHOXYANTHRACHINON (CZECH) see DNY800
1,5-DIMETHOXYANTHRAQUINONE see DNY800
1-5,6-DIMETHOXYAPORPHINE see NOE500
1,2-DIMETHOXY-6a-β-APORPHINE see NOE500
(R)-1,2-DIMETHOXYAPORPHINE see NOE500

β-(2,5-DIMETHOXYPHENYL)-β-HYDROXYISOPROPYLAMINE HYDROCHLORIDE
  see MDW000
2-(2,5-DIMETHOXYPHENYL)ISOPROPYLAMINE see DOK600
β-(2,5-DIMETHOXYPHENYL)ISOPROPYLAMINE HYDROCHLORIDE see DOJ800
DIMETHOXYPHENYLMETHANE see DOG700
1-((3,4-DIMETHOXYPHENYL)METHYL)-6,7-DIMETHOXYISOQUINOLINE see
  PAH000
((3,4-DIMETHOXYPHENYL)METHYL)HYDRAZINE see VIK150
DIMETHOXYPHENYLMETHYLSILANE see DOH400
1,1-DIMETHOXY-2-PHENYLPROPANE see HII600
1-(3,4-DIMETHOXYPHENYL)-1-PROPANONE see PMX600
1-(3,4-DIMETHOXYPHENYL)-2-PROPANONE see VIK300
1-(3,4-DIMETHOXYPHENYL)-2-PROPENE see AGE250
3-(4,6-DIMETHOXY-α-PHENYL-m-TOLUOYL)-PROPIONIC ACID SODIUM SALT
  see DOA875
2,2-DI-(p-METHOXYPHENYL)-1,1,1-TRICHLOROETHANE see MEI450
DI(p-METHOXYPHENYL)-TRICHLOROMETHYL METHANE see MEI450
DIMETHOXYPHOSPHINE OXIDE see DSG600
((DIMETHOXYPHOSPHINOTHIOYL)THIO)BUTANEDIOIC ACID DIETHYL ESTER
  see MAK700
2-DIMETHOXYPHOSPHINOTHIOYLTHIOMETHYL-4,6-DIAMINO-s-TRIAZINE see
  ASD000
2-((DIMETHOXYPHOSPHINYL)OXY)-1H-BENZ(d,e)ISOQUINOLINE-1,3(2H)-DIONE
  see DOL400
3-((DIMETHOXYPHOSPHINYL)OXY)-2-BUTENOIC ACID METHYL ESTER see
  MQR750
(E)-3-((DIMETHOXYPHOSPHINYL)OXY)-2-BUTENOIC ACID 1-PHENYLETHYL ES-
  TER (9CI) see COD000
3-(DIMETHOXYPHOSPHINYLOXY)-N,N-DIMETHYL-cis-CROTONAMIDE see
  DGQ875
3-(DIMETHOXYPHOSPHINYLOXY)-N,N-DIMETHYLISOCROTONAMIDE see
  DGQ875
3-(DIMETHOXYPHOSPHINYLOXY)-N-METHYL-N-METHOXY-cis-CROTONAMIDE
  see DOL800
3-(DIMETHOXYPHOSPHINYLOXY)N-METHYL-cis-CROTONAMIDE see MRH209
((DIMETHOXYPHOSPHINYL)THIO)ACETIC ACID ETHYL ESTER see DRB600
((DIMETHOXYPHOSPHINYL)THIO)-BUTANEDIOIC ACID DIETHYL ESTER (9CI)
  see OPK250
1,1-DIMETHOXYPROPANE see DOM200
2,2-DIMETHOXYPROPANE see DOM400
3,3-DIMETHOXYPROPENE see DOM600
1,2-DIMETHOXY-4-PROPENYLBENZENE see IKR000
3′,4′-DIMETHOXYPROPIOPHENONE see PMX600
3,4-DIMETHOXYPROPIOPHENONE see PMX600
N-(3,6-DIMETHOXY-4-PYRIDAZINYL)SULFANILAMIDE see DON700
N¹-(4,6-DIMETHOXYPYRIMIDIN-2-YL)SULFANILAMIDE see SNI500
N¹-(2,6-DIMETHOXY-4-PYRIMIDINYL)SULFANILAMIDE see SNN300
N′-(5,6-DIMETHOXY-4-PYRIMIDYL)SULFANILAMIDE see AIE500
4,7-DIMETHOXY-6-(2-PYRROLIDINYLETHOXY)-5-CINNAMOYLBENZOFURAN
  MALEATE see DON000
2,6-DIMETHOXYQUINOL see DON200
2′,5′-DIMETHOXYSTILBENAMINE see DON400
2,5-DIMETHOXY-4′-STILBENAMINE see DON400
4-(2,5-DIMETHOXY)STILBENAMINE see DON400
2,3-DIMETHOXYSTRYCHNIDIN-10-ONE see BOL750
2,3-DIMETHOXYSTRYCHNINE see BOL750
DIMETHOXY STRYCHNINE (DOT) see BOL750
DIMETHOXYSULFADIAZINE see SNN300
2,4-DIMETHOXY-6-SULFANILAMIDO-1,3-DIAZINE see SNN300
3,6-DIMETHOXY-4-SULFANILAMIDOPYRIDAZINE see DON700
2,6-DIMETHOXY-4-SULFANILAMIDOPYRIMIDINE see SNN300
DIMETHOXYTETRAETHYLENE GLYCOL see PBO500
DIMETHOXYTETRAGLYCOL see PBO500
2,5-DIMETHOXYTETRAHYDROFURAN see DON800
(s-(4*,S*))-6,7-DIMETHOXY-3-(5,6,7,8-TETRAHYDRO-4-METHOXY-6-METHYL-1,3-
  DIOXOLO(4,5-g)ISOQUINOLIN-5-YL)-1(3H)-ISOBENZOFURANONE, N-OXIDE,
  HYDROCHLORIDE see NBP300
3′,5′-DIMETHOXY-3,4′,5,7-TETRAHYDROXYFLAVYLIUM ACID ANION see
  MAO750
DI(METHOXYTHIOCARBONYL) DISULFIDE see DUN600
DIMETHOXY-2,2,2-TRICHLORO-1-N-BUTYRYLOXY-ETHYLPHOSPHINE OXIDE
  see BPG000
DIMETHOXY-2,2,2-TRICHLORO-1-HYDROXY-ETHYL-PHOSPHINE OXIDE see
  TIQ250
3,3′-DIMETHOXYTRIPHENYLMETHANE-4,4′-BIS(1′′-AZO-2′′-NAPHTHOL)) see
  DOO400
6,7-DIMETHOXY-1-VERATRYLISOQUINOLINE see PAH000
6,7-DIMETHOXY-1-VERATRYLISOQUINOLINE-3-CARBOXYLIC ACID SODIUM
  SALT see PAG750
6,7-DIMETHOXY-1-VERATRYLISOQUINOLINE HYDROCHLORIDE see PAH250
16,17-DIMETHOXYVIOLANTHRONE see JAT000
DIMETHOXYVIOLANTHRONE see JAT000
DIMETHPRAMIDE see DUO300
DIMETHPYRINDENE MALEATE see FMU409
DIMETHULENE see DRV000

DIMETHWLEN see DRV000
DIMETHYL see EDZ000, SDF000
DIMETHYLACETAL see DOO600
N,N-DIMETHYLACETAMIDE see DOO800
DIMETHYLACETAMIDE see DOO800
1,1-DIMETHYL-3-(p-ACETAMIDOPHENYL)TRIAZENE see DUI000
2,4-DIMETHYLACETANILIDE see ABO740
2′,4′-DIMETHYLACETANILIDE see ABO740
2,6-DIMETHYLACETANILIDE see ABP760
3,4-DIMETHYLACETANILIDE see ABP770
3′,4′-DIMETHYLACETANILIDE see ABP770
DIMETHYLACETIC ACID see IJU000
N,N-DIMETHYLACETOACETAMIDE see DOP000
2′,4′-DIMETHYLACETOACETANILIDE see OOI100
DIMETHYLACETONE see DJN750
DIMETHYLACETONE AMIDE see DOO800
DIMETHYLACETONITRILE see IJX000
N,N-DIMETHYL-β-ACETOXY β-PHENYLETHYLAMINE see ABN700
DIMETHYLACETYLENE see COC500
DIMETHYLACETYLENECARBINOL see MHX250
DIMETHYL ACETYLENEDICARBOXYLIC ACID see DOP400
DIMETHYLACETYLENYLCARBINOL see MHX250
O,S-DIMETHYLACETYLPHOSPHOROAMIDOTHIOATE see DOP600
DIMETHYL ACID PHOSPHITE see DSG600
N,N-DIMETHYLACRYLAMIDE see DOP800
3,3-DIMETHYL-ACRYLATE de 2,4-DINITRO-6-(1-METHYLPROPYLE) PHENYLE
  (FRENCH) see BGB500
(E)-2,3-DIMETHYLACRYLIC ACID see TGA700
3,3-DIMETHYLACRYLIC ACID see MHT500
β,β-DIMETHYLACRYLIC ACID see MHT500
trans-2,3-DIMETHYLACRYLIC ACID see TGA700
trans-α-β-DIMETHYLACRYLIC ACID see TGA700
3,3-DIMETHYLACRYLIC ACID 2-sec-BUTYL-4,5-DINITROPHENYL ESTER see
  BGB500
cis-α,β-DIMETHYL ACRYLIC ACID, GERANIOL ESTER see GDO000
DIMETHYL ADIPATE see DOQ300
DIMETHYLAETHANOLAMIN (GERMAN) see DOY800
N-(5-(1,1-DIMETHYLAETHYL)-1,3,4-THIADIAZOL-2-YL)-N,N′-DIMETHYLHARNS-
  TOFF (GERMAN) see BSN000
O,O-DIMETHYL-O-(2-AETHYLTHIO-AETHYL) MONOTHIOPHOSPHAT (GERMAN)
  see DAO800
DIMETHYL ALDEHYDE see DOO600
4-(N,4-DIMETHYL-I-ALLOISOLEUCINE)-8-(N,4-DIMETHYL-I-ALLOISOLEUCINE)-QUI-
  NOMYCIN A see QQS075
3,3-DIMETHYLALLYL ACETATE see DOQ350
γ,γ-DIMETHYLALLYL ACETATE see DOQ350
DIMETHYLALLYL ACETATE see DOQ350
3,3-DIMETHYLALLYL ALCOHOL see MHU110
γ,γ-DIMETHYLALLYL ALCOHOL see MHU110
DIMETHYLALLYL ALCOHOL see MHU110
2-(3,3-DIMETHYLALLYL)CYCLAZOCINE see DOQ400
2-(3,3-DIMETHYLALLYL-5,9-DIMETHYL-2′-HYDROXYBENZOMORPHAN see DOQ400
2-(3,3-DIMETHYLALLYL)-5-ETHYL-2′-HYDROXY-9-METHYL-6,7-BENZOMORPHAN
  see DOQ600
2-(3,3-DIMETHYLALLYL)-2′,2′-HYDROXY-5,9-DIMETHYL-6,7-BENZOMORPHAN
  see DOQ800
DIMETHYLALUMINUM CHLORIDE see DOQ700
DIMETHYLALUMINUM HYDRIDE see DOQ750
DIMETHYLAMIDE ACETATE see DOO800
DIMETHYLAMID KYSELINY CHLORMRAVENCI see DQY950
DIMETHYLAMIDOETHOXYPHOSPHORYL CYANIDE see EIF000
DIMETHYLAMINE, anhydrous (DOT) see DOQ800
DIMETHYLAMINE, aqueous solution (DOT) see DOQ800
DIMETHYLAMINE, solution (DOT) see DOQ800
DIMETHYLAMINE see DOQ800
DIMETHYLAMINE BENZHYDRYL ESTER HYDROCHLORIDE see BAU750
DIMETHYLAMINE BORANE see DOR200
4-DIMETHYLAMINE m-CRESYL METHYLCARBAMATE see DOR400
DIMETHYLAMINE with DIBORANE (1:1) see DOX200
DIMETHYLAMINE HYDROCHLORIDE see DOR600
4-DIMETHYLAMINEPYRIDINE see DQB600
DIMETHYLAMINE SALT of 2,4-D see DFY800
DIMETHYLAMINE SALTS of mixed POLYCHLOROBENZOIC ACIDS see
  PJQ000
DIMETHYLAMINE-2,3,6-TRICHLOROBENZOATE see DOR800
4-(DIMETHYLAMINE)-3,5-XYLYL-N-METHYLCARBAMATE see DOS000
(DIMETHYLAMINO)ACETIC ACID HYDROCHLORIDE see MPI100
2-DIMETHYLAMINOACETONITRILE (DOT) see DOS200
DIMETHYLAMINOACETONITRILE see DOS200
(DIMETHYLAMINO)ACETYLENE see DOS300
N′,N′-DIMETHYL-4′-AMINO-N-ACETYL-N-MONOMETHYL-4-AMINOAZOBENZENE
  see DPQ200
DIMETHYLAMINOETHANOL (GERMAN) see DOY800
β-DIMETHYLAMINO-AETHYL-BENZHYDRYL-AETHER (GERMAN) see BBV500

N-(2′-DIMETHYLAMINOAETHYL)-(o-BENZYLPHENOL)-AETHER HYDROCHLORID (GERMAN) see DPD400
N-(4-((1-(DIMETHYLAMINO)-AETHYLIDEN)AMINO)PHENYL)-2-METHOXYACETAM-ID-HYDROCHLORID (GERMAN) see DPF200
N,N-DIMETHYL-β-AMINOAETHYL-ISOTHIURONIUM DIHYDROCHLORID (GERMAN) see NNL400
N-DIMETHYLAMINO-AETHYL-N-p-METHOXY-BENZYL-α-AMINO-PYRIDIN-MALEAT (GERMAN) see WAK000
5-(DIMETHYLAMINOAETHYL-OXYIMINO)-5H-DIBENZO(a,d)CYCLOHEPTA-1,4-DIENHYDROCHLORID (GERMAN) see DPH600
DIMETHYLAMINO-ANALGESINE see DOT000
9-(p-DIMETHYLAMINOANILINO)ACRIDINE see DOS800
4-(DIMETHYLAMINO)ANTIPYRINE see DOT000
DIMETHYLAMINOANTIPYRINE see DOT000
4-(DIMETHYLAMINO)ANTIPYRINE mixed with SODIUM NITRITE (1:1) see DOT200
1-(((((4-DIMETHYLAMINO)-1,4,4A,5,5A,6,11,12A-OCTAHYDRO-3,6,10,12,12A-PENTAHYDROXY-6-METHYL-1,11-DIOXO-2-NAPHTHACE-NYL)CARBONYL)AMINO)METHYL)-I-PROLINE see PMI000
p-DIMETHYLAMINOAZOBENZEN (CZECH) see DOT300
2′,3-DIMETHYL-4-AMINOAZOBENZENE see AIC250
N,N-DIMETHYL-4-AMINOAZOBENZENE see DOT300
4-(N,N-DIMETHYLAMINO)AZOBENZENE see DOT300
4-DIMETHYLAMINOAZOBENZENE see DOT300
N,N-DIMETHYL-p-AMINOAZOBENZENE see DOT300
p-DIMETHYLAMINOAZOBENZENE see DOT300
DIMETHYLAMINOAZOBENZENE see DOT300
4-DIMETHYLAMINOAZOBENZENE AMINE-N-OXIDE see DTK600
4′-DIMETHYLAMINOAZOBENZENE-2-CARBOXYLIC ACID see CCE500
p-(DIMETHYLAMINO)AZOBENZENE-o-CARBOXYLIC ACID see CCE500
N,N-DIMETHYLAMINOAZOBENZENE-N-OXIDE see DTK600
4-DIMETHYLAMINOAZOBENZENE-4′-SULPHONIC ACID SODIUM SALT see MND600
4-DIMETHYLAMINOAZOBENZOL see DOT300
p-DIMETHYLAMINO-AZOBENZOL (GERMAN) see DOT300
DIMETHYLAMINOAZOBENZOL see DOT300
DIMETHYLAMINOAZOPHENE see DOT000
4-(DIMETHYLAMINO) BENZALDEHYDE see DOT400
p-(DIMETHYLAMINO)BENZALDEHYDE see DOT400
1-(4-DIMETHYLAMINOBENZAL)INDENE see DOT600
p-(DIMETHYLAMINO)BENZAL-5-RHODANINE see DOT800
5-(p-DIMETHYLAMINOBENZAL)RHODANINE see DOT800
p-DIMETHYLAMINOBENZALRHODANINE see DOT800
3,4-DIMETHYLAMINOBENZENE see XNS000
(DIMETHYLAMINO)BENZENE see DQF800
p-DIMETHYLAMINOBENZENE-1-AZO-1-NAPHTHALENE see DSU600
p-DIMETHYLAMINOBENZENEAZO-1-NAPHTHALENE see DSU600
p-DIMETHYLAMINOBENZENE-1-AZO-2-NAPHTHALENE see DSU800
5(4-DIMETHYLAMINOBENZENEAZO)TETRAZOLE see DOU000
4-DIMETHYLAMINOBENZENECARBONAL see DOT400
p-DIMETHYLAMINOBENZENEDIAZONIUM CHLOROZINCATE (6CI) see DCP300
4-(DIMETHYLAMINO)BENZENEDIAZONIUM TRICHLOROZINCATE(1-) see DCP300
p-(DIMETHYLAMINO)BENZENEDIAZONIUM TRICHLOROZINCATE (7CI) see DCP300
p-DIMETHYLAMINOBENZENE DIAZO SODIUM SULFONATE see DOU600
p-DIMETHYLAMINOBENZENEDIAZOSODIUM SULPHONATE see DOU600
p-(DIMETHYLAMINO)BENZENEDIAZOSULFONATE see DOU600
p-DIMETHYLAMINOBENZENEDIAZOSULFONIC ACID, SODIUM SALT see DOU600
4-DIMETHYLAMINOBENZENEDIAZOSULFONIC ACID, SODIUM SALT see DOU600
p-(DIMETHYLAMINO)BENZENEDIAZOSULPHONATE see DOU600
p-(DIMETHYLAMINO)BENZENEDIAZOSULPHONIC ACID, SODIUM SALT see DOU600
4-DIMETHYLAMINOBENZENEDIAZOSULPHONIC ACID, SODIUM SALT see DOU600
p-DIMETHYLAMINO BENZOIC ACID see DOU650
p-DIMETHYLAMINOBENZOIC ACID, OCTYL ESTER see AOI500
p-DIMETHYLAMINOBENZOIC ACID, PENTYL ESTER see AOI500
p-DIMETHYLAMINOBENZOLDIAZOSULFONAT (NATRIUMSALZ) (GERMAN) see DOU600
4,4′-DIMETHYLAMINOBENZOPHENONIMIDE see IBB000
5-(p-DIMETHYLAMINOBENZOYLIDENE)RHODANINE see DOT800
5-DIMETHYLAMINO-3-BENZOYLINDOLE see DOU700
p-DIMETHYLAMINOBENZYLIDEN-1,2-BENZ-9-METHYL-ACRIDINE see DQC200
p-DIMETHYLAMINOBENZYLIDEN-3,4-BENZ-9-METHYLACRIDINE see DQC000
p-DIMETHYLAMINOBENZYLIDENE-3,4,5,6-DIBENZ-9-METHYLACRIDINE see DOV000
(4-DIMETHYLAMINOBENZYLIDENE)INDENE see DOT600
p-DIMETHYLAMINOBENZYLIDENE RHODAMINE see DOT800
2′,3-DIMETHYL-4-AMINOBIPHENYL see BLV250
3,2′-DIMETHYL-4-AMINOBIPHENYL see BLV250
3,3′-DIMETHYL-4-AMINOBIPHENYL see DOV200
4-DIMETHYLAMINOBIPHENYL see BGF899

4-(DIMETHYLAMINO)-3-BIPHENYLOL see DOV400
4-(DIMETHYLAMINO)-α,α-BIS(4-(DIMETHYLAMINO)PHENYL)-BENZENEMETHA-NOL (9CI) see TJK000
4-DIMETHYLAMINO-1,1-BIS((3,4-(METHYLENEDIOXY)PHENOXY)METHYL)-1-BU-TANOL, METHYLCARBAMATE (ester), CITRATE see DOV800
N-DIMETHYL AMINO-β-CARBAMYL PROPIONIC ACID see DQD400
3-((((DIMETHYLAMINO)CARBONYL)AMINO)PHENYL-1,1-DIMETHYLE-THYL)CARBAMATE see DUM800
N,N-DIMETHYLAMINOCARBONYL CHLORIDE see DQY950
(DIMETHYLAMINO)CARBONYL CHLORIDE see DQY950
3-((DIMETHYLAMINO)CARBONYL)OXY)-1-METHYL-PYRIDINIUM BROMIDE) see MDL600
3-(((DIMETHYLAMINO)CARBONYL)OXY)-1-METHYL-PYRIDINIUM (9CI) see PPI800
p-DIMETHYLAMINO-CARVACROLDIMETHYLURETHANE METHIODIDE see DOW875
1-(N,N-DIMETHYLAMINO)-3-(p-CHLOROPHENYL-3-α-PYRIDYL)PROPANE MALE-ATE see TAI500
3-β-(DIMETHYLAMINO)CON-5-ENINE-DIHYDROBROMIDE see DOX000
3-β-(DIMETHYLAMINO)CON-5-ENINE HYDROCHLORIDE see CNH660
4-DIMETHYLAMINO-3-CRESYL METHYLCARBAMATE see DOR400
1-DIMETHYLAMINO-3-CYANO-3-PHENYL-4-METHYLHEXANE HYDROCHLORIDE see DOX100
DIMETHYLAMINOCYANPHOSPHORSAEUREAETHYLESTER (GERMAN) see EIF000
N,N-DIMETHYLAMINOCYCLOHEXANE see DRF709
(DIMETHYLAMINO)CYCLOHEXANE see DRF709
7-DIMETHYLAMINO-6-DEMETHYL-6-DEOXYTETRACYCLINE see MQW250
DIMETHYLAMINODIBORANE see DOX200
4-(DIMETHYLAMINO)-1,2-DIHYDRO-1,5-DIMETHYL-2-PHENYL-3H-PYRAZOL-3-ONE see DOT000
1-(DIMETHYLAMINO)-2-((DIMETHYLAMINO)METHYL)-2-BUTANOL BENZO-ATE,(ESTER) see AHI250
4-(DIMETHYLAMINO)-3,5-DIMETHYLPHENOL METHYLCARBAMATE (ESTER) see DOS000
4-(DIMETHYLAMINO)-3,5-DIMETHYLPHENYL ESTER, METHYLCARBAMIC ACID see DOS000
4-(DIMETHYLAMINO)-3,5-DIMETHYLPHENYL-N-METHYLCARBAMATE see DOS000
4-DIMETHYLAMINO-2,3-DIMETHYL-1-PHENYL-3-PYRAZOLIN-5-ONE see DOT000
4-DIMETHYLAMINO-2,3-DIMETHYL-1-PHENYL-5-PYRAZOLONE see DOT000
3-DIMETHYLAMINO-1,2-DIMETHYLPROPYL p-AMINOBENZOATE HYDROCHLO-RIDE see AIT750
2-(DIMETHYLAMINO)-5,6-DIMETHYL-4-PYRIMIDINYLDIMETHYLCARBAMATE see DOX600
3,2′-DIMETHYL-4-AMINODIPHENYL see BLV250
3,3′-DIMETHYL-4-AMINODIPHENYL see DOV200
(3S,6S)-(−)-6-(DIMETHYLAMINO)-4,4-DIPHENYL-3-HEPTANOL ACETATE (ester) HYDROCHLORIDE see ACQ690
d-6-(DIMETHYLAMINO)-4,4-DIPHENYL-3-HEPTANONE see DBE100
l-6-(DIMETHYLAMINO)-4,4-DIPHENYL-3-HEPTANONE see MDO775
(s)-6-(DIMETHYLAMINO)-4,4-DIPHENYL-3-HEPTANONE see DBE100
dl-6-DIMETHYLAMINO-4,4-DIPHENYL-3-HEPTANONE HYDROCHLORIDE see MDP750
1-6-DIMETHYLAMINO-4,4-DIPHENYL-3-HEPTANONE HYDROCHLORIDE see MDP250
6-DIMETHYLAMINO-4,4-DIPHENYL-3-HEPTANONE HYDROCHLORIDE see MDP000
6-(DIMETHYLAMINO)-4,4-DIPHENYL-3-HEPTANONE dl-MIXTURE see MDO760
p,p-DIMETHYLAMINODIPHENYLMETHANE see MJN000
α-4-DIMETHYLAMINO-1,2-DIPHENYL-3-METHYL-2-BUTANOL PROPIONATE see PNA250
α-(+)-4-DIMETHYLAMINO-1,2-DIPHENYL-3-METHYL-2-BUTANOL PROPIONATE ESTER see DAB879
2-(DIMETHYLAMINO)-N-(1,3-DIPHENYL-1H-PYRAZOL-5-YL) PROPANAMIDE see PAM500
4-(DIMETHYLAMINO)-2,2-DIPHENYLVALERAMIDE see DOY400
2-DIMETHYLAMINO ETHANETHIOL HYDROCHLORIDE see DOY600
2-(DIMETHYLAMINO)ETHANOL see DOY800
β-DIMETHYLAMINOETHANOL see DOY800
N,N-DIMETHYLAMINOETHANOL see DOY800
N-DIMETHYLAMINOETHANOL see DOY800
DIMETHYLAMINOETHANOL see DOY800
2-DIMETHYLAMINOETHANOL-p-ACETAMIDOBENZOATE see DOZ000
2-DIMETHYLAMINOETHANOL ACETATE see DOZ100
DIMETHYLAMINOETHANOL ACETATE see DOZ100
2-(DIMETHYLAMINO)ETHANOL BITARTRATE see DPA000
2-DIMETHYLAMINOETHANOL-4-N-BUTYLAMINOBENZOATE HYDROCHLORIDE see TBN000
β-DIMETHYLAMINOETHANOL DIPHENYLMETHYL ETHER see BBV500
2-(DIMETHYLAMINO)ETHANOL METHACRYLATE see DPG600
1-(DIMETHYLAMINOETHOXYACETAMIDO)ADAMANTANE HYDROCHLORIDE see AEF500

11-(3-DIMETHYLAMINOPROPYLIDENE)-6,11-DIHYDRODIBENZ(b,e)OXIPIN see DYE409

9-(3-DIMETHYLAMINOPROPYLIDENE)-10,10-DIMETHYL-9,10-DIHYDROANTHRACENE HYDROCHLORIDE see TDL000

2,2'-(3-DIMETHYLAMINOPROPYLIMINO)DIBENZYL see DLH600

N-(γ-DIMETHYLAMINOPROPYL)IMINODIBENZYL see DLH600

N-(3-DIMETHYLAMINOPROPYL)IMINODIBENZYL HYDROCHLORIDE see DLH630

5-DIMETHYLAMINO-6-PROPYL-5H-INDENO(5,6-d)-1,3-DIOXOLE HYDROCHLORIDE see DPY600

10-(3-DIMETHYLAMINOPROPYL)-2-METHOXYPHENOTHIAZINE see MFK500

10-((3-(DIMETHYLAMINO)PROPYL)-2-METHOXY)PHENOTHIAZINE, MALEATE see MFK750

17-β-((3-(DIMETHYLAMINO)-PROPYL)METHYLAMINO)ANDROST-5-EN-3-β-OL DIHYDROCHLORIDE see ARX800

10-(3-(DIMETHYLAMINO)PROPYL)-1-NITRO-9-ACRIDANONE HYDROCHLORIDE see NFW460

10-(3-(DIMETHYLAMINO)PROPYL)-1-NITRO-9(10H)-ACRIDINONE MONOHYDROCHLORIDE (9CI) see NFW460

o-(2-DIMETHYLAMINO)PROPYL)OXIME-5H-DIBENZO(a,d)CYCLOHEPTEN-5-ONE MONOHYDROCHLORIDE see LHX498

10-(2-(DIMETHYLAMINO)PROPYL)PHENOTHIAZINE see DQA400

10-(3-(DIMETHYLAMINO)PROPYL)PHENOTHIAZINE see DQA600

10-(3-DIMETHYLAMINOPROPYL)PHENOTHIAZINE-3-ETHYLONE see ABH500

10-(2-DIMETHYLAMINOPROPYL)PHENOTHIAZINE HYDROCHLORIDE see PMI750

10-(γ-DIMETHYLAMINO-N-PROPYL)PHENOTHIAZINE HYDROCHLORIDE see PMI500

10-(3-(DIMETHYLAMINO)PROPYL)PHENOTHIAZINE HYDROCHLORIDE see PMI500

N-(2-DIMETHYLAMINOPROPYL-1)PHENOTHIAZINE HYDROCHLORIDE see PMI750

(DIMETHYLAMINO-2-PROPYL-10-PHENOTHIAZINE HYDROCHLORIDE (FRENCH) see DQA400

10-(2-(DIMETHYLAMINO)PROPYL)PHENOTHIAZINE MONOHYDROCHLORIDE see PMI750

1-(10-(3-(DIMETHYLAMINO)PROPYL)PHENOTHIAZIN-2-YL)-1-BUTANONE MALEATE see BTA125

1-(10-(3-(DIMETHYLAMINO)PROPYL)-10H-PHENOTHIAZIN-2-YL)ETHANONE see ABH500

10-(3-DIMETHYLAMINOPROPYL)PHENOTHIAZIN-3-YLMETHYL KETONE see ABH500

10-(3-(DIMETHYLAMINO)PROPYL)PHENOTHIAZIN-2-YL METHYL KETONE MALEATE (1:1) see AAF750

10-(3-(DIMETHYLAMINO)PROPYL)PHENOTHIAZIN-2-YL MORPHOLINOMETHYL KETONE see DQA710

10-(2-(DIMETHYLAMINO)PROPYL)PHENOTHIAZIN-2-YL MORPHOLINOMETHYL KETONE see DQA700

1-(10-(2-DIMETHYLAMINOPROPYL)-PHENOTHIAZIN-2-YL)-1-PROPANONE MALEATE see IDA500

1-(10-(3-(DIMETHYLAMINO)PROPYL)PHENOTHIAZIN-2-YL)-1-PROPANONE MALEATE see PMX500

1-(2-(DIMETHYLAMINO)PROPYL)-2-(PHENYLACETYL)PYRROLE CITRATE see BEL525

α-(2-(DIMETHYLAMINO)PROPYL)-α-PHENYLBENZENEACETAMIDE see DOY400

10-(2-DIMETHYLAMINOPROPYL)-2-PROPIONYLPHENOTHIAZINE MALEATE see IDA500

((4-DIMETHYLAMINOPROPYLPYRIDO (3,2b) BENZOTHIAZINE)) HYDROCHLORIDE see DYB800

2-(3-DIMETHYLAMINOPROPYL)-3a,4,7,7a-TETRAHYDRO-4,7-ETHANOISOINDOLINE DIMETHIODIDE see DQB309

10-(2-DIMETHYLAMINOPROPYL)-9-THIA-1,10-DIAZAANTHRACENE HYDROCHLORIDE see AEG625

N-(3-DIMETHYLAMINOPROPYL)THIOCARBAMINSAEURE-S-AETHYLESTER-HYDROCHLORID (GERMAN) see EIH500

2'-((3-(DIMETHYLAMINO)PROPYL)THIO)CINNAMANILIDE HYDROCHLORIDE see CMP900

n-(2-((3-(DIMETHYLAMINO)PROPYL)THIO)PHENYL)-3-PHENYL-2-PROPENAMIDE MONOHYDROCHLORIDE see CMP900

DIMETHYLAMINO-N-PROPYL-THIOPHENYLPYRIDYLAMINE see DYB600

10-(3-(DIMETHYLAMINO)PROPYL)-2-(TRIFLUOROMETHYL) PHENOTHIAZINE see TKL000

4-DIMETHYLAMINOPYRIDINE see DQB600

γ-(DIMETHYLAMINO)PYRIDINE see DQB600

p-DIMETHYLAMINOPYRIDINE see DQB600

2-(DIMETHYLAMINO) RESERPILINATE see DQB800

2-(DIMETHYLAMINO) RESERPILIN-24-OIC ACID ETHYL ESTER see DQB800

DIMETHYLAMINORHODANBENZOL see TFH500

4-DIMETHYLAMINOSTILBEN (GERMAN) see DUB800

N,N-DIMETHYL-4-AMINOSTILBENE see DUB800

cis-4-DIMETHYLAMINOSTILBENE see DUC200

trans-4-DIMETHYLAMINOSTILBENE see DUC000

4-DIMETHYLAMINO-trans-STILBENE see DUC000

trans-p-(DIMETHYLAMINO)STILBENE see DUC000

DIMETHYLAMINOSTOVAINE see AHI250

12-(p-DIMETHYLAMINO)STYRYLBENZ(a)ACRIDINE see DQC200

7-(p-(DIMETHYLAMINO)STYRYL)BENZ(c)ACRIDINE see DQC000

2-(4-DIMETHYLAMINOSTYRYL)BENZOTHIAZOLE see DQC400

2-(p-(DIMETHYLAMINO)STYRYL)BENZOTHIAZOLE see DQC400

14-(p-(DIMETHYLAMINO)STYRYL)DIBENZ(a,j)ACRIDINE see DOV000

4-(p-(DIMETHYLAMINO)STYRYL)-6,8-DIMETHYLQUINOLINE see DQC600

2-(4-N,N-DIMETHYLAMINOSTYRYL)QUINOLINE see DQD000

4-(4-DIMETHYLAMINOSTYRYL)QUINOLINE see DQD000

4-(p-(DIMETHYLAMINO)STYRYL)QUINOLINE see DQD000

4-(p-(DIMETHYLAMINO)STYRYL)QUINOLINE MONOHYDROCHLORIDE see DQD200

N-(DIMETHYLAMINO)SUCCINAMIC ACID see DQD400

DIMETHYLAMINOSUCCINAMIC ACID see DQD400

N-DIMETHYLAMINO-SUCCINAMIDSAEURE (GERMAN) see DQD400

O-(4-((DIMETHYLAMINO)SULFONYL)PHENYL) O,O-DIMETHYL PHOSPHOROTHIOATE see FAB600

(DIMETHYLAMINO)-TERMINATED see SDF000

DIMETHYLAMINO-4-THIOCYANOBENZENE see TFH500

4-DIMETHYLAMINOTHIOCYANOBENZENE see TFH500

p-N,N-DIMETHYLAMINOTHIOCYANOBENZENE see TFH500

p-(DIMETHYLAMINO)THIOCYANOBENZENE see TFH500

p-DIMETHYLAMINOTHYMOLDIMETHYLURETHANE METHIODIDE see DQD500

4-((4-(DIMETHYLAMINO)-m-TOLYL)AZO)-2-PICOLINE-1-OXIDE see DQD600

4-((4-(DIMETHYLAMINO)-o-TOLYL)AZO)-2-PICOLINE-1-OXIDE see DQD800

4-((4-(DIMETHYLAMINO)-m-TOLYL)AZO)-3-PICOLINE-1-OXIDE see DQE000

4-((4-(DIMETHYLAMINO)-o-TOLYL)AZO)-3-PICOLINE-1-OXIDE see DQE200

5-((4-(DIMETHYLAMINO)-m-TOLYL)AZO)QUINOLINE see DQE400

5-((4-(DIMETHYLAMINO)-o-TOLYL)AZO)QUINOLINE see DQE600

5-DIMETHYLAMINO-4-TOLYL METHYLCARBAMATE see DQE800

4-(DIMETHYLAMINO)-m-TOLYL METHYLCARBAMATE see DOR400

S,S'-(2-(DIMETHYLAMINO)TRIMETHYLENE)BIS(THIOCARBAMATE) HYDROCHLORIDE see BHL750

DIMETHYLAMINOTRIMETHYLSILANE see DQE900

4-DIMETHYLAMINOTRIPHENYLMETHAN (GERMAN) see DRQ000

4-DIMETHYLAMINOTRIPHENYLMETHANE see DRQ000

3-DIMETHYLAMINO-1,1,2-TRIS(4-METHOXYPHENYL)-1-PROPENE HYDROCHLORIDE see TNJ750

5-DIMETHYLAMINO-1,2,3-TRITHIANE HYDROGENOXALATE see TFH750

4-DIMETHYLAMINO-3,5-XYLENOL see DQF000

4-(DIMETHYLAMINO)-3,5-XYLENOL METHYLCARBAMATE (ESTER) see DOS000

4-((4-(DIMETHYLAMINO)-2,3-XYLYL)AZO)PYRIDINE-1-OXIDE see DQF200

4-((4-(DIMETHYLAMINO)-2,5-XYLYL)AZO)PYRIDINE-1-OXIDE see DQF400

4-((4-(DIMETHYLAMINO)-3,5-XYLYL)AZO)PYRIDINE-1-OXIDE see DQF600

4-(DIMETHYLAMINO)-3,5-XYLYL ESTER METHYLCARBAMIC ACID see DOS000

4-DIMETHYLAMINO-3,5-XYLYL METHYLCARBAMATE see DOS000

4-(N,N-DIMETHYLAMINO)-3,5-XYLYL N-METHYLCARBAMATE see DOS000

4-DIMETHYLAMINO-3,5-XYLYL-N-METHYLCARBAMATE see DOS000

DIMETHYLAMMONIUM CHLORIDE see DOR600

DIMETHYLAMMONIUM 2,4-DICHLOROPHENOXYACETATE see DFY800

N-((2-DIMETHYLAMMONIUM)ETHYL)-4,5,6,7-TETRACHLOROISOINDOLINIUM DIMETHOCHLORIDE see CDY000

DIMETHYLAMMONIUM PERCHLORATE see DQF650

DI(4-METHYL-2-AMYL) MALEATE see DKP400

2,3-DIMETHYLANILINE see XMJ000

2,4-DIMETHYLANILINE see XMS000

2,5-DIMETHYLANILINE see XNA000

2,6-DIMETHYLANILINE see XNJ000

3,4-DIMETHYLANILINE see XNS000

3,5-DIMETHYLANILINE see XOA000

N,N-DIMETHYLANILINE see DQF800

N-DIMETHYL-ANILINE (OSHA) see DQF800

DIMETHYLANILINE see XMA000

N,N-DIMETHYL-p-ANILINEDIAZOSULFONIC ACID SODIUM SALT see DOU600

2,4-DIMETHYLANILINE HYDROCHLORIDE see XOJ000

2,5-DIMETHYLANILINE HYDROCHLORIDE see XOS000

N,N-DIMETHYLANILINE METHIODIDE see TMB750

2-(2,6-DIMETHYLANILINO)-5,6-DIHYDRO-4H-1,3-THIAZINE see DMW000

6,12-DIMETHYLANTHANTHRENE see DQG000

9,10-DIMETHYLANTHRACENE see DQG200

3-(10,10-DIMETHYL(10H)-ANTHRACENYLIDENE)-N,N-DIMETHYL-1-PROPANAMINE (9CI) see AEG875

3-(10,10-DIMETHYL-9(10H)-ANTHRACENYLIDENE)-N,N-DIMETHYL-1-PROPANAMINE HYDROCHLORIDE see TDL000

DIMETHYL ANTHRANILATE (FCC) see MGQ250

DIMETHYLANTIMONY CHLORIDE see DQG400

DIMETHYLARSENIC ACID see HKC000

DIMETHYLARSENIC ACID SODIUM SALT TRIHYDRATE see HKC550

DIMETHYLARSINAT SODNY see HKC500

DIMETHYLARSINE see DQG600

DIMETHYLARSINIC ACID see HKC000

DIMETHYLARSINIC ACID SODIUM SALT TRIHYDRATE see HKC550

DIMETHYL ARSINIC SULFIDE see DQG700

((DIMETHYLARSINO)OXY)SODIUM-As-OXIDE see HKC500

α,α-DIMETHYL-α'-CARBOBUTOXY-DIHYDRO-γ-PYRONE see BRT000
O,O-DIMETHYL-O-(2-CARBOMETHOXY-1-METHYLVINYL) PHOSPHATE see MQR750
DIMETHYL-1-CARBOMETHOXY-1-PROPEN-2-YL PHOSPHATE see MQR750
DIMETHYL CARBONATE see MIF000
DIMETHYLCELLOSOLVE see DOE600
N,N-DIMETHYLCETYLAMINE see HCP525
DIMETHYLCETYLAMINE see HCP525
DIMETHYL-CHLORMETHYL-ETHOXYSILAN (CZECH) see DRC400
O,O-DIMETHYL-O-3-CHLOR-4-NITROFENYLTIOFOSFAT (CZECH) see MIJ250
O,O-DIMETHYL-O-(3-CHLOR-4-NITROPHENYL)-MONOTHIOPHOSPHAT (GERMAN) see MIJ250
1,10-DIMETHYL-2-CHLORO-5,6-BENZACRIDINE see CGH250
1,10-DIMETHYL-2-CHLORO-7,8-BENZACRIDINE (FRENCH) see DRB800
8,12-DIMETHYL-9-CHLOROBENZ(a)ACRIDINE see CGH250
7,11-DIMETHYL-10-CHLOROBENZ(c)ACRIDINE see DRB800
O,O-DIMETHYL-O-(6-CHLOROBICYCLO(3.2.0)HEPTADIEN-1,5-YL)PHOSPHATE see HBK700
DIMETHYL 2-CHLORO-2-DIETHYLCARBAMOYL-1-METHYLVINYL PHOSPHATE see FAB400
O,O-DIMETHYL O-(2-CHLORO-2-(N,N-DIETHYLCARBAMOYL)-1-METHYLVINYL) PHOSPHATE see FAB400
DIMETHYLCHLOROETHER see CIO250
DIMETHYL(2-CHLOROETHYL)AMINE see CGW000
DIMETHYL(2-CHLOROETHYL)AMINE HYDROCHLORIDE see DRC000
DIMETHYL-β-CHLOROETHYLAMINE HYDROCHLORIDE see DRC000
DIMETHYLCHLOROFORMAMIDE see DQY950
2,2-DIMETHYL-4-(CHLOROMETHYL)-1,3-DIOXA-2-SILACYCLOPENTANE see CIL775
DIMETHYLCHLOROMETHYLETHOXYSILANE see DRC400
O,O-DIMETHYL O-2-CHLORO-4-NITROPHENYL PHOSPHOROTHIOATE see NFT000
O,O-DIMETHYL-O-(3-CHLORO-4-NITROPHENYL) PHOSPHOROTHIOATE see MIJ250
DIMETHYL-2-CHLORO-4-NITROPHENYLTHIONOPHOSPHATE see DRC500
DIMETHYL-3-CHLORO-4-NITROPHENYL THIONOPHOSPHATE see MIJ250
DIMETHYL-2-CHLORONITROPHENYL THIOPHOSPHATE see NFT000
O,O-DIMETHYL-O-(3-CHLORO-4-NITROPHENYL) THIOPHOSPHATE see MIJ250
α,α-DIMETHYL-o-CHLOROPHENETHYLAMINE HYDROCHLORIDE see DRC600
α,α-DIMETHYL-p-CHLOROPHENETHYLAMINE HYDROCHLORIDE see ARW750
N-DIMETHYL-4-(p-CHLOROPHENOXY-1,1'-DIMETHYLACETATE)BUTYRAMIDE see LGK200
N-DIMETHYL-4-(1,4'-CHLOROPHENOXY-1,1'-DIMETHYLACETATE)BUTYRAMIDE see LGK200
N-DIMETHYL-4-(1,4'-CHLOROPHENOXYISOBUTYRATE)BUTYRAMIDE see LGK200
N-DIMETHYL-4-(p-CHLOROPHENOXYISOBUTYRATE)BUTYRAMIDE see LGK200
N,N-DIMETHYL-p-((m-CHLOROPHENYL)AZO)ANILINE see DRC800
N,N-DIMETHYL-p-((o-CHLOROPHENYL)AZO)ANILINE see DRD000
N,N-DIMETHYL-p-((p-CHLOROPHENYL)AZO)ANILINE see CGD250
DIMETHYL-p-CHLOROPHENYLTHIOMETHYL DITHIOPHOSPHATE see MQH750
1,1-DIMETHYL-3-(p-CHLOROPHENYL)UREA see CJX750
DIMETHYL CHLOROTHIOPHOSPHATE (DOT) see DTQ600
DIMETHYLCHLORTHIOFOSAT (CZECH) see DTQ600
O,O-DIMETHYL-O-2-CHLOR-1-(2,4,5-TRICHLORPHENYL)-VINYL-PHOSPHAT (GERMAN) see TBW100
1,3-DIMETHYLCHOLANTHRENE see DRE000
15,20-DIMETHYLCHOLANTHRENE see DRE000
16:20-DIMETHYLCHOLANTHRENE see DRD800
2,3-DIMETHYLCHOLANTHRENE see DRD800
3,6-DIMETHYLCHOLANTHRENE see DRD850
(+)-o,o'-DIMETHYLCHONDROCURARINE DIIODIDE see DUM000
1,11-DIMETHYLCHRYSENE see DRE400
1,2-DIMETHYLCHRYSENE see DRE200
4,5-DIMETHYLCHRYSENE see DRE600
5,11-DIMETHYLCHRYSENE see DRF000
5,6-DIMETHYLCHRYSENE see DRE800
5,7-DIMETHYLCHRYSENE see DRE800
DIMETHYL CITRACONATE see DRF200
2,2-DIMETHYL-7-COUMARANYL-N-METHYLCARBAMATE see CBS275
4,6-DIMETHYLCOUMARIN see DRF400
DIMETHYLCYANAMIDE see DRF600
DIMETHYLCYANOARSINE see COL750
O,O-DIMETHYL-O-(4-CYANO-PHENYL)-MONOTHIOPHOSPHAT (GERMAN) see COQ399
O,O-DIMETHYL-O-4-CYANOPHENYL PHOSPHOROTHIOATE see COQ399
O,O-DIMETHYL-O-p-CYANOPHENYL PHOSPHOROTHIOATE see COQ399
O,O-DIMETHYL-O-4-CYANOPHENYL THIOPHOSPHATE see COQ399
N,N-DIMETHYLCYCLOHEXANAMINE see DRF709
1,3-DIMETHYLCYCLOHEXANE see DRG000
1,4-DIMETHYLCYCLOHEXANE see DRG200
1,2-DIMETHYLCYCLOHEXANE (DOT) see DRF800
cis-1,2-DIMETHYLCYCLOHEXANE see DRF800
m-DIMETHYLCYCLOHEXANE see DRG000
o-DIMETHYLCYCLOHEXANE see DRF800

trans-1,2-DIMETHYLCYCLOHEXANE see DRG400
(±)N,α-DIMETHYL-CYCLOHEXANEETHANAMINE see PNN400
α,N-DIMETHYLCYCLOHEXANEETHYLAMINE see PNN400
N,α-DIMETHYLCYCLOHEXANEETHYLAMINE see PNN400
N,α-DIMETHYLCYCLOHEXANEETHYLAMINE HYDROCHLORIDE see PNN300
DIMETHYL-3-CYCLOHEXENE-1-CARBOXALDEHYDE see LJE200
1,5-DIMETHYL-5-(1-CYCLOHEXENYL)BARBITURIC ACID see ERD500
1,5-DIMETHYL-5-CYCLOHEXENYL-1'-BARBITURIC ACID, SODIUM SALT see ERE000
N,N-DIMETHYLCYCLOHEXYLAMINE (DOT) see DRF709
DIMETHYLCYCLOHEXYLAMINE see DRF709
N,N-DIMETHYL-N-CYCLOHEXYLMETHYLAMINE see CPR500
N,α-DIMETHYLCYCLOPENTANEETHYLAMINE HYDROCHLORIDE see CPV609
11,17-DIMETHYL-15H-CYCLOPENTA(a)PHENANTHRENE see DRH200
12,17-DIMETHYL-15H-CYCLOPENTA(a)PHENANTHRENE see DRH400
3,4-DIMETHYL-1,2-CYCLOPENTENOPHENANTHRENE see DRH800
N-DIMETHYLCYSTEAMINE HYDROCHLORIDE see DOY600
β,β-DIMETHYLCYSTEINE see MCR750
DIMETHYLCYSTEINE see MCR750
9,10-DIMETHYL-DBA see DRI800
3,5-DIMETHYL-4-DIALLYLAMINOPHENYL-N-METHYLCARBAMATE see DBI800
3,3'-DIMETHYL-4,4'-DIAMINODIPHENYLMETHANE see MJO250
DIMETHYLDIAMINODIPHENYLMETHANE see MJO000
N,N'-DIMETHYLDIAMINOETHANE see DRI600
asym-DIMETHYL-3,7-DIAMINOPHENAZATHIONIUM CHLORIDE see DUG700
N,N-DIMETHYL-1,3-DIAMINOPROPANE see AJQ100
O,O-DIMETHYL- S-(4,6-DIAMINO-s-TRIAZIN-2-YLMETHYL)PHOSPHORODITHIOATE see ASD000
DIMETHYLDIAMINOXANTHENYL CHLORIDE see ADK000
3,3'-DIMETHYLDIAN see IPK000
DIMETHYLDIARSINE SULFIDE see DQG700
1,5-DIMETHYL-1,5-d-DIAZAUNDECAMETHYLENE POLYMETHOBROMIDE see HCV500
1,1-DIMETHYLDIAZENIUM PERCHLORATE see DRI700
2,3-DIMETHYL-1,4-DIAZINE see DTU400
2,5-DIMETHYL-1,4-DIAZINE see DTU600
9,10-DIMETHYL-1,2,5,6-DIBENZANTHRACENE see DRI800
7,14-DIMETHYLDIBENZ(a,h)ANTHRACENE see DRI800
N,N-DIMETHYL-5H-DIBENZO(a,d)CYCLOHEPTENE-Δ⁵,γ-PROPYLAMINE see PMH600
N,N-DIMETHYL-5H-DIBENZO(a,d)CYCLOHEPTENE-Δ⁵,⁷-PROPYLAMINE HYDROCHLORIDE see DPX800
N,N-DIMETHYLDIBENZO(b,e)THIEPIN-Δ¹¹(⁶ᴴ),ʸPROPYLAMINE see DYC875
N,N-DIMETHYLDIBENZO(b,e)THIEPIN-Δ¹¹(⁶ᴴ),ʸ-PROPYLAMINE HYDROCHLORIDE see DPY200
4,9-DIMETHYL-2,3,5,6-DIBENZOTHIOPHENTHRENE see DRJ000
N,N-DIMETHYLDIBENZ(b,e)OXEPIN-Δ¹¹(⁶ᴴ)-γ-PROPYLAMINE see DYE409
N,N-DIMETHYLDIBENZ(b,e)OXEPIN-Δ¹¹(⁶ᴴ)-PROPYLAMINE HYDROCHLORIDE see AEG750
1,1-DIMETHYLDIBORANE see DRJ200
1,2-DIMETHYLDIBORANE see DRJ400
O,O-DIMETHYL-O-(1,2-DIBROMO-2,2-DICHLOROETHYL)PHOSPHATE see NAG400
DIMETHYL-1,2-DIBROMO-2,2-DICHLOROETHYL PHOSPHATE (OSHA) see NAG400
2,5-DIMETHYL-2,5-DI(t-BUTYLPEROXY)HEXANE see DRJ800
2,5-DIMETHYL-2,5-DI(tert-BUTYLPEROXY)HEXANE see DRJ800
2,5-DIMETHYL-2,5-DI(tert-BUTYLPEROXY)HEXYNE-3 see DRJ825
DIMETHYL-1,3-DI(CARBOMETHOXY)-1-PROPEN-2-YL PHOSPHATE see SOY000
O,O-DIMETHYL-O-(2,5-DICHLOR-4-BROMPHENYL)-THIONOPHOSPHAT (GERMAN) see BNL250
O,O-DIMETHYL-O-(2,5-DICHLOR-4-JODPHENYL)-MONOTHIOPHOSPHAT (GERMAN) see IEN000
O,O-DIMETHYL-O-(2,5-DICHLOR-4-JODPHENYL)-THIONOPHOSPHAT (GERMAN) see IEN000
O,O-DIMETHYL-O-(2,5-DICHLORO-4-BROMOPHENYL)PHOSPHOROTHIOATE see BNL250
O,O-DIMETHYL-O-(2,5-DICHLORO-4-BROMOPHENYL) THIOPHOSPHATE see BNL250
O,O-DIMETHYL-O-2,2-DICHLORO-1,2-DIBROMOETHYL PHOSPHATE see NAG400
DIMETHYL-2,2-DICHLOROETHENYL PHOSPHATE see DGP900
DIMETHYL-1,1'-DICHLOROETHER see BIK000
O,O-DIMETHYL-O-2,5-DICHLORO-4-IODOPHENYL THIOPHOSPHATE see IEN000
O,O-DIMETHYL S-(2,5-DICHLOROPHENYLTHIO)METHYL PHOSPHORODITHIOATE see MNO750
1,1-DIMETHYL-3-(3,4-DICHLOROPHENYL)UREA see DXQ500
DIMETHYLDICHLOROSILANE (DOT) see DFE259
DIMETHYLDICHLOROSTANNANE see DUG825
DIMETHYLDICHLOROTIN see DUG825
DIMETHYL-2,2-DICHLOROVINYL PHOSPHATE see DGP900
O,O-DIMETHYL-O-2,2-DICHLOROVINYL PHOSPHATE see DGP900
DIMETHYL DICHLOROVINYL PHOSPHATE see DGP900

O,O-DIMETHYL DICHLOROVINYL PHOSPHATE see DGP900
DIMETHYL-DICHLORSILAN see DFE259
O,O-DIMETHYL-O-(2,2-DICHLOR-VINYL)-PHOSPHAT (GERMAN) see DGP900
DIMETHYLDIDECYLAMMONIUM CHLORIDE see DGX200
2,5-DIMETHYL-1,2,5,6-DIEPOXYHEX-3-YNE see DRK400
DIMETHYL-DIETHOXYSILAN (CZECH) see DHG000
DIMETHYLDIETHOXYSILANE (DOT) see DHG000
DIMETHYL DIETHYLAMIDO-1-CHLOROCROTONYL (2) PHOSPHATE see FAB400
N,N-DIMETHYL-p-((3,4-DIETHYLPHENYL)AZO)ANILINE see DJB400
2,6-DIMETHYL-1,1-DIETHYLPIPERIDINIUM BROMIDE see DRK600
trans-4,4'-DIMETHYL-α-α'-DIETHYLSTILBENE see DRK500
N,N-DIMETHYL-2,5-DIFLUORO-p-(2,5-DIFLUOROPHENYLAZO)ANILINE see DRK800
N,N-DIMETHYL-p-(2,5-DIFLUOROPHENYLAZO)ANILINE see DKH000
N,N-DIMETHYL-p-(3,4-DIFLUOROPHENYLAZO)ANILINE see DRL000
N,N-DIMETHYL-p-(3,5-DIFLUOROPHENYLAZO)ANILINE see DKH100
N,N-DIMETHYL-3',4'-DIFLUORO-4-(PHENYLAZO)BENZENEAMINE see DRL000
N,N-DIMETHYLDIGUANIDE see DQR600
9:10-DIMETHYL-9-10-DIHYDRO-1,2-BENZANTHRACENE-9,10-OXIDE see DQL800
2,2-DIMETHYL-2,3-DIHYDROBENZOFURAN-7-YL ESTER, METHYLCARBAMIC ACID see CBS275
2,2-DIMETHYL-2,3-DIHYDRO-7-BENZOFURANYL-N-METHYLCARBAMATE see CBS275
11,17-DIMETHYL-16,17-DIHYDRO-15H-CYCLOPENTA(a)PHENANTHRENE see DLI200
7,11-DIMETHYL-15,16-DIHYDROCYCLOPENTA(a)PHENANTHREN-17-ONE see DLI300
2,5-DIMETHYL-2,5-DIHYDROFURAN-2,5-ENDO PEROXIDE see DUL589
2,5-DIMETHYL-2,5-DIHYDROPEROXYHEXANE, >82% with water (DOT) see DSE800
5,5-DIMETHYL-DIHYDRORESORCINOL-N,N-DIMETHYLCARBAMAT (GERMAN) see DRL200
5,5-DIMETHYLDIHYDRORESORCINOL DIMETHYLCARBAMATE see DRL200
5,5-DIMETHYL-4,5-DIHYDRO-3-RESORCYL-DIMETHYL-CARBAMAT (GERMAN) see DRL200
1,3-DIMETHYL-7-(2,3-DIHYDROXYPROPYL)XANTHINE see DNC000
DIMETHYL DIKETONE see BOT500
2,6-DIMETHYL-3,5-DIMETHOXYCARBONYL-4-(o-DIFLUOROMETHOXYPHENYL)-1,4-DIHYDROPYRIDINE see RSZ375
DIMETHYL-α-(DIMETHOXYPHOSPHINYL)-p-CHLOROBENZYL PHOSPHATE see CMU875
DIMETHYL 3-(DIMETHOXYPHOSPHINYLOXY)GLUTACONATE see SOY000
2',3'-DIMETHYL-4-DIMETHYLAMINOAZOBENZENE see DUN800
2,4'-DIMETHYL-4-DIMETHYLAMINOAZOBENZENE see DRL400
N,N-DIMETHYL-4-DIMETHYLAMINO-3-ISOPROPYLPHENYL ESTER METHIODIDE, CARBAMIC ACID see HJZ000
3,3-DIMETHYL-4-(DIMETHYLAMINO)-4-(p-METHOXYPHENYL)BUTYL p-METHOXYPHENYL KETONE see DRL460
3,3-DIMETHYL-4-(DIMETHYLAMINO)-4-(o-METHOXYPHENYL)BUTYL o-METHOXYPHENYL KETONE see DRL450
2,3-DIMETHYL-8-(DIMETHYLAMINOMETHYL)-7-METHOXYCHROMONE HYDROCHLORIDE see DRL600
3-keto-1,5-DIMETHYL-4-DIMETHYLAMINO-2-PHENYL-2,3-DIHYDROPYRAZOLE see DOT000
1,5-DIMETHYL-4-DIMETHYLAMINO-2-PHENYL-3-PYRAZOLONE see DOT000
2,3-DIMETHYL-4-DIMETHYLAMINO-1-PHENYL-5-PYRAZOLONE see DOT000
9,9-DIMETHYL-10-(3-(DIMETHYLAMINO)PROPYL)ACRIDAN see DNW700
9,9-DIMETHYL-10-DIMETHYLAMINOPROPYLACRIDAN HYDROGEN TARTRATE see DRM000
9,9-DIMETHYL-10-(3-DIMETHYLAMINO)PROPYLACRIDINE TARTRATE see DRM000
5,6-DIMETHYL-2-DIMETHYLAMINO-4-PYRIMIDINYLDIMETHYLCARBAMATE see DOX600
6,8-DIMETHYL-(4-p-(DIMETHYLAMINO)STYRYL)QUINOLINE see DQC600
3,3-DIMETHYL-4-(DIMETHYLAMINO)-4-(m-TOLYL)BUTYL m-TOLYL KETONE see DRM100
3,3-DIMETHYL-4-(DIMETHYLAMINO)-4-(o-TOLYL)BUTYL o-TOLYL KETONE see DRM110
3,3-DIMETHYL-4-(DIMETHYLAMINO)-4-(p-TOLYL)BUTYL p-TOLYL KETONE see DRM120
3',4'-DIMETHYL-4-DIMETHYLAMINOZOBENZENE see DUO000
O,O-DIMETHYL-O-(2-DIMETHYL-CARBAMOYL-1-METHYL-VINYL)PHOSPHAT (GERMAN) see DGQ875
O,O-DIMETHYL-O-(N,N-DIMETHYLCARBAMOYL-1-METHYLVINYL) PHOSPHATE see DGQ875
3,4-DIMETHYL-4-(3,4-DIMETHYL-5-ISOXAZOLYAZO)-ISOXAZOLIN-5-ONE see DRM600
O,O-DIMETHYL-O-(3,5-DIMETHYL-4-METHYLTHIOPHENYL) PHOSPHOROTHIOATE see DST200
O,O-DIMETHYL-O-(1,4-DIMETHYL-3-OXO-4-AZA-PENT-1-ENYL)FOSFAAT (DUTCH) see DGQ875
O,O-DIMETHYL-O-(1,4-DIMETHYL-3-OXO-4-AZA-PENT-1-ENYL)PHOSPHATE see DGQ875

N,N-DIMETHYL-p-(2',3'-DIMETHYLPHENYLAZO)ANILINE see DUN800
N,N-DIMETHYL-p-(3',4'-DIMETHYLPHENYLAZO)ANILINE see DUO000
3-DIMETHYL-1,2-DIMETHYLPROPYL p-AMINOBENZOATE HYDROCHLORIDE see AIT750
N,N-DIMETHYL-4-(4'-(2',5'-DIMETHYLPYRIDYL-1'-OXIDE)AZO)ANILINE see DPP600
N,N-DIMETHYL-4-(4'-(2',6'-DIMETHYLPYRIDYL-1'-OXIDE)AZO)ANILINE see DPP800
N,N-DIMETHYL-4-(4'-(3',5'-DIMETHYLPYRIDYL-1'-OXIDE)AZO)ANILINE see DPP709
2,5-DIMETHYL-1-(5-(2,5-DIMETHYLPYRROLIDINO)-2,4-PENTADIENYLIDENE) PYRROLIDINIUM CHLORIDE SESQUIHYDRATE see DRM800
O,O-DIMETHYL-O-(p-(N,N-DIMETHYLSULFAMOYL)PHENYL)PHOSPHOROTHIOATE see FAB600
3,4-DIMETHYL-2,6-DINITRO-N-(1-ETHYLPROPYL)ANILINE see DRN200
DIMETHYL-DI-NITROSO-AETHYLENDIAMIN (GERMAN) see DVE400
1,6-DIMETHYL-1,6-DINITROSOBIUREA see DRN400
DIMETHYLDINITROSOETHYLENEDIAMINE see DVE400
2,5-DIMETHYL-1,4-DINITROSOPIPERAZINE see DRN800
2,5-DIMETHYLDINITROSOPIPERAZINE see DRN800
2,6-DIMETHYLDINITROSOPIPERAZINE see DRO000
DIMETHYLDIOCTADECYLAMMONIUM CHLORIDE see DXG625
2,2-DIMETHYL-1,3-DIOXA-6-AZA-2-SILACYCLOOCTANE-6-ETHANOL see SDH670
DIMETHYL-1,1-DIOXA-2,8-HYDROXYETHYL-5 SILA-1 AZA-5 CYCLOOCTANE (FRENCH) see SDH670
4,4-DIMETHYLDIOXANE-1,3 see DVQ400
DIMETHYL-p-DIOXANE (DOT) see DRO800
DIMETHYL DIOXANE see DRO800
2,2-DIMETHYL-1,3-DIOXANE-4,6-DIONE see DRP200
2,2-DIMETHYL-m-DIOXANE-4,6-DIONE see DRP200
DIMETHYLDIOXANES (DOT) see DRO800
2,6-DIMETHYL-m-DIOXAN-4-OL ACETATE see ABC250
2,6-DIMETHYL-m-DIOXAN-4-YL ACETATE see ABC250
2,2-DIMETHYL-4,6-DIOXO-m-DIOXANE see DRP200
2,2-DIMETHYL-1,3-DIOXOLAN see DRP400
2,2-DIMETHYL-1,3-DIOXOLANE-4-METHANOL see DVR600
2-(4,5-DIMETHYL-1,3-DIOXOLAN-2-YL)PHENYL-N-METHYLCARBAMATE see DRP600
N,N-DIMETHYL-2,2-DIPHENYLACETAMIDE see DRP800
N,N-DIMETHYL-α,α-DIPHENYLACETAMIDE see DRP800
N,N-DIMETHYLDIPHENYLACETAMIDE see DRP800
2',3'-DIMETHYL-2-DIPHENYLAMINECARBOXYLIC ACID see XQS000
cis-N,N-DIMETHYL-2-(p-(1,2-DIPHENYL-1-BUTENYL)PHENOXY)ETHYLAMINE see NOA600
N,N-DIMETHYL-2-(p-(1,2-DIPHENYL-1-BUTENYL)PHENOXY)ETHYLAMINE CITRATE see DRP875
3,3'-DIMETHYL-4,4'-DIPHENYLDIAMINE see TGJ750
3,3'-DIMETHYLDIPHENYL-4,4'-DIAMINE see TGJ750
(R) (−)-N,N-DIMETHYL-1,2-DIPHENYLETHYLAMINE HYDROCHLORIDE see DWA600
3,3'-DIMETHYLDIPHENYLMETHANE-4,4'-DIISOCYANATE see MJN750
N,N-DIMETHYL-4-(DIPHENYLMETHYL)ANILINE see DRQ000
α-N,N-DIMETHYL-3,4-DIPHENYL-2-METHYL-3-PROPIONOXY-1-BUTYLAMINE see PNA250
1,2-DIMETHYL-3,5-DIPHENYL-1-H-PYRAZOLIUM METHYL SULFATE see ARW000
2,2'-DIMETHYLDIPROPYLINITROSOAMINE see DRQ200
1,1'-DIMETHYL-4,4'-DIPYRIDINIUM-DICHLORID see PAJ000
1',1'-DIMETHYL-4,4'-DIPYRIDINIUM DI(METHYLSULFATE) see PAJ250
4,4'-DIMETHYLDIPYRIDYL DICHLORIDE see PAJ000
1,1'-DIMETHYL-4,4'-DIPYRIDYLIUM CHLORIDE see PAJ000
N,N'-DIMETHYL-4,4'-DIPYRIDYLIUM DICHLORIDE see PAJ000
DIMETHYLDISULFIDE see DRQ400
N,N-DIMETHYL-4,4-DI-2-THIENYL-3-BUTEN-2-AMINE HYDROCHLORIDE see TLQ250
N,1-DIMETHYL-3,3-DI-2-THIENYL-N-ETHYLALLYLAMINE HYDROCHLORIDE see EIJ000
o,o-DIMETHYL DITHIOBIS(THIOFORMATE) see DUN600
DIMETHYLDITHIOCARBAMATE ZINC SALT see BJK500
DIMETHYLDITHIOCARBAMIC ACID COPPER SALT see CNL500
DIMETHYLDITHIOCARBAMIC ACID with DIMETHYLAMINE (1:1) see DRQ600
DIMETHYLDITHIOCARBAMIC ACID DIMETHYL AMINE SALT see DRQ600
N,N-DIMETHYLDITHIOCARBAMIC ACID DIMETHYLAMINOMETHYL ESTER see DRQ650
DIMETHYLDITHIOCARBAMIC ACID DIMETHYLAMMONIUM SALT see DRQ600
DIMETHYLDITHIOCARBAMIC ACID, IRON(3+) SALT see FAS000
DIMETHYLDITHIOCARBAMIC ACID, IRON SALT see FAS000
DIMETHYLDITHIOCARBAMIC ACID, LEAD SALT see LCW000
DIMETHYLDITHIOCARBAMIC ACID, SODIUM SALT see SGM500
N,N-DIMETHYLDITHIOCARBAMIC ACID S-TRIBUTYLSTANNYL ESTER see TID150
DIMETHYLDITHIOCARBAMIC ACID, ZINC SALT see BJK500

N,N-DIMETHYL-DITHIOCARBAMINSAEURE-DIMETHYLAMINOMETHYL-ESTER see DRQ650

O,O-DIMETHYLDITHIOFOSFORECNAN SODNY (CZECH) see PHI250

2,4-DIMETHYL-1,3-DITHIOLANE-2-CARBOXALDEHYDE O-((METHYLAMI-NO)CARBONYL)OXIME see DRR000

2,4-DIMETHYL-1,3-DITHIOLANE-2-CARBOXALDEHYDE O-(METHYLCARBA-MOYL)OXIME see DRR000

O,O-DIMETHYLDITHIOPHOSPHATE see PHH500

O,O-DIMETHYLDITHIOPHOSPHATE DIETHYLMERCAPTOSUCCINATE see MAK700

DIMETHYLDITHIOPHOSPHORIC ACID see PHH500

O,O-DIMETHYL DITHIOPHOSPHORIC ACID see PHH500

DIMETHYLDITHIOPHOSPHORIC ACID N-METHYLBENZAZIMIDE ESTER see ASH500

O,O-DIMETHYL DITHIOPHOSPHORYLACETIC ACID-N-METHYL-N-FORMYLAM-IDE see DRR200

O,O-DIMETHYLDITHIOPHOSPHORYLACETIC ACID-N-MONOMETHYLAMIDE SALT see DSP400

O,O-DIMETHYL-DITHIOPHOSPHORYLESSIGSAEURE MONOMETHYLAMID (GERMAN) see DSP400

(O,O-DIMETHYLDITHIOPHOSPHORYLPHENYL)ACETIC ACID ETHYL ESTER see DRR400

DIMETHYL DIXANTHOGEN see DUN600

N,N-DIMETHYL-2-(DI-2,6-XYLYLMETHOXY)ETHYLAMINE HYDROCHLORIDE see DRR500

DIMETHYL DIZENEDICARBOXYLATE see DQH509

2,5-DIMETHYL-DNPZ see DRN800

2,6-DIMETHYL-DNPZ see DRO000

N,N-DIMETHYL-1-DODECANAMINE see DRR800

2,6-DIMETHYLDODECA-2,6,8-TRIEN-10-ONE see DRR700

7,11-DIMETHYL-4,6,10-DODECATRIEN-3-ONE see DRR700

N,N-DIMETHYLDODECYLAMINE see DRR800

N,N-DIMETHYLDODECYLAMINE ACETATE see DRS000

DIMETHYLDODECYLAMINE ACETATE see DRS000

DIMETHYLDODECYLAMINE HYDROCHLORIDE mixed with SODIUM NITRITE (7:8) see SIS150

N,N-DIMETHYLDODECYLAMINE OXIDE see DRS200

DIMETHYLDODECYLAMINE-N-OXIDE see DRS200

N,N-DIMETHYL-DODECYLAMINOXID (CZECH) see DRS200

N,N-DIMETHYLDODECYLBETAINE see LBU200

N,N-DIMETHYL-N-DODECYLGLYCINE see LBU200

N,N-DIMETHYL-n-DODECYL(2-HYDROXY-3-CHLOROPROPYL)AMMONIUM CHLORIDE see DRS400

N,N-DIMETHYL-n-DODECYL(3-HYDROXYPROPENYL) AMMONIUM CHLORIDE see DRS600

3:4-DIMETHYLENE-1:2-BENZANTHRACENE see AAF000

8:9-DIMETHYLENE-1:2-BENZANTHRACENE see BAW000

DIMETHYLENEDIAMINE see EEA500

DIMETHYLENE DIISOTHIOCYANATE see ISK000

DIMETHYLENE GLYCOL see BOT000

DIMETHYLENEIMINE see EJM900

DIMETHYLENE OXIDE see EJN500

N,N-DIMETHYLENEOXIDEBIS(PYRIDINIUM-4-ALDOXIME) DICHLORIDE see BGS250

DIMETHYLENIMINE see EJM900

N,N-DIMETHYLENOXID-BIS-(PYRIDINIUM-4-ALDOXIM)-DICHLORID (GERMAN) see BGS250

exo-1,2-cis-DIMETHYL-3,6-EPOXYHEXAHYDROPHTHALIC ANHYDRIDE see CBE750

6,7-DIMETHYLESCULETIN see DRS800

O,S-DIMETHYL ESTER AMIDE of AMIDOTHIOATE see DTQ400

O,O-DIMETHYLESTER KYSELINY CHLORTHIOFOSFORECNE (CZECH) see DTQ600

DIMETHYLESTER KYSELINY FOSFORITE (CZECH) see DSG600

DIMETHYLESTER KYSELINY MALEINOVE see DSL800

DIMETHYLESTER KYSELINY SIROVE (CZECH) see DUD100

DIMETHYLESTER KYSELINY TEREFTALOVE (CZECH) see IML000

DIMETHYL ESTER PHOSPHORIC ACID ESTER with METHYL 3-HYDROXY-CROTONATE see MQR750

O,O-DIMETHYL ESTER PHOSPHOROTHIOIC ACID-S-ESTER with 1,2-BIS(METHOXYCARBONYL)ETHANETHIOL see OPK250

O,O-DIMETHYL ESTER PHOSPHOROTHIOIC ACID-S-ESTER with ETHYL MER-CAPTOACETATE see DRB600

N,N'-DIMETHYL-1,2-ETHANEDIAMINE see DRI600

N,N'-DIMETHYLETHANEDIAMINE see DRI600

N,N-DIMETHYLETHANETHIOAMIDE see DUG450

7,14-DIMETHYL-7,14-ETHANODIBENZ(a,b)ANTHRACENE-15,16-DICARBOXYLIC ACID see DRT000

1,1-DIMETHYLETHANOL see BPX000

DIMETHYLETHANOLAMINE (DOT) see DOY800

N,N-DIMETHYLETHANOLAMINE see DOY800

DIMETHYLETHANOLAMINE see DOY800

DIMETHYL ETHER (DOT) see MJW500

DIMETHYL ETHER HYDROQUINONE see DOA400

(DIMETHYL ETHER)OXODIPEROXO CHROMIUM(VI) see DRT089

DIMETHYLETHER RESORCINOL see REF025

DIMETHYLETHER of d-TUBOCURARINE IODIDE see DUM000

6-α,21-DIMETHYLETHISTERONE see DRT200

3,7-DIMETHYL-3-(1-ETHOXYETHOXY)-1,6-OCTADIENE see ELZ050

2',6'-DIMETHYL-2-(2-ETHOXYETHYLAMINO)ACETANILIDE see DRT400

2',6'-DIMETHYL-2-(2-ETHOXYETHYLAMINO)ACETANILIDE HYDROCHLORIDE see DRT600

N,N-DIMETHYL-p-((3-ETHOXYPHENYL)AZO)ANILINE see DRU000

DIMETHYLETHOXYPHENYLSILANE see DRU200

DIMETHYL ETHYL ALLENOLIC ACID METHYL ETHER see DRU600

1,1-DIMETHYLETHYLAMINE see BPY250

N,N-DIMETHYL-4'-ETHYL-4-AMINOAZOBENZENE see EOI500

(±)-1-((1,1-DIMETHYLETHYL)AMINO)-3-(2,3-DIMETHYLPHENOXY)-2-PROPANOL HYDROCHLORIDE see XGA500

5-(2-((1,1-DIMETHYLETHYL)AMINO)-1-HYDROXYETHYL)-1,3-BENZENEDIOL see TAN100

5-(2-((1,1-DIMETHYLETHYL)AMINO)-1-HYDROXYETHYL)-1,3-BENZENEDIOL, SULFATE (2:1) (SALT) see TAN250

1-(4-(3-((1,1-DIMETHYLETHYL)AMINO)-2-HYDROXYPROPOXY)-2-BENZOFURA-NYL)ETHA NONE see ACN300

1-(7-(3-((1,1-DIMETHYLETHYL)AMINO)-2-HYDROXYPROPOXY)-2-BENZOFURA-NYL)ETHA NONE see ACN320

2-(3-((1,1-DIMETHYLETHYL)AMINO)-2-HYDROXYPROPOXY)-BENZONITRILE HY-DROCHLORIDE see BON400

5-(3-((1,1-DIMETHYLETHYL)AMINO)-2-HYDROXYPROPOXY)-3,4-DIHYDRO-2(1H)-QUINOLINONE see CCL800

5-(3-((1,1-DIMETHYLETHYL)AMINO)-2-HYDROXYPROPOXY)-1,2,3,4-TETRAHY-DRO-2,3-NAPHTHALENEDIOL see CNR675

α-1-(((1,1-DIMETHYLETHYL)AMINO)METHYL)-4-HYDROXY-1,3-BENZENEDIMETH-ANOL see BQF500

$α^6$-(((1,1-DIMETHYLETHYL)AMINO)METHYL)-3-HYDROXY-2,6-PYRIDINEDIMETHA-NOL DIHYDROCHLORIDE see POM800

2-((1,1-DIMETHYLETHYL)AZO)-5-METHYL-2-HEXANOL see BQI300

1,4-DIMETHYL-7-ETHYLAZULENE see DRV000

4-(1,1-DIMETHYLETHYL)-1,2-BENZENEDIOL see BSK000

4-(1,1-DIMETHYLETHYL)BENZENEPROPANAL see BMK300

DI-N-METHYL ETHYL CARBAMATE see EII500

DIMETHYLETHYLCARBINOL see PBV000

2-(1,1-DIMETHYLETHYL)CYCLOHEXANOL ACETATE see BQW490

2-(1,1-DIMETHYLETHYL)-4,6-DINITROPHENOL see DRV200

2-(1,1-DIMETHYLETHYL)-4,6-DINITROPHENOL ACETATE see DVJ400

1-(1,1-DIMETHYLETHYL)-4,4-DIPHENYLPIPERIDINE see BOM530

DIMETHYLETHYLENE see BOW500

N,N'-DIMETHYLETHYLENEDIAMINE see DRI600

sym-DIMETHYLETHYLENEDIAMINE see DRI600

N,N-DIMETHYLETHYLENEUREA see EJA100

1,1-DIMETHYLETHYL ETHER see EHA550

DIMETHYL ETHYL HEXADECYL AMMONIUM BROMIDE see EKN500

1,1-DIMETHYLETHYL HYDROPEROXIDE see BRM250

DIMETHYLETHYL-β-HYDROXYETHYLAMMONIUM SULFATE DIBUTYLURETHAN see DDW000

DIMETHYL-ETHYL-β-HYDROXYETHYL-AMMONIUM-SULFATE-DI-n-BUTYLCARBA-MATE see DDW000

DIMETHYLETHYL(m-HYDROXYPHENYL)AMMONIUM CHLORIDE see EAE600

O,O-DIMETHYL-2-ETHYLMERCAPTOETHYL THIOPHOSPHATE see TIR250

O,O-DIMETHYL-O-ETHYLMERCAPTOETHYL THIOPHOSPHATE see DAO800

O,O-DIMETHYL 2-ETHYLMERCAPTOETHYL THIOPHOSPHATE, THIONO ISO-MER see DAO800

4-(1,1-DIMETHYLETHYL)-α-METHYLBENZENEPROPANAL see LFT000

6-(1,1-DIMETHYLETHYL)-3-METHYL-2,4-DINITROPHENYL ACETATE see BRU750

2-(1,1-DIMETHYLETHYL)-5-METHYLPHENOL see BQV600

N,N-DIMETHYL-p-(4'-ETHYL-3'-METHYLPHENYLAZO)ANILINE see EPU000

N,N-DIMETHYL-p-(3'-ETHYL-4'-METHYLPHENYLAZO)ANILINE see EPT500

4-(1,1-DIMETHYLETHYL)-N-(1-METHYLPROPYL)-2,6-DINITROBENZENAMINE see BQN600

α,α-DIMETHYLETHYL NITRITE see BRV760

1,1-DIMETHYL-3-ETHYL-3-NITROSOUREA see DRV600

3,5-DIMETHYL-5-ETHYLOXAZOLIDINE-2,4-DIONE see PAH500

4-(1,1-DIMETHYLETHYL)PHENOL see BSE500

2-(4-(1,1-DIMETHYLETHYL)PHENOXY)CYCLOHEXYL 2-PROPYNYL ESTER, SULFUROUS ACID see SOP000

2-(4-(1,1-DIMETHYLETHYL)PHENOXY)CYCLOHEXYL 2-PROPYNYL SULFITE see SOP000

4-(3-(4-(1,1-DIMETHYLETHYL)PHENOXY)-2-HYDROXYPROPOXY)BENZOIC ACID see BSE750

N,N-DIMETHYL-p-(3'-ETHYLPHENYLAZO)ANILINE see EOI000

N,N-DIMETHYL-p-((4-ETHYLPHENYL)AZO)ANILINE see EOI500

N,N-DIMETHYL-p((m-ETHYLPHENYL)AZO)ANILINE see EOI000

N,N-DIMETHYL-p((o-ETHYLPHENYL)AZO)ANILINE see EIF450

N,N-DIMETHYL-3'-ETHYL-4-(PHENYLAZO)BENZENAMINE see EOI000

S-(4-(1,1-DIMETHYLETHYL)PHENYL)-o-ETHYL ETHYLPHOSPHONODITHIOATE see BSG000

α-(4-(1,1-DIMETHYLETHYL)PHENYL)-4-(HYDROXYDIPHENYLMETHYL)-1-PIPERI-DINEBUTANOL see TAI450

N,N-DIMETHYLIMIDODICARBONIMIDIC DIAMIDE MONOHYDROCHLORIDE see DQR800
DIMETHYLIMIPRAMINE see DSI709
DIMETHYLIMIPRAMINE HYDROCHLORIDE see DLS600
N,N-DIMETHYL-4-(6'-1H-INDAZYLAZO)ANILINE see DSI800
N,N-DIMETHYL-p-(6-INDAZYLAZO)ANILINE see DSI800
N,N-DIMETHYL-β-3-INDOLYLETHYLAMINE SULFOSALICYLATE see DPG000
N,N-DIMETHYL-α-INDOLYLIDENE-p-TOLUIDINE see DOT600
DIMETHYLIODOARSINE see DSI889
1,3-DIMETHYL-α-IONONE see COW780
DIMETHYLIONONE see COW780
2,8-DIMETHYL-6-ISOBUTYLNONANOL-4 see DSJ200
DIMETHYL ISOPHTHALATE see IML000
DIMETHYLISOPROPANOLAMINE see DPT800
α-α-DIMETHYL-m-ISOPROPENYL BENZYL ISOCYANATE see IKG800
1,4-DIMETHYL-7-ISOPROPYLAZULENE see DSJ800
DIMETHYL-5-(1-ISOPROPYL-3-METHYLPYRAZOLYL)CARBAMATE see DSK200
N,N-DIMETHYL-4-(4'-ISOQUINOLINYLAZO)ANILINE see DPO800
N,N'-DIMETHYL-4-(5'-ISOQUINOLINYLAZO)ANILINE see DPP000
N,N-DIMETHYL-4-(7'-ISOQUINOLINYLAZO)ANILINE see DPP200
N,N-DIMETHYL-4-(5'-ISOQUINOLYL-2'-OXIDE)AZOANILINE see DPP400
1,3-DIMETHYLISOTHIOUREA see DSK900
3,4-DIMETHYLISOXALE-5-SULFANILAMIDE see SNN500
3,4-DIMETHYLISOXALE-5-SULPHANILAMIDE see SNN500
3,5-DIMETHYLISOXAZOLE see DSK950
N'-(3,4-DIMETHYL-5-ISOXAZOLYL)SULFANILAMIDE see SNN500
N'-(3,4-DIMETHYL-5-ISOXAZOLYL)SULFANILAMIDE LITHIUM SALT see DSL000
N'-(3,4-DIMETHYL-5-ISOXAZOLYL)SULFANILAMIDE SODIUM SALT see DSL200
N'-(3,4)DIMETHYLISOXAZOL-5-YL-SULPHANILAMIDE see SNN500
N'-(3,4-DIMETHYL-5-ISOXAZOLYL)SULPHANILAMIDE see SNN500
DIMETHYLKARBAMOYLCHLORID see DQY950
DIMETHYLKETAL see ABC750
DIMETHYLKETENE see DSL289
DIMETHYLKETOL see ABB500
DIMETHYL KETONE see ABC750
N,N-DIMETHYLLAURYLAMINE see DRR800
DIMETHYL LAURYLBENZENE AMMONIUM BROMIDE see BEO000
DIMETHYLLAURYLBETAINE see LBU200
DIMETHYL LEAD DIACETATE see DSL400
N,N-DIMETHYL-4-((3',5'-LUTIDYL-1'-OXIDE)AZO)ANILINE see DPP709
DIMETHYLMAGNESIUM see DSL600
DIMETHYL MALEATE see DSL800
α,β-DIMETHYLMALEIC ANHYDRIDE see DSM000
DIMETHYLMALEIC ANHYDRIDE see DSM000
DIMETHYL MALONATE see DSM200
DIMETHYL MANGANESE see DSM289
DIMETHYL MERCURY see DSM450
DIMETHYLMESCALINE see DOE200
N,N-DIMETHYLMETHANAMINE OXIDE, DIHYDRATE see TLE250
DIMETHYLMETHANE see PMJ750
O,O-DIMETHYL-O-2-METHOXYCARBONYL-1-METHYL-VINYL-PHOSPHAT (GERMAN) see MQR750
DIMETHYL 2-METHOXYCARBONYL-1-METHYLVINYL PHOSPHATE see MQR750
DIMETHYL METHOXYCARBONYLPROPENYL PHOSPHATE see MQR750
DIMETHYL (1-METHOXYCARBOXYPROPEN-2-YL)PHOSPHATE see MQR750
1,1-DIMETHYL-5-METHOXY-3-(DITHIEN-2-YLMETHYLENE)PIPERIDINIUM BROMIDE see TGB160
2',6'-DIMETHYL-2-(2-METHOXYETHYLAMINO)-PROPIONANILIDE see MEN250
2,6-DIMETHYL-N-(2-METHOXYETHYL)CHLOROACETANILIDE see DSM500
2,3-DIMETHYL-7-METHOXY-8-(MORPHOLINOMETHYL)CHROMONE HYDROCHLORIDE see DSM600
α,α-DIMETHYL-2-(6-METHOXYNAPHTHYL)PROPIONIC ACID see DRU600
2,6-DIMETHYL-2-METHOXY-4-NITROSO-MORPHOLINE see NKO600
3,7-DIMETHYL-7-METHOXY-1-OCTANAL see DSM800
3,7-DIMETHYL-7-METHOXY-2-OCTANOL see DLR300
N,N-DIMETHYL-p-(2-METHOXYPHENYLAZO)ANILINE see DSN000
N,N-DIMETHYL-p-(3-METHOXYPHENYLAZO)ANILINE see DSN200
N,N-DIMETHYL-p-(4-METHOXYPHENYLAZO)ANILINE see DSN400
3,3-DIMETHYL-1-p-METHOXYPHENYLTRIAZENE see DSN600
2',6'-DIMETHYL-2-(2-METHOXYPROPYLAMINO)ACETANILIDE HYDROCHLORIDE see DSN800
2',6-DIMETHYL-2-(2-METHOXYPROPYLAMINO)-PROPIONANILIDE PERCHLORATE see MEN500
1,9-DIMETHYL-7-METHOXY-9H-PYRIDO(3,4-b)INDOLE HYDROCHLORIDE see ICU100
3,7-DIMETHYL-9-(4-METHOXY-2,3,6-TRIMETHYLPHENYL)-2,4,6,8-NONANETETRAENOIC ACID ETHYL ESTER see EMJ500
all-trans-3,7-DIMETHYL-9-(4-METHOXY-2,3,6-TRIMETHYLPHENYL)-2,4,6,8-NONA-TETRAENOIC ACID see REP400
2,2-DIMETHYL-4-(N-METHYLAMINOCARBOXYLATO)-1,3-BENZODIOXOLE see DQM600
1,3-DIMETHYL-5-(METHYLAMINO)-4-PYRAZOLYL o-FLUOROPHENYL KETONE see DSO500

(1,3-DIMETHYL-5-(METHYLAMINO)-1H-PYRAZOL-4-YL)(2-FLUOROPHENYL)METHANONE see DSO500
N,N-DIMETHYL-α-METHYLBENZYLAMINE see DSO800
2,2-DIMETHYL-4-(N-METHYLCARBAMATO)-1,3-BENZODIOXOLE see DQM600
O,O-DIMETHYL METHYLCARBAMOYLMETHYL PHOSPHORODITHIOATE see DSP400
O,O-DIMETHYL-O-(2-N-METHYLCARBAMOYL-1-METHYL-VINYL-FOSFAAT (DUTCH) see MRH209
O,O-DIMETHYL-O-(2-N-METHYLCARBAMOYL-1-METHYL-)-VINYL-PHOSPHAT (GERMAN) see MRH209
O,O-DIMETHYL-O-(2-N-METHYLCARBAMOYL-1-METHYL-VINYL) PHOSPHATE see MRH209
N,N-DIMETHYL-α-METHYLCARBAMOYLOXYIMINO-α-(METHYLTHIO)ACETAMIDE see DSP600
N',N'-DIMETHYL-N-((METHYLCARBAMOYL)OXY)-1-METHYLTHIOOXAMIMIDIC ACID see DSP600
N',N'-DIMETHYL-N-((METHYLCARBAMOYL)OXY)-1-THIOOXAMIMIDIC ACID METHYL ESTER see DSP600
O,O-DIMETHYL-O-(1-METHYL-2-CARBOXY-α-PHENYLETHYL)VINYL PHOSPHATE see COD000
O,O-DIMETHYL O-(1-METHYL-2-CARBOXYVINYL) PHOSPHATE see MQR750
O,O-DIMETHYL-O-(1-METHYL-2-CHLOR-2-N,N-DIAETHYL-CARBAMOYL)-VINYL-PHOSPHAT see FAB400
(O,O-DIMETHYL-O-(1-METHYL-2-CHLORO-2-DIETHYLCARBAMOYL-VINYL) PHOSPHATE) see FAB400
2,6-DIMETHYL-1-((2-METHYLCYCLOHEXYL)CARBONYL)PIPERIDINE see DSP650
6,6-DIMETHYL-2-METHYLENEBICYCLO(3.1.1)HEPTANE see POH750
N,N'-DIMETHYL-4,4'-METHYLENEDIANILINE see MJO000
DIMETHYLMETHYLENE-p,p'-DIPHENOL see BLD500
DIMETHYL METHYL MALEATE see DRF200
O,O-DIMETHYL-O-4-(METHYLMERCAPTO)-3-METHYLPHENYL PHOSPHOROTHIOATE see FAQ900
O,O-DIMETHYL-p-4-(METHYLMERCAPTO)-3-METHYLPHENYL THIOPHOSPHATE see FAQ900
O,O-DIMETHYL O-(4-METHYLMERCAPTOPHENYL)PHOSPHATE see PHD250
(E)-DIMETHYL 1-METHYL-3-(METHYLAMINO)-3-OXO-1-PROPENYL see MRH209
DIMETHYL-1-METHYL-2-(METHYLCARBAMOYL)VINYLPHOSPHATE, cis PHOSPHATE see MRH209
O,O-DIMETHYL-O-(3-METHYL-4-METHYLMERCAPTOPHENYL)PHOSPHOROTHIOATE see FAQ900
N,N-DIMETHYL-2-METHYL-4-(4'-(2'-METHYLPYRIDYL)-1'-OXIDE)AZO)ANILINE see DQD600
N,N'DIMETHYL-3-METHYL-4-(4'-(2'-METHYLPYRIDYL-1'OXIDE)AZO)ANILINE see DQD800
O,O-DIMETHYL-O-(3-METHYL-4-METHYLTHIO-FENYL)-MONOTHIOFOSFAAT (DUTCH) see FAQ900
O,O-DIMETHYL-O-(3-METHYL-4-METHYLTHIOPHENYL)-MONOTHIOPHOSPHAT (GERMAN) see FAQ900
O,O-DIMETHYL-O-3-METHYL-4-METHYLTHIOPHENYL PHOSPHOROTHIOATE see FAQ900
O,O-DIMETHYL-O-(3-METHYL-4-METHYLTHIO-PHENYL)-THIONOPHOSPHAT (GERMAN) see FAQ900
O,O-DIMETHYL-O-(3-METHYL-4-NITROFENYL)-MONOTHIOFOSFAAT (DUTCH) see DSQ000
O,O-DIMETHYL-O-(3-METHYL-4-NITRO-PHENYL)-MONOTHIOPHOSPHAT (GERMAN) see DSQ000
O,O-DIMETHYL O-(3-METHYL-4-NITROPHENYL)PHOSPHORATE see PHD750
O,O-DIMETHYL-O-(3-METHYL-4-NITROPHENYL) PHOSPHOROTHIOATE see DSQ000
DIMETHYL-3-METHYL-4-NITROPHENYLPHOSPHOROTHIONATE see DSQ000
O,O-DIMETHYL-O-(3-METHYL-4-NITROPHENYL) THIOPHOSPHATE see DSQ000
N-DIMETHYL-1-METHYL-N'-(1-NITRO-9-ACRIDINYL)-1,2-ETHANEDIAMINE DIHYDROCHLORIDE see NFW435
α-4-DIMETHYL-α-(4-METHYL-3-PENTENYL)-3-CYCLOHEXENE-1-METHANOL see BGO775
N,N-DIMETHYL-p-(2'-METHYLPHENYLAZO)ANILINE see DUH800
N,N-DIMETHYL-p-(3'-METHYLPHENYLAZO)ANILINE see DUH600
N,N-DIMETHYL-4-((2-METHYLPHENYL)AZO)BENZENAMINE see DUH800
N,N-DIMETHYL-4-((3-METHYLPHENYL)AZO)BENZENAMINE see DUH600
N,N-DIMETHYL-4-((4-METHYLPHENYL)AZO)BENZENAMINE see DUH400
N,N-DIMETHYL-2-((α-METHYL-α-PHENYLBENZYL)OXY)ETHYLAMINE see DSQ600
N,N-DIMETHYL-2-(α-METHYL-α-PHENYLBENZYLOXY)ETHYLAMINE see DSQ600
N,N-DIMETHYL-2-((o-METHYL-α-PHENYL-BENZYL)OXY)-ETHYLAMINE CITRATE see DPH000
N,N-DIMETHYL-2-(o-METHYL-α-PHENYLBENZYLOXY)ETHYLAMINE HYDROCHLORIDE see OJW000
DIMETHYL-cis-1-METHYL-2-(1-PHENYLETHOXYCARBONYL)VINYL PHOSPHATE see COD000
3,3-DIMETHYL-6-(((5-METHYL-3-PHENYL)-4-ISOXAZOLECARBOXAMIDE)-7-OXO)-4-THIA-1-AZABICYCLO(3.2.0)HEPTANE see DSQ800
DIMETHYL-5-(3-METHYL-1-PHENYLPYRAZOLYL) CARBAMATE see PPQ625
3,3-DIMETHYL-1-(m-METHYLPHENYL)TRIAZENE see DSR200

3,3-DIMETHYL-1-(o-METHYLPHENYL)TRIAZENE see MNT500

DIMETHYL METHYLPHOSPHONATE see DSR400

O,O-DIMETHYL-O-(3-METHYL) PHOSPHOROTHIOATE see DSQ000

N,N-DIMETHYL-9-(3-(4-METHYL-1-PIPERANIZLY)PROPYLIDENE)-9H-THIOXAN-THENE-2-SULFONAMIDE, (Z)- see NBP500

N,N-DIMETHYL-9-(3-(4-METHYL-1-PIPERAZINYL)PROPYLIDENE)THIAXANTHENE-2-SULFONAMIDE see NBP500

N,N-DIMETHYL-9-(3-(4-METHYL-1-PIPERAZINYL)PROPYLIDENE)THIOXANTHENE-2-SULFONAMIDE see NBP500

DIMETHYL-3-(2-METHYL-1-PROPENYL)CYCLOPROPANECARBOXYLATE see BEP500

(+)-2,2-DIMETHYL-3-(2-METHYLPROPENYL)-CYCLOPROPANECARBOXYLIC ACID-(E)-,ESTER with (+)- see AFR750

2,2-DIMETHYL-3-(2-METHYLPROPENYL)CYCLOPROPANECARBOXYLIC ACID ETHYL ESTER see EHM100

(+)-(Z)-2,2-DIMETHYL-3-(2-METHYLPROPENYL)-CYCLOPROPANECARBOXYLIC ACID ESTER with 2-ALLYL-4-HYDROXY-3-METHYL-2-CYCLOPENTEN-ONE see AFR500

2,2-DIMETHYL-3-(2-METHYLPROPYL)CYCLOPROPANECARBOXYLIC ACID-p-(METHOXYMETHYL)BENZYL ESTER see MNG525, MBV700

1,3-DIMETHYL-4-(p-((p-((1-METHYLPYRIDINIUM-4-YL)AMINO)PHENYL)CARBAMOYL)ANILINOQUINOLINIUM, DIBROMIDE see DSR600

1,8-DIMETHYL-4-((p-((p-((1-METHYLPYRIDINIUM-4-YL)AMINO)PHENYL)CARBAMOYL)ANILINO)QUINOLINIUM)DI-p-TOLUENES see DSS000

1,6-DIMETHYL-4-((p-((p-((1-METHYLPYRIDINIUM-4-YL)AMINO)PHENYL) CARBAM-OYL)ANILINO)QUINOLINIUM) DI-p-TOLUENESUL see DSR800

N,N-DIMETHYL-4-((2-METHYL-4-PYRIDINYL)AZO)BENZENAMINE-N-OXIDE see DSS200

N,N-DIMETHYL-4-(2-METHYL-4-PYRIDYLAZO)ANILINE-N-OXIDE see DSS200

N,N-DIMETHYL-4-(4'-METHYLPYRIDYL-1'-OXIDE)AZO)ANILINE see DSS200

N,N-DIMETHYL-4-(4'-(3'-METHYLPYRIDYL-1'-OXIDE)AZO)ANILINE see MOY750

N,N'-DIMETHYL-4-(4'-(2'-METHYLPYRIDYL-1-OXIDE)AZO)-o-TOLUIDINE see DQD800

N,N-DIMETHYL-4-(5-(3'-METHYLQUINOLYL)AZO)ANILINE see MJF500

N,N-DIMETHYL-4-(5'-(6'-METHYLQUINOLYL)AZO)ANILINE see MJF750

N,N-DIMETHYL-4-(5'-(7'-METHYLQUINOLYL)AZO)ANILINE see DPQ400

N,N-DIMETHYL-4-(5'-(8'-METHYLQUINOLYL)AZO)ANILINE see MJG000

N,N-DIMETHYL-2'-METHYLSTILBENAMINE see TMF750

O,O-DIMETHYL-o-(4-(METHYLSULFONYL)-m-TOLYL) PHOSPHOROTHIOATE see DSS800

3,3-DIMETHYL-1-(METHYLTHIO)-2-BUTANONE-o-((METHYLAMI-NO)CARBONYL)OXIME see DAB400

3,3-DIMETHYL-1-(METHYLTHIO)-2-BUTANONE OXIME see DSS900

O,O-DIMETHYL-O-(4-METHYLTHIO-3-METHYLPHENYL) PHOSPHOROTHIOATE see FAQ900

3,5-DIMETHYL-4-(METHYLTHIO)PHENOL METHYLCARBAMATE see DST000

3,5-DIMETHYL-4-METHYL-THIOPHENYL-N-CARBAMAT (GERMAN) see DST000

O-(3,5-DIMETHYL-4-(METHYLTHIO)PHENYL)-O,O-DIETHYL ESTER PHOSPHO-ROTHIOIC ACID see DJR800

O-(3,5-DIMETHYL-4-(METHYLTHIO)PHENYL)-O,O-DIETHYL PHOSPHOROTH-IOATE see DJR800

O-(3,5-DIMETHYL-4-(METHYLTHIO)PHENYL)-O,O-DIMETHYL PHOSPHOROTH-IOATE see DST200

3,5-DIMETHYL-4-METHYLTHIOPHENYL-N-METHYLCARBAMATE see DST000

DIMETHYL-p-(METHYLTHIO)PHENYL PHOSPHATE see PHD250

O,O-DIMETHYL-O-(4-(METHYLTHIO)-m-TOLYL) PHOSPHOROTHIOATE see FAQ900

O,O-DIMETHYL-o-((4-METHYLTHIO)-m-TOLYL) PHOSPHOROTHIOATE SULFONE see DSS800

O,O-DIMETHYL-O-4-(METHYLTHIO)-3,5-XYLYL PHOSPHOROTHIOATE see DST200

DIMETHYL MONOSULFATE see DUD100

(9-α,13-α,14-α)-3,17-DIMETHYLMORPHINAN PHOSPHATE see MLP250

3,17-DIMETHYL-9-α,13-α,14-α-MORPHINAN PHOSPHATE see MLP250

2,6-DIMETHYLMORPHOLINE see DST600

O,O-DIMETHYL MORPHOLINOCARBONYLMETHYL PHOSPHORODITHIOATE see MRU250

DIMETHYL S-(MORPHOLINOCARBONYLMETHYL) PHOSPHOROTHIOLOTHION-ATE see MRU250

DIMETHYLMORPHOLINOPHOSPHONATE see DST800

DIMETHYL MORPHOLINOPHOSPHORAMIDATE see DST800

DIMETHYLMYLERAN see DSU000

N,N-DIMETHYL-N'-(1-ANTHRACHINONYL)FORMAMIDINIUMCHLORID (GERMAN) see APK750

N,N-DIMETHYL-1-NAPHTHYLAMINE see DSU400

N,N-DIMETHYL-α-NAPHTHYLAMINE see DSU400

α-DIMETHYLNAPHTHYLAMINE see DSU400

DIMETHYL-α-NAPHTHYLAMINE see DSU400

N,N-DIMETHYL-4(2'-NAPHTHYLAZO)ANILINE see DSU800

N,N-DIMETHYL-p-(1-NAPHTHYLAZO)ANILINE see DSU600

N,N-DIMETHYL-p-(2-(1-NAPHTHYL)VINYL)ANILINE see DSV000

N,N-DIMETHYL-N'-BENZYL-N'-(α-PYRIDYL)ETHYLENEDIAMINE see TMP750

N,N'-DIMETHYL-N,N'-BIS(3-(3',4',5'-TRIMETHOXYBENZOX-Y)PROPYL)ETHYLENEDIAMINE DIHYDROCHLORIDE see HFF500

N,N-DIMETHYL-N'-(4-CHLOROPHENYL)UREA see CJX750

3,3'-DIMETHYL-N,N'-DIACETYLBENZIDINE see DRI400

N,N-DIMETHYL-p-(N',N'-DIMETHYLCARBAMOYLOXY)PHENETHYLAMINE, HY-DROCHLORIDE see DQX800

N,N'-DIMETHYL-N,N'-DINITROOXAMIDE see DRN300

N,N'-DIMETHYL-N,N'-DINITROSOETHYLENEDIAMINE see DVE400

N,N'-DIMETHYL-N,N'-DINITROSO-1,2-HYDRAZINEDICARBOXAMIDE see DRN400

N,N'-DIMETHYL-N,N'-DINITROSOOXAMIDE see DRN600

N,N'-DIMETHYL-N,N'-DINITROSO-1,3-PROPANEDIAMINE see DRO200

N,N'-DIMETHYL-N,N'-DIPHENYLUREA see DRB200

N,N-DIMETHYL-N'-ETHYL-N'-1-NAPHTHYLETHYLENEDIAMINE see DRV500

N,N-DIMETHYL-N'-ETHYL-N'-2-NAPHTHYLETHYLENEDIAMINE see DRV550

N,N-DIMETHYL-N'-ETHYL-N'-PHENYLETHYLENEDIAMINE see DRV850

DIMETHYLNITRAMIN (GERMAN) see DSV200

DIMETHYLNITRAMINE see DSV200

N',N'-DIMETHYL-N²-(1-NITRO-9-ACRIDINYL)-1,2-PROPANEDIAMINE DIHYDRO-CHLORIDE see NFW435

DIMETHYLNITROAMINE see DSV200

1,4-DIMETHYL-2-NITROBENZENE see NMS520

2,6-DIMETHYLNITROBENZENE see NMS500

α,N-DIMETHYL-N-NITROSOBENZYLAMINE see NKW000

3,3-DIMETHYL-1-NITRO-1-BUTYNE see DSV289

O,O-DIMETHYL-O-4-NITRO-3-CHLOROPHENYL THIOPHOSPHATE see MIJ250

O,O-DIMETHYL-p-NITRO-m-CHLOROPHENYL THIOPHOSPHATE see MIJ250

DIMETHYL-p-NITROFENYLESTER KYSELINY FOSFORECNE (CZECH) see PHD500

O,O-DIMETHYL-O-p-NITROFENYLESTER KYSELINY THIOFOSFORECNE (CZECH) see MNH000

O,O-DIMETHYL-O-(4-NITROFENYL)-MONOTHIOFOSFAAT (DUTCH) see MNH000

4,6-DIMETHYL-2-(5-NITRO-2-FURYL)PYRIMIDINE see DSV400

1,2-DIMETHYL-5-NITRO-1H-IMIDAZOLE see DSV800

4,5-DIMETHYL-2-NITROIMIDAZOLE see DSW500

1,2-DIMETHYL-5-NITROIMIDAZOLE see DSV800

DIMETHYLNITROMETHANE see NIY000

O,O-DIMETHYL-O-(4-NITRO-3-METHYLPHENYL)THIOPHOSPHATE see DSQ000

N,N-DIMETHYL-p-((m-NITROPHENYL)AZO)ANILINE see DSW600

N,N-DIMETHYL-p-((o-NITROPHENYL)AZO)ANILINE see DSW800

O,O-DIMETHYL-O-(4-NITRO-PHENYL)-MONOTHIOPHOSPHAT (GERMAN) see MNH000

DIMETHYL p-NITROPHENYL MONOTHIOPHOSPHATE see MNH000

DIMETHYL-4-NITROPHENYL PHOSPHATE see PHD500

DIMETHYL-p-NITROPHENYL PHOSPHATE see PHD500

O,O-DIMETHYL-O-(4-NITROPHENYL) PHOSPHOROTHIOATE see MNH000

O,O-DIMETHYL-O-(p-NITROPHENYL) PHOSPHOROTHIOATE see MNH000

DIMETHYL 4-NITROPHENYL PHOSPHOROTHIONATE see MNH000

O,O-DIMETHYL-O-(4-NITROPHENYL)-THIONOPHOSPHAT (GERMAN) see MNH000

O,O-DIMETHYL-O-(p-NITROPHENYL)-THIONOPHOSPHAT (GERMAN) see MNH000

DIMETHYL-p-NITROPHENYL THIONPHOSPHATE see MNH000

O,O-DIMETHYL-O-p-NITROPHENYL THIOPHOSPHATE see MNH000

DIMETHYL p-NITROPHENYL THIOPHOSPHATE see MNH000

3,3-DIMETHYL-1-(p-NITROPHENYL)TRIAZENE see DSX400

2,3-DIMETHYL-4-NITROPYRIDINE-1-OXIDE see DSX800

2,5-DIMETHYL-4-NITROPYRIDINE-1-OXIDE see DSY000

DIMETHYLNITROSAMIN (GERMAN) see NKA600

N,N-DIMETHYLNITROSAMINE see NKA600

DIMETHYLNITROSAMINE see NKA600

DIMETHYLNITROSOAMINE see NKA600

DIMETHYL-p-NITROSOANILINE (DOT) see DSY600

N,N-DIMETHYL-p-NITROSOANILINE see DSY600

N,N-DIMETHYL-4-NITROSOBENZENAMINE see DSY600

α,N-DIMETHYL-N-NITROSOBENZYLAMINE see NKW000

N,m-DIMETHYL-N-NITROSOBENZYLAMINE see NKS000

N,o-DIMETHYL-N-NITROSOBENZYLAMINE see NKR500

N,p-DIMETHYL-N-NITROSOBENZYLAMINE see NKS500

3,2'-DIMETHYL-4-NITROSOBIPHENYL see DSY800

DIMETHYLNITROSOHARNSTOFF (GERMAN) see DTB200

1,2-DIMETHYLNITROSOHYDRAZINE see DSY889

N,O-DIMETHYL-N-NITROSOHYDROXYLAMINE see DSZ000

2,6-DIMETHYL-4-NITROSOMORPHOLINE cis and trans mixture (2:1) see DTA050

2,6-DIMETHYLNITROSOMORPHOLINE see DTA000

2,6-DIMETHYL-N-NITROSOMORPHOLINE see DTA000

DIMETHYLNITROSOMORPHOLINE see DTA000

N,N'-DIMETHYL-N-NITROSO-N'-PHENYLUREA see DTN875

DIMETHYL(p-NITROSOPHENYL)AMINE see DSY600

3,5-DIMETHYL-1-NITROSOPIPERAZINE see NKA850

3,5-DIMETHYL-1-NITROSOPIPERIDINE see DTA600

cis-3,5-DIMETHYL-1-NITROSOPIPERIDINE see DTA690

trans-3,5-DIMETHYL-1-NITROSOPIPERIDINE see DTA700

3,6-DIMETHYL-OCTYN-4-DIOL-(3,6) see DTF850
N,N-DIMETHYLOKTADECYLAMIN (CZECH) see DTC400
DIMETHYLOL DIHYDROXYETHYLENE UREA see DTG000
DIMETHYLOLGLYOXALUREA see DTG000
N,N-DIMETHYLOL-2-METHOXYETHYL CARBAMATE see DTG200
1,1-DIMETHYLOL-1-NITROETHANE see NHO500
DIMETHYLOLPROPANE see DTG400
DIMETHYLOLPROPANE DIACRYLATE see DUL200
DIMETHYLOL-TETRAKIS-BUTOXYMETHYLMELAMIN (CZECH) see BHB500
1,3-DIMETHYLOLUREA see DTG700
2,3-DIMETHYL-7-OXABICYCLO(2.2.1)HEPTANE-2,3-DICARBOXYLIC ANHYDRIDE
   see CBE750
6,10-DIMETHYL-3-OXA-9-UNDECENAL see CMT300
4,4-DIMETHYLOXAZOLIDINE see DTG750
DIMETHYL OXAZOLIDINE see DTG750
5,5-DIMETHYLOXAZOLIDINE-2,4-DIONE see PMO250
5,5-DIMETHYL-2,4-OXAZOLIDINEDIONE see PMO250
DIMETHYLOXAZOLIDINEDIONE see PMO250
N¹-(4,5-DIMETHYL-2-OXAZOLYL)-SULFANILAMIDE see AIE750
3,3-DIMETHYLOXETANE see EBQ500
3,3-DIMETHYL-2-OXETANONE see DTH000
3,3-DIMETHYL-2-OXETHANONE see DTH000
5,5-DIMETHYL-3-(2-(OXIRANYLMETHOXY)PROPYL)-1-(OXIRANYLMETHYL)-2,4-
   IMIDAZ OLIDINEDIONE see DTH100
N,N-DIMETHYL-3-OXOBUTANAMIDE see DOP000
N-(1,1-DIMETHYL-3-OXOBUTYL)ACRYLAMIDE see DTH200
N-(1,1-DIMETHYL-3-OXOBUTYL)-2-PROPENAMIDE see DTH200
(5,5-DIMETHYL-3-OXO-CYCLOHEX-1-EN-YL)-N,N-DIMETHYL-CARBAMAAT
   (DUTCH) see DRL200
(5,5-DIMETHYL-3-OXO-CYCLOHEX-1-EN-YL)-N,N-DIMETHYL-CARBAMAT (GER-
   MAN) see DRL200
5,5-DIMETHYL-3-OXO-1-CYCLOHEXEN-1-YL DIMETHYLCARBAMATE see
   DRL200
5,5-DIMETHYL-3-OXOCYCLOHEX-1-ENYL DIMETHYLCARBAMATE see DRL200
3-(2-(3,5-DIMETHYL-2-OXOCYCLOHEXYL)-2-HYDROXYETHYL)GLUTARIMIDE
   see CPE750
1,6-DIMETHYL-4-OXO-1,6,7,8,9,9a-HEXAHYDRO-4H-PYRIDO(1,2-a)PYRIMIDINE-3-
   CARBOXAMIDE see CDM500
3,7-DIMETHYL-1-(5-OXOHEXYL)-1H,3H-PURIN-2,6-DIONE see PBU100
3,7-DIMETHYL-1-(5-OXOHEXYL)XANTHINE see PBU100
DIMETHYLOXOHEXYLXANTHINE see PBU100
3,3-DIMETHYL-7-OXO-6-(2-PHENYLACETAMIDO)-4-THIA-1-AZABICY-
   CLO(3.2.0)HEPTANE-2-CARBOXYLIC ACID compounded with EPHEDRINE
   (1:1) see PAQ120
2-(2,2-DIMETHYL-1-OXOPROPYL)-1H-INDENE-1,3(2H)-DIONE see PIH175
endo-8,8-DIMETHYL-3-((1-OXO-2-PROPYLPENTYL)OXY)-8-AZONIABICY-
   CLO(3.2.1)OCTANE BROMIDE see LJS000
DIMETHYLOXOSTANNANE see DTH400
2,2-DIMETHYL-4-OXYMETHYL-1,3-DIOXOLANE see DVR600
α,γ-DIMETHYL-α-OXYMETHYL GLUTARALDEHYDE see DTH600
1-(2,5-DIMETHYLOXYPHENYLAZO)-2-NAPHTHOL see DOK200
DIMETHYLOXYQUINAZINE see AQN000
O,O-DIMETHYL-1-OXY-2,2,2-TRICHLOROETHYL PHOSPHONATE see TIQ250
N,N-DIMETHYLPALMITAMIDE see DTH700
DIMETHYLPALMITYLAMINE see HCP525
DIMETHYL PARANITROPHENYL THIONOPHOSPHATE see DTH800
DIMETHYL PARAOXON see PHD500
DIMETHYL PARATHION see MNH000
2,3-DIMETHYL-1-PENTANOL see DTI400
2,3-DIMETHYLPENTANOL see DTI400
2,4-DIMETHYL-3-PENTANONE see DTI600
S,S-DIMETHYLPENTASULFUR HEXANITRIDE see DTI709
DI(4-METHYL-2-PENTYL) MALEATE see DKP400
3,5-DIMETHYLPERHYDRO-1,3,5-THIADIAZIN-2-THION (CZECH, GERMAN) see
   DSB200
DIMETHYL PEROXIDE see DTJ000
DIMETHYLPEROXYCARBONATE see DTJ159
16,16-DIMETHYL-trans-Δ²-PGE1 METHYL ESTER see CDB775
N,N¹-DIMETHYLPHAEANTHINE DIIODIDE see TDX835
1,4-DIMETHYLPHENANTHRENE see DTJ200
α,α-DIMETHYLPHENETHYL ACETATE see BEL750, DQQ375
α,α-DIMETHYLPHENETHYL ALCOHOL see DQQ200
α,α-DIMETHYLPHENETHYL ALCOHOL ACETATE see BEL750
α,α-DIMETHYLPHENETHYL ALCOHOL PROPIONATE see DQQ400
α,α-DIMETHYLPHENETHYLAMINE see DTJ400
N,α-DIMETHYLPHENETHYLAMINE HYDROCHLORIDE see DBA800
(−)-N-α-DIMETHYLPHENETHYLAMINE HYDROCHLORIDE see MDQ500
α,α-DIMETHYLPHENETHYL BUTYRATE see BEL850
2,3-DIMETHYLPHENOL see XKJ000
2,4-DIMETHYLPHENOL see XKJ500
2,5-DIMETHYLPHENOL see XKS000
2,6-DIMETHYLPHENOL see XLA000
3,4-DIMETHYLPHENOL see XLJ000
3,5-DIMETHYLPHENOL see XLS000
3,6-DIMETHYLPHENOL see XKS000

4,5-DIMETHYLPHENOL see XLJ000
4,6-DIMETHYLPHENOL see XKJ500
DIMETHYLPHENOL see XKA000
3,4-DIMETHYLPHENOL METHYLCARBAMATE see XTJ000
3′,3″-DIMETHYLPHENOLPHTHALEIN see CNX400
2,8-DIMETHYLPHENOSAFRANINE see GJI400
N,N-DIMETHYL-10H-PHENOTHIAZINE-10-PROPANAMINE see DQA600
N,N-DIMETHYL-3-PHENOTHIAZINESULFONAMIDE see DTK300
5-(2,5-DIMETHYLPHENOXY)-2,2-DIMETHYLPENTANOIC ACID (9CI) see
   GCK300
5-((3,5-DIMETHYLPHENOXY)METHYL)-2-OXAZOLIDINONE see XVS000
(2-(2,6-DIMETHYLPHENOXY)PROPYL)TRIMETHYLAMMONIUM CHLORIDE
   MONOHYDRATE see TLQ000
2-(2,6-DIMETHYLPHENOXY)-N,N,N-TRIMETHYL-ETHANAMINIUM BROMIDE (9CI)
   see XSS900
2-(2,6-DIMETHYLPHENOXY)-N,N,N-TRIMETHYL-1-PROPANAMINIUM HYDRATE
   see TLQ000
α,α-DIMETHYLPHENRTHYL BUTYRATE see DQQ380
N,N-DIMETHYL-2-(2-PHENYLACETAMIDO)ACETAMIDE see PEB775
2,3-DIMETHYLPHENYLAMINE see XMJ000
2,4-DIMETHYLPHENYLAMINE see XMS000
2,5-DIMETHYLPHENYLAMINE see XNA000
3,4-DIMETHYLPHENYLAMINE see XNS000
3,5-DIMETHYLPHENYLAMINE see XOA000
N,N-DIMETHYLPHENYLAMINE see DQF800
DIMETHYLPHENYLAMINE see DQF800, XMA000
2-((2,3-DIMETHYLPHENYL)AMINO)BENZOIC ACID see XQS000
2-((2,6-DIMETHYLPHENYLAMINO)-4H-5,6-DIHYDRO-1,3-THIAZINE see DMW000
1,5-DIMETHYL-2-PHENYL-4-AMINOPYRAZOLINE see AIB300
N-(2,3-DIMETHYLPHENYL)ANTHRANILIC ACID see XQS000
N-(2,6-DIMETHYLPHENYL)-2-AZABICYCLO(2.2.2)OCTANE-3-CARBOXAMIDE
   MONOHYDROCHLORIDE (9CI) see EAV100
2,3-DIMETHYL-4-PHENYLAZOANILINE see DTL000
N,N-DIMETHYL-p-PHENYLAZOANILINE see DOT300
N,N-DIMETHYL-p-PHENYLAZOANILINE-N-OXIDE see DTK600
N,N-DIMETHYL-4-PHENYLAZO-o-ANISIDINE see DTK800
N,N-DIMETHYL-4-(PHENYLAZO)BENZAMINE see DOT300
2,3-DIMETHYL-4-(PHENYLAZO)BENZENAMINE see DTL000
N,N-DIMETHYL-4-(PHENYLAZO)BENZENAMINE see DOT300
4-((2,4-DIMETHYLPHENYL)AZO)-3-HYDROXY-2,7-NAPHTHALENEDISULFONIC
   ACID, DISODIUM SALT see FMU070
4-((2,4-DIMETHYLPHENYL)AZO)-3-HYDROXY-2,7-NAPHTHALENEDISULPHONIC
   ACID, DISODIUM SALT see FMU070
1-((2,4-DIMETHYLPHENYL)AZO)-2-NAPHTHALENOL see XRA000
N,N-DIMETHYL-4-(PHENYLAZO)-m-TOLUIDINE see TLE750
N,N-DIMETHYL-4-(PHENYLAZO)-o-TOLUIDINE see MJF000
N,N-DIMETHYL-α-PHENYLBENZENEACETAMIDE see DRP800
(R)-N,N-DIMETHYL-α-PHENYLBENZENEETHANAMINE, HYDROCHLORIDE see
   DWA600
2-DI(N-METHYL-N-PHENYL-tert-BUTYL-CARBAMOYLMETHYL)AMINOETHANOL
   see DTL200
N-(N′-(2,6-DIMETHYLPHENYL)CARBAMOYLMETHYL)IMINODIACETIC ACID see
   LFO300
N-(2,6-DIMETHYLPHENYLCARBAMOYLMETHYL)-IMINODIACETIC ACID see
   LFO300
DIMETHYLPHENYLCARBINOL see DTN100
N-(2,6-DIMETHYLPHENYL)-5,6-DIHYDRO-4H-1,3-THIAZIN-2-AMINE see DMW000
N-(2,6-DIMETHYLPHENYL)-5,6-DIHYDRO-4H-1,3-THIAZINE-2-AMINE (9CI) see
   DMW000
N′-(2,4-DIMETHYLPHENYL)-N-(((2,4-DIMETHYLPHENYL)IMINO)METHYL)-N-ME-
   THYLME THANIMIDAMIDE see MJL250
1,3-DIMETHYL-3-PHENYL-2,5-DIOXOPYRROLIDINE see MLP800
DIMETHYL-4,4′-o-PHENYLENE-BIS-(3-THIOALLOPHANATE) see PEX500
N,N-DIMETHYL-p-PHENYLENEDIAMINE see DTL600, DTL800
DIMETHYL-p-PHENYLENEDIAMINE see DTL600, DTL800
N,N-DIMETHYL-p-PHENYLENEDIAMINE HEMISULFATE see DTM200
N,N-DIMETHYL-p-PHENYLENEDIAMINE MONOHYDROCHLORIDE see DTM400
1,1-DIMETHYL-2-PHENYLETHANAMINE see DTJ400
1,1-DIMETHYL-2-PHENYLETHANOL see DQQ200
(E)-N,N,-DIMETHYL-4-(2-PHENYLETHENYL)BENZENAMINE see DUC000
α,α-DIMETHYL-β-PHENYLETHYLAMINE see DTJ400
DIMETHYLPHENYLETHYL CARBINOL see BEC250
DIMETHYLPHENYLETHYLCARBINYL ACETATE see MNT000
o,p-DIMETHYL-β-PHENYLETHYLHYDRAZINE DIHYDROGEN SULFATE see
   DTM600
β-(2,4-DIMETHYLPHENYL)ETHYLHYDRAZINE DIHYDROGEN SULPHATE see
   DTM600
DIMETHYLPHENYLETHYNYLTHALLIUM see DTM800
N,N-DIMETHYL-3-PHENYL-1-INDANAMINE see DRX400
2,4-DIMETHYLPHENYLMALEIMIDE see DTN000
2,4-DIMETHYL-N-PHENYLMALEIMIDE see DTN000
DIMETHYLPHENYLMETHANOL see DTN100
N-(2,6-DIMETHYLPHENYL)-N-(METHOXYACETYL)-DL-ALANINE METHYL ESTER
   see MDM100

N-(2,6-DIMETHYLPHENYL)-N-(METHOXYACETYL)-ALANINE METHYL ESTER see MDM100
3,4-DIMETHYLPHENYL-N-METHYLCARBAMATE see XTJ000
3,5-DIMETHYLPHENYL-N-METHYLCARBAMATE see DTN200
N,N-DIMETHYL''-(PHENYLMETHYL)GUANIDINE SULPHATE (2:1) see BFW250
N,N-DIMETHYL-3((1-PHENYLMETHYL)-1H-INDAZOL-3-YL)OXY-1-PROPANAMINE HYDROCHLORIDE see BBW500
4-(DIMETHYLPHENYLMETHYL)PHENOL see COF400
N-(2,6-DIMETHYLPHENYL)-1-METHYL-2-PIPERIDINECARBOXAMIDE-MONOHY-DROCHLORIDE see CBR250
5,5-DIMETHYL-2-PHENYLMORPHOLINE see DTN775
3,4-DIMETHYL-2-PHENYLMORPHOLINE BITARTRATE see DKE800
3,4-DIMETHYL-2-PHENYLMORPHOLINEHYDROCHLORIDE see DTN800
1,3-DIMETHYL-3-PHENYL-1-NITROSOUREA see DTN875
N-(2,6-DIMETHYLPHENYL)-N'-(1-METHYL-2-PYRROLIDINYLIDENE)UREA see XGA725
N-(2,4-DIMETHYLPHENYL)-3-OXOBUTANAMIDE see OOI100
DIMETHYLPHENYLPHOSPHINE see DTN896
1,1-DIMETHYL-4-PHENYLPIPERAZINE IODIDE see DTO000
1,1-DIMETHYL-4-PHENYLPIPERAZINIUM IODIDE see DTO000
1,1-DIMETHYL-4-PHENYLPIPERIDINIUM IODIDE see DTO100
1,3-DIMETHYL-4-PHENYL-4-PIPERIDINOL PROPIONATE (ESTER) see NEA500
1,3-DIMETHYL-4-PHENYL-4-PIPERIDINOL, PROPIONATE, HYDROCHLORIDE see NEB000
dl-1,3-DIMETHYL-4-PHENYL-4-PIPERIDINOL PROPIONATE HYDROCHLORIDE see NEB000
α-1,3-DIMETHYL-4-PHENYL-4-PIPERIDINYL PROPIONATE see NEA500
(±)-1,3-DIMETHYL-4-PHENYL-4-PIPERIDYL ESTER PROPIONIC ACID HYDRO-CHLORIDE see NEB000
1,3-DIMETHYL-4-PHENYL-4-PIPERIDYL PROPIONATE HYDROCHLORIDE see NEB000
1,1-DIMETHYL-3-PHENYL-1-PROPANOL see BEC250
1,1-DIMETHYL-3-PHENYLPROPANOL see BEC250
1,3-DIMETHYL-4-PHENYL-4-PROPIONOXYPIPERIDINE see NEA500
α-1,3-DIMETHYL-4-PHENYL-4-PROPIONOXYPIPERIDINE see NEA500
(±)-α-1,3-DIMETHYL-4-PHENYL-4-PROPIONOXYPIPERIDINE HYDROCHLORIDE see NEB000
α,α-DIMETHYL-Δ-PHENYLPROPYL ALCOHOL see BEC250
(1,1-DIMETHYL-3-PHENYLPROPYL)ESTER ACETIC ACID see MNT000
2,3-DIMETHYL-1-PHENYL-3-PYRAZOLIN-5-ONE see AQN000
2,3-DIMETHYL-1-PHENYL-5-PYRAZOLONE see AQN000
N,N-DIMETHYL-2-(1-PHENYL-1-(2-PYRIDINYL)ETHOXY)ETHANAMINE (9CI) see DYE500
Z-(±)-2,6-DIMETHYL-α-PHENYL-α-(2-PYRIDYL)-1-PIPERIDINEBUTANOL HYDRO-CHLORIDE see PJA170
(±)-cis-2,6-DIMETHYL-α-PHENYL-α-2-PYRIDYL-1-PIPERIDINEBUTANOL MONO-HYDROCHLORIDE see PJA170
N,N-DIMETHYL-3-PHENYL-3-(2-PYRIDYL)PROPYLAMINE see TMJ750
1,3-DIMETHYL-3-PHENYL-PYRROLIDIN-2,5-DIONE see MLP800
3,4-trans-2,2-DIMETHYL-3-PHENYL-4-p-(β-PYRROLIDINOETHOXY)PHENYL-7-ME-THOXYCHROMAN HCl see CCW750
1,2-DIMETHYL-3-PHENYL-3-PYRROLIDINOL PROPIONATE (ester) see DTO200
1,2-DIMETHYL-3-PHENYL-3-PYRROLIDYL PROPIONATE see DTO200
1-(2,4-DIMETHYL-5-PHENYL-1H-PYRROL-3-YL)ETHANONE see DTO300
N,2-DIMETHYL-2-PHENYLSUCCINIMIDE see MLP800
N,N-DIMETHYL-2-(α-PHENYL-o-TOLOXY)ETHYLAMINE DIHYDROGEN CITRATE see DTO600
N,N-DIMETHYL-2-(α-PHENYL-o-TOLOXY)ETHYLAMINE HYDROCHLORIDE see DTO800
3,3-DIMETHYL-1-PHENYL-1-TRIAZENE see DTP000
3,3-DIMETHYL-1-PHENYLTRIAZENE see DTP000
1,1-DIMETHYL-3-PHENYLUREA see DTP400
DIMETHYL PHOSPHATE see PHC800
DIMETHYL PHOSPHATE of 2-CHLORO-N,N-DIETHYL-3-HYDROXYCROTONAM-IDE see FAB400
DIMETHYL PHOSPHATE ESTER with 2-CHLORO-N-ETHYL-3-HYDROXYCRO-TONAMIDE see DTP600
DIMETHYL PHOSPHATE ESTER with 2-CHLORO-N-METHYL-3-HYDROXYCRO-TONAMIDE see DTP800
DIMETHYLPHOSPHATE ESTER with 3-HYDROXY-N,N-DIMETHYL-cis-CROTO-NAMIDE see DGQ875
DIMETHYL PHOSPHATE ESTER of 3-HYDROXY-N-METHYL-cis-CROTONAMIDE see MRH209
DIMETHYL PHOSPHATE-3-HYDROXY-CROTONIC ACID, p-CHLOROBENZYL ESTER see CEQ500
DIMETHYL PHOSPHATE of 3-HYDROXY-N,N-DIMETHYL-cis-CROTONAMIDE see DGQ875
DIMETHYL PHOSPHATE of 3-HYDROXY-N-METHYL-cis-CROTONAMINE see MRH209
DIMETHYL PHOSPHATE of α-METHYLBENZYL-3-HYDROXY-cis-CROTONATE see COD000
DIMETHYL PHOSPHINE see DTQ089
DIMETHYL PHOSPHITE see DSG600
DIMETHYL PHOSPHODITHIONATE see PHH500
DIMETHYL PHOSPHONATE see DSG600

DIMETHYLPHOSPHORAMIDOCYANIDIC ACID, ETHYL ESTER see EIF000
O,S-DIMETHYL PHOSPHORAMIDOTHIOATE see DTQ400
DIMETHYL PHOSPHOROCHLORIDOTHIOATE (DOT) see DTQ600
O,O-DIMETHYLPHOSPHOROCHLORIDOTHIOATE see DTQ600
DIMETHYL PHOSPHORODITHIOATE see PHH500
O,O-DIMETHYL PHOSPHORODITHIOATE see PHH500
O,O-DIMETHYL PHOSPHORODITHIOATE N-FORMYL-2-MERCAPTO-N-METHY-LACETAMIDE-S-ESTER see DRR200
S-(O,O-DIMETHYLPHOSPHORODITHIOATE) of N-(2-MERCAPTOE-THYL)ETHYLCARBAMATE see EMC000
N-((O,O-DIMETHYLPHOSPHORODITHIOYL)ETHYL)ACETAMIDE see DOP200
O,O-DIMETHYL PHOSPHOROTHIOATE-O,O-DIESTER with 4,4'-THIODIPHENOL see TAL250
O,O-DIMETHYL PHOSPHOROTHIOATE-O-ESTER with 4-HYDROXY-m-ANISONI-TRILE see DTQ800
DIMETHYL PHOSPHOROUS ACID see DSG600
5-(O,O-DIMETHYLPHOSPHORYL)-6-CHLOROBICYCLO(3.2.0)HEPTA-1,5-DIEN see HBK700
DIMETHYL PHTHALATE see DTR200
(O,O-DIMETHYL-PHTHALIMIDIOMETHYL-DITHIOPHOSPHATE) see PHX250
O,O-DIMETHYL S-(N-PHTHALIMIDOMETHYL) DITHIOPHOSPHATE /S O,O-DI-METHYL S-PHTHALIMIDOMETHYL PHOSPHORODITHIOATE see PHX250
1,4-DIMETHYLPIPERAZINE see LIQ500
2,5-DIMETHYLPIPERAZINE see DTR400
N,N'-DIMETHYLPIPERAZINE see LIQ500
α,4-DIMETHYL-1-PIPERAZINEACETIC ACID-6-CHLORO-o-TOLYL ESTER DIHY-DROCHLORIDE see FAC185
α,4-DIMETHYL-1-PIPERAZINEACETIC ACID-2,6-DIISOPROPYLPHENYL ESTER DIHYDROCHLORIDE see FAC160
2-β,16-β-(4'-DIMETHYL-1'-PIPERAZINO)-3-α,17-β-DIACETOXY-5-α-ANDROSTANE 2BR see PII250
2,6-DIMETHYLPIPERIDINE see LIQ550
2-(2,6-DIMETHYLPIPERIDINO)-2',6'-ACETOXYLIDIDE HYDROCHLORIDE see DTR800
2,4'-DIMETHYL-3-PIPERIDINOPROPIOPHENONE see TGK200
2,4'-DIMETHYL-3-PIPERIDINOPROPIOPHENONE HYDROCHLORIDE see MRW125
N,N-DIMETHYL-4-PIPERIDYLIDENE-1,1-DIPHENYLMETHANE METHYLSULFATE see DAP800
DIMETHYLPOLYSILOXANE see DTR850
6,17-DIMETHYLPREGNA-4,6-DIENE-3,20-DIONE see MBZ100
1,2-DIMETHYLPROPANAMINE see AOE200
2,2-DIMETHYLPROPANE see NCH000
2,2-DIMETHYLPROPANE, other than pentane and isopentane (DOT) see NCH000
N,N-DIMETHYL-1,3-PROPANEDIAMINE see AJQ100
DIMETHYL PROPANEDIOATE see DSM200
2,2-DIMETHYL-1,3-PROPANEDIOL see DTG400
2,2-DIMETHYL-1,3-PROPANEDIOL DIACRYLATE see DUL200
2,2'-((2,2-DIMETHYL-1,3-PROPANEDIYL)BIS(OXYMETHYLENE))BISOXIRANE see NCI300
2,2-DIMETHYLPROPANOIC ACID see PJA500
2,2-DIMETHYLPROPANOIC ACID-3-(2-(ETHYLAMINO)-1-HYDROXYETH-YL)PHENYL ESTER HYDROCHLORIDE see EGC500
(±)-2,2-DIMETHYL-PROPANOIC ACID-4-(1-HYDROXY-2-(METHYLAMINO)ETHYL)-1,2-PHENYLENE ESTER, HYDROCHLORIDE see DWP559
2,2-DIMETHYLPROPANOIC ACID ISOOCTADECYL ESTER see ISC550
2,2-DIMETHYL-PROPANOIC ACID-2-OXO-2-PHENYLETHYL ESTER (9CI) see PCV350
2,2-DIMETHYLPROPANOYL CHLORIDE see DTS400
1,1-DIMETHYLPROPARGYL ALCOHOL see MHX250
α-α-DIMETHYLPROPARGYL ALCOHOL see MHX250
N,N-DIMETHYL-2-PROPENAMIDE see DOP800
N,N-DIMETHYL-N-2-PROPENYL-2-PROPEN-1-AMINIUM CHLORIDE HOMOPOLY-MER (9CI) see DTS500
3,3-DIMETHYL-β-PROPIOLACTONE see DTH000
DIMETHYL PROPIOLACTONE see DTH000
N,N-DIMETHYLPROPIONAMIDE see DTS600
2,2-DIMETHYLPROPIONIC ACID see PJA500
α,α-DIMETHYLPROPIONIC ACID see PJA500
2,2-DIMETHYLPROPIONYL CHLORIDE see DTS400
2-(2,2-DIMETHYL-3-PROPIONYL)-1-METHYL-3-(METHYLETHE-NYL)CYCLOPENTENE see MJW300
2,6-DIMETHYL-4-PROPOXY-BENZOIC ACID 2-METHYL-2-(1-PYRROLIDI-NYL)PROPYLESTER see DTS625
2,6-DIMETHYL-4-PROPOXY-BENZOIC ACID 2-METHYL-2-(1-PYRROLIDI-NYL)PROPYL ESTER see UAG050
2,6-DIMETHYL-4-PROPOXY-BENZOIC ACID 2-(1-PYRROLIDINYL)PROPYL ES-TER HYDROCHLORIDE see UAG075
1,2-DIMETHYLPROPYLAMINE see AOE200
7,12-DIMETHYL-8-PROPYL-BENZ(a)ANTHRACENE see PNI500
4-(1,1-DIMETHYLPROPYL)CYCLOHEXANONE see AOH750
N,N-DIMETHYL-1,3-PROPYLENEDIAMINE see AJQ100
DIMETHYLPROPYLENEUREA see DSE489
1,1-DIMETHYLPROPYL HYDROPEROXIDE see PBX325

5-(3-DIMETHYLPROPYLIDENE)DIBENZO(a,d)(1,4)CYCLOHEPTADIENE see EAH500

p-(1,1-DIMETHYLPROPYL)PHENOL see AON000

p-(α,α-DIMETHYLPROPYL)PHENOL see AON000

N,N-DIMETHYL-p-((p-PROPYLPHENYL)AZO)ANILINE see DTT400

trans-(±)-3-(1,3-α-DIMETHYL-4-α-PROPYL-4-β-PIPERIDINYL)PHENOL HYDROCHLORIDE see PIB700

(±)-3-(1,3-DIMETHYL-4-PROPYL-4-PIPERIDINYL)PHENOL HYDROCHLORIDE see PIB700

m-(1,2-DIMETHYL-3-PROPYL-3-PYRROLIDINYL)PHENOL see DSI000

1,1-DIMETHYLPROPYNOL see MHX250

(−)-N,α-DIMETHYL-N-2-PROPYNYLBENZENEETHANAMINE HYDROCHLORIDE see DAZ125

N-(1,1-DIMETHYLPROPYNYL)-3,5-DICHLOROBENZAMIDE see DTT600

(±)-N,α-DIMETHYL-N-2-PROPYNYLPHENETHYLAMINE HYDROCHLORIDE see DAZ118

(+)-N,α-DIMETHYL-N-2-PROPYNYLPHENETHYLAMINE HYDROCHLORIDE see DAZ120

16,16-DIMETHYL-trans-Δ²-PROSTAGLANDIN E1 METHYL ESTER see CDB775

3,4-DIMETHYLPROTOCATECHUIC ACID see VHP600

3,5-DIMETHYL-4H-PYRAN-4-ONE-2-METHOXY-6-(TETRAHYDRO-4-(β-METHYL-p-NITROCINNAMYLIDENE)-2-FURYL) see DTU200

2,3-DIMETHYLPYRAZINE see DTU400

2,5-DIMETHYLPYRAZINE see DTU600

2,6-DIMETHYLPYRAZINE see DTU800

3,5-DIMETHYLPYRAZOLE see DTU850

2,4-DIMETHYLPYRIDINE see LIY990

2,6-DIMETHYLPYRIDINE see LJA010

3,4-DIMETHYLPYRIDINE see LJB000

α-α′-DIMETHYLPYRIDINE see LJA010

α-γ-DIMETHYLPYRIDINE see LIY990

2,6-DIMETHYLPYRIDINE-N-OXIDE see DTV089

2,6-DIMETHYLPYRIDINE-1-OXIDE-4-AZO-p-DIMETHYLANILINE see DPP800

N,N-DIMETHYL-3-(1-(2-PYRIDINYL)ETHYL)-1H-INDENE-2-ETHANAMINE (Z)-2-BUTENEDIOATE (1:1) see FMU409

N¹-(4,6-DIMETHYL-2-PYRIDINYL)SULFANILAMIDE, MONOSODIUM SALT see SJW500

5,11-DIMETHYL-6H-PYRIDO(4,3-b)CARBAZOLE see EAI850

5,11-DIMETHYL-6H-PYRIDO(4,3-b)CARBAZYL-9-OL see HKH000

1,4-DIMETHYL-5H-PYRIDO(4,3-b)INDOL-3-AMINE see TNX275

1,4-DIMETHYL-5H-PYRIDO(4,3-b)INDOL-3-AMINE ACETATE see AJR500

1,4-DIMETHYL-5H-PYRIDO(4,3-b)INDOL-3-AMINE MONOACETATE see AJR500

N,N-DIMETHYL-4-(3′-PYRIDYLAZO)ANILINE see POP750

N,N-DIMETHYL-p-(3-PYRIDYLAZO)ANILINE see POP750

(3,3-DIMETHYL-1-(m-PYRIDYL-N-OXIDE))TRIAZENE see DTV200

S-(4,6-DIMETHYL-2-PYRIMIDINYL)-O,O-DIETHYL PHOSPHORODITHIOATE see DTV400

N¹-(4,6-DIMETHYL-2-PYRIMIDINYL)SULFANILAMIDE see SNJ000

N¹-(2,6-DIMETHYL-4-PYRIMIDINYL)SULFANILAMIDE see SNJ350

(N¹-(4,6-DIMETHYL-2-PYRIMIDINYL)SULFANILAMIDO) SODIUM see SJW500

6,8-DIMETHYL-PYRIMIDO(5,4-e)-1,2,4-TRIAZINE-5,7(6H,8H)-DIONE see FBP300

6,8-DIMETHYLPYRIMIDO(5,4-e)-as-TRIAZINE-5,7(6H,8H)-DIONE see FBP300

N-(4,6-DIMETHYL-2-PYRIMIDYL)SULFANILAMIDE see SNJ000

1,3-DIMETHYL PYROGALLATE see DOJ200

20-(2,4-DIMETHYL-1H-PYRROLE-3-CARBOXYLATE) BATRACHOTOXININ A see BAR750

20-α-(2,4-DIMETHYL-1H-PYRROLE-3-CARBOXYLATE) BETRACHOTOXININ A see BAR750

3,4-DIMETHYLPYRROLIDINE ETHANOL see HKR550

2,4-DIMETHYLPYRROL-3-YL METHYL KETONE see ACI500

2,5-DIMETHYLPYRROL-3-YL METHYL KETONE see ACI550

N-(2,5-DIMETHYL-1H-PYRROL-1-YL)-6-(4-MORPHOLINYL)-3-PYRIDAZINAMINE HYDROCHLORIDE see MBV735

N,N-DIMETHYL-4-(4′-QUINOLYLAZO)ANILINE see DTY200

N,N-DIMETHYL-4-(5′-QUINOLYLAZO)ANILINE see DPQ800

N,N-DIMETHYL-4-(6′-QUINOLYLAZO)ANILINE see DPR000

N,N-DIMETHYL-p-(5′-QUINOLYLAZO)ANILINE see DPQ800

N,N-DIMETHYL-4-(5′-QUINOLYLAZO)-m-TOLUIDINE see DQE400

DIMETHYLQUINOLYL METHYLSULFATE UREA see PJA120

N,N-DIMETHYL-4-((4′-QUINOLYL-1′-OXIDE)AZO)ANILINE see DTY400

N,N′-DIMETHYL-4-((6′-QUINOLYL-1′-OXIDE)AZO)ANILINE see DPR400

N,N-DIMETHYL-4-((5′-QUINOLYL-I′-OXIDE)AZO)ANILINE see DPR200

3,3-DIMETHYL-1-(3-QUINOLYL)TRIAZENE see DTY600

2,3-DIMETHYLQUINOXALINE see DTY700

N,N-DIMETHYL-p-(6-QUINOXALINYLAZO)ANILINE see DUA400

N,N-DIMETHYL-p-(6-QUINOXALYAZO)ANILINE see DUA400

N,N-DIMETHYL-p-(5-QUINOXALYLAZO)ANILINE see DUA200

7,8-DIMETHYL-10-d-RIBITYLISOALLOXAZINE see RIK000

6,7-DIMETHYL-9-d-RIBITYLISOALLOXAZINE see RIK000

7,8-DIMETHYL-10-(d-RIBO-2,3,4,5-TETRAHYDROXYPENTYL)-4a,5-DIHYDROISOALLOXAZINE see DUA600

7,8-DIMETHYL-10-(d-RIBO-2,3,4,5-TETRAHYDROXYPENTYL)ISOALLOXAZINE see RIK000

O,O-DIMETHYL-S-(2-ACETAMIDOETHYL) ESTER PHOSPHORODITHIOIC ACID see DOP200

O,O-DIMETHYL-S-(2-(ACETYLAMINO)ETHYL) DITHIOPHOSPHATE see DOP200

O,O-DIMETHYL-S-(2-ACETYLAMINOETHYL) PHOSPHORODITHIOATE see DOP200

O,O-DIMETHYL-S-(2-AETHYLSULFINYL-AETHYL)-THIOLPHOSPHAT (GERMAN) see DAP000

O,O-DIMETHYL-S-(2-AETHYLSULFONYL-AETHYL)-THIOLPHOSPHAT (GERMAN) see DAP600

O,O-DIMETHYL-S-(2-AETHYLTHIO-AETHYL)-DITHIO PHOSPHAT (GERMAN) see PHI500

O,O-DIMETHYL-S-(2-AETHYLTHIO-AETHYL)-MONOTHIOPHOSPHAT (GERMAN) see DAP400

N,N-DIMETHYLSALICYLAMIDE see DUA800

DIMETHYL SALICYLATE see MLH800

O,O-DIMETHYL-S-(BENZAZIMINOMETHYL) DITHIOPHOSPHATE see ASH500

O,O-DIMETHYL-S-(1,2,3-BENZOTRIAZINYL-4-KETO)METHYL PHOSPHORODITHIOATE see ASH500

O,O-DIMETHYL-S-(1,2-BIS(ETHOXYCARBONYL)ETHYL)DITHIOPHOSPHATE see MAK700

O,O-DIMETHYL-S-1,2-BIS(ETHOXYCARBONYL)ETHYL PHOSPHOROTHIOATE see OPK250

O,O-DIMETHYL-S-(CARBETHOXY)METHYL PHOSPHOROTHIOLATE see DRB600

O,O-DIMETHYL-S-(1-CARBOETHOXYBENZYL) DITHIOPHOSPHATE see DRR400

O,O-DIMETHYL-S-CARBOETHOXYMETHYL THIOPHOSPHATE see DRB600

O,O-DIMETHYL-S-(CARBONYLMETHYLMORPHOLINO) PHOSPHORODITHIOATE see PHI500

O,O-DIMETHYL-S-p-CHLOROPHENYL PHOSPHOROTHIOATE see FOR000

O,O-DIMETHYL-S-(p-CHLOROPHENYLTHIOMETHYL)PHOSPHORODITHIOATE see MQH750

O,O-DIMETHYL-S-(4,6-DIAMINO-1,3,5-TRIAZINYL-2-METHYL) DITHIOPHOSPHATE see ASD000

O,O-DIMETHYL-S-(4,6-DIAMINO-1,3,5-TRIAZIN-2-YL)METHYL PHOSPHORODITHIOATE see ASD000

O,O-DIMETHYL-S-(4,6-DIAMINO-1,3,5-TRIAZIN-2-YL)METHYL PHOSPHOROTHIOLOTHIONATE see ASD000

O,O-DIMETHYL-S-1,2-(DICARBAETHOXYAETHYL)-DITHIOPHOSPHAT (GERMAN) see MAK700

O,O-DIMETHYL-S-(1,2-DICARBETHOXYETHYL) DITHIOPHOSPHATE see MAK700

O,O-DIMETHYL-S-(1,2-DICARBETHOXYETHYL)PHOSPHORODITHIOATE see MAK700

O,O-DIMETHYL-S-(1,2-DICARBETHOXY)ETHYL PHOSPHOROTHIOATE see OPK250

O,O-DIMETHYL-S-(1,2-DICARBETHOXYETHYL) THIOTHIONOPHOSPHATE see MAK700

O,O-DIMETHYL-S-1,2-DI(ETHOXYCARBAMYL)ETHYL PHOSPHORODITHIOATE see MAK700

O,O-DIMETHYL-S-(3,4-DIHYDRO-4-KETO-1,2,3-BENZOTRIAZINYL-3-METHYL) DITHIOPHOSPHATE see ASH500

O,O-DIMETHYL-S-1,2-DIKARBETOXYLETHYLDITIOFOSFAT (CZECH) see MAK700

DIMETHYL SELENATE see DUB000

DIMETHYL SELENIDE see DUB200

DIMETHYLSELENIUM see DUB200

N,N-DIMETHYLSEROTONIN see DPG109

O,O-DIMETHYL-S-α-ETHOXY-CARBONYLBENZYL PHOSPHORODITHIOATE see DRR400

O,O-DIMETHYL-S-(5-ETHOXY-1,3,4-THIADIAZOLINYL-3-METHYL)DITHIOPHOSPHATE see DRU400

O,O-DIMETHYL-S-(5-ETHOXY-1,3,4-THIADIAZOL-2(3H)-ONYL-(3)-METHYL)DITHIOPHOSPHATE see DRU400

O,O-DIMETHYL-S-(5-ETHOXY-1,3,4-THIADIAZOL-2(3H)-ONYL-(3)-METHYL)PHOSPHORODITHIOATE see DRU400

O,O-DIMETHYL-S-(2-ETHSULFONYLETHYL)PHOSPHOROTHIOATE see DAP600

DIMETHYL-S-(2-ETHSULFONYLETHYL)THIOPHOSPHATE see DAP600

O,O-DIMETHYL-S-(2-ETHTHIOETHYL)PHOSPHOROTHIOATE see DAP400

DIMETHYL-S-(2-ETHTHIOETHYL)THIOPHOSPHATE see DAP400

O,O-DIMETHYL-S-(2-ETHTHIONYLETHYL) PHOSPHOROTHIOATE see DAP000

DIMETHYL-S-(2-ETHTHIONYLETHYL) THIOPHOSPHATE see DAP000

O,O-DIMETHYL-S-(N-ETHYLCARBAMOYLMETHYL) DITHIOPHOSPHATE see DNX600

O,O-DIMETHYL-S-(N-ETHYLCARBAMOYLMETHYL) PHOSPHORODITHIOATE see DNX600

O,O-DIMETHYL-S-(2-ETHYLMERCAPTOETHYL) DITHIOPHOSPHATE see PHI500

O,O-DIMETHYL-S-ETHYLMERCAPTOETHYL THIOPHOSPHATE see DAP400

O,O-DIMETHYL-S-ETHYLMERCAPTOETHYL THIOPHOSPHATE, THIOLO ISOMER see DAP400

O,O-DIMETHYL-S-2-ETHYLMERKAPTOETHYLESTER KYSELINY DITHIOFOSFORECNE (CZECH) see PHI500

O,O-DIMETHYL-S-(2-ETHYLSULFINYL-ETHYL)-MONOTHIOFOSFAAT (DUTCH) see DAP000

O,O-DIMETHYL-S-(2-(ETHYLSULFINYL)ETHYL) PHOSPHOROTHIOATE see DAP000

O,O-DIMETHYL-S-(2-ETHYLSULFINYL)ETHYL THIOPHOSPHATE see DAP000

(2,6-DIMETHYL-4-TERTIARYBUTYL-3-HYDROXYPHENYL)METHYLIMIDAZOLINE
  HYDROCHLORIDE see AEX000
DIMETHYL 2,3,5,6-TETRACHLOROTEREPHTHALATE see TBV250
DIMETHYL TETRACHLOROTEREPHTHALATE see TBV250
N,N-DIMETHYL-N-TETRADECYLBENZENEMETHANAMINIUM, CHLORIDE (9CI)
  see TCA500
7,12-DIMETHYL-1,2,3,4-TETRAHYDROBENZ(a)ANTHRACENE see TCP600
7,12-DIMETHYL-8,9,10,11-TETRAHYDROBENZ(a)ANTHRACENE see DUF000
1,11-DIMETHYL-1,2,3,4-TETRAHYDROCHRYSENE see DUF200
2,6-DIMETHYL-2,3,5,6-TETRAHYDRO-4H-1,4-OXAZINE see DST600
DIMETHYL TETRAHYDROPHTHALATE see DUF400
2,6-DIMETHYL TETRAHYDRO-1,4-PYRONE see TCQ350
DIMETHYLTETRAHYDROPYRONE see TCQ350
3,5-DIMETHYL-1,2,3,5-TETRAHYDRO-1,3,5-THIADIAZINETHIONE-2 see DSB200
3,5-DIMETHYL-1,3,5-2H-TETRAHYDROTHIADIAZINE-2-THIONE see DSB200
3,5-DIMETHYLTETRAHYDRO-1,3,5-2H-THIADIAZINE-2-THIONE see DSB200
3,5-DIMETHYLTETRAHYDRO-1,3,5-THIADIAZINE-2-THIONE see DSB200
3,5-DIMETHYLTETRAHYDRO-2H-1,3,5-THIADIAZINE-2-THIONE see DSB200
4,4'-(2,3-DIMETHYLTETRAMETHYLENE)DIPYROCATECHOL see NBR000
2,2'-DIMETHYLTETRANDRINIUM DIIODIDE see TDX835
3,6-DIMETHYL-1,2,4,5-TETRAOXANE see DUF800
DIMETHYLTHALLIUM FULMINATE see DUG00
DIMETHYLTHALLIUM-N-METHYLACETOHYDROXAMATE see DUG089
DIMETHYLTHIAMBUTENE HYDROCHLORIDE see TLQ250
2,4-DIMETHYLTHIAZOLE see DUG200
DIMETHYLTHIENYLCETONE see DUG425
DIMETHYLTHIOACETAMID see DUG450
N,N-DIMETHYLTHIOACETAMIDE see DUG450
DIMETHYLTHIOACETAMIDE see DUG450
N,N'-DIMETHYLTHIOCARBAMIDE see DSK900
DIMETHYLTHIOCARBAMIDE see DSK900
m,N-DIMETHYLTHIOCARBANILIC ACID-o-2 NAPHTHYL ESTER see TGB475
2,2'-DIMETHYLTHIOCARBANILIDE see DXP600
DIMETHYLTHIOMETHYLPHOSPHATE see DUG500
2,2-DIMETHYL-3-THIOMORPHOLINONE see DUG600
2,2-DIMETHYL-3-THIOMORPHOLONE see DUG600
DIMETHYLTHIONINE see DUG700
3,5-DIMETHYL-2-THIONOTETRAHYDRO-1,3,5-THIADIAZINE see DSB200
O,O-DIMETHYLTHIOPHOSPHORIC ACID, p-CHLOROPHENYL ESTER see
  MQH750
1,3-DIMETHYLTHIOUREA see DSK900
sym-DIMETHYLTHIOUREA see DSK900
DIMETHYLTIN BIS(DIBUTYLDITHIOCARBAMATE) see BIW500
DIMETHYL-TIN BIS(ISOOCTYLTHIOGLYCOLLATE) see BKK500
DIMETHYLTIN DIBROMIDE see DUG800
DIMETHYLTIN DICHLORIDE see DUG825
DIMETHYLTIN DIFLUORIDE see DKH200
DIMETHYLTIN DINITRATE see DUG889
DIMETHYLTIN FLUORIDE see DKH200
DIMETHYLTIN OXIDE see DTH400
DIMETHYL-m-TOLUIDINE see TLG100
N,N-DIMETHYL-o-TOLUIDINE see DUH200
DIMETHYL-o-TOLUIDINE see DUH200
DIMETHYL-p-TOLUIDINE see TLG150
DIMETHYLTOLUTHIONINE CHLORIDE see AJP250
N,N-DIMETHYL-4-(p-TOLYLAZO)ANILINE see DUH400
N,N-DIMETHYL-p-(m-TOLYLAZO)ANILINE see DUH600
N,N-DIMETHYL-p-((o-TOLYL)AZO)ANILINE see DUH800
N,N-DIMETHYL-2-(α-(p-TOLYL)BENZYLOXY)ETHYLAMINE HYDROCHLORIDE
  see TGJ475
N,N-DIMETHYL-N-(4-TOLYL)-N-(DICHLOROFLUOR-METHYLTHIO)SULFAMIDE
  see DFL400
3,3-DIMETHYL-1-(m-TOLYL)TRIAZENE see DSR200
3,3-DIMETHYL-1-(o-TOLYL)TRIAZENE see MNT500
N,N'-DIMETHYLTRIAZENE see DUI709
4'-(3,3-DIMETHYL-1-TRIAZENO)ACETANILIDE see DUI000
4'-DIMETHYLTRIAZENOACETANILIDE see DUI000
4'-(3-(3,3-DIMETHYL-1-TRIAZENO)-9-ACRIDINYLAMI-
  NO)METHANESULFONANILIDE see DUI200
p-(3,3-DIMETHYLTRIAZENO)BENZAMIDE see CCC325
5-(3,3-DIMETHYL-1-TRIAZENO)IMIDAZOLE-4-CARBOXAMIDE see DAB600
5-(3,3-DIMETHYLTRIAZENO)IMIDAZOLE-4-CARBOXAMIDE see DAB600
5-(DIMETHYLTRIAZENO)IMIDAZOLE-4-CARBOXAMIDE see DAB600
4-(3,3-DIMETHYL-1-TRIAZENO)IMIDAZOLE-5-CARBOXAMIDE see DAB600
4-(5)-(3,3-DIMETHYL-1-TRIAZENO)IMIDAZOLE-5(4)-CARBOXAMIDE see DAB600
4-(DIMETHYLTRIAZENO)IMIDAZOLE-5-CARBOXAMIDE see DAB600
(DIMETHYLTRIAZENO)IMIDAZOLECARBOXAMIDE see DAB600
3-(3',3'-DIMETHYLTRIAZENO)PYRIDINE-N-OXIDE see DTV200
3-(3',3'-DIMETHYLTRIAZENO)-PYRIDIN-N-OXID (GERMAN) see DTV200
4-(3,3-DIMETHYL-1-TRIAZENYL)BENZAMIDE see CCC325
p-(3,3-DIMETHYL-1-TRIAZENYL)BENZAMIDE see CCC325
5-(3,3-DIMETHYL-1-TRIAZENYL)-1H-IMIDAZOLE-4-CARBOXAMIDE see DAB600
N-(4-(3,3-DIMETHYL-1-TRIAZENYL)PHENYL)ACETAMIDE see DUI000
1,3-DIMETHYL-1-TRIAZINE see DUI709

O,O-DIMETHYL-(2,2,2-TRICHLOOR-1-HYDROXY-ETHYL)-FOSFONAAT (DUTCH)
  see TIQ250
O,O-DIMETHYL-(2,2,2-TRICHLOR-1-HYDROXY-AETHYL)PHOSPHONAT (GER-
  MAN) see TIQ250
O,O-DIMETHYL 2,2,2-TRICHLORO-1-(N-BUTYRYLOXY)ETHYLPHOSPHONATE
  see BPG000
DIMETHYL-2,2,2-TRICHLORO-1-HYDROXYETHYLPHOSPHONATE see TIQ250
O,O-DIMETHYL-2,2,2-TRICHLORO-1-HYDROXYETHYL PHOSPHONATE see
  TIQ250
DIMETHYLTRICHLOROHYDROXYETHYL PHOSPHONATE see TIQ250
3,5-DIMETHYL-1-(TRICHLOROMETHYLMERCAPTO)PYRAZOLE see DUJ000
O,O-DIMETHYL-O-2,4,5-TRICHLOROPHENYL PHOSPHOROTHIOATE see
  RMA500
O,O-DIMETHYL-O-(2,4,5-TRICHLOROPHENYL)THIOPHOSPHATE see RMA500
DIMETHYL TRICHLOROPHENYL THIOPHOSPHATE see RMA500
3,3-DIMETHYL-1-(2,4,6-TRICHLOROPHENYL)-TRIAZINE see TJA000
DIMETHYL-3,5,6-TRICHLORO-2-PYRIDYL PHOSPHATE see PHE250
DIMETHYL-3,5,6-TRICHLOROPYRIDYL PHOSPHATE see PHE250
O,O-DIMETHYL-O-(3,5,6-TRICHLORO-2-PYRIDYL)PHOSPHOROTHIOATE see
  CMA250
O,O-DIMETHYL-O-(2,4,5-TRICHLORPHENYL)-THIONOPHOSPHAT(GERMAN) see
  RMA500
2,6-DIMETHYL-4-TRIDECYLMORPHOLINE see DUJ400
N,N-DIMETHYL-2-(TRIFLUOROMETHYL)-10H-PHENOTHIAZINE-10-PROPANA-
  MINE see TKL000
1,1-DIMETHYL-3-(3-TRIFLUOROMETHYLPHENYL)UREA see DUK800
2',4'-DIMETHYL-5-((TRIFLUOROMETHYL)SULFONAMIDO)ACETANILIDE see
  DUK000
N-(2,4-DIMETHYL-5-(((TRIFLUOROMETH-
  YL)SULFONYL)AMINO)PHENYL)ACETAMIDE see DUK000
N,N-DIMETHYL-p-(2,4,6-TRIFLUOROPHENYLAZO)ANILINE see DUK200
1,1-DIMETHYL-3-(α,α,α-TRIFLUORO-m-TOLYL) UREA see DUK800
9-cis-3,7-DIMETHYL-9-(2,6,6-TRIMETHYL-1-CYCLOHEXEN-1-YL)-2,4,6,8-NONATE-
  TRAENAL see VSK975
3,7-DIMETHYL-9-(2,6,6-TRIMETHYL-1-CYCLOHEXEN)-1-YL-2,4,6,8-NONATET-
  RAENOIC ACID see VSK950
3,7-DIMETHYL-9-(2,6,6-TRIMETHYL-1-CYCLOHEXEN-1-YL)-2,4,6,8-NONATET-
  RAEN-1-OL see VSK600
2,2-DIMETHYLTRIMETHYLENE ACRYLATE see DUL200
2,2-DIMETHYLTRIMETHYLENE ESTER ACRYLIC ACID see DUL200
DIMETHYLTRIMETHYLENE GLYCOL see DTG400
3,3-DIMETHYLTRIMETHYLENE OXIDE see EBQ500
β,β-DIMETHYLTRIMETHYLENE OXIDE see EBQ500
N,N-DIMETHYL-4-(3,4,5-TRIMETHYLPHENYL)AZOANILINE see DUL400
N,N-DIMETHYL-4-((3,4,5-TRIMETHYLPHENYL)AZO)BENZENAMINE see DUL400
1,2-DIMETHYL-2-TRIMETHYLSILYLHYDRAZINE see DUL500
DIMETHYLTRIMETHYLSILYLPHOSPHINE see DUL550
1,4-DIMETHYL-2,3,7-TRIOXABICYCLO[2.2.1]HEPT-5-ENE see DUL589
(3,5-DIMETHYL-1,2,4-TRIOXOLANE) see BOX825
N,N-DIMETHYL-1,2,3-TRITHIAN-5-AMINE, ETHANEDIOATE (1:1) see TFH750
N,N-DIMETHYL-1,2,3-TRITHIAN-5-AMINE HYDROGENOXALATE see TFH750
N,N-DIMETHYL-1,2,3-TRITHIAN-5-YLAMMONIUM HYDROGEN OXALATE see
  TFH750
N,N-DIMETHYLTRYPTAMINE see DPF600
o,o'-DIMETHYLTUBOCURARINE see DUL800
o,o-DIMETHYLTUBOCURARINE see DUL800
DIMETHYL TUBOCURARINE see DUL800
DIMETHYL TUBOCURARINE IODIDE see DUM000
α,3-DIMETHYLTYROSINE METHYL ESTER HYDROCHLORIDE see DUM100
6,10-DIMETHYL-UNDECA-5,9-DIEN-2-ONE see GDE400
2,6-DIMETHYLUNDECA-2,6,8-TRIENE-10-ONE see POH525
6,10-DIMETHYL-3,5,9-UNDECATRIEN-2-ONE see POH525
1,1-DIMETHYLUREA see DUM150
1,3-DIMETHYLUREA see DUM200
N,N'-DIMETHYLUREA see DUM200
sym-DIMETHYLUREA see DUM200
DIMETHYLUREA and SODIUM NITRITE see DUM400
p-N,N-DIMETHYLUREIDOAZOBENZENE see DUM600
m-(3,3-DIMETHYLUREIDO)PHENYL-tert-BUTYL CARBAMATE see DUM800
β,β-DIMETHYLVINYL CHLORIDE see IKE000
DIMETHYLVINYLETHINYL-p-HYDROXYPHENYLMETHANE see DUN400
DIMETHYL(VINYL)ETHYNYLCARBINOL see MKM300
α-(2,2-DIMETHYLVINYL)-α-ETHYNYL-p-CRESOL see DUN400
1,5-DIMETHYL-1-VINYL-4-HEXEN-1-OL BENZOATE see LFZ000
1,5-DIMETHYL-1-VINYL-4-HEXEN-1-OL CINNAMATE see LGA000
1,5-DIMETHYL-1-VINYL-4-HEXEN-1-YL-o-AMINOBENZOATE see APJ000
1,5-DIMETHYL-1-VINYL-4-HEXEN-1-YL BENZOATE see LFZ000
1,5-DIMETHYL-1-VINYL-4-HEXEN-1-YL CINNAMATE see LGA000
1,5-DIMETHYL-1-VINYL-4-HEXENYL ESTER, ISOBUTYRIC ACID see LGB000
DIMETHYL VIOLOGEN see PAI990
DIMETHYL VIOLOGEN CHLORIDE see PAJ000
DIMETHYL XANTHIC DISULFIDE see DUN600
1,3-DIMETHYLXANTHINE see TEP000
1,7-DIMETHYLXANTHINE see PAK300
3,7-DIMETHYLXANTHINE see TEO500

3-((1,3-DIMETHYLXANTHIN-7-YL)METHYL)-5-METHYL-1,2,4-OXADIAZOLE see CDM575

DIMETHYLXANTHOGEN DISULFIDE see DUN600

3,3-DIMETHYL-1-XENYL-TRIAZENE see BGL500

N,N-DIMETHYL-p-(2,3,XYLYLAZO)ANILINE see DUN800

N,N-DIMETHYL-p-(3,4-XYLYLAZO)ANILINE see DUO000

2,2-DIMETHYL-5-(2,5-XYLYLOXY)VALERIC ACID see GCK300

DIMETHYL YELLOW see DOT300

DIMETHYL YELLOW-N,N-DIMETHYLANILINE see DOT300

DIMETHYLZINC see DUO200

DIMETHYLZINN-S,S'-BIS(ISOOCTYLTHIOGLYCOLAT) (GERMAN) see BKK500

DIMETHYOXYDOPAMINE see DOE200

2,6-DIMETHYPYRIDINE see LJA010

10,11-DIMETHYSTRYCHNINE see BOL750

5-(DIMETILAMINOETILOSIMINO-5H-DIBENZO(a,d)CICLOEPTA-1,4-DIENE) CLORIDRATO (ITALIAN) see DPH600

(4-DIMETILAMINO-3-METIL-FENIL)-N-METIL-CARBAMMATO (ITALIAN) see DOR400

5-(3-DIMETILAMINOPROPILIDEN)-5H-DIBENZO-(a,d)-CICLOPENTENE (ITALIAN) see PMH600

N-(γ-DIMETILAMINOPROPIL)-IMINODIBENZILE CLORIDRATO (ITALIAN) see DLH630

9-(3-DIMETILAMINOPROPYLIDEN)-10,10-DIMETIL-9,10-DIIDROANTHRACENE (ITALIAN) see AEG875

DIMETILAN see DQZ000

DIMETILANE see DQZ000

2,5-DIMETILBENZOCHINONE (1:4) (ITALIAN) see XQJ000

O,O-DIMETIL-O-(1,4-DIMETIL-3-OXO-4-AZA-PENT-1-ENIL)-FOSFATO (ITALIAN) see DGQ875

2,6-DIMETIL-EPTAN-4-ONE (ITALIAN) see DNI800

O,O-DIMETIL-O-(2-ETILTIO-ETIL)-MONOTIOFOSFATO (ITALIAN) see DAO800

2,6-DIMETILFENILICO DELL'ACIDO α-N-METILPIPERAZINOBUTIRRICO IDOCLORIDRAT (ITALIAN) see FAC130

DIMETILFORMAMIDE (ITALIAN) see DSB000

O,O-DIMETIL-O-(2-N-METILCARBAMOIL-1-METIL-VINIL)-FOSFATO (ITALIAN) see MRH209

O,O-DIMETIL-O-(3-METIL-4-METILTIO-FENIL)-MONOTIOFOSFATO (ITALIAN) see FAQ900

O,O-DIMETIL-O-(3-METIL-4-NITRO-FENIL)-MONOTIOFOSFATO (ITALIAN) see DSQ000

O,O-DIMETIL-O-(4-NITRO-FENIL)-MONOTIOFOSFATO (ITALIAN) see MNH000

(5,5-DIMETIL-3-OXO-CICLOES-1-EN-IL)-N,N-DIMETIL-CARBAMMATO (ITALIAN) see DRL200

3,5-DIMETIL-PERIDRO-1,3,5-THIADIAZIN-2-TIONE (ITALIAN) see DSB200

O,O-DIMETIL-S-(2-ETILTIO-ETIL)-MONOTIOFOSFATO (ITALIAN) see DAP400

O,O-DIMETIL-S-(2-ETIL-SOLFINIL-ETIL)-MONOTIOFOSFATO (ITALIAN) see DAP000

O,O-DIMETIL-S-(ETILTIO-ETIL)-DITIOFOSFATO (ITALIAN) see PHI500

O,O-DIMETIL-S-(N-FORMIL-N-METIL-CARBAMOIL-METIL)-DITIOFOSFATO (ITALIAN) see DRR200

O,O-DIMETIL-S-(N-METIL-CARBAMOIL-METIL)-DITIOFOSFATO (ITALIAN) see DSP400

O,O-DIMETIL-S-(N-METIL-CARBAMOIL)-METIL-MONOTIOFOSFATO (ITALIAN) see DNX800

O,O-DIMETIL-S-((2-METOSSI-1,3,4-(4H)-TIADIZAOL-5-ON-4-IL)-METIL)-DITIFOSFATO (ITALIAN) see DSO000

O,O-DIMETIL-S-((MORFOLINO-CARBONIL)-METIL)-DITIOFOSFATO (ITALIAN) see MRU250

DIMETILSOLFATO (ITALIAN) see DUD100

O,O-DIMETIL-S-((4-OXO-3H-1,2,3-BENZOTRIAZIN-3-IL)-METIL)-DITIOFOSFATO (ITALIAN) see ASH500

DIMETIL-m-TOLUIDINA see TLG100

DIMETIL-p-TOLUIDINA see TLG150

O,O-DIMETIL-(2,2,2-TRICLORO-1-IDROSSI-ETIL)-FOSFONATO (ITALIAN) see TIQ250

DIMETINA see BEM500

DIMETINDENE MALEATE see FMU409

DIMETOL see FON200

DIMETON see DSP400

3,3'-DIMETOSSIBENZODINA (ITALIAN) see DCJ200

DIMETOX see TIQ250

DIMETPRAMIDE see DUO300

DIMETRIDAZOLE see DSV800

DIMETYLFORMAMIDU (CZECH) see DSB000

O,O-DIMETYL-O-p-NITROFENYLFOSFAT (CZECH) see PHD500

DIMEVAMIDE see DOY400

DIMEVUR see DSP400

DIMEXAN see DUN600

DIMEXANO see DUN600

DIMEXIDE see DUD800

DIMEZATHINE see SNJ000

DIMID see DRP800

DIMIDIN see DUO350

DIMILIN see CJV250

DIMIPRESSIN see DLH600, DLH630

DIMITAN see BIE500

DIMITE see BIN000

DIMITRON see CMR100

DIMITRONAL see CMR100

DIMO see DLW600

DIMONOCLOROACETILAJMALINA CLORIDRATO (ITALIAN) see AFH275

DIMORLIN see AFJ400

DIMORPHOLAMINE see DUO400

DIMORPHOLINE DISULFIDE see BKU500

DIMORPHOLINETHIURAM DISULFIDE see MRR090

DIMORPHOLINIUM HEXACHLOROSTANNATE see DUO500

DIMORPHOLINO DISULFIDE see BKU500

1,5-DIMORPHOLINO-3-(1-NAPHTHYL)-PENTANE see DUO600

DIMP see DNQ875

DIMPEA see DOE200

DIMPYLATE see DCM750

DIM-SA see DNV800

DIMYRCETOL see DUO800

DIN 2.4602 see CNA750

DIN 2.4964 see CNA750

DINA see NFW000

DINACORYL see DJS200

DINACRIN see ILD000

DINAPACRYL see BGB500

DINAPHTAZIN (GERMAN) see DUP000

3,4,5,6-DINAPHTHACARBAZOLE see DCY000

1,2,5,6-DINAPHTHACRIDINE see DCS400

3,4,6,7-DINAPHTHACRIDINE see DCS600

DINAPHTHAZINE see DUP000

16H-DINAPHTHO(2,3-a:2',3'-i)CARBAZOLE-5,10,15,17-TETRAONE, 6,9-DIBENZAMIDO- see DUP100

16H-DINAPHTHO(2,3-a:2',3'-i)CARBAZOLE-5,10,15,17-TETRAONE, 6,9-DIBENZAMIDO-1-METHOXY- see CMU800

2,2'-DINAPHTHOL see BGC100

DINAPHTHO(2,3-a:2',3'-i)NAPHTH(2',3':6,7)INDOLO(2,3-c)CARBAZOLE-5,10,15,17,22,24-HEXAONE, 16,23-DIHYDRO- see CMU770

DINAPHTHO(1,2,3-cd:3',2',1'-lm)PERYLENE-5,10-DIONE see DCU800

DI-(1-NAPHTHOYL)PEROXIDE see DUP200

DI-β-NAPHTHYLDIIMIDE see ASN750

DI-β-NAPHTHYL-p-PHENYLDIAMINE see NBL000

N,N-DI-β-NAPHTHYL-p-PHENYLENEDIAMINE see NBL000

DI-β-NAPHTHYL-p-PHENYLENEDIAMINE see NBL000

sym-DI-β-NAPHTHYL-p-PHENYLENEDIAMINE see NBL000

N,N'-DI(α-(1-NAPHTHYL)PROPIONYLOXY-2-ETHYL)PIPERAZINE DIHYDROCHLORIDE see NAD000

DINARKON see DLX400

DINATE see DXE600

DINATRIUM-AETHYLENBISDITHIOCARBAMAT (GERMAN) see DXD200

DINATRIUM-(3,6-EPOXY-CYCLOHEXAAN-1,2-DICARBOXYLAAT) (DUTCH) see DXD000

DINATRIUM-(3,6-EPOXY-CYCLOHEXAN-1,2-DICARBOXYLAT) (GERMAN) see DXD000

DINATRIUM-(N,N'-AETHYLEN-BIS(DITHIOCARBAMAT)) (GERMAN) see DXD200

DINATRIUM-(N,N'-ETHYLEEN-BIS(DITHIOCARBAMAAT)) (DUTCH) see DXD200

DINATRIUMPYROPHOSPHAT (GERMAN) see DXF800

DINDEVAN see PFJ750

DINEODYMIUM TRIOXIDE see NCC000

DINEVAL see PFJ750

DINEX see CPK500

DINEZIN see DII200

DINGSABLCH, LEAF EXTRACT see KCA100

DINICKEL TRIOXIDE see NDH500

DINIL see PFA860

DINILE see SNE000

DINIOBIUM PENTAOXIDE see NEA050

DINIOBIUM PENTOXIDE see NEA050

DINITOLMID see DUP300

DINITOLMIDE see DUP300

DINITRAMINE see CNE500

2,4-DINITRANILINE see DUP600

DINITRATE de DIETHYLENE-GLYCOL (FRENCH) see DJE400

1,3-DINITRATO-2,2-BIS(NITRATOMETHYL)PROPANE see PBC250

DINITRATODIOXOURANIUM, HEXAHYDRATE see URS000

2,2'-DINITRATO-N-NITRODI-ETHYLAMINE see NFW000

DINITRILE of ISOPHTHALIC ACID see PHX550

2,3-DINITRILO-1,4-DITHIA-ANTHRAQUINONE see DLK200

2,3-DINITRILO-1,4-DITHIOANTHRACHINON (GERMAN) see DLK200

DINITRO-3 see BRE500

DINITRO see BRE500

DINITROAMINE see CNE500

4,6-DINITRO-2-AMINOPHENOL see DUP400

2,4-DINITROANILIN (GERMAN) see DUP600

2,4-DINITROANILINA (ITALIAN) see DUP600

2,4-DINITROANILINE see DUP600

2,4-DINITROANISOL see DUP800

2,4-DINITROANISOLE see DUP800
α-DINITROANISOLE see DUP800
1,5-DINITRO-9,10-ANTHRACENEDIONE see DUQ000
1,5-DINITROANTHRACHINON (CZECH) see DUQ000
1,5-DINITROANTHRAQUINONE see DUQ000
2,4-DINITROBENZENAMIME see DUP600
1,2-DINITROBENZENE see DUQ400
1,3-DINITROBENZENE see DUQ200
2,4-DINITROBENZENE see DUQ200
DINITROBENZENE, solution (DOT) see DUQ180
m-DINITROBENZENE see DUQ200
o-DINITROBENZENE see DUQ400
p-DINITROBENZENE see DUQ600
DINITROBENZENE see DUQ180
2,4-DINITROBENZENESULFENYL CHLORIDE see DUR200
2,4-DINITROBENZENESULFONIC ACID see DUR400
4,6-DINITROBENZOFURAZAN-N-OXIDE see DUR500
1,3-DINITROBENZOL see DUQ200
DINITROBENZOL, solid (DOT) see DUQ180
5,7-DINITRO-1,2,3-BENZOXADIAZOLE see DUR800
4,4'-DINITROBIFENYL (CZECH) see DUS000
4,4'-DINITROBIPHENYL see DUS000
2,4-DINITRO-6-BROMANILIN (CZECH) see DUS200
2,4-DINITRO-6-BROMOANILINE see DUS200
2,3-DINITRO-2-BUTENE see DUS400
4,6-DINITRO-2-sec.BUTYLFENOL (CZECH) see BRE500
4,6-DINITRO-2-sec.BUTYLFENOLATE AMMONY (CZECH) see BPG250
2,4-DINITRO-6-sec-BUTYLFENYLESTER KYSELINY OCTOVE (CZECH) see ACE500
2,4-DINITRO-6-sek.BUTYL-ISOPROPYLPHENYLCARBONAT (GERMAN) see CBW000
4,6-DINITRO-2-sec-BUTYLPHENOL see BRE500
2,4-DINITRO-6-sec-BUTYLPHENOL see BRE500
2,4-DINITRO-6-tert-BUTYLPHENOL see DRV200
4,6-DINITRO-o-sec-BUTYLPHENOL see BRE500
DINITROBUTYLPHENOL see BRE500
4,6-DINITRO-2-sec-BUTYLPHENOL AMMONIUM SALT see BPG250
4,6-DINITRO-o-sec-BUTYLPHENOL AMMONIUM SALT see BPG250
DINITROBUTYLPHENOL-2,2',2''-NITRILOTRIETHANOL SALT see BRE750
2,4-DINITRO-6-sek.BUTYL-PHENYLACETAT (GERMAN) see ACE500
4,6-DINITRO-2-s-BUTYLPHENYL ACETATE see ACE500
4,6-DINITRO-2-sec-BUTYLPHENYL β,β-DIMETHYLACRYLATE see BGB500
2,4-DINITRO-6-sec-BUTYLPHENYL ISOPROPYL CARBONATE see CBW000
2,4-DINITRO-6-tert-BUTYLPHENYL METHANESULFONATE see DUS500
2,4-DINITRO-6-sec-BUTYLPHENYL-2-METHYLCROTONATE see BGB500
4,6-DINITRO-2-(2-CAPRYL)PHENYL CROTONATE see AQT500
4,6-DINITRO-2-CAPRYLPHENYL CROTONATE see AQT500
2,4-DINITROCHLORBENZEN-6-SULFONAN SODNY (CZECH) see CJC000
2,6-DINITRO-4-CHLOROANILINE see CGL500
2,4-DINITRO-1-CHLOROBENZENE see CGM000
2,4-DINITROCHLOROBENZENE see CGM000
1,3-DINITRO-4-CHLOROBENZENE see CGM000
DINITROCHLOROBENZENE (DOT) see CGL750
DINITROCHLOROBENZENE see CGL750
DINITROCHLOROBENZOL see CGM000
2,4-DINITRO-1-CHLORO-NAPHTHALENE see DUS600
3,5-DINITRO-4-CHLORO-α,α,α-TRIFLUOROTOLUENE see CGM225
4,5-DINITROCHRYSAZIN see DMN400
2,4-DINITRO-o-CRESOL see DUS700
3,5-DINITRO-o-CRESOL see DUT000
4,6-DINITRO-o-CRESOL see DUS700
DINITRO-o-CRESOL see DUS700
2,6-DINITRO-p-CRESOL see DUT600
3,5-DINITRO-p-CRESOL see DUT200
DINITRO-p-CRESOL see DUT600
DINITROCRESOL see DUS700
4,6-DINITRO-o-CRESOL AMMONIUM SALT see DUT800
4,6-DINITRO-o-CRESOL DIETHYLAMINE SALT see DUU000
4,6-DINITRO-o-CRESOL METHYLAMINE (1:1) see DUU200
4,6-DINITRO-o-CRESOL MORPHOLINE (1:1) see DUU400
4,6-DINITRO-o-CRESOLO (ITALIAN) see DUS700
3,5-DINITRO-o-CRESOL SODIUM SALT see DUU600
4,6-DINITRO-o-CRESOL SODIUM SALT see DUU600
DINITRO-o-CRESOL SODIUM SALT see DUU600, SGP550
2,4-DINITRO-6-CYCLOHEXYLPHENOL see CPK500
DINITROCYCLOHEXYLPHENOL (DOT) see CPK500
4,6-DINITRO-o-CYCLOHEXYLPHENOL see CPK500
DINITRO-o-CYCLOHEXYLPHENOL see CPK500
DINITROCYCLOHEXYLPHENOL see CPK500
DINITROCYCLOPENTYLPHENOL see CQB250
DINITRODENDTROXAL see DUS700
N,N'-DINITRO-1,2-DIAMINOETHANE see DUU800
3,3'-DINITRO-4,4'-DIFLUORODIPHENYL SULFONE see DKH250
DINITRODIGLICOL (ITALIAN) see DJE400
DINITRODIGLYKOL (CZECH) see DJE400

2,4-DINITRODIPHENYLAMINE see DUV100
3',5'-DINITRO-4'-(DI-n-PROPYLAMINO)ACETOPHENONE see DUV400
2,6-DINITRO-N,N-DIPROPYL-4-(TRIFLUOROMETHYL)BENZENAMINE see DUV600
2,6-DINITRO-N,N-DI-N-PROPYL-α,α,α-TRIFLURO-p-TOLUIDINE see DUV600
2,2'-DINITRO-5,5'-DITHIODIBENZOESAEURE see DUV700
2,2'-DINITRO-5,5'-DITHIODIBENZOIC ACID see DUV700
1,1-DINITROETHANE see DUV710
1,2-DINITROETHANE see DUV720
1,1-DINITROETHANE (dry) (DOT) see DUV710
N,N'-DINITROETHANEDIAMINE see DUU800
N,N'-DINITROETHYLENEDIAMINE see DUU800
2,5-DINITRO-N-(1-ETHYLPROPYL)-3,4-XYLIDINE see DRN200
2,3-DINITROFENOL see DUY900
2,6-DINITROFENOL see DVA200
3,4-DINITROFENOL see DVA400
2,4-DINITROFENOL (DUTCH) see DUZ000
DINITROFENOLO (ITALIAN) see DUZ000
2,4-DINITROFENYLHYDRAZIN (CZECH) see DVC400
3,7-DINITROFLUORANTHENE see DUW100
3,9-DINITROFLUORANTHENE see DUW120
4,12-DINITROFLUORANTHENE see DUW120
2,7-DINITROFLUORENE see DUW200
2,4-DINITRO-1-FLUOROBENZENE see DUW400
2,4-DINITROFLUOROBENZENE see DUW400
2,2-DINITRO-2-FLUOROETHANOL see FHW000
DINITROGEN MONOXIDE see NGU000
DINITROGEN TETRAFLUORIDE see TCI000
DINITROGEN TETROXIDE, liquefied (DOT) see NGU500
DINITROGEN TETROXIDE see NGU500
DINITROGLICOL (ITALIAN) see EJG000
1,3-DINITROGLYCERIN see GGA200
DINITROGLYCOL see EJG000
2,4-DINITRO-p-HYDROXYDIPHENYLAMINE see DUW500
3,5-DINITRO-2-HYDROXYTOLUENE see DUS700
1,3-DINITRO-2-IMIDAZOLIDINONE see DUW503
1,3-DINITRO-2-IMIDAZOLIDONE see DUW503
2,4-DINITRO-6-ISOBROPYL-m-CRESOL see DVG200
2,6-DINITRO-4-ISOPROPYLPHENOL see IOX000
4,6-DINITRO-o-KRESOL (CZECH) see DUS700
4,6-DINITROKRESOL (DUTCH) see DUS700
4,6-DINITRO-o-KRESYLESTER KYSELINY OCTOVE (CZECH) see AAU250
DINITROL see DUS700
2,4-DINITROMESITYLENE see DUW505
DINITROMETHANE see DUW507
2,6-DINITRO-3-METHOXY-4-tert-BUTYLTOLUENE see BRU500
2,6-DINITRO-4-METHYLANILINE see DVI100
3,5-DINITRO-2-METHYLBENZENEDIAZONIUM-4-OXIDE see DUX509
2,5-DINITRO-3-METHYLBENZOIC ACID see DUX560
2,4-DINITRO-3-METHYL-6-tert-BUTYLPHENYLACETAT (GERMAN) see BRU750
2,4-DINITRO-3-METHYL-6-tert-BUTYLPHENYL ACETATE see BRU750
DINITROMETHYL CYCLOHEXYLTRIENOL see DUS700
N,N'-DINITRO-N-METHYL-1,2-DIAMINOETHANE see DUX600
DINITRO(1-METHYLHEPTYL)PHENYL CROTONATE see AQT500
2,4-DINITRO-6-(1-METHYLHEPTYL)PHENYL CROTONATE see AQT500
2,4-DINITRO-6-METHYLPHENOL see DUS700
2,4-DINITRO-6-METHYLPHENOL SODIUM SALT see DUU600
2,4-DINITRO-6-(1-METHYL-PROPYL)PHENOL (FRENCH) see BRE500
4,6-DINITRO-2-(1-METHYL-N-PROPYL)PHENOL see BRE500
2,4-DINITRO-1-NAFTOL see DUX800
1,5-DINITRONAPHTHALENE see DUX700
2,4-DINITRO-1-NAPHTHOL see DUX800
2-4 DINITRO-α-NAPHTOL see DUX800
2-4 DINITRO-α-NAPHTOL (FRENCH) see DUX800
3,5-DINITRO-N,N'-BIS(2,4,6-TRINITROPHENYL)-2,6-PYRIDINEDIAMINE see PQC525
3,5-DINITRO-N⁴,N⁴-DIPROPYLSULFANILAMIDE see OJY100
2,4-DINITRO-6-(2-OCTYL)PHENYL CROTONATE see AQT500
2,6-DINITRO-4-PERCHLORYLPHENOL see DUY200
2,4-DINITROPHENETOLE see DUY400
2,3-DINITROPHENOL see DUY900
2,4-DINITROPHENOL see DUZ000
2,5-DINITROPHENOL see DVA000
2,6-DINITROPHENOL see DVA200
3,4-DINITROPHENOL see DVA400
3,5-DINITROPHENOL see DVA600
α-DINITROPHENOL see DUZ000
β-DINITROPHENOL see DVA200
γ-DINITROPHENOL see DVA000
DINITROPHENOL see DUY600
2,4-DINITROPHENOL SODIUM SALT see DVA800
DINITROPHENOL SOLUTIONS (UN 1599) (DOT) see DUY600
DINITROPHENOL, wetted with not <15% water, by weight (UN 1320) (DOT) see DUY600

DINITROPHENOL, dry or wetted with <15% water, by weight (UN 0076) (DOT) see DUY600
2,4-DINITROPHENYLACETYL CHLORIDE see DVB200
5-((2,4-DINITROPHENYL)AMINO)-2-(PHENYLAMINO)BENZENESULFONIC ACID MONOSODIUM SALT see SGP500
4,6-DINITROPHENYL-2-sec-BUTYL-3-METHYL-2-BUTENONATE see BGB500
2,4-DINITROPHENYL-DIMETHYL-DITHIOCARBAMATE see DVB850
2,4-DINITROPHENYL-2,4-DINITRO-6-sec-BUTYLPHENYL CARBONATE see DVC200
2,4-DINITROPHENYL ETHER of MORPHINE see DVC800
2,4-DINITROPHENYLHYDRAZINE see DVC400
2,4-DINITROPHENYLHYDRAZINIUMPERCHLORATE see DVC600
o-(2,4-DINITROPHENYL)HYDROXYLAMINE see DVC700
DINITROPHENYLMETHANE see DVG600
2,4-DINITROPHENYLMETHYL ETHER see DUP800
2,4-DINITROPHENYLMORPHINE HYDROCHLORIDE see DVC800
DI-p-NITROPHENYL PHOSPHATE see BLA600
2,4-DINITROPHENYL THIOCYANATE see DVF800
2,2-DINITRO-1,3-PROPANEDIOL see DVD000
2,2-DINITRO-1-PROPANOL see DVD200
2,2-DINITROPROPANOL see DVD200
1,3-DINITROPYRENE see DVD400
1,6-DINITROPYRENE see DVD600
1,8-DINITROPYRENE see DVD800
DINITROPYRENE see DVD400, DVD600, DVD800
4,6-DINITROQUINOLINE-1-OXIDE see DVE000
4,7-DINITROQUINOLINE-1-OXIDE see DVE200
2,4-DINITRORESORCINOL (heavy metal salts of) (dry) (DOT) see DVF300
2,4-DINITRO-RHODANBENZOL (GERMAN) see DVF800
3,5-DINITROSALICYLIC ACID see HKE600
1,4-DINITROSOBENZENE see DVE260
p-DINITROSOBENZENE see DVE260
1,4-DINITROSOBENZENE HOMOPOLYMER see PJR500
p-DINITROSOBENZENE POLYMERS see PJR500
DINITROSO-2,5-DIMETHYLPIPERAZINE see DRN800
1,4-DINITROSO-2,6-DIMETHYLPIPERAZINE see DRO000
N,N'-DINITROSO-2,6-DIMETHYLPIPERAZINE see DRO000
DINITROSO-2,6-DIMETHYLPIPERAZINE see DRO000
DINITROSODIMETHYLPROPANEDIAMINE see DRO200
DINITROSOHOMOPIPERAZINE see DVE600
N,4-DINITROSO-N-METHYLANILINE see MJG750
N,N'-DINITROSO-N,N'-DIETHYLETHYLENEDIAMINE see DJB800
N,N'-DINITROSO-N,N'-DIMETHYLETHYLENEDIAMINE see DVE400
N,N'-DINITROSO-N,N'-DIMETHYLOXAMID (GERMAN) see DRN600
N,N'-DINITROSO-N,N'-DIMETHYL-1,3-PROPANEDIAMINE see DRO200
3,4-DI-N-NITROSOPENTAMETHYLENETETRAMINE see DVF400
3,7-DI-N-NITROSOPENTAMETHYLENETETRAMINE see DVF400
N,N-DINITROSOPENTAMETHYLENETETRAMINE see DVF400
DINITROSOPENTAMETHYLENETETRAMINE see DVF400
DI(N-NITROSO)-PERHYDROPYRIMIDINE see DVF000
N',N³-DI(NITROSO)PENTAMETHYLENETETRAMINE see DVF400
DINITROSOPIPERAZIN (GERMAN) see DVF200
1,4-DINITROSOPIPERAZINE see DVF200
DINITROSOPIPERAZINE see DVF200
N,N'-DINITROSOPIPERAZINE see DVF200
DINITROSORBIDE see CCK125
2,4-DINITROSO-m-RESORCINOL see DVF300
3,7-DINITROSO-1,3,5,7-TETRAAZABICYCLO[3.3.1]NONANE see DVF400
4,4'-DINITRO-2,2'-STILBENEDISULFONIC ACID see DVF600
DINITROSTILBENEDISULFONIC ACID see DVF600
2,4-DINITROTHIOCYANATOBENZENE see DVF800
2,4-DINITRO-1-THIOCYANOBENZENE see DVF800
2,4-DINITROTHIOCYANOBENZENE see DVF800
2,4-DINITROTHIOPHENE see DVG000
2,6-DINITROTHYMOL see DVG200
DINITROTHYMOL 1-2-4 (FRENCH) see DVG200
3,5-DINITRO-o-TOLUAMIDE see DUP300
2,3-DINITROTOLUENE see DVG800
2,4-DINITROTOLUENE see DVH000
2,5-DINITROTOLUENE see DVH200
2,6-DINITROTOLUENE see DVH400
3,4-DINITROTOLUENE see DVH600
3,5-DINITROTOLUENE see DVH800
DINITROTOLUENE see DVG600
DINITROTOLUENES, molten (DOT) see PGN100
DINITROTOLUENES, liquid or solid (DOT) see DVG600
3,5-DINITRO-o-TOLUIDINE see AJR750
2,6-DINITRO-p-TOLUIDINE see DVI100
2,4-DINITROTOLUOL see DVH000
4,6-DINITRO-1,2,3-TRICHLOROBENZENE see DVI600
2,6-DINITRO-4-TRIFLUORMETHYL-N,N-DIPROPYLANILIN (GERMAN) see DUV600
2.4'-DINITRO-4-TRIFLUOROMETHYL-DIPHENYL ETHER see NIX000
2,4-DINITRO-1,3,5-TRIMETHYLBENZENE (DOT) see DUW505
sym-DINITROXYDIETHYLNITRAMINE see NFW000

α-α'-DI-(NITROXY)METHYL ETHER (DOT) see OPQ200
DINKUM OIL see EQQ000
DINOBUTON see CBW000
DINOC see DUS700, DUU600
DINOCTON-6 see DVI800
DINOCTON-O see DVI800
DINOFEN see CBW000
DINOLEINE see DNF600
DINONYL-1,2-BENZENEDICARBOXYLATE see DVJ000
DI-n-NONYL PHTHALATE see DVJ000
DINOPOL 235 see OEU000
DINOPOL NOP see DVL600
DINOPROST see POC500
DINOPROST METHYL ESTER see DVJ100
DINOPROSTONE see DVJ200
DINOPROST TROMETHAMINE (USDA) see POC750
18,19-DINOR-17-α-PREGN-4-EN-3-ONE, 13-ETHYL-17-HYDROXY- see NNE600
DINOSEB see BRE500
DINOSEB-ACETATE see ACE500
DINOSEB (AMINE) see BPG250
DINOSEBE (FRENCH) see BRE500
DINOSEBE ACETATE see ACE500
DINOSEB METHACRYLATE see BGB500
DINOSOL see SNN300
DINOTERB see DRV200
DINOTERB ACETATE see DVJ400
DINOVEX see DAL600
DINOXOL see DAA800, TAA100
DINOZOL 50 see DUT800
DINOZOL see DUT800
DINTOIN see DKQ000
DINTOINA see DNU000
DINULCID see OLM300
DINURANIA see DUS700
DINYL see PFA860
DIOCID see DVJ500
DIOCIDE see DVJ500
DIOCTADECYL 3,3'-THIODIPROPIONATE see DXG700
DIOCTADECYL THIODIPROPIONATE see DXG700
DIOCTANOL-2-PHTHALATE see BLB750
DIOCTANOYL PEROXIDE see CBF705
DIOCTLYN see DJL000
DIOCTYL ADIPATE see AEO000
DIOCTYLAL see DJL000
DIOCTYLAMINE see DVJ600
DIOCTYL AZELATE see BJQ500
DIOCTYL-o-BENZENEDICARBOXYLATE see DVL600
DIOCTYLBIS(LAUROYLOXY)STANNANE see DVJ800
DIOCTYLBIS(NONYLOXYMALEOYLOXY)STANNANE see DEI800
DIOCTYLDIDODECANOYLOXYSTANNANE see DVJ800
DIOCTYLDI(LAUROYLOXY)STANNANE see DVJ800
2,2-DIOCTYL-1,3-DIOXA-2-STANNA-7-THIADECAN-4,10-DIONE see DVN909
2,2-DIOCTYL-1,3,2-DIOXASTANNEPIN-4,7-DIONE see DVK200
4,4'-DIOCTYLDIPHENYLAMINE see DVK400
DIOCTYL ESTER of SODIUM SULFOSUCCINATE see DJL000
DIOCTYL ESTER of SODIUM SULFOSUCCINIC ACID see DJL000
DIOCTYL ETHER see OEY000
DIOCTYL(ETHYLENEDIOXYBIS(CARBONYLMETHYLTHIO))STANNANE see DVN400
DIOCTYL FUMARATE see DVK600
DIOCTYLISOPENTYLPHOSPHINE OXIDE see DVK709
DIOCTYL ISOPHTHALATE see BJQ750
DI-n-OCTYL KETONE see OFE020
DIOCTYL KETONE see OFE020
DI-N-OCTYL MALEATE see DVK800
"DIOCTYL" MALEATE see BJR000
DIOCTYL MALEATE see DVK800
DIOCTYL-MEDO FORTE see DJL000
N,N-DIOCTYL-1-OCTANAMINE see DVL000
2,2-DIOCTYL-1,3,2-OXATHIASTANNOLANE-5-OXIDE see DVL200
DIOCTYLOXOSTANNANE see DVL400
n-DIOCTYL PHTHALATE see DVL600
DI-sec-OCTYL PHTHALATE see DVL700
DIOCTYL PHTHALATE see DVL600, DVL700
DIOCTYL(1,2-PROPYLENEDIOXYBIS(MALEOYLDIOXY))STANNANE see DVL800
DIOCTYL SEBACATE see BJS250
DIOCTYL SODIUM SULFOSUCCINATE (FCC) see DJL000
DI-n-OCTYL SODIUM SULFOSUCCINATE see SOD300
DIOCTYLSTANNIUM DICHLORIDE see DVN300
(Z,Z)-4,4'-((DIOCTYLSTANNYLENE)BIS(OXY))BIS(4-OXO-2-BUTANOIC ACID DI-ISOOCTYL ESTER see BKL000
DIOCTYLSTANNYLENE MALEATE see DVK200
DIOCTYL SULFOSUCCINATE SODIUM SALT see DJL000
DIOCTYLTHIOACETOXYSTANNANE see DVL200
DIOCTYLTHIOXOSTANNANE see DVM000

DI-N-OCTYLTIN BIS(BUTYL MALEATE) see BHK500
DI-n-OCTYLTIN BIS(BUTYL MERCAPTOACETATE) see DVM200
DI-n-OCTYLTIN BIS(DODECYL MERCAPTIDE) see DVM400
DI-n-OCTYLTIN BIS(2-ETHYLHEXYL MALEATE) see DVM600
DI-n-OCTYLTIN BIS(2-ETHYLHEXYL) MERCAPTOACE-TATE see DVM800
DIOCTYLTINBIS(ISOOCTYL MALEATE) see BKL000
DIOCTYLTIN-S,S'-BIS(ISOOCTYL MERCAPTOACETATE) see BKK750
DIOCTYLTIN BIS(ISOOCTYL MERCAPTOACETATE) see BKK750
DIOCTYLTIN BIS(ISOOCTYL THIOGLYCOLATE) see BKK750
DIOCTYL-TIN BIS(ISOOCTYLTHIOGLYCOLLATE) see BKK750
DI-n-OCTYLTIN BIS(LAURYLTHIOGLYCOLATE) see DVN000
DI-n-OCTYLTIN-1,4-BUTANEDIOL-BIS-MERCAPT-OACETATE see DVN200
DI-n-OCTYLTINDICHLORIDE see DVN300
DIOCTYLTIN DICHLORIDE see DVN300
DI-n-OCTYLTIN DIISOOCTYL THIOGLYCOLATE see BKK750
DI-n-OCTYLTIN DILAURATE see DVJ800
DIOCTYLTIN DILAURATE see DVJ800
DI-n-OCTYLTIN DIMONOBUTYLMALEATE see BHK500
DI-n-OCTYLTIN DI(1,2-PROPYLENEGLYCOLMALEATE) see DVL800
DI-n-OCTYLTIN ETHYLENEGLYCOL DITHIOGLYCOLATE see DVN400
DI-N-OCTYLTIN-2-ETHYLHEXYLDIMERCAPTOETHANOATE see DVM800
DI-N-OCTYLTIN-ITHIOGLYCOLIC ACID 2-ETHYLHEXYL ESTER see DVM800
DI-n-OCTYLTIN MALEATE see DVK200
DIOCTYLTIN MALEATE see DVK200
DI-n-OCTYLTIN MERCAPTIDE see DVN600
DIOCTYLTIN MERCAPTIDE see DVN600
DI-n-OCTYLTIN β-MERCAPTOPROPIONATE see DVN800
DIOCTYLTIN-β-MERCAPTOPROPIONATE see DVN800
DI-n-OCTYLTIN OXIDE see DVL400
DIOCTYLTIN OXIDE see DVL400
DI-n-OCTYLTIN SULFIDE see DVM000
DIOCTYLTIN-3,3'-THIODIPROPIONATE see DVN909
DI-n-OCTYLTIN THIOGLYCOLATE see DVL200
DIOCTYLTIN THIOGLYCOLATE see DVL200
DI-n-OCTYL-ZINN AETHYLENGLYKOL-DITHIOGLYKOLAT (GERMAN) see DVN400
DI-n-OCTYL-ZINN-BIS(2-AETHYLHEXYLMALEINAT) (GERMAN) see DVM600
DI-n-OCTYL-ZINN-BIS(LAURYL-THIOGLYKOLAT) (GERMAN) see DVN000
DI-n-OCTYL-ZINN-1,4-BUTANDIOL-BIS-MERCAPTOACETAT (GERMAN) see DVN200
DI-n-OCTYL-ZINN DICHLORID see DVN300
DI-n-OCTYL-ZINN-DI-ISOOCTYLTHIOGLYKOLAT (GERMAN) see BKK750
DI-n-OCTYL-ZINN DILAURAT (GERMAN) see DVJ800
DI-N-OCTYLZINN-DIMONOBUTYLMALEINAT (GERMAN) see BHK500
DI-n-OCTYLZINN-DIMONOMETHYLMALEINAT (GERMAN) see BKO500
DI-n-OCTYL-ZINN-DI-(1,2-PROPYLENGLYKOLMALEINAT)(GERMAN) see DVL800
DI-n-OCTYLZINN MALEINAT see DVK200
DI-n-OCTYL-ZINN β-MERCAPTOPROPIONAT (GERMAN) see DVN800
DI-n-OCTYL-ZINN OXYD (GERMAN) see DVL400
DI-n-OCTYL-ZINN THIOGLYKOLAT (GERMAN) see DVL200
DIOCYDE see DVO000
DIODON see DNG400
DIODONE see DNG400
DIODOQUIN see DNF600
DIODOXYLIN see DNF600
DIODRAST see DNG400
DIOFORM see DFI210
DIOGYN see EDO000
DIOGYN B see EDP000
DIOGYNETS see EDO000
DIOKAN see DVQ000
DIOKSAN (POLISH) see DVQ000
DIOKSYNY (POLISH) see TAI000
DIOKTYLESTER SULFOJANTARANU SODNEHO see SOD300
DIOLAMINE see DHF000
DIOLANDRONE see AOO475
DIOLANE see HFP875
DIOLENE see IPU000
DIOL-EPOXIDE-1 see DMO500
DIOL-EPOXIDE 2 see DMO800
anti-DIOLEPOXIDE see DVO175
DIOLICE see CNU750
DIOLOSTENE see AOO475
DIOMEDICONE see DJL000
DI-ON see DXQ500
DIONE 21-ACETATE see RKP000
2-4-DIONE-1,3-DIAZASPIRO(4.5)DECANE see DVO600
DIONIN see ENK000, ENK500
DIONINE see ENK000
DIONINE HYDROCHLORIDE see DVO700
DIONIN HYDROCHLORIDE see DVO700
DIONONE see DMH400
DIONONYL PHTHALATE see DVJ000
DIOPAL see SBH500
DIOPHYLLIN see TEP500

DIORTHOTOLYLGUANIDINE see DXP200
DIOSPYROL see DVO809
DIOSPYROS VIRGINIANA see PCP500
1,4-DIOSSAN-2,3-DIYL-BIS(O,O-DIETIL-DITIOFOSFATO) (ITALIAN) see DVQ709
DIOSSANO-1,4 (ITALIAN) see DVQ000
1,4-DIOSSIBENZENE (ITALIAN) see QQS200
2,4-DIOSSI-5-DIAZOPIRIMIDINA (ITALIAN) see DCQ600
DIOSSIDONE see BRF500
DIOSUCCIN see DJL000
DIOTHANE see DVO819
DIOTHANE HYDROCHLORIDE see DVV500
DIOTHENE see PJS750
DIOTILAN see DJL000
DIOVAC see DJL000
DIOVOCYCLIN see EDR000
DIOVOCYLIN see EDR000
DIOXAAN-1,4 (DUTCH) see DVQ000
1,4-DIOXAAN-2,3-DIYL-BIS(O,O-DIETHYL-DITHIOFOSFAAT) (DUTCH) see DVQ709
5-(1,4-DIOXA-8-AZASPIRO(4.5)DEC-8-YLMETHYL)-3-ETHYL-6,7-DIHYDRO-2-METHYL-INDOL-4(5H)-ONE see AFH550
DIOXABENZOFOS see MEC250
2,3-DIOXABICYCLO(2.2.2)OCT-5-ENE, 1-ISOPROPYL-4-METHYL- see ARM500
DIOXACARB see DVS000
1,6-DIOXACYCLOHEPTADECAN-17-ONE see OKW110
1,7-DIOXACYCLOHEPTADECAN-17-ONE see OKW100
1,8-DIOXACYCLOHEPTADECAN-9-ONE see OLE100
1,3-DIOXACYCLOHEXANE see DVP600
1,3-DIOXACYCLOPENTANE see DVR800
6,9-DIOXA-3,12-DIAZATETRADECANEDIOIC ACID, 3,12-BIS(CARBOXYMETHYL)-(9CI) see EIT000
(R*,S*)-3,14-DIOXA-2,15-DITHIA-6,11-DIAZEHEXADECANE-8,9-DIOL, 2,2,15,15-TETRAOXIDE (9CI) see LJD500
3,6-DIOXADODECANOL-1 see HFN000
1-DIOXADROL HYDROCHLORIDE see LFG100
d-DIOXADROL HYDROCHLORIDE see DVP400
2,5-DIOXAHEXANE see DOE600
p-DIOXAN (CZECH) see DVQ000
DIOXAN-1,4 (GERMAN) see DVQ000
m-DIOXAN see DVP600
2,3-p-DIOXANDITHIOL S,S-BIS(O,O-DIETHYL PHOSPHORODITHIOATE) see DVQ709
1,4-DIOXAN-2,3-DIYL-BIS(O,O-DIAETHYL-DITHIOPHOSPHAT) (GERMAN) see DVQ709
1,4-DIOXAN-2,3-DIYL-BIS(O,O-DIETHYLPHOSPHOROTHIOLOTHIONATE) see DVQ709
1,4-DIOXAN-2,3-DIYL-O,O,O',O'-TETRAETHYL DI(PHOSPHORODITHIOATE) see DVQ709
1,3-DIOXANE see DVP600
1,4-DIOXANE (MAK) see DVQ000
p-DIOXANE see DVQ000
DIOXANE see DVQ000
2,3-p-DIOXANE-S,S-BIS(O,O-DIETHYLPHOSPHOROITHIOATE) see DVQ709
1,3-DIOXANE, 2-BUTYL-4,4,6-TRIMETHYL- see BSR600
m-DIOXANE-4,4-DIMETHYL see DVQ400
cis-2,3-p-DIOXANEDITHIOL-S,S-BIS(O,O-DIETHYLPHOSPHORODITHIOATE) see DVQ600
p-DIOXANE-2,3-DITHIOL-S,S-DIESTER with O,O-DIETHYL PHOSPHORODITH-IOATE see DVQ709
p-DIOXANE-2,3-DIYL ETHYL PHOSPHORODITHIOATE see DVQ709
1,3-DIOXANE, 2-(PHENYLMETHYL)-4,4,6-TRIMETHYL- see PFR400
DIOXANNE (FRENCH) see DVQ000
3,6-DIOXAOCTANE-1,8-DIOL see TJQ000
2,7-DIOXAPYRENE-1,3,6,8-TETRONE see NAP300
1,3-DIOXA-2-SILACYCLOPENTANE, 4-(CHLOROMETHYL)-2,2-DIMETHYL- see CIL775
2,4-DIOXASPIRO(5.5)UNDEC-8-ENE, 3-(3-CYCLOHEXENYL)- see CPD650
1,3,2-DIOXASTANNEPIN-4,7-DIONE, 2,2-DIBUTYL- see DEJ100
4,10-DIOXATETRACYCLO(5.4.0$^{3,5}$.0$^{1,7}$.0$^{9,11}$)UNDECANE see BFY750
4,9-DIOXATETRACYCLO(5.4.0.0$^{3,5}$.0$^{8,10}$)UNDECANE see BFY750
1,3,2-DIOXATHIANE-2,2-DIOXIDE (9CI) see TLR750
1,3,2-DIOXATHIOLANE-2-OXIDE (9CI) see COV750
DIOXATHION see DVQ709
DIOXATRINE see BBU800
cis-1,4-DIOXENEDIOXETANE see DVQ759
1,4-DI-N-OXIDE 2,3-BIS(OXYMETHYL)QUINOXLINE see DVQ800
1,4-DI-N-OXIDE of DIHYDROXYMETHYLQUINOXALINE see DVQ800
2,2-DIOXIDE-1,3,2-DIOXATHIOLANE see EJP000
4,4-DIOXIDE-1,4-OXATHIANE see DVR000
1,4-DIOXIDE-2,3-QUINOXALINEDIMETHANOL see DVQ800
1,1-DIOXIDETETRAHYDROTHIOFURAN see SNW500
1,1-DIOXIDETETRAHYDROTHIOPHENE see SNW500
DIOXIDE of VITAVAX see DLV200
DIOXIDIN see DVQ800
DIOXIDINE see DVQ800

DIOXIME-p-BENZOQUINONE see DVR200
DIOXIME-2,5-CYCLOHEXADIENE-1,4-DIONE see DVR200
DIOXIME-1,4-CYCLOHEXADIENEDIONE see DVR200
DIOXIN (herbicide contaminant) see TAI000
DIOXINE see TAI000
DIOXIN (bactericide) (OBS.) see ABC250
DIOXITOL see CBR000
DIOXOAMINOPYRINE see DVU100
9,10-DIOXOANTHRACENE see APK250
p-DIOXOBENZENE see HIH000
2,2'-((1,4-DIOXO-1,4-BUTANEDIYL)BIS(OXY)BIS(N,N,N-TRIMETHYLETHANAMINI-UM) DICHLORIDE see HLC500
2,2'-((1,4-DIOXO-1,4-BUTANEDIYL)BIS(OXY))BIS(N,N,N-TRIMETHYLETHANAMINI-UM) see CMG250
1,1'-((1,4-DIOXO-1,4-BUTANEDIYL)BIS(OXY-2,1-ETHANEDIYL))BIS-QUINOLINIUM DIIODIDE see KGK400
1,1',1''-(3,6-DIOXO-1,4-CYCLOHEXADIENE-1,2,4-TRIYL)TRISAZIRIDINE see TND000
2,6-DIOXO-5-DIAZOPYRIMIDINE see DCQ600
DIOXODICHLOROCHROMIUM see CML125
3,3'-(DIOXODIGERMOXANYLENE) DIPROPANOIC ACID see CCF125
DIOXO-9,9-(DIMETHYLAMINO-3-METHYL-2-PROPYL)-10-PHENOTHIAZINE (FRENCH) see AFL750
5,5-DIOXO-10-(2-(DIMETHYLAMINO)PROPYL)PHENOTHIAZINE HYDROCHLO-RIDE see DVT400
3,5-DIOXO-1,2-DIPHENYL-4-N-BUTYLPYRAZOLIDENE see BRF500
3,5-DIOXO-1,2-DIPHENYL-4-N-BUTYLPYRAZOLIDIN SODIUM see BOV750
2,2'-DIOXO-DI-N-PROPYLNITROSAMINE see NJN000
DI-OXO-DI-N-PROPYLNITROSAMINE see NJN000
2,4-DIOXO-5-FLUORO-N-HEXYL-3,4-DIHYDRO-1(2H)-PYRIMIDINECARBOXAM-IDME see CCK630
2,4-DIOXO-5-FLUORO-N-HEXYL-1,2,3,4-TETRAHYDRO-1-PYRIMIDINECARBOX-AMIDE see CCK630
2,3-DIOXOINDOLINE see ICR000
1,3-DIOXOLAN see DVR800
DIOXOLAN see DVR600
1,4-DIOXOLAN-2,5-DIYLDIMETHYLENEBIS(NITROMERCURY) see BKZ000
1,3-DIOXOLANE see DVR800
DIOXOLANE (DOT) see DVR600
1,3-DIOXOLANE-2-ACETIC ACID, 2-METHYL-, ETHYL ESTER see EFR100
1,3-DIOXOLANE, 4-(CHLOROMETHYL)-2-(o-NITROPHENYL)- see CIQ400
1,3-DIOXOLANE, 2-(1,2-DICHLOROETHYL)-4-METHYL- see DFJ500
1,3-DIOXOLANE, 2-(2,6-DIMETHYL-1,5-HEPTADIENYL)- see CMS324
1,3-DIOXOLANE-4-METHANOL see DVR909
1,3-DIOXOLANE, 4-METHYL-2-PENTYL- see MNM450
2-(1,3-DIOXOLANE-2-YL)PHENYL N-METHYLCARBAMATE see DVS000
1,3-DIOXOLAN-2-ONE see GHM000
1,3-DIOXOLAN-2-ONE, 4-(CHLOROMETHYL)- see CBW400
1,3-DIOXOLAN-2-ONE, 4-ETHYL- see BOT200
α-((1,3-DIOXOLAN-2-YLMETHOXY)IMINO)BENZENEACETONITRILE see OKS200
7-(1,3-DIOXOLAN-2-YLMETHYL)THEOPHYLLINE see TEQ175
((1,3-DIOXOLAN-4-YL)METHYL)TRIMETHYLAMMONIUM IODIDE see FMX000
2-(1,3-DIOXOLAN-2-YL)PHENYL-N-METHYLCARBAMAT see DVS000
o-(1,3-DIOXOLAN-2-YL)PHENYL METHYLCARBAMATE see DVS000
1,3-DIOXOL-4-EN-2-ONE see VOK000
DIOXOLONE-2 see GHM000
1,3-DIOXOL-2-ONE see VOK000
(1,3)DIOXOLO(4,5-j)PYRROLO(3,2,1-de)PHENANTHRIDINIUM, 4,5-DIHYDRO-2-HYDROXY-(9CI) see LJB800
2,6-DIOXO-4-METHYL-4-ETHYLPIPERIDINE see MKA250
3-(2-(1,3-DIOXO-2-METHYLINDANYL))GLUTARIMIDE see DVS100
DIOXONE see DJT400
2,2'-DIOXO-N-NITROSODIPROPYLAMINE see NJN000
3,5-DIOXO-1-PHENYL-2-(p-HYDROXYPHENYL)-4-N-BUTYLPYRAZOLIDENE see HNI500
3-(2-(1,3-DIOXO-2-PHENYLINDANYL))GLUTARIMIDE see DVS300
3-(2-(1,3-DIOXO-2-PHENYL-4,5,6,7-TETRAHYDRO-4,7-DITHIAINDA-NYL))GLUTARIMIDE see DVS400
1,3-DIOXOPHTHALAN see PHW750
1,3-DIOXO-5-PHTHALANCARBOXYLIC ACID see TKV000
2,6-DIOXO-3-PHTHALIMIDOPIPERIDINE see TEH500
2-(2,6-DIOXOPIPERIDEN-3-YL) PHTHALIMIDINE see DVS600
2-(2-6-DIOXO-3-PIPERIDINYL)1H-ISOINDOLE-1,3(2H)-DIONE see TEH500
(s)-2-(2,6-DIOXO-3-PIPERIDINYL)-1H-ISOINDOLE-1,3(2H)-DIONE see TEH520
N-(2,6-DIOXO-3-PIPERIDYL)PHTHALIMIDE see TEH500
DIOXOPROMETHAZINE HYDROCHLORIDE see DVT400
N,N-DI(2-OXOPROPYL)NITROSAMINE see NJN000
2,2'-DIOXOPROPYL-N-PROPYLNITROSAMINE see NJN000
2,6-DIOXOPURINE see XCA000
1,3-DIOXO-2-(3-PYRIDYLMETHYLENE)INDAN see DVT459
2,4-DIOXOPYRIMIDINE see UNJ800
2,4-DIOXOTETRAHYDROQUINAZOLINE see QEJ800
3,5-DIOXO-2,3,4,5-TETRAHYDRO-1,2,4-TRIAZINE RIBOSIDE see RJA000
2,4-DIOXOTHIAZOLIDINE see TEV500
1,1-DIOXOTHIOLAN see SNW500

DIOXOTHIOLAN see SNW500
DIOXYAMINOPYRINE see DVU100
DIOXYANTHRANOL see APH250
1,4-DIOXYANTHRAQUINONE (RUSSIAN) see DMH000
2,6-DIOXY-8-AZAPURINE see THT350
1,4-DIOXYBENZENE see QQS200
m-DIOXYBENZENE see REA000
o-DIOXYBENZENE see CCP850
p-DIOXYBENZENE see HIH000
3,3'-DIOXYBENZIDINE see DMI400
1,4-DIOXY-BENZOL (GERMAN) see QQS200
DIOXYBIS(2,2'-DI-tert-BUTYLBUTANE see DVT500
DIOXYBUTADIENE see BGA750
3-β,14-DIOXY-CARDEN-(20:22)-OLID (GERMAN) see DMJ000
3-α,7-β-DIOXYCHOLANIC ACID see DMJ200
DIOXYDE de BARYUM (FRENCH) see BAO250
DIOXYDEMETON-S-METHYL see DAP600
3,3'-(DIOXYDICARBONYL)DIPROPIONIC ACID see SNC500
2,4-DIOXY-3,3-DIETHYL-5-METHYLPIPERIDINE see DNW400
3-β,14-DIOXY-DIGEN-(20:22)-OLID (GERMAN) see DMJ000
DIOXYDINE see DVQ800
4,4'-DIOXYDIPHENYLSULFIDE see TFJ000
1,4-DIOXYETHYLAMINO-5,8-DIOXYANTHRAQUINONE see DMM400
N,N-DIOXYETHYLANILINE see BKD500
DIOXYETHYLENE ETHER see DVQ000
DIOXYGENYL TETRAFLUOROBORATE see DVT800
DIOXYMETHYLENE-PROTOCATECHUIC ALDEHYDE see PIW250
1,3-DIOXY-2-NICOTINSAEURENITRIL-TETRAZOL (GERMAN) see DND900
1,4-DI-p-OXYPHENYL-2,3-DI-ISONITRILO-1,3-BUTADIENE see DVU000
DI(p-OXYPHENYL)-2,4-HEXADIENE see DAL600
2-(3,4-DIOXYPHENYL)TETRAHYDRO-1,4-OXAZIN (GERMAN) see MRU100
7-(2,3-DIOXYPROPYL)THEOPHYLLINE see DNC000
DIOXYPYRAMIDON see DVU100
DIOZOL see BRF500
DIP see DNS200
DIPA see DNL600, DNM200, DNM400
DIPAC see DNN900
DIPALLADIUM TRIOXIDE see DVU200
DIPAM see DCK759
DIPAN see DVX200
DIPANE see WAK000
DIPANOL see MCC250
DIPAR see PDF250
DIPARAANISYL-MONOPHENETHYL-GUANIDIN-HYDROCHLORID (GERMAN) see PDN500
DI-PARALEN see CFF500
DIPARALENE see CFF500
DI-PARALENE-2-HYDROCHLORIDE see CDR000
DIPARALENE HYDROCHLORIDE see CDR250
DIPARCOL see DHF600, DII200
DIPAV see PAH250
DIPAXIN see DVV600
DIPEGYL see NCR000
DIPEL see BAC040
DIPELARGONYL PEROXIDE see NNA100
DIPENICILLINA-G-ALLUMINIO-SULFAMETOSSIPIRIDAZINA (ITALIAN) see DVU300
DIPENICILLIN-G-ALUMINIUM-SULPHAMETHOXYPYRIDAZINE see DVU300
DIPENINBROMID (GERMAN) see DGW600
DIPENTAMETHYLENETHIURAM TETRASULFIDE see TEF750
DI(PENTANOYLOXY)DIBUTYLSTANNANE see DEA600
DIPENTENE see MCC250
DIPENTENE DIOXIDE see LFV000
DIPENTYLAMINE see DCH200
DIPENTYL ETHER see PBX000
2,5-DI-tert-PENTYLHYDROQUINONE see DCH400
DIPENTYL KETONE see ULA000
DIPENTYL LEAD DIACETATE see DVU600
DIPENTYL MALEATE see MAL000
DI-n-PENTYLNITROSAMINE see DCH600
DIPENTYLNITROSAMINE see DCH600
DIPENTYLOXOSTANNANE see DVV000
DIPENTYL PHENOL see DCH800
2,4-DI-tert-PENTYLPHENOL see DCI000
2-(2,4-DI-tert-PENTYLPHENOXY)BUTYRIC ACID see DVV109
DIPENTYLTIN DICHLORIDE see DVV200
DIPENTYLTIN OXIDE see DVV000
DIPEPTIDE SWEETENER see ARN825
2,6-DIPERCHLORYL-4,4'-DIPHENOQUINONE see DVV400
DIPERDON HYDROCHLORIDE see DVV500
DIPERFLUOROBUTYL ETHER see PCG755
DIPERODON HYDROCHLORIDE see DVV500
DIPEROXYTEREPHTHALIC ACID see DVV550
DIPHACIL see DHX800
DIPHACIN see DVV600

DIPHACINONE see DVV600
DIPHACYL see DHX800
DIPHANTINE see BBV500
DIPHANTOIN see DKQ000
DIPHANTOINE SODIUM see DNU000
M-DIPHAR see MAS500
DIPHEBUZOL see BRF500
DIPHEDAL see DKQ000
DIPHEDAN see DNU000
DIPHEMANIL see DAP800
DIPHEMANIL METHYLSULFATE see DAP800
DIPHENACIN see DVV600
DIPHENADIONE see DVV600
DIPHENALDEHYDE see DVV800
DIPHENAMID see DRP800
DIPHENAMIDE see DRP800
DIPHENAMIZOLE see PAM500
DIPHENAN (pharmaceutical) see PFR325
DIPHENAN see PFR325
DIPHENANE see PFR325
DIPHENATE see DNU000
DIPHENATIL see DAP800
DIPHENATRILE see DVX200
DIPHENAZINE DIHYDROCHLORIDE see BLF250
DIPHENCHLOXAZINE HYDROCHLORIDE see DVW000
DIPHENETHYLAMINE, o,o′-DIMETHOXY-α-α′-DIMETHYL-, compounded with LACTIC ACID see BKP200
DIPHENHYDRINATE see DYE600
DIPHENIDOL see DWK200
DIPHENIN see DNU000
DIPHENINE see DKQ000
DIPHENINE SODIUM see DNU000
DIPHENMANIL METHYLSULFATE see DAP800
DIPHENMETHANIL see DAP800
DIPHENMETHANIL METHYLSULFATE see DAP800
o-DIPHENOL see CCP850
1,5-DIPHENOXYANTHRAQUINONE see DVW100
N,N′-DI(3-PHENOXY-2-HYDROXYPROPYL)ETHYLENEDIAMINE DIHYDROCHLORIDE see IAG700
DIPHENOXYLATE HYDROCHLORIDE see LIB000
1,3-DIPHENOXY-2-PROPANOL see DVW600
DIPHENPYRALINE HYDROCHLORIDE see DVW700
DIPHENTHANE 70 see MJM500
DIPHENTOIN see DKQ000, DNU000
DIPHENYL (OSHA) see BGE000
DIPHENYLACETAMIDE see PDX500
N,N′-DIPHENYLACETAMIDINE see DVW750
α,α-DIPHENYLACETIC ACID see DVW800
DIPHENYLACETIC ACID see DVW800
DIPHENYLACETIC ACID 2-(BIS(2-HYDROXYETHYL)AMINO)ETHYL ESTER HYDROCHLORIDE see DVW900
DIPHENYLACETIC ACID, 2-(DIETHYLAMINO)ETHYL ESTER see DHX800, THK000
DIPHENYLACETIC ACID DIETHYLAMINOETHYL ESTER see DHX800, THK000
DIPHENYLACETIC ACID, ESTER with 3-QUINUCLIDINOL, HYDROGEN SULFATE (2:1) DIHYDRATE see QVJ000
DIPHENYLACETIC ACID-1-ETHYL-3-PIPERIDYL ESTER HYDROCHLORIDE see EOY000
1,1-DIPHENYL ACETONE see MJH910
DIPHENYLACETONITRILE see DVX200
2,2-DIPHENYLACETOPHENONE see TMS100
ω,ω-DIPHENYLACETOPHENONE see TMS100
DIPHENYLACETOPHENONE see TMS100
6-DIPHENYLACETOXY-1-AZABICYCLO(3.2.1)OCTANE HYDROCHLORIDE see ARX500
DIPHENYLACETYLDIETHYLAMINOETHANOL see DHX800, THK000
2-DIPHENYLACETYL-1,3-DIKETOHYDRINDENE see DVV600
2-(DIPHENYLACETYL)INDAN-1,3-DIONE see DVV600
2-DIPHENYLACETYL-1,3-INDANDIONE see DVV600
2-(DIPHENYLACETYL)-1H-INDENE-1,3(2H)-DIONE see DVV600
2,3-DIPHENYLACRYLONITRILE see DVX600
α,β-DIPHENYLACRYLONITRILE see DVX600
3-(2,2-DIPHENYLAETHYL)-5-(2-PIPERIDINOAETHYL)-1,2,4-OXADIAZOL (GERMAN) see LFJ000
DIPHENYLAMIDE see DRP800
DIPHENYLAMIN-(β-DIAETHYLAMINOAETHYL)CARBAMIDTHIOESTER (GERMAN) see PDC875
N,N-DIPHENYLAMINE see DVX800
DIPHENYLAMINE see DVX800
DIPHENYLAMINE, 4-AMINO see PFU500
DIPHENYLAMINE, p-AMINO- see PFU500
DIPHENYLAMINE-2-CARBOXYLIC ACID see PEG500
DIPHENYLAMINECHLORARSINE see PDB000
DIPHENYLAMINECHLOROARSINE (DOT) see PDB000
DIPHENYLAMINE, 2,4-DINITRO- see DUV100

DIPHENYLAMINE, HEXANITRO- see HET500
DIPHENYLAMINE HYDROGEN SULFATE see DVY000
DIPHENYLAMINE, 4-(PHENYLAZO)- see PEI800
DIPHENYLAMINE SULFATE see DVY000
DIPHENYLAN see DKQ000
N,N-DIPHENYLANILINE see TMQ500
DIPHENYLAN SODIUM see DNU000
DIPHENYLARSINOUS ACID see DVY100
DIPHENYLARSINOUS CHLORIDE see CGN000
1,2-DIPHENYLBENZENE see TBC640
1,4-DIPHENYLBENZENE see TBC750
m-DIPHENYLBENZENE see TBC620
p-DIPHENYLBENZENE see TBC750
DIPHENYLBENZENE see TBD000
DIPHENYLBIS(PHENYLTHIO)STANNANE see DVY800
DIPHENYLBIS(PHENYLTHIO)TIN see DVY800
DIPHENYL BLACK see PFU500
DIPHENYL BLUE 2B see CMO000
DIPHENYL BLUE 3B see CMO250
DIPHENYL BLUE BLACK GHS see CMN800
DIPHENYL BLUE BLACK MBH see CMN800
DIPHENYL BLUE G see CMO600
DIPHENYL BRILLIANT BLUE see CMO500
DIPHENYL BROWN BS see CMO820
DIPHENYL BROWN 3GT see CMO825
DIPHENYL BROWN PT see CMO810
DIPHENYL BROWN TB see CMO820
DIPHENYLBUTAZONE see BRF500
1,3-DIPHENYL-2-BUTEN-1-ONE see MPL000
(Z)-2-(p-(1,2-DIPHENYL-1-BUTENYL)-PHENOXY)-N,N-DIMETHYLETHYLAMINE see NOA600
1,2-DIPHENYL-4-BUTYL-3,5-DIKETOPYRAZOLIDINE CALCIUM SALT see PEO750
1,2-DIPHENYL-4-BUTYL-3,5-DIOXOPYRAZOLIDINE see BRF500
1,1-DIPHENYL-2-BUTYNYL-N-CYCLOHEXYLCARBAMATE see DVY900
DIPHENYLCARBAMIC ACID-2-(DIETHYLAMINO)ETHYL ESTER HYDROCHLORIDE see DHY000
DIPHENYLCARBAMOTHIOIC ACID-S-(2-(DIETHYLAMINO)ETHYL) ESTER HYDROCHLORIDE see PDC875
DIPHENYL CARBINOL see HKF300
DIPHENYL CARBONATE see DVZ000
4-DIPHENYLCARBOXYLIC ACID see PEL600
DIPHENYLCHLOORARSINE (DUTCH) see CGN000
DIPHENYLCHLOROARSINE (DOT) see CGN000
DIPHENYL-(2-CHLOROPHENYL)-1-IMIDAZOYLMETHANE see MRX500
DIPHENYLCHROMIUM see BGY700
DIPHENYLCHROMIUM(III) IODIDE see BGY720
α,β-DIPHENYLCINNAMONITRILE see TMQ250
DIPHENYL CRESOL PHOSPHATE see TGY750
DIPHENYL CRESYL PHOSPHATE see TGY750
DIPHENYL-α-CYANOMETHANE see DVX200
DIPHENYL DARK GREEN B see CMO830
DIPHENYL DARK GREEN BN see CMO830
DIPHENYL DEEP BLACK G see AQP000
2,4′-DIPHENYLDIAMINE see BGF109
1,2-DIPHENYLDIAZENE see ASL250
DIPHENYLDIAZENE see ASL250
1,1′-DIPHENYLDIAZOMETHANE see DVZ100
DIPHENYLDIBROMOTIN see DDN200
DIPHENYL DICHLOROSILANE (DOT) see DFF000
DIPHENYLDICHLORO TIN DIPYRIDINE complex see DWA400
DIPHENYLDIIMIDE see ASL250
2,2-DIPHENYL-4-DIISOPROPYLAMINOBUTYRAMIDE METHIODIDE see DAB875
DIPHENYL-α,β-DIKETONE see BBY750
DIPHENYLDIKETOPYRAZOLIDINE see DWA500
2,2-DIPHENYL-N,N-DIMETHYLACETAMIDE see DRP800
1,2-DIPHENYL-1-(DIMETHYLAMINO)ETHANE see DWA600
1,1-DIPHENYL-1-(DIMETHYLAMINOISOPROPYL)BUTANONE-2 see IKZ000
1,3-DIPHENYL-5-(2-DIMETHYLAMINOPROPIONAMIDO)PYRAZOLE see PAM500
1,1-DIPHENYL-1-(β-DIMETHYLAMINOPROPYL)BUTANONE-2 HYDROCHLORIDE see MDP000
α,α-DIPHENYL-γ-DIMETHYLAMINOVALERAMIDE see DOY400
α,α-DIPHENYL-γ-DIMETHYLAMINOVALERAMIDE HYDROCHLORIDE see ALR250
3,3-DIPHENYL-3-DIMETHYLCARBAMOYL-1-PROPYNE see DWA700
DIPHENYLDIOXOBUTYLPYRAZOLIDINE-BUTAZOLIDINE-SODIUM see BOV750
d-2-(2,2-DIPHENYL-1,3-DIOXOLAN-4-YL)PIPERIDINE HYDROCHLORIDE see DVP400
1-2-(2,2-DIPHENYL-1,3-DIOXOLAN-4-YL)PIPERIDINE HYDROCHLORIDE see LFG100
1,2-DIPHENYL-3,5-DIOXOPYRAZOLIDIN (GERMAN) see DWA500
1,4-DIPHENYL-3,5-DIOXO-PYRAZOLIDIN (GERMAN) see DWL600
DIPHENYL mixed with DIPHENYL OXIDE see PFA860
DIPHENYL DISULFIDE see PEW250
DIPHENYLE CHLORE, 42% de CHLORE (FRENCH) see PJM500

DIPHENYLE CHLORE, 54% de CHLORE (FRENCH) see PJN000
4,4'-DIPHENYLENEDIAMINE see BBX000
DIPHENYLENE DIOXIDE see DDA800
DIPHENYLENEIMINE see CBN000
DIPHENYLENEMETHANE see FDI100
DIPHENYLENE SULFIDE see TES300
DIPHENYLENIMIDE see CBN000
DIPHENYLENIMINE see CBN000
1,3-DIPHENYL-1,3-EPIDIOXY-1,3-DIHYDROISOBENZOFURAN see DWA800
1,2-DIPHENYLETHANE see BFX500
1,2-DIPHENYLETHANEDIONE see BBY750
DIPHENYLETHANOLAMINE HYDROCHLORIDE see DWB000
1,2-DIPHENYLETHANONE see PEB000
trans-1,2-DIPHENYLETHENE see SLR100
trans-DIPHENYLETHENE see SLR100
DIPHENYL ETHER see PFA850
2,2-DIPHENYL-2-ETHOXYACETIC ACID (2-(DIMETHYLAMINO)ETHYL) ESTER HYDROCHLORIDE see DPE200
1,2-DIPHENYLETHYLAMINE HYDROCHLORIDE see DWB300
1,2-DIPHENYLETHYLENE see SLR000
(E)-1,2-DIPHENYLETHYLENE see SLR100
α-β-DIPHENYLETHYLENE see SLR000
trans-1,2-DIPHENYLETHYLENE see SLR100
trans-α-β-DIPHENYLETHYLENE see SLR100
N,N'-DIPHENYLETHYLENEDIAMINE see DWB400
DIPHENYL-2-ETHYLHEXYL PHOSPHATE see DWB800
1-(2-(3-(2,2-DIPHENYLETHYL)-1,2,4-OXADIAZOL-5-YL)ETHYL)PIPERIDINE MONO-HYDROCHLORIDE see LFJ000
4,4-DIPHENYL-1-ETHYLPIPERIDINE MALEATE see DWB875
N-(1,2-DIPHENYLETHYL)-2-(PYRROLIDINYL)ACETAMIDE HYDROCHLORIDE see DWC000
3,3-DIPHENYL-2-ETHYL-1-PYRROLINE see DWC050
3,3-DIPHENYL-3-(ETHYLSULFONYL)-N,N,1-TRIMETHYLPROPYLAMINE HYDRO-CHLORIDE see DWC100
DIPHENYL FAST 5BL SUPRA I RED see CMO885
DIPHENYL FAST BROWN BRL see CMO750
DIPHENYL FAST BROWN F see CMO820
DIPHENYL FAST RED B see CMO870
DIPHENYL FAST RED 5BL see CMO885
DIPHENYL FAST RED 5BLN see CMO885
DIPHENYL FAST RED F see CMO870
α-α-DIPHENYLGLYCOLIC ACID see BBY990
DIPHENYLGLYCOLIC ACID see BBY990
DIPHENYLGLYCOLIC ACID 2-(DIETHYLAMINO)ETHYL ESTER see DHU900
DIPHENYLGLYCOLLIC ACID-2-(DIETHYLAMINO)ETHYL ESTER HYDROCHLO-RIDE see BCA000
DIPHENYLGLYOXAL see BBY750
DIPHENYLGLYOXAL PEROXIDE see BDS000
DIPHENYL GREEN BB see CMO840
DIPHENYL GREEN BY see CMO840
DIPHENYL GREEN C see CMO840
DIPHENYL GREEN GPD see CMO840
DIPHENYL GREEN KG see CMO840
DIPHENYL GREEN MB see CMO840
1,3-DIPHENYLGUANIDINE see DWC600
N,N'-DIPHENYLGUANIDINE see DWC600
DIPHENYLGUANIDINE see DWC600
sym-DIPHENYLGUANIDINE HYDROCHLORIDE see DWC625
2,2-DIPHENYL-4-N-HEXAMETHYLENIMINOBUTYRAMIDE METHIODIDE see BTA325
5,5-DIPHENYLHYDANTOIN see DKQ000
DIPHENYLHYDANTOIN see DKQ000
DIPHENYLHYDANTOINE (FRENCH) see DKQ000
DIPHENYLHYDANTOIN and PHENOBARBITAL see DWD000
5,5-DIPHENYLHYDANTOIN SODIUM see DNU000
DIPHENYLHYDANTOIN SODIUM see DNU000
DIPHENYLHYDRAMINE see BBV500
DIPHENYLHYDRAMINE HYDROCHLORIDE see BAU750
1,2-DIPHENYLHYDRAZINE see HHG000
sym-DIPHENYLHYDRAZINE see HHG000
DIPHENYLHYDROXYACETIC ACID see BBY990
DIPHENYLHYDROXYARSINE see DVY100
3,3-DIPHENYL-3-HYDROXYPROPIONIC ACID DIETHYLAMINOETHYL ESTER HYDROCHLORIDE see DWD200
2,2-DIPHENYL-3-HYDROXYPROPIONIC ACID LACTONE see DWI400
5,5-DIPHENYLIMIDAZOLIDIN-2,4-DIONE see DKQ000
5,5-DIPHENYL-2,4-IMIDAZOLIDINEDIONE see DKQ000
5,5-DIPHENYL-2,4-IMIDAZOLIDINE-DIONE, MONOSODIUM SALT see DNU000
DIPHENYLINE see BGF109
3,4-DIPHENYL-1-ISOBUTYLPYRAZOLE-5-ACETIC ACID SODIUM SALT see DWD400
4,4-DIPHENYL-N-ISOPROPYLCYCLOHEXYLAMINE HYDROCHLORIDE see SDZ000
1,2-DIPHENYL-4-(γ-KETOBUTYL)-3,5-PYRAZOLIDINEDIONE see KGK000
DIPHENYL KETONE see BCS250

DIPHENYL KETOXIME see BCS400
8,10-DIPHENYL LOBELIONOL see LHY000
DIPHENYLMERCURY see DWD800
DIPHENYLMETHAN-4,4'-DIISOCYANAT (GERMAN) see MJP400
4,4'-DIPHENYLMETHANEBISMALEIMIDE see BKL800
DIPHENYLMETHANEBISMALEIMIDE see BKL800
2,4'-DIPHENYLMETHANEDIAMINE see MJP750
4,4'-DIPHENYLMETHANEDIAMINE see MJQ000
4,4'-DIPHENYLMETHANE DIISOCYANATE see MJP400
DIPHENYLMETHANE 4,4'-DIISOCYANATE (DOT) see MJP400
p,p'-DIPHENYLMETHANE DIISOCYANATE see MJP400
DIPHENYL METHANE DIISOCYANATE see MJP400
DIPHENYLMETHANE-4,4'-DIISOCYANATE-TRIMELLIC ANHYDRIDE-ETHOMID HT POLYMER see TKV000
4,4'-DIPHENYLMETHANEDIMALEIMIDE see BKL800
DIPHENYLMETHANOL see HKF300
DIPHENYLMETHANONE see BCS250
DIPHENYLMETHANONE OXIME see BCS400
2-(DIPHENYLMETHOXY)ACETAMIDOXIME HYDROGEN MALEATE see DWE200
2-(DIPHENYLMETHOXY)-N,N-DIMETHYL-ETHANAMINE HYDROCHLORIDE see BAU750
2-(DIPHENYLMETHOXY)-N,N-DIMETHYLETHYLAMINE see BBV500
2-DIPHENYLMETHOXY-N,N-DIMETHYLETHYLAMINE HYDROCHLORIDE see BAU750
endo-3-(DIPHENYLMETHOXY)-8-ETHYL-8-AZABICYCLO(3.2.1)OCTANE see DWE800
3-α-(DIPHENYLMETHOXY)-8-ETHYLNORTROPANE see DWE800
3-(DIPHENYLMETHOXY)-8-ETHYLNORTROPANE see DWE800
4-(DIPHENYLMETHOXY)-1-METHYLPIPERIDINE see LJR000
4-(DIPHENYLMETHOXY)-1-METHYLPIPERIDINE CHLOROTHEOPHYLLINE see DWF000
3-α-(DIPHENYLMETHOXY)-1-α-H,5-α-H-TROPANE see BDI000
3-DIPHENYLMETHOXYTROPANE MESYLATE see TNU000
3-DIPHENYLMETHOXYTROPANE METHANESULFONATE see TNU000
DIPHENYLMETHYL ALCOHOL see HKF300
DIPHENYLMETHYL BROMIDE (DOT) see BNG750
DIPHENYL METHYL BROMIDE, solution (DOT) see BNH000
DIPHENYLMETHYLCYANIDE see DVX200
α-(DIPHENYLMETHYLENE)BENZENEACETIC ACID see TMQ250
2-DIPHENYLMETHYLENEBUTYLAMINE HYDROCHLORIDE see DWF200
3-(DIPHENYLMETHYLENE)-1,1-DIETHYL-2-METHYLPYRROLIDINIUM BROMIDE see PAB750
4-(DIPHENYLMETHYLENE)-1,1-DIMETHYLPIPERIDINIUM METHYLSULFATE see DAP800
3-(DIPHENYLMETHYLENE)-1-ETHYL-2-METHYLPYRROLIDINE ETHYL BROMIDE see PAB750
(DIPHENYLMETHYLENE)HYDROXYLAMINE see BCS400
1-(DIPHENYLMETHYL)-4-(2-(2-HYDROXYETHOXY)ETHYL)PIPERAZINE see BBV750
1-DIPHENYLMETHYL-4-METHYLPIPERAZINE see EAN600
(±)-1-DIPHENYLMETHYL-4-METHYLPIPERAZINE HYDROCHLORIDE see MAX275
(+)-2,2-DIPHENYL-3-METHYL-4-MORPHOLINOBUTYRYLPYRROLIDINE see AFJ400
2-(3,3-DIPHENYL-3-(5-METHYL-1,3,4-OXADIAZOL-2-YL)PROPYL)-2-AZABICY-CLO(2.2.2)OCTANE see DWF700
2-(2-((4-(DIPHENYLMETHYL)-1-PIPERAZINYL)ETHOXY)ETHANOL HYDROXYDIE-THYLPHENAMINE see BBV750
1-(3-(4-(DIPHENYLMETHYL)-1-PIPERAZINYL)PROPYL)-2-BENZIMIDAZOLINONE see OMG000
1-(3-(4-(DIPHENYLMETHYL)-1-PIPERAZINYL)PROPYL)-1,3-DIHYDRO-2H-BENZIMI-DAZOL-2-ONE see OMG000
4-DIPHENYLMETHYLPIPERIDINE see DWF865
4,4-DIPHENYL-1-METHYLPIPERIDINE MALEATE see DWF869
5,5-DIPHENYL-3-(3-(2-METHYLPIPERIDINO)PROPYL)-2-THIOHYDANTOIN HY-DROCHLORIDE see DWF875
5,5-DIPHENYL-3-(3-(2-METHYLPIPERIDINO)PROPYL)-2-THIOHYDANTOIN MONO-HYDROCHLORIDE see DWF875
α-2,2-DIPHENYL-4-(1-METHYL-2-PIPERIDYL)-1,3-DIOXOLANE HYDROCHLORIDE see DWG600
β-2,2-DIPHENYL-4-(1-METHYL-2-PIPERIDYL)-1,3-DIOXOLANE HYDROCHLORIDE see DWH000
α-2,2-DIPHENYL-4-(1-METHYL-2-PIPERIDYL)-1,3-DIOXOLANE METHYLIODIDE see DWH200
3,3-DIPHENYL-2-METHYL-1-PYRROLINE see DWH500
4-DIPHENYLMETHYLTROPYLTROPINIUM BROMIDE see PEM750
4-DIPHENYLMETHYL-dl-TROPYLTROPINIUM BROMIDE see PEM750
4,4-DIPHENYL-6-MORPHOLINO-3-HEPTANONE HYDROCHLORIDE see DWH600
4,4-DIPHENYL-6-MORPHOLINO-3-HEXANONE HYDROCHLORIDE see DWH800
2,3-DIPHENYL-3H-NAPHTHO(1,2-d)TRIAZOLIUM CHLORIDE see DWH875
DIPHENYLNITROSAMIN (GERMAN) see DWI000
N,N-DIPHENYLNITROSAMINE see DWI000
DIPHENYLNITROSAMINE see DWI000

DIPHENYL N-NITROSOAMINE see DWI000
o-DIPHENYLOL see BGJ250
2,2-DI(4-PHENYLOL)PROPANE see BLD500
2,5-DIPHENYLOXAZOLE see DWI200
4,5-DIPHENYL-2-OXAZOLEPROPANOIC ACID see OLW600
4,5-DIPHENYL-2-OXAZOLEPROPIONIC ACID see OLW600
N-(4,5-DIPHENYLOXAZOL-2-YL)DIETHANOLAMINE MONOHYDRATE see
    BKB250
2,2'-((4,5-DIPHENYL-2-OXAZOLYL)IMINO)-DIETHANOLMONOHYDRATE see
    BKB250
3,3-DIPHENYL-2-OXETANONE see DWI400
DIPHENYL OXIDE see PFA850
DIPHENYLOXIDE-4,4'-DISULFOHYDRAZIDE (DOT) see OPE000
DIPHENYL-4-γ-OXO-γ-BUTRIC ACID see BGL250
1,2-DIPHENYL-4-(3'-OXOBUTYL)-3,5-DIOXOPYRAZOLIDINE see KGK000
DIPHENYLOXOSTANNANE, POLYMER see DWO600
DIPHENYL PENTACHLORIDE see PAV600
1,5-DIPHENYL-1,4-PENTAZDIENE see DWI800
N,N'-DIPHENYL-p-PHENYLENEDIAMINE see BLE500
DIPHENYL-p-PHENYLENEDIAMINE see BLE500
1,2-DIPHENYL-4-(2'-PHENYLSULFINETHYL)-3,5-PYRAZOLIDINEDIONE see
    DWM000
5,5-DIPHENYL-1-PHENYLSULFONYLHYDANTOIN see DWJ300
5,5-DIPHENYL-1-(PHENYLSULFONYL)-2,4-IMIDAZOLIDINEDIONE (9CI) see
    DWJ300
1,2-DIPHENYL-4-PHENYLTHIOETHYL-3,5-PYRAZOLIDINEDIONE see DWJ400
α,α-DIPHENYL-1-PIPERIDINEBUTANOL see DWK200
α,α-DIPHENYL-1-PIPERIDINEBUTANOL HYDROCHLORIDE see CCR875
α,α-DIPHENYL-2-PIPERIDINEMETHANOL see DWK400
α,α-DIPHENYL-2-PIPERIDINEMETHANOL HYDROCHLORIDE see PII750
α,α-DIPHENYL-1-PIPERIDINEMETHANOL HYDROCHLORIDE see PIY750
α,α-DIPHENYL-1-PIPERIDINEPROPANOL HYDROCHLORIDE see PMC250
α,α-DIPHENYL-1-PIPERIDINEPROPANOL METHANESULFONATE (ester) see
    PMC275
DIPHENYL-PIPERIDINO-AETHYL-ACETAMID-BROMMETHYLAT (GERMAN) see
    RDA375
1,1-DIPHENYL-3-N-PIPERIDINOBUTANOL-1 HYDROCHLORIDE see ARN700
2,2-DIPHENYL-4-(4-PIPERIDINO-4-CARBAMOYLPIPERIDINO)BUTYRONITRILE see
    PJA140
1,1-DIPHENYL-3-PIPERIDINO-1-PROPANOL HYDROCHLORIDE see PMC250
5,5-DIPHENYL-3-(3-PIPERIDINOPROPYL)-2-THIOHYDANTOIN HYDROCHLORIDE
    see DWK500
1-2,2-DIPHENYL-4-(2-PIPERIDYL)-1,3-DIOXOLANE HYDROCHLORIDE see
    LFG100
d-2,2-DIPHENYL-4-(2-PIPERIDYL)-1,3-DIOXOLANE HYDROCHLORIDE see
    DVP400
1,1-DIPHENYL-3-(1-PIPERIDYL)-1-PROPANOL HYDROCHLORIDE see PMC250
1,3-DIPHENYL-1-PROPEN-3-ONE see CDH000
α,α-DIPHENYL-β-PROPIOLACTONE see DWI400
1,2-DIPHENYL-2-PROPIONOXY-3-METHYL-4-DIMETHYLAMINOBUTANE see
    PNA250
(+)-1,2-DIPHENYL-2-PROPIONOXY-3-METHYL-4-DIMETHYLAMINOBUTANE HY-
    DROCHLORIDE see PNA500
3-(3,3-DIPHENYLPROPYLAMINO)PROPYL-3',4',5'-TRIMETHOXYBENZOATE HY-
    DROCHLORIDE see DWK700
3-(N-d,d-DIPHENYLPROPYL-N-METHYL)AMINOPROPAN-1-OL HYDROCHLORIDE
    see DWK900
N-(3,3-DIPHENYLPROPYL)-α-METHYLPHENETHYLAMINE see PEV750
N-(3,3-DIPHENYLPROPYL)-α-METHYLPHENETHYLAMINE LACTATE see
    DWL200
1,1-DIPHENYL-2-PROPYN-1-OL CYCLOHEXANECARBAMATE see DWL400
1,1-DIPHENYL-2-PROPYNYL-N-CYCLOHEXYLCARBAMATE see DWL400
1,1-DIPHENYL-2-PROPYNYL ESTER CYCLOHEXANECARBAMIC ACID see
    DWL400
1,1-DIPHENYL-2-PROPYNYL-N-ETHYLCARBAMATE see DWL500
1,1-DIPHENYL-2-PROPYNYL 1-PYRROLIDINECARBOXYLATE see DWL525
DIPHENYLPYRALIN-8-CHLOR-THEOPHYLLINAT (GERMAN) see PIZ250
DIPHENYLPYRALINE see LJR000
DIPHENYLPYRALINE HYDROCHLORIDE see DVW700
DIPHENYLPYRALINE TEOCLATE see PIZ250
1,2-DIPHENYL-3,5-PYRAZOLIDINEDIONE see DWA500
1,4-DIPHENYL-3,5-PYRAZOLIDINEDIONE see DWL600
1,3-DIPHENYL-5-PYRAZOLONE see DWL800
DIPHENYLPYRAZONE see DWM000
DIPHENYLPYRILENE see LJR000
3,3-DIPHENYL-3-(PYRROLIDINE-CARBONYLOXY)-1-PROPYNE see DWL525
α,α-DIPHENYL-3-QUINUCLIDINEMETHANOL HYDROCHLORIDE see DWM400
DIPHENYL RED 8B see CMO880
DIPHENYL RED B see CMO870
DIPHENYL RED 3BS see CMO875
DIPHENYL SCARLET 3BS see CMO875
DIPHENYL SELENIDE see PGH250
DIPHENYLSELENONE see DWM600
DIPHENYL SKY BLUE 6B see CMO500
DIPHENYLSTANNANE see DWM800

DIPHENYL SULFIDE see PGI500
DIPHENYL SULFONE see PGI750
DIPHENYL SULFOXIDE see DWN000
1-((DIPHENYLSULFOXIMIDO)METHYL)-1-METHYLPYRROLIDINIUM BROMIDE
    see HAK200
DIPHENYL SULPHONE see PGI750
3,3',4,4'-DIPHENYLTETRAMINE see BGK500
DIPHENYLTHIOCARBAMIC ACID-S-(2-(DIETHYLAMINO)ETHYL ESTER HYDRO-
    CHLORIDE see PDC875
DIPHENYLTHIOCARBAMIC ACID-S-(2-(DIETHYLAMINO)ETHYL) ESTER see
    PDC850
N,N'-DIPHENYLTHIOCARBAMIDE see DWN800
sym-DIPHENYLTHIOCARBAMIDE see DWN800
DIPHENYL THIOCARBAZIDE see DWN400
DIPHENYLTHIOCARBAZONE see DWN200
1,5-DIPHENYL-3-THIOCARBOHYDRAZIDE see DWN400
DIPHENYL THIOETHER see PGI500
5,5-DIPHENYL-2-THIOHYDANTOIN see DWN600
DIPHENYLTHIOLACETIC ACID-2-DIETHYLAMINOETHYL ESTER HYDROCHLO-
    RIDE see DHY400
1,3-DIPHENYLTHIOUREA see DWN800
1,1-DIPHENYL-2-THIOUREA see DWO000
1,3-DIPHENYL-2-THIOUREA see DWN800
DIPHENYLTHIOUREA see DWN800
N,N'-DIPHENYLTHIOUREA see DWN800
sym-DIPHENYLTHIOUREA see DWN800
DIPHENYLTIN see DWM800
DIPHENYLTIN DIBROMIDE see DDN200
DIPHENYLTIN DICHLORIDE see DWO400
DIPHENYLTIN DIHYDRIDE see DWM800
DIPHENYLTIN OXIDE POLYMER see DWO600
DIPHENYL TOLYL PHOSPHATE see TGY750
1,3-DIPHENYLTRIAZENE see DWO800
5,6-DIPHENYL-as-TRIAZIN-3-OL see DWO875
3,5-DIPHENYL-1H-1,2,4-TRIAZOLE see DWO950
3,5-DIPHENYL-1,2,4-TRIAZOLE see DWO950
3,5-DIPHENYL-s-TRIAZOLE see DWO950
2,2-DIPHENYL-1,1,1-TRICHLOROETHANE see DWP000
DIPHENYLTRICHLOROETHANE see DAD200
1,3-DIPHENYLUREA see CBM250
N,N'-DIPHENYLUREA see CBM250
sym-DIPHENYLUREA see CBM250
DIPHER see EIR000
DIPHERGAN see PMI750
DI-PHETINE see DKQ000, DNU000
DIPHEXAMIDE METHIODIDE see BTA325
DIPHONE see SOA500
DIPHOSGEN see TIR920
DIPHOSGENE see PGX000, TIR920
DIPHOSPHANE see DWP229
1,2-DIPHOSPHINOETHANE see DWP250
DIPHOSPHOPYRIDINE NUCLEOTIDE see CNF390
DIPHOSPHORIC ACID, DISODIUM SALT see DXF800
DIPHOSPHORIC ACID TETRAETHYL ESTER see TCF250
DIPHOSPHORIC ACID, TETRAMETHYL ESTER see TDV000
DIPHOSPHORUS PENTOXIDE see PHS250
DIPHOSPHORUS TRIOXIDE see PHT500
DIPHTHERIA TOXIN see DWP300
DIPHYL see PFA860
DIPHYLLIN see DNC000
DIPHYLLIN-3,4-o-DIMETHYL XYLOSIDE see CMV375
DIPICRYLAMINE (DOT) see HET500
DIPICRYLAMINE see HET500
DIPICRYLOXAMIDE see HET700
DIPIDOLOR see PJA140
DIPIGYL see NCR000
DIPIKRYLAMIN see HET500
DIPIN see BJC250
DIPINE see BJC250
DIPIPERAL see FHG000
2-β,16-β-DIPIPERIDINO-5-α-ANDROSTAN-3-α,17-β-DIOL DIPIVALATE HYDR see
    DWP500
2,2',2'',2'''-(4,8-DIPIPERIDINOPYRIMIDO(5,4-d)PYRIMIDINE-2,6-DIYLDINITRI-
    LO)TETRAETHANOL see PCP250
DIPIPERON see FHG000
DIPIPERONE see FHG000
DIPIRARTRIL-TROPICO see DUD800
DIPIRIN see DOT000
DIPIRITRAMIDE see PJA140
DIPIVEFRINE HYDROCHLORIDE see DWP559
DIPIVEFRIN HYDROCHLORIDE see DWP559
DIPLIN see DMZ000
DIPLOSAL see SAN000
DIPN see DNB200
DIPOFENE see DCM750

DIPO-SAFT see BFC750
DIPOTASSIUM 4,4'-BIS(4-PHENYL-1,2,3-TRIAZOL-2-YL)STILBENE-2,2'-DISULFO-
NATE see BLG100
DIPOTASSIUM 4,4'-BIS(4-PHENYL-1,2,3-TRIAZOL-2-YL)STILBENE-2,2'-SULFO-
NATE see BLG100
DIPOTASSIUM CHLORAZEPATE see CDQ250
DIPOTASSIUM CHROMATE see PLB250
DIPOTASSIUM CLORAZEPATE see CDQ250
DIPOTASSIUM CYCLOOCTATETRAENE see DWP900
DIPOTASSIUM DIAZIRINE-3,3-DICARBOXYLATE see DWP950
DIPOTASSIUM DICHLORIDE see PLA500
DIPOTASSIUM DICHROMATE see PKX250
DIPOTASSIUM-meso-N,N-DISULFO-3,4-DIPHENYLHEXANE see SPA650
DIPOTASSIUM ETHYLENEDIAMINETETRAACETATE see EEB100, EJA250
DIPOTASSIUM MONOCHROMATE see PLB250
DIPOTASSIUM MONOPHOSPHATE see PLQ400
DIPOTASSIUM NICKEL TETRACYANIDE see NDI000
DIPOTASSIUM OXALATE see PLN300
DIPOTASSIUM PERSULFATE see DWQ000
DIPOTASSIUM PHOSPHATE see PLQ400
DIPOTASSIUM SELENITE see SBO100
DIPOTASSIUM TETRACHLOPALLADATE see PLN750
DIPOTASSIUM TETRACYANONICKELATE see NDI000
DIPOTASSIUM TRICHLORONITROPLATINATE see PLN050
DIPPEL'S OIL see BMA750
DIPPING ACID see SOI500
DIPRAM see DGI000
DIPRAMID see DAB875
DIPRAMIDE see DAB875
DIPRAZINE see DQA400
DIPRIVAN see DNR800
DIPROFILLIN see DNC000
DIPROFILLINE see DNC000
DIPRON see SNM500
DIPROPANEDIOL DIBENZOATE see DWS800
DIPROPANOIC ACID GERMANIUM SESQUIOXIDE see CCF125
DIPROPARGYL ETHER see POA500
DI-2-PROPENYLAMINE see DBI600
DI-2-PROPENYL ESTER, 1,2-BENZENEDICARBOXYLIC ACID see DBL200
DI-2-PROPENYL ISOPHTHALATE see IMK000
DI-2-PROPENYL PHOSPHONITE see DBL000
N-N-DI-2-PROPENYL-2-PROPEN-1-AMINE see THN000
5,5-DI-2-PROPENYL-2,4,6(1H,3H,5H)-PYRIMIDINETRIONE (9CI) see AFS500
DIPROPETRYN see EPN500
DIPROPETRYNE see EPN500
DIPROPHYLLIN see DNC000
DIPROPHYLLINE see DNC000
DIPROPIONATE BECLOMETHASONE see AFJ625
DIPROPIONATE d'OESTRADIOL (FRENCH) see EDR000
DIPROPIONATO de ESTILBENE (SPANISH) see DKB000
p,p'-DIPROPIONOXY-trans-α,β-DIETHYLSTILBENE see DKB000
DIPROPIONYL PEROXIDE, >28% in solution (DOT) see DWQ800
DIPROPIONYL PEROXIDE see DWQ800
1,1-DIPROPOXYETHANE see AAG850
4,5-DIPROPOXY-2-IMIDAZOLIDINONE see DWQ850
DIPROPYL ACETAL see AAG850
DIPROPYLACETAMIDE see PNX600
DIPROPYLACETATE SODIUM see PNX750
N-DIPROPYLACETIC ACID see PNR750
DIPROPYLACETIC ACID see PNR750
DIPROPYLACETIC ACID CALCIUM SALT see CAY675
DI-n-PROPYL ADIPATE see DWQ875
DIPROPYL ADIPATE see DWQ875
DI-n-PROPYLAMINE see DWR000
n-DIPROPYLAMINE see DWR000
DIPROPYLAMINE see DWR000
1-DIPROPYLAMINOACETYLINDOLINE see DWR200
4-(DI-N-PROPYLAMINO)-3,5-DINITRO-1-TRIFLUOROMETHYLBENZENE see
DUV600
4-((DIPROPYLAMINO)SULFONYL)BENZOIC ACID see DWW000
DIPROPYLCARBAMIC ACID ETHYL ESTER see DWT400
DIPROPYLCARBAMOTHIOIC ACID-S-ETHYL ESTER see EIN500
DIPROPYLDIAZENE 1-OXIDE see ASP500
DIPROPYL-2,2'-DIHYDROXYAMINE see DNL600
9,10-DI-n-PROPYL-9-10-DIHYDROXY-9,10-DIHYDRO-1,2,5,6-DIBENZANTHRA-
CENE see DLK600
N,N-DI-N-PROPYL-2,6-DINITRO-4-TRIFLUOROMETHYLANILINE see DUV600
N³,N³-DIPROPYL-2,4-DINITRO-6-TRIFLUOROMETHYL-m-PHENYLENEDIAMINE
see DWS200
DI-n-PROPYL DISELENIDE see PNI850
α,α'-DIPROPYLENEDINITRILODI-o-CRESOL see DWS400
DIPROPYLENE GLYCOL see DWS500, OQM000
DIPROPYLENE GLYCOL BUTYL ETHER see DWS600
DIPROPYLENE GLYCOL DIACRYLATE see DWS650
DIPROPYLENE GLYCOL DIBENZOATE see DWS800

DIPROPYLENE GLYCOL DIPELARGONATE see DWT000
DIPROPYLENE GLYCOL METHYL ETHER see DWT200
DIPROPYLENE GLYCOL MONOMETHYL ETHER see DWT200
DIPROPYLENE GLYCOL, 3,3,5-TRIMETHYLCYCLOHEXYL ETHER see TLO600
DIPROPYLENETRIAMINE see AIX250
DI-n-PROPYLESSIGSAURE (GERMAN) see PNR750
DIPROPYL ETHER see PNM000
N,N-DI-n-PROPYL ETHYL CARBAMATE see DWT400
1-(N,N-DIPROPYLGLYCYL)INDOLINE see DWR200
DI-N-PROPYL-ISOCINCHOMERONATE (GERMAN) see EAU500
DIPROPYL ISOCINCHOMERONATE see EAU500
DIPROPYL KETONE see DWT600
DI-n-PROPYL MALEATE-ISOSAFROLE CONDENSATE see PNP250
DIPROPYL MERCURY see DWU000
DIPROPYL METHANE see HBC500
DI-n-PROPYL 6,7-METHYLENEDIOXY-3-METHYL-1,2,3,4-TETRAHYDRONA-
PHTHALENE see PNP250
DI-n-PROPYL-3-METHYL-6,7-METHYLENEDIOXY-1,2,3,4-TETRAHYDRONA-
PHTHALENE-1,2-DICARBOXYLATE see PNP250
S,S-DIPROPYL METHYLPHOSPHONOTRITHIOATE see DWU200
O,O-DI-n-PROPYL-O-(4-METHYLTHIOPHENYL)PHOSPHATE see DWU400
DI-n-PROPYLNITROSAMINE see NKB700
α-DIPROPYLNITROSAMINE METHYL ETHER see DWU800
DIPROPYLNITROSOAMINE see NKB700
DIPROPYL OXIDE see PNM000
DIPROPYLOXOSTANNANE see DWV000
DIPROPYL PEROXIDE see DWV200
DI-n-PROPYL PEROXYDICARBONATE see DWV400
DI-n-PROPYL PHTHALATE see DWV500
DIPROPYL PHTHALATE see DWV500
N,N-DIPROPYL-1-PROPANAMINE see TMY250
DIPROPYL 2,5-PYRIDINEDICARBOXYLATE see EAU500
DI-N-PROPYL SUCCINATE see DWV800
DIPROPYL SUCCINATE see DWV800
4-(DIPROPYLSULFAMOYL)BENZOIC ACID see DWW000
p-(DIPROPYLSULFAMOYL)BENZOIC ACID see DWW000
p-(DIPROPYLSULFAMOYL)BENZOIC ACID SODIUM SALT see DWW200
p-(DIPROPYLSULFAMYL)BENZOIC ACID see DWW000
p-(DI-N-PROPYLSULFAMYL)BENZOIC ACID SODIUM SALT see DWW200
DIPROPYL-5,6,7,8-TETRAHYDRO-7-METHYLNAPHTHO(2,3-d)-1,3-DIOXOLE-5,6-
DICARBOXYLATE see PNP250
N,N-DIPROPYLTHIOCARBAMIC ACID-S-ETHYL ESTER see EIN500
DIPROPYLTHIOCARBAMIC ACID-S-PROPYL ESTER see PNI750
DI-n-PROPYLTIN BISMETHANESULFONATE see DWW400
DIPROPYLTIN CHLORIDE see DFF400
DI-n-PROPYLTIN DICHLORIDE see DFF400
DIPROPYLTIN DICHLORIDE see DFF400
DIPROPYLTIN OXIDE see DWV000
N,N-DIPROPYL-4-TRIFLUOROMETHYL-2,6-DINITROANILINE see DUV600
DIPROPYL ZINC see DWW500
DI(2-PROPYNYL) ETHER see POA500
DIPROSONE see BFV765
DIPROSTRON see EDR000
DIPROZIN see DQA400
DIPTERAX see TIQ250
DIPTEREX 50 see TIQ250
DIPTEREX see TIQ250
DIPTEVUR see TIQ250
DIPTHAL see DNS600
DIPYRIDAMINE see PCP250
DIPYRIDAMOL see PCP250
DIPYRIDAMOLE see PCP250
DIPYRIDAN see PCP250
4,4'-DIPYRIDINE see BGO600
DIPYRIDINESODIUM see DWW600
DIPYRIDO(2,3-D,2,3-K)PYRENE see NAU500
DIPYRIDO(1,2-a:3',2'-d)IMIDAZOL-2-AMINE see DWW700
DIPYRIDO(1,2-a:3',2'-d)IMIDAZOLE, 2-AMINO-, HYDROCHLORIDE see AJS225
DIPYRIDO(1,2-a:3',2'-d)IMIDAZOLE, 2-AMINO-6-METHYL-, HYDROCHLORIDE
see AKS275
DIPYRIDO(2,3-d,2,3-1)PYRENE see NAU000
2,2'-DIPYRIDYL see BGO600
4,4-DIPYRIDYL see BGO600
4,4'-DIPYRIDYL see BGO600
α,α'-DIPYRIDYL see BGO500
γ,γ'-DIPYRIDYL see BGO600
DIPYRIDYLDIHYDRATE see PAI995
DIPYRIDYL HYDROGEN PHOSPHATE see DWX000
DI-3-PYRIDYLMERCURY see DWX200
1,2-DI-3-PYRIDYL-2-METHYL-1-PROPANONE see MCJ370
DIPYRIDYL PHOSPHATE see DWX000
DIPYRIN see DOT000
DIPYROXIME see TLQ500
cis-DIPYRROLIDINEDICHLOROPLATINUM(II) see DEU200
1,4-DIPYRROLIDINYL-2-BUTYNE see DWX600

DI-1H-PYRROL-2-YL KETONE see PPY300
DIPYUDAMINE see PCP250
DIQUAT see DWX800
DIQUAT DIBROMIDE see DWX800
DIQUAT DICHLORIDE see DWY000
DI-QUINOL see DNF600
1,3-DIQUINOLIN-6-YLUREA BISMETHOSULFATE see PJA120
DIRALGAN see TKG000
DIRAM A see ANG500
DIRAME see PMX250
DIRAX see AQN635, IDJ500
DIRCA PALUSTRIS see LEF100
DIRECT ARTIFICIAL SILK BLACK G see CMN240
DIRECT BLACK 17 see CMN230
DIRECT BLACK 19 see CMN240
DIRECT BLACK 38 see AQP000
DIRECT BLACK 3 see AQP000
DIRECT BLACK A see AQP000
DIRECT BLACK BH see CMN800
DIRECT BLACK BRN see AQP000
DIRECT BLACK CX see AQP000
DIRECT BLACK CXR see AQP000
DIRECT BLACK E see AQP000
DIRECT BLACK EW see AQP000
DIRECT BLACK EX see AQP000
DIRECT BLACK FR see AQP000
DIRECT BLACK GAC see AQP000
DIRECT BLACK GREEN see CMO830
DIRECT BLACK GW see AQP000
DIRECT BLACK GX see AQP000
DIRECT BLACK GXR see AQP000
DIRECT BLACK JET see AQP000
DIRECT BLACK META see AQP000
DIRECT BLACK METHYL see AQP000
DIRECT BLACK N see AQP000
DIRECT BLACK RX see AQP000
DIRECT BLACK SD see AQP000
DIRECT BLACK WS see AQP000
DIRECT BLACK Z see AQP000
DIRECT BLUE 14 see CMO250
DIRECT BLUE 15 see CMO500
DIRECT BLUE 22 see CMO600
DIRECT BLUE 2 see CMN800
DIRECT BLUE 6 see CMO000
DIRECT BLUE BLACK BH see CMN800
DIRECT BLUE BR see CMO600
DIRECT BLUE 10G see CMO500
DIRECT BLUE HH see CMO500
DIRECT BLUE MRW see CMO600
DIRECT BLUE RW see CMO600
DIRECT BLUE RWN see CMO600
DIRECT BRILLIANT GREEN BB see CMO840
DIRECT BRILLIANT GREEN C see CMO840
DIRECT BRILLIANT GREEN CBM see CMO840
DIRECT BROWN 1:2 see CMO810
DIRECT BROWN 154 see CMO825
DIRECT BROWN 31 see CMO820
DIRECT BROWN 95 see CMO750
DIRECT BROWN 1A see CMO810
DIRECT BROWN 3B see CMO820
DIRECT BROWN B see CMO820
DIRECT BROWN BR see PEY000
DIRECT BROWN BS see CMO820
DIRECT BROWN BSB see CMO820
DIRECT BROWN 5C see CMO810
DIRECT BROWN CGN see CMO810
DIRECT BROWN CMD see CMO825
DIRECT BROWN D3Y see CMO825
DIRECT BROWN FS see CMO820
DIRECT BROWN 5G see CMO810
DIRECT BROWN 5GR see CMO825
DIRECT BROWN 2GS see CMO810
DIRECT BROWN TRB see CMO820
DIRECT DARK BLUE BH see CMN800
DIRECT DARK GREEN A see CMO830
DIRECT DARK GREEN B see CMO830
DIRECT DARK GREEN BF see CMO830
DIRECT DARK GREEN BG see CMO830
DIRECT DARK GREEN MB see CMO830
DIRECT DARK GREEN S see CMO830
DIRECT DARK GREEN SUPRA see CMO830
DIRECT DARK GREEN WS see CMO830
DIRECT DEEP BLACK E see AQP000
DIRECT DEEP BLACK EAC see AQP000
DIRECT DEEP BLACK EA-CF see AQP000

DIRECT DEEP BLACK E EXTRA see AQP000
DIRECT DEEP BLACK EW see AQP000
DIRECT DEEP BLACK EX see AQP000
DIRECT DEEP GREEN A see CMO830
DIRECT DIAZO BLACK see CMN800
DIRECT DIAZO BLACK C see CMN800
DIRECT DIAZO BLACK N see CMN800
DIRECT DIAZO BLACK S see CMN800
DIRECT FAST BLACK G see CMN240
DIRECT FAST BLACK GU see CMN240
DIRECT FAST BLACK SA see CMN240
DIRECT FAST BROWN BP see CMO820
DIRECT FAST BROWN TSN see CMO820
DIRECT FAST BROWN TWC see CMO820
DIRECT FAST PURPURINE 8B see CMO880
DIRECT FAST RED 5B see CMO885
DIRECT FAST RED B see CMO870
DIRECT FAST RED 8BL see CMO885
DIRECT FAST RED F see CMO870
DIRECT FAST RED FN see CMO870
DIRECT FAST RED FR see CMO870
DIRECT FAST RED G see CMO870
DIRECT FAST RED MF see CMO870
DIRECT FAST RED 2S see CMO885
DIRECT FAST SCARLET 3B see CMO875
DIRECT FAST VIOLET N see CMP000
DIRECT GREEN 6 see CMO840
DIRECT GREEN see CMO840
DIRECT GREEN A see CMO840
DIRECT GREEN 2B see CMO840
DIRECT GREEN BN see CMO840
DIRECT GREEN BP see CMO840
DIRECT GREEN BX see CMO840
DIRECT GREEN MB see CMO840
DIRECT GREEN WAC see CMO830
DIRECT LIGHTFAST RED 2S see CMO885
DIRECT LIGHT RED 4B see CMO885
DIRECT LIGHT RED 8B see CMO885
DIRECT LIGHT RED M 8BL see CMO885
DIRECT NAVY BLUE BH see CMN800
DIRECT PURE BLUE see CMO500
DIRECT PURE BLUE M see CMO500
DIRECT RAYON BLACK KSG see CMN240
DIRECT RED 1 see CMO870
DIRECT RED 28 see SGQ500
DIRECT RED 39 see CMO875
DIRECT RED 81 see CMO885
DIRECT RED 8BS see CMO880
DIRECT RED F see CMO870
DIRECT RED FR see CMO870
DIRECT RED Kh see CMO870
DIRECT RED M see CMO870
DIRECT RED MN see CMO870
DIRECT ROSE MN see CMO870
DIRECT SCARLET 3BS see CMO875
DIRECT SKY BLUE A see CMO500
DIRECT VIOLET C see CMP000
DIRECT YELLOW MTZ see CMP050
DIRECT YELLOW TZ see CMP050
DIREKTAN see NCQ900
DIREMA see CFY000
DIREN see UVJ450
DIRESORCYL SULFIDE see BJE500
DIREXIODE see DNF600
DIREX 4L see DXQ500
DIREZ see DEV800
DIRIAN see BOL325
DIRIDONE see PDC250, PEK250
DIRIMAL see OJY100
DIRNATE see SNL840
DIRONYL see DLR150
DIROX see HIM000
DIRUBIDIUM CARBONATE see RPB200
DIRUBIDIUM MONOCARBONATE see RPB200
DISADINE see PKE250
DISALCID see SAN000
N,N'-DISALICYLIDENE-1,2-PROPANEDIAMINE see DWS400
DISALICYLALPROPYLENEDIIMINE see DWS400
DISALICYLIC ACID see SAN000
N,N'-DISALICYLIDENE-1,2-DIAMINOPROPANE see DWS400
N,N'-DISALICYLIDENE ETHYLENEDIAMINE see DWY200
DISALUNIL see CFY000
DISALYL see SAN000
DISAN see DNO800
DISATABS TABS see VSK600

DISCOLITE see FMW000
DISDOLEN see PHA575
DISELENIDE, BIS(2,2-DIETHOXYETHYL)- see BJA200
DISELENIDE, DIBUTYL-(9CI) see BRF550
DISELENIDE, DIPROPYL-(9CI) see PNI850
DISELENIUM DICHLORIDE see SBS500
α,α′-DISELENOBIS-o-ACETOTOLUIDIDE see DWY400
4,4′-DISELENOBIS(2-AMINOBUTYRIC ACID) see SBU710
2,2′-DISELENOBIS(N-PHENYLACETAMIDE) see DWY600
3,3′-DISELENODIALANINE see DWY800, SBP600
p,p′-DISELENODIANILINE see DWZ000
β,β′-DISELENODIPROPIONIC ACID, SODIUM SALT see DWZ100
DISELENO SALICYLIC ACID see SBU150
DI-SEPTON see DAO500
DISETIL see DXH250
DISFLAMOLL TKP see TNP500
DISFLAMOLL TOF see TNI250
DISILANE see DXA000
DISILOXANE, 1,3-BIS(3-AMINOPROPYL)-1,1,3,3-TETRAMETHYL- see OPC100
DISILVER ACETYLIDE SILVER NITRATE see SDJ025
DISILVER CYANAMIDE see DXA500
DISILVER KETENIDE see DXA600
DISILVER OXALATE see SDU000
DISILVER OXIDE see SDU500
DISILVER PENTATIN UNDECAOXIDE see DXA800
DISILYN see BEN000
DISIPAL see MJH900
DISIPAL HYDROCHLORIDE see OJW000
DI-SIPIDIN see ORU500
DISODIUM ADENOSINE 5′-TRIPHOSPHATE see AEM100
DISODIUM ADENOSINE TRIPHOSPHATE see AEM100
DISODIUM ANTHRAQUINONE-1,5-DISULFONATE see DLJ700
DISODIUM ARSENATE see ARC000
DISODIUM ARSENATE, HEPTAHYDRATE see ARC250
DISODIUM ARSENIC ACID see ARC000
DISODIUM ATP see AEM100
DISODIUM AUROTHIOMALATE see GJC000
DISODIUM-4,4′-BIS((4-AMINO-6-(2-HYDROXYETHYL)AMINO-s-TRIAZIN-2-YL)AMINO)-2,2′-STILBENDISULFON see DXB400
DISODIUM-4,4′-BIS((4-ANILINO-6-METHOXY-s-TRIAZIN-2-YL)AMINO)STILBENE-2,2′-DISULFONATE see BGW100
DISODIUM 4,4′-BIS((4-ANILINO-6-MORPHOLINO-1,3,5-TRIAZIN-2-YL)AMINO)STILBENE-2,2′-DISULFONATE see CMP200
DISODIUM-4,4′-BIS((4,6-DIANILINO-1,3,5-TRIAZIN-2-YL)AMINO)STILBENE-2,2′-DISULFONATE see DXB450
DISODIUM-4,4′-BIS(2-SULFOSTYRYL)BIPHENYL see TGE150
DISODIUM BROMOSULFOPHTHALEIN see HAQ600
DISODIUM CALCIUM EDTA see CAR780
DISODIUM CALCIUM ETHYLENEDIAMINETETRAACETATE see CAR780
DISODIUM CARBONATE see SFO000
DISODIUM-N-(3-(CARBOXYMETHYLTHIOMERCURI)-2-METHOXYPROPYL)-α-CAMPHORAMATE see TFK270
DISODIUM CHROMATE see DXC200
DISODIUM CHROMOGLYCATE see CNX825
DISODIUM CINNAMYLIDENE BISULFITE derivative of SULFAPYRIDINE see DXF400
DISODIUM CITRATE see DXC400
DISODIUM CROMOGLICATE see CNX825
DISODIUM CROMOGLYCATE see CNX825
DISODIUM DEXAMETHASONE PHOSPHATE see DAE525
DISODIUM DIACID ETHYLENEDIAMINETETRAACETATE see EIX500
DISODIUM-3,3′-DIAMINO-4,4′-DIHYDROXYARSENOBENZENE-N-DIMETHYLENE-SULFONATE see SNR000
DISODIUM-3,3′-DIAMINO-4,4′-DIHYDROXYARSENOBENZENE-N,N′-DIMETHY-LENEBISULFITE see SNR000
DISODIUM p,p′-DIAMINODIPHENYLSULFONE-N,N′-DIGLUCOSE SULFONATE see AOO800
DISODIUM-2,7-DIBROM-4-HYDROXY-MERCURI-FLUORESCEIN see MCV000
DISODIUM-2′,7′-DIBROMO-4′-(HYDROXYMERCURY)FLUORESCEIN see MCV000
DISODIUM DICHROMATE see SGI000
DISODIUM DIFLUORIDE see SHF500
DISODIUM DIHYDROGEN ATP see AEM100
DISODIUM DIHYDROGEN ETHYLENEDIAMINETETRAACETATE see EIX500
DISODIUM DIHYDROGEN(ETHYLENEDINITRILO)TETRAACETATE see EIX500
DISODIUM DIHYDROGEN-(1-HYDROXYETHYLIDENE)DIPHOSPHONATE see DXD400
DISODIUM DIHYDROGEN PYROPHOSPHATE see DXF800
DISODIUM-1,3-DIHYDROXY-1,3-BIS-(aci-NITROMETHYL)-2,2,4,4-TETRAMETHYL-CYCLO BUTANE see DXC600
DISODIUM (2,4-DIMETHYLPHENYLAZO)-2-HYDROXYNAPHTHALENE-3,6-DISUL-PHONATE see FMU070
DISODIUM (2,4-DIMETHYLPHENYLAZO)-2-HYDROXYNAPHTHALENE-3,6-DISUL-FONATE see FMU070
DISODIUM DIOXIDE see SJC500

DISODIUM DIPHOSPHATE see DXF800
DISODIUM DISULFITE see SII000
DISODIUM-4,4′-DISULFOXYDIPHENYL-(2-PYRIDYL)METHANE see SJJ175
DISODIUM EDATHAMIL see EIX500
DISODIUM EDETATE see EIX500
DISODIUM EDTA (FCC) see EIX500
DISODIUM-3,6-ENDOXOHEXAHYDROPHTHALATE see DXD000
DISODIUM EOSIN see BNH500
DISODIUM-3,6-EPOXYCYCLOHEXANE-1,2-DICARBOXYLATE see DXD000
DISODIUM ETHANOL-1,1-DIPHOSPHONATE see DXD400
DISODIUM ETHYDRONATE see DXD400
DISODIUM ETHYLENE-1,2-BISDITHIOCARBAMATE see DXD200
DISODIUM ETHYLENEBIS(DITHIOCARBAMATE) see DXD200
DISODIUM ETHYLENEDIAMINETETRAACETATE see EIX500
DISODIUM ETHYLENEDIAMINETETRAACETIC ACID see EIX500
DISODIUM (ETHYLENEDINITRILO)TETRAACETATE see EIX500
DISODIUM (ETHYLENEDINITRILO)TETRAACETIC ACID see EIX500
DISODIUM ETIDRONATE see DXD400
DISODIUM FLUOROPHOSPHATE see DXD600
DISODIUM FOSFOMYCIN see DXF600
DISODIUM FOSFOMYCIN HYDRATE see FOL200
DISODIUM FUMARATE see DXD800
DISODIUM GLYCYRRHIZIN see DXD875
DISODIUM GLYCYRRHIZINATE see DXD875
DISODIUM-5′-GMP see GLS800
DISODIUM GMP see GLS800
DISODIUM-5′-GUANYLATE see GLS800
DISODIUM GUANYLATE (FCC) see GLS800
DISODIUM-5′-GUANYLATE mixed with DISODIUM 5′-INOSINATE (1:1) see RJF400
(2-)-DISODIUM HEXAFLUOROSILICATE see DXE000
DISODIUM HEXAFLUOROSILICATE see DXE000
DISODIUM HYDROGEN ARSENATE see ARC000
DISODIUM HYDROGEN CITRATE see DXC400
DISODIUM HYDROGEN NITRILOTRIACETATE see DXF000
DISODIUM HYDROGEN ORTHOARSENATE see ARC000
DISODIUM HYDROGEN PHOSPHATE see SJH090
DISODIUM-6-HYDROXY-3-OXO-9-XANTHENE-o-BENZOATE see FEW000
DISODIUM-5,5′-((2-HYDROXYTRIMETHYLENE)DIOXY)-BIS(4-OXO-4H-1-BENZO-PYRAN-2-CARBOXYLATE) see CNX825
DISODIUM-3-HYDROXY-4-((2,4,5-TRIMETHYLPHENYL)AZO)-2,7-NAPHTHALEN-EDISULFONIC ACID see FAG018
DISODIUM-3-HYDROXY-4-((2,4,5-TRIMETHYLPHENYL)AZO)-2,7-NAPHTHALEN-EDISULPHONIC ACID see FAG018
DISODIUM-3-HYDROXY-4-((2,4,5-TRIMETHYLPHENYL)AZO)-2,7-NAPHTHALEN-EDISULFONATE see FAG018
DISODIUM-3-HYDROXY-4-((2,4,5-TRIMETHYLPHENYL)AZO)-2,7-NAPHTHALEN-EDISULPHONATE see FAG018
DISODIUM IMINODIACETATE see DXE200
DISODIUM IMP see DXE500
DISODIUM INDIGO-5,5-DISULFONATE see FAE100
DISODIUM-5′-INOSINATE see DXE500
DISODIUM INOSINATE see DXE500
DISODIUM-5′-INOSINATE mixed with DISODIUM 5′-GUANYLATE (1:1) see RJF400
DISODIUM INOSINE-5′-MONOPHOSPHATE see DXE500
DISODIUM INOSINE-5′-PHOSPHATE see DXE500
DISODIUM LATAMOXEF see LBH200
DISODIUM METASILICATE see SJU000
DISODIUM METHANEARSENATE see DXE600
DISODIUM METHANEARSONATE see DXE600
DISODIUM METHOTREXATE see MDV600
DISODIUM METHYLARSENATE see DXE600
DISODIUM METHYLARSONATE see DXE600
DISODIUM MOLYBDATE see DXE800
DISODIUM MOLYBDATE DIHYDRATE see DXE875
DISODIUM MONOFLUOROPHOSPHATE see DXD600
DISODIUM MONOHYDROGEN ARSENATE see ARC000
DISODIUM MONOHYDROGEN PHOSPHATE see SJH090
DISODIUM MONOMETHYLARSONATE see DXE600
DISODIUM MONOSELENIDE see SJT000
DISODIUM MONOSILICATE see SJU000
DISODIUM MONOXIDE see SIN500
N,N′-DISODIUM N,N′-DIMETHOXYSULFONYLDIAMIDE see DXC800
DISODIUM NITRILOTRIACETATE see DXF000
DISODIUM NITROPRUSSIDE DIHYDRATE see SIW500
DISODIUM NITROSYLPENTACYANOFERRATE see SIU500
DISODIUM OCTABORATE, TETRAHYDRATE see DXF200
DISODIUM ORTHOPHOSPHATE see SJH090
DISODIUM-7-OXABICYCLO(2.2.1)HEPTANE-2,3-DICARBOXYLATE see DXD000
DISODIUM OXIDE see SIN500
DISODIUM PEROXIDE see SJC500
DISODIUM PEROXYDICARBONATE see SJB400
DISODIUM-2-(p-(γ-PHENYLPROPYLAMINO)BENZENESULFONAMIDO) PYRIDINE see DXF400

DISODIUM PHOSPHATE see SJH090
DISODIUM PHOSPHONOMYCIN see DXF600
DISODIUM PHOSPHONOMYCIN HYDRATE see FOL200
DISODIUM PHOSPHORIC ACID see SJH090
DISODIUM PHOSPHOROFLUORIDATE see DXD600
DISODIUM PREDNISOLONE 21-PHOSPHATE see PLY275
DISODIUM PROTOPORPHYRIN see DXF700
DISODIUM PYROPHOSPHATE see DXF800
DISODIUM PYROSULFITE see SII000
DISODIUM-5′-RIBONUCLEOTIDE see RJF400
DISODIUM (−)-(1R,2S)-(1,2-EPOXYPROPYL)PHOSPHONATE HYDRATE see FOL200
DISODIUM SALT of EDTA see EIX500
DISODIUM SALT of ENDOTHALL see DXD000
DISODIUM SALT of 1-INDIGOTIN-S,S′-DISULPHONIC ACID see FAE100
DISODIUM SALT of 7-OXABICYCLO(2.2.1)HEPTANE-2,3-DICARBOXYLIC ACID see DXD000
DISODIUM SALT of 2-(4-SULPHO-1-NAPHTHYLAZO)-1-NAPHTHOL-4-SULPHONIC ACID see HJF500
DISODIUM SALT of 1-p-SULPHOPHENYLAZO-2-NAPHTHOL-6-SULPHONIC ACID see FAG150
DISODIUM SALT of 1-(2,4-XYLYLAZO)-2-NAPHTHOL-3,6-DISULFONIC ACID see FMU070
DISODIUM SALT of 1-(2,4-XYLYLAZO)-2-NAPHTHOL-3,6-DISULPHONIC ACID see FMU070
DISODIUM SELENATE see DXG000
DISODIUM SELENITE see SJT500
DISODIUM SEQUESTRENE see EIX500
DISODIUM SILICOFLUORIDE see DXE000
DISODIUM 2-(4-STYRYL-3-SULFOPHENYL)-7-SULFO-2H-NAPHTHO(1,2-d)TRIAZOLE see DXG025
DISODIUM SUCCINATE see SJW100
DISODIUM SULBENICILLIN see SNV000
DISODIUM SULFATE see SJY000
DISODIUM SULFITE see SJZ000
DISODIUM α-SULFOBENZYLPENICILLIN see SNV000
DISODIUM SULFOBENZYLPENICILLIN see SNV000
DISODIUM-2-(4-SULFO-1-NAPHTHYLAZO)-1-NAPHTHOL-4-SULFONATE see HJF500
DISODIUM-2-(4-SULPHO-1-NAPHTHYLAZO)-1-NAPHTHOL-4-SULPHONATE see HJF500
DISODIUM l-(+)-TARTRATE see BLC000
DISODIUM TARTRATE see BLC000
DISODIUM TETRABORATE see DXG035
DISODIUM TETRACEMATE see EIX500
DISODIUM TETRAOXATUNGSTATE (2-) see SKN500
DISODIUM TETRAOXOTUNGSTATE (2-) see SKN500
DISODIUM-5-TETRAZOLAZOCARBOXYLATE see DXG050
DISODIUM TUNGSTATE see SKN500
DISODIUM VERSENATE see EIX500
DISODIUM VERSENE see EIX500
DISOFEN see DNG000
DISOLFURO DI TETRAMETILTIOURAME (ITALIAN) see TFS350
DISOMAR see DXE600
DISOMER MALEATE see DXG100
2,4-D ISOOCTYL ESTER see ILO000
DISOPHENOL see DNG000
2,4-D ISOPROPYL ESTER see IOY000
DISOPYRAMIDE see DNN600
DISOQUIN see DNF600
DISORLON see CCK125
DISOTAT see DNM400
DISPADOL see DAM700
DISPAL see AHE250
DISPAMIL see PAH250
DISPARICIDA see ABX500
DISPARLURE see ECB200
DISPASOL M see PKB500
DISPERGATOR NF see BLX000
DISPERMINE see PIJ000
DISPERSE BLUE 110 see BKP500
DISPERSE BLUE 56 see CMP070
DISPERSE BLUE 78 see BKP500
DISPERSE BLUE 7 see DMM400
DISPERSE BLUE GREEN see DMM400
DISPERSE BLUE K see MGG250
DISPERSE BLUE NO 1 see TBG700
DISPERSE BORDEAUX S see CMP080
DISPERSE BRILLIANT PINK see DBX000
DISPERSE BRILLIANT ROSE see DBX000
DISPERSED BLUE 12195 see FAE000
DISPERSED ORANGE 11348 see FAG150
DISPERSED VIOLET 12197 see FAG120
DISPERSE DYE FAST YELLOW 4K see CMP090
DISPERSED YELLOW 12116 see FAG150

DISPERSE FAST PINK B see AKE250
DISPERSE FAST VIOLET B see AKP250
DISPERSE FAST YELLOW G see AAQ250
DISPERSE FAST YELLOW 2K see DUW500
DISPERSE FAST YELLOW 4K see CMP090
DISPERSE MB-61 see TFC600
DISPERSE ORANGE see AKP750
DISPERSE PINK Zh see AKO350
DISPERSE POLYESTER PINK 2S see AKI750
DISPERSE RED 11 see DBX000
DISPERSE RED 13 see CMP080
DISPERSE RED 15 see AKE250
DISPERSE RED 25 see AKE250
DISPERSE RED-4 see AKO350
DISPERSE RED 60 see AKI750
DISPERSER NF see BLX000
DISPERSE ROSE Zh see AKO350
DISPERSE VIOLET K see DBP000
DISPERSE VIOLET 2S see DBY700
DISPERSE VIOLET 4S see AKP250
DISPERSE YELLOW 3 see AAQ250
DISPERSE YELLOW 7 see CMP090
DISPERSE YELLOW G see AAQ250
DISPERSE YELLOW R see DUW500
DISPERSE YELLOW STABLE 2K see DUW500
DISPERSE YELLOW 6Z see MEB750
DISPERSE YELLOW Z see AAQ250
DISPERSING AGENT NF see BLX000
DISPERSIVE blue-green see DMM400
DISPERSIVE YELLOW 3T see AAQ250
DISPERSOL ACA see BLX000
DISPERSOL BLUE B-R see CMP070
DISPERSOL FAST CRIMSON B see CMP080
DISPERSOL FAST YELLOW A see DUW500
DISPERSOL FAST YELLOW G see AAQ250
DISPERSOL ORANGE D-G see AKE250
DISPERSOL PRINTING YELLOW A see DUW500
DISPERSOL PRINTING YELLOW G see AAQ250
DISPERSOL RED B 2B see AKI750
DISPERSOL RED B 3B see DBX000
DISPERSOL RED PP see SBC500
DISPERSOL RUBINE B see CMP080
DISPERSOL VIOLET B see AKP250
DISPERSOL YELLOW A-G see AAQ250
DISPERSOL YELLOW B-A see DUW500
DISPERSOL YELLOW PP see PEJ500
DISPEX C40 see ADW200
DISPHEX see PKE250
DISPHOLIDUS TYPHUS VENOM see DXG150
DISPIRO(CYCLOHEXANE-1,2′(3′H)-QUINAZOLINE-4′,1‴(4a′H)-CYCLOHEXANE), 5′,6′,7′,8′-TETRAHYDRO- see TCQ600
3,6-DI(SPIROCYCLOHEXANE)TETRAOXANE see DXG200
DISPRANOL see BIK500
DISRUPT see ECB200
DISSENTEN see LIH000
DISSOLVANT APV see DJD600
DISTACLOR see CCR850
DISTAKAPS V-K see PDT750
DISTAMICINA A (ITALIAN) see SLI300
DISTAMINE see PAP550
DISTAMYCIN A see SLI300
DISTAMYCIN A HYDROCHLORIDE see DXG600
DISTANNANE, HEXAISOPROPYL- see HDY100
DISTANNATHIANE, HEXABUTYL-(9CI) see HCA700
DISTANNATHIANE, HEXAKIS(PHENYLMETHYL)-(9CI) see BLK750
DISTANNOXANE, BIS(1,3-DITHIOCYANATO-1,1,3,3-TETRABUTYL)- see BJM700
DISTANNOXANE, 1,3-BIS(2,4,5-TRICHLOROPHENOXY)-1,1,3,3-TETRABUTYL- see OPE100
DISTANNOXANE, HEXAETHYL- see HCX050
DISTANNOXANE, HEXAOCTYL- see HEW100
DISTANNOXANE, 1,1,1,3,3,3-HEXAPROPYL- see BLT300
DISTANNTHIANE, HEXABUTYL- see HCA700
DISTANNTHIANE, HEXAETHYL- see HCX100
DISTANNTHIANE, HEXAOCTYL- see HEW150
DISTANNTHIANE, HEXAPROPYL- see HEW200
DISTAQUAINE V see PDT500
DISTAQUAINE V-K see PDT750
DISTAVAL see TEH500
DISTAXAL see TEH500
DISTEARIN see OAV000
DISTEARYL DIMETHYLAMMONIUM CHLORIDE see DXG625
DISTEARYL 3,3′-THIODIPROPIONATE see DXG700
DISTEARYL β,β′-THIODIPROPIONATE see DXG700
DISTEARYL β-THIODIPROPIONATE see DXG700
DISTEARYL THIOPROPIONATE see DXG700

DISTERYL see CDT250
DISTESOL see EHP000
DISTESSOL see EHP000
DISTHENE see AHF500
DISTIGMINE BROMIDE see DXG800
DISTILBENE see DKA600, DKB000
DISTILLATES (COAL TAR) see CMY900
DISTILLATES (PETROLEUM), ACID-TREATED HEAVY NAPHTHENIC (9CI) see MQV760
DISTILLATES (PETROLEUM), ACID-TREATED LIGHT NAPHTHENIC (9CI) see MQV770
DISTILLATES (PETROLEUM), ACID-TREATED LIGHT PARAFFINIC (9CI) see MQV775
DISTILLATES (PETROLEUM), HEAVY CATALYTIC CRACKED see DXG810
DISTILLATES (PETROLEUM), HEAVY NAPHTHENIC (9CI) see MQV780
DISTILLATES (PETROLEUM), HEAVY PARAFFINIC (9CI) see MQV785
DISTILLATES (PETROLEUM), HYDRODESULFURIZED MIDDLE see DXG820
DISTILLATES (PETROLEUM), HYDROTREATED (mild) HEAVY NAPHTHENIC (9CI) see MQV790
DISTILLATES (PETROLEUM), HYDROTREATED (mild) HEAVY PARAFFINIC (9CI) see MQV795
DISTILLATES (PETROLEUM), HYDROTREATED (mild) LIGHT NAPHTHENIC (9CI) see MQV800
DISTILLATES (PETROLEUM), HYDROTREATED (mild) LIGHT PARAFFINIC (9CI) see MQV805
DISTILLATES (PETROLEUM), HYDROTREATED MIDDLE see DXG830
DISTILLATES (PETROLEUM), LIGHT CATALYTIC CRACKED see DXG840
DISTILLATES (PETROLEUM), LIGHT NAPHTHENIC (9CI) see MQV810
DISTILLATES (PETROLEUM), LIGHT PARAFFINIC (9CI) see MQV815
DISTILLATES (PETROLEUM), SOLVENT-DEWAXED HEAVY NAPHTHENIC (9CI) see MQV820
DISTILLATES (PETROLEUM), SOLVENT-DEWAXED HEAVY PARAFFINIC (9CI) see MQV825
DISTILLATES (PETROLEUM), SOLVENT-DEWAXED LIGHT NAPHTHENIC (9CI) see MQV835
DISTILLATES (PETROLEUM), SOLVENT-DEWAXED LIGHT PARAFFINIC (9CI) see MQV840
DISTILLATES (PETROLEUM), SOLVENT-REFINED (mild) HEAVY NAPHTHENIC (9CI) see MQV845
DISTILLATES (PETROLEUM), SOLVENT-REFINED (mild) HEAVY PARAFFINIC (9CI) see MQV850
DISTILLATES (PETROLEUM), SOLVENT-REFINED (mild) LIGHT NAPHTHENIC (9CI) see MQV852
DISTILLATES (PETROLEUM), SOLVENT-REFINED (mild) LIGHT PARAFFINIC (9CI) see MQV855
DISTILLATES (PETROLEUM), STRAIGHT-RUN MIDDLE see GBW000
DISTILLED LIME OIL see OGO000
DISTILLED MUSTARD see BIH250
DISTIVIT (B12 PEPTIDE) see VSZ000
DISTOBRAM see TGI250
DISTOKAL see HCI000
DISTOL 8 see EIV000
DISTOPAN see HCI000
DISTOPIN see HCI000
DISTOVAL see TEH500
DISTRANEURIN see CHD750
DISTYLIN see DMD000
DISUL see CNW000
1,3-DI(4-SULFAMOYLPHENYL)TRIAZENE see THQ900
1,3-DISULFAMYL-4,5-DICHLOROBENZENE see DEQ200
DISULFAN see DXH250
DISULFATON see DXH325
DISULFATOZIRCONIC ACID see ZTJ000
DISULFIDE, BENZYL METHYL see MHN350
DISULFIDE, BIS(MORPHOLINOTHIOCARBONYL) see MRR090
DISULFIDE, BIS(2-NITRO-α-α-α-TRIFLUORO-p-TOLYL) see BLA800
DISULFIDE DIBENZYL see DXH200
DISULFIDE DIPHENYL see PEW250
DISULFIDE, DI-2-PROPENYL (9CI) see AGF300
DISULFINE BLUE VN see ADE500
DISULFIRAM see DXH250
DISULFIRAM mixed with SODIUM NITRITE see SIS200
2,5-DISULFO-1-AMINOBENZENE see AIE000
2,5-DISULFOANILINE see AIE000
1,5-DISULFOANTHRAQUINONE see APK625
1,8-DISULFOANTHRAQUINONE see APK635
2,2'-DISULFOBENZIDINE see BBX500
3,5-DISULFOCATECHOL DISODIUM SALT see DXH300
DISULFO-meso-4,4-DIPHENYLHEXANE DIPOTASSIUM see SPA650
4,8-DISULFO-2-NAPHTHALAMINE see ALH250
2,2'-DISULFO-4,4'-STILBENEDIAMINE DISODIUM SALT see FCA200
DISULFOTON see DXH325
DISULFOTON DISULIDE see OQS000
DISULFOTON SULFOXIDE see OQS000
DISULFURAM see DXH250

DISULFUR DIBROMIDE see DXH350
DISULFUR DICHLORIDE see SON510
DISULFUR DINITRIDE see DXH400
DISULFURE de TETRAMETHYLTHIOURAME (FRENCH) see TFS350
DISULFUR HEPTAOXIDE see DXH600
DISULFUROUS ACID, DISODIUM SALT see SII000
DISULFUR PENTOXYDICHLORIDE see PPR500
DISULFURYL CHLORIDE see PPR500
DISULFURYL DIAZIDE see DXH800
DISUL-Na see CNW000
DISULONE see SOA500
DISULPHINE BLUE AN see ERG100
DISULPHINE BLUE VN 150 see ADE500
DISULPHINE LAKE BLUE AN see ERG100
DISULPHINE LAKE BLUE EG see FMU059
DISULPHINE VN see ADE500
DISULPHURAM see DXH250
DISULPHURIC ACID see SOI520
DISUL-SODIUM see CNW000
DISYNCRAM see MPE250
DISYNCRAN see MDT500, MPE250
DISYNFORMON see EDV000
DI-SYSTON see DXH325
DISYSTON SULFOXIDE see OQS000
DISYSTOX see DXH325
DI-TAC see DXE600
DITAK see UVJ450
DITALLOW DIMETHYL AMMONIUM CHLORIDE see QAT550
DITAVEN see DKL800
DITAZOL MONOHYDRATE see BKB250
DITEFTIN see FQJ100
1,2-DI(5-TETRAZOLYL)HYDRAZINE see DXI000
1,3-DI(5-TETRAZOYL)TRIAZENE see DXI200
DITHALLIUM CARBONATE see TEJ000
DITHALLIUM OXIDE see TEL040
DITHALLIUM(1+) SULFATE see TEM000
DITHALLIUM SULFATE see TEM000
DITHALLIUM TRIOXIDE see TEL050
DITHANE D-14 see DXD200
DITHANE M-45 see DXI400
DITHANE A-4 see DUQ600
DITHANE A-40 see DXD200
DITHANE M 22 see MAS500
DITHANE M 22 SPECIAL see MAS500
DITHANE R-24 see BPU000
DITHANE S60 see DXI400
DITHANE SPC see DXI400
DITHANE STAINLESS see ANZ000
DITHANE ULTRA see DXI400
DITHANE Z see EIR000
1,4-DITHIAANTHRAQUINONE-2,3-DICARBONITRILE see DLK200
1,4-DITHIAANTHRAQUINONE-2,3-DINITRILE see DLK200
1,4-DITHIACYCLOHEXANE see DXI550
2,4-DITHIA-1,3-DIOXANE-2,2,4,4-TETRAOXIDE see DXI500
1,4-DITHIANE see DXI550
p-DITHIANE see DXI550
DITHIANON see DLK200
DITHIANONE see DLK200
4,5-DITHIA-1,7-OCTADIENE see AGF300
1,3,2-DITHIARSENOLANE, 2-(p-BIS(2-CHLOROETHYL)AMINOPHENYL)- see BHW300
1,3,2-DITHIARSENOLANE, 2-(p-(DIETHYLAMINO)PHENYL)- see DIP100
3,6-DITHIA-3,4,5,6-TETRAHYDROPHTHALIMIDE see DLK700
2,6-DITHIA-1,3,5,7-TETRAZATRICYCLO(3.3.1.1³·⁷)DECANE-2,2,6,6-TETROXIDE see TDX500
DITHIAZANINE see DXI600
DITHIAZANINE IODIDE see DJT800
DITHIAZANIN IODIDE see DJT800
DITHIAZININE see DJT800
(R,S)-α-(1-((3,3-DI-3-THIENYLALLYL)AMINO)ETHYL))BENZYL ALCOHOL (+)-(α)-HYDROCHLORIDE see DXI800
(+)-α-(1-(93,3-DI-3-THIENYLALLYL)AMINO)ETHYL)BENZYL ALCOHOL HYDRO-CHLORIDE see DXI800
1-((3,3-DI-2-THIENYL-1-METHYL)ALLYL)PYRROLIDINE HYDROCHLORIDE see DXJ100
3-(DI(2-THIENYL)METHYLENE)-5-METHYLDECAHYDROQUINOLIZANIUM BRO-MIDE see TGF075
3-(DI-2-THIENYLMETHYLENE)-1-METHYLPIPERIDINE see BLV000
3-(DI-2-THIENYLMETHYLENE)-1-METHYLPIPERIDINE CITRATE see ARP875
3-(DI-2-THIENYLMETHYLENE)-5-METHYL-trans-QUINOLIZIDINIUM BROMIDE see TGF075
3,3-DI-2-THIENYL-N,N,1-TRIMETHYLALLYLAMINE HYDROCHLORIDE see TLQ250
1,3-DITHIETAN-2-YLIDENE PHOSPHORAMIDIC ACID DIETHYL ESTER see DHH200

DITHIO see SOD100
2,2′-DITHIOBIS(N-(1-ADAMANTYLMETHYL)ACETAMIDINE) DIHYDROCHLORIDE
 HEMIHYDRATE see DXJ400
2,2′-DITHIOBISANILINE see DXJ800
O,O-DITHIO-BIS-ANILINE see DXJ800
2′,2′′′-DITHIOBISBENZANILIDE see BDK800
2,2′-DITHIOBIS(BENZOIC ACID) see BHM300
2,2′-DITHIOBIS(BENZOTHIAZOLE) see BDE750
1,1′-DITHIOBIS(N,N-DIETHYLTHIOFORMAMIDE) see DXH250
2,2′-DITHIOBIS(N,N-DIMETHYLETHYLAMINE) DIHYDROCHLORIDE see BJG100
α,α′-DITHIOBIS(DIMETHYLTHIO)FORMAMIDE see TFS350
1,1′-DITHIOBIS(N,N-DIMETHYLTHIO)FORMAMIDE see TFS350
2,2′-DITHIOBIS(ETHYLAMINE) see MCN500
2,2′-DITHIO-BIS-(ETHYLAMINE) DIHYDROCHLORIDE see CQJ750
3,3′-DITHIOBIS(METHYLENE)BIS(5-HYDROXY-6-METHYL-4-PYRIDINEMETHANOL)
 DIHYDROCHLORIDE see BMB000
4,4′-DITHIOBIS(MORPHOLINE) see BKU500
DITHIOBISMORPHOLINE see BKU500
3,3′-DITHIOBIS(6-NITROBENZOIC ACID) see DUV700
DITHIOBIS(THIOFORMIC ACID) O,O-DIBUTYL ESTER) see BSS550
DITHIOBIS(THIOFORMIC ACID)-o,o-DIETHYL ESTER see BJU000
2,5-DITHIOBIUREA see BLJ250
DITHIOBIURET see DXL800
DITHIOCARB see SGJ000, SGJ500
DITHIOCARBAMATE see SGJ000
DITHIOCARBAMOYLHYDRAZINE see MLJ500
DITHIOCARBANILIC ACID ETHYL ESTER see EOK550
DITHIOCARBONIC ACID-o-BUTYL ESTER POTASSIUM SALT see PKY850
DITHIOCARBONIC ACID-o-sec-BUTYL ESTER SODIUM SALT see DXM000
DITHIOCARBONIC ACID O-ISOPROPYL ESTER POTASSIUM SALT see
 IRG050
DITHIOCARBONIC ACID O-PENTYL ESTER POTASSIUM SALT see PKV100
DITHIOCARBONIC ANHYDRIDE see CBV500
DITHIOCARBOXYMETHYL-p-CARBAMIDOPHENYLARSENOUS OXIDE see
 DXM100
DITHIOCARBOXYPHENYL-p-CARBAMIDOPHENYLARSENOUS OXIDE see
 ARK800
DI-μ-(THIOCYANATODI-n-BUTYLSTANNYLOXO)BIS(THIOCYANATODI-n-BUTYL-
 TIN) see BJM700
DITHIODEMETON see DXH325
β,β′-DITHIODIALANINE see CQK325
2,2′-DITHIODIANILINE see DXJ800
2′,2′′′-DITHIODIBENZANILIDE see BDK800
2,2′-DITHIODIBENZOESAEURE see BHM300
2,2′-DITHIODIBENZOIC ACID see BHM300
N,N′-(DITHIODICARBONOTHIOYL)BIS(N-METHYLMETHANAMINE) see TFS350
4,4′-(DITHIODICARBONOTHIOYL)BISMORPHOLINE see MRR090
2,2-DITHIODIETHANOL see DXM600
1,1′-DITHIODIETHYLENEBIS(3-(2-(CHLOROETHYL)-3-NITROSOUREA)) see
 BIF625
DITHIODIGLYCOL see DXM600
3,3′-DITHIODIMETHYLENEBIS(5-HYDROXY-6-METHYL-4-PYRIDINEMETHANOL)
 DIHYDROCHLORIDE HYDRATE see BMB000
5,5′-DITHIODIMETHYLENEBIS(2-METHYL-3-HYDROXY-4-HYDROXYMETHYLPYRI-
 DINE)DIHYDROCHLORIDE HYDRATE see BMB000
4,4′-DITHIODIMORPHOLINE see BKU500
N,N-DITHIODIMORPHOLINE see BKU500
N,N′-(DITHIODI-2,1-PHENYLENE)BISBENZAMIDE see BDK800
DITHIODIPHOSPHORIC ACID, TETRAETHYL ESTER see SOD100
1,4-DITHIOERYTHRITOL see DXN350
DITHIOERYTHRITOL see DXN350
DITHIOETHYLENEGLYCOL see EEB000
1,1′-DITHIOFORMAMIDINE DIHYDROCHLORIDE see DXN400
DITHIOFOS see SOD100
1,2-DITHIOGLYCEROL see BAD750
DITHIOGLYCEROL see BAD750
DITHIOGLYCOL see EEB000
DITHIOGLYCOLYL p-ARSENOBENZAMIDE see TFA350
DITHIOHYDANTOIN see IAT100
DITHIOLANE see DXN600
1,3-DITHIOLANE, 2-ACETYL-2-METHYL- see ACR050
DITHIOLANE IMINOPHOSPHATE see DXN600
1,2-DITHIOLANE-3-PENTANAMIDE (9CI) see DXN709
1,3-DITHIOLANE-2-THIONE see EJQ100
1,2-DITHIOLANE-3-VALERAMIDE see DXN709
1,2-DITHIOLANE-3-VALERIC ACID see DXN800
1,3-DITHIOLAN-2-YLIDENE-PHOSPHORAMIDOTHIOIC ACID-O,O-DIETHYL ES-
 TER see DXN600
1,3-DITHIOLAN-2-YLIDENE-PHOSPHORAMIDOTHIOIC ACID DIETHYL ESTER
 see DXN600
4-((5-(1,2-DITHIOLAN-3-YL)-1-OXOPENTYL)AMINO)BUTANOIC ACID see
 LGK100
5-(1,2-DITHIOLAN-3-YL)VALERIC ACID see DXN800
1,3-DITHIOLIUM PERCHLORATE see DXN850
1,2-DITHIOLPROPANE see PML300

1,3-DITHIOL-2-YLIDENE-PROPANEDIOIC ACID BIS(1-METHYLETHYL) ESTER
 see MAO275
DITHIO-METHANEARSONOUS ACID BIS(ANHYDROSULFIDE) with DIMETHYL-
 DITHIOCARBAMIC ACID see USJ075
DITHIOMETON (FRENCH) see PHI500
4,4′-DITHIOMORPHOLINE see BKU500
DITHION see DXO000, SOD100
DITHIONE see DXO000, SOD100
DITHIONIC ACID see SOI520
DI(THIONOCARBOMETHOXY) DISULFIDE see DUN600
DITHIONOUS ACID, ZINC SALT (1:1) see ZGJ100, ZIJ100
6,8-DITHIOOCTANOIC ACID see DXN800
DITHIOOXALDIIMIDIC ACID see DXO200
DITHIOOXAMIDE see DXO200
DITHIOPHOS see SOD100
DITHIOPHOSPHATE de O,O-DIETHYLE et de (4-CHLORO-PHENYL) THIOME-
 THYLE (FRENCH) see TNP250
DITHIOPHOSPHATE de O,O-DIETHYLE et de S-(2-ETHYLTHIO-ETHYLE)
 (FRENCH) see DXH325
DITHIOPHOSPHATE de O,O-DIETHYLE et d'ETHYLTHIOMETHYLE (FRENCH)
 see PGS000
DITHIOPHOSPHATE de-O,O-DIETHYLE et de S(2,5-DICHLOROPHENYL)
 THIOMETHYLE (FRENCH) see PDC750
DITHIOPHOSPHATE de O,O-DIETHYLE et de S-N-METHYL-N-CARBOETHOXY
 CARBAMOYLMETHYLE (FRENCH) see DJI000
DITHIOPHOSPHATE de O,O-DIMETHYLE et de S-((4,6-DIAMINO-1,3,5-TRIA-
 ZINE-2-YL)-METHYLE) (FRENCH) see ASD000
DITHIOPHOSPHATE de O,O-DIMETHYLE et de S-(1,2-DICARBOETHOXYE-
 THYLE) (FRENCH) see MAK700
DITHIOPHOSPHATE de O,O-DIMETHYLE et de S-(2-ETHYLTHIO-ETHYLE)
 (FRENCH) see PHI500
DITHIOPHOSPHATE de O,O-DIMETHYLE et de S-((MORPHOLINOCARBO-
 NYLE)-METHYLE) (FRENCH) see MRU250
DITHIOPHOSPHATE de O,O-DIMETHYLE et de S(-N-METHYLCARBAMOYL-ME-
 THYLE) (FRENCH) see DSP400
DI(THIOPHOSPHORIC) ACID, TETRAETHYL ESTER see SOD100
DITHIOPHOSPHORSAEURE-O-AETHYL-S,S-DIPHENYLESTER (GERMAN) see
 EIM000
2,3-DITHIOPROPANOL see BAD750
DITHIOPROPYLTHIAMINE see DXO300
DITHIOPROPYLTHIAMINE HYDROCHLORIDE see DXO400
DITHIOPYROPHOSPHATE de TETRAETHYLE (FRENCH) see SOD100
DITHIOSYSTOX see DXH325
DITHIOTEP see SOD100
DITHIOTEREPHTHALIC ACID see DXO600
1,4-DITHIOTHREITOL see DXO800
d-1,4-DITHIOTHREITOL see DXO800
dl-1,4-DITHIOTHREITOL see DXO775
dl-DITHIOTHREITOL see DXO775
rac-DITHIOTHREITOL see DXO775
DITHIOTHREITOL see DXO775
DITHIOTRIMETHYLENEGLYCOL see PML350
DITHIOXAMIDE see DXO200
DITHIZON see DWN200
DITHIZONE see DWN200
DITHRANOL, 10-PROPIONYL- see PMW760
DITIAMINA see EIR000
DITILIN see CMG250, HLC500
DITILINE see CMG250, HLC500
DITILIN IODIDE see BJI000
DITIOVIT see DXO300
DITOIN see DNU000
DITOINATE see DKQ000
4,4′-DI-o-TOLUIDINE see TGJ750
1,4-DI-p-TOLUIDINOANTHRAQUINONE see BLK000
DI-o-TOLUYLTHIOUREA see DXP600
DITOLYLETHANE see DXP000
1,3-DI-o-TOLYLGUANIDINE see DXP200
DI-o-TOLYLGUANIDINE see DXP200
N,N′-DI-o-TOLYL-p-PHENYLENE DIAMINE see DXP400
DI-o-TOLYLTHIOUREA see DXP600
DITRAN see DXP800
DITRANIL see RDP300
DI-TRAPEX see ISE000, MLC000
DITRAZIN see DIW200
DITRAZIN CITRATE see DIW200
DITRAZINE see DIW200
DITRAZINE BASE see DIW000
DITRAZINE CITRATE see DIW200
1,3-DI(TRICHLOROMETHYL)BENZENE see BLL825
DITRIDECYLAMINE see DXQ000
DITRIDECYL PHTHALATE see DXQ200
DI(TRI-(2,2-DIMETHYL-2-PHENYLETHYL)TIN)OXIDE see BLU000
DITRIFON see TIQ250
DITRIPENTAT see CAY500

DMPS see DNU860
DMPT see DTP000
DMPTP see TFD500
DMS-70 see DUD800
DMS-90 see DUD800
DMS see DNV800, DUD100, DUD400, TFP000
DMSA see DNV800, DQD400
DMSO see DUD800
DMSP see FAQ800
D 65MT see XAJ000
DMT see DPF600
DMTP see FAQ900
DMTP (JAPAN) see DSO000
DMTT see DSB200
DMU see DTG700, DXQ500
DN 289 see BRE500
DNA see DUP600
DNAASE see EDK650, PAF575
DNA DEPOLYMERASE see EDK650, PAF575
DNA ENDONUCLEASE see EDK650, PAF575
DNA NUCLEASE see EDK650, PAF575
DNASE I see EDK650, PAF575
DNASW see EDK650, PAF575
DNBP see BRE500
DNBP AMMONIUM SALT see BPG250
DNBS see SNA500
DNCB see CGM000
DNDMP see DRO000
DN DRY MIX No. 1 see CPK500
DN-DRY MIX No.2 see DUS700
DN DUST No. 12 see CPK500
2,4-DNFB see DUW400
1,3-DNG see GGA200
DNOC AMMONIUM SALT see DUT800
DNOCHP see CPK500
DNOC-SODIUM see SGP550
DNOC SODIUM SALT see DUU600
DNOK (CZECH) see DUS700
DNOK-ACETAT (CZECH) see AAU250
DNOP see DVL600
DNOSBP see BRE500
2,4-DNP see DUZ000
2,5-DNP see DVA000
DNP see BIN500, DNG000
DNPC see DUT600
DNPD see NBL000
2,4-DNPH see DVC400
DNPMT see DVF400
DNPOH see DVD200
DNPT see DVF400
DNPZ see DVF200
DNRB see DVF800
DNSBP see BRE500
2,3-DNT see DVG800
2,4-DNT see DVH000
2,5-DNT see DVH200
2,6-DNT see DVH400
3,4-DNT see DVH600
3,5-DNT see DVH800
DNT see DVH000
DNTB see DVF800
DNTBP see DRV200
DNTP see PAK000
DO 9 see DXY000
DO 14 see SOP000
DOA see AEO000, AOO500
DOBANIC ACID 83 see LBU100
DOBANIC ACID JN see LBU100
DOBENDAN see CCX000
DOBESILATE CALCIUM see DMI300
DOBETIN see VSZ000
DOBO see TEP500
DOBREN see EPD500
DOBUTAMINE HYDROCHLORIDE see DXS375
DOBUTREX see DXS375
DOCA see DAQ800
DOCA ACETATE see DAQ800
DOC-AC see DAQ800
DOC ACETATE see DAQ800
DOCEMINE see VSZ000
DOCIBIN see VSZ000
DOCIGRAM see VSZ000
DOCITON see ICB000, ICC000
DOCTAMICINA see CDP250

DOCUSATE SODIUM see DJL000
DODAT see DAD200
DODDLE-DO (PUERTO RICO) see CAK325
DODECABEE see VSZ000
DODECACARBONYLDIVANADIUM see DXS400
DODECACARBONYLTRIIRON see DXS600
DODECACHLOROOCTAHYDRO-1,3,4-METHENO-2H-CYCLOBU-
    TA(c,d)PENTALENE see MQW500
1,1a,2,2,3,3a,4,5,5,5a,5b,6-DODECACHLOROOCTAHYDRO-1,3,4-METHENO-1H-
    CYCLOBUTA(c,d)PENTALENE see MQW500
DODECACHLOROPENTACYCLO(3,2,2,0$^{2,6}$,0$^{3,9}$,0$^{5,10}$)DECANE see MQW500
DODECACHLOROPENTACYCLODECANE see MQW500
8,10-DODECADIEN-1-OL, (E,E)- see CNG760
(E,E)-8,10-DODECADIEN-1-OL see CNG760
8E,10E-DODECADIEN-1-OL see CNG760
trans-8,trans-10-DODECADIEN-1-OL see CNG760
DODECAHYDRODIPHENYLAMINE see DGT600
DODECAHYDRO-7,14-METHANO-2H,6H-DIPYRIDO(1,2-a:1′,2′-e)(1,5)DIAZOCINE
    see SKX500
DODECAHYDROPHENYLAMINE NITRITE see DGU200
γ-DODECALACTONE see OES100
Δ-DODECALACTONE see DXS700, HBP450
1,12′-DODECAMETHYLENEDIAMINE see DXW800
DODECAMETHYLPENTASILOXANE see DXS800
n-DODECAN (GERMAN) see DXT200
1-DODECANAL see DXT000
DODECANAMINE ACETATE see DXW050
1-DODECANAMINE (9CI) see DXW000
DODECANAMINE HYDROCHLORIDE see DKW100
1-DODECANAMINE, HYDROCHLORIDE (9CI) see DKW100
1-DODECANAMINIUM, N-(CARBOXYMETHYL)-N,N-DIMETHYL-, HYROXIDE, in-
    ner salt see LBU200
DODECANE see DXT200
1,12-DODECANEDIAMINE see DXW800
1,12′-DODECANEDIAMINE see DXW800
DODECANENITRILE see DXT400
1-DODECANETHIOL see LBX000
tert-DODECANETHIOL see DXT800
DODECANOIC ACID see LBL000
DODECANOIC ACID BENZYL ESTER see BEU750
DODECANOIC ACID, CADMIUM SALT (9CI) see CAG775
DODECANOIC ACID-2,3-DIHYDROXYPROPYL ESTER see MRJ000
DODECANOIC ACID, 2-THIOCYANATOETHYL ESTER see LBO000
1-DODECANOL see DXV600
n-DODECANOL see DXV600
1-DODECANOL ACETATE see DXV400
DODECANOL ACETATE see DXV400
DODECANOL, ETHOXYLATE see DXY000
DODECANOL-ETHYLENE OXIDE (9.5 moles) CONDENSATE see DXY000
DODECANOLIDE-1,4 see OES100
1-DODECANOL, 2-OCTYL- see OEW100
DODECANOL, POLYETHOXYLATED see DXY000
1-((n-DODECANOYLOXY)METHYL)-3-METHYLIMIDAZOLIUM CHLORIDE see
    HMG500
3-(DODECANOYLOXYMETHY)-1-METHYL-1H-IMIDAZOLIUM CHLORIDE see
    MKR150
DODECANOYL PEROXIDE see LBR000
DODECAN-1-YL ACETATE see DXV400
1,6,10-DODECATRIEN-3-OL, 3,7,11-TRIMETHYL-, ACETATE, (S-(Z))- see
    NCN800
4,6,10-DODECATRIEN-3-ONE, 7,11-DIMETHYL- see DRR700
DODECATRIETHYLAMMONIUM BROMIDE see DXU200
5,7,11-DODECATRIYN-1-OL see DXU250
DODECAVITE see VSZ000
2-DODECENAL see DXU280
DODECENE EPOXIDE see DXU400
9a-DODECENOATE see DXU600
(Z)-7-DODECEN-1-OL see DXU800
(Z)-7-DODECEN-1-OLACETATE see DXU830
7-DODECEN-1-OL, ACETATE, (Z)- see DXU830
9-DODECEN-1-OL, ACETATE, (Z)- see GJU050
(Z)-7-DODECENYL ACETATE see DXU830
(Z)-9-DODECENYL ACETATE see GJU050
cis-7-DODECENYL ACETATE see DXU830
cis-9-DODECENYL ACETATE see GJU050
DODECENYLSUCCINIC ANHYDRIDE see DXV000
DODECOIC ACID see LBL000
DODECONIUM see HEF200
n-DODECYL ACETATE see DXV400
DODECYL ACETATE see DXV400
n-DODECYL ALCOHOL see DXV600
DODECYL ALCOHOL see DXV600
DODECYL ALCOHOL ACETATE see DXV400
DODECYL ALCOHOL CONDENSED with 23 MOLES ETHYLENE OXIDE see
    LBU000

DODECYL ALCOHOL CONDENSED with 4 MOLES ETHYLENE OXIDE see LBS000
DODECYL ALCOHOL CONDENSED with 7 MOLES ETHYLENE OXIDE see LBT000
DODECYL ALCOHOL, ETHOXYLATED see DXY000
DODECYL ALCOHOL, HYDROGEN SULFATE, SODIUM SALT see SIB600
1-DODECYL ALDEHYDE see DXT000
1-DODECYLAMINE see DXW000
n-DODECYLAMINE see DXW000
DODECYLAMINE see DXW000
1-DODECYLAMINE ACETATE see DXW050
DODECYLAMINE, ACETATE see DXW050
n-DODECYLAMINE HYDROCHLORIDE see DKW100
DODECYLAMINE, HYDROCHLORIDE see DKW100
1-DODECYL-4-AMINOQUINALDINIUM ACETATE see AJS750
N-DODECYL-4-AMINOQUINALDINIUM ACETATE see AJS750
n-DODECYLAMMONIUM CHLORIDE see DKW100
DODECYLAMMONIUM CHLORIDE see DKW100
DODECYL AMMONIUM SULFATE see SOM500
DODECYLBENZENE see PEW500
DODECYL BENZENE SODIUM SULFONATE see DXW200
DODECYL BENZENESULFONATE see DXW400
DODECYLBENZENESULFONIC ACID (DOT) see LBU100
n-DODECYLBENZENESULFONIC ACID see LBU100
DODECYLBENZENESULFONIC ACID, compd. with 2,2',2''-NITRILO-TRIS(ETHANOL) (1:1) see TJK800
DODECYLBENZENESULFONIC ACID SODIUM SALT see DXW200
DODECYLBENZENESULFONIC ACID TRIETHANOLAMINE SALT see TJK800
DODECYLBENZENESULPHONATE, SODIUM SALT see DXW200
DODECYLBENZENESULPHONIC ACID see LBU100
DODECYLBENZENSULFONAN SODNY (CZECH) see DXW200
DODECYLBENZYL CHLORIDE see DXW600
DODECYLBETAINE see LBU200
DODECYLBIS(AMINOETHYL)GLYCINE HYDROCHLORIDE see DYA850
N-DODECYL-N,N-BIS(2-HYDROXYETHYL)BENZENEMETHANAMINIUM CHLORIDE see BDJ600
DODECYLDIAMINE see DXW800
N-DODECYLDIMETHYLAMINE see DRR800
DODECYLDIMETHYLAMINE see DRR800
N-DODECYLDIMETHYLAMINE OXIDE see DRS200
DODECYLDIMETHYLAMINE OXIDE see DRS200
(DODECYLDIMETHYLAMMONIO)ACETATE see LBU200
N-DODECYL-N,N-DIMETHYLBENZENEMETHANAMINIUM BROMIDE see BEO000
DODECYL DIMETHYL BENZYLAMMONIUM CHLORIDE see BEM000
DODECYLDIMETHYLBETAINE see LBU200
DODECYLDIMETHYL(2-PHENOXYETHYL)AMMONIUM BROMIDE see DXX000
DODECYL-DI(β-OXYAETHYL)-BENZYL-AMMONIUMCHLORID see BDJ600
1,12'-DODECYLENEDIAMINE see DXW800
DODECYLESTER KYSELINY GALLOVE see DXX200
1-DODECYL-1-ETHYLPIPERIDINIUM BROMIDE see DXX100
DODECYL GALLATE see DXX200
N-DODECYLGUANIDINACETAT (GERMAN) see DXX400
N-DODECYLGUANIDINE ACETATE see DXX400
DODECYLGUANIDINE ACETATE see DXX400
DODECYLGUANIDINE ACETATE with SODIUM NITRITE (3:5) see DXX600
DODECYLGUANIDINE HYDROCHLORIDE see DXX800
1-DODECYLHEXAHYDRO-1H-AZEPINE-1-OXIDE see DXX875
1-DODECYLHEXAMETHYLENIMINE-N-OXIDE see DXX875
α-DODECYL-ω-HYDROXY-POLYOXYETHYLENE see DXY000
2-DODECYLISOQUINOLINIUM BROMIDE see LBW000
1-DODECYL MERCAPTAN see LBX000
m-DODECYL MERCAPTAN see LBX000
DODECYL MERCAPTAN see LBX000
tert-DODECYLMERCAPTAN see DXT800
tert. DODECYLMERKAPTAN (CZECH) see DXT800
DODECYL METHACRYLATE see DXY200
3-DODECYL-1-METHYL-2-PHENYLBENZIMIDAZOLIUM FERRICYANIDE see TNH750
1-DODECYL-3-METHYL-2-PHENYL-1H-BENZIMIDAZOLIUM, HEXACYANOFER-RATE(III) see TNH750
DODECYL-2-METHYL-2-PROPENOATE see DXY200
4-DODECYLMORPHOLINE-4-OXIDE see DXY300
4-DODECYLMORPHOLINE-N-OXIDE see DXY300
DODECYL NITRATE see NEE000
2-(DODECYLOXY)ETHANOL HYDROGEN SULFATE SODIUM SALT see DYA000
2-(2-(2-(DODECYLOXY)ETHOXY)ETHOXY)ETHANESULFONIC ACID, SODIUM SALT see SIC000
2-(2-(2-(DODECYLOXY)ETHOXY)ETHOXY)ETHANOL HYDROGEN SULFATE SODIUM SALT see SIC000
N-(2-DODECYLOXYETHYL)-N-METHYL-2-(PYRROLIDINYL)ACETAMIDE HYDRO-CHLORIDE see DXY400
DODECYLPHENOL (mixed isomers) see DXY600
DODECYL PHENYLMERCURI SULFIDE see DYA400

1-DODECYLPIPERIDINE-1-OXIDE see DXY700
1-DODECYLPIPERIDINE-N-OXIDE see DXY700
DODECYL-POLYAETHYLENOXYD-AETHER (GERMAN) see DXY000
DODECYL POLY(OXYETHYLENE)ETHER see DXY000
1-DODECYLPYRIDINIUM CHLORIDE see DXY725, LBX050
N-DODECYLPYRIDINIUM CHLORIDE see DXY725, LBX050
DODECYLPYRIDINIUM CHLORIDE see DXY725, LBX050
1-DODECYL-2-PYRROLIDINONE see DXY750
N-DODECYLPYRROLIDINONE see DXY750
N-DODECYLSARCOSINE SODIUM SALT see DXZ000
DODECYL SODIUM ETHOXYSULFATE see DYA000
DODECYL SODIUM SULFATE see SIB600
DODECYL SODIUM SULFOACETATE see SIB700
DODECYL SULFATE see MRH250
DODECYL SULFATE, SODIUM SALT see SIB600
DODECYLSULFURIC ACID see MRH250
n-DODECYL THIOCYANATE see DYA200
tert-DODECYLTHIOL see DXT800
(DODECYLTHIO)PHENYLMERCURY see DYA400
DODECYL-p-TOLYL TRIMETHYL AMMONIUM CHLORIDE see DYA600
DODECYLTRICHLOROSILANE see DYA800
DODECYLTRIMETHYLAMMONIUM BROMIDE see DYA810
6-DODECYL-2,2,4-TRIMETHYL-1,2-DIHYDROQUINOLINE see SAU480
DODEMORFE (FRANCE) see COX000
DODEX see VSZ000
DODGUADINE see DXX400
DODICIN HYDROCHLORIDE see DYA850
DODINE see DXX400
DODINE ACETATE see DXX400
DODINE, mixture with GLYODIN see DXX400
DODINE with SODIUM NITRITE (3:5) see DXX600
1,4-DOEA-5,8-DAPFA (RUSSIAN) see DMM400
DOF see DVK600
DOFSOL see VSK600
DOG HOBBLE see DYA875
DOG LAUREL see DYA875
DOGMATIL see EPD500
DOGMATYL see EPD500
DOG PARSLEY see FMU200
DOG POISON see FMU200
DOGQUADINE see DXX400
DOISYNOESTROL see BIT000
DOISYNOLIC ACID see DYB000
DOJYOPICRIN see CKN500
DOKIRIN see BLC250
DOKTACILLIN see AIV500
DOL see BBQ750
DOLAN, combustion products see ADX750
DOLADENE see DJM800, XCJ000
DOLAFLUX see HGM100
DOLANTAL see DAM700
DOLANTIN see DAM700
DOLANTIN HYDROCHLORIDE see DAM700
DOLANTIN-N-OXIDE HYDROCHLORIDE see DYB250
DOLANTOL see DAM700
DOLAREN see DAM700
DOLARGAN see DAM700
DOLCOL see PIZ000
DOLCO MOUSE CEREAL see SMN500
DOLCONTRAL see DAM600, DAM700
DOLCYMENE see CQI000
DOLEAN pH 8 see ADA725
DOLENAL see DAM700
DOLENE see DAB879, PNA500
DOLENOL see DAM700
DOLEN-PUR see HCD250
DOLESTAN see BAU750
DOLESTINE see DAM700
DOLGIN see IIU000
DOL GRANULE see BBQ500
DOLICUR see DUD800
DOLIGUR see DUD800
DOLIN see DAM700
DOLIPOL see BSQ000
DOLIPRANE see HIM000
DOLISINA see DII200
DOLKWAL AMARANTH see FAG020
DOLKWAL BRILLIANT BLUE see FAE000
DOLKWAL ERYTHROSINE see FAG040
DOLKWAL INDIGO CARMINE see FAE100
DOLKWAL ORANGE SS see TGW000
DOLKWAL PONCEAU 3R see FAG018
DOLKWAL SUNSET YELLOW see FAG150
DOLKWAL TARTRAZINE see FAG140
DOLKWAL YELLOW AB see FAG130

DOLKWAL YELLOW OB see FAG135
DOLLS EYES see BAF325
DOLMIX see BBQ750
DOLOBID see DKI600
DOLOBIL see DKI600
DOLOBIS see DKI600
DOLOCAP see PNA500
DOLOCHLOR see CKN500
DOLOCONEURASE see DYB300
DOLOGAL see DAM700
DOLOMIDE see SAH000
DOLOMITE see CAO000
DOLONEURINE see DAM700
DOLONIL see PDC250
DOLOPETHIN see DAM700
DOLOPHINE see MDO750, MDP000
d-DOLOPHINE HYDROCHLORIDE see MDP240
DOLOPHINE HYDROCHLORIDE see MDP000, MDP750
DOLOPHIN HYDROCHLORIDE see MDP750
DOLOSAL see DAM600, DAM700
DOLOVIN see IDA000
DOLOXENE see DAB879, DYB400, PNA500
DOLPHINE see MDP000
DOLSIMA see DII200
DOLSIN see DAM600
DOLVANOL see DAM700
DOM see BJR000
DOMAGK'S T.B.1 CONTEBEN see FNF000
DOMAIN see SNJ350
DOMALIUM see DCK759
DOMAR see CAT775, PIH100
DOMARAX see IPU000
DOMATOL 88 see AMY050
DOMATOL see AMY050
DOMESTROL see DKA600
DOMF see MCV000
DOMICAL see EAI000
DOMICILLIN see SEQ000
DOMINAL see DYB600
DOMINAL HYDROCHLORIDE see DYB800
DOMOSO see DUD800
DOMPERIDONE see DYB875
DOMUCOR see ELH600
DON see DCQ400
DONAXINE see DYC000
DONMOX see AAI250
DONOPON see SBG500
DONOREST see CKI750
DONOVAN'S SOLUTION see ARI500
DOOJE see HBT500
DOMAKOL see FNF000
DOP see DVL700
l-DOPA see DNA200
(−)-DOPA see DNA200
DOPAFLEX see DNA200
l-DOPA HYDROCHLORIDE see DYC200
DOPAL see DNA200
DOPAMET see DNA800, MJE780
l-DOPA METHYL ESTER see DYC300
DOPAMINE see DYC400
DOPAMINE CHLORIDE see DYC600
DOPAMINE HYDROCHLORIDE see DYC600
DOPAN see DYC700
DOPANE see DYC700
DOPARKINE see DNA200
DOPASOL see DNA200
DOPATEC see MJE780
DOPEGYT see DNA800, MJE780
DOPIDRIN see DBA800
DOPN see NJN000
DOPRAM see ENL100, SLU000
DOPRIN see DNA200
DOPTAEC see DNA800
DORAL see VSZ100
DORANTAMIN see WAK000
DORAPHEN see PNA500
DORBANE see DMH400
DORBANEX see DMH400
DORCOSTRIN see DAQ800
DOREVANE see IDA500
DORICO see ERD500
DORICO SOLUBLE see ERE000
DORIDEN see DYC800
DORIDEN-SED see DYC800
DORINAMIN see BBW500

DORISUL see SNN300
DORLOTYN see AMX750
DORLYL see PKQ059
DORM see AFS500
DORMABROL see MQU750
DORMAL see CDO000
DORMALLYL see AFS500
DORMATE see MBW750
DORME see PMI750
DORMETHAN see DBE200
DORMIDIN see EQL000
DORMIGENE see BNP750
DORMIGOA see QAK000
DORMIN see TEO250
DORMIPHEN see EQL000
DORMIRAL see EOK000
DORMITURIN see BNK000
DORMODOR see DAB800
DORMOGEN see QAK000
DORMONAL see BAG000
DORMONE see DAA800
DORMOSAN see EQL000
DORMUPHAR see SKS700
DORMUTIL see QAK000
DORMWELL see SKS700
DORMYTAL see AMX750
DORNASE see EDK650, PAF575
DORNAVA see EDK650, PAF575
DORNAVAC see EDK650, PAF575
DORNWAL see ALY250
DORNWALL see ALY250
DORSACAINE see OPI300
DORSACAINE HYDROCHLORIDE see OPI300
DORSEDIN see QAK000
DORSIFLEX see MFD500
DORSILON see MFD500
DORSITAL see NBT500
DORSULFAN see SNN500
DORVICIDE A see BGJ750
DORVON see SMQ500
DORVON FR 100 see SMQ500
DORYL (PHARMACEUTICAL) see CBH250
DOS see BJS250
DOSAFLO see MQR225
DOSAGRAN see MQR225
DOSANEX see MQR225
DOSANEX FL see MQR225
DOSANEX MG see MQR225
DOSEGRAN see AFL750
DOSULEPIN see DYC875
DOSULEPIN CHLORIDE see DPY200
DOSULEPIN HYDROCHLORIDE see DPY200
D.O.T. see DUP300
DOTAN see CDY299
DOTC see DVN300
DOTG see BKK750
DOTG ACCELERATOR see DXP200
DOTHEIPIN HYDROCHLORIDE see DPY200
DOTHIEPIN see DYC875
DOTMENT 324 see AHE250
DOTRIACONTANE see DYC900
DOTYCIN see EDH500
DOUBLE STRENGTH see TIX500
DOVENIX see HLJ500
DOVIP see FAB600
DOW 209 see SMR000
DOW 360 see SMQ500
DOW 456 see SMQ500
DOW 665 see SMQ500
DOW 860 see SMQ500
DOW 874 see CGW300
DOW 1329 see DGD800
DOW 1683 see SMQ500
DOWANOL see CBR000
DOWANOL 33B see PNL250
DOWANOL-50B see DWT200
DOWANOL 62B see TNA000
DOWANOL DB see DJF200
DOWANOL DE see CBR000
DOWANOL DM see DJG000
DOWANOL DPM see DWT200
DOWANOL EB see BPJ850
DOWANOL EE see EES350
DOWANOL EIPAT see INA500
DOWANOL EM see EJH500

DOWANOL EP see PER000
DOWANOL EPH see PER000
DOWANOL PM see PNL250
DOWANOL (R) PMA GLYCOL ETHER ACETATE see PNL265
DOWANOL PM GLYCOL ETHER see PNL250
DOWANOL PPH GLYCOL ETHER see PNL300
DOWANOL TE see EFL000
DOWANOL TMAT see TJQ750
DOWANOL TPM see MEV500
DOWCC 132 see COD850
DOWCHLOR see CDR750
DOWCIDE 1 see BGJ250
DOWCIDE 7 see PAX250
DOWCIDE 1 ANTIMICROBIAL see BGJ250
DOWCO 109 see BQU500
DOWCO 118 see DGD800
DOWCO 133 see DYD200
DOWCO 139 see DOS000
DOWCO 159 see DYD400
DOWCO 160 see DYD600
DOWCO 161 see EOQ500
DOWCO-163 see CLP750
DOWCO 169 see PEV500
DOWCO 177 see DYD800
DOWCO 179 see CMA100
DOWCO-183 see DYE200
DOWCO 184 see CEG550
DOWCO 186 see HON000
DOWCO 187 see AGU500
DOWCO-213 see CQH650
DOWCO 217 see CMA250
DOWCO 233 see TJE890
DOWCO 290 see DGJ100
DOWCO 356 see TJK100
DOW CORNING 200 see HEE000
DOW CORNING 345 see DAF350
DOW CORNING 346 see PJR000
DOW CORNING 345 FLUID see DAF350
DOW-CORNING 200 FLUID-LOT No. AA-4163 see DUB600
DOW CORNING SILICONE FLUID and FLUOROHYDROCARBON see XGA000
DOW DEFOLIANT see SFU500
DOW DORMANT FUNGICIDE see SJA000
DOWELL L 37 see NEI100
DOW ET 14 see RMA500
DOW ET 57 see RMA500
DOWFLAKE see CAO750
DOWFROST see PML000
DOWFROTH 250 see MKS250
DOWFUME 40 see EIY500
DOWFUME see MHR200
DOWFUME EB-5 see DYE400
DOWFUME EDB see EIY500
DOWFUME MC-2 SOIL FUMIGANT see MHR200
DOWFUME N see DGG000
DOWFUME W-8 see EIY500
DOW GENERAL see BRE500
DOW GENERAL WEED KILLER see BRE500
DOWICIDE 2 see TIV750
DOWICIDE 4 see CHN500
DOWICIDE 6 see TBT000
DOWICIDE 7 see PAX250
DOWICIDE 9 see CFI750
DOWICIDE 31 see SFV500
DOWICIDE see BGJ750
DOWICIDE A see BGJ750
DOWICIDE A & A FLAKES see BGJ750
DOWICIDE B see SKK500, TIV750
DOWICIDE EC-7 see PAX250
DOWICIDE G see PAX250
DOWICIDE G-ST see SJA000
DOWICIDE Q see CEG550
DOWICIDE 2S see TIW000
DOWICIL 75 see CEG550
DOWICIL 100 see CEG550
DOWIZID A see BGJ750
DOW LATEX 612 see SMR000
DOW LATEX 874 see CGW300
DOWLEX see PKF750
DOWLEX FILM see PJS750
DOW MCP AMINE WEED KILLER see CIR250
DOW MX 5514 see SMQ500
DOW MX 5516 see SMQ500
DOWMYCIN E see EDJ500
DOW PENTACHLOROPHENOL DP-2 ANTIMICROBIAL see PAX250
DOWPEN V-K see PDT750

DOW-PER see PCF275
DOWPON see DGI400, DGI600
DOWPON M see DGI400
DOW SEED DISINFECTANT No. 5 see TBO500
DOW SELECTIVE see BPG250
DOW SELECTIVE WEED KILLER see BRE500
DOW SODIUM TCA INHIBITED see TII500
DOW SODIUM TCA SOLUTION see TII250
DOWSPRAY 17 see CPK500
DOWTHERM 209 see PNL250
DOWTHERM see PFA860
DOWTHERM A see PFA860
DOWTHERM E see DEP600
DOWTHERM SR 1 see EJC500
DOW-TRI see TIO750
DOWZENE DHC see PIK000
DOXAL see TMJ150
DOXAPRAM see ENL100
DOXAPRAM HYDROCHLORIDE HYDRATE see SLU000
DOXCIDE 50 see CDW450
DOXEPHRIN see DBA800
DOXEPIN see DYE409
DOXEPIN HYDROCHLORIDE see AEG750
DOXERGAN see AFL750
DOX HYDROCHLORIDE see HKA300
DOXICICLINA (ITALIAN) see DYE425
DOXIFLURIDINE see DYE415
DOXIGALUMICINA see HGP550
DOXINATE see DJL000
DOXIUM see DMI300
DOXO see DAQ800
DOXOL see DJL000
DOXORUBICIN see AES750, HKA300
DOXORUBICIN HYDROCHLORIDE see HKA300
DOXYCYCLINE see DYE425
DOXYCYCLINE HYCLATE see HGP550
DOXYCYCLINE HYDROCHLORIDE see HGP550
DOXYFED see DBA800, MDT600
DOXY-II see HGP550
DOXYLAMINE see DYE500
DOXYLAMINE SUCCINATE (1:1) see PGE775
DOXYLAMINE SUCCINATE see PGE775
DOXYPYRROMYCIN see DAY835
DOXY-TABLINEN see HGP550
DOZAR see DPJ400
2,4-DP see DGB000
2-(2,4-DP) see DGB000
2,2-DPA see DGI600
n-DPA see PNR750
DPA see DGI000, DVX800, HET500
D-P-A INJECTION see PAG200
DPA SODIUM see PNX750
DPBS see DFY400
DPC see DXY725, LBX050
DPD see DJV200
D & P DOUBLE O CRABGRASS KILLER see PLC250
DPE-HCl see DWD200
DPF-1 see DYE550
DPF see DYE550
DPG see DWC600
DPG ACCELERATOR see DWC600
2,4-D PGBE see DFZ000
DPH see DKQ000, DNU000
DPID see DLH600
DPMA see DMB200
DPN see CNF390, NKB700
DPNA see NKB700
DPP see PAK000, PEK250
DPPD see BLE500
2,4-D PROPYLENE GLYCOL BUTYL ETHER ESTER see DFZ000
DPS see PGI750
3,5-DPT see DWO950
DPT see DCA600
DPTA see DCB100
DPX 1108 see ANG750
DPX 1410 see DSP600
DP X 1410 see MME809
DPX 3217 see COM300
DPX 3674 see HFA300
DPX 3217M see COM300
DQDA 1868 see PJS750
DQUIGARD see DGP900
DQV-K see PDT750
DQWA 0355 see PJS750
DR-15771 see SBE500

DRABET see BSQ000
DRACONIC ACID see AOU600
DRACYLIC ACID see BCL750
DRAGIL-P see SLL000
DRAGOCAL see CAS750
DRAKEOL see MQV750, MQV875
DRALZINE see HGP500
DRAMAMIN see DYE600
DRAMAMINE see DYE600
DRAMARIN see DYE600
DRAMYL see DYE600
DRAPEX 4.4 see FAB920
DRAPOLENE see AFP250
DRAPOLEX see BBA500
DRASIL 507 see CNH125
DRAT see CJJ000
DRAWIN 755 see MPU250
DRAWINOL see CBW000
DRAZA see DST000
DRAZINE see PDV700
DRAZOXOLON see MLC250
DRAZOXOLONE see MLC250
DRB see DVF800
DRC-714 see BIM000
DRC 1339 see CLK215
DRC-1,339 see CLK230
DRC 3340 see DTN200
DRC 3341 see MIB750
DRC 3345 see PFS350
DRC-4575 see AMU125
DREFT see SIB600
DRENAMIST see VGP000
DRENE see SON000
DRENOBYL see DAL000
DRENOL see CFY000
DRENUSIL-R see RDK000
DREWMULSE POE-SMO see PKL100
DREWMULSE TP see OAV000
DREWMULSE V see OAV000
DREWSORB 60 see SKV150
DREXEL see DXQ500
DREXEL DEFOL see SFS000
DREXEL DIURON 4L see DXQ500
DREXEL DSMA LIQUID see DXE600
DREXEL METHYL PARATHION 4E see MNH000
DREXEL PARATHION 8E see PAK000
DREXEL-SUPER P see DMC600
DRIBAZIL see DYE700
DRICOL see AHL500
DRI-DIE PESTICIDE 67 see SCH000
DRIDOL see DYF200
DRIERITE see CAX500
DRI-KIL see RNZ000
DRILL TOX-SPEZIAL AGLUKON see BBQ500
DRINALFA see DBA800, MDT600
DRINOX see AFK250, HAR000
DRINUPAL HYDROCHLORIDE see CKB500
DRIOL see DYE700
DRIOL-LABAZ see DYE700
DRISDOL see VSZ100
DRISTAN INHALER see PNN400
DRI-TRI see SJH200
DROCODE see DKW800
DROCTIL see ERE200
DROGENIL see FMR050
DROLEPTAN see DYF200
DROMETRIZOLE see HML500
DROMILAC see DBV400
DROMISOL see DUD800
levo-DROMORAN see LFG000
racemic DROMORAN see MKR250
DROMORAN see MKR250
DROMORAN-HYDROBROMIDE see HMH500
DROMORAN HYDROBROMIDE see MDV250
l-DROMORAN TARTRATE see DYF000
DROMOSTAT see TKU675
DROMYL see DYE600
DRONACTIN see PCI500
DRONCIT see BGB400
DROPCILLIN see BDY669
DROPERIDOL see DYF200
DROP LEAF see SFS000
DROPP see TEX600
DROPRENILAMINE HYDROCHLORIDE see CPP000
DROPSPRIN see SAH000

DROTEBANOL see ORE000
DROXAROL see BPP750
DROXARYL see BPP750
DROXOL see MDQ250
DROXOLAN see DAQ400
DRP 859025 see TND500
DRUMULSE AA see OAV000
DRUPINA 90 see BJK500
DRW 1139 see ALA500
DRY AND CLEAR see BDS000
DRY ICE see CBU250
DRY ICE (UN 1845) (DOT) see CBU250
DRYISTAN see BBV500
DRYLIN see TKX000
DRYLISTAN see BBV500
DRY MIX No. 1 see CPK500
DRYOBALANOPS CAMPHOR see BMD000
DRYPTAL see CHJ750
DS-36 see SNL800
DS-15647 see DAB400
DS 18302 see CBW000
DS see DKB110
DSDP see DJA400
DSE see DXD200
D 58SI see DLU900
DS-M-1 see DBD750
DSMA LIQUID see DXE600
DS-Na see SFW000
2,4-D SODIUM SALT see SGH500
DSP see SJH090
DSPT see THQ900
DSS see DJL000, SOA500
DS SUBSTANDE see AJX500
DST 50 see SMR000
DST see DME000
DSTDP see DXG700
DSTP see DXG700
DT see DNC000, DWP000, TFX790
DTA see DCB100, DLK200
D-40TA see CKL250
DTAS see MJK750
DTB see DXL800
DTBN see DEF090
DTBP see BSC750
DTBT see BLN750
DTDP see DXQ200
DTE see DXN350
DTHYD see TFX790
DTIC see DAB600
DTIC-DOME see DAB600
DTMC see BIO750
DTPA see DJG800
DTPA CALCIUM TRISODIUM SALT see CAY500
DTP HYDROCHLORIDE see DXO400
N-D1-TRIMETHYL-3,3-DI-2-THIENYLALLYLAMINE HYDROCHLORIDE see TLQ250
DTS see DNV800
D-DTT see DXO800
DTT see DXO775, DXO800
DU-717 see CIT625
DU 21220 see RLK700
DU 112307 see CJV250
DU see DCQ600
DU-A 1 see SCQ000
DU-A 2 see SCQ000
DU-A 3 see SCQ000
DU-A 4 see SCQ000
DU-A 3C see SCQ000
DUAL see MQQ450
DUAMINE see MCH525
DUASYN ACID YELLOW RRT see SGP500
DUATOK see TEX250
DUAZOMYCIN B see ASO510, ASO501
DUBIMAX see NAE000
DUBORIMYCIN see DAC300
DUBOS CRUDE CRYSTALS see TOG500
DUBRONAX see SOA500
DUCKALGIN see SEH000
DUCOBEE see VSZ000
O-DUE see TAG750
DUFALONE see BJZ000
DUGERASE see AIT250
DUGRO see MMN750
DUKERON see TIO750
DUKSEN see DCK759

DULCIDOR see GFA000
DULCINE see EFE000
DULCITOLDIEPOXIDE see DCI600
DULCOLAN see PPN100
DULCOLAX see PPN100
DUL-DUL (PUERTO RICO) see CAK325
DULL 704 see PJY500
DULSIVAC see DJL000
DULZOR-ETAS see SGC000
DUMASIN see CPW500
DUMBCANE see DHB309
DUMB PLANT see DHB309
DUMITONE see SOA500
DUMOCYCIN see TBX250
DUMOGRAN see MPN500
DUNCAINE see DHK400
DUNCAINE HYDROCHLORIDE see DHK600
DUNERYL see EOK000
DUNKELGELB see PEJ500
DUODECANE see DXT200
DUODECIBIN see VSZ000
DUODECYL ALCOHOL see DXV600
DUODECYLIC ACID see LBL000
DUODECYLIC ALDEHYDE see DXT000
DUOGASTRONE see CBO500
DUO-KILL see COD000, DGP900
DUOLAX see DMH400
DUOLIP see TEQ500
DUOLUTON see NNL500
DUOMYCIN see CMA750
DUOSAN (pesticide) see MAP300
DUOSAN see MAP300
DUOSOL see DJL000
DUO-STREPTOMYCIN see SLY200
DUOTRATE see PBC250
DUPHAPEN see PAQ200
DUPHAR see CKM000
DUPICAL see OBW100
DUPLEX RED LAKE CD 20-5925 see CMS150
DUPLEX TOLUIDINE RED L 20-3140 see MMP100
DUPONOL 80 see OFU200
DUPONOL see SIB600
DU PONT 326 see DGD600
DU PONT 732 see BQT750
DU PONT 1991 see BAV575
DU PONT HERBICIDE 326 see DGD600
DU PONT HERBICIDE 732 see BQT750
DU PONT HERBICIDE 976 see BMM650
DU PONT INSECTICIDE 1179 see MDU600
DU PONT INSECTICIDE 1519 see DVS000
DU PONT INSECTICIDE 1642 see MPS250
DU PONT PC CRABGRASS KILLER see PLC250
DU PONT WK see DXY000
DUPRENE see PJQ050
DURABIOTIC see BFC750
DURABOL see DYF450
DURABOLIN see DYF450
DURA CLOFIBRAT see ARQ750
DURAD see TNP500
DURAD MP280ᴿ HYDRAULIC FLUID see DYF500
DURA-ESTRADIOL see EDS100
DURAFUR BLACK R see PEY500
DURAFUR BLACK RC see PEY650
DURAFUR BROWN see ALL750
DURAFUR BROWN MN see DBO400
DURAFUR BROWN 2R see ALL750
DURAFUR BROWN RB see ALT250
DURAFUR DEVELOPER C see CCP850
DURAFUR DEVELOPER D see NAW500
DURAFUR DEVELOPER G see REA000
DURALUTON see HNT500
DURAMAX see ADA725
DURAMITE see CAT775
DURAN see DXQ500
DURANATE EXP-D 101 see PKO750
DURANIT see SMR000
DURANITRAT see CCK125
DURAN MP280ᴿ see DYF500
DURANOL BLUE GREEN B see DMM400
DURANOL BLUE TR see CMP070
DURANOL BRILLIANT BLUE B see MGG250
DURANOL BRILLIANT BLUE CB see TBG700
DURANOL BRILLIANT BLUE VIOLET BR see DBY700
DURANOL BRILLIANT RED T 2B see AKI750
DURANOL BRILLIANT VIOLET B see AKP250

DURANOL BRILLIANT VIOLET BR see DBY700
DURANOL BRILLIANT YELLOW G see MEB750
DURANOL ORANGE G see AKP750
DURANOL PRINTING BLUE GREEN B see DMM400
DURANOL RED 2B see AKE250
DURANOL RED X3B see DBX000
DURANOL VIOLET WR see DBP000
DURANTA REPENS see GIW200
DURA-PENITA see BFC750
DURAPHOS see MQR750
DURAPROST see OLW600
DURASORB see DUD800
DURA-TAB S.M. AMINOPHYLLINE see TEP500
DURATION see AEX000
DURATOX see DAP400, MIW100
DURAVOS see DGP900
DURAX see CPI250
DURAZOL BLUE 5G see CMO600
DURAZOL GRAY B see CMN230
DURAZOL RED 2B see CMO885
DURAZOL RED 2BP see CMO885
DURCAL 10 see CAT775
DUR-EM 204 see GGR200
DURETHAN BK see PJY500
DURETTER see FBN100
DUREX see CBT750
DURFAX 80 see PKL100
DURGACET RUBINE B see CMP080
DURGACET YELLOW G see AAQ250
DURGASOL SCARLET GG SALT see DEO295
DURINDONE ORANGE R see CMU815
DURINDONE ORANGE RP see CMU815
DURINDONE PRINTING BLACK BL see CMU320
DURINDONE PRINTING ORANGE R see CMU815
DURINDONE PRINTING RED B see DNT300
DURINDONE RED B see DNT300
DURINDONE RED BP see DNT300
DUROCHROME BLUE OCG see CMP880
DUROCHROME CYANINE G see CMP880
DUROCHROME FAST CYANINE 6BN see CMP880
DUROFERON see FBN100
DUROFOL P see PKQ059
DUROID 5870 see TAI250
DUROLAX see PPN100
DUROMINE see DTJ400
DURONITRIN see TJL250
DUROPENIN see BFC750
DUROPROCIN see MDU300
DUROSPERSE YELLOW G see AAQ250
DUROTOX see PAX250
DUROX see AKO500
DURSBAN see CMA100
DURSBAN F see CMA100
DURSBAN METHYL see CMA250
DURTAN 60 see SKV150
DUSICNAN BARNATY (CZECH) see BAN250
DUSICNAN CERITY (CZECH) see CDB000
DUSICNAN KADEMNATY (CZECH) see CAH250
DUSICNAN VAPENATY (CZECH) see CAU250
DUSICNAN ZINECNATY (CZECH) see NEF500
DUSICNAN ZIRKONICITY (CZECH) see ZSA000
DUSITAN DICYKLOHEXYLAMINU (CZECH) see DGU200
DUSITAN SODNY (CZECH) see SIQ500
DUSOLINE see CMD750
DUSORAN see CMD750
DUSPAR 125B see HKS400
DU-SPREX see DER800
DUST M see RAF100
DUS-TOP see MAE250
DUTCH LIQUID see EIY600
DUTCH OIL see EIY600
DUTCH-TREAT see HKC500
DU-TER see HON000
DUTOM see SKQ400
DUVADILAN see VGA300, VGF000
DUVALINE see PPH050
DUVILAX BD 20 see AAX250
DUVOID see HOA500
DV-17 see DBC575
DV-79 see AJV500
DV 400 see PKB500
DV 714 see LEF400
DV see DAL600
DVEDEG (RUSSIAN) see DJE600
D3-VIGANTOL see CMC750

DW-61 see FCB100
DW 62 see DNV000, DNV200
DW3418 see BLW750
DWARF BAY see LAR500
DWARF LAUREL see MRU359
DWARF PINE NEEDLE OIL see PIH400
DWARF POINCIANA see CAK325
DWELL see EFK000
DWUBROMOETAN (POLISH) see EIY500
DWUCHLOROCZTEROFLUOROETAN (POLISH) see DGL600
DWUCHLORODWUETYLOWY ETER (POLISH) see DFJ050
DWUCHLORODWUFLUOROMETAN (POLISH) see DFA600
2,4-DWUCHLOROFENOKSYOCTOSY KWAS (POLISH) see DAA800
DWUCHLOROFLUOROMETAN (POLISH) see DFL000
DWUCHLOROPROPAN (POLISH) see PNJ400
DWUETYLOAMINA (POLISH) see DHJ200
DWUETYLOWY ETER (POLISH) see EJU000
DWUFENYLOGUANIDYNA (POLISH) see DWC600
DWUMETHYLOFORMAMID (POLISH) see DSB000
DWUMETYLOANILINA (POLISH) see DQF800
3,5-DWUMETYLOIZOKSAZOLU see DSK950
3,5-DWUMETYLOPIRAZOLU see DTU850
DWUMETYLOSULFOTLENKU (POLISH) see MAO250
DWUMETYLOWY SIARCZAN (POLISH) see DUD100
DWU-β-NAFTYLO-p-FENYLODWUAMINA (POLISH) see NBL000
DWUNITROBENZEN (POLISH) see DUQ200
DWUNITRO-o-KREZOL (POLISH) see DUS700
3,3′-DWUOKSYBENZYDYNA (POLISH) see DMI400
DWUSIARCZEK DWUBENZOTIAZYLU (POLISH) see BDE750
DX see AES750
DXEWMULSE POE-SML see PKG000
DXG see BSS550
DXM 100 see PJS750
DXMS see SOW000
(D)-XYLOSE see XQJ300
DYALL see PJS750
DYANACIDE see ABU500
DYAZIDE see CFY000
DYBAR see DTP400
DYCARB see DQM600
DYCHOLIUM see SGD500
DYCILL see DGE200
DYCLOCAINUM see BPR500
DYCLONE HYDROCHLORIDE see BPR500
DYCLONINE HYDROCLORIDE see BPR500
DYCLOTHANE see BPR500
DYDELTRONE see PMA000
DYDROGESTERONE and HYDROXYPROGESTERONE (9:10) see PMA250
DYE EVANS BLUE see BGT250
DYE FD AND C RED No. 4 see FAG050
DYE FDC RED 2 see FAG020
DYE FD&C RED No. 3 see FAG040
DYE FD & C RED No. 4 see FAG050
DYE FD & C YELLOW LAKE 6 see FAG150
DYE FDC YELLOW LAKE 6 see FAG150
DYE FD and C YELLOW NO. 5 see FAG140
DYE FDC YELLOW NO. 6 see FAG150
DYE FD & C YELLOW NO. 6 see FAG150
DYE GS see ALL750
DYE ORANGE No. 1 see FAG010
DYE QUINOLINE YELLOW see CMM510
DYE RED RASPBERRY see FAG020
DYESTROL see DKA600
DYE SUNSET YELLOW see FAG150
DYETONE see SFG000
DYFLOS see IRF000
DYFONATE see FMU045
DYGRATYL see DME300
DYKANOL see PJL750
DYKOL see DAD200
DYLAMON see BBV500
DYLAN see PJS750
DYLAN SUPER see PJS750
DYLAN WPD 205 see PJS750
DYLARK 111 see SEA500
DYLARK 230 see SEA500
DYLARK 231 see SEA500
DYLARK 232 see SEA500
DYLARK 238 see SEA500
DYLARK 250 see SMQ500
DYLARK 332 see SEA500
DYLENE 8 see SMQ500
DYLENE 9 see SMQ500
DYLENE see SMQ500
DYLENE 8G see SMQ500

DYLEPHRIN see VGP000
DYLITE F 40 see SMQ500
DYLITE F 40L see SMQ500
DYLOX see TIQ250
DYLOX-METASYSTOX-R see TIQ250
DYMADON see HIM000
DYMANTHINE see DTC400
DYMEL 22 see CFX500
DYMELOR see ABB000
DYMEX see ABH000
DYMID see DRP800
DYNA-CARBYL see CBM750
DYNACORYL see DJS200
DYNADUR see PKQ059
DYNALIN INJECTABLE see TET800
DYNAMICARDE see DJS200
DYNAMITE see DYG000
DYNAMONE see AES650
DYNAMUTILIN see TET800
DYNAPEN see DGE200
DYNAPHENIL see AOB500
DYNAPRIN see DLH600
DYNARSAN see ABX500
DYNASTEN see PAN100
DYNA-ZINA see DLH600, DLH630
DYNAZONE see NGE500
DYNEL see ADY250
DYNEL NYGL see ADY250
DYNERIC see CMX700
DYNEX see DXQ500
DYNH see PJS750
DYNK 2 see PJS750
DYNOMIN MM 9 see MCB050
DYNOMIN MM 75 see MCB050
DYNOMIN MM 100 see MCB050
DYNOMIN UI 16 see UTU500
DYNOMIN UM 15 see UTU500
DYNONE see EIH500
DYNOSOL see DUU600
DYNOTHEL see SKJ300
DYNOVAS see PAH250
DYODIN see DNF600
DYPERTANE COMPOUND see RDK000
DYP-97 F see LBR000
DYPHONATE see FMU045
DYPHYLLINE see DNC000
DYPNONE see MPL000
DYPRIN see MDT740
DYREN see UVJ450
DYRENE see DEV800
DYRENE 50W see DEV800
DYRENIUM see UVJ450
DYREX see TIQ250
DYSEDON see AFL750
DYSPNE-INHAL see VGP000
DYSPROSIUM see DYG400
DYSPROSIUM CHLORIDE see DYG600
DYSPROSIUM CITRATE see DYG800
DYSPROSIUM NITRATE see DYH100
DYSPROSIUM(III) NITRATE HEXAHYDRATE (1:3:6) see DYH000
DYSPROSIUM TRINITRATE see DYH100
DYTAC see UVJ450
DYTHERM X 214 see SEA500
DYTHOL see CMD750
DYTOL M-83 see OEI000
DYTOL S-91 see DAI600
DYTOL E-46 see OAX000
DYTOL F-11 see HCP000
DYTOL J-68 see DXV600
DYTRANSIN see IJG000
DYVON see TIQ250
DYZOL see DCM750
DZ see MDH500
DZhp-4K see CMP090


E¹ see EDV000
E² see EDO000
E2 see PJY100
E-3 see ELF500
E6 see PKO500
E-48 see DJV600
E 62 see PKQ059
E 102 see FAG140
E-103 see BSN000

E 104 see CMM510
E 110 see FAG150
E-111 see PNN300
E 125 see CMP620
E 127 see FAG040
E 130 see IBV050
E 131 see ADE500, CMM062
E 132 see FAE100
E 140 see CKN000
E 141 see DIS600
E 142 see ADF000
E 151 see BMA000
E 158 see DAP600
E-212 see VGZ000
E-236 see DUJ400
(±)-E-250 see DAZ118
(+)-E-250 see DAZ120
(−)-E-250 see DAZ125
E 261 see DTS500
E-298 see MRU080
E393 see SOD100
E 534 see PKB100
583E see POC750
E 600 see NIM500
E-646 see EAV700
686E see SMQ500
E 736 see TFD000
E-1059 see DAO500
E 1059 see DAO600
E 1440 see CCU250
E-2663 see CML835
E 3314 see HAR000
E 7256 see LBU100
EA 166 see GLS700
E-733-A see CQH000
EA 2277 see QVA000
EA 3547 see DDE200
EAA see EFS000
EAB see EOH500
EACA see AJD000
EACA KABI see AJD000
E-73 ACETATE see ABN000
EACS see AJD000
EAGLE GERMANTOWN see CBT750
EA-1 HYDROCHLORIDE see NCL300
EAK see ODI000
EAMN see ENR500
α-EARLEINE see GHA050
EARTHCIDE see PAX000
EARTH GALL see FAB100
EARTHNUT OIL see PAO000
EASEPTOL see HJL000
EASTBOND M 5 see PMP500
EASTER FLOWER see EQX000
EASTERN COTTONMOUTH VENOM see EAB200
EASTERN DIAMOND-BACK RATTLESNAKE VENOM see EAB225
EASTERN STATES DUOCIDE see WAT200
EAST INDIAN COPAIBA see GME000
EAST INDIAN LEMONGRASS OIL see LEG000
EAST INDIAN SANDALWOOD OIL see OHG000
EASTMAN 910 see MIQ075
EASTMAN 1334 see KHU050
EASTMAN 7663 see DJT800
EASTMAN 910 ADHESIVE see MIQ075
EASTMAN BLUE BNN see MGG250
EASTMAN FAST BLUE B-GLF see CMP060
EASTMAN INHIBITOR DHPB see DMI600
EASTMAN INHIBITOR HPT see HEK000
EASTMAN INHIBITOR RMB see HNH500
EASTMAN 910 MONOMER see MIQ075
EASTONE YELLOW GN see AAQ250
EASTOZONE 31 see BJT500
EASTOZONE 33 see BJL000
EASTOZONE see BJL000
EASY OFF-D see TIG250
EATAN see DLY000
EAZAMINE see DII200
EB 80 see PAT850
EB-382 see MGC200
EB see BGT250, EGP500
EBECRYL 110 see PER250
E-D-BEE see EIY500
EBERPINE see RDK000
EBERSPINE see RDK000
EBI see ISK000

EBIS see ISK000
EBNA see NKD500
EBNS see EHC800
EBONTA see EIT000
EBRANTIL see USJ000
EBS see FAG040
EBSERPINE see RDK000
EBUCIN see CAS750
EBZ see EDP000
E.C. 1.1.3.4 see GFG100
E.C. 3.1.1.3. see GGA800
E.C. 3.1.4.5 see EDK650, PAF575
E.C. 3.4.4.5 see CML850
E.C. 3.4.4.6 see CML850
E.C. 3.4.4.7 see EAG875
E.C. 3.2.1.18 see NCQ200
E.C. 3.2.1.23 see GAV100
E.C. 3.4.21.1 see CML850
E.C. 3.4.21.5 see TFU800
E.C. 3.4.4.13 see TFU800
E.C. 3.4.4.16 see BAC000
E.C. 3.4.4.24 see BMO000
E.C. 3.4.2.1.11 see EAG875
E.C. 3.4.21.14 see BAC000
E.C. 3.4.21.15 see SBI860
ECARAZINE see TGJ150
ECARAZINE HYDROCHLORIDE see TGJ150
E. CARINATUS VENOM see EAD600
ECARLATE GN see CMP620
ECATOX see PAK000
ECBOLINE see EDC565
ECBOLINE ETHANESULFONATE see EDC575
ECCOTHAL see TEM000
β-ECDYSONE see HKG500
β-ECDYSTERONE see HKG500
ECDYSTERONE see HKG500
ECF see EHK500
ECGONINE, METHYL ESTER, BENZOATE (ESTER) see CNE750
ECH see EAZ500
ECHIMIDINE see EAC500
ECHINOMYCIN see EAD500
ECHINOMYCIN A see EAD500
ECHIS CARINATUS VENOM see EAD600
ECHIS COLORATA VENOM see EAD650
ECHIS COLORATUS VENOM see EAD650
ECHIUM PLANTAGINEUM see VQZ675
ECHIUM VULGARE see VQZ675
ECHLOMEZOL see EFK000
ECHODIDE see TLF500
ECHOTHIOPHATE see TLF500
ECHOTHIOPHATE IODIDE see TLF500
ECHUJETIN see DMJ000
ECIPHIN see EAW000
ECLERIN see BOV825
ECLORIL see CDO500
ECLORION see HBT500
ECM see ADA725
ECODOX see HGP550
E. COLI 0111: B4 LPS see EDK750
ECOLID see CDY000
ECOLID CHLORIDE see CDY000
E. COLI ENDOTOXIN see EDK700
ECONAZOLE NITRATE see EAE000
ECONOCHLOR see CDP250
ECOPRO see TAL250
ECOSTATIN see EAE000
ECOTHIOPATE IODIDE see TLF500
ECOTHIOPHATE IODIDE see TLF500
ECOTRIN see ADA725
ECP see DFK600
ECPN see NKE000
ECTIBAN see AHJ750
ECTIDA see EHP000
ECTILURAN see TEH500
ECTILUREA see EHP000
ECTODEX see MJL250
ECTON see EHP000
ECTORAL see RMA500
ECTRIN see FAR100
ECTYDA see EHP000
ECTYLCARBAMIDE see EHP000
ECTYLUREA see EHO700, EHP000
ECTYN see EHP000
ECUANIL see MQU750
ECYLERT see PAP000

ECZECIDIN see CHR500
ED see DFH200, EQJ100
EDA 200 see EIV750
EDA see DCN800
3',4'-Et2-DAB see DJB400
EDATHAMIL see EIX000
EDATHAMIL CALCIUM DISODIUM see CAR780
EDATHAMIL DISODIUM see EIX500
EDATHAMIL MONOSODIUM FERRIC SALT see EJA379
EDATHANIL TETRASODIUM see EIV000
EDB-85 see EIY500
EDB see EIY500
EDBPHA see EIV100
EDC see EIY600
EDCO see MHR200
EDDHA see EIV100
EDDO see EAI600
EDDP see EIM000
EDECRIL see DFP600
EDECRIN see DFP600
EDECRINA see DFP600
EDEMEX see BDE250
EDEMOX see AAI250
EDEN see LFK000
EDENAL see MQU750
EDETAMIN see CAR780
EDETAMINE see CAR780
EDETATE CALCIUM see CAR780
EDETATE DISODIUM see EIX500
EDETATE SODIUM see EIV000
EDETATE TRISODIUM see TNL250
EDETIC ACID see EIX000
EDETIC ACID CALCIUM DISODIUM SALT see CAR780
EDETIC ACID TETRASODIUM SALT see EIV000
EDHPA see EIV100
EDICOL AMARANTH see FAG020
EDICOL BLUE CL 2 see FAE000
EDICOL PONCEAU RS see FMU070
EDICOL SUPRA 10B see CMS231
EDICOL SUPRA BLACK BN see BMA000
EDICOL SUPRA BLUE E6 see FMU059
EDICOL SUPRA BLUE VR see ADE500
EDICOL SUPRA 10BS see CMS231
EDICOL SUPRA CARMOISINE WS see HJF500
EDICOL SUPRA ERYTHROSINE A see FAG040
EDICOL SUPRA GERANINE 2G see CMM300
EDICOL SUPRA GERANINE 2GS see CMM300
EDICOL SUPRA GREEN B see ADF000
EDICOL SUPRA PONCEAU 4R see FMU080
EDICOL SUPRA PONCEAU SX see FAG050
EDICOL SUPRA ROSE B see FAG070
EDICOL SUPRA TARTRAZINE N see FAG140
EDICOL SUPRA YELLOW FC see FAG150
EDIFENPHOS see EIM000
EDION see TLP750
EDIPHENPHOS see EIM000
EDIPOSIN see VBK000
EDISOL M see MIF760
EDISTIR RB 268 see SMR000
EDISTIR RB see SMQ500
EDIWAL see NNL500
EDPA HYDROCHLORIDE see DWF200
EDRIZAR see MJL250
EDROFURADENE see EAE500
EDROPHONIUM BROMIDE see TAL490
EDROPHONIUM CHLORIDE see EAE600
EDRUL see EAE675
EDTA (chelating agent) see EIX000
EDTA ACID see EIX000
EDTACAL see CAR780
EDTA CALCIUM DISODIUM SALT see CAR780
d'E.D.T.A. DISODIQUE (FRENCH) see EIX500
EDTA, DISODIUM SALT see EIX500
EDTA FERRIC SODIUM SALT see IGX875
EDTA, SODIUM SALT see EIV000
EDTA TETRASODIUM SALT see EIV000
EDTA TRISODIUM SALT see TNL250
EDTA TRISODIUM SALT (TRIHYDRATE) see TNL500
E-103-E see MQQ050
EEC No. E924 see PKY300
EECPE see QFA250
EEC SERIAL No. 124 see CMP250
EEDDKK see EIJ500
EENA see ELG500
EENKAPTON (DUTCH) see PDC750

EEREX GRANULAR WEED KILLER see BMM650
EEREX WATER SOLUBLE CONCENTRATE WEED KILLER see BMM650
EF 10 see SCK600
EFACIN see NCQ900
EF CORLIN see CNS750
EFED see TMU250
EFEDRIN see EAW000
EFERON see PEC250
EFEROX see LFG050
EFFEMOLL DOA see AEO000
EFFISAX see MOV500
EFFLUDERM (free base) see FMM000
EFFORTIL see EGE500
EFFROXINE see DBA800
EFFUSAN see DUS700
EFH see EKL250
EFIRAN 99 see TIR800
EFLORAN see MMN250
EFLORNITHINE see DKH875
EFLORNITHINE HYDROCHLORIDE see EAE775
EFLOXATE see ELH600
EFO-DINE see PKE250
EFRICEL see SPC500
EFROXINE see MDT600
EFSIOMYCIN see FMR500
EFTAPAN see ECU550
EFTOLON see AIF000
EFUDEX see FMM000
EFUDIX see FMM000
EFURANOL see DLH630
EFV 250/400 see IGK800
EGACID ORANGE GG see FAG010
EGACID RED G see CMM300
EGDME see DOE600
EGDN see EJG000
EGF-UROGASTRONE see UVJ475
EGGOBESIN see PNN300
EGG YELLOW A see FAG140
EGITOL see HCI000
EGLONYL see EPD500
EGM see EJH500
EGME see EJH500
EGPEA see PNA225
EGRI M 5 see CAT775
EGTA see EIT000
EGYPTIAN RATTLEPOD see SBC550
EGYT 201 see BAV250
EGYT 341 see GJS200
EGYT 739 see CAY875
EGYT-1050 see DAJ400
EH2 see DGQ200
EH 121 see TNX000
EHBN see ELE500
EHDP see HKS780
EHEN see ELG500
EHRLICH 5 see ARL000
EHRLICH 594 see ABX500
EHRLICH 606 see SAP500
EHRLICH'S REAGENT see DOT400
EI-103 see BSN000
EI-1642 see MPS250
EI-12880 see DSP400
EI-18706 see DNX600
EI 38,555 see CMF400
EI 47031 see PGW750
EI 47300 see DSQ000
EI-47470 see DHH400
EI 52160 see TAL250
EI see EJM900
2,2,3,3,4,4,5,5,6,6,7,7,8,8,9,9,10,10,11,11-EICOSAFLUOROUNDECANOIC ACID
    see EAE875
EICOSAFLUOROUNDECANOIC ACID see EAE875
11-H-EICOSAFLUORUNDEKANSAEURE (GERMAN) see EAE875
ω-H-EICOSAFLUORUNDEKANSAEURE (GERMAN) see EAE875
EICOSANOIC ACID see EAF000
EICOSANYL DIMETHYL BENZYLAMMONIUM CHLORIDE see BEM250
EINALON S see CLY500
EIRENAL see PGA750
EISENDEXTRAN (GERMAN) see IGS000
EISENDIMETHYLDITHIOCARBAMAT (GERMAN) see FAS000
EISEN-III-HYDROXID-POLYMALTOSE (GERMAN) see IHA000
EISENOXYD see IHD000
EISEN(III)-TRIS(N,N-DIMETHYLDITHIOCARBAMAT) (GERMAN) see FAS000
EITDRONATE DISODIUM see DXD400
EJIBIL see EGQ000

EK 54 see DUU600
EK 1700 see PEY250
EKAFLUVIN see KDK000
EKAGOM CBS see CPI250
EKAGOM TB see TFS350
EKAGOM TEDS see DXH250
EKALUX see DJY200
EKATIN see PHI500
EKATIN AEROSOL see PHI500
EKATINE-25 see PHI500
EKATIN ULV see PHI500
EKATOX see PAK000
EKAVYL SD 2 see PKQ059
EK 1108GY-A see TAI250
EKILAN see MFD500
EKKO CAPSULES see DKQ000
EKOA (HAWAII) see LED500
EKOMINE see MGR500
EKTAFOS see DGQ875
EKTASOLVE de ACETATE see CBQ750
EKTASOLVE DB see DJF200
EKTASOLVE DB ACETATE see BQP500
EKTASOLVE DIB see DJF800
EKTASOLVE EB see BPJ850
EKTASOLVE EB ACETATE see BPM000
EKTASOLVE EE see EES350
EKTASOLVE EE ACETATE SOLVENT see EES400
EKTASOLVE EIB see IIP000
EKTASOLVE EP see PNG750
EKTEBIN see PNW750
EKTYLCARBAMID see EHP000
EKVACILLIN see SLJ000, SLJ050
EL-103 see BSN000
EL-119 see OJY100
EL-161 see ENE500
EL 222 see FAK100
EL 228 see CHJ300
EL-291 see MQC000
EL 400 see BNL250
EL 2289 see CHJ300
EL 4049 see MAK700
ELAIOMYCIN see EAG000
ELALDEHYDE see PAI250
ELAMOL see TGJ250
ELAN see PGG350
ELANCOBAN see MRE225
ELANCOLAN see DUV600
ELANIL see EAH500
ELAOL see DEH200
ELASIOMYCIN see EAG500
ELASTASE see EAG875
ELASTOFIX ACS see MCB050
ELASTONON see AOA250, BBK000
ELASTOPAR see MJG750
ELASTOPAX see MJG750
ELASTOZONE 31 see BJT500
ELASTOZONE 34 see PFL000
ELASZYM see EAG875
ELATERICIN A see EAH100
α-ELATERIN see COE250
ELAVIL see EAH500, EAI000
ELAVIL HYDROCHLORIDE see EAI000
ELAYL see EIO000
ELBANIL see CKC000
ELBENYL ORANGE A-3RD see SGP500
ELBRUS see CGA000
ELCACID MILLING FAST RED RS see CMM330
ELCEMA F 150 see CCU150
ELCEMA G 250 see CCU150
ELCEMA P 050 see CCU150
ELCEMA P 100 see CCU150
ELCIDE 75 see MDI000
ELCORIL see CDO500
EL-CORTELAN SOLUBLE see HHR000
ELCOSINE see SNJ350
ELCOZINE CHRYSOIDINE Y see PEK000
ELCOZINE RHODAMINE B see FAG070
ELCOZINE RHODAMINE 6GDN see RGW000
ELDADRYL see BAU750
ELDEPRYL see DAZ125
ELDERBERRY see EAI100
ELDERFIELD PYRIMIDINE MUSTARD see DYC700
ELDESINE see VGU750
ELDEZOL see NGE500
ELDIATRIC C see CAK500

ELDODRAM see DYE600
ELDOPAL see DNA200
ELDOPAQUE see HIH000
ELDOQUIN see HIH000
ELDRIN see RSU000
ELEAGOL see SCA750
ELECOR see PEV750
ELECTRO-CF 11 see TIP500
ELECTRO-CF 12 see DFA600
ELECTRO-CF 22 see CFX500
ELECTROCORUNDUM see EAL100
ELECTRONIC E-2 see HIC000
ELEMENTAL SELENIUM see SBO500
ELEMI see EAI500
ELEMI OIL see EAI500
ELEN see IBW500
ELENIUM see LFK000, MDQ250
ELEPHANT'S EAR see CAL125, EAI600
ELEPHANT TRANQUILIZER see AOO500
ELEPSINDON see DKQ000
ELESANT see DLR300
ELESTOL see CLD000
ELEUDRON see TEX250
ELEVAN see DOZ000
ELEX 334 see EAI800
ELF see CBT750
ELFANEX see RDK000
ELFAN 4240 T see SON000
ELFAN WA SULPHONIC ACID see LBU100
ELFTEX see CBT750
ELGACID ORANGE 2G see FAG010
ELGETOL 318 see BRE500
ELGETOL see BRE500, DUS700, DUU600
ELIAMINA RED 8BL see CMO885
ELICIDE see MDI000
ELIMIN see DLV000
ELIMOCLAVIN see EAJ000
ELINTAAL see ELZ050
ELIPOL see DUS700
ELIPTEN see AKC600
ELITE FAST RED BG see NAO600
ELITE FAST RED G see CMM320
ELITE FAST RED GRS see CMM320
ELITE FAST RED R see NAO600
ELITE FAST RED RS see NAO600
ELITONE see DJS200
ELIXICON see TEP000
ELIXIR of VITRIOL see SOI510
ELIXOPHYLLIN see TEP000
ELIXOPHYLLINE see TEP000
ELJON FAST ORANGE G see CMS145
ELJON FAST SCARLET PV EXTRA see MMP100
ELJON FAST SCARLET RN see MMP100
ELJON LAKE RED C see CHP500
ELJON MADDER see DMG800
ELJON PINK TONER see RGW000
ELJON RUBINE BS see CMS148
ELJON YELLOW BG see DEU000
ELKAPIN see MNG000
ELKON FAST YELLOW GR see CMS208
ELKOSIL see SNJ350
ELKOSIN see SNJ350
ELKOSINE see SNJ350
ELLIPTICINE see EAI850
ELLIPTISINE see EAI850
ELLSYL see MHJ500
ELMASIL see AMY050
ELMEDAL see BRF500
ELMER'S GLUE ALL see AAX250
ELOBROMOL see DDJ000
ELOCRON see DVS000
ELON see MGJ750
ELON WORT see CCS650
ELORINE see CPQ250
ELORINE CHLORIDE see EAI875
ELORINE SULFATE see TJG225
EL P.E.T.N. see PBC250
ELPI see ARQ750
ELPON see PMP500
ELRODORM see DYC800
ELSAN see DRR400
ELSERPINE see RDK000
ELSYL see MHJ500
ELTEX 6037 see PJS750
ELTEX see PJS750

ELTEX A 1050 see PJS750
ELTREN see DXY725, LBX050
ELTRIANYL see TKX000
ELTROXIN see LFG050
ELVACITE see PKB500
ELVANOL 50-42 see PKP750
ELVANOL 52-22 see PKP750
ELVANOL 70-05 see PKP750
ELVANOL 71-30 see PKP750
ELVANOL 90-50 see PKP750
ELVANOL 522-22 see PKP750
ELVANOL see PKP750
ELVANOL 73125G see PKP750
ELVARON see DFL200
ELYMOCLAVIN see EAJ000
ELYMOCLAVINE see EAJ000
ELYSION see AOO500
ELYZOL see MMN250
ELZOGRAM see CCS250
EM 12 see DVS600
EM 136 see OOM300
EM 255 see EAJ100
EM 923 see DFY400
EM see EDH500
EMAFORM see CHR500
EMAGRIN see SPC500
EMALEX 103 see PJT300
EMALEX 115 see PJT300
EMAL T see SON000
EMANAY ATOMIZED ALUMINUM POWDER see AGX000
EMANAY ZINC DUST see ZBJ000
EMANAY ZINC OXIDE see ZKA000
EMANDIONE see PFJ750
EMANIL see DAS000
EMANON 4115 see PJY100
EMAR see ZKA000
EMASOL 41S see SKV170
EMATHLITE see KBB600
EMAZOL RED B see EAJ500
EMB see TGA500
EMBACETIN see CDP250
EMBADOL see TFK300
EMBAFUME see MHR200
EMBANOX see BQI000
EMBARIN see ZVJ000
EMBARK see DUK000
EMBARK PLANT GROWTH REGULATOR see DUK000
EMBATHION see EEH600
EMBAY 8440 see BGB400
EMBECHINE see BIE500
EMBELIC ACID see EAJ600
EMBELIN see EAJ600
EMBEQUIN see DNF600
EMBERLINE see EAJ600
EMB-FATOL see EDW875
EMBICHIN 7 see NOB700
EMBICHIN see BIE250, BIE500
EMBICHIN HYDROCHLORIDE see BIE500
EMBIKHINE see BIE500
EMBINAL see BAG250
EMBIOL see VSZ000
EMBITOL see EAK000
EMBONIC ACID see PAF100
EMBRAMINE HYDROCHLORIDE see BMN250
EMBUTAL see NBU000
EMBUTOX see DGA000, EAK500
EMBUTOX KLEAN-UP see DGA000
EMC see CHC500
EMCEPAN see CIR250
EMCOL CA see OAV000
EMCOL DS-50 CAD see HKJ000
EMCOL E-607 see LBD100
EMCOL H-2A see PJY100
EMCOL H 31A see PJY100
EMCOL-IM see IQN000
EMCOL-IP see IQW000
EMCOL MSK see OAV000
EMCOL O see GGR200
EMCOL PS-50 RHP see SLL000
EMD 9806 see SDZ000
EMDABOL see TFK300
EMDABOLIN see TFK300
EMEDAN see BSM000
EMERALD GREEN see BAY750, COF500
EMERESSENCE 1150 see EJQ500

EMERESSENCE 1160 see PER000
EMEREST 2301 see OHW000
EMEREST 2314 see IQN000
EMEREST 2316 see IQW000
EMEREST 2325 see BSL600
EMEREST 2350 see EJM500
EMEREST 2381 see SLL000
EMEREST 2400 see OAV000
EMEREST 2401 see OAV000
EMEREST 2646 see PJY100
EMEREST 2660 see PJY100
EMEREST 2801 see OHW000
EMERGIL see FMO129
EMERICID see LIF000
EMERSAL 6400 see SIB600
EMERSAL 6434 see SON000
EMERSAL 6465 see TAV750
EMERSOL 120 see SLK000
EMERSOL 140 see PAE250
EMERSOL 143 see PAE250
EMERSOL 210 see OHU000
EMERSOL 213 see OHU000
EMERSOL 6321 see OHU000
EMERSOL 233LL see OHU000
EMERSOL 221 LOW TITER WHITE OLEIC ACID see OHU000
EMERSOL 220 WHITE OLEIC ACID see OHU000
EMERY 655 see MSA250
EMERY 2218 see MJW000
EMERY 2219 see OHW000
EMERY 2310 see OHW000
EMERY 5703 see BKD500
EMERY 5709 see DHF400
EMERY 5711 see TGT000
EMERY 5712 see DMT800
EMERY 5714 see HKS100
EMERY 5715 see CJU200
EMERY 5717 see CJU200
EMERY 5770 see EHJ600
EMERY 5791 see MCN250
EMERY 6705 see PER000
EMERY 6802 see PJW500
EMERY see EAL100
EMERY OLEIC ACID ESTER 2221 see GGR200
EMERY OLEIC ACID ESTER 2301 see OHW000
EMERY X-88-R see DWT000
EMESIDE see ENG500
EMETE-CON see BDW000
EMETHIBUTIN HYDROCHLORIDE see EIJ000
EMETIC HOLLY see HGF100
EMETICON see BDW000
EMETIC WEED see CCJ825
EMETINE see EAL500
(−)-EMETINE see EAL500
EMETINE ANTIMONY IODIDE see EAM000
EMETINE BISMUTH IODIDE see EAM500
EMETINE with BISMUTH(III) TRIIODIDE see EAM500
1-EMETINE DIHYDROCHLORIDE see EAN000
EMETINE, DIHYDROCHLORIDE see EAN000
(−)-EMETINE DIHYDROCHLORIDE see EAN000
EMETINE DIHYDROCHLORIDE TETRAHYDRATE see EAN500
EMETINE HYDROCHLORIDE see EAN000
EMETINE TRIIODOBISMUTH(III) see EAM500
EMETIQUE (FRENCH) see AQG250
EMETIRAL see PMF250
EMETREN see CDP250
EMFAC 1202 see NMY000
EMI-CORLIN see HHR000
EMID 6511 see BKE500
EMID 6541 see BKE500
EMINEURINA see CHD750
EMIPHEROL see VSZ450
EMISAN 6 see MEP250
EMISOL 50 see AMY050
EMISOL see AMY050
EMISOL F see AMY050
EMMATOS see MAK700
EMMATOS EXTRA see MAK700
EMMI see EME050
EMOCICLINA see VSZ000
EMODIN see IIU000, MQF250
EMODOL see MQF250
EMO-NIK see NDN000
EMOQUIL see EAN600
EMOREN see DTL200
EMORFAZONE see EAN700

EMORHALT see AJV500
EMOTIVAL see CFC250
EMP see EDT100, EME100
EMPAL see CIR250
EMPECID see MRX500
EMPG see ENC000
EMPILAN 2848 see EJM500
EMPILAN BP 100 see PJY100
EMPILAN BQ 100 see PJY100
EMPIRIN see ADA725
EMPIRIN COMPOUND see ABG750, ARP250
EMPLETS POTASSIUM CHLORIDE see PLA500
EMPP see EAV700
EMQ see SAV000
EMRITE 6009 see GGR200
EMS see EMF500
EMSORB 2500 see SKV100
EMSORB 2502 see SKV170
EMSORB 2505 see SKV150
EMSORB 2510 see MRJ800
EMSORB 2515 see SKV000
EMSORB 6900 see PKL100
EMSORB 6907 see SKV195
EMSORB 6915 see PKG000
EMT 25,299 see MDV500
EMTAL 596 see TAB750
EMTEXATE see MDV500
EMTRYL see DSV800
EMTRYLVET see DSV800
EMTRYMIX see DSV800
EMTS see EME500
EMULGATOR 8972 see SKV170
EMULGEN 100 see DXY000
EMULGEN 210 see PJT300
EMUL P.7 see OAV000
EMULPHOR A see PJY100
EMULPHOR ON-870 see PJW500
EMULPHOR UN-430 see PJY100
EMULPHOR VN 430 see PJY100
EMULSAMINE BK see DAA800
EMULSAMINE E-3 see DAA800
EMULSIFIABLE OIL see MQV855
EMULSIFIER No. 104 see SIB600
EMULSION 212 see FQU875
EMULSIPHOS 440/660 see SJH200
EMULSOV O EXTRA P see EAN800
E-MYCIN see EDH500
EMYRENIL see OOG000
EN 237 see EAL100
EN 313 see EEI025
EN-1530 see NAH000
EN 1627 see DPE000
EN 1639 see CQF099
EN 1939 see CQF099
EN-15304 see NAH000
EN 18133 see EPC500
EN-28,450 see AGT250
ENADEL see CMY525
EN-1733A HYDROCHLORIDE see MRB250
ENDOTAL see DXD000
ENALAPRIL MALEATE see EAO100
ENALLYNYMAL SODIUM see MDU500
ENAMEL WHITE see BAP000
ENANTHAL see HBB500
ENANTHALDEHYDE see HBB500
ENANTHIC ACID see HBE000
ENANTHIC ALCOHOL see HBL500
ENANTHOLE see HBB500
ENANTHONE see TJJ300
ENANTHOTOXIN see EAO500
ENANTHYLIC ACID see HBE000
ENANTHYLIC ETHER see EKN050
ENARMON see TBG000
ENAVEN see EGS000
ENAVID see EAP000
ENBU see ENT000
ENC see ENT500
ENCEPHALARTOS HILDEBRANDTII see EAQ000
ENCETROP see NNE400
ENCORDIN see EAQ050
ENCORTON see PLZ000
ENCYPRATE see EGT000
ENDAK see MQT550
ENDAK MITE see MQT550
ENDECRIL see DFP600

ENDEP see EAI000
ENDIARON see DEL300
ENDIEMALUM see DJO800
ENDISON see DCV800
ENDOBIL see IFP800
ENDOBION see NCR000
ENDOCEL see EAQ750
ENDOCID see EAS000
ENDOCIDE see EAS000
ENDOCISTOBIL see BGB315
ENDODAN see EJQ000
ENDODEOXYRIBONUCLEASE I see EDK650, PAF575
ENDOD, EXTRACT see EAQ100
ENDO E see VSZ450
6,14-ENDOETHENO-7-(2-HYDROXY-2-PENTYL)-TETRAHYDRO-ORIPAVINE HY-
    DROCHLORIDE see EQO500
ENDOFOLLICOLINA D.P. see EDR000
ENDOFOLLICULINA see EDV000
ENDOGRAFIN see BGB315
ENDOGRAPHIN see BGB315
ENDOKOLAT see PPN100
ENDOLAT see DAM700
2,5-ENDOMETHYLENE CYCLOHEXANECARBOXYLIC ACID, ETHYL ESTER
    (mixed formyl isomers) see NNG500
(2,5-ENDOMETHYLENECYCLOHEXYLMETHYL)AMINE see AKY750
cis-3,6-ENDOMETHYLENE-Δ⁴-TETRAHYDROPHTHALIC ACID DIMETHYL ESTER
    see DRB400
ENDOMETHYLENETETRAHYDROPHTHALIC ACID, N-2-ETHYLHEXYL IMIDE
    see OES000
esc ENDONUCLEASE I see PAF575
3,6-ENDOOXOHEXAHYDROPHTHALIC ACID see EAR000
ENDOPANCRINE see IDF300
ENDOPITUITRINA see ORU500
ENDOSAN see BGB500
ENDOSOL see EAQ750
ENDOSULFAN see EAQ750
ENDOSULPHAN see EAQ750
ENDOX see EAT600
ENDOXAN see CQC650
ENDOXANA see CQC500
ENDOXANAL see CQC650
ENDOXAN-ASTA see CQC500
ENDOXAN MONOHYDRATE see CQC500
ENDOXAN R see CQC500
3,6-ENDOXOHEXAHYDROPHTHALIC ACID see EAR000
3,6-ENDOXOHEXAHYDROPHTHALIC ACID DISODIUM SALT see DXD000
ENDRATE see EIX000
ENDRATE DISODIUM see EIX500
ENDRATE TETRASODIUM see EIV000
ENDREX see EAT500
ENDRIN see EAT500
ENDRINE (FRENCH) see EAT500
ENDROCID see EAT600
ENDROCIDE see EAT600
ENDURACIDIN see EAT800
ENDURACIDIN A see EAT810
ENDURANCE see DWS200
ENDURON see MIV500
ENDURONYL see RDF000
ENDUXAN see CQC500
ENDYDOL see ADA725
ENDYL see TNP250
E.N.E. see ENX500
ENE 11183 B see EAT600
ENELFA see HIM000
ENERIL see HIM000
ENERZER see IKC000
ENFENEMAL see ENB500
ENFLURANE see EAT900
ENGLISH HOLLY see HGF100
ENGLISH IVY see AFK950
ENGLISH RED see IHD000
ENDOTHAL see DXD000, EAR000
ENDOTHALL see EAR000
ENDOTHAL-NATRIUM (DUTCH) see DXD000
ENDOTHAL-SODIUM see DXD000
ENDOTHAL TECHNICAL see EAR000
ENDOTHAL WEED KILLER see DXD000
ENHEPTIN see ALQ000
ENHEPTIN A see ABY900
ENHEXYMAL see ERD500, ERE000
ENHEXYMAL NFN see ERE000
ENDOTHION see EAS000
ENHYDRINA SCHISTOSA VENOM see EAU000
ENIACID BLACK IVS see FAB830

ENIACID BLACK SH see FAB830
ENIACID BRILLIANT RUBINE 3B see HJF500
ENIACID FUCHSINE BN see CMS231
ENIACID LIGHT RED 3G see CMM300
ENIACID METANIL YELLOW GN see MDM775
ENIACID ORANGE I see FAG010
ENIACID SUNSET YELLOW see FAG150
ENIACROMO YELLOW G see SIT850
ENIALIT LIGHT RED RL see MMP100
ENIAL ORANGE I see PEJ500
ENIAL RED IV see SBC500
ENIAL YELLOW 2G see DOT300
ENIAMETHYL ORANGE see MND600
ENIANIL BLACK CN see AQP000
ENIANIL BLUE 2BN see CMO000
ENIANIL BLUE RW see CMO600
ENIANIL BROWN 2GS see CMO810
ENIANIL DARK GREEN BG see CMO830
ENIANIL FAST RED F see CMO870
ENIANIL GREEN B see CMO840
ENIANIL GREEN BBN see CMO840
ENIANIL PURE BLUE AN see CMO500
ENIAZOL BLUE BLACK BHN see CMN800
ENICOL see CDP250
ENIDE see DRP800
ENIDRAN see DYE700
ENIDREL see CFZ000, EAP000
ENILOCONAZOL (SP) see FPB875
ENIPRESSER see RDK000
ENJAY CD 460 see PMP500
ENJI see CNF050
ENKALON see NOH000
ENKEFAL see DNU000
ENKELFEL see DKQ000
ENNG see ENU000
ENOCITABINE see EAU075
ENORDEN see MQU750
ENOVID see EAP000
ENOVID-E see EAP000
ENOVIT see DJV000
ENOVIT M see PEX500
ENOXACIN see EAU100
ENOXACIN HYDRATE (2:3) see EAU150
ENOXAPARIN see HAQ500
hmi ENDOTOXIN, phenol water extract see HAB710
ENDOTOXIN, klp see EAS260
ENDOTOXIN, AEROMONAS HYDROPHILA A₃ see AFT500
Δ-ENDOTOXIN, from BACILLUS THURINGIENSIS see EAS230
ENDOTOXIN, BACT. AERTRYCKE see EAS200
ENDOTOXIN, BACTEROIDES FRAGILIS see BAC390
ENDOTOXIN, BRUCELLA MELITENSIS see BOL600
ENDOTOXIN, ESCHERICHIA COLI see EDK700
ENDOTOXIN see EAS100
hmi ENDOTOXIN, NaCl-citrate extract see HAB700
ENDOTOXIN, PSEUDOMONAS AERUGINOSA see POH620
ENDOTOXIN, VIBRIO CHOLERAE see VJZ100
ENOXOLONE see GIE000
ENPHENEMAL see ENB500
ENPROMATE see DWL400
ENQUIK see UTU600
ENRADIN see EAT800
ENRADINE see EAT810
ENRAMYCIN see EAT800, EAT810
ENRICHED SUPERPHOSPHATE see SOV500
ENRUMAY see WAK000
E.N.S. see ENX500
ENS see EPI300
ENSEAL see PLA500
ENSIGN see CMF350
ENSODORM see EOK000
ENSTAMINE HYDROCHLORIDE see DCJ850
ENSTAR see POB000
ENSURE see EAQ750
ENS-ZEM WEEVIL BAIT see DXE000
ENT 5 see LBO000
ENT 6 see BPL250
ENT 9 see BRT000
17-ENT see ABU000
ENT 38 see PDP250
ENT 54 see ADX500
ENT 92 see IHZ000
ENT 114 see DYA200
ENT 123 see VHZ000
ENT 133 see RNZ000
ENT 154 see DUS700

ENT 157 see CPK500
ENT 262 see DTR200
ENT-337 see DED500
ENT 375 see EKV000
ENT 666 see SNA500
ENT 884 see COF500
ENT 988 see BJK500
ENT 1025 see HCP100
ENT 1,122 see BRE500
ENT 1,501 see DXE000
ENT 1,506 see DAD200
ENT 1,656 see EIY600
ENT 1,716 see MEI450
ENT 1,860 see PCF275
ENT 2,435 see NDR500
ENT 3,424 see NDN000
ENT 3,776 see DFT000
ENT 3,797 see TBO500
ENT 4,225 see BIM500
ENT 4,504 see DFJ050
ENT 4,585 see CJR500
ENT 4,705 see CBY000
ENT 7,543 see POO100
ENT 7,796 see BBQ500
ENT 8,184 see OES000
ENT 8286 see BRP250
ENT 8,420 see DGG000
ENT 8,538 see DAA800
ENT 8,601 see BBP750
ENT 9,232 see BBQ000
ENT 9,233 see BBR000
ENT 9,234 see BFW500
ENT 9,624 see BIN000
ENT 9,735 see CDV100
ENT 9,932 see CDR750
ENT 14,250 see PIX250
ENT 14,611 see ASL250
ENT 14,689 see FAS000
ENT 14,874 see EIR000
ENT 14,875 see MAS500
ENT 15,108 see PAK000
ENT 15,152 see HAR000
ENT 15,208 see CKE750, NCM700
ENT 15,266 see PNP250
ENT 15,349 see EIY500
ENT 15,406 see PNJ400
ENT 15,748 see HBL100
ENT 15,949 see AFK250
ENT 16,087 see NIM500
ENT 16,225 see DHB400
ENT 16,273 see SOD100
ENT 16,275 see BGC750
ENT 16,358 see CJT750
ENT 16,391 see KEA000
ENT 16,436 see DXX400
ENT 16,519 see SOP500
ENT 16,634 see ISA000
ENT 16,894 see TED500
ENT 17,034 see MAK700
ENT 17,035 see NFT000
ENT 17,251 see EAT500
ENT 17,291 see OCM000
ENT 17,292 see MNH000
ENT 17,295 see DAO600
ENT 17,470 see DFK600
ENT 17,510 see AFR250
ENT 17,588 see PPQ625
ENT 17,591 see EAU500
ENT 17,596 see BHJ500
ENT 17,798 see EBD700
ENT 17,941 see CKI625
ENT 17,956 see CNU750
ENT 18,060 see CKC000
ENT 18,066 see BON250
ENT 18,544 see PDK000
ENT 18,596 see DER000
ENT 18,771 see TCF250
ENT 18,861 see MIJ250
ENT 18,862 see DAO800, MIW100
ENT 18,870 see DMC600
ENT 19,060 see DSK200
ENT 19,109 see BJE750
ENT 19,244 see IKO000
ENT 19,442 see TBC500
ENT 19,507 see DCM750

ENT 27,341 see MDU600
ENT 27,346 see DOP200
ENT 27,349 see CFC500
ENT 27,350 see MDX250
ENT 27,351 see MLX000
ENT 27,357 see DTP800
ENT 27,358 see DTP600
ENT 27,389 see DVS000
ENT 27,394 see DJY200
ENT 27,396 see DTQ400
ENT 27,407 see MPG250
ENT 27,408 see IEN000
ENT 27,410 see DRP600
ENT 27,411 see MPS250
ENT 27,438 see DGA200
ENT 27,474 see BEP500
ENT 27,488 see BAT750
ENT 27,520 see CMA250
ENT 27,521 see PHE250
ENT 27,552 see IOS000
ENT 27,566 see DSO200
ENT 27,567 see CJJ250, CJJ500
ENT 27,572 see FAK000
ENT 27,625 see DOL800
ENT 27,635 see CLJ875
ENT 27,696 see DRR000
ENT 27,738 see BLU000
ENT 27,766 see DOX600
ENT 27,822 see DOP600
ENT 27,851 see DAB400
ENT 27,910 see DED000
ENT 27,967 see MJL250
ENT 27,989 see MKA000
ENT 28,009 see HON000
ENT 28,344 see PIZ499
ENT 29,054 see CJV250
ENT 29,118 see EPC175
ENT 31,472 see MJK500
ENT 31,560 see BQT600
ENT 32,833 see AAR500
ENT 33,266 see DXU830
ENT 33,335 see HFX000
ENT 33,348 see HFE520
ENT 34,872 see CNG760
ENT 34,886 see ECB200
ENT 35,349 see TJF400
ENT 50,003 see TNK250
ENT 50,107 see BJC250
ENT 50,146 see RDK000
ENT 50,172 see HEF500
ENT 50,324 see EJM900
ENT 50,434 see AQG250
ENT 50,439 see BIA250
ENT 50,698 see DYC700
ENT 50,787 see BGX775
ENT 50,825 see MLY000
ENT 50,838 see TDQ225
ENT 50,852 see HEJ500
ENT 50,882 see HEK000
ENT 50,909 see AGU500
ENT 51,253 see BGY500
ENT 51,254 see MGE100
ENT 51,762 see NNF000
ENT 51,799 see MJQ500
ENT 51,904 see TLR250
ENT 61,241 see ACM750
ENT 61,969 see DNU850
ENT 70,459 see EQD000
ENT 70,460 see KAJ000
ENT 70,531 see POB000
ENT 27,300-A see CQI500
ENTACYL see HEP000
ENT AI3-29261 see AFK000
ENTEPAS see AMM250
ENTERAMINE see AJX500
ENTERICIN see ADA725
ENTERO-BIO FORM see CHR500
ENTERO-EXOTOXIN, CHOLERA see CMC800
ENTEROMYCETIN see CDP250
ENTEROPHEN see ADA725
ENTEROQUINOL see CHR500
ENTEROSALICYL see SJO000
ENTEROSALIL see SJO000
ENTEROSARINE see ADA725
ENTEROSEDIV see TEH500

ENTEROSEPT see DNF600
ENTEROSEPTOL see CHR500
ENTEROTOXIN, CHOLERA see CMC800
ENTEROTOXIN, CLOSTRIDIUM PERFRINGENS, TYPE A see CMY220
ENTEROTOXON see NGG500
ENTERO-VIOFORM see CHR500
ENTEROZOL see CHR500
ENTERUM LOCORTEN see CHR500
ENTEX see FAQ900
ENT 27,386GC see DRR400
ENT 27,699GC see DIN800
ENTHOHEX see PCY300
ENTIZOL see MMN250
ENTOBEX see PCY300
ENTOMOXAN see BBQ500
ENTPROL see QAT000
ENTRA see TMX775
ENTRAMIN see ALQ000
ENTROKIN see CHR500
ENTRONON see PCY300
ENTROPHEN see ADA725
ENTSUFON see TMN490
ENTUSIL see SNN500
ENT 24,980-X see DJA400
ENT 25,545-X see OAN000
ENT 25,552-X see CDR750
ENT 25,554-X see MNO750
ENT 25,555-X see DJA200
ENT 25,595-X see DQZ000
ENT 25,602-X see COD850
ENT 25,700-X see HCK000
ENT 27,395-X see CQH650
ENTYDERMA see AFJ625
ENU see ENV000, NKE500
ENUCLEN see BBA500
ENVERT 171 see DAA800
ENVERT DT see DAA800
ENVERT-T see TAA100
ENVIOMYCIN SULFATE see VQZ100
ENZACTIN see THM500
ENZAMIN see BBW500
ENZAPROST see POC500
ENZAPROST F see POC500
ENZEON see CML850
ENZOPRIDE see CNF390
ENZOSE see DBI800
EO 122 see EAV100
E.O. see EJN500
EO 5A see IGK800
EOC see EEO000
EOCT see COD475
EOSIN see BMO250, BNK700
EOSIN BLUE see ADG250
EOSINE see BMO250, BNH500
EOSINE B see BNK700
EOSINE BLUE see ADG250
EOSINE BLUISH see ADG250
EOSINE FA see BNK700
EOSINE LAKE RED Y see BNK700
EOSINE SODIUM SALT see BNH500
EOSINE YELLOWISH see BNH500
EOSIN GELBLICH (GERMAN) see BNH500
EP 30 see PAX250
E 66P see PKQ059
EP-145 see ECO500
EP 160 see PKP750
EP-185 see DNI200
EP 201 see ECB000
EP-205 see BJN250
EP-206 see VOA000
EP 316 see CQI500
EP-332 see DSO200
EP-333 see CJJ250
EP 333 see CJJ500
EP-411 see PFC750
EP-452 see MEG250
EP 453 see OMY925
EP-475 see EEO500
EP-1086 see TKS000
EP 1463 see AAX250
EPAL 6 see HFJ500
EPAL 8 see OEI000
EPAL 10 see DAI600
EPAL 12 see DXV600
EPAL 16NF see HCP000

E-PAM see DCK759
EPAMIN see DKQ000, DNU000
EPANUTIN see DKQ000, DNU000
EPAREN see DFL200
EPASMIR '5' see DKQ000
EPATIOL see MCI375
EPC (the plant regulator) see CBL750
EPDANTOINE SIMPLE see DKQ000
EP-161E see ISE000
EPE see EAV500
EPELIN see DKQ000, DNU000
EPERISONE HYDROCHLORIDE see EAV700
EPHEDRAL see EAW000
EPHEDRATE see EAW000
EPHEDREMAL see EAW000
EPHEDRIN see EAW000
dl-EPHEDRINE see EAW100
d-psi-EPHEDRINE see POH000
d-EPHEDRINE see EAW200
l-EPHEDRINE see EAW000
l-(+)-EPHEDRINE see EAW200
l(−)-EPHEDRINE see EAW000
psi-EPHEDRINE see POH000
EPHEDRINE see EAW000
(±)-EPHEDRINE see EAW100
EPHEDRINE, (±)- see EAW100
EPHEDRINE, (+)- see EAW200
(+)-EPHEDRINE see EAW200
dl-EPHEDRINE HYDROCHLORIDE see EAX500
d-EPHEDRINE HYDROCHLORIDE see EAW995
l-EPHEDRINE, HYDROCHLORIDE see EAW500
l-EPHEDRINE HYDROCHLORIDE see EAX000
EPHEDRINE HYDROCHLORIDE see EAW500, EAX000
(−)-EPHEDRINE HYDROCHLORIDE see EAX000
EPHEDRINE PENICILLIN see PAQ120
dl-EPHEDRINE PHOSPHATE (ESTER) see EAY175
d-EPHEDRINE PHOSPHATE (ESTER) see EAY150
l-EPHEDRINE PHOSPHATE (ESTER) see EAY075
1-EPHEDRINE SULFATE see EAY500
EPHEDRITAL see EAW000
EPHEDROL see EAW000
EPHEDROSAN see EAW000
EPHEDROTAL see EAW000
EPHEDSOL see EAW000
EPHENDRONAL see EAW000
EPHERON see PEC250
EPHETONIN see EAX500
EPHETONINE see EAX500
EPHININE HYDROCHLORIDE see EAZ000
EPHIRSULPHONATE see CJT750
EPHORRAN see DXH250
EPHOXAMIN see EAW000
EPHYNAL see VSZ450
EPIB see ARQ750
EPIBENZALIN see DLY000
EPIBLOC see CDT750
EPIBROMHYDRIN see BNI000
EPIBROMOHYDRIN (DOT) see BNI000
EPIBROMOHYDRINE see BNI000
EPICHLOORHYDRINE (DUTCH) see EAZ500
EPICHLORHYDRIN (GERMAN) see EAZ500
EPICHLORHYDRINE (FRENCH) see EAZ500
α-EPICHLOROHYDRIN see EAZ500
(dl)-α-EPICHLOROHYDRIN see EAZ500
EPICHLOROHYDRIN see EAZ500
EPICHLOROHYDRYNA (POLISH) see EAZ500
EPICHLOROPHYDRIN see EAZ500
EPICHOLESTANOL see EBA100
EPICLASE see PEC250
EPI-CLEAR see BDS000
EPICLORIDRINA (ITALIAN) see EAZ500
EPICUR see MQU750
EPICURE DDM see MJQ000
EPICURE NMA see NAC000
EPIDEHYDROCHOLESTERIN see EBA100
EPIDERMOL see ACR300
EPIDIAN 5 see IPO000
3,17-EPIDIHYDROXYESTRATRIENE see EDO000
3,17-EPIDIHYDROXYOESTRATRIENE see EDO000
EPIDIONE see TLP750
1,4-EPIDIOXY-1,4-DIHYDRO-6,6-DIMETHYLFULVENE see EBA600
EPIDONE see TLP750
EPIDORM see EOK000
EPIDOSIN see VBK000
4'-EPIDOXORUBICIN see EBB100

EPIDOZIN see VBK000
EPIDROPAL see ZVJ000
EPI-DX see EBB100
EPIFEN see FMS875
EPIFENYL see DKQ000, DNU000
EPIFLUOROHYDRIN see EBU000
EPIFOAM see HHQ800
EPIFRIN see VGP000
EPIHYDAN see DKQ000, DNU000
EPIHYDRIN ALCOHOL see GGW500
EPIHYDRINALDEHYDE see GGW000
EPIHYDRINAMINE, N,N-DIETHYL- see GGW800
EPIHYDRINE ALDEHYDE see GGW000
EPIKURE DDM see MJQ000
EPILAN see DKQ000, MKB250
EPILAN-D see DNU000
EPILANTIN see DKQ000, DNU000
EPILEO PETIT MAL see ENG500
EPILIM see PNR750
EPILIN see PNX750
EPIMID see MNZ000
EPINAL see AGN000
EPINAT see DKQ000, DNU000
EPINELBON see DLY000
EPINEPHRAN see VGP000
1-EPINEPHRINE see VGP000
(R)-EPINEPHRINE see VGP000
dl-EPINEPHRINE see EBB500
d-EPINEPHRINE see AES250
EPINEPHRINE racemic see EBB500
EPINEPHRINE see VGP000
1-EPINEPHRINE (synthetic) see VGP000
(−)-EPINEPHRINE see VGP000
1-EPINEPHRINE-d-BITARTRATE see AES000
EPINEPHRINE-d-BITARTRATE see AES000
l-EPINEPHRINE BITARTRATE see AES000
EPINEPHRINE BITARTRATE see AES000
(−)-EPINEPHRINE BITARTRATE see AES000
1-EPINEPHRINE CHLORIDE see AES500
EPINEPHRINE CHLORIDE see AES500
1-EPINEPHRINE HYDROCHLORIDE see AES500
dl-EPINEPHRINE HYDROCHLORIDE see AES625
(±)-EPINEPHRINE HYDROCHLORIDE see AES625
(−)-EPINEPHRINE HYDROCHLORIDE see AES500
EPINEPHRINE HYDROGEN TARTRATE see AES000
EPINEPHRINE ISOPROPYL HOMOLOG see DMV600
1-EPINEPHRINE TARTRATE see AES000
EPINOVAL see DJU200
EPI-PEVARYL see EAE000
EPIPREMNUM AUREUM see PLW800
EPIRENAMINE see VGP000
EPIRENAN see VGP000
EPI-REZ 508 see BLD750
EPI-REZ 510 see BLD750
EPI-REZ 508 mixed with ERR 4205 (1:1) see OPI200
EPIRIZOLE see MCH550
EPIROTIN see BBW500
EPIRUBICIN see EBB100
EPISED see DKQ000
EPISEDAL see EOK000
EPITELIOL see VSK600
EPITHELONE see ACR300
2,3-EPITHIOANDROSTAN-17-OL see EBD500
2-α,3-α-EPITHIO-5-α-ANDROSTAN-17-β-OL see EBD500
2,2-EPITHIO-17-((1-METHOXYCYCLOPENTYL)OXY)-ANDROSTANE (2-α,3-α,5-
    α,17-β) see MCH600
4,5-EPITHIOVALERONITRILE see EBD600
2-α,3-α-EPITHIO-17-β-YL 1-METHOXYCYCLOPENTYL ETHER see MCH600
EPITIOSTANOL see EBD500
EPITOPIC see DKJ300
EPITRATE see VGP000
EPL see EDN000
EPN see EBD700
EPOBRON see IIU000
EPOCAN see BMB000
EPOCELIN see EBE100
EPODYL see TJQ333
EPOK U 9048 see UTU500
EPOK U 9192 see MCB050
EPOLENE C 10 see PJS750
EPOLENE C 11 see PJS750
EPOLENE C see PJS750
EPOLENE E 10 see PJS750
EPOLENE E 12 see PJS750
EPOLENE E see PJS750

EPOLENE M 5K see PMP500
EPOLENE N see PJS750
EPON 562 see EBF000
EPON 820 see EBF500
EPON 828 see BLD750, IPO000
EPON 1001 see EBG000
EPON 1007 see EBG500
EPON 828 mixed with ERR 4205 (1:1) see OPI200
EPONOC B see TGX550
EPONTHOL see PMM000
EPORAL see SOA500
(3,6-EPOSSI-CICLOESAN-1,2-DICARBOSSILATO) DISODICO (ITALIAN) see DXD000
EPOSTANE see EBH400
EPOSTAR EPS-S see MCB050
EPOXIDE-201 see ECB000
EPOXIDE 269 see LFV000
EPOXIDE A see BLD750
EPOXIDE ERLA-0510 see EBH500
1,2-EPOXYAETHAN (GERMAN) see EJN500
(E)-1-α-2-α-EPOXYBENZ(c)ACRIDINE-3-α-4-β-DIOL see EBH850
(Z)-1-β,2-β-EPOXYBENZ(c)ACRIDINE-3-α-4-β-DIOL see EBH875
1,2-EPOXYBUTANE see BOX750
1,4-EPOXYBUTANE see TCR750
EPOXYBUTANE see BOX750
3,4-EPOXY-1-BUTENE see EBJ500
1,2-EPOXYBUTENE-3 see EBJ500
2,3-EPOXYBUTYRIC ACID BUTYL ESTER see EBK000
2,3-EPOXYBUTYRIC ACID, ETHYL ESTER see EJS000
1,2-EPOXYBUTYRONITRILE see EBK500
3,4-EPOXYBUTYRONITRILE see EBK500
EPOXYCARYOPHYLLENE see CCN100
4,9-EPOXYCEVANE-3-α,4-β,12,14,16-β,17,20-HEPTOL see EBL000
4,9-EPOXYCEVANE-3-β,4-β,7-α,14,15-α,16-β,20-HEPTOL see EBL500
4,9-EPOXYCEVANE-3,4,12,14,16,17,20-HEPTOL 3-(3,4-DIMETHOXYBENZOATE) see VHU000
1,2-EPOXY-3-CHLOROPROPANE see EAZ500
5,6-α-EPOXY-5-α-CHOLESTAN-3-β-OL see EBM000
5-α,6-α-EPOXYCHOLESTANOL see EBM000
EPOXYCHOLESTEROL see EBM000
1,2-EPOXYCYCLOHEXANE see CPD000
3,6-EPOXY-CYCLOHEXANE 1,2-CARBOXYLATE DISODIQUE (FRENCH) see DXD000
3,6-endo-EPOXY-1,2-CYCLOHEXANEDICARBOXYLIC ACID see EAR000
β-(3,4-EPOXYCYCLOHEXYL)ETHYLTRIMETHOXYSILANE see EBO000
EPOXYCYCLOHEXYLETHYL TRIMETHOXY SILANE see EBO000
3,4-EPOXYCYCLOHEXYLMETHYL 3,4-EPOXYCYCLOHEXANE CARBOXYLATE see EBO050
(3-α,4-β)-12,13-EPOXY-4,15-DIACETATE-TRICHOTHEC-9-ENE-3,4,15-TRIOL see AOP250
12,13-EPOXY-4-β,15-DIACETOXY-3-α,8-α-DIHYDROXYTRICHOTHEC-9-ENE see NCK000
12,13-EPOXY-4-β,15-DIAZETOXY-3-α-HYDROXY-TRICHOTHEC-9-ENE see AOP250
5,6-EPOXY-5,6-DIHYDROBENZ(a)ANTHRACENE see EBP000
7,8-EPOXY-7,8-DIHYDROBENZO(a)PYRENE see BCV750
(−)-EPOXYDIHYDROCARYOPHYLLENE see CCN100
5,6-EPOXY-5,6-DIHYDRODIBENZ(a,h)ANTHRACENE see EBP500
15,20-EPOXY-15,30-DIHYDRO-12-HYDROXYSENECIONAN-11,16-DIONE see JAK000
EPOXYDIHYDROLINALOOL, mixed isomers see LFY500
5,6-EPOXY-5,6-DIHYDRO-7-METHYLBENZ(A) ANTHRACENE see MGZ000
11,12-EPOXY-11,12-DIHYDRO-3-METHYLCHOLANTHRENE see MIL500
9,10-EPOXY-9,10-DIHYDROPHENANTHRENE see PCX000
(4-α-5-α-17-β)-4,5-EPOXY-3,17-DIHYDROXY-4,17-DIMETHYLANDROST- 2-ENE-2-CARBONITRILE see EBH400
4,5-α-EPOXY-3,14-DIHYDROXY-17-METHYLMORPHINAN-6-ONE HYDROCHLO-RIDE see ORG100
1,3-EPOXY-2,2-DIMETHYLPROPANE see EBQ500
endo-2,3-EPOXY-7,8-DIOXABICYCLO(2.2.2)OCT-5-ONE see EBQ550
1,2-EPOXYDODECANE see DXU400
3,4-EPOXY-2,5-ENDOMETHYLENECYCLOHEXANECARBOXYLIC ACID, ETHYL ESTER see ENZ000
1,2-EPOXY-4-(EPOXYETHYL)CYCLOHEXANE see VOA000
1,2-EPOXY-7,8-EPOXYOCTANE see DHD800
4,5-EPOXY-2-(2,3-EPOXYPROPYL)VALERIC ACID, METHYL ESTER see MJD000
1,2-EPOXYETHANE see EJN500
EPOXYETHANE see EJN500
1,2-EPOXY-3-ETHOXYPROPANE see EKM200
1,2-EPOXY-3-ETHOXYPROPANE (DOT) see EKM200
7-(EPOXYETHYL)-BENZ(a)ANTHRACENE see ONC000
1,2-EPOXYETHYLBENZENE see EBR000
EPOXYETHYLBENZENE (8CI) see EBR000
α-EPOXYETHYL-1,3-BENZODIOXOLE-5-METHANOL see HOE500

2-(1,2-EPOXYETHYL)-5,6-EPOXYBENZENE see EBR500
2-(α,β-EPOXYETHYL)-5,6-EPOXYBENZENE see EBR500
1-EPOXYETHYL-3,4-EPOXYCYCLOHEXANE see VOA000
2,3,-EPOXY-2-ETHYLHEXANAMIDE see OLU000
α-(EPOXYETHYL)-p-METHOXYBENZYL ALCOHOL see HKI075
3-(1,2-EPOXYETHYL)-7-OXABICYCLO(4.1.0)HEPTANE see VOA000
3-(EPOXYETHYL)-7-OXABICYCLO(4.1.0)HEPTANE see VOA000
4-(1,2-EPOXYETHYL)-7-OXABICYCLO(4.1.0)HEPTANE see VOA000
4-(EPOXYETHYL)-7-OXABICYCLO(4.1.0)HEPTANE see VOA000
1-EPOXYETHYLPYRENE see ONG000
2′,3′-EPOXYEUGENOL see EBT500
1,2-EPOXY-3-FLUOROPROPANE see EBU000
2-β,3-β,6-β,7-α)-2,3-EPOXY-GRAYANOTOXANE-5,6,7,10,16-PENTOL-6-ACETATE see LJE100
EPOXYGUAIENE see EBU100
EPOXY HARDENER ZZL-0814 see EBU500
EPOXY HARDENER ZZL-0816 see EBV000
EPOXY HARDENER ZZL-0822 see EBV100
EPOXY HARDENER ZZL-0854 see EBV500
EPOXY HARDENER ZZLA-0334 see EBW000
EPOXYHEPTACHLOR see EBW500
1,2-EPOXYHEXADECANE see EBX500
5-β,20-EPOXY-1,2-α,4,7-β,10-β,13-α-HEXAHYDROXY-TAX-11-EN-9-ONE 4,10-DI-ACETATE 2-BENZOATE 13-ESTER with (2R,3S)-N-BENZOYL-3-PHENYLI-SOSERINE see TAH775
14,15-β-EPOXY-3-β-HYDROXY-5-β-BUFA-20,22-DIENOLIDE see BOM650
4,5-EPOXY-3-HYDROXY-N-METHYLMORPHINAN see DKX600
4,5-α-EPOXY-3-HYDROXY-17-METHYLMORPHINAN-6-ONE HYDROCHLORIDE see DNU300
4,5-α-EPOXY-3-HYDROXY-17-METHYLMORPHINAN-6-ONE-N-OXIDE TARTRATE see EBY500
(2-α-4-α-5-α-17-β)-4,5-EPOXY-17-HYDROXY-3-OXOANDROSTANE-2-CARBONI-TRILE see EBY600
4-α-5-EPOXY-17-β-HYDROXY-3-OXO-5-α-ANDROSTANE-2-α-CARBONITRILE see EBY600
5,6-EPOXY-3-HYDROXY-p-TOLUQUINONE see TBF325
12,13-EPOXY-4-HYDROXYTRICHOTHEC-9-EN-8-ONE CROTONATE see TJE750
4,5-EPOXY-3-HYDROXYVALERIC ACID-β-LACTONE see ECA000
1,2-EPOXY-3-ISOPROPOXYPROPANE see IPD000
1,4-EPOXY-p-MENTHANE see IKC100
1,8-EPOXY-p-MENTHANE see CAL000
EPOXYMETHAMINE BROMIDE see SBH500
4,5α-EPOXY-3-METHOXY-17-METHYLMORPHINAN-6-ONE see OOI000
11,12-EPOXY-3-METHYLCHOLANTHRENE see ECA500
3,4-EPOXY-6-METHYLCYCLOHEXANECARBOXYLIC ACID,ALLYL ESTER see AGF500
3,4-EPOXY-6-METHYLCYCLOHEXENECARBOXYLIC ACID (3,4-EPOXY-6-ME-THYLCYCLOHEXYLMETHYL) ESTER see ECB000
3,4-EPOXY-6-METHYLCYCLOHEXYLMETHYL-3,4-EPOXY-6-METHYLCYCLOHEX-ANECARBOXYLATE see ECB000
4,5-EPOXY-2-METHYLCYCLOHEXYLMETHYL-4,5-EPOXY-2-METHYLCYCLOHEX-ANECARBOXYLATE see ECB000
3,4-EPOXY-6-METHYLCYCLOHEXYLMETHYL-3′,4′-EPOXY-6′-METHYLCYCLO-HEXANE CARBOXYLATE see ECB000
4-(1,2-EPOXY-1-METHYLETHYL)-1-METHYL-7-OXABICYCLO(4.1.0)HEPTANE see LFV000
α-β-EPOXY-β-METHYLHYDROCINNAMIC ACID, ETHYL ESTER see ENC000
4,5-α-EPOXY-17-METHYLMORPHINAN-3-OL see DKX600
cis-7,8-EPOXY-2-METHYLOCTADECANE see ECB200
2,3-EPOXY-2-METHYLPENTANE see ECC600
1,2-EPOXY-3(p-NITROPHENOXY)-PROPANE see NIN050
5,6-EPOXY-2-NORBORNANECARBOXYLIC ACID, ETHYL ESTER see ENZ000
cis-9,10-EPOXYOCTADECANOATE see ECD500
cis-9,10-EPOXYOCTADECANOIC ACID see ECD500
9,10-EPOXYOCTADECANOIC ACID BUTYL ESTER see BRH250
1,2-EPOXYOCTANE see ECE000
EPOXYOLEIC ACID see ECD500
12,13-EPOXY-4-((1-OXO-2-BUTENYL)OXY)TRICHOTHEC-9-EN-8-ONE see TJE750
2,3-EPOXY-4-OXO-7,10-DODECADIENAMIDE see ECE500
(2R,3S)-2,3-EPOXY-4-OXO-7E,10E-DODECADIENAMIDE see ECE500
(2S)(3R)-2,3-EPOXY-4-OXO-7,10-DODECADIENOYLAMIDE see ECE500
4,5-EPOXY-2-PENTENAL see ECE550
EPOXYPERCHLOROVINYL see TBQ275
1,2-EPOXY-3-PHENOXYPROPANE see PFH000
EPOXYPIPERAZINE see DHE100
2,3-EPOXY-1-PROPANAL see GGW000
2,3-EPOXYPROPANAL see GGW000
1,2-EPOXYPROPANE see PNL600
1,3-EPOXYPROPANE see OMW000
2,3-EPOXYPROPANE see PNL600
EPOXYPROPANE see PNL600
2,3-EPOXY-1-PROPANOL see GGW500
2,3-EPOXYPROPANOL see GGW500
2,3-EPOXY-1-PROPANOL (OSHA) see GGW500

2,3-EPOXY-1-PROPANOL ACRYLATE see ECH500
2,3-EPOXY-1-PROPANOL METHACRYLATE see ECI000
2,3-EPOXY-1-PROPANOL OLEATE see ECJ000
2,3-EPOXY-1-PROPANOL STEARATE see SLK500
2,3-EPOXYPROPIONALDEHYDE see GGW000
2,3-EPOXY PROPIONALDEHYDE OXIME see ECG100
4-(2,3-EPOXYPROPOXY)BUTANOL see ECG200
2-(N-(2-(2,3-EPOXYPROPOXY)ETHYL)ANILINO)ETHANOL see MRI500
N-(4-(2,3-EPOXYPROPOXY)PHENYL)BENZYLAMINE see ECG500
p-(G-(2,3-EPOXYPROPOXY)-N-PHENYLBENZYLAMINE see ECG500
3-(2,3-EPOXYPROPOXY)PROPYLTRIMETHOXYSILANE see ECH000
2,3-EPOXYPROPYL ACRYLATE see ECH500
2,3-EPOXYPROPYL BUTYL ETHER see BRK750
2,3-EPOXYPROPYL CHLORIDE see EAZ500
N-(2,3-EPOXYPROPYL)DIETHYLAMINE see GGW800
2,3-EPOXYPROPYL ESTER ACRYLIC ACID see ECH500
2,3-EPOXYPROPYL ESTER METHACRYLIC ACID see ECI000
2,3-EPOXYPROPYL ESTER of OLEIC ACID see ECJ000
2,3-EPOXYPROPYL ESTER of STEARIC ACID see SLK500
2,3-EPOXYPROPYL ISOPROPYL ETHER see IPD000
2,3-EPOXYPROPYL METHACRYLATE see ECI000
4-(2,3-EPOXYPROPYL)-2-METHOXYPHENOL see EBT500
2,3-EPOXYPROPYL NITRATE see ECI600
2,3-EPOXYPROPYL OLEATE see ECJ000
3-(2,3-EPOXYPROPYLOXY)-2,2-DINITROPROPYL AZIDE see ECJ100
2,3-EPOXYPROPYLPHENYL ETHER see PFH000
(1R,2S)(−)-(1,2-EPOXYPROPYL)PHOSPHONIC ACID DISODIUM SALT see
    DXF600
N-(2,3-EPOXYPROPYL)-PHTHALIMIDE see ECK000
2,3-EPOXYPROPYL STEARATE see SLK500
(2,3-EPOXYPROPYL)TRIMETHYLAMMONIUM CHLORIDE see GGY200
3,12-EPOXY-12H-PYRANO(4,3-j)-1,2-BENZODIOXEPIN, DECAHYDRO-10-ME-
    THOXY-3,6,9-TRIMETHYL-, (3-α-5a-β,6-β,8a-β,9-α-12-β,12aR)-, (+)- see
    ARL425
EPOXY RESIN ERL-2795 see ECL000
EPOXY RESINS, CURED see ECL500
EPOXY RESINS, UNCURED see ECM500
9,10-EPOXYSTEARIC ACID see ECD500
cis-9,10-EPOXYSTEARIC ACID see ECD500
9,10-EPOXYSTEARIC ACID ALLYL ESTER see ECO500
9,10-EPOXYSTEARIC ACID-2-ETHYLHEXYL ESTER see EKV500
α,β-EPOXYSTYRENE see EBR000
EPOXYSTYRENE see EBR000
3,4-EPOXYSULFOLANE see ECP000
cis-1-β,2-β-EPOXY-1,2,3,4-TETRAHYDROBENZO(c)PHENANTHRENE-3-α-4-β-
    DIOL see BCS103
trans-1-α-2-α-EPOXY-1,2,3,4-TETRAHYDROBENZO(c)PHENANTHRENE-3-α,4-β-
    DIOL see BCS105
trans-1-β,2-β-EPOXY-1,2,3,4-TETRAHYDROBENZO(c)PHENANTHRENE-3-β, 4-α-
    DIOL see BCS110
9,10-EPOXY-9,10,11,12-TETRAHYDROBENZO(e)PYRENE see ECQ150
9,10-EPOXY-7,8,9,10-TETRAHYDROBENZO(a)PYRENE see ECQ100
3,4-EPOXY-1,2,3,4-TETRAHYDROCHRYSENE see ECQ200
(+)-(E)-3,4-EPOXY-1,2,3,4-TETRAHYDRO-CHRYSENEDIOL see DMS000
1,2-EPOXY-1,2,3,4-TETRAHYDROPHENANTHRENE see ECR259
3,4-EPOXY-1,2,3,4-TETRAHYDROPHENANTHRENE see ECR500
( ±)-3-α,4-α-EPOXY-1,2,3,4-TETRAHYDRO-1-β,2-α-PHENANTHRENEDIOL see
    ECS000
12,13-EPOXY-3-α,4-β,8-α,15-TETRAHYDROXYTRICHOTHEC-9-ENE-8-ISOVALER-
    ATE see DAD600
12,13-EPOXY-3,4,7,15-TETRAHYDROXYTRICHOTHEC-9-EN-8-ONE see NMV000
(3-α,4-β,7-α)-12,13-EPOXY-3,4,7,15-TETRAHYDROXYTRICHOTHEC-9-EN-8-ONE
    see NMV000
2,3-EPOXYTETRAMETHYLENE SULFONE see ECP000
1,2-EPOXY-3-(TOLYLOXY)PROPANE see TGZ100
1,2-EPOXY-4,4,4-TRICHLOROBUTANE see ECT500
EPOXY-1,1,2-TRICHLOROETHANE see ECT600
1,2-EPOXY-3,3,3-TRICHLOROPROPANE see TJC250
12,13-EPOXY-TRICHOTHEC-9-ENE-3-α,15-TETROL 15-ACETATE, 8-ISOVALER-
    ATE see THI250
6-β,7-β-EPOXY-3-α-TROPANYL S-(−)-TROPATE see SBG000
EPOXYTROPINE TROPATE see SBG000
EPOXYTROPINE TROPATE METHYLBROMIDE see SBH500
EPOXYTROPINE TROPATE METHYLNITRATE see SBH000
1,2-EPOXY-4-VINYLCYCLOHEXANE see VNZ000
EPRAZIN see POL500
EPRAZINONE DIHYDROCHLORIDE see ECU550
EPRAZINONE HYDROCHLORIDE see ECU550
EPROFIL see TEX000
EPROLIN see VSZ450
EPROZINOL see EQY600
EPROZINOL DIHYDROCHLORIDE see ECU600
EPSAMON see AJD000
EPSICAPRON see AJD000
EPSILAN see VSZ450

EPSOM SALTS see MAJ250, MAJ500
EPSYLON KAPROLAKTAM (POLISH) see CBF700
EPT 500 see IHC550
EPT see EQP000
EPTAC 1 see BJK500
EPTACLORO (ITALIAN) see HAR000
1,4,5,6,7,8,8-EPTACLORO-3a,4,7,7a-TETRAIDRO-4,7-endo-METANO-INDENE
    (ITALIAN) see HAR000
EPTAL see DKQ000
EPTAM see EIN500, EPC150
EPTANI (ITALIAN) see HBC500
EPTAN-3-ONE (ITALIAN) see EHA600
EPTAPUR see CKF750
EPTC see EIN500
EPTOIN see DKQ000, DNU000
E-PVC see PKQ059
EQ see SAV000
EQUAL see ARN825
EQUANIL SUSPENSION see MQU750
EQUIBRAL see MDQ250
EQUI BUTE see BRF500
EQUIGEL see DGP900
EQUIGYNE see ECU750
EQUILASE see CCP525
EQUILENIN see ECV000
EQUILENINA (SPANISH) see ECV000
EQUILENIN BENZOATE see ECV500
EQUILENINE see ECV000
EQUILIN see ECW000
EQUILIN BENZOATE see ECW500
EQUILIN SODIUM SULFATE see ECW520
EQUILIN, SULFATE, SODIUM SALT (6CI) see ECW520
EQUILIUM see MQU750, PGE000
EQUINE CYONIN see SCA750
EQUINE GONADOTROPHIN see SCA750
EQUINE GONADOTROPIN see SCA750
EQUINIL see MQU750
EQUINO-ACID see TIQ250
EQUINO-AID see TIQ250
EQUIPERTINE see ECW600
EQUIPOISE see CJR909
EQUIPROXEN see MFA500
EQUIPUR see VLF000
EQUISETIC ACID see ADH000
EQUITDAZIN see MHC750
EQUIZOLE see TEX000
ER 115 see CFY750
ER5461 see CQG250
ERABUTOXINA see ECX000
ERADE see ORU000
ERADEX see CMA100
ERALDIN see ECX100
ERALON see ILD000
ERAMIDE see CDR250
ERAMIN see HGD000
ERANTIN see PNA500
ERASE see HKC000
ERASOL see BIE500
ERASOL HYDROCHLORIDE see BIE500
ERASOL-IDO see BIE500
ERAZIDON see ORU000
ERBAPLAST see CDP250
ERBAPRELINA see TGD000
ERBITOX see DUT800
ERBIUM CHLORIDE see ECX500
ERBIUM CITRATE see ECY000
ERBIUM(III) NITRATE (1:3) see ECY500
ERBIUM(III) NITRATE, HEXAHYDRATE (1:3:6) see ECZ000
ERBIUM TRICHLORIDE see ECX500
ERBN see PBK000
ERBOCAIN see PDU250
ERBON see PBK000
ERCEFUROL see DGQ500
ERCEFURYL see DGQ500
ERCO-FER see FBJ100
ERCOFERRO see FBJ100
ERCORAX see HKR500
ERCOTINA see HKR500
ERE 1359 see REF000
EREBILE see DAL000
E-RETINAL see VSK985
ERGADENYLIC ACID see AOA125
ERGAM see EDC500
ERGAMINE see HGD000
ERGATE see EDC500

ERGENYL see PNX750
ERGOATETRINE see LJL000
ERGOBASINE see LJL000
ERGOCALCIFEROL see VSZ100
ERGOCHROME AA (2,2')-5-β,6-α,10-β-5',6'-α,10'-β see EDA500
ERGOCORNIN see EDA600
ERGOCORNINE see EDA600
α-ERGOCRYPTINE see EDB100
ERGOCRYPTINE see EDB100
ERGOKLININE see LJL000
α-ERGOKRYPTINE see EDB100
ERGOLINE-8-ACETAMIDE, 6-ETHYL-, (8-β)-, (R-(R*,R*))-2,3-DIHYDROXYBU-
    TANEDIOATE (1:1) see EJS100
ERGOLINE-8-ACETAMIDE, 6-METHYL-, (8-β)-, (R-(R*,R*))-2,3-DIHYDROXYBU-
    TANEDIOATE (2:1) see MJV500
ERGOLINE-8-ACETAMIDE, 6-(2-PROPENYL)-, (8-β)-, (R-(R*,R*))-2,3-DIHYDROX-
    YBUTANEDIOATE see PMR800
ERGOLINE-8-ACETAMIDE, 6-PROPYL-, (8-β)-, (R-(R*,R*))-2,3-DIHYDROXYBU-
    TANEDIOATE (2:1) see PNL800
ERGOLINE-8-CARBAMIC ACID, 6-PROPYL-, ETHYL ESTER, (8S)- see EPC115
ERGOLINE-8-β-PROPIONAMIDE, α-ACETYL-6-METHYL- see ACR100
ERGOLINE-8-PROPIONAMIDE, 6-ALLYL-α-CYANO- see AGC200
ERGOLINE-8-PROPIONAMIDE, 2-CHLORO-α-CYANO-6-METHYL- see CFF100
ERGOLINE-8-PROPIONAMIDE, α-CYANO-2,6-DIMETHYL- see COM075
ERGOLINE-8-PROPIONAMIDE, α-CYANO-6-ISOBUTYL- see COP600
ERGOLINE-8-β-PROPIONAMIDE, N-ETHYL-6-METHYL-α-(METHYLSULFONYL)-
    see EMW100
ERGOLINE-8-β-PROPIONITRILE, 6-METHYL-α-(4-METHYL-1-PIPERAZINYLCAR-
    BONYL)- see MLR400
ERGOLINE-8-β-PROPIONITRILE, 6-METHYL-α-(1-PYRROLIDINYLCARBONYL)-
    see MPC300
ERGOMAR see EDC500
ERGOMETRINE see LJL000
ERGOMETRINE ACID MALEATE see EDB500
ERGOMETRINE MALEATE see EDB500
ERGONOVINE see LJL000
ERGONOVINE, MALEATE (1:1) (SALT) see EDB500
ERGOPLAST ADC see DGT500
ERGOPLAST AdDO see AEO000
ERGOPLAST.FDC see DGV700
ERGOPLAST FDO see DVL700
ERGORONE see VSZ100
α-ERGOSINE METHANESULFONATE see EDB200
ERGOSINE METHANESULFONATE see EDB200
ERGOSINE MONOMETHANESULFONATE see EDB200
ERGOSTAT see EDC500
ERGOSTEROL, activated see VSZ100
ERGOSTEROL, irradiated see VSZ100
ERGOT see EDB500
ERGOTAMAN-3',6',18-TRIONE, 12'-HYDROXY-2',5'-BIS(1-METHYLETHYL)-, (5'-
    α)- see EDA600
ERGOTAMAN-3',6',18-TRIONE, 12'-HYDROXY-2'-METHYL-5'-(2-METHYLPRO-
    PYL)-, (5'-α)-, MONOMETHANESULFONATE (salt) see EDB200
ERGOTAMINE see EDC000
ERGOTAMINE BITARTRATE see EDC500
ERGOTAMINE TARTRATE see EDC500
ERGOTARTRATE see EDC500
ERGOTERM OTGO see DVN600
ERGOTERM TGO see EDC560
ERGOTIDINE see HGD000
ERGOTOCINE see LJL000
ERGOTOXIN see EDC565
ERGOTOXINE see EDC565
ERGOTOXINE ETHANESULFONATE see EDC575
ERGOTOXINE ETHANESULPHONATE see EDC575
ERGOTOXINE ETHANSULFONATE see EDC575
ERGOTRATE see EDB500, LJL000
ERGOTRATE MALEATE see EDB500
ERIAMYCIN see EDC600
ERIBUTAZONE see BRF500
ERIDAN see DCK759
ERIE BLACK B see AQP000
ERIE BLACK BF see AQP000
ERIE BLACK GAC see AQP000
ERIE BLACK GXOO see AQP000
ERIE BLACK JET see AQP000
ERIE BLACK NUG see AQP000
ERIE BLACK RXOO see AQP000
ERIE BRILLIANT BLACK S see AQP000
ERIE BROWN 3GN see CMO825
ERIE CONGO 4B see SGQ500
ERIE FAST BROWN B see CMO820
ERIE FAST RED FD see CMO870
ERIE FIBRE BLACK VP see AQP000
ERIE GREEN GPD see CMO840

ERIE GREEN MT see CMO840
ERIE GREEN TCM see CMO840
ERIE GREEN WT see CMO830
ERIE SCARLET 3B see CMO875
ERIE VIOLET 3R see CMP000
ERINA see MQU750
ERINIT see PBC250
ERINITRIT see SIQ500
ERIO ANTHRACENE BRILLIAN BLUE B see CMM090
ERIO ANTHRACENE BRILLIANT BLUE RFF see CMM080
ERIO ANTHRACENE FAST BLUE BB see CMM090
ERIOBOTRYA JAPONICA see LIH200
ERIO BRILLIANT BLUE V see ADE500
ERIOCHROMAL YELLOW G see SIT850
ERIOCHROME BLUE BLACK see CMP880
ERIOCHROME BLUE BLACK 2B see CMP880
ERIOCHROME BLUE BLACK B see CMP880
ERIOCHROME BLUE BLACK BC see CMP880
ERIOCHROME BLUE BLACK 2BP see CMP880
ERIOCHROME BLUE BLACK BSS see CMP880
ERIOCHROME BLUE BLACK 2G see CMP880
ERIOCHROME YELLOW 2G see SIT850
ERIOCHROME YELLOW GS see SIT850
ERIO FAST BLUE BRL see CMM070
ERIO FAST BLUE BS see CMM090
ERIO FAST YELLOW AE see SGP500
ERIO FAST YELLOW AEN see SGP500
ERIO FAST YELLOW RL see CMM759
ERIO FLOXINE 2G see CMM300
ERIO FLOXINE 2GN see CMM300
ERIOGLAUCINE see ADE500, FMU059
ERIOGLAUCINE G see FAE000
ERIOGLAUCINE SUPRA see ADE500
ERIO GLAUCINE X see ERG100
ERIO GLAUCINE XS see ERG100
ERIO GREEN S see ADF000
ERION see ARQ250, CFU750
ERIONITE see EDC650
ERIONYL BLUE BFF see CMM100
ERIONYL BLUE E-B see CMM090
ERIONYL BLUE E-BFF see CMM100
ERIONYL BLUE E-RFF see CMM080
ERIONYL RED G see CMM320
ERIONYL RED RS see CMM330
ERIONYL YELLOW E-AEN see SGP500
ERIO ORANGE II see CMM220
ERIOSIN BLUE BLACK B see FAB830
ERIOSIN FAST BLUE B see CMM090
ERIOSIN FAST BLUE BFF see CMM100
ERIOSIN FAST BLUE RFF see CMM080
ERIOSIN RHODAMINE B see FAG070
ERIOSKY BLUE see FMU059
ERIO TARTRAZINE see FAG140
ERIO YELLOW AEN see SGP500
ERISIMIN DIHYDRATE see HAO000
ERISPAN see FDB100
ERITRONE see VSZ000
ERITROXILINA see CNE750
ERIZIMIN see HAN800
ERIZOMYCIN see PPI775
ERL-2774 see BLD750
ERL-2795 see ECL000
ERL-4221 see EBO050
ERLA-2270 see VOA000
ERLA-2271 see VOA000
ERMALONE see MJE760
ERMETRINE see LJL000
EROCYANINE 540 see EDD500
EROINA see HBT500
ERR 4205 see BJN250
ERR 4205 mixed with ARALDITE 6010 (1:1) see OPI200
ERR 4205 mixed with EPI-REZ 508 (1:1) see OPI200
ERR 4205 mixed with EPON 828 (1:1) see OPI200
ERROLON see CHJ750
ERSERINE see PIA500
ERTALON 6SA see PJY500
ERTILEN see CDP250
ERTRON see VSZ100
ERTUBAN see ILD000
ERYCIN see EDH500
ERYCORBIN see SAA025
ERYCYTOL see VSZ000
ERYPAR see EDJ500
ERYSAN see BIF250
ERYSIMIN see HAN800

ERYSIMIN DIHYDRATE see HAO000
ERYSIMOTOXIN see HAN800
ERYSIMUPICRONE see SMM500
ERYSIMUPIKRON see SMM500
ERYSODINE HYDROCHLORIDE see EDE000
ERYSOPINE HYDROCHLORIDE see EDE500
d-ERYTHORBIC ACID see SAA025
ERYTHORBIC ACID see SAA025
ERYTHRALINE HYDROBROMIDE see EDF000
ERYTHRENE see BOP500
ERYTHRITOL ANHYDRIDE see BGA750, DHB800
ERYTHRITOL, 1,4-DITHIO- see DXN350
ERYTHROCIN see EDH500
ERYTHROCIN STEARATE see EDJ500
ERYTHROGRAN see EDH500
ERYTHROGUENT see EDH500
ERYTHROHYCIN GLUCEPTATE see EDI500
β-ERYTHROIDINE see EDG500
β-ERYTHROIDINE HYDROCHLORIDE see EDH000
ERYTHROMYCIN see EDH500
ERYTHROMYCIN A see EDH500
ERYTHROMYCIN GLUCOHEPTONATE (1:1) see EDI500
ERYTHROMYCIN HYDROCHLORIDE see EDJ000
ERYTHROMYCIN OCTADECANOATE (salt) see EDJ500
ERYTHROMYCIN STEARATE see EDJ500
ERYTHROMYCIN STEARIC ACID SALT see EDJ500
ERYTHROSIN see FAG040
ERYTHROSINE B-FO (BIOLOGICAL STAIN) see FAG040
ERYTHROTIN see VSZ000
ERYTROXYLIN see CNE750
ES 902 see PPN000
ESACHLOROBENZENE (ITALIAN) see HCC500
ESACICLONATO see SHL500
ESAIDRO-1,3,5-TRINITRO-1,3,5-TRIAZINA (ITALIAN) see CPR800
ESAMETILENTETRAMINA (ITALIAN) see HEI500
ESAMETINA see HEA000
ESAMETONIO IODURO (ITALIAN) see HEB000
ESANI (ITALIAN) see HEN000
ESANITRODIFENILAMINA (ITALIAN) see HET500
ESANTENE see HFG550
ESAPROPIMATO see POA250
ESBATAL see BFW250
ESBECYTHRIN see DAF300
ESBERICARD see EDK600
ESBIOL see AFR750
ESBIOL CONCENTRATE 90% see AFR750
ESBRITE 2 see SMQ500
ESBRITE 4 see SMQ500
ESBRITE 8 see SMQ500
ESBRITE 4-62 see SMQ500
ESBRITE see SMQ500
ESBRITE G 10 see SMQ500
ESBRITE G-P 2 see SMQ500
ESBRITE 500HM see SMQ500
ESBRITE LBL see SMQ500
ESCAMBIA 2160 see PKQ059
ESCASPERE see RDK000
ESCHERICHIA COLI ENDONUCLEASE I see EDK650
ESCHERICHIA COLI ENDOTOXIN see EDK700
ESCHERICHIA COLI LIPOPOLYSACCHARIDE see EDK750
α-ESCIN see EDL000
β-ESCIN see EDL500
ESCIN see EDK875
ESCINA (ITALIAN) see EDK875
β-ESCINIC ACID see EDL000
ESCIN TRIETHANOLAMINE SALT see EDM500
ESCIN, SODIUM SALT see EDM000
ESCLAMA see NHH000
ESCOPARONE see DRS800
ESCOREZ 7404 see SMQ500
ESCORPAL see PDC850, PDC875
ESCULETIN DIMETHYL ETHER see DRS800
ESDRAGOL see AFW750
ESELIN see DIS600
ESEN see PHW750
ESERINE see PIA500
ESERINE SALICYLATE see PIA750
ESERINE SULFATE see PIB000
ESERINE SULPHATE see PIB000
ESEROLEIN, METHYLCARBAMATE (ESTER) see PIA500
ESERPINE see RDK000
ESFAR see CPJ000
ESFAR CALCIUM see CPJ250
ESGRAM see PAJ000
ESICLENE see FNK040

ESIDREX see CFY000
ESIDRIX see CFY000
ESILGAN see CKL250
ESJAYDIOL see AOO475
ESKABARB see EOK000
ESKACILLIN V see PDT500
ESKACILLIN see BFD000
ESKADIAZINE see PPP500
ESKALIN V see VRA700, VRF000
ESKALITH see LGZ000
ESKALON 100 see CAT775
ESKASERP see RDK000
ESKAZINE see TKK250
ESKAZINE DIHYDROCHLORIDE see TKK250
ESKEL see AHK750
ESKIMON 11 see TIP500
ESKIMON 12 see DFA600
ESKIMON 22 see CFX500
ESMAIL see CGA000
ESMARIN see HII500
ESOBARBITALE (ITALIAN) see ERD500
E 39 SOLUBLE see BDC750
ESOMID CHLORIDE see HEA500
ESOPHOTRAST see BAP000
ESOPIN see MDL000
ESORB see VSZ450
ESP see DSK600
ESPADOL see CLW000
ESPARIN see DQA600
ESPASMO GASIUM see DAB750
ESPECTINOMICINA DIHYDROCHLORIDE PENTAHYDRATE see SKY500
ESPENAL see DXH250
ESPERAL see DXH250
ESPERAN see OLW400
ESPERFOAM FR see MKA500
ESPEROX 10 see BSC500
ESPEROX 24M see BSC600
ESPEROX 31M see BSD250
ESPHYGMOGENINA see VGP000
ESPIGA de AMOR (PUERTO RICO) see CAK325
ESPIGELIA (CUBA) see PIH800
ESPINOMYCIN A see MBY150
ESPIRAN see DAI200
ESPIRITIN see LAG010
ESPRIL see BET000
ESQUINON see BGX750
ESSENCE of MIRBANE see NEX000
ESSENCE of MYRBANE see NEX000
ESSENCE of NIOBE see EGR000
ESSENCE OF NIOBE see MHA750
ESSENCE of ROSE see RNA000
ESSENTIAL OIL see CNT350
ESSENTIAL OIL of ACORUS CALAMUS Linn. see OGL020
ESSENTIAL OIL of CYMBOPOGON NARDUS see CMT000
ESSENTIAL OIL from MYRTLE see OGU000
ESSENTIAL PHOSPHOLIPIDS see EDN000
ESSEX 1360 see HLB400
ESSEX see CBT750
ESSEX GUM 1360 see HLB400
ESSIGESTER (GERMAN) see EFR000
ESSIGSAEURE (GERMAN) see AAT250
ESSIGSAEUREANHYDRID (GERMAN) see AAX500
ESSO FUNGICIDE 406 see CBG000
ESSO HERBICIDE 10 see BQZ000
EST see OMY925
ESTABAL see BFW250
ESTABEX S see BOU550
ESTABEX U 18 see DVK200
ESTANE 5703 see UVA000
ESTAR see CMY800, PKF750
ESTASIL see MQU750
ESTAZOLAM see CKL250
ESTER 25 see NIM500
ESTER d'ACIDE BENZILIQUE et DU-1-METHYLSULFATE de 1,1-DIMETHYL-(2-HYDROXY-METHYL)PIPERIDINIUM see CDG250
ESTER del ACIDO BENCILICO del-1,1-DIMETIL-2-OXIMETIL-PIPERIDINIO-METIL-SULFATO (SPANISH) see CDG250
ESTERCIDE T-2 and T-245 see TAA100
S-ESTER with O,O-DIMETHYL PHOSPHOROTHIOATE see MAK700
ESTER DWETYLOAMINOETYLOWSKY KWASU DWUFENYLOOCTOWEGO (POLISH) see THK000
ESTER DWUETYLOAMINOETYLOWY KWASU DWUFENYLOOCTOWEGO see DHX800
ESTERE CIANOACETICO see EHP500

ESTERE ETOSSIMETILICO dell' ACIDO N-(2,6-DICLORO-m-TO-
LIL)ANTRANILICO (ITALIAN) see DGN000
ESTERE ISOAMILICO dell'ACIDO α-(N-(PIRROLIDINOETIL))-AMINOFENILACETI-
CO (ITALIAN) see CBA375
S-ESTER of (2-MERCAPTOETHYL)TRIMETHYLAMMONIUM IODIDE with O,O-DI-
ETHYL PHOSPHOROTHIOATE see TLF500
O-ESTER-p-NITROPHENOL with O-ETHYL PHENYL PHOSPHONOTHIOATE see
EBD700
ESTERON 44 see IOY000
ESTERON 99 see DAA800
ESTERON see DAA800
ESTERON 76 BE see DAA800
ESTERON 245 BE see TAA100
ESTERON BRUSH KILLER see DAA800, TAA100
ESTERON 99 CONCENTRATE see DAA800
ESTERONE see EDV000
ESTERONE FOUR see DAA800
ESTERON 44 WEED KILLER see DAA800
ESTEROQUINONE LIGHT BLUE 4JL see DMM400
ESTEROQUINONE LIGHT PINK RLL see AKO350
ESTEROQUINONE LIGHT YELLOW 4JL see AAQ250
α-ESTER PALMITIC ACID with D-threo-(−)-2,2-DICHLORO-N-(β-HYDROXY-α-
(HYDROXYMETHYL)-p-NITROPHENETHYL)ACETAMIDE see CDP700
ESTERS see EDN500
ESTER SULFONATE see CJT750
ESTERTRICHLOROSTANNANE see TIM500
ESTEVE see BRF500
ESTIBOGLUCONATO SODICO see AQH800, AQI250
ESTILBEN see DKA600, DKB000
ESTILBIN see DKB000
ESTIMULEX see DBA800
ESTINERVAL see PFC750
ESTOCINE see DPE200
ESTOL 103 see IQW000
ESTOL 603 see OAV000
ESTOL 1550 see DJX000
ESTOMYCIN see NCF500
ESTON see DSK600, QFA250
ESTONATE see DAD200
ESTON-B see EDP000
ESTONE see DAA800
ESTONE YELLOW GN see AAQ250
ESTONMITE see CJT750
ESTONOX see CDV100
ESTOSTERIL see PCL500
ESTOTSIN see DPE200
ESTOX see DSK600
ESTRACYT see EDT100
ESTRACYT HYDRATE see EDN600
17-α-ESTRADIOL see EDO500
ESTRADIOL-17-β see EDO000
17-β-ESTRADIOL see EDO000
3,17-β-ESTRADIOL see EDO000
α-ESTRADIOL see EDO000
β-ESTRADIOL see EDO000
cis-ESTRADIOL see EDO000
d-3,17-β-ESTRADIOL see EDO000
d-ESTRADIOL see EDO000
ESTRADIOL see EDO000
ESTRADIOL-17-β-3-BENZOATE see EDP000
ESTRADIOL-17-β-BENZOATE see EDP000
17-β-ESTRADIOL BENZOATE see EDP000
17-β-ESTRADIOL-3-BENZOATE see EDP000
β-ESTRADIOL-3-BENZOATE see EDP000
ESTRADIOL-3-BENZOATE see EDP000
β-ESTRADIOL BENZOATE see EDP000
ESTRADIOL BENZOATE see EDP000
ESTRADIOL-17-BENZOATE-3,n-BUTYRATE see EDP500
ESTRADIOL BENZOATE mixed with PROGESTERONE (1:14 moles) see
EDQ000
ESTRADIOL-3-BENZOATE mixed with PROGESTERONE (1:14 moles) see
EDQ000
ESTRADIOL-17-CAPRYLATE see EDQ500
ESTRADIOL-17-β 3-CYCLOPENTYL ETHER see QFA250
17-β-ESTRADIOL DIPROPIONATE see EDR000
β-ESTRADIOL-3,17-DIPROPIONATE see EDR000
3,17-β-ESTRADIOL DIPROPIONATE see EDR000
ESTRADIOL-3,17-DIPROPIONATE see EDR000
β-ESTRADIOL DIPROPIONATE see EDR000
ESTRADIOL DIPROPIONATE see EDR000
17-β-ESTRADIOL MONOBENZOATE see EDP000
ESTRADIOL MONOBENZOATE see EDP000
ESTRADIOL MUSTARD see EDR500
ESTRADIOL PHOSPHATE POLYMER see EDS000
ESTRADIOL POLYESTER with PHOSPHORIC ACID see EDS000

ESTRADIOL 17-β-VALERATE see EDS100
ESTRADIOL-17-VALERATE see EDS100
ESTRADIOL VALERATE see EDS100
ESTRADIOL VALERIANATE see EDS100
ESTRADURIN see EDS000, PJR750
ESTRAGARD see DAL600
ESTRAGON OIL see TAF700
ESTRALDINE see EDO000
ESTRALUTIN see HNT500
ESTRAMUSTINE PHOSPHATE DISODIUM see EDT100
ESTRAMUSTINE PHOSPHATE DISODIUM HYDRATE see EDN600
ESTRAMUSTINE PHOSPHATE SODIUM see EDT100
ESTRAMUSTINE PHOSPHATE SODIUM HYDRATE see EDN600
α-ESTRA-1,3,5,7,9-PENTANE-3,17-DIOL see EDT500
β-ESTRA-1,3,5,7,9-PENTANE-3,17-DIOL see EDU000
ESTRATAB see ECU750
1,3,5,7-ESTRATETRAEN-3-OL-17-ONE see ECW000
ESTRA-1,3,5(10),7-TETRAEN-17-ONE, 3-HYDROXY-, HYDROGEN SULFATE
SODIUM SALT (8CI) see ECW520
ESTRA-1,3,5(10),7-TETRAEN-17-ONE, 3-(SULFOOXY)-, SODIUM SALT see
ECW520
(17-β)-ESTRA-1,3,5(10)-TRIEN-3,17-DIOL, 3-METHOXY-17-(3,3,3-TRIFLUORO-1-
PROPYNYL) see TKH050
ESTRA-1,3,5(10)-TRIENE-2,4-D2-3,17-DIOL, (17-β)- see DHA425
17-β-ESTRA-1,3,5(10)-TRIENE-3,17-DIOL see EDO000
ESTRA-1,3,5(10)-TRIENE-3,17-α-DIOL see EDO500
1,3,5-ESTRATRIENE-3,17-α-DIOL see EDO500
ESTRA-1,3,5(10)-TRIENE-3,17-β-DIOL see EDO000
1,3,5-ESTRATRIENE-3,17-β-DIOL see EDO000
ESTRA-1,3,5(10)-TRIENE-3,17-DIOL (17-β)-3-BENZOATE see EDP000
ESTRA-1,3,5(10)-TRIENE-3,17-β-DIOL, 3-BENZOATE see EDP000
1,3,5(10)-ESTRATRIENE-3,17-β-DIOL 3-BENZOATE see EDP000
ESTRA-1,3,5(10)-TRIENE-3,17-β-DIOL-17-BENZOATE-3-n-BUTYRATE see
EDP500
ESTRA-1,3,5(10)-TRIENE-3,17-DIOL 3-(BIS(2-CHLOROETHYL)CARBAMATE)17-
DISODIUM PHOSPHATE see EDT100
ESTRA-1,3,5(10)-TRIENE-3,17-DIOL (17-β)-DIPROPIONATE see EDR000
1,3,5(10)-ESTRATRIENE-3,17-β-DIOL DIPROPIONATE see EDR000
ESTRA-1,3,5(10)-TRIENE-3,17-DIOL, 11-β-ETHYL- see EJT575
ESTRA-1,3,5(10)-TRIENE-3,17-β-DIOL, 4-FLUORO-, (17-β)-(9CI) see FIA500
ESTRA-1,3,5(10)-TRIENE-3,17-β-DIOL-17-OCTANOATE see EDQ500
(17-β)-ESTRA-1,3,5(10)-TRIENE-3,17-DIOL-17-PENTANOATE (9CI) see EDS100
(17-β)-ESTRA-1,3,5(10)-TRIENE-3,17-DIOL POLYMER with PHOSPHORIC ACID
see EDS000
(16-α,17-β)-ESTRA-1,3,5(10)-TRIENE-3,16,17-TRIOL see EDU500
ESTRA-1,3,5(10)-TRIENE-3,16-α,17-β-TRIOL see EDU500
1,3,5-ESTRATRIENE-3-β,16-α,17-β-TRIOL see EDU500
1,3,5(10)-ESTRATRIEN-3-OL-17-ONE see EDV000
1,3,5-ESTRATRIEN-3-OL-17-ONE see EDV000
Δ-1,3,5-ESTRATRIEN-3-β-OL-17-ONE see EDV000
ESTRA-1,3,5(10)-TRIEN-17-ONE, 3-(SULFOXY)-, SODIUM SALT (9CI) see
EDV600
ESTRA-1,3,5(10)-TRIETNE-3,17-β-DIOL-3-BENZOATE mixed with PROGESTER-
ONE (1:14 moles) see EDQ000
ESTRATRIOL see EDU500
ESTRAVEL see EDS100
ESTRELLA DEL NORTE (CUBA) see ROU450
ESTREPTOCIDA see SNM500
ESTRIFOL see ECU750
ESTRIL see DKA600
ESTRIN see EDV000
16-α,17-β-ESTRIOL see EDU500
3,16-α,17-β-ESTRIOL see EDU500
ESTRIOL see EDU500
ESTRIOLO (ITALIAN) see EDU500
ESTROATE see ECU750
ESTROBEN see DKB000
ESTROBENE see DKA600, DKB000
ESTROCON see ECU750
ESTRODIENOL see DAL600
ESTROFOL see PKF750
ESTROGEN see DKA600, EEH500
ESTROGENIN see DKB000
ESTROGENS, CONJUGATES see PMB000
ESTROICI see EDR000
ESTROL see EDV000
ESTROMED see ECU750
ESTROMENIN see DKA600
ESTRON see EDV000
ESTRONA (SPANISH) see EDV000
ESTRONE see EDV000
ESTRONE-A see EDV000
ESTRONE BENZOATE see EDV500
ESTRONE, HYDROGEN SULFATE, SODIUM SALT see EDV600
ESTRONE SODIUM SULFATE see EDV600

ESTRONE SULFATE SODIUM see EDV600
ESTRONE-3-SULFATE SODIUM SALT see EDV600
ESTRONE SULFATE SODIUM SALT see EDV600
ESTRONEX see EDR000
ESTROPAN see ECU750
ESTRORAL see DAL600
ESTROSEL see DGP900
ESTROSOL see DGP900
ESTROSTILBEN see DKB000
ESTROSYN see DKA600
ESTROVIS 4000 see QFA250
ESTROVIS see QFA250
ESTROVISTER see QFA250
ESTROVITE see EDO000
ESTRU BENZYLOWEGO KWASU NIKOTYNOWEGO see NCR040
ESTRUGENONE see EDV000
ESTRUSOL see EDV000
ESTULIC see GKU300
ESTYRENE 4-62 see SMQ500
ESTYRENE AS see ADY500
ESTYRENE G 15 see SMQ500
ESTYRENE G 20 see SMQ500
ESTYRENE G-P 4 see SMQ500
ESTYRENE H 61 see SMQ500
ESTYRENE 500SH see SMQ500
ESUCOS see MKQ000
ESZ see IHL000
ET 14 see RMA500
ET 57 see RMA500
ET 67 see CGW300
ET-394 see THW750
ET 495 see TNR485
ETs see EHG100
ETABENZARONE see EDV700
ETABETACIN see EDW000
ETABUS see DXH250
ETACRINIC ACID see DFP600
ETAFENONE HYDROCHLORIDE see DHS200
ETAIN (TETRACHLORURE d') (FRENCH) see TGC250
ETAKRINIC ACID see DFP600
ETAMBRO see TCC000
ETAMBUTOL see EDW875
ETAMICAN see VSZ450
ETAMIDE see EEM000
ETAMINAL SODIUM see NBU000
ETAMINOPHYLLINE see CNR125
ETAMIPHYLLIN see CNR125
ETAMIPHYLLINE see CNR125
ETAMIPHYLLINE HYDROCHLORIDE see DIH600
ETAMIPHYLLIN HYDROCHLORIDE see DIH600
ETAMON CHLORIDE see TCC250
ETAMSYLATE see DIS600
ETANAUTINE see BBV500
ETANOLAMINA (ITALIAN) see EEC600
ETANOLO (ITALIAN) see EFU000
ETANTIOLO (ITALIAN) see EMB100
ETAPERAZIN see CJM250
ETAPERAZINE see CJM250
ETAVIT see VSZ450
ETAZIN see BQC250
ETAZINE see BQC250
ETC see EPY600
ETCHLORVINOLO see CHG000
ETEAI see EDW300
ETEM see EJQ000
ETERE ETILICO (ITALIAN) see EJU000
E TETRAETHYLPYRONIN see DIS200
ETH see EEI000, EPQ000
ETHAANTHIOL (DUTCH) see EMB100
ETHACRIDINE LACTATE see EDW500
ETHACRYNIC ACID see DFP600
ETHAL see HCP000
ETHALFLURALIN see ENE500
ETHALFLURLIN see ENE500
ETHAL LA-X see DXY000
ETHAMBUTOL see TGA500
ETHAMBUTOL DIHYDROCHLORIDE see EDW875
ETHAMBUTOL HYDROCHLORIDE see EDW875
ETHAMBUTOL mixed with SODIUM NITRITE (1:1) see SIS500
ETHAMIDE see DWW000
ETHAMINAL see NBT500
ETHAMINAL SODIUM see NBU000
ETHAMON DS see EDX000
ETHAMOXYTRIPHETOL see DHS000
ETHAMSYLATE see DIS600

ETHAN see PII500
ETHANAL see AAG250
ETHANAL OXIME see AAH250
ETHANAMIDE see AAI000
ETHANAMINE see EFU400
ETHANAMINIUM, 2-(ACETYLOXY)-N,N,N-TRIMETHYL-, BROMIDE (9CI) see CMF260
ETHANAMINIUM, 2-(ACETYLTHIO)-N,N,N-TRIMETHYL-, IODIDE (9CI) see ADC300
ETHANAMINIUM, 2-CHLORO-N,N,N-TRIMETHYL-, CHLORIDE (9CI) see CMF400
ETHANAMINIUM, 2-((CYCLOHEXYLHYDROXYPHENYLACETYL)OXY)-N,N-DIE-THYL-N-METHYL-, BROMIDE (9CI) see ORQ000
ETHANAMINIUM, N-(4-((4-(DIETHYLAMINO)PHENYL)(5-HYDROXY-2,4-DISULFO-PHENYL)METHYLENE)-2,5-CYCLOHEXADIEN-1-YLIDENE)-N-ETHYL-, HY-DROXIDE, inner salt, CALCIUM SALT (2:1) see CMM062
ETHANAMINIUM, N,N-DIETHYL-N-METHYL-2-((9H-XANTHEN-9-YLCARBO-NYL)OXY)-, BROMIDE (9CI) see XCJ000
ETHANAMINIUM, 2-HYDRAZINO-N,N,N-TRIMETHYL-2-OXO-, CHLORIDE (9CI) see GEQ500
ETHANAMINIUM, 2-HYDROXY-N,N,N-TRIS(2-HYDROXYETHYL)-, HYDROXIDE (9CI) see TCB500
ETHANAMINIUM, N,N,N-TRIETHYL-, IODIDE (9CI) see TCC750
ETHANAMINIUM, N,N,N-TRIETHYL-, PERCHLORATE see TCD002
ETHANAMINIUM, N,N,N-TRIMETHYL-, IODIDE (9CI) see EQC600
ETHANAMINIUM, N,N,N-TRIMETHYL-2-(1-OXOPROPOXY)-, IODIDE (9CI) see PMW750
ETHANE see EDZ000
ETHANE, 1,2-BIS(2,3-EPOXYPROPOXY)- see EEA600
ETHANECARBOXYLIC ACID see PMU750
ETHANE, 2-CHLORO-2-(DIFLUOROMETHOXY)-1,1,1-TRIFLUORO- (9CI) see IKS400
ETHANE, CHLOROPENTAFLUORO-, mixt. with CHLORODIFLUOROMETHANE see FOO510
ETHANEDIAL DIOXIME see EEA000
ETHANEDIAMIDE see OLO000
1,2-ETHANEDIAMINE see EEA500
1,2-ETHANEDIAMINE, DIHYDROCHLORIDE see EIW000
1,2-ETHANEDIAMINE, N-METHYL-(9CI) see MJW100
1,2-ETHANEDIAMINE, N,N'-DIMETHYL-(9CI) see DRI600
ETHANEDIAMINE, N,N'-DINITRO-(9CI) see DUU800
ETHANE, 1,2-DIAMINO-, COPPER COMPLEX see DBU800
ETHANE, 1,2-DIAZIDO- see DCL600
ETHANE, 1,2-DIBROMOTETRAFLUORO- see FOO525
ETHANE, 1,2-DIBROMO-1,1,2,2-TETRAFLUORO-(9CI) see FOO525
1,2-ETHANEDICARBOXYLIC ACID see SMY000
ETHANE DICHLORIDE see EIY600
ETHANE, 1,1-DICHLORO-1-FLUORO- see FOO550
ETHANEDINITRILE see COO000
ETHANE, 1,1-DINITRO- see DUV710
ETHANE, 1,2-DINITRO- see DUV720
ETHANEDIOIC ACID see OLA000
ETHANEDIOIC ACID, COPPER(2+) SALT (1:1) see CNN755
ETHANEDIOIC ACID DIAMMONIUM SALT see ANO750
ETHANEDIOIC ACID, DIPOTASSIUM SALT (9CI) see PLN300
ETHANEDIOIC ACID, DISILVER(1+) SALT see SDU000
ETHANEDIOIC ACID, TIN(2+) SALT (1:1) (9CI) see TGE250
1,2-ETHANEDIOL see EJC500
1,2-ETHANEDIOL DIACETATE see EJD759
1,2-ETHANEDIOL DIGLYCIDYL ETHER see EEA600
1,2-ETHANEDIOL DIMETHACRYLATE see BKM250
ETHANEDIOL DIMETHACRYLATE see BKM250
1,2-ETHANEDIOL DIMETHANESULFONATE (9CI) see BKM125
ETHANEDIOL DINITRATE see EJG000
1,2-ETHANEDIOL DIPROPANOATE (9CI) see COB260
1,2-ETHANEDIOL, MONOACETATE see EJI000
1,2-ETHANEDIOL, 1-PHENYL- see SMQ100
1,2-ETHANEDIOL, PHENYL- see SMQ100
1,2-ETHANEDIOL, 1,1,2,2-TETRAPHENYL- see TEA600
ETHANEDIONIC ACID see OLA000
ETHANEDITHIOAMIDE see DXO200
1,2-ETHANEDITHIOL see EEB000
1,2-ETHANEDITHIOL, CYCLIC S,S-ESTER with PHOSPHONODITHIOIMIDOCAR-BONIC ACID P,P-DIETHYL ESTER see PGW750
1,2-ETHANEDITHIOL, CYCLIC ESTER with P,P-DIETHYL PHOSPHONODI-THIOIMIDOCARBONATE see PGW750
ETHANEDIUREA see EJC100
N,N'-1,2-ETHANEDIYLBIS(N-BUTYL-4-MORPHOLINECARBOXAMIDE) see DUO400
1,2-ETHANEDIYLBIS(CARBAMODITHIOATO)(2−)-MANGANESE see MAS500
((1,2-ETHANEDIYLBIS(CARBAMODITHIOATO))(2-)ZINC see EIR000
1,2-ETHANEDIYLBIS(CARBAMODITHIOATO) (2-)-S,S'-ZINC see EIR000
1,2-ETHANEDIYLBISCARBAMODITHIOIC ACID DISODIUM SALT see DXD200
1,2-ETHANEDIYLBISCARBAMODITHIOIC ACID MANGANESE COMPLEX see MAS500

1,2-ETHANEDIYLBISCARBAMODITHIOIC ACID, MANGANESE(2+) SALT (1:1) see MAS500

1,2-ETHANEDIYLBISCARBAMODITHIOIC ACID, ZINC COMPLEX see EIR000

1,2-ETHANEDIYLBISCARBAMOTHIOIC ACID, ZINC SALT see EIR000

N,N′-1,2-ETHANEDIYLBIS(N-(CARBOXYMETHYL))GLYCINE see EIX000

N,N′-1,2-ETHANEDIYLBIS(N-(CARBOXYMETHYL)GLYCINE), DIPOTASSIUM SALT see EEB100

N,N′-1,2-ETHANEDIYLBIS(N-(CARBOXYMETHYL)GLYCINE) DISODIUM SALT see EIX500

N,N′-1,2-ETHANEDIYLBIS(N-(CARBOXYMETHYL)GLYCINE TETRASODIUM SALT see EIV000

N,N′-1,2-ETHANEDIYLBIS(N-CARBOXYMETHYL)GLYCINE, TRISODIUM SALT see TNL250

1,2-ETHANEDIYLBISMANEB, MANGANESE(2+) SALT (1:1) see MAS500

1,1′-(1,2-ETHANEDIYLBIS(OXY))BIS-BUTANE see DDW400

2,2′-(1,2-ETHANEDIYLBIS(OXY))BISETHANOL see TJQ000

2,2′-(1,2-ETHANEDIYLBIS(OXYMETHYLENE))BISOXIRANE see EEA600

1,2-ETHANEDIYLBIS(TRIS(2-CYANOETHYL)PHOSPHONIUM DIBROMIDE see EIU000

N,N′′-1,2-ETHANEDIYLBISUREA see EJC100

(R)-2,2′-(1,2-ETHANEDIYLDIIMINO)BIS-1-BUTANOL see TGA500

1,1′-(1,2-ETHANEDIYLDIIMINO)BIS(3-(4-METHOXYPHENOXY)-2-PROPANOL, DI-METHANESULFONATE (salt) see MQY125

1,2-ETHANEDIYL DIMETHANESULFONATE see BKM125

1,2-ETHANEDIYL DIMETHANESULFONATE see BKM125

1,2-ETHANEDIYL ESTER CARBAMIMIDOTHIOIC ACID DIHYDROBROMIDE see EJA000

ETHANE HEXACHLORIDE see HCI000

ETHANE HEXAMERCARBIDE see BKH125, EEB500

ETHANE, HEXANITRO- see HET675

ETHANEHYDRAZONIC ACID see ACM750

ETHANE-1-HYDROXY-1,1-DIPHOSPHONATE see HKS780

ETHANE-1-HYDROXY-1,1-DIPHOSPHONIC ACID DISODIUM SALT see DXD400

ETHANE-1-HYDROXY-1,1-DIPHOSPHONIC ACID, TETRAPOTASSIUM SALT see TEC250

ETHANE-1-HYDROXY-1,1-DIPHOSPHONIC ACID, TRATRASODIUM SALT see TEE250

ETHANE-1-HYDROXY-1,1-DIPHOSPHONIC ACID, TRISODIUM SALT see TNL750

ETHANENITRILE see ABE500

ETHANE PENTACHLORIDE see PAW500

ETHANEPEROXOIC ACID see PCL500

ETHANEPEROXOIC ACID, 1,1-DIMETHYLETHYL ESTER see BSC250

ETHANESELENOL, 2-AMINO-, HYDROCHLORIDE see AJS900

ETHANE-1,1′-SULFINYLBIS see EPI500

ETHANESULFONIC ACID, 2-AMINO- see TAG750

ETHANESULFONIC ACID, 2-HYDROXY-, AMMONIUM SALT see ANL100

ETHANESULFONYL CHLORIDE see EEC000

ETHANE, 1,1,2,2-TETRACHLORO-1,2-DIFLUORO- see TBP050

ETHANETHIOAMIDE see TFA000

ETHANETHIOAMIDE, N,N-DIMETHYL-(9CI) see DUG450

ETHANETHIOIC ACID see TFA500

ETHANETHIOL see EMB100

ETHANETHIOLIC ACID see TFA500

ETHANE TRICHLORIDE see TIN000

ETHANE, 1,1,1-TRIFLUORO- see TJY900

1,1,1-ETHANETRIOL DIPHOSPHONATE see HKS780

ETHANE, compressed (UN 1035) (DOT) see EDZ000

ETHANE, refrigerated liquid (UN 1961) (DOT) see EDZ000

ETHANIMIDOTHIOIC ACID, 2-(DIMETHYLAMINO)-N-HYDROXY-2-OXO-, METHYL ESTER see OLT100

ETHANIMIDOTHIOIC ACID, N-HYDROXY-, METHYL ESTER see MPS270

ETHANIMINIUM-N-(6-(DIETHYLAMINO)-3H-XANTHEN-3-YLIDENE)-N-ETHYL CHLORIDE see DIS200

6,10-ETHANO-5-AZONIASPIRO(4.5)DECAN-8-OL CHLORIDE BENZILATE see BCA375

1H-4,9a-ETHANOCYCLOHEPTA(c)PYRAN-7-CARBOXYLIC ACID, 4a-(ACETY-LOXY)-3-(4-CARBOXY-1,3-PENTADIENYL)3,4,4a,5,6,9-HEXAHYDRO-3-METH-YL-1-OXO-, 7-METHYL ESTER, (3-α(1E,3E),4-α-4a-α-9a-α)-(-)- (9CI) see POH600

3,8a-ETHANO-8aH-1-BENZOPYRAN-2(3H)-ONE, HEXAHYDRO-3,5,5-TRIMETHYL- see LAQ100

ETHANOIC ACID see AAT250

ETHANOIC ACID, ETHENYL ESTER see VLU250

ETHANOIC ANHYDRIDE see AAX500

ETHANOL (MAK) see EFU000

N-ETHANOLACETAMIDE see HKM000

β-ETHANOLAMINE see EEC600

ETHANOLAMINE, solution (DOT) see EEC600

ETHANOLAMINE see EEC600

ETHANOLAMINE CHLORIDE see EEC700

ETHANOLAMINE-N,N-DIACETIC ACID see HKM500

ETHANOLAMINE HYDROCHLORIDE see EEC700

ETHANOLAMINE PERCHLORATE see HKM175

ETHANOLAMINE PHOSPHATE see EED000

ETHANOLAMINE SALT of 5,2′-DICHLORO-4′-NITROSALICYCLICANILIDE see DFV600

ETHANOL, 1-AMINO-(8CI,9CI) see AAG500

ETHANOL, 2-AMINO-, HYDROCHLORIDE see EEC700

ETHANOL, 2-((4-AMINO-2-NITROPHENYL)AMINO)- see ALO750

ETHANOL, 2-AMINO-1-PHENYL- see HNF000

ETHANOL, 2-AZIDO-, NITRATE (ester) see ASF800

ETHANOL-2-(2-BUTOXYETHOXY) THIOCYANATE see BPL250

ETHANOL, 2-CHLORO-, ACETATE see CGO600

ETHANOL, 2-((4-((2-CHLORO-4-NITROPHENYL)AZO)PHENYL)ETHYLAMINO)- see CMP080

ETHANOL, 2-(2-(4-(p-CHLORO-α-PHENYLBENZYL)-1-PIPERAZINYL)ETHOXY)-, DIHYDROCHLORIDE see HOR470

ETHANOL, 2-(2-(2-(4-(p-CHLORO-α-PHENYLBENZYL)-1-PIPERAZI-NYL)ETHOXY)ETHOXY)-, DIMALEATE see HHK100

ETHANOL, 2,2′-((3-CHLOROPHENYL)IMINO)BIS- see CJU200

ETHANOL, 2,2′-(1,2-ETHANEDIYLBIS(OXY))BIS-, DINITRATE see TJQ500

ETHANOLETHYLENE DIAMINE see AJW000

ETHANOL, 2,2′-(ETHYLIMINO)BIS-(9CI) see ELP000

ETHANOL, 2-(N-ETHYL-m-TOLUIDINO)- see HKS100

ETHANOL, 2-(HEPTYLOXY)- see HBN250

ETHANOL, 2-(2-(HEPTYLOXY)ETHOXY)- see HBP275

ETHANOL,2,2′-IMINODI-,3,5-DIIODO-4-OXO-1(4H)-PYRIDINEACETATE (salt) see DNG400

ETHANOL,2,2′-IMINODI- with 3,5-DIIODO-4-OXO-1(4H)-PYRIDINEACETIC ACID (1:1) see DNG400

ETHANOLISOPROPYLAMINE see INN400

ETHANOL, 2-(ISOPROPYLAMINO)- see INN400

ETHANOLMERCURY BROMIDE see EED600

ETHANOL, 2-(2-(2-METHOXYETHOXY)ETHOXY)-, ACETATE (8CI,9CI) see AAV500

ETHANOL, 2-METHOXY-, OLEATE see MEP750

ETHANOL, 2-METHOXY-, PHOSPHATE (3:1) see TLA600

ETHANOL, 2-((1-METHYLETHYL)AMINO)- (9CI) see INN400

ETHANOL, 2,2′-(METHYLIMINO)BIS- see MKU250

ETHANOL, 2,2′,2′′-NITRILOTRIS-, TRINITRATE (ester), PHOSPHATE (1:2)(SALT) (9CI) see TJL250

ETHANOL, 2-(((5-NITRO-2-FURANYL)METHYLENE)AMINO)-, N-OXIDE (9CI) see HKW450

ETHANOL, 2-((5-NITROFURFURYLIDENE)AMINO)-, N-OXIDE see HKW450

ETHANOL, 2-NITRO-, NITRATE (ester) see NFY570

ETHANOL, 1-PHENYL- see PDE000

ETHANOL, 2-PHENYL- see PDD750

ETHANOL 200 PROOF see EFU000

ETHANOL, 2,2′,2′′-(PROPYLIDYNETRIS(METHYLENEOXY))TRI-, TRIACRYLATE see TLX100

4-ETHANOLPYRIDINE see POR500

ETHANOL, 2,2′,2′′,2′′′-SILANETETRAYLTETRAKIS- see TDH100

ETHANOL SOLUTIONS (UN 1170) (DOT) see EFU000

ETHANOLSULFONIC ACID see HKI500

ETHANOL THALLIUM(1+) SALT see EEE000

1-ETHANOL-2-THIOL see MCN250

ETHANOL, 2,2,2-TRICHLORO-, DIHYDROGEN PHOSPHATE see TIO800

ETHANOL-2-(2,4,5-TRICHLOROPHENOXY)-, 2,2-DICHLOROPROPIONATE see PBK000

ETHANONE, 1-(4-AMINO-5-(2-CHLOROPHENYL)-2-METHYL-1H-PYRROL-3-YL)- see AJI300

ETHANONE, 1-(4-AMINO-5-(p-CHLOROPHENYL)-2-METHYL-1H-PYRROL-3-YL)- see AJI330

ETHANONE, 1-(4-AMINO-5-(3,4-DICHLOROPHENYL)-2-METHYL-1H-PYRROL-3-YL)- see AJM600

ETHANONE, 1-(4-AMINO-5-(3,4-DIMETHOXYPHENYL)-2-METHYL-1H-PYRROL-3-YL)- see AJO800

ETHANONE, 1-(4-AMINO-1,2-DIMETHYL-5-PHENYL-1H-PYRROL-3-YL)- see AJR100

ETHANONE, 1-(4-AMINO-1-ETHYL-2-METHYL-5-PHENYL-1H-PYRROL-3-YL)- see AKA600

ETHANONE, 1-(4-AMINO-5-(3-FLUOROPHENYL)-2-METHYL-1H-PYRROL-3-YL)- see AKC550

ETHANONE, 1-(4-AMINO-5-(4-FLUOROPHENYL)-2-METHYL-1H-PYRROL-3-YL)- see AKC560

ETHANONE, 1-(4-AMINO-5-(4-METHOXY-3-METHYLPHENYL)-2-METHYL-1H-PYRROL-3-YL)- see AKO100

ETHANONE, 1-(4-AMINO-5-(4-METHOXYPHENYL)-2-METHYL-1H-PYRROL-3-YL)- see AKO430

ETHANONE, 1-(4-AMINO-5-(2-METHOXYPHENYL)-2-METHYL-1H-PYRROL-3-YL)- see AKO450

ETHANONE, 1-(4-AMINO-5-(3-METHOXYPHENYL)-2-METHYL-1H-PYRROL-3-YL)- see AKO400

ETHANONE, 1-(4-AMINO-2-METHYL-5-(2-METHYLPHENYL)-1H-PYRROL-3-YL)- see AKT800

ETHANONE, 1-(4-AMINO-2-METHYL-5-(3-METHYLPHENYL)-1H-PYRROL-3-YL)- see AKT830

ETHANONE, 1-(4-AMINO-2-METHYL-5-(4-METHYLPHENYL)-1H-PYRROL-3-YL)- see AKT850

ETHANONE, 1-(4-AMINO-2-METHYL-5-PHENYL-1H-PYRROL-3-YL)-, MONOHY-
DROCHLORIDE see ALA300
ETHANONE, 1-(9-AZABICYCLO(4.2.1)NON-2-EN-2-YL)-, (1R)- see AOO120
ETHANONE, 1-(2-BENZOFURANYL)-(9CI) see ACC100
ETHANONE, 1-(2H-1-BENZOPYRAN-3-YL)-(9CI) see BCS500
ETHANONE, 1-(1,1'-BIPHENYL)-4-YL-(9CI) see MHP500
ETHANONE, 1-(1,1'-BIPHENYL)-4-YL-2-((4-(DICHLOROACETYL)PHENYL)AMINO)-
2-HYDROXY- see BGL400
1-ETHANONE, 1-(3-BROMOADAMANTYL)-2-DIAZO- see EEE025
ETHANONE, 1-(1-BROMO-3-ADAMANTYL)-2-ETHOXY- see BMT100
ETHANONE, 1-(1-BROMO-3-ADAMANTYL)-2-HYDROXY- see BMT130
ETHANONE, 2-BROMO-1-(4-CHLOROPHENYL)- see CJJ100
1-ETHANONE, 1-(3-CHLOROADAMANTYL)-2-DIAZO- see CEE800
ETHANONE, 1-(1-CHLORO-3-ADAMANTYL)-2-ETHOXY- see CEE825
ETHANONE, 1-(1-CHLORO-3-ADAMANTYL)-2-HYDROXY- see CEE850
ETHANONE, 2-CHLORO-1-(9H-FLUOREN-2-YL)- see CEC700
ETHANONE, 1-(4-CHLORO-3-NITROPHENYL)-2-ETHOXY-2-((4-(METHYLTHI-
O)PHENYL)AMINO)- see CJD630
ETHANONE, 2-CHLORO-1-PHENYL- see CEA750
ETHANONE, 1-(3-CHLOROPHENYL)-2-((4-(DICHLOROACETYL)PHENYL)AMINO)-
2-HYDROXY- see CJU300
ETHANONE, 2-DIAZO-1-PHENYL-(9CI) see PET800
ETHANONE, 2-((4-(DIBROMOACETYL)PHENYL)AMINO)-2-ETHOXY-1-(4-NITRO-
PHENYL)- see DDJ850
ETHANONE, 2-((4-(DICHLOROACETYL)PHENYL)AMINO)-2-ETHOXY-1-(4-NITRO-
PHENYL)- see DEN820
ETHANONE, 2-((4-(DICHLOROACETYL)PHENYL)AMINO)-2-HYDROXY-1-(4-ME-
THOXYPHENYL)- see DEN840
ETHANONE, 2-((4-(DICHLOROACETYL)PHENYL)AMINO)-2-HYDROXY-1-(4-ME-
THYLPHENYL)- see DEN860
ETHANONE, 2-((4-(DICHLOROACETYL)PHENYL)AMINO)-2-HYDROXY-1-(4-PHE-
NOXYPHENYL)- see DEN880
ETHANONE, 2-((4-(DICHLOROACETYL)PHENYL)AMINO)-2-HYDROXY-1-(4-(PHE-
NYLTHIO)PHENYL)- see DEN910
ETHANONE, 2-((4-(DICHLOROACETYL)PHENYL)AMINO)-2-HYDROXY-1-PHE-
NYL- see DEN900
ETHANONE, 1-(4,7-DIMETHOXY-2-BENZOFURANYL)-(9CI) see DOA810
ETHANONE, 1-(6,7-DIMETHOXY-2-BENZOFURANYL)-(9CI) see DOA815
ETHANONE, 1-(5,8-DIMETHOXY-2H-1-BENZOPYRAN-3-YL)-(9CI) see DOA820
ETHANONE, 1-(7,8-DIMETHOXY-2H-1-BENZOPYRAN-3-YL)-(9CI) see DOA830
ETHANONE, 1-(DIMETHYLAMINO)-2-METHYL-5-PHENYL-1H-PYRROL-3-YL)-
see DPL950
ETHANONE, 1-(10-(2-(DIMETHYLAMINO)PROPYL)-10H-PHENOTHIAZIN-2-YL)-
(9CI) see AAF800
ETHANONE, 1-(7-(3-((1,1-DIMETHYLETHYL)AMINO)-2-HYDROXYPROPOXY)-2-
BENZOFURANYL)- see ACN320
ETHANONE, 1-(4-(3-((1,1-DIMETHYLETHYL)AMINO)-2-HYDROXYPROPOXY)-2-
BENZOFURANYL)- see ACN300
ETHANONE, 1-(2,4-DIMETHYL-5-PHENYL-1H-PYRROL-3-YL)- see DTO300
ETHANONE, 1-(2,5-DIMETHYL-3-THIENYL)- see DUG425
ETHANONE, 1-(2-ETHYL-7-(2-HYDROXY-3-((1-METHYLE-
THYL)AMINO)PROPOXY)-4-BENZOFURANYL)- see ELI600
ETHANONE, 1-(3-ETHYL-5,6,7,8-TETRAHYDRO-5,5,8,8-TETRAMETHYL-2-NA-
PHTHALENYL)(9CI) see ACL750
ETHANONE, 1-(9H-FLUOREN-2-YL) see FEI200
ETHANONE, 2-HYDROXY-1,2-DIPHENYL- see BCP250
ETHANONE, 2-HYDROXY-1-(1-IODO-3-ADAMANTYL)- see HME100
ETHANONE, 1-(7-(2-HYDROXY-3-((1-METHYLETHYL)AMINO)PROPOXY)-2-BEN-
ZOFURANYL)-(9CI) see HLK600
ETHANONE, 1-(4-HYDROXY-2-METHYL-5-PHENYL-1H-PYRROL-3-YL)- see
HMN100
ETHANONE, 1-(7-(2-HYDROXY-3-((1-METHYLPROPYL)AMINO)PROPOXY)-2-BEN-
ZOFURANYL)- see ACN310
ETHANONE, 2-HYDROXY-1-(1-PHENYL-3-ADAMANTYL)- see HMK150
ETHANONE, 1-(2-HYDROXYPHENYL)-(9CI) see HIN500
ETHANONE, 1-(10-(3-(4-HYDROXY-1-PIPERIDINYL)PROPYL)-10H-PHENOTHI-
AZIN-2-YL)- see HNT200
ETHANONE, 1-(8-HYDROXY-5-QUINOLINYL)-(9CI) see HOE200
ETHANONE, 1-(5-METHOXY-2-BENZOFURANYL)-(9CI) see MEC330
ETHANONE, 1-(6-METHOXY-2-BENZOFURANYL)-(9CI) see MEC300
ETHANONE, 1-(7-METHOXY-2-BENZOFURANYL)-(9CI) see MEC320
ETHANONE, 1-(5-METHOXY-2H-1-BENZOPYRAN-3-YL)-(9CI) see MEC340
ETHANONE, 1-(5-METHOXY-1H-INDOL-3-YL)-2-(4-PIPERIDINYL)-, MONOHYDRO-
CHLORIDE (9CI) see MES900
ETHANONE, 1-(4-METHOXY-2-METHYL-5-PHENYL-1H-PYRROL-3-YL)- see
MEX275
ETHANONE, 1-(7-METHOXYNAPHTHO(2,1-b)FURAN-2-YL)- see ACQ790
ETHANONE, 1-(4-METHOXYPHENYL)-(9CI) see MDW750
ETHANONE, 1-(10-(3-(4-METHOXY-1-PIPERIDINYL)PROPYL)-10H-PHENOTHI-
AZIN-2-YL)- see MFJ200
ETHANONE, 1-(4-METHYLPHENYL)-(9CI) see MFW250
ETHANONE, 1-(2-METHYL-5-PHENYL-1H-PYRROL-3-YL)- see MNY800
ETHANONE, 1-(5-METHYL-2-THIENYL)- see MPR300
ETHANONE, 1-(1-NAPHTHALENYL)-(9CI) see ABC475
ETHANONE, 1-(2-NITRONAPHTHO(2,1-b)FURAN-7-YL)- see NHR100

ETHANONE, 1-(4-NITROPHENYL)- see NEL600
ETHANONE, 1-(3-NITROPHENYL)-(9CI) see NEL500
ETHANONE, 1-(5-NITRO-2-THIENYL)-(9CI) see NML100
ETHANONE, 1-(3-PHENOXYPHENYL)-(9CI) see PDR200
ETHANONE, 1-PHENYL-, OXIME see ABH150
ETHANONE, 1-(2-PYRIDINYL)-(9CI) see PPH200
ETHANONE, 1,2,2-TRIPHENYL- see TMS100
ETHANOX 330 see TMJ000
ETHANOX 701 see DEG100
ETHANOX 736 see TFD000
ETHANOX see EEH600
ETHANOXYTRIPHETOL see DHS000
ETHANOYL CHLORIDE see ACF750
ETHAPERAZINE see CJM250
ETHAPHENE see HLV500
ETHAVAN see EQF000
ETHAVERINE HYDROCHLORIDE see PAH260
ETHAZATE see BJC000
ETHAZOLE (FUNGICIDE) see EFK000
ETHBENZAMIDE see EEM000
ETHCHLOROVYNOL see CHG000
ETHCHLORVINYL see CHG000
ETHCLORVYNOL see CHG000
ETHEFON see CDS125
ETHEL see CDS125
ETHENE see EIO000
ETHENE, 1,1-DICHLORO-, POLYMER with CHLOROETHENE (9CI) see
CGW300
ETHENE, 1,1-DIFLUORO- see VPP000
2,3'-(1,2-ETHENEDIYL)BIS(5-AMINOBENZENESULFONIC ACID) DISODIUM
SALT see FCA200
(E)-1,1'-(1,2-ETHENEDIYL)BISBENZENE see SLR100
1,1'-(1,2-ETHENEDIYL)DIBENZENE see SLR000
ETHENE, FLUORO- see VPA000
ETHENE, NITRO-, HOMOPOLYMER see NFY560
ETHENE OXIDE see EJN500
ETHENE POLYMER see PJS750
ETHENETETRACARBONITRILE see EEE500
ETHENE, TRIBROMO- see THV100
ETHENE, TRIFLUORO(TRIFLUOROMETHOXY)- see TKG800
ETHENOL HOMOPOLYMER (9CI) see PKP750
ETHENONE see KEU000
6,14-endo-ETHENOTETRAHYDROORIPAVINE, 7-α-(1-HYDROXY-1-METHYLBU-
TYL)- see EQO450
1-ETHENOXY-2-ETHYLHEXANE see ELB500
ETHENYL ACETATE see VLU250
1-ETHENYLAZIRIDINE see VLZ000
α-ETHENYL-1-AZIRIDINEETHANOL (9CI) see VMA000
ETHENYLBENZENE see SMQ000
ETHENYLBENZENE HOMOPOLYMER see SMQ500
ETHENYLBENZENE POLYMER with 1,3-BUTADIENE see SMR000
4-ETHENYL-BENZENESULFONIC ACID SODIUM SALT, HOMOPOLYMER (9CI)
see SJK375
5-ETHENYLBICYCLO(2.2.1)HEPT-2-ENE see VQA000
9-ETHENYL-9H-CARBAZOLE see VNK100
N-ETHENYLCARBAZOLE see VNK100
4-ETHENYL-1-CYCLOHEXENE see CPD750
1-ETHENYLCYCLOHEXENE see VNZ990
ETHENYL ETHANOATE see VLU250
S-ETHENYL-dl-HOMOCYSTEINE see VLU200
S-ETHENYL-l-HOMOCYSTEINE see VLU210
6-ETHENYL-6-(METHOXYETHOXY)-2,5,7,10-TETRAOXA-6-SILAUNDECANE (9CI)
see TNJ500
4-ETHENYL-2-METHOXYPHENOL see VPF100
1-ETHENYL-4-METHYLBENZENE see VQK700
5-ETHENYL-2-METHYLPYRIDINE see MQM500
1-(ETHENYLOXY) BUTANE see VMZ000
1-(ETHENYLOXY)DECANE see EEE200
2-(ETHENYLOXY)ETHANOL see EJL500
ETHENYLOXYETHENE see VOP000
1-(2-(ETHENYLOXY)ETHOXY)BUTANE see VMU000
2-(2-(ETHENYLOXY)ETHOXY)ETHANOL see DJG400
2-(ETHENYLOXY)ETHYL 2-METHYL-2-PROPENOATE see VQA150
4-ETHENYLPHENOL ACETATE see ABW550
1-ETHENYL PYRENE see EEF000
4-ETHENYL PYRENE see EEF500
2-ETHENYLPYRIDINE see VQK560
4-ETHENYLPYRIDINE see VQK590
ETHENYLPYRIDINE 1-OXIDE HOMOPOLYMER see PKQ100
1-ETHENYL-2-PYRROLIDINONE see EEG000
1-ETHENYL-2-PYRROLIDINONE HOMOPOLYMER see PKQ250
1-ETHENYL-2-PYRROLIDINONE HOMOPOLYMER compounded with IODINE
see PKE250
1-ETHENYL-2-PYRROLIDINONE POLYMER with ETHENYLBENZENE see
VQK595

1-ETHENYL-2-PYRROLIDINONE POLYMERS see PKQ250
1-ETHENYLSILATRANE see VQK600
ETHENYLTRIMETHOXYSILANE see TLD000
ETHENZAMID see EEM000
ETHENZAMIDE see EEM000
ETHEPHON see CDS125
ETHER see EJU000
ETHER, ALLYL GLYCERYL see AGP500
ETHER, BIS(4-CHLOROBUTYL) see OPE040
ETHER, BIS(2-CHLORO-1-METHYLETHYL), mixed with 2-CHLORO-1-METHYLE-
   THYL-(2-CHLOROPROPYL)ETHER (7:3) see BIK100
ETHER, BIS(2-CYANOETHYL) see OQQ000
ETHER, BIS(2-(2-ETHOXYETHOXY)ETHYL) see PBO250
ETHER BIS-14-HYDROXY-IMINOMETHYLOPYRIDINE-(1)-METYLODICHLORIDE
   (POLISH) see BGS250
ETHER, BIS(2-(2-METHOXYETHOXY)ETHYL) see PBO500
ETHER, BUTYL 2,3-EPOXYPROPYL see BRK750
ETHER, tert-BUTYL ETHYL see EHA550
ETHER, BUTYL GLYCIDYL see BRK750
ETHER BUTYLIQUE (FRENCH) see BRH750
ETHER, BUTYL METHYL see BRU780
ETHER CHLORATUS see EHH000
ETHER, 4-CHLOROPHENYL (4'-CHLORO-2'-NITRO)PHENYL see CJD600
ETHER, 1-CHLORO-2,2,2-TRIFLUOROETHYL DIFLUOROMETHYL see IKS400
ETHER CYANATUS see PMV750
ETHER, DECYL METHYL see MIW075
ETHER, DECYL VINYL (6Cl,7Cl,8Cl) see EEE200
ETHER DICHLORE (FRENCH) see DFJ050
ETHER, DICYCLOPENTYL see CQB275
ETHER, DI-n-HEPTYL- see HBO000
ETHER, DI-n-PENTYL- see PBX000
ETHER, DI-n-PROPYL- see PNM000
ETHER ETHYLBUTYLIQUE (FRENCH) see EHA500
ETHER ETHYLIQUE (FRENCH) see EJU000
ETHER, ETHYL PHENYL see PDM000
ETHER, HEXACHLOROPHENYL see CDV175
ETHER HYDROCHLORIC see EHH000
ETHERIN see PJS750
ETHER ISOPROPYLIQUE (FRENCH) see IOZ750
ETHER METHYLIQUE MONOCHLORE (FRENCH) see CIO250
ETHER, METHYL PHENYL see AOX750
ETHER, METHYL PROPYL see MOU830
ETHER MONOETHYLIQUE de l'ETHYLENE-GLYCOL (FRENCH) see EES350
ETHER MONOETHYLIQUE de l'HYDROQUINONE see EFA100
ETHER MONOMETHYLIQUE de l'ETHYLENE-GLYCOL (FRENCH) see EJH500
ETHER MURIATIC see EHH000
ETHEROL E see PJS750
ETHERON see PKL500
ETHERON SPONGE see PKL500
ETHEROPHENOL see IHX400
ETHER, PENTACHLOROPHENYL see PAW250
ETHER, PHENYL m-TOLYL- see MNV770
ETHERS see EEG500
ETHERSULFONATE see CJT750
ETHER, TRIFLUOROMETHYL TRIFLUOROVINYL see TKG800
ETHEVERSE see CDS125
ETHIBI see EDW875
ETHICON PTFE see TAI250
ETHIDE see DFU000
ETHIDIUM BROMIDE see DBV400
ETHIDIUM CHLORIDE see HGI000
ETHIENOCARB see DFE469
ETHIMIDE see EPQ000
ETHINA see EPQ000
ETHINAMATE see EEH000
ETHINAMIDE see EPQ000
ETHINE see ACI750
ETHINODIOL see EQJ100
ETHINODIOL DIACETATE see EQJ500
ETHINONE see GEK500
ETHINYLBENZENE see PEB750
2-ETHINYLBUTANOL-2 see EQL000
1-ETHINYLCYCLOHEXYL CARBAMATE see EEH000
1-ETHINYLCYCLOHEXYL CARBONATE see EEH000
17-α-ETHINYL-3,17-DIHYDROXY-Δ^{1,3,5}-ESTRATRIENE see EEH500
17-α-ETHINYL-3,17-DIHYDROXY-Δ^{1,3,5}-OESTRATRIENE see EEH500
17-α-ETHINYLESTRADIOL see EEH500
17-α-ETHINYL-17-β-ESTRADIOL see EEH500
17-ETHINYLESTRADIOL see EEH500
17-ETHINYL-3,17-ESTRADIOL see EEH500
ETHINYL ESTRADIOL see EEH500
17-α-ETHINYLESTRADIOL 3-CYCLOPENTYL ETHER see QFA250
ETHINYL ESTRADIOL and DIMETHISTERONE see DNX500
ETHINYLESTRADIOL and MEDROXYPROGESTERONE ACETATE see POF275
17-α-ETHINYL ESTRADIOL 3-METHYL ETHER see MKB750

ETHINYLESTRADIOL-3-METHYL ETHER see MKB750
ETHINYLESTRADIOL-3-METHYL ETHER and NORETHYNODRED (1:50) see
   EAP000
ETHINYL ESTRADIOL and NORETHINDRONE ACETATE see EEH520
ETHINYLESTRADIOL mixed with dl-NORGESTREL see NNL500
ETHINYL ESTRADIOL mixed with NORGESTREL see NNL500
17-α-ETHINYL-ESTRA(5,10)ENEOLONE see EEH550
17-ETHINYL-5(10)-ESTRAENEOLONE see EEH550
17-α-ETHINYLESTRA-4-EN-17-β-OL-3-ONE see NNP500
17-α-ETHINYLESTRA-1,3,5(10)-TRIENE-3,17-β-DIOL see EEH500
Δ^4-17-α-ETHINYLESTREN-17-β-OL see NNV000
ETHINYLESTRENOL see NNV000
17-α-ETHINYL-5,10-ESTRENOLONE see EEH550
ETHINYLESTRIOL see EEH500
17-α-ETHINYL-13-β-ETHYL-17-β-HYDROXY-4-ESTREN-3-ONE see NNQ520
17-α-ETHINYL-17-β-HYDROXYESTR-4-ENE see NNV000
17-α-ETHINYL-17-β-HYDROXY-Δ 4-ESTREN-3-ONE see NNP500
17-α-ETHINYL-17-β-HYDROXY-Δ^{5(10)}-ESTREN-3-ONE see EEH550
17-α-ETHINYL-17-β-HYDROXYOESTR-4-ENE see NNV000
ETHINYLMETHYLETHYLCARBINOL see EQL000
17-α-ETHINYL-19-NORTESTOSTERONE see NNP500
17-α-ETHINYL-Δ^{5,10-19}-NORTESTOSTERONE see EEH550
17-α-ETHINYL-19-NORTESTOSTERONE-17-β-ACETATE see ABU000
17-α-ETHINYL-19-NORTESTOSTERONE ACETATE see ABU000
17-α-ETHINYL-19-NORTESTOSTERONE ENANTHATE see NNQ000
17-ETHINYL-3,17-OESTRADIOL see EEH500
ETHINYLOESTRADIOL see EEH500
ETHINYLOESTRADIOL mixed with LYNOESTRENOL see EEH575
17-α-ETHINYL OESTRADIOL-3-METHYL ETHER see MKB750
ETHINYLOESTRADIOL-3-METHYL ETHER see MKB750
ETHINYL OESTRADIOL mixed with NORETHISTERONE ACETATE see
   EEH520
ETHINYLOESTRADIOL mixed with NORGESTREL see NNL500
ETHINYL-OESTRANOL see EEH500
ETHINYLOESTRANOL see NNV000
17-α-ETHINYLOESTRA-1,3,5(10)-TRIENE-3,17-β-DIOL see EEH500
17-α-ETHINYL-Δ^{1,3,5(10)}OESTRATRIENE-3,17-β-DIOL see EEH500
Δ^4-17-α-ETHINYLOESTREN-17-β-OL see NNV000
ETHINYL OESTRENOL see NNV000
ETHINYLOESTRIOL see EEH500
17-ETHINYLTESTOSTERONE see GEK500
ETHINYLTESTOSTERONE see GEK500
ETHINYL TRICHLORIDE see TIO750
ETHION, dry see CMA100
ETHIODAN see ELQ500
ETHIOFENCARB see EPR000
ETHIOFOS see AMD000
ETHIOL see EEH600
ETHIOLACAR see MAK700
ETHION see EEH600
ETHIONIAMIDE see EPQ000
ETHIONIN see EEI000
dl-ETHIONINE see EEI000
l-ETHIONINE see AKB250
ETHIONINE see AKB250, EEI000
(±)-ETHIONINE see EEI000
ETHIOPHENCARP see EPR000
ETHIRIMOL see BRI750
ETHISTERONE see GEK500
ETHISTERONE and DIETHYLSTILBESTROL see EEI050
ETHLON see PAK000
ETHMOSINE see EEI025
ETHMOZINE see EEI025
ETHNINE see TCY750
ETHOATE METHYL see DNX600
ETHOBROM see ARW250, THV000
ETHOCAINE see AIT250
ETHOCEL 150 see EHG100
ETHOCEL 890 see EHG100
ETHOCEL see EHG100
ETHOCEL E7 see EHG100
ETHOCEL E50 see EHG100
ETHOCEL MED see EHG100
ETHOCEL N7 see EHG100
ETHOCEL N10 see EHG100
ETHOCEL N200 see EHG100
ETHOCEL STD see EHG100
ETHOCHLORVYNOL see CHG000
ETHODAN see EEH600
ETHODIN see EDW500
ETHODRYL see DIW000
ETHODRYL CITRATE see DIW200
ETHODUOMEEN, HYDROFLUORIDE see EEI100
ETHODUOMEEN see EEI060
ETHOFAT O 15 see PJY100

ETHOFAT O see PJY100
ETHOGLUCID see TJQ333
ETHOGLUCIDE see TJQ333
ETHOHEPTAZINE CITRATE see MIE600
ETHOHEXADIOL see EKV000
ETHOL see HCP000
ETHOMEEN C/15 see EEJ000
ETHOMEEN S/12 see EEJ500
ETHOMEEN S/15 see EEK000
ETHOMEEN T/15 see EEK050
ETHONE see ENY500
ETHONE, 2-CHLORO-1,2-DIPHENYL-(9CI) see CFJ100
ETHOPHYLLINE see TEP500
ETHOPIAN see EDW875
ETHOPROP see EIN000
ETHOPROPAZINE see DIR000
ETHOPROPHOS see EIN000
ETHOSALICYL see EEM000
ETHOSPERSE CL20 see PJT300
ETHOSPERSE LA-4 see DXY000
ETHOSUCCIMIDE see ENG500
ETHOSUCCINIMIDE see ENG500
ETHOSUXIDE see ENG500
ETHOSUXIMIDE see ENG500
ETHOTOIN see EOL100
ETHOVAN see EQF000
ETHOXENE see VMA000
ETHOXOL 20 see PJW500
2-ETHOXYACETANILIDE see ABG350
2'-ETHOXYACETANILIDE see ABG350
3'-ETHOXYACETANILIDE see ABG250
3-ETHOXYACETANILIDE see ABG250
4-ETHOXYACETANILIDE see ABG750
m-ETHOXYACETANILIDE see ABG250
p-ETHOXYACETANILIDE see ABG750
ETHOXY ACETATE see EES400
4-ETHOXYACETOACETANILIDE see AAZ000
4'-ETHOXYACETOACETANILIDE see AAZ000
ETHOXY ACETYLENE see EEL000
3-ETHOXY ACROLEIN DIETHYL ACETAL see TJM500
2-ETHOXYANILINE see PDK819
3-ETHOXYANILINE see PDK800
4-ETHOXYANILINE see PDK890
m-ETHOXYANILINE see PDK800
o-ETHOXYANILINE see PDK819
p-ETHOXYANILINE see PDK890
4-ETHOXYANILINE HYDROCHLORIDE see PDL750
6-ETHOXY-m-ANOL see IRY000
4-ETHOXYBENZALDEHYDE see EEL500
p-ETHOXYBENZALDEHYDE see EEL500
ETHOXYBENZALDEHYDE see EEL500
2-ETHOXYBENZAMIDE see EEM000
o-ETHOXYBENZAMIDE see EEM000
3-ETHOXYBENZENAMINE see PDK800
4-ETHOXYBENZENAMINE see PDK890
4-ETHOXYBENZENAMINE HYDROCHLORIDE see PDL750
ETHOXYBENZENE see PDM000
o-ETHOXYBENZOIC ACID (1-CARBOXYETHYLIDENE)HYDRAZIDE see RSU450
o-ETHOXY-BENZOIL-IDRAZONE DELL'ACIDO PIRUVICO (ITALIAN) see RSU450
o-ETHOXY-BENZOYL-HYDRAZONE of PYRUVIC ACID see RSU450
4-(p-ETHOXYBENZOYL)PYRIDINE see EEN600
8-ETHOXYCAFFEINE see EEO000
3-(ETHOXYCARBONYLAMINO)PHENOL see ELE600
m-(ETHOXYCARBONYLAMINO)PHENOL see ELE600
3-ETHOXYCARBONYLAMINOPHENYL-N-PHENYLCARBAMATE see EEO500
N-(ETHOXYCARBONYL)AZIRIDINE see ASH750
S-α-ETHOXYCARBONYLBENZYL-O,O-DIMETHYL PHOSPHORODITHIOATE see DRR400
S-α-ETHOXYCARBONYLBENZYL DIMETHYL PHOSPHOROTHIOLOTHIONATE see DRR400
ETHOXYCARBONYLDIAZOMETHANE see DCN800
ETHOXY CARBONYL DIGOXIN see EEP000
ETHOXYCARBONYLETHYLENE see EFT000
N-ETHOXYCARBONYLETHYLENEIMINE see ASH750
ETHOXYCARBONYL-1-ETHYLENIMINE see ASH750
ETHOXYCARBONYL HYDRAZIDE see EHG000
1-(ETHOXYCARBONYL) HYDRAZINE see EHG000
N-(ETHOXYCARBONYL) HYDRAZINE see EHG000
(ETHOXYCARBONYL) HYDRAZINE see EHG000
7-ETHOXYCARBONYL-4-HYDROXYMETHYL-6,8-DIMETHYL-1(2H)-PHTHALAZI-NONE see PHV725
4-ETHOXYCARBONYL-1-(2-HYDROXY-3-PHENOXYPROPYL) 4-PHENYLPIPERI-DINE HYDROCHLORIDE see EEQ000
ETHOXYCARBONYLMETHYL BROMIDE see EGV000

S-((ETHOXYCARBONYL)METHYLCARBAMOYL)METHYL-O,O-DIETHYL PHOS-PHORODITHIOATE see DJI000
N-ETHOXYCARBONYL-N-METHYLCARBAMOYLMETHYL-O,O-DIETHYL PHOS-PHORODITHIOATE see DJI000
S-(N-ETHOXYCARBONYL-N-METHYLCARBAMOYLMETHYL)-DIETHYL PHOS-PHORODITHIOATE see DJI000
(2-(ETHOXYCARBONYL)-1-METHYL)ETHYL CARBONIC ACID-p-IODOBENZYL ESTER see EEQ100
N-(ETHOXYCARBONYL)-3-(4-MORPHOLINYL)SYDNONE IMINE see MRN275
N-(N-((S)-1-ETHOXYCARBONYL-3-PHENYLPROPYL)-l-ALANYL)-N-(INDAN-2-YL)GLYCINE HYDROCHLORIDE see DAM315
(S)-1-(N-(1-(ETHOXYCARBONYL)-3-PHENYLPROPYL)-l-ALANYL)-l-PROLINE MA-LEATE see EAO100
N-((S)-1-ETHOXYCARBONYL-3-PHENYLPROPYL)-l-ALANYL-l-PROLINE MALEATE see EAO100
S-ETHOXYCARBONYLTHIAMINE HYDROCHLORIDE see EEQ500
ETHOXY CHLOROMETHANE see CIM000
ETHOXY CLEVE'S ACID see AJU500
2-ETHOXY-6,9-DIAMINOACRIDINE LACTATE see EDW500
2-ETHOXY-6,9-DIAMINOACRIDINE LACTATE HYDRATE see EDW500
2-ETHOXY-6,9-DIAMINOACRIDINIUM LACTATE see EDW500
ETHOXY DIETHYL ALUMINUM see EER000
ETHOXY DIGLYCOL see CBR000
11-ETHOXY-15,16-DIHYDRO-17-CYCLOPENTA(a)PHENANTHREN-17-ONE see EER400
2-ETHOXY-3,4-DIHYDRO-1,2-PYRAN see EER500
2-ETHOXY-3,4-DIHYDRO-2H-PYRAN see EER500
2-ETHOXY DIHYDROPYRAN see EER500
2-ETHOXY-2,3-DIHYDRO-γ-PYRAN see EER500
6-ETHOXY-1,2-DIHYDRO-2,2,4-TRIMETHYLQUINOLINE see SAV000
ETHOXYDIISOBUTYLALUMINUM see EES000
3'-ETHOXY-4-DIMETHYLAMINOAZOBENZENE see DRU000
1-ETHOXY-3,7-DIMETHYL-2,6-OCTADIENE see GDG100
8-ETHOXY-2,6-DIMETHYLOCTENE-2 see EES100
ETHOXYDIPHENYLACETIC ACID see EES300
ETHOXYETHANE see EJU000
2-ETHOXYETHANOL see EES350
2-ETHOXYETHANOL ACETATE see EES400
2-ETHOXYETHANOL, ESTER with ACETIC ACID see EES400
ETHOXY ETHENE see EQF500
4-(1-ETHOXYETHENYL)-3,3,5,5-TETRAMETHYLCYCLOHEXANONE see EES370
1-(2-ETHOXYETHOXY)-BUTANE see BPK750
2-(2-ETHOXYETHOXY)ETHANOL see CBR000
2-(2-ETHOXYETHOXY)ETHANOL ACETATE see CBQ750
1-ETHOXY-2-(β-ETHOXYETHOXY)ETHANE see DIW800
2-(2-(2-ETHOXYETHOXY)ETHOXY)ETHANOL see EFL000
(2-(1-ETHOXYETHOXY)ETHYL)BENZENE see EES380
3-(2-ETHOXY)ETHOXY-2-PROPANOL see EET000
2-ETHOXY-ETHYLACETAAT (DUTCH) see EES400
2-ETHOXYETHYL ACETATE see EES400
β-ETHOXYETHYL ACETATE see EES400
ETHOXYETHYL ACETATE see EES400
2-ETHOXYETHYL ACRYLATE see ADT500
ETHOXYETHYL ACRYLATE see ADT500
2-(2-ETHOXYETHYLAMINO)-2',6'-ACETOXYLIDIDE see DRT400
2-(2-ETHOXYETHYLAMINO)-2',6'-ACETOXYLIDIDE HYDROCHLORIDE see DRT600
2-(2-ETHOXYETHYLAMINO)-2'-METHYL-PROPIONANILIDE see EES500
2-(2-ETHOXYETHYLAMINO)-o-PROPIONOTOLUIDIDE see EES500
2-ETHOXYETHYLE, ACETATE de (FRENCH) see EES400
1,1'-((1-ETHOXYETHYL)ETHANEDIYLIDENE)BIS(3-THIOSEMICARBAZIDE) see KFA100
ETHOXYETHYL ETHER of PROPYLENE GLYCOL see EET000
(1-ETHOXYETHYL)GLYOXAL BIS(THIOSEMICARBAZONE) see KFA100
1-ETHOXYETHYLIDENEMALONONITRILE see EET100
(1-ETHOXYETHYLIDENE)MALONONITRILE see EET100
(1-ETHOXYETHYLIDENE)PROPANEDINITRILE see EET100
2-ETHOXYETHYL-2-METHOXYETHYLETHER see DJE800
2-ETHOXYETHYL-2-PROPENOATE see ADT500
2-ETHOXYETHYL-2-(VINYLOXY)ETHYL ETHER see DJF000
ETHOXYETHYNE see EEL000
ETHOXYFORMIC ANHYDRIDE see DIX200
3-ETHOXYHEXANAL DIETHYL ACETAL see TJM000
(Z)-1-ETHOXY-1-(3-HEXENYLOXY)ETHANE see EKS100
1-ETHOXY-1-HEXYLOXYETHANE see EKS120
4-ETHOXY-3-HYDROXYBENZALDEHYDE see IKO100
3-ETHOXY-4-HYDROXYBENZALDEHYDE see EQF000
4'-ETHOXY-3-HYDROXYBUTYRANILIDE see HJS850
N-(2-ETHOXY-3-HYDROXYMERCURIPROPYL)BARBITAL see EET500
1-ETHOXY-2-HYDROXY-4-PROPENYLBENZENE see IRY000
2-(3-ETHOXY-1-INDANYLIDENE)-1,3-DINDANDIONE see BGC250
1-ETHOXY-3-ISOPROPOXY-2-PROPANOL see EET600
1-ETHOXY-3-ISOPROPOXYPROPAN-2-OL see EET600
ETHOXYKARBONYLMETHYL-METHYLESTER KYSELINY FTALOVE (CZECH) see MOD000

1-ETHOXY-4-(1-KETO-2-HYDROXYETHYL)-NAPHTHALENE SUCCINATE see SNB500
ETHOXYLATED LAURYL ALCOHOL see DXY000
ETHOXYLATED OCTYL PHENOL see GHS000
ETHOXYLATED SORBITAN MONOOLEATE see PKL100
1-ETHOXY-1-LINALYLOXETHANE see ELZ050
ETHOXYMETHANE see EMT000
(ETHOXYMETHOXY)CYCLODODECANE see FMV100
10-ETHOXYMETHYL-1:2-BENZANTHRACENE see EEV000
7-ETHOXY-12-METHYLBENZ(a)ANTHRACENE see MJX000
7-(ETHOXYMETHYL)BENZ(a)ANTHRACENE see EEV000
4-ETHOXY-2-METHYL-3-BUTYN-2-OL see EEV100
ETHOXY METHYL CHLORIDE see CIM000
ETHOXYMETHYL-N-(2,6-DICHLORO-m-TOLYL)ANTHRANILATE see DGN000
ETHOXYMETHYLENEMALONIC ACID, ETHYL ESTER see EEV200
ETHOXYMETHYLENE MALONONITRILE see EEW000
2-((ETHOXY((1-METHYLETHYL)AMINO)PHOSPHINOTHIOYL)OXY)BENZOIC ACID 1-METHYLETHYL ESTER see IMF300
1-(ETHOXYMETHYL)-2-METHOXYBENZENE see EMF600
7-ETHOXY METHYL-12-METHYL BENZ(a)ANTHRACENE see EEX000
2-ETHOXY-N-METHYL-N-(2-(METHYLPHENETHYLAMINO)ETHYL)-2,2-DIPHENY-LACETAMIDE HYDROCHLORIDE see CBQ625
4-ETHOXY-2-METHYL-5-MORPHOLINO-3(2H)-PYRIDAZINONE see EAN700
4-ETHOXY-2-METHYL-5-(4-MORPHOLINYL)-3(2H)-PYRIDAZINONE see EAN700
(ETHOXYMETHYL)OXIRANE see EKM200
2-ETHOXY-2-METHYLPROPANE see EHA550
N-ETHOXYMORPHOLINO DIAZENIUM FLUOROBORATE see EEX500
O-ETHOXY-N-(3-MORPHOLINOPROPYL)BENZAMIDE see EEY000
2-ETHOXYNAPHTHALENE see EEY500
6-(2-ETHOXY-1-NAPHTHAMIDO)PENICILLIN SODIUM see SGS500
1-ETHOXY-4-NITROBENZENE see NID000
ETHOXY-4-NITROPHENOXYPHENYLPHOSPHINE SULFIDE see EBD700
N-(4-ETHOXY-3-NITRO)PHENYLACETAMIDE see NEL000
ETHOXY-4-NITROPHENYLOXYETHYLPHOSPHINEOXIDE see ENQ000
3-ETHOXY-2-OXOBUTYRALDEHYDE BIS(THIOSEMICARBAZONE) see KFA100
ETHOXYPHAS see DIX000
3-(4-(β-ETHOXYPHENETHYL)-1-PIPERAZINYL)-2-METHYL-1-PHENYL-1-PROPA-NONE DIHYDROCHLORIDE see ECU550
3-(4-(β-ETHOXYPHENETHYL)-1-PIPERAZINYL)-2-METHYLPROPIOPHENONE see PFB000
2-(4-(β-ETHOXYPHENETHYL)-1-PIPERAZINYLMETHYL)PROPIOPHENONE DIHY-DROCHLORIDE see ECU550
4-ETHOXYPHENOL see EFA100
p-ETHOXYPHENOL see EFA100
2-((2-ETHOXYPHENOXY)METHYL)MORPHOLINE see VKA875
2-((2-ETHOXYPHENOXY)METHYL)MORPHOLINE HYDROCHLORIDE see VKF000
2-((o-ETHOXYPHENOXY)METHYL)MORPHOLINE HYDROCHLORIDE see VKF000
2-(2-ETHOXYPHENOXYMETHYL)TETRAHYDRO-1,4-OXAZINE see VKA875
2-(2-ETHOXYPHENOXYMETHYL)TETRAHYDRO-1,4-OXAZINE HYDROCHLORIDE see VKF000
N-(2-ETHOXYPHENYL)ACETAMIDE see ABG350
N-(4-ETHOXYPHENYL)ACETAMIDE see ABG750
N-p-ETHOXYPHENYLACETAMIDE see ABG750
N-(3-ETHOXYPHENYL)ACETAMIDE (9CI) see ABG250
N-(4-ETHOXYPHENYL)ACETOHYDROXAMIC ACID see HIN000
α-ETHOXY-α-PHENYL-BENZENEACETIC ACID (9CI) see EES300
2-(4-ETHOXYPHENYL)-1,3-BIS(4-METHOXYPHENYL)GUANIDINE HYDROCHLO-RIDE see PDN500
S-2-((4-(p-ETHOXYPHENYL)BUTYL)AMINO)ETHYL THIOSULFATE see EFC000
1-(p-ETHOXYPHENYL)-1-DIETHYLAMINO-3-METHYL-3-PHENYLPROPANE HY-DROCHLORIDE see PEF500
4-ETHOXY-7-PHENYL-3,5-DIOXA-6-AZA-4-PHOSPHAOCT-6-ENE-8-NITRILE-4-SUL-FIDE see BAT750
3-(4-(2-ETHOXY-2-PHENYLETHYL)-1-PIPERAZINYL)-2-METHYL-1-PHENYL-1-PRO-PANONE see PFB000
3-(4-(2-ETHOXY-2-PHENYLETHYL)-1-PIPERAZINYL)-2-METHYL-1-PHENYL-1-PRO-PANONE DIHYDROCHLORIDE see ECU550
N-(4-ETHOXYPHENYL)-N-HYDROXYACETAMIDE see HIN000
3-(4-ETHOXYPHENYL)-2-METHYL-4(3H)-QUINAZOLINONE see LID000
N-(4-ETHOXYPHENYL)-3'-NITROACETAMIDE see NEL000
p-ETHOXYPHENYL 2-PYRIDYLKETONE see EFC700
p-ETHOXYPHENYL 3-PYRIDYLKETONE see EFC800
p-ETHOXYPHENYL 4-PYRIDYL KETONE see EEN600
4-ETHOXYPHENYLUREA see EFE000
N-(4-ETHOXYPHENYL)UREA see EFE000
p-ETHOXYPHENYLUREA see EFE000
ETHOXYPHOS see DIX000
4-ETHOXY-β-(1-PIPERIDYL)PROPIOPHENONE HYDROCHLORIDE see EFE500
1-ETHOXYPROPANE see EPC125
3-ETHOXY-1,2-PROPANEDIOL see EFF000
3-ETHOXY-1-PROPANOL see EFG000
1-ETHOXY-2-PROPANOL see EFF500
ETHOXYPROPANOL see EJV000

ETHOXY PROPIONALDEHYDE see EFG500
ETHOXYPROPIONIC ACID see EFH000
3-ETHOXYPROPIONIC ACID, ETHYL ESTER see EJV500
ETHOXYPROPIONIC ACID, ETHYL ESTER see EJV500
β-ETHOXYPROPIONITRILE see EFH500
ETHOXYPROPYLACRYLATE see EFI000
ETHOXYPROPYL ESTER ACRYLIC ACID see EFI000
1-ETHOXY-2-PROPYNE see EFJ000
4'-ETHOXY-2-PYRROLIDINYLACETANILIDE HYDROCHLORIDE see PPU750
ETHOXYQUIN (FCC) see SAV000
ETHOXYQUINE see SAV000
2-ETHOXY-1(2H)-QUINOLINECARBOXYLIC ACID, ETHYL ESTER see EFJ500
3-ETHOXYSALICYLALDEHYDE see NOC100
1-ETHOXYSILATRANE see EFJ600
ETHOXYSILATRANE see EFJ600
5-ETHOXY-3-TRICHLOROMETHYL-1,2,4-THIADIAZOLE see EFK000
ETHOXYTRIETHYLENE GLYCOL see EFL000
ETHOXYTRIGLYCOL see EFL000
6-ETHOXY-2,2,4-TRIMETHYL-1,2-DIHYDROQUINOLINE see SAV000
ETHOXYTRIMETHYLSILANE see EFL500
ETHRANE see EAT900
ETHREL see CDS125
ETHRIL see EDJ500
ETHRIOL see HDF300
ETHYBENZTROPINE see DWE800
ETHYL 703 see DEA100, FAB000
ETHYL 736 see TFD000
ETHYLAC see BDF250
ETHYLACETAAT (DUTCH) see EFR000
N-ETHYLACETAMIDE see EFM000
ETHYLACETAMIDE see EFM000
d-N-ETHYLACETAMIDE CARBANILATE see CBL500
2-(2-(3-(N-ETHYLACETAMIDO)-2,4,6-TRIIODOPHENOXY)ETHOXY)ACETIC ACID SODIUM SALT see EFM500
2-(2-(3-(N-ETHYLACETAMIDO)-2,4,6-TRIIODOPHENOXY)ETHOXY)-2-PHENYL ACETIC ACID SODIUM SALT see EFN500
2-(2-(3-(N-ETHYLACETAMIDO)-2,4,6-TRIIODOPHENOXY)ETHOXY)PROPIONIC ACID SODIUM SALT see EFO000
N-ETHYLACETANILIDE see EFQ500
ETHYLACETANILIDE see EFQ500
ETHYL ACETATE see EFR000
ETHYLACETIC ACID see BSW000
ETHYL ACETIC ESTER see EFR000
ETHYL ACETOACETATE (FCC) see EFS000
ETHYL ACETOACETATE ETHYLENE KETAL see EFR100
ETHYL ACETONE see PBN250
N-ETHYL-p-ACETOPHENETIDIDE see EOD500
N-ETHYL-N-(ACETOXYMETHYL)NITROSAMINE see ENR500
ETHYL ACETOXYMETHYLNITROSAMINE see ENR500
ETHYL ACETYL ACETATE see EFS000
ETHYL ACETYLACETONATE see EFS000
ETHYL 2-ACETYLCAPRYLATE see EKS150
ETHYL ACETYLENE see EFS500
ETHYL ACETYLENE, INHIBITED see EFS500
2-ETHYL-4-ACETYL-7-(2-HYDROXY-3-ISOPROPYLAMINOPROPOX-Y)BENZOFURAN see ELI600
ETHYL 2-ACETYLOCTANOATE see EKS150
ETHYL 3-ACETYLPROPIONATE see EFS600
ETHYLACRYLAAT (DUTCH) see EFT000
ETHYL ACRYLATE see EFT000
ETHYL ADIPATE see AEP750
ETHYLADRIANOL see EGE500
ETHYLAENE GLYCOL FORMAL see DVR800
ETHYLAKRYLAT (CZECH) see EFT000
ETHYLAL see EFT500
ETHYLALCOHOL (DUTCH) see EFU000
ETHYL ALCOHOL, anhydrous see EFU000
ETHYL ALCOHOL see EFU000
ETHYL ALCOHOL SOLUTIONS (UN 1170) (DOT) see EFU000
ETHYL ALCOHOL THALLIUM (I) see EEE000
ETHYL ALDEHYDE see AAG250
1-ETHYL-3-ALLYL-6-AMINOURACIL see AFW500
ETHYL ALUMINUM DICHLORIDE see EFU050
ETHYL ALUMINUM DIIODIDE see EFU100
ETHYLALUMINUM SESQUICHLORIDE see TJP775
ETHYLAMINE see EFU400
ETHYLAMINE with BORON FLUORIDE (1:1) see EFU500
ETHYLAMINE, 2-(α-(p-CHLOROPHENYL)-α-METHYLBENZYLOXY)-N,N-DIETHYL see CKE000
ETHYLAMINE-2-(DIPHENYLMETHOXY)-N,N-DIMETHYL, compound with 8-CHLOROTHEOPHYLLINE (1:1) see DYE600
ETHYLAMINE, 2,2'-DITHIOBIS(N,N-DIMETHYL), DIHYDROCHLORIDE see BJG100
ETHYLAMINE HYDROCHLORIDE see EFW000
ETHYLAMINE, β-HYDROXY-β-PHENYL- see HNF000

ETHYLAMINE, 2-MERCAPTO-, HYDROCHLORIDE see MCN750
ETHYLAMINE (UN 1036) (DOT) see EFU400
ETHYLAMINE, aqueous solution with not <50% but not >70% ethylamine (UN 2270) (DOT) see EFU400
N-ETHYL-4-AMINOAZOBENZENE see EOH500
4-(ETHYLAMINOAZOBENZENE) see EOH500
N-ETHYLAMINOBENZENE see EGK000
ETHYL-4-AMINOBENZOATE see EFX000
ETHYL-o-AMINOBENZOATE see EGM000
ETHYL-p-AMINOBENZOATE see EFX000
ETHYL AMINOBENZOATE see EFX000
ETHYL-m-AMINOBENZOATE METHANE SULFONATE see EFX500
ETHYL m-AMINOBENZOATE, METHANESULFONIC ACID SALT see EFX500
ETHYL 2-AMINO-6-BENZYL-4,5,6,7-TETRAHYDROTHIENO(2,3-c)PYRIDINE-3-CARBOXYLATE HYDROCHLORIDE see TGE165
ETHYL-2-AMINO-6-BENZYL-3-THIENO(2,3-c)PYRIDINECARBOXYLATE HYDRO-CHLORIDE see AJU000
2-ETHYLAMINO-4-DIETHYLAMINO-6-CHLORO-s-TRIAZINE see TJL500
2-ETHYLAMINOETHANOL see EGA500
2-(ETHYLAMINO)ETHANOL see EGA500
2-(2-(ETHYLAMINO)ETHYL)-2-METHYL-1,3-BENZODIOXOLE HYDROCHLORIDE see EGC000
ETHYL-N-(2-AMINO-6-(4-FLUOROPHENYLMETHYLAMINO)PYRIDIN-3-YL)CARBAMATE MALEATE see FMP100
ETHYL-N-(2-AMINO-6-(4-FLUOR-PHENYLMETHYLAMINO)PYRIDIN-3-YL)CARBAMAT MALEAT (GERMAN) see FMP100
dl-(±)-3-(2-ETHYLAMINO-1-HYDROXYETHYL)PHENYL PIVALATE HYDROCHLO-RIDE see EGC500
(2-(((ETHYLAMINO)IMINOMETHYL)THIO)ETHYL)TRIMETHYL AMMONIUM BRO-MIDE HYDROBROMIDE see EQN700
2-(((ETHYLAMINO)IMINOMETHYL)THIO)-N,N,N-TRIMETHYLETHANAMINIUM BRO-MIDE, MONOHYDROBROMIDE see EQN700
6-ETHYLAMINO-4-ISOPROPYLAMINO-2-METHOXY-1,3,5-TRIAZINE see EGD000
4-ETHYLAMINO-6-ISOPROPYLAMINO-2-METHOXY-s-TRIAZINE see EGD000
2-ETHYLAMINO-4-ISOPROPYLAMINO-6-METHOXY-s-TRIAZINE see EGD000
2-ETHYLAMINO-4-ISOPROPYLAMINO-6-METHYLMERCARPO-s-TRIAZINE see MPT500
2-ETHYLAMINO-4-ISOPROPYLAMINO-6-METHYLTHIO-1,3,5-TRIAZINE see MPT500
2-ETHYLAMINO-4-ISOPROPYLAMINO-6-METHYLTHIO-s-TRIAZINE see MPT500
2-ETHYLAMINO-4-METHYL-5-n-BUTYL-6-HYDROXYPYRIMIDINE see BRI750
α-((ETHYLAMINO)METHYL)-m-HYDROXYBENZYL ALCOHOL see EGE500
α-((ETHYLAMINO)METHYL)-m-HYDROXYBENZYL ALCOHOL 2,2-DIMETHYL-PROPIONATE HYDROCHLORIDE see EGC500
dl-(±)α-(ETHYLAMINOMETHYL)-3'-HYDROXYBENZYL ALCOHOL 3-(2,2-DIME-THYLPROPIONATE)HCl see EGC500
α-((ETHYLAMINO)METHYL)-m-HYDROXYBENZYL ALCOHOL HYDROCHLORIDE see EGF000
S-(2-(ETHYLAMINO-2-OXOETHYL))-O,O-DIMETHYL PHOSPHORODITHIOATE see DNX600
ETHYL-1-(p-AMINOPHENETHYL)-4-PHENYLISONIPECOTATE see ALW750
3-ETHYL-3-(p-AMINOPHENYL)-2,6-DIOXOPIPERIDINE see AKC600
ETHYL-p-AMINOPHENYL KETONE see AMC000
2-ETHYLAMINO-3-PHENYL-NORCAMPHANE HYDROCHLORIDE see EOM000
N-((3-(ETHYLAMINO)PROPOXY)METHYL)DIPHENYLAMINE see EGG000
ETHYLAMINOPROPYLDIPHENYLAMINOCARBINOL HYDROCHLORIDE see EGG000
2-ETHYLAMINO-1,3,4-THIADIAZOLE see EGI000
2-ETHYLAMINOTHIADIAZOLE see EGI000
2-ETHYL-3-(3-AMINO-2,4,6-TRIIODOPHENYL)PROPIONIC ACID see IFY100
ETHYL AMMONIUM CHLORIDE see EFW000
ETHYLAMPHETAMINE see EGI500
ETHYL-n-AMYLCARBINOL see OCY100
ETHYLAMYLCARBINOL see OCY100
ETHYL sec-AMYL KETONE see EGI750
ETHYL AMYL KETONE see EGI750, ODI000
2-(1-ETHYLAMYLOXY)ETHANOL see EGJ000
2-(2-(1-ETHYLAMYLOXY)ETHOXY)ETHANOL see EGJ500
ETHYLAN see DJC000, IPS500
ETHYLAN A3 see PJY100
ETHYLAN A6 see PJY100
ETHYLAN CP see GHS000
2-ETHYL ANILINE see EGK500
2-ETHYLANILINE see EGK500
4-ETHYLANILINE see EGL000
2-ETHYLANILINE (DOT) see EGK500
N-ETHYLANILINE see EGK000
p-ETHYLANILINE see EGL000
ETHYLANILINE see EGK000
3-(N-ETHYLANILINO)PROPIONITRILE see EHQ500
ETHYL-p-ANISATE (FCC) see AOV000
ETHYL ANISATE see AOV000
ETHYLAN MLD see BKE500
2-ETHYL-9,10-ANTHRACENEDIONE see EGL500
ETHYL ANTHRANILATE see EGM000

2-ETHYLANTHRAQUINONE see EGL500
2-ETHYL-9,10-ANTHRAQUINONE see EGL500
ETHYL ANTIOXIDANT 736 see TFD000
ETHYL ANTIOXIDANT 703 see DEA100, FAB000
ETHYL APOVINCAMINATE see EGM100
ETHYL APOVINCAMIN-22-OATE see EGM100
ETHYLARSONOUS DICHLORIDE see DFH200
ETHYL AZIDE see EGM500
ETHYL AZIDOFORMATE see EGN000
ETHYL-2-AZIDO-2-PROPENOATE see EGN100
ETHYL-1-AZIRIDINECARBOXYLATE see ASH750
ETHYL AZIRIDINECARBOXYLATE see ASH750
ETHYL AZIRIDINOCARBOXYLATE see ASH750
ETHYL-1-AZIRIDINYLCARBOXYLATE see ASH750
ETHYL AZIRIDINYLFORMATE see ASH750
ETHYLBARBITAL see BAG000
ETHYLBENATROPINE see DWE800
9-ETHYL-3,4-BENZACRIDINE see BBM500
7-ETHYLBENZ(c)ACRIDINE see BBM500
10-ETHYL-1,2-BENZANTHRACENE see EGO500
5-ETHYL-1,2-BENZANTHRACENE see EGO000
12-ETHYLBENZ(a)ANTHRACENE see EGP000
7-ETHYLBENZ(a)ANTHRACENE see EGO500
8-ETHYLBENZ(a)ANTHRACENE see EGO000
ETHYLBENZEEN (DUTCH) see EGP500
2-ETHYLBENZENAMINE see EGK500
N-ETHYLBENZENAMINE see EGK000
N-ETHYLBENZENAMINO see EGK000
ETHYL BENZENE see EGP500
ETHYL BENZENEACETATE see EOH000
α-ETHYLBENZENEACETIC ACID-2-((2-DIETHYLAMINO)ETHOXY)ETHYL ESTER CITRATE see BOR350
α-ETHYLBENZENEMETHANOL see EGQ000
ETHYL BENZOATE see EGR000
1-(7-ETHYLBENZOFURAN-2-YL)-2-tert-BUTYLAMINO-1-HYDROXYETHANE HY-DROCHLORIDE see BQD000
2-ETHYL-3-BENZOFURANYL p-HYDROXYPHENYL KETONE see BBJ500
ETHYLBENZOL see EGP500
5-ETHYLBENZO(c)PHENANTHRENE see EGS500
2-ETHYLBENZOXAZOLE see EGR500
ETHYL BENZOYL ACETATE see EGR600
ETHYL-o-BENZOYL-3-CHLORO-2,6-DIMETHOXY-BENZOHYDROXIMATE see BCP000
ETHYL-N-BENZOYL-N-(3,4-DICHLOROPHENYL)-2-AMINOPROPIONATE see EGS000
2-ETHYL-3:4-BENZPHENANTHRENE see EGS500
ETHYLBENZTROPINE see DWE800
ETHYL BENZYL ACETOACETATE see EFS000
α-ETHYLBENZYL ALCOHOL see EGQ000
ETHYLBENZYLBARBITURIC ACID see BEA500
ETHYL-N-BENZYLCYCLOPROPANECARBAMATE see EGT000
ETHYL-N-BENZYL-N-CYCLOPROPYLCARBAMATE see EGT000
ETHYL-1-(2-BENZYLOXYETHYL)-4-PHENYLPIPERIDINE-4-CARBOXYLATE see BBU625
1,2-ETHYL BIS-AMMONIUM PERCHLORATE see EGT500
ETHYL (BIS(1-AZIRIDINYL)PHOSPHINYL)CARBAMATE see EHV500
ETHYLBIS(2-CHLOROETHYL)AMINE see BID250
ETHYLBIS(β-CHLOROETHYL)AMINE see BID250
ETHYLBIS(β-CHLOROETHYL)AMINE HYDROCHLORIDE see EGU000
ETHYL BISCOUMACETATE see BKA000
2-ETHYL-1,3-BIS(DISMETHYLAMINO)-2-PROPANOL BENZOATE see AHI250
ETHYL BIS(4-HYDROXY-3-COUMARINYL)ACETATE see BKA000
ETHYL BIS(4-HYDROXYCOUMARINYL)ACETATE see BKA000
ETHYLBIS(2-HYDROXYETHYL)AMINE see ELP000
S-ETHYL BIS(2-METHYLPROPYL)CARBAMOTHIOATE see EID500
ETHYL BORATE (DOT) see BMC250
ETHYL BROMACETATE see EGV000
ETHYL BROMIDE see EGV400
ETHYL-α-BROMOACETATE see EGV000
ETHYL BROMOACETATE see EGV000
N-ETHYL-N-o-BROMOBENZYL-N,N-DIMETHYLAMMONIUM TOSYLATE see BMV750
ETHYL BROMOPHOS see EGV500
2-ETHYLBUTANAL see DHI000
2-ETHYL-1-BUTANAMINE see EHA000
ETHYL BUTANOATE see EHE000
2-ETHYL BUTANOIC ACID see DHI400
2-ETHYL-1-BUTANOL see EGW000
2-ETHYLBUTANOL-1 see EGW000
2-ETHYLBUTANOL see EGW000
2-ETHYL-1-BUTANOL, SILICATE see TCD250
2-ETHYL-2-BUTENAL see EHO000
2-ETHYL-1-BUTENE see EGW500
2-ETHYL-1-BUTENE-1-ONE see DJN700
ETHYL (E)-2-BUTENOATE see COB750

3-ETHYLBUTINOL see EQL000
2-(2-ETHYLBUTOXY)ETHANOL see EGX000
3-(2-ETHYLBUTOXY)PROPIONIC ACID see EGY000
3-(2-ETHYLBUTOXY)PROPIONITRILE see EGY500
ETHYLBUTYLACETALDEHYDE see BRI000
ETHYL BUTYLACETATE (DOT) see EHF000
2-ETHYLBUTYLACRYLATE see EGZ000
2-ETHYLBUTYL ALCOHOL see EGW000
2-ETHYLBUTYLAMINE see EHA000
5-ETHYL-5-N-BUTYLBARBITURIC ACID see BPF500
ETHYL-N,N-BUTYLCARBAMATE see EHA100
ETHYL BUTYLCARBAMATE see EHA100
ETHYLBUTYLCETONE (FRENCH) see EHA600
2-ETHYLBUTYL ESTER, ACRYLIC ACID see EGZ000
2-ETHYLBUTYLESTER KYSELINY AKRYLOVE see EGZ000
ETHYL BUTYL ETHER see EHA500
ETHYL tert-BUTYL ETHER see EHA550
ETHYLBUTYLKETON (DUTCH) see EHA600
ETHYL BUTYL KETONE see EHA600
ETHYL-N-BUTYLNITROSAMINE see EHC000
ETHYL-tert-BUTYLNITROSAMINE see NKD500
ETHYL N-BUTYL-N-NITROSOSUCCINAMATE see EHC800
ETHYL tert-BUTYL OXIDE see EHA550
2-ETHYL-2-BUTYL-1,3-PROPANEDIOL see BKH625
2-ETHYLBUTYL SILICATE see EHC900
α-ETHYL-α',sec-BUTYLSTILBENE see EHD000
α-ETHYL-β-sec-BUTYLSTILBENE see EHD000
3-ETHYLBUTYNOL see EQL000
α-ETHYLBUTYRALDEHYDE see DHI000
2-ETHYLBUTYRALDEHYDE (DOT,FCC) see DHI000
ETHYL BUTYRALDEHYDE (DOT) see DHI000
ETHYL BUTYRALDEHYDE see DHI000
ETHYL BUTYRATE (DOT,FCC) see EHE000
ETHYL n-BUTYRATE see EHE000
α-ETHYLBUTYRIC ACID see DHI400
2-ETHYLBUTYRIC ACID (FCC) see DHI400
2-ETHYLBUTYRIC ALDEHYDE see DHI000
γ-ETHYLBUTYROLACTONE see HDY600
γ-ETHYL-n-BUTYROLACTONE see HDY600
ETHYL CADMATE see BJB500
ETHYL CAPRATE see EHE500
ETHYL CAPRINATE see EHE500
α-ETHYLCAPROALDEHYDE see BRI000
ETHYL CAPROATE see EHF000
α-ETHYLCAPROIC ACID see BRI250
2-ETHYLCAPROYL CHLORIDE see EKO600
ETHYL CAPRYLATE see ENY000
ETHYL CARBAMATE see UVA000
N-ETHYLCARBAMIC ACID-3-DIMETHYLAMINOPHENYL ESTER, METHOSUL-
  FATE see HNQ500
ETHYLCARBAMIC ACID 1,1-DIPHENYL-2-PROPYNYL ESTER see DWL500
ETHYLCARBAMIC ACID, ETHYL ESTER see EJW500
N-ETHYLCARBAMIC ACID-3-(TRIMETHYLAMMONIO)PHENYL ESTER, METHYL-
  SULFATE see HNQ500
ETHYLCARBAMIC ESTER of m-OXYPHENYLTRIMETHYLAMMONIUM METHYL-
  SULFATE see HNQ500
ETHYL CARBAMIMONIOTHIOATE HYDROGEN SULFATE see ELY550
d-(−)-1-(ETHYLCARBAMOYL)ETHYL PHENYLCARBAMATE see CBL500
2-(N-ETHYL CARBAMOYLHYDROXYMETHYL)FURAN see EHF500
S-(N-ETHYLCARBAMOYLMETHYL) DIMETHYL PHOSPHORODITHIOATE see
  DNX600
(3-(N-ETHYLCARBAMOYLOXY)PHENYL)TRIMETHYLAMMONIUM METHYLSUL-
  FATE see HNQ500
ETHYL CARBANILATE see CBL750
ETHYL CARBAZATE see EHG000
ETHYL CARBAZINATE see EHG000
ETHYL CARBINOL see PND000
ETHYL CARBITOL see CBR000
ETHYL-Δ-CARBOETHOXYVALERATE see AEP750
ETHYL CARBONATE see DIX200
ETHYL CARBONAZIDATE see EGN000
ETHYLCARBONYL PIPERAZINE see EHG050
1-(p-ETHYLCARBOXYPHENYL)-3,3-DIMETHYLTRIAZENE see CBP250
N-ETHYL-N-(3-CARBOXYPROPYL)NITROSOAMINE see NKE000
ETHYL CELLOSOLVE see EES350
ETHYL CELLOSOLVE ACETAAT (DUTCH) see EES400
ETHYLCELLULOSE see EHG100
ETHYL CETAB see EKN500
ETHYLCHLOORFORMIAAT (DUTCH) see EHK500
ETHYL CHLORACETATE see EHG500
ETHYL CHLORIDE see EHH000
ETHYL-α-CHLOROACETATE see EHG500
ETHYL CHLOROACETATE see EHG500
ETHYL 4-CHLOROACETOACETATE see EHH100
ETHYL γ-CHLOROACETOACETATE see EHH100

ETHYL 2-CHLOROACRYLATE see EHH200
ETHYL α-CHLOROACRYLATE see EHH200
ETHYLCHLOROBENZENE see EHH500
ETHYL CHLOROCARBONATE (DOT) see EHK500
ETHYL 4-CHLORO-α-(4-CHLOROPHENYL)-α-HYDROXYBENZENEACETATE see
  DER000
ETHYL CHLOROETHANOATE see EHG500
ETHYL-β-CHLOROETHYLAMINE HYDROCHLORIDE see CGX250
7-(2-(ETHYL-2-CHLOROETHYL)AMINOETHYLAMINO)BENZ(c)ACRIDINE DIHY-
  DROCHLORIDE see EHI500
7-(3-(ETHYL-2-(CHLOROETHYLAMINO)PROPYLAMINO))BENZ(c)ACRIDINE DIHY-
  DROCHLORIDE see EHJ000
9-(3-(ETHYL(2-CHLOROETHYL)AMINO)PROPYLAMINO)-6-CHLORO-2-METHOX-
  YACRIDINE DIHYDROCHLORIDE see ADJ875
9-((3-ETHYL-2-CHLOROETHYL)AMINOPROPYLAMINO)-4-METHOXYACRIDINE
  DIHYDROCHLORIDE see EHJ500
ETHYL(CHLOROETHYL)ANILINE see EHJ600
ETHYL-β-CHLOROETHYLETHYLENIMONIUM PICRYLSULFONATE see EHK000
1-ETHYL-1-(β-CHLOROETHYL)ETHYLENIMONIUM PICRYLSULFONATE see
  EHK000
ETHYL(2-CHLOROETHYL)ETHYLENIMONIUM PICRYLSULFONATE see EHK000
ETHYL-β-CHLOROETHYL-β-HYDROXYETHYLAMINE PICRYLSULFONATE see
  EHK300
ETHYL-N-(β-CHLOROETHYL)-N-NITROSOCARBAMATE see CHF500
ETHYL-2-CHLOROETHYL SULFIDE see CGY750
ETHYL-β-CHLOROETHYL SULFIDE see CGY750
ETHYL-7-CHLORO-5-(o-FLUOROPHENYL)-2,3-DIHYDRO-2-OXO-1H-1,4-BENZODI-
  AZEPINE-3-CARBOXYLATE see EKF600
ETHYL CHLOROFORMATE see EHK500
7-ETHYL-10-CHLORO-11-METHYLBENZ(c)ACRIDINE see EHL000
ETHYL 4-(4-CHLORO-2-METHYLPHENOXY)BUTYLATE see EHL500
ETHYL-4-(4-CHLORO-2-METHYLPHENOXY)BUTYRATE see EHL500
ETHYL 4-CHLORO-3-OXOBUTANOATE see EHH100
ETHYL-2-(p-CHLOROPHENOXY)ISOBUTYRATE see ARQ750
ETHYL-α-(4-CHLOROPHENOXY)ISOBUTYRATE see ARQ750
ETHYL-α-p-CHLOROPHENOXYISOBUTYRATE see ARQ750
ETHYL-p-CHLOROPHENOXYISOBUTYRATE see ARQ750
ETHYL CHLOROPHENOXYISOBUTYRATE see ARQ750
ETHYL 2-(4-CHLOROPHENOXY)-2-METHYLPROPIONATE see ARQ750
ETHYL 2-(p-CHLOROPHENOXY)-2-METHYLPROPIONATE see ARQ750
ETHYL-α-(4-CHLOROPHENOXY)-α-METHYLPROPIONATE see ARQ750
ETHYL-α-(p-CHLOROPHENOXY)-α-METHYLPROPIONATE see ARQ750
ETHYL CHLOROTHIOFORMATE (DOT) see CLJ750
ETHYL-β-CHLOROVINYLETHYNYL CARBINOL see CHG000
ETHYLCHLORVYNOL see CHG000
20-ETHYLCHOLANTHRENE see EHM000
3-ETHYL-CHOLANTHRENE see EHM000
3-ETHYLCHOLANTHRENE see EHM000
ETHYL CHRYSANTHEMATE see EHM100
ETHYL CHRYSANTHEMUMATE see EHM100
ETHYL CINNAMATE (FCC) see EHN000
ETHYL-trans-CINNAMATE see EHN000
ETHYL CITRAL see EHN500
ETHYL CITRATE see TJP750
ETHYL CLOFIBRATE see ARQ750
2-ETHYLCROTONALDEHYDE see EHO000
ETHYL CROTONATE (DOT) see COB750
ETHYLCROTONATE see COB750
ETHYL (E)-CROTONATE see COB750
ETHYL CROTONATE see EHO200
ETHYL trans-CROTONATE see COB750
N-ETHYL-o-CROTONOTOLUIDIDE see EHO500
N-ETHYL-o-CROTONOTOLUIDINE see EHO500
trans-1-(2-ETHYLCROTONOYL)UREA see EHO700
(α-ETHYL-cis-CROTONYL)CARBAMIDE see EHP000
2-ETHYLCROTONYLUREA see EHP000
2-ETHYL-cis-CROTONYLUREA see EHP000
cis-(2-ETHYLCROTONYL) UREA see EHP000
2-ETHYL-trans-CROTONYLUREA see EHO700
ETHYL CYANIDE see PMV750
ETHYL CYANOACETATE see EHP500
ETHYL 1-(3-CYANO-3,3-DIPHENYLPROPYL)-4-PHENYLISONIPECOTATE MONO-
  HYDROCHLORIDE see LIB000
ETHYL CYANOETHANOATE see EHP500
N-ETHYL-N-(2-CYANOETHYL)ANILINE see EHQ500
N-ETHYL-N,β-CYANOETHYLANILINE see EHQ500
o-ETHYL-o-4-CYANOPHENYL PHENYLPHOSPHOROTHIOATE see CON300
ETHYL CYCLOBUTANE see EHR000
5-ETHYL-5-(1'-CYCLOHEPTENYL)-BARBITURIC ACID see COY500
1-ETHYLCYCLOHEXANOL see EHR500
5-ETHYL-5-CYCLOHEXENYLBARBITURIC ACID see TDA500
ETHYLCYCLOHEXYL ACETATE see EHS000
N-ETHYL(CYCLOHEXYL)AMINE see EHT000
N-ETHYL-CYCLOHEXYLAMINE see EHT000
S-ETHYL CYCLOHEXYLETHYLTHIOCARBAMATE see EHT500

ETHYLENEBISDITHIOCARBAMATE MANGANESE see MAS500
N,N'-ETHYLENE BIS(DITHIOCARBAMATE MANGANEUX) (FRENCH) see MAS500
N,N'-ETHYLENE BIS(DITHIOCARBAMATE de SODIUM) (FRENCH) see DXD200
ETHYLENEBIS(DITHIOCARBAMATO) MANGANESE see MAS500
ETHYLENE BIS(DITHIOCARBAMATO)ZINC see EIR000
ETHYLENEBIS(DITHIOCARBAMIC ACID) DISODIUM SALT see DXD200
ETHYLENEBIS(DITHIOCARBAMIC ACID) MANGANESE SALT see MAS500
ETHYLENEBIS(DITHIOCARBAMIC ACID MANGANESE ZINC COMPLEX (8CI) see DXI400
ETHYLENEBIS(DITHIOCARBAMIC ACID) MANGANOUS SALT see MAS500
ETHYLENEBIS(DITHIOCARBAMIC ACID) NICKEL(II) SALT see EIR500
ETHYLENEBIS(DITHIOCARBAMIC ACID), ZINC SALT see EIR000
N,N'-ETHYLENE BIS(3-FLUOROSALICYLIDENEIMINATO)COBALT(II) see EIS000
N,N'-ETHYLENEBIS(2-(o-HYDROXYPHENYL)GLYCINE) see EIV100
ETHYLENEBIS(IMINODIACETIC ACID) DISODIUM SALT see EIX500
ETHYLENEBIS(IMINODIACETIC ACID) TETRASODIUM SALT see EIV000
ETHYLENEBISISOTHIOCYANATE see ISK000
ETHYLENE BIS(MERAPTOACETATE) see MCN000
ETHYLENE BIS(METHANESULFONATE) see BKM125
2,2'-(ETHYLENEBIS(NITROSOIMINO))BISBUTANOL see HMQ500
(ETHYLENEBIS(OXYETHYLENENITRILO)TETRAACETIC ACID see EIT000
1,1'-ETHYLENEBIS(PYRIDINIUM)BROMIDE see EIT100
N,N'-ETHYLENEBIS(SALICYLIDENEIMINATO)COBALT(II) see BLH250
ETHYLENE BIS(THIOGLYCOLATE) see MCN000
2,2'-ETHYLENE-BIS-(2-THIOPSEUDOUREA), DIHYDROBROMIDE see EJA000
ETHYLENE BISTHIURAM MONOSULFIDE see EJQ000
ETHYLENE-BIS-THIURAMMONO-SULFIDE see ISK000
ETHYLENEBIS(TRIS(2-CYANOETHYL))PHOSPHONIUM BROMIDE see EIU000
1,1'-ETHYLENEBISUREA see EJC100
ETHYLENE BRASSYLATE see EJQ500
ETHYLENE BROMIDE see EIY500
ETHYLENE BROMOACETATE see BHD250
ETHYLENEBROMOHYDRIN see BNI500
ETHYLENE CARBONATE see GHM000
ETHYLENE CARBONIC ACID see GHM000
ETHYLENECARBOXAMIDE see ADS250
ETHYLENECARBOXYLIC ACID see ADS750
ETHYLENE CHLORIDE see EIY600
ETHYLENE CHLOROBROMIDE see CES500
ETHYLENE CHLOROFORMATE see EIQ000
ETHYLENE CHLOROHYDRIN see EIU800
ETHYLENE, CHLOROTRIFLUORO-, POLYMERS (8CI) see KDK000
ETHYLENE CYANIDE see SNE000
ETHYLENE CYANOHYDRIN see HGP000
ETHYLENE DIACRYLATE see EIP000
1,2-ETHYLENEDIAMINE see EEA500
ETHYLENE-DIAMINE (FRENCH) see EEA500
ETHYLENEDIAMINE (OSHA) see EEA500
ETHYLENEDIAMINEACETIC ACID TRISODIUM SALT see TNL250
ETHYLENEDIAMINE, N-(5-CHLORO-2-THENYL)-N',N'-DIMETHYL-N-2-PYRIDYL- see CHY250
N,N'-ETHYLENEDIAMINEDIACETIC ACID TETRASODIUM SALT see EIV000
ETHYLENEDIAMINEDICHLORIDE PLATINUM (II) see DFJ000
ETHYLENEDIAMINE, N,N-DIETHYL-N'-PHENYL- see DJV300
ETHYLENEDIAMINE-DI(2-HYDROXYPHENYL)ACETIC ACID see EIV100
ETHYLENEDIAMINE-DI(o-HYDROXYPHENYL)ACETIC ACID see EIV100
ETHYLENEDIAMINE, N,N-DIISOPROPYL- see DNP700
ETHYLENEDIAMINEDINITRATE see EIV700
ETHYLENE DIAMINE DIPERCHLORATE (DOT) see EGT500
ETHYLENEDIAMINE ETHOXYLATE see EIV750
ETHYLENEDIAMINE ETHYLENE OXIDE ADDUCT see EIV750
ETHYLENEDIAMINE HYDROCHLORIDE see EIW000
ETHYLENEDIAMINE, N-METHYL- see MJW100
ETHYLENEDIAMINE-N,N'-BIS(2-HYDROXYPHENYLACETIC ACID) see EIV100
ETHYLENEDIAMINE, N,N'-DIBENZYL- see BHB300
ETHYLENEDIAMINE, N,N'-DIBENZYL-, DIACETATE see DDF800
ETHYLENEDIAMINE, N,N'-DIMETHYL- see DRI600
ETHYLENEDIAMINE, N,N'-DIPHENYL- see DWB400
ETHYLENEDIAMINE-N,N,N',N'-TETRAACETIC ACID see EIX000
ETHYLENEDIAMINE PERCHLORATE see EGT500
ETHYLENEDIAMINETETRAACETATE see EIX000
ETHYLENEDIAMINETETRAACETATE DISODIUM SALT see EIX500
ETHYLENEDIAMINETETRAACETIC ACID see EIX000
ETHYLENEDIAMINETETRAACETIC ACID, CALCIUM DISODIUM CHELATE see CAR780
ETHYLENEDIAMINETETRAACETIC ACID, DISODIUM SALT see EIX500
ETHYLENEDIAMINE TETRAACETIC ACID, IRON(III) SALT see HIA000
ETHYLENEDIAMINETETRAACETIC ACID, TETRASODIUM SALT see EIV000
ETHYLENEDIAMINETETRAACETIC ACID, TRISODIUM SALT see TNL250
ETHYLENEDIAMINETETRAACETONITRILE see EJB000
ETHYLENEDIAMINE, compounded with THEOPHYLLINE (1:2) see TEP500
ETHYLENEDIAMMONIUM CHLORIDE see EIW000
1,2-ETHYLENE DIBROMIDE see EIY500

(E)1,2-ETHYLENEDICARBOXYLIC ACID see FOU000
cis-1,2-ETHYLENEDICARBOXYLIC ACID see MAK900
trans-1,2-ETHYLENEDICARBOXYLIC ACID see FOU000
trans-1,2-ETHYLENEDICARBOXYLIC ACID DIMETHYL ESTER see DSB600
ETHYLENEDICESIUM see EIY550
1,2-ETHYLENE DICHLORIDE see EIY600
ETHYLENE DICHLORIDE see EIY600
ETHYLENE, 1,1-DICHLORO-, POLYMER with CHLOROETHYLENE see CGW300
ETHYLENE DICYANIDE see SNE000
ETHYLENE DIGLYCIDYL ETHER see EEA600
ETHYLENE DIGLYCOL see DJD600
ETHYLENE DIGLYCOL MONOETHYL ETHER see CBR000
ETHYLENE DIGLYCOL MONOMETHYL ETHER see DJG000
ETHYLENE DIHYDRATE see EJC500
(+)-2,2'-(ETHYLENEDIIMINO)DI-1-BUTANOL see TGA500
N,N'-ETHYLENE DIIMINO DI(o-CRESOL) see DWY200
ETHYLENE DIISOTHIOCYANATE see ISK000
ETHYLENE DIISOTHIOUREA DIHYDROBROMIDE see EJA000
ETHYLENE DIISOTHIOURONIUM DIBROMIDE see EJA000
α-ETHYLENE DIMERCAPTAN see EEB000
ETHYLENE DIMERCAPTAN see EEB000
ETHYLENE DIMETHANESULFONATE see BKM125
ETHYLENE DIMETHANESULPHONATE see BKM125
ETHYLENE DIMETHYL ETHER see DOE600
ETHYLENEDINITRAMINE see DUU800
ETHYLENE DINITRATE see EJG000
(ETHYLENEDINITRILO)TETRAACETATE DIPOTASSIUM SALT see EEB100, EJA250
((ETHYLENEDINITRILO)TETRAACETATO(2-))-COBALTATE(2-) COBALT(2+) SALT see DGQ400
((ETHYLENEDINITRILO)TETRAACETATO)-FERATE(1-), SODIUM see EJA379
ETHYLENEDINITRILOTETRAACETIC ACID see EIX000
(ETHYLENEDINITRILO)-TETRAACETIC ACID see TNL500
(ETHYLENEDINITRILO)TETRAACETIC ACID CADMIUM(II) COMPLEX see CAF750
(ETHYLENEDINITRILO)TETRAACETIC ACID COPPER(II) COMPLEX see CNL750
(ETHYLENEDINITRILO)-TETRAACETIC ACID DISODIUM SALT see EIX500
(ETHYLENEDINITRILO)TETRA ACETIC ACID, LEAD(II) COMPLEX see LDD000
(ETHYLENEDINITRILO)TETRA ACETIC ACID, MERCURY(II) COMPLEX see MDB250
(ETHYLENEDINITRILO)TETRA ACETIC ACID NICKEL(II) COMPLEX see EJA500
(ETHYLENEDINITRILO)TETRAACETONITRILE see EJB000
1,1',1'',1'''-(ETHYLENEDINITRILO)TETRA-2-PROPANOL see QAT000
ETHYLENEDINITROAMINE see DUU800
6,6'-(ETHYLENEDIOXY)BIS(4-AMINOQUINALDINE) DIHYDROCHLORIDE see EJB100
ETHYLENEDIOXYBIS(ETHYLENEAMINO)TETRAACETIC ACID see EIT000
2,2'-ETHYLENEDIOXYDIETHANOL see TJQ000
2,2'-ETHYLENEDIOXYDIETHANOL DIACETATE see EJB500
2,2'-(ETHYLENEDIOXY)DI(ETHYL ACETATE) see EJB500
2,2'-(ETHYLENEDIOXY)DI(ETHYL 2-ETHYLBUTYRATE) see TJQ250
2,2'-ETHYLENEDIOXYETHANOL see TJQ000
16,17-ETHYLENEDIOXYVIOLANTHRONE see CMU500
ETHYLENE DIPERCHLORATE see EJC000
1,1'-ETHYLENEDIPYRIDINIUM DIBROMIDE see EIT100
1,1'-ETHYLENE-2,2'-DIPYRIDINIUM DICHLORIDE see DWY000
1,1-ETHYLENE 2,2-DIPYRIDYLIUM DIBROMIDE see DWX800
1,1'-ETHYLENE-2,2'-DIPYRIDYLIUM DIBROMIDE see DWX800
ETHYLENE DIPYRIDYLIUM DIBROMIDE see DWX800
2,2-ETHYLENEDITHIODIPSEUDOUREA DIHYDROBROMIDE see EJA000
ETHYLENE DITHIOGLYCOL see EEB000
ETHYLENEDITHIOL see EEB000
1,1'-ETHYLENEDIUREA see EJC100
ETHYLENEDIUREA see EJC100
2,2'-(1,2-ETHYLENEDIYL)BIS(5-AMINOBENZENESULFONIC ACID) see FCA100
1,2-ETHYLENEDIYLBIS(CARBAMODITHIOATO)MANGANESE see MAS500
ETHYLENE EPISULFIDE see EJP500
ETHYLENE EPISULPHIDE see EJP500
ETHYLENE FLUORIDE see ELN500
ETHYLENE, FLUORO-(8CI) see VPA000
ETHYLENE FORMATE see EJF000
ETHYLENE GLYCOL see EJC500
ETHYLENE GLYCOL ACETATE see EJD759, EJI000
ETHYLENE GLYCOL ACRYLATE see ADV250
ETHYLENE GLYCOL BIS(β-AMINOETHYL ETHER)-N,N'-TETRAACETIC ACID see EIT000
ETHYLENE GLYCOL BIS(2-AMINOETHYL ETHER)-N,N,N',N'-TETRAACETIC ACID see EIT000
ETHYLENE GLYCOL BIS(β-AMINOETHYL ETHER)TETRAACETATE see EIT000
ETHYLENE GLYCOL BIS(AMINOETHYL ETHER)TETRAACETATE see EIT000
ETHYLENE GLYCOL BIS(2-AMINOETHYL ETHER)TETRAACETIC ACID see EIT000
ETHYLENE GLYCOL BIS(BROMOACETATE) see BHD250

ETHYLENE GLYCOL, BISCHLOROFORMATE see EIQ000
ETHYLENE GLYCOL BIS(CHLOROMETHYL)ETHER see BIJ250
ETHYLENE GLYCOL BIS(2,3-EPOXY-2-METHYLPROPYL) ETHER see EJD000
ETHYLENE GLYCOL-BIS-(2-HYDROXYETHYL ETHER) see TJQ000
ETHYLENE GLYCOL BIS(METHACRYLATE) see BKM250
ETHYLENE GLYCOL BIS(THIOGLYCOLATE) see MCN000
ETHYLENE GLYCOL-n-BUTYL ETHER see BPJ850
ETHYLENE GLYCOL CARBONATE see GHM000
ETHYLENE GLYCOL, CHLOROHYDRIN see EIU800
ETHYLENE GLYCOL, CYCLIC CARBONATE see GHM000
ETHYLENE GLYCOL, CYCLIC SULFATE see EJP000
ETHYLENE GLYCOL DIACETATE see EJD759
ETHYLENE GLYCOL DIACRYLATE see EIP000
ETHYLENE GLYCOL DIALLYL ETHER see EJE000
ETHYLENE GLYCOL, DIBUTOXYTETRA see TCE350
ETHYLENE GLYCOL, DIBUTYL see DDW400
ETHYLENE GLYCOL DI(CHLOROFORMATE) see EIQ000
ETHYLENE GLYCOL DI(2,3-EPOXY-2-METHYLPROPYL)ETHER see EJD000
ETHYLENE GLYCOL DIETHYL ETHER see EJE500
ETHYLENE GLYCOL DIFORMATE see EJF000
ETHYLENE GLYCOL DIGLYCIDYL ETHER see EEA600
ETHYLENE GLYCOL DIHYDROXYDIETHYL ETHER see TJQ000
ETHYLENE GLYCOL DIMETHACRYLATE see BKM250
ETHYLENE GLYCOL DIMETHYL ETHER see DOE600
ETHYLENE GLYCOL DINITRATE see EJG000
ETHYLENE GLYCOL DINITRATE mixed with NITROGLYCERIN (1:1) see
  NGY500
ETHYLENE GLYCOL DIPROPIONATE (8CI) see COB260
ETHYLENE GLYCOL ETHYL ETHER see EES350
ETHYLENE GLYCOL ETHYL ETHER ACETATE see EES400
ETHYLENE GLYCOL-N-HEXYL ETHER see HFT500
ETHYLENE GLYCOLIDE (2,3-EPOXY-2-METHYLPROPYL)ETHER see EJD000
ETHYLENE GLYCOL ISOPROPYL ETHER see INA500
ETHYLENE GLYCOL MALEATE see EJG500
ETHYLENE GLYCOL METHACRYLATE see EJH000
ETHYLENE GLYCOL METHYL ETHER see EJH500
ETHYLENE GLYCOL METHYL ETHER ACETATE see EJJ500
ETHYLENE GLYCOL MONOACETATE see EJI000
ETHYLENE GLYCOL MONOACRYLATE see ADV250
ETHYLENE GLYCOL MONOBENZYL ETHER see EJI500
ETHYLENE GLYCOL MONOBUTYL ETHER (MAK, DOT) see BPJ850
ETHYLENE GLYCOL MONO-sec-BUTYL ETHER see EJJ000
ETHYLENE GLYCOL MONOBUTYL ETHER ACETATE (MAK) see BPM000
ETHYLENE GLYCOL MONOETHYL ETHER (DOT) see EES350
ETHYLENE GLYCOL MONOETHYL ETHER see EES350
ETHYLENE GLYCOL MONOETHYL ETHER ACETATE (MAK, DOT) see
  EES400
ETHYLENE GLYCOL MONOETHYL ETHER ACRYLATE see ADT500
ETHYLENE GLYCOL MONOETHYL ETHER PROPENOATE see ADT500
ETHYLENE GLYCOL MONOHEPTYL ETHER see HBN250
ETHYLENE GLYCOL, MONO-2,4-HEXADIENE ETHER see HCT500
ETHYLENE GLYCOL MONOHEXYL ETHER see HFT500
ETHYLENE GLYCOL, MONO(HYDROGEN MALEATE) see EJG500
ETHYLENE GLYCOL MONOISOBUTYL ETHER see IIP000
ETHYLENE GLYCOL, MONOISOPROPYL ETHER see INA500
ETHYLENE GLYCOL, MONOMETHACRYLATE see EJH000
ETHYLENE GLYCOL MONOMETHYL ETHER (MAK, DOT) see EJH500
ETHYLENE GLYCOL MONOMETHYL ETHER ACETATE see EJJ500
ETHYLENE GLYCOL MONOMETHYL ETHER ACRYLATE see MEM250,
  MIF750
ETHYLENE GLYCOL MONOMETHYL ETHER OLEATE see MEP750
ETHYLENE GLYCOL MONO-2-METHYLPENTYL ETHER see EJK000
ETHYLENE GLYCOL MONOMETHYLPENTYL ETHER see EJK000
ETHYLENE GLYCOL MONOPHENYL ETHER see PER000
ETHYLENEGLYCOL MONOPHENYL ETHER PROPIONATE see EJK500
ETHYLENE GLYCOL-MONO-n-PROPYL ETHER see PNG750
ETHYLENE GLYCOL-MONO-PROPYL ETHER see PNG750
ETHYLENE GLYCOL MONOPROPYL ETHER ACETATE see PNA225
ETHYLENE GLYCOL, MONOSTEARATE see EJM500
ETHYLENE GLYCOL MONO-2,6,8-TRIMETHYL-4-NONYL ETHER see EJL000
ETHYLENE GLYCOL MONOVINYL ETHER see EJL500
ETHYLENE GLYCOL PHENYL ETHER see PER000
ETHYLENE GLYCOL STEARATE see EJM500
ETHYLENE GLYCOL VINYL ETHER see EJL500
ETHYLENE HEXACHLORIDE see HCI000
ETHYLENE HOMOPOLYMER see PJS750
ETHYLENEIMINE see EJM900
ETHYLENE IMINE, INHIBITED (DOT) see EJM900
1-ETHYLENEIMINO-2-HYDROXY-3-BUTENE see VMA000
ETHYLENE MERCAPTOACETATE see MCN000
ETHYLENE METHACRYLATE see BKM250
ETHYLENE MONOCHLORIDE see VNP000
1,8-ETHYLENE NAPHTHALENE see AAE750
1,8-ETHYLENENAPHTHALENE see AAF275
N,N-ETHYLENE-N',N'-DIMETHYLUREA see EJA100

ETHYLENE NITRATE see EJG000
ETHYLENE, NITRO-, POLYMERS see NFY560
ETHYLENENITROSOUREA see NKL000
N,N-ETHYLENE-N'-METHYLUREA see EJN400
ETHYLENE OXIDE see EJN500
ETHYLENE OXIDE, mixed with CARBON DIOXIDE see EJO000
ETHYLENE OXIDE and CARBON DIOXIDE MIXTURES (DOT) see EJO000
ETHYLENE OXIDE CYCLIC TETRAMER see COD475
ETHYLENE OXIDE, ETHYL- see BOX750
ETHYLENE (OXYDE d') (FRENCH) see EJN500
1-ETHYLENEOXY-3,4-EPOXYCYCLOHEXANE see VOA000
ETHYLENE OZONIDE see EJO500
1,4-ETHYLENEPIPERAZINE see DCK400
ETHYLENE POLYMER see PJS750
ETHYLENE POLYMERS (8CI) see PJS750
ETHYLENE PROPIONATE see COB260
ETHYLENESUCCINIC ACID see SMY000
ETHYLENE SULFATE see EJP000
ETHYLENE SULFIDE see EJP500
1,2-ETHYLENE SULFITE see COV750
ETHYLENE SULFITE see COV750
ETHYLENE SULPHIDE see EJP500
ETHYLENE TEREPHTHALATE POLYMER see PKF750
ETHYLENE TETRACHLORIDE see PCF275
ETHYLENETHIOCARBAMYL SULFIDE see EJQ000
1,3-ETHYLENE-2-THIOUREA see IAQ000
N,N'-ETHYLENETHIOUREA see IAQ000
ETHYLENE THIOUREA see IAQ000
ETHYLENETHIOUREA mixed with SODIUM NITRITE see IAR000
l'ETHYLENE THIOUREE (FRENCH) see IAQ000
ETHYLENE THIURAM MONOSULFIDE see EJQ000
ETHYLENE THIURAM MONOSULPHIDE see EJQ000
ETHYLENE, TRIBROMO- see THV100
ETHYLENE TRICHLORIDE see TIO750
ETHYLENE TRITHIOCARBONATE see EJQ100
ETHYLENE UNDECANE DICARBOXYLATE see EJQ500
1,3-ETHYLENE UREA see IAS000
ETHYLENE UREA see IAS000
ETHYLENGLYCOL MONOVINYL ESTER (RUSSIAN) see EJL500
ETHYLENGLYKOLDIGLYCIDYLETHER see EEA600
ETHYLENGLYKOLDINITRAT (CZECH) see EJG000
ETHYLENIMINE see EJM900
1-ETHYLENIMINO-2-HYDROXYBUTENE see VMA000
1-N-ETHYLEPHEDRINE HYDROCHLORIDE see EJR500
ETHYL-2,3-EPOXYBUTYRATE see EJS000
ETHYL-α,β-EPOXYHYDROCINNAMATE see EOK600
ETHYL α,β-EPOXY-β-METHYLHYDROCINNAMATE see ENC000
ETHYL 2,3-EPOXY-3-METHYL-3-PHENYLPROPIONATE see ENC000
ETHYL-α,β-EPOXY-α-PHENYLPROPIONATE see EOK600
(8-β)-6-ETHYLERGOLINE-8-ACETAMIDE TARTRATE see EJS100
ETHYL ESTER of N-ACETYL-dl-SARCOLYSYL-l-PHENYLALANINE see ACD250
ETHYL ESTER of N-ACETYL-dl-SARCOSYLYL-dl-VALINE see ARM000
ETHYL ESTER-1-CYSTEINE HYDROCHLORIDE (9CI) see EHU600
ETHYL ESTER l-CYSTEINE HYDROCHLORIDE (9CI) see MBX800
ETHYL ESTER of 1,2,5,6-DIBENZANTHRACENE-endo-α,β-SUCCINO GLYCINE
  see EJT000
ETHYL ESTER of 4,4'-DICHLOROBENZILIC ACID see DER000
ETHYL ESTER of DIMETHYLDITHIOCARBAMIC ACID see EIJ500
ETHYL ESTER of O,O-DIMETHYLDITHIOPHOSPHORYL α-PHENYL ACETATE
  ACID see DRR400
ETHYL ESTER of 2,3-EPOXY-3-PHENYLBUTANOIC ACID see ENC000
O-ETHYLESTER KYSELINY DICHLORTHIOFOSFORECNE (CZECH) see
  MRI000
ETHYLESTER KYSELINY DUSITE see ENN000
ETHYLESTER KYSELINY N-ETHYL-N-NITROSOKARBAMINOVE see NKE500
ETHYLESTER KYSELINY FLUOROCTOVE see EKG500
ETHYLESTER KYSELINY GALLOVE see EKM100
ETHYLESTER KYSELINY KROTONOVE see EHO200
ETHYLESTER KYSELINY KYANOCTOVE see EHP500
ETHYLESTER KYSELINY METHYLKARBAMINOVE see EMQ500
ETHYLESTER KYSELINY MLECNE see LAJ000
ETHYLESTER KYSELINY ORTHOMRAVENCI (CZECH) see ENY500
ETHYL ESTER of METHANESULFONIC ACID see EMF500
ETHYL ESTER of 3-METHYLCHOLANTHRENE-endo-α,β-SUCCINOGLYCINE
  see EJT500
ETHYL ESTER of METHYLNITROSO-CARBAMIC ACID see MMX250
ETHYL ESTER of METHYLSULFONIC ACID see EMF500
ETHYL ESTER of METHYLSULPHONIC ACID see EMF500
ETHYL ESTER of MONOACETIC ACID see MFW100
11-β-ETHYLESTRADIOL see EJT575
11-β-ETHYLESTRA-1,3,5(10)-TRIENE-3,17-β-DIOL see EJT575
N-ETHYL-ETHANAMINE see DHJ200
N-ETHYL-ETHANAMINE HYDROCHLORIDE (9CI) see DIS500
ETHYL ETHANOATE see EFR000
ETHYL ETHER see EJU000

ETHYL ETHER of 10-(β-MORPHOLYLPROPIONYL)PHENTHIAZINECARBAMINO ACID HYDROCHLORIDE see EEI025
ETHYL ETHER OF PROPYLENE GLYCOL see EJV000
11-β-ETHYL-17-α-ETHINYLESTRADIOL see EJV400
ETHYL-β-ETHOXYPROPIONATE see EJV500
3-ETHYL-2-(5-(3-ETHYL-2-BENZOTHIAZOLINYLIDENE)-1,3-PENTADIE-NYL)BENZOTHIAZOLIUM IODIDE see DJT800
ETHYL-N-ETHYL CARBAMATE see EJW500
S-ETHYL-N-ETHYL-N-CYCLOHEXYLTHIOLCARBAMATE see EHT500
α-ETHYL-6-(3-ETHYL-1,5-DIMETHYL-4-OXO-1,5-HEPTADIENYL)-N-(1,8,14,15,18,21,27-HEPTAAZA-21-HYDROXY-7-(1-HYDROXYETHYL-2,6,9.16,19,22-HEXAOXO-4-ISOPROPYL-20- (METHOXYMETHYL)-17,18-DI-METHYL-5-OXATRICYCLO(21.4.0.0(sup 10,15)HEPTACOSAN-3-6L)TETRAHYDRO-α-2-DIHYDROXY-5-METHYL-2H-PYRAN-2-ACETAMIDE see ASH425
3-ETHYL-5-(4,4-ETHYLENEDIOXYPIPERIDINO-1-METHYL)-6,7-DIHYDRO-2-ME-THYLINDOL-4(5H)-ONE see AFH550
ETHYL ETHYLENE OXIDE see BOX750
2-ETHYL-N-(2-ETHYLHEXYL)-1-HEXANAMINE see DJA800
ETHYL-6-(ETHYL(2-HYDROXYPROPYL)AMINO)-3-PYRIDAZINECARBAZATE see CAK275
ETHYL-2-(6(ETHYL(2-HYDROXYPROPYL)AMINO)-3-PYRIDAZI-NYL)HYDRAZINECARBOXYLATE see CAK275
ETHYL N-ETHYLNITROSOCARBAMATE see NKE500
ETHYL 3-ETHYL-4-OXO-5-PIPERIDINO-Δ²·°-THIAZOLIDINEACETATE see EOA500
ETHYL (Z)-(3-ETHYL-4-OXO-5-PIPERIDINOTHIAZOLIDIN-2-YLIDENE)ACETATE see EOA500
5-ETHYL-5-(1-ETHYLPROPYL)BARBITURIC ACID see EJY000
5-ETHYL-5-(1-ETHYLPROPYL)2,4,6(1H,3H,5H)-PYRIMIDINETRIONE see EJY000
1-ETHYL-4-(p-(p-((p-((1-ETHYLPYRIDINIUM-4-YL)AMINO)-2-AMINOPHE-NYL)CARBAMOYL)CINNAMAMIDO)ANILINO)PYRIDINIUM, see EJZ000
1-ETHYL-4-(p-(p-((1-ETHYLPYRIDINIUM-4-YL)AMINO)BENZAMIDO)QUINOLINIUM DIBROMIDE see EKA000
1-ETHYL-4-(p-((p-((1-ETHYLPYRIDINIUM-4-YL)AMINO)PHENYL)CARBAMOYL)ANILINO)QUINOLINIUM, DIBROMIDE see EKA500
1-ETHYL-4-(p-(p-((p-((1-ETHYLPYRIDINIUM-4-YL)AMINO)PHENYL)CARBAMOYL)CINNAMAMIDO)ANILINO)PYRIDINIUM, DI-p- see EKB000
1-ETHYL-4-(p-((p-((1-ETHYLPYRIDINIUM-4-YL)PHENYL)CARBAMOYL)ANILINO)QUINOLINIUM), DI-p-TOLUENE SULFO-NATE see EKC500
1-ETHYL-6-(p-(p-((1-ETHYLQUINOLINIUM-6-YL)CARBAMOYL)BENZAMIDO)BENZAMIDO)QUINOLINIUM, DI-p-TOLUENE SULFONATE see EKD500
1-ETHYL-7-((p-(p-((1-ETHYLQUINOLINIUM-7-YL)CARBAMOYL)BENZAMIDO)BENZAMIDO)QUINOLINIUM), DI-p-TOLUENE SULFONATE see EKD000
ETHYLETHYNE see EFS500
13-ETHYL-17-α-ETHYNYLGON-4-EN-17-β-OL-3-ONE see NNQ500, NNQ520
(±)-13-ETHYL-17-α-ETHYNYL-17-HYDROXYGON-4-EN-3-ONE see NNQ500
13-ETHYL-17-α-ETHYNYL-17-β-HYDROXY-4-GONEN-3-ONE see NNQ520, NNQ500
dl-13-β-ETHYL-17-α-ETHYNYL-17-β-HYDROXYGON-4-EN-3-ONE see NNQ500
dl-13-β-ETHYL-17-α-ETHYNYL-19-NORTESTOSTERONE see NNQ500
ETHYL ETRINOATE see EMJ500
ETHYLEX GUM 2020 see HLB400
ETHYL (E,Z)-2,4-DECADIENOATE see EHV100
ETHYL FLAVONE-7-OXYACETATE see ELH600
ETHYL-7-FLAVONOXYACETATE see ELH600
ETHYL FLAVONYL-7-OXYACETATE see ELH600
ETHYL FLAVON-7-YLOXYACETATE see ELH600
ETHYL FLUCLOZEPATE see EHV600
ETHYL FLUORIDE (DOT) see FIB000
ETHYL FLUOROACETATE see EKG500
ETHYL-10-FLUORODECANOATE see EKI000
ETHYL-ω-FLUORODECANOATE see EKI000
1-ETHYL-6-FLUORO-1,4-DIHYDRO-4-OXO-7-(1-PIPERAZINYL)-1,8-NAPHTHYRI-DINE-3-CARBOXYLIC ACID see EAU100
1-ETHYL-6-FLUORO-1,4-DIHYDRO-4-OXO-7-(1-PIPERAZINYL)-3-QUINOLINECAR-BOXYLIC ACID see BAB625
ETHYL-9-FLUORONONANECARBOXYLATE see EKI000
ETHYL-8-FLUORO OCTANOATE see EKK500
ETHYL-ω-FLUOROOCTANOATE see EKK500
ETHYL-p-FLUOROPHENYL SULFONE see FLG000
ETHYL FLUOROSULFATE see EKK550
N-ETHYLFORMAMIDE see EKK600
ETHYLFORMAMIDE see EKK600
ETHYL FORMATE see EKL000
ETHYLFORMIAAT (DUTCH) see EKL000
ETHYLFORMIC ACID see PMU750
ETHYL FORMIC ESTER see EKL000
1-ETHYL-1-FORMYLHYDRAZINE see EKL250
N-ETHYL-N-FORMYLHYDRAZINE see EKL250

6-ETHYL-7-FORMYL-1,1,4,4-TETRAMETHYL-1,2,3,4-TETRAHYDRONAPHTHAL-ENE see FNK200
ETHYL FUMARATE see DJJ800
5-ETHYL-2(5H)-FURANONE see EKL500
ETHYL FUROATE see EKM000
ETHYL GALLATE see EKM100
ETHYL GERANYL ETHER see GDG100
α-ETHYL GLYCEROL ETHER see EFF000
ETHYL GLYCIDYL ETHER see EKM200
ETHYL GLYCOLATE see EKM500
ETHYLGLYKOLACETAT (GERMAN) see EES400
ETHYL GLYME see EJE500
ETHYL GREEN see BAY750
ETHYL-p-(6-GUANIDINOHEXANOYLOXY) BENZOATE METHANESULFONATE see GAD400
ETHYL GUSATHION see EKN000
ETHYL GUTHION see EKN000
ETHYL 10-HENDECENOATE see EQD200
ETHYL n-HEPTANOATE see EKN050
ETHYL HEPTANOATE see EKN050
ETHYL HEPTOATE see EKN050
ETHYL HEPTYLATE see EKN050
(3-ETHYL-N-HEPTYL)METHYLCARBINOL see ENW500
1-ETHYL-1-HEPTYLPIPERIDINIUM BROMIDE see EKN100
ETHYLHEXABITAL see TDA500
ETHYL HEXADECYL DIMETHYL AMMONIUM BROMIDE see EKN500
1-ETHYL-1-HEXADECYLPIPERIDINIUM BROMIDE see EKN600
ETHYL HEXAFLUORO-2-BROMOBUTYRATE see EKO000
sec-ETHYL HEXAHYDRO-1H-AZEPINE-1-CARBOTHIOATE see EKO500
5-ETHYLHEXAHYDRO-4,6-DIOXO-5-PHENYLPHYIMIDINE see DBB200
5-ETHYLHEXAHYDRO-5-PHENYLPYRIMIDINE-4,6-DIONE see DBB200
2-ETHYLHEXALDEHYDE see BRI000
ETHYLHEXALDEHYDE (DOT) see BRI000
ETHYL-1-HEXAMETHYLENEIMINECARBOTHIOLATE see EKO500
S-ETHYL-1-HEXAMETHYLENEIMINOTHIOCARBAMATE see EKO500
S-ETHYL-N-HEXAMETHYLENETHIOCARBAMATE see EKO500
2-ETHYLHEXANAL see BRI000
2-ETHYLHEXANE-1,3-DIOL see EKV000
2-ETHYLHEXANEDIOL-1,3 see EKV000
2-ETHYL-1,3-HEXANEDIOL see EKV000
ETHYL HEXANEDIOL see EKV000
ETHYL HEXANOATE (FCC) see EHF000
2-ETHYLHEXANOIC ACID see BRI250
2-ETHYLHEXANOIC ACID CHLORIDE see EKO600
2-ETHYLHEXANOIC ACID, 2-ETHYLHEXYL ESTER see EKW000
2-ETHYLHEXANOIC ACID, VINYL ESTER see VOU000
2-ETHYL-1-HEXANOL see EKQ000
2-ETHYLHEXANOL see EKQ000
2-ETHYL-1-HEXANOL ESTER with DIPHENYL PHOSPHATE see DWB800
2-ETHYL-1-HEXANOL HYDROGEN PHOSPHATE see BJR750
2-ETHYL-1-HEXANOL HYDROGEN SULFATE, SODIUM SALT see TAV750
2-ETHYL-1-HEXANOL PHOSPHATE see TNI250
2-ETHYL-1-HEXANOL SILICATE see EKQ500
2-ETHYL-1-HEXANOL SULFATE SODIUM SALT see TAV750
2-ETHYLHEXANOYL CHLORIDE see EKO600
((2-ETHYLHEXANOYL)OXY)TRIBUTYLSTANNANE see TID250
2-ETHYLHEXANYL ACETATE see OEE000
2-ETHYL-2-HEXENAL see EKR000
2-ETHYLHEXENAL see EKR000
2-ETHYL-1-HEXENE see EKR500
2-ETHYL HEXENE-1 see EKR500
2-ETHYL-2-HEXENOIC ACID see EKS000
ETHYL cis-3-HEXENYL ACETAL see EKS100
2-ETHYLHEXOIC ACID see BRI250
2-ETHYLHEXOIC ACID, VINYL ESTER see VOU000
ETHYL HEXYL ACETAL see EKS120
2-ETHYLHEXYL ACETATE see OEE000
β-ETHYLHEXYL ACETATE see OEE000
ETHYL-2-HEXYL ACETOACETATE see EKS150
ETHYL α-HEXYLACETOACETATE see EKS150
2-ETHYLHEXYL ACRYLATE see ADU250
2-ETHYLHEXYL ALCOHOL see EKQ000
2-ETHYL HEXYLAMINE see EKS500
2-ETHYLHEXYL-3-AMINOPROPYL ETHER see ELA000
5-ETHYL-5-HEXYLBARBITURIC ACID SODIUM SALT see EKT500
N-(2-ETHYLHEXYL)BICYCLO-(2,2,1)-HEPT-5-ENE-2,3-DICARBOXIMIDE see OES000
2-ETHYLHEXYL-1-CHLORIDE see EKU000
2-ETHYLHEXYL-6-CHLORIDE see EKU100
2-ETHYLHEXYL DIPHENYL ESTER PHOSPHORIC ACID see DWB800
2-ETHYLHEXYL DIPHENYLPHOSPHATE see DWB800
ETHYL HEXYLENE GLYCOL see EKV000
2-ETHYLHEXYL-9,10-EPOXYOCTADECANOATE see EKV500
2-ETHYLHEXYL EPOXYSTEARATE see EKV500
2-ETHYLHEXYLESTER KYSELINY 2-ETHYLKAPRONOVE see EKW000

2-ETHYLHEXYL ETHANOATE see OEE000
2-ETHYLHEXYL-2-ETHYLHEXANOATE see EKW000
2-ETHYLHEXYL FUMARATE see DVK600
2-ETHYLHEXYL GLYCIDYL ETHER see GGY100
N-(2-ETHYLHEXYL)-3-HYDROXYBUTYRAMIDE HYDROGEN SUCCINATE see BPF825
N-2-ETHYLHEXYLIMIDEENDOMETHYLENETETRAHYDROPHTHALIC ACID see OES000
2-ETHYLHEXYLMALEINAN DI-N-BUTYLCINICITY (CZECH) see BJR250
2-ETHYLHEXYL MERCAPTOACETATE see EKW300
N-(2-ETHYLHEXYL)-5-NORBORNENE-2,3-DICARBOXIMIDE see OES000
2-ETHYLHEXYL OCTADECANOATE see OFU300
N-2-ETHYLHEXYL-β-OXYBUTYRAMIDE SEMISUCCINATE see BPF825
2-((2-ETHYLHEXYL)OXY)ETHANOL see EKX500
2-(2-ETHYLHEXYLOXY)ETHANOL see EKX500
4-(2-ETHYLHEXYLOXY)-2-HYDROXYBENZOPHENONE see EKY000
(((2-ETHYLHEXYL)OXY)METHYL)OXIRANE see GGY100
3-((2-ETHYLHEXYL)OXY)PROPANENITRILE see EKZ000
3-(2-ETHYLHEXYLOXY)PROPIONITRILE see EKZ000
2-ETHYLHEXYLOXYPROPYLAMINE see ELA000
3-((2-ETHYLHEXYL)OXY)PROPYLAMINE see ELA000
2-ETHYLHEXYL PALMITATE see OFE100
2-ETHYLHEXYL PHTHALATE see DVL700
ETHYLHEXYL PHTHALATE see DVL700
1-ETHYL-1-HEXYLPIPERIDINIUM BROMIDE see ELA600
2-ETHYLHEXYL-2-PROPENOATE see ADU250
5-ETHYL-5-HEXYL-2,4,6-(1H,3H,5H)-PYRIMIDINETRIONE MONOSODIUM SALT see EKT500
2-ETHYLHEXYL SALICYLATE see ELB000
2-ETHYLHEXYL SEBACATE see BJS250
2-ETHYLHEXYL SODIUM SULFATE see TAV750
2-ETHYLHEXYL STEARATE see OFU300
2-ETHYLHEXYL SULFATE see ELB400
2-ETHYLHEXYL SULFOSUCCINATE SODIUM see DJL000
2-(2-ETHYLHEXYL)-3a,4,7,7a-TETRAHYDRO-4,7-METHANO-1H-ISOINDOLE-1,3(2H)-DIONE see OES000
5-ETHYL-5-HEXYL-2-THIOBARBITURIC ACID SODIUM SALT see SHN275
2-ETHYLHEXYL THIOGLYCOLATE see EKW300
2-ETHYLHEXYL VINYL ETHER see ELB500
S-ETHYL-HOMOCYSTEINE see EEI000
S-ETHYL-dl-HOMOCYSTEINE see EEI000
S-ETHYL-l-HOMOCYSTEINE see AKB250
ETHYL HYDRATE see EFU000
ETHYLHYDRAZINE HYDROCHLORIDE see ELC000
ETHYL HYDRIDE see EDZ000
ETHYLHYDROCUPREINE see HHR700
ETHYLHYDROCUPREINE HYDROCHLORIDE see ELC500
ETHYL HYDROGEN ADIPATE see ELC600
ETHYL(HYDROGEN CYSTEINATO)MERCURY see EME000
ETHYL HYDROGEN PEROXIDE see ELD000
ETHYL HYDROPEROXIDE see ELD000
ETHYL HYDROPERSULFIDE see EEB000
ETHYL HYDROSULFIDE see EMB100
ETHYL HYDROXIDE see EFU000
1-ETHYL-3-(HYDROXYACETYL)INDOLE see ELD100
4'-ETHYL-4-HYDROXYAZOBENZENE see ELD500
ETHYL-o-HYDROXYBENZOATE see SAL000
ETHYL-p-HYDROXYBENZOATE see HJL000
2-ETHYL-4'-HYDROXY-3-BENZOYLBENZOFURAN see BBJ500
2-ETHYL-3-(p-HYDROXYBENZOYL)BENZOFURAN see BBJ500
ETHYL-2 (HYDROXY-4 BENZOYL)-3 BENZOFURANNE see BBJ500
ETHYL-2-HYDROXY-2,2-BIS(4-CHLOROPHENYL)ACETATE see DER000
N-ETHYL-N-(4-HYDROXYBUTYL)NITROSOAMINE see ELE500
ETHYL-N-HYDROXYCARBAMATE see HKQ025
ETHYL 3-HYDROXYCARBANILATE see ELE600
ETHYL m-HYDROXYCARBANILATE see ELE600
ETHYL-m-HYDROXYCARBANILATE CARBANILATE (ESTER) see EEO500
α-ETHYL-1-HYDROXYCYCLOHEXANEACETIC ACID see COW700
N-ETHYL-3-HYDROXY-N,N-DIMETHYL-BENZENAMINIUM BROMIDE (9CI) see TAL490
N-ETHYL-3-HYDROXY-N,N-DIMETHYLBENZENAMINIUM CHLORIDE (9CI) see EAE600
(±)-13-ETHYL-17-HYDROXY-18,19-DINOR-17-α-PREGN-4-EN-20-YN-3-ONE see NNQ500
N-ETHYL-2-((HYDROXYDIPHENYLACETYL)OXY)-N,N-DIMETHYLETHANAMINIUM CHLORIDE see ELF500
16-ETHYL-17-HYDROXYESTER-4-EN-3-ONE see ELF100
16-β-ETHYL-17-β-HYDROXYESTER-4-EN-3-ONE ACETATE see ELF110
16-β-ETHYL-17-β-HYDROXY-4-ESTREN-3-ONE see ELF100
16-β-ETHYL-17-β-HYDROXYESTR-4-EN-3-ONE see ELF100
2-(N-ETHYL-N-2-HYDROXYETHYLAMINO)ETHANOL see ELP000
1-ETHYL-1-(2-HYDROXYETHYL)AZIRIDINIUM SALT with 2,4,6-TRINITROBENZENESULFONIC ACID see ELG000
1-ETHYL-1-(2-HYDROXYETHYL)AZIRIDINIUM-2,4,6-TRINITROBENZENESULFONATE see ELG000

ETHYL (2-HYDROXYETHYL)DIMETHYLAMMONIUM BENZILATE CHLORIDE see ELF500
ETHYL(2-HYDROXYETHYL)DIMETHYLAMMONIUM CHLORIDE BENZILATE see ELF500
ETHYL(2-HYDROXYETHYL)DIMETHYL-AMMONIUM SULFATE (SALT), BIS(DIBUTYLCARBAMATE) see DDW000
ETHYL-β-HYDROXYETHYLETHYLENIMONIUM PICRYLSULFONATE see ELG000
ETHYL(2-HYDROXYETHYL)ETHYLENIMONIUM PICRYLSULFONATE see ELG000
1-ETHYL-1-(β-HYDROXYETHYL)ETHYLENIMONIUM PICRYLSULFONATE see ELG000
ETHYL-2-HYDROXYETHYLNITROSAMINE see ELG500
N-ETHYL-N-HYDROXYETHYLNITROSAMINE see ELG500
ETHYL-2-HYDROXYETHYL SULFIDE see EPP500
ETHYL-2-HYDROXYETHYL THIOETHER see EPP500
N-ETHYL-N-(2-HYDROXYETHYL)-m-TOLUIDINE see HKS100
ETHYL-7-HYDROXYFLAVONE see ELH600
ETHYL 2-HYDROXYISOBUTYRATE see ELH700
ETHYL α-HYDROXYISOBUTYRATE see ELH700
ETHYL-4-HYDROXY-3-METHOXYBENZOATE see EQE500
(8-α,9R)-1-ETHYL-9-HYDROXY-6'-METHOXYCINCHONAN-1-IUM IODIDE see QIS000
ETHYL-2-(HYDROXYMETHYL)ACRYLATE see ELI500
ETHYL-α-(HYDROXYMETHYL)ACRYLATE see ELI500
1-(2-ETHYL-7-(2-HYDROXY-3-((1-METHYLETHYL)AMINO)PROPOXY)-4-BENZOFURANYL) ETHANONE see ELI600
α-ETHYL-β-(HYDROXYMETHYL)-1-METHYL-IMIDAZOLE-5-BUTYRIC ACID, γ-LACTONE see PIF000
1-ETHYL-7-HYDROXY-2-METHYL-1,2,3,4,4a,9,10,10a-OCTAHYDROPHENANTHRENE-2-CARBOXYLIC ACID see DYB000
1-ETHYL-3-HYDROXY-1-METHYL-PIPERIDINIUM BROMIDE BENZILATE see PJA000
2-ETHYL-2-(HYDROXYMETHYL)-1,3-PROPANEDIOL, CYCLIC PHOSPHATE (1:1) see ELJ500
2-ETHYL-2-(HYDROXYMETHYL)-1,3-PROPANEDIOL TRIACRYLATE see TLX175
2-ETHYL-2-HYDROXYMETHYL-1,3-PROPANEDIOL TRIMETHACRYLATE see TLX250
ETHYL 2-HYDROXY-2-METHYLPROPANOATE see ELH700
ETHYL-4-HYDROXY-3-MORPHOLINOMETHYLBENZOATE see ELK000
5-ETHYL-2'-HYDROXY-2(N)-(3-METHYL-2-BUTENYL)-9-METHYL-6,7-BENZOMORPHAN see DOQ600
β-ETHYL-β-HYDROXYPHENETHYL CARBAMATE see HNJ000
β-ETHYL-β-HYDROXYPHENETHYL CARBAMIC ACID ESTER see HNJ000
ETHYL (3-HYDROXYPHENYL)CARBAMATE see ELE600
ETHYL N-(3-HYDROXYPHENYL)CARBAMATE see ELE600
ETHYL(m-HYDROXYPHENYL)DIMETHYLAMMONIUM BROMIDE (8CI) see TAL490
ETHYL(m-HYDROXYPHENYL)DIMETHYLAMMONIUM CHLORIDE see EAE600
ETHYL-p-HYDROXYPHENYL KETONE see ELL500
ETHYL (4-(m-HYDROXYPHENYL)-1-METHYL)-4-PIPERIDYL KETONE see KFK000
ETHYL 2-HYDROXYPROPIONATE see LAJ000
ETHYL α-HYDROXYPROPIONATE see LAJ000
3-(6-(ETHYL-(2-HYDROXYPROPYL)AMINO)PYRIDAZIN-3-YL)CARBAZIC ACID ETHYL ESTER see CAK275
2-(6-ETHYL(2-HYDROXYPROPYL)AMINO)-3-PYRIDAZINYL)-HYDRAZINECARBOXYLIC ACID ETHYL ESTER see CAK275
(S)-4-ETHYL-4-HYDROXY-1H-PYRANO(3',4':6,7)INDOLIZINO(1,2-b)QUINOLINE-3,14(4H,12H)-DIONE see CBB870
2-ETHYL-3-HYDROXY-4H-PYRAN-4-ONE see EMA600
α-ETHYL-3-HYDROXY-2,4,6-TRIIODOHYDROCINNAMIC ACID see IFZ800
α-ETHYL-β-(3-HYDROXY-2,4,6-TRIIODOPHENYL)PROPIONIC ACID see IFZ800
ETHYL HYPOCHLORITE see ELM500
ETHYLIC ACID see AAT250
3-ETHYLIDENE-3,4,5,6,9,11,13,14,14A,14B-DECAHYDRO-6-HYDROXY-5,6-DIMETHYL(1,6)DIOXACYCLODODECINO(2,3,4-GH)-PYRROLIZINE-2,7-DIONE see IDG000
ETHYLIDENE ACETONE see PBR500
5-ETHYLIDENEBICYCLO(2.2.1)HEPT-2-ENE see ELO500
N,N'-ETHYLIDENE-BIS(ETHYL CARBAMATE) see ELO000
1,1'-(ETHYLIDENEBIS(OXY))BISBUTANE see DDT400
ETHYLIDENE BROMIDE see DDN800
ETHYLIDENE CHLORIDE see DFF809
ETHYLIDENE DIBROMIDE see DDN800
ETHYLIDENEDICARBAMIC ACID, DIETHYL ESTER see ELO000
ETHYLIDENE DICHLORIDE see DFF809
ETHYLIDENE DIETHYL ETHER see AAG000
ETHYLIDENE DIFLUORIDE see ELN500
5-ETHYLIDENEDIHYDRO-2(3H)-FURANONE see HLF000
ETHYLIDENE DIMETHYL ETHER see DOO600
ETHYLIDENE DINITRATE see ELN600
ETHYLIDENE DIURETHAN see ELO000
ETHYLIDENE FLUORIDE see ELN500
ETHYLIDENE-2(5H)-FURANONE see MOW500

ETHYLIDENE GYROMITRIN see AAH000
ETHYLIDENEHYDROXYLAMINE see AAH250
trans-15-ETHYLIDENE-12-β-HYDROXY-4,12-α,13-β-TRIMETHYL 8-OXO-4,8 SEC-OSENEC-1-ENINE see DMX200
ETHYLIDENELACTIC ACID see LAG000
5-ETHYLIDENE-2-NORBORNENE see ELO500
ETHYLIDENE NORBORNENE see ELO500
ETHYLIDICHLORARSINE see DFH200
ETHYLIDICHLOROARSINE (DOT) see DFH200
ETHYLIMINE see EJM900
2,2'-(ETHYLIMINO)BISETHANOL see ELP000
N-ETHYL-2,2'-IMINODIETHANOL see ELP000
2,2'-(ETHYLIMINO)DIETHANOL see ELP000
ETHYL IODIDE see ELP500
ETHYL IODOACETATE see ELQ000
ETHYLIODOMETHYLARSINE see ELQ100
ETHYL 10-(p-IODOPHENYL)UNDECANOATE see ELQ500
ETHYL-10-(p-IODOPHENYL)UNDECYLATE see ELQ500
5-ETHYL-5-ISOAMYLBARBITURIC ACID see AMX750
5-ETHYL-5-ISOAMYLMALONYL UREA see AMX750
ETHYL ISOBUTANOATE see ELS000
ETHYL ISOBUTENOATE see MIP800
ETHYLISOBUTYLMETHANE, ISOHEPTANE see MKL250
ETHYLISOBUTYRATE (DOT) see ELS000
ETHYL ISOBUTYRATE see ELS000
ETHYL ISOCYANATE (DOT) see ELS500
ETHYL ISOCYANATE see ELS500
ETHYL ISOCYANIDE see ELT000
2-ETHYLISOHEXANOL see ELT500, EMZ000
ETHYL ISONICOTINATE see ELU000
2-ETHYLISONICOTINIC ACID THIOAMIDE see EPQ000
α-ETHYLISONICOTINIC ACID THIOAMIDE see EPQ000
2-ETHYLISONICOTINIC THIOAMIDE see EPQ000
α-ETHYLISONICOTINOYLTHIOAMIDE see EPQ000
ETHYL ISONITRILE see ELT000
5-ETHYL-5-ISOPENTYLBARBITURIC ACID see AMX750
ETHYLISOPENTYLBARBITURIC ACID see AMX750
5-ETHYL-5-ISOPENTYLBARBITURIC ACID SODIUM SALT see AON750
O-ETHYL-O-(2-ISOPROPOXY-CARBONYL)-PHENYL ISOPROPYLPHOSPHORAMI-DOTHIOATE see IMF300
5-ETHYL-5-ISOPROPYLBARBITURIC ACID see ELX000
ETHYL ISOPROPYLBARBITURIC ACID see ELX000
ETHYL ISOPROPYL FLUOROPHOSPHONATE see ELX100
ETHYL ISOPROPYLIDENE ACETATE see MIP800
4-ETHYL-3-ISOPROPYL-4-METHYL-1-OXACYCLOBUTAN-2-ONE see DSH400
ETHYL ISOPROPYLNITROSAMINE see ELX500
5-ETHYL-5-ISOPROPYL-2-THIOBARBITURIC ACID SODIUM SALT see SHY500
ETHYLISOTHIAMIDE see EPQ000
ETHYL ISOTHIOCYANATOACETATE see ELX530
ETHYL ISOTHIOCYANOACETATE see ELX530
2-ETHYLISOTHIONICOTINAMIDE see EPQ000
α-ETHYLISOTHIONICOTINAMIDE see EPQ000
S-ETHYLISOTHIOURONIUM HYDROGEN SULFATE see ELY550
S-ETHYLISOTHIURONIUM DIETHYL PHOSPHATE see CBI675
S-ETHYLISOTHIURONIUM METAPHOSPHATE see ELY575
ETHYL ISOVALERATE (FCC) see ISY000
ETHYLISOVANILLIN see IKO100
ETHYLJODID see ELP500
ETHYL 4-KETOVALERATE see EFS600
ETHYL KETOVALERATE see EFS600
N-ETHYL-N-2-KYANETHYLANILIN see EHQ500
d-N-ETHYLLACTAMIDE CARBANILATE (ESTER) see CBL500
ETHYL LACTATE (DOT,FCC) see LAJ000
ETHYL LAEVULINATE see EFS600
ETHYL LAURATE see ELY700
ETHYL LEVULATE see EFS600
ETHYL LINALOOL see ELZ000
ETHYLLINALYL ACETAL see ELZ050
ETHYL LINALYL ACETATE see HGI585
ETHYLLITHIUM see ELZ100
ETHYL LOFLAZEPATE see EKF600
ETHYL MAGNESIUM IODIDE see EMA000
ETHYL MALEATE see DJO200
N-ETHYLMALEIMIDE see MAL250
ETHYL MALONATE see EMA500
ETHYL MALTOL see EMA600
ETHYL MANDELATE see EMB000
ETHYLMERCAPTAAN (DUTCH) see EMB100
ETHYL MERCAPTAN see EMB100
ETHYL-2-MERCAPTOACETATE see EMB200
ETHYL-α-MERCAPTOACETATE see EMB200
ETHYL MERCAPTOACETATE see EMB200
ETHYL(MERCAPTOACETATO(2)-O,S)-MERCURATE(1)-POTASSIUM see PLE750
ETHYL MERCAPTOACETIC ACID see EMB200

ETHYL (2-MERCAPTOETHYL) CARBAMATE S-ESTER with O,O-DIMETHYL PHOSPHORODITHIOATE see EMC000
β-ETHYLMERCAPTOETHYL DIMETHYL THIONOPHOSPHATE see DAO800
2-ETHYLMERCAPTOMETHYLPHENYL-N-METHYLCARBAMATE see EPR000
2-ETHYLMERCAPTO-10-(3-(1-METHYL-4-PIPERAZINYL)PROPYL)PHENOTHIAZINE DIMALEATE see TEZ000
3-ETHYLMERCAPTO-10-(1'-METHYLPIPERAZINYL-4'-PROPYL)PHENOTHIAZINE DIMALEATE see TEZ000
ETHYL MERCAPTOPHENYLACETATE-O,O-DIMETHYL PHOSPHOROCITHIOATE see DRR400
9-ETHYL-6-MERCAPTOPURINE see EMC500
ETHYLMERCURIC ACETATE see EMD000
ETHYLMERCURIC CHLORIDE see CHC500
ETHYLMERCURIC CYSTEINE see EME000
ETHYLMERCURICHLORENDIMIDE see EME050
ETHYLMERCURIC PHOSPHATE see BJT250, EME100
N-(ETHYLMERCURI)-1,4,5,6,7,7-HEXACHLOROBICYCLO(2.2.1)HEPT-5-ENE-2,3-DICARBOXIMIDE see EME050
N-ETHYLMERCURI-3,4,5,6,7,7-HEXACHLORO-3,6-ENDOMETHYLENE-1,2,3,6-TETRAHYDROPHTHALIMIDE see EME050
N-ETHYLMERCURI-N-PHENYL-p-TOLUENESULFONAMIDE see EME500
N-ETHYLMERCURI-1,2,3,6-TETRAHYDRO-3,6-ENDOMETHANO-3,4,5,6,7,7-HEXA-CHLOROPHTHALIMIDE see EME050
o-(ETHYLMERCURITHIO)BENZOIC ACID SODIUM SALT see MDI000
ETHYLMERCURITHIOSALICYLIC ACID SODIUM SALT see MDI000
N-(ETHYLMERCURI)-p-TOLUENESULFONANILIDE see EME500
N-(ETHYLMERCURI)-p-TOLUENESULPHONANILIDE see EME500
ETHYLMERCURY CHLORIDE see CHC500
ETHYLMERCURY PHOSPHATE see BJT250, EME100
ETHYLMERCURY p-TOLUENESULFANILIDE see EME500
ETHYLMERCURY-p-TOLUENE SULFONAMIDE see EME500
ETHYLMERCURY-p-TOLUENESULFONANILIDE see EME500
ETHYLMERKAPTAN (CZECH) see EMB100
β-ETHYLMERKAPTOETHANOL (CZECH) see EPP500
β-ETHYLMERKAPTOETHYLCHLORID (CZECH) see CGY750
ETHYLMERKURIACETAT see EMD000
ETHYL METHACRYLATE see EMF000
ETHYL METHACRYLATE, INHIBITED (DOT) see EMF000
ETHYL METHANESULFONATE see EMF500
ETHYLMETHANESULFONATO-CNU see CHF250
ETHYL METHANESULPHONATE see EMF500
ETHYL METHANOATE see EKL000
ETHYL METHANSULFONATE see EMF500
ETHYL METHANSULPHONATE see EMF500
ETHYLMETHIAMBUTENE HYDROCHLORIDE see EIJ000
7-ETHYL-5-METHOXY-BENZ(a)ANTHRACENE see MEN750
ETHYL-4-METHOXYBENZOATE see AOV000
ETHYL-p-METHOXYBENZOATE see AOV000
ETHYL 2-METHOXYBENZYL ETHER see EMF600
ETHYL o-METHOXYBENZYL ETHER see EMF600
2'-ETHYL-2-(2-METHOXY BUTYLAMINO) PROPIONANILIDE see EMG000
ETHYL O-(o-(METHOXYCARBONYL)BENZOYL)GLYCOLATE see MOD000
ETHYL o-(METHOXYCARBONYL)BENZOYLOXYACETATE see MOD000
2'-ETHYL-3-(2-METHOXYETHYL)AMINOBUTYRANILIDE HYDROCHLORIDE see EMG500
2'-ETHYL-4-(2-METHOXYETHYL)AMINOBUTYRANILIDE HYDROCHLORIDE see EMH000
2'-ETHYL-3-(2-METHOXYETHYL)AMINO-3-METHYLBUTYRANILIDE CYCLAMATE see EMH500
2'-ETHYL-3-(2-METHOXYETHYL)AMINO-3-METHYLBUTYRANILIDE CYCLOHEX-ANE SULFAMATE see EMH500
2'-ETHYL-2-(2-METHOXYETHYLAMINO)PROPIONANILIDE see EMI000
2'-ETHYL-2-(2-METHOXYETHYLAMINO)-PROPIONANILIDE HYDROCHLORIDE see EMI500
N-ETHYL-6-METHOXY-N'-(1-METHYLETHYL)-1,3,5-TRIAZINE-2,4-DIAMINE see EGD000
3-ETHYL-4-(p-METHOXYPHENYL)-2-METHYL-3-CYCLOHEXENE-1-CARBOXYLIC ACID see CBO625
ETHYL all-trans-9-(4-METHOXY-2,3,6-TRIMETHYLPHENYL)-3,7-DIMETHYL-2,4,6,8-NONATETRAENOATE see EMJ500
ETHYL-2-METHYLACRYLATE see EMF000
ETHYL-α-METHYL ACRYLATE see EMF000
4'-ETHYL-N-METHYL-4-AMINOAZOBENZENE see EOJ000
4-ETHYLMETHYLAMINOAZOBENZENE see ENB000
N-ETHYL-N-METHYL-p-AMINOAZOBENZENE see ENB000
p-ETHYLMETHYLAMINOAZOBENZENE see ENB000
3-ETHYLMETHYLAMINO-1,1-DI(2'-THIENYL)BUT-1-ENE HYDROCHLORIDE see EIJ000
α-(1-(ETHYLMETHYLAMINO)ETHYL)BENZYL ALCOHOL HYDROCHLORIDE (−) see EJR500
4-((((4-ETHYL-4-METHYL)AMINO)PHENYL)AZO)PYRIDINE 1-OXIDE see MKB500
2-ETHYL-6-METHYLANILINE see MJY000
ETHYL METHYL ARSINE see EMK600
7-ETHYL-9-METHYLBENZ(c)ACRIDINE see EMM000
7-ETHYL-12-METHYLBENZ(a)ANTHRACENE see EMM500

12-ETHYL-7-METHYLBENZ(a)ANTHRACENE see EMN000
2-ETHYL-6-METHYL-BENZENAMINE see MJY000
1-ETHYL-2-METHYLBENZENE see EPS500
1-ETHYL-4-METHYLBENZENE see EPT000
o-ETHYL METHYLBENZENE see EPS500
p-ETHYLMETHYLBENZENE see EPT000
ETHYL-p-METHYL BENZENESULFONATE see EPW500
ETHYL (E)-2-METHYL-2-BUTENOATE see TGA800
5-ETHYL-5-(1-METHYL-1-BUTENYL)BARBITURATE see EMO500
5-ETHYL-5-(1-METHYL-1-BUTENYL)BARBITURIC ACID see EMO500
5-ETHYL-5-(1-METHYL-2-BUTENYL)BARBITURIC ACID see EMO875
5-ETHYL-5-(1-METHYL-1-BUTENYL)BARBITURIC ACID SODIUM SALT see
   VKP000
5-ETHYL-5-(1-METHYL-1-BUTENYL)-2,4,6(1H,3H,5H)-PYRIMIDINETRIONE see
   EMO500
5-ETHYL-5-(1-METHYL-1-BUTENYL)-2,4,6(1H,3H,5H)-PYRIMIDINETRIONE SODI-
   UM SALT see VKP000
5-ETHYL-5-(1-METHYLBUTYL)BARBITURIC ACID see NBT500
5-ETHYL-5-(3-METHYLBUTYL)BARBITURIC ACID see AMX750
5-ETHYL-5-(3-METHYLBUTYL)BARBITURIC ACID, SODIUM DERIVATIVE see
   AON750
5-ETHYL-5-(1-METHYLBUTYL)BARBITURIC ACID SODIUM SALT see NBU000
R(+)-5-ETHYL-5-(1-METHYLBUTYL)BARBITURIC ACID SODIUM SALT see
   PBS500
S(−)-5-ETHYL-5-(1-METHYLBUTYL)BARBITURIC ACID SODIUM SALT see
   PBS750
ETHYL 2-METHYLBUTYL KETOXINE see EMP550
5-ETHYL-5-(1-METHYLBUTYL)MALONYLUREA see NBT500
5-ETHYL-5-(1-METHYLBUTYL)-2,4,6(1H,3H,5H)-PYRIMIDINETRIONE (9CI) see
   NBT500
5-ETHYL-5-(1-METHYLBUTYL)-2,4,6(1H,3H,5H)-PYRIMIDINETRIONE MONOSODI-
   UM SALT (9CI) see NBU000
5-ETHYL-5-(1-METHYLBUTYL)-2-THIOBARBITURIC ACID see PBT250
5-ETHYL-5-(1-METHYLBUTYL)-2-THIOBARBITURIC ACID MONOSODIUM see
   PBT500
ETHYL 2-METHYLBUTYRATE see EMP600
ETHYL-N-METHYLCARBAMATE see EMQ500
ETHYL METHYLCARBAMATE see EMQ500
ETHYLMETHYL CARBINOL see BPW750
ETHYL METHYL CETONE (FRENCH) see MKA400
3-ETHYL-7-METHYL-9-α-(4′-CHLOROBENZOYLOXY)-3,7-DIAZABICY-
   CLO(3.3.1)NONAME HYDROCHLORIDE see YGA700
ETHYL-2-METHYL-4-CHLOROPHENOXYACETATE see EMR000
ETHYL (E)-2-METHYLCROTONATE see TGA800
ETHYL 3-METHYLCROTONATE see MIP800
ETHYL α-METHYLCROTONATE see MIP800, TGA800
ETHYL METHYL 1,4-DIHYDRO-2,6-DIMETHYL-4-(m-NITROPHENYL)-3,5-PYRIDI-
   NEDICARBOXYLATE see EMR600
1-ETHYL-7-METHYL-1,4-DIHYDRO-1,8-NAPHTHYRIDINE-4-ONE-3-CARBOXYLIC
   ACID see EID000
1-ETHYL-7-METHYL-1,4-DIHYDRO-1,8-NAPHTHYRIDIN-4-ONE-3-CARBOXYLIC
   ACID see EID000
4′-ETHYL-2-METHYL-4-DIMETHYLAMINOAZOBENZENE see EMS000
2-ETHYL-2-METHYL-1,3-DIOXOLANE see EIO500
4-ETHYL-4-METHYL-2,6-DIOXOPIPERIDINE see MKA250
1-ETHYL-6,7-METHYLENEDIOXY-4(1H)-OXOCINNOLINE-3-CARBOXYLIC ACID
   see CMS130
1-ETHYL-6,7-METHYLENEDIOXY-4-QUINOLONE-3-CARBOXYLIC ACID see
   OOG000
ETHYL METHYLENE PHOSPHORODITHIOATE see EEH600
ETHYL METHYL ETHER (DOT) see EMT000
ETHYL METHYL ETHER see EMT000
5-ETHYL-5-(1-METHYLETHYL)-2,4,6(1H,3H,5H)-PYRIMIDINETRIONE see ELX000
3-ETHYL-3-METHYLGLUTARIMIDE see MKA250
β-ETHYL-β-METHYLGLUTARIMIDE see MKA250
7-ETHYL-2-METHYL-4-HENDECANOL SULFATE SODIUM SALT see EMT500
7-ETHYL-2-METHYL-4-HEXADECANOL SULFATE SODIUM SALT see SIO000
ETHYLMETHYLKETON (DUTCH) see MKA400
ETHYL METHYL KETONE (DOT) see MKA400
ETHYL METHYL KETONE AZINE see EMT600
ETHYL METHYL KETONE OXIME see EMU500
ETHYL METHYL KETONE PEROXIDE see MKA500
ETHYL-METHYLKETONOXIM see EMU500
ETHYL METHYL KETOXIME see EMU500
ETHYL 2-METHYLLACTATE see ELH700
O-ETHYL-O-(4-(METHYLMERCAPTO)PHENYL)-S-N-PROPYLPHOSPHOROTHION-
   OTHIOLATE see SOU625
2-ETHYL-6-METHYL-1-N-(2-METHOXY-1-EMTHYLETHYL)CHLOROACETANILIDE
   see MQQ450
1-ETHYL-2-METHYL-7-METHOXY-1,2,3,4-TETRAHYDROPHENANTHRYL-2-CAR-
   BOXYLIC ACID see BIT000
N-ETHYL-6-METHYL-α-(METHYLSULFONYL)ERGOLINE-8-β-PROPIONAMIDE see
   EMW100
ETHYL-3-METHYL-4-(METHYLTHIO)PHENYL(1-METHYLE-
   THYL)PHOSPHORAMIDATE see FAK000

O-ETHYL-O-METHYL-O-p-NITROFENYLESTER KYSELINY THIOFOSFORECNE
   see ENI175
ETHYL (2-(2-METHYL-5-NITRO-1-IMIDAZOLYL)ETHYL)SULFONE see TGD250
ETHYLMETHYLNITROSAMINE see MKB000
4-ETHYL-1-METHYLOCTYLAMINE see EMY000
2-ETHYL-1-(3-METHYL-1-OXO-2-BUTENYL)PIPERIDINE see EMY100
1-ETHYL-7-METHYL-4-OXO-1,4-DIHYDRO-1,8-NAPHTHYRIDINE-3-CARBOXYLIC
   ACID see EID000
ETHYL (Z)-(3-METHYL-4-OXO-5-PIPERIDINO-THIAZOLIDIN-2-YLIDENE)ACETATE
   see MNG000
2-ETHYL-4-METHYL-1-PENTANOL see ELT500, EMZ000
2-ETHYL-4-METHYLPENTANOL see ELT500, EMZ000
5-ETHYL-5-(1-METHYL-1-PENTENYL)BARBITURIC ACID see ENA000
ETHYL METHYL PEROXIDE see ENA500
N-ETHYL-N-METHYL-p-(PHENYLAZO)ANILINE see ENB000
5-ETHYL-1-METHYL-5-PHENYLBARBITURIC ACID see ENB500
5-ETHYL-N-METHYL-5-PHENYLBARBITURIC ACID see ENB500
N-ETHYLMETHYLPHENYLBARBITURIC ACID see ENB500
ETHYL METHYLPHENYLGLYCIDATE see ENC000
5-ETHYL-1-METHYL-5-PHENYLHYDANTOIN see ENC500
5-ETHYL-3-METHYL-5-PHENYLHYDANTOIN see MKB250
5-ETHYL-3-METHYL-5-PHENYL-2,4(3H,5H)-IMIDAZOLEDIONE see MKB250
5-ETHYL-3-METHYL-5-PHENYLIMIDAZOLIDIN-2,4-DIONE see MKB250
ETHYL-1-METHYL-4-PHENYLISONIPECOTATE see DAM600
ETHYL-1-METHYL-4-PHENYLISONIPECOTATE HYDROCHLORIDE see DAM700
ETHYL-1-METHYL-4-PHENYLPIPERIDINE-4-CARBOXYLATE see DAM600
ETHYL-1-METHYL-4-PHENYLPIPERIDINE-4-CARBOXYLATE HYDROCHLORIDE
   see DAM700
ETHYL-1-METHYL-4-PHENYLPIPERIDYL-4-CARBOXYLATE HYDROCHLORIDE
   see DAM700
5-ETHYL-1-METHYL-5-PHENYL-2,4,6(1H,3H,5H)-PYRIMIDINETRIONE see
   ENB500
α-ETHYL-1-METHYL-α-PHENYL-3-PYRROLIDINEMETHANOL PROPIONATE FU-
   MARATE see ENC600
α-ETHYL-4-METHYL-1-PIPERAZINEACETIC ACID-2,6-DIETHYLPHENYL ESTER
   DIHYDROCHLORIDE see FAC165
α-ETHYL-4-METHYL-1-PIPERAZINEACETIC ACID MESITYL ESTER DIHYDRO-
   CHLORIDE see FAC195
α-ETHYL-4-METHYL-1-PIPERAZINEACETIC ACID-2,6-XYLYL ESTER HYDRO-
   CHLORIDE see FAC130
5-ETHYL-2-METHYLPIPERIDINE see END000
3-ETHYL-6-METHYLPIPERIDINE see END000
4-ETHYL-4-METHYL-2,6-PIPERIDINEDIONE see MKA250
N-ETHYL-2-(METHYLPIPERIDINO)-N-(1-PHENOXY-2-PROPYL)ACETAMIDE HY-
   DROCHLORIDE see EOG500
4′-ETHYL-2-METHYL-3-PIPERIDINOPROPIOPHENONE HYDROCHLORIDE see
   EAV700
N-ETHYL-2-(2-METHYLPIPERIDINO)-N-(1-(2,4-XYLYLOXY)-2-PROPYL) ACET-
   AMIDE HYDROCHLORIDE see END500
ETHYL-2-METHYLPROPANOATE see ELS000
ETHYL-2-METHYL-2-PROPENOATE see EMF000
N-ETHYL-N-(2-METHYL-2-PROPENYL)-2,6-DINITRO-4-(TRIFLUOROMETH-
   YL)BENZENAMINE see ENE500
ETHYL-2-METHYLPROPIONATE see ELS000
5-ETHYL-5-(1-METHYLPROPYL)BARBITURATE see BPF000
5-ETHYL-5-(1-METHYLPROPYL)BARBITURIC ACID see BPF000
5-ETHYL-5-(1-METHYLPROPYL)BARBITURIC ACID SODIUM SALT see BPF250
α-ETHYL-4-(2-METHYLPROPYL)BENZENEACETIC ACID see IJH000
1-ETHYL-1-METHYLPROPYL CARBAMATE see ENF000
5-ETHYL-5-(1-METHYLPROPYL)-2,4,6(1H,3H,5H)-PYRIMIDINETRIONE (9CI) see
   BPF000
5-ETHYL-5-(1-METHYLPROPYL)-2,4,6(1H,3H,5H)-PYRIMIDINETRIONE MONOSO-
   DIUM SALT see BPF250
1-ETHYL-1-METHYL-2-PROPYNYL CARBAMATE see MNM500
2-ETHYL-3-METHYLPYRAZINE see ENF200
5-ETHYL-2-METHYLPYRIDINE see EOS000
3-ETHYL-6-METHYLPYRIDINE see EOS000
3-ETHYL-3-METHYL-2,5-PYRROLIDINE-DIONE see ENG500
3-ETHYL-3-METHYLPYRROLIDINE-2,5-DIONE see ENG500
2-ETHYL-2-METHYLSUCCINIMIDE see ENG500
α-ETHYL-α-METHYLSUCCINIMIDE see ENG500
ETHYLMETHYLTHIAMBUTENE HYDROCHLORIDE see EIJ000
O-ETHYL-O-(4-(METHYLTHIO)PHENYL)PHOSPHORODITHIOIC ACID-S-PROPYL
   ESTER see SOU625
O-ETHYL-O-(4-(METHYLTHIO)PHENYL) S-PROPYL PHOSPHORODITHIOATE
   see SOU625
ETHYLMETHYLTHIOPHOS see ENI175
ETHYL-4-(METHYLTHIO)-m-TOLYL ISOPROPYL PHOSPHOR AMIDATE see
   FAK000
O-ETHYL-O-(4-METHYLTHIO-m-TOLYL) METHYLPHOSPHORAMIDOTHIOATE
   see ENI500
N-ETHYL-α-METHYL-m-(TRIFLUOROMETHYL)PHENETHYLAMINE see ENJ000
N-ETHYL-α-METHYL-m-TRIFLUOROMETHYLPHENETHYLAMINE see PDM250
N-ETHYL-α-METHYL-m-(TRIFLUOROMETHYL)PHENETHYLAMINE HYDROCHLO-
   RIDE see PDM250

3-ETHYL-5-METHYL-1,2,4-TRIOXOLANE see PBQ300
7-ETHYL-2-METHYL-4-UNDECANOL SULFATE SODIUM SALT see EMT500
2-ETHYL-3-METHYLVALERAMIDE see ENJ500
ETHYLMETHYL VALERAMIDE see ENJ500
ETHYL MONOBROMOACETATE see EGV000
ETHYL MONOCHLORACETATE see EHG500
ETHYL MONOCHLOROACETATE see EHG500
ETHYL MONOIODOACETATE see ELQ000
ETHYL MONOSULFIDE see EPH000
3-o-ETHYLMORPHINE see ENK000
ETHYLMORPHINE see ENK000
o-ETHYLMORPHINE HYDROCHLORIDE see DVO700
ETHYLMORPHINE HYDROCHLORIDE see DVO700, ENK500
ETHYL MORPHINE HYDROCHLORIDE DIHYDRATE see ENK500
4-ETHYLMORPHOLINE see ENL000
N-ETHYLMORPHOLINE see ENL000
1-(N-p-ETHYLMORPHOLINE)-5-NITROIMIDAZOLE see NHH000
1-ETHYL-4-(2-MORPHOLINOETHYL)-3,3-DIPHENYL-2-PYRROLIDINONE see
    ENL100
1-ETHYL-4-(2-MORPHOLINOETHYL)-3,3-DIPHENYL-2-PYRROLIDINONE HYDRO-
    CHLORIDE HYDRATE see SLU000
ETHYL 1-(2′-MORPHOLINOETHYL)-4-PHENYLPIPERIDINE-4-CARBOXYLATE DI-
    HYDROCHLORIDE see MRN675
ETHYL 10-(3-MORPHOLINOPROPIONYL)PHENOTHIAZINE-2-CARBAMATE HY-
    DROCHLORIDE see EEI025
9-ETHYL-6-MP see EMC500
ETHYL MYRISTATE see ENL850
ETHYL 1-NAPHTHALENEACETATE see ENL900
ETHYL-β-NAPHTHOLATE see EEY500
ETHYL 1-NAPHTHYLACETATE see ENL900
ETHYL-2-NAPHTHYL ETHER see EEY500
ETHYL-β-NAPHTHYL ETHER see EEY500
N,N′-ETHYL-N,N′-(β-CHLOROETHYL)ETHYLENEDIAMINE DIHYDROCHLORIDE
    see BIC500
N-ETHYL-N′-2-INDANYL-N′-PHENYL-1,3-PROPANEDIAMINE HYDROCHLORIDE
    see DBA475
3-ETHYLNIRVANOL see MKB250
ETHYL NITRATE see ENM500
ETHYL NITRILE see ABE500
ETHYL NITRITE see ENN000
ETHYL NITRITE SOLUTIONS (DOT) see ENN000
ETHYL NITROACETATE see ENN100
ETHYL-p-NITROBENZOATE see ENO000
ETHYL NITROBENZOATE, PARA ESTER see ENO000
ETHYL (4-NITROBENZOYL)ACETATE see ENO100
ETHYL (p-NITROBENZOYL)ACETATE see ENO100
N-ETHYL-N-NITROETHANAMINE (9CI) see DJS500
O-ETHYL-O-((4-NITROFENYL)-FENYL)-MONOTHIOFOSFONAAT (DUTCH) see
    EBD700
ETHYLNITROLIC ACID see NHY100
ETHYL 4-NITRO-β-OXOBENZENEPROPANOATE see ENO100
O-ETHYL O-(4-NITROPHENYL)BENZENETHIONOPHOSPHONATE see EBD700
ETHYL-p-NITROPHENYL BENZENETHIONOPHOSPHONATE see EBD700
ETHYL-p-NITROPHENYL BENZENETHIOPHOSPHATE see EBD700
ETHYL-p-NITROPHENYL BENZENETHIOPHOSPHONATE see EBD700
ETHYL p-NITROPHENYL ETHYLPHOSPHATE see NIM500
ETHYL-4-NITROPHENYL ETHYLPHOSPHONATE see ENQ000
ETHYL-p-NITROPHENYLPENTYLPHOSPHONATE see ENQ500
O-ETHYL-O-(4-NITROPHENYL) PHENYLPHOSPHONOTHIOATE see EBD700
ETHYL-p-NITROPHENYL PHENYLPHOSPHONOTHIOATE see EBD700
O-ETHYL-O-p-NITROPHENYL PHENYLPHOSPHONOTHIOLATE see EBD700
O-ETHYL-O-p-NITROPHENYL PHENYLPHOSPHOROTHIOATE see EBD700
ETHYL-p-NITROPHENYL THIONOBENZENEPHOSPHATE see EBD700
ETHYL-p-NITROPHENYL THIONOBENZENEPHOSPHONATE see EBD700
3-ETHYL-4-NITROPYRIDINE-1-OXIDE see ENR000
2-ETHYL-4-NITROQUINOLINE-1-OXIDE see NGA500
2-(ETHYLNITROSAMINO)ETHANOL see ELG500
(ETHYLNITROSAMINO)METHYL ACETATE see ENR500
4-((ETHYLNITROSAMINO)METHYL)PYRIDINE see NLH000
4-((ETHYLNITROSOAMINO)-1-BUTANOL see ELE500
4-(ETHYLNITROSOAMINO)BUTYRIC ACID see NKE000
5-(N-ETHYL-N-NITROSO)AMINO-3-(5-NITRO-2-FURYL)-s-TRIAZOLE see ENS000
ETHYLNITROSOANILINE see NKD000
N-ETHYL-N-NITROSOBENZENAMINE see NKD000
N-ETHYL-N-NITROSOBENZYLAMINE see ENS500
N-ETHYL-N-NITROSOBIURET see ENT000
ETHYLNITROSOBIURET see ENT000
N-ETHYL-N-NITROSO-tert-BUTANAMINE see NKD500
N-ETHYL-N-NITROSOBUTYLAMINE see EHC000
N-ETHYL-N-NITROSOCARBAMIC ACID ETHYL ESTER see NKE500
ETHYLNITROSOCARBAMIC ACID, ETHYL ESTER see NKE500
N-ETHYL-N-NITROSOCARBAMIDE see ENV000
ETHYLNITROSOCYANAMIDE see ENT500
N-ETHYL-N-NITROSO-ETHANAMINE see NJW500
N-ETHYL-N-NITROSOETHENAMINE see NKF000

N-ETHYL-N-NITROSOETHENYLAMINE see NKF000
N-ETHYL-N-NITROSO-N′-NITROGUANIDINE see ENU000
ETHYL 4-NITROSO-1-PIPERAZINECARBOXYLATE see NJQ000
1-ETHYL-1-NITROSOUREA see ENV000
N-ETHYL-N-NITROSO-UREA see ENV000
ETHYLNITROSOUREA see ENV000
N-ETHYL-N-NITROSOURETHAN see NKE500
ETHYL NITROSOURETHAN see NKE500
N-ETHYL-N-NITROSOURETHANE see NKE500
ETHYL NITROSOURETHANE see NKE500
N-ETHYL-N-NITROSOVINYLAMINE see NKF000
1-ETHYL-3-(5-NITRO-2-THIAZOLYL) UREA see ENV500
N-ETHYL-N′-(3-METHYL-2-THIAZOLIDINYLIDENE)UREA see ENH000
N-ETHYL-N′-NITRO-N-NITROSOGUANIDINE see ENU000
N-ETHYL-N′-(5-NITRO-2-THIAZOLYL)UREA see ENV500
ETHYL NONANOATE see ENW000
5-ETHYL-2-NONANOL see ENW500
5-ETHYL-3-NONEN-2-ONE see ENX000
ETHYL NONYLATE see ENW000
1-ETHYL-1-NONYLPIPERIDINIUM BROMIDE see ENX100
ETHYLNORADRENALINE HYDROCHLORIDE see ENX500
ETHYL NORADRIANOL see EGE500
α-ETHYLNOREPINEPHRINE HYDROCHLORIDE see ENX500
ETHYL NOREPINEPHRINE HYDROCHLORIDE see ENX500
(−)-N-ETHYL-NORHYOSCINE-METHOBROMIDE see ONI000
N-ETHYLNORPHENYLEPHRINE see EGE500
11-β-ETHYL-19-NOR-17-α-PREGNA-1,3,5(10)-TRIEN-20-YNE-3,17-DIOL see
    EJV400
N-ETHYL-NORSCOPOLAMINEMETHOBROMIDE see ONI000
ETHYLNORSUPRARENIN HYDROCHLORIDE see ENX500
16-β-ETHYL-19-NORTESTOSTERONE see ELF100
N-ETHYLNORTROPINE BENZHYDRYL ETHER see DWE800
ETHYL n-OCTADECANOATE see EPF700
ETHYL OCTADECANOATE see EPF700
1-ETHYL-1-OCTADECYLPIPERIDINIUM BROMIDE see ENX875
1-ETHYL-1,2,3,4,4a,9,10,10a-OCTAHYDRO-7-HYDROXY-2-METHYL-2-PHENAN-
    THRENECARBOXYLIC ACID see DYB000
ETHYL OCTANOATE see ENY000
ETHYL OCTYLATE see ENY000
1-ETHYL-1-OCTYLPIPERIDINIUM BROMIDE see ENY100
ETHYL OENANTHATE see EKN050
ETHYL OENANTHYLATE see EKN050
ETHYLOLAMINE see EEC600
1-(β-ETHYLOL)-2-METHYL-5-NITRO-3-AZAPYRROLE see MMN250
ETHYL ORTHOFORMATE see ENY500
ETHYL ORTHOSILICATE see EPF550
ETHYL OXALATE (DOT) see DJT200
ETHYL OXALATE see DJT200
ETHYL-3-OXATRICYCLO-(3.2.1.0²,⁴)OCTANE-6-CARBOXYLATE see ENZ000
1-ETHYL-3-(2-OXAZOLYL)UREA see EOA000
ETHYLOXIRANE see BOX750
ETHYL β-OXOBENZENEPROPANOATE see EGR600
ETHYL-3-OXOBUTANOATE see EFS000
ETHYL-3-OXOBUTYRATE see EFS000
ETHYL 3-OXOBUTYRATE ETHYLENE KETAL see EFR100
ETHYL 4-OXOPENTANOATE see EFS600
(Z)-2-(3-ETHYL-4-OXO-5-PIPERIDINO-2-THIAZOLIDINYLIDENE) ACETIC ACID see
    EOA500
cis-2-(3-ETHYL-4-OXO-5-PIPERIDINO-2-THIAZOLIDINYLIDENE)ACETIC ACID see
    EOA500
2-(3-ETHYL-4-OXO-5-PIPERIDINO-2-THIAZOLIDINYLIDENE)ACETIC ACID ETHYL
    ESTER see EOB000
(3-ETHYL-4-OXO-5-(1-PIPERIDINYL)-2-THIAZOLIDINYLIDENE)ACETIC ACID ETH-
    YL ESTER see EOA500
ETHYL 4-OXOVALERATE see EFS600
4-ETHYLOXYPHENOL see EFA100
ETHYL PARABEN see HJL000
ETHYL PARAOXON see NIM500
ETHYL PARASEPT see HJL000
S-ETHYL PARATHION see DJT000
ETHYL PARATHION see PAK000
ETHYL PELARGONATE see ENW000
ETHYL PENTABORANE (9) see EOB100
1-ETHYL-1-PENTADECYLPIPERIDINIUM BROMIDE see EOB200
3-ETHYL-2-PENTANOL see EOB300
3-ETHYL-3-PENTANOL see TJP550
5-ETHYL-5-PENTYLBARBITURIC ACID see PBS250
2-((1-ETHYLPENTYL)OXY)ETHANOL see EGJ000
2-(((1-ETHYLPENTYL)OXY)ETHOXY)ETHANOL see EGJ500
ETHYL PERCHLORATE see EOD000
ETHYL PEROXYCARBONATE see DJU600
ETHYLPHENACEMIDE see PFB350
ETHYL PHENACETATE see EOH000
ETHYLPHENACETIN see EOD500
2-ETHYLPHENOL see PGR250

ETHYNYLOESTRADIOL METHYL ETHER see MKB750
17-α-ETHYNYL-1,3,5(10)-OESTRATRIENE-3,17-β-DIOL see EEH500
17-α-ETHYNYLOESTRA-1,3,5(10)-TRIENE-3,17-β-DIOL see EEH500
17-ETHYNYLOESTRA-1,3,5(10)-TRIENE-3,17-β-DIOL see EEH500
17-α-ETHYNYL-1,3,5-OESTRATRIENE-3,17-β-DIOL see EEH500
17-α-ETHYNYLOESTRENOL see NNV000
17-α-ETHYNYLOESTR-4-EN-17-β-OL see NNV000
17-α-ETHYNYLTESTOSTERONE see GEK500
ETHYNYLTESTOSTERONE see GEK500
ETHYNYL VINYL SELENIDE see EQN500
ETHYONOMIDE see EPQ000
ETICOL see NIM500
ETIDRONIC ACID see HKS780
ETIFELMIN HYDROCHLORIDE see DWF200
ETIFOXIN see CGQ500
ETIFOXINE see CGQ500
ETIL ACRILATO (ITALIAN) see EFT000
ETILACRILATULUI (ROMANIAN) see EFT000
ETILAMINA (ITALIAN) see EFU400
ETILBENZENE (ITALIAN) see EGP500
ETILBUTILCHETONE (ITALIAN) see EHA600
ETIL CLOROCARBONATO (ITALIAN) see EHK500
ETIL CLOROFORMIATO (ITALIAN) see EHK500
ETILE (ACETATO di) (ITALIAN) see EFR000
ETILE (FORMIATO di) (ITALIAN) see EKL000
ETILEFRINE see EGE500
ETILEFRINE HYDROCHLORIDE see EGF000
ETILEFRINE PIVALATE HYDROCHLORIDE see EGC500
N,N'-ETILEN-BIS(DITIOCARBAMMATO) di MANGANESE (ITALIAN) see MAS500
N,N'-ETILEN-BIS(DITIOCARBAMMATO) di SODIO (ITALIAN) see DXD200
ETILENE (OSSIDO di) (ITALIAN) see EJN500
ETILENIMINA (ITALIAN) see EJM900
ETILEN-XANTISAN TABL. see TEP500
ETILFEN see EOK000
ETILMERCAPTANO (ITALIAN) see EMB100
O-ETIL-O-((4-NITRO-FENIL)-FENIL)-MONOTIOFOSFONATO (ITALIAN) see EBD700
S-2-ETIL-SULFINIL-1-METIL-ETIL-O,O-DIMETIL-MONOTIOFOSFATO see DSK600
ETIMID see EPQ000
ETIMIDIN see EQI600
ETIN see EDC500
ETINAMATE see EEH000
ETIOCHOLANE-17-β-OL-3-ONE see HJB100
ETIOCHOLAN-17-β-OL-3-ONE see HJB100
ETIOCIDAN see EPQ000
ETIOL see MAK700
ETIONAMID see EPQ000
ETIONIZINA see EPQ000
ETIZOLAM see EQN600
ETM (heterocycle) see EJQ000
ETM see EJQ000
ETMA see EQN700
ETMOZIN see EEI025
ETMT see EFK000
ETO see EJN500
ETOCIL see EEM000
ETOCLOFENE see DGN000
ETODROXIZINE DIMALEATE see HHK100
ETODROXYZINE DIMALEATE see HHK100
ETOFEN see DGN000
ETOFENAMATE see HKK000
ETOFYLLINCLOFIBRAT (GERMAN) see TEQ500
ETOFYLLINE see HLC000
ETOFYLLINE CLOFIBRATE see TEQ500
ETOGLUCID see TJQ333
ETOGYN see MBZ100
ETOKSYETYLOWY ALKOHOL (POLISH) see EES350
ETOMAL see ENG500
ETOMIDATE see HOU100
ETOMIDE HYDROCHLORIDE see CBQ625
ETOMIDOLINE see AHP125
ETONITAZENE HYDROCHLORIDE see EQN750
ETOPERIDONE see EQO000
ETOPOSIDE see EAV500
ETOPROPEZINA see DIR000
7-α-ETORPHINE see EQO450
ETORPHINE see EQO450
(−)-ETORPHINE see EQO450
ETORPHINE HYDROCHLORIDE see EQO500
ETOSALICIL see EEM000
ETOSALICYL see EEM000
2-ETOSSIETIL-ACETATO (ITALIAN) see EES400
ETOSUXIMIDA see ENG500
ETOVAL see BPF500
ETOZOLINE see MNG000

ETP see EPQ000, EQP000
ETs (POLYSACCHARIDE) see EHG100
ETRENOL see HGO500
ETRETIN see REP400
ETRETINATE see EMJ500
ETRIDIAZOLE see EFK000
ETRIOL see HDF300
ETROFLEX see GKK000
ETROFOL see CKF000
ETROFOLAN see MIA250
ETROLENE see RMA500
ETROPRES see RCA200
ETROZOLIDINA see HNI500
ETRUSCOMICINA see LIN000
ETRUSCOMYCIN see LIN000
ETRYPTAMINE ACETATE see AJB250
ETTRIOL see HDF300
ETU see IAQ000
ETYBENZATROPINE see DWE800
ETYDION see TLP750
ETYLENU TLENEK (POLISH) see EJN500
ETYLOAMINA (POLISH) see EFU400
ETYLOBENZEN (POLISH) see EGP500
ETYLON see TCC000
ETYLOWY ALKOHOL (POLISH) see EFU000
ETYLU BROMEK (POLISH) see EGV400
ETYLU CHLOREK (POLISH) see EHH000
ETYLU KRZEMIAN (POLISH) see EPF550
E. TYPHOSA LIPOPOLYSACCHARIDE see EQP100
EU-1806 see NAE100
EU 4200 see TNR485
EUBASIN see PPO000
EUBASINUM see PPO000
EUBINE see DLX400
β-EUCAINE see EQP500
EUCAINE B see EQP500
EUCALMYL see FLU000
EUCALYPTOL (FCC) see CAL000
EUCALYPTOLE see CAL000
EUCALYPTUS CITRIODORA OIL, ACETYLATED see OHC100
EUCALYPTUS OIL see EQQ000
EUCALYPTUS OIL, ACETYLATED see OHC100
EUCALYPTUS REDUNCA TANNIN see MSC000
EUCANINE GB see TGL750
EUCAST see EQQ100
EUCHESSINA see PDO750
EUCHEUMA SPINOSUM GUM see CCL250
EUCHRYSINE see BJF000
EUCISTEN see EID000
EUCISTIN see PDC250
EUCLIDAN see EQQ100
EUCLORINA see CDP000
EUCODAL see DLX400
EUCORAN see DJS200
EUCTAN see TGJ885
EUCUPIN DIHYRDOCHLORIDE see EQQ500
EUDATINE see BEX500
EUDEMINE INJECTION see DCQ700
EUDESMA-3,11(13)-DIEN-12-OIC ACID see EQR000
EUFIN see DIX200
EUFLAVINE see DBX400, XAK000
EUFODRIANL see MDT600
EUGENIA JAMBOS Linn., extract excluding roots see SPF200
EUGENIC ACID see EQR500
EUGENOL see EQR500
1,3,4-EUGENOL ACETATE see EQS000
EUGENOL ACETATE see EQS000
EUGENOL FORMATE see EQS100
1,3,4-EUGENOL METHYL ETHER see AGE250
EUGENOL OXIDE see EBT500
EUGENOL PHENYLACETATE see AGL000
EUGENO-2',3'-OXIDE see EBT500
EUGENYL ACETATE see EQS000
EUGENYL FORMATE see EQS100
EUGENYL METHYL ETHER see AGE250
EUGENYL PHENYLACETATE see AGL000
EUGLUCAN see CEH700
EUGLUCON 5 see CEH700
EUGLUCON see CEH700
EUGLYKON see CEH700
EUGYON see NNL500
EUHAEMON see VSZ000
EUHYPNOS see CFY750
EUKAIN B see EQP500
EUKALYPTUS OEL (GERMAN) see EQQ000

EUKODAL see DLX400
EUKRATON see MKA250
EUKUPIN DIHYDROCHLORIDE see EQQ500
EUKYSTOL see CLY500
EULAVA SM see MAG250
EULAXAN see PPN100
EULICIN see EQS500
EULISSIN A see DAF800
EULIXINE see DAF800
EUMICTON see AAI250
EUMIDRINA see MGR500
EUMIN see MMN250
EUMOTOL see CCD750
EUMOVATE see CMW400
EUMYCETIN see EQT000
EUMYDRIN see MGR500
EUNASIN see BAV000
EUNATROL see OIA000
EUNERPAN see FKI000
EUNOCTAL see AMX750
EUNOCTIN see DLY000
EUONYMUS EUROPAEUS see ERA309
EUPAREN see DFL200
EUPARENE see DFL200
EUPHODRIN see DBA800
EUPHORBIA ABYSSINICA LATEX see EQT500
EUPHORBIA CANARIENSIS LATEX see EQU000
EUPHORBIA CANDELABRIUM LATEX see EQU500
EUPHORBIA ESULA LATEX see EQV000
EUPHORBIA GRANDIDENS LATEX see EQV500
EUPHORBIA LATHYRIS LATEX see EQW000
EUPHORBIA OBOVALIFOLIA LATEX see EQW500
EUPHORBIA POINSETTIS BUIST see EQX000
EUPHORBIA PULCHERRIMA WILLD. see EQX000
EUPHORBIA SERRATA LATEX see EQX500
EUPHORBIA TIRUCALLI LATEX see EQY000
EUPHORIA WULFENII LATEX see EQY500
EUPHORIN see AHK750, CBL750
EUPHOZID see ILE000
EUPHYLLIN see TEP500
EUPHYLLINE see TEP500
EUPLACID see EHP000
EUPNERON see ECU600, EQY600
EUPRACTONE see PMO250
EUPRAMIN see DLH600, DLH630
EUPREX see TLG000
EURAZYL see EQZ000
EURECEPTOR see TAB250
EURECOR see CCK125
EUREKENE see PNX750
EURESOL see RDZ900
EUREX see EHT500
EURINOL see HII500
EUROCERT AMARANTH see FAG020
EUROCERT AZORUBINE see HJF500
EUROCERT COCHINEAL RED A see FMU080
EUROCERT ORANGE FCF see FAG150
EUROCERT SCARLET GN see CMP620
EUROCERT TARTRAZINE see FAG140
EURODIN see CKL250
EURODOPA see DNA200
EUROGALE see MKP500
EUROPEAN HOLLY see HGF100
EUROPEAN MISTLETOE see ERA100
EUROPEAN SPINDLE TREE see ERA309
EUROPEN see MGR500
EUROPHAN see PKQ059
EUROPIC CHLORIDE see ERA500
EUROPIUM CHLORIDE see ERA500
EUROPIUM CITRATE see ERB000
EUROPIUM EDETATE see ERB500
EUROPIUM NITRATE see ERC550
EUROPIUM(III) NITRATE, HEXAHYDRATE (1:3:6) see ERC000
EUROPIUM(II) SULFIDE see ERC500
EUROPIUM TRINITRATE see ERC550
EUROTIN (A) see ERC600
EURPHYLLIN see TEP500
EUSAL see EEM000
EUSAPRIM see SNK000, TKX000
EUSCOPOL see HOT500
EUSMANID see BFW250
EUSPIRAN see IMR000
EUSTIDIL see DFH600
EUSTIGMIN see NCL100
EUSTIGMIN BROMIDE see POD000

EUSTIGMINE see NCL100
EUSTIGMIN METHYLSULFATE see DQY909
EUSTROPHINUM see SMN000
EUTAGEN see DLX400
EUTANOL G see OEW100
EUTENSIN see CHJ750
EUTHATAL see NBU000
EUTHYROX see LFG050
EUTIMOX see PMI250
EUTONYL see MOS250
EUTRIT see XPJ000
EUUFILIN see TEP500
EUVERNIL see SNQ550
EUVESTIN see DKB000
EUVIFOR see NNE400
EVABLIN see BGT250
EVAC-Q-KIT see PDO750
EVAC-Q-KWIK see PDO750
EVAC-U-GEN see PDO750
EVADYNE see BPT750
EVALON see ILD000
EVANOL see SJJ175
EVANS BLUE DYE see BGT250
EVASPIRINE see PGG350
EVASPRINE see PGG350
EVAU-SUPER see SFS000
EVE see EQF500
EVENING TRUMPET FLOWER see YAK100
EVENTIN see PNN300
EVENTIN HYDROCHLORIDE see PNN300
EVERCYN see HHS000
EVERGREEN CASSENA see HGF100
EVERLASTING FLOWER OIL see HAK500
EVERNINOMICIN D see ERC800
EVERNINOMYCIN-B see ERD000
EVEX see ECU750, EDV600
EVION see VSZ450
EVIPAL see ERD500
EVIPAL SODIUM see ERE000
EVIPAN see ERD500
EVIPAN SODIUM see ERE000
EVIPLAST 80 see DVL700
EVIPLAST 81 see DVL700
EVISECT see TFH750
EVISEKT see TFH750
EVITAMINUM see VSZ450
EVOLA see DEP800
EVONOGENIN see DMJ000
EVRONAL see SBM500
EVRONAL SODIUM see SBN000
EVRRONAL see SBN000
EWEISS see BAP000
EX 4355 see DLS600
EX10781 see BAV000
EX 10-781 see MHJ500
EXACTHIN see AES650
EXACT-S see TFP000
EXACYL see AJV500
EXADRIN see VGP000
EXAGAMA see BBQ500
EXAL see VLA000
EXALAMIDE see HFS759
EXAPROLOL HYDROCHLORIDE see ERE100
EXCELSIOR see CBT750
EXD see BJU000
EXDOL see HIM000
EXELMIN see PIJ500
EXHAUST GAS see CBW750
EXHORAN see DXH250
EXHORRAN see DXH250
EXIPROBEN SODIUM see ERE200
EXITELITE see AQF000
EXLAN, combustion products see ADX750
EXLUTEN see NNV000
EXLUTION see NNV000
EXLUTON see NNV000
EXLUTONA see NNV000
EXMIGRA see EDC500
EXMIN see AHJ750
EXNA see BDE250
EXOFENE see HCL000
EXOLAN see APH500
EXON 450 see AAX175
EXON 454 see AAX175
EXON 605 see PKQ059

EXONAL see FLZ050
EXOSALT see BDE250
EXOTHERM see TBQ750
EXOTHERM TERMIL see TBQ750
EXOTHION see EAS000
EXOTOXIN, BACILLUS THURINGIENSIS MORRISONI see BAC140
EXOTOXIN, CLOSTRIDIUM PERFRINGENS see CMY170
EXOTOXIN see BAC125
EXOTOXIN A, PSEUDOMONAS AERUGINOSA see POH650
β-EXOTOXIN (BACILLUS THURINGIENSIS) see BAC125
EXOTOXIN, PSEUDOMONAS AERUGINOSA see POH630
EXOTOXIN, PSEUDOMONAS PSEUDOMALLEI see POH660
EXP 126 see AJU625
EXP 338 see AMQ500
EXP 999 see MQR000
EXP 105-1 see AED250
EXP-105-1 see TJG250
EXPANDEX see DBD700
EXPANSIN see CMV000
EXPERIMENTAL CHEMOTHERAPEUTANT 1,207 see NNH000
EXPERIMENTAL FUNGICIDE 341 see GII000
EXPERIMENTAL FUNGICIDE 5223 see DXX400
EXPERIMENTAL FUNGICIDE 224 (UNION CARBIDE) see ZJA000
EXPERIMENTAL HERBICIDE 2 see DGQ200
EXPERIMENTAL HERBICIDE 732 see BQT750
EXPERIMENTAL INSECTICIDE 711 see IKO000
EXPERIMENTAL INSECTICIDE 4049 see MAK700
EXPERIMENTAL INSECTICIDE 4124 see NFT000
EXPERIMENTAL INSECTICIDE 7744 see CBM750
EXPERIMENTAL INSECTICIDE 12008 see DJN600
EXPERIMENTAL INSECTICIDE 12,880 see DSP400
EXPERIMENTAL INSECTICIDE 52160 see TAL250
EXPERIMENTAL INSECTICIDE S-4087 see CON300
EXPERIMENTAL NEMATOCIDE 18,133 see EPC500
EXPERIMENTAL TICK REPELLENT 3 see AEO750
EXPLOSION ACETYLENE BLACK see CBT750
EXPLOSION BLACK see CBT750
EXPLOSIVE D see ANS500
EXPLOSIVES, HIGH see ERF000
EXPLOSIVES, LOW see ERF500
EXPLOSIVES, PERMITTED see ERG000
EXP. MITICIDE No. 7 see BRP250
EXPONCIT see NNW500
EXPORSAN see DNO800
EXSEL see SBR000
EXSICATED SODIUM SULFITE see SJZ000
EXSICCATED FERROUS SULFATE see FBN100
EXSICCATED FERROUS SULPHATE see FBN100
EXSICCATED SODIUM PHOSPHATE see SJH090
EXTACOL see PGA750
EXT D and C BLUE No. 3 see ERG100
EXT D and C GREEN NO. 1 see NAX500
EXT. D&C ORANGE No. 3 see FAG010
EXT D and C RED NO. 11 see CMM300
EXT. D&C RED No. 15 see FAG018
EXT. D and C RED NO. 7 see SEH475
EXT. D&C YELLOW No. 10 see FAG135
EXT D&C YELLOW No. 1 see MDM775
EXT. D&C YELLOW No. 9 see FAG130
EXTENCILLINE see BFC750
EXTENDED ZINC INSULIN SUSPENSION see LEK000
EXTENICILLINE see BFC750
EXTERMATHION see MAK700
EXTERNAL BLUE 1 see BJI250
EXTHRIN see AFR250
EXTRACT D&C ORANGE No. 4 see TGW000
EXTRACT D&C RED No. 14 see XRA000
EXTRACT D&C RED No. 10 see HJF500
EXTRACT of JAMAICA GINGER see JBA000
EXTRACTS (PETROLEUM), HEAVY NAPHTHENIC DISTILLATE SOLVENT (9CI) see MQV857
EXTRACTS (PETROLEUM), HEAVY PARAFFINIC DISTILLATE SOLVENT (9CI) see MQV859
EXTRACTS (PETROLEUM), LIGHT NAPHTHENIC DISTILLATE SOLVENT (9CI) see MQV860
EXTRACTS (PETROLEUM), LIGHT PARAFFINIC DISTILLATE SOLVENT (9CI) see MQV862
EXTRACTS (PETROLEUM), RESIDUAL OIL SOLVENT (9CI) see MQV863
EXTRA FINE 200 SALT see SFT000
EXTRA FINE 325 SALT see SFT000
EXTRAMYCIN see APY500
EXTRANASE see BMO000
EXTRA-PLEX see DLB400
EXTRAR see DUS700
EXTRAX see RNZ000

EXTREMA see AQF000, EPF550, GDY000, SCQ500
EXTREN see ADA725
EXTREX P 60 see PJY100
EXTRINSIC FACTOR see VSZ000
EXURATE see DDP200
EYE BRIGHT see CCJ825
EYE-CORT see HHQ800
EYEULES see ARR000
E-Z-OFF see MAE000
E-Z-OFF D see BSH250
E-Z-PAQUE see BAP000

F 1 see DEA100, FAB000, WCJ000
F 3 see KDK000
F 6 see NCP875
11 F see KDK000
F 12 see DFA600
12 F see KDK000
F 13 see CLR250
F 14 see CBY250
16 F see ARM266
F 22 see CFX500
F-33 see FLJ000
F 44 see SCK600
F-112 see TBP050
F 114 see FOO509
F-115 see CJI500
F 125 see SCK600
F-139 see BPG000
F-150 see FPI000
F 156 see HEL500
190 F see ABX500
F 190 see ABX500
F-400 see DAB840
F461 see DLV200
F 735 see CCC500
F 849 see AKR500
F 933 see BCI500
F 1162 see SNM500
1167 F see DJV300
1262 F see BGO000
F 1262 see BGO000
1358F see SOA500
F 1358 see SOA500
1399 F see SNY500
2249F see FMX000
F 2387 see DVX600
F 2559 see PDD300
F 2966 see DXI400
Fi 5853 see APP500
F 6066 see FBP100
F 7771 see HDW100
23F203 see PJS750
F 1 (antioxidant) see FAB000
F 1 (complexon) see EIX500
F 10 (pesticide) see MAS500
FA see FMV000, FPQ900
2-FAA see FDR000
2,7-FAA see BGP250
Fa 100 see EQR500
F-diAA see DBF200
FAA see FDR000, FFF000, FIC000
FABANTOL see PMM000
FABIANOL see APT000
FABRITONE PE see PJS750
FABT (CZECH) see ALV500
FAC 20 see IOT000
FAC 5273 see PIX250
FA-Ca see FQR100
FAC see FAS700, IOT000
FACCLA see PKU600
FACCULA see PKU600
FACTITIOUS AIR see NGU000
FACTOR II (VITAMIN) see VSZ000
FACTOR PP see NCR000
FACTOR S see VSU100
FACTOR S (vitamin) see VSU100
FAD see RIF100
FADORMIR see QAK000
FAECLA see PKU600
FAECULA see PKU600
FAFT see FQQ400
FAGINE see CMF800
FAIR 30 see DMC600

FAIR 85 see DAI800, ODE000
FAIR PS see DMC600
FAIRY BELLS see FOM100
FAIRY CAP see FOM100
FAIRY GLOVE see FOM100
FAIRY LILY see RBA500
FAIRY THIMBLES see FOM100
FALAPEN see BFD000
FALICAIN see PNB250
FALICAINE HYDROCHLORIDE see PNB250
FALIGRUEN see CNK559
FALISAN see MEP250
FALITHION see DSQ000
FALITIRAM see TFS350
FALKITOL see HCI000
FALL see SFS000
FALL CROCUS see CNX800
FALSE ACACIA see GJU475
FALSE BITTERSWEET see AHJ875
FALSE HELLEBORE see FAB100
FALSE KOA (HAWAII) see LED500
FALSE MISTLETOE see MQW525
FALSE PARSLEY see FMU200
FALSE SYCAMORE see CDM325
FAM see MME809
FAMFOS see FAB400, FAB600
FAMID see DVS000
FAMODIL see FAB500
FA MONOMER see FPQ900
FAMOPHOS see FAB600
FAMOPHOS WARBEX see FAB600
FAMOTIDINE see FAB500
FAMPHOS see FAB600
FAMPHUR see FAB600
FANAL PINK B see RGW000
FANAL PINK GFK see RGW000
FANAL RED 25532 see RGW000
FANASIL see AIE500
FANFOS see FAB600
FANFT see NGM500
FANNOFORM see FMV000
FANODORMO see TDA500
FANTERRIN see HOH500
F 1 (ANTIOXIDANT) see DEA100
FANZIL see AIE500
FAP see FPT100
FARBRUSS see CBT750
FAREDINA see TEY000
FARENAL see TMJ150
FARGAN see DQA400, PMI750
FAR-GO see DNS600
FARIAL see IBR200
FARINEX 100 see SLJ500
FARINGOSEPT see AHI875
FARLUTIN see MCA000
FARMA 939 see BAT000
FARMA see BAT000
FARMAMID see SNQ000
FARMCO see DAA800
FARMCO ATRAZINE see ARQ725
FARMCO DIURON see DXQ500
FARMCO FENCE RIDER see TAA100
FARMCO PROPANIL see DGI000
FARMICETINA see CDP250
FARMIGLUCIN see APP500
FARMIN 80 see OBC000
FARMINOSIDIN see APP500
FARMISERINE see CQH000
FARMITALIA 204/122 see MFN500
FARMORUBICIN see EBB100
FARMOTAL see PBT250, PBT500
FARNESOL see FAB800
FARNESYL ALCOHOL see FAB800
FAROLITO (CUBA) see JBS100
FARTOX see PAX000
FAS see FAF000
FAS-CILE see MQU750
FASCIOLIN see CBY000, HCI000
FASCO-TERPENE see CDV100
FASCO WY-HOE see CKC000
FASERTON see AHE250
FASIGIN see TGD250
FASIGYN see TGD250
FASINEX see CFL200
FAST ACID BLUE RL see ADE750

FAST ACID LIGHT BLUE BRL see CMM100
FAST ACID MAGENTA see CMS231
FAST ACID MAGENTA B see CMS231
FASTBALLS see BBK500
FAST BENZIDENE ORANGE YB 3 see CMS145
FAST BLUE B BASE see DCJ200
FAST BLUE R SALT see PFU500
FAST BROWN SALT RR see MIH500
FAST CHROME CYANINE 6B see CMP880
FAST CHROME CYANINE G see CMP880
FAST CORINTH BASE B see BBX000
FAST CRIMSON GR see CMM300
FAST DARK BLUE BASE R see TGJ750
FAST DISPERSE YELLOW 2K see DUW500
FAST DRIMSON GR see CMM300
FAST GARNET GBC BASE see AIC250
FAST GREEN see AFG500
FAST GREEN FCF see FAG000
FAST GREEN JJO see BAY750
FAST LIGHT YELLOW E see SGP500
FASTOGEN BLUE FP-3100 see DNE400
FASTOGEN BLUE SH-100 see DNE400
FASTOGEN GREEN Y see CMS140
FASTOGEN GREEN 2YK see CMS140
FAST OIL ORANGE see PEJ500
FAST OIL ORANGE II see XRA000
FAST OIL ORANGE T see CMP600
FAST OIL RED B see SBC500
FAST OIL SCARLET III see OHA000
FAST OIL YELLOW see AIC250
FAST OIL YELLOW B see DOT300
FAST OIL YELLOW 2G see CMP600
FAST OIL YELLOW G see CMP600
FASTOLITE RED 8BL see CMO885
FASTONA ORANGE G see CMS145
FASTONA RED B see MMP100
FASTONA SCARLET RL see MMP100
FASTONA SCARLET YS see MMP100
FAST ORANGE see PEJ500
FAST ORANGE BASE GR see NEO000
FAST ORANGE BASE JR see NEO000
FAST ORANGE G see CMS145
FAST ORANGE GC BASE see CEH675
FAST ORANGE GR BASE see NEO000
FAST ORANGE O BASE see NEO000
FAST ORANGE RD OIL see CEG800
FAST ORANGE RD SALT see CEG800
FAST ORANGE R SALT see NEN500
FAST ORANGE SALT RD see CEG800
FAST ORANGE SALT RDA see CEG800
FAST ORANGE SALT RDN see CEG800
FAST PURPURINE see CMO880
FAST RED A see MMP100
FAST RED AB see MMP100
FAST RED A (PIGMENT) see MMP100
FAST RED B see NEQ000
FAST RED BASE B see NEQ000
FAST RED BASE GG see NEO500
FAST RED BASE 3GL SPECIAL see KDA050
FAST RED BASE 2J see NEO500
FAST RED BASE 3JL see KDA050
FAST RED BASE TR see CLK220
FAST RED BB see SBC500
FAST RED B BASE see NEQ000
FAST RED BB BASE see AOX250
FAST RED 5CT BASE see CLK220
FAST RED 5CT SALT see CLK235
FAST RED F see CMO870
FAST RED 2G BASE see NEO500
FAST RED GG BASE see NEO500
FAST RED 3GL BASE see KDA050
FAST RED 3GL SPECIAL BASE see KDA050
FAST RED J see MMP100
FAST RED JE see MMP100
FAST RED KB AMINE see CLK225
FAST RED KB BASE see CLK225
FAST RED KB SALT see CLK225
FAST RED KB SALT SUPRA see CLK225
FAST RED KBS SALT see CLK225
FAST RED MP BASE see NEO500
FAST RED 5NA BASE see NEQ000
FAST RED 2NC BASE see KDA050
FAST RED P BASE see NEO500
FAST RED R see MMP100, OHA000
FAST RED SALT TR see CLK235

FAST RED SALT TRA see CLK235
FAST RED SALT TRN see CLK235
FAST RED SG BASE see NMP500
FAST RED SGG BASE see DEO295
FAST RED TR11 see CLK220
FAST RED TR see CLK220
FAST RED TR BASE see CLK220
FAST RED TRO BASE see CLK220
FAST RED TR SALT see CLK235
FAST SCARLET BASE B see NBE500
FAST SCARLET BASE G see NMP500
FAST SCARLET BASE J see NMP500
FAST SCARLET G see NMP500
FAST SCARLET G BASE see NMP500
FAST SCARLET GC BASE see NMP500
FAST SCARLET G SALT see NMP500
FAST SCARLET J SALT see NMP500
FAST SCARLET M4NT BASE see NMP500
FAST SCARLET R see NEQ500
FAST SCARLET S see CMM320
FAST SCARLET T BASE see NMP500
FAST SCARLET TR BASE see CLK200
FAST SILK YELLOW SH see CMM759
FAST SPIRIT YELLOW AAB see PEI000
FAST SULON BLACK BN see FAB830
FASTUM see BDU500
FASTUSOL BLUE 9GLP see CMO650
FASTUSOL RED 4BA-CF see CMO885
FAST WHITE see LDY000
FAST WOOL BLUE R see ADE750
FAST YELLOW AB see CMM758
FAST YELLOW AT see AIC250
FAST YELLOW B see AIC250
FAST YELLOW EXTRA SPECIALLY PURE see CMM758
FAST YELLOW GC BASE see CEH670
FAST YELLOW S see CMM758
FAST YELLOW S EXTRA SPECIALLY PURE see CMM758
FAST YELLOW Y see CMM758
FATAL see TBV250
FAT BROWN 2G see NBG500
FAT BROWN 2R see NBG500
FAT BROWN RR see NBG500
FATOLIAMID see EPQ000
FAT ORANGE A see CMP600
FAT ORANGE G see CMP600
FAT ORANGE GS see CMP600
FAT ORANGE II see TGW000
FAT ORANGE RG see CMP600
FAT PONCEAU R see SBC500
FAT RED 2B see SBC500
FAT RED 7B see EOJ500
FAT RED B see SBC500
FAT RED BB see SBC500
FAT RED BG see CMS242
FAT RED (BLUISH) see OHA000
FAT RED BS see SBC500
FAT RED G see CMS242, OHA000
FAT RED HRR see OHA000
FAT RED R see OHA000
FAT RED RS see CMS242, OHA000
FAT RED TS see SBC500
FAT RED (YELLOWISH) see XRA000
FAT SCARLET 2G see XRA000
FAT SCARLET LB see OHA000
FATSCO ANT POISON see ARD750
FAT SOLUBLE GREEN ANTHRAQUINONE see BLK000
FAT SOLUBLE RED S see CMS242
FAT SOLUBLE RED ZH see OHA000
FATTY ACID, LANOLIN, ISO-PR ESTERS see IPS500
FATTY ACID, TALL OIL, EPOXIDIZED, OCTYL ESTER see FAB920
FAT VICTORIA YELLOW D see CMP600
FAT YELLOW see DOT300
FAUSTAN see DCK759
FAVISTAN see MCO500
FB/2 see DWX800
F-13B1 see TJY100
FB 217 see PJS750
F-114B2 see FOO525
FB 5097 see MRX500
FBA 52 see DPJ800
FBA 1420 see PMM000
FBA 1500 see MCA100
FBA 4503 see PMX250
FBC CMPP see CIR500
FBHC see BBQ750

FC-1 see CMB125
5-FC see FHI000
FC 12 see DFA600
FC 14 see CBY250
FC-21 see DFL000
FC 22 see CFX500
FC 31 see CHI900
FC 43 see HAS000
FC 47 see HAS000
FC 112 see TBP050
FC 113 see TJE100
FC 114 see FOO509
FC 123 see TJY500
FC133a see TJE100
FC 133a see TJY175
FC142b see CFX250
FC143a see TJY900
FC-143 see ANP625
FC 152a see ELN500
FC 402 see FAC050
FC 403 see XSS375
FC-410 see FAC060
FC 448 see MMA525
FC 455 see DHM309
FC 457 see FAC100
FC 4/58 see CBA375
FC 480 see FAC130
FC 590 see FAC150
FC 591 see FAC155
FC 642 see FAC157
FC 646 see FAC160
FC 650 see FAC163
FC 651 see FAC165
FC 652 see FAC166
FC 657 see FAC170
FC 659 see FAC175
FC 660 see FAC179
FC 668 see FAC185
FC 676 see DHI875
FC 681 see FAC195
FC-1318 see OBO000
FC 3001 see SOX400
FC 4648 see PKQ059
FC 114B2 see FOO525
FC-C 318 see CPS000
FCDR see FHO000
FC-MY 5450 see SMQ500
FCNU see CPL750
F-COL see FHH100
F-CORTEF see FHH100
FCdR see FHO000
FD-1 see BLH325
12FD see KDK000
FDA 0101 see SHF500
FDA 0109 see BHT250
FDA 0121 see ALL250
FDA 0345 see BIF750
FDA 1446 see AFR250
FDA 1541 see EIN500
FDA 1725 see TNX400
FDA 1902 see MCA100
FDA see DGB600
2 FDBP see MQY400
FD&C ACID RED 32 see FAG080
FD&C BLUE No. 1 see FAE000
FD&C BLUE No. 2 see FAE100
FDC GREEN 1 see FAE950
FD&C GREEN No. 1 see FAE950
FD&C GREEN No. 2 see FAF000
FD&C GREEN No. 3 see FAG050
FD&C GREEN No. 2-ALUMINUM LAKE see FAF000
FD and C NO. 6 see FAG150
FD & C NO. 6 see FAG150
FDC ORANGE I see FAG010
FD&C ORANGE No. 1 see FAG010
FD&C RED No. 1 see FAG018
FD&C RED No. 2 see FAG020
FD&C RED No. 3 see FAG040
FD&C RED No. 4 see FAG050
FD&C RED No. 19 see FAG070
FD&C RED No. 32 see FAG080
FD&C RED No. 2-ALUMINIUM LAKE see FAG020
FD & C RED No. 4-ALUMINIUM LAKE see FAG050
FD&C VIOLET No. 1 see FAG120
FD and C YELLOW 6 see FAG150

FD and C YELLOW LAKE NO. 6 see FAG150
FD and C YELLOW No. 10 see CMM510
FD&C YELLOW No. 3 see FAG130
FD&C YELLOW No. 4 see FAG135
FD&C YELLOW No. 5 see FAG140
FD and C YELLOW NO. 6 see FAG150
FDC YELLOW NO. 6 see FAG150
FD&C YELLOW No. 6 see FAG150
FD & C YELLOW NO. 6 ALUMINIUM LAKE see FAG150
FD & C YELLOW NO. 5 TARTRAZINE see FAG140
Fe-DEXTRAN see IGS000
FDN see DRP800
F3DThd see TKH325
FDUR see DAR400
FEBANTEL see BKN250
FEBRILIX see HIM000
FEBRININA see DOT000
FEBRO-GESIC see HIM000
FEBROLIN see HIM000
FEBRON see DOT000
FEBRUARY DAPHNE see LAR500
FEBUPROL see BPP250
FECAMA see DGP900
FECTO see CBT750
FECTRIM see SNK000, TKX000
FEDACIN see NGE500
FEDAL-UN see PCF275
FEDRIN see EAW000
FEEN-A-MINT GUM see PDO750
FEENO see PDP250
FEGLOX see DWX800
FEIDMIN 5 see TES800
FEINALMIN see DLH630
FEKABIT see PLA250
FELACRINOS see DAL000
FELAN see EKO500
FELAZINE see PFC750
FELBEN see BAU750
FELDENE see FAJ100
FELICAIN (GERMAN) see PNB250
FELICUR see EGQ000
FELISON see DAB800
FELITROPE see EGQ000
FELIXYN see CKE750
α-FELLANDRENE see MCC000
FELLING ZINC OXIDE see ZKA000
FELLOZINE see PMI750
FELMANE see FMQ000
FELONWORT see CCS650
FELSULES see CDO000
FELUREA see PEC250
FELVITEN see AOO490
FEMA 3174 see HKC575
FEMA 3186 see MJH905
FEMA 3230 see POM000
FEMA 3251 see PPH200
FEMACOID see ECU750
FEMADOL see PNA500
FEMALE WATER DRAGON see WAT300
FEMAMIDE see FAJ150
FEMA No. 2003 see AAG250
FEMA No. 2005 see MDW750
FEMA No. 2006 see AAT250
FEMA No. 2007 see THM500
FEMA No. 2008 see ABB500
FEMA No. 2009 see ABH000
FEMA No. 2011 see AEN250
FEMA No. 2026 see AGC500
FEMA No. 2031 see AGH250
FEMA No. 2032 see AGA500
FEMA No. 2033 see AGI500
FEMA No. 2034 see AGJ250
FEMA No. 2037 see AGM500
FEMA No. 2045 see ISV000
FEMA No. 2055 see IHO850
FEMA No. 2058 see IHP100
FEMA No. 2060 see IHP400
FEMA No. 2061 see AOG500
FEMA No. 2063 see AOG600
FEMA No. 2069 see IHS000
FEMA No. 2075 see IHU100
FEMA No. 2082 see AON350
FEMA No. 2085 see ITB000
FEMA No. 2086 see PMQ750
FEMA No. 2097 see AOX750

FEMA No. 2098 see AOY400
FEMA No. 2099 see MED500
FEMA No. 2109 see ARN000
FEMA No. 2127 see BAY500
FEMA No. 2134 see BCS250
FEMA No. 2135 see BDX000
FEMA No. 2137 see BDX500
FEMA No. 2138 see BCM000
FEMA No. 2140 see BED000
FEMA No. 2141 see IJV000
FEMA No. 2142 see BEG750
FEMA No. 2149 see BFD400
FEMA No. 2151 see BFJ750
FEMA No. 2152 see ISW000
FEMA No. 2159 see BMD100
FEMA No. 2160 see IHX600
FEMA No. 2170 see MKA400
FEMA No. 2174 see BPU750
FEMA No. 2175 see IIJ000
FEMA No. 2178 see BPW500
FEMA No. 2179 see IIL000
FEMA No. 2183 see BQI000
FEMA No. 2184 see BFW750
FEMA No. 2186 see BQM500
FEMA No. 2187 see BSW500
FEMA No. 2188 see BRQ350
FEMA No. 2190 see BQP000
FEMA No. 2193 see IIQ000
FEMA No. 2203 see DTC800
FEMA No. 2209 see BBA000
FEMA No. 2210 see IJF400
FEMA No. 2213 see IJN000
FEMA No. 2218 see ISX000
FEMA No. 2219 see BSU250
FEMA No. 2220 see IJS000
FEMA No. 2221 see BSW000
FEMA No. 2222 see IJU000
FEMA No. 2223 see TIG750
FEMA No. 2224 see CAK500
FEMA No. 2229 see CBA500
FEMA No. 2245 see CCM000
FEMA No. 2249 see CCM100, CCM120
FEMA No. 2252 see CCN000
FEMA No. 2286 see CMP969
FEMA No. 2288 see CMP975
FEMA No. 2293 see CMQ730
FEMA No. 2294 see CMQ740
FEMA No. 2295 see API750
FEMA No. 2299 see CMR500
FEMA No. 2301 see CMR850
FEMA No. 2302 see CMR800
FEMA No. 2306 see CMS750
FEMA No. 2307 see CMS845
FEMA No. 2309 see CMT250
FEMA No. 2311 see AAU000
FEMA No. 2312 see CMT600
FEMA No. 2313 see CMT900
FEMA No. 2314 see CMT750
FEMA No. 2316 see CMU100
FEMA No. 2341 see COE500
FEMA No. 2356 see CQI000
FEMA No. 2361 see DAF200
FEMA No. 2362 see DAG000, DAG200
FEMA No. 2365 see DAI600
FEMA No. 2366 see DAI350
FEMA No. 2370 see BOT500
FEMA No. 2371 see BEO250
FEMA No. 2375 see EMA500
FEMA No. 2376 see DJY600
FEMA No. 2377 see SNB000
FEMA No. 2379 see DKV150
FEMA No. 2381 see HHR500
FEMA No. 2391 see DTE600
FEMA No. 2392 see DQQ375
FEMA No. 2393 see DQQ200
FEMA No. 2394 see DQQ380
FEMA No. 2401 see DXS700
FEMA No. 2414 see EFR000
FEMA No. 2415 see EFS000
FEMA No. 2418 see EFT000
FEMA No. 2420 see AOV000
FEMA No. 2421 see EGM000
FEMA No. 2422 see EGR000
FEMA No. 2426 see DHI000
FEMA No. 2427 see EHE000

FEMA No. 2428 see ELS000
FEMA No. 2429 see DHI400
FEMA No. 2430 see EHN000
FEMA No. 2432 see EHE500
FEMA No. 2433 see EJN500
FEMA No. 2434 see EKL000
FEMA No. 2437 see EKN050
FEMA No. 2439 see EHF000
FEMA No. 2440 see LAJ000
FEMA No. 2441 see ELY700
FEMA No. 2443 see EMP600
FEMA No. 2444 see ENC000
FEMA No. 2445 see ENL850
FEMA No. 2447 see ENW000
FEMA No. 2449 see ENY000
FEMA No. 2452 see EOH000
FEMA No. 2454 see EOK600
FEMA No. 2456 see EPB500
FEMA No. 2458 see SAL000
FEMA No. 2463 see ISY000
FEMA No. 2464 see EQF000
FEMA No. 2465 see CAL000
FEMA No. 2467 see EQR500
FEMA No. 2468 see IKQ000
FEMA No. 2469 see EQS000
FEMA No. 2470 see AAX750
FEMA No. 2475 see AGE250
FEMA No. 2476 see IKR000
FEMA No. 2478 see FAB800
FEMA No. 2489 see FPQ875
FEMA No. 2497 see DSD775, FQT000
FEMA No. 2507 see DTD000
FEMA No. 2509 see DTD800
FEMA No. 2511 see GDE800
FEMA No. 2512 see GDE825
FEMA No. 2514 see GCY000
FEMA No. 2516 see GDM400
FEMA No. 2517 see GDM450
FEMA No. 2539 see HBA550
FEMA No. 2540 see HBB500
FEMA No. 2544 see MGN500
FEMA No. 2545 see EHA600
FEMA No. 2548 see HBL500
FEMA No. 2557 see HEM000
FEMA No. 2559 see HEU000
FEMA No. 2560 see HFA525
FEMA No. 2562 see HFD500
FEMA No. 2563 see HFE000
FEMA No. 2565 see HFI500
FEMA No. 2567 see HFJ500
FEMA No. 2569 see HFO500
FEMA No. 2583 see CMS850
FEMA No. 2585 see HJV700
FEMA No. 2588 see RBU000
FEMA No. 2593 see ICM000
FEMA No. 2594 see IFW000
FEMA No. 2595 see IFX000
FEMA No. 2615 see DXT000
FEMA No. 2617 see DXV600
FEMA No. 2633 see LFU000
FEMA No. 2635 see LFX000
FEMA No. 2636 see LFY100
FEMA No. 2638 see LFZ000
FEMA No. 2640 see LGB000
FEMA No. 2642 see LGA050
FEMA No. 2645 see LGC100
FEMA No. 2665 see MCF750, MCG000, MCG250
FEMA No. 2667 see MCE250, MCG275
FEMA No. 2668 see MCG500, MCG750
FEMA No. 2670 see AOT530
FEMA No. 2672 see MFF580
FEMA No. 2677 see MFW250
FEMA No. 2681 see MGP000
FEMA No. 2682 see APJ250
FEMA No. 2683 see MHA750
FEMA No. 2684 see MNT075
FEMA No. 2685 see PDE000
FEMA No. 2690 see MIP750
FEMA No. 2697 see MIO000
FEMA No. 2698 see MIO500
FEMA No. 2700 see HMB500
FEMA No. 2707 see MKK000
FEMA No. 2718 see MGQ250
FEMA No. 2719 see MLL600
FEMA No. 2723 see ABC500

FEMA No. 2729 see MND275
FEMA No. 2731 see HFG500
FEMA No. 2733 see MHA500
FEMA No. 2743 see COU500
FEMA No. 2745 see MPI000
FEMA No. 2749 see MQI550
FEMA No. 2762 see MRZ150
FEMA No. 2770 see DTD200
FEMA No. 2780 see NMV780
FEMA No. 2781 see CNF250
FEMA No. 2782 see NMW500
FEMA No. 2788 see NNB400
FEMA No. 2789 see NNB500
FEMA No. 2797 see OCO000
FEMA No. 2798 see OCE000
FEMA No. 2800 see OEI000
FEMA No. 2802 see ODG000
FEMA No. 2806 see OEG000
FEMA No. 2809 see OEY100
FEMA No. 2841 see ABX750
FEMA No. 2842 see PBN250
FEMA No. 2856 see MCC000
FEMA No. 2857 see PFB250
FEMA No. 2858 see PDD750
FEMA No. 2862 see PDF750
FEMA No. 2866 see PDI000
FEMA No. 2868 see PDK200
FEMA No. 2871 see PDF775
FEMA No. 2873 see PDS900
FEMA No. 2874 see BBL500
FEMA No. 2876 see PDX000
FEMA No. 2878 see PDY850
FEMA No. 2885 see HHP050
FEMA No. 2886 see COF000
FEMA No. 2887 see HHP000
FEMA No. 2890 see HHP500
FEMA No. 2902 see PIH250
FEMA No. 2903 see POH750
FEMA No. 2911 see PIW250
FEMA No. 2922 see IRY000
FEMA No. 2923 see PMT750
FEMA No. 2926 see INE100
FEMA No. 2930 see PNE250
FEMA No. 2962 see MCE750
FEMA No. 2980 see CMT250
FEMA No. 2981 see DTF400, RHA000
FEMA No. 3006 see OHG000
FEMA No. 3007 see SAU400
FEMA No. 3045 see TBD500
FEMA No. 3047 see TBE250
FEMA No. 3053 see TBE600
FEMA No. 3060 see TCU600
FEMA No. 3073 see MNR250
FEMA No. 3075 see THA250
FEMA No. 3082 see TJJ400
FEMA No. 3091 see UJA800
FEMA No. 3092 see UJJ000
FEMA No. 3095 see ULJ000
FEMA No. 3097 see UNA000
FEMA No. 3101 see VAQ000
FEMA No. 3102 see ISU000
FEMA No. 3103 see VAV000
FEMA No. 3107 see VFK000
FEMA No. 3126 see ADA350
FEMA No. 3135 see DAE450
FEMA No. 3149 see EIL100
FEMA No. 3155 see ENF200
FEMA No. 3164 see HAV450
FEMA No. 3183 see MEX350
FEMA No. 3213 see NNA300
FEMA No. 3237 see TDV725
FEMA No. 3244 see TME270
FEMA No. 3264 see DAI360
FEMA No. 3271 see DTU400
FEMA No. 3272 see DTU600
FEMA No. 3273 see DTU800
FEMA No. 3289 see HBI800
FEMA No. 3291 see BOV000
FEMA No. 3302 see MFN285
FEMA No. 3309 see MOW750
FEMA No. 3317 see NMV760
FEMA No. 3326 see ABC750
FEMA No. 3354 see HFM600
FEMA No. 3386 see PPS250
FEMA No. 3497 see HFE550

FEMA No. 3498 see ISZ000
FEMA No. 3499 see HFR200
FEMA No. 3558 see MLA250
FEMA No. 3559 see MCB750
FEMA No. 3565 see DKV175
FEMA No. 3581 see OCY100
FEMA No. 3583 see OEG100
FEMA No. 3632 see PDF790
FEMANTHREN GOLDEN YELLOW see DCZ000
FEMERGIN see EDC500
FEMEST see ECU750
FEMESTRAL see EDO000
FEMESTRONE see EDP000
FEMESTRONE INJECTION see EDV000
FEM H see ECU750
FEMIDYN see EDV000
FEMMA see ABU500
FEMOGEN see ECU750, EDO000
FEMOGEX see EDS100
FEMOXETINE HYDROCHLORIDE see FAJ200
FEMPROPAZINE see DIR000
FEMULEN see EQJ500
FENAB see TIY500
FENAC see TIY500
FENACAINE see BJO500
FENACEMID see PEC250
FENACEMIDE see PEC250
FENACET AST BLUE FF see MGG250
FENACET BLUE G see TBG700
FENACETEAMIDE see PEC250
FENACET FAST PINK B see AKE250
FENACET FAST PINK 3BE see DBX000
FENACET FAST PINK RF see AKO350
FENACET FAST RUBINE B see CMP080
FENACET FAST TURQUOISE B see DMM400
FENACET FAST VIOLET 6B see AKP250
FENACET FAST VIOLET B see DBY700
FENACET FAST VIOLET 5R see DBP000
FENACET FAST YELLOW G see AAQ250
FENACET FAST YELLOW 2R see DUW500
FENACETIL-KARBAMIDE see PEC250
FENACETINA see ABG750
FENACET YELLOW G see AAQ250
FENACILIN see PDT500
FENACTIL see CKP250
FENADONE see MDP750
FENAFOR RED PB see CMM330
FENAFOR RED PG see CMM320
FENAGLICODOLO see CKE750
FENAKROM RED W see SEH475
FENAKROM YELLOW R see SIT850
FENAKRON BLUE BLACK EB see CMP880
FENAKTYL see CKP250
FENALAC RED FKB EXTRA see NAY000
FENALAMIDE see FAJ150
FENALAN BLUE B see CMM070
FENALAN YELLOW E see SGP500
FEN-ALL see TIK500
FENALLYMAL see AGQ875
FENALUZ RED 4B see CMO885
FENAM see DRP800
FENAMIC ACID see PEG500
FENAMIDE see FAJ150
FENAMIN see ARQ725
FENAMIN BLACK E see AQP000
FENAMIN BLACK GR see CMN240
FENAMIN BLUE 2B see CMO000
FENAMIN BLUE RW see CMO600
FENAMIN BROWN PBL see CMO820
FENAMIN DIAZO BLACK D see CMN230
FENAMINE see AMY050, ARQ725
FENAMIN FAST RED F see CMO870
FENAMIN GREEN A see CMO840
FENAMIN GREEN B see CMO840
FENAMIN GREEN G see CMO840
FENAMIN GREEN M see CMO830
FENAMIN NAVY BLUE H see CMN800
FENAMINOSULF see DOU600
FENAMIN SCARLET 3B see CMO875
FENAMIN SKY BLUE see CMO500
FENAMIPHOS see FAK000
FENAMIZOL HYDROCHLORIDE see DCA600
FENAN BLUE BCS see DFN425
FENAN BLUE RSN see IBV050
FENANTHREN BLUE BC see DFN425

FENANTHREN BLUE BD see DFN425
FENANTHREN BLUE RS see IBV050
FENANTHREN BRILLIANT ORANGE GR see CMU820
FENANTHREN BRILLIANT VIOLET 2R see DFN450
FENANTHREN BRILLIANT VIOLET 4R see DFN450
FENANTHREN BROWN BR see CMU770
FENANTHRENE RED BROWN 5R see CMU800
FENANTHREN OLIVE R see DUP100
FENANTHREN RUBINE R see CMU825
FENANTOIN see DKQ000, DNU000
FENARIMOL see FAK100
FENAROL see CJR909, CKF500
FENARSONE see CBJ000
FENARTIL see BRF500
FENASAL see DFV400
FENASPARATE see SAN600
FENATE see IGS000
FENATROL see ARQ725, TIY500
FENAVAR see AMY050
FENAZAFLOR see DGA200
FENAZEPAM see PDB350
FENAZIL see DQA400, PMI750
FENAZO BLUE BLACK see FAB830
FENAZO BLUE SR see ADE750
FENAZO BLUE XF see ADE500
FENAZO BLUEXG see ERG100
FENAZO BLUE XI see FAE000
FENAZO BLUE XR see FMU059
FENAZO BLUE XV see ADE500
FENAZO EOSINE XG see BNH500, BNK700
FENAZO GREEN 7G see FAF000
FENAZO GREEN L see FAE950
FENAZO LIGHT BLUE AC see CMM090
FENAZO LIGHT BLUE RA see CMM080
FENAZO ORANGE see CMM220
FENAZO RED B see CMM300
FENAZO RED C see HJF500
FENAZO RED FG see CMM325
FENAZO SCARLET 2R see FMU070
FENAZO SCARLET 3R see FMU080
FENAZOXINE see FAL000
FENAZOXINE HYDROCHLORIDE see NBS500
FENAZO YELLOW M see MDM775
FENAZO YELLOW T see FAG140
FENBITAL see EOK000
FENBUFEN see BGL250
FENBUTATIN OXIDE see BLU000
FENCAL see ARB750
FENCAMFAMINE HYDROCHLORIDE see EOM000
FENCAMINE HYDROCHLORIDE see FAM000
FENCARBAMIDE see PDC850
FENCARBAMIDE HYDROCHLORIDE see PDC875
FENCAROL see DWM400
FENCHEL OEL (GERMAN) see FAP000
FENCHLOORFOS (DUTCH) see RMA500
FENCHLORFOS see RMA500
FENCHLORFOSU (POLISH) see RMA500
FENCHLOROPHOS see RMA500
FENCHLORPHOS see RMA500
α-FENCHOL see TLW000
endo-FENCHOL see TLW000
FENCHON (GERMAN) see TLW250
d(+)-FENCHONE see FAM300
d-FENCHONE see FAM300
(+)-FENCHONE see FAM300
FENCHONE see TLW250
FENCHYL ACETATE see FAO000
α-FENCHYL ALCOHOL see TLW000
FENCLOR 42 see PJL750
FENCLOR see PJL750
FENCLOZIC ACID see CKK250
FENDON see HIM000
FENDOSAL see FAO100
FENDOZAL see FAO100
FENELZIN see PFC750
FENERGAN see DQA400, PMI750
FENESTERIN see CME250
FENESTREL see FAO200
FENESTRIN see CME250
FENETAZINA see DQA400
FENETHAZINE HYDROCHLORIDE see DPI000
β-FENETHYLALKOHOL see PDD750
FENETHYLLINE HYDROCHLORIDE see CBF825
FENETICILLINE see PDD350
*p*-FENETIDIN see PDK890

1-FENETILBIGUANIDE CLORIDRATO (ITALIAN) see PDF250
N′-β-FENETILFORMAMIDINILIMINOUREA (ITALIAN) see PDF000
8-N-FENETIL-1-OXA-2-OXO-3,8-DIAZASPIRO-(4,5)-DECANO CLORIDRATO (ITALIAN) see DAI200
FENFLUORAMINE HYDROCHLORIDE see PDM250
FENFLURAMINE see ENJ000
FENFORMINA see PDF000
FENHYDREN see PFJ750
FENIBUT see PEE500
FENIBUTAZONA see BRF500
FENIBUT HYDROCHLORIDE see GAD000
FENIBUTOL see BRF500
FENICOL see CDP250, EGQ000
FENIDANTOIN 'S' see DKQ000
FENIDIN see DTP400
FENIDRONE see OPK300
FENIGAM see PEE500
FENIGAMA see PEE500
FENIGAM HYDROCHLORIDE see GAD000
2-FENIL-2-(p-AMINOFENIL)PROPIONAMMIDE (ITALIAN) see ALX500
2-FENIL-5-BROMO-INDANDIONE (ITALIAN) see UVJ400
FENILBUTINE see BRF500
FENILDICLOROARSINA (ITALIAN) see DGB600
FENILEP see PEC250
FENILFAR see SPC500
4-FENIL-4-FORMILPIPERIDINA (ITALIAN) see PFY100
FENILIDINA see BRF500
FENILIDRAZINA (ITALIAN) see PFI000
FENILIN see PFJ750
FENILISOPROPILIDRAZINA see PDN000
4-FENIL-α-METILFENILACETATO-γ-PROPILSOLFONATO SALE SODICO (ITALIAN) see PFR350
FENILOR see DDS600
FENILPRENAZONE see PEW000
2-FENILPROPANO (ITALIAN) see COE750
FENILPROPANOLAMINA (ITALIAN) see NNM000
FENIPENTOL see BQJ500
FENISED see PEC250
FENISTIL see FMU409
FENISTIL-RETARD see FMU409
FENITOIN see DNU000
FENITOX see DSQ000
FENITROOXON see PHD750
FENITROOXONE see PHD750
FENITROTHION see DSQ000
FENITROTHION OXON see PHD750
FENITROTION (HUNGARIAN) see DSQ000
FENITROXON see PHD750
FENIZON (FRENCH) see CJR500
FENKAROL see DWM400
FENMEDIFAM see MEG250
FENNEL OIL see FAP000
FENNOSAN see QPA000
FENNOSAN B 100 see DSB200
FENOBARBITAL see EOK000
FENOBCARB see MOV000
FENOBOLIN see DYF450
FENOBUCARB see MOV000
FENOCYCLIN see BIT000
FENOCYCLINE see BIT000
FENOFIBRIC ACID see CJP750
FENOFLURAZOLE see DGA200
FENOL (DUTCH, POLISH) see PDN750
FENOLFTALEIN see PDO750
FENOLIPUNA see PDO800
FENOLO (ITALIAN) see PDN750
FENOLOVO see HON000
FENOLOVO ACETATE see ABX250
FENOPHOSPHON see EPY000
FENOPRAIN see PMJ525
FENOPROFEN CALCIUM DIHYDRATE see FAP100
FENOPROFEN CALCIUM SALT DIHYDRATE see FAP100
FENOPROFEN SODIUM see FAQ000
FENOPROMIN see AOA250
FENOPRON see FAP100
FENOPROP see TIX500
FENORMONE see TIX500
FENOSMOLIN see PDP100
FENOSPEN see PDT500
FENOSTENYL see PEC250
FENOSTIL see FMU409
FENOSUCCIMIDE see MNZ000
FENOTEROL BROMIDE see FAQ100
FENOTEROL HYDROBROMIDE see FAQ100
FENOTHIAZINE (DUTCH) see PDP250

FENOTIAZINA (ITALIAN) see PDP250
FENOTONE see BRF500
FENOVERM see PDP250
FENOX see SPC500
FENOXEDIL see BPM750
FENOXEDIL HYDROCHLORIDE see BPM750
FENOXYBENZAMIN see DDG800
2-FENOXYETHANOL (CZECH) see PER000
FENOXYL CARBON N see DUZ000
FENOXYPEN see PDT500, PDT750
FENOZAFLOR see DGA200
FENPENTADIOL see CKG000
FENPHOSPHORIN see MEC250
FENPIVERIMIUM BROMIDE see RDA375
FENPROBAMATO see PGA750
FENPROPANAGE see DAB825
FENPROPATHRIN see DAB825
FENPROPAZINA see DIR000
FENSON see CJR500
FENSPIRIDE see DAI200
FENSPIRIDE HYDROCHLORIDE see DAI200
FENSULFOTHION see FAQ800
FENTAL see FLZ050
FENTANEST see PDW500, PDW750
FENTANIL see PDW500
FENTANYL see PDW500
FENTANYL CITRATE see PDW750
FENTAZIN see CJM250
FENTHION see FAQ900
FENTHION SULFONE see DSS800
FENTHIURAM see FAQ930
FENTHOATE see DRR400
FENTIAPRIL see FAQ950
FENTIAZAC see CKI750
FENTIAZAC CALCIUM SALT see CAP250
FENTIAZIN see PDP250
FENTIN ACETAAT (DUTCH) see ABX250
FENTIN ACETAT (GERMAN) see ABX250
FENTIN ACETATE see ABX250
FENTIN CHLORIDE see CLU000
FENTINE ACETATE (FRENCH) see ABX250
FENTIN HYDROXIDE see HON000
FENTIURAM see FAQ930
FENTRINOL see AHL500
FENUGREEK ABSOLUTE see FAR000
FENULON see DTP400
FENURAL see PEC250
FENUREA see PEC250
FENURON see DTP400
FENURONE see PEC250
FENVALERATE see FAR100
2-FENYL-5-AMINOBENZTHIAZOL (CZECH) see ALV500
1-FENYL-4-AMINO-5-CHLOR-6-PYRIDAZINON (CZECH) see PEE750
1-FENYL-3-AMINOPYRAZOLIN (CZECH) see ALX750
FENYLBUTAZON see BRF500
FENYL-CELLOSOLVE (CZECH) see PER000
FENYLDICHLORARSIN see DGB600
1-FENYL-4,5-DICHLOR-6-PYRIDAZINON (CZECH) see DGE800
1-FENYL-3,3-DIETHYLTRIAZEN (CZECH) see PEU500
1-FENYL-3,3-DIMETHYLTRIAZIN (CZECH) see DTP000
m-FENYLENDIAMIN (CZECH) see PEY000
FENYLENODWUAMINA (POLISH) see PEY500
FENYLEPSIN see DKQ000
FENYLESTER KYSELINY CHLORMRAVENCI (CZECH) see CBX109
FENYLESTER KYSELINY OCTOVE see PDY750
FENYLESTER KYSELINY SALICYLOVE see PGG750
1-FENYL-1,2-ETHANDIOL see SMQ100
1-FENYLETHANOL see PDE000
β-FENYLETHANOL see PDD750
4-(1-FENYLETHYL)FENOL see PFD400
FENYLETTAE see EOK000
N-FENYL-p-FENYLENDIAMIN see PFU500
FENYLFOSFIN see PFV250
FENYL-GLYCIDYLETHER (CZECH) see PFH000
FENYLGLYCOL see SMQ100
FENYLHIST see BAU750
FENYLHYDRAZID KYSELINY OCTOVE see ACX750
FENYLHYDRAZINE (DUTCH) see PFI000
FENYL-α-HYDROXYBENZYLKETON see BCP250
FENYL-β-HYDROXYETHYLSULFON (CZECH) see BBT000
N-FENYLIMID KYSELINY MALEINOVE (CZECH) see PFL750
FENYLISOKYANAT see PFK250
FENYLKYANID see BCQ250
FENYLMERCURIACETAT (CZECH) see ABU500
FENYLMERCURICHLORID (CZECH) see PFM500

FENYLMERKURINITRAT see MCU750
FENYL-METHYLKARBINOL see PDE000
1-FENYL-3-METHYL-2-PYRAZOLIN-5-ON see NNT000
N-FENYL-N'-CYKLOHEXYL-p-FENYLENDIAMIN (CZECH) see PET000
2-FENYLOTIOMOCZNIK (POLISH) see DWN800
1-FENYLPIPERAZIN see PFX000
2-FENYL-PROPAAN (DUTCH) see COE750
1-FENYL-3-PYRAZOLON (CZECH) see PGE250
FENYLSILAN see SDX250
FENYLSILATRAN (CZECH) see PGH750
FENYL-TRIFLUORSILAN see PGO500
FENYPRIN see DBA800
FENYRAMIDOL see PGG350
FENYRAMIDOL HYDROCHLORIDE see PGG355
FENYRIPOL see PGG350
FENYRIPOL HYDROCHLORIDE see FAR200
FENYTAN see PEC250
FENYTOINE see DKQ000, DNU000
FENZAFLOR see DGA200
FENZEN (CZECH) see BBL250
FEOJECTIN see IHG000
FEOSOL see FBN100, FBO000
FEOSPAN see FBN100
FEOSTAT see FBJ100
FEPRAZONE see PEW000
FEPRONA see FAP100
FERBAM 50 see FAS000
FERBAM see FAS000
FERBAM, IRON SALT see FAS000
FERBECK see FAS000
FERDEX 100 see IGS000
FERGON see FBK000
FERGON PREPARATIONS see FBK000
FER-IN-SOL see FBN100, FBO000
FERISAN see EJA379
FERKETHION see DSP400
FERLUCON see FBK000
FERMATE FERBAM FUNGICIDE see FAS000
FERMENICIDE LIQUID see SOH500
FERMENICIDE POWDER see SOH500
FERMENTATION ALCOHOL see EFU000
FERMENTATION AMYL ALCOHOL see IHP000
FERMENTATION BUTYL ALCOHOL see IIL000
FERMIDE see TFS350
FERMINE see DTR200
FERMOCIDE see FAS000
FERNACOL see TFS350
FERNASAN see TFS350
FERNESTA see BQZ000, DAA800
FERNEX see DIN600
FERNIDE see TFS350
FERNIMINE see DAA800
FERNISOLONE see PMA000
FERNISONE see HHQ800
FERNOS see DOX600
FERNOXENE see SGH500
FERNOXONE see DAA800
FERODIN SL see FAS100
FERO-GRADUMET see FBN100, FBO000
FEROTON see FBJ100
FER PENTACARBONYLE (FRENCH) see IHG500
FERRADOW see FAS000
FERRALYN see FBN100
FERRATE(1-), (GLYCINATO-N,O)(SULFATO(2-)-O',O')-, HYDROGEN, (T-4)-(9CI) see FBD500
FERRATE(4-), HEXACYANO-, TETRAPOTASSIUM see TEC500
FERRATE(4-), HEXAKIS(CYANO-C)-, IRON(3+) (3:4), (OC-6-11)-(9CI) see IGY000
FERRATE(4-), HEXAKIS(CYANO-C)-, TETRAPOTASSIUM, (OC-6-11)- see TEC500
FERRATE(2-), PENTAKIS(CYANO-C)NITROSYL-, DISODIUM, DIHYDRATE (OC-6-22)-, (9CI) see SIW500
FERRIAMICIDE see MQW500
FERRIC ACETYLACETONATE see IGL000
FERRIC AMMONIUM CITRATE see FAS700
FERRIC AMMONIUM CITRATE, GREEN see FAS700
FERRIC AMMONIUM OXALATE (DOT) see ANG925
FERRIC AMMONIUM OXALATE see ANG925
FERRIC ARSENATE, solid (DOT) see IGN000
FERRIC ARSENITE, solid (DOT) see IGO000
FERRIC ARSENITE, BASIC see IGO000
FERRIC CHLORIDE see FAU000
FERRIC CHLORIDE HEXAHYDRATE see FAW000
FERRIC CHLORIDE (UN 1733) (DOT) see FAU000
FERRIC CHLORIDE, solution (UN 2582) (DOT) see FAU000

FERRIC CHOLINE CITRATE see FBC100
FERRIC DEXTRAN see IGS000
FERRIC DIMETHYLDITHIOCARBAMATE see FAS000
FERRIC FERROCYANIDE see IGY000
FERRIC FLUORIDE see FAX000
FERRIC HEXACYANOFERRATE (II) see IGY000
FERRIC HYDROXIDE NITRILOTRIPROPIONIC ACID COMPLEX see FAY000
FERRIC NITRATE (DOT) see FAY200, IHB900
FERRIC NITRATE see FAY200, IHB900
FERRIC NITRATE, NONAHYDRATE see IHC000
FERRIC NITRILOTRIACETATE see IHC100
FERRIC NITROSODIMETHYL DITHIOCARBAMATE and TETRAMETHYL THIURAM DISULFIDE see FAZ000
FERRICON see ROF300
FERRIC OXIDE see IHD000
FERRIC OXIDE, SACCHARATED see IHG000
FERRIC SACCHARATE IRON OXIDE (MIX.) see IHG000
FERRIC SODIUM EDETATE see EJA379
FERRIC SODIUM EDTA see EJA379
FERRIC SODIUM GLUCONATE COMPLEX see IHK000
FERRIC SODIUM PYROPHOSPHATE see SHE700
FERRIC TRIACETYLACETONATE see IGL000
FERRIC TRICHLORIDE HEXAHYDRATE see FAW000
FERRICYANURE de TRI(1-DODECYL-2-PHENYL-3-METHYL)-1,3-BENZIMIDAZO-LIUM (FRENCH) see TNH750
FERRIDEXTRAN see IGS000
FERRIGEN see IGT000
FERRIHEXACYANOFERRATE see IGY000
FERRITIN see FBB000
FERRIVENIN see IHG000
FERRLECIT see FBB100
FERROACTINOLITE see FBG200
FERROANTHOPHYLLITE see ARM264
FERROCENE see FBC000
FERROCENE, ACETYL- see ABA750
FERROCHOLINATE see FBC100
FERROCHROME (exothermic) see FBD000
exothermic FERROCHROME see FBD000
FERROCHROME see FBD000
FERROCHROMIUM see FBD000
FERROCIN see IGY000
FERROCYANIDES see FBD100
FERRODEXTRAN see IGS000
FERROFLUKIN 75 see IGS000
FERROFOS 509 see NEI100
FERROFOS 510 see HKS780
FERROFUME see FBJ100
FERROGLUCIN see IGS000
FERROGLUKIN 75 see IGS000
FERROGLYCINE SULFATE see FBD500
FERROGLYCINE SULFATE COMPLEX see FBD500
FERRO-GRADUMET see FBN100
FERRO LEMON YELLOW see CAJ750
FERROLIP see FBC100
exothermic FERROMANGANESE (DOT) see FBE000
FERROMANGANESE (exothermic) see FBE000
FERRON see IEP200
FERRONAT see FBJ100
FERRONE see FBJ100
FERRONICUM see FBK000
FERRONORD see FBD500
FERRO ORANGE YELLOW see CAJ750
FERROSANOL see FBD500
FERROSILICON see FBG000, IHJ000
FERROSILICON, containing more than 30% but less than 90% SILICON (DOT) see FBG000
FERROSULFAT (GERMAN) see FBN100
FERROSULFATE see FBN100
FERROTEMP see FBJ100
FERRO-THERON see FBN100
FERROTREMOLITE see FBG200
FERROTSIN see IGY000
FERROUS see FBN000
FERROUS ACETATE see FBH000
FERROUS ARSENATE (DOT) see IGM000
FERROUS ARSENATE, solid (DOT) see IGM000
FERROUS ASCORBATE see FBH050
FERROUS CARBONATE see FBH100
FERROUS CHLORIDE see FBI000
FERROUS CHLORIDE, solid (NA 1759) (DOT) see FBI000
FERROUS CHLORIDE, solution (NA 1760) (DOT) see FBI000
FERROUS CHLORIDE TETRAHYDRATE see FBJ000
FERROUS FERRITE see IHG100
FERROUS FUMARATE see FBJ100
FERROUS GLUCONATE see FBK000

FERROUS GLUCONATE DIHYDRATE see FBL000
FERROUS GLUTAMATE see FBM000
FERROUS ION see FBN000
FERROUS LACTATE see LAL000
FERROUS SULFATE (FCC) see FBO000
FERROUS SULFATE see FBN100
FERROUS SULFATE HEPTAHYDRATE see FBO000
FERROVAC E see IGK800
FERROVANADIUM DUST see FBP000
FERRO YELLOW see CAJ750
FERRUGO see IHD000
FERRUM see FBJ100, FBP050
FERSAMAL see FBJ100
FERSOLATE see FBN100
FERTENE see PJS750
FERTILYSIN see BIX250
FERTIRAL see LIU360
FERTODUR see FBP100
FERULIC ACID see FBP200
trans-FERULIC ACID see FBP200
FERVENULIN see FBP300
FERVENULINE see FBP300
FES see ZAT000
FESOFOR see FBO000
FESOTYME see FBO000
FETID NIGHTSHADE see HAQ100
FETT see FNK200
FETTER BUSH see DYA875
FETTERBUSH see FBP520
FETTORANGE B see XRA000
FETTORANGE R see PEJ500
FETTPONCEAU G see OHA000
FETTROT see OHA000
FETTSCHARLACH see OHA000
FETTSCHARLACH LB see OHA000
FEUILLES CRABE (HAITI) see SLJ650
FEVER TWIG see AHJ875
FF 106 see IJH000
FF see FQN000
FFB 32 see PHB500
F-5-FU see FLZ050
F 151 FUMARATE see MDP800
FG 834 see SMQ500
FG 4963 see FAJ200
FG 5111 see FKI000
FH 099 see DTL200
FH 122-A see NNQ500
FHCH see BBQ750
FHD-3 see BOF750
F.I 106 see AES750
FI 106 see HKA300
FI 1163 see LIN000
F.I. 58-30 see EPQ000
FI 6120 see DLH200
F.I. 6145 see PIW000
FI 6146 see BTA325
FI6339 see DAC000
FI 6341 see FDB000
FI 6714 see NDM000
FI 6804 see HKA300
FIBERGLASS see FBQ000
FIBERS, REFRACTORY CERAMIC see RCK725
FIBER V see PKF750
FIBORAN see FBP850
FIBRALEM see ARQ750
FIBRASET TC see UTU500
FIBRE BLACK VF see AQP000
FIBRENE C 400 see TAB750
FIBRINOGENASE see TFU800
FIBROTAN see PFN000
FIBROUS CROCIDOLITE ASBESTOS see ARM275
FIBROUS GLASS see FBQ000
FIBROUS GLASS DUST (ACGIH) see FBQ000
FIBROUS GRUNERITE see ARM250
FIBROUS TREMOLITE see ARM280
FICAM see DQM600
FICHLOR 91 see TIQ750
FICIN see FBS000
FI CLOR 71 see DGN200
FI CLOR 91 see TIQ750
FI CLOR 60S see SGG500
FICUS PROTEASE see FBS000
FICUS PROTEINASE see FBS000
FIDDLE FLOWER see SDZ475
FIDDLE-NECK see TAG250

FIELD GARLIC see WBS850
FIETIN see FBW000
FIGUIER MAUDIT MARRON (HAITI) see BAE325
FIGWORT see FBS100
F III (sugar fraction) see FAB010
FILARIOL see EGV500
FILARSEN see DFX400
FILIGRANA (CUBA) see LAU600
FILMERINE see SIQ500
FILORAL see CMF500
FILTEX WHITE BASE see CAT775
FILTRASORB 200 see CBT500
FILTRASORB 400 see CBT500
FILTRASORB see CBT500
FIMALENE see ILD000
FINA, combustion products see ADX750
FINAVEN see ARW000
FINDOLAR see GGS000
FINE GUM HES see SFO500
FINEMEAL see BAP000
FINIMAL see HIM000
FINISH EN see DTG700
FINLEPSIN see DCV200
FINNCARB 6002 see CAT775
FINQUEL see EFX500
FINTIN ACETATO (ITALIAN) see ABX250
FINTINE HYDROXYDE (FRENCH) see HON000
FINTIN HYDROXID (GERMAN) see HON000
FINTIN HYDROXYDE (DUTCH) see HON000
FINTIN IDROSSIDO (ITALIAN) see HON000
FINTROL see AQM000
FIORINAL see ABG750
FIR BALSAM ABSOLUTE see FBS200
FIRE DAMP see MDQ750
FIREMASTER BP-6 see FBU000
FIREMASTER FF-1 see FBU509
FIREMASTER T23P-LV see TNC500
FIRMACEF see CCS250
FIRMATEX RK see DTG000
FIRMAZOLO see AIF000
FIRMOTOX see POO250
FIR NEEDLE OIL, SIBERIAN see FBV000
FIRON see FBJ100
FISCHER'S SOLUTION see FBV100
FISETHOLZ see FBW000
FISETIN see FBW000
FISH BERRY see PIE500
FISH POISON see HGL575
FISHTAIL PALM see FBW100
FISH-TOX see RNZ000
FISIOQUENS see EEH575
FISONS B25 see CEW500
FISONS NC 2964 see DSO000
FISONS NC 5016 see DGA200
FISSUCAIN see BQA010
FITIOS see DNX600
FITIOS B/77 see DNX600
FITROL see KBB600
FITROL DESICCATE 25 see KBB600
FIXANOL BLACK E see AQP000
FIXANOL BLUE 2B see CMO000
FIXANOL BLUE BH see CMN800
FIXANOL BROWN LF see CMO820
FIXANOL C see HCP800
FIXANOL GREEN BN see CMO840
FIXANOL ORANGE BROWN X see CMO810
FIXANOL RED FS see CMO870
FIXANOL VIOLET N see CMP000
FIXAPRET CP see DTG000
FIXATIVE IS see CNH125
FIXER IS see CNH125
FIXIERER P see THM900
FIXOL see CMS850
6FK see HCZ000
FK 235 see NDY650
FK 749 see EBE100
FK 1160 see CJH750
F KLOT see AJP250
FL-1039 see MCB550
FL see MQR225
FLAC see ARB750
FLACAVON R see TNC500
FLACETHYLE see ELH600
FLAGECIDIN see AOY000
FLAGEMONA see MMN250

FLAGESOL see MMN250
FLAGIL see MMN250
FLAGYL see MMN250
FLAKE WHITE see BKW100
FLAMARIL see HNI500
FLAMAZINE see SNI425
FLAMENCO see TGG760
FLAMINGO FLOWER see APM875
FLAMINGO LILY see APM875
FLAMMEX AP see TNC500
FLAMOLIN MF 15711 see PJS750
FLAMRUSS see CBT750
FLAMULA (CUBA) see CMV390
FLAMYCIN see CMA750
FLANARIL see HNI500
FLANOGEN ELA see CCL250
FLAROXATE HYDROCHLORIDE see FCB100
FLAVACRIDINUM HYDROCHLORICUM see DBX400
4-FLAVANONE see FBW150
FLAVANONE see FBW150
FLAVASPIDIC ACID see FBY000
FLAVASPIDSAEURE (GERMAN) see FBY000
FLAVAXIN see RIK000
FLAVAZONE see NGE500
FLAVENSOMYCIN see FBZ000
FLAVIN see XAK000
FLAVIN ADENIN DINUCLEOTIDE see RIF100
FLAVIN ADENINE DINUCLEOTIDE see RIF100
FLAVINAT see RIF100
FLAVINE see DBX400, XAK000
FLAVINE-ADENINE DINUCLEOTIDE see RIF100
FLAVINE ADENOSINE DIPHOSPHATE see RIF100
FLAVIOFORM see DBX400
FLAVIPIN see DBX400
FLAVISEPT see DBX400
FLAVISPIDIC ACID BB see FBY000
FLAVITAN see RIF100
FLAVITAN see RIF100
FLAVITROL see EDW500
FLAVOFUNGIN (1:10) see FCA000
FLAVOMYCELIN see LIV000
FLAVOMYCIN see MRA250
FLAVONE see PER700
7-FLAVONE ETHYL HYDROXYACETATE see ELH600
FLAVONE-7-ETHYLOXYACETATE see ELH600
FLAVONE, 3-HYDROXY- see HLC600
FLAVONIC ACID see FCA100
FLAVONIC ACID DISODIUM SALT see FCA200
7-FLAVONOXYACETIC ACID ETHYL ESTER see ELH600
FLAVOPHOSPHOLIPOL see MRA250
FLAVOSAN see XAK000
FLAVOXATE HYDROCHLORIDE see FCB100
FLAVUMYCIN B see FCC000
FLAVUROL see MCV000
FLAVYLIUM, 3,4′,5,7-TETRAHYDROXY-3′,5′-DIMETHOXY-, CHLORIDE see MAO600
FLAXEDIL see PDD300
FLAX OLIVE see LAR500
FLEBOCORTID see HHR000
FLECK-FLIP see TIO750
FLECTOL A see TLP500
FLECTOL H see TLP500
FLECTOL H, POLYMER see PJQ750
FLECTOL PASTILLES see TLP500
FLEET-X see TLM050
FLEET-X-DV-99 see TLM000
FLEUR SUREAU (CANADA, HAITI) see EAI100
FLEXAL see IPU000
FLEXAMINE G see BLE500
FLEXARTAL see IPU000
FLEXARTEL see IPU000
FLEXAZONE see BRF500
FLEXERIL see DPX800
FLEXIBAN see DPX800
FLEXIBLE COLLODION see CCU250
FLEXICHEM see CAX350
FLEXICHEM CS see CAX350
FLEXILON see AJF500
FLEXIMEL see DVL700
FLEXIN see AJF500
FLEXOL A 26 see AEO000
FLEXOL DOP see DVL700
FLEXOL EPO see FCC100
FLEXOL 4GO see FCD512
FLEXOL GPE see PJC500
FLEXOL PLASTICIZER 810 see FCN050

FLEXOL PLASTICIZER CC-55 see FCN000
FLEXOL PLASTICIZER DIP see ILR100
FLEXOL PLASTICIZER DOP see DVL700
FLEXOL PLASTICIZER 3GO see FCD560
FLEXOL PLASTICIZER TCP see TNP500
FLEXOL TOF see TNI250
FLEXO RED 482 see RGW000
FLEXZONE 3C see PFL000
FL-G 5 see KDK000
FL-G 35 see KDK000
FL-G 100 see KDK000
FL-G 330 see KDK000
FLIBOL E see TIQ250
FLIEGENTELLER see TIQ250
FLINDIX see CCK125
FLINT see SCI500, SCJ500
FLIT 406 see CBG000
FLOBACIN see OGI300
FLOCOOL 180 see SJC500
FLOCOR see PKQ059
FLOCTAFENINE see TKG000
FLOGAR see OLM300
FLOGENE see CKI750
FLOGHENE see HNI500
FLOGICID see BPP750
FLOGINAX see MFA500
FLOGISTIN see HNI500
FLOGITOLO see HNI500
FLOGOCID N PLASTIGEL see BPP750
FLOGODIN see HNI500
FLOGORIL see HNI500
FLOGOS see SOX875
FLOGOSTOP see HNI500
FLO-MOR see PAI000
FLOMORE see BSQ750
FLOMOXEF see FCN100
FLOMOXEF SODIUM see FCN100
FLOPIRINA see HNI500
FLOPROPION see TKP100
FLOPROPIONE see TKP100
FLO PRO T SEED PROTECTANT see TFS350
FLO PRO V SEED PROTECTANT see CCC500
FLOR de ADONIS (CUBA) see PCU375
FLORALOZONE see EIJ600
FLORALTONE see GEM000, TKQ250
FLORANE see HBP425
FLORAQUIN see DNF600
FLOR de BARBERO (CUBA) see AFQ625
FLOR de CAMARON (MEXICO) see CAK325
FLOR de CULEBRA (PUERTO RICO) see APM875
FLORDIMEX see CDS125
FLOREL see CDS125
FLORES MARTIS see FAU000
FLORIDA ARROWROOT see CNH789
FLORIDA HOLLY see PCB300
FLORIDIN see TEY000
FLORIDINE see SHF500
FLORIMYCIN see VQZ000
FLORINEF see FHH100
FLORIPONDIO (PUERTO RICO) see AOO825
FLORITE R see CAW850
FLOROCID see SHF500
FLOROL see PGR250
FLOROMYCIN see VQZ000
FLORONE (ITALIAN) see XQJ000
FLOROPIPAMIDE see FHG000
FLOROPIPETON see FLN000
FLOROPRYL see IRF000
FLOROXENE see TKB250
FLOR del PERU see YAK350
FLOSIN see IDA400
FLOSINT see IDA400
FLOTHENE see PJS750
FLOU see POF500
FLOUVE OIL see FDA000
FLOVACIL see DKI600
FLOWER FENCE see CAK325
FLOWER of PARADISE see HMX600
FLOWERS of ANTIMONY see AQF000
FLOWERS of SULPHUR (DOT) see SOD500
FLOWERS of ZINC see ZKA000
FLOXACILLIN SODIUM see FDA100
FLOXACILLIN SODIUM MONOHYDRATE see CHJ000
FLOXAPEN see CHJ000
FLOXAPEN SODIUM see FDA100

FLOXIN see OGI300
FLOXURIDIN see DAR400
FLOXURIDINE see DAR400
FLOZENGES see SHF500
FLUALAMIDE see FDA875
FLUANISON see HAF400
FLUANISONE see HAF400
FLUANISONE HYDROCHLORIDE see FDA880
FLUANXOL see FMO129
FLUATE see TIO750
FLUAZIFOP-BUTYL see FDA885
FLUBE see BLX000
FLUBENDAZOLE see FDA887
FLUCHLORALIN see FDA900
FLUCINAR see SPD500
FLUCLOXACILLIN SODIUM see FDA100
FLUCLOXACILLIN SODIUM MONOHYDRATE see CHJ000
FLUCLOXACILLIN SODIUM SALT see FDA100
FLUCORT see SPD500
FLUCYTOSINE see FHI000
FLUDERMA see FDB000
FLUDEX see IBV100
FLUDIAZEPAM see FDB100
FLUDILAT see BAV250
FLUDROCORTISONE see FHH100
FLUDROCORTONE see FHH100
FLUE DUST, ARSENIC CONTAINING see ARE500
FLUE GAS see CBW750
FLUENETIL see FDB200
FLUENYL see FDB200
FLUFENAMIC ACID see TKH750
FLUFENAMINSAURE (GERMAN) see TKH750
FLUGENE 22 see CFX500
FLUGERIL see FMR050
FLUGEX 12B1 see BNA250
FLUIBIL see CDL325
FLUID-EXTRACT of JAMAICA GINGER U.S.P. see JBA000
FLUIFORT see CBR675
FLUIMUCETIN see ACH000
FLUIMUCIL see ACH000
FLUITRAN see HII500
FLUKOIDS see CBY000
FLUMAMINE see DQR600
FLUMARK see EAU100
FLUMEN see CLH750
FLUMESIL see BEQ625
FLUMETHIAZIDE see TKG750
FLUMICIL see ACH000
FLUMOPERONE HYDROCHLORIDE see TKK750
FLUNARIZINE DIHYDROCHLORIDE see FDD080
FLUNARIZINE HYDROCHLORIDE see FDD080
FLUNIGET see DKI600
FLUNISOLIDE see FDD085
FLUNITRAZEPAM see FDD100
FLUOBORIC ACID (DOT) see FDD125, HHS600
FLUOBORIC ACID see FDD125, HHS600
FLUOBRENE see FOO525
FLUOCINOLIDE see FDD150
FLUOCINOLONE 16,17-ACETONIDE see SPD500
FLUOCINOLONE ACETONIDE see SPD500
FLUOCINOLONE ACETONIDE 21-ACETATE see FDD150
FLUOCINOLONE ACETONIDE ACETATE see FDD150
FLUOCINONIDE see FDD150
FLUODROCORTISONE see FHH100
FLUOHYDRISONE see FHH100
FLUOHYDROCORTISONE see FHH100
FLUO-KEM see TAI250
FLUOMETURON see DUK800
FLUOMINE see EIS000
FLUOMINE DUST see EIS000
FLUON see TAI250
FLUOOXENE see TKB250
FLUOPERAZINE see TKE500, TKK250
FLUOPERIDOL see FLU000
FLUOPHOSGENE see CCA500
FLUOPHOSPHORIC ACID DI(DIMETHYLAMIDE) see BJE750
FLUOPHOSPHORIC ACID, DIETHYL ESTER see DJJ400
FLUOPHOSPHORIC ACID, DIISOPROPYL ESTER see IRF000
FLUOPHOSPHORIC ACID, DIMETHYL ESTER see DSA800
FLUOR (DUTCH, FRENCH, GERMAN, POLISH) see FEZ000
FLUORACETATO di SODIO (ITALIAN) see SHG500
5-FLUORACIL (GERMAN) see FMM000
FLUORAKIL 100 see FFF000
FLUORAL see SHF500
FLUORAL HYDRATE see TJZ000

3',6'-FLUORANDIOL see FEV000
3,6-FLUORANDIOL see FEV000
FLUORANE 114 see FOO509
4-FLUORANILIN see FFY000
FLUORANTHENE see FDF000
N-FLUORANTHEN-3-YLACETAMIDE see AAK400
N-3-FLUORANTHENYLACETAMIDE see AAK400
FLUORAQUIN see DNF600
o-FLUORBENZOESAEURE (GERMAN) see FGH000
5-FLUOR-DESOXYCYTIDIN (GERMAN) see FHO000
2-FLUORENAMINE see FDI000
FLUOREN-2-AMINE see FDI000
FLUOREN-9-AMINE, N-(2-CHLOROETHYL)-N-ETHYL-, HYDROCHLORIDE see
   FEE100
9H-FLUOREN-9-AMINE, N-(2-CHLOROETHYL)-N-ETHYL-, HYDROCHLORIDE
   (9CI) see FEE100
9H-FLUORENE see FDI100
FLUORENE see FDI100
2-FLUORENEAMINE see FDI000
9-FLUORENECARBOXYLATE-3-QUINUCLIDINOL HYDROCHLORIDE see
   FDK000
FLUORENE-9-CARBOXYLIC ACID-2-(DIETHYLAMINO)ETHYL ESTER see
   CBS250
FLUORENE-9-CARBOXYLIC ACID-3-QUINUCLIDINYL ESTER see FDK000
FLUORENE, 2-(CHLOROACETYL)- see CEC700
2,7-FLUORENEDIAMINE see FDM000
FLUORENE-2,7-DIAMINE see FDM000
9H-FLUORENE, 1,9-DIMETHYL- see DRY100
1,1'-(9H-FLUORENE-2,7-DIYL)BIS(2-(DIETHYLAMINO)ETHANONE) DIHYDRO-
   CHLORIDE TRIHYDRATE see FDN000
9H-FLUOREN-9-ONE see FDO000
FLUOREN-9-ONE see FDO000
9-FLUORENONE see FDO000
FLUORENO(9,1-gh)QUINOLINE see FDP000
N-FLUOREN-1-YL ACETAMIDE see FDQ000
1-FLUORENYLACETAMIDE see FDQ000
N-FLUOREN-2-YL ACETAMIDE see FDR000
2-FLUORENYLACETAMIDE see FDR000
N-9H-FLUOREN-3-YL ACETAMIDE see FDS000
N-FLUOREN-3-YL ACETAMIDE see FDS000
3-FLUORENYL ACETAMIDE see FDS000
N-FLUOREN-4-YLACETAMIDE see ABY000
N-1-FLUORENYLACETAMIDE see FDQ000
N-2-FLUORENYLACETAMIDE see FDR000
N-3-FLUORENYL ACETAMIDE see FDS000
N-4-FLUORENYLACETAMIDE see ABY000
1-FLUORENYL ACETHYDROXAMIC ACID see FDT000
FLUORENYL-2-ACETHYDROXAMIC ACID see HIP000
3-FLUORENYL ACETHYDROXAMIC ACID see FDU000
N-(FLUOREN-2-YL)ACETOHYDROXAMIC ACETAMIDE see ABL000
N-(FLUOREN-3-YL)ACETOHYDROXAMIC ACETATE see ABO250
N-(FLUOREN-4-YL)ACETOHYDROXAMIC ACETATE see ABO500
N-FLUOREN-1-YL ACETOHYDROXAMIC ACID see FDT000
N-FLUOREN-2-YL ACETOHYDROXAMIC ACID see HIP000
N-FLUOREN-3-YL ACETOHYDROXAMIC ACID see FDU000
N-2-FLUORENYL ACETOHYDROXAMIC ACID see HIP000
N-FLUOREN-2-YLACETOHYDROXAMIC ACID, COBALT(2+) COMPLEX see
   FDU875
N-FLUOREN-2-YL ACETOHYDROXAMIC ACID, COPPER(2+) COMPLEX see
   HIP500
N-FLUOREN-2-YL ACETOHYDROXAMIC ACID, IRON(3+) COMPLEX see
   HIQ000
N-FLUOREN-2-YL ACETOHYDROXAMIC ACID, MANGANESE(2+) COMPLEX
   see HIR000
N-FLUOREN-2-YL ACETOHYDROXAMIC ACID, NICKEL(2+) COMPLEX see
   HIR500
N-FLUOREN-2-YL ACETOHYDROXAMIC ACID, POTASSIUM SALT see HIS000
N-FLUOREN-2-YL ACETOHYDROXAMIC ACID SULFATE see FDV000
N-FLUOREN-2-YLACETOHYDROXAMIC ACID, ZINC COMPLEX see ZHJ000
N-FLUOREN-2-YL BENZAMIDE see FDX000
N-(2-FLUORENYL)BENZAMIDE see FDX000
N-9H-FLUOREN-2-YL-BENZAMIDE (9CI) see FDX000
N-FLUOREN-1-YL BENZOHYDROXAMIC ACID see FDY000
N-FLUOREN-2-YL BENZOHYDROXAMIC ACID see FDZ000
N-(2-FLUORENYL)BENZOHYDROXAMIC ACID see FDZ000
N-FLUOREN-2-YL BENZOHYDROXAMIC ACID ACETATE see ABO750
N,N'-FLUOREN-2,7-YLBISACETAMIDE see BGP250
2,7-FLUORENYLBISACETAMIDE see BGP250
N-FLUOREN-1-YLDIACETAMIDE see DBF000
N-FLUOREN-2-YLDIACETAMIDE see DBF200
2-FLUORENYLDIACETAMIDE see DBF200
N-1-FLUORENYLDIACETAMIDE see DBF000
N-2-FLUORENYLDIACETAMIDE see DBF200
2-FLUORENYLDIMETHYLAMINE see DPJ600
N,N'-FLUOREN-2,5-YLENEBISACETAMIDE see BGR250

N-4-(4'-FLUORO)BIPHENYLACETAMIDE see FKZ000
4'-FLUORO-4-BIPHENYLAMINE see AKC500
N-(4'-FLUORO-4-BIPHENYLYL)ACETAMIDE see FKZ000
2-(2-FLUORO-4-BIPHENYLYL)PROPIONIC ACID see FLG100
FLUOROBISISOPROPYLAMINO-PHOSPHINE OXIDE see PHF750
5-FLUORO-1,3-BIS(TETRAHYDRO-2-FURANYL)-2,4(1H,3H)-PYRIMIDINEDIONE
    see BLH325
FLUOROBIS(TRIFLUOROMETHYL)PHOSPHINE see FGW100
FLUOROBLASTIN see FMM000
1-FLUORO-2-BROMOBENZENE see FGX000
1-FLUORO-3-BROMOBENZENE see FGY000
6-FLUORO-7-BROMOMETHYLBENZ(a)ANTHRACENE see BNQ250
4-FLUOROBUTYL BROMIDE see FHA000
4-FLUOROBUTYL CHLORIDE see FHB000
4-FLUOROBUTYL IODIDE see FHC000
3-FLUOROBUTYL ISOCYANATE see FHC200
4-FLUOROBUTYL THIOCYANATE see FHD000
4-FLUORO-BUTYRIC ACID-2-CHLOROETHYL ESTER see CGZ000
4-FLUOROBUTYRIC ACID METHYL ESTER see MKE000
4-FLUOROBUTYRONITRILE see FHF000
γ-FLUOROBUTYRONITRILE see FHF000
FLUOROBUTYROPHENONE see FHG000
FLUOROCARBON 113 see FOO000
FLUOROCARBON 114 see FOO509
FLUOROCARBON-115 see CJI500
FLUOROCARBON-12 see DFA600
FLUOROCARBON-22 see CFX500
FLUOROCARBON 1211 see BNA250
FLUOROCARBON FC142b see CFX250
FLUOROCARBON FC143a see TJY900
FLUOROCARBON FC 43 see HAS000
FLUOROCARBON No. 11 see TIP500
FLUOROCHLOROCARBON LIQUID see FHH000
5-FLUORO-7-CHLOROMETHYL-12-METHYLBENZ(a)ANTHRACENE see FHH025
FLUOROCHROME see MCV000
FLUOROCORTISONE see FHH100
4-FLUORO-CROTONIC ACID METHYL ESTER see MKE250
2-FLUORO-2'-CYANODIETHYL ETHER see CON500
5-FLUOROCYSTOSINE see FHI000
5-FLUOROCYTOSINE see FHI000
10-FLUORODECYL CHLORIDE see FHN000
5-FLUORO-2'-DEOXYCYTIDINE see FHO000
5-FLUORODEOXYCYTIDINE see FHO000
5-FLUORO-2'-DEOXYURIDINE see DAR400
5-FLUORO-2-DEOXYURIDINE see DAR400
β-5-FLUORO-2'-DEOXYURIDINE see DAR400
5-FLUORODEOXYURIDINE see DAR400
FLUORODEOXYURIDINE see DAR400
4-FLUORO-1,2:5,6-DIBENZANTHRACENE see FHP000
6-FLUORODIBENZ(a,h)ANTHRACENE see FHP000
FLUORODICHLOROMETHANE see DFL000
FLUORODIFEN see NIX000
9-α-FLUORO-11-β,21-DIHYDROXY-16-α-ISOPROYLIDENEDIOXY-1,4-PREGNADI-
    ENE, 3,20-DIONE see AQX500
9-α-FLUORO-11-β,17-β-DIHYDROXY-17-α-METHYL-4-ANDROSTENE-3-ONE see
    AOO275
FLUORO-9-α DIHYDROXY-11-β,17-β METHYL-17-α ANDROSTENE-4 ONE-3
    (FRENCH) see AOO275
9-FLUORO-11-β-,17-β-DIHYDROXY-17-METHYLANDROST-4-EN-3-ONE see
    AOO275
FLUORODIISOPROPYL PHOSPHATE see IRF000
2-FLUORO-4-DIMETHYLAMINOAZOBENZENE see FHQ000
2'-FLUORO-4-DIMETHYLAMINOAZOBENZENE see FHQ010
3'-FLUORO-4-DIMETHYLAMINOAZOBENZENE see FHQ100
4'-FLUORO-4-DIMETHYLAMINOAZOBENZENE see DSA000
4'-FLUORO-N,N-DIMETHYL-4-AMINOAZOBENZENE see DSA000
m-FLUORODIMETHYLAMINOAZOBENZENE see FHQ100
4'-FLUORO-p-DIMETHYLAMINOAZOBENZENE see DSA000
2'-FLUORO-4-DIMETHYLAMINOSTILBENE see FHU000
4'-FLUORO-4-DIMETHYLAMINOSTILBENE see FHV000
3-FLUORO-2,10-DIMETHYL-5,6-BENZACRIDINE see FHR000
10-FLUORO-9,12-DIMETHYLBENZ(a)ACRIDINE see FHR000
11-FLUORO-7,12-DIMETHYLBENZ(a)ANTHRACENE see DRZ000
1-FLUORO-7,12-DIMETHYLBENZ(a)ANTHRACENE see FHS000
4-FLUORO-7,12-DIMETHYLBENZ(a)ANTHRACENE see DRY400
5-FLUORO-7,12-DIMETHYLBENZ(a)ANTHRACENE see DRY600
8-FLUORO-7,12-DIMETHYLBENZ(a)ANTHRACENE see DRY800
4'-FLUORO-N,N-DIMETHYL-p-PHENYLAZOANILINE see DSA000
2'-FLUORO-N,N-DIMETHYL-4-STILBENAMINE see FHU000
4'-FLUORO-N,N-DIMETHYL-4-STILBENAMINE see FHV000
1,2,4-FLUORODINITROBENZENE see DUW400
1-FLUORO-2,4-DINITROBENZENE see DUW400
1-FLUORO-1,1-DINITRO-2-BUTENE see FHV300
2-FLUORO-1,1-DINITROETHANE see FHV800
2-FLUORO-2,2-DINITROETHANOL see FHW000

2-FLUORO-2,2-DINITROETHYLAMINE see FHX000
FLUORO DINITROMETHANE see FHY000
FLUORO DINITROMETHYL AZIDE see FHZ000
1-FLUORO-1,1-DINITRO-2-PHENYLETHANE see FHZ200
12-FLUORO DODECANO NITRILE see FIA000
2,7-FLUOROENEDIAMINE see FDM000
4-FLUOROESTRADIOL see FIA500
4-FLUOROESTRA-1,3,5-(10)-TRIENE-3,17-β-DIOL see FIA500
FLUOROETHANE see FIB000
FLUOROETHANOIC ACID see FIC000
2-FLUOROETHANOL see FIE000
β-FLUOROETHANOL see FIE000
FLUOROETHANOL see FID000
2-FLUOROETHANOL, PHOSPHITE (3:1) see PHO250
FLUOROETHENE see VPA000
FLUOROETHYL see HDC000
β-FLUOROETHYL-N-(β-CHLOROETHYL)-N-NITROSOCARBAMATE see FIH000
2-FLUOROETHYL CHLOROFORMATE see FIH100
1-FLUOROETHYL-3-CYCLOHEXYL-1-NITROSOUREA see CPL750
FLUOROETHYLENE see VPA000
FLUOROETHYLENE OZONIDE see FIK875
2-FLUOROETHYL ESTER DIPHENYLACETIC ACID see FIP999
2-FLUOROETHYL FLUOROACETATE see FIM000
β-FLUOROETHYL FLUOROACETATE see FIM000
2-FLUORO ETHYL-γ-FLUORO BUTYRATE see FIN000
β-FLUOROETHYL-γ-FLUOROBUTYRATE see FIN000
2-FLUOROETHYL-5-FLUOROHEXOATE see FIO000
(8R)-8-(2-FLUOROETHYL)-3-α-HYDROXY-1-α-H,5-α-H-TROPANIUM BROMIDE
    BENZILATE H2O see FMR300
β-FLUOROETHYLIC ESTER of XENYLACETIC ACID see FIP999
2-FLUOROETHYL IODIDE see FIQ000
2-FLUOROETHYL-N-METHYL-N-NITROSOCARBAMATE see FIS000
1-(2-FLUOROETHYL)-1-NITROSO-UREA see NKG000
1-FLUORO-2-FAA see FFL000
3-FLUORO-2-FAA see FFM000
4-FLUORO-2-FAA see FFN000
5-FLUORO-2-FAA see FFO000
6-FLUORO-2-FAA see FFP000
7-FLUORO-2-FAA see FIT200
8-FLUORO-2-FAA see FFQ000
FLUOROFLEX see TAI250
N-(7-FLUOROFLUORENE-2-YL)ACETAMIDE see FFG000
N-(1-FLUOROFLUOREN-2-YL)ACETAMIDE see FFL000
N-(3-FLUOROFLUOREN-2-YL)ACETAMIDE see FFM000
N-(4-FLUOROFLUOREN-2-YL)ACETAMIDE see FFN000
N-(5-FLUOROFLUOREN-2-YL)ACETAMIDE see FFO000
N-(6-FLUOROFLUOREN-2-YL)ACETAMIDE see FFP000
N-(8-FLUOROFLUOREN-2-YL)ACETAMIDE see FFQ000
7-FLUORO-2-N-(FLUORENYL)ACETHYDROXAMIC ACID see FIT200
N-(7-FLUORO-2-FLUORENYL)ACETOHYDROXAMIC ACID see FIT200
7-FLUORO-N-(FLUOREN-2-YL)ACETOHYDROXAMIC ACID see FIT200
4'-FLUORO-4-(8-FLUORO-2,3,4,5-TETRAHYDRO-1H-PYRIDO(4,3-b)INDOL-2-
    YL)BUTYROPHENONE HYDROCHLORIDE see FIW000
FLUOROFORM see CBY750
FLUOROFORMYL FLUORIDE see CCA500
FLUOROFORMYLON see FDB000
FLUOROFUR see FLZ050
1-FLUOROHEPTANE see FIX000
7-FLUOROHEPTANONITRILE see FIY000
7-FLUOROHEPTYLAMINE see FIZ000
1-FLUOROHEXANE see FJA000
FLUOROHEXANE see FJA000
5-FLUORO-1-HEXYLCARBAMOYL-URACIL see CCK630
9-α-FLUOROHYDROCORTISONE see FHH100
7-FLUORO-N-HYDROXY-N-2-ACETYLAMINOFLUORENE see FIT200
4-FLUORO-3-HYDROXY-BUTANETHIOIC ACID METHYL ESTER see MKF000
γ-FLUORO-β-HYDROXY-BUTYRIC ACID METHYL ESTER see MKE750
4-FLUORO-2-HYDROXYBUTYRIC ACID SODIUM SALT see SHI000
4'-FLUORO-4-(4-HYDROXY-4-(4'-CHLOROPHE-
    NYL)PIPERIDINO)BUTYROPHENONE see CLY500
9-α-FLUORO-17-HYDROXYCORTICOSTERONE see FHH100
9-α-FLUORO-11-β-HYDROXY-17-METHYLTESTOSTERONE see AOO275
9-α-FLUORO-16-α-HYDROXYPREDNISOLONE see AQX250
9-α-FLUORO-16-HYDROXYPREDNISOLONE ACETONIDE see AQX500
4'-FLUORO-4-(4-HYDROXY-4-p-TOLYLPIPERIDINO)BUTYROPHENONE, HYDRO-
    CHLORIDE see MNN250
4'-FLUORO-4-(4-HYDROXY-4-(α,α,α-TRIFLUORO-m-TOL-
    YL)PIPERIDINO)BUTYROPHENONE see TKK500
4-FLUORO-4,4-IDROSSI-4-(m-TRIFLUOROMETIL-FENIL)-PIPERIDINO-BUTIRRO-
    FENONE (ITALIAN) see TKK500
1-FLUOROIMINOHEXAFLUOROPROPANE see FJI500
2-FLUOROIMINOHEXAFLUOROPROPANE see FJI510
4-FLUORO-3-IODOBENZOIC ACID-3-(DIBUTYLAMINO)PROPYL ESTER, HYDRO-
    CHLORIDE see HHM500
1-FLUORO-2-IODOETHANE see FIQ000

FLUOROISOPROPOXYMETHYLPHOSPINE OXIDE see IPX000
9-α-FLUORO-16-α-17-α-ISOPROPYLEDENE DIOXY PREDNISOLONE see AQX500
9-α-FLUORO-16-α-17-α-ISOPROPYLIDENEDIOXY-Δ-1-HYDROCORTISONE see AQX500
FLUOROLON 3 see KDK000
FLUOROLON 4 see TAI250
FLUOROLUBE 2000 see KDK000
FLUOROLUBE 300/140 see KDK000
FLUOROLUBE FS 5 see KDK000
FLUOROLUBE GR 470 see KDK000
FLUOROLUBE S 30 see KDK000
FLUOROMAR see TKB250
FLUOROMETHANE see FJK000
4′-FLUORO-4-(4-(o-METHOXYPHENYL)-1-PIPERAZINYL)BUTYROPHENONE see HAF400
4′-FLUORO-4-(4-(o-METHOXYPHENYL)-1-PIPERAZINYL)BUTYROPHENONE HYDROCHLORIDE see FDA880
3-FLUORO-10-METHYL-5,6-BENZACRIDINE see MKD750
1-FLUORO-10-METHYL-7,8-BENZACRIDINE see MKD500
3-FLUORO-10-METHYL-7,8-BENZACRIDINE (FRENCH) see MKD250
10-FLUORO-12-METHYLBENZ(a)ACRIDINE see MKD750
3′-FLUORO-10-METHYL-1,2-BENZANTHRACENE see FJO000
6-FLUORO-10-METHYL-1,2-BENZANTHRACENE see FJQ000
7-FLUORO-10-METHYL-1,2-BENZANTHRACENE see FJR000
10-FLUORO-7-METHYLBENZ(a)ANTHRACENE see FJR000
2-FLUORO-7-METHYLBENZ(a)ANTHRACENE see FJN000
3-FLUORO-7-METHYLBENZ(a)ANTHRACENE see FJO000
6-FLUORO-7-METHYLBENZ(a)ANTHRACENE see FJP000
9-FLUORO-7-METHYLBENZ(a)ANTHRACENE see FJQ000
1-FLUORO-2-METHYLBENZENE see FLZ100
4-FLUORO-7-METHYL-6H-(1)BENZOTHIOPYRANO(4,3-b)QUINOLINE see FJT000
2-FLUORO-α-METHYL(1,1′-BIPHENYL)-4-ACETIC ACID see FLG100
2-FLUORO-α-METHYL-4-BIPHENYLACETIC ACID see FLG100
2-FLUORO-α-METHYL-(1,1′-BIPHENYL)-4-ACETIC ACID 1-(ACETYLOXY)ETHYL ESTER see FJT100
2-FLUORO-3-METHYLCHOLANTHRENE see FJU000
6-FLUORO-3-METHYLCHOLANTHRENE see FJV000
9-FLUORO-3-METHYLCHOLANTHRENE see FJW000
11-FLUORO-5-METHYLCHRYSENE see FKD000
12-FLUORO-5-METHYLCHRYSENE see FKE000
1-FLUORO-5-METHYLCHRYSENE see FJY000
6-FLUORO-5-METHYLCHRYSENE see FKA000
7-FLUORO-5-METHYLCHRYSENE see FKB000
9-FLUORO-5-METHYLCHRYSENE see FKC000
Δ¹-9-α-FLUORO-16-α-METHYLCORTISOL see SOW000
FLUOROMETHYL CYANIDE see FFJ000
9-α-FLUORO-17-α-METHYL-11-β,17-DIHYDROXY-4-ANDROSTEN-3-ONE see AOO275
cis-5-FLUORO-2-METHYL-1-((4-METHYLSULFINYL)PHENYL)METHYLENE))-1H-INDENE-3-ACETIC ACID see SOU550
2-FLUORO-N-METHYL-N-1-NAPHTHALENYLACETAMIDE see MME809
2-FLUORO-N-METHYL-N-1-NAPHTHYLACETAMIDE see MME809
p-FLUORO-N-METHYL-N-NITROSOANILINE see FKF800
4′-FLUORO-4-(4-METHYLPIPERIDINO)BUTYROPHENONE HYDROCHLORIDE see FKI000
9-α-FLUORO-16-α-METHYLPREDNISOLONE see SOW000
9-α-FLUORO-16-β-METHYLPREDNISOLONE see BFV750
6-α-FLUORO-16-α-METHYLPREDNISOLONE-21-ACETATE see PAL600
9-α-FLUORO-16-β-METHYL- 1,4-PREGNADIENE-11-β,17-α,21-TRIOL-3,20-DIONE see BFV750
9-α-FLUORO-16-α-METHYL-1,4-PREGNADIENE-11-β,17-α,21-TRIOL-3,20-DIONE see SOW000
4-α-FLUORO-16-α-METHYL-11-β,17,21-TRIHYDROXYPREGNA-1,4-DIENE-3,20-DIONE see SOW000
FLUOROMETHYL(1,2,2-TRIMETHYLPROPOXY)PHOSPHINE OXIDE see SKS500
4-FLUORO-3-NITROANILINE see FKK100
1-FLUORO-4-NITROBENZENE see FKL000
4-FLUORONITROBENZENE see FKL000
p-FLUORONITROBENZENE see FKL000
3-FLUORO-4-NITROQUINOLINE-1-OXIDE see FKO000
8-FLUORO-4-NITROQUINOLINE-1-OXIDE see FKP000
FLUORONIUM PERCHLORATE see FKQ000
9-FLUORONONYL PHENYL KETONE see FKQ100
8-FLUOROOCTANOIC ACID, ETHYL ESTER see EKK500
8-FLUOROOCTYL BROMIDE see FKS000
8-FLUOROOCTYL CHLORIDE see FKT000
8-FLUOROOCTYL PHENYL KETONE see FKT050
5-FLUOROOROTATE see TCQ500
5-FLUOROOROTIC ACID see TCQ500
4′-FLUORO-4-(4-OXO-1-PHENYL-1,3,8-TRIAZASPIRO(4,5)DECAN-8-YL)-BUTYROPHENONE see SLE500
FLUOROPAK 80 see TAI250
5-FLUOROPENTYLAMINE see FFV000
5-FLUOROPENTYL THIOCYANATE see FFX000

2-FLUOROPHENOL see FKT100
4-FLUOROPHENOL see FKV000
o-FLUOROPHENOL see FKT100
7-(2-(4-(4-FLUOROPHENOXY)-3-HYDROXY-1-BUTENYL)-3,5-DIHYDROXYCYCLO-PENTYL)-(1-α-(Z),2-β-(1E,3S*),3-α,5-α)-5-HEPTENOIC ACID see IAD100
2-FLUORO-N-PHENYLACETAMIDE see FFH000
2′-FLUORO-4′-PHENYLACETANILIDE see FKX000
4′-(m-FLUOROPHENYL)ACETANILIDE see FKY000
4′-(p-FLUOROPHENYL)ACETANILIDE see FKZ000
4-FLUOROPHENYLACETONITRILE see FLC000
p-FLUOROPHENYLACETONITRILE see FLC000
p-FLUOROPHENYLAMINE see FFY000
p-((p-FLUOROPHENYL)AZO)-N,N-DIMETHYLANILINE see DSA000
4-((4-FLUOROPHENYL)AZO)-N,N-DIMETHYLBENZENAMINE see DSA000
p-FLUOROPHENYL ETHYL SULFONE see FLG000
3-FLUORO-4-PHENYLHYDRATROPIC ACID see FLG100
1-(4-FLUOROPHENYL)-4-(4-HYDROXY-4-(4-METHYLPHENYL)-1-PIPERIDINYL)-1-BUTANONE HYDROCHLORIDE see MNN250
1-(4-FLUOROPHENYL)-4-(4-HYDROXY-4-(3-(TRIFLUOROMETHYL)PHENYL)-1-PIPERIDINYL)-1-BUTANONE see TKK500
4-(p-FLUOROPHENYL)-1-ISOPROPYL-7-METHYL-2(1H)-QUINAZOLINONE see THH350
3-FLUOROPHENYL ISOTHIOCYANATE see ISL000
4-FLUOROPHENYL ISOTHIOCYANATE see ISM000
m-FLUOROPHENYL ISOTHIOCYANATE see ISL000
p-FLUOROPHENYL ISOTHIOCYANATE see ISM000
4-FLUOROPHENYLLITHIUM see FLH100
(±)-α-(p-FLUOROPHENYL)-4-(o-METHOXYPHENYL)-1-PIPERAZINEBUTANOL see HAH000
dl-1-(4-FLUOROPHENYL)-4-(1-(4-(2-METHOXY-PHENYL))-PIPERAZINYL)BUTANOL see HAH000
4-(4-FLUOROPHENYL)-7-METHYL-1-(1-METHYLETHYL)-2(1H)-QUINAZOLINONE see THH350
8-(4-p-FLUORO PHENYL-4-OXOBUTYL)-2-METHYL-2,8-DIAZASPI-RO(4.5)DECANE-1,3-DIONE see FLJ000
8-(4-(4-FLUOROPHENYL)-4-OXOBUTYL)-1-PHENYL-1,3,8-TRIAZASPI-RO(4.5)DECAN-4-ONE see SLE500
1-(1-(4-(4-FLUOROPHENYL)-4-OXOBUTYL)-4-PIPERIDINYL)-1,3-DIHYDRO-2H-BENZIMIDAZOL-2-ONE see FLK100
1-(1-(4-(p-FLUOROPHENYL-4-OXOBUTYL))-1,2,3,6-TETRAHYDRO-4-PYRIDYL)-2-BENZIMIDAZOLINONE see DYF200
1-(4-FLUOROPHENYL)-4-(4-PHENYL-1-PIPERAZINYL)-1-BUTANONE see FLL000
4′-FLUORO-4-(1-(4-PHENYL)PIPERAZINO)BUTYROPHENONE see FLL000
1-(4-FLUOROPHENYL)-4-(4-(2-PYRIDINYL)-1-PIPERAZINYL)-1-BUTANONE see FLU000
FLUOROPHOSGENE see CCA500
FLUOROPHOSPHORIC ACID, anhydrous see PHJ250
4′-FLUORO-4-(4-N-PIPERIDINO-4-CARBAMIDOPIPERIDINO)BUTYROPHENONE see FHG000
p-FLUORO-γ-(4-PIPERIDINO-4-CARBAMOYLPIPERIDINO)BUTYROPHENONE see FHG000
4′-FLUORO-4-(4-PIPERIDINO-4-PROPIONYLPIPERIDINO)BUTYROPHENONE see FLN000
FLUOROPLAST 3 see CLQ750, KDK000
FLUOROPLAST 4 see TCH500
FLUOROPLAST F 3 see KDK000
FLUOROPLAST 3P see KDK000
FLUOROPLEX see FMM000
3-FLUOROPROPENE see AGG500
2-FLUORO-2-PROPEN-1-OL see FLQ000
3-FLUOROPROPIONIC ACID see FLR000
ω-FLUOROPROPIONIC ACID see FLR000
3-FLUOROPROPYL ISOCYANATE see FLR100
FLUOROPRYL see IRF000
2-FLUOROPYRIDINE see FLT100
4′-FLUORO-4-(4-(2-PYRIDYL)-1-PIPERAZINYL)BUTYROPHENONE see FLU000
5-FLUORO-2,4(1H,3H)-PYRIMIDINEDIONE see FMM000
5-FLUORO-2,4-PYRIMIDINEDIONE see FMM000
5-FLUORO-4(1H)-PYRIMIDINONE see FMO100
4′-FLUORO-4-(n-(4-PYRROLIDINAMIDO)-4-m-TOLYPIPERIDI-NO)BUTYROPHENONE see FLV000
4′-FLUORO-4-(1-PYRROLIDINYL)BUTYROPHENONE MALEATE see FGW000
FLUOROSILICIC ACID see SCO500
4′-FLUORO-4-STILBENAMINE see FLY000
2-FLUORO-4-STILBENYL-N,N-DIMETHYLAMINE see FHU000
4′-FLUORO-4-STILBENYL-N,N-DIMETHYLAMINE see FHV000
FLUOROSUFONIC ACID (DOT) see FLZ000
FLUOROSULFONATES see FLY100
m-FLUOROSULFONYLBENZENESULFONYL CHLORIDE see FLY200
FLUOROSULFONYL CHLORIDE see SOT500
FLUOROSULFURIC ACID see FLZ000
FLUOROSULFURYL HYPOFLUORITE see FFB000
FLUOROTANE see HAG500
mu-FLUOROTETRAFLUORODISTANNATE(1-), SODIUM see SJA500

5-FLUORO-1,2,3,6-TETRAHYDRO-2,6-DIOXO-4-PYRIMIDINECARBOXYLIC ACID see TCQ500
5-FLUORO-1-(TETRAHYDRO-2-FURANYL)-2,4(1H,3H)-PYRIMIDINEDIONE see FLZ050
5-FLUORO-1-(TETRAHYDRO-2-FURANYL)-2,4-PYRIMIDINEDIONE see FLZ050
5-FLUORO-1-(TETRAHYDROFURAN-2-YL)URACIL see FLZ050
5-FLUORO-1-(TETRAHYDRO-3-FURYL)URACIL see FLZ050
9-α-FLUORO-11-β,16-α,17-α,21-TETRAHYDROXYPREGNA-1,4-DIENE-3,20-DIONE see AQX250
9-α-FLUORO-11-β,16-α,17,21-TETRAHYDROXY-1,4-PREGNADIENE-3,20-DIONE see AQX250
9-α-FLUORO-11-β,16-α,17,21-TETRAHYDROXYPREGNA-1,4-DIENE-3,20-DIONE see AQX250
6-α-FLUORO-11-β,16-α,17,21-TETRAHYDROXYPREGNA-1,4-DIENE,-3,20-DIONE, CYCLIC 16,17-ACETAL with ACETONE see FDD085
FLUOROTHENE see KDK000
2-FLUOROTHIOPYRANO(4,3-b)BENZ(e)INDOLE see FGJ000
4-FLUOROTHIOPYRANO(4,3-b)BENZ(e)INDOLE see FGL000
4'-FLUORO-4-(4-(2-THIOXOBENZIMIDAZOL-1-YL)PIPERIDINO)-BUTYROPHENONE see TGB175
FLUOROTHYL see HDC000
2-FLUOROTOLUENE see FLZ100
o-FLUOROTOLUENE see FLZ100
p-FLUOROTOLUENE see FMC000
FLUOROTRIBUTYLSTANNANE see FME000
FLUOROTRICHLOROMETHANE (OSHA) see TIP500
2-FLUOROTRICYCLOQUINAZOLINE see FMF000
3-FLUORO-TRICYCLOQUINAZOLINE see FMG000
4-FLUORO-4'-TRIFLUOROMETHYLBENZOPHENONE GUANYLHYDRAZONE HYDROCHLORIDE see FMH000
9-FLUORO-11-β,17,21-TRIHYDROXY-16-β-METHYLPREGNA-1,4-DIENE-3,20-DIONE see BFV750
9-α-FLUORO-11-β,17,21-TRIHYDROXY-16-β-METHYLPREGNA-1,4-DIENE- 3,20-DIONE see BFV750
9-α-FLUORO-11-β,17-α,21-TRIHYDROXY-16-α-METHYLPREGNA-1,4-DIENE-3,20-DIONE see SOW000
9-FLUORO-11-β,17,21-TRIHYDROXY-16-α-METHYLPREGNA-1,4-DIENE-3,20-DIONE see SOW000
9-FLUORO-11-β,17,21-TRIHYDROXY-16-β-METHYLPREGNA-1,4-DIENE-3,20-DIONE, 21-(DIHYDROGEN PHOSPHATE), DISODIUM SALT see BFV770
9-FLUORO-11-β,17,21-TRIHYDROXY-16-α-METHYLPREGNA-1,4-DIENE-3,20-DIONE-21-(DIHYDROGEN PHOSPHATE) DISODIUM SALT see DAE525
9-FLUORO-11-β,17,21-TRIHYDROXY-16-α-METHYLPREGNA-1,4-DIENE-3,20-DIONE-17,21-DIPROPIONATE see DBC500
9-FLUORO-11-β,17,21-TRIHYDROXY-16-β-METHYLPREGNA-1,4-DIENE-3,20-DIONE, 17,21-DIPROPIONATE see BFV765
9-FLUORO-11-β,17,21-TRIHYDROXY-16-α-METHYLPREGNA-1,4-DIENE-3,20-DIONE-21-(HYDROGEN SULFATE), MONOSODIUM SALT see DBC550
9-FLUORO-11-β,17,21-TRIHYDROXY-16-α-METHYLPREGNA-1,4-DIENE-3,20-DIONE, 21-ISONICOTINATE see DBC510
9-FLUORO-11-β,17,21-TRIHYDROXY-16-α-METHYLPREGNA-1,4-DIENE-3,20-DIONE-17-VALERATE see DBC575
9-FLUORO-11-β,17,21-TRIHYDROXY-16-β-METHYLPREGNA-1,4-DIENE-3,20,DIONE-17-VALERATE see VCA000
9-FLUORO-11-β,17,21-TRIHYDROXYPREGN-4-ENE-3,20-DIONE see FHH100
9-α-FLUORO-11-β,17-α,21-TRIHYDROXY-4-PREGNENE-3,20-DIONE see FHH100
FLUOROTRINITROMETHANE see FMI000
3-FLUORO-1,2,4-TRIOXOLANE see FIK875
FLUOROTRIPHENYLSTANNANE see TMV850
FLUOROTROJCHLOROMETAN (POLISH) see TIP500
3-FLUOROTYROSIN see FMJ000
3-FLUOROTYROSINE see FMJ000
m-FLUOROTYROSINE see FMJ000
5-FLUOROURACIL see FMM000
FLUOROURACIL see FMM000
5-FLUOROURACIL-2'-DEOXYRIBOSIDE see DAR400
5-FLUOROURACIL DEOXYRIBOSIDE see DAR400
5-FLUOROURIDINE see FMN000
5-FLUOROVALERONITRILE see FMO000
FLUOROWODOR (POLISH) see HHU500
FLUOROXENE see TKB250
4-FLUORPHENYL-1-ISOPROPYL-7-METHYL-2(1H)-CHINAZOLINON (GERMAN) see THH350
FLUORPLAST 4 see TAI250
5-FLUORPROPYRIMIDINE-2,4-DIONE see FMM000
FLUORSPAR see CAS000
FLUORTHYRIN see FMJ000
3-FLUORTYROSIN (GERMAN) see FMJ000
5-FLUORURACIL (GERMAN) see FMM000
FLUORURE de BORE (FRENCH) see BMG700
FLUORURE de N,N'-DIISOPROPYLE PHOSPHORODIAMIDE (FRENCH) see PHF750
FLUORURE de N,N,N',N'-TETRAMETHYLE PHOSPHORO-DIAMIDE (FRENCH) see BJE750
FLUORURE de POTASSIUM (FRENCH) see PLF500

FLUORURES ACIDE (FRENCH) see FEZ000
FLUORURE de SODIUM (FRENCH) see SHF500
FLUORURE de SULFURYLE (FRENCH) see SOU500
FLUORURE de THIONYLE (FRENCH) see TFL250
FLUORURI ACIDI (ITALIAN) see FEZ000
FLUORURIDINE DEOXYRIBOSE see DAR400
FLUORWASSERSTOFF (GERMAN) see HHU500
FLUORWATERSTOF (DUTCH) see HHU500
FLUORXENE see TKB250
FLUORYL see CBY750
FLUOSILICATE de ALUMINUM (FRENCH) see THH000
FLUOSILICATE de AMMONIUM (FRENCH) see COE000
FLUOSILICATE de MAGNESIUM (FRENCH) see MAG250
FLUOSILICATE de SODIUM see DXE000
FLUOSILICIC ACID see SCO500
FLUOSOL 43 see HAS000
FLUOSOL-DA 20% see FMO050
FLUOSTIGMINE see IRF000
FLUOSULFONIC ACID (DOT) see FLZ000
FLUOTESTIN see AOO275
FLUOTHANE see HAG500
FLUOTITANATE de POTASSIUM (FRENCH) see PLI000
FLUOTRACEN see DMF800
FLUOVITIF see SPD500
FLUOXIDINE see FMO100
FLUOXIMESTERONE see AOO275
FLUOXYDINE see FMO100
FLUOXYMESTERONE see AOO275
FLUOXYMESTRONE see AOO275
FLUOXYPREDNISOLONE see AQX250
(α,β)-FLUPENTHIXOL see FMO129
cis-(Z)-FLUPENTHIXOL see FMO129
FLUPENTHIXOL see FMO129
FLUPENTHIXOLE see FMO129
FLUPENTIXOL see FMO129
FLUPENTIXOL DIHYDROCHLORIDE see FMO150
FLUPENTIXOL HYDROCHLORIDE see FMO150
FLUPHENAMIC ACID see TKH750
FLUPHENAZINE see TJW500
FLUPHENAZINE DIHYDROCHLORIDE see FMP000
FLUPHENAZINE ENANTHATE see PMI250
FLUPHENAZINE HYDROCHLORIDE see FMP000
FLUPIRTINE MALEATE see FMP100
FLUPIRTIN-MALEAT (GERMAN) see FMP100
FLUPROQUAZONE see THH350
FLURACIL see FMM000
FLURA-GEL see SHF500
FLURAZEPAM see FMQ000
FLURAZEPAM DIHYDROCHLORIDE see DAB800
FLURAZEPAM HYDROCHLORIDE see DAB800
FLURAZEPAM MONOHYDROCHLORIDE see FMQ100
FLURAZOLE see BEG300
FLURBIPROFEN see FLG100
FLURBIPROFEN AXETIL see FJT100
FLURCARE see SHF500
FLURENTIXOL see FMO129
FLURI see EMM000
FLURIDONE see FMQ200
FLURIL see FMM000
4'-FLURO-4-(1,2,4,4a,5,6-HEXAHYDRO-3H-PYRANZINO(1,2-A)QUINOLIN-3-YL)-BUTYROPHENONE 2HCl see CNH500
FLUROTHYL see HDC000
FLUROXENE see TKB250
FLUSTERON see AOO275
FLUTAMIDE see FMR050
FLUTAZOLAM see FMR075
FLUTESTOS see AOO275
FLUTONE see AQX500
FLUTOPRAZEPAM see FMR100
FLUTRA see HII500
FLUTROPIUM BROMIDE HYDRATE see FMR300
FLUVIN see CFY000
FLUVOMYCIN see FMR500
FLUXANXOL see FMO129
FLUXEMA see POD750
FLUX MAAG see NDN000
FLY BAIT GRITS see SOY000
FLY-DIE see DGP900
FLY FIGHTER see DGP900
FLYPEL see DKC800
F 3M see KDK000
13FM see KDK000
FM 100 see LFN000
FM 510 see PJS750
FMA (analytical reagent) see FEV100

FMA see ABU500, FEV100
FMC 249 see AFR250
FMC-1240 see EEH600
FMC 5273 see PIX250
FMC 5462 see EAQ750
FMC 5488 see CKM000
FMC 9044 see BGB500
FMC-9102 see MQQ250
FMC 10242 see CBS275
FMC 11092 see DUM800
FMC-16388 see PNV750
FMC 17370 see BEP500
FMC 30980 see RLF350
FMC 33297 see AHJ750
FMC 41655 see AHJ750
FMC 45497 see RLF350
FMC 45806 see RLF350
FM-NTS see CCU250
4-F-3NA see FKK100
F-6-NDBA see NJN300
F-NORSTEARANTHRENE see TCJ500
FNT see NDY500
FO see TCQ500
FOA see TCQ500
FOBEX see BCA000
FOCUSAN see TGB475
FOGARD see FMR700
FOGARD S see FMR700
FOIN ABSOLUTE see FMS000
FOIN COUPE see FMS000
FOLACIN see FMT000
FOLAN RED B see CMM330
FOLAN YELLOW G see CMM320
FOLATE see FMT000
FOLBEX see DER000
FOLBEX SMOKE-STRIPS see DER000
FOLCID see CBF800
FOLCODAL see CMR100
FOLCODINE see TCY750
FOLCYSTEINE see FMT000
FOLEDRIN see FMS875
FOLETHION see DSQ000
FOLEX see TIG250
FOLIANDRIN see OHQ000
FOLIC ACID see FMT000
FOLIC ACID, 4-AMINO- see AMG750
FOLIDOL see PAK000
FOLIDOL M see MNH000
FOLIGAN see ZVJ000
FOLIKRIN see EDV000
FOLIMAT see DNX800
FOLINERIN see OHQ000
FOLINEVIN see OHQ000
FOLIONE see MND275
FOLIPEX see EDV000
FOLISAN see EDV000
FOLKS GLOVE see FOM100
FOLLESTRINE see EDV000
FOLLICORMON see EDP000
FOLLICULAR HORMONE see EDV000
FOLLICULAR HORMONE HYDRATE see EDU500
FOLLICULIN see EDV000
FOLLICULINE BENZOATE see EDV000
FOLLICUNODIS see EDV000
FOLLICYCLIN P see EDR000
FOLLIDIENE see DAL600, DKA600
FOLLIDRIN see EDP000, EDV000
FOLLINYL see NNL500
FOLLORMON see DAL600
FOLOSAN see PAX000, TBR750
FOLOSAN DB-905 FUMITE see TBS000
FOLPAN see TIT250
FOLPET see TIT250
FOLSAN see TBS000
FOMAC 2 see PAX000
FOMAC see HCL000
FOM-Ca HYDRATE see CAW376
FOMINOBEN HYDROCHLORIDE see FMU000
FOM-Na see DXF600
p-FOMOCAINE see PDU250
FOMOCAINE see PDU250
FOMREZ SUL-3 see DBF800
FOMREZ SUL-4 see DDV600
FONATOL see DKA600
FONAZINE MESYLATE see FMU039

FONDAREN see DRP600
FONOFOS see FMU045
FONOLINE see MQV750
FONTARSAN see ARL000
FONTARSOL see DFX400
FONTEGO see BON325
FONTILIX see MQQ050
FONTILIZ see MQQ050
FONURIT see AAI250
FONZYLANE see BOM600
FOOD BLUE 1 see FMU059
FOOD BLUE 2 see FAE000
FOOD BLUE 3 see ADE500
FOOD BLUE 4 see IBV050
FOOD BLUE DYE No. 1 see FAE000
FOODCOL SUNSET YELLOW FCF see FAG150
FOOD DYE RED No. 104 see ADG250, CMM000
FOOD GREEN 1 see FAE950
FOOD GREEN 2 see FAF000
FOOD GREEN S see ADF000
FOOD RED 2 see CMP620, FAG020
FOOD RED 4 see FAG050
FOOD RED 5 see HJF500
FOOD RED 6 see FMU080
FOOD RED 7 see FMU080
FOOD RED 9 see FAG020
FOOD RED 14 see FAG040
FOOD RED 15 see FAG070
FOOD RED 16 see CMS242
FOOD RED COLOR No. 105, SODIUM SALT see RMP175
FOOD RED No. 101 see FMU070
FOOD RED No. 102 see FMU080
FOOD RED No. 104 see ADG250
FOOD RED No. 105, SODIUM SALT see RMP175
FOOD YELLOW 13 see CMM510
FOOD YELLOW 2 see CMM758
FOOD YELLOW 3 see FAG150
FOOD YELLOW 4 see FAG140
FOOD YELLOW 5 see FAG140
FOOD YELLOW 6 see FAG150
FOOD YELLOW NO. 4 see FAG140
FOOL'S CICELY see FMU200
FOOL'S PARSLEY see FMU200
FOPIRTOLINA (SPANISH) see FMU225
FOPIRTOLINE HYDROCHLORIDE see FMU225
FORAAT (DUTCH) see PGS000
FORALAMIN FUMARATE see MDP800
FORALMINE FUMARATE see MDP800
FORANE 22 see CFX500
FORANE see IKS400
FORAPIN see MCB525
FORDIURAN see BON325
FORE see DXI400
FOREDEX 75 see DAA800
FORENOL see NDX500
FORHISTAL MALEATE see FMU409
FORIOD see TDE750
FORIT see ECW600
FORLEX see MLC000
FORLIN see BBQ500
FORMAGENE see PAI000
FORMAL see MAK700, MGA850
FORMALDEHYD (CZECH, POLISH) see FMV000
FORMALDEHYDE, solution (DOT) see FMV000
FORMALDEHYDE see FMV000
FORMALDEHYDE-ANILINE COPOLYMER see FMW330
FORMALDEHYDE BIS(β-CHLOROETHYL) ACETAL see BID750
FORMALDEHYDE COPOLYMER with UREA see UTU500
FORMALDEHYDE CYANOHYDRIN see HIM500
FORMALDEHYDE CYCLODODECYL ETHYL ACETAL see FMV100
FORMALDEHYDE CYCLODODECYL METHYL ACETAL see FMV200
FORMALDEHYDE DIMETHYLACETAL see MGA850
FORMALDEHYDE HYDROSULFITE see FMW000
FORMALDEHYDE-MELAMINE CONDENSATE see MCB050
FORMALDEHYDE-MELAMINE COPOLYMER see MCB050
FORMALDEHYDE-MELAMINE POLYMER see MCB050
FORMALDEHYDE-MELAMINE RESIN see MCB050
FORMALDEHYDE OXIDE POLYMER see FMW300
FORMALDEHYDE, POLYMER with BENZENAMINE see FMW330
FORMALDEHYDE SODIUM BISULFITE ADDUCT see FMW000
FORMALDEHYDE SODIUM SULFOXYLATE see FMW000
FORMALDEHYDESULFOXYLIC ACID SODIUM SALT see FMW000
FORMALDEHYDE-UREA CONDENSATE see UTU500
FORMALDEHYDE-UREA COPOLYMER see UTU500
FORMALDEHYDE-UREA POLYMER see UTU500

FORMALDEHYDE-UREA PRECONDENSATE see UTU500
FORMALDEHYDE-UREA PREPOLYMER see UTU500
FORMALDEHYDE-UREA RESIN see UTU500
FORMAL FAST BLACK 2B see CMN240
FORMAL GLYCOL see DVR800
FORMAL HYDRAZINE see FNN000
FORMALIN 40 see FMV000
FORMALIN (DOT) see FMV000
FORMALIN see FMV000
FORMALINA (ITALIAN) see FMV000
FORMALINE (GERMAN) see FMV000
FORMALINE BLACK C see AQP000
FORMALIN-LOESUNGEN (GERMAN) see FMV000
FORMALIN-MELAMINE COPOLYMER see MCB050
FORMALIN-UREA COPOLYMER see UTU500
FORMALITH see FMV000
FORMAL-γ-TRIMETHYLAMMONIUM PROPANEDIOL see FMX000
FORMAMIDE see FMY000
FORMAMIDE, N-(1,1'-BIPHENYL)-4-YL-N-HYDROXY- see HLE650
FORMAMIDE, N-ETHYL- see EKK600
FORMAMIDE, N,N'-(DITHIOBIS(2-(2-HYDROXYETHYL)-1-METHYLVINY-
   LENE))BIS(N-((4-AMINO-2-METHYL-5-PYRIMIDINYL)METHYL)- see TES800
FORMAMIDE, N,N'-(1,4-PIPERAZINEDIYLBIS(2,2,2-TRICHLOROETHYLI-
   DENE))BIS-(8CI,9CI) see TKL100
FORMAMIDOBENZENE see FNJ000
FORMAMINE see HEI500
FORMANILIDE see FNJ000
FORMARIN see MNM500
FORMATRIX see ECU750
3-FORMAZANTHIOL, 1,5-DIPHENYL- see DWN200
FORMEBOLONE see FNK040
FORMETANATE HYDROCHLORIDE see DSO200
FORMHYDRAZID (GERMAN) see FNN000
FORMHYDRAZIDE see FNN000
FORMHYDROXAMIC ACID see FMZ000
FORMHYDROXAMSAEURE (GERMAN) see FMZ000
FORMIATE de METHYLE (FRENCH) see MKG750
FORMIATE de PROPYLE (FRENCH) see PNM500
FORMIC ACID see FNA000
FORMIC ACID, ALLYL ESTER see AGH000
FORMIC ACID AMMONIUM SALT see ANH500
FORMIC ACID, AZIDO-, tert-BUTYL ESTER see BQI250
FORMIC ACID, AZIDODITHIO- see ASF750
FORMIC ACID, 1-BUTYLHYDRAZIDE see BRK100
FORMIC ACID, CALCIUM SALT see CAS250
FORMIC ACID, CHLORO-, 2-CHLOROETHYL ESTER see CGU199
FORMIC ACID, CHLORO-, FLUOREN-9-YLMETHYL ESTER see FEI100
FORMIC ACID, CHLORO-, 2-FLUOROETHYL ESTER see FIH100
FORMIC ACID, CHLORO-, OXYDIETHYLENE ESTER see OPO000
FORMIC ACID, CHLORO-, TRICHLOROMETHYL ESTER see TIR920
FORMIC ACID, CHLORO-, 2,4,6-TRICHLOROPHENYL ESTER see TIY800
FORMIC ACID, CINNAMYL ESTER see CMR500
FORMIC ACID, CITRONELLYL ESTER see CMT750
FORMIC ACID-3,7-DIMETHYL-6-OCTEN-1-YL ESTER see CMT750
FORMIC ACID, DITHIOBIS(THIO-, O,O-DIBUTYL ESTER see BSS550
FORMIC ACID, ETHYL ESTER see EKL000
FORMIC ACID, 1-ETHYLHYDRAZIDE see EKL250
FORMIC ACID, GERANIOL ESTER see GCY000
FORMIC ACID, HEPTYL ESTER see HBO500
FORMIC ACID, HEXYL ESTER see HFQ100
FORMIC ACID, HYDRAZIDE see FNN000
FORMIC ACID, ISOBUTYL ESTER see IIR000
FORMIC ACID, ISOPENTYL ESTER see IHS000
FORMIC ACID, ISOPROPYL ESTER see IPC000
FORMIC ACID, METHYLHYDRAZIDE see FNW000
FORMIC ACID, METHYLPENTYLIDENEHYDRAZIDE see PBJ875
FORMIC ACID (2-(4-METHYL-2-THIAZOLYL))HYDRAZIDE see FNB000
FORMIC ACID, NERYL ESTER see FNC000
FORMIC ACID, OCTYL ESTER see OEY100
FORMIC ACID, 1-PROPYLHYDRAZIDE see PNM650
FORMIC ACID, compounded with QUININE (1:1) see QIS300
FORMIC ALDEHYDE see FMV000
FORMIC ANAMMONIDE see HHS000
FORMIC BLACK BA see AQP000
FORMIC BLACK C see AQP000
FORMIC BLACK CW see AQP000
FORMIC BLACK MTG see AQP000
FORMIC BLACK TG see AQP000
FORMIC ETHER see EKL000
FORMIC HYDRAZIDE see FNN000
FORMIC 2-(4-(5-NITROFURYL)-2-THIAZOLYL)HYDRAZIDE see NDY500
FORMILOXIN see FND100
FORMILOXINE see FND100
FORMIN see HEI500
FORMIN K-K see MCB050

FORMOCORTAL see FDB000
FORMOHYDRAZIDE see FNN000
FORMOL see FMV000
FORMOLA 40 see DAA800
FORMOMALENIC THALLIUM see TEM399
FORMONITRILE see HHS000
FORMOPAN see FMW000
FORMOSA CAMPHOR see CBA750
FORMOSA CAMPHOR OIL see CBB500
FORMOSE OIL of CAMPHOR see CBB500
FORMOSULFACETAMIDE see SNP500
FORMOSULFATHIAZOLE see TEX250
FORMOTEROL FUMARATE DIHYDRATE see FNE100
FORMOTHION see DRR200
FORMPARANATE see FNE500
FORMULA 40 see DFY800
FORMVAR 1285 see AAX250
4'-FORMYLACETANILIDE THIOSEMICARBAZONE see FNF000
2-FORMYLAMINOFLUORENE see FEF000
2-FORMYLAMINO-4-(5-NITRO-2-FURYL)THIAZOLE see NGM500
N-FORMYLANILINE see FNJ000
FORMYLANILINE see FNJ000
6-FORMYLANTHANTHRENE see FNK000
4A-FORMYL-1,4,4A,5A,6,9,9A,9B-OCTAHYDRODIBENZOFURAN see BHJ500
7-FORMYLBENZ(c)ACRIDINE see BAX250
4-FORMYLBENZALDEHYDE see TAN500
p-FORMYLBENZALDEHYDE see TAN500
7-FORMYLBENZO(c)ACRIDINE see BAX250
2-FORMYLBENZOIC ACID see FNK010
o-FORMYLBENZOIC ACID see FNK010
4-FORMYLBENZONITRILE see COK250
p-FORMYLBENZONITRILE see COK250
6-FORMYLBENZO(a)PYRENE see BCT250
2-FORMYL-3:4-BENZPHENANTHRENE see BCS000
O-FORMYLCEFAMANDOLE SODIUM see FOD000
4-FORMYLCYCLOHEXENE see FNK025
5-FORMYL-1,2:3,4-DIBENZOPYRENE see DCY800
5-FORMYL-3,4:8,9-DIBENZOPYRENE see DCY600
5-FORMYL-3,4:9,10-DIBENZOPYRENE see BCQ750
FORMYLDIENOLONE see FNK040
3-FORMYL-DIGITOXIGENIN see FNK050
3-12-FORMYL-DIGOXIGENIN see FNK075
2-FORMYL-3,4-DIHYDRO-2H-PYRAN see ADR500
N-FORMYLDIMETHYLAMINE see DSB000
p-FORMYLDIMETHYLANILINE see DOT400
2-FORMYL-6,6-DIMETHYLBICYCLO(3.1.1)HEPT-2-ENE see FNK150
N-FORMYLETHYLAMINE see EKK600
α-FORMYLETHYLBENZENE see COF000
FORMYLETHYLTETRAMETHYLTETRALIN see FNK200
N-FORMYL-N-2-FLUORENYLHYDROXYLAMINE see FEG000
16-FORMYL-GITOXIN see GES100
5-FORMYLGUAIACOL see FNM000
6-FORMYLGUAIACOL see VFP000
FORMYLHYDRAZIDE see FNN000
N-FORMYLHYDRAZINE see FNN000
FORMYLHYDRAZINE see FNN000
2-(2-FORMYLHYDRAZINO)-4-(5-NITRO-2-FURYL)THIAZOLE see NDY500
N-FORMYL HYDROXYAMINOACETIC ACID see FNO000
N-FORMYL-N-HYDROXYGLYCINE see FNO000
N-FORMYLHYDROXYLAMINE see FMZ000
2-FORMYL-11-α-HYDROXY-Δ¹-METHYLTESTOSTERONE see FNK040
FORMYLIC ACID see FNA000
3-FORMYLINDOLE see FNO100
1-FORMYLISOQUINOLINE THIOSEMICARBAZONE see IRV300
N-FORMYLJERVINE see FNP000
S-(2-(FORMYLMETHYLAMINO)-2-OXOETHYL)-O,O-DIMETHYLPHOSPHORODITH-
   IOATE see DRR200
2-FORMYL-17-α-METHYLANDROSTA-1,4-DIENE-11-α,17-β-DIOL-3-ONE see
   FNK040
6-FORMYL-12-METHYLANTHANTHRENE see FNQ000
7-FORMYL-11-METHYLBENZ(c)ACRIDINE see FNS000
7-FORMYL-9-METHYLBENZ(c)ACRIDINE see FNR000
7-FORMYL-12-METHYLBENZ(a)ANTHRACENE see FNT000
12-FORMYL-7-METHYLBENZ(a)ANTHRACENE see MGY500
S-(N-FORMYL-N-METHYLCARBAMOYLMETHYL)-O,O-DIMETHYL PHOSPHORO-
   DITHIOATE see DRR200
N-FORMYL-N-METHYLCARBAMOYLMETHYL-O,O-DIMETHYL PHOSPHORO-
   DITHIOATE see DRR200
S-(N-FORMYL-N-METHYLCARBAMOYLMETHYL) DIMETHYL PHOSPHORO-
   THIOLOTHIONATE see DRR200
5-FORMYL-10-METHYL-3,4:8,9-DIBENZOPYRENE see FNV000
5-FORMYL-8-METHYL-3,4:9,10-DIBENZOPYRENE see FNU000
2-FORMYL-5-METHYLFURAN see MKH550
1-FORMYL-1-METHYLHYDRAZINE see FNW000
N-FORMYL-N-METHYLHYDRAZINE see FNW000

1-FORMYL-1-METHYL-4-(4-METHYLPENTYL)-3-CYCLOHEXENE see VIZ150
4-FORMYL-4'-METHYL-1,1'-(OXYDIMETHYLENE)DIPYRIDINIUM, DICHLORIDE
    OXIME see MBZ000
N-FORMYL-N-METHYL-p-(PHENYLAZO)ANILINE see FNX000
2-FORMYL-1-METHYLPYRIDINIUM CHLORIDE OXIME see FNZ000
2-FORMYL-1-METHYLPYRIDINIUM IODIDE OXIME see POS750
2-FORMYL-1-METHYLPYRIDINIUM METHANESULFONATE OXIME see PLX250
2-FORMYL-N-METHYLPYRIDINIUM OXIME CHLORIDE see FNZ000
2-FORMYL-N-METHYLPYRIDINIUM OXIME IODIDE see POS750
2-FORMYL-N-METHYLPYRIDINIUM OXIME METHANESULFONATE see PLX250
4-FORMYLMONOMETHYLAMINOAZOBENZENE see FNX000
N-FORMYLMORFOLIN see FOA100
4-FORMYLMORPHOLINE see FOA100
N-FORMYLMORPHOLINE see FOA100
1-FORMYLNAPHTHALENE see NAJ000
N-FORMYL-N'-(3',4'-DICHLORPHENYL)-2,2,2-TRICHLORACETALDEHYDAM
    (GERMAN) see CDP750
3-FORMYLNITROBENZENE see NEV000
p-FORMYLNITROBENZENE see NEV500
(6R-(6-α,7-β(R)))-7-(((FORMYLOXY)PHENYLACE-TYL)AMINO)-3-(((1-METHYL-1H-
    TETRAZOL-5-YL)THIO)METHYL)-8-OXO-5-THIA-1-A see FOD000
FORMYLOXYTRIBENZYLSTANNANE see FOE000
(FORMYLOXY)TRIS(PHENYLMETHYL)STANNANE see FOE000
2-FORMYLPHENOL see SAG000
4-FORMYLPHENOL see FOF000
o-FORMYLPHENOL see SAG000
p-FORMYLPHENOL see FOF000
N-FORMYLPIPERIDIN (GERMAN) see FOH000
1-FORMYLPIPERIDINE see FOH000
3-FORMYLPYRIDINE see NDM100
4-FORMYLPYRIDINE see IKZ100
2-FORMYLPYRIDINE THIOSEMICARBAZONE see PIB925
2-FORMYLQUINOXALINE-1,4-DIOXIDE CARBOMETHOXYHYDRAZONE see
    FOI000
4-FORMYLRESORCINOL see REF100
N-FORMYL-l-p-SARCOLYSIN see BHX250
2-FORMYLTHIOPHENE see TFM500
α-FORMYLTHIOPHENE see TFM500
1-FORMYL-3-THIOSEMICARBAZIDE see FOJ000
FORMYL TRICHLORIDE see CHJ500
8-FORMYL-1,6,7-TRIHYDROXY-5-ISOPROPYL-3-METHYL-2,2'-BISNAPHTHALENE
    see GJM000
1-FORMYL-3,5,6-TRIMETHYL-3-CYCLOHEXENE and 1-FORMYL-2,4,6-TRIMETH-
    YL-3-CYCLOHEXENE see IKJ000
FORMYL VIOLET S4BN see FAG120
FOROMACIDIN see SLC000
FORON BRILLIANT RED E 2BL see AKI750
FOROTOX see TIQ250
FORPEN see BFD000
FORPHENICINOL see FOJ100
FORRON see TAA100
FORSTAN see ORU000
FORST U 46 see TAA100
FOR-SYN see BEP500
FORTALGESIC see DOQ400
FORTALIN see DOQ400
FORTECORTIN see SOW000
FORTEX see TAA100
FORTHION see MAK700
FORTIFICAR see SFX000
FORTIFLEX 6015 see PJS750
FORTIFLEX A 60/500 see PJS750
FORTIGRO see FOI000
FORTIMICIN A see FOK000
FORTIMICIN A SULFATE see FOL000
FORTION NM see DSP400
FORTODYL see VSZ100
FORTOMBRINE-N see AAN000
FORTRACIN see BAC250
FORTRACIN (BACITRACIN-MD) see BAC260
FORTRAL see DOQ400
FORTROL see BLW750
FORTURF see TBQ750
FOSAMINE AMMONIUM see ANG750
FOSAZEPAM see FOL100
FOSCHLOR see TIQ250
FOSCHLOREM (POLISH) see TIQ250
FOSCHLOR R-50 see TIQ250
FOSCOLIC ACID see PGW600
FOSDRIN see MQR750
FOSFAKOL see NIM500
FOS-FALL "A" see BSH250
FOSFAMID see DSP400
FOSFAMIDON see FAB400
FOSFAMIDONE see FAB400

FOSFAZIDE see ILE000
FOSFERMO see PAK000
FOSFESTROL see DKA200
FOSFESTROL TETRASODIUM see TEE300
FOSFEX see PAK000
FOSFIVE see PAK000
FOSFOCINA see PHA550
FOSFOMYCIN see PHA550
FOSFOMYCIN CALCIUM HYDRATE see CAW376
FOSFOMYCIN-Ca HYDRATE see CAW376
FOSFOMYCIN DISODIUM see DXF600
FOSFOMYCIN DISODIUM HYDRATE see FOL200
FOSFOMYCIN DISODIUM SALT see DXF600
FOSFOMYCIN SODIUM HYDRATE see FOL200
FOSFOMYCIN SODIUM SALT see DXF600
FOSFON D see THY500
FOSFONO 50 see EEH600
FOSFONOMYCIN see PHA550
FOSFORAN TROJ-(1,3-DWUCHLOROIZOPROPYLOWY) (POLISH) see FQU875
FOSFORO BIANCO (ITALIAN) see PHP010
FOSFORO(PENTACHLORURO di) (ITALIAN) see PHR500
FOSFORO(TRICLORURO di) (ITALIAN) see PHT275
FOSFOROWODOR (POLISH) see PGY000
FOSFOROXYCHLORID see PHQ800
FOSFORPENTACHLORIDE (DUTCH) see PHR500
FOSFORTHIOCHLORID see TFO000
FOSFORTRICHLORIDE (DUTCH) see PHT275
FOSFORYN TROJETYLOWY (CZECH) see TJT800
FOSFORYN TROJMETYLOWY (CZECH) see TMD500
FOSFORZUUROPLOSSINGEN (DUTCH) see PHB250
FOSFOSAL see PHA575
FOSFOTHION see MAK700
FOSFOTION see MAK700
FOSFOTOX see DSP400
FOSFURI di ALLUMINIO (ITALIAN) see AHE750
FOSFURI di MAGNESIO (ITALIAN) see MAI000
FOSGEEN (DUTCH) see PGX000
FOSGEN (POLISH) see PGX000
FOSGENE (ITALIAN) see PGX000
FOSOVA see PAK000
FOSPIRATE see PHE250
FOSPIRATE METHYL see PHE250
FOSSIL FLOUR see SCH000
FOSTEN see PAF550
FOSTERN see PAK000
FOSTEX see BDS000
FOSTER GRANT 834 see SMQ500
FOSTHIETAN see DHH200
FOSTION see IOT000
FOSTION MM see DSP400
FOSTOX see PAK000
FOSTRIL see HCL000
FOSVEL see LEN000
FOSVEX see TCF250
FOSZFAMIDON see FAB400
FOUADIN see AQH500
FOUMARIN see ABF500
FOURAMIEN 2R see ALL750
FOURAMINE see TGL750
FOURAMINE BA see DBO400
FOURAMINE BROWN AP see PPQ500
FOURAMINE D see PEY500
FOURAMINE EG see ALT500
FOURAMINE ERN see NAW500
FOURAMINE J see TGL750
FOURAMINE OP see ALT000
FOURAMINE P see ALT250
FOURAMINE PCH see CCP850
FOURAMINE RS see REA000
FOURAMINE STD see TGM400
FOURINE DS see PEY650
FOURNEAU 190 see ABX500
FOURNEAU 309 see BAT000
FOURNEAU 933 see BCI500
FOURNEAU 1162 see SNM500
FOURNEAU 2268 see MJH250
FOURNEAU 2987 see DII200
FOURNEAU see BAT000
FOURRINE 1 see PEY500
FOURRINE 36 see ALL750
FOURRINE 57 see NEM480
FOURRINE 64 see PEY650
FOURRINE 65 see ALT500
FOURRINE 68 see CCP850
FOURRINE 76 see DBO400

FOURRINE 79 see REA000
FOURRINE 81 see CEG625
FOURRINE 84 see ALT250
FOURRINE 88 see AHQ250
FOURRINE 93 see DUP400
FOURRINE 94 see TGL750
FOURRINE 99 see NAW500
FOURRINE A see AHQ250
FOURRINE BROWN PR see NEM480
FOURRINE BROWN PROPYL see NEM480
FOURRINE BROWN 2R see ALL750
FOURRINE D see PEY500
FOURRINE EG see ALT500
FOURRINE ERN see NAW500
FOURRINE M see TGL750
FOURRINE P BASE see ALT250
FOURRINE PG see PPQ500
FOURRINE 4R see DUP400
FOURRINE SLA see DBO400
FOURRINE SO see CEG625
FOUR THOUSAND FORTY-NINE see MAK700
FOVANE see BDE250
FOWLER'S SOLUTION see FOM050
FOXGLOVE see DKL200, FOM100
FOY see GAD400
FOZALON see BDJ250
FP 4 see PJS750
FP 70 see FLG100
FP-83 see FJT100
FPA see FHG000
FPF 1002 see TAB250
FPL 670 see CNX825
FPL see PIV650
FR 2 see TNG750
FR 28 see DXG035
FR-33 see FLJ000
FR 222 see MKA500
FR 300 see PAU500
FR 860 see HAQ500
FR 1138 see DDQ400
FR-1923 see NMV480
FR 3068 see TGA600
FR-13,479 see EBE100
FR 34235 see NDY650
FRABEL see HNI500
FRACINE see NGE500
FRACTION AB see CKP250
FRADEMICINA see LGC200
FRADIOMYCIN SULFATE see NCG000
FRAESEOL see ENC000
FRAGAROL see NBJ000
FRAGIVIX see BBJ500
FRAILECILLO (CUBA) see CNR135
FRAMBINONE see RBU000
FRAMED see BJP000
FRAMYCETIN see NCF000
FRAMYCETIN SULFATE see NCD550
FRAMYCIN SULFATE see NCD550
FRANCILLADE (HAITI) see CAK325
FRANGULA EMODIN see MQF250
FRANKINCENSE GUM see OIM000
FRANKINCENSE OIL see OIM025
FRANKLIN see CAO000
FRANOCIDE see DIW200
FRANOZAN see DIW200
FRANROZE see FLZ050
FRAQUINOL see NCD550
FRATOL see SHG500
FRAXINELLONE see FOM200
FRAXINUS JAPONICA Blume, bark extract see FON100
FRAXIPARIN see HAQ500
FRAZALON see AOO300
FREE ACID see SLW475
FREE BENZYLPENICILLIN see BDY669
FREE COCONUT OIL see CNR000
FREE HISTAMINE see HGD000
FREEMANS WHITE LEAD see LDY000
FREESIOL see FON200
FREEURIL see BDE250
FREKAPHYLLIN see HLC000
FREKVEN see ICC000
FRENACTIL see FGU000, FLK100
FRENACTYL see FGU000, FLK100
FRENANTOL see ELL500
FRENASMA see CNX825

FRENCH GREEN see COF500
FRENCH JASMINE see COD675
FRENCH ROSE ABSOLUTE see RMP000
FRENOGASTRICO see DJM800, XCJ000
FRENOHYPON see ELL500
FRENOLON DIFUMARATE see MDU750
FRENOLYSE see AJV500
FRENQUEL HYDROCHLORIDE see PIY750
FRENTIROX see MCO500
FREON 11 see TIP500
FREON 12 see DFA600
FREON 13 see CLR250
FREON 14 see CBY250
FREON 21 see DFL000
FREON 22 see CFX500
FREON 23 see CBY750
FREON 30 see MJP450
FREON 31 see CHI900
FREON 41 see FJK000
FREON 112 see TBP050
FREON 113 see FOO000
FREON 114 see FOO509
FREON 115 see CJI500
FREON 123 see TJY500
FREON 133a see TJY175
FREON 141 see FOO550
FREON 142b see CFX250
FREON 142 see CFX250
FREON 152 see ELN500
FREON 253 see TJY200
FREON 500 see DFB400
FREON 502 see FOO510
FREON 503 see FOO515
FREON see CFX500
FREON 12B1 see BNA250
FREON 13B1 see TJY100
FREON 12-B2 see DKG850
FREON 114B2 see FOO525
FREON C-318 see CPS000
FREON F-12 see DFA600
FREON F-23 see CBY750
FREON FT see TJE100
FREON MF see TIP500
FREON R 112 see TBP050
FREON 113TR-T see FOO000
FRESENIUS D 6 see CCU150
FRESMIN see VSZ000
FREUND'S ADJUVANT see FOO600
FRIAR'S CAP see MRE275
FRIDERON see RBF100
FRIGEN 11 see TIP500
FRIGEN 12 see DFA600
FRIGEN 22 see CFX500
FRIGEN 113a see FOO000
FRIGEN 114 see FOO509
FRIGEN see CFX500
FRIGIDERM see FOO509
FRIJOLILLO (MEXICO) see NBR800
FRISIUM see CIR750
FROBEN see FLG100
FRP 53 see PAU500
FRUCOTE see BPY000
FRUCTOFURANOSE, TETRANICOTINATE see TDX860
FRUCTONE see EFR100
FRUCTOSE (FCC) see LFI000
FRUCTUS PIPERIS LONGI see PIV650
FRUITDO see BLC250
FRUITONE see NAK000, NAK500
FRUITONE A see TAA100
FRUITONE CPA see CJQ300
FRUITONE T see TIX500
FRUIT RED A EXTRA YELLOWISH GEIGY see HJF500
FRUIT RED A GEIGY see FAG020
FRUIT SALAD PLANT see SLE890
FRUIT SUGAR see LFI000
FRUMIN AL see DXH325
FRUSEMIDE see CHJ750
FRUSEMIN see CHJ750
FRUSID see CHJ750
FRUSTAN see DCK759
FRUTABS see LFI000
FS 74 see SCK600
F3T see TKH325
FT 8 see BEQ625
FT see FPY000

FTA see FQQ100
FTAALZUURANHYDRIDE (DUTCH) see PHW750
F1-TABS see SHF500
FTAFLEX DIBA see DNH125
FTALAN see TIT250
FTALOPHOS see PHX250
FTALOWY BEZWODNIK (POLISH) see PHW750
FTBG see FMH000
F3TDR see TKH325
FTIVAZID see VEZ925
FTIVAZIDE see VEZ925
FTORAFUR see FLZ050
FTORIN see FOO875
FTORIN (PHARMACEUTICAL) see FOO875
FTORLON 3 see KDK000
FTORLON 4 see TAI250
FTORLON F3 see KDK000
FTORLON 3M see KDK000
FTOROPLAST 3 see KDK000
FTOROPLAST 4 see TAI250
FTOROPLAST 3P see KDK000
FTOROTAN (RUSSIAN) see HAG500
F-2 TOXIN see ZAT000
FT mixture with URACIL (1:4) see UNJ810
5-FU see FMM000
FUBERIDATOL see FQK000
FUBERIDAZOLE see FQK000
FUBERISAZOL see FQK000
FUBRIDAZOLE see FQK000
FUCHSIN (basic) see MAC500
p-FUCHSIN see RMK020
FUCHSIN see MAC250
FUCHSINE see MAC250
FUCHSINE A see MAC250
FUCHSINE BASE see MAC500
FUCHSINE CS see MAC250
FUCHSINE DR-001 see RMK020
FUCHSINE G see MAC250
FUCHSINE HF BASE see MAC500
FUCHSINE HO see MAC250
FUCHSINE N see MAC250
FUCHSINE RTN see MAC250
FUCHSINE SBP see MAC250
FUCHSINE SPC see RMK020
FUCHSINE Y see MAC250
FUCIDINA see SHK000
FUCIDINE see SHK000
FUCLASIN see BJK500
FUCLASIN ULTRA see BJK500
5-FUDR see DAR400
FUDR see DAR400
FUEL OIL, pyrolyzate see FOP100
FUEL OIL see FOP000
FUEL OIL NO.2. (DOT) see DHE800
FUEL OIL, RESIDUAL see FOP050
FUELS, DIESEL see DHE900, FOP000
FUGACILLIN see CBO250
FUGEREL see FMR050
FUGILIN see FOZ000
FUGOA see NNW500
FUGU POISON see FOQ000
FUJI HEC-BL 20 see HKQ100
FUJITHION see FOR000
FUKI-NO-TOH (JAPANESE) see PCR000
FUKINOTOXIN (neutral) see PCQ750
FUKINOTOXIN see PCQ750
FUKLASIN see BJK500
FUKLASIN ULTRA see FAS000
FULAID see FLZ050
FULCIN see GKE000
FULCINE see GKE000
FULFEEL see FLZ050
FUL-GLO see FEW000
FULL RANGE CATALYTIC REFORMED NAPHTHA see NAQ510
FULLSAFE see TKH750
FULMINATE of MERCURY (dry) (DOT) see MDC000
FULMINATE of MERCURY see MDC000
FULMINATES see FOS000
FULMINATING MERCURY (DOT) see MDC000
FULMINATING SILVER (DOT) see SDR000
FULMINIC ACID see FOS050
FULSIX see CHJ750
FULUMINOL see FOS100
FULUVAMIDE see CHJ750
6-FULVENOSELONE see FOS300

FULVICAN GRISACTIN see GKE000
FULVICIN see GKE000
FULVINA see GKE000
FULVINE see FOT000
FULVISTATIN see GKE000
FUMADIL B see FOZ000
FUMAFER see FBJ100
FUMAGILLIN see FOZ000
FUMAGON see DDL800
FUMAR-F see FBJ100
FUMARIC ACID see FOU000
FUMARIC ACID, DIBUTYL ESTER see DEC600
FUMARIC ACID DIHEXYL ESTER see DKP000
FUMARIC ACID DIISOPROPYL ESTER see DNQ200
FUMARIC ACID, DIMETHYL ESTER see DSB600
FUMARIC ACID DINITRILE see FOX000
FUMARIC ACID, MAGNESIUM SALT see MAF750
FUMARIC ACID, METHYL- see MDI250
FUMARINE see FOW000
FUMARONITRILE see FOX000
FUMAROYL CHLORIDE see FOY000
FUMARSAEUREDINITRIL see FOX000
FUMARYLCHLORID (CZECH) see FOY000
FUMARYL CHLORIDE see FOY000
FUMAZONE see DDL800
FUMED SILICA see SCH000
FUMED SILICON DIOXIDE see SCH000
FUMETOBAC see NDN000
FUMETTE see MDR750
FUMIDIL see FOZ000
FUMIGANT-1 (OBS.) see MHR200
FUMIGRAIN see ADX500
FUMILAT A see MJL250
FUMING LIQUID ARSENIC see ARF500
FUMING SULFURIC ACID see SOI520
FUMIRON see FBJ100
FUMITOXIN see AHE750
FUMO-GAS see EIY500
FUNDAL 500 see CJJ250
FUNDAL see CJJ250
FUNDAL SP see CJJ500
FUNDASOL see BAV575
FUNDEX see CJJ250
FUNDUSCEIN see FEW000
FUNGACETIN see THM500
FUNGAFLOR see FPB875
FUNGAREST see KFK100
FUNGICHROMIN, HYDRATE see FPC000
FUNGICHROMIN see FPC000
FUNGICHTHOL see IAD000
FUNGICIDE 1991 see BAV575
FUNGICIDE FX see MJM500
FUNGICLOR see PAX000
FUNGIFEN see PAX250
FUNGIFOS see MIE250
FUNGILIN see AOC500
FUNGILON see TNH750
FUNGIMAR see CNO000
FUNGINEX see TKL100
FUNGINON see FPC100
FUNGISONE see AOC500
FUNGISTOP see TGB475
FUNGITOX see PEX500
FUNGITOX OR see ABU500
FUNGIVIN see GKE000
FUNGIZONE see AOC500
FUNGOCIN see BAB750
FUNGOL see ZIA000
FUNGOL B see SHF500
FUNGONIT GF 2 see ZIA000
FUNGO-POLYCID see CDY325
FUNGORAL see KFK100
FUNGOSTOP see BJK500
FUNGUS BAN TYPE II see CBG000
FUNICOLOSIN see FPD000
FUQUA see FPD100
5-FUR see FMN000
FUR see FMN000
FURAN, 2-(BROMOMETHYL)TETRAHYDRO- see TCS550
FURACILLIN see NGE500
FURACIN see SPC500
FURACINETTEN see NGE500
FURACOCCID see NGE500
FURACORT see NGE500
FURACYCLINE see NGE500

FURADAN see CBS275
FURADANTIN see NGE000
FURAN, 2,5-DIETHYLTETRAHYDRO- see DKB165
FURADONIN see NGE000
FURADROXYL see FPE100
FURAN, 2-(N-ETHYLCARBAMOYLHYDROXYMETHYL)- see EHF500
FURAFLUOR see FLZ050
FURAL see FPQ875
FURALAZIN see FPF000
2-FURALDEHYDE see FPQ875
2-FURALDEHYDE AZINE see FPH000
2-FURALDEHYDE, 2,3:4,5-BIS(2-BUTENYLENE)TETRAHYDRO- see BHJ500
2-FURALDEHYDE, 5-METHYL- see MKH550
FURALDON see NGE500
FURALE see FPQ875
FURALTADONE see FPI000
l-FURALTADONE HYDROCHLORIDE see FPI150
FURALTADONE HYDROCHLORIDE see FPI100
FURAMETHRIN see POD875
FURAMON see FPY000
FURAMON IODIDE see FPY000
FURAN see FPK000
FURANACE-10 see NDY400
FURANACE see NDY400
3-FURANACROLEIN, 2-METHYL- see FPX050
2-FURANACRYLIC ACID see FPK050
2-FURANALDEHYDE see FPQ875
FURAN-2-AMIDOXIME see FPK100
2-FURANCARBINOL see FPU000
2-FURANCARBONAL see FPQ875
2-FURANCARBOXALDEHYDE see FPQ875
2-FURANCARBOXALDEHYDE, (2-FURANYLMETHYLENE)HYDRAZONE (9CI) see FPH000
α-FURANCARBOXYLIC ACID see FQF000
FURAN-α-CARBOXYLIC ACID METHYL ESTER see MKH600
2,5-FURANDIONE see MAM000
2,5-FURANDIONE, DIHYDRO-3-(TETRAPROPENYL)-(9CI) see TEC600
2,5-FURANDIONE, 3-(DODECENYL)DIHYDRO- see DXV000
2,5-FURANDIONE, POLYMER with ETHENYLBENZENE (9CI) see SEA500
FURANEOL see HKC575
FURANIDINE see TCR750
FURANIUM see FEW000
2-FURANMETHANETHIOL see FPM000
2-FURANMETHANOL see FPU000
2-FURANMETHANOL, α-(N-ETHYLCARBAMOYL)- see EHF500
2-FURANMETHYLAMINE see FPW000
FURAN-OFTENO see NGE500
FURANOL see FPY000
3-FURANOL, TETRAHYDRO- see HOI200
2(3H)-FURANONE, DIHYDRO-5-(3-HEXENYL)-5-METHYL-, (Z)- see MIW060
2(3H)-FURANONE, DIHYDRO-5-OCTYL- see OES100
3(2H)-FURANONE, 2,5-DIMETHYL-4-HYDROXY- see HKC575
2(3H)-FURANONE, 5-ETHYLDIHYDRO- see HDY600
2(3H)-FURANONE, 5-HEPTYLDIHYDRO- see HBN200
2-FURANPROPIONIC ACID, TETRAHYDRO-α-(1-NAPHTHYLMETHYL)-, 2-(DIETHYLAMINO)ETHYL ESTER, OXALATE (1:1) see NAE100
FURANTHRIL see CHJ750
FURANTHRYL see CHJ750
FURANTOIN see NGE000
FURANTRIL see CHJ750
2-(2-FURANYL)-1H-BENZIMIDAZOLE see FQK000
5-(3-FURANYL)-5-HYDROXY-2-PENTANONE see IGF325
N-(2-FURANYLMETHYL)-N',N'-DIMETHYL-N-2-PYRIDINYL-1,2-ETHANEDIAMINE FUMARATE see MDP800
1-(3-FURANYL)-4-METHYL-1-PENTANONE see PCI750
1-(3-FURANYL)-2,4-PENTANEDIOL see IGF200
1-(3-FURANYL)-1,4-PENTANEDIONE see IGF300
3-FURANYLPHENYLMETHANONE see FQO050
FURAN, 2,2'-(OXYBIS(METHYLENE))BIS- see FPX025
FURAN, 2,2'-(OXYDIMETHYLENE)DI-(6CI,8CI) see FPX025
FURAPLAST see NGE500
FURAPYRIMIDONE see FPO100
FURASEPTYL see NGE500
FURAN, TETRAHYDRO-2-(BROMOMETHYL)- see TCS550
FURAN, TETRAHYDRO-2-HEPTYL- see HBP425
FURATOL see SHG500
FURATONE see BKH500
FURATONE-S see BKH500
FURAXONE see NGG500
FURAZABOL see AOO300
FURAZOL see NGG500
FURAZOLIDON see NGG500
FURAZOLIDONE (USDA) see NGG500
FURAZOLIN see FPI000
FURAZOLINE see FPI000

FURAZON see NGG500
FURAZONE see NGE500
FURAZOSIN see AJP000
FURAZOSIN HYDROCHLORIDE see FPP100
FUR BLACK 41867 see PEY500
FUR BROWN 41866 see PEY500
FURCELLERAN GUM see FPQ000
FURESIS see CHJ750
FURESOL see NGE500
FURETHIDINE see FPQ100
2-FURFURAL see FPQ875
FURFURAL see FPQ875
FURFURAL-ACETONE ADDUCT see FPQ900
1:1 FURFURAL-ACETONE MONOMER see FPQ900
FURFURAL-ACETONE MONOMER see FPQ900
FURFURAL ACETONE MONOMER FA see FPQ900
FURFURAL ALCOHOL see FPU000
FURFURALDAZINE see FPH000
FURFURALDEHYDE see FPQ875
FURFURALDEHYDE AZINE see FPH000
FURFURALE (ITALIAN) see FPQ875
FURFURAL OXIME see FPR000
FURFURAMIDE see FPS000
FURFURAN see FPK000
FURFURIN see NGE500
FURFUROL see FPQ875
FURFUROLATSETONOVYI MONOMER FA see FPQ900
FURFUROLE see FPQ875
FURFURYLACETONE see FPT000
6-FURFURYLADENINE see FPT100
N⁶-FURFURYLADENINE see FPT100
N-FURFURYLADENINE see FPT100
FURFURYL ALCOHOL see FPU000
FURFURYL ALCOHOL PHOSPHATE (3:1) see FPV000
2-FURFURYLALKOHOL (CZECH) see FPU000
FURFURYLAMINE see FPW000
6-(FURFURYLAMINO)PURINE see FPT100
N⁶-(FURFURYLAMINO)PURINE see FPT100
FURFURYL AZINE see FPH000
FURFURYL-BIS(2-CHLOROETHYL)AMINE HYDROCHLORIDE see FPX000
FURFURYL-BIS(β-CHLOROETHYL)AMINE HYDROCHLORIDE see FPX000
FURFURYL ETHER see FPX025
FURFURYLIDINE-2-PROPANAL see FPX050
FURFURYL MERCAPTAN see FPM000
N-(2-FURFURYL)-N-(2-PYRIDYL)-N',N'-DIMETHYLETHYLENEDIAMINE FUMARATE see MDP800
1-FURFURYLPYRROLE see FPX100
N-(2-FURFURYL)PYRROLE see FPX100
N-FURFURYL PYRROLE see FPX100
FURFURYLTRIMETHYLAMMONIUM IODIDE see FPY000
FURIDAZOL see FQK000
FURIDAZOLE see FQK000
FURIDIAZINE see NGI500
FURIDON see NGG500
2,2'-FURIL see FPZ000
α-FURIL see FPZ000
FURIL see FPZ000
2-FURIL-METANALE (ITALIAN) see FPQ875
α-FURILMONOXIME see FQB000
FURITON see NGB700
FURLOE see CKC000
FURLOE 4EC see CKC000
FURMETHANOL see FPI000
FURMETHIDE see FPY000
FURMETHONOL see FPI000, FPI150
FURMETONOL see FPI000
FURMITHIDE IODIDE see FPY000
FURNAL see CBT750
FURNEX see CBT750
FURNEX N 765 see CBT750
FUROBACTINA see NGE000
2H-FURO(2,3-h)(1)BENZOPYRAN-2-ONE see FQC000
FURO(5',4',7,8)COUMARIN see FQC000
FURODAN see CBS275
FUROFUTRAN see FLZ050
2-FUROIC ACID see FQF000
α-FUROIC ACID see FQF000
FUROIC ACID see FQE000
2-FUROIC ACID, METHYL ESTER see MKH600
FUROIN see FQI000
α-FUROLE see FPQ875
FUROLE see FPQ875
FUROSEDON see CHJ750
FUROSEMID see CHJ750
FUROSEMIDE see CHJ750

FUROSEMIDE "MITA" see CHJ750
FUROTHIAZOLE see FQJ000
FUROVAG see NGG500
FUROX see NGG500
FUROXAL see NGG500
FUROXANE see NGG500
FUROXONE SWINE MIX see NGG500
2-FUROYL AZIDE see FQJ025
FUROYL CHLORIDE see FQJ050
N-FUROYL-N'-n-BUTYLHARNSTOFF (GERMAN) see BRK250
(FUROYLOXY)TRIETHYL PLUMBANE see TJS750
2-(4-(2-FUROYL)PIPERAZIN-1-YL)-4-AMINO-6,7-DIMETHOXYQUINAZOLINE see
    AJP000
2-(4-(2-FUROYL)PIPERAZIN-1-YL)-4-AMINO-6,7-DIMETHOXYQUINAZOLINE HY-
    DROCHLORIDE see FPP100
FUROZOLIDINE see NGG500
FURPIRINOL see NDY400
FURPYRINOL see NDY400
FURRO D see PEY500
FURRO EG see ALT500
FURRO ER see NAW500
FURRO L see DBO000
FURRO P BASE see ALT250
FURRO 4R see DUP400
FURRO SLA see DBO400
FURSEMID see CHJ750
FURSEMIDE see CHJ750
FURSULTIAMIN see FQJ100
FURSULTIAMINE see FQJ100
FURTHRETHONIUM IODIDE see FPY000
FURTRETHONIUM IODIDE see FPY000
FURTRIMETHONIUM IODIDE see FPY000
FUR YELLOW see PEY500
3-(α-FURYL-β-ACETYLAETHYL)-4-HYDROXYCUMARIN (GERMAN) see ABF500
3-(1-FURYL-3-ACETYLETHYL)-4-HYDROXYCOUMARIN see ABF500
FURYL ALCOHOL see FPU000
FURYLAMIDE see FQN000
2-(2'-FURYL)-BENZIMIDAZOLE see FQK000
2-(2-FURYL)BENZIMIDAZOLE see FQK000
4-(2-FURYL)-2-BUTANONE see FPT000
2-FURYLCARBINOL see FPU000
α-FURYLCARBINOL see FPU000
FURYLFURAMIDE see FQN000
2-FURYL α-HYDROXYFURFURYL KETONE see FQI000
5-(3-FURYL)-5-HYDROXY-2-PENTANONE see IGF325
1-(3-FURYL)-4-HYDROXYPENTANONE see FQL100
1-(β-FURYL)-4-HYDROXYPENTANONE see FQL100
2-FURYL p-HYDROXYPHENYL KETONE see FQL200
β-FURYL ISOAMYL KETONE see PCI750
2-FURYLISOPROPYLAMINE SULFATE see FQM000
β-(2-FURYL)ISOPROPYLAMINE SULFATE see FQM000
2-FURYLMERCURIC CHLORIDE see CHK000
2-FURYLMERCURY CHLORIDE see CHK000
2-FURYL-METHANAL see FPQ875
(2-FURYL)METHANOL see FPU000
1-(2-FURYL)METHYLAMINE see FPW000
1-(3-FURYL)-4-METHYL-1-PENTANONE see PCI750
α-2-FURYL-5-NITRO-2-FURANACYRLAMIDE see FQN000
2-(2-FURYL)-3-(5-NITRO-2-FURYL)ACRYLAMIDE see FQN000
2-(2-FURYL)-3-(5-NITRO-2-FURYL)ACRYLIC ACID AMIDE see FQN000
α-(FURYL)-β-(5-NITRO-2-FURYL)ACRYLIC AMIDE see FQN000
1-(3-FURYL)-1,4-PENTANEDIONE see IGF300
3-FURYL PHENYL KETONE see FQO050
N-(4-(2-FURYL)-2-THIAZOLYL)ACETAMIDE see FQQ100
N-(4-(2-FURYL)-2-THIAZOLYL)FORMAMIDE see FQQ400
FUSARENONE X see FQR000
FUSAREX see TBR750, TBS000
FUSARIC ACID see BSI000
FUSARIC ACID-Ca see FQR100
FUSARIC ACID CALCIUM SALT see FQR100
FUSARINIC ACID see BSI000
FUSARIOTOXIN T 2 see FQS000
FUSARIUM TOXIN see ZAT000
FUSED BORAX see DXG035
FUSED BORIC ACID see BMG000
FUSED QUARTZ see SCK600
FUSED SILICA see SCK600
FUSELEX see SCK600
FUSELEX RD 120 see SCK600
FUSELEX RD 40-60 see SCK600
FUSELEX ZA 30 see SCK600
FUSELOEL (GERMAN) see FQT000
FUSEL OIL see FQT000
FUSEL OIL, REFINED (FCC) see FQT000
FUSID see CHJ750

FUSIDATE SODIUM see SHK000
FUSIDIC ACID see FQU000
FUSIDIN see SHK000
FUSIDINE see FQU000
FUSILADE see FDA885
FUSIN see SHK000
FUSSOL see FFF000
FUSTEL see FBW000
FUSTET see FBW000
FUTRAFUL see FLZ050
FUTRAMINE D see PEY500
FUTRAMINE EG see ALT500
FUTRICAN see CDY325
FUVACILLIN see NGE500
FUXAL see SNN300
FUZIDIN see FQU000
F 3 (VINYL POLYMER) see KDK000
FW 50 see WCJ000
FW 293 see BIO750
FW 734 see DGI000
FW 925 see DFT800
FW 200 (mineral) see WCJ000
FWH 399 see TNV625
FWH 429 see TNV625
FX 703 see FMO150
FYCOL 8 see CNK559
FYDALIN see BNK000
FYDE see FMV000
FYFANON see MAK700
FYROL CEF see CGO500
FYROL FR 2 see FQU875
FYROL FR2 see TNG750
FYROL HB32 see TNC500
FYRQUEL 150 see TNP500
FYSIOQUENS see EEH575
FYTIC ACID see PIB250
FYTOLAN see CNK559

G 0 see HDG000
G 1 see HIM000
G 3 see MCB050
G 4 see MJM500
G-11 see HCL000
G 50 see EHG100
G-52 see GAA100
G 130 see DTN775
Ge 132 see CCF125
G 200 see EHG100
G 301 see DCM750
G 316 see PJC000
G 338 see DER000
G 347 see DST200
G 475 see COY500
G 821 see MCB050
Go 919 see EOA500
G 996 see CDS125
G-1029 see AIH000
G-2129 see PJY000
G 3063 see GAC000
G 3707 see DXY000
G 3802 see PJT300
G 3816 see PJT300
G 3820 see PJT300
γ-6480 see BOV000
G 14744 see DWA500
G 22150 see DLH630
G 22355 see DLH600, DLH630
G-23350 see ABF750
G 23992 see DER000
G 24,163 see PNH750
G-24480 see DCM750
G-24622 see DJX400
G-25804 see TBO500
G 27202 see HNI500
G 27365 see DJA200
G 27901 see TJL500
G-29288 see MQH750
G 30027 see ARQ725
G-31435 see MFL250
G 32883 see DCV200
G 33040 see DCV800
G 33182 see CLY600
G 34161 see BKL250
G 34360 see INR000

G 34586 see CDV000
G 35020 see DLS600
GA 242 see CDF400
G-2130A see DXY000
GA-10832 see CQG250
GA 56 (enzyme) see GGA800
GA see EIF000, GEM000
GABA see PIM500
GABACET see NNE400
p-GABA HYDROCHLORIDE see GAD000
GABAIL see GCU050
GABBROMICINA see APP500
GABBROMYCIN see APP500, NCF500
GABBROPAS see AMM250
GABBRORAL see APP500
GABBROROL see APP500
GABEXATE MESILATE see GAD400
GABEXATE MESYLATE see GAD400
GABOB see AKF375
GABOMADE see AKF375
GABRITE see UTU500
GADEXYL see MQU750
GADOLINIA see GAP000
GADOLINIUM see GAF000
GADOLINIUM CHLORIDE see GAH000
GADOLINIUM CITRATE see GAJ000
GADOLINIUM(III) NITRATE (1:3) see GAL000
GADOLINIUM(III) NITRATE, HEXAHYDRATE (1:3:6) see GAN000
GADOLINIUM(3+) OXIDE see GAP000
GADOLINIUM(III) OXIDE see GAP000
GADOLINIUM OXIDE see GAP000
GADOLINIUM SESQUIOXIDE see GAP000
GADOLINIUM TRICHLORIDE see GAH000
GADOLINIUM TRIOXIDE see GAP000
GAFCOL EB see BPJ850
GAFCOTE see PKQ059
GAG ROOT see CCJ825
GAIMAR see RLU000
GAINEX see PEL250
GALACTARIC ACID see GAR000
GALACTASOL see GLU000
GALACTASOL A see SLJ500
GALACTICOL see DDJ000
β-d-GALACTOPYRANOSIDE, (3-β)-SOLANID-5-EN-3-YL O-6-DEOXY-α-l-MANNO-
   PYRANOSYL-(1-2)-O-(β-d-GLUCOPYRANOSYL-(1-3))-, HYDROCHLORIDE see
   SKS100
GALACTOQUIN see GAX000
GALACTOSACCHARIC ACID see GAR000
d-GALACTOSAMINE HYDROCHLORIDE see GAT000
d-GALACTOSE see GAV000
GALACTOSE see GAV000
β-GALACTOSIDASE see GAV100
4-(β-d-GALACTOSIDO)-d-GLUCOSE see LAR000
GALACTURONIC ACID with α-(6-METHOXY-4-QUINOLYL)-5-VINYL-2-QUINUCLI-
   DINEMETHANOL see GAX000
GALAN de DIA (CUBA) see DAC500
GALAN de NOCHE (CUBA) see DAC500
GALANTHAMINE HYDROBROMIDE see GBA000
GALANTHUS NIVALIS see SED575
GALATONE see GBB500
GALATUR see DPX200
GALBANUM OIL see GBC000
GALECRON see CJJ250
GALECRON SP see CJJ500
GALENA see LDZ000
GALFER see FBJ100
GALIPAN see BAV000
GALLACETOPHENONE see TKN250
GALLALDEHYDE-3,5-DIMETHYL ETHER see DOF600
GALLAMINE see PDD300
GALLIA see GBS050
GALLIC ACID see GBE000
GALLIC ACID, DODECYL ESTER see DXX200
GALLIC ACID, ETHYL ESTER see EKM100
GALLIC ACID, LAURYL ESTER see DXX200
GALLIC ACID, PROPYL ESTER see PNM750
GALLIMYCIN see EDJ500
GALLITO (CUBA) see SBC550
GALLIUM see GBG000
GALLIUM ARSENIDE see GBK000
GALLIUM(3+) CHLORIDE see GBM000
GALLIUM CHLORIDE see GBM000
GALLIUM CITRATE see GBO000
GALLIUM COMPOUNDS see GBO500
GALLIUM MONOARSENIDE see GBK000

GALLIUM-NICKEL ALLOY see NDD500
GALLIUM(III) NITRATE (1:3) see GBS000
GALLIUM NITRATE see GBS000
GALLIUM OXIDE see GBS050
GALLIUM SESQUIOXIDE see GBS050
GALLIUM SULFATE see GBS100
GALLIUM TRIOXIDE see GBS050
GALLOCHROME see MCV000
GALLODESOXYCHOLIC ACID see CDL325
GALLOGAMA see BBQ500
GALLOTANNIC ACID see TAD750
GALLOTANNIN see TAD750
GALLOTOX see ABU500, PFO550
GALLOXON see DFH600
GALOFAK see BDY669
GALOPERIDOL see CLY500
GALOXANE see DFH600
GALOXOLIDE see GBU000
GALOZONE see CCL250
GALSEPTIL see SNQ000
GALVANISONE see FLL000
GALVATOL 1-60 see PKP750
GAMACID see BBQ500
GAMAPHEX see BBQ500
GAMAQUIL see PGA750
GAMAREX see PIM500
GAMASERPIN see RDK000
GAMASOL 90 see DUD800
GAMEFAR see RHZ000
GAMENE see BBQ500
GAMIBETAL see AKF375
GAMISO see BBQ500
GAMMA-COL see BBQ500
GAMMACORTEN see SOW000
GAMMAHEXA see BBQ500
GAMMAHEXANE see BBQ500
GAMMALIN see BBQ500
GAMMALON see PIM500
GAMMA OH see HJS500
GAMMAPHOS see AMD000
GAMMASERPINE see RDK000
GAMMEXANE see BBP750
GAMMOPAZ see BBQ500
GAMONIL see CBM750, IFZ900
GAMOPHENE see HCL000
GANCIDIN (unpurified) see GBU600
GANEAKE see ECU750
GANEX P 804 see PKQ250
GANGESOL see HII500
GANGLIOSTAT see HEA000
GANIDAN see AHO250
GANJA see CBD750
GANLION see MKU750
GANOCIDE see MLC250
GANOZAN see CHC500
GANPHEN see PMI750
GANSIL see CDP000
GANTANOL see SNK000
GANTAPRIN see TKX000
GANTRIM see TKX000
GANTRISINE see SNN500
GANU see CHA000
GARAMYCIN see GCO000, GCS000
GARANTOSE see BCE500
GARBANCILLO (CUBA) see GIW200
GARDCIDE see RAF100
GARDENAL SODIUM see SID000
GARDENIOL II see SMP600
GARDENOL see SMP600
GARDENTOX see DCM750
GARDEPANYL see EOK000
GARDIQUAT 1450 see AFP250
GARDIQUAT 1250AF see QAT520
GARDONA see RAF100
GARDOPRIM see BQB000
GARGET see PJJ315
GARGON see TFQ275
GARLIC see WBS850
GARLIC OIL see GBU800
GARLIC POWDER see GBU850
GARLON see TJE890
GARMIAN see BOV825
GARNITAN see DGD600
GAROLITE SA see CAT775
GAROX see BDS000

GARRATHION see TNP250
GARVOX see DQM600
GAS-FURNACE BLACK see CBT750
GASHOUSE TANKAGE see SKZ000
GAS OIL (DOT) see DHE800
GAS OIL see DHE800, GBW000
GAS OILS (petroleum), hydrodesulfurized heavy vacuum see GBW010
GAS OILS (petroleum), light vacuum see GBW025
GAS OILS (PETROLEUM), heavy vacuum see GBW005
GASOLINE (100–130 octane) see GCA000
GASOLINE (115–145 octane) see GCC000
GASOLINE see GBY000
GASOLINE ENGINE EXHAUST CONDENSATE see GCE000
GASOLINE ENGINE EXHAUST "TAR" see GCE000
GASOLINE, UNLEADED see GCE100
GASTER see FAB500
GASTEX see CBT750
GASTOMAX see CDQ500
GASTRACID see PEK250
GASTRIDENE see BGC625
GASTRIDIN see FAB500
GASTRINIDE see TEH500
GASTRIPON see PEM750
GASTRODYN see GIC000
GASTROGRAFIN see AOO875
GASTROMET see TAB250
GASTRON see DJM800, XCJ000
GASTROPIDIL see CBF000
GASTROSEDAN see DJM800, XCJ000
GASTROTELOS see COI750
GASTROTEST see PEK250
GASTROTOPIC see BGC625
GASTROZEPIN see GCE500
GATALONE see GBB500
GATINON see MHM500
GAULTHERIA OIL, ARTIFICIAL see MPI000
GB 94 see BMA625
GB see IPX000
GBH see BDD000
GBL see DWT600
GBS see SEG800
GBX see CDB770
GC 928 see CJR500
GC-1106 see HCL500
GC-2466 see MRV000
GC-2996 see CJY000
GC 3707 see SOY000
GC 4072 see CDS750
GC 6936 see ABX250
GC 7787 see HDA500
GC 8993 see CLU000
GC 10000 see SCQ000
GC 3944-3-4 see PAX000
GCP 5126 see DED000
G-CURE see ADW200
GD see SKS500
GEA 6414 see CLK325
GEABOL see DAL300
GEARPHOS see MNH000, PAK000
GEASTIGMOL see GCE600
GEASTIMOL see GCE600
GEBUTOX see BRE500
GECHLOREERDEDIFENYL (DUTCH) see PJM500
GEDEX see SMQ500
GEFARNATE see GDG200
GEFARNIL see GDG200
GEFARNYL see GDG200
GEIGY-444E see TBO500
GEIGY 338 see DER000
GEIGY 2747 see PET250
GEIGY 12968 see DRU400
GEIGY 13005 see DSO000
GEIGY 19258 see DRL200
GEIGY 22870 see DQZ000
GEIGY 24480 see DCM750
GEIGY 27,692 see BJP000
GEIGY 30,027 see ARQ725
GEIGY 30,044 see BJP250
GEIGY 30494 see MNO750
GEIGY 32,293 see EGD000
GEIGY 32883 see DCV200
GEIGY-BLAU 536 see BGT250
GEIGY G-23611 see DSK200
GEIGY G-27365 see DJA200
GEIGY G-28029 see PDC750

GEIGY G-29288 see MQH750
GEIGY G.S. 14254 see BQC250
GEIGY GS-19851 see IOS000
GEISSOSPERMINE see GCG300
GELACILLIN see BDY669
GELAN I see EIF000
GELATIN see PCU360
GELATINE see PCU360
GELATINE DYNAMITE see DYG000
GELATIN-EPINEPHRINE see AES500
GELATIN FOAM see PCU360
GELATINS see PCU360
L-GELB 2 see FAG140
L-GELB 3 see CMM510
GELBER PHOSPHOR (GERMAN) see PHP010
GELBIN see CAP500
GELBIN YELLOW ULTRAMARINE see CAP750
GELBORANGE-S see FAG150
GELCARIN see CCL250
GELCARIN HMR see CCL250
GELDANAMYCIN see GCI000
G-ELEVEN see HCL000
GELFOAM see PCU360
GEL II see SHF500
GELIOMYCIN see HAL000
GELOCATIL see HIM000
GELOSE see AEX250
GELOZONE see CCL250
GELSEMIN see GCK000
GELSEMINE see GCK000
GELSEMIUM SEMPERVIRENS see YAK100
GELSTAPH see SLJ000, SLJ050
GELTABS see VSZ100
GELUCYSTINE see CQK325
GELUTION see SHF500
GELVA CSV 16 see AAX250
GELVATOL 1-30 see PKP750
GELVATOL 1-90 see PKP750
GELVATOL 3-91 see PKP750
GELVATOL 20-30 see PKP750
GELVATOL 2090 see PKP750
GELVATOL see PKP750
GEMCADIOL see TDO260
GEMEPROST see CDB775
GEMFIBROZIL see GCK300
GEMONIL see DJO800
GEMONIT see DJO800
GENACORT see CNS750
GENACRON YELLOW G see AAQ250
GENACRYL ORANGE G see CMM764
GENACRYL PINK G see CMM768
GENAMIN DSAC see DXG625
GENAMIN 16R302D see HCP525
GENAZO RED KB SOLN see CLK225
GENDRIV 162 see GLU000
GENEP EPTC see EIN500
GENERAL CHEMICALS 1189 see KEA000
GENERAL CHEMICALS 3707 see SOY000
GENERAL CHEMICALS 8993 see CLU000
GENET ABSOLUTE see GCM000
GENETRON 11 see TIP500
GENETRON 12 see DFA600
GENETRON 13 see CLR250
GENETRON 21 see DFL000
GENETRON 22 see CFX500
GENETRON-23 see CBY750
GENETRON 100 see ELN500
GENETRON 101 see CFX250
GENETRON 112 see TBP050
GENETRON 113 see FOO000
GENETRON 114 see FOO509
GENETRON 115 see CJI500
GENETRON 133a see TJY175
GENETRON 142b see CFX250
GENETRON 316 see FOO509
GENETRON 1113 see CLQ750
GENETRON 1112A see DFA300
GENETRONE 1112A see DFA300
GENIPHENE see CDV100
GENIPIN see GCM300
GENISIS see ECU750
GENITE 883 see CJT750
GENITE see DFY400
GENITHION see PAK000
GENITOL see DFY400

GENITOX see DAD200
GENITRON BSH see BBS300
GENO-CRISTAUZ GREMY see TBF500
GENOFACE see TIL500
GENOGRIS see NNE400
GENOL see MGJ750
GENOPHYLLIN see TEP500
GENOPTIC see GCS000
GENOPTIC S.O.P. see GCS000
GENOTHERM see PKQ059
GENOXAL see CQC500, CQC650
GENOZYM see CMX700
GENPROPATHRIN see DAB825
GENSALATE SODIUM see GCU050
GENTALPIN see GCU050
GENTAMICIN C²b see MQS579
GENTAMICIN C2b see MQS579
GENTAMICIN C COMPLEX see GCO200
GENTAMYCIN see GCO000
GENTAMYCIN-CREME (GERMAN) see GCO000
GENTAMYCIN SULFATE see GCS000
GENTASOL see GCU050
GENTERSAL see AOR500
GENTIANAE SCABRAE RADIX (LATIN) see RSZ675
GENTIANAVIOLETT see AOR500
GENTIAN VIOLET see AOR500
GENTIAVERM see AOR500
GENTICID see AOR500
GENTIDOL see GCU050
GENTIMON see ALB000
GENTINATRE see GCU050
GENTIOLETTEN see AOR500
GENTISAN see GCU050
GENTISATE see GCU000
GENTISATE SODIUM see GCU050
GENTISIC ACID see GCU000
GENTISIC ACID, MONOSODIUM SALT see GCU050
GENTISOD see GCU050
GENTRAN see DBD700
GENTRON 142B see CFX250
GENU see CCL250, PAO150
GENUGEL see CCL250
GENUGEL CJ see CCL250
GENUGOL RLV see CCL250
GENUINE ACETATE CHROME ORANGE see LCS000
GENUINE ORANGE CHROME see LCS000
GENUINE PARIS GREEN see COF500
GENUVISCO J see CCL250
GENVIS see SLJ500
GEOCARB 50EC see MOV000
GEOFOS see DHH200
GEOMYCIN see HOH500
GEON 135 see AAX175
GEON 222 see CGW300
GEON 652 see CGW300
GEON see PJR000, PKQ059
GEON LATEX 151 see PKQ059
GEOPEN see CBO250
GEOTRICYN see MCH525
GERANALDEHYDE see GCU100
GERANIAL see GCU100
GERANIC ACID see GCW000
GERANINE 2GS see CMM300
GERANIOL (FCC) see DTD000
GERANIOL ACETATE see DTD800
GERANIOL ALCOHOL see DTD000
GERANIOL BUTYRATE see GDE810
GERANIOL CROTONATE see DTE000
GERANIOL EXTRA see DTD000
GERANIOL FORMATE see GCY000
GERANIOL TETRAHYDRIDE see DTE600
GERANIUM CRYSTALS see PFA850
GERANIUM LAKE N see FAG070
GERANIUM OIL see GDA000
GERANIUM OIL ALGERIAN TYPE see GDA000
GERANIUM OIL BOURBON see GDC000
GERANIUM OIL, EAST INDIAN TYPE see PAE000
GERANIUM OIL MOROCCAN see GDE000
GERANIUM OIL, TURKISH TYPE see PAE000
GERANIUM THUNBERGII Sieb. et Zucc., extract see GDE300
GERANONITRILE see GDM000
GERANOXY ACETALDEHYDE see GDM100
GERANYL ACETATE (FCC) see DTD800
GERANYL ACETONE see GDE400
GERANYL ALCOHOL see DTD000

GERANYL BENZOATE see GDE800
GERANYL BUTANOATE see GDE810
GERANYL-2-BUTENOATE see DTE000
GERANYL n-BUTYRATE see GDE810
GERANYL BUTYRATE see GDE810, GDE825
GERANYL CAPROATE see GDG000
GERANYL CROTONATE see DTE000
GERANYL ETHYL ETHER see GDG100
GERANYL FARNESYL ACETATE see GDG200
GERANYL FORMATE (FCC) see GCY000
GERANYL HEXANOATE see GDG000
GERANYLINALOOL see GDG300
GERANYL ISOBUTYRATE see GDI000
GERANYL ISOVALERATE see GDK000
GERANYL NITRILE see GDM000
GERANYL OXYACETALDEHYDE see GDM100
GERANYL PHENYLACETATE see GDM400
GERANYL PROPIONATE see GDM450
GERANYL TIGLATE see GDO000
GERANYL α-TOLUATE see GDM400
GERASTOP see ARQ750
GERBITOX see BAV000
GERFIL see PMP500
GERHARDITE see CNN000
GERISON see SNM500
GERLACH 1396 see NBJ100
GERMAIN'S see CBM750
GERMALGENE see TIO750
GERMALL 11 see IAS100
GERMALL 115 see GDO800
GERMA-MEDICA see HCL000
GERMANATE(2-), BIS(2-CARBOXYLATOETHYL)TRIOXODI-, DIHYDROGEN (9CI)
    see CCF125
GERMAN CHAMOMILE OIL see CDH500
GERMANE (DOT) see GEI100
GERMANE, TETRAPROPYL- see TED650
GERMANIA see GEC000
GERMANIC ACID see GEC000
GERMANIC OXIDE (crystalline) see GDS000
GERMANIN see BAT000
GERMANIUM, metal powder see GDU000
GERMANIUM see GDU000
GERMANIUM BROMIDE see GDW000
GERMANIUM CHLORIDE see GDY000
GERMANIUM COMPOUNDS see GEA000
GERMANIUM, (l-CYSTEINE)TETRAHYDROXY- see CQK100
GERMANIUM DIOXIDE see GEC000
GERMANIUM ELEMENT see GDU000
GERMANIUM HYDRIDE see GEI100
GERMANIUM OXIDE see GEC000
GERMANIUM OXIDE (GeO₂) see GEC000
GERMANIUM(II) SULFIDE see GEI000
GERMANIUM TETRABROMIDE see GDW000
GERMANIUM TETRACHLORIDE see GDY000
GERMANIUM TETRAHYDRIDE see GEI100
3,3'-(GERMANOIC ANHYDRIDE) DIPROPANOIC ACID see CCF125
GERMERIN (GERMAN) see VHF000
GERMICICLIN see MDO250
GERMIN see EBL500
GERMINE see EBL500
GERMINOL see BBA500
GERMISAN see PFO250
GERMITOL see BBA500
GERNEBCIN see TGI250
GEROBIT see DBA800
GERODYL see DWK400
GERONTINE see DCC400
GERONTINE TETRAHYDROCHLORIDE see GEK000
GEROSTOP see GEK200
GEROT-EPILAN see MKB250
GEROT-EPILAN-D see DKQ000
GEROVIT see DBA800
GEROVITAL see AIL750
GEROX see SLW500
GERTLEY BORATE see SFF000
GERVOT see MDT600
GESADURAL see BJP250
GESAFID see DAD200
GESAFLOC see TJL500
GESAFRAM 50 see MFL250
GESAFRAM see MFL250
GESAGARD see BKL250
GESAGRAM see MFL250
GESAKUR see PNH750
GESAMIL see PMN850

GESAPON see DAD200
GESAPRIM see ARQ725
GESARAN see BJP000, INQ000
GESAREX see DAD200
GESAROL see DAD200
GESATAMIN see EGD000
GESATOP see BJP000
GESFID see MQR750
GESOPRIM see ARQ725
GESTEROL L.A. see HNT500
GESTID see MQR750
GESTONORONE CAPROATE see GEK510
GESTONORONE CAPRONATE see GEK510
GESTORAL see GEK500
GESTRONOL CAPROATE see GEK510
GESTRONOL HEXANOATE see GEK510
GESTYL see SCA750
GETTYSOLVE-B see HEN000
GETTYSOLVE-C see HBC500
GEUM ELATUM (Royle) Hook. f., extract see GEK600
GEVILON see GCK300, SHL500
GF 58 see TNJ500
GF see DVR909
GFX-E see GEK875
GFX-ES see GEK880
GH 20 see PKP750
GH see MJM500
GHA 331 see AHC000
G 3063 HYDROCHLORIDE see GAC000
G-I see AOO375
GIACOSIL HYDROCHLORIDE see LFK200
GIALLO CROMO (ITALIAN) see LCR000
GIANT MILKWOOD see COD675
GIARDIL see NGG500
GIARLAM see NGG500
GIATRICOL see MMN250
GIBBERELLIC ACID see GEM000
GIBBERELLIN see GEM000
GIBBREL see GEM000
GIBS see CAX500
GIB-SOL see GEM000
GIB-TABS see GEM000
GICHTEX see ZVJ000
GIE see DVR600
GIEGY GS-13798 see MPG250
GIFBLAAR POISON see FIC000
GIGANTIN see CMV000
GIHITAN see DCK759
GILDER'S WHITING see CAT775
GILEMAL see CEH700
GILOTHERM OM 2 see TBD000
GILSONITE see GEO000
GILUCARD see RDK000
GILUCOR NITRO see NGY000
GILURYTMAL see AFH250
GILUTENSIN see DWF200
GILVOCARCIN V see GEO200
GIMID see DYC800
GINANDRIN see AOO410
GINARSOL see ABX500
GINBEY (DOMINICAN REPUBLIC) see SLJ650
GINDARINE HYDROCHLORIDE see GEO600
GINDARIN HYDROCHLORIDE see GEO600
GINEFLAVIR see MMN250
GINGER OIL see GEQ000
GINGERONE see VFP100
GINGICAIN M see TBN000
GINGILLI OIL see SCB000
GINSENG see GEQ400
GINSENG, ROOT EXTRACT see GEQ425
GINSENG ROOT-NEUTRAL SAPONINS see GEQ425
GINSENGWURZEL, EXTRACT (GERMAN) see GEQ425
GINSENOSIDE RB1 see PAF450
GIQUEL see HKR500
GIRACID see PDC250
GIRARD REAGENT T see GEQ500
GIRARD'S REAGENT T see GEQ500
GIRARD T REAGENT see GEQ500
GIRL see CNE750
GIROSTAN see TFQ750
GITALIN see GES000
GITALOXIGENIN + DIGITALOSE (GERMAN) see VIZ200
GITALOXIGENIN-TRIDIGITOXOSID (GERMAN) see GES100
GITALOXIN see GES100
GITALOXIN-16-FORMATE see GES100

GITHAGENIN see GMG000
GITOFORMATE see FND100
GITOXIGENIN-3-o-MONODIGITALOSIDE see SMN275
GITOXIGENIN-TRIDIGITOXOSID (GERMAN) see GEU000
GITOXIGENIN TRIDIGITOXOXIDE-16-FORMATE see GES100
GITOXIN see GEU000
GITOXIN PENTAACETATE see GEW000
GITOXOGENIN-d-DIGITALOSID (GERMAN) see SMN275
GIUBA BLACK D see CMN230
GL 02 see PKP750
GL 03 see PKP750
G.L. 102 see EIT100
G.L. 105 see PBH100
GL 2487 see DLO880
GLACIAL ACETIC ACID see AAT250
GLACIAL ACRYLIC ACID see ADS750
GLANDUBOLIN see EDV000
GLANDUCORPIN see PMH500
GLANIL see CMR100
GLARUBIN see GEW700
GLASS see FBQ000
GLASS FIBERS see FBQ000
GLAUCARUBIN see GEW700
d-GLAUCINE see TDI475
(+)-GLAUCINE see TDI475
GLAUCINE see TDI475
s-(+)-GLAUCINE see TDI475
GLAUCINE HYDROCHLORIDE see TDI500
dl-GLAUCINE PHOSPHATE see TDI600
dl-GLAUCINPHOSPHAT (GERMAN) see TDI600
GLAUMEBA see GEW700
GLAUPAX see AAI250
GLAURAMINE see IBB000
GLAUVENT see TDI475
GLAXORIDIN see TEY000
GLAZD PENTA see PAX250
GLEEM see SHF500
GLENTONIN-RETARD see CCK125
GLIANIMON see FGU000, FLK100
GLIANIMON MITE see FLK100
GLIBENCLAMIDE see CEH700
GLIBORNURIDE see GFY100
GLIBUTIDE see BQL000
GLICLAZIDE see DBL700
GLICOL MONOCLORIDRINA (ITALIAN) see EIU800
GLICOSIL see CPR000
GLIKOCEL TA see SFO500
GLIMID see DYC800
GLIODIN see GIO000
GLIOTOXIN see ARO250
GLIPORAL see BQL000
GLIRICIDIAL SEPIUM see RBZ000
GLISEMA see CKK000
GLISOLAMIDE see DBE885
GLISOXEPID see PMG000
GLISOXEPIDE see PMG000
GLO 5 see PKP750
GLOBENICOL see CDP250
GLOBOCICLINA see MDO250
GLOBOID see ADA725
GLOBULARIACITRIN see RSU000
GLOGAL see HNI500
GLOMAX see KBB600
GLONOIN see NGY000
GLONSEN see SHK800
GLORIOSA LILY see GEW800
GLORIOSA ROTHSCHILDIANA see GEW800
GLORIOSA SUPERBA see GEW800
GLOROUS see CDP250
GLORY LILY see GEW800
GLOSSO STERANDRYL see MPN500
GLOVER see LCF000
GLOXAZON see KFA100
GLOXAZONE see KFA100
GLUBORID see GFY100
GLUCAGON see GEW875
GLUCAL see CAS750
GLUCAZIDE see GBB500
GLUCID see BCE500
GLUCIDORAL see BSM000
GLUCINIUM see BFO750
GLUCINUM see BFO750
d-GLUCITOL see SKV200
GLUCITOL see SKV200

d-GLUCITOL, 1-DEOXY-1-(METHYLAMINO)-, 3,5-BIS(ACETYLAMINO)-2,4,6-TRI-IODOBENZOATE (SALT) see AOO875
d-GLUCITOL, 1,1'-(SULFONYLBIS(4,1-PHENYLENEIMINO))BIS(1-DEOXY-1-SUL-FO)-, DISODIUM SALT (9CI) see AOO800
GLUCOBASIN see ARQ500
GLUCOBIOGEN see CAS750
GLUCOCHLORAL see GFA000
α-d-GLUCOCHLORALOSE see GFA000
GLUCOCHLORALOSE see GFA000
GLUCODIGIN see DKL800
GLUCO-FERRUM see FBK000
GLUCOFREN see BSM000
GLUCOFURANOSE, 1:2,5:6-DI-O-ISOPROPYLIDENE-, α-D- see DVO100
GLUCOGITOXIN see DAD650
GLUCOLIN see GFG000
GLUCONATE de CALCIUM (FRENCH) see CAS750
GLUCONATO di CALCIO see CAS750
GLUCONATO di SODIO (ITALIAN) see SHK800
d-GLUCONIC ACID, CALCIUM SALT (2:1) (9CI) see CAS750
d-GLUCONIC ACID, CYCLIC ESTER with ANTIMONIC ACID (H$_6$Sb$_2$O$_9$) (2:1),TRISODIUM SALT, NONAHYDRATE see AQH800
d-GLUCONIC ACID, CYCLIC ESTER with ANTIMONIC ACID (H8Sb2O9) (2:1),TRISODIUM SALT, NONAHYDRATE see AQI250
GLUCONIC ACID, compd. with 1,1'-HEXAMETHYLENE BIS(5-(p-CHLOROPHE-NYL)BIGUANIDE) (2:1), D-(8CI) see CDT250
d-GLUCONIC ACID, MONOPOTASSIUM SALT (9CI) see PLG800
d-GLUCONIC ACID, 2,4:2',4'-O-(OXYDISTIBYLIDYNE)BIS-, Sb,Sb'-DIOXIDE, TRI-SODIUM SALT, NONAHYDRATE see AQH800
GLUCONIC ACID POTASSIUM SALT see PLG800
GLUCONIC ACID SODIUM SALT see SHK800
GLUCONSAN K see PLG800
GLUCOPHAGE see DQR600, DQR800
GLUCOPHAGE LA 6023 see DQR600
GLUCOPROSCILLARIDIN A see GFC000
α-d-GLUCOPYRANOSIDE, β-d-FRUCTOFURANOSYL O-α-d-GALACTOPYRANO-SYL-(1-6)- see RBA100
α-D-GLUCOPYRANOSIDE, METHYL-, TETRANITRATE see MKI125
GLUCOPYRANOSIDE, PHENYL-, β-D- see PDO300
β-D-GLUCOPYRANOSIDE, PHENYL- see PDO300
α-D-GLUCOPYRANOSIDE, 1,3,4,6-TETRA-O-ACETYL-β-D-FRUCTOFURANOSYL-, TETRAACETATE (9CI) see OAF100
4-(α-d-GLUCOPYRANOSIDO)-α-GLUCOPYRANOSE see MAO500
α-D-GLUCOPYRANOSIDURONIC ACID, (3-β,20-β)-20-CARBOXY-11-OXO-30-NO-ROLEAN-12-EN-3-YL 2-O-β-D-GLUCOPYRANURONOSYL-, AMMONIATE see GIE100
β-D-GLUCOPYRANOSIDURONIC ACID, (METHYL-ONN-AZOXY)METHYL- see MGS700
10-GLUCOPYRANOSYL-1,8-DIHYDROXY-3-(HYDROXYMETHYL)-9(10H)-ANTHRA-CENONE see BAF825
α-d-GLUCOPYRANOSYL β-d-FRUCTOFURANOSIDE see SNH000
3-β-((α-d-GLUCOPYRANOSYL)OXY)-14-HYDROXY-19-OXO-BUFA-4,20,22-TRIEN-OLIDE see AGW625
2-(β-d-GLUCOPYRANOSYLOXY)ISOBUTYRONITRILE see GFC100
2-(β-d-GLUCOPYRANOSYLOXY)-2-METHYLPROPANENITRILE see GFC100
(d-GLUCOPYRANOSYLTHIO)GOLD see ART250
β-D-GLUCOPYRANURONIC ACID, 1-DEOXY-1-(9H-FLUOREN-2-YLHYDROX-YAMINO)- see HLE750
GLUCOSACCHARONIC ACID see SAA025
d-GLUCOSE, anhydrous see GFG000
d-GLUCOSE see GFG000
GLUCOSE see GFG000
GLUCOSE AERODEHYDROGENASE see GFG100
GLUCOSE LIQUID see GFG000
β-d-GLUCOSE OXIDASE see GFG100
GLUCOSE OXIDASE see GFG100
GLUCOSE-RINGER'S SOLUTION (23.3%) see GFG200
GLUCOSE-RINGER'S SOLUTION (29.2%) see GFG205
(α-d-GLUCOSIDO)-β-d-FRUCTOFURANOSIDE see SNH000
4-(α-d-GLUCOSIDO)-d-GLUCOSE see MAO500
GLUCOSTIBAMINE SODIUM see NCL000
GLUCOSTIMIDINE SODIUM see NCL000
GLUCOSULFONE see AOO800
GLUCOSULFONE SODIUM see AOO800
N-d-GLUCOSYL(2)-N'-NITROSOMETHYLHARNSTOFF (GERMAN) see SMD000
N-d-GLUCOSYL-(2)-N'-NITROSOMETHYLUREA see SMD000
β-d-GLUCOSYLOXYAZOXYMETHANE see COU000
(1-d-GLUCOSYLTHIO)GOLD see ART250
α-d-GLUCOTHIOPYRANOSE see GFK000
GLUCOXY see GFM000
GLUCURON see GFM000
GLUCURONE see GFM000
GLUCURONIC ACID, 1-DEOXY-1-(2-FLUORENYLHYDROXYAMINO)- see HLE750
d-GLUCURONIC ACID-γ-LACTONE see GFM000

GLUCURONIC ACID-γ-LACTONE see GFM000
d-GLUCURONIC ACID LACTONE see GFM000
GLUCURONIC ACID LACTONE see GFM000
N-GLUCURONIDE of N-HYDROXY-2-AMINOFLUORENE see HLE750
d-GLUCURONOLACTONE see GFM000
GLUCURONOLACTONE see GFM000
GLUCURONOSAN see GFM000
GLUDIASE see GFM200
GLUEOPHOGE see DQR600
GLUKRESIN see GGS000
GLUMAL see AGX125
GLUMAMYCIN see AOB875
GLUMIN see GFO050
GLU-P-2 see DWW700
GLUPAN see TEH500
GLUPAX see AAI250
GLU-P-I see AKS250
GLURONAZID see GBB500
GLURONAZIDE see GBB500
GLUSATE see GFO000
GLUSIDE see BCE500
GLUTACID see GFO000
GLUTACYL see MRL500
α-GLUTAMIC ACID see GFO000
GLUTAMIC ACID, dl- see GFM300
l-GLUTAMIC ACID see GFO000
(±)-GLUTAMIC ACID see GFM300
GLUTAMIC ACID see GFO000
GLUTAMIC ACID-5-AMIDE see GFO050
GLUTAMIC ACID AMIDE see GFO050
GLUTAMIC ACID, N-(p-(((2-AMINO-3,4,5,6,7,8-HEXAHYDRO-4-OXO-6-PTERIDI-NYL)METHYL)AMINO) BENZOYL)-, L- see TCR400
l-GLUTAMIC ACID, 5-(2-(4-CARBOXYPHENYL)HYDRAZIDE) see GFO100
dl-GLUTAMIC ACID (9CI) see GFM300
GLUTAMIC ACID, N-(p-(((2,4-DIAMINO-6-PTERIDI-NYL)METHYL)METHYLAMINO)BENZOYL)-, DISODIUM SALT, L-(+)- see MDV600
GLUTAMIC ACID, IRON (2+) SALT (1:1) see FBM000
l-GLUTAMIC ACID, MAGNESIUM SALT (1:1), HYDROBROMIDE see MAG050
l-GLUTAMIC ACID, MONOPOTASSIUM SALT see MRK500
GLUTAMIC ACID, N-NITROSO-N-PTEROYL-, l- see NKG450
GLUTAMIC ACID, SODIUM SALT see MRL500
d-GLUTAMIENSUUR see GFO000
γ-GLUTAMINE see GFO050
GLUTAMINE see GFO050
GLUTAMINE, N-(1-CARBOXY-2-(METHYLENECYCLOPROPYL)ETHYL)- (7CI) see HOW100
l-GLUTAMINE (9CI, FCC) see GFO050
l-GLUTAMINE,N,N'-(SELENOBIS(THIO(1-(((CARBOXYMETH-YL)AMINO)CARBONYL)-2,1-ET HANEDIYL)))BIS- see SBU500
GLUTAMINE, N,N'-((SELENODITHIO)BIS(1-((CARBOXYMETH-YL)CARBAMOYL)ETHYLENE))DI-, L- see SBU500
l-GLUTAMINIC ACID see GFO000
GLUTAMINIC ACID see GFO000
GLUTAMINOL see GFO000
GLUTAMMATO MONOSODICO (ITALIAN) see MRL500
N²-(γ-l-(+)-GLUTAMYL)-4-CARBOXYPHENYLHYDRAZINE see GFO100
1-(l-α-GLUTAMYL)-2-ISOPROPYLHYDRAZINE see GFO200
GLUTANON see TEH500
GLUTARAL see GFQ000
GLUTARALDEHYD (CZECH) see GFQ000
GLUTARALDEHYDE see GFQ000
GLUTARDIALDEHYDE see GFQ000
GLUTARIC ACID see GFS000
GLUTARIC ACID DINITRILE see TLR500
GLUTARIC ANHYDRIDE see GFU000
GLUTARIC DIALDEHYDE see GFQ000
GLUTARIMIDE, 2-(p-AMINOPHENYL)-2-ETHYL- see AKC600
GLUTARIMIDE, 3-(1,3-DIOXO-2-METHYLINDAN-2-YL)- see DVS100
GLUTARIMIDE, 3-(1,3-DIOXO-2-PHENYLINDAN-2-YL)- see DVS300
GLUTARIMIDE, 3-(5,7-DIOXO-6-PHENYL-2,3,6,7-TETRAHYDRO-5H-CYCLOPEN-TA-p-DITHIIN-6-YL)- see DVS400
GLUTARODINITRILE see TLR500
GLUTARONITRILE see TLR500
GLUTARYL DIAZIDE see GFU200
GLUTATHIMID see DYC800
GLUTATHIONE (reduced) see GFW000
GLUTATHIONE see GFW000
GLUTATIOL see GFW000
GLUTATIONE see GFW000
GLUTATON see GFO000
GLUTAVENE see MRL500
GLUTETHIMID see DYC800
GLUTETHIMIDE see DYC800
GLUTETIMIDE see DYC800

GLUTIDE see GFW000
GLUTINAL see GFW000
GLUTOFIX 600 see HKQ100
GLUTRIL see GFY100
GLYBENZCYCLAMIDE see CEH700
GLYBIGID see BRA625
GLYBURIDE see CEH700
GLYBUTAMIDE see BSM000
GLYBUZOLE see GFM200
dl-GLYCERALDEHYDE see GFY200
GLYCERALDEHYDE, (±)- see GFY200
GLYCERIN, anhydrous see GGA000
GLYCERIN, synthetic see GGA000
dl-GLYCERIC ALDEHYDE see GFY200
GLYCERIN see GGA000
GLYCERIN 1-ALLYL ETHER see AGP500
GLYCERIN DIACETATE see GGA100
GLYCERINE see GGA000
GLYCERINE DIACETATE see GGA100
GLYCERINE MONOOLEATE see GGR200
GLYCERINE TRIACETATE see THM500
GLYCERINE TRIPROPIONATE see TMY000
GLYCERINFORMALE see DVR909
GLYCERIN GUAIACOLATE see RLU000
GLYCERIN-α-MONOCHLORHYDRIN see CDT750
GLYCERINMONOGUAIACOL ETHER see RLU000
GLYCERIN MONOOLEATE see GGR200
GLYCERIN MONOSTEARATE see OAV000
GLYCERINTRINITRATE (CZECH) see NGY000
GLYCERITE see TAD750
GLYCERITOL see GGA000
GLYCEROL see GGA000
GLYCEROL-1-ACETATE see GGO000
GLYCEROLACETONE see DVR600
GLYCEROL α-ALLYL ETHER see AGP500
GLYCEROL-1-p-AMINOBENZOATE see GGQ000
GLYCEROL-α-CHLOROHYDRIN see CDT750
GLYCEROL CHLOROHYDRIN see CDT750
GLYCEROL DIACETATE see DBF600, GGA100
GLYCEROL-α,γ-DIBROMOHYDRINE see DDR800
GLYCEROL-α,β-DICHLOROHYDRINE see DGG600
GLYCEROL α,γ-DICHLOROHYDRIN see DGG400
sym-GLYCEROL DICHLOROHYDRIN see DGG400
GLYCEROL DIMETHYLKETAL see DVR600
GLYCEROL-1,3-DINITRATE see GGA200
GLYCEROL, 1,3-DINITRATE see GGA200
GLYCEROL EPICHLORHYDRIN see EAZ500
GLYCEROL ESTER HYDROLASE see GGA800
GLYCEROL-α-ETHYL ETHER see EFF000
GLYCEROL FORMAL see DVR909
GLYCEROL GUAIACOLATE see RLU000
GLYCEROL-α-(o-METHOXYPHENYL)ETHER see RLU000
GLYCEROL-1-MONOACETATE see GGO000
GLYCEROL-α-MONOACETATE see GGO000
GLYCEROL MONOACETATE see GGO000
GLYCEROL-α-MONOCHLOROHYDRIN (DOT) see CDT750
GLYCEROL-α-MONOGUAIACOL ETHER see RLU000
GLYCEROL MONO(2-METHOXYPHENYL)ETHER see RLU000
GLYCEROL MONOOCTADECYL ETHER see GGA915
GLYCEROL MONOOLEATE see GGR200
GLYCEROL MONOSTEARATE see OAV000
GLYCEROL, NITRIC ACID TRIESTER see NGY000
GLYCEROL OLEATE see GGR200
GLYCEROL PHENYL ETHER DIACETATE see PFF300
GLYCEROL TRIACETATE see THM500
GLYCEROL TRIBROMOHYDRIN see GGG000
GLYCEROL TRIBUTYRATE see TIG750
GLYCEROL TRICAPROATE see GGK000
GLYCEROL TRICAPRYLATE see TMO000
GLYCEROL TRICHLOROHYDRIN see TJB600
GLYCEROL (TRI(CHLOROMETHYL))ETHER see GGI000
GLYCEROL TRIHEXANOATE see GGK000
GLYCEROLTRINITRAAT (DUTCH) see NGY000
GLYCEROL(TRINITRATE de) (FRENCH) see NGY000
GLYCEROL TRINITRATE see NGY000
GLYCEROL TRIOCTANOATE see TMO000
GLYCEROPLUMBONITRATE see LDH000
GLYCERYL ACETATE see GGO000
GLYCERYL-p-AMINOBENZOATE see GGQ000
GLYCERYL-α-CHLOROHYDRIN see CDT750
GLYCERYL-1,3-DIACETATE see DBF600
GLYCERYL DIACETATE see GGA100
1,3-GLYCERYL DINITRATE see GGA200
GLYCERYL-α,γ-DIPHENYL ETHER see DVW600
GLYCERYL GUAIACOLATE see RLU000

GLYCERYLGUAIACOLATE CARBAMATE see GKK000
α-GLYCERYL GUAIACOLATE ETHER see RLU000
GLYCERYL GUAIACYL ETHER see RLU000
GLYCERYLGUAJACOL-CARBAMAT see GKK000
GLYCERYL MONOACETATE see GGO000
GLYCERYL MONOLAURATE see MRJ000
GLYCERYL MONOOLEATE see GGR200
GLYCERYL MONOSTEARATE see OAV000
GLYCERYL NITRATE see NGY000
GLYCERYL OLEATE see GGR200
GLYCERYL-o-TOLYL ETHER see GGS000
GLYCERYL TRIACETATE see THM500
GLYCERYL TRIBROMOHYDRIN see GGG000
GLYCERYL TRICAPROATE see GGK000
GLYCERYL TRICHLOROHYDRIN see TJB600
GLYCERYL TRINITRATE see NGY000
GLYCERYL TRIOCTANOATE see TMO000
GLYCERYL TRIPROPIONATE see TMY000
GLYCIDAL see GGW000
GLYCIDALDEHYDE see GGW000
GLYCIDE see GGW500
GLYCIDOL see GGW500
GLYCIDOL OLEATE see ECJ000
GLYCIDOL STEARATE see SLK500
γ-GLYCIDOXYPROPYLTRIMETHOXYSILANE see ECH000
GLYCIDYL ACRYLATE see ECH500
GLYCIDYL ALCOHOL see GGW500
GLYCIDYLALDEHDYE see GGW000
GLYCIDYL BUTYL ETHER see BRK750
N-GLYCIDYL DIETHYL AMINE see GGW800
GLYCIDYLDIETHYLAMINE see GGW800
GLYCIDYL ESTER of DODECANOIC ACID see TJI500
GLYCIDYL ESTER of HEXANOIC ACID see GGY000
GLYCIDYL 2-ETHYLHEXYL ETHER see GGY100
GLYCIDYL ISOPROPYL ETHER see IPD000
GLYCIDYL LAURATE see LBM000
GLYCIDYL METHACRYLATE see ECI000
GLYCIDYL-α-METHYL ACRYLATE see ECI000
GLYCIDYL METHYLPHENYL ETHER see TGZ100
GLYCIDYL 4-NITROPHENYL ETHER see NIN050
GLYCIDYL p-NITROPHENYL ETHER see NIN050
GLYCIDYL OCTADECANOATE see SLK500
GLYCIDYL OCTADECENOATE see ECJ000
GLYCIDYL OLEATE see ECJ000
GLYCIDYL PHENYL ETHER see PFH000
N-GLYCIDYLPHTHALIMIDE see ECK000
GLYCIDYL PROPENATE see ECH500
GLYCIDYL STEARATE see SLK500
GLYCIDYL-TRIMETHYL-AMMONIUM CHLORIDE see GGY200
GLYCINE see GHA000
GLYCINE, N-BENZOYL- see HGB300
GLYCINE BETAINE see GHA050
GLYCINE, N,N-BIS(CARBOXYMETHYL)-(9CI) see AMT500
GLYCINE, N,N-BIS(CARBOXYMETHYL)-, DISODIUM SALT (9CI) see DXF000
GLYCINE, N,N-BIS(CARBOXYMETHYL)-, SODIUM SALT (9CI) see SEP500
GLYCINE, N-(CARBOXYMETHYL)-N-((9,10-DIHYDRO-3,4-DIHYDROXY-9,10-DI-
    OXO-2-ANTHRACENY L))- see AFM400
GLYCINE, N-(2,3-DIHYDRO-1H-INDEN-2-YL)-N-(N-(1-(ETHOXYCARBONYL)-3-
    PHENYLPROPYL)-I-A LANYL)-, MONOHYDROCHLORIDE, (S)- see DAM315
GLYCINE, N,N-DIHYDROXYETHYL- see DMT500
GLYCINE, N,N-DIMETHYL-, HYDROCHLORIDE see MPI100
GLYCINE, DODECYLDIMETHYLBETAINE (6CI) see LBU200
GLYCINE, ETHYL ESTER, HYDROCHLORIDE see GHA100
GLYCINE, N-METHYL-N-NITROSO- see NLR500
GLYCINE MUSTARD see GHE000
GLYCINE, N,N'-1,2-CYCLOHEXANEDIYLBIS(N-(CARBOXYMETHYL))- (9CI) see
    CPB120
GLYCINE, N,N'-ETHYLENEBIS(2-(o-HYDROXYPHENYL))- see EIV100
GLYCINE, N,N'-(2-HYDROXY-1,3-PROPANEDIYL)BIS(N-(CARBOXYMETHYL)-
    (9CI) see DCB100
GLYCINE NITROGEN MUSTARD see GHE000
GLYCINE, N-(p-NITROPHENYL)- see NIN100
GLYCINE, SODIUM SALT see GHG000
GLYCINOL see EEC600
GLYCIRENAN see VGP000
GLYCOALKALOID EXTRACT from POTATO BLOSSOMS see PLW550
GLYCOCOIL HYDROCHLORIDE see GHK000
GLYCOCOLL BETAINE see GHA050
GLYCODIAZINE SODIUM SALT see GHK200
GLYCODINE see TCY750
GLYCO-FLAVINE see DBX400
GLYCOGEN see GHK300
GLYCOHYDROCHLORIDE see GHK000
GLYCOL see EJC500
GLYCOL ALCOHOL see EJC500

GLYCOLANILIDE see GHK500
GLYCOL BIS(HYDROXYETHYL) ETHER see TJQ000
GLYCOL BIS(MERCAPTOACETATE) see MCN000
GLYCOL BROMIDE see EIY500
GLYCOL BROMOHYDRIN see BNI500
GLYCOL BUTYL ETHER see BPJ850
GLYCOL CARBONATE see GHM000
GLYCOL CHLOROHYDRIN see EIU800
GLYCOL CYANOHYDRIN see HGP000
GLYCOL DIACETATE see EJD759
GLYCOL DIBROMIDE see EIY500
GLYCOL DICHLORIDE see EIY600
GLYCOL, DIETHOXYTETRAETHYLENE see PBO250
GLYCOL DIFORMATE see EJF000
GLYCOL DIGLYCIDYL ETHER see EEA600
GLYCOL DIMERCAPTOACETATE see MCN000
GLYCOL DIMETHACRYLATE see BKM250
GLYCOL DIMETHYL ETHER see DOE600
GLYCOLDINITRAAT (DUTCH) see EJG000
GLYCOL DINITRATE see EJG000
GLYCOL (DINITRATE DE) (FRENCH) see EJG000
GLYCOL ETHER see DJD600
GLYCOL ETHER de ACETATE see CBQ750
GLYCOL ETHER DB see DJF200
GLYCOL ETHER DB ACEATATE see BQP500
GLYCOL-ETHERDIAMINETETRAACETIC ACID see EIT000
GLYCOL ETHER EB see BPJ850
GLYCOL ETHER EB ACETATE see BPJ850
GLYCOL ETHER EE see EES350
GLYCOL ETHER EE ACETATE see EES400
GLYCOL ETHER EM see EJH500
GLYCOL ETHER EM ACETATE see EJJ500
GLYCOL ETHER PM see PNL250
GLYCOL ETHERS see GHN000
GLYCOL ETHYLENE ETHER see DVQ000
GLYCOL ETHYL ETHER see DJD600, EES350
GLYCOL FORMAL see DVR800
GLYCOLIC ACID see GHO000
GLYCOLIC ACID, 2,2-DI-2-THIENYL-, 6,6,9-TRIMETHYL-9-AZABICY-
CLO(3.3.1)NON-3-YL ESTER, HCl see MBV100
GLYCOLIC ACID, ETHYL ESTER, METHYL PHTHALATE see MOD000
GLYCOLIC ACID, PHENYL- see MAP000
GLYCOLIC ACID PHENYL ETHER see PDR100
GLYCOLIC ACID, 2-THIO- see TFJ100
GLYCOLIC ACID, THIO- see TFJ100
GLYCOLIC NITRILE see HIM500
GLYCOLIXIR see GHA000
GLYCOLLIC ACID PHENYL ETHER see PDR100
GLYCOL METHACRYLATE see EJH000
GLYCOLMETHYL ETHER see EJH500
GLYCOL MONOACETATE see EJI000
GLYCOL-MONOACETIN see EJI000
GLYCOL MONOBUTYL ETHER see BPJ850
GLYCOL MONOBUTYL ETHERACETATE see BPM000
GLYCOLMONOCHLOORHYDRINE (DUTCH) see EIU800
GLYCOL MONOCHLOROHYDRIN see EIU800
GLYCOL MONOETHYL ETHER see EES350
GLYCOL MONOETHYL ETHER ACETATE see EES400
GLYCOL MONOMETHACRYLATE see EJH000
GLYCOL MONOMETHYL ETHER see EJH500
GLYCOL MONOMETHYL ETHER ACETATE see EJJ500
GLYCOL MONOMETHYL ETHER ACRYLATE see MEM250, MIF750
GLYCOL MONOPHENYL ETHER see PER000
GLYCOL MONOSTEARATE see EJM500
GLYCOLMUL SOC see SKV170
GLYCOLONITRILE see HIM500
GLYCOLONITRILE ACETATE see COP750
(GLYCOLOYLOXY)TRIBUTYLSTANNANE see GHQ000
GLYCOLPYRAMIDE see GHR609
GLYCOLS, POLYETHYLENE, (ALKYLIMINO)DIETHYLENE ETHER, MONOFAT-
TY ACID ESTER see PKE550
GLYCOLS, POLYETHYLENE, MONOCHOLESTERYL ETHER see PKE700
GLYCOLS, POLYETHYLENE, MONOHEXADECYL ETHER (8CI) see PJT300
GLYCOLS, POLYETHYLENE, MONOLAURATE (8CI) see PJY000
GLYCOLS, POLYETHYLENE, MONO(NONYLPHENYL) ETHER see NND500
GLYCOLS, POLYETHYLENE, MONO((1,1,3,3-TETRAMETHYLBUTYL)PHENYL)
ETHER see GHS000
GLYCOLS, POLYETHYLENE POLYPROPYLENE, MONOBUTYL ETHER (non-
ionic) see GHY000
GLYCOL STEARATE see EJM500
GLYCOL SULFATE see EJP000
GLYCOL SULFITE see COV750
GLYCOMONOCHLORHYDRIN see EIU800
GLYCOMUL O see SKV100
GLYCOMUL P see MRJ800

GLYCOMUL S see SKV150
GLYCOMUL SOC see SKV170
GLYCOMUL SOC SPECIAL see SKV170
GLYCON S-70 see SLK000
GLYCON DP see SLK000
GLYCONIAZIDE see GBB500
GLYCONITRILE see HIM500
GLYCONORMAL see GHK200
GLYCON RO see OHU000
GLYCON TP see SLK000
GLYCON WO see OHU000
GLYCOPHEN see GIA000
GLYCOPHENE see GIA000
GLYCOPYRROLATE see GIC000
GLYCOPYRROLATE BROMIDE see GIC000
GLYCOPYRRONIUM BROMIDE see GIC000
GLYCOSPERSE L-20 see PKG000
GLYCOSPERSE O-20 see PKL100
GLYCOSPERSE TO-20 see TOE250
GLYCOSPERSE TS 20 see SKV195
GLYCOSPERSE L-20X see PKG000, PKL000
GLYCO STEARIN see HKJ000
GLYCOTUSS see RLU000
GLYCOXALINEDICARBOXYLIC ACID see IAM000
GLYCURONE see GFM000
GLYCYCLAMIDE see CPR000
2-GLYCYLAMINOFLUORENE see AKC000
GLYCYLBETAINE see GHA050
GLYCYRON see GIG000
18-β-GLYCYRRHETIC ACID see GIE000
GLYCYRRHETIC ACID see GIE000
GLYCYRRHETIN see GIE000
18-β-GLYCYRRHETINIC ACID see GIE000
α-GLYCYRRHETINIC ACID see GIE000
β-GLYCYRRHETINIC ACID see GIEO5O
GLYCYRRHETINIC ACID see GIE000
GLYCYRRHETINIC ACID GLYCOSIDE see GIG000
18-β-GLYCYRRHETINIC ACID HYDROGEN SUCCINATE DISODIUM SALT see
CBO500
GLYCYRRHETINIC ACID HYDROGEN SUCCINATE DISODIUM SALT see
CBO500
GLYCYRRHIZA see LFN300
GLYCYRRHIZAE (LATIN) see LFN300
GLYCYRRHIZA EXTRACT see LFN300
18-β-GLYCYRRHIZIC ACID see GIG000
GLYCYRRHIZIC ACID see GIG000
GLYCYRRHIZIC ACID, AMMONIUM SALT see GIE100
GLYCYRRHIZIC ACID (8CI) see GIG000
β-GLYCYRRHIZIN see GIG000
GLYCYRRHIZIN see GIG000
GLYCYRRHIZINA see LFN300
GLYCYRRHIZINIC ACID see GIG000
GLYCYRRHIZINIC ACID DISODIUM SALT see DXD875
GLYECINE A see TFI500
GLYESTRIN see ECU750
GLYFERRO see FBD500
GLYFYLLIN see DNC000
GLYKOKOLAN SODNY (CZECH) see SHT000
GLYKOKOLLBETAIN see GHA050
GLYKOKOLLBETAIN-CHLORID see CCH850
GLYKOLDINITRAT (GERMAN) see EJG000
GLYME-3 see TKL875
GLYME see DOE600
GLYMIDINE SODIUM SALT see GHK200
GLYMOL see MQV750
GLYODEX 3722 see CBG000
GLYODIN see GII000, GIO000
GLYODIN ACETATE see GII000
GLYOTOL see GGS000
GLYOXAL see GIK000
GLYOXAL, DIMETHYL- see BOT500
GLYOXAL, DIOXIME see EEA000
GLYOXALIDIN see IAL000
GLYOXALIN see IAL000
GLYOXALINE see IAL000
GLYOXIDE see GII000, GIO000
GLY-OXIDE see HIB500
GLYOXIDE DRY see GII000
GLYOXIME see EEA000
GLYOXYLIC ACID see GIQ000
GLYOXYLIC ACID, METHYL ESTER see MKI550
GLYPED see THM500
GLYPHOSATE see PHA500
GLYPHOSINE see BLG250

GLYPHYLLIN see DNC000
GLYPHYLLINE see DNC000
GLYSANOL B see ART250
GLYSOLETTEN see EOK000
GLYTAC see GGY200
GLYTAC A 100 see GGY200
GM 3 see MCB050
GM 4 see MCB050
GM 14 see PKP750
G-MAC see GGY200
G-M-F see DVR200
GMI see DLS600
5'-GMP see GLS750
GMP see GLS750
5'-GMP DISODIUM SALT see GLS800
GMP DISODIUM SALT see GLS800
GMP SODIUM SALT see GLS800
GM SULFATE see GCS000
GNATROL see BAC130
GNOSCOPINE see NBP275
GNS see GEQ425
GO-80 see FBV100
GO 186 see SHL500
GO-1261 see EIH000
GO 10213 see SAY950
GODALAX see PPN100
GOE 1734 see DBQ125
GOE 687 (GERMAN) see MNG000
GOHSENOL see PKP750
GOHSENOL AH 22 see PKP750
GOHSENOL GH 17 see PKP750
GOHSENOL GH 23 see PKP750
GOHSENOL GH 20 see PKP750
GOHSENOL GH see PKP750
GOHSENOL GL 03 see PKP750
GOHSENOL GL 05 see PKP750
GOHSENOL GL 08 see PKP750
GOHSENOL GM 14 see PKP750
GOHSENOL GM 94 see PKP750
GOHSENOL GM 14L see PKP750
GOHSENOL KH 17 see PKP750
GOHSENOL NH 17 see PKP750
GOHSENOL NH 18 see PKP750
GOHSENOL NH 26 see PKP750
GOHSENOL NH 20 see PKP750
GOHSENOL NH 05 see PKP750
GOHSENOL NK 114 see PKP750
GOHSENOL NL 05 see PKP750
GOHSENOL NM 14 see PKP750
GOHSENYL E 50 Y see AAX250
D-GOITRIN see VQA100
(R)-GOITRIN see VQA100
GOLARSYL see ACN250
1721 GOLD see CNI000
GOLD see GIS000
GOLD(I) ACETYLIDE see GIT000
GOLDBALLS see FBS100
GOLD BOND see CCP250
GOLD BRONZE see CNI000
GOLD(III) CHLORIDE see GIW176
GOLD CHLORIDE see GIW176
GOLD CHLORIDE SODIUM see GIZ100
GOLD COMPOUNDS see GIW179
GOLD(I) CYANIDE see GIW189
GOLD DUST see CNE750
GOLDEN ANTIMONY SULFIDE see AQF500
GOLDEN BROWN RK-FQ see CMP250
GOLDEN CEYLON CREEPER see PLW800
GOLDEN CHAIN see GIW195
GOLDEN DEWDROP see GIW200
GOLDEN HUNTER'S ROBE see PLW800
GOLDEN HURRICANE LILY see REK325
GOLDEN POTHOS see PLW800
GOLDEN RAIN see GIW300
GOLDEN SHOWER see GIW300
GOLDEN SPIDER LILY see REK325
GOLDEN YELLOW see DCZ000, DUX800
GOLDEN YELLOW RUAF see CMM758
GOLD FLAKE see GIS000
GOLD(III) HYDROXIDE-AMMONIA see GIX300
GOLD LEAF see GIS000
GOLD(I) NITRIDE-AMMONIA see GIY300
GOLD(III) NITRIDE TRIHYDRATE see GIZ000
GOLD ORANGE see MND600

GOLD ORANGE MP see DOU600
GOLD POWDER see GIS000
GOLDQUAT 276 see PAJ000
GOLD SATINOBRE see LDS000
GOLD SODIUM CHLORIDE see GIZ100
GOLD SODIUM THIOMALATE see GJC000
GOLD SODIUM THIOSULFATE see GJE000
GOLD SODIUM THIOSULFATE DIHYDRATE see GJG000
GOLD THIOGLUCOSE see ART250
GOLD TRICHLORIDE see GIW176
GOLDWEED see FBS100
GOLTIX see ALA500
GOMBARDOL see SNM500
GOMELIN see BAC040
GOMISIN A see SBE450
GONACRINE see DBX400, XAK000
GONADORELIN see LIU360
GONADORELIN DIACETATE see LIU380
GONADYL see SCA750
GONADOTRAPHON FSH see SCA750
GONADOTROPIN-RELEASING FACTOR see LIU360
GONADOTROPIN RELEASING HORMONE see LIU360
GONA-1,3,5(10)-TRIENE-3,16-α-17-β-TRIOL, 13-ETHYL- see HGI700
GONA-1,3,5(10)-TRIENE-3,16,17-TRIOL, 13-ETHYL-, (16-α-17-β)-(9CI) see HGI700
GONDAFON see GHK200
GONOCRIN see XAK000
GONOSAN see GJI250
GONTOCHIN see CLD000
GONTOCHIN PHOSPHATE see CLD250
GONYAULAX TOXIC DIHYDROCHLORIDE see SBA500
GOOD-RITE see PJR000
GOOD-RITE GP 264 see DVL700
GOOD-RITE K 37 see ADW200
GOOD-RITE K 702 see ADW200
GOOD-RITE K-700 see ADW200
GOOD-RITE K727 see ADW200
GOOD-RITE NIX see SIA000
GOOD-RITE WS 801 see ADW200
GOODRITE 1800X73 see SMR000
GOOLS see MBU550
GOOSEBERRY TOMATO see JBS100
GOPHACIDE see BIM000
GOPHER BAIT see SMN500
GOPHER-GITTER see SMN500
GORDOLOBO YERBA (MEXICO) see RBA400
GORDONA see RAF100
GORDON'S MECOMEC see CLO200
GORE-TEX see TAI250
GORMAN see SCA750
GOSHTAM see CNT350
GOSLING see PAM780
GOSSYPIMINE see GJI400
GOSSYPINE see CMF800
GOSSYPLURE see GJK000
GOSSYPLURE H.F. see GJK000
GOSSYPOL see GJM000
(+)-GOSSYPOL see GJM025
GOSSYPOL ACETATE see GJM259
GOSSYPOL ACETIC ACID see GJM259
GOSSYPOSE see RBA100
GOTAMINE TARTRATE see EDC500
GOTA de SANGRE (CUBA) see PCU375
GOTHNION see ASH500
GOUGEROTIN see ARP000
GOUT STALK see CNR135
G 2 (OXIDE) see AHE250
GOYL see ABX500
GP-121 see AOO500
GP 130 see DTN775
2100 GP see PJS750
GP 33679 see DPX400
GP 38383 see IBP309
GP 45840 see DEO600
GP-40-66:120 see HCD250
GP-AMIN see MFB000
GP 7I see SCK600
GP 11I see SCK600
GPKh see HAR000
G 3802POE see PJT300
G 3804POE see PJT300
G 50 (POLYSACCHARIDE) see EHG100
GR-23 see GFG200
GR-29 see GFG205
GR 2/234 see AFK875

GR 33040 see DCV800
GR see GLS000
de GRAAFINA see EDP000
GRAAFINA see EDP000
GRACET VIOLET 2R see DBP000
GRAFESTROL see DKA600
GRAHAM'S SALT see SII500
GRAIN ALCOHOL see EFU000
GRAINES D'EGLISE (GUADELOUPE) see RMK250
GRAINS de LIN PAYS (HAITI) see LED500
GRAIN SORGHUM HARVEST-AID see SFS000
GRAMAXIN see CCS250
GRAMEVIN see DGI400, DGI600
GRAMICIDIN see GJO000
GRAMICIDIN A see GJO025
GRAMIN see DYC000
GRAMINE see DYC000
GRAMINIC ACID see GJQ100
GRAMISAN see MEP250
GRAMIXEL see PAJ000
GRAMONOL see PAJ000
GRAMOXON see PAJ000
GRAMOXONE see PAJ000
GRAMOXONE D see PAJ000
GRAMOXONE DICHLORIDE see PAJ000
GRAMOXONE METHYL SULFATE see PAJ250
GRAMOXONE S see PAI990, PAJ000
GRAMOXONE W see PAJ000
GRAMPENIL see AIV500
GRAMURON see PAJ000
GRANALINO (DOMINICAN REPUBLIC) see LED500
GRANATICIN see GJS000
GRANATICIN A see GJS000
GRANDAXIN see GJS200
GRANEX O see SFS000
GRANMAG see MAH500
GRANOSAN see CHC500
GRANOSAN M see EME100, EME500
GRANOX NM see HCC500
GRANOX PPM see CBG000
GRANULAR ZINC see ZBJ000
GRANULATED SUGAR see SNH000
GRANULIN see ACR300
GRANUREX see BRA250
GRANUTOX see PGS000
GRAPE BLUE A GEIGY see FAE100
GRAPEFRUIT OIL see GJU000
GRAPEFRUIT OIL, coldpressed see GJU000
GRAPEFRUIT OIL, expressed see GJU000
GRAPEMONE see GJU050
GRAPE SUGAR see GFG000
GRAPHITE see CBT500
GRAPHITE SYNTHETIC (ACGIH,OSHA) see CBT500
GRAPHLOX see HCE500
GRAPHOL see MGJ750
GRAPHTOL BLUE RL see IBV050
GRAPHTOL ORANGE GP see CMS145
GRAPHTOL RED A-4RL see MMP100
GRAPHTOL YELLOW A-HG see DEU000
GRAPHTOL YELLOW GXS see CMS212
GRAPHTOL YELLOW RGS see CMS208
GRASAL BRILLIANT RED B see SBC500
GRASAL BRILLIANT RED G see OHA000
GRASAL BRILLIANT YELLOW see DOT300
GRASAL YELLOW see FAG130
GRASAN BRILLIANT RED B see SBC500
GRASAN BROWN DT NEW see NBG500
GRASAN ORANGE 3R see XRA000
GRASAN ORANGE R see PEJ500
GRASCIDE see DGI000
GRASEX see CDN550
GRASLAN see BSN000
GRASOL BLUE 2GS see TBG700
GRASOL VIOLET R see DBP000
GRASOL YELLOW RSF see CMP600
GRASSLAND WEEDKILLER see BAV000
GRATIBAIN see OKS000
GRATUS STROPHANTHIN see OKS000
GRAVIDOX see PPK250
GRAVINOL see DYE600
GRAVISTAT see NNL500
GRAVOCAIN see DHK400
GRAVOCAIN HYDROCHLORIDE see DHK600
GRAVOL see DYE600
GRAY ACETATE see CAL750

GRAY AMBER see AHJ000
GRAYANOTOXANE-3,5,6,10,14,16-HEXOL 14 ACETATE see AOO375
GRAYANOTOXIN I see AOO375
GREAT BLUE LOBELIA see CCJ825
GREEN 5 see ADF000
1724 GREEN see FAG000
11091 GREEN see BLK000
11661 GREEN see CMJ900
12078 GREEN see ADF000
GREEN BS see ADF000
GREEN CHROME OXIDE see CMJ900
GREEN CHROMIC OXIDE see CMJ900
GREEN CINNABAR see CMJ900
GREEN CROSS COUCH GRASS KILLER see TII500
GREEN CROSS CRABGRASS KILLER see PLC250
GREEN CROSS WARBLE POWDER see RNZ000
GREEN-DAISEN M see DXI400
GREEN DENSIC see SCQ000
GREEN DENSIC GC 800 see SCQ000
GREENHARTEN see HLY500
GREEN HELLEBORE see FAB100, VIZ000
GREEN HYDROQUINONE see QFJ000
GREEN LILY see GJU460
GREEN LOCUST see GJU475
GREEN MAMBA VENOM see GJU500
GREEN NICKEL OXIDE see NDF500
GREEN No. 2 see BLK000
GREEN No. 203 see FAF000
GREENOCKITE see CAJ750
GREEN OIL see APG500, COD750
GREEN ROUGE see CMJ900
GREEN S see ADF000
GREEN SEAL-8 see ZKA000
GREEN VITRIOL see FBN100
GREEN VITROL see FBO000
GRELAN see AMK250
GRENADE see GJU600
GRENADES, empty primed (NA0349) (DOT) see GJU600
GRENADES, hand or rifle, with bursting charge (UN0284, UN0285, UN0292, UN0293) (DOT) see GJU600
GRENADES, practice, hand or rifle (UN0452, UN0110, UN0318, UN0372) (DOT) see GJU600
GREOSIN see GKE000
GRESFEED see GKE000
G 3 (RESIN) see MCB050
G-RESINS see PJS750
GREX see PJS750
GREX PP 60-002 see PJS750
GREY ARSENIC see ARA750
GREY NICKER see CAK325
Gn-RH see LIU360
GRICIN see GKE000
GRIFFEX see ARQ725
GRIFFIN MANEX see MAS500
GRIFFITH'S ZINC WHITE see LHX000
GRIFOMIN see TEP500
GRIFULVIN see GKE000
GRILON see NOH000, PJY500
GRILONIT RV 1806 see BOS100
GRIPENIN see CBO250
GRIPPEX see TEH500
GRISACTIN see GKE000
GRISCOFULVIN see GKE000
GRISEFULINE see GKE000
GRISEIN see GJU800
GRISEMIN see CEX250
GRISEO see GKE000
(+)-GRISEOFULVIN see GKE000
GRISEOFULVIN-FORTE see GKE000
GRISEOFULVINUM see GKE000
GRISEOLUTEIN B see GJW000
GRISEOMYCIN see GJY000
GRISEORUBIN COMPLEX see GJY100
GRISEORUBIN I HYDROCHLORIDE see GKA000
GRISEOVIRIDIN see GKC000
GRISETIN see GKE000
GRISIN see CEX250, GJU800
GRISOFULVIN see GKE000
GRISOL see TCF250
GRISOLEN see PJS750
GRISOVIN see GKE000
GRIS-PEG see GKE000
GR-M see PJQ050
GROCEL see GEM000
GROCO 2 see OHU000

GROCO 4 see OHU000
GROCO 54 see SLK000
GROCO 5810 see BSL600
GROCO see LGK000
GROCO 5L see OHU000
GROCOLENE see GGA000
GROCOR 5500 see OAV000
GROCOR 6000 see OAV000
GROFAS see QSA000
GROSAFE see CBT500
GROTAN B see THR820
GROTAN BK see THR820
GROUND HEMLOCK see YAK500
GROUNDNUT OIL see PAO000
GROUND RYANIA SPECISA(VAHL) STEMWOOD (ALKOLOID RYANODINE) see RSZ000
GROUNDSEL see RBA400
GROUND VOCLE SULPHUR see SOD500
L-GRUEN No. 1 see CKN000
L-GRUEN No. 1 (GERMAN) see CKN000
GRUMEX see BQW490
GRUNDIER ARBEZOL see PAX250
GRUNERITE see GKE900
GRYSIO see GKE000
GRYZBOL see DVF800
GRZYBOL see DVF800
GS 6 see IGK800
GS 015 see BMB125
GS-95 see TEZ000
G 339 S see CAX350
GS-3065 see DYE425
GS 6244 see FOI000
GS 13528 see THR600
GS 13529 see BQB000
GS-13,798 see MPG250
GS 14259 see BQC500
GS 15254 see BQC250
GS-16068 see EPN500
GS 34360 see INR000
GSH see GFW000
G-STROPHANTHIN see OKS000
G 52 SULFATE see GAA120
GT41 see BOT250
GT-1012 see DNB000
GT 2041 see BOT250
GT see PCU360
GTG see ART250
GTN see NGY000
GUABENXANE see GKG300
GUACALOTE AMARILLO (CUBA) see CAK325
GUACAMAYA (CUBA) see CAK325
GUACIS (MEXICO) see LED500
GUAIAC ACETATE see GKG400
GUAIAC GUM see GLY100
m-GUAIACOL see REF050
GUAIACOL see GKI000
GUAIACOLGLICERINETERE see RLU000
GUAIACOL GLYCERYL ETHER see RLU000
GUAIACOL GLYCERYL ETHER CARBAMATE see GKK000
GUAIACURANE see RLU000
GUAIACWOOD ACETATE see GKG400
GUAIAC WOOD OIL see GKM000
GUAIACYL GLYCERYL ETHER see RLU000
GUAIACYLPROPANE see MFM750
GUAIA-1(5),7(11)-DIENE see GKO000
GUAIAMAR see RLU000
GUAIANESIN see RLU000
s-GUAIAZULENE see DSJ800
GUAICOL see GKI000
β-GUAIENE see GKO000
GUAIENE see GKO000
GUAIFENESIN see RLU000
GUAIMERCOL see GKO500
GUAIOL ACETATE see GKG400
GUAIPHENESINE see RLU000
GUAJACOL-GLYCERINAETHER (GERMAN) see RLU000
GUAJACOL-α-GLYCERINETHER see RLU000
GUAMIDE see AHO250
GUANABENZ see GKO750
GUANABENZ ACETATE see GKO750
GUANATOL HYDROCHLORIDE see CKB500
GUANAZODINE see GKO800
GUANAZODINE SULFATE MONOHYDRATE see CAY875
GUANAZOL see AJO500
GUANAZOLE see DCF200

GUANAZOLO see AJO500
GUANERAN see AKY250
GUANETHIDINE see GKQ000
GUANETHIDINE BISULFATE see GKS000
GUANETHIDINE MONOSULFATE see GKU000
GUANETHIDINE SULFATE see GKS000
GUANFACINE HYDROCHLORIDE see GKU300
GUANICAINE see PDN500
GUANICIL see AHO250
GUANIDAN see AHO250
GUANIDINE see GKW000
GUANIDINE, AMINO-, HYDROCHLORIDE see AKC800
GUANIDINE CARBONATE see GKW100
GUANIDINE CHLORIDE see GKY000
GUANIDINE, CYANO-, METHYLMERCURY deriv. see MLF250
GUANIDINE, 1,3-DIPHENYL-, MONOHYDROCHLORIDE see DWC625
GUANIDINE HYDROCHLORIDE see GKY000
GUANIDINE, N-METHYL-N′-NITRO-N-NITROSO-(9CI) see MMP000
GUANIDINE, MONOHYDROCHLORIDE see GKY000
GUANIDINE MONONITRATE see GLA000
GUANIDINE MONOPHOSPHATE see GLS750
GUANIDINE NITRATE (DOT) see GLA000
GUANIDINE, NITRO- see NHA500
GUANIDINE, N′-NITRO-N-NITROSO-N-PENTYL- see NLC500
GUANIDINE, SULFANILYL- see AHO250
GUANIDINE THIOCYANATE see TFF250
GUANIDINIUM CARBONATE see GKW100
GUANIDINIUM CHLORIDE see GKY000
GUANIDINIUM DICHROMATE see GLB100
GUANIDINIUM HYDROCHLORIDE see GKY000
GUANIDINIUM NITRATE see GLB300
GUANIDINIUM PERCHLORATE see GLC000
GUANIDINIUM THIOCYANATE see TFF250
p-GUANIDINOBENZOIC ACID 4-ALLYL-2-METHOXYPHENYL ESTER see MDY300
p-GUANIDINOBENZOIC ACID 4-METHYL-2-OXO-2H-1-BENZOPYRAN-7-YL ESTER see GLC100
p-GUANIDINOBENZOIC ACID p-NITROPHENYL ESTER see NIQ500
1-(2-GUANIDINOETHYL)HEPTAMETHYLENIMINE see GKQ000
N-(2-GUANIDINO ETHYL)HEPTAMETHYLENIMINE SULFATE see GKU000
1-(2-GUANIDINOETHYL)OCTAHYDROAZOCINE see GKQ000
1-(2-GUANIDINOETHYL)OCTAHYDROAZOCINE SULFATE (2:1) see GKS000
GUANIDINO-6-METHYL-1,4-BENZODIOXANE see GKG300
6-GUANIDINOMETHYL 1,4-BENZODIOXANE SULFATE see GKG300
GUANIDINO METHYL-6-BENZODIOXANNE-1,4 (FRENCH) see GKG300
GUANINE see GLI000
GUANINE DEOXYRIBOSIDE see DAR800
GUANINE-3-N-OXIDE see GLK100
GUANINE-7-N-OXIDE see GLM000
GUANINE-3-N-OXIDE HEMIHYDROCHLORIDE see GLO000
GUANINE, 9-β-D-RIBOFURANOSYL- see GLS000
GUANINE RIBOSIDE see GLS000
GUANIOL see DTD000
GUANOSINE see GLS000
GUANOSINE 5′-MONOPHOSPHATE see GLS750
GUANOSINE MONOPHOSPHATE see GLS750
GUANOSINE 5′-MONOPHOSPHORIC ACID see GLS750
GUANOSINE 5′-PHOSPHATE see GLS750
GUANOTHIAZON see AHI875
GUANOXYFEN SULFATE see GLS700
GUANTLET see CHJ300
GUANYL DISULFIDE DIHYDROCHLORIDE see DXN400
GUANYL HYDRAZINE see AKC750
GUANYLHYDRAZINE HYDROCHLORIDE see AKC800
5′-GUANYLIC ACID see GLS750
GUANYLIC ACID see GLS750
GUANYLIC ACID SODIUM SALT see GLS800
1-GUANYL-4-NITROSAMINOGUANYLTETRAZENE see TEF500
GUANYL NITROSAMINOGUANYLTETRAZENE (dry) (DOT) see TEF500
GUANYL NITROSAMINOGUANYLTETRAZENE, wetted or tetrazene, wetted with not <30% water or (DOT) see TEF500
GUANYL NITROSAMINO GUANYL TETRAZENE see TEF500
GUAR see GLU000
GUARAN see GLU000
GUARANINE see CAK500
GUAR FLOUR see GLU000
GUAR GUM see GLU000
GUASTIL see EPD500
GUATEMALA LEMONGRASS OIL see LEH000
GUAVA see GLW000
GUAYIGA (DOMINICAN REPUBLIC) see CNH789
GUBERNAL see AGW000
GUDAKHU (INDIA) see SED400
GUESAPON see DAD200
GUESAROL see DAD200

GUIACOL-GLICERILETERE MONOCARBAMMATO see GKK000
GUICITRINA see AIV500
GUICITRINE see AIV500
GUIDAZIDE see GBB500
GUIGNER'S GREEN see CMJ900
GUINDA (PUERTO RICO) see APM875
GUINEA GREEN see FAE950
GUINEA GREEN B see FAE950
GUINEA GREEN BA see FAE950
GULF S-15126 see DED000
l-GULITOL see SKV200 ·
GULITOL see SKV200
GULLIOSTIN see PCP250
GUM see PJR000
GUM ARABIC see AQQ500
GUM CAMPHOR see CBA750
GUM CARRAGEENAN see CCL250
GUM CHON 2 see CCL250
GUM CHROND see CCL250
GUM CYAMOPSIS see GLU000
GUM GHATTI see GLY000
GUM GUAIAC see GLY100
GUM GUAR see GLU000
GUM NAFKACRYSTAL see ROH900
GUM OPIUM see OJG000
GUM OVALINE see AQQ500
GUM SENEGAL see AQQ500
GUM STERCULIA see KBK000, SLO500
GUM TRAGACANTH see THJ250
GUNACIN see GMC000
GUNCOTTON see CCU250
GUNPOWDER see ERF500, PLL750
GUNPOWDER, granular or as a meal (UN 0027) (DOT) see ERF500, PLL750
GUNPOWDER, compressed (UN 0028) (DOT) see ERF500, PLL750
GUNPOWDER, in pellets (UN 0028) (DOT) see ERF500, PLL750
GURJUN BALSAM see GME000
GURONSAN see GFM000
GUSATHION see ASH500
GUSATHION A see EKN000
GUSERVIN see GKE000
GUSTAFSON CAPTAN 30-DD see CBG000
GUTHION (DOT) see ASH500
GUTHION (ETHYL) see EKN000
GUTRON see MQT530
GUTTAGENA see PKQ059
GUTTALAX see SJJ175
α-2-GUTTIFERIN see GME300
β-GUTTIFERIN see CBA125
α-GUTTIFERIN (9CI) see GME300
G1V GARD DXN see ABC250
GYKI 11679 see RFU800
GYKI 14166 see PEE100
GYN see MAG000
GYNAESAN see EDU500
GYN-ANOVLAR see EEH520
GYNECLORINA see CDP000
GYNECORMONE see EDP000
GYNEFOLLIN see DAL600
GYNE-LOTRIMIN see MRX500
GYNE-MERFEN see MDH500
GYNERGEN see EDC500
GYNERGON see EDO000
GYNESTREL see EDO000
GYNFORMONE see EDP000
GYNOCHROME see MCV000
GYNOESTRYL see EDO000
GYNOFON see ABY900
GYNOKHELLAN see AHK750
GYNOLETT see DKB000
GYNONLAR 21 see EEH520
GYNO-PEVARYL see EAE000
GYNOPHARM see DKA600
GYNOPLIX see ABX500
GYNOVLAR see EQM500
GY-PHENE see CDV100
GYPSINE see LCK000, LCK100
GYPSOGENIN see GMG000
GYPSOPHILASAPOGENIN see GMG000
GYPSOPHILASAPONIN see GMG000
GYPSUM see CAX750
GYPSUM STONE see CAX750
GYRANE see GMG100
GYROMITRIN see AAH000
GYRON see DAD200

H-22 see BSG300
H-28 see BSG250
H-33 see BPP250
H-34 see HAR000
H 46 see AHC000
H-69 see DTN200
H-88 see TKF250
H 95 see CKF750
H115 see CDY325
H 133 see DER800
H 224 see BRA625
H 321 see DST000
H-365 see ELL500
454H see SMQ500
H-490 see ENG500
H 520 see RDK000
Hg 532 see PGA750
H 610 see EOM000
H-899 see AKG500
H 940 see ENG500
H 990 see AEX000
H4 099 see DTL200
H 1032 see MHJ500
H3 111 see DBC510
H 1313 see DER800
H 1672 see PEF750
2903 H see CDS275
H 3292 see DNN600
H 3452 see FBP100
H 4007 see MCH535
H-4723 see CIR750
H 56/28 see AGW250
H 59/64 see DUM100
H 8717 see PMN250
H 93/26 see MQR144
96H60 see DFH600
cis-H 102.09 see ZBA500
''H'' see HBT500
HA 106 see TLG000
HA see HHQ800
70H-2AAF see HIK500
HAAS see SMT500
HABA de SAN ANTONIO (PUERTO RICO) see CAK325
HABILLA (PUERTO RICO) see YAG000
HACHE UNO SUPER see FDA885
HACHI-SUGAR see SGC000
H ACID see AKH000
cis-HAD see DGU709
HADACIDIN see FNO000
HADACIDINE see FNO000
HADACIN see FNO000
2,4-HADIYNYLENE CHLOROFORMATE see HAB600
HAEMATITE see HAO875
HAEMODAN see MGC350
HAEMOFORT see FBO000
HAEMOPHILUS INFLUENZAE, ENDOTOXIN, phenol water extract see HAB710
HAEMOPHILUS INFLUENZAE, ENDOTOXIN, hypertonic NaCl citrate extract see HAB700
HAEMOSTASIN see VGP000
HAFFKININE see ARQ250
HAFNIUM, wet with not less than 25% water (DOT) see HAC000
HAFNIUM see HAC000
HAFNIUM CHLORIDE see HAC800
HAFNIUM CHLORIDE OXIDE see HAD500
HAFNIUM CHLORIDE OXIDE OCTAHYDRATE see HAD000
HAFNIUM DICYCLOPENTADIENE DICHLORIDE see HAE500
HAFNIUM OXYCHLORIDE see HAD500
HAFNIUM OXYCHLORIDE OCTAHYDRATE see HAD000
HAFNIUM POWDER, wetted with not <25% water (UN 1326) (DOT) see HAC000
HAFNIUM POWDER, dry (UN 2545) (DOT) see HAC000
HAFNIUM TETRACHLORIDE see HAC800
HAFNOCENE DICHLORIDE see HAE500
HAIARI see RNZ000
HAIMASED see SIA500
HAIROXAL see PII100
HAIRY see HBT500
HAITIN see HON000
HAKUENKA CC see CAT775
HAKUENKA R 06 see CAT775
HALAMID see CDP000
HALANE see DFE200
HALAR 200 see KDK000
HALARSOL see DFX400
HALAZONE see HAF000

HALBMOND see BAU750
HALCIDERM see HAF300
HALCIMAT see HAF300
HALCINONIDE see HAF300
HALCION see THS800
HALCORT see HAF300
HALDI RHIZOME EXTRACT see COF850
HALDOL see CLY500
HALDRONE see PAL600
HALEITE see DUU800
HALF-CYSTEINE see CQK000
HALF-CYSTINE see CQK000
HALF-MUSTARD GAS see CGY750
HALF MUSTARD GAS see CHC000
HALF-MYLERAN see EMF500
HALF SULFUR MUSTARD see CHC000
HALIDO see BAV250
HALIDOR see POD750
HALITE see SFT000
HALIZAN see TDW500
HALKAN see DYF200
HALLOYSITE see HAF375
HALLTEX see SEH000
HALOANISONE see HAF400
HALOANISONE COMPOSITUM see FDA880
HALOCARBON 11 see TIP500
HALOCARBON 14 see CBY250
HALOCARBON 23 see CBY750
HALOCARBON 112a see TBP000
HALOCARBON 112 see TBP050
HALOCARBON 113 see FOO000
HALOCARBON 114 see FOO509
HALOCARBON 115 see CJI500
HALOCARBON 152A see ELN500
HALOCARBON 1132A see VPP000
HALOCARBON C-138 see CPS000
HALOCARBON OIL 11-14 see KDK000
HALOCARBON OIL 11-21 see KDK000
HALOCARBON OIL 13-21 see KDK000
HALOCARBON 13/UCON 13 see CLR250
HALOFANTRINE HYDROCHLORIDE see HAF500
HALOFANTRINO (SPANISH) see HAF500
HALOFLEX 202 see ADW200
HALOFLEX 208 see ADW200
HALOG see HAF300
HALOMICIN see HAF825
HALOMYCETIN see CDP250
HALON 14 see CBY250
HALON 1001 see MHR200
HALON 1011 see CES650
HALON 1202 see DKG850
HALON 1211 see BNA250
HALON 1301 see TJY100
HALON 2001 see EGV400
HALON 2402 see FOO525
HALON 10001 see MKW200
HALON see DFA600, KDK000
HALONISON see HAF400
HALON (POLYMER) see KDK000
HALON TFEG 180 see TAI250
HALOPERIDOL see CLY500
HALOPERIDOL DECANOATE see HAG300
HALOPREDONE ACETATE see HAG325
HALOPROGIN see IFA000
HALOPROPANE see BOF750
HALOPYRAMINE see CKV625
HALOSTEN see CLY500
HALOTAN see HAG500
HALOTESTIN see AOO275
HALOTEX see IFA000
HALOTHANE see HAG500
HALOWAX 1014 see HCK500
HALOWAX see TBR000, TIT500
HALOXAZOLAM see HAG800
HALOXON see DFH600
HALSAN see HAG500
HALSO 99 see CLK100
HALTRON 22 see CFX500
HALTS see BAF250
HALVIC 223 see PKQ059
HALVISOL see HAH000
HAMAMELIS see WCB000
HAMIDOP see DTQ400
HAMILTON RED see CHP500
HAMOVANNID see HFG550

HAMP-ENE 100 see EIV000
HAMP-ENE 215 see EIV000
HAMP-ENE 220 see EIV000
HAMP-ENE ACID see EIX000
HAMP-ENE Na4 see EIV000
HAMP-EX ACID see DJG800
HAMP-OL ACID see HKS000
HAMPSHIRE see IBH000
HAMPSHIRE GLYCINE see GHA000
HAMPSHIRE NTA see SIP500
HAMPSHIRE NTA ACID see AMT500
HAMYCIN see HAH800
HANA see HMX600
HANANE see BJE750
HANCOCK YELLOW 10010 see DEU000
HANQ see HIX200
HANSACOR see DJS200
HANSAMID see NCR000
HANSA RED B see MMP100
HANSA RED G see MMP100
HANSA SCARLET RB see MMP100
HANSA SCARLET RN see MMP100
HANSA SCARLET RNC see MMP100
HANSOLAR see SNY500
HAOS see HLN100
HAPA-B SULFATE see SBE800
HAPLOS see EOK000
HAPPY DUST see CNE750
HAPTOCIL see PPO000
HAQ see BKB300
HARE-RID see SMN500
HARMALOL HYDROCHLORIDE see HAI300
HARMAN see MPA050
HARMANE see MPA050
HARMAR see PNA500
HARMINE see HAI500
HARMONE B 79 see DFN425
HARMONIN see MQU750
HARMONYL see RDF000
HAROWAX L 9 see GGR200
HARRICAL see CCK125
HARRY see HBT500
HARTOL see MQU750
HART'S HORN see MBU825
HARTSHORN PLANT see PAM780
HARVAMINE see WAK000
HARVATRATE see MGR500
HARVEN see SGD000
HARVEST-AID see SFS000
HARWAX A see HOG000
HAS (GERMAN) see HLB400
HASACH see CBD750
HASETHROL see PBC250
HASHISH see CBD750
HASTELLOY C see CNA750
HASTINGS CARMINE 2G see CMM300
HATCOL 200 see TJR600
HATCOL DIBP see DNJ400
HATCOL DOP see DVL700
HAURYMELLIN see DQR800
HAVERO-EXTRA see DAD200
HAVIDOTE see EIX000
HAVOC see TAC800
HAY ABSOLUTE see FMS000
HAYNES 25 see CNA750
HAYNES ALLOY NUMBER 25 see CNA750
HAYNES STELLITE 21 see CNA750
HAYNON see CLX300
HAZODRIN see MRH209
HB[17] see HHQ825
H.17B see HHQ825
HB-218 see AKM125
HB 419 see CEH700
HB see HEA000
m-HBA see HJI100
HBB see HCA500
HBBN see HJQ350
HBC see HEF500
HBF 386 see AEA750, AEB000
HBK see HAI600
9,10-H2 B(e)P see DKU400
HBP see HHQ850
HC-3 see HAQ000
HC-064 see NGB700
292 HC see PIK400

336 HC see PIK375
HC 1281 see TIK500
1352 HC see DHY200
HC 2072 see NIM500
HC7901 see DUG500
HC see CNS750
HCA see HCL500, HHQ800
HCA WEEDKILLER see HCL500
HCB see HCC500
HCBD see HCD250
HC BLUE 1 see BKF250
HC BLUE No. 2 see HKN875
α-HCC see HJV000
HCCH see BBP750, BBQ500
HCCPD see HCE500
HCDD see HAJ500
HCE see EBW500
HCFU see CCK630
HC 20-511 FUMARATE see KGK200
HCG-004 see CJP750
α-HCH see BBQ000
β-HCH see BBR000
γ-HCH see BBQ500
HCH see BBQ500
HCL SALZ des p-AMINO-BENZOESAEURE-DIMETHYLAMINO-AETHYL-ESTER
  (GERMAN) see DPC200
HCl SALZ des p-AMINO-SALICYLSAEURE-DIAETHYLAMINOAETHYLESTER
  (GERMAN) see AMM750
HCl,1; SALZ des p-AMINO-SALICYLSAEURE-DIMETHYLAMINOAETHYL-ESTER
  (GERMAN) see AMN000
HCl SALZ des p-AMINO-SALICYLSAEURE-DIMETHYLAMINOAETHYL-ESTER
  (GERMAN)
HCl,1; SALZ DES p,N,N-
  BUTYLAMINOSALICYLSAEUREDIAETHYLAMINOAETHYLESTER (GERMAN)
  see BQH250
HCPHCl SALZ DES p,N,N-
  BUTYLAMINOSALICYLSAEUREDIAETHYLAMINOAETHYLESTER (GERMAN)
  see ABD000, HCL000
HC RED NO. 3 see ALO750
HCS 3260 see CDR750
HCT see HGO500
HCTZ see CFY000
H. CYANOCINCTUS VENOM see SBI900
HCZ see CFY000
HD 419 see CEH700
HD AMARANTH B see FAG020
HDEHP see BJR750
H.D. EUTANOL see OBA000
HDMTX see MDV500
HDO see HEP500
HD OLEYL ALCOHOL 70/75 see OBA000
HD OLEYL ALCOHOL 80/85 see OBA000
HD OLEYL ALCOHOL 90/95 see OBA000
HD OLEYL ALCOHOL CG see OBA000
HD PONCEAU 4R see FMU080
HD SUNSET YELLOW FCF see FAG150
HD SUNSET YELLOW FCF SUPRA see FAG150
HD TARTRAZINE see FAG140
HD TARTRAZINE SUPRA see FAG140
HDTMP see HEQ600
HDU see HEF500
(2-HDYROXYETHYL)TRIMETHYLAMMONIUM see CMF000
17-β-HDYROXY-17-α-METHYL-5-α-ANDROSTANO(2,3-c)FURAZAN see AOO300
HE 166 see DUS500
HEADACHE WEED see CMV390
HEALON see HGN600
HEART-OF-JESUS see CAL125
HEARTS see AOB250, BBK500
HEAT see HAJ700
HEAT PRE see MJG750
HEAVENLY BLUE see DJO000
HEAVY CATALYTICALLY CRACKED DISTILLATE see DXG810
HEAVY CATALYTIC CRACKED NAPHTHA see NAQ520
HEAVY CATALYTIC REFORMED NAPHTHA see NAQ530
#6 HEAVY FUEL OILS see HAJ750
HEAVY NAPHTHENIC DISTILLATE see MQV780
HEAVY NAPHTHENIC DISTILLATE SOLVENT EXTRACT see MQV857
HEAVY NAPHTHENIC DISTILLATES (PETROLEUM) see MQV780
HEAVY OIL see CMY825
HEAVY PARAFFINIC DISTILLATE see MQV785
HEAVY PARAFFINIC DISTILLATE, SOLVENT EXTRACT see MQV859
HEAVY THERMAL CRACKED NAPHTA see NAQ580
HEAVY VACUUM DISTILLATE see GBW005
HEAVY VACUUM GAS OIL (PETROLEUM) see GBW005
HEAZLEWOODITE see NDJ500

HEBABIONE HYDROCHLORIDE see PPK500
HEBANIL see CKP500
HEBARAL see EKT500
HEB-CORT see CNS750
HEBUCOL see COW700
HEC see HKQ100
HEC-AL 5000 see HKQ100
HECLOTOX see BBQ500
HECNU see CHB750
HECNU-MS see HKQ300
HECTOGRAPH VIOLET SR see AOR500
HECTO VIOLET R see AOR500
HEDAPUR M 52 see CIR250
HEDERA CANARIENSIS see AFK950
HEDERA HELIX see AFK950
HEDEX see HIM000
HEDGE PLANT see PMD550
HED-HEPARIN see HAQ550
HEDIONDA (PUERTO RICO) see CNG825
HEDIONDILLA (PUERTO RICO) see LED500
HEDIONE see HAK100
HEDOLIT see DUS700
HEDONAL see DGB000, MOU500
HEDONAL DP see DGB000
HEDONAL MCPP see CIR500, CLO200
HEDONAL (The herbicide) see DAA800
HEDP see HKS780
HEDTA see HKS000
HEDULIN see PFJ750
HEEDTA see HKS000
HEF-2 see JDA100
HEF-3 see JDA125
HE-HK-52 see HAK200
HEKBILIN see CDL325
HEKSAN (POLISH) see HEN000
HEKSOGEN (POLISH) see CPR800
HEKTALIN see VGP000
HELAKTYN BLACK DN see CMS227
HELAKTYN PURE BLUE FR see CMS228
HELANTHRENE BLACK BL see CMU320
HELANTHRENE BLUE BC see DFN425
HELANTHRENE BROWN GR see CMU770
HELANTHRENE YELLOW see DCZ000
HELENALIN see HAK300
HELFERGIN see AAE500, DPE000
HEL-FIRE see BRE500
HELFO DOPA see DNA200
HELFOSERPIN see RDK000
HELIACID LIGHT YELLOW 4R see SGP500
HELIANE ORANGE RF see CMU815
HELIANE RED 5B see DNT300
HELIANTHINE see MND600
HELIANTHINE B see MND600
HELIANTHRENE BLUE RS see IBV050
HELICHRYSUM OIL see HAK500
HELICOCERIN see ECE500
HELICON see ADA725
HELIC YELLOW GW see DEU000
HELINDON ORANGE R see CMU815
HELINDON RED BB see DNT300
HELIO FAST GREEN GN see CMS140
HELIO FAST GREEN GT see CMS140
HELIO FAST RED BN see MMP100
HELIO FAST RED RL see MMP100
HELIO FAST RED RN see MMP100
HELIO FAST YELLOW GRF see CMS208
HELIO FAST YELLOW GRN see CMS208
HELIOGEN see CDP000
HELIOGEN BLUE 6470 see IBV050
HELIOGEN GREEN 9360 see CMS140
HELIOGEN GREEN 6G see CMS140
HELIOGEN GREEN 6GA see CMS140
HELIOGEN GREEN 8GA see CMS140
HELIOMYCIN see HAL000
HELION RED 8B see CMO885
HELIOPAR see CLD000
HELIO RED RL see MMP100
HELIO RED TONER see MMP100
HELIO RED TONER LCLL see CHP500
HELIOSTABLE BRILLIANT PINK B EXTRA see RGW000
HELIOTRIDINE ESTER with LASIOCARPUM and ANGELIC ACID see LBG000
HELIOTRINE see HAL500
HELIOTRON see HAL500
HELIOTROPIN see PIW250
HELIOTROPIN ACETAL see PIZ499

HELIOTROPIUM SUPINUM L. see HAM000
HELIOTROPIUM TERNATUM see SAF500
HELIOTROPYL ACETATE see PIX000
HELIOTROPYL ACETONE see MJR250
HELIUM see HAM500
HELIUM, compressed (UN 1046) (DOT) see HAM500
HELIUM, refrigerated liquid (cryogenic liquid) (UN 1963) (DOT) see HAM500
HELLEBORE see CMG700
HELLEBORUS NIGER see CMG700
HELLEBRIGENIN-GLUCO-RHAMNOSID (GERMAN) see HAN600
HELLEBRIN see HAN600
HELLEGRIGENIN GLUCORHAMNOSIDE see HAN600
HELLIPIDYL see AMM250
HELMATAC see BQK000
HELMET FLOWER see MRE275
HELMETINA see PDP250
HELMIRANE see DFH600
HELMIRON see DFH600
HELMIRONE see DFH600
HELMOX see COH250
HELODERMA SUSPECTUM VENOM see HAN625
HELOTHION see SOU625
HELOXY WC68 see NCI300
HELVETICOSID (GERMAN) see HAN800
HELVETICOSIDE see HAN800
HELVETICOSIDE DIHYDRATE see HAO000
HEMACHATUS HAEMACHATES VENOM see HAO500
HEMACHATUS HAEMACHATUS VENOM see HAO500
HEMATITE see HAO875
HEMATOPORPHYRIN see PHV275
HEMATOPORPHYRIN MERCURY DISODIUM SALT see HAP000
HEMATOXYLIN see HAP500
HEMEL see HEJ500
HEMICHOLINE see HAQ000
HEMICHOLINIUM-3 see HAQ000
HEMICHOLINIUM-3-BROMIDE see HAQ000
HEMICHOLINIUM BROMIDE see HAQ000
HEMICHOLINIUM-3-DIBROMIDE see HAQ000
HEMICHOLINIUM DIBROMIDE see HAQ000
HEMIMELLITENE see TLL500
HEMINEVRIN see CHD750
HEMISINE see VGP000
HEMITON see CMX760
HEMLOCK OIL see SLG650
HEMLOCK WATER DROPWORT see WAT315
HEMO-B-DOZE see VSZ000
HEMOCAPROL see AJD000
HEMOCOAGULASE see CMY725
HEMODAL see MMD500
HEMODESIS see PKQ250
HEMODEX see DBD700
HEMODEZ see PKQ250
HEMOFURAN see NGE500
HEMOMIN see VSZ000
HEMOPAR see AJD000
HEMOSTASIN see VGP000
HEMOSTYPTANON see EDU500
HEMOTON see FBJ100
HEMOTROPE see PEU000
HEMPA see HEK000
HENBANE see HAQ100
HENDECANAL see UJJ000
HENDECANALDEHYDE see UJJ000
HENDECANE see UJS000
HENDECANOIC ACID see UKA000
HENDECANOIC ALCOHOL see UNA000
1-HENDECANOL see UNA000
2-HENDECANONE see UKS000
HENDECENAL see ULJ000
10-HENDECEN-1-YL ACETATE see UMS000
HENDECYL ALCOHOL see UNA000
n-HENDECYLENIC ALCOHOL see UNA000
10-HENEDECENOIC ACID see ULS000
HENEICOSAFLUORO-1-IODODECANE see IEU075
HENKEL'S COMPOUND see TBC450
HENNA see HMX600
HENNOLETTEN see EOK000
HENU see HKW500
HEOD see DHB400
HEPACHOLINE see CMF750
HEPADIAL see DOK400
HEPAGON see VSZ000
HEPALIDINE see TEV000
HEPAR CALCIS see CAY000
HEPAREGENE see TEV000

α-HEPARIN see HAQ500
HEPARIN see HAQ500
HEPARINATE see HAQ500
HEPARINIC ACID see HAQ500
HEPARINOID see SEH450
HEPARIN SODIUM see HAQ550
HEPARIN SULFATE see HAQ500
HEPARLIPON see DXN800
HEPARTEST see HAQ600
HEPARTESTABROME see HAQ600
HEPATHROM see HAQ550
HEPATION see MAO275
HEPATOSULFALEIN see HAQ600
HEPAVIS see VSZ000
HEPCOVITE see VSZ000
HEPERAL see DXG600
HEPIN see AJD000
HEPINOID see SEH450
HEPORAL see AOO490
HEPT see TCF250
HEPTABARB see COY500
HEPTABARBITAL see COY500
HEPTABARBITONE see COY500
HEPTABARBUM see COY500
HEPTACAINE see PIO750
HEPTACHLOOR (DUTCH) see HAR000
1,4,5,6,7,8,8-HEPTACHLOOR-3a,4,7,7a-TETRAHYDRO-4,7-endo-METHANO-IN-DEEN (DUTCH) see HAR000
HEPTACHLOR (technical grade) see HAR500
HEPTACHLOR see HAR000
HEPTACHLORE (FRENCH) see HAR000
HEPTACHLOR EPOXIDE (USDA) see EBW500
2,2,5-endo,6-exo,8,9,10-HEPTACHLOROBORNANE see THH575
3,4,5,6,7,8,8a-HEPTACHLORODICYCLOPENTADIENE see HAR000
3,4,5,6,7,8,8-HEPTACHLORODICYCLOPENTADIENE see HAR000
1,4,5,6,7,8,8-HEPTACHLORO-2,3-EPOXY-2,3,3a,4,7,7a-HEXAHYDRO-4,7-METHA-NOINDENE see EBW500
1,4,5,6,7,8,8-HEPTACHLORO-2,3-EPOXY-3a,4,7,7a-TETRAHYDRO-4,7-METHA-NOINDAN see EBW500
1,4,5,6,7,8,8-HEPTACHLORO-3a,4,5,6,7,7a-HEXAHYDRO-4,7-METHANO-1H-IN-DENE see SCD500
2,3,4,5,6,7,7-HEPTACHLORO-1a,1b,5,5a,6,6a-HEXAHYDRO-2,5-METHANO-2H-IN-DENO(1,2-b)OXIRENE see EBW500
1,4,5,6,7,8,8-HEPTACHLORO-3a,4,7,7a-TETRAHYDRO-4,7-ENDOMETHANOIN-DENE see HAR000
1,4,5,6,7,10,10-HEPTACHLORO-4,7,8,9,-TETRAHYDRO-4,7-ENDOMETHYLENEIN-DENE see HAR000
1,4,5,6,7,8,8a-HEPTACHLORO-3a,4,7,7a-TETRAHYDRO-4,7-METHANOINDANE see HAR000
1(3a),4,5,6,7,8,8-HEPTACHLORO-3a(1),4,7,7a-TETRAHYDRO-4,7-METHANOIN-DENE see HAR000
1,4,5,6,7,8,8-HEPTACHLORO-3a,4,7,7a-TETRAHYDRO-4,7-METHANOINDENE see HAR000
1,4,5,6,7,8,8-HEPTACHLORO-3a,4,7,7a-TETRAHYDRO 4,7-METHANOINDENE (technical grade) see HAR500
1,4,5,6,7,8,8-HEPTACHLORO-3a,4,7,7a-TETRAHYDRO-4,7-METHANOL-1H-IN-DENE see HAR000
1,4,5,6,7,8,8-HEPTACHLORO-3a,4,7,7,7a-TETRAHYDRO-4,7-METHYLENE IN-DENE see HAR000
1,4,5,6,7,8,8-HEPTACHLOR-3a,4,7,7,7a-TETRAHYDRO-4,7-endo-METHANO-IN-DEN (GERMAN) see HAR000
HEPTACOSAFLUOROTRIBUTYLAMINE see HAS000
HEPTADECAFLUORO-1-IODOOCTANE see PCH325
n-HEPTADECANE see HAS100
HEPTADECANE see HAS100
1-HEPTADECANECARBOXYLIC ACID see SLK000
HEPTADECANOIC ACID see HAS500
HEPTADECANOIC ACID, 16-METHYL-, ISOPROPYL ESTER see IPS450
HEPTADECANOL (mixed primary isomers) see HAT000
9-HEPTADECANONE see OFE020
2,8,10-HEPTADECATRIENE-4,6-DIYNE-1,14-DIOL see EAO500
8,10,12-HEPTADECATRIENE-4,6-DIYNE-1,14-DIOL, (E,E,E)-(−)- see CMN000
2-(8-HEPTADECENYL)-2-IMIDAZOLINE-1-ETHANOL see AHP500
n-HEPTADECOIC ACID see HAS500
HEPTADECYL CYANIDE see SLL500
2-HEPTADECYL-4,5-DIHYDRO-1H-IMIDAZOLYL MONOACETATE see GII000
2-HEPTADECYL GLYOXALIDINE see GIO000
2-HEPTADECYL GLYOXALIDINE ACETATE see GII000
2-HEPTADECYL-1-HYDROXYETHYLIMIDAZOLINE see HAT500
n-HEPTADECYLIC ACID see HAS500
2-HEPTADECYL-2-IMIDAZOLINE see GIO000
2-HEPTADECYL-2-IMIDAZOLINE ACETATE see GII000
2-HEPTADECYL-2-IMIDAZOLINE-1-ETHANOL see HAT500
HEPTADECYL-TRIMETHYLAMMONIUM METHYLSULFAT (CZECH) see HAU500
HEPTADECYLTRIMETHYLAMMONIUM METHYLSULFATE see HAU500

2,4-HEPTADIENAL see HAV450
HEPTADIENAL-2,4 see HAV450
trans,trans-2,4-HEPTADIENAL see HAV450
1,6-HEPTADIYNE see HAV500
HEPTADONE see MDO750
HEPTADON HYDROCHLORIDE see MDP750
HEPTADORM see COY500
HEPTAFLUORJODPROPAN see HAY300
HEPTAFLUORMASELNAN STRIBRNY (CZECH) see HAW000
HEPTAFLUOROBUTANOIC ACID, SILVER SALT see HAW000
1,1-H,H-HEPTAFLUOROBUTANOL see HAW100
2,2,3,3,4,4,4-HEPTAFLUOROBUTANOL see HAW100
2,2,3,3,4,4,4-HEPTAFLUOROBUTYRAMIDE see HAX000
HEPTAFLUOROBUTYRIC ACID see HAX500
HEPTAFLUOROBUTYRIC ACID, ETHYL ESTER see HAY000
HEPTAFLUOROBUTYRYL HYPOCHLORITE see HAY059
HEPTAFLUOROBUTYRYL HYPOFLUORITE see HAY100
HEPTAFLUOROBUTYRYL NITRATE see HAY200
HEPTAFLUOROIODOPROPANE see HAY300
HEPTAFLUOROISOBUTYLENE METHYL ETHER see HAY500
2-HEPTAFLUOROPROPYL-1,3,4-DIOXAZOLONE see HAY600
HEPTAFLUOROPROPYL HYPOFLUORITE see HAY650
HEPTAFLUR see CDF400
HEPTAGRAN see HAR000
2,3,3′,4,4′5,7-HEPTAHYDROXYFLAVAN see HBA259
γ-HEPTALACTONE see HBA550
n-HEPTALDEHYDE see HBB500
HEPTALDEHYDE see HBB500
HEPTALDEHYDE METHYLANTHRANILATE, Schiff's base see HBO700
HEPTAMAL see COY500
HEPTAMETHYLENEIMINE mixed with SODIUM NITRITE see SIT000
HEPTAMINE see HBM490
HEPTAMUL see HAR000
HEPTAN (POLISH) see HBC500
HEPTANAL see HBB500
1-HEPTANAMINE see HBL600
2-HEPTANAMINE see HBM490
2-HEPTANAMINE SULFATE (2:1) see AKD600
HEPTANDIOIC ACID see PIG000
n-HEPTANE see HBC500
HEPTANE see HBC500
HEPTANE, 1-BROMO- see HBN100
1-HEPTANECARBOXYLIC ACID see OCY000
HEPTANE, 1-CHLORO-5-METHYL- see EKU100
1,7-HEPTANEDICARBOXYLIC ACID see ASB750
HEPTANEDICARBOXYLIC ACID see ASB750
HEPTANEDINITRILE see HBD000
1,7-HEPTANEDIOIC ACID see PIG000
HEPTANE-1,7-DIOIC ACID see PIG000
HEPTANEDIOIC ACID see PIG000
HEPTANEN (DUTCH) see HBC500
HEPTANE, 1,1′-OXYBIS-(9CI) see HBO000
1-HEPTANETHIOL see HBD500
1,1,1,3,5,5,5-HEPTANITROPENTANE see HBD650
HEPTANOIC ACID see HBE000
HEPTANOIC ACID, ISOBUTYL ESTER see IIS000
HEPTANOIC ACID, METHYL ESTER see MKK100
HEPTANOIC ACID, 2-METHYLPROPYL ESTER see IIS000
HEPTANOIC ACID, ester with TESTOSTERONE see TBF750
1-HEPTANOL see HBL500
2-HEPTANOL see HBE500
HEPTANOL-2 see HBE500
3-HEPTANOL see HBF000
n-HEPTANOL-1 (FRENCH) see HBL500
n-HEPTANOL-1 see HBL500
2-HEPTANOL, 2,6-DIMETHYL- see FON200
3-HEPTANOL, 6-(DIMETHYLAMINO)-4,4-DIPHENYL-, ACETATE (ester), HYDRO-
    CHLORIDE, (3S,6S)-(-)- see ACQ690
3-HEPTANOL, 6-(DIMETHYLAMINO)-4,4-DIPHENYL-, ACETATE (ester), (3S,6S)-
    (-)- see ACQ666
HEPTANOL, FORMATE see HBO500
HEPTANOLIDE-1,4 see HBA550
HEPTANOLIDE-4,1 see HBA550
HEPTANON see MDO750
HEPTAN-3-ON (DUTCH, GERMAN) see EHA600
2-HEPTANONE see MGN500
3-HEPTANONE see EHA600
HEPTAN-3-ONE see EHA600
4-HEPTANONE see DWT600
HEPTAN-4-ONE see DWT600
4-HEPTANONE, 3-BENZYL-(8CI) see BEN800
3-HEPTANONE, 6-(DIMETHYLAMINO)-4,4-DIPHENYL-, (±)- see MDO760
3-HEPTANONE, 5-METHYL- see EGI750
3-HEPTANONE, 5-METHYL-, OXIME see EMP550
4-HEPTANONE, 3-(PHENYLMETHYL)- see BEN800

1-HEPTANONE, 1-(4-PYRIDYL)- see HBF600
17-β-HEPTANOYLOXY-19-NOR-17-α-PREGNEN-20-YNONE see NNQ000
4-HEPTANOYLPYRIDINE see HBF600
HEPTANYL ACETATE see HBL000
HEPTARINOID see SEH450
HEPTA-1,3,5-TRIYNE see HBI725
2,4-HEPTDAIENAL see HAV450
HEPTEDRINE see HBM490
cis-4-HEPTEN-1-AL see HBI800
2-HEPTENAL, (E)- see HBI770
4-HEPTENAL see HBI800
(E)-2-HEPTEN-1-AL see HBI770
1-h-HEPTENE see HBJ000
2-HEPTENE see HBJ500
3-HEPTENE (mixed isomers) see HBK350
n-HEPTENE see HBJ000
1-HEPTENE-4,6-DIYNE see HBK450
2-HEPTENOIC ACID see HBK500
5-HEPTEN-2-OL, 6-METHYL-2-(4-METHYL-3-CYCLOHEXEN-1-YL)- see BGO775
5-HEPTEN-2-ONE, 6-METHYL- see MKK000
HEPTENOPHOS see HBK700
HEPTENYL ACROLEIN see DAE450
6-HEPTEN-4-YN-3-OL, 1-(PROPYLTHIO)-2,3,6-TRIMETHYL- see TME260
HEPTHLIC ACID see HBE000
HEPTOCOAGULASE see CMY725
n-HEPTOIC ACID see HBE000
1-HEPTYL ACETATE see HBL000
n-HEPTYL ACETATE see HBL000
HEPTYL ACETATE see HBL000
HEPTYL ACRYLATE see HBL100
HEPTYL ALCOHOL see HBL500
1-HEPTYLAMINE see HBL600
2-HEPTYLAMINE see HBM490
3-HEPTYLAMINE see HBM500
n-HEPTYLAMINE see HBL600
HEPTYLAMINE see HBL600
HEPTYLAMINE, 1-METHYL- see OEK010
2-HEPTYLAMINE SULFATE see AKD600
8-HEPTYLBENZ(a)ANTHRACENE see HBN000
5-n-HEPTYLBENZ(1:2)BENZANTHRACENE see HBN000
n-HEPTYL BROMIDE see HBN100
HEPTYL BROMIDE see HBN100
HEPTYL BUTANOATE see HBN150
HEPTYL BUTYRATE see HBN150
γ-n-HEPTYLBUTYROLACTONE see HBN200
γ-HEPTYLBUTYROLACTONE see HBN200
HEPTYL CARBINOL see OEI000
HEPTYL CELLOSOLVE see HBN250
2-n-HEPTYL CYCLOPENTANONE see HBN500
α-HEPTYL CYCLOPENTANONE see HBN500
HEPTYLDICHLORARSINE see HBN600
1-HEPTYLENE see HBJ000
HEPTYLENE see HBK350
HEPTYL ETHER see HBO000
HEPTYL FORMATE see HBO500
HEPTYL HYDRAZINE see HBO600
HEPTYL HYDRIDE see HBC500
HEPTYL 4-HYDROXYBENZOATE see HBO650
HEPTYL p-HYDROXYBENZOATE see HBO650
n-HEPTYLIC ACID see HBE000
HEPTYLIDENE METHYL ANTHRANILATE see HBO700
HEPTYL KETONE see HBO790
n-HEPTYLMERCAPTAN see HBD500
HEPTYL MERCAPTAN see HBD500
n-HEPTYL METHANOATE see HBO500
HEPTYL METHYL KETONE see NMY500
HEPTYLMETHYLNITROSAMINE see HBP000
1-HEPTYL-1-NITROSOUREA see HBP250
n-HEPTYL NITROSOUREA see HBP250
2-HEPTYLOXYCARBANILIC ACID-2-(1-PIPERIDINYL)ETHYL ESTER HYDRO-
    CHLORIDE see PIO750
2-(HEPTYLOXY)ETHANOL see HBN250
2-(2-(HEPTYLOXY)ETHOXY)ETHANOL see HBP275
p-(HEPTYLOXY)PHENOL see HBP285
4-(HEPTYLOXY)PHENOL (9CI) see HBP285
(2-(HEPTYLOXY)PHENYL)CARBAMIC ACID-2-(1-PIPERIDINYL)ETHYL ESTER
    HYDROCHLORIDE see PIO750
N-(2-(HEPTYLOXYPHENYLCARBAMOYLOXY)ETHYL)PIPERIDINIUM CHLORIDE
    see PIO750
HEPTYL PARABEN see HBO650
2-HEPTYLPHENOL see HBP350
o-n-HEPTYLPHENOL see HBP350
o-HEPTYLPHENOL see HBP350
HEPTYL 4-PYRIDYL KETONE see HBF600
2-HEPTYL-TETRAHYDROFURAN see HBP425

1-α,2-α,3-β,4-α,5-α,6-β-HEXACHLOROCYCLOHEXANE see BBQ500
1-α,2-α,3-β,4-α,5-β,6-β-HEXACHLOROCYCLOHEXANE see BBQ000
1-α,2-β,3-α,4-β,5-α,6-β-HEXACHLOROCYCLOHEXANE see BBR000
α-HEXACHLOROCYCLOHEXANE see BBQ000
β-HEXACHLOROCYCLOHEXANE see BBR000
α-1,2,3,4,5,6-HEXACHLOROCYCLOHEXANE (MAK) see BBQ000
β-1,2,3,4,5,6-HEXACHLOROCYCLOHEXANE (MAK) see BBR000
γ-HEXACHLOROCYCLOHEXANE (MAK) see BBQ500
1,2,3,4,5,6-HEXACHLOROCYCLOHEXANE (mixture of isomers) see BBQ750
HEXACHLOROCYCLOHEXANE, delta and epsilon mixture see HCE400
HEXACHLOROCYCLOHEXANE see BBP750
Δ-1,2,3,4,5,6-HEXACHLOROCYCLOHEXANE see BFW500
Δ-HEXACHLOROCYCLOHEXANE see BFW500
Δ-HEXACHLOROCYCLOHEXANE mixed with epsilon-HEXACHLOROCYCLO-
    HEXANE see HCE400
1,2,3,4,5,6-HEXACHLOROCYCLOHEXANE, γ-ISOMER see BBQ500
1,2,3,4,5,5-HEXACHLORO-1,3-CYCLOPENTADIENE see HCE500
HEXACHLORO-1,3-CYCLOPENTADIENE see HCE500
HEXACHLOROCYCLOPENTADIENE (ACGIH,DOT,OSHA) see HCE500
HEXACHLOROCYCLOPENTADIENE see HCE500
1,2,3,4,5,5-HEXACHLORO-1,3-CYCLOPENTADIENE DIMER see MQW500
HEXACHLOROCYCLOPENTADIENEDIMER see MQW500
1,2,3,4,7,8-HEXACHLORODIBENZO-p-DIOXIN see HCF000
1,2,3,6,7,8-HEXACHLORODIBENZO-p-DIOXIN see HAJ500
HEXACHLORODIBENZO-p-DIOXIN see HAJ500
1,2,3,7,8,9-HEXACHLORODIBENZO-p-DIOXIN mixed with 1,2,3,6,7,8-HEXA-
    CHLORODIBENZO-p-DIOXIN see HCF500
1,2,3,6,7,8-HEXACHLORODIBENZO-p-DIOXIN mixed with 1,2,3,7,8,9-HEXA-
    CHLORODIBENZO-p-DIOXIN see HCF500
1,2,3,6,7,8-HEXACHLORODIBENZODIOXIN mixed with PENTACHLORO ISO-
    MERS and HEPTACHLORO ISOMERS 96.8%:1.97%:1.23% see HCG000
2,2′,3,3′,5,5′-HEXACHLORO-6,6′-DIHYDROXYDIPHENYLMETHANE see HCL000
HEXACHLORODIPHENYL ETHER see CDV175
HEXACHLORO DIPHENYL OXIDE see CDV175
HEXACHLORODISILANE see HCH500
HEXACHLOROEPOXYOCTAHYDRO-endo,endo-DIMETHANONAPHTHALENE
    see EAT500
HEXACHLOROEPOXYOCTAHYDRO-endo,exo-DIMETHANONAPHTHALENE see
    DHB400
1,1,1,2,2,2-HEXACHLOROETHANE see HCI000
HEXACHLOROETHANE see HCI000
HEXACHLOROETHYLENE see HCI000
1,4,5,6,7,7-HEXACHLORO-N-(ETHYLMERCURI)-5-NORBORNENE-2,3-DICARBOXI-
    MIDE see EME050
HEXACHLOROFEN (CZECH) see HCL000
1,2,3,4,10,10-HEXACHLORO-1,4,4a,5,8,8a-HEXAHYDRO-1,4,5,8-DIMETHANO-
    NAPHTHALENE see AFK250
1,2,3,4,10,10-HEXACHLORO-1,4,4a,5,8,8a-HEXAHYDRO-1,4,5,8-endo,endo-DI-
    METHANONAPHTHALENE see IKO000
1,2,3,4,10,10-HEXACHLORO-1,4,4a,5,8,8a-HEXAHYDRO-1,4-endo-exo-5,8-DI-
    METHANONAPHTHALENE see AFK250
1,2,3,4,10,10-HEXACHLORO-1,4,4a,5,8,8a-HEXAHYDRO-1,4-endo,endo-5,8-DI-
    METHANONAPHTHALENE see IKO000
HEXACHLOROHEXAHYDRO-endo-exo-DIMETHANONAPHTHALENE see
    AFK250
1,2,3,4,10,10-HEXACHLORO-1,4,4a,5,8,8a-HEXAHYDRO-exo-1,4,-endo-5,8-DI-
    METHANONAPHTHALENE see AFK250
5,6,7,8,9,9-HEXACHLORO-1,4,4a,5,8,8a-HEXAHYDRO-1,4:5,8-DIMETHANO-
    PHTHALAZINE-2-OXIDE see HCI475
HEXACHLOROHEXAHYDROMETHANO 2,4,3-BENZODIOXATHIEPIN-3-OXIDE
    see EAQ750
6,7,8,9,10,10-HEXACHLORO-1,5,5a,6,9,9a-HEXAHYDRO-6,9-METHANO-2,4,3-
    BENZODIOXATHIEPIN-3-OXIDE see EAQ750
6,7,8,9,10,10-HEXACHLORO-1,5,5a,6,9,9a-HEXAHYDRO-6,9-METHANO-3-METH-
    YL-2,4-BENZODIOXEPIN see HCK000
6,7,8,9,10,10-HEXACHLORO-1,5,5a,6,9,9a-HEXAHYDRO-3-METHYL-6,9-METHA-
    NO-2,4-BENZDIOXEPIN see HCK000
HEXACHLOROIRIDATE(2−) DIHYDROGEN, (OC-6-11) see HIA500
HEXACHLOROMELAMINE see TNG275
HEXACHLORONAPHTHALENE see HCK500
1,4,5,6,7,7-HEXACHLORO-5-NORBORNENE-2,3-DICARBOXIMIDE see CDS100
1,4,5,6,7,7-HEXACHLORO-5-NORBORNENE-2,3-DICARBOXYLIC ACID see
    CDS000
1,4,5,6,7,7-HEXACHLORO-5-NORBORNENE-2,3-DIMETHANOL CYCLIC SULFITE
    see EAQ750
3,4,5,6,9,9-HEXACHLORO-1a,2,2a,3,6,6a,7,7a-OCTAHYDRO-2,7:3,6-DIMETHA-
    NONAPHTH(2,3-b)OXIRENE see EAT500
3,4,5,6,9,9-HEXACHLORO-1a,2,2a,3,6,6a,7,7a-OCTAHYDRO-2,7:3,6-DIMETHA-
    NONAPHTH(2,3-b)OXIRENE see DHB400
HEXACHLOROPHANE see HCL000
HEXACHLOROPHEN see HCL000
HEXACHLOROPHENE (DOT) see HCL000
HEXACHLOROPHENE see HCL000, MRM000
HEXACHLOROPLATINATE(2) DIPOTASSIUM see PLR000
HEXACHLOROPLATINIC(IV) ACID see CKO750

HEXACHLOROPLATINIC ACID see CKO750
HEXACHLOROPLATINIC(4+) ACID, HYDROGEN- see CKO750
1,1,1,3,3,3-HEXACHLORO-2-PROPANONE see HCL500
HEXACHLORO-2-PROPANONE see HCL500
HEXACHLOROPROPENE see HCM000
HEXACHLOROPROPYLENE see HCM000
4,5,6,7,8,8-HEXACHLORO-3a,4,7,7a-TETRAHYDRO-4,7-METHANOINDENE see
    HCN000
N,N,N′,N′,N′′,N′′-HEXACHLORO-1,3,5-TRIAZINE-2,4,6-TRIAMINE see TNG275
α,α,α,α′,α′,α′-HEXACHLOROXYLENE see BLL825
α,α′-HEXACHLOROXYLENE see HCM500
α,α,α,α′,α′,α′-HEXACHLORO-m-XYLENE see BLL825
α,α′-HEXACHLORO-m-XYLENE see BLL825
α,α,α,α′,α′,α′-HEXACHLORO-p-XYLENE see HCM500
4,5,6,7,8,8-HEXACHLOR-Δ¹,⁵-TETRAHYDRO-4,7-METHANOINDEN see HCN000
HEXACID 698 see HEU000
HEXACID 898 see OCY000
HEXACID 1095 see DAH400
HEXACID C-7 see HBE000
HEXACID C-9 see NMY000
HEXACO BLUE VRS see ADE500
HEXACOL ACID YELLOW G see CMM758
HEXACOL BLACK PN see BMA000
HEXACOL BLUE VRS see ADE500
HEXACOL BRILLIANT BLUE A see FAE000
HEXACOL CARMOISINE see HJF500
HEXACOL ERYTHROSINE BS see FAG040
HEXACOL GREEN S see ADF000
HEXACOL OIL ORANGE SS see TGW000
HEXACOL OIL YELLOW GG see CMP600
HEXACOL PONCEAU MX see FMU070
HEXACOL PONCEAU 4R see FMU080
HEXACOL PONCEAU SX see FAG050
HEXACOL RED 10B see CMS231
HEXACOL RED 2G see CMM300
HEXACOL RHODAMINE B EXTRA see FAG070
HEXACOL SUNSET YELLOW FCF see FAG150
HEXACOL SUNSET YELLOW FCF SUPRA see FAG150
HEXACOL SUNSET YELLOW FCP see FAG150
HEXACOL SUNSET YELLOW F & F SUPRA see FAG150
HEXACOL TARTRAZINE see FAG140
HEXACOSE see HCS500
HEXACYANOFERRATE(3) TRIPOTASSIUM see PLF250
HEXACYANOTRIS(3-DODECYL-1-METHYL)-2-PHENYLBENZIMIIMIDAZOLINIUM
    FERRATE (3−) see TNH750
HEXACYCLEN TRISULFATE see HCN100
HEXACYCLONAS see SHL500
HEXACYCLONATE SODIUM see SHL500
7,11-HEXADECADIEN-1-OL, ACETATE see GJK000
HEXADECADROL see SOW000
HEXADECAFLUOROHEPTANE see PCH000
HEXADECAFLUORO-1-NONANOL see HCO000
2,2,3,3,4,4,5,5,6,6,7,7,8,8,9,9-HEXADECAFLUORONONANOL see HCO000
HEXADECANAMIDE, N,N-DIMETHYL- see DTH700
1-HEXADECANAMINE see HCO500
1-HEXADECANAMINE HYDROFLUORIDE (9CI) see CDF400
1-HEXADECANAMINIUM, N-(CARBOXYMETHYL)-N,N-DIMETHYL-, HYDROXIDE,
    inner salt see CDF450
n-HEXADECANE see HCO600
HEXADECANE see HCO600
HEXADECANOIC ACID see PAE250
HEXADECANOIC ACID, 2-ETHYLHEXYL ESTER (9CI) see OFE100
HEXADECANOIC ACID, HEXADECYL ESTER see HCP700
HEXADECANOIC ACID, ISOPROPYL ESTER see IQW000
1-HEXADECANOL see HCP000
HEXADECAN-1-OL see HCP000
n-HEXADECANOL see HCP000
HEXADECANOL see HCP000
1-HEXADECANOL, ACETATE see HCP100
1,16-HEXADECANOLACTONE see OKU000
1-HEXADECANOL, HYDROGEN SULFATE, SODIUM SALT see HCP900
HEXADECANOLIDE see OKU000
1,6,10,14-HEXADECATETRAEN-3-OL, 3,7,11,15-TETRAMETHYL-, (E,E)- see
    GDG300
11-HEXADECENAL, (Z)- see HCP050
(Z)-11-HEXADECENAL see HCP050
HEXADECENE EPOXIDE see EBX500
2-HEXADECEN-1-OL, 3,7,11,15-TETRAMETHYL-, (R-(R*,R*-(E)))-(9CI) see
    PIB600
n-HEXADECOIC ACID see PAE250
HEXADECYL ACETATE see HCP100
n-HEXADECYL ALCOHOL see HCP000
HEXADECYL ALCOHOL see HCP000
N-HEXADECYLAMINE see HCO500
HEXADECYLAMINE, N,N-DIMETHYL- see HCP525

HEXADECYLAMINE HYDROFLUORIDE see CDF400
HEXADECYLBETAINE see CDF450
HEXADECYL CYCLOPROPANECARBOXYLATE see HCP500
HEXADECYLDIMETHYLAMINE see HCP525
HEXADECYL 2-ETHYLHEXANOATE see HCP550
HEXADECYLIC ACID see PAE250
HEXADECYLMALEINAN DI-n-BUTYLCINICITY (CZECH) see DEI600
HEXADECYL NEODECANOATE see HCP600
HEXADECYL PALMITATE see HCP700
HEXADECYLPOLY(ETHYLENEOXY) ETHANOL see PJT300
HEXADECYLPYRIDINE BROMIDE see HCP800
1-HEXADECYLPYRIDINIUM BROMIDE see HCP800
N-HEXADECYLPYRIDINIUM BROMIDE see HCP800
HEXADECYLPYRIDINIUM BROMIDE see HCP800
1-HEXADECYL-PYRIDINIUM BROMIDE mixture with CHLORO(2-HYDROXYETH-
     YL)MERCURY see DVJ500
1-HEXADECYLPYRIDINIUM CHLORIDE see CCX000
n-HEXADECYLPYRIDINIUM CHLORIDE see CCX000
HEXADECYLPYRIDINIUM CHLORIDE see CCX000
1-HEXADECYLPYRIDINIUM CHLORIDE MONOHYDRATE see CDF750
HEXADECYL SODIUM SULFATE see HCP900
HEXADECYLTRICHLOROSILANE see HCQ000
HEXADECYLTRIETHYLAMMONIUM BROMIDE see CDF500
(1-HEXADECYL)TRIMETHYLAMMONIUM BROMIDE see HCQ500
N-HEXADECYL-N,N,N-TRIMETHYLAMMONIUM BROMIDE see HCQ500
N-HEXADECYLTRIMETHYLAMMONIUM BROMIDE see HCQ500
HEXADECYLTRIMETHYLAMMONIUM BROMIDE see HCQ500
HEXADECYLTRIMETHYLAMMONIUM PENTACHLOROPHENOL see TLN150
ω-h-HEXADEKAFLUORNONANOL-1 (GERMAN) see HCO000
HEXADENOL see HCS500
HEXADERM RED MRG see CMM325
2,4-HEXADIENAL see SKT500
HEXA-2,4-DIENAL see SKT500
1,4-HEXADIENE see HCR000
1,5-HEXADIENE see HCR500
HEXA-1,5-DIENE see HCR500
HEXADIENE see HCQ600
1,3-HEXADIENE-5-YNE see HCS100
HEXADIENIC ACID see SKU000
2,4-HEXADIENOIC ACID see SKU000
HEXADIENOIC ACID see SKU000
trans-trans-2,4-HEXADIENOIC ACID see SKU000
2,4-HEXADIENOIC ACID POTASSIUM SALT see PLS750
2,4-HEXADIENOL see HCS500
2,4-HEXADIEN-1-OL see HCS500
2,4-HEXADIEN-1-OL ACETATE see AAU750
2,4-HEXADIENOL BUTANOATE see HCS600
2,4-HEXADIENYL ACETATE see AAU750
2,4-HEXADIENYL BUTYRATE see HCS600
2,4-HEXADIENYL ISOBUTYRATE see HCS700
2,4-HEXADIENYL 2-METHYLPROPANOATE see HCS700
2-(2,4-HEXADIENYLOXY)ETHANOL see HCT500
1,5-HEXADIEN-3-YNE see HCU500
HEXADIMETHRINE BROMIDE see HCV500
HEXADIONA see DBB200
1,5-HEXADIYNE see HCV850
2,4-HEXADIYNE-1,6-DIOIC ACID see HCV875
1,5-HEXADIYNE-3-ONE see HCV880
HEXADRIN see CQC650, EAT500
HEXADROL see SOW000
HEXAETHYLBENZENE see HCX000
HEXAETHYLDILEAD see TJS000
1,1,1,3,3,3-HEXAETHYLDISTANNOXANE see HCX050
HEXAETHYLDISTANNOXANE see HCX050
1,1,1,3,3,3-HEXAETHYLDISTANNTHIANE see HCX100
HEXAETHYLDISTANNTHIANE see HCX100
HEXAETHYLTETRAFOSFAT see HCY000
HEXAETHYL TETRAPHOSPHATE, liquid or solid (DOT) see HCY000
HEXAETHYL TETRAPHOSPHATE see HCY000
HEXAETHYLTRIALUMINUM TRITHIOCYANATE see HCY500
HEXAFEN see HCL000
HEXAFERB see FAS000
HEXAFLUORENIUM DIBROMIDE see HEG000
HEXAFLUOROACETIC ANHYDRIDE see TJX000
HEXAFLUOROACETONE see HCZ000
HEXAFLUOROACETONE HYDRATE see HDA000
HEXAFLUOROACETONE SESQUIHYDRATE see HDE500
HEXAFLUORO ACETONE TRIHYDRATE see HDA500
HEXAFLUOROBENZENE see HDB000
1,1,1,4,5,5-HEXAFLUORO-2-CHLORO-2-BUTENE see CHK750
HEXAFLUORODICHLOROBUTENE see HDB500
HEXAFLUORODIETHYL ETHER see HDC000
HEXAFLUORO FERRATE (3-) TRIAMMONIUM SALT see ANI000
HEXAFLUOROGLUTARIC ACID DIETHYL ESTER see DJK100
HEXAFLUOROGLUTARONITRILE see HDC300

HEXAFLUOROGLUTARYL DIHYPOCHLORITE see HDC425
3,3,3,4,4,4-HEXAFLUOROISOBUTYLENE see HDC450
HEXAFLUOROISOBUTYLENE see HDC450
HEXAFLUOROISOBUTYRIC ACID METHYL ESTER see MKK750
HEXAFLUOROISOPROPANOL see HDC500
HEXAFLUOROKIESELSAEURE (GERMAN) see SCO500
HEXAFLUOROKIEZELZUUR (DUTCH) see SCO500
4,4,4,4',4',4'-HEXAFLUORO-N-NITROSODIBUTYLAMINE see NJN300
HEXAFLUOROPENTANEDIOIC ACID DIETHYL ESTER see DJK100
HEXAFLUOROPHOSPHORIC ACID see HDE000
1,1,1,3,3,3-HEXAFLUORO-2-PROPANOL see HDC500
HEXAFLUORO-2-PROPANONE HYDRATE see HDA000
HEXAFLUORO-2-PROPANONE SESQUIHYDRATE see HDE500
HEXAFLUOROPROPENE see HDF000
HEXAFLUOROPROPYLENE (DOT) see HDF000
HEXAFLUOROSILICATE(2-) DIHYDROGEN see SCO500
HEXAFLUOROSILICATE (2-1) LEAD(II) SALT DIHYDRATE see LDG000
HEXAFLUOROSILICATE(2-) MAGNESIUM (1:1) see MAF600
HEXAFLUOROSILICATE (2−), NICKEL see NDD000
HEXAFLUOROSILICATE(2-) STRONTIUM see SMJ000
HEXAFLUORO-SILICATE(2−), THALLIUM see TEK250
HEXAFLUORO VANADATE (3-) TRIAMMONIUM SALT see ANI500
α,α,α,α,α,α-HEXAFLUORO-3,5-XYLIDINE see BLO250
1-(α,α,α,α',α',α'-HEXAFLUORO-3,5-XYLYL)-4-METHYL-3-THIO-SEMICARBAZIDE
     see BLP325
HEXAFLUORURE de SOUFRE (FRENCH) see SOI000
HEXAFLUOSILICIC ACID see SCO500
HEXAFLURATE see PLH500
HEXAFLURONIUM BROMIDE see HEG000
HEXAFORM see HEI500
HEXAFUNGIN see HDF100
HEXAGLYCERINE see HDF300
HEXAHYDRO-3A,7A-DIMETHYL-4,7-EPOXYISOBENZOFURAN-1,3-DIONE see
     CBE750
1,5A,6,9,9A,9B-HEXAHYDRO-4A(4H)-DIBENZOFURANCARBOXALDEHYDE see
     BHJ500
2,3,5,6,7,8-HEXAHYDRO-9-AMINO-1H-CYCLOPENTA(b)QUINOLINE HYDRO-
     CHLORIDE HYDRATE see AKD775
HEXAHYDROANILINE see CPF500
HEXAHYDROANILINE HYDROCHLORIDE see CPA775
HEXAHYDRO-1H-AZEPINE see HDG000
HEXAHYDROAZEPINE see HDG000
HEXAHYDRO-2H-AZEPIN-2-ONE see CBF700
HEXAHYDRO-2-AZEPINONE see CBF700
HEXAHYDRO-2H-AZEPIN-2-ONE HOMOPOLYMER see PJY500
N-(((HEXAHYDRO-1H-AZEPIN-1-YL)-AMINO)CARBONYL)-4-METHYLBENZENE-
     SULFONAMIDE see TGJ500
((2-HEXAHYDRO-1-AZEPINYL)ETHYL)GUANIDINE SULFATE (2:1) see HDG600
(+)-6-(((HEXAHYDRO-1H-AZEPIN-1-YL)METHYLENE)AMINO)-3,3-DIMETHYL-7-
     OXO-4-THIA-1-AZABICYCLO(3.2.0)HEPTANE-2-CARBOXYLIC ACID HY-
     DROXYMETHYL ESTER, PIVALATE (ester), MONOHYDROCHLORIDE see
     MCB550
3-(3-(HEXAHYDRO-1H-AZEPIN-1-YL)PROPOXY)-1-(α,α,α-TRIFLUORO-m-TOLYL)-
     1H-PYRAZOLO(3,4-b)PYRIDINE MONOHYDROCHLORIDE see ITD100
1-(HEXAHYDRO-1H-AZEPIN-1-YL)-3-(p-TOLYLSULFONYL)UREA see TGJ500
1-(HEXAHYDRO-1-AZEPINYL)-3-p-TOLYLSULFONYLUREA see TGJ500
(2-(HEXAHYDRO-1(2H)-AZOCINYL)ETHYL) GUANIDINE HYDROGEN SULFATE
     see GKU000
(2-(HEXAHYDRO-1(2H)-AZOCINYL)ETHYL)GUANIDINE SULFATE see GKS000
1,2,4,5,6,7-HEXAHYDROBENZ(e)ACEANTHRYLENE see HDH000
HEXAHYDROBENZENAMINE see CPF500
HEXAHYDROBENZENE see CPB000
HEXAHYDROBENZOIC ACID see HDH100
HEXAHYDROBENZOIC ACID AMIDE see CPB050
HEXAHYDROBENZYL ALCOHOL see HDH200
HEXAHYDROCARQUEJENE see MCE275
HEXAHYDROCRESOL see MIQ745
N-(((HEXAHYDROCYCLOPENTA(c)PYRROL-2(1H)-YL)AMINO)CARBONYL)-4-
     METHYL-BENZENESULFONAMIDE see DBL700
1-(HEXAHYDROCYCLOPENTA(c)PYRROL-2(1H)-YL)-3-(p-TOLYLSULFO-
     NYL)UREA see DBL700
HEXAHYDRO-p-CYMENE HYDROPEROXIDE see IQE000
HEXAHYDRO-1,8-DIAMINO-4,7-METHANOINDAN see DCF600
HEXAHYDRO-1,4-DIAZEPINE see HGI900
HEXAHYDRO-1,4-DIAZINE see PIJ000
1,2,3,4,12,13-HEXAHYDRODIBENZ(a,h)ANTHRACENE see HDK000
1a-β,1b,5a6,6a7a-β-HEXAHYDRO-1b-α,6-β-DIHYDROXY-8a-ISOPROPENYL-6a-α-
     METHYL-SPIRO(2,5-METHANO-7H-OXIRENO(3,4)CYCLOPENT(1,2-d)OXEPIN-
     7,2'-OXIRAN)-3(2aH)-ONE see TOE175
(−)-2,3,4,5,6,7-HEXAHYDRO-1,4-DIMETHYL-1,6-METHANO-1H-4-BENZAZONIN-
     10-OL HYDROBROMIDE see SLI200
1,2,3,4,5,6-HEXAHYDRO-6,11-DIMETHYL-3-(3-METHYL-2-BUTENYL)-2,6-METHA-
     NO-3-BENZAZOCIN-8-OL HCl see PBP300
1,2,3,4,5,6-HEXAHYDRO-6,11-DIMETHYL-3-(3-METHYL-2-BUTENYL)-2,6-METHA-
     NO-3-BENZAZOCINE see DOQ400

HEXAHYDRO-1,4-DINITROSO-1H-1,4-DIAZEPINE see DVE600
HEXAHYDRO-1-DODECYL-1H-AZEPINE-1-OXIDE see DXX875
(4aRS,5RS,9bRS)-2,3,4,4a,5,9b-HEXAHYDRO-2-ETHYL-7-METHYL-5-p-TOLYL-1H-INDENO(1,2-c)PYRIDINE HYDROCHLORIDE see CNH300
2,2,4,4,6,6-HEXAHYDRO-2,2,4,4,6,6-HEXAKIS(1-AZIRIDINYL)-1,3,5,2, 4,6-TRIAZA-TRIPHOSPHORINE see AQO000
1,3,4,6,7,8-HEXAHYDRO-4,6,6,7,8,8-HEXAMETHYL-CYCLOPENTA-γ-2-BENZO-PYRAN see GBU000
1,2,3a,4,5,9b-HEXAHYDRO-8-HYDROXY-3-METHYL-9b-PROPYL-3H-BENZ(e)INDOLE, HYDROCHLORIDE see HDO500
1,2,3,3a,8,8a-HEXAHYDRO-5-HYDROXY-1,3a,8-TRIMETHYL-PYRROLO(2,3-b)INCOLE METHYLCARBAMATE (ester), (3aS-cis)-, SULFATE (2:1) see PIB000
1,3,4,6,7,11b-HEXAHYDRO-3-ISOBUTYL-9,10-DIMETHOXY-2H-BEN-ZO(a)QUINOLIZIN- 2-ONE see TBJ275
4-HEXAHYDROISONICOTINIC ACID see ILG100
exo-HEXAHYDRO-4,7-METHANOINDAN see TLR675
3-(5-(3a,4,5,6,7,7a-HEXAHYDRO-4,6-METHANOINDANYL))-1,1-DIMETHYLUREA see HDP500
3-(HEXAHYDRO-4,7-METHANOINDAN-5-YL)-1,1-DIMETHYLUREA see HDP500
1-(5-(3a,4,5,6,7,7a-HEXAHYDRO-4,7-METHANOINDANYL))-3,3-DIMETHYLUREA see HDP500
1-(3a,4,5,6,7,7a-HEXAHYDRO-4,7-METHANO-5-INDANYL)-3,3-DIMETHYLUREA see HDP500
3a,4,5,6,7,7a-HEXAHYDRO-4,7-METHANO-1H-INDEN-6-OL ACETATE see DLY400
1,2,3,4,5,6-HEXAHYDRO-1,5-METHANO-8H-PYRIDO(1,2-A)(1,5)DIAZOCIN-8-ONE see CQL500
1,2,3,4,5,6-HEXAHYDRO-6-METHYLAZEPINO(4,5-b)INDOLE HYDROCHLORIDE see HDQ500
HEXAHYDRO-1-METHYL-2H-AZEPIN-2-ONE see MHY750
6,7,8,9,10,12b-HEXAHYDRO-3-METHYL CHOLANTHRENE see HDR500
1,2,3,4,10,14b-HEXAHYDRO-2-METHYLDIBENZO(c,f)PYRAZINO(1,2-a)AZEPINE see MQS220
1,2,3,4,10,14b-HEXAHYDRO-2-METHYLDIBENZO(c,f)PYRAZINO(1,2-a)AZEPINE HYDROCHLORIDE see BMA625
dl-1,3,4,9,10,10A-HEXAHYDRO-11-METHYL-2H-10,4A-IMINOETHANOPHENAN-THREN-6-OL see MKR250
(+)-cis-1,3,4,9,10,10A-HEXAHYDRO-11-METHYL-2H-10,4a-IMINOETHANOPHE-NANTHREN-6-OL see DBE800
1,2,3,4,5,6-HEXAHYDRO-1-(2'-METHYL-3'-(N-METHYLAMINO)PROPYL)-1-BENZA-ZOCINE HYDROCHLORIDE see HDS200
HEXAHYDROMETHYLPHENOL see MIQ745
HEXAHYDRO-1-NITROSO-1H-AZEPINE see NKI000
HEXAHYDRO-2-OXO-1,4-CYCLOHEXANEDIMETHANOL see BKH325
HEXAHYDRO-3,6-endo-OXYPHTHALIC ACID see EAR000
HEXAHYDROPHENOL see CPB750
1,3,6,7,8,9-HEXAHYDRO-5-PHENYL-2H-(1)BENZOTHIENO(2,3-e)-1,4-DIAZEPIN-2-ONE see BAV625
HEXAHYDROPHENYL ETHYL ACETATE see EHS000
HEXAHYDROPHENYLETHYL ALCOHOL see CPL250
HEXAHYDROPICOLINIC ACID see HDS300
HEXAHYDROPYRAZINE see PIJ000
HEXAHYDROPYRIDINE see PIL500
HEXAHYDROTHYMOL see MCF750
HEXAHYDROTOLUENE see MIQ740
HEXAHYDRO-1,3,5-s-TRIAZINE see HDV500
HEXAHYDRO-1,3,5-TRIETHYL-s-TRIAZINE see HDW000
HEXAHYDRO-3,5,5-TRIMETHYL-3,8a-ETHANO-8aH-1-BENZOPYRAN-2(3H)-ONE see LAQ100
HEXAHYDRO-1,3,5-TRIMETHYL-s-TRIAZINE see HDW100
HEXAHYDRO-1,3,5-TRINITROSO-1,3,5-TRIAZINE see HDV500
HEXAHYDRO-1,3,5-TRINITROSO-s-TRIAZINE see HDV500
HEXAHYDRO-1,3,5-TRINITRO-1,3,5-TRIAZIN see CPR800
HEXAHYDRO-1,3,5-TRINITRO-1,3,5-TRIAZIN (GERMAN) see CPR800
HEXAHYDRO-1,3,5-TRINITRO-1,3,5-TRIAZINE see CPR800
HEXAHYDRO-1,3,5-TRIS(DIMETHYLAMINOPROPYL)-s-TRIAZINE see TNH250
HEXAHYDRO-1,3,5-TRIS(HYDROXYETHYL)TRIAZINE see THR820
HEXAHYDRO-1,3,5-TRIS(1-OXO-2-PROPENYL)-1,3,5-TRIAZINE see THM900
2,2',3,3,3',3'-HEXAHYDROXY(2,2' -BI INDAN)-1,1'-DIONE see HHI000
2-β,3-β,14,20,22,25-HEXAHYDROXY-5-β-CHOLET-7-EN-6-ONE see HKG500
3,3',4',5,5',7-HEXAHYDROXYFLAVYLIUM ACID ANION see DAM400
HEXAHYDROXYLAMINECOBALT(III) NITRATE see HDX500
HEXA(HYDROXYMETHYL)MELAMINE see HDY000
HEXAISOBUTYLDITIN see HDY100
HEXAKIS(μ-ACETATO-O:O')-μ⁴-OXOTETRABERYLLIUM see BFT500
HEXAKIS(μ-ACETATO)-μ⁴-OXOTETRABERYLLIUM see BFT500
2,2,4,4,6,6-HEXAKIS(1-AZIRIDINYL)CYCLOTRIPHOSPHAZA-1,3,5-TRIENE see AQO000
2,2,4,4,6,6-HEXAKIS(1-AZIRIDINYL)-2,2,4,4,6,6-HEXAHYDRO-1,3,5,2,4,6-TRIAZA-TRIPHOSPHORINE see AQO000
HEXAKIS-(1-AZIRIDINYL)PHOSPHONITRILE see AQO000
HEXAKIS(AZIRIDINYL)PHOSPHOTRIAZINE see AQO000
HEXAKIS(DIHYDROGEN PHOSPHATE) MYO-INOSITOL see PIB250
HEXAKIS(β,β-DIMETHYLPHENETHYL)DISTANNOXANE see BLU000

HEXAKIS(HYDROXYMETHYL)MELAMINE see HDY000
HEXAKIS(HYDROXYMETHYL)-1,3,5-TRIAZINE-2,4,6-TRIAMINE see HDY000
HEXAKIS-METHOXYMETHYLMELAMIN (CZECH) see HDY500
HEXAKIS(METHOXYMETHYL)MELAMINE see HDY500
HEXAKIS(METHOXYMETHYL)-s-TRIAZINE-2,4,6-TRIAMINE see HDY500
N,N,N',N',N'',N''-HEXAKIS(METHOXYMETHYL)-1,3,5-TRIAZINE-2,4,6-TRIAMINE see HDY500
HEXAKIS(2-METHYL-2-PHENYLPROPYL)DISTANNOXANE see BLU000
HEXAKIS(PYRIDINE)IRON(II) TRIDECACARBONYLTETRAFERRATE(2-) see HEY000
HEXAKOSE see HCS500
γ-HEXALACTONE see HDY600
HEXALAN RED B see CMS231
HEXALAN RED 2G see CMM300
HEXALDEHYDE (DOT) see HEM000
HEXALIN see CPB750
HEXAMARIUM see DXG800
HEXAMETAPHOSPHATE, SODIUM SALT see SHM500
HEXAMETAPOL see HEK000
HEXAMETHIONIUM BROMIDE see HEA000
HEXAMETHIONIUM CHLORIDE see HEA500
HEXAMETHONIUM BROMIDE see HEA000
HEXAMETHONIUM CHLORIDE see HEA500
HEXAMETHONIUM DIBROMIDE see HEA000
HEXAMETHONIUM DICHLORIDE see HEA500
HEXAMETHONIUM DIIODIDE see HEB000
HEXAMETHONIUM IODIDE see HEB000
HEXA(METHOXYMETHYL)MELAMINE see HDY500
HEXAMETHYLBENZENE see HEC000
HEXAMETHYL-BICYCLO(2.2.0)HEXA-2,5-DIEN (GERMAN) see HEC500
HEXAMETHYLBICYCLO(2.2.0)HEXA-2,5-DIENE see HEC500
N,N,N,N',N',N'-HEXAMETHYL-1,10-DECANEDIAMINIUM DIBROMIDE see DAF600
N,N,N,N',N',N'-HEXAMETHYL-1,10-DECANEDIAMINIUM DIIODIDE see DAF800
HEXAMETHYLDIPLATINUM see HED000
HEXAMETHYLDISILANE see HED425
HEXAMETHYLDISILAZANE see HED500
HEXAMETHYLDISILOXANE see HEE000
HEXAMETHYLDISTANNANE see HEE500
HEXAMETHYLDITIN see HEE500
HEXAMETHYLENAMINE see HEI500
HEXAMETHYLENDIISOKYANAT see DNJ800
HEXAMETHYLEN-1,6-(N-DIMETHYLCARBODESOXYMETHYL)AMMONIUM DI-CHLORIDE see HEF200
HEXAMETHYLENE see CPB000
HEXAMETHYLENEAMINE see HEI500
N,N'-HEXAMETHYLENEBIS-1-AZIRIDINECARBOXAMIDE see HEF500
HEXAMETHYLENEBIS((CARBOXYMETHYL)DIMETHYLAMMONIUM), DICHLO-RIDE, DIDODECYL ESTER see HEF200
1,1'-HEXAMETHYLENEBIS(5-(p-CHLOROPHENYL)BIGUANIDE) see BIM250
1,1'-HEXAMETHYLENEBIS(5-(p-CHLOROPHENYL)BIGUANIDE) DIACETATE see CDT125
1,1'-HEXAMETHYLENEBIS(5-(p-CHLOROPHENYL)BIGUANIDE)DIGLUCONATE see CDT250
N,N'-HEXAMETHYLENEBIS(2,2-DICHLORO-N-ETHYLACETAMIDE) see HEF300
HEXAMETHYLENEBIS(DIMETHYL-9-FLUORENYLAMMONIUM BROMIDE) see HEG000
HEXAMETHYLENEBIS(DITHIOCARBAMIC ACID) DISODIUM SALT see HEF400
1,6-HEXAMETHYLENEBIS(ETHYLENEUREA) see HEF500
HEXAMETHYLENEBIS(ETHYLENEUREA) see HEF500
HEXAMETHYLENEBIS(FLUOREN-9-YLDIMETHYLAMMONIUM BROMIDE) see HEG000
HEXAMETHYLENE BIS(9-FLUORENYL DIMETHYLAMMONIUM)DIBROMIDE see HEG000
HEXAMETHYLENEBIS(TRIMETHYLAMMONIUM) BROMIDE see HEA000
HEXAMETHYLENE(BISTRIMETHYLAMMONIUM)CHLORIDE see HEA500
HEXAMETHYLENEBIS(TRIMETHYLAMMONIUM) DIBENZENESULFONATE see HEG100
HEXAMETHYLENEBIS(TRIMETHYLAMMONIUM IODIDE) see HEB000
1,6-HEXAMETHYLENEDIAMINE see HEO000
HEXAMETHYLENEDIAMINE see HEO000
HEXAMETHYLENEDIAMINE ADIPATE (1:1) see HEG120
1,6-HEXAMETHYLENEDIAMINE DIHYDROCHLORIDE see HEO100
HEXAMETHYLENEDIAMINE DIHYDROCHLORIDE see HEO100
HEXAMETHYLENEDIAMINE MONOADIPATE see HEG120
HEXAMETHYLENEDIAMINE SEBACATE see HEG130
HEXAMETHYLENEDIAMINE, solution (UN 1783) (DOT) see HEO000
HEXAMETHYLENEDIAMINE, solid (UN 2280) (DOT) see HEO000
HEXAMETHYLENEDIAMMONIUM ADIPATE see HEG120
HEXAMETHYLENEDIAMMONIUM SEBACATE see HEG130
HEXAMETHYLENEDIETHYLENEUREA see HEF500
1,6-HEXAMETHYLENE DIISOCYANATE see DNJ800
HEXAMETHYLENE-1,6-DIISOCYANATE see DNJ800
HEXAMETHYLENE DIISOCYANATE (DOT) see DNJ800
HEXAMETHYLENE DIISOCYANATE see DNJ800

HEXAMETHYLENE DIISOCYANATE POLYMER see HEG300
HEXAMETHYLENE DIISOCYANATE TRIMER see HEG300
HEXAMETHYLENEDIOL see HEP500
HEXAMETHYLENE GLYCOL see HEP500
HEXAMETHYLENE IMINE (DOT) see HDG000
HEXAMETHYLENEIMINE-3,5-DINITROBENZOATE see ACI250
1-(2-HEXAMETHYLENEIMINOETHYL)-2-OXOCYCLOHEXANECARBOXYLIC ACID
    BENZYL ESTER HYDROCHLORIDE see HEI000
HEXAMETHYLENE ISOCYANATE POLYMER see HEG300
HEXAMETHYLENETETRAAMINE see HEI500
HEXAMETHYLENETETRAMINE see HEI500
HEXAMETHYLENE TETRAMINE TETRAIODIDE see HEI650
HEXAMETHYLENETETRAMMONIUM TETRAPEROXOCHROMATE(V) see
    HEJ000
HEXAMETHYLENETRIPEROXYDIAMINE see DCK700
HEXAMETHYLENIMINE see HDG000
2-(β-HEXAMETHYLENIMINOAETHYL)CYCLOHEXANON-2-CARBONSAUREBEN-
    ZYLESTER-HYDROCHLORIDE (GERMAN) see HEI000
HEXAMETHYLENTETRAMIN (GERMAN) see HEI500
HEXAMETHYLERBIUM-HEXAMETHYLETHYLENEDIAMINE LITHIUM COMPLEX
    see HEJ350
2,4,6,8,9,10-HEXAMETHYLHEXAAZA-1,3,5,7-TETRAPHOSPHAADAMANTANE
    see HEJ375
2,3,3,4,4,5-HEXAMETHYL-2-HEXANETHIOL see DXT800
N,N,N,N′,N′,N′-HEXAMETHYL-1,6-HEXANEDIAMINIUM DIBROMIDE see HEA000
N,N,N,N′,N′,N′-HEXAMETHYL-1,6-HEXANEDIAMINIUM DICHLORIDE see
    HEA500
N,N,N,N′,N′,N′-HEXAMETHYL-1,6-HEXANEDIAMINIUM DIIODIDE see HEB000
1,1,2,3,3,6-HEXAMETHYLINDAN-5-YL METHYL KETONE see HEJ400
HEXAMETHYLMELAMINE see HEJ500
HEXAMETHYL METHYLOLMELAMINE see HDY500
HEXAMETHYLOLMELAMIN (CZECH) see HDY000
HEXAMETHYLOLMELAMINE see HDY000
HEXAMETHYLOL-MELAMIN-HEXA-METHYLAETHER see HDY500
HEXAMETHYLPARAOSANILINE CHLORIDE see AOR500
HEXAMETHYLPHOSPHORAMIDE see HEK000
HEXAMETHYLPHOSPHORIC ACID TRIAMIDE (MAK) see HEK000
HEXAMETHYLPHOSPHORIC TRIAMIDE see HEK000
N,N,N,N,N,N-HEXAMETHYLPHOSPHORIC TRIAMIDE see HEK000
HEXAMETHYLPHOSPHOROTRIAMIDE see HEK000
HEXAMETHYLPHOSPHOTRIAMIDE see HEK000
HEXAMETHYLRHENIUM see HEK550
HEXAMETHYL-p-ROSANILINE CHLORIDE see AOR500
HEXAMETHYL-p-ROSANILINE HYDROCHLORIDE see AOR500
N,N,N′,N′,N″,N″-HEXAMETHYLSILANETRIAMINE see TNH300
HEXAMETHYLSILAZANE see HED500
2,6,10,15,19,23-HEXAMETHYLTETRACOSANE see SLG700
N,N,N′,N′,N″,N″-HEXAMETHYL-1,3,5-TRIAZINE-2,4,6-TRIAMINE see HEJ500
HEXAMETHYL-1,3,5-TRITHIANE see HEL000
2,2,4,4,6,6-HEXAMETHYLTRITHIANE see HEL000
HEXAMETHYL-s-TRITHIANE see HEL000
HEXAMETHYL VIOLET see AOR500
HEXAMETON see HEA000
HEXAMETON CHLORIDE see HEA500
HEXAMIC ACID see CPQ625
HEXAMID see HEL500
HEXAMIDINE (the antispasmodic) see DBB200
HEXAMIDINE see DBB200
HEXAMINE (DOT) see HEI500
HEXAMITE see TCF250
HEXAMOL SLS see SIB600
1-HEXANAL see HEM000
HEXANAL see ERE000, HEM000
HEXANAL, 3,5,5-TRIMETHYL- see ILJ100
HEXANAMIDE see HEM500
1-HEXANAMINE see HFK000
HEXANAPHTHENE see CPB000
HEXANASTAB see ERE000
HEXANASTAB ORAL see ERD500
HEXANATE see TIX250
HEXANATE D see HCP600
HEXANATRIUMTETRAPOLYPHOSPHAT (GERMAN) see HEY500
HEXANE (DOT) see HEN000
n-HEXANE see HEN000
1-HEXANECARBOXYLIC ACID see HBE000
HEXANEDIAMIDE (9CI) see AEN000
1,6-HEXANEDIAMINE see HEO000
1,6-HEXANEDIAMINE, DIHYDROCHLORIDE see HEO100
HEXANEDIAMINE DIHYDROCHLORIDE see HEO100
1,6-HEXANEDIAMINE, N,N′-DICARBOXYMETHYL-N,N′-DIMETHYL-, DIMETHO-
    CHLORIDE, DIDODECYL ESTER see HEF200
1,6-HEXANEDIAMINE, N,N,N′,N′-TETRAMETHYL- see TDR100
1,6-HEXANEDIAMINIUM, N,N′-BIS[2-(DODECYLOXY)-2-OXOETHYL]-N,N,N′,N′-
    TETRAMETHYL-, DICHLORIDE see HEF200

1,6-HEXANEDIAMINIUM, N,N,N,N′,N′,N′-HEXAMETHYL-, DIBENZENESULFO-
    NATE (9CI) see HEG100
HEXANE, 1,6-DIISOCYANATO-, HOMOPOLYMER (9CI) see HEG300
HEXANE, 2,5-DIMETHYL-, 2,5-DIHYDROPEROXIDE see DSE800
HEXANEDINITRILE see AER250
1,6-HEXANEDIOIC ACID see AEN250
HEXANEDIOIC ACID, BIS[2-(2-BUTOXYETHOXY)ETHYL] ESTER (9CI) see
    DDT500
HEXANEDIOIC ACID, BIS(2-ETHYLHEXYL) ESTER see AEO000
HEXANEDIOIC ACID, BIS(2-(HEXYLOXY)ETHYL)ESTER (9CI) see AEQ000
HEXANEDIOIC ACID, BIS(3-METHYLBUTYL) ESTER see AEQ500
HEXANEDIOIC ACID, BIS(1-METHYLETHYL) ESTER see DNL800
HEXANEDIOIC ACID BIS(4-METHYL-7-OXABICYCLO(4.1.0)HEPT-3-YL)METHYL
    ESTER see AEN750
HEXANEDIOIC ACID–DIBUTYL ESTER see AEO750
HEXANEDIOIC ACID, DICYCLOHEXYL ESTER (9CI) see DGT500
HEXANEDIOIC ACID DIHYDRAZIDE see AEQ250
HEXANEDIOIC ACID DINITRILE see AER250
HEXANEDIOIC ACID, DIOCTYL ESTER see AEO000
HEXANEDIOIC ACID-DI-2-PROPENYL ESTER see AEO500
HEXANEDIOIC ACID ETHENYL METHYL ESTER see MQL000
HEXANEDIOIC ACID, compd. with 1,6-HEXANEDIAMINE (1:1) (9CI) see
    HEG120
HEXANEDIOIC ACID, OCTAFLUORO-, DIETHYL ESTER see DJT100
HEXANEDIOIC ACID, compound with PIPERAZINE (1:1) see HEP000
HEXANEDIOIC ACID, POLYMER with 1,4-BUTANEDIOL and 1,1′-METHYLENE-
    BIS(4-ISOCYANATOBENZENE) see PKM250
HEXANEDIOIC ACID, POLYMER with 1,3-ETHANEDIOL and 1,1′-METHYLENE-
    BIS(4-ISOCYANATOBENZENE) see PKL750
1,2-HEXANEDIOL see HFP875
1,6-HEXANEDIOL see HEP500
2,5-HEXANEDIOL see HEQ000
α-ω-HEXANEDIOL see HEP500
ω-HEXANEDIOL see HEP500
1,6-HEXANEDIOL DIACRYLATE see HEQ100
1,6-HEXANEDIOL DIISOCYANATE see DNJ800
2,5-HEXANEDIOL DIMETHYLSULFONATE see DSU000
2,3-HEXANEDIONE see HEQ200
2,5-HEXANEDIONE see HEQ500
2,3-HEXANEDIONE, 5-METHYL- see MKL300
N,N′-1,6-HEXANEDIYLBIS-1-AZIRIDINECARBOXAMIDE see HEF500
3,3′-(1,6-HEXANEDIYLBIS-((METHYLIMINO)CARBONYL)OXY)BIS(1-METHYLPYRI-
    DINIUMDIBROMIDE) see DXG800
(1,6-HEXANEDIYLBIS(NITRILOBIS(METHYLENE)))TETRAKISPHOSPHONIC ACID
    see HEQ600
(1,6-HEXANEDIYLBIS(NITRILOBIS(METHYLENE)))TETRAKISPHOSPHONIC ACID
    POTASSIUM SALT see HEQ610
HEXANE, 1-(1-ETHOXYETHOXY)- see EKS120
1,2,3,4,5,6-HEXANEHEXOL see HER000
6-HEXANELACTAM see CBF700
HEXA-NEMA see DFK600
HEXANEN (DUTCH) see HEN000
HEXANENITRILE see HER500
HEXANES (FCC) see HEN000
1-HEXANETHIOL see HES000
1,2,6-HEXANETRIOL see HES500
HEXANE-1,2,6-TRIOL see HES500
HEXANETRIOL-1,2,6 see HES500
HEXANHYDROPYRIDINE HYDROCHLORIDE see HET000
HEXANICIT see HFG550
HEXANICOTINOYL INOSITOL see HFG550
HEXANICOTOL see HFG550
HEXANITROBENZENE see HET350
HEXANITROCOBALTATE(3) TRIPOTASSIUM see PLI750
2,2′,4,4′,6,6′-HEXANITRODIFENYLAMIN see HET500
HEXANITRODIFENYLAMINE (DUTCH) see HET500
2,2′,4,4′,6,6′-HEXANITRODIPHENYLAMINE see HET500
2,4,6,2′,4′,6′-HEXANITRODIPHENYLAMINE see HET500
HEXANITRODIPHENYLAMINE (FRENCH) see HET500
HEXANITRODIPHENYLAMINE see HET500
HEXANITRODIPHENYLSULFIDE see BLR750
HEXANITROETHANE see HET675
HEXANITROL see MAW250
2,2′,4,4′,6,6′-HEXANITROOXANILIDE see HET700
HEXANITROOXANILIDE see HET700
HEXANOESTROL see DLB400
n-HEXANOIC ACID see HEU000
HEXANOIC ACID see HEU000
HEXANOIC ACID, BIS(2-ETHOXYETHYL) ESTER see BJO225
HEXANOIC ACID, BUTYL ESTER see BRK900
HEXANOIC ACID, 2-ETHYL-, HEXADECYL ESTER see HCP550
HEXANOIC ACID, 3-HEXENYL ESTER, (Z)- see HFE510
HEXANOIC ACID, HEXYL ESTER see HFQ500
HEXANOIC ACID, ISOBUTYL ESTER see IIT000
HEXANOIC ACID, METHYL ESTER see MHY700

HEXANOIC ACID, 2-METHYLPROPYL ESTER see IIT000
HEXANOIC ACID, MONOETHYL ESTER (9CI) see ELC600
HEXANOIC ACID, PENTYL ESTER see PBW450
HEXANOIC ACID, 1,2,3-PROPANETRIYL ESTER (9CI) see GGK000
HEXANOIC ACID, 3,5,5-TRIMETHYL-, ALLYL ESTER see AGU400
HEXANOIN, TRI-(6CI,7CI,8CI) see GGK000
1-HEXANOL see HFJ500
sec-HEXANOL (DOT) see EGW000
n-HEXANOL see HFJ500
HEXANOL see HFJ500
6-HEXANOLACTONE see LAP000
γ-HEXANOLACTONE see HDY600
tert-HEXANOL CARBAMATE see ENF000
tert-HEXANOL (9CI, DOT) see HFJ600
2-HEXANOL, 2-((1,1-DIMETHYLETHYL)AZO)-5-METHYL- see BQI300
1-HEXANOL, 2-ETHYL-, HYDROGEN PHOSPHATE, SODIUM SALT (7CI) see TBA750
1-HEXANOL, 2-ETHYL-, PHOSPHATE, SODIUM SALT (6CI) see TBA750
HEXANOLIDE-1,4 see HDY600
1,6-HEXANOLIDE see LAP000
3-HEXANOL, 3-METHYL- see MKL350
HEXANON see CPC000
2-HEXANONE see HEV000
HEXANONE-2 see HEV000
3-HEXANONE see HEV500
HEXANONE ISOXIME see CBF700
3-HEXANONE, 2-METHYL- see MKL400
1-HEXANONE, 1-(4-PYRIDYL)- see HEW050, PBX800
HEXANONISOXIM (GERMAN) see CBF700
1,4,7,10,13,16-HEXANOXACYCLOOCTADECANE see COD500
p-HEXANOYLANILINE see AJD250
1-HEXANOYLAZIRIDINE see HEW000
HEXANOYL CHLORIDE, 2-ETHYL- see EKO600
HEXANOYLETHYLENEIMINE see HEW000
17-α-HEXANOYLOXY-19-NOR-4-PREGNENE-3,20-DIONE see GEK510
17-α-HEXANOYLOXYPREGN-4-ENE-3,20-DIONE see HNT500
4-HEXANOYLPYRIDINE see HEW050, PBX800
1,1,1,3,3,3-HEXAOCTYLDISTANNOXANE see HEW100
HEXAOCTYLDISTANNOXANE see HEW100
1,1,1,3,3,3-HEXAOCTYLDISTANNTHIANE see HEW150
HEXAOCTYLDISTANNTHIANE see HEW150
13,16,21,24-HEXAOXA-1,10-DIAZABICYCLO-(8,8,8)-HEXACOSANE see LFQ000
5,8,11,13,16,19-HEXAOXATRICOSANE (9CI) see BHK750
1,1,1,3,3,3-HEXAPHENYLDISTANNTHIANE see BLT250
HEXAPLAS M/1B see DNJ400
HEXAPLAS M/B see DEH200
HEXAPLAS M/O see ILR100
HEXAPLIN see RDK000
HEXAPROMIN see AJV500
1,1,1,3,3,3-HEXAPROPYLDISTANNOXANE see BLT300
1,1,1,3,3,3-HEXAPROPYLDISTANNTHIANE see HEW200
HEXAPROPYLDISTANNTHIANE see HEW200
HEXAPROPYMATE see POA250
HEXAPROPYNATE see POA250
HEXA-3-PYRIDINECARBOXYLATE-myo-INOSITOL (9CI) see HFG550
HEXAPYRIDINEIRON(II) TRIDECACARBONYL TETRAFERRATE(2·) see HEY000
HEXASODIUM TETRAPHOSPHATE see HEY500
HEXASODIUM TETRAPOLYPHOSPHATE see HEY500
HEXASTAT see HEJ500
HEXASUL see SOD500
HEXATHANE see EIR000
HEXATHIDE see HEB000
HEXATHIR see TFS350
HEXATOX see BBQ500
1,3,5-HEXATRIENE see HEZ000
1,3,5-HEXATRIYNE see HEZ375
HEXATRON see AJV500
HEXATYPE BROWN N see NBG500
HEXATYPE CARMINE B see EOJ500
HEXAUREA CHROMIC CHLORIDE see HEZ800
HEXAUREACHROMIUM(III) NITRATE see HFA000
HEXAUREAGALLIUM(III) PERCHLORATE see HFA225
HEXAVIBEX see PPK500
HEXAVIN see CBM750
HEXAZANE see PIL500
HEXAZINONE see HFA300
HEXAZIR see BJK500
HEXEMAL see TDA500
trans-2-HEXEN-1-AL see HFA525
2-HEXENAL see HFA500
HEX-2-ENAL see HFA500
2-HEXENAL, (E)- see PNC500
3-(Z)-HEXENAL see HFA515
3-HEXENAL, (Z)- see HFA515
HEX-2-EN-1-AL see HFA500

HEXENAL (barbiturate) see ERD500
cis-3-HEXENAL see HFA515
HEXENAL see ERD500
trans-2-HEXENAL see PNC500
trans-2-HEXENAL DIETHYL ACETAL see HFA600
trans-2-HEXENAL DIMETHYL ACETAL see HFA620
5-HEXENAL, 2,6-EPOXY- see ADR500
2-HEXEN-1-AL, 2-ISOPROPYL-5-METHYL- see IKM100
2-HEXEN-1-AL, 5-METHYL-2-(1-METHYLETHYL)- see IKM100
HEXENAL SODIUM see ERE000
1-HEXENE see HFB000
2-HEXENE see HFB500
HEXENE see HFB000
2-HEXENE, 1,1-DIETHOXY-, (E)- see HFA600
2-HEXENE, 1,1-DIMETHOXY-, (E)- see HFA620
3-HEXENE DINITRILE see HFC000
3-HEXENE, 1-(1-ETHOXYETHYL)-, (Z)- see EKS100
HEXENE-OL see HCS500
2-HEXENE-4-ONE see HFE200
trans-2-HEXENE OZONIDE see HFC500
4-HEXENE-1-YNE-3-OL see HFF000
4-HEXENE-1-YNE-3-ONE see HFF300
3-HEXENOIC ACID see HFD000
4-HEXENOIC ACID, 2-ACETYL-5-HYDROXY-3-OXO-, Δ-LACTONE, SODIUM derivative see SGD000
2-HEXENOL see HFD500
2-HEXEN-1-OL, (E)- see HFD500
β-γ-HEXENOL see HFE000
cis-3-HEXEN-1-OL (FCC) see HFE000
trans-2-HEXEN-1-OL (FCC) see HFD500
cis-3-HEXENOL see HFE000
HEXENOL see HCS500
trans-2-HEXENOL see HFD500
2-HEXEN-1-OL ACETATE see HFE100
3-HEXEN-1-OL, ACETATE, (Z)- see HFE150
cis-3-HEXENOL ACETATE see HFE150
γ-HEXENOLACTONE see PAJ500
3-HEXEN-OL, 2-AMINOBENZOATE, (Z)- see HFE300
3-HEXEN-1-OL, FORMATE, (Z)- see HFE516
2-HEXEN-5,1-OLIDE see PAJ500
D''-HEXENOLLACTONE see PAJ500
4-HEXEN-1-OL, 5-METHYL-2-(1-METHYLETHENYL)-, ACETATE see LCA100
2-HEXEN-1-OL, PROPANOATE, (E)- see HFE700
3-HEXEN-1-OL, PROPANOATE (Z)- see HFE650
2-HEXEN-4-ONE see HFE200
d-erythro-HEX-2-ENONIC ACID, γ-LACTONE see SAA025
2-HEXEN-1-YL-ACETATE see HFE100
2-HEXENYL ACETATE see HFE100
(E)-2-HEXENYL ACETATE see HFE100
HEX-2-ENYL ACETATE see HFE100
(Z)-3-HEXENYL ACETATE see HFE150
cis-3-HEXENYL ACETATE see HFE150
trans-2-HEXENYL ACETATE see HFE100
3-HEXEN-1-YL 2-AMINOBENZOATE see HFE300
cis-3-HEXENYL ANTHRANILATE see HFE300
cis-3-HEXENYL BENZOATE see HFE500
cis-3-HEXENYL CAPROATE see HFE510
2-(2-HEXENYL CYCLOPENTANONE) see HFE513
HEXENYL CYCLOPENTANONE see HFE513
3-HEXENYL ESTER, BENZOIC ACID (Z)- see HFE500
cis-3-HEXENYL ETHANOATE see HFE150
cis-3-HEXENYL ETHYL ACETAL see EKS100
cis-3-HEXENYL FORMATE see HFE516
cis-3-HEXENYL HEXANOATE see HFE510
β,γ-HEXENYL ISOBUTANOATE see HFE520
cis-3-HEXENYL ISOBUTYRATE see HFE520
cis-3-HEXENYL ISOVALERATE (FCC) see ISZ000
β,γ-HEXENYL METHANOATE see HFE516
cis-3-HEXENYL-2-METHYL-trans-2-BUTENOATE see HFE710
cis-3-HEXENYL 2-METHYLBUTYRATE see HFE550
cis-HEXENYL OCYACETALDEHYDE see HFE600
cis-3-HEXENYL PENTANOATE see HFE800
cis-3-HEXENYL PHENYLACETATE see HFE625
2-HEXENYL PROPANOATE see HFE700
β,γ-HEXENYL PROPANOATE see HFE650
cis-3-HEXENYL PROPIONATE see HFE650
trans-2-HEXENYL PROPIONATE see HFE700
β,γ-cis-HEXENYL SALICYLATE see SAJ000
cis-3-HEXENYL SALICYLATE see SAJ000
cis-3-HEXENYL TIGLATE see HFE710
β,γ-HEXENYL α-TOLUATE see HFE625
cis-3-HEXENYL VALERATE see HFE800
4-HEXEN-1-YN-3-OL see HFF000
4-HEXEN-1-YN-3-ONE see HFF300
HEXERMIN see PPK500

HEXERMIN P see PII100
meso-HEXESTROL see DLB400
HEXESTROL see DLB400
HEXESTROL DIPHOSPHATE SODIUM see SHN150
HEXETHAL SODIUM see EKT500
HEXICIDE see BBQ500
HEXIDE see HCL000
HEXILMETHYLENAMINE see HEI500
HEXMETHYLPHOSPHORAMIDE see HEK000
HEXOBARBITAL see ERD500
HEXOBARBITAL Na see ERE000
HEXOBARBITAL SODIUM see ERE000
HEXOBARBITONE see ERD500
HEXOBARBITONE Na see ERE000
HEXOBARBITONE SODIUM see ERE000
HEXOBENDINE DIHYDROCHLORIDE see HFF500
HEXOBION see PPK500
HEXOCYCLIUM METHYLSULFATE see HFG400
HEXOESTROL see DLB400
HEXOGEEN (DUTCH) see CPR800
HEXOGEN see CPR800
HEXOGEN (Explosive) see CPR800
HEXOGEN 5W see CPR800
HEXOGEN, desensitized (UN 0483) (DOT) see CPR800
HEXOGEN, wetted (UN 0072) (DOT) see CPR800
n-HEXOIC ACID see HEU000
1,6-HEXOLACTAM see CBF700
HEXOLITE see CPR800
HEXOLITE, dry or wetted with <15% water, by weight (UN 0118) (DOT) see CPR800
HEXON (CZECH) see HFG500
HEXON CHLORIDE see HEA500
HEXONE see HFG500
HEXONE CHLORIDE see HEA500
HEXONIUM DIBROMIDE see HEA000
HEXONIUM DIIODIDE see HEB000
HEXOPAL see HFG550
HEXOPHENE see HCL000
HEXOPRENALINE DIHYDROCHLORIDE see HFG600
HEXOPRENALINE SULFATE see HFG650
HEXOSAN see HCL000
2-HEXOXYACETALDEHYDE DIMETHYLACETAL see HFG700
β-HEXOXYACETALDEHYDE DIMETHYLACETAL see HFG700
HEXOXYACETALDEHYDE DIMETHYLACETAL see HFG700
3-HEXOXY-1-(2′-CARBOXYPHENOXY)-PROPANOL-(2) SODIUM SALT see ERE200
n-HEXOXYETHOXYETHANOL see HFN000
HEXYCLAN see BBQ750
HEXYL (GERMAN, DUTCH) see HET500
1-HEXYL ACETATE see HFI500
n-HEXYL ACETATE (FCC) see HFI500
sec-HEXYL ACETATE see HFJ000
HEXYL ACETATE see HFI500
β-HEXYLACROLEIN see NNA300
N-HEXYL ACRYLATE see ADV000
HEXYL ACRYLATE see ADV000
n-HEXYL ALCOHOL see HFJ500
sec-HEXYL ALCOHOL see EGW000
HEXYL ALCOHOL see HFJ500
tert-HEXYL ALCOHOL see HFJ600
HEXYL ALCOHOL, ACETATE see HFI500
N-HEXYLAMINE see HFK000
HEXYLAMINE see HFK000
HEXYLAN see BBP750
5-n-HEXYL-1,2-BENZANTHRACENE see HFL000
8-HEXYL-BENZ(a)ANTHRACENE see HFL000
4-HEXYL-1,3-BENZENEDIOL see HFV500
n-HEXYLBENZOATE see HFL500
HEXYL BENZOATE see HFL500
HEXYL BROMIDE see HFM500
n-HEXYL n-BUTANOATE see HFM700
n-HEXYL BUTANOATE see HFM700
HEXYL BUTANOATE see HFM700
n-HEXYL 2-BUTENOATE see HFM600
HEXYL-2-BUTENOATE see HFM600
1-HEXYL BUTYRATE see HFM700
n-HEXYL BUTYRATE see HFM700
HEXYL BUTYRATE see HFM700
γ-N-HEXYL-γ-BUTYROLACTONE see HKA500
HEXYLCAINE HYDROCHLORIDE see COU250
HEXYL CAPROATE see HFQ500
n-HEXYL CAPRYLATE see HFS600
HEXYL CAPRYLATE see HFS600
1-HEXYLCARBAMOYL-5-FLUOROURACIL see CCK630
n-HEXYL CARBITOL see HFN000

HEXYL CARBITOL see HFN000
n-HEXYL CARBORANE see HFN500
N-HEXYL CELLOSOLVE see HFT500
2-HEXYL-4-CHLOROPHENOL see CHL500
α-HEXYLCINNAMALDEHYDE (FCC) see HFO500
HEXYL CINNAMALDEHYDE see HFO500
α-HEXYLCINNAMIC ALDEHYDE see HFO500
HEXYL CINNAMIC ALDEHYDE see HFO500
HEXYL CROTONATE see HFM600
2-n-HEXYLCYCLOPENTANONE see HFO600
2-HEXYLCYCLOPENTANONE see HFO600
2-n-HEXYL-2-CYCLOPENTEN-1-ONE see HFO700
2-HEXYLDECANOIC ACID see HFP500
2-HEXYLDECANSAEURE (GERMAN) see HFP500
HEXYLDICARBADODECABORANE(12) see HFN500
HEXYLDICHLORARSINE see HFP600
5-HEXYLDIHYDRO-2(3H)-FURANONE see HKA500
4-HEXYL-1,3-DIHYDROXYBENZENE see HFV500
HEXYL 2,2-DIMETHYLPROPANOATE see HFP700
HEXYLENE see HFB000
HEXYLENE GLYCOL see HFP875
HEXYLENE GLYCOL DIACETATE see HFQ000
cis-β,γ-HEXYLENIC ALDEHYDE see HFA515
HEXYLENIC ALDEHYDE see HFA500
HEXYL ETHANOATE see HFI500
N-HEXYL ETHER see DKO800
HEXYL ETHER see DKO800
n-HEXYL FORMATE see HFQ100
HEXYL FORMATE see HFQ100
HEXYL FUMARATE see DKP000
n-HEXYL HEXANOATE see HFQ500
HEXYL HEXOATE see HFQ500
n-HEXYL ISOBUTANOATE see HFQ550
HEXYL ISOBUTANOATE see HFQ550
1-HEXYL ISOBUTYRATE see HFQ550
n-HEXYL ISOBUTYRATE see HFQ550
HEXYL ISOBUTYRATE see HFQ550
n-HEXYL ISOPENTANOATE see HFQ575
HEXYL ISOVALERATE see HFQ575
HEXYL KETONE see TJJ300
HEXYL MANDELATE see HFR000
HEXYL MERCAPTAN see HES000
n-HEXYLMERCURIC BROMIDE see HFR100
HEXYLMERCURIC BROMIDE see HFR100
HEXYL MERCURY BROMIDE see HFR100
HEXYL 3-METHYL BUTANOATE see HFQ575
n-HEXYL trans-2-METHYL-2-BUTENOATE see HFX000
HEXYL 2-METHYLBUTYRATE see HFR200
2-HEXYL-4-METHYL-1,3-DIOXOLANE see HFR500
HEXYL NEOPENTANOATE see HFP700
1-HEXYL-1-NITROSOUREA see HFS500
HEXYL OCTANOATE see HFS600
2-n-HEXYLOXYBENZAMIDE see HFS759
2-(HEXYLOXY)BENZAMIDE see HFS759
o-HEXYLOXYBENZAMIDE see HFS759
2-(HEXYLOXY)ETHANOL see HFT500
2-((2-HEXYLOXY)ETHOXY)ETHANOL see HFN000
o-(3-(HEXYLOXY)-2-HYDROXYPROPOXY)BENZOIC ACID MONOSODIUM SALT see ERE200
4′-(HEXYLOXY)-3-PIPERIDINOPROPIOPHENONE HYDROCHLORIDE see HFU500
4-HEXYLOXY-β-(1-PIPERIDYL)PROPIOPHENONE HYDROCHLORIDE see HFU500
2-HEXYLPHENOL see HFU600
o-HEXYLPHENOL see HFU600
α-n-HEXYL-β-PHENYLACROLEIN see HFO500
2-(1-HEXYL-3-PIPERIDYL)ETHYL ESTER, BENZOIC ACID HYDROCHLORIDE see HFV000
1-HEXYL PROPANOATE see HFV100
HEXYL PROPANOATE see HFV100
HEXYL-2-PROPENOATE see ADV000
1-HEXYL PROPIONATE see HFV100
HEXYL PROPIONATE see HFV100
HEXYLRESORCIN (GERMAN) see HFV500
4-HEXYLRESORCINE see HFV500
4-n-HEXYLRESORCINOL see HFV500
4-HEXYLRESORCINOL see HFV500
p-HEXYLRESORCINOL see HFV500
HEXYLRESORCINOL see HFV500
3-HEXYL-7,8,9,10-TETRAHYDRO-6,6,9-TRIMETHYL-6H-DIBENZO(B,D)PYRAN-1-OL see HGK500
HEXYLTHIOCARBAM see EHT500
3-(5-(HEXYLTHIO)PENTYL)THIAZOLIDINE HYDROCHLORIDE see HFW500
n-HEXYL TIGLINATE see HFX000
HEXYL TIGLATE see HFX000

HEXYLTRICHLOROSILANE see HFX500
n-HEXYL VINYL SULFONE see HFY000
3-HEXYNE, 2,5-DIMETHYL-2,5-DI(t-BUTYLPEROXY)- see DRJ825
HEXYNE-3-DIOL-2,5 see HFY500
3-HEXYNE-2,5-DIOL see HFY500
1-HEXYN-3-OL see HFZ000
HEXYNOL see HGA000
HEYDEFLON see TAI250
HEYDEN 768 see TGC300
HF 10 see SMQ500
HF 55 see SMQ500
HF 77 see SMQ500
HF 191 see DGV700
HF 264 see BRJ325
HF-1854 see CMY650
HF 1927 see DCW600, DCW800
HF-2333 see HOU059
HF3170 see DCS200
HFA see FIC000
H-35-F 87 (BVM) see DSQ000
HFCB see HDB500
HFDB 4201 see PJS750
HFE see HDC000
HF-2159 HYDROCHLORIDE see CIT000
HFIP see HDC500
81723 HFU see TET800
HG-203 see CHX250
H.G. BLENDING see SFT000
HGG-12 see HGA050
HGI see BBQ500
HH 102 see SMQ500
HH-197 see BOR350
H. HAEMACHATES VENOM see HAO500
HHDN see AFK250
Hg-HEMATOPORPHYRIN-Na see HAP000
HHI 11 see SMQ500
HI-6 see HGE900
HIADELON see PII100
HI-ALAZIN see TGB475
HI-A-VITA see VSK600
HIBANIL see CKP250, CKP500
HIBAWOOD OIL see HGA100
HIBERNA see DQA400
HIBERNAL see CKP250, CKP500
HIBERNON HYDROCHLORIDE see HGA500
HIBERNYL see TCY750
HIBESTROL see DKA600
HIBICLENS see CDT250
HIBICON see BEG000
HIBIDIL see CDT250
HIBISCOLIDE see OKW110
HIBISCRUB see CDT250
HIBISCUS MANIHOT Linn., extract see HGA550
HIBITANE see BIM250, CDT250
HIBITANE DIACETATE see CDT125
HIBROM see NAG400
HICAL-2 see HGB000
HI-CAL 3 see JDA125
HICHILLOS see KGK000
HICO CCC see CMF400
HICOFOR PR see MCB050
HICOPHOR PR see MCB000
HIDA see LFO300
HIDACHROME YELLOW 2G see SIT850
HIDACIAN see COH250
HIDACIANN see COH250
HIDACID AMARANTH see FAG020
HIDACID AZO RUBINE see HJF500
HIDACID AZURE BLUE see FMU059
HIDACID BLUE A see ERG100
HIDACID BLUE AF see ERG100
HIDACID BLUE V see ADE500
HIDACID BROMO ACID REGULAR see BNK700
HIDACID DIBROMO FLUORESCEIN see BNH500, BNK700
HIDACID EMERALD GREEN see FAE950
HIDACID FAST CRIMSON see CMM300
HIDACID FAST SCARLET 3R see FMU080
HIDACID FLUORESCEIN see FEV000
HIDACID METANIL YELLOW see MDM775
HIDACID ORANGE II see CMM220
HIDACID SCARLET 2R see FMU070
HIDACID URANINE see FEW000
HIDACID WOOL GREEN see ADF000
HIDACO BRILLIANT CRYSTAL VIOLET see AOR500
HIDACO BRILLIANT GREEN see BAY750

HIDACO CRYSTAL VIOLET see AOR500
HIDACO MALACHITAE GREEN BASE see AFG500
HIDACO METHYLENE BLUE SALT FREE see BJI250
HIDACO OIL ORANGE see PEJ500
HIDACO OIL RED see SBC500
HIDACO OIL YELLOW see AIC250
HIDACO SAFRANINE see GJI400
HIDACO VICTORIA BLUE R see VKA600
HIDAN see DKQ000
HIDANTILO see DKQ000
HIDANTINA SENOSIAN see DKQ000
HIDANTINA VITORIA see DKQ000
HIDANTOMIN see DKQ000
HIDAZID TARTRAZINE see FAG140
HI-DERATOL see VSZ100
HIDRALAZIN see HGP495, HGP500
HIDRANIZIL see ILD000
HIDRIL see CFY000
HIDRIX see HOO500
HIDROCHLORTIAZID see CFY000
HIDRO-COLISONA see CNS750
HIDROESTRON see EDP000
HIDROMEDIN see DFP600
HIDRORONOL see CFY000
HIDROTIAZIDA see CFY000
N-(trans-4-HIDROXICICLOHEXIL)-(2-AMINO-3,5-DIBROMOBENCIL)AMINA (SPAN-
    ISH) see AHJ250
HIDROXIFENAMATO see HNJ000
HIDROXITEOFILLINA see DNC000
HIDRULTA see ILD000
HI-DRY see TCE250
HI-ENTEROL see CHR500
HIERBA de SANTIAGO (MEXICO) see RBA400
HIESTRONE see EDV000
HI-FAX 1900 see PJS750
HI-FAX 4401 see PJS750
HI-FAX 4601 see PJS750
HI-FAX see PJS750
HI-FLASH NAPHTHA see NAI500
HIFOL see BIO750
HIGH BELIA see CCJ825
HIGILITE see AHC000
HIGOSAN see MEP250
HIGUERETA CIMARRONA see CNR135
HIGUERETA (CUBA, PUERTO RICO) see CCP000
HIGUERILLA (MEXICO) see CCP000
HIGUEROXYL DELABARRE see FBS000
HI-JEL see BAV750
HIKIZIMYCIN see APF000
HILDAN see EAQ750
HILDIT see DAD200
HILITE 60 see DGN200
HILLS-OF-SNOW see HGP600
HILONG see CFZ000
HILTHION see MAK700
HILTHION 25WDP see MAK700
HILTONIL FAST BLUE B BASE see DCJ200
HILTONIL FAST ORANGE GR BASE see NEO000
HILTONIL FAST ORANGE R BASE see NEN500
HILTONIL FAST RED B BASE see NEQ000
HILTONIL FAST RED 3GL BASE see KDA050
HILTONIL FAST RED KB BASE see CLK225
HILTONIL FAST SCARLET 2G BASE see DEO295
HILTONIL FAST SCARLET G BASE see NMP500
HILTONIL FAST SCARLET GC BASE see NMP500
HILTONIL FAST SCARLET G SALT see NMP500
HILTOSAL FAST BLUE B SALT see DCJ200
HILTOSAL FAST ORANGE RD SALT see CEG800
HILTOSAL FAST SCARLET 2G SALT see DEO295
HINDAMINE SCARLET GG see DEO295
HINDASOL BLUE B SALT see DCJ200
HINDASOL RED TR SALT see CLK235
H. INFLUENZAE ENDOTOXIN, phenol water extract see HAB710
H. INFLUENZAE ENDOTOXIN, NaCl-citrate extract see HAB700
HINOKITIOL see IRR000
HINOKITOL see IRR000
HINOSAN see EIM000
HINSALU see RQU300
HIOHEX CHLORIDE see HEA500
HIOXYL see HIB000
HIP see COF250
HIPERCILINA see BFD000
HIPHYLLIN see DNC000
HIPNAX see DLY000
HIPOFTALIN see HGP495, HGP500

HI-POINT 90 see MKA500
HI-POINT 180 see MKA500
HI-POINT PD-1 see MKA500
HIPOSERPIL see RDK000
HIPPEASTRUM (VARIOUS SPECIES) see AHI635
HIPPOBROMA LONGIFLORA see SLJ650
HIPPODIN see HGB200
HIPPOMANE MANCINELLA see MAO875
HIPPOPHAIN see AJX500
HIPPURAN see HGB200
HIPPURIC ACID see HGB300
HIPPURIC ACID SODIUM SALT see SHN500
HIPPUZON see TEH500
HIPSAL see DLY000
HIPTAGENIC ACID see NIY500
HI-PYRIDOXIN see PII100
HIRATHIOL see IAD000
HIRSUTIC ACID N see HGB500
HISERPIA see RDK000
HISHIREX 502 see PKQ059
HISINDAMONE A see CDY000
HISMANAL see ARP675
HISPACET FAST YELLOW G see AAQ250
HISPACID BRILLIANT SCARLET 3RF see FMU080
HISPACID FAST BLUE R see ADE750
HISPACID FAST CARMOISINE G see CMM300
HISPACID FAST YELLOW T see FAG140
HISPACID GREEN GB see FAE950
HISPACID ORANGE 1 see FAG010
HISPACID ORANGE AF see CMM220
HISPACID YELLOW MG see MDM775
HISPACROM BLUE BG see CMP880
HISPACROM YELLOW 2G see SIT850
HISPACROM YELLOW 2GR see SIT850
HISPADIAZO BLACK D see CMN230
HISPALIT FAST SCARLET RN see MMP100
HISPALUZ RED 8BL see CMO885
HISPAMIN BLACK EF see AQP000
HISPAMIN BLUE 2B see CMO000
HISPAMIN BLUE 3BX see CMO250
HISPAMIN BLUE RW see CMO600
HISPAMIN CONGO 4B see SGQ500
HISPAMIN FAST BLACK CG see CMN240
HISPAMIN FAST BROWN NZ see CMO820
HISPAMIN FAST RED FN see CMO870
HISPAMIN GREEN B see CMO840
HISPAMIN GREEN WT see CMO830
HISPAMIN PURE YELLOW T2G see CMP050
HISPAMIN SKY BLUE 3B see CMO500
HISPAMIN VIOLET 3R see CMP000
HISPAVIC 229 see PKQ059
HISPERSE YELLOW G see AAQ250
HISPRIL see LJR000
HISPRIL HYDROCHLORIDE see DVW700
HISTABID see SPC500
HISTABUTYZINE DIHYDROCHLORIDE see BOM250
HISTACAP see WAK000
HISTADUR see CLX300, TAI500
HISTADUR DURA-TABS see TAI500
HISTADYL see TEO250
HISTADYL HYDROCHLORIDE see DPJ400
HISTAFED see DPJ400
HISTAGLOBIN see HGC400
HISTALEN see TAI500
HISTALON see WAK000
HISTAMETHINE see HGC500
HISTAMETHIZINE see HGC500
HISTAMETIZINE see HGC500
HISTAMETIZYNE see HGC500
HISTAMINE see HGD000
HISTAMINE ACID PHOSPHATE see HGE000
HISTAMINE DICHLORIDE see HGD500
HISTAMINE DIHYDROCHLORIDE see HGD500
HISTAMINE DIPHOSPHATE see HGE000
HISTAMINE HYDROCHLORIDE see HGE500
HISTAMINE PHOSPHATE (1:2) see HGE000
HISTAMINOS see ARP675
HISTAN see WAK000
HISTANTIN see CFF500
HISTANTINE see CFF500
HISTANTINE DIHYDROCHLORIDE see CDR000
HISTAPAN see TAI500
HISTAPYRAN see WAK000
HISTARGAN see DQA400
HISTASAN see WAK000

HISTASPAN see CLD250
HISTATEX see DBM800
HISTAXIN see BBV500
l-HISTIDINE, N-β-ALANYL- see CCK665
HISTIDYL see DPJ400
β-HISTINE see HGE820
HISTOCARB see CBJ000
HISTOSTAB see PDC000
HISTRYL see LJR000
HISTYN see LJR000
HISTYRENE S 6F see SMR000
HI-STYROL see SMQ500
HITACERAM SC 101 see SCQ000
HIVOL-44 see DAA800
HI-YIELD DESICCANT H-10 see ARB250
HIZAROCIN see CPE750
HIZEX 5000 see PJS750
HIZEX 5100 see PJS750
HIZEX see PJS750
HIZEX 3000B see PJS750
HIZEX 3300F see PJS750
HIZEX 7000F see PJS750
HIZEX 7300F see PJS750
HIZEX 1091J see PJS750
HIZEX 1291J see PJS750
HIZEX 1300J see PJS750
HIZEX 2100J see PJS750
HIZEX 2200J see PJS750
HIZEX 2100LP see PJS750
HIZEX 5100LP see PJS750
HIZEX 6100P see PJS750
HIZEX 3000S see PJS750
HIZEX 3300S see PJS750
HIZEX 5000S see PJS750
HJ 6 see HGE900
HK-141 see BAW500
HK 256 see LFJ000
H.K. FORMULA No. K. 7117 see FMU059
HL 267 see DGW600
HL-331 see ABU500
HL 2153 see DPD400
HL 2197 see CEP675
HL 2447 see DFV400
HL 8700 see PMI750
HL-8731 see MQF750
HLR 4219 see MJQ260
HLR 4448 see MJQ260
HLS 831 see BSQ000
HM 100 see MCB050
HMB see HFR100
7-HMBA see BBH250
HMBD see HLX900, HLX925
HMD see PAN100
HMDA see HEO000
HMDI see DNJ800
HMDS see HED500
HMF see ORG000
HMG see HMC000
HMGA see HMC000
HMM see HEJ500
7-HM-12-MBA see HMF000
12-HM-7-MBA see HMF500
HMP see SHM500
HMPA see HEK000
HMPT see HEK000
HMT see HEI500
HMX (dry or unphlegmatized) (DOT) see CQH250
HMX see CQH250
HMX, wetted (UN 0226) (DOT) see CQH250
beta HMY see CQH250
HN1 see BID250
HN 1 see CGW000
HN2.HCl see BIE500
HN2 see BIE250
HNl●HCl see EGU000
HN$_2$ AMINE OXIDE see CFA500
HN1 CHLOROHYDRIN see EHK300
HNED CERVENAVA OSTANTHRENOVA 5 RF see CMU800
HNED KYPOVA 1 see CMU770
HNED KYPOVA 25 see CMU800
HNED OSTANTHRENOVA BR see CMU770
HNED ROZPOUSTEDLOVA 1 see NBG500
HNED SUDAN RR see NBG500
HN1 HYDROCHLORIDE see EGU000
HN2 HYDROCHLORIDE see BIE500

HN₂ OXIDE HYDROCHLORIDE see CFA750
HN₂ OXIDE MUSTARD see CFA500
HNT see HHD500
HNU see HKW500
HO 11513 see TMK000
3-HO-AAF see HIO875
HOCA see DXH250
HOCH see FMV000
HODAG GMS see OAV000
HODAG PSML-20 see PKG000
HODAG SMS see SKV150
HODAG SSO see SKV170
HODAG SVO 9 see PKL100
HODOSTIN see DQY909
HODSON see DSK800
HOE 118 see PDW250
HOE 280 see OGI300
HOE 296 see BAR800
HOE 766 see LIU420
HOE 893 see PAP225
HOE 933 see CAZ125
HOE 984 see NMV725
HOE 2,671 see EAQ750
HOE 2747 see CKD500
HOE 2784 see BGB500
HOE 2810 see DGD600
HOE-2824 see ABX250
HOE 2872 see CLU000
HOE 2904 see ACE500
HOE 2982 see HBK700
HOE 17411 see MHC750
HOE 33258 see MOD500
HOE 36801 see CGQ500
HOE 766A see LIU420
HOECHST 1082 see MDP000
HOECHST 10720 see KFK000
HOECHST 10,820 see MDP750
12494 HOECHST see RDA375
33258 HOECHST see MOD500
HOECHST DYE 33258 see MOD500
HOECHST PA 190 see PJS750
HOECHST WAX PA 520 see PJS750
HOE 893D see PAP225, PAP230
HOE 2960 OJ see THT750
HOE 2982 OJ see HBK700
HOG see AOO500, PDC890
HOG APPLE see MBU800
HOG BUSH see CDM325
HOGGAR see BNK000
HOGGAR N see PGE775
HOG PHYSIC see CCJ825
HOG'S POTATO see DAE100
HOJA GRANDE (CUBA) see APM875
HOKKO-MYCIN see SLW500
HOKMATE see FAS000
HOLBAMATE see MQU750
HO LEAF OIL see HGF000
HOLIN see EDU500
HOLLICHEM HQ 3300 see BBA625
HOLLY see HGF100
HOLMIUM see HGF500
HOLMIUM CHLORIDE see HGG000
HOLMIUM CITRATE see HGG500
HOLMIUM NITRATE see HGH100
HOLMIUM(III) NITRATE, HEXAHYDRATE (1:3:6) see HGH000
HOLMIUM TRINITRATE see HGH100
HOLOCAINE see BJO500
HOLODORM see QAK000
HOLOPAN see SBH500
HOLOXAN see IMH000
HOMANDREN (amps) see TBG000
HOMANDREN see MPN500
HOMAPIN see MDL000
HOMATROMIDE see MDL000
(±)-HOMATROPINE BROMIDE see HGH150
HOMATROPINE HYDROBROMIDE see HGH150
HOMATROPINE METHYLBROMIDE see MDL000
HOMBITAN see TGG760
HOME HEATING OIL No.2 see DHE800
#2 HOME HEATING OILS see DHE800, HGH200
HOMIDIUM BROMIDE see DBV400
HOMIDIUM CHLORIDE see HGI000
HOMOANISIC ACID see MFE250
HOMOCAL D see CAT775
HOMOCHLORCYCLIZINE see HGI100

HOMOCHLORCYCLIZINE DIHYDROCHLORIDE see CJR809
HOMOCHLORCYCLIZINE HYDROCHLORIDE see HGI200
HOMOCHLOROCYCLIZINE see HGI100
HOMOCHLOROCYCLIZINE DIHYDROCHLORIDE see CJR809
HOMOCHLOROCYCLIZINE HYDROCHLORIDE see HGI200
HOMOCLOMINE see HGI100
HOMOCODEINE see TCY750
HOMODAMON see HGI100
18-HOMO-ESTRIOL see HGI700
HOMOGUAIACOL see MEK250
HOMOHARRINGTONINE see HGI575
HOMOLINALYL ACETATE see HGI585
HOMOLLE'S DIGITALIN see DKN400
HOMOMENTHOL see TLO500
HOMOMYRETENOL see DTB800
HOMONEURINE CHLORIDE see HGI600
18-HOMO-OESTRIOL see HGI700
HOMOOLAN see HIM000
HOMOPIPERAZINE see HGI900
HOMOPIPERIDINE see HDG000
HOMOPROLINE see HDS300
HOMORESTAR see HGI100
HOMOSALICYLIC ACID see CNX625
HOMOSTERONE see TBF500
3-HOMOTETRA HYDRO CANNIBINOL see HGK500
HOMOTRYPTAMINE HYDROCHLORIDE see AME750
HOMOVERATRIC ACID see HGK600
HOMOVERATRONITRILE see VIK100
HOMOVERATRYLAMINE see DOE200
HON see AKG500
HONEY DIAZINE see PPP500
HONEY YELLOW 3GNT see CMO810
HONG KIEN see ABU500
HONVAN see DKA200
HONVAN TETRASODIUMTETRASODIUM SALT, (E)- see TEE300
HOOKER HRS-16 see DAE425
HOOKER HRS-1422 see DNS200
HOOKER HRS 1654 see DAE425
HOOKER NO. 1 CHRYSOTILE ASBESTOS see ARM268
HOPA see CAT175
HOPANTENATE CALCIUM see CAT125
HOPANTENATE CALCIUM HEMIHDYRATE see CAT175
HOPCIDE see CKF000
HOPCIN see MOV000
HOP EXTRACT EI see HGK725
HOPFENEXTRAKTE EI see HGK725
HORACE VERNET'S BLUE see CNQ000
HORBADOX see DRN200
HORFEMINE see DKB000
HORMALE see MPN500
HORMATOX see DGB000
HORMEX ROOTING POWDER see ICP000
HORMIN see DFY800
HORMIT see SGH500
HORMOCEL-2CCC see CMF400
HORMODIN see ICP000
HORMOESTROL see DLB400
HORMOFEMIN see DAL600
HORMOFLAVEINE see PMH500
HORMOFOLLIN see EDV000
17-HORMOFORIN see AOO450
HORMOFORT see HNT500
HORMOGYNON see EDP000
HORMOLUTON see PMH500
HORMOMED see EDU500
HORMONIN see EDU500
HORMONISENE see CLO750
HORMOSLYR 500T see TAH900
HORMOTESTON see TBG000
HORMOTUHO see CIR250
HORMOVARINE see EDV000
HORSE see HBT500
HORSE BLOB see MBU550
HORSE CHESTNUT see HGL575
HORSE GOLD see FBS100
HORSE HEAD A-410 see TGG760
HORSE NICKER see CAK325
HORSE POISON (JAMAICA) see SLJ650
HORTENSIA (CUBA) see HGP600
HORTFENICOL see CDP250
HORTICULTURAL SPRAY OIL see MQV855
HORTOCRITT see EJQ000
HOSALON see DSK800
HOSDON GRANULE see DSK800
HOSTACAIN see BQA750

HOSTACAINE see BQA750
HOSTACAINE HYDROCHLORIDE see BQA750
HOSTACILLIN see PAQ200
HOSTACORTIN see PLZ000, PMA000
HOSTACYCLIN see TBX000
HOSTADUR see PKF750
HOSTAFLEX VP 150 see AAX175
HOSTAFLON see KDK000, TAI250
HOSTAFLON C see KDK000
HOSTAGINAN see PEV750
HOSTALEN see PJS750
HOSTALEN GD 620 see PJS750
HOSTALEN GD 6250 see PJS750
HOSTALEN GF 4760 see PJS750
HOSTALEN GF 5750 see PJS750
HOSTALEN GM 5010 see PJS750
HOSTALEN GUR see PJS750
HOSTALEN HDPE see PJS750
HOSTALEN PP see PMP500
HOSTALIT see PKQ059
HOSTALIVAL see NMV725
HOSTANOX SE 2 see DXG700
HOSTANOX VP-SE 2 see DXG700
HOSTAPERM GREEN 8G see CMS140
HOSTAPERM ORANGE GR see CMU820
HOSTAPERM VAT ORANGE GR see CMU820
HOSTAPERM YELLOW GR see CMS208
HOSTAPERM YELLOW GT see CMS212
HOSTAPERM YELLOW GTT see CMS212
HOSTAPHAN see PKF750
HOSTAPON T see SIY000
HOSTAQUICK see ABU500, HBK700
HOSTATHERM PINK FBL see AKI750
HOSTATHION see THT750
HOSTAVAT BRILLIANT ORANGE GR see CMU820
HOSTAVAT GOLDEN YELLOW see DCZ000
HOSTAVAT ORANGE R see CMU815
HOSTAVIK (RUSSIAN) see HBK700
HOSTETEX L-PEC see TIK250
HOSTYREN N 4000 see SMQ500
HOSTYREN N 7001 see SMQ500
HOSTYREN N see SMQ500
HOSTYREN N 4000V see SMQ500
HOSTYREN S see SMQ500
HOT PEPPER see PCB275
HOURBESE see DKE800
HOWFLEX GBP see DDY400, DJD700
HOX 1901 see EPR000
H.P. 34 see SAH000
HP 129 see FAO100
H.P. 165 see BPG750
H.P. 206 see PMY750
H.P. 209 see EEM000
H.P. 216 see HFS759
HP 1275 see PDV700
HP 1325 see HGL600
HP see PHV275
HPA see EPI300
HPC see HNT500, OPK300
β-HPN see HGP000
HPOP see HNX500
HPP see ZVJ000
2-HPPN see NLM500
β-HPPN see NLM500
HPT see HEK000
HQ-275 see HGL630
HR 376 see CIR750
HR 756 see CCR950
HR 930 see FOL100
HRS-16 see DAE425
HRS 860 see CEP000
HRS 1276 see MQW500
HRS 1654 see DAE425
HRS 1655 see HCE500
HRS 16A see DAE425
HRW 13 see SLJ500
HS 3 see HGL650
H 3S see IHC550
HS 4 see PPC000
HS 21 see CNA750
HS 25 see CNA750
HS 55 see AFM375
HS 592 see FOS100
HS-119-1 see PEE750
HS see HGW500

HSP 2986 see SDZ000
HSR-902 see TGF075
H-STADUR see CLD250
5-HT see AJX500
HT 88 see SMQ500
HT 91-1 see SMQ500
HT 972 see MJQ000
5-HTA see AJX500
HT 88A see SMQ500
H. TERNATUM see SAF500
HT-F 76 see SMQ500
HTH see HOV500
5-HTP see HOO100
l-5-HTP see HOA600, HOO000
HTP see HCY000
HT-2 TOXIN see THI250
HUBBUCK'S WHITE see ZKA000
HUBER see CBT750
HUELE de NOCHE (MEXICO) see DAC500
HUEVO de GATO (CUBA) see JBS100
HUILE d'ANILINE (FRENCH) see AOQ000
HUILE de CAMPHRE (FRENCH) see CBA750
HUILE de FUSEL (FRENCH) see FQT000
HUILE H50 see TFM250
HULS P 6500 see PMP500
HUMAN SPERM see HGM000
HUMATIN see APP500, NCF500
HUMEDIL see BCP650
HUMENEGRO see CBT750
HUMIC ACID, SODIUM SALT see HGM100
HUMIDIN see HGM500
HUMIFEN NBL 85 see BLX000
HUMIFEN WT 27G see DJL000
HUMINSAURE NATRIUM see HGM100
HUMULIN I see IDF325
HUMYCIN see NCF500
HUMYCIN SULFATE see APP500
HUNDRED PACE SNAKE VENOM see HGM600
HUNGARIAN CHAMOMILE OIL see CDH500
HUNGAZIN see ARQ725
HUNGAZIN DT see BJP000
HUNGAZIN PK see ARQ725
HUNGER WEED see FBS100
HUNTER'S ROBE see PLW800
HURA CREPITANS see SAT875
HURRICANE PLANT see SLE890
HUSEPT EXTRA see CLW000
HUSTODIL see RLU000
HUSTOSIL see RLU000
HVA 2 see BKL750
HVA-2 CURING AGENT see BKL750
HW 4 see CQH250
HW 920 see DXQ500
HWA 153 see HLF500
HX 3/5 see CCU250
HX 752 see BLG400
HX-868 see HGN000
HXR see IDE000
HY 951 see TJR000
HYACINTH see ZSS000
HYACINTH ABSOLUTE see HGN500
HYACINTHAL see COF000
HYACINTH BASE see BOF000
HYACINTH BODY see EES380
HYACINTHIN see BBL500
HYACINTH-OF-PERU see SLH200
HYADRINE see BBV500
HYADUR see DUD800
HYALURONIC ACID, SODIUM SALT see HGN600
HYAMINE 1622 see BBU750, BEN000
HYAMINE 2389 see DYA600
HYAMINE 3500 see AFP250, QAT520
HYAMINE see BEN000
HYAMINE 10X see MHB500
HYANILID see SAH500
HYASORB see BFD000
HYBAR X see UNJ800
HYBERNAL see CKP500
HYCANTHON see LIM000
HYCANTHONE see LIM000
HYCANTHONE MESYLATE see HGO500
HYCANTHONE METHANESULFONATE see HGO500
HYCANTHONE METHANESULPHONATE see HGO500
HYCANTHONE MONOMETHANESULPHONATE see HGO500
HYCAR see PJR000

HYCAR LX 407 see SMR000
HY-CHLOR see HOV500
HYCHOTINE see CJR909
HYCLORATE see ARQ750
HYCLORITE see SHU500
HYCORACE see HHR000
HYCORTOLE ACETATE see HHQ800
HYCOZID see ILD000
NHYD see NKJ000
HYDAN (antiseptic) see DFE200
HYDAN see DFE200
HYDANTAL see DKQ000
HYDANTIN SODIUM see DNU000
HYDANTOIN, BIS(2,3-EPOXYPROPYL)-5-ETHYL-5-METHYL- see ECI200
HYDANTOIN, 1,3-DIBROMO-5,5-DIMETHYL- see DDI900
HYDANTOIN, 5-ETHYL-5-PHENYL-, (−)- see EOL050
HYDANTOIN, 1-NITRO- see NHE100
HYDANTOIN, 1-((5-(p-NITROPHENYL)FURFURYLIDENE)AMINO)- see DAB845
HYDANTOIN SODIUM see DNU000
HYDELTRA see PMA000
HYDELTRASOL see PLY275
HYDELTRONE see PMA000
HYDERGIN see DLL400
HYDERGINE see DLL400
HYDOUT see DXD000, EAR000
HYDOXIN see PPK250
HYDRACETIN see ACX750
HYDRACILLIN see PAQ200
HYDRACRYLIC ACID β-LACTONE see PMT100
HYDRACRYLONITRILE see HGP000
HYDRACRYLONITRILE ACRYLATE see ADT111
HYDRAL 705 see AHC000
HYDRAL see CDO000
HYDRALAZINE see HGP495
HYDRALAZINE CHLORIDE see HGP500
HYDRALAZINE HYDROCHLORIDE see HGP500
HYDRALAZINE MONOHYDROCHLORIDE see HGP500
HYDRAL de CHLORAL see CDO000
HYDRALIN see CPB750
HYDRALLAZINE see HGP495
HYDRALLAZINE HYDROCHLORIDE see HGP500
HYDRAM see EKO500
HYDRAMYCIN see HGP550
HYDRANGEA see HGP600
HYDRANGEA MACROPHYLLA see HGP600
HYDRAPHEN see PFN000
HYDRAPRESS see HGP500
HYDRARGAPHEN see PFN000
HYDRARGYRUM BIJODATUM (GERMAN) see MDD000
HYDRASTINE HYDROCHLORIDE see HGQ500
HYDRASTININE see HGR000
HYDRASTIS CANADENSIS L., ROOT EXTRACT see HGR500
HYDRATED ALUMINA see AHC250
HYDRATED LIME see CAT225
HYDRATROP ALDEHYDE see COF000
HYDRATROPIC ACETATE see PGB750
HYDRATROPIC ALCOHOL see HGR600
HYDRATROPIC ALDEHYDE see COF000
HYDRATROPYL ACETATE see PGB750
HYDRATROPYL ALCOHOL see HGR600
HYDRAZID see ILD000
HYDRAZIDE BSG see BBS300
l-HYDRAZIDE CYSTEINE see CQK125
HYDRAZID KYSELINY MALEINOVE see DMC600
HYDRAZINE, anhydrous (DOT) see HGS000
HYDRAZINE see HGS000
HYDRAZINE AQUEOUS SOLUTIONS with >64% hydrazine, by weight (DOT) see HGS000
HYDRAZINE AQUEOUS SOLUTIONS, with not >64% hydrazine, by weight (DOT) see HGU500
HYDRAZINE AZIDE see HGS100
HYDRAZINE, AZIDO- see HGS100
HYDRAZINE-BENZENE see PFI000
HYDRAZINE-BENZENE and BENZIDINE SULFATE see BBY300
HYDRAZINE BISBORANE see HGT500
HYDRAZINE, 1-(p-BROMOPHENYL)-, HYDROCHLORIDE see BNW550
HYDRAZINECARBOHYDRAZONOTHIOIC ACID see TFE250
HYDRAZINECARBOTHIOAMIDE see TFQ000
HYDRAZINECARBOTHIOAMIDE, 2,2′-(1-(1-ETHOXYETHYL)-1,2-ETHANEDIYLIDENE)BIS see KFA100
HYDRAZINECARBOTHIOAMIDE, 2-PHENYL-(9CI) see PGM750
HYDRAZINECARBOTHIOAMIDE, N-PHENYL-(9CI) see PGN000
HYDRAZINECARBOXALDEHYDE see FNN000
HYDRAZINE CARBOXAMIDE see HGU000

HYDRAZINECARBOXAMIDE MONOHYDROCHLORIDE see SBW500
HYDRAZINECARBOXIMIDAMIDE see AKC750
HYDRAZINECARBOXIMIDAMIDE HYDROCHLORIDE see AKC800
HYDRAZINECARBOXYLIC ACID, ETHYL ESTER see EHG000
HYDRAZINE, 1-(2-(o-CHLOROPHENOXY)ETHYL)-, HYDROGEN SULFATE (1:1) see CJP500
HYDRAZINE, 1-(2-(o-CHLOROPHENOXY)ETHYL)-, SULFATE (1:1) see CJP500
HYDRAZINE DICARBONIC ACID DIAZIDE (DOT) see DCL200
1,2-HYDRAZINEDICARBOTHIOAMIDE see BLJ250
HYDRAZINE, (2,5-DICHLOROPHENYL)- see DGC850
HYDRAZINE DIFLUORIDE see HGU100
HYDRAZINE, DIHYDROFLUORIDE see HGU100
HYDRAZINE, 1,1-DI-2-PROPENYL-(9CI) see DBK100
HYDRAZINE, HEPTYL- see HBO600
HYDRAZINE HYDRATE, with not >64% hydrazine, by weight (DOT) see HGU500
HYDRAZINE HYDRATE see HGU500
HYDRAZINE, HYDROCHLORIDE see HGV000
HYDRAZINE HYDROGEN SULFATE see HGW500
HYDRAZINE, 1-(o-METHOXYPHENETHYL)-, SULFATE (1:1) see MFG200
HYDRAZINE, N-((2-METHYLCYCLOPROPYL)CARBONYL)- see MIV300
HYDRAZINE, 1-(α-METHYLPHENETHYL)-2-PHENETHYL- see MNP400
HYDRAZINE, 1-(o-METHYLPHENETHYL)-, SULFATE (1:1) see MNU100
HYDRAZINE, 1-(p-METHYLPHENETHYL)-, SULFATE (1:1) see MNU150
HYDRAZINE, (1-METHYL-2-PHENOXYETHYL)-, (Z)-2-BUTENEDIOATE (1:1) (9CI) see PDV700
HYDRAZINE, 1-(2-(o-METHYLPHENOXY)ETHYL)-, HYDROGEN SULFATE (1:1) see MNR100
HYDRAZINE, (1-METHYL-2-PHENOXYETHYL)-, MALEATE see PDV700
HYDRAZINE, (2-(4-METHYLPHENYL)ETHYL)-, SULFATE (1:1) (9CI) see MNU150
HYDRAZINE MONOBORANE see HGV500
HYDRAZINE MONOCHLORIDE see HGV000
HYDRAZINE MONOSULFATE see HGW500
HYDRAZINE PERCHLORATE (DOT) see HGV600
HYDRAZINE, PERCHLORATE see HGV600
HYDRAZINE, 1-(1-PHENOXY-2-PROPYL)-, MALEATE see PDV700
HYDRAZINE PROPANEMETHANE SULFONATE see HGW000
HYDRAZINE SELENATE see HGW100
HYDRAZINE, SELENATE see HGW100
HYDRAZINE SULFATE (1:1) see HGW500
HYDRAZINE SULPHATE see HGW500
HYDRAZINE, TRIFLUOROSTANNITE see HHA100
HYDRAZINE YELLOW see FAG140
HYDRAZINIUM CHLORATE see HGX000
HYDRAZINIUM CHLORIDE see HGV000
HYDRAZINIUM CHLORITE see HGX500
HYDRAZINIUM DIPERCHLORATE see HGY000
HYDRAZINIUM HYDROGENSELENATE see HGY500
HYDRAZINIUM MONOCHLORIDE see HGV000
HYDRAZINIUM NITRATE see HGZ000
HYDRAZINIUM PERCHLORATE see HHA000
HYDRAZINIUM SULFATE see HGW500
HYDRAZINIUM TRIFLUOROSTANNITE see HHA100
2-HYDRAZINO-4-(4-AMINOPHENYL)THIAZOLE see HHB000
2-HYDRAZINO-4-(p-AMINOPHENYL)THIAZOLE see HHB000
HYDRAZINOBENZENE see PFI000
4-HYDRAZINOBENZOIC ACID see HHB100
p-HYDRAZINOBENZOIC ACID see HHB100
2-HYDRAZINOBENZOTHIAZOLE see HHB500
3-HYDRAZINO-6-(N,N-BIS-(2-HYDROXYETHYL)AMINO) PYRIDAZINE DIHYDROCHLORIDE see BKB500
(S)-α-HYDRAZINO-3,4-DIHYDROXY-α-METHYL-BENZENEPROPANOIC ACID (9CI) see CBQ500
(−)-l-α-HYDRAZINO-3,4-DIHYDROXY-α-METHYLHYDROCINNAMIC ACID MONOHYDRATE see CBQ529
S(−)-α-HYDRAZINO-3,4-DIHYDROXY-α-METHYLHYDROCINNAMIC ACID MONOHYDRATE see CBQ529
2-HYDRAZINOETHANOL see HHC000
3-HYDRAZINO-6-((2-HYDROXYPROPYL)METHYLAMINO)PYRIDAZINE DIHYDROCHLORIDE see HHD000
HYDRAZINO-α-METHYLDOPA see CBQ500
p-HYDRAZINONITROBENZENE see NIR000
2-HYDRAZINO-4-(5-NITRO-2-FURANYL)THIAZOLE see HHD500
2-HYDRAZINO-4-(5-NITRO-2-FURYL)THIAZOLE see HHD500
2-HYDRAZINO-4-(4-NITROPHENYL)THIAZOLE see HHE000
1-HYDRAZINO-2-PHENYLETHANE see PFC500
1-HYDRAZINO-2-PHENYLETHANE HYDROGEN SULPHATE see PFC750
2-HYDRAZINO-1-PHENYLPROPANE see PDN000
1-HYDRAZINOPHTHALAZINE see HGP495
HYDRAZINOPHTHALAZINE see HGP495
1-HYDRAZINOPHTHALAZINE HYDROCHLORIDE see HGP500
1-HYDRAZINOPHTHALAZINE MONOHYDROCHLORIDE see HGP500
4-HYDRAZINO-2-THIOURACIL see HHF500
HYDRAZOBENZEN (CZECH) see HHG000

HYDRAZOBENZENE see HHG000
HYDRAZODIBENZENE see HHG000
HYDRAZODICARBONSAEUREABIS(METHYLNITROSAMID) (GERMAN) see
 DRN400
HYDRAZODICARBOXYLIC ACID BIS(METHYLNITROSAMIDE) see DRN400
HYDRAZODIFORMIC ACID see DKJ600
HYDRAZOETHANE see DJL400
HYDRAZOIC ACID see HHG500
HYDRAZOMETHANE see DSF600, MKN000
HYDRAZONIUM SULFATE see HGW500
2,2'-HYDRAZONODIETHANOL see HHH000
HYDRAZYNA (POLISH) see HGS000
HYDREA see HOO500
HYDREL see HHH100
HYDRIDES see HHH500
HYDRIN-2 see HHQ800
HYDRINDANTIN, anhydrous see HHI000
HYDRIODIC ACID see HHI500
HYDRIODIC ACID, solution (UN 1787) (DOT) see HHI500
HYDRIODIC ETHER see ELP500
HYDRIODIDE-ENTROL see CHR500
HYDRITE see KBB600
HYDRO-AQUIL see CFY000
HYDROAZODICARBOXYBIS(METHYLNITROSAMIDE) see DRN400
HYDROAZOETHANE see DJL400
HYDROBIS(2-METHYLPROPYL)ALUMINUM see DNI600
HYDROBROMIC ACID see HHJ000
HYDROBROMIC ACID MONOAMMONIATE see ANC250
HYDROBROMIC ACID SOLUTION, >49% hydrobromic acid (UN 1788) (DOT)
 see HHJ000
HYDROBROMIC ACID SOLUTION, not >49% hydrobromic acid (UN 1788)
 (DOT) see HHJ000
HYDROBROMIC ETHER see EGV400
HYDROCARB 60 see CAT775
HYDROCARBON GAS see HHJ500
HYDROCARBON GASES, COMPRESSED, N.O.S. (UN 1964) (DOT) see
 HHJ500
HYDROCARBON GASES, LIQUEFIED, N.O.S. (UN 1965) (DOT) see HHJ500
HYDROCARBON GASES MIXTURES, COMPRESSED, N.O.S. (UN 1964)
 (DOT) see HHJ500
HYDROCARBON GASES MIXTURES, LIQUEFIED, N.O.S. (UN 1965) (DOT)
 see HHJ500
HYDROCELLULOSE see HHK000
HYDROCERIN see CMD750
HYDROCHINON (CZECH, POLISH) see HIH000
HYDROCHLORBENZETHYLAMINE DIMALEATE see HHK100
HYDROCHLORIC ACID see HHL000
HYDROCHLORIC ACID DICARBOXIDE see DEK000
HYDROCHLORIC ACID DIMETHYLAMINE see DOR600
HYDROCHLORIC ACID, mixed with NITRIC ACID (3:1) see HHM000
HYDROCHLORIC ACID, solution (UN 1789) (DOT) see HHL000
HYDROCHLORIC ETHER see EHH000
H102/09 HYDROCHLORIDE see ZBA525
R 18553 HYDROCHLORIDE see LIH000
L-67 HYDROCHLORIDE see CMS250
d-8955 HYDROCHLORIDE see DXI800
HYDROCHLORIDE see HHL000
l-HYDROCHLORIDE ARGININE see AQW000
HYDROCHLORIDE of DI-n-BUTYLAMINOPROPYL-3-IODO-4-FLUOROBENZOATE
 see HHM500
HYDROCHLORIDE DIETHYLAMINE see DIS500
HYDROCHLORIDE OF 1-PHENYLCYCLOPENTANECARBOXYLIC ACID DIE-
 THYLAMINOETHYL ESTER see PET250
HYDROCHLORID SALZ des p-N-n-BUTYLAMINO-BENZOESAURE-DIAETHY-
 LAMINOAETHYLESTERS (GERMAN) see BQA000
HYDROCHLOROAURIC ACID, SODIUM SALT see GIZ100
HYDROCHLOROFLUOROCARBON 142b see CFX250
HYDROCHLOROUS ACID, SODIUM SALT, PENTAHYDRATE see SHU525
HYDROCHLORTHIAZID see CFY000
HYDROCINNAMALDEHYDE see HHP000
HYDROCINNAMALDEHYDE, p-ISOPROPYL-α-METHYL-, DIETHYL ACETAL see
 COU510
HYDROCINNAMALDEHYDE, p-ISOPROPYL-α-METHYL-, DIMETHYL ACETAL
 see COU525
HYDROCINNAMIC ACID, α-AMINO- see PEC750
HYDROCINNAMIC ALCOHOL see HHP050
HYDROCINNAMIC ALDEHYDE see HHP000
HYDROCINNAMIQUE NITRILE (FRENCH) see HHP100
HYDROCINNAMONITRILE see HHP100
HYDROCINNAMYL ACETATE see HHP500
HYDROCINNAMYL ALCOHOL see HHP050
HYDROCINNAMYL CINNAMATE see PGB800
HYDROCINNAMYL FORMATE see HHQ000
HYDROCINNAMYL ISOBUTYRATE see HHQ500
HYDROCINNAMYL PROPIONATE see HHQ550

HYDROCODIN see DKW800
HYDROCODONE see OOI000
HYDROCODONE BITARTRATE see DKX050
HYDROCONQUININE see HIG500
11-β-HYDROCORTISONE see CNS750
Δ¹-HYDROCORTISONE see PMA000
HYDROCORTISONE-21-ACETATE see HHQ800
HYDROCORTISONE-17-BUTYRATE see HHQ825
HYDROCORTISONE BUTYRATE see HHQ825
HYDROCORTISONE-17-BUTYRATE-21-PROPIONATE see HHQ850
HYDROCORTISONE 17-α-BUYTRATE see HHQ825
HYDROCORTISONE BUYTRATE PROPIONATE see HHQ850
HYDROCORTISONE FREE ALCOHOL see CNS750
HYDROCORTISONE-21-SODIUM SUCCINATE see HHR000
HYDROCORTISONE SODIUM SUCCINATE see HHR000
HYDROCORTISYL see CNS750
HYDROCORTONE see CNS750
HYDROCOUMARIN see HHR500
HYDROCUPREINE ETHYL ESTER HYDROCHLORIDE see ELC500
HYDROCUPREINE ETHYL ETHER see HHR700
HYDROCYANIC ACID see HHS000
HYDROCYANIC ACID, aqueous solutions <5% HCN (NA 1613) (DOT) see
 HHS000
HYDROCYANIC ACID, ION(CN¹⁻) see COI500
HYDROCYANIC ACID, POTASSIUM SALT see PLC500
HYDROCYANIC ACID (PRUSSIC), unstabilized (DOT) see HHS000
HYDROCYANIC ACID, SODIUM SALT see SGA500
HYDROCYANIC ACID, aqueous solutions not >20% hydrocyanic acid (UN
 1613) (DOT) see HHS000
HYDROCYANIC ETHER see PMV750
HYDROCYCLIN see HOI000
HYDROCYCLINE see MCH525
HYDRODARCO see CBT500
HYDRODELTALONE see PMA000
HYDRODELTISONE see PMA000
HYDRODESUFURIZED MIDDLE DISTILLATE see DXG820
HYDRODESULFURIZED HEAVY VACUUM GAS OIL see GBW010
HYDRODESULFURIZED KEROSINE (PETROLEUM) see KEK110
HYDRODIISOBUTYLALUMINUM see DNI600
HYDRODIURETIC see CFY000
HYDRO-DIURIL see CFY000
HYDROFLUOBORIC ACID see FDD125, HHS600
HYDROFLUORIC ACID see HHU500
HYDROFLUORIC ACID, solution, not >60% strength (UN 1790) (DOT) see
 HHU500
HYDROFLUORIC ACID, solution, >60% strength (UN 1790) (DOT) see
 HHU500
HYDROFLUORIDE see HHU500
HYDROFLUORIDE-1927 WANDER see DCW800
HYDROFLUOSILICIC ACID see SCO500
HYDROFOL see PAE250
HYDROFOL ACID 1255 see LBL000
HYDROFOL ACID 1495 see MSA250
HYDROFOL ACID 1655 see SLK000
HYDROFOL ACID 200 see HOG000
α-HYDROFORMAMINE CYANIDE see HHW000
HYDROFURAMIDE see FPS000
HYDROFURAN see TCR750
HYDROGEN, compressed (DOT) see HHW500
HYDROGEN, refrigerated liquid (DOT) see HHW500
HYDROGEN (DOT) see HHW500
HYDROGEN see HHW500
HYDROGEN ANTIMONIDE see SLQ000
HYDROGEN ARSENIDE see ARK250
HYDROGENATED MDI see MJM600
HYDROGENATED TERPHENYLS see HHW800
HYDROGEN AZIDE see HHG500
HYDROGEN BROMIDE (ACGIH,OSHA,MAK) see HHJ000
HYDROGEN BROMIDE, anhydrous (UN 1048) (DOT) see HHJ000
HYDROGEN CARBOXYLIC ACID see FNA000
HYDROGEN CHLORIDE see HHX000
HYDROGEN CHLORIDE, anhydrous (UN 1050) (DOT) see HHL000
HYDROGEN CHLORIDE, refrigerated liquid (UN 2186) (DOT) see HHL000
HYDROGEN CYANAMIDE see COH500
HYDROGEN CYANIDE (ACGIH,OSHA) see HHS000
HYDROGEN CYANIDE see HHS000
HYDROGEN CYANIDE, anhydrous, stabilized (UN 1051) (DOT) see HHS000
HYDROGEN CYANIDE, anhydrous, stabilized, absorbed in a porous inert
 material (UN 1614) (DOT) see HHS000
HYDROGEN DIMETHYL PHOSPHITE see DSG600
HYDROGEN DIOXIDE see HIB000
HYDROGEN DISULFIDE see HHZ000
HYDROGENE SULFURE (FRENCH) see HIC500
(HYDROGEN(ETHYLENEDINITRILO)TETRAACETATO)IRON see HIA000
HYDROGEN FLUORIDE, anhydrous (UN 1052) (DOT) see HHU500

HYDROGEN HEXACHLOROIRIDATE (4+) see HIA500
HYDROGEN HEXACHLOROPLATINATE(4+) see CKO750
HYDROGEN HEXAFLUOROPHOSPHATE see HDE000
HYDROGEN HEXAFLUOROSILICATE see SCO500
HYDROGEN IODIDE see HHI500
HYDROGEN IODIDE, anhydrous (UN 2197) (DOT) see HHI500
HYDROGEN ISOTHIOCYANATE see ISF100
HYDROGEN NITRATE see NED500
HYDROGEN OXALATE of AMITON see AMX825
HYDROGEN PEROXIDE, 8% to 20% see HIB005
HYDROGEN PEROXIDE, 30% see HIB010
HYDROGEN PEROXIDE, solution, 30% see HIB010
HYDROGEN PEROXIDE, 90% see HIB000
HYDROGEN PEROXIDE, solution, 8% to 20% (DOT) see HIB005
HYDROGEN PEROXIDE, stabilized with >60% hydrogen peroxide (DOT) see HIB000
HYDROGEN PEROXIDE CARBAMIDE see HIB500
HYDROGEN PEROXIDE & PEROXYACETIC ACID MIXT., with acids, H2O not >5% peroxyacetic acid (DOT) see PCL500
HYDROGEN PEROXIDE, aqueous solutions with >40%, not >60% hydrogen peroxide (UN 2014) see HIB010
HYDROGEN PEROXIDE, aqueous solutions with not <8% but <20% hydrogen peroxide (UN 2984) (DOT) see HIB010
HYDROGEN PEROXIDE with UREA (1:1) see HIB500
HYDROGEN PHOSPHIDE see PGY000
HYDROGEN POTASSIUM FLUORIDE see PKU250
HYDROGEN SELENIDE, anhydrous (DOT) see HIC000
HYDROGEN SELENIDE see HIC000
21-(HYDROGEN SUCCINATE)CORTISOL, MONOSODIUM SALT see HHR000
HYDROGEN SULFIDE see HIC500
HYDROGEN SULFITE see HIC600
HYDROGEN SULFITE SODIUM see SFE000
HYDROGEN SULFURIC ACID see HIC500
HYDROGEN TETRAFLUOROBORATE see FDD125, HHS600
HYDROGEN TRISULFIDE see HID000
α-HYDRO-ω-HYDROXY-POLY(OXY-1,2-ETHANEDIYL) see PJT500
α-HYDRO-omega-HYDROXYPOLY(OXY-1,2-ETHANEDIYL) see PJT000
HYDROL see DBI800
HYDROLIN see SHR500
HYDROLIT see FMW000
HYDROLOSE see MIF760
HYDROL SW see PKF500
HYDRO-MAG MA see MAG750
HYDROMAGNESITE see MAC650
HYDROMEDIN see DFP600
HYDROMIREX see MRI750
HYDROMORPHONE see DLW600
HYDROMORPHONE HYDROCHLORIDE see DNU300
HYDROMOX R see RDK000
HYDRONITRIC ACID see HHG500
HYDRONOL see HID350
HYDRONSAN see GBB500
HYDROOT see SOI500
HYDROPERIT see HIB500
HYDROPEROXIDE see HIB000
HYDROPEROXIDE, ACETYL see PCL500
HYDROPEROXIDE, ETHYL see ELD000
6-β-HYDROPEROXYCHOLEST-4-EN-3-ONE see HID500
6-β-HYDROPEROXY-4-CHOLESTEN-3-ONE see HID500
6-β-HYDROPEROXY-Δ⁴CHOLESTEN-3-ONE see HID500
1-HYDROPEROXY-3-CYCLOHEXENE see HIE000
1-HYDROPEROXYCYCLOHEX-3-ENE see HIE000
1-HYDROPEROXYCYCLOHEXYL-1-HYDROXYCYCLOHEXYL PEROXIDE see CPC300
4-HYDROPEROXYCYCLOPHOSPHAMIDE see HIF000
HYDROPEROXYDE de BUTYLE TERTIAIRE (FRENCH) see BRM250
HYDROPEROXYDE de CUMENE (FRENCH) see IOB000
HYDROPEROXYDE de CUMYLE (FRENCH) see IOB000
4-HYDROPEROXYIFOSFAMIDE see HIE550
N-(HYDROPEROXYMETHYL)-N-NITROSOPROPYLAMINE see HIE570
2-HYDROPEROXY-2-METHYLPROPANE see BRM250
HYDROPEROXY-N-NITROSODIBUTYLAMINE see HIE600
1-(HYDROPEROXY)-N-NITROSODIMETHYLAMINE see HIE700
4-HYDROPEROXYPHOSPHAMIDE see HIF000
2-HYDROPEROXYPROPANE see IPI000
4-HYDROPEROXY-4-VINYL-1-CYCLOHEXENE see HIF575
1-HYDROPEROXY-1-VINYLCYCLOHEX-3-ENE see HIF575
HYDROPHENOL see CPB750
HYDROPHIS CYANOCINCTUS VENOM see SBI900
HYDROPHIS ELEGANS (AUSTRALIA) VENOM see HIG000
HYDROPRENE see EQD000
HYDROPRES see RDK000
HYDROPRES KA see RDK000
HYDROQUINIDINE see HIG500
HYDROQUINOL see HIH000

HYDROQUINOLE see HIH000
α-HYDROQUINONE see HIH000
HYDROQUINONE, liquid or solid (DOT) see HIH000
m-HYDROQUINONE see REA000
o-HYDROQUINONE see CCP850
p-HYDROQUINONE see HIH000
HYDROQUINONE see HIH000
HYDROQUINONE, compounded with p-BENZOQUINONE see QFJ000
HYDROQUINONE BENZYL ETHER see AEY000
HYDROQUINONE, BROMO- see BNL260
HYDROQUINONE CALCIUM SULFONATE see DMI300
HYDROQUINONECARBOXYLIC ACID see GCU000
HYDROQUINONE, 2,5-DI-tert-BUTYL- see DEC800
HYDROQUINONE MONOBENZYL ETHER see AEY000
HYDROQUINONE MONOETHYL ETHER see EFA100
HYDROQUINONE MONOMETHYL ETHER see MFC700
HYDROQUINONE MUSTARD see BHB750
HYDROQUINONE, PHENYL- see BGG250
HYDRO-RAPID see CHJ750
HYDRORETROCORTIN see PMA000
HYDRORUBEANIC ACID see DXO200
HYDROSALURIC see CFY000
HYDROSARPAN see AFG750
HYDROSCINE HYDROBROMIDE see HOT500
HYDROSILICOFLUORIC ACID see SCO500
HYDROSORBIC ACID see HFD000
HYDROSULFITE ANION see HIC600
HYDROSULFITE AWC see FMW000
HYDROTHAL-47 see EAR000
HYDROTHIADENE see HII000
HYDROTHIDE see CFY000
HYDROTHOL see DXD000
HYDROTREATED (mild) HEAVY NAPHTHENIC DISTILLATE see MQV790
HYDROTREATED (mild) HEAVY NAPHTHENIC DISTILLATES (PETROLEUM) see MQV790
HYDROTREATED (mild) HEAVY PARAFFINIC DISTILLATE see MQV795
HYDROTREATED KEROSENE see KEK100
HYDROTREATED (mild) LIGHT NAPHTHENIC DISTILLATE see MQV800
HYDROTREATED (mild) LIGHT NAPHTHENIC DISTILLATES (PETROLEUM) see MQV800
HYDROTREATED (mild) LIGHT PARAFFINIC DISTILLATE see MQV805
HYDROTREATED NAPHTHA see NAI500
HYDROTRICHLOROTHIAZIDE see HII500
HYDROTRICINE see TOG500
HYDROTROPALDEHYDE DIMETHYL ACETAL see HII600
HYDROTROPIC ALDEHYDE DIMETHYL ACETAL see HII600
HYDROXAMIC ACID, BENZSULFO- see BDW100
HYDROXINE see CJR909
HYDROXINE YELLOW L see FAG140
HYDROXON see HNT500
N-HYDROXY-AABP see ACD000
N-HYDROXY-AAF see HIP000
trans-4′-HYDROXY-AAS see HIK000
12-HYDROXYABIETIC ACID BIS(2-CHLOROETHYL)AMINE see HIJ000
12-HYDROXYABIETIC ACID BIS(2-CHLOROETHYL)AMINE SALT see HIJ000
N-HYDROXY-4-ACETAMIDOBIPHENYL see ACD000
N-4-(N-HYDROXYACETAMIDO)BIPHENYL see ACD000
N-HYDROXY-4-ACETAMIDODIPHENYL see ACD000
1-HYDROXY-2-ACETAMIDOFLUORENE see HIJ400
N-HYDROXY-2-ACETAMIDOFLUORENE see HIP000
2-(N-HYDROXYACETAMIDO)FLUORENE see HIP000
2-(N-HYDROXYACETAMIDO)NAPHTHALENE see NBD000
2-(N-HYDROXYACETAMIDO)PHENANTHRENE see PCZ000
HYDROXY-2-(β-(2′-ACETAMIDOPHENYL))ETHYLNAPHTHAMIDE see HIJ500
4-(N-HYDROXYACETAMIDO)STILBENE see SMT000
trans-4′-HYDROXY-4-ACETAMIDOSTILBENE see HIK000
trans-N-HYDROXY-4-ACETAMIDOSTILBENE see SMT500
HYDROXY(o-ACETAMINOBENZOATO)MERCURY SODIUM SALT see SJP500
7-HYDROXY-2-ACETAMINOFLUORENE see HIK500
2′-HYDROXYACETANILIDE see HIL000
3-HYDROXYACETANILIDE see HIL500
4′-HYDROXYACETANILIDE see HIM000
4-HYDROXYACETANILIDE see HIM000
m-HYDROXYACETANILIDE see HIL500
o-HYDROXYACETANILIDE see HIL000
p-HYDROXYACETANILIDE see HIM000
HYDROXYACETIC ACID see GHO000
HYDROXYACETIC ACID ETHYL ESTER see EKM500
HYDROXYACETIC ACID, MONOSODIUM SALT see SHT000
HYDROXYACETONE see ABC000
2-HYDROXYACETONITRILE see HIM500
HYDROXYACETONITRILE see HIM500
N-HYDROXY-p-ACETOPHENETIDIDE see HIN000
2′-HYDROXYACETOPHENONE see HIN500
4′-HYDROXYACETOPHENONE see HIO000

o-HYDROXYACETOPHENONE see HIN500
p-HYDROXYACETOPHENONE see HIO000
N-HYDROXY-4-ACETYLAMINOBIBENZYL see PDI250
N-HYDROXY-4-ACETYLAMINOBIPHENYL see ACD000
3-HYDROXY-N-ACETYL-2-AMINOFLUORENE see HIO875
N-HYDROXY-N-ACETYL-2-AMINOFLUORENE see HIP000
N-HYDROXY-2-ACETYLAMINOFLUORENE see HIP000
N-HYDROXY-2-ACETYLAMINOFLUORENE, COBALTOUS CHELATE see
　　FDU875
N-HYDROXY-2-ACETYLAMINOFLUORENE, CUPRIC CHELATE see HIP500
N-HYDROXY-2-ACETYLAMINOFLUORENE, FERRIC CHELATE see HIQ000
N-HYDROXY-2-ACETYLAMINOFLUORENE-o-GLUCURONIDE see HIQ500
N-HYDROXY-2-ACETYLAMINOFLUORENE, MANGANOUS CHELATE see
　　HIR000
N-HYDROXY-2-ACETYLAMINOFLUORENE, NICKELOUS CHELATE see HIR500
N-HYDROXY-2-ACETYLAMINOFLUORENE, POTASSIUM SALT see HIS000
N-HYDROXY-2-ACETYLAMINOFLUORENE, ZINC CHELATE see ZHJ000
N-HYDROXY-2-ACETYLAMINONAPHTHALENE see NBD000
N-HYDROXY-2-ACETYLAMINOPHENANTHRENE see PCZ000
N-HYDROXY-4-ACETYLAMINOSTILBENE see SMT000
trans-N-HYDROXY-4-ACETYLAMINOSTILBENE see SMT500
trans-N-HYDROXY-4-ACETYLAMINOSTILBENE CUPRIC CHELATE see SMU000
2-(HYDROXYACETYL)INDOLE see HIS100
5-(HYDROXYACETYL)INDOLE see HIS110
2-(2-HYDROXYACETYL)-1-METHYLINDOLE see HIS120
2-(HYDROXYACETYL)-1-METHYLINDOLE see HIS120
3-(HYDROXYACETYL)-1-METHYLINDOLE see HIS140
3-(2-HYDROXYACETYL)-2-METHYLINDOLE see HIS150
2-(HYDROXYACETYL)-3-METHYLINDOLE see HIS130
2-HYDROXYACLACINOMYCIN A see HIS300
3-(2-HYDROXY-1-ADAMANTYL)-N-METHYLPROPLYAMINE HYDROCHLORIDE
　　see MGK750
N-HYDROXYADENINE see HIT000
6-N-HYDROXYADENOSINE see HIT100
6-HYDROXYADENOSINE see HIT100
N⁶-HYDROXYADENOSINE see HIT100
N-HYDROXYADENOSINE see HIT100
2-HYDROXYADIPALDEHYDE see HIT500
α-HYDROXYADIPALDEHYDE see HIT500
HYDROXYADIPALDEHYDE see HIT500
2-(2-HYDROXYAETHOXY)AETHYLESTER der FLUTENAMINSAEURE (GERMAN)
　　see HKK000
4-HYDROXYAFLATOXIN B1 see AEW000
β-HYDROXYALANINE see SCA355
(−)-3-HYDROXY-N-ALLYLMORPHINAN see AGI000
l-3-HYDROXY-N-ALLYL MORPHINAN see AGI000
1-HYDROXY-4-AMINOANTHRAQUINONE see AKE250
2-HYDROXY-4-AMINOBENZOIC ACID see AMM250
N-HYDROXY-4-AMINOBIPHENYL see BGI250
4-HYDROXYAMINOBIPHENYL see BGI250
β-HYDROXY-α-AMINOBUTYRIC ACID see AKF375
4-HYDROXY-3-AMINODIPHENYL HYDROCHLORIDE see AIW250
3-HYDROXY-4-AMINODIPHENYL SULPHATE see AKF250
5-HYDROXY-3-(β-AMINOETHYL)INDOLE see AJX500
N-HYDROXY-2-AMINOFLUORENE see HIU500
N-HYDROXY-3-AMINOFLUORENE see FEH000
N-HYDROXY-2-AMINOFLUORENE N-GLUCURONIDE see HLE750
2-HYDROXY-4-AMINO-5-FLUOROPYRIMIDINE see FHI000
3-HYDROXY-5-AMINOMETHYLISOXAZOLE see AKT750
3-HYDROXY-5-AMINOMETHYLISOXAZOLE-AGARIN see AKT750
4-(HYDROXYAMINO)-5-METHYLQUINOLINE-1-OXIDE see HIV000
4-(HYDROXYAMINO)-6-METHYLQUINOLINE-1-OXIDE see HIV500
4-(HYDROXYAMINO)-7-METHYLQUINOLINE-1-OXIDE see HIW000
4-(HYDROXYAMINO)-8-METHYLQUINOLINE-1-OXIDE see HIW500
N-HYDROXY-1-AMINONAPHTHALENE see HIX000
N-HYDROXY-2-AMINONAPHTHALENE see NBI500
2-HYDROXYAMINO-1,4-NAPHTHOQUINONE see HIX200
3-HYDROXY-4-AMINONITROBENZENE see ALO000
4-(HYDROXYAMINO)-6-NITROQUINOLINE-1-OXIDE see HIX500
4-(HYDROXYAMINO)-7-NITROQUINOLINE-1-OXIDE see HIY000
N-(2-(HYDROXYAMINO)-2-OXOETHYL)-2-THIOPHENECARBOXAMIDE see
　　TEN725
dl-α-HYDROXY-β-AMINOPROPYLBENZENE see NNM500
α-HYDROXY-β-AMINOPROPYLBENZENE HYDROCHLORIDE see PMJ500
6-HYDROXYAMINOPURINE see HIT000
4-(HYDROXYAMINO)-6-QUINOLINECARBOXYLIC ACID-1-OXIDE see CCF750
4-(HYDROXYAMINO)QUINOLINE-1-OXIDE see HIY500
4-(HYDROXYAMINO)QUINOLINE-1-OXIDE HYDROCHLORIDE see HIZ000
4-(HYDROXYAMINO)QUINOLINE-1-OXIDE, HYDROCHLORIDE see HIZ000
4-(HYDROXYAMINO)QUINOLINE-1-OXIDE MONOHYDROCHLORIDE see HIZ000
4-HYDROXYAMINOQUINOLINE-1-OXIDE MONOHYDROCHLORIDE see HIZ000
trans-N-HYDROXY-4-AMINOSTILBENE see HJA000
HYDROXYAMMONIUM CHLORIDE see HLN000
17-β-HYDROXY-5-β-ANDROSTAN-3-ONE see HJB100
3-β-HYDROXY-5-ANDROSTEN-17-ONE see AOO450

17-HYDROXY-(17-β)-ANDROST-4-EN-3-ONE see TBF500
17-β-HYDROXY-4-ANDROSTEN-3-ONE see TBF500
17-β-HYDROXYANDROST-4-EN-3-ONE see TBF500
7-β-HYDROXYANDROST-4-EN-3-ONE see TBF500
17-β-HYDROXY-Δ⁴-ANDROSTEN-3-ONE see TBF500
3-β-HYDROXYANDROST-5-EN-17-ONE ESTER with SODIUM SULFATE DIHY-
　　DRATE see DAL030
3-β-HYDROXY-5-ANDROSTEN-17-ONE SODIUM SULFATE DIHYDRATE see
　　SJK410
2-HYDROXYANILINE see ALT000
3-HYDROXYANILINE see ALT500
4-HYDROXYANILINE see ALT250
o-HYDROXYANILINE see ALT000
p-HYDROXYANILINE see ALT250
2-HYDROXY-m-ANISALDEHYDE see VFP000
4-HYDROXY-m-ANISALDEHYDE see VFK000
3-HYDROXY-p-ANISALDEHYDE see FNM000
3-HYDROXYANISIC ACID see HJC000
2-HYDROXY-m-ANISIC ACID see HJB500
3-HYDROXY-m-ANISIC ACID see HJB500
4-HYDROXY-m-ANISIC ACID see VFF000
2-HYDROXYANISOLE see GKI000
3-HYDROXYANISOLE see REF050
m-HYDROXYANISOLE see REF050
o-HYDROXYANISOLE see GKI000
1-HYDROXY-9,10-ANTHRACENEDIONE see HJE000
1-HYDROXYANTHRACHINON (CZECH) see HJE000
3-HYDROXYANTHRANILIC ACID see AKE750
3-HYDROXYANTHRANILIC ACID METHYL ESTER see HJC500
3-(3-HYDROXYANTHRANILOYL)ALANINE see HJD000
3-(3-HYDROXYANTHRANILOYL)-l-ALANINE see HJD500
3-HYDROXY-ANTHRANILSAEURE (GERMAN) see AKE750
1-HYDROXYANTHRAQUINONE see HJE000
1-HYDROXY-9,10-ANTHRAQUINONE see HJE000
4-HYDROXY-1-ANTHRAQUINONYLAMINE see AKE250
N-(4-HYDROXY-1-ANTHRAQUINONYL)-4-METHYLANILINE see HOK250
N-(4-HYDROXY-1-ANTHRAQUINONYL)-p-TOLUIDINE see HOK000
4-HYDROXY-3-ARSANILIC ACID see HJE500
4-HYDROXY-m-ARSANILIC ACID see HJE500
2-HYDROXY-p-ARSANILIC ACID see HJE400
trans-N-HYDROXY-ASS see SMT500
HYDROXYATHYLSTARKE (GERMAN) see HLB400
3-HYDROXY-8-AZAXANTHINE see HJE575
14-HYDROXYAZIDOMORPHINE see HJE600
4-HYDROXYAZOBENZENE see HJF000
p-HYDROXYAZOBENZENE see HJF000
4-HYDROXY-3,4'-AZODI-1-NAPHTHALENESULFONIC ACID, DISODIUM SALT
　　see HJF500
4-HYDROXY-3,4'-AZODI-1-NAPHTHALENESULPHONIC ACID, DISODIUM SALT
　　see HJF500
2-HYDROXY-1,1'-AZONAPHTHALENE-3,6,4'-TRISULFONIC ACID TRISODIUM
　　SALT see FAG020
4'-HYDROXY-2,3'-AZOTOLUENE see HJG000
2-HYDROXYBENZALDEHYDE see SAG000
4-HYDROXYBENZALDEHYDE see FOF000
o-HYDROXYBENZALDEHYDE see SAG000
p-HYDROXYBENZALDEHYDE see FOF000
o-HYDROXYBENZALDEHYDE OXIME see SAG500
2-HYDROXYBENZAMIDE see SAH000
N-HYDROXYBENZAMIDE see BCL500
o-HYDROXYBENZAMIDE see SAH000
o-HYDROXYBENZANILIDE see SAH500
3-HYDROXY-1,2-BENZANTHRACENE see BBI000
5-HYDROXYBENZ(a)ANTHRACENE see BBI000
HYDROXYBENZENE see PDN750
4-HYDROXYBENZENEACETIC ACID see HNG600
α-HYDROXYBENZENEACETIC ACID 2-(2-ETHOXYETHOXY)ETHYL ESTER see
　　HJG100
α-HYDROXY-BENZENEACETIC ACID 3,3,5-TRIMETHYLCYCLOHEXYL ESTER
　　(9CI) see DNU100
p-HYDROXYBENZENEARSONIC ACID see PDO250
4-HYDROXYBENZENEDIAZONIUM-3-CARBOXYLATE see HJH000
N-HYDROXYBENZENESULFONANILIDE see HJH100
HYDROXYBENZENESULFONIC ACID see HJH500
p-HYDROXY-BENZENESULFONIC ACID MERCURY DERIVATIVE, DISODIUM
　　SALT see MCU500
p-HYDROXYBENZENESULFONIC ACID ZINC SALT see ZIJ300
o-HYDROXYBENZHYDRAZIDE see HJL100
2-HYDROXYBENZHYDROXAMIC ACID see SAL500
3-HYDROXYBENZISOTHIAZOL-S,S-DIOXIDE see BCE500
2-HYDROXY-1,3,2-BENZODIOXASTIBOLE see HJI000
2-HYDROXYBENZOHYDRAZIDE see HJL100
2-HYDROXYBENZOHYDROXAMIC ACID see SAL500
o-HYDROXYBENZOHYDROXAMIC ACID see SAL500
2-HYDROXYBENZOIC ACID see SAI000

3-HYDROXYBENZOIC ACID see HJI100
4-HYDROXYBENZOIC ACID see SAI500
m-HYDROXYBENZOIC ACID see HJI100
o-HYDROXYBENZOIC ACID see SAI000
p-HYDROXYBENZOIC ACID see SAI500
2-HYDROXYBENZOIC ACID BUTYL ESTER see BSL250
p-HYDROXYBENZOIC ACID BUTYL ESTER see BSC000
p-HYDROXYBENZOIC ACID BUTYL ESTER, SODIUM SALT see HJI500
p-HYDROXYBENZOIC ACID ETHYL ESTER see HJL000
p-HYDROXYBENZOIC ACID HEPTYL ESTER see HBO650
2-HYDROXYBENZOIC ACID HYDRAZIDE see HJL100
o-HYDROXYBENZOIC ACID HYDRAZIDE see HJL100
2-HYDROXYBENZOIC ACID METHYL ESTER see MPI000
o-HYDROXYBENZOIC ACID, METHYL ESTER see MPI000
p-HYDROXYBENZOIC ACID METHYL ESTER see HJL500
2-HYDROXYBENZOIC ACID MONOAMMONIUM SALT see ANT600
2-HYDROXYBENZOIC ACID MONOSODIUM SALT see SJO000
4-HYDROXYBENZOIC ACID ((5-NITRO-2-FURANYL)METHYLENE)HYDRAZIDE
    see DGQ500
p-HYDROXYBENZOIC ACID (5-NITROFURFURYLIDENE)HYDRAZIDE see
    DGQ500
4-HYDROXYBENZOIC ACID PROPYL ESTER see HNU500
p-HYDROXYBENZOIC ACID PROPYL ESTER see HNU500
p-HYDROXYBENZOIC ACID, PROPYL ESTER, SODIUM DERIVATIVE see
    PNO250
p-HYDROXYBENZOIC ACID, SODIUM SALT see HJM000
2-HYDROXYBENZOIC ACID STRONTIUM SALT (2:1) see SML500
p-HYDROXYBENZOIC ETHYL ESTER see HJL000
o-HYDROXYBENZOIC SODIUM SALT see SJO000
2-HYDROXYBENZOIC-5-SULFONIC ACID see SOC500
4-HYDROXYBENZONITRILE see HJN000
p-HYDROXYBENZONITRILE see HJN000
4-HYDROXY-2H-1-BENZOPYRAN-2-ONE see HJY000
11-HYDROXYBENZO(a)PYRENE see BCY750
12-HYDROXYBENZO(a)PYRENE see BCZ000
10-HYDROXYBENZO(a)PYRENE see BCY500
2-HYDROXYBENZO(a)PYRENE see BCX000
3-HYDROXYBENZO(a)PYRENE see BCX250
5-HYDROXYBENZO(a)PYRENE see BCX500
6-HYDROXYBENZO(a)PYRENE see BCX750
7-HYDROXYBENZO(a)PYRENE see BCY000
9-HYDROXYBENZO(a)PYRENE see BCY250
HYDROXYBENZOPYRIDINE see QPA000
1-HYDROXY-1H-BENZOTRIAZOLE AMMONIUM SALT see HJN600
1-HYDROXY-1H-BENZOTRIAZOLE HYDRATE see HJN650
1-HYDROXYBENZOTRIAZOLE HYDRATE see HJN650
N-HYDROXYBENZOTRIAZOLE HYDRATE see HJN650
2-HYDROXYBENZOXAZOLE see BDJ000
N-HYDROXY-2-BENZOYLAMINOFLUORENE see FDZ000
HYDROXY-4 BENZOYL-2-FURANNE see FQL200
N-o-HYDROXYBENZOYLGLYCINE see SAN200
N-(2-HYDROXYBENZOYL)-GLYCINE (9CI) see SAN200
2-HYDROXYBENZOYLHYDRAZIDE see HJL100
o-HYDROXYBENZOYLHYDRAZIDE see HJL100
2-HYDROXYBENZOYLHYDRAZINE see HJL100
o-HYDROXYBENZOYLHYDRAZINE see HJL100
(HYDROXY-4 BENZOYL)-3 MESITYL-2 BENZOFURANNE see HNK700
(HYDROXY-4 BENZOYL)-4 MESITYL-2 BENZOFURANNE see HNK800
2-(p-HYDROXYBENZOYL)PYRIDINE see HJN700
8-HYDROXY-3,4-BENZPYRENE see BCX250
p-HYDROXYBENZYL ACETONE see RBU000
5-((p-HYDROXYBENZYLIDENE)AMINO)-3-METHYLISOTHIAZOLO(5,4-
    d)PYRIMIDINE-4,6(5H,7H)-DIONE see IGE100
5-((4'-HYDROXYBENZYLIDENOIMINO)-3-METHYLISOTHIAZOLO(5,4-
    d)PYRIMIDINE-(7H)-4,6)-DIONE see IGE100
α-HYDROXYBENZYL PHENYL KETONE see BCP250
(p-HYDROXYBENZYL)TARTARIC ACID see HJO500
2-HYDROXYBIFENYL (CZECH) see BGJ250
2-HYDROXYBIPHENYL see BGJ250
4-HYDROXYBIPHENYL see BGJ500
o-HYDROXYBIPHENYL see BGJ250
p-HYDROXYBIPHENYL see BGJ500
N-HYDROXY-N-4-BIPHENYLACETAMIDE see ACD000
N-HYDROXY-4-BIPHENYLBENZAMIDE see HJP500
2-HYDROXYBIPHENYL SODIUM SALT see BGJ750
N-HYDROXY-4-BIPHENYLYLBENZAMIDE see HJP500
HYDROXY-4-BIPHENYLYLBENZENESULFONAMIDE see BGN000
HYDROXYBIS(2-(p-CHLOROPHENOXY)ISOBUTYRIC ACID) ALUMINUM see
    AHA150
12'-HYDROXY-2',5'-α-BIS(1-METHYLETHYL)ERGOTAMAN-3',6',18-TRIONE see
    EDA600
2-HYDROXY-4,6-BIS(NITROAMINO)-1,3,5-TRIAZINE see HJP575
1-((1R)-2-endo-HYDROXY-3-endo-BORNYL)-3-(p-TOLYLSULFONYL)UREA see
    GFY100
3-HYDROXYBUTANAL see AAH750

1-HYDROXYBUTANE see BPW500
2-HYDROXYBUTANE see BPW750
HYDROXYBUTANEDIOIC ACID see MAN000
4-HYDROXYBUTANOIC ACID LACTONE see BOV000
3-HYDROXYBUTANOIC ACID-β-LACTONE see BSX000
4-HYDROXYBUTANOIC ACID MONOLITHIUM SALT see LHM800
3-HYDROXY-2-BUTANONE see ABB500
2-HYDROXY-3-BUTENENITRILE see HJQ000
1-(2-HYDROXYBUT-1-ENYL)AZIRIDINE see VMA000
o-(2-HYDROXY-3-(tert-BUTYLAMINO)PROPOXY)BENZONITRILE HYDROCHLO-
    RIDE see BON400
1-HYDROXY-4-tert-BUTYLBENZENE see BSE500
4-HYDROXYBUTYLBUTYLNITROSAMINE see HJQ350
1-HYDROXY-3-BUTYL HYDROPEROXIDE see HJQ500
4-((4-HYDROXYBUTYL)NITROSAMINO)BUTYRIC ACID METHYL ESTER ACE-
    TATE (ESTER) see MEI000
4-HYDROXYBUTYL(2-PROPENYL)NITROSAMINE see HJS400
3-HYDROXYBUTYRALDEHYDE see AAH750
β-HYDROXYBUTYRALDEHYDE see AAH750
γ-HYDROXYBUTYRATE SODIUM SALT see HJS500
γ-HYDROXYBUTYRIC ACID CYCLIC ESTER see BOV000
3-HYDROXYBUTYRIC ACID-2-HEPTYL ESTER see MKM750
3-HYDROXYBUTYRIC ACID LACTONE see BSX000
4-HYDROXYBUTYRIC ACID γ-LACTONE see BOV000
HYDROXYBUTYRIC ACID LACTONE see BSX000
β-HYDROXYBUTYRIC ACID-p-PHENETIDIDE see HJS850
4-HYDROXYBUTYRIC ACID SODIUM SALT see HJS500
γ-HYDROXYBUTYROLACTONE see BOV000
3-HYDROXY-p-BUTYROPHENETIDIDE see HJS850
2-HYDROXYCAMPHANE see BMD000
HYDROXYCARBAMIC ACID ETHYL ESTER see HKQ025
HYDROXYCARBAMINE see HOO500
m-HYDROXYCARBANILIC ACID ETHYL ESTER see ELE600
m-HYDROXYCARBANILIC ACID METHYL ESTER m-METHYLCARBANILATE
    see MEG250
3-HYDROXY-4-CARBOXYANILINE see AMM250
cis-4-HYDROXY-CCNU see CHA500
trans-4-HYDROXY-CCNU see CHB000
HYDROXYCELLULOSE see CCU150
8-HYDROXY-CHINOLIN (GERMAN) see QPA000
8-HYDROXY-CHINOLIN-SULFAT (GERMAN) see QPS000
2-HYDROXY-5-CHLORO-N-(2-CHLORO-4-NITROPHENYL)BENZAMIDE see
    DFV400
4-(4-HYDROXY-4'-CHLORO-4-PHENYLPIPERIDINO)-4'-FLUOROBUTYROPHE-
    NONE see CLY500
2-(3-(4-HYDROXY-4-p-CHLOROPHENYLPIPERIDINO)-PROPYL)-3-METHYL-7-
    FLUOROCHROMONE see HJU000
HYDROXYCHLOROQUINE see PJB750
3-α-HYDROXYCHOLANIC ACID see LHW000
3-α-HYDROXY-5-β-CHOLANIC ACID see LHW000
(3-α,5-β)-3-HYDROXY-CHOLAN-24-OIC ACID see LHW000
(3-α,5-β)-3-HYDROXYCHOLAN-24-OIC ACID, MONOSODIUM SALT see SIC500
3-α-HYDROXY-5-β-CHOLAN-24-OIC ACID, MONOSODIUM SALT see SIC500
1-α-HYDROXYCHOLECALCIFEROL see HJV000
1-HYDROXYCHOLECALCIFEROL see HJV000
HYDROXYCHOLECALCIFEROL see HJV000
3-β-HYDROXYCHOLESTANE see DKW000
3-β-HYDROXYCHOLEST-5-ENE see CMD750
6-HYDROXYCHOLEST-4-EN-3-ONE see HJV500
3-β-HYDROXYCHOLEST-5-EN-7-ONE (8CI) see ONO000
(3-β)3-HYDROXYCHOLEST-5-EN-7-ONE (9CI) see ONO000
3-HYDROXYCINCHOPEN see OPK300
HYDROXYCINCHOPHENE see OPK300
HYDROXYCINE see CJR909
4-HYDROXYCINNAMIC ACID see CNU825
4'-HYDROXYCINNAMIC ACID see CNU825
p-HYDROXYCINNAMIC ACID see CNU825
trans-2-HYDROXYCINNAMIC ACID see CNU850
trans-o-HYDROXYCINNAMIC ACID see CNU850
o-HYDROXYCINNAMIC ACID LACTONE see CNV000
7-HYDROXYCITRONELLAL see CMS850
HYDROXYCITRONELLAL (FCC) see CMS850
HYDROXYCITRONELLAL DIMETHYL ACETAL see HJV700, LBO050
HYDROXYCITRONELLAL DMA see LBO050
HYDROXYCITRONELLAL-INDOLE (SCHIFF BASE) see ICS100
HYDROXYCITRONELLA METHYL ETHER see DSM800
HYDROXYCITRONELLOL see DTE400
HYDROXYCITRONELLYLIDENE-INDOLE see ICS100
HYDROXY CNU METHANESULPHONATE see HKQ300
14-β-HYDROXYCODEINONE see HJX500
14-HYDROXYCODEINONE see HJX500
HYDROXYCODEINONE see HJX500
HYDROXYCOPPER(II) GLYOXIMATE see HJX625
17-HYDROXYCORTICOSTERONE see CNS750
17-α-HYDROXYCORTICOSTERONE ACETATE see HHQ800

17-HYDROXY-CORTICOSTERONE 21-ACETATE see HHQ800
11-β-HYDROXYCORTISONE see CNS750
4-HYDROXYCOUMARIN see HJY000
(HYDROXY-4 COUMARINYL 3)-3 PHENYL-3 (BROMO-4 BIPHENYLYL-4)-1 PROPANOL-1 (FRENCH) see BMN000
(E)-3-HYDROXY-CROTONIC ACID α-METHYLBENZYL ESTER, DIMETHYL PHOSPHATE see COD000
3-HYDROXYCROTONIC ACID METHYL ESTER DIMETHYL PHOSPHATE see MQR750
1-HYDROXYCUMENE see DTN100
(4-HYDROXY-o-CUMENYL)TRIMETHYLAMMONIUM CHLORIDE, METHYLCARBAMATE see HJY500
(4-HYDROXY-o-CUMENYL)TRIMETHYLAMMONIUM IODIDE, DIMETHYLCARBAMATE see HJZ000
3-HYDROXY-CYCLOBUT-3-ENE-1,2-DIONE see MRE250
1-HYDROXY-CYCLOBUT-1-ENE-3,4-DIONE see MRE250
1-HYDROXYCYCLOHEPTANECARBONITRILE see HKA000
3-HYDROXYCYCLOHEXADIEN-1-ONE see REA000
HYDROXYCYCLOHEXANE see CPB750
1-HYDROXY-CYCLOHEXANECARBONITRILE see HKA000
2-HYDROXYCYCLOHEXANE-1,1,3,3-TETRAMETHANOL TETRAESTER with NICOTINIC ACID see CME675
N-(trans-4-HYDROXYCYCLOHEXYL)-(2-AMINO-3,5-DIBROMOBENZYL)-AMINE see AHJ250
α-(1-HYDROXYCYCLOHEXYL)BUTYRIC ACID see COW700
4-HYDROXYCYCLOPHOSPHAMIDE see HKA200
(—)-3-HYDROXY-N-CYCLOPROPYLMETHYLMORPHINAN see CQG750
2-HYDROXY-p-CYMENE see CCM000
3-HYDROXY-p-CYMENE see TFX810
14-HYDROXYDAUNOMYCIN see AES750
14'-HYDROXYDAUNOMYCIN see AES750
14-HYDROXYDAUNORUBICINE see AES750
HYDROXYDAUNORUBICIN HYDROCHLORIDE see HKA300
2-HYDROXYDECALIN see DAF000
HYDROXYDECANOIC ACID-γ-LACTONE see HKA500
17-α-HYDROXY-11-DEHYDROCORTICOSTERONE see CNS800
17-HYDROXY-11-DEHYDROCORTICOSTERONE see CNS800
1'-HYDROXY-2',3'-DEHYDROESTRAGOLE see HKA700
5-HYDROXY-α-6-DEOXYTETRACYCLINE see DYE425
HYDROXYDE de POTASSIUM (FRENCH) see PLJ500
HYDROXYDE de SODIUM (FRENCH) see SHS000
HYDROXYDE de TETRAMETHYLAMMONIUM (FRENCH) see TDK500
HYDROXYDE de TRIPHENYL-ETAIN (FRENCH) see HON000
3-HYDROXY-4,15-DIACETOXY-8-(3-METHYLBUTYRYLOXY)-12,13-EPOXY-Δ⁹-TRICHOTHECENE see FQS000
3-HYDROXYDIAZEPAM see CFY750
HYDROXYDIAZEPAM see CFY750
4-HYDROXY-3,5-DIBROMOBENZONITRILE see DDP000
4-HYDROXY-3,5-DI-tert-BUTYLTOLUENE see BFW750
2-HYDROXY-3-(3,3-DICHLOROALLYL)-1,4-NAPHTHOQUINONE see DFN500
2-HYDROXY-3,5-DICHLOROPHENYL SULPHIDE see TFD250
HYDROXYDICHLOROQUINALDINE see CLC500
o-HYDROXY-N,N-DIETHYLBENZAMIDE see DJY400
β-HYDROXYDIETHYL SULFIDE see EPP500
HYDROXY(3-(5,5-DIETHYL-2,4,6-TRIOXO-(1H,3H,5H)-PYRIMIDINO)-2-ETHOXYPROPYL)MERCURY see EET500
HYDROXY(3-(5,5-DIETHYL-2,4,6-TRIOXO-(1H,3H,5H)-PYRIMIDINO)-2-ISOPROPOXYPROPYL)MERCURY see INE000
HYDROXY(3-(5,5-DIETHYL-2,4,6-TRIOXO-(1H,3H,5H)-PYRIMIDINO)-2-METHOXYPROPYL)MERCURY see MES250
p-HYDROXYDIFENYLAMIN (CZECH) see AOT000
2-(HYDROXY)-5-(2,4-DIFLUOROPHENYL)BENZOIC ACID see DKI600
14-HYDROXYDIHYDROCODEINONE see PCG500
14-HYDROXYDIHYDROCODEINONE HYDROCHLORIDE see DLX400
HYDROXYDIHYDROCYCLOPENTADIENE see HKB200
11-HYDROXY-15,16-DIHYDROCYCLOPENTA(a)PHENANTHRACEN-17-ONE ACETATE (ESTER) see ABN250
1-HYDROXY-13-DIHYDRODAUNOMYCIN see DAC300
14-HYDROXYDIHYDRO-6-β-THEBAINOL 4-METHYL ETHER see ORE000
4-HYDROXY-3,5-DIIODOBENZONITRILE see HKB500
3-(4-(4-HYDROXY-3,5-DIIODOPHENOXY)-3,5-DIIODOPHENYL)ALANINE see TFZ300
o-(4-HYDROXY-3,5-DIIODOPHENYL)-3,5-DIIODO-TYROSINE see TFZ300
o-(4-HYDROXY-3,5-DIIODOPHENYL)-3,5-DIIODO-d-TYROSINE MONOSODIUM SALT see SKJ300
β-HYDROXY-3,5-DIIODOPHENYL)-α-PHENYLPROPIONIC ACID see PDM750
8-HYDROXY-5,7-DIIODOQUINOLINE see DNF600
4-HYDROXY-3,5-DIMETHOXYBENZOIC ACID see SPE700
4-HYDROXY-3,5-DIMETHOXY-BENZOIC ACID ETHYL CARBONATE, ester with METHYL RESERPATE see RCA200
β-HYDROXY-β-(2,5-DIMETHOXYPHENYL)-ISOPROPYLAMINE HYDROCHLORIDE see MDW000
18-α-HYDROXY-11, 17-α-DIMETHOXY-3-β, 20-α-YOHIMBAN-16-β-CARBOXYLI see HKB600
HYDROXYDIMETHYLARSINE OXIDE see HKC000

HYDROXYDIMETHYLARSINE OXIDE, SODIUM SALT see HKC500
HYDROXYDIMETHYLARSINE OXIDE, SODIUM SALT TRIHYDRATE see HKC550
1-HYDROXY-2,4-DIMETHYLBENZENE see XKJ500
4-HYDROXY-N-DIMETHYLBUTYRAMIDE-4-CHLOROPHENOXY-ISOBUTYRATE see LGK200
3-HYDROXYDIMETHYL CROTONAMIDE DIMETHYL PHOSPHATE see DGQ875
3-HYDROXY-N,N-DIMETHYL-cis-CROTONAMIDE DIMETHYL PHOSPHATE see DGQ875
3-HYDROXY-5,5-DIMETHYL-2-CYCLOHEXEN-1-ONE DIMETHYLCARBAMATE see DRL200
2'-HYDROXY-5,9-DIMETHYL-2-(3,3-DIMETHYLALLYL)-6,7-BENZOMORPHAN see DOQ400
dl-2'-HYDROXY-5,9-DIMETHYL-2-(3,3-DIMETHYLALLYL)-6,7-BENZOMORPHAN see DOQ400
17-β-HYDROXY-7-α,17-DIMETHYLESTR-4-EN-3-ONE see MQS225
(7-α,17-β)-17-HYDROXY-7,17-DIMETHYL-ESTR-4-EN-3-ONE (9CI) see MQS225
4-HYDROXY-2,5-DIMETHYL-3(2H)FURANONE see HKC575
7-HYDROXY-3,7-DIMETHYLOCTAN-1-AL see CMS850
7-HYDROXY-3,7-DIMETHYL OCTANAL see CMS850
7-HYDROXY-3,7-DIMETHYL OCTANAL:ACETAL see HJV700
7-HYDROXY-3,7-DIMETHYLOCTAN-1-OL see DTE400
3-HYDROXY-4,5-DIMETHYLOL-α-PICOLINE see PPK250
3-HYDROXY-4,5-DIMETHYLOL-α-PICOLINE HYDROCHLORIDE see PPK500
p-HYDROXY-N,α-DIMETHYLPHENETHYLAMINE see FMS875
2'-HYDROXY-5,9-DIMETHYL-2-PHENETHYL-6,7-BENZOMORPHAN HYDROBROMIDE see PMD325
3-HYDROXY-1,1-DIMETHYLPIPERIDINIUM BROMIDE BENZILATE see CBF000
5-HYDROXY-4,6-DIMETHYL-3-PYRIDINEMETHANOL HYDROCHLORIDE see DAY825
4-HYDROXY-3,5-DIMETHYL-1,2,4-TRIAZOLE see HKD550
4-HYDROXY-N,N-DIMETHYLTRYPTAMINE see HKE000
5-HYDROXY-N,N-DIMETHYLTRYPTAMINE see DPG109
1-HYDROXY-2,4-DINITROBENZENE see DUZ000
4-HYDROXY-3,5-DINITROBENZENEARSONIC ACID see HKE500
2-HYDROXY-3,5-DINITROBENZOIC ACID see HKE600
β-(2-HYDROXY-3,5-DINITROPHENYL)BUTANE ACETATE see ACE500
2-HYDROXY-3,5-DINITROPYRIDINE see HKE700
HYDROXYDIONE see VJZ000
HYDROXYDIONE SODIUM see VJZ000
HYDROXYDIONE SUCCINATE see VJZ000
2-HYDROXYDIPHENYL see BGJ250
4-HYDROXYDIPHENYL see BGJ500
o-HYDROXYDIPHENYL see BGJ250
p-HYDROXYDIPHENYL see BGJ500
HYDROXYDIPHENYLACETIC ACID see BBY990
3-((HYDROXYDIPHENYLACETYL)OXY)-1,1-DIMETHYLPIPERIDINIUM BROMIDE see CBF000
2-(((HYDROXYDIPHENYLACETYL)OXY)METHYL)-1,1-DIMETHYLPIPERIDINIUM METHYL SULFATE (SALT) see CDG250
4-((HYDROXYDIPHENYLACETYL)OXY)-1,1,2,2,6-PENTAMETHYLPIPERIDINIUM CHLORIDE (β FORM) see BCB000
4-((HYDROXYDIPHENYLACETYL)OXY)-1,1,2,2,6-PENTAMETHYLPIPERIDINIUM CHLORIDE (α FORM) see BCB250
4-HYDROXYDIPHENYLAMINE see AOT000
p-HYDROXYDIPHENYLAMINE see AOT000
p-HYDROXYDIPHENYLAMINE ISOPROPYL ETHER see HKF000
4-HYDROXYDIPHENYLDIMETHYLMETHANE see COF400
4,4'-HYDROXY-γ,Δ-DIPHENYL-β,Δ-HEXADIENE see DAL600
HYDROXYDIPHENYLMETHANE see HKF300
α-HYDROXYDIPHENYLMETHANE-β-DIMETHYLAMINOETHYL ETHER HYDROCHLORIDE see BAU750
2-HYDROXYDIPHENYL SODIUM see BGJ750
2-HYDROXYDIPHENYL, SODIUM SALT see BGJ750
1-HYDROXY-1,1-DIPHOSPHONOETHANE see HKS780
N-HYDROXYDITHIOCARBAMIC ACID see HKF600
5-HYDROXYDODECANOIC ACID LACTONE see HBP450
5-HYDROXYDODECANOIC ACID Δ-LACTONE see HBP450
6-HYDROXYDOPAMINE HYDROCHLORIDE see HKG000
N-HYDROXY-EAB see HKY000
20-HYDROXYECDYSONE see HKG500
9-HYDROXYELLIPTICIN see HKH000
9-HYDROXYELLIPTICINE see HKH000
HYDROXY-9 ELLIPTICINE (FRENCH) see HKH000
p-HYDROXYEPHEDRINE see HKH500
3-β-HYDROXY-14,15-β-EPOXY-5-β-BUFA-20,22-DIENOLIDE see BOM650
3-HYDROXY-1,2-EPOXYPROPANE see GGW500
16-α-HYDROXYESTRADIOL see EDU500
4-HYDROXY-17-β-ESTRADIOL see HKH850
4-HYDROXYESTRADIOL see HKH850
17-β-HYDROXY-ESTRA-4-EN-3-ONE-17-PHENYLPROPIONATE see DYF450
1'-HYDROXYESTRAGOLE see HKI000
1'-HYDROXY-ESTRAGOLE-2',3'-OXIDE see HKI075
3-HYDROXYESTRA-1,3,5(10),6,8-PENATEN-17-ONE see ECV000
3-HYDROXYESTRA-1,3,5,7,9-PENTAEN-17-ONE BENZOATE see ECV500

3-HYDROXYESTRA-1,3,5(10),7-TETRAEN-17-ONE see ECW000
3-HYDROXYESTRA-1,3,5(10),7-TETRAEN-17-ONE BENZOATE see ECW500
3-HYDROXYESTRA-1,3,5(10)-TRIEN-17-ONE see EDV000
3-HYDROXYESTRA-1,3,5(10)-TRIEN-17-ONE BENZOATE see EDV500
17-β-HYDROXY-ESTR-4-ENE-3-ONE-3-PHENYLPROPIONATE see DYF450
HYDROXYESTRIN BENZOATE see EDP000
1-HYDROXYETHANECARBOXYLIC ACID see LAG000
1-HYDROXYETHANEDIPHOSPHONIC ACID see HKS780
HYDROXYETHANEDIPHOSPHONIC ACID see HKS780
2-HYDROXYETHANESULFONIC ACID see HKI500
2-HYDROXYETHANESULFONIC ACID AMMONIUM SALT see ANL100
2-HYDROXY-1-ETHANETHIOL see MCN250
HYDROXYETHANOIC ACID see GHO000
HYDROXY ETHER see EES350
HYDROXY ETHER see EES350
HYDROXY ETHER see EES350
17-β-HYDROXY-17-α-ETHINYL-5(10)-ESTREN-3-ONE see EEH550
4-HYDROXY-3-ETHOXYBENZALDEHYDE see EQF000
2-(2-HYDROXYETHOXY)ETHYL ESTER STEARIC ACID see HKJ000
2-(2-HYDROXYETHOXY)ETHYL PERCHLORATE see HKJ500
10-(3-(4-HYDROXYETHOXYETHYL-1-PIPERAZINYL)-2-METHYLPRO-
  PYL)PHENOTHIAZINE see MKQ000
2-(2-HYDROXYETHOXY)ETHYL-N-(α,α,α-TRIFLUORO-m-TOLYL)ANTHRANILATE
  see HKK000
(2-HYDROXYETHOXY)METHOXYMETHANE see MET000
β-HYDROXYETHYLACETAMIDE see HKM000
N-(2-HYDROXYETHYL)ACETAMIDE see HKM000
N-β-HYDROXYETHYLACETAMIDE see HKM000
HYDROXYETHYL ACETAMIDE see HKM000
2-HYDROXYETHYL ACETATE see EJI000
2-HYDROXYETHYL ACETOACETATE ACRYLATE see AAY750
1-(2-HYDROXYETHYL)-4-(3-(2-ACETYL-10-PHENOTHIAZYL)PROPYL)PIPERAZINE
  see ABG000
2-HYDROXYETHYL ACRYLATE see ADV250
β-HYDROXYETHYL ACRYLATE see ADV250
HYDROXYETHYL ACRYLATE see ADV250
N-HYDROXYETHYLAMID KYSELINY 4-AMINOTOLUEN-2-SULFONOVE (CZECH)
  see AKG000
2-HYDROXYETHYLAMINE see EEC600
β-HYDROXYETHYLAMINE see EEC600
2-HYDROXYETHYLAMINIUM PERCHLORATE see HKM175
N-HYDROXY-N-ETHYL-4-AMINOAZOBENZENE see HKY000
2-HYDROXYETHYLAMINODIACETIC ACID see HKM500
7-(3-(N-(2-HYDROXYETHYL)AMINO)-2-HYDROXYPROPYL)THIOPHYLLINE NICO-
  TINATE see XCS000
4-HYDROXYETHYLAMINO-1-METHYLAMINOANTHRAQUINONE see MGG250
2-(β-HYDROXYETHYLAMINOMETHYL)-1,4-BENZODIOXANE HYDROCHLORIDE
  see HKN500
4-(2-HYDROXYETHYL)AMINO-3-NITROANILINE see ALO750
2,2'-((4-((2-HYDROXYETHYL)AMINO-3-NITROPHENYL)IMINO))BISETHANOL see
  HKN875
2,2'-((4-((2-HYDROXYETHYL)AMINO)-3-NITROPHENYL)IMINO)DIETHANOL see
  HKN875
4-(2-HYDROXYETHYLAMINO)-2-(5-NITRO-2-THIENYL)QUINAZOLINE see
  HKO000
2-HYDROXY-1-ETHYLAZIRIDINE see ASI000
β-HYDROXY-1-ETHYLAZIRIDINE see ASI000
N-(2-HYDROXYETHYL)AZIRIDINE see ASI000
N-(β-HYDROXYETHYL)AZIRIDINE see ASI000
10-β-HYDROXYETHYL-1:2-BENZANTHRACENE see BBG500
β-HYDROXYETHYLBENZENE see PDD750
N-HYDROXY ETHYL CARBAMATE see HKQ025
(2-HYDROXYETHYL)CARBAMIC ACID,γ-LACTONE see OMM000
1-(((2-HYDROXYETHYL)CARBAMOYL)METHYL)PYRIDINIUM CHLORIDE LAU-
  RATE (ESTER) see LBD100
2-HYDROXYETHYL CELLULOSE see HKQ100
HYDROXYETHYL CELLULOSE see HKQ100
2-HYDROXYETHYL CELLULOSE ETHER see HKQ100
HYDROXYETHYL CELLULOSE ETHER see HKQ100
1-(2-HYDROXYETHYL)-3-(2-CHLOROETHYL)-3-NITROSOUREA see CHB750
2-HYDROXYETHYL-2-CHLOROETHYL SULFIDE see CHC000
1-(2-HYDROXYETHYL)-4-(3-(2-CHLORO-10-PHENOTHIAZI-
  NYL)PROPYL)PIPERAZINE see CJM250
HYDROXYETHYL CNU see CHB750
HYDROXYETHYL CNU METHANESULFONATE see HKQ300
2-HYDROXYETHYL CYCLOHEXANECARBOXYLATE see HKQ500
1-HYDROXY-α-ETHYLCYCLOHEXYLACETIC ACID see COW700
N-(2-HYDROXYETHYL)CYCLOHEXYLAMINE see CPG125
N-(2-HYDROXYETHYL)DIETHYLENETRIAMINE see HKR000
N-(HYDROXYETHYL)DIETHYLENETRIAMINE see HKR000
(2-HYDROXYETHYL)DIISOPROPYLMETHYLAMMONIUMBROMIDE XANTHENE-9-
  CARBOXYLATE see HKR500
β-HYDROXYETHYLDIMETHYLAMINE see DOY800
N-2-HYDROXYETHYL-3,4-DIMETHYLAZOLIDIN see HKR550
(2-HYDROXYETHYL)DIMETHYL(3-STEARAMIDOPROPYL)-AMMONIUM PHOS-
  PHATE (1:1) (SALT) see CCP675
N-HYDROXYETHYLENEDIAMINETRIACETIC ACID see HKS000

2-HYDROXYETHYL ESTER MALEIC ACID see EJG500
2-HYDROXYETHYL ESTER METHACRYLIC ACID see EJH000
2-HYDROXYETHYL ESTER STEARIC ACID see EJM500
N-(2-HYDROXYETHYL)-1,2-ETHANEDIAMINE see AJW000
HYDROXYETHYL ETHER CELLULOSE see HKQ100
N-(2-HYDROXYETHYL)ETHYLENEDIAMINE see AJW000
N-(β-HYDROXYETHYL)ETHYLENEDIAMINE see AJW000
N-(β-HYDROXYETHYLETHYLENEDIAMINE)-N,N',N'-TRIACETIC ACID see
  HKS000
(N-HYDROXYETHYLETHYLENEDINITRILO)TRIACETIC ACID see HKS000
N-HYDROXYETHYL ETHYLENE IMINE see ASI000
1-(2-HYDROXYETHYL)ETHYLENIMINE see ASI000
N-(2-HYDROXYETHYL)ETHYLENIMINE see ASI000
N-HYDROXYETHYL-N-ETHYL-m-TOLUIDINE see HKS100
1-(2-HYDROXYETHYL)-2-HEPTADECENYLGLYOXALIDINE see AHP500
1-HYDROXYETHYL-2-HEPTADECENYLGLYOXALIDINE see AHP500
1-(2-HYDROXYETHYL)-2-HEPTADECENYL-2-IMIDAZOLINE see AHP500
1-(2-HYDROXYETHYL)-2-N-HEPTADECENYL-2-IMIDAZOLINE see AHP500
2-HYDROXY-3-ETHYLHEPTANOIC ACID see HKS300
2-HYDROXY-4-(2'-ETHYLHEXOXY)BENZOFENON (CZECH) see EKY000
β-HYDROXYETHYLHYDRAZINE see HHC000
N-(2-HYDROXYETHYL)HYDRAZINE see HHC000
HYDROXYETHYL HYDRAZINE see HHC000
N-HYDROXYETHYL-N-2-HYDROXYALKYLAMINE see HKS400
2-HYDROXYETHYL (2-HYDROXYETHYL)CARBAMATE see HKS550
2-(1-HYDROXYETHYL)-7-(2-HYDROXY-3-ISOPROPYLAMINOPROPOX-
  Y)BENZOFURAN see HKS600
o-(2-HYDROXYETHYL)HYDROXYLAMINE see HKS775
5-(2-HYDROXYETHYL)-3-((4-HYDROXY-2-METHYL-5-PYRIMIDINYL)METHYL)-4-
  METHYL-THIAZOLIUM see ORS200
1-HYDROXYETHYLIDENE-1,1-DIPHOSPHONIC ACID see HKS780
(1-HYDROXYETHYLIDENE)DIPHOSPHONIC ACID DISODIUM SALT see
  DXD400
(1-HYDROXYETHYLIDENE)DIPHOSPHONIC ACID, TETRAPOTASSIUM SALT
  see TEC250
(1-HYDROXYETHYLIDENE)DIPHOSPHONIC ACID, TETRASODIUM SALT see
  TEE250
(1-HYDROXYETHYLIDENE)DIPHOSPHONIC ACID, TRISODIUM SALT see
  TNL750
3-(1-HYDROXYETHYLIDENE)-6-METHYL-2H-PYRAN-2,4(3H)-DIONE, SODIUM
  SALT see SGD000
2,2'-((2-HYDROXYETHYL)IMINO BIS(N-(α,α-DIMETHYLPHENETHYL))-N-METHYL-
  ACETAMIDE see DTL200
2,2'-((2-HYDROXYETHYL)IMINO)BIS(N-(1,1-DIMETHYL-2-PHENYLETHYL))-N-ME-
  THYLACETAMIDE) see DTL200
(2-HYDROXYETHYL)IMINODIACETIC ACID see HKM500
(N-HYDROXYETHYL)ISOPROPYLAMINE see INN400
β-HYDROXYETHYL ISOPROPYL ETHER see INA500
N-β-HYDROXYETHYL-N-β-KYANETHYLANILIN (CZECH) see CPJ500
2-HYDROXYETHYL MERCAPTAN see MCN250
2-HYDROXYETHYLMERCURY(II) NITRATE see HKT200
2-HYDROXYETHYL METHACRYLATE see EJH000
β-HYDROXYETHYL METHACRYLATE see EJH000
HYDROXYETHYL METHACRYLATE see EJH000
2-(N-2-HYDROXYETHYL-N-METHYLAMINO)ETHANOL see MKU250
7-(2-HYDROXYETHYL)-12-METHYLBENZ(a)ANTHRACENE see HKU000
N-HYDROXYETHYL-α-METHYLBENZYLAMINE see HKU500
1-HYDROXYETHYL METHYL KETONE see ABB500
1-(2-HYDROXY-1-ETHYL)-2-METHYL-5-NITROIMIDAZOLE see MMN250
1-(2-HYDROXYETHYL)-2-METHYL-5-NITROIMIDAZOLE see MMN250
1-(β-HYDROXYETHYL)-2-METHYL-5-NITROIMIDAZOLE see MMN250
1-HYDROXYETHYL-2-METHYL-5-NITROIMIDAZOLE see MMN250
3-(2-HYDROXYETHYL)-3-METHYL-1-PHENYLTRIAZENE see HKV000
N-(2-HYDROXYETHYL)MORPHOLINE see MRQ500
N-β-HYDROXYETHYLMORPHOLINE see MRQ500
N-(β-HYDROXYETHYL)MORPHOLINE HYDROCHLORIDE see HKW300
N-(2-HYDROXYETHYL)NICOTINAMIDE NITRATE (ESTER) see NDL800
N-(2-HYDROXYETHYL)-3-NITROBENZYLIDENIMINE N-OXIDE see HKW345
N-(2-HYDROXYETHYL)-4-NITROBENZYLIDENIMINE N-OXIDE see HKW350
1-(2-HYDROXYETHYL)-3-((5-NITROFURFURYLIDENE)AMINO)-2-IMIDAZOLIDI-
  NONE see EAE500
N-(2-HYDROXYETHYL)-α-(5-NITRO-2-FURYL)NITRONE see HKW450
N(sup 1)-(2-HYDROXYETHYL)-2-NITRO-p-PHENYLENEDIAMINE see ALO750
3-((2-HYDROXYETHYL)NITROSAMINO)-1,2-PROPANEDIOL see NBR100
1-((2-HYDROXYETHYL)NITROSAMINO)-2-PROPANOL see HKW475
1-(2-HYDROXYETHYL)-1-NITROSOUREA see HKW500
2-HYDROXYETHYL-n-OCTYL SULFIDE see OGA000
1-HYDROXYETHYL PERACETATE see HKX600
1-HYDROXYETHYL PEROXYACETATE see HKX600
N-(2-HYDROXYETHYL)PHENYLAMINE see AOR750
N-HYDROXY-N-ETHYL-p-(PHENYLAZO) ANILINE see HKY000
β-HYDROXYETHYL PHENYL ETHER see PER000
4-(2-HYDROXYETHYL)-α-PHENYL-1-PIPERAZINEACETIC ACID-2,6-XYLYL ES-
  TER HYDROCHLORIDE see FAC155
1-(2-HYDROXYETHYL)PIPERAZINE see HKY500

N-(β-HYDROXYETHYL)PIPERAZINE see HKY500
5-(γ-(β-HYDROXYETHYLPIPERAZINO)PROPYL)-5H-DIBENZO(b,f)AZEPINE DIHY-
    DROCHLORIDE see IDF000
10-(3-(2-HYDROXYETHYL)PIPERAZINOPROPYL)-2-(TRIFLUOROMETHYL) PHE-
    NOTHIAZINE see TJW500
γ-(4-(β-HYDROXYETHYL)PIPERAZIN-1-YL)PROPYL-2-CHLOROPHENOTHIAZINE
    see CJM250
5-(3-(4-(2-HYDROXYETHYL)-1-PIPERAZINYL)PROPYL)-5H-DIBENZ(b,f)AZEPINE
    see DCV800
1-(10-(3-(4-(2-HYDROXYETHYL)-1-PIPERAZINYL)PROPYL)-10H-PHENOTHIAZIN-2-
    YL)ETHANONE see ABG000
10-(3-(4-(2-HYDROXYETHYL)-1-PIPERAZINYL)PROPYL)PHENOTHIAZIN-2-YL
    METHYL KETONE see ABG000
10-(3-(4-(2-HYDROXYETHYL)PIPERAZINYL)PROPYL)-2-TRIFLUOROMETHYLPHE-
    NOTHIAZINE DIHYDROCHLORIDE see FMP000
10-(3-(4-(2-HYDROXYETHYL)-1-PIPERAZINYL)PROPYL)-2-(TRIFLUOROMETH-
    YL)PHENOTHIAZINE see TJW500
1-(2-HYDROXYETHYL)PIPERIDINE see HKY600
N-(2-HYDROXYETHYL)PIPERIDINE see HKY600
N-(β-HYDROXYETHYL)PIPERIDINE see HKY600
N-(HYDROXYETHYL)PIPERIDINE see HKY600
1-(2-HYDROXYETHYL)PIPERIDINE HYDROCHLORIDE see PIN275
N-(2-HYDROXYETHYL)PIPERIDINE HYDROCHLORIDE see PIN275
N-(β-HYDROXYETHYL)PIPERIDINE HYDROCHLORIDE see PIN275
N-(HYDROXYETHYL)PIPERIDINE HYDROCHLORIDE see PIN275
10-(3-(4-(2-HYDROXYETHYL)PIPERIDINO)PROPYL)PHENOTHIAZIN-2-YL METHYL
    KETONE see PII500
9-(2-HYDROXYETHYL)-2-(4-PIPERONYL-1-PIPERAZINYL)-9H-PURINE DIHYDRO-
    CHLORIDE see PIX750
4-(2-HYDROXYETHYL)PYRIDINE see POR500
4-(β-HYDROXYETHYL)PYRIDINE see POR500
4-(1-HYDROXYETHYL)PYRIDINE-N-OXIDE see HLB350
3-HYDROXY-2-ETHYL-4-PYRONE see EMA600
1-(2-HYDROXYETHYL)-2-PYRROLIDINONE see HLB380
N-(2-HYDROXYETHYL)-2-PYRROLIDONE see HLB380
2-(2-HYDROXYETHYL)QUINOLINE see QNJ000
2-HYDROXYETHYL STARCH see HLB400
o-(2-HYDROXYETHYL)STARCH see HLB400
o-(HYDROXYETHYL)STARCH see HLB400
HYDROXYETHYL STARCH see HKQ100, HLB400
2-HYDROXYETHYL STARCH ETHER see HLB400
β-HYDROXYETHYL SULFIDE see TFI500
HYDROXYETHYLSULFONIC ACID see HKI500
4′-(2-HYDROXYETHYLSULFONYL)ACETANILIDE see HLB500
p′-((4-(5-((2-HYDROXYETHYL)SULFONYL)-2-METHOXYPHENYL)AZO)-5-HY-
    DROXY-3-METHYLPYRAZOL-1-YL)-3-CHLORO-5-METHYL-BENZENESULFON-
    IC ACID, HYDROGEN SULFATE (ESTER), DISODIUM SALT see RCZ000
7-(2′-HYDROXYETHYL)THEOPHYLLINE see HLC000
7-(2-HYDROXYETHYL)THEOPHYLLINE see HLC000
7-(β-HYDROXYETHYL)THEOPHYLLINE see HLC000
7-(HYDROXYETHYL)THEOPHYLLINE see HLC000
HYDROXYETHYLTHEOPHYLLINE see HLC000
4-(α-HYDROXYETHYL)TOLUENE see TGZ000
N-β-HYDROXYETHYL-o-TOLUIDINO see TGT000
3-(2-HYDROXYETHYL)-1-(3-(TRIFLUOROMETHYL)PHENYL)-2,4(1H,3H)-QUINAZO-
    LINEDIONE see TKF250
(2-HYDROXYETHYL)TRIMETHYLAMMONIUM CHLORIDE see CMF750
(2-HYDROXYETHYL)TRIMETHYLAMMONIUM CHLORIDE ACETATE see
    ABO000
(2-HYDROXYETHYL)TRIMETHYLAMMONIUM CHLORIDE CARBAMATE see
    CBH250
(2-HYDROXYETHYL)TRIMETHYLAMMONIUM CHLORIDE SUCCINATE see
    HLC500
(2-HYDROXYETHYL)TRIMETHYLAMMONIUM IODIDE PROPIONATE see
    PMW750
(2-HYDROXYETHYL)TRIMETHYLAMMONIUM SALICYLATE see CMG000
(2-HYDROXYETHYL)TRIMETHYLAMMONIUM with THEOPHYLLINE see
    CMF500
2-HYDROXYETHYL VINYL ETHER see EJL500
17-β-HYDROXY-17-α-ETHYNYL-4-ANDROSTER-3-ONE see GEK500
N-HYDROXY-2-FAA see HIP000
N-HYDROXY-3-FAA see FDU000
N-HYDROXY-FABP see HLE650
7-HYDROXY-3H-FENOXAZIN-3-ON-10-OXID see HNG500
3-(4′-HYDROXYFENYL)AMINOKARBAZOL see CBN100
o-HYDROXYFENYLMERKURICHLORID see CHW675
14-HYDROXYFILIPIN see FPC000
3-HYDROXYFLAVONE see HLC600
N-HYDROXY-1-FLUORENYL ACETAMIDE see FDT000
9-HYDROXY-2-FLUORENYLACETAMIDE see ABY150
N-(1-HYDROXY-2-FLUORENYL)ACETAMIDE see HIJ400
N-(3-HYDROXY-2-FLUORENYL)ACETAMIDE see HIO875
N-(9-HYDROXYFLUOREN-2-YL)ACETAMIDE see ABY150
N-HYDROXY-3-FLUORENYL ACETAMIDE see FDU000
7-HYDROXY-N-2-FLUORENYLACETAMIDE see HIK500

N-HYDROXY-N-(2-FLUORENYL)ACETAMIDE see HIP000
2-N-HYDROXYFLUORENYL ACETAMIDE, CUPRIC CHELATE see HIP500
N-HYDROXY-N-2-FLUORENYL ACETAMIDE, CUPRIC CHELATE see HIP500
N-(7-HYDROXYFLUOREN-2-YL)ACETOHYDROXAMIC ACID see HLE450
N-HYDROXY-1-FLUORENYL BENZAMIDE see FDY000
N-HYDROXY-N-2-FLUORENYLBENZAMIDE see FDZ000
N-HYDROXY-2-FLUORENYLBENZENESULFONAMIDE see HLE500
N-HYDROXY-N-FLUORENYLBENZENESULFONAMIDE see HLE500
N-HYDROXY-7-FLUORO-2-ACETYLAMINOFLUORENE see FIT200
N-HYDROXY-4-FORMYLAMINOBIPHENYL see HLE650
N-HYDROXY-N-2-FORMYLAMINOFLUORENE see FEG000
4-HYDROXY-4H-FURO(3,2-C)PYRAN-2(6H)-ONE see CMV000
N-HYDROXY-N-GLUCURONOSYL-2-AMINOFLUORENE see HLE750
3-HYDROXYGLUTACONIC ACID, DIMETHYL ESTER, DIMETHYL PHOSPHATE
    see SOY000
7-HYDROXYGUANINE see GLM000
1-HYDROXYHEPTANE see HBL500
2-HYDROXYHEPTANE see HBE500
3-HYDROXYHEPTANE see HBF000
4-HYDROXYHEPTANOIC ACID LACTONE see HBA550
4-HYDROXYHEPTANOIC ACID, γ-LACTONE see HBA550
16-HYDROXYHEXADECANOIC ACID LACTONE see OKU000
1-HYDROXY-2,4-HEXADIENE see HCS500
4-HYDROXYHEXA-2,4-DIENOIC ACID LACTONE see MOW500
1-HYDROXYHEXANE see HFJ500
2-HYDROXYHEXANEDIAL see HIT500
4-HYDROXYHEXANOIC ACID LACTONE see HDY600
6-HYDROXYHEXANOIC ACID LACTONE see LAP000
4-HYDROXYHEX-2-ENOIC ACID LACTONE see EKL500
5-HYDROXY-2-HEXENOIC ACID LACTONE see PAJ500
4-HYDROXYHEX-4-ENOIC ACID LACTONE see HLF000
7-(5-HYDROXYHEXYL)-3-METHYL-1-PROPYLXANTHINE see HLF500
9-((6-HYDROXYHEXYL)OXY)NONANOIC ACID omicron-LACTONE see OLE100
1-HYDROXY-3-N-HEXYL-6,6,9-TRIMETHYL-7,8,9,10-TETRAHYDRO-6-DIBENZO-
    PYRAN see HGK500
o-HYDROXYHIPPURIC ACID see SAN200
5-HYDROXYHYDRINDENE see IBU000
trans-p-HYDROXY HYDROCINNAMIC ACID-4-(AMINOMETHYL)CYCLOHEXANE
    CARBOXYLATE see CDF375
o-HYDROXY-HYDROCINNAMIC ACID-Δ-LACTONE see HHR500
1-HYDROXY-1′-HYDROPEROXYDICYCLOHEXYL PEROXIDE see CPC300
1-HYDROXY-1-HYDROPEROXYDICYCLOHEXYL PEROXIDE see CPC300
HYDROXYHYDROQUINONE see BBU250
2-HYDROXYHYDROQUINONE TRIACETATE see HLF600
HYDROXYHYDROQUINONE TRIACETATE see HLF600
HYDROXY(6-HYDROXY-2,7-DIIODO-3-OXO-9-(o-SULFOPHENYL)-3H-XANTHEN-5-
    YL)MERCURY DISODIUM SALT see SIF500
3-(2-HYDROXY-2-(5-HYDROXY-3,5-DIMETHYL-2-OXOCYCLOHEX-
    YL)ETHYL)GLUTARIMIDE see SMC500
4-(2-HYDROXY-2-(5-HYDROXY-3,5-DIMETHYL-2-OXOCYCLOHEXYL)ETHYL)-2,6-
    PIPERIDINEDIONE see SMC500
7-(2-HYDROXY-3-((2-HYDROXYETHYL)METHYLAMINO)PROPYL)THEOPHYLLINE,
    compound with NICOTINIC ACID see XCS000
N-HYDROXY-N-(7-HYDROXY-2-FLUORENYL)ACETAMIDE see HLE450
2′-HYDROXY-5′-(1-HYDROXY-2-(ISOPROPYLAMINO)ETHYL)-METHANESULFON-
    ANILIDE MONOHYDROCHLORIDE see SKW500
4-HYDROXY-5-(HYDROXYMERCURI)-2-PHENYLPENTANOIC ACID, SODIUM
    SALT see SJQ000
(p-((2-HYDROXY-3-HYDROXYMERCU-
    RI)PROPYL)CARBAMOYL)PHENOXYACETIC ACID see HLP500
(m-((2-HYDROXY-3-HYDROXYMERCURI)PROPYL))PHENOXYACETIC ACID see
    HLP000
(p-(2-HYDROXY-3-HYDROXYMERCURI)PROPYL)PHENOXY))ACETIC ACID, SO-
    DIUM SALT see CCF500
(2-HYDROXY-3-HYDROXYMERCURI)PROPYL(PHENYL)MALONIC ACID SODIUM
    SALT see SJQ500
(o-((2-HYDROXY-3-(HYDROXYMERCU-
    RY)PROPYL)CARBAMOYL)PHENOXY)ACETIC ACID see NCM800
2′-(HYDROXY-5′-(1-HYDROXY-2-(p-METHOXYPHENE-
    THYL)AMINO)PROPYL)METHANESULFONANILIDE HCl see MDM000
4-HYDROXY-3-HYDROXYMETHYL-α-((tert-BUTYLAMINO)METHYL)BENZYL AL-
    COHOL see BQF500
17-β-HYDROXY-2-HYDROXYMETHYLENE-17-α-METHYL-3-ANDROSTANONE
    see PAN100
17-β-HYDROXY-2-(HYDROXYMETHYLENE)-17-METHYL-5-α-ANDROSTAN-3-ONE
    see PAN100
17-β-HYDROXY-2-(HYDROXYMETHYLENE)-17-α-METHYL-5-α-ANDROSTAN-3-
    ONE see PAN100
17-HYDROXY-2-(HYDROXYMETHYLENE)-17-METHYL-5-α-17-β-ANDROSTAN-3-
    ONE see PAN100
8-HYDROXY-5-(1-HYDROXY-2-((1-METHYLETHYL)AMINO)BUTYL)-2(1H)-QUINOLI-
    NONE see PME600
3-HYDROXY-5-(HYDROXYMETHYL)-2-METHYLISONICOTINALDEHYDE, HYDRO-
    CHLORIDE see VSU000
5-HYDROXY-2-(HYDROXYMETHYL)-4H-PYRAN-4-ONE see HLH500

5-HYDROXY-2-(HYDROXYMETHYL)-4H-PYRAN-4-ONE SODIUM SALT see HLH000

5-HYDROXY-2-(HYDROXYMETHYL)-4-PYRONE see HLH500

3-HYDROXY-4-(2-HYDROXY-1-NAPHTHYLAZO)-1-NAPHTHALENESULFONIC ACID, SODIUM see HLI100

HYDROXY(4-HYDROXY-3-NITROPHENYL)MERCURY see HLO400

1-3-HYDROXY-2-(3-HYDROXY-1-OCTENYL)-5-OXOCYCLOPENTANEHEPTANOIC ACID see POC350

3-HYDROXY-2-(3-HYDROXY-1-OCTENYL)-5-OXOCYCLOPENTANEHEPTANOIC ACID see POC350

7-(3-HYDROXY-2-(3-HYDROXY-1-OCTENYL)-5-OXOCYCLOPENTYL)-5-HEPTEN-OIC ACID see DVJ200

erythro-p-HYDROXY-α-(1-((p-HYDROXYPHENETHYL)AMINO)ETHYL)BENZYL ALCOHOL HYDROCHLORIDE see RLK700

(±)-N-(2-(HYDROXY-3-(4-HYDROXYPHENOX-Y)PROPYLAMINO)ETHYL)MORPHOLINE-4-CA RBOXAMIDE FUMARATE see XAH000

HYDROXY(2-HYDROXYPROPYL)PHENYLARSINE OXIDE see HNX600

1-HYDROXYIMIDAZOLE-2-CARBOXALDOXIME-3-OXIDE see HLI300

1-HYDROXYIMIDAZOL-N-OXIDE see HLI325

2-HYDROXYIMINOACETOPHENONE see ILH000

5-HYDROXYIMINOBARBITURIC ACID see AFU000

1-HYDROXYIMINOBUTANE see IJT000

(HYDROXYIMINO)CYCLOHEXANE see HLI500

2-(HYDROXYIMINOMETHYL)-1-METHYLPYRIDINIUM CHLORIDE see FNZ000

2-HYDROXYIMINOMETHYL-1-METHYLPYRIDINIUM IODIDE see POS750

2-((HYDROXYIMINO)METHYL)-1-METHYLPYRIDINIUM METHANESULFONATE see PLX250

2-HYDROXYIMINOMETHYL-N-METHYLPYRIDINIUM METHANESULPHONATE see PLX250

8-HYDROXYINDENO(1,2,3-cd)PYRENE see IBZ100

5-HYDROXYINDOLEACETIC ACID see HLJ000

5-HYDROXYINDOLYLACETIC ACID see HLJ000

2-HYDROXYINIPRAMINE see DPX400

m-HYDROXYIODOBENZENE see IEV010

4-HYDROXY-3-IODO-5-NITROBENZONITRILE see HLJ500

l-3-(4-(4-HYDROXY-3-IODOPHENOXY)-3,5-DIIODOPHENYL)ALANINE see LGK050

O-(4-HYDROXY-3-IODOPHENYL)-3,5-DIIODO-l-TYROSINE see LGK050

8-HYDROXY-7-IODOQUINOLINE SULFONATE see IEP200

8-HYDROXY-7-IODO-5-QUINOLINESULFONIC ACID see IEP200

8-HYDROXY-7-IODOQUINOLINESULFONIC ACID see IEP200

HYDROXYISOBUTYLISOPROPYLARSINE OXIDE see IPS100

2-HYDROXY-3-ISOBUTYL-6-(1-METHYLPROPYL)PYRAZINE 1-OXIDE see ARO000

α-HYDROXYISOBUTYRONITRILE see MLC750

5-(1-HYDROXY-2-ISOPROPYLAMINO)BUTYL-8-HYDROXYCARBOSTYRIL HYDROCHLORIDE see PME600

4'-(1-HYDROXY-2-(ISOPROPYLAMINO)ETHYL)METHANESULFOANILIDE HYDROCHLORIDE see CCK250

4'-(1-HYDROXY-2-ISOPROPYLAMINO)ETHYL)METHANESULFONANILIDE MONO-HYDROCHLORIDE see CCK250

4-HYDROXY-α-ISOPROPYLAMINOMETHYLBENZYL ALCOHOL see HLK000

p-HYDROXY-α-ISOPROPYLAMINOMETHYLBENZYL ALCOHOL see HLK000

3-HYDROXY-α-ISOPROPYLAMINOMETHYLBENZYL ALCOHOL HYDROCHLORIDE see HLK500

N-(4-(2-HYDROXY-3-(ISOPROPYLAMINO)PROPOXY)ACETAMIDE (9CI) see ECX100

4'-(2-HYDROXY-3-(ISOPROPYLAMINO)PROPOXY)ACETANILIDE see ECX100

7-(2-HYDROXY-3-(ISOPROPYLAMINO)PROPOXY)-2-BENZOFURANYL METHYL KETONE see HLK600

8-(2-HYDROXY-3-ISOPROPYLAMINO)PROPOXY-2H-1-BENZOPYRAN see HLK800

4-(2-HYDROXY-3-ISOPROPYLAMINOPROPOXY)-INDOLE see VSA000

7-(2-HYDROXY-3-(ISOPROPYLAMINO)PROPOXY)-α-METHYL-2-BENZOFURAN-METHANOL see HKS600

2-(p-2-HYDROXY-3-(ISOPROPYLAMINO)PROPOXY)PHENYL)ACETAMIDE see TAL475

2-HYDROXY-4-ISOPROPYL-2,4,6-CYCLOHEPTATRIEN-1-ONE see IRR000

2-HYDROXY-5-ISOPROPYL-2,4,6-CYCLOHEPTATRIEN-1-ONE see TFV750

β-HYDROXY-N-ISOPROPYL-3,4-DICHLOROPHENETHYLAMINE see DFN400

HYDROXYISOPROPYLMERCURY see IPW000

(2-(4-HYDROXY-2-ISOPROPYL-5-METHYLPHENOXY)ETHYL)DIMETHYLAMINE see DAD850

(2-(4-HYDROXY-2-ISOPROPYL-5-METHYLPHENOXY)ETHYL)METHYLAMINE see DAD500

12-HYDROXY-13-ISOPROPYLPODOCARPA-7,13-DIEN-15-OIC ACID BIS(2-CHLO-ROETHYL)AMINE SALT see HIJ000

3-α-HYDROXY-8-ISOPROPYL-1-α-H,5-α-H-TROPANIUM BROMIDE (±)-TROPATE see IGG000

(8r)-3-α-HYDROXY-8-ISOPROPYL-1-α-H,5-α-H-TROPIUMBROMIDE-(±)-TROPATE see IGG000

HYDROXYISOXAZOLE see HLM000

3-HYDROXY-17-KETO-ESTRA-1,3,5-TRIENE see EDV000

4-HYDROXY-2-KETO-4-METHYLPENTANE see DBF750

3-HYDROXY-17-KETO-OESTRA-1,3,5-TRIENE see EDV000

3-HYDROXYKYNURENINE see HJD000

l-3-HYDROXYKYNURENINE see HJD500

3-HYDROXY-l-KYNURENINE see HJD500

HYDROXYKYNURENINE see HJD000

HYDROXYLAMINE see HLM500

HYDROXYLAMINE, N-ACETYL-N-(7-IODO-2-FLUORENYL)-O-MYRISTOYL- see MSB100

HYDROXYLAMINE CHLORIDE (1:1) see HLN000

HYDROXYLAMINE CHLORIDE see HLN000

HYDROXYLAMINE, HYDRIODIDE see HLM600

HYDROXYLAMINE HYDROCHLORIDE see HLN000

HYDROXYL AMINE IODIDE (DOT) see HLM600

HYDROXYLAMINE, N-METHYL-N-(4-QUINOLINYL)-, 1-OXIDE see HLX550

HYDROXYLAMINE NEUTRAL SULFATE see OLS000

HYDROXYLAMINE SULFATE (2:1) see OLS000

HYDROXYLAMINE SULFATE see OLS000

HYDROXYLAMINE-O-SULFONIC ACID see HLN100

HYDROXYLAMINESULFONIC ACID see HLN100

4-HYDROXYLAMINOBIPHENYL see BGI250

4-(HYDROXYLAMINO)-2-METHYL-QUINOLINE, 1-OXIDE see MKR000

6-N-HYDROXYLAMINOPURINE see HIT000

6-HYDROXYLAMINOPURINE see HIT000

6-HYDROXYLAMINOPURINE RIBOSIDE see HIT100

N⁶-HYDROXYLAMINOPURINE RIBOSIDE see HIT100

HYDROXYLAMMONIUM CHLORIDE see HLN000

HYDROXYLAMMONIUM PHOSPHINATE see HLN500

HYDROXYLAMMONIUM SULFATE see OLS000

O-HYDROXYLAMMONIUM SULFONATE see HLN100

o-HYDROXYLBENZHYDRAZIDE see HJL100

(2-HYDROXYLIMINOBUTANE CHLORIDE) see BOV625

HYDROXYLIZARIC ACID see TKN750

13-HYDROXYLUPANINE-2-PYRROLE CARBOXYLIC ACID ESTER see CAZ125

HYDROXYUREA see HOO500

N-HYDROXY-MAB see HLV000

1-HYDROXY-MA-144-N1 see RHA125

1-HYDROXY MA144 S1 see APV000

(E)-1-HYDROXY-MC-9,10-DIHYDRODIOL see DLT600

6-HYDROXY-2-MERCAPTOPYRIMIDINE see TFR250

o-(N-3-HYDROXYMERCURI-2-HYDROXYETHOXYPROPYLCARBAMYL)PHENOXY-ACETIC ACID, SODIUM SALT see HLO300

5-(3-HYDROXYMERCURI-2-METHOXYPROPYL) BARBITURIC ACID SODIUM SALT see HLU000

o-((3-HYDROXYMERCURI-2-METHOXYPROPYL)CARBAMOYL)PHENOXYACETIC ACID MONOSODIUM SALT see SIH500

N-((3-(HYDROXYMERCURI)-2-METHOXYPROPYL)-CARBAMOYL)SUCCINAMIC ACID see MFC000

8-(γ-HYDROXYMERCURI-β-METHOXYPROPYL)-3-COUMARINCARBOXYLICACID THEOPHYLLINE SODIUM see MCS000

8-(3-(HYDROXYMERCURI)-2-METHOXYPROPYL)-2-OXO-2H-1-BENZOPYRAN-3-CARBOXYLIC ACID SODIUM SALT COMPOUND with THEOPHYLLINE (1:1) see MCS000

N-(γ-HYDROXYMERCURI-β-METHOXYPROPYL)SALICYLAMIDE-o-ACETIC ACID SODIUM SALT see SIH500

2-HYDROXYMERCURI-3-NITROBENZOIC ACID-1,2-CYCLIC ANHYDRIDE see NHY500

HYDROXYMERCURI-o-NITROPHENOL see HLO400

o-(HYDROXYMERCURI)PHENOL see HLO500

HYDROXYMERCURIPROPANOLAMIDE of p-CARBOXYPHENOXYACETIC ACID see HLP500

HYDROXYMERCURIPROPANOLAMIDE of m-CARBOXYPHENOXYACETIC ACID see HLP000

HYDROXYMERCURIPROPANOLAMIDE of o-CARBOXYPHENOXYACETIC ACID see NCM800

(5-(HYDROXYMERCURI)-2-THIENYL)MERCURY ACETATE see HLQ000

2-HYDROXYMESITYLENE see MDJ740

p-HYDROXYMETHAMPHETAMINE see FMS875

N-HYDROXY-METHANAMINE see MKQ875

HYDROXYMETHANESULFINIC ACID SODIUM SALT see FMW000

4'-HYDROXY-3'-METHOXYACETOPHENONE see HLQ500

4-HYDROXY-3-METHOXYALLYLBENZENE see EQR500

1-HYDROXY-2-METHOXY-4-ALLYLBENZENE see EQR500

4-HYDROXY-3-METHOXY-4-AMINOAZOBENZENE see HLR000

N-HYDROXY-3-METHOXY-4-AMINOAZOBENZENE see HLR000

3-HYDROXY-4'-METHOXY-4-AMINODIPHENYL see HLR500

3-HYDROXY-4'-METHOXY-4-AMINODIPHENYL HYDROCHLORIDE see AKN250

2-HYDROXY-3-METHOXYBENZALDEHYDE see VFP000

3-HYDROXY-4-METHOXYBENZALDEHYDE see VFK000

4-HYDROXY-3-METHOXYBENZALDEHYDE see FNM000

1-HYDROXY-2-METHOXYBENZENE see GKI000

6-HYDROXY-7-METHOXY-5-BENZOFURANACRYLIC ACID Δ-LACTONE see XDJ000

2-HYDROXY-3-METHOXYBENZOIC ACID see HJB500

4-HYDROXY-3-METHOXYBENZOIC ACID see VFF000

3-HYDROXY-4-METHOXYBENZOIC ACID see HJC000

2-HYDROXY-4-(1-METHYLETHYL)-2,4,6-CYCLOHEPTATRIEN-1-ONE see IRR000
(HYDROXYMETHYL)ETHYLENE ACETATE see DBF600
N-(1-(HYDROXYMETHYL)ETHYL)-d-LYSERGOMIDE see LJL000
N-(α-(HYDROXYMETHYL)ETHYL)-d-LYSERGOMIDE see LJL000
12′-HYDROXY-2′-(1-METHYLETHYL)-5′-α-(2-METHYLPROPYL)ERGOTAMAN-
  3′,6′,18-TRIONE see EDB100
5-HYDROXYMETHYLFURALDEHYDE see ORG000
2-HYDROXYMETHYLFURAN see FPU000
5-(HYDROXYMETHYL)-2-FURANCARBOXALDEHYDE see ORG000
4-HYDROXY-5-METHYL-2(5H)-FURANONE see MPQ500
5-(HYDROXYMETHYL)FURFURAL see ORG000
HYDROXYMETHYLFURFUROLE see ORG000
3-HYDROXY-3-METHYLGLUTARIC ACID see HMC000
β-HYDROXY-β-METHYLGLUTARIC ACID see HMC000
3-HYDROXY-1-METHYLGUANINE see HMC500
3-HYDROXY-7-METHYLGUANINE see HMD000
3-HYDROXY-9-METHYLGUANINE see HMD500
3-HYDROXYMETHYL-n-HEPTAN-4-OL see EKV000
HYDROXYMETHYL HYDROPEROXIDE see HME000
5-HYDROXY-2-METHYL-1H-IMIDAZOLE-4-CARBOXAMIDE see MID800
3-HYDROXYMETHYLINDOLE see ICP100
3-HYDROXY-1-METHYL-5,6-INDOLINEDIONE see AES639
HYDROXYMETHYL 2-INDOLYL KETONE see HIS100
7-(3-((2-HYDROXY-3-((2-METHYLINDOL-4-
  YL)OXY)PROPYL)AMINO)BUTYL)THEOPHYLLINE see TAL560
HYDROXYMETHYLINITRILE see HIM500
HYDROXYMETHYL 1-IODO-3-ADAMANTYL KETONE see HME100
6-HYDROXYMETHYL-2-ISOPROPYLAMINOMETHYL-7-NITRO-1,2,3,4-TETRAHY-
  DROQUINOLINE see OLT000
3-HYDROXY-1-METHYL-4-ISOPROPYLBENZENE see TFX810
3-HYDROXY-5-METHYLISOXAZOLE see HLM000
2-HYDROXY-2-METHYLMALONATE see MPN250
HYDROXYMETHYLMERCURY see MLG000
7-HYDROXYMETHYL-12-METHYLBENZ(a)ANTHRACENE see HMF000
12-HYDROXYMETHYL-7-METHYLBENZ(a)ANTHRACENE see HMF500
1-HYDROXYMETHYL-2-METHYLDITMIDE-2-OXIDE see HMG000
1-(HYDROXYMETHYL)-3-METHYLIMIDAZOLINIUM CHLORIDE, DODECANOATE
  see HMG500
3-(HYDROXYMETHYL)-1-METHYLIMIDAZOLINIUM CHLORIDE LAURATE (ES-
  TER) see HMG600
HYDROXYMETHYL 1-METHYL-2-INDOLYL KETONE see HIS120
HYDROXYMETHYL 3-METHYL-2-INDOLYL KETONE see HIS130
2-(HYDROXYMETHYL)-2-(METHYLPENTYL) BUTYLCARBAMATE CARBAMATE
  see MOV500
2-(HYDROXYMETHYL)-2-METHYLPENTYL ESTER, CARBAMATE, BUTYL CAR-
  BAMIC ACID see MOV500
HYDROXYMETHYL METHYL PEROXIDE see HMH000
5-HYDROXYMETHYL-6-METHYL-2,4(1H,3H)-PYRIMIDINEDIONE see HMH300
5-HYDROXYMETHYL-4-METHYLURACIL see HMH300
5-HYDROXYMETHYL-6-METHYLURACIL see HMH300
(±)-3-HYDROXY-N-METHYLMORPHINAN see MKR250
d-3-HYDROXY-N-METHYLMORPHINAN see DBE800
dl-3-HYDROXY-N-METHYLMORPHINAN see MKR250
(−)-3-HYDROXY-N-METHYLMORPHINAN see LFG000
l-3-HYDROXY-N-METHYLMORPHINAN BITARTRATE see DYF000
dl-3-HYDROXY-N-METHYLMORPHINAN HYDROBROMIDE see HMH500,
  MDV250
d-3-HYDROXY-N-METHYLMORPHINAN TARTRATE see DBE825
l-3-HYDROXY-N-METHYLMORPHINAN TARTRATE see DYF000
(−)-3-HYDROXY-N-METHYLMORPHINAN TARTRATE DIHYDRATE see MLZ000
(+)-3-HYDROXY-N-METHYLMORPHINAN TARTRATE HYDRATE see MMA000
5-HYDROXY-2-METHYL-1,4-NAPHTHALENEDIONE see PJH610
3-HYDROXY-2-METHYL-1,4-NAPHTHOQUINONE see HMI000
5-HYDROXY-2-METHYL-1,4-NAPHTHOQUINONE see PJH610
2-HYDROXY-3-METHYL-1,4-NAPHTHOQUINONE see HMI000
o-HYDROXYMETHYLNITROBENZENE see NFM550
3-HYDROXYMETHYL-1-((3-(5-NITRO-2-FURYL)ALLYLIDENE)AMINO)HYDANTOIN
  see HMI500
6-HYDROXYMETHYL-2-(2-(5-NITRO-2-FURYL)VINYL)-PYRIDINE see NDY400
2-HYDROXYMETHYL-2-NITROPROPANE-1,3-DIOL see HMJ500
2-HYDROXYMETHYLNORCAMPHANE see NNH000
17-β-HYDROXY-18-METHYL-19-NOR-17-α-PREGN-4-EN-20-YN-3-ONE see
  NNQ500, NNQ520
3-HYDROXY-4-METHYL-7-OXABICYCLO(4.1.0)HEPT-3-ENE-2,5-DIONE see
  TBF325
(1R-cis)-3-HYDROXY-4-METHYL-7-OXABICYCLO(4.1.0)HEPT-3-ENE-2,5-DIONE
  see TBF325
7-HYDROXY-4-METHYL-2-OXO-2H-1-BENZOPYRAN see MKP500
3-(HYDROXYMETHYL)-8-OXO-7-(2-(4-PYRIDYLTHIO)ACETAMIDO)-5-THIA-1-AZA-
  BICYCLO(4.2.0)OCT-2-ENE-2-CARBOXYLIC ACID, ACETATE (ESTER) see
  CCX500
3-(HYDROXYMETHYL)-8-OXO-7-(2-(4-PYRIDYLTHIO)ACETAMIDO)-5-THIA-1-AZA-
  BICYCLO(4,2,0)OCT-2-ENE-2-CARBOX see HMK000
3-HYDROXY-3-METHYLPENTANEDIOIC ACID see HMC000
4-HYDROXY-4-METHYL-PENTAN-2-ON (GERMAN, DUTCH) see DBF750

4-HYDROXY-4-METHYLPENTANONE-2 see DBF750
4-HYDROXY-4-METHYL PENTAN-2-ONE see DBF750
4-HYDROXY-4-METHYL-2-PENTANONE see DBF750
4-HYDROXY-4-METHYL-2-PENTANONE PEROXIDE see HMK050
4-HYDROXY-α-(1-((1-METHYL-2-PHENOXYETHYL)AMINO)ETHYL)-BENZENEM-
  ETHANOL HYDROCHLORIDE see VGA300
p-HYDROXY-N-(1-METHYL-2-PHENOXYETHYL)NOREPHEDRINE see VGF000
HYDROXYMETHYL 1-PHENYL-3-ADAMANTYL KETONE see HMK150
HYDROXYMETHYLPHENYLARSINE OXIDE see HMK200
N-(4-((2-HYDROXY-5-METHYLPHENYL)AZO)PHENYL)ACETAMIDE see AAQ250
2-(2-HYDROXY-5-METHYLPHENYL)BENZOTRIAZOLE see HML500
p-HYDROXY-α-(1-((1-METHYL-3-PHENYLPROPYL)AMINO)ETHYL)BENZYL ALCO-
  HOL HYDROCHLORIDE see DNU200
5-(1-HYDROXY-2-((1-METHYL-3-PHENYLPROPYL)AMINO)ETHYL)SALICYLAMIDE
  HYDROCHLORIDE see HMM500
2-(HYDROXYMETHYL)-6-PHENYL-3(2H)-PYRIDAZINONE see HMN000
2-HYDROXYMETHYL-6-PHENYL-3-PYRIDAZONE see HMN000
1-(4-HYDROXY-2-METHYL-5-PHENYL-1H-PYRROL-3-YL)ETHANONE see
  HMN100
3-HYDROXY-3-METHYL-1-PHENYLTRIAZENE see HMP000
3-HYDROXY-2-METHYL-5-((PHOSPHONOOXY)METHYL)-4-PYRIDINECARBOXAL-
  DEHYDE see PII100
17-HYDROXY-6-METHYLPREGNA-4,6-DIENE-3,20-DIONE ACETATE see VTF000
17-HYDROXY-6-METHYLPREGNA-4,6-DIENE-3,20-DIONE ACETATE mixed with
  19-NOR-17-α-PREGNA-1,3,5(10)-TRIEN-2-YNE-3,17-DIOL see MCA500
17-α-HYDROXY-6-α-METHYLPREGN-4-ENE-3,20-DIONE ACETATE see MCA000
17-HYDROXY-6-α-METHYLPREGN-4-ENE-3,20-DIONE ACETATE see MCA000
17-α-HYDROXY-6-α-METHYLPROGESTERONE ACETATE see MCA000
1-HYDROXYMETHYLPROPANE see IIL000
HYDROXYMETHYL-PROPANEDIOIC ACID (9CI) see MPN250
N-(HYDROXYMETHYL)-2-PROPENAMIDE see HLU500
2-HYDROXY-2-METHYLPROPIONITRILE see MLC750
1-(HYDROXYMETHYL)PROPYLAMIDE of 1-METHYL-(+)-LYSERGIC ACID HY-
  DROGEN MALEATE see MQP500
d-N,N′-(1-HYDROXYMETHYLPROPYL)ETHYLENEDINITROSAMINE see HMQ500
(6-α,17-β)-17-HYDROXY-6-METHYL-17-(1-PROPYNYL)-ANDROST-4-EN-3-ONE
  see DRT200
17-β-HYDROXY-6-α-METHYL-17-(1-PROPYNYL)ANDROST-4-EN-3-ONE see
  DRT200
4-HYDROXY-6-METHYL-2H-PYRAN-2-ONE see THL800
3-HYDROXY-2-METHYL-4H-PYRAN-4-ONE see MAO350
2-(HYDROXYMETHYL)PYRIDINE see POR800
3-(HYDROXYMETHYL)PYRIDINE see NDW510
4-(HYDROXYMETHYL)PYRIDINE see POR810
5-HYDROXY-6-METHYL-3,4-PYRIDINEDICARBINOL HYDROCHLORIDE see
  PPK500
5-HYDROXY-6-METHYL-3,4-PYRIDINEDIMETHANOL see PPK250
5-HYDROXY-6-METHYL-3,4-PYRIDINEDIMETHANOL HYDROCHLORIDE see
  PPK500
3-HYDROXY-1-METHYLPYRIDINIUM BROMIDE DIMETHYLCARBAMATE (ES-
  TER) see MDL600
3-HYDROXY-1-METHYLPYRIDINIUM BROMIDE HEXAMETHYLENE-
  BIS(METHYLCARBAMATE) see DXG800
3-HYDROXY-1-METHYL-PYRIDINIUM DIMETHYLCARBAMATE (ester) see
  PPI800
4-HYDROXY-2-METHYL-N-2-PYRIDINYL-2H-THIENO(2,3-e)-1,2-THIAZINE-3-CAR-
  BOXAMIDE 1,1-DIOXIDE see TAL485
4-HYDROXY-2-METHYL-N-(2-PYRIDYL)-2H-1,2-BENZOTHIAZIN-3-CARBOXYAMID-
  1,1-DIOXID (GERMAN) see FAJ100
4-HYDROXY-2-METHYL-N-(2-PYRIDYL)-2H-1,2-BENZOTHIAZINE-3-CARBOXAM-
  IDE-1,1-DIOXIDE see FAJ100
3-HYDROXY-2-METHYL-4-PYRONE see MAO350
3-HYDROXY-2-METHYL-γ-PYRONE see MAO350
8-HYDROXY-1-METHYLQUINOLINIUM METHYLSULFATE DIMETHYLCARBA-
  MATE see DQY400
4-HYDROXY-1-METHYL-2-QUINOLONE see HMR500
12-HYDROXY-4-METHYL-4,8-SECOSENECIONAN-8,11,16-TRIONE see DMX200
2-HYDROXYMETHYLTETRAHYDROPYRAN see MDS500
2-HYDROXY-4-(METHYLTHIO)BUTANOIC ACID see HMR600
1-(HYDROXYMETHYL)-2,8,9-TRIOXA-5-AZA-1-SILABICYCLO(3.3.3) UNDECA-
  NEMETHACRYLATE see HMS500
3-α-HYDROXY-8-METHYL-1-α-H,5-α-H-TROPANIUM BROMIDE MANDELATE see
  MDL000
3-α-HYDROXY-8-METHYL-1-α-H,5-H-TROPANIUM BROMIDE 2-PROPYLVALER-
  ATE see LJS000
3-α-HYDROXY-8-METHYL-1-α,5-α-H-TROPANIUM BROMIDE (±)TROPATE see
  MGR250
3-α-HYDROXY-8-METHYL-1-α-H,5-α-H-TROPANIUM NITRATE (±)-TROPATE (ES-
  TER) see MGR500
3-HYDROXY-α-METHYL-l-TYROSINE see DNA800
3-HYDROXY-α-METHYL-l-TYROSINE-1-(2,2-DIMETHYL-1-OXOPROPOXY)ETHYL
  ESTER PHOSPHATE HYDRATE see HMS875
4-HYDROXY-3-MORPHOLINOMETHYLBENZOIC ACID ETHYL ESTER see
  ELK000
HYDROXYMYCIN see NCF500

N-HYDROXY-N-MYRISTOYL-2-AMINOFLUORENE see HMU000
5-HYDROXY-1,4-NAFTOCHINON see WAT000
2-HYDROXYNAPHTHALDEHYDE see HMU200
β-HYDROXYNAPHTHALDEHYDE see HMU200
1-HYDROXYNAPHTHALENE see NAW500
2-HYDROXYNAPHTHALENE see NAX000
α-HYDROXYNAPHTHALENE see NAW500
β-HYDROXYNAPHTHALENE see NAX000
2-HYDROXY-1-NAPHTHALENECARBOXALDEHYDE see HMU200
3-HYDROXY-2-NAPHTHALENECARBOXYLIC ACID see HMX520
2-HYDROXY-1,4-NAPHTHALENEDIONE see HMX600
6-HYDROXY-2-NAPHTHALENESULFONIC ACID see HMU500
2-HYDROXY-6-NAPHTHALENESULFONIC ACID see HMU500
N-HYDROXY-N-2-NAPHTHALENYLACETAMIDE see NBD000
4-((4-HYDROXY-1-NAPHTHALENYL)AZO)BENZENESULPHONIC ACID, MONO-
    SODIUM SALT see FAG010
N-HYDROXYNAPHTHALIMIDE, DIETHYL PHOSPHATE see HMV000
N-HYDROXYNAPHTHALIMIDE-O,O-DIETHYL PHOSPHOROTHIOATE see
    NAQ500
3-HYDROXY-2-NAPHTHAMIDE see HMW500
2-HYDROXYNAPHTHENESULFONIC ACID see HMX000
2-HYDROXY-1-NAPHTHOIC ACID see HMX500
3-HYDROXY-2-NAPHTHOIC ACID see HMX520
2-HYDROXY-1,4-NAPHTHOQUINONE see HMX600
5-HYDROXY-1,4-NAPHTHOQUINONE see WAT000
2-HYDROXYNAPHTHOQUINONE see HMX600
N-HYDROXY-1-NAPHTHYLAMINE see HIX000
N-HYDROXY-2-NAPHTHYLAMINE see NBI500
2-HYDROXY-1-NAPHTHYLAMINE HYDROCHLORIDE see ALK000
1-HYDROXY-2-NAPHTHYLAMINE HYDROCHLORIDE see ALK250
p-((4-HYDROXY-1-NAPHTHYL)AZO)BENZENESULFONIC ACID, MONOSODIUM
    SALT see FAG010
p-((2-HYDROXY-1-NAPHTHYL)AZO)BENZENESULFONIC ACID SODIUM SALT
    see CMM220
p-((4-HYDROXY-1-NAPHTHYL)AZO)BENZENESULPHONIC ACID, MONOSODI-
    UM SALT see FAG010
p-((4-HYDROXY-1-NAPHTHYL)AZO)BENZENESULPHONIC ACID, SODIUM SALT
    see FAG010
N-HYDROXYNAPHTHYLIMIDE DIETHYL PHOSPHATE see HMV000
3-HYDROXY-4-(1-NAPHTHYLOXY)BUTYRAMIDOXIME HYDROCHLORIDE see
    NAC500
N-HYDROXY-N,N′-DIACETYLBENZIDINE see HKB000
p-HYDROXYNEOZONE see NBF500
4-HYDROXY-3-NITROANILINE see NEM480
2-HYDROXY-4-NITROANILINE see ALO000
2-HYDROXY-5-NITROANILINE see NEM500
2-HYDROXYNITROBENZENE see NIE500
3-HYDROXYNITROBENZENE see NIE010
4-HYDROXYNITROBENZENE see NIF000
m-HYDROXYNITROBENZENE see NIE010
4-HYDROXY-3-NITROBENZENEARSONIC ACID see HMY000
4-HYDROXY-3-NITROBENZENESULFONYL CHLORIDE see HMY050
4-HYDROXY-3-NITROBENZOIC ACID METHYL ESTER see HMY075
2-HYDROXY-5-NITROMETANILIC ACID see HMY500
4-HYDROXY-3-NITROPHENYLARSONIC ACID see HMY000
4-HYDROXY-3-(1-(4-NITROPHENYL)-3-OXOBUTYL)-2H-1-BENZOPYRAN-2-ONE
    see ABF750
3-(HYDROXYNITROSOAMINO)-l-ALANINE (9CI) see AFH750
N-HYDROXY-N-NITROSO-BENZENAMINE, AMMONIUM SALT see ANO500
2-HYDROXY-3-NITROSO-2,7-NAPHTHALENEDISULFONIC ACID DISODIUM
    SALT see HNB000
trans-4-HYDROXY-1-NITROSO-l-PROLINE see HNB500
3-HYDROXYNITROSOPYRROLIDINE see NLP700
8-HYDROXY-5-NITROSOQUINOLINE see NLR000
HYDROXY No. 253 see NNC500
4-HYDROXYNONANOIC ACID, γ-LACTONE see CNF250
4-HYDROXY-2-NONENAL see HNB600
4-HYDROXYNONENAL see HNB600
1-m-HYDROXYNOREPHEDRINE see HNC000
m-HYDROXY NOREPHEDRINE see HNB875
HYDROXYNOREPHEDRINE see HNB875
17-HYDROXY-19-NOR-17-α-PREGNA-4,20-DIEN-3-ONE see NCI525
17-HYDROXY-19-NOR-4-PREGNENE-3,20-DIONE HEXANOATE see GEK510
17-HYDROXY-19-NORPREGN-4-ENE-3,20-DIONE HEXANOATE see GEK510
17-HYDROXY-19-NOR-17-α-PREGN-4-EN-20-YN-3-ONE see NNP500
(17-α)-17-HYDROXY-19-NORPREGN-4-EN-20-YN-3-ONE see NNP500
17-β-HYDROXY-19-NORPREGN-4-EN-20-YN-3-ONE see NNP500
17-HYDROXY(17-α)-19-NORPREGN-5(10)-EN-20-YN-3-ONE see EEH550
17-HYDROXY-19-NOR-17-α-PREGN-5(10)-EN-20-YN-3-ONE see EEH550
(17-α)-17-HYDROXY-19-NORPREGN-5(10)-EN-20-YN-3-ONE see EEH550
17-β-HYDROXY-19-NOR-17-α-PREGN-4-EN-20-YN-3-ONE ACETATE see
    ABU000
17-HYDROXY-19-NOR-17-α-PREGN-4-EN-20-YN-3-ONE ACETATE see ABU000
17-α-HYDROXY-19-NORPROGESTERONE CAPROATE see GEK510
13-HYDROXY-9,11-OCTADECADIENOIC METHYL ESTER see MKR500

12-HYDROXY-cis-9-OCTADECENOIC ACID see RJP000
1-HYDROXYOCTANE see OEI000
5-HYDROXYOCTANOIC ACID LACTONE see OCE000, ODE300
7-(2-(3-HYDROXY-1-OCTENYL)-5-OXO-3-CYCLOPENTEN-1-YL)-5-HEPTENOIC
    ACID see MCA025
4-HYDROXY-5-OCTYL-2(5H)FURANONE see HND000
2-HYDROXY-4-(OCTYLOXY)BENZOPHENONE see HND100
(2-HYDROXY-4-(OCTYLOXY)PHENYL)PHENYLMETHANONE see HND100
16-α-HYDROXYOESTRADIOL see EDU500
3-HYDROXY-OESTRA-1,3,5(10)-TRIEN-17-ONE see EDV000
3-HYDROXY-1,3,5(10)-OESTRATRIEN-17-ONE see EDV000
2-HYDROXY-4-OKTYLOXYBENZOFENON see HND100
16-HYDROXY-11-OXAHEXADECANOIC ACID, ω-LACTONE see OKW100
16-HYDROXY-12-OXAHEXADECANOIC ACID, ω-LACTONE see OKW110
γ-HYDROXY-β-OXOBUTANE see ABB500
(11-β)-11-HYDROXY-17-(1-OXOBUTOXY)-21-(1-OXOPROPOXY)-PREGN-4-ENE-
    3,20-DIONE see HHQ850
1-HYDROXY-4-OXO-2,5-CYCLOHEXADIENE-1-SULFONIC ACID compound with
    DIETHYLAMINE see DIS600
5-β-HYDROXY-19-OXODIGITOXIGENIN see SMM500
6α-HYDROXY-3-OXO-11-EPIISOEUSANTONA-1,4-DIENIC ACID, γ-LACTONE see
    SAU500
6α-HYDROXY-3-OXO-EUDESMA-1,4-DIEN-12-OIC ACID, γ-LACTONE (11S)-(−)-
    see SAU500
4-HYDROXY-3-(3-OXO-1-FENYL-BUTYL) CUMARINE (DUTCH) see WAT200
5-HYDROXY-4-OXO-NORVALINE see AKG500
Δ-HYDROXY-γ-OXO-l-NORVALINE see AKG500
3-β-HYDROXY-23-OXO-OLEAN-12-EN-28-OIC ACID see GMG000
3-β-HYDROXY-11-OXOOLEAN-12-EN-30-OIC ACID see GIE000
3-β-HYDROXY-11-OXO-18-α-OLEAN-12-EN-30-OIC ACID see GIE05O
3-β-HYDROXY-11-OXO-OLEAN-12-EN-30-OIC ACID, HYDROGEN SUCCINATE
    see BGD000
3-β-HYDROXY-11-OXOOLEAN-12-EN-30-OIC ACID HYDROGEN SUCCINATE
    DISODIUM SALT see CBO500
4-HYDROXY-3-(3-OXO-1-PHENYLBUTYL)-2H-1-BENZOPYRAN-2-ONE POTASSI-
    UM SALT see WAT209
4-HYDROXY-3-(3-OXO-1-PHENYLBUTYL)-2H-1-BENZOPYRAN-2-ONE SODIUM
    SALT (9CI) see WAT220
4-HYDROXY-3-(3-OXO-1-PHENYL-BUTYL)-CUMARIN (GERMAN) see WAT200
HYDROXYOXOPHENYL IODANIUM PERCHLORATE see HND375
(HYDROXY)(OXO)(PHENYL)-LAMBDA³-IODANIUM PERCHLORATE see IFS000
15-HYDROXY-9-OXO-PROSTA-5,10-13-TRIEN-1-OIC ACID, (5Z,13E,15S)- (9CI)
    see MCA025
3-HYDROXY-2-OXOPURINE see HOB000
trans-6-(10-HYDROXY-6-OXO-1-UNDECENYL)-μ-LACTONE, RESORCYLIC ACID
    see ZAT000
6-(10-HYDROXY-6-OXO-trans-1-UNDECENYL)-β-RESORCYCLIC ACID-N-LAC-
    TONE see ZAT000
2′-HYDROXYPELARGIDENOLON 1522 see MRN500
3-HYDROXY-2-PENTANEDIOIC ACID, DIMETHYL ESTER, DIMETHYL PHOS-
    PHATE see SOY000
4-HYDROXYPENTANOIC ACID LACTONE see VAV000
4-HYDROXYPENT-3-ENOIC ACID LACTONE see AOO750, MKH250
4-HYDROXY-2-PENTENOIC ACID γ-LACTONE see MKH500
S-3-HYDROXY-4-PENTENONITRILE see COP400
N-HYDROXYPHENACETIN see HIN000
HYDROXYPHENAMATE see HNJ000
N-HYDROXY-N-2-PHENANTHRENYL-ACETAMIDE (9CI) see PCZ000
N-HYDROXY-4′-PHENETHYLACETANILIDE see PDI250
β-HYDROXYPHENETHYL ALCOHOL-α-CARBAMATE see PFJ000
4-HYDROXYPHENETHYLAMINE see TOG250
p-HYDROXY-β-PHENETHYLAMINE see TOG250
β-HYDROXYPHENETHYLAMINE see HNF000
p-HYDROXYPHENETHYLAMINE see TOG250
2-(β-HYDROXYPHENETHYLAMINO)PYRIDINE see PGG350
2-(β-HYDROXY-β-PHENETHYLAMINO)-PYRIMIDINE HYDROCHLORIDE see
    FAR200
β-HYDROXYPHENETHYL CARBAMATE see PFJ000
(−)-2-(6-(β-HYDROXYPHENETHYL)-1-METHYL-2-PIPERIDYL)-ACETOPHENONE
    HYDROCHLORIDE see LHZ000
(m-HYDROXYPHENETHYL)TRIMETHYLAMMONIUM PICRATE see HNG000
p-HYDROXYPHENETOLE see EFA100
2-HYDROXYPHENOL see CCP850
3-HYDROXYPHENOL see REA000
m-HYDROXYPHENOL see REA000
o-HYDROXYPHENOL see CCP850
p-HYDROXYPHENOL see HIH000
7-HYDROXY-3H-PHENOXAZIN-3-ONE-10 OXIDE see HNG500
1-HYDROXY-2-PHENOXYETHANE see PER000
(2-HYDROXYPHENOXY)PHENYLMERCURY (8CI) see PFO750
2-HYDROXY-N-PHENYLACETAMIDE see GHK500
N-(4-HYDROXYPHENYL)ACETAMIDE see HIM000
m-HYDROXYPHENYL ACETATE see RDZ900
(4-HYDROXYPHENYL)ACETIC ACID see HNG600
α-HYDROXYPHENYLACETIC ACID see MAP000

(p-HYDROXYPHENYL)ACETIC ACID see HNG600
p-HYDROXYPHENYLACETIC ACID see HNG700
HYDROXYPHENYLACETONITRILE see MAP250
2-HYDROXY-2-PHENYLACETOPHENONE see BCP250
α-HYDROXY-α-PHENYLACETOPHENONE see BCP250
β-(4-HYDROXYPHENYL)ACRYLIC ACID see CNU825
p-HYDROXYPHENYLACRYLIC ACID see CNU825
l-β-(p-HYDROXYPHENYL)ALANINE see TOG300
α-(4-HYDROXYPHENYL)-β-AMINOETHANE see TOG250
1-(3′-AMINOPHENYL)-2-AMINOETHANOL see AKT000
1-(m-HYDROXYPHENYL)-2-AMINOETHANOL see AKT000
1-(p-HYDROXYPHENYL)-2-AMINOETHANOL see AKT250
l-1-(m-HYDROXYPHENYL)-2-AMINO-1-PROPANOL-d-HYDROGEN TARTRATE
    see HNC000
4-HYDROXYPHENYLARSENOUS ACID see HNG800
p-HYDROXYPHENYLARSONIC ACID see PDO250
(4-HYDROXYPHENYL)ARSONIC ACID polymer with FORMALDEHYDE see
    BCJ150
α-HYDROXY-α-PHENYLBENZENEACETIC ACID see BBY990
α-HYDROXY-α-PHENYLBENZENEACETIC ACID-2-(DIETHYLAMINO)ETHYL ES-
    TER see DHU900
N-HYDROXYPHENYLBENZENESULFONAMIDE see HJH100
3-HYDROXYPHENYL BENZOATE see HNH500
2-(o-HYDROXYPHENYL)BENZOXAZOLE see HNI000
p-HYDROXYPHENYL BENZYL ETHER see AEY000
3-α-HYDROXY-8-(p-PHENYLBENZYL)-1-α-H,5-α-H-TROPANIUM BROMIDE, (±)-
    TROPATE see PEM750
4-(4-HYDROXYPHENYL)-2-BUTANONE see RBU000
1-(p-HYDROXYPHENYL)-3-BUTANONE see RBU000
4-(p-HYDROXYPHENYL)-2-BUTANONE (FCC) see RBU000
4-(p-HYDROXYPHENYL)-2-BUTANONE ACETATE see AAR500
p-HYDROXYPHENYLBUTAZONE see HNI500
1-(p-HYDROXYPHENYL)-2-BUTYLAMINOETHANOL see BQF250
2-HYDROXY-2-PHENYLBUTYL CARBAMATE see HNJ000
3-HYDROXY-2-PHENYLCINCHONINIC ACID see OPK300
4-(p-HYDROXYPHENYL)-2,5-CYCLOHEXADIEN-1-ONE SODIUM SALT see
    HNJ500
(3-HYDROXYPHENYL)DIETHYLMETHYLAMMONIUM BROMIDE see HNK000
(m-HYDROXYPHENYL)DIETHYLMETHYLAMMONIUM IODIDE, DIMETHYLCAR-
    BAMATE see HNK500
(m-HYDROXYPHENYL)DIETHYLMETHYLAMMONIUM METHOSULFATE ME-
    THYLCARBAMATE see HNK550
2-(p-HYDROXYPHENYL)-5,7-DIHYDROXYCHROMONE see CDH250
4-(4-(4-HYDROXYPHENYL)-2,3-DIISOCYANO-1,3-BUTADIENYL)-1,2-BENZENE-
    DIOL see XCS680
(−)-(R)-1-(p-HYDROXYPHENYL)-2-((3,4-DIMETHOXYPHENE-
    THYL)AMINO)ETHANOL see DAP850
3-HYDROXYPHENYLDIMETHYLETHYLAMMONIUM BROMIDE see TAL490
(4-HYDROXY-m-PHENYLENE)BIS(ACETATOMERCURY) see HNK575
m-HYDROXYPHENYLETHANOLAMINE see AKT000
p-HYDROXYPHENYLETHANOLAMINE see AKT250
m-HYDROXYPHENYLETHANOLETHYLAMINE see EGE500
1-(4-HYDROXYPHENYL)ETHANONE see HIO000
N-HYDROXY-N-(4-(2-PHENYLETHENYL)PHENYL)ACETAMIDE see SMT000
2-(p-HYDROXYPHENYL)ETHYLAMINE see TOG250
4-HYDROXYPHENYLETHYLAMINE see TOG250
β-HYDROXY-β-PHENYLETHYLAMINE see HNF000
β-HYDROXYPHENYLETHYLAMINE see TOG250
1-(3′-HYDROXYPHENYL)-2-ETHYLAMINOETHANOL see EGE500
2-(β-(HYDROXYPHENYL)ETHYLAMINOMETHYL)TETRALONE HYDROCHLORIDE
    see HAJ700
2-(β-HYDROXY-β-PHENYL-ETHYL-AMINO)-PYRIMIDINE CHLORHYDRATE
    (FRENCH) see FAR200
2-HYDROXY-2-PHENYLETHYL CARBAMATE see PFJ000
β-HYDROXY-β-PHENYLETHYL DIMETHYLAMINE see DSH700
1-(3-HYDROXYPHENYL)-1-HYDROXY-2-AMINOETHANE see AKT000
1-(4-HYDROXYPHENYL)-1-HYDROXY-2-BUTYLAMINOETHANE see BQF250
(2-HYDROXYPHENYL)HYDROXYMERCURY see HLO500
7-HYDROXY-4-PHENYL-3-(4-HYDROXYPHENYL)COUMARIN see HNK600
1-(3-HYDROXYPHENYL)-2-ISOPROPYLAMINOETHANOL HYDROCHLORIDE see
    HLK500
β-(p-HYDROXYPHENYL)ISOPROPYLMETHYLAMINE see FMS875
(2R,4R)-2-(o-HYDROXYPHENYL)-3-(3-MERCAPTOPROPIONYL)-4-THIAZOLIDINE-
    CARBOX YLIC ACID see FAQ950
o-HYDROXYPHENYLMERCURIC CHLORIDE see CHW675
HYDROXYPHENYLMERCURY see PFN100
p-HYDROXYPHENYL 2-MESITYLBENZOFURAN-3-YL KETONE see HNK700
p-HYDROXYPHENYL 2-MESITYLBENZOFURAN-4-YL KETONE see HNK800
1-1-(m-HYDROXYPHENYL)-2-METHYLAMINOETHANOL see NCL500
1-(4-HYDROXYPHENYL)-2-METHYLAMINOETHANOL see HLV500
p-HYDROXYPHENYLMETHYLAMINOETHANOL see HLV500
l-1-(m-HYDROXYPHENYL)-2-METHYL-AMINOETHANOL HYDROCHLORIDE see
    SPC500
(+−)-1-(4-HYDROXYPHENYL)-2-METHYLAMINOETHANOL TARTRATE see
    SPD000

1-(p-HYDROXYPHENYL)-2-METHYLAMINOPROPANE see FMS875
α-(p-HYDROXYPHENYL)-β-METHYLAMINOPROPANE see FMS875
1-(4-HYDROXYPHENYL)-2-METHYLAMINOPROPANOL see HKH500
p-HYDROXYPHENYLMETHYLAMINOPROPANOL see HKH500
2-HYDROXYPHENYL METHYLCARBAMATE see HNK900
o-HYDROXYPHENYL METHYLCARBAMATE see HNK900
1-(3-HYDROXYPHENYL)-N-METHYLETHANOLAMINE see NCL500
1-(4-HYDROXYPHENYL)-N-METHYLETHANOLAMINE see HLV500
4-(m-HYDROXYPHENYL)-1-METHYLISONIPECOTINOYL METHYL KETONE see
    HNK950
o-HYDROXYPHENYL METHYL KETONE see HIN500
p-HYDROXYPHENYL METHYL KETONE see HIO000
1-(4-HYDROXYPHENYL)-2-(1-METHYL-2-PHENOXYETHYLAMINO)PROPANOL
    see VGF000
1-(p-HYDROXYPHENYL)-2-(1′-METHYL-2′-PHENOXY)ETHYLAMINOPROPANOL-1
    HYDROCHLORIDE see VGA300
1-(p-HYDROXYPHENYL)-2-(1′-METHYL-2′-PHENOXYETHYLAMINO)PROPANOL-2-
    HYDROCHLORIDE see VGF000
α-(4-HYDROXYPHENYL)-β-METHYL-4-(PHENYLMETHYL)-1-PIPERIDINEETHANOL
    (9CI) see IAG600
1-p-HYDROXYPHENYL-2-(1′-METHYL-3′-PHENYLPROPYLAMINO)-1-PROPANOL
    HYDROCHLORIDE see DNU200
(±)-4-(2-((3-(4-HYDROXYPHENYL)-1-METHYLPROPYL)AMINO)ETHYL)-1,2-BEN-
    ZENEDIOL HYDROCHLORIDE see DXS375
4-(2-((3-(p-HYDROXYPHENYL)-1-METHYLPRO-
    PYL)AMINO)ETHYL)PYROCATECHOL HYDROCHLORIDE see DXS375
N-p-HYDROXYPHENYL-2-NAPHTHYLAMINE see NBF500
p-HYDROXYPHENYL-2-NAPHTHYLAMINE see NBF500
N-p-HYDROXYPHENYL-β-NAPHTHYLAMINE see NBF500
p-HYDROXYPHENYL-β-NAPHTHYLAMINE see NBF500
1-HYDROXY-1-PHENYLPENTANE see BQJ500
1-(p-HYDROXYPHENYL)-2-PHENYL-4-BUTYL-3,5-PYRAZOLIDINEDIONE see
    HNI500
1-p-HYDROXYPHENYL-2-PHENYL-3,5-DIOXO-4-N-BUTYLPYRAZOLIDINE see
    HNI500
1-(p-HYDROXYPHENYL)-2-(3′-PHENYLTHIOPROPYLAMINO)-1-PROPANOL HY-
    DROCHLORIDE see HNL100
1-(4-HYDROXYPHENYL)-1-PROPANONE see ELL500
p-HYDROXYPHENYL-1-PROPANONE see ELL500
N-HYDROXY-3-PHENYL-2-PROPENAMIDE (9CI) see CMQ475
(E)-3-(2-HYDROXYPHENYL)-2-PROPENOIC ACID see CNU850
3-(4-HYDROXYPHENYL)-2-PROPENOIC ACID see CNU825
3-((1-HYDROXY-1-PHENYL-2-PROPYL)METHYLAMINO)PROPIONITRILE see
    CCX600
(p-HYDROXYPHENYL) 2-PYRIDYL KETONE see HJN700
p-HYDROXYPHENYLPYRUVIC ACID see HNL500
3-HYDROXY-2-PHENYL-4-QUINOLINECARBOXYLIC ACID see OPK300
N-(4-HYDROXYPHENYL)SALICYLAMIDE see DYE700
N-(p-HYDROXYPHENYL)SALICYLAMIDE see DYE700
p-HYDROXYPHENYLSALICYLAMIDE see DYE700
HYDROXYPHENYL SALICYLAMIDE see DYE700
4-HYDROXYPHENYLSULFONIC ACID see HNL600
4-(3-HYDROXY-3-PHENYL-3-(2-THIENYL)PROPYL)-4-METHYLMORPHOLINIUM-
    IODIDE see HNM000
4-HYDROXY-α-(1-((3-(PHENYLTHIO)PROPYL)AMINO)ETHYL)-BENZENEMETHA-
    NOL HYDROCHLORIDE see HNL100
p-HYDROXY-α-(1-((3-(PHENYLTHIO)PROPYL)AMINO)ETHYL)BENZYL ALCOHOL
    HYDROCHLORIDE see HNL100
2-((N-(m-HYDROXYPHENYL)-p-TOLUIDINO)METHYL)-2-IMIDAZOLINE see
    PDW400
(m-HYDROXYPHENYL)TRIMETHYLAMMONIUM BROMIDE DIMETHYLCARBA-
    MATE see POD000
3-HYDROXYPHENYLTRIMETHYLAMMONIUM BROMIDE DIMETHYLCARBAMIC
    ESTER see POD000
(m-HYDROXYPHENYL)TRIMETHYLAMMONIUM CHLORIDE, METHYLCARBA-
    MATE see HNN000
(m-HYDROXYPHENYL)TRIMETHYLAMMONIUM DIMETHYLCARBAMATE (ester)
    see NCL100
(m-HYDROXYPHENYL)TRIMETHYLAMMONIUM IODIDE see HNN500
(m-HYDROXYPHENYL)TRIMETHYLAMMONIUM IODIDE, BENZYLCARBAMATE
    see HNO000
(m-HYDROXYPHENYL)TRIMETHYLAMMONIUM IODIDE, METHYLCARBAMATE
    see HNO500
(p-HYDROXYPHENYL)TRIMETHYLAMMONIUM IODIDE, METHYLCARBAMATE
    see HNP000
(m-HYDROXYPHENYL)TRIMETHYLAMMONIUM METHYLSULFATE BENZYLCAR-
    BAMATE see BED500
(m-HYDROXYPHENYL)TRIMETHYLAMMONIUM METHYLSULFATE, CARBA-
    MATE see HNP500
(m-HYDROXYPHENYL)TRIMETHYLAMMONIUM METHYL SULFATE DIMETHYL-
    CARBAMATE see DQY909
(3-HYDROXYPHENYL)TRIMETHYLAMMONIUM METHYL SULFATE DIMETHYL-
    CARBAMIC ESTER see DQY909
(m-HYDROXYPHENYL)TRIMETHYLAMMONIUM METHYLSULFATE DIETHYL-
    CARBAMATE see HNQ000

(m-HYDROXYPHENYL)TRIMETHYLAMMONIUM METHYLSULFATE, ETHYLCAR-BAMATE see HNQ500

(m-HYDROXYPHENYL)TRIMETHYLAMMONIUM METHYLSULFATE, METHYL-CARBAMATE see HNR000

(m-HYDROXYPHENYL)TRIMETHYLAMMONIUM METHYLSULFATE METHYLPHE-NYLCARBAMATE see HNR500

(m-HYDROXYPHENYL)TRIMETHYLAMMONIUM METHYLSULFATE, PENTAME-THYLENECARBAMATE see HNS000

(m-HYDROXYPHENYL)TRIMETHYLAMMONIUM METHYLSULFATE, PHENYL-CARBAMATE see HNS500

N-HYDROXYPHTHALIMIDE see HNS600

3-HYDROXY-2-PICOLINE-4,5-DIMETHANOL see PPK250

3-(1-HYDROXY-2-PIPERIDINOETHYL)-5-PHENYLISOXAZOLE see PCJ325

3-(1-HYDROXY-2-PIPERIDINOETHYL)-5-PHENYLISOXAZOLE CITRATE (2:1) see PCJ325

3-(1-HYDROXY-2-PIPERIDINOETHYL)-5-PHENYLISOXAZOLE CITRATE see HNT075

2-HYDROXY-N-(3-(m-(PIPERIDINOMETHYL)PHENOXY)PROPYL)ACETAMIDE AC-ETATE (ester) HYDROCHLORIDE see HNT100

10-(3-(4-HYDROXYPIPERIDINO)PROPYL)PHENOTHIAZINE-2-CARBONITRILE see PIW000

10-(3-(4-HYDROXYPIPERIDINO)PROPYL)PHENOTHIAZIN-2-YL METHYL KETONE see HNT200

HYDROXYPOLYETHOXYDODECANE see DXY000

3-HYDROXYPREGNANE-11,20-DIONE see AFK875

3-α-HYDROXY-5-α-PREGNANE-11,20-DIONE see AFK875

3-α-HYDROXY-5-α-PREGNANE-11,20-DIONE mixed with 3-α,21-DIHYDROXY-5-α-PREGNANE-11,20-DIONE 21-ACETATE (3:1) see AFK500

21-HYDROXYPREGNANE-3,20-DIONE SODIUM HEMISUCCINATE see VJZ000

21-HYDROXY-5-β-PREGNANE-3,20-DIONE SODIUM HEMISUCCINATE see VJZ000

21-HYDROXY-5-β-PREGNANE-3,20-DIONE, SODIUM SALT, HEMISUCCINATE see VJZ000

21-HYDROXYPREGN-4-ENE-3,20-DIONE see DAQ600

21-HYDROXYPREGN-4-ENE-3,20-DIONE-21-ACETATE see DAQ800

17-HYDROXYPREGN-4-ENE-3,20-DIONE HEXANOATE see HNT500

17α,21-HYDROXYPREGN-4-ENE-3,11,20-TRIONE see CNS800

17-HYDROXY-17-α-PREGN-4-EN-20-YN-3-ONE see GEK500

m-HYDROXYPROCAINE see DHO600

HYDROXYPROCAINE see DHO600

21-HYDROXYPROGESTERONE see DAQ600

17-α-HYDROXYPROGESTERONE CAPROATE see HNT500

17-α-HYDROXY PROGESTERONE-N-CAPROATE see HNT500

HYDROXYPROGESTERONE CAPROATE see HNT500

17-α-HYDROXYPROGESTERONE HEXANOATE see HNT500

m-HYDROXYPROPADRINE see HNB875

1-HYDROXYPROPANE see PND000

2-HYDROXY-1,2,3-PROPANECARBOXYLIC ACID TERBIUM (3+) SALT (1:1) see TAM500

((2-HYDROXY-1,3-PROPANE-DIYL)BIS(NITRILOBIS(METHYLENE)))TETRAKISPHOSPHONIC ACID see DYE550

3-HYDROXYPROPANENITRILE see HGP000

3-HYDROXY-1-PROPANESULFONIC ACID γ-SULTONE see PML400

3-HYDROXY-1-PROPANESULPHONIC ACID SULTONE see PML400

2-HYDROXY-1,2,3-PROPANETRICARBOXYLIC ACID see CMS750

2-HYDROXY-1,2,3-PROPANETRICARBOXYLIC ACID COPPER(2+) SALT (1:2) (9CI) see CNK625

2-HYDROXY-1,2,3-PROPANETRICARBOXYLIC ACID EBRIUM(3+) salt (1:1) see ECY000

2-HYDROXY-1,2,3-PROPANETRICARBOXYLIC ACID MONOSODIUM SALT see MRL000

2-HYDROXY,1,2,3-PROPANETRICARBOXYLIC ACID, TRIETHYL ESTER see TJP750

2-HYDROXY-1,2,3-PROPANETRISCARBOXYLIC ACID CERIUM(3+) SALT (1:1) (9CI) see CCZ000

2-HYDROXYPROPANNITRIL see LAQ000

2-HYDROXYPROPANOIC ACID see LAG000

(S)-2-HYDROXYPROPANOIC ACID see LAG010

2-HYDROXYPROPANOIC ACID, BUTYL ESTER see BRR600

2-HYDROXYPROPANOIC ACID CALCIUM SALT see CAT600

2-HYDROXYPROPANOIC ACID MONOSODIUM SALT see LAM000

1-HYDROXY-2-PROPANONE see ABC000

3-HYDROXY-2-PROPENAL SODIUM SALT see MAN700

3-HYDROXYPROPENE see AFV500

2-HYDROXYPROPIONIC ACID see LAG000

α-HYDROXYPROPIONIC ACID see LAG000

(S)-2-HYDROXYPROPIONIC ACID see LAG010

3-HYDROXYPROPIONIC ACID LACTONE see PMT100

2-HYDROXYPROPIONITRILE see LAQ000

3-HYDROXYPROPIONITRILE see HGP000

β-HYDROXYPROPIONITRILE see HGP000

N-HYDROXY-N-2-PROPIONYLAMINO FLUORENE see FEO000

l-5α-HYDROXYPROPIONYLAMINO-2,4,6-TRIIODOISOPHTHALIC ACID DI(1,3-DI-HYDROXY-2-PROPYLAMIDE) see IFY000

4-HYDROXYPROPIOPHENONE see ELL500

p-HYDROXYPROPIOPHENONE see ELL500

HYDROXYPROPIOPHENONE see ELL500

2-(2-(2-HYDROXYPROPOXY)PROPOXY)-1-PROPANOL see TMZ000

O-(2-HYDROXYPROPYL)-1-ACETYLBENZOCYCLOBUTENE OXIME see HNT550

2-HYDROXYPROPYL ACRYLATE see HNT600

β-HYDROXYPROPYL ACRYLATE see HNT600

HYDROXY PROPYL ALGINATE see PNJ750

2-HYDROXYPROPYLAMINE see AMA500

3-HYDROXYPROPYLAMINE see PMM250

2′-(2-HYDROXY-3-(PROPYLAMINO)PROPOXY)-3-PHENYLPROPIOPHENONE HY-DROCHLORIDE see PMJ525

(3-HYDROXYPROPYL)BENZENE see HHP050

α-HYDROXYPROPYLBENZENE see EGG000

p-HYDROXYPROPYL BENZOATE see HNU500

HYDROXYPROPYL CELLULOSE see HNV000

1-(3-HYDROXYPROPYL)-CNU see CHC250

1-(3-HYDROXYPROPYL)-3,7-DIMETHYLXANTHINE see HNY500

1-(3-HYDROXYPROPYL)-3,7-DIMETHYLXANTHINE see HNZ000

(2-HYDROXY-1,3-PROPYLENEDIAMINE)-N,N,N′,N′-TETRAMETHYLENEPHOS-PHORIC ACID see DYE550

HYDROXYPROPYL ETHER of CELLULOSE see HNV000

2-HYDROXYPROPYL METHACRYLATE see HNV500

β-HYDROXYPROPYL METHACRYLATE see HNV500

HYDROXYPROPYL METHYLCELLULOSE see HNX000

2-HYDROXYPROPYLMETHYLNITROSAMINE see NKU500

2-HYDROXYPROPYL 2-METHYL-2-PROPENOATE see HNV500

3-((2-HYDROXYPROPYL)NITROSAMINO)-1,2-PROPANEDIOL see NOC400

1-(((2-HYDROXYPROPYL)NITROSO)AMINO)ACETONE see HNX500

1-((2-HYDROXYPROPYL)NITROSOAMINO)-2-PROPANONE see HNX500

1-(2-HYDROXYPROPYL)-1-NITROSOUREA see NKO400

1-(3-HYDROXYPROPYL)-1-NITROSOUREA see NKK000

2-HYDROXYPROPYL PHENYL ARSINIC ACID see HNX600

N-(2-HYDROXYPROPYL)-1,2-PROPANEDIAMINE see HNX800

β-HYDROXYPROPYLPROPYLNITROSAMINE see NLM500

(2-HYDROXYPROPYL)PROPYLNITROSOAMINE see NLM500

HYDROXYPROPYL STARCH see HNY000

1-(2-HYDROXYPROPYL)THEOBROMINE see HNY500

1-(3-HYDROXYPROPYL)THEOBROMINE see HNZ000

1-(β-HYDROXYPROPYL)THEOBROMINE see HNY500

7-(2-HYDROXYPROPYL)THEOPHYLLINE see HOA000

7-(β-HYDROXYPROPYL)THEOPHYLLINE see HOA000

β-HYDROXYPROPYLTHEOPHYLLINE see HOA000

HYDROXYPROPYLTHEOPHYLLINE see HOA000

(3-HYDROXYPROPYL)TRIMETHOXYSILANE METHACRYLATE see TLC250

(2-HYDROXYPROPYL)TRIMETHYLAMMONIUM CHLORIDE see MIM300

(2-HYDROXYPROPYL)TRIMETHYLAMMONIUMCHLORIDE ACETATE see ACR000

(2-HYDROXYPROPYL)TRIMETHYLAMMONIUM CHLORIDE CARBAMATE see HOA500

4-(1-(HYDROXY)PROPYOXY)BENZENEARSONIC ACID see PMQ000

N-HYDROXY-1H-PURIN-6-AMINE (9CI) see HIT000

3-HYDROXYPURIN-2(3H)-ONE see HOB000

4′-HYDROXYPYRAZOLOL(3,4-d)PYRIMIDINE see ZVJ000

4-HYDROXY-3,4-PYRAZOLOPYRIMIDINE see ZVJ000

4-HYDROXY-1H-PYRAZOLO(3,4-d)PYRIMIDINE see ZVJ000

4-HYDROXYPYRAZOLO(3,4-d)PYRIMIDINE see ZVJ000

4-HYDROXYPYRAZOLYL(3,4-d)PYRIMIDINE see ZVJ000

8-HYDROXYPYRENE-1,3,6-TRISULFONIC ACID SODIUM SALT see TNM000

8-HYDROXY-1,3,6-PYRENETRISULFONIC ACID TRISODIUM SALT see TNM000

6-HYDROXY-3(2H)-PYRIDAZINONE see DMC600

6-HYDROXY-3-(2H)-PYRIDAZINONE DIETHANOLAMINE see DHF200

2-HYDROXYPYRIDINE see OOO200

3-HYDROXYPYRIDINE see PPH025

4-HYDROXYPYRIDINE see HOB100

β-HYDROXYPYRIDINE see PPH025

γ-HYDROXYPYRIDINE see HOB100

1-HYDROXY-2-(1H)-PYRIDINETHIONE see HOB500

1-HYDROXY-2-PYRIDINETHIONE see HOB500

1-HYDROXY-2-(1H)-PYRIDINETHIONE SODIUM SALT see HOC000

(1-HYDROXY-2-PYRIDINETHIONE), SODIUM SALT, TECH see MCQ750

5-(α-HYDROXY-α-2-PYRIDYLBENZYL)-7-(α-2-PYRIDYLBENZYLIDENE)-5-NOR-BORNENE-2,3-DICARBOXIMIDE see NNF000

4-HYDROXY-2(1H)-PYRIMIDINETHIONE see TFR250

8-HYDROXYQUINALDIC ACID see HOE000

HYDROXYQUINOL see BBU250

N-HYDROXY-4-QUINOLINAMINE-1-OXIDE MONOHYDROCHLORIDE see HIZ000

4-HYDROXYQUINOLINAMINE-1-OXIDE MONOHYDROCHLORIDE see HIZ000

8-HYDROXYQUINOLINE see QPA000

8-HYDROXYQUINOLINE COPPER COMPLEX see BLC250

8-HYDROXYQUINOLINE SULFATE see QPS000

1-(8-HYDROXY-5-QUINOLINYL)ETHANONE see HOE200

8-HYDROXY-5-QUINOLYL METHYL KETONE see HOE200

3-HYDROXYQUINUCLIDINE see QUJ990

14-HYDROXY-3-β-(RHAMNOSYLOXY)BUFA-4,20,22-TRIENOLIDE see POB500

5-HYDROXY-1-β-d-RIBOFURANOSYL-1H-IMIDAZOLE-4-CARBOXAMIDE see BMM000
1'-HYDROXYSAFROLE see BCJ000
1'-HYDROXYSAFROLE-2',3'-OXIDE see HOE500
4-HYDROXYSALICYLALDEHYDE see REF100
4'-HYDROXYSALICYLANILIDE see DYE700
p'-HYDROXYSALICYLANILIDE see DYE700
5-HYDROXYSALICYLATE SODIUM see GCU050
5-HYDROXYSALICYLIC ACID see GCU000
14-β-HYDROXY-3-β-SCILLOBIOSIDOBUFA-4,20,22-TRIENOLIDE see GFC000
3-HYDROXY-16,7-SECOESTRA-1,3,5(10)-TRIEN-17-OIC ACID see DYB000
12-HYDROXYSENECIONAN-11,16-DIONE see ARS500
(15E)-12-HYDROXY-SENECIONAN-11,16-DIONE (9CI) see IDG000
HYDROXYSENKIRKINE see HOF000
5-HYDROXY-1(N-SODIO-5-TETRAZOLYLAZO)TETRAZOLE see HOF500
3-HYDROXY-SPIRO(8-AZONIABICYCLO(3.2.1)OCTANE-8,1'-PYRROLIDINIUM CHLORIDE) BENZILATE see BCA375
3-α-HYDROXY-SPIRO(1-α-H,5-α-H-NORTROPANE-8,1'-PYRROLIDINIUM) CHLORIDE BENZILATE see KEA300
12-HYDROXYSTEARIC ACID see HOG000
12-HYDROXYSTEARIC ACID, METHYL ESTER see HOG500
HYDROXYSTREPTOMYCIN see SLX500
4-HYDROXYSTYRENE see VQA200
p-HYDROXYSTYRENE see VQA200
(E)-4'-(p-HYDROXYSTYRYL) ACETANILIDE see HIK000
α-HYDROXYSUCCINIC ACID see MAN000
HYDROXYSUCCINIC ACID see MAN000
4-HYDROXY-3-((4-SULFO-1-NAPHTHALENYL)AZO)-1-NAPHTHALENESULFONIC ACID, DISODIUM SALT see HJF500
3-HYDROXY-4-((4-SULFO-1-NAPHTHALENYL)AZO)-2,7-NAPHTHLENEDISULFON-IC ACID, TRISODIUM SALT see FAG020
3-HYDROXY-4-((4-SULFO-1-NAPHTHALENYL)AZO)-2,7-NAPHTHALENEDISULFONIC ACID, TRISODIUM SALT see FAG020
6-HYDROXY-5-((p-SULFOPHENYL)AZO)-2-NAPHTHALENESULFONIC ACID, DI-SODIUM SALT see FAG150
6-HYDROXY-5-((4-SULFOPHENYL)AZO)-2-NAPHTHALENESULFONIC ACID, DI-SODIUM SALT see FAG150
4-HYDROXY-3-((5-SULFO-2,4-XYLYL)AZO)-1-NAPHTHALENESULFONIC ACID, DISODIUM SALT see FAG050
3-HYDROXY-4-((4-SULPHO-1-NAPHTHALENYL)AZO)-2,7-NAPHTHALENEDISUL-PHONIC ACID, TRISODIUM SALT see FAG020
3-HYDROXY-4-((4-SULPHO-1-NAPHTHYL)AZO)-2,7-NAPHTHALENEDISULPHONIC ACID, TRISODIUM SALT see FAG020
6-HYDROXY-5-((p-SULPHOPHENYL)AZO)-2-NAPHTHALENESULPHONIC ACID, DISODIUM SALT see FAG150
6-HYDROXY-5-((4-SULPHOPHENYL)AZO)-2-NAPHTHALENESULPHONIC ACID, DISODIUM SALT see FAG150
4-HYDROXY-3-((5-SULPHO-2,4-XYLYL)AZO)-1-NAPHTHALENESULPHONIC ACID, DISODIUM SALT see FAG050
5-HYDROXYTETRACYCLINE see HOH500
5-HYDROXYTETRACYCLINE HYDROCHLORIDE see HOI000
N-HYDROXY-N-TETRADECANOYL-2-AMINOFLUORENE see HMU000
m-HYDROXYTETRAETHYLDIAMINOTRIPHENYLCARBINOL ANHYDRIDE DISUL-FONIC ACID CALCIUM SALT see CMM062
3-HYDROXYTETRAHYDROFURAN see HOI200
4-HYDROXY-3-(1,2,3,4-TETRAHYDRO-1-NAFTYL)-4-CUMARINE (DUTCH) see EAT600
4-HYDROXY-3-(1,2,3,4-TETRAHYDRO-1-NAPHTHALENYL)-2H-1-BENZOPYRAN-2-ONE (9CI) see EAT600
4-HYDROXY-3-(1,2,3,4-TETRAHYDRO-1-NAPHTHYL)CUMARIN see EAT600
6-HYDROXY-2,5,7,8-TETRAMETHYLCHROMAN-2-CARBOXYLIC ACID see TNR625
7-HYDROXY-3,4-TETRAMETHYLENECOUMARIN-O,O-DIETHYL THIOPHOSP-HATE see DXO000
3-HYDROXY-2,2,5,5-TETRAMETHYLTETRAHYDRO-3-FURYL METHYL KETONE see HOI400
14-HYDROXY-6-β-THEBAINOL 4-METHYL ETHER see ORE000
7-HYDROXYTHEOPHYLLIN (GERMAN) see HOJ000
7-HYDROXYTHEOPHYLLINE see HOJ000
9-(2-HYDROXYTHEOXYMETHYL)GUANINE see AEC700
6-HYDROXY-2-THIO-8-AZAPURINE see HOJ100
4-HYDROXYTOLUENE see CNX250
α-HYDROXYTOLUENE see BDX500
m-HYDROXYTOLUENE see CNW750
o-HYDROXYTOLUENE see CNX000
p-HYDROXYTOLUENE see CNX250
HYDROXYTOLUENE see BDX500
α-HYDROXY-α-TOLUIC ACID see MAP000
1-HYDROXY-4-(p-TOLUIDINO)ANTHRAQUINONE see HOK000
HYDROXYTOLUOLE (GERMAN) see CNW500
2-(m-HYDROXY-N-p-TOLYLANILINOMETHYL)-2-IMIDAZOLINE see PDW400
4'-((6-HYDROXY-m-TOLYL)AZO)ACETANILIDE see AAQ250
2-HYDROXY-3-o-TOLYLOXYPROPYL-1-CARBAMATE see CBK500
(3-HYDROXY-p-TOLYL)TRIMETHYLAMMONIUM CHLO-RIDE,METHYLCARBAMATE see HOL000

3-HYDROXY-4β,8-α-15-TRIACETOXY-12,13-EPOXYTRICHOTHEC-9-ENE see ACS500
4-HYDROXY-3H-o-TRIAZOLO(4,5-d)PYRIMIDINE-5,7(4H-6H)-DIONE see HJE575
(HYDROXY)TRIBUTYLSTANNANE see TID500
HYDROXYTRIBUTYLSTANNANE-4,4-DIMETHYLOCTANOATE see TIF250
HYDROXYTRIBUTYLSTANNANE, SULFATE (2:1) see TIF600
β-HYDROXYTRICARBALLYLIC ACID see CMS750
1-HYDROXY-2,2,2-TRICHLOROETHYL DIETHYL PHOSPHONITE see EPY600
1-HYDROXY-2,2,2-TRICHLOROETHYLPHOSPHONIC ACID DIMETHYL ESTER see TIQ250
2'-HYDROXY-2,4,4'-TRICHLORO-PHENYLETHER see TIQ000
2-HYDROXYTRIETHYLAMINE see DHO500
(S)-N-(3-HYDROXY-9,10,11-TRIMETHOXY-5H-DIBENZO(a,c)CYCLOHEPTEN-5-YL)-ACETAMIDE (9CI) see ACG250
3-HYDROXY-N,N,N-TRIMETHYLBENZENAMINIUM IODIDE see HNN500
3-HYDROXY-N,N,N-TRIMETHYLBENZENEETHANAMINIUM PICRATE see HNG000
((2-HYDROXYTRIMETHYLENE)DINITRILO)TETRAACETIC ACID see DCB100
2-HYDROXY-N,N,N-TRIMETHYLETHANAMINIUM see CMF000
2-HYDROXY-N,N,N-TRIMETHYLETHANAMINIUM SALT with 3,7-DIHYDRO-1,3-DI-METHYLPURINE-2,6-DIONE see CMF500
2-HYDROXY-N,N,N-TRIMETHYLETHANAMINIUM SALT with 2-HYDROXYBEN-ZOIC ACID (1:1) see CMG000
3-HYDROXY-2,2,4-TRIMETHYL-3-PENTENOIC ACID, β-LACTONE see IPL000
3-HYDROXY-4-((2,4,5-TRIMETHYLPHENYL)AZO)-2,7-NAPHTHALENEDISULFONIC ACID, DISODIUM SALT see FAG018
3-HYDROXY-4-((2,4,5-TRIMETHYLPHENYL)AZO)-2,7-NAPHTHALENEDISULPHON-IC ACID, DISODIUM SALT see FAG018
2-HYDROXY-N,N,N-TRIMETHYL-1-PROPANAMINIUM CHLORIDE see MIM300
3-HYDROXY-N,N,5-TRIMETHYLPYRAZOLE-1-CARBOXAMIDE DIMETHYLCARBA-MATE (ESTER) see DQZ000
HYDROXYTRIMETHYLSTANNANE see TMI250
2-HYDROXY-1,3,5-TRINITROBENZENE see PID000
3-HYDROXY-2,4,6-TRINITROPHENOL see SMP500
HYDROXYTRIPHENYLSILANE see HOM300
HYDROXYTRIPHENYLSTANNANE see HON000
HYDROXYTRIPHENYLTIN see HON000
3-β-HYDROXY-1-α-H,5-α-H-TROPANE-2-β-CARBOXYLIC ACID METHYL ESTER, BENZOATE see CNE750
3-β-HYDROXY-1-α-H,5-α-H-TROPANE-2-β-CARBOXYLIC ACID METHYL ESTER, BENZOATE (ESTER), HYDROCHLORIDE see CNF000
3-HYDROXYTROPOLONE see HON500
5-HYDROXYTRYPTAMINE see AJX500
5-HYDROXYTRYPTAMINE CREATININE SULFATE see AJX750
5-HYDROXYTRYPTAMINE CREATININE SULFATE MONOHYDRATE see AJX750
5-HYDROXYTRYPTOPHAN see HOA575
(±)-5-HYDROXYTRYPTOPHAN see HOA575
5-HYDROXYTRYPTOPHAN see HON800
(±)-5-HYDROXYTRYPTOPHAN see HON800
5-HYDROXYTRYPTOPHAN see HOO100
dl-5-HYDROXYTRYPTOPHAN see HOA575
dl-5-HYDROXYTRYPTOPHAN see HOA575, HON800
5-HYDROXY-l-TRYPTOPHAN see HOA600
l-5-HYDROXYTRYPTOPHAN see HOA600, HOO000
5-HYDROXY-l-TRYPTOPHAN see HOO000
HYDROXYTRYPTOPHAN see HOO100
5-HYDROXYTRYPTOPHANE see HOO100
3-HYDROXYTYRAMINE see DYC400
3-HYDROXYTYRAMINE HYDROCHLORIDE see DYC600
m-HYDROXYTYRAMINE HYDROCHLORIDE see DYC600
3-HYDROXY-l-TYROSINE see DNA200
l-o-HYDROXYTYROSINE see DNA200
3-HYDROXY-l-TYROSINE HYDROCHLORIDE see DYC200
4-HYDROXYUNDECANOIC ACID LACTONE see HBN200
5-HYDROXYUNDECANOIC ACID LACTONE see UKJ000
4-HYDROXYUNDECANOIC ACID, γ-LACTONE see HBN200
N-HYDROXYUREA see HOO500
HYDROXYUREA see HOO500
N-HYDROXYURETHAN see HKQ025
N-HYDROXYURETHANE see HKQ025
3-HYDROXYURIC ACID see HOO875
4-HYDROXYVALERIC ACID LACTONE see VAV000
17-HYDROXY-17-α-VINYL-4-ESTREN-3-ONE see NCI525
1-α-HYDROXYVITAMIN D3 see HJV000
9-HYDROXYXANTHENE see XBJ000
3-HYDROXYXANTHINE see HOP000, HOQ500
7-HYDROXYXANTHINE see HOP259
1-HYDROXYXANTHINE DIHYDRATE see HOQ000
3-HYDROXYXANTHINE HYDRATE see HOQ500
3-HYDROXY-4-(2,4-XYLYLAZO)-3,7-NAPHTHALENEDISULFONIC ACID, DISODI-UM SALT see FMU070
3-HYDROXY-4-(2,4-XYLYLAZO)-3,7-NAPHTHALENEDISULPHONIC ACID, DISO-DIUM SALT see FMU070
3-(4-HYDROXY-3,5-XYLYL)-2-METHYL-4(3H)-QUINAZOLINONE see DSH800

17-HYDROXYYOHIMBAN-16-CARBOXYLIC ACID METHYL ESTER see YBJ000
17-α-HYDROXY-20-α-YOHIMBAN-16-β-CARBOXYLIC ACID, METHYL ESTER,
   HYDROCHLORIDE see YCA000
o-HYDROXYZIMTSAEURE-LACTON (GERMAN) see CNV000
HYDROXYZINE see CJR909
HYDROXYZINE DIHYDROCHLORIDE see HOR470
HYDROXYZINE HYDROCHLORIDE see HOR470, VSF000
HYDROXYZINE PAMOATE see HOR500
HYDRURE de LITHIUM (FRENCH) see LHH000
HYDURA see HOO500
HYFLAVIN see RIK000
HYGROMIX-8 see AQB000
HYGROMULL see UTU500
HYGROMYCIN B (USDA) see AQB000
HYGROSTATIN see HOS500
HYGROTON see CLY600
HYGROTON-RESERPINE see RDK000
HYKOLEX see DAL000
HYLEMOX see EEH600
HYLENE M50 see MJP400
HYLENE-T see TGM740
HYLENE T see TGM750
HYLENE TCPA see TGM750
HYLENE TLC see TGM750
HYLENE TM-65 see TGM750
HYLENE TM see TGM750, TGM800
HYLENE TRF see TGM750
HYLENTA see BFD000
HYLITE LF see DTG000
HYLUTIN see HNT500
HYMECROMONE see MKP500
HYMECROMONE SODIUM see HMB000
HYMENOCALLIS (VARIOUS SPECIES) see BAR325
HYMENOXON see HOT200
HYMENOXONE see HOT200
HYMEXAZOL see HLM000
HYMINAL see QAK000
HYMORPHAN see DLW600, DNU300
HYONIC PE-250 see PKF500
HYOSAN see MJM500
HYOSCIN-N-BUTYLBROMID (GERMAN) see SBG500
HYOSCIN-N-BUTYL BROMIDE see SBG500
HYOSCINE see SBG000
(−)-HYOSCINE see SBG000
HYOSCINE BROMIDE see HOT500
HYOSCINE BUTOBROMIDE see SBG500
HYOSCINE-N-BUTYL BROMIDE see SBG500
HYOSCINE BUTYL BROMIDE see SBG500
HYOSCINE F HYDROBROMIDE see HOT500
1-HYOSCINE HYDROBROMIDE see HOT500
HYOSCINE HYDROBROMIDE see HOT500
(−)-HYOSCINE HYDROBROMIDE see HOT500
HYOSCINE METHYL BROMIDE see SBH500
1-HYOSCYAMINE see HOU000
dl-HYOSCYAMINE see ARR000
HYOSCYAMINE see HOU000
(−)-HYOSCYAMINE see HOU000
HYOSCYAMINE METHYLBROMIDE see MGR250
dl-HYOSCYAMINE METHYLNITRATE see MGR500
HYOSCYAMINE SULFATE see HOT600
(−)-HYOSCYAMINE SULFATE see HOT600
HYOSCYAMUS NIGER see HAQ100
HYOSCYINE HYDROBROMIDE see HOT500
HYOSOL see SBG000
dl-HYOSYAMINE METHYLNITRATE see MGR500
HYOZID see ILD000
HYPAQUE 60 see AOO875
HYPAQUE 13.4 see AOO875
HYPAQUE see SEN500
HYPAQUE CYSTO see AOO875
HYPAQUE M 30 see AOO875
HYPAQUE MEGLUMINE see AOO875
HYPAQUE SODIUM see SEN500
HYPCOL see QAK000
HYPERAN see HFS759
HYPERAZIN see HGP500
HYPERBUTAL see BPF500
HYPERCAL B see RDK000
HYPERNEPHRIN see VGP000
HYPERNIC EXTRACT see LFT800
HYPEROL see HIB500
HYPERPAX see DNA800, MJE780
HYPERSIN see BFW250
HYPERSTAT see DCQ700
HYPERTANE FORTE see RDK000

HYPERTENAIN see MAW250
HYPERTENSAN see RDK000
HYPERTENSIN see AOO900
HYPERTONALUM see DCQ700
HY-PHI 1055 see OHU000
HY-PHI 1088 see OHU000
HY-PHI 1199 see SLK000
HY-PHI 2066 see OHU000
HY-PHI 2088 see OHU000
HY-PHI 2102 see OHU000
HYPHYLLINE see DNC000
HYPNODIN see HOU059
HYPNOGEN see EOK000
HYPNOGENE see BAG000
HYPNOMIDATE see HOU100
HYPNON see DLV000
HYPNONE see ABH000
HYPNOREX see LGZ000
HYPNORM see HAF400
HYPNOSTAN see PBT500
HYPNO-TABLINETTEN see EOK000
HYPO see SKI000, SKI500
HYPOCHLORITES see HOU500
HYPOCHLORITE SOLUTION containing >7% available CHLORINE by /S HY-
   POCHLORITE SOLUTIONS with >5% but <16% available chlorine (DOT)
   see SHU500
HYPOCHLORITE SOLUTIONS with 16% or more available chlorine by wt.
   (UN 1791) see SHU500
HYPOCHLOROUS ACID see HOV000
HYPOCHLOROUS ACID, CALCIUM SALT see HOV500
HYPOCHLOROUS ACID, POTASSIUM SALT see PLI250, PLK000
HYPODERMACID see TIQ250
HYPOGLYCIN see MJP500
HYPOGLYCIN A see MJP500
HYPOGLYCIN B see HOW100
HYPOGLYCINE A see MJP500
HYPOGLYCINE B see HOW100
HYPONITROUS ACID see HOW500
HYPONITROUS ACID ANHYDRIDE see NGU000
α-HYPOPHAMINE see ORU500
HYPOPHENON see ELL500
HYPOPHOSPHOROUS ACID see PGY250
HYPOPHTHALIN see HGP495, HGP500
HYPOPHYSEAL GROWTH HORMONE see PJA250
HYPOPRESOL see OJD300
HYPORENIN see VGP000
HYPOS see HGP500
HYPOTHIAZIDE see CFY000
HYPOTROL see SBM500, SBN000
HYPOXANTHINE see DMC000
HYPOXANTHINE NUCLEOSIDE see IDE000
HYPOXANTHINE RIBONUCLEOSIDE see IDE000
HYPOXANTHINE-d-RIBOSIDE see IDE000
HYPOXANTHINE RIBOSIDE see IDE000
HYPOXANTHOSINE see IDE000
HYPROVAL-PA see HNT500
HYPTOR BASE see QAK000
HYRE see RIK000
HYSCO see HOT500
HYSCYLENE P see PDX000
HYSONE-A see HHQ800
HYSSOP OIL see HOX000
HYSTEROL see HOX100
HYSTRENE 80 see SLK000
HYSTRENE 8016 see PAE250
HYSTRENE 9014 see MSA250
HYSTRENE 9512 see LBL000
HYTAKEROL see DME300
HYTONE LOTION see CNS750
HYTOX see MIA250
HYTRIN see TEF700
HYVAR see BMM650, BNM000
HYVAREX see BMM650
HYVAR X see BMM650
HYVAR X BROMACIL see BMM650
HYVAR X WEED KILLER see BMM650
HYVERMECTIN see ITD875
HYZYD see ILD000


I 677 see OLK200
I-1431 see CQH000
IA-4 see CFP750
I 337A see CDP250
IA 887 see NCP875

IA see IDZ000
IAA see ICN000
IAB see AOC500
I ACID see AKI000
IAMBOLEN see PKF750
IA-4 N-OXIDE see CFQ000
IA-PRAM see DLH630
IARACTAN see TAF675
IdB-1027 see COI750
IBA see ICP000
IBATRAN see ACE000
IBD 78 see MGL500
IBD see CCK125
IBDU see IIV000
IBENZMETHYZINE see PME250
IBENZMETHYZINE HYDROCHLORIDE see PME500
IBENZMETHYZIN HYDROCHLORIDE see PME500
IBERIN see MPN100
IBIFUR see FPI000
IBINOLO see TAL475
IBIODORM see SKS700
IBIODRAL see BBV500
IBIOFURAL see NGE500
IBIOSUC see SGC000
IBIOTON see TAI500
IBIOTYZIL see BCA000
IBIOZEDRINE see AOB250
IBN see IJD000
IBOGAMINE-18-CARBOXYLIC ACID, METHYL ESTER, HYDROCHLORIDE see
    CNS200
IBOGAMINE-18-CARBOXYLIC ACID, METHYL ESTER, MONOHYDROCHLO-
    RIDE (9CI) see CNS200
IBOMAL see QCS000
IBOTENIC ACID see AKG250
IBOTENSAURE (GERMAN) see AKG250
IBP see BKS750, DIU800
IBUFEN see IIU000
IBUFENAC see IJG000
IBUNAC see IJG000
IBUPROCIN see IIU000
IBUPROFEN see IIU000
IBUPROFEN GUAIACOL ESTER see IJJ000
IBUPROFEN PICONOL see IAB000
IBYLCAINE HYDROCHLORIDE see IAC000
IBZ see PME500
I-C 26 see DWC100
IC 6002 see PIW000
ICE-NUCLEATION-ACTIVE PSEUDOMONAS SYRINGAE see IAC100
ICG see CCK000
ICHDEN see IAD000
ICHTAMMON see IAD000
ICHTHADONE see IAD000
ICHTHALUM see IAD000
ICHTHAMMOL see IAD000
ICHTHAMMONIUM see IAD000
ICHTHIUM see IAD000
ICHTHOSAN see IAD000
ICHTHOSAURAN see IAD000
ICHTHOSULFOL see IAD000
ICHTHYMALL see IAD000
ICHTHYNAT see IAD000
ICHTHYOL see IAD000
ICHTHYOPON see IAD000
ICHTHYSALLE see IAD000
ICI 350 see AJH129
ICI 543 see PMP500
ICI 3435 see SNI500
ICI 28257 see ARQ750
ICI 32525 see SNL800
ICI-32865 see TJQ333
I.C.I. 33,828 see MLJ500
ICI 35868 see DNR800
ICI 38174 see INT000
ICI 43823 see BPI300
ICI 45520 see ICB000, ICC000
ICI 46,474 see NOA600, TAD175
ICI 46638 see DMZ000
ICI 48213 see FBP100
ICI 50172 see ECX100
ICI 51426 see CQH625
ICI 54,450 see CKK250
ICI-58834 see VKA875
ICI 58,834 see VKF000
ICI 59118 see RCA375
ICI 66,082 see TAL475

racemic-ICI 79,939 see IAD100
ICI 79,939 see IAD100
ICI 118587 see XAH000
ICI 123215 see LIU420
ICI 146814 see GJU600
ICI 156834 DISODIUM see CCS371
ICI-EP 5850 see MQD750
ICIG 770 see EAI850
ICIG 772 see MEL775
I.C.I.G. 1105 see RFU600
ICIG 1109 see CGV250
ICIG 1110 see CHD250
I.C.I.G. 1163 see ROF200
I.C.I.G. 1325 see BIF625
I.C.I. HYDROCHLORIDE see INT000
ICI 54856 METHYL ESTER see MIO975
ICIPEN see PDT750
ICI-PP 557 see AHJ750
ICI-PP 563 see GJU600
ICN-1229 see RJA500
ICORAL B see HNB875
ICOSAHYDRO DIBENZO(b,k)(1,4,7,10,13,16)HEXAOXACYCLOOCTADECIN see
    DGV100
ICR 10 see QDS000
ICR-48b see DFH000
ICR-125 see QCS875
ICR 170 see ADJ875
ICR 180 see CGS750
ICR 217 see CGX750
ICR 220 see BHZ000
ICR 290 see CGX500
ICR 292 see EHJ000
ICR 311 see EHI500
ICR 340 see IAE000
ICR 342 see CFA250
ICR 368 see CGY000
ICR 372 see CGS500
ICR 377 see EHJ500
ICR 394 see CHH000
ICR 395 see CGR500
ICR 442 see CKU250
ICR-450 see BIJ750
ICR 451 see CIG250
ICR 486 see CFB500
ICR 498 see BNO750
ICR 502 see BNR000
ICR 506 see BNO500
ICR-25A see CLD500
ICRF-159 see PIK250
ICRF 159 see RCA375
ICTALIS SIMPLE see DKQ000
ID 540 see FDB100
ID-622 see IAG300
ID-1229 see FGQ000
IDA see IBH000
IDALENE see MOV500
IDANTOIL see DNU000
IDANTOIN see DKQ000
IDANTOINAL see DNU000
IDARAC see TKG000
ID 480 DIHYDROCHLORIDE see DAB800
IDEXUR see DAS000
IDMT see IOW500
IDOCYL NOVUM see SJO000
IDOMETHINE see IDA000
IDONOR see PJB500
IDOSERP see RDK000
IDOXENE see DAS000
IDOXURIDIN see DAS000
IDOXURIDINE see DAS000
IDPN see BIQ500
IDRAGIN see ADA725
IDRALAZINA (ITALIAN) see HGP495
IDRAZIDE DELL'ACIDO ISONICOTINICO see ILD000
IDRAZIL see ILD000
IDRAZINA SOLFATO (ITALIAN) see HGW500
2-IDRAZINO-6-(N,N-BIS(2-IDROSSIETIL)-AMINO)-PIRIDAZINA CLORIDRATO (ITAL-
    IAN) see BKB500
3-IDRAZINO-6-(N-(2-IDROSSIPROPIL)METILAMINO)PIRIDAZINA DICLORIDRATO
    (ITALIAN) see HHD000
IDRIANOL see SPC500
IDROBUTAZINA see HNI500
IDROCHINONE (ITALIAN) see HIH000
IDROESTRIL see DKA600
IDROGENO SOLFORATO (ITALIAN) see HIC500

IDROGESTENE see HNT500
IDROPEROSSIDO di CUMENE (ITALIAN) see IOB000
IDROPEROSSIDO di CUMOLO (ITALIAN) see IOB000
IDROSSIDO DI STAGNO TRIFENILE (ITALIAN) see HON000
1′,1-(2-IDROSSIETIL)4-(3-(2-CLORO-10-FENOTIAZIL)PROPILPIPERAZINA (ITALIAN) see CJM250
(+−)-1-(4-IDROSSIFENIL)-2-METILAMINOETANOLO TARTRATO (ITALIAN) see SPD000
4-IDROSSI-4-METIL-PENTAN-2-ONE (ITALIAN) see DBF750
d-N,N′-(1-IDROSSIMETIL PROPIL)-ETILENDINITROSAMINA (ITALIAN) see HMQ500
4-IDROSSI-3-(3-OXO-)-(FENIL-BUTIL)-CUMARINE (ITALIAN) see WAT200
l-5α-IDROSSIPROPIONILAMINO-2,4,6-TRIIODOISOFTAL-DI(1,3-DIIDROSSI-2-PROPILAMIDE) see IFY000
4-IDROSSI-3-(1,2,3,4-TETRAIDRO-1-NAFTIL)CUMARINA (ITALIAN) see EAT600
IDROSSIZINA see CJR909
IDROTIADENE see HII000
IDROTIAZIDE see CFY000
IDRYL see FDF000
IDSOSERP see RDK000
IDU see DAS000
IDUCHER see DAS000
IDULEA see DAS000
IDULIAN see DLV800
IDUOCULOS see DAS000
IDUR see DAS000
IDURIDIN see DAS000
IEM-1-15 see CEX250
IEM 455 see CJN750
IERGIGAN see DQA400
IEROIN see HBT500
IF (fumigant) see TBV750
IFC-45 see CGG500
IFC see CBM000
IFENEC see EAE000
IFENPRODIL see IAG600
IFENPRODIL TARTRATE (2:1) see IAG625
IFENPRODIL l-(+)-TARTRATE see IAG625
IFENPRODIL TARTRATE see IAG625
IFIBRIUM see LFK000
IFOSFAMID see IMH000
IFOSFAMIDE see IMH000
IF ROM 203 see IAG700
IGE (OSHA) see IPD000
IGE see IPD000
IGELITE F see PKQ059
IGEPAL CA-63 see PKF500
IGEPAL CA see GHS000
IGEPAL CO-630 see PKF000
IGEPAL GAS see IAH000
IGEPON T-33 see SIY000
IGEPON T-43 see SIY000
IGEPON T 51 see SIY000
IGEPON T-71 see SIY000
IGEPON T-73 see SIY000
IGEPON T-77 see SIY000
IGEPON TE see SIY000
IGIG 929 see HKH000
IGNAZIN see AFH250
IGNOTINE see CCK665
IGROSIN see SIH500
IGROTON see CLY600
IH 773B see FDA885
II-C-2 see DOQ400
IIH see ILE000
IKACLOMIN see CMX700
IKADA RHODAMINE B see FAG070
IKhS 1 see CGW300
IKTEROSAN see PGG000
IKURIN see ANU650
IL 6001 see DLH200
ILBION see FNF000
ILETIN see IDF300
ILETIN U 40 see LEK000
ILEX AQUIFOLIUM see HGF100
ILEX OPACA see HGF100
ILEX VOMITORIA see HGF100
ILIADIN see AEX000
ILIDAR see AGD500, ARZ000
ILIDAR BASE see ARZ000
ILIDAR PHOSPHATE see AGD500
ILITIA see VSZ450
ILIXATHIN see RSU000
ILLOXOL see DHB400
ILLUDINE S see LiO600

ILLUDIN S see LIO600
IL-6302 MESYLATE see FMU039
ILOPAN see PAG200
ILOTYCIN see EDH500
ILOTYCIN HYDROCHLORIDE see EDJ000
IM see DLH600
IMADYL see CCK800
IMAGON see CLD000
IMAVATE see DLH630
IMAVEROL see FPB875
IMAZALIL see FPB875
IMBARAL see SOU550
IMBRILON see IDA000
IMC 3950 see SAZ000
IMD 760 see ARX800
IMESONAL see SBM500, SBN000
IMET 3393 see CQM750
IMETRO see IBP200
IMFERON see IGS000
IMI 115 see TGF250
IMIDA-LAB see TEH500
IMIDALIN see BBW750
IMIDALINE HYDROCHLORIDE see BBJ750
IMIDAMINE see PDC000
IMIDAN see PHX250
IMIDAN (PEYTA) see TEH500
7H-IMIDAZO(4,5-D)PYRIMIDINE see POJ250
IMIDAZOL see IAL000
IMIDAZOLE see IAL000
IMIDAZOLE-4-ACRYLIC ACID see UVJ440
5-IMIDAZOLEACRYLIC ACID see UVJ440
IMIDAZOLEACRYLIC ACID see UVJ440
5-IMIDAZOLECARBOXAMIDE, 4-AMINO- see AKK250
α-β-IMIDAZOLECARBOXYLIC ACID see IAM000
4,5-IMIDAZOLEDICARBOXYLATE see IAM000
IMIDAZOLE-4,5-DICARBOXYLIC ACID see IAM000
IMIDAZOLE, 1-(2-(2,4-DICHLOROPHENYL)-2-((2,4-DICHLOROPHENYL)METHOXY)ETHYL)-(9CI) see MQS550
1H-IMIDAZOLE, 4,5-DIHYDRO-2-PHENYL- see PFJ300
1H-IMIDAZOLE-4-ETHANAMINE see HGD000
1H-IMIDAZOLE-4-ETHANAMINE PHOSPHATE (1:2) see HGE000
4-IMIDAZOLEETHYLAMINE see HGD000
IMIDAZOLE-4-ETHYLAMINE see HGD000
5-IMIDAZOLEETHYLAMINE see HGD000
1H-IMIDAZOLE, 1-(2-(ETHYLSULFONYL)ETHYL)-2-METHYL-5-NITRO- see TGD250
IMIDAZOLE-2-HYDROXYBENZOATE see IAM100
IMIDAZOLE MUSTARD see IAN000
IMIDAZOLE, 4-NITRO- see NHG100
IMIDAZOLEPYRAZOLE see IAN100
IMIDAZOLE with SALICYLIC ACID see IAM100
IMIDAZOLE-2-THIOL see IAO000
IMIDAZOL-2-HYDROXYBENZOAT (GERMAN) see IAM100
2,4-IMIDAZOLIDINEDIONE, 1,3-DIBROMO-5,5-DIMETHYL-(9CI) see DDI900
2,4-IMIDAZOLIDINEDIONE, 5,5-DIMETHYL- see DSF300
2,4-IMIDAZOLIDINEDIONE, 5,5-DIMETHYL-3-(2-(OXIRANYLMETHOXY)PROPYL)-1-(OXIRANYLMETHYL)- see DTH100
2,4-IMIDAZOLIDINEDIONE, 5-ETHYL-5-METHYL-1,3-BIS(OXIRANYLMETHYL)- see ECI200
2,4-IMIDAZOLIDINEDIONE, 5-ETHYL-5-PHENYL-, (R)- (9CI) see EOL050
2,4-IMIDAZOLIDINEDIONE, 1-NITRO- see NHE100
2,4-IMIDAZOLIDINEDIONE, 1-(((5-(4-NITROPHENYL)-2-FURANYL)METHYLENE)AMINO)-, SODIUM SALT see DAB840
2-IMIDAZOLIDINETHIONE see IAQ000
2-IMIDAZOLIDINETHIONE mixed with SODIUM NITRITE see IAR000
2-IMIDAZOLIDINONE see IAS000
4-IMIDAZOLIDINONE, 1-ACETYL-2-THIOXO- see ADC750
2-IMIDAZOLIDINONE, 1,3-DINITRO- see DUW503
2-IMIDAZOLIDINONE, 1-(1-METHYL-5-NITRO-1H-IMIDAZOL-2-YL)-3-(METHYLSULFONYL)- see SAY950
IMIDAZOLIDINYL UREA 11 see IAS100
2-IMIDAZOLIDONE see IAS000
2-IMIDAZOLINE see IAT000
IMIDAZOLINE see IAT000
2-IMIDAZOLINE, 2-(2,6-DICHLOROANILINO)- see DGB500
IMIDAZOLINE-2,4-DITHIONE see IAT100
2-IMIDAZOLINE, 2-METHYL-1-(3,4,5-TRIMETHOXYBENZOYL)- see IAT200
2-IMIDAZOLINE, 2-PHENYL- see PFJ300
N-(2-IMIDAZOLINE-2-YL)-N-(4-INDANYL)AMINE MONOHYDROCHLORIDE see IBR200
N-(2-IMIDAZOLINE-2-YL)-N-(4-INDANYL)AMIN-MONOHYDROCHLORID (GERMAN) see IBR200
1H-IMIDAZOLIUM, 4,5-DIHYDRO-1-(CARBOXYMETHYL)-1-(2-HYDROXYETHYL)-2-UNDECYL-, HYDROGEN SULFATE (salt), MONOSODIUM SALT see AOC275

2-(4-IMIDAZOLYL)ETHYLAMINE see HGD000
2-IMIDAZOL-4-YL-ETHYLAMINE see HGD000
β-IMIDAZOLYL-4-ETHYLAMINE see HGD000
3-(1H-IMIDAZOL-4-YL)-2-PROPENOIC ACID see UVJ440
1H-IMIDAZO(2,1-f)PURINE-2,4(3H,6H)-DIONE,7,8-DIHYDRO-8-ALLYL-1,3-DIMETH-YL-(8Cl) see KHU100
4-IMIDAZO(1,2-a)PYRIDIN-2-YL-α-METHYLBENZENEACETIC ACID see IAY000
2-(4-IMIDAZO(1,2-a)PYRIDIN-2-YL)PHENYL)PROPIONIC ACID see IAY000
2-(p-(2-IMIDAZO(1,2-a)PYRIDYL)PHENYL)PROPIONIC ACID see IAY000
IMIDAZO(5,1-d)-1,2,3,5-TETRAZINE-8-CARBOXAMIDE, 3-(2-CHLOROETHYL)-3,4-DIHYDRO-4-OXO- see MQY110
IMIDAZO(2,1-β)THIAZOLE MONOHYDROCHLORIDE see LFA020
IMIDENE see TEH500
IMIDIN see NAH500
IMIDOBENZYLE see DLH600, DLH630
4,4'-(IMIDOCARBONYL)BIS(N,N-DIMETHYLAMINE) MONOHYDROCHLORIDE see IBA000
4,4'-(IMIDOCARBONYL)BIS(N,N-DIMETHYLANILINE) see IBB000
1,1'-IMIDODIACETONITRILE see IBB100
IMIDODICARBONIC DIAMIDE, N,N',2-TRIS(6-ISOCYANATOHEXYL)- see TNJ300
IMIDODICARBONIC DIHYDRAZIDE (9 Cl) see IBC000
IMIDODICARBONIMIDIC DIAMIDE, N-(2-METHYLPHENYL)-(9Cl) see TGX550
IMIDODICARBOXYLIC ACID, DIHYDRAZIDE see IBC000
3,3'-IMIDODI-1-PROPANOL, DIMETHANESULFONATE (ester), HYDROCHLO-RIDE see YCJ000
IMIDODISULFURIC ACID, AMMONIUM SALT see ANK650
IMIDOL see DLH630
IMIDOLE see PPS250
IMILANYLE see DLH630
IMINAZOLE see IAL000
IMINO-1,1'-BIANTHRAQUINONE see IBI000
IMINOBIS(ACETIC ACID) see IBH000
2,2'-IMINOBISACETONITRILE see IBB100
1,1'-IMINOBIS(4-AMINO-9,10-ANTHRACENEDIONE) see IBD000
1,1'-IMINOBIS(4-AMINOANTHRAQUINONE) see IBD000
1,1'-IMINOBIS-9,10-ANTHRACENEDIONE see IBI000
1,1'-IMINOBIS(4-BENZAMIDOANTHRAQUINONE) see IBJ000
4,5'-IMINOBIS(4-BENZAMIDOANTHRAQUINONE) see IBE000
2,2'-IMINOBISETHANOL see DHF000
2,2'-IMINOBISETHYLAMINE see DJG600
4,4'-IMINOBISPHENOL see IBJ100
3,3'-IMINOBISPROPANENITRILE see BIQ500
3,3'-IMINOBIS-1-PROPANOL DIMETHANESULFONATE (ester), 4-METHYLBEN-ZENESULFONATE (salt) see IBQ100
3,3'-IMINOBIS(PROPYLAMINE) see AIX250
IMINOBIS(PROPYLAMINE) see AIX250
4-IMINO-2,5-CYCLOHEXADIEN-1-ONE see BDD500
4,4'-((4-IMINO-2,5-CYCLOHEXADIEN-1-YLIDENE)METHYLENE)DIANILINE MONO-HYDROCHLORIDE-o-TOLUIDINE see RMK020
2,2'-IMINODIACETIC ACID see IBH000
IMINODIACETIC ACID see IBH000
IMINODIACETONITRILE see IBB100
1,1'-IMINODIANTHRAQUINONE see IBI000
N,N'-(IMINODI-4,1-ANTHRAQUINONYLENE)BISBENZAMIDE see IBJ000
IMINODIBENZYL see DKY800
IMINODIETHANOIC ACID see IBH000
2,2'-IMINODIETHANOL see DHF000
2,2'-IMINODI-ETHANOL with 1,2-DIHYDRO-3,6-PYRIDAZINEDIONE (1:1) see DHF200
2,2'-IMINODI-N-NITROSOETHANOL see NKM000
IMINODIOCTAN SODNY (CZECH) see DXE200
4,4'-IMINODIPHENOL see IBJ100
1,1'-IMINODI-2-PROPANOL see DNL600
3,3'-IMINODIPROPIONITRILE see BIQ500
β,β'-IMINODIPROPIONITRILE see BIQ500
IMINO-β,β'-DIPROPIONITRILE see BIQ500
β,β-IMINODIPROPIONITRILE see BIQ500
IMINODIPROPYL DIMETHANESULFONATE 4-TOLUENESULPHONATE see IBQ100
5-IMINO-1,2,4-DITHIAZOLIDINE-3-THIONE see IBL000
(2-((IMINO(METHYLAMINO)METHYL)THIO)ETHYL)TRIETHYLAMMONIUM BRO-MIDE HYDROBROMIDE see MRU775
(2-IMINO-5-PHENYL-4-OXAZOLIDINONATO(2-))DIAQUOMAGNESIUM see PAP000
2-IMINO-5-PHENYL-4-OXAZOLIDINONE see IBM000
2-IMINO-5-PHENYL-4-OXAZOLIDINONE MAGNESIUM CHELATE see PAP000
IMINOPHOSPHATE see DXN600
2(3H)-IMINO-9-β-D-RIBOFURANOSYL-9H-PURIN-6(1H)-ONE see GLS000
2-IMINOTHIAZOLIDINE see TEV600
IMINOUREA see GKW000
IMINOUREA HYDROCHLORIDE see GKY000
IMIPHOS see BGY000
IMIPRAMINA (ITALIAN) see DLH600, DLH630
IMIPRAMINE see DLH600, DLH630
IMIPRAMINEDEMETHYL HYDROCHLORIDE see DLS600

IMIPRAMINE HYDROCHLORIDE see DLH630
IMIPRAMINE MONOHYDROCHLORIDE see DLH630
IMIPRAMINE-N-OXIDE HYDROCHLORIDE see IBP000
IMIPRIN see DLH600, DLH630
IMIZIN see DLH600
IMIZINUM see DLH600
IMMENOCTAL see SBM500, SBN000
IMMENOX see SBM500
IMMETROPAN see IBP200
IMMORTELLE see HAK500
IMODIUM see LIH000
IMOL S 140 see TNP500
IMOTRYL see BBW500
IMOVANCE see ZUA450
IMOVANE see ZUA450
5'-IMP see IDE200
IMP see IDE200
o-IMPC see PMY300
5'-IMP DISODIUM SALT see DXE500
IMP DISODIUM SALT see DXE500
IMPEDEX see SJL500
IMPERATOR see RLF350
IMPERATORIN see IHR300
IMPERIAL GREEN see COF500
IMPERON FIXER T see TND250
IMPERVOTAR see CMY800
IMPF see IPX000
IMP HYDROCHLORIDE see DLH630
IMPINGEMENT BLACK see CBT750
IMPIRAMINE-N-OXIDE see IBP309
IMPOSIL see IGS000
IMPRAMINE see DLH600
IMPROMEN see BNU725
IMPROMIDINE HYDROCHLORIDE see IBQ075
IMPROMIDINE TRIHYDROCHLORIDE see IBQ075
IMPROSULFAN-p-TOLUENESULFONATE see IBQ100
IMPROSULFAN TOSILATE see IBQ100
IMPROSULFAN TOSYLATE see IBQ100
IMPROVED WILT PRUF see PKQ059
IMPRUVOL see BFW750
IMP SODIUM SALT see DXE500
IMPY see IAN100
IMS see IPY000
IMUGAN see CDP750
IMURAN see ASB250
IMUREK see ASB250
IMUREL see ASB250
IMUTEX see IAL000
IMVITE I.G.B.A. see BAV750
IMWITOR 191 see OAV000
IMWITOR 900K see OAV000
IN-117 see HEG000
IN 399 see MKW000
IN 511 see PGG350, PGG355
IN 836 see FAR200
INACID see IDA000
INACILIN see AOD000
INACTIVE LIMONENE see MCC250
INAKOR see ARQ725
INALONE O see AFJ625
INALONE R see AFJ625
INAMIL see AIF000
INAMYCIN see NOB000, SMB000
INAPPIN see DYF200
INAPSIN see DYF200
INBESTAN see FOS100
INBUTON see BSM000
INCASAN see IBQ300
INCAZANE see IBQ300
INCIDOL see BDS000
2-(4-(3-INCOLYLMETHYL)-1-PIPERAZINYL)-QUINOLINE DIMALEATE see ICZ100
INCORTIN see CNS800, CNS825
INCRECEL see CMF400
INDACRINONE see IBQ400
INDAN, 4,6-DINITRO-1,1,3,3,5-PENTAMETHYL- see MRU300
INDAFLEX see IBV100
INDALCA AG see GLU000
INDALCA AG-BV see GLU000
INDALCA AG-HV see GLU000
INDALONE see BRT000
INDAMOL see IBV100
INDANAL see CFH825, CMV500
INDANAZOLIN (GERMAN) see IBR200
INDANAZOLINE see IBR200
INDANAZOLINE HYDROCHLORIDE see IBR200

1,3-INDANDIONE see IBS000
INDAN-5-OL see IBU000
5-INDANOL see IBU000
2-INDANONE see IBV000
INDANTHREN BLUE see IBV050
INDANTHREN BLUE BC see DFN425
INDANTHREN BLUE BCA see DFN425
INDANTHREN BLUE BCS see DFN425
INDANTHREN BLUE GP see IBV050
INDANTHREN BLUE GPT see IBV050
INDANTHREN BLUE RPT see IBV050
INDANTHREN BLUE RS see IBV050
INDANTHREN BLUE RSN see IBV050
INDANTHREN BLUE RSP see IBV050
INDANTHREN BRILLIANT BLUE R see IBV050
INDANTHREN BRILLIANT ORANGE GR see CMU820
INDANTHREN BRILLIANT VIOLET 4R see DFN450
INDANTHREN BRILLIANT VIOLET RR see DFN450
INDANTHREN BRONZE BR see CMU770
INDANTHREN BROWN BR see CMU770
INDANTHREN BROWN GR see CMU770
INDANTHRENE see IBV050
INDANTHRENE BLUE see IBV050
INDANTHRENE BLUE BC see DFN425
INDANTHRENE BLUE BCF see DFN425
INDANTHRENE BLUE GP see IBV050
INDANTHRENE BLUE RP see IBV050
INDANTHRENE BLUE RS see IBV050
INDANTHRENE BLUE RSA see IBV050
INDANTHRENE BLUE RSN see IBV050
INDANTHRENE BRILLIANT ORANGE GR see CMU820
INDANTHRENE BRILLIANT ORANGE GRP see CMU820
INDANTHRENE BRILLIANT VIOLET 4R see DFN450
INDANTHRENE BRILLIANT VIOLET RR see DFN450
INDANTHRENE BROWN BR see CMU770
INDANTHRENE GOLDEN YELLOW see DCZ000
INDANTHRENE NAVY BLUE G see CMU500
INDANTHRENE OLIVE R see DUP100
INDANTHRENE PRINTING BLACK BL see CMU320
INDANTHRENE RED BROWN 5RF see CMU800
INDANTHRENE REDDISH BROWN 5RF see CMU800
INDANTHRENE RUBINE R see CMU825
INDANTHREN GREY M see CMU475
INDANTHREN GREY MG see CMU475
INDANTHREN NAVY BLUE TRR see CMU750
INDANTHREN OLIVE R see DUP100
INDANTHREN PRINTING BLUE FRS see IBV050
INDANTHREN PRINTING BLUE KRS see IBV050
INDANTHREN PRINTING VIOLET F 4R see DFN450
INDANTHREN RED BROWN 5RF see CMU800
INDANTHREN RUBINE R see CMU825
INDANTHREN RUBINE RS see CMU825
INDANTHRONE see IBV050
1,2,3-INDANTRIONE-2-HYDRATE see DMV200
1,2,3-INDANTRIONE MONOHYDRATE see DMV200
2-INDANYLAMINE HYDROCHLORIDE see AKL000
INDAPAMIDE see IBV100
INDAR see BPU000
2-(p-(2H-INDAZOL-2-YL)PHENYL)PROPIONIC ACID see IBW100
INDECAINIDE HYDROCHLORIDE see IBW400
INDELOXAZINE HYDROCHLORIDE see IBW500
INDEMA see PFJ750
INDENE see IBX000
1H-INDENE-1,3(2H)-DIONE see IBS000
INDENE TRIPROPYLAMINE see IBY000
INDENOLOL HYDROCHLORIDE see IBY600, ICA000
13H-INDENO(1,2-1)PHENANTHRENE see DCX000
INDENO(1,2,3-cd)PYRENE see IBZ000
INDENO(1,2,3-cd)PYRENE-1,2-OXIDE see DLK750
INDENO(1,2,3-cd)PYREN-8-OL see IBZ100
4-(1H-INDEN-1-YLIDENEMETHYL)-N,N-DIMETHYLBENZENAMINE see DOT600
4-(1-H-INDEN-1-YLIDENEMETHYL)-N-METHYL-N-NITROSOBENZENAMINE see MMR750
1-(4(or 7)-INDENYLOXY)-3-(ISOPROPYLAMINO)-2-PROPANOL HYDROCHLORIDE see IBY600
(±)-1-(7-INDENYLOXY)-3-ISOPROPYLAMINOPROPAN-2-OL HYDROCHLORIDE see ICA000
2-(7-INDENYLOXYMETHYL)MORPHOLINE HYDROCHLORIDE see IBW500
2-((1H-INDEN-7-YLOXY)METHYL)MORPHOLINE HYDROCHLORIDE see IBW500
INDEPENDENCE RED see MMP100
INDERAL see ICB000
INDERAL HYDROCHLORIDE see ICC000
INDEREX see ICC000
INDEROL see ICC000
INDI see ICD100

INDIA see ICI100
INDIAN APPLE see MBU800
INDIAN BERRY see PIE500
INDIAN BLACK DRINK see HGF100
INDIAN CANNABIS see CBD750
INDIAN COBRA VENOM see ICC700
INDIAN GUM see AQQ500, GLY000
INDIAN HEMP see CBD750
INDIAN LABURNUM see GIW300
INDIAN LAUREL see MBU780
INDIAN LICORICE SEED see AAD000
INDIAN LILAC see CDM325
INDIAN PINK see CCJ825, PIH800
INDIAN POKE see FAB100, VIZ000
INDIAN POLK see PJJ315
INDIAN RED see IHD000, LCS000
INDIAN SAVIN TREE (JAMAICA) see CAK325
INDIAN TOBACCO see CCJ825
INDIAN WALNUT see TOA275
INDIA RUBBER see ROH900
INDIA RUBBER VINE see ROU450
INDICAN (POTASSIUM SALT) see ICD000
INDICINE-N-OXIDE see ICD100
INDIGENE BLACK D see CMN230
INDIGENOUS PEANUT OIL see PAO000
INDIGO BLUE 2B see CMO000
INDIGO CARMINE see FAE100
INDIGO CARMINE (BIOLOGICAL STAIN) see FAE100
INDIGO CARMINE DISODIUM SALT see FAE100
INDIGO EXTRACT see FAE100
INDIGO-KARMIN (GERMAN) see FAE100
5,5'-INDIGOTIN DISULFONIC ACID see FAE100
INDIGOTINE see FAE100
INDIGOTINE DISODIUM SALT see FAE100
INDION see PFJ750
INDISAN see IKA000
INDISULFAT (GERMAN) see ICJ000
INDIUM see ICF000
INDIUM ACETYLACETONATE see ICG000
INDIUM CHLORIDE see ICK000
INDIUM CITRATE see ICH000
INDIUM NITRATE see ICI000
INDIUM (3+) OXIDE see ICI100
INDIUM (III) OXIDE see ICI100
INDIUM OXIDE see ICI100
INDIUM SESQUIOXIDE see ICI100
INDIUM SULFATE see ICJ000
INDIUM TRICHLORIDE see ICK000
INDIUM TRIOXIDE see ICI100
INDOBLACK GR see CMN240
INDOBLOC see ICC000
INDO BLUE B-I see DFN425
INDO BLUE WD 279 see DFN425
INDOCID see IDA000
INDOCYANINE GREEN see CCK000, ICL000
INDOCYBIN see PHU500
INDOFAST ORANGE OV 5983 see CMU820
INDOFAST VIOLET LAKE see DFN450
INDOKLON see HDC000
INDOL (GERMAN) see ICM000
3-INDOLACETONITRILE see ICW000
INDOLACIN see CMP950
β-INDOLAETHYLAMIN-CHLORHYDRAT (GERMAN) see AJX250
α-INDOLAETHYLAMIN SALZSAEURE (GERMAN) see LIU100
INDOLAPRIL HYDROCHLORIDE see ICL500
INDOLE see ICM000
INDOLE, 3-ACETATO- see IDA600
1H-INDOLE-3-ACETIC ACID see ICN000
β-INDOLE-3-ACETIC ACID see ICN000
3-INDOLEACETIC ACID see ICN000
β-INDOLEACETIC ACID see ICN000
1H-INDOLE-3-ACETONITRILE see ICW000
INDOLE-3-ACETONITRILE see ICW000
INDOLEACETONITRILE see ICW000
INDOLE-3-ACRYLIC ACID see ICO000
INDOLE-3-ALANINE see TNX000
INDOLE-3-ALDEHYDE see FNO100
INDOLE-3-(2-AMINOBUTYL) ACETATE see AJB250
INDOLE, 3-(2-AMINOETHYL)-5-METHOXY- see MFS400
INDOLE, 5-BENZYLOXY-3-ISONIPECOTOYL- see BFC200
1H-INDOLE-3-BUTANOIC ACID see ICP000
INDOLE BUTYRIC see ICP000
γ-(INDOLE-3)-BUTYRIC ACID see ICP000
3-INDOLEBUTYRIC ACID see ICP000
β-INDOLEBUTYRIC ACID see ICP000

INDOLE BUTYRIC ACID see ICP000
INDOLE-3-CARBALDEHYDE see FNO100
INDOLE-3-CARBINOL see ICP100
INDOLE-5-CARBONITRILE, 3-ACETYL- see COP550
INDOLE-5-CARBONITRILE, 3-(2-METHYLPROPIONYL)- see COP525
INDOLE-3-CARBOXALDEHYDE see FNO100
INDOLE-3-CARBOXALDEHYDE, 2-(m-AMINOPHENYL)-, 4-(m-TOLYL)-3-THIO-
    SEMICARBAZONE see ALW900
1H-INDOLE-3-CARBOXALDEHYDE (9CI) see FNO100
INDOLE-2,3-DIONE see ICR000
1H-INDOLE-3-ETHANAMINE see AJX000
INDOLE ETHANOL see ICS000
INDOLE-3-ETHYLAMINE HYDROCHLORIDE see AJX250
α-INDOLEETHYLAMINE HYDROCHLORIDE see LIU100
β-INDOLE-ETHYLAMINE HYDROCHLORIDE see AJX250
INDOLE, 3-(HYDROXYACETYL)-1-METHYL see HIS140
INDOLE, 3-(HYDROXYACETYL)-2-METHYL see HIS150
INDOLE-3-METHANOL see ICP100
1H-INDOLE-3-METHANOL (9CI) see ICP100
INDOLENE see ICS100
1H-INDOLE-3-PROPIONIC ACID see ICS200
β-INDOLEPROPIONIC ACID see ICS200
INDOLEPROPIONIC ACID see ICS200
INDOLE, 1-PROPIONYL- see ICW100
INDOLE-3-PROPYLAMINE HYDROCHLORIDE see AME750
1H-INDOLE-2-SULFONIC ACID, 5-((AMINOCARBONYL)HYDRAZONO)-2,3,5,6-
    TETRAHYDRO-1-METHYL-6-OXO-, MONOSODIUM SALT, TRIHYDRATE see
    AER666
INDOL-3-ETHYLAMINE see AJX000
INDOLIN see BBW500
INDOLINE, 5-ACETYL- see ACO320
2,3-INDOLINEDIONE see ICR000
INDOLINE, 2-METHYLENE-1,3,3-TRIMETHYL- see TLU200
3-INDOLINONE, 5-BROMO-2-(9-CHLORO-3-OXONAPHTHO(1,2-b)THIEN-2(3H)-
    YLIDENE)- see CMU320
3H-INDOLIUM, 2-(p-((2-CHLOROETHYL)METHYLAMINO)STYRYL)-1,3,3-TRIMETH-
    YL-, CHLORIDE see CMM768
3H-INDOLIUM, 2-(2-(2-METHYLINDOL-3-YL)VINYL)-1,3,3-TRIMETHYL-, CHLO-
    RIDE see CMM764
3H-INDOLIUM, 1,3,3-TRIMETHYL-2-(3-(1,3,3-TRIMETHYL-2-INDOLINYLI-
    DENE)PROPENYL)-, CHLORIDE see CMM765
INDOL-N-METHYLHARMINE HYDROCHLORIDE see ICU100
INDOL-3-OL, ACETATE (ester) (8CI) see IDA600
1H-INDOL-3-OL, ACETATE (ester) (9CI) see IDA600
INDOL-3-OL, HYDROGEN SULFATE (ESTER), POTASSIUM SALT see ICD000
INDOL-3-OL, POTASSIUM SULFATE see ICD000
Δ-INDOLYBUTYLAMINE HYDROCHLORIDE see AJB500
α-INDOL-3-YL-ACETIC ACID see ICN000
3-INDOLYLACETIC ACID see ICN000
INDOLYL-3-ACETIC ACID see ICN000
β-INDOLYLACETIC ACID see ICN000
INDOLYLACETIC ACID see ICN000
3-INDOLYLACETONITRILE see ICW000
INDOLYLACETONITRILE see ICW000
3-INDOLYLACRYLIC ACID see ICO000
1-β-3-INDOLYLALANINE see TNX000
β-INDOLYLALDEHYDE see FNO100
ω-3-INDOLYLAMYLAMINE ADIPINATE see ALR500
4-(INDOL-3-YL)BUTYRIC ACID see ICP000
γ-(INDOL-3-YL)BUTYRIC ACID see ICP000
INDOLYL-3-BUTYRIC ACID see ICP000
4-(3-INDOLYL)BUTYRIC ACID see ICP000
4-(INDOLYL)BUTYRIC ACID see ICP000
3-INDOLYL-γ-BUTYRIC ACID see ICP000
γ-(3-INDOLYL)BUTYRIC ACID see ICP000
3-INDOLYLCARBINOL see ICP100
3-INDOLYLETHANOL see ICS000
2-(3-INDOLYL)ETHYLAMINE see AJX000
β-3-INDOLYLETHYLAMINE HYDROCHLORIDE see AJX250
INDOL-1-YL ETHYL KETONE see ICW100
N-(2-INDOL-3-YLETHYL)NICOTINAMIDE see NDW525
N-(2-(3-INDOLYL)ETHYL)-NICOTINAMIDE see NDW525
2-INDOLYL METHOXYMETHYL KETONE see ICW200
(±)-1-(3-INDOLYL)-2-METHYLAMINOETHANOL see ICY000
1-(INDOLYL-3)-2-METHYLAMINOETHANOL-1 RACEMATE see ICY000
INDOL-3-YL METHYL KETONE see ICW100
1-(3'-INDOLYLMETHYL)-4-(2''-QUINOLYL)PIPERAZINE DIMALEATE see ICZ100
INDOLYL-3-MORPHOLINOMETHYL KETONE see ICZ150
1-(4-INDOLYLOXY)-3-(ISOPROPYLAMINO)-2-PROPANOL see VSA000
1-(1H-INDOL-4-YLOXY)-3-((1-METHYLETHYL)AMINO)-2-PROPANOL see VSA000
INDOLYL-3-PIPERIDINOMETHYL KETONE see ICZ200
INDOL-3-YL POTASSIUM SULFATE see ICD000
3-(3-INDOLYL)PROPANOIC ACID see ICS200
3-(1-H-INDOL-3-YL)-2-PROPENOIC ACID see ICO000
γ-3-INDOLYLPROPYLAMINE HYDROCHLORIDE see AME750

INDOL-3-YL SULFATE, POTASSIUM SALT see ICD000
INDO MAROON LAKE RV 6666 see CMU825
INDOMECOL see IDA000
INDOMED see IDA000
INDOMETHACIN see IDA000
INDOMETHAZINE see IDA000
INDOMETICINA (SPANISH) see IDA000
INDON see PFJ750
INDONAPHTHENE see IBX000
INDOPAN see AME500
INDOPHENOL, 2,6-DICHLORO-, SODIUM SALT see SGG650
INDOPROFEN see IDA400
INDOPTIC see IDA000
INDO-RECTOLMIN see IDA000
INDORM see IDA500
INDO-TABLINEN see IDA000
INDOXAMIC ACID see OLM300
INDOXINE KL see CMN800
INDOXYL-O-ACETATE see IDA600
INDOXYLACETATE see IDA600
INDUSTRENE 105 see OHU000
INDUSTRENE 205 see OHU000
INDUSTRENE 206 see OHU000
INDUSTRENE 4516 see PAE250
INDUSTRENE 5016 see SLK000
INERTEEN see PJL750
INETOL see INS000, INT000
INEXIT see BBQ500
INF 1837 see TKH750
INF 3355 see XQS000
INF 4668 see DGM875
INFAMIL see HNI500
INFERNO see DJA400
INFILTRINA see DUD800
INFLAMASE see PLY275
INFLAMEN see BMO000
INFLATINE see LHY000
INFLAZON see IDA000
INFRON see VSZ100
INFUSORIAL EARTH see DCJ800
INGALAN (RUSSIAN) see DFA400
INGALAN see DFA400
INGENANE HEXADECANOATE see IDB000
INHALAN see DFA400
INH-G see GBB500
INHG-SODIUM see IDB100
INHIBINE see HIB000
INHIBISOL see MIH275
INHISTON see TMK000
'INIA (HAWAII) see CDM325
INICARDIO see DJS200
INIPROL see PAF550
INITIATING EXPLOSIVE DIAZODINITROPHENOL (DOT) see DUR800
INITIATING EXPLOSIVE LEAD MONONITRORESORCINATE (DOT) see LDP000
INITIATING EXPLOSIVE NITROSOGUANIDINE see NKH000
INITIATING EXPLOSIVE PENTAERYTHRITE TETRANITRATE (DOT) see
    PBC250
INJERTO (TEXAS, MEXICO) see MQW525
INKASAN see IBQ300
INK BERRY see PJJ315
INK RED JSN see CMM300
INNOVAN see DYF200
INNOVAR see DYF200
INNOXALON see EID000
INO see IDE000
INOCOR see AOD375
INOLIN see IDD100
INDOTONER BLUE B 79 see DFN425
INOPHYLLINE see TEP500
INOPSIN see DYF200
INOSIE see IDE000
β-INOSINE see IDE000
INOSINE see IDE000
INOSINE, 2-AMINO- see GLS000
INOSINE-5'-MONOPHOSPHATE see IDE200
INOSINE-5'-MONOPHOSPHATE DISODIUM see DXE500
INOSINE-5'-MONOPHOSPHORIC ACID see IDE200
INOSINE-5'-PHOSPHATE see IDE200
INOSINIC ACID see IDE200
5'-INOSINIC ACID, DISODIUM SALT, mixed with DISODIUM-5'-GUANYLATE
    (1:1) see RJF400
5'-INOSINIC ACID, HOMOPOLYMER complex with 5'-CYTIDYLIC ACID HOMO-
    POLYMER (1:1) see PJY750
INOSIN-5'-MONOPHOSPHATE DISODIUM see DXE500
INOSITHEXAPHOSPHORSAEURE (GERMAN) see PIB250

meso-INOSITOL HEXANICOTINATE see HFG550
m-INOSITOL HEXANICOTINATE see HFG550
myo-INOSITOL HEXANICOTINATE see HFG550
INOSITOL HEXANICOTINATE see HFG550
INOSITOL HEXAPHOSPHATE see PIB250
INOSITOL NIACINATE see HFG550
INOSITOL NICOTINATE see HFG550
INOSTRAL see CNX825
INOTREX see DXS375
INOVAL see DYF200
INOVITAN PP see NCR000
INPC see PMS825
INSANE ROOT see HAQ100
INSARIOTOXIN see FQS000
INSECTICIDE 1,179 see MDU600
INSECTICIDE-NEMATICIDE 1410 see DSP600
INSECTICIDE No. 4049 see MAK700
INSECTICIDE No. 497 see DHB400
INSECTOPHENE see EAQ750
INSECT POWDER see POO250
INSECT REPELLENT 448 see PES000
6-12-INSECT REPELLENT see EKV000
INSIDON see DCV800
INSIDON DIHYDROCHLORIDE see IDF000
INSOLUBLE SACCHARINE see BCE500
INSOMNOL see EQL000
INSOM-RAPIDO see CAV000
IN-SONE see PLZ000
INSPIR see ACH000
INSULAMIN see BOM750, BQL000
INSULAMINA see DNA200
INSULAR see IDF300
INSULATARD see IDF325
INSULIN see IDF300
INSULIN INJECTION see IDF300
INSULIN LENTE see LEK000
INSULIN NOVO LENTE see LEK000
INSULIN PROTAMINE ZINC see IDF325
INSULIN RETARD RI see IDF325
INSULIN ZINC COMPLEX see LEK000
INSULIN ZINC PROTAMINATE see IDF325
INSULIN ZINC PROTAMINE see IDF325
INSULIN ZINC SUSPENSION see LEK000
INSULTON see MKB250
INSULYL see IDF300
INSULYL-RETARD see IDF325
INSUMIN see DAB800
INTAL see CNX825
INTALBUT see BRF500
INTALPRAM see DLH600, DLH630
INTEBAN SP see IDA000
INTEGERRIMINE see IDG000
INTEGRIN see ECW600
INTENKORDIN see CBR500
INTENSAIN see CBR500
INTENSAIN HYDROCHLORIDE see CBR500
INTENSE BLUE see FAE100
INTERCAIN see BQA010
INTERCHEM ACETATE BLUE B see MGG250
INTERCHEM ACETATE BORDEAUX B see CMP080
INTERCHEM ACETATE DEVELOPED BLACK see DPO200
INTERCHEM ACETATE FAST PINK DNA see AKO350
INTERCHEM ACETATE GREEN BLUE ALF see DMM400
INTERCHEM ACETATE PINK 3B see DBX000
INTERCHEM ACETATE PINK BLF see AKE250
INTERCHEM ACETATE VIOLET 6B see AKP250
INTERCHEM ACETATE VIOLET R see DBP000
INTERCHEM ACETATE YELLOW G see AAQ250
INTERCHEM DIRECT BLACK Z see AQP000
INTERCHEM DISPERSE YELLOW GH see AAQ250
INTERCHEM HISPERSE GREEN BLUE ALFH see DMM400
INTERCHEM HISPERSE PINK BH see AKE250
INTERFLO see PJS750
INTERKELLIN see AHK750
INTERNATIONAL ORANGE 2221 see LCS000
INTEROX see HIB010
INTERPINA see RDK000
INTERTULLE FUCIDIN see SHK000
INTEXAN LB-50 see AFP250
INTEXAN SB-85 see DTC600
INTEXSAN CPC see CCX000
INTEXSAN LQ75 see LBW000
INTOCOSTRIN see TOA000
INTOCOSTRINE see COF750
INTOLEX see BEQ625

INTRABILIX see EQC000
INTRABLIX see BGB315
INTRACID GREEN F see FAE950
INTRACID PURE BLUE L see FAE000
INTRACID PURE BLUE V see ADE500
INTRACORT see HHR000
INTRADERM TYROTHRICIN see TOG500
INTRADEX see DBD700
INTRAMYCETIN see CDP250
INTRANEFRIN see VGP000
INTRANYL ORANGE T-4R see SGP500
INTRAPERSE YELLOW GBA see AAQ250
INTRASPERSE YELLOW GBA EXTRA see AAQ250
INTRASPORIN see TEY000
INTRASTIGMINA see NCL100
INTRATHION see PHI500
INTRATION see PHI500
INTRAVAL see PBT250
INTRAVAL SODIUM see PBT500
INTRAVAT BLUE GF see DFN425
INTRAZONE RED BR see CMM330
INTROMENE see HII500
INTROPIN see DYC600
INVENOL see BSM000
INVERSAL see AHI875
INVERSINE see VIZ400
INVERSINE HYDROCHLORIDE see MQR500
INVERTON 245 see TAA100
INVISI-GARD see PMY300
IOB 82 see AFQ575
IOCARMATE MEGLUMINE see IDJ500
IOCARMIC ACID see TDQ230
IOCARMIC ACID DI-N-METHYLGLUCAMINE SALT see IDJ500
IODAIRAL see HGB200
IODAMIDE 380 see IDJ600
IODAMIDE see AAI750
IODAMIDE MEGLUMINE see IDJ600
IODAMIDE METHYLGLUCAMINE see IDJ600
IODATES see IDJ700
IODE (FRENCH) see IDM000
IODEIKON see TDE750
IODENTEROL see CHR500
IODIC ACID see IDK000
IODIC ACID, SODIUM SALT see SHV500
IODIC ACIODIC ACID, POTASSIUM SALT see PLK250
IODIDES see IDL000
IODINE see IDM000
IODINE AZIDE (dry) (DOT) see IDN000
IODINE(I) AZIDE see IDN000
IODINE AZIDE see IDN000
IODINEBENZOL see IEC500
IODINE BROMIDE see IDN200
IODINE CHLORIDE see IDS000
IODINE CRYSTALS see IDM000
IODINE CYANIDE see COP000
IODINE DIOXIDE TRIFLUORIDE see IDP000
IODINE DIOXYGEN TRIFLUORIDE see IDP000
IODINE HEPTAFLUORIDE see IDQ000
IODINE ISOCYANATE see IDR000
IODINE MONOCHLORIDE see IDS000
IODINE(V) OXIDE see IDS300
IODINE PENTAFLUORIDE see IDT000
IODINE(III) PERCHLORATE see IDU000
IODINE PHOSPHIDE see PHA000
IODINE SUBLIMED see IDM000
IODINE TRIACETATE see IDV000
IODIO (ITALIAN) see IDM000
IODIPAMIDE MEGLUMINE see BGB315
IODIPAMIDE MEGLUMINE SALT see BGB315
IODIPAMIDE METHYLGLUCAMINE SALT see BGB315
2-IODOACETAMIDE see IDW000
α-IODOACETAMIDE see IDW000
IODOACETAMIDE see AOC500, IDW000
4'-IODOACETANILIDE see IDY000
4-IODOACETANILIDE see IDY000
p-IODOACETANILIDE see IDY000
IODOACETATE see IDZ000
IODOACETATE SODIUM SALT see SIN000
IODOACETIC ACID see IDZ000
IODOACETIC ACID ETHYL ESTER see ELQ000
IODOACETIC ACID SODIUM SALT see SIN000
(IODOACETOXY)TRIBUTYLSTANNANE see TID750
(IODOACETOXY)TRIPROPYLSTANNANE see TNB500
N-(IODOACETYL)-3-AZABICYCLO(3.2.2)NONANE see IDZ100
1-IODOACETYL-α-α-DIPHENYL-4-PIPERIDINEMETHANOL see IDZ200

IODOACETYLENE see IDZ400
IODOALPHIONIC ACID see PDM750
3-IODOANILINE see IEB000
4-IODOANILINE see IEC000
m-IODOANILINE see IEB000
p-IODOANILINE see IEC000
IODOAZIDE see IDN000
4-IODOBENZENAMINE see IEC000
IODOBENZENE see IEC500
4-IODOBENZENEDIAZONIUM-2-CARBOXYLATE see IED000
o-IODOBENZOIC ACID see IEE000
p-IODOBENZOIC ACID see IEE025
o-IODOBENZOIC ACID SODIUM SALT see IEE050
p-IODOBENZOIC ACID SODIUM SALT see IEE100
o-IODOBENZOIC ACID TRIBUTYLSTANNYL ESTER see TIE000
p-IODOBENZOIC ACID TRIBUTYLSTANNYL ESTER see TIE250
N-(2-IODOBENZOYL)-GLYCIN MONOSODIUM SALT (9CI) see HGB200
N-p-IODOBENZOYL-N′,N′,N′,N′-DIETHYLENETRIAMIDE of PHOSPHORIC ACID
    see BGY140
(4-IODOBENZOYLOXY)TRIBUTYLSTANNANE see TIE250
(o-IODOBENZOYLOXY)TRIPROPYLSTANNANE see IEF000
IODOBIL see PDM750
1-IODO-1,3-BUTADIYNE see IEG000
1-IODOBUTANE see BRQ250
2-IODOBUTANE see IEH000
IODOCHLORHYDROXYQUINOL see CHR500
IODOCHLORHYDROXYQUINOLINE see CHR500
7-IODO-5-CHLORO-8-HYDROXYQUINOLINE see CHR500
7-IODO-5-CHLOROXINE see CHR500
5-IODO-2′-DEOXYURIDINE see DAS000
5-IODODEOXYURIDINE see DAS000
IODODIMETHYLARSINE see IEI000
4-IODO-3,5-DIMETHYLISOXAZOLE see IEI600
IODOENTEROL see CHR500
IODOETHANE see ELP500
2-(2-IODOETHYL)-1,3-DIOXOLANE see IEL700
(2-IODOETHYL)TRIMETHYLAMMONIUM IODIDE see IEM300
3-IODO-2-FAA see IEO000
IODOFENOPHOS see IEN000
N-(3-IODO-2-FLUORENYL)ACETAMIDE see IEO000
IODOFORM see IEP000
IODOGNOST see TDE750
IODOHEPTAFLUOROPROPANE see HAY300
IODOHIPPURA see HGB200
o-IODOHIPPURATE SODIUM see HGB200
IODOHIPPURATE SODIUM see HGB200
4-(3-IODO-4-HYDROXYPHENOXY)-3,5-DIIODOPHENYLALANINE see LGK050
7-IODO-8-HYDROXYQUINOLINE-5-SULFONIC ACID see IEP200
IODOMETANO (ITALIAN) see MKW200
IODOMETHANE see MKW200
IODOMETHANESULFONIC ACID SODIUM SALT see SHX000
1-IODO-3-METHYLBUTANE see IHU200
IODOMETHYLMAGNESIUM see MLE250
7-IODOMETHYL-12-METHYLBENZ(a)ANTHRACENE see IER000
1-IODO-2-METHYLPROPANE see IIV509
2-IODO-2-METHYLPROPANE see TLU000
IODOMETHYLTRIMETHYLARSONIUM IODIDE see IET000
IODOMETHYLZINC see MQP250
1-IODOOCTANE see OFA100
IODOPACT see AAN000
IODOPANIC ACID see IFY100
IODOPANOIC ACID see IFY100
IODOPAQUE see AAN000
1-IODOPENTANE see IET500
1-IODO-3-PENTEN-1-YNE see IEU000
1-IODOPERFLUORODECANE see IEU075
1-IODOPERFLUOROOCTANE see PCH325
IODOPHENE see IEU100
IODOPHENE SODIUM see TDE750
3-IODOPHENOL see IEV010
4-IODOPHENOL see IEW000
m-IODOPHENOL see IEV010
p-IODOPHENOL see IEW000
1-IODO-3-PHENYL-2-PROPYNE see IEX000
IODOPHOS see IEN000
IODOPHTHALEIN SODIUM see TDE750
1-IODOPROPANE see PNO750
2-IODOPROPANE see IPS000
3-IODO-1-PROPENE see AGI250
3-IODOPROPENE see AGI250
3-IODOPROPIONIC ACID see IEY000
(IODOPROPIONYLOXY)TRIBUTYLSTANNANE see TIE500
3-IODOPROPYLENE see AGI250
3-IODOPROPYNE see IEZ800
3-IODO-2-PROPYNYL-2,4,5-TRICHLOROPHENYL ETHER see IFA000

IODOPYRACET see DNG400
IODOQUINOL see DNF600
IODORAYORAL see TDE750
IODOSOBENZENE see IFC000
IODOSOBENZENE DIACETATE see IFD000
IODOSYLBENZENE see IFE000
4-IODOSYLTOLUENE see IFE875
2-IODOSYLVINYL CHLORIDE see IFE879
5-IODO-2-THIOURACIL, SODIUM SALT see SHX500
IODO(p-TOLYL)MERCURY see IFL000
11-IODO-10-UNDECINIC ACID see IFO700
IODO-UNDECINIC ACID see IFO700
11-IODO-10-UNDECYNOIC ACID see IFO700
5-IODOURACIL see IFP000
5-IODOURACIL DEOXYRIBOSIDE see DAS000
IODOXAMATE MEGLUMINE see IFP800
IODOXAMIC ACID MEGLUMINE SALT see IFP800
IODTETRAGNOST see TDE750
IODURE d′ETHYL-TRIMETHYL-AMMONIUM see EQC600
IODURE de MERCURE (FRENCH) see MDC750
IODURE de METHYLE (FRENCH) see MKW200
l′IODURE de α-THIENYL-1 PHENYL-1 N-METHYL MORPHOLINIUM-3 PROPA-
    NOL-1 (FRENCH) see HNM000
IODURIL see SHW000
IODURON B see DNG400
4-IODYLANISOLE see IFQ775
IODYLBENZENE PERCHLORATE see IFS000
4-IODYL TOLUENE see IFS350
2-IODYLVINYL CHLORIDE see IFS385
3-IODOTETRAHYDROTHIOPHENE-1,1-DIOXIDE see IFG000
IOFENDYLATE see ELQ500
IOGLUCOMIDE see IFS400
IMAKOL see AFL750
IOMESAN see DFV400
IOMEX see IFT100
IOMEZAN see DFV400
6,3-IONENE BROMIDE see IFT300
IONET S-80 see SKV100
IONET MO-400 see PJY100
IONET S 60 see SKV150
IONOL 6 see MHR050
α-IONOL see IFT400
β-IONOL see IFT500
IONOL (antioxidant) see BFW750
IONOL see BFW750
α-IONONE see IFW000
β-IONONE see IFX000
IONONE see IFV000
IONOX 100 see IFX200
IONOX 100 ANTIOXIDANT see IFX200
α-IONYL ACETATE see IFX300
4-IODOTOLUENE see IFK509
IOP see AES000
IOPAMIDOL see IFY000
IOPAMIRON see IFY000
IOPANOIC ACID see IFY100
IOPEZITE see PKX250
IOPHENDYLATE see ELQ500
IOPHENOXIC ACID see IFZ800
IOPODATE SODIUM see SKM000
IOPRAMINE HYDROCHLORIDE see IFZ900
IOPRONIC ACID see IGA000
IOPYRACIL see DNG400
IOQUIN SUSPENSION see DNF600
IODOTRIBUTYLSTANNANE see IFM000
IODOTRIMETHYLSTANNANE see IFN000
IODOTRIMETHYLTIN see IFN000
IODOTRIPHENYLSTANNANE see IFO000
IODOTRIPROPYLSTANNANE see TNB250
IOSALIDE see JDS200
IOTALAMATE de METHYLGLUCAMINE (FRENCH) see IGC000
IOTHALAMATE MEGLUMINE see IGC000
IOTHALAMATE METHYLGLUCAMINE see IGC000
IOTHALAMATE METHYLGLUCAMINE SALT see IGC000
IOTOX see HKB500
IOTROXATE MEGLUMINE see IGD075
IOTROXATE METHYLGLUCMINE SALT see IGD075
IOTROXIC ACID see IGD100
IOTROXINSAEURE (GERMAN) see IGD100
IOXAGLIC ACID see IGD200
IOXYNIL see HKB500
IOXYNIL, LITHIUM SALT see DNF400
IOXYNIL OCTANOATE see DNG200
IP-10 see IGE100
IP-82 see IIU000

o-IPA see INA200
IPA see DMV600, IMJ000
IPABUTONA see HNI500
IPAMIX see IBV100
IPANER see DAA800
IPD see YCJ000
IPDI see IMG000
IP-1,2-DIOL see DLE400
IPECAC SYRUP see IGF000
IPECACUANHA see IGF000
IPE-TOBACCO WOOD see HLY500
IPHOSPHAMIDE see IMH000
IPN see ILE000, PHX550
IPNO see IBP309
IPNOFIL see QAK000
IPO 8 see TBW100
IPODATE SODIUM see SKM000
IPOGLICONE see BSQ000
IPOGNOX 89 see TFD500
IPOLAB see HMM500
IPOLINA see HGP500
1,4-IPOMEADIOL see IGF200
IPOMEANIN see IGF300
IPOMEANINE see IGF300
4-IPOMEANOL see FQL100
IPOMEANOL see FQL100, IGF325
IPORES see RCA200
IPOTENSIVO see MBW750
IPPC see CBM000
IPRAL see ELX000
IPRAL SODIUM see NBU000
IPRATROPIUMBROMID (GERMAN) see IGG000
IPRATROPIUM BROMIDE see IGG000
IPRAZID see ILE000
IPRINDOLE see DPX200
IPROBENFOS see BKS750
IPROCLOZIDE see CJN250
IPRODIONE see GIA000
IPROGEN see DLH630
IPROHEPTINE HYDROCHLORIDE see IGG300
IPRONIAZID see ILE000
IPRONIAZID DIHYDROCHLORIDE see IGG600
IPRONIAZID HYDROCHLORIDE see IGG600
IPRONIAZID PHOSPHATE see IGG700
IPRONID see ILE000
IPRONIDAZOLE (USDA) see IGH000
IPRONIN see ILE000
IPROPLATIN see IGG775
IPROPRAN see IGH000
IPROVERATRIL see IRV000
IPROVERATRIL HYDROCHLORIDE see VHA450
IPROX see CMY650
IPSILON see AJD000
IPSOFLAME see BRF500
IPSOTIAN see MQU750
IPT (GERMAN) see INP100
I.P.Z. see IDF325
IQ 1 see IRV300
IQ DIHYDROCHLORIDE see AKT620
IR 125 see CCK000
IRADICAV see SHF500
IRAGEN RED L-U see FAG070
IRALDEINE see IFV000
IRAMIL see DLH600, DLH630
IRAX see PJS750
IRC 453 see CIN750
IRC-50 ARVIN see VGU700
IRCON see FBJ100
IRENAL see ELX000, PLO500
IRENAT see PLO500
IRENE see DLS600
IRETIN see AQQ750, AQR000
IRG see AKY250
IRGACHROME ORANGE OS see LCS000
IRGALITE 1104 see CBT500
IRGALITE BRONZE RED CL see BNH500, BNK700
IRGALITE FAST RED P4R see MMP100
IRGALITE FAST SCARLET RND see MMP100
IRGALITE ORANGE P see CMS145
IRGALITE ORANGE PG see CMS145
IRGALITE ORANGE PX see CMS145
IRGALITE RED 4B see CMS155
IRGALITE RED C see CMS150
IRGALITE RED CBN see CHP500
IRGALITE RED PV2 see MMP100

IRGALITE RED RNPX see MMP100
IRGALITE RUBINE PB see CMS155
IRGALITE SCARLET RB see MMP100
IRGALITE YELLOW BAW see CMS208
IRGALITE YELLOW BAWX see CMS208
IRGALITE YELLOW BO see DEU000
IRGALITE YELLOW BR see CMS212
IRGALITE YELLOW BRE see CMS212
IRGALON see EIV000
IRGANOX PS 802 see DXG700
IRGANOX PS 800 see TFD500
IRGAPLAST ORANGE G see CMS145
IRGAPLAST YELLOW IRS see CMS208
IRGAPYRIN see IGI000
IRGAPYRINE see IGI000
IRGASAN see TIQ000
IRGASAN BS-200 see TBV000
IRGASAN DP300 see TIQ000
IRGSTAB T 4 see DEJ100
IRGASTAB T 150 see DEJ100
IRGASTAB T 290 see DEJ100
IRIDIL see HNI500
IRIDIUM see IGJ000
IRIDIUM(IV) CHLORIDE see IGJ499
IRIDIUM CHLORIDE see IGJ300
IRIDIUM MURIATE see IGJ300
IRIDIUM TETRACHLORIDE see IGJ499
IRIDOCIN see EPQ000
IRIDOZIN see EPQ000
IRIFAN see TDA500
IRIS see OHC130
IRIS ABSOLUTE see OHC130
IRISH GUM see CCL250
IRISH MOSS EXTRACT see CCL250
IRISH MOSS GELOSE see CCL250
IRISOL BASE see HOK000
IRISONE see IFV000
IRISONE ACETATE see CNS825
IRITONE see IGJ600
IRIUM see SIB600
IRMIN see DLH600
IRON, (N,N-BIS(CARBOXYMETHYL)GLYCINATO(3-)-N,O,O',O'')-, (T-4)-(9CI) see IHC100
IROCAINE see AIT250
IRON, CARBONYL (FCC) see IGK800
IRON, ELECTROLYTIC see IGK800
IRON, ELEMENTAL see IGK800
IROINI see HBT500
IRO-JEX see IGS000
IROMIN see FBK000
IRON(2+) see FBN000
IRON see IGK800
IRON(2+) ACETATE see FBH000
IRON(II) ACETATE see FBH000
IRON ACETYLACETONATE see IGL000
IRON(III) AMMONIUM CITRATE see FAS700
IRON(III) ARSENATE (1:1) see IGN000
IRON(II) ARSENATE (3:2) see IGM000
IRON ARSENATE (DOT) see IGM000
IRON(III)-o-ARSENITE PENTAHYDRATE see IGO000
IRON(II) ASCORBATE see FBH050
IRONATE see FBO000
IRON BIS(CYCLOPENTADIENE) see FBC000
IRON BLACK see IHC550
IRON BLUE see IGY000
IRON(III) BROMIDE see IGQ000
IRON(II) BROMIDE see IGP000
IRON CARBIDE see IGQ750
IRON CARBOHYDRATE COMPLEX see IGT000
IRON(II) CARBONATE see FBH100
IRON CARBONYL see IHG500, NMV740
IRON(II) CHLORIDE (1:2) see FBI000
IRON(III) CHLORIDE see FAU000
IRON CHLORIDE see FAU000
IRON(3+) CHLORIDE HEXAHYDRATE see FAW000
IRON(III), CHLORIDE HEXAHYDRATE see FAW000
IRON(2+) CHLORIDE TETRAHYDRATE see FBJ000
IRON (II) CHLORIDE TETRAHYDRATE see FBJ000
IRON CHLORIDE TETRAHYDRATE see FBJ000
IRON CHOLINE CITRATE COMPLEX see FBC100
IRON CHROMITE see CMI500
IRON COMPOUNDS see IGR499
IRON CYANIDE see IGY000
IRON DEXTRAN see IGS000
IRON-DEXTRAN COMPLEX see IGS000

IRON DEXTRAN GLYCEROL GLYCOSIDE see IGU000
IRON DEXTRAN INJECTION see IGS000
IRON-DEXTRIN COMPLEX see IGT000
IRON DEXTRIN INJECTION see IGT000
IRON DIACETATE see FBH000
IRON DICHLORIDE see FBI000
IRON DICHLORIDE TETRAHYDRATE see FBJ000
IRON DICYCLOPENTADIENYL see FBC000
IRON DIMETHYLDITHIOCARBAMATE see FAS000
IRON DISULFIDE see IGV000
IRON DUST see IGW000
α-IRONE see IGW500
IRON(III)-EDTA SODIUM see IGX875
IRON (Fe 2+) see FBN000
IRON(3+) FERROCYANIDE see IGY000
IRON (III) FERROCYANIDE see IGY000
IRON complex with N-FLUOREN-2-YL ACETOHYDROXAMIC ACID see HIQ000
IRON FLUORIDE see FAX000
IRON FUMARATE see FBJ100
IRON GLUCONATE see FBK000
IRON(III) HEXACYANOFERRATE(4) see IGY000
IRON HYDROGENATED DEXTRAN see IGS000
IRON(+3) HYDROXIDE COMPLEX with NITRILO-TRI-PROPIONIC ACID see FAY000
IRON(III)HYDROXIDE-POLYMALTOSE see IHA000
IRON(II,III) OXIDE see IHC550
IRON(II) ION see FBN000
IRON(2+) LACTATE see LAL000
IRON(II) MALEATE see IHB675
IRON MANGANESE ZINC OXIDE see IHB677
IRON MONOSULFATE see FBN100
IRON NICKEL SULFIDE see NDE500
IRON NITRATE see FAY200, IHB900
IRON (III) NITRATE, ANHYDROUS see FAY200, IHB900
IRON(III) NITRATE, NONAHYDRATE (1:3:9) see IHC000
IRON NITRILOTRIACETATE see IHC100
IRON-NITRILOTRIACETATE CHELATE see IHC100
IRON(3+) NTA see IHC100
IRON ORE see HAO875
IRONORM INJECTION see IGS000
IRON(III) OXIDE see IHD000
IRON(II) OXIDE see IHC500
IRON OXIDE see IHD000, IHG100
IRON OXIDE, spent see IHG100
IRON OXIDE, CHROMIUM OXIDE, and NICKEL OXIDE FUME see IHE000
IRON OXIDE FUME see IHF000
IRON OXIDE RED see IHD000
IRON OXIDE RED 130B see IHG100
IRON OXIDE, SACCHARATED see IHG000
IRON PENTACARBONYL see IHG500
IRON-POLYSACCHARIDE COMPLEX see IHH000
IRON-POLY(SORBITOL-GLUCONIC ACID) COMPLEX see IHH300
IRON PROTOCHLORIDE see FBI000
IRON PROTOSULFATE see FBN100
IRON PYRITES see IGV000
IRON SACCHARATE see IHG000
IRON SESQUIOXIDE see IHD000
IRON SESQUIOXIDE HYDRATED see LFW000
IRON-SILICON see IHJ000
IRON SODIUM GLUCONATE see IHK000
IRON SORBITEX see IHL000
IRON-SORBITOL see IHK100
IRON SORBITOL CITRATE see IHL000
IRON-SORBITOL-CITRIC ACID see IHL000
IRON SPONGE, spent obtained from coal gas purification (DOT) see IHG100
IRON SUGAR see IHG000
IRON(II) SULFATE (1:1) see FBN100
IRON(II) SULFATE (1:1), HEPTAHYDRATE see FBO000
IRON(III) SULFIDE see IHN050
IRON(II) SULFIDE see IHN000
IRON SULFIDE see IGV000
IRON TRICHLORIDE see FAU000
IRON TRICHLORIDE HEXAHYDRATE see FAW000
IRON TRIFLUORIDE see FAX000
IRON TRINITRATE see FAY200, IHB900
IRON VITRIOL see FBN100
IRON VITROL see FBO000
IROQUINE see CLD000
IRON, REDUCED (FCC) see IGK800
IROSPAN see FBN100
IROSUL see FBN100, FBO000
IRON, TRIS(5,6-DIHYDRO-5,6-DIOXO-2-NAPHTHALENESULFONIC ACID-5-OXI-MATO)-, TRISODIUM SALT see NAX500
IROX (GADOR) see FBK000
IRRADIATED ERGOSTA-5,7,22-TRIEN-3-β-OL see VSZ100

IRRATHENE R see PJS750
IRTRAN 1 see MAF500
IRTRAN 3 see CAS000
IRTRAN 6 see CAJ800
IS 401 see DNM400
1205 I.S. see MHD300
1530 I.S. see MRR125
1665 I.S. see BKS825
2341 I.S. see BIQ500
2406 I.S. see IBB100
2466 I.S. see MHQ750
2723 I.S. see PFY100
ISACONITINE see PIC250
ISADRINE see IMR000
dl-ISADRINE HYDROCHLORIDE see IQS500
ISADRINE-HYDROCHLORIDE see IMR000
ISALIZINA see BET000
ISAMIN see WAK000
ISAROL see AIF000
ISATEX see ACD500
ISATIC ACID LACTAM see ICR000
ISATIDINE see RFU000
ISATIN see ICR000
ISATINIC ACID ANHYDRIDE see ICR000
ISAZOFOS see PHK000
ISAZOPHOS see PHK000
ISCEON 22 see CFX500
ISCEON 113 see FOO000
ISCEON 122 see DFA600
ISCEON 131 see TIP500
ISCHELIUM see DLL400
ISCOBROME see MHR200
ISCOBROME D see EIY500
ISCOVESCO see DKA600
ISDIN see CCK125
I-SEDRIN see EAW000
ISEPAMICIN SULFATE see SBE800
ISETHION see GFW000
ISETHIONIC ACID see HKI500
ISF 2001 see TNP275
ISF 2469 see CAK275
ISF 2508 see CFJ000
ISICAINA see DHK400
ISICAINE HYDROCHLORIDE see DHK600
ISIDRINA see ILD000
ISINDONE see IDA400
ISLACTID see AES650
ISLANDITOXIN see COW750
ISMAZIDE see ILD000
ISMELIN see GKQ000, GKS000
ISMELIN SULFATE see GKS000
ISMICETINA see CDP250
ISMIPUR see POK000
5-ISMN see ISC500
ISMOTIC see HID350
ISOACETOPHORONE see IMF400
ISOADRENALINE HYDROCHLORIDE see AMB000
ISOALLOXAZINE-ADENINE DINUCLEOTIDE see RIF100
ISOAMIDONE II see IKZ000
ISOAMINILE CITRATE see PCK639
ISOAMINILE CYCLAMATE see IHO200, PCK669
ISOAMYCIN see BBK000
ISOAMYL ACETATE see IHO850
ISOAMYL ALCOHOL see IHP000, IHP010
ISOAMYL ALDEHYDE see MHX500
ISOAMYL ALKOHOL (CZECH) see IHP000
ISO-AMYLALKOHOL (GERMAN) see IHP000
ISOAMYL BENZOATE see IHP100
ISOAMYL BENZYL ETHER see BES500
ISOAMYL BROMIDE see BNP250
ISOAMYL BUTANOATE see IHP400
ISOAMYL BUTYLATE see IHP400
ISOAMYL-n-BUTYRATE see IHP400
ISOAMYL BUTYRATE see IHP400
ISOAMYL CAPROATE see IHU100
ISOAMYL CAPRYLATE see IHP500
ISOAMYL CINNAMATE see AOG600
ISOAMYL CYANIDE see MNJ000
ISOAMYLDICHLORARSINE see IHQ100
ISOAMYLDICHLOROARSINE see IHQ100
ISOAMYL α-(N-(β-DIETHYLAMINOETHYL))-AMINOPHENYLACETATE see NOC000
ISOAMYL N-(β-DIETHYLAMINOETHYL)-α-AMINOPHENYLACETATE see NOC000

ISOAMYL 5,6-DIHYDRO-7,8-DIMETHYL-4,5-DIOXO-4H-PYRANO(3,2-c)QUINOLINE-2-CARBOXYLATE see IHR200
ISOAMYLENE see IHP150
ISOAMYLENE ALCOHOL see PBL000
β-ISOAMYLENE OXIDE see TLY175
8-ISOAMYLENOXYPSORALEN see IHR300
4-(β-ISOAMYLENYL)-1,2-DIPHENYL-3,5-PYRAZOLIDINEDIONE see PEW000
ISOAMYL ETHANOATE see IHO850
5-ISCAMYL-5-ETHYLBARBITURIC ACID see AMX750
ISOAMYLETHYLBARBITURIC ACID see AMX750
5-ISOAMYL-5-ETHYLBARBITURIC ACID, SODIUM DERIVATIVE see AON750
ISOAMYL FORMATE see IHS000
ISOAMYL GERANATE see IHS100
ISOAMYL HEXANOATE see IHU100
ISOAMYLHYDRIDE see EIK000
ISOAMYL HYDROCUPREINE DIHYDROCHLORIDE see EQQ500
ISOAMYL IODIDE see IHU200
ISOAMYL ISOVALERATE (FCC) see ITB000
ISOAMYL METHANOATE see IHS000
ISOAMYL METHYL KETONE see MKW450
ISO-AMYL NITRATE see ILW100
ISOAMYLNITRILE see ITD000
ISOAMYL NITRITE see IMB000
ISOAMYL OCTANOATE see IHP500
ISOAMYLOL see IHP000
ISOAMYL 3-PENTYL PROPENATE see AOG600
ISOAMYL PHENYLACETATE see IHV000
ISOAMYL PHENYLAMINOACETATE HYDROCHLORIDE see PCV750
ISOAMYL PHENYLETHYL ETHER see IHV050
ISOAMYL POTASSIUM XANTHATE see PLK580
ISOAMYL PROPIONATE see AON350
1-ISOAMYL THEOBROMINE see IHX200
ISOANETHOLE see AFW750
trans-ISOASARONE see IHX400
d-ISOASCORBIC ACID see SAA025
ISOASCORBIC ACID see SAA025
ISOBAC 20 see HCL000, MRM000
ISOBAMATE see IPU000
ISOBARB see NBU000
ISOBENZAMIC ACID see AMU750
ISOBENZAN see OAN000
1,3-ISOBENZOFURANDIONE see PHW750
1,3-ISOBENZOFURANDIONE, 4,5,6,7-TETRACHLORO-(9CI) see TBT150
1,3-ISOBENZOFURANDIONE, 3a,4,7,7a-TETRAHYDROMETHYL- see MPP000
1(3H)-ISOBENZOFURANONE, 3,3-BIS(4-HYDROXYPHENYL)- see PDO750
1(3H)-ISOBENZOFURANONE, 3-BUTYLIDENE-(9CI) see BRQ100
1(3H)-ISOBENZOFURANONE, 3-(3-FURANYL)-3a,4,5,6-TETRAHYDRO-3a,7-DI-METHYL-, (3R-cis)-(9CI) see FOM200
ISOBERGAMATE see IHX450
ISO-BID see CCK125
ISOBIDE see HID350
ISOBIND 100 see PKB100
ISO-BNU see IJF000
dl-ISOBORNEOL see IHY000
ISOBORNEOL see IHY000
ISOBORNEOL METHYL ETHER see IHX500
ISOBORNEOL THIOCYANATOACETATE see IHZ000
ISOBORNYL ACETATE see IHX600
ISOBORNYL ALCOHOL see IHY000
ISOBORNYL METHYL ETHER see IHX500
ISOBORNYL THIOCYANATOACETATE see IHZ000
ISOBORNYL THIOCYANOACETATE see IHZ000
ISOBROMYL see BNP750
ISOBUTANAL see IJS000
ISOBUTANDIOL-2-AMINE see ALB000
ISOBUTANE (DOT) see MOR750
ISOBUTANE see MOR750
ISOBUTANE MIXTURES (DOT) see MOR750
ISOBUTANOL (DOT) see IIL000
ISOBUTANOL-2-AMINE see IIA000
ISOBUTANOLAMINE see IIA000
ISOBUTAZINA see HNI500
ISOBUTENAL see MGA250
ISOBUTENE see IIC000
ISOBUTENYL CHLORIDE see CIU750
ISOBUTENYL METHYL KETONE see MDJ750
ISOBUTIL see HNI500
ISOBUTINYL-N-(3-CHLORPHENYL)-CARBAMAT (GERMAN) see CEX250
2-ISOBUTOXYETHANOL see IIP000
2-ISOBUTOXY-2-ETHOXY-2-ETHANOL see DJF800
2-(2-ISOBUTOXYETHOXY)ETHANOL see DJF800
N-ISOBUTOXYMETHYLACRYLAMIDE see IIE100
2-ISOBUTOXYNAPHTHALENE see NBJ000
2-ISOBUTOXY TETRAHYDROPYRAN see III000
ISOBUTYL ACETATE see IIJ000

N-ISOBUTYL-N-(ACETOXYMETHYL)NITROSAMINE see ABR125
ISOBUTYL ACRYLATE, inhibited (DOT) see IIK000
ISOBUTYL ACRYLATE see IIK000
ISOBUTYL ADIPATE see DNH125
ISOBUTYL ALCOHOL see IIL000
ISOBUTYL ALDEHYDE (DOT) see IJS000
ISOBUTYLALDEHYDE see IJS000
ISOBUTYLALKOHOL (CZECH) see IIL000
ISO-BUTYLALLYLBARBITURIC ACID see AGI750
ISOBUTYLALLYLBARTURIC ACID see AGI750
ISOBUTYLAMINE see IIM000
2-ISOBUTYLAMINOETHANOL HYDROCHLORIDE ACID SALT, p-AMINOBENZO-IC ACID ESTER see IAC000
2-(ISOBUTYLAMINO)ETHYL-p-AMINOBENZOATE HYDROCHLORIDE see IAC000
2-(ISOBUTYLAMINO)-2-METHYL-1-PROPANOL BENZOATE (ESTER), HYDRO-CHLORIDE see IIM100
2-(ISOBUTYLAMINO)-2-METHYL-1-PROPANOL BENZOATE HYDROCHLORIDE see IIM100
3-(3-(ISOBUTYLAMINO)PROPYL)-2-(3,4-METHYLENEDIOXYPHENYL)-4-THIAZOLI-DINONE HYDROCHLORIDE see WBS855
ISOBUTYLBENZENE see IIN000
ISOBUTYLBIS(2-CHLOROETHYL)AMINE HYDROCHLORIDE see BIE750
ISOBUTYLBIS(β-CHLOROETHYL)AMINE HYDROCHLORIDE see BIE750
ISOBUTYL BROMIDE see BNR750
ISOBUTYL BUTANOATE see BSW500
3-ISOBUTYL-6-sec-BUTYL-2-HYDROXYPYRAZINE-1-OXIDE see ARO000
ISOBUTYL BUTYRATE (FCC) see BSW500
ISOBUTYL CAPROATE see IIT000
ISOBUTYLCARBINOL see IHP000
(2-(p-(5-(ISOBUTYLCARBOZMOYL)-2-OCTYLOXYBENZAMI-DO)BENZAMIDO)ETHYL)-TRIETHYLAMMONIUM IODIDE see IIO750
ISOBUTYL CELLOSOLVE see IIP000
ISOBUTYL CHLORIDE see CIU500
ISOBUTYL CINNAMATE see IIQ000
ISOBUTYL CYANOACETATE see IIQ100
2-((1-ISOBUTYL-3,5-DIMETHYLHEXYL)OXY)ETHANOL see EJL000
1-ISOBUTYL-3,4-DIPHENYLPYRAZOLE-5-ACETIC ACID SODIUM SALT see DWD400
ISOBUTYLDIUREA see IIV000
ISOBUTYLENE (DOT) see IIC000
ISOBUTYLENE CHLORIDE see IIQ200
ISOBUTYLENEDIUREA see IIV000
ISOBUTYLESTER KYSELINY METHAKRYLOVE see IIY000
ISOBUTYLESTER KYSELINY MRAVENCI see IIR000
ISOBUTYLESTER KYSELINY OCTOVE see IIJ000
ISO-BUTYL FORMATE see IIR000
ISOBUTYL FORMATE see IIR000
ISOBUTYL-2-FURANPROPIONATE see IIR100
ISOBUTYL FURYLPROPIONATE see IIR100
ISOBUTYLGLYCEROL, NITRO- see HMJ500
ISOBUTYL HEPTANOATE see IIS000
ISOBUTYL HEPTYLATE see IIS000
ISOBUTYL HEXANOATE see IIT000
4-ISOBUTYLHYDRATROPIC ACID see IIU000
p-ISOBUTYLHYDRATROPIC ACID see IIU000
ISOBUTYL-o-HYDROXYBENZOATE see IJN000
1,1'-ISOBUTYLIDENEBISUREA see IIV000
ISOBUTYLIDENEDIUREA see IIV000
ISOBUTYL IODIDE see IIV509
ISOBUTYLISOBUTYRATE (DOT) see IIW000
ISOBUTYL ISOBUTYRATE see IIW000
ISOBUTYL ISOPENTANOATE see ITA000
ISOBUTYL ISOVALERATE see ITA000
ISOBUTYLJODID see IIV509
ISOBUTYL KETONE see DNI800
ISOBUTYL LINALOL see IIW100
ISOBUTYL MERCAPTAN see IIX000
ISOBUTYL-α-METHACRYLATE see IIY000
ISOBUTYL METHACRYLATE see IIY000
2-(ISOBUTYL-3-METHYLBUTOXY)ETHANOL see IJA000
2-(2-(1-ISOBUTYL-3-METHYLBUTOXY)ETHOXY)ETHANOL see IJB000
ISOBUTYL METHYL CARBINOL see MKW600
ISOBUTYL-METHYLKETON (CZECH) see HFG500
ISOBUTYL METHYL KETONE see HFG500
ISOBUTYL METHYL KETONE PEROXIDE see IJB100
ISOBUTYLMETHYLMETHANOL see MKW600
5-ISOBUTYL-5-(METHYLTHIOMETHYL)BARBITURIC ACID SODIUM SALT see SHY000
ISOBUTYL 2-NAPHTHYL ETHER see NBJ000
ISOBUTYL β-NAPHTHYL ETHER see NBJ000
ISOBUTYL NITRITE see IJD000
1-ISO-BUTYL-1-NITROSOUREA see IJF000
N-ISOBUTYL-N-NITROSOUREA see IJF000
ISOBUTYL ORTHOVANADATE see VDK000

ISOBUTYL PHENYLACETATE see IJF400
4-ISOBUTYLPHENYLACETIC ACID see IJG000
(p-ISOBUTYLPHENYL)ACETIC ACID see IJG000
2-(4-ISOBUTYLPHENYL)BUTYRIC ACID see IJH000
2-(4-ISOBUTYLPHENYL)PROPANOIC ACID see IIU000
2-(p-ISOBUTYLPHENYL)PROPIONIC ACID see IIU000
α-(4-ISOBUTYLPHENYL)PROPIONIC ACID see IIU000
α-p-ISOBUTYLPHENYLPROPIONIC ACID see IIU000
2-(p-ISOBUTYLPHENYL)PROPIONIC ACID-o-METHOXYPHENYL ESTER see IJJ000
2-(p-ISOBUTYLPHENYL)PROPIONIC ACID 2-PYRIDYLMETHYL ESTER see IAB000
ISOBUTYL POTASSIUM XANTHATE see PLK600
ISOBUTYL-2-PROPENOATE see IIK000
ISOBUTYL PROPENOATE see IIK000
ISOBUTYL PROPIONATE (DOT) see PMV250
2-ISOBUTYLQUINOLINE see IJM000
α-ISOBUTYLQUINOLINE see IJM000
ISOBUTYL SALICYLATE see IJN000
ISOBUTYL STEARATE see IJN100
ISOBUTYLSULFHYDRATE see IJO000
p-ISOBUTYL-α-TOLUIC ACID see IJG000
ISOBUTYLTRIMETHYLETHANE see TLY500
ISOBUTYL VINYL ETHER see IJQ000
p-ISOBUTYOXYBENZOIC ACID-3-(2′-METHYLPIPERIDINO)PROPYL ESTER see IJR000
ISOBUTYRALDEHYD (CZECH) see IJS000
ISOBUTYRALDEHYDE see IJS000
ISOBUTYRALDEHYDE, OXIME see IJT000
ISOBUTYRIC ACID (DOT) see IJU000
ISOBUTYRIC ACID see IJU000
ISOBUTYRIC ACID, BENZYL ESTER see IJV000
ISOBUTYRIC ACID, CINNAMYL ESTER see CMR750
ISOBUTYRIC ACID, 3,7-DIMETHYL-2,6-OCTADIENYL ESTER, (Z)-(8CI) see NCO200
ISOBUTYRIC ACID, ETHYL ESTER see ELS000
ISOBUTYRIC ACID, 2,4-HEXADIENYL ESTER (8CI) see HCS700
ISOBUTYRIC ACID, HEXYL ESTER see HFQ550
ISOBUTYRIC ACID, ISOBUTYL ESTER see IIW000
ISOBUTYRIC ACID, 1-ISOPROPYL-2,2-DIMETHYLTRIMETHYLENE ESTER see TLZ000
ISOBUTYRIC ACID, METHYL ESTER see MKX000
ISOBUTYRIC ACID, 2-PHENOXYETHYL ESTER (6CI,8CI) see PDS900
ISOBUTYRIC ACID-3-PHENYLPROPYL ESTER see HHQ500
ISOBUTYRIC ACID, p-TOLYL ESTER see THA250
ISOBUTYRIC ALDEHYDE see IJS000
ISOBUTYRIC ANHYDRIDE see IJW000
ISOBUTYRONE see DTI600
ISOBUTYRONITRILE see IJX000
3-ISOBUTYRYL-2-ISOPROPYLPYRAZOLO(1,5-a)PYRIDINE see KDA100
ISOCAINE see IJZ000
ISOCAINE-ASID see AIT250
ISOCAINE BASE see PIV750
ISOCAINE-HEISLER see AIT250
ISOCAMPHANYL CYCLOHEXANOL (mixed isomers) see IKA000
ISOCAMPHOL see IHY000
ISOCAMPHYL CYCLOHEXANOL (mixed isomers) see IKA000
ISOCAPROALDEHYDE see MQJ500
ISOCAPRONITRILE see MNJ000
ISOCARAMIDINE SULFATE see IKB000
ISOCARB see PMY300
ISOCARBONAZID see IKC000
ISOCARBOSSAZIDE see IKC000
ISOCARBOXAZID see IKC000
ISOCARBOXAZIDE see IKC000
ISOCARBOXYZID see IKC000
ISOCETYL STEARATE see IKC050
ISOCHINOL see DNX400
ISOCHLOORTHION (DUTCH) see NFT000
ISOCHRYSENE see TMS000
ISOCID see ILD000
ISOCIL see BNM000
ISOCILLIN see PDT750
ISOCINEOLE see IKC100
cis-ISOCITRATO-(1,2-DIAMINO-CYCLOHEXAN)-PLATIN(II) (GERMAN) see PGQ275
ISO-CORNOX see CIR500
ISOCOTIN see ILD000
ISOCROTYL CHLORIDE see IKE000
ISOCUMENE see IKG000
ISOCYANATE 580 see PKB100
ISOCYANATE de METHYLE (FRENCH) see MKX250
ISOCYANATES see IKG349
ISOCYANATOCYCLOHEXANE see CPN500
ISOCYANATOETHANE see ELS500

2-ISOCYANATOETHYL METHACRYLATE see IKG700
β-ISOCYANATOETHYL METHACRYLATE see IKG700
ISO-CYANATOMETHANE see MKX250
1-(1-ISOCYANATO-1-METHYLETHYL)-3-(1-METHYLETHENYL)BENZENE see IKG800
3-ISOCYANATOMETHYL-3,5,5-TRIMETHYLCYCLOHEXYLISOCYANATE see IMG000
1-ISOCYANATOPROPANE see PNP000
ISOCYANIC ACID, ALLYL ESTER see AGJ000
ISOCYANIC ACID, BENZ(a)ANTHRACEN-7-YL ESTER see BBJ250
ISOCYANIC ACID, 3-BROMOALLYL ESTER see BMT150
ISOCYANIC ACID, BUTYL ESTER see BRQ500
ISOCYANIC ACID-2-CHLOROETHYL ESTER see IKH000
ISOCYANIC ACID, 6-CHLOROHEXYL ESTER see CHL250
ISOCYANIC ACID-m-CHLOROPHENYL ESTER see CKA750
ISOCYANIC ACID-p-CHLOROPHENYL ESTER see CKB000
ISOCYANIC ACID, CYCLOHEXYL ESTER see CPN500
ISOCYANIC ACID-3,4-DICHLOROPHENYL ESTER see IKH099
ISOCYANIC ACID, DIESTER with 1,6-HEXANEDIOL see DNJ800
ISOCYANIC ACID, 3,3′-DIMETHYL-4,4′-BIPHENYLENE ESTER see DQS000
ISOCYANIC ACID, ESTER with DI-o-TOLUENEMETHANE see MJN750
ISOCYANIC ACID, ETHYL ESTER see ELS500
ISOCYANIC ACID, 4-FLUOROBUTYL ESTER see FHC200
ISOCYANIC ACID, 3-FLUOROPROPYL ESTER see FLR100
ISOCYANIC ACID, HEXAHYDRO-4,7-METHANOINDAN-1,8-YLENE ESTER see TJG500
ISOCYANIC ACID, HEXAMETHYLENE ESTER see DNJ800
ISOCYANIC ACID, HEXAMETHYLENE ESTER, POLYMER with 1,4-BUTANE-DIOL see PKO750
ISOCYANIC ACID, HEXAMETHYLENE ESTER, POLYMERS see HEG300
ISOCYANIC ACID, m-ISOPROPENYL-α-α-DIMETHYL BENZYL ESTER see IKG800
ISOCYANIC ACID, METHYLENEDI-p-PHENYLENE ESTER, POLYMER with 1,4-BUTANEDIOL see PKP000
ISOCYANIC ACID, METHYL ESTER see MKX250
ISOCYANIC ACID, 4-METHYL-m-PHENYLENE ESTER see TGM750
ISOCYANIC ACID, METHYLPHENYLENE ESTER see TGM740, TGM750
ISOCYANIC ACID-1,5-NAPHTHYLENE ESTER see NAM500
ISOCYANIC ACID, OCTADECYL ESTER see OBG000
ISOCYANIC ACID, PHENYL ESTER see PFK250
ISOCYANIC ACID, PROPYL ESTER see PNP000
ISOCYANIC ACID, TRIESTER with 1,3,5-TRIS(6-HYDROXYHEXYL)BIURET see TNJ300
ISOCYANIC ACID, (m-TRIFLUOROMETHYLPHENYL) ESTER see TKJ250
ISOCYANIC ACID, p-XYLYLENE ESTER see XSS260
ISOCYANIDE see COI500
ISOCYANIDES see IKH339
ISOCYANOAMIDE see IKH669
ISOCYANOETHANE see ELT000
ISOCYANURIC ACID see THS000
ISOCYANURIC ACID, DICHLORO- see DGN200
ISOCYANURIC ACID, DICHLORO-, POTASSIUM SALT see PLD000
ISOCYANURIC ACID TRIALLYL ESTER see THS100
ISOCYANURIC CHLORIDE see TIQ750
ISOCYANURIC DICHLORIDE see DGN200
ISOCYCLEX see IOE000
ISOCYCLOCITRAL see IKJ000
ISODECANOL see IKK000
ISODECYL ACRYLATE see IKL000
ISODECYL ALCOHOL see IKK000
ISODECYL ALCOHOL ACRYLATE see IKL000
ISODECYL METHACRYLATE see IKM000
ISODECYL PROPENOATE see IKL000
ISODEMETON see DAP200
ISODIAZOMETHANE see IKH669
ISODIENESTROL see DAL600
ISODIHYDROLAVANDULOL see IQH000
ISODIHYDROLAVANDULYL ALDEHYDE see IKM100
ISODILAN see IJG000
ISODIN see CFZ000
ISODINE see PKE250
ISODIPHENYLBENZENE see TBC620
ISODIPRENE see CCK500
ISODONAZOLE NITRATE see IKN200
ISODORMID see IQX000
ISODRIN see IKO000
ISODRINE see FMS875
ISODRINUM see FMS875
ISODUR see IIV000
ISODURENE see TDM500
β-ISODURYLIC ACID see IKN300
ISOENDOXAN see IMH000
d-ISOEPHEDRINE see POH000
ISOESTRAGOLE see PMQ750
ISOETHADIONE see PAH500

ISOETHYLVANILLIN see IKO100
ISOEUGENOL see IKQ000
ISOEUGENOL ACETATE see AAX750
ISOEUGENOL AMYL ETHER see AOK000
1,3,4-ISOEUGENOL METHYL ETHER see IKR000
ISOEUGENYL ACETATE (FCC) see AAX750
ISOEUGENYL METHYL ETHER see IKR000
ISOFEDROL see EAW000, EAY500
ISOFENPHOS see IMF300
ISOFLAV BASE see DBN600
ISOFLUOROPHATE see IRF000
ISOFLURANE see IKS400
ISOFLUROPHATE see IRF000
ISOFORON see IMF400
ISOFORONE (ITALIAN) see IMF400
ISOFOSFAMIDE see IMH000
ISOFTALODINITRIL (CZECH) see PHX550
ISOGAMMA ACID see AKI000
ISOGLAUCON see CMX760
ISOHARRINGTONINE see IKS500
ISOHEXANAL see MQJ500
ISOHEXANE see IKS600
ISOHEXANOIC ACID (mixed isomers) see IKT000
ISOHEXYL ALCOHOL see AOK750
ISOHOL see INJ000
ISOHOMOGENOL see IKR000
ISOINDOLE-1,3-DIONE see PHX000
1,3-ISOINDOLEDIONE see PHX000
1H-ISOINDOLE-1,3(2H)-DIONE, 2-(CYCLOHEXYLTHIO)- see CPQ700
1H-ISOINDOLE-1,3(2H)-DIONE,2-(2,6-DIOXO-3-PIPERIDINYL)-, (R)- see TEH510
1H-ISOINDOLE-1,3(2H)-DIONE, 3-α-4,7,7-α-TETRAHYDRO-(9CI) see TDB100
1H-ISOINDOLE-5-SULFONAMIDE, 6-CHLORO-2,3-DIHYDRO-1,3-DIOXO- see
   CGM500
ISOINDOLINE, 2-(2-DIMETHYLAMINOETHYL)-4,5,6,7-TETRACHLORODIMETHO
   CHLORIDE see CDY000
1,3-ISOINDOLINEDIONE see PHX000
5-ISOINDOLINESULFONAMIDE, 6-CHLORO-1,3-DIOXO- see CGM500
ISOINOKOSTERONE see HKG500
ISOJASMONE see IKV000
ISO-K see BDU500
ISOKET see CCK125
ISOL see HFP875
ISOLAN see DSK200
ISOLANE (FRENCH) see DSK200
ISOLANID see LAU000
ISOLANIDE see LAU000
ISOLASALOCID A see IKW000
ISOL BENZIDINE FAST YELLOW GRX see CMS208
ISOL BENZIDINE YELLOW G see DEU000
ISOL BENZIDINE YELLOW GO see CMS212
ISOL BENZIDINE YELLOW GRX 2548 see CMS208
ISOL BONA RED N 5R BARIUM SALT see CMS148
ISOL BONA RED NR BARIUM SALT see CMS148
ISOL BONA RUBINE BK see CMS155
ISOL BONA RUBINE BKS see CMS155
ISOL BONA RUBINE KBK see CMS155
ISOL DIARYL YELLOW GRF see CMS208
I-ISOLEUCINE (FCC) see IKX000
ISOLEUCINE see IKX000
ISOL FAST RED HB see MMP100
ISOL FAST RED RN2B see MMP100
ISOL FAST RED RNB see MMP100
ISOL FAST RED RN2G see MMP100
ISOL FAST RED RNG see MMP100
ISOL LAKE RED LCL see CMS150
ISOL LAKE RED LCR see CMS150
ISOL LAKE RED LCS 12527 see CHP500
ISOL LAKE RED LCT see CMS150
ISOL RUBY BK see CMS155
ISOL RUBY BKS see CMS155
ISOL TOLUIDINE RED HB see MMP100
ISOL TOLUIDINE RED RN2B see MMP100
ISOL TOLUIDINE RED RNB see MMP100
ISOL TOLUIDINE RED RN2G see MMP100
ISOL TOLUIDINE RED RNG see MMP100
ISOLYN see ILD000
ISOMACK see CCK125
ISOMALIC ACID see MPN250
ISOMEBUMAL see EJY000
ISOMELAMINE see MCB000
ISOMELIN see GKS000
ISOMENTHONE see IKY000
ISOMEPROBAMATE see IPU000
β-ISOMER see BBR000
ISOMERIC CHLORTHION see NFT000

ISOMETASYSTOX see DAP400
ISOMETASYSTOX SULFONE see DAP600
ISOMETHADONE see IKZ000
ISOMETHEPTENE see ILK000
ISOMETHEPTENE HYDROCHLORIDE see ODY000
ISOMETHYLSYSTOX see DAP400
ISOMETHYLSYSTOX SULFONE see DAP600
ISOMETHYLSYSTOX SULFOXIDE see DAP000
ISOMIN see TEH500
ISOMYL see AMX750
ISOMYN see BBK000
ISOMYST see IQN000
ISOMYTAL see AMX750
ISONAL see AFT000, ENB500
ISONAL (ROUSSEL) see ENB500
ISONAPHTHOIC ACID see NAV500
ISONAPHTHOL see NAX000
ISONATE see MJP400
ISONATE 390P see PKB100
ISONEX see ILD000
ISONIACID see ILD000
ISONIAZIDE see ILD000
ISONIAZID SODIUM METHANESULFONATE see SIJ000
ISONICAZIDE see ILD000
ISONICO see ILD000
ISONICOTAN see ILD000
ISONICOTINALDEHYDE see IKZ100
ISONICOTINAMIDE PENTAAMMINE RUTHENIUM(II) PERCHLORATE see
   ILB150
ISONICOTINHYDRAZID see ILD000
ISONICOTINIC ACID see ILC000
ISONICOTINIC ACID, ETHYL ESTER see ELU000
ISONICOTINIC ACID HYDRAZIDE see ILD000
ISONICOTINIC ACID-2-ISOPROPYLHYDRAZIDE see ILE000
ISONICOTINIC ACID, 2-ISOPROPYLHYDRAZIDE, PHOSPHATE see IGG700
ISONICOTINIC ACID, SODIUM SALT see ILF000
ISONICOTINIC ACID, VANILLYLIDENEHYDRAZIDE see VEZ925
ISONICOTINIC ALDEHYDE see IKZ100
N-ISONICOTINOYL-N'-(β-N-BENZYLCARBOXAMIDOETHYL)HYDRAZINE see
   BET000
2-(2-ISONICOTINOYLHYDRAZINO)-d-GLUCOPYRANURONIC ACID SODIUM
   SALT DIHYDRATE see IDB100
4-(ISONICOTINOYLHYDRAZONE)PIMELIC ACID see ILG000
1-ISONICOTINOYL-2-ISOPROPYLHYDRAZINE see ILE000
N-ISONICOTINOYL-N'-GLUCURONID-HYDRAZIN-NATRIUMSALZ DIHYDRAT
   (GERMAN) see IDB100
N-ISONICOTINOYL-N'-GLUCURONSAEURE-γ-LACTON-HYDRAZON (GERMAN)
   see GBB500
ISONICOTINSAEUREHYDRAZID see ILD000
ISONICOTINYL HYDRAZIDE see ILD000
1-ISONICOTINYL-2-ISOPROPYLHYDRAZINE see ILE000
ISONIDE see ILD000
ISONIKAZID see ILD000
ISONIN see ILD000
ISONIPECAINE see DAM600
ISONIPECAINE HYDROCHLORIDE see DAM700
ISONIPECOTIC ACID see ILG100
ISONIPECOTIC ACID, 1-(2-HYDROXY-3-PHENOXYPROPYL)-4-PHENYL-, ETHYL
   ESTER, HYDROCHLORIDE see EEQ000
3-ISONIPECOTYLINDOLE see ILG200
ISONIRIT see ILD000
ISONITOX see MIW250
ISONITROPROPANE see NIY000
ISONITROSOACETOPHENONE see ILH000
5-ISONITROSOBARBITURIC ACID see AFU000
β-ISONITROSOPROPANE see ABF000
ISONITROSOPROPIOPHENONE see ILI000
ISONIXIN see XLS300
ISONIXINE see XLS300
ISONIZIDE see ILD000
ISONONYL ACETATE see TLT500
ISONONYL ALCOHOL see ILJ000
ISONONYLALDEHYDE see ILJ100
ISONORENE see DMV600
ISONYL see ILK000
ISOOCTADECANOIC ACID, 1-METHYLETHYL ESTER see IPS450
ISOOCTANE (DOT) see TLY500
ISOOCTANOL see ILL000
ISOOCTYL ALCOHOL see ILL000
ISOOCTYL ALCOHOL (2,4-DICHLOROPHENOXY)ACETATE see ILO000
2-ISOOCTYL AMINE see ILM000
ISOOCTYL-2,4-DICHLOROPHENOXYACETATE see ILO000
ISOOCTYL ESTER, MERCAPTOACETATE ACID see ILR000
ISOOCTYL MERCAPTOACETATE see ILR000

ISOOCTYL PHTHALATE see ILR100
p-ISOOCTYLPOLYOXYETHYLENEPHENOL FORMALDEHYDE POLYMER see TDN750
ISOOCTYL THIOGLYCOLATE see ILR000
ISOPAL see IQW000
ISOPARATHION see DJT000
ISOPELLETIERINE see PAO500
ISOPENTALDEHYDE see MHX500
ISOPENTANE (DOT) see EIK000
ISOPENTANOIC ACID (DOT) see ISU000
ISOPENTANOIC ACID, PHENYLMETHYL ESTER see ISW000
ISOPENTANOL see IHP000
ISOPENTENE see IHP150
ISOPENTENES (DOT) see IHP150
ISOPENT-2-ENYL ACETATE see DOQ350
4-(2-ISOPENTENYL)-1,2-DIPHENYL-3,5-PYRAZOLIDINEDIONE see PEW000
5-(2-ISOPENTENYL)-5-ISOPROPYL-1-METHYLBARBITURIC ACID see ILU000
8-ISOPENTENYLOXYPSORALENE see IHR300
ISOPENTYL ACETATE see IHO850
ISOPENTYL ALCHOL, ESTER with N-(2-(DIETHYLAMINO)ETHYL)-2-PHENYL-GLYCINE see NOC000
ISOPENTYL ALCOHOL see IHP000
ISOPENTYL ALCOHOL ACETATE see IHO850
ISOPENTYL ALCOHOL, FORMATE see IHS000
ISOPENTYL ALCOHOL, NITRATE see ILW100
ISOPENTYL ALCOHOL NITRITE see IMB000
ISOPENTYL BENZOATE see IHP100
ISOPENTYL BROMIDE see BNP250
ISOPENTYL BUTANOATE see IHP400
ISOPENTYL BUTYRATE see IHP400
ISOPENTYL FORMATE see IHS000
ISOPENTYL-n-HEXANOATE see IHU100
ISOPENTYL HEXANOATE see IHU100
ISOPENTYL IODIDE see IHU200
ISOPENTYL ISOVALERATE see ITB000
ISOPENTYL METHYL KETONE see MKW450
ISOPENTYL NITRATE see ILW100
ISOPENTYL NITRITE see IMB000
3-((ISOPENTYL)NITROSOAMINO)-2-BUTANONE see MHW350
ISOPENTYL OCTANOATE see IHP500
2-(ISOPENTYLOXY)-4-METHYL-1-PENTANOL see IJA000
ISOPENTYLPHENYLACETATE see IHV000
ISOPENTYL-2-PHENYLGLYCINATE HYDROCHLORIDE see PCV750
1-ISOPENTYL-THEOBROMINE see IHX200
ISOPERTHIOCYANIC ACID see IBL000
ISOPESTOX see PHF750
ISOPHANE INSULIN see IDF325
ISOPHANE INSULIN INJECTION see IDF325
ISOPHANE INSULIN SUSPENSION see IDF325
ISOPHEN (pesticide) see CBW000
ISOPHEN see CBW000, DBA800, MDT600
ISOPHENERGAN see DQA400
ISOPHENICOL see CDP250
ISOPHENPHOS see IMF300
ISOPHORONE see IMF400
ISOPHORONE DIAMINE DIISOCYANATE see IMG000
ISOPHORONEDIISOCYANATE, solution, 70%, by weight (DOT) see IMG000
ISOPHORONE DIISOCYANATE see IMG000
ISOPHOSPHAMIDE see IMH000
ISOPHRIN see NCL500
ISOPHRINE see SPC500
ISOPHRIN HYDROCHLORIDE see SPC500
ISOPHTHALALDEHYDES (FRENCH) see IMI000
ISOPHTHALALDEHYDE see IMI000
ISOPHTHALIC ACID see IMJ000
ISOPHTHALIC ACID CHLORIDE see IMO000
ISOPHTHALIC ACID, DIALLYL ESTER see IMK000
ISOPHTHALIC ACID, DIALLYL ESTER (6Cl,7Cl,8Cl) see IMK000
ISOPHTHALIC ACID DICHLORIDE see IMO000
ISOPHTHALIC ACID, DIETHYL ESTER see IMK100
ISOPHTHALIC ACID, DIMETHYL ESTER see IML000
ISOPHTHALODINITRILE see PHX550
ISOPHTHALONITRILE see PHX550
3,3'-(ISOPHTHALOYLBIS(IMINO-p-PHENYLENECARBONYLIMINO))BIS(1-ETHYL-PYRIDINIUM), DI-p-TOLUENESULFONATE see IMM000
ISOPHTHALOYL CHLORIDE see IMO000
ISOPHTHALOYL DICHLORIDE see IMO000
ISOPHTHALYL DICHLORIDE see IMO000
ISOPIRINA see INM000
ISOPLAIT see VGA300
ISO PPC see PMS825
ISOPRAL see IMQ000
ISOPRENALINE see DMV600
ISOPRENALINE CHLORIDE see IMR000
dl-ISOPRENALINE HYDROCHLORIDE see IQS500

racemic ISOPRENALINE HYDROCHLORIDE see IQS500
ISOPRENALINE HYDROCHLORIDE see IMR000
(±)-ISOPRENALINE HYDROCHLORIDE see IQS500
dl-ISOPRENALINE SULFATE see IRU000
(±)-ISOPRENALINE SULFATE see IRU000
ISOPRENE see IMS000, MNH750
ISOPRENE, INHIBITED (DOT) see IMS000
ISOPROCARB see MIA250
ISOPROCARBE see MIA250
ISOPROCIL (FRENCH) see BNM000
ISOPROMEDOL see IMT000
ISOPROMETHAZINE see DQA400
ISOPROPAMIDE see DAB875
ISOPROPAMIDE IODIDE see DAB875
ISOPROPANETHIOL see IMU000
ISOPROPANOL (DOT) see INJ000
ISOPROPANOLAMINE see AMA500
ISOPROPANOLAMINES, mixed see IMV000
ISOPROPANOLAMINES see IMV000
ISOPROPENE CYANIDE see MGA750
ISOPROPENIL-BENZOLO (ITALIAN) see MPK250
ISOPROPENYL ACETATE (DOT) see MQK750
ISOPROPENYL ACETATE see MQK750
ISOPROPENYL-BENZEEN (DUTCH) see MPK250
ISOPROPENYLBENZENE see MPK250
ISOPROPENYL-BENZOL (GERMAN) see MPK250
ISOPROPENYL CARBINOL see IMW000
4-ISOPROPENYL-1-CYCLOHEXENE-1-CARBOXALDEHYDE see DKX100
4-ISOPROPENYL-1-CYCLOHEXENE-1-CARBOXALDEHYDE, ANTI-OXIME see PCJ000
4-ISOPROPENYL-CYCLOHEX-1-ENE-1-METHANOL see PCI550
ISOPROPENYLESTER KYSELINY OCTOVE see MQK750
2'-ISOPROPENYL-2-(2-METHOXYETHYLAMINO)PROPRIONANILIDE OXALATE see IMX000
(+)-4-ISOPROPENYL-1-METHYLCYCLOHEXENE see LFU000
ISOPROPENYLNITRILE see MGA750
ISOPROPHYL METHYLPHOSPHONOFLUORIDATE see IPX000
1-ISOPROPILBIGUANIDE CLORIDRATO (ITALIAN) see IOF100
(2-ISOPROPILCROTONIL)UREA (ITALIAN) see PMS825
ISOPROPILE (ACETATO di) (ITALIAN) see INE100
ISOPROPIL-N-FENIL-CARBAMMATO (ITALIAN) see CBM000
N-ISOPROPILFTALIMMIDE see IRG000
ISOPROPILIDRAZIDE dell'ac. p-CLORO-FENOSSIACETICO (ITALIAN) see CJN250
(1-ISOPROPIL-3-METIL-1H-PIRAZOL-5-IL)-N,N-DIMETIL-CARBAMMATO (ITALIAN) see DSK200
o-ISOPROPOXYANILINE see INA200
(3)-O-2-ISOPROPOXY-CARBONYL-1-METHYLVINYL-O-METHYL ETHYLPHOS-PHORAMIDOTHIOATE see MKA000
11-ISOPROPOXY-15,16-DIHYDRO-17-CYCLOPENTA(a)PHENANTHREN-17-ONE see INA400
4-ISOPROPOXYDIPHENYLAMINE see HKF000
p-ISOPROPOXYDIPHENYLAMINE see HKF000
2-ISOPROPOXYETHANOL see INA500
N-(2-ISOPROPOXY-3-HYDROXYMERCURIPROPYL)BARBITAL see INE000
(ISOPROPOXYMETHYL)OXIRANE see IPD000
ISOPROPOXYMETHYLPHORYL, FLUORIDE see IPX000
2-ISOPROPOXYNITROBENZENE see NIR550
o-ISOPROPOXYNITROBENZENE see NIR550
N-(4-ISOPROPOXYPHENYL)ANILINE see HKF000
2-ISOPROPOXYPHENYL-N-METHYLCARBAMATE see PMY300
o-ISOPROPOXYPHENYL-N-METHYLCARBAMATE see PMY300
2-ISOPROPOXYPHENYL N-METHYLCARBAMATE, nitrosated see PMY310
o-ISOPROPOXYPHENYL METHYLCARBAMATE see PMY300
o-ISOPROPOXYPHENYL N-METHYL-N-NITROSOCARBAMATE see PMY310
o-ISOPROPOXYPHENYL METHYLNITROSOCARBAMATE see PMY310
2-ISOPROPOXYPROPANE see IOZ750
ISOPROPYDRIN see DMV600
ISOPROPYLACETAAT (DUTCH) see INE100
ISOPROPYLACETAT (GERMAN) see INE100
ISOPROPYL (ACETATE d') (FRENCH) see INE100
ISOPROPYL ACETATE see INE100
ISOPROPYLACETIC ACID see ISU000
ISOPROPYL ACETIC ACID, BENZYL ESTER see ISW000
ISOPROPYLACETONE see HFG500
ISOPROPYL-N-ACETOXY-N-PHENYLCARBAMATE see ING400
ISOPROPYL ACID PHOSPHATE solid see PHE500
2-ISOPROPYL ACRYLALDEHYDE OXIME see ING509
N-ISOPROPYLACRYLAMIDE see INH000
ISOPROPYL ACRYLAMIDE see INH000
ISOPROPYL ADIPATE see DNL800

ISOPROPYLADRENALINE see DMV600
ISOPROPYL ALCOHOL see INJ000
ISOPROPYL ALCOHOL, TITANIUM(4+) SALT see IRN200
ISO-PROPYLALKOHOL (GERMAN) see INJ000
ISOPROPYLALLYLAZETYLKARBAMID (GERMAN) see IQX000
ISOPROPYLALLYLBARBITURIC ACID see AFT000
ISOPROPYLAMID KYSELINY AKRYLOVE see INH000
ISOPROPYLAMINE see INK000
ISOPROPYLAMINE HYDROCHLORIDE see INL000
4-(2-ISOPROPYLAMINE-1-HYDROXYETHYL)METHANESULFOANILIDE HYDRO-
 CHLORIDE see CCK250
1-ISOPROPYLAMINE-3-(1-NAPHTHYLOXY)-2-PROPANOL see ICB000
4-(ISOPROPYLAMINO)ANTIPYRINE see INM000
ISOPROPYLAMINOANTIPYRINE see INM000
ISOPROPYLAMINO-4-AZIDO-6-METHYLTHIO-1,3,5-TRIAZIN (GERMAN) see
 ASG250
1-ISOPROPYLAMINO-3-(2-CYCLOHEXYLPHENOXY)-2-PROPANOL HYDRO-
 CHLORIDE see ERE100
4-ISOPROPYLAMINO-2,3-DIMETHYL-1-PHENYL-3-PYRAZOLIN-5-ONE see
 INM000
4-ISOPROPYLAMINODIPHENYLAMINE see PFL000
2-ISOPROPYLAMINOETHANOL see INN400
N-ISOPROPYLAMINOETHANOL see INN400
ISOPROPYLAMINOETHANOL see INN400
2-(ISOPROPYLAMINO)ETHANOL HYDROCHLORIDE see INN500
ISOPROPYLAMINO ETHANOL HYDROCHLORIDE see INN500
ISOPROPYLAMINO-O-ETHYL-(4-METHYLMERCAPTO-3-METHYLPHE-
 NYL)PHOSPHATE see FAK000
4-(2-ISOPROPYLAMINO-1-HYDROXYAETHYL)METHANESULFONALID HYDRO-
 CHLORID (GERMAN) see CCK250
1-(ISOPROPYLAMINO)-2-HYDROXY-3-(o-(ALLYLOXY)PHENOXY)PROPANE see
 CNR500
ISOPROPYLAMINOHYDROXYETHYLMETHANESULFONALIDE HYDROCHLO-
 RIDE see CCK250
1-(1-(2-(3-ISOPROPYLAMINO-2-HYDROXYPROPOXY)-3,6-DICHLOROPHE-
 NYL)VINYL)-1H-IMIDAZOLE HCl see DFM875
4-((3-ISOPROPYLAMINO)-2-HYDROXYPROPOXY)-3-METHOXY-BENZOIC ACID
 METHYL ESTER see VFP200
7-(ISOPROPYLAMINOISOPROPYL)THEOPHYLLINE HYDROCHLORIDE see
 INP100
(±)-1-(ISOPROPYLAMINO)-3-(p-(2-METHOXYETHYL)PHENOXY)-2-PROPANOL
 HEMI-I-TARTRATE see MQR150
1-ISOPROPYLAMINO-3-(p-(2-METHOXYETHYL)PHENOXY)-2-PROPANOL TAR-
 TRATE see SBV500
1-(ISOPROPYLAMINO)-3-(p-(2-METHOXYETHYL)PHENOXY)-2-PROPANOL TAR-
 TRATE (2:1) see MQR150
2-ISOPROPYLAMINO-4-(3-METHOXYPROPYLAMINO)-6-METHYLTHIO-1,3,5-
 TRIAZIN (GERMAN) see INQ000
2-ISOPROPYLAMiNO-4-(3-METHOXYPROPYLAMINO)-6-METHYLTHIO-s-TRIAZINE
 see INQ000
4-ISOPROPYLAMINO-6-(3'-METHOXYPROPYLAMINO)-2-METHYTHIO-1,3,5-TRIAZ-
 IN (GERMAN) see INQ000
2-ISOPROPYLAMINO-4-METHYLAMINO-6-METHYLMERCAPTO-s-TRIAZINE see
 INR000
2-ISOPROPYLAMINO-4-METHYLAMINO-6-METHYLTHIO-1,3,5-TRIAZINE see
 INR000
α-(ISOPROPYLAMINOMETHYL)-3,4-DIHYDROXYBENZYL ALCOHOL HYDRO-
 CHLORIDE see IMR000
ISOPROPYLAMINOMETHYL-3,4-DIHYDROXYPHENYL CARBINOL see DMV600
7-(2-ISOPROPYLAMINO-2-METHYLETHYL)THEOPHYLLINE HYDROCHLORIDE
 see INP100
α-(ISOPROPYLAMINO)METHYL)-2-NAPHTHALENE-METHANOL see INS000
α-((ISOPROPYLAMINO)METHYL)NAPHTHALENE-METHANOL, HYDROCHLO-
 RIDE see INT000
α-(ISOPROPYLAMINOMETHYL)-4-NITROBENZYL ALCOHOL see INU000
2-((ISOPROPYLAMINO)METHYL)-7-NITRO-1,2,3,4-TETRAHYDRO-6-QUINOLINEM-
 ETHANOL see OLT000
α-(ISOPROPYLAMINOMETHYL)PROTOCATECHUYL ALCOHOL see DMV600
1-ISOPROPYLAMINO-3-(1-NAPHTHOXY)-PROPAN-2-OL HYDROCHLORIDE see
 ICC000
1-(ISOPROPYLAMINO)-3-(α-NAPHTHOXY)-2-PROPANOL HYDROCHLORIDE see
 ICC000
2-ISOPROPYLAMINO-1-(2-NAPHTHYL)ETHANOL see INS000
2-ISOPROPYLAMINO-1-(NAPHTH-2-YL)ETHANOL see INS000
2-ISOPROPYLAMINO-1-(2-NAPHTHYL)ETHANOL HYDROCHLORIDE see INT000
1-ISOPROPYLAMINO-3-(1-NAPHTHYLOXY)-2-PROPANOL see ICB000
(±)-1-(ISOPROPYLAMINO)-3-(1-NAPHTHYLOXY)-2-PROPANOL HYDROCHLO-
 RIDE see PNB800
1-(ISOPROPYLAMINO)-3-(1-NAPHTHYLOXY)PROPAN-2-OL HYDROCHLORIDE
 see ICC000
1-(ISOPROPYLAMINO)-3-(1-NAPTHYLOXY)-2-PROPANOL HYDROCHLORIDE
 see ICC000
ISOPROPYLAMINOPHENAZON see INM000
ISOPROPYLAMINOPHENAZONE see INM000

(±)-1-(ISOPROPYLAMINO)-3-(o-PHENOXYPHENOXY)-2-PROPANOL HYDRO-
 CHLORIDE see INU200
4-ISOPROPYLAMINO-1-PHENYL-2,3-DIMETHYL-3-PYRAZOLIN-5-ONE see
 INM000
9-((3-(ISOPROPYLAMINO)PROPYL)AMINO)-1-NITROACRIDINE DIHYDROCHLO-
 RIDE see INW000
7-(2-(ISOPROPYLAMINO)PROPYL)-THEOPHYLLINE MONOHYDROCHLORIDE
 see INP100
2-ISOPROPYL ANILINE see INW100
N-ISOPROPYLANILINE see INX000
o-ISOPROPYLANILINE see INW100
ISOPROPYLANTIPYRIN see INY000
4-ISOPROPYLANTIPYRINE see INY000
ISOPROPYLANTIPYRINE see INY000
ISOPROPYLANTIPYRINE with ETHENZAMIDE and CAFFEINE MONOHYDRATE
 see INY100
ISOPROPYLARTERENOL see DMV600
ISOPROPYLARTERENOL HYDROCHLORIDE see IMR000
4-ISOPROPYLBENZALDEHYDE see COE500
p-ISOPROPYLBENZALDEHYDE see COE500
5-ISOPROPYL-1:2-BENZANTHRACENE see INZ000
6-ISOPROPYL-1:2-BENZANTHRACENE see IOA000
8-ISOPROPYLBENZ(a)ANTHRACENE see INZ000
9-ISOPROPYLBENZ(a)ANTHRACENE see IOA000
ISOPROPYLBENZEEN (DUTCH) see COE750
ISOPROPYL BENZENE see COE750
p-ISOPROPYLBENZENECARBOXALDEHYDE see COE500
ISOPROPYLBENZENE HYDROPEROXIDE see IOB000
ISOPROPYLBENZENE PEROXIDE see DGR600
ISOPROPYL BENZOATE see IOD000
p-ISOPROPYLBENZOIC ACID-2-(DIETHYLAMINO)ETHYL ESTER HYDROCHLO-
 RIDE see IOI600
ISOPROPYL-BENZOL (GERMAN) see COE750
ISOPROPYLBENZOL see COE750
p-ISOPROPYLBENZONITRILE see IOD050
3-ISOPROPYL-2,1,3-BENZOTHIADIAZINON-(4)-2,2-DIOXID (GERMAN) see
 MJY500
3-ISOPROPYL-1H-2,1,3-BENZOTHIADIAZIN-4(3H)-ONE-2,2-DIOXIDE see MJY500
N-ISOPROPYL-2-BENZOTHIAZOLE SULFONAMIDE see IOE000
2-ISOPROPYL-3:4-BENZPHENANTHRENE see IOF000
4-ISOPROPYLBENZYL ACETATE see IOF050
p-ISOPROPYLBENZYL ALCOHOL see CQI250
1-ISOPROPYLBIGUANIDE HYDROCHLORIDE see IOF100
N-ISOPROPYLBIGUANIDE HYDROCHLORIDE see IOF100
ISOPROPYLBIPHENYL see IOF200
ISOPROPYL-BIS(β-CHLOROETHYL)AMINE see IOF300
ISOPROPYL BIS(β-CHLOROETHYL)AMINE HYDROCHLORIDE see IPG000
ISOPROPYL BORATE see IOI000
5-ISOPROPYL-5-BROMALLYLBARBITURIC ACID see QCS000
ISOPROPYL BROMIDE see BNY000
5-ISOPROPYL-5-(2-BROMOALLYL)BARBITUATE see QCS000
3-ISOPROPYL-5-BROMO-6-METHYLURACIL see BNM000
ISOPROPYLCAINE HYDROCHLORIDE see IOI600
ISOPROPYL CARBAMATE see IOJ000
ISOPROPYLCARBAMIC ACID, ESTER with 2-(HYDROXYMETHYL)-2-METHYL-
 PENTYL CARBAMATE see IPU000
ISOPROPYLCARBAMIC ACID, ETHYL ESTER see IPA000
2-(p-ISOPROPYL CARBAMOYL BENZYL)-1-METHYLHYDRAZINE see PME250
1-(p-ISOPROPYLCARBAMOYLBENZYL)-2-METHYLHYDRAZINE HYDROCHLO-
 RIDE see PME500
2-(p-(ISOPROPYLCARBAMOYL)BENZYL)-1-METHYLHYDRAZINE HYDROCHLO-
 RIDE see PME500
1-ISOPROPYL CARBAMOYL-3-(3,5-DICHLOROPHENYL)-HYDANTOIN see
 GIA000
ISOPROPYL CARBANILATE see CBM000
ISOPROPYL CARBANILIC ACID ESTER see CBM000
ISOPROPYLCARBINOL see IIL000
4-ISOPROPYLCATECHOL see IOK000
ISOPROPYL CELLOSOLVE see INA500
ISOPROPYL CHLORIDE see CKQ000
N-ISOPROPYL-2-CHLOROACETANILIDE see CHS500
N-ISOPROPYL-α-CHLOROACETANILIDE see CHS500
ISOPROPYL-3-CHLOROCARBANILATE see CKC000
ISOPROPYL-m-CHLOROCARBANILATE see CKC000
ISOPROPYL CHLOROCARBONATE see IOL000
ISOPROPYL CHLOROFORMATE see IOL000
ISOPROPYL CHLOROMETHANOATE see IOL000
ISOPROPYL-3-CHLOROPHENYLCARBAMATE see CKC000
o-ISOPROPYL-N-(3-CHLOROPHENYL)CARBAMATE see CKC000
ISOPROPYL-N-(3-CHLOROPHENYL)CARBAMATE see CKC000
ISOPROPYL-4-CHLOROPHENYL KETONE see IOL100
ISOPROPYL-N-(3-CHLORPHENYL)-CARBAMAT (GERMAN) see CKC000
20-ISOPROPYLCHOLANTHRENE see ION000
3-ISOPROPYLCHOLANTHRENE see ION000
ISOPROPYL CINNAMATE see IOO000

ISOPROPYLMETHYLPYRIMIDYL DIETHYL THIOPHOSPHATE see DCM750
ISOPROPYLMORPHOLINE see IQM000
ISOPROPYL MYRISTATE see IQN000
2-ISOPROPYLNAPHTHALENE see IQN100
N′-ISOPROPYL-N,N′-DIMETHYL-1,3-PROPANE-DIAMINE see IOW000
N-ISOPROPYL-N′-FENYL-p-FENYLENDIAMIN (CZECH) see PFL000
ISOPROPYL NITRATE see IQP000
ISOPROPYL NITRILE see IJX000
ISOPROPYL NITRITE see IQQ000
ISOPROPYL o-NITROPHENYL ETHER see NIR550
1-ISOPROPYL-1-NITROSOUREA see NKO425
1-ISOPROPYLNORADRENALINE see DMV600
N-ISOPROPYLNORADRENALINE see DMV600
ISOPROPYL NORADRENALINE see DMV600
dl-N-ISOPROPYLNORADRENALINE HYDROCHLORIDE see IQS500
dl-ISOPROPYLNORADRENALINE HYDROCHLORIDE see IQS500
8-ISOPROPYLNORATROPINE METHOBROMIDE see IGG000
N-ISOPROPYLNORATROPINIUM BROMOMETHYLATE see IGG000
dl-ISOPROPYLNOREPINEPHRINE HYDROCHLORIDE see IQS500
ISOPROPYLNOREPINEPHRINE-HYDROCHLORIDE see IMR000
N-ISOPROPYL-N′-PHENYL-p-PHENYLENEDIAMINE see PFL000
N-ISOPROPYLOCTADECYLAMINE see IQT000
ISOPROPYLOCTADECYLAMINE see IQT000
ISOPROPYL OILS see IQU000
ISOPROPYL ORTHOTITANATE see IRN200
3-ISOPROPYLOXYPROPYLENE OXIDE see IPD000
ISOPROPYL PALMITATE see IQW000
(2-ISOPROPYL-4-PENTENOYL)UREA see IQX000
ISOPROPYL PERCARBONATE see DNR400
ISOPROPYL PEROXYDICARBONATE see DNR400
ISOPROPYLPHENAZONE see INY000
2-ISOPROPYLPHENOL see IQX100
3-ISOPROPYLPHENOL see IQX090
4-ISOPROPYLPHENOL see IQZ000
m-ISOPROPYLPHENOL see IQX090
o-ISOPROPYLPHENOL see IQX100
p-ISOPROPYLPHENOL see IQZ000
m-ISOPROPYLPHENOL-N-METHYLCARBAMATE see COF250
o-ISOPROPYLPHENOL METHYLCARBAMATE see MIA250
4-ISOPROPYL PHENYLACETALDEHYDE see IRA000
p-ISOPROPYLPHENYLACETALDEHYDE see IRA000
ISOPROPYL-N-PHENYL-CARBAMAT (GERMAN) see CBM000
ISOPROPYL-N-PHENYLCARBAMAT (GERMAN) see PMS825
o-ISOPROPYL-N-PHENYL CARBAMATE see CBM000
ISOPROPYL-N-PHENYLCARBAMATE see CBM000
ISOPROPYL PHENYLCARBAMATE see CBM000
3-ISOPROPYLPHENYL METHYLCARBAMATE see COF250
2-ISOPROPYLPHENYL N-METHYLCARBAMATE see MIA250
m-ISOPROPYLPHENYL-N-METHYLCARBAMATE see COF250
o-ISOPROPYLPHENYL N-METHYLCARBAMATE see MIA250
m-ISOPROPYLPHENYL METHYLCARBAMATE see COF250
ISOPROPYL 3-PHENYLPROPENOATE see IOO000
ISOPROPYL-N-PHENYLURETHAN (GERMAN) see CBM000
ISOPROPYL PHOSPHONATE see DNQ600
ISOPROPYLPHOSPHORAMIDOTHIOIC ACID-O-2,4-DICHLOROPHENYL-O-METH-
  YL ESTER see DGD800
ISOPROPYL-PHOSPHORAMIDOTHIOIC ACID O-ETHYL O-(2-ISOPROPOXYCAR-
  BONYLPHENYL) ESTER see IMF300
N-ISOPROPYLPHOSPHORAMIDOTHIOIC ACID-O-(2,4,5-TRICHLOROPHE-
  NYL)ESTER see DYD200
ISOPROPYL PHOSPHORIC ACID see PHE500
ISOPROPYL PHOSPHOROFLUORIDATE see IRF000
N-ISOPROPYLPHTHALIMIDE see IRG000
13-ISOPROPYLPODOCARPA-7,13-DIEN-15-OIC ACID see AAC500
13-ISOPROPYLPODOCARPA-8,11,13-TRIEN-15-OIC ACID see DAK400
ISOPROPYL POTASSIUM XANTHATE see IRG050
ISOPROPYL PYROPHOSPHATE see TDF750
1-ISOPROPYL-2-PYRROLIDINONE see IRG100
N-ISOPROPYLPYRROLIDINONE see IRG100
6-ISOPROPYL QUINOLINE see IRL000
p-ISOPROPYL QUINOLINE see IRL000
ISOPROPYL QUINOLINE see IRL000
ISOPROPYL-S see IOF300
ISOPROPYL SALICYLATE O-ESTER with O-
  ETHYLISOPROPYLPHOSPHORAMIDOTHIOATE see IMF300
ISOPROPYL-S HYDROCHLORIDE see IPG000
O-ISOPROPYL SODIUM DITHIOCARBONATE see SIA000
ISOPROPYL STEARATE see IRL100
ISOPROPYL SULFATE see DNO900
N-ISOPROPYL TEREPHTHALAMIC ACID see IRN000
ISOPROPYL TETRADECANOATE see IQN000
S-2-ISOPROPYLTHIOETHYL-O,O-DIMETHYL PHOSPHORODITHIOATE see
  DSK800
ISOPROPYLTHIOL see IMU000

(ISOPROPYLTHIO)-METHANETHIOL-S-ESTER with O,O-DIETHYL PHOSPHORO-
  DITHIOATE see DJN600
erythro-p-(ISOPROPYLTHIO)-α-((1-(OCTYLAMINO)ETHYL)BENZYL ALCOHOL
  see SOU600
erythro-1-(4-ISOPROPYLTHIOPHENYL)-2-n-OCTYLAMINOPROPANOL see
  SOU600
ISOPROPYL TIGLATE see IRN100
ISOPROPYL TITANATE(IV) see IRN200
p-ISOPROPYLTOLUENE see CQI000
4-ISOPROPYL-2-(α,α,α-TRIFLUORO-m-TOLYL)MORPHOLINE see IRP000
N-ISOPROPYL-4-(3,4,5-TRIMETHOXYCINNAMOYL)-1-PIPERAZINEACETAMIDE
  MALEATE see IRQ000
4-ISOPROPYL-2,6,7-TRIOXA-1-ARSABICYCLO(2.2.2)OCTANE see IRQ100
β-ISOPROPYLTROPOLON see IRR000
4-ISOPROPYLTROPOLONE see IRR000
1-ISOPROPYLUREA see IRR100
N-ISOPROPYLUREA see IRR100
ISOPROPYLUREA see IRR100
ISOPROPYLUREA and SODIUM NITRITE see SIS675
ISOPROPYL VINYL ETHER see IRS000
ISOPROPYLXANTHIC ACID, SODIUM SALT see SIA000
ISOPROPYLXANTHOGENAN SODNY see SIA000
l-ISOPROTERENOL see DMV600
ISOPROTERENOL see DMV600
racemic ISOPROTERENOL HDYROCHLORIDE see IQS500
dl(±)-ISOPROTERENOL HYDROCHLORIDE see IQS500
dl-ISOPROTERENOL HYDROCHLORIDE see IQS500
ISOPROTERENOL HYDROCHLORIDE see IMR000
(±)-ISOPROTERENOL HYDROCHLORIDE see IQS500
dl-ISOPROTERENOL SULFATE see IRU000
(±)-ISOPROTERENOL SULFATE see IRU000
ISOPROTERNOL MONOHYDROCHLORIDE see IMR000
ISOPSORALIN see FQC000
ISOPTIN see IRV000, VHA450
ISOPTO CARBACHOL see CBH250
ISOPTO-CARPINE see PIF250
ISOPTO CETAMIDE see SNP500
ISOPTO FENICOL see CDP250
ISOPTO-HYDROCORTISONE see HHQ800
ISOPTO HYOSCINE see SBG000
ISOPULEGOL (FCC) see MCE750
ISO-PUREN see CCK125
ISOPURINE see POJ250
ISOPYRIN see INM000
ISOPYRINE see INM000, INY000
ISOQUINALDEHYDE THIOSEMICARBAZONE see IRV300
ISOQUINOLINE see IRX000
ISOQUINOLINE, 1-(3,4-DIETHOXYBENZYL)-6,7-DIETHOXY-, HYDROCHLORIDE
  see PAH260
ISOQUINOLINE, 1-((3,4-DIETHOXYPHENYL)METHYL)-6,7-DIETHOXY-, HYDRO-
  CHLORIDE see PAH260
ISOQUINOLINE, 6,7-DIMETHOXY-1-VERATRYL-, HYDROCHLORIDE see
  PAH250
2-(1-ISOQUINOLINYLMETHYLENE)-HYDRAZINECARBOTHIOAMIDE (9CI) see
  IRV300
1-(2-ISOQUINOLYL)-5-METHYL-4-PYRAZOLYL METHYL KETONE see IRX100
ISORBID see CCK125
ISORDIL see CCK125
ISORDIL TEMBIDS see CCK125
ISOREN see CLY600
ISORENIN see DMV600
ISORETINENE a see VSK975
ISOSAFROEUGENOL see IRY000
ISOSAFROLE see IRZ000
ISOSAFROLE-n-OCTYLSULFOXIDE see ISA000
ISOSAFROLE, OCTYL SULFOXIDE see ISA000
ISOSCOPIL see HOT500
(+)-d-ISOSORBIDE see HID350
ISOSORBIDE see HID350
ISOSORBIDE DINITRATE see CCK125
ISOSORBIDE 5-MONONITRATE see ISC500
ISOSORBIDE 5-NITRATE see ISC500
ISOSTEARYL NEOPENTANOATE see ISC550
ISOSTENASE see CCK125
ISOSUMITHION see MKC250
ISO SYSTOX SULFOXIDE see ISD000
ISOTACTIC POLYPROPYLENE see PMP500
ISOTAZIN see DIR000
ISOTEBEZID see ILD000
ISOTENIC ACID see AKG250
ISOTENSE see RCA200
ISOTHAN see LBW000
ISOTHAZINE see DIR000
ISOTHIAZINE see DIR000

ISOTHIN see EPQ000
ISOTHIOATE see DSK800
ISOTHIOCYANATE d'ALLYLE (FRENCH) see AGJ250
ISOTHIOCYANATE de METHYLE (FRENCH) see ISE000
1-ISOTHIOCYANATE-NAPHTHALENE see ISN000
ISOTHIOCYANATOBENZENE see ISQ000
(2-ISOTHIOCYANATOETHYL)BENZENE see ISP000
ISOTHIOCYANATOMETHANE see ISE000
1-ISOTHIOCYANATONAPHTHALENE see ISN000
N-(3-ISOTHIOCYANATOPHENYL)ACETAMIDE see ISG000
3-ISOTHIOCYANATO-1-PROPENE see AGJ250
(ISOTHIOCYANATO)TRIMETHYLSTANNE see ISF000
ISOTHIOCYANATOTRIMETHYLTIN see ISF000
(ISOTHIOCYANATO)TRIPROPYLSTANNANE see TNB750
ISOTHIOCYANIC ACID (DOT) see ISF100
ISOTHIOCYANIC ACID see ISF100
ISOTHIOCYANIC ACID-m-ACETAMIDOPHENYL ESTER see ISG000
ISOTHIOCYANIC ACID BENZYL ESTER see BEU250
ISOTHIOCYANIC ACID,-p-CHLOROBENZYL ESTER see CEQ750
ISOTHIOCYANIC ACID, p-CHLOROPHENYL ESTER see ISH000
ISOTHIOCYANIC ACID, CYCLOHEXYL ESTER see ISJ000
ISOTHIOCYANIC ACID, ETHYLENE ESTER see ISK000
ISOTHIOCYANIC ACID-m-FLUOROPHENYL ESTER see ISL000
ISOTHIOCYANIC ACID-p-FLUOROPHENYL ESTER see ISM000
ISOTHIOCYANIC ACID, METHYL ESTER see ISE000
ISOTHIOCYANIC ACID, 3-(METHYLSULFINYL)PROPYL ESTER see MPN100
ISOTHIOCYANIC ACID-1-NAPHTHYL ESTER see ISN000
ISOTHIOCYANIC ACID-m-NITROPHENYL ESTER see ISO000
ISOTHIOCYANIC ACID, PHENETHYL ESTER see ISP000
ISOTHIOCYANIC ACID-p-PHENYLENE ESTER see PFA500
ISOTHIOCYANIC ACID, PHENYL ESTER see ISQ000
ISOTHIOINDIGO see DNT300
ISOTHIONIC ACID see HKI500
ISOTHIOUREA see ISR000
ISOTHIOURONIUM CHLORIDE, BENZYL see BEU500
ISOTHIPENDYL HYDROCHLORIDE see AEG625
ISOTHYMOL see CCM000, IQJ000
ISOTIAMIDA see EPQ000
ISOTIOCIANATO di METILE (ITALIAN) see ISE000
ISOTONIL see DRM000
ISOTOX see BBQ500
ISOTRATE see CCK125
ISOTRETINOIN see VSK955
ISOTRICYCLOQUINAZOLINE NITRATE see TBI750
ISOTRIMETHYLTETRAHYDRO BENZYL ALCOHOL see ISR100
ISOTRON 11 see TIP500
ISOTRON 12 see DFA600
ISOTRON 22 see CFX500
ISOUREA see USS000
ISOURETHANE see PKL500
ISOVAL see BNP750
ISOVALERAL see MHX500
ISOVALERALDEHYDE see MHX500
8-ISOVALERATE see FQS000
ISO-VALERIANATE de BORNYLE see HOX100
ISOVALERIANIC AICD see ISU000
ISOVALERIC ACID see ISU000
ISOVALERIC ACID, ALLYL ESTER see ISV000
ISOVALERIC ACID, BENZYL ESTER see ISW000
ISOVALERIC ACID, 2-BORNYL ESTER (7CI,8CI) see HOX100
ISOVALERIC ACID, BUTYL ESTER see ISX000
ISOVALERIC ACID, CINNAMYL ESTER (6CI,7CI,8CI) see PEE200
(E)-ISOVALERIC ACID-3,7-DIMETHYL-2,6-OCTADIENYL ESTER see GDK000
ISOVALERIC ACID, (4,7-DIMETHYL-1,6-OCTADIEN-3-YL) ESTER see LGC000
ISOVALERIC ACID-8-ESTER with 3-FORMAMIDO-N-(7-HEXYL-8-HYDROXY-4,9-
   DIMETHYL-2,6-DIOXO-1,5-DIOXONAN-3-YL)SALICYLAMIDE see DUO350
ISOVALERIC ACID, ETHYL ESTER see ISY000
(Z)-ISOVALERIC ACID-3-HEXENYL see ISZ000
ISOVALERIC ACID, HEXYL ESTER (8CI) see HFQ575
ISOVALERIC ACID, ISOBUTYL ESTER see ITA000
ISOVALERIC ACID, ISOPENTYL ESTER see ITB000
ISOVALERIC ACID, METHYL ESTER see ITC000
ISOVALERIC ALDEHYDE see MHX500
ISOVALERONE see DNI800
ISOVALERONITRILE see ITD000
ISOVANILLIC ACID see HJC000
ISOVANILLIN see FNM000
ISOVANILLINE see FNM000
ISOVITAMIN C see SAA025
ISOXAL see PCJ325
ISOXAMIN see SNN500
ISOXANTHINE see XCA000
ISOXATHION see DJV600
ISOXAZOLE, 3,5-DIMETHYL- see DSK950
ISOXAZOLIDINE, 2-(3,4,5-TRIETHYOXYBENZOYL)- see ITD025

2-ISOXAZOLIDINYL 3,4,5-TRIETHOXYPHENYL KETONE see ITD025
ISOXSUPRINE see VGF000
ISOXSUPRINE HYDROCHLORIDE see VGA300
ISOXSUPRIN HYDROCHLORIDE see VGA300
ISOZIDE see ILD000
ISPENORAL see PDT750
ISPHAMYCIN see CMB000
ISRAVIN see DBX400
ISTIN see DMH400
ISTONYL see DRM000
ISUPREL see DMV600, IMR000
ISUPREL HYDROCHLORIDE see IMR000
ISUPREN see DMV600
ISZILIN see IDF300
IT 40 see SMQ500
IT 931 see DLK200
IT 3456 see CDT000
DL-111-IT see EOL600
ITA-104 see PJB500
ITA 312 see LII400
ITACHIGARDEN see HLM000
ITALCHIN see CFU750
ITALIAN ARUM see ITD050
ITAMID see PJY500
ITAMIDONE see DOT000
ITAMO REAL (PUERTO RICO) see SDZ475
ITAMYCIN see HAL000
ITCH WEED see FAB100
ITF 182 see IAM100
ITF 611 see TKF699
ITF 1016 see ITD100
ITINEROL see HGC500
ITIOCIDE see EPQ000
ITOBARBITAL see AGI750
ITOPAZ see EEH600
ITROP see IGG000
ITRUMIL SODIUM see SHX500
ITRUMIL SODIUM SALT see SHX500
ITURAN see NGE000
I.U. 7 see CAL075
5-IUDR see DAS000
IUDR see DAS000
IUGLON see WAT000
IVALON see FMV000, PKP750
IVAUGAN see CFY000
IVE see IJQ000
IVERMECTIN see ITD875
IVERSAL see AHI875
IVERTOL see AHI875
IVIRON see IHG000
IVORAN see DAD200
IVORIT see EJM500
IVORY see ITE000
IVOSIT see ACE500
IVY see AFK950
IVY ARUM see PLW800
IVY BUSH see MRU359
IXODEX see DAD200
IXOTEN see TNT500
IXPER 25M see MAH750
IYLOMYCIN see ITF000
IYOMYCIN see ITG000
IZ 914 see DYC875
IZADRIN see IMR000
IZOACRIDINA see AHS500
IZOFORON (POLISH) see IMF400
o-IZOPROPOKSYANILINA (POLISH) see INA200
o-IZOPROPOKSYNITROBENZEN (CZECH) see NIR550
IZOPROPYLOWY ETER (POLISH) see IOZ750
IZOPTIN see VHA450
IZOPTIN HYDROCHLORIDE see VHA450
IZOSYSTOX (CZECH) see DAP200
IZSAB see LEK000


J3 see BPP750
J 164 see HKQ100
J 242 see AHG000
J 400 see PMP500
J 820 see MCB050
JACINTO de PERU (CUBA) see SLH200
JACK BEAN UREASE see UTU550
JACK WILSON CHLORO 51 (oil) see CKC000
JACOBINE see JAK000, SBX500
JACODINE HYDROCHLORIDE see SBX525

JACUTIN see BBP750, BBQ500
JADE GREEN BASE see JAT000
JADO see CBT500
JA-FA IPM see IQN000
JA-FA IPP see IQW000
JAFFNA TOBACCO see BFW135
JAGUAR 6000 see GLU000
JAGUAR see GLU000
JAGUAR A 20 B see GLU000
JAGUAR A 20D see GLU000
JAGUAR A 40F see GLU000
JAGUAR GUM A-20-D see GLU000
JAGUAR No. 124 see GLU000
JAGUAR PLUS see GLU000
JAIKIN see SFF000
JALAN see EKO500
JAMAICA GINGER EXTRACT see JBA000
JAMAICAN WALNUT see TOA275
JAMBERRY see JBS100
JAMESTOWN WEED see SLV500
JANIMINE see DLH630
JANTARAN SODNY see SJW100
JANUPAP see HIM000
JANUS GREEN B see DHM500
JANUS GREEN V see DHM500
JAPAN AGAR see AEX250
JAPAN CAMPHOR see CBA750
JAPANESE AUCUBA see JBS050
JAPANESE BEAD TREE see CDM325
JAPANESE CAMPHOR see CBB250
JAPANESE CAMPHOR OIL see CBB500
JAPANESE LANTERN PLANT see JBS100
JAPANESE LAUREL see JBS050
JAPANESE MEDLAR see LIH200
JAPANESE OIL of CAMPHOR see CBB500
JAPANESE PLUM see LIH200
JAPANESE POINSETTIA see SDZ475
JAPAN ISINGLASS see AEX250
JAPAN LACQUER see JCA000
JAPAN OIL TREE see TOA275
JAPANOL BLACK BHK see CMN800
JAPANOL BRILLIANT BLUE RWL (6CI) see CMO600
JAPANOL FAST BLACK D see CMN230
JAPANOL FAST RED 1 see CMO870
JAPANOL FAST RED F see CMO870
JAPANOL VIOLET J see CMP000
JAPAN RED 104 see ADG250
JAPAN RED 201 see CMS155
JAPAN RED 203 see CMS150
JAPAN RED 219 see CMS160
JAPAN RED NO. 203 see CMS150
JAPAN RED NO. 505 see CMO885
JAPAN YELLOW 203 see CMM510
JASAD see ZBJ000
JASMINALDEHYDE see AOG500
JASMIN de NUIT (HAITI) see DAC500
JASMOLIN I or II see POO250
cis-JASMONE see JCA100
(Z)-JASMONE see JCA100
JASMONE see JCA100
JATRONEURAL see TKE500, TKK250
JATROPHA CATHARTICA see CNR135
JATROPHA CURCAS see CNR135
JATROPHA GOSSYPIIFOLIA see CNR135
JATROPHA INTEGERRIMA see CNR135
JATROPHA MACRORHIZA see CNR135
JATROPHA MULTIFIDA see CNR135
JATROPHA PODAGRICA see CNR135
JATROPUR see UVJ450
JAUNE AB see FAG130
JAUNE de BEURRE (FRENCH) see DOT300
JAUNE OB see FAG135
JAUNE ORANGE S see FAG150
JAUNE de QUINOLEINE see CMM510
JAUNE SOLEIL see FAG150
JAVA AMARANTH see FAG020
JAVA CHROME BLUE BLACK BN see CMP880
JAVA CHROME YELLOW GT see SIT850
JAVA METANIL YELLOW G see MDM775
JAVA NAPHTOL RED G see CMM300
JAVA ORANGE I see FAG010
JAVA ORANGE II see CMM220
JAVA PONCEAU 2R see FMU070
JAVA RUBINE N see HJF500
JAVA SCARLET 3R see FMU080

JAVA UNICHROME YELLOW GT see SIT850
JAVEX see SHU500
JAVILLO (DOMINICAN REPUBLIC, PUERTO RICO) see SAT875
JAYANTI, extract see SCB100
JAYFLEX DTDP see DXQ200
JAYSOL see EFU000
JAYSOL S see EFU000
J.B. 305 see EOY000
JB-323 see PJA000
JB 329 see DXP800
JB 336 see MON250
JB 340 see CBF000
JB 516 see PDN000, PDN250
JB 8181 see DLS600
177 J.D. see AJD000
JECTOFER see IHK100
JEFFAMINE AP22 see FMW330
JEFFAMINE AP27 see FMW330
JEFFAMINE AP-20 see MJQ000
JEFFERSOL DB see DJF200
JEFFERSOL EB see BPJ850
JEFFERSOL EE see EES350
JEFFERSOL EM see EJH500
JEFFOX see PJT000, PJT200, PKI500
JEFFOX OL 2700 see MKS250
JELLIN see SPD500
JENACAINE see AIL750
JENACILLIN O see PAQ200
JEN-DIRIL see CFY000
JERUSALEM OAK see SAF000
JERVINE see JCS000
JERVINE, 11-DEOXO-12-β,13-α-DIHYDRO-11-β-HYDROXY- see DAQ002
JESTRYL see CBH250
JESUIT'S BALSAM see CNH792
JET BEAD see JDA075
JET FUEL HEF-2 see JDA100
JET FUEL HEF-3 see JDA125
JET FUEL JP-4 see JDA135
JET FUELS see JDJ000
JETRIUM see AFJ400
JETRIUM R see AFJ400
JEW BUSH see SDZ475
JEWELER'S ROUGE see IHD000
J 2Fp see GLU000
JF 5705F see RLF350
JICAMA de AQUA (CUBA) see YAG000
JICAMILLA (MEXICO, TEXAS) see CNR135
JICAMO (MEXICO) see YAG000
JIFFY GROW see ICP000
JILKON HYDROBROMIDE see GBA000
JIMBAY BEAN (BAHAMAS) see LED500
JIMSON WEED see SLV500
JISC 3108 see AGX000
JISC 3110 see AGX000
JL 130 see COW700
J-LIBERTY see MDQ250
JM 8 see CCC075
JM-28 see IGG775
JMC 45498 see DAF300
JN-21 see CQH000
JODAIROL see HGB200
JODAMID (GERMAN) see AAI750
4-JODBENZOESAEURE see IEE025
2-JODBUTAN see IEH000
JODCYAN see COP000
JODDEOXIURIDIN see DAS000
JODETHAN see ELP500
JODFENPHOS see IEN000
JOD (GERMAN, POLISH) see IDM000
o-JODHIPPURSAEURE NATRIUM (GERMAN) see HGB200
JODID SODNY see SHW000
JOD-METHAN (GERMAN) see MKW200
1-JOD-3-METHYLBUTAN see IHU200
1-JOD-2-METHYLPROPAN see IIV509
JODOBIL see PDM750
1-JODOKTAN see OFA100
JODOMIRON see AAI750
JODOPAX see AAN000
1-JODPENTAN see IET500
3-JODPHENOL see IEV010
JODPHOSPHONIUM (GERMAN) see PHA000
JOGEN see JDJ100
JOHNKOLOR see TKQ250
JOJOBA Liquid WAX see JDJ300
JOJOBA OIL see JDJ300

JOLIPEPTIN see JDS000
JOLT see EIN000
JOMYBEL see JDS200
JONIT see PFA500
JONNIX see SNQ500
JONQUIL see DAB700
JON-TROL see DXE600
JOOD (DUTCH) see IDM000
JOODMETHAAN (DUTCH) see MKW200
JOPAGNOST see IFY100
JORCHEM 400 ML see PJT000
JORTAINE see GHA050
JOSAMINA see JDS200
JOSAMYCIN see JDS200
JOSCINE see SBG500
JOY POWDER see HBT500
JP-10 see TLR675
JP 61 see AJU625
JP 428 HYDROCHLORIDE see DAI200
J SOFT C 4 see DTC600
J-SUL see SNN500
JUCA see CCO680
JUDEAN PITCH see ARO500
JUDOLOR see FQJ100
JUGLANE see WAT000
JUGLON see WAT000
JUGLONE see WAT000
JULIN'S CARBON CHLORIDE see HCC500
JULODIN see CKL250
JUMBEE BEADS (VIRGIN ISLANDS) see RMK250
JUMBLE BEAD see AAD000
JUMEX see DAZ125
JUMP-AND-GO (BAHAMAS) see LED500
JUNGER FUSTIK see FBW000
JUNIPEN see LID100
JUNIPENE see LID100
JUNIPER BERRY OIL see JEA000
JUNIPERIC ACID LACTONE see OKU000
JUNIPER OIL see JEA000
JUNIPER TAR see JEJ000
JUNIPERUS COMMUNIS auct. non. Linn., extract excluding roots see JEJ100
JUNIPERUS COMMUNIS Linn. var. SAXATILIS Pallas, extract excluding roots see JEJ100
JUNIPERUS PHOENICEA OIL see JEJ200
JUNLON 110 see ADW200
JURIMER AC 10H see ADW200
JURIMER AC 10P see ADW200
JUSONIN see SFC500
JUSQUIAME (CANADA) see HAQ100
JUSTAMIL see AIE750
JUVASON see PLZ000
JUVASTIGMIN see NCL100
JUVENIMICIN A3 see RMF000
JUVOCAINE see AIT250
JZF see BLE500


K 0 see UTU500
06K see AJI250
K-9 see MEC250
K 17 see TEH500, UTU500
K 52 see OHU000
K 55E see SMR000
K-117 see NCP875
K2$_{20}$ see VTA650
K 250 see CAT775
K 257 see CBT500
K-300 see CQH000
K-315 see THJ750
K 351 see NEA100
K-373 see DAP700
K385 see UTU500
K 525 see SMQ500
K6-30 see ARM268
K 653 see HGW000
K 694 see CDR550
K-708 see AAE625
K 1875 see NCM700
K 1900 see NHH000
K-1902 see PPW000
K 2680 see AHP125
K3917 see CFY750
K 4277 see IDA400
K 4710 see KFK000
K 6451 see CJT750

K 8870 see UTU500
K-10033 see BPP250
K 121-02 see MCB050
K 22023 see DGD800
K-30052 see EGC500
K 411-02 see UTU500
K 421-01 see MCB050
K 421-02 see MCB050
K 421-05 see MCB050
K 423-02 see MCB050
K62-105 see LEN000
K 31 (pharmaceutical) see CME675
K25 (polymer) see PKQ250
KABAT see KAJ000
KABIKINASE see SLW450
KADMIUM (GERMAN) see CAD000
KADMIUMCHLORID (GERMAN) see CAE250
KADMIUMSTEARAT (GERMAN) see OAT000
KADMU TLENEK (POLISH) see CAH500
KADOX-25 see ZKA000
KAERGONA see MMD500
KAFAR COPPER see CNI000
KAFIL SUPER see RLF350
KAFOCIN see CCR890
KAISER CHEMICALS 11 see FOO000
KAISER CHEMICALS 12 see DFA600
KAISER NCO 20 see PKB100
KAKEN see CPE750
KAKERBIN see PAK250
KAKO BLUE B SALT see DCJ200
KAKODYLAN DODNY see HKC500
KAKO RED B BASE see NEQ000
KAKO RED TR BASE see CLK220
KAKO SCARLET GG SALT see DEO295
KAKO TARTRAZINE see FAG140
KALCIT see CAD800
KALEX see EIV000
KALGAN see PFC750
K'-ALGILINE see SEH000
KALITABS see PLA500
KALIUMARSENIT (GERMAN) see PKV500
KALIUM-BETA see PLG800
KALIUMCARBONAT (GERMAN) see PLA000
KALIUMCHLORAAT (DUTCH) see PLA250
KALIUMCHLORAT (GERMAN) see PLA250
KALIUMCYANAT (GERMAN) see PLC250
KALIUM-CYANID (GERMAN) see PLC500
KALIUMDICHROMAT (GERMAN) see PKX250
KALIUMHYDROXID (GERMAN) see PLJ500
KALIUMHYDROXYDE (DUTCH) see PLJ500
KALIUMNITRAT (GERMAN) see PLL500
KALIUMPERMANGANAAT (DUTCH) see PLP000
KALIUMPERMANGANAT (GERMAN) see CAV250, PLP000
KALKHYDRATE see CAT225
KALLIKREIN-TRYPSIN INACTIVATOR see PAF550
KALLOCRYL K see PKB500
KALLODENT CLEAR see PKB500
KALMETTUMSOMNIFERUM see GFA000
KALMIA ANGUSTIFOLIA see MRU359
KALMIA LATIFOLIA see MRU359
KALMIA MICROPHYLLA see MRU359
KALMOCAPS see LFK000, MDQ250
KALMUS OEL (GERMAN) see OGK000
KALO see ARB750, EAI600
KALODIL see DNM400
KALOMEL (GERMAN) see MCW000
KALPREN see CAS750
KALPUR TE see THR820
KALTOSTAT see CAM200
KALZIUMARSENIAT (GERMAN) see ARB750
KALZIUMZYKLAMATE (GERMAN) see CAR000
KAM 1000 see SLK000
KAM 2000 see SLK000
KAM 3000 see SLK000
KAMALIN see KAJ500
KAMANI (HAWAII) see MBU780
KAMAVER see CDP250
KAMBAMINE RED TR see CLK220
KAMBAMINE SCARLET GG BASE see DEO295
KAMFOCHLOR see CDV100
KAMILLENOEL (GERMAN) see DRV000
KAMILLEN OEL see CBA200
KAMILLENOEL see CDH500
KAMPFER (GERMAN) see CBA750
KAMPOSAN see CDS125

KAMPSTOFF "LOST" see BIH250
KAMYCIN see KAM000, KAV000
KAMYNEX see KAM000, KAV000
KANABRISTOL see KAM000, KAV000
KANACEDIN see KAM000, KAV000
KANAMICINA (ITALIAN) see KAL000
KANAMYCIN see KAL000
KANAMYCIN A see KAL000
KANAMYCIN A SULFATE see KAM000, KAV000
KANAMYCIN B see BAU270
KANAMYCIN B SULFATE see KBA100
KANAMYCIN B, SULFATE (1:1) (SALT) see KBA100
KANAMYCIN MONOSULFATE see KAM000, KAV000
KANAMYCIN SULFATE see KAM000, KAV000
KANAMYCIN SULFATE (1:1) SALT see KAV000
KANAMYTREX see KAL000, KAM000, KAV000
KANAQUA see KAM000, KAV000
KANASIG see KAM000, KAV000
KANATROL see KAM000, KAV000
KANDISET see BCE500
KANECHLOR 300 see PJL750, PJO500
KANECHLOR 400 see PJL750, PJO750
KANECHLOR 500 see PJP000, PJP250
KANECHLOR see PJL750
KANECHLOR C see PJP250
KANEKALON see ADY250
KANEKROL 500 see PAV600
KANENDOMYCIN see BAU270
KANENDOMYCIN SULFATE see KBA100
KANEPAR see TIY500
KANESCIN see KAM000, KAV000
KANICIN see KAM000
KANNASYN see KAM000, KAV000
KANNIT see XPJ000
KANO see KAM000, KAV000
KANOCHOL see DYE700
KANONE see MMD500
KANTEC see MAO275
KANTREX see KAL000, KAM000, KAV000
KANTREXIL see KAM000, KAV000
KANTROX see KAM000, KAV000
KANZO (JAPANESE) see LFN300
KAOCHLOR see PLA500
KAOLIN see KBB600
KAON see PLG800
KAON-CI see PLA500
KAON ELIXIR see PLG800
KAOPAOUS see KBB600
KAOPHILLS-2 see KBB600
KAPPAXAN see MMD500
e-KAPROLAKTAM (CZECH) see CBF700
KAPROLIT see PJY500
KAPROLON see PJY500
KAPROMIN see PJY500
KAPRON see NOH000, PJY500
KAPRONAN DI-N-BUTYLCINICITY (CZECH) see BJX750
KAPRONAN DICYKLOHEXYLAMINU (CZECH) see DGU400
KAPRYLAN DI-N-BUTYLCINICITY (CZECH) see BLB250
KAPSOLAT see MEP750
KAPTAN see CBG000
KAPTAX see BDF000
KARAMATE see DXI400
KARAYA GUM see KBK000
KARBAM BLACK see FAS000
KARBAMOL see UTU500
KARBAMOL B/M see UTU500
KARBAM WHITE see BJK500
KARBANIL see PFK250
KARBARYL (POLISH) see CBM750
KARBARYL see CBM750
KARBASPRAY see CBM750
KARBATION see SIL550, VFU000
KARBATOX 75 see CBM750
KARBATOX see CBM750
KARBATOX ZAWIESINOWY see CBM750
KARBOFOS see MAK700
KARBOKROMEN (RUSSIAN) see CBR500
5-KARBOKSIMETIL-3-p-TOLIL-TIAZOLIDIN-2,4-DION-2-ACETOFENONHIDRAZON
    (CZECH) see CCH800
KARBORAFIN see CDI000
KARBOSEP see CBM750
1-(4'-KARBOXYLAMIDOFENYL)-3,3-DIMETHYLTRIAZENU (CZECH) see
    CCC325
1-(4'-KARBOXYLAMIDOPHENYL)-3,3-DIMETHYLTRIAZEN (GERMAN) see
    CCC325

KARBROMAL see BNK000
KARBUTILATE see DUM800
KARCON see MMD500
KARDIAMID see DJS200
KARDIN see TJL250
KARDONYL see DJS200
KAREON see MMD500
KARIDIUM see SHF500
KARIGEL see SHF500
KARION see SKV200
KARI-RINSE see SHF500
KARLAN see RMA500
KARMESIN see HJF500
KARMEX see DXQ500
KARMEX DIURON HERBICIDE see DXQ500
KARMEX DW see DXQ500
KARMEX MONURON HERBICIDE see CJX750
KARMEX W. MONURON HERBICIDE see CJX750
KARMINOMYCIN see KBU000
KARMINOMYCIN HYDROCHLORIDE see KCA000
KARNOZZN see CCK665
KARO KAROUNDE ABSOLUTE see KCA050
KARPHOS see DJV600
KARSAN see FMV000
KARSULPHAN see SNR000
KARTRYL see BNK000
KARWINSKIA HUMBOLDTIANA see BOM125
KASAL see SEM305
KASEBON see CBW000
KASH, LEAF EXTRACT see KCA100
KASHMIRJA see CNT350
KASIMID see BBW750
KASSIA OEL (GERMAN) see CCO750
KASTAM see CNT350
KASTONE see HIB010
KASUGAMYCIN see AEB500
KASUGAMYCIN HYDROCHLORIDE see KCK000
KASUGAMYCIN MONOHYDROCHLORIDE see KCK000
KASUGAMYCIN PHOSPHATE see KCU000
KASUGAMYCIN SULFATE see KDA000
KASUMIN see KCK000
KAT 256 see CMW500
KATAGRIPPE see EEM000
KATALYSIN see AKK750
KATAMINE AB see AFP250, DTC600
KATANOL C12 see BDJ600
KATAPYRIN see AFW500
KATCHUNG OIL see PAO000
KATEXOL 300 see KDA025
KATHON LP PRESERVATIVE see OFE000
KATHON SP 70 see OFE000
KATHRO see CMD750
KATINE see NNM510
KATIV-G see MMD500
KATIV N see VTA000
KATLEX see CHJ750
KATONIL see CHX250
KATORIN see PLG800
KATOVIT HYDROCHLORIDE see PNS000
KATRON see BCA000, PDN000
KATRONIAZID see PDN000
KAURAMIN 542 see MCB050
KAURAMIN 650 see MCB050
KAURAMIN 700 see MCB050
KAURAMIN 782 see MCB050
KAURESIN K244 see UTU500
KAURIT 420 see UTU500
KAURIT 285 FL see UTU500
KAURITIL see CNK559
KAURIT M 70 see MCB050
KAURIT S see DTG700
KAUTSCHIN see MCC250
KAVAIN see GJI250
(+)-KAVAIN see GJI250
KAWAIN see GJI250
KAYACYL BLUE BR see CMM070
KAYACYL PURE BLUE FGA see ERG100
KAYACYL SKY BLUE R see CMM080
KAYAFUME see MHR200
KAYAHARD MCD see NAC000
KAYAKU ACID BRILLIANT SCARLET 3R see FMU080
KAYAKU ALIZARINE SKY BLUE R see CMM080
KAYAKU AMARANTH see FAG020
KAYAKU BLUE B BASE see DCJ200
KAYAKU CONGO RED see SGQ500

KAYAKU DIRECT see CMO000
KAYAKU DIRECT BLACK BH see CMN800
KAYAKU DIRECT BRILLIANT BLUE RW see CMO600
KAYAKU DIRECT DARK GREEN B see CMO830
KAYAKU DIRECT DEEP BLACK EX see AQP000
KAYAKU DIRECT DEEP BLACK GX see AQP000
KAYAKU DIRECT DEEP BLACK S see AQP000
KAYAKU DIRECT FAST BLACK D see CMN230
KAYAKU DIRECT FAST RED F see CMO870
KAYAKU DIRECT GREEN B see CMO840
KAYAKU DIRECT LEATHER BLACK EX see AQP000
KAYAKU DIRECT SCARLET 3B see CMO875
KAYAKU DIRECT SKY BLUE 5B see CMO500
KAYAKU DIRECT SPECIAL BLACK AAX see AQP000
KAYAKU FAST RED 3GL BASE see KDA050
KAYAKU FOOD COLOUR YELLOW NO. 4 see FAG140
KAYAKU MORDANT YELLOW GG see SIT850
KAYAKU RED B BASE see NEQ000
KAYAKU SCARLET G BASE see NMP500
KAYAKU SCARLET GG BASE see DEO295
KAYAKU TARTRAZINE see FAG140
KAYALON FAST BLUE BR see TBG700
KAYALON FAST BLUE FN see MGG250
KAYALON FAST RUBINE B see CMP080
KAYALON FAST VIOLET BB see AKP250
KAYALON FAST VIOLET BR see DBY700
KAYALON FAST YELLOW G see AAQ250
KAYALON FAST YELLOW 4R see CMP090
KAYALON FAST YELLOW RR see DUW500
KAYALON POLYESTER BLUE EBL-E see CMP070
KAYALON POLYESTER YELLOW 4R-E see CMP090
KAYALON POLYESTER YELLOW RF see CMP090
KAYANOL MILLING RED PG see CMM320
KAYANOL MILLING RED RS see CMM330
KAYANOL RED PG see CMM320
KAYANOL RED RS see NAO600
KAYARUS BLACK G see CMN240
KAYARUS BLACK G CONC. see CMN240
KAYASET YELLOW G see AAQ250
KAYAZINON see DCM750
KAYAZOL see DCM750
KAY CIEL see PLA500
KAYDOL see MQV750, MQV875
KAYEXALATE see SJK375
KAYKLOT see MMD500
KAYLITE see PKQ059
KAYQUINONE see MMD500
KAYTRATE see PBC250
KAYTWO see VTA650
KAYVISYN see AKX500
KAZOE see SFA000
KB-16 see MIH250
KB-53 see CCP875
KB 95 see BCP650
KB 101 see HAQ500
KB 227 see RGP450
KB-227 see TCY260
KB 227 see THJ825
KB-509 see FMR100
KB-944 see DIU500
KB 1585 see LEJ500
KBM 1003 see TLD000
KB (POLYMER) see SMQ500
K-BRITE see SHR500
KBT-1585 see LEJ500
KC-400 see PJO750
KC-404 see KDA100
KC-500 see PJP000
KC 9147 see MQA000
KCA ACETATE CRIMSON B see CMP080
KCA ACETATE FAST YELLOW G see AAQ250
KCA ACID MILLING YELLOW M see CMM759
KCA FOODCOL AMARANTH A see FAG020
KCA FOODCOL SUNSET YELLOW FCF see FAG150
KCA FOODCOL TARTRAZINE PF see FAG140
KCA METHYL ORANGE see MND600
KCA SILK RED G see CMM320
KCA TARTRAZINE PF see FAG140
KD 83 see DXG625
KD-136 see HAG300
KDM see BAU270
KE see CNS800
KEBILIS see CDL325
KEBUZONE see KGK000
KEDACILLIN see SNV000

KEDAVON see TEH500
KEESTAR see SLJ500
KEFENID see BDU500
KEFGLYCIN see CCR890
KEFLEX see ALV000
KEFLIN see SFQ500
KEFLODIN see TEY000
KEFORAL see ALV000
KEFZOL see CCS250
KEIMSTOP see CBL750
KELACID see AFL000
KELAMERAZINE see ALF250
KELCO GEL LV see SEH000
KELCOLOID see PNJ750
KELCOSOL see SEH000
KELENE see EHH000
KEL-F 3 see KDK000
KEL-F 81 see KDK000
KEL-F 90 see KDK000
KEL-F 200 see KDK000
KEL-F 95/5 see KDK000
KEL-F 6061 see KDK000
KEL-F see KDK000
KELFIZIN see MFN500
KELGIN see SEH000
KELGUM see SEH000
KELICORIN see AHK750
KELINCOR see AHK750
KELOFORM see EFX000
KEL-S see KDK000
KELSET see SEH000
KELSIZE see SEH000
KELTANE see BIO750
KELTEX see SEH000
KELTHANE (DOT) see BIO750
p,p'-KELTHANE see BIO750
KELTHANE DUST BASE see BIO750
KELTHANETHANOL see BIO750
KELTONE see SEH000
KEMADRINE see CPQ250
KEMAMIDE S see OAR000
KEMAMINE 9902D see DTC400
KEMAMINE P690 see DXW000
KEMAMINE P 989 see OHM700
KEMAMINE P990 see OBC000
KEMAMINE Q 9702C see QAT550
KEMAMINE QSML2 see QAT550
KEMAMINE S 190 see KDU100
KEMATE see DEV800
KEMDAZIN see MHC750
KEMESTER 105 see OHW000
KEMESTER 115 see OHW000
KEMESTER 205 see OHW000
KEMESTER 213 see OHW000
KEMI see ICC000
KEMICETINE see CDP250
KEMICETINE SUCCINATE see CDP725
KEMIKAL see CAT225
KEMITHAL see TES500
KEMITRACIN 10 see BAC260
KEMODRIN see DBA800
KEMOLATE see PHX250
KEMOVIRAN see MKW250
KENACHROME BLACK 6B see CMP880
KENACHROME BLUE 2R see HJF500
KENACORT see AQX250
KENACORT-A see AQX500
KENALOG see AQX500
KENAPON see DGI400
KENDALL'S COMPOUND E see CNS800
KENDALL'S COMPOUND F see CNS750
KENGSHENGMYCIN see AEB500
KENROX 106 see PBP200
KEOBUTANE-JADE see KGK000
KEPHALIS see EES370
KEPHRINE see MGC350
KEPHTON see VTA000
KEPINOL see TKX000
KEPMPLEX 100 see EIV000
KEPONE see KEA000
KEPTAN see KEA300
KER 710 see EAL100
KERALYT see SAI000
KERAPHEN see TDE750
KERASALICYL see SJO000

KERB see DTT600
KERECID see DAS000
KERLONE see KEA350
KERMAC 600W (MINERAL SEAL OIL) see DXG830
KERNECHTROT see AJQ250
KEROCAINE see AIT250
KEROPUR see BAV000
KEROSAL see SJO000
KEROSENE see KEK000
KEROSENE (PETROLEUM), hydrotreated see KEK100
KEROSINE (petroleum) see KEK000
KEROSINE see KEK000
KEROSINE, HYDRODESULFURIZED see KEK110
KESELAN see CLY500
KESSAR see TAD175
KESSCO 40 see OAV000
KESSCO BSC see BSL600
KESSCOCIDE see ISN000
KESSCOFLEX see BHK000
KESSCOFLEX BS see BSL600
KESSCOFLEX MCP see DOF400
KESSCOFLEX TRA see THM500
KESSCO ICS see IKC050
KESSCO ISOPROPYL see IQW000
KESSCO ISOPROPYL MYRISTATE see IQN000
KESSCOMIR see IQN000
KESSOBAMATE see MQU750
KESSODANTEN see DKQ000
KESSODRATE see CDO000
KESTREL (Pesticide) see AHJ750
KESTRIN see ECU750
KESTRONE see EDV000
KETAJECT see CKD750
KETALAR see CKD750
KETALGIN see MDO000
KETALGIN HYDROCHLORIDE see MDP750
KETAMAN see HKR500
KETAMINE see CKD750, KEK200
KETAMINE HYDROCHLORIDE see CKD750
KETANEST see CKD750
KETANRIFT see ZVJ000
KETASET see CKD750
KETASON see KGK000
KETAVET see CKD750
KETAZONE see KGK000
KETENE see KEU000
KETENE DIMER see KFA000
KETHOXAL-BIS-THIOSEMICARBAZIDE see KFA100
KETJENBLACK EC see CBT750
3-(3-KETO-7-α-ACETYLTHIO-17-β-HYDROXY-4-ANDROSTEN-17-α-YL)PROPIONIC
    ACID LACTONE see AFJ500
4-KETOAMYLTRIMETHYLAMMONIUM IODIDE see TLY250
KETOBEMIDONE see KFK000
4-KETOBENZOTRIAZINE see BDH000
β-KETOBUTYRANILIDE see AAY000
KETOCAINE HYDROCHLORIDE see DNN000
KETOCHOL see DAL000
KETOCHOLANIC ACID see DAL000
7-KETOCHOLESTEROL see ONO000
KETOCONAZOL see KFK100
KETOCONAZOLE see KFK100
KETOCYCLOHEPTANE see SMV000
KETOCYCLOPENTANE see CPW500
KETODERM see KFK100
KETODESTRIN see EDV000
2-KETO-3-ETHOXY-BUTYRALDEHYDE-BIS(THIOSEMICARBAZONE) see KFA100
KETO-ETHYLENE see KEU000
KETOGAZE see MGC350
3-KETO-I-GULOFURANOLACTONE see ARN000
KETOHEPTAMETHYLENE see SMV000
KETOHEXAMETHYLENE see CPC000
2-KETOHEXAMETHYLENIMINE see CBF700
KETOHYDROXY-ESTRATRIENE see EDV000
KETOHYDROXYESTRIN see EDV000
KETOHYDROXYESTRIN BENZOATE see EDV500
KETOHYDROXYOESTRIN see EDV000
2,3-KETOINDOLINE see ICR000
KETOISDIN see KFK100
KETOLAR see CKD750
KETOLE see ICM000
KETOLIN-H see CJU250
γ-KETO-β-METHOXY-Δ-METHYLENE-Δ°-HEXENOIC ACID see PAP750
15-KETO-20-METHYLCHOLANTHRENE see MIM250
KETONE, 1-ADAMANTYLDIAZO METHYL see DAB807

KETONE, 2-AMINO-5-BENZIMIDAZOLYL PHENYL see AIH000
KETONE, (4-AMINO-5-(o-CHLOROPHENYL)-2-METHYLPYRROL-3-YL) METHYL
    see AJI300
KETONE, (4-AMINO-5-(p-CHLOROPHENYL)-2-METHYLPYRROL-3-YL) METHYL
    see AJI330
KETONE, (4-AMINO-5-(3,4-DICHLOROPHENYL)-2-METHYLPYRROL-3-YL) METH-
    YL see AJM600
KETONE, (4-AMINO-5-(3,4-DIMETHOXYPHENYL)-2-METHYLPYRROL-3-YL)
    METHYL see AJO800
KETONE, (4-AMINO-1,2-DIMETHYL-5-PHENYLPYRROL-3-YL) METHYL see
    AJR100
KETONE, 5-AMINO-1,3-DIMETHYLPYRAZOL-4-YL o-FLUOROPHENYL see
    AJR400
KETONE, (4-AMINO-1-ETHYL-2-METHYL-5-PHENYLPYRROL-3-YL) METHYL see
    AKA600
KETONE, (4-AMINO-5-(p-FLUOROPHENYL)-2-METHYLPYRROL-3-YL) METHYL
    see AKC560
KETONE, (4-AMINO-5-(m-FLUOROPHENYL)-2-METHYLPYRROL-3-YL) METHYL
    see AKC550
KETONE, (4-AMINO-5-(2-METHOXYPHENYL)-2-METHYLPYRROL-3-YL) METHYL
    see AKO450
KETONE, (4-AMINO-5-(p-METHOXYPHENYL)-2-METHYLPYRROL-3-YL) METHYL
    see AKO430
KETONE, (4-AMINO-5-(m-METHOXYPHENYL)-2-METHYLPYRROL-3-YL) METHYL
    see AKO400
KETONE, (4-AMINO-5-(4-METHOXY-m-TOLYL)-2-METHYLPYRROL-3-YL) METH-
    YL see AKO100
KETONE, (4-AMINO-2-METHYL-5-PHENYLPYRROL-3-YL) METHYL, MONOHY-
    DROCHLORIDE see ALA300
KETONE, (4-AMINO-2-METHYL-5-(m-TOLYL)PYRROL-3-YL) METHYL see
    AKT830
KETONE, (4-AMINO-2-METHYL-5-(o-TOLYL)PYRROL-3-YL) METHYL see
    AKT800
KETONE, (4-AMINO-2-METHYL-5-(p-TOLYL)PYRROL-3-YL) METHYL see
    AKT850
KETONE, (4-AMINOPIPERIDINO)METHYL INDOL-3-YL see AMA100
KETONE, p-ANISYL 3-PYRIDYL see MED100
KETONE, p-ANISYL 4-PYRIDYL see MFH930
KETONE, 3-AZABICYCLO(3.2.2)NONYL CHLOROMETHYL see CEC100
KETONE, 3-AZABICYCLO(3.2.2)NONYL IODOMETHYL see IDZ100
KETONE, 1-AZIRIDINYL 3-(BIS(2-CHLOROETHYL)AMINO)-p-TOLYL see
    BHQ760
KETONE, 2-BENZOFURANYL METHYL see ACC100
KETONE, 2H-1-BENZOPYRAN-3-YL METHYL see BCS500
KETONE, (1,2,4-BENZOTRIAZIN-3-YL)METHYL 1-PYRROLIDINYL see PPS600
KETONE, 1-BENZYL-2-INDOLYL HYDROXYMETHYL- see BES300
KETONE, 5-BENZYLOXY-3-INDOLYL 4-PIPERIDYL see BFC200
KETONE, 4-BENZYLPIPERAZINYL β-(p-CHLOROPHENYL)PHENETHYL see
    BFE770
KETONE, BENZYL(4-PYRIDYL), THIOSEMICARBAZONE see BFH100
KETONE, BICYCLO(4.2.0)OCTA-1,3,5-TRIEN-7-YL BENZYL see BFZ100
KETONE, BICYCLO(4.2.0)OCTA-1,3,5-TRIEN-7-YL BENZYL, OXIME see
    BFZ110
KETONE, BICYCLO(4.2.0)OCTA-1,3,5-TRIEN-7-YL BUTYL see BCH800
KETONE, BICYCLO(4.2.0)OCTA-1,3,5-TRIEN-7-YL METHYL see BFZ120
KETONE, BICYCLO(4.2.0)OCTA-1,3,5-TRIEN-7-YL METHYL, O-ACETYLOXIME
    see BFZ130
KETONE, BICYCLO(4.2.0)OCTA-1,3,5-TRIEN-7-YL METHYL, O-ALLYLOXIME
    see BFZ140
KETONE, BICYCLO(4.2.0)OCTA-1,3,5-TRIEN-7-YL METHYL, O-BUTYLOXIME
    see BFZ150
KETONE, BICYCLO(4.2.0)OCTA-1,3,5-TRIEN-7-YL METHYL, O-(2-HYDROXY-
    PROPYL)OXIME see HNT550
KETONE, BICYCLO(4.2.0)OCTA-1,3,5-TRIEN-7-YL METHYL, O-METHYLOXIME
    see MFX550
KETONE, BICYCLO(4.2.0)OCTA-1,3,5-TRIEN-7-YL METHYL, OXIME see
    BFZ160
KETONE, BICYCLO(4.2.0)OCTA-1,3,5-TRIEN-7-YL PENTYL, OXIME see
    BFZ170
KETONE, BICYCLO(4.2.0)OCTA-1,3,5-TRIEN-7-YL PHENYL see BFZ180
KETONE, 4-BIPHENYL ETHYL see BGM100
KETONE, 1-(1,1'-BIPHENYL)-4-YL-2-((4-(DICHLOROACETYL)PHENYL)AMINO)-2-
    HYDROXY- see BGL400
KETONE, m-(BIS(2-CHLOROETHYL)AMINO)PHENYL PIPERIDINO see BHP150
KETONE, 3-(BIS(2-CHLOROETHYL)AMINO)-p-TOLYL MORPHOLINO- see
    BHR400
KETONE, 3-(BIS(2-CHLOROETHYL)AMINO)-p-TOLYL PIPERIDYL- see BIA100
KETONE, 3-BROMO-1-ADAMANTYL DIAZOMETHYL see EEE025
KETONE, BROMOMETHYL 4-(DIPHENYLHYDROXYMETHYL)PIPERIDINO see
    BMS300
KETONE, 7-(3-(sec-BUTYLAMINO)-2-HYDROXYPROPOXY)-2-BENZOFURANYL
    METHYL see ACN310
KETONE, 4-(3-(tert-BUTYLAMINO)-2-HYDROXYPROPOXY)-2-BENZOFURANYL
    METHYL see ACN300

KETONE, METHOXYMETHYL 1-METHYL-2-INDOLYL see MDX300
KETONE, METHOXYMETHYL 3-METHYL-2-INDOLYL- see MDX310
KETONE, (4-METHOXY-2-METHYL-5-PHENYLPYRROL-3-YL) METHYL see MEX275
KETONE, 7-METHOXYNAPHTHO(2,1-b)FURAN-2-YL METHYL see ACQ790
KETONE, 4-(o-METHOXYPHENYL)-1-PIPERAZINYLMETHYL 3-PYRIDYL- see KGK130
KETONE, 4-(p-METHOXYPHENYL)PIPERAZINYL 3,4,5-TRIMETHOXYPHENYL see MFH770
KETONE, 4-(o-METHOXYPHENYL)PIPERAZINYL 3,4,5-TRIMETHOXYPHENYL see MFH760
KETONE, p-METHOXYPHENYL 2-PYRIDYL see MFH900
KETONE, (p-METHOXYPHENYL) 3-PYRIDYL see MED100
KETONE, (p-METHOXYPHENYL) 4-PYRIDYL see MFH930
KETONE, 10-(3-(4-METHOXYPIPERIDINO)PROPYL)PHENOTHIAZIN-2-YL METHYL see MFJ200
KETONE, 2-METHYLCYCLOHEXYL 4-METHYLPIPERIDINO see MLM700
KETONE, 2-METHYLCYCLOPROPYL HYDRAZINO see MIV300
KETONE, 2-METHYL-2-IMIDAZOLIN-1-YL 3,4,5-TRIMETHOXYPHENYL- see IAT200
KETONE, METHYL ISOAMYL see MKW450
KETONE, METHYL 2-METHYL-1,3-DITHIOLAN-2-YL see ACR050
KETONE, METHYL 5-METHYL-3-(5-NITRO-2-FURYL)-4-ISOXAZOLYL see MLP300
KETONE, METHYL 10-(3-(4-METHYL-1-PIPERAZINYL)PROPYL)PHENOTHIAZIN-2-YL see ACR500
KETONE, METHYL 5-METHYL-1-(2-QUINOLYL)-4-PYRAZOLYL see MLW600
KETONE, METHYL 5-METHYL-1-(2-QUINOXALINYL)-4-PYRAZOLYL see MLW630
KETONE, METHYL 5-METHYL-2-THIENYL see MPR300
KETONE, METHYL 10-(3-MORPHOLINOPROPYL)PHENOTHIAZIN-2-YL see MMA600
KETONE, METHYL 3-(5-NITRO-2-FURYL)-5-PHENYL-4-ISOXAZOLYL see MMJ955
KETONE, METHYL 2-NITRONAPHTHO(2,1-b)FURAN-7-YL see NHR100
KETONE, METHYL (4-NITRO-2-PYRROLYL) see ACT300
KETONE, METHYL (5-NITRO-2-PYRROLYL) see ACT330
KETONE, METHYL (5-NITRO-2-THIENYL) see NML100
KETONE-METHYL-5-OXO-5H-(1)BENZOPYRANO(2,3-b)PYRIDYL see ACU125
KETONE METHYL PHENYL see ABH000
KETONE, (2-METHYL-5-PHENYLPYRROL-3-YL) METHYL see MNY800
KETONE, 2-METHYLPIPERIDINO 2-NAPHTHYL see MOG600
KETONE, METHYL 10-(3-PIPERIDINOPROPYL)PHENOTHIAZIN-2-YL see MOL300
KETONE, METHYL 10-(3-PIPERIDINOPROPYL)PHENOXAZIN-2-YL see MOL400
KETONE, 1-METHYL-3-PIPERIDYL PIPERIDINO see MOG650
KETONE, METHYL 2-PYRIDYL see PPH200
KETONE, METHYL 4-PYRIDYL see ADA365
KETONE, METHYL 1,4,5,6-TETRAHYDRO-2-METHYLCYCLOPENTA(b)PYRROL-3-YL see MPO800
KETONE, (α-METHYL-m-TRIFLUOROMETHYLPHENETHYLAMINOMETHYL) PIPERIDINO see MQE100
KETONE, METHYL 2,4,5-TRIMETHYLPYRROL-3-YL see ADE050
KETONE, MORPHOLINO(1,2,3,4-TETRAHYDRO-9-ACRIDINYL) see MRR760
KETONE, PENTYL 4-PYRIDYL see HEW050, PBX800
KETONE, 1-PHENETHYL-3-PIPERIDYL PIPERIDINO see PDI550
KETONE, 4-PHENYL-1-PIPERAZINYLMETHYL-3-PYRIDYL- see NDW520
KETONE, 4-PHENYLPIPERAZINYL 3,4,5-TRIMETHOXYPHENYL see PFX600
KETONE, PHENYL (1-PIPERIDINOCYCLOHEXYL) see PFY200
KETONE, PHENYL 2-PYRIDYL see PGE760
KETONE, PHENYL 3-PYRIDYL see PGE765
KETONE, PHENYL 4-PYRIDYL see PGE768
KETONE, PHENYL PYRROL-2-YL see PGF900
KETONE, PIPERIDINO 3-PIPERIDYL see PIR200
KETONE, PIPERIDINO(1,2,3,4-TETRAHYDRO-9-ACRIDINYL) see PIU100
KETONE PROPANE see ABC750
KETONE, PROPYL 4-PYRIDYL see PNV755
KETONE, 4-(2-PYRIDYL)PIPERAZINYL 3,4,5-TRIMETHOXYPHENYL see TKY300
KETONE, 4-(2-PYRIMIDYL)PIPERAZINYL 3,4,5-TRIMETHOXYPHENYL see PPP550
KETONES see KGA000
KETONE, 4-(2-THIAZOLYL)PIPERAZINYL 3,4,5-TRIMETHOXYPHENYL see TEX220
KETONE, 4-(m-TOLYL)PIPERAZINYL 3,4,5-TRIMETHOXYPHENYL see THF300
KETONE, 4-(p-TOLYL)PIPERAZINYL 3,4,5-TRIMETHOXYPHENYL see THF310
4-KETONIRIDAZOLE see KGA100
KETONOX see MKA500
KETOPENTAMETHYLENE see CPW500
4-KETOPENTANOIC ACID BUTYL ESTER see BRR700
KETOPHENYLBUTAZONE see KGK000
KETOPROFEN see BDU500
KETOPROFEN SODIUM see KGK100
KETOPRON see BDU500
β-KETOPROPANE see ABC750
1-KETOPROPIONALDEHYDE see PQC000

2-KÉTOPROPIONALDEHYDE see PQC000
α-KETOPROPIONALDEHYDE see PQC000
α-KETOPROPIONIC ACID see PQC100
1-(2-KETO-2-(3'-PYRIDYL)ETHYL)-4-(2'-CHLOROPHENYL)PIPERAZINE see KGK120
1-(2-KETO-2-(3'-PYRIDYL)ETHYL)-4-(2'-METHOXYPHENYL)PIPERAZINE see KGK130
1-(2-KETO-2-(3'-PYRIDYL)ETHYL)-4-(PHENYL)PIPERAZINE see NDW520
2-KETOPYRROLIDINE-1-YLACETAMIDE see NNE400
2-KETO-4-QUINAZOLINONE see QEJ800
4-KETOSTEARIC ACID see KGK150
I-3-KETOTHREOHEXURONIC ACID LACTONE see ARN000
KETOTIFEN FUMARATE see KGK200
8-KETOTRICYCLO(5.2.1.0$^{2,6}$)DECANE see OPC000
2-KETO-1,7,7-TRIMETHYLNORCAMPHANE see CBA750
4-KETOVALERIC ACID see LFH000
γ-KETOVALERIC ACID see LFH000
KEUTEN see DEE600
KEVADON see TEH500
KEY-SERPINE see RDK000
KEY-TUSSCAPINE see NOA000
KEY-TUSSCAPINE HYDROCHLORIDE see NOA500
2KF see DOR800
KF-868 see KGK300
KF 994 see OCE100
KF 995 see DAF350
KF-1820 see DOQ400
KF 2 (HERBICIDE) see DOR800
K-FLEBO see PKV600
K-FLEX DP see DWS800
KFS see UTU500
K-GRAN see PLA000
KH 360 see TGG760
KHAINI (INDIA) see SED400
KHALADON 22 see CFX500
KHAROPHEN see ABX500
KHE 0145 see MIA250
KHIMCOCCID see RLK890
KHIMCOECID see RLK890
KHIMKOKTSID see RLK890
KHIMKOKTSIDE see RLK890
KHINALIZARIN see TDD000
KHINGAMIN see CLD250
KHINOTILIN see KGK400
KHLADON 113 see FOO000
KHLADON 744 see CBU250
KHLADON 114B2 see FOO525
KHLORAKON see BEG000
KHLORIDIN see TGD000
KHLORTRIANIZEN see CLO750
KHLOTAZOL see CMB675
KHOMECIN see ZJS300
KHOMEZIN see ZJS300
KHP 2 see AHE250
KHROMOLAN see CMH300
K-IAO see PLG800
KIATRIUM see DCK759
KID KILL see MRU359
KIDNEY BEAN TREE see WCA450
KIDOLINE see VGP000
KIEFERNADEL OEL (GERMAN) see PIH500
KIESELGUHR see DCJ800
KIESELSAEURE (GERMAN) see SCL000
KIEZELFLUORWATERSTOFZUUR (DUTCH) see SCO500
K III see DUS700
KIKUTHRIN see PMN700
KILDIP see DGB000
KILEX 3 see BSQ750
KILEX see CNK559
KILL-ALL see SEY500
KILLAX see TCF250
KILL COW see PJJ300
KILLEEN see CCL250
KILL KANTZ see AQN635
KILMAG see ARB750
KILMITE 40 see TCF250
KILOSEB see BRE500
KILPROP see CIR500
KILRAT see ZLS500
KILSEM see CIR250
KILVAL see MJG500
KIMAVOXYL see CBK500
KINADION see VTA000
KINAVOSYL see GGS000
KINEKS see AKO500

KINETIN see FPT100
KINETIN (PLANT HORMONE) see FPT100
KINEX see AKO500
KINGCOT see CCU150
KINGCUP see MBU550
KING'S GOLD see ARI000
KING'S GREEN see COF500
KING'S YELLOW see ARI000
KING'S YELLOW see LCR000
KINIDIN DURETTER see QFS100
KINIDIN DURULES see QFS100
KINILENTIN see QFS100
KINNIKINNIK see CCJ825
KINOPRENE see POB000
KINOTOMIN see FOS100
KIPCA see MMD500
KIRESUTO B see EIX500
KIRESUTO NTB see DXF000
KIRKSTIGMINE BROMIDE see POD000
KIRKSTIGMINE METHYL SULFATE see DQY909
α-KIRONDRIN see GEW700
β-KIRONDRIN see GEW700
KIR RICHTER see PAF550
KITASAMYCIN see SLC000
KITASAMYCIN A3 see JDS200
KITASAMYCIN TARTRATE see LEX000
KITAZIN see DIU800
KITAZIN L see BKS750
KITAZIN P see BKS750
KITINE see RDK000
KITON BLUE A see ERG100
KITON CRIMSON 2R see HJF500
KITON FAST YELLOW A see SGP500
KITON GREEN F see FAE950
KITON GREEN FC see FAE950
KITON GREEN S see ADF000
KITON ORANGE II see CMM220
KITON ORANGE MNO see MDM775
KITON PONCEAU R see FMU070
KITON PURE BLUE L see FMU059
KITON PURE BLUE V see ADE500
KITON PURE BLUE V.FQ see ADE500
KITON RED 2G see CMM300
KITON RED G see CMM300
KITON RUBINE S see FAG020
KITON SCARLET 4R see FMU080
KITON YELLOW EXTRA see CMM758
KITON YELLOW MS see MDM775
KITON YELLOW T see FAG140
KIVATIN see HKR500
KIWAM (INDIA) see SED400
KIWI LUSTR 277 see BGJ250
K. IXINA see CAC500
KL-001 see BMN750
KL 255 see BQB250
(—)-KL 255 see BQB250
KL 373 see BGD500
KLAVI KORDAL see NGY000
KLEBCIL see KAM000, KAV000
KLEER-LOT see AMY050
KLEESALZ (GERMAN) see OLE000
KLEGECELL see PKQ059
KLERAT see TAC800
KLIMANOSID see RDK000
KLIMORAL see EDU500
KLINE see BBK250
KLINGTITE see NAK500
KLINIT see XPJ000
KLION see MMN250
KLOBEN see BRA250
KLOBEN NEBURON see BRA250
KLOFIRAN see ARQ750
K-LOR see PLA500
KLORALFENAZON see SKS700
KLORAMIN see BIE500, CDP000
KLORAMINE-T see CDP000
KLOREX see SFS000
KLORINOL see TIX000
KLOROCIN see SHU500
KLOROKIN see CLD000
KLORPROMAN see CKP500
KLORPROMEX see CKP500
KLORT see MQU750
KLOT see AJP250
KLOTRIX see PLA500

KLOTTONE see MMD500
KLT 40 see PKF750
KLUCEL see HNV000
KM 2 see UTU500
4K-2M see CIR250
KM 200 see THR820
KM-208 see BAC175
KM-1146 see KGU100
KM (the antibiotic) see KAL000
KM see KAL000, SMQ500
KMH see DMC600
KM 2 (POLYMER) see UTU500
KM (POLYMER) see SMQ500
KMTS 212 see SFO500
K1-N see PLK500
KN 320 see IHC550
KNEE PINE OIL see PIH400
KNITTEX ASL see DTG700
KNITTEX LE see DTG000
KNITTEX TC see UTU500
KNITTEX TS see UTU500
Kh15N55M16 see CNA750
Kh15N55M16V see CNA750
KNOCKBAL see BSG300
KNOCKMATE see FAS000
KNOLL H75 see FMS875
KNOLLIDE see PLK500
KO 7 see EAL100
KO 08 see SCR400
KO 1173 see MQR775
KOA-HAOLE (HAWAII) see LED500
KOAXIN see MMD500
KOBALT CHLORID (GERMAN) see CNB599
KOBALT-EDTA (GERMAN) see DGQ400
KOBALT (GERMAN, POLISH) see CNA250
KOBALT HISTIDIN (GERMAN) see BJY000
KOBAN see EFK000
KOBU see PAX000
KOBUTOL see PAX000
KOCHINEAL RED A FOR FOOD see FMU080
KOCIDE see CNM500, SOD500
KO 1366-CL see BON400
KODAFLEX see TIG750
KODAFLEX DBS see DEH600
KODAFLEX DIBP see DNJ400
KODAFLEX DOA see AEO000
KODAFLEX DOP see DVL700
KODAFLEX DOTP see BJS500
KODAFLEX TOTM see TJR600
KODAFLEX TRIACETIN see THM500
KODAK LR 115 see CCU250
KODAK SILVER HALIDE SOLVENT HS-103 see BGT500
KODOCYTOCHALASIN-1 see PAM775
KOE 1366 CHLORIDE see BON400
KOE 1173 HYDROCHLORIDE see MQR775
KOFFEIN (GERMAN) see CAK500
KOHLENDIOXYD (GERMAN) see CBU250
KOHLENDISULFID (SCHWEFELKOHLENSTOFF) (GERMAN) see CBV500
KOHLENMONOXID (GERMAN) see CBW750
KOHLENOXYD (GERMAN) see CBW750
KOHLENSAEURE (GERMAN) see CBU250
KOJIC ACID see HLH500
KOKAIN see CNE750
KOKAN see CNE750
KOKAYEEN see CNE750
KOKOTINE see BBQ500
KOLALES HALOMTANO (GUAM) see RMK250
KOLCHAMIN see MIW500
KOLCHICIN see MIW500
KOLI (HAWAII) see CCP000
KOLKAMIN see MIW500
KOLKLOT see MMD500
2,4,6-KOLLIDIN see TME272
KOLLIDON see PKQ250
KOLOFOG see SOD500
KOLOSPRAY see SOD500
KOLPHOS see PAK000
KOLPON see EDV000
KOLTON see PIZ250
KOLTONAL see PIZ250
KOMBE-STROPHANTHIN see SMN000
KOMBETIN see SMN000
KOMEEN see DBU800
KOMPLEXON I see AMT500
KOMPLEXON IV see CPB120

KOMPLXON see EIV000
KONAKION see VTA000
KONDREMUL see MQV750
KONESSIN DIHYDROBROMIDE see DOX000
KONESTA see TII250
KONLAX see DJL000
KONTRAST-U see SHX000
KOOLMONOXYDE (DUTCH) see CBW750
KOOLSTOFDISULFIDE (ZWAVELKOOLSTOF) (DUTCH) see CBV500
KOOLSTOFOXYCHLORIDE (DUTCH) see PGX000
KOOST see CNT350
KOOT see CNT350
KOPFUME see EIY500
KOP KARB see CNJ750
KOPLEN 2 see SMQ500
KOPLEX AQUATIC HERBICIDE see DBU800
KOP MITE see DER000
KOPOLYMER BUTADIEN STYRENOVY (CZECH) see SMR000
KOPREZ 87-110 see UTU500
KOPROL see PDO750
KOPROSTERIN (GERMAN) see DKW000
KOPSOL see DAD200
KOP-THIODAN see EAQ750
KOP-THION see MAK700
KORAD see PKB500
KORBUTONE see AFJ625
KORDIAMIN see DJS200
KOREON see CMK415, NBW000
KORGLYKON see CNH780
KORIUM see MJM500
KORLAN see RMA500
KORLANE see RMA500
KORMOGRIZEIN see GJU800
KORODIL see CCK125
KOROSEAL see PKQ059
KOROSTAN RED G see CMM325
KORO-SULF see SNN500
KORUM see HIM000
KORUND see EAL100
KOSATE see DJL000
KOSMINK see CBT750
KOSMOBIL see CBT750
KOSMOLAK see CBT750
KOSMOS see CBT750
KOSMOTHERM see CBT750
KOSMOVAR see CBT750
KOST see CNT350
KOSTIL see ADY500
KOTAMITE see CAT775
K-OTHRIN see DAF300
KOTION see DSQ000
KOTOL see BBQ750
KOTORAN see KHK000
KOW see HOG000
KP 2 see PAX000
KP 140 see BPK250
KP 201 see DGV700
KPB see KGK000
KPE see KHK100
K PHENETHICILLIN see PDD350
K-PIN see PIB900
K. PNEUMONIAE ENDOTOXIN see EAS260
K 17 (POLYMER) see UTU500
K-PRENDE-DOME see PLA500
K PREPARATION see BJU000
06K-QUINONE see AJI250
KR 492 see MKK500
KR 2537 see SMQ500
KRAMERIA IXINA see CAC500
KRAMERIA TRIANDRA see RGA000
KRASTEN 1.4 see SMQ500
KRASTEN 052 see SMQ500
KRASTEN SB see SMQ500
KRATEDYN see EAW000
KREBON see BQL000
KRECALVIN see DGP900, PHC750
KREDAFIL 150 EXTRA see CAT775
KREDAFIL RM 5 see CAT775
KREGASAN see TFS350
KRENITE see ANG750
KRENITE BRUSH CONTROL AGENT see ANG750
KRENITE (OBS.) see DUS700, DUU600
KREOSOL see MEK250
KRESAMONE see DUS700
KRESIDIN see MGO750

o-KRESOL (GERMAN) see CNX000
m-KRESOL see CNW750
p-KRESOL see CNX250
KRESOLE (GERMAN) see CNW500
KRESOLEN (DUTCH) see CNW500
o-KRESOL-GLYCERINAETHER (GERMAN) see GGS000
KRESONIT E see DUT800
KRESOXYPROPANDIOL see GGS000
KREZAMON see DUT800
KREZIDINE see MGO750
KREZOL (POLISH) see CNW500
KREZONE see CIR250
KREZONIT E see DUT800
KREZONITE see DUU600
KREZOTOL 50 see DUS700
KRINO B 15 see DNM400
KRINOCORTS see DAQ800
KRIPLEX see DEO600
KRIPTIN see WAK000
KRISOLAMINE see DBP000
KRISTALLOSE see SJN700
KRISTALL-VIOLETT see AOR500
KRMD 58 see MAE000
KRO 1 see SMR000
KROKYDOLITH (GERMAN) see ARM275
KROLOR ORANGE RKO 786D see MRC000
KROMAD see KHU000
KROMFAX SOLVENT see TFI500
KROMON GREEN B see CLK235
KROMON HELIO FAST RED see MMP100
KROMON HELIO FAST RED YS see MMP100
KROMON LAKE ORANGE TONER see CMM220
KROMON LAKE RED C see CMS150
KROMON ORANGE G see CMS145
KROMON PERMANENT RED 4B see CMS155
KROMON YELLOW GXR see CMS208
KROMON YELLOW MTB see DEU000
KRONISOL see BHK000
KRONITEX see TNP500
KRONITEX KP-140 see BPK250
KRONITEX TOF see TNI250
KRONOS TITANIUM DIOXIDE see TGG760
KROTENAL see DXH250
KROTILINE see DAA800
KROTONALDEHYD (CZECH) see COB250
KROTYLCHLORID see CEU825
KROVAR II see BMM650
KRUMKIL see ABF500
KRYOGENIN see CBL000
KRYOLITH (GERMAN) see SHF000
KRYPTOCUR see LIU360
KRYPTOCYANINE IODIDE see KHU050
KRYSID see AQN635
KRZEWOTOKS see BSQ750
KS 11 see UTU500
KS 35 see UTU500
KhS 596 see CGW300
KS 1300 see CAT775
KS 1675 see DBA600
KS 4B see DEJ100
KS 68M see UTU500
KS-M 0.3P see UTU500
K-STROPHANTHIDIN see SMM500
K-STROPHANTHIN-$\alpha$ see CQH750
K-STROPHANTHIN-$\beta$ see SMN002
KSYLEN (POLISH) see XGS000
KT 35 see CNK559
KT 136 see KHU100
K-THROMBYL see MMD500
KTS (PHARMACEUTICAL) see KFA100
KU 5-3 see EAL100
KUBACRON see HII500
KUBARSOL see ABX500
KU 13-032-C see DFL200
KUE 13032c see DFL200
KUEMMEL OIL (GERMAN) see CBG500
KUKUI (HAWAII, GUAM) see TOA275
KULU 40 see CAT775
KUMADER see WAT200
KUMIAI see MIB750
KUMORAN see BJZ000
KUMULUS see SOD500
KUPAOA (HAWAII) see DAC500
KUPFERCARBONAT (GERMAN) see CNJ750
KUPFEROXYCHLORID (GERMAN) see CNK559

KUPFEROXYDUL (GERMAN) see CNO000
KUPFERRON (CZECH) see ANO500
KUPFERSULFAT (GERMAN) see CNP250
KUPFERSULFAT-PENTAHYDRAT (GERMAN) see CNP500
KUPFERVITRIOL (GERMAN) see CNP500
KUPRABLAU see CNM500
KUPRATSIN see EIR000
KUPRICOL see CNK559
KUPRIKOL see CNK559
KUPROTSIN see ZJS300
KUR see CNT350
KURALON VP see PKP750
KURAN see TIX500
KURARE OM 100 see AAX250
KURARE POVAL 1700 see PKP750
KURARE PVA 205 see PKP750
KURATE POVAL 120 see PKP750
KURCHICINE see KHU136
KURDUMANA, root extract see CNH750
KUREHALON A0 see CGW300
KUROMATSUEN see LID100
KUROMATSUENE see LID100
KURON see TIX500, TIX750
KUROSAL see TIX500
KUSA-TOHRU see SFS000
KUSATOL see SFS000
KUSHTHA see CNT350
KUSNARIN see EID000
KUSTA see CNT350
KUTH see CNT350
KUTROL see UVJ475
K-VITAN see MMD500
KW-066 see PNX750
KW 110 see AGX125
KW-125 see AES750
KW-1062 see MQS579
KW-1070 see FOK000
KW-3149 see FDD080
KW-4354 see OMG000
06K-50W see AJI250
KW-5338 see DYB875
KWAS BENZYDYNODWUKAROKSYLOWY (POLISH) see BFX250
KWAS 2,4-DWUCHLOROFENOKSYOCTOWY see DAA800
KWAS DWUMETYLO-DWUTIOFOSFOROWY see PHH500
KWAS METANIOWY (POLISH) see FNA000
KWASU 2,4-DWUCHLOROFENOKSYOCTOWEGO see DAA800
KWD 2019 see TAN100, TAN250
KWELL see BBQ500
KWELLS see HOT500
KWIETAL see QCS000
KWIK (DUTCH) see MCW250
KWIK-KIL see SMN500
KWIKSAN see ABU500
KWIT see EEH600
KW-2-LE-T see LFA020
KYAMEPROMAZINE MALEATE see COS899
KYANACETHYDRAZID see COH250
KYANID SODNY (CZECH) see SGA500
KYANID STRIBRNY (CZECH) see SDP000
KYANITE see AHF500
2-KYANMETHYLBENZIMIDAZOL (CZECH) see BCC000
KYANOSTRIBRNAN DRASELNY (CZECH) see PLS250
KYANURCHLORID (CZECH) see TJD750
KYLAR see DQD400
KYNEX see AKO500
KYOCRISTINE see LEZ000
KYONATE see PLV750
KYPCHLOR see CDR750
KYPFOS see MAK700
KYPMAN 80 see MAS500
KYPTHION see PAK000
KYPZIN see EIR000
KYSELINA ADIPOVA (CZECH) see AEN250
KYSELINA AKRYLOVA see ADS750
KYSELINA AMIDOSULFONOVA (CZECH) see SNK500
KYSELINA-4-AMINOANISOL-3-SULFONOVA see AIA500
KYSELINA 4-AMINOAZOBENZEN-3,4'-DISULFONOVA (CZECH) see AJS500
KYSELINA p-AMINOBENZOOVA see AIH600
KYSELINA 1-AMINO-2-ETHOXYNAFTALEN-6-SULFONOVA (CZECH) see AJU500
KYSELINA-3-AMINO-4-METHOXYBENZOOVA (CZECH) see AIA250
KYSELINA 1-AMINO-8-NAFTOL-3,6-DISULFONOVA (CZECH) see AKH000
KYSELINA 1-AMINO-8-NAFTOL-4-SULFONOVA (CZECH) see AKH750
KYSELINA 2-AMINO-5-NAFTOL-7-SULFONOVA (CZECH) see AKI000
KYSELINA-p-AMINOSALICYLOVA (CZECH) see AMM250

KYSELINA ANILIN-2,5-DISULFONOVA (CZECH) see AIE000
KYSELINA ANILIN-3-SULFONOVA (CZECH) see SNO000
KYSELINA 4,4'-AZO-BIS-(4-KYANVALEROVA) see ASL500
KYSELINA BENZIDIN-2,2'-DISULFONOVA (CZECH) see BBX500
KYSELINA BENZOOVA (CZECH) see BCL750
KYSELINA BROMOCTOVA see BMR750
KYSELINA C (CZECH) see ALH250
KYSELINA CEROMSALICYLOVA (CZECH) see BHA000
KYSELINA 2-CHLOR-6-AMINOFENOL-4-SULFONOVA (CZECH) see AJH500
KYSELINA o-CHLORBENZOOVA (CZECH) see CEL250
KYSELINA CHLOROCTOVA see CEA000
KYSELINA 4-CHLORO-3-NITROBENZOOVA (CZECH) see CJC500
KYSELINA-2-CHLORO-4-NITROBENZOOVA (CZECH) see CJC250
KYSELINA 2-CHLOR-4-TOLUIDIN-5-SULFONOVA (CZECH) see AJJ250
KYSELINA CITRAZINOVA see DMV400
KYSELINA CITRONOVA (CZECH) see CMS750
KYSELINA CLEVE (CZECH) see ALI250
KYSELINA 1,2-CYKLOHEXYLENDIAMINTETRAOCTOVA see CPB120
KYSELINA-2,4-DIAMINOBENZENSULFONOVA (CZECH) see PFA250
KYSELINA 2,4-DICHLORFENOXYOCTOVA see DAA800
KYSELINA DICHLORISOKYANUROVA (CZECH) see DGN200
KYSELINA 2,5-DICHLOR-4-(3'-METHYL-5'-PYRAZOLON-1'-YL)BENZENSULFONOVA (CZECH) see DFQ200
KYSELINA DICHLOROCTOVA see DEL000
KYSELINA 3,6-DICHLORPIKOLINOVA see DGJ100
KYSELINA O,O-DIETHYLDITHIOFOSFORECNA (CZECH) see PHG500
KYSELINA DI-(2-ETHYLHEXYL)FOSFORECNA see BJR750
KYSELINA 2,3-DIHYDROXYBUTANDIOVA see TAF750
KYSELINA 3,5-DIKARBOXYBENZENSULFONOVA (CZECH) see SNU500
KYSELINA O,O-DIMETHYLDITHIOFOSFORCNA (CZECH) see PHH500
KYSELINA-2,4-DINITROBENZENSULFONOVA (CZECH) see DUR400
KYSELINA-4,4'-DINITROSTILBEN-2,2'-DISULFONOVA (CZECH) see DVF600
KYSELINA DUSICNE see NED500
KYSELINA DUSITE see NMR000
KYSELINA 3,6-ENDOMETHYLEN-3,4,5,6,7,7-HEXACHLOR-Δ⁴-TETRAHYDROFTALOVA (CZECH) see CDS000
KYSELINA ETHOXY-CLEVE-1,6 (CZECH) see AJU500
KYSELINA 3-(2-ETHYLBUTOXY)PROPIONOVA see EGY000
KYSELINA-1,3-FENYLENDIAMIN-4-SULFONOVA (CZECH) see PFA250
KYSELINA 2-FENYL-2-HYDROXYETHANOVA see MAP000
KYSELINA FUMAROVA (CZECH) see FOU000
KYSELINA GLYOXYLOVA see GIQ000
KYSELINA H (CZECH) see AKH000
KYSELINA HET (CZECH) see CDS000
KYSELINA HYDROXYBUTANDIOVA (CZECH) see MAN000
KYSELINA 3-HYDROXY-2-NAFTOOVA see HMX520
KYSELINA 12-HYDROXY-9-OKTADECENOVA see RJP000
KYSELINA 2-HYDROXYPROPANOVA see LAG000
KYSELINA ISOFTALOVA (CZECH) see IMJ000
KYSELINA ISOMASELNA see IJU000
KYSELINA JABLECNA (CZECH) see MAN000
KYSELINA JANTAROVA see SMY000
KYSELINA KAKODYLOVA see HKC000
KYSELINA KOCHOVA (CZECH) see ALI750
KYSELINA KYANUROVA (CZECH) see THS000
KYSELINA MANDLOVA see MAP000
KYSELINA MERKAPTOOCTOVA see TFJ100
KYSELINA MESAKONOVA (CZECH) see MDI250
KYSELINA METANILOVA (CZECH) see SNO000
KYSELINA METHAKRYLOVA see MDN250
KYSELINA METHANSULFONOVA (CZECH) see MDR250
KYSELINA 4-METHOXYBENZOOVA see AOU600
KYSELINA 2-METHYLAMINO-5-NAFTOL-7-SULFONOVA (CZECH) see HLX000
KYSELINA N-METHYLANTHRANILOVA (CZECH) see MGQ000
KYSELINA METHYLARSONOVA see MGQ530
KYSELINA-N-METHYL-I (CZECH) see HLX000
KYSELINA MLECNA (CZECH) see LAG000
KYSELINA MUKOCHLOROVA see MRU900
KYSELINA-2-NAFTOL-1-SULFONOVA (CZECH) see HMX000
KYSELINA 2-NAFTYLAMIN-1,5-DISULFONOVA (CZECH) see ALH000
KYSELINA-2-NAFTYLAMIN-4,8-DISULFONOVA (CZECH) see ALH250
KYSELINA-2-NAFTYLAMIN-1-SULFONOVA (CZECH) see ALH750
KYSELINA-1-NAFTYLAMIN-6-SULFONOVA (CZECH) see ALI250
KYSELINA 2-NAFTYLAMIN-3,6,8-TRISULFONOVA (CZECH) see ALI750
KYSELINA NITRILOTRIOCTOVA see AMT500
KYSELINA 6-NITRO-2-AMINOFENOL-4-SULFONOVA (CZECH) see PHH500
KYSELINA-4-NITRO-2-AMINOFENOL-6-SULFONOVA (CZECH) see HMY500
KYSELINA 4-NITRO-4'-AMINOSTILBEN-2,2'-DISULFONOVA (CZECH) see ALP750
KYSELINA NITROBENZEN-m-SULFONOVA (CZECH) see NFB500
KYSELINA-p-NITROBENZOOVA (CZECH) see CCI250
KYSELINA N-(4-NITROFENYL)OXAMOVA see NHX100
KYSELINA-4-NITROTOLUEN-2-SULFONOVA (CZECH) see MMH250
KYSELINA OXALANILOVA see OLW000
KYSELINA PEROXYOCTOVA see PCL500

KYSELINA PIKROVA see PID000
KYSELINA PROPIONOVA see PMU750
KYSELINA RICINOLOVA see RJP000
KYSELINA-S-(8-CHLORMETHYL-1-NAFTYL)THIOGLYKOLOVA (CZECH) see CIQ000
KYSELINA STAVELOVA (CZECH) see OLA000
KYSELINA SULFAMINOVA (CZECH) see SNK500
KYSELINA SULFANILOVA see SNN600
KYSELINA 1-SULFOMETHYL-2-NAFTYLAMIN-6-SULFONOVA (CZECH) see AMP000
KYSELINA SULFO-TOBIAOVA (CZECH) see ALH000
KYSELINA TERFTALOVA (CZECH) see TAN750
KYSELINA 1,2,5,6-TETRAHYDROBENZOOVA see CPC650
KYSELINA-β,β′-THIODIPROPIONOVA (CZECH) see BHM000
KYSELINA THIOGLYKOLOVA see TFJ100
KYSELINA THIOOCTOVA see TFA500
KYSELINA TOBIASOVA (CZECH) see ALH750
KYSELINA p-TOLUENESULFONOVA (CZECH) see TGO000
KYSELINA-4-TOLUIDIN-3-SULFONOVA (CZECH) see AKQ000
KYSELINA 2-TOLUIDIN-4-SULFONOVA (CZECH) see AMT000
KYSELINA-3-TOLUIDIN-6-SULFONOVA (CZECH) see AMT250
KYSELINA o-TOSYL-H (CZECH) see AKH500
KYSELINA TRICHLOISOKYANUROVA (CZECH) see TIQ750
KYSELINA 2,3,6-TRICHLORBENZOOVA DIMETHYLAMONNA SUL see DOR800
KYSELINA TRICHLOROCTOVA see TII250
KYSELINA TRIFLUOROCTOVA see TKA250
KYSELINA VINNA see TAF750
KYSELINA WOLFRAMOVA (CZECH) see TOD000
KYSLICNIK DI-n-AMYLCINICITY (CZECH) see DVV000
KYSLICNIK DI-n-BUTYLCINICITY (CZECH) see DEF400
KYSLICNIK DIISOAMYLCINICITY (CZECH) see DNL400
KYSLICNIK DIISOBUTYLCINICITY (CZECH) see DNJ000
KYSLICNIK DIISOPROPYLCINICITY (CZECH) see DNR200
KYSLICNIK DI-N-PROPYLCINICITY (CZECH) see DWV000
KYSLICNIK TRI-N-BUTYLCINICITY (CZECH) see BLL750
KYURINETT see YCJ200
K-ZINC see ZKA000
KZ 3M see SCQ000
KZ 5M see SCQ000
KZ 7M see SCQ000


L-2 see BIQ250
L16 see AGX000
84L see DIW000
L-99 see CDG250
L-105 see CCS635
L 195 see UTU500
L-310 see LGK000
L343 see IOT000
L-395 see DSP400
L-561 see DRR400
L. 1633 see DEE600
L 1718 see DYE700
L-01748 see DJT800
L 1811 see DJT400
L-2103 see POA250
L2214 see DDP200
L 2329 see EID200
L. 3428 see AJK750
L-5103 see RKP000
L 6150 see BKB500
L 6504 see MCB050
L 8580 see MNB250
L 10499 see PEU650
L 109-65 see MCB050
L 121-60 see MCB050
L 12717 see CKL325
L-36352 see DUV600
Le-100 see EIF000
LA 01 see CCU150
LA 1 see DLY000
LA 1221 see PEU000
LA 6023 see DQR600
LA see DXY000
LA96A see PPN100
LAAM see ACQ666
LAAM HYDROCHLORIDE see ACQ690
LA'AU-'AILA (HAWAII) see CCP000
LABA see LGK100
2329 LABAZ see EID200
LABAZ see AJK750
LABAZENE see PNX750
LABDANOL see IIQ000
LABDANUM OIL see LAC000

LABETALOL HYDROCHLORIDE see HMM500
LABICAN see MDQ250
LABILITE see MAP300
LABITON see DHF600
LABOPAL see DKQ000
LABOR-NR 2683 see DHR800
LABROCOL see HMM500
LABRODA see TKP100
LABRODAX see TKP100
LABRODAX SUPANATE see TKP100
LABURNUM see GIW195
LABURNUM ANAGYROIDES see GIW195
LABYRIN see CMR100
LAC-43 see BOO000
LACHESIN see ELF500
LACHESINE CHLORIDE see ELF500
LAC LSP-1 see LIJ000
LACO see PPN100
LACOLIN see LAM000
LACQREN 506 see SMQ500
LACQREN 550 see SMQ500
LACQTEN 1020 see PJS750
LACQUER DILUENT see ROU000
LACQUER ORANGE V see TGW000
LACQUER ORANGE V 3G see CMP600
LACQUER ORANGE VG see PEJ500
LACQUER ORANGE VR see XRA000
LACQUER RED V see SBC500
LACQUER RED V3B see EOJ500
LACQUER RED 2G see CMS242
LACQUER RED VS see SBC500
LACQUERS see LAD000
LACQUERS, NITROCELLULOSE see LAE000
LACQUER YELLOW T see CMM758
LACRETIN see FOS000
LACRIMIN see OPI300
LACTASE see GAV100
LACTATE d'ETHYLE (FRENCH) see LAJ000
dl-LACTIC ACID see LAG000
(S)-(+)-LACTIC ACID see LAG010
(S)-LACTIC ACID see LAG010
d-LACTIC ACID see LAG010
l-(+)-LACTIC ACID see LAG010
LACTIC ACID see LAG000
(+)-LACTIC ACID see LAG010
LACTIC ACID, ANTIMONY SALT see AQE250
LACTIC ACID, BERYLLIUM SALT see LAH000
LACTIC ACID, BUTYL ESTER see BRR600
LACTIC ACID, BUTYL ESTER, BUTYRATE see BQP000
LACTIC ACID, CADMIUM SALT see CAG750
LACTIC ACID, ETHYL ESTER see LAJ000
LACTIC ACID, IRON(2+) SALT (2:1) see LAL000
LACTIC ACID, LEAD(2+) SALT (2:1) see LDL000
LACTIC ACID LITHIUM SALT see LHL000
LACTIC ACID, MAGNESIUM SALT see LAL100
LACTIC ACID, METHYL ESTER see MLC600
LACTIC ACID, 2-METHYL-, ETHYL ESTER see ELH700
LACTIC ACID, MONOSODIUM SALT see LAM000
LACTIC ACID SODIUM SALT see LAM000
LACTIC ACID, TRIS(2-HYDROXYETHYL)(PHENYLMERCURI)AMMONIUM derivative see TNI500
LACTIC ACID, ion(1−), TRIS((2-HYDROXYETHYL)PHENYLMERCURIO)AMMONIUM see TNI500
LACTIC ACID, ZIRCONIUM SALT (3:1) see ZRJ000
LACTIC ACID, ZIRCONIUM SALT (4:1) see ZRS000
LACTIN see LAR000
LACTOBACILLUS LACTIS DORNER FACTOR see VSZ000
LACTOBARYT see BAP000
LACTOBIOSE see LAR000
LACTOCAINE see AIT250
LACTOFLAVIN see RIK000
LACTOFLAVINE see RIK000
ε-LACTONE HEXANOIC ACID see LAP000
LACTONITRILE see LAQ000
LACTOSCATONE see LAQ100
d-LACTOSE see LAR000
LACTOSE see LAR000
LACUMIN see MOQ250
LADAKAMYCIN see ARY000
LADIE'S THIMBLES see FOM100
LADOGAL see DAB830
LADY LAUREL see LAR500
LAE-32 see LJI000
LAETRILE see LAS000
LAEVORAL see LFI000

LAEVOSAN see LFI000
LAEVOXIN see LFG050
LAEVULIC ACID see LFH000
LAEVULINIC ACID see LFH000
LAGOSIN see FPC000
LAGRIMAS de MARIA see CAL125
LAIDLOMYCIN see DAK000
LAI (HAITI) see WBS850
LAKANA, MIKINOLIA-HIHIU (HAWAII) see LAU600
LAKE BLUE AFX see ERG100
LAKE BLUE B BASE see DCJ200
LAKE FAST BLUE BS see IBV050
LAKE FAST BLUE GGS see IBV050
LAKE ORANGE A see CMM220
LAKE ORANGE II YS see CMM220
LAKE PONCEAU see FMU070
LAKE RED C 18958 see CMS150
LAKE RED C see CHP500, CMS150
LAKE RED CY see CMS150
LAKE RED KB BASE see CLK225
LAKE RED 4R see MMP100
LAKE RED 4RII see MMP100
LAKE SCARLET G BASE see NMP500
LAKE SCARLET GG BASE see DEO295
LAKE YELLOW see FAG140
LAKE YELLOW GA see CMS212
LALOI (HAITI) see AGV875
LAM see PPT500
LAMAR see FLZ050
LAMBAST see CFW750
LAMBDAMYCIN see CDK250
LAMBETH see PMP500
LAMB KILL see MRU359
LAMBRATEN see AJT250
LAMBRIL see BBW750
LAMBROL see FDB200, FIP999
LAMDIOL see EDO000
LAMIDON see IIU000
LAMITEX see SEH000
LAMIUM ALBUM LINN., EXTRACT see LAS500
LAMORYL see GKE000
LAMPIT see NGG000
LAMPTEROL see LIO600
LAMURAN see AFG750
LANACORT see HHQ800
LANADIGENIN see DKN300
LANADIN see TIO750
LANALENE L see IPS500
LANALENE P see IPS500
LANALENE S see IPS500
LANAPERL BLUE B see CMM070
LANAPERL FAST RED 3G see CMM320
LANAPERL RED G see CMM320
LANAPERL YELLOW BROWN GT see SGP500
LANASYN GREEN BL see CMM200
LANATOSID A (GERMAN) see LAT000
LANATOSID B (GERMAN) see LAT500
LANATOSID C (GERMAN) see LAU000
LANATOSIDE A see LAT000
LANATOSIDE B see LAT500
LANATOSIDE C see LAU000
LANATOSIDES see LAU400
LANATOXIN see DKL800
LANAZINE see DBA800
LANCOL see OBA000
LANDALGINE see AFL000
LANDAMYCINE see RIP000
LANDISAN see MEO750
LANDOCAINE see BQA010
LAND PLASTER see CAX750
LANDRIN see TMC750, TMD000
LANDRIN, NITROSO DERIVATIVE see NLY500
LANDRUMA see NDX500
LANESTA see CDV700
LANESTA L see IPS500
LANESTA P.S. see IPS500
LANETTE WAX-S see SIB600
LANEX see DUK800
LANGFORD see KBB600
LANGORAN see CCK125
LANI-ALI'I (HAWAII) see AFQ625
LANIAZID see ILD000
LANICOR see DKN400
LANIRAPID see MJD300
LANITOP see MJD300

LANNAGOL LF see PJY100
LANNATE see MDU600
LANODOXIN see DNF600
LANOL see CMD750
LANOPHYLLIN see TEP000
LANOSTABIL see LAU400
LANOXIN see DKN400
LANTANA see LAU600
LANTANA CAMARA see LAU600
LANTHANA see LBA100
LANTHANACETAT (GERMAN) see LAW000
LANTHANIA (La2O3) see LBA100
LANTHANUM see LAV000
LANTHANUM ACETATE see LAW000
LANTHANUM AMMONIUM NITRATE see ANL500
LANTHANUM CHLORIDE see LAX000
LANTHANUM DIHYDRIDE see LAY499
LANTHANUM EDETATE see LAZ000
LANTHANUM HYDRIDE see LBC000
LANTHANUM NITRATE see LBA000
LANTHANUM(3+) OXIDE see LBA100
LANTHANUM(III) OXIDE see LBA100
LANTHANUM OXIDE see LBA100
LANTHANUM SESQUIOXIDE see LBA100
LANTHANUM(III) SULFATE (2:3) see LBB000
LANTHANUM SULFATE see LBB000
LANTHANUM TRIACETATE see LAW000
LANTHANUM TRIHYDRIDE see LBC000
LANTHANUM TRIOXIDE see LBA100
LANTOSIDE see LAU400
LANVIS see AMH250
LAPACHIC ACID see HLY500
LAPACHOL see HLY500
LAPACHOL WOOD see HLY500
LAPAQUIN see CLD000
LAPAV see PAH250
LAPEMIS HARDWICKII VENOM see SBI910
LAPLEN see CGW300
(+)-LAPPACONITINE see LBD000
LAPPACONITINE see LBD000
LAPYRIUM CHLORIDE see LBD100
LARAHA see LBE000
LARD FACTOR see VSK600
LAREX see UTU500
LARGACTIL see CKP250
LARGACTIL MONOHYDROCHLORIDE see CKP500
LARGACTILOTHIAZINE see CKP250
LARGACTYL see CKP250
LARGAKTYL see CKP500
LARIXIC ACID see MAO350
LARIXIN see ALV000
LARIXINIC ACID see MAO350
LAROCAINE see DNY000
LARODON see INY000
LARODOPA see DNA200
LAROXIL see EAH500
LAROXYL see EAH500
LARTEN see MQU750
LARVACIDE see CKN500
LARVATROL see BAC040
LAS see AFO500, CMV325
LASALOCID see LBF500
LASERDIL see CCK125
LASEX see CHJ750
LASIOCARPINE see LBG000
LASIX see CHJ750
LAS-Mg see LGF800
LAS, MAGNESIUM SALT see LGF800
LAS-Na see LGF825
LASODEX see TEP500
LASSO see CFX000
LAS, SODIUM SALT see LGF825
LATAMOXEF SODIUM see LBH200
LATEX see PJR000
LATEXOL FAST BLUE SD see IBV050
LATEXOL FAST ORANGE J see CMS145
LATEXOL FAST YELLOW JR see CMS208
LATEXOL SCARLET R see CHP500
LATEX SVKh see CGW300
LATHANOL see SIB700
LATHANOL-LAL 70 see SIB700
LATHANOL LAL see SIB700
LATHYRUS ODORATUS, SEEDS see SOZ000
LATIBON see DII200
LATICATOXIN see LBI000

LATICAUDA SEMIFASCIATA VENOM see LBI000
LATKA 666 see BBP750
LATKA 7744 see CBM750
LATRODECTUS M. MACTANS VENOM see BLW500
LATSCHENKIEFEROEL see PIH400
LATUSATE see AFY500
LATYL BLUE BCN see CMP070
LATYL CERISE N see AKI750
LAUDICON see DLW600
LAUDOCAINE see BQA010
LAUDRAN DI-n-BUTYLCINICITY (CZECH) see DDV600
LAUGHING GAS see NGU000
LAURAMIDE DEA see BKE500
LAUREL CAMPHOR see CBA750
LAUREL LEAF OIL see BAT500, LBK000
LAURELWOOD see MBU780
LAURETH see DXY000
LAURIC ACID see LBL000
LAURIC ACID, BARIUM CADMIUM SALT see BAI770
LAURIC ACID, CADMIUM SALT (2:1) see CAG775
LAURIC ACID, DIBUTYLSTANNYLENE derivative see DDV600
LAURIC ACID, DIBUTYLSTANNYLENE SALT see DDV600
LAURIC ACID DIETHANOLAMIDE see BKE500
LAURIC ACID-2,3-EPOXYPROPYL ESTER see LBM000
LAURIC ACID ESTER with 2-HYDROXYETHYL THIOCYANATE see LBO000
LAURIC ACID, METHYL ESTER see MLC800
LAURIC ACID, SODIUM SALT see LBN000
LAURIC ACID, 2-THIOCYANATOETHYL ESTER see LBO000
LAURIC ALCOHOL see DXV600
LAURIC DIETHANOLAMIDE see BKE500
LAURIER ROSE (HAITI) see OHM875
LAURINAMINE see DXW000
LAURINE see CMS850
LAURINE DIMETHYL ACETAL see LBO050
LAURINIC ALCOHOL see DXV600
LAURODIN see AJS750
LAUROLINIUM ACETATE see AJS750
LAUROLITSINE see LBO100
LAUROMACROGOL 400 see DXY000
LAURONITRILE see DXT400
LAUROSCHOLTZINE see TKX700
LAUROSTEARIC ACID see LBL000
LAUROTETANIN see LBO200
(+)-LAUROTETANINE see LBO200
LAUROTETANINE see LBO200
LAUROX see LBR000
1-LAUROYLAZIRIDINE see LBQ000
N-(LAUROYLCOLAMENOFORMYLMETHYL)PYRIDINIUM CHLORIDE see LBD100
LAUROYL DIETHANOLAMIDE see BKE500
LAUROYLETHYLENEIMINE see LBQ000
(LAUROYLOXY)TRIBUTYLSTANNANE see TIE750
LAUROYL PEROXIDE see LBR000
LAUROYL PEROXIDE, TECHNICALLY PURE (DOT) see LBR000
LAURUS NOBILIS OIL see OGQ150
LAURYDOL see LBR000
LAURYL 24 see DXV600
LAURYL ACETATE see DXV400
LAURYL ALCOHOL (FCC) see DXV600
LAURYL ALCOHOL CONDENSED with 23 MOLES ETHYLENE OXIDE see LBU000
LAURYL ALCOHOL CONDENSED with 4 MOLES ETHYLENE OXIDE see LBS000
LAURYL ALCOHOL EO (23) see LBU000
LAURYL ALCOHOL EO (4) see LBS000
LAURYL ALCOHOL EO (7) see LBT000
LAURYL ALCOHOL, ETHOXYLATED see DXY000
n-LAURYL ALCOHOL, PRIMARY see DXV600
LAURYL ALDEHYDE (FCC) see DXT000
n-LAURYLAMINE see DXW000
LAURYLAMINE see DXW000
LAURYLAMINE HYDROCHLORIDE see DKW100
LAURYLAMMONIUM HYDROCHLORIDE see DKW100
LAURYL AMMONIUM SULFATE see SOM500
LAURYLBENZENESULFONIC ACID see LBU100
LAURYLBETAIN see LBU200
LAURYL-N-BETAINE see LBU200
LAURYLBETAINE see LBU200
LAURYL DIETHANOLAMIDE see BKE500
LAURYLDIETHYLENETRIAMINE see LBV000
N-LAURYLDIMETHYLAMINE see DRR800
LAURYLDIMETHYLAMINE see DRR800
LAURYLDIMETHYLAMINE OXIDE see DRS200
LAURYLDIMETHYLAMMONIOACETATE see LBU200
LAURYLDIMETHYLBETAINE see LBU200

LAURYLESTER KYSELINY DUSICNE (CZECH) see NEE000
LAURYLESTER KYSELINYMETHAKRYLOVE (CZECH) see DXY200
LAURYL GALLATE see DXX200
LAURYLGUANIDINE ACETATE see DXX400
LAURYLISOQUINOLINIUM BROMIDE see LBW000
m-LAURYL MERCAPTAN see LBX000
LAURYL MERCAPTAN see LBX000
LAURYL METHACRYLATE see DXY200
LAURYL-N-METHYLSARCOSINE see LBU200
LAURYLNITRAT (CZECH) see NEE000
LAURYL NITRATE see NEE000
LAURYL POLYETHYLENE GLYCOL ETHER see DXY000
1-LAURYLPYRIDINIUM CHLORIDE see DXY725, LBX050
LAURYLPYRIDINIUM CHLORIDE see DXY725, LBX050
LAURYL RHODANATE see DYA200
LAURYL SODIUM SULFATE see SIB600
LAURYL SULFATE see MRH250
LAURYL SULFATE AMMONIUM SALT see SOM500
LAURYL SULFATE, SODIUM SALT see SIB600
LAURYL SULFURIC ACID see MRH250
LAURYL THIOCYANATE see DYA200
LAURYL 3,3'-THIODIPROPIONATE see TFD500
LAUSIT see IDA000
LAUTHSCHES VIOLETT (GERMAN) see AKK750
LAUXTOL see PAX250
LAUXTOL A see PAX250
LAV see CMY800
LAVAMENTHE see LBX100
LAVANDIN ABSOLUTE see LCA000
LAVANDIN BENZOL ABSOLUTE see LCA000
LAVANDIN OIL see LCA000
LAVANDULYL ACETATE see LCA100
LAVATAR see CMY800
LAVENDEL OEL (GERMAN) see LCD000
LAVENDEL OEL see LCC000
LAVENDER ABSOLUTE see LCC000
LAVENDER OIL see LCD000
LAVENDER OIL, SPIKE see SLB500
LAVOFLAGIN see ABY900
LAVSAN see PKF750
LAWN-KEEP see DAA800
LAWSONE see HMX600, WAT000
LAWSONITE see PKF750
LAXADIN see PPN100
LAXAGEN see ACD500
LAXAGETTEN see ACD500, PPN100
LAXANIN N see PPN100
LAXANORM see DMH400
LAXANS see PPN100
LAXANTHREEN see DMH400
LAXESIN see ELF500
LAXIDOGOL see SJJ175
LAXINATE see DJL000
LAXIPUR see DMH400
LAXIPURIN see DMH400
LAXOBERAL see SJJ175
LAXOBERON see SJJ175
LAXOGEN see PDO750
LAXOREX see PPN100
LA XVII see BMN750
LAYOR CARANG see AEX250
LAZETA see CMR100
LAZO see CFX000
LB-46 see VSA000
LB 502 see CHJ750
(L)-BC-2605 see CQF079
LBF DISULFIDE see PAG150
LBI see LEF200
LB-ROT 1 see FAG040
LC 44 see FMO129
LC-80 see CCK660
'L,' CARPSERP see RDK000
L-N-(3-CHLORO-1-METHYLACETONYL)-p-TOLUENESULFONAMIDE see THH360
LCR see LEY000
LD 400 see PJS750
LD 600 see PJS750
LD-813 see LCE000
L.D. 3055 see IBP200
LDA see BKE500
LDE see BKE500
LDPE 4 see PJS750
LD RUBBER RED 16913 see CHP500
29060 LE see VLA000
LEA-COV see SHF500

LEAD see LCF000
LEAD(2+) ACETATE see LCV000
LEAD(II) ACETATE see LCV000
LEAD ACETATE see LCG000, LCV000
LEAD(IV) ACETATE AZIDE see LCG500
LEAD ACETATE, BASIC see LCH000
LEAD ACETATE BROMATE see LCI000
LEAD ACETATE-LEAD BROMITE see LCI600
LEAD ACETATE(II), TRIHYDRATE see LCJ000
LEAD ACETATE TRIHYDRATE see LCJ000
LEAD ACID ARSENATE see LCK000
LEAD ARSENATE, solid (DOT) see LCK000
LEAD ARSENATE see ARC750, LCK000, LCK100
LEAD ARSENATE (standard) see LCK000
LEAD(II) ARSENITE see LCL000
LEAD ARSENITES (DOT) see LCL000
LEAD AZIDE (dry) (DOT) see LCM000
LEAD AZIDE, wetted with not <20% water or mixture of alcohol and water,
  by weight (DOT) see LCM000
LEAD(II) AZIDE see LCM000
LEAD(IV) AZIDE see LCN000
LEAD BOTTOMS see LDY000
LEAD BROMATE see LCO000
LEAD BROWN see LCX000
LEAD(2+) CARBONATE see LCP000
LEAD CARBONATE see LCP000
LEAD(2+) CHLORIDE see LCQ000
LEAD(II) CHLORIDE see LCQ000
LEAD CHLORIDE see LCQ000
LEAD(II) CHLORITE see LCQ300
LEAD CHROMATE(VI) see LCR000
LEAD CHROMATE see LCR000
LEAD CHROMATE, BASIC see LCS000
LEAD CHROMATE MOLYBDATE SULFATE RED see MRC000
LEAD CHROMATE OXIDE (MAK) see LCS000
LEAD CHROMATE, RED see LCS000
LEAD CHROMATE, SULPHATE and MOLYBDATE see LDM000
LEAD COMPOUNDS see LCT000
LEAD CYANIDE (DOT) see LCU000
LEAD(II) CYANIDE see LCU000
LEAD DIACETATE see LCV000
LEAD DIACETATE TRIHYDRATE see LCJ000
LEAD DIBASIC ACETATE see LCV000
LEAD DIBASIC PHOSPHITE see LCV100
LEAD DICHLORIDE see LCQ000
LEAD DIFLUORIDE see LDF000
LEAD DIMETHYLDITHIOCARBAMATE see LCW000
LEAD DINITRATE see LDO000
LEAD DIOXIDE see LCX000
LEAD DIPERCHLORATE see LDS499
LEAD DIPHENYL ACID PROPIONATE see LCZ000
LEAD DIPHENYL NITRATE see LDA000
LEAD DIPICRATE see LDA500
LEAD DISODIUM EDTA see LDB000
LEAD DISODIUM ETHYLENEDINITRILOTETRACETATE see LDB000
LEAD DROSS (DOT) see LDY000
LEAD DROSS see LDC000
LEAD(II) EDTA COMPLEX see LDD000
LEAD(2-), ((ETHYLENEDINITRILO)TETRAACETATO)-, DISODIUM see LDB000
LEAD FLAKE see LCF000
LEAD FLUOBORATE see LDE000
LEAD FLUORIDE (DOT) see LDF000
LEAD(II) FLUORIDE see LDF000
LEAD(II) FLUOROSILICATE see LDG000
LEAD GLYCERONITRATE see LDH000
LEAD HYPONITRITE see LDI000
LEAD HYPOPHOSPHITE see LDJ000
LEAD IMIDE see LDK000
LEAD LACTATE see LDL000
LEAD-MOLYBDENUM CHROMATE see LDM000
LEAD MONONITRORESORCINATE (DRY) (DOT) see LDP000
LEAD MONOSUBACETATE see LCH000
LEAD MONOXIDE see LDN000
LEAD NAPHTHENATE see NAS500
LEAD(II) NITRATE (1:2) see LDO000
LEAD(2+) NITRATE see LDO000
LEAD(II) NITRATE see LDO000
LEAD NITRATE see LDO000
LEAD NITRORESORCINATE see LDP000
LEAD(II) OLEATE (1:2) see LDQ000
LEAD ORTHOPHOSPHATE see LDU000
LEAD ORTHOPLUMBATE see LDS000
LEAD(II) OXIDE see LDN000
LEAD(IV) OXIDE see LCX000
LEAD OXIDE see LDN000

LEAD OXIDE BROWN see LCX000
LEAD OXIDE PHOSPHONATE, HEMIHYDRATE see LCV100
LEAD OXIDE RED see LDS000
LEAD OXIDE YELLOW see LDN000
LEAD(2+) PERCHLORATE see LDS499
LEAD PERCHLORATE, solid or solution (DOT) see LDS499
LEAD(II) PERCHLORATE see LDS499
LEAD PERCHLORATE see LDS499
LEAD(II) PERCHLORATE, HEXAHYDRATE (1:2:6) see LDT000
LEAD PEROXIDE (DOT) see LCX000
LEAD(2+) PHOSPHATE see LDU000
LEAD(II) PHOSPHATE (3:2) see LDU000
LEAD PHOSPHATE (3:2) see LDU000
LEAD PHOSPHATE see LDU000
LEAD(II) PHOSPHINATE see LDJ000
LEAD PHOSPHITE, dibasic (DOT) see LCV100
LEAD PICRATE (dry) (DOT) see PID100
LEAD PICRATE see PID100
LEAD POTASSIUM THIOCYANATE see LDV000
LEAD PROTOXIDE see LDN000
LEAD S2 see LCF000
LEAD SCRAP see LDC000
LEAD SILICATE see LDW000
LEAD STEARATE see LDX000
LEAD STYPHNATE (dry) (DOT) see LEE000
LEAD STYPHNATE, wetted or lead trinitroresorcinate, wetted with not <20%
  water or mixt. (DOT) see LEE000
LEAD STYPHNATE see LEE000
LEAD SUBACETATE see LCH000
LEAD(II) SULFATE (1:1) see LDY000
LEAD SULFATE, solid, containing more than 3% free acid (DOT) see
  LDY000
LEAD SULFIDE see LDZ000
LEAD SUPEROXIDE see LCX000
LEAD(II) TARTRATE (1:1) see LEA000
LEAD TETRACETATE see LEB000
LEAD TETRACHLORIDE see LEC000
LEAD, TETRAETHYL- see TCF000
LEAD TETRAOXIDE see LDS000
LEAD(II) THIOCYANATE see LEC500
LEAD TITANATE see LED000
LEAD TITANATE ZIRCONATE see LED100
LEAD TITANIUM ZIRCONIUM OXIDE see LED100
LEAD TREE see LED500
LEAD TRINITRORESORCINATE see LEE000
LEAD(II) TRINITROSOBENZENE-1,3,5-TRIOXIDE see LEF000
LEAD(II) TRINITROSOPHLOROGLUCINOLATE see LEF000
LEAD TRIPROPYL see TNA500
LEAD ZIRCONATE TITANATE see LED100
LEAD ZIRCONIUM TITANATE see LED100
LEAD ZIRCONIUM TITANIUM OXIDE see LED100
LEAF ALCOHOL see HFE000
LEAF ALDEHYDE see HFA500
LEAF DROP see SFS500
LEAF GREEN see CMJ900
LEALGIN COMPOSITUM see CLY500
LEANDIN see COH250
LEATHER BLUE G see ADE500
LEATHER BUSH see LEF100
LEATHER FAST RED B see CMM330
LEATHER FLOWER see CMV390
LEATHER GREEN B see FAE950
LEATHER GREEN SF see FAF000
LEATHER ORANGE EXTRA see CMM220
LEATHER ORANGE HR see PEK000
LEATHER PURE BLUE HB see BJI250
LEATHER RED G see CMM300
LEATHER RED HT see GJI400
LEATHERWOOD see LEF100
LEAVER WOOD see LEF100
LEBAYCID see FAQ900
LEBON 15 HYDROCHLORIDE see DYA850
LE CAPTANE (FRENCH) see CBG000
LECASOL see FOS100
LECITHIN-BOUND IODINE see LEF200
LECITHIN IODIDE see LEF200
LECTOPAM see BMN750
LECTRAPEL see DTS500
LEDAKRIN see LEF300, NFW500
LEDCLAIR see CAR780
LEDERCILLIN VK see PDT750
LEDERFEN see BGL250
LEDERKYN see AKO500
LEDERLE AA223 see AFI625
LEDERMYCIN see MIJ500

LEDERMYCIN HYDROCHLORIDE see DAI485
LE DINITROCRESOL-4,6 (FRENCH) see DUS700
LEDON 11 see TIP500
LEDON 12 see DFA600
LEDON 114 see FOO509
LEDOSTEN see DJT400
LEFEBAR see EOK000
LEGENTIAL see GCU050
LEGUMEX see CLN750
LEGUMEX D see DGA000
LEGUMEX DB see CIR250
LEGUMEX EXTRA see BAV000
LEGURAME see CBL500
LEHYDAN see DKQ000
LEINOLEIC ACID see LGG000
LEIOPLEGIL see LEF400
LEIOPYRROLE see LEF400
LEIPZIG YELLOW see LCR000
LEIVASOM see TIQ250
LEKAMIN see TNF500
LEKUTHERM 2159 see DKM500
LEKUTHERM X 100 see DKM500
LEMAC 1000 see AAX250
LEMBROL see DCK759
LEMISERP see RDK000
LEMOFLUR see SHF500
LEMOL 5-88 see PKP750
LEMOL 5-98 see PKP750
LEMOL 12-88 see PKP750
LEMOL 16-98 see PKP750
LEMOL 24-98 see PKP750
LEMOL 30-98 see PKP750
LEMOL 51-98 see PKP750
LEMOL 60-98 see PKP750
LEMOL 75-98 see PKP750
LEMOL see PKP750
LEMOL GF-60 see PKP750
LEMON AJAX see AFG625
LEMON CHROME see BAK250
LEMONENE see BGE000
LEMONGRAS OEL (GERMAN) see LEG000
LEMONGRASS OIL EAST INDIAN see LEG000
LEMONGRASS OIL WEST INDIAN see LEH000
LEMONILE see LEH100
LEMON OIL, desert type, coldpressed see LEI025
LEMON OIL, distilled see LEI030
LEMON OIL see LEI000
LEMON OIL, COLDPRESSED (FCC) see LEI000
LEMON OIL, EXPRESSED see LEI000
LEMONOL see DTD000
LEMON PETITGRAIN OIL see LEJ000
LEMON YELLOW see BAK250, LCR000
LEMON YELLOW A see FAG140
LEMON YELLOW A GEIGY see FAG140
LEMON YELLOW ZN 3 see CMM510
LEMORAN see DYF000
LENAMPICILLIN HYDROCHLORIDE see LEJ500
LENAMYCIN see ACJ250
LENDINE see BBQ500
LENDORM see LEJ600
LENDORMIN see LEJ600
LENETRAN see MFD500
LENETRANAT see MFD500
LENETRAN TAB see MFD500
LENGUNA de VACA (CUBA, PUERTO RICO) see APM875
LENITRAL see NGY000
LENOCYCLINE see HOH500
LENOREMYCIN see LEJ700
LENOTAN see BAV350
LENTAC see AKO500
LENTE see LEK000
LENTE INSULIN see LEK000
LENTIN see CBH250
LENTINAN see LEK100
LENTINE (FRENCH) see CBH250
LENTISQUE ABSOLUTE see MBU777
LENTIZOL see EAI000
LENTOCILLIN see BFC750
LENTOPENIL see BFC750
LENTOTRAN see MDQ250
LENTOX see BBQ500
LEO 72a see BHO250
LEO 1727 see NNX600
LEO 640 HYDROCHLORIDE see IFZ900
LEOMYPEN see BFC750

LEOPENTAL see PBT500
LEOSTESIN see DHK400
LEOSTESIN HYDROCHLORIDE see DHK600
LEOSTIGMINE BROMIDE see POD000
LEOSTIGMINE METHYL SULFATE see DQY909
LEPARGYLIC ACID see ASB750
LEPASEN see SEP000
LEPENIL see MQU750
LEPETOWN see MQU750
LEPHEBAR see EOK000
LEPIMIDIN see DBB200
LEPINAL see EOK000
LEPITOIN see DKQ000, DNU000
LEPITOIN SODIUM see DNU000
LEPONEX see CMY650
LEPOTEX see CMY650
LEPSIN see DKQ000
LEPSIRAL see DBB200
LEPTAMIN see DJS200
LEPTANAL see DYF200, PDW750
LEPTODACTYLINE PICRATE see HNG000
LEPTOFEN see DYF200
LEPTON see DJT400
LEPTOPHOS see LEN000
LEPTOPHOS PHENOL see LEN050
LEPTRYL see LEO000
LERBEK see CMX850
LERCIGAN see DQA400
LERENOX see BDD000
LERGIGAN see DQA400, PMI750
LERGINE see CPQ250
LERGINE CHLORIDE see EAI875
LERGITIN see BEM500
LERGOTRILE MESYLATE see LEP000
LERTUS see BDU500
LESAN see DOU600
LESCOPINE BROMIDE see SBH500
LESSER CELANDINE see FBS100
LESSER HEMLOCK see FMU200
LESTEMP see HIM000
LETHALAIRE G-52 see TCF250
LETHALAIRE G-54 see PAK000
LETHALAIRE G-57 see SOD100
LETHALAIRE G-58 see CJT750
LETHALAIRE G-59 see OCM000
LETHANE 60 see LBO000
LETHANE 384 see BPL250
LETHANE see BPL250
LETHANE (special) see LEQ000
LETHANE 384 REGULAR see BPL250
LETHELMIN see PDP250
LETHIDROME see AFT500
LETHOX see TNP250
LETHURIN see TIO750
LETIDRONE see AFT500
LETTER see LFG050
LETUSIN see PNA250
LETYL see MQU750
LEUCAENA LEUCOCEPHAIA see LED500
LEUCARSONE see CBJ000
LEUCETHANE see UVA000
LEUCIDIL see BCA000
LEUCIN (GERMAN) see LES000
ε-LEUCINE see AJD000
dl-LEUCINE see LER000
l-LEUCINE see LES000
LEUCINE see LES000
LEUCINE, N-CARBOXY-, N-BENZYL 1-VINYL ESTER see CBR215
l-LEUCINE, N-((PHENYLMETHOXY)CARBONYL)-, ETHENYL ESTER see
    CBR215
LEUCINOCAINE see LET000
LEUCINOCAINE MESYLATE see LEU000
LEUCINOCAINE METHANESULFONATE see LEU000
LEUCO-4 see AEH000
LEUCO-1,4-DIAMINOANTHRAQUINONE see DBO600
LEUCOGEN see ARN800
LEUCOHARMINE see HAI500
LEUCOINDOPHENOL see IBJ100
LEUCOL see QMJ000
LEUCOLINE see IRX000, QMJ000
LEUCOMYCIN see SLC000
LEUCOMYCIN A3 see JDS200
LEUCOMYCIN A6 see LEV025
LEUCOMYCIN B see LEW000
LEUCOMYCIN TARTRATE see LEX000

LEUCOPARAFUCHSIN see THP000
LEUCOPARAFUCHSINE see THP000
LEUCOPIN see CMV000
LEUCOQUINIZARIN see LEX200
LEUCOSOL GOLDEN YELLOW see DCZ000
LEUCOSULFAN see BOT250
1,4,5,8-LEUCOTETRAOXYANTHRAQUINONE see TDD250
LEUCOTHANE see UVA000
LEUCOTHOE (VARIOUS SPECIES) see DYA875
LEUCOVYL PA 1302 see AAX175
LEUKAEMOMYCIN C see DAC000
LEUKAEMOMYCIN D see DAC300
LEUKANOL NF see BLX000
LEUKERAN see CDO500, POK000
LEUKERSAN see CDO500
LEUKICHTHOL see IAD000
LEUKO-1,4-DIAMINOANTHRACHINON (CZECH) see DBO600
LEUKOL see QMJ000
LEUKOMYAN see CDP250
LEUKOMYCIN A6 see LEV025
LEUKORAN see CDO500
LEUNA M see CIR250
LEUPEPTIN Ac-LL see LEX400
LEUPURIN see POK000
LEUROCRISTINE see LEY000
LEUROCRISTINE SULFATE (1:1) see LEZ000
LEVADONE see MDP250
(— )-LEVALLORPHAN see AGI000
l-LEVALLORPHAN TARTRATE see LIH300
LEVALLORPHAN TARTRATE see LIH300
LEVALLORPHINE TARTRATE see LIH300
LEVAMISOLE see LFA000, LFA020
LEVAMISOLE HYDROCHLORIDE see LFA020
LEVANIL see EHP000
LEVANOL RED GG see CMM330
LEVANOX GREEN GA see CMJ900
LEVANOX RED 130A see IHD000
LEVANOX WHITE RKB see TGG760
LEVANXENE see CFY750
LEVANXOL see CFY750
LEVARGIN see AQW000
LEVARTERENOL see NNO500
LEVARTERENOL BITARTRATE see NNO699
LEVATROM see ARQ750
LEVAXIN see LFG050
LEVEDRINE see BBK750
LEV HYDROCHLORIDE see LFA020
LEVIGATED CHALK see CAT775
LEVIL see EHP000
LEVIUM see DCK759
LEVOARTERENOL see NNO500
LEVOGLUTAMID see GFO050
LEVOGLUTAMIDE see GFO050
LEVOMEPATE HYDROCHLORIDE see LFC000
LEVOMEPROMAZINE see MCI500
LEVOMETHADONE see MDO775
LEVOMETHORPHAN HYDROBROMIDE see LFD200
LEVOMYCETIN see CDP250
LEVOMYCETIN HEMISUCCINATE see CDP725
LEVOMYCETIN SUCCINATE see CDP725
LEVOMYCIN see EAD500
LEVOMYSOL HYDROCHLORIDE see LFA020
LEVONORADRENALINE see NNO500
LEVONOREPINEPHRINE see NNO500
LEVONORGESTREL see NNQ525
LEVOPHACETOPERANE HYDROCHLORIDE see LFO000
LEVOPHACETOPERAN HYDROCHLORIDE see LFO000
LEVOPHED see NNO500
LEVOPHED HYDROCHLORIDE see NNP000
LEVOPROMAZINE see MCI500
LEVORENIN see VGP000
LEVORIN see LFF000
LEVOROXINE see LFG050
LEVORPHAN see LFG000
LEVORPHANOL see LFG000
LEVORPHANOL TARTRATE see DYF000
LEVORPHAN TARTRATE see DYF000
LEVOSAN see LFG100
LEVOTHROID see LFG050
LEVOTHYL see MDO775, MDP250
LEVOTHYROX see LFG050
LEVOTHYROXINE see TFZ275
LEVOTHYROXINE SODIUM see LFG050
LEVOTOMIN see MCI500
LEVOXADROL HYDROCHLORIDE see LFG100

LEVUGEN see LFI000
LEVULIC ACID see LFH000
LEVULINIC ACID see LFH000
LEVULINIC ACID, BUTYL ESTER see BRR700
LEVULINIC ACID, ETHYL ESTER see EFS600
LEVULINIC ACID, TRIPHENYLSTANNYL ESTER see TMW250
LEVULOSE see LFI000
LEWISITE see CLV000
LEWISITE (ARSENIC COMPOUND) see CLV000
LEWISITE II see BIQ250
LEWISITE I OXIDE see DEW000
LEWIS-RED DEVIL LYE see SHS000
LEXATOL see SPC500
LEXIBIOTICO see ALV000
LEXOMIL see BMN750
LEXONE see MQR275
LEXOTAN see BMN750
LEXOTANIL see BMN750
LEY-CORNOX see BAV000
LEYMIN see BAV000
LEYSPRAY see CIR250
LEYTOSAN see ABU500, PFP500
4LF see KDK000
LF 62 see POA250
LFA 2043 see GIA000
LFP 83 see FJT100
LG 61 see DJR700
LG 50043 see TKX250
L.G. 11,457 HYDROCHLORIDE see DHS200
LGYCOSPERSE S-20 see PKL030
LH 3012 see ZMA000
LH see ILE000
L'HEMISULFATE de GUANIDINO METHYL-6-BENZODIOXANNE-1,4 (FRENCH)
    see GKG300
LH RELEASING FACTOR see LIU360
LH-RELEASING HORMONE see LIU360
LH-RF see LIU360
LH-RH see LIU360
LHRH see LIU360
LH-RH see LIU380
LHRH DIACETATE see LIU380
LHRH DIACETATE TETRAHYDRATE see LIU400
LH-RH/FSH-RH see LIU360
LH 30/Z see ZMA000
LIANE BON GARCON (HAITI) see CMV390
LIATRIS see DAJ800
LIATRIX OLEORESIN see DAJ800
LIBAVIUS FUMING SPIRIT see TGC250
LIBERETAS see DCK759
LIBEXIN see LFJ000
LIBIOLAN see MQU750
LIBRATAR see CDQ500
LIBRAX see LFK000
LIBRININ see LFK000
LIBRITABS see LFK000
LIBRIUM see LFK000, MDQ250
LIBRIUM HYDROCHLORIDE see MDQ250
LICABILE HYDROCHLORIDE see LFK200
LICARAN HYDROCHLORIDE see LFK200
LICAREOL ACETATE see LFY100
LICHENIC ACID see FOU000
LICHENIFORMIN A see LFL500
LICHENOL see IRL000
LICHTGRUEN (GERMAN) see FAF000
LICIDRIL see DPE000
LICORICE see LFN300
LICORICE COMPONENT FM 100 see LFN000
LICORICE EXTRACT see LFN300
LICORICE ROOT see LFN300
LICORICE ROOT EXTRACT see LFN300
LICORICE VINE see RMK250
LIDA-MANTLE see DHK400
LIDAMYCIN CREME see NCG000
LIDANAR see MON750
LIDANIL see MON750
LIDENAL see BBQ500
LIDEPRAN HYDROCHLORIDE see LFO000
LIDOCAINE see DHK400
LIDOCAINE HYDROCHLORIDE see DHK600
LIDOFENIN see LFO300
LIDOL see DAM600, DAM700
LIDONE see MRB250
LIENOMYCIN see LFP000
LIFEAMPIL see AIV500, AOD125
LIFENE see MNZ000

LIFRIL see FLZ050
LIGAND 222 see LFQ000
LIGHT CAMPHOR OIL see CBB500
LIGHT CATALYTICALLY CRACKED DISTILLATE see DXG840
LIGHT CATALYTICALLY CRACKED NAPHTHA see NAQ540
LIGHT CATALYTIC REFORMED NAPHTHA see NAQ550
LIGHT ESTER PO-A see PER250
LIGHT FAST YELLOW ES see SGP500
LIGHT GAS OIL see GBW025
LIGHT GREEN FCF YELLOWISH see FAF000
LIGHT GREEN LAKE see FAF000
LIGHT GREEN N see AFG500
LIGHTHOUSE CHROME BLACK 6B see CMP880
LIGHTHOUSE CHROME BLUE 2R see HJF500
LIGHT NAPHTHENIC DISTILLATE see MQV810
LIGHT NAPHTHENIC DISTILLATE, SOLVENT EXTRACT see MQV860
LIGHT NAPHTHENIC DISTILLATES (PETROLEUM) see MQV810
LIGHT OIL of CAMPHOR see CBB500
LIGHT ORANGE CHROME see LCS000
LIGHT PARAFFINIC DISTILLATE see MQV815
LIGHT PARAFFINIC DISTILLATE, SOLVENT EXTRACT see MQV862
LIGHT RED see IHD000
LIGHT SF YELLOWISH (BIOLOGICAL STAIN) see FAF000
LIGHT SPAR see CAX750
LIGHT STRAIGHT-RUN NAPHTHA see NAQ560
LIGHT YELLOW JB see DEU000
LIGHT YELLOW JBR see CMS208
LIGHT YELLOW JBV see CMS212
LIGHT YELLOW JBVT see CMS212
LIGNASAN see EME100
LIGNASAN FUNGICIDE see BJT250
LIGNASAN-X see BJT250
LIGNOCAINE see DHK400
LIGNOCAINE HYDROCHLORIDE see DHK600
LIGNYL ACETATE see DTC000
LIGROIN see PCT250
LIGUSTRUM VULGARE see PMD550
LIDOTHESIN HYDROCHLORIDE see DHK600
LIHOCIN see CMF400
LIKUDEN see GKE000
LILACILLIN see SNV000
LILAILA see CDM325
LILAS (HAITI, DOMINICAN REPUBLIC) see CDM325
LILAS de NUIT (HAITI) see DAC500
LILIA-O-KE-AWAWA (HAWAII) see LFT700
LILIAL see LFT000
LILIAL-METHYLANTHRANILATE, Schiff's base see LFT100
LILLY 1516 see DQA400
LILLY 01516 see DQA400
LILLY 22113 see AGL875
LILLY 22451 see MDU500
LILLY 22641 see RDF000
LILLY 34,314 see DRP800
LILLY 36,352 see DUV600
LILLY 37231 see LEZ000
LILLY 38253 see SFQ500
LILLY 39435 see CCR890
LILLY 40602 see TEY000
LILLY-68618 see MRX000
LILLY 69323 see FAP100
LILLY 99094 see VGU750
LILLY 99638 see PAG075
LILLY COMPOUND LY133314 see TMP175
LILLY 99638 HYDRATE see CCR850
LILO see PDO750
LILYAL see LFT000
LILYL ALDEHYDE see CMS850
LILY OF THE FIELD see PAM780
LILY-OF-THE-VALLEY see LFT700
LILY-OF-THE-VALLEY BUSH see FBP520
LIMARSOL MALAGRIDE see ABX500
LIMAS see LGZ000
LIMAWOOD EXTRACT see LFT800
LIMBIAL see CFZ000
LIMBUX see CAT225
LIME see CAU500
LIME ACETATE see CAL750
LIME, BURNED see CAU500
LIME CHLORIDE see HOV500
LIMED ROSIN see CAW500
LIME MILK see CAT225
LIME-NITROGEN (DOT) see CAQ250
LIME OIL, distilled (FCC) see OGO000
LIME OIL see OGO000
LIME PYROLIGNITE see CAL750

LIMESTONE (FCC) see CAO000
LIME, UNSLAKED (DOT) see CAU500
LIME WATER see CAT225
LIMIT see DKE800
1-LIMONENE see MCC500
(−)-LIMONENE (FCC) see MCC500
(+)-R-LIMONENE see LFU000
dl-LIMONENE see MCC250
d-(+)-LIMONENE see LFU000
d-LIMONENE see LFU000
LIMONENE see MCC250
LIMONENE DIOXIDE see LFV000
LIMONENE OXIDE see CAL000
LIMONITE see LFW000
LIMONIUM NASHII see MBU750
LIN 1418 see EPD100
LINADRYL HYDROCHLORIDE see LFW300
LINALOL see LFX000
LINALOL ACETATE see LFY100
p-LINALOOL see LFY000
LINALOOL see LFX000
LINALOOL ACETATE see LFY100
LINALOOL ISOBUTYRATE see LGB000
LINALOOL OXIDE see LFY500
LINALOOL TETRAHYDRIDE see LFY510
LINALYL ACETATE see LFY100
LINALYL ALCOHOL see LFX000
LINALYL-o-AMINOBENZOATE see APJ000
LINALYL ANTHRANILATE see APJ000
LINALYL BENZOATE see LFZ000
LINALYL CINNAMATE see LGA000
LINALYL FORMATE see LGA050
LINALYL ISOBUTYRATE see LGB000
LINALYL ISOVALERATE see LGC000
LINALYL PHENYLACETATE see LGC050
LINALYL PROPIONATE see LGC100
LINALYL α-TOLUATE see LGC050
LINAMARIN see GFC100
LINAMIN see FQJ100
LINAMPHETA see AOB250, TEP500
LINARIS see TKX000
LINARODIN see MDW750
LINASEN see EOK000
LINAXAR see PFJ000
LINCOCIN see LGC200, LGD000, LGE000
LINCOCIN HYDROCHLORIDE HYDRATE see LGC200
LINCOLCINA see LGD000
LINCOLNENSIN see LGD000
LINCOMYCIN, HYDROCHLORIDE, HEMIHYDRATE see LGC200
LINCOMYCIN, HYDROCHLORIDE HYDRATE see LGC200
LINCOMYCIN see LGD000
LINCOMYCINE (FRENCH) see LGD000
LINCOMYCIN HYDROCHLORIDE see LGE000
LINCOMYCIN HYDROCHLORIDE MONOHYDRATE see LGC200
LINCTUSSAL see DEE600
LINDAGRAIN see BBQ500
LINDAN see DGP900
α-LINDANE see BBQ000
β-LINDANE see BBR000
LINDANE (ACGIH, DOT, USDA) see BBQ500
Δ-LINDANE see BFW500
LINDATOX see EEM000
LINDEROL see NCQ820
LIN-DIBENZANTHRACENE see PAV000
LINDOL see TNP500
LINDRIDE see MPP000
LINEAR ALKYLBENZENE SULFONATE see AFO500
LINEAR ALKYLBENZENE SULFONATE, MAGNESIUM SALT see LGF800
LINEAR ALKYLBENZENE SULFONATE, SODIUM SALT see LGF825
LINEAR ALKYLBENZENE SULPHONATE see AFO500
LINE RIDER see TAA100
LINESTRENOL see NNV000
LINEVOL 79 see LGF875
LINEVOL 7-9 see LGF875
LINEX 4L see DGD600
LINFADOL see AFJ400
LINFOLIZIN see CDO500
LINFOLYSIN see CDO500
LINGEL see BRF500
LINGRAINE see EDC500
LINGRAN see EDC500
LINGUSORBS see PMH500
LIN-NAPHTHOANTHRACENE see PAV000
LINODIL see HFG550
9,12-LINOLEIC ACID see LGG000

LINOLEIC ACID see LGG000
LINOLEIC ACID mixed with OLEIC ACID see LGI000
LINOLEIC DIETHANOLAMIDE see BKF500
LINOLENATE HYDROPEROXIDE see PCN000
LINOLENIC ACID see OAX100
LINOLENIC HYDROPEROXIDE see PCN000
(LINOLEOYLOXY)TRIBUTYLSTANNANE see LGJ000
LINORMONE see CIR250
LINOROX see DGD600
LINSEED OIL see LGK000
LINTON see CLY500
LINTOX see BBQ500
LINUREX see DGD600
LINURON (herbicide) see DGD600
LINURON see DGD600
LIOMYCIN see HGP550
LIONOGEN BLUE R see IBV050
LIONOL YELLOW FG 1310 see CMS208
LIONOL YELLOW GGR see CMS212
LION'S BEARD see PAM780
LION'S MOUTH see FOM100
LIORESAL see BAC275
LIOTHYRONIN see LGK050
I-LIOTHYRONINE see LGK050
LIOTHYRONINE see LGK050
LIOXIN see VFK000
LIPAL 400-DL see PJY100
LIPAL 4LA see DXY000
LIPAL 20-OA see PJW500
LIPAL 30W see PJY100
LIPAMID see ARQ750
α-LIPAMIDE see DXN709
LIPAMIDE see DXN709
4-(dl-α-LIPAMIDO)BUTYRIC ACID see LGK100
LIPAN see DUS700
LIPARITE see CAS000
LIPARON see DPA000
LIPASE see GGA800
LIPASE AP6 see LGK150
LIPAVIL see ARQ750
LIPAVLON see ARQ750
LIPAZIN see GGA800
LIPENAN see LGK200
LIPFEN see FJT100
LIPHADIONE see CJJ000
LIPID CRIMSON see SBC500
LIPIDE 500 see ARQ750
LIPIDSENKER see ARQ750
LIPOACIN see DXN709
LIPOAMID see DXN709
α-LIPOAMIDE see DXN709
LIPOAMIDE see DXN709
LIPOCLIN see CMV700
LIPOCOL L-4 see DXY000
LIPOCOL O-n20 see PJW500
LIPOCOL C2 see PJT300
LIPOCTON see DXN709
LIPO-DIAZINE see PPP500
LIPO EGMS see EJM500
LIPOFACTON see ARQ750
LIPO-FLURBIPROFEN AXETIL see FJT100
LIPOGLUTAREN see HMC000
LIPO GMS 410 see OAV000
LIPO GMS 450 see OAV000
LIPO GMS 600 see OAV000
LIPO-HEPIN see HAQ500
α-LIPOIC ACID see DXN800
LIPOIC ACID see DXN800
α-LIPOIC ACID AMIDE see DXN709
LIPOICIN see DXN709
LIPO-LEVAZINE see PPP500
LIPO-LUTIN see PMH500
LIPOMID see ARQ750
LIPONEURINA see DXO300
α-LIPONIC ACID see DXN800
LIPONORM see ARQ750
α-LIPONSAEURE (GERMAN) see DXN800
LIPOPILL see DTJ400
LIPOPOLYSACCHARIDE see PPQ600
LIPOPOLYSACCHARIDE, from B. ABORTUS Bang. see LGK350
LIPOPOLYSACCHARIDE, ESCHERICHIA COLI see EDK750, LGK300
LIPOREDUCT see ARQ750
LIPORIL see ARQ750
LIPOSAN see DXN800
LIPOSID see ARQ750

LIPOSOMAL FLURBIPROFEN AXETIL see FJT100
LIPOSORB L-20 see PKG000
LIPOSORB O-20 see PKL100, SKV100
LIPOSORB S-20 see PKL030, SKV150
LIPOSORB O see SKV100
LIPOSORB P see MRJ800
LIPOSORB S see SKV150
LIPOSORB SQ0 see SKV170
LIPOTHION see DXN800
LIPOTRIL see CMF750
LIPOVAL A see ARW800
LIPOZYME see DXN709
LIPPIA CITRIODORA OIL see VIK500
LIPRIN see ARQ750
LIPRINAL see ARQ750
LIPTAN see IIU000
LIPUR see GCK300
LIQUACILLIN see BDY669
LIQUADIAZINE see PPP500
LIQUAEMIN see HAQ500
LIQUAEMIN SODIUM see HAQ550
LIQUAGESIC see HIM000
LIQUAMYCIN see TBX000
LIQUAMYCIN INJECTABLE see HOI000
LIQUAMYCIN LA 200 see HOH500
LIQUA-TOX see WAT200
LIQUEFIED CARBON DIOXIDE see LGL000
LIQUEFIED PETROLEUM GAS (DOT) see IIC000
LIQUEFIED PETROLEUM GAS see LGM000
LIQUEMIN see HAQ500, HAQ550
LIQUIBARINE see BAP000
LIQUIDAMBAR STYRACIFLUA see SOY500
LIQUID BRIGHT PLATINUM see PJD500
LIQUID CAMPHOR see CBB500
LIQUID CARBONIC GAS see LGL000
LIQUID DERRIS see RNZ000
LIQUID ETHYENE see EIO000
LIQUIDOW see CAO750
LIQUID PITCH OIL see CMY825
LIQUID ROSIN see TAC000
LIQUIGEL see AHC000
LIQUIMETH see MDT750
LIQUIPHENE see ABU500
LIQUIPRIN see SAH000
LIQUIPRON see LGM200
LIQUI-SAN see MLH000
LIQUI-STIK see NAK500
LIQUITAL see EOK000
LIQUOPHYLLINE see TEP000
LIQUORICE see GIG000
LIRANOL see DQA600
LIRANOX see CIR500
LIRCAPYL see PFB350
LIRIO (SPANISH) see BAR325, SLB250
LIRIO see AHI635
LIRIO CALA (SPANISH) see CAY800
LIROBETAREX see CJX750
LIRO CIPC see CKC000
LIROHEX see TCF250
LIROMAT see OPK250
LIROMATIN see ABX250
LIRONOX see BQZ000
LIROPON see DGI400
LIROPREM see PAX250
LIROSTANOL see ABX250
LIROTAN see EIR000
LIROTHION see PAK000
LISACORT see PLZ000
LISERGAN see ABH500
LISERGAN HYDROCHLORIDE see DPI000
LISKONUM see LGZ000
LISOLIPIN see SKJ300
LISSAMINE AMARANTH AC see FAG020
LISSAMINE BLUE AR see CMM070
LISSAMINE FAST YELLOW AE see SGP500
LISSAMINE FAST YELLOW AES see SGP500
LISSAMINE GREEN B see ADF000
LISSAMINE GREEN BN see ADF000
LISSAMINE GREEN G see FAE950
LISSAMINE LAKE GREEN SF see FAF000
LISSAMINE RED 2G see CMM300
LISSAMINE TURQUOISE VN see ADE500
LISSAMINE ULTRA BLUE AR see CMM070
LISSAMINE YELLOW AE see SGP500
LISSATAN AC see BLX000

LISSENPHAN see GGS000
LISSOLAMINE see HCQ500
LISTENON see HLC500
LISTICA see HNJ000
LISULFEN see AKO500
LISURIDE HYDROGEN MALEATE see LJE500
LITAC see ADY500
LITALER see HOO500
LITEX CA see SMR000
LITHALURE see LHQ100
LITHAMIDE see LGT000
LITHANE see LGZ000
LITHARGE see LDN000
LITHARGE YELLOW L-28 see LDN000
LITHIC ACID see UVA400
LITHICARB see LGZ000
LITHIDRONE see AFT500
LITHINATE see LGZ000
LITHIUM see LGO000
LITHIUM ACETATE see LGO100
LITHIUM ACETYLIDE see LGP875
LITHIUM ACETYLIDE COMPLEXED with ETHYLENEDIAMINE see LGQ000
LITHIUM ACETYLIDE-ETHYLENEDIAMINE COMPLEX see LGQ000
LITHIUM ALANATE see LHS000
LITHIUM ALUMINOHYDRIDE see LHS000
LITHIUM ALUMINUM HYDRIDE (DOT) see LHS000
LITHIUM ALUMINUM HYDRIDE, ETHEREAL (DOT) see LHS000
LITHIUM ALUMINUM TETRAHYDRIDE see LHS000
LITHIUM ALUMINUM TRI-tert-BUTOXYHYDRIDE see LGS000
LITHIUM AMIDE see LGT000
LITHIUM AMIDE, POWDERED see LGT000
LITHIUM ANTIMONIOTHIOMALATE see LGU000
LITHIUM ANTIMONY THIOMALATE see LGU000
LITHIUM ANTIMONY THIOMALATE NONAHYDRATE see AQE500
LITHIUM AZIDE see LGV000
LITHIUM BENZENEHEXOXIDE see LGV700
LITHIUM BIS(TRIMETHYLSILYL)AMIDE see LGX000
LITHIUM BOROHYDRIDE (DOT) see LHT000
LITHIUM BROMIDE see LGY000
LITHIUM CARBONATE (2:1) see LGZ000
LITHIUM CARBONATE see LGZ000
LITHIUM CHLORIDE see LHB000
LITHIUM CHLOROACETYLIDE see LHC000
LITHIUM CHLOROETHYNIDE see LHC000
LITHIUM CHROMATE(VI) see LHD099
LITHIUM CHROMATE see LHD000
LITHIUM DIAZOMETHANIDE see LHE450
LITHIUM DIETHYL AMIDE see LHE475
LITHIUM-2,2-DIMETHYLTRIMETHYLSILYL HYDRAZIDE see LHE525
LITHIUM FERRO SILICON see LHK000
LITHIUM FLUORIDE see LHF000
LITHIUM FLUORURE (FRENCH) see LHF000
LITHIUM-1-HEPTYNIDE see LHF625
LITHIUM HYDRIDE see LHH000
LITHIUM HYDRIDE (UN 1414) (DOT) see LHH000
LITHIUM HYDRIDE, fused solid (UN 2805) (DOT) see LHH000
LITHIUM HYDROXIDE, monohydrate or lithium hydroxide, solid (DOT) see LHI100
LITHIUM HYDROXIDE, solution (DOT) see LHI100
LITHIUM HYDROXIDE see LHI100
LITHIUM HYDROXIDE (Li(OH)) (9CI) see LHI100
LITHIUM γ-HYDROXYBUTYRATE see LHM800
LITHIUM HYDROXYBUTYRATE see LHM800
LITHIUM HYPOCHLORITE see LHJ000
LITHIUM HYPOCHLORITE COMPOUND, dry, containing more than 39% available chlorine (DOT) see LHJ000
LITHIUM IRON SILICON see LHK000
LITHIUM LACTATE see LHL000
LITHIUM METAL (DOT) see LGO000
LITHIUM METAL, IN CARTRIDGES (DOT) see LGO000
LITHIUM MONOBROMIDE see LGY000
LITHIUM NITRIDE see LHM000
LITHIUM-4-NITROTHIOPHENOXIDE see LHM750
LITHIUM OCTADECANOATE see LHQ100
LITHIUM OXYBUTYRATE see LHM800
LITHIUM PENTAMETHYLTITANATE-BIS(2,2'-BIPYRIDINE) see LHM850
LITHIUM PERCHLORATE see LHM875
LITHIUM PERCHLORATE TRIHYDRATE see LHN000
LITHIUM PEROXIDE see LHO000
LITHIUMRHODANID see TFF500
LITHIUM SILICON see LHP000
LITHIUM STEARATE see LHQ100
LITHIUM SULFATE (2:1) see LHR000
LITHIUM SULFOCYANATE see TFF500
LITHIUM SULPHATE see LHR000

LITHIUM TETRAAZIDOALUMINATE see LHR650
LITHIUM TETRAAZIDOBORATE see LHR675
LITHIUM TETRAHYDROALUMINATE see LHS000
LITHIUM TETRAHYDROBORATE see LHT000
LITHIUM TETRAMETHYLBORATE see LHT400
LITHIUM THIOCYANATE see TFF500
LITHOBID see LGZ000
LITHOCHOLIC ACID see LHW000
LITHOFOR BROWN A see NBG500
LITHOGRAPHIC STONE see CAO000
LITHOL see IAD000
LITHOL FAST SCARLET RN see MMP100
LITHOLITE see LHQ100
LITHOL RUBIN B see CMS155
LITHOL RUBINE see CMS155
LITHOL RUBINE BNA see CMS155
LITHOL SCARLET K 3700 see CMS148
LITHONATE see LGZ000
LITHOPONE see LHX000
LITHOSOL ORANGE R BASE see NMP500
LITHOSOL RED C see CMS150
LITHOSOL RED CLM see CMS150
LITHOTABS see LGZ000
LITICON see DOQ400
LITLURE A see FAS100
LITLURE B see TBX300
LITMOMYCIN see GJS000
LITSOEINE see LBO200
LIV 1176 see DVK200
LIVAZONE see FNF000
LIVER STARCH see GHK300
LIVERT see EKS120
LIVIATIN see DYE425
LIVICLINA see CCS250
LIVIDOMYCIN see LHX350
LIVIDOMYCIN B see DAS600
LIVODYMYCIN see LHX350
LIVONAL see EGQ000
LIXIL see BON325
L.J. 48 see MBX800
L.J. 206 see CBR675
LJ 206 see CCH125
LK 36 see HDY500
LL 1530 see NAC500
LL 1656 see BOM600
L 2642-LABAZ see EDV700
LLD FACTOR see VSZ000
LLONCEFAL see TEY000
LM-16 see CMS125
LM 91 see CJJ000
LM-209 see MCJ400
LM 280 see IDJ500
LM 433 see GKG300
LM-637 see BMN000
LM-2717 see CIR750
LM 2910 see LHX498
LM 2916 see LHX500
LM 2917 see LHX510
LM 2918 see LHX515
LM 2930 see LHX600
LM 22070 see DWD400
LMB 357 see MCB050
LM SEED PROTECTANT see MLH000
LOBAK see CKF500
LOBAMINE see MDT740
LO-BAX see HOV500
LOBELIA, INDIAN TOBACCO see LHY000
LOBELIA INFLATA see CCJ825
LOBELIA NICOTIANIFOLIA Roth ex R. & S., extract see LHX700
LOBELIA SIPHILITICA see CCJ825
LOBELINE see LHY000
LOBELINE HYDROCHLORIDE see LHZ000
(−)-LOBELINE HYDROCHLORIDE see LHZ000
LOBELIN HYDROCHLORIDE see LHZ000
LOBELLOA CARDINALIS see CCJ825
LOBENZARIT DISODIUM see LHZ600
LOBENZARIT SODIUM see LHZ600
LOBETRIN see ARQ750
LOBNICO see LHY000
LOBULANTINA see SCA750
LOBUTEROL see BQE250
LOCALYN see SPD500
LOCOID see HHQ825
LOCRON EXTRA see AHA000
LOCULA see SNQ000

LOCUNDIEZIDE see RFZ000
LOCUNDIOSIDE see RFZ000
LOCUST BEAN GUM see LIA000
LOCUTURINE see MPA050
LODESTONE YELLOW YB-57 see DEU000
LODIBON see DII200
LODITAC see OMY700
LODOPIN see ZUJ000
LODOSIN see CBQ500
LODOSYN see CBQ500
LOFEPRAMINE see DLH630
LOFEPRAMINE HYDROCHLORIDE see IFZ900
LOFETENSIN see LIA400
LOFETENSIN HYDROCHLORIDE see LIA400
LOFEXIDINE HYDROCHLORIDE see LIA400
LOFTYL see BOM600
LOGGERHEAD WEED (BARBADOS) see PIH800
LOHA see IGK800
LOISOL see TIQ250
LOKARIN see DPE200
LOKUNDJOSID (GERMAN) see RFZ000
LOKUNDJOSIDE see RFZ000
LabelOL see HMM500
LOLITOL see FON200
LOMADES see BDJ600
LOMAPECT see LFJ000
LOMAR D see BLX000
LOMAR LS see BLX000
LOMAR PW see BLX000
LOMBRICERA (PUERTO RICO) see PIH800
LOMBRICERO (CUBA) see APM875
LOMBRISTOP see TEX000
LO MICRON TALC 1 see TAB750
LO MICRON TALC, BC 1621 see TAB750
LO MICRON TALC USP, BC 2755 see TAB750
LOMIDIN see DBL800
LOMIDINE see DBL800
LOMIDINE ISOETHIONATE see DBL800
LOMOTIL see LIB000
LOMPER see MHL000
LOMUDAL see CNX825
LOMUDAS see CNX825
LOMUPREN see DMV600
LOMUSTINE see CGV250
LOMYCIN see GJY000
LON 41 see DTM600
LON 798 see GKU300
LON-954 see PFV000
S-LON see PKQ059
LONACOL see EIR000
LONAMIN see DTJ400
LONARID see HIM000
LONAVAR see AOO125
LONDOMIN see RCA200
LONDOMYCIN see MDO250
LONDON PURPLE see LIC000
LONETHYL see LID000
LONETIL see LID000
LONGACILIAN see BFC750
LONGANOCT see BPF500
LONGATIN see NOA000
LONGATIN HYDROCHLORIDE see NOA500
LONGICIL see BFC750
LONGIFENE see BOM250, HGC500
LONGIFOLEN see LID100
d-LONGIFOLENE see LID100
(+)-LONGIFOLENE see LID100
LONGIFOLENE (6CI) see LID100
β-LONGILOBINE see RFP000
LONGIN see AKO500
LONG PEPPER see PCB275
LONGUM see MFN500
LONIDAMINE see DEL200
LONIN 3 see BHO500
LONITEN see DCB000
LONOCOL M see MAS500
LONOMYCIN, MONOSODIUM SALT see LIG000
LONOMYCIN see LIF000
LONOMYCIN, SODIUM SALT see LIG000
LONTREL 3 see DGJ100
LONTREL see DGJ100
LONZA G see PKQ059
LONZAINE 16S see CDF450
LOOPLURE see DXU830
LOOPLURE INHIBITOR see DXU800

LO/OVRAL see NNL500
LOPEMID see LIH000
LOPEMIN see LIH000
LOPERAMIDE HYDROCHLORIDE see LIH000
LOPERYL see LIH000
LOPHINE see TMS750
LOPID see GCK300
LOPIRIN see MCO750
LOPRAMINE HYDROCHLORIDE see IFZ900
LOPREMONE see TNX400
LOPRESOR see SBV500
LOPRESS see HGP500
LOPRESSOR see SBV500
LOPURIN see ZVJ000
LOQUAT see LIH200
LORAMET see MLD100
LORAMINE AMB 13 see GHA050
L. ORANGE Z2010 see FAG150
LORAX see CFC250
LORAZEPAM see CFC250
LORDS-AND-LADIES see ITD050
LORETIN see IEP200
LOREX see DGD600
LORFAN see AGI000, LIH300
LORFAN TARTRATE see LIH300
LORIDIN see TEY000
LORINAL see CDO000
LORMETAZEPAM see MLD100
LORMIN see CBF250
LORO see DYA200
LOROL 20 see OEI000
LOROL 22 see DAI600
LOROL 24 see HCP000
LOROL 28 see OAX000
LOROL see DXV600
LOROL THIOCYANATE see DYA200
LOROMISIN see CDP250
LOROTHIDOL see TFD250
LOROX see BNM000, DGD600
LOROXIDE see BDS000
LOROX LINURON WEED KILLER see DGD600
LORPHEN see TAI500
LORPOX see BAR800
LORSBAN see CMA100
LORSILAN see CFC250
LOSANTIN see HOV500
N-LOST (GERMAN) see BIE250
N-LOST see BIE500
S-LOST see BIH250
N-LOST-PHOSPHORSAEUREDIAMID (GERMAN) see PHA750
LOSUNGSMITTEL APV see CBR000
LOTRIFEN see CKL325
LOTRIMIN see MRX500
(−)-LOTUCAINE HYDROCHLORIDE see TGJ625
LOTURINE see MPA050
LOTUSATE see AFY500
LOUISIANA LOBELIA see CCJ825
LOVAGE see PMD550
LOVAGE OIL see LII000
LOVE BEAD see RMK250
LOVENOX see HAQ500
LOVISCOL see CBR675
LOVOSA see SFO500
LOVOZAL see DGA200
LOW BELIA see CCJ825
LOWESERP see RDK000
LOWETRATE see PBC250
LOWPSTRON see CHJ750
LOXACOR see LIA400
LOXACOR HYDROCHLORIDE see LIA400
LOXANOL 95 see OBA000
LOXANOL K see HCP000
LOXANOL M see OBA000
LOXAPINE see DCS200
LOXIOL G 10 see GGR200
LOXIOL G 21 see HOG000
LOXON see DFH600
LOXOPROFEN SODIUM DIHYDRATE see LII300
LOXURAN see DIW200
LOXYNIL (GERMAN) see HKB500
LOZILUREA see LII400
LOZOL see IBV100
LPA 39 see SEA500
LPC see DXY725, LBX050
L.P.G. (OSHA, ACGIH) see LGM000

LPG see BFC750, LGM000
LPG ETHYL MERCAPTAN 1010 see EMB100
LPS see EDK700, PPQ600
LPT see PKB500
LR 115 see CCU250
LR-529 see CFO750
L RED Z 3000 see CMP620
LRF see LIU360
LRH see LIU360
LS 121 see NAE100
LS 1028E see SMQ500
LS 1727 see NNX600
L.S. 3394 see BLL750
LS 4442 see CLU000
LS 061A see SMQ500
LSD-25 see DJO000
d-LSD see DJO000
LSD see DJO000
LS 519 DIHYDROCHLORIDE see GCE500
LSD-25-PYRROLIDATE see LJM000
LSD TARTRATE see LJG000
LSM-775 see LJJ000
LSP 1 see LIJ000
L-T3 see LGK050
LTAN see DMH400
1,4,5,8-LTOA (RUSSIAN) see TDD250
LT-SOR see CMY240
LU 1631 see MQS100
LUBALIX see CMY525
LUBERGAL see AGQ875, EOK000
LUBRICATING OIL see LIK000
LUBRICATING OIL (mainly mineral) see LIK000
LUBRICATING OIL, CYLINDER see LIK000
LUBRICATING OIL, MOTORS see LIK000
LUBRICATING OIL, SPINDLE see LIK000
LUBRICATING OIL, TURBINE see LIK000
LUBROL 12A9 see DXY000
LUBROL W see PJT300
LUCALOX see AHE250
LUCAMIDE see EEM000
LUCANTHON see DHU000
LUCANTHONE see DHU000
LUCANTHONE HYDROCHLORIDE see SBE500
LUCANTHONE METABOLITE see LIM000
LUCANTHONE MONOHYDROCHLORIDE see SBE500
LUCEL (polysaccharide) see SFO500
LUCEL see CMA500
LUCENSOMYCIN see LIN000
LUCIDOL see BDS000
LUCIDOL DELTAX see MKA500
LUCIDRIL see AAE500
LUCIDRYL see DPE000
LUCIDRYL HYDROCHLORIDE see AAE500
LUCIJET see LIN400
LUCITE see PKB500
LUCKNOMYCIN see LIN600
LUCKY NUT see YAK350
LUCOFEN see ARW750
LUCOFENE see ARW750
LUCOFLEX see PKQ059
LUCORTEUM ORAL see GEK500
LUCORTEUM SOL see PMH500
LUCOSIL see MPQ750
LUCOVYL PE see PKQ059
LUDIGOL F,60 see NFC500
LUDIOMIL see LIN800, MAW850
LUDOX see SCH000
LUEH 6 see BGS250
LUETOCRIN DEPOT see HNT500
LUFYLLIN see DNC000
LUH[6] see BGS250
LUH6 see BGS250
LUH[6]-Cl2 see BGS250
LUH6-CHLORIDE see BGS250
LUH6-DINITRAT (GERMAN) see OPY000
LULAMIN see TEH500, TEO250
LULIBERIN see LIU360
LULLAMIN see DPJ400, TEO250
LUMBANG (GUAM) see TOA275
LUMBRICAL see PIJ000
LUMEN see EOK000
LUMESETTES see EOK000
LUMILAR 100 see PKF750
LUMINAL see EOK000
LUMINAL SODIUM see SID000

LUMIRELAX see GKK000
LUMIRROR see PKF750
LUMNITZERA RACEMOSA Willd., extract excluding roots see LIN850
LUMOFRIDETTEN see EOK000
LUNACOL see ZJS300
LUNAMIN see POA250
LUNAMYCIN see LIO600
LUNAR CAUSTIC see SDS000
LUNARINE HYDROCHLORIDE see LIP000
LUNETORON see BON325
LUNIPAK see DAB800
LUNIS see FDD085
LUPANIN see LIQ000
d-LUPANINE see LIQ000
(+)-LUPANINE see LIQ000
LUPANINE see LIQ000
LUPAREEN see PMP500
LUPERCO see BDS000, DGR600
LUPERCO CST see BIX750
LUPEROX see DGR600
LUPEROX FL see BDS000
LUPEROX 500R see DGR600
LUPEROX 500T see DGR600
LUPERSOL 8 see BSC600
LUPERSOL 11 see BSD250
LUPERSOL 70 see BSC250
LUPERSOL see MKA500
LUPERSOL DDA 30 see MKA500
LUPERSOL DDM see MKA500
LUPERSOL DNF see MKA500
LUPERSOL DSW see MKA500
LUPERSOL Δ-X see MKA500
LUPERSOL 228Z see ACG300
LUPETAZINE see LIQ500
2,6-LUPETIDINE see LIQ550
LUPETIDINE see LIQ550
LUPIN see LIT000
LUPINIDINE see SKX500
LUPINUS see LIT000
LUPINUS ANGUSTIFOLIUS, seed alkaloid mixture see LIT000
LUPOLEN 1010H see PJS750
LUPOLEN 1810H see PJS750
LUPOLEN 6011H see PJS750
LUPOLEN 4261A see PJS750
LUPOLEN 6042D see PJS750
LUPOLEN 1800H see PJS750
LUPOLEN KR 1032 see PJS750
LUPOLEN KR 1051 see PJS750
LUPOLEN KR 1257 see PJS750
LUPOLEN 6011L see PJS750
LUPOLEN L 6041D see PJS750
LUPOLEN N see PJS750
LUPOLEN 1800S see PJS750
LUPRANATE M 10 see PKB100
LUPRANATE M 70 see PKB100
LUPRANATE M 20S see PKB100
LUPRINATE M 20 see PKB100
LUPROSIL see PMU750
β-LUPULIC ACID see LIU000
LUPULON see LIU000
LUPULONE see LIU000
LURAFIX BLUE FFR see MGG250
LURAMIN see EOK000
LURAN see ADY500
LURAZOL BLACK BA see AQP000
LURAZOL ORANGE E see CMM220
LURGO see DSP400
LURIDE see SHF500
LURIDINE see CMF800
LUSIL see SNM500
LUSMIT see TFD500
LUSMIT SS see DXG700
LUSTRAN see ADY500
LUSTREX see SMQ500
LUSTREX H 77 see SMQ500
LUSTREX HH 101 see SMQ500
LUSTREX HP 77 see SMQ500
LUSTREX HT 88 see SMQ500
LUTALYSE see POC750
LUTAMIN see LIU100
LUTATE see HNT500
LUTEAL HORMONE see PMH500
LUTEINIZING HORMONE-RELEASING FACTOR see LIU360
LUTEINIZING HORMONE-RELEASING HORMONE see LIU360

LUTEINIZING HORMONE-RELEASING HORMONE, DIACETATE (SALT) see LIU380
LUTEINIZING HORMONE-RELEASING HORMONE, DIACETATE, TETRAHYDRATE see LIU400
LUTEINIZING HORMONE-RELEASING HORMONE (PIG), 6-(O-(1,1-DIMETHYLETHYL)-d-SERINE)-9-(N-ETHYL-l-PROLINAMIDE)-10- see LIU420
LUTEINIZING HORMONE-RELEASING HORMONE, (d-SER(TBU)$^6$-EA$^{10}$)- see LIU420
LUTEOCRIN see HNT500
LUTEOHORMONE see PMH500
LUTEOSAN see PMH500
LUTEOSKYRIN see LIV000
(—)-LUTEOSKYRIN see LIV000
LUTEOSTIMULIN see LIU360
LUTETIA FAST BLUE RS see IBV050
LUTETIA FAST RED 3R see MMP100
LUTETIA FAST SCARLET RF see MMP100
LUTETIA FAST SCARLET RJN see MMP100
LUTETIA ORANGE J see CMS145
LUTETIA ORANGE 3JR see CMM220
LUTETIA RED C see CMS150
LUTETIA RED CLN see CHP500
LUTETIUM CHLORIDE see LIW000
LUTETIUM CITRATE see LIX000
LUTETIUM(III) NITRATE (1:3) see LIY000
LUTEX see PMH500
2,4-LUTIDINE see LIY990
2,6-LUTIDINE see LJA010
3,4-LUTIDINE see LJB000
α-α′-LUTIDINE see LJA010
LUTIDON ORAL see GEK500
LUTINYL see CBF250
LUTOCYCLIN see PMH500
LUTOCYLOL see GEK500
LUTOFAN see PKQ059
LUTOGAN see DRT200
LUTO-METRODIOL see EQJ500
LUTOPRON see HNT500
LUTOSAN see DRT200
LUTOSOL see INJ000
LUTRELEF see LIU360
LUTROL-9 see EJC500
LUTROL see EAE675, EIM000, PJT000
LUTROMONE see PMH500
LUVATRENE see MNN250
LUVIPAL 066 see MCB050
LUVISKOL see PKQ250
LUWIPAL 012 see MCB050
LUXAN BLACK R see PFU500
LUXISTELM see PHI500
LUXOMNIN see AGQ875
LUXON see DFH600
261LV see DTS500
bacL VITAMIN H1 see AIH600
LVM see LHX350
LW 3170 see DCS200
LX 14-0 see CQH250
LXON see DFH600
LY 61017 see BGE125
LY 127935 see LBH200
LY133314 see TMP175
LY150720 see PIB700
LYCANOL see GHK200
LYCEDAN see AOA125
LYCINE see GHA050
LYCOBETAINE see LJB800
LYCOID DR see GLU000
LYCOPERSICIN see THG250
LYCORANIUM, 1,2,3,3a,6,7,12b,12c-OCTADEHYDRO-2-HYDROXY- see LJB800
LYCORCIDINOL see LJC000
LYCOREMINE HYDROBROMIDE see GBA000
LYCORICIDIN-A see LJC000
LYCORICIDINOL see LJC000
LYCORIS AFRICANA see REK325
LYCORIS RADIATA see REK325
LYCORIS SQUAMIGERA see REK325
LYCURIM see LJD500
LYDOL see DAM700
LYE see PLJ500
LYE (DOT) see SHS000
LYE, solution see SHS500
LYGODIUM FLEXUOSUM (LINN.) SWARTZ., EXTRACT see LJD600
LYGOMME CDS see CCL250
LYKURIM see LJD500

LYMECYCLINE see MRV250
LYMPHCHIN see AOQ875
LYMPHOCIN see AOQ875
LYMPHOQUIN see AOQ875
LYMPHOSCAN see AQL500
LYNAMINE see AHK750
LYNDIOL see LJE000
LYNENOL see NNV000
LYNESTRENOL see NNV000
LYNESTRENOL mixed with MESTRANOL see LJE000
LYNESTROL see EEH550
LYNESTROL mixed with MESTRANOL see LJE000
LYNOESTRENOL see NNV000
LYNOESTRENOL mixed with ETHINYLOESTRADIOL see EEH575
LYNOESTRENOL mixed with MESTRANOL see LJE000
LYOBEX see NOA000
LYOBEX HYDROCHLORIDE see NOA500
LYOFIX CH see MCB050
LYOFIX CHN see MCB050
LYOFIX CHN-ZA see MCB050
LYOFIX MLF see MCB050
LYOGLYCOGEN see GHK300
LYONIATOXIN see LJE100
LYONIOL A see LJE100
LYOPECT see CNG250
LYOPHRIN see AES000, VGP000
LYOVAC COSMEGEN see AEB000
LYP 97 see LBR000
LYPOARAN see DXN709
LYRAL see LJE200
LYSALGO see XQS000
LYSANXIA see DAP700
LYSENYL see LJE500
LYSENYL BIMALEATE see LJE500
LYSENYL HYDROGEN MALEATE see LJE500
LYSERGAMID see DJO000
LYSERGAURE DIETHYLAMID see DJO000
LYSERGIC ACID DIETHYLAMIDE-25 see DJO000
d-LYSERGIC ACID DIETHYLAMIDE see DJO000, LJF000
LYSERGIC ACID DIETHYLAMIDE, 1-ISOMER see LJF000
d-LYSERGIC ACID DIETHYLAMIDE TARTRATE see LJG000
LYSERGIC ACID DIETHYLAMIDE TARTRATE see LJG000
d-LYSERGIC ACID DIMETHYLAMIDE see LJH000
LYSERGIC ACID ETHYLAMIDE see LJI000
d-LYSERGIC ACID-1-HYDROXYMETHYLETHYLAMIDE see LJL000
d-LYSERGIC ACID MONOETHYLAMIDE see LJI000
d-LYSERGIC ACID MORPHOLIDE see LJJ000
LYSERGIC ACID MORPHOLIDE see LJJ000
d-LYSERGIC ACID-l,2-PROPANOLAMIDE see LJL000
LYSERGIC ACID PROPANOLAMIDE see LJL000
LYSERGIC ACID PYROLIDE see LJM000
d-LYSERGIC ACID PYRROLIDIDE see LJM000
LYSERGIDE see DJO000
LYSERGSAEUREDIAETHYLAMID see DJO000
LYS (HAITI) see SLB250
LYSILAN see HGI100
l-LYSINE ACETATE see LJM800
dl-LYSINE ACETYLSALICYLATE see ARP125
dl-LYSINE ACETYLSALICYLIC ACID SALT see ARP125
l-LYSINE HYDROCHLORIDE see LJO000
l-LYSINE, MONOACETATE see LJM800
dl-LYSINE MONO(2-(ACETYLOXY)BENZOATE see ARP125
l-LYSINE MONOHYDROCHLORIDE see LJO000
LYSINE MONOHYDROCHLORIDE see LJO000
N-LYSINOMETHYLTETRACYCLINE see MRV250
LYSIVANE see DIR000
LYSOCELLIN see LJP000
LYSOCOCCINE see SNM500
LYSOFON see AGW750
LYSOFORM see FMV000
LYSOL see LJP500
LYSOSTAPHIN see LJQ000
LYSSIPOLL see LJR000
LYSTENON see HLC500
LYSTHENONE see HLC500
LYSURON see ZVJ000
LYTECA SYRUP see HIM000
LYTHIDATHION see DRU400
LYTISPASM see LJS000
LYTRON 810 see SEA500
LYTRON 812 see SEA500
LYTRON 822 see SEA500
LYTRON 5202 see SMR000

01M see AGD250
M-2H see BPF825
M-2 see CAR875
M 2 see UTU500
M 3 see MCB050
M-14 see RKZ000
M-30 see DAE695
M-32 see DAE700
M 40 see CIR250
M 60 see UTU500
M 70 see UTU500
M-71 see CNW125
M 73 see DFV600
M-74 see DXH325
M 76 see MCB050
M 81 see PHI500
M-101 see TFY000
M-108 see TAL575
M 140 see CDR750
M 176 see DUD800
M 410 see CDR750
M 551 see PFB350
M 6/42 see RSU450
M 1028 see IFA000
M 13/20 see PKP750
M 1703 see IOV000
M 2060 see FIP999
M 3/158 see DAP600
M 3180 see CQH650
M-4209 see CBT250
M-4212 see MAB050, MQX775
M 4888 see CKB500
M 5055 see CDV100
5512-M see PDC000
M5943 see DGD100
M-6029 see BOO630
M-9500 see TND500
M-12210 see MAB055
M73101 see EAN700
MDL-035 see THG700
MDL-899 see MBV735
MA-110 see EHP000
MA-117 see EHO700
MA-1214 see DXV600
MA 1291 see QWJ500
MA 1337 see CMX840
M-4365A2 see RMF000
MA see DAL300, MNQ500
MA300A see DEJ100
MA-144A2 see TAH675
MAA see AFI850, MGQ530
1-MA-4-AA see AKP250
MAAC see HFJ000
MAA SODIUM SALT see DXE600
MA144 B2 see TAH650
MAB see MNR500
MABA see AIH500
MABERTIN see CFY750
MABLIN see BOT250
MABUTEROL see MAB300
MA 100 (CARBON) see CBT500
MACASIROOL see CHJ750
MACBAL see DTN200
MACBECIN II see MAB400
MACE (lachrymator) see CEA750
MACE OIL see OGQ100
MACHETE (herbicide) see CFW750
MACHETE see CFW750
MACHETTE see CFW750
MACH-NIC see NDN000
MACKREAZID see COH250
MACLEYINE see FOW000
MACOCYN see HOH500
MACODIN see MDP750
MACQUER'S SALT see ARD250
MACRABIN see VSZ000
MACROCYCLON see TDN750
MACRODANTIN see NGE000
MACRODIOL see EDO000
MACROGOL 1000 see PJT250
MACROGOL 4000 see PJT750
MACROGOL 400 BPC see EJC500
MACROGOL OLEATE 600 see PJY100
MACROL see EDO000
MACROLEX BLUE FR see BKP500
MACROMOMYCIN see MAB750

MACRONDRAY see DAA800
MACROPAQUE see BAP000
MACROSE see DBD700
MACULOTOXIN see FOQ000
MAD see AOO475
MADAGASCAR LEMONGRASS OIL see LEH000
MADAME FATE (JAMAICA) see SLJ650
MADAR see CGA500
MADECASSOL see ARN500
MADEIRA IVY see AFK350
MADHURIN see SJN700
MADIOL see AOO475, MQU750
MADRESELVA (MEXICO) see YAK100
MADRIBON see SNN300
MADRIGID see SNN300
MADRINE see DBA800
MADRIQID see SNN300
MADROX see MEU300
MADROXIN see SNN300
MADROXINE see SNN300
MADURIT 152 see MCB050
MADURIT MS see MCB050
MADURIT MW 111 see MCB050
MADURIT MW 112 see MCB050
MADURIT MW 150 see MCB050
MADURIT MW 161 see MCB050
MADURIT MW 166 see MCB050
MADURIT MW 392 see MCB050
MADURIT MW 484 see MCB050
MADURIT MW 559 see MCB050
MADURIT MW 630 see MCB050
MADURIT MW 815 see MCB050
MADURIT MW 909 see MCB050
MADURIT 5238N see MCB050
MADURIT OP see MCB050
MADURIT TN see MCB050
MADURIT VMW 3113 see MCB050
MADURIT VMW 3114 see MCB050
MADURIT VMW 3151 see MCB050
MADURIT VMW 3163 see MCB050
MADURIT VMW 3284 see MCB050
MADURIT VMW 3399 see MCB050
MADURIT VMW 3489 see MCB050
MADURIT VMW 3494 see MCB050
MADURIT VMW 3490 see MCB050
MADURIT VMW 3819 see MCB050
MADURIT VMW 3822 see MCB050
1-MA-40EAA (RUSSIAN) see MGG250
MAFATATE see MAC000
MAFENIDE ACETATE see MAC000
MAFU see DGP900
MAG see MRF000
MAGADI SODA see SJT750
MAGBOND see BAV750
MAGCAL see MAH500
MAGCHEM 100 see MAH500
MAGECOL see CBT750
MAGENTA see MAC250
MAGENTA BASE see MAC500
MAGENTA DP see MAC250
MAGENTA E see MAC250
MAGENTA G see MAC250
MAGENTA I see MAC250
MAGENTA PN see MAC250
MAGENTA POWDER N see MAC250
MAGENTA S see MAC250
MAGENTA SUPER FINE see MAC250
MAGIC LILY see REK325
MAGIC METHYL see MKG250
MAGISTERY OF BISMUTH see BKW100
MAGK'S T.B.1 CONTEBEN see FNF000
MAGLITE see MAH500
MAGMASTER see MAC650
MAGNACIDE H see ADR000
MAGNAMYCIN see CBT250
MAGNAMYCIN A see CBT250
MAGNAMYCIN HYDROCHLORIDE see MAC600
MAGNESIA see MAH500
MAGNESIA ALBA see MAC650
MAGNESIA MAGMA see MAG750
MAGNESIA USTA see MAH500
MAGNESIA WHITE see CAX750
MAGNESIO (ITALIAN) see MAC750
MAGNESITE see MAC650
MAGNESIUM see MAC750

MAGNESIUM ACETATE see MAD000
MAGNESIUM ALLOYS, powder (UN 1418) (DOT) see MAC750
MAGNESIUM ALLOYS with >50% magnesium in pellets, turnings or ribbons (UN 1869) (DOT) see MAC750
MAGNESIUM ALUMINUM PHOSPHIDE (DOT) see AHD250
MAGNESIUM ARSENATE see ARD000
MAGNESIUM ARSENATE PHOSPHOR see ARD000
MAGNESIUM AUREOLATE see MAD050
MAGNESIUM BIS(2,3-DIBROMOPROPYL)PHOSPHATE see MAD100
MAGNESIUMBIS(2,3-DIBROMOPROPYL)PHOSPHATE see MAD100
MAGNESIUM, BIS(2-HYDROXYPROPANOATO-O(1),O(2))-, (T-4)-(9CI) see LAL100
MAGNESIUM, BIS(LACTATO)-(8CI) see LAL100
MAGNESIUM BORIDE see MAD250
MAGNESIUM(II) CARBONATE (1:1) see MAC650
MAGNESIUM CARBONATE see MAC650
MAGNESIUM CARBONATE, PRECIPITATED see MAC650
MAGNESIUM CHLORATE see MAE000
MAGNESIUM CHLORIDE see MAE250
MAGNESIUM CHLORIDE HEXAHYDRATE see MAE500
MAGNESIUM CLIPPINGS see MAC750
MAGNESIUM COMPOUNDS see MAE750
MAGNESIUM DIACETATE see MAD000
MAGNESIUM DICHLORATE see MAE000
MAGNESIUM DIPROPYLACETATE see MAK275
MAGNESIUM DROSS, wet or hot (DOT) see MAF000
MAGNESIUM DROSS (HOT) see MAF000
MAGNESIUM FLUORIDE see MAF500
MAGNESIUM FLUOROSILICATE (DOT) see MAF600
MAGNESIUM FLUORURE (FRENCH) see MAF500
MAGNESIUM FLUOSILICATE see MAF600, MAG250
MAGNESIUMFOSFIDE (DUTCH) see MAI000
MAGNESIUM FUMARATE see MAF750
MAGNESIUM GLUCONATE see MAG000
MAGNESIUM GLUTAMATE HYDROBROMIDE see MAG050
MAGNESIUM GOLD PURPLE see GIS000
MAGNESIUM GRANULES, coated particle size not <149 microns (UN 2950) (DOT) see MAC750
MAGNESIUM HEXAFLUOROSILICATE see MAG250
MAGNESIUM HYDRATE see MAG750
MAGNESIUM HYDRIDE see MAG500
MAGNESIUM HYDROXIDE see MAG750
MAGNESIUM LACTATE see LAL100
MAGNESIUM LINEAR ALKYLBENZENE SULFONATE see LGF800
MAGNESIUM(II) NITRATE (1:2) see MAH000
MAGNESIUM NITRATE (DOT) see MAH000
MAGNESIUM OXIDE see MAH500
MAGNESIUM OXIDE FUME (ACGIH) see MAH500
MAGNESIUM-5-OXO-2-PYRROLIDINECARBOXYLATE see PAN775
MAGNESIUM PELLETS see MAC750
MAGNESIUM PEMOLINE see PAP000
MAGNESIUM PERCHLORATE see PCE000
MAGNESIUM PEROXIDE, solid (DOT) see MAH750
MAGNESIUM PEROXIDE see MAH750
MAGNESIUM PHOSPHATE, DIBASIC see MAH775
MAGNESIUM PHOSPHATE, TRIBASIC see MAH780
MAGNESIUM PHOSPHIDE see MAI000
MAGNESIUM PHTHALOCYANINE see MAI250
MAGNESIUM POTASSIUM-I-ASPARTATE see MAI600
MAGNESIUM POTASSIUM ASPARTATE see MAI600
MAGNESIUM POWDERED see MAC750
MAGNESIUM RIBBONS see MAC750
MAGNESIUM SALT of AUREOLIC ACID see MAD050
MAGNESIUM SILICATE HYDRATE see MAJ000
MAGNESIUM SILICOFLUORIDE see MAG250
MAGNESIUM STEARATE (ACGIH) see MAJ030
MAGNESIUM STEARATE see MAJ030
MAGNESIUM SULFATE (1:1) see MAJ250
MAGNESIUM SULFATE and CALOMEL (8:5) see CAZ000
MAGNESIUM SULFATE HEPTAHYDRATE see MAJ500
MAGNESIUM SULPHATE see MAJ250
MAGNESIUM TETRAHYDROALUMINATE see MAJ775
MAGNESIUM THIOSULFATE see MAK000
MAGNESIUM THIOSULFATE HEXAHYDRATE see MAK250
MAGNESIUM TURNINGS (DOT) see MAC750
MAGNESIUM, powder (UN 1418) (DOT) see MAC750
MAGNESIUM (UN 1869) (DOT) see MAC750
MAGNESIUM VALPROATE see MAK275
MAGNETIC 70, 90, and 95 see SOD500
MAGNETIC BLACK see IHC550
MAGNETIC OXIDE see IHC550
MAGNETITE see IHC550
MAGNEZU TLENEK (POLISH) see MAH500
MAGNIFLOC 509C see MCB050
MAGNOFENYL see OPK300

MAGNOLIONE see OOO100
MAGNOPHENYL see OPK300
MAGNOSULF see MAK000
MAGOX 85 see MAH500
MAGOX 90 see MAH500
MAGOX 95 see MAH500
MAGOX 98 see MAH500
MAGOX see MAG750, MAH500
MAGOX OP see MAH500
MAGRACROM YELLOW GG see SIT850
MAGRON see MAE000
MAGSALYL see SJO000
MAH see DMC600
MAHOGANY see MAK300
MAIKOHIS see FOS100
MAINTAIN 3 see DMC600
MAINTAIN A see CDT000
MAINTAIN CF125 see CDT000
MAIORAD see TGF175
MAIPEDOPA see DNA200
MAIS BOUILLI (HAITI) see GIW200
MAITANSINE see MBU820
MAITENIN see TGD125
MAIZENA see SLJ500
MAJEPTYL see TFM100
MAJOL PLX see SFO500
MAJSOLIN see DBB200
MAKI see BMN000
MALACHITE see CNJ750
MALACHITE GREEN G see BAY750
MALACHITE GREEN OXALATE see MAK600
MALACID see TGD000
MALACIDE see MAK700
MALAFOR see MAK700
MALAGRAN see MAK700
MALAKILL see MAK700
MALAMAR 50 see MAK700
MALAMAR see MAK700
MALAMINE, HEXAKIS(METHOXYMETHYL) see HDY500
MALA MUJER (MEXICO) see CNR135
MALANGA CARA de CHIVO (CUBA) see EAI600
MALANGA (CUBA) see XCS800
MALANGA DEUX PALLES (HAITI) see EAI600
MALANGA ISLENA (CUBA) see EAI600
MALANGA de JARDIN (CUBA) see EAI600
MALANGA TREPADORA (CUBA) see PLW800
MALAOXON see OPK250
MALAOXONE see OPK250
MALAPHELE see MAK700
MALAPHOS see MAK700
MALAQUIN see CLD000
MALAREN see CLD000
MALAREX see CLD000
MALARICIDA see CFU750
MALARIDINE see PPQ650
MALASOL see MAK700
MALASPRAY see MAK700
MALATHION see MAK700
MALATHION-O-ANALOG see OPK250
MALATHION ULV CONCENTRATE see MAK700
MALATHIOZOO see MAK700
MALATHON see MAK700
MALATHYL LV CONCENTRATE & ULV CONCENTRATE see MAK700
MALATION (POLISH) see MAK700
MALATOL see MAK700
MALATOX see MAK700
MALAYAN CAMPHOR see BMD000
MALAZIDE see DMC600
MALCOTRAN see MDL000
MALDISON see MAK700
MALDOCIL see TEN750
MALEALDEHYDIC ACID, DICHLORO-4-OXO-, (Z)-(9CI) see MRU900
MALEATE ACIDE de l'ACETYL-3-DIMETHYLAMINO-3-PROPYL-10-PHENOTHI-AZINE (FRENCH) see AAF750
MALEATE de 2-((p-CHLOROPHENYL)-2-(PYRIDYL)HYDROXY METH-YL)IMIDAZOLINE (FRENCH) see SBC800
MALEATE de CINEPAZIDE (FRENCH) see VGK000
MALEATE de CINPROPAZIDE see IRQ000
MALEIC ACID see MAK900
MALEIC ACID ANHYDRIDE (MAK) see MAM000
MALEIC ACID, DIALLYL ESTER see DBK200
MALEIC ACID, DIBUTLY ESTER see DED600
MALEIC ACID DI(1,3-DIMETHYLBUTYL) ESTER see DKP400
MALEIC ACID, DIETHYL ESTER see DJO200
MALEIC ACID DIHEXYL ESTER see DKP400

MANGANESE, (4-AMINOBUTANOATO-N,O)(N-(2,4-DIHYDROXY-3,3-DIMETHYL-1-OXOBUTYL)-β-ALANINATO)- see MAQ600
MANGANESE γ-AMINOBUTYRATOPANTOTHENATE see MAQ600
MANGANESE BINOXIDE see MAS000
MANGANESE (BIOSSIDO di) (ITALIAN) see MAS000
MANGANESE (BIOXYDE de) (FRENCH) see MAS000
MANGANESE(II) BIS(ACETYLIDE) see MAQ780
MANGANESE, BIS(I-HISTIDINATO-N,O)-, TETRAHYDRATE see BJX800
MANGANESE, BIS(I-HISTIDINATO)-, TETRAHYDRATE see BJX800
MANGANESE, BIS(5-SULFO-8-QUINOLINOLATO)- see BKJ275
MANGANESE BLACK see MAS000
MANGANESE(II) CHLORIDE (1:2) see MAR000
MANGANESE(II) CHLORIDE TETRAHYDRATE see MAR250
MANGANESE COMPOUNDS see MAR500
MANGANESE CYCLOPENTADIENYL TRICARBONYL see CPV000
MANGANESE DIACETATE see MAQ000
MANGANESE DIACETATE TETRAHYDRATE see MAQ250
MANGANESE DICHLORIDE see MAR000
MANGANESE DIMETHYL DITHIOCARBAMATE see MAR750
MANGANESE DINITRATE see MAS900
MANGANESE (DIOSSIDO di) (ITALIAN) see MAS000
MANGANESE DIOXIDE see MAS000
MANGANESE (DIOXYDE de) (FRENCH) see MAS000
MANGANESE EDTA COMPLEX see MAS250
MANGANESE, ((1,2-ETHANEDIYLBIS(CARBAMODITHIOATO))(2-))-, mixed with DIMETHYL (1,2-PHENYLENEBIS(IMINOCARBONOTHIOYL))BIS(CARBAMATE) see MAP300
MANGANESE ETHYLENE-1,2-BISDITHIOCARBAMATE see MAS500
MANGANESE(II) ETHYLENEBIS(DITHIOCARBAMATE) see MAS500
MANGANESE(II) ETHYLENE DI(DITHIOCARBAMATE) see MAS500
MANGANESE complex with N-FLUOREN-2-YL ACETOHYDROXAMIC ACID see HIR000
MANGANESE(II) FLUORIDE see MAS750
MANGANESE FLUORIDE see MAS750
MANGANESE FLUORURE (FRENCH) see MAS750
MANGANESE GREEN see MAT250
MANGANESE MANGANATE see MAT500
MANGANESE, (METHYLCYCLOPENTADIENYL)TRICARBONYL- see MAV750
MANGANESE MONOXIDE see MAT250
MANGANESE(2+) NITRATE see MAS900
MANGANESE NITRATE (DOT) see MAS900
MANGANESE(II) NITRATE see MAS900
MANGANESE NITRATE see MAS900
MANGANESE (II) NITRATE, ANHYDROUS see MAS900
MANGANESE(III) OXIDE see MAT500
MANGANESE(II) OXIDE see MAT250
MANGANESE(IV) OXIDE see MAS000
MANGANESE(VII) OXIDE see MAT750
MANGANESE OXIDE see MAS000, MAU800
MANGANESE(II) PERCHLORATE see MAT899
MANGANESE PERCHLORATE HEXAHYDRATE see MAU000
MANGANESE PEROXIDE see MAS000
MANGANESE SESQUIOXIDE see MAT500
MANGANESE(II) SULFATE (1:1) see MAU250
MANGANESE(II) SULFATE TETRAHYDRATE (1:1:4) see MAU750
MANGANESE(II) SULFIDE see MAV000
MANGANESE SUPEROXIDE see MAS000
MANGANESE(II) TELLURIDE see MAV250
MANGANESE, TETRACARBONYL(TRIFLUOROMETHYLTHIO)-, dimer see TBN200
MANGANESE(II) TETRAHYDROALUMINATE see MAV500
MANGANESE TETROXIDE see MAU800
MANGANESE TRICARBONYL METHYLCYCLOPENTADIENYL see MAV750
MANGANESE TRIFLUORIDE see MAW000
MANGANESE TRIOXIDE see MAT500
MANGANESE(2+), TRIS(OCTAMETHYLDIPHOSPHORAMIDE-Op,Op')-, (OC-6-11)-, DIPERCHLORATE see TNK400
MANGANESE ZINC BERYLLIUM SILICATE see BFS750
MANGANESE ZINC FERRATE see IHB677
MANGANESE ZINC FERRITE see IHB677
MANGANIC OXIDE see MAT500
MANGAN(II)-(N,N'-AETHYLEN-BIS(DITHIOCARBAMAT)) (GERMAN) see MAS500
MANGAN NITRIDOVANY (CZECH) see MAP750
MANGANOMANGANIC OXIDE see MAU800
MANGANOUS ACETATE see MAQ000
MANGANOUS ACETATE TETRAHYDRATE see MAQ250
MANGANOUS ACETYLACETONATE see MAQ500
MANGANOUS CHLORIDE see MAR000
MANGANOUS CHLORIDE TETRAHYDRATE see MAR250
MANGANOUS DIMETHYLDITHIOCARBAMATE see MAR750
MANGANOUS DINITRATE see MAS900
MANGANOUS NITRATE see MAS900
MANGANOUS OXIDE see MAT250
MANGANOUS SULFATE see MAU250

MANGAN-ZINK-AETHYLENDIAMIN-BIS-DITHIO-CARBAMAT (GERMAN) see DXI400
MAN-GRO see MAU250
MANHEXIN see MAW250
MANICOLE see MAW250
MANIHOT ESCULENTA see CCO680
MANINIL see CEH700
MANIOC see CCO680
MANIOKA see CCO680
MANIOL see CCR875
MANITE see MAW250
MANNA SUGAR see HER000
MANNEX see MAW250
MANNITE see HER000
MANNIT-LOST (GERMAN) see MAW500
MANNIT-MUSTARD (GERMAN) see MAW500
d-MANNITOL see HER000
MANNITOL see HER000
d-MANNITOL BUSULFAN see BKM500
MANNITOL HEXANITRATE (dry) (DOT) see MAW250
d-MANNITOL HEXANITRATE see MAW250
MANNITOL HEXANITRATE see MAW250
MANNITOL HEXANITRATE, wetted with not <40% water, by weight or mixture (NA 0133) (DOT) see MAW250
MANNITOL MUSTARD DIHYDROCHLORIDE see MAW750
MANNITOL MYLERAN see BKM500
MANNITOL NITROGEN MUSTARD see MAW500
MANNITRIN see MAW250
MANNOGRANOL see BKM500
MANNOMUSTINE see MAW500
MANNOMUSTINE DIHYDROCHLORIDE see MAW750
MANNOPYRANOSIDE, STROPHANTHIDIN-3 6-DEOXY-, α-l- see CNH780
MANNOSULFAN see MAW800
MANOLENE 6050 see PJS750
MANOSEB see DXI400
MANOXAL OT see DJL000
MANRO PTSA 65 E see TGO000
MANRO PTSA 65 H see TGO000
MANRO PTSA 65 LS see TGO000
MANSIL see OLT000
MAN'S MOTHERWORT see CCP000
MANTA see KAJ000
MANTADAN see AED250
MANTHELINE see DJM800, XCJ000
MANUCOL see SEH000
MANUCOL DM see SEH000
MANUFACTURED IRON OXIDES see IHD000
MANUTEX see SEH000
MANZANA (SPANISH) see AQP875
MANZANILLO (CUBA, PUERTO RICO, DOMINICAN REPUBLIC) see MAO875
MANZATE 200 see DXI400, MAS500
MANZATE see MAS500
MANZATE D see MAS500
MANZATE MANEB FUNGICIDE see MAS500
MANZEB see DXI400, MAS500
MANZIN 80 see DXI400
MANZIN see MAS500
MAOA see QAK000
MAOH see MKW600
MAOLATE see CJQ250
MAO-REM see PFC750
4M2AP see ALC250
MAPHARSAL see ARL000
MAPHARSEN see ARL000
MAPHARSIDE see ARL000
MAPHENIDE ACETATE see MAC000
MAPLE AMARANTH see FAG020
MAPLE BRILLIANT BLUE FCF see FMU059
MAPLE ERYTHROSINE see FAG040
MAPLE INDIGO CARMINE see FAE100
MAPLE LACTONE see HMB500
MAPLE PONCEAU 3R see FAG018
MAPLE SUNSET YELLOW FCF see FAG150
MAPLE TARTRAZOL YELLOW see FAG140
MAPO see TNK250
MAPOLOSE M25 see MIF760
MAPOLOSE 60SH50 see MIF760
MAPOSOL see SIL550, VFU000
MAPP (OSHA) see MFX600
MAPRENAL 980 see MCB050
MAPRENAL MF 590 see MCB050
MAPRENAL MF 650 see MCB050
MAPRENAL MF 915 see MCB050
MAPRENAL MF 910 see MCB050
MAPRENAL MF 927 see MCB050

MAPRENAL MF 929 see MCB050
MAPRENAL MF 920 see MCB050
MAPRENAL MF 904 see MCB050
MAPRENAL MF 980 see MCB050
MAPRENAL MF 900 see MCB050
MAPRENAL MF see MCB050
MAPRENAL MP 500 see MCB050
MAPRENAL NPX see MCB050
MAPRENAL RT-MF 650 see MCB050
MAPRENAL TTX see MCB050
MAPRENAL VMF 52/7 see MCB050
MAPRENAL VMF 3655 see MCB050
MAPRENAL VMF 3925 see MCB050
MAPRENAL VMF 3935 see MCB050
MAPRENAL VMF see MCB050
MAPROFIX 563 see SIB600
MAPROFIX TLS 65 see SON000
MAPROFIX WAC-LA see SIB600
MAPROTILINE see LIN800
MAPROTILINE HYDROCHLORIDE see MAW850
MAPTC see CMP900
MAQBARL see DTN200
MAQUAT MC 1412 see QAT520
MARALATE see MEI450
MARANHIST see WAK000
MARANTA see SLJ500
MARANYL F 114 see PJY500
MARAPLAN see IKC000
MARASMIC ACID see MAW875
MARATHON see DWS200
MARAVILLA (MEXICO) see CAK325
MARAZINE see EAN600
MAR BATE see MQU750
MARBLE see CAO000
MARBLEWHITE 325 see CAT775
MARBON 9200 see SMR000
MARBORAN see MKW250
MARCAIN see BOO000
MARCELLOMYCIN see MAX000
MARCOPHANE see BGY000
MARETIN see HMV000
MAREVAN (SODIUM SALT) see WAT220
MAREX see HGC500
MAREZINE see EAN600
MAREZINE HYDROCHLORIDE see MAX275
MARFIL see CAT775
MARFOTOKS see DJI000
MARGARIC ACID see HAS500
MARGONIL see MQU750
MARGONOVINE see LJL000
MARGOSA OIL see NBS300
MARIA GRANDE (PUERTO RICO) see MBU780
MARICAINE see DHK400
MARIHUANA see CBD750
MARIJUANA see CBD750
MARIJUANA, SMOKE RESIDUE see CBD760
MARIMET 45 see CAW850
MARINCO H see MAG750
MARINDININ see DLR000
MARINEURINA see DXO300
MARISAN see CMR100
MARISILAN see AIV500
MARITUS YELLOW see DUX800
MARJORAM OIL, SPANISH see MBU500
MARKOFANE see BGY000
MARK PEP 24 see COV800
MARKS 4-CPA see CJN000
MARKURE UL2 see DEJ100
MARLATE see MEI450
MARLEX 9 see PJS750
MARLEX 50 see PJS750
MARLEX 60 see PJS750
MARLEX 960 see PJS750
MARLEX 6003 see PJS750
MARLEX 6009 see PJS750
MARLEX 6015 see PJS750
MARLEX 6050 see PJS750
MARLEX 6060 see PJS750
MARLEX 9400 see PMP500
MARLEX EHM 6001 see PJS750
MARLEX M 309 see PJS750
MARLEX TR 704 see PJS750
MARLEX TR 880 see PJS750
MARLEX TR 885 see PJS750
MARLEX TR 906 see PJS750

MARLIPAL 1217 see DXY000
MARLON AS 3 see LBU100
MARLOPHEN 820 see PKF500
MARMAG see MAH500
MARMELOSIN see IHR300
MARMER see DXQ500
MARNITENSION SIMPLE see RDK000
MAROMERA (CUBA) see RBZ400
MAROXOL-50 see DUZ000
MARPLAN see IKC000
MARPLON see IKC000
MARRONNIER (CANADA) see HGL575
MARSALID see ILE000
MARS BROWN see IHD000
MARSH GAS see MDQ750
MARSH MARIGOLD see MBU550
MARSH ROSEMARY EXTRACT see MBU750
MARSILID see ILE000
MARSIN see MNV750
MARS RED see IHD000
MARSTHINE see FOS100
MARTRATE-45 see PBC250
MARUCOTOL see AAE500
MARUNGUEY (PUERTO RICO) see CNH789
MARVEX see DGP900
MARVINAL see PKQ059
MARYGIN-M see DAB875
MARZIN see DXI400
MARZINE see EAN600
MARZULENE S see MBU775
MA144 S2 see APV000
MAS see MGQ750
MASCHITT see CFY000
MASDIOL see AOO475
MASELNAN DI-n-BUTYLCINICITY (CZECH) see DDX600
MASENATE see TBG000
MASENONE see MPN500
MASEPTOL see HJL500
MASHERI (INDIA) see SED400
MASLETINE see FOS100
MASOTEN see TIQ250
MASSICOT see LDN000
MASSICOTITE see LDN000
MASTESTONA see MPN500
MASTIC ABSOLUTE see MBU777
MASTIC (RESIN) see MBU777
MASTIPHEN see CDP250
MASTWOOD see MBU780
MA 144T1 see DAY835
MATACIL see DOR400
MATANOL see AMK250
MATAVEN see ARW000
MATCH see MLJ500
MATENON see MQS225
MATILIT CM 2L see SEA500
MATO AZUL (PUERTO RICO) see CAK325
MATO de PLAYA (PUERTO RICO) see CAK325
MATRICARIA CAMPHOR see CBA750
MATRIGON see DGJ100
MATROMYCIN see OHO200
MATSUTAKE ALCOHOL (JAPANESE) see ODW000
MATTING ACID (DOT) see SOI500
MATULANE see PME250, PME500
MAURYLENE see PMP500
MAVISERPIN see RDK000
MAXATASE see BAC000
MAXICHEM 1DTM see MCB050
MAXIDEX see SOW000
MAXIFEN see AOD000
MAXIPEN see PDD350
MAXITATE see MAW250
MAXOLON see AJH000
MAXULVET see SNN300
MA 144 Y see ADG425
MAY APPLE see MBU800
MAY & BAKER S-4084 see COQ399
MAY BLOB see MBU550
MAYCOR see CCK125
MAYFLOWER see LFT700
MAYSANINE see MBU820
MAYSERPINE see RDK000
MAYT see MBU820
MAYTANSINE see MBU820
MAYTANSINOL ISOVALERATE see APE529
MAYTENIN see TGD125

2-MAYTHIC ACID see NAV500
MAY THORN see MBU825
MAYVAT BROWN BR see CMU770
MAYVAT GOLDEN YELLOW see DCZ000
MAYVAT OLIVE AR see DUP100
MAZATICOL HYDROCHLORIDE see MBV100
MAZEPTYL see TFM100
MAZIDE 30 see DHF200
MAZINDOL see MBV250
MAZOTEN see TIQ250
M+B 695 see PPO000
M+B 760 see TEX250
M & B 800 see DBL800
MB 2878 see EAK500
MB-3000 see BKL800
M&B 3046 see CLO000
M & B 4620 see TDS250
M&B 8873 see HKB500
MB 9057 see SNQ500
MB 10064 see DDP000
M&B 11,461 see DNG200
M & B 39565 see MQY110
MB see MHR200
7-MBA see MGW750
MB 2050A see PBT000
M&B 17,803A see AAE125
MBA see BIE250
7-MBA-3,4-DIHYDRODIOL see MBV500
MBA HYDROCHLORIDE see BIE500
MBAO see CFA500
MBAO HYDROCHLORIDE see CFA750
MBBA see CQK600
MBBH 766 see CMP200
MBC see BAV575, MHC750
MBCP see LEN000
M & B 782 DIHYDROCHLORIDE see DBM400
MBDZ see MHL000
MBH see PME500
MBK see HEV000
MBMA see MJQ250
MBNA see MHW500
MBOCA see MJM200
MBOT see MJO250
MB PYRETHROID see MBV700, MNG525
MBR 6168 see DRR000
MBR 8251 see TKF750
MBR 12325 see DUK000
MBR 3092-42 see MBV710
MBT see BDF000
MBTH see MHJ250
MBTS see BDE750
MBTS RUBBER ACCELERATOR see BDE750
MBX see MHR200
MBY see MHX250
6-MC see MIP750
2M-4C see CIR250
MC 338 see NIW500
MC 474 see DJI000
MC 1053 see CBW000
MC 1108 see DVJ400
MC 1415 see PMB250
MC 1478 see NIW500
MC 1488 see BRU750
MC 1945 see DVI800
MC 2188 see CDY299
MC 2303 see CBK500, GGS000
MC 3199 see BCP685
MC6897 see DQM600
MeC see MCB050
3-MCA see MIJ750
MCA-600 see BDG250
MeCsAc see EJJ500
MCA see CEA000
MCA-11,12-EPOXIDE see ECA500
MCA-11,12-OXIDE see ECA500
M. CARINICAUDUS DUMERILII VENOM see MQT000
4-(MCB) see CLN750
MCB see CEJ125
MCC see DEV600
Me-CCNU see CHD500
MC DEFOLIANT see MAE000
(E)-MC 11,12-DIHYDRODIOL see DML800
M.C. DUMERILLI VENOM see MQT000
MCE see EIF000
MCF see MIG000

MCH 52 see UTU500
M 4212 (pesticide) (9CI) see MAB050
MCI-C54875 see DXY000
MCIT see CEX275
MCM-JR-4584 see FLK100
MCN-485 see AJF500
MCN 1025 see NNF000
MCN 2559 see TGJ850
MCN-3113 see XGA725
MCN 2783-21-98 see ZUA300
MCNAMEE see KBB600
MCNEIL 481 see DQU200
MCN-JR-2498 see TKK500
MCN-JR-3345 see FHG000
MCN-JR-4584 see FGU000
MCN-JR-4749 see DYF200
MCN-JR-8299 see TDX750
MCN-JR-4263-49 see PDW750
MCO 8000 see MIF760
M1 (COPPER) see CNI000
M2 (COPPER) see CNI000
2M-4CP see CIR500
MCP 875 see AFJ400
MC 4000 cP see MIF760
MCP see CIR250
MCPA see CIR250
MCPABN see MEI000
MCPA DIETHANOLAMINE SALT see MIH750
MCPA-ETHYL see EMR000
MCPA SODIUM SALT see SIL500
MCPB see CLN750, CLO000
MCPB-ETHYL see EHL500
MCP-BUTYRIC see CLN750
4-(MCPD) see CLO000
MCPEE see EMR000
2-MCPP see CIR500
MCPP-D-4 see CIR500
MCPP see CIR500
MCPP 2,4-D see CIR500
MCPP-K-4 see CIR500
MCPP POTASSIUM SALT see CLO200
MCR see MAB750
MC 20000S see MIF760
MC-T see CAT775
MCT see CPV000, TMO000
MCZ NITRATE see MQS560
10-MD see MEK700
MD 141 see DIS600
516 MD see CMR100
MD 2028 see HAF400
2028 MD see HAF400
5579 MD see DAB875
MD 67350 see VGK000
68111 M.D. see IRQ000
MD 780515 see MEW800
l-(α-MD) see DNA800
MDA 150 see FMW330
MDA 220 see FMW330
MDA see MJQ000, MJQ775
3'-MDAB see DUH600
2-MeDAB see TLE750
MDAB see DUH600
MDBA see MEL500
MD BACITRACIN see BAC260
MDBCP see MBV720
MDEA see MKU250
bu-MDI see BRC500
pr-MDI see DPY600
MDI see MJP400
MDI-CR 100 see PKB100
MDI-CR 200 see PKB100
MDI-CR 300 see PKB100
MDI-CR see PKB100
M 141 DIHYDROCHLORIDE PENTAHYDRATE see SKY500
MDI-PC see DGE200
MDS see DBD750
2-ME see MCN250
ME 277 see PAM000
ME-1700 see BIM500
ME 3625 see DFD000
MEA 610 see NAC000
MEA see AJT250, EEC600, MCN750
MEAD JOHNSON 1999 see CCK250
MEADOW BRIGHT see MBU550
MEADOW GARLIC see WBS850

MEADOW GREEN see COF500
MEADOW ROSE LEEK see WBS850
MEADOW SAFFRON see CNX800
MEA HYDROCHLORIDE see EEC700
MEARLMAID see GLI000
MEASURIN see ADA725
MEAVERIN see CBR250
MEB 6447 see CJO250
MEB see MAS500
MEBACID see ALF250
MEBALLYMAL see SBM500
MEBALLYMAL SODIUM see SBN000
MEBANAZINE OXALATE see MBV750
MEBANAZINE SULPHATE see MHO000
MEBARAL see ENB500
MnEBD see MAS500
MEBENDAZOLE (USDA) see MHL000
MEBENVET see MHL000
MEBEREL see ENB500
MEBHYDROLIN NAPADISYLATE see MBW100
MEBICAR see MBW250
MEBICAR-A see MBW500
MEBICHLORAMINE see BIE500
MEBR see MHR200
ME4 BROMINAL see DDP000
MEBRON see MCH550
MEBROPHENHYDRAMINE see BMN250
MEBROPHENHYDRAMINE HYDROCHLORIDE see BMN250
MEBRYL see BMN250
MEBUBARBITAL see NBT500
MEBUBARBITAL SODIUM see NBU000
MEBUMAL NATRIUM see NBU000
MEBUMAL SODIUM see NBU000
MEBUTAMATE see MBW750
MEBUTINA see MBW750
MECADOX see FOI000
MECALMIN see CCR875
MECAMILAMINA (ITALIAN) see VIZ400
MECAMINE see VIZ400
MECAMINE HYDROCHLORIDE see MQR500
MECAMYLAMINE see VIZ400
MECAMYLAMINE HYDROCHLORIDE see MQR500
MECARBAM see DJI000
MECARBENIL see MBW780
MECAROL see EQL000
MECB see DJG000
ME-CCNU see CHD250
MECHLORETHAMINE see BIE250
MECHLORETHAMINE HYDROCHLORIDE see BIE500
MECHLORETHAMINE OXIDE see CFA500
MECHLORETHAMINE OXIDE HYDROCHLORIDE see CFA750
MECHOLYL see ACR000
MECHOTHANE see HOA500
MECICLIN see DAI485
MECINARONE see MBX000
MECLASTINE HYDROGEN FUMARATE see FOS100
MECLIZINE see HGC500
MECLIZINE DIHYDROCHLORIDE see MBX250
MECLIZINE HYDROCHLORIDE see MBX500
MECLOFENAMATE SODIUM see SIF425
MECLOFENAMIC ACID see DGM875
MECLOFENOXANE see DPE000
MECLOFENOXATE HYDROCHLORIDE see AAE500
MECLOMEN see SIF425
MECLOPHENAMIC ACID see DGM875
MECLOPHENOXATE see DPE000
MECLOZINE see HGC500
MECLOZINE HYDROCHLORIDE see MBX500
MECOBALAMIN see VSZ050
MECODIN see MDO750
MECODRIN see BBK000
MECOMEC see CIR500
MECOPEOP see CIR500
MECOPER see CIR500
MECOPEX see CIR500, CLO200
MECOPROP see CIR500
MECOPROP POTASSIUM SALT see CLO200
MECOTURF see CIR500
MECPROP see CIR500
MECRAMINE see AJT250
MECRILAT see MIQ075
MECROTHENE F see PJS750
MECRYL see CFU750
MECRYLATE see MIQ075
MECS see EJH500

MECYSTEINE HYDROCHLORIDE see MBX800
MEDAMYCIN see TBX250
MEDAPAN see COY500
MEDARON see NGG500
MEDARSED see BPF000
MEDAZEPAM see CGA000
MEDAZEPAM HYDROCHLORIDE see MBY000
MEDAZEPOL see CGA000
MEDEMANOL see MAW250
MEDEMYCIN see MBY150
MEDFALAN see SAX200
MEDIAMID see DJS200
MEDIAMYCETINE see CDP250
MEDIBEN see MEL500
MEDICAINE see BQA010
MEDI-CALGON see SHM500
MEDICEL see AKO500
MEDICON see DBE200
MEDIDRYL see BBV500
MEDIFENAC see AGN000
MEDIFLAVIN see DBX400
MEDIFLOR FC 43 see HAS000
MEDIFURAN see FPI000
MEDIGOXIN see MJD300
MEDIHALER-EPI see AES000, VGP000
MEDIHALER-TETRACAINE see BQA010
MEDILLA see MKP500
MEDINAL see BAG250
MEDINOTERB ACETATE see BRU750
MEDIQUIL see IPU000
MEDIREX see IJG000
MEDITERRANEAN BAY OIL see OGQ150
MEDITRENE see IEP200
MEDIUM BLUE see IBV050
MEDIUM BLUE EMBL see ADE750
MEDLEXIN see ALV000
MEDOMET see DNA800, MJE780
MEDOMIN see COY500
MEDOMINE see COY500
α-MEDOPA see MJE780
MEDOPAQUE see HGB200
MEDOPREN see DNA800, MJE780
2,4-MEDP see MBY500
4,4-MEDP see MBZ000
MEDPHALAN see SAX200
MEDROGESTERONE see MBZ100
MEDROGESTONE see MBZ100
MEDROGLUTARIC ACID see HMC000
MEDROL see MOR500
MEDROL ACETATE see DAZ117
MEDROL DOSEPAK see MOR500
MEDRONE see MOR500
MEDROXYPROGESTERONE ACETATE see MCA000
MEDROXYPROGESTERONE ACETATE and ETHINYLESTRADIOL see POF275
MEDULLIN see MCA025
MEE see EDP000
MEETHOBALM see BQA010
MEFEDINA see DAM700
MEFENAL see SNJ350
MEFENAMIC ACID see XQS000
MEFENOXALONA see MFD500
MEFENOXALONE see MFD500
MEFENSINA see GGS000
MEFLUIDIDE see DUK000
MEFOXIN see CCS510
MEFOXITIN see CCS510
MEFRUSID see MCA100
MEFRUSIDE see MCA100
MEFURINA see AHK750
M.E.G. see EJC500
MEGABA see AJC625
MEGABION see VSZ000
MEGABION (JAPANESE) see AOO475
MEGA BT see BAC040
MEGACE see VTF000
MEGACILLIN ORAL see PDT750
MEGACILLIN SUSPENSION see BFC750
MEGACILLIN TABLETS see BFD000
MEGACORT see DAE525
MEGADIURIL see CFY000
MEGALOMICIN A see MCA250
MEGALOMYCIN-A see MCA250
MEGALOVEL see VSZ000
MEGAMYCINE see MDO250
MEGAPHEN see CKP250, CKP500

MEGASEDAN see CGA000
MEGATOX see FFF000
MEGEPTIL see TFM100
MEGESTROL ACETATE see VTF000
MEGESTROL ACETATE 4 mg, ETHINYLOESTRADIOL 50 µg see MCA500
MEGESTROL ACETATE + ETHINYLOESTRADIOL see MCA500
MEGESTRYL ACETATE see VTF000
MEGIMIDE see MKA250
MEGLUMINE AMIDOTRIZOATE see AOO875
MEGLUMINE CONRAY see IGC000
MEGLUMINE DIATRIZOATE see AOO875
MEGLUMINE IOCARMATE see IDJ500
MEGLUMINE IODAMIDE see IDJ600
MEGLUMINE IODIPAMIDE see BGB315
MEGLUMINE IODOXAMATE see IFP800
MEGLUMINE IOTHALAMATE see IGC000
MEGLUMINE IOTROXATE see IGD075
MEGLUMINE ISOTHALAMATE see IGC000
MEGLUMINE SODIUM IODAMIDE see MCA775
6-ME-GLU-P-2 see AKS250
MEGLUTOL see HMC000
MEGUAN see DQR800
MEHENDI see HMX600
MEHP see MRI100
4-ME-I see MKU000
MEICELIN see CCS365
MEISEI SCARLET GG SALT see DEO295
MEISEI TERYL DIAZO BLACK CR see DPO200
MEISEI TERYL DIAZO BLUE HR see DCJ200
MEITO MY 30 see GGA800
MEK see MKA400
MEKAMINE see VIZ400
MEKAMIN HYDROCHLORIDE see MQR500
MEK-OXIME see EMU500
MEKP (OSHA) see MKA500
MEK PEROXIDE see MKA500
MELABON see ABG750
MELADININ see XDJ000
MELADININE see XDJ000
MELADUR MS 80 see MCB050
MELAFORM 45 see MCB050
MELAFORM 150 see MCB050
MELAFORM see MCB050
MELAFORM E 45 see MCB050
MELAFORM E 55 see MCB050
MELAFORM E 50 see MCB050
MELAFORM M 45S, see MCB050
MELAFORM WM6 see MCB050
MELAFORM WM 100 see MCB050
MELALEUCA ALTERNIFOLIA OIL see TAI150
MELALIT see MCB050
MELAMINE 20 see MCB050
MELAMINE 366 see MCB050
MELAMINE see MCB000
MELAMINE, polymer with FORMALDEHYDE see MCB050
MELAMINE-FORMALDEHYDE CONDENSATE see MCB050
MELAMINE-FORMALDEHYDE COPOLYMER see MCB050
MELAMINE-FORMALDEHYDE POLYMER see MCB050
MELAMINE-FORMALDEHYDE RESIN see MCB050
MELAMINE-FORMOL COPOLYMER see MCB050
MELAMINE, HEXACHLORO-(6Cl,7Cl,8Cl) see TNG275
MELAMINE, POLYMER with FORMALDEHYDE (8Cl) see MCB050
MELAMINE RESIN see MCB050
MELAMINKYANURAT (CZECH) see THO750
MELAMIN-N,N′,N″-TRIMETHYLSULFONSAURES NATRIUM (GERMAN) see THS500
MELAN 11 see UTU500
MELAN 15 see MCB050
MELAN 20 see MCB050
MELAN 22 see MCB050
MELAN 23 see MCB050
MELAN 26 see MCB050
MELAN 27 see MCB050
MELAN 28 see MCB050
MELAN 29 see MCB050
MELAN 125 see MCB050
MELAN 220 see MCB050
MELAN 243 see MCB050
MELAN 245 see MCB050
MELAN 287 see MCB050
MELAN 445 see MCB050
MELAN 523 see MCB050
MELAN 620 see MCB050
MELAN 630 see MCB050
MELAN 2000 see MCB050

MELAN 8000 see MCB050
MELAN 21A see MCB050
MELAN 28A see MCB050
MELAN 284A see MCB050
MELAN BLACK see BMA000
MELAN 28D see MCB050
MELANILINE see DWC600
MELANOL LP20T see SON000
MELANOMYCIN see MCB100
MELANOSPORIN see MCB250
MELANTHERINE BH see CMN800
MELANTHERINE BHX see CMN800
MELAN X 28 see MCB050
MELAN X 65 see MCB050
MELAN X 71 see MCB050
MELAPRET P see MCB050
MELAROM 3 see MCB050
MELARSONYL POTASSIUM SALT see PLK800
MELASIL K 1 see MCB050
MELASIL K 2 see MCB050
MELASIL K3 see MCB050
MELASIL K 1S see MCB050
MELASIL U 1 see MCB050
MELASIL U 2 see MCB050
MELASIL U see MCB050
MELATONIN see MCB350
MELATONINE see MCB350
MELBEX see MRX000
MELBIN see DQR600
MELDIAN see CKK000
MELDONE see CNU750
MELERIL see MOO250
MELETIN see QCA000
MELEX see MQR760
MEL-F see MCB050
MELIA AZEDARACH see CDM325
MELIFORM see PKF750
MELILOTAL see MFW250
MELILOTIN see HHR500
MELILOTOL see HHR500
MELIN see RSU000
MELINEX see PKF750
MELINITE see PID000
MELIPAN see MCB500
MELIPAX see CDV100
MELIPRAMIN see DLH600, DLH630
MELIPRAMINE see DLH600, DLH630
MELIPRAMINE HYDROCHLORIDE see DLH630
MELIPRAMIN HYDROCHLORIDE see DLH630
MEL-IRON A see MCB050
MELITASE see CKK000
MELITOSE see RBA100
MELITOXIN see BJZ000
MELITRACEN see AEG875
MELITRACENE see AEG875
MELITRACENE HYDROCHLORIDE see TDL000
MELITRACEN HYDROCHLORIDE see TDL000
MELITRIOSE see RBA100
MELITTIN see MCB525
MELITTIN-I see MCB525
MELLARIL see MOO250
MELLARIL HYDROCHLORIDE see MOO500
MELLERETTE see MOO250
MELLERETTEN see MOO250
MELLERIL see MOO250
MELLINESE see CKK000
MELLITE 131 see DEH650
MELLITE 825 see DVK200
MELLOSE see MIF760
MELOCHIA TOMENTOSA see BAR500
MELOCOTON (SPANISH) see AQP890
MELOGEL see SLJ500
MELOLAK B see MCB050
MELOLAK B-II see MCB050
MELOLAM 285 see MCB050
MELOLAM see MCB050
MELOPAS AMP 1 see MCB050
MELOPAS 183GF see MCB050
MELOPAS N 37601 see MCB050
MELOPLAST B see MCB050
MELOXINE see XDJ000
MELPHALAN (RUSSIAN) see BHV000
MELPHALAN see PED750
MELPHALAN HYDROCHLORIDE see BHV250
MELPREX see DXX400

MELSEDIN see MDT250
MELSEDIN BASE see QAK000
MELSMON see MCB535
MELSOMIN see QAK000
MELTROL see PDF250
MELUNA see SLJ500
MEL W see PLK800
MELYSIN see MCB550
MEMA see MEO750, MLF250
MEMC see MEP250
MEMCOZINE see SNN300
ME-MDA see MJO250
MEMINE see TCY750
MEMMI see MLF500
MEMPA see HEK000
MENADIOL see MMC250
MENADION see MMD500
MENADIONE see MMD500
MENADION-NATRIUM-BISULFIT TRIHYDRAT (GERMAN) see MMD750
MENAGEN see EDV000
MENAPHTAM see CBM750
MENAPHTHON see MMD500
MENAPHTONE see MMD500
MENAQUINONE-4 see VTA650
MENAQUINONE K₄ see VTA650
MENATENSINA see RCA200
MENATETRENONE see VTA650
MENAZON see ASD000
dd-MENCS see MLC000
MENDEL see MQU750
MENDI see HMX600
MENDIAXON see MKP500
MENDON see CDQ250
MENDRIN see EAT500
MENESIA see MAG000
MENEST see ECU750
MENETYL see EJR500
MENFORMON see EDV000
MENHYDRINATE see DYE600
MENICHLOPHOLAN see DFD000
MENIPHOS see MQR750
MENISPERMUM CANADENSE see MRN100
MENITE see MQR750
MENOCIL see ALX250
MENOGAROL see MCB600
MENOGEN see ECU750
MENOMYCIN see MRA250
MENONASAL see TBN000
MENOQUENS see MCA500
MENOSTILBEEN see DKA600
MENOTAB see ECU750
MENOTROL see ECU750
MENSISO see APY500
MENTA-BAL see ENB500
MENTHA CITRATA OIL see BFN990
p-MENTHA-1,8-DIEN-7-AL see DKX100
p-MENTHA-1,3-DIENE see MLA250
p-MENTHA-1,4-DIENE see MCB750
p-MENTHA-1,5-DIENE see MCC000
1,8(9)-p-MENTHADIENE see MCC250
(S)-(−)-p-MENTHA-1,8-DIENE see MCC500
d-p-MENTHA-1,8-DIENE see LFU000
p-MENTHA-1,8-DIENE see LFU000, MCC250
o-1,4-MENTHADIENE see MCB700
l-p-MENTHA-6,8-DIEN-2-OL see MKY250
p-MENTHA-1,8-DIEN-7-OL see PCI550
p-MENTHA-6,8-DIEN-2-OL, PROPIONATE see MCD000
1-6,8(9)-p-MENTHADIEN-2-ONE see CCM120
d-p-MENTHA-6,8,(9)-DIEN-2-ONE see CCM100
6,8(9)-p-MENTHADIEN-2-ONE see MCD250
(R)-(−)-p-MENTHA-6,8-DIEN-2-ONE see CCM120
p-MENTHA-6,8-DIEN-2-ONE see MCD250
1-p-MENTHA-6(8,9)-DIEN-2-YL ACETATE see CCM750
MENTHADIENYL FORMIATE see IHX450
(−)-trans-2-p-MENTHA-1,8-DIEN-3-YL-5-PENTYLRESORCINOL see CBD599
1-p-MENTHA-6,8(9)-DIEN-2-YL PROPIONATE see MCD000
p-MENTHANE-1,8-DIAMINE see MCD750
MENTHANE DIAMINE see MCD750
p-MENTHANE, 1,4-EPOXY- see IKC100
p-MENTHANE-8-HYDROPEROXIDE see MCE000
p-MENTHANE HYDROPEROXIDE see MCE000
p-MENTHANE HYDROPEROXIDE, TECHNICALLY PURE see IQE000
p-MENTHAN-3-OL see MCF750
p-MENTHAN-8-OL see MCE100
dl-3-p-MENTHANOL see MCG000

p-MENTHAN-8-OL ACETATE see DME400
l-p-MENTHAN-3-ONE see MCG275
(Z)-p-MENTHAN-3-ONE see IKY000
p-MENTHAN-3-ONE racemic see MCE250
o-1-MENTHENE see MCE275
p-MENTH-8-EN-1-OL see TBD775
8-p-MENTHEN-2-OL see DKV150
p-MENTH-8-EN-3-OL see MCE750
1-p-MENTHEN-4-OL see TBD825
p-MENTH-1-EN-8-OL see TBD500
t-MENTH-1-EN-8-OL see TBD775
8(9)-p-MENTHEN-3-OL see MCE750
p-MENTH-8-EN-2-OL, ACETATE see DKV160
p-MENTH-1-EN-8-OL (8CI) see TBD750
p-MENTH-1-EN-8-OL, FORMATE (mixed isomers) see TBE500
8-p-MENTHEN-2-ONE see DKV175
p-MENTH-8-EN-2-ONE see DKV175
p-MENTH-1-EN-3-ONE see MCF250
d-p-MENTH-4(8)-EN-3-ONE see MCF500
4(8)-p-MENTHEN-3-ONE see MCF500
p-MENTH-4(8)-EN-3-ONE see MCF500
p-MENTH-4(8)-EN-3-ONE, (R)-(+)- see POI615
p-MENTH-8-EN-2-YL ACETATE see DKV160
p-MENTH-1-EN-8-YL ISOBUTYRATE see MCF515
MENTHENYL KETONE see MCF525
1-(p-MENTHEN-6-YL)-1-PROPANONE see MCF525
MENTHEN-1-YL-8 PROPIONATE see TBE600
3-p-MENTHOL see MCG000
MENTHOL racemique (FRENCH) see MCG000
dl-MENTHOL see MCG000
l-MENTHOL see MCF750, MCG250
MENTHOL racemic see MCG000
MENTHOL see MCF750
MENTHOL, ACETATE (8CI) see MCG500
MENTHOL ACETOACETATE see MCG850
l-MENTHONE (FCC) see MCG275
p-MENTHONE see MCG275
MENTHONE, racemic see MCE250
MENTHONE see MCG275
trans-MENTHONE see MCG275
1-p-MENTH-3-YL ACETATE see MCG750
l-p-MENTH-3-YL ACETATE see MCG750
l-MENTHYL ACETATE (FCC) see MCG750
dl-MENTHYL ACETATE see MCG500
MENTHYL ACETATE racemic see MCG500
MENTHYL ACETATE see MCG500
(−)-MENTHYL ACETATE see MCG750
(−)-MENTHYL ACETOACETATE see MCG850
MENTHYL ACETOACETATE see MCG850
(−)-MENTHYL ALCOHOL see MCG250
p-MENTH-3-YL ESTER-dl-ACETIC ACID see MCG500
(-)-MENTHYL PHENYLACETATE see MCG910
MEOBAL see XTJ000
MEOBAL, NITROSATED (JAPANESE) see XTS000
MEONAL see BPF500
MEONINE see MDT740
MEOTHRIN see DAB825
MEP see EOS000
MEPACRINE see ARQ250
MEPACRINE DIHYDROCHLORIDE see CFU750
MEPACRINE HYDROCHLORIDE see CFU750
MEPADIN see DAM700
MEPAMTIN see MQU750
ME-PARATHION see MNH000
MEPARFYNOL CARBAMATE see MNM500
MEPATON see MNH000
MEPAVLON see MQU750
MEPAZIN see MOQ250
MEPAZINE BASE see MOQ250
MEPEDYL see PIZ250
MEPENICYCLINE see MCH525
MEPENTAMATE see MNM500
MEPENTAMATO see EQL000, MNM500
MEPENTIL see EQL000
MEPENZOLATE see CBF000
MEPENZOLATE BROMIDE see CBF000
MEPERIDIDE see FLV000
MEPERIDINE see DAM600
MEPERIDINE HYDROCHLORIDE see DAM700
MEPHABUTAZONE see BRF500
MEPHACYCLIN see TBX250
MEPHADRYL see BBV500
MEPHANAC see CIR250
MEPHASERPIN see RDK000
MEPHATE see GGS000

MEPHEDAN see GGS000
MEPHEDINE see DAM700
MEPHELOR see GGS000
MEPHENAMINE HYDROCHLORIDE see OJW000
MEPHENAMIN HYDROCHLORIDE see OJW000
MEPHENAMINIC ACID see XQS000
MEPHENESIN CARBAMATE see CBK500
MEPHENON see MDP750
MEPHENOXALONE see MFD500
MEPHENSIN see GGS000
MEPHENYTOIN see MKB250
MEPHEXAMIDE see DIB600, MCH250
MEPHOBARBITAL see ENB500
MEPHOBARBITONE see ENB500
MEPHOSAL see GGS000
MEPHOSFOLAN see DHH400
MEPHSON see GGS000
MEPHYTAL see ENB500
MEPHYTON see VTA000
MEPIBEN see LJR000
MEPICYCLINE PENICILLINATE see MCH525
MEPIOSINE see MQU750
MEPIPRAZOLE DIHYDROCHLORIDE see MCH535
MEPIPRAZOLE HYDROCHLORIDE see MCH535
MEPIRESERPATE HYDROCHLORIDE see MDW100
MEPIRIZOL see MCH550
MEPISERATE HYDROCHLORIDE see MDW100
MEPITIOSTANE see MCH600
dl-MEPIVACAINE see SBB000
MEPIVACAINE see SBB000
dl-MEPIVACAINE HYDROCHLORIDE see CBR250
MEPIVACAINE HYDROCHLORIDE see CBR250
MEPIVASTESIN see CBR250
MEPOSED see MQU750
MEP (Pesticide) see DSQ000
MEPRANIL see MQU750
MEPRIN (detoxicant) see MCI375
MEPRIN see MCI375
MEPRO see CIR500
MEPROBAM see MQU750
MEPROBAMAT (GERMAN) see MQU750
MEPROBAMATE see MQU750
MEPROBAMATO (ITALIAN) see MQU750
MEPROCOMPREN see MQU750
MEPROCON CMC see MQU750
MEPRODIL see MQU750
MEPRODINE (GERMAN) see NOE550
MEPROFEN see BDU500
MEPROLEAF see MQU750
MEPROMAZINE see MCI500
MEPROSAN see MQU750
MEPROSCILLARIN see MCI750
MEPROTABS see MQU750
MEPROZINE see MQU750
MEPRYLCAINE HYDROCHLORIDE see OJI600
MEPTIN see PME600
MEPTOX see MNH000
MEPTRAN see MQU750
MEPYRAMIN (GERMAN) see WAK000
MEPYRAMINE HYDROCHLORIDE see MCJ250
MEPYRAMINE MALEATE see DBM800
MEPYRAMINE 7-THEOPHYLLINE ACETATE see MCJ300
MEPYRAPONE see MCJ370
MEPYREN see WAK000
MEQUIN see QAK000
MEQUINOL see MFC700
MEQUITAZINE see MCJ400
MER 25 see DHS000
MER 29 see TMP500
MER-41 see CMX700
MERABITOL see AFH250
MERACTINOMYCIN see AEB000
MERADAN see AJU625
MERADANE see AJU625
MERAKLON see PMP500
MERALEIN DISODIUM see SIF500
MERALEN see TKH750
MERALLURIDE see MFC000, TEQ000
MERALLURIDE SODIUM see SIG000
MERAMEC M 25 see IHC550
MERANTINE BLUE AF see ERG100
MERANTINE BLUE EG see FAE000
MERANTINE BLUE V see CMM062
MERANTINE BLUE VF see ADE500
MERANTINE GREEN G see FAE950

MERANTINE GREEN SF see FAF000
MERATRAN see DWK400
MERBAPHEN see CCG500
MERBENTUL see CLO750
MERBROMIN see MCV000
MERCALEUKIN see POK000
MERCAMINE see AJT250
MERCAMINE DISULFIDE see MCN500
MERCAPTAMINE see AJT250
MERCAPTAN AMYLIQUE (FRENCH) see PBM000
MERCAPTAN METHYLIQUE (FRENCH) see MLE650
MERCAPTAN METHYLIQUE PERCHLORE (FRENCH) see PCF300
MERCAPTANS see MCJ500
MERCAPTAZOLE see MCO500
2-MERCAPTOACETANILIDE see MCK000
α-MERCAPTOACETANILIDE see MCK000
MERCAPTOACETATE see TFJ100
(MERCAPTOACETATO(2-)-O,S)METHYL-MERCURATE(1-), SODIUM see SIM000
2-MERCAPTOACETIC ACID see TFJ100
α-MERCAPTOACETIC ACID see TFJ100
MERCAPTOACETIC ACID see TFJ100
MERCAPTOACETIC ACID ACETATE see ACQ250
MERCAPTOACETIC ACID, DIESTER with DITHIO-p-UREIDOBENZENEARSO-
   NOUS ACID see CBI250
MERCAPTOACETIC ACID ETHYL ESTER see EMB200
MERCAPTOACETIC ACID-2-ETHYLHEXYL ESTER see EKW300
MERCAPTOACETIC ACID METHYL ESTER see MLE750
MERCAPTOACETIC ACID, SODIUM-BISMUTH SALT see BKX750
MERCAPTOACETIC ACID SODIUM SALT see SKH500
MERCAPTOACETONITRILE see MCK300
4-(MERCAPTOACETYL)MORPHOLINE O,O-DIMETHYL PHOSPHORODITHIOATE
   see MRU250
β-MERCAPTOAETHYLAMIN CHLORHYDRAT see MCN750
β-MERCAPTOALANINE see CQK000
6-MERCAPTO-2-AMINOPURINE see AMH250
2-MERCAPTO-5-AMINO-1,3,4-THIADIAZOLE see AKM000
4-MERCAPTOANILINE see AIF750
o-MERCAPTOANILINE see AIF500
p-MERCAPTOANILINE see AIF750
2-MERCAPTOBARBITURIC ACID see MCK500
7-MERCAPTOBENZ(a)ANTHRACENE see BBH750
2-MERCAPTOBENZIMIDAZOLE see BCC500
2-MERCAPTOBENZIMIDAZOLE ZINC SALT (2:1) see ZIS000
MERCAPTOBENZIMIDAZOLE ZINC SALT see ZIS000
o-MERCAPTOBENZOESAEURE (GERMAN) see MCK750
o-MERCAPTOBENZOIC ACID see MCK750
o-MERCAPTOBENZOIC ACID, DIESTER with DITHIO-p-UREIDOBENZENEARSO-
   NOUS ACID see TFD750
2-MERCAPTOBENZOIMIDAZOLE see BCC500
MERCAPTOBENZOIMIDAZOLE see BCC500
2-MERCAPTOBENZOTHIAZOLE see BDF000
MERCAPTOBENZOTHIAZOLE see BDF000
2-MERCAPTOBENZOTHIAZOLEDISULFIDE see BDE750
2-MERCAPTOBENZOTHIAZOLE SODIUM DERIVATIVE see SIG500
2-MERCAPTOBENZOTHIAZOLE SODIUM SALT see SIG500
2-MERCAPTOBENZOTHIAZOLE ZINC SALT see BHA750
2-MERCAPTOBENZOTHIAZYLDISULFIDE see BDE750
2-MERCAPTOBENZOXAZOLE see MCK900
(MERCAPTOBUTANEDIOATO(1-))GOLD DISODIUM SALT see GJC000
(2-MERCAPTOCARBAMOYL)DI-ACETANILIDE see MCL500
MERCAPTOCYCLOPENTANE see CPW300
MERCAPTODIACETIC ACID see MCM750
MERCAPTO DI-ACETIC ACID, ETHYLENE ESTER see MCN000
MERCAPTODIMETHUR (DOT) see DST000
1-MERCAPTODODECANE see LBX000
2-MERCAPTOETHANESULFONIC ACID MONOSODIUM SALT see MDK875
2-MERCAPTOETHANOL see MCN250
β-MERCAPTOETHANOL see MCN250
MERCAPTOETHANOL see MCN250
(2-MERCAPTOETHYL)AMINE see AJT250
β-MERCAPTOETHYLAMINE see AJT250
β-MERCAPTOETHYLAMINE DISULFIDE see MCN500
2-MERCAPTOETHYLAMINE HYDROCHLORIDE see MCN750
β-MERCAPTOETHYLAMINE HYDROCHLORIDE see MCN750
MERCAPTOETHYLAMINE HYDROCHLORIDE see MCN750
2-MERCAPTOETHYLAMINE (OXIDIZED) see MCN500
N-(2-MERCAPTOETHYLBENZENESULFONAMIDE)-S-(O,O-DIISOPROPYL PHOS-
   PHORODITHIOATE) see DNO800
(2-MERCAPTOETHYL)CARBAMIC ACID, ETHYL ESTER, S-ESTER with O,O-DI-
   METHYL PHOSPHORODITHIOATE see EMC000
N-(2-MERCAPTOETHYL)DIMETHYLAMINE HYDROCHLORIDE see DOY600
2-MERCAPTOETHYL TRIMETHOXY SILANE see MCO250
(2-MERCAPTOETHYL)TRIMETHYLAMMONIUM S-ESTER with O,O'-DIETHYL-
   PHOSPHOROTHIOATE see PGY600
(2-MERCAPTOETHYL)TRIMETHYLAMMONIUM IODIDE ACETATE see ADC300

(2-MERCAPTOETHYL)TRIMETHYLAMMONIUM IODIDE S-ESTER with O,O-DIE-
  THYL PHOSPHOROTHIOATE see TLF500
MERCAPTOFOS (RUSSIAN) see DAO500
1-MERCAPTOGLYCEROL see MRM750
6-MERCAPTOGUANINE see AMH250
6-MERCAPTOGUANOSINE see TFJ500
2-MERCAPTO-4-HYDROXY-6-METHYLPYRIMIDINE see MPW500
2-MERCAPTO-4-HYDROXY-6-N-PROPYLPYRIMIDINE see PNX000
2-MERCAPTO-4-HYDROXYPYRIMIDINE see TFR250
2-MERCAPTOIMIDAZOLE see IAO000
2-MERCAPTOIMIDAZOLINE see IAQ000
MERCAPTOMERIN SODIUM see TFK270
(MERCAPTOMETHYL)BENZENE see TGO750
3-(MERCAPTOMETHYL)-1,2,3-BENZOTRIAZIN-4(3H)-ONE-O,O-DIMETHYL PHOS-
  PHORODITHIOATE see ASH500
3-(MERCAPTOMETHYL)-1,2,3-BENZOTRIAZIN-4(3H)-ONE-O,O-DIMETHYL PHOS-
  PHORODITHIOATE-S-ESTER see ASH500
(trans)-2-MERCAPTOMETHYLCYCLOBUTYLAMINE HYDROCHLORIDE see
  AKL750
2-MERCAPTO-1-METHYLIMIDAZOLE see MCO500
N-(2-MERCAPTO-2-METHYL-1-OXOPROPYL)-l-CYSTEINE see MCO775
N-(2-MERCAPTO-2-METHYL-1-OXOPROPYL)-l-CYSTEINE SODIUM SALT see
  SAY875
1-(d-3-MERCAPTO-2-METHYL-1-OXOPROPYL)-l-PROLINE (S,S) see MCO750
1-(3-MERCAPTO-2-METHYL-1-OXOPROPYL)-l-PROLINE see MCO750
N-(MERCAPTOMETHYL)PHTHALIMIDE S-(O,O-DIMETHYL PHOSPHORODITH-
  IOATE) see PHX250
N-(2-MERCAPTO-2-METHYLPROPANOYL)-l-CYSTEINE see MCO775
N-(2-MERCAPTO-2-METHYLPROPANOYL)-l-CYSTEINE SODIUM SALT see
  SAY875
1-((2S)-3-MERCAPTO-2-METHYLPROPIONYL)-l-PROLINE see MCO750
2-(MERCAPTOMETHYL)PYRIDINE see POR790
2-MERCAPTO-6-METHYL-4-PYRIMIDONE see MPW500
2-MERCAPTO-6-METHYLPYRIMID-4-ONE see MPW500
MERCAPTOMETHYLTRIETHOXYSILANE see MCP000
2-MERCAPTO-N-(2-METHOXYETHYL)-ACETAMIDE S-ESTER with O,O-DI-
  METHYL PHOSPHORODITHIOATE see AHO750
2-MERCAPTONAPHTHALENE see NAP500
β-MERCAPTONAPHTHALENE see NAP500
5-MERCAPTO-1-PHENYLTETRAZOLE see PGJ750
5-MERCAPTO-3-PHENYL-2H-1,3,4-THIADIAZOLE-2-THIONE see MCP500
MERCAPTOPHOS see DAO600, FAQ900
2-MERCAPTOPROPANE see IMU000
3-MERCAPTO-1,2-PROPANEDIOL see MRM750
1-MERCAPTO-2,3-PROPANEDIOL see MRM750
α-MERCAPTOPROPANOIC ACID see TFK250
β-MERCAPTOPROPANOIC ACID see MCQ000
3-MERCAPTOPROPANOL see PML500
2-MERCAPTOPROPIONIC ACID see TFK250
3-MERCAPTOPROPIONIC ACID see MCQ000
α-MERCAPTOPROPIONIC ACID see TFK250
MERCAPTOPROPIONIC ACID, DIBUTYLTIN SALT see DEJ200
(2-MERCAPTOPROPIONYL)GLYCINE see MCI375
α-MERCAPTOPROPIONYLGLYCINE see MCI375
N-(2-MERCAPTOPROPIONYL)GLYCINE see MCI375
MERCAPTOPROPIONYLGLYCINE see MCI375
2-MERCAPTO-6-PROPYLPYRIMID-4-ONE see PNX000
2-MERCAPTO-6-PROPYL-4-PYRIMIDONE see PNX000
3-MERCAPTOPROPYLTRIMETHOXYSILANE see TLC000
γ-MERCAPTOPROPYLTRIMETHOXYSILANE see TLC000
6-MERCAPTOPURIN see POK000
MERCAPTOPURIN (GERMAN) see POK000
6-MERCAPTOPURINE see POK000
6-MERCAPTOPURINE MONOHYDRATE see MCQ100
6-MERCAPTOPURINE 3-N-OXIDE see MCQ250
MERCAPTOPURINE-3-N-OXIDE see MCQ250
6-MERCAPTOPURINE 3-N-OXIDE MONOHYDRATE see OMY800
MERCAPTOPURINE RIBONUCLEOSIDE see MCQ500
6-MERCAPTOPURINE RIBOSIDE see MCQ500
2-MERCAPTOPYRIDINE see TFP300
2-MERCAPTOPYRIDINE MONOXIDE see MCQ700
2-MERCAPTOPYRIDINE-N-OXIDE SODIUM SALT see MCQ750
2-MERCAPTO-4-PYRIMIDINOL see TFR250
2-MERCAPTO-4-PYRIMIDONE see TFR250
2-MERCAPTOPYRIMID-4-ONE see TFR250
2-MERCAPTOQUINOLINE see QOJ100
MERCAPTOSUCCINIC ACID see MCR000
MERCAPTOSUCCINIC ACID ANTIMONATE(III) HEXALITHIUM SALT see
  LGU000
MERCAPTOSUCCINIC ACID DIETHYL ESTER see MAK700
MERCAPTOSUCCINIC ACID, GOLD SODIUM SALT see GJC000
MERCAPTOSUCCINIC ACID-S-ANTIMONY DERIVATIVE LITHIUM SALT see
  LGU000
MERCAPTOSUCCINIC ACID, THIOANTIMONATE(III), DILITHIUM SALT see
  LGU000

p-MERCAPTO SULFADIAZINE see MCR250
7-MERCAPTO-1,3,4,6-TETRAZAINDENE see POK000
2-MERCAPTOTHIAZOLINE see TFS250
MERCAPTOTHION see MAK700
MERCAPTOTION (SPANISH) see MAK700
α-MERCAPTOTOLUENE see TGO750
m-MERCAPTOTOLUENE see TGO800
o-MERCAPTOTOLUENE see TGP000
p-MERCAPTOTOLUENE see TGP250
3-MERCAPTO-1H-1,2,4-TRIAZOLE see THT000
d,3-MERCAPTOVALINE see MCR750
dl-α-MERCAPTOVALINE see PAP500
d-MERCAPTOVALINE see MCR750
3-MERCAPTO-dl-VALINE (9CI) see PAP500
(R)-MERCAPTURIC ACID see ACH000
MERCAPTURIC ACID see ACH000
MERCAPURIN see POK000
MERCARDAN see SIG000
MERCATE 5 see SAA025
MERCAZOLYL see MCO500
MERCHLORATE see MEP250
MERCHLORETHANAMINE see BIE500
MERCK 261 see DTS500
MERCKOGEN 6000 see AAX250
MERCLORAN see CHX250
MERCOL 25 see DXW200
MERCORAL see CHX250
MERCUFENOL CHLORIDE see CHW675
MERCUHYDRIN see TEQ000
MERCUMATILIN SODIUM see MCS000
MERCUPURIN see MCV750
MERCURAM see TFS350
MERCURAMIDE see SIH500
MERCURAN see MEO750
MERCURANINE see MCV000
MERCURATE(1-), ACETATOPHENYL-, AMMONIUM SALT see PFO550
MERCURE (FRENCH) see MCW250
MERCURETIN see SIG000
MERCURHYDRIN SODIUM see SIG000
MERCURIACETATE see MCS750
MERCURIALIN see MGC250
2,2′-MERCURIBIS(6-ACETOXYMERCURI-4-NITRO)ANILINE see MCS250
MERCURIBIS(DIETHYL(2,2-DIMETHYL-4-DITHIOCARBOXYAMI-
  NO))BUTYLAMMONIUM DICHLORIDE see MCS500
MERCURIBIS-o-NITROPHENOL see MCS600
MERCURIC ACETATE see MCS750
MERCURIC AMMONIUM CHLORIDE, solid see MCW500
MERCURIC ARSENATE see MDF350
MERCURIC BASIC SULFATE see MDG000
MERCURIC BENZOATE, solid (DOT) see MCX500
MERCURIC BENZOATE see MCX500
MERCURIC BROMIDE see MCY000
MERCURIC BROMIDE, solid see MCY000
MERCURIC CHLORIDE (DOT) see MCY475
MERCURIC CHLORIDE, AMMONIATED see MCW500
MERCURIC CHLORIDE-1,4-OXATHIANE see OMA000
MERCURIC CYANIDE, solid (DOT) see MDA250
MERCURIC DIACETATE see MCS750
MERCURIC-8,8-DICAFFEINE see MCT000
MERCURIC DINAPHTHYLMETHANE DISULPHONATE see MCT250
MERCURIC IODIDE see MDD000
MERCURIC IODIDE, solid see MDD000
MERCURIC IODIDE, solution see MDD000
MERCURIC IODIDE, RED see MDD000
MERCURIC LACTATE see MDD500
MERCURIC NITRATE see MDF000
MERCURIC OLEATE, solid (DOT) see MDF250
MERCURIC OXIDE, solid (DOT) see MCT500
MERCURIC OXIDE see MCT500
MERCURIC OXIDE, RED see MCT500
MERCURIC OXIDE, YELLOW see MCT500
MERCURIC OXYCYANIDE, solid (desensitized) (DOT) see MDA500
MERCURIC OXYCYANIDE see MDA500
MERCURIC PEROXYBENZOATE see MCT750
MERCURIC POTASSIUM CYANIDE (DOT) see PLU500
MERCURIC POTASSIUM CYANIDE, solid (DOT) see PLU500
MERCURIC POTASSIUM IODIDE, solid (DOT) see NCP500
MERCURIC POTASSIUM IODIDE see NCP500
MERCURIC SALICYLATE, solid (DOT) see MCU000
MERCURIC SALICYLATE see MCU000
MERCURIC SUBSULFATE, solid see MDG000
MERCURIC SULFATE, solid see MDG500
MERCURIC SULFOCYANATE see MCU250
MERCURIC SULFOCYANIDE see MCU250
MERCURIC SULFOCYANTE, solid (DOT) see MCU250

MERCURIC THIOCYANATE, solid (DOT) see MCU250
MERCURIC THIOCYANATE see MCU250
MERCURIDIACETALDEHYDE see BJW800
N,N'-MERCURIDIANILINE see DCJ000
3,3'-MERCURIDI-2-PROPYN-1-OL see BKJ250
MERCURIDISALICYLIC ACID, DISODIUM SALT see SJR000
MERCURI-HEMATOPORPHYRIN DISODIUM SALT see HAP000
MERCURIO (ITALIAN) see MCW250
MERCURIPHENOLDISULFONATE SODIUM see MCU500
MERCURIPHENYL ACETATE see ABU500
MERCURIPHENYL CHLORIDE see PFM500
MERCURIPHENYL NITRATE see MCU750
MERCURISALICYLIC ACID see MCU000
MERCURITAL see SIH500
MERCUROCHLORID (DUTCH) see MCW000
MERCUROCHROME see MCV000
MERCUROCHROME-220 SOLUBLE see MCV000
MERCUROCOL see MCV000
MERCUROL see MCV250
MERCUROME see MCV000
MERCUROPHAGE see MCV000
MERCUROPHEN see MCV500
MERCUROPHYLLINE see MCV750
MERCUROTHIOLATE see MDI000
MERCUROUS ACETATE, solid (DOT) see MDE250
MERCUROUS ACETATE see MDE250
MERCUROUS AZIDE (DOT) see MCX000
MERCUROUS BROMIDE, solid (DOT) see MCX750
MERCUROUS CHLORIDE see MCW000, MCY300
MERCUROUS GLUCONATE, solid (DOT) see MDC500
MERCUROUS GLUCONATE see MDC500
MERCUROUS IODIDE see MDC750
MERCUROUS NITRATE, solid (DOT) see MDE750
MERCUROUS OXIDE, BLACK, solid (DOT) see MDF750
MERCUROUS SULFATE, solid (DOT) see MDG250
MERCURY see MCW250
MERCURY(2+) ACETATE see MCS750
MERCURY(II) ACETATE see MCS750
MERCURY ACETATE see MCS750, MDE250
MERCURY (II) ACETATE, PHENYL- see ABU500
MERCURY, (ACETATO-O)PHENYL-, AMMONIATE see PFO550
MERCURY, ACETOXY(2-METHOXYETHYL)- see MEO750
MERCURY, ACETOXYPHENYL- see ABU500
MERCURY ACETYLIDE (DOT) see MCW349
MERCURY(II) ACETYLIDE see MCW350
MERCURY ACETYLIDE see MCW349
MERCURY AMIDE CHLORIDE see MCW500
MERCURY AMINE CHLORIDE see MCW500
MERCURY AMMONIATED see MCW500
MERCURY(II) AZIDE see MCX250
MERCURY(I) AZIDE see MCX000
MERCURY AZIDE see MCX000
MERCURY(II) BENZOATE see MCX500
MERCURY BICHLORIDE see MCY475
MERCURY BINIODIDE see MDD000
MERCURY, BIS(ACETATO)(mu-(3',6'-DIHYDROXY-2',7'-FLUORANDIYL))DI- see FEV100
MERCURY BIS(CHLOROACETYLIDE) see MCX600
MERCURY, BIS(4-HYDROXY-3-NITROPHENYL)- see MCS600
MERCURY BISULFATE see MDG500
MERCURY(I) BROMATE see MCX700
MERCURY(I) BROMIDE (1:1) see MCX750
MERCURY(II) BROMIDE (1:2) see MCY000
MERCURY(II) BROMIDE COMPLEX with TRIS(2-ETHYLHEXYL) PHOSPHITE see MCY250
MERCURY, BROMOHEXYL see HFR100
MERCURY, BROMO(2-HYDROXYETHYL)-, compound with AMMONIA (1:0.8 moles) see BNL275
MERCURY, (3-(α-CARBOXY-o-ANISAMIDO)-2-HYDROXYPROPYL)HYDROXY- see NCM800
MERCURY, (3-(α-CARBOXYMETHOXYPROPYL)HYDROXY) MONOSODIUM SALT, COMPOUNDED with THEOPHYLLINE (1:1) see MDH750
MERCURY(II) CHLORIDE see MCY475
MERCURY(I) CHLORIDE see MCW000
MERCURY CHLORIDE see MCY300
MERCURY(II) CHLORIDE COMPLEX with TRIS(2-ETHYLHEXYL) PHOSPHITE see MCY500
MERCURY(II) CHLORITE see MCY755
MERCURY (E)-CHLORO(2-(3-BROMOPROPIONAMIDO)CYCLOHEXYL) see CET000
MERCURY, CHLORO(2-HYDROXYPHENYL)- see CHW675
MERCURY COMPOUNDS, INORGANIC see MCZ000
MERCURY COMPOUNDS, ORGANIC see MDA000
MERCURY(I) CYANAMIDE see MDA100
MERCURY(II) CYANATE see MDA150

MERCURY(II) CYANIDE see MDA250
MERCURY CYANIDE OXIDE see MDA500
MERCURY-O,O-DI-n-BUTYL PHOSPHORODITHIOATE see MDA750
MERCURY DICHLORITE see MCY755
MERCURY, DIISOAMYL- see DNL200
MERCURY, DIMETHYL see DSM450
MERCURY(II) aci-DINITROMETHANIDE see MDA800
MERCURY DITHIOCYANATE see MCU250
MERCURY(II) EDTA COMPLEX see MDB250
MERCURY(II) FLUOROACETATE see MDB500
MERCURY(II) FORMOHYDROXAMATE see MDB775
MERCURY(II) FULMINATE see MDC000
MERCURY FULMINATE, wetted with not <20% water, or mixture (UN 0135) (DOT) see MDC000
MERCURY(I) GLUCONATE see MDC500
MERCURY, HYDROXY(4-HYDROXY-3-NITROPHENYL)- see HLO400
MERCURY, HYDROXY(NITRATO)DIPHENYLDI- see MDH500
MERCURY, HYDROXYPHENYL- see PFN100
MERCURY, (4-HYDROXY-m-PHENYLENE)BIS(ACETATO)- see HNK575
MERCURY, HYDROXYPHENYL-, cmpd. with NITRATOPHENYLMERCURY (1:1) see MDH500
MERCURY IODIDE (DOT) see MDC750
MERCURY IODIDE, solution (DOT) see MDC750
MERCURY(II) IODIDE see MDD000
MERCURY(I) IODIDE see MDC750
MERCURY(II) IODIDE (solution) see MDD250
MERCURYL ACETATE see MCS750
MERCURY(2+) LACTATE see MDD500
MERCURY, METALLIC (DOT) see MCW250
MERCURY METHYLCHLORIDE see MDD750
MERCURY(II) METHYLNITROLATE see MDE000
MERCURY, (2-METHYL-5-NITROPHENOLATO(2-)-C⁶,O')-(9CI) see NHK900
MERCURY MONOACETATE see MDE250
MERCURY MONOCHLORIDE see MCW000
MERCURY-2-NAPHTHALENEDIAZONIUM TRICHLORIDE see MDE500
MERCURY(I) NITRATE (1:1) see MDE750
MERCURY(II) NITRATE (1:2) see MDF000
MERCURY NITRATE see MDF000
MERCURY NITRIDE see TKW000
MERCURY(II) 5-NITROTETRAZOLIDE see MDF100
MERCURY NUCLEATE, solid (DOT) see MCV250
MERCURY, (9-OCTADECENOATO-O)PHENYL-, (Z)-(9CI) see PFP100
MERCURY OLEATE see MDF250
MERCURY, (OLEATO)PHENYL- see PFP100
MERCURY(II) ORTHOARSENATE see MDF350
MERCURY(II) OXALATE see MDF500
MERCURY(II) OXIDE see MCT500
MERCURY(I) OXIDE see MDF750
MERCURY OXIDE SULFATE see MDG000
MERCURY OXYCYANIDE see MDA500
MERCURY(II) PERCHLORATE see MDG200
MERCURY PERCHLORIDE see MCY475
MERCURY PERNITRATE see MDF000
MERCURY(II) PEROXYBENZOATE see MCT750
MERCURY PERSULFATE see MDG500
MERCURY(II) POTASSIUM IODIDE see NCP500
MERCURY PROTOCHLORIDE see MCW000
MERCURY PROTOIODIDE see MDC750
MERCURY SALICYLATE see MCU000
MERCURY and SODIUM PHENOLSULFONATE see MCU500
MERCURY SUBCHLORIDE see MCY300
MERCURY SUBSALICYLATE see MCU000
MERCURY(II) SULFATE (1:1) see MDG500
MERCURY(I) SULFATE see MDG250
MERCURY, SULFATOBIS(METHYL- see BKS810
MERCURY(II) SULFIDE see MDG750
MERCURY TETRAVANADATE see MDH000
MERCURY THIOCYANATE (DOT) see MCU250
MERCURY(II) THIOCYANATE see MCU250
MERCURY(I) THIONITROSYLATE see MDH250
MERCURY, (((p-TOLYL)SULFAMOYL)IMINO)BIS(METHYL- see MLH100
MERCURY ZINC CHROMATE COMPLEX see ZJA000
MERCUSAL see SIH500
MERCUTAL see PFO750
MERCUZANTHIN see MCV750
MEREPRINE see PGE775
MEREX see KEA000
MERFALAN see BHT750
MERFAMIN see MDI000
MERFAZIN see PFM500
MERFEN see PFP250
MERFEN-STYLI see MDH500
MERGAMMA see ABU500
MERGE see MRL750

MERIAN see AIF000
MERIDIL see MNQ000
MERILID see CHX250
MERINAX see POA250
MERIT see WBJ700
MERITAL see NMV725
MERITIN see DDT300
MERIZONE see BRF500
MERKAPOL P see MCB050
MERKAPOL PG see MCB050
MERKAPTOBENZIMIDAZOL (CZECH) see BCC500
2-MERKAPTOBENZOTIAZOL see BDF000
2-MERKAPTOBENZTHIAZOL see BDF000
5-MERKAPTO-3-FENYL-1,3,4-THIADIAZOL-2-THION DRASELNY (CZECH) see MCP500
2-MERKAPTOIMIDAZOLIN (CZECH) see IAQ000
6-MERKAPTOPURIN, MONOHYDRAT see MCQ100
MERKAZIN see BKL250
MERMETH see SNJ000
MERN see POK000
MEROCYANINE 540 see EDD500
MERODICEIN see SIF500
MERONIDAL see MMN250
MEROPENIN see PDT500
o-MEROPHAN see BHT250
MEROPHAN see BHT250
MEROXYL see AOR500
MEROXYLAN see AOR500
MEROXYLAN-WANDER see AOR500
MEROXYL-WANDER see AOR500
MERPACYL BLUE SK see CMM090
MERPAN see CBG000
o-MERPHALAN see BHT750
MERPHALAN see BHT750
MERPHALAN HYDROCHLORIDE see BHV000
MERPHEN see MDH500
MERPHENE see MDH500
MERPHENYL NITRATE see MCU750, MDH500
MERPHOS see TIG250
MERPHYLLIN see HAP000
MERPHYRIN see HAP000
MERPOL see EJN500
MERQUAT 100 see DTS500
MERRILLITE see ZBJ000
MERSALIN see SIH500
MERSALYL see SIH500
MERSALYL THEOPHYLLINE see MDH750
MERSALYL with THEOPHYLLINE see SAQ000
MERSOLITE 1 see PFN100
MERSOLITE 2 see PFM500
MERSOLITE 7 see MCU750
MERSOLITE 8 see ABU500
MERSOLITE see ABU500
MERTEC see TEX000
MERTESTATE see TBF500
MERTHIOLATE see MDI000
MERTHIOLATE SALT see MDI000
MERTHIOLATE SODIUM see MDI000
MERTIONIN see MDT740
MERTORGAN see MDI000
MERURAN see MDI200
MERVAMINE see DJL000
MERVAN see AGN000
MERVAN ETHANOLAMINE SALT see MDI225
MERXIN see CCS510
MERZONIN SODIUM see MDI000
MESA see TGE300
MESACONATE see MDI250
MESACONIC ACID see MDI250
MESAMATE see MRL750
MESAMATE CONCENTRATE see MRL750
MESANOLON see MJE760
MESANTOIN see MKB250
MESATON see NCL500
MESCAL BEAN see NBR800
MESCALINE see MDI500
MESCALINE ACID SULFATE see MDJ000
MESCALINE HYDROCHLORIDE see MDI750
MESCALINE SULFATE see MDJ000
MESCOPIL see SBH500
MESECLAZONE see CIL500
MESENTOL see ENG500
MESEREIN see MDJ250
MESIDICAINE HYDROCHLORIDE see DHL800
MESIDIN (CZECH) see TLG500

MESIDINE see TLG500
MESIDINE HYDROCHLORIDE see TLH000
MESITOIC ACID see IKN300
MESITOL see MDJ740
MESITYL ALCOHOL see MDJ740
MESITYLAMINE see TLG500
MESITYLAMINE HYDROCHLORIDE see TLH000
2-MESITYLBENZOFURAN-3-YL (p-METHOXYPHENYL) KETONE see AOY300
MESITYLENE see TLM050
MESITYLENE, 2,4-DINITRO- see DUW505
α¹,α³-MESITYLENEDIOL, 2-HYDROXY-(7CI,8CI) see HLV100
MESITYLENE, 2,4,6-TRINITRO- see TMI800
MESITYLENIC ACID see MDJ748
MESITYL ESTER of 1-PIPERIDINEACETIC ACID HYDROCHLORIDE see FAC175
MESITYLOXID (GERMAN) see MDJ750
MESITYL OXIDE see MDJ750
MESITYL OXIDE (1-PHTHALAZINYL)HYDRAZONE see DQU400
MESITYLOXYDE (DUTCH) see MDJ750
2-MESITYLOXYDIISOPROPYLAMINE HYDROCHLORIDE see MDK000
N-(2-MESITYLOXYETHYL)-N-METHYL-2-(2-METHYLPIPERIDINO)ACETAMIDE HYDROCHLORIDE see MDK250
N-(2-MESITYLOXYETHYL)-N-METHYL-2-(MORPHOLINE)ACETAMIDE HYDROCHLORIDE see MRT250
(1-MESITYLOXY-2-PROPYL)-N-METHYLCARBAMIC ACID-2-(DIETHYLAMINO)ETHYL ESTER, HYDROCHLORIDE see MDK500
N-(1-MESITYLOXY-2-PROPYL)-N-METHYL-2-(2-METHYLPIPERIDINO) ACETAMIDE HYDROCHLORIDE see MDK750
MESNA see MDK875
MESNUM see MDK875
MESOCAINE HYDROCHLORIDE see DHL800
MESOFOLIN see MDV000
MESOKAIN HYDROCHLORIDE see DHL800
MESOMILE see MDU600
MESONEX see HFG550
MESOPIN see MDL000
MESORANIL see ASG250
MESORIDAZINE see MON750
MESOTAL see HFG550
MESOXALONITRILE see MDL250
MESOXALONITRILE, (m-CHLOROPHENYL)HYDRAZONE (8CI) see CKA550
MESOXALYLCARBAMIDE see AFT750
MESOXALYLCARBAMIDE MONOHYDRATE see MDL500
MESOXALYLUREA see AFT750
MESOXALYLUREA MONOHYDRATE see MDL500
MESPAFIN see HGP550
MESTALONE see MJE760
MESTANOLONE see MJE760
MESTENEDIOL see AOO475
MESTERONE see MPN500
MESTINON see MDL600
MESTRANOL see MKB750
MESTRANOL mixed with ANAGESTONE ACETATE (1:10) see AOO000
MESTRANOL mixed with CHLOROETHYNYL NORGESTREL (1:20) see CHI750
MESTRANOL mixed with ETHYNERONE (1:20) see EQJ000
MESTRANOL mixed with ETHYNODIOL see EQK100
MESTRANOL mixed with ETHYNODIOL DIACETATE see EQK010
MESTRANOL mixed with LYNESTRENOL see LJE000
MESTRANOL mixed with LYNESTROL see LJE000
MESTRANOL mixed with NORETHINDRONE see MDL750
MESTRANOL mixed with NORETHISTERONE see MDL750
MESTRANOL mixed with NORETHYNODREL see EAP000
MESTRANOL mixed with NORGESTREL see NNR000
MESTRENOL see MKB750
MESULFA see ALF250
MESUPRINE see MDM000
MESUPRINE HYDROCHLORIDE see MDM000
MESURAL see LFK000
MESUROL see DST000
MESUXIMIDE see MLP800
MESYLITH see CLD000
META see TDW500
METAARSENIC ACID see ARB000
METABARBITAL see DJO800
META BLACK see AQP000
METABOLITE C see SOA500
METABOLITE I see HNI500
METACAINE see EFX500
METACARDIOL see AKT000
METACE see CLO750
METACEN see IDA000
METACETALDEHYDE see TDW500
METACETONE see DJN750
METACETONIC ACID see PMU750

METACHLOR see CFX000
METACHLORPHENPROP see CJQ300
METACHROME RED F see CMO870
METACHROME YELLOW see SIT850
METACHROME YELLOW RA see SIT850
METACIDE see MNH000
METACIL see MPW500
METACIN see ORQ000
METACLOPROMIDE see AJH000
METACORTANDRACIN see PLZ000
METACORTANDRALONE see PMA000
METACRATE see MIB750
METADEE see VSZ100
METADELPHENE see DKC800
METADIAZINE see PPO750
METADIAZOL BROWN 450 see CMO825
METADOMUS see MDO250
METAFOS see MNH000, SII500
METAFUME see MHR200
METAHYDRIN see HII500
METAHYDROXYPROCAINE see DHO600
METAISOSEPTOX see DAP400
METAISOSYSTOX see DAP400
METAISOSYSTOX-SOLFON 20 315 see DAP600
METAISOSYSTOXSULFOXIDE see DAP000
METAKRYLAN METYLU (POLISH) see MLH750
METALAXIL see MDM100
METALAXYL see MDM100
METALCAPTASE see MCR750, PAP550
METALDEHYD (GERMAN) see TDW500
METALDEHYDE (DOT) see TDW500
METALDEIDE (ITALIAN) see TDW500
METALKAMATE see BTA250
METALLIBURE see MLJ500
METALLIC ARSENIC see ARA750
METALLIC OSMIUM see OKE000
METAM see SIL550
METAMFETAMINA see DBA800
METAMID see PJY500
METAMIDOFOS ESTRELLA see DTQ400
METAMIN see TJL250
METAMINE see TJL250
METAMITON see ALA500
METAMITRON (GERMAN) see ALA500
METAMPHETAMIN see DBA800
METAMPHETAMINE HYDROCHLORIDE see MDT600
METAM-SODIUM (DUTCH, FRENCH, GERMAN, ITALIAN) see VFU000
METANA ALUMINUM PASTE see AGX000
METANABOL see DAL300
METANDIENON see DAL300
METANDIENONE see DAL300
METANDIENONUM see DAL300
METANDIOL see AOO475
METANDREN see MPN500
METANDRIOL see AOO475
METANDROSTENOLON see DAL300
METANDROSTENOLONE see DAL300
METANEPHRIN see VGP000
METANEX see CBT750
METANFETAMINA see DBA800
METANICOTINE see MDM750
METANILAMIDE see MDM760
METANILAN SODNY (CZECH) see AIF250
METANILE YELLOW O see MDM775
METANILIC ACID see SNO000
METANIL YELLOW 1955 see MDM775
METANIL YELLOW see MDM775
METANIL YELLOW C see MDM775
METANIL YELLOW E see MDM775
METANIL YELLOW EXTRA see MDM775
METANIL YELLOW F see MDM775
METANIL YELLOW G see MDM775
METANIL YELLOW GRIESBACH see MDM775
METANIL YELLOW K see MDM775
METANIL YELLOW KRSU see MDM775
METANIL YELLOW M3X see MDM775
METANIL YELLOW O see MDM775
METANIL YELLOW PL see MDM775
METANIL YELLOW S see MDM775
METANIL YELLOW SUPRA P see MDM775
METANIL YELLOW VS see MDM775
METANIL YELLOW WS see MDM775
METANIL YELLOW Y see MDM775
METANIL YELLOW YK see MDM775
METANIN see CPQ250

METANITE see MGR500
METANOLO (ITALIAN) see MGB150
METANTIOLO (ITALIAN) see MLE650
METANTYL see DJM800, XCJ000
METAOKSEDRIN see SPC500
METAOXEDRIN see NCL500, SPC500
METAOXEDRINUM see SPC500
METAOXON see PHD750
METAPHEN see NHK900
METAPHENYLALANINE MUSTARD see BHU500
METAPHENYLENEDIAMINE see PEY000
METAPHOR see MNH000
METAPHOS see MNH000
METAPHOSPHORIC ACID, CALCIUM SODIUM SALT see CAX260
METAPHYLLIN see TEP500
METAPHYLLINE see TEP500
METAPIRAZONE see AIB300
METAPLEXAN see MCJ400
METAPLEX NO see PKB500
METAPREL see MDM800
METAPROTERENOL see DMV800
METAPROTERENOL SULFATE see MDM800
METAQUALON see QAK000
METAQUEST A see EIX000
METAQUEST B see EIX500
METAQUEST C see EIV000
METARADRINE see HNB875
1-METARAMINOL see HNB875
(–)-METARAMINOL see HNB875
METARAMINOL see HNB875
METARAMINOL BITARTRATE see HNC000
METARAMINOL TARTRATE (1:1) see HNC000
METARSENOBILLON see SNR000
METARTRIL see IDA000
METASAP XX see AHH825
METASILICIC ACID see SCL000
METASOL 30 see ABU500
METASOL see MLH000
METASOL P-6 see PFO000
METASOL TK-100 see TEX000
METASON see TDW500
METASQUALENE see TMP500
METASTENOL see DAL300
METASYMPATOL see NCL500
METASYNEPHRINE see NCL500
METASYSTEMOX see DAP000
METASYSTOX see MIW100
METASYSTOX FORTE see DAP400
METASYSTOX-R see DAP000
METASYSTOX-S see DSK600
METATENSIN see RDK000
METATHION, S-METHYL ISOMER see MKC250
METATHIONE see DSQ000
METATION see DSQ000
META TOLUYLENE DIAMINE see TGL750
METATOLYLENEDIAMINE DIHYDROCHLORIDE see DCE000
METATSIN see ORQ000
METATYL see MGJ750
d,l-METATYROSINE see TOG275
METAUPON see OHU000
METAUPON PASTE see SIY000
METAXALONE see XVS000
METAXAN see DJM800, XCJ000
METAXANIN see MDM100
METAXITE see ARM268
METAXON see CIR250
METAXONE see SIL500
METAZALONE see XVS000
METAZIN see HDY500, SNJ000
METAZINE see HDY500
METAZIN 6U see MCB050
METAZOL see MLH000
METAZOLO see MCO500
METAZOLONE see XVS000
METEBANYL see ORE000
METELILACHLOR see MQQ450
METENDIOL see AOO475
METENIX see ZAK300
METENOLONE ACETATE see PMC700
METEPA see TNK250
METERAZIN MALEATE see PMF250
METERFER see FBJ100
METERFOLIC see FBJ100
METET see MDN100
METFENOSSIDIOLO see RLU000

METFORMIN see DQR600
METFORMIN HYDROCHLORIDE see DQR800
"METH" see MDQ500
METHAANTHIOL (DUTCH) see MLE650
METHABOL see PAN100
METHACETONE see DJN750
METHACHLOR see CFX000
METHACHLORPHENPROP see CFC750
METHACHOLINE CHLORIDE see ACR000
METHACHOLINIUM CHLORIDE see ACR000
METHACIDE see TGK750
METHACIN see ORQ000
METHACON see DPJ400
METHACRALDEHYDE (DOT) see MGA250
METHACROLEIN see MGA250
METHACROLEIN DIMER see DLI600
METHACRYLALDEHYDE DIMER see DLI600
METHACRYLATE de BUTYLE (FRENCH) see MHU750
METHACRYLATE de METHYLE (FRENCH) see MLH750
METHACRYL CHLORIDE see MDN899
METHACRYLIC ACID (ACGIH,OSHA) see MDN250
METHACRYLIC ACID, inhibited (DOT) see MDN250
METHACRYLIC ACID see MDN250
METHACRYLIC ACID, ALLYL ESTER see AGK500
METHACRYLIC ACID AMIDE see MDN500
METHACRYLIC ACID ANHYDRIDE see MDN699
METHACRYLIC ACID CHLORIDE see MDN899
METHACRYLIC ACID-3,4-DICHLOROANILIDE see DFO800
METHACRYLIC ACID, DIESTER with TETRAETHYLENE GLYCOL see TCE400
METHACRYLIC ACID, DIESTER with TRIETHYLENE GLYCOL see MDN510
METHACRYLIC ACID, 2-(DIETHYLAMINO)ETHYL ESTER see DIB300
METHACRYLIC ACID DODECYL ESTER see DXY200
METHACRYLIC ACID, 2-HYDROXYPROPYL ESTER see HNV500
METHACRYLIC ACID, ISOBUTYL ESTER see IIY000
METHACRYLIC ACID, ISODECYL ESTER see IKM000
METHACRYLIC ACID LAURYL ESTER see DXY200
METHACRYLIC ACID, METHYL ESTER (MAK) see MLH750
METHACRYLIC ACID METHYL ESTER POLYMERS see PKB500
METHACRYLIC ACID, PHENETHYL ESTER (7CI,8CI) see PFP600
METHACRYLIC ACID, 2-(VINYLOXY)ETHYL ESTER see VQA150
METHACRYLIC ALDEHYDE see MGA250
METHACRYLIC AMIDE see MDN500
METHACRYLIC ANHYDRIDE see MDN699
METHACRYLIC CHLORIDE see MDN899
METHACRYLONITRILE, inhibited see MGA750
γ-METHACRYLOXYPROPYLTRIMETHOXYSILANE see TLC250
METHACRYLOYL ANHYDRIDE see MDN699
α-METHACRYLOYL CHLORIDE see MDN899
METHACRYLOYL CHLORIDE see MDN899
METHACRYLOYLOXYETHYL ISOCYANATE see IKG700
METHACRYLSAEUREBUTYLESTER (GERMAN) see MHU750
METHACRYLSAEUREMETHYL ESTER (GERMAN) see MLH750
METHACRYLSAEUREMETHYL ESTER see MLH750
METHACRYLYL CHLORIDE see MDN899
METHACYCLINE HYDROCHLORIDE see MDO250
METHACYCLINE MONOHYDROCHLORIDE see MDO250
METHADON see MDO760
6s-METHADONE see DBE100
dl-METHADONE see MDO760
d-METHADONE see DBE100
l-(+)-METHADONE see DBE100
l-METHADONE see MDO775
(+)-METHADONE see DBE100
METHADONE see MDO750
(±)-METHADONE see MDO760
METHADONE see MDO760
s-(+)-METHADONE see DBE100
(−)-METHADONE see MDO775
dl-METHADONE HYDROCHLORIDE see MDP750
d-METHADONE HYDROCHLORIDE see MDP240
l-METHADONE HYDROCHLORIDE see MDP250
racemic METHADONE HYDROCHLORIDE see MDP750
METHADONE HYDROCHLORIDE see MDP000
(±)-METHADONE HYDROCHLORIDE see MDP750
METHADRENE see MJV000
METHAFORM see ABD000
METHAFRONE see AHK750
METHAFURILEN FUMARATE see MDP800
METHAFURYLENE FUMARATE see MDP800
METHAKRYLALDEHYD see MGA250
METHAKRYLOXYMETHYLSILATRAN (CZECH) see HMS500
METHALLIBURE see MLJ500
METHALLYL ALCOHOL (DOT) see IMW000
α-METHALLYL CHLORIDE see CEV250, CIU750
γ-METHALLYL CHLORIDE see CEU825

METHALLYL CHLORIDE see CIU750
1-METHALLYL-3-METHYL-6-AMINOTETRAHYDROPYRIMIDINEDIONE see AKL625
METHAMBUCAINE HYDROCHLORIDE see AJA500
METHAMBUTOXYCAINE HYDROCHLORIDE see AJA500
METHAM DIHYDRATE see SIL550
METHAMIDOPHOS see DTQ400
METHAMIN see HEI500
METHAMINOACETOCATECHOL see MGC350
METHAMINODIAZEPINE HYDROCHLORIDE see MDQ250
METHAMINODIAZEPOXIDE see LFK000
METHAMINODIAZEPOXIDE HYDROCHLORIDE see MDQ250
METHAMPHETAMINE see DBB000
(+)-METHAMPHETAMINE CHLORIDE see MDT600
dl-METHAMPHETAMINE HYDROCHLORIDE see DAR100
d-METHAMPHETAMINE HYDROCHLORIDE see MDT600
l-METHAMPHETAMINE HYDROCHLORIDE see MDQ500
METHAMPHETAMINE HYDROCHLORIDE see DBA800, MDT600
(+)-METHAMPHETAMINE HYDROCHLORIDE see MDT600
METHAMPHETAMINIUM CHLORIDE see MDT600
METHAM SODIUM see VFU000
METHANABOL see AOO475
METHANAL see FMV000
METHANAMIDE see FMY000
METHANAMINE (9CI) see MGC250
METHANAMINE, N-METHYL-, 2,3,6-TRICHLOROBENZOATE see DOR800
METHANAMINE, N-NITRO-(9CI) see NHN500
METHANAMINE, compd. with TRINITROMETHANE (1:1) see MGC300
METHANAMINIUM, 1-CARBOXY-N,N,N-TRIMETHYL-, CHLORIDE see CCH850
METHANAMINIUM NITRATE see MGN150
METHANDIENONE see DAL300
METHANDIOL see AOO475
METHANDRIOL see AOO475
METHANDROLAN see AOO475
METHANDROLONE see DAL300
METHANDROSTENOLONE see DAL300
METHANE see MDQ750
METHANEARSONIC ACID see MGQ530
METHANEARSONIC ACID DIMERCURY SALT see DNW000
METHANE BASE see MJN000
METHANE, BIS(4-CHLOROPHENYL)- see BIM800
METHANEBIS(N,N′-(5-UREIDO-2,4-DIKETOTETRAHYDROIMIDAZOLE)-N,N-DIMETHYLOL) see GDO800
METHANE BORONIC ANHYDRIDE-PYRIDINE COMPLEX see MDQ800
METHANE, BROMOTRIPHENYL- see TNP600
METHANECARBONITRILE see ABE500
METHANECARBOTHIOLIC ACID see TFA500
METHANECARBOXAMIDE see AAI000
METHANECARBOXYLIC ACID see AAT250
METHANE, CHLOROTRIFLUORO-, mixt. with TRIFLUOROMETHANE (9CI) see FOO515
METHANE, CYANO- see ABE500
METHANE, DIAZODIPHENYL- see DVZ100
METHANEDICARBOXYLIC ACID see CCC750
METHANEDICARBOXYLIC ACID, DIETHYL ESTER see EMA500
METHANE DICHLORIDE see MJP450
METHANE, DINITRO- see DUW507
METHANEDIOL, DINITRATE see MJU150
METHANEDISULFONIC ACID, CALCIUM SALT (1:1) see CAT700
METHANEDITHIOL-S,S-DIESTER with O,O-DIETHYL ESTER PHOSPHORODITHIOIC ACID see EEH600
METHANEPEROXOIC ACID see PCM500
METHANE, PHENYL- see TGK750
METHANESULFONIC ACID see MDR250
METHANESULFONIC ACID CHLOROETHYL ESTER see CHC750
METHANESULFONIC ACID, COMPOUND with 2-(DIETHYLAMINO)-4-METHYLPENTYL PAMINOBENZOATE (1:1) see LEU000
METHANESULFONIC ACID, METHYLENE ESTER see MJQ500
METHANESULFONIC ACID-1-METHYLETHYL ESTER see IPY000
METHANESULFONIC ACID, 2-PROPENYL ESTER (9CI) see AGK750
METHANESULFONIC ACID, SILVER(1+) SALT see SDR500
METHANESULFONIC ACID, SILVER SALT see SDR500
METHANESULFONIC ACID TETRAMETHYLENE ESTER see BOT250
METHANESULFONYL FLUORIDE see MDR750
1-(3-(2-(METHANESULFONYL)PHENOTHIAZIN-10-YL)PROPYL)-4-PIPERIDINECARBOXAMIDE see MPM750
METHANESULPHONIC ACID ETHYL ESTER see EMF500
METHANESULPHONIC ACID METHYL ESTER see MLH500
METHANESULPHONYL FLUORIDE see MDR750
METHANETELLUROL see MDR775
METHANE, TETRABROMIDE see CBX750
METHANE, TETRABROMO- see CBX750
METHANE TETRACHLORIDE see CBY000
METHANE, TETRAFLUORO- see CBY250
METHANE TETRAMETHYLOL see PBB750

N,N'-METHANETETRAYLBIS-2-PROPANAMINE see DNO400
N,N'-METHANE TETRAYLBIS(1,1,1-TRIMETHYLSILANAMINE) see BLQ950
1,1',1'',1'''-METHANETHETRAYLTETRAKISBENZENE see TEA750
METHANETHIOL see MLE650
METHANETHIOL, PHENYL- see TGO750
METHANE TRICHLORIDE see CHJ500
METHANE, TRIFLUORO-, mixt. with CHLOROTRIFLUOROMETHANE see FOO515
METHANE, TRIMETHYLOLNITRO- see HMJ500
METHANE, compressed (UN 1971) (DOT) see MDQ750
METHANE, refrigerated liquid (cryogenic liquid) (UN 1972) (DOT) see MDQ750
METHANIDE see DJM800, XCJ000
METHANIMIDAMIDE, N'-(2,4-DIMETHYLPHENYL)-N-(((2,4-DIMETHLPHE-NYL)IMINO)METHYL)-N-METHYL- see MJL250
1,4-METHANOAZULENE, DECAHYDRO-9-METHYLENE-4,8,8-TRIMETHYL-, (1S-(1-α-3a-β,4-α-8a-β))- see LID100
1,4-METHANOAZULENE, DECAHYDRO-4,8,8-TRIMETHYL-9-METHYLENE-, (1S,3aR,4S,8aS)-(+)-(8CI) see LID100
1H-3a,7-METHANOAZULENE, OCTAHYDRO-6-METHOXY-3,6,8,8-TETRA-METHYL-,(3R-(3-α-3a-β, 6-α-7-β,8aα-)- see CCR525
1H-3-α-7-METHANOAZULEN-6-OL, OCTAHYDRO-3,6,8,8-TETRAMETHYL-, FOR-MATE, (3R-(3-α-3a-β,6-α-7-β,8aα-))- see CCR524
2,6-METHANO-3-BENZAZOCIN-8-OL-3-ALLYL-6-ETHYL-1,2,3,4,5,6-HEXAHYDRO-11-METHYL see AGG250
4,7-METHANOBENZOTRITHIOLE, HEXAHYDRO- see TNO300
METHANOIC ACID see FNA000
4,7-METHANOINDAN, 2,2,4,5,6,7,8,8-OCTACHLORO-3a,4,7,7a-TETRAHYDRO- see CDR575
4,7-METHANO-1H-INDENE-2-CARBOXALDEHYDE, OCTAHYDRO-5-METHOXY- see OBW200
4,7-METHANOINDENE-6-CARBOXYLIC ACID, 3a,4,5,6,7,7a-HEXAHYDRO-, ETH-YL ESTER see TJG600
4,7-METHANOINDENE, 4,5,6,7,8,8-HEXACHLORO-3a,4,7,7a-TETRAHYDRO- see HCN000
4,7-METHANO-1H-INDENE, 2,2,4,5,6,7,8,8-OCTACHLORO-2,3,3a,4,7,7a-HEXAH-YDRO- (9CI) see CDR575
4,7-METHANOISOBENZOFURAN-1,3-DIONE, 3a,4,7,7a-TETRAHYDROMETHYL- see NAC000
METHANOL see MGB150
METHANOLACETONITRILE see HGP000
N-METHANOLACRYLAMIDE see HLU500
METHANOL, BENZYL- see PDD750
METHANOL, CYCLOHEXYL- see HDH200
METHANOL, (METHYL-ONN-AZOXY)-, BENZOATE (ester) (9CI) see MGS925
METHANOL, OXIRANYL- see GGW500
METHANOL, OXYDI-, DINITRATE see OPQ200
METHANOL, SODIUM SALT see SIK450
2-METHANOL TETRAHYDROPYRAN see MDS500
METHANOL, TRICHLORO-, CHLOROFORMATE see TIR920
4,7-METHANO-2,3,8-METHENOCYCLOPENT(a)INDENE, DODECAHYDRO-, ste-reoisomer see DLJ500
1,4-METHANONAPHTHALENE-5,8-DIONE, 1,4,4a,8a-TETRAHYDRO- see TCU650
METHANONE, (2-AMINO-1H-BENZIMIDAZOL-5-YL)PHENYL- see AIH000
METHANONE, (2-AMINO-5-CHLOROPHENYL)PHENYL- see AJE400
METHANONE, (5-AMINO-1,3-DIMETHYL-1H-PYRAZOL-4-YL)(2-FLUOROPHENYL)- see AJR400
METHANONE, (3,5-DIBROMO-4-HYDROXYPHENYL)(2-(2,4,6-TRIMETHYLPHE-NYL)-3-BENZOFURANYL)- see DDP300
METHANONE, (4,5-DICHLORO-1H-PYRROL-2-YL)(2,6-DIHYDROXYPHENYL)- see POK400
METHANONE, (4-(2-(DIETHYLAMINO)ETHOXY)PHENYL)(2-ETHYL-3-BENZOFU-RANYL)-(9CI) see EDV700
METHANONE, (4-(2-(DIETHYLAMINO)ETHOXY)PHENYL)(2-(2,4,6-TRIMETHYL-PHENYL)-3-BENZOFURA NYL)-, HCl see DHR900
METHANONE, (2,3-DIHYDRO-1,4-BENZODIOXIN-6-YL)(4-FLUOROPHENYL)- see DKT500
METHANONE, (1,3-DIMETHYL-5-(METHYLAMINO)-1H-PYRAZOL-4-YL)(2-FLUO-ROPHENYL)- see DSO500
METHANONE, DIPHENYL-, OXIME (9CI) see BCS400
METHANONE, 9H-FLUOREN-2-YL(4-METHOXYPHENYL)- see MEC600
METHANONE, 9H-FLUOREN-2-YLPHENYL- see FEM100
METHANONE, 3-FURFANYLPHENYL-(9CI) see FQO050
METHANONE, (4-HYDROXY-3,5-DIIODOPHENYL)(2-(2,4,6-TRIMETHYLPHENYL)-3-BENZOFURANYL)- see DNF550
METHANONE, (2-HYDROXY-4-(OCTYLOXY)PHENYL)PHENYL-(9CI) see HND100
METHANONE, (4-HYDROXYPHENYL)(2-(2,4,6-TRIMETHYLPHENYL)-4-BENZOFU-RANYL)- see HNK800
METHANONE, (4-HYDROXYPHENYL)(2-(2,4,6-TRIMETHYLPHENYL)-3-BENZOFU-RANYL)- see HNK700
METHANONE, (4-METHOXYPHENYL)(2-(2,4,6-TRIMETHYLPHENYL)-3-BENZOFU-RANYL)- see AOY300
METHANONE, PHENYL-1H-PYRROL-2-YL-(9CI) see PGF900

6,9-METHANO-8H-PYRIDO(1',2':1,2)AZEPINO(4,5-b)INDOLE-6(6aH)-CARBOXYL-IC ACID, 7,8,9,10,12,13- HEXAHYDRO-6-ETHYL-13a-HYDROXY-, METHYL ESTER, HYDROCHLORIDE see CNS200
4,7-METHANOSELENOPHENE, OCTAHYDRO-2,2,3,3-TETRAFLUORO- see TCI100
o-(1,4-METHANO-1,2,3,4-TETRAHYDRO-6-NAPHTHYL)-N-METHYL-N-(m-TOLYL)-THIOCARBAMATE see MQA000
METHANTHELINE BROMIDE see DJM800, XCJ000
METHANTHELINIUM BROMIDE see XCJ000
METHANTHINE BROMIDE see XCJ000
METHANTHIOL (GERMAN) see MLE650
METHAPHENILENE HYDROCHLORIDE see DCJ850
METHAPHOXIDE see TNK250
METHAPYRAPONE see MCJ370
METHAPYRILENE see DPJ200, TEO250
METHAPYRILENE HYDROCHLORIDE see DPJ400
METHAPYRILENE HYDROCHLORIDE (L.A.) see DPJ400
METHAPYRILENE HYDROCHLORIDE (S.A.) see DPJ400
METHAPYRILENE mixed with SODIUM NITRITE (1:2) see MDT000
METHAQUALONE see QAK000
METHAQUALONE HYDROCHLORIDE see MDT250
METHAQUALONEINONE see QAK000
METHAR see DXE600
METHARBITAL see DJO800
METHARBITONE see DJO800
METHARBUTAL see DJO800
METHARSINAT see DXE600
METHASAN see BJK500
β-METHASONE-17-VALERATE see VCA000
METHASQUIN see DBY500
METHAZATE see BJK500
METHAZINE see IDA000
METHAZOLE see BGD250
METHAZONIC ACID see NEJ600
METHCAINE HYDROCHLORIDE see IJZ000
METHDILAZINE see MPE250
METHDILAZINE HYDROCHLORIDE see MDT500
METHEDRINE see DBA800, DBB000, MDT600
METHEDRINE HYDROCHLORIDE see DBA800, MDT600
METHEGRIN see MJV750
METHELINA see DJM800, XCJ000
METHENAMIC ACID see XQS000
METHENAMINE see HEI500
METHENOLONE ACETATE see PMC700
N,N'-METHENYL-o-PHENYLENEDIAMINE see BCB750
METHENYL TRIBROMIDE see BNL000
METHENYL TRICHLORIDE see CHJ500
METHERGINE see PAM000
METHETOIN see ENC500
METHEXENYL see ERD500
METHEXENYL SODIUM see ERE000
METHIACIL see MPW500
METHIAMAZOLE see MCO500
METHIAZIC ACID see MNQ500
METHIAZINIC ACID see MNQ500
METHIDATHION see DSO000
METHILANIN see MDT740
METHIOCARB see DST000
METHIOCIL see MPW500
METHIODAL SODIUM see SHX000
METHIODIDE of N-BENZYLURETHANE of 3-DIMETHYLAMINOPHENOL see HNO000
METHIODIDE of N-METHYLURETHANE of 3-DIETHYLAMINOPHENOL see MID250
METHIODIDE of N-METHYLURETHANE of 3-DIMETHYLAMINOPHENOL see HNO500
METHIODIDE of N-METHYLURETHANE of 4-DIMETHYLAMINOPHENOL see HNP000
METHIONAL see MPV400, TET900
METHIONAMINE see ACQ275
dl-METHIONINE see MDT740
d-METHIONINE see MDT730
l-METHIONINE see MDT750
l-(−)-METHIONINE see MDT750
(±)-METHIONINE see MDT740
METHIONINE see MDT750
METHIONINE, N-ACETYL-, dl- see ACQ270
dl-METHIONINE, N-ACETYL-(9CI) see ACQ270
l-METHIONINE, N-(N-(3-(p-FLUOROPHENYL)-l-ANANYL)-3-(m-(BIS(2-CHLOROE-THYL)AMINO)PHENYL)-l-ALANYL)-, ETHYL ESTER, HYDROCHLORIDE see POI550
METHIONINE HYDROXY ANALOG see HMR600
dl-METHIONINE SULFOXIDE see ALF600
METHIONINE SULFOXIDE see ALF600
dl-METHIONINE-dl-SULFOXIMINE see MDU100

METHIONINE SULFOXIMINE see MDU100
METHIOPLEGIUM see TKW500
METHISAZONE see MKW250
METHIUM CHLORIDE see HEA500
METHIXENE HYDROCHLORIDE see THL500
METHOCARBAMOL see GKK000
METHOCEL 10 see MIF760
METHOCEL 15 see MIF760
METHOCEL 181 see MIF760
METHOCEL 400 see MIF760
METHOCEL 4000 see MIF760
METHOCEL A see MIF760
METHOCEL CHG see MIF760
METHOCEL 400CPS see MIF760
METHOCEL 4000CPS see MIF760
METHOCEL HG see HNX000
METHOCEL MC 25 see MIF760
METHOCEL MC4000 see MIF760
METHOCEL MC 8000 see MIF760
METHOCEL MC see MIF760
METHOCEL SM 100 see MIF760
METHOCHLOPRAMIDE see AJH000
METHOCHLORIDE of N-METHYLURETHANE of 3-DIMETHYLAMINOPHENOL
   see HNN000
METHOCILLIN-S see SLJ000
METHOCROTOPHOS see DOL800
METHODICHLOROPHEN see MQR100
METHOFADIN see MDU300
METHOFAZINE see MDU300
METHOFLURANE see DFA400
METHOGAS see MHR200
METHOHEXITAL SODIUM see MDU500
METHOHEXITONE SODIUM see MDU500
METHOIDAL SODIUM see SHX000
METHOIN see MKB250
METHOLCARB see MIB750
METHOLENE 2095 see MHY650
METHOLENE 2218 see MJW000
METHOLENE 2296 see MLC800
METHOMYL see MDU600
METHOPHENAZATE ACID FUMARATE see MDU750
METHOPHENAZINE DIFUMARATE see MDU750
METHOPHOLINE see MDV000
METHOPHYLLINE see TEP500
METHOPIRAPONE see MCJ370
METHOPLAIN see DNA800, MJE780
METHOPRENE see KAJ000
METHOPROMAZINE MALEATE see MFK750
METHOPROPTRYNE see INQ000
METHOPTERIN see MDV500
METHOPYRAPONE see MCJ370
METHOPYRININE see MCJ370
METHOPYRONE see MCJ370
METHOQUINE see CFU750
METHORATE HYDROBROMIDE see DBE200
d-METHORPHAN see DBE150
Δ-METHORPHAN see DBE150
d-METHORPHAN HYDROBROMIDE see DBE200
METHORPHINAN see MKR250
METHORPHINAN HYDROBROMIDE see MDV250
METHOSCOPYLAMINE BROMIDE see SBH500
METHOSERPEDINE see MEK700
METHOSERPIDINE see MEK700
METHOSTAN see AOO475
METHOTEXTRATE see MDV500
METHOTREXATE see MDV500
METHOTREXATE DISODIUM SALT see MDV600
METHOTREXATE SODIUM see MDV750
METHOTRIMEPRAZINE see MCI500
METHOXA-DOME see XDJ000
METHOXADONE see MFD500
METHOXAMINE HYDROCHLORIDE see MDW000
METHOXANE see DFA400
METHOXCIDE see MEI450
METHOXERPATE HYDROCHLORIDE see MDW100
METHOXIPHENADRIN HYDROCHLORIDE see OJY000
METHOXO see MEI450
METHOXOLONE see XVS000
METHOXONE see CIR250, CIR500, SIL500
METHOXONE M see CLO200
METHOXSALEN see XDJ000
α-METHOXYACETALDEHYDE see MDW250
METHOXYACETALDEHYDE see MDW250
1-METHOXY-2-ACETAMIDOFLUORENE see MER000
2-METHOXYACETIC ACID see MDW275

METHOXYACETIC ACID see MDW275
2-METHOXYACETOACETANILIDE see ABA500
2′-METHOXYACETOACETANILIDE see ABA500
o-METHOXYACETOACETANILIDE see ABA500
4-METHOXYACETOFENON see MDW750
1-METHOXYACETONE see MDW300, MFL100
METHOXYACETONE see MDW300, MFL100
4-METHOXYACETOPHENONE see MDW750
4′-METHOXYACETOPHENONE see MDW750
p-METHOXYACETOPHENONE see MDW750
1-METHOXY-2-ACETOXYPROPANE see PNL265
METHOXYACETYL CHLORIDE see MDW780
METHOXYACETYLENE see MDX000
2-(METHOXYACETYL)INDOLE see ICW200
(METHOXYACETYL)METHYLCARBAMIC ACID-o-ISOPROPOXYPHENYL ESTER
   see MDX250
2-(METHOXYACETYL)-1-METHYLINDOLE see MDX300
2-(METHOXYACETYL)-3-METHYLINDOLE see MDX310
5-METHOXY-N-ACETYLTRYPTAMINE see MCB350
4′-METHOXYACRYLOPHENONE see ONW100
p-METHOXYACRYLOPHENONE see ONW100
2-METHOXY-AETHANOL (GERMAN) see EJH500
2-METHOXYAETHYLACETAT (GERMAN) see EJJ500
1-METHOXY-AETHYL-AETHYLNITROSAMIN (GERMAN) see MEO500
1-METHOXY-AETHYL-METHYLNITROSAMIN (GERMAN) see MEP500
METHOXYAETHYLQUECKSILBERCHLORID see MEP250
METHOXYAETHYLQUECKSILBERSILIKAT (GERMAN) see MEP000
p-METHOXYALLYLBENZENE see AFW750
2-METHOXY-4-ALLYLPHENOL see EQR500
2′-METHOXY-4′-ALLYLPHENYL 4-GUANIDINOBENZOATE see MDY300
2-METHOXY-1-AMINOBENZENE see AOV900
2-METHOXY-1-AMINOBENZENE HYDROCHLORIDE see AOX250
4-METHOXY-2-AMINOBENZOTHIAZOLE see AKM750
2-METHOXY-4-AMINO-5-CHLORO-N-β-(DIETHYLAMINOETHYL)BENZAMIDE DI-
   HYDROCHLORIDE MONOHYDRATE see MQQ300
2-METHOXY-3-AMINODIBENZOFURAN see MDZ000
3-METHOXY-4-AMINODIPHENYL see MEA000
3-METHOXY-2-AMINODIPHENYLENE OXIDE see AKN500
2-METHOXY-4-AMINO-5-HYDROXYMETHYLPYRIMIDINE see AKO750
6-METHOXY-8-(4-AMINO-1-METHYLBUTYLAMINO)QUINOLINE see PMC300
1-METHOXY-2-AMINONAPHTHALENE see MFA250
4-METHOXY-4-AMINO-2-PENTANOL see MEA250
3-METHOXY-4-AMINOSTILBENE see MFP500
4-METHOXYAMPHETAMINE HYDROCHLORIDE see MEA500
dl,4-METHOXYAMPHETAMINE HYDROCHLORIDE see MEA500
2-METHOXYANILINE see AOV900
3-METHOXYANILINE see AOV890
4-METHOXYANILINE see AOW000
o-METHOXYANILINE see AOV900
p-METHOXYANILINE see AOW000
2-METHOXYANILINE HYDROCHLORIDE see AOX250
o-METHOXYANILINE HYDROCHLORIDE see AOX250
2-METHOXYANILINIUM NITRATE see MEA600
3-METHOXYANISOLE see REF025
METHOXYAZOXYMETHANOLACETATE see MEA750
4-METHOXYBENZALACETONE see MLI400
p-METHOXYBENZALACETONE see MLI400
2-METHOXYBENZALDEHYDE see AOT525
4-METHOXYBENZALDEHYDE see AOT530
6-METHOXYBENZALDEHYDE see AOT525
p-METHOXYBENZALDEHYDE (FCC) see AOT530
o-METHOXYBENZALDEHYDE see AOT525
2-METHOXYBENZAMIDE see AOT750
o-METHOXYBENZAMIDE see AOT750
10-METHOXY-1,2-BENZANTHRACENE see MEB500
3-METHOXY-1,2-BENZANTHRACENE see MEB000
5-METHOXY-1,2-BENZANTHRACENE see MEB250
5-METHOXY-1,2-BENZ(a)ANTHRACENE see MEB250
5-METHOXY-BENZ(a)ANTHRACENE see MEB000
7-METHOXY-BENZ(a)ANTHRACENE see MEB500
8-METHOXY-BENZ(a)ANTHRACENE see MEB250
3-METHOXY-7H-BENZ(de)ANTHRACEN-7-ONE see MEB750
3-METHOXYBENZANTHRONE see MEB750
2-METHOXYBENZENAMINE see AOV900
3-METHOXYBENZENAMINE see AOV890
4-METHOXYBENZENAMINE see AOW000
2-METHOXYBENZENAMINE HYDROCHLORIDE see AOX250
METHOXYBENZENE see AOX750
4-METHOXYBENZENEACETIC ACID see MFE250
p-METHOXYBENZENEACETONITRILE see MFF000
4-METHOXYBENZENEAMINE see AOW000
2-METHOXYBENZENEAMINE HYDROCHLORIDE see AOX250
2-METHOXYBENZENECARBOXALDEHYDE see AOT525
4-METHOXY-1,3-BENZENEDIAMINE see DBO000
4-METHOXY-1,3-BENZENEDIAMINE SULFATE (1:1) see DBO400

4-METHOXY-1,3-BENZENEDIAMINE SULFATE see DBO400
2-METHOXY-1,4-BENZENEDIAMINE SULFATE see MEB820
4-METHOXY-1,3-BENZENEDIAMINE SULPHATE see DBO400
4-METHOXYBENZENEMETHANOL see MED500
2-METHOXYBENZENETHIOL see MFQ300
o-METHOXYBENZENETHIOL see MFQ300
2-METHOXY-4H-1,2,3-BENZODIOXAPHOSPHORINE-2-SULFIDE see MEC250
1-(5-METHOXY-2-BENZOFURANYL)ETHANONE see MEC330
1-(6-METHOXY-2-BENZOFURANYL)ETHANONE see MEC300
1-(7-METHOXY-2-BENZOFURANYL)ETHANONE see MEC320
5-METHOXY-2-BENZOFURANYL METHYL KETONE see MEC330
6-METHOXY-2-BENZOFURANYL METHYL KETONE see MEC300
7-METHOXY-2-BENZOFURANYL METHYL KETONE see MEC320
o-METHOXYBENZOHYDRAZIDE see AOV250
2-METHOXYBENZOIC ACID see MPI000
3-METHOXYBENZOIC ACID see AOU500
4-METHOXYBENZOIC ACID see AOU600
m-METHOXYBENZOIC ACID see AOU500
o-METHOXYBENZOIC ACID see MPI000
p-METHOXYBENZOIC ACID see AOU600
2-METHOXYBENZOIC ACID HYDRAZIDE see AOV250
4-METHOXYBENZOIC ACID HYDRAZIDE see AOV500
o-METHOXYBENZOIC ACID HYDRAZIDE see AOV250
p-METHOXYBENZOIC ACID HYDRAZIDE see AOV500
o-METHOXYBENZOIC ACID METHYL ESTER see MLH800
p-METHOXYBENZOIC HYDRAZIDE see AOV500
7-METHOXY-2H-1-BENZOPYRAN-2-ONE see MEK300
1-(5-METHOXY-2H-1-BENZOPYRAN-3-YL)ETHANONE see MEC340
5-METHOXY-2H-1-BENZOPYRAN-3-YL METHYL KETONE see MEC340
6-METHOXYBENZO(a)PYRENE see MEC500
METHOXYBENZOYL CHLORIDE see AOY250
2-(4'-METHOXYBENZOYL)FLUORENE see MEC600
2-METHOXYBENZOYL HYDRAZIDE see AOV250
4-METHOXYBENZOYL HYDRAZIDE see AOV500
o-METHOXYBENZOYLHYDRAZIDE see AOV250
2-METHOXYBENZOYLHYDRAZINE see AOV250
4-METHOXYBENZOYLHYDRAZINE see AOV500
(p-METHOXYBENZOYL)HYDRAZINE see AOV500
N-(4-METHOXY)BENZOYLOXYPIPERIDINE see MED000
3-(4-METHOXYBENZOYL)PYRIDINE see MED100
4-(4-METHOXYBENZOYL)PYRIDINE see MFH930
8-METHOXY-3,4-BENZPYRENE see MED250
p-METHOXYBENZYL ACETATE see AOY400
4-METHOXYBENZYLACETONE see MFF580
4-METHOXYBENZYL ALCOHOL see MED500
p-METHOXYBENZYL ALCOHOL see MED500
p-METHOXYBENZYL BUTYRATE see MED750
p-METHOXYBENZYL CYANIDE see MFF000
2-METHOXYBENZYL ETHYL ETHER see EMF600
o-METHOXYBENZYL ETHYL ETHER see EMF600
p-METHOXYBENZYL FORMATE see MFE250
4-METHOXYBENZYLIDENEACETONE see MLI400
p-METHOXYBENZYLIDENEACETONE see MLI400
4-METHOXYBENZYL METHYL KETONE see AOV875
p-METHOXYBENZYL METHYL KETONE see AOV875
N-(p-METHOXYBENZYL)-N',N'-DIMETHYL-N-2-PYRIDYLETHYLENEDIAMINE see WAK000
N-p-METHOXYBENZYL-N',N'-DIMETHYL-N-α-PYRIDYLETHYLENEDIAMINE see WAK000
N-p-METHOXYBENZYL-N',N'-DIMETHYL-N-2-PYRIMIDINYLETHYLENE DIAMINE HYDROCHLORIDE see RDU000
N-p-METHOXYBENZYL-N',N'-DIMETHYL-N-α-PYRIDYLETHYLENEDIAMINE MALEATE see DBM800
p-METHOXYBENZYL PHENYLACETATE see APE000
2-METHOXY BIPHENYL see PEG000
4-METHOXYBIPHENYL see PEG250
p-METHOXYBIPHENYL see PEG250
3-METHOXYBIPHENYLAMINE see MEA000
1-((4-METHOXY(1,1'-BIPHENYL)-3-YL)METHYL)PYRROLIDINE see MEF400
2-METHOXY-4,6-BIS(ETHYLAMINO)-s-TRIAZINE see BJP250
2-METHOXY-4,6-BIS(ISOPROPYLAMINO)-1,3,5-TRIAZINE see MFL250
2-METHOXY-4,6-BIS(ISOPROPYLAMINO)-s-TRIAZINE see MFL250
4-METHOXY-α-(BIS(4-METHOXYPHENYL)METHYLENE)-N,N-DIMETHYLBENZENEETHANAMINE HCl see TNJ750
exo-2-METHOXYBORNANE see IHX500
2-METHOXYBROMOBENZENE see BMT400
4-METHOXYBROMOBENZENE see AOY450
o-METHOXYBROMOBENZENE see BMT400
p-METHOXYBROMOBENZENE see AOY450
1-METHOXYBUTANE see BRU780
α-METHOXYBUTANE see BRU780
3-METHOXY BUTANOIC ACID see MEF500
4-((1-METHOXYBUTOXY)(5-VINYL-2-QUINUCLIDINYL)METHYL)-6-QUINOLINOL DIHYDROCHLORIDE see EQQ500
3-METHOXYBUTYL ACETATE see MHV750

2-METHOXY-4-sec-BUTYLAMINO-6-AETHYLAMINO-s-TRIAZIN (GERMAN) see BQC250
2-METHOXY-4-tert-BUTYLAMINO-6-AETHYLAMINO-s-TRIAZIN (GERMAN) see BQC500
3-METHOXYBUTYLESTER KYSELINY OCTOVE see MHV750
4-METHOXY-2-tert-BUTYLPHENOL see BRN000
3-METHOXY BUTYRALDEHYDE see MEF750
3-METHOXYBUTYRIC ACID see MEF500
2-(METHOXY-CARBONYLAMINO)-BENZIMIDAZOL see MHC750
2-(METHOXYCARBONYLAMINO)-BENZIMIDAZOLE see MHC750
3-METHOXYCARBONYLAMINOPHENYL N-3'-METHYLPHENYLCARBAMATE see MEG250
2-(METHOXYCARBONYL)ANILINE see APJ250
METHOXYCARBONYL CHLORIDE see MIG000
METHOXYCARBONYLETHYLENE see MGA500
3-METHOXYCARBONYL-N-(3'-METHYLPHENYL)-CARBAMAT see MEG250
(2-METHOXYCARBONYL-1-METHYL-VINYL)-DIMETHYL-FOSFAAT (DUTCH) see MQR750
(2-METHOXYCARBONYL-1-METHYL-VINYL)-DIMETHYL-PHOSPHAT (GERMAN) see MQR750
2-METHOXYCARBONYL-1-METHYLVINYL DIMETHYLPHOSPHATE see MQR750
1-METHOXYCARBONYL-1-PROPEN-2-YL DIMETHYL PHOSPHATE see MQR750
3-METHOXYCARBONYL PROPEN-2-YL TRIFLUOROMETHANE SULFONATE see MEH775
N-(3-METHOXYCARBONYLPROPYL)-N-(1-ACETOXYBUTYL)NITROSAMINE see MEI000
(1-METHOXYCARBOXYPROPEN-2-YL)PHOSPHORIC ACID, DIMETHYL ESTER see MQR750
p,p'-METHOXYCHLOR see MEI450
METHOXYCHLOR see MEI450
METHOXYCHLOR mixed with DIAZINON see MEI500
2-METHOXY-6-CHLORO-9-(4-BIS(2-CHLOROETHYL)AMINO-1-METHYLBUTYLAMINO)ACRIDINE DIHYDROCHLORIDE see QDS000
2-METHOXY-6-CHLORO-9-(3-(2-CHLOROETHYL) MERCAPTO PROPYLAMINO) ACRIDINE HYDROCHLORIDE see CFA250
2-METHOXY-6-CHLORO-9-(4-DIETHYLAMINO-1-METHYLBUTYLAMINO)ACRIDINEDIHYDROCHLORIDE see CFU750
2-METHOXY-6-CHLORO-9-DIETHYLAMINOPENTYLAMINOACRIDINE see ARQ250
2-METHOXY-6-CHLORO-9-(3-(ETHYL-2-CHLOROETHYL)AMINOPROPYLAMINO)ACRIDINE DIHYDROCHLORIDE see QDS000, ADJ875
2-METHOXY-5-CHLOROPROCAINAMIDE see AJH000
5-METHOXYCHRYSENE see MEJ500
6'-METHOXYCINCHONAN-9-OL see QFS000
(8-α,9R)-6'-METHOXYCINCHONAN-9-OL see QHJ000
6'-METHOXYCINCHONAN-9-OL DIHYDROCHLORIDE see QIJ000
(9S)-6'-METHOXYCINCHONAN-9-OL SULFATE (1:1) (SALT) see QFS100
(9S)-6'-METHOXYCINCHONAN-9-OL SULFATE (2:1) (SALT) see QHA000
6-METHOXYCINCHONINE see QHJ000
o-METHOXY CINNAMALDEHYDE see MEJ750
cis-2-METHOXYCINNAMIC ACID see MEJ775
cis-o-METHOXYCINNAMIC ACID see MEJ775
o-METHOXYCINNAMIC ALDEHYDE see MEJ750
5-(p-METHOXY-CINNAMOYL)-4,7-DIMETHOXY-6-DIMETHYLAMINOETHOXYBENZOFURAN see MBX000
METHOXYCITRONELLAL METHYL ETHER see DSM800
7-METHOXYCOUMARIN see MEK300
2-METHOXY-p-CRESOL see MEK250
β-(2-METHOXY-5-CYCLOHEXYLBENZOYL)PROPIONIC ACID SODIUM SALT see MEK350
1-METHOXYCYCLOPROPANE see CQE750
METHOXYCYCLOPROPANE see CQE750
3-METHOXY-DBA see MEK750
9-METHOXY-DBA see MEL000
METHOXY-DDT see MEI450
1-METHOXYDECANE see MIW075
10-METHOXYDESERPIDINE see MEK700
10-METHOXY-11-DESMETHOXYRESERPINE see MEK700
3-METHOXY-1,2:5,6-DIBENZANTHRACENE see MEK750
9-METHOXY-1,2,5,6-DIBENZANTHRACENE see MEL000
5-METHOXYDIBENZ(a,h)ANTHRACENE see MEK750
7-METHOXYDIBENZ(a,h)ANTHRACENE see MEL000
2-METHOXY-3,6-DICHLOROBENZOIC ACID see MEL500
(3-METHOXY-4-((N,N-DIETHYLCARBAMIDO)METHOXY)PHENYL)ACETIC ACID n-PROPYL ESTER see PMM000
METHOXYDIGLYCOL see DJG000
11-METHOXY-15,16-DIHDYROCYCLOPENTA(a)PHENANTHREN-17-ONE see DLR200
5-METHOXY-2-(DIMETHOXYPHOSPHINYLTHIOMETHYL)PYRONE-4 see EAS000
2'-METHOXY-4-DIMETHYLAMINOAZOBENZENE see DSN000
3'-METHOXY-4-DIMETHYLAMINOAZOBENZENE see DSN200
3-METHOXY-4-DIMETHYLAMINOAZOBENZENE see DTK800
4'-METHOXY-4-DIMETHYLAMINOAZOBENZENE see DSN400

11-METHOXY-17-METHYL-15H-CYCLOPENTA(a)PHENANTHRENE see MEV000
3-METHOXY-17-METHYL 15H-CYCLOPENTA(a)PHENANTHRENE see MEU750
1-METHOXY-1-METHYL-3-(3,4-DICHLOROPHENYL)UREA see DGD600
6-METHOXY-11-METHYL-15,16-DIHYDRO-17H-CYCLOPENTA(a) PHENAN-
  THREN-17-ONE see MEV250
19-METHOXY-1,2-(METHYLENEDIOXY)-6a-α-APORPHIN-11-OL see HLT000
(2-(2-METHOXY METHYL ETHOXY)METHYL ETHOXY)PROPANOL see
  MEV500
METHOXYMETHYL ETHYL NITROSAMINE see MEV750
p-METHOXY-α-METHYLHYDROCINNAMALDEHYDE see MLJ050
4-METHOXYMETHYL-5-HYDROXY-6-METHYL-3-PYRIDINEMETHANOL HYDRO-
  CHLORIDE see MEY000
2-METHOXY-10-(2-METHYL-3-(4-HYDROXYPIPERIDI-
  NO)PROPYL)PHENOTHIAZINE see LEO000
7-METHOXYMETHYL-12-METHYLBENZ(a)ANTHRACENE see MEW000
METHOXYMETHYL 1-METHYL-2-INDOLYL KETONE see MDX300
METHOXYMETHYL 3-METHYL-2-INDOLYL KETONE see MDX310
METHOXYMETHYL METHYL KETONE see MDW300, MFL100
METHOXYMETHYL-METHYLNITROSAMIN (GERMAN) see MEW250
METHOXYMETHYL METHYLNITROSAMINE see MEW250
6-METHOXY-1-METHYL-4-((p-((1-METHYLPYRIDINIUM-4-
  YL)AMINO)PHENYL)CARBAMOYL)ANILINO)QUINOLINIUM, DI-p- see
  MEW750
8-METHOXY-1-METHYL-4-((p-((1-METHYLPYRIDINIUM-4-
  YL)AMINO)PHENYL)CARBAMOYL)ANILINO)QUINOLINIUM, DIBROMI see
  MEW500
4-METHOXY-5-METHYL-6-(7-METHYL-8-(TETRAHYDRO-3,4-DIHYDROXY-2,4,5-
  TRIMETHYL-2-FURANYL)-1,3,5,7-OCTATETRAENYL)-2H-PYRAN-2-ONE see
  CMS500
3-METHOXY-17-METHYLMORPHINAN HYDROBROMIDE see LFD200
(±)-3-METHOXY-17-METHYLMORPHINAN HYDROBROMIDE see RAF300
3-METHOXY-17-METHYL-9-α,13-α,14-α-MORPHINAN HYDROBROMIDE see
  DBE200
d-3-METHOXY-N-METHYLMORPHINAN HYDROBROMIDE see DBE200
(+)-6-METHOXY-α-METHYL-2-NAPHTHALENEACETIC ACID see MFA500
(S)-6-METHOXY-α-METHYL-2-NAPHTHALENEACETIC ACID see MFA500
l-(-)-6-METHOXY-α-METHYL-2-NAPHTHALENEACETIC ACID SODIUM SALT see
  NBO550
4-METHOXY-N-METHYLNAPHTHALIMIDE see MEW760
α-(METHOXYMETHYL)-2-NITRO-1H-IMIDAZOLE-1-ETHANOL see NHH500
α-(METHOXYMETHYL)-2-NITROIMIDAZOLE-1-ETHANOL see NHH500
α-(METHOXYMETHYL)-2-NITRO-1H-IMIDAZOLE-1-ETHANOL (9CI) see NHH500
7-METHOXY-1-METHYL-2-NITRONAPHTHO(2,1-b)FURAN see MEW775
N-METHOXY-N-METHYLNONANAMIDE see EQS500
3-METHOXY-5-METHYL-4-OXO-2,5-HEXADIENOIC ACID see PAP750
3-((4-(5-(METHOXYMETHYL)-2-OXO-3-OXAZOLIDI-
  NYL)PHENOXY)METHYL)BENZONITRILE see MEW800
N-(7-METHOXY-3-METHYL-4-OXO-2-PHENYL-4H-CHROMEN-8-YL)METHYL-N,N-
  DIMETHYLAMINE see DNV000
5-METHOXY-2-METHYL-1-(1-OXO-3-PHENYL-2-PROPENYL)-1H-INDOLE-3-ACETIC
  ACID (9CI) see CMP950
4-METHOXY-4-METHYL-2-PENTANONE see MEX250
4-METHOXY-4-METHYLPENTAN-2-ONE (DOT) see MEX250
dl-p-METHOXY-α-METHYL-PHENETHYLAMINE HYDROCHLORIDE see MEA500
2-METHOXY-4-METHYLPHENOL see MEK250
4-β-METHOXY-1-METHYL-4-α-PHENYL-3-α,5-α-PROPANOPIPERIDINE HYDRO-
  GEN CITRATE see ARX150
1-(4-METHOXY-2-METHYL-5-PHENYL-1H-PYRROL-3-YL)ETHANONE see
  MEX275
2-METHOXY-2-METHYLPROPANE see MHV859
N-(METHOXYMETHYL)-2-PROPENAMIDE see MEX300
2-METHOXY-3(5)-METHYLPYRAZINE see MEX350
2-(3-METHOXY-5-METHYLPYRAZOL-2-YL)-4-METHOXY-6-METHYLPYRIMIDINE
  see MCH550
7-METHOXY-1-METHYL-9H-PYRIDO(3,4-b)INDOLE see HAI500
4-METHOXYMETHYLPYRIDOXINE HYDROCHLORIDE see MEY000
4-METHOXYMETHYLPYRIDOXOL HYDROCHLORIDE see MEY000
1-(4-METHOXY-6-METHYL-2-PYRIMIDINYL)-3-METHYL-5-METHOXYPYRAZOLE
  see MCH550
p-METHOXY-β-METHYLSTYRENE see PMQ750
6-METHOXY-1-METHYL-1,2,3,4-TETRAHYDRO-β-CARBOLINE see MLJ100
N-(4-(METHOXYMETHYL)-1-(2-(2-THIENYL)ETHYL)-4-PIPERIDINYL)-N-PHENYL-
  PROPIONAMIDE CITRATE see SNH150
N-(4-(METHOXYMETHYL)-1-(2-(2-THIENYL)ETHYL)-4-PIPER-
  IDYL)PROPIONANILIDE CITRATE see SNH150
4-METHOXY-6-METHYL-1,3,5-TRIAZIN-2-AMINE see MGH800
3-METHOXY-4-MONOMETHYLAMINOAZOBENZENE see MNS000
METHOXYN see DBA800
METHOXYNAL see AAE500
2-METHOXY-1,4-NAPHTHALENEDIONE see MEY800
4-(6-METHOXY-2-NAPHTHALENYL)-2-BUTANONE see MFA300
2-METHOXY-1,4-NAPHTHOQUINONE see MEY800
2-METHOXYNAPHTHOQUINONE see MEY800
3-(4-METHOXY-1-NAPHTHOYL)PROPIONIC ACID see MEZ300
β-(1-METHOXY-4-NAPHTHOYL)-PROPIONSAEURE (GERMAN) see MEZ300

1-METHOXY-2-NAPHTHYLAMINE see MFA000
1-METHOXY-2-NAPHTHYLAMINE HYDROCHLORIDE see MFA250
4-(6-METHOXY-2-NAPHTHYL)-2-BUTANONE see MFA300
(+)-2-(METHOXY-2-NAPHTHYL)-PROPIONIC ACID see MFA500
(+)-2-(METHOXY-2-NAPHTHYL)-PROPIONSAEURE see MFA500
d-2-(6'-METHOXY-2'-NAPHTHYL)-PROPIONSAEURE see MFA500
4-METHOXY-2-NITROANILIN (CZECH) see MFB000
4-METHOXY-2-NITROANILINE see MFB000
2-METHOXY-4-NITROANILINE see NEQ000
2-METHOXY-5-NITROANILINE see NEQ500
2-METHOXY-4-NITROBENZENAMINE see NEQ000
2-METHOXY-5-NITROBENZENAMINE see NEQ500
1-METHOXY-2-NITROBENZENE see NER000
2-METHOXYNITROBENZENE see NER000
1-METHOXY-4-NITROBENZENE see NER500
4-METHOXYNITROBENZENE see NER500
p-METHOXYNITROBENZENE see NER500
3-METHOXY-2-(2-NITRO-1-IMIDAZOLYL)-1-PROPANOL see DBA700
7-METHOXY-2-NITRONAPHTHO(2,1-b)FURAN see MFB400
8-METHOXY-6-NITROPHENANTHOL-(3,4-d)-1,3-DIOXOLE-5-CARBOXYLIC ACID
  see AQY250
8-METHOXY-6-NITROPHENANTHRO(3,4-d)-1,3-DIOXOLE-5-CARBOXYLIC ACID
  SODIUM SALT see AQY125
4-METHOXY-2-NITROPHENYLTHIOCYANATE see MFB500
N-METHOXY-N-NITROSOMETHYLAMINE see DSZ000
1-METHOXY-N-NITROSO-N-PROPYLPROPYLAMINE see DWU800
3-METHOXY-17-α-19-NORPREGNA-K,3,5(10)-TRIEN-20-YN-17-OL see MKB750
11-β-METHOXY-19-NOR-17-α-PREGNA-1,3,5(10)-TRIEN-20-YNE-3,17-DIOL see
  MRU600
3-METHOXY-19-NOR-17-α-PREGNA-1,3,5(10)-TRIEN-10-YN-17-OL see MKB750
(17-α)-3-METHOXY-19-NORPREGNA-1,3,5(10)-TRIEN-20-YN-17-OL see MKB750
d-threo-METHOXY-3-(1-OCTENYL-O,N,N-AZOXY)-2-BUTANOL see EAG000
METHOXYOXIMERCURIPROPYLSUCCINYL UREA see MFC000
S-5-METHOXY-4-OXOPYRAN-2-YLMETHYL DIMETHYL PHOSPHOROTHIOATE
  see EAS000
S-((5-METHOXY-2-OXO-1,3,4-THIADIAZOL-3(2H)-YL)METHYL)-O,O-DIMETHYL
  PHOSPHORODITHIOATE see DSO000
METHOXYPECTIN see PAO150
2-METHOXY-1-(PENTYLOXY)-4-(1-PROPENYL)-BENZENE see AOK000
METHOXYPHENAMINE HYDROCHLORIDE see OJY000
METHOXYPHENAMINIUM CHLORIDE see OJY000
6-METHOXY-1-PHENAZINOL 5,10-DIOXIDE see HLT100
p-METHOXYPHENETHYLAMINE see MFC500
1-(p-METHOXYPHENETHYL)HYDRAZINE HYDROGEN SULFATE see MFC600
1-(p-METHOXYPHENETHYL)-HYDRAZINE SULFATE (1:1) see MFC600
4-(β-METHOXYPHENETHYL)-α-PHENYL-1-PIPERAZINEPROPANOL DIHYDRO-
  CHLORIDE see ECU600
2-METHOXYPHENOL see GKI000
3-METHOXYPHENOL see REF050
4-METHOXYPHENOL see MFC700
m-METHOXYPHENOL see REF050
o-METHOXYPHENOL see GKI000
p-METHOXYPHENOL see MFC700
METHOXYPHENOTHIAZINE see MCI500
1-(3-(2-METHOXYPHENOTHIAZIN-10-YL)-2-METHYLPROPYL)-4-PIPERIDINOL see
  LEO000
2-(p-METHOXYPHENOXY)-N-(2-(DIETHYLAMINO)ETHYL)ACETAMIDE see
  MCH250, DIB600
METHOXYPHENOXYDIOL see RLU000
N-(2-o-METHOXYPHENOXYETHYL)-6,7-DIMETHOXY-3,4-DIHYDRO-2-(1H)-ISOQUI-
  NOLINE CARBOXAMIDINE HBr see SBA875
(2-(p-METHOXYPHENOXY)ETHYL)HYDRAZINE HYDROCHLORIDE see MFD250
3-(2-METHOXYPHENOXY)-1-GLYCERYL CARBAMATE see GKK000
3-(o-METHOXYPHENOXY)-2-HYDROXYPROPYL CARBAMATE see GKK000
3-(o-METHOXYPHENOXY)-2-HYDROXYPROPYL-1-NICOTINATE see HLT300
(3R-trans)-3-((4-METHOXYPHENOXY)METHYL)-1-METHYL-4-PHENYLPIPERIDINE
  HYDROCHLORIDE see FAJ200
5-(o-METHOXYPHENOXYMETHYL)-2-OXAZOLIDINONE see MFD500
5-(o-METHOXYPHENOXYMETHYL)-2-OXAZOLIDONE see MFD500
3-(o-METHOXYPHENOXY)-1,2-PROPANEDIOL see RLU000
3-o-METHOXYPHENOXYPROPANE 1:2-DIOL see RLU000
3-(o-METHOXYPHENOXY)-1,2-PROPANEDIOL 1-CARBAMATE see GKK000
4-METHOXYPHENYLACETIC ACID see MFE250
p-METHOXYPHENYLACETIC ACID see MFE250
p-METHOXYPHENYLACETONE see AOV875
4-METHOXYPHENYLACETONITRILE see MFF000
p-METHOXYPHENYLACETONITRILE see MFF000
β-(o-METHOXYPHENYL)ACROLEIN see MEJ750
o-METHOXYPHENYLAMINE see AOV900
p-METHOXYPHENYLAMINE see AOW000
o-METHOXYPHENYLAMINE HYDROCHLORIDE see AOX250
N-(p-METHOXYPHENYL)-1-AZIRIDINECARBOXAMIDE see MFF250
2-METHOXY-4-PHENYLAZOANILINE see MFF500
4-((p-METHOXYPHENYL)AZO)-o-ANISIDINE see DNY400
N-(2-METHOXY-4-(PHENYLAZO)PHENYL)HYDROXYLAMINE see HLR000

2-METHOXYPHENYL BROMIDE see BMT400
4-METHOXYPHENYL BROMIDE see AOY450
o-METHOXYPHENYL BROMIDE see BMT400
p-METHOXYPHENYL BROMIDE see AOY450
4-(p-METHOXYPHENYL)-2-BUTANONE see MFF580
4-p-METHOXYPHENYL-2-BUTANONE see MFF580
p-METHOXYPHENYLBUTANONE see MFF580
4-(p-METHOXYPHENYL)-3-BUTEN-2-ONE see MLI400
p-METHOXYPHENYL-N-CARBAMOYLAZIRIDINE see MFF250
(E)-o-METHOXY-α-PHENYL-CINNAMIC ACID see MFG600
α-p-METHOXYPHENYL-α-DI-n-BUTYLAMINOACETAMIDE see DDT300
3-p-METHOXYPHENYL-5-DIETHYLAMINOETHYL-1,2,4-OXADIAZOLE see CPN750
2-(p-(6-METHOXY-2-PHENYL-3,4-DIHYDRO-1-NAPH-THYL)PHENOXY)TRIETHYLAMINE HYDROCHLORIDE see MFF625
1-(m-METHOXYPHENYL)-2-DIMETHYLAMINOMETHYLCYCLOHEXAN-1-OL HY-DROCHLORIDE see THJ750
trans-1-(m-METHOXYPHENYL)-2-DIMETHYLAMINOMETHYLCYCLOHEXAN-1-OL HYDROCHLORIDE see THJ750
1-(p-METHOXYPHENYL)-3,3-DIMETHYLTRIAZENE see DSN600
2-(p-METHOXYPHENYL)-3,3-DIPHENYLACRYLONITRILE see MFF750
α-(p-METHOXYPHENYL)-β,β-DIPHENYLACRYLONITRILE see MFF750
5-(p-METHOXYPHENYL)-1,2-DITHIOCYCLOPENTEN-3-THIONE see AOO490
5-(p-METHOXYPHENYL)-3H-1,2-DITHIOLE-3-THIONE see AOO490
5-(4-METHOXYPHENYL)-3H-1,2-DITHIOLE-3-THIONE (9CI) see AOO490
4-METHOXY-m-PHENYLENEDIAMINE see DBO000
p-METHOXY-m-PHENYLENEDIAMINE see DBO000
4-METHOXY-m-PHENYLENEDIAMINE SULFATE see DBO400
4-METHOXY-m-PHENYLENEDIAMINE SULPHATE see DBO400
p-METHOXY-m-PHENYLENEDIAMINE SULPHATE see DBO400
p-METHOXYPHENYLETHYLAMINE see MFC500
p-METHOXY-β-PHENYLETHYLHYDRAZINE DIHYDROGEN SULFATE see MFC600
o-METHOXY-β-PHENYLETHYLHYDRAZINE DIHYDROGEN SULFATE see MFG200
1-(2-METHOXY-2-PHENYL)ETHYL-4-(2-HYDROXY-3-METHOXY-3-PHE-NYL)PROPYLPIPERAZINE DIHYDROCHLORIDE see MFG250
1-(2-METHOXY-2-PHENYLETHYL)-4-(3-HYDROXY-3-PHENYLPRO-PYL)PIPERAZINE DIHYDROCHLORIDE see ECU600
4-(2-METHOXY-2-PHENYLETHYL)-α-PHENYL-1-PIPERAZINEPROPANOL (9CI) see EQY600
o-METHOXYPHENYL GLYCERYL ETHER see RLU000
2-(p-(6-METHOXY-2-PHENYLINDEN-3-YL)PHENOXY)TRIETHYLAMINE HYDRO-CHLORIDE see MFG260
2-(p-(6-METHOXY-2-PHENYL-3-INDENYL)PHENOXY)TRIETHYLAMINE HYDRO-CHLORIDE see MFG260
β-(o-METHOXYPHENYL)ISO-p-TROPYLMETHYLAMINEHYDROCHLORIDE see OJY000
α-(2-METHOXYPHENYL)-β-METHYLAMINOPROPANEHYDROCHLORIDE see OJY000
9-β-METHOXY-9-α-PHENYL-3-METHYL-3-AZABICYCLO(3.3.1)NONANE CITRATE see ARX150
2-(4-METHOXYPHENYL)-3-(1-METHYLETHYL)-3H-NAPHTH(1,2-d)IMIDAZOLE see THG700
4-METHOXYPHENYL METHYL KETONE see MDW750
p-METHOXYPHENYL METHYL KETONE see MDW750
1-(p-METHOXYPHENYL)-3-METHYL-3-NITROSOUREA see MFG400
1-(4-METHOXYPHENYL)-3-METHYL TRIAZENE see MFG510
1-(2-(p-α-(p-METHOXYPHENYL)-β-NITROSTY-RYL)PHENOXY)ETHYL)PYRROLIDINE CITRATE (1:1) see NHP500
1-(2-(p-α-(p-METHOXYPHENYL)-β-NITROSTY-RYL)PHENOXY)ETHYL)PYRROLIDINE MONOCITRATE see NHP500
trans-3-(o-METHOXYPHENYL)-2-PHENYLACRYLIC ACID see MFG600
2-((3-METHOXYPHENYLPIPERAZINO)-PROPYL)-3-METHYL-7-METHOXYCHRO-MONE see MFH000
4-(4-(o-METHOXYPHENYL)-1-PIPERAZINYL)-p-FLUOROBUTYROPHENONE see HAF400
6-(3-(4-(o-METHOXYPHENYL)-1-PIPERAZINYL)PROPYLAMINO)-1,3-DIMETHYLU-RACIL see USJ000
4-(p-METHOXYPHENYL)PIPERAZINYL 3,4,5-TRIMETHOXYPHENYL KETONE see MFH770
4-(o-METHOXYPHENYL)PIPERAZINYL 3,4,5-TRIMETHOXYPHENYL KETONE see MFH760
1-(p-METHOXYPHENYL)-2-PROPANONE see AOV875
3-(2-METHOXYPHENYL)-2-PROPENAL see MEJ750
1-(p-METHOXYPHENYL)PROPENE see PMQ750
(Z)-3-(2-METHOXYPHENYL)-2-PROPENOIC ACID (9CI) see MEJ775
1-(4-METHOXYPHENYL)-2-PROPEN-1-ONE see ONW100
p-METHOXYPHENYL 2-PYRIDYL KETONE see MFH900
p-METHOXYPHENYL 3-PYRIDYL KETONE see MED100
p-METHOXYPHENYL 3-PYRIDYL KETONE see MFH930
(2-(p-METHOXYPHENYLTHIO)ETHYL)HYDRAZINE MALEATE see MFJ000
1-(p-METHOXYPHENYL)-4-(3,4,5-TRIMETHOXYBENZOYL)PIPERAZINE see MFH770

1-(o-METHOXYPHENYL)-4-(3,4,5-TRIMETHOXYBENZOYL)PIPERAZINE see MFH760
(4-METHOXYPHENYL)(2-(2,4,6-TRIMETHYLPHENYL)-3-BENZOFURA-NYL)METHANONE see AOY300
5-(p-METHOXYPHENYL)TRITHIONE see AOO490
4-METHOXY-6-(β-PHENYLVINYL)-5,6-DIHYDRO-α-PYRONE see GJI250
p-METHOXYPHENYL VINYL KETONE see ONW100
10-(3-(4-METHOXYPIPERIDINO)PROPYL)PHENOTHIAZIN-2-YL METHYL KETONE see MFJ200
METHOXY POLYETHYLENE GLYCOL 350 see MFJ750
METHOXY POLYETHYLENE GLYCOL 550 see MFK000
METHOXY POLYETHYLENE GLYCOL 750 see MFK250
2-METHOXYPROMAZINE see MFK500
METHOXYPROMAZINE MALEATE see MFK750
1-METHOXYPROPANE see MOU830
α-METHOXY PROPANE see MOU830
METHOXYPROPANEDIOL see RLU000
3-METHOXYPROPANENITRILE see MFL750
3-METHOXYPROPANNITRIL see MFL750
3-METHOXY-1-PROPANOL see MFL000
1-METHOXY-2-PROPANOL see PNL250
1-METHOXY-2-PROPANONE see MDW300, MFL100
METHOXY-2-PROPANONE see MDW300, MFL100
METHOXYPROPAZINE see MFL250
2-METHOXYPROPENE see MFL300
(E)-1-METHOXY-4-(1-PROPENYL)BENZENE see PMR250
1-METHOXY-4-PROPENYLBENZENE see PMQ750
1-METHOXY-4-(2-PROPENYL)BENZENE see AFW750
4-METHOXYPROPENYLBENZENE see PMQ750
2-METHOXY-4-PROP-2-ENYLPHENOL see EQR500
2-METHOXY-4-(2-PROPENYL)PHENOL see EQR500
2-METHOXY-4-PROPENYLPHENOL see IKQ000
2-METHOXY-4-PROPENYLPHENYL ACETATE see AAX750
3-METHOXYPROPIONIC ACID METHYL ESTER see MFL400
β-METHOXYPROPIONIC ACID, METHYL ESTER see MFL400
3-METHOXYPROPIONITRILE see MFL750
2-(2-(2-METHOXYPROPOXY)PROPOXY) PROPANOL see TNA000
3-METHOXYPROPYLAMINE see MFM000
2-(2-METHOXYPROPYLAMINO)-2',6'-ACETOXYLIDIDE HYDROCHLORIDE see DSN800
1-METHOXY-4-PROPYLBENZENE see PNE250
2-METHOXY-4-PROPYLPHENOL see MFM750
1-METHOXYPROPYLPROPYLNITROSAMIN (GERMAN) see DWU800
1-METHOXYPROPYLPROPYLNITROSAMINE see DWU800
9-(2-METHOXY-4-(PROPYLSULFONAMIDO)ANILINO)-4-ACRIDINECARBOXAMIDE HYDROCHLORIDE see MFN000
(2-METHOXYPROPYL)UREA, MERCURY COMPLEX see CHX250
3-METHOXYPROPYNE see MFN250
5-METHOXY PSORALEN see MFN275
8-METHOXYPSORALEN see XDJ000
9-METHOXYPSORALEN see XDJ000
2-METHOXYPYRAZINE see MFN285
3-METHOXYPYRAZINE SULFANILAMIDE see MFN500
N¹-(3-METHOXY-2-PYRAZINYL)SULFANILAMIDE see MFN500
N¹-(6-METHOXY-3-PYRIDAZINYL)SULFANILAMIDE see AKO500
4-METHOXYPYRIDOXINE see MFN600
METHOXYPYRIMAL SODIUM see MFO000
N¹-(5-METHOXY-2-PYRIMIDINYL)SULFANILAMIDE, SODIUM SALT see MFO000
4-METHOXY-7H-PYRIMIDO(4,5-b)(1,4)THIAZIN-6-AMINE MONOHYDROCHLO-RIDE see THG600
S-((5-METHOXY-4H-PYRON-2-YL)-METHYL)-O,O-DIMETHYL-MONOTHIOFOSFAAT (DUTCH) see EAS000
S-((5-METHOXY-4H-PYRON-2-YL)-METHYL)-O,O-DIMETHYL-MONOTHIOPHOSP-HAT (GERMAN) see EAS000
S-(5-METHOXY-4-PYRON-2-YLMETHYL) DIMETHYL PHOSPHOROTHIOLATE see EAS000
METHOXY-5-PYRROLIDINO-2'-ETHYL-3-INDOLE see MFO250
5-METHOXY-3-(2-PYRROLIDINOETHYL)INDOLE see MFO250
3-METHOXY-4-PYRROLIDINYLMETHYLDIBENZOFURAN see MFO500
N⁴-(6-METHOXY-8-QUINOLINYL)-1,4-PENTANEDIAMINE PHOSPHATE (1:2) (9CI) see PMC310
1-(6-METHOXY-4-QUINOLYL)-3-(3-VINYL-4-PIPERIDYL)-1-PROPANONE OXALATE see QFJ300
α-(6-METHOXY-4-QUINOLYL)-5-VINYL-2-QUINUCLIDINEMETHANOL see QFS000
α-(6-METHOXY-4-QUINOLYL)-5-VINYL-2-QUINUCLIDINEMETHANOL GALACTU-RONATE (salt) see GAX000
α-(6-METHOXY-4-QUINOYL)-5-VINYL-2-QUINCLIDINEMETHANOL see QHJ000
3-METHOXYSALICYLALDEHYDE see VFP000
3-METHOXYSALICYLIC ACID see HJB500
METHOXY SIMAZINE see BJP250
3-METHOXY-4-STILBENAMINE see MFP500
p-METHOXYSTYRYL METHYL KETONE see MLI400
5-METHOXYSULFADIAZINE SODIUM see MFO000
6-METHOXY-3-SULFANILAMIDOPYRIDAZINE see AKO500
3-METHOXY-2-SULFAPYRAZINE see MFN500

4-METHOXY-2-SULFOANILINE see AIA500
2-METHOXYTETRACHLOROPHENOL see TBQ290
2-METHOXY-3,4,5,6-TETRACHLOROPHENOL see TBQ290
6-METHOXY-1,2,3,4-TETRAHYDRO-β-CARBOLINE HYDROCHLORIDE see MFT250
6-METHOXY-1,2,3,4-TETRAHYDRO-9H-PYRIDO(3,4-B)INDOLE HYDROCHLORIDE see MFT250
5-METHOXY-1,2,3,4-THIATRIAZOLE see MFQ250
2-METHOXYTHIOPHENOL see MFQ300
4-METHOXYTOLUENE see MGP000
p-METHOXYTOLUENE see MGP000
4-METHOXY-m-TOLUIDINE see MFQ500, MGO750
2-METHOXY-3,5,6-TRICHLOROBENZOIC ACID see TIK000
8-METHOXYTRICYCLO(5.2.2.1)DECANE-4-CARBOXALDEHYDE see OBW200
2-METHOXYTRICYCLOQUINAZOLINE see MFQ750
3-METHOXYTRICYCLOQUINAZOLINE see MFR000
METHOXY TRIETHYLENE GLYCOL VINYL ETHER see MFR250
METHOXYTRIGLYCOL see TJQ750
METHOXYTRIGLYCOL ACETATE see AAV500
METHOXYTRIMEPRAZINE see MCI500
exo-2-METHOXY-1,7,7-TRIMETHYLBICYCLO(2.2.1)HEPTANE see IHX500
(E,E)-11-METHOXY-3,7,11-TRIMETHYL-2,4-DODECANDIENOATE see KAJ000
exo-2-METHOXY-1,7,7-TRIMETHYLNORBORNANE see IHX500
(all-E)-9-(4-METHOXY-2,3,6-TRIMETHYLPHENYL)-3,7-DIMETHYL-2,4,6,8-NONA-TETRAENOIC ACID see REP400
1-METHOXY-3,4,5-TRIMETHYL PYRAZOLE-N-OXIDE see MFR775
2-METHOXY-1,3,5-TRINITROBENZENE see TMK300
5-METHOXYTRYPTAMINE see MFS400
6-METHOXYTRYPTAMINE see MFS500
METHOXYTRYPTAMINE see MFS400
5-METHOXYTRYPTAMINE HYDROCHLORIDE see MFT000
6-METHOXYTRYPTOLINE HYDROCHLORIDE see MFT250
N-METHOXYURETHANE see MEO000
p-METHOXY-α-VINYLBENZYL ALCOHOL see HKI000
p-METHOXY-α-VINYLBENZYL ALCOHOL ACETATE (ester) see ABN725
1-METHOXY-2-(VINYLOXY)ETHANE see VPZ000
2-METHOXY-4-VINYLPHENOL see VPF100
6-METHOXY-α-(5-VINYL-2-QUINUCLIDINYL)-4-QUINOLINEMETHANOL see QFS000
METHRAZONE see PEW000
METHSCOPOLAMINE BROMIDE see SBH500
METHSUXIMIDE see MLP800
METHULOSE see MIF760
METHURAL see DTG700
METHURIN (RUSSIAN) see DTG700
12-METHYBENZ(a)ANTHRACENE-7-METHANOL see HMF000
METHYBOL see MJE760
METHYCHLOTHIAZIDE see MIV500
METHYCLOTHIAZIDE see MIV500
METHYCOBAL see VSZ050
METHYCYCLOTHIAZIDE see MIV500
METHYL-E-600 see PHD500
METHYL ABIETATE see MFT500
METHYLACETAAT (DUTCH) see MFW100
METHYLACETALDEHYDE see PMT750
N-METHYLACETAMIDE see MFT750
METHYLACETAMIDE see MFT750
2-(2-(3-(N-METHYLACETAMIDO)-2,4,6-TRIIODOPHENOXY)ETHOXY)ACETIC ACID SODIUM SALT see MFU000
2-(2-(3-(N-METHYLACETAMIDO)-2,4,6-TRIIODOPHENOXY)ETHOXY)BUTYRIC ACID SODIUM SALT see MFU250
2-(2-(3-(N-METHYLACETAMIDO)-2,4,6-TRIIODOPHENOXY)ETHOXY)-2-PHENYL-ACETIC ACID SODIUM SALT see MFU500
2-METHYLACETANILIDE see ABJ000
2'-METHYLACETANILIDE see ABJ000
3-METHYLACETANILIDE see ABI750
3'-METHYLACETANILIDE see ABI750
4-METHYLACETANILIDE see ABJ250
4'-METHYLACETANILIDE see ABJ250
N-METHYLACETANILIDE see MFW000
m-METHYLACETANILIDE see ABI750
o-METHYLACETANILIDE see ABJ000
p-METHYLACETANILIDE see ABJ250
METHYLACETAPHOS see DRB600
METHYLACETAT (GERMAN) see MFW100
METHYL ACETATE see MFW100
METHYL ACETIC ACID see PMU750
METHYLACETIC ANHYDRIDE see PMV500
2'-METHYLACETOACETANILIDE see ABA000
METHYLACETOACETATE see MFX250
METHYL ACETONE (DOT) see MKA400
4'-METHYL ACETOPHENONE see MFW250
p-METHYL ACETOPHENONE see MFW250
METHYL ACETOPHOS see DRB600
METHYLACETOPYRONONE see MFW500

METHYL ACETOXON see DRB600
METHYL-β-ACETOXYETHYL-β-CHLOROETHYLAMINE see MFW750
METHYLACETOXYMALONONITRILE see MFX000
METHYL(ACETOXYMETHYL)NITROSAMINE see AAW000
METHYL-12-ACETOXY-9-OCTADECENOATE see MFX750
METHYL-12-ACETOXYOLEATE see MFX750
6-METHYL-17-α-ACETOXYPREGNA-4,6-DIENE-3,20-DIONE see VTF000
6-α-METHYL-17-α-ACETOXYPREGN-4-ENE-3,20-DIONE see MCA000
6-α-METHYL-17-α-ACETOXYPROGESTERONE see MCA000
METHYL ACETYLACETATE see MFX250
METHYL ACETYLACETONATE see MFX250
3-METHYL-4-ACETYLAMINOBIPHENYL see PEB250
1-METHYL-4-ACETYLBENZENE see MFW250
O-METHYL-1-ACETYLBENZOCYCLOBUTENE OXIME see MFX550
METHYLACETYL CHOLINE see ACR000
β-METHYLACETYLCHOLINE CHLORIDE see ACR000
METHYL ACETYLENE see MFX590
METHYL ACETYLENEDICARBOXYLATE see DOP400
METHYL ACETYLENE-PROPADIENE MIXTURE see MFX600
METHYLACETYLENE and PROPADIENE MIXTURES, stabilized (DOT) see MFX600
METHYL ACETYL RICINOLEATE see MFX750
2-METHYLACROLEIN see MGA250
α-METHYLACROLEIN see MGA250
β-METHYLACROLEIN see COB250
β-METHYL ACROLEIN see COB260
METHYLACRYLAAT (DUTCH) see MGA500
METHYLACRYLALDEHYDE see MGA250
2-METHYL ACRYLALDEHYDE OXIME see MGA275
2-METHYLACRYLAMIDE see MDN500
N-METHYLACRYLAMIDE see MGA300
METHYL-ACRYLAT (GERMAN) see MGA500
METHYL ACRYLATE see MGA500
METHYL ACRYLATE, INHIBITED (DOT) see MGA500
3-METHYLACRYLIC ACID see COB500
α-METHYLACRYLIC ACID see MDN250
β-METHYLACRYLIC ACID see COB500
2-METHYLACRYLIC ACID DODECYL ESTER see DXY200
α-METHYL ACRYLIC AMIDE see MDN500
α-METHYLACRYLONITRILE see MGA750
β-METHYLACRYLONITRILE see COC300
METHYLACRYLONITRILE see MGA750
α-METHYL-1-ADAMANTANEMETHYLAMINE HYDROCHLORIDE see AJU625
1,10-(N-METHYL-N-(1'-ADAMANTYL)AMINO)DECANE DIIODOMETHYLATE see DAE500
N⁶-METHYLADENOSINE mixed with SODIUM NITRITE (1:4) see SIS650
N-METHYLADENOSINE mixed with SODIUM NITRITE (1:4) see SIS650
METHYL ADIPATE see DOQ300
N-METHYLADRENALINE see MJV000
4-METHYLAESCULETIN see MJV800
METHYLAETHYLNITROSAMIN (GERMAN) see MKB000
METHYLAL see MGA850
2-METHYLALANINE see MGB000
α-METHYLALANINE see MGB000
METHYL ALCOHOL see MGB150
METHYL ALDEHYDE see FMV000
1-METHYL-2-ALDOXIMINOPYRIDINIUM CHLORIDE see FNZ000
1-METHYL-2-ALDOXIMINOPYRIDINIUM IODIDE see POS750
METHYLALKOHOL (GERMAN) see MGB150
1-METHYL-2-(p-ALLOPHANOYLBENZYL)HYDRAZINE HYDROBROMIDE see MKN750
(±)-2-(p-((2-METHYLALLYL)AMINO)PHENYL)PROPIONIC ACID see MGC200
2-METHYL-ALLYLCHLORID (GERMAN) see CIU750
1-METHYLALLYL CHLORIDE see CEV250
2-METHYLALLYL CHLORIDE see CIU750
α-METHYLALLYL CHLORIDE see CEV250
β-METHYLALLYL CHLORIDE see CIU750
METHYL ALLYL CHLORIDE (DOT) see CIU750
γ-METHYLALLYL CHLORIDE see CEU825
1-METHYL-5-ALLYL-5-(1-METHYL-2-PENTYNYL)BARBITURIC ACID SODIUM SALT see MDU500
METHYLALLYLNITROSAMIN (GERMAN) see MMT500
METHYLALLYLNITROSAMINE see MMT500
dl-3-METHYL-3-ALLYL-4-PROPIONOXYPIPERIDINE HYDROCHLORIDE see AGV890
1-α-METHYLALLYLTHIOCARBAMOYL-2-METHYLTHIOCARBAMOYLHYDRAZINE see MLJ500
N-((1-METHYLALLYL)THIOCARBAMOYL)-N'-(METHYLTHIOCARBA-MOYL)HYDRAZINE see MLJ500
METHYL ALUMINIUM SESQUIBROMIDE see MGC225
METHYL ALUMINIUM SESQUICHLORIDE see MGC230
METHYL ALUMINUM SESQUIBROMIDE see MGC225
METHYL ALUMINUM SESQUICHLORIDE see MGC230
METHYLAMID KYSELINY-2,4-5-TRICHLORBENZEN-SULFONOVE (CZECH) see MQC250

METHYLAMINE (ACGIH,OSHA) see MGC250
METHYLAMINE see MGC250
(2-METHYLAMINE-1-HYDROXYETHYL)METHANESULFONANILIDE METHANE-
   SULFONATE see AHL500
METHYLAMINEN (DUTCH) see MGC250
METHYLAMINE NITROFORM (DOT) see MGC300
METHYLAMINE, compd. with TRINITROMETHANE see MGC300
METHYLAMINE, aqueous solution (UN 1235) (DOT) see MGC250
METHYLAMINE, anhydrous (UN 1061) (DOT) see MGC250
4-METHYLAMINOACETOCATECHOL see MGC350
4-METHYLAMINOACETOPYROCATECHOL see MGC350
9-(p-(METHYLAMINO)ANILINO)ACRIDINE HYDROBROMIDE see MGD000
4-METHYL-2-AMINOANISOLE see MGO750
METHYLAMINOANTIPYRINE SODIUM METHANESULFONATE see AMK500
4'-METHYL-4-AMINOAZOBENZENE see TGV750
N-METHYL-4-AMINOAZOBENZENE see MNR500
4-(METHYLAMINO)AZOBENZENE see MNR500
N-METHYL-p-AMINOAZOBENZENE see MNR500
2-METHYL-1-AMINOBENZENE see TGQ750
1-METHYL-2-AMINOBENZENE see TGQ750
N-METHYLAMINOBENZENE see MGN750
(METHYLAMINO)BENZENE see MGN750
4-METHYLAMINOBENZENE-1,3-BIS(SULFONYL AZIDE) see MGD100
2-METHYL-1-AMINOBENZENE HYDROCHLORIDE see TGS500
1-METHYL-2-AMINOBENZENE HYDROCHLORIDE see TGS500
METHYL-N-(4-AMINOBENZENESULFONYL)CARBAMATE see SNQ500
3-(METHYLAMINO)-2,1-BENZISOTHIAZOLE see MGD200
3-(METHYLAMINO)-2,1-BENZISOTHIAZOLE HYDROCHLORIDE see MGD210
METHYL 2-AMINOBENZOATE see APJ250
METHYL o-AMINOBENZOATE see APJ250
METHYL p-AMINOBENZOATE see AIN150
N-METHYL-2-AMINOBENZOIC ACID see MGQ000
2-(METHYLAMINO)BENZOIC ACID see MGQ000
N-METHYL-o-AMINOBENZOIC ACID see MGQ000
o-(METHYLAMINO)BENZOIC ACID see MGQ000
4-METHYL-2-AMINOBENZOTHIAZOLE see AKQ500
6-METHYL-2-AMINOBENZOTHIAZOLE see MGD500
3-METHYL-4-AMINOBIPHENYL see MGE000
METHYLAMINO-BIS(1-AZIRIDINYL)PHOSPHINE OXIDE see MGE100
1-METHYLAMINO-4-BROMANTHRACHINON see BNN550
1-(METHYLAMINO)-4-BROMOANTHRAQUINONE see BNN550
N-((METHYLAMINO)CARBONYL)-N-(((METHYLAMI-
   NO)CARBONYL)OXY)ACETAMIDE see CBG075
2-(METHYLAMINO)-2-(2-CHLOROPHENYL)CYCLOHEXANONE see KEK200
METHYLAMINOCOLCHICIDE see MGF000
4-(METHYLAMINO)-o-CRESOL, HYDROGEN SULFATE (2:1) see MGF250
4-METHYLAMINO CRESOL-2-SULFATE see MGF250
N-METHYLAMINODIGLYCOL see MKU250
4-METHYLAMINO-1,5-DIMETHYL-2-PHENYL-3-PYRAZOLONE SODIUM METH-
   ANESULFONATE see AMK500
2-METHYL-4-AMINODIPHENYL see MGF500
3-METHYL-4-AMINODIPHENYL see MGE000
4'-METHYL-4-AMINODIPHENYL see MGF750
2-METHYLAMINOETHANOL see MGG000
β-(METHYLAMINO)ETHANOL see MGG000
N-METHYLAMINOETHANOL see MGG000
1-METHYLAMINO-4-ETHANOLAMINOANTHRAQUINONE see MGG250
1-METHYLAMINOETHANOLCATHECHOL HYDROCHLORIDE see AES500
m-METHYLAMINOETHANOLPHENOL see NCL500
p-METHYLAMINOETHANOLPHENOL see HLV500
m-METHYLAMINOETHANOLPHENOL HYDROCHLORIDE see SPC500
2-(METHYLAMINO)ETHYLAMINE see MJW100
S-(R,R)-d-(α-(1-METHYLAMINO)ETHYL)BENZENEMETHANOL HYDROCHLORIDE
   see POH250
(R-(R*,S*))-α-(1-(METHYLAMINO)ETHYL)BENZENEMETHANOL HYDROCHLO-
   RIDE see EAX000
1-α-(1-METHYLAMINOETHYL)BENZYL ALCOHOL see EAW000
α-(1-(METHYLAMINO)ETHYL)BENZYL ALCOHOL see POH000
(−)-α-(1-METHYLAMINOETHYL)BENZYL ALCOHOL see EAW000
d-(α-(1-METHYLAMINO)ETHYL)BENZYL ALCOHOL HYDROCHLORIDE see
   POH250
dl-α-(1-(METHYLAMINO)ETHYL) BENZYL ALCOHOL HYDROCHLORIDE see
   EAX500
1-α-(1-(METHYLAMINO)ETHYL)BENZYL ALCOHOL SULFATE see EAY500
d-α-(1-METHYLAMINO)ETHYL)BENZYL PHOSPHATE see EAY150
dl-α-(1-(METHYLAMINO)ETHYL)BENZYL PHOSPHATE see EAY175
l-α-(1-(METHYLAMINO)ETHYL)BENZYL PHOSPHATE see EAY075
α-(1-METHYLAMINOETHYL)-p-HYDROXYBENZYL ALCOHOL see HKH500
2-METHYL-3-(β-AMINOETHYL)-5-METHOXYBENZOFURAN see MGH750
N-METHYLAMINOETHYL-2-METHYLBENZHYDRYL ETHER HYDROCHLORIDE
   see TGJ250
2-(2-(METHYLAMINO)ETHYL)PYRIDINE see HGE820
4-(2-METHYLAMINOETHYL)PYROCATECHOL HYDROCHLORIDE see EAZ000
2-METHYLAMINOFLUORENE see FEI500
2-METHYL-6-AMINOHEPTANE see ILM000

2-METHYLAMINO-HEPTANE, HYDROCHLORIDE see NCL300
1-METHYLAMINO-4-(β-HYDROXYETHYLAMINO)ANTHRAQUINONE see
   MGG250
3'-(2-(METHYLAMINO)-1-HYDROXYETHYL)METHANESULFONANILIDE METH-
   ANESULFONATE see AHL500
2-METHYL-4-AMINO-1-HYDROXYNAPHTHALENE see AKX500
β-METHYLAMINO-α-(4-HYDROXYPHENYL)ETHYL ALCOHOL see HLV500
2-METHYLAMINOISOCAMPHANE see VIZ400
3-METHYLAMINOISOCAMPHANE see VIZ400
3-METHYLAMINOISOCAMPHANE HYDROCHLORIDE see MQR500
2-METHYLAMINOISOOCTANE HYDROCHLORIDE see ODY000
2-METHYL-4-AMINO-6-METHOXY-s-TRIAZINE see MGH800
2-METHYLAMINO METHYL BENZOATE see MGQ250
1-METHYLAMINOMETHYLDIBENZO(b,c)BICYCLO(2,2,2)OCTADIENE HYDRO-
   CHLORIDE see BCH750
6-METHYLAMINO-2-METHYLHEPTENE see ILK000
METHYLAMINO-METHYLHEPTENE HYDROCHLORIDE see ODY000
METHYLAMINOMETHYL(4-HYDROXYPHENYL)CARBINOL see HLV500
2-(6-(METHYLAMINO)-3-(METHYLIMINO)-3H-XANTHEN-9-YL)BENZOIC ACID
   MONOPERCHLORATE see RGW100
2-METHYLAMINO-4-METHYLTHIO-6-ISOPROPYLAMINO-1,3,5-TRIAZINE see
   INR000
2-METHYL-4-AMINO-1-NAPHTHOL see AKX500
o-METHYL-2-AMINO-1-NAPHTHOL HYDROCHLORIDE see MFA250
4-METHYLAMINO-N-NITROSOAZOBENZENE see MMY250
1-METHYLAMINO-4-OXYETHYLAMINOANTHRAQUINONE (RUSSIAN) see
   MGG250
3-(α-METHYLAMINOPHENETHYL)PHENOL HYDROCHLORIDE see MGJ600
N-(α-METHYLAMINOPHENETHYL)PHENOL HYDROCHLORIDE see MGJ600
3-METHYL-4-AMINOPHENOL see AKZ000
METHYL-p-AMINOPHENOL SULFATE see MGJ750
p-METHYLAMINOPHENOLSULFATE see MGJ750
p-(METHYLAMINO)PHENOL SULFATE (2:1) (SALT) see MGJ750
6-METHYL-2-(p-AMINO PHENYL) see ALX000
METHYLAMINOPHENYLDIMETHYLPYRAZOLONE METHANESULFONATE SODI-
   UM see AMK500
1-2-METHYLAMINO-1-PHENYLPROPANOL see EAW000
d-psi-2-METHYLAMINO-1-PHENYL-1-PROPANOL see POH000
METHYL ((4-AMINOPHENYL)SULFONYL)CARBAMATE see SNQ500
3-METHYL-4-AMINO-6-PHENYL-1,2,4-TRIAZIN(4H)-ON (GERMAN) see ALA500
2-METHYL-2-AMINOPROPANE HYDROCHLORIDE see MGJ800
1-(3-METHYLAMINOPROPYL)-2-ADAMANTANOL HYDROCHLORIDE see
   MGK750
1-(3-METHYLAMINOPROPYL)DIBENZO(b,e)BICYCLO(2.2.2)OCTADIENE HYDRO-
   CHLORIDE see MAW850
5-(3-METHYLAMINOPROPYL)-5H-DIBENZO(a,d)CYCLOHEPTENE see DDA600
5-(3-METHYLAMINOPROPYL)-5H-DIBENZO(a,d)CYCLOHEPTENE HYDROCHLO-
   RIDE see POF250
9-(γ-METHYLAMINOPROPYL)-9,10-DIHYDRO-9,10-ETHANOANTHRACENE HY-
   DROCHLORIDE see MAW850
5-(3-METHYLAMINO)PROPYLIDENE)DIBENZO(a,e)CYCLOHEPTA(1,5)DIENE see
   NNY000
4-(3'-METHYLAMINOPROPYLIDENE)-9,10-DIHYDRO-4H-BEN-
   ZO(4,5)CYCLOHEPTA(1,2-b)THIOPHEN see MGL500
5-(3-METHYLAMINOPROPYLIDENE)-10,11-DIHYDRO-5H-DIBEN-
   ZO(a,d)CYCLOHEPTENE see NNY000
METHYLAMINOPROPYLIMINODIBENZYL see DSI709
N-(γ-METHYLAMINOPROPYL)IMINODIBENZYL HYDROCHLORIDE see DLS600
p-(3-METHYLAMINOPROPYL)PHENOL see FMS875
METHYLAMINOPTERIN see MDV500
METHYL-4-AMINO-2-PYRIDINE see ALC250
4-METHYL-2-AMINOPYRIDINE see ALC250
1-METHYL-3-AMINO-5H-PYRIDO(4,3-b)INDOLE see ALD500
N-METHYL-4-AMINO-1,2,5-SELENADIAZOLE-3-CARBOXAMIDE see MGL600
2-METHYL-4-AMINOSTILBENE see MPJ250
3-METHYL-4-AMINOSTILBENE see MPJ500
3-β-METHYLAMINO-2,2,3-TRIMETHYLBICYCLO(2.2.1)HEPTANE see VIZ400
2-METHYLAMINO-2,3,3-TRIMETHYLNORBORANE see VIZ400
2-METHYLAMINO-2,3,3-TRIMETHYLNORBORNANE see VIZ400
METHYLAMMONIUM CHLORITE see MGN000
METHYLAMMONIUM NITRATE see MGN150
METHYLAMMONIUM PERCHLORATE see MGN250
N-METHYLAMPHETAMINE see DBB000
METHYLAMPHETAMINE see DBB000
N-METHYLAMPHETAMINE HYDROCHLORIDE see MDT600
d-METHYLAMPHETAMINE HYDROCHLORIDE see MDT600
METHYLAMPHETAMINE HYDROCHLORIDE see DBA800, MDT600
2-METHYLAMYL ACETATE see MGN300
METHYL AMYL ACETATE (DOT) see HFJ000
METHYLAMYL ACETATE see HFJ000
METHYLAMYL ALCOHOL see AOK750
METHYL AMYL ALCOHOL see MKW600
METHYL AMYL CARBINOL see HBE500
METHYL-AMYL-CETONE (FRENCH) see MGN500
2-METHYLAMYLESTER KYSELINY OCTOVE see MGN300

METHYL AMYL KETONE (DOT) see MGN500
METHYL n-AMYL KETONE see MGN500
METHYLAMYLNITROSAMIN (GERMAN) see AOL000
METHYL-N-AMYLNITROSAMINE see AOL000
METHYLAMYLNITROSAMINE see AOL000
17-α-METHYL-5-α-ANDROSTANO(2,3-c)(1,2,5)OXADIAZOL-17-β-OL see
  AOO300
17-METHYL-5-α-ANDROSTANO(2,3-c)(1,2,5)OXADIAZOL-17-β-OL see AOO300
METHYLANDROSTENDIOL see AOO475
2-METHYLANILINE see TGQ750
3-METHYLANILINE see TGQ500
4-METHYLANILINE see TGR000
N-METHYLANILINE (ACGIH,DOT) see MGN750
N-METHYLANILINE see MGN750
m-METHYLANILINE see TGQ500
o-METHYLANILINE see TGQ750
p-METHYLANILINE see TGR000
METHYLANILINE see MGN750
2-METHYLANILINE HYDROCHLORIDE see TGS500
4-METHYLANILINE HYDROCHLORIDE see TGS750
o-METHYLANILINE HYDROCHLORIDE see TGS500
METHYLANILINE and SODIUM NITRITE (1.2:1) see MGO000
N-METHYLANILINE mixed with SODIUM NITRITE (1:35) see MGO250
2-(N-METHYLANILINO)ETHANOL see MKQ250
N-METHYLANILIN UND NATRIUMNITRIT (GERMAN) see MGO250
METHYL o-ANISATE see MLH800
METHYL-p-ANISATE see AOV750
5-METHYL-o-ANISIDINE see MGO750
2-METHYL-p-ANISIDINE see MGO500
p-METHYL ANISOLE see MGP000
METHYLANTALON see MJE760
6-METHYLANTHANTHRENE see MGP250
9-METHYLANTHRACENE see MGP750
METHYL ANTHRANILATE (FCC) see APJ250
N-METHYLANTHRANILIC ACID see MGQ000
N-METHYLANTHRANILIC ACID, METHYL ESTER see MGQ250
2-METHYL-1-ANTHRAQUINONYLAMINE see AKP750
METHYLANTIFEBRIN see MFW000
METHYL APHOXIDE see TNK250
METHYLARSENIC ACID see MGQ530
METHYLARSENIC ACID, SODIUM SALT see MRL750
METHYLARSENIC DIMETHYL DITHIOCARBAMATE see USJ075
METHYLARSENIC SULFIDE see MGQ750
METHYL ARSINE-BIS(DIMETHYLDITHIOCARBAMATE) see USJ075
METHYLARSINE DICHLORIDE see DFP200
METHYLARSINE DIIODIDE see MGQ775
METHYLARSINE SULFIDE see MGQ750
METHYLARSINIC ACID see MGQ530
METHYLARSINIC SULFIDE see MGQ750
METHYLARSINIC SULPHIDE see MGQ750
METHYLARSONIC ACID see MGQ530
METHYLARSONOUS DICHLORIDE see DFP200
METHYLARTERENOL see VGP000
METHYL ASPARTYLPHENYLALANATE see ARN825
1-METHYL N-l-α-ASPARTYL-l-PHENYLALANINE see ARN825
METHYLATROPINE BROMIDE see MGR250
N-METHYLATROPINE NITRATE see MGR500
METHYL ATROPINE NITRATE see MGR500
8-METHYLATROPINIUM BROMIDE see MGR250
METHYLATROPINIUM BROMIDE see MGR250
8-METHYLATROPINIUM NITRATE see MGR500
N-METHYLATROPINIUM NITRATE see MGR500
8-METHYL-8-AZABICYCLO(3.2.1)OCTAN-3-OL BENZOATE (ester), HYDRO-
  CHLORIDE, exo- see TNS200
2-METHYLAZACYCLOPROPANE see PNL400
3-METHYL-3-AZAPENTANE-1,5-BIS(ETHYLDIMETHYLAMMONIUM) BROMIDE
  see MKU750
METHYL AZIDE see MGR750
METHYL-2-AZIDOBENZOATE see MGR800
METHYLAZINPHOS see ASH500
2-METHYLAZIRIDINE see PNL400
N-METHYL-1-AZIRIDINECARBOXAMIDE see EJN400
METHYL-AZOXY-BUTANE see MGS500
METHYLAZOXYMETHANOL see HMG000
METHYLAZOXYMETHANOL ACETATE see MGS750
METHYLAZOXYMETHANOL-β-d-GLUCOSIDE see COU000
METHYLAZOXYMETHANOL GLUCOSIDE see COU000
METHYLAZOXYMETHANOL-β-D-GLUCOSIDURONIC ACID see MGS700
METHYLAZOXYMETHYL ACETATE see MGS750
METHYLAZOXYMETHYL BENZOATE see MGS925
METHYLAZOXYMETHYLESTER KYSELINY OCTOVE (CZECH) see MGS750
METHYLAZOXYOCTANE see MGT000
METHYL-AZULENO(5,6,7-c,d)PHENALENE see MGT250
METHYL-B₁₂ see VSZ050
1-METHYLBARBITAL see DJO800

N-METHYLBARBITAL see DJO800
METHYLBARBITAL see DJO800
METHYLBEN see HJL500
12-METHYL BENZ(A)ANTHRACENE-7-ETHANOL see HKU000
6-METHYL-11H-BENZ(bc)ACEANTHRYLENE see MLN500
3-METHYLBENZ(j)ACEANTHRYLENE see DAL200
12-METHYLBENZ(e)ACEPHENANTHRYLENE see MGT420
3-METHYLBENZ(e)ACEPHENANTHRYLENE see MGT400
7-METHYLBENZ(e)ACEPHENANTHRYLENE see MGT410
8-METHYLBENZ(e)ACEPHENANTHRYLENE see MGT415
9-METHYL-1,2-BENZACRIDINE see MGU500
9-METHYL-3,4-BENZACRIDINE see MGT500
10-METHYL-5,6-BENZACRIDINE see MGU500
10-METHYL-7,8-BENZACRIDINE (FRENCH) see MGT500
12-METHYLBENZ(a)ACRIDINE see MGU500
7-METHYLBENZ(c)ACRIDINE see MGT500
11-METHYLBENZ(c)ACRIDINE-7-CARBOXALDEHYDE see FNS000
9-METHYLBENZ(c)ACRIDINE-7-CARBOXALDEHYDE see FNR000
7-METHYLBENZ(c)ACRIDINE 3,4-DIHYDRODIOL see MGU550
α-METHYL-α-BENZALACETONE see MNS600
N-METHYLBENZAMIDE see MHA250
10-METHYL-1,2-BENZANTHRACEN (GERMAN) see MGW750
1′-METHYL-1,2-BENZANTHRACENE see MGU750
10′-METHYL-1,2-BENZANTHRACENE see MGW750
2′-METHYL-1,2-BENZANTHRACENE see MGV000
3-METHYL-1,2-BENZANTHRACENE see MGV750
4′-METHYL-1:2-BENZANTHRACENE see MGV500
4-METHYL-1,2-BENZANTHRACENE see MGW000
5-METHYL-1,2-BENZANTHRACENE see MGW250
6-METHYL-1,2-BENZANTHRACENE see MGW500
7-METHYL-1,2-BENZANTHRACENE see MGW740
8-METHYL-1:2-BENZANTHRACENE see MGX250
9-METHYL-1,2-BENZANTHRACENE see MGX500
11-METHYLBENZ(a)ANTHRACENE see MGX250
12-METHYLBENZ(a)ANTHRACENE see MGX500
1-METHYLBENZ(a)ANTHRACENE see MGU750
10-METHYLBENZ(a)ANTHRACENE see MGW740
2-METHYLBENZ(a)ANTHRACENE see MGV000
3-METHYLBENZ(a)ANTHRACENE see MGV250
4-METHYLBENZ(a)ANTHRACENE see MGV500
5-METHYLBENZ(a)ANTHRACENE see MGV750
6-METHYLBENZ(a)ANTHRACENE see MGW000
7-METHYLBENZ(a)ANTHRACENE see MGW750
8-METHYLBENZ(a)ANTHRACENE see MGW250
9-METHYLBENZ(a)ANTHRACENE see MGW500
10-METHYL-1,2-BENZANTHRACENE-5-CARBONAMIDE see MGX750
7-METHYLBENZ(a)ANTHRACENE-10-CARBONITRILE see MGY250
7-METHYLBENZ(a)ANTHRACENE-8-CARBONITRILE see MGY000
7-METHYLBENZ(a)ANTHRACENE-12-CARBOXALDEHYDE see MGY500
12-METHYLBENZ(a)ANTHRACENE-7-CARBOXALDEHYDE see FNT000
7-METHYLBENZ(a)ANTHRACENE-12-METHANOL see HMF500
12-METHYLBENZ(a)ANTHRACENE-7-METHANOL ACETATE (ESTER) see
  ABR250
12-METHYLBENZ(a)ANTHRACENE-7-METHANOL BENZOATE (ESTER) see
  BDQ250
7-METHYLBENZ(a)ANTHRACENE-5,6-OXIDE see MGZ000
7-METHYLBENZ(a)ANTHRACEN-8-YL CARBAMIDE see MGX750
S-(12-METHYL-7-BENZ(a)ANTHRYLMETHYL)HOMOCYSTEINE see MHA000
N-METHYLBENZAZIMIDE, DIMETHYLDITHIOPHOSPHORIC ACID ESTER see
  ASH500
METHYLBENZEDRIN see DBA800
METHYL 1H-BENZEMEDAZOL-2-YLCARBAMATE see MHC750
N-METHYLBENZENAMIDE see MHA250
2-METHYLBENZENAMINE see TGQ750
3-METHYLBENZENAMINE see TGQ500
4-METHYLBENZENAMINE see TGR000
N-METHYLBENZENAMINE see MGN750
m-METHYLBENZENAMINE see TGQ500
o-METHYLBENZENAMINE see TGQ750
p-METHYLBENZENAMINE see TGR000
2-METHYLBENZENAMINE HYDROCHLORIDE see TGR500, TGS500
4-METHYLBENZENAMINE HYDROCHLORIDE see TGS750
o-METHYLBENZENAMINE HYDROCHLORIDE see TGS500
5-METHYL-1,3-BENZENDIOL see MPH500
METHYLBENZENE see TGK750
METHYL BENZENEACETATE see MHA500
2-METHYLBENZENECARBONITRILE see TGT500
METHYL BENZENECARBOXYLATE see MHA750
2-METHYL-1,2-BENZENEDIAMINE see TGM100
3-METHYL-1,2-BENZENEDIAMINE see TGY800
4-METHYL-1,2-BENZENEDIAMINE see TGM250
4-METHYL-1,3-BENZENEDIAMINE see TGL750
2-METHYL-1,4-BENZENEDIAMINE see TGM000, TGM400
2-METHYL-1,4-BENZENEDIAMINE DIHYDROCHLORIDE see DCE200
2-METHYL-1,4-BENZENEDIAMINE SULFATE see DCE600

N-METHYL-N-BENZYLPROPYNYLAMINE see MOS250
METHYLBICYCLO(2.2.1)HEPTENE-2,3-DICARBOXYLIC ANHYDRIDE ISOMERS
  see NAC000
1-METHYLBIGUANIDE HYDROCHLORIDE see MHP400
2-METHYL-2,2'-BIOXIRANE (9CI) see DHD200
4-METHYL-1,1'-BIPHENYL see MJH905
p-METHYLBIPHENYL see MJH905
α-METHYL-(1,1'-BIPHENYL)-4-ACETIC ACID 3-SULFOPROPYL ESTER SODIUM
  SALT see PFR350
4'-METHYLBIPHENYLAMINE see MGF750
METHYL 4-BIPHENYLYL KETONE see MHP500
METHYLBIS(3-AMINOPROPYL)AMINE see BGU750
N-METHYL-BIS-CHLORAETHYLAMIN (GERMAN) see BIE250
γ(1-METHYL-5-BIS(β-CHLORAE-
  THYL)AMINOBENZIMIDAZOLYL)BUTTERSAEUREHYDROCHLORID(GERMAN)
  see CQM750
METHYL-BIS-(β-CHLORAETHYL)-AMIN-N-OXYD-HYDROCHLORID (GERMAN)
  see CFA750
N-METHYL-BIS-β-CHLORETHYLAMINE HYDROCHLORIDE see BIE500
γ-(1-METHYL-5-BIS(β-CHLOROAETHYL) AMINOBENZIMIDAZOYL) BUTTERSAU-
  ERHYDROCHLORID(GERMAN) see CQM750
N-METHYL-BIS(β-CHLOROETHYL)AMINE see BIE250
METHYLBIS(β-CHLOROETHYL)AMINE see BIE250
N-METHYL-BIS(2-CHLOROETHYL)AMINE (MAK) see BIE250
N-METHYLBIS(2-CHLOROETHYL)AMINE HYDROCHLORIDE see BIE500
METHYLBIS(2-CHLOROETHYL)AMINE HYDROCHLORIDE see BIE500
METHYLBIS(β-CHLOROETHYL)AMINE HYDROCHLORIDE see BIE500
METHYL-BIS(β-CHLOROETHYL)AMINE OXIDE see CFA500
METHYLBIS(β-CHLOROETHYL)AMINE N-OXIDE see CFA500
N-METHYLBIS(2-CHLOROETHYL)AMINE-N-OXIDE HYDROCHLORIDE see
  CFA750
METHYLBIS(β-CHLOROETHYL)AMINE-N-OXIDE HYDROCHLORIDE see CFA750
6-METHYL-5-(BIS(2-CHLOROETHYL)AMINO)URACIL see DYC700
4-METHYL-5-(BIS(β-CHLOROETHYL)AMINO)URACIL see DYC700
METHYL-BIS(2-CHLOROETHYL-MERCAPTOETHYL)AMINE HYDROCHLORIDE
  see MHQ500
METHYLBIS(β-CHLOROETHYLTHIOETHYL)AMINE HYDROCHLORIDE see
  MHQ500
METHYL BIS(β-CYANOETHYL)AMINE see MHQ750, MHQ750
7-METHYLBISDEHYDRODOISYNOLIC ACID see BIT000
METHYLBIS(DIMETHYLDITHIOCARBAMOYLTHIO)ARSINE see USJ075
METHYLBIS(2-HYDROXYETHYL)AMINE see MKU250
2-METHYL-4,5-BIS(HYDROXYMETHYL)-3-HYDROXYPYRIDINE see PPK250
N-METHYL-BIS-(2-KYANETHYL)AMIN see MHQ750
7-METHYL-2,3: 9,10-BIS(METHYLENEDIOXY)-7,13a-SECOBERBIN-13a-ONE see
  FOW000
N-METHYL-N,N-BIS(3-METHYLSULFONYLOXYPROPYL)AMINE 4,4'-BIPHENYLDI-
  SULFONATE see MHQ775
METHYLBISMUTH OXIDE see MHR000
4-METHYL-2,6-BIS(1-PHENYLETHYL)PHENOL see MHR050
N-METHYL-BIS(2,4-XYLYLIMINOMETHYL)AMINE see MJL250
β-METHYLBIVINYL see IMS000
METHYL BORATE see TLN000
1-METHYL-BP see MHG250
5-METHYL-BP see MHH200
METHYLBROMID (GERMAN) see MHR200
dl-METHYLBROMIDE see MDL000
METHYL BROMIDE see MHR200
METHYL BROMIDE and ETHYLENE DIBROMIDE MIXTURE, liquid (DOT) see
  BNM750
METHYL α-BROMOACETATE see MHR250
METHYL BROMOACETATE see MHR250
2-METHYLBROMOBENZENE see BOG260
3-METHYLBROMOBENZENE see BOG300
m-METHYLBROMOBENZENE see BOG300
METHYL-4-BROMOBENZENEDIAZOATE see MHR750
METHYL 4-BROMO-2-BUTENOATE see MHR790
METHYL 4-BROMOCROTONATE see MHR790
METHYL γ-BROMOCROTONATE see MHR790
METHYL BROMOCROTONATE see MHR790
O-METHYL-O-(4-BROMO-2,5-DICHLOROPHENYL)PHENYL THIOPHOSPHONATE
  see LEN000
METHYL-(BROMOMERCURI)FORMATE see MHS250
1-METHYL-7-BROMOMETHYLBENZ(a)ANTHRACENE see BNQ750
1-METHYL-3-(p-BROMOPHENYL)-1-NITROSOUREA see BNX125
1-METHYL-3-(p-BROMOPHENYL)UREA see MHS375
1-METHYL-3-(p-BROMOPHENYL)UREA mixed with SODIUM NITRITE see
  SIQ675
METHYL 1-BROMOVINYL KETONE see MHS400
1-METHYL-3-(p-BROMPHENYL)HARNSTOFF see MHS375
1-METHYL-3-(p-BROMPHENYL)-1-NITROSOHARNSTOFF (GERMAN) see
  BNX125
N-METHYLBUTABARBITAL see BRJ250
2-METHYLBUTADIENE see IMS000
2-METHYL-1,3-BUTADIENE (DOT) see IMS000

2-METHYL-1-BUTANAL see MJX500
2-METHYLBUTANAL see MJX500
3-METHYLBUTANAL see MHX500
2-METHYLBUTANAL-4 see MHX500
α-METHYLBUTANAL see MJX500
3-METHYL-2-BUTANAMINE see AOE200
2-METHYLBUTANE see EIK000
2-METHYL-1,4-BUTANEDIOL see MHS500
2-METHYLBUTANE SECONDARY MONONITRILE see ITD000
METHYL n-BUTANOATE see MHY000
3-METHYLBUTANOIC ACID see ISU000
3-METHYLBUTANOIC ACID, BUTYL ESTER see ISX000
3-METHYLBUTANOIC ACID, ETHYL ESTER see ISY000
2-METHYLBUTANOIC ACID, n-HEXYL ESTER see HFR200
3-METHYLBUTANOIC ACID, METHYL ESTER see ITC000
3-METHYL-BUTANOIC ACID 2-PHENYLETHYL ESTER see PDF775
3-METHYLBUTANOIC ACID, PHENYLETHYL ESTER see ISW000
3-METHYLBUTANOIC ACID 3-PHENYL-2-PROPENYL ESTER see PEE200
3-METHYLBUTANOIC ACID, 2-PROPENYL ESTER see ISV000
endo-3-METHYLBUTANOIC ACID 1,7,7-TRIMETHYLBICYCLO(2.2.1)HEPT-2-YL
  ESTER see HOX100
2-METHYL BUTANOL-1 see MHS750
3-METHYLBUTAN-1-OL see IHP000
2-METHYL BUTANOL-2 see PBV000
2-METHYL-2-BUTANOL see PBV000
2-METHYLBUTANOL see MHS750
3-METHYLBUTAN-3-OL see PBV000
3-METHYL BUTANOL see IHP000
2-METHYL-4-BUTANOL see IHP000
3-METHYL-1-BUTANOL (CZECH) see IHP000
3-METHYL-1-BUTANOL NITRATE see ILW100
3-METHYLBUTANOL NITRITE see IMB000
3-METHYL-2-BUTANONE see MLA750
2-METHYL BUTAN-2-ONE (DOT) see MLA750
2-METHYL-1-BUTENE see MHT000
3-METHYL-1-BUTENE see MHT250
2-METHYLBUTENE see IHP150
METHYLBUTENE see IHP150
2-METHYL-2-BUTENEDIOIC ACID see CMS320
trans-2-METHYL-2-BUTENEDIOIC ACID see MDI250
cis-METHYLBUTENEDIOIC ACID see CMS320
(Z)-2-METHYL-2-BUTENEDIOIC ACID (9CI) see CMS320
cis-2-METHYL-2-BUTENEDIOIC ACID, DIMETHYL ESTER see DRF200
METHYL trans-2-BUTENOATE see COB825
3-METHYL-2-BUTENOIC ACID see MHT500
trans-2-METHYL-2-BUTENOIC ACID see TGA700
2-METHYLBUTENOIC ACID-7-((2,3-DIHYDROXY-2-(1-METHYLETHYL)-1-OXOBU-
  TOXY)METHYL)-2,3,5,7a-TETRAHYDRO-1H-PYRROLIZIN-1-YL ESTER see
  SPB500
(E)-2-METHYL-2-BUTENOIC ACID ETHYL ESTER see TGA800
3-METHYL-2-BUTENOIC ACID ETHYL ESTER see MIP800
2-METHYL-2-BUTENOIC ACID, HEXYL ESTER see HFX000
3-METHYL-2-BUTEN-1-OL see MHU110
2-METHYL-3-BUTEN-2-OL see MHU100
3-METHYL-1-BUTEN-3-OL see MHU100
3-METHYL-BUTEN-(1)-OL-(3) (GERMAN) see MHU100
METHYLBUTENOL see MHU100
3-METHYL-3-BUTEN-2-ON (GERMAN) see MKY500
2-METHYL-1-BUTEN-3-ONE see MKY500
3-METHYL-2-BUTENYL ACETATE see DOQ350
3-METHYL-2-BUTENYL BENZOATE see MHU150
4-(3-METHYL-2-BUTENYL)-1,2-DIPHENYL-3,5-PYRAZOLIDINEDIONE see
  PEW000
3-(3-METHYL-2-BUTENYL)-1,2,3,4,5,6-HEXAHYDRO-6,11-DIMETHYL-2,6-METHA-
  NO-3-BENZAZOCIN-8-OL see DOQ400
9-((3-METHYL-2-BUTENYL)OXY)-7H-FURO(3,2-g)(1)BENZOPYRAN-7-ONE see
  IHR300
5-(1-METHYL-1-BUTENYL)-5-PROPYLBARBITURIC ACID see MHU200
3-METHYL-2-BUTENYL SALICYLATE see PMB600
2-METHYL-1-BUTEN-3-YNE see MHU250
3-METHYL-3-BUTEN-1-YNYLTRIETHYLLEAD see MHU500
3-METHYL-BUTIN-(1)-OL-(3) (GERMAN) see MHX250
(2-(3-METHYLBUTOXY)ETHYL)BENZENE see IHV050
3'-METHYLBUTTERGELB (GERMAN) see DUH600
3-METHYL-1-BUTYL ACETATE see IHO850
1-METHYLBUTYL ACETATE see AOD735
3-METHYLBUTYL ACETATE see IHO850
2-METHYL-BUTYLACRYLAAT (DUTCH) see MHU750
2-METHYL-BUTYLACRYLAT (GERMAN) see MHU750
2-METHYL BUTYLACRYLATE see MHU750
N-(METHYL) BUTYL AMINE see MHV000
N-METHYL-n-BUTYLAMINE see MHV000
METHYLBUTYLAMINE see MHV000

METHYLCARBAMIC ACID, TRIMETHYLPHENYL ESTER see TMC750
METHYLCARBAMIC ACID-3,4-XYLYL ESTER see XTJ000
METHYLCARBAMIC ESTER of α-3-HYDROXYPHENYLETHYLDIMETHYLAMINE HYDROCHLORIDE see MIC250
METHYLCARBAMIC ESTER of 3-OXYPHENYLDIMETHYLAMINE HYDROCHLORIDE see MID000
METHYLCARBAMIC ESTER of OXYPHENYLMETHYLDIETHYLAMMONIUM IODIDE see MID250
METHYLCARBAMIC ESTER of 3-OXYPHENYLTRIMETHYLAMMONIUM METHYLSULFATE see HNR000
METHYLCARBAMIC ESTER of p-OXYPHENYLTRIMETHYLAMMONIUM IODIDE see HNP000
2-METHYL-4-CARBAMOYL-5-HYDROXYIMIDAZOLE see MID800
S-METHYLCARBAMOYLMETHYL-O,O-DIMETHYL PHOSPHORODITHIOATE see DSP400
o-METHYLCARBAMOYL-2-METHYLPROPENEALDOXIME see MBW780
(4-METHYLCARBAMOYLOXY-o-CUMENYL)TRIMETHYLAMMONIUM CHLORIDE see HJY500
(3-(N-METHYLCARBAMOYLOXY)PHENYL)ALLYLDIMETHYLAMMONIUM BROMIDE see TGH680
(3-(N-METHYLCARBAMOYLOXY)PHENYL)DIETHYLMETHYLAMMONIUM CHLORIDE see TGH665
(3-(N-METHYLCARBAMOYLOXY)PHENYL)DIETHYLMETHYL-AMMONIUM IODIDE see MID250
(3-(N-METHYLCARBAMOYLOXY)PHENYL)DIMETHYLPROPYLAMMONIUM BROMIDE see TGH675
((p-METHYLCARBAMOYLOXY)PHENYL)TRIETHYLAMMONIUM IODIDE see HNP000
(3-(METHYLCARBAMOYLOXY)PHENYL)TRIMETHYLAMMONIUM CHLORIDE see HNN000
(4-(N-METHYLCARBAMOYLOXY)PHENYL)TRIMETHYLAMMONIUM IODIDE see HNP000
(3-(METHYLCARBAMOYLOXY)PHENYL)TRIMETHYLAMMONIUM IODIDE see HNO500
(3-(N-METHYLCARBAMOYLOXY)PHENYL)TRIMETHYLAMMONIUM METHYLSULFATE see HNR000
(3-(N-METHYLCARBAMOYLOXY)PHENYL)TRIMETHYL-ARSONIUM IODIDE see MID900
METHYL N-(CARBAMOYLOXY)THIOACETIMIDATE see MPS250
(4-METHYLCARBAMOYLOXY-o-TOLYL)ALLYLDIMETHYLAMMONIUM IODIDE see TGH685
(3-(METHYLCARBAMOYLOXY)-p-TOLYL)TRIMETHYLAMMONIUM BISULFATE see TGH655
(3-(METHYLCARBAMOYLOXY)-p-TOLYL)TRIMETHYLAMMONIUM METHYLSULFATE see TGH650
o-METHYLCARBANILIC ACID-N-ETHYL-3-PIPERDINYL ESTER see MIE250
1-METHYL-1-CARBETHOXY-4-PHENYL HEXAMETHYLENIMINE CITRATE see MIE600
1-METHYL-4-CARBETHOXY-4-PHENYLPIPERIDINE HYDROCHLORIDE see DAM700
METHYL CARBETHOXYSYRINGOYL RESERPATE see RCA200
METHYLCARBINOL see EFU000
METHYL CARBITOL see DJG000
METHYL CARBITOL ACETATE see MIE750
3-METHYL-4-CARBOLINE see MPA050
2-METHYL-β-CARBOLINE see MPA050
METHYL-4-CARBOMETHOXY BENZOATE see DUE000
METHYL CARBONATE see MIF000
METHYLCARBONYL FLUORIDE see ACM000
METHYLCARBOPHENOTHION see MQH750
17-β-(1-METHYL-3-CARBOXYPROPYL)ETHIOCHOLAN-3-α-OL see LHW000
17-β-(1-METHYL-3-CARBOXYPROPYL)-ETIOCHOLANE-3-α,12-α-DIOL see DAQ400
17-β-(1-METHYL-3-CARBOXYPROPYL)ETIOCHOLANE-3-α,7-β-DIOL see DMJ200
N-METHYL-N-(3-CARBOXYPROPYL)NITROSAMINE see MIF250
3-METHYLCATECHOL see DNE000
METHYLCATECHOL see GKI000
METHYL-CCNU see CHD250
trans-METHYL-CCNU see CHD250
METHYL CEDRYL ETHER see CCR525
METHYL CELLOSOLVE (OSHA, DOT) see EJH500
METHYL CELLOSOLVE ACETATE (OSHA, DOT) see EJJ500
METHYL CELLOSOLVE ACRYLATE see MEM250, MIF750
METHYL CELLOSOLVEAT OLEATE see MEP750
METHYL CELLOSOLYE ACETAAT (DUTCH) see EJJ500
METHYL CELLULOSE see MIF760
METHYL CELLULOSE-A see MIF760
METHYL CELLULOSE ETHER see MIF760
METHYL CENTRALITE see DRB200
4-METHYLCHALCONE see MIF762
β-METHYLCHALCONE see MPL000
p-METHYLCHALCONE see MIF762
METHYL CHAVICOL see AFW750
METHYL CHEMOSEPT see HJL500

6-METHYL-CHINOXALIN-2,3-DITHIOL-CYCLO-CARBONAT (GERMAN) see ORU000
METHYLCHLOORFORMIAT (DUTCH) see MIG000
METHYL-2-CHLORAETHYLNITROSAMIN (GERMAN) see CIQ500
METHYLCHLORID (GERMAN) see MIF765
METHYL CHLORIDE see MIF765
METHYL CHLORIDE–METHYLENE CHLORIDE MIXTURE see CHX750
METHYL CHLOROACETATE (DOT) see MIF775
METHYL CHLOROACETATE see MIF775
METHYL γ-CHLOROACETOACETATE see MIH600
METHYL-2-CHLOROACRYLATE see MIF800
METHYL-α-CHLOROACRYLATE see MIF800
4-METHYL-2-CHLOROANILINE see CLK210
2-METHYL-4-CHLOROANILINE see CLK220
4-METHYL-3-CHLOROANILINE HYDROCHLORIDE see CLK230
2-METHYL-4-CHLOROANILINE HYDROCHLORIDE see CLK235
2-(2'-METHYL-3'-CHLORO)ANILINONICOTINIC ACID see CMX770
1-METHYL-2-CHLOROBENZENE see CLK100
2-METHYLCHLOROBENZENE see CLK100
o-METHYL-o-2-CHLORO-4-tert-BUTYLPHENYL-N-METHYLAMIDOPHOSPHATE see COD850
METHYL CHLOROCARBONATE see MIG000
METHYL-2-CHLORO-3-(4-CHLOROPHENYL)PROPIONATE see CFC750
2-METHYL-5-CHLORO-2-(N,N-DIETHYLAMINOETHOXYETHYL)-1,3-BENZODIOXOLE see CFO750
METHYL-β-CHLOROETHYLAMINE HYDROCHLORIDE see MIG250
1-METHYL-1-(β-CHLOROETHYL)ETHYLENIMONIUM see MIG800
METHYL-β-CHLOROETHYL-ETHYLENIMONIUM PICRYLSULFONATE see MIG850
1-METHYL-1-(β-CHLOROETHYL)ETHYLENIMONIUM PICRYLSULFONATE see MIG850
METHYL-β-CHLOROETHYL-β-HYDROXYETHYLAMINE HYDROCHLORIDE see MIH000
METHYL(2-CHLOROETHYL)NITROSAMINE see CIQ500
METHYL-N-(β-CHLOROETHYL)-N-NITROSOCARBAMATE see MIH250
4-METHYL-5-(β-CHLOROETHYL)THIAZOLE see CHD750
5-METHYL-3-(2-CHLORO-6-FLUOROPHENYL)-4-ISOXAZOLYLPENICILLIN SODIUM see FDA100
METHYL CHLOROFORM see MIH275
METHYL CHLOROFORMATE (DOT) see MIG000
METHYL-2-CHLORO-9-HYDROXYFLUORENE-9-CARBOXYLATE see CDT000
1-METHYL-5-CHLOROINDOLINE METHYLBROMIDE see MIH300
METHYL CHLOROMETHYL ETHER, anhydrous (DOT) see CIO250
METHYLCHLOROMETHYL ETHER (DOT) see CIO250
METHYL-6-(((2-CHLORO-4-NITRO)PHENYL)AZO)-m-ANISIDINE see MIH500
METHYL 4-CHLORO-3-OXOBUTANOATE see MIH600
3-METHYL-4-CHLOROPHENOL see CFE250
2-METHYL-4-CHLOROPHENOXYACETIC ACID see CIR250
(2-METHYL-4-CHLOROPHENOXY)ACETIC ACID, DIETHANOLAMINE SALT see MIH750
(2-METHYL-4-CHLOROPHENOXY)ACETIC ACID, SODIUM SALT see SIL500
2-METHYL-4-CHLOROPHENOXYBUTYRIC ACID see CLN750
4-(2-METHYL-4-CHLOROPHENOXY)BUTYRIC ACID see CLN750
γ-2-METHYL-4-CHLOROPHENOXYBUTYRIC ACID see CLN750
4-(2-METHYL-4-CHLOROPHENOXY)BUTYRIC ACID, SODIUM SALT see CLO000
2-(2-METHYL-4-CHLOROPHENOXY)PROPIONIC ACID see CIR500
α-(2-METHYL-4-CHLOROPHENOXY)PROPIONIC ACID see CIR500
2-METHYL-4-CHLOROPHENOXY-α-PROPIONIC ACID see CIR500
N-(2-METHYL-3-CHLOROPHENYL)ANTHRANILIC ACID see CLK325
2-METHYL-3-(4'-CHLOROPHENYL)CHINAZOLON-(4) (GERMAN) see MII250
N'-(2-METHYL-4-CHLOROPHENYL)-N,N-DIMETHYLFORMAMIDINE see CJJ250
3-METHYL-4-((o-CHLOROPHENYL)HYDRAZONE)-4,5-ISOXAZOLEDIONE see MLC250
3-METHYL-4-(o-CHLOROPHENYLHYDRAZONO)-5-ISOXAZOLONE see MLC250
1-((2-METHYL-4-CHLOROPHENYL)METHYL)-1H-INDAZOLE-3-CARBOXYLIC ACID see TGJ875
1-((2-METHYL-4-CHLOROPHENYL)METHYL)-INDAZOLE-3-CARBOXYLIC ACID see TGJ875
2-METHYL-4-(p-CHLOROPHENYL)-2,4-PENTANEDIOL see CKG000
METHYL-2-(4-(p-CHLOROPHENYL)PHENOXY)-2-METHYLPROPIONATE see MIO975
2-METHYL-3-(4-CHLOROPHENYL)-4(3H)-QUINAZOLINONE see MII250
METHYL-4-CHLOROPHENYL SULFIDE see CKG500
METHYL-p-CHLOROPHENYL SULFIDE see CKG500
METHYL-4-CHLOROPHENYL SULFONE see CKG750
METHYL-4-CHLOROPHENYL SULFOXIDE see CKH000
METHYL CHLOROPHOS see TIQ250
METHYLCHLOROPINDOL see CMX850
METHYL-2-CHLOROPROPIONATE (DOT) see CKT000
2-METHYL-6-CHLORO-4-QUINAZOLINONE see MII750
METHYLCHLOROTHIAZIDE see MIV500
METHYLCHLOROTHION see MIJ250
4-(2-METHYL-4-CHLORPHENOXY)-BUTTERSAEURE (GERMAN) see CLN750
2-METHYL-4-CHLORPHENOXYESSIGSAEURE (GERMAN) see CIR250

2-(2-METHYL-4-CHLORPHENOXY)-PROPIONSAEURE (GERMAN) see CIR500
N'-(2-METHYL-4-CHLORPHENYL)-FORMAMIDIN-HYDROCHLORID (GERMAN)
   see CJJ250
METHYLCHLORPINDOL see CMX850
METHYL CHLORPYRIFOS see CMA250
METHYLCHLORTETRACYCLINE see MIJ500
22-METHYLCHOLANTHRENE see MIK250
20-METHYLCHOLANTHRENE see MIJ750
3-METHYLCHOLANTHRENE see MIJ750
4-METHYLCHOLANTHRENE see MIK250
5-METHYLCHOLANTHRENE see MIK000
METHYLCHOLANTHRENE see MIJ750
20-METHYLCHOLANTHRENE CHOLEIC ACID see MIK500
3-METHYLCHOLANTHRENE COMPOUND with PICRIC ACID (1:1) see MIL750
3-METHYLCHOLANTHRENE COMPOUND with 1,3,5-TRINITROBENZENE (1:1)
   see MIM000
(E)-3-METHYLCHOLANTHRENE-11,12-DIHYDRODIOL see DML800
cis-3-METHYLCHOLANTHRENE-1,2-DIOL see MIK750
3-METHYLCHOLANTHRENE-11,12-EPOXIDE see ECA500
3-METHYLCHOLANTHRENE-2-ONE see MIL250
3-METHYLCHOLANTHRENE-11,12-OXIDE see MIL500
20-METHYLCHOLANTHRENE PICRATE see MIL750
20-METHYLCHOLANTHRENE-TRINITROBENZENE see MIM000
3-METHYL-1-CHOLANTHRENOL see HMA000
3-METHYLCHOLANTHREN-1-OL see HMA000
3-METHYLCHOLANTHREN-2-OL see HMA500
3-METHYLCHOLANTHREN-1-ONE see MIM250
20-METHYLCHOLANTHREN-15-ONE see MIM250
3-METHYLCHOLANTHREN-2-ONE see MIL250
20-METHYLCHOLANTHRYLENE see DAL200
3-METHYLCHOLANTHRYLENE see DAL200
β-METHYLCHOLINE CHLORIDE see MIM300
β-METHYLCHOLINE CHLORIDE CARBAMINOYL see HOA500
β-METHYLCHOLINE CHLORIDE URETHAN see HOA500
1-METHYLCHRYSENE see MIM500
2-METHYLCHRYSENE see MIM750
3-METHYLCHRYSENE see MIN000
4-METHYLCHRYSENE see MIN250
5-METHYLCHRYSENE see MIN500
6-METHYLCHRYSENE see MIN750
α-METHYLCINNAMALDEHYDE see MIO000
METHYL CINNAMATE see MIO500
α-METHYLCINNAMIC ALDEHYDE see MIO000
METHYL CINNAMIC ALDEHYDE see MIO000
METHYL CINNAMYLATE see MIO500
α-METHYLCINNIMAL see MIO000
METHYLCISTOX see DAO800
METHYL CLOFENAPATE see MIO975
METHYLCLOTHIAZIDE see MIV500
METHYLCOBALAMIN see VSZ050
METHYL COBALAMINE see VSZ050
METHYLCOLCHAMINONE see MGF000
METHYLCOLCHICINE see MIW500
METHYL COPPER see MIP250
4-METHYLCOUMARIN see MIP500
6-METHYLCOUMARIN see MIP750
7-METHYLCOUMARIN see MIP775
6-METHYLCOUMARINIC ANHYDRIDE see MIP750
6-METHYL-m-CRESOL see XKS000
METHYL α-CROTONATE see COB825
METHYL CROTONATE see COB825
METHYL trans-CROTONATE see COB825
(Z)-3-((Z)-2-METHYLCROTONATE)4,9-EPOXYCEVANE-3-β,4-β,12,14,16-β,17,20-
   HEPTOL see CDG000
(E)-2-METHYLCROTONIC ACID see TGA700
3-METHYLCROTONIC ACID see MHT500
β-METHYLCROTONIC ACID see MHT500
trans-2-METHYLCROTONIC ACID see TGA700
3-METHYLCROTONIC ACID 2-sec-BUTYL-4,6-DINITROPHENYL ESTER see
   BGB500
3-METHYLCROTONIC ACID, ETHYL ESTER see MIP800
γ-METHYL-α,β-CROTONOLACTONE see MKH500
γ-METHYL-β,γ-CROTONOLACTONE see MKH250
p-METHYL-CUMENE see CQI000
METHYL CYANIDE see ABE500
METHYL 2-CYANOACETATE see MIQ000
METHYL CYANOACETATE see MIQ000
METHYL 2-CYANOACRYLATE see MIQ075
METHYL α-CYANOACRYLATE see MIQ075
METHYL CYANOACRYLATE see MIQ075
9-METHYL-10-CYANO-1,2-BENZANTHRACENE see MIQ250
4-METHYLCYANOBENZENE see TGT750
METHYL CYANOCARBAMATE DIMER see MIQ350
METHYL CYANOETHANOATE see MIQ000
METHYL β-CYANOETHYL ETHER see MFL750

o-METHYLCYCLIZINE DIHYDROCHLORIDE see MIQ725
1-METHYLCYCLODODECYL METHYL ETHER see MEU300
N-METHYLCYCLOHEXANAMINE see MIT000
METHYLCYCLOHEXANE see MIQ740
4-METHYL-1,3-CYCLOHEXANEDIAMINE see DBY000
1-METHYL-2,4-CYCLOHEXANEDIAMINE see DBY000
(±)-α-METHYLCYCLOHEXANEETHYLAMINE HYDROCHLORIDE see CPG625
METHYLCYCLOHEXANOL (ACGIH,DOT,OSHA) see MIQ745
METHYLCYCLOHEXANOL see MIQ745
METHYL CYCLOHEXANOLS, Fp not >60.5 degrees C (DOT) see MIQ745
2-METHYL-CYCLOHEXANON (GERMAN, DUTCH) see MIR500
1-METHYLCYCLOHEXAN-2-ONE see MIR500
2-METHYLCYCLOHEXANONE see MIR500
3-METHYLCYCLOHEXANONE see MIR600
4-METHYLCYCLOHEXANONE see MIR625
METHYL-3-CYCLO-HEXANONE-1 (FRENCH) see MIR600
METHYL-4-CYCLO-HEXANONE-1 (FRENCH) see MIR625
o-METHYLCYCLOHEXANONE see MIR500
METHYLCYCLOHEXANONE see MIR250
6-METHYL-3-CYCLOHEXENE-1-CARBOXALDEHYDE see MPO000
2-METHYL-4-CYCLOHEXENE-1-CARBOXALDEHYDE see MPO000
1-((6-METHYL-3-CYCLOHEXEN-1-YL)CARBONYL)PIPERIDINE see MIS500
(METHYL-3-CYCLOHEXENYL)METHANOL see MIS750
N-METHYL-5-CYCLOHEXENYL-5-METHYLBARBITURIC ACID see ERD500
N-METHYLCYCLOHEXYLAMINE see MIT000
METHYLCYCLOHEXYLAMINE see MIT000
S-2-((4-(4-METHYLCYCLOHEXYL)BUTYL)AMINO)ETHYL THIOSULFATE see
   MIT250
METHYL CYCLOHEXYLFLUOROPHOSPHONATE see MIT600
METHYLCYCLOHEXYLNITROSAMIN (GERMAN) see NKT500
METHYLCYCLOHEXYLNITROSAMINE see NKT500
N-METHYL-N-(2-CYCLOHEXYL-2-PHENYL-1,3-DIOXOLAN-4-YL-METHYL)-PIPERI-
   DINIUM IODIDE see OLW400
3-METHYL-1-CYCLOPENTADECANONE see MIT625
3-METHYLCYCLOPENTADECANONE see MIT625
2-METHYLCYCLOPENTADIENYL MANGANESETRICARBONYL see MAV750
2-METHYLCYCLOPENTADIENYL MANGANESE TRICARBONYL (ACGIH) see
   MAV750
METHYLCYCLOPENTADIENYL MANGANESE TRICARBONYL (OSHA) see
   MAV750
METHYLCYCLOPENTADIENYL MANGANESE TRICARBONYL see MAV750
METHYL CYCLOPENTANE (DOT) see MIU500
METHYLCYCLOPENTANE see MIU500
3-METHYLCYCLOPENTANE-1,2-DIONE see HMB500
17-METHYL-15H-CYCLOPENTA(a)PHENANTHRENE see MIU750
METHYL CYCLOPENTENOLONE (FCC) see HMB500
10-METHYL-1,2-CYCLOPENTENOPHENANTHRENE see MIV250
METHYLCYCLOPROPANECARBONYLHYDRAZINE see MIV300
METHYL CYCLOPROPYL ETHER see CQE750
METHYLCYCLOTHIAZIDE see MIV500
3-METHYLCYKLOHEXANONPEROXID (CZECH) see BKR250
METHYLCYKLOPENTADIENTRIKARBONYLMANGANIUM see MAV750
METHYL CYSTEINE HYDROCHLORIDE see MBX800
2-METHYL-DAB see TLE750
3'-METHYL-DAB see DUH600
3-METHYL-4-DAB see MJF000
METHYL-DBCP see MBV720
N-METHYL-N-DEACETYLCOLCHICINE see MIW500
(E)-N-(2-METHYLDECAHYDROISOQUINOL-5-YL)-3,4,5-TRIMETHOXY-BENZAMIDE
   see DAE700
trans-N-(2-METHYLDECAHYDROISOQUINOL-5-YL)-3,4,5-TRIMETHOXYBENZAM-
   IDE see DAE700
α-METHYL DECALACTONE see MIW050
2-METHYL-1-DECANAL see MIW000
METHYL n-DECANOATE see MHY650
METHYL DECANOATE see MHY650
4-METHYLDECANOLIDE see MIW050
4-METHYL-cis-DECENE γ-LACTONE see MIW060
METHYL n-DECYL ETHER see MIW075
METHYL DECYL ETHER see MIW075
6-METHYL-6-DEHYDRO-17-α-ACETOXYPROGESTERONE see VTF000
6-METHYL-6-DEHYDRO-17-α-ACETYLPROGESTERONE see VTF000
6-METHYL-6-DEHYDRO-17-METHYLPROGESTERONE see MBZ100
N-METHYLDEMECOLCINE see MIW500
O-METHYLDEMETON see DAO800
METHYL DEMETON see MIW100
METHYL DEMETON METHYL see MIW250
METHYL-DEMETON-O see DAO800
METHYL DEMETON-O-SULFOXIDE see DAP000
METHYL DEMETON THIOESTER see DAP400
β-d-METHYL-2-DEOXY-2-(3-METHYL-3-NITROSOUREIDO)GLUCOPYRANOSIDE
   see MPJ800
5-METHYLDEOXYURIDINE see TFX790
N-METHYL-N-DESACETYLCOLCHICINE see MIW500
METHYL DIACETOACETATE see MIW750

2-METHYLDIACETYLBENZIDINE see MIX000
METHYLDIAETHYLCARBINOL (GERMAN) see MNJ100
N-METHYLDIAMINOETHANE see MJW100
METHYLDIAZENE see MIX250
METHYL DIAZEPINONE see DCK759
METHYL DIAZINE see MIX250
3-METHYLDIAZIRINE see MIX500
METHYL DIAZOACETATE see MIX750
N-METHYLDIAZOACETYLGLYCINE AMIDE see DCO200
10-METHYL-1,2:5,6-DIBENZACRIDINE see MIY200
10-METHYL-3,4,5,6-DIBENZACRIDINE see MIY250
9-METHYL-3,4,5,6-DIBENZACRIDINE see MIY000
14-METHYLDIBENZ(a,h)ACRIDINE see MIY200
14-METHYLDIBENZ(a,j)ACRIDINE see MIY250
7-METHYLDIBENZ(c,h)ACRIDINE see MIY000
2'-METHYL-1:2:5:6-DIBENZANTHRACENE see MIY500
3'-METHYL-1:2:5:6-DIBENZANTHRACENE see MIY750
4-METHYL-1,2,5,6-DIBENZANTHRACENE see MIZ000
10-METHYLDIBENZ(a,c)ANTHRACENE see MJA000
2'-METHYLDIBENZ(a,h)ANTHRACENE see MIY500
3-METHYLDIBENZ(a,h)ANTHRACENE see MIY750
6-METHYL DIBENZ(a,h)ANTHRACENE see MIZ000
N-METHYL-3:4:5:6-DIBENZCARBAZOLE see MJA250
N-METHYL-7H-DIBENZO(c,g)CARBAZOLE see MJA250
5-METHYL-DIBENZO(b,def)CHRYSENE see MJA500
12-METHYL DIBENZO(def,mno)CHRYSENE see MGP250
10-METHYLDIBENZO(def,p)CHRYSENE see MJB000
6-METHYLDIBENZO(def,mno)CHRYSENE-12-CARBOXALDEHYDE see FNQ000
14-METHYLDIBENZO(b,def)CHRYSENE-7-CARBOXALDEHYDE see FNV000
1-METHYL-4-(5-DIBENZO(a,e)CYCLOHEPTATRIENYLIDENE)PIPERIDINE HYDRO-
    CHLORIDE see PCI250
N-METHYL-5H-DIBENZO(a,d)CYCLOHEPTENE-5-PROPYLAMINE HYDROCHLO-
    RIDE see POF250
7-METHYLDIBENZO(h,rst)PENTAPHENE see MJA750
5-METHYL-1,2,3,4-DIBENZOPYRENE see MJB000
2'-METHYL-1,2:4,5-DIBENZOPYRENE see MMD000
3'-METHYL-1,2:4,5-DIBENZOPYRENE see MMD250
5-METHYL-3,4:8,9-DIBENZOPYRENE (FRENCH) see MJA500
7-METHYL-1:2:3:4-DIBENZPYRENE see MJB250
5-METHYL-3,4,9,10-DIBENZPYRENE (FRENCH) see MHE250
METHYLDIBORANE see MJB300
2-METHYL-1,2-DIBROMO-3-CHLOROPROPANE see MBV720
N-METHYL-DIBROMOMALEINIMIDE see MJB500
4-METHYL-2,6-DI-terc. BUTYLFENOL (CZECH) see BFW750
4-METHYL-2,6-DI-tert-BUTYLPHENOL see BFW750
METHYL DI-tert-BUTYLPHENOL see BFW750
METHYLDICHLORARSINE see DFP200
METHYL DICHLOROACETATE (DOT) see DEM800
METHYLDICHLOROARSINE (DOT) see DFP200
O-METHYL-O-2,5-DICHLORO-4-BROMOPHENYL PHENYLTHIOPHOSPHONATE
    see LEN000
METHYL-3,4-DICHLOROCARBANILATE see DEV600
N-METHYL-2,2'-DICHLORODIETHYLAMINE see BIE250
N-METHYL-2,2'-DICHLORODIETHYLAMINE HYDROCHLORIDE see BIE500
N-METHYL-2,2'-DICHLORODIETHYLAMINE-N-OXIDE HYDROCHLORIDE see
    CFA750
METHYL DICHLOROETHANOATE see DEM800
METHYLDI(2-CHLOROETHYL)AMINE see BIE250
N-METHYL-DI-2-CHLOROETHYLAMINE HYDROCHLORIDE see BIE500
METHYLDI(2-CHLOROETHYL)AMINE HYDROCHLORIDE see BIE500
METHYLDI(β-CHLOROETHYL)AMINE HYDROCHLORIDE see BIE500
N-METHYL-DI-2-CHLOROETHYLAMINE-N-OXIDE see CFA500
METHYLDI(2-CHLOROETHYL)AMINE-N-OXIDE HYDROCHLORIDE see CFA750
N-METHYLDICHLOROMALEINIMIDE see DFP800
METHYL-N-(3,4-DICHLOROPHENYL) CARBAMATE see DEV600
METHYL α,4-DICHLOROPHENYLPROPANOATE see CFC750
METHYL DICHLOROSILANE (DOT) see DFS000
METHYL-DICHLORSILAN (CZECH) see DFS000
N-METHYLDICYCLOHEXYLAMINE see MJC750
N-METHYL-N-(β-DICYCLOHEXYLAMINOETHYL)PIPERIDINE BROMIDE see
    MJC775
METHYL DIEPOXYDIALLYLACETATE see MJD000
N-METHYLDIETHANOLAMINE see MKU250
METHYLDIETHANOLAMINE see MKU250
N-METHYLDIETHANOLIMINE see MKU250
2-METHYL-4-DIETHYLAMINOPENTAN-5-OL p-AMINOBENZOATE see LEU000
5-METHYL-7-DIETHYLAMINO-s-TRIAZOLO-(1,5-a)PYRIMIDINE see DIO200
2-METHYL-N,N-DIETHYLBENZAMIDE see DKD000
3-METHYL-N,N-DIETHYLBENZAMIDE see DKC800
1-METHYL-4-DIETHYLCARBAMOYLPIPERAZINE CITRATE see DIW200
1-METHYL-4-DIETHYLCARBAMYLPIPERAZINE see DIW000
METHYLDIETHYLCARBINOL see MNJ100
METHYL DIETHYL CARBINOLURETHAN see ENF000
METHYL DIFLUOROPHOSPHITE see MJD275
4'''-β-METHYLDIGOXIN see MJD300

4'''-o-METHYLDIGOXIN see MJD300
4'''-METHYLDIGOXIN see MJD300
β-METHYLDIGOXIN see MJD300
METHYLDIGOXIN see MJD300
4'''-o-METHYLDIGOXIN ACETONE (2:1) see MJD500
METHYL-DIHYDROARTEMISININE see ARL425
6-METHYL-3,4-DIHYDRO-2H-1-BENZOTHIOPYRAN-7-SULFONAMIDE 1,1-DIOX-
    IDE see MQQ050
trans-3-METHYL-11,12-DIHYDROCHOLANTHRENE-11,12-DIOL see DML800
trans-3-METHYL-9,10-DIHYDROCHOLANTHRENE-9,10-DIOL see MJD610
11-METHYL-15,16-DIHYDRO-17H-CYCLOPENTA(a)PHENANTHRENE see
    MJD750
11-METHYL-15,16-DIHYDRO-17H-CYCLOPENTA(a)PHENANTHREN-17-ONE see
    MJE500
3-METHYL-2,3-DIHYDRO-9H-ISOXAZOLO(3,2-b)QUINAZOLIN-9-ONE see
    MJE250
METHYL DIHYDROJASMONATE see HAK100
11-METHYL-15,16-DIHYDRO-17-OXOCYCLOPENTA(a)PHENANTHRENE see
    MJE500
2-(8-METHYL-10,11-DIHYDRO-11-OXODIBENZ(b,f)OXEPIN-2-YL)PROPIONIC see
    DLI650
1-METHYL-1,6-DIHYDROPICOLINALDEHYDE OXIME HYDROCHLORIDE see
    PMJ550
METHYLDIHYDROPYRAN see MJE750
N-METHYL-1,6-DIHYDROPYRIDINE-2-CARBALDOXIME HYDROCHLORIDE see
    PMJ550
1-METHYL-1,6-DIHYDRO-2-PYRIDINECARBOXYALDEHYDE OXIME HYDRO-
    CHLORIDE see PMJ550
17-α-METHYLDIHYDROTESTOSTERONE see MJE760
METHYLDIHYDROTESTOSTERONE see MJE760
2-METHYL-1,3-DIHYDROXYBENZENE see MPH400
(3-(2R,3S))-4-METHYL-2,3-DIHYDROXY-2-(3-METHYLBUTYL)BUTANEDIOATE
    (ESTER) CEPHALOTAXINE see IKS500
METHYL (±)-11-α-16-DIHYDROXY-16-METHYL-9-OXOPROST-13-EN-1-OATE see
    MJE775
l-α-METHYL-3,4-DIHYDROXYPHENYLALANINE see DNA800
l-(−)-α-METHYL-β-(3,4-DIHYDROXYPHENYL)ALANINE see DNA800
α-METHYL-l-3,4-DIHYDROXYPHENYLALANINE see DNA800, MJE780
α-METHYL-β-(3,4-DIHYDROXYPHENYL)-l-ALANINE see DNA800
3-METHYL-3,4-DIHYDROXY-4-PHENYL-BUTIN-1 (GERMAN) see DMX800
3-METHYL-3,4-DIHYDROXY-4-PHENYL-1-BUTYNE see DMX800
METHYL-(β-(3,4-DIHYDROXY PHENYL ETHYL) AMINE HYDROCHLORIDE see
    EAZ000
METHYLDIIODOARSINE see MGQ775
4-METHYL-1,2-DIMERCAPTOBENZENE see TGN000
1-METHYL-2,6-DI-(p-METHOXYPHENETHYL)PIPERIDINE ETHANESULFONATE
    see MJE800
METHYL-3-(DIMETHOXYPHOSPHINYLOXY)CROTONATE see MQR750
5-METHYL-10-β-DIMETHYLAMINOAETHYL-10,11-DIHYDRO-11-OXO-5-DIBEN-
    ZO(b,e)(1,4)DIAZEPIN see DCW800
3'-METHYL-4-DIMETHYLAMINOAZOBENZEN (CZECH) see DUH600
2'-METHYL-4-DIMETHYLAMINOAZOBENZENE see DUH800
2-METHYL-4-DIMETHYLAMINOAZOBENZENE see TLE750
3'-METHYL-4-DIMETHYLAMINOAZOBENZENE see DUH600
3-METHYL-4-DIMETHYLAMINOAZOBENZENE see MJF000
4'-METHYL-4-DIMETHYLAMINOAZOBENZENE see DUH400
2-METHYL-N,N-DIMETHYL-4-AMINOAZOBENZENE see DUH800, TLE750
3'-METHYL-N,N-DIMETHYL-4-AMINOAZOBENZENE see DUH600
M'-METHYL-p-DIMETHYLAMINOAZOBENZENE see DUH600
o'-METHYL-p-DIMETHYLAMINOAZOBENZENE see DUH800
p'-METHYL-p-DIMETHYLAMINOAZOBENZENE see DUH400
3'-METHYLDIMETHYLAMINOAZOBENZOL (GERMAN) see DUH600
3-METHYL-4-DIMETHYLAMINO-2,2-DIPHENYLBUTYRAMIDE see DOY400
α-METHYL-β-(N-(2-DIMETHYLAMINOETHYL)-NPHENYL)AMINO-Δ^{α,β}BUTENOLID
    see MJF250
3-METHYL-4-(N-(2-DIMETHYLAMINOETHYL)-N-PHENYLAMINO)-2(5H) FURA-
    NONE see MJF250
METHYL 2-(DIMETHYLAMINO)-N-HYDROXY-2-OXOETHANIMIDOTHIOATE see
    OLT100
METHYL-2-(DIMETHYLAMINO)-N-(((METHYLAMINO)CARBONYL)OXY)-2-OXOE-
    THANIMIDOTHIOATE see DSP600
2-METHYL-5'-(p-DIMETHYLAMINOPHENYLAZO)QUINOLINE see DQE600
2'-METHYL-5'-(p-DIMETHYLAMINOPHENYLAZO)QUINOLINE see DPQ600
3-METHYL-5'-(p-DIMETHYLAMINOPHENYLAZO)QUINOLINE see DQE400
3'-METHYL-5'-(p-DIMETHYLAMINOPHENYLAZO)QUINOLINE see MJF500
6'-METHYL-5'-(p-DIMETHYLAMINOPHENYLAZO)QUINOLINE see MJF750
7'-METHYL-5'-(p-DIMETHYLAMINOPHENYLAZO)QUINOLINE see DPQ400
8'-METHYL-5'-(p-DIMETHYLAMINOPHENYLAZO)QUINOLINE see MJG000
4-METHYL-3-DIMETHYLAMINOPHENYL ESTER-N-METHYLCARBAMIC ACID see
    DQE800
2'-METHYL-4-DIMETHYLAMINOSTILBENE see TMF750
3'-METHYL-4-DIMETHYLAMINOSTILBENE see TMG000
4'-METHYL-4-DIMETHYLAMINOSTILBENE see TMG250
METHYL-4-DIMETHYLAMINO-3,5-XYLYL CARBAMATE see DOS000

METHYL-4-DIMETHYLAMINO-3,5-XYLYL ESTER of CARBAMIC ACID see DOS000
o-METHYLDIMETHYLANILINE see DUH200
METHYL-2,2-DIMETHYLBICYCLO(2.2.1)HEPTANE-3-CARBOXYLATE see MHY600
METHYL-1,1-DIMETHYLBUTANON(3)-NITROSAMIN (GERMAN) see MMX750
METHYL-1-(DIMETHYLCARBAMOYL)-N-(METHYLCARBAMOYLOX-Y)THIOFORMIMIDATE see DSP600
S-METHYL-1-(DIMETHYLCARBAMOYL)-N-((METHYLCARBA-MOYL)OXY)THIOFORMIMIDATE see DSP600
METHYL 1,1-DIMETHYLETHYL ETHER see MHV859
2-METHYL-3-(3,5-DIMETHYL-4-HYDROXYPHENYL)-3,4-DIHYDROQUINAZOLIN-4-ONE see DSH800
(E)2-METHYL-3,7-DIMETHYL-2,6-OCTADIENYL ESTER PROPANOIC ACID see GDI000
METHYL 2-(((((4,6-DIMETHYL-2-PYRIMIDIN-YL)AMINO)CARBONYL)AMINO)SULFONYL)BENZOATE see SNW550
N-METHYL-O,O-DIMETHYLTHIOLOPHOSPHORYL-5-THIA-3-METHYL-2-VALERAM-IDE see MJG500
4-METHYL-2,6-DINITROANILINE see DVI100
2-METHYL-3,5-DINITROBENZAMIDE see DUP300
4-METHYL-2,6-DINITROBENZENAMINE see DVI100
2-METHYL-3,5-DINITROBENZENAMINE see AJR750
4-METHYL-1,2-DINITROBENZENE see DVH600
2-METHYL-1,3-DINITROBENZENE see DVH400
2-METHYL-1,4-DINITROBENZENE see DVH200
1-METHYL-2,4-DINITROBENZENE see DVH000
1-METHYL-3,5-DINITRO-BENZENE see DVH800
METHYLDINITROBENZENE see DVG600
1-METHYL-2,3-DINITRO-BENZENE (9CI) see DVG800
2-METHYL-4,6-DINITROPHENOL see DUS700
2-METHYL-4,6-DINITROPHENOL, AMMONIUM SALT see DUT800
2-METHYLDINITROPHENOL SODIUM SALT see SGP550
2-METHYL-4,6-DINITROPHENOL SODIUM SALT see DUU600
N-METHYL-N,4-DINITROSOANILINE see MJG750
N-METHYL-N,p-DINITROSOANILINE see MJG750
N-METHYL-N,4-DINITROSOBENZENAMINE see MJG750
2-METHYLDINITROSOPIPERAZINE see MQH000
N-METHYL-N,N-DIOCTYL-1-OCTANAMINIUM CHLORIDE see MQH000
((2-METHYL-1,3-DIOXALAN-4-YL)METHYL)TRIMETHYLAMMONIUM IODIDE see MJH250
2-METHYL-1,3-DIOXOLANE see MJH775
METHYLDIOXOLANE see MJH775
2-METHYL-1,3-DIOXOLANE-2-ACETIC ACID ETHYL ESTER see EFR100
((2-METHYL-1,3-DIOXOLAN-4-YL)METHYL)TRIMETHYLAMMONIUM CHLORIDE see MJH800
6-METHYL-1,11-DIOXY-2-NAPHTHACENECARBOXAMIDE see TBX000
2-METHYLDIPHENHYDRAMINE see MJH900
o-METHYLDIPHENHYDRAMINE see MJH900
4-METHYLDIPHENYL see MJH905
p-METHYLDIPHENYL see MJH905
4'-METHYL-2-DIPHENYLAMINECARBOXYLIC ACID see TGV000
3-METHYLDIPHENYL ETHER see MNV770
METHYL DIPHENYL ETHER see PEG000
METHYL DIPHENYLMETHYL KETONE see MJH910
(±)-γ-METHYL-α,α-DIPHENYL-1-PYRROLIDINEPROPANOL HYDROCHLORIDE see ARN700
2-METHYL-1,2-DI-3-PYRIDINYL-1-PROPANONE (9CI) see MCJ370
6-METHYL DIPYRIDO(1,2-a:3',2'-d)IMIDAZOL-2-AMINE see AKS250
2-METHYL-1,2-DIPYRIDYL-(3'-1-OXOPROPANE) DITARTRATE see MJJ000
2-METHYL-1,2-DI-3-PYRIDYL-1-PROPANONE TARTRATE (1:2) see MJJ000
N-METHYL-3,6-DITHIA-3,4,5,6-TETRAHYDROPHTHALIMIDE see MJJ250
1-METHYL-3-(DI-2-THIENYLMETHYLENE)PIPERIDINE CITRATE see ARP875
1-(1-METHYL-3,3-DI-2-THIENYL-2-PROPENYL)PIPERIDINE HYDROCHLORIDE see MJJ500
N-METHYLDITHIOCARBAMATE de SODIUM (FRENCH) see VFU000
N-METHYLDITHIOCARBAMIC ACID, SODIUM SALT see VFU000
METHYLDITHIOCARBAMIC ACID, SODIUM SALT see VFU000
METHYL DITHIOCARBANILATE see MJK500
METHYLDITHIOCYANATOARSINE HOMOPOLYMER see MJK750
METHYLDITHIOKARBAMAN SODNY DIHYDRAT see SIL550
(4-METHYL-1,3-DITHIOLAN-2-YLIDENE)PHOSPHORAMIDIC ACID, DIETHYL ES-TER see DHH400
6-METHYL-1,3-DITHIOLO(4,5-b)QUINOXALIN-2-ONE see ORU000
METHYL DIVINYL ACETYLENE see MJL000
2-METHYL-1,3-DI(2,4-XYLYLIMINO)-2-AZAPROPANE see MJL250
2-METHYL-DNPZ see MJH000
METHYL DODECANOATE see MLC800
METHYL DODECYLATE see MLC800
METHYL DODECYL BENZYL AMMONIUM CHLORIDE see DYA600
α-METHYL-I-DOPA see DNA800
l-α-METHYLDOPA see DNA800
METHYLDOPA see DNA800
l-α-METHYLDOPAHYDRAZINE see CBQ500
N-METHYLDOPAMINE HYDROCHLORIDE see EAZ000

METHYL DOPA SESQUIHYDRATE see MJE780
METHYL DURSBAN see CMA250
METHYL-E 605 see MNH000
METHYLE (ACETATE de) (FRENCH) see MFW100
METHYL E-CROTONATE see COB825
O,O-METHYLEEN-BIS-(4-CHLOORFENOL) see MJM500
METHYLEEN-S,S'-BIS(O,O-DIETHYL-DITHIOFOSFAAT) (DUTCH) see EEH600
3,3'-METHYLEEN-BIS-(4-HYDROXY-CUMARINE) (DUTCH) see BJZ000
METHYLE (FORMIATE de) (FRENCH) see MKG750
METHYL ENANTHATE see MKK100
S,S'-METHYLEN-BIS(O,O-DIAETHYL-DITHIOPHOSPHAT) (GERMAN) see EEH600
3,3'-METHYLEN-BIS-(4-HYDROXY-CUMARIN) (GERMAN) see BJZ000
METHYLENDIC ANHYDRIDE see NAC000
3,4-METHYLENDIOXY-6-PROPYLBENZYL-n-BUTYL-DIAETHYLENGLYKOLAETH-ER (GERMAN) see PIX250
METHYLENDIRHODANID (CZECH, GERMAN) see MJT500
METHYLENE ACETONE see BOY500
METHYLENEAMINOACETONITRILE see HHW000
endo-α-METHYLENEBENZENEACETIC ACID 8-METHYL-8-AZABICY-CLO(3.2.1)OCT-3-YL ESTER see AQO250
METHYLENE BICHLORIDE see MJP450
2,2'-METHYLENEBIPHENYL see FDI100
N,N'-METHYLENEBIS(ACRYLAMIDE) see MJL500
METHYLENEBISACRYLAMIDE see MJL500
METHYLENEBIS(2-AMINO-BENZOIC ACID) DIMETHYL ESTER (9CI) see MJQ250
METHYLENEBIS(4-AMINOCYCLOHEXANE) see MJQ260
N,N'-METHYLENEBIS(2-AMINO-1,3,4-THIADIAZOLE) see MJL750
2,4'-METHYLENEBIS(ANILINE) see MJP750
4,4'-METHYLENEBISANILINE see MJQ000
METHYLENEBIS(ANILINE) see MJQ000
4,4'-METHYLENEBIS(BENZENEAMINE) see MJQ000
2,2'-METHYLENEBIS(6-tert-BUTYL-p-CRESOL) see MJO500
2,2'-METHYLENEBIS(6-tert-BUTYL-4-ETHYLPHENOL) see MJN250
4,4'-METHYLENEBIS(2-CARBOMETHOXYANILINE) see MJQ250
4,4'-METHYLENE BIS(2-CHLOROANILINE) see MJM200
4,4'-METHYLENE(BIS)-CHLOROANILINE see MJM200
p,p'-METHYLENEBIS(α-CHLOROANILINE) see MJM200
4,4'-METHYLENEBIS(o-CHLOROANILINE) see MJM200
METHYLENE-4,4'-BIS(o-CHLOROANILINE) see MJM200
p,p'-METHYLENEBIS(o-CHLOROANILINE) see MJM200
4,4'-METHYLENE-BIS(2-CHLOROANILINE) HYDROCHLORIDE see MJM250
4,4'-METHYLENEBIS-2-CHLOROBENZENAMINE see MJM200
2,2'-METHYLENEBIS(4-CHLOROPHENOL) see MJM500
4,4'-METHYLENEBIS(CYCLOHEXANAMINE) see MJQ260
4,4'-METHYLENEBIS(CYCLOHEXYLAMINE) see MJQ260
METHYLENE BIS(4-CYCLOHEXYLISOCYANATE) see MJM600
METHYLENE BIS(4-CYCLOHEXYLISOCYANATE) (ACGIH,OSHA) see MJM600
METHYLENE-S,S'-BIS(O,O-DIAETHYL-DITHIOPHOSPHAT) (GERMAN) see EEH600
METHYLENEBIS(DIETHYLPHOSPHONATE) see MJT100
6,6'-METHYLENEBIS(1,2-DIHYDRO-2,2,4-TRIMETHYLQUINOLINE) see MRU760
4,4'-METHYLENEBIS(N,N-DIMETHYL)BENZENAMINE see MJN000
2,2'-METHYLENEBIS(4-ETHYL-6-tert-BUTYLPHENOL) see MJN250
N,N-METHYLENE-BIS(ETHYL CARBAMATE) see MJT750
3,3'-METHYLENEBIS(4-HYDROXY-1,2-BENZOPYRONE) see BJZ000
3,3'-METHYLENEBIS(4-HYDROXYCOUMARIN) see BJZ000
3,3'-METHYLENE-BIS(4-HYDROXYCOUMARINE) (FRENCH) see BJZ000
4,4'-METHYLENEBIS(3-HYDROXY-2-NAPHTHOIC ACID) see PAF100
4,4'-METHYLENEBIS(3-HYDROXY-2-NAPHTHOIC ACID) with TRIS-(p-AMINO-PHENYL)METHYLIUM SALT (1:2) see TNC750
4,4'-METHYLENEBIS(3-HYDROXY-2-NAPHTHOIC ACID) with TRIS-(p-AMINO-PHENYL)CARBONIUM SALT (1:2) see TNC725
4,4'-METHYLENEBIS(3-HYDROXY-2-NAPHTHOIC ACID) with TRIS-(p-AMINO-PHENYL)CARBONIUM SALT (1:2) see TNC750
4,4'-METHYLENEBIS(3-HYDROXY-2-NAPHTHOIC ACID) with TRIS-(p-AMINO-PHENYL)METHYLIUM SALT (1:2) see TNC725
4,4'-METHYLENEBIS(3-HYDROXY-2-NAPTHOIC ACID ESTER) with 2-(2-(4-(p-CHLORO-α-PHENYLBENZYL)-1-PIPERAZINYL)-ETHOXY)ETHANOL see HOR500
1,1-METHYLENEBIS(4-ISOCYANATOBENZENE) see MJP400
METHYLENEBIS(4-ISOCYANATOBENZENE) see MJP400
5,5'-METHYLENEBIS(2-ISOCYANATO)TOLUENE see MJN750
METHYLENE BIS(METHANESULFONATE) see MJQ500
4,4'-METHYLENEBIS(2-METHYLANILINE) see MJO250
4,4'-METHYLENEBIS(N-METHYLANILINE) see MJO000
METHYLENE-BIS(METHYL ANTHRANILATE) see MJQ250
4,4'-METHYLENEBIS(2-METHYLBENZENAMINE) see MJO250
4,4'-METHYLENEBIS(N-METHYLBENZENAMINE) see MJO000
2,2''-METHYLENEBIS(4-METHYL-6-tert-BUTYLPHENOL) see MJO500
4,4'-METHYLENE BIS(N,N'-DIMETHYLANILINE) see MJN000
N,N''-METHYLENEBIS(N'-(1-(HYDROXYMETHYL)-2,5-DIOXO-4-IMIDAZOLIDI-NYL)UREA) see GDO800
METHYLENE BIS(NITRAMINE) see MJO775

METHYLENE-BIS-ORTHOCHLOROANILINE see MJM200
2,2'-METHYLENEBIS OXIRANE (9CI) see DHE000
1,1'-(METHYLENEBIS(OXY))BIS(4-CHLORO)BENZENE see NCM700
1,1'-(METHYLENEBIS(OXY)BIS(2-CHLOROETHANE)) see BID750
4,4'-METHYLENEBISPHENOL see BKI250
METHYLENEBIS(4-PHENYLENE ISOCYANATE) see MJP400
METHYLENEBIS(p-PHENYLENE ISOCYANATE) see MJP400
4,4'-METHYLENEBIS(PHENYL ISOCYANATE) see MJP400
METHYLENEBIS(4-PHENYL ISOCYANATE) see MJP400
p,p'-METHYLENEBIS(PHENYL ISOCYANATE) see MJP400
METHYLENEBIS(p-PHENYL ISOCYANATE) see MJP400
METHYLENE BISPHENYL ISOCYANATE see MJP400
4,4'-METHYLENEBIS(PHENYLMALEIMIDE) see BKL800
2,2'-METHYLENEBIS(3,4,6-TRICHLOROPHENOL) see HCL000
METHYLENE BLUE (medicinal) see BJI250
METHYLENE BLUE see BJI250
METHYLENE BLUE A see BJI250
METHYLENE BLUE BB see BJI250
METHYLENE BLUE BB ZINC FREE see BJI250
METHYLENE BLUE CHLORIDE see BJI250
METHYLENE BLUE CHLORIDE (biological stain) see BJI250
METHYLENE BLUE D see BJI250
METHYLENE BLUE I (medicinal) see BJI250
METHYLENE BLUE NF (medicinal) see BJI250
METHYLENE BLUE POLYCHROME see BJI250
METHYLENE BLUE USP (medicinal) see BJI250
METHYLENE BLUE USP XII (medicinal) see BJI250
METHYLENE BROMIDE see DDP800
(4-(2-METHYLENEBUTYRYL)-2,3-DICHLOROPHENOXY)ACETIC ACID see
    DFP600
METHYLENEBUTYRYL PHENOXYACETIC ACID see DFP600
METHYLENE CHLORIDE see MJP450
METHYLENE CHLOROBROMIDE see CES650
1,2-α-METHYLENE-6-CHLORO-Δ⁶-17-α-HYDROXYPROGESTERONE ACETATE
    see CQJ500
1,2-α-METHYLENE-6-CHLORO-PREGNA-4,6-DIENE-3,20-DIONE 17-α-ACETATE
    see CQJ500
1,2-α-METHYLENE-6-CHLORO-Δ⁴·⁶-PREGNADIENE-17-α-OL-3,20-DIONE 17-α-AC-
    ETATE see CQJ500
1,2-α-METHYLENE-6-CHLORO-Δ⁴·⁶-PREGNADIENE-17-α-OL-3,20-DIONE ACE-
    TATE see CQJ500
METHYLENE CYANIDE see MAO250
2-METHYLENE-CYCLOPENTAENE-3-ONO-4,5-EPOXY-4,5-DIMETHYL-1-CARBOX-
    YLIC ACID see MJU500
2-METHYLENECYCLOPROPANEALANINE see MJP500
2-METHYLENECYCLOPROPANYLALANINE see MJP500
β-(METHYLENECYCLOPROPYL)ALANINE see MJP500
N,N'-METHYLENEDIACRYLAMIDE see MJL500
2,4'-METHYLENEDIANILINE see MJP750
4,4'-METHYLENEDIANILINE see MJQ000
4,4'-METHYLENEDIANILINE (ACGIH) see MJQ000
p,p'-METHYLENEDIANILINE see MJQ000
METHYLENEDIANILINE see MJQ000
4,4'-METHYLENEDIANILINE DIHYDROCHLORIDE see MJQ100
p,p'-METHYLENEDIANILINE DIHYDROCHLORIDE see MJQ100
METHYLENEDIANTHRANILIC ACID DIMETHYL ESTER see MJQ250
1',9-METHYLENE-1,2:5,6-DIBENZANTHRACENE see DCR600
METHYLENE DIBROMIDE see DDP800
METHYLENE DICHLORIDE see MJP450
4,4'-METHYLENEDICYCLOHEXANAMINE see MJQ260
4,4'-METHYLENEDICYCLOHEXANEAMINE see MJQ260
4,4'-METHYLENEDICYCLOHEXYLAMINE see MJQ260
3,4-METHYLENE-DIHYDROXYBENZALDEHYDE see PIW250
METHYLENE DIIODIDE see DNF800
METHYLENE DIISOCYANATE see DNK100
METHYLENEDILITHIUM see MJQ325
METHYLENE DIMETHANESULFONATE see MJQ500
METHYLENE DIMETHYL ETHER see MGA850
4,4'-METHYLENEDIMORPHOLINE see MJQ750
METHYLENEDINAPHTHALENESULFONIC ACID BISPHENYLMERCURI SALT
    see PFN000
3,3'-METHYLENEDI-2-NAPHTHALENESULFONIC ACID, MERCURY SALT see
    MCT250
METHYLENE DINITRATE see MJU150
1,5-METHYLENE-3,7-DINITROSO-1,3,5,7-TETRAAZACYCLOOCTAINE see
    DVF400
1,5-METHYLENE-3,7-DINITROSO-1,3,5,7-TETRAAZACYCLOOCTANE see
    DVF400
3,4-METHYLENEDIOXY-ALLYBENZENE see SAD000
1,2-METHYLENEDIOXY-4-ALLYLBENZENE see SAD000
3,4-METHYLENEDIOXY-AMPHETAMINE see MJQ775
METHYLENEDIOXYAMPHETAMINE see MJQ775
3,4-(METHYLENEDIOXY)BENZALACETONE see MJR250
3,4-METHYLENEDIOXYBENZALDEHYDE see PIW250
1,2-METHYLENEDIOXYBENZENE see MJR000

METHYLENEDIOXYBENZENE see MJR000
3,4-METHYLENEDIOXYBENZYL ACETATE see PIX000
3,4-METHYLENEDIOXYBENZYL ACETONE see MJR250
2-(4-(3,4-METHYLENEDIOXYBENZYL)PIPERAZINO)PYRIMIDINE see TNR485
1-(3,4-METHYLENEDIOXYBENZYL)-4-(2-PYRIMIDYL)PIPERAZINE see TNR485
1,2-(METHYLENEDIOXY)-4-(3-BROMO-1-PROPENYL)BENZENE see BOA750
3,4-METHYLENEDIOXY-N,α-DIMETHYL-β-PHENYLETHYLAMINE HYDROCHLO-
    RIDE see MJR500
3,4-METHYLENEDIOXY-α-ETHYL-β-PHENYLETHYLAMINE see MJR750
1,2-METHYLENEDIOXY-4-(1-HYDROXYALLYL)BENZENE see BCJ000
3,4-METHYLENEDIOXY-α-METHYL-β-PHENYLETHYLAMINE HYDROCHLORIDE
    see MJS750
1,2-(METHYLENEDIOXY)-4-(2-(OCTYLSULFINYL)PROPYL)BENZENE see ISA000
3,4-METHYLENEDIOXYPHENOL see MJU000
2-(3,4-(METHYLENEDIOXY)PHENOXY)-1-((3,4-(METHYLENEDIOX-
    Y)PHENOXY)METHYL)ETHYLAMINE see MJS250
1-(5-(3,4-(METHYLENEDIOXY)PHENOXY)-4-((3,4-(METHYLENEDIOX-
    Y)PHENOXY)METHYL)PENTYLPIPERIDINE) CITRATE see MJS500
1-(3,4-METHYLENEDIOXYPHENYL)-2-AMINOPROPANE see MJS750
4-(3,4-(METHYLENEDIOXY)PHENYL)-3-BUTEN-2-ONE see MJR250
3,4-METHYLENEDIOXY-β-PHENYLETHYLAMINE HYDROCHLORIDE see
    MJT000
3,4-METHYLENEDIOXY-1-PROPENYL BENZENE see IRZ000
1,2-METHYLENEDIOXY-4-PROPENYLBENZENE see IRZ000
1,2-(METHYLENEDIOXY)-4-PROPYLBENZENE see DMD600
3,4-METHYLENEDIOXY-6-PROPYLBENZYL-n-BUTYL DIETHYLENEGLYCOL
    ETHER see PIX250
(3,4-METHYLENEDIOXY-6-PROPYLBENZYL)(BUTYL)DIETHYLENE GLYCOL
    ETHER see PIX250
4,4'-METHYLENE DIPHENOL see BKI250
4,4'-METHYLENEDIPHENYL DIISOCYANATE see MJP400
1,1'-(METHYLENEDI-4,1-PHENYLENE)BIS-1H-PYRROLE-2,5-DIONE see BKL800
METHYLENEDI-p-PHENYLENE DIISOCYANATE see MJP400
4,4'-METHYLENEDIPHENYLENE ISOCYANATE see MJP400
METHYLENE DI(PHENYLENE ISOCYANATE) (DOT) see MJP400
METHYLENEDI-p-PHENYLENE ISOCYANATE see MJP400
4,4'-METHYLENEDIPHENYL ISOCYANATE see MJP400
METHYLENEDI(PHOSPHONIC ACID) TETRAETHYL ESTER see MJT100
METHYLENE DITHIOCYANATE see MJT500
4,4'-METHYLENE DI-o-TOLUIDINE see MJO250
METHYLENE DIURETHAN see MJT750
METHYLENE ETHER of OXYHYDROQUINONE see MJU000
N-METHYLENE GLYCINONITRILE see HHW000
METHYLENE GLYCOL see FMV000
METHYLENE GLYCOL DINITRATE see MJU150
METHYLENE GREEN see MJU250
METHYLENE IODIDE see DNF800
METHYLENEMAGNESIUM see MJU350
METHYLENEMELAMINE POLYCONDENSATE see MCB050
(1S-cis)-5-METHYLENE-6-(1-METHYLETHENYL)-2-CYCLOHEXEN-1-OL see
    CCL109
3-METHYLENE-7-METHYL-1,6-OCTADIENE see MRZ150
3-METHYLENE-7-METHYLOCTAN-7-YL ACETATE see DLX100
3-METHYLENE-7-METHYL-1-OCTEN-7-OL see MLO250
3-METHYLENE-7-METHYL-1-OCTEN-7-YL ACETATE see AAW500
S,S'-METHYLENE O,O,O',O'-TETRAETHYL PHOSPHORODITHIOATE see
    EEH600
4-METHYLENE-2-OXETANONE see KFA000
METHYLENE OXIDE see FMV000
2-METHYLENE-3-OXO-CYCLOPENTANECARBOXYLIC ACID see SAX500
2-METHYLENE-3-OXO-1-CYCLOPENTANECARBOXYLIC ACID compounded
    with NICOTINIC HYDRAZIDE see SAY000
2-METHYLENE-3-OXOCYCLOPENTANECARBOXYLIC ACID, SODIUM SALT
    see SJS500
2,5-METHYLENE-6-PROPYL-3-CYCLOHEXENECARBOXALDEHYDE see
    CML620
METHYLENESUCCINYLOXYBIS(TRIBUTYLSTANNANE) see BLL500
2,5-endo-METHYLENE-Δ³-TETRAHYDROBENZYL ACRYLATE see BFY250
8-METHYLENE-4,11,11-(TRIMETHYL)BICYCLO(7.2.0)UNDEC-4-ENE see
    CCN000
2-METHYLENE-1,3,3-TRIMETHYLINDOLINE see TLU200
METHYLENIMINOACETONITRILE see HHW000
METHYLENIUM CERULEUM see BJI250
METHYLENOMYCIN A see MJU500
METHYLEPHEDRIN (GERMAN) see MJU750
METHYLEPHEDRINE see MJU750
N-METHYLEPHEDRINE HYDROCHLORIDE see DPD200
l-METHYLEPHEDRINE HYDROCHLORIDE see DPD200
METHYLEPHEDRINE HYDROCHLORIDE see DPD200
dl-METHYLEPHENDRINE HYDROCHLORIDE see MNN350
N-METHYLEPINEPHRINE see MJV000
METHYL-18-EPIRESERPATE METHYL ETHER HYDROCHLORIDE see MQR200
3-METHYL-11,12-EPOXYCHOLANTHRENE see ECA500
6-METHYL-3,4-EPOXYCYCLOHEXYLMETHYL-6-METHYL-3,4-EPOXYCYCLOHEX-
    ANE CARBOXYLATE see ECB000

METHYLERGOBASINE see PAM000
METHYLERGOBASINE MALEATE see MJV750
METHYLERGOBREVIN see PAM000
(8-β)-6-METHYLERGOLINE-8-ACETAMIDE TARTRATE (2:1) see MJV500
1-((5R,8S,10R)-6-METHYL-8-ERGOLINYL)-3,3-DIETHYLUREA see DLR100
1-((5R,8S,10R)-6-METHYL-8-ERGOLINYL)-3,3-DIETHYLUREA HYDROGEN MALE-
ATE see DLR150
METHYLERGOMETRIN see PAM000
METHYLERGOMETRINE see PAM000
METHYLERGOMETRINE MALEATE see MJV750
METHYLERGONOVIN see PAM000
METHYLERGONOVINE see PAM000
METHYLERGONOVINE MALEATE see MJV750
4-METHYLESCULETIN see MJV800
METHYLESCULETIN see MJV800
4-METHYLESCULETOL see MJV800
METHYL ESTER FLUOROSULFURIC ACID see MKG250
METHYL ESTER of GABA see AJC625
METHYL ESTER of p-HYDROXYBENZOIC ACID see HJL500
METHYLESTER KISELINY OCTOVE (CZECH) see MFW100
METHYLESTER KYSELINY ANTHRANILOVE see APJ250
METHYLESTER KYSELINY BENZOOVE see MHA750
METHYLESTER KYSELINY BROMOCTOVE see MHR250
METHYLESTER KYSELINY 2-CHLOR-5-NITROBENZOOVE see CJC515
METHYLESTER KYSELINY-2-CHLOR-5-NITROBENZOOVE (CZECH) see
CJD625
METHYLESTER KYSELINY CHLOROCTOVE see MIF775
METHYLESTER KYSELINY 4-FLUORMASELNE see MKE000
METHYLESTER KYSELINY FOSFORITE (CZECH) see PGZ950
METHYLESTER KYSELINY METHAKRYLOVE see MLH750
METHYLESTER KYSELINY 3-METHOXYPROPIONOVE see MFL400
METHYLESTER KYSELINY ORTHOMRAVENCI (CZECH) see TLX600
METHYLESTER KYSELINY p-TOLUENSULFONOVE (CZECH) see MLL250
METHYL ESTER of METHANESULFONIC ACID see MLH500
METHYL ESTER of METHANESULPHONIC ACID see MLH500
METHYL ESTER of SERPENTINIC ACID see SCA475
METHYL ESTER STEARIC ACID see MJW000
METHYL ESTER STREPONIGRIN see SMA500
METHYL ESTER of WOOD ROSIN, partially hydrogenated (FCC) see MFT500
METHYL ESTER of WOOD ROSIN see MFT500
METHYLE (SULFATE de) (FRENCH) see DUD100
N-METHYL-1,2-ETHANEDIAMINE see MJW100
N-METHYLETHANEDIAMINE see MJW100
1,1'-(METHYLETHANEDILIDENEDINITRILO)BIGUANIDINE DIHYDROCHLORIDE
DIHYDRATE see MKI000
((((1-METHYL-1,2-ETHANEDIYL)BIS(CARBAMODITHIOATO))(2−))ZINC HOMO-
POLYMER see ZMA000
4,4'-(1-METHYL-1,2-ETHANEDIYL)BIS-2,6-PIPERAZINEDIONE see PIK250
1-METHYLETHANETHIOL see IMU000
N-METHYLETHANOANTHRACENE-9-(10H)-METHYLAMINE HYDROCHLORIDE
see BCH750
N-METHYL-9,10-ETHANOANTHRACENE-9(10H)-PROPANAMINE HYDROCHLO-
RIDE see MAW850
METHYL ETHANOATE see MFW100
N-METHYLETHANOLAMINE see MGG000
METHYLETHENE see PMO500
4-(1-METHYLETHENYL)-1-CYCLOHEXENE-1-CARBOXALDEHYDE (9CI) see
DKX100
5-(1-METHYLETHENYL)-β,β,2-TRIMETHYL-1-CYCLOPENTENE-1-PROPANOL
PROPANOATE see MJW300
METHYL ETHER see MJW500
METHYL ETHER of PROPYLENE GLYCOL (α) see MJW750
METHYL ETHERTRIMETHYLCOLCHICINIC ACID-l-TARTRATE see DBA175
METHYL ETHER TRIMETHYLCOLCHICINIC ACID-d-TARTRATE, HYDRATE see
DBA175
8-METHYL ETHER of XANTHURENIC ACID see HLT500
11-β-METHYL-17-α-ETHINYLESTRADIOL see MJW875
METHYL ETHOXOL see EJH500
2-(1-METHYLETHOXY)-BENZENAMINE see INA200
9-METHYL-10-ETHOXYMETHYL-1,2-BENZANTHRACENE see EEX000, MJX000
((1-METHYLETHOXY)METHYL)OXIRANE see IPD000
2-(1-METHYLETHOXY)PHENOL METHYLCARBAMATE see PMY300
2-(1-METHYLETHOXY)PHENYL (METHOXYACETYL)METHYLCARBAMATE see
MDX250
2-(1-(METHYLETHOXY)PHENYL METHYL(METHYLTHIO)ACETYL)CARBAMATE
see MLX000
METHYLETHYLACETALDEHYDE see MJX500
METHYLETHYLACETYLENYLCARBINOL see EQL000
1-METHYLETHYLAMINE see INK000
4-(METHYLETHYL)AMINOAZOBENZENE see ENB000
N-METHYL-4'-ETHYL-p-AMINOAZOBENZENE see EOJ000
N-METHYL-N-ETHYL-p-AMINOAZOBENZENE see ENB000
4-(((1-METHYLETHYL)AMINO)CARBONYL)-BENZOIC ACID (9CI) see IRN000
α-(((1-METHYLETHYL)AMINO)METHYL)-2-NAPHTHALENEMETHANOL, HYDRO-
CHLORIDE see INT000

α-(((1-METHYLETHYL)AMINO)METHYL)-4-NITROBENZENEMETHANOL see
INU000
(±)-1-((1-METHYLETHYL)AMINO)-3-(1-NAPHTHALENYLOXY)-2-PROPANOL (9CI)
see PNB790
1-((1-METHYLETHYL)AMINO)-3-(1-NAPHTHALENYLOXY)-2-PROPANOL HYDRO-
CHLORIDE see ICC000
2-METHYLETHYLAMINO-1-PHENYL-1-PROPANOL HYDROCHLORIDE see
EJR500
1-((1-METHYLETHYL)AMINO)-3-(2-(2-PROPENYLOXY)PHENOXY)-2-PROPANOL
see CNR500
1-((1-METHYLETHYL)AMINO)-3-(2-(2-PROPENYL)PHENOXY)-2-PROPANOL see
AGW250
9-(3-((1-METHYLETHYL)AMINO)PROPYL)-9H-FLUORENE-9-CARBOXAMIDE HY-
DROCHLORIDE see IBW400
2-METHYL-6-ETHYL ANILINE see MJY000
3-METHYL-10-ETHYL-7,8-BENZACRIDINE (FRENCH) see EMM000
10-METHYL-3-ETHYL-7,8-BENZACRIDINE (FRENCH) see MJY250
3-METHYL-10-ETHYLBENZ(c)ACRIDINE see EMM000
7-METHYL-9-ETHYLBENZ(c)ACRIDINE see MJY250
4-(1-METHYLETHYL)-BENZALDEHYDE (9CI) see COE500
2-(1-METHYLETHYL)BENZENAMINE see INW100
4-METHYLETHYLBENZENE see EPT000
o-METHYLETHYLBENZENE see EPS500
p-METHYLETHYLBENZENE see EPT000
4-(1-METHYLETHYL)-1,2-BENZENEDIOL see IOK000
4-(1-METHYLETHYL)BENZENEMETHANOL ACETATE see IOF050
4-(1-METHYLETHYL)BENZONITRILE see IOD050
3-(1-METHYLETHYL)-1H-2,1,3-BENZOTHIAZAIN-4(3H)-ONE-2,2-DIOXIDE see
MJY500
1-METHYLETHYL 4-BROMO-α-(4-BROMOPHENYL)-α-HYDROXYBENZENEACE-
TATE see IOS000
METHYLETHYLBROMOMETHANE see BMX750
(1-METHYLETHYL)CARBAMIC ACID 2-(((AMINOCARBONYL)OXY)METHYL)-2-
METHYLPENTYL ESTER see IPU000
METHYLETHYLCARBINOL see BPW750
1-METHYLETHYL-4-CHLORO-α-(4-CHLOROPHENYL)-α-HYDROXYBENZENEACE-
TATE see PNH750
4-(1-METHYLETHYL)CYCLOHEXADIENE-1-ETHANOL FORMATE see IHX450
4-(1-METHYLETHYL)CYCLOHEXADIENE-1-ETHYL FORMATE see IHX450
6-(1-METHYLETHYL)-2-DECAHYDRO-NAPHTHALENOL see IOX400
4-METHYL-4-ETHYL-2,6-DIOXOPIPERIDINE see MKA250
METHYLETHYLENE see PMO500
1-METHYLETHYLENE CARBONATE see CBW500
N-METHYLETHYLENEDIAMINE see MJW100
METHYLETHYLENE GLYCOL see PML000
METHYL ETHYLENE OXIDE see PNL600
4-METHYLETHYLENETHIOUREA see MJZ000
N-METHYLETHYLENEUREA see EJN400
2-METHYLETHYLENIMINE see PNL400
METHYLETHYLENIMINE see PNL400
METHYL ETHYL ETHER (DOT) see EMT000
1-METHYLETHYL-2-((ETHOXY((1-METHYLE-
THYL)AMINO)PHOSPHINOTHIOYL)OXY)BENZOATE see IMF300
(E)-1-METHYLETHYL-3-((ETHYLAMINO)METHOXYPHOSPHINOTHIOYL)OXY-2-
BUTENOATE see MKA000
1-(METHYLETHYL)-ETHYL 3-METHYL-4-(METHYLTHIO)PHENYL PHOSPHORA-
MIDATE see FAK000
1-METHYLETHYL-2-(1-ETHYLPROPYL)-4,6-DINITROPHENYL CARBONATE see
CBW000
METHYLETHYLETHYNYLCARBINOL see EQL000
3-METHYL-3-ETHYLGLUTARIMIDE see MKA250
β-METHYL-β-ETHYLGLUTARIMIDE see MKA250
2-(1-METHYLETHYL)HYDRAZIDE 4-PYRIDINECARBOXYLIC ACID DIHYDRO-
CHLORIDE (9CI) see IGG600
4,4'-(1-METHYLETHYLIDENE)BIS(2,6-DICHLOROPHENOL) see TBO750
4,4'-(1-METHYLETHYLIDENE)BIS(2-METHYLPHENOL) see IPK000
2,2'-((1-METHYLETHYLIDENE)BIS(4,1-PHENYLENEOXYMETHY-
LENE))BISOXIRANE see BLD750
N-METHYLETHYLIDENEDIAMINE see MJW100
METHYLETHYL KETAZINE see EMT600
METHYL ETHYL KETONE see MKA400
METHYL ETHYL KETONE AZINE see EMT600
METHYL ETHYL KETONE HYDROPEROXIDE see MKA500
METHYL ETHYL KETONE KETAZINE see EMT600
METHYL ETHYL KETONE PEROXIDE, in solution with >9% by weight active
oxygen (DOT) see MKA500
METHYL ETHYL KETONE PEROXIDE see MKA500
METHYL ETHYL KETONE SEMICARBAZONE see MKA750
METHYLETHYLKETONHYDROPEROXIDE see MKA500
METHYL ETHYL KETOXIME see EMU500
METHYLETHYLMETHANE see BOR500
2-METHYL-3-ETHYL-4-p-METHOXYPHENYL-Δ³-CYCLOHEXENE CARBOXYLIC
ACID see CBO625
N-(1-METHYLETHYL)-4-((2-METHYLHYDRAZINO)METHYL)BENZAMIDE MONO-
HYDROCHLORIDE see PME500

1-METHYLETHYL-2-(1-METHYLPROPYL)-4,5-DINITROPHENYLESTER CARBONIC ACID see CBW000
O-ETHYL O-p-NITROPHENYL THIOPHOSPHATE see ENI175
N,N-METHYLETHYLNITROSAMINE see MKB000
METHYLETHYLNITROSAMINE see MKB000
N-(1-METHYLETHYL)-N-NITROSO-UREA (9CI) see NKO425
N-(1-METHYLETHYL)-N'-(1-NITRO-9-ACRIDINYL)-1,3-PROPANEDIAMINE DIHYDROCHLORIDE see INW000
1-METHYLETHYL OCTADECANOATE see IRL100
1-METHYL-4-ETHYLOCTYLAMINE see EMY000
METHYLETHYLOLAMINE see MGG000
2-(1-METHYLETHYL)PHENOL see IQX100
4-(1-METHYLETHYL)PHENOL see IQZ000
1-METHYL-5-ETHYL-5-PHENYLBARBITURIC ACID see ENB500
2-METHYL-3-ETHYL-4-PHENYL-4-CYCLOHEXENE CARBOXYLIC ACID see FAO200
2-METHYL-3-ETHYL-4-PHENYL-Δ⁴-CYCLOHEXENECARBOXYLIC ACID see FAO200
METHYLETHYL PHENYLETHYL CARBINYL ACETATE see PFR300
3-METHYL-5-ETHYL-5-PHENYLHYDANTOIN see MKB250
2-(1-METHYLETHYL)PHENYL METHYLCARBAMATE see MIA250
1-METHYLETHYL 3-PHENYL-2-PROPENOATE see IOO000
1-METHYLETHYL 3-PHENYLPROPENOATE see IOO000
(1-METHYLETHYL)PHOSPHORAMIDOTHIOIC ACID O-(2,4-DICHLOROPHENYL)-O-METHYL ESTER see DGD800
N-(1-METHYLETHYL)-2-PROPANAMINE see DNM200
N-(1-METHYLETHYL)-2-PROPENAMIDE see INH000
6-METHYL-3-ETHYLPYRIDINE see EOS000
2-METHYL-5-ETHYLPYRIDINE see EOS000
2-METHYL-5-ETHYLPYRIDINE (DOT) see EOS000
METHYL ETHYL PYRIDINE (DOT) see EOS000
N-METHYL-N-ETHYL-4-(4'-(PYRIDYL-1'OXIDE)AZO)ANILINE see MKB500
3-METHYL-3-ETHYLPYRROLIDINE-2,5-DIONE see ENG500
γ-METHYL-γ-ETHYL-SUCCINIMIDE see ENG500
METHYLETHYLTHIOFOS see ENI175
(R*,S*)-4-((1-METHYLETHYL)THIO)-α-(1-OCTYLAMINO)ETHYL)BENZENEMETHANOL see SOU600
METHYLETHYLTHIOPHOS see ENI175
O-METHYL-O-ETHYL-O-2,4,5-TRICHLOROPHENYL THIOPHOSPHATE see TIR250
4-(1-METHYLETHYL)-2-(3-(TRIFLUOROMETHYL)PHENYL)MORPHOLINE see IRP000
(1-METHYLETHYL)UREA see IRR100
1-(METHYLETHYL)UREA and SODIUM NITRITE see SIS675
3-METHYLETHYNYLESTRADIOL see MKB750
18-METHYL-17-α-ETHYNYL-19-NORTESTOSTERONE see NNQ500, NNQ520
3-METHYLETHYNYLOESTRADIOL see MKB750
METHYL EUGENOL (FCC) see AGE250
S-METHYL FENITROOXON see MKC250
S-METHYL FENITROTHION see MKC250
2-METHYLFLUORANTHENE see MKC500
3-METHYLFLUORANTHENE see MKC750
METHYL FLUORIDE see FJK000
METHYL FLUOROACETATE see MKD000
10-METHYL-3-FLUORO-5,6-BENZACRIDINE see MKD750
7-METHYL-11-FLUOROBENZ(c)ACRIDINE see MKD500
7-METHYL-9-FLUOROBENZ(c)ACRIDINE see MKD250
7-METHYL-2-FLUOROBENZ(a)ANTHRACENE see FJN000
METHYL N-(5-(p-FLUOROBENZOYL)-2-BENZIMIDAZOLYL)CARBAMATE see FDA887
METHYL-4-FLUOROBUTYRATE see MKE000
METHYL-γ-FLUOROBUTYRATE see MKE000
METHYL-ω-FLUOROBUTYRATE see MKE000
METHYL-γ-FLUOROCROTONATE see MKE250
16-α-METHYL-9-α-FLUORO-1-DEHYDROCORTISOL see SOW000
METHYLFLUOROFORM see TJY900
16-α-METHYL-9-α-FLUORO-Δ¹-HYDROCORTISONE see SOW000
METHYL-γ-FLUORO-β-HYDROXYBUTYRATE see MKE750
17-α-METHYL-9-α-FLUORO-11-β-HYDROXYTESTERONE see AOO275
METHYL-γ-FLUORO-β-HYDROXYTHIOLBUTYRATE see MKF000
METHYLFLUOROPHOSPHORIC ACID, ISOPROPYL ESTER see IPX000
METHYL-FLUORO-PHOSPHORYLCHOLINE see MKF250
16-α-METHYL-9-α-FLUOROPREDNISOLONE see SOW000
16-α-METHYL-6-α-FLUORO-PREDNISOLONE-21-ACETATE see PAL600
16-α-METHYL-9-α-FLUORO-1,4-PREGNADIENE-11-β,17-α,21-TRIOL-3,20-DIONE see SOW000
METHYL FLUOROSULFATE see MKG250
METHYL FLUOROSULFONATE see MKG250
16-α-METHYL-9-α-FLUORO-11-β,17-α,21-TRIHYDROXYPREGNA-1,4-DIENE-3,20-DIONE see SOW000
METHYLFLUORPHOSPHORSAEURECHOLINESTER (GERMAN) see MKF250
METHYLFLUORPHOSPHORSAEUREISOPROPYLESTER (GERMAN) see IPX000
METHYLFLUORPHOSPHORSAEUREPINAKOLYLESTER (GERMAN) see SKS500
METHYL FLUORSULFONATE see MKG250
METHYLFLURETHER see EAT900

N-METHYLFORMAMIDE see MKG500
METHYLFORMAMIDE see MKG500
METHYL FORMATE see MKG750
METHYLFORMIAAT (DUTCH) see MKG750
METHYLFORMIAT (GERMAN) see MKG750
METHYL FORMYL see DOO600
N-METHYL-N-FORMYLHYDRAZINE see FNW000
N-METHYL-N-FORMYL HYDRAZONE of ACETALDEHYDE see AAH000
1-METHYL-2-FORMYLPYRIDINIUM CHLORIDE OXIME see FNZ000
METHYL FOSFERNO see MNH000
METHYLFUMARIC ACID see MDI250
5-METHYL-2-FURALDEHYDE see MKH550
2-METHYLFURAN see MKH000
METHYLFURAN see MKH000
5-METHYL-2-FURANCARBOXALDEHYDE see MKH550
METHYL 2-FURANCARBOXYLATE see MKH600
3-METHYL-2,5-FURANDIONE see CMS322
5-METHYL-2(3H)-FURANONE see MKH250
5-METHYL-2(5H)-FURANONE see MKH500
5-METHYL-2-FURFURAL see MKH550
5-METHYL FURFURAL see MKH550
5-METHYLFURFURALDEHYDE see MKH550
METHYL 2-FUROATE see MKH600
METHYL FUROATE see MKH600
2-METHYL-3-(2-FURYL)ACROLEIN see FPX050
2-METHYL-3-FURYLACROLEIN see FPX050
α-METHYLFURYLACROLEIN see FPX050
α-METHYL-β-FURYLACROLEIN see FPX050
5-(5-METHYL-2-FURYL)-1-PHENYL-3-(2(PIPERIDINO)ETHYL)-1H-PYRAZOLINE HYDROCHLORIDE see MKH750
2-METHYL-3-(2-FURYL)PROPENAL see FPX050
METHYL GAG see MKI000
METHYL GALLATE see MKI100
6'-N-METHYLGENTAMICIN C¹a see MQS579
METHYL GLUCAMINE BILIGRAFIN see BGB315
METHYLGLUCAMINE DIATRIXOATE see AOO875
METHYLGLUCAMINE-3,5-DIIODO-4-PYRIDONE-N-ACETATE see DNG400
METHYL GLUCAMINE IODIPAMIDE see BGB315
METHYLGLUCAMINE IOTALAMATE see IGC000
METHYLGLUCAMINE IOTHALAMATE see IGC000
N-METHYLGLUCAMINE and SULFALENE see SNJ400
N-METHYLGLUCAMINE and SULFAMONOMETOXINE see SNL830
α-METHYLGLUCOSIDE TETRANITRATE see MKI125
3-METHYL GLUTARALDEHYDE see MKI250
α-METHYLGLYCEROL TRINITRATE see MKI300
METHYL GLYCOL see EJH500, PML000
METHYL GLYCOL ACETATE see EJJ500
METHYL GLYCOL MONOACETATE see EJJ500
METHYLGLYKOL (GERMAN) see EJH500
METHYLGLYKOLACETAT (GERMAN) see EJJ500
METHYLGLYOXAL see PQC000
METHYL GLYOXYLATE see MKI550
4-METHYLGUAIACOL see MEK250
p-METHYLGUAIACOL see MEK250
METHYLGUANIDIN (GERMAN) see MKI750
METHYLGUANIDINE see MKI750
METHYLGUANIDINE mixed with SODIUM NITRITE (1:1) see MKJ000
METHYL GUTHION see ASH500
METHYLHARNSTOFF and NATRIUMNITRIT (GERMAN) see MQJ250
METHYL HEPTANETHIOL see MKJ250
METHYL HEPTANOATE see MKK100
5-METHYL-3-HEPTANONE see EGI750
3-METHYL-5-HEPTANONE see EGI750
5-METHYL-3-HEPTANONE (OSHA) see ODI000
5-METHYL-3-HEPTANONE OXIME see EMP550
6-METHYL-5-HEPTEN-2-ONE see MKK000
METHYL HEPTENONE see MKK000
METHYL HEPTINE CARBONATE see MND275
METHYL HEPTOATE see MKK100
1-METHYLHEPTYLAMINE see OEK010
2-METHYL-2-HEPTYLAMINE see ILM000
6-METHYL-2-HEPTYLAMINE see ILM000
METHYL n-HEPTYLATE see MKK100
(6-(1-METHYL-HEPTYL)-2,4-DINITRO-FENYL)-CROTONAAT (DUTCH) see AQT500
(6-(1-METHYL-HEPTYL)-2,3-DINITRO-PHENYL)-CROTONAT (GERMAN) see AQT500
2-(1-METHYLHEPTYL)-4,6-DINITROPHENYL CROTONATE see AQT500
6-METHYL-2-HEPTYLHYDRAZINE see MKK250
6-METHYL-2-HEPTYL-ISOPROPYLIDENHYDRAZIN (GERMAN) see MKK500
6-METHYL-2-HEPTYLISOPROPYLIDENHYDRAZINE see MKK500
METHYL HEPTYL KETONE see NMY500
METHYLHEPTYLNITROSAMIN (GERMAN) see HBP000
METHYLHESPERIDIN see MKK600
METHYLHEXABARBITAL see ERD500

METHYLHEXABITAL see ERD500
3-β-14-METHYLHEXADECANOATE-CHOLEST-5-EN-3-OL see CCJ500
2-METHYL-1,5-HEXADIENE-3-YNE see MJL000
METHYL HEXAFLUOROISOBUTYRATE see MKK750
N-METHYLHEXAHYDRO-2-PICOLINIC ACID, 2,6-DIMETHYLANILIDE see SBB000
2-METHYLHEXANE see MKL250
5-METHYL-2,3-HEXANEDIONE see MKL300
METHYL n-HEXANOATE see MHY700
METHYL HEXANOATE see MHY700
3-METHYL-3-HEXANOL see MKL350
3-METHYL-HEXANOL-(3) see MKL350
5-METHYL-2-HEXANONE see MKW450
2-METHYL-3-HEXANONE see MKL400
2-METHYL-5-HEXANONE see MKW450
5-METHYLHEXAN-2-ONE (DOT) see MKW450
2-METHYL-5-HEXEN-3-YN-2-OL see MKM300
METHYL HEXOATE see MHY700
METHYLHEXYLACETALDEHYDE see MNC175
1-METHYLHEXYLAMINE see HBM490
1-METHYLHEXYLAMINE SULFATE see AKD600
METHYL HEXYLATE see MHY700
METHYL HEXYL KETONE (FCC) see ODG000
1-METHYLHEXYL-β-OXYBUTYRATE see MKM750
METHYLHOMATROPINE BROMIDE see MDL000
8-METHYLHOMOTROPINIUM BROMIDE see MDL000
α-((N-METHYL-N-HOMOVERATRYL)-γ-AMINOPROPYL)-3,4-DIMETHOXYPHENYLACETONITRILE see IRV000
METHYL HYDANTOIN see MKB250
1-METHYL HYDRAZINE see MKN000
METHYLHYDRAZINE (DOT) see MKN000
METHYL HYDRAZINE see MKN000
METHYLHYDRAZINE HYDROCHLORIDE see MKN250
METHYL HYDRAZINE SULFATE see MKN500
METHYLHYDRAZINIUM NITRATE see MRJ250
4-((2-METHYLHYDRAZINO)METHYL)-N-ISOPROPYLBENZAMIDE see PME250
p-(N'-METHYLHYDRAZINOMETHYL)-N-ISOPROPYLBENZAMIDE HYDROCHLORIDE see PME500
(α-(2-METHYLHYDRAZINO)-p-TOLUOYL)UREA, MONOHYDROBROMIDE see MKN750
METHYL HYDRIDE see MDQ750
Δ¹-6-α-METHYLHYDROCORTISONE see MOR500
METHYL HYDROGEN SILOXANE see TDJ100
METHYL HYDROPEROXIDE see MKO000
N-METHYL-N-(HYDROPEROXYMETHYL)NITROSAMINE see HIE700
2-METHYLHYDROQUINONE see MKO250
METHYL-p-HYDROQUINONE see MKO250
METHYLHYDROQUINONE see MKO250
p-METHYL HYDROTROPALDEHYDE see THD750
METHYLHYDROXAMIC ACID see ABB250
METHYL HYDROXIDE see MGB150
METHYL N-HYDROXYACETIMIDOTHIOATE see MPS270
4-(N-METHYLHYDROXYAMINO)-QUINOLINE-1-OXIDE see HLX550
17-α-METHYL-17-β-HYDROXY-1,4-ANDROSTADIEN-3-ONE see DAL300
METHYL-3-HYDROXYANTHRANILATE see HJC500
1-METHYL-4-HYDROXYBENZENE see CNX250
METHYL-o-HYDROXYBENZOATE see MPI000
METHYL p-HYDROXYBENZOATE see HJL500
2-METHYL-3-HYDROXY-4,5-BIS(HYDROXYMETHYL)PYRIDINE see PPK250
2-METHYL-3-HYDROXY-4,5-BIS(HYDROXYMETHYL)PYRIDINE HYDROCHLORIDE see PPK500
N-METHYL-N-(4-HYDROXYBUTYL)NITROSAMINE see MKP000
METHYL-3-HYDROXYBUTYRATE see MKP250
METHYL m-HYDROXYCARBANILATE, m-METHYLCARBANILATE see MEG250
1-METHYL-4-HYDROXY-2-CHINOLON (CZECH) see HMR500
4-METHYL-7-HYDROXYCOUMARIN see MKP500
4-METHYL-7-HYDROXY COUMARIN DIETHOXYTHIOPHOSPHATE see PKT000
2-METHYL-5-HYDROXY-8-(β-DIAETHYLAMINO-AETHOXY)FURANO-6,7:2'3'-CHROMON HYDROCHLORIDE see AHP375
2-METHYL-3-HYDROXY-4,5-DIHYDROXYMETHYL-PYRIDIN (GERMAN) see PPK250
2-METHYL-3-HYDROXY-4,5-DI(HYDROXYMETHYL)PYRIDINE see PPK250
METHYL N-HYDROXYETHANIMIDOTHIOATE see MPS270
(2-METHYL-3-(1-HYDROXYETHOXYETHYL-4-PIPERAZINYL)PROPYL)-10-PHENOTHIAZINE see MKQ000
METHYL(β-HYDROXYETHYL)AMINE see MGG000
N-METHYL-N-β-HYDROXYETHYLANILINE see MKQ250
N-METHYL-N-HYDROXYETHYLANILINE see MKQ250
1-METHYL-1-(β-HYDROXYETHYL)ETHYLENAMMONIUM PICRYLSULFONATE see MKQ500
METHYL-β-HYDROXYETHYL-ETHYLENIMONIUM PICRYLSULFONATE see MKQ500
2-METHYL-3-(2-HYDROXYETHYL)-4-NITROIMIDAZOLE see MMN250
2-METHYL-1-(2-HYDROXYETHYL)-5-NITROIMIDAZOLE see MMN250
METHYL-2-HYDROXYETHYLNITROSAMINE see NKU350

METHYL-2-HYDROXYETHYLNITROSOAMINE see NKU350
2-METHYL-3-HYDROXY-4-FORMYL-5-HYDROXYMETHYLPYRIDINE HYDROCHLORIDE see VSU000
1-METHYL-2-HYDROXYIMINOMETHYLPYRIDINIUM IODIDE see POS750
1-METHYL-3-HYDROXY-4-ISOPROPYLBENZENE see TFX810
N-METHYL-4-HYDROXYKARBOSTYRIL (CZECH) see HMR500
METHYLHYDROXYLAMINE see MKQ875
2-METHYL-4-HYDROXYLAMINOQUINOLINE 1-OXIDE see MKR000
5-METHYL-4-HYDROXYLAMINOQUINOLINE-1-OXIDE see HIV000
6-METHYL-4-HYDROXYLAMINOQUINOLINE-1-OXIDE see HIV500
7-METHYL-4-HYDROXYLAMINOQUINOLINE-1-OXIDE see HIW000
8-METHYL-4-HYDROXYLAMINOQUINOLINE-1-OXIDE see HIW500
METHYL-o-(4-HYDROXY-3-METHOXYCINNAMOYL)RESERPATE see MKR100
7-METHYL-12-HYDROXYMETHYLBENZ(a)ANTHRACENE see HMF500
17-α-METHYL-2-HYDROXYMETHYLENE-17-HYDROXY-5-α-ANDROSTAN-3-ONE see PAN100
1-METHYL-3-(HYDROXYMETHYL)IMIDAZOLIUM CHLORIDE LAURATE see MKR150
4-METHYL-5-HYDROXYMETHYLURACIL see HMH300
N-METHYL-3-HYDROXYMORPHINAN see MKR250
2-METHYL-3-HYDROXY-1,4-NAPHTHOQUINONE see HMI000
2-METHYL-5-HYDROXY-1,4-NAPHTHOQUINONE see PJH610
3-METHYL-4-HYDROXY-1-NAPHTHYLAMINE see AKX500
METHYL HYDROXYOCTADECADIENOATE see MKR500
2-METHYL-2-p-HYDROXYPHENYLBUTANE see AON000
METHYL-p-HYDROXYPHENYL KETONE see HIO000
METHYL (4-(m-HYDROXYPHENYL)-1-METHYL)-4-PIPERIDYL KETONE see HNK950
N-METHYL-1-(3-HYDROXYPHENYL)-2-PHENYLETHYLAMINE HYDROCHLORIDE see MGJ600
α-METHYL-ω-HYDROXYPOLY(OXY(METHYL-1,2-ETHANEDIYL)) see MKS250
6-α-METHYL-17-α-HYDROXYPROGESTERONE ACETATE see MCA000
6-METHYL-17-α-HYDROXY-Δ⁶-PROGESTERONE ACETATE see VTF000
6-METHYL-4-HYDROXYPYRON-(2) see THL800
2-METHYL-3-HYDROXY-4-PYRONE see MAO350
3-METHYL-4-HYDROXY-3-PYRROLIN-2-ONE AMMONIUM SALT see MKS750
METHYL-12-HYDROXYSTEARATE see HOG500
METHYL-3-β-HYDROXY-1-α-H,5-α-H-TROPANE-2-β-CARBOXYLATE BENZOATE (ESTER) see CNE750
N-METHYLHYOSCINE BROMIDE see SBH500
N-METHYLHYOSCINE METHYL SULFATE see DAB750
(−)-α-METHYLHYOSCYAMINE HYDROCHLORIDE see LFC000
METHYL HYPOCHLORITE see MKT000
N,N'-METHYLIDENEBISACRYLAMIDE see MJL500
4,4',4''-METHYLIDYNETRIANILINE see THP000
4,4',4''-METHYLIDYNETRISBENZENEAMINE see THP000
1,1',1'-(METHYLIDYNETRIS(OXY))TRIS(ETHANE) see ENY500
1-METHYLIMIDAZOLE see MKT500
2-METHYLIMIDAZOLE see MKT750
4-METHYLIMIDAZOLE see MKU000
1-METHYLIMIDAZOLE-2-THIOL see MCO500
4-METHYL-2-IMIDAZOLIDINETHIONE see MJZ000
2-METHYL-3-(Δ²-IMIDAZOLINYLMETHYL)BENZO(b)THIOPHENE HYDROCHLORIDE see MMH250
N-METHYLIMIDODICARBONIMIDIC DIAMIDE MONOHYDROCHLORIDE see MHP400
2,2'-(METHYLIMINO)BIS(N-ETHYL-N,N-DIMETHYLETHANAMINIUM) DIBROMIDE see MKU750
3,3'-(METHYLIMINO)BIS-1-PROPANOL DIMETHANESULFONATE (ESTER), (1,1'-BIPHENYL)-4,4'-DISULFONATE (1:1) (SALT) see MHQ775
N-METHYL-2,2'-IMINODIETHANOL see MKU250
2,2'-(METHYLIMINO)DIETHANOL see MKU250
N-METHYLIMINODIETHANOL see MKU250
METHYLIMINODIETHANOL see MKU250
((METHYLIMINO)DIETHYLENE)BIS(ETHYLDIMETHYLAMMONIUM BROMIDE) see MKU750
N,N'-((METHYLIMINO)DIMETHYLIDYNE)DI-2,4-XYLIDINE see MJL250
N-METHYL-3,3'-IMINODIPROPIONITRILE see MHQ750
2-METHYL-1,3-INDANDIONE see MKV500
α-METHYL-β-INDOLAETHYLAMINE (GERMAN) see AME500
3-METHYL-1H-INDOLE see MKV750
3-METHYLINDOLE see MKV750
β-METHYLINDOLE see MKV750
N-METHYLINDOLE-2,3-DIONE THIOSEMICARBAZONE see MKW250
1-METHYLINDOLE-2,3-DIONE-3-(THIOSEMICARBAZONE) see MKW250
α-METHYL-β-INDOLEETHYLAMINE see AME500
4-(N-METHYL-3-INDOLYLETHYL)PYRIDINIUM HYDROCHLORIDE see MKW000
4-(1-METHYL-3-INDOLYLETHYL)PYRIDINE HYDROCHLORIDE see MKW000
4-(2-(1-METHYL-3-INDOLYL)ETHYL)-1-(3-(TRIMETHYLAMMONIO)PROPYL)PYRIDINIUM DICHLORIDE see MKW100
METHYL IODIDE see MKW200
N-METHYLISATIN-3-(THIOSEMICARBAZONE) see MKW250
N-METHYLISATIN THIOSEMICARBAZONE see MKW250
METHYLISOAMYL ACETATE see HFJ000
METHYL ISOAMYL KETONE see MKW450

METHYL ISOAMYL KETOXIME see MKW500
N-METHYL-dl-ISOBORNYLAMINE HYDROCHLORIDE see MQR500
METHYL ISOBUTENYL KETONE see MDJ750
METHYL ISOBUTYL CARBINOL see AOK750, MKW600
METHYLISOBUTYL CARBINOL see MKW600
METHYLISOBUTYLCARBINOL ACETATE see HFJ000
METHYLISOBUTYLCARBINYL ACETATE see HFJ000
METHYL-ISOBUTYL-CETONE (FRENCH) see HFG500
METHYLISOBUTYLKETON (DUTCH, GERMAN) see HFG500
METHYL ISOBUTYL KETONE (ACGIH, DOT) see HFG500
METHYL ISOBUTYL KETONE PEROXIDE, in solution with >9% by weight
    active oxygen (DOT) see IJB100
METHYL ISOBUTYL KETOXIME see MKW750
METHYL ISOBUTYRATE see MKX000
N-METHYL-2-ISOCAMPHANAMINE see VIZ400
METHYLISOCYANAAT (DUTCH) see MKX250
METHYL ISOCYANAT (GERMAN) see MKX250
METHYL ISOCYANATE, solutions (DOT) see MKX250
METHYL ISOCYANATE see MKX250
METHYL ISOCYANIDE see MKX500
METHYL ISOCYANOACETATE see MKX575
METHYL ISOEUGENOL (FCC) see IKR000
METHYLISOMIN see DBA800
METHYLISOMYN see MDT600
METHYL ISONITRILE see MKX500
METHYLISOOCTENYLAMINE see ILK000
METHYL ISOPENTANOATE see ITC000
1-METHYL-4-ISOPROPENYLCYCLOHEXAN-3-OL see MCE750
1-METHYL-4-ISOPROPENYL-1-CYCLOHEXENE see MCC250
1-METHYL-4-ISOPROPENYL-6-CYCLOHEXEN-2-OL see MKY250
1-1-METHYL-4-ISOPROPENYL-6-CYCLOHEXEN-2-ONE see CCM120
d-1-METHYL-4-ISOPROPENYL-6-CYCLOHEXEN-2-ONE see CCM100
Δ-1-METHYL-4-ISOPROPENYL-6-CYCLOHEXEN-2-ONE see MCD250
METHYL ISOPROPENYL KETONE see MKY500
METHYL ISOPROPENYL KETONE INHIBITED (DOT) see MKY500
N-METHYL-2-ISOPROPOXYPHENYLCARBAMATE see PMY300
1-METHYL-4-ISOPROPYLBENZENE see CQI000
p-METHYLISOPROPYL BENZENE see CQI000
1-METHYL-2-p-(ISOPROPYLCARBAMOYL)BENZOHYDRAZINE HYDROCHLO-
    RIDE see PME500
1-(METHYL-2-(-ISOPROPYLCARBAMOYL)BENZYL)HYDRAZINE see PME250
1-METHYL-2-(p-ISOPROPYLCARBAMOYLBENZYL)HYDRAZINE HYDROCHLO-
    RIDE see PME500
1-METHYL-4-ISOPROPYLCYCLOHEXADIENE-1,3 see MLA250
1-METHYL-4-ISOPROPYL-1,3-CYCLOHEXADIENE see MLA250
2-METHYL-5-ISOPROPYL-1,3-CYCLOHEXADIENE see MCC000
1-METHYL-4-ISOPROPYLCYCLOHEXADIENE-1,4 see MCB750
1-METHYL-4-ISOPROPYLCYCLOHEXANE-8-OL see MCE100
6-METHYL-3-ISOPROPYLCYCLOHEXANOL see DKV150
1-METHYL-4-ISOPROPYL-1-CYCLOHEXEN-3-ONE see MCF250
α-METHYL-p-ISOPROPYLHYDROCINNAMALDEHYDE see COU500
α-METHYL-p-ISOPROPYL HYDROCINNAMIC ALDEHYDE DIETHYL ACETAL
    see COU510
α-METHYL-p-ISOPROPYLHYDROCINNAMIC ALDEHYDE DIMETHYL ACETAL
    see COU525
1-METHYL-4-ISOPROPYLIDENE-3-CYCLOHEXANONE see MCF500
METHYL ISOPROPYL KETONE see MLA750
5-METHYL-2-ISOPROPYL-1-PHENOL see TFX810
2-METHYL-5-ISOPROPYLPHENOL see CCM000
N-METHYL-3-ISOPROPYLPHENYL CARBAMATE see COF250
N-METHYL-m-ISOPROPYLPHENYL CARBAMATE see COF250
(3-METHYL-5-ISOPROPYLPHENYL)-N-METHYLCARBAMAT (GERMAN) see
    CQI500
3-METHYL-5-ISOPROPYLPHENYL-N-METHYLCARBAMATE see CQI500
2-METHYL-3-(p-ISOPROPYLPHENYL)PROPIONALDEHYDE see COU500
5-METHYL-2-ISOPROPYL-3-PYRAZOLYL DIMETHYLCARBAMATE see DSK200
METHYL(5-ISOPROPYL-N-(p-TOLYL)-o-TOLUENESULFONAMIDO)MERCURY see
    MLB000
METHYLISOPSEUDOIONONE see TMJ100
METHYL ISOSYSTOX see DAP400
METHYLISOTHIOCYANAAT (DUTCH) see ISE000
METHYL-ISOTHIOCYANAT (GERMAN) see ISE000
METHYL ISOTHIOCYANATE (DOT) see ISE000
d-d-METHYLISOTHIOCYANATE see MLC000
S-METHYLISOTHIURONIUM SULFATE see MPV790
METHYLISOVALERATE (DOT) see ITC000
METHYL ISOVALERATE see ITC000
5-METHYL-3-ISOXAZOLECARBOXYLIC ACID-2-BENZYLHYDRAZIDE see
    IKC000
3-METHYL-4,5-ISOXAZOLEDIONE-4-((2-CHLOROPHENYL)HYDRAZONE) see
    MLC250
N'-(5-METHYL-3-ISOXAZOLE)SULFANILAMIDE see SNK000
5-METHYL-3-ISOXAZOLOL see HLM000
5-METHYL-3(2H)-ISOXAZOLONE see HLM000
N'-(5-METHYL-3-ISOXAZOLYL)SULFANILAMIDE see SNK000

N'-(5-METHYLISOXAZOL-3-YL)SULPHANILAMIDE see SNK000
N'-(5-METHYL-3-ISOXAZOLYL)SULPHANILAMIDE see SNK000
METHYLIUM, TRIS(4-AMINOPHENYL)-, 4,4'-METHYLENEBIS(3-HYDROXY-
    NAPHTHOATE) (2:1) see TNC725
METHYLJODID (GERMAN) see MKW200
METHYLJODIDE (DUTCH) see MKW200
METHYLKARBITOLACETAT see MIE750
METHYL KETONE see ABC750
METHYLKYANID see ABE500
METHYL LACTATE see MLC600
2-METHYLLACTIC ACID ETHYL ESTER see ELH700
2-METHYLLACTONITRILE see MLC750
METHYL LAURATE see MLC800
METHYL LAURINATE see MLC800
N-METHYLLAUROTETANINE see TKX700
α-METHYL-L-3,4-DIHYDROXYPHENYLALANINE see MJE780
METHYL LEADATE see LCW000
METHYLLITHIUM see MLD000
METHYL-LOMUSTINE see CHD250
N-METHYLLORAZEPAM see MLD100
METHYLLORAZEPAM see MLD100
N-METHYL-LOST see BIE250
1-METHYL-LUMILYSERGOL-8-(5-BROMONICOTINATE)-10-METHYL ETHER see
    NDM000
1-METHYLLYSERGIC ACID BUTANOLAMIDE see MLD250
METHYLLYSERGIC ACID BUTANOLAMIDE see MLD250
1-METHYLLYSERGIC ACID ETHYLAMIDE see MLD500
d-1-METHYL LYSERGIC ACID MONOETHYLAMIDE see MLD500
METHYLMAGNESIUM BROMIDE (ethyl ether solution) see MLE000
METHYL MAGNESIUM BROMIDE in ETHYL ETHER (DOT) see MLE000
METHYLMAGNESIUM IODIDE see MLE250
METHYL MALEATE see DSL800
METHYLMALEIC ACID see CMS320
METHYLMALEIC ACID, DIMETHYL ESTER see DRF200
2-METHYLMALEIC ANHYDRIDE see CMS322
3-METHYLMALEIC ANHYDRIDE see CMS322
α-METHYLMALEIC ANHYDRIDE see CMS322
METHYLMALEIC ANHYDRIDE see CMS322
METHYL MALONATE see DSM200
METHYLMERCAPTAAN (DUTCH) see MLE650
METHYL MERCAPTAN see MLE650
METHYL-2-MERCAPTOACETATE see MLE750
METHYLMERCAPTOACETATE see MLE750
3-METHYLMERCAPTOANILINE see ALX100
2-METHYLMERCAPTO-4,6-BIS(ISOPROPYLAMINO)-s-TRIAZINE see BKL250
4-METHYLMERCAPTO-3,5-DIMETHYLPHENYL N-METHYLCARBAMATE see
    DST000
2-METHYLMERCAPTO-4-ETHYLAMINO-6-ISOPROPYLAMINO-s-TRIAZINE see
    MPT500
METHYL-MERCAPTOFOS TEOLOVY see DAP400
1-METHYL-2-MERCAPTOIMIDAZOLE see MCO500
2-METHYLMERCAPTO-4-ISOPROPYLAMINO-6-ETHYLAMINO-s-TRIAZINE see
    MPT500
METHYLMERCAPTO-4-ISOPROPYLAMINO-6-METHYLAMINO-s-TRIAZINE see
    INR000
4-METHYLMERCAPTO-3-METHYLPHENYL DIMETHYL THIOPHOSPHATE see
    FAQ900
2-METHYLMERCAPTO-10-((2-N-METHYL-2-PIPERIDYL)ETHYL)PHENOTHIAZINE
    see MOO250
2-METHYLMERCAPTO-10-(2-(N-METHYL-2-PIPERIDYL)ETHYLPHENOTHIAZINE
    HYDROCHLORIDE see MOO500
4-METHYLMERCAPTOPHENOL see MPV300
METHYLMERCAPTOPHOS see DAO800
METHYL-MERCAPTOPHOS see MIW100
d-2-METHYL-3-MERCAPTOPROPANOYL-l-PROLINE see MCO750
3-(METHYLMERCAPTO)PROPIONALDEHYDE see MPV400, TET900
β-(METHYLMERCAPTO)PROPIONALDEHYDE see MPV400, TET900
METHYLMERCAPTOPROPIONIC ALDEHYDE see MPV400, TET900
6-METHYLMERCAPTOPURINE RIBONUCLEOSIDE see MPU000
6-METHYLMERCAPTOPURINE RIBOSIDE see MPU000
1-METHYL-5-MERCAPTO-1,2,3,4-TETRAZOLE see MPQ250
4-METHYLMERCAPTO-3,5-XYLYL METHYLCARBAMATE see DST000
N-METHYLMERCURI-BIS-p-TOLUENSULFONAMID (CZECH) see BLK250
METHYLMERCURIC CHLORIDE see MDD750
METHYLMERCURIC CYANOGUANIDINE see MLF250
METHYLMERCURIC DICYANDIAMIDE see MLF250
METHYLMERCURICHLORENDIMIDE see MLF500
METHYLMERCURIC HYDROXIDE see MLG000
METHYLMERCURIC SULFATE see BKS810
S-(METHYLMERCURIC)THIOGLYCOLIC ACID SODIUM SALT see SIM000
N-(METHYLMERCURI)-1,4,5,6,7,7-HEXACHLOROBICYCLO(2.2.1)HEPT-5-ENE-2,3-
    DICARBOXIMIDE see MLF500
8-(METHYLMERCURIOXY)QUINOLINE see MLH000
METHYLMERCURIPENTACHLORFENOLAT (CZECH) see MLG250

N-METHYLMERCURI-1,2,3,6-TETRAHYDRO-3,6-ENDOMETHANO-3,4,5,6,7,7-HEX-
ACHLOROPHTHALIMIDE see MLF500
N-METHYLMERCURI-1,2,3,6-TETRAHYDRO-3,6-METHANO-3,4,5,6,7,7-HEXA-
CHLOROPHTHALIMIDE see MLF500
METHYLMERCURY see MLF550
METHYLMERCURY(I) CATION see MLF550
METHYLMERCURY CHLORIDE see MDD750
METHYL-MERCURY(1+) (9CI) see MLF550
METHYLMERCURY DICYANDIAMIDE see MLF250
METHYLMERCURY DIMERCAPTOPROPANOL see MLF750
METHYLMERCURY HYDROXIDE see MLG000
METHYLMERCURY β-HYDROXYQUINOLATE see MLH000
METHYLMERCURY 8-HYDROXYQUINOLINATE see MLH000
METHYLMERCURY ION(1+) see MLF550
METHYLMERCURY ION see MLF550
METHYLMERCURY OXINATE see MLH000
METHYLMERCURY OXYQUINOLINATE see MLH000
METHYLMERCURY PENTACHLOROPHENATE see MLG250
METHYLMERCURY PERCHLORATE see MLG500
METHYLMERCURY PROPANEDIOLMERCAPTIDE see MLG750
METHYLMERCURY QUINOLINOLATE see MLH000
METHYL-MERCURY TOLUENESULPHAMIDE see MLH100
METHYLMERKURIDIKYANDIAMID see MLF250
α-METHYLMESCALINE see MLH250
METHYL MESYLATE see MLH500
METHYLMETHACRYLAAT (DUTCH) see MLH750
METHYL-METHACRYLAT (GERMAN) see MLH750
METHYL METHACRYLATE see MLH750
METHYL METHACRYLATE HOMOPOLYMER see PKB500
METHYL METHACRYLATE MONOMER, INHIBITED (DOT) see MLH750
METHYL METHACRYLATE POLYMER see PKB500
METHYL METHACRYLATE RESIN see PKB500
N-METHYLMETHANAMINE see DOQ800
N-METHYLMETHANAMINE with BORANE (1:1) see DOR200
N-METHYLMETHANAMINE HYDROCHLORIDE see DOR600
METHYLMETHANE see EDZ000
METHYL METHANESULFONATE see MLH500
METHYL METHANESULPHONATE see MLH500
METHYL METHANOATE see MKG750
METHYL METHANSULFONAT (GERMAN) see MLH500
METHYL METHANSULFONATE see MLH500
METHYL METHANSULPHONATE see MLH500
2-METHYL-4-METHOXYANILINE see MGO500
4-METHYL-1-METHOXYBENZENE see MGP000
METHYL 2-METHOXYBENZOATE see MLH800
METHYL o-METHOXYBENZOATE see MLH800
METHYL-p-METHOXYBENZOATE see AOV750
1-METHYL-7-METHOXY-β-CARBOLINE see HAI500
METHYL-3-METHOXY CARBONYLAZOCROTONATE see MLI350
METHYL-p-METHOXYCINNAMYLKETONE see MLI400
3-METHYL-7-METHOXY-8-(DIMETHYLAMINO-METHYL)-FLAVONE HYDROCHLO-
RIDE see DNV200
METHYL-2-METHOXY-5-N-DIMETHYLTRYPTAMINE see MLI750
2-METHYL-5-METHOXY-N-DIMETHYLTRYPTAMINE see MLI750
(E)-4-METHYL-5-METHOXY-7-HYDROXY-6-(5-CARBOXY-3METHYLPENT-2-EN-1-
YL)PHTHALIDE see MRX000
METHYL-3-METHOXY-4-HYDROXY STYRYL KETONE see MLI800
9-METHYL-10-METHOXYMETHYL-1,2-BENZANTHRACENE see MEW000
METHYL(METHOXYMETHYL)NITROSAMINE see MEW250
N-METHYL-4-METHOXYNAPHTHALIMIDE see MEW760
2-METHYL-3-(p-METHOXYPHENYL)PROPANAL see MLJ050
N-METHYL-5-METHOXY-3-PIPERIDYLIDENEDITHIENYLMETHANE METHOBROM-
IDE see TGB160
2-(3-METHYL-5-METHOXY-1-PYRAZOLYL)-4-METHOXY-6-METHYLPYRIMIDINE
see MCH550
METHYL p-METHOXYSTYRYL KETONE see MLI400
1-METHYL-6-METHOXY-1,2,3,4-TETRAHYDRO-β-CARBOLINE see MLJ100
3-METHYL-8-METHOXY-3H,1,2,5,6-TETRAHYDROPYRAZINO-(1,2,3-ab)-β-CAR-
BOLINE HYDROCHLORIDE see IBQ300
METHYL-α-METHYLACRYLATE see MLH750
1-METHYL-6-(1-METHYLALLYL)-2,5-DITHIOBIUREA see MLJ500
1-METHYL-6-(1-METHYLALLYL)DITHIOBIUREA see MLJ500
N-METHYL-2'-METHYL-4-AMINOAZOBENZENE see MPY250
N-METHYL-3'-METHYL-4-AMINOAZOBENZENE see MPY000
N-METHYL-4'-METHYL-4-AMINOAZOBENZENE see MPY500
N-METHYL-2'-METHYL-p-AMINOAZOBENZENE see MPY250
N-METHYL-3-METHYL-p-AMINOAZOBENZENE see MLY250
N-METHYL-3'-METHYL-p-AMINOAZOBENZENE see MPY000
N-METHYL-4'-METHYL-p-AMINOAZOBENZENE see MPY500
METHYL METHYLAMINOBENZOATE see MGQ250
2-METHYL-2-(METHYLAMINO)-1,3-BENZODIOXOLE HYDROCHLORIDE see
MLJ750
(( METHYL N-((METHYLAMINO)CARBONYL)OXY)ETHANIMIDO)THIOATE see
MDU600
2-METHYL-4-(2-(METHYLAMINO)ETHOXY)-5-ISOPROPYL-PHENOL see DAD500

2-METHYL-4-(2-(METHYLAMINO)ETHOXY)-5-(1-METHYLETHYL)-PHENOL see
DAD500
2-METHYL-6-METHYLAMINO-2-HEPTENE see ILK000
N-METHYL-4'-(p-METHYLAMINOPHENYLAZO)ACETANILIDE see MLK750
N-METHYL-N-(4-((4-(METHYLAMINO)PHENYL)AZO)PHENYLACETAMIDE) see
MLK750
6-METHYL-8-METHYLAMINO-s-TRIAZOLO(4,3-b)PYRIDAZINE see MLK800
METHYL-6-METHYL-AMINO-8-s-TRIAZOLO(4,3b)PYRIDAZINE (FRENCH) see
MLK800
METHYL-N-METHYL ANTHRANILATE see MGQ250
METHYL-4-METHYLBENZENESULFONATE see MLL250
METHYL-p-METHYLBENZENESULFONATE see MLL250
METHYL-4-METHYLBENZOATE see MPX850
METHYL-p-METHYLBENZOATE see MPX850
4-METHYL-1-(2-(2-METHYL-1,3-BENZODIOXOL-2-YL)ETHYL) PIPERAZINE HY-
DROCHLORIDE see MLL500
1-METHYL-5-(4-METHYLBENZOYL)-PYRROLE-2-ACETIC ACID see TGJ850
1-METHYL-5-(4-METHYLBENZOYL)-1H-PYRROLE-2-ACETIC ACID SODIUM
SALT see SKJ340
METHYL 1-(α-METHYLBENZYL)IMIDAZOLE-5-CARBOXYLATE see MQQ500
METHYL 2-METHYLBUTANOATE see MLL600
METHYL-3-METHYLBUTANOATE see ITC000
METHYL 2-METHYLBUTYRATE see MLL600
METHYL-3-METHYLBUTYRATE see ITC000
((2-METHYL-5-METHYLCARBAMOYLOXY)PHENYL)TRIMETHYLAMMONIUM ME-
THYLSULFATE see TGH660
S-METHYL N-[METHYLCARBAMOYLOXY]THIOACETIMIDATE see MDU600
METHYL-N-((METHYLCARBAMOYL)OXY)THIOACETIMIDATE see MDU600
cis-1-METHYL-2-METHYL CARBAMOYL VINYL PHOSPHATE see MRH209
METHYL (E)-2-METHYLCROTONATE see MPW700
METHYL α-METHYLCROTONATE see MPW700
METHYL trans-2-METHYLCROTONATE see MPW700
4-METHYL-1-((6-METHYL-3-CYCLOHEXEN-1-YL)CARBONYL)PIPERIDINE see
MLL660
3-METHYL-1-((6-METHYL-3-CYCLOHEXEN-1-YL)CARBONYL)PIPERIDINE see
MLL655
2-METHYL-1-((6-METHYL-3-CYCLOHEXEN-1-YL)CARBONYL)PIPERIDINE see
MLL650
6-METHYL-2-(4-METHYL-3-CYCLOHEXEN-1-YL)-5-HEPTEN-2-OL see BGO775
2-METHYL-1-((2-METHYLCYCLOHEXYL)CARBONYL)PIPERIDINE see MLM500
3-METHYL-1-((2-METHYLCYCLOHEXYL)CARBONYL)PIPERIDINE see MLM600
4-METHYL-1-((2-METHYLCYCLOHEXYL)CARBONYL)PIPERIDINE see MLM700
10-METHYL-1',9-METHYLENE-1,2-BENZANTHRACENE see MLN500
α-METHYL-3,4-(METHYLENEDIOXY)PHENETHYLAMINE see MJQ775
α-METHYL-3,4-METHYLENEDIOXYPHENETHYLAMINE HYDROCHLORIDE see
MJS750
1-METHYL-2-(3,4-METHYLENEDIOXYPHENYL)ETHYL OCTYL SULFOXIDE see
ISA000
d-2-METHYL-5-(1-METHYLENENYL)-CYCLOHEXANONE see DKV175
7-METHYL-3-METHYLENE-1,6-OCTADIENE see MRZ150
2-METHYL-6-METHYLENE-2-OCTANOL ACETATE (ESTER) see DLX100
2-METHYL-6-METHYLENE-7-OCTEN-2-OL see MLO250
2-METHYL-6-METHYLENE-7-OCTEN-2-OL ACETATE see AAW500
2-METHYL-6-METHYLENE-7-OCTEN-2-YL ACETATE see AAW500
1-METHYL-4-(1-METHYLETHENYL)CYCLOHEXANOL see TBD775
(R)-1-METHYL-4-(1-METHYLETHENYL)-CYCLOHEXENE see LFU000
1-METHYL-4-(1-METHYLETHENYL)-(S)-CYCLOHEXENE see MCC500
2-METHYL-5-(1-METHYLETHENYL)-2-CYCLOHEXEN-1-OL ACETATE see
CCM750
2-METHYL-5-(1-METHYLETHENYL)-2-CYCLOHEXEN-1-OL PROPIONATE see
MCD000
(S)-2-METHYL-5-(1-METHYLETHENYL)-2-CYCLOHEXEN-1-ONE see CCM100
(R)-2-METHYL-5-(1-METHYLETHENYL)-2-CYCLOHEXEN-1-ONE (9CI) see
CCM120
(1R-trans)-2-(3-METHYL-6-(1-METHYLETHENYL)-2-CYCLOHEXEN-1-YL)-5-PEN-
TYL-1,3-BENZENEDIOL see CBD599
2-METHYL-5-(1-METHYLETHENYL)CYCLOHEXYL ACETATE see DKV160
5-METHYL-2-(1-METHYLETHENYL)-4-HEXEN-1-OL ACETATE see LCA100
(1S-1-α,4-α,5-α)-4-METHYL-1-(1-METHYLETHYL)-BICYCLO(3.1.0)HEXAN-3-ONE
see TFW000
1-METHYL-4-(1-METHYLETHYL)-1,4-CYCLOHEXADIENE see MCB700
5-METHYL-2-(1-METHYLETHYL)-CYCLOHEXANOL (1-α,2-β,5-α) see MCG000
(1R-(1-α,2-β,5-α))-5-METHYL-2-(1-METHYLETHYL)CYCLOHEXANOL see
MCG250
5-METHYL-2-(1-METHYLETHYL)CYCLOHEXANOL see MCF750
(R-(1α,2β,5α))-5-METHYL-2-(1-METHYLETHYL)-CYCLOHEXANOL ACETATE
(9CI) see MCG750
(Z)-5-METHYL-2-(1-METHYLETHYL)CYCLOHEXANONE see IKY000
5-METHYL-2-(1-METHYLETHYL)CYCLOHEXANONE see IKY000
trans-5-METHYL-2-(1-METHYLETHYL)-CYCLOHEXANONE see MCG275
1-METHYL-2-(1-METHYLETHYL)-1-CYCLOHEXENE, DIDEHYDRO deriv. see
MCE275
4-METHYL-1-(1-METHYLETHYL)-3-CYCLOHEXEN-1-OL (9CI) see TBD825
3-METHYL-6-(1-METHYLETHYL)-2-CYCLOHEXEN-1-ONE see MCF250

1-METHYL-4-(1-METHYLETHYL)-2,3-DIOXABICYCLO(2.2.2)OCT-5-ENE see ARM500

6-METHYL-N-(1-METHYLETHYL)-2-HEPTANAMINE HYDROCHLORIDE see IGG300

5-METHYL-2-(1-METHYLETHYL)-2-HEXEN-1-OL see IQH000

1-METHYL-4-(1-METHYLETHYLIDENE)CYCLOHEXENE see TBE000

1-METHYL-2-(1-METHYLETHYL)-5-NITRO-1H-IMIDAZOLE see IGH000

1-METHYL-4-(1-METHYLETHYL)-7-OXABICYCLO(2.2.1)HEPTANE see IKC100

5-METHYL-2-(1-METHYLETHYL)PHENOL see TFX810

3-METHYL-5-(1-METHYLETHYL)PHENOLMETHYLCARBAMATE see CQI500

1-METHYL-4-(1-METHYLETHYL)-2-(1-PROPENYL)BENZENE see VIP100

2-METHYL-2-(α-METHYLHEXYLAMINO)PROPYL-p-AMINOBENZOATE HYDROCHLORIDE see MLO750

METHYL 3-METHYLHEXYL KETONE see MLO800, MND075

METHYL(4-METHYL-N-((4-METHYLPHENYL)SULFONYL)BENZENESULFONAMIDATO-N)-MERCURY see BLK250

d-3-METHYL-N-METHYLMORPHINAN PHOSPHATE see MLP250

METHYL 5-METHYL-3-(5-NITRO-2-FURYL)-4-ISOXAZOLYL KETONE see MLP300

1-METHYL-4-(4-METHYLPENTYL)-3-CYCLOHEXENE-1-CARBOXALDEHYDE see VIZ150

2-METHYL-2-(2-(N-METHYL-N-PHENETHYLAMINO)ETHYL)-1,3-BENZODIOXOLE HYDROCHLORIDE see MLP750

2-METHYL-4-((2-METHYLPHENYL)AZO)BENZENAMINE see AIC250

1-((2-METHYL-4-((2-METHYLPHENYL)AZO)PHENYL)AZO)-2-NAPHTHALENOL see SBC500

N-METHYL-2-((o-METHYL-α-PHENYLBENZYL)OXY)ETHYLAMINE HYDROCHLORIDE see TGJ250

1-METHYL-4-(o-METHYL-α-PHENYLBENZYL)PIPERAZINE DIHYDROCHLORIDE see MIQ725

METHYL (3-METHYLPHENYL)CARBAMOTHIOIC ACID, o-2-NAPHTHALENYL ESTER see TGB475

3-METHYL-2-((1-METHYL-2-PHENYL-1H-INDOL-3-YL)AZO)THIAZOLIUM CHLORIDE see CMM770

N-METHYL-2-((2-METHYLPHENYL)PHENYLMETHOXY)ETHANAMINE HYDROCHLORIDE see TGJ250

1-METHYL-4-((2-METHYLPHENYL)PHENYLMETHYL)PIPERAZINE DIHYDROCHLORIDE (9CI) see MIQ725

2-METHYL-3-(2-METHYLPHENYL)-4-QUINAZOLINONE see QAK000

2-METHYL-3-(2-METHYLPHENYL)-4(3H)-QUINAZOLINONE see QAK000

N-METHYL-α-METHYLPHENYLSUCCINIMIDE see MLP800

N-METHYL-α-METHYL-α-PHENYLSUCCINIMIDE see MLP800

3-METHYL-1-(4-METHYLPHENYL)TRIAZENE see MQB250

6-METHYL-α-(4-METHYL-1-PIPERAZINYLCARBONYL)ERGOLINE-8-β-PROPIONITRILE see MLR400

2-METHYL-11-(4-METHYL-1-PIPERAZINYL)-DIBENZO(b,f)(1,4)THIAZEPINE see MQQ000

2-METHYL-2-(2-(4-METHYL-1-PIPERAZINYL)ETHYL)-1,3-BENZODIOXOLE HYDROCHLORIDE see MLL500

METHYL 10-(3-(4-METHYL-1-PIPERAZINYL)PROPYL)PHENOTHIAZIN-2-YL KETONE see ACR500

5-METHYL-3-(4-METHYL-1-PIPERAZINYL)-5H-PYRIDAZINO(3,4-b)(1,4)BENZOXAZINE HYDROCHLORIDE see ASC250

5-METHYL-4-(4-METHYL-1-PIPERAZINYL)THIENO(2,3-d)PYRIMIDINE HYDROCHLORIDE see MLR500

N-METHYL-2-(2-METHYLPIPERIDINO)-N-(2-PHENOXYETHYL)ACETAMIDE HYDROCHLORIDE see MLS250

N-METHYL-2-(2-METHYLPIPERIDINO)-N-(2-(o-TOLYLOXY)ETHYL)-ACETAMIDE HYDROCHLORIDE see MLS750

METHYL-2-METHYL-2-PROPENOATE see MLH750

α-METHYL-4-(2-METHYLPROPYL)BENZENEACETIC ACID see IIU000

α-METHYL-4-(2-METHYLPROPYL)BENZENEACETIC ACID 2-PYRIDINYLMETHYL ESTER see IAB000

2-METHYL-N-(2-METHYLPROPYL)-1-PROPANAMINE see DNH400

1-METHYL-4-((p-(p-(1-METHYLPYRIDINIUM-4-YL)AMINO)BENZAMIDO)ANILINO)QUINOLINIUM), DI-p-TOLUENESULFONATE see MLT250

1-METHYL-4-(((4-(3-((4-((1-METHYLPYRIDINIUM-4-YL)AMINO)PHENYL)AMINO)3-OXO-1-PROPENYL)PHENYL)AMINO)Q see MLT500

1-METHYL-4-((p-((p-((1-METHYLPYRIDINIUM-4-YL)AMINO)PHENYL)CARBAMOYL)ANILINO)QUINOLINIUM), DIBROMIDE see MLU000

1-METHYL-4-((p-((p-((1-METHYLPYRIDINIUM-4-YL)AMINO)PHENYL)CARBAMOYL)ANILINO)-7-NITROQUINOLINIUM), DI-p-TO see MLT750

1-METHYL-4-(p-(p-((1-METHYLPYRIDINIUM-4-YL)AMINO)STYRYL)ANILINO)QUINOLIUM DIBROMIDE see MLU250

1-METHYL-4-((p-((p-((1-METHYLPYRIDINIUM-4-YL)PHENYL)CARBAMOYL)ANILINO)QUINOLINIUM), DI-p-TOLUENE SULFONATE see MLU750

1-METHYL-6-((p-(p-((1-METHYLQUINOLINIUM-6-YL)CARBAMOYL)BENZAMIDO)BENZAMIDO)QUINOLINIUM), DI-p-TOLUENE SULFONA see MLW250

METHYL 5-METHYL-1-(2-QUINOLYL)-4-PYRAZOLYL KETONE see MLW600

METHYL 5-METHYL-1-(2-QUINOXALINYL)-4-PYRAZOLYL KETONE see MLW630

METHYL α-METHYLSTYRYL KETONE see MNS600

2-METHYL-2-(METHYLSULFINYL)PROPANAL O-((METHYLAMINO)CARBONYL)OXIME see MLW750

2-METHYL-2-(METHYLSULFINYL)PROPIONALDEHYDE O-(METHYLCARBAMOYL)OXIME see MLW750

2-METHYL-2-(METHYLSULFONYL)PROPANAL-o-((METHYLAMINO)CARBONYL)OXIME see AFK000

2-METHYL-2-(METHYLSULFONYL)PROPIONALDEHYDE-o-(METHYLCARBAMOYL)OXIME see AFK000

METHYL((METHYLTHIO)ACETYL)CARBAMIC ACID-o-ISOPROPOXYPHENYL ESTER see MLX000

3-METHYL-4-METHYLTHIOPHENOL see MLX750

2-METHYL-2-(METHYLTHIO)PROPANAL-O-((METHYLAMINO)CARBONYL)OXIME see CBM500

2-METHYL-2-(METHYLTHIO)PROPANAL-o-((METHYLNITROSOAMINO)CARBONYL)OXIME see NJJ500

2-METHYL-2-(METHYLTHIO)PROPANAL OXIME see MLX800

2-METHYL-2-(METHYLTHIO)PROPIONALDEHYDE-O-(METHYLCARBAMOYL)OXIME see CBM500

2-METHYL-2-(METHYLTHIO)PROPIONALDEHYDE-o-((METHYLNITROSO)CARBAMOYL) OXIME see NJJ500

2-METHYL-2-(METHYLTHIO)PROPIONALDEHYDE OXIME see CBM500

2-METHYL-2-METHYLTHIO-PROPIONALDEHYD-O-(N-METHYL-CARBAMOYL)-OXIM (GERMAN) see CBM500

METHYL METIRAM see MLX850

METHYLMITOMYCIN see MLY000

N-METHYLMITOMYCIN C see MLY000

METHYL MONOBROMOACETATE see MHR250

METHYL MONOCHLORACETATE see MIF775

METHYL MONOCHLOROACETATE see MIF775

3-METHYL-4-MONOMETHYLAMINOAZOBENZENE see MLY250

3′-METHYL-4-MONOMETHYLAMINOAZOBENZENE see MPY000

N-METHYLMONOTHIOSUCCINIMIDE see MLY500

(+−)-17-METHYLMORPHINAN-3-OL HYDROBROMIDE see HMH500

(−)-17-METHYLMORPHINAN-3-OL TARTRATE DIHYDRATE see MLZ000

(+)-17-METHYLMORPHINAN-3-OL TARTRATE HYDRATE see MMA000

METHYLMORPHINE see CNF500

N-METHYL MORPHINE CHLORIDE see MRP000

N-METHYLMORPHINIUM CHLORIDE see MRP000

4-METHYLMORPHOLINE see MMA250

METHYLMORPHOLINE (DOT) see MMA250

N-METHYL MORPHOLINE see MMA250

α-METHYL-4-MORPHOLINEACETIC ACID-2,6-XYLYL ESTER HYDROCHLORIDE see MMA525

1-METHYL-MORPHOLINO-3-PHTHALIMIDO-GLUTARIMIDE see MRU080

METHYL 10-(3-MORPHOLINOPROPYL)PHENOTHIAZIN-2-YL KETONE see MMA600

6-METHYL-MP-RIBOSIDE see MPU000

METHYL MUSTARD OIL see ISE000

N-METHYL-N′-(p-ACETYLPHENYL)-N-NITROSOUREA see MFX725

METHYLNAFTALEN see MMB500

5-METHYLNAFTOYLBENZIMIDAZOL (CZECH) see MHC000

N-METHYL-1-NAFTYL-CARBAMAAT (DUTCH) see CBM750

METHYL NAMATE see SGM500

2-METHYL-1,4-NAPHTHALENDIONE see MMD500

1-METHYLNAPHTHALENE see MMB750

2-METHYLNAPHTHALENE see MMC000

α-METHYLNAPHTHALENE see MMB750

β-METHYLNAPHTHALENE see MMC000

METHYLNAPHTHALENE see MMB500

2-METHYL-1,4-NAPHTHALENEDIOL see MMC250

2-METHYL-1,4-NAPHTHALENEDIONE see MMD500

2-METHYL-1,4-NAPHTHOCHINON (GERMAN) see MMD500

2-METHYL-1,4-NAPHTHOCHINON-NATRIUM-BISULFIT TRIHYDRAT (GERMAN) see MMD750

5-METHYLNAPHTHO(1,2,3,4-def)CHRYSENE see MMD000

6-METHYLNAPHTHO(1,2,3,4-def)CHRYSENE see MMD250

2-METHYL-1,4-NAPHTHOHYDROQUINONE see MMC250

METHYLNAPHTHOHYDROQUINONE see MMC250

2-METHYL-1,4-NAPHTHOQUINOL see MMC250

2-METHYL-1,4-NAPHTHOQUINONE see MMD500

3-METHYL-1,4-NAPHTHOQUINONE see MMD500

2-METHYL-1,4-NAPHTHOQUINONE, SODIUM BISULFITE, TRIHYDRATE see MMD750

3-METHYL-2-NAPHTHYLAMINE see MME500

3-METHYL-2-NAPHTHYLAMINE HYDROCHLORIDE see MME750

N-METHYL-1-NAPHTHYL-CARBAMAT (GERMAN) see CBM750

N-METHYL-1-NAPHTHYL CARBAMATE see CBM750

N-METHYL-α-NAPHTHYLCARBAMATE see CBM750

N-METHYL NAPHTHYLCARBAMATE see MME800

N-METHYL-N-(1-NAPHTHYL)FLUOROACETAMIDE see MME809

METHYL 1-NAPHTHYL KETONE see ABC475

METHYL-2-NAPHTHYL KETONE see ABC500

METHYL α-NAPHTHYL KETONE see ABC475
α-METHYL NAPHTHYL KETONE see ABC475
β-METHYL NAPHTHYL KETONE see ABC500
METHYL-β-NAPHTHYL KETONE (FCC) see ABC500
N-METHYL-N-(1-NAPHTHYL)MONOFLUOROACETAMIDE see MME809
N-METHYL-α-NAPHTHYLURETHAN see CBM750
N-METHYL-N'-BENZHYDRYLPIPERAZINE see EAN600
N-METHYL-N'-(4-CHLOROBENZHYDRYL)PIPERAZINE DIHYDROCHLORIDE see CDR000
N-METHYL-N'-(p-CHLOROPHENYL)-N-NITROSOUREA see MMW775
2-METHYL-N,N'-DIACETYLBENZIDINE see MIX000
METHYL-N',N'-DIMETHYL-N-((METHYLCARBAMOYL)OXY)-1-THIOOXAMIMIDATE see DSP600
METHYL NICOTINATE see NDV000
METHYL NIRAN see MNH000
METHYL NITRAMINE (dry) (DOT) see NHN500
METHYLNITRAMINE see NHN500
METHYL NITRATE see MMF500
2-METHYL-1-NITRATODIMERCURIO-2-NITRATOMERCURIO PROPANE see MMF600
1-METHYLNITRAZEPAM see DLV000
METHYL NITRITE see MMF750
1-METHYL-4-NITRO-5-(2'-AMINO-6'-PURINYL)MERCAPTOIMIDAZIDE see AKY250
4-METHYL-3-NITROANILINE see NMP000
6-METHYL-3-NITROANILINE see NMP500
N-METHYL-4-NITROANILINE see MMF800
2-METHYL-5-NITROANILINE see NMP500
2-METHYL-1-NITRO-9,10-ANTHRACENEDIONE see MMG000
2-METHYL-1-NITROANTHRAQUINONE see MMG000
2-METHYLNITROBENZENE see NMO525
3-METHYLNITROBENZENE see NMO500
4-METHYLNITROBENZENE see NMO550
m-METHYLNITROBENZENE see NMO500
o-METHYLNITROBENZENE see NMO525
p-METHYL NITROBENZENE see NMO550
2-METHYL-5-NITRO-BENZENEAMINE see NMP500
9-METHYL-ω-(p-NITROBENZENEAZO)-3,4-BENZACRIDINE see NIK500
METHYL-2-NITROBENZENE DIAZOATE see MMH000
4-METHYL-3-NITROBENZENE SULFONIC ACID see MMH400
2-METHYL-5-NITROBENZENESULFONIC ACID see MMH250
2-METHYL-5-NITROBENZENESULFONYL CHLORIDE see MMH500
1-METHYL-2-NITROBENZIMIDAZOLE see MMH740
3-METHYL-2-NITROBENZOYL CHLORIDE see MMI250
3-METHYL-4-NITRO-2-BUTEN-1-YL ACETATE see MMI650
3-METHYL-4-NITRO-1-BUTEN-3-YL ACETATE see MMI640
1-METHYL-7-NITRO-5-(2-FLUOROPHENYL)-3H-1,4-BENZODIAZEPIN-2(1H)-ONE see FDD100
4-METHYL-1-((5-NITROFURFURYLIDENE)AMINO)-2-IMIDAZOLIDINONE see MMJ000
3-METHYL-4-(5'-NITROFURYLIDENE-AMINO)-TETRAHYDRO-4H-1,4-THIAZINE-1,1-DIOXIDE see NGG000
5-METHYL-3-(5-NITRO-2-FURYL)ISOXAZOLE see MMJ950
N-METHYL-3-(5-NITRO-2-FURYL)-N-NITROSO-1H-1,2,4-TRIAZOL-5-AMINE see MMU000
METHYL 3-(5-NITRO-2-FURYL)-5-PHENYL-4-ISOXAZOLYL KETONE see MMJ955
5-METHYL-3-(5-NITRO-2-FURYL)PYRAZOLE see MMJ960
2-METHYL-4-(5-NITRO-2-FURYL)THIAZOLE see MMJ975
N-(1-METHYL-3-(5-NITRO-2-FURYL)-1H-1,2,4-TRIAZOL-5-YL)ACETAMIDE see MMK000
N-(1-METHYL-3-(5-NITRO-2-FURYL)-s-TRIAZOL-5-YL-ACETAMIDE see MMK000
N-METHYL-N-NITROGLYCINE see NJH500
1-METHYL-3-NITROGUANIDINE mixed with SODIUM NITRITE (1:1) see MML500
1-METHYL-3-NITROGUANIDINIUM NITRATE see MML550
1-METHYL-3-NITROGUANIDINIUM PERCHLORATE see MML575
1-METHYL-2-NITROIMIDAZOLE see MML750
2-METHYL-5-NITROIMIDAZOLE-1-ETHANOL see MMN250
1-METHYL-5-NITRO-1H-IMIDAZOLE-2-METHANOL CARBAMATE ESTER see MMN750
1-METHYL-5-NITROIMIDAZOLE-2-METHANOL CARBAMATE (ESTER) see MMN750
4-((E)-2-(1-METHYL-5-NITRO-1H-IMIDAZOL-2-YL)-AETHENYL)-2-PYRIMIDINAMIN (GERMAN) see TJF000
4-((E)-2-(1-METHYL-5-NITRO-1H-IMIDAZOL-2-YL)-ETHENYL)-2-PYRIMIDINAMINE see TJF000
6-(1'-METHYL-4'-NITRO-5'-IMIDAZOLYL)-MERCAPTOPURINE see ASB250
METHYLNITROIMIDAZOLYLMERCAPTOPURINE see ASB250
1-(1-METHYL-5-NITRO-1H-IMIDAZOL-2-YL)-3-(METHYLSULFONYL)-2-IMIDAZOLI-DINO NE see SAY950
6-((1-METHYL-4-NITRO-1H-IMIDAZOL-5-YL)THIO)-1H-PURINE see ASB250
6-(1-METHYL-4-NITROIMIDAZOL-5-YLTHIO)PURINE see ASB250
6-((1-METHYL-4-NITROIMIDAZOL-5-YL)THIO)PURINE see ASB250
6-(1-METHYL-p-NITRO-5-IMIDAZOLYL)-THIOPURINE see ASB250

6-(METHYL-p-NITRO-5-IMIDAZOLYL)-THIOPURINE see ASB250
METHYLNITROLIC ACID see NHY250
3-METHYL-4-NITRO-1-(p-NITROPHENYL)-2-PYRAZOLIN-5-ONE see PIE000
1-METHYL-3-NITRO-1-NITROSOGUANIDINE see MMP000
METHYLNITRONITROSOGUANIDINE see MMP000
5-METHYL-2-NITRO-7-OXA-8-MERCURABICYCLO(4.2.0)OCTA-1,3,5-TRIENE see NHK900
4-METHYL-2-NITROPHENOL see NFU500
2-METHYL-4-NITROPHENOL see NFV010
1-((4-METHYL-2-NITROPHENYL)AZO)-2-NAPHTHALENOL see MMP100
1-METHYL-7-NITRO-5-PHENYL-1,3-DIHYDRO-2H-1,4-BENZODIAZEPIN-2-ONE see DLV000
3-METHYL-4-NITROPHENYL DIMETHYL PHOSPHATE see PHD750
METHYL 3-NITROPHENYL KETONE see NEL500
METHYL-p-NITROPHENYL KETONE see NEL600
METHYLNITROPHOS see DSQ000
2-METHYL-2-NITROPROPANE see NFQ500
2-METHYL-2-NITROPROPANE-1,3-DIOL see NHO500
2-METHYL-2-NITRO-PROPANOL NITRATE see MMP200
N-(2-METHYL-2-NITROPROPYL)-p-NITROSOANILINE see NHK800
N-(2-METHYL-2-NITROPROPYL)-4-NITROSOBENZAMINE see NHK800
2-METHYL-4-NITROPYRIDINE-1-OXIDE see MMP500
3-METHYL-4-NITROPYRIDINE-1-OXIDE see MMP750
2-METHYL-4-NITROQUINOLINE-1-OXIDE see MMQ250
3-METHYL-4-NITROQUINOLINE-1-OXIDE see MMQ500
5-METHYL-4-NITROQUINOLINE-1-OXIDE see MMQ750
6-METHYL-4-NITROQUINOLINE-1-OXIDE see MMR000
7-METHYL-4-NITROQUINOLINE-1-OXIDE see MMR250
8-METHYL-4-NITROQUINOLINE-1-OXIDE see MMR500
2-METHYL-4-NITROQUINOLINE N-OXIDE see MMQ250
1-(4-N-METHYL-N-NITROSAMINOBENZYLIDENE)INDENE see MMR750
1-(METHYLNITROSAMINO)-2-BUTANONE see MMR800
4-(METHYLNITROSAMINO)-2-BUTANONE see MMR810
1-N-METHYL-N-NITROSAMINO-1-DEOXY-d-GLUCITOLE see DAS400
METHYLNITROSAMINOMETHYL-d3 ESTER ACETIC ACID see MMS000
2-METHYLNITROSAMINO-2-METHYLPENTANON(4) (GERMAN) see MMX750
3-(METHYLNITROSAMINO)-1,2-PROPANEDIOL see NMV450
3-METHYLNITROSAMINOPROPIONITRILE see MMS200
6-(METHYLNITROSAMINO)PURINE see MMT250
2-(METHYLNITROSAMINO)PYRIDINE see NKQ000
γ-(METHYLNITROSAMINO)-3-PYRIDINEBUTYRALDEHYDE see MMS250
4-(N-METHYL-N-NITROSAMINO)-4-(3-PYRIDYL)BUTANAL see MMS250
4-(N-METHYL-N-NITROSAMINO)-1-(3-PYRIDYL)-1-BUTANONE see MMS500
4-(4-N-METHYL-N-NITROSAMINOSTYRYL)QUINOLINE see MMS750
METHYLNITROSOACETAMID (GERMAN) see MMT000
N-METHYL-N-NITROSOACETAMIDE see MMT000
METHYLNITROSOACETAMIDE see MMT000
1-METHYL-1-NITROSOACETYLUREA see ACR400
N-METHYL-N-NITROSOADENINE see MMT250
N⁶-(METHYLNITROSO)ADENOSINE see MMT300
N-METHYL-N-NITROSOALLYLAMINE see MMT500
2-(N-METHYL-N-NITROSO)AMINOACETONITRILE see MMT750
4-(METHYLNITROSOAMINO)BUTYRIC ACID see MIF250
1-(N-METHYL-N-NITROSOAMINO)-1-DEOXY-d-GLUCITOL see DAS400
1-N-METHYL-N-NITROSOAMINO-1-DESOXY-d-GLUCIT (GERMAN) see DAS400
2-(METHYLNITROSOAMINO)ETHANOL see NKU350
α-(1-(N-METHYL-N-NITROSOAMINO)ETHYL)BENZYL ALCOHOL see NKC000
5-(N-METHYL-N-NITROSO)AMINO-3-(5-NITRO-2-FURYL)-s-TRIAZOLE see MMU000
2-(N-METHYL-N-NITROSOAMINO)-1-PHENYL-1-PROPANOL see NKC000
1-(METHYLNITROSOAMINO)-2-PROPANOL see NKU500
1-(METHYLNITROSOAMINO)2-PROPANONE see NKV000
N-METHYL-N-NITROSO-2-AMINOPYRIDINE see NKQ000
γ-(METHYLNITROSOAMINO)-3-PYRIDINEBUTANAL see MMS250
4-(N-METHYL-N-NITROSOAMINO)-4-(3-PYRIDYL)-1-BUTANONE see MMS500
N-METHYL-N-NITROSOANILINE see MMU250
N-METHYL-N-NITROSOBENZAMIDE see MMU500
N-METHYL-N-NITROSOBENZENAMINE see MMU250
1-METHYL-2-NITROSOBENZENE see NLW500
o-METHYLNITROSOBENZENE see NLW500
2-METHYL-N-NITROSO-BENZIMIDAZOLE CARBAMATE and SODIUM NITRITE (1:1) see SIQ700
N-METHYL-N-NITROSOBENZYLAMINE see MHP250
1-METHYL-1-NITROSOBIURET see MMV000
N-METHYL-N-NITROSOBIURET see MMV000
1-METHYL-1-NITROSO-3-(p-BROMOPHENYL)UREA see BNX125
N-METHYL-N-NITROSOBUTYLAMINE see MHW500
METHYLNITROSOCARBAMIC ACID-2,3-DIHYDRO-2,3-DIMETHYL-7-BENZOFURA-NYL ESTER see NJQ500
N-METHYL-N-NITROSOCARBAMIC ACID, ETHYL ESTER see MMX250
METHYLNITROSOCARBAMIC ACID-α-(ETHYLTHIO)-o-TOLYL ESTER see MMV750
METHYLNITROSOCARBAMIC ACID o-ISOPROPOXYPHENYL ESTER see PMY310
METHYL-NITROSOCARBAMIC ACID-1-NAPHTHYL ESTER see NBJ500

N-METHYL-N-NITROSOCARBAMIC ACID-m-3-PENTYLPHENYL ESTER see PBX750
N-METHYL-N-NITROSOCARBAMIC ACID, PHENYL ESTER see NKV500
N-METHYL-N-NITROSOCARBAMIC ACID, TRIMETHYLPHENYL ESTER see NLY500
METHYLNITROSOCARBAMIC ACID 3,4-XYLYL ESTER see XTS000
N⁵-(METHYLNITROSOCARBAMOYL)-l-ORNITHINE see MQY325
N⁵-(N-METHYL-N-NITROSOCARBAMOYL)-l-ORNITHINE see MQY325
Nᴬ-(N-METHYL-N-NITROSOCARBAMOYL)-l-ORNITHINE see MQY325
N-((METHYLNITROSOCARBAMOYL)OXY)-2-METHYLTHIOACETIMIDIC ACID see NKX000
1-METHYL-1-NITROSO-3-(p-CHLOROPHENYL)UREA see MMW775
METHYLNITROSOCYANAMIDE see MMX000
N-METHYL-N-NITROSOCYCLOHEXYLAMINE see NKT500
N-METHYL-N-NITROSODECYLAMINE see MMX200
1-METHYL-N-NITROSODIETHYLAMINE see ELX500
N-METHYL-N-NITROSO-ETHAMINE see MKB000
N-METHYL-N-NITROSO-ETHENYLAMINE see NKY000
N-METHYL-N-NITROSOETHYLAMINE see MKB000
N-METHYL-N-NITROSOETHYLCARBAMATE see MMX250
N-METHYL-N-NITROSO-β-d-GLUCOSAMINE see MMX500
N-METHYL-N-NITROSO-β-d-GLUCOSYLAMIN (GERMAN) see MMX500
N-METHYL-N-NITROSO-β-d-GLUCOSYLAMINE see MMX500
N-METHYL-N-NITROSOGLYCINE see NLR500
N-METHYL-N-NITROSO-HARNSTOFF (GERMAN) see MNA750
METHYLNITROSO-HARNSTOFF (GERMAN) see MNA750
N-METHYL-N-NITROSOHEPTYLAMINE see HBP000
N-METHYL-N-NITROSO-1-HEXANAMINE see NKU400
N-METHYL-N-NITROSOHEXYLAMINE see NKU400
N-METHYL-N-NITROSOLAURYLAMINE see NKU000
N-METHYL-N-NITROSOMETHANAMINE see NKA600
4-METHYL-4-N-(NITROSOMETHYLAMINO)-2-PENTANONE see MMX750
2-METHYL-4-NITROSOMORPHOLINE see NKU550
N-METHYL-N-NITROSO-N'-ACETYLUREA see ACR400
N-METHYL-N-NITROSO-N'-(2-BENZOTHIAZOLYL)-HARNSTOFF (GERMAN) see NKR000
N-METHYL-N-NITROSO-N'-(2-BENZOTHIAZOLYL)-UREA see NKR000
N-METHYL-N-NITROSO-N'-CARBAMOYLUREA see MMV000
N-METHYL-N-NITROSONITROGUANIDIN see MMP000
1-METHYL-1-NITROSO-3-NITROGUANIDINE see MMP000
N-METHYL-N-NITROSO-N'-NITROGUANIDINE see MMP000
N-METHYL-N-NITROSO-N'-PHENYLUREA see MMY500
N-METHYL-N-NITROSOOCTYLAMINE see NKU590
2-METHYL-3-NITROSO-1,3-OXAZOLIDINE see NKU600
5-METHYL-3-NITROSO-1,3-OXAZOLIDINE see NKU875
N-METHYL-N-NITROSO-4-OXO-4-(3-PYRIDYL)BUTYL AMINE see MMS500
N-METHYL-N-NITROSOPENTYLAMINE see AOL000
N-METHYL-N-NITROSOPHENETHYLAMINE see MNU250
N-METHYL-N-NITROSO-4-(PHENYLAZO)ANILINE see MMY250
N-METHYL-N-NITROSO-1-PHENYLETHYLAMINE see NKW000
METHYL-(4-NITROSOPHENYL)NITROSAMINE see MJG750
1-METHYL-1-NITROSO-3-PHENYLUREA see MMY500
1-METHYL-4-NITROSOPIPERAZINE see NKW500
N'-METHYL-N-NITROSOPIPERAZINE see NKW500
2-METHYLNITROSOPIPERIDINE see NLI000
3-METHYLNITROSOPIPERIDINE see MMY750
4-METHYLNITROSOPIPERIDINE see MMZ000
3-METHYL-1-NITROSO-4-PIPERIDONE see MMZ800
N-METHYL-N-NITROSO-1-PROPANAMINE see MNA000
N-METHYL-N-NITROSO-2-PROPEN-1-AMINE see MMT500
N-METHYL-N-NITROSOPROPIONAMIDE see MNA250
METHYLNITROSO-PROPIONAMIDE see MNA250
METHYL-NITROSOPROPIONSAEUREAMID (GERMAN) see MNA250
METHYLNITROSOPROPIONYLUREA see MNA250
N-METHYL-N-NITROSO-1H-PURIN-6-AMINE see MMT250
N-METHYL-N-NITROSO-4-(2-(4-QUINOLINYL)ETHENYL)BENZENAMINE see MMS750
METHYLNITROSO-p-TOLUENESULFONAMIDE see THE500
1-METHYL-1-NITROSO-3-(p-TOLYL)UREA see MNA650
N-METHYL-N-NITROSOUNDECYLAMINE see NKX500
1-METHYL-1-NITROSOUREA see MNA750
N-METHYL-N-NITROSOUREA see MNA750
METHYLNITROSOUREA see MNA750
METHYLNITROSOUREE (FRENCH) see MNA750
METHYLNITROSOURETHAN (GERMAN) see MMX250
N-METHYL-N-NITROSO-URETHANE see MMX250
METHYLNITROSOURETHANE see MMX250
N-METHYL-N-NITROSOVINYLAMINE see NKY000
4-METHYL-5-(5-NITRO-2-4H-1,2,4-TRIAZOL-3-AMINE) see AKY000
2-METHYL-N-(4-NITRO-3-(TRIFLUOROMETHYL)PHENYL)PROPANAMIDE (9CI) see FMR050
1-METHYL-2-NITRO-5-VINYL-1H-IMIDAZOLE see MNB250
N-METHYL-N'-(p-METHOXYPHENYL)-N-NITROSOUREA see MFG400
N-METHYL-N-(5-(N'-METHYLANILINO)-2,4-PENTADIENYLIDENE) ANILINIUM CHLORIDE see MLL000

N-METHYL-N'-(p-METHYLPHENYL)-N-NITROSOUREA see MNA650
N-METHYL-N'-NITRO-N-NITROSOGUANIDINE see MMP000
N⁶-METHYL-N⁶-NITROSO-1H-PURIN-6-AMINE see MMT250
N⁶-METHYL-N⁶-NITROSO-9b-d-RIBOFURANOSYL-9H-PURIN-6-AMINE see MMT300
7-o-METHYLNOGAROL see MCB600
3-METHYL-2(3)-NONENENITRILE see MNB500
METHYL-2-NONENOATE see MNB750
METHYL n-NONYL ACETALDEHYDE see MQI550
METHYLNONYLACETALDEHYDE see MQI550
METHYL NONYL ACETALDEHYDE DIMETHYL ACETAL see MNB600
METHYL NONYL ACETIC ALDEHYDE see MQI550
METHYL NONYLENATE see MNB750
METHYL-n-NONYL KETONE see UKS000
METHYL NONYL KETONE see UKS000
1-METHYL-7-(NONYLOXY)-9H-PYRIDO(3,4-b)INDOLE HYDROCHLORIDE see NNC100
METHYL-2-NONYNOATE see MNC000
α-METHYLNORADRENALINE HYDROCHLORIDE see AMB000
N-METHYLNORAPORMORPHINE HYDROCHLORIDE see AQP500
N-METHYL-NORDOCEINE see CNF500
N-METHYLNOREPHEDRINE see EAW000
7(R)-o-METHYLNORGAROL see MCB600
1-METHYLNORHARMAN see MPA050
11-β-METHYL-19-NOR-17-α-PREGNA-1,3,5(10)-TRIEN-20-YNE-3,17-DIOL see MJW875
N-METHYL-N'-PHENYL-N-NITROSOUREA see MMY500
N-METHYL-N'-PHENYL THIOUREA see MOA500
N-METHYL-N'-2,4-XYLYL-N-(N-2,4-XYLYLFORMIMIDOYL)FORMAMIDINE see MJL250
METHYL OCTADECANOATE see MJW000
METHYL-9-OCTADECENOATE see OHW000
METHYL (Z)-9-OCTADECENOATE see OHW000
METHYL cis-9-OCTADECENOATE see OHW000
2-METHYLOCTANAL see MNC175
α-METHYLOCTANAL see MNC175
2-METHYLOCTANE see MNC250
METHYL OCTANOATE see MHY800
METHYL OCTANOATE and METHYL DECANOATE see MND000
7-METHYL-1-OCTANOL see ILJ000
3-METHYL-3-OCTANOL see MND050
3-METHYLOCTAN-3-OL see MND050
5-METHYL-2-OCTANONE see MLO800, MND075
3-METHYL-1-OCTEN-OL see MND100
METHYLOCTENYLAMINE see ILK000
METHYL 2-OCTINATE see MND275
METHYL OCTINE CARBONATE see MNC000
METHYL OCTYL ACETALDEHYDE see MIW000
METHYLOCTYLDIAZENE 1-OXIDE see MGT000
METHYL n-OCTYL KETONE see OFE050
METHYL OCTYL KETONE see OFE050
METHYL OCTYNE CARBONATE see MNC000
METHYL 2-OCTYNOATE see MND275
METHYL OENANTHYLATE see MKK100
METHYLOL see MGB150
N-METHYLOLACRYLAMIDE see HLU500
N'-METHYLOL-o-CHLORTETRACYCLINE see MND500
N-METHYLOL DIMETHYLPHOSPHONOPROPIONAMIDE see MND550
METHYL OLEATE see OHW000
N-METHYL-N-OLEOYLTAURINE SODIUM SALT see SIY000
3-METHYLOLPENTANE see EGW000
METHYLOLPROPANE see BPW500
METHYLOLUREA RESIN see UTU500
METHYL-4-OMBELLIFERONE SODEE (FRENCH) see HMB000
(METHYL-ONN-AZOXY)METHANOL see HMG000
(METHYL-ONN-AZOXY)METHANOL, ACETATE (ester) see MGS750
(METHYL-ONN-AZOXY)METHYL-β-d-GLUCOPYRANOSIDE see COU000
(METHYL-ONN-AZOXY)METHYL-β-D-GLUCOPYRANOSIDURONIC ACID see MGS700
METHYL ORANGE see MND600
METHYL ORANGE B see MND600
METHYL ORTHOFORMATE see TLX600
METHYL ORTHOSILICATE (DOT) see MPI750
METHYL ORTHOSILICATE see MPI750
9-METHYL-3-OXA-9-AZATRICYCLO(3.3.1.0²⁴)NONAN-7-OL TROPATE (ester) see SBG000
4-METHYL-7-OXABICYCLO(4.1.0)HEPTANE-3-CARBOXYLIC ACID, ALLYL ESTER see AGF500
2-(3-(5-METHYL-1,3,4-OXADIAZOL-2-YL)-3,3-DIPHENYLPROPYL)-2-AZABICYCLO(2.2.2)OCTANE see DWF700
N-METHYLOXAZEPAM see CFY750
METHYLOXAZEPAM see CFY750
3-METHYL-2-OXAZOLIDONE see MND750
4-METHYL-2-OXETANONE see BSX000
METHYL OXIRANE see PNL600

METHYL PHENIDYL ACETATE see MNQ000
METHYLPHENIDYLACETATE HYDROCHLORIDE see RLK000
4-METHYLPHENISOPROPYLAMINE SULFATE see AQQ000
1-METHYLPHENOBARBITAL see ENB500
N-METHYLPHENOBARBITAL see ENB500
METHYLPHENOBARBITAL see ENB500
N-METHYLPHENOBARBITOL see ENB500
METHYLPHENOBARBITONE see ENB500
2-METHYLPHENOL see CNX000
3-METHYLPHENOL see CNW750
4-METHYLPHENOL see CNX250
m-METHYLPHENOL see CNW750
o-METHYLPHENOL see CNX000
p-METHYLPHENOL see CNX250
4-METHYLPHENOL METHYL ETHER see MGP000
3-METHYLPHENOL SODIUM SALT see SJP000
4-METHYLPHENOL, SODIUM SALT see SIM100
10-METHYLPHENOTHIAZINE-2-ACETIC ACID see MNQ500
(10-METHYL-2-PHENOTHIAZINYL)ACETIC ACID see MNQ500
N-METHYL-3-PHENOTHIAZINYLACETIC ACID see MNQ500
2-(2-(4-(2-METHYL-3-PHENOTHIAZIN-10-YLPROPYL)-1-PIPERAZI-
    NYL)ETHOXY)ETHANOL DIHYDROCHLORIDE see DXR800
2-(2-(4-(2-METHYL-3-PHENOTHIAZIN-10-YLPROPYL)-1-PIPERAZI-
    NYL)ETHOXY)ETHANOL see MKQ000
1-METHYL-3-PHENOXYBENZENE see MNV770
(±)-α-METHYL-3-PHENOXYBENZENEACETIC ACID CALCIUM SALT DIHY-
    DRATE see FAP100
1-(2-(o-METHYLPHENOXY)ETHYL)HYDRAZINE HYDROGEN SULFATE see
    MNR100
(1-METHYL-2-PHENOXYETHYL)HYDRAZINE MALEATE see PDV700
(1-METHYL-2-PHENOXYETHYL)HYDRAZINIUM see PDV700
((METHYLPHENOXY)METHYL)OXIRANE see TGZ100
3-(2-METHYLPHENOXY)-1,2-PROPANEDIOL see GGS000
3-(2-METHYLPHENOXY)-1,2-PROPANEDIOL 1-CARBAMATE see CBK500
α-METHYLPHENSUXIMIDE see MLP800
α-METHYL PHENYLACETALDEHYDE see COF000
4-METHYLPHENYL ACETATE see MNR250
METHYL PHENYLACETATE (FCC) see MHA500
p-METHYLPHENYL ACETATE see MNR250
METHYL(2-PHENYLAETHYL)NITROSAMIN (GERMAN) see MNU250
N-METHYLPHENYLAMINE see MGN750
METHYLPHENYLAMINE see MGN750
2-((4-METHYLPHENYL)AMINO)BENZOIC ACID see TGV000
2-METHYL-4-PHENYLANILINE see MGF500
N-(4-METHYLPHENYL)ANTHRANILIC ACID see TGV000
N-(p-METHYLPHENYL)ANTHRANILIC ACID see TGV000
METHYLPHENYLARSINIC ACID see HMK200
METHYLPHENYLARSONIC ACID see HMK200
N-METHYL-p-(PHENYLAZO)ANILINE see MNR500
N-METHYL-4-(PHENYLAZO)-o-ANISIDINE see MNS000
4-((4-METHYLPHENYL)AZO)BENZENAMINE see TGV750
4-METHYL-6-(PHENYLAZO)-1,3-BENZENEDIAMINE see CMM760
1-((2-METHYLPHENYL)AZO)-2-NAPHTHALENAMINE see FAG135
1-(2-METHYLPHENYL)AZO-2-NAPHTHALENAMINE see FAG135
1-((2-METHYLPHENYL)AZO)-2-NAPHTHALENOL see TGW000
1-(2-METHYLPHENYL)AZO-2-NAPHTHYLAMINE see FAG135
N-METHYL-N-(p-(PHENYLAZO)PHENYL)HYDROXYLAMINE see HLV000
METHYLPHENYLBARBITURIC ACID see ENB500
5-METHYL-7-PHENYL-1:2-BENZACRIDINE see MNS250
7-METHYL-9-PHENYLBENZ(c)ACRIDINE see MNS250
2-METHYL-N-PHENYLBENZAMIDE see MNS500
4-METHYL-N-PHENYLBENZENESULFONAMIDE see TGN600
4-METHYLPHENYL BENZOATE see TGX100
o-METHYLPHENYL BROMIDE see BOG260
2-METHYL-4-PHENYL-2-BUTANOL see BEC250
2-METHYL-4-PHENYL-2-BUTANOL ACETATE see MNT000
3-METHYL-4-PHENYL-3-BUTEN-2-ONE see MNS600
2-METHYL-4-PHENYL-2-BUTYL ACETATE see MNT000
2-METHYL-1-PHENYL-3-BUTYNE-1,2-DIOL see DMX800
3-(METHYLPHENYL)CARBAMIC ACID 3-((METHOXYCARBO-
    NYL)AMINO)PHENYL ESTER see MEG250
METHYLPHENYLCARBAMIC ESTER OF 3-OXYPHENYLTRIMETHYLAMMONIUM
    METHYLSULFATE see HNR500
N-METHYL-4-PHENYL-4-CARBETHOXYPIPERIDINE see DAM600
N-METHYL-4-PHENYL-4-CARBETHOXYPIPERIDINE HYDROCHLORIDE see
    DAM700
METHYLPHENYLCARBINOL see PDE000
METHYLPHENYLCARBINOL ACETATE see SMP600
METHYL PHENYLCARBINYL ACETATE see MNT075
METHYL PHENYL CARBINYL ACETATE see SMP600
METHYLPHENYLCARBINYL PROPIONATE see PFR000
1-METHYL-4-PHENYL-4-CARBOETHOXYPIPERIDINE HYDROCHLORIDE see
    DAM700
1-METHYL-5-PHENYL-7-CHLORO-1,3-DIHYDRO-2H-1,4-BENZODIAZEPIN-2-ONE
    see DCK759

1-METHYL-3-PHENYL-5-CHLOROIMIDAZO(4,5-b)PYRIDIN-2-ONE see MNT100
p-METHYLPHENYLDIAZONIUM FLUOROBORATE see TGM450
METHYLPHENYLDICHLOROSILANE (DOT) see DFQ800
2-METHYL-9-PHENYL-2,3-DIHYDRO-1-PYRIDINDENE HYDROBROMIDE see
    NOE525
METHYLPHENYLDIMETHOXYSILANE see DOH400
1-(o-METHYLPHENYL)-3,3-DIMETHYL-TRIAZEN (GERMAN) see MNT500
1-(2-METHYLPHENYL)-3,3-DIMETHYLTRIAZENE see MNT500
1-(3-METHYLPHENYL)-3,3-DIMETHYLTRIAZENE see DSR200
1-(m-METHYLPHENYL)-3,3-DIMETHYLTRIAZENE see DSR200
1-(o-METHYLPHENYL)-3,3-DIMETHYL-TRIAZENE see MNT500
2-(METHYLPHENYL)-1,3-DIOXAN-5-OL (mixed isomers) see TGK500
METHYLPHENYL DIPHENYL PHOSPHATE see TGY750
METHYL PHENYLDITHIOCARBAMATE see MJK500
1-METHYL-2-PHENYL-3-DODECYLBENZIMIDAZOLINIUM FERROCYANIDE see
    TNH750
1-METHYL-2-PHENYL-3-m-DODECYLBENZIMIDAZOLIUM HEXACYANOFERRATE
    see TNH750
1,1'-(4-METHYL-1,3-PHENYLENE)BIS-1H-PYRROLE-2,5-DIONE see TGY770
4-METHYL-m-PHENYLENEDIAMINE see TGL750
2-METHYL-p-PHENYLENEDIAMINE see TGM000
METHYLPHENYLENEDIAMINE see TGL500
2-METHYL-p-PHENYLENEDIAMINE SULPHATE see DCE600
4-METHYL-PHENYLENE DIISOCYANATE see TGM750
METHYL-m-PHENYLENE DIISOCYANATE see TGM740
2-METHYL-m-PHENYLENE ESTER, ISOCYANIC ACID see TGM800
4-METHYL-PHENYLENE ISOCYANATE see TGM750
2-METHYL-m-PHENYLENE ISOCYANATE see TGM800
METHYLPHENYLENE ISOCYANATE see TGM740
1-(4-(METHYLPHENYL))ETHANOL see TGZ000
1-(p-METHYLPHENYL)ETHANOL see TGZ000
METHYL PHENYL ETHER see AOX750
α-METHYL-β-PHENYLETHYL ACETATE see ABU800
β-METHYLPHENYLETHYL ACETATE see PGB750
α-METHYL PHENYLETHYL ALCOHOL see HGR600
METHYL(2-PHENYLETHYL)ARSINIC ACID see MNU050
N-METHYL-5-PHENYL-5-ETHYLBARBITAL see ENB500
1-METHYL-5-PHENYL-5-ETHYLBARBITURIC ACID see ENB500
4-(1-METHYL-1-PHENYLETHYL)-2,6-BIS-(1-PIPERIDINYLMETHYL)PHENOL DIHY-
    DROBROMIDE see BHR750
as-METHYLPHENYLETHYLENE see MPK250
METHYL PHENYLETHYL ETHER see PFD325
METHYL PHENYLETHYL ETHER see PFD325
3-METHYL-5,5-PHENYLETHYLHYDANTOIN see MKB250
(2-(2-METHYLPHENYL)ETHYL)-HYDRAZINE see MNO775
o-METHYL-β-PHENYLETHYLHYDRAZINE DIHYDROGEN SULFATE see
    MNU100
p-METHYL-β-PHENYLETHYLHYDRAZINE DIHYDROGEN SULFATE see
    MNU150
(1-METHYL-2-PHENYLETHYL)-HYDRAZINEIUM CHLORIDE see PDN250
METHYL-2-PHENYLETHYL KETONE see PDF800
METHYL PHENYLETHYL KETONE see PDF800
METHYL-PHENYLETHYL-NITROSAMINE see MNU250
N-(1-METHYL-2-PHENYLETHYL)-γ-PHENYLBENZENEPROPANAMINE see
    PEV750
3-METHYL-3-PHENYLGLYCIDIC ACID ETHYL ESTER see ENC000
4-METHYLPHENYLHYDRAZINE HYDROCHLORIDE see MNU500
4-METHYLPHENYL 2-HYDROXYBENZOATE see THD850
N-(2-METHYLPHENYL)-HYDROXYLAMINE see THA000
N-(2-METHYLPHENYL)IMIDODICARBONIMIDIC DIAMIDE see TGX550
2,2'-((2-METHYLPHENYL)IMINO)BISETHANOL see DMT800
2,2'-((3-METHYLPHENYL)IMINO)BISETHANOL see DHF400
N'-(5-METHYL-3-PHENYL-1-INDOLYL)-N,N,N'-TRIMETHYLETHYLENEDIAMINE
    HYDROCHLORIDE see MNU750
1-METHYL-4-PHENYLISONIPECOTIC ACID, ETHYL ESTER see DAM600
1-METHYL-4-PHENYLISONIPECOTIC ACID ETHYL ESTER HYDROCHLORIDE
    see DAM700
1-METHYL-4-PHENYLISONIPECOTIC ACID ETHYL ESTER-1-OXIDE HYDRO-
    CHLORIDE see DYB250
N-METHYL-β-PHENYLISOPROPYLAMIN (GERMAN) see DBB000
N-METHYL-β-PHENYLISOPROPYLAMINE see DBB000
dl-N-METHYL-β-PHENYLISOPROPYLAMINE HYDROCHLORIDE see DAR100
l-N-METHYL-β-PHENYLISOPROPYLAMINE HYDROCHLORIDE see MDQ500
N-METHYL-β-PHENYLISOPROPYLAMINHYDROCHLORID (GERMAN) see
    DBA800
5-METHYL-3-PHENYLISOXAZOLE-4-CARBOXYLIC ACID see MNV000
5-METHYL-3-PHENYL-4-ISOXAZOLYL-PENICILLIN see DSQ800
5-METHYL-3-PHENYL-4-ISOXAZOLYL PENICILLIN, SODIUM see MNV250
METHYL PHENYL KETONE see ABH000
4-METHYLPHENYLMERCAPTAN see TGP250
p-METHYLPHENYLMERCAPTAN see TGP250
3-METHYLPHENYL N-METHYLCARBAMATE see MIB750
m-METHYLPHENYL METHYLCARBAMATE see MIB750
(4-METHYLPHENYL)METHYL CHLORIDE see MHN300
1-(1-METHYL-2-(2-(PHENYLMETHYL)PHENOXY)ETHYL)PIPERIDINE PHOSPHATE
    (9CI) see PJA130

10-(2-(4-METHYL-1-PIPERAZINYL)ETHYL)PHENOTHIAZINE see MOE250
8-(((4-METHYL-1-PIPERAZINYL)IMINO)METHYL)RIFAMYCIN SV see RKP000
3-(4-METHYLPIPERAZINYLIMINOMETHYL)-RIFAMYCIN SV see RKP000
8-(4-METHYLPIPERAZINYLIMINOMETHYL) RIFAMYCIN SV see RKP000
2-(4-METHYL-1-PIPERAZINYL)-10-METHYL-3,4-DIAZAPHENOXAZINDIHYDRO-CHLORID (GERMAN) see ASC250
6-(4-METHYL-1-PIPERAZINYL)MORPHANTHRIDINE see HOU059
N-METHYL-PIPERAZINYL-N'-AETHYL-PHENOTHIAZIN (GERMAN) see MOE250
N-METHYL-PIPERAZINYL-N'-PROPYL-PHENOTHIAZIN (GERMAN) see PCK500
N-(γ-(4'-METHYLPIPERAZINYL-1')PROPYL)-3-CHLOROPHENOTHIAZINE see PMF500
N-(3-(4-METHYL-1-PIPERAZINYL)PROPYL)PHENOTHIAZINE see PCK500
10-(3-(4-METHYL-1-PIPERAZINYL)PROPYL)-10H-PHENOTHIAZINE (9CI) see PCK500
1-(10-(3-(4-METHYL-1-PIPERAZINYL)PROPYL)PHENOTHIAZIN-2-YL)-1-BUTANONE DIMALEATE see BSZ000
10-(3-(4-METHYL-1-PIPERAZINYL)PROPYL)-2-(TRIFLUOROMETHYL) PHENOTHIAZINE see TKE500
10-(3-(4-METHYL-1-PIPERAZINYL)PROPYL)-2-TRIFLUOROMETHYLPHENOTHIAZINE DIHYDROCHLORIDE see TKK250
4-(4-METHYL-1-PIPERAZINYL)-5,6,7,8-TETRAHYDRO-(1)-BENZOTHIENO(2,3-d) PYRIMIDINE HYDROCHLORIDE see MOF750
4-(4-METHYL-1-PIPERAZINYL)THIENO(2,3-d) PYRIMIDINE HYDROCHLORIDE see MOG000
2-(4-METHYL-1-PIPERAZINYL)-11-(p-TOLYL)-10,11-DIHYDROPYRIDAZINO(3,4-b)(1,4)BENZOXAZEPINE see MOG250
3-METHYLPIPERIDINE see MOH000
4-METHYLPIPERIDINE see MOH250
1-METHYLPIPERIDINE (DOT) see MOG500
N-METHYLPIPERIDINE see MOG500
1-METHYL-PIPERIDINE-4-CARBONSAURE-O,O-XYLIDID HYDROCHLORID (GERMAN) see MQO250
(1-METHYL-dl-PIPERIDINE-2-CARBOXYLIC ACID)-2,6-DIMETHYLANILIDE HYDROCHLORIDE see CBR250
γ-(4-METHYLPIPERIDINE)-p-FLUOROBUTYROPHENONE HYDROCHLORIDE see FKI000
2-METHYLPIPERIDINE β-NAPHTHOAMIDE see MOG600
2-METHYL-1-PIPERIDINEPROPANOL BENZOATE HYDROCHLORIDE see IJZ000
1-METHYL-3-(PIPERIDINOCARBONYL)PIPERIDINE see MOG650
β-4-METHYLPIPERIDINOETHYL BENZOATE HYDROCHLORIDE see MOI250
2-METHYL-2-(2-PIPERIDINOETHYL)-1,3-BENZODIOXOLE HYDROCHLORIDE see MOI500
N-(1-METHYL-2-PIPERIDINOETHYL)-N-2-PYRIDYLPROPIONAMIDE FUMARATE see PMX250
β-METHYL-4-PIPERIDINOPHENETHYLAMINE DIHYDROCHLORIDE see MOJ500
2-METHYL-2-(4-PIPERIDINOPHENYL)ETHYLAMINE DIHYDROCHLORIDE see MOJ500
2-METHYL-1-PIPERIDINOPROPANOL, BENZOATE see PIV750
METHYL-4-(3-PIPERIDINOPROPIONYLAMINO)SALICYLATE, METHIODIDE see MOK000
γ-3-METHYLPIPERIDINOPROPYL-p-AMINOBENZOATE HYDROCHLORIDE see MOK500
(2-METHYLPIPERIDINO)PROPYL BENZOATE see PIV750
3-(2-METHYLPIPERIDINO)PROPYL BENZOATE HYDROCHLORIDE see IJZ000
γ-(2-METHYLPIPERIDINO)PROPYL BENZOATE HYDROCHLORIDE see IJZ000
dl-(2-METHYLPIPERIDINO)PROPYL BENZOATE HYDROCHLORIDE see IJZ000
METHYL 10-(3-PIPERIDINOPROPYL)PHENOTHIAZIN-2-YL KETONE see MOL300
METHYL 10-(3-PIPERIDINOPROPYL)PHENOXAZIN-2-YL KETONE see MOL400
2-METHYL-3-PIPERIDINOPYRAZINE MONOSULFATE see MOM750
2-METHYL-3-PIPERIDINOPYRAZINE SULFATE see MOM750
2-METHYL-3-PIPERIDINO-1-p-TOLYLPROPAN-1-ONE see TGK200
2-METHYL-3-PIPERIDINO-1-p-TOLYLPROPAN-1-ONE HYDROCHLORIDE see MRW125
N-(1-METHYL-2-(1-PIPERIDINYL)ETHYL)-N-2-PYRIDINYLPROPANAMIDE-(E)-2-BUTENEDIOATE (1:1) see PMX250
1-METHYL-4-PIPERIDYL-p-AMINOBENZOATE HYDROCHLORIDE see MON000
N-METHYLPIPERIDYL-(4)-BENZHYDRYLAETHER SALZSAUREN SALZE (GERMAN) see LJR000
N-METHYL-3-PIPERIDYL BENZILATE see MON250
N-METHYL-3-PIPERIDYL BENZILATE METHOBROMIDE see CBF000
1-METHYL-4-PIPERIDYL BENZOATE HYDROCHLORIDE see MON500
N-METHYL-3-PIPERIDYLDIPHENYLGLYCOLATE METHOBROMIDE see CBF000
9-(4'-(N-METHYLPIPERIDYLENE))THIOXANTHENE MALEATE see MOP000
1-METHYL-3-PIPERIDYL ESTER METHOBROMIDE BENZILIC ACID see CBF000
10-(2-(1-METHYL-2-PIPERIDYL)ETHYL)-2-METHYLSULFINYL PHENOTHIAZINE see MON750
10-(2-(1-METHYL-2-PIPERIDYL)ETHYL)-2-(METHYLTHIO)PHENOTHIAZINE see MOO250
10-(2-(1-METHYL-2-PIPERIDYL)ETHYL)-2-METHYLTHIOPHENOTHIAZINE HYDROCHLORIDE see MOO500
(1-METHYL-4-PIPERIDYLIDENE)-9-ANTHROL-9,10-DIHYDRO-10-HYDROCHLORIDE see WAJ000

1-METHYL-3-PIPERIDYLIDENEDI(2-THIENYL)METHANE see BLV000
9-(1-METHYL-4-PIPERIDYLIDENE)THIOXANTHENE see MOO750
9-(N-METHYL-PIPERIDYLIDEN-4)THIOXANE MALEATE see MOP000
9-(1-METHYL-PIPERIDYL-(2)-METHYL)-CARBAZOL (GERMAN) see MOP500
9-(METHYL-2-PIPERIDYL)METHYLCARBAZOLE see MOP500
10-(1-METHYLPIPERIDYL-3-METHYL)PHENOTHIAZINE see MOQ250
10-(1-METHYL-3-PIPERIDYL)METHYL PHENOTHIAZINE see MOQ250
(N-METHYL-3-PIPERIDYL)METHYLPHENOTHIAZINE see MOQ250
9-(1-METHYL-3-PIPERIDYLMETHYL)THIAXANTHENE HYDROCHLORIDE see THL500
9-((N-METHYL-3-PIPERIDYL)METHYL)-THIOXANTHENHYDROCHLORID (GERMAN) see THL500
d-1-METHYL-3-PIPERIDYL-dl-α-PHENYLCYCLOHEXANEGLYCOLATE HYDROCHLORIDE see PMS800
1-METHYL-3-PIPERIDYL PIPERIDINO KETONE see MOG650
γ-(2-METHYLPIPERIDYL)PROPYL BENZOATE see PIV750
(+—)-γ-(2-METHYLPIPERIDYL)PROPYL BENZOATE HYDROCHLORIDE see IJZ000
METHYL PIRIMIPHOS see DIN800
METHYL POTASSIUM see MOR250
6-α-METHYLPREDNISOLONE see MOR500
METHYLPREDNISOLONE see MOR500
METHYLPREDNISOLONE 21-ACETATE see DAZ117
6-α-METHYLPREDNISOLONE ACETATE see DAZ117
6-α-METHYLPREDNISOLONE ACETATE see DAZ117
METHYLPREDNISOLONE ACETATE see DAZ117
6-α-METHYLPREDNISOLONE SODIUM SUCCINATE see USJ100
METHYLPREDNISOLONE SODIUM SUCCINATE see USJ100
16-β-METHYL-1,4-PREGNADIENE-9-α-FLUORO-11-β,17-α,21-TRIOL- 3,20-DIONE see BFV750
6-METHYL-Δ⁴˒⁶-PREGNADIEN-17-α-OL-3,20-DIONE ACETATE see VTF000
6-α-METHYL-4-PREGNENE-3,20-DION-17-α-OL ACETATE see MCA000
METHYLPROMAZINE see AFL500
METHYLPROPAMINE see DBA800
2-METHYL-1-PROPANAL see IJS000
2-METHYLPROPANAL see IJS000
2-METHYL-1-PROPANAL OXIME see IJT000
N-METHYLPROPANAMIDE see MOS900
2-METHYLPROPANE see MOR750
2-METHYLPROPANENITRILE see IJX000
α-METHYLPROPANENITRILE see IJX000
2-METHYL-2-PROPANETHIOL see MOS000
2-METHYLPROPANETHIOL see IIX000
METHYL PROPANOATE see MOT000
(Z)-2-METHYLPROPANOIC ACID 3,7-DIMETHYL-2,6-OCTADIENYL ESTER see NCO200
2-METHYLPROPANOIC ACID 2,4-HEXADIENYL ESTER see HCS700
2-METHYLPROPANOIC ACID 1-METHYL-1-(4-METHYL-3-CYCLOHEXEN-1-YL)ETHYL ESTER see MCF515
2-METHYL-PROPANOIC ACID-3-PHENYL-2-PROPENYL ESTER see CMR750
2-METHYLPROPAN-1-OL see IIL000
2-METHYL-1-PROPANOL see IIL000
2-METHYL-2-PROPANOL see BPX000
2-METHYL PROPANOL see IIL000
N-METHYL-N-PROPARGYLBENZYLAMINE see MOS250
N-METHYL-N-PROPARGYL-3-(2,4-DICHLOROPHENOXY)PROPYLAMINE HYDROCHLORIDE see CMY000
METHYL PROPARGYL ETHER see MFN250
2-METHYLPROPENAL (CZECH) see MGA250
2-METHYLPROPENAMIDE see MDN500
METHYL PROPENATE see MGA500
2-METHYLPROPENE see IIC000
2-METHYL-2-PROPENE-1,1-DIOL DIACETATE see AAW250
2-METHYLPROPENENITRILE see MGA750
2-METHYL-1-PROPENE-1-ONE see DSL289
METHYL-2-PROPENOATE see MGA500
METHYL PROPENOATE see MGA500
2-METHYLPROPENOIC ACID see MDN250
2-METHYL-2-PROPENOIC ACID ANHYDRIDE (9CI) see MDN699
2-METHYL-2-PROPENOIC ACID CHLORIDE see MDN899
2-METHYL-2-PROPENOIC ACID 2-(DIMETHYLAMINO)-1-((DIMETHYLAMINO)METHYL)ETHYL ESTER (9CI) see BJI125
2-METHYL-2-PROPENOIC ACID, ETHYL ESTER see EMF000
2-METHYL-2-PROPENOIC ACID METHYL ESTER see MLH750
2-METHYL-2-PROPENOIC ACID METHYL ESTER HOMOPOLYMER see PKB500
2-METHYL-2-PROPENOIC ACID METHYL ESTER, POLYMER with TRIBUTYL(2-METHYL-1-OXO-2-PROPENYL)OXY)STANNANE see OIY000
2-METHYL-2-PROPENOIC ACID METHYL ESTER, POLYMER with TRIBUTYL(92-METHYL-1-OXO-2-PROPENYL)OXY)STANNANE and ((2-METHYL-1-OXO-2-PROPENYL)OXY)TRIPOPYLSTANNANE see OIW000
2-METHYL-2-PROPENOIC ACID-2-METHYLPROPYL ESTER see IIY000
2-METHYL-2-PROPENOIC ACID-3-(TRIMETHOXYSILYL)PROPYL ESTER see TLC250
2-METHYL-2-PROPEN-1-OL see IMW000

1-METHYL PROPENOL see MQL250
2-METHYL-2-PROPENOYL CHLORIDE see MDN899
2-METHYLPROPENYL CHLORIDE see MDN899
METHYL PROPENYL KETONE see PBR500
METHYL PROPIOLATE see MOS875
2-METHYLPROPIONALDEHYDE see IJS000
N-METHYLPROPIONAMIDE see MOS900
METHYL PROPIONATE see MOT000
2-METHYLPROPIONIC ACID see IJU000
α-METHYLPROPIONIC ACID see IJU000
N-METHYLPROPIONIC ACID AMIDE see MOS900
2-METHYLPROPIONIC ACID, ETHYL ESTER see ELS000
2-METHYLPROPIONITRILE see IJX000
N-METHYLPROPIONSAEUREAMID see MOS900
3-(2-METHYLPROPIONYL)-5-INDOLECARBONITRILE see COP525
METHYL 5-n-PROPOXY-2-BENZIMIDAZOLE CARBAMATE see OMY700
N-((2-METHYLPROPOXY)METHYL)-2-PROPENAMIDE see IIE100
2-METHYL-1-PROPYL ACETATE see IIJ000
2-METHYLPROPYL ACETATE see IIJ000
METHYLPROPYLACETIC ACID see MQJ750
Z-METHYLPROPYL ACRYLATE see IIK000
2-METHYLPROPYL ALCOHOL see IIL000
1-METHYLPROPYLAMINE see BPY000
(2-METHYLPROPYL)-p-AMINOBENZOATE see MOT750
2-METHYL-2-(PROPYLAMINO)-1-PROPANOL BENZOATE (ester), HYDROCHLO-
RIDE see OJI600
2-METHYL-2-(PROPYLAMINO)-1-PROPANOL BENZOATE HYDROCHLORIDE
see OJI600
METHYLPROPYLARSINIC ACID see MOT800
METHYL PROPYLATE see MOT000
12-METHYL-7-PROPYLBENZ(a)ANTHRACENE see MOU250
4-(2-METHYLPROPYL)BENZENEACETIC ACID see IJG000
2-METHYLPROPYL BUTYRATE see BSW500
METHYL PROPYL CARBINOL see PBM750
METHYLPROPYLCARBINOL CARBAMATE see MOU500
METHYL-PROPYL-CETONE (FRENCH) see PBN250
3-(1-METHYLPROPYL)-6-CHLOROPHENYL METHYLCARBAMATE see MOU750
2-(1-METHYLPROPYL)CYCLOHEXANONE see MOU800
METHYL PROPYL DIKETONE see HEQ200
6-(1-METHYL-PROPYL)-2,4-DINITROFENOL (DUTCH) see BRE500
(6-(1-METHYL-PROPYL)-2,4-DINITRO-FENYL)-3,3-DIMETHYL ACRYLAAT
(DUTCH) see BGB500
2-(1-METHYLPROPYL)-4,6-DINITROPHENOL see BRE500
2-(1-METHYL-N-PROPYL) 4,6-DINITROPHENOL AMMONIUM SALT see BPG250
2-(1-METHYL-N-PROPYL)-4,6-DINITROPHENOL TRIETHANOLAMINE SALT see
BRE750
2-(1-METHYLPROPYL)-4,6-DINITROPHENYL ACETATE see ACE500
2-(1-METHYLPROPYL)-4,6-DINITROPHENYL-β,β-DIMETHACRYLATE see
BGB500
(6-(1-METHYL-PROPYL)-2,4-DINITRO-PHENYL)-3,3-DIMETHYL ACRYLAT (GER-
MAN) see BGB500
2-(1-METHYL-2-PROPYL)-4,6-DINITROPHENYL ISOPROPYLCARBONATE see
CBW000
β-METHYLPROPYL ETHANOATE see IIJ000
2-METHYL-2-PROPYLETHANOL see AOK750
METHYL n-PROPYL ETHER see MOU830
METHYL PROPYL ETHER see MOU830
2-METHYLPROPYL HEXANOATE see IIT000
N,N''-(2-METHYLPROPYLIDENE)BISUREA (9CI) see IIV000
2-METHYLPROPYL ISOBUTYRATE see IIW000
2-METHYLPROPYL ISOVALERATE see ITA000
METHYL PROPYL KETONE (ACGIH, DOT) see PBN250
METHYL-n-PROPYL KETONE see PBN250
2-METHYLPROPYL METHACRYLATE see IIY000
2-METHYLPROPYL-3-METHYLBUTYRATE see ITA000
METHYL-N-PROPYLNITROSAMINE see MNA000
METHYLPROPYLNITROSOAMINE see MNA000
N-(2-METHYLPROPYL)-N-NITROSOUREA see IJF000
8-METHYL-3-(2-PROPYLPENTANOYLOXY)TROPINIUM BROMIDE see LJS000
2-(1-METHYLPROPYL)PHENYL METHYLCARBAMATE see MOV000
m-(1-METHYLPROPYL)PHENYLMETHYLCARBAMATE see BSG250
2-METHYL-2-PROPYL-1,3-PROPANEDIOL BUTYLCARBAMATE CARBAMATE
see MOV500
2-METHYL-2-PROPYL-1,3-PROPANEDIOL CARBAMATE ISOPROPYLCARBA-
MATE see IPU000
2-METHYL-2-N-PROPYL-1,3-PROPANEDIOL DICARBAMATE see MQU750
2-METHYLPROPYLPROPANOIC ACID-2-METHYLPROPYL ESTER (9CI) see
IIW000
5-(2-METHYLPROPYL)-5-(2-PROPENYL)-2,4,6(1H,3H,5H)-PYRIMIDINETRIONE
(9CI) see AGI750
5-(1-METHYLPROPYL)-5-(2-PROPENYL)-2,4,6(1H,3H,5H)-PYRIMIDINETRIONE
(9CI) see AFY500
2-METHYLPROPYL PROPIONATE see PMV250
METHYL 5-(PROPYLTHIO)-2-BENZIMIDAZOLECARBAMATE see VAD000

2-METHYL-2-PROPYLTRIMETHYLENE BUTYLCARBAMATE CARBAMATE see
MOV500
2-METHYL-2-PROPYLTRIMETHYLENE CARBAMATE see MQU750
3-METHYL-5-PROPYL-1,2,4-TRIOXOLANE see HFC500
METHYL PROPYNOATE see MOS875
N-METHYL-N-2-PROPYNYLBENZYLAMINE see MOS250
N-METHYL-N-(2-PROPYNYL)BENZYLAMINE HYDROCHLORIDE see BEX500
1-METHYL-2-PROPYNYL-m-CHLOROCARBANILATE see CEX250
1-METHYL-2-PROPYNYL-m-CHLOROPHENYLCARBAMATE see CEX250
1-METHYLPROPYNYL 3-CHLOROPHENYLCARMATE see CEX250
1-METHYLPROPYNYL ESTER of 3-CHLOROPHENYLCARBAMIC ACID see
CEX250
2-METHYL-5-(2-PROPYNYL)-3-FURYLMETHYL-cis-trans-CHRYSANTHEMATE see
PMN700
6-α-METHYL-17-(1-PROPYNYL)TESTOSTERONE see DRT200
6-α-METHYL-17-α-PROPYNYLTESTOSTERONE see DRT200
15(R)-METHYLPROSTAGLANDIN E2 see AQT575
15(R)-METHYLPROSTAGLANDIN E2 see AQT575
15(s)-15-METHYL-PROSTAGLANDIN E2 see MOV800
15(S)-15-METHYL-PROSTAGLANDIN F2-α see CCC100
15(S)-METHYLPROSTAGLANDIN F2-α see CCC100
15-METHYLPROSTAGLANDIN F2-α see CCC100
(15S)-15-METHYLPROSTAGLANDIN F2-α see CCC100
15(s)15-METHYL-PROSTAGLANDIN-F2-α-METHYL ESTER see MLO300
15-METHYL-PROSTAGLANDIN-F2-α-METHYL ESTER see MLO300
15(S)15-METHYL PROSTAGLANDIN F2-α TROMETHAMINE see CCC110
N-METHYL-4-PROTOADAMANTANEAMINE HYDROCHLORIDE see MOW000
N-METHYL-4-PROTOADAMANTANEMETHANAMINE MALEATE see MOW250
METHYL PROTOANEMONIN see MOW500
METHYLPROTOCATECHUALDEHYDE see VFK000
2-METHYLPYRAZINE see MOW750
3(5)-METHYLPYRAZOLE see MOX010
3-METHYLPYRAZOLE see MOX010
4-METHYLPYRAZOLE see MOX000
5-METHYLPYRAZOLE see MOX010
3-METHYL-2-PYRAZOLIN-5-ONE see MOX100
3-METHYL-PYRAZOLON-(5) see MOX100
3-METHYLPYRAZOLYL-5-DIETHYLPHOSPHATE see MOX250
METHYLPYRAZOLYL DIETHYLPHOSPHATE see MOX250
5-METHYL-1H-PYRAZOL-3-YL DIMETHYLCARBAMATE see DQZ000
1-METHYLPYRENE see MOX875
3-METHYLPYRENE see MOX875
2-METHYLPYRIDINE see MOY000
3-METHYLPYRIDINE see PIB920
4-METHYLPYRIDINE see MOY250
α-METHYLPYRIDINE see MOY000
N-METHYLPYRIDINE-2-ALDOXIME IODIDE see POS750
2-METHYLPYRIDINE-4-AZO-p-DIMETHYLANILINE see MOY500
N-METHYL-2-PYRIDINEETHANAMINE see HGE820
N-METHYL-2-PYRIDINEETHANAMINE DIMETHANESULFONATE see BFV350
3-METHYLPYRIDINE-1-OXIDE see MOY550
4-METHYLPYRIDINE 1-OXIDE see MOY790
2-METHYLPYRIDINE-1-OXIDE-4-AZO-p-DIMETHYLANILINE see DSS200
3-METHYLPYRIDINE-1-OXIDE-4-AZO-p-DIMETHYL-ANILINE see MOY750
1-METHYL-2-PYRIDINIUM ALDOXIME CHLORIDE see FNZ000
N-METHYLPYRIDINIUM-2-ALDOXIME IODIDE see POS750
1-METHYLPYRIDINIUM-2-ALDOXIME METHANESULFONATE see PLX250
N-METHYLPYRIDINIUM-2-ALDOXIME METHANESULPHONATE see PLX250
1-METHYLPYRIDINIUM CHLORIDE see MOY875
N-METHYLPYRIDINIUM CHLORIDE see MOY875
METHYL PYRIDINIUM CHLORIDE see MOY875
N-METHYLPYRIDINIUM CHLORIDE-2-ALDOXIME see FNZ000
1-METHYLPYRIDINIUM IODIDE see MOZ000
N-METHYLPYRIDINIUM IODIDE see MOZ000
N-METHYLPYRIDINIUM METHANE SULFONATE-2-ALDOXIME see PLX250
1-METHYL-2(1H)-PYRIDINONE see MPA075
1-METHYL-9H-PYRIDO(3,4-b)INDOLE see MPA050
1-METHYL-2(1H)-PYRIDONE see MPA075
N-METHYLPYRIDONE see MPA075
α⁴-O-METHYLPYRIDOXOL see MFN600
α-((4-METHYL-2-PYRIDYLAMINO)METHYL)BENZYL ALCOHOL HYDROCHLO-
RIDE see MPA100
N-METHYL-N-β-(2-PYRIDYL)ETHYLAMINE see HGE820
9-(2-(2-METHYLPYRIDYL-5)ETHYL)-3,6-DIMETHYL-1,2,3,4-TETRAHYDRO-γ-CAR-
BOLINE 2HCl see DNU875
METHYL 2-PYRIDYL KETONE see PPH200
METHYL-3-PYRIDYL KETONE see ABI000
METHYL 4-PYRIDYL KETONE see ADA365
METHYL-β-PYRIDYL KETONE see ABI000
METHYL PYRIDYL KETONE see ABI000
1-METHYL-2-(3-PYRIDYL)PYRROLE see NDX300
1-METHYL-2-(3-PYRIDYL)PYRROLIDINE see NDN000
l-1-METHYL-2-(3-PYRIDYL)-PYRROLIDINE SULFATE see NDR500
METHYLPYRIMAL see ALF250
6-METHYL-2,4(1H,3H)-PYRIMIDINEDIONE see MQI750

N¹-(4-METHYL-2-PYRIMIDINYL)SULFANILAMIDE see ALF250
N¹-(4-METHYL-2-PYRIMIDINYL)SULFANILAMIDE SODIUM SALT see SJW475
3-METHYLPYROCATECHOL see DNE000
2-METHYL PYROMECONIC ACID see MAO350
METHYL PYROMUCATE see MKH600
METHYL PYROPHOSPHATE see TDV000
1-METHYLPYRROLE see MPB000
1-METHYLPYRROLE-2,3-DIMETHANOL see MPB175
1-METHYLPYRROLIDINE see MPB250
1-METHYL-2-PYRROLIDINONE see MPF200
N-METHYL-2-PYRROLIDINONE see MPF200
1-METHYL-5-PYRROLIDINONE see MPF200
N-METHYLPYRROLIDINONE see MPF200
3-(N-METHYLPYRROLIDINO)PYRIDINE see NDN000
6-METHYL-α-(1-PYRROLIDINYLCARBONYL)ERGOLINE-8-β-PROPIONITRILE see MPC300
1-(1-METHYL-2-PYRROLIDINYLIDENE)-3-(2,6-XYLYL)UREA see XGA725
3-(1-METHYL-2-PYRROLIDINYL)INDOLE see MPD000
3-(1-METHYL-3-PYRROLIDINYL)INDOLE see MPD250
10-((1-METHYL-3-PYRROLIDINYL)METHYL)-PHENOTHIAZINE see MPE250
10-((1-METHYL-3-PYRROLIDINYL)METHYL)PHENOTHIAZINE, HYDROCHLORIDE see MDT500
α-1-(1-METHYL-3-PYRROLIDINYL)-1-PHENYLPROPYL PROPIONATE FUMARATE see ENC600
3-(1-METHYL-2-PYRROLIDINYL)PYRIDINE see NDN000
(S)-3-(1-METHYL-2-PYRROLIDINYL)PYRIDINE (9CI) see NDN000
(S)-3-(1-METHYL-2-PYRROLIDINYL)PYRIDINE (R-(R,R))-2,3-DIHYDROXYBU-TANEDIOATE (1:2) see NDS500
(S)-3-(1-METHYL-2-PYRROLIDINYL)PYRIDINE SULFATE (2:1) see NDR500
N-(N'-METHYLPYRROLIDINIUMMETHYL)-2,2-DIPHENYL SULFOXIMIDE BRO-MIDE see HAK200
1-METHYL-2-PYRROLIDONE see MPF200
N-METHYL-2-PYRROLIDONE see MPF200
N-METHYLPYRROLIDONE see MPF200
METHYLPYRROLIDONE see MPF200
1-METHYL-2-(PYRROLID-2-YL)METHYL PHENYLCYCLOHEXYL GLYCOLLATE METHOBROMIDE see IBP200
1-METHYL-3-PYRROLIDYL-α-PHENYLCYCLOPENTANEGLYCOLATE HYDRO-CHLORIDE see AEY400
I-3-(1-METHYL-2-PYRROLIDYL)PYRIDINE see NDN000
(−)-3-(1-METHYL-2-PYRROLIDYL)PYRIDINE see NDN000
I-3-(1-METHYL-2-PYRROLIDYL)PYRIDINE SULFATE see NDR500
3-(1-METHYL-2-PYRROLYL)PYRIDINE see NDX300
3-(1-METHYL-1H-PYRROL-2-YL)-PYRIDINE (9CI) see NDX300
METHYL-QUECKSILBER-TOLUOLSULFAMID see MLH100
2-METHYL-4-QUINAZOLINONE see MPF500
2-METHYL-4(3H)-QUINAZOLINONE see MPF500
3-METHYL-4-QUINAZOLINONE see MPF750
2-METHYLQUINAZOLONE see MPF500
METHYLQUINAZOLONE HYDROCHLORIDE see MDT250
2-METHYLQUINOLINE see QEJ000
6-METHYLQUINOLINE see MPF800
p-METHYLQUINOLINE see MPF800
8-(METHYLQUINOLYL)-N-METHYL CARBAMATE see MPG250
2-METHYL-1,4-QUINONE see MHI250
6-METHYL-2,3-QUINOXALINE DITHIOCARBONATE see ORU000
6-METHYL-2,3-QUINOXALINEDITHIOL CYCLIC CARBONATE see ORU000
6-METHYL-2,3-QUINOXALINEDITHIOL CYCLIC DITHIOCARBONATE see ORU000
6-METHYL-QUINOXALINE-2,3-DITHIOLCYCLOCARBONATE see ORU000
METHYL RED see CCE500
METHYL RESERPATE see MPH300
METHYL RESERPATE-4-ETHOXYCARBONYL-3,5-DIMETHOXYBENZOIC ACID ESTER see RCA200
METHYL RESERPAT ESTER of SYRINGIG ACID ETHYL CARBONATE see RCA200
METHYL RESERPATE 3,4,5-TRIMETHOXYBENZOIC ACID see RDK000
METHYL RESERPATE 3,4,5-TRIMETHOXYBENZOIC ACID ESTER see RDK000
METHYL RESERPATE 3,4,5-TRIMETHOXYCINNAMIC ACID ESTER see TLN500
METHYL RESERPINOLATE see MPH300
2-METHYLRESORCIN see MPH400
2-METHYLRESORCINOL see MPH400
5-METHYLRESORCINOL see MPH500
5-METHYLRESORCINOL ORCINOL see MPH500
METHYLRHODANID (GERMAN) see MPT000
3-METHYLRHODANINE see MPH750
6-METHYL-9-RIBOFURANOSYLPURINE-6-THIOL see MPU000
3-METHYLRODANIN see MPH750
METHYLROSANILINCHLORID see AOR500
METHYLROSANILINE see TJK000
METHYLROSANILINE CHLORIDE see AOR500
METHYLROSANILINUM CHLORATUM see AOR500
METHYL SALICYLATE see MPI000
METHYLSALICYLATE METHYL ETHER see MLH800

3-METHYLSALYCILIC ACID see CNX625
N-METHYLSARCOSINE HYDROCHLORIDE see MPI100
METHYLSCOPOLAMINE BROMIDE see SBH500
METHYLSCOPOLAMINE HYDROBROMIDE see SBH500
N-METHYLSCOPOLAMINE METHOSULFATE see DAB750
N-METHYLSCOPOLAMINE METHYL SULFATE see DAB750
METHYLSCOPOLAMINE METHYL SULFATE see DAB750
METHYL SCOPOLAMINE NITRATE see SBH000
N-METHYLSCOPOLAMMONIUM BROMIDE see SBH500
METHYLSCOPOLAMMONIUM METHYLSULFATE see DAB750
METHYL SELENAC see SBQ000
METHYL SELENIDE see DUB200
METHYL SELENIUM see DUB200
METHYLSELENO-2-BENZOIC ACID see MPI200
o-(METHYLSELENO)BENZOIC ACID see MPI200
o-(METHYLSELENO)BENZOIC ACID SODIUM SALT see MPI205
(METHYLSELENO)TRIS(DIMETHYLAMINO)PHOSPHONIUM IODIDE see MPI225
METHYLSENFOEL (GERMAN) see ISE000
METHYLSERGIDE BIMALEATE see MQP500
1-METHYLSILACYCLOPENTA-2,4-DIENE see MPI600
METHYLSILANE see MPI625
METHYLSILATRAN see MPI650
1-METHYLSILATRANE see MPI650
METHYLSILATRANE see MPI650
METHYL SILICATE 51 see TDJ100
METHYL SILICATE see MPI750, TDJ100
METHYL SILICONE see PJR000
METHYLSILVER see MPI800
O-METHYLSINAPIC ACID see CMQ100
O-METHYL-S-(4-METHYLPHENYL) ETHYLPHOSPHONODITHIOATE see EOO000
METHYL SODIUM see MPJ000
METHYL STEARATE see MJW000
METHYL STIBINE see MPJ100
2-METHYL-4-STILBENAMINE see MPJ250
3-METHYL-4-STILBENAMINE see MPJ500
METHYL STREPONIGRIN see SMA500
β-METHYLSTREPTOZOTOCIN see MPJ800
α-METHYLSTYREEN (DUTCH) see MPK250
α-METHYL STYRENE see MPK250
p-METHYLSTYRENE see VQK700
METHYLSTYRENE see VQK650
α-METHYL-STYROL (GERMAN) see MPK250
METHYL trans-STYRYL KETONE see BAY275
METHYL STYRYLPHENYL KETONE see MPL000
6-METHYL-7-SULFAMIDO-THIOCHROMAN-1,1-DIOXIDE see MQQ050
5-METHYL-3-SULFANILAMIDOISOXAZOLE see SNK000
5-METHYL-2-SULFANILAMIDO-1,3,4-THIADIAZOLE see MPQ750
METHYL SULFANILYL CARBAMATE see SNQ500
METHYL SULFATE (DOT) see DUD100
METHYLSULFAZIN see ALF250
METHYL SULFIDE (DOT) see TFP000
METHYL SULFIDE compound with BORANE (1:1) see MPL250
2-((METHYLSULFINYL)ACETYL)PYRIDINE see MPL500
METHYLSULFINYL ETHYLTHIAMINE DISULFIDE see MPL600
METHYLSULFINYLMETHANE see DUD800
METHYL SULFOCYANATE see MPT000
METHYLSULFONAL see BJT750
9-(p-(METHYLSULFONAMIDO)ANILINO)-3-ACRIDINE CARBAMIC ACID METHYL ESTER see MPL750
N-(9-(p-(METHYLSULFONAMIDO)ANILINO)ACRIDIN-3-YL)ACETAMIDE METHANE-SULFONATE see MPM000
METHYLSULFONIC ACID, ETHYL ESTER see EMF500
N-(9-((4-((METHYLSULFONYL)AMINO)PHENYL)AMINO)-3-ACRIDI-NYL)ACETAMIDE METHANESULFONATE see MPM000
2-METHYLSULFONYL-10-(3-(4'-CARBAMOYLPIPERIDINO)PROPYL)-PHENOTHI-AZINE see MPM750
2-METHYLSULFONYL-10-(3-(4-CARBAMOYLPIPERIDI-NO)PROPYL)PHENOTHIAZINE see MQR000
METHYLSULFONYL CHLORAMPHENICOL see MPN000
2-METHYLSULFONYL-o-(N-METHYL-CARBAMOYL)-BUTANON-(3)-OXIM (GER-MAN) see SOB500
1-METHYLSULFONYL-3-(1-METHYL-5-NITRO-2-IMIDAZOLYL)-2-IMIDAZOLIDI-NONE see SAY950
1,4-(METHYLSULFONYLOXYETHYLAMINO)-1,4-DIDEOXY-ERYTHRIOLDIMETHYL-SULFONATE see LJD500
((METHYLSULFONYL)OXY)TRIBUTYLSTANNANE see TIF000
((METHYLSULFONYL)OXY)TRIPHENYLSTANNANE see TMW500
1-(3-(2-(METHYLSULFONYL)PHENOTHIAZIN-10-YL)PROPYL)ISONIPECOTAMIDE see MQR000
1-(3-(2-(METHYLSULFONYL)-10H-PHENOTHIAZIN-10-YL)PROPYL)-4-PIPERIDINE CARBOXAMIDE see MQR000
1-(3-(2-(METHYLSULFONYL)PHENOTHIAZIN-10-YL)PROPYL)-4-PIPERIDINE CAR-BOXAMIDE see MQR000
2-(4-(METHYLSULFONYL)PHENYL)-IMIDAZO(1,2-a)PYRIDINE see ZUA200

METHYL SULFOXIDE see DUD800
5-METHYL-3-SULPHANIL-AMIDOISOXAZOLE see SNK000
METHYL SULPHIDE see TFP000
3-METHYLSULPHINYLPROPYLISOTHIOCYANATE see MPN100
METHYLSULPHONAL see BJT750
METHYLSYSTOX see DAO800
METHYL SYSTOX see MIW100
METHYL TARTRONIC ACID see MPN250
o-(METHYLTELLURO)BENZOIC ACID see MPN275
METHYLTERT-BUTYL KETONE see DQU000
17-METHYLTESTOSTERON see MPN500
17-α-METHYLTESTOSTERONE see MPN500
17-METHYLTESTOSTERONE see MPN500
METHYLTESTOSTERONE see MPN500
Δ'-17-α-METHYLTESTOSTERONE see DAL300
Δ'-17-METHYLTESTOSTERONE see DAL300
2-METHYL-3,4,5,6-TETRABROMOPHENOL see TBJ500
2-METHYL-3,3,4,5-TETRAFLUORO-2-BUTANOL see MPN750
10-METHYL-1,2-TETRAHYDRO-1,2:5,6-BENZACRIDINE see MPO400
10-METHYL-1,2-TETRAHYDRO-1,2:7,8-BENZACRIDINE see MPO390
2-METHYL-1,2,3,6-TETRAHYDROBENZALDEHYDE see MPO000
4-METHYL-1',2',3',4'-TETRAHYDRO-1,2-BENZANTHRACENE see MPO250
6-METHYL-1,2,3,4-TETRAHYDROBENZ(a)ANTHRACENE see MPO250
14-METHYL-8,9,10,11-TETRAHYDRODIBENZ(a,h)ACRIDINE see MPO400
7-METHYL-1,2,3,4-TETRAHYDRODIBENZ(c,h)ACRIDINE see MPO390
6-METHYL-2,3,5,7-TETRAHYDRO-6H-p-DITHIINO-(2,3-C)PYRROLE-5,7-DIONE see MJJ250
2-METHYLTETRAHYDROFURAN (CZECH) see MPO500
METHYLTETRAHYDROFURAN see MPO500
METHYL TETRAHYDRO-5-HYDROXY-4-METHYL-3-FURYL KETONE see BMK290
N-METHYL-1,2,3,4-TETRAHYDROISOQUINOLINE HYDROCHLORIDE see MPO750
METHYL 1,4,5,6-TETRAHYDRO-2-METHYLCYCLOPENTA(b)PYRROL-3-YL KETONE see MPO800
METHYL-1,2,5,6-TETRAHYDRO-1-METHYLNICOTINATE see AQT750
METHYL-1,2,5,6-TETRAHYDRO-1-METHYLNICOTINATE, HYDROBROMIDE see AQU000
2-METHYL-2-(4-(1,2,3,4-TETRAHYDRO-1-NAPHTHALE-NYL)PHENOXY)PROPANOIC ACID see MCB500
2-METHYL-2-(4-(1,2,3,4-TETRAHYDRO-1-NAPHTHYL)PHENOXY)PROPANOIC ACID see MCB500
2-METHYL-2-(p-(1,2,3,4-TETRAHYDRO-1-NAPHTHYL)PHENOXY)PROPIONIC ACID see MCB500
α-METHYL-α-(p-1,2,3,4-TETRAHYDRONAPHTH-1-YLPHENOXY)PROPIONIC ACID see MCB500
N-METHYL-Δ-TETRAHYDRONICOTINIC ACID METHYL ESTER see AQT750
METHYLTETRAHYDROPHTHALIC ANHYDRIDE see MPP000
N-METHYLTETRAHYDROPYRIDINE-β-CARBOXYLIC ACID METHYL ESTER see AQT750
N-METHYLTETRAHYDROPYRROLE see MPB250
2-METHYL-3-(3,7,11,15-TETRAMETHYL-2,6,10,14-HEXADECATETRAENYL)-1,4-NAPHTHOQUINONE see VTA650
2-METHYL-3-(3,7,11,15-TETRAMETHYL-2-HEXADECENYL)-1,4-NAPHTHALENE-DIONE see VTA000
2-METHYL-3-trans-TETRAMETHYL-1,4-NAPHTHQUINONE see VTA650
N-METHYL-N,2,4,6-TETRANITROANILINE see TEG250
N-METHYL-N,2,4,6-TETRANITROBENZENAMINE see TEG250
1-METHYL-1H-TETRAZOLE-5-THIOL see MPQ250
α-METHYL TETRONIC ACID see MPQ250
METHYL 5-(2-THENOYL)-2-BENZIMIDAZOLECARBAMATE see OJD100
METHYLTHEOBROMIDE see CAK500
1-METHYLTHEOBROMINE see CAK500
7-METHYLTHEOPHYLLINE see CAK500
N'-(5-METHYL-1,3,4-THIADIAZOL-2-YL)-SULFANILAMIDE see MPQ750
N-METHYL-3-THIA-2-METHYL-VALERAMID DER O,O-DIMETHYLTHIOLPHOS-PHORSAEURE (GERMAN) see MJG500
METHYL 3-THIANAPHTHENYL KETONE see MPQ600
2-METHYLTHIAZOLIDINE see MPR000
METHYL-2-THIAZOLIDINE (FRENCH) see MPR000
N-(3-METHYL-2-THIAZOLIDINYLIDENE)NICOTINAMIDE see MPR250
METHYL-2 Δ-2 THIAZOLINE see DLV900
α-METHYL-4-(2-THIENYLCARBONYL)BENZENEACETIC ACID see TEN750
METHYL (5-(2-THIENYLCARBONYL)-1H-BENZIMIDAZOL-2-YL)CARBAMATE see OJD100
METHYLTHIENYLCETONE see MPR300
METHYLTHIOACETALDEHYDE-o-(CARBAMOYL)OXIME see MPS250
1-(METHYLTHIO)ACETALDEHYDE OXIME see MPS270
2-METHYLTHIO-ACETALDEHYD-O-(METHYLCARBAMOYL)-OXIM (GERMAN) see MDU600
1-(METHYLTHIO)ACETALDOXIME see MPS270
METHYL(THIOACETAMIDO) MERCURY see MPS300
METHYL THIOACETOHYDROXAMATE see MPS270
l-γ-METHYLTHIO-α-AMINOBUTYRIC ACID see MDT750
4-(METHYLTHIO)ANILINE see AMS675

m-(METHYLTHIO)ANILINE see ALX100
3-(METHYLTHIO)BENZENAMINE see ALX100
4-(METHYLTHIO)BENZENAMINE see AMS675
(METHYLTHIO)BENZENE see TFC250
2-METHYLTHIOBENZIMIDAZOLE see MPS500
2-METHYLTHIO-4,6-BIS(ISOPROPYLAMINO)-s-TRIAZINE see BKL250
6-METHYLTHIOCHROMAN-7-SULFONAMIDE 1,1-DIOXIDE see MQQ050
4-(METHYLTHIO)-m-CRESOL see MLX750
4-(METHYLTHIO)-m-CRESOL-O-ESTER with O-ETHYL METHYLPHOSPHORAMI-DOTHIOATE see ENI500
METHYL THIOCYANATE see MPT000
4-METHYLTHIO-3,5-DIMETHYLPHENYL METHYLCARBAMATE see DST000
2-(METHYLTHIO)-ETHANETHIOL-O,O-DIMETHYL PHOSPHOROTHIOATE see MIW250
2-(METHYLTHIO)-ETHANETHIOL-S-ESTER with O,O-DIMETHYL PHOSPHO-ROTHIOATE see MIW250
2-(METHYLTHIO)ETHANOL ACRYLATE see MPT250
2-METHYLTHIOETHYL ACRYLATE see MPT250
2-METHYLTHIO-4-ETHYLAMINO-6-tert-BUTYLAMINO-s-TRIAZINE see BQC750
2-(2-METHYLTHIOETHYLAMINO)ETHYLGUANIDINE SULFATE see MPT300
2-METHYLTHIO-4-ETHYLAMINO-6-ISOPROPYLAMINO-s-TRIAZINE see MPT500
7-(2-METHYLTHIOETHYL)-THEOPHYLLINE see MDN100
7-(β-METHYLTHIOETHYL)THEOPHYLLINE see MDN100
METHYLTHIOGLYCOLATE see MLE750
6-METHYLTHIOINOSINE see MPU000
METHYLTHIOINOSINE see MPU000
2-METHYLTHIO-4-ISOPROPYLAMINO-6-METHYLAMINO-s-TRIAZINE see INR000
METHYLTHIOKYANAT see MPT000
METHYLTHIOMETHANE see TFP000
3-(METHYLTHIO)-o-((METHYLAMINO)CARBONYL)OXIME-2-BUTANONE see MPU250
2-(METHYLTHIO)-4-(METHYLAMINO)-6-(ISOPROPYLAMINO)-s-TRIAZINE see INR000
2-METHYLTHIO-o-(N-METHYLCARBAMOYL)-BUTANONOXIM-3 (GERMAN) see MPU250
8-β-((METHYLTHIO)METHYL)-6-PROPYLERGOLINE METHANESULFONATE see MPU500
METHYLTHIONINE CHLORIDE see BJI250
METHYLTHIONIUM CHLORIDE see BJI250
METHYL THIOPHANATE see PEX500
METHYLTHIOPHANATE-MANEB mixture see MAP300
2-METHYLTHIOPHENE see MPV000
3-METHYLTHIOPHENE see MPV250
2-METHYLTHIOPHENOL see TGP000
3-METHYLTHIOPHENOL see TGO800
4-(METHYLTHIO)PHENOL see MPV300
4-METHYLTHIOPHENOL see TGP250
m-METHYLTHIOPHENOL see TGO800
o-METHYLTHIOPHENOL see TGP000
p-(METHYLTHIO)PHENOL see MPV300
p-METHYLTHIOPHENOL see TGP250
4-METHYLTHIOPHENYLDIMETHYL PHOSPHATE see PHD250
METHYLTHIOPHOS see MNH000
3-(METHYLTHIO)PROPANAL see MPV400
3-(METHYLTHIO)PROPIONALDEHYDE see MPV400, TET900
β-(METHYLTHIO)PROPIONALDEHYDE see MPV400, TET900
2-METHYLTHIO-PROPIONALDEHYD-O-(METHYLCARBAMOYL)-OXIM (GERMAN) see MDU600
2-METHYL-2-THIOPSEUDOUREA SULFATE see MPV790
S-METHYLTHIOPSEUDOUREA SULFATE see MPV790
2-METHYL-2-THIOPSEUDOUREA SULFATE (1:1) SODIUM SALT see MPV800
6-(METHYLTHIO)PURINE RIBONUCLEOSIDE see MPU000
6-METHYLTHIOPURINE RIBOSIDE see MPU000
6-METHYL-2-THIO-2,4-(1H3H)PYRIMIDINEDIONE see MPW500
4-METHYL-2-THIOURACIL see MPW500
6-METHYL-2-THIOURACIL see MPW500
6-METHYLTHIOURACIL see MPW500
METHYLTHIOURACIL see MPW500
1-METHYLTHIOUREA see MPW600
METHYL THIOUREA see MPW600
S-METHYLTHIOURONIUM SULFATE see MPV790
1-METHYL-3-(THIOXANTHEN-9-YLMETHYL)-1-PIPERIDINE HYDROCHLORIDE see THL500
METHYLTHIOXOARSINE see MGQ750
4-(METHYLTHIO)-3,5-XYLENOL-O-ESTER with O,O-DIETHYL PHOSPHOROTH-IOATE see DJR800
4-(METHYLTHIO)-3,5-XYLENOL METHYLCARBAMATE see DST000
4-(METHYLTHIO)-3,5-XYLYL METHYLCARBAMATE see DST000
METHYL THIRAM see TFS350
METHYL THIURAMDISULFIDE see TFS350
O-METHYLTHYMOL see MPW650
METHYL THYMOL ETHER see MPW650
METHYL THYMYL ETHER see MPW650
METHYL TIGLATE see MPW700
METHYLTIN TRICHLORIDE see MQC750

METHYL-4-TOLUATE see MPX850
METHYL-α-TOLUATE see MHA500
METHYL-p-TOLUATE see MPX850
o-METHYLTOLUENE see XHJ000
p-METHYLTOLUENE see XHS000
METHYL TOLUENE see XGS000
METHYL TOLUENE-4-SULFONATE see MLL250
METHYL-p-TOLUENESULFONATE see MLL250
α-METHYL-α-TOLUIC ALDEHYDE see COF000
6-METHYL-m-TOLUIDINE see XNA000
4-METHYL-o-TOLUIDINE see XMS000
5-METHYL-o-TOLUIDINE see XNA000
2-METHYL-p-TOLUIDINE see XMS000
6-METHYL-m-TOLUIDINE HYDROCHLORIDE see XOS000
4-METHYL-o-TOLUIDINE HYDROCHLORIDE see XOJ000
5-METHYL-o-TOLUIDINE HYDROCHLORIDE see XOS000
2-METHYL-p-TOLUIDINE HYDROCHLORIDE see XOJ000
1-METHYL-5-(p-TOLUOYL)-2-PYRROLEACETIC ACID see SKJ340
1-METHYL-5-p-TOLUOYL-PYRROLE-2-ACETIC ACID see TGJ850
2-METHYL-3-o-TOLYL-6-AMINO-CHINAZOLINON-4 (GERMAN) see AKM125
N-METHYL-p-(m-TOLYLAZO)ANILINE see MPY000
N-METHYL-p-(o-TOLYLAZO)ANILINE see MPY250
N-METHYL-p-(p-TOLYLAZO)ANILINE see MPY500
2'-METHYL-4'-(o-TOLYLAZO)OXANILIC ACID see OLG000
N-METHYL-2-(α-(2-TOLYLBENZYL)OXY)ETHYLAMINE HYDROCHLORIDE see
    TGJ250
8-METHYL-3-(α-(o-TOLYL)BENZYLOXY)TROPANIUM IODIDE see MPZ000
1-METHYL-4-α-o-TOLYLBENZYL)PIPERAZINE DIHYDROCHLORIDE see MIQ725
N-METHYL-N-(m-TOLYL)CARBAMOTHIOIC ACID-(1,2,3,4-TETRAHYDRO-1,4-
    METHANONAPHTHALEN-6-YL) ESTER see MQA000
METHYL 3-(m-TOLYLCARBAMOYLOXY)PHENYLCARBAMATE see MEG250
METHYL-p-TOLYLCARBINOL see TGZ000
2-METHYL-3-o-TOLYL-4(3H)-CHINAZOLINON (GERMAN) see QAK000
2-METHYL-3-o-TOLYL-4(3H)-CHINAZOLONE see QAK000
2-METHYL-3-TOLYLCHINAZOLON-4 HYDROCHLORIDE (GERMAN) see
    MDT250
2-METHYL-3-(o-TOLYL)-3,4-DIHYDRO-4-QUINAZOLINONE see QAK000
2-METHYL-3-(o-TOLYL)-3,4-DIHYDRO-4-(QUINAZOLINONE) see QAK000
METHYL-p-TOLYL ETHER see MGP000
METHYL-p-TOLYL KETONE see MFW250
2-METHYL-3-TOLYL-4-OXYBENZDIAZINE see QAK000
3-METHYL-1-p-TOLYL-PYRAZOLIN-5-ONE see THA300
2-METHYL-3-o-TOLYL-4(3H)-QUINAZOLINONE see QAK000
2-METHYL-3-o-TOLYL-4(3H)-QUINAZOLINONE HYDROCHLORIDE see MDT250
2-METHYL-3-(2-TOLYL)QUINAZOL-4-ONE see QAK000
2-METHYL-3-o-TOLYL-4-QUINAZOLONE see QAK000
2-METHYL-3-(o-TOLYL)-4-QUINAZOLONE HYDROCHLORIDE see MDT250
2-METHYL-3-o-TOLYL-6-SULFAMYL-7-CHLORO-1,2,3,4-TETRAHYDRO-4-QUINA-
    ZOLINONE see ZAK300
3-METHYL-5-(p-TOLYLSULFONYL)-1,2,4-THIADIAZOLE see MQB100
3-METHYL-1-(p-TOLYL)-TRIAZENE see MQB250
METHYL-p-TOSYLATE see MLL250
METHYL TOSYLATE see MLL250
METHYLTRETAMINE see TNK000
METHYLTRIACETOXYSILANE see MQB500
5-(3-METHYL-1-TRIAZENO)IMIDAZOLE-4-CARBOXAMIDE see MQB750
5-METHYL-1,2,4-TRIAZOLE(3,4-b)BENZOTHIAZOLE see MQC000
5-METHYL-s-TRIAZOLO(3,4-b)BENZOTHIAZOLE see MQC000
2'-METHYL-1,2:4,5:8,9-TRIBENZOPYRENE see MJA750
METHYLTRICAPRYLYLAMMONIUM CHLORIDE see MQH000
METHYL TRICHLORIDE see CHJ500
METHYL TRICHLOROACETATE see MQC150
N-METHYL-2,4,5-TRICHLOROBENZENESULFONAMIDE see MQC250
METHYLTRICHLOROMETHANE see MIH275
METHYLTRICHLOROSILANE see MQC500
METHYLTRICHLOROSTANNANE see MQC750
METHYLTRICHLOROTIN see MQC750
METHYL-TRICHLORSILAN (CZECH) see MQC500
α-METHYLTRICYCLO(3.3.1.1³·⁷)DECANE-1-METHANAMINE HYDROCHLORIDE
    see AJU625
1-METHYLTRICYCLOQUINAZOLINE see MQD000
3-METHYLTRICYCLOQUINAZOLINE see MQD250
4-METHYLTRICYCLOQUINAZOLINE see MQD500
METHYLTRIETHOXYSILANE see MQD750
METHYL TRIFLUORIDE see CBY750
5-METHYL-2-TRIFLUOROMETHYLOXAZOLIDINE see MQE000
(α-METHYL-m-TRIFLUOROMETHYLPHENETHYLAMINOMETHYL) PIPERIDINO
    KETONE see MQE100
8-METHYL-3-(α-(α,α,α-TRIFLUORO-o-TOLYL)BENZYLOXY)TROPANIUM IODIDE
    see MQF000
METHYL-3,3,3-TRIFLUORO-2-(TRIFLUOROMETHYL)PROPIONATE see MKK750
METHYL TRIFLUOROVINYL ETHER see MQF200
6-METHYL-1,3,8-TRIHYDROXYANTHRAQUINONE see MQF250
METHYL 3,4,5-TRIHYDROXYBENZOATE see MKI100
1-METHYL-3,7,9-TRIHYDROXY-6H-DIBENZO(b,d)PYRAN-6-ONE see AGW476

METHYL 3,4,5-TRIMETHOXYBENZOATE see MQF300
METHYL-18-o-(3,4,5-TRIMETHOXYCINNAMOYL RESERPATE see TLN500
α-METHYL-3,4,5-TRIMETHOXYPHENETHYLAMINE HYDROCHLORIDE see
    TKX500
METHYLTRIMETHOXYSILANE see MQF500
(2-METHYL-2-(2-(TRIMETHYLAMMONIO)ETHOXY)ETHYL)DIETHYLMETHYL AM-
    MONIUM DIIODIDE see MQF750
METHYLTRIMETHYLENE GLYCOL see BOS500
4-METHYL TRIMETHYLENE SULFITE see MQG500
METHYL 2,4,5-TRIMETHYLPYRROL-3-YL KETONE see ADE050
N-METHYL-N-TRIMETHYLSILYLTRIFLUOROACETAMIDE see MQG750
1-METHYL-2,4,5-TRINITROBENZENE see TMN400
3-METHYL-2,4,6-TRINITROPHENOL see TML500
METHYLTRIOCTYLAMMONIUM CHLORIDE see MQH000
4-METHYL-2,6,7-TRIOXA-1-ARSABICYCLO(2.2.2)OCTANE see MQH100
1-METHYL-2,8,9-TRIOXO-5-AZA-1-SILABICYCLO(3.3.3)UNDECANE see MPI650
3-METHYL-1,2,4-TRIOXOLANE see PMP250
1-METHYL-4-(3,3,3-TRIS(p-CHLOROPHENYL)PROPIONYLPIPERAZINE) MONO-
    HYDROCHLORIDE see MQH250
METHYLTRIS(2-ETHYLHEXYLOXYCARBONYLMETHYLTHIO)STANNANE see
    MQH500
METHYL TRITHION see MQH750
8-METHYLTROPINIUM BROMIDE 2-PROPYLVALERATE see LJS000
4-METHYL-TROPOLONE see MQI000
α-METHYLTRYPTAMINE see AME500
l-5-METHYLTRYPTOPHAN see MQI250
METHYL TUADS see TFS350
α-METHYL-m-TYRAMINE see AMF500
4-METHYLUMBELLIFERON (CZECH) see MKP500
4-METHYLUMBELLIFERONE see MKP500
β-METHYLUMBELLIFERONE see MKP500
METHYLUMBELLIFERONE see MEK300
4-METHYLUMBELLIFERONE-O,O-DIETHYL THIOPHOSPHATE see PKT000
4-METHYLUMBELLIFERONE 4-GUANIDINOBENZOATE see GLC100
METHYL-4-UMBELLIFERONE SODIUM see HMB000
2-METHYLUNDECANAL see MQI550
METHYL 10-UNDECENOATE see ULS400
METHYL 9-UNDECENOATE see ULS400
METHYL UNDECYLENATE see ULS400
4-METHYLURACIL see MPW500, MQI750
5-METHYLURACIL see TFX800
6-METHYLURACIL see MQI750
1-METHYLUREA see MQJ000
N-METHYLUREA see MQJ000
METHYLUREA see MQJ000
METHYL UREA and SODIUM NITRITE see MQJ250
N-METHYL URETHAN see EMQ500
METHYLURETHAN see MHZ000
N-METHYLURETHANE see EMQ500
METHYLURETHANE see EMQ500, MHZ000
N-METHYLURETHANE of HYDROCHLORIDE of 2-DIMETHYLAMINO-p-CRESOL
    see DPL900
N-METHYLURETHANE of HYDROCHLORIDE of 3-DIMETHYLAMINOPHENOL
    see MID000
N-METHYLURETHANE HYDROCHLORIDE of 6-DIMETHYLAMINO-o-4-XYLENOL
    see MIA775
4-METHYL VALERALDEHYDE see MQJ500
METHYL n-VALERATE see VAQ100
METHYL VALERATE see VAQ100
METHYL VALERIANATE see VAQ100
2-METHYLVALERIC ACID see MQJ750
α-METHYLVALERIC ACID see MQJ750
2-METHYL-1,5-VALERODINITRILE see MQK000
4-METHYLVALERONITRILE see MNJ000
4-o-METHYLVANILLIN see VHK000
METHYLVANILLIN see VHK000
trans-8-METHYL-N-VANILLYL-6-NONEAMIDE see CBF750
METHYLVINYL ACETATE see MQK750
METHYL VINYL ADIPATE see MQL000
METHYL VINYL CARBINOL see MQL250
METHYL-VINYL-CETONE (FRENCH) see BOY500
METHYL VINYL ETHER see MQL750
METHYLVINYLKETON (GERMAN) see BOY500
METHYL VINYL KETONE see BOY500, MQM100
METHYLVINYLNITROSAMIN (GERMAN) see NKY000
METHYLVINYLNITROSAMINE see NKY000
2-METHYL-5-VINYLPYRIDINE see MQM500
METHYL VINYL SULFONE see MQM750
2-METHYL-5-VINYL TETRAZOLE see MQM775
4-METHYL-5-VINYL THIAZOLE see MQM800
METHYL VIOLET see MQN025
METHYL VIOLET 10B see AOR500
METHYL VIOLET 2B see MQN025
METHYL VIOLET 6B see MQN000
METHYL VIOLET BB see MQN025

METHYL VIOLET 10BD see AOR500
METHYL VIOLET 10BK see AOR500
METHYL VIOLET 10BN see AOR500
METHYL VIOLET 5BNO see AOR500
METHYL VIOLET 10BNS see AOR500
METHYL VIOLET 10BO see AOR500
METHYL VIOLET 5BO see AOR500
METHYL VIOLET CARBINOL see MQN250
METHYL VIOLET CARBINOL BASE see MQN250
METHYL VIOLET FN see MQN025
METHYL VIOLET N see MQN025
METHYL-VIOLETT (GERMAN) see MQN025
METHYLVIOLETT see AOR500
METHYL VIOLOGEN (2+) see PAI990
METHYLVIOLOGEN see PAJ000
METHYL VIOLOGEN DICHLORIDE see PAJ000
METHYL VIOLOGEN (REDUCED) see PAJ000
3-METHYLXANTHINE see MQN500
1-METHYL-N-(2,6-XYLYL)ISONIPECOTAMIDE HYDROCHLORIDE see MQO250
1-METHYL-2-(2,6-XYLYLOXY)-ETHYLAMINE HYDROCHLORIDE see MQR775
N-METHYL-N-((1-(3,5-XYLYLOXY)-2-PROPYL)CARBAMIC ACID-2-DIETHYLAMI-
   NO)ETHYL ESTER, HYDROCHLORIDE see MQO500
N-METHYL-N-(1-(3,5-XYLYLOXY)-2-PROPYL)CARBAMIC ACID-2-(2-METHYLPI-
   PERIDINO)ETHYL ESTER HYDROCHLORIDE see MQO750
N-METHYL-N-(1-(3,5-XYLYLOXY)-2-PROPYL)CARBAMIC ACID-2-(PYRROLIDI-
   NYL)ETHYL ESTER, HCl see MQP000
METHYL YELLOW see DOT300
METHYL ZIMATE see BJK500
METHYLZINC IODIDE see MQP250
METHYL ZINEB see BJK500, ZMA000
METHYL ZIRAM see BJK500
METHYPHENYLMETHANOL see PDE000
METHYPROLON see DNW400
METHYPRYLON see DNW400
METHYRIMOL see BRD000
METHYSERGID see MLD250
METHYSERGIDE see MLD250
METHYSERGIDE DIMALEATE see MQP500
METIAPINE see MQQ000
METIAZIC ACID see MNQ500
METIAZINIC ACID see MNQ500
METICORTELANE see PMA100
METICORTELONE see PMA000
METICORTELONE ACETATE see SOV100
METICORTELONE SOLUBLE see PMA100
METICRAN see MQQ050
METICRANE see MQQ050
METI-DERM see PMA000
METIDIONE see AOO475
METIFEX see EDW500
METIFONATE see TIQ250
METIGUANIDE see DQR800
METILACRILATO (ITALIAN) see MGA500
METILAMIL ALCOHOL (ITALIAN) see MKW600
METILAMINE (ITALIAN) see MGC250
2-METIL-6-AMINO-EPTANO (ITALIAN) see ILM000
1-METILBIGUANIDE CLORIDRATO (ITALIAN) see MHP400
3-METIL-BUTANOLO (ITALIAN) see IHP000
METIL CELLOSOLVE (ITALIAN) see EJH500
2-METILCICLOESANONE (ITALIAN) see MIR500
METILCLOROFORMIATO (ITALIAN) see MIG000
METILCLORPINDOL see CMX850
N-METIL-N-(β-DICICLOESILAMINOETIL)PIPERIDINIO BROMURO (ITALIAN) see
   MJC775
METILDIGOXIN see MJD300, MJD500
METILDIGOXINA (SPANISH) see MJD300
2'-METIL-3'-DIMETILAMINO-PROPIL-5-IMINODIBENZILE (ITALIAN) see DLH200
METILDIOLO see AOO475
N-METIL-DITIOCARBAMMATO di SODIO (ITALIAN) see VFU000
METILE (ACETATO di) (ITALIAN) see MFW100
METILENBIOTIC see MDO250
O,O-METILEN-BIS(4-CLORO-FENOLO) see MJM500
3,3'-METILEN-BIS(4-IDROSSI-CUMARINA) (ITALIAN) see BJZ000
4,4-METILENE-BIS-o-CLOROANILINA (ITALIAN) see MJM200
(6-(1-METIL-EPITL)-2,4-DINITRO-FENIL)-CROTONATO (ITALIAN) see AQT500
METILESTER del ACIDO BISDEHIDROISYNOLICO (SPANISH) see BIT000
7-METILETER del ACIDO BISDEHIDRODOISYNOLICO see BIT030
METILETILCHETONE (ITALIAN) see MKA400
METIL (FORMIATO di) (ITALIAN) see MKG750
METILISOBUTILCHETONE (ITALIAN) see HFG500
METIL ISOCIANATO (ITALIAN) see MKX250
METILMERCAPTANO (ITALIAN) see MLE650
METILMERCAPTOFOSOKSID see DAP000
METIL METACRILATO (ITALIAN) see MLH750

α-METIL-β-(2-METILENE-4,5-DIIDROIMIDAZOLIL)BENZOTIOFANE CLORIDRATO
   (ITALIAN) see MHJ500
N-METIL-1-NAFTIL-CARBAMMATO (ITALIAN) see CBM750
N-(4-(β-(5-METILOSSAZOL-3-CARBOSSAMIDO)-ETIL)-BENZENESOLFONIL)-N'-CI-
   CLOESIL-UREA see DBE885
METILPARATION (HUNGARIAN) see MNH000
4-METILPENTAN-2-OLO (ITALIAN) see MKW600
4-METILPENTAN-2-ONE (ITALIAN) see HFG500
4-METIL-3-PENTEN-2-ONE (ITALIAN) see MDJ750
1-(N-METIL-PIPERIDIL-4')-3-FENIL-4-BENZIL-PIRAZOLONE-5 (ITALIAN) see
   BCP650
3-γ-(α-METIL-1-PIPERIDINO)PROPIL-5,5-DIFENILTIOIDANTOINA CLORIDRATO
   (ITALIAN) see DWF875
(6-(1-METIL-PROPIL)-2,4-DINITRO-FENIL)-3,3-DIMETIL-ACRILATO (ITALIAN) see
   BGB500
6-(1-METIL-PROPIL)-2,4-DINITRO-FENOLO (ITALIAN) see BRE500
α-METIL-STIROLO (ITALIAN) see MPK250
2-METIL-2-TIOMETIL-PROPIONALDEID-O-(N-METIL-CARBAMOIL)-OSSIMA (ITAL-
   IAN) see CBM500
6-METIL-TIOURACILE (ITALIAN) see MPW500
METILTRIAZOTION see ASH500
METINDOL see IDA000
METIONE see MDT740
d-METIONIEN (AUSTRALIAN) see MDT730
METIPREGNONE see MCA000
METIPRILONE see DNW400
METIRAM see MQQ250
METISAZONUM see MKW250
METIZOL see MCO500
METIZOLIN see BAV000
METIZOLINE HYDROCHLORIDE see MHJ500
METMERCAPTURON see DST000
METOBROMURON see PAM785
METOCARBAMOL see GKK000
METOCARBAMOLO see GKK000
METOCHLOPRAMIDE see AJH000
METOCLOL see AJH000
METOCLOPRAMIDE DIHYDROCHLORIDE MONOHYDRATE see MQQ300
METOCRYST see AOO475
METOFANE see DFA400
METOFENINA see GKK000
METOFOLINE see MDV000
2-METOKSY-4-ALLILOFENOL (POLISH) see EQR500
METOKSYCHLOR (POLISH) see MEI450
METOKSYETYLOWY ALKOHOL (POLISH) see EJH500
METOLACHLOR see MQQ450
METOLAZONE see ZAK300
METOLCARB see MIB750
METOLOSE MC 8000 see MIF760
METOLOSE 60SH see MIF760
METOLOSE 60SH400 see MIF760
METOLOSE SM 15 see MIF760
METOLOSE SM 100 see MIF760
METOLOSE SM 4000 see MIF760
METOLQUIZOLONE see QAK000
METOMEGA CHROME YELLOW GM see SIT850
METOMIDATE see MQQ500
METOMIDATE HYDROCHLORIDE see MQQ750
METOMIL (ITALIAN) see MDU600
METON see HEA500
METOPIMAZINE see MQR000
METOPIRON see MCJ370, MJJ000
METOPIRONE see MCJ370
METOPIRONE DITARTRATE see MJJ000
METOPRINE see MQR100
METOPROLOL see MQR144
(±)-METOPROLOL see MQR144
METOPROLOL HEMITARTRATE see MQR150
METOPROLOL TARTRATE see MQR150, SBV500
METOPRYL see MOU830
METOPYRONE see MCJ370
METOQUINE see CFU750
METOSERPATE HYDROCHLORIDE see MDW100, MQR200
(2-METOSSICARBONIL-1-METIL-VINIL)-DIMETIL-FOSFATO (ITALIAN) see
   MQR750
2-METOSSIETANOLO (ITALIAN) see EJH500
2-METOSSIETILACETATO (ITALIAN) see EJJ500
S-((5-METOSSI-4H-PIRON-2-IL)-METIL)-O,O-DIMETIL-MONOTIOFOSFATO (ITAL-
   IAN) see EAS000
METOSSIPROPANDIOLO see RLU000
METOSYN see FDD150
METOTHYRINE see MCO500
METOX see CEP000, MEI450
METOXADONE see MFD500
METOXAL see SNK000

METOXFLURAN see DFA400
METOXIDON see SNN300
METOXIFLURAN see DFA400
METOXON see CKC000
METOXURON see MQR225
METOXYDE see HJL500
4-o-METPA see MQR250
METRAMAK see DJA400
METRACTYL see MQU750
METRAMAC see DJA400
1-METRAMINOL BITARTRATE see HNC000
d-(—)-METRAMINOL BITARTRATE see HNC000
METRAMINOL BITARTRATE see HNC000
(—)-METRAMINOL (+)-BITARTRATE see HNC000
METRANIL see PBC250
METRASPRAY see BQA010
METRIBEN see TIK000
METRIBUZIN see MQR275
METRIFONATE see TIQ250
METRIPHONATE see TIQ250
METRISONE see MOR500
METRIZAMIDE see MQR300
METRIZOATE see MQR350
METRIZOIC ACID see MQR350
METRODIOL see EQJ500
METRODIOL DIACETATE see EQJ500
METROGEN RED FORMER KB SOLN see CLK225
METROGESTONE see MBZ100
METRON see MNH000
METRONE see MPN500
METRONIDAZ see MMN250
METRONIDAZOL see MMN250
METRONIDAZOLO see MMN250
METROPINE see MGR500
METRORAT see DBE200
METRO TALC 4604 see TAB750
METRO TALC 4608 see TAB750
METRO TALC 4609 see TAB750
METROXEDRINE see SPC500
METSO 20 see SJU000
METSO BEADS 2048 see SJU000
METSO BEADS, DRYMET see SJU000
METSO PENTABEAD 20 see SJU000
MET-SPAR see CAS000
METSUCCIMIDE see MLP800
METURAL see DTG700
METYCAINE see PIV750
METYLAL (POLISH) see MGA850
METYLENO-BIS-FENYLOIZOCYJANIAN see DNJ800
METYLENU CHLOREK (POLISH) see MJP450
METYLESTER KYSELINY SALICYLOVE (CZECH) see MPI000
METYLFENEMAL see ENB500
METYLOAMINA (POLISH) see MGC250
METYLOCYKLOHEKSAN (POLISH) see MIQ740
METYLOCYKLOHEKSANOL (POLISH) see MIQ745
METYLOCYKLOHEKSANON (POLISH) see MIR250
METYLOETYLOKETON (POLISH) see MKA400
METYLOHYDRAZYNA (POLISH) see MKN000
METYLOIZOBUTYLOKETON (POLISH) see HFG500
1-METYLO-2-MERKAPTOIMIDAZOLEM (POLISH) see MCO500
N-METYLO-N'-NITRO-N-NITROZOGUANIDYNY see MMP000
METYLOPARATION (POLISH) see MNH000
METYLOPROPYLOKETON (POLISH) see PBN250
7-(β-METYLOTIOETYLO)-TEOFILINA (POLISH) see MDN100
METYLOWY ALKOHOL (POLISH) see MGB150
METYLPARATION (CZECH) see MNH000
METYLU BROMEK (POLISH) see MHR200
METYLU CHLOREK (POLISH) see MIF765
METYLU JODEK (POLISH) see MKW200
METYNA see ENB500
METYPRAPONE BITARTRATE see MJJ000
METYRAPON see MCJ370
METYRAPONE see MCJ370
METYRAPONE DITARTRATE see MJJ000
MEV see MNB250
MEVASINE see VIZ400
MEVASIN HYDROCHLORIDE see MQR500
MEVINFOS (DUTCH) see MQR750
MEVINPHOS see MQR750
MEXACARBATE (DOT) see DOS000
MEXAMINE HYDROCHLORIDE see MFT000
MEXAZOLAM see MQR760
MEXENE see BJK500
MEXEPHENAMIDE see DIB600, MCH250
MEXICA FLAME LEAF see EQX000

MEXICAN BREADFRUIT see SLE890
MEXICAN FLAME TREE see EQX000
MEXICAN FLOWER PLANT see EQX000
MEXICAN HUSK TOMATO see JBS100
MEXICO WEED see CCP000
MEXIDE see RNZ000
MEXIDEX see SOW000
MEXILETINE HYDROCHLORIDE see MQR775
MEXITIL see MQR775
MEXOCINE see DAI485, MIJ500
MEXOLAMINE see CPN750
MEXPECTIN see PAO150
MEXYL see ABX500
MEYPRALGIN R/LV see SEH000
MEZARONIL see ASG250
MEZATON see NCL500, SPC500
R(—)-MEZATON see NCL500
MEZCALINE see MDI500
MEZCALINE SULFATE see MDJ000
MEZCLINE see MDI500
MEZENE see BJK500
MEZEPAN see CGA000
MEZEREIN see MDJ250
MEZEREUM see LAR500
MEZIDINE see TLG500
MEZINEB see ZMA000
MEZINIUM METHYL SULFATE see MQS100
MEZLOCILLIN see MQS200
MEZOLIN see IDA000
MEZOTOX see DFT800
MEZURON see ASG250
MF 1 see UTU500
MF 004 see MCB050
MF 009 see MCB050
MF 17 see UTU500
MF 27 see UTU500
MF-344 see EFK000
MF 565 see MAP300
MF 598 see MAP300
MF 910 see MCB050
MF see UTU500
MFA see FIC000, MKD000, TLX175
MFAS-R 100P see MCB050
M.F. FULVIUS VENOM see CNR150
MFH see FNW000
MFI see IPX000
MFI-PC see FDA100
MFM see TLX175
MFNA see MME809
M 60 (FORMALDEHYDE POLYMER) see UTU500
MFP 8 see MCB050
MFPS 1 see UTU500
MF RESIN see UTU500
M. FULVIUS FULVIUS VENOM see CNR150
MG 46 see LGK200
M.G. 8823 see ERE100
MG 8926 see CPP000
MG 18037 see BQX000
MG 18370 see DTJ400
MG 18415 see AJK000
MG 18512 see CPP750
MG 18570 see DTJ400
M.G. 18590 see TEY600
M.G.18755 see IBW100
h-MG see CGY750
M.G. 8948-2HCl see BJW000
M.G. 18001-3HCl see BJV750
M7-GIFTKOERNER see TEM000
MGK 11 see BHJ500
MGK-264 see OES000
MGK 326 see EAU500
MGK DIETHYLTOLUAMIDE see DKC800
MGK DOG AND CAT REPELLENT see UKS000
MGK REPELLENT 11 see BHJ500
MGK REPELLENT-326 see EAU500
MGK REPELLENT 874 see OGA000
MH 30 see DMC600
MH-40 see DMC600
MH-532 see PGA750
MH see DMC600
M.H. see HAP000
MH see MKK600
MHA ACID see HMR600
MHA-FA see HMR600
MH 36 BAYER see DMC600

MHOROMER see EJH000
MHP see NKU500
MHSK see MLI800
683 M HYDROCHLORIDE see OOE100
3-MI see MKV750
MI 85 see AQN750
217 MI see TLF500
N-6-MI see NKI000
MIA see IDZ000
MIADONE see MDP750
MIAK see MKW450
MIALEX see CKK250
MIANESINA see GGS000
MIANSERIN see MQS220
MIANSERINE see MQS220
MIANSERINE HYDROCHLORIDE see BMA625
MIANSERIN HYDROCHLORIDE see BMA625
MIANSERYNA see MQS220
MIARSENOL see NCJ500
MIAZOLE see IAL000
MIBC see MKW600
MIBK see HFG500
MIBOLERON see MQS225
MIBOLERONE see MQS225
MIBT see MKW250
3-MIC see MKW600
MIC see ISE000, MKW600, MKX250
MICA see MQS250
MICA SILICATE see MQS250
MICASIN see CDS500, CKL500
MICHLER'S BASE see MJN000
MICHLER'S HYDRIDE see MJN000
p,p'-MICHLER'S HYDROL see TDO750
p,p'-MICHLER'S KETONE see MQS500
MICHLER'S KETONE see MQS500
MICHLER'S METHANE see MJN000
MICHROME No. 226 see AJQ250
MICIDE see EIR000
MICOCHLORINE see CDP250
MICOFENOLICO ACIDO (SPANISH) see MRX000
MICOFUME see DSB200
MICOFUR see NGC000
MICOL see HCQ500
MICONAZOLE see MQS550
MICONAZOLE NITRATE see MQS560
MICREST see DKA600
MICROCAL 160 see CAW850
MICROCAL ET see CAW850
MICROCARB see CAT775
MICRO-CEL see CAW850
MICRO-CEL A see CAW850
MICRO-CEL B see CAW850
MICRO-CEL C see CAW850
MICRO-CEL E see CAW850
MICRO-CEL T26 see CAW850
MICRO-CEL T38 see CAW850
MICRO-CEL T41 see CAW850
MICRO-CEL T see CAW850
MICROCETINA see CDP250
MICRO-CHECK 12 see CBG000
MICRO-CHEK 11 see OFE000
MICRO-CHEK SKANE see OFE000
MICROCID see GFG100
MICROCILLIN see CBO250
MICROCOCCIN see MQS565
MICROCOCCIN P1 see MQS565
MICROCOCCIN P, 13',19'-DIDEHYDRO-19'-DEOXY-28,44-DIHYDRO-44-HY-
    DROXY- see MQS565
MICROCOP see CNK559
MICROCRYSTALLINE QUARTZ see SCK600
MICRO DDT 75 see DAD200
MICRODIOL see EDO000
MICRO DRY see AHA000
MICROEST see DKA600
MICROFLOTOX see SOD500
MICROGRIT WCA see AHE250
MICROGYNON see NNL500
MICRO-LEX GREEN 5B see BLK000
MICROLYSIN see CKN500
MICROMIC CR 16 see CAT775
MICROMICIN see MQS579
MICROMYA see CAT775
MICROMYCIN see MQS579
MICRONASE see CEH700
MICRONEX see CBT750

MICRONOMICIN SULFATE see MQS600
MICRONOMYCIN SULFATE see MQS600
MICRO-PEN see PAQ200
MICROPENIN see MNV250
MICROSETILE BLUE EB see TBG700
MICROSETILE BLUE FF see MGG250
MICROSETILE DIAZO BLACK G see DPO200
MICROSETILE ORANGE RA see AKP750
MICROSETILE PINK BN see AKE250
MICROSETILE RUBINE 2B see CMP080
MICROSETILE VIOLET B see AKP250
MICROSETILE VIOLET 3R see DBP000
MICROSETILE YELLOW GR see AAQ250
MICROSETILE YELLOW 2R see DUW500
MICROSUL see MPQ750
MICROTAN PIRAZOLO see AIF000
MICROTEX LAKE RED CR see CHP500
MICROTHENE 510 see PJS750
MICROTHENE 710 see PJS750
MICROTHENE 704 see PJS750
MICROTHENE see PJS750
MICROTHENE FN 510 see PJS750
MICROTHENE FN 500 see PJS750
MICROTHENE MN 754-18 see PJS750
MICROTIN see SFQ500
MICROTRIM see TKX000
MICROWHITE 25 see CAT775
MICROZUL see CJJ000
MICRURUS ALLENI YATESI VENOM see MQS750
MICRURUS CARINICAUDUS DUMERILII VENOM see MQT000
MICRURUS FULVIUS FULVIUS VENOM see CNR150
MICRURUS FULVIUS VENOM see MQT100
MICRURUS MIPARTIUS HERTWIGI VENOM see MQT250
MICRURUS NIGROCINCTUS VENOM see MQT500
MICTINE see AFW500
MIDARINE see HLC500
MIDAZOLAM see MQT525
MIDECAMYCIN see LEV025, MBY150
MIDECAMYCIN A₁ see MBY150
MIDELID see BGC625
MIDETON FAST RED VIOLET R see DBP000
MI 85 DI see ASA000
MIDICEL see AKO500
MIDIKEL see AKO500
MIDLON RED PG see CMM320
MIDLON RED PRS see CMM330
MIDLON YELLOW PROPYL see CMM759
MIDODRINE see MQT530
(±)-MIDODRINE HYDROCHLORIDE see MQT530
MIDONE see DBB200
MIDOXIN see HGP550
MIDRONAL see CMR100
MIEDZ (POLISH) see CNM000
MIEDZIAN 50 see CNK559
MIEDZIAN see CNK559
MIELUCIN see BOT250
MIERENZUUR (DUTCH) see FNA000
MIFUROL see CCK630
MI-GEE see DNF800
MIGHTY 150 see NAJ500
MIGLYOL 812 NEUTRAL OIL see CBF710
MIGLYOL 810 NEUTRAL OIL see CBF710
MIGRISTENE see FMU039
MIGUGEN see EFJ600
MIH see PME250
MIH HYDROCHLORIDE see PME500
MIIKE 20 see CBT750
MIK see HFG500
MIKACION BRILLIANT BLUE RS see CMS228
MIKACION BRILLIANT RED 5BS see PMF540
MIKAMETAN see IDA000
MIKAMYCIN see VRF000
MIKAMYCIN B see VRA700
MIKAMYCIN IA see VRA700
MIKEDIMIDE see MKA250
MIKELAN see MQT550
MIKEPHOR TB see CMP200
MIKETAZOL DEVELOPER ONS see HMX520
MIKETHREN BRILLIANT ORANGE GR see CMU820
MIKETHRENE BLUE BC see DFN425
MIKETHRENE BLUE BCS see DFN425
MIKETHRENE BLUE RSN see IBV050
MIKETHRENE BRILLIANT BLUE R see IBV050
MIKETHRENE BROWN BR see CMU770
MIKETHRENE BROWN GR see CMU770

MIKETHRENE GOLD YELLOW see DCZ000
MIKETHRENE GREY K see IBJ000
MIKETHRENE GREY M see CMU475
MIKETHRENE GREY MG see CMU475
MIKETHRENE MARINE BLUE G see CMU500
MIKETHRENE OLIVE R see DUP100
MIKETHRENE ORANGE GR see CMU820
MIKETHRENE ORANGE R see CMU815
MIKETHRENE RED BROWN 5RF see CMU800
MIKETHREN NAVY BLUE FRA see VGP100
MIKETON BRILLIANT BLUE B see MGG250
MIKETON FAST BLUE see TBG700
MIKETON FAST PINK FF 3B see DBX000
MIKETON FAST PINK RL see AKO350
MIKETON FAST TURQUOISE BLUE G see DMM400
MIKETON FAST VIOLET B see DBY700
MIKETON FAST YELLOW G see AAQ250
MIKETON POLYESTER BLUE FBL see CMP070
MIKETON POLYESTER PINK RL see AKO350
MIKETON POLYESTER RED FB see AKI750
MIKETON POLYESTER YELLOW 5R see CMP090
MIKETORIN see EAI000
MIKIPALAOA (HAWAII) see CNG825
MIKROKALCIT see CAD800
MIKROLOUR see PJS750
MIKRONAL S 40 see MCB050
MIL see MQU000
MILAXEN see HEG000
MILBAM see BJK500
MILBAN see BJK500
MILBAN F see TFD500
MIL-B-4394-B see CES650
MILBEDOCE see VSZ000
MILBEMYCIN D see MQT600
MILBEX see CKL500
MILBEX mixed with PHOSALON (2:1) see MQT750
MILBEX mixed with PHOSALONE (2:1) see MQT750
MILBOL 49 see BBQ500
MILBOL see BIO750
MILCHSAEURE (GERMAN) see LAG000
MIL-COL see MLC250
MILCURB see BRD000, BRI750
MILCURB SUPER see BRI750
MILDIOMYCIN see MQU000
MILDMEN see LFK000, MDQ250
MILD MERCURY CHLORIDE see MCW000
MIL-DU-RID see BGJ750
MILEPSIN see DBB200
MILESTROL see DKA600
MILEZIN see MCI500
MILFARON see CDP750
MILGO see BRI750
MILGO E see BRI750
MILID see BGC625
MILIDE see BGC625
MILK ACID see LAG000
MILK OF LIME see CAT225
MILK OF MAGNESIA see MAG750
MILK SUGAR see LAR000
MILK WHITE see LDY000
MIL L 7808 see MQU250
MIL L 17535 see MQU500
MILLER P.C. WEEDKILLER see PLC250
MILLER'S FUMIGRAIN see ADX500
MILLICORTEN see SOW000
MILLING BRILLIANT SCARLET GN see CMM320
MILLING FAST RED B see CMM330
MILLING FAST RED G see CMM320
MILLING FAST RED GL see CMM320
MILLING FAST RED PG see CMM320
MILLING FAST RED R see NAO600
MILLING FAST RED RS see NAO600
MILLING RED A see CMM325
MILLING RED B see CMM330
MILLING RED BB see CMM330
MILLING RED J see CMM320
MILLING RED PRX see NAO600
MILLING RED SWB see CMM330
MILLING RED SWG see CMM320
MILLING SCARLET DH see CMM325
MILLING SCARLET 2G see CMM325
MILLING SCARLET G see CMM320
MILLING SCARLET R see CMM325
MILLING YELLOW 3G see CMM759
MILLING YELLOW 3J see CMM759

MILLING YELLOW RX see CMM759
MILLIONATE 300 see PKB100
MILLIONATE MR 100 see PKB100
MILLIONATE MR 200 see PKB100
MILLIONATE MR 340 see PKB100
MILLIONATE MR 300 see PKB100
MILLIONATE MR 400 see PKB100
MILLIONATE MR 500 see PKB100
MILLIONATE MR see PKB100
MILLIPHYLLINE see CNR125
MILLON'S BASE ANHYDRIDE see DNW200
MILLOPHYLLINE see CNR125
MILMER see BLC250
MILOCEP see MQQ450
MILOGARD see PMN850
MILONTIN see MNZ000
MILORI BLUE see IGY000
MILPREM see ECU750, MQU750
MILPREX see DXX400
MILSAR see CKL500
MILSTEM see BRI750
MILSTEM SEED DRESSING see BRI750
MILTANN see MQU750
MILTON see SHU500
MILTOWN see MQU750
MILTOX see EIR000, ZJS300
MILTOX SPECIAL see EIR000
MIMEDRAN see PIK625
MIMOSA ABSOLUTE see MQV000
MIMOSA TANNIN see MQV250
MIMOSA Z see CMP050
MINACIDE see CQI500
MINAPHIL see TEP500
MINCARD see AFW500
MINERAL DUSTS see MQV500
MINERAL FIRE RED 5DDS see MRC000
MINERAL FIRE RED 5GS see MRC000
MINERAL GREEN see COF500
MINERAL NAPHTHA see BBL250
MINERAL OIL see MQV750
MINERAL OIL, PETROLEUM CONDENSATES, VACUUM TOWER see MQV755
MINERAL OIL, PETROLEUM DISTILLATES, ACID-TREATED HEAVY PARAFFIN-
    IC (severe solvent-refining and/or hydrotreatment) see MQV765
MINERAL OIL, PETROLEUM DISTILLATES, ACID-TREATED HEAVY NAPH-
    THENIC (mild or no solvent-refining or hydrotreatment) see MQV760
MINERAL OIL, PETROLEUM DISTILLATES, ACID-TREATED LIGHT NAPH-
    THENIC (mild or no solvent-
MINERAL OIL, PETROLEUM DISTILLATES, ACID-TREATED LIGHT PARAFFIN-
    IC (mild or no solvent-refining or hydrotreatment) see MQV775
MINERAL OIL, PETROLEUM DISTILLATES CATALYTIC DEWAXED LIGHT PA-
    RAFFINIC (mild or no solvent-refining or hydrotreatment) see MQV870
MINERAL OIL, PETROLEUM DISTILLATES CATALYTIC DEWAXED HEAVY
    NAPHTENIC (mild or no solvent-refining or hydrotreatment) see MQV865
MINERAL OIL, PETROLEUM DISTILLATES CATALYTIC DEWAXED LIGHT
    NAPHTHENIC (mild or no solvent-refining or hydrotreatment) see MQV867
MINERAL OIL, PETROLEUM DISTILLATES CATALYTIC DEWAXED HEAVY
    PARAFFINIC (mild or no solvent-refining or hydrotreatment) see MQV868
MINERAL OIL, PETROLEUM DISTILLATES, HEAVY NAPHTHENIC see
    MQV780
MINERAL OIL, PETROLEUM DISTILLATES, HEAVY PARAFFINIC see MQV785
MINERAL OIL, PETROLEUM DISTILLATES, HYDROTREATED (mild) HEAVY
    NAPHTHENIC see MQV790
MINERAL OIL, PETROLEUM DISTILLATES, HYDROTREATED (mild) HEAVY
    PARAFFINIC see MQV795
MINERAL OIL, PETROLEUM DISTILLATES, HYDROTREATED (mild) LIGHT PA-
    RAFFINIC see MQV805
MINERAL OIL, PETROLEUM DISTILLATES, HYDROTREATED (mild) LIGHT
    NAPHTHENIC see MQV800
MINERAL OIL, PETROLEUM DISTILLATES, LIGHT NAPHTHENIC see MQV810
MINERAL OIL, PETROLEUM DISTILLATES, LIGHT PARAFFINIC see MQV815
MINERAL OIL, PETROLEUM DISTILLATES, SOLVENT-DEWAXED HEAVY PA-
    RAFFINIC (mild or no solvent-refining or hydrotreatment) see MQV825
MINERAL OIL, PETROLEUM DISTILLATES, SOLVENT-DEWAXED HEAVY
    NAPHTHENIC (mild or no solvent-refining or hydrotreatment) see MQV820
MINERAL OIL, PETROLEUM DISTILLATES, SOLVENT-DEWAXED LIGHT
    NAPHTHENIC (mild or no solvent-refining or hydrotreatment) see MQV835
MINERAL OIL, PETROLEUM DISTILLATES, SOLVENT-DEWAXED LIGHT PA-
    RAFFINIC (mild or no solve
MINERAL OIL, PETROLEUM DISTILLATES, SOLVENT-REFINED (mild) HEAVY
    NAPHTHENIC see MQV845
MINERAL OIL, PETROLEUM DISTILLATES, SOLVENT-REFINED (mild) HEAVY
    PARAFFINIC see MQV850
MINERAL OIL, PETROLEUM DISTILLATES, SOLVENT-REFINED (mild) LIGHT
    PARAFFINIC see MQV855

MINERAL OIL, PETROLEUM DISTILLATES, SOLVENT-REFINED (mild) LIGHT NAPHTHENIC see MQV852
MINERAL OIL, PETROLEUM EXTRACTS, HEAVY NAPHTHENIC DISTILLATE SOLVENT see MQV857
MINERAL OIL, PETROLEUM EXTRACTS, HEAVY PARAFFINIC DISTILLATE SOLVENT see MQV859
MINERAL OIL, PETROLEUM EXTRACTS, LIGHT NAPHTHENIC DISTILLATE SOLVENT see MQV860
MINERAL OIL, PETROLEUM EXTRACTS, LIGHT PARAFFINIC DISTILLATE SOLVENT see MQV862
MINERAL OIL, PETROLEUM EXTRACTS, RESIDUAL OIL SOLVENT see MQV863
MINERAL OIL, PETROLEUM RESIDUAL OILS, ACID-TREATED see MQV872
MINERAL OIL, WHITE (FCC) see MQV750
MINERAL OIL, WHITE see MQV875
MINERAL ORANGE see LDS000
MINERAL PITCH see ARO500
MINERAL RED see LDS000
MINERAL WHITE see CAX750
MINETOIN see DKQ000, DNU000
MINGIT see DFP600
MINIDRIL see NNL500
MINIHIST see DBM800, WAK000
MINILYN see EEH575
MINIMYCIN see OMK000
MINIPLANOR see ZVJ000
MINIUM see LDS000
MINIUM NON-SETTING RL-95 see LDS000
MINOALEUIATIN see TLP750
MINOCIN see MQW100
MINOCYCLIN see MQW250
MINOCYCLINE see MQW250
MINOCYCLINE CHLORIDE see MQW100
MINOCYCLINE HYDROCHLORIDE see MQW100
MINOPHAGEN A see AQW000
MINORAN see MEK700
MINORLAR see EEH520
MINOSSIDILE (ITALIAN) see DCB000
MINOVLAR see EEH520
MINOXIDIL see DCB000
MINOZINAN see MCI500
MINPROG see POC350
MINTACO see NIM500
MINTACOL see NIM500
MINTAL see NBU000
MINTEZOL see TEX000
MINT-O-MAG see MAG750
MINTUSSIN see MGR250
MIN-U-GEL 200 see PAE750
MIN-U-GEL 400 see PAE750
MIN-U-GEL FG see PAE750
MINURIC see DDP200
MINUS see DKE800, SEH000
MINUSIN see NNW500
MINZIL see CLH750
MINZOLUM see TEX000
MIOARTRINA see IPU000
MIOBLOCK see PAF625
MIOCURIN see RLU000
MIODAR see PGG350
MIOFILIN see TEP500
MIO 40GN see IHG100
MIOLAXENE see GKK000
MIOLISODAL see IPU000
MIOLISODOL see IPU000
MIONAL see EAV700
MIO-PRESSIN see RDK000
MIORATRINA see IPU000
MIORELAX see RLU000
MIORIL see IPU000
MIORILAS see GKK000
MIORILAX see CKF500
MIORIODOL see IPU000
MIO-SED see CKF500
MIOSTAT see CBH250
MIOTICOL see DNR309
MIOTINE see MIC250
MIOTISAL see NIM500
MIOTISAL A see NIM500
MIOTOLON see AOO300
MIOWAS see GKK000
MIPAFOX see PHF750
MIPAX see DTR200
MIPC see MIA250
MIPCIN see MIA250

MIPCINE see MIA250
MI-PILO OPHTH SOL see PIF250
MIPK see MLA750
MIPSIN see MIA250
MeIQ see AJQ600
MIRACIL D see DHU000, SBE500
MIRACIL D HYDROCHLORIDE see SBE500
MIRACLE see DAA800
MIRACOL see SBE500
MIRADOL see EPD500
MIRAL see PDN000, PHK000
MIRAL 10 G see PHK000
MIRAMEL see EQL000
MIRAMID WM 55 see PJY500
MIRANOL C2M-SF CONC see AOC250
MIRANOL MHT see AOC275
MIRAPRONT see DTJ400
MIRASON 9 see PJS750
MIRASON 16 see PJS750
MIRASON M 15 see PJS750
MIRASON M 50 see PJS750
MIRASON M 68 see PJS750
MIRASON NEO 23H see PJS750
MIRATHEN 1313 see PJS750
MIRATHEN 1350 see PJS750
MIRATHEN see PJS750
MIRBANE 850 see MCB050
MIRBANE MR2 see MCB050
MIRBANE OIL see NEX000
MIRBANE SM 607 see MCB050
MIRBANE SM 800 see MCB050
MIRBANE SM 850 see MCB050
MIRBANE SU 118K see UTU500
MIRBANIL see EPD500
MIRCOL see MCJ400
MIRCOSULFON see PPP500
MIREX see MQW500
MIRLON see NOH000
MIROISTONIL see DRM000
MIROMORFALIL see NAG500
MIROPROFEN see IAY000
MIROSERINA see CQH000
MIROTIN see MNZ000
MIRREX MCFD 1025 see PKQ059
MISASIN see CKL500
MISCLERON see ARQ750
MISHERI (INDIA) see SED400
MISHRI (INDIA) see SED400
MISODINE see DBB200
MISOLYNE see DBB200
MISONIDAZOLE see NHH500
MISOPROSTOL see MJE775
MISPICKEL see ARJ750
MISTABRON see MDK875
MISTABRONCO see MDK875
MISTLETOE (AMERICAN) see MQW525
MISTRON FROST P see TAB750
MISTRON RCS see TAB750
MISTRON 2SC see TAB750
MISTRON STAR see TAB750
MISTRON SUPER FROST see TAB750
MISTRON VAPOR see TAB750
MISTURA C see CBH250
MISULBAN see BOT250
MISULVAN see EPD500
MISURAN see MSC100
MIT see ISE000
MITABAN see MJL250
MITAC see MJL250
MITACIL see DOR400
MITANOLINE see PMC250
MIT-C see AHK500
MITC see ISE000
MITENON see MMD500
MITEXAN see MDK875
MITHRACIN see MQW750
MITHRAMYCIN, MAGNESIUM SALT see MAD050
MITHRAMYCIN see MQW750
MITHRAMYCIN A see MQW750
MITICIDE K-101 see CJT750
MITIGAN see BIO750
MITION see CKM000
MITIS GREEN see COF500
MITOBROMOL see DDP600
MITOBRONITOL see DDP600

MITO-C see AHK500
MITOCHROMIN see MQX000
MITOCIN-C see AHK500
MITOCROMIN see MQX000
MITOLAC see DDJ000
MITOLACTOL see DDJ000
MITOMALCIN see MQX250
MITOMEN see CFA500, CFA750
MITOMIN see CFA500
MITOMYCIN see AHK500
MITOMYCIN A see MQX500
MITOMYCIN B see MQX750
MITOMYCIN-C see AHK500
MITOMYCINUM see AHK500
MITONAFIDE see MQX775
MITOSTAN see BOT250
MITOTANE see CDN000
MITOX see CEP000
MITOXAN see CQC500, CQC650
MITOXANA see IMH000
MITOXANTHRONE see MQY090
MITOXANTRONE see MQY090
MITOXANTRONE HYDROCHLORIDE see MQY100
MITOXINE see BIE500
MITOZOLOMIDE see MQY110
MITRAMYCIN see MQW750
MITRONAL see CMR100
MITSUI ACID PURE BLUE VX see CMM062
MITSUI ALIZARINE B see DMG800
MITSUI ALIZARINE RED S see SEH475
MITSUI ANTHRACENE BROWN see TKN500
MITSUI AURAMINE O see IBA000
MITSUI BLUE B BASE see DCJ200
MITSUI BRILLIANT GREEN G see BAY750
MITSUI CHROME BLUE BLACK BC see CMP880
MITSUI CHROME YELLOW GG see SIT850
MITSUI CONGO RED see SGQ500
MITSUI CRYSTAL VIOLET see AOR500
MITSUI DIRECT BLACK BH see CMN800
MITSUI DIRECT BLACK D see CMN230
MITSUI DIRECT BLACK EX see AQP000
MITSUI DIRECT BLACK GX see AQP000
MITSUI DIRECT BLUE 2BN see CMO000
MITSUI DIRECT BLUE RW see CMO600
MITSUI DIRECT BRILLIANT SCARLET 8B see CMO880
MITSUI DIRECT DARK GREEN BX see CMO830
MITSUI DIRECT FAST RED F see CMO870
MITSUI DIRECT GREEN BC see CMO840
MITSUI DIRECT SCARLET 3BX see CMO875
MITSUI DIRECT SKY BLUE 5B see CMO500
MITSUI METANIL YELLOW see MDM775
MITSUI METHYLENE BLUE see BJI250
MITSUI MILLING SCARLET G see CMM320
MITSUI NYLON FAST SKY BLUE B see CMM090
MITSUI RED B BASE see NEQ000
MITSUI RED 3GL BASE see KDA050
MITSUI RED TR BASE see CLK220
MITSUI RHODAMINE see RGW000
MITSUI RHODAMINE BX see FAG070
MITSUI RHODAMINE 6GCP see RGW000
MITSUI SAFRANINE see GJI400
MITSUI SCARLET G BASE see NMP500
MITSUI TARTRAZINE see FAG140
MITUSI SCARLET GG BASE see DEO295
MIVIZON see FNF000
MIXTURE of p-METHENOLS see TBD500
MIZODIN see DBB200
MIZOLIN see DBB200
MIZORIBINE see BMM000
MJ 505 see PGG350, PGG355
MJ 1992 see SKW000, SKW500
MJ 1998 see HLX500
MJ 1999 see CCK250
MJ 5022 see MPE250
MJ 5190 see AHL500
MJ 10061 see DDP200
MJ 4309-1 see OPK000
MJF 9325 see IMH000
MJF-12264 see FLZ050
MJ 1999 HYDROCHLORIDE see CCK250
MK₄ see VTA650
MK-33 see GEW700
MK 56 see POL500
MK 75 see NCW100
MK 125 see SOW000

MK 141 see PCI500
MK 142 see MQY125
MK 184 see TAF675
MK-188 see RBF100
(±)-MK 196 see IBQ400
MK 196 see IBQ400
MK 231 see SOU550
MK 240 see DDA600
MK 351 see DNA800
MK-351 see MJE780
MK 360 see TEX000
MK-366 see BAB625
MK 421 see EAO100
MK 486 see CBQ500
MK-595 see DFP600
MK 647 see DKI600
MK 665 see CHP750
MK-872 see HMS875
MK 933 see ITD875
MK 950 see TGB185
MK-955 see PHA550
MK. B51 see DNA800
MK.B51 see MJE780
MK-142 DIMETHANESULFONATE see MQY125
2M-4KH see CIR250
MKH 52 see UTU500
2M 4KHP see CIR500
2M-4KH SODIUM SALT see SIL500
MK 421 MALEATE see EAO100
ML 21 see MCB050
ML 045 see MCB050
ML 97 see FAB400, PGX300
ML 133 see MCB050
ML 630 see MCB050
ML 1024 see TEQ500
ML 3120 see MCB050
MLA-74 see MLD500
ML 33F see SKV100
ML 55F see SKV100
MLO 81-125 see KDK000
MLO 83-322 see KDK000
MLO-5277 (ORGANO-SILICATE) see MQY250
MLT see MAK700
MM 83 see MCB050
MM 4462 see AOP250
MM 14151 see CMV250
MM see BKM500
MMA see MGQ250
3M MBR 6168 see DRR000
MMC see AHK500, MDD750
MMD see MLF250
MME see MFC700, MLH750
M 3 (MELAMINE POLYMER) see MCB050
MMH see MKN000
4-MMPD see DBO000
4-MMPD SULPHATE see DBO400
MMS see MLH500
MMT see MAV750
MMTs-BTR see MIF760
MMTP see MLX750
MM 160V30M see MCB050
ω-MMYCIN see TBX000
MN-1695 see MQY300
6-MNA see MMT250
MNA see MMU250, NEN500
MNA DIMETHYL ACETAL see MNB600
MNB see MMU500
MNBK see HEV000
MNC see MMX000
MN-CELLULOSE see CCU150
MNCO see MQY325
MNE see NDM000
MNFA see MME809
MNG see MMP000
McN-JR 4263 see PDW750
McN-JR-6238 see PIH000
McN-JR-16,341 see PAP250
Me₂NMOR see DTA000
MNNG see MMP000
MNPA see MFA500
MNPN see MMS200
MNQ see MMD500
MNT see NMO500
MNU see MMX250, MNA750
MNU and CYCLOPHOSPHAMIDE (2:1) see CQC600

MO 338 see NIW500
MO 709 see DRM000
MO see NIW500
MOB see MES000
M-2-OB see MMR800
M-3-OB see MMR810
MOBAM see BDG250
MOBAM PHENOL see BDG250
MOBAN see MRB250
MOBAY MRS see PKB100
MOBENOL see BSQ000
MOBILAN see IDA000
MOBILAT see MQY350
MOBILAWN see DFK600
MOBIL DBHP see DEG800
MOBIL MC-A-600 see BDG250
MOBIL V-C 9-104 see EIN000
MOBUTAZON see MQY400
MOBUZON see MQY400
MOCA see MJM200
MOCAP see EIN000
MOCK AZALEA see DBA450
MOCTYNOL see GGS000
MODALINE SULFATE see MOM750
MODANE see DMH400
MODANE SOFT see DJL000
MODECCIN see MRA000
MODECCIN TOXIN see MRA000
MODENOL see RDK000
MODERAMIN see MFD500
MODERIL see TLN500
MODIFIER 113-63 see MRA100
MODIRAX see POA250
MODITEN see FMP000, TJW500
MODITEN ENANTHATE see PMI250
MODITEN-RETARD see PMI250
MODOCOLL 1200 see SFO500
MODR ALIZARINOVA CISTA B (CZECH) see AIY750
MODR BRILANTNI ALIZARINOVA BRL (CZECH) see DKR400
MODR BRILANTNI OSTAZINOVA S-R (CZECH) see DGN400
MODR DISPERZNI 56 see CMP070
MODRENAL see EBY600
MODR FRALOSTANOVA 3G (CZECH) see DNE400
MODR KYPOVA 18 see VGP100
MODR KYPOVA 4 see IBV050
MODR KYSELA 1 see ADE500
MODR KYSELA 41 see CMM070
MODR KYSELA 78 see CMM090
MODR KYSELA 129 see CMM100
MODR METHYLENOVA (CZECH) see BJI250
MODR MIDLONOVA STALA ER (CZECH) see TMA750
MODR NAMORNICKA OSTANTHRENOVA G (CZECH) see CMU500
MODR NAMORNICKA OSTANTHRENOVA RA (CZECH) see CMU750
MODR OSTACETOVA LG (CZECH) see BND250
MODR OSTACETOVA LR see BNC800
MODR OSTACETOVA P3R (CZECH) see MGG250
MODR OSTACETOVA SE-LB (CZECH) see DBT000
MODR PIGMENT 60 see IBV050
MODR POTRAVINARSKA 3 see ADE500
MODR POTRAVINARSKA 4 see IBV050
MODR PRIMA 15 see CMO500
MODR REAKTIVNI 4 see CMS228
MODULEX see CBT750
MOEBIQUIN see DNF600
MOENOMYCIN see MRA250
MOENOMYCIN A see MRA250
MOF see DFA400
MOFEBUTAZONE see MQY400
MOGADAN see DLY000
MOGETON GRANULE see AJI250
MOGUL see CBT750
MOGUL L see CBT750
MOHEPTAN see MDP750
MOHICAN RED A-8008 see CHP500
MOLACCO see CBT750
MOLANTIN P see MRA300
MOLASSES ALCOHOL see EFU000
MOLATOC see DJL000
MOLCER see DJL000
MOLDAMIN see BFC750
MOLDCIDIN B see FPC000
MOLDEX see HJL500
MOLECULAR CHLORINE see CDV750
MOLECULAR SIEVE 13X with 14.6% DI-n-BUTYLAMINE see MRA750
MOLE DEATH see SMN500

MOLEVAC see PQC500
MOLINATE see EKO500
MOLINDONE HYDROCHLORIDE see MRB250
MOLINILLO (PUERTO RICO) see SAT875
MOLIPAXIN see CKJ000, THK880
MOL-IRON see FBO000
MOLIVATE see CMW400
MOLLAN O see DVL700
MOLLINOX see QAK000
MOLLUSCICIDE BAYER 73 see DFV600
MOLMATE see EKO500
MOLOFAC see DJL000
MOLOL see MQV750
MOLSIDOLAT see MRN275
MOLSIDOMINE see MRN275
MOLTEN ADIPIC ACID see AEN250
MOLURAME see BJK500
MOLYBDATE see MRC250
MOLYBDATE, CALCIUM see CAT750
MOLYBDATE ORANGE see MRC000
MOLYBDATE ORANGE Y 786D see MRC000
MOLYBDATE ORANGE YE 421D see MRC000
MOLYBDATE ORANGE YE 698D see MRC000
MOLYBDATE RED see MRC000
MOLYBDATE RED AA3 see MRC000
MOLYBDENITE see MRD750
MOLYBDEN RED see MRC000
MOLYBDENUM see MRC250
MOLYBDENUM AZIDE PENTACHLORIDE see MRC500
MOLYBDENUM AZIDE TRIBROMIDE see MRC600
MOLYBDENUM BORIDE see MRC650
MOLYBDENUM-COBALT-CHROMIUM ALLOY see VSK000
MOLYBDENUM COMPOUNDS see MRC750
MOLYBDENUM DIAZIDE TETRACHLORIDE see MRD000
MOLYBDENUM DIOXIDE see MRD250
MOLYBDENUM-LEAD CHROMATE see LDM000
MOLYBDENUM ORANGE see LDM000
MOLYBDENUM(IV) OXIDE see MRD250
MOLYBDENUM(VI) OXIDE see MRE000
MOLYBDENUM OXIDE see MRD250
MOLYBDENUM PENTACHLORIDE see MRD500
MOLYBDENUM RED see MRC000
MOLYBDENUM SULFIDE see MRD750
MOLYBDENUM, TRICARBONYL(1,3,5-CYCLOHEPTATRIENE)- see COY100
MOLYBDENUM TRIOXIDE see MRE000
MOLYBDIC ACID DIAMMONIUM SALT see ANM750
MOLYBDIC ACID, DISODIUM SALT see DXE800
MOLYBDIC ACID ($H_2MoO_4$), CALCIUM SALT (1:1) see CAT750
MOLYBDIC ANHYDRIDE see MRE000
MOLYBDIC SULFIDE see MRD750
MOLYBDIC TRIOXIDE see MRE000
MOLYKOTE 522 see TAI250
MOLYKOTE see MRD750
MOMENTOL see TKX000
MOMENTUM see HIM000
MOMETINE see CJL500
MOMORDICA BALSAMINA see FPD100
MOMORDICA CHARANTIA see FPD100
MOMORDIQUE A FUEILLES de VIGNE see FPD100
MON 0573 see PHA500
MON 2139 see PHA500
MONACETYLFERROCENE see ABA750
MONACRIN see AHS500, AHS750
MONACRIN HYDROCHLORIDE see AHS750
MONAGYL see MMN250
MONALIDE see CGL250
MONAM see SIL550
MONAMID 150-LW see BKE500
MONAQUEST see DJG800
MONARCH see CBT750, ZVJ000
MONARGAN see ABX500
MONASIRUP see PMQ750
MONASPOR see CCS550
MONASTRAL FAST GREEN 3Y see CMS140
MONASTRAL FAST GREEN 6Y see CMS140
MONASTRAL FAST GREEN 3YA see CMS140
MONASTRAL FAST GREEN 6YA see CMS140
MONASTRAL GREEN Y-GT 805D see CMS140
MONATE see MRL750
MONAWET MD 70E see DJL000
MONAWET MM 80 see DKP800
MONAZAN see MQY400
MONCIDE see HKC000
MONDUR E 429 see PKB100
MONDUR E 441 see PKB100

MONDUR E 541 see PKB100
MONDUR MR 200 see PKB100
MONDUR MR see PKB100
MONDUR MRS 10 see PKB100
MONDUR MRS see PKB100
MONDUR P see PFK250
MONDUR-TD-80 see TGM740
MONDUR TD-80 see TGM750
MONDUR-TD see TGM740
MONDUR TD see TGM750
MONDUR TDS see TGM750
MONELAN see MRE225
MONELGIN see OAV000
MONENSIC ACID see MRE225
MONENSIN, MONOSODIUM SALT (9CI) see MRE230
MONENSIN (USDA) see MRE225
MONENSIN A see MRE225
MONENSIN SODIUM see MRE230
MONENSIN SODIUM SALT see MRE230
MONEX see BJL600
MONGARE see MJK750
MONHYDRIN see PMJ500
MONILIFORMIN see MRE250
MONISTAT see MQS550
MONITAN see PKL100
MONITOR see DTQ400
MONKEY PISTOL see SAT875
MONKEY'S DINNER BELL see SAT875
MONKIL WP see MGQ750
MONKSHOOD see MRE275
1-MONOACETIN see GGO000
α-MONOACETIN see GGO000
MONOACETIN see GGO000
MONOACETYL GLYCERINE see GGO000
MONOACETYLHYDRAZINE see ACM750
MONOACETYL 4-HYDROXYAMINOQUINOLINE see QQA000
MONOACETYLISOSPIRAMYCIN see MRE500
MONOACETYLSPIRAMYCIN see MRE750
MONOAETHANOLAMIN (GERMAN) see EEC600
MONOALLYLAMINE see AFW000
MONOALLYLUREA see AGV000
MONOALUMINUM PHOSPHATE see PHB500
MONOAMMONIUM CARBONATE see ANB250
MONOAMMONIUM l-GLUTAMATE see MRF000
MONOAMMONIUM GLUTAMATE see MRF000
MONOAMMONIUM GLYCYRRHIZINATE see GIE100
MONOAMMONIUM SULFAMATE see ANU650
MONOAMMONIUM SULFATE see ANJ500
MONOAMMONIUM SULFIDE see ANJ750
MONOAMMONIUM SULFITE see ANB600
MONOAMYLAMINE see PBV505
MONOAZO see MDM775
MONOBASIC racemic AMPHETAMINE PHOSPHATE see AOB500
MONOBASIC CHROMIUM SULFATE see NBW000
MONOBASIC CHROMIUM SULPHATE see NBW000
MONOBASIC LEAD ACETATE see LCH000
MONOBASIC dl-α-METHYLPHENETHYLAMINE PHOSPHATE see AOB500
MONOBENZALPENTAERYTHRITOL see MRF200
MONOBENZONE see AEY000
MONOBENZYL-p-AMINOPHENOL HYDROCHLORIDE see MRF250
MONOBENZYL ETHER HYDROQUINONE see AEY000
MONOBENZYL HYDROQUINONE see AEY000
MONOBOR-CHLORATE see SFS500
MONOBROMESSIGSAEURE (GERMAN) see BMR750
MONOBROMOACETIC ACID see BMR750
MONOBROMOACETONE see BNZ000
MONOBROMOBENZENE see PEO500
MONOBROMODIFLUOROPHOSPHINE SULFIDE see TFO500
MONOBROMOETHANE see EGV400
MONOBROMOGLYCEROL see MRF275
2-MONOBROMOISOVALERYLUREA see BNP750
MONOBROMOISOVALERYLUREA see BNP750
MONOBROMOMETHANE see MHR200
MONOBUTILAMINA see BPX750
MONOBUTYL see MQY400
MONO-n-BUTYLAMINE see BPX750
MONOBUTYLAMINE see BPX750
MONOBUTYL DIPHENYL SODIUM MONOSULFONATE see AQU750
MONOBUTYL GLYCOL ETHER see BPJ850
MONO-tert-BUTYLHYDROQUINONE see BRM500
MONOBUTYL PHOSPHITE see MRF500
MONOBUTYLTIN TRICHLORIDE see BSO750
MONOBUTYLTIN TRILAURATE see BSO750
MONOCAINE HYDROCHLORIDE see IAC000
MONOCALCIUM ARSENITE see CAM500

MONOCALCIUM CARBONATE see CAT775
MONOCALCIUM DISODIUM EDTA see CAR780
MONOCALCIUM PHOSPHATE see CAW110
MONOCARBAMIDE DIHYDROGEN SULFATE see UTU600
MONOCARBETHOXYHYDRAZINE see EHG000
MONOCHLOORAZIJNZUUR (DUTCH) see CEA000
MONOCHLOORBENZEEN (DUTCH) see CEJ125
MONOCHLORACETIC ACID see CEA000
MONOCHLORACETONE see CDN200
MONOCHLORAMIDE see CDO750
MONOCHLORAMINE see CDO750
MONOCHLORBENZENE see CEJ125
MONOCHLORBENZOL (GERMAN) see CEJ125
MONOCHLORESSIGSAEURE (GERMAN) see CEA000
MONOCHLORETHANE see EHH000
MONOCHLORHYDRIN see CDT750
MONOCHLORHYDRINE du GLYCOL (FRENCH) see EIU800
MONOCHLORIMIPRAMINE see CDU750
MONOCHLOROACETALDEHYDE see CDY500
MONOCHLOROACETIC ACID see CEA000
MONOCHLOROACETIC ACID METHYL ESTER see MIF775
MONOCHLOROACETONE, inhibited (DOT) see CDN200
MONOCHLOROACETONE, stabilized (DOT) see CDN200
MONOCHLOROACETONE, unstabilized (DOT) see CDN200
MONOCHLOROACETONE see CDN200
MONOCHLOROACETONITRILE see CDN500
MONOCHLOROACETYL CHLORIDE see CEC250
MONOCHLOROAMINE see CDO750
MONOCHLOROAMMONIA see CDO750
α-MONOCHLOROANTHRAQUINONE see CEI000
MONOCHLOROBENZENE see CEJ125
MONOCHLOROCYCLOHEXANE see CPI400
MONOCHLORODIFLUOROMETHANE see CFX500
MONOCHLORODIMETHYL ETHER (MAK) see CIO250
MONOCHLORODIPHENYL OXIDE see MRG000
MONOCHLOROETHANOIC ACID see CEA000
2-MONOCHLOROETHANOL see EIU800
MONOCHLOROETHENE see VNP000
MONOCHLOROETHYLENE (DOT) see VNP000
MONOCHLOROETHYLENE OXIDE see CGX000
α-MONOCHLOROHYDRIN see CDT750
MONOCHLOROHYDRIN see CDT750
MONOCHLOROISOTHYMOL see CEX275
MONOCHLOROMETHANE see MIF765
MONOCHLOROMETHYL CYANIDE see CDN500
MONO-CHLORO-MONO-BROMO-METHANE see CES650
MONOCHLOROMONOFLUOROMETHANE see CHI900
MONOCHLOROPENTAFLUOROETHANE (DOT) see CJI500
MONOCHLOROPHENYLETHER see MRG000
p-MONOCHLOROPHENYL PHENYL SULFONE see CKI625
MONOCHLOROPOLYOXYETHYLENE see PJU250
β-MONOCHLOROPROPIONIC ACID see CKS500
MONOCHLOROSULFURIC ACID see CLG500
MONOCHLOROTETRAFLUOROETHANE (DOT) see CLH000
MONOCHLOROTRIFLUOROETHYLENE see CLQ750
MONOCHLOROTRIFLUOROMETHANE (DOT) see CLR250
MONOCHORIA VAGINALIS (Burm. f.) Presl., extract see MRG100
MONOCHROME YELLOW MG see SIT850
MONOCHROMIUM OXIDE) see CMK000
MONOCHROMIUM TRIOXIDE see CMK000
MONOCIL 40 see MRH209
MONOCIL see MRH209
"MONOCITE" METHACRYLATE MONOMER see MLH750
MONOCLAIR see CCK125
MONOCLOROBENZENE (ITALIAN) see CEJ125
MONOCOBALT OXIDE see CND125
MONOCOPPER MONOSULFIDE see CNQ000
MONOCORTIN see PAL600
MONOCRATILIN see MRH000
MONOCRESYL DIPHENYL PHOSPHATE see TGY750
MONOCRON see MRH209
MONOCROTALINE see MRH000
MONOCROTALINE, 3,8-DIDEHYDRO- see DAL350
MONOCROTOPHOS (ACGIH,OSHA) see MRH209
MONOCROTOPHOS see ASN000, MRH209
MONOCYANOACETIC ACID see COJ500
MONOCYCLOHEXYLTIN ACID see MRH212
MONODEHYDROSORBITOL MONOOLEATE see SKV100
MONODEMETHYLIMIPRAMINE see DSI709
MONO(2,3-DIBROMOPROPYL)AMMONIUM PHOSPHATE see MRH214
MONO-DIGITOXID (GERMAN) see DKL875
MONODION see VTA000
MONODODECYLAMINE see DXW000
MONODODECYL ESTER SULFURIC ACID see MRH250
MONODORM see BPF500

MONODRAL see PBS000
MONODRAL BROMIDE see PBS000
MONODRIN see MRH209
MONOESTER with 4-BUTYL-4-(HYDROXYMETHYL)-1,2-DIPHENYL-SUCCINIC ACID 3,5-PYRAZOLIDINEDIONE see SOX875
MONOETHANEAMINE BENZOATE see MRH300
MONOETHANOLAMINE see EEC600
MONOETHANOLAMINE HYDROCHLORIDE see EEC700
MONOETHANOLETHYLENEDIAMINE see AJW000
MONOETHYL ADIPATE see ELC600
MONOETHYLADIPIC ACID ESTER see ELC600
N-MONOETHYLAMIDE of O,O-DIMETHYLDITHIOPHOSPHORYLACETIC ACID see DNX600
MONOETHYLAMINE, anhydrous (DOT) see EFU400
MONOETHYLAMINE (DOT) see EFU400
2-N-MONOETHYLAMINOETHANOL see EGA500
MONOETHYLAMMONIUM CHLORIDE see EFW000
MONOETHYLDICHLOROTHIOPHOSPHATE see MRI000
MONOETHYLENEDIUREA see EJC100
MONOETHYLENE GLYCOL see EJC500
MONOETHYLENE GLYCOL DIMETHYL ETHER see DOE600
MONOETHYL ETHER of DIETHYLENE GLYCOL see CBR000
MONOETHYL HEXANEDIOATE see ELC600
MONO(2-ETHYLHEXYL)PHTHALATE see MRI100
MONOETHYLHEXYL PHTHALATE see MRI100
MONO(2-ETHYLHEXYL)SULFATE SODIUM SALT see TAV750
MONOETHYLISOSELENOURONIUMBROMIDE-HYDROBROMIDE see AJY000
MONOETHYLTIN TRICHLORIDE see EPS000
MONOFLUORAZIJNZUUR (DUTCH) see FIC000
MONOFLUORESSIGSAEURE (GERMAN) see FIC000
MONOFLUORESSIGSAEURE, NATRIUM (GERMAN) see SHG500
MONOFLUORETHANOL see FID000
MONOFLUOROACETAMIDE see FFF000
MONOFLUOROACETATE see FIC000
MONOFLUOROACETIC ACID see FIC000
MONOFLUOROETHANE see FIB000
MONOFLUOROETHANOL see FID000
MONOFLUOROETHYLENE see VPA000
MONOFLUOROPHOSPHORIC ACID, anhydrous see PHJ250
MONOFLUOROTRICHLOROMETHANE see TIP500
MONOFURACIN see NGE500
MONOGERMANE see GEI100
MONOGLYCEROL-p-AMINOBENZOATE see GGQ000
MONOGLYCERYL OLEATE see GGR200
MONOGLYCIDYL ETHER of N-PHENYLDIETHANOLAMINE see MRI500
MONO-GLYCOCOARD see DKL800
MONOGLYME see DOE600
MONO-N-HEXYLAMINE see HFK000
8-MONOHYDRO MIREX see MRI750
MONOHYDROXYBENZENE see PDN750
MONO-2-(2-HYDROXYETHOXY)ETHYLESTER KYSELINY MALEINOVE (CZECH) see MAL500
MONO(HYDROXYETHYL) ESTER MALEIC ACID see EJG500
MONOHYDROXY-MERCURI-DI-IODORESORCIN-SULPHONPHTHALEIN DISODIUM see SIF500
MONOHYDROXYMETHANE see MGB150
MONO-(2-HYDROXYPROPYL)ESTER KYSELINY MALEINOVE (CZECH) see MAL750
MONOIODOACETAMIDE see IDW000
MONOIODOACETATE see IDZ000
MONOIODOACETIC ACID see IDZ000
MONOIODOMETHANESULFONIC ACID, SODIUM SALT see SHX000
MONOIODURO di METILE (ITALIAN) see MKW200
MONOISOBUTYLAMINE see IIM000
MONO-ISO-PROPANOLAMINE see AMA500
N-MONOISOPROPYLAMIDE of O,O-DIETHYLDITHIOPHOSPHORYLACETIC ACID see IOT000
MONOISOPROPYLAMINE see INK000
MONOISOPROPYLAMINE HYDROCHLORIDE see INL000
MONOISOPROPYLAMINOETHANOL see INN400
4-MONOISOPROPYLAMINO-1-PHENYL-2,3-DIMETHYL-5-PYRAZOLONE see INM000
MONOISOPROPYLBIPHENYL see IOF200
MONOISOPROPYL ETHER of ETHYLENE GLYCOL see INA500
MONO-KAY see VTA000
1-MONOLAURIN see MRJ000
MONOLAURYL DIMETHYLAMINE see DRR800
MONOLINURON see CKD500
MONOLITE FAST BLUE 3R see IBV050
MONOLITE FAST BLUE 3RD see IBV050
MONOLITE FAST BLUE 2RV see DFN425
MONOLITE FAST BLUE RV see IBV050
MONOLITE FAST BLUE 2RVSA see DFN425
MONOLITE FAST BLUE SRS see IBV050
MONOLITE FAST ORANGE G see CMS145

MONOLITE FAST ORANGE GA see CMS145
MONOLITE FAST SCARLET CA see MMP100
MONOLITE FAST SCARLET GSA see MMP100
MONOLITE FAST SCARLET RB see MMP100
MONOLITE FAST SCARLET RBA see MMP100
MONOLITE FAST SCARLET RN see MMP100
MONOLITE FAST SCARLET RNA see MMP100
MONOLITE FAST SCARLET RNV see MMP100
MONOLITE FAST SCARLET RT see MMP100
MONOLITE FAST YELLOW GLV see CMS208
MONOLITE RED CN see CMS150
MONOLITE YELLOW GL see CMS208
MONOLITE YELLOW GLA see CMS208
MONOLITE YELLOW GT see DEU000
MONOLITHIUM ACETYLIDE-AMMONIA see MRJ125
MONOMER FA see FPQ900
MONOMER MG-1 see EJH000
MONOMETHACRYLIC ETHER of ETHYLENE GLYCOL see EJH000
N-MONOMETHYLACETAMIDE see MFT750
N-MONOMETHYLAMIDE of O,O-DIMETHYLDITHIOPHOSPHORYLACETIC ACID see DSP400
MONOMETHYLAMINE see MGC250
MONOMETHYL-AMINOAETHANOL (GERMAN) see MGG000
4-MONOMETHYLAMINOAZOBENZENE see MNR500
p-MONOMETHYLAMINOAZOBENZENE see MNR500
N-MONOMETHYLAMINOETHANOL see MGG000
MONOMETHYLAMINOETHANOL see MGG000
N-MONOMETHYL-2-AMINOFLUORENE see FEI500
2-MONOMETHYLAMINOFLUORENE see FEI500
N-MONOMETHYLANILINE see MGN750
MONOMETHYL ANILINE (OSHA) see MGN750
MONOMETHYLANILINE see MGN750
MONOMETHYLARSINIC ACID see MGQ530
10-MONOMETHYLBENZO(a)PYRENE see MHH500
MONOMETHYL ETHER of ETHYLENE GLYCOL see EJH500
MONOMETHYL ETHER HYDROQUINONE see MFC700
MONOMETHYLFORMAMIDE see MKG500
MONOMETHYLFOSFIT (CZECH) see PGZ950
MONOMETHYL GUANIDIN (GERMAN) see MKI750
MONOMETHYLGUANIDINE see MKI750
MONOMETHYL HYDRAZINE see MKN000
MONOMETHYLHYDRAZINE NITRATE see MRJ250
MONOMETHYLMALEIC ANHYDRIDE see CMS322
MONOMETHYL MERCURY CHLORIDE see MDD750
MONOMETHYLOLACRYLAMIDE see HLU500
MONOMETHYLTIN TRICHLORIDE see MQC750
MONOMYCIN see MRJ600
MONOMYCIN A see NCF500
MONONICKEL MONOSULFIDE see NDL100
MONONITROCHLOROBENZENE see CJA950
MONONITRONAPHTHALENE see NHP990
MONONITROSOPIPERAZINE see MRJ750
MONOOCTADECYLAMINE see OBC000
MONOOCTADECYL ETHER of GLYCEROL see GGA915
MONO-n-OCTYLTIN TRICHLORIDE see OGG000
MONO-n-OCTYL-TIN-TRIS-(2-ETHYLHEXYLMERCAPTOACETATE) see OGI000
MONO-N-OCTYL-ZINN-TRICHLORID (GERMAN) see OGG000
MONOOLEIN see GGR200
MONOOLEOYLGLYCEROL see GGR200
MONOPALMITATE SORBITAN see MRJ800
MONOPEN see BFD000
MONOPENTEK see PBB750
MONOPERACETIC ACID see PCL500
MONOPEROXY SUCCINIC ACID see MRK000
MONOPHEN see PFC750
MONOPHENOL see PDN750
MONOPHENYLBUTAZONE see MQY400
MONOPHENYLUREA see PGP250
MONOPHOR see AOB500
MONOPHOS see AOB500
MONOPHYLLINE see HOA000
MONOPLEX DBS see DEH600
MONOPLEX DCP see BLB750
MONOPLEX DOA see AEO000
MONOPLEX DOS see BJS250
MONOPOTASSIUM ARSENATE see ARD250
MONOPOTASSIUM DIHYDROGEN ARSENATE see ARD250
MONOPOTASSIUM aci-1-DINITROETHANE see MRK250
MONOPOTASSIUM l-GLUTAMATE (FCC) see MRK500
MONOPOTASSIUM GLUTAMATE see MRK500
MONOPOTASSIUM 2-HYDROXYPROPANOATE ACID see PLK650
MONOPOTASSIUM PHOSPHATE see PLQ405
MONOPOTASSIUM SALT of ACETYLENEDICARBOXYLIC ACID see ACJ500
MONOPOTASSIUM SULFATE see PKX750
MONOPOTASSIUM TARTRATE see PKU600

MONOPRIM see TKZ000
MONO-N-PROPYLAMINE see PND250
MONOPROPYLENE GLYCOL see PML000
MONOPROPYL ETHER of ETHYLENE GLYCOL see PNG750
MONOPYRROLE see PPS250
MONORHEUMETTEN see MQY400
MONOSAN see DAA800
MONOSILANE see SDH575
MONOSIZER W710L see TJR600
MONOSODIOGLUTAMMATO (ITALIAN) see MRL500
MONOSODIUM ACETYLIDE see MRK609
MONOSODIUM ACID METHANEARSONATE see MRL750
MONOSODIUM ACID METHARSONATE see MRL750
MONOSODIUM-β-AMINOETHYL THIOPHOSPHATE see AKB500
MONOSODIUM (4-AMINOPHENYL)ARSONATE see ARA500
MONOSODIUM ARSENATE see ARD600
MONOSODIUM BARBITURATE see MRK750
MONOSODIUM CARBONATE see SFC500
MONOSODIUM CEFAZOLIN see CCS250
MONOSODIUM CEROXITIN see CCS510
MONOSODIUM CITRATE see MRL000
MONOSODIUM CYANURATE see SGB550
MONOSODIUM DIHYDROGEN CITRATE see MRL000
MONOSODIUM DIHYDROGEN PHOSPHATE see SJH100
MONOSODIUM 2,5-DIHYDROXYBENZOATE see GCU050
MONOSODIUM-5-ETHYL-5-(1-METHYLBUTYL) THIOBARBITURATE see PBT500
MONOSODIUM FERRIC EDTA see EJA379
MONOSODIUM FLUCLOXACILLIN see FDA100
MONOSODIUM GLUCONATE see SHK800
α-MONOSODIUM GLUTAMATE see MRL500
MONOSODIUM-l-GLUTAMATE (FCC) see MRL500
MONOSODIUM GLUTAMATE see MRL500
MONOSODIUM GLYCOLATE see SHT000
MONOSODIUM METHANEARSONATE see MRL750
MONOSODIUM METHANEARSONIC ACID see MRL750
MONOSODIUM METHYLARSONATE see MRL750
MONOSODIUM NOVOBIOCIN see NOB000
MONOSODIUM PHOSPHATE see SJH100
MONOSODIUM SALT of 2,2′-METHYLENE BIS(3,4,6-TRICHLOROPHENOL) see MRM000
MONOSODIUM-2-SULFANILAMIDOPYRIMIDINE see MRM250
MONOSODIUM-2-SULFANILAMIDOTHIAZOLE see TEX500
MONOSODIUM THYROXINE see LFG050
MONOSODIUM URATE see SKO575
MONOSORB XP-4 see SJH100
MONOSPAN see SBG500
MONOSTEARIN see OAV000
MONOSTEARYL TRIMETHYL AMMONIUM CHLORIDE see TLW500
MONOSTEOL see SLL000
MONOSULFUR DICHLORIDE see SOG500
MONOTARD see LEK000
MONOTEN see PFC750
MONOTHIOBENZOIC ACID see TFC550
MONOTHIOETHYLENEGLYCOL see MCN250
α-MONOTHIOGLYCEROL see MRM750
MONOTHIOGLYCEROL see MRM750
MONOTHIOSUCCINIMIDE see MRN000
MONO-THIURAD see BJL600
MONOTHIURAM see BJL600
MONOTRICHLOR-AETHYLIDEN-α-GLUCOSE (GERMAN) see GFA000
MONOUREA SULFURIC ACID ADDUCT see UTU600
MONOVAR see NNQ500
MONOVERIN see SDZ000
MONOVINYL PHOSPHATE see VQA400
MONOXONE see SFU500
β-MONOXYNAPHTHALENE see NAX000
MONSANTO CP-16226 see OAL000
MONSANTO CP-19699 see CON300
MONSANTO CP-40294 see MOB699
MONSANTO CP-40507 see MOB750
MONSANTO CP-43858 see TIP750
MONSANTO CP 47114 see DSQ000
MONSANTO CP-48985 see CFC500
MONSANTO CP-49674 see DOP200
MONSANTO CP 51969 see BNL250
MONSTERA DELICIOSA see SLE890
MONSUR see CBM750
MONTAN 80 see SKV100
MONTANE 40 see MRJ800
MONTANE 60 see SKV150
MONTANE 83 see SKV170
MONTANOA TOMENTOSA, leaf extract, crude see ZTS600
MONTANOA TOMENTOSA, leaf extract, semi-purified see ZTS625
MONTANOX 80 see PKL100
MONTAR see HKC000, PJL750

MONTECATINI L-561 see DRR400
MONTHYBASE see EJM500
MONTHYLE see EJM500
MONTMORILLONITE see BAV750
MONTREL see COD850
MONTROSE PROPANIL see DGI000
MONUREX see CJX750
MONURON see CJX750
MONURON-TCA see CJY000
MONUROX see CJX750
MONURUON see CJX750
MONUURON see CJX750
MONZAOMYCIN see TAB300
MONZET see USJ075
MOON see GGA000
MOONSEED see MRN100
MOOSEWOOD see LEF100
8-MOP see XDJ000
MOP see NKV000
MOPA see MFE250
MOPARI see DGP900
MOPAZIN see MFK500
MOPAZINE see MFK500
MOPERONE CHLORHYDRATE see MNN250
MOPERONE HYDROCHLORIDE see MNN250
MOPLEN see PMP500
MOPLEN RO-QG 6015 see PJS750
MOPOL M see MRD750
MOPOL S see MRD750
MOQUIZONE see MRN250
MOQUIZONE HYDROCHLORIDE see PCJ350
MORAMIDE see AFJ400
MORBOCID see FMV000
MORBUSAN see ENB500
MORDANT YELLOW 3R see SIU000
MOREPEN see AOD125
MORESTAN see ORU000
MORESTANE see ORU000
MORESTIN see EDV600
MORFAMQUAT see BJK750
MORFAX see BDF750
MORFINA (ITALIAN) see MRO500
MORFLEX 510 see TJR600
MORFOTHION (DUTCH) see MRU250
MORFOXONE see BJK750
MORIAL see MRN275
MORIN see MRN500
MORINGA OLEIFERA Lamk., extract excluding roots see MRN550
MORIPERAN see AJH000
MORISYLYTE CITRATE see MRN600
MORNIDINE see CJL500
MOROCIDE see BGB500
MORONAL see NOH500
MOROSAN see DCK759
MORPAN CBP see HCP800
MORPAN T see TCB200
MORPHACETIN see HBT500
MORPHACTIN see CDT000
MORPHANQUAT DICHLORIDE see BJK750
MORPHERIDINE see MRN675
MORPHERIDINE DIHYDROCHLORIDE see MRN675
MORPHIA see MRO500
MORPHINA see MRO500
MORPHINAN, 3-METHOXY-17-PHENETHYL-, TARTRATE, (−)- see MRN700
MORPHINAN-6-α-OL, 7,8-DIDEHYDRO-4,5-α-EPOXY-3-METHOXY-17-METHYL-, PHOSPHATE (1:1) see CNG500
MORPHINAN-6-α-OL, 7,8-DIDEHYDRO-4,5-α-EPOXY-3-METHOXY-17-METHYL-, SULFATE (2:1) (salt) see CNG750
MORPHINAN-6-ONE, 3,14-DIHYDROXY-4,5-α-EPOXY-17-METHYL-, HYDROCHLORIDE see ORG100
MORPHINAN-6-ONE, 4,5-EPOXY-3,14-DIHYDROXY-17-METHYL-, HYDROCHLORIDE, (5-α)- (9CI) see ORG100
MORPHINAN-6-ONE, 4,5-α-EPOXY-3,14-DIHYDROXY-17-(2-PROPENYL)- see NAG550
MORPHINAN-6-ONE, 4,5-α-EPOXY-3-METHOXY-17-METHYL-, TARTRATE (1:1) see DKX050
MORPHINAN, 6,7,8,14-TETRADEHYDRO-4,5-α-EPOXY-3,6-DIMETHOXY-17-METHYL-, HYDROCHLORIDE see TEN100
MORPHINE see MRO500
(−)-MORPHINE see MRO500
MORPHINE CHLORHYDRATE see MRO750
MORPHINE CHLORIDE see MRO750
MORPHINE DIACETATE see HBT500
MORPHINE HYDROCHLORIDE see MRO750
MORPHINE METHOCHLORIDE see MRP000

MORPHINE METHYLCHLORIDE see MRP000
MORPHINE-3-METHYL ETHER see CNF500
MORPHINE MONOMETHYL ETHER see CNF500
MORPHINE SULFATE see MRP250
MORPHINE SULPHATE see MRP250
MORPHINISM see MRO500
MORPHINUM see MRO500
MORPHIUM see MRO500
MORPHOLINE see MRP750
MORPHOLINE, N-AMINOPROPYL- see AMF250
MORPHOLINE, AQUEOUS MIXTURE (DOT) see MRP750
MORPHOLINE, 4-(3-(BIS(2-CHLOROETHYL)AMINO)-p-TOLUOYL)- see BHR400
MORPHOLINE, compounded with BORANE (1:1) see MRQ250
MORPHOLINEBORANE see MRQ250
MORPHOLINE, 4-BUTYL- see BRV100
4-MORPHOLINECARBOXALDEHYDE see FOA100
4-MORPHOLINECARBOXAMIDE, N-(2-((2-HYDROXY-3-(4-HYDROXYPHENOX-Y)PROPYL)AMINO)ETHYL)-, (E)-2-BUTENEDIOATE (2:1) (salt) see XAH000
MORPHOLINE, 4-(1-CYCLOPENTEN-1-YL)- see CPY800
MORPHOLINE, 4-DECANOYL- see CBF725
MORPHOLINE, 2,6-DIMETHYL-4-NITROSO-, (Z)- see NKA695
MORPHOLINE, 2,6-DIMETHYL-N-NITROSO-, (cis)- see NKA695
MORPHOLINE DISULFIDE see BKU500
MORPHOLINE, 4,4′-(DITHIODICARBONOTHIOYL)BIS-(9CI) see MRR090
4-MORPHOLINEETHANAMINE see AKA750
MORPHOLINE ETHANOL see MRQ500
4-MORPHOLINEETHANOL, HYDROCHLORIDE see HKW300
MORPHOLINE HYDROCHLORIDE see MRQ600
MORPHOLINE, N-METHYL- see MMA250
MORPHOLINE, 4-((MORPHOLINOTHIOCARBONYL)THIO)- see OPQ100
MORPHOLINE, 4-((4-MORPHOLINYLTHIO)THIOXOMETHYL)-(9CI) see OPQ100
4-MORPHOLINENONYLIC ACID see MRQ750
MORPHOLINE, 4-(1-OXODECYL)-(9CI) see CBF725
MORPHOLINE SALICYLATE see SAI100
MORPHOLINE, compounded with SALICYLIC ACID (1:1) see SAI100
MORPHOLINE and SODIUM NITRITE (1:1) see SIN675
4-MORPHOLINE SULFENYL CHLORIDE see MRR075
4-MORPHOLINETHIOCARBONYL DISULFIDE see MRR090
MORPHOLINIUM, HEXACHLOROSTANNATE(2-) (2:1) see DUO500
MORPHOLINIUM, (3-INDOLYLMETHYLENE)-, HEXACHLOROSTANNATE(2-) (2:1) see BKJ500
MORPHOLINIUM PERCHLORATE see MRR100
1-MORPHOLINOACETYL-3-PHENYL-2,3-DIHYDRO-4(1H)-QUINAZOLINONE HY-DROCHLORIDE see PCJ350
N-MORPHOLINO-β-(2-AMINOMETHYLBENZODIOXAN)-PROPIONAMIDE see MRR125
MORPHOLINOBENZENE see PFS750
MORPHOLINO-2-BENZOTHIAZOLYL DISULFIDE see BDF750
MORPHOLINOCARBONYLACETONITRILE see MRR750
4-MORPHOLINOCARBONYL-2,3-TETRAMETHYLENEQUINOLINE see MRR760
MORPHOLINO-CNU see MRR775
(1-MORPHOLINOCYCLOPENTENE) see CPY800
MORPHOLINODAUNOMYCIN see MRR850
3′-MORPHOLINO-3′-DEAMINODAUNORUBICIN see MRT100
MORPHOLINODISULFIDE see BKU500
2-(MORPHOLINODITHIO)BENZOTHIAZOLE see BDF750
β-MORPHOLINOETHYL BENZHYDRYL ETHER HYDROCHLORIDE see LFW300
3-O-(2-MORPHOLINOETHYL)MORPHINE see TCY750
β-MORPHOLINOETHYLMORPHINE see TCY750
N-2-MORPHOLINOETHYL-5-NITROIMIDAZOLE see NHH000
MORPHOLINOETHYL NORPETHIDINE DIHYDROCHLORIDE see MRN675
1-(2-MORPHOLINOETHYL)-4-PHENYLISONIPECOTIC ACID ETHYL ESTER DI-HYDROCHLORIDE see MRN675
N-(1-(MORPHOLINOMETHYL)-2,6-DIOXO-3-PIPERIDYL)PHTHALIMIDE see MRU080
2-(MORPHOLINO)-N-METHYL-N-(2-MESITYLOXYETHYL)ACETAMIDE HYDRO-CHLORIDE see MRT250
I-5-(MORPHOLINOMETHYL)-3-((5-NITROFURFURYLIDENE)AMINO)-2-OXAZOLIDI-NONEHYDROCHLORIDE see FPI150
5-MORPHOLINOMETHYL-3-(5-NITRO-2-FURFURYLIDINE-AMINO)-2-OXAZOLIDI-NONE see FPI000
5-MORPHOLINOMETHYL-3-(5-NITROFURFURYLIDINE)AMINO-2-OXAZOLIDINONE HYDROCHLORIDE see FPI100
1-MORPHOLINOMETHYLTHALIDOMIDE see MRU080
4-MORPHOLINO-2-(5-NITRO-2-THIENYL)QUINAZOLINE see MRU000
N-MORPHOLINO NONANAMIDE see MRQ750
MORPHOLINOPHOSPHONIC ACID DIMETHYL ESTER see DST800
1-(6-MORPHOLINO-3-PYRIDAZINYL)-2-(1-(tert-BUTOXYCARBONYL)-2-PROPYLI-DENE)HYDRAZINE see RFU800
3-MORPHOLINOSYDNONE IMINE HYDROCHLORIDE see MRU075
MORPHOLINO(7,8,9,10-TETRAHYDRO-11-(6H-CYCLOHEPTA(b)QUINOLINYL)) KETONE see MRU077
MORPHOLINO-THALIDOMIDE see MRU080
2-(MORPHOLINOTHIO)BENZOTHIAZOLE see BDG000
4-((MORPHOLINOTHIOCARBONYL)THIO)MORPHOLINE see OPQ100

MORPHOLIN SALICYLAT see SAI100
N-MORPHOLINYL-2-BENZOTHIAZOLYL DISULFIDE see BDF750
4-MORPHOLINYL-2-BENZOTHIAZYL DISULFIDE see BDF750
3-(2-(4-MORPHOLINYL)ETHYL)MORPHINE see TCY750
MORPHOLINYLETHYLMORPHINE see TCY750
1-(2-N-MORPHOLINYLETHYL)-5-NITROIMIDAZOLE see NHH000
1-((2-MORPHOLINYL)ETHYL)-4-PHENYL-4-PIPERIDINECARBOXYLIC ACID ETH-YL ESTER DIHYDROCHLORIDE see MRN675
MORPHOLINYLMERCAPTOBENZOTHIAZOLE see BDG000
4-MORPHOLINYLPHOSPHONIC ACID DIMETHYL ESTER see DST800
3-((6-(4-MORPHOLINYL)-3-PYRIDAZINYL)HYDRAZONO)BUTANOIC ACID 1,1-DIMETHYL ETHYL ESTER see RFU800
4-(2-MORPHOLINYL)PYROCATECHOL see MRU100
2-(4-MORPHOLINYLTHIO)BENZOTHIAZOLE see BDG000
4-((4-MORPHOLINYLTHIO)THIOXOMETHYL)MORPHOLINE see OPQ100
3-MORPHOLYLAETHYLMORPHIN (GERMAN) see TCY750
MORPHOTHION see MRU250
MORROCID see BGB500
MORSODREN see MLF250
MORSYDOMINE see MRN275
MORTON EP-227 see MLF250
MORTON EP-316 see CQI500
MORTON EP332 see DSO200
MORTON EP 333 see CJJ500
MORTON SOIL DRENCH see MLF250
MORTON SOIL-DRENCH-C see MLF250
MORTON WP-161E see ISE000
MORTOPAL see TCF250
MORYL see CBH250
MOS-708 see BDG250
MOSANON see SEH000
MOSATIL see CAR780
MOSCARDA see MAK700
MOSCHUS KETONE see MIT625
MOSE see MRU255
MOSKENE see MRU300
MOSS GREEN see COF500
MOSTEN see PMP500
MOSYLAN see DNG400
5-MOT see MFS400
MOTAZOMIN see MRN275
MOTH BALLS (DOT) see NAJ500
MOTHER-IN-LAW PLANT see CAL125
MOTHER-IN-LAW'S TONGUE PLANT see DHB309
MOTH FLAKES see NAJ500
MOTIAX see FAB500
MOTILIUM see DYB875
MOTILYN see PAG200
MOTIORANGE R see PEJ500
MOTIROT G see XRA000
MOTIROT 2R see OHA000
MOTOLON see QAK000
MOTOR BENZOL see BBL250
MOTOR SPIRIT (DOT) see GBY000
MOTOX see CDV100
MOTRIN see IIU000
MOTTENHEXE see HCI000
MOULDRITE A256 see UTU500
MOUNTAIN GREEN see COF500
MOUNTAIN LAUREL see MRU359
MOUNTAIN TOBACCO see AQY500
MOUS-CON see ZLS000
MOUSE-NOTS see SMN500
MOUSE PAK see WAT200
MOUSE-RID see SMN500
MOUSE-TOX see SMN500
MOVINYL 100 see PKQ059
MOVINYL 114 see AAX250
MOWIOL see PKP750
MOWIOL N 30-88 see PKP750
MOWIOL N 50-98 see PKP750
MOWIOL N 70-98 see PKP750
MOXADIL see AOA095
MOXALACTAM DISODIUM see LBH200
MOXAM see LBH200
MOXESTROL see MRU600
MOXIE see MEI450
MOXISYLYTE HYDROCHLORIDE see TFY000
MOXNIDAZOLE see MRU750
MOXONE see DAA800
MOZAMBIN see QAK000
8-MP see XDJ000
MP 655 see CHE750
MP 1023 see AAN000
MP 12-50 see TAB750

MP 25-38 see TAB750
MP 45-26 see TAB750
MP 1 (refractory) see EAL100
MP see POK000
3MPA see MCQ000
3-MPA see MFM000
MPA see DAZ117
3-MPC see MNM500
M.P. CHLORCAPS T.D. see TAI500
MPCM see XTJ000
MPE see MPU500
MPF 2 see UTU500
MPG see MRK500
15M-PGF2-α see MLO300
15-M3-PGF2-α see CCC100
M 4212 (PHARMACEUTICAL) see MQX775
MPI-PC see DSQ800
MPI-PENICILLIN see DSQ800
MPK 90 see PKQ250
MPK see PBN250
MPN see MNA000, NKV100
MPNU see MMY500
2'-MePO4' see DSS200
M 2 (POLYMER) see UTU500
M 70 (POLYMER) see UTU500
M 76 (POLYMER) see MCB050
MPP see FAQ900
1-MPPN see DWU800
MPS see USJ100
MPT see MPX850
MR 1 see MCB050
MR 56 see BOO630
MR 67 see MCB050
MR 84 see SCK600
MR 200 see PKB100
MR 231 see MCB050
MR 2000 see PKB100
MR see MPH300
MRAVENCAN DI-n-BUTYLCINICITY (CZECH) see DDZ000
MRAVENCAN SODNY see SHJ000
MRAVENCAN TRIBENZYLCINICITY (CZECH) see FOE000
MRAVENCAN VAPENATY (CZECH) see CAS250
MRC 910 see GIA000
MRD 108 see DWK400
MRL 41 see CMX700
MROWCZAN ETYLU (POLISH) see EKL000
MR 1 (RESIN) see MCB050
MS 1 see ASM050
MS 001 see MCB050
3-MS see CNX625
MS 21 see MCB050
MS 33 see SKV150
MS 53 see SNK000
MS-222 see EFX500
MS. 752 see SBE500
MS 1053 see DJI000
MS-1112 see BFV760
MS 1143 see DJI000
MS-4101 see FMR075
MS-5075 see FBP850
MS-ANTIGEN 40 see MRU755
MS-BENZANTHRONE see BBI250
MSC-102824 see AQN000
MS 1 (CATALYST) see ASM050
MS 33F see SKV150
MSF see MDR750
MSG see MRL500
MSK-C see CAT775
MSMA see MRL750
MSMED see ECU750
MSP 100F see MCB050
MS-R 100S see MCB050
MSTFA see MQG750
MSZYCOL see BBQ500
MT-45 see MRU757
MT-141 see CCS365
MT 14-411 see CMY525
MT see HOA000
MTs see MIF760
MTB 51 see DJM800, XCJ000
MTB see HNY500
MTBE see MHV859
MTBHQ see BRM500
MTD see DTQ400, TGL750
MTDQ see MRU760

MTEAI see MRU775
M.T.F. see DUV600
MTIC see MQB750
MTIQ see MPO750
MTMC see MIB750
M.T. MUCORETTES see MPN500
MTN see TGT250
MTQ see QAK000
MTQ HYDROCHLORIDE see MDT250
MTU see MPW500
MTX see MDV500
MTX DISODIUM see MDV600
MTX SODIUM see MDV750
MUCAESTHIN see BQA010
MUCAINE see DTL200
MUCALAN see IHO200
MUCIC ACID see GAR000
MUCICLAR see CBR675
MUCIDRIL see DPE000
MUCIDRINA see VGP000
MUCINOL see AOO490
MUCITUX see ECU550
MUCOCHLORIC ACID see MRU900
MUCOCHLORIC ANHYDRIDE see MRV000
MUCOCIS see CBR675
MUCODYNE see CBR675
MUCOFLUID see MDK875
MUCOLASE see CBR675
MUCOLEX see CBR675
MUCOLYSIN see MCI375
MUCOLYTICUM see ACH000
MUCOLYTICUM LAPPE see ACH000
MUCOMYCIN see MRV250
MUCOMYST see ACH000
MUCONOMYCIN A see MRV500
MUCOPOLYSACCHARIDE, POLYSULFURIC ACID ESTER see MRV525
MUCOPOLYSACCHARIDIPOLY SCHWEFELSAEUREESTER (GERMAN) see MRV525
MUCOPRONT see CBR675
MUCORAMA see PMJ500
MUCOSOLVAN see AHJ500
MUCOSOLVIN see ACH000
MUCOXIN see DTL200
MUCUNA MONOSPERMA DC. ex Wight (extract excluding roots) see MRV600
MUDAR see COD675
MUGAN see CBM750
MUGB see GLC100
MUGUET (CANADA) see LFT700
MUIRAMID see AAI250
MULDAMINE see MRV750
MUL F 66 see PKL750
MULHOUSE WHITE see LDY000
MULSIFEROL see VSZ100
MULSOPAQUE see ELQ500
MULTACID YELLOW 3R see SGP500
MULTAMAT see DQM600
MULTERGAN see MRW000
MULTERGAN METHYL SULFATE see MRW000
MULTEZIN see MRW000
MULTICHLOR see CDP000
MULTICUER BROWN MPH see SGP500
MULTIFLEX MM see CAT775
MULTIFUGE CITRATE see PIJ500
MULTIN see HIM000
MULTIPROP see CDT000
MUNDISAL see CMG000
MU OIL TREE see TOA275
MURACIL see MPW500
MURATOX see DJI000
MURCIL see MDQ250
MUREL see VBK000
MUREX see GFA000
MURFOS see PAK000
MURFOTOX see DJI000
MURFULVIN see GKE000
MURIATE of PLATINUM see PJE000
MURIATIC ACID see HHL000
MURIATIC ETHER see EHH000
MURIOL see CJJ000
MURITAN see DEQ000
MUROTOX see DJI000
MUROX see CHJ300
MURPHOTOX see DJI000
MURUTOX see DJI000

MURVESCO see CJR500
MURVIN see CBM750
MUSARIL see CFG750
MUSCALM see MRW125
MUSCALURE see TJF400
MUSCAMONE see TJF400
MUSCARIN see MRW250
dl-MUSCARINE see MRW250
MUSCARINE see MRW250
MUSCIMOL see AKT750
MUSCLE ADENYLIC ACID see AOA125
MUSCONE see MIT625
MUSCULAMINE see DCC400
MUSCULAMINE TETRAHYDROCHLORIDE see GEK000
MUSCULARON see IHH000
MUSETTAMYCIN see APV000
MUSHROOM AMNITA RUBESCENS TOXIN see AHI500
MUSHROOMS see MRW269
MUSK see MRW250
MUSK 36A see ACL750
MUSK AMBRETTE see BRU500, OKU100
MUSK AMBRETTE (NATURAL) see OKU100
MUSKARIN see MRW250
MUSKEL see CKF500
MUSKEL-TRANCOPAL see CKF500
MUSK NATURAL see OKU100
MUSKONE see MIT625
MUSK R 1 see OKW100
MUSK-T see EJQ500
MUSK TIBETENE see MRW272
MUSK XYLENE (DOT) see TML750
MUSK XYLENE see TML750
MUSK XYLOL see TML750
MUSQUASH POISON see WAT325
MUSQUASH ROOT see WAT325
MUSSEL POISON DIHYDROCHLORIDE see SBA500
N-MUSTARD (GERMAN) see BIE500
MUSTARD CHLOROHYDRIN see CHC000
MUSTARD GAS see BIH250
MUSTARD GAS SULFONE see BIH500
MUSTARD HD see BIH250
MUSTARD OIL see AGJ250
MUSTARD SULFONE see BIH500
MUSTARD VAPOR see BIH250
MUSTARGEN see BIE250, BIE500
MUSTARGEN HYDROCHLORIDE see BIE500
MUSTINE see BIE250
MUSTINE HYDROCHLOR see BIE500
MUSTINE HYDROCHLORIDE see BIE500
MUSTRON see CFA750
MUSUET SYNTHETIC see CMS850
MUSUETTINE PRINCIPLE see CMS850
MUTABASE see DCQ700
MUTAGEN see BIE250
MUTAMYCIN see AHK500
MUTAMYCIN (MITOMYCIN for INJECTION) see AHK500
MUTHESA see DTL200
MUTHMANN'S LIQUID see ACK250
MUTOXIN see DAD200
MUZOLIMINE see EAE675
MV see MAK275
MV 119A see DLK200
MVEEG (RUSSIAN) see EJL500
MVNA see NKY000
MW 30 see MCB050
M 33W see MCB050
MX 40 see MCB050
2M-4X see SIL500
MX 705 see MCB050
MX 4500 see SMQ500
MX 5514 see SMQ500
MX 5516 see SMQ500
MX 5517-02 see SMQ500
MXDA see XHS800
2M-4XP see CIR325
MY/68 see FAF000
MY 33-7 see TGJ625
MY 41-6 see AJC000
MY-5116 see IHR200
MYACINE see NCD550
MYACYNE see NCE000
MYALEX see CKK250
MYAMBUTOL see EDW875
MYANIL see GGS000
MYARSENOL see SNR000

MYASUL see AKO500
MYBASAN see ILD000
MY-B-DEN see AOA125
MYBORIN see MRW275
MYCAIFRADIN SULFATE see NCG000
MYCANDEN see IFA000
MYCARDOL see PBC250
MYCELAX see MRX500
MYCELEX see MRX500
MYCHEL see CDP250
MYCIFRADIN see NCE000
MYCIFRADIN-N see NCG000
MYCIGIENT see NCG000
MYCILAN see IFA000
MYCINAMICIN 1 see MRW500
MYCINAMICINS II see MRW750
MYCINOL see CDP250
MYCIVIN see LGC200
1-MYCOBACIDIN see CCI500
MYCOBACIDIN see CMP885
MYCOBACTYL see GBB500
MYCOBAN see SJL500
MYCOBUTOL see EDW875
MYCOCURAN see GGS000
MYCOFARM see BFD250
MYCOHEPTIN see MRW800
MYCOHEPTYNE see MRW800
MYCOIN see CMV000
MYCOLUTEIN see DTU200
MYCOPHENOLIC ACID see MRX000
MYCOPHYT see PIF750
MYCO-POLYCID see CDY325
MYCOSHIELD TMQTHC 20 see HOH500
MYCOSPOR see BGA825
MYCOSPORIN see MRX500
MYCOSTATIN 20 see NOH500
MYCOSTATIN see NOH500
MYCOTICIN (1:1) see MRY000
MYCOTICIN see MRY000
MYCOTOXIN F2 see ZAT000
MYCOZOL see TEX000
MYCRONIL see BJK500
MYDECAMYCIN see MBY150
MYDETON see TGK200
MYDFRIN see SPC500
MYDOCALM see MRW125, TGK200
MYDRIAL see AOA250
MYDRIASIN see MGR250
MYDRIATIN see NNM000
MYDRIATINE see NNN000, PMJ500
MYEBROL see DDP600
MYELOBROMOL see DDP600
MYELOLEUKON see BOT250
MYELOTRAST see SHX000, TDQ230
MYELOTRAST DI-N-METHYLGLUCAMINE SALT see IDJ500
MYGAL see BET000
MYKOSTIN see VSZ100
MYLAR see PKF750
MYLAXEN see HEG000
MYLEPSIN see DBB200
MYLEPSINUM see DBB200
MYLERAN see BOT250
MYLIS see DAL040
MYLODORM see AMX750
MYLOFANOL see PGG000
MYLON (CZECH) see DSB200
MYLONE 85 see DSB200
MYLONE see DSB200
MYLOSAR see ARY000
MYLOSUL see AKO500
MYNOSEDIN see IIU000
MYOCAINE see RLU000
MYOCHOLINE see HOA500
MYOCHRYSINE see GJC000
MYOCOL see AEH750
MYOCON see NGY000
MYOCORD see TAL475
MYOCRISIN see GJC000
MYODETENSINE see GGS000
MYODIGIN see DKL800
MYODIL see ELQ500
MYODYL see ELQ500
MYOFER 100 see IGS000
MYOFLEXINE see CDQ750
MYOGLYCERIN see NGY000

MYO-INOSISTOL HEXAKISPHOSPHATE see PIB250
MYO-INOSITOL HEXAPHOSPHATE see PIB250
MYOLASTAN see CFG750
MYOLAX see GGS000
MYOLAXENE see GKK000
MYOLYSEEN see PIL550
MYOMYCIN B see MRY100
MYOMYCIN SULFATE see MRY250
MYOPAN see GGS000
MYOPLEGINE see HLC500
MYOPONE see WBJ700
MYOPORUM LAETUM see MRY600
MYORDIL see TNJ750
MYORELAX see RLU000
MYOREXON see CCK125
MYOSALVARSAN see SNR000
MYOSCAINE see RLU000
MYOSEROL see GGS000
MYOSPAZ see PFJ000
MYOSTHENINE see VGP000
MYOSTIBIN see AQH800, AQI250
MYOSTON see AOA125
MYOTOLON see AOO300
MYOTRATE "10" see PBC250
MYOTRIPHOS see ARQ500
MYOXANE see GGS000
MYPROZINE see PIF750
MYRABOLAM TANNIN see MRZ100
MYRAFORM see PKQ059
MYRCENE see MRZ150
MYRCENOL see MLO250
MYRCENYL ACETATE see AAW500
MYRCIA OIL see BAT500, LBK000
MYRICA CERIFERA see WBA000
MYRICIA OIL see BAT500
MYRINGACAINE DROPS see CHW675
MYRISIIC ALCOHOL see TBY250
MYRISTALDEHYDE see TBX500
MYRISTAN DI-n-BUTYLCINICITY (CZECH) see BLH309
MYRISTICA see NOG000
MYRISTIC ACID see MSA250
MYRISTIC ACID, BUTYL ESTER see MSA300
MYRISTIC ACID, ISOPROPYL ESTER see IQN000
MYRISTIC ACID, SODIUM SALT see SIN900
MYRISTIC ALDEHYDE see TBX500
MYRISTICA OIL see NOG500
MYRISTICIN see MSA500
9-MYRISTOYL-1,7,8-ANTHRACENETRIOL see MSA750
10-MYRISTOYL-1,8,9-ANTHRACENETRIOL see MSA750
1-MYRISTOYLAZIRIDINE see MSB000
MYRISTOYLETHYLENEIMINE see MSB000
N-MYRISTOYLOXY-AAF see ACS000
N-MYRISTOYLOXY-AAIF see MSB100
N-MYRISTOYLOXY-N-ACETYL-2-AMINOFLUORENE see ACS000
N-MYRISTOYLOXY-N-ACETYL-2-AMINO-7-IODOFLUORENE see MSB100
N-MYRISTOYLOXY-N-MYRISTOYL-2-AMINOFLUORENE see MSB250
MYRISTYL ALCOHOL (mixed isomers) see TBY500
MYRISTYL-γ-PICOLINIUM CHLORIDE see MSB500
MYRISTYL STEARATE see TCB100
MYRISTYL SULFATE, SODIUM SALT see SIO000
MYRISTYLTRIMETHYLAMMONIUM BROMIDE see TCB200
MYRITICALORIN see RSU000
MYRITICOALORIN see RSU000
MYRITOL 318 see CBF710
MYROBALANS TANNIN see MSB750
MYRRH OIL see MSB775
MYRTAN TANNIN see MSC000
MYRTENAL see FNK150
(+)-MYRTENYL ACETATE see MSC050
MYRTENYL ACETATE see MSC050
MYRTLE OIL see OGU000
MYSEDON see DBB200
MYSOLINE see DBB200
MYSONE see HHQ800
MYSORITE see ARM262
MYSTECLIN-F see AOC500
MYSTER GRASS see DAE100
MYSTERIA see CNX800
MYSTOX WFA see BGJ750
MYSURAN see MSC100
MYSURAN CHLORIDE see MSC100
MYTAB see TCB200
MYTELASE see MSC100
MYTELASE CHLORIDE see MSC100
MYTOMYCIN see AHK500

MYTRATE see VGP000
MYVAK see VSK900
MYVAX see VSK900
MYVIZONE see FNF000
MYVPACK see VSK600
MYXIN see HLT100
MYXOVIROMYCIN see AHN625


N 5 see EHG100
N-9 see NNB300
N 34 see CAT775
N 50 see UTU500
N 68 see BRS000
Ni 270 see NCW500
N-399 see PEM750
N 521 see DSB200
N-553 see MRW125
N 642 see PIR100
N 714 see TAF675
N 715 see THL500
N-746 see CLX250
N 869 see SIL550
N 2038 see ZJS300
N 2404 see CJD650
N 2790 see FMU045
N 3051 see BSG000
N 4328 see EOO000
N 4548 see MOB250
N 7001 see AEG875
N 7009 see FMO129, FMO150
168N15 see SMQ500
NA see EID000, NBE500
NA-22 see IAQ000
NA-53 see BCP000
NA 97 see PAF625
NA-101 see TDX000
NA-872 see AHJ250
NA 872 see AHJ500
N-1544A see PFC750
NAA 800 see NAK500
NAAM see NAK000
NAB 365 see VHA350
NAB see NJK150
NABAC see HCL000
NABADIAL see AOO475
NABAM see DXD200
NABAME (FRENCH) see DXD200
NAB 365Cl see VHA350
NABOLIN see MPN500
NABOR BRILLIANT PINK 28 see CMM768
NABOR ORANGE G see CMM764
NABUMETONE see MFA300
NAC see ACH000, CBM750
NACARAT A EXPORT see HJF500
NACCANOL NR see DXW200
NACCONATE-100 see TGM740
NACCONATE 300 see MJP400
NACCONATE 400 see BBP000
NACCONATE H 12 see MJM600
NACCONATE 1OO see TGM750
NACCONOL LAL see SIB700
NACCONOL 98SA see LBU100
NACELAN BLUE CBG see DMM400
NACELAN BLUE G see TBG700
NACELAN BLUE KLT see MGG250
NACELAN FAST YELLOW CG see AAQ250
NACELAN PINK 3B see DBX000
NACELAN PINK B see AKE250
NACELAN VIOLET 4B see AKP250
NACELAN VIOLET 4R see DBP000
NACIMYCIN see RKK000
NACLEX see BDE250
NACM-CELLULOSE SALT see SFO500
NACRICLASINE see LJC000
NAC-TB see ACH000
NACYCLYL see EDR000
β-NAD see CNF390
NAD+ see CNF390
NAD see CNF390, NAK000
NADEINE see DKW800
NA-DESOXYCHOLAT (GERMAN) see SGE000
NADIC METHYL ANHYDRIDE see NAC000
NADIDE see CNF390
NADISAL see SJO000

NADISAN see BSM000
NADIZAN see BSM000
NADOLOL see CNR675
NADONE see CPC000
NADOXOLOL HYDROCHLORIDE see NAC500
NADOZONE see BRF500
β-NADP see CNF400
NADP see CNF400
NAD PHOSPHATE see CNF400
NADROTHYRON D see SKJ300
NAEPAINE see PBV750
Na-AESCINAT see EDM000
NAFCILLIN SODIUM SALT see SGS500
NAFEEN see SHF500
NAFENOIC ACID see MCB500
NAFENOPIN see MCB500
NAFIVERINE DIHYDROCHLORIDE see NAD000
NAFKA see ROH900
NAFKA CRYSTAL GUM see ROH900
NAFKA KRISTALGOM see ROH900
NAFOXIDINE see NAD500
NAFOXIDINE HYDROCHLORIDE see NAD750
NAFRINE see AEX000
NAFRONYL see NAE000
NAFRONYL OXALATE see NAE100
NAFTALEN (POLISH) see NAJ500
NAFTALIN-BUTIL-SOLFONATO (ITALIAN) see NBS700
NAFTALOFOS see HMV000
NAFTIDROFURYL see NAE000
NAFTIDROFURYL OXALATE see NAE100
β-NAFTILAMINA (ITALIAN) see NBE500
1-NAFTILAMINA (SPANISH) see NBE700
1-NAFTIL-TIOUREA (ITALIAN) see AQN635
NAFTIPRAMIDE see IOU000
NAFTIZIN see NAH550
2-NAFTOL (DUTCH) see NAX000
β-NAFTOL (DUTCH) see NAX000
2-NAFTOLO (ITALIAN) see NAX000
β-NAFTOLO (ITALIAN) see NAX000
NAFTOPEN see SGS500
α-NAFTYLAMIN (CZECH) see NBE700
β-NAFTYLAMIN (CZECH) see NBE500
2-NAFTYLAMIN-5,7-DISULFONAN SODNY (CZECH) see ALH500
1-NAFTYLAMINE (DUTCH) see NBE700
2-NAFTYLAMINE (DUTCH) see NBE500
4-(2-NAFTYLAMINO)FENOL see NBF500
1-NAFTYLESTER KYSELINY METHYLKARBAMINOVE see CBM750
α-NAFTYL-N-METHYLKARBAMAT see CBM750
β-NAFTYLOAMINA (POLISH) see NBE500
1-NAFTYLTHIOUREUM (DUTCH) see AQN635
NAFTYPRAMIDE see IOU000
NAFUSAKU see NAK500
NAGANOL see BAT000
NAGARMOTHA OIL see NAE505
NAGARMUSTA OIL see NAE505
NAGARSE see BAC000
NAGASE (BACILLUS SUBTILIS) see NAE507
NAGENT see GCU050
NAGENTIS see GCU050
NAGRAVON see VSZ000
NAH 80 see SHO500
NAH see NCQ900
Na III HYDROCHLORIDE see DWF200
NAHP see NJJ950
NA 65 HYDROCHLORIDE see PIV500
NAILAMIDE YELLOW BROWN E-L see SGP500
NAIRIT see PJQ050
NAIXAN see MFA500
NAJA FLAVA VENOM see NAE510
NAJA HAJE ANNULIFERA VENOM see NAE512
NAJA HAJE VENOM see NAE515
NAJA MELANOLEUCA VENOM see NAE875
NAJA NAJA ATRA VENOM see NAF000
NAJA NAJA KAOUTHIA VENOM see NAF200
NAJA NAJA NAJA VENOM see ICC700
NAJA NAJA SIAMENSIS VENOM see NAF250
NAJA NIGRICOLLIS VENOM see NAG000
NAJA NIVEA VENOM see NAG200
NAKED LADY LILY see AHI635
NAKO BROWN R see ALT250
NAKO FAST GREY BL see CMU320
NAKO H see PEY500
NAKO TEG see ALT500
NAKO TGG see REA000
NAKO TMT see TGL750

NAKO TRB see NAW500
NAKO TSA see DBO400
NAKO YELLOW EGA see ALT000
NAKVA see HII500
NALADOR see SOU650
NALCAMINE G-13 see AHP500
NALCAST see SCK600
NALCO 680 see AHG000
NALCOAG see SCH000
NALCON 240 see BMS250
NALCON 243 see DSB200
NALCROM see CNX825
NALDE see AHL500
NALED see NAG400
NALFLOC 636 see ADW200
NALFON see FAP100
NALGESIC see FAP100
NALIDIC ACID see EID000
NALIDICRON see EID000
NALIDIXIC ACID see EID000
NALIDIXIN see EID000
NALINE HYDROCHLORIDE see NAG500
NALITUCSAN see EID000
NALKIL see BMM650
NALLINE see AFT500
NALORFINA see AFT500
NALORPHINE see AFT500
NALORPHINE HYDROCHLORIDE see NAG500
NALORPHINIUM see AFT500
NALOX see MMN250
NALOXIPHAN see AGI000
l-NALOXONE see NAG550
NALOXONE see NAG550
NALOXONE HYDROCHLORIDE see NAH000
NALTREXONE see CQF099
NALTROPINE see AGM250
NALUTORAL see GEK500
NALUTRON see PMH500
NAM see NCR000
NAMATE see DXE600
NAMEKIL see TDW500
NaAMP see AEM750
NAMPHEN see XQS000
NAMURON see TDA500
5-NAN see NEJ500
NANA-HONUA (HAWAII) see AOO825
NANAOMYCIN A see RMK200
NANCHOR see RMA500
NANDERVIT-N see NCR000
NANDROLIN see DYF450
NANDROLONE PHENPROPIONATE see DYF450
NANDROLONE PHENYLPROPIONATE see DYF450
NANDRON see EAN700
NANI-ALI'I (HAWAII) see AFQ625
NANI-O-HILO (HAWAII) see PCB300
NANKAI ACID ORANGE I see FAG010
NANKER see RMA500
NANKOR see RMA500
NANM see AFT500
NANOFIN see LIQ550
NANOPHYN see LIQ550
NANOPLAST FB 101 see MCB050
NANSA 1042P see LBU100
NANSA SSA see LBU100
NAOP see AGM125
NAOTIN see NCQ900
NAPA see HIM000
NAPACETIN see IIU000
NAPCLOR-G see SJA000
NAPENTAL see NBU000
NAPHAZOLINE see NAH500
NAPHAZOLINE HYDROCHLORIDE see NCW000
NAPHAZOLINE NITRATE see NAH550
NAPHCON see NCW000
NAPHCON FORTE see NCW000
NAPHID see NAR000
NAPHTAMINE BLUE 2B see CMO000, CMO250
NAPHTAMINE BLUE 10G see CMO500
NAPHTAMINE BLUE RW see CMO600
NAPHTAMINE FAST RED F see CMO870
NAPHTAMINE GREEN B see CMO840
NAPHTAMINE VIOLET N see CMP000
NAPHTHA, hydrotreated see NAI500
NAPHTHA see NAI500, ROU000
NAPHTHACAINE HYDROCHLORIDE see NAH800

NAPHTHALOXIMIDODIETHYL THIOPHOSPHATE see NAQ500
α-NAPHTHALTHIOHARNSTOFF (GERMAN) see AQN635
NAPHTHAMINE DARK GREEN B see CMO830
NAPHTHANE see DAE800
NAPHTHANIL BLUE B BASE see DCJ200
NAPHTHANIL RED B BASE see NEQ000
NAPHTHANIL RED 3G BASE see KDA050
NAPHTHANIL SCARLET 2G BASE see DEO295
NAPHTHANIL SCARLET G BASE see NMP500
NAPHTHANTHRACENE see BBC250
NAPHTHANTHRONE see BBI250
Δ^{5,7,9}-NAPHTHANTRIENE see TCX500
NAPHTHA (PETROLEUM), CATALYTIC REFORMED see NAQ510
NAPHTHA (PETROLEUM), HEAVY CATALYTIC CRACKED see NAQ520
NAPHTHA (PETROLEUM), HEAVY CATALYTIC REFORMED see NAQ530
NAPHTHA (PETROLEUM), LIGHT CATALYTIC CRACKED see NAQ540
NAPHTHA (PETROLEUM), LIGHT CATALYTIC REFORMED see NAQ550
NAPHTHA (PETROLEUM), LIGHT STRAIGHT-RUN see NAQ560
NAPHTHA (PETROLEUM), SWEETENED see NAQ570
NAPHTHA (PETROLEUM), THERMAL CRACKED see NAQ580
1,2 NAPHTHAQUINONE see NBA000
NAPHTHA SAFETY SOLVENT see SLU500
NAPHTHA, petroleum (UN1255) (DOT) see NAI500
NAPHTHA, solvent (UN1256) (DOT) see NAI500
NAPHTHA (UN2553) (DOT) see NAI500
NAPHTHAZINE GREEN S see ADF000
NAPHTHAZINE ROSE 2G see CMM300
NAPHTHENATE de COBALT (FRENCH) see NAR500
NAPHTHENE see NAJ500
NAPHTHENIC ACID see NAR000
NAPHTHENIC ACID ALUMINUM SALT see NAR100
NAPHTHENIC ACID, COBALT SALT see NAR500
NAPHTHENIC ACID, COPPER SALT see NAS000
NAPHTHENIC ACID, LEAD SALT see NAS500
NAPHTHENIC ACID, PHENYLMERCURY SALT see PFP000
NAPHTHENIC ACIDS, ALUMINUM SALT see NAR100
NAPHTHENIC ACID, ZINC SALT see NAT000
NAPHTHENIC BASE LUBE STOCK see MQV845
NAPHTHENIC OILS (PETROLEUM), CATALYTIC DEWAXED HEAVY (9CI) see MQV865
NAPHTHENIC OILS (PETROLEUM), CATALYTIC DEWAXED LIGHT (9CI) see MQV867
NAPHTH(2′,3′:6,7)INDOLO(2,3-c)DINAPHTHO(2,3-a:2′,3′-i)CARBAZOLE-5,10,15,17,22,24-HEXONE see CMU770
NAPHTHIOMATE T see TGB475
1,4-NAPHTHIONIC ACID see ALI000
NAPHTHIONIC ACID see ALI000
NAPHTHIPRAMIDE see IOU000
NAPHTHISEN see NAH550
NAPHTHIZEN see NAH550
NAPHTHIZINE see NAH500
NAPHTHOCAINE HYDROCHLORIDE see NAH800
NAPHTHO(1,2,3,4-def)CHRYSENE see NAT500
α-β-NAPHTHO-2,3-DIPHENYL-TRIAZOLIUM CHLORID (GERMAN) see DWH875
NAPHTHO(1,8-gh:4,5-g′h′)DIQUINOLINE see NAU000
NAPHTHO(1,8-gh:5,4-g′h′)DIQUINOLINE see NAU500
NAPHTHOELAN NAVY BLUE see PFU500
NAPHTHOELAN RED B BASE see NEQ000
α-NAPHTHOFLAVONE see NBI100
β-NAPHTHOFLAVONE see NAU525
1-NAPHTHOIC ACID see NAV490
2-NAPHTHOIC ACID see NAV500
α-NAPHTHOIC ACID see NAV490
β-NAPHTHOIC ACID see NAV500
2-NAPHTHOIC ACID, 3-HYDROXY- see HMX520
2-NAPHTHOIC ACID, 4,4′-METHYLENEBIS(3-HYDROXY)-, compounded with (E)-1,4,5,6-TETRAHYDRO-1-METHYL-2-(2-(2-THIENYL)VINYL)PYRIMIDINE (1:1) see POK575
1-NAPHTHOL see NAW500
2-NAPHTHOL see NAX000
α-NAPHTHOL see NAW500
β-NAPHTHOL see NAX000
NAPHTHOL see NAW000
1,8-NAPHTHOLACTAM see NAX100
NAPHTHOLACTAM see NAX100
2-NAPHTHOL, 5-AMINO- see ALJ500
NAPHTHOL B see NAX000
NAPHTHOL B.O.N. see HMX520
2-NAPHTHOL-3-CARBOXYLIC ACID see HMX500
2-NAPHTHOL, DECAHYDRO-, ACETATE see DAF100
2-NAPHTHOL, DECAHYDRO-, FORMATE see DAF150
2-NAPHTHOL-3,6-DISULFONIC ACID SODIUM SALT see ROF300
2-NAPHTHOL ETHYL ETHER see EEY500
β-NAPHTHOL ETHYL ETHER see EEY500
NAPHTHOL GREEN see NAX500

NAPHTHOL GREEN B see NAX500
β-NAPHTHOL ISOBUTYL ETHER see NBJ000
1-NAPHTHOL N-METHYLCARBAMATE see CBM750
α-NAPHTHOL ORANGE see FAG010
β-NAPHTHOL ORANGE see CMM220
NAPHTHOL ORANGE see CMM220, FAG010
2-NAPHTHOL ORANGE II see CMM220
NAPHTHOL RED B 20-7575 see NAY000
NAPHTHOL RED B see FAG020, NAY000
NAPHTHOL RED DEEP 10459 see NAY000
NAPHTHOL RED D TONER 35-6001 see NAY000
2-NAPHTHOL-6-SULFONIC ACID see HMU500
β-NAPHTHOL-6-SULFONIC ACID see HMU500
β-NAPHTHOLSULFONIC ACID S see HMU500
1-NAPHTHOL, 1,2,3,4-TETRAHYDRO- see TDI350
NAPHTHOL YELLOW see DUX800
3H-NAPHTHO(2,1-b)PYRAN-2-CARBOXYLIC ACID, 3-OXO-, ETHYL ESTER see OOI200
4H-NAPHTHO(1,2-b)PYRAN-4-ONE, 2-PHENYL- see NBI100
1H-NAPHTHO(2,1-b)PYRAN-1-ONE, 3-PHENYL- see NAU525
NAPHTHOPYRIN see NAY500
NAPHTHO(2,3-f)QUINOLINE see NAZ000
1,2-NAPHTHOQUINONE see NBA000
1,4-NAPHTHOQUINONE see NBA500
α-NAPHTHOQUINONE see NBA500
β-NAPHTHOQUINONE see NBA000
1,4-NAPHTHOQUINONE, 2-HYDROXY- see HMX600
1,4-NAPHTHOQUINONE, 5-HYDROXY- see WAT000
1,4-NAPHTHOQUINONE, 8-HYDROXY- see WAT000
1,4-NAPHTHOQUINONE, 2-METHOXY- see MEY800
β-NAPHTHOQUINONE-4-SULFONATE SODIUM SALT see DLK000
NAPHTHORESORCINOL see NAN000
NAPHTHOSOL FAST RED KB BASE see CLK225
NAPHTHOSTYRIL see NAX100
NAPHTHO(1,2-e)THIANAPHTHENO(3,2-b)PYRIDINE see BCF500
NAPHTHO(2,1-e)THIANAPHTHENO(3,2-b)PYRIDINE see BCF750
α-NAPHTHOTHIOUREA see AQN635
2H-NAPHTHO(1,2-d)TRIAZOLE, 2-(4-STYRYL-3-SULFOPHENYL)-7-SULFO-, DI-SODIUMSALT see DXG025
NAPHTHO(1,2-d)TRIAZOLE-7-SULFONIC ACID,2-(4-(2-PHENYLETHENYL)-3-SUL-FOPHENYL)-, DISODIUM see DXG025
4-(2H-NAPHTHO(1,2-d)TRIAZOL-2-YL)-2-STILBENESULFONIC ACID SODIUM SALT see TGE155
2-NAPHTHOXYACETIC ACID see NBJ100
β-NAPHTHOXYACETIC ACID see NBJ100
1-(2-NAPHTHOYL)-AZIRIDINE see NBC000
β-NAPHTHOYLETHYLENEIMINE see NBC000
1-NAPHTHYLACETAMIDE see NAK000
α-NAPHTHYLACETAMIDE see NAK000
α-NAPHTHYLACETIC see NAK500
1-NAPHTHYLACETIC ACID see NAK500
α-NAPHTHYLACETIC ACID see NAK500
NAPHTHYLACETIC ACID see NAK500
N-2-NAPHTHYLACETOHYDROXAMIC ACID see NBD000
α-(1-NAPHTHYL)ACETONITRILE see NBD500
α-NAPHTHYL ACETONITRILE see NBD500
β-NAPHTHYL ALCOHOL see NAX000
1-NAPHTHYLALDEHYDE see NAJ000
α-NAPHTHYLALDEHYDE see NAJ000
1-NAPHTHYLAMIN (GERMAN) see NBE700
2-NAPHTHYLAMIN (GERMAN) see NBE500
β-NAPHTHYLAMIN (GERMAN) see NBE500
1-NAPHTHYLAMINE see NBE700
2-NAPHTHYLAMINE see NBE500
6-NAPHTHYLAMINE see NBE500
α-NAPHTHYLAMINE see NBE700
β-NAPHTHYLAMINE see NBE500
NAPHTHYLAMINE BLUE see CMO250
2-NAPHTHYLAMINE-4,8-DISULFONIC ACID see ALH250
β-NAPHTHYLAMINE-4,8-DISULFONIC ACID see ALH250
2-NAPHTHYLAMINE-6,8-DISULFONIC ACID see NBE850
β-NAPHTHYLAMINEDISULFONIC ACID see ALH250
2-NAPHTHYLAMINE-1-d-GLUCOSIDURONIC ACID see ALL000
1-NAPHTHYLAMINE HYDROCHLORIDE see NBF000
α-NAPHTHYLAMINE HYDROCHLORIDE see NBF000
2-NAPHTHYLAMINE MUSTARD see NBE500
NAPHTHYLAMINE MUSTARD see BIF250
2-NAPHTHYLAMINE-1-SULFONIC ACID see ALH750
5-NAPHTHYLAMINE-2-SULFONIC ACID see ALI250
1-NAPHTHYLAMINE-4-SULFONIC ACID see ALI000
1-NAPHTHYLAMINE-6-SULFONIC ACID see ALI250
2-NAPHTHYLAMINE-8-SULFONIC ACID see ALI300
α-NAPHTHYLAMINE-p-SULFONIC ACID see ALI000
4-(2-NAPHTHYLAMINO)PHENOL see NBF500
p-(2-NAPHTHYLAMINO)PHENOL see NBF500

N-(1-NAPHTHYL)ANILINE see PFT250
N-(2-NAPHTHYL)ANILINE see PFT500
4-(1-NAPHTHYLAZO)-2-NAPHTHYLAMINE see NBG000
4-(1-NAPHTHYLAZO)-m-PHENYLENEDIAMINE see NBG500
5-(β-NAPHTHYLAZO)-2,4,6-TRIAMINOPYRIMIDINE see NBO500
2-NAPHTHYLBIS(2-CHLOROETHYL)AMINE see BIF250
β-NAPHTHYL-BIS-(β-CHLOROETHYL)AMINE see BIF250
α-NAPHTHYLCARBOXALDEHYDE see NAJ000
α-NAPHTHYLCARBOXYLIC ACID see NAV490
1-NAPHTHYL CHLOROCARBONATE see NBH200
1-NAPHTHYL CHLOROFORMATE see NBH200
α-NAPHTHYL CHLOROFORMATE see NBH200
β-NAPHTHYL-DI-(2-CHLOROETHYL)AMINE see BIF250
α-NAPHTHYLENEACETIC ACID see NAK500
1,2-(1,8-NAPHTHYLENE)BENZENE see FDF000
1,5-NAPHTHYLENEDIAMINE see NAM000
1,5-NAPHTHYLENE DISULFONIC ACID see AQY400
NAPHTHYLENEETHYLENE see AAF275
NAPHTHYLENE YELLOW see DUX800
α-NAPHTHYLESSIGSAEURE (GERMAN) see NAK500
N-(1-NAPHTHYL)ETHYLENEDIAMINE DIHYDROCHLORIDE see NBH500
α-NAPHTHYLFLAVONE see NBI100
O-(2-NAPHTHYL)GLYCOLIC ACID see NBJ100
β-NAPHTHYL HYDROXIDE see NAX000
1-NAPHTHYLHYDROXYLAMINE see HIX000
2-NAPHTHYLHYDROXYLAMINE see NBI500
N-1-NAPHTHYLHYDROXYLAMINE see HIX000
β-NAPHTHYL ISOBUTYL ETHER see NBJ000
(2-NAPHTHYL)-1-ISOPROPYLAMINOETHANOL see INS000
NAPHTHYLISOPROTERENOL see INS000
NAPHTHYLISOPROTERENOL HYDROCHLORIDE see INT000
1-NAPHTHYL ISOTHIOCYANATE see ISN000
α-NAPHTHYL ISOTHIOCYANATE see ISN000
2-NAPHTHYL MERCAPTAN see NAP500
β-NAPHTHYL MERCAPTAN see NAP500
1-NAPHTHYL METHYLCARBAMATE see CBM750
α-NAPHTHYL METHYLCARBAMATE see CBM750
1-NAPHTHYL N-METHYLCARBAMATE see CBM750
α-NAPHTHYL N-METHYLCARBAMATE see CBM750
2-(1-NAPHTHYLMETHYL)-2-IMIDAZOLINE see NAH500
2-(α-NAPHTHYLMETHYL)-IMIDAZOLINE see NAH500
α-NAPHTHYLMETHYL IMIDAZOLINE see NAH500
2-(1-NAPHTHYLMETHYL)IMIDAZOLINE HYDROCHLORIDE see NCW000
2-(1-NAPHTHYLMETHYL)-2-IMIDAZOLINE HYDROCHLORIDE see NCW000
1-NAPHTHYL-N-METHYL-KARBAMAT see CBM750
1-NAPHTHYL METHYL KETONE see ABC475
2-NAPHTHYL METHYL KETONE see ABC500
α-NAPHTHYL METHYL KETONE see ABC475
β-NAPHTHYL METHYL KETONE see ABC500
1-NAPHTHYL METHYLNITROSOCARBAMATE see NBJ500
1-NAPHTHYL-N-METHYL-N-NITROSOCARBAMATE see NBJ500
2-NAPHTHYL N-METHYL-N-(3-TOLYL)THIONOCARBAMATE see TGB475
o-2-NAPHTHYL m,N-DIMETHYLTHIOCARBANILATE see TGB475
N-1-NAPHTHYL-N,N',N'-TRIETHYLETHYLENEDIAMINE see NBO515
β-NAPHTHYL ORANGE see CMM220
(2-NAPHTHYLOXY)ACETIC ACID see NBJ100
(2-(α-NAPHTHYLOXY)ETHYL)HYDRAZINE MALEATE see NBK000
4-(α-NAPHTHYLOXY)-3-HYDROXYBUTYRAMIDOXIME HYDROCHLORIDE see NAC500
4-α-NAPHTHYLOXY-3-HYDROXY-BUTYRAMIDOXIM-HYDROCHLORID (GERMAN) see NAC500
1-(1-NAPHTHYLOXY)-2-HYDROXY-3-ISOPROPYLAMINOPROPANE HYDROCHLORIDE see ICC000
4,4'-(3-(1-NAPHTHYL)-1,5-PENTAMETHYLENE)DIMORPHOLINE see DUO600
2-NAPHTHYLPHENYLAMINE see PFT500
β-NAPHTHYLPHENYLAMINE see PFT500
2-NAPHTHYL-p-PHENYLENEDIAMINE see NBL000
α-NAPHTHYLPHTHALAMIC ACID SODIUM SALT see SIO500
N-1-NAPHTHYLPHTHALAMIC ACID SODIUM SALT see SIO500
1-NAPHTHYL-N-PROPYL-N-NITROSOCARBAMATE see NBM500
α-NAPHTHYL RED see PEJ600
β-NAPHTHYLSULFONIC ACID see NAP000
3-(1-NAPHTHYL)-2-TETRAHYDROFURFURYLPROPIONIC ACID 2-(DIETHYLAMINO)ETHYL ESTER see NAE000
1-NAPHTHYL THIOACETAMIDE see NBN000
α-NAPHTHYLTHIOCARBAMIDE see AQN635
1-NAPHTHYL-THIOHARNSTOFF (GERMAN) see AQN635
2-NAPHTHYL THIOL see NAP500
1-(1-NAPHTHYL)-2-THIOUREA see AQN635
N-(1-NAPHTHYL)-2-THIOUREA see AQN635
α-NAPHTHYLTHIOUREA see AQN635
α-NAPHTHYLTHIOUREA (DOT) see AQN635
1-NAPHTHYL THIOUREA (MAK) see AQN635
1-NAPHTHYL-THIOUREE (FRENCH) see AQN635
5-β-NAPHTHYL-2:4:6-TRIAMINOAZOPYRIMIDINE see NBO500

4-(1-NAPHTHYLVINYL)-PYRIDINE HYDROCHLORIDE see NBO525
4-(2-(1-NAPHTHYL)VINYL)PYRIDINE HYDROCHLORIDE see NBO525
NAPHTHYPRAMIDE see IOU000
NAPHTOCARD ORANGE II see CMM220
NAPHTOCARD RED 2G see CMM300
NAPHTOCARD YELLOW O see FAG140
NAPHTOELAN FAST RED 3GL BASE see KDA050
NAPHTOELAN FAST SCARLET G BASE see NMP500
NAPHTOELAN FAST SCARLET G SALT see NMP500
NAPHTOELAN MITSUI SCARLET GG SALT see DEO295
NAPHTOELAN ORANGE R BASE see NEN500
NAPHTOELAN RED GG BASE see NEO500
2-NAPHTOL (FRENCH) see NAX000
β-NAPHTOL (GERMAN) see NAX000
NAPHTOL AS-KG see TGR000
NAPHTOL AS-KGLL see TGR000
2-NAPHTOL-6-SULFOSAURE (GERMAN) see HMU500
NAPHTO(1,2-c-d-e)NAPHTACENE (FRENCH) see BCP750
NAPHTOX see AQN635
NAPHTYZIN see NAH550
NAPHURIDE see BAT000
NAPHYL-1-ESSIGSAEURE (GERMAN) see NAK500
NAPOLONE see MIF760
NAPOTON see LFK000, MDQ250
NAPRINOL see HIM000
NAPROPION see SJL500
NAPROSINE see MFA500
NAPROSYN see MFA500
NAPROXEN see MFA500
NAPROXEN SODIUM see NBO550
NAPRUX see MFA500
NAPTALAM SODIUM see SIO500
NAPTHOLROT S see FAG020
NAQUA see HII500
NAQUIVAL see RDK000
NARAMYCIN see CPE750
NARCAN see NAH000
NARCEOL see MNR250
NARCICLASINE see LJC000
NARCISO (CUBA, MEXICO) see DAB700
NARCISSUS see DAB700
NARCISSUS ABSOLUTE see NBP000
NARCISSUS POETICUS see DAB700
NARCISSUS PSEUDONARCISSUS see DAB700
NARCOGEN see TIO750
NARCOLAN see ARW250, THV000
NARCOLO see AFJ400
NARCOMPREN see NOA000
NARCOMPREN HYDROCHLORIDE see NOA500
NARCOSAN see ERD500
NARCOSAN SOLUBLE see ERE000
NARCOSINE see NOA000
NARCOSINE HYDROCHLORIDE see NOA500
NARCOSINE, HYDROCHLORIDE (8CI) see NOA500
NARCOTANE see HAG500
NARCOTANN NE-SPOFA (RUSSIAN) see HAG500
NARCOTILE see EHH000
1-α-NARCOTINE see NOA000
NARCOTINE see NBP275, NOA000
1-α-NARCOTINE HYDROCHLORIDE see NOA500
NARCOTINE N-OXIDE HYDROCHLORIDE see NBP300
NARCOTUSSIN see NOA000
NARCOZEP see FDD100
NARCYLEN see ACI750
NARDELZINE see PFC750
NARDIL see PFC500, PFC750
NAREA see HDP500
NARGOLINE see NDM000
NARIGIX see EID000
NARITHERACIN see VGZ000
NARKOLAN see ARW250, THV000
NARKOSOID see TIO750
NARLENE see BQU500
NARPHEN see PMD325
NARSIS see CGA000
NASALIDE see FDD085
NASDOL see TBG000
NASIVINE see ORA100
NASMIL see CNX825
NASOL see EAW000
NASS (IRAN) see SED400
NASTENON see PAN100
NASTYN see EHP000
NASWAR (PAKISTAN and AFGHANISTAN) see SED400
NA TA see TII250

NATA see TII500
NATACYN see PIF750
NATAMYCIN see PIF750
NATASOL FAST RED TR SALT see CLK235
NATERETIN see BEQ625
NATHULANE see PME500
NAT-333 HYDROCHLORIDE see DAI200
NATICARDINA see GAX000
NATIL see DNU100
NATIONAL 120-1207 see AAX250
NATIVE CALCIUM SULFATE see CAX750
NaATP see AEM250
NATRASCORB INJECTABLE see ARN000
NATREEN see BCE500, SGC000
NATRILIX see IBV100
NATRINAL see BAG250
NATRIN HERBICIDE see SKL000
NATRIONEX see AAI250
NATRI-PAS see SEP000
NATRIPHENE see BGJ750
NATRIUM see SEE500
NATRIUMACETAT (GERMAN) see SEG500
NATRIUMALUMINUMFLUORID (GERMAN) see SHF000
NATRIUMANTIMONYLTARTRAT (GERMAN) see AQI750
NATRIUMARSENIT (GERMAN) see ARJ500
NATRIUMAZID (GERMAN) see SFA000
NATRIUMAZIDE (DUTCH) see SFA000
NATRIUMBARBITALS (GERMAN) see BAG250
NATRIUMBICHROMAAT (DUTCH) see SGI000
NATRIUMCHLORAAT (DUTCH) see SFS000
NATRIUMCHLORAT (GERMAN) see SFS000
NATRIUMCHLORID (GERMAN) see SFT000
NATRIUM CITRICUM (GERMAN) see DXC400
NATRIUMDEHYDROCHOLAT (GERMAN) see SGD500
NATRIUM-2,4-DICHLORPHENOXYATHYLSULFAT (GERMAN) see CNW000
NATRIUMDICHROMAAT (DUTCH) see SGI000
NATRIUMDICHROMAT (GERMAN) see SGI000
NATRIUM-3-α,12-α-DIHYDROXYCHOLANAT (GERMAN) see SGE000
NATRIUMFLUORACETAAT (DUTCH) see SHG500
NATRIUMFLUORACETAT (GERMAN) see SHG500
NATRIUM FLUORIDE see SHF500
NATRIUMGLUTAMINAT (GERMAN) see MRL500
NATRIUMHEXAFLUOROALUMINATE (GERMAN) see SHF000
NATRIUM-HUMAT see HGM100
NATRIUMHYDROXID (GERMAN) see SHS000
NATRIUMHYDROXYDE (DUTCH) see SHS000
NATRIUMHYPOPHOSPHIT (GERMAN) see SHV000
NATRIUM-O-ISOPROPYLDITHIOKARBONAT see SIA000
NATRIUMJODAT (GERMAN) see SHV500
NATRIUMJODID (GERMAN) see SHW000
NATRIUMMALAT (GERMAN) see MAN250
NATRIUM-N-METHYL-DITHIOCARBAMAAT (DUTCH) see VFU000
NATRIUM-N-METHYL-DITHIOCARBAMAT (GERMAN) see VFU000
NATRIUMMETHYLDITHIOCARBAMAT (GERMAN) see SIL550
NATRIUMMOLYBDAT (GERMAN) see DXE800
NATRIUMNICOTINAT (GERMAN) see NDW500
NATRIUMNITRAT UND N-METHYLANILIN (GERMAN) see MGO250
NATRIUM NITRIT (GERMAN) see SIQ500
NATRIUMOXALAT (GERMAN) see SIY500
NATRIUMPERCHLORAAT (DUTCH) see PCE750
NATRIUMPERCHLORAT (GERMAN) see PCE750
NATRIUMPHOSPHAT (GERMAN) see SJH090
NATRIUMPROPIONAT see SJL500
NATRIUMPYROPHOSPHAT see TEE500
NATRIUMRHODANID (GERMAN) see SIA500
NATRIUMSALZ DER 2,2-DICHLORPROPIONSAEURE see DGI600
NATRIUMSELENIAT (GERMAN) see DXG000
NATRIUMSELENIT (GERMAN) see SJT500
NATRIUMSILICOFLUORID (GERMAN) see DXE000
NATRIUMSULFAT (GERMAN) see SJY000
NATRIUMSULFID (GERMAN) see SJZ000
NATRIUMTARTRAT (GERMAN) see SKB000
NATRIUMTRICHLOORACETAAT (DUTCH) see TII500
NATRIUMTRICHLORACETAT (GERMAN) see TII500
NATRIUMTRIPOLYPHOSPHAT (GERMAN) see SKN000
NATRIUMZYKLAMATE (GERMAN) see SGC000
NATRIURAN see CLY600
NATROSOL 250 see HKQ100
NATROSOL see HKQ100
NATROSOL 250G see HKQ100
NATROSOL 250H see HKQ100
NATROSOL 300H see HKQ100
NATROSOL 250HHP see HKQ100
NATROSOL 250HHR see HKQ100
NATROSOL 250HR see HKQ100

NATROSOL 250H4R see HKQ100
NATROSOL 250HX see HKQ100
NATROSOL 240JR see HKQ100
NATROSOL 150L see HKQ100
NATROSOL 180L see HKQ100
NATROSOL 250L see HKQ100
NATROSOL LR see HKQ100
NATROSOL 250M see HKQ100
NATROSOL 250MH see HKQ100
NATULAN see PME250, PME500
NATULANAR see PME500
NATULAN HYDROCHLORIDE see PME500
NATURAL CALCIUM CARBONATE see CAO000
NATURAL GASOLINE (DOT) see GBY000
NATURAL GAS, compressed (with high methane content) (UN 1971) (DOT) see MDQ750
NATURAL GAS, refrigerated liquid (cryogenic liquid) (with high methane content) (UN 1972) (DOT) see MDQ750
NATURAL IRON OXIDES see IHD000
NATURAL LEAD SULFIDE see LDZ000
NATURAL MUSK AMBRETTE see OKU100
NATURAL RED OXIDE see IHD000
NATURAL RUBBER see ROH900
NATURAL WINTERGREEN OIL see MPI000
NATURETIN see BEQ625
NATURINE see BEQ625
NATYL see PCP250
NAUCAINE see AIT250
NAUGARD DSTDP see DXG700
NAUGARD TJB see DWI000
NAUGARD TKB see NKB500
NAUGATUCK D-014 see SOP000
NAUGATUCK DET see DKC800
NAULI "GUM" see PMQ750
NAUROCTIL see MCI500
NAUSEN see BBV500
NAUSIDOL see CJL500
NAUTAZINE see EAN600
NAVADYL see NOC000
NAVANE see NBP500
NAVARON see NBP500
NAVICALM see HGC500
NAVIDREX see CPR750
NAVIDRIX see CPR750
NAVINON BLUE BC see DFN425
NAVINON BLUE RSN see IBV050
NAVINON BLUE RSN REDDISH SPECIAL see IBV050
NAVINON BRILLIANT BLUE RCL see DFN425
NAVRON see FFF000
NAVY BLUE EMBL see CMN800
NAXAMIDE see IMH000
NAXEN see MFA500
NAXOFEM see NHH000
NAXOGIN see NHH000
NAXOL see CPB750
NAXYN see MFA500
NAYPER B and BO see BDS000
NB2B see CMO000
NBBA see NJO150
NBHA see HJQ350
N.B. MECOPROP see CIR500
NBN see BRV500
NBS 706 see SMQ500
NBT see DVF800
NC5 see HER500
NC12 see TJH750
NC 26 see BHO250
NC 100 see IGK800
NC 123 see MON750
NC 150 see PDC250, PEK250
NC-262 see DSP400
N 714C see TAF675
NC 1667 see TJL500
NC-2962 see DRU400
NC 3363 see DGO400
NC 5016 see DGA200
NCC 45 see CAT775
NCE of H. INFLUENZAE see HAB700
Na-CEMMIX see BLX000
NC11F see FIA000
NCI 1136 see DBA175
NCI 143-418 see RKA000
NCI 145-604 see RKK000
NCI-C00044 see AFK250
NCI-C00055 see AJM000

NCI-C00066 see ASH500
NCI-C00077 see CBG000
NCI-C00099 see CDR750
NCI-C00102 see TBQ750
NCI-C00113 see DGP900
NCI-C00124 see DHB400
NCI-C00135 see DSP400
NCI-C00157 see EAT500
NCI-C00168 see TBW100
NCI-C00180 see HAR000
NCI-C00191 see KEA000
NCI-C00204 see BBQ500
NCI-C00215 see MAK700
NCI-C00226 see PAK000
NCI-C00237 see PIB900
NCI-C00259 see CDV100
NCI-C00260 see HON000
NCI-C00395 see DVQ709
NCI-C00408 see DER000
NCI-C00419 see PAX000
NCI-C00420 see DFT800
NCI-C00431 see DFV600
NCI-C00442 see DUV600
NCI-C00453 see CDO250
NCI-C00464 see DAD200
NCI-C00475 see BIM500
NCI-C00486 see BIO750
NCI-C00497 see MEI450
NCI-C00500 see DDL800
NCI-C00511 see EIY600
NCI-C00522 see EIY500
NCI-C00533 see CKN500
NCI-C00544 see DOS000
NCI-C00555 see BIM750
NCI-C00566 see EAQ750
NCI-C00588 see FAB400
NCI-C00920 see EJC500
NCI-C01445 see NEI000
NCI-C01478 see LBG000
NCI-C01514 see AES750
NCI-C01536 see ADR750
NCI-C01547 see YCJ000
NCI-C01558 see CME250
NCI-C01569 see ARY000
NCI-C01570 see EDR500
NCI-C01592 see BOT250
NCI-C01605 see EAN500
NCI-C01616 see IAN000
NCI-C01627 see PIK250
NCI-C01638 see IMH000
NCI-C01649 see TFQ750
NCI-C01661 see DDG800
NCI-C01672 see PDC250
NCI-C01683 see TGD000
NCI-C01694 see EPQ000
NCI-C01707 see TFI000
NCI-C01718 see SOA500
NCI-C01729 see TNX000
NCI-C01730 see API500
NCI-C01741 see PDF000
NCI-C01785 see POL500
NCI-C01810 see PME500
NCI-C01821 see DSY600
NCI-C01832 see DCE600, TGM400
NCI-C01843 see NMP500
NCI-C01854 see HHG000
NCI-C01865 see DVH000
NCI-C01876 see AIB000
NCI-C01887 see AJT750
NCI-C01901 see AKP750
NCI-C01912 see NFD500
NCI-C01923 see MMG000
NCI-C01934 see NEQ500
NCI-C01945 see NES500
NCI-C01956 see NHQ000
NCI-C01967 see NEJ500
NCI-C01978 see NEL000
NCI-C01989 see DBO400
NCI-C01990 see MJN000
NCI-C02006 see MQS500
NCI-C02017 see PGN250
NCI-C02028 see DBF800
NCI-C02039 see CEH680
NCI-C02040 see CLK215
NCI-C02051 see CLK225

NCI-C02073 see EFE000
NCI-C02084 see SIQ500
NCI-C02095 see AEN000
NCI-C02108 see AAI000
NCI-C02119 see USS000
NCI-C02142 see HEM500
NCI-C02153 see THG000
NCI-C02175 see DCJ400
NCI-C02186 see TMH750
NCI-C02200 see SMQ000
NCI-C02211 see NMC100
NCI-C02222 see ALL750
NCI-C02233 see PFU500
NCI-C02244 see NKB500
NCI-C02299 see TLG250
NCI-C02302 see TGL750
NCI-C02335 see TGS500
NCI-C02368 see CLK235
NCI-C02551 see CAH500
NCI-C02653 see HCL000
NCI-C02664 see PJN000
NCI-C02686 see CHJ500
NCI-C02697 see ARP250
NCI-C02711 see CAJ750
NCI-C02722 see TGC000
NCI-C02733 see CAK500
NCI-C02766 see AMT500
NCI-C02799 see FMV000
NCI-C02813 see PIX250
NCI-C02824 see ISA000
NCI-C02835 see SGJ000
NCI-C02846 see CJX750
NCI-C02857 see EPJ000
NCI-C02868 see DJC000
NCI-C02880 see DWI000
NCI-C02891 see LCW000
NCI-C02904 see TIW000
NCI-C02915 see PFT500
NCI-C02926 see ASL250
NCI-C02937 see CAQ250
NCI-C02959 see DXH250
NCI-C02960 see CMF400
NCI-C02971 see MNH000
NCI-C02982 see MGO750
NCI-C02993 see MGO500
NCI-C03009 see BLJ250
NCI-C03010 see DOU600
NCI-C03021 see NAM000
NCI-C03032 see TBR250
NCI-C03043 see AJV250
NCI-C03054 see DFE200
NCI-C03065 see ALQ000
NCI-C03076 see NIY500
NCI-C03134 see EFU000
NCI-C03167 see SMD000
NCI-C03190 see ADL500
NCI-C03258 see ANO500
NCI-C03269 see DEU000
NCI-C03270 see TNC500
NCI-C03281 see NBH500
NCI-C03292 see CFK125
NCI-C03305 see CJY120
NCI-C03316 see CEG625
NCI-C03327 see TGJ500
NCI-C03361 see BBX000
NCI-C03372 see IAQ000
NCI-C03474 see ASB250
NCI-C03485 see CDO500
NCI-C03510 see API750
NCI-C03521 see BDH250
NCI-C03554 see TBQ100
NCI-C03565 see DCZ000
NCI-C03598 see BFW750
NCI-C03601 see PHW750
NCI-C03612 see BBO500
NCI-C03656 see DDA800
NCI-C03667 see DAC800
NCI-C03678 see OAJ000
NCI-C03689 see DVQ000
NCI-C03703 see HCF500
NCI-C03714 see TAI000
NCI-C03736 see AOQ000, BBL000
NCI-C03747 see AOX250
NCI-C03758 see AOX500
NCI-C03770 see CEC000

NCI-C03781 see TMD250
NCI-C03792 see ENV500
NCI-C03805 see BNK000
NCI-C03816 see DKC400
NCI-C03827 see DQD400
NCI-C03838 see CIV000
NCI-C03849 see SHX000
NCI-C03850 see DVR200
NCI-C03861 see LHW000
NCI-C03907 see PIC000
NCI-C03918 see DQJ200
NCI-C03930 see PEY650
NCI-C03941 see ALL500
NCI-C03952 see NNT000
NCI-C03963 see NEM480
NCI-C03974 see TNL250, TNL500
NCI-C03985 see DGG950
NCI-C04126 see DTH000
NCI-C04137 see DBN200
NCI-C04159 see BKF250
NCI-C04240 see TGG760
NCI-C04251 see TGF250
NCI-C04502 see DGW200
NCI-C04535 see DFF809
NCI-C04546 see TIO750
NCI-C04557 see DMF000
NCI-C04568 see IEP000
NCI-C04579 see TIN000
NCI-C04580 see PCF275
NCI-C04591 see CBV500
NCI-C04604 see HCI000
NCI-C04615 see AGB250
NCI-C04626 see MIH275
NCI-C04637 see TIP500
NCI-C04671 see MDV500
NCI-C04682 see AEB000
NCI-C04693 see DAC000
NCI-C04706 see AHK500
NCI-C04717 see DAB600
NCI-C04728 see AQQ750
NCI-C04739 see ADA000
NCI-C04740 see CGV250
NCI-C04762 see DDP600
NCI-C04773 see BIF750
NCI-C04784 see MPU000
NCI-C04795 see DDJ000
NCI-C04819 see DCF200
NCI-C04820 see BIA250
NCI-C04831 see HOO500
NCI-C04842 see VKZ000
NCI-C04853 see PED750
NCI-C04864 see LEY000
NCI-C04875 see DFO000
NCI-C04886 see POK000
NCI-C04897 see PLZ000
NCI-C04900 see CQC650
NCI-C04922 see CQN000
NCI-C04933 see CDN000
NCI-C04944 see BHT750
NCI-C04955 see CHD250
NCI-C05970 see REA000
NCI-C06008 see TAB750
NCI-C06111 see BDX500
NCI-C06155 see BQQ750
NCI-C06224 see EHH000
NCI-C06360 see BEE375
NCI-C06428 see MQW500
NCI-C06462 see SFA000
NCI-C06508 see BDX000
NCI-C07272 see TGK750
NCI-C08628 see OPK250
NCI-C08640 see CBM500
NCI-C08651 see FAQ900
NCI-C08662 see CNU750
NCI-C08673 see DCM750
NCI-C08684 see DEV800
NCI-C08695 see DUK800
NCI-C08991 see ARM250, ARM280
NCI-C09007 see ARM275
NCI-C50011 see BCP250
NCI-C50033 see SBT000
NCI-C50044 see BII250
NCI-C50055 see DUE000
NCI-C50077 see PMO500
NCI-C50088 see EJN500

NCI-C50099 see PNL600
NCI-C50102 see MJP450
NCI-C50124 see PDN750
NCI-C50135 see EIU800
NCI-C50146 see OPM000
NCI-C50157 see RDK000
NCI-C50168 see CNU000
NCI-C50191 see SIB600
NCI-C50204 see TAV750
NCI-C50226 see ZAT000
NCI-C50259 see HEJ500
NCI-C50260 see DBR400
NCI-C50282 see BGE325
NCI-C50317 see DCE400
NCI-C50351 see BGJ250
NCI-C50362 see HER000
NCI-C50384 see EFT000
NCI-C50395 see GLU000
NCI-C50419 see LIA000
NCI-C50442 see BJK500
NCI-C50453 see EQR500
NCI-C50464 see AGJ250
NCI-C50475 see AEX250
NCI-C50533 see TGM750
NCI-C50544 see WCB000
NCI-C50602 see BOP500
NCI-C50613 see AMW000
NCI-C50635 see BLD500
NCI-C50646 see CBF700
NCI-C50657 see DBL200
NCI-C50668 see MJP400
NCI-C50680 see MLH750
NCI-C50715 see MCB000
NCI-C50737 see CQK600
NCI-C50748 see AQQ500
NCI-C52459 see TBQ000
NCI-C52733 see DVL700
NCI-C52904 see NAJ500
NCI-C53634 see HCA500
NCI-C53781 see AAQ250
NCI-C53792 see CHP500
NCI-C53849 see HJF500
NCI-C53894 see PAW500
NCI-C53907 see FAG150
NCI-C53929 see PEJ500
NCI-C54262 see VPK000
NCI-C54375 see BEC500
NCI-C54386 see AEO000
NCI-C54546 see SBW000
NCI-C54557 see AQP000
NCI-C54568 see CMO750
NCI-C54579 see CMO000
NCI-C54604 see MJQ100
NCI-C54626 see MRC000
NCI-C54660 see DHF200
NCI-C54706 see FEW000
NCI-C54717 see ISV000
NCI-C54728 see DTD800
NCI-C54739 see RMK020
NCI-C54740 see DST800
NCI-C54751 see TNI250
NCI-C54773 see DSG600
NCI-C54808 see ARN000
NCI-C54819 see IKE000
NCI-C54820 see CIU750
NCI-C54831 see TIQ250
NCI-C54842 see PMK000
NCI-C54853 see EES350
NCI-C54886 see CEJ125
NCI-C54897 see HKN875
NCI-C54900 see TBG700
NCI-C54911 see SGP500
NCI-C54922 see ALO750
NCI-C54933 see PAX250
NCI-C54944 see DEP600
NCI-C54955 see DEP800
NCI-C54966 see REF000
NCI-C54977 see EBR000
NCI-C54988 see TCF000
NCI-C54999 see CPD750
NCI-C55005 see CPC000
NCI-C55050 see TDI000
NCI-C55061 see TDH750
NCI-C55072 see CDS000
NCI-C55107 see CEA750

NCI-C55118 see CEQ600
NCI-C55129 see DRS200
NCI-C55130 see BNL000
NCI-C55141 see PNJ400
NCI-C55152 see AQF000
NCI-C55163 see CCP250
NCI-C55174 see DHF000
NCI-C55185 see GEO000
NCI-C55196 see NGE000
NCI-C55209 see OLA000
NCI-C55210 see RNZ000
NCI-C55221 see SHF500
NCI-C55232 see XGS000
NCI-C55243 see BND500
NCI-C55254 see CFK500
NCI-C55265 see TAI500
NCI-C55276 see BBL250
NCI-C55287 see PAU500
NCI-C55298 see QPA000
NCI-C55301 see POP250
NCI-C55323 see BKE500
NCI-C55345 see DFX800
NCI-C55367 see BPX000
NCI-C55378 see PAX250
NCI-C55425 see GFQ000
NCI-C55436 see DDS000
NCI-C55447 see MKA500
NCI-C55458 see AJL500
NCI-C55470 see WCJ000
NCI-C55481 see EGV400
NCI-C55492 see PEO500
NCI-C55505 see SEM000
NCI-C55516 see DDQ400
NCI-C55527 see BOX750
NCI-C55538 see EBX500
NCI-C55549 see GGW500
NCI-C55550 see TEO250
NCI-C55561 see TBX250
NCI-C55572 see LFU000
NCI-C55583 see NKM000
NCI-C55594 see MHZ000
NCI-C55607 see HCE500
NCI-C55618 see IMF400
NCI-C55630 see IMR000
NCI-C55641 see SPC500
NCI-C55652 see EAY500
NCI-C55663 see AES500
NCI-C55674 see EDJ500
NCI-C55685 see PDE000
NCI-C55696 see SNC000
NCI-C55709 see CDP250
NCI-C55710 see AOB250
NCI-C55721 see DNA800
NCI-C55743 see PBC250
NCI-C55765 see DKQ000
NCI-C55776 see PJD000
NCI-C55787 see HFV500
NCI-C55798 see PDO750
NCI-C55801 see HIM000
NCI-C55812 see MIP750
NCI-C55823 see GEM000
NCI-C55834 see HIH000
NCI-C55845 see QQS200
NCI-C55856 see CCP850
NCI-C55867 see MCD250
NCI-C55878 see BOV000
NCI-C55889 see HAP500
NCI-C55890 see HHR500
NCI-C55903 see XDJ000
NCI-C55925 see CFY000
NCI-C55936 see CHJ750
NCI-C55947 see TDY250
NCI-C55958 see NEM500
NCI-C55969 see AOR500
NCI C55970 see ALO000
NCI-C55992 see NIF000
NCI-C56031 see DFI210
NCI C56042 see UVJ450
NCI-C56064 see NGE500
NCI-C56075 see BAU750
NCI-C56086 see AOD125
NCI-C56097 see DWW000
NCI-C56100 see BFC750
NCI-C56111 see CMP969
NCI-C56122 see RGW000

NCI-C56133 see BAY500
NCI-C56144 see IDA000
NCI-C56155 see TMN490
NCI-C56166 see BCK250
NCI-C56177 see FPQ875
NCI-C56188 see XNJ000
NCI-C56199 see EID000
NCI-C56202 see FPK000
NCI-C56213 see ABC250
NCI-C56224 see FPU000
NCI-C56235 see IPU000
NCI-C56246 see QFS000
NCI-C56257 see HKR500
NCI-C56279 see COB260
NCI-C56280 see MNQ000
NCI-C56291 see BSU250
NCI-C56304 see PFL000
NCI-C56315 see DQV250
NCI-C56326 see AAG250
NCI-C56337 see BJL000
NCI-C56348 see DTC800
NCI-C56359 see PAH250
NCI-C56360 see DBB200
NCI-C56382 see BIE500
NCI-C56393 see EGP500
NCI-C56406 see VQK650
NCI-C56417 see BMC000
NCI-C56428 see DQF800
NCI-C56439 see IPD000
NCI-C56440 see HCZ000
NCI-C56451 see PKL500
NCI-C56462 see MRH000
NCI-C56473 see HOH500
NCI-C56484 see OGQ100
NCI-C56508 see HMY000
NCI-C56519 see BDF000
NCI-C56520 see MNH250
NCI-C56531 see BRF500
NCI-C56542 see SGB550
NCI-C56553 see BRV500
NCI-C56564 see CBF750
NCI-C56575 see CAL000
NCI-C56586 see CHP250
NCI-C56597 see SNH000
NCI-C56600 see SNJ000
NCI-C56611 see TBO780
NCI-C56633 see TKV000
NCI-C56644 see HOO100
NCI-C56655 see PAX250
NCI-C56666 see AGH150
NCI-C56762 see DSR400
NCI-C60015 see COG000
NCI-C60026 see AHP000
NCI-C60048 see DJX000
NCI-C60066 see MKQ875
NCI-C60071 see MRL750
NCI-C60082 see NEX000
NCI-C60093 see PFJ250
NCI-C60102 see QCJ000
NCI-C60106 see QCA000
NCI-C60117 see TAJ000
NCI-C60128 see CGO500
NCI-C60139 see VOA000
NCI-C60162 see FCA100
NCI-C60173 see MCY475
NCI-C60184 see PAD500
NCI-C60195 see PEC500
NCI-C60208 see VPP000
NCI-C60219 see PGX000
NCI-C60220 see TJB600
NCI-C60231 see CEA000
NCI-C60242 see TIK500
NCI-C60286 see PKL100
NCI-C60297 see CNV000
NCI-C60311 see CNA250
NCI-C60322 see DTG000
NCI-C60333 see HLU500
NCI-C60344 see NDK500
NCI-C60355 see KDA050
NCI-C60366 see MMP100
NCI-C60377 see NAY000
NCI-C60388 see NER000
NCI-C60399 see MCW250
NCI-C60402 see EEA500
NCI-C60413 see DER000

NEMALITH (GERMAN) see NBT000
NEMAMORT see BII250
NEMANAX see DDL800
NEMAPAN see TEX000
NEMAPAZ see DDL800
NEMAPHOS see EPC500
NEMASET see DDL800
NEM-A-TAK see DHH200
NEMATOCIDE see DDL800, EPC500
NEMATOLYT see PAG500
NEMATOX see DDL800
NEMAZENE see PDP250
NEMAZINE see PDP250
NEMAZON see DDL800
NEMBUSEN see GGS000
NEMBU-SERPIN see RDK000
NEMBUTAL see NBT500
NEMBUTAL CALCIUM see CAV000
NEMBUTAL SODIUM see NBU000
NEMEROL see DAM600
NEMICIDE see LFA020
NEMISPOR see DXI400
NENDRIN see EAT500
NENESIN see BNK000
NEO see TEH500
NEOABIETIC ACID see NBU800
NEOANTERGAN see WAK000
NEOANTERGAN HYDROCHLORIDE see MCJ250
NEOANTERGAN MALEATE see DBM800
NEOANTERGAN PHOSPHATE see NBV000
NEOANTICID see CAT775
NEOANTIMOSAN see AQH500
NEO-ANTITENSOL see RDK000
NEO-ANTITERSOL see RDK000
NEOARSOLUIN see NCJ500
NEOARSPHENAMINE see NCJ500
NEOASYCODILE see DXE600
NEO-ATOMID see ARQ750
NEO-AVAGAL see SBH500
NEOBAN see CEW500
NEOBAR see BAP000
NEOBARENE see NBV100
NEOBEE M-5 see CBF710
NEOBEE O see CBF710
NEO-BENODINE see TGJ475
NEOBIOTIC see NCG000
NEOBOR see SFF000
NEOBRETTIN see NCD550
NEOBRIDAL see WAK000
NEOCAINE see AIL750, AIT250
NEO-CALMA see CJR909
NEOCARBORANE see NBV100
NEOCARCINOSTATIN see NBV500
NEO-CARDIAMINE see GCE600
NEOCARZINOSTATIN see NBV500
NEOCARZINOSTATIN K see NBV500
NEO-CEBICURE see SAA025
NEOCHIN see CLD000
NEOCHROMIUM see NBW000
NEOCID see DAD200
NEOCIDOL see DCM750
NEO-CLEANER see SHU500
NEOCLYM see FBP100
NEOCOCCYL see SNM500
NEO-CODEMA see CFY000
NEOCOMPENSAN see PKQ250
NEO-COROVAS see PBC250
NEOCROSEDIN see EHP000
NEOCRYL A-1038 see ADW200
NEO-CULTOL see MQV750
NEOCYCLINE see TBX000
NEOCYCLOHEXIMIDE see CPE750
NEODALIT see DCW800
NEODECANOIC ACID see NBW500
NEODECANOIC ACID, HEXADECYL ESTER see HCP600
NEODELPREGNIN see MCA500
NEO-DEMA see CLH750
NEO-DEVOMIT see EAN600
NEODICOUMARIN see BKA000
NEODICOUMAROL see BKA000
NEODICUMARINUM see BKA000
NEODORM (NEW) see NBT500
NEODRENAL see DMV600
NEODRINE see DBA800
NEODYMACETAT (GERMAN) see NBX500

NEODYMIA see NCC000
NEODYMIUM see NBX000
NEODYMIUM ACETATE see NBX500
NEODYMIUM CHLORIDE see NBY000
NEODYMIUM(III) NITRATE (1:3) see NCB000
NEODYMIUM(III) NITRATE, HEXAHYDRATE (1:3:6) see NCB500
NEODYMIUM(3+) OXIDE see NCC000
NEODYMIUM(III) OXIDE see NCC000
NEODYMIUM OXIDE see NCC000
NEODYMIUM SESQUIOXIDE see NCC000
NEODYMIUM TRIACETATE see NBX500
NEODYMIUM TRIOXIDE see NCC000
NEO-EPININE see DMV600
NEO-ERGOTIN see EDC500
NEOESERINE BROMIDE see POD000
NEOESERINE METHYL SULFATE see DQY909
NEO-ESTRONE see ECU750
NEO-FARMADOL see HNI500
NEO-FAT 8 see OCY000
NEO-FAT 10 see DAH400
NEO-FAT 12 see LBL000
NEO-FAT 18-61 see SLK000
NEO-FAT 90-04 see OHU000
NEO-FAT 92-04 see OHU000
NEO-FAT 18-S see SLK000
NEOFEMERGEN see LJL000
NEOFEN see HNI500
NEO-FERRUM see IHG000
NEOFIX FP see CNH125
NEOFLUMEN see CFY000
NEOFOLLIN see EDS100
NEO-FULCIN see GKE000
NEOGEL see CBO500
NEO GERM-I-TOL see AFP250
NEO-GILURYTMAL see DNB000
NEOGLAUCIT see IRF000
NEOGYNON see NNL500
NEOHETRAMINE see NCD500
NEOHETRAMINE HYDROCHLORIDE see RDU000
NEOHEXANE (DOT) see DQT200
NEO-HIBERNEX see DQA600
NEO-HOMBREOL see TBG000
NEO-HOMBREOL-M see MPN500
NEOHYDRAZID see COH250
NEOHYDRIN see CHX250
NEOISCOTIN see SIJ000
NEO-ISTAFENE see HGC500
NEOL see DTG400
NEOLAMIN see TES800
NEOLEXINA see ALV000
NEOLIN see BFC750
NEOLITE F see CAT775
NEOLOID see CCP250
NEOLUTIN see HNT500
NEO-MANTLE CREME see NCG000
NEOMCIN see NCE000
NEOMETANTYL see HKR500
NEOMETHIODAL see DNG400
NEOMIX see NCD550, NCG000
NEOMYCIN see NCE000
NEOMYCIN A see NCE500
NEOMYCIN B see NCF000
NEOMYCIN B SULFATE (SALT) see NCD550
NEOMYCIN C see NCF300
NEOMYCIN E see NCF500
NEOMYCINE SULFATE see NCG000
NEOMYCIN SULFATE see NCG000
NEOMYCIN SULPHATE see NCG000
NEOMYSON G HYDROCHLORIDE see UVA150
NEON see NCG500
NEO-NACLEX see BEQ625
NEONAL see BPF500
NEO-NAVIGAN see DYE600
NEONICOTINE see AON875
NEONIKOTIN see AON875
NEO-NILOREX see DKE800
NEONOL see NCG700
NEONOL AF-14 see NCG715
NEONOL 2B1317-12 see NCG725
NEONYL BLUE 4R see CMM100
NEONYL RED 2B see CMO870
NEONYL SCARLET R see CMM320
NEO-OESTRANOL 1 see DKA600
NEO-OESTRANOL II see DKB000
NEO-ORMONAL see AOO275

NEOOXEDRINE see SPC500
NEO-OXYPATE see PQC500
NEOPANTANOYL CHLORIDE see DTS400
NEOPAX see SJJ175
NEOPELLIS see TFD250
NEOPENTANE see NCH000
NEOPENTANETETRAYL NITRATE see PBC250
1,1',1'',1'''-(NEOPENTANE TETRAYLTETRAOXY)TETRAKIS(2,2,2-TRICHLOROE-
    THANOL) see NCH500
NEOPENTANOIC ACID see PJA500
NEOPENTYLENE GLYCOL see DTG400
NEOPENTYL GLYCOL see DTG400
NEOPENTYL GLYCOL DIACRYLATE see DUL200
NEOPENTYL GLYCOL DIGLYCIDYL ETHER see NCI300
NEOPEPULSAN see HKR500
NEOPHARMEDRINE see DBA800
NEOPHEDAN see PEC250
NEOPHENAL see PEC250
NEOPHRYN see SPC500
NEOPHYILINE see TEP500
NEOPHYLLIN see DNC000
NEOPHYLLINE see DNC000
NEOPHYLLIN M see DNC000
NEOPLATIN see PJD000
NEOPOLEN see PJS750
NEOPOLEN 30N see PJS750
NEOPRENE see NCI500, PJQ050
NEOPROGESTIN see NCI525
NEOPROMA see MFK500
NEOPROSERINE see NCI550
NEOPROTOVERATRIN see NCI600
NEOPROTOVERATRINE see NCI600
NEOPSICAINE HYDROCHLORIDE see NCJ000
NEORAM BLU see CNK559
NEORESERPAN see RCA200
NEORESTAMIN see TAI500
NEORON see IOS000
NEO-RONTYL see BEQ625
NEOSABENYL see CJU250
NEO-SALICYL see SJO000
NEOSALVARSAN see NCJ500
NEOSAR see CQC650
NEO-SCABICIDOL see BBQ500
NEOSEDYN see TEH500
NEOSEPT see HCL000
NEOSEPTAL CL see SHU500
NEOSERFIN see RDK000
NEOSERINE BROMIDE see POD000
NEO-SERP see RDK000
NEOSETILE BLUE EB see TBG700
NEOSETILE PINK BN see AKE250
NEOS-HIDANTOINA see DKQ000
NEO SILOX D see SFT500
NEO-SINEFRINA see SPC500
NEO-SKIODAN see DNG400
NEOSLOWTEN see RDK000
NEOSOLANIOL see NCK000
NEOSOLANIOL MONOACETATE see ACS500
NEOSONE D see MFA250
NEOSORB ND see KDK000
NEO-SPECTRA see CBT750
NEO SPECTRA II see CBT750
NEOSPIRAMYCIN see NCK500
NEOSPIRAN see GCE600
NEOSTAM see NCL000
NEOSTAM STIBAMINE GLUCOSIDE see NCL000
NEOSTEN see AGN000
NEOSTENE see AOO475
NEOSTENOVASAN see DNC000
NEOSTERON see AOO475
NEOSTIBOSAN see SLP600
NEOSTIGMETH see DQY909
NEOSTIGMINE see NCL100
NEOSTIGMINE BROMIDE see POD000
NEOSTIGMINE METHOSULFATE see DQY909
NEOSTIGMINE METHYL BROMIDE see POD000
NEOSTIGMINE METHYL SULFATE see DQY909
NEOSTIGMINE MONOMETHYLSULFATE see DQY909
NEOSTON see AGN000
NEOSTREPAL see SNN300
NEOSTREPSAN see TEX250
NEOSULF see NCD550
NEOSUPRANOL HYDROCHLORIDE see NCL300
NEO-SUPRIMAL see HGC500
NEO-SUPRIMEL see HGC500

NEOSYDYN see TEH500
NEOSYMPATOL see SPC500
NEOSYNEPHRINE see NCL500, SPC500
NEOSYNEPHRINE HYDROCHLORIDE see SPC500
NEOSYNESINE see SPC500
NEO-TENEBRYL see DNG400
NEO-TESTIS see TBF500
NEOTEX see CBT750
NEOTHESIN see PIV750
NEOTHESIN HYDROCHLORIDE see IJZ000
NEOTHRAMYCIN see NCM275
NEOTHRAMYCIN A see TCP000
NEOTHYL see MOU830
NEOTHYLLINE see DNC000
NEOTIBIL see FNF000
NEOTILINA see DNC000
NEOTIZIDE see SIJ000
NEO-TIZIDE SODIUM SALT see SIJ000
NEOTONZIL see MCI500
NEOTOPSIN see PEX500
NEO-TRAN see MQU750
NEOTRAN see NCM700
NEO-TRIC see MMN250
NEON, compressed (UN 1065) (DOT) see NCG500
NEON, refrigerated liquid (cryogenic liquid) (UN 1913) (DOT) see NCG500
NEO-VASOPHYLINE see DNC000
NEOVITAMIN A ACID see VSK955
NEOVLETTA see NNL500
NEOXAZOL see SNN500
NEOXIN see ILD000
NEOZEPAM see DLY000
NEOZEX 45150 see PJS750
NEOZEX 4010B see PJS750
NEOZINE see MCI500
NEO-ZINE see MNV750
NEOZONE A see PFT250
NEOZONE D see PFT500
NEPHELOR see GGS000
NEPHENTINE see MQU750
NEPHIS see EIY500
NEPHOCARP see TNP250
NEPHRAMIDE see AAI250
NEPHRIDINE see VGP000
NEPRESOL see OJD300
NEPRESOLIN see OJD300
NEPRESSOL see OJD300
NEPTAL see NCM800
NEPTUNE BLUE BRA CONCENTRATION see FMU059
NEPTUNIUM see NCN500
NERA see ZLS200
NERACID see CBG000
NERA EMULZE (CZECH) see ZLS200
NERAL see DTC800
NERAN BRILLIANT GREEN G see FAE950
NEREB see MAS500
NERIINE DIHYDRBROMIDE see DOX000
NERIODIN see DEO600
NERIOL see OHQ000
NERIOLIN see OHQ000
NERIOSTENE see OHQ000
NERISONA see DKF130
NERISONE see DKF130
NERIUM OLEANDER see OHM875
NERKOL see DGP900
NEROBIL see DYF450
NEROBIOLIL see DYF450
NEROBOL see DAL300
NEROBOLETTES see DAL300
NEROL (FCC) see DTD200
NEROL ACETATE (6CI) see NCO100
NEROLI BIGARADE OIL, TUNISIAN see NCO000
NEROLIDYL ACETATE see NCN800
NEROLIN see EEY500
NEROLINE see EEY500
NEROLIN FRAGAROL see NBJ000
NEROLIN II see EEY500
NEROLIN NEW see EEY500
NEROLI OIL, ARTIFICAL see APJ250
NEROLI OIL, TUNISIAN see NCO000
NERONE see MCF525
NERPRUN (CANADA) see MBU825
NERUSIL see NCQ100
NERVACTON see BCA000
NERVANAID B ACID see EIX000
NERVANAID B LIQUID see EIV000

NERVANID B see EIV000
NERVATIL see BCA000
NERVILON see DXO300
NERVOSETON see BAG250
NERVOTON see DOZ000
NERYL ACETATE see NCO100
NERYL FORMATE see FNC000
NERYL ISOBUTYRATE see NCO200
NERYL ISOVALERIANATE see NCO500
NERYL-β-METHYL BUTYRATE see NCO500
NERYL PROPIONATE see NCP000
NESDONAL see PBT250
NESDONAL SODIUM see PBT500
NESOL see MCC250
NESONTIL see CFZ000
NESPOR see MAS500
NESSLER REAGENT see NCP500
NESTON see MDT740
NESTYN see EHP000
NETAGRONE 600 see DAA800
NETAL see BNL250
NETALID see INS000
NETAZOL see CIR250
NETH see INS000
NETHALIDE see INS000
NETHALIDE HYDROCHLORIDE see INT000
NETHAMINE HYDROCHLORIDE see EJR500
NETILLIN see NCP550
NETILMICIN see SBD000
NETILMICIN SULFATE see NCP550
NETOCYD see DJT800
NETROMYCIN see NCP550
NETROPSIN see NCP875
NETROPSIN HYDROCHLORIDE see NCQ000
NETSUSARIN see DOT000
NEU see ENV000, NKE500
NEUCHLONIC see DLY000
NEUCOCCIN see FMU080
NEUFIL see DNC000
NEULACTIL see PIW000
NEULEPTIL see PIW000
NEUMOLISINA see ELC500
NEUPERM GFN see DTG000
NEURACEN see BEG000
NEURACTIL see MCI500
NEURAKTIL see BCA000
NEURALEX see NCQ100
α-NEURAMINIDASE see NCQ200
NEURAMINIDASE see NCQ200
NEURAXIN see GKK000
NEURAZINE see CKP500
NEURIDINE see DCC400
NEURIDINE TETRAHYDROCHLORIDE see GEK000
NEURIN see VQR300
NEURINE see VQR300
NEURIPLEGE see CLY750
NEUROBARB see EOK000
NEUROBENZIL see BCA000
NEUROCAINE see CNE750
NEUROCIL see MCI500
NEURODYN see TEH500
NEUROLEPTONE see BCA000
NEURONIKA see ADA725
NEUROPROCIN see EHP000
NEUROSEDIN see TEH500
NEUROSTAIN see NCQ500
NEUROTONE see RLU000
NEUROTOXIN, CLOSTRIDIUM BOTULINUM see BMM292
NEUROTRAST see ELQ500
NEUROZINA see CJR909
NEURVITA see DXO300
NEURYL see SHL500
NEUSTAB see FNF000
NEUT see SFC500
NEUTHION see GFW000
NEUTRAFIL see DNC000
NEUTRAFILLINA see DNC000
NEUTRAL ACRIFLAVINE see DBX400, XAK000
NEUTRAL AMMONIUM CHROMATE see ANF500, NCQ550
NEUTRAL AMMONIUM FLUORIDE see ANH250
NEUTRAL BERBERINE SULFATE see BFN625
NEUTRAL POTASSIUM CHROMATE see PLB250
NEUTRAL RED see AJQ250
NEUTRAL RED CHLORIDE see AJQ250
NEUTRAL RED PG see CMM320

NEUTRAL RED W see AJQ250
NEUTRAL SAPONINS of PANAX GINSENG ROOT see GEQ425
NEUTRAL SODIUM CHROMATE see DXC200
NEUTRAL VERDIGRIS see CNI250
NEUTRAPHYLLIN see DNC000
NEUTRAPHYLLINE see DNC000
NEUTRAZYME see SIB600
NEUTROFLAVINE see XAK000
NEUTRONYX 605 see PKF500
NEUTRONYX 600 see PKF000
NEUTRONYX 622 see GHS000
NEUTRORMONE see AOO475
NEUTROSEL NAVY BN see DCJ200
NEUTROSEL RED TRVA see CLK235
NEUTROSTERON see AOO475
NEUTROXANTINA see DNC000
NEUVITAN see GEK200
NEUWIED GREEN see COF500
NEVANAID-B POWDER see TNL250
NEVAX see DJL000
NEVIGRAMON see EID000
NEVIN (mixture) see SNL850
NEVIN see ILD000, SNL850
NEVRACTEN see CDQ250
NEVRITON see DXO300
NEVRODYN see TEH500
NEW BLITANE see ZJS300
NEW COCCIN see FMU080
NEWCOL 60 see SKV150
NEW GREEN see COF500
NEW IMPROVED CERESAN see BJT250
NEW IMPROVED GRANOSAN see BJT250
NEWLASE see NCQ600
NEW MILSTEM see BRI750
NEW PATENT BLUE A-CE EXTRA see CMM062
NEW PATENT BLUE EXTRA PURE A see CMM062
NEWPHRINE see SPC500
NEW PINK BLUISH GEIGY see FAG040
NEWPOL LB3000 see BRP250
NEW PONCEAU 4R see FMU070
NEWTOL see XPJ000
NEW VICTORIA GREEN EXTRA I see AFG500
NEXAGAN see EGV500
NEXION 40 see BNL250
NEXION see BNL250
NEXIT see BBQ500
NEXOVAL see CKC000
NF 6 see TGI250
NF 44 see PEX500
NF 64 see NGB700
NF 246 see NDY000
NF 260 see FPI000
NF-260 see FPI150
NF-269 see FPI100
NF 323 see NDY400
NF 1010 see EAE500
β-NF see NAU525
NF (dispersant) see BLX000
NF 35 (fungicide) see DJV000
NF see BLX000, NGE500
NF-A see BLX000
NFHAA see FNO000
NF-902 HYDROCHLORIDE see FPI100
NFIP see NGI800
NFN see DLU800
NaFPAK see SHF500
Na FRINSE see SHF500
NFT see FPF000
NG-180 see NGG500
NG see NGY000
NGAI CAMPHOR see NCQ820
N(sup 1)-GUANYLSULFANILAMIDE see AHO250
NH 18 see PKP750
NH 188 see RDU000
NHA see NBI500
NHC see HFN500
NH-LOST see BHN750
NHMI see OBY000
NHU see HKQ025
NIA 249 see AFR250
NIA 2,995 see DFO600
NIA 4556 see DFO800
NIA 5273 see PIX250
NIA 5462 see EAQ750
NIA 5488 see CKM000

NIA-5767 see EAS000
NIA 5996 see DER800
NIA 9044 see BGB500
NIA 9102 see MQQ250
NIA-9241 see BDJ250
NIA 10242 see CBS275
NIA 11092 see DUM800
NIA 16,388 see PNV750
NIA 17170 see BEP500
NIA-18739 see BEP750
NIA 33297 see AHJ750
NIACEVIT see NCR000
NIACIDE see FAS000
NIACIN see NCQ900
NIACINAMIDE see NCR000
NIACIN BENZYL ESTER see NCR040
NIAGARA 1,137 see BIT250
NIAGARA 1240 see EEH600
NIAGARA 4512 see SKQ400
NIAGARA 4556 see DFO800
NIAGARA 5006 see DER800
NIAGARA 5,462 see EAQ750
NIAGARA 5767 see EAS000
NIAGARA 5943 see AIX000
NIAGARA 5,996 see DER800
NIAGARA 9044 see BGB500
NIAGARA 9241 see BDJ250
NIAGARA BLUE see CMO250
NIAGARA BLUE 2B see CMO000
NIAGARA BLUE 4B see CMO500
NIAGARA BLUE RW see CMO600
NIAGARAMITE see SOP500
NIAGARA P.A. DUST see NDN000
NIAGARA SKY BLUE see CMO500
NIAGARA-STIK see NAK500
NIAGARATHAL see DXD000
NIAGARATRAN see CJT750
NIAGARIL see BEQ625
NIAGATHAL see TBT150
NIAGRA 10242 see CBS275
NIA 2995J see DEV600
NIALAMIDE see BET000
NIALATE see EEH600
NIALK see TIO750
NIAMID see BET000
NIAMIDAL see BET000
NIAMIDE see NCR000
NIAMINE see DJS200
NIAPROOF 4 see SIO000
NIA PROOF 08 see TAV750
NIAQUITIL see BET000
NIAX AFPI see PKB100
NIAX CATALYST AL see BJH750
NIAX CATALYST ESN see NCS000
NIAX FLAME RETARDANT 3 CF see CGO500
NIAX ISOCYANATE TDI see TGM740
NIAX POLYOL LG-168 see NCT000
NIAX POLYOL LHT-42 see NCT500
NIAX TDI see TGM750, TGM800
NIAX TDI-P see TGM750
NIAX TRIOL 6000 see NCV500
NIAZOL see NCW000
NIBAL see PMC700
NIBREN WAX see TIT500
NIBROL see TEH500
NIBUFIN see NIM000
NICACID see NCQ900
NICAMETATE CITRATE see EQQ100
NICAMETATE DIHYDROGEN CITRATE see EQQ100
NICAMIDE see DJS200, NCR000
NICAMIN see NCQ900
NICAMINA see NCR000
NICAMINDON see NCR000
NICANGIN see NCQ900
NICARB see NCW100
NICARBAZIN see NCW100
NICARBAZINE see NCW100
NICARDIPINE HYDROCHLORIDE see PCG550
NICASIR see NCR000
NICAZIDE see ILD000
NICEL see MIF760
NICELATE see EID000
N. I. CERESAN see EME100
NICERGOLIN (GERMAN) see NDM000
NICERGOLINE see NDM000

NICERITROL see NCW300
NICETAMIDE see DJS200
NICETHAMIDE see DJS200
NICHEL (ITALIAN) see NCW500
NICHEL TETRACARBONILE (ITALIAN) see NCZ000
NICHOLIN see CMF350
NICIZINA see ILD000
NICKEL 270 see NCW500
NICKEL see NCW500
NICKEL(II) ACETATE (1:2) see NCX000
NICKEL(II) ACETATE TETRAHYDRATE see NCX500
NICKEL ACETATE TETRAHYDRATE see NCX500
NICKEL ALLOY, Ni,Be see NCY000
NICKEL AMMONIUM SULFATE see NCY050
NICKEL ANTIMONIDE see NCY100
NICKEL ARSENIDE see NCY110
NICKEL ARSENIDE (As$_8$-Ni$_{11}$) see NDJ400
NICKEL ARSENIDE (As$_2$-Ni$_5$) see NDJ399
NICKEL ARSENIDE SULFIDE see NCY125
NICKEL BISCYCLOPENTADIENE see NDA500
NICKEL(2+), BIS(N,N'-BIS(2-AMINOETHYL)-1,2-ETHANEDIAMINE-N,N',N$^\#$)-, (T-4)-
    TETRAOXOTUNGSTATE(2-) (1:1) see BLN100
NICKEL, BIS(TRIETHYLENETETRAMINE)TUNGSTATO- see BLN100
NICKEL BISTRIPHENYLPHOSPHINE DITHIOCYANATE see BLS500
NICKEL BLACK see NDE010
NICKEL BOROFLUORIDE see NDC000
NICKEL(II) CARBONATE (1:1) see NCY500
NICKEL CARBONATE HYDROXIDE see NCY600
NICKEL, (CARBONATO(2-))TETRAHYDROXYTRI- see NCY600
NICKEL CARBONYL see NCZ000
NICKEL CARBONYLE (FRENCH) see NCZ000
NICKEL(II) CHLORIDE (1:2) see NDH000
NICKEL CHLORIDE see NDH000
NICKEL(II) CHLORIDE HEXAHYDRATE (1:2:6) see NDA000
NICKEL CHROMATE see NDA100
NICKEL CHROMITE see DXA100
NICKEL CHROMIUM OXIDE see NDA100
NICKEL COMPLEX with N-FLUOREN-2-YL ACETOHYDROXAMIC ACID see
    HIR500
NICKEL, COMPOUND with pi-CYCLOPENTADIENYL (1:2) see NDA500
NICKEL COMPOUNDS see NDB000
NICKEL CYANIDE (DOT) see NDB500
NICKEL CYANIDE (solid) see NDB500
NICKEL DIACETATE TETRAHYDRATE see NCX500
NICKEL DIBUTYLDITHIOCARBAMATE see BIW750
NICKEL DIFLUORIDE see NDC500
NICKEL DIHYDROXIDE see NDE000
NICKEL DINITRITE see NDG550
NICKEL DISULFIDE see NDB875
NICKEL (DUST) see NCW500
NICKEL(II) EDTA COMPLEX see EJA500
NICKEL(II) FLUOBORATE see NDC000
NICKEL(II) FLUORIDE (1:2) see NDC500
NICKEL FLUOROBORATE see NDC000
NICKEL(II) FLUOSILICATE (1:1) see NDD000
NICKEL-GALLIUM ALLOY see NDD500
NICKEL(III) HYDROXIDE see NDE010
NICKEL(II) HYDROXIDE see NDE000
NICKELIC HYDROXIDE see NDE010
NICKELIC OXIDE see NDH500
NICKEL IRON CHROMITE BLACK SPINEL see CMS135
NICKEL IRON SULFIDE see NDE500
NICKEL-IRON SULFIDE MATTE see NDE500
NICKEL(II) ISODECYL ORTHOPHOSPHATE (3:2) see NDF000
NICKEL MONOANTIMONIDE see NCY100
NICKEL MONOARSENIDE see NCY110
NICKEL MONOSELENIDE see NDF400
NICKEL MONOSULFATE HEXAHYDRATE see NDL000
NICKEL MONOSULFIDE see NDL100
NICKEL MONOXIDE see NDF500
NICKEL(II) NITRATE (1:2) see NDG000
NICKEL NITRATE see NDG000
NICKEL(II) NITRATE, HEXAHYDRATE (1:2:6) see NDG500
NICKEL(2+) NITRATE, HEXAHYDRATE see NDG500
NICKEL NITRITE see NDG550
NICKELOCENE see NDA500
NICKELOUS ACETATE see NCX000
NICKELOUS ACETATE TETRAHYDRATE see NCX500
NICKELOUS CARBONATE see NCY500
NICKELOUS CHLORIDE see NDH000
NICKELOUS FLUORIDE see NDC500
NICKELOUS HYDROXIDE see NDE000
NICKELOUS OXIDE see NDF500
NICKELOUS SULFATE see NDK500
NICKELOUS SULFIDE see NDL100

NICKELOUS TETRAFLUOROBORATE see NDC000
NICKEL(II) OXIDE (1:1) see NDF500
NICKEL OXIDE (MAK) see NDF500
NICKEL OXIDE see NDH500
NICKEL OXIDE, IRON OXIDE, and CHROMIUM OXIDE FUME see IHE000
NICKEL OXIDE PEROXIDE see NDH500
NICKEL PARTICLES see NCW500
NICKEL(2+) PERCHLORATE, HEXAHYDRATE see NDJ000
NICKEL PEROXIDE see NDH500
NICKEL POTASSIUM CYANIDE see NDI000
NICKEL PROTOXIDE see NDF500
NICKEL REFINERY DUST see NDI500
NICKEL(2+) SALT PERCHLORIC ACID HEXAHYDRATE see NDJ000
NICKEL SELENIDE see NDF400, NDJ475
NICKEL SELENIDE (3:2) CRYSTALLINE see NDJ475
NICKEL SESQUIOXIDE see NDH500
NICKEL SPONGE see NCW500
NICKEL SUBARSENIDE see NDJ399, NDJ400
NICKEL SUBSELENIDE see NDJ475
NICKEL SUBSULFIDE see NDJ500
NICKEL SUBSULPHIDE see NDJ500
NICKEL (II) SULFAMATE see NDK000
NICKEL SULFARSENIDE see NCY125
NICKEL(2+)SULFATE(1:1) see NDK500
NICKEL(II) SULFATE (1:1) see NDK500
NICKEL SULFATE(1:1) see NDK500
NICKEL(II) SULFATE see NDK500
NICKEL SULFATE see NDK500
NICKEL(II) SULFATE HEXAHYDRATE (1:1:6) see NDL000
NICKEL (II) SULFATE HEXAHYDRATE see NDL000
NICKEL SULFATE HEXAHYDRATE see NDL000
NICKEL(2+) SULFIDE see NDL100
NICKEL(II) SULFIDE see NDL100
NICKEL SULFIDE see NDB875, NDJ500, NDL100
α-NICKEL SULFIDE (3:2) CRYSTALLINE see NDJ500
NICKEL SULPHATE HEXAHYDRATE see NDL000
NICKEL SULPHIDE see NDJ500
NICKEL TELLURIDE see NDL425
NICKEL TETRACARBONYL see NCZ000
NICKEL TETRACARBONYLE (FRENCH) see NCZ000
NICKEL(II) TETRAFLUOROBORATE see NDC000
NICKEL-TITANATE see NDL500
NICKEL TITANIUM OXIDE see NDL500
NICKEL TRIOXIDE see NDH500
NICKEL TRITADISULPHIDE see NDJ500
NICLOFEN see DFT800
NICLOFOLAN see DFD000
NICLOSAMIDE see DFV400, DFV600
NICO-400 see NCQ900
NICO see NCQ900
NICOBID see NCQ900
NICOBION see NCR000
NICOCAP see NCQ900
NICOCIDE see NDN000
NICOCIDIN see NCQ900
NICOCODINE see CNG250
NICOCRISINA see NCQ900
NICODAN see NCQ900
NICODEL see PCG550
NICODELMINE see NCQ900
NICO-DUST see NDN000
NICOFORT see NCR000
NICO-FUME see NDN000
NICOGEN see NCR000
NICOLANE see NOA000
NICOLAR see NCQ900
NICOLEN see NGG500
NICOLIN see CMF350
NICOMETH see NDV000
NICOMIDOL see NCR000
NICOMOL see CME675
NICONACID see NCQ900
NICONAT see NCQ900
NICONAZID see NCQ900
NICOR see DJS200
NICORANDIL see NDL800
NICORDAMIN see DJS200
NICORINE see DJS200
NICOROL see CHJ750, NCQ900
NICORYL see DJS200
NICOSAN 2 see NCR000
NICOSIDE see NCQ900
NICO-SPAN see NCQ900
NICOSYL see NCQ900
NICOTA see NCR000

NICOTAMIDE see NCR000
NICOTAMIN see NCQ900
NICOTENE see NCQ900
NICOTERGOLINE see NDM000
NICOTIANA ATTENUATA see TGI100
NICOTIANA GLAUCA see TGI000, TGI100
NICOTIANA LONGIFLORA see TGI100
NICOTIANA RUSTICA see TGI100
NICOTIANA TABACUM see TGH725, TGI100
NICOTIBINA see ILD000
NICOTIL see NCQ900
NICOTILAMIDE see NCR000
NICOTILILAMIDO see NCR000
NICOTINA (ITALIAN) see NDN000
NICOTINALDEHYDE see NDM100
NICOTINAMIDE-ADENINE DINUCLEOTIDE see CNF390
NICOTINAMIDE, N-(2-INDOL-3-YLETHYL)- see NDW525
2-NICOTINAMIDOETHYL NITRATE see NDL800
NICOTINAMIDOMETHYLAMINOPYRAZOLONE see AQN500
NICOTINE, liquid (DOT) see NDN000
NICOTINE, solid (DOT) see NDN000
l-NICOTINE see NDN000
NICOTINE see NDN000
(—)-NICOTINE see NDN000
NICOTINE ACID see NCQ900
NICOTINE ACID AMIDE see NCR000
NICOTINE ACID TARTRATE see NDS500
NICOTINEALDEHYDE see NDM100
NICOTINE ALKALOID see NDN000
NICOTINEAMIDE ADENINE DINUCLEOTIDE see CNF390
NICOTINE BITARTRATE see NDS500
NICOTINE, COMPOUND, with NICKEL(II)-o-BENZOYL BENZOATE TRIHYD-
    RATE (2:1) see BHB000
NICOTINE, 1′-DEMETHYL-1′-NITROSO-, 1-OXIDE see NLD525
1-NICOTINE HYDROCHLORIDE see NDP500
NICOTINE HYDROCHLORIDE, solution (DOT) see NDP400
NICOTINE HYDROCHLORIDE (d,l) see NDP400
d-NICOTINE HYDROCHLORIDE see NDQ000
NICOTINE HYDROCHLORIDE see NDP400
(+)-NICOTINE HYDROCHLORIDE see NDQ000
NICOTINE HYDROGEN TARTRATE see NDS500
(—)-NICOTINE HYDROGEN TARTRATE see NDS500
NICOTINE MONOSALICYLATE see NDR000
NICOTINE, 1′-NITROSO-1′-DEMETHYL- see NLD500
NICOTINE SALICYLATE (DOT) see NDR000
NICOTINE SULFATE, solid or solution (DOT) see NDR500
NICOTINE SULFATE see NDR500
NICOTINE TARTRATE (1:2) see NDS500
NICOTINE TARTRATE (DOT) see NDS500
NICOTINE TARTRATE see NDS500
NICOTINIC ACID see NCQ900
NICOTINIC ACID AMIDE see NCR000
NICOTINIC ACID, BENZYL ESTER see NCR040
NICOTINIC ACID, BUTYL ESTER see NDT500
NICOTINIC ACID-7,8-DIDEHYDRO-4,5-α-EPOXY-3-METHOXY-17-METHYLMOR-
    PHINAN-6-α-YL ESTER see CNG250
NICOTINIC ACID DIETHYLAMIDE see DJS200
NICOTINIC ACID, 3-(2,6-DIMETHYLPIPERIDINO)PROPYL ESTER, HYDROCHLO-
    RIDE see NDU000
NICOTINIC ACID, ESTER with CODEINE see CNG250
NICOTINIC ACID, HYDRAZIDE see NDU500
NICOTINIC ACID-2-HYDROXY-3-(o-METHOXYPHENOXY)PROPYL ESTER see
    HLT300
NICOTINIC ACID, METHYL ESTER see NDV000
NICOTINIC ACID, 3-(2-METHYLPIPERIDINO)PROPYL ESTER, HYDROCHLORIDE
    see NDW000
NICOTINIC ACID NITRILE see NDW515
NICOTINIC ACID, SODIUM SALT see NDW500
NICOTINIC ACID-1,2,5,6-TETRAHYDRO-1-METHYL-, METHYL ESTER, HYDRO-
    CHLORIDE see AQU250
NICOTINIC ALCOHOL see NDW510
NICOTINIC ALDEHYDE see NDM100
NICOTINIC AMIDE see NCR000
NICOTINIPCA see NCQ900
NICOTINONITRILE see NDW515
NICOTINOYL HYDRAZINE see NCQ900, NDU500
1-NICOTINOYLMETHYL-4-PHENYL-PIPERAZINE see NDW520
N-NICOTINOYLTRYPTAMIDE see NDW525
NICOTINSAEURE (GERMAN) see NCQ900
NICOTINSAEUREAMID (GERMAN) see NCR000
NICOTINYL ALCOHOL see NDW510
NICOTINYLHYDRAZIDE see NDU500
NICOTION see EPQ000
NICOTOL see NCR000
NICOTYLAMIDE see NCR000

3,2′-NICOTYRINE see NDX300
β-NICOTYRINE see NDX300
NICOTYRINE see NDX300
NICOULINE see RNZ000
NICOUMALONE see ABF750
NICOVASAN see NCQ900
NICOVASEN see NCQ900
NICOVEL see NCQ900, NCR000
NICOVIT see NCR000
NICOVITOL see NCR000
NICOXIN see NCW100
NICOZIDE see ILD000
NICOZYMIN see NCR000
NICRAZIN see NCW100
NICRAZINE see NCW100
NICYL see NCQ900
NIDA see MMN250
NIDANTIN see OOG000
NIDORAL see CCY000
NIDRAFUR see NGB700
NIDRAFUR P see NGB700
NIDRAN see ALF500
NIDRANE see BEG000
NIDRAZID see ILD000
NIDROXYZONE see FPE100
NIEA see HKW475
NIERALINE see VGP000
NIESYMETRYCZNA DWU METYLOHYDRAZYNA (POLISH) see DSF400
NIFEDIN see AEC750
NIFEDIPINE see AEC750
NIFELAT see AEC750
NIFENALOL see INU000
NIFLAN see PLX400
NIFLEX see SAV000
NIFLUMIC ACID see NDX500
NIFLURIL see NDX500
NIFOS T see TCF250
NIFTHOLIDE see FMR050
NIFTOLIDE see FMR050
NIFULIDONE see NGG500
NIFURADENE see NDY000
NIFURAN see NGG500
NIFURATRONE see HKW450
NIFURDAZIL see EAE500
NIFURHYDRAZONUM see NGB700
NIFUROXAZID see DGQ500
NIFUROXAZIDE see DGQ500
NIFUROXIME see NGC000
NIFURPIPONE see NDY350
NIFURPIPONE ACETATE see NDY360
NIFURPIPONE DIHYDROCHLORIDE see NDY370
NIFURPIPONE HYDROCHLORIDE see NDY380
NIFURPIPONE MALEATE (1:1) see NDY390
NIFURPIRINOL see NDY400
NIFURTHIAZOLE see NDY500
NIFURTIMOX see NGG000
NIFUZON see NGE500
NIGALAX see PPN100
NIGERICIN SODIUM SALT see SIO788
NIGHT BLOOMING JESSAMINE see DAC500
NIGHTCAPS see PAM780
NIGHTSAGE (JAMAICA) see YAK300
NIGHTSHADE see DAD880
NIGLYCON see NGY000
NIGRIN see SMA000
NIH 204 see TKZ000
NIH 3127 see SBE500
NIH 7574 see BBU625
NIH 7672 see MDV000
NIH 7958 see DOQ400
NIH 7981 see COV500
NIH 7519 HYDROBROMIDE see PMD325
NIH-4185 HYDROCHLORIDE see DJP500
NIH-5145 HYDROCHLORIDE see EIJ000
NIH 7440 HYDROCHLORIDE see AGV890
NIHONTHRENE BLUE BC see DFN425
NIHONTHRENE BLUE RSN see IBV050
NIHONTHRENE BRILLIANT BLUE RCL see DFN425
NIHONTHRENE BRILLIANT BLUE RP see IBV050
NIHONTHRENE BRILLIANT VIOLET 4R see DFN450
NIHONTHRENE BRILLIANT VIOLET RR see DFN450
NIHONTHRENE BROWN BR see CMU770
NIHONTHRENE BROWN GR see CMU770
NIHONTHRENE FAST ORANGE R see CMU815
NIHONTHRENE GOLDEN YELLOW see DCZ000

NIHONTHRENE GREY M see CMU475
NIHONTHRENE NAVY BLUE G see CMU500
NIHONTHRENE OLIVE R see DUP100
NIHONTHRENE RED BB see CMU825
NIHYDRAZON see NGB700
NIHYDRAZONE see NGB700
NIKAFLOC D 1000 see CNH125
NIKALAC 031 see MCB050
NIKALAC MS 001 see MCB050
NIKALAC MS 11 see MCB050
NIKALAC MS 21 see MCB050
NIKALAC MS 40 see MCB050
NIKALAC MW 12 see MCB050
NIKALAC MW 22 see MCB050
NIKALAC MW 30 see MCB050
NIKALAC MW 40 see MCB050
NIKALAC MW 12LF see MCB050
NIKALAC MW 10LF see MCB050
NIKALAC MW 30M see MCB050
NIKALAC MX 032 see MCB050
NIKALAC MX 45 see MCB050
NIKALAC MX 40 see MCB050
NIKALAC MX 054 see MCB050
NIKALAC MX 65 see MCB050
NIKALAC MX 430 see MCB050
NIKALAC MX 485 see MCB050
NIKALAC MX 705 see MCB050
NIKALAC MX 706 see MCB050
NIKALAC MX 750 see MCB050
NIKARDIN see DJS200
NIKARESIN S 176 see MCB050
NIKARESIN S 260 see MCB050
NIKARESIN S 305 see MCB050
NIKARESIN S 306 see MCB050
NIKA-TEMP see PKQ059
NIKAVINYL SG 700 see PKQ059
NIKETAMID see DJS200
NIKETHAROL see DJS200
NIKETHYL see DJS200
NIKETILAMID see DJS200
NIKION see BEQ625
NIKKELTETRACARBONYL (DUTCH) see NCZ000
NIKKOL BC 30 see PJT300
NIKKOL BC 40 see PJT300
NIKKOL BC see PJT300
NIKKOL BC 15TX see PJT300
NIKKOL BC 20TX see PJT300
NIKKOL BL see DXY000
NIKKOL IPIS see IPS450
NIKKOL MYO 2 see PJY100
NIKKOL MYO 10 see PJY100
NIKKOL OTP 70 see DJL000
NIKKOL S.C.S see HCP900
NIKKOL SO 10 see SKV100
NIKKOL SO-15 see SKV100
NIKKOL SO 15 see SKV170
NIKKOL SO-30 see SKV100
NIKKOL SP10 see MRJ800
NIKKOL SS 30 see SKV150
NIKKOL TO see PKL100
NIKORIN see DJS200
NIKO-TAMIN see NCR000
NIKOTIN (GERMAN) see NDN000
NIKOTINSAEUREAMID (GERMAN) see NCR000
NIKOTINSULFAT see NDR500
NIKOTYNA (POLISH) see NDN000
NILACID see ABX500
NILE BLUE see AJN250
NILE BLUE A see AJN250
NILE BLUE AX see AJN250
NILE BLUE BASE see AJN250
NILE BLUE CHLORIDE see AJN250
NILE BLUE HYDROCHLORIDE see AJN250
NILERGEX HYDROCHLORIDE see AEG625
NILHISTIN see DCJ850
NILODIN see DHU000, SBE500
NILOX see BPF000
NILOX PBNA see PFT500
NILSTAT see NOH500
NILUDIPINE see NDY600
NILVADIPINE see NDY650
NILVERM see TDX750
NILVEROM see TDX750
NIMBLE WEED see PAM780
NIMCO CHOLESTEROL BASE H see CMD750

NIMERGOLINE see NDM000
NIMETAZEPAM see DLV000
NIMITEX see TAL250
NIMITOX see TAL250
NIMODIPINE see NDY700
NIM OIL see NBS300
NIMORAZOLE see NHH000
NIMORAZOLO see NHH000
NIMOTOP see NDY700
NIMROD see BRJ000
NIMROD T see BRJ000
NIMUSTINE see NDY800
NIMUSTINE HYDROCHLORIDE see ALF500
NINCALUICOLFLASTINE see VKZ000
NINHYDRIN see DMV200
NINHYDRIN HYDRATE see DMV200
NINOL 4821 see BKE500
NINOL AA62 see BKE500
NINOL AA-62 EXTRA see BKE500, LBL000
NIOBATE, OCTAPOTASSIUM see PLL250
NIOBE OIL see MHA750
NIOBIA see NEA050
NIOBIUM see NDZ000
NIOBIUM CHLORIDE see NEA000
NIOBIUM(5+) OXIDE see NEA050
NIOBIUM(V) OXIDE see NEA050
NIOBIUM OXIDE see NEA050
NIOBIUM PENTACHLORIDE see NEA000
NIOBIUM PENTAOXIDE see NEA050
NIOBIUM PENTOXIDE see NEA050
NIOBIUM POTASSIUM FLUORIDE see PLN500
NIOBIUM POTASSIUM OXIDE see PLL250
NIOCINAMIDE see NCR000
NIOFORM see CHR500
NIOGEN ES 160 see PJY100
NIOI-PEPA (HAWAII) see PCB275
NIOMIL see DQM600
NIONATE see FBK000
NIONG see NGY000
NIOPAM see IFY000
NIOZYMIN see NCR000
NIP see DFT800
NIPA 49 see PNM750
NIPABUYL see BSC000
NIPAGALLIN A see EKM100
NIPAGALLIN LA see DXX200
NIPAGALLIN P see PNM750
NIPAGIN see HJL500
NIPAGIN A see HJL000
NIPAHEPTYL see HBO650
NIPAM see INH000
NIPA NO. 48 see EKM100
NIPANTIOX 1-F see BQI000
NIPAR S-20 see NIY000
NIPAR S-20 SOLVENT see NIY000
NIPAR S-30 SOLVENT see NIY000
NIPASOL see HNU500
NIPAXON see NOA000
NIPAZIN A see HJL000
NIPELLEN see NCQ900
NIPEON A 21 see PKQ059
NIPERYT see PBC250
NIPERYTH see PBC250
NIPHANOID see BQA010
NIPLEN see ILD000
NIPODAL see PMF500
NIPOL 407 see SMR000
NIPOL 576 see PKQ059
NIPPAS see SEP000
NIPPON BLUE BB see CMO000
NIPPON DARK GREEN B see CMO830
NIPPON DEEP BLACK see AQP000
NIPPON DEEP BLACK GX see AQP000
NIPPON DIRECT SKY BLUE see CMO500
NIPPON GREEN B see CMO840
NIPPON KAGAKU SAFRANINE GK see GJI400
NIPPON KAGAKU SAFRANINE T see GJI400
NIPPON PURPURINE 8B see CMO880
NIPRADILOL see NEA100
NIPRIDE see SIU500
NIPRIDE DIHYDRATE see SIW500
NIPSAN see DCM750
NIQUETAMIDA see DJS200
NIRAN see CDR750, PAK000
NIRATIC HYDROCHLORIDE see LFA020

NIRATIC-PURON HYDROCHLORIDE see LFA020
NIRAZIN see NCW100
NIRIDAZOLE see NML000
NIRIT see DVF800
NIROSAN see TDY120
(R)-NIRVANOL see EOL050
l-NIRVANOL see EOL050
NIRVANOL see EOL000
(−)-NIRVANOL see EOL050
NIRVOTIN see PGE000
NISENTIL see NEA500, NEB000
NISENTIL HYDROCHLORIDE see NEB000
NISETAMIDE see DJS200
NISIDANA see DCV800
NISOLONE see SOV100
NISOTIN see EPQ000
NISPERO DEL JAPON (CUBA, PUERTO RICO) see LIH200
NISSAN AMINE AB see OBC000
NISSAN AMINE BB see DXW000
NISSAN ANON BL see LBU200
NISSAN CATION AB see TLW500
NISSAN CATION M2-100 see TCA500
NISSAN CATION S2-100 see DTC600
NISSAN DIAPION S see SIY000
NISSAN DIAPON T see SIY000
NISSAN NONION OP 83 see SKV170
NISSAN NONION OP 83RAT see SKV170
NISSAN NONION P 213 see PJT300
NISSAN NONION P 210 see PJT300
NISSAN NONION P 220 see PJT300
NISSAN NONION P 208 see PJT300
NISSAN NONION PP40 see MRJ800
NISSAN NONION PP 40R see MRJ800
NISSAN NONION SP 60 see SKV150
NISSOCAINE see AIL750
NISSOL EC see MME809
NISY see DRB400
NITARSONE see NIJ500
NITAZOL see ABY900
NITAZOLE see ABY900
NITEBAN see ILD000
NITER see PLL500
NITHIAMIDE see ABY900
NITHIAZID see ENV500
NITHIAZIDE see ENV500
NITICID see CHS500
NITICOLIN see CMF350
NITOBANIL see FLZ050
NITOFEN see DFT800
NITOL see BIE500
NITOL "TAKEDA" see BIE500
NITOMAN see TBJ275
NITORA see NGY000
NITRADOR see DUS700
NITRADOS see DLY000
NITRAFEN see DFT800
NITRALAMINE HYDROCHLORIDE see NEC000
NITRALDONE see FPI000
NITRAMIDE see NEG000
NITRAMIN see ALQ000
NITRAMINE see ALQ000, TEG250
NITRAMINE, METHYL- see NHN500
NITRAMINOACETIC ACID see NEC500, NGY700
NITRAN see DUV600
4-NITRANILINE see NEO500
m-NITRANILINE see NEN500
p-NITRANILINE see NEO500
NITRANITOL see MAW250
NITRANOL see TJL250
NITRAPHEN see DFT800
NITRAPYRIN (ACGIH) see CLP750
NITRATE d'AMYLE (FRENCH) see AOL250
NITRATE d'ARGENT (FRENCH) see SDS000
NITRATE de BARYUM (FRENCH) see BAN250
NITRATE MERCUREUX (FRENCH) see MDE750
NITRATE MERCURIQUE (FRENCH) see MDF000
NITRATE PHENYL MERCURIQUE see MDH500
NITRATE de PLOMB (FRENCH) see LDO000
NITRATE de PROPYLE NORMAL (FRENCH) see PNQ500
NITRATES see NED000
NITRATE de SODIUM (FRENCH) see SIO900
NITRATE de STRONTIUM (FRENCH) see SMK000
NITRATE de ZINC (FRENCH) see ZJJ000
NITRATINE see SIO900
NITRATION BENZENE see BBL250

NITRAZEPAM see DLY000
NITRAZOL CF EXTRA see NEO500
NITRE see PLL500
NITRE CAKE see SEG800
NITRENDIPINE see EMR600
NITRENPAX see DLY000
NITRETAMIN see TJL250
NITRETAMIN PHOSPHATE see TJL250
NITRIC ACID, over 40% (DOT) see NED500
NITRIC ACID other than red fuming with not >70% nitric acid (DOT) see NED500
NITRIC ACID other than red fuming with >70% nitric acid (DOT) see NED500
NITRIC ACID see NED500
NITRIC ACID, ALUMINUM(3+) SALT see AHD750
NITRIC ACID, ALUMINUM SALT see AHD750
NITRIC ACID, ALUMINUM SALT, NONAHYDRATE (8CI,9CI) see AHD900
NITRIC ACID, AMMONIUM LANTHANUM SALT see ANL500
NITRIC ACID, AMMONIUM SALT see ANN000
NITRIC ACID, ANHYDRIDE with PEROXYACETIC ACID see PCL750
NITRIC ACID, BARIUM SALT see BAN250
NITRIC ACID, BERYLLIUM SALT see BFT000
NITRIC ACID, BISMUTH(3+) SALT see BKW250
NITRIC ACID, CADMIUM SALT see CAH000
NITRIC ACID, CADMIUM SALT, TETRAHYDRATE see CAH250
NITRIC ACID, CALCIUM SALT (8CI,9CI) see CAU000
NITRIC ACID, CALCIUM SALT, TETRAHYDRATE see CAU250
NITRIC ACID, CERIUM(3+) SALT (8CI, 9CI) see CDB000
NITRIC ACID, CERIUM(3+) SALT, HEXAHYDRATE see CDB250
NITRIC ACID, CESIUM SALT see CDE250
NITRIC ACID, CHROMIUM (3+) SALT see CMJ600
NITRIC ACID, COBALT(2+) SALT see CNC500
NITRIC ACID, COPPER(2+) SALT TRIHYDRATE see CNN000
NITRIC ACID, DODECYL ESTER see NEE000
NITRIC ACID DYSPROSIUM(3+) SALT HEXAHYDRATE see DYH000
NITRIC ACID, ERBIUM (3+) SALT see ECY500
NITRIC ACID, ERBIUM (3+) SALT, HEXAHYDRATE see ECZ000
NITRIC ACID, ETHYL ESTER see ENM500
NITRIC ACID, EUROPIUM(3+) SALT, HEXAHYDRATE see ERC000
NITRIC ACID, FUMING (DOT) see NEE500
NITRIC ACID, GADOLINIUM(3+) SALT see GAL000
NITRIC ACID, GALLIUM(3+) SALT see GBS000
NITRIC ACID, HOLMIUM(3+) SALT, HEXAHYDRATE see HGH000
NITRIC ACID, IRON(3+) SALT see FAY200, IHB900
NITRIC ACID, IRON (3+) SALT, NONAHYDRATE see IHC000
NITRIC ACID, ISOPROPYL ESTER see IQP000
NITRIC ACID, LANTHANUM AMMONIUM SALT see ANL500
NITRIC ACID, LEAD(2+) SALT see LDO000
NITRIC ACID, LUTETIUM(3+) SALT see LIY000
NITRIC ACID, MAGNESIUM SALT (2:1) see MAH000
NITRIC ACID, MANGANESE(2+) SALT see MAS900
NITRIC ACID, MERCURY(II) SALT see MDF000
NITRIC ACID, MERCURY(I) SALT see MDE750
NITRIC ACID METHYL ESTER see MMF500
NITRIC ACID, NEODYMIUM SALT see NCB000
NITRIC ACID, NEODYMIUM (3+) SALT, HEXAHYDRATE see NCB500
NITRIC ACID, NICKEL(II) SALT see NDG000
NITRIC ACID, NICKEL(2+) SALT, HEXAHYDRATE see NDG500
NITRIC ACID, PHENYLMERCURY SALT see MCU750
NITRIC ACID, POTASSIUM SALT see PLL500
NITRIC ACID, PRASEODYMIUM(3+) SALT see PLY250
NITRIC ACID, PROPYL ESTER see PNQ500
NITRIC ACID, RED FUMING (DOT) see NEE500
NITRIC ACID (RED FUMING) see NEE500
NITRIC ACID, SAMARIUM(3+) SALT, HEXAHYDRATE see SAT000
NITRIC ACID, SILVER(1+) SALT see SDS000
NITRIC ACID, SODIUM SALT see SIO900
NITRIC ACID, STRONTIUM SALT see SMK000
NITRIC ACID, THALLIUM(1+) SALT see TEK750
NITRIC ACID, THORIUM(4+) SALT see TFT500
NITRIC ACID, THULIUM(3+) SALT, HEXAHYDRATE see TFX250
NITRIC ACID TRIESTER OF GLYCEROL see NGY000
NITRIC ACID (WHITE FUMING) see NEF000
NITRIC ACID, YTTERBIUM(3+) SALT see YDS800
NITRIC ACID, YTTERBIUM(3+) SALT, HEXAHYDRATE see YEA000
NITRIC ACID, YTTRIUM(3+) SALT see YFJ000
NITRIC ACID, YTTRIUM(3+)SALT, HEXAHYDRATE see YFS000
NITRIC ACID, ZINC SALT see ZJJ000
NITRIC ACID, ZINC SALT, HEXAHYDRATE see NEF500
NITRIC AMIDE see NEG000
NITRIC ETHER see ENM500
NITRIC OXIDE see NEG100
NITRIC OXIDE and NITROGEN TETROXIDE MIXTURES see NGT500
NITRIDAZOLE see NML000
NITRIDES see NEH000

NITRILE ACRILICO (ITALIAN) see ADX500
NITRILE ACRYLIQUE (FRENCH) see ADX500
NITRILE ADIPICO (ITALIAN) see AER250
NITRILES see NEH500
NITRILE TRICHLORACETIQUE (FRENCH) see TII750
NITRIL KISELINY DIETHYLAMINOOCTOVE (CZECH) see DHJ600
NITRIL KYSELINY-o-CHLORBENZOOVE (CZECH) see CEM000
NITRIL KYSELINY p-CHLORBENZOOVE (CZECH) see CEM250
NITRIL KYSELINY ISOFTALOVE (CZECH) see PHX550
NITRIL KYSELINY MALONOVE (CZECH) see MAO250
NITRIL KYSELINY MANDLOVE (CZECH) see MAP250
NITRIL KYSELINY STEAROVE (CZECH) see SLL500
NITRIL KYSELINY TEREFTALOVE (CZECH) see BBP250
NITRIL KYSELINY β,β'-THIODIPROPIONOVE (CZECH) see DGS600
NITRIL KYSELINY m-TOLUYLOVE (CZECH) see TGT250
NITRIL KYSELINY p-TOLUYLOVE (CZECH) see TGT750
NITRIL MASTNE KYSELINY S (CZECH) see AFQ000
NITRILOACETIC ACID TRISODIUM SALT MONOHYDRATE see NEI000
NITRILOACETONITRILE see COO000
NITRILOMALONAMIDE see COJ250
NITRILOTRIACETIC ACID see AMT500
NITRILOTRIACETIC ACID, DISODIUM SALT see DXF000
NITRILOTRIACETIC ACID SODIUM SALT see SEP500
NITRILOTRIACETIC ACID, TRISODIUM SALT see SIP500
NITRILOTRIACETIC ACID TRISODIUM SALT MONOHYDRATE see NEI000
NITRILO-2,2',2''-TRIETHANOL see TKP500
2,2',2''-NITRILOTRIETHANOL see TKP500
(2,2',2''-NITRILOTRIETHANOL)PHENYLMERCURY(1+) LACTATE (salt) see TNI500
2,2',2''-NITRILOTRIETHANOL TRINITRATE PHOSPHATE see TJL250
NITRILOTRIMETHANEPHOSPHONIC ACID see NEI100
NITRILOTRIMETHYLENEPHOSPHONIC ACID see NEI100
NITRILOTRIMETHYLPHOSPHONIC ACID see NEI100
NITRILOTRIMETHYLPHOSPHONIC ACID ZINC complex TRISODIUM TETRAHYDRATE see ZJJ200
1,1',1''-NITRILOTRI-2-PROPANOL see NEI500
NITRILOTRISILANE see TNJ250
(NITRILOTRIS(METHYLENE))TRISPHOSPHONIC ACID see NEI100
NITRILOTRIS(METHYLPHOSPHONIC ACID) see NEI100
NITRIMIDAZINE see NHH000
NITRIN see NGY000
NITRINE see NGY000
NITRINE-TDC see NGY000
N-NITRISO-N-(2,3-DIHYDROXYPROPYL)-N-(2-HYDROXYETHYL)AMINE see NBR100
NITRITE see DVF800
NITRITES see NEJ000
NITRITE de SODIUM (FRENCH) see SIQ500
NITRITO see NGR500
NITRITO D'AMILE see ILW100
5-NITROACENAPHTHENE see NEJ500
5-NITROACENAPHTHYLENE see NEJ500
5-NITROACENAPTHENE see NEJ500
2-NITROACETALDEHYDE OXIME see NEJ600
2-NITRO-4-ACETAMINOFENETOL (CZECH) see NEL000
4'-NITROACETANILIDE see NEK000
4-NITROACETANILIDE see NEK000
p-NITROACETANILIDE see NEK000
2-NITRO-p-ACETANISIDIDE see NEK100
5-NITRO-2-ACETILAMINOTIAZOLO see ABY900
3-NITROACETOFENON see NEL500
3-NITRO-p-ACETOPHENETIDE see NEL000
3-NITRO-p-ACETOPHENETIDIDE see NEL000
5-NITRO-p-ACETOPHENETIDIDE see NEL000
3'-NITRO-p-ACETOPHENETIDIN see NEL000
2'-NITROACETOPHENONE see NEL450
3'-NITROACETOPHENONE see NEL500
4'-NITROACETOPHENONE see NEL600
m-NITROACETOPHENONE see NEL500
o-NITROACETOPHENONE see NEL450
p-NITROACETOPHENONE see NEL600
NITRO ACID 100 percent see HMY000
NITRO ACID SULFITE see NMJ000
2,2'-((2-(8-NITRO-9-ACRIDINYLAMINO)ETHYL)DIETHANOL HYDROCHLORIDE see NFW350
4'-(3-NITRO-9-ACRIDINYLAMINO)METHANESULFONANILIDE see NEM000
N-(4-((3-NITRO-9-ACRIDINYL)AMINO)PHENYL)METHANESULFONAMIDE see NEM000
N'-(1-NITRO-9-ACRIDINYL)-N,N,1-TRIMETHYL-1,2-ETHANEDIAMINE HYDROCHLORIDE see NFW435
β-NITROALCOHOL see NFY550
NITROALKANES see NEM300
m-NITROAMINOBENZENE see NEN500
4-NITRO-2-AMINOFENOL (CZECH) see NEM500
p-NITROAMINOFENOL (POLISH) see NEM500

4'-NITRO-4-AMINO-3-HYDROXYDIPHENYL HYDROCHLORIDE see AKI500
4'-NITRO-4-AMINO-3-HYDROXYDIPHENYL HYDROGEN CHLORIDE see AKI500
5-NITRO-2-AMINOPHENOL see ALO000
2-NITRO-4-AMINOPHENOL see NEM480
3-NITRO-6-AMINOPHENOL see ALO000
p-NITRO-o-AMINOPHENOL see NEM500
o-NITRO-p-AMINOPHENOL see NEM480
2-(N-NITROAMINO)PYRIDINE-N-OXIDE see NEM600
5-N-NITROAMINOTETRAZOLE see NEN000
5-NITRO-2-AMINOTHIAZOLE see ALQ000
4-NITRO-2-AMINOTOLUENE see NMP500
2-NITRO-4-AMINOTOLUENE see NMP000
4-NITROAMINO-1,2,4-TRIAZOLE see NEN300
p-NITROANILINA see NEO500
2-NITROANILINE see NEO000
3-NITROANILINE see NEN500
N-NITROANILINE see NEO510
m-NITROANILINE see NEN500
o-NITROANILINE see NEO000
p-NITROANILINE see NEO500
4-NITROANILINE, 2,6-DICHLORO- see RDP300
p-NITROANILINE MERCURY(II) derivative see NEP000
4-NITROANILINE-2-SULFONIC ACID see NEP500
4-NITROANILINIUM PERCHLORATE see NEP600
4-NITRO-o-ANISIDINE see NEQ000
5-NITRO-o-ANISIDINE see NEQ500
p-NITROANISOL see NER500
2-NITROANISOLE see NER000
4-NITROANISOLE see NER500
o-NITROANISOLE see NER000
p-NITROANISOLE see NER500
2-NITROANTHRACENE see NES000
1-NITRO-9,10-ANTHRACENEDIONE see NET000
1-NITROANTHRACHINON (CZECH) see NET000
4-NITROANTHRANILIC ACID see NES500
1-NITROANTHRAQUINONE see NET000
α-NITROANTHRAQUINONE see NET000
3-NITRO-3-AZAPENTANE-1,5-DIISOCYANATE see NHI500
5-NITROBARBITURIC ACID see NET550
2-NITROBENZALDEHYDE see NEU500
3-NITROBENZALDEHYDE see NEV000
4-NITROBENZALDEHYDE see NEV500
m-NITROBENZALDEHYDE see NEV000
o-NITROBENZALDEHYDE see NEU500
p-NITROBENZALDEHYDE see NEV500
p-NITROBENZAMIDE see NEV525
NITROBENZEEN (DUTCH) see NEX000
NITROBENZEN (POLISH) see NEX000
3-NITROBENZENAMINE see NEN500
4-NITROBENZENAMINE see NEO500
NITROBENZENE, liquid (DOT) see NEX000
NITROBENZENE see NEX000
4-NITROBENZENEACETIC ACID see NII510
4-NITROBENZENEACETONITRILE see NIJ000
p-NITROBENZENEACETONITRILE see NIJ000
4-NITROBENZENEARSONIC ACID see NIJ500
o-NITROBENZENEARSONIC ACID see NEX500
m-NITROBENZENEBORONIC ACID see NEY500
m-NITROBENZENECARBOXYLIC ACID see NFG000
4-NITRO-1,2-BENZENEDIAMINE see ALL500
2-NITRO-1,4-BENZENEDIAMINE see ALL750
4-NITROBENZENEDIAZONIUM AZIDE see NEZ000
3-NITROBENZENEDIAZONIUM CHLORIDE see NEZ100
4-NITROBENZENEDIAZONIUM NITRATE see NFA000
3-NITROBENZENEDIAZONIUM PERCHLORATE see NFA500
m-NITROBENZENE DIAZONIUM PERCHLORATE (DOT) see NFA500
2-NITROBENZENEDIAZONIUM SALTS see NFA600
4-NITROBENZENEDIAZONIUM SALTS see NFA625
2-NITROBENZENEMETHANOL see NFM550
p-NITROBENZENESULFONAMIDE see NFB000
3-NITROBENZENESULFONIC ACID see NFB500
m-NITROBENZENESULFONIC ACID see NFB500
m-NITROBENZENESULFONIC ACID, SODIUM SALT see NFC500
NITROBENZEN-m-SULFONAN SODNY (CZECH) see NFC500
5-NITRO-1H-BENZIMIDAZOLE see NFD500
2-NITROBENZIMIDAZOLE see NFC700
6-NITRO-BENZIMIDAZOLE see NFD500
2-NITRO-1H-BENZIMIDAZOLE (9CI) see NFC700
3-NITROBENZOIC ACID see NFG000
4-NITROBENZOIC ACID see CCI250
m-NITROBENZOIC ACID see NFG000
p-NITROBENZOIC ACID see CCI250
4-NITROBENZOIC ACID CHLORIDE see NFK100
p-NITROBENZOIC ACID CHLORIDE see NFK100
p-NITROBENZOIC ACID HYDRAZIDE see NFH000

p-NITROBENZOIC ACID, PIPERIDINO ESTER see NFL500
NITROBENZOL, liquid (DOT) see NEX000
NITROBENZOL (DOT) see NEX000
3-NITROBENZONITRILE see NFH500
4-NITROBENZONITRILE see NFI010
m-NITROBENZONITRILE see NFH500
p-NITROBENZONITRILE see NFI010
5-NITROBENZOTRIAZOL (DOT) see NFJ000
5-NITRO-1H-BENZOTRIAZOLE see NFJ000
6-NITRO-1H-BENZOTRIAZOLE see NFJ000
5-NITROBENZOTRIAZOLE see NFJ000
3-NITROBENZOTRIFLUORIDE see NFJ500
m-NITROBENZOTRIFLUORIDE (DOT) see NFJ500
7-NITRO-3H-2,1-BENZOXAMERCUROL-3-ONE see NHY500
2-NITROBENZOYL CHLORIDE see NFL000
4-NITROBENZOYL CHLORIDE see NFK100
m-NITROBENZOYL CHLORIDE see NFK500
o-NITROBENZOYL CHLORIDE see NFL000
p-NITROBENZOYL CHLORIDE see NFK100
3-NITROBENZOYL NITRATE see NFL100
N-(4-NITRO)BENZOYLOXYPIPERIDINE see NFL500
6-NITROBENZ(a)PYRENE see NFM500
2-NITROBENZYL ALCOHOL see NFM550
4-NITROBENZYL ALCOHOL see NFM560
o-NITROBENZYL ALCOHOL see NFM550
2-NITROBENZYL BROMIDE see NFM700
4-NITROBENZYL BROMIDE see NFN000
p-NITROBENZYL BROMIDE see NFN000
p-NITROBENZYL CHLORIDE see NFN400
4-NITRO-BENZYL-CYANID (GERMAN) see NIJ000
4-NITROBENZYL CYANIDE see NIJ000
p-NITROBENZYLCYANIDE see NIJ000
p-NITROBENZYLIDENEACETOPHENONE see NFS505
2-((m-NITROBENZYLIDENE)AMINO)ETHANOL N-OXIDE see HKW345
2-((p-NITROBENZYLIDENE)AMINO)ETHANOL N-OXIDE see HKW350
m-NITROBENZYLIDENE-1,2-BENZ-9-METHYLACRIDINE see NMF000
o-NITROBENZYLIDENE-1,2-BENZ-9-METHYLACRIDINE see NMF500
p-NITROBENZYLIDENE-1,2-BENZ-9-METHYLACRIDINE see NMG000
m-NITROBENZYLIDENE-3,4-BENZ-9-METHYLACRIDINE see NMD500
o-NITROBENZYLIDENE-3,4-BENZ-9-METHYLACRIDINE see NME000
p-NITROBENZYLIDENE-3,4-BENZ-9-METHYLACRIDINE see NME500
1-(p-NITROBENZYL)-2-NITROIMIDAZOLE see NFO700
1-(p-NITROBENZYL)PYRIDINE see NFO750
o-NITROBIPHENYL see NFP500
p-NITROBIPHENYL see NFQ000
4'-NITRO-4-BIPHENYLAMINE see ALM000
4-NITROBIPHENYL ESTER see NIU000
2-NITRO-1,1-BIS(p-CHLOROPHENYL)PROPANE see BIN500
tert-NITROBUTANE see NFQ500
2-NITRO-2-BUTENE see NFR500
1-NITRO-3-BUTENE see NFR000
4-NITROCALONE see NFS505
N-NITROCARBAMIDE see NMQ500
NITROCARBOL see NHM500
NITRO CARBO NITRATE see NFS500
NITROCARDIOL see TJL250
NITROCELLULOSE E950 see CCU250
NITROCELLULOSE with water not <25% water, by weight (UN 2555) (DOT) see CCU250
NITROCELLULOSE with alcohol not <25% alcohol by weight, and not >12.6% nitrogen (UN 2556) (DOT) see CCU250
NITROCELLULOSE with plasticizing not <18% plasticizing substance, by weight (UN 2557) (DOT) see CCU250
NITROCELLULOSE, solution, flammable with not >12.6% nitrogen, by weight (UN 2059) (DOT) see CCU250
NITROCELLULOSE, unmodified or plasticized with <18% plasticizing substance (UN 0341) (DOT) see CCU250
NITROCELLULOSE, wetted with not <25% alcohol, by weight (UN 0342) (DOT) see CCU250
NITROCELLULOSE, plasticized with not <18% plasticizing substance, by weight (UN 0343) (DOT) see CCU250
NITROCELLULOSE, dry or wetted with <25% water (or alcohol), by weight (UN 0340) (DOT) see CCU250
4-NITROCHALCONE see NFS505
4-NITROCHINOLIN N-OXID (SWEDISH) see NJF000
p-NITROCHLOORBENZEEN (DUTCH) see NFS525
NITROCHLOR see DFT800
4-NITRO-2-CHLOROANILINE see CJA175
2-NITRO-4-CHLOROANILINE see KDA050
1-NITRO-5-CHLOROANTHRAQUINONE see CJA250
m-NITROCHLOROBENZENE, solid (DOT) see CJB250
p-NITROCHLOROBENZENE solid (DOT) see NFS525
m-NITROCHLOROBENZENE see CJB250
o-NITROCHLOROBENZENE see CJB750
p-NITROCHLOROBENZENE see NFS525

NITROCHLOROBENZENE see CJA950
p-NITROCHLOROBENZOL (GERMAN) see NFS525
3-NITRO-4-CHLOROBENZOTRIFLUORIDE see NFS700
NITROCHLOROFORM see CKN500
p-NITRO-m-CHLOROPHENYL DIMETHYL THIONOPHOSPHATE see MIJ250
p-NITRO-o-CHLOROPHENYL DIMETHYL THIONOPHOSPHATE see NFT000
3-NITRO-4-CHLORO-α,α,α-TRIFLUOROTOLUENE see NFS700
6-NITROCHRYSENE see NFT400
p-NITROCINNAMALDEHYDE see NFT425
4-NITROCINNAMIC ACID see NFT440
p-NITROCINNAMIC ACID see NFT440
p-NITROCLOROBENZENE (ITALIAN) see NFS525
NITRO COMPOUNDS see NFT459
NITRO COMPOUNDS of AROMATIC HYDROCARBONS see NFT500
NITROCOTTON see CCU250
4-NITRO-m-CRESOL see NFV000
4-NITRO-o-CRESOL see NFV010
2-NITRO-p-CRESOL see NFU500
NITROCRESOLAMINE see DBY700
NITROCYCLOHEXANE see NFV500
4-NITRO-1,2-DIAMINOBENZENE see ALL500
2-NITRO-1,4-DIAMINOBENZENE see ALL750
NITRO-p-DICHLOROBENZENE see DFT400
4′-NITRO-2,4-DICHLORODIPHENYL ETHER see DFT800
N-NITRODIETHANOLAMINE DINITRATE see NFW000
NITRODIETHANOLAMINE DINITRATE see NFW000
N-NITRODIETHYLAMINE see DJS500
1-NITRO-9-(DIETHYLAMINOETHYLAMINE)-ACRIDINE DIHYDROCHLORIDE see NFW100
1-NITRO-9-(DIETHYLAMINOPROPYLAMINE)-ACRIDINE DIHYDROCHLORIDE see NFW200
4-NITRODIFENYLAMIN (CZECH) see NFY000
4-NITRODIFENYLAMIN-2-SULFONAN SODYN (CZECH) see AOS500
4-NITRODIFENYLETHER (CZECH) see NIU000
1-NITRO-9-(2-DIHYDROXYETHYLAMINO-ETHYLAMINO)-ACRIDINE HYDROCHLORIDE see NFW350
N-NITRODIMETHYLAMINE see DSV200
1-NITRO-9-(DIMETHYLAMINE)-ACRIDINE DIHYDROCHLORIDE see NFW400
1-NITRO-9-(DIMETHYLAMINO)-ACRIDINE HYDROCHLORIDE see NFW425
2′-NITRO-4-DIMETHYLAMINOAZOBENZENE see DSW800
3′-NITRO-4-DIMETHYLAMINOAZOBENZENE see DSW600
1-NITRO-10-(DIMETHYLAMINOETHYL)-9-ACRIDONE HYDROCHLORIDE see NFW430
1-NITRO-9-((2-DIMETHYLAMINO)-1-METHYLETHYLAMINO)-ACRIDINE DIHYDROCHLORIDE see NFW435
1-NITRO-9-(5-DIMETHYLAMINOPENTYLAMINO)-ACRIDINE DIHYDROCHLORIDE see NFW450
1-NITRO-10-(3-DIMETHYLAMINOPROPYL)-ACRIDONE HYDROCHLORIDE see NFW460
1-NITRO-14-(DIMETHYLAMINOPROPYL)-ACRIDONE HYDROCHLORIDE see NFW460
1-NITRO 10-(3-DIMETHYLAMINOPROPYL)-ACRIDON HYDROCHLORIDE see NFW460
1-NITRO-9-(3-DIMETHYLAMINOPROPYLAMINE)ACRIDINE-N¹⁰-OXIDE DIHYDROCHLORIDE see CAB125
1-NITRO-9-(3-DIMETHYLAMINOPROPYLAMINE)-ACRIDINE-N-OXIDE DIHYDROCHLORIDE see NFW470
1-NITRO-9-(3′-DIMETHYLAMINOPROPYLAMINO)-ACRIDINE see NFW500
1-NITRO-9-(3-DIMETHYLAMINOPROPYLAMINO)-ACRIDINE DIHYDROCHLORIDE see LEF300
NITRODIMETHYLBENZENE see NMS000
1-NITRO-3-(2,4-DINITROPHENYL)UREA see NFX500
2-NITRODIPHENYL see NFP500
4-NITRODIPHENYL see NFQ000
o-NITRODIPHENYL see NFP500
4-NITRODIPHENYLAMINE see NFY000
p-NITRODIPHENYLAMINE see NFY000
2-NITRODIPHENYL ETHER see NIT500
4-NITRODIPHENYL ETHER see NIU000
p-NITRODIPHENYL ETHER see NIU000
N-NITRO-DMA see DSV200
4-NITRODRACYLIC ACID see CCI250
NITRO-DUR see NGY000
NITRODURAN see TJL250
dl-1-(2-NITRO-3-EMTHYLPHENOXY)-3-tert-BUTYLAMINO-PROPAN-2-OL see BQF750
NITROETAN (POLISH) see NFY500
NITROETHANE see NFY500
2-NITROETHANOL see NFY550
NITROETHYLENE POLYMER see NFY560
NITROETHYL NITRATE (DOT) see NFY570
4-NITRO-2-ETHYLQUINOLINE-N-OXIDE see NGA500
NITROFAN see DUS700
NITRO FAST GREEN GB see BLK000
NITRO FAST YELLOW SL see CMS245

NITROFEN see DFT800
NITROFENE (FRENCH) see DFT800
4-NITROFENOL (DUTCH) see NIF000
m-NITROFENOL see NIE010
o-NITROFENOL see NIE500
1-p-NITROFENYL-3,3-DIMETHYLTRIAZEN (CZECH) see DSX400
4-NITRO-1,3-FENYLENDIAMIN see NIM550
3-NITROFLUORANTHENE see NGA700
4-NITROFLUORANTHENE see NGA700
2-NITROFLUORENE see NGB000
4-NITROFLUOROBENZENE see FKL000
p-NITROFLUOROBENZENE see FKL000
NITROFORM see TMM500
5-NITRO-2-FURALDEHYDE ACETYLHYDRAZONE see NGB700
5-NITRO-2-FURALDEHYDE-2-(2-HYDROXYETHYL)SEMICARBAZONE see FPE100
5-NITRO-2-FURALDEHYDE OXIME see NGC000
5-NITROFURALDEHYDE SEMICARBAZIDE see NGE500
6-NITROFURALDEHYDE SEMICARBAZIDE see NGE500
5-NITRO-2-FURALDEHYDE SEMICARBAZONE see NGE500
5-NITRO-2-FURALDEHYDE THIOSEMICARBAZONE see NGC400
5-NITRO-2-FURALDOXIME see NGC000
5-NITRO-2-FURANACRYLAMIDE see NGL500
5-NITROFURAN-2-ALDEHYDE SEMICARBAZONE see NGE500
5-NITRO-2-FURANCARBOXALDEHYDE SEMICARBAZONE see NGE500
NITROFURANTOIN see NGE000
6-(2-(5-NITRO-2-FURANYL)ETHENYL)-2-PYRIDINEMETHANOL (9CI) see NDY400
1-(((5-NITRO-2-FURANYL)METHYLENE)AMINO)-2-IMIDAZOLIDINONE see NDY000
3-(((5-NITRO-2-FURANYL)METHYLENE)AMINO)-2-OXAZOLIDINONE see NGG500
1-((5-NITROFURANYL-2)METHYLENEAMINO)TETRAHYDROPYRIMIDONE-2-ONE see FPO100
((5-NITRO-2-FURANYL)METHYLENE)HYDRAZIDEACETIC ACID (9CI) see NGB700
2((5-NITRO-2-FURANYL)METHYLENE)HYDRAZINECARBOXAMIDE see NGE500
5-(5-NITRO-2-FURANYL)-1,3,4-THIADIAZOL-2-AMINE see NGI500
N-(4-(5-NITRO-2-FURANYL)-2-THIAZOLYL)ACETAMIDE see AAL750
2-(4-(5-NITRO-2-FURANYL)-2-THIAZOLYL)-HYDRAZINECARBOXALDEHYDE see NDY500
NITROFURAZOLIDONE see NGG500
NITROFURAZOLIDONUM see NGG500
NITROFURAZONE see NGE500
5-NITROFURFURALACETYLHYDRAZONE see NGB700
3-(5′-NITROFURFURALAMINO)-2-OXAZOLIDONE see NGG500
5-NITROFURFURAL SEMICARBAZONE see NGE500
2-(5-NITRO-2-FURFURYLIDENE)AMINOETHANOL N-OXIDE see HKW450
N-(5-NITROFURFURYLIDENE)-1-AMINOHYDANTOIN see NGE000
1-((5-NITROFURFURYLIDENE)AMINO)HYDANTOIN see NGE000
N-(5-NITROFURFURYLIDENE)-1-AMINOHYDANTOIN see NGE000
1-((5-NITROFURFURYLIDENE)AMINO)-2-IMIDAZOLIDINONE see NDY000
N-(5-NITROFURFURYLIDENEAMINO)-2-IMIDAZOLIDINONE see NDY000
N-(5-NITRO-2-FURFURYLIDENE)-1-AMINO-2-IMIDAZOLIDONE see NDY000
1-((5-NITROFURFURYLIDENE)AMINO)-2-METHYLTETRAHYDRO-1,4-THIAZINE-4,4-DIOXIDE see NGG000
4-((5-NITROFURFURYLIDENE)AMINO)-3-METHYLTHIOMORPHOLINE-1,1-DIOXIDE see NGG000
N-(5-NITRO-2-FURFURYLIDENE)-3-AMINOOXAZOLIDINE-2-ONE see NGG500
N-(5-NITRO-2-FURFURYLIDENE)-3-AMINO-2-OXAZOLIDONE see NGG500
3-((5-NITROFURFURYLIDENE)AMINO)-2-OXAZOLIDONE see NGG500
(5-NITRO-2-FURFURYLIDENEAMINO)UREA see NGE500
(NITRO-5′ FURFURYLIDENE-2′) HYDROXY-4 BENZHYDRAZIDE (FRENCH) see DGQ500
N-(6-(5-NITROFURFURYLIDENEMETHYL)-1,2,4-TRIAZIN-3-YL)IMINODIMETHANOL see BKH500
NITROFURMETHONE see FPI000
NITROFURMETON see FPI000
NITROFUROXIME see NGC000
NITROFUROXON see NGG500
3-(5-NITRO-2-FURYL)ACRYLAMIDE see NGL500
5-NITRO-2-FURYLACRYLAMIDE see NGL500
5-(5-NITRO-2-FURYL)-2-AMINO-1,3,4-THIADIAZOLE see NGI500
2-(5-NITRO-2-FURYL)-5-AMINO-1,3,4-THIADIAZOLE see NGI500
3-((5-NITROFURYLIDENE)AMINO)-2-OXAZOLIDONE see NGG500
3-(5-NITRO-2-FURYL)-IMIDAZO(1,2-a)PYRIDINE see NGI800
(5-NITRO-2-FURYL) METHYL KETONE see ACT250
((3-(5-NITRO-2-FURYL)-1-(2-(5-NITRO-2-FU-RYL)VINYL)ALLYIDENE)AMINO)GUANIDINE see PAF500
N-(3-(5-NITRO-2-FURYL)-6H-1,2,4-OXADIAZINYL)ACETAMIDE see AAL500
5-(5-NITRO-2-FURYL)-1,3,4-OXADIAZOLE-2-OL see NGK000
N-((3-(5-NITRO-2-FURYL)-1,2,4-OXADIAZOLE-5-YL)METHYL)ACETAMIDE see NGK500
3-(5-NITRO-2-FURYL)-2-PHENYLACRYLAMIDE see NGL000
3-(5-NITRO-2-FURYL)-2-PHENYL-2-PROPENAMIDE see NGL000
3-(5-NITRO-2-FURYL)-2-PROPENAMIDE see NGL500

N-(5-(5-NITRO-2-FURYL)-1,3,4-THIADIAZOL-2-YL)ACETAMIDE see FQJ000
4-(5-NITRO-2-FURYL)THIAZOLE see NGM400
N-(4-(5-NITRO-2-FURYL)-2-THIAZOLYL)ACETAMIDE see AAL750
N-(4-(5-NITRO-2-FURYL)THIAZOL-2-YL)ACETAMIDE see AAL750
N-(4-(5-NITRO-2-FURYL)-2-THIAZOLYL)FORMAMID (GERMAN) see NGM500
N-(4-(5-NITRO-2-FURYL)-2-THIAZOLYL)FORMAMIDE see NGM500
(4-(5-NITRO-2-FURYL)THIAZOL-2-YL)HYDRAZONOACETONE see NGN000
N-(4-(5-NITRO-2-FURYL)-2-THIAZOLYL)-2,2,2-TRIFLUOROACETAMIDE see
   NGN500
N,N'-(6-(5-NITRO-2-FURYL)-s-TRIAZINE-2,4-DIYL)BISACETAMIDE see DBF400
3-(5-NITRO-2-FURYL)-1H-1,2,4-TRIAZOL-5-AMINE see ALM750
N-(3-(5-NITRO-2-FURYL)-s-TRIAZOL-5-YL)-N-NITROSOETHYLAMINE see ENS000
6-(5-NITRO-2-FURYLVINYL)-3-(DIHYDROXYDIMETHYLAMINO)-1,2,4-TRIAZENE
   see BKH500
6-(2-(5-NITRO-2-FURYL)VINYL-2-PYRIDINE-METHANOL see NDY400
N-(6-(2-(5-NITRO-2-FURYL)VINYL)-1,2,4-TRIAZIN-3-YL)IMINODIMETHANOL see
   BKH500
((6-(2-(5-NITRO-2-FURYL)VINYL)-as-TRIAZIN-3-YL)IMINO)DIMETHANOL see
   BKH500
NITROGEN see NGP500
NITROGEN BROMIDE see BMP250
NITROGEN CHLORIDE see CDW000, NGQ500
NITROGEN CHLORIDE DIFLUORIDE see NGR000
NITROGEN DIOXIDE see NGR500
NITROGEN DIOXIDE, DI- see NGU500
NITROGEN FLUORIDE see NGW000
NITROGEN FLUORIDE OXIDE see NGS500
NITROGEN GAS see NGP500
NITROGEN GLUCOSIDE of SODIUM-p-AMINOPHENYLSTIBONATE see
   NCL000
NITROGEN HALF MUSTARD see CGW000
NITROGEN IODIDE see IDN000, NGW500
NITROGEN LIME see CAQ250
NITROGEN MONOXIDE see NEG100
NITROGEN MONOXIDE, mixed with NITROGEN TETROXIDE see NGT500
NITROGEN MUSTARD see BIE250
NITROGEN MUSTARD HYDROCHLORIDE see BIE500
NITROGEN MUSTARD-N-OXIDE see CFA500, CFA750
NITROGEN MUSTARD OXIDE see CFA500, CFA750
NITROGEN MUSTARD-N-OXIDE HYDROCHLORIDE see CFA750
NITROGENOL see HCP800
NITROGEN OXIDE see NGU000
NITROGEN OXIDES mixed with OZONE (47%:53%) see ORY000
NITROGEN OXYCHLORIDE see NMH000
NITROGEN OXYFLUORIDE see NMH500
NITROGEN PEROXIDE see NGR500
NITROGEN SELENIDE see SBT100
NITROGEN TETROXIDE see NGU500
NITROGEN TETROXIDE-NITRIC OXIDE MIXTURE see NGT500
NITROGEN TRIBROMIDE HEXAAMMONIATE see NGV500
NITROGEN TRICHLORIDE (DOT) see NGQ500
NITROGEN TRICHLORIDE see NGQ500
NITROGEN TRIFLUORIDE see NGW000
NITROGEN TRIIODIDE see NGW500
NITROGEN TRIIODIDE-AMMONIA see NGX000
NITROGEN TRIIODIDE-SILVER AMIDE see NGX500
NITROGEN, compressed (UN 1066) (DOT) see NGP500
NITROGEN, refrigerated liquid (cryogenic liquid) (UN 1977) (DOT) see
   NGP500
NITROGLICERINA (ITALIAN) see NGY000
NITROGLICERYNA (POLISH) see NGY000
N-NITROGLICIN see NGY700
NITROGLYCERIN, liquid, not desensitized (DOT) see NGY000
NITROGLYCERIN see NGY000
NITROGLYCERINE see NGY000
NITROGLYCERIN mixed with ETHYLENE GLYCOL DINITRATE (1:1) see
   NGY500
NITROGLYCERIN, SPIRITS OF see NGY000
NITROGLYCERIN, solution in alcohol, with not >1% nitroglycerin (UN 1204)
   (DOT) see NGY000
NITROGLYCERIN, desensitized, not <40% non-volatile water insoluble phleg-
   matizer (UN 0143) (DOT) see NGY000
NITROGLYCERIN, solution in alcohol, with >1% but not >10% nitroglycerin
   (UN 0144) (DOT) see NGY000
NITROGLYCERIN, solution in alcohol, with >1% but not >5% nitroglycerin
   (UN 3064) (DOT) see NGY000
NITROGLYCEROL see NGY000
N-NITROGLYCINE see NGY700
NITROGLYCOL see EJG000
NITROGLYKOL (CZECH) see EJG000
NITROGLYN see NGY000
NITROGRANULOGEN see BIE500
NITROGRANULOGEN HYDROCHLORIDE see BIE500
4-NITROGUAIACOL see NHA000
2-NITROGUANIDINE see NHA500

α-NITROGUANIDINE see NHA500
NITROGUANIDINE see NHA500
NITROGUANIDINE, wetted (UN 1336) (DOT) see NHA500
NITROGUANIDINE, dry or wetted with <20% water, by weight (UN 0282)
   (DOT) see NHA500
NITROGUANIL see NIJ400
2-NITRO-2-HEPTENE see NHB000
3-NITRO-2-HEPTENE see NHB500
2-NITRO-2-HEXENE see NHD000
3-NITRO-3-HEXENE see NHE000
1-NITROHYDANTOIN see NHE100
NITROHYDRENE see NHE500
NITROHYDROCHLORIC ACID, diluted (DOT) see HHM000
NITROHYDROCHLORIC ACID (DOT) see HHM000
2-NITRO-1-HYDROXYBENZENE-4-ARSONIC ACID see HMY000
3-NITRO-4-HYDROXYBENZENEARSONIC ACID see HMY000
6-NITRO-4-HYDROXYLAMINOQUINOLINE-1-OXIDE see HIX500
7-NITRO-4-HYDROXYLAMINOQUINOLINE-1-OXIDE see HIY000
4-NITRO-5-HYDROXYMERCURIORTHOCRESOL see NHK900
2-NITRO-2-(HYDROXYMETHYL)-1,3-PROPANEDIOL see HMJ500
3-NITRO-4-HYDROXYPHENYLARSENOUS ACID see NHE600
3-NITRO-4-HYDROXYPHENYLARSONIC ACID see HMY000
5-NITRO-8-HYDROXYQUINOLINE see NHF500
6-NITRO IA see NHK600
2-NITRO-1H-IMIDAZOLE see NHG000
2-NITROIMIDAZOLE see NHG000
4-NITRO-1H-IMIDAZOLE (9CI) see NHG100
1-NITRO-2-IMIDAZOLIDONE see NKL000
N-NITRO-2-IMIDAZOLIDONE see NKL000
4-(2-(5-NITROIMIDAZOL-1-YL)ETHYL)MORPHOLINE see NHH000
1-(2-NITRO-1-IMIDAZOLYL)-3-METHOXY-2-PROPANOL see NHH500
1-(2-NITROIMIDAZOL-1-YL)-3-METHOXYPROPAN-2-OL see NHH500
3-(2-NITRO-1H-IMIDAZOL-1-YL)-1,2-PROPANEDIOL see NHI000
3-(2-NITROIMIDAZOL-1-YL)-1,2-PROPANEDIOL see NHI000
2,2'-(NITROIMINO)BISETHANOL, DINITRATE (ESTER) (9CI) see NFW000
2,2'-NITROIMINOBIS(ETHYLNITRATE) see NFW000
2,2'-NITROIMINODIETHANOL NITRATE see NFW000
NITROIMINODIETHYLENEDIISOCYANIC ACID see NHI500
3,3'-NITROIMINODIPROPIONIC ACID see NHI600
2,2'-(NITROIMINO)ETHANOL DINITRATE see NFW000
4-NITROINDANE see NHJ000
5-NITROINDANE see NHJ009
6-NITRO-ISATOIC ANHYDRIDE see NHK600
NITRO ISOBUTANE TRIOL TRINITRATE (DOT) see NHK650
NITROISOBUTYL GLYCERYL TRINITRATE see NHK650
NITROISOBUTYL GLYCOL TRINITRATE see NHK650
NITROISOPROPANE see NIY000
1-NITRO-9-(3-ISOPROPYLAMINOPROPYLAMINE)-ACRIDINE DIHYDROCHLORIDE
   see INW000
1-NITRO-9-(3-ISOPROPYLAMINOPROPYLAMINO)-ACRIDINE DIHYDROCHLORIDE
   see INW000
NITRO KLEENUP see DUZ000
NITROL see NGY000, NHK800
NITROLAN see NGY000
NITRO-LENT see NGY000
NITROLETTEN see NGY000
NITROLIME see CAQ250
NITROLINGUAL see NGY000
NITROLOWE see NGY000
NITROL (PROMOTER) see NHK800
NITROMANNITE (dry) (DOT) see MAW250
NITRO MANNITE see MAW250
NITROMANNITE, wetted with not <40% water, by weight or mixture (NA
   0133) (DOT) see MAW250
NITROMANNITOL see MAW250
NITROMEL see NGY000
NITROMERSOL see NHK900
NITROMERSOL SOLUTION see NHK900
NITROMESITYLENE see NHM000
NITROMETAN (POLISH) see NHM500
N-NITROMETHANAMINE see NHN500
NITROMETHANE see NHM500
5-NITRO-2-METHOXYANILINE see NEQ500
3-NITRO-6-METHOXYANILINE see NEQ500
2-NITRO-7-METHOXYNAPHTHO(2,1-b)FURAN see MFB400
N-NITROMETHYLAMINE see NHN500
5-NITRO-2-METHYLANILINE see NMP500
3-NITRO-4-METHYLANILINE see NMP000
3-NITRO-6-METHYLANILINE see NMP500
1-NITRO-2-METHYLANTHRAQUINONE see MMG000
((α-NITROMETHYL)-o-CHLOROBENZYLTHIO)ETHYLAMINE HYDROCHLORIDE
   see NEC000
3-(5-NITRO-1-METHYL-2-IMIDAZOLYL)-METHYLENE-AMINO-5-MORPHOLINO-
   METHYL-2-OXAZOLIDONE HCI see MRU750
1-((2-NITRO-4-METHYLPHENYL)AZO)-2-NAPHTHOL see MMP100

2-NITRO-2-METHYL-1,3-PROPANEDIOL see NHO500
2-NITRO-2-METHYL-1-PROPANOL see NHP000
2-NITRO-2-METHYLPROPANOL NITRATE (DOT) see MMP200
NITROMIDINE see TJF000
NITROMIFENE CITRATE see NHP500
NITROMIM see CFA750
NITROMIN see CFA500
NITROMIN HYDROCHLORIDE see CFA750
NITROMIN IDO see ALQ000
NITROMURIATIC ACID (DOT) see HHM000
NITRON see CCU250
1-NITRONAPHTHALENE see NHQ000
2-NITRONAPHTHALENE see NHQ500
α-NITRONAPHTHALENE see NHQ000
β-NITRONAPHTHALENE see NHQ500
NITRONAPHTHALENE (DOT) see NHP990
NITRONAPHTHALENE see NHP990
5-NITRONAPHTHALENE ETHYLENE see NEJ500
2-NITRONAPHTHO(2,1-b)FURAN see NHQ950
1-(2-NITRONAPHTHO(2,1-b)FURAN-7-YL)ETHANONE see NHR100
3-NITRO-2-NAPHTHYLAMINE see NHR500
NITRONET see NGY000
NITRONG see NGY000
2-NITRO-1-(p-NITROBENZYL)IMIDAZOLE see NFO700
2-NITRO-1-(4-NITROPHENOXY)-4-(TRIFLUOROMETHYL)BENZENE see NIX000
N'-NITRO-N-NITROSO-N-METHYLGUANIDINE see MMP000
NITRONIUM TETRAFLUOROBORATE(1−) see NHS500
NITRON LAVSAN see PKF750
NITRON (NITROCELLULOSE) see CCU250
3-NITRO-N',N',N⁴-TRIS(2-HYDROXYETHYL)-p-PHENYLENEDIAMINE see HKN875
2-NITRO-2-NONENE see NHT000
3-NITRO-3-NONENE see NHU000
5-NITRO-4-NONENE see NHV500
NITRON (POLYESTER) see PKF750
2-NITRO-2-OCTENE see NHW000
3-NITRO-2-OCTENE see NHW500
3-NITRO-3-OCTENE see NHX000
4'-NITROOXANILIC ACID see NHX100
1-NITRO-1-OXIMINOETHANE see NHY100
NITROOXIMINOMETHANE see NHY250
7-NITRO-3-OXO-3H-2,1-BENZOXAMERCUROLE see NHY500
5-NITRO-N-(2-OXO-3-OXAZOLIDINYL)-2-FURANMETHANIMINE see NGG500
NITROPENTA see PBC250
NITROPENTAERYTHRITE see PBC250
NITROPENTAERYTHRITOL see PBC250
1-NITROPENTANE see AOL500
2-NITRO-2-PENTENE see NHZ000
3-NITRO-2-PENTENE see NIA000
3-NITROPERCHLORYLBENZENE see NIA700
p-NITROPEROXYBENZOIC ACID see NIB000
NITROPHEN see DFT800
NITROPHENE see DFT800
p-NITROPHENETOL (GERMAN) see NID000
p-NITROPHENETOLE see NID000
2-NITROPHENOL see NIE500
3-NITROPHENOL see NIE600
4-NITROPHENOL see NIF000
m-NITROPHENOL (DOT) see NIE010
p-NITROPHENOL (DOT) see NIF000
o-NITROPHENOL see NIE500
p-NITROPHENOL ACETATE see ABS750
3-(α-(p-NITROPHENOL)-β-ACETYLETHYL)-4-HYDROXYCOUMARIN see ABF750
NITROPHENOLARSONIC ACID see HMY000
P-NITROPHENOL, ESTER with DIETHYL PHOSPHATE see NIM500
p-NITROPHENOL, O-ESTER with O,O-DIETHYLPHOSPHOROTHIOATE see
    PAK000
p-NITROPHENOL, MERCURY(II) SALT see NIG500
p-NITROPHENOL SODIUM SALT see SIV600
p-NITROPHENOL TIN(IV) SALT see NIH000
p-NITROPHENOL ZINC SALT see NIH500
(4-NITROPHENOXY)ACETIC ACID see NII000
p-NITROPHENOXYACETIC ACID see NII000
1-NITRO-4-PHENOXYBENZENE (9CI) see NIU000
((4-NITROPHENOXY)METHYL)OXIRANE see NIN050
p-NITROPHENOXYTRIBUTYLTIN see NII200
N-(4-NITROPHENYL)ACETAMIDE see NEK000
4-NITROPHENYL ACETATE see ABS750
p-NITROPHENYL ACETATE see ABS750
2-(p-NITROPHENYL)ACETIC ACID see NII510
(4-NITROPHENYL)ACETIC ACID see NII510
(p-NITROPHENYL)ACETIC ACID see NII510
4-NITROPHENYLACETONITRILE see NIJ000
p-NITROPHENYLACETONITRILE see NIJ000
2-NITROPHENYLACETYL CHLORIDE see NIJ200
NITROPHENYL ACETYLENE see NIJ300

3-(α-(4'-NITROPHENYL)-β-ACETYLETHYL)-4-HYDROXYCOUMARIN see ABF750
3-(α-p-NITROPHENYL-β-ACETYLETHYL)-4-HYDROXYCOUMARIN see ABF750
NITROPHENYLACETYLETHYL-4-HYDROXYCOUMARINE see ABF750
3-(4-NITROPHENYL)ACRYLIC ACID see NFT440
1-p-NITROPHENYL-3-AMIDINOUREA HYDROCHLORIDE see NIJ400
m-NITROPHENYLAMINE see NEN500
p-NITROPHENYLAMINE see NEO500
((4-NITROPHENYL)AMINO)OXOACETIC ACID see NHX100
4-NITROPHENYLARSONIC ACID see NIJ500
p-NITROPHENYLARSONIC ACID see NIJ500
4-((p-NITROPHENYL)AZO)DIPHENYLAMINE see NIK000
7-((p-NITROPHENYLAZO))METHYLBENZ(c)ACRIDINE see NIK500
4-NITRO-N-PHENYLBENZENAMINE see NFY000
1-(p-NITROPHENYL)BIGUANIDE HYDROCHLORIDE see NIL500
N-(2-NITROPHENYL)-1,3-DIAMINOETHANE see NIL625
p-NITROPHENYLDI-N-BUTYLPHOSPHINATE see NIM000
p-NITROPHENYLDIBUTYLPHOSPHINATE see NIM000
d-(−)-threo-1-p-NITROPHENYL-2-DICHLORACETAMIDO-1,3-PROPANEDIOL see
    CDP250
d-threo-1-(p-NITROPHENYL)-2-(DICHLOROACETYLAMINO)-1,3-PROPANEDIOL
    see CDP250
p-NITROPHENYL DIETHYLPHOSPHATE see NIM500
7-NITRO-5-PHENYL-2,3-DIHYDRO-1H-1,4-BENZODIAZEPIN-2-ONE see DLY000
4-(2'-NITROPHENYL)-2,6-DIMETHYL-3,5-DICARBOMETHOXY-1,4-DIHYDRO-
    PYRIDINE see AEC750
p-NITROPHENYLDIMETHYLTHIONOPHOSPHATE see MNH000
1-(p-NITROPHENYL-3,3-DIMETHYL-TRIAZEN (GERMAN) see DSX400
1-(4-NITROPHENYL)-3,3-DIMETHYLTRIAZENE see DSX400
1-(p-NITROPHENYL)-3,3-DIMETHYL-TRIAZENE see DSX400
4-NITRO-1,2-PHENYLENEDIAMINE see ALL500
4-NITRO-1,3-PHENYLENEDIAMINE see NIM550
2-NITRO-1,4-PHENYLENEDIAMINE see ALL750
o-NITRO-p-PHENYLENEDIAMINE (MAK) see ALL750
4-NITRO-m-PHENYLENEDIAMINE see NIM550
4-NITRO-o-PHENYLENE-DIAMINE see ALL500
p-NITRO-o-PHENYLENEDIAMINE see ALL500
2-NITRO-p-PHENYLENEDIAMINE see ALL750
NITRO-p-PHENYLENEDIAMINE see ALL750
p-NITROPHENYL ESTER of DIETHYLPHOSPHINIC ACID see DJW200
1-(2-NITROPHENYL)ETHANONE see NEL450
1-(3-NITROPHENYL)ETHANONE see NEL500
p-NITROPHENYL ETHYLBUTYLPHOSPHONATE see NIN000
p-NITROPHENYL ETHYL PENTYLPHOSPHONATE see ENQ500
1-((5-(p-NITROPHENYL)FURFURYLIDENE)AMINO)HYDANTOIN SODIUM see
    DAB840
p-NITROPHENYL GLYCIDYL ETHER see NIN050
NITROPHENYL GLYCIDYL ETHER see NIN050
N-(p-NITROPHENYL)GLYCINE see NIN100
4-NITROPHENYL 4-GUANIDINOBENZOATE see NIQ500
p-NITROPHENYL-p'-GUANIDINOBENZOATE see NIQ500
p-NITROPHENYL p-GUANIDINO-BENZOATE see NIQ500
2-(p-NITROPHENYL)HYDRAZIDEACETIC ACID see NIQ550
2-(p-NITROPHENYL)HYDRAZIDE FORMIC ACID see NIQ600
2-NITROPHENYLHYDRAZINE see NIQ800
4-NITROPHENYLHYDRAZINE see NIR000
(o-NITROPHENYL)HYDRAZINE see NIQ800
p-NITROPHENYLHYDRAZINE see NIR000
N-(4-NITROPHENYL)-IMIDODICARBONIMIDIC DIAMIDE MONOHYDROCHLORIDE
    see NIL500
o-NITROPHENYL ISOPROPYL ETHER see NIR550
3-NITROPHENYL ISOTHIOCYANATE see ISO000
m-NITROPHENYL ISOTHIOCYANATE see ISO000
o-NITROPHENYL MERCURY ACETATE see NIS000
o-NITROPHENYL MERCURY (ACETATO) see NIS000
o-NITROPHENYL METHYL ETHER see NER000
p-NITROPHENYL METHYL KETONE see NEL600
p-NITROPHENYL-2-NITRO-4-(TRIFLUOROMETHYL) PHENYL ETHER see
    NIX000
N-(4-NITROPHENYL)-N'-(3-PYRIDINYLMETHYL)UREA see PPP750
N-(4-NITROPHENYL)OXAMIC ACID see NHX100
2-NITROPHENYL PHENYL ETHER see NIT500
4-NITROPHENYL PHENYL ETHER see NIU000
p-NITROPHENYL PHENYLETHER see NIU000
O-(4-NITROPHENYL) O-PHENYLMETHYL PHOSPHONOTHIOATE see MOB699
3-(4-NITROPHENYL)-1-PHENYL-2-PROPEN-1-ONE see NFS505
3-(4-NITROPHENYL)PROPENOIC ACID see NFT440
2-NITROPHENYL SULFONYL DIAZOMETHANE see NIW300
4-(4-NITROPHENYL)THIAZOLE see NIW400
p-NITROPHENYL-2,4,6-TRICHLOROPHENYL ETHER see NIW500
p-NITROPHENYL-α,α,α-TRIFLUORO-2-NITRO-p-TOLYL ETHER see NIX000
NITROPHOS see DSQ000
NITROPONE C see BRE500
NITROPORE OBSH see BBS300, OPE000
1,1'-(2-NITROPORPYLIDENE)BIS(4-CHLOROBENZENE) see BIN500
1-NITROPROPANE see NIX500

2-NITROPROPANE see NIY000
β-NITROPROPANE see NIY000
NITROPROPANE (DOT) see NIY000
NITROPROPANE see NIY000
2-NITROPROPENE see NIY200
3-NITROPROPIONIC ACID see NIY500
β-NITROPROPIONIC ACID see NIY500
NITROPRUSSIATE de SODIUM see SIW500
NITROPRUSSIDNATRIUM (GERMAN) see SIU500
NITROPRUSSIDNATRIUM see SIW500
1-NITROPYRENE see NJA000
3-NITROPYRENE see NJA000
4-NITROPYRENE see NJA100
4-NITROPYRIDINE-1-OXIDE see NJA500
4-NITROPYRIDINE-N-OXIDE see NJA500
5-NITRO-2,4,6(1H,3H,5H)-PYRIMIDINETRIONE see NET550
4-NITROQUINALDINE-N-OXIDE see MMQ250
2-NITROQUINOLINE see NJB500
8-NITROQUINOLINE see NJB500
4-NITROQUINOLINE-6-CARBOXYLIC ACID-1-OXIDE see CCG000
4-NITRO-6-QUINOLINECARBOXYLIC ACID-1-OXIDE see CCG000
4-NITROQUINOLINE-1-OXIDE see NJF000
4-NITROQUINOLINE-N-OXIDE see NJF000
5-NITRO-8-QUINOLINOL see NHF500
NITRORECTAL see NGY000
4-NITROSALICYLIC ACID see NJF100
p-NITROSALICYLIC ACID see NJF100
4-NITRO-SALICYLSAEURE see NJF100
NITROSAMINES see NJH000
N-NITROSAMINO DIACETONITRIL (GERMAN) see NKL500
N-NITROSARCOSINE see NJH500
N-NITROSAZETIDINE see NJL000
NITRO-SIL see AMY500
NITROSIMINODIACETONITRILE see NKL500
N-NITROSOACETANILIDE see NJI700
N-NITROSO-N-(1-ACETOXYMETHYL)BUTYLAMINE see BRX500
N-NITROSO-N-(ACETOXYMETHYL)-N-ISOBUTYLAMINE see ABR125
N-NITROSO-N-(ACETOXY)METHYL-N-METHYLAMINE see AAW000
N-NITROSO-N-(1-ACETOXYMETHYL)PROPYL AMINE see PNR250
NITROSO-N-(1-ACETOXYMETHYL)TRIDEUTEROMETHYLAMINE see MMS000
NITROSO-4-ACETYL-3,5-DIMETHYLPIPERAZINE see NJI850
N-NITROSOAETHYLAETHANOLAMIN (GERMAN) see ELG500
NITROSOAETHYLDIMETHYLHARNSTOFF see DRV600
NITROSOALDICARB see NJJ500
N-NITROSOALLYL-2,3-DIHYDROXYPROPYLAMINE see NJY500
N-NITROSOALLYLETHANOLAMINE see NJJ875
N-NITROSOALLYL-2-HYDROXYPROPYLAMINE see NJJ950
NITROSO-ALLYL-2-HYDROXYPROPYLAMINE see NJJ950
N-NITROSO-N-ALLYL-N-(2-HYDROXYPROPYL)AMINE see NJJ950
N-NITROSOALLYLMETHYLAMINE see MMT500
N-NITROSOALLYL-2-OXOPROPYLAMINE see AGM125
NITROSOALLYLUREA see NJK000
N-NITROSOAMINODIETHANOL see NKM000
4-(NITROSOAMINO-N-METHYL)-1-(3-PYRIDYL)-1-BUTANONE see MMS500
1-NITROSOANABASINE see NJK150
N-NITROSOANABASINE see NJK150
N'-NITROSOANABASINE see NJK150
N-(p-NITROSOANILINOMETHYL)-2-NITROPROPANE see NHK800
N-NITROSOAZACYCLOHEPTANE see NKI000
N-NITROSOAZACYCLONONANE see OCA000
N-NITROSOAZACYCLOOCTANE see OBY000
1-NITROSOAZACYCLOTRIDECANE see NJK500
NITROSO-AZETIDIN (GERMAN) see NJL000
1-NITROSOAZETIDINE see NJL000
N-NITROSOAZETIDINE see NJL000
NITROSOAZETIDINE see NJL000
NITROSO-BAYGON see PMY310
1-NITROSO-4-BENZOYL-3,5-DIMETHYLPIPERAZINE see NJL850
N-NITROSO-4-BENZOYL-3,5-DIMETHYLPIPERAZINE see NJL850
N-NITROSOBENZTHIAZURON see NKR000
N-NITROSOBENZYLMETHYLAMINE see MHP250
NITROSOBENZYLUREA see NJM000
N-NITROSOBIS(ACETOXYMETHYL)AMINE see NKM500
N-NITROSOBIS(2-ACETOXYPROPYL)AMINE see NJM500
NITROSOBIS(2-CHLOROETHYL)AMINE see BIF500
NITROSOBIS(2-CHLOROPROPYL)AMINE see DFW000
N-NITROSOBIS(2-ETHOXYETHYL)AMINE see BJO250
N-NITROSOBIS(2-HYDROXYETHYL)AMINE see NKM000
N-NITROSOBIS(2-HYDROXYPROPYL)AMINE see DNB200
N-NITROSOBIS(2-METHOXYETHYL)AMINE see BKO000
N-NITROSOBIS(2-OXOBUTYL)AMINE see NKL500
N-NITROSOBIS(2-OXOPROPYL)AMINE see NJN000
N-NITROSO-BIS-(4,4,4-TRIFLUORO-n-BUTYL)AMINE see NJN300
N-NITROSO-1-BUTYLAMINO-2-PROPANONE see BRY000
N-NITROSO-N-BUTYLBUTYRAMIDE see NJO150

N-NITROSO-N-(BUTYL-N-BUTYROLACTONE)AMINE see NJO200
N-NITROSO-N-BUTYL-N-(3-CARBOXYPROPYL)AMINE see BQQ250
N-NITROSO-N-BUTYLETHYLAMINE see EHC000
N-NITROSO-tert-BUTYLETHYLAMINE see NKD500
N-NITROSO-n-BUTYL-(4-HYDROXYBUTYL)AMINE see HJQ350
N-NITROSO-N-BUTYLMETHYLAMINE see MHW500
N-NITROSO-N-BUTYL-N-PENTYLAMINE see BRY250
N-NITROSO-N-BUTYLPENTYLAMINE see BRY250
N-NITROSO-4-tert-BUTYLPIPERIDINE see BRZ200, NJO300
N-NITROSOBUTYLUREA see BSA250
NITROSO-sec-BUTYLUREA see NJO500
N-NITROSO-N-BUTYROXY-BUTYLAMINE see NJO150
N-NITROSO-N-(1-BUTYROXYMETHYL)METHYL AMINE see BSX500
NITROSO-BUX-TEN see NJP000, PBX750
N-NITROSOCARBARYL see NBJ500
N-NITROSO-N-(3-CARBOETHOXYPROPIONYL)BUTYLAMINE see EHC800
NITROSOCARBOFURAN see NJQ500
N-NITROSO-N-(3-CARBOXYPROPYL)ETHYLAMINE see NKE000
NITROSOCHLOROETHYLDIETHYLUREA see NJR000
1-NITROSO-1-(2-CHLOROETHYL)-3,3-DIMETHYLUREA see NJR500
NITROSOCHLOROETHYLDIMETHYLUREA see NJR500
1-NITROSO-1-(2-CHLOROETHYL)UREA see CHE750
N-NITROSO-2-CHLOROETHYLUREA see CHE750
N-NITROSOCIMETIDINE see NJS300
NITROSOCIMETIDINE see NJS300
N-NITROSO COMPOUNDS see NJT550
NITROSO COMPOUNDS see NJT500
NITROSOCYCLOHEXYLUREA see NJV000
1'-NITROSO-1'-DEMETHYLNICOTINE see NLD500
N-NITROSODIACETONITRILE see NKL500
N-NITROSODIAETHANOLAMIN (GERMAN) see NKM000
N-NITROSODIAETHYLAMIN (GERMAN) see NJW500
N-NITROSODIALLYL AMINE see NJV500
N-NITROSO-3,4-DIBROMOPIPERIDINE see DDQ800
N-NITROSODI-n-BUTYLAMINE (MAK) see BRY500
N-NITROSODIBUTYLAMINE see BRY500
N-NITROSO-DI-sec-BUTYLAMINE see NJW000
NITROSODI-sec-BUTYLAMINE see NJW000
N-NITROSO-2,2'-DICHLORODIETHYLAMINE see BIF500
N-NITROSO-3,4-DICHLOROPIPERIDINE see DFW200
N-NITROSODI(CYANOMETHYL)AMINE see NKL500
N-NITROSODIETHANOLAMINE (MAK) see NKM000
N-NITROSODIETHYLAMINE see NJW500
NITROSODIETHYLAMINE see NJW500
NITROSO-1,1-DIETHYL-3-METHYLUREA see DJP600
N-NITROSODIFENYLAMIN (CZECH) see DWI000
p-NITROSODIFENYLAMIN (CZECH) see NKB500
N-NITROSO-3,6-DIHYDROOXAZIN-1,2 (GERMAN) see NJX000
N-NITROSO-3,6-DIHYDRO-1,2-OXAZINE see NJX000
1-NITROSO-5,6-DIHYDROURACIL see NJY000
N-NITROSO-2,3-DIHYDROXYPROPYLALLYLAMINE see NJY500
N-NITROSO-N,N-DI(2-HYDROXYPROPYL)AMINE see DNB200
N-NITROSODIHYDROXYPROPYLETHANOLAMINE see NBR100
N-NITROSO-2,3-DIHYDROXYPROPYL-2-HYDROXYETHYLAMINE see NBR100
N-NITROSO-2,3-DIHYDROXYPROPYL-2-HYDROXYPROPYLAMINE see NOC400
N-NITROSODIHYDROXYPROPYL-2-OXOPROPYLAMINE see NJY550
NITROSO-DIHYDROXYPROPYLOXOPROPYLAMINE see NJY550
N-NITROSODI-ISO-BUTYLAMINE see DRQ200
N-NITROSODIISOBUTYLAMINE see DRQ200
NITROSODIISOBUTYLAMINE see DRQ200
N-NITROSODIISOPROPYLAMINE see NKA000
N-NITROSODIMETHYLAMINE see NKA600
NITROSODIMETHYLAMINE see NKA600
4-NITROSODIMETHYLANILINE see DSY600
p-NITROSODIMETHYLANILINE (DOT) see DSY600
p-NITROSO-N,N-DIMETHYLANILINE see DSY600
N-NITROSO-2,2'-DIMETHYLDI-n-PROPYLAMINE see DRQ200
NITROSO-1,1-DIMETHYL-3-ETHYLUREA see DRV600
N-NITROSO-2,6-DIMETHYLMORPHOLINE see DTA000
cis-N-NITROSO-2,6-DIMETHYLMORPHOLINE see NKA695
cis-NITROSO-2,6-DIMETHYLMORPHOLINE see NKA695
NITROSO-2,6-DIMETHYLMORPHOLINE see DTA000
trans-N-NITROSO-2,6-DIMETHYLMORPHOLINE see NKA700
trans-NITROSO-2,6-DIMETHYLMORPHOLINE see NKA700
1-NITROSO-3,5-DIMETHYLPIPERAZINE see NKA850
N-NITROSO-3,5-DIMETHYLPIPERAZINE see NKA850
NITROSO-3,5-DIMETHYLPIPERAZINE see NKA850
N-NITROSO-2,6-DIMETHYLPIPERIDINE see DTA400
N-NITROSO-3,5-DIMETHYLPIPERIDINE see DTA600
NITROSO-3,5-DIMETHYLPIPERIDINE cis-isomer see DTA690
NITROSO-3,5-DIMETHYLPIPERIDINE trans-isomer see DTA700
N-NITROSODIMETHYLUREA see DTB200
NITROSODIMETHYLUREA see DTB200
N-NITROSO-N,N-DI(2-OXYPROPYL)AMINE see NJN000
N-NITROSODIPENTYLAMINE see DCH600

N-NITROSODI-n-PENTYLAMINE see DCH600
4-NITROSODIPHENYLAMINE see NKB500
N-NITROSODIPHENYLAMINE see DWI000
p-NITROSODIPHENYLAMINE see NKB500
NITROSODIPHENYLAMINE see DWI000
N-NITROSODI-i-PROPYLAMINE (MAK) see NKA000
N-NITROSODI-N-PROPYLAMINE see NKB700
N-NITROSO-N-DIPROPYLAMINE see NKB700
N-NITROSODIPROPYLAMINE see NKB700
N-NITROSODODECAMETHYLENEIMINE see NJK500
N-NITROSODODECAMETHYLENIMINE see NJK500
NITROSODODECAMETHYLENIMINE see NJK500
N-NITROSOEPHEDRINE see NKC000
N-NITROSO-3,4-EPOXYPIPERIDINE see NKC300
NITROSOETHANECARBAMONITRILE see ENT500
N-NITROSOETHANOLISOPROPANOLAMINE see HKW475
NITROSOETHIOFENCARB see MMV750
NITROSOETHOXYETHYLAMINE see NKC500
N-NITROSO-N-ETHYL ANILINE see NKD000
NITROSOETHYLANILINE see NKD000
N-NITROSO-N-ETHYLBENZYLAMIN (GERMAN) see ENS500
N-NITROSO-N-ETHYLBENZYLAMINE see ENS500
N-NITROSO-N-ETHYL BIURET see ENT000
N-NITROSOETHYL-N-BUTYLAMINE see EHC000
N-NITROSOETHYL-tert-BUTYLAMINE see NKD500
N-NITROSO-ETHYL(3-CARBOXYPROPYL)AMINE see NKE000
1-NITROSO-1-ETHYL-3,3-DIMETHYLUREA see DRV600
NITROSOETHYLDIMETHYLUREA see DRV600
N-NITROSOETHYLENETHIOUREA see NKK500
N-NITROSOETHYLETHANOLAMINE see ELG500
N-NITROSOETHYL-2-HYDROXYETHYLAMINE see ELG500
N-NITROSO-N-ETHYL-N-(2-HYDROXYETHYL)AMINE see ELG500
N-NITROSOETHYLISOPROPYLAMINE see ELX500
N-NITROSOETHYLMETHYLAMINE see MKB000
N-NITROSOETHYLPHENYLAMINE (MAK) see NKD000
NITROSOETHYLUREA see ENV000
N-NITROSO-N-ETHYLURETHAN see NKE500
NITROSOETHYLURETHAN see NKE500
NITROSO-N-ETHYLURETHANE see NKE500
N-NITROSO-N-ETHYLVINYLAMINE see NKF000
N-NITROSOETHYLVINYLAMINE see NKF000
N-NITROSOFENYLHYDROXYLAMIN AMONNY (CZECH) see ANO500
2-NITROSOFLUORENE see NKF500
NITROSOFLUOROETHYLUREA see NKG000
NITROSOFOLIC ACID see NKG450, NLP000
NITROSOGUANIDIN (GERMAN) see NKH000
NITROSOGUANIDINE (DOT) see MMP000
N-NITROSOGUANIDINE see NKH000
NITROSOGUANIDINE see NKH000
N-NITROSOGUVACINE see NKH500
N-NITROSOGUVACOLINE see NKH500
NITROSOGUVACOLINE see NKH500
N-NITROSOHEPTAMETHYLENEIMINE see OBY000
NITROSOHEPTAMETHYLENEIMINE see OBY000
NITROSO-HEPTAMETHYLENIMIN (GERMAN) see OBY000
N-NITROSO-4,4,4,4',4',4'-HEXAFLUORODIBUTYLAMINE see NJN300
N-NITROSOHEXAHYDROAZEPINE see NKI000
N-NITROSOHEXAMETHYLENEIMINE see NKI000
NITROSOHEXAMETHYLENIMINE see NKI000
NITROSO-n-HEXYLMETHYLAMINE see NKU400
NITROSO-N-HEXYLUREA see HFS000
NITROSO HYDANTOIC ACID see NKI500
1-NITROSOHYDANTOIN see NKJ000
1-NITROSO-1-HYDROXYETHYL-3-CHLOROETHYLUREA see NKJ050
1-NITROSO-1-(2-HYDROXYETHYL)-3-(2-CHLOROETHYL)UREA see NKJ050
1-NITROSO-1-(2-HYDROXYETHYL)UREA see HKW500
NITROSO-2-HYDROXYETHYLUREA see HKW500
N-NITROSOHYDROXYETHYLUREA see HKW500
N-NITROSOHYDROXYLAMINE see HOW500
N-NITROSO-3-HYDROXYPIPERIDINE see NLK000
N-NITROSOHYDROXYPROLINE see HNB500
1-NITROSO-1-HYDROXYPROPYL-3-CHLOROETHYLUREA see NKJ100
1-NITROSO-1-(2-HYDROXYPROPYL)-3-(2-CHLOROETHYL)UREA see NKJ100
N-NITROSO-(2-HYDROXYPROPYL)-(2-HYDROXYETHYL)AMINE see HKW475
N-NITROSO(2-HYDROXYPROPYL)(2-OXOPROPYL)AMINE see HNX500
N-NITROSO-2-HYDROXY-n-PROPYL-n-PROPYLAMINE see NLM500
NITROSO-3-HYDROXYPROPYLUREA see NKK000
N-NITROSO-2-HYDROXY-N-PROPYLUREA see NKO400
NITROSO-2-HYDROXY-N-PROPYLUREA see NKO400
NITROSO-3-HYDROXY-N-PROPYLUREA see NKK000
N-NITROSO-3-HYDROXYPYRROLIDINE see NLP700
N-NITROSOIMIDAZOLIDINETHIONE see NKK500
1-NITROSOIMIDAZOLIDINONE see NKL000
1-NITROSO-2-IMIDAZOLIDINONE see NKL000
N-NITROSO-IMIDAZOLIDON (GERMAN) see NKL000

N-NITROSOIMIDAZOLIDONE see NKL000
2,2'-(NITROSOIMINO)BISACETONITRILE see NKL500
1,1'-(NITROSOIMINO)BIS-2-BUTANONE see NKL300
2,2'-(NITROSOIMINO)BISETHANOL see NKM000
(NITROSOIMINO)DIACETONE see NJN000
2,2'-(N-NITROSOIMINO)DIACETONITRILE see NKL500
NITROSOIMINO DIETHANOL see NKM000
N-NITROSO-2,2'-IMINODIETHANOLDIACETATE see NKM500
N-NITROSO-1,1'-IMINODI-2-PROPANOL see DNB200
1,1'-NITROSOIMINODI-2-PROPANOL see DNB200
1,1-(N-NITROSOIMINO)DI-2-PROPANOL, DIACETATE see NJM500
N-NITROSOINDOLIN (GERMAN) see NKN000
1-NITROSOINDOLINE see NKN000
N-NITROSOINDOLINE see NKN000
N-NITROSO-ISO-BUTYLUREA see IJF000
NITROSOISOPROPANOL-ETHANOLAMINE see HKW475
NITROSOISOPROPANOLUREA see NKO400
NITROSOISOPROPYLUREA see NKO425
NITROSO-LANDRIN see NLY500
NITROSO-METHOMYL see NKX000
N-NITROSO-2-METHOXY-2,6-DIMETHYLMORPHOLINE see NKO600
1-NITROSOMETHOXYETHYLUREA see NKO900
NITROSO-2-METHOXYETHYLUREA see NKO900
N-NITROSOMETHOXYMETHYLAMINE see DSZ000
N-NITROSO-N-METHOXYMETHYLMETHYLAMINE see MEW250
N-NITROSO-N-METHYLACETAMIDE see MMT000
N-NITROSO-N-METHYL-N-ACETOXYMETHYLAMINE see AAW000
N-NITROSOMETHYLALLYLAMINE see MMT500
NITROSOMETHYLALLYLAMINE see MMT500
N-NITROSOMETHYLAMINACETONITRIL (GERMAN) see MMT750
N-NITROSOMETHYLAMINOACETONITRILE see MMT750
N-NITROSO-4-METHYLAMINOAZOBENZENE see MMY250
N-NITROSO-4-METHYLAMINOAZOBENZOL (GERMAN) see MMY250
N-NITROSO-N-METHYL-2-AMINOPYRIDINE see NKQ000
2-NITROSOMETHYLAMINOPYRIDINE see NKQ000
4-(N-NITROSO-N-METHYLAMINO)-4-(3-PYRIDYL)BUTANAL see MMS250
4-(N-NITROSO-N-METHYLAMINO)-1-(3-PYRIDYL)-1-BUTANONE see MMS500
3-(N-NITROSOMETHYLAMINO)SULFOLAN (GERMAN) see NKQ500
N-NITROSOMETHYLAMINOSULFOLANE see NKQ500
N-NITROSO-N-METHYL-N-AMYLAMINE see AOL000
N-NITROSO-N-METHYLANILINE see MMU250
NITROSOMETHYLANILINE see MMU250
N-NITROSOMETHYLBENZYLAMINE see MHP250
N-NITROSO-N-(2-METHYLBENZYL)-METHYLAMIN (GERMAN) see NKR500
N-NITROSO-N-(3-METHYLBENZYL)-METHYLAMIN (GERMAN) see NKS000
N-NITROSO-N-(4-METHYLBENZYL)-METHYLAMIN (GERMAN) see NKS500
N-NITROSO-N-(2-METHYLBENZYL)METHYLAMINE see NKR500
N-NITROSO-N-(3-METHYLBENZYL)METHYLAMINE see NKS000
N-NITROSO-N-(4-METHYLBENZYL)METHYLAMINE see NKS500
1-NITROSO-1-METHYL-3,3-BIS-(2-CHLOROETHYL)UREA see NKT000
NITROSOMETHYLBIS(CHLOROETHYL)UREA see NKT000
N-NITROSO-N-METHYLBIURET see MMV000
N-NITROSO-N-METHYL-N-n-BUTYL-1-d²-AMINE see NKT100
N-NITROSOMETHYL-N-BUTYLAMINE see MHW500
NITROSOMETHYL-d³-n-BUTYLAMINE see NKT105
N-NITROSO-N-METHYLCARBAMIDE see MNA750
N-NITROSOMETHYL-3-CARBOXYPROPYLAMINE see MIF250
N-NITROSOMETHYL-2-CHLOROETHYLAMINE see CIQ500
N-NITROSO-N-METHYLCYCLOHEXYLAMINE see NKT500
N-NITROSOMETHYLCYCLOHEXYLAMINE see NKT500
NITROSOMETHYL-n-DECYLAMINE see MMX200
NITROSOMETHYLDIAETHYLHARNSTOFF see DJP600
1-NITROSO-1-METHYL-3,3-DIETHYLUREA see DJP600
NITROSOMETHYLDIETHYLUREA see DJP600
N-NITROSOMETHYL-(2,3-DIHYDROXYPROPYL)AMINE see NMV450
N-NITROSO-N-METHYL-N-DODECYLAMIN (GERMAN) see NKU000
N-NITROSO-N-METHYL-N-DODECYLAMINE see NKU000
NITROSOMETHYL-n-DODECYLAMINE see NKU000
N-NITROSOMETHYLETHANOLAMINE see NKU350
N-NITROSOMETHYLETHYLAMINE (MAK) see MKB000
N-NITROSO-N-METHYL-4-FLUOROANILINE see FKF800
N-NITROSOMETHYLGLYCINE see NLR500
N-NITROSO-N-METHYL-HARNSTOFF (GERMAN) see MNA750
N-NITROSO-N-METHYLHEPTYLAMINE see HBP000
NITROSOMETHYL-n-HEXYLAMINE see NKU400
N-NITROSO-N-METHYL-(4-HYDROXYBUTYL)AMINE see MKP000
N-NITROSOMETHYL-(2-HYDROXYETHYL)AMINE see NKU350
N-NITROSOMETHYL-2-HYDROXYPROPYLAMINE see NKU500
N-NITROSOMETHYLMETHOXYAMINE see DSZ000
N-NITROSO-N-METHYL-o-METHYLHYDROXYLAMIN (GERMAN) see DSZ000
N-NITROSO-N-METHYL-o-METHYL-HYDROXYLAMINE see DSZ000
N-NITROSO-2-METHYLMORPHOLINE see NKU550
NITROSO-2-METHYLMORPHOLINE see NKU550
N-NITROSO-N-METHYL-N'-(2-BENZOTHIAZOLYL)UREA see NKR000
NITROSOMETHYLNEOPENTYLAMINE see NKU570

N-NITROSO-N-METHYLNITROGUANIDINE see MMP000
N-NITROSO-N-METHYL-4-NITROSO-ANILINE see MJG750
NITROSO-N-METHYL-n-NONYLAMINE see NKU580
N-NITROSOMETHYL-n-NONYLAMINE see NKU580
NITROSOMETHYL-n-NONYLAMINE see NKU580
N-NITROSO-N-METHYL-n-OCTYLAMINE see NKU590
N-NITROSO-N-METHYL-n-OCTYLAMINE see NKU590
NITROSOMETHYL-n-OCTYLAMINE see NKU590
N-NITROSO-2-METHYL-1,3-OXAZOLIDINE see NKU600
NITROSO-2-METHYL-1,3-OXAZOLIDINE see NKU600
N-NITROSO-5-METHYL-1,3-OXAZOLIDINE see NKU875
NITROSO-5-METHYL-1,3-OXAZOLIDINE see NKU875
NITROSO-5-METHYLOXAZOLIDONE see NKU875
N-NITROSOMETHYL(2-OXOBUTYL)AMINE see MMR800
N-NITROSOMETHYL(3-OXOBUTYL)AMINE see MMR810
N-NITROSOMETHYL-2-OXOPROPYLAMINE see NKV000
NITROSOMETHYL-N-PENTYLAMINE see AOL000
N-NITROSOMETHYLPENTYLNITROSAMINE see NKV100
N-NITROSOMETHYLPHENYLAMINE (MAK) see MMU250
N-NITROSO-N-METHYLPHENYLCARBAMATE see NKV500
NITROSOMETHYLPHENYLCARBAMATE see NKV500
N-NITROSO-N-METHYL-(1-PHENYL)-ETHYLAMIN (GERMAN) see NKW000
N-NITROSO-N-METHYL-1-(1-PHENYL)ETHYLAMINE see NKW000
N-NITROSO-N-METHYL-2-PHENYLETHYLAMINE see MNU250
NITROSOMETHYLPHENYLUREA see MMY500
1-NITROSO-4-METHYLPIPERAZINE see NKW500
S(+)-N-NITROSO-2-METHYL-PIPERIDIN (GERMAN) see NLJ000
R(−)-N-NITROSO-2-METHYL-PIPERIDIN (GERMAN) see NLI500
S(+)-N-NITROSO-2-METHYLPIPERIDINE see NLJ000
R(−)-N-NITROSO-2-METHYLPIPERIDINE see NLI500
NITROSO-3-METHYL-4-PIPERIDONE see MMZ800
NITROSOMETHYL-N-PROPYLAMINE see MNA000
NITROSOMETHYLPROPYLAMINE see MNA000
N-NITROSO-N-METHYL-n-TETRADECYLAMINE see NKW800
NITROSO-2-METHYLTHIOPROPIONALDEHYDE-o-METHYL CARBAMOYL-OXIME see NKX000
N-NITROSO-N-METHYL-4-TOLYLSULFONAMIDE see THE500
NITROSOMETHYL-2-TRIFLUOROETHYLAMINE see NKX300
NITROSOMETHYLUNDECYLAMINE see NKX500
1-NITROSO-1-METHYLUREA see MNA750
N-NITROSO-N-METHYLUREA see MNA750
NITROSOMETHYLUREA see MNA750
NITROSOMETHYL UREA see MOA500
NITROSOMETHYLURETHAN (GERMAN) see MMX250
N-NITROSO-N-METHYLURETHANE see MMX250
NITROSOMETHYLURETHANE see MMX250
N-NITROSOMETHYLVINYLAMINE see NKY000
N-NITROSOMORPHOLIN (GERMAN) see NKZ000
4-NITROSOMORPHOLINE see NKZ000
N-NITROSOMORPHOLINE (MAK) see NKZ000
NITROSOMORPHOLINE see NKZ000
NITROSO-NAC see NBJ500
1-NITROSO-2-NAFTOL see NLB000
α-NITROSO-β-NAFTOL see NLB000
1-NITROSONAPHTHALENE see NLA000
2-NITROSONAPHTHALENE see NLA500
2-NITROSO-1-NAPHTHOL see NLB500
1-NITROSO-2-NAPHTHOL see NLB000
α-NITROSO-β-NAPHTHOL see NLB000
NITROSO-β-NAPHTHOL see NLB000
N-NITROSO-N'-CARBAETHOXYPIPERAZIN (GERMAN) see NJQ000
N-NITROSO-N'-CARBETHOXYPIPERAZINE see NJQ000
1-NITROSO-2-NITROAMINO-2-IMIDAZOLINE see NLB700
1-NITROSO-3-NITRO-1-BUTYLGUANIDINE see NLC000
1-NITROSO-3-NITRO-1-PENTYLGUANIDINE see NLC500
3-NITROSO-1-NITRO-1-PROPYLGUANIDINE see NLD000
N-NITROSO-N-(NITROSOMETHYL)PENTYLAMINE see NKV100
NITROSONIUM BISULFITE see NMJ000
N-NITROSO-N'-METHYLPIPERAZIN (GERMAN) see NKW500
N-NITROSO-N'-METHYLPIPERAZINE see NKW500
N-NITROSO-N'-NITRO-N-BUTYLGUANIDINE see NLC000
N'-NITROSONORNICOTINE see NLD500
N'-NITROSONORNICOTINE-1-N-OXIDE see NLD525
N-NITROSO-N,N',N'-TRIMETHYLHYDRAZINE see NLY000
N-NITROSOOCTAMETHYLENEIMINE see OCA000
1-NITROSO-1-OCTYLUREA see NLD800
N-NITROSO-O,N-DIAETHYLHYDROXYLAMIN (GERMAN) see NKC500
N-NITROSO-O,N-DIETHYLHYDROXYLAMINE see NKC500
3-NITROSO-7-OXA-3-AZABICYCLO(4.1.0)HEPTANE see NKC300
N-NITROSOOXAZOLIDIN (GERMAN) see NLE000
N-NITROSO-1,3-OXAZOLIDINE see NLE000
3-NITROSOOXAZOLIDINE see NLE000
N-NITROSOOXAZOLIDINE see NLE000
NITROSOOXAZOLIDONE see NLE000
N-NITROSO-N-(2-OXOBUTYL)BUTYLAMINE see BSB500

N-NITROSO-N-(3-OXOBUTYL)BUTYLAMINE see BSB750
N-NITROSO(2-OXOBUTYL)(2-OXOPROPYL)AMINE see NLE400
NITROSO-(2-OXOPROPYL)-N-BUTYLAMINE see BRY000
N-NITROSO-2-OXOPROPYL-2,3-DIHYDROXYPROPYLAMINE see NJY550
N-NITROSO-2-OXO-N-PROPYL-N-PROPYLAMINE see ORS000
N-NITROSO-N-PENTYLCARBAMIC ACID-ETHYL ESTER see AOL750
N-NITROSO-N-PENTYL-(4-HYDROXYBUTYL)AMINE see NLE500
1-NITROSO-1-PENTYLUREA see PBX500
N-NITROSOPERHYDROAZEPINE see NKI000
N-NITROSOPHENACETIN see NLE550
2-NITROSOPHENOL see NLF300
4-NITROSOPHENOL see NLF200
p-NITROSOPHENOL see NLF200
NITROSOPHENOL see NLF200
4-NITROSO-N-PHENYLANILINE see NKB500
N-NITROSO-N-PHENYLANILINE see DWI000
p-NITROSO-N-PHENYLANILINE see NKB500
4-NITROSO-N-PHENYLBENZENAMINE see NKB500
N-NITROSOPHENYLHYDROXYLAMIN AMMONIUM SALZ (GERMAN) see ANO500
N-NITROSOPHENYLHYDROXYLAMINE AMMONIUM SALT see ANO500
N-NITROSO-N-(PHENYLMETHYL)UREA see NJM000
1-NITROSO-4-PHENYLPIPERIDINE see PFT600
N-NITROSO-4-PHENYLPIPERIDINE see PFT600
NITROSO-4-PHENYLPIPERIDINE see PFT600
N-NITROSO-N-PHENYLUREA see NLG500
NITROSOPHENYLUREA see NLG500
N-NITROSO-4-PICOLYLETHYLAMINE see NLH000
1-NITROSOPIPECOLIC ACID see NLH500
N-NITROSOPIPECOLIC ACID see NLH500
1-NITROSO-2-PIPECOLINE see NLI000
1-NITROSO-3-PIPECOLINE see MMY750
1-NITROSO-4-PIPECOLINE see MMZ000
S(+)-N-NITROSO-α-PIPECOLINE see NLJ000
R(−)-N-NITROSO-α-PIPECOLINE see NLI500
1-NITROSOPIPERAZINE see MRJ750
N-NITROSOPIPERAZINE see MRJ750
N-NITROSO-PIPERIDIN see NLJ500
NITROSOPIPERIDIN see NLJ500
1-NITROSOPIPERIDINE see NLJ500
N-NITROSOPIPERIDINE see NLJ500
N-NITROSO-Δ²-PIPERIDINE see NLU480
N-NITROSO-Δ³-PIPERIDINE see NLU500
1-NITROSO-2-PIPERIDINECARBOXYLIC ACID see NLH500
N-NITROSO-3-PIPERIDINOL see NLK000
NITROSO-3-PIPERIDINOL see NLK000
NITROSO-4-PIPERIDINOL see NLK500
N-NITROSO-4-PIPERIDINOL see NLK500
N-NITROSO-4-PIPERIDINONE see NLL000
NITROSO-4-PIPERIDINONE see NLL000
1-NITROSO-4-PIPERIDONE see NLL000
NITROSO-4-PIPERIDONE see NLL000
1-NITROSO-I-PROLINE see NLL500
N-NITROSO-I-PROLINE see NLL500
N-NITROSO-N-2-PROPENYLUREA see NJK000
NITROSOPROPOXUR see PMY310
1-(NITROSOPROPYLAMINO)-2-PROPANOL see NLM500
1-(NITROSOPROPYLAMINO)-2-PROPANONE see ORS000
N-NITROSO-N-PROPYL-1-BUTANAMINE see PNF750
N-NITROSO-N-PROPYLBUTYLAMINE see PNF750
N-NITROSO-N-PROPYLCARBAMIC ACID ETHYL ESTER see PNR500
N-NITROSO-N-PROPYLCARBAMIC ACID-1-NAPHTHYL ESTER see NBM500
N-NITROSO-N-PROPYL-(4-HYDROXYBUTYL)AMINE see NLN000
N-NITROSO-N-PROPYL-1-PROPANAMINE see NKB700
N-NITROSO-N-PROPYLPROPANAMINE see NKB700
N-NITROSO-N-PROPYLPROPIONAMIDE see NLN500
N-NITROSO-N-PROPYL-PROPRIONAMID (GERMAN) see NLN500
N-NITROSO-N-PROPYLUREA see NLO500
NITROSO-N-PROPYLUREA see NLO500
NITROSOPROPYLUREA see NLO500
N-NITROSO-N-PTEROYL-I-GLUTAMIC ACID see NLP000
1-NITROSOPYRENE see NLP375
N-NITROSO-2-(3'-PYRIDYL)PIPERIDINE see NJK150
1-NITROSO-2-(3-PYRIDYL)PYRROLIDINE see NLD500
N-NITROSOPYRROLIDIN (GERMAN) see NLP500
1-NITROSOPYRROLIDINE see NLP500
N-NITROSO-2-PYRROLIDINE see NLP600
N-NITROSOPYRROLIDINE see NLP500
NITROSOPYRROLIDINE see NLP480
1-NITROSO-3-PYRROLIDINOL see NLP700
1-NITROSO-2-PYRROLIDINONE see NLP600
3-(1-NITROSO-2-PYRROLIDINYL)PYRIDINE see NLD500
NITROSO-3-PYRROLIN (GERMAN) see NLQ000
N-NITROSO-3-PYRROLINE see NLQ000
4-NITROSOQUINOLINE-1-OXIDE see NLQ500
5-NITROSO-8-QUINOLINOL see NLR000

NITROZAN K see MJG750
NITROZELL RETARD see NGY000
NITROZONE see NGE500
NITRUMON see BIF750
NITRYL CHLORIDE see NMT000
NITRYL FLUORIDE see NMT500
NITRYL HYPOFLUORITE see NMU000
NITRYL KWASU NIKOTYNOWEGO see NDW515
NITRYL PERCHLORATE see NMU500
NITTO ACID RED PG see CMM320
NITTO DIRECT SKY BLUE 5B see CMO500
NIVACHINE see CLD000
NIVALENOL see NMV000
NIVALENOL-4-O-ACETATE see FQR000
NIVALIN see GBA000
NIVAQUINE B see CLD000
NIVELONA see DAP800
NIVEMYCIN see NCE000
NIVITIN see SKV200
NIVOCILIN see HGP550
NIX see SIA000
NIXON E/C see EHG100
NIXON N/C see CCU250
NIX-SCALD see SAV000
NIXYN see XLS300
NIXYN HERMES see XLS300
NIZOFENONE FUMARATE see NMV400
NIZORAL see KFK100
NIZOTIN see EPQ000
NK 5 see KHU050
NK 136 see DJT800
NK 171 see EAV500
NK-381 see LEX400
NK 421 see BFV300
NK 631 see BLY770, PCB000
NK 711 see LEN000
NK-843 see NGY000
NK 1006 see BAU270
NK-1158 see DWU400
NK 2272 see EDD500
NK 5 (DYE) see KHU050
NK ESTER A-TMPT see TLX175
NK ESTER 3G see MDN510
NK FASTER see MCB050
NKK 105 see MAO275
NLPD see PHA750
N-2-M see CCK665
NM 11 see PKP750
NM 14 see PKP750
NMA see MMU250, NAC000
1-N-2-MA (RUSSIAN) see MMG000
NMBA-d2 see NKT100
NMBA-d3 see NKT105
NMBA see MHW500
NMC 50 see CBM750
Na MCPA see SIL500
NMDDA see NKU000
NMDHP see NMV450
NMEA see MKB000
NMH see MNA750
NMHP see NKU500
NMNA see NKU570
NMO see CFA500
NMOP see NKV000
NMOR see NKZ000
NMP see MPF200
NMU see MNA750
NMUM see MMX250
NMUT see MMX250
NMVA see NKY000
NNA see MMS250
N. NAJA NAJA VENOM see ICC700
NNDG see DQR600
N. NIGRICOLLIS VENOM see NAG000
NNK see MMS500
No. 48-80 see PMB800
NOBACID see SAN000
NOBECUTAN see TFS350
NOBEDON see HIM000
NOBEDORM see QAK000
NOBFELON see IIU000
NOBFEN see IIU000
NOBGEN see IIU000
NOBILEN see TFR250
NOBITOCIN S see ORU500

NOBLEN see PMP500
NOBRIUM see CGA000
NO BUNT 40 see HCC500
NO BUNT 80 see HCC500
NO BUNT see HCC500
NO BUNT LIQUID see HCC500
NOCA see SLL000
NOCARDICIN A see APT750
NOCARDICIN COMPLEX see NMV480
NOCBIN see DXH250
NOCCELER BG see TGX550
NOCCELER CZ see CPI250
NOCODAZOLE see OJD100
No. 49 CONCENTRATED BENZIDINE YELLOW see DEU000
NO. 56 CONC. PERMANENT ORANGE G see CMS145
No. 3 CONC. SCARLET see CHP500
No. 55 Conc. PALE YELLOW SF see CMS212
NOCRAC 224 see PJQ750
NOCRAC NS 5 see MJN250
NOCRAC NS 6 see MJO500
NOCTAL see QCS000
NOCTAMID see MLD100
NOCTAN see DNW400
NOCTAZEPAM see CFZ000
NOCTEC see CDO000
NOCTENAL see QCS000
NOCTILENE see QAK000
NOCTIVANE SODIUM see ERE000
NOCTOSEDIV see TEH500
NOCTOSOM see FMQ000
NOCTOVANE see ERD500
NODAPTON see GIC000
NO-DHU see NJY000
NO-DOZ see CAK500
NO-ETU see NKK500
NOFLAMOL see PJL750
NO. 59 FORTHFAST BENZIDINE YELLOW see CMS145
NO. 2 FORTHFAST SCARLET see MMP100
NOGALAMYCIN see NMV500
NOGAL de LA INDIA (CUBA) see TOA275
NOGALOMYCIN see NMV500
NOGEDAL see DPH600
NOGEST see MCA000, POF275
NOGOS see DGP900
NOGRAM see EID000
NOHDABZ see HKB000
NOHFAA see HIP000
NOHO-MALIE (HAWAII) see YAK350
NOIGEN 160 see DXY000
NOIR BRILLANT BN (FRENCH) see BMA000
NOISETTE des GRANDS-FONDS (GUADELOUPE) see TOA275
NOIX de BANCOUL (GUADELOUPE) see TOA275
NOIX des MOLLUQUES (GUADELOUPE) see TOA275
NOKHEL see AHP375
NOLEPTAN see FMU000
NOLTAM see TAD175
NOLTRAN see CMA250
NOLUDAR see DNW400
NOLVADEX see TAD175
NOLVASAN see BIM250
NOMATE CHOKEGARD see HCP050
NOMATE PBW see GJK000
NOMERSAN see TFS350
NOMETIC see DWK200
NOMETINE see CJL500
No. 907 METRO TALC see TAB750
NOMIFENSIN see NMV700, NMV725
NOMIFENSINE see NMV700
NOMIFENSINE HYDROGEN MALEATE see NMV725
NOMIFENSINE MALEATE see NMV725
NOMIFENSIN HYDROGEN MALEATE see NMV725
NONABROMOBIPHENYL see NMV735
NONACARBONYL DIIRON see NMV740
NONACHLAZINE see NMV750
NONADECAFLUORO-n-DECANOIC ACID see PCG725
NONADECAFLUORODECANOIC ACID see PCG725
2,6-NONADIENAL see NMV760
trans-2,cis-6-NONADIENAL see NMV760
trans,cis-2,6-NONADIENAL see NMV760
2,6-NONADIENENITRILE, 3,7-DIMETHYL- see LEH100
2-trans-6-cis-NONADIEN-1-OL see NMV780
NONADIENOL see NMV780
trans,cis-2,6-NONADIENOL see NMV780
1,6-NONADIEN-3-OL, 3,7-DIMETHYL-, ACETATE see HGI585
γ-NONALACTONE (FCC) see CNF250

1-NONALDEHYDE see NMW500
NONALOL see NNB500
1,4-NONALOLIDE see CNF250
NONANE see NMX000
1-NONANECARBOXYLIC ACID see DAH400
4-NONANECARBOXYLIC ACID see NMX500
NONANEDIOIC ACID see ASB750
NONANEDIOTIC ACID, DIHEXYL ESTER see ASC000
n-NONANENITRILE see OES300
NONANENITRILE see OES300
NONANOIC ACID see NMY000
NONANOIC ACID, ETHYL ESTER see ENW000
NONANOIC ACID OXYDI-3,1-PROPANEDIYL ESTER (9CI) see DWT000
NONANOIC ACID OXYDIPROPYLENE ESTER see DWT000
NONANOIC ACID, TRIBUTYLSTANNYL ESTER see TIF500
1-NONANOL see NMW500
NONAN-1-OL see NNB500
1-NONANOL see NNB500
2-NONANONE see NMY500
NONAN-2-ONE see NMY500
NONAN-5-ONE see NMZ000
5-NONANONE see NMZ000
n-NONANONITRILE see OES300
NONANONITRILE see OES300
NONANOPHENONE, 9-FLUORO- see FKT050
1-NONANOYLAZIRIDINE see NNA000
NONANOYLETHYLENEIMINE see NNA000
4-NONANOYLMORPHOLINE see MRQ750
(NONANOYLOXY)TRIBUTYLSTANNANE see TIF500
NONANOYL PEROXIDE see NNA100
cis-6-NONEN-1-AL see NNA325
NON-2-ENAL see NNA300
2-NONENAL see NNA300
(Z)-6-NONENAL see NNA325
6-NONENAL, (Z)- see NNA325
2-NONEN-1-AL see NNA300
trans-2-NONENAL (FCC) see NNA300
cis-6-NONENAL see NNA325
2-NONENENITRILE, 3-METHYL- see MNB500
2-NONENOIC ACID, METHYL ESTER see MNB750
2-NONEN-4,6,8-TRIYN-1-AL see NNB000
α-NONENYL ALDEHYDE see NNA300
NONEX 25 see PJY100
NONEX 30 see PJY100
NONEX 52 see PJY100
NONEX 64 see PJY100
NONEX 411 see HKJ000
NON-FER-AL see CAT775
NONFLAMIN see TGE165
NONFLEX RD see PJQ750
NONIDET P40 see GHS000
NONION 06 see PJY100
NONION HS 206 see GHS000
NONION O2 see PJY100
NONION O4 see PJY100
NONION OP80R see SKV100
NONION P 208 see PJT300
NONION P 210 see PJT300
NONION PP40 see MRJ800
NONION SP 60 see SKV150
NONION SP 60R see SKV150
NONISOL 200 see PJY100
NONISOLD see CNH125
n-NONOIC ACID see NMY000
NONOX CL see NBL000
NONOX D see PFT500
NONOX DCP see IPK000
NONOX DPPD see BLE500
NONOX TBC see BFW750
NONOXYNOL-9 see NNB300
NONOXYNOL see NND500
NONOX ZA see PFL000
NONYL ACETATE see NNB400
n-NONYL ALCOHOL see NNB500
sec-NONYL ALCOHOL see DNH800
NONYL ALCOHOL see NNB500
1-NONYL ALDEHYDE see NMW500
1-NONYLAMINE, N-METHYL-N-NITROSO- see NKU580
NONYLCARBINOL see DAI600
o-NONYL-HARMOL HYDROCHLORIDE see NNC100
n-NONYLIC ACID see NMY000
NONYL METHYL KETONE see UKS000
4-NONYLPHENOL see NNC510
NONYL PHENOL (mixed isomers) see NNC500
NONYLPHENOL, POLYOXYETHYLENE ETHER see NND500

NONYLPHENOXYPOLY(ETHYLENEOXY)ETHANOL see NNB300
NONYLPHENOXYPOLYETHYLENEOXY ETHANOL-IODINE COMPLEX see NND000
NONYL PHENYL POLYETHYLENE GLYCOL see NND500
NONYL PHENYL POLYETHYLENE GLYCOL ETHER see NND500
NONYLTRICHLOROSILANE see NNE000
2-NONYN-1-AL DIMETHYLACETAL see NNE100
2-NONYNAL, DIMETHYL ACETAL see NNE100
2-NONYNAL DIMETHYLACETAL see NNE100
2-NONYNOIC ACID, METHYL ESTER see MNC000
No. 156 ORANGE CHROME see LCS000
No. 177 ORANGE LAKE see CMM220
NOOTRON see NNE400
NOOTROPIL see NNE400
NOOTROPYL see NNE400
NO-Pip see NLJ500
NO-Pro see NLL500
NOPALCOL 1-0 see PJY100
NOPALCOL 4-0 see PJY250
NOPALCOL 6-0 see PJY100
NOPALCOL 6-L see PJY000
NOPALMATE see PLH500
NOPCAINE see PAQ200
NOPCOCIDE see TBQ750
NOPCO 2272-R see NNE500
NOPCOTE C 104 see CAX350
4-NOPD see ALL500
NO-PEST see DGP900
NO-PEST STRIP see DGP900
NOPIL see TKX000
NOPINEN see POH750
NOPINENE see POH750
NOPOL see DTB800
NOPOL ACETATE see DTC000
NOPOL (POLYMER) see PJS750
NOPOL (TERPENE) see DTB800
NO-PRESS see MBW750
NOPTIL see EOK000
NOPYL ACETATE see DTC000
NO-PYR see NLP500
NORACYCLINE see LJE000
(−)-NORADREC see NNO500
NORADRENALIN see NNO500
NORADRENALINA (ITALIAN) see NNO500
d-(−)-NORADRENALINE see NNO500
l-NORADRENALINE see NNO500
NORADRENALINE see NNO500
(−)-NORADRENALINE see NNO500
(−)-NORADRENALINE ACID TARTRATE see NNO699
l-NORADRENALINE BITARTRATE see NNO699
(−)-NORADRENALINE BITARTRATE see NNO699
(−)-NORADRENALINE BITARTRATE MONOHYDRATE see NNO699
dl-NORADRENALINE HYDROCHLORIDE see NNP050
NORADRENALINE HYDROCHLORIDE see NNP000
(±)-NORADRENALINE HYDROCHLORIDE see NNP050
NORADRENALINE HYDROGEN TARTRATE see NNO699
NORADRENALINE TARTRATE (1:1) see NNO699
l-NORADRENALINE TARTRATE see NNO699
(−)-NORADRENALINE TARTRATE see NNO699
NOR-ADRENALIN HYDROCHLORIDE see NNP000
NORADRENLINE see NNO500
NORAL ALUMINUM see AGX000
NORAL EXTRA FINE LINING GRADE see AGX000
NORAL INK GRADE ALUMINUM see AGX000
NORAL NON-LEAFING GRADE see AGX000
NOR-AM EP 332 see DSO200
NOR-AM EP 333 see CJJ500
NORAMITRIPTYLINE see NNY000
NORAMIUM M 2SH15 see QAT550
NORAMIUM M 2SH see QAT550
NORAM O see OHM700
NORANAT see IBV100
NORANDROLONE PHENYLPROPIONATE see DYF450
NORANDROSTENOLONE PHENYLPROPIONATE see DYF450
NORANTIPYRINE see NNT000
NORARTRINAL see NNO500
NORBILAN see TGZ000
NORBOLDINE see LBO100
(+)-NORBOLDINE see LBO100
NORBOLETHONE see NNE600
NORBORAL see BSM000
NORBORMIDE see NNF000
2,5-NORBORNADIENE see NNG000
NORBORNADIENE see NNG000
endo-2-NORBORNANECARBOXYLIC ACID ETHYL ESTER see NNG500

2-NORBORNANEMETHANOL see NNH000
2-NORBORNANONE see NNK000
2-NORBORNANONE, 1,3,3-TRIMETHYL-, (1R,4S)-(+)- see FAM300
2-NORBORNENE see NNH500
trans-5-NORBORNENE-2,3-DICARBONYL CHLORIDE see NNI000
cis-5-NORBORNENE-2,3-DICARBOXYLIC ACID DIMETHYL ESTER see DRB400
5-NORBORNENE-2,3-DICARBOXYLIC ACID, 1,4,5,6,7,7-HEXACHLORO-, DIBU-
    TYL ESTER (8CI) see CDS025
5-NORBORNENE-2,3-DICARBOXYLIC ANHYDRIDE, 1,4,5,6,7,7-HEXACHLORO-
    see CDS050
5-NORBORNENE-2-METHANOL ACRYLATE see BFY250
5-NORBORNENE-2-METHYLOLACRYLATE see BFY250
α-5-NORBORNEN-2-YL-α-PHENYL-PIPERIDINE PROPANOL HYDROCHLORIDE
    see BGD750
NORBORNYLENE see NNH500
NORBUTORPHANOL TARTRATE see NNJ600
NORCAIN see EFX000
NORCAMPHANE see EOM000
2-NORCAMPHANE METHANOL see NNH000
NORCAMPHENE see NNH500
NORCAMPHOR see NNK000
NORCOZINE see CKP500
20-NORCROTALANAN-11,15-DIONE, 3,8-DIDEHYDRO-14,19-DIHYDRO-12,13-DI-
    HYDROXY-, (13-α-14-α)- see DAL350
NORCURON see VGU075
NORDEN see AKT250
NORDETTE see NNL500
NORDHAUSEN ACID (DOT) see SOI500
NORDIALEX see DBL700
NORDIAZEPAM see CGA500
NORDICOL see EDO000
NORDICORT see HHR000
NORDIHYDROGUAIARETIC ACID see NBR000
NORDIHYDROGUAIRARETIC ACID see NBR000
NORDIMAPRIT see NNL400
NORDIOL-28 see NNL500
NORDIOL see NNL500
NORDOPAN see BIA250
NOREA see HDP500
NORENOL see AKT000
NOREPHEDRANE see AOB250, BBK000
dl-NOREPHEDRINE see NNM500
d-NOR-psi-EPHEDRINE see NNM510
NOR-psi-EPHEDRINE see NNM510
psi-NOREPHEDRINE see NNM510
(−)-NOREPHEDRINE see NNM000
dl-NOREPHEDRINE HYDROCHLORIDE see PMJ500
d-NOREPHEDRINE HYDROCHLORIDE see NNO000
l-NOREPHEDRINE HYDROCHLORIDE see NNN500
NOREPHEDRINE HYDROCHLORIDE see NNN000
(−)-NOREPHEDRINE HYDROCHLORIDE see NNN500, NNO000
l-NOREPINEPHRINE see NNO500
NOREPINEPHRINE see NNO500
(−)-NOREPINEPHRINE see NNO500
l-NOREPINEPHRINE BITARTRATE see NNO699
(−)-NOREPINEPHRINE BITARTRATE see NNO699
dl-NOREPINEPHRINE HYDROCHLORIDE see NNP050
NOREPINEPHRINE HYDROCHLORIDE see NNP000
(±)-NOREPINEPHRINE HYDROCHLORIDE see NNP050
(−)-NOREPINEPHRINE TARTRATE see NNO699
NOREPIRENAMINE see NNO500
NORES see HDP500
NORETHANDROL see EAP000
NORETHINDRONE-17-ACETATE see ABU000
NORETHINDRONE ACETATE 3-CYCLOPENTYL ENOL ETHER see QFA275
NORETHINDRONE ACETATE and ETHINYLESTRADIOL see EEH520
NORETHINDRONE mixed with ETHYNYLESTRADIOL see EQM500
NORETHINDRONE mixed with MESTRANOL see MDL750
NORETHINODREL see EEH550
19-NOR-ETHINYL-4,5-TESTOSTERONE see NNP500
19-NOR-ETHINYL-5,10-TESTOSTERONE see EEH550
NORETHINYNODREL see EEH550
19-NORETHISTERONE see NNP500
19-NORETHISTERONE ACETATE see ABU000
NORETHISTERONE ACETATE mixed with ETHINYL OESTRADIOL see
    EEH520
NORETHISTERONE ENANTHATE see NNQ000
NORETHISTERONE mixed with ETHINYL OESTRADIOL (60:1) see EQM500
NORETHISTERONE mixed with MESTRANOL see MDL750
NORETHISTERONE OENANTHATE see NNQ000
NORETHYNODRAL see EEH550
19-NORETHYNODREL see EEH550
NORETHYNODREL see EEH550
NORETHYNODREL and ETHINYLESTRADIOL-3-METHYL ETHER (50:1) see
    EAP000

NORETHYNODREL mixed with MESTRANOL see EAP000
19-NOR-17-α-ETHYNYLANDROSTEN-17-β-OL-3-ONE see NNP500
19-NOR-17-α-ETHYNYL-17-β-HYDROXY-4-ANDROSTEN-3-ONE see NNP500
19-NOR-17-α-ETHYNYLTESTOSTERONE see NNP500
19-NORETHYNYLTESTOSTERONE ACETATE see ABU000
NORETHYSTERONE ACETATE see ABU000
NOREX see CJQ000
NORFENEFRINE HYDROCHLORIDE see NNT100
NORFLEX see DPH000
NORFLOXACIN see BAB625
NORFORMS see ABU500
NORGAMEM see TEV000
NORGE SALTPETER see CAU000
α-NORGESTREL see NNQ500
dl-NORGESTREL see NNQ500
d(−)-NORGESTREL see NNQ500
d-NORGESTREL see NNQ500, NNQ520
d(−)-NORGESTREL see NNQ520
l-NORGESTREL see NNQ525
(±)-NORGESTREL see NNQ500
NORGESTREL see NNQ500
dl-NORGESTREL mixed with ETHINYLESTRADIOL see NNL500
NORGESTREL mixed with ETHINYL ESTRADIOL see NNL500
NORGESTREL mixed with MESTRANOL see NNR000
NORGINE see AFL000
NORGLAUCIN see NNR200
d-N-NORGLAUCINE see NNR200
NORGLAUCINE see NNR200
NORHARMAN see NNR300
NOR-HN2 see BHO250
NOR-HN2 HYDROCHLORIDE see BHO250
NORHOMOEPINEPHRINE HYDROCHLORIDE see AMB000
NORIDIL see UVJ450
NORIDYL see UVJ450
NORIGEST see NNQ000
NORILGAN-S see SNN500
NORIMIPRAMINE see DSI709
NORIMYCIN V see CDP250
NORINYL-1 see MDL750
NORISODRINE see DMV600
NORISODRINE HYDROCHLORIDE see IMR000
NORIT see CBT500
NORKEL see AHK750
NORKOOL see EJC500
NORLESTRIN see EEH520
NORLEUCAMINE see PBV505
ε-NORLEUCINE see AJD000
NOR-LOST HYDROCHLORID (GERMAN) see BHO250
NORLUTATE see ABU000
NORLUTIN see NNP500
NORLUTINE ACETATE see ABU000
NORLUTIN ENANTHATE see NNQ000
NORMABRAIN see NNE400
NORMADATE see HMM500
NORMALIP see ARQ750
NORMAL LEAD ACETATE see LCV000
NORMAL LEAD ORTHOPHOSPHATE see LDU000
NORMAL SUPERPHOSPHATE see SOV500
NORMASTIGMIN see DQY909, NCL100
NORMAT see ARQ750
NORMERSAN see TFS350
NORMETHYL EX4442 see POF250
NORMETOL see AKT000
NORMI-NOX see QAK000
NORMISON see CFY750
NORMOCYTIN see VSZ000
NORMODYNE see HMM500
NORMO-LEVEL see NNR400
NORMOLIPOL see ARQ750
NORMONSON see CHG000
NORMORESCINA see TLN500
NORMORPHINONE, N-ALLYL-7,8-DIHYDRO-14-HYDROXY-, (−)- see NAG550
NORMOSAN see CHG000
NORMOSON see CHG000
NORMOSTEROL see NNR400
NORMUSCONE see CPU250
NORNICOTIN see NNR500
NORNICOTINE see NNR500
d-NORNICOTINE HYDROCHLORIDE see NNS500
NORNICOTINE (+)-HYDROCHLORIDE see NNS500
NORNICOTYRINE see PQB500
NOR-NITROGEN MUSTARD see BHN750
NORNITROGEN MUSTARD HYDROCHLORIDE see BHO250
NOROCAINE see AIL750
NORODIN see DBA800

NORODIN HYDROCHLORIDE see MDT600
NOROX BZP-250 see BDS000
19-NOR-P see NNT500
NORPACE see RSZ600
NORPHEN see AKT250
NORPHENAZONE see NNT000
NORPHENYLEPHRINE see AKT000
(±)-NORPHENYLEPHRINE HYDROCHLORIDE see NNT100
2-NORPINENE-2-CARBOXALDEHYDE, 6,6-DIMETHYL- see FNK150
NORPLANT see NNQ525
NORPRAMIN see DLS600
NORPRAZEPAM see CGA500
19-NOR-17-α-PREGNA-1,3,5(10)-TRIEN-2-YNE-3,17-DIOL see EEH500
(17-α)-19-NORPREGNA-1,3,5(10)-TRIEN-20-YNE-3,17,DIOL see EEH500
19-NORPREGN-4-ENE-3,20-DIONE see NNT500
(3-β,17-α)-19-NORPREGN-4-EN-20-YNE-3,17-DIOL DIACETATE see EQJ500
19-NOR-17-α-PREGN-4-EN-20-YNE-3-β,17-DIOL DIACETATE mixed with 3-METHYOXY-19-NOR-17-α-PREGNA-1,3,5(10)-TRIEN-20-YN-17-OL see EQK010
10-NOR-17-α-PREGN-4-EN-20-YNE-3-β,17-DIOL mixed with 3-METHOXY-17-α-19-NORPREGNA-1-3-5(10)-TRIEN-20-YN-17-OL see EQK100
19-NOR-17-α-PREGN-4-EN-20-YN-17-OL see NNV000
(17-α)-19-NORPREGN-4-EN-20-YN-17-OL see NNV000
19-NOR-17-α-PREGN-4-EN-20-YN-3-ONE, 17-HYDROXY-, mixed with 19-NOR-17-α-PREGNA-1,3,5(10)-TRIEN-2-YNE-3,17-DIOL (60:1) see EQM500
NOR-PRESS 25 see HGP500
NOR-PROGESTELEA see NCI525
19-NORPROGESTERONE see NNT500
d-NORPSEUDOEPHEDRINE see NNM510
NORPSEUDOEPHEDRINE, (+)- see NNM510
(−)-NORPSEUDOEPHEDRINE see NNV500
d-NORPSEUDOEPHEDRINE HYDROCHLORIDE see NNW500
(+)-NORPSEUDOEPHEDRINE HYDROCHLORIDE see NNW500
1-NOR-PSI-EPHEDRIN (GERMAN) see NNV500
NORSTEARANTHRENE see TCJ500
NORSULFASOL see TEX250
NORSULFAZOLE see TEX250
NORSYMPATHOL see AKT250
NORSYNEPHRINE see AKT000, AKT250
NORTEC see CDO000
19-NORTESTOSTERONE-17-N-(2-CHLOROETHYL)-N-NITROSOCARBAMATE see NNX600
19-NORTESTOSTERONE PHENYLPROPIONATE see DYF450
NORTHERN COPPERHEAD VENOM see NNX700
NORTIMIL see DLS600
NORTRIPTYLINE see NNY000
NORURON see HDP500
NORVAL see BMA625, DJL000
NORVALAMINE see BPX750
NORVALINE, 3-METHYL- see IKX000
NORVALINE, 4-METHYL- see LES000
NORVEDAN see CKI750
NORVINISTERONE see NCI525
NORVINYL see PKQ059
NORVINYL P 6 see AAX175
19-NOR-17-α-VINYLTESTOSTERONE see NCI525
NORWAY SALTPETER see CAU000
NORWEGIAN SALTPETER see CAU000
NORZETAM see NNE400
NO SCALD see DVX800
NOSCAPAL see NOA000
NOSCAPALIN see NOA000
NOSCAPINE see NBP275, NOA000
NOSCAPINE HYDROCHLORIDE see NOA500
NOSCOSED see CLD250
NOSIM see CCK125
NOSOPHENE SODIUM see TDE750
NOSPAN see MOV500
NOSPASM see IPU000
NOSTAL see EHP000, QCS000
NOSTEL see CHG000
NOSTIN see EHP000
NOSTRAL see QCS000
NOSYDRAST see DNG400
NOTANDRON see AOO475
NOTANDRON-DEPOT see AOO475
NOTARAL see BFD000, CFZ000
NOTATIN see GFG100
NOTECHIS SCUTATUS VENOM see ARV550
NOTENQUIL see ABH500
NOTENSIL see AAF750, ABH500
NOTESIL see ABH500
NOTEZINE see DIW000
NOTOMYCIN A1 see CNV500
NOURALGINE see SEH000

NOURITHION see PAK000
NOURYSET 200 see AGD250
NOVACRYSIN see GJG000
NOVADELOX see BDS000
NOVADEX see NOA600
NOVADOX see HGP550
NOVADRAL see AKT000, NNT100
NOVAFED see POH000, POH250
NOVAKOL see FNF000
NOVALICHIN see NOA700
NOVALLYL see AFS500
NOVALON YELLOW 2GN see AAQ250
NOVAMIDON see DOT000
NOVAMIN see APT000, DYE600, PMF500
NOVAMINE see DYE600
NOVAMONT 2030 see PMP500
NOVANTOINA see DKQ000, DNU000
NOVA-PHENO see EOK000
NOVARSAN see NCJ500
NOVARSENOBENZOL see NCJ500
NOVARSENOBILLON see NCJ500
NOVASMASOL see MDM800
NOVASUROL see CCG500
NOVATEC JUO 80 see PJS750
NOVATEC JVO 80 see PJS750
NOVATHION see DSQ000
NOVATONE see MDW750
NOVATRIN see MDL000
NOVATROPINE see MDL000
NOVECYL see SAH000
NOVEDRIN HYDROCHLORIDE see EJR500
NOVEGE see PBK000
NOVEMBICHIN see NOB700
NOVEMBIKHIN see NOB700
NOVERIL see DCW800
NOVERME see MHL000
NOVERYL see DCW800
NOVESIN see OPI300
NOVESINE see OPI300
NOVICET see VLF000
NOVICODIN see DKW800
NOVID see ADA725
NOVIDIUM CHLORIDE see HGI000
NOVIDORM see THS800
NOVIGAM see BBQ500
NOVISMUTH see BKW100
NOVIZIR see EIR000
NOVOBIOCIN, MONOSODIUM SALT see NOB000
NOVOBIOCIN see SMB000
NOVOBIOCIN MONOSODIUM see NOB000
NOVOBIOCIN, SODIUM derivative see NOB000
NOVOCAINAMIDE see AJN500
NOVOCAIN-CHLORHYDRAT (GERMAN) see AIT250
NOVOCAINE see AIL750
NOVOCAINE AMIDE see AJN500
NOVOCAINE HYDROCHLORIDE see AIT250
NOVOCAIN HYDROCHLORID (GERMAN) see AIT250
NOVOCAL see CAS750
NOVOCAMID see AJN500
NOVOCAMID HYDROCHLORIDE see PME000
NOVOCEBRIN HYDROCHLORIDE see DXI800
NOVOCHLOROCAP see CDP250
NOVOCILLIN see BFD250
NOVOCOLIN see DAL000
NOVOCONESTRON see ECU750
NOVODIL see DNU100
NOVODIPHENYL see DNU000
NOVODOLAN see TKG000
NOVODRIN see DMV600
NOVOEMBICHIN see NOB700
NOVOHEPARIN see HAQ500
NOVOHETRAMIN see RDU000
NOVOL see OBA000
NOVOLEN see PMP500
NOVOL KETONE see NOB800
NOVOL POE 20 see PJW500
NOVOMAZINA see CKP250
NOVOMYCETIN see CDP250
NOVON 712 see PKQ059
NOVON see PBK000
NOVONAL see DJU200
NOVONIDAZOL see MMN250
NOVOPHENICOL see CDP250
NOVOPHENYL see BRF500
NOVOPHONE see SOA500

NOVO-R see SMB000
NOVOSAXAZOLE see SNN500
NOVOSCABIN see BCM000
NOVOSED see MDQ250
NOVOSERIN see CQH000
NOVOSERPINA see RCA200
NOVOSIR N see EIR000
NOVOSPASMIN see NOC000
NOVOVANILLIN see NOC100
NOVOX see BSC500
NOVYDRINE see BBK000
2-NOXA see NBJ100
NOXA see NBJ100
NOXABEN see DGE200
NOXAL see DXH250
NOXFISH see RNZ000
NOXIPTILINE HYDROCHLORIDE see DPH600
NOXIPTILIN HYDROCHLORID (GERMAN) see DPH600
NOXIPTYLINE HYDROCHLORIDE see DPH600
NOXODYN see TEH500
NOXOKRATIN see EQL000
NOXYLIN see UTU500
NOXYRON see DYC800
1-NP see NIX500
2-NP see ALL750
NP 2 see NCW500
2-NP see NIY000
NP-9 see NNB300
NP 212 see CJR909
NP 246 see CJN750
NPA-3 see SIO500
3N4HPA see HMY000
NPA see DNB000, SIO500
NPAB see PNC925
NPA, SODIUM SALT see SIO500
NPD see TED500
NPG see DTG400
NPGB see NIQ500
NPH 83 see MJG500
NPH-1091 see BDJ250
NPH ILETIN see IDF325
NPH 50 INSULIN see IDF325
NPH INSULIN see IDF325
N-N-PIP see NLJ500
NPIP see NLJ500
NPOH see DVD200
NPP see DYF450
2-NPPD see ALL750
NPRO see NLL500
NaPSt see SMQ500
NPU see NLO500
N-N-PYR see NLP500
NPYR see NLP500
4-NQO see NJF000
NR.C 2294 see DNA800
NRDC 104 see BEP500
NRDC 107 see BEP750
NRDC 119 see RDZ875
NRDC 149 see RLF350
NRDC 160 see RLF350
NRDC 161 see DAF300
NRDC 166 see RLF350
NRL-18-B see TNI000
NS 11 see DTG000
Ni 0901-S see NCW500
NS 100 (carbonate) see CAT775
NS (carbonate) see CAT775
NS 200 (filler) see CAT775
NSAR see NLR500
NSC-185 see CPE750
NSC 339 see DVF200
NSC 423 see DAA800
NSC 729 see PIB925
NSC 739 see AMG750
NSC-740 see MDV500
NSC 741 see HHQ800
NSC-742 see ASA500
NSC 743 see POJ500
NSC 744 see CKV500
NSC 746 see UVA000
NSC-749 see AJO500
NSC-750 see BOT250
NSC 751 see EEI000
NSC-752 see AMH250
NSC 753 see POJ250

NSC 755 see POK000
NSC-756 see THT350
NSC 757 see CNG830
NSC 759 see BCB750
NSC 762 see BIE250, BIE500
NSC-763 see DUD800
NSC 1026 see AJK250
NSC-1390 see ZVJ000
NSC 1393 see POM600
NSC 1532 see DUZ000
NSC 1895 see DCF200
NSC-2066 see TEP000
NSC 2083 see AMM250
NSC-2100 see NGE500
NSC-2101 see HMY000
NSC 2107 see NGE000
NSC 2666 see DDV800
NSC-2752 see FOU000
NSC 3051 see MKG500
NSC 3053 see AEB000
NSC-3055 see AEI000
NSC-3058 see BDH250
NSC 3060 see PKV500
NSC-3061 see TGD000
NSC 3063 see DAY825
NSC 3069 see CDP250
NSC-3070 see DKA600
NSC 3072 see SFA000
NSC 3073 see FMT000
NSC-3088 see CDO500
NSC 3094 see DTP000
NSC 3095 see IBC000
NSC 3096 see MIW500
NSC 3138 see DOO800
NSC 3233 see TNI300
NSC-3424 see QDJ000
NSC-3425 see THR750
NSC 4170 see MMD500
NSC 4320 see SMN002
NSC-4729 see TES000
NSC 4730 see EGI000
NSC 4911 see MCQ500
NSC 5088 see SAL500
NSC 5159 see CDK250
NSC 5230 see EQA000
NSC 5340 see AHL000
NSC 5354 see PEY250
NSC 5356 see DSB000
NSC-5366 see NBP275
NSC 5366 see NOA000
NSC-6091 see SOA500
NSC-6396 see TFQ750
NSC-6470 see NDY000
NSC-6738 see DGP900
NSC 7365 see DCQ400
NSC-7571 see AHS750
NSC-7760 see POS750
NSC 7764 see LAQ000
NSC-7778 see MES000
NSC 8028 see IIM000
NSC 8194 see EQL500
NSC 8260 see MGA250
NSC-8491 see THY100
NSC 8652 see DSF300
NSC-8806 see BHV250, PED750
NSC 8819 see ADR000
NSC-9166 see TBG000
NSC 9169 see HOI000
NSC-9324 see AFS500
NSC 9369 see MMP000
NSC 9659 see ILD000
NSC-9698 see MAW750
NSC-9701 see MPN500
NSC-9704 see PMH500
NSC 9706 see TND500
NSC 9717 see TND250
NSC-9895 see EDO000
NSC 10023 see PLZ000
NSC-10107 see CFA500, CFA750
NSC-10108 see CLO750
NSC 10483 see CNS750
NSC 10815 see AMX750
NSC-10873 see BHN750
NSC 10873 see BHO250
NSC 11595 see PNW800

NSC-11905 see HLY500
NSC-12165 see AOO275
NSC-12169 see EDU500
NSC 13002 see DOS800
NSC-13252 see CMB000
NSC 13875 see HEJ500
NSC 14083 see SLW500
NSC-14210 see BHT750
NSC 14574 see SBE500
NSC 14575 see EMC500
NSC 15193 see DAR600
NSC-15432 see EEH550
NSC 15747 see BFL125
NSC-15780 see AOD500
NSC-16498 see BIA000
NSC-16895 see LGZ000
NSC-17118 see CLD500
NSC-17262 see BDC750
NSC-17591 see TBF750
NSC-17592 see HNT500
NSC 17661 see GHE000
NSC 17663 see BHN500
NSC-17777 see PGG355
NSC 18016 see CHC750
NSC-18268 see AEA750
NSC 18321 see BHB750
NSC-18429 see AOQ875
NSC-18439 see BIC600
NSC-19477 see HEG000
NSC 19488 see BRS500
NSC-19494 see MQR100
NSC-19893 see FMM000
NSC-19962 see CBO625
NSC-19987 see MOR500
NSC-20264 see AOA125
NSC-20526 see THW750
NSC-20527 see BOD600
NSC 21206 see ALL250
NSC 21402 see AKM000
NSC-22314 see TCF000
NSC 22420 see DCP400
NSC 23519 see DCQ600
NSC-23890 see DSU000
NSC-23892 see BCC250
NSC 23909 see MNA750
NSC 24559 see MQW750
NSC 24639 see NLC000
NSC 24818 see PJJ225
NSC 24890 see NKE500
NSC-24970 see BMZ000
NSC-25154 see BHJ250
NSC-25855 see TEV000
NSC 25958 see NLB700
NSC-26154 see AJD000
NSC-26198 see PAN100
NSC 26271 see CQC500, CQC650
NSC 26805 see EMF500
NSC-26806 see VMA000
NSC-26812 see AQO000
NSC 26980 see AHK500
NSC-27640 see DAR400
NSC 28693 see MRH000
NSC-29215 see TND000
NSC 29421 see BDG325
NSC-29422 see TFJ500
NSC-29630 see DFO000
NSC-29863 see GKS000
NSC-30152 see PMO250
NSC-30211 see TNF500
NSC 30622 see SBX525
NSC 31083 see AEA109
NSC-31712 see TCQ500
NSC 32065 see HOO500
NSC 32074 see RJA000
NSC-32606 see BOP750
NSC 32743 see ABN000
NSC 32946 see MKI000
NSC-33531 see DMH600
NSC-33659 see LBD100
NSC-33669 see EAL500, EAN000
NSC 33674 see NLD000
NSC-34372 see DFH000
NSC 34391 see KHU050
NSC-34462 see BIA250
NSC 34533 see GKE000

NSC-34652 see MKB250
NSC 34699 see NLC500
NSC-35051 see BHU750, SAX200
NSC 35770 see CMX700
NSC-36354 see DBA175
NSC 37095 see EHV500
NSC 37448 see PDT250
NSC-37538 see BKM500
NSC-37725 see NNV000
NSC 38191 see ENU000
NSC 38270 see OIU499
NSC 38721 see CDN000
NSC-38887 see AKY250
NSC-39069 see TFU500
NSC-39084 see ASB250
NSC 39147 see SMC500
NSC-39470 see BFV750
NSC 39661 see DAS000
NSC 40774 see MPU000
NSC-40902 see AOO500
NSC 42076 see BER500
NSC-42722 see DAL300
NSC 44185 see EAM500
NSC 44690 see DIS200
NSC 45383 see SMA000
NSC-45384 see SMA500
NSC-45388 see DAB600
NSC 45403 see ENV000
NSC-45463 see DOY600
NSC-45624 see SKI500
NSC-46015 see TCW500
NSC 47547 see CHE750
NSC 47842 see VKZ000
NSC-49171 see SAN000
NSC 49842 see VLA000
NSC-50256 see MLH500
NSC-50364 see MMN250
NSC-50413 see AQY250
NSC-50982 see BHW500
NSC 51143 see IAN100
NSC-52695 see CQN000
NSC-52760 see DAK200
NSC-52947 see PAB500
NSC 53396 see AEA250
NSC 56408 see TNY500
NSC-56410 see MLY000
NSC 56654 see ASO510
NSC-57199 see BHT250
NSC-58404 see DCO800
NSC-58514 see CMK650
NSC-58775 see DLY000
NSC 59729 see SKX000
NSC-60195 see MQG500
NSC 60380 see CMA250
NSC 60520 see DGH400
NSC-62209 see BIF250
NSC 62579 see DJB800
NSC 62580 see DRO200
NSC-62939 see TKX250
NSC 63346 see DQD200
NSC-63701 see VGZ000
NSC 63878 see AQQ750, AQR000
NSC 64375 see BDW000
NSC-64393 see SLC000
NSC-65104 see SBZ000
NSC-65346 see SAU000
NSC 65426 see PML250
NSC-66847 see TEH500
NSC-67239 see THM750
NSC-67574 see LEY000
NSC 67574 see LEZ000
NSC-68075 see TEH250
NSC-68626 see ADA000
NSC 69187 see MEL775
NSC-69536 see MLJ500
NSC-69811 see MKW250
NSC 69856 see NBV500
NSC-69945 see PHA750
NSC-70731 see LGD000
NSC-70762 see TGJ500
NSC-70845 see NMV500
NSC-71045 see HKQ025
NSC-71261 see TFJ250
NSC-71423 see VTF000
NSC-71795 see EAI850

NSC-71936 see CMQ725
NSC 71964 see PCU425
NSC-72005 see TIL500
NSC 73438 see NKL000
NSC 74437 see DHE100
NSC 75520 see TKH325
NSC-76098 see ARW750
NSC-77213 see PME250, PME500
NSC 77471 see MQX000
NSC 77517 see MKN750
NSC-77518 see DCK759
NSC 78409 see BIH325
NSC-78559 see DAB800
NSC 78572 see HIZ000
NSC 79019 see PQC000
NSC-79037 see CGV250
NSC-79389 see ARQ750
NSC-80087 see DOT600
NSC-81430 see CQJ500
NSC-82116 see KFA100
NSC-82151 see DAC000, DAC200
NSC-82174 see EID000
NSC-82196 see IAN000
NSC-82261 see GCS000
NSC-82699 see TKH750
NSC 83265 see TNR475
NSC-83629 see HKQ025
NSC 83653 see AED250
NSC 84241 see NAP300
NSC-84423 see HGI000
NSC 84954 see CFH500
NSC 84963 see AMN300
NSC 85598 see SMD000
NSC-85998 see SMD000
NSC-86078 see SMM500
NSC 87418 see IKH000
NSC 87419 see CPN500
NSC 87974 see CPL750
NSC 88104 see CFH750
NSC-88106 see CHE500
NSC 89936 see JAK000
NSC-89945 see DMX200
NSC-91523 see ICB000, ICC000
NSC 91726 see CGV000
NSC-92338 see CBF250
NSC-93150 see AQR500
NSC 93169 see MGL600
NSC-94100 see DDP600
NSC 94600 see CBB870
NSC-95441 see CHD250
NSC-95466 see CGW250
NSC 100880 see CBB870
NSC-100880 see CBB875
NSC-101983 see MMR750
NSC-101984 see MMS750
NSC 102627 see YCJ000
NSC-102816 see ARY000
NSC 103548 see CHF000
NSC 104469 see CME250
NSC-104800 see DDJ000
NSC-104801 see CQK600, SIK000
NSC 105023 see RQF350
NSC-105388 see FPC000
NSC-106563 see BFW250
NSC-106568 see TKZ000
NSC-106959 see CBQ625
NSC-106962 see SKM000
NSC-107429 see COV500
NSC-107430 see DOQ400
NSC-107434 see SKO500
NSC-107654 see CFC000
NSC-108034 see HNJ000
NSC-108160 see AEG750
NSC-109229 see ARN800
NSC 109723 see TNT500
NSC-109724 see IMH000
NSC-110326 see CNH800
NSC-110364 see OOG000
NSC-110430 see CBR500
NSC-110431 see MIV500
NSC-110432 see DFA400
NSC-111071 see CBO250
NSC 111180 see ACH000
NSC 112259 see EDR500
NSC-113233 see MQX250

NSC 113619 see AAE500
NSC 113926 see RKP000
NSC-114649 see FCB100
NSC-114650 see DNV200
NSC 114900 see DLH630
NSC-114901 see DLS600
NSC-115944 see EAT900
NSC-119875 see PJD000
NSC-119876 see TBO776
NSC-120949 see PJY750
NSC 122023 see VBZ000
NSC 122053 see PCV775
NSC-122402 see LJD500
NSC-122758 see VSK950
NSC-122819 see EQP000
NSC 122870 see DBY500
NSC-123127 see AES750
NSC-125973 see TAH775
NSC 126771 see DFN500
NSC-129185 see MRX000
NSC-129943 see PIK250
NSC-130044 see TEZ000
NSC 132313 see DCI600
NSC-132319 see ICD100
NSC-133099 see RKA000
NSC-133129 see DCI800
NSC-134434 see LIM000
NSC-135758 see PIK075
NSC 137679 see SBU700
NSC 138780 see FQS000
NSC-139105 see THR500
NSC-140117 see IBQ100
NSC-141537 see AOP250
NSC 141540 see EAV500
NSC 141549 see ADL500, ADL750
NSC 141633 see HGI575
NSC 141634 see IKS500
NSC-143019 see MJL750
NSC 143504 see CNH625
NSC-143969 see TGB000
NSC-145668 see COW900
NSC-145669 see DLX300
NSC-148958 see FLZ050
NSC-150014 see HGW500
NSC-153858 see MBU820
NSC-154020 see TJE870
NSC 154890 see CNR250
NSC-164011 see ROZ000
NSC-165563 see BOL500
NSC 166100 see POC000
NSC-172112 see SLD900
NSC 176319 see QOJ250
NSC-177023 see LFA020
NSC 178248 see CLX000
NSC-180024 see CCK625
NSC 181815 see HIF000
NSC 182986 see ASK875
NSC 190935 see CJJ250
NSC 190945 see DOP200
NSC 190947 see CFC500
NSC 190948 see MDX250
NSC 190949 see MLX000
NSC 190955 see DTP800
NSC 190956 see DTP600
NSC 190978 see DRR400
NSC 190981 see DVS000
NSC 190986 see DJY200
NSC 190987 see DTQ400
NSC 190997 see MPG250
NSC 190998 see IEN000
NSC 191000 see DRP600
NSC 191001 see MPS250
NSC 191025 see DGA200
NSC 192965 see SLD800
NSC 194814 see DAD040
NSC 195022 see BEP500
NSC 195058 see PHE250
NSC 195087 see IOS000
NSC 195102 see CJJ500
NSC 195106 see FAK000
NSC 195154 see DOL800
NSC 195164 see CLJ875
NSC 217697 see MRU760
NSC-218321 see PBT100
NSC-224131 see PGY750

NSC 226080 see RBK000
NSC 238159 see OJD100
NSC-239717 see CHB000
NSC 239724 see CHA500
NSC-241240 see CCC075
NSC-245382 see ALF500
NSC-246131 see TJX350
NSC 247516 see PCC000
NSC 249992 see ADL750
NSC-253272 see CBG075
NSC 253947 see CHA750
NSC-256439 see DAN000
NSC 259968 see BMK325
NSC-261036 see DBA700
NSC 261036 see NHI000
NSC-261037 see NHH500
NSC 262666 see BIC325
NSC-267703 see QBS000
NSC-269148 see MCB600
NSC 271674 see CCJ350
NSC 271675 see DEU115
NSC 279836 see MQY090
NSC 281278 see SNS100
NSC-286193 see RJF500
NSC 287513 see BKB300
NSC 294895 see CHB750
NSC-294896 see CHF250
NSC-296934 see TBC450
NSC-296961 see AMD000
NSC 298223 see APT375
NSC 299195 see DBE875
NSC 300288 see MQX775
NSC 301739 see MQY100
NSC 311056 see SLE875
NSC-322921 see MOD500
NSC-325319 see DHA300
NSC 330500 see MAB400
NSC-339004 see CMA600
NSC 353451 see MQY110
NSC 403169 see ADR750
NSC-405124 see TIQ750
NSC-407347 see MQB750
NSC 409425 see DRN400
NSC-409962 see BIF750
NSC 521778 see FNO000
NSC-525334 see NDY500
NSC-526062 see DVP400
NSC 526417 see EAD500
NSC 527017 see AOC500
NSC-528986 see AIV500
NSC 616586 see PGQ500, PGQ750
NSC-1461711 see AFQ575
NSC A15920 see AEA250
NSC B-2992 see MQX250
NSC D 254157 see CHA000, CLX000
NSC-280594 HYDRATE see TJE875
NSC 5366 HYDROCHLORIDE see NOA500
NSC-762 HYDROCHLORIDE see BIE500
NSD 256927 see IGG775
NTs 62 see CCU250
NTs 218 see CCU250
NTs 222 see CCU250
NTs 539 see CCU250
NTs 542 see CCU250
Ni 4303T see NCW500
NT see NLV500
NTA-194 see CJH750
NTA see AMT500, SIP500
NTA SODIUM HYDRATE see NEI000
NTD 2 see BHL750
NTG see NGY000
NTL see SBD000
NTL SULFATE see NCP550
NTM see DTR200
NTN-8629 see DGC800
NTO (DOT) see NMP620
NTO see NMP620
NTOI see NML000
NTPA see NOC400
19NTPP see DYF450
NTPP see DYF450
NTS 40 see PJT300
NU-1196 see NEA500, NEB000
NU-1932 see NOE550
NU 2017 see ABV250

NU-2121 see NDW510
NU 2206 see MDV250, MKR250
NU-2221 see DDF200
NU 2222 see TKW500
NUARIMOL see CHJ300
NUARSOL see ARA500
NU-BAIT II see MDU600
NUBIAN YELLOW TB see PEJ600
NUBILON ORANGE R see CMM220
NUCES NUCISTAE see NOG000
NUCHAR 722 see CDI000
NUCHAR see CBT500
NUCIDOL see DCM750
NUCIFERIN see NOE500
1-NUCIFERINE see NOE500
NUCIFERINE see NOE500
(—)-NUCIFERINE see NOE500
NUCIN see WAT000
NUCLEAR FAST RED see AJQ250
NUCLEASE, DEOXYRIBO- see EDK650, PAF575
NUCLEOCARDYL see AEH750
NUCLON ARGENTINIAN see FNF000
NUC SILICONE VS 7158 see DAF350
NUC SILICONE VS 7207 see OCE100
NUCTALON see CKL250
NUDRIN see MDU600
NUEZ de la INDIA (PUERTO RICO) see TOA275
NUEZ NOGAL (PUERTO RICO) see TOA275
NUFENOXOLE see DWF700
NUFLUOR see SHF500
NUGESTORAL see GEK500
NU-1326 HYDROBROMIDE see NOE525
NU 1604 HYDROCHLORIDE see TEQ700
NUISANCE DUSTS and AEROSOLS see NOF000
NUJOL see MQV750
NULANS see CFY500
NU-LAWN WEEDER see DDP000
NULLAPON B see EIV000
NULLAPON BF-78 see EIV000
NULLAPON BF ACID see EIX000
NULLAPON BFC CONC see EIV000
NULOGYL see NHH000
NULSA see BGC625
NUMAL see AFT000
NU MAN see MPN500
NU-MANESE see MAT250
NUMBER 2 BURNER FUEL see DHE800
NUMBER 2 FUEL OIL see DHE800
NUMOQUIN see HHR700
NUMOQUIN HYDROCHLORIDE see ELC500
NUMORPHAN HYDROCHLORIDE see ORG100
NUODEX V 1525 see DEJ100
NUOPLAZ see DXQ200
NUOPLAZ DOP see DVL700
NUPERCAINAL see DDT200
NUPERCAINE see DDT200
NUPERCAINE HYDROCHLORIDE see NOF500
NUPRIN see AIE750
NURAN see PCI250
NUREDAL see BET000
NURELLE see TIV750
NU REXFORM see LCK100
NUSYN-NOXFISH see PIX250
NUTINAL see BCA000
NUTMEG see NOG000
NUTMEG OIL see NOG500
NUTMEG OIL, EAST INDIAN see NOG500
NU-TONE see NAK500
NUTRASWEET see ARN825
NUTRIFOS STP see SJH200
NUTROP see SBH500
NUTROSE see SFQ000
NUVA see DGP900
NUVACRON 20 see MRH209
NUVACRON see MRH209
NUVANOL see DSQ000
NUVANOL N see IEN000
NUVAPEN see AIV500
NUVELBI V.C.A. see DXO300
NUX MOSCHATA see NOG000
NUX VOMICA see SMN500
NUX-VOMICA TREE see NOG800
N 4000V see SMQ500
NVC 9025 see PJS750
NVP see NBO525

NYACOL 830 see SCH000
NYACOL 1430 see SCH000
NYACOL see SCH000
NYACOL A 1530 see AQF000
NYAD 10 see WCJ000
NYAD 325 see WCJ000
NYA G see WCJ000
NYANTHRENE BLUE BFP see DFN425
NYANTHRENE BRILLIANT VIOLET 4R see DFN450
NYANTHRENE BROWN RB see CMU770
NYANTHRENE GOLDEN YELLOW see DCZ000
NYANTHRENE OLIVE R see DUP100
NYANTHRENE RED G 2B see CMU825
NYANZA BLUE RW see CMO600
NYANZA FAST RED FA see CMO870
NYAZIN see BET000
NYCOLINE see PJT200
NYCO Liquid RED GF see RGW000
NYCOR 200 see WCJ000
NYCOR 300 see WCJ000
NYCOTON see CDO000
NYCTAL see BNK000
NYDRAN see BEG000
NYDRANE see BEG000
NYDRAZID see ILD000
NY IV-34-1 see TEH250
NYLIDRIN HYDROCHLORIDE see DNU200
NYLMERATE see ABU500
NYLOCROM YELLOW 3R see SGP500
NYLOMINE ACID BLUE B-B see CMM070
NYLOMINE ACID RED C-R see CMO870
NYLOMINE ACID RED P4B see HJF500
NYLOMINE ACID SCARLET C-R see CMM320
NYLOMINE ACID SCARLET P-R see CMM320
NYLOMINE ACID YELLOW B-RD see SGP500
NYLOMINE BLUE A 2B see CMM090
NYLOMINE BLUE A 3R see CMM100
NYLON-6 see PJY500
NYLON see NOH000
NYLON 66 SALT see HEG120
NYLON 610 SALT see HEG130
NYLOQUINONE BLUE 2J see TBG700
NYLOQUINONE BLUE 4J see DMM400
NYLOQUINONE BORDEAUX B see CMP080
NYLOQUINONE LIGHT YELLOW 4JL see AAQ250
NYLOQUINONE ORANGE JR see AKP750
NYLOQUINONE PINK B see AKO350
NYLOQUINONE PURE BLUE see MGG250
NYLOQUINONE VIOLET R see DBP000
NYLOQUINONE YELLOW 4J see AAQ250
NYLOQUINONE YELLOW 2R see DUW500
NYLOSAN GREEN F-BL see CMM200
NYLOSAN YELLOW E-3R see SGP500
NYMCEL S see SFO500
NYSCAPS see WAK000
NYSCONITRINE see NGY000
NYSTAN see NOH500
NYSTATIN see NOH500
NYSTATINE see NOH500
NYSTAVESCENT see NOH500
NYTAL see TAB750
NZ see CAT775


O 250 see SKV100
O-2857 see BDE500
OA-A 1102 see CAT775
OAAT see AIC250
OAP see HED500
OBB see OAH000
OBDZ see OMY700
OBELINE PICRATE see ANS500
OBEPAR see DKE800
OBESIN see PNN300
OBESITABS see AOB500
OBIDOXIME CHLORIDE see BGS250
OBIDOXIME DICHLORIDE see BGS250
OBIDOXIME HYDROCHLORIDE see BGS250
OBLEVIL see EQL000
OBLITEROL see EMT500
m-OBLIVON see EQL000
OBLIVON C see MNM500
OBLIVON CARBAMATE see MNM500
OBOB see NLE400
OBPA see OMY850

OBRACINE see TGI500
OBRAMYCIN see TGI250
OBSH see OPE000
OBSTON see DJL000
OC 1085 see MQT550
OCDD see OAJ000
OCENOL see OBA000
OCEOL see OBA000
OCHRATOXIN see OAD090
OCHRATOXIN A see CHP250
OCHRE see IHD000
OCI 56 see SGG500
OCIMENE see DTF200
OCIMENOL see DTD400
OCIMUM BASILICUM OIL see BAR250
OCNA see CJA200
OCOTEA CYMBARUM OIL see OAF000
OCPA see CEH750
OCPNA see CJA175
OCRITEN see DAQ800
OCTA see CPB120
OCTAACETYLSUCROSE see OAF100
OCTABENZONE see HND100
ar,ar,ar,ar,ar',ar',ar',ar'-OCTABROMO-1,1'-BIPHENYL see OAH000
OCTABROMOBIPHENYL see OAH000
OCTABROMODIPHENYL see OAH000
OCTACAINE HYDROCHLORIDE see MLO750
OCTACARBONYLDICOBALT see CNB500
1,2,4,5,6,7,8,8-OCTACHLOOR-3a,4,7,7a-TETRAHYDRO-4,7-endo-METHANO-IN-
    DAAN (DUTCH) see CDR750
OCTACHLOR see CDR750
OCTACHLOROCAMPHENE see CDV100
1,2,3,4,6,7,8,9-OCTACHLORODIBENZODIOXIN see OAJ000
OCTACHLORODIBENZO(b,e)(1,4)DIOXIN see OAJ000
OCTACHLORODIBENZO-p-DIOXIN see OAJ000
OCTACHLORODIBENZODIOXIN see OAJ000
OCTACHLORODIHYDRODICYCLOPENTADIENE see CDR750
OCTACHLORODIPROPYLETHER see OAL000
1,2,4,5,6,7,8,8-OCTACHLORO-2,3,3a,4,7,7a-HEXAHYDRO-4,7-METHANO-1H-IN-
    DENE see CDR750
1,2,4,5,6,7,8,8-OCTACHLORO-2,3,3a,4,7,7a-HEXAHYDRO-4,7-METHANOINDENE
    see CDR750
1,3,4,5,6,8,8-OCTACHLORO-1,3,3a,4,7,7a-HEXAHYDRO-4,7-METHANOISOBEN-
    ZOFURAN see OAN000
OCTACHLORO-HEXAHYDRO-METHANOISOBENZOFURAN see OAN000
1,2,4,5,6,7,8,8-OCTACHLORO-3a,4,7,7a-HEXAHYDRO-4,7-METHYLENE INDANE
    see CDR750
OCTACHLORO-4,7-METHANOHYDROINDANE see CDR750
1,2,4,5,6,7,8,8-OCTACHLORO-4,7-METHANO-3a,4,7,7a-TETRAHYDROINDANE
    see CDR750
OCTACHLORO-4,7-METHANOTETRAHYDROINDANE see CDR750
1,3,4,5,6,7,10,10-OCTACHLORO-4,7-endo-METHYLENE-4,7,8,9-TETRAHYDRO-
    PHTHALAN see OAN000
OCTACHLORONAPHTHALENE see OAP000
1,3,4,5,6,7,8,8-OCTACHLORO-2-OXA-3a,4,7,7a-TETRAHYDRO-4,7-METHANOIN-
    DENE see OAN000
1,2,4,5,6,7,8,8-OCTACHLORO-3a,4,7,7a-TETRAHYDRO-4,7-METHANOINDAN
    see CDR750
2,2,4,5,6,7,8,8-OCTACHLORO-3a,4,7,7a-TETRAHYDRO-4,7-METHANOINDAN
    see CDR575
1,2,4,5,6,7,8,8-OCTACHLORO-3a,4,7,7a-TETRAHYDRO-4,7-METHANOINDANE
    see CDR750
1,2,4,5,6,7,10,10-OCTACHLORO-4,7,8,9-TETRAHYDRO-4,7-METHYLENEINDANE
    see CDR750
1,2,4,5,6,7,8,8-OCTACHLOR-3a,4,7,7a-TETRAHYDRO-4,7-endo-METHANO-IN-
    DAN (GERMAN) see CDR750
OCTACIDE 264 see OES000
9,12-OCTADECADIENOIC ACID see LGG000
cis-9,cis-12-OCTADECADIENOIC ACID see LGG000
cis,cis-9,12-OCTADECADIENOIC ACID see LGG000
OCTADECAFLUORODECAHYDRONAPHTHALENE see PCG700
OCTADECAFLUOROOCTANE see OAP100
OCTADECANAMINE ACETATE see OAP300
1-OCTADECANAMINIUM, N,N-DIMETHYL-N-OCTADECYL-, CHLORIDE (9CI)
    see DXG625
OCTADECANE, 7,8-EPOXY-2-METHYL-, cis-(8CI) see ECB200
OCTADECANENITRILE see SLL500
OCTADECANNAMIDE see OAR000
OCTADECANOIC ACID see SLK000
OCTADECANOIC ACID, ALUMINUM SALT see AHH825
OCTADECANOIC ACID, AMMONIUM SALT see ANU200
OCTADECANOIC ACID, BARIUM CADMIUM SALT (4:1:1) (9CI) see BAI800
OCTADECANOIC ACID, BARIUM SALT (9CI) see BAO825
OCTADECANOIC ACID, BUTYL ESTER (9CI) see BSL600
OCTADECANOIC ACID, CADMIUM SALT see OAT000

OCTANE, 1,1,1,2,2,3,3,4,4,5,5,6,6,7,7,8,8-HEPTADECAFLUORO-8-IODO-(9CI) see PCH325
OCTANE, 1-IODO- see OFA100
OCTANENITRILE see OCW100
OCTANE-1-NNO-AZOXYMETHANE see MGT000
OCTANE, OCTADECAFLUORO- see OAP100
tert-OCTANETHIOL see MKJ250, OFE030
OCTAN ETOKSYETYLU (POLISH) see EES400
OCTAN ETYLU (POLISH) see EFR000
OCTAN FENYLRTUTNATY (CZECH) see ABU500
OCTANIL see ILK000
OCTAN KOBALTNATY (CZECH) see CNA500
OCTAN MANGANATY (CZECH) see MAQ000
OCTAN MEDNATY (CZECH) see CNI250
OCTAN METYLU (POLISH) see MFW100
OCTANOIC ACID see OCY000
OCTANOIC ACID, 2-ACETYL-, ETHYL ESTER see EKS150
OCTANOIC ACID ALLYL ESTER see AGM500
OCTANOIC ACID, CADMIUM SALT (2:1) see CAD750
OCTANOIC ACID, 2,2-DIMETHYL-, HEXADECYL ESTER see HCP600
OCTANOIC ACID, ETHYL ESTER see ENY000
OCTANOIC ACID, HEXYL ESTER see HFS600
OCTANOIC ACID, 5-HYDROXY-, Δ-LACTONE see ODE300
OCTANOIC ACID, 5-HYDROXY-, LACTONE (6CI) see ODE300
OCTANOIC ACID, ISOPENTYL ESTER see IHP500
OCTANOIC ACID, METHYL ESTER see MHY800
OCTANOIC ACID, METHYL ESTER mixed with METHYL DECANOATE see MND000
OCTANOIC ACID, PENTADECAFLUORO- see PCH050
OCTANOIC ACID, 1,2,3-PROPANETRIYL ESTER see TMO000
OCTANOIC ACID-2-PROPENYL ESTER see AGM500
OCTANOIC ACID, PROPYL ESTER see PNR600
OCTANOIC ACID, SODIUM SALT see SIX600
OCTANOIC ACID, p-TOLYL ESTER see THB000
OCTANOIC ACID TRIGLYCERIDE see TMO000
OCTANOIC ACID, ZINC SALT (2:1) see ZEJ000
OCTANOIC/DECANOIC ACID TRIGLYCERIDE see CBF710
2-OCTANOL see OCY090
OCTANOL-3 see OCY100
3-OCTANOL see OCY100
D-n-OCTANOL see OCY100
1-OCTANOL (FCC) see OEI000
OCTANOL, mixed isomers see ODE000
n-OCTANOL see OEI000
OCTANOL see OEI000
1-OCTANOL ACETATE see OEG000
2-OCTANOL, 8,8-DIMETHOXY-2,6-DIMETHYL-(9CI) see LBO050
3-OCTANOL, 3,7-DIMETHYL- see LFY510
2-OCTANOL, 3,7-DIMETHYL-7-METHOXY- see DLR300
OCTANOLIDE-1,4 see OCE000
5-OCTANOLIDE see ODE300
3-OCTANOL, 3-METHYL- see MND050
2-OCTANONE see ODG000
3-OCTANONE see ODI000
2-OCTANONE, 5-METHYL- see MLO800, MND075
OCTANONITRILE see OCW100
n-OCTANOYL PEROXIDE (DOT) see CBF705
2-OCTANOYL-1,2,3,4-TETRAHYDROISOQUINOLINE see ODO000
OCTAN PROPYLU (POLISH) see PNC250
OCTAN WINYLU (POLISH) see VLU250
n-OCTANYL ACETATE see OEG000
OCTAN ZINECNATY (CZECH) see ZCA000
OCTATENSIN see GKQ000
OCTATENZINE see GKQ000
1,3,7-OCTATRIEN-5-YNE see ODQ300
OCTATROPINE METHYLBROMIDE see LJS000
2-OCTENE, 8-ETHOXY-2,6-DIMETHYL- see EES100
1-OCTEN-3-OL see ODW000
1-OCTEN-3-OL ACETATE see ODW028
6-OCTEN-1-OL, 3,7-DIMETHYL- see DTF410
7-OCTEN-2-OL, 2,6-DIMETHYL-8-(1H-INDOL-1-YL)- see ICS100
1-OCTEN-3-OL, 3-METHYL- see MND100
OCTENYL ACETATE see ODW028
OCTHILINONE see OFE000
OCTIC ACID see OCY000
OCTILIN see OEI000
OCTIN see ILK000
OCTIN HYDROCHLORIDE see ODY000
OCTINUM see ILK000
OCTOCLOTHEPIN see ODY100
OCTOCLOTHEPINE see ODY100
OCTODRINE see ILM000
OCTOGEN see CQH250
OCTOGEN, wetted with not <15% water, by weight (UN 0226) (DOT) see CQH250

OCTOGEN, desensitized (UN 0483) (DOT) see CQH250
n-OCTOIC ACID see OCY000
OCTOIL see DVL700
OCTOIL S see BJS250
OCTON see ILK000
m-OCTOPAMINE see AKT000
OCTOPAMINE see AKT250
dl-m-OCTOPAMINE HYDROCHLORIDE see NNT100
OCTOTIAMINE see GEK200
OCTOWY ALDEHYD (POLISH) see AAG250
OCTOWY BEZWODNIK (POLISH) see AAX500
OCTOWY KWAS (POLISH) see AAT250
OCTOXINOL see PKF500
OCTOXYNOL 3 see PKF500
OCTOXYNOL 9 see PKF500
OCTOXYNOL see PKF500
OCTSETAN see SNQ000
1-OCTYL ACETATE see OEG000
3-OCTYL ACETATE see OEG100
n-OCTYL ACETATE see OEG000
OCTYL ACETATE see OEE000, OEG000
β-OCTYL ACROLEIN see DXU280
OCTYL ACRYLATE see ADU250
OCTYL ADIPATE see AEO000
OCTYL ALCOHOL see OEI000
OCTYL ALCOHOL ACETATE see OEG000
OCTYL ALCOHOL, NORMAL-PRIMARY see OEI000
n-OCTYL ALDEHYDE see OCO000
2-OCTYLAMINE see OEK010
N-OCTYLAMINE see OEK000
1-OCTYLAMINE, N-METHYL-N-NITROSO- see NKU590
N-n-OCTYLATROPINE BROMIDE see OEM000
N-n-OCTYL-ATROPINIUMBROMID (GERMAN) see OEM000
8-OCTYLATROPINIUM BROMIDE see OEM000
N-n-OCTYLATROPINIUM BROMIDE see OEM000
N-OCTYLATROPINIUM BROMIDE see OEM000
4-n-OCTYLBENZOIC ACID see OEQ000
p-OCTYLBENZOIC ACID see OEQ000
N-OCTYLBICYCLO-(2.2.1)-5-HEPTENE-2,3-DICARBOXIMIDE see OES000
N-OCTYL BICYCLOHEPTENE DICARBOXIMIDE see OES000
n-OCTYL BROMIDE see BNU000
γ-n-OCTYL-γ-n-BUTYROLACTONE see OES100
OCTYL CARBINOL see NNB500
1-OCTYL CYANIDE see OES300
n-OCTYL CYANIDE see OES300
OCTYL CYANIDE see OES300
n-OCTYL n-DECYL PHTHALATE see OEU000
OCTYL DECYL PHTHALATE see OEU000
OCTYL-DIMETHYL-p-AMINOBENZOIC ACID see AOI500
OCTYL-DIMETHYL-BENZYLAMMONIUM CHLORIDE see OEW000
2-OCTYLDODECANOL see OEW100
OCTYLDODECANOL see OEW100
2-OCTYLDODECYL ALCOHOL see OEW100
OCTYLENE EPOXIDE see ECE000
OCTYLENE GLYCOL see EKV000
OCTYL EPOXYTALLATE see FAB920
n-OCTYL ESTER of 3,4,5-TRIHYDROXYBENZOIC ACID see OFA000
OCTYL ETHER see OEY000
n-OCTYL FORMATE see OEY100
OCTYL FORMATE see OEY100
OCTYL GALLATE see OFA000
N-OCTYL-o-HYDROXYBENZOATE see SAJ500
γ-OCTYL-β-HYDROXY-Δ^a,β-BUTENOLID (GERMAN) see HND000
n-OCTYLIC ACID see OCY000
1-n-OCTYL IODIDE see OFA100
1-OCTYL IODIDE see OFA100
n-OCTYL IODIDE see OFA100
OCTYL IODIDE see OFA100
n-OCTYLISOSAFROLE SULFOXIDE see ISA000
2-OCTYL-4-ISOTHIAZOLIN-3-ONE see OFE000
2-OCTYL-3(2H)-ISOTHIAZOLONE see OFE000
OCTYL KETONE see OFE020
tert-OCTYLMERCAPTAN (DOT) see OFE030
tert-OCTYLMERCAPTAN see MKJ250
OCTYL METHYL KETONE see OFE050
n-OCTYL NITROSUREA see NLD800
OCTYL-OCTADECYL DIMETHYL ETHYLBENZYL AMMONIUM CHLORIDES see BBA500
cis-3-OCTYL-OXIRANEOCTANOIC ACID see ECD500
4'-(OCTYLOXY)-3-PIPERIDINOPROPIOPHENONE HYDROCHLORIDE see OFG000
4-OCTYLOXY-β-(1-PIPERIDYL)PROPIOPHENONE HYDROCHLORIDE see OFG000
OCTYL PALMITATE see OFE100
OCTYLPEROXIDE see OFI000

p-tert-OCTYLPHENOL see TDN500
OCTYL PHENOL see OFK000
OCTYL PHENOL CONDENSED with 12-13 MOLES ETHYLENE OXIDE see
   PKF500
OCTYL PHENOL EO (16) see OFS000
OCTYLPHENOL EO (10) see OFQ000
OCTYLPHENOL EO (3) see OFM000
OCTYL PHENOL EO (5) see OFO000
OCTYL PHENOL condensed with 3 MOLES ETHYLENE OXIDE see OFM000
OCTYL PHENOL condensed with 5 MOLES ETHYLENE OXIDE see OFO000
OCTYL PHENOL condensed with 8–10 MOLES ETHYLENE OXIDE see
   OFQ000
OCTYL PHENOL condensed with 16 MOLES ETHYLENE OXIDE see OFS000
p-tert-OCTYLPHENOXYETHOXYETHYLDIMETHYLBENZYL AMMONIUM CHLO-
   RIDE see BEN000
p-tert-OCTYLPHENOXYPOLYETHOXYETHANOL see PKF500
OCTYLPHENOXYPOLY(ETHOXYETHANOL) see GHS000
tert-OCTYLPHENOXYPOLY(ETHOXYETHANOL) see GHS000
OCTYLPHENOXYPOLY(ETHYLENEOXY)ETHANOL see GHS000
tert-OCTYLPHENOXY POLY(OXYETHYLENE)ETHANOL see GHS000
n-OCTYL PHTHALATE see DVL600
OCTYL PHTHALATE see DVL600, DVL700
4′-OCTYL-3-PIPERIDINOPROPIOPHENONE HYDROCHLORIDE see PIY000
1-OCTYL-2-PYRROLIDINONE see OFU100
N-OCTYLPYRROLIDINONE see OFU100
N-OCTYLPYRROLIDONE see OFU100
OCTYL SALICYLATE see SAJ500
OCTYL SEBACATE see BJS250
OCTYL SODIUM SULFATE see OFU200
OCTYL STEARATE see OFU300
2-(OCTYLTHIO)ETHANOL see OGA000
OCTYLTRICHLOROSILANE see OGE000
OCTYLTRICHLOROSTANNANE see OGG000
OCTYLTRIS(2-ETHYLHEXYLOXYCARBONYLMETHYL- THIO)STANNANE see
   OGI000
4-OCTYN-3,6-DIOL, 3,6-DIMETHYL- see DTF850
OCTYNECARBOXYLIC ACID, METHYL ESTER see MNC000
1-OCTYN-3-OL see OGI050
OC-U-MID see SNQ000
OCUSEPTINE see SJL500
OCUSOL see SPC500
OCU-VINC see VLF000
ODA see SJN700
ODB see DEP600, EDP000
ODCB see DEP600
ODISTON see DCK000
ODODIBORANE see OGI100
OE3 see EDU500
OE 7 see EDJ500
OEDROPHANIUM see TAL490
OEKOLP see DKA600
OENANTHAL see HBB500
OENANTHALDEHYDE see HBB500
OENANTHE CROCATA see WAT315
OENANTHIC ACID see HBE000
OENANTHIC ALDEHYDE see HBB500
OENANTHIC ETHER see EKN050
OENANTHOL see HBB500
OENANTHOTOXIN see EAO500
OENANTHYLIC ACID see HBE000
OENETHYL HYDROCHLORIDE see NCL300
OESTERGON see EDO000
OESTRADIOL-17-α see EDO500
OESTRADIOL-17-β see EDO000
3,17-β-OESTRADIOL see EDO000
α-OESTRADIOL see EDO000
β-OESTRADIOL see EDO000
cis-OESTRADIOL see EDO000
d-3,17-β-OESTRADIOL see EDO000
d-OESTRADIOL see EDO000
OESTRADIOL see EDO000
17-β-OESTRADIOL-3-BENZOATE see EDP000
β-OESTRADIOL-3-BENZOATE see EDP000
OESTRADIOL-3-BENZOATE see EDP000
β-OESTRADIOL BENZOATE see EDP000
OESTRADIOL BENZOATE see EDP000
17-β-OESTRADIOL DIPROPIONATE see EDR000
3,17-β-OESTRADIOL DIPROPIONATE see EDR000
OESTRADIOL-3,17-DIPROPIONATE see EDR000
β-OESTRADIOL DIPROPIONATE see EDR000
OESTRADIOL DIPROPIONATE see EDR000
OESTRADIOL MONOBENZOATE see EDP000
OESTRADIOL MUSTARD see EDR500
OESTRADIOL PHOSPHATE POLYMER see EDS000
OESTRADIOL POLYESTER with PHOSPHORIC ACID see EDS000

OESTRADIOL R see EDO000
OESTRAFORM (BDH) see EDP000
OESTRASID see DAL600
17-β-OESTRA-1,3,5(10)-TRIENE-3,17-DIOL see EDO000
OESTRA-1,3,5(10)-TRIENE-3,17-β-DIOL see EDO000
1,3,5(10)-OESTRATRIENE-3,17-β-DIOL 3-BENZOATE see EDP000
(16-α,17-β)-OESTRA-1,3,5(10)-TRIENE-3,16,17-TRIOL see EDU500
OESTRA-1,3,5(10)-TRIENE-3,16-α,17-β-TRIOL see EDU500
1,3,5-OESTRATRIENE-3-β-3,16-α,17-β-TRIOL see EDU500
1,3,5(10)-OESTRATRIEN-3-OL-17-ONE see EDV000
1,3,5-OESTRATRIEN-3-OL-17-ONE see EDV000
Δ-1,3,5-OESTRATRIEN-3-β-OL-17-ONE see EDV000
OESTRATRIOL see EDU500
OESTRILIN see ECU750
OESTRIN see EDV000
16-α,17-β-OESTRIOL see EDU500
3,16-α,17-β-OESTRIOL see EDU500
OESTRIOL see EDU500
OESTRODIENE see DAL600
OESTRODIENOL see DAL600
OESTRO-FEMINAL see ECU750
OESTROFORM see EDV000
OESTROGENINE see DKA600
OESTROGLANDOL see EDO000
OESTROGYNAEDRON see DKB000
OESTROGYNAL see EDO000
OESTROL VETAG see DKA600
OESTROMENIN see DKA600
OESTROMENSIL see DKA600
OESTROMENSYL see DKA600
OESTROMIENIN see DKA600
OESTROMON see DKA600
OESTRONBENZOAT (GERMAN) see EDV500
OESTRONE see EDV000
OESTRONE-3-SULPHATE SODIUM SALT see EDV600
OESTROPAK MORNING see ECU750
OESTROPEROS see EDV000
OESTRORAL see DAL600
OETs see HKQ100
OF see PMI250
OFF see DKC800
OFFITRIL see HNI500
OFF-SHOOT-O see MND000
OFHC Cu see CNI000
OFLOCET see OGI300
OFLOCIN see OGI300
OFLOXACIN see OGI300
OFLOXACINE see OGI300
OFNACK see POP000
OFNA-PERL SALT RRA see CLK235
OFTALENT see CDP250
OFTALFRINE see SPC500
OFTANOL see IMF300
OFUNACK see POP000
OG 1 see SEH000
OGEEN 515 see OAV000
OGEEN GRB see OAV000
OGEEN M see OAV000
OGINEX see FBP100
OGOSTAL see CBF675
3-OHAA see AKE750
'OHAI-ALI'I (HAWAII) see CAK325
'OHAI (HAWAII) see SBC550
'OHAI-KE'OKE'O (HAWAII) see SBC550
'OHAI-'ULA'ULA (HAWAII) see SBC550
OHB see SAH000
OH-BBN see HJQ350
1-α-OH-CC see HJV000
cis-4-OH-CCNU see CHA500
trans-2-OH-CCNU see CHA750
trans-4-OH CCNU see CHB000
4-OH-CP see HKA200
1-α-OH-D³ see HJV000
4-OH-E2 see HKH850
17-β-OH-ESTRADIOL see EDO000
4-OH-ESTRADIOL see HKH850
OHIO 347 see EAT900
7-OHM-MBA see HMF000
7-OHM-12-MBA see HMF000
17-β-OH-OESTRADIOL see EDO000
OHRIC see DGF000
1-α-OH VITAMIN D3 see HJV000
OIL of ABIES ALBA see AAC250
OIL of ACORUS CALAMUS Linn. see OGL020
OIL de ACORUS CALAMUS (SPANISH) see OGL020

OIL of ALLSPICE see PIG740
OIL of AMERICAN WORMSEED see CDL500
OIL of ANISE see AOU250
OIL of ANISEED see PMQ750
OIL of ARBOR VITAE see CCQ500
OIL of ARGEMONE mixed with OIL of MUSTARD see OGS000
OIL, ARTEMISIA see ARL250
OIL of BAY see BAT500, LBK000
OIL of BERGAMOT, coldpressed see BFO000
OIL of BERGAMOT, rectified see BFO000
OIL, BITTER ALMOND see BLV500
OIL BLUE see CNQ000
OIL of CALAMUS, GERMAN see OGK000
OIL of CALAMUS, SPANISH see OGL020
OIL of CAMPHOR RECTIFIED see CBB500
OIL CAMPHOR SASSAFRASSY see CBB500
OIL of CAMPHOR WHITE see CBB500
OIL of CARAWAY see CBG500
OIL of CARDAMON see CCJ625
OIL of CASSIA see CCO750
OIL CEDAR see CCR000
OIL of CEDAR LEAF see CCQ500
OIL of CELERY see OGL100
OIL of CHENOPODIUM see CDL500
OIL of CHINESE CINNAMON see CCO750
OIL of CINNAMON, CEYLON see CCO750
OIL of CINNAMON see CCO750
OIL of CITRONELLA see CMT000
OIL CITRUS RETICULATA see OHC150
OIL of CLOVE see CMY475
OIL of CORIANDER see CNR735
OIL-DRI see AHF500
OIL of EUCALYPTUS see EQQ000
OIL of FENNEL see FAP000
OIL of FUR see AAC250
OIL-FURNACE BLACK see CBT750
OIL GARLIC see AGS250
OIL GAS see OGM000
OIL of GERANIUM see GDA000
OIL GERANIUM REUNION see GDC000
OIL of GRAPEFRUIT see GJU000
OIL GREEN see CMJ900
OIL of HARTSHORN see BMA750
OIL of HEERABOL-MYRRH see MSB775
OIL, HIBAWOOD see HGA100
OIL of JUNIPER BERRY see JEA000
OIL of LABDANUM see LAC000
OIL of LAVANDIN, ABRIAL TYPE see LCA000
OIL of LAVANDIN see LCA000
OIL of LAVENDER see LCC000, LCD000
OIL of LEMON, desert type, coldpressed see LEI025
OIL of LEMON, distilled see LEI030
OIL of LEMON see LEI000
OIL of LEMONGRASS, EAST INDIAN see LEG000
OIL of LEMONGRASS, WEST INDIAN see LEH000
OIL of LIME, distilled see OGO000
OIL of MACE see OGQ100
OIL MANDARIN see OHC150
OIL of MARJORAM, SPANISH see MBU500
OIL of MEDITERRANEAN BAY see OGQ150
OIL of MIRBANE (DOT) see NEX000
OIL MIST, MINERAL (OSHA, ACGIH) see MQV750
OIL of MOUNTAIN PINE see PIH400
OIL of MUSTARD, artificial see AGJ250
OIL of MUSTARD, EXPRESSED mixed with OIL of ARGEMONE see OGS000
OIL of MYRBANE see NEX000
OIL of MYRCIA see BAT500, LBK000
OIL of MYRISTICA see NOG500
OIL of MYRTLE see OGU000
OIL of NUTMEG, expressed see OGQ100
OIL of NUTMEG see NOG500
OIL OF BASIL see BAR250
OIL OF CIVET see OGM100
OIL OF CUBEB see COE175
OIL OF MUSCATEL see OGQ200
OIL OF NIOBE see MHA750
OIL OF ROSE see RNA000
OIL OF ROSE BLOSSOM see RNA000
OIL OF ROSE BULGARIAN see RNA000, RNF000
OIL OF ROSE MOROCCAN see RNK000
OIL OF SPEARMINT see SKY000
OIL OF SPIKE LAVENDER see SLB500
OIL OF THYME see TFX500
OIL OF VETIVER see VJU000
OIL OF VITRIOL (DOT) see SOI500

OIL of ONION see OJD200
OIL of ORANGE see OGY000
OIL ORANGE see PEJ500
OIL ORANGE 4G see CMP600
OIL ORANGE G see CMP600
OIL ORANGE KB see XRA000
OIL ORANGE MO see CMP600
OIL ORANGE MON see CMP600
OIL ORANGE MON EXTRA see CMP600
OIL ORANGE N EXTRA see XRA000
OIL ORANGE O'PEL see TGW000
OIL ORANGE 2R see XRA000
OIL ORANGE R see XRA000
OIL ORANGE SS see TGW000
OIL ORANGE X see XRA000
OIL ORANGE XO see XRA000
OIL of ORIGANUM see OJO000
OIL of PALMA CHRISTI see CCP250
OIL of PALMAROSA see PAE000
OIL of PARSLEY see PAL750
OIL of PELARGONIUM see GDA000
OIL of PIMENTA see PIG740
OIL PIMENTA BERRIES see PIG740
OIL of PIMENTA LEAF see PIG730
OIL of PIMENTO see PIG740
OIL of PINE see PIH750
OIL PINK see CMS242
OIL RED 3 see SBC500
OIL RED 7 see SBC500
OIL RED 47 see SBC500
OIL RED 113 see CMS242
OIL RED 282 see SBC500
OIL RED 6566 see OHA000
OIL RED see CMS242, OHA000
OIL RED A see SBC500
OIL RED APT see SBC500
OIL RED AS see OHA000
OIL RED 2B see SBC500
OIL RED 3B see OHA000, SBC500
OIL RED 4B see SBC500
OIL RED B see OHA000
OIL RED BB see SBC500
OIL RED BS see SBC500, WAT000
OIL RED D see SBC500
OIL RED ED see SBC500
OIL RED F see SBC500
OIL RED 3G see OHA000
OIL RED G see OHA000
OIL RED GO see SBC500
OIL RED GRO see XRA000
OIL RED IV see SBC500
OIL RED O see OHA000, XRA000
OIL RED OG see CMS242
OIL RED PEL see SBC500
OIL RED RC see SBC500
OIL RED RO see XRA000
OIL RED RR see SBC500
OIL RED S see SBC500
OIL RED TAX see SBC500
OIL RED XO see FAG080, XRA000
OIL RED ZD see SBC500
OIL of ROSE GERANIUM see GDA000
OIL ROSE GERANIUM see GDC000
OIL ROSE GERANIUM ALGERIAN see GDA000
OIL of RUE see OHE000
OILS, ALLSPICE see PIG740
OILS, AMYRIS see WBJ650
OIL of SANDALWOOD, EAST INDIAN see OHG000
OILS, ANGELICA ROOT see AOO760
OIL of SANTAL see OHG000
OIL of SASSAFRAS see OHI000
OIL of SASSAFRAS BRAZILIAN see OAF000
OILS, BASIL see BAR250
OIL SCARLET 48 see SBC500
OIL SCARLET 371 see XRA000
OIL SCARLET 389 see CMS242
OIL SCARLET see OHA000, SBC500, XRA000
OIL SCARLET APYO see XRA000
OIL SCARLET AS see OHA000
OIL SCARLET BL see XRA000
OIL SCARLET 6G see XRA000
OIL SCARLET G see OHA000
OIL SCARLET L see XRA000
OIL SCARLET YS see XRA000
OILS, CARROT see CCL750

OILS, CEDAR LEAF see CCQ500
OILS, CELERY see OGL100
OILS, CHAMOMILE, GERMAN see CDH500
OILS, CINNAMON see CCO750
OILS, CINTONELLA see CMT000
OILS, CLARY SAGE see OGQ200
OILS, CLOVE see CMY475
OILS, CLOVE LEAF see CMY500
OILS, CLOVE STEM see CMY510
OILS, CORIANDER see CNR735
OILS, COSTUS see CNT350
OILS, CUBEB see COE175
OILS, CUMIN see COF325
OILS, EUCALYPTUS, E. CITRIODORA, ACETYLATED see OHC100
OILS, GUAIACWOOD, ACETATES see GKG400
OIL of SHADDOCK see GJU000
OIL-SHALE PYROLYSE LAC LSP-1 see LIJ000
OIL of SILVER FIR see AAC250
OIL of SILVER PINE see AAC250
OILS, JUNIPER see JEA000
OILS, JUNIPERUS PHOENICEA see JEJ200
OILS, KARO-KAROUNDE see KCA050
OILS, LIME see OGO000
OILS, MINT, MENTHA CITRATA see BFN990
OILS, NIM see NBS300
OIL SOLUBLE ANILINE YELLOW see PEI000
OIL SOLUBLE RED S see CMS242
OIL-SOL. YELLOW ZH see CMP600
OILS, ORRIS see OHC130
OILS, PALM see PAE500
OILS, PETITGRAIN see OHC150
OILS, PIMENTA see PIG740
OILS, PINE see PIH750
OILS, SANDALWOOD see OHG000
OILS, SPRUCE see SLG650
OILS, SWEET BAY see OGQ150
OILS, SWEET BIRCH see SOY100
OILS, TEA-TREE see TAI150
OILS, VERBENA see VIK500
OIL of SWEET FLAG see OGK000
OIL of SWEET ORANGE see OGY000
OILS, WHEAT GERM see WBJ700
OIL of THUJA see CCQ500
OIL THUJA see CCQ500
OIL of TURPENTINE see TOD750
OIL of TURPENTINE, RECTIFIED see TOD750
OIL VERMILION see CMS242
OIL VERMILION LP see CMS242
OIL VIOLET see EOJ500
OIL VIOLET IRS see HOK000
OIL VIOLET R see DBP000
OIL of WHITE CEDAR see CCQ500
OIL of WINTERGREEN see MPI000
OIL YELLOW 21 see AIC250
OIL YELLOW 2681 see AIC250
OIL YELLOW see AIC250, DOT300
OIL YELLOW A see AIC250, FAG130
OIL YELLOW AAB see PEI000
OIL YELLOW AT see AIC250
OIL YELLOW C see AIC250
OIL YELLOW G EXTRA see CMP600
OIL YELLOW GG see CMP600
OIL YELLOW HA see OHK000
OIL YELLOW I see AIC250
OIL YELLOW OB see FAG135
OIL YELLOW OPS see OHK000
OIL YELLOW 2R see AIC250
OIL YELLOW SIS see CMS245
OIL YELLOW T see AIC250
OJO de CANGREJO (PUERTO RICO) see RMK250
OK 7 see PJY100
OK 622 see PAJ000
OKASA-MASCUL see TBG000
OKITEN G 23 see PJS750
OKO see DGP900
OK PRE-GEL see SLJ500
OKSAZIL see MSC100
OKSISYKLIN see HOH500
OKTADECYLAMIN (CZECH) see OBC000
ω-H-OKTAFLUORPENTANOL-1 (GERMAN) see OBS800
OKTAMETHYLCYKLOTETRASILOXAN see OCE100
OKTAMETHYLENDIKYANID see SBK500
OKTAN (POLISH) see OCU000
OKTANEN (DUTCH) see OCU000
terc. OKTANTHIOL see MKJ250, OFE030

OKTATERR see CDR750
OKTOGEN see CQH250
OKTYLESTER KYSELINY GALLOVE see OFA000
p-terc.OKTYLFENOL (CZECH) see TDN500
OKULTIN M see CIR250
OKY-046 SODIUM see SHV100
OL 27-400 see CQH100
OLAMINE see EEC600
OLATE FLAKES see OIA000
OLCADIL see CMY525
OLD 01 see ADW200
OLDHAMITE see CAY000
OLDREN see ZAK300
OLEAL ORANGE R see PEJ500
OLEAL ORANGE SS see TGW000
OLEAL RED BB see SBC500
OLEAL RED G see CMS242
OLEAL YELLOW 2G see DOT300
OLEAL YELLOW RE see OHK000
OLEAMINE see OHM700
'OLEANA (HAWAII) see OHM875
OLEANDER see OHM875
OLEANDOMYCIN HYDROCHLORIDE see OHO000
OLEANDOMYCIN MONOHYDROCHLORIDE see OHO000
OLEANDOMYCIN PHOSPHATE see OHO200
OLEANDRIN see OHQ000
OLEANDRINE see OHQ000
OLEATE of MERCURY see MDF250
OLEFIANT GAS see EIO000
OLEFINS see OHS000
α-OLEFIN SULFONATE see AFN500
α-OLEFIN SULPHONATE see AFN500
OLEIC ACID see OHU000
OLEIC ACID, compd. with BUTYLAMINE (1:1) see OIA200
OLEIC ACID, BUTYL ESTER see BSB000
OLEIC ACID GLYCEROL MONOESTER see GGR200
OLEIC ACID GLYCIDYL ESTER see ECJ000
OLEIC ACID, 12-HYDROXY- see RJP000
OLEIC ACID, LEAD(2+) SALT (2:1) see LDQ000
OLEIC ACID LEAD SALT see LDQ000
OLEIC ACID mixed with LINOLEIC ACID see LGI000
cis-OLEIC ACID, METHYL ESTER see OHW000
OLEIC ACID MONOGLYCERIDE see GGR200
OLEIC ACID POLY(OXYETHYLENE) ESTER see PJY100
OLEIC ACID, POTASSIUM SALT see OHY000
OLEIC ACID, SODIUM SALT see OIA000
OLEIC ACID, ZINC SALT see ZJS000
OLEINAMINE see OHM700
OLEJ NAPEDOWY III see DHE900, FOP000
OLEOAKARITHION see TNP250
OLEOCUIVRE see CNO000
OLEOFAC see IOT000
OLEOFOS 20 see PAK000
OLEOGESAPRIM see ARQ725
OLEOL see OBA000
OLEOMYCETIN see CDP250
OLEO NORDOX see CNO000
OLEOPARAPHENE see PAK000
OLEOPARATHION see PAK000
OLEOPHOSPHOTHION see MAK700
OLEORESIN TUMERIC see TOD625
OLEOSUMIFENE see DSQ000
OLEOVITAMIN A see VSK600
OLEOVITAMIN D3 see CMC750
OLEOVITAMIN D see VSZ100
OLEOVOFOTOX see MNH000
OLEOX 5 see PJY100
1-OLEOYLAZIRIDINE see OIC000
OLEOYLETHYLENEIMINE see OIC000
OLEOYLGLYCEROL see GGR200
OLEOYLMETHYLTAURINE SODIUM SALT see SIY000
OLEPAL I see PJY100
OLEPAL III see PJY100
OLEPTAN see FMU000
OLETAC 100 see PMP500
OLETETRIN see TBX000
OLEUM see SOI520
OLEUM ABIETIS see PIH750
OLEUM SINAPIS VOLATILE see AGJ250
OLEUM TIGLII see COC250
OLEYL ALCOHOL see OBA000
OLEYL ALCOHOL EO (10) see OIG040
OLEYL ALCOHOL EO (2) see OIG000
OLEYL ALCOHOL EO (20) see PJW500
OLEYL ALCOHOL condensed with 2 moles ETHYLENE OXIDE see OIG000

OLEYL ALCOHOL condensed with 10 moles ETHYLENE OXIDE see OIG040
OLEYL ALCOHOL condensed with 20 MOLES ETHYLENE OXIDE see PJW500
OLEYLAMIN (GERMAN) see OHM700
OLEYL AMINE see OHM700
OLEYLAMINE-HF see OII200
OLEYLAMINE HYDROFLUORIDE see OII200
OLEYLAMINHYDROFLUORID (GERMAN) see OII200
OLEYLMONOGLYCERIDE see GGR200
OLEYLPOLYOXETHYLENE GLYCOL ETHER see OIK000
OLIBANUM GUM see OIM000
OLIBANUM OIL see OIM025
OLICINE see GGR200
OLIGOMYCIN A, mixed with OLIGOMYCIN B see OIS000
OLIGOMYCIN B, mixed with OLIGOMYCIN A see OIS000
OLIMPEN see MCH525
OLIN 53139 see HEF500
'OLINANA (HAWAII) see OHM875
OLIN MATHIESON 2,424 see EFK000
OLIN MO. 2174 see AQO000
OLIO DI CROTON (ITALIAN) see COC250
OLITENSOL see BBW750
OLITREF see DUV600
OLIVE AR, ANTHRIMIDE see IBJ000
OLIVE OIL see OIQ000
OLIVE R BASE see IBJ000
OLIVOMITSIN see OIS000, OIU499
OLIVOMYCIN see OIS000, OIU499
OLIVOMYCIN A see OIU499
OLIVOMYCIN D see OIU499
OLIVOMYCINE see OIU499
OLIVOMYCIN, SODIUM SALT see CMK750
OLIV OSTANTHRENOVY R see DUP100
'OLIWA (HAWAII) see OHM875
OLOSED see MNM500
OLOTHORB see PKL100
OLOW (POLISH) see LCF000
OLPISAN see PAX000
OLTITOX see CBM750
OM 518 see MFD500
OM-1455 see LIN400
OM-1463 see TIP750
OM-1563 see ZMJ000
OM 2424 see EFK000
OM 53139 see HEF500
OMACIDE 24 see HOC000
OMADINE see HOB500, MCQ700
OMADINE ZINC see ZMJ000
OMAFLORA see HHC000
OMAHA see LCF000
OMAHA & GRANT see LCF000
OMAIN see MIW500
OMAINE see MIW500
OMAIT see SOP000
OMAL see TIW000
OMCA see TJW500
OMCHLOR see DFE200
OMEGA 127 see MKP500
OMEGA CHROME BLUE BLACK B see CMP880
OMEGA CHROME BLUE FB see HJF500
7-OMEN see MCB600
OMETHOAT see DNX800
OMETHOATE see DNX800
OMF 59 see PHA750
OM-HYDANTOINE see DKQ000
OM-HYDANTOINE SODIUM see DNU000
OMI see DEA100, FAB000
OMIFIN see CMX700
OMITE see SOP000
OMNIBON see SNN300
OMNI-PASSIN see DJT800
OMNIPEN see AIV500
OMNIPEN-N see SEQ000
OMNIZOLE see TEX000
OMNYL see QAK000
O³-(2-MORPHOLINOETHYL)MORPHINE see TCY750
OMP-1 see OIW000
OMP-2 see OIY000
OMP-4 see OJA000
OMP-5 see OJC000
OMPA see OCM000
OMPACIDE see OCM000
OMPATOX see OCM000
OMPAX see OCM000
OMPERAN see EPD500

OMS 2 see FAQ900
OMS 14 see DGP900
OMS-15 see COF250
OMS-29 see CBM750
OMS-32 see MIA250
OMS-33 see PMY300
OMS 43 see DSQ000
OMS-47 see DOS000
OMS-93 see DST000
OMS 115 see DGD800
OMS-174 see CGI500
OMS 468 see AFR250
OMS 570 see EAQ750
OMS-597 see TMD000
OMS-658 see BNL250
OMS-659 see EGV500
OMS-708 see BDG250
OMS-771 see CBM500
OMS 834 see MRH209
OMS-0971 see CMA100
OMS 1075 see DRR400
OMS-1092 see IOV000
OMS-1155 see CMA250
OMS-1206 see BEP500
OMS-1211 see IEN000
OMS 1325 see FAB400
OMS 1328 see CDS750
OMS 1342 see CLJ875
OMS 1804 see CJV250
OMSAT see TKX000
OMT see SIY000
OMTAN see OAN000
OMYA see CAT775
OMYA BLH see CAT775
OMYACARB F see CAT775
OMYALENE G 200 see CAT775
OMYALITE 90 see CAT775
ONAONA-IAPANA (HAWAII) see DAC500
ONCB see CJB750
ONCO-CARBIDE see HOO500, TGN250
ONCODAZOLE see OJD100
ONCOSTATIN see AEA500
ONCOSTATIN K see AEB000
ONCOTEPA see TFQ750
ONCOTIOTEPA see TFQ750
ONCOVEDEX see TND000
ONCOVIN see LEY000, LEZ000
ONDENA see DAC200
ONDOGYNE see FBP100
ONDONID see FBP100
ONECIDE see FDA885
ONECIDE EC see FDA885
ONE-IRON see FBJ100
ONEX see CJT750
ONGROVIL S 165 see PKQ059
ONION see WBS850
ONION OIL see OJD200
ONION TREE see WBS850
ONKOTIN see DBD700
ONO 802 see CDB775
ONOZUKA P 500 see CCU150
ONQUININ see PAF550
ONT see NMO525
ONTOSEIN see OJM400
ONTRACIC 800 see MFL250
ONTRACK 8E see MQQ450
ONTRACK see MFL250
ONTRACK-WE-2 see MFL250
ONYX see SCI500, SCJ500
ONYX BTC (ONYX OIL & CHEM CO) see AFP250
ONYXIDE 200 see THR820
ONYXIDE 3300 see BBA625
ONYXIDE (ONYX OIL & CHEM CO) see AFN750
ONYXOL 345 see BKE500
OOS see TAB750
OP 1062 see GHS000
OPACIN see TDE750
OPALON 400 see AAX175
OPALON see PKQ059
OPARENOL see DNG400
OPC 1085 see MQT550
OPC 2009 see PME600
OPC-8212 see DLF700
OPC-13013 see CMP825
OPE-3 see OFM000

OPE 30 see PKF500
OPERIDINE see DAM700
OPERTIL see ECW600
OPHIOBOLIN see CNF109
OPHIOBOLIN A see CNF109
OPHIOPHAGUS HANNAH VENOM see ARV125
OPHTALMOKALIKAN see KAV000
OPHTALMOKALIXAN see KAM000
OPHTHALAMIN see VSK600
OPHTHALMADINE see DAS000
OPHTHAZIN see OJD300
OPHTHOCILLIN see DVU000
OPIAN see NBP275, NOA000
OPIAN HYDROCHLORIDE see NOA500
OPIANINE see NOA000
OPIANINE HYDROCHLORIDE see NOA500
OPILON HYDROCHLORIDE see TFY000
OPINSUL see AKO500
OPIPRAMOL see DCV800
OPIPRAMOL DIHYDROCHLORIDE see IDF000
OPIUM see OJG000
OPLOSSINGEN (DUTCH) see FMV000
OPOBALSAM see BAF000
OPOSIM see ICC000
OPP see BGJ250
OPP-Na see BGJ750
OPP-SODIUM see BGJ750
OPRAMIDOL see DCV800
OPREN see OJI750
OPSB see BRP250
OP-SULFA 30 see SNP500
OPTAL see PND000
OPTALIDON see AGI750
OPTANOL FAST SCARLET GN see CMM320
OPTANOL RED R see NAO600
OPTANOL SCARLET GS see CMM325
OPTANOL YELLOW R see CMM759
OPTEF see CNS750
OPTENYL see PAH250
OP-THAL-ZIN see ZNA000
OPTHEL-S see SNP500
OPTHOCHLOR see CDP250
OPTICAL BLEACH 13-61 see OOI200
OPTIDASE see CCP525
OPTIMIL see MDT250
OPTIMINE see DLV800
OPTIMYCIN see MDO250
OPTINOXAN see QAK000
OPTIPEN see PDD350
OPTIPHYLLIN see TEP000
OPTISULIN LONG see IDF300
OPTIVAL see PLY275
OPTOCHIN see HHR700
OPTOCHIN HYDROCHLORIDE see ELC500
OPTOCHINIDIN see QFS100
OPTOCIL see SCK600
OPTOCIL (QUARTZ) see SCK600
OPTOQUINE see HHR700
OPTOQUINHYDROCHLORIDE see ELC500
8-OQ see QPA000
OR 1191 see FAB400, PGX300
OR-1549 see AJP125
OR-1550 see AJL875
OR-1556 see AJN375
OR-1578 see AJO625
ORABET see BSQ000
ORABILEX see EQC000
ORACAINE HYDROCHLORIDE see OJI600
ORACEF see ALV000
ORACET RED 3B see AKE250
ORACET SAPPHIRE BLUE G see TBG700
ORACET VIOLET B see AKP250
ORACET VIOLET BN see AKP250
ORACET VIOLET 2R see DBP000
α-ORACILLIN see PDD350
ORACILLIN see PDT500
ORACIL-VK see PDT750
ORACON see DNX500
ORACONAL see MCA500
ORADEXON see SOW000
ORADIAN see CDR550, CKK000
ORADIL see CLY600
ORAFLEX see OJI750
ORAFURAN see NGE000
ORAGEST see MCA000

ORAGRAFIN-SODIUM see SKM000
ORAHEXAL see CDT250
ORALCID see ABX500
ORALIN see BSQ000
ORALITH ORANGE PG see CMS145
ORALITH RED P4R see MMP100
ORALOPEN see PDD350
ORALSONE see HHR000
ORALSTERONE see AOO275
ORA-LUTIN see GEK500
ORAMID see SAH000
L-ORANGE 2 see FAG150
ORANGE 3 see MND600
1333 ORANGE see FAG010
11550 ORANGE see CMM220
ORANGE A l'HUILE see PEJ500
ORANGE BASE CIBA II see NEO000
ORANGE BASE IRGA I see NEN500
ORANGE BASE IRGA II see NEO000
ORANGE CHROME see LCS000
ORANGE CRYSTALS see ABC500
ORANGE EXTRA N see CMM220
ORANGE EXTRA P see CMM220
ORANGE G see CMS145
ORANGE GC BASE see CEH675
ORANGE I see FAG010
ORANGE II see CMM220
ORANGE IIC see CMM220
ORANGE III see MND600
ORANGE II for LAKES see CMM220
ORANGE IIP see CMM220
ORANGE II R see FAG150
ORANGE IIS see CMM220
ORANGE IISM see CMM220
ORANGE II SPECIAL FOR LACQUER see CMM220
ORANGE INSOLUBLE OLG see PEJ500, XRA000
ORANGE INSOLUBLE RR see XRA000
ORANGE LEAD see LDS000
ORANGE LEAF OIL, BITTER see OHC150
ORANGE LEAF WATER, ABSOLUTE see OHC150
ORANGE NITRATE CHROME see LCS000
ORANGE No. 205 see CMM220
ORANGE OIL, coldpressed (FCC) see OGY000
ORANGE OIL see OGY000, PDD750
ORANGE OIL KB see XRA000
ORANGE PAL see FAG150
ORANGE PEL see PEJ500
ORANGE RESENOLE No. 3 see PEJ500
ORANGE RGL CONC. SPECIALLY PURE see FAG150
ORANGE 3R SOLUBLE IN GREASE see TGW000
ORANGES see BBK500
ORANGE SALT CIBA II see NEO000
ORANGE SALT IRGA II see NEO000
ORANGE SALT NRD see CEG800
ORANGE 2 SODIUM SALT see CMM220
ORANGE SOLUBLE A l'HUILE see PEJ500
ORANGE TONER GRT see CMM220
ORANGE Y see CMM220, CMS145
ORANGE YA see CMM220
ORANGE YELLOW S see FAG150
ORANGE YELLOW S.AF see FAG150
ORANGE YELLOW S.FQ see FAG150
ORANGE YZ see CMM220
ORANGE ZH see CMM764
ORANIL see BSM000
ORANIXON see GGS000
ORANYL see BSM000
ORANZ BRILANTNI OSTAZINOVA S-2R (CZECH) see DGO000
ORANZ KYPOVA 5 see CMU815
ORANZ KYSELA 7 see CMM220
ORANZ POTRAVINARSKA 3 see CMP600
ORANZ ROZPOUSTEDLOVA 1 see CMP600
ORANZ ZASADITA 2 (CZECH) see DBP999
ORAP see PIH000
ORAPEN see PDT750
ORARSAN see ABX500
ORASECRON see NNL500
ORASONE see PLZ000
ORASPOR see CCS530
ORASTHIN see ORU500
ORASULIN see BSM000
ORATESTIN see AOO275
ORA-TESTRYL see AOO275
ORATRAST see BAP000
ORATREN see PDT500

ORATROL see DEQ200
ORAVIRON see MPN500
ORAVUE see IGA000
ORBAMIN GREEN B see CMO840
ORBENIN SODIUM see SLJ050
ORBENIN SODIUM HYDRATE see SLJ000
ORBICIN see DCQ800, PAG050
ORBINAMON see NBP500
ORBON see OAV000
ORCA 100-2 see MCB050
ORCANON see MPW500
ORCED see DGN200
ORCHARD BRAND ZIRAM see BJK500
ORCHIOL see TBG000
ORCHISTIN see TBG000
ORCIN see MPH500
ORCINOL see MPH500
ORCIPRENALINE see DMV800
ORCIPRENALINE SULFATE see MDM800
ORDIMEL see ABB000
ORDINARY AZOXYBENZENE see ASO750
ORDINARY LACTIC ACID see LAG000
ORDRAM see EKO500
OREGON HOLLY see HGF100
OREMET see TGF250
ORESOL see RLU000
ORESON see RLU000
ORESTOL see DKB000
ORETIC see CFY000
ORETON-F see TBF500
ORETON-M see MPN500
ORETON METHYL see MPN500
ORETON PROPIONATE see TBG000
OREZAN see BSQ000
ORF 2166 see CBO625
ORF 3858 see FAO200
ORF 5513 see BJG150
ORF 18489 see OGI300
ORFENADRINA see DPH000
ORG 485-50 see NNV000
ORGA-414 see AMY050
ORGAMETIL see NNV000
ORGAMETRIL see NNV000
ORGAMETROL see NNV000
ORGAMIDE see PJY500
ORGANEX see CAK500
ORGANIC GLASS E 2 see PKB500
ORGANIL 644 see MAP300
ORGANOL BORDEAUX B see EOJ500
ORGANOL BROWN 2R see NBG500
ORGANOL FAST GREEN J see BLK000
ORGANOL ORANGE see PEJ500
ORGANOL ORANGE 2J see CMP600
ORGANOL ORANGE 2R see TGW000
ORGANOL 2R see NBG500
ORGANOL RED B see SBC500
ORGANOL RED BS see OHA000
ORGANOL SCARLET see OHA000
ORGANOL VERMILION see CMS242
ORGANOL YELLOW 25 see AIC250
ORGANOL YELLOW see PEI000
ORGANOL YELLOW ADM see DOT300
ORGANOMETALS see OJM000
ORGANON'S DOCA ACETATE see DAQ800
ORGASEPTINE see SNM500
ORGASTYPTIN see EDU500
ORGATRAX see HOR470
ORG GB 94 see BMA625
ORG NA 97 see PAF625
ORG NC 45 see VGU075
ORGOTEIN see OJM400
ORGOTEINS see OJM400
ORICUR see CHX250
ORIDIN see OJV500
ORIENTAL BERRY see PIE500
ORIENT BASIC MAGENTA see MAC250
ORIENT OIL ORANGE PS see PEJ500
ORIENT OIL RED OG see CMS242
ORIENT OIL RED RR see SBC500
ORIENT OIL YELLOW GG see DOT300
ORIENTOMYCIN see CQH000
ORIENT WATER BLACK 100L see CMN240
ORIENT WATER BLACK 200L see CMN240
ORIENT WATER PINK 2 see ADG250

ORIFUNGAL M see KFK100
ORIGANUM OIL see OJO000
ORIMETEN see AKC600
ORIMON see PDP250
ORINASE see BSQ000
ORINAZ see BSQ000
ORION BLUE 3B see CMO250
ORIPAVINE, 6,14-endo-ETHYLENETETRAHYDRO-7-(1-HYDROXY-1-METHYLBU-
   TYL)-(7CI) see EQO450
ORISUL see AIF000
ORISULF see AIF000
ORIZON 805 see PJS750
ORIZON see PJS750
ORLON, combustion products see ADX750
ORLUTATE see ABU000
ORMETEIN see OJM400
ORNAMENTAL WEED see AJM000
ORNID see BMV750
ORNIDAZOLE see OJS000
dl-ORNITHINE, 2-(DIFLUOROMETHYL)-, MONOHYDROCHLORIDE see EAE775
ORNITROL see ARX800
OROBETINA see DXO300
OROFAR see BDJ600
OROLEVOL see MQU750
ORONOL see ART250
OROPUR see OJV500
OROSOMYCIN see TFC500
OROTIC ACID see OJV500
OROTIC ACID mixed with CHOLESTEROL and CHOLIC ACID (2:2:1) see
   OJV525
OROTONIN see OJV500
OROTSAURE (GERMAN) see OJV500
OROTURIC see OJV500
OROTYL see OJV500
OROVERMOL see TDX750
OROXIN see ALV000
OROXINE see LFG050
ORPHENADINE see MJH900
ORPHENADRIN see MJH900
ORPHENADRINE see MJH900
ORPHENADRINE CITRATE see DPH000
ORPHENADRINE HYDROCHLORIDE see OJW000
ORPHENOL see BGJ750
ORPIMENT see ARI000
ORQUISTERONE see TBF500
ORRIS see OHC130
ORRIS ABSOLUTE see OHC130
ORRIS CONCRETE see OHC130
ORRIS CONCRETE OIL see OHC130
ORRIS EXTRACT see OHC130
ORRIS Liquid see OHC130
ORRIS OIL see OHC130
ORRIS RESIN see OHC130
ORRIS RESINOID see OHC130
ORRIS ROOT EXTRACT see OHC130
ORRIS ROOT OIL see OHC130
ORSILE see BEQ625
ORSIN see PEY500
ORTAL SODIUM see EKT500
ORTEDRINE see BBK000
ORTHAMINE see PEY250
ORTHANILIC ACID see SNO100
ORTHENE-755 see DOP600
ORTHENE see DOP600
ORTHESIN see EFX000
ORTHO 4355 see NAG400
ORTHO 5353 see BTA250
ORTHO-5655 see MOU750
ORTHO 5865 see CBF800
ORTHO 9006 see DTQ400
ORTHO 12420 see DOP600
l'ORTHO, p-AMINONITROPHENOL, SEL SODIQUE (FRENCH) see ALO500
ORTHOARSENIC ACID see ARB250
ORTHOARSENIC ACID HEMIHYDRATE see ARC500
ORTHOBENZOIC ACID, cyclic 7,8,10a-ESTER with 5,6-EPOXY-
   4,5,6,6a,7,8,9,10,10a,10b-DECAHYDRO-3a,4,7,8,10a-PENTAHYDROXY-5-(HY-
   DROXYMETHYL)-8-ISOPROPENYL-2,10-DIMETHYLBEN Z(e)AZULEN-3(3aH)-
   ONE see DAB850
ORTHOBORIC ACID see BMC000
ORTHOCAINE see AKF000
ORTHO C-1 DEFOLIANT & WEED KILLER see SFS000
ORTHOCHLOROPARANISIDINE see CEH750
ORTHOCHROME YELLOW GGW see SIT850
ORTHOCIDE see CBG000
ORTHOCRESOL see CNX000

ORTHODERM see AKF000
ORTHODIAZINE see OJW200
ORTHODIBROM see NAG400
ORTHODIBROMO see NAG400
ORTHODICHLOROBENZENE see DEP600
ORTHODICHLOROBENZOL see DEP600
ORTHO N-4 DUST see NDN000
ORTHO N-5 DUST see NDN000
ORTHO EARWIG BAIT see DXE000
ORTHOFORM see AKF000
ORTHOFORMIC ACID, ETHYL ESTER see ENY500
ORTHOFORMIC ACID, TRIETHYL ESTER see ENY500
ORTHOFORMIC ACID, TRIMETHYL ESTER see TLX600
ORTHO GRASS KILLER see CBM000
ORTHOHYDROXYBENZOIC ACID see SAI000
ORTHOHYDROXYDIPHENYL see BGJ250
ORTHOIODIN see HGB200
ORTHO-KLOR see CDR750
ORTHO L10 DUST see LCK000, LCK100
ORTHO L40 DUST see LCK000
ORTHO-LM APPLE SPRAY see MLH000
ORTHO LM CONCENTRATE see MLH000
ORTHO LM SEED PROTECTANT see MLH000
ORTHO MALATHION see MAK700
ORTHO MC see MAE000
ORTHO-MITE see SOP500
ORTHOMRAVENCAN ETHYLNATY (CZECH) see ENY500
ORTHOMRAVENCAN METHYLNATY (CZECH) see TLX600
ORTHONAL see QAK000
ORTHO-NOVUM see MDL750
ORTHO PARAQUAT CL see PAJ000
ORTHO P-G BAIT see COF500
ORTHOPHALTAN see TIT250
ORTHOPHENANTHROLINE see PCY250
ORTHOPHENYLALANINE MUSTARD see BHT250
ORTHOPHENYLPHENOL see BGJ250
ORTHOPHOS see PAK000
ORTHO PHOSPHATE DEFOLIANT see BSH250
ORTHOPHOSPHORIC ACID see PHB250
ORTHOPHOSPHORUS ACID see PGZ899
ORTHOSAN MB see DTC600
ORTHOSERPINA see RDK000
ORTHOSIL see SJU000
ORTHOTELLURIC ACID see TAI750
ORTHOTOLUIC ACID see TGQ000
ORTHO-TOLUOL-SULFONAMID (GERMAN) see TGN250
ORTHOTRAN see CJT750
ORTHOVANILLINE see VFP000
ORTHO WEEVIL BAIT see DXE000
ORTHOXENOL see BGJ250
ORTHOXINE HYDROCHLORIDE see OJY000
ORTIN see TJL250
ORTISPORINA see ALV000
ORTIZON see HIB500
ORTODRINEX HYDROCHLORIDE see OJY000
ORTOL see GGS000
ORTONAL see QAK000
ORTRAN see DOP600
ORTRIL see DOP600
ORTUDUR see PKQ059
ORUDIS see BDU500
ORUVAIL see BDU500
ORVAGIL see MMN250
ORVINYLCARBINOL see AFV500
ORVUS WA PASTE see SIB600
ORYZAEMATE see PMD800
ORYZALIN see OJY100
ORYZANIN see TES750
ORYZANINE see TES750
OS 30 see PJT300
OS 55 see PJT300
OS 1836 see CLV375
OS 1897 see DDL800
OS 2046 see MQR750
OS-3966-A see RMK200
OSACYL see AMM250
OSAGE ORANGE see MRN500
OSAGE ORANGE CRYSTALS see MRN500
OSAGE ORANGE EXTRACT see MRN500
OSALMID see DYE700
OSALMIDE see DYE700
OSARSAL see ABX500
OSARSOLE see ABX500
OSBAC see MOV000
OSBON AC see PCL500

OS-CAL see CAT775
OSCINE see SBG000
OSCOPHEN see ARP250
O. SCUTELLATUS (AUSTRALIA) VENOM see ARV500
OSDMP see DJR700
OSIREN see AFJ500
OSIROL see DLR300
OSMIC ACID see OKK000
OSMITROL see HER000
OSMIUM see OKE000
OSMIUM HEXAFLUORIDE see OKG000
OSMIUM(IV) OXIDE see OKI000
OSMIUM(VIII) OXIDE see OKK000
OSMIUM TETROXIDE see OKK000
OSMOFERRIN see IHK000
OSMOSOL EXTRA see PND000
OSNERVAN see CPQ250
OSOCIDE see CBG000
OSPEN see PDT500
OSPENEFF see PDT750
OS-40 (PHOSPHATE ESTER) see OKK500
OSSALIN see SHF500
OSSIAMINA see CFA750
OSSIAN see OOG000
OSSICHLORIN see CFA750
OSSIDO di MESITILE (ITALIAN) see MDJ750
OSSIN see SHF500
OSTACET BRILLIANT RED E-LB see AKI750
OSTACET YELLOW P2G see AAQ250
OSTAMER see CDV625
OSTANTHREN BLUE BCL see DFN425
OSTANTHREN BLUE BCS see DFN425
OSTANTHREN BLUE RS see IBV050
OSTANTHREN BLUE RSN see IBV050
OSTANTHREN BLUE RSZ see IBV050
OSTANTHREN BROWN BR see CMU770
OSTANTHRENE BLUE RS see IBV050
OSTANTHRENE BROWN BR see CMU770
OSTANTHRENE ORANGE GR see CMU820
OSTANTHREN GREY M see CMU475
OSTANTHREN OLIVE R see DUP100
OSTANTHREN ORANGE GR see CMU820
OSTANTHREN REDDISH BROWN 5RF see CMU800
OSTAZIN BLACK H-N see CMS227
OSTAZIN BRILLIANT BLUE S-R see CMS228
OSTAZIN BRILLIANT RED S 5B see PMF540
OSTELIN see VSZ100
OSTENOL see TLG000
OSTENSIN see TLG000
OSTEOBOND SURGICAL BONE CEMENT see PKB500
OSTREOGRYCIN see VRF000
OSTREOGRYCIN B see VRA700
OSVARSAN see ABX500
OSWEGO ORANGE X 2065 see CMS145
OSYRITRIN see RSU000
OSYROL see AFJ500, DLR300
OT see HLC000
OTACRIL see DFP600
OTAHEITE WALNUT (VIRGIN ISLANDS) see TOA275
OTAMOL OKO200
OTB see ONI000
OTBE (FRENCH) see BLL750
OTC see HOH500
OTERBEN see BSQ000
OTETRYN see HOI000
OTIFURIL see FPI000
OTILONIUM BROMIDE see OKO400
OTOBIOTIC see NCG000
OTODYNE see EQQ500
OTOFURAN see NGE500
OTOKALIXIN see KAM000, KAV000
OTOPHEN see CDP250
OTOS see OPQ100
OTRACID see CJT750
OTRIVINE HYDROCHLORIDE see OKO500
OTRIVIN HYDROCHLORIDE see OKO500
OTS 11 see DVM800
OTsS 14 see PJT300
OTS 15 see DVN400
OTsS 20 see PJT300
OTS see TGN250
OTTAFACT see CFE250
OTTANE (ITALIAN) see OCU000
OTTASEPT see CLW000
OTTASEPT EXTRA see CLW000

OTTILONIO BROMURO (ITALIAN) see OKO400
1,2,4,5,6,7,8,8-OTTOCHLORO-3A,4,7,7A-TETRAIDRO-4,7-endo-METANO-INDA-NO (ITALIAN) see CDR750
OTTOMETIL-PIROFOSFORAMMIDE (ITALIAN) see OCM000
OTTO of ROSE see RNA000, RNF000
OTTO ROSE see RNA000
OUABAGENIN-I-RHAMNOSID (GERMAN) see OKS000
OUABAGENIN-I-RHAMNOSIDE see OKS000
OUABAIN see OKS000
OUABAINE see OKS000
OU-B see CBT500
OUBAIN see OKS000
OUDENONE see OKS100
OUDENONE SODIUM SALT see OKS150
OUPLATE see CNT350
OURARI see COF750
OUTFLANK see AHJ750
OUTFLANK-STOCKADE see AHJ750
OUTFOX see CQI750
(OV₁) see HNK600
OVABAN see VTF000
OVADOFOS see DSQ000
OVADZIAK see BBQ500
OVAHORMON see EDO000
OVAHORMON BENZOATE see EDP000
OVANON see LJE000
OVARELIN see LIU360
OVARIOSTAT (FRENCH) see LJE000
OVASTEROL see EDO000
OVASTEROL-B see EDP000
OVASTEVOL see EDO000
OVATRAN see CJT750
OVEST see ECU750
OVESTERIN see EDU500
OVESTIN see EDU500
OVESTINON see EDU500
OVESTRION see EDU500
OVETTEN see MNM500
OVEX see CJT750, EDP000, EDV000
OVIDON see NNL500
OVIFOLLIN see EDV000
OVIN see DNX500
OVISOT see ABO000, CMF250
OVITELMIN see MHL000
OVOCHLOR see CJT750
OVOCICLINA see EDO000
OVOCYCLIN see EDO000
OVOCYCLIN BENZOATE see EDP000
OVOCYCLIN DIPROPIONATE see EDR000
OVOCYCLINE see EDO000
OVOCYCLIN M see EDP000
OVOCYCLIN-MB see EDP000
OVOCYCLIN-P see EDR000
OVOCYLIN see EDO000
OVOSTAT 1375 see EEH575
OVOSTAT see EEH575
OVOSTAT E see EEH575
OVOTOX see CJT750
OVOTRAN see CJT750
OVRAL 21 see NNL500
OVRAL 28 see NNL500
OVRAL see NNL500
OVRAN see NNL500
OVRANETT see NNL500
O-V STATIN see NOH500
OVULEN 50 see EQJ500
OVULEN see EQK100
OWISPOL GF see SMQ500
OX see CFZ000
OXAALZUUR (DUTCH) see OLA000
10H-9-OXAANTHRACENE see XAT000
5-OXA-1-AZABICYCLO(4.2.0)OCT-2-ENE-2-CARBOXYLIC ACID, 7-(2-((DIFLUO-ROMETHYL)THIO)ACETAMIDO)-3-(((1-(2-HYDROXYETHYL)-1H-TETRAZOL-5-YL)THIO)METHYL)-7-METHOXY-8-OXO-, SODIUM SALT, (6R,7R)- see FCN100
1-OXA-4-AZACYCLOHEXANE see MRP750
1-OXA-3-AZAINDENE see BDI500
OXABEL see MNV250
OXABETRINIL see OKS200
7-OXABICYCLO(4.1.0)HEPTANE see CPD000
7-OXABICYCLO(4.1.0)HEPTANE-3-CARBOXYLIC ACID, 7-OXABICY-CLO(4.1.0)HEPT-3-YLMETHYL ESTER see EBO050
7-OXABICYCLO(2.2.1)HEPTANE-2,3-DICARBOXYLIC ACID see EAR000
7-OXABICYCLO(2.2.1)HEPTANE, 1-ISOPROPYL-4-METHYL-(6CI) see IKC100

7-OXABICYCLO(2.2.1)HEPTANE, 1-METHYL-4-(1-METHYLETHYL)-(9CI) see IKC100
OXACILLIN see DSQ800
OXACILLIN-AMPICILLIN MIXTURE see AOC875
OXACILLIN SODIUM SALT see MNV250
OXACYCLOBUTANE see OMW000
OXACYCLOHEPTADECAN-2-ONE see OKU000
OXACYCLOHEPTADEC-8-EN-2-ONE, (Z)- see OKU100
OXACYCLOPENTADIENE see FPK000
OXACYCLOPENTANE see TCR750
OXACYCLOPROPANE see EJN500
4H-1,3,5-OXADIAZINE-4-THIONE, TETRAHYDRO-3,5-DIMETHYL- see TCQ275
(1,2,3)OXADIAZOLO(5,4-d)PYRIMIDIN-5(4H)-ONE see DCQ600
8-OXA-3,5-DITHIA-4-STANNAHEPTADECANOIC ACID, 4,4-DIBUTYL-7-OXO-, NONYLESTER (9CI) see DEH650
OXAF see BHA750
OXAFLOZANE see IRP000
OXAFLOZANO (SPANISH) see IRP000
OXAFLUMAZINE DISUCCINATE see OKW000
OXAFLUMINE see OKW000
OXAFURADENE see NDY000
3-OXA-1-HEPTANOL see BPJ850
11-OXAHEXADECANOLIDE see OKW100
12-OXAHEXADECANOLIDE see OKW110
10-OXAHEXADECANOLIDE see OLE100
OXAINE see DTL200
8-OXA-9-KETOTRICYCLO(5.3.1.0²·⁶)UNDECANE see OOK000
OXALAMIDE see OLO000
OXALATES see OKY000
OXALDIN see OGI300
OXALIC ACID see OLA000
OXALIC ACID, COPPER(2+) SALT (1:1) see CNN755
OXALIC ACID DIAMIDE see OLO000
OXALIC ACID, DIAMMONIUM SALT see ANO750
OXALIC ACID, DIETHYL ESTER see DJT200
OXALIC ACID DINITRILE see COO000
OXALIC ACID, DIPOTASSIUM SALT see PLN300
OXALIC ACID DISILVER SALT see SDU000
OXALIC ACID, DISODIUM SALT see SIY500
OXALIC ACID, MONOPOTASSIUM SALT see OLE000
OXALIC ACID SILVER SALT (1:2) see SDU000
OXALIC ACID, TIN(2+) SALT (1:1) (8CI) see TGE250
OXALID see HNI500
OXALIDE see OLE100
OXALONITRILE see COO000
OXALSAEURE (GERMAN) see OLA000
(OXALYBIS(IMINOETHYLENE)BIS((o-CHLOROBENZYL)DIETHYLAMMONIUM)) DI-CHLORIDE see MSC100
2,2'-OXALYDIFURAN OXIME see FQB000
OXALYL-o-AMINOAZOTOLUENE see OLG000
4'-OXALYLAMINO-2,3'-DIMETHYLAZOBENZENE see OLG000
OXALYL CHLORIDE see OLI000
OXALYL CYANIDE see COO000
OXALYL FLUORIDE see OLK000
l-4-OXALYSINE see OLK200
OXALYSINE see OLK200
7-OXA-8-MERCURABICYCLO(4.2.0)OCTA-1,3,5-TRIENE, 5-METHYL-2-NITRO- see NHK900
1,4-OXAMERCURANE see OLM000
OXAMETACIN see OLM300
OXAMETACINE see OLM300
OXAMETHACIN see OLM300
OXAMIC ACID, PHENYL- see OLW000
d-OXAMICINA (ITALIAN) see CQH000
OXAMID (CZECH) see OLO000
OXAMIDE see OLO000
OXAMIDE, N,N'-BIS(2-HYDROXYETHYL)DITHIO- see BKD800
OXAMIDE, N,N'-DIBENZYLDITHIO- see DDF700
OXAMIDE, N,N'-DIPICRYL- see HET700
OXAMIMIDIC ACID see OLO000
OXAMIZIL see MSC100
OXAMMONIUM see HLM500
OXAMMONIUM HYDROCHLORIDE see HLN000
OXAMMONIUM SULFATE see OLS000
OXAMNIQUINE see OLT000
d-OXAMYCIN see CQH000
OXAMYL see DSP600
OXAMYL OXIME see OLT100
OXAN 600 see ARQ750
OXANAL FAST RED SW see SEH475
OXANAL YELLOW T see FAG140
OXANAMIDE see OLU000
OXANDROLONE see AOO125
OXANE see EJN500
OXANILIC ACID see OLW000

OXANTHRENE see DDA800
3-OXAPENTANEDIOIC ACID see ONQ100
3-OXAPENTANE-1,5-DIOL see DJD600
3-OXA-1,5-PENTANEDIOL see DJD600
OXAPHENAMID see DYE700
OXAPHENAMIDE see DYE700
OXAPIUM IODIDE see OLW400
OXAPRIM see TKX000
OXAPRO see OLW600
OXAPROPANIUM IODIDE see FMX000
OXAPROZIN see OLW600
OXARSANILIC ACID see PDO250
OXASULFA see AIE750
6-OXA-3-THIABICYCLO(3.1.0)HEXANE-3,3-DIOXIDE see ECP000
1,4-OXATHIANE see OLY000
OXATHIANE see OLY000
p-OXATHIANE-4,4-DIOXIDE see DVR000
1,4-OXATHIANE compound with MERCURIC CHLORIDE see OMA000
1,3,2-OXATHIASTANNOLANE, 2,2-DIBUTYL- see DEF150
1,2-OXATHIOLANE-2,2-DIOXIDE see PML400
OXATIMIDE see OMG000
OXATOMIDA see OMG000
OXATOMIDE see OMG000
5-OXATRICYCLO(8.2.0.0⁴·⁶)DODECANE, 4,12,12-TRIMETHYL-9-METHYLENE-, (1R,4R,6R,10S)- see CCN100
3-OXATRICYCLO(3.2.1.0²·⁴)OCTANE-6-CARBOXYLIC ACID, ETHYL ESTER see ENZ000
2-H-1,3,2-OXAZAPHOSPHORINANE see CQC650
OXAZEPAM see CFZ000
OXAZIL see MSC100
OXAZIMEDRINE see PMA750
OXAZINOMYCIN see OMK000
OXAZOCILLIN see DSQ800
OXAZOLAM see OMK300
OXAZOLAZEPAM see OMK300
OXAZOLIDIN see HNI500
OXAZOLIDINE A see DTG750
2-OXAZOLIDINETHIONE, 5-ETHENYL-, (R)-(9CI) see VQA100
2-OXAZOLIDINETHIONE, 5-VINYL-, (R)- see VQA100
OXAZOLIDINE, 3-(3,4,5-TRIETHOXYBENZOYL)- see OMM100
OXAZOLIDIN-GEIGY see HNI500
2-OXAZOLIDINONE see OMM000
OXAZOLIDINONE see MFD500
3-OXAZOLIDINYL 3,4,5-TRIETHOXYPHENYL KETONE see OMM100
OXAZOLIDONE see OMM000
OXAZYL see MSC100
OXCORD see AEC750
OXEBETRINIL see OKS200
m-OXEDRINE see NCL500, SPC500
p-OXEDRINE see HLV500
OXEDRINE see HLV500
(−)-m-OXEDRINE see NCL500
OXEDRINE TARTRATE see SPD000
OXELADIN see DHQ200
OXELADIN CITRATE see OMS400
OXELADINE CITRATE see OMS400
OXENDOLONE see ELF100
2-OXEPANONE (8CI, 9CI) see LAP000
OXEPINAC see OMU000
OXEPINACO (SPANISH) see OMU000
OXEPIN-3,6-ENDOPEROXIDE see EBQ550
OXETACAINE see DTL200
OXETAN see OMW000
OXETANE see OMW000
OXETHACAINA (ITALIAN) see DTL200
OXETHAZINE see DTL200
OXIAMIN see IDE000
OXIARSOLAN see ARL000
OXIBENDAZOLE see OMY700
OXIBUTININA HYDROCHLORIDE see OPK000
OXIBUTOL see HNI500
OXICOB see CNK559
OXIDASE GLUCOSE see GFG100
OXIDATE LE see MHA750
OXIDATION BASE 22 see ALL750
OXIDATION BASE 25 see NEM480
OXIDATION BASE 12A see DBO400
OXIDATION BASE 10A see PEY650
NNN-1-N-OXIDE see NLD525
OXIDE of CHROMIUM see CMJ900
N¹⁰-OXIDE-1-NITRO-9-(3-DIMETHYLAMINOPROPYLAMINO)-DIHYDROCHLORIDE ACRIDINE see CAB125
N-OXIDE de PHENAZINE (FRENCH) see PDB750
3-N-OXIDE PURIN-6-THIOL MONOHYDRATE see OMY800
10,10′-OXIDIPHENOXARSINE see OMY850

OXIDIZED I-CYSTEINE see CQK325
α,β-OXIDOETHANE see EJN500
OXIDOETHANE see EJN500
1,8-OXIDO-p-MENTHANE see CAL000
OXI-FENIBUTOL see HNI500
OXIFENON see ORQ000
OXIFENYLBUTAZON see HNI500
OXIFLAVIL see ELH600
OXIKON see DLX400
OXILAPINE see DCS200
OXILORPHAN see CQF079
α-OXIME BENZOIN see BCP500
OXIME COPPER see BLC250
OXIMES see OMY899
OXIMETHOLONUM see PAN100
OXIMETOLONA see PAN100
2-OXIMINO-3-BUTANONE see OMY910
OXIMINO OXAMYL see OLT100
1-OXINDENE see BCK250
OXINE see QPA000
OXINE COPPER see BLC250
OXINE CUIVRE see BLC250
OXINE SULFATE see QPS000
OXINOFEN see OPK300
p-OXINOZON see NBF500
2-OXI-PROPYL-PROPYLNITROSAMIN (GERMAN) see ORS000
OXIRAAN (DUTCH) see EJN500
OXIRANE see EJN500
OXIRANE, 2,2′-(1,4-BUTANEDIYL)BIS-(9CI) see DHD800
OXIRANE, 2,2′-(1,4-BUTANEDIYLBIS(OXYMETHYLENE))BIS-(9CI) see BOS100
OXIRANE-CARBOXALDEHYDE see GGW000
OXIRANE CARBOXALDEHYDE OXIME see ECG100
OXIRANECARBOXYLIC ACID, 3-(((3-METHYL-1-(((3-METHYLBU-TYL)AMINO)CARBONYL)BUTYL)AMINO) CARBONYL)-, ETHYL ESTER, (2S- see OMY925
OXIRANE, 2-DECYL-3-(5-METHYLHEXYL)-, cis- see ECB200
OXIRANE, 2-(3,5-DICHLOROPHENYL)-2-(2,2,2-TRICHLOROETHYL)- see TJK100
OXIRANE, 2,2′-(1,2-ETHANEDIYLBIS(OXYMETHYLENE))BIS-(9CI) see EEA600
OXIRANE, (ETHOXYMETHYL)-(9CI) see EKM200
OXIRANE, (((2-ETHYLHEXYL)OXY)METHYL)-(9CI) see GGY100
OXIRANEMETHANAMINIUM, N,N,N-TRIMETHYL-, CHLORIDE (9CI) see GGY200
OXIRANEMETHANOL NITRATE see ECI600
OXIRANE, ((1-METHYLETHOXY)METHYL)-(9CI) see IPD000
OXIRANE ((METHYLPHENOXY)METHYL) (9CI) see TGZ100
OXIRANE, ((4-NITROPHENOXY)METHYL)-(9CI) see NIN050
OXIRANE, 2,2′-(2,5,8,11-TETRAOXADODECANE-1,12-DIYL)BIS- (9CI) see TJQ333
7-OXIRANYLBENZ(a)ANTHRACENE see ONC000
6-OXIRANYLBENZO(a)PYRENE see ONE000
OXIRANYLMETHYL ESTER of OCTADECANOIC ACID see SLK500
OXIRANYLMETHYL ESTER of 9-OCTADECENOIC ACID see ECJ000
2-(OXIRANYLMETHYL)-1H-ISOINDOLE-1,3(2H)-DIONE see ECK000
N-(OXIRANYLMETHYL)-N-PHENYL-OXIRANEMETHANAMINE (9CI) see DKM120
3-OXIRANYL-7-OXABICYCLO(4.1.0)HEPTENE see VOA000
1-OXIRANYLPYRENE see ONG000
OXISURAN see MPL500
OXITETRACYCLIN see HOH500
OXITOL see EES350
OXITOSONA-50 see PAN100
OXITROPIUM BROMIDE see ONI000
OXIURAN see AOR500
OXIVOR see CNK559
OXLOPAR see HOI000
OXO see TAB750
3′-(3-OXO-7-α-ACETYLTHIO-17-β-HYDROXYANDROST-4-EN-17-β-YL)PROPIONIC ACID LACTONE see AFJ500
7-OXOBENZ(de)ANTHRACENE see BBI250
2-OXOBENZIMIDAZOLE see ONI100
5-OXO-5H-BENZO(E)ISOCHROMENO(4,3-b)INDOLE see DLY800
2-OXO-1,2-BENZOPYRAN see CNV000
γ-OXO(1,1′-BIPHENYL)-4-BUTANOIC ACID see BGL250
OXOBIS(SULFATO(2-)-O)-ZIRCONATE(2-), DISODIUM (9CI) see ZTS100
OXOBOI see OOG000
2-OXOBORNANE see CBA750
3-OXO-BUTANOIC ACID BUTYL ESTER see BPV250
3-OXOBUTANOIC ACID ETHYL ESTER see EFS000
3-OXOBUTANOIC ACID METHYL ESTER see MFX250
1-(1-OXOBUTYL)AZIRIDINE see BSY000
4-(3-OXOBUTYL)-1,2-DIPHENYL-3,5-PYRAZOLIDINEDIONE see KGK000
γ-OXO-α-BUTYLENE see BOY000
p-(3-OXOBUTYL)PHENYL ACETATE see AAR500
trans-pi-OXOCAMPHOR see VSF400
19-OXO-CARDOGENEN-(20:22)-TRIOL(3-β,5,14) (GERMAN) see SMM500
4-OXO-2-(β-CHLOROETHYL)-2,3-DIHYDROBENZO-1,3-OXAZINE see CDS250

3-OXO-N-(2-CHLOROPHENYLBUTANAMIDE) see AAY600
7-OXOCHOLESTEROL see ONO000
2-OXOCHROMAN see HHR500
((4-OXO-2,5-CYCLOHEXADIEN-1-YLIDENE)AMINO)GUANIDINE THIOSEMICAR-
 BAZONE see AHI875
OXODIACETIC ACID see ONQ100
N-(6-OXO-6H-DIBENZO(b,d)PYRAN-1-YL)ACETAMIDE see BCH250
5-OXO-5,13-DIHYDROBENZO(E)(2)BENZOPYRANO(4,3-b)INDOLE see DLY800
OXODIOCTYLSTANNANE see DVL400
OXODIPEROXOPYRIDINE CHROMIUM-N-OXIDE see ONU100
α-OXODIPHENYLMETHANE see BCS250
OXODIPHENYLSTANNANE, POLYMER see DWO600
OXODISILANE see ONW000
OXODOLIN see CLY600
1'-OXOESTRAGOLE see ONW100
9-OXO-2-FLUORENYLACETAMIDE see ABY250
N-(9-OXO-2-FLUORENYL)ACETAMIDE see ABY250
3-OXO-I-GULOFURANOLACTONE see ARN000
4-OXOHEPTANEDIOIC ACID, ISONICOTINOYL HYDRAZONE see ILG000
(17-β)17-((1-OXOHEPTYL)OXY)-ANDROST-4-EN-3-ONE see TBF750
2-OXOHEXAMETHYLENIMINE see CBF700
1-(5-OXOHEXYL)-3,7-DIMETHYLXANTHINE see PBU100
17-((1-OXOHEXYL)OXY)PREGN-4-ENE-3,20-DIONE see HNT500
1-(5-OXOHEXYL)THEOBROMINE see PBU100
(E)-7-OXO-3-β-HYDROXY-14-α-METHYL-8-β-PODOCARPANE-Δ¹³-α-ACETIC ACID-
 2-(DIMETHYLAMINO)ETHYL ESTER HYDROCHLORIDE see CCO675
4-OXO-4H-IMIDAZO(4,5-D)-v-TRIAZINE see ARY500
2-(3-OXO-1-INDANYLIDENE)-1,3-INDANDIONE see ONY000
2-OXO-3-ISOBUTYL-9,10-DIMETHOXY-1,3,4,6,7,11-β-HEXAHYDRO-2H-BENZO-
 QUIN OLIZINE see TBJ275
8-OXO-8H-ISOCHROMENO(4',3':4,5)PYRROLO(2,3-f)QUINOLINE see OOA000
p-(1-OXO-2-ISOINDOLINYL)-HYDRATROPIC ACID see IDA400
2-(p-(1-OXO-2-ISOINDOLINYL)PHENYL)-PROPIONIC ACID see IDA400
α-(4-(1-OXO-2-ISO-INDOLINYL)-PHENYL)-PROPIONIC ACID see IDA400
OXOLAMINA CLORIDRATO (ITALIAN) see OOE100
OXOLAMINE see OOC000
OXOLAMINE CITRATE see OOE000
OXOLAMINE HYDROCHLORIDE see OOE100
OXOLANE see TCR750
OXOLE see FPK000
OXOLINIC ACID see OOG000
OXOMEMAZINE see AFL750
OXOMETHANE see FMV000
6-OXO-3-METHOXY-N-METHYL-4,5-EPOXYMORPHINAN see OOI000
1-OXO-2-(p-((α-METHYL)CARBOXYMETHYL)PHENYL)ISOINDOLINE see IDA400
3-OXO-N-(2,4-METHYLPHENYL)BUTANAMIDE see OOI100
3-OXO-3H-NAPHTHO(2,1-b)PYRAN-2-CARBOXYLIC ACID ETHYL ESTER see
 OOI200
OXONIUM, TRIETHYL-, TETRAFLUOROBORATE(1-) see TJL600
(2R-(2-α,3-α(4E,7E)))-3-(1-OXO-4,7-NONADIENYL)OXIRANECARBOXAMIDE see
 ECE500
3-(1-OXO-4,7-NONADIENYL)OXIRANECARBOXAMIDE see ECE500
5-OXONONANE see NMZ000
(E)-12-OXO-10-OCTADECENOIC ACID, METHYL ESTER see MNF250
12-OXO-trans-10-OCTADECENOIC ACID, METHYL ESTER see MNF250
α-1-(OXOOCTADECYL)-ω-HYDROXYPOLY(OXY-1,2-ETHANEDIYL) see PJW750
9-OXO-8-OXATRICYCLO(5.3.1.0²·⁶)UNDECANE see OOK000
8-OXOPENTADECANE see HBO790
4-OXOPENTANOIC ACID see LFH000
OXOPHENARSINE see OOK100
OXOPHENARSINE HYDROCHLORIDE see ARL000
β-OXO-α-PHENYLBENZENEPROPANENITRILE see OOK200
((4-OXO-2-PHENYL-4H-1-BENZOPYRAN-7-YL)OXY)ACETIC ACID ETHYL ESTER
 see ELH600
3-OXO-N-PHENYLBUTANAMIDE see AAY000
(12-β(E,E))-12-((1-OXO-5-PHENYL-2,4-PENTADIENYL)OXY)-DAPHNETOXIN see
 MDJ250
(17-β)-17-(1-OXO-3-PHENYLPROPOXY)-ESTR-4-EN-3-ONE (9CI) see DYF450
N-(2-OXO-3-PIPERIDYL)PHTHALIMIDE see OOM300
5-OXO-I-PROLYL-I-HISTIDYL-I-PROLINAMIDE TARTRATE see POE050
2-OXOPROPANAL see PQC000
OXOPROPANEDINITRILE see OOM400
OXOPROPANEDINITRILE (CARBONYL DICYANIDE) see MDL250
2-OXOPROPANOIC ACID see PQC100
2-OXOPROPIONIC ACID see PQC100
17-(1-OXOPROPOXY)-(17-β)-ANDROST-4-EN-3-ONE see TBG000
p-(2-OXOPROPOXY)BENZENEARSONIC ACID see OOO000
3-(2-OXOPROPYL)-2-PENTYLCYCLOPENTANONE see OOO100
1-OXOPROPYLPROPYLNITROSAMIN (GERMAN) see NLN500
1-OXOPROPYLPROPYLNITROSAMINE see NLN500
2-OXO-PROPYL-PROPYLNITROSAMINE see ORS000
(2-OXOPROPYL)PROPYLNITROSOAMINE see ORS000
2-OXOPYRIDINE see OOO200
2-OXOPYRROLIDINE see PPT500
2-OXO-1-PYRROLIDINEACETAMIDE see NNE400

2-OXO-PYRROLIDINE ACETAMIDE see NNE400
5-OXO-2-PYRROLIDINECARBOXYLIC ACID MAGNESIUM SALT (2:1) see
 PAN775
2'-OXOPYRROLIDINO-1-PYRROLIDINO-4-BUTYNE see OOY000
2-OXOPYRROLIDIN-1-YLACETAMIDE see NNE400
OXOSILANE see OOS000
2-OXOSPARTEINE see LIQ000
OXOSULFATOVANADIUM PENTAHYDRATE see VEZ100
OXOSUMITHION see PHD750
4-OXO-2,2,6,6-TETRAMETHYLPIPERIDINE see TDT770
1-4-OXO-2-THIAZOLIDINEHEXANOIC ACID see CCI500
4-OXO-2-THIONOTHIAZOLIDINE see RGZ550
OXOTREMORIN see OOY000
OXOTREMORINE see OOY000
8-OXOTRICYCLO(5.2.1.0²·⁶)DECANE see OPC000
2,2'-(2-OXO-TRIMETHYLENE)BIS(1,1-DIMETHYLPYRROLIDINIUM) DIBENZENE-
 SUFLONATE see COG500
6-OXOUNDECANE see ULA000
4-OXOVALERIC ACID see LFH000
β-(γ-OXOVALEROYL)FURAN see IGF300
(4-OXOVALERYLOXY)TRIPHENYLSTANNANE see TMW250
22-OXOVINCALEUKOBLASTINE see LEY000
OXPENTIFYLLINE see PBU100
OXPRENOLOL see AGW000, CNR500
OXPRENOLOL HYDROCHLORIDE see THK750
OXSORALEN see XDJ000
OXTRIMETHYLLINE see CMF500
OXTRIPHYLLINE see CMF500
OXUCIDE see PIJ500
OXURASIN see HEP000
OXY-5 see BDS000
OXY-10 see BDS000
p-OXYACETOPHENONE see HIO000
OXY ACID BLACK BASE see PFU500
β-OXYAETHYL-MORPHOLIN (GERMAN) see MRQ500
OXYAETHYLTHEOPHYLLIN (GERMAN) see HLC000
OXYAMINE see CFA750
3-OXYANTHRANILIC ACID see AKE750
p-OXYBENZALDEHYDE see FOF000
OXYBENZENE see PDN750
p-OXYBENZOESAEUREAETHYLESTER (GERMAN) see HJL000
p-OXYBENZOESAEUREHEPTYLESTER see HBO650
p-OXYBENZOESAEUREMETHYLESTER (GERMAN) see HJL500
p-OXYBENZOESAEUREPROPYLESTER (GERMAN) see HNU500
p-OXYBENZOESAURE (GERMAN) see SAI500
OXYBENZONE see MES000
OXYBENZOPYRIDINE see QPA000
2,2'-OXYBISACETIC ACID see ONQ100
OXYBISACETIC ACID see ONQ100
OXYBIS(4-AMINOBENZENE) see OPM000
OXYBIS((3-AMINOPROPYL)DIMETHYLSILANE) see OPC100
4,4'-OXYBISANILINE see OPM000
p,p'-OXYBIS(ANILINE) see OPM000
4,4'-OXYBISBENZENAMINE see OPM000
p,p'-OXYBISBENZENE DISULFONYLHYDRAZIDE see OPE000
OXYBISBENZENESULFONIC ACID DIHYDRAZIDE see OPE000
OXYBIS(BENZENESULFONYL HYDRAZIDE) see OPE000
p,p'-OXYBIS(BENZENESULFONYL) HYDRAZINE see OPE000
1,1'-OXYBIS(BUTANE) see BRH750
2,2'-OXYBISBUTANE see BRH760, OPE030
4,4'-OXYBIS(2-CHLOROANILINE) see BGT000
4,4'-OXYBIS(2-CHLORO-BENZENAMINE) see BGT000
1,1'-OXYBIS(4-CHLOROBUTANE) see OPE040
1,1'-OXYBIS(1-CHLOROETHANE) see BID000
1,1'-OXYBIS(2-CHLORO)ETHANE see DFJ050
1,1'-OXYBIS(2-(2-CHLOROETHYL)THIOETHANE see DFK200
OXYBIS(CHLOROMETHANE) see BIK000
2,2'-OXYBIS(1-CHLOROPROPANE) see BII250
OXYBIS(DIBUTYL(2,4,5-TRICHLOROPHENOXY)TIN) see OPE100
5,5'-OXYBIS(3,4-DICHLORO-2(5H))-FURANONE see MRV000
4,4'-OXYBIS(2,3-DICHLORO-4-HYDROXYCROTONIC ACID)-DI-γ-LACTONE see
 MRV000
OXYBIS(N,N-DIMETHYLACETAMIDETRIPHENYLSTI- BONIUM) DIPERCHLORATE
 see OPG000
1,1'-OXYBISETHANE see EJU000
1,1'-(OXYBIS(2,1-ETHANEDIYLOXY))BISBUTANE see DDW200
3,3'-(OXYBIS(2,1-ETHANEDIYLOXY))BIS-1-PROPANAMINE see DJD800
(OXYBIS(2,1-ETHANEDIYLOXY))BIS-PROPANAMINE see EBV100
2,2'-OXYBISETHANOL see DJD600
1,1'-OXYBISETHENE see VOP000
2,2'-(OXYBIS(ETHYLENEOXY))DIETHANOL see TCE250
1,1'-OXYBIS(2-ETHYLHEXANE) see DJK600
1,1'-OXYBISHEPTANE see HBO000
1,1'-OXYBISHEXANE see DKO800
OXYBISMETHANE see MJW500

(1,1'-(OXYBIS(METHYLENE))BIS(4-(1,1-DIMETHYLETHYL))-PYRIDINIUM, DICHLO-
RIDE (9CI) see SAB800
2,2'-(OXYBIS(METHYLENE))BISFURAN see FPX025
1,1'-OXYBIS(METHYLENE))BIS(4-(HYDROXYIMINO)METHYL)PYRIDINIUM DI-
CHLORIDE see BGS250
2,2'-OXYBIS-6-OXABICYCLO-(3.1.0)HEXANE see BJN250
2,2'-OXYBIS(6-OXABICYCLO(3.1.0)HEXANE) mixed with 2,2-BIS(p-(2,3-EPOXY-
PROPOXY)PHENYL)PROPANE see OPI200
1,1'-OXYBIS(2,3,4,5,6-PENTABROMOBENZENE (9CI) see PAU500
1,1'-OXYBISPENTANE see PBX000
10-10' OXYBISPHENOXYARSINE see OMY850
1,1'-OXYBISPROPANE see PNM000
3,3'-OXYBIS(1-PROPENE) see DBK000
OXYBIS(TRIBUTYLTIN) see BLL750
OXYBIS(TRIMETHYLSILANE) see HEE000
OXYBULAN see PFR325
OXYBUPROCAINE HYDROCHLORIDE see OPI300
OXYBUTANAL see AAH750
β-OXYBUTTERSAEURE-p-PHENETIDID see HJS850
OXYBUTYNIN CHLORIDE see OPK000
OXYBUTYRIC ALDEHYDE see AAH750
OXYCAINE see DHO600
OXYCARBON SULFIDE see CCC000
OXYCARBOPHOS see OPK250
OXYCARBOXIN see DLV200
OXYCARBOXINE see DLV200
OXY CHEK 114 see MJO500
o-OXYCHINOLIN (GERMAN) see QPA000
OXYCHINOLIN see QPA000
OXYCHLORID FOSFORECNY see PHQ800
OXYCHLORUE DE CUIVRE see CNK559
OXYCHLORURE CHROMIQUE (FRENCH) see CML125
OXYCHOLIN see DAL000
OXYCIL see SFS000
OXYCINCHOPHEN see OPK300
OXYCLOR see CNK559
OXYCLOZANID see DMZ000
OXYCLOZANIDE see DMZ000
OXYCODEINONE see PCG500
OXYCODONE HYDROCHLORIDE see DLX400
OXYCODON HYDROCHLORIDE see DLX400
OXYCOLOR see AOR500
OXYCON see DLX400
OXYCUR see CNK559
OXY DBCP see DDL800
OXYDE d'ALLYLE et de GLYCIDYLE (FRENCH) see AGH150
OXYDE de BARYUM (FRENCH) see BAO000
OXYDE de CALCIUM (FRENCH) see CAU500
OXYDE de CARBONE (FRENCH) see CBW750
OXYDE de CHLORETHYLE (FRENCH) see DFJ050
OXYDE d'ETHYLE (FRENCH) see EJU000
OXYDE de MERCURE (FRENCH) see MCT500
OXYDE de MESITYLE (FRENCH) see MDJ750
OXYDEMETONMETHYL see DAP000
OXYDEMETON-METILE (ITALIAN) see DAP000
OXYDE NITRIQUE (FRENCH) see NEG100
OXYDEPROFOS see DSK600
OXYDE de PROPYLENE (FRENCH) see PNL600
OXYDE de TRIBUTYLETAIN see BLL750
2,2'-OXYDIACETIC ACID see ONQ100
OXYDIACETIC ACID see ONQ100
OXYDIACETIC ACID, DISODIUM SALT see SIZ000
OXYDIAMINE BROWN 3GN see CMO810
4,4'-OXYDIANILINE see OPM000
p,p'-OXYDIANILINE see OPM000
OXYDIANILINE see OPM000
OXYDIAZEPAM see CFY750
OXYDIAZOL see BGD250
1,1'-OXYDI-4-CHLOROBUTANE see OPE040
2,2'-OXYDIETHANOL see DJD600
OXYDIETHANOLIC ACID see ONQ100
N-OXYDIETHYL-2-BENZOTHIAZOLSULFENAMID (CZECH) see BDF750
OXYDIETHYLENE ACRYLATE see ADT250
N-(OXYDIETHYLENE)BENZOTHIAZOLE-2-SULFENAMIDE see BDG000
OXYDIETHYLENE BIS(CHLOROFORMATE) see OPO000
OXYDIETHYLENE CHLOROFORMATE see OPO000
OXYDIETHYLENE DIACRYLATE see ADT250
OXYDIETHYLENEDICARBONIC ACID DIALLYL ESTER see AGD250
3,3'-(2,2'-OXYDIETHYLENEDIOXYBISACETAMIDO)BIS(2,4,6-TRIIODOBENZOIC
ACID) see IGD100
(N,N'-OXYDIETHYLENEDIOXYDIETHYLENE)BIS(TRIETHYLAMMONIUM IODIDE)
see OPQ000
N-OXYDIETHYLENE THIOCARBAMYL-N-OXYDIETHYLENE SULFENAMIDE see
OPQ100
OXYDIFORMIC ACID DIETHYL ESTER see DIZ100

β-(4-OXY-3,5-DIJOD-PHENYL)-α-PHENYL-PROPIONSAEURE (GERMAN) see
PDM750
OXYDIMETHANOL DINITRATE see OPQ200
1,1'-OXYDIMETHYLENE BIS(4-tert-BUTYLPYRIDINIUM CHLORIDE) see SAB800
3,3'-(OXYDIMETHYLENEBIS(CARBONYLIMINO))BIS(2,4,6-TRIIODOBENZOIC
ACID DISODIUM SALT) see BGB350
1,1'-(OXYDIMETHYLENE)BIS(4-FORMYLPYRIDINIUM)DICHLORIDE DIOXIME see
BGS250
1,1'-(OXYDIMETHYLENE)BIS(4-FORMYLPYRIDINIUM) DINITRATE DIOXIME see
OPY000
1,1'-(OXYDIMETHYLENE)BIS(4-FORMYLPYRIDINIUM) DIOXIME DICHLORIDE
see BGS250
OXYDIMETHYLQUINAZINE see AQN000
4,4'-OXYDIPHENOL see OQI000
p,p'-OXYDIPHENOL see OQI000
4,4'-OXYDIPHENYLAMINE see OPM000
p-OXYDIPHENYLAMINE see AOT000
OXYDI-p-PHENYLENEDIAMINE see OPM000
1,1'-OXYDI-2-PROPANOL see OQM000
3,3'-OXYDI-1-PROPANOL DIBENZOATE see DWS800
OXYDIPROPANOL PHOSPHITE (3:1) see OQO000
3,3'-OXYDIPROPIONITRILE see OQQ000
β,β'-OXYDIPROPIONITRILE see OQQ000
OXYDISULFOTON see OQS000
N-OXYD-LOST see CFA500, CFA750
N-OXYD-MUSTARD see CFA500
OXYDOL see HIB000
OXYDRENE see DBA800
OXYEPHEDRINE see HKH500
OXYETHYLATEED TERTIARY OCTYL-PHENOL-FORMALDEHYDE POLYMER
see TDN750
OXYETHYLENATED DODECYL ALCOHOL see DXY000
OXYETHYLENATED HEXADECYL ALCOHOL see PJT300
OXYETHYLIDENEDIPHOSPHONIC ACID see HKS780
1-(β-OXYETHYL)-2-METHYL-5-NITROIMIDAZOLE see MMN250
OXYETHYLTHEOPHYLLINE see HLC000
OXYFED see DBA800
OXYFENAMATE see HNJ000
OXYFENON see ORQ000
OXYFLAVIL see ELH600
OXYFUME 12 see EJN500
OXYFUME 20 see EJO000
OXYFUME 30 see EJO000
OXYFUME see EJN500
OXYFURADENE see NDY000
β-OXY-GABA see AKF375
OXYGEN see OQW000
OXYGEN DIFLUORIDE see ORA000
OXYGEN FLUORIDE see ORA000
OXYGERON see VLF000
OXYGEN, compressed (UN 1072) (DOT) see OQW000
OXYGEN, refrigerated liquid (cryogenic liquid) (UN 1073) (DOT) see OQW000
OXYHYDROCHINON (GERMAN) see BBU250
OXYHYDROQUINONE see BBU250
OXYJECT 100 see HOI000
OXYKODAL see DLX400
OXYKON see DLX400
OXYLAN see DKQ000, PFR325
OXYLITE see BDS000
OXYMAG see MAH500
OXY MBC see SFS500
OXYMEMAZINE see AFL750
OXYMETAZOLINE see ORA100
OXYMETAZOLINE CHLORIDE see AEX000
OXYMETAZOLINE HYDROCHLORIDE see AEX000
OXYMETEBANOL see ORE000
OXYMETHALONE see PAN100
OXYMETHAZOLINE see ORA100
OXYMETHEBANOL see ORE000
OXYMETHENOLONE see PAN100
OXYMETHOLONE see PAN100
OXY-2-METHOXY-3-BENZALDEHYDE (FRENCH) see VFP000
OXY-3-METHOXY-4 BENZALDEHYDE (FRENCH) see FNM000
OXYMETHUREA see DTG700
OXYMETHYLENE see FMV000
5-OXYMETHYLFURFUROLE see ORG000
OXYMETOZOLINE see ORA100
OXYMORPHINONE HYDROCHLORIDE see ORG100
OXYMORPHONE HYDROCHLORIDE see ORG100
OXYMURIATE OF POTASH see PLA250
OXYMYCIN see CQH000, HOH500
OXYMYKOIN see HOH500
β-OXYNAPHTHOIC ACID see HMX500
OXYNEURINE see GHA050
OXY-NH2 see CFA500

OXYOZYL see AOR500
OXYPAAT see HEP000
OXYPANAMINE see ORI300
OXYPANGAM see DNM400
OXYPARATHION see NIM500
OXYPERTIN see ECW600
OXYPERTINE see ECW600
OXYPHENALON see RBU000
OXYPHENAMATE see HNJ000
OXYPHENBUTAZONE see HNI500
OXYPHENIC ACID see CCP850
OXYPHENON see ORQ000
OXYPHENONIUM see ORQ000
OXYPHENONIUM BROMIDE see ORQ000
OXYPHENYLBUTAZONE see HNI500
1-(p-OXYPHENYL)-2-METHYLAMINOPROPAN (GERMAN) see FMS875
α-(p-OXYPHENYL)-β-METHYL AMINOPROPANE see FMS875
3-OXYPHENYL TRIMETHYLAMMONIUM IODIDE see HNN500
OXYPHIONFOS see DSK600
OXYPHYLLINE see HLC000
OXYPHYLLINE (AMIDO) see HLC000
OXYPROCAIN see DHO600
OXYPROCAINE see DHO600
p-OXYPROPIOPHENONE see ELL500
2-OXY-1,3-PROPYLENEDIAMINE-N,N,N′,N′-TETRAMETHYLENEPHOSPHONIC ACID see DYE550
β-OXYPROPYLPROPYLNITROSAMINE see ORS000
γ-(γ-OXYPROPYL)-THEOBROMIN (GERMAN) see HNZ000
γ-OXYPROPYLTHEOBROMIN (GERMAN) see HNZ000
β-OXYPROPYLTHEOBROMINE see HNY500
β-OXYPROPYLTHEOPHYLLIN see HOA000
OXYPROPYLTHEOPHYLLINE see HOA000
OXYPSORALEN see XDJ000
OXYPYRRONIUM see IBP200
OXYPYRRONIUM BROMIDE see IBP200
8-OXYQUINOLINE see QPA000
OXYQUINOLINE see QPA000
OXYQUINOLINE SULFATE see QPS000
OXYQUINOLINOLEATE de CUIVRE (FRENCH) see BLC250
5-OXYRESORCINOL see PGR000
OXYRITIN see RSU000
OXYSTIN see ORU500
OXYSULFATOVANADIUM see VEZ000
OXYTERRACIN see HOH500
OXYTERRACINE see HOH500
OXYTERRACYNE see HOH500
OXYTETRACYCLINE see HOH500
OXYTETRACYCLINE AMPHOTERIC see HOH500
OXYTETRACYCLINE HYDROCHLORIDE see HOI000
OXYTHANE see NCM700
OXYTHEONYL see HLC000
OXYTHIAMIN see ORS200
OXYTHIAMINE see ORS200
OXYTHIOQUINOX see ORU000
OXYTOCIC see EDB500
OXYTOCIN see ORU500
OXYTOL ACETATE see EES400
m-OXYTOLUENE see CNW750
o-OXYTOLUENE see CNX000
p-OXYTOLUENE see CNX250
OXYTRIL see HKB500
OXYTRIL M see DDP000
OXYTROPIUM BROMIDE see ONI000
OXYURANUS SCUTELLATUS (AUSTRALIA) VENOM see ARV500
OXYUREA see HOO500, TGN250
OXY WASH see BDS000
OXYZINE see PIJ500
OXYZIN (TABL.) see HEP000
OZAGREL see SHV100
OZIDE see ZKA000
OZLO see ZKA000
OZOCERITE see ORU900
OZOKERITE see ORU900
OZON (POLISH) see ORW000
OZONE see ORW000
OZONE mixed with NITROGEN OXIDES (53%:47%) see ORY000
OZONIDES see ORY499
OZP 9 see DXB450


P07 see PKO500
P 12 see MNV250
P-25 see SLJ000
P-33 see CBT750
P-40 see DXG000

P 48 see CMV400
P-50 see AIV500
P 55 see VJZ000
P68 see CBT750
134 P see MLK800
P 1-60 see LED100
P-165 see ASA500
P 237 see BTA000
P 241 see PEO750
P 253 see LJR000
P-267 see BPR500
P 271 see PAB000
P 284 see DWF000
P 301 see HNJ000
P-314 see DHR000
P-329 see DPU800
P 391 see MOQ250
P-398 see ACX500
Ph. 458 see SCA525
501 P see AOO800
P 527 see MOE250
P 652 see PDU250
P 887 see ILE000
P 1011 see DGE200
P-1108 see DVJ400
P 1133 see BET000
P 1134 see COS500
P 1142 see PDN000
P1250 see CBT750
P 1297 see BEQ500
P 1393 see BDE250
P 1488 see BRU750
P1496 see RBF100
P 1531 see PFC750
P-2292 see ADA000
P 2647 see BCL250
P-5048 see DRT200
P-5307 see PDY870
6020P see PJS750
P-7138 see NDY400
P 71-0129 see FAO100
P 286 (contrast medium) see IGD200
P 11H see ADW200
Pt-05 see DFJ000
Pt-09 see TBO776
Pt-93 see PLW200
PA 93 see SMB000
PA 94 see CQH000
PA 130 see PJS750
PA 144 see MQW750
PA 190 see PJS750
PA 520 see PJS750
PA 560 see PJS750
PA 6 (polymer) see PJY500
PA see PAP750, PEG500
PAA-25 see ADW200
PAA see BBL500, PJK350
PAA-701 DIHYDROCHLORIDE see BFX125
PA'AILA (HAWAII) see CCP000
PA 114B see VRA700
PAB see SEO500
PABA see AIH600
PABANOL see AIH600
PABAVJT see SEO500
PABESTROL see DKA600, DKB000
PABIALGIN see IGI000
PABRACORT see HHQ800
PABS see SNM500
PABYRIN see HAQ500
PACAMINE HYDROCHLORIDE see NCL300
PACATAL see MOQ250
PACATAL BASE see MOQ250
PACEMO see HIM000
PACETYN see EHP000
PACHAK see CNT350
6-β,7-α,9-α,11-α-PACHYCARPINE see SKX500
PACHYCARPINE see PAB250
PACHYRHIZUS EROSUS see YAG000
PACIENCIA see DAB700
PACIENX see CFZ000
PACIFAN see NBU000
PACINOL see TJW500
PACITANE see BBV000
PACITRAN see DCK759, MQR200
PACM 20 see MJQ260

PACTAMYCIN see PAB500
PAD 522 see PJS750
PADAN see BHL750
PADARYL see BQW000
PADISAL see MRW000
PADOPHENE see PDP250
PADRIN see PAB750
PAE see INN500
PAEONIA MOUTAN see PAC000
PAEONIFLORIN see PAC000
PAEONOL see PAC250
PAEONY ROOT see PAC000
PAFE see PKE550
PAG see NNR400
PAGANO-COR see DHS200
PAGITANE HYDROCHLORIDE see CQH500
PAGODA TREE see NBR800
PAIDAZOLO see AIF000
PA'INA (HAWAII) see JBS100
PAIN de COULEUVRE (CANADA) see BAF325
PAINTERS NAPHTHA see PCT250
PAINT WHITE see BKW100
PAISAJE (PUERTO RICO) see PGQ285
PAISLEY POLYMER see PMP500
PAKA (HAWAII) see TGI100
PAKHTARAN see DUK800
PAL see PEC750
PALA see PGY750
PALACET SCARLET B see CMP080
PALACET YELLOW GN see AAQ250
PALACOS see PKB500
PALACRIN see CFU750
PALAFER see FBJ100
PALAFUGE see PFR325
PALANIL BLUE 7G see DMM400
PALANIL BLUE R see CMP070
PALANILCARRIER A see PAC500
PALANIL PINK RF see AKO350
PALANIL RED BF see AKI750
PALANIL VIOLET 3B see DBY700
PALANIL VIOLET 6R see DBX000
PALANIL YELLOW G see AAQ250
PALANIL YELLOW 5R see CMP090
PALANIL YELLOW 5RX see CMP090
PALANTHRENE BLUE BC see DFN425
PALANTHRENE BLUE BCA see DFN425
PALANTHRENE BLUE GPT see IBV050
PALANTHRENE BLUE GPZ see IBV050
PALANTHRENE BLUE RPT see IBV050
PALANTHRENE BLUE RPZ see IBV050
PALANTHRENE BLUE RSN see IBV050
PALANTHRENE BRILLIANT BLUE R see IBV050
PALANTHRENE BRILLIANT ORANGE GR see CMU820
PALANTHRENE BROWN BR see CMU770
PALANTHRENE GOLDEN YELLOW see DCZ000
PALANTHRENE NAVY BLUE G see CMU500
PALANTHRENE NAVY BLUE RB see CMU750
PALANTHRENE OLIVE R see DUP100
PALANTHRENE ORANGE R see CMU815
PALANTHRENE PRINTING BLUE KRS see IBV050
PALANTHRENE RED G 2B see CMU825
PALAPENT see NBU000
PALASH SEED EXTRACT see BOV800
PALATINOL A see DJX000
PALATINOL AH see DVL700
PALATINOL BB see BEC500
PALATINOL C see DEH200
PALATINOL IC see DNJ400
PALATINOL M see DTR200
PALATONE see MAO350
PALAVALE see EAE000
PALE ORANGE CHROME see LCS000
PALESTROL see DKA600
PALETA de PINTOR (PUERTO RICO) see CAL125
PALFADONNA see AFJ400
PALFIUM see AFJ400
PALIN see PIZ000
PALINUM see TDA500
PALIUROSIDE see RSU000
PALLADATE(2-), TETRACHLORO-, DIAMMONIUM (8CI) see ANE750
PALLADATE(2-), TETRACHLORO-, DIAMMONIUM, (SP-4-1)-(9CI) see ANE750
PALLADIUM see PAD250
PALLADIUM(2+) ACETATE see PAD300
PALLADIUM(II) ACETATE see PAD300
PALLADIUM ACETATE (Pd(OAc)₂) (7CI) see PAD300

PALLADIUM(2+) CHLORIDE see PAD500
PALLADIUM CHLORIDE see PAD500
PALLADIUM CHLORIDE, DIHYDRIDE see PAD750
PALLADIUM DIACETATE see PAD300
PALLADOUS ACETATE see PAD300
PALLADOUS CHLORIDE see PAD500
PALLETHRINE see AFR250
PALLICID see ABX500
PALMA CHRISTI (HAITI) see CCP000
PALMAROSA OIL see PAE000
PALMATINE HYDROXIDE see PAE100
PALMATINIUM HYDROXIDE see PAE100
PALM BUTTER see PAE500
PALMITA see PGA750
PALMITA de JARDIN (PUERTO RICO) see CNH789
PALMITIC ACID see PAE250
PALMITIC ACID, 2-ETHYLHEXYL ESTER see OFE100
PALMITIC ACID, HEXADECYL ESTER see HCP700
PALMITYL ACETATE see HCP100
PALMITYL ALCOHOL see HCP000
PALMITYLAMINE see HCO500
PALMITYLDIMETHYLAMINE see HCP525
PALMITYL PALMITATE see HCP700
PALMO-ARA-C see AQS875
PALM OIL see PAE500
PALO GUACO (CUBA) see CDH125
PALOHEX see HFG550
PALO de NUEZ (PUERTO RICO) see TOA275
PALOPAUSE see ECU750
PALOSEIN see OJM400
PAL-P see PII100
PALTET see TBX250
PALUDRINE see CKB250
PALUDRINE DIHYDROCHLORIDE see PAE600
PALUDRINE HYDROCHLORIDE see CKB500
PALUSIL HYDROCHLORIDE see CKB500
PALYGORSCITE see PAE750
PALYGORSKIT (GERMAN) see PAE750
PALYTHOATOXIN see PAE875
PALYTOXIN see PAF000
PA 11M see ADW200
PAM (CZECH) see POS750
l-PAM see PED750
PAM see ALW500
PAMA see PGV250
PAMACEL YELLOW G-3 see AAQ250
PAMACYL see AMM250
PAMAQUIN see RHZ000
PAMAZONE see CJR909
2-PAM CHLORIDE see FNZ000
PAMEION see PAH250
PAMINE see SBH500
PAMINE BROMIDE see SBH500
2-PAM IODIDE see POS750
PAMISAN see ABU500
PAMISYL see AMM250
PAMISYL SODIUM see SEP000
2-PAM METHANESULFONATE see PLX250
PAMN see PNR250
PAMOIC ACID see PAF100
PAMOLYN see OHU000
PAMOSOL 2 FORTE see EIR000
PAMOVIN see PQC500
PAN see BFW120, PCL750
PANA see PFT250
PANACAINE see PDU250
PANACEF see CCR850
PANACELAN see POC500
PANACID see PAF250
PANACIDE see MJM500
PANADOL see HIM000
PANADON see PAG200
PANALDINE see TGA525
PANAM see CBM750
PANAMIDIN DIHYDROCHLORIDE see DBM400
PANAX see GEQ400
PANAX GINSENG, ROOT EXTRACT see GEQ425
PANAX SAPONIN E see PAF450
PANAZONE see PAF500
PANCAL see CAU750
PANCALMA see MQU750
PANCID see SNN500
PANCIL see OFE000
PANCODINE see DLX400
PANCORAL see BQJ500

PANCREATIC BASIC TRYPSIN INHIBITOR see PAF550
PANCREATIC DEOXYRIBONUCLEASE see EDK650, PAF575
PANCREATIC DORNASE see EDK650, PAF575
PANCREATIC TRYPSIN INHIBITOR see PAF550
PANCREATIC TRYPSIN INHIBITOR (KUNITZ) see PAF550
PANCREATIN see PAF600
PANCREATOPEPTIDASE E see EAG875
PANCURONIUM BROMIDE see PAF625
PANCURONIUM BROMIDE HYDRATE see PAF630
PANCURONIUM DIBROMIDE HYDRATE see PAF630
PANDEX see PKM250
PANDIGAL see LAU400
PANDRINOX see MLF250
PANDUROL see BHD250
PANESTIN see TBG000
PANETS see HIM000
PANEX see HIM000
PANFLAVIN see DBX400, XAK000
PANFORMIN see BQL000
PANFUNGOL see KFK100
PANFURAN see FPF000
PANFURAN-S see BKH500
PANGUL see TEH500
PANHIBIN see SKS600
PANIMYCIN see PAG050
PANINI-'AWA'AWA (HAWAII) see AGV875
PANITHAL see SAH000
PANLANAT see LAU400
PANMYCIN see TBX000
PANMYCIN HYDROCHLORIDE see TBX250
PANO-DRENCH 4 see MLF250
PANODRIN A-13 see MLF250
PANOFEN see HIM000
PANOGEN 15 see MLF250
PANOGEN 43 see MLF250
PANOGEN see MEO750, MLF250
PANOGEN M see MEO750
PANOGEN METOX see MEO750
PANOGEN PX see MLF250
PANOGEN TURF FUNGICIDE see MLF250
PANOGEN TURF SPRAY see MLF250
PANOLID see DWE800
PANORAL see CCR850, PAG075
PANORAL HYDRATE see CCR850
PANORAM 75 see TFS350
PANORAM D-31 see DHB400
PANOSINE see MMD500
PANOSPRAY 30 see MLF250
PANOXYL see BDS000
PANPARNIT see PET250
PANRONE see FNF000
PANSOIL see EFK000
PANTALGINE see DAM700
PANTAS see HKR500
PANTASOTE R 873 see PKQ059
PANTELMIN see MHL000
d-PANTETHINE see PAG150
PANTETHINE see PAG150
PANTETINA see PAG150
PANTHELINE see HKR500
d(+)-PANTHENOL (FCC) see PAG200
d-PANTHENOL see PAG200
PANTHENOL see PAG200
PANTHER CREEK BENTONITE see BAV750
PANTHESIN see LEU000
PANTHESINE see LEU000
PANTHION see PAK000
PANTHODERM see PAG200
PANTHOJECT see CAU750
PANTHOLIC-L see IDE000
PANTHOLIN see CAU750
PANTOBROMINO see HNY500
PANTOCAINE see BQA010
PANTOCAINE HYDROCHLORIDE see TBN000
PANTOCID see HAF000
PANTOCRIN see PAG225
PANTOGAM see CAT125
PANTOL see PAG200
PANTOLAX see HLC500
PANTOMICINA see EDH500
PANTOMIN see PAG150
PANTONSILETTEN see DBX400
PANTOPAQUE see ELQ500
PANTOPRIM see TKX000
PANTOSEDIV see TEH500

PANTOSIN see PAG150
PANTOTHENATE CALCIUM see CAU750
PANTOTHENATE de ZINC (FRENCH) see ZKS000
PANTOTHENIC ACID, CALCIUM SALT see CAU750
(+)-PANTOTHENIC ACID, CALCIUM SALT see CAU750
PANTOTHENIC ACID, ZINC SALT see ZKS000
d-PANTOTHENOL see PAG200
PANTOTHENOL see PAG200
d(+)-PANTOTHENYL ALCOHOL see PAG200
d-PANTOTHENYL ALCOHOL see PAG200
PANTOTHENYL ALCOHOL see PAG200
PANTOVERNIL see CDP250
PANTOZOL 1 see COD000
PAN-TRANQUIL see MQU750
PANURIN see CFY000
PAN-W-29 see MOZ000
PANWARFIN see WAT220
PAP-1 see AGX000
PAP see ALT250, DRR400, PDC250
PAPAIN see PAG500
PAPANERINE see PAH000
PAPAO-APAKA (GUAM) see EAI600
PAPAO-ATOLONG (GUAM) see EAI600
PAPAVARINE CHLORHYDRATE see PAH250
PAPAVERINA (ITALIAN) see PAH000
PAPAVERIN CARBOXYLIC ACID, SODIUM SALT see PAG750
PAPAVERINE see PAH000
PAPAVERINE CHLOROHYDRATE see PAH250
PAPAVERINE HYDROCHLORIDE see PAH250
PAPAVERINE MONOHYDROCHLORIDE see PAH250
PAPAVERIN-HCL (GERMAN) see PAH250
PAPAYOTIN see PAG500
PAPER BLACK BA see AQP000
PAPER BLACK T see AQP000
PAPER BLUE R see AOR500
PAPER DEEP BLACK C see AQP000
PAPER RED HRR see FMU070
PAPER SCARLET 3BX see CMO875
PAPETHERINE HYDROCHLORIDE see PAH260
PAP H see PAH250
PAPI 20 see PKB100
PAPI 27 see PKB100
PAPI 135 see PKB100
PAPI 580 see PKB100
PAPI 901 see PKB100
PAPI see PKB100
PAPOOSE ROOT see BMA150
PAPP see AMC000
PAPTHION see DRR400
PARA see PEY500
PARAAMINODIPHENYL see AJS100
PARABAR 441 see BFW750
PARABEN see HJL500, HNU500
PARABENCIL see PFR325
PARABENCILFENOL see PFR325
PARABIS see MJM500
PARABROMODYLAMINE MALEATE see BNE750
PARABROMOTOLUENE see BOG255
PARABUTYRALDEHYDE see TNC100
PARACAIN see AIT250
PARACETALDEHYDE see PAI250
PARACETAMOLE see HIM000
PARACETAMOLO (ITALIAN) see HIM000
PARACETANOL see HIM000
PARACETOPHENETIDIN see ABG750
PARACHLORAMINE see HGC500
PARACHLOROCIDUM see DAD200
PARACHLOROPHENOL see CJK750
PARACIDE see DEP800
PARACODIN see DKW800
PARACODINE see DKW800
PARA-COL see PAJ000
PARACORT see PLZ000
PARACORTOL see PMA000
PARACOTOL see PMA000
PARACRESYL ACETATE see MNR250
PARACRESYL ISOBUTYRATE see THA250
PARA CRYSTALS see DEP800
d-PARACURARINE CHLORIDE see TOA000
PARACYMENE see CQI000
PARACYMOL see CQI000
PARADERIL see RNZ000
PARADI see DEP800
PARADIAZINE see POL490
PARADICHLORBENZOL (GERMAN) see DEP800

PARADICHLOROBENZENE see DEP800
PARADICHLOROBENZOL see DEP800
PARA-DIEN see DAL600
PARADIONE see PAH500
PARADISE TREE see CDM325
(0)-PARADOL see VFP100
PARADONE BLUE RC see DFN425
PARADONE BLUE RS see IBV050
PARADONE BRILLIANT BLUE R see IBV050
PARADONE BRILLIANT ORANGE GR see CMU820
PARADONE BRILLIANT ORANGE GR NEW see CMU820
PARADONE DARK BLUE RFW see VGP100
PARADONE GOLDEN YELLOW see DCZ000
PARADONE GREY M see CMU475
PARADONE GREY MG see CMU475
PARADONE NAVY BLUE G see CMU500
PARADONE OLIVE GREEN B see ALT000
PARADONE OLIVE R see DUP100
PARADONE PRINTING BLACK BLSF see CMU320
PARADONE PRINTING BLACK TLSF see CMU320
PARADONE PRINTING BLUE FRS see IBV050
PARADONE RED BROWN 2RD see CMU770
PARADORMALENE see TEO250
PARADOW see DEP800
PARADUST see PAK000
PARAESIN see BQH250
PARAFFIN see PAH750
PARAFFIN HYDROCARBONS see PAH770
PARAFFIN OIL see MQV750
PARAFFIN OILS (PETROLEUM), CATALYTIC DEWAXED HEAVY (9CI) see MQV868
PARAFFIN OILS (PETROLEUM), CATALYTIC DEWAXED LIGHT (9CI) see MQV870
PARAFFIN WAX see PAH750
PARAFFIN WAXES and HYDROCARBON WAXES, CHLORINATED (C23, 43% CHLORINE) see PAH810
PARAFFIN WAXES and HYDROCARBON WAXES, CHLORINATED (C12, 60% CHLORINE) see PAH800
PARAFFIN WAX FUME (ACGIH) see PAH750
PARAFLEX see CDQ750
PARAFLU see TKH750
PARAFORM see FMV000
PARAFORMALDEHYDE see PAI000
PARAFORMALDEHYDE-UREA POLYMER see UTU500
PARAFORMALDEHYDE-UREA RESIN see UTU500
PARAFORSN see PAI000
PARAFUCHSIN (GERMAN) see RMK020
PARAGLAS see PKB500
PARAGLYINE see MOS250
PARAHEXYL see HGK500
PARAHYDROXYBENZALDEHYDE see FOF000
PARAISO (MEXICO) see CDM325
PARAL see PAI250
PARALACTIC ACID see LAG010
PARALDEHYD (GERMAN) see PAI250
PARALDEHYDE see PAI250
PARALDEIDE (ITALIAN) see PAI250
PARALERGIN see ARP675
PARALEST see BBV000
PARALKAN see BBA500
PARALYTIC SHELLFISH POISON DIHYDROCHLORIDE see SBA500
PARA M see AKE250
PARA-MAGENTA see RMK020
PARAMAL see DBM800
PARAMANDELIC ACID see MAP000
PARAMAR see PAK000
PARAMEL DC see CNH125
PARAMENTHANE HYDROPEROXIDE see IQE000
PARAMETADIONE see PAH500
PARAMETHADIONE see PAH500
PARAMETHASONE 21-ACETATE see PAL600
PARAMETHAZONE ACETATE see PAL600
PARAMETHYL PHENOL see CNX250
PARAMIBE see DDS600
PARAMICINA see APP500
PARAMID see AKO500
PARAMIDIN see BQW825
PARAMIDINE see BQW825
PARAMID SUPRA see AKO500
PARAMINE BLACK B see AQP000
PARAMINE BLACK BH see CMN800
PARAMINE BLACK E see AQP000
PARAMINE BLUE 2B see CMO000
PARAMINE BLUE 3B see CMO250
PARAMINE FAST RED F see CMO870

PARAMINE FAST SCARLET 3B see CMO875
PARAMINE FAST VIOLET N see CMP000
PARAMINE GREEN B see CMO840
PARAMINE GREEN BN see CMO840
PARAMINOL see AIH600
PARAMINOPROPIOPHENONE see AMC000
PARAMINYL see WAK000
PARAMINYL MALEATE see DBM800
PARAMORFAN see DNU310
PARAMORPHAN see DLW600
PARAMORPHINE see TEN000
PARAMOTH see DEP800
PARAMYCIN see AMM250
PARANAPHTHALENE see APG500
PARANATE see AIH600
PARANEPHRIN see VGP000
PARANIT ETHANE DISULFONATE see CBG250
PARANITROFENOL (DUTCH) see NIF000
PARANITROFENOLO (ITALIAN) see NIF000
PARANITROPHENOL (FRENCH, GERMAN) see NIF000
PARANITROSODIMETHYLANILIDE see DSY600
PARANOL see ALT250
PARANOL FAST RED 8BL see CMO885
PARA-NONYL PHENOL see NNC510
PARANUGGETS see DEP800
PARA ORANGE see FAG150
PARAOXON see NIM500
PARAOXONE see NIM500
PARAOXON-METHYL see PHD500
PARAPAN see HIM000
PARA-PAS see AMM250
PARAPEST M-50 see MNH000
PARAPHENOLAZO ANILINE see PEI000
PARA PHENYL BENZOIC ACID see PEL600
PARAPHENYLEN-DIAMINE see PEY500
PARAPHOS see PAK000
PARAPLEX P 543 see PKB500
PARAPLEX RG-2 (60%) see PAI750
PARAQUAT (ACGIH) see PAJ000
PARAQUAT see PAI990
PARAQUAT BIS(METHYL SULFATE) see PAJ250
PARAQUAT CHLORIDE see PAJ000
PARAQUAT CL see PAJ000
PARAQUAT DICATION see PAI990
PARAQUAT, DICHLORIDE see PAJ000
PARAQUAT DICHLORIDE see PAJ000
PARAQUAT DIHYDRIDE see PAI995
PARAQUAT DIMETHYL SULFATE see PAJ250
PARAQUAT DIMETHYL SULPHATE see PAJ250
PARAQUAT I see PAJ250
PARAQUAT ION see PAI990
PARAROSANILINE see RMK020
PARAROSANILINE CHLORIDE see RMK020
PARAROSANILINE HYDROCHLORIDE see RMK020
PARAROSANILINE, N,N,N',N',N'',N''-HEXAMETHYL-, CHLORIDE see AOR500
PARASAL see AMM250
PARASALICIL see AMM250
PARASALINDON see AMM250
PARASAN see BCA000
PARASCORBIC ACID see PAJ500
PARASEPT see BSC000, HJL500, HNU500
PARASOL see CNM500
PARASORBIC ACID see PAJ500
(+)-PARASORBINSAEURE (GERMAN) see PAJ500
PARASPAN see SBH500
PARASPEN see HIM000
PARASTARIN see EJM500
PARASULFONDICHLORAMIDO BENZOIC ACID see HAF000
PARASYMPATOL see HLV500
PARATAF see MNH000
PARATHENE see PAK000
PARATHESIN see EFX000
PARATHIAZINE see PAJ750
PARATHION, liquid (DOT) see PAK000
PARATHION and compressed gas mixture (DOT) see PAK230
M-PARATHION see MNH000
PARATHION see PAK000
PARATHION-ETHYL see PAK000
PARATHION METHYL see MNH000
PARATHION-METILE (ITALIAN) see MNH000
PARATHION S see DJS800
PARATHORMONE see PAK250
PARATHYRIN see PAK250
PARATHYROID HORMONE see PAK250
PARATOX see MNH000

PARAWET see PAK000
PARAXANTHINE see PAK300
PARAXENOL see BGJ500
PARAXIN see CDP250
PARAXIN SUCCINATE see CDP725
PARAZENE see DEP800
PARAZINE see PIJ500
PARAZONE see FNF000
PARBENDAZOLE see BQK000
PARBOCYL-REV see SJO000
PARCIDOL see DIR000
PARCLAY see KBB600
PARCLOID see PKQ059
PARDA see DNA200
PARDIDOL see DIR000
PARDISOL see DIR000
PARDROYD see PAN100
PAREDRINOL see FMS875
PARENTERAL see CJR909
PARENTRACIN see BAC250
PARENZYME see TNW000
PARENZYMOL see TNW000
PAREPHYLLIN see CNR125
PAREST see MDT250, QAK000
PAR ESTRO see ECU750
PAREZ 607 see MCB050
PAREZ 613 see MCB050
PAREZ 707 see MCB050
PARFENAC see BPP750
PARFENAL see BPP750
PARFEZINE see DIR000
PARFURAN see NGE000
PARGITAN see BBV000
PARGLYAMINE see MOS250
PARGONYL see NCF500
PARGYLINE see MOS250
PARGYLINE HYDROCHLORIDE see BEX500
PARICINA see APP500
PARIDINE RED LCL see CHP500
PARIDOL see HJL500
PARIS GREEN see COF500
PARIS RED see LDS000
PARIS VIOLET R see MQN025
PARIS YELLOW see LCR000
PARITOL see SEH450
PARKE DAVIS CI-628 see NHP500
PARKEMED see XQS000
PARKIBLEU see CMO250
PARKIN see DIR000
PARKINSAN see BBV000
PARKIPAN see CMO250
PARKISOL see DIR000
PARKOPAN see BBV000, PAL500
PARKOPHYLLIN see TEP000
PARKOSED see BNK000
PARKOTAL see EOK000
PAR KS-12 see PMC250
PARLEF see TKH750
PARLIF see TKH750
PARLODEL see BNB325
PARLODION see CCU250
PARMAL see WAK000
PARMATHASONE ACETATE see PAL600
PARMAVERT see NNE100
PARMETOL see CFE250
PARMIDIN see PPH050
PARMIDINE see PPH050
PARMIDINE R see PPH050
PARMINAL see QAK000
PARMOL see HIM000
PARMONE see NAK500
PARNATE see PET500, PET750
PAROL see CFE250, MQV750, MQV875
PAROLEINE see MQV750
PAROMOMYCIN see NCF500
PAROMOMYCIN SULFATE see APP500
PAROXAN see NIM500
PAROXON see ELL500
PAROXYL see ABX500
PAROXYPROPIONE see ELL500
PAROZONE see SHU500
PARPANIT see PET250
PARPHEZEIN see DIR000
PARPON see BCA000
PARROT GREEN see COF500

PARRYCOP see CNK559
PARSAL see AJC000
PARSALMIDE see AJC000
PARSIDOL see DIR000
PARSITAN see DIR000
PARSLEY APIOL see AGE500
PARSLEY CAMPHOR see AGE500
PARSLEY HERB OIL (FCC) see PAL750
PARSLEY OIL see PAL750
PARSLEY SEED OIL (FCC) see PAL750
PARTEL see DJT800
PARTERGIN see PAM000
PARTEROL see DME300
PARTHENICIN see PAM175
PARTHENIN see PAM175
PARTOCON see ORU500
PARTREX see TBX250
PARTUSISTEN see FAQ100
PARVOLEX see ACH000
PARZATE see DXD200, EIR000
PARZONE see DKW800
PAS see AMM250
PASA see AMM250
PASADE see SEP000
PASALIN see PAM500
PASALON see AMM250
PASALON-RAKEET see SEP000
PASARA see AMM250
PAS-C see AMM250
PASCO see ZBJ000, ZKA000
PASCORBIC see AMM250
PASEM see AMM250
PASEPTOL see HNU500
PASEXON 100T see SHK800
PAS-H 10 see DTS500
PASILLA (PUERTO RICO) see CDM325
PASK see AMM250
PASMED see AMM250
PASNAL see SEP000
PASNODIA see AMM250
PASOLAC see AMM250
PASOTOMIN see PMF250
PASPALIN see PAM775
PASPALIN P I see PAM775
PASQUE FLOWER see PAM780
PASSIFLORAE INCARNATAE EXTRACTUM see PAM782
PASSIFLORA EXTRACT see PAM782
PASSIFLORA INCARNATA, EXTRACT see PAM782
PASSIFLORIN see MPA050
PASSIONFLOWER EXTRACT see PAM782
PASSODICO see SEP000
PAT see PKE600
PATAP see BHL750
PATCHOULI OIL see PAM783
PATCHOULY OIL see PAM783
PATCHUK see CNT350
PATENTBLAU V see ADE500
PATENT BLUE see ADE500
PATENT BLUE A see ERG100
PATENT BLUE AE see FMU059
PATENT BLUE AF see ERG100
PATENT BLUE V see ADE500, CMM062
PATENT BLUE V CARMINE BLUE V see CMM062
PATENT BLUE VF see ADE500
PATENT BLUE VF-CF see ADE500
PATENT BLUE VF SPECIAL see ADE500
PATENT BLUE VS see ADE500
PATENT GREEN see COF500
PATHCLEAR see PAJ000
PATHOCIDIN see AJO500
PATHOCIDINE see AJO500
PATHOCIL see DGE200
PATHOCLON see DXN709
PATHOMYCIN see APY500
PATORAN see PAM785
PATRICIN see VRF000
PATRINOSIDE see PAM789
PATRINOSIDE-AGLYCONE see PAM800
PATROVINE see DHX800, THK000
PATTINA V 82 see PKQ059
PATTONEX see PAM785
PATULIN see CMV000
PAUCIMYCIN see NCF500
PAUSITAL see CKE750
P. AUSTRALIS VENOM see ARV000

PAVABID see PAH250
PAVAGRANT see PAH250
PAVAKEY see PAH250
PAVASED see PAH250
PAVATEST see PAH250
PAVATRIN see CBS250
PAVATRINE see FDK000
PAVATRINEAT see CBS250
PAVISOID see PAN100
PAVULON see PAF625, PAF630
PAXAREL see ACE000
PAXATE see DCK759
PAXILON see BGD250
PAXISTIL see CJR909
PAXISYN see DLY000
PAXITAL see MOQ250
PAXYL see TAF675
PAY-OFF see DRN200
PAYZE see BLW750
PAYZONE see PAF500
PAZITAL see CGA000
PB-106 see PIX800
P 2010B see PJS750
P-4657-B see NBP500
PB see PIX250, POQ250
PBA, DIMETHYLAMINE SALT see PJQ000
PBAN-560 see PAN250
PBAN-560 (degassed) see PAN500
PBB see FBU000, FBU509, PJL335
PB 89 CHLORIDE see FMU000
PBDZ see BQK000
PB-89 HYDROCHLORIDE see FMU000
PBK 1 see DTS500
PBNA see PFT500
PBQI see BDD500
PB-1,6-QUINONE see BCU500
PBS see SID000
PBX(AF) 108 see CPR800
PBXW 108(E) see CPR800
PBZ see TMP750
PC 1 see BQJ500
PC 222 see CKD250
PC 603 see CJN250
PC-1421 see PII500
1PC6115 see MCB050
PC see PGP250
PCA see MCR750, PEE750, TDV300
PC ALCOHOL DF see PAN750
PCB 54 see TBO600
PCB see PJL750, PJL800, PJM000, PJM250, PJM500, PJM750, PJN000,
  PJN250, PJN500, PJN750, PJO000, PJO250, PJO500, PJO750, PJP000,
  PME250
PCBs see PJL750
PCB HYDROCHLORIDE see PME500
PCBS see CJR500
PCC see CDV100, PIN225
P.C. 80 CRABGRASS KILLER see PLC250
PCDF see PJP100
PCEO see TBQ275
PCHO see PAI250
PCI see CJR500
PCL see HCE500
PCM see PCF300
PCMC see CFE250
PCMH see PAN775
PCMX see CLW000
PCNB see PAX000
PCNU see CGW250
PCON see KDA050
PCONA see KDA050
PCP (anesthetic) see PDC890
PCP see PAX250
PCPA see CJN000
PCPBS see CJR500
PCPCBS see CJT750
PCP HYDROCHLORIDE see AOO500
PCPI see CKB000
PCT see PJP250
PCTP see PAY500
PD 93 see PAF250
PD 71627 see AJR400
PD 73093 see DSO500
PD 109394 see AAI100
p-PD HCl see PEY650
P-D see BBO500

2-N-p-PDA see ALL750
p-PDA HCl see PEY650
P.D.A.B. see DOT300
PDB see DEP800
PDC see DGG800
PDCB see DEP800
PDD see PGT250
PDD 6040I see CJV250
PDMT see DTP000
m-PDN see PHX550
p-PDN see BBP250
PDP see PDC250
PDQ see CLN750
PDT see DTP000
PDTA-Sb see AQI500
PDU see DTP400
PE-043 see PAN800
PE 512 see PJS750
PE 617 see PJS750
PE see PBB750
PE 60A see LED100
β-PEA see PDD750
PEA see PDD750
PEACE PILL see AOO500
PEACH see AQP890
PEACH ALDEHYDE see HBN200, UJA800
PEACHES see AOB250
PEACH LACTONE see HBN200
PEACOCK BLUE X-1756 see FMU059
PEA FLOWER LOCUST see GJU475
PEANUT OIL see PAO000
PEARL ASH see PLA000, PLA250
PEARLPUSS see CCL250
PEARL STEARIC see SLK000
PEARLY GATES see DJO000
PEAR OIL see AOD725, IHO850
PEARSALL see AGY750
PEB1 see DAD200
PEBC see PNF500
PEBULATE see PNF500
PECAN SHELL POWDER see PAO000
PECATAL see MOQ250
PECAZINE see MOQ250
PECNON see KGK000
PECTA-DIAZINE, suspension see PPP500
PECTALGINE see SEH000
PECTAMOL see DHQ200
PECTIN see PAO150
PECTINATE see PAO150
PECTINIC ACID see PAO150
PECTINS see PAO150
PECTITE see MBX800
PECTOLIN see TCY750
PECTOX see CBR675
PED see LBT000
PEDIAFLOR see SHF500
PEDIDENT see SHF500
PEDILANTHUS TITHYMALOIDES see SDZ475
PEDINEX (FRENCH) see CPK500
PEDIPEN see PDT750
PEDRACZAK see BBQ500
PEDRIC see HIM000
PE 60E see LED100
PEERACID ORANGE II see CMM220
PEERAMINE BLACK E see AQP000
PEERAMINE BLACK GXOO see AQP000
PEERAMINE BRIGHT YELLOW MN see CMP050
PEERAMINE CONGO RED see SGQ500
PEERLESS see CBT750, KBB600
PEG 200 see PJT200
PEG 300 see PJT225
PEG 400 see PJT230
PEG 600 see PJT240
PEG 1000 see PJT250
PEG 1500 see PJT500
PEG 4000 see PJT750
PEG 6000 see PJU000
PEGANONE see EOL100
PEG 200MO see PJY100
PEG 600MO see PJY100
PEG 1000MO see PJY100
PEG-9 NONYL PHENYL ETHER see PKF000
PEGOANONE see EOL100
PEG-9 OCTYL PHENYL ETHER see PKF500
PEG-6 OLEATE see PJY100

PEG-20 OLEATE see PJY100
PEG-32 OLEATE see PJY100
PEG-20 OLEYL ETHER see PJW500
PEGOSPERSE 400MO see PJY100
PEGOTERATE see PKF750
PEHA see PBD000
PEHANORM see TEM500
PELADOW see CAO750
PELAGOL BA see DBO400
PELAGOL CD see PEY650
PELAGOL D see PEY500
PELAGOL DA see DBO000
PELAGOL DR see PEY500
PELAGOL EG see ALT500
PELAGOL 3GA see ALT000
PELAGOL GREY see DBO400
PELAGOL GREY C see CCP850
PELAGOL GREY CD see PEY650
PELAGOL GREY D see PEY500
PELAGOL GREY GG see ALT000
PELAGOL GREY J see TGL750
PELAGOL GREY L see DBO000
PELAGOL GREY P BASE see ALT250
PELAGOL GREY RS see REA000
PELAGOL GREY SLA see DBO400
PELAGOL J see TGL750
PELAGOL L see DBO000
PELAGOL P BASE see ALT250
PELAGOL SLA see DBO400
PELA PUERCO (DOMINICAN REPUBLIC) see DHB309
PELARGIC ACID see NMY000
PELARGIDENON 1449 see CDH250
PELARGOL see DTE600
PELARGON (RUSSIAN) see NMY000
PELARGONE see OFE020
PELARGONIC ACID see NMY000
PELARGONIC ALCOHOL see NNB500
PELARGONIC ALDEHYDE see NMW500
PELARGONIC MORPHOLIDE see MRQ750
PELARGONITRILE see OES300
PELARGONIUM OIL see GDA000
PELARGONONITRILE see OES300
PELARGONOYL PEROXIDE see NNA100
PELARGONYL PEROXIDE, technically pure (DOT) see NNA100
PELARGONYL PEROXIDE see NNA100
PELASPAN 333 see SMQ500
PELASPAN ESP 109s see SMQ500
PELAZID see ILD000
PELENTAN see BKA000
PELIDORM see BNK000
PELIKAN C 11/1431a see CBT500
PELLAGRAMIN see NCQ900
PELLAGRA PREVENTIVE FACTOR see NCQ900
PELLAGRIN see NCQ900
PELLCAFS see BBK500
PELLCAP see BBK500
PELLCAPS see BBK500
PELLETEX see CBT750
(R)-(−)-PELLETIERINE see PAO500
PELLETIERINE see PAO500
PELLIDOL see ACR300
PELLIDOLE see ACR300
PELLON 2506 see PMP500
PELLUGEL see CCL250
PELMIN see NCR000
PELMINE see NCR000
PELONIN see NCQ900
PELONIN AMIDE see NCR000
PELSON see DLY000
PELT-44 see PEX500
PELTAR see MAP300
PELTOL BR see PFU500
PELTOL BR II see PFU500
PELTOL D see PEY500
PELT SOL see DJV000
PELUCES see CLY500
PELVIRAN see DNG400
PEMAL see ENG500
PEMALIN see ENG500
PEMOLINE MAGNESIUM see PAP000
PEMOLINE MAGNESIUM CHELATE see PAP000
PEMOLIN and MAGNESIUM HYDROXIDE see PAP000
PEMPIDINA TARTRATO (ITALIAN) see PAP110
PEMPIDINE HYDROCHLORIDE see PAP100
PEMPIDINE TARTRATE see PAP110

PEN 100 see PJS750
PEN 200 see PDD350
PEN A see AOD125
PEN-A-BRASIVE see BFD250
PENADUR see BFC750
PENADUR L-A see BFC750
PENAGEN see PDT750
PENALEV see BFD000
d-PENAMINE see MCR750
PEN A/N see SEQ000
PENAR see DRS000
PENATIN see CMV000, GFG100
PENBAR see NBU000
PENBRISTOL see AIV500
PENBRITIN see AIV500
PENBRITIN PAEDIATRIC see AIV500
PENBRITIN-S see SEQ000
PENBRITIN SYRUP see AIV500
PENBROCK see AIV500
PENBUTOLOL see PAP225
(−)-PENBUTOLOL see PAP225
PENBUTOLOL SULFATE see PAP230
PENCAL see ARB750
PENCARD see PBC250
PENCHLOROL see PAX250
PENCIL GREEN SF see FAF000
PENCILLIC ACID see PAP750
PENCOGEL see CCL250
PENCOMPREN see PDT750
PENDEPON see BFC750
PENDEROL see PMB250
PEN-DI-BEN see BFC750
PENDIMETHALIN see DRN200
PENDIOMID see MKU750
PENDIOMIDE BROMIDE see MKU750
PENDIOMIDE DIBROMIDE see MKU750
PENDITAN see BFC750
PENDURAN see BFC750
PENETECK see MQV750, MQV875
PENETRACYNE see MCH525
PENFLURIDOL see PAP250
PENFORD 260 see HLB400
PENFORD 280 see HLB400
PENFORD 290 see HLB400
PENFORD GUM 380 see SLJ500
PENFORD P 208 see HLB400
PENGITOXIN see GEW000
PENGLOBE see BAB250
PENIALMEN see SEQ000
PENICIDIN see CMV000
(S)-PENICILLAMIN see MCR750
PENICILLAMIN see MCR750
dl-PENICILLAMINE see PAP500
d-PENICILLAMINE see MCR750
PENICILLAMINE see MCR750
d-PENICILLAMINE HYDROCHLORIDE see PAP550
PENICILLAMINE HYDROCHLORIDE see PAP550
PENICILLANIC ACID 1,1-DIOXIDE SODIUM SALT see PAP600
PENICILLANIC ACID DIOXIDE SODIUM SALT see PAP600
PENICILLANIC ACID SULFONE SODIUM SALT see PAP600
PENICILLIN, compounded with 9-AMINOACRIDINE see AHT000
PENICILLIC ACID see PAP750
PENICILLIN see PAQ000
PENICILLIN AT see AGK250
PENICILLIN BT see BRS250
PENICILLIN compounded with CHOLINE CHLORIDE see PAQ060
PENICILLIN-CHOLINESTER CHLORID (GERMAN) see PAQ060
PENICILLIN G see BDY669
PENICILLIN G BENETHAMINE see PAQ100
PENICILLIN G EPHEDRINE SALT see PAQ120
PENICILLIN-G, MONOSODIUM SALT see BFD250
PENICILLIN G, compounded with N,N′-DIBENZYLETHYLENEDIAMINE (2:1) see BFC750
PENICILLIN G POTASSIUM see BFD000
PENICILLIN G POTASSIUM SALT see BFD000
PENICILLIN G PROCAINE see PAQ200
PENICILLIN G SALT of N,N′-DIBENZYLETHYLENEDIAMINE see BFC750
PENICILLIN G, SODIUM see BFD250
PENICILLIN G, SODIUM SALT see BFD250
PENICILLIN O see AGK250
PENICILLIN P-12 see DSQ800, MNV250
PENICILLIN PHENOXYMETHYL see PDT500
PENICILLIN POTASSIUM PHENOXYMETHYL see PDT750
PENICILLIN V see PDT500
PENICILLIN V POTASSIUM see PDT750

PENICILLIN V POTASSIUM SALT see PDT750
PENICILLIUM ROQUEFORTI TOXIN see PAQ875
PENICIN see PCU500
PENICLINE see AIV500
PENIDURAL see BFC750
PENIDURE see BFC750
PENILARYN see BFD250
PENILENTE see BFC750
PENILTETRA see MCH525
PENIMEPICYCLINE see MCH525
PENIN see PCU500
PENISEM see BFD000
PENITE see SEY500
PENITRACIN see BAC250
PENITREM A see PAR250
PENIZILLIN (GERMAN) see PAQ000
PENNAC see SGF500
PENNAC CBS see CPI250
PENNAC CRA see IAQ000
PENNAC MBT POWDER see BDF000
PENNAC MS see BJL600
PENNAC TBBS see BQK750
PENNAC ZT see BHA750
PENNAMINE see DAA800
PENNAMINE D see DAA800
PENNCAP-M see MNH000
PENNFLOAT M see LBX000
PENNFLOAT S see LBX000
PENNSALT TD-72 see DJI000
PENN SALT TD-183 see TBV750
PENNSALT TD 5032 see HEE500
PENNWALT C-4852 see DSQ000
PENNWHITE see SHF500
PENNYROYAL OIL see PAR500
PENNZONE B see DEI000
PENNZONE E see DKC400
PENOCTONIUM BROMIDE see PAR600
PEN-ORAL see PDT500
PENOTRANE see PFN000
PENOXALINE see DRN200
PENPHENE see CDV100, TBV750
PENRECO see MQV750
PENSIG see PDD350
PEN-SINT see DGE200
PENSTAPHOCID see MNV250
PENSYN see AOD125
PENTA see PAX250
PENTA-o-ACETYLGITOXIN see GEW000
PENTAACETYLGITOXIN see GEW000
PENTAAMMINEAQUACOBALT(III) CHLORATE see PAR750
PENTAAMMINEPYRAZINERUTHENIUM(II) PERCHLORATE see PAR799
PENTAAMMINEPYRIDINERUTHENIUM(II) PERCHLORATE see PAS829
PENTAAMMINETHIOCYANATOCOBALT(III) PERCHLORATE see PAS859
PENTAAMMINETHIOCYANATORUTHENIUM(II) PERCHLORATE see PAS879
PENTAAZAACENAPHTHYLENE-5'-PHOSPHATE ESTER MONOHYDRATE see TJE875
PENTAAZACENTOPTHYLENE see TJE870
1,4,7,10,13-PENTAAZATRIDECANE see TCE500
1,3,5,5,5-PENTAAZIRIDINO-1-THIA-2,4,6-TRIAZA-3,5-DIPHOSPHORINE-1-OXIDE see SED700
PENTABARBITAL SODIUM see NBU000
PENTABARBITONE see NBT500
PENTABORANE(11) see PAT799
PENTABORANE(9) see PAT750
PENTABORANE (ACGIH,DOT,OSHA) see PAT750
PENTABORANE(9)DIAMMONIATE see DCF725
PENTABROMFENOL see PAU250
2,3,4,5,6-PENTABROMOETHYLBENZENE see PAT850
PENTABROMOETHYLBENZENE see PAT850
PENTABROMOPHENOL see PAU250
PENTABROMOPHENYL ETHER see PAU500
PENTABROMO PHOSPHORANE see PHR250
PENTABROMO PHOSPHORUS see PHR250
PENTAC see DAE425
PENTACAINE see PPW000
PENTACAINE HYDROCHLORIDE see PPW000
PENTACARBONYLIRON see IHG500
PENTACENE see PAV000
PENT-ACETATE see AOD725
PENTACHLOORETHAAN (DUTCH) see PAW500
PENTACHLOORFENOL (DUTCH) see PAX250
PENTACHLORAETHAN (GERMAN) see PAW500
PENTACHLORETHANE (FRENCH) see PAW500
PENTACHLORIN see DAD200
PENTACHLORNITROBENZOL (GERMAN) see PAX000

PENTACHLOROACETONE see PAV225
2',3',4',5',6'-PENTACHLOROACETOPHENONE see PAV250
PENTACHLOROACETOPHENONE see PAV250
2,3,4,5,6-PENTACHLOROANISOLE see MNH250
PENTACHLOROANISOLE see MNH250
PENTACHLOROANTIMONY see AQD000
PENTACHLOROBENZENE see PAV500
PENTACHLORO-BENZENETHIOL see PAY500
PENTACHLOROBIPHENYL see PAV600
PENTACHLOROBUTANE see PAV750
2,2,3,4,4-PENTACHLORO-3-BUTENOIC ACID see PAV775
PENTACHLORO-3-BUTENOIC ACID see PAV775
1,2,3,7,8-PENTACHLORODIBENZO-p-DIOXIN see PAW000
2,3,4,7,8-PENTACHLORODIBENZOFURAN see PAW100
3,3',5,5',6-PENTACHLORO-2,2'-DIHYDROXYBENZANILIDE see DMZ000
3,5,6,3',5'-PENTACHLORO-2,2'-DIHYDROXYBENZANILIDE see DMZ000
PENTACHLORODIPHENYL see PAV600
PENTACHLORO DIPHENYL OXIDE see PAW250
PENTACHLOROETHANE see PAW500
PENTACHLOROFENOL see PAX250
2',2'',4',4'',5-PENTACHLORO-4-HYDROXY-ISOPHTHALANILIDE see PAW600
3,3',5,5',6-PENTACHLORO-2'-HYDROXYSALICYLANILIDE see DMZ000
PENTACHLOROMETHOXYBENZENE see MNH250
PENTACHLORONAPHTHALENE see PAW750
PENTACHLORONITROBENZENE see PAX000
PENTACHLOROPHENATE see PAX250
PENTACHLOROPHENATE SODIUM see SJA000
2,3,4,5,6-PENTACHLOROPHENOL see PAX250
PENTACHLOROPHENOL (GERMAN) see PAX250
PENTACHLOROPHENOL see PAX250
PENTACHLOROPHENOL, DOWICIDE EC-7 see PAX250
PENTACHLOROPHENOL, DP-2 see PAX250
PENTACHLOROPHENOL, SODIUM SALT see SJA000
PENTACHLOROPHENOL, SODIUM derivative mixed with TETRACHLOROPHE-
   NOL SODIUM derivative (4:1) see PAX750
PENTACHLOROPHENOL, TECHNICAL see PAX250
PENTACHLOROPHENOXY SODIUM see SJA000
PENTACHLOROPHENYL CHLORIDE see HCC500
PENTACHLOROPHENYL METHYL ETHER see MNH250
1,1,2,2,3-PENTACHLOROPROPANE see PAY000
1,1,1,3,3-PENTACHLOROPROPANONE see PAV225
1,1,1,3,3-PENTACHLORO-2-PROPANONE see PAV225
1,1,2,3,3-PENTACHLORO-1-PROPENE see PAY200
1,1,2,3,3-PENTACHLOROPROPENE see PAY200
1,1,2,3,3-PENTACHLOROPROPYLENE see PAY200
2,3,4,5,6-PENTACHLOROPYRIDINE see PAY250
PENTACHLOROTHIOFENOLAT ZINECNATY (CZECH) see BLC500
PENTACHLOROTHIOPHENOL see PAY500
PENTACHLORTHIOFENOL (CZECH) see PAY500
PENTACHLORURE d'ANTIMOINE (FRENCH) see AQD000
PENTACIN see CAY500
PENTACINE see CAY500
PENTACLOROETANO (ITALIAN) see PAW500
PENTACLOROFENOLO (ITALIAN) see PAX250
PENTACON see PAX250
PENTAC WP see DAE425
PENTACYANONITROSYLFERRATE BARIUM see PAY600
PENTACYANONITROSYLFERRATE COBALT see PAY610
PENTACYANONITROSYLFERRATE ZINC see ZJJ400
PENTADECAFLUORO-n-OCTANOIC ACID see PCH050
PENTADECAFLUOROOCTANOIC ACID see PCH050
1-(2,2,3,3,4,4,5,5,6,6,7,7,8,8,8-PENTADECAFLUOROOCTYL)PYRIDINIUM TRI-
   FLUOROMETHANESULFONATE see MBV710
1-PENTADECANAMINE see PBA000
N-PENTADECANE see PAY750
1-PENTADECANECARBOXYLIC ACID see PAE250
7-PENTADECANECARBOXYLIC ACID see HFP500
PENTADECANOIC ACID see PAZ000
PENTADECAN-8-ONE see HBO790
8-PENTADECANONE see HBO790
PENTADECYCLIC ACID see PAZ000
1-PENTADECYLAMINE see PBA000
n-PENTADECYLAMINE see PBA000
PENTADECYLAMINE see PBA000
(E)-1,3-PENTADIENE see PBA250
trans-1,3-PENTADIENE see PBA250
1,3-PENTADIENE-1-CARBOXALDEHYDE see SKT500
1,3-PENTADIENE-1-CARBOXYLIC ACID see SKU000
1:4-PENTADIENE DIOXIDE see DHE000
2,4-PENTADIENOIC ACID, 5-(4a-(ACETYLOXY)-3,4,4a,5,6,9-HEXAHYDRO-3,7-
   DIMETHYL-1-OXO-1H-4,9a-ETHANOCYCLOHEPTA(c)PYRAN-3-YL)-2-METHYL-,
   (3-α(2E,4E),4-α-9a-α)-(−)- (9CI) see POH550
1,4-PENTADIEN-3-ONE, 1,5-BIS(4-AZIDOPHENYL)-(9CI) see BGW720
1,3-PENTADIYNE see PBB000
1,3-PENTADIYN-1-YL COPPER see PBB229

1,3-PENTADIYN-1-YL SILVER see PBB449
PENTADORM see EQL000
PENTAERYTHRITE see PBB750
PENTAERYTHRITE TETRANITRATE, dry (DOT) see PBC250
PENTAERYTHRITE TETRANITRATE (DOT) see PBC250
PENTAERYTHRITE TETRANITRATE, with not less than 7% wax (DOT) see
  PBC250
PENTAERYTHRITE TETRANITRATE, desensitized, wet (DOT) see PBC250
PENTAERYTHRITE TETRANITRATE see PBC250
PENTAERYTHRITOL see PBB750
PENTAERYTHRITOL CHLOROL see NCH500
PENTAERYTHRITOL DIBROMIDE see DDQ400
PENTAERYTHRITOL DIBROMOHYDRIN see DDQ400
PENTAERYTHRITOL DICHLOROHYDRIN see ALB000
PENTAERYTHRITOL, TETRAACETATE see NNR400
PENTAERYTHRITOL TETRABENZOATE see PBC000
PENTAERYTHRITOL TETRANICOTINATE see NCW300
PENTAERYTHRITOL TETRANITRATE, diluted see PBC250
PENTAERYTHRITOL TETRANITRATE see PBC250
PENTAERYTHRITOL TRIACRYLATE see PBC750
PENTAETHYLENEHEXAMINE see PBD000
PENTAFIN see PBC250
2,3,4,5,6-PENTAFLUOROANILINE see PBD250
PENTAFLUOROANILINE see PBD250
PENTAFLUOROANTIMONY see AQF250
PENTAFLUOROBENZOIC ACID see PBD275
1,2,3,4,5-PENTAFLUOROBICYCLO(2.2.0)HEXA-2,5-DIENE see PBD300
PENTAFLUOROCHLOROBENZENE see PBE100
PENTAFLUOROGUANIDINE see PBD500
PENTAFLUOROIODINE see IDT000
PENTAFLUOROPHENOL see PBD750
PENTAFLUOROPHENYLALUMINUM DIBROMIDE see PBE000
PENTAFLUOROPHENYLAMINE see PBD250
PENTAFLUOROPHENYL CHLORIDE see PBE100
PENTAFLUOROPHENYLLITHIUM see PBE250
2,2,3,3,3-PENTAFLUORO-1,1-PROPANEDIOL see PBE500
2,2,3,3,3-PENTAFLUORO-1-PROPANOL see PBE750
PENTAFLUOROPROPIONIC ACID see PBF000
PENTAFLUORO-PROPIONIC ACID METHYL ESTER see PCH350
PENTAFLUOROPROPIONIC ACID SILVER SALT see PBF250
PENTAFLUOROPROPIONYL FLUORIDE see PBF300
PENTAFLUOROPROPIONYL HYPOCHLORITE see PBF400
PENTAFLUOROPROPIONYL HYPOFLUORITE see PBF500
PENTAFLUOROSULFUR PEROXYACETATE see PBG100
PENTAFLUORPROPIONAN STRIBRNY (CZECH) see PBF250
PENTAFORMYL DIGOXIN see DKN875
PENTAFORMYLGITOXIN see FND100
PENTAGEN see PAX000
PENTAGIN see DOQ400
PENTAGIT see GEW000
3,3',4',5,7-PENTAHYDROXYFLAVANONE see DMD000
3,3',4',5',7-PENTAHYDROXYFLAVANONE see RLP000
2',3,4',5,7-PENTAHYDROXYFLAVONE see MRN500
2',4',3,5,7-PENTAHYDROXYFLAVONE see MRN500
3,3',4',5,7-PENTAHYDROXYFLAVONE see QCA175
3,5,7,2',4'-PENTAHYDROXYFLAVONE see MRN500
3,5,7,3',4'-PENTAHYDROXYFLAVONE see QCA000
3,3',4',5,7-PENTAHYDROXYFLAVONE-3-l-RHAMNOSIDE see QCJ000
3,3',4',5,7-PENTAHYDROXYFLAVONE-3-(o-RHAMNOSYLGLUCOSIDE) see
  RSU000
3,3',4',5,7-PENTAHYDROXYFLAVONE-3-RUTINOSIDE see RSU000
3,5,7,2',4'-PENTAHYDROXYFLAVONOL see MRN500
3,3'4',5,7-PENTAHYDROXYFLAVYLIUM CHLORIDE see COI750
3,3',4',5,7-PENTAHYDROXY-5'-METHOXYFLAVYLIUM ACID ANION see
  PCU000
2-(1,2,3,4,5-PENTAHYDROXY)-N-PENTYLTHIAZOLIDINE see TEW000
PENTAHYDROXY-TIGLIADIENONE-MONOACETATE(C)MONOMYRISTATE(B) see
  PGV000
PENTAKAIN see PPW000
PENTA-KIL see PAX250
PENTAKIS(N²-ACETYL-l-GLUTAMINATO)TETRAHYDROXYTRIALUMINUM see
  AGX125
PENTAL see NBU000
γ-PENTALACTONE see VAV000
PENTALIN see PAW500
PENTALINIUM TARTRATE see PBT000
PENTAMETHAZENE DIBROMIDE see MKU750
PENTAMETHAZINE see MKU750
1,1'-PENTAMETHLENEBIS-PYRIDINIUM DIBROMIDE see PBH100
N,N-3',4',5'-PENTAMETHYLAMINOAZOBENZENE see DUL400
PENTAMETHYLBENZYL-p-ROSANILINE CHLORIDE see MQN000
PENTAMETHYLDIETHYLENETRIAMINE see PBG500
N,N,N',N',N''-PENTAMETHYLDIETHYLENETRIAMINE see PBG500
1,1,3,3,5-PENTAMETHYL-4,6-DINITROINDANE see MRU300

1,5-PENTAMETHYLEN-BIS-N-AETHYLPYRROLIDINIUM)-DIBROMID (GERMAN)
  see PAB000
1,5-PENTAMETHYLEN-BIS(N-sec-BUTYLPIPERIDINIUM)-DIBROMID (GERMAN)
  see PBG725
1,5-PENTAMETHYLEN-BIS-(N-n-BUTYLPIPERIDINIUM)-DIBROMID (GERMAN) see
  PBG850
1,5-PENTAMETHYLEN-BIS-(PIPERIDINIUM)-DIHYDROBROMID (GERMAN) see
  PBI200
PENTAMETHYLENE see CPV750
1,1'-PENTAMETHYLENEBIS(sec-BUTYLPIPERIDINIUM)DIBROMIDE see PBG725
1,1'-PENTAMETHYLENEBIS(3-(2-CHLOROETHYL))-3-NITROSOUREA see
  PBG750
1,1'-PENTAMETHYLENEBIS(1-ETHYLPIPERIDINIUM)DIBROMIDE see PBG850
1,1'-PENTAMETHYLENEBIS(1-ETHYL-PYRROLIDINIUM DIBROMIDE) see
  PAB000
1,1'-PENTAMETHYLENEBIS(1-METHYLPIPERIDINIUM)DIBROMIDE see PBH075
1,1'-PENTAMETHYLENEBIS(1-METHYLPYRROLIDINIUM HYDROGEN TAR-
  TRATE) see PBT000
PENTAMETHYLENE-1,5-BIS(1'-METHYLPYRROLIDINIUM TARTRATE) see
  PBT000
1,1'-PENTAMETHYLENEBIS(1-METHYLPYRROLIDINIUM TARTRATE) see
  PBT000
1,1'-PENTAMETHYLENEBIS(PYRIDINIUM BROMIDE) see PBH100
N,N-PENTAMETHYLENECARBAMIC ACID-3-DIMETHYLAMINOPHENYL ESTER,
  METHOSULFATE see HNS000
PENTAMETHYLENECARBAMIC ACID-m-(TRIMETHYLAMMONIO)PHENYL ES-
  TER, METHYLSULFATE see HNS000
PENTAMETHYLENECARBAMIC ESTER of 3-OXYPHENYLTRIMETHYLAMMONI-
  UM METHYLSULFATE see HNS000
(3-(PENTAMETHYLENECARBAMOYLOXY)PHENYL)TRIMETHYLAMMONIUM ME-
  THYLSULFATE see HNS000
1,5-PENTAMETHYLENEDIAMINE see PBK500
PENTAMETHYLENEDIAMINE see PBK500
PENTAMETHYLENE DIIODIDE see PBH125
p,p'-(PENTAMETHYLENE-DIOXY)BIS-BENZAMIDINE see DBM000
1,1'-(PENTAMETHYLENEDIOXY)BIS(3-CHLORO-2-PROPANOL) see PBH150
4,4'-(PENTAMETHYLENEDIOXY)DIBENZAMIDINE see DBM000
p,p'-(PENTAMETHYLENEDIOXY)DIBENZAMIDINE see DBM000
p,p'-(PENTAMETHYLENEDIOXY)DIBENZAMIDINE BIS(β-HYDROXYETHANESUL-
  FONATE) see DBL800
(N,N'-PENTAMETHYLENE DIOXYDIETHYLENE)BIS(DIMETHYLETHYL AMMO-
  NIUM IODIDE) see PBH500
(N,N'-PENTAMETHYLENE DIOXYDIETHYLENE)BIS(TRIETHYL AMMONIUM IO-
  DIDE) see PBI000
1,1'-PENTAMETHYLENEDIPIPERIDINE DIHYDROBROMIDE see PBI200
PENTAMETHYLENEDITHIOCARBAMATE see PIY500
PENTAMETHYLENE GLYCOL see PBK750
PENTAMETHYLENEIMINE see PIL500
PENTAMETHYLENETETRAZOL see PBI500
PENTAMETHYLENE-1,5-TETRAZOLE see PBI500
1,5-PENTAMETHYLENETETRAZOLE see PBI500
N,N-PENTAMETHYLENEUREA see PIL525
N,N,N',N',3-PENTAMETHYL-N,N'-DIETHYL-3-AZAPENTYLENE-1,5-DIAMMONIUM
  DIBROMIDE see MKU750
1,2,2,6,6-PENTAMETHYLPIPERIDINE see PBJ000
1,2,2,6,6-PENTAMETHYLPIPERIDINE HYDROCHLORIDE see PAP100
1,2,2,6,6-PENTAMETHYLPIPERIDINE TARTRATE see PAP110
PENTAMETHYLTANTALUM see PBJ600
2,2,4,6,6-PENTAMETHYLTETRAHYDRO PYRIMIDINE see PBJ750
PENTAMIDINE see DBM000
PENTAMIDINE DIISETHIONATE see DBL800
PENTAMIDINE ISETHIONATE see DBL800
PENTAMINE see MKU750
PENTAMON see MKU750
PENTAMYCIN see FPC000
PENTAN (POLISH) see PBK250
n-PENTANAL see VAG000
PENTANAL see VAG000
PENTANAL, N-FORMYL-N-METHYLHYDRAZONE see PBJ875
PENTANAL, 4-METHYL-(9CI) see MQJ500
PENTANAL METHYLFORMYLHYDRAZONE see PBJ875
1-PENTANAMINE see PBV505
PENTANATE see PBK000
PENTANDIOIC ACID see GFS000
PENTANE (ITALIAN) see PBK250
n-PENTANE see PBK250
tert-PENTANE see NCH000
PENTANE, 1-BROMO- see AOF800
2-PENTANECARBOXYLIC ACID see MQJ750
3-PENTANECARBOXYLIC ACID see DHI400
1,5-PENTANEDIAL see GFQ000
1,5-PENTANEDIAMINE see PBK500
2,4-PENTANEDIAMINE, 2-METHYL- see MNI525
1,5-PENTANEDICARBOXYLIC ACID see PIG000
PENTANE, 1,5-DIIODO- see PBH125

PENTANEDINITRILE see TLR500
1,5-PENTANEDIOIC ACID see GFS000
PENTANEDIOIC ACID see GFS000
PENTANE-1,5-DIOL see PBK750
1,5-PENTANEDIOL see PBK750
2,4-PENTANEDIOL see PBL000
PENTANEDIOL-2,4 see PBL000
(2,4-PENTANEDIONATO-O,O')PHENYLMERCURY see PBL250
1,5-PENTANEDIONE see GFQ000
2,3-PENTANEDIONE see PBL350
2,4-PENTANEDIONE (FCC) see ABX750
PENTANEDIONE see ABX750
2,4-PENTANEDIONE, NICKEL(II) DERIVATIVE see PBL500
2,4-PENTANEDIONE PEROXIDE see PBL600
2,4-PENTANEDIONE, PHENYLMERCURIC SALT see PBL250
2,4-PENTANEDIONE, ZIRCONIUM COMPLEX see PBL750
4,4'-(1,5-PENTANEDIYLBIS(OXY))BIS-BENZENECARBOXIMIDAMIDE, (9CI) see DBM000
PENTANE, 1-IODO- see IET500
PENTANEN (DUTCH) see PBK250
PENTANENITRILE (9CI) see VAV300
PENTANE, 1,1'-OXYBIS-(9CI) see PBX000
1,2,3,4,5-PENTANEPENTOL see RIF000
1-PENTANETHIOL see PBM000
2-PENTANETHIOL, 2,4,4-TRIMETHYL- see MKJ250, OFE030
2,3,4,5,6-PENTANITROANILINE see PBM100
PENTANITROANILINE (dry) (DOT) see PBM100
PENTANITROANILINE see PBM100
PENTANOCHLOR see SKQ400
n-PENTANOIC ACID see VAQ000
PENTANOIC ACID see VAQ000
tert-PENTANOIC ACID see PJA500
PENTANOIC ACID, 4,4'-AZOBIS(4-CYANO)- (9CI) see ASL500
PENTANOIC ACID, 2,2-DIPHENYL, 2-(N,N-DIETHYLAMINO)ETHYL ESTER, HY-DROCHLORIDE see PBM500
PENTANOIC ACID, 3-HEXENYL ESTER, (Z)- see HFE800
PENTANOIC ACID, METHYL ESTER (9CI) see VAQ100
PENTANOIC ACID, 4-OXO-, BUTYL ESTER (9CI) see BRR700
PENTANOIC ACID, 4-OXO-, ETHYL ESTER (9CI) see EFS600
PENTANOIC ACID, 2-PROPYL-, MAGNESIUM SALT see MAK275
PENTAN-1-OL see AOE000
PENTANOL-1 see AOE000
PENTANOL-2 see PBM750
2-PENTANOL see PBM750
PENTAN-3-OL see IHP010
3-PENTANOL see IHP010
PENTANOL-3 see IHP010
N-PENTANOL see AOE000
tert-PENTANOL see PBV000
1-PENTANOL ACETATE see AOD725
2-PENTANOL, ACETATE see AOD735
2-PENTANOL CARBAMATE see MOU500
2-PENTANOL, 3-ETHYL- see EOB300
3-PENTANOL, 3-ETHYL- see TJP550
4-PENTANOLIDE see VAV000
3-PENTANOL, 3-METHYL-1-PHENYL- see PFR200
2-PENTANONE see PBN250
3-PENTANONE see DJN750
PENTANONE-3 see DJN750
PENTANONE, 1-BENZOCYCLOBUTYL- see BCH800
PENTANONE, 1-BENZOCYCLOBUTYL-, OXIME see BFZ170
1-PENTANONE, 1-BICYCLO(4.2.0)OCTA-1,3,5-TRIEN-7-YL- see BCH800
1-PENTANONE, 1-BICYCLO(4.2.0)OCTA-1,3,5-TRIEN-7-YL-, OXIME see BFZ170
2-PENTANONE, 5-(DIETHYLAMINO)- see NOB800
2-PENTANONE, 4-HYDROXY-1-PHENYL-5,5,5-TRIFLUORO-4-(TRIFLUOROMETH-YL)- see TKB285
2-PENTANONE, 4-METHYL-, PEROXIDE see IJB100
1-PENTANONE, 1-(4-PYRIDYL)- see VBA100
N-PENTAN-4-ONE-N,N,N-TRIMETHYLAMMONIUM IODIDE see TLY250
3-PENTANONE, 1-(2,6,6-TRIMETHYL-2-CYCLOHEXEN-1-YL)- see TLO530
PENTANTIN see DAM700
PENTANTIN HYDROCHLORIDE see DAM700
PENTANYL see PDW750
1,4,7,10,13-PENTAOXACYLOPENTADECANE see PBO000
3,6,9,12,15-PENTAOXAHEPTADECANE see PBO250
2,5,8,11,14-PENTAOXAPENTADECANE see PBO500
PENTAPHEN see AON000
PENTAPHENATE see SJA000
PENTAPHENE HYDROCHLORIDE see PET250
PENTAPOTASSIUM TRIPHOSPHATE see PLW400
PENTAPYRROLIDIUM BITARTRATE see PBT000
PENTASILOXANE, DODECAMETHYL- see DXS800
PENTASILVER DIAMIDOPHOSPHATE see PBO800
PENTASILVER DIIMIDOTRIPHOSPHATE see PBP000
PENTASILVER ORTHODIAMIDOPHOSPHATE see PBP100

PENTASODIUM COLISTINMETHANESULFONATE see SFY500
PENTASODIUM NITRILOTRIS(METHYLENEPHOSPHONATE) see PBP200
PENTASODIUM TRIPHOSPHATE see SKN000
PENTASOL see AOE000, PAX250
PENTASULFURE de PHOSPHORE (FRENCH) see PHS000
PENTAZOCINE see DOQ400
PENTAZOCINE HYDROCHLORIDE see PBP300
PENTECH see DAD200
PENTEK see PBB750
2-PENTENAL see PBP500
4-PENTENAL see PBP750
2-PENTENAL, 4,5-EPOXY- see ECE550
1-PENTENE see PBQ000
2-PENTENE see PBQ250
2-PENTENE, 2-METHYL- see MNK100
(Z)-2-PENTENENITRILE see PBQ275
2-PENTENENITRILE, (Z)- see PBQ275
cis-2-PENTENENITRILE see PBQ275
trans-2-PENTENE OZONIDE see PBQ300
4-PENTENOIC ACID see PBQ750
4-PENTEN-1-OL see PBR000
3-PENTEN-2-ONE see PBR500
1-PENTEN-3-ONE see PBR250
1-PENTEN-3-ONE, 1-(2,6,6-TRIMETHYL-2-CYCLOHEXEN-1-YL)-2-METHYL- see COW780
cis-2-PENTENONITRILE see PBQ275
4-PENTENONITRILE, 3-HYDROXY-, S- see COP400
2-PENTEN-4-YN-3-OL see PBR750
PENTESTAN-80 see PBC250
PENTETATE TRISODIUM CALCIUM see CAY500
PENTETRATE UNICELLES see PBC250
PENTHAMIL see CAY500, DJG800
PENTHAZINE see PDP250
PENTHIENATE BROMIDE see PBS000
PENTHIOBARBITAL see PBT250
PENTHIOBARBITAL SODIUM see PBT500
PENTHRANE see DFA400
PENTICORT see COW825
PENTID see BFD000
PENTIDS see BFD000
PENTIFORMIC ACID see HEU000
PENTILEN see CFU750
PENTILIUM see PBT000
PENTIN C see MNM500
R-PENTINE see CPU500
PENTINIMID see ENG500
PENTITOL see RIF000
PENTOBARBITAL see NBT500, PBS250
PENTOBARBITAL CALCIUM see CAV000
R(+)-PENTOBARBITAL SODIUM see PBS500
PENTOBARBITONE SODIUM see NBU000
PENTOBARBITURATE see NBT500
PENTOBARBITURIC ACID see NBT500
PENTOFRAN see DSI709
PENTOFURYL see DGQ500
PENTOKSIL see HMH300
PENTOLE see CPU500
PENTOLINIUM BITARTRATE see PBT000
PENTOLINIUM DITARTRATE see PBT000
PENTOLINIUM TARTRATE see PBT000
PENTOLITE, dry or wetted with <15% water, by weight (DOT) see PBT050
PENTOLITE see PBT050
PENTOMID see MKU750
PENTONAL see NBU000
PENTOSTAM see AQH800, AQI250
PENTOSTATIN see PBT100
PENTOTHAL see PBT250
PENTOTHAL SODIUM see PBT500
PENTOTHIOBARBITAL see PBT250
PENTOXIFYLLIN see PBU100
PENTOXIFYLLINE see PBU100
PENTOXIL see HMH300
PENTOXIPHYLLIUM see PBU100
PENTOXYL see HMH300
1-PENTOXY-2-METHOXY-4-PROPENYLBENZENE see AOK000
PENTOXYPHYLLINE see PBU100
PENTOYL see EAN700
PENTRAN see DFA400
PENTRANE see DFA400
PENTRATE see PBC250
PENTREX see AIV500
PENTREXL see AIV500
PENTRIOL see PBC250
PENTRYATE 80 see PBC250
PENTYDORM see EQL000

PENTYL see NBU000
1-PENTYL ACETATE see AOD725
2-PENTYL ACETATE see AOD735
n-PENTYL ACETATE see AOD725
PENTYL ACETATE see AOD725
sec-PENTYL ALCOHOL see PBM750
PENTYL ALCOHOL see AOE000
tert-PENTYL ALCOHOL see PBV000
1-PENTYLALLYL ACETATE see ODW028
1-PENTYLAMINE see PBV505
PENTYLAMINE (mixed isomers) see PBV500
n-PENTYLAMINE see PBV505
PENTYLAMINE see PBV505
2,N-PENTYLAMINOETHYL-p-AMINOBENZOATE see PBV750
8-PENTYLBENZ(a)ANTHRACENE see AOE750
tert-PENTYLBENZENE see AOF000
n-PENTYL BENZOATE see PBV800
PENTYL BENZOATE see PBV800
4-n-PENTYLBENZOIC ACID see PBW000
p-PENTYLBENZOIC ACID see PBW000
PENTYLBIPHENYL see AOF500
1-PENTYL BROMIDE see AOF800
n-PENTYL BROMIDE see AOF800
PENTYL BROMIDE see AOF800
PENTYL BUTYRATE see AOG000
PENTYLCANNABICHROMENE see PBW400
PENTYL CAPROATE see PBW450
3-PENTYLCARBINOL see EGW000
sec-PENTYLCARBINOL see EGW000
PENTYLCARBINOL see HFJ500
PENTYL CHLORIDE see PBW500
α-PENTYLCINNAMALDEHYDE see AOG500
α-PENTYL CINNAMYL ACETATE see AOG750
PENTYLCYCLOHEXANOL ACETATE see AOI000
4-tert-PENTYLCYCLOHEXANONE see AOH500
PENTYLCYCLOPENTANONEPROPANONE see OOO100
2-PENTYL-2-CYCLOPENTEN-1-ONE see PBW600
PENTYLDICHLOROARSINE see AOI200
PENTYLENE see AOI800
1,5-PENTYLENE GLYCOL see PBK750
PENTYLENETETRAZOL see PBI500
PENTYL ESTER PHOSPHORIC ACID see PBW750
PENTYL ETHER see PBX000
n-PENTYL FORMATE see AOJ500
PENTYL FORMATE see AOJ500
PENTYLFORMIC ACID see HEU000
PENTYL HEXANOATE see PBW450
n-PENTYLHYDRAZINE HYDROCHLORIDE see PBX250
tert-PENTYL HYDROPEROXIDE see PBX325
N-PENTYL-N-(4-HYDROXYBUTYL)NITROSAMINE see NLE500
2-PENTYLIDENECYCLOHEXANONE see PBX350
PENTYLIDENE GYROMITRIN see PBJ875
1-PENTYL IODIDE see IET500
n-PENTYL IODIDE see IET500
PENTYL IODIDE see IET500
PENTYL KETONE see ULA000
PENTYL MERCAPTAN see PBM000
2-PENTYL-3-METHYL-2-CYCLOPENTEN-1-ONE see DLQ600
PENTYL NITRITE see AOL500
1-PENTYL-3-NITRO-1-NITROSOGUANIDINE see NLC500
4-(PENTYLNITROSAMINO)-1-BUTANOL see NLE500
n-PENTYLNITROSOUREA see PBX500
m-PENTYLOXYCARBANILIC ACID, trans-2-(1-PYRROLIDINYL)CYCLOHEXYL ES-
   TER HYDROCHLORIDE see PPW000
(3-(PENTYLOXY)PHENYL)CARBAMIC ACID, 2-(1-PYRROLIDINYL)CYCLOHEXYL
   ESTER, HCl, (E)- see PPW000
PENTYL PENTYLAMINE see DCH200
2-PENTYLPHENOL see AOM325
o-(sec-PENTYL) PHENOL see AOM500
o-PENTYLPHENOL see AOM325
p-(sec-PENTYL) PHENOL see AOM750
p-PENTYLPHENOL see AOM250
p-tert-PENTYLPHENOL see AON000
m-(3-PENTYL)PHENYL-N-METHYL-N-NITROSOCARBAMATE see PBX750
PENTYL 4-PYRIDYL KETONE see HEW050, PBX800
PENTYLTRICHLOROSILANE see PBY750
PENTYLTRIETHOXYSILANE see PBZ000
PENTYLTRIMETHYLAMMONIUM IODIDE see TMA500
3-PENTYL-6,6,9-TRIMETHYL-6a,7,8,10a-TETRAHYDRO-6H-DIBENZO(b,d)PYRAN-
   1-OL see TCM250
PENTYMAL see AMX750
PENTYMALUM see AMX750
1-PENTYNE see PCA250
2-PENTYNE see PCA300
m-PENTYNOL see EQL000

PENTYREST see EQL000
PEN V see PDT500
PEN-VEE see PDT500
PEN-VEE-K see PDT750
PEN-VEE-K POWDER see PDT750
PENVIKAL see PDT750
PEN-V-K POWDER see PDT750
PENWAR see PAX250
PENZAL N 300 see PAQ200
PENZYLPENICILLIN SODIUM SALT see BFD250
PEONIA (CUBA) see RMK250
PEP 211 see PJS750
PEP see BLY770, EDS000, PCC000, PJR750
PEPCID see FAB500
PEPCIDINE see FAB500
PEPDUL see FAB500
PEPLEOMYCIN see BLY770
PEPLEOMYCIN SULFATE see PCB000
PEPLOMYCIN see BLY770
PEPPER BUSH see DYA875
PEPPERMINT CAMPHOR see MCF750
PEPPERMINT OIL see PCB250
PEPPER PLANTS see PCB275
PEPPER-ROOT see FAB100
PEPPER TREE see PCB300
PEPROSAN see CNK559
PEPSTATIN see PCB750
PEPSTATIN A see PCB750
PEPTAZIN BAFD see BDK800
PEPTICHEMIO see PCC000
PEPTISANT 10 see BDK800
PEPTON 22 see BDK800
PER-ABRODIL see DNG400
PERACETIC ACID see PCL500
PERACON see PCK639
PERAGAL ST see PKQ250
PERAGIT see BBV000
PERANDREN see TBF500, TBG000
PERATOX see PAX250
PERAWIN see PCF275
PERAZIL see CDR000, CDR500
PERAZIL DIHYDROCHLORIDE see CDR000
PERAZINE see PCK500
PERAZINE MALEATE see PCC475
PERBENZOATE de BUTYLE TERTIAIRE (FRENCH) see BSC500
PERBENZOIC ACID see PCM000
PERBROMYL FLUORIDE see PCC750
PERBUNAN C see PJQ050
PERBUTYL H see BRM250
PERCAIN see NOF500
PERCAINE HYDROCHLORIDE see NOF500
PERCA ORANGE GR see CMM220
PERCAPYL see CHX250
PERCARBAMIDE see HIB500
PERCELINE OIL see HCP550
PERCHLOORETHYLEEN, PER (DUTCH) see PCF275
PERCHLOR see PCF275
PERCHLORAETHYLEN, PER (GERMAN) see PCF275
PERCHLORATE ACID, LEAD SALT, HEXAHYDRATE see LDT000
PERCHLORATE de MAGNESIUM (FRENCH) see PCE000
PERCHLORATES see PCD000
PERCHLORATE de SODIUM (FRENCH) see PCE750
PERCHLORETHYLENE see PCF275
PERCHLORETHYLENE, PER (FRENCH) see PCF275
PERCHLORIC ACID, >72% acid by weight (DOT) see PCD250
PERCHLORIC ACID see PCD250
PERCHLORIC ACID, AMMONIUM SALT see PCD500
PERCHLORIC ACID, BARIUM SALT●3H$_2$O see PCD750
PERCHLORIC ACID, CHROMIUM(3+) SALT see CMI300
PERCHLORIC ACID, COBALT(II) SALT, HEXAHYDRATE see CND900
PERCHLORIC ACID, COPPER(II) SALT, DIHYDRATE (8CI, 9CI) see CNO500
PERCHLORIC ACID, ETHYL ESTER see EOD000
PERCHLORIC ACID, LITHIUM SALT, TRIHYDRATE see LHN000
PERCHLORIC ACID, MAGNESIUM SALT see PCE000
PERCHLORIC ACID, MANGANESE(2+) SALT, HEXAHYDRATE see MAU000
PERCHLORIC ACID, MANGANESE(2+) SALT, compounded with 3 mols. of
   OCTAMETHYLPYROPHOSPHORAMIDE see TNK400
PERCHLORIC ACID, NICKEL(II) SALT compounded with OCTAMETHYL PY-
   ROPHOSPHORAMIDE see PCE250
PERCHLORIC ACID, SODIUM SALT see PCE750
PERCHLORIC ACID, SODIUM SALT, MONOHYDRATE see SJB450
PERCHLORIC ACID, TRICHLOROMETHYL ESTER see TIS750
PERCHLORIC ACID, not >50% acid, by weight (UN 1802) (DOT) see
   PCD250

PERCHLORIC ACID, >50% but not >72% acid, by weight (UN 1873) (DOT) see PCD250
PERCHLORIC ACID, ZINC SALT, HEXAHYDRATE see ZKS100
PERCHLORIDE of MERCURY see MCY475
PERCHLORMETHYLMERKAPTAN (CZECH) see PCF300
PERCHLOROBENZENE see HCC500
PERCHLOROBUTADIENE see HCD250
PERCHLORO-2-CYCLOBUTENE-1-ONE see PCF250
PERCHLOROCYCLOPENTADIENE see HCE500
PERCHLORODIHOMOCUBANE see MQW500
PERCHLOROETHANE see HCI000
PERCHLOROETHYLENE see PCF275
PERCHLOROMELAMINE see TNG275
PERCHLOROMETHANE see CBY000
PERCHLOROMETHYL MERCAPTAN see PCF300
PERCHLORON see HOV500
PERCHLOROPENTACYCLO(5.2.1.0$^{2,6}$.0$^{3,9}$.0$^{5,8}$)DECANE see MQW500
PERCHLOROPENTACYCLODECANE see MQW500
PERCHLOROPYRIMIDINE see TBU250
PERCHLOROTHIOPHENE see TBV750
PERCHLORURE d'ANTIMOINE (FRENCH) see AQD000
PERCHLORURE de FER see FAU000
PERCHLORYLBENZENE see PCF500
PERCHLORYL FLUORIDE see PCF750
PERCHLORYL HYPOFLUORITE see FFD000
PERCHLORYL PERCHLORATE see PCF775
1-PERCHLORYLPIPERIDINE see PCG000
PERCHROMATES see PCG250
PERCIN see ILD000
PERCLENE see PCF275
PERCLOROETILENE (ITALIAN) see PCF275
PERCLUSON see CMW750
PERCLUSONE see CMW750
PERCOBARB see ABG750, PCG500
PERCOCCIDE see ALF250
PERCODAN see ABG750, PCG500
PERCODAN HYDROCHLORIDE see DLX400
PERCOL 1697 see DTS500
PERCOLATE see PHX250
PERCORAL see DJS200
PERCORTEN see DAQ800
PERCOSOLVE see PCF275
PERCOTOL see DAQ800
PERCUTACRINE see PMH500
PERCUTACRINE ANDROGENIQUE see TBF500
PERCUTATRINE OESTROGENIQUE ISCOVESCO see DKA600
PERCUTINA see SPD500
PERDIPINE see PCG550
PEREBRAL see DNU100
PEREGRINA (PUERTO RICO, CUBA, US) see CNR135
PEREMESIN see HGC500
PEREQUIL see MQU750
PERFECTA see MQV750
PERFECTHION see DSP400
PERFENAZINA (ITALIAN) see CJM250
PERFLUIDONE see TKF750
PERFLUORIDE see FEX875
PERFLUOROACETIC ACID see TKA250
PERFLUOROACETIC ANHYDRIDE see TJX000
PERFLUOROACETYL CHLORIDE see TJX500
PERFLUOROADIPIC ACID DINITRILE see PCG600
PERFLUOROADIPINIC ACID DINITRILE see PCG600
PERFLUOROADIPONITRILE see PCG600
PERFLUOROAMMONIUM OCTANOATE see ANP625
PERFLUOROBENZOIC ACID see PBD275
PERFLUOROBUT-2-ENE see OBO000
PERFLUORO-2-BUTENE (DOT) see OBO000
PERFLUORO-tert-BUTYL PEROXYHYPOFLUORITE see PCG650
PERFLUOROCAPRYLIC ACID see PCH050
PERFLUOROCTANOIC ACID see PCH050
PERFLUOROCYCLOBUTANE see CPS000
PERFLUORODECAHYDRONAPHTHALENE see PCG700
PERFLUORODECALIN see PCG700
PERFLUORO-N-DECANOIC ACID see PCG725
PERFLUORODECANOIC ACID see PCG725
PERFLUORODECYL IODIDE see IEU075
PERFLUORO-n-DIBUTYL ETHER see PCG755
PERFLUORODIBUTYLETHER see PCG755
PERFLUOROETHENE see TCH500
PERFLUORO ETHER see PCG760
PERFLUOROETHYLENE see TCH500
PERFLUOROFORMAMIDINE see PCG775
PERFLUOROGLUTARIC ACID DINITRILE see HDC300
PERFLUOROGLUTARONITRILE see HDC300
PERFLUORO-n-HEPTANE see PCH000

PERFLUOROHEPTANE see PCH000
PERFLUOROHEPTANECARBOXYLIC ACID see PCH050
PERFLUOROHEXYL IODIDE see PCH100
PERFLUORO HYDRAZINE see TCI000
PERFLUORO-1-IODODODECANE see IEU075
PERFLUORO-1-IODOOCTANE see PCH325
PERFLUOROISOBUTYLENE (ACGIH) see OBM000
PERFLUOROMETHANE see CBY250
PERFLUOROMETHOXYPROPIONIC ACID METHYL ESTER see PCH275
PERFLUOROMETHYLCYCLOHEXANE see PCH290
PERFLUORO-tert-NITROSOBUTANE see PCH300
n-PERFLUOROOCTANE see OAP100
PERFLUOROOCTANE see OAP100
PERFLUOROOCTANOIC ACID see PCH050
PERFLUOROOCTYL IODIDE see PCH325
PERFLUOROPROPENE see HDF000
PERFLUOROPROPIONIC ACID, METHYL ESTER see PCH350
PERFLUOROPROPYLENE see HDF000
PERFLUOROSUCCINIC ACID see TCJ000
PERFLUOROTOLUENE see PCH500
PERFLUOROTRIBUTYLAMINE see HAS000
PERFMID see BSN000
PERGACID VIOLET 2B see FAG120
PERGANTENE see SHF500
PERGITRAL see PBC250
PERGLOTTAL see NGY000
PER-GLYCERIN see LAM000
PERGOLIDE MESYLATE see MPU500
PERHEXILINE MALEATE see PCH800
PERHYDRIT see HIB500
PERHYDROAZEPINE see HDG000
2-PERHYDROAZEPINONE see CBF700
PERHYDROGERANIOL see DTE600
PERHYDROL see HIB000
PERHYDROL-UREA see HIB500
PERHYDRONAPHTHALENE see DAE800
PERIACTIN see PCI500
PERIACTIN HYDROCHLORIDE see PCI250
PERIACTINOL see PCI250, PCI500
PERIACTIN SYRUP see PCI250
PERICHLOR see NCH500
PERICHLORAL see NCH500
PERICHTHOL see IAD000
PERICIAZINE see PIW000
PERICLASE see MAH500
PERICLOR see NCH500
PERICYAZINE see PIW000
PERIDAMOL see PCP250
PERIDEX see CDT250
PERIDEX-LA see PBC250
PERI-DINAPHTHALENE see PCQ250
PERIETHYLENENAPHTHALENE see AAF275
PERIFUNAL see PJB500
PERILAX see GKK000, PPN100
PERILENE see PCQ250
PERILITON BRILLIANT PINK R see AKO350
PERILLA ALCOHOL see PCI550
PERILLA ALDEHYDE see DKX100
PERILLA FRUTESCENS (Linn.) Britt., extract see PCI600
PERILLA FRUTESCENS OIL see PCJ100
PERILLA KETONE see PCI750
PERILLAL see DKX100
PERILLALDEHYDE see DKX100
1-PERILLALDEHYDE-α-ANTIOXIME see PCJ000
PERILLA OCIMOIDES Linn., extract see PCI600
PERILLA OIL see PCJ100
PERILLARTINE see PCJ000
PERILLA SUGAR see PCJ000
PERILLOL see PCI550
PERILLYL ALCOHOL see PCI550
PERILLYL ALDEHYDE see DKX100
PERIMETAZINE see LEO000
PERIMETHAZINE see LEO000
2(1H)-PERINAPHTHAZOLONE see NAX100
trans-PERINONE see CMU820
O-PERIODIC ACID see PCJ250
PERIODIN see PLO500
PERIPHERINE see BBW750
PERIPHERMIN see ACR300
PERIPHETOL see BOV825
PERISOLOL see NDL800
PERISOXAL CITRATE see PCJ325
PERISTIL see PCJ350
PERISTON see PKQ250
PERITOL see PCI250

PERITRATE see PBC250
PERITYL see PBC250
PERK see PCF275
PERKADOX SE 8 see CBF705
PERKE see BBK500
PERKLONE see PCF275
PERLAPINE see HOU059
PERLATAN see EDV000
PERLITE see PCJ400
PERLITON BLUE B see TBG700
PERLITON BLUE FFR see MGG250
PERLITON BLUE GREEN B see DMM400
PERLITON ORANGE 3R see AKP750
PERLITON PINK 3B see AKE250
PERLITON RED VIOLET FFB see DBX000
PERLITON RUBINE 4B see CMP080
PERLITON VIOLET B see DBY700
PERLITON VIOLET 3R see DBP000
PERLITON YELLOW G see AAQ250
PERLITON YELLOW RR see DUW500
PERLON see NOH000
PERLUTEX see MCA000
PERM-A-CHLOR see TIO750
PERMACIDE see PAX250
PERMAFRESH 183 see DTG000
PERMAFRESH 477 see DTG700
PERMAFRESH SW see TCJ900
PERMAGARD see PAX250
PERMAGEL see PAE750
PERMAGEN YELLOW see CMS212
PERMAGEN YELLOW GA see CMS212
PERMA KLEER see DJG800
PERMA KLEER 50 ACID see EIX000
PERMA KLEER 50 CRYSTALS see EIV000
PERMA KLEER 50 CRYSTALS DISODIUM SALT see EIX500
PERMA KLEER TETRA CP see EIV000
PERMA KLEER 50, TRISODIUM SALT see TNL250
PERMANAX 45 see PJQ750
PERMANAX TQ see PJQ750
PERMANENT ORANGE G see CMS145
PERMANENT ORANGE G EXTRA see CMS145
PERMANENT RED 4B see CMS155
PERMANENT RED BBa see CMS148
PERMANENT RED BB see CMS148
PERMANENT RED F 6R see CMS155
PERMANENT RED 4R see MMP100
PERMANENT WHITE see BAP000, ZKA000
PERMANENT YELLOW see BAK250
PERMANENT YELLOW G see CMS212
PERMANENT YELLOW GHG see DEU000
PERMANENT YELLOW GR01 see CMS208
PERMANENT YELLOW GR see CMS208
PERMANENT YELLOW 2K see DUW500
PERMANENT YELLOW LIGHT see CMS212
PERMANGANATE of POTASH (DOT) see PLP000
PERMANGANATE de POTASSIUM (FRENCH) see PLP000
PERMANGANATES see PCJ500
PERMANGANATE de SODIUM (FRENCH) see SJC000
PERMANGANIC ACID AMMONIUM SALT see PCJ750
PERMANGANIC ACID, BARIUM SALT see PCK000
PERMANGANIC ACID(HMnO₄, CALCIUM SALT (8CI,9CI) see CAV250
PERMANGANIC ACID, SODIUM SALT see SJC000
PERMAPEN see BFC750
PERMASAN see PAX250
PERMASEAL see PJH500
PERMATOX DP-2 see PAX250
PERMATOX PENTA see PAX250
PERMEK N see MKA500
PERMETHRIN (USDA) see AHJ750
PERMETRINA (PORTUGUESE) see AHJ750
PERMETRIN (HUNGARIAN) see AHJ750
PERMICORT see CNS750
PERMINAL FC-P see CNH125
PERMITAL see DYE600
PERMITE see PAX250
PERMITIL see TJW500
PERMITIL HYDROCHLORIDE see FMP000
PERMONID see DKX600
PERMONOSULFAMIC ACID see HLN100
PERNAEMON see VSZ000
PERNAEVIT see VSZ000
PERNAMBUCO EXTRACT see LFT800
PERNAZINE see PCK500
PERNETTYA (various species) see PCK600
PERNIPURON see VSZ000

PERNITHRENE BLUE BC see DFN425
PERNITHRENE BLUE RS see IBV050
PERNITHRENE OLIVE R see DUP100
PERNOCTON see BOR000
PERNOSTON see BOR000
PERNOX see CLY500
PERNSATOR-WIRKSTOFF see PNN300
PEROCAN CITRATE see PCK639
PEROCAN CYCLAMATE see PCK669
PEROLYSEN see PBJ000
PERONE 30 see HIB010
PERONE 35 see HIB010
PERONE 50 see HIB010
PERONE see HIB000
PERONIAS (PUERTO RICO) see RMK250
PEROPAL see THT500
PEROPYRENE see DDC400
PEROSIN see EIR000
PEROSSIDO di BENZOILE (ITALIAN) see BDS000
PEROSSIDO di BUTILE TERZIARIO (ITALIAN) see BSC750
PEROSSIDO di IDROGENO (ITALIAN) see HIB000
PEROXAN see HIB000
PEROXIDE see HIB000
PEROXIDE, ACETYL CYCLOHEXYLSULFONYL see ACG300
PEROXIDE, BIS(1-OXONONYL) see NNA100
PEROXIDE, BIS(1-OXOOCTYL) (9CI) see CBF705
PEROXIDE, BIS(1-OXOPROPYL) see DWQ800
PEROXIDE, 1-HYDROPEROXYCYCLOHEXYL 1-HYDROXYCYCLOHEXYL see CPC300
PEROXIDE, OCTANOYL see CBF705
PEROXIDE, (PHENYLENEBIS(1-METHYLETHYLIDENE))BIS(1,1-DIMETHYLE-THYL)- see BHL100
PEROXIDE, (PHENYLENEDIISOPROPYLIDENE)BIS(tert-BUTYL- see BHL100
PEROXIDES, INORGANIC see PCL000
PEROXIDES, ORGANIC see PCL250
PEROXIDE, (1,1,4,4-TETRAMETHYL-1,4-BUTANEDIYL)BIS((1,1-DIMETHYLETHYL) see DRJ800
PEROXIDE, (1,1,4,4-TETRAMETHYLTETRAMETHYLENE)BIS(tert-BUTYL see DRJ800
1,4-PEROXIDO-p-MENTHENE-2 see ARM500
PEROXOACETIC ACID see PCL500
PEROXOMONOPHOSPHORIC ACID see PCN500
PEROXOMONOSULFURIC ACID see PCN750
PEROXYACETIC ACID, >43% and with >6% hydrogen peroxide (DOT) see PCL500
PEROXYACETIC ACID see PCL500
PEROXYACETYL NITRATE see PCL750
PEROXYACETYL PERCHLORATE see PCL775
PEROXYBENZOIC ACID see PCM000
4-PEROXY-CPA see HIF000
PEROXYDE de BARYUM (FRENCH) see BAO250
PEROXYDE de BENZOYLE (FRENCH) see BDS000
PEROXYDE de BUTYLE TERTIAIRE (FRENCH) see BSC750
PEROXYDE d'HYDROGENE (FRENCH) see HIB000
PEROXYDE de LAUROYLE (FRENCH) see LBR000
PEROXYDE de PLOMB (FRENCH) see LCX000
PEROXYDICARBONATE d'ISOPROPYLE see DNR400
PEROXYDICARBONIC ACID, BIS(2-ETHYLHEXYL) ESTER see DJK800
PEROXYDICARBONIC ACID, BIS(1-METHYLETHYL) ESTER see DNR400
PEROXYDICARBONIC ACID, DIBENZYL ESTER see DDH200
PEROXYDICARBONIC ACID, DIBUTYL ESTER see BSC800
PEROXYDICARBONIC ACID, DICYCLOHEXYL ESTER see DGV650
PEROXYDICARBONIC ACID, DIETHYL ESTER see DJU600
PEROXYDICARBONIC ACID, DI(2-ETHYLHEXYL) ESTER see DJK800
PEROXYDICARBONIC ACID DIPROPYL ESTER see DWV400
PEROXYDICARBONIC ACID, DISODIUM SALT see SJB400
PEROXYDISULFIRIC ANHYDRIDE see DXH600
PEROXYDISULFURIC ACID DIPOTASSIUM SALT see DWQ000
PEROXYDISULFURYL DIFLUORIDE see PCM250
PEROXYFORMIC ACID see PCM500
PEROXYFUROIC ACID see PCM550
PEROXYHEXANOIC ACID see PCM750
PEROXYISOBUTYRIC ACID, tert-BUTYL ESTER see BSC600
PEROXYLINOLEIC ACID, SODIUM SALT see SIC250
PEROXYLINOLENIC ACID see PCN000
PEROXYMONOPHOSPHORIC ACID see PCN500
PEROXYMONOSULFURIC ACID see PCN750
PEROXYNITRIC ACID see PCO000
PEROXYPROPIONIC ACID see PCO100
PEROXYPROPIONYL NITRATE see PCO150
PEROXYPROPIONYL PERCHORATE see PCO175
PEROXYSULFURIC ACID, POTASSIUM SALT see PLP750
PEROXYTRIFLUOROACETIC ACID see PCO250
PERPARINE HYDROCHLORIDE see PAH260
PERPARIN HYDROCHLORIDE see PAH260

PERPERINE HYDROCHLORIDE see PAH260
PERPHENAZIN see CJM250
PERPHENAZINE see CJM250
PERPHENAZINE DIHYDROCHLORIDE see PCO500
PERPHENAZINE HYDROCHLORIDE see PCO750
PERPHENAZINE MALEATE see PCO850
PER-RADIOGRAPHOL see DNG400
PERSADOX see BDS000
PERSAMINE see DLH630
PERSANTIN see PCP250
PERSANTINAT see ARQ750
PERSANTINE see PCP250
PERSEC see PCF275
PERSIAN BERRY see MBU825
PERSIAN BERRY LAKE see MQF250
PERSIAN LILAC see CDM325
PERSIAN ORANGE see CMM220
PERSIAN ORANGE LAKE see CMM220
PERSIAN ORANGE X see CMM220
PERSIAN RED see LCS000
PERSIA-PERAZOL see DEP800
PERSICOL see HBN200
PERSIMMON see PCP500
PERSISTEN see DJT400
PERSISTOL see TND500
PERSKLERAN see RDK000
PERSPEX see PKB500
PERSULFATE d'AMMONIUM (FRENCH) see ANR000
PERSULFATE de SODIUM (FRENCH) see SJE000
PERSULFEN see SNN300
PERTESTIS see TBF600
PERTHANE see DJC000
PERTHIOXANTHATE, TRICHLOROMETHYL ALLYL see TIR750
PERTHIOXANTHATE, TRICHLOROMETHYL METHYL see TIS500
PERTOFRAM see DLH630
PERTOFRAN see DLS600, DSI709
PERTOFRANE see DLS600, DSI709
PERTOXIL see CMW500
PERU BALSAM see PCP750
PERU BALSAM OIL see PCQ000
PERUSCABIN see BCM000
PERUVIAN BALSAM see BAE750
PERUVIAN JACINTH see SLH200
PERUVIAN MASTIC TREE see PCB300
PERUVOSID see EAQ050
PERUVOSIDE see EAQ050
PERVAGAL see HKR500
PERVAL see VLF000
PERVERTIN see DBB000
PERVINCAMINE see VLF000
PERVITIN see DBA800, MDT600
PERVONE see VLF000
PERYCIT see NCW300
PERYLENE see PCQ250
PES 100 see PJS750
PES 200 see PJS750
PESTAN see DJI000
PESTMASTER see EIY500
PESTMASTER EDB-85 see EIY500
PESTMASTER (OBS.) see MHR200
PESTON XV see PHF750
PESTOX 3 see OCM000
PESTOX 14 see BJE750
PESTOX 15 see PHF750
PESTOX 101 see NIM500
PESTOX see OCM000
PESTOX III see OCM000
PESTOX IV see BJE750
PESTOX PLUS see PAK000
PESTOX XIV see BJE750
PESTOX XV see PHF750
PETA see PBC750
PETASITENINE (neutral) see PCQ750
PETASITENINE see PCQ750
PETASITES JAPONICUS MAXIM see PCR000
PETEHA see PNW750
PETE-PETE (HAITI) see RBZ400
PETERPHYLLIN see TEP500
PETERSILIENSAMEN OEL (GERMAN) see PAL750
PETHIDINE CHLORIDE see DAM700
PETHIDINETER see DAM600
PETHIDOINE see DAM600
PETHION see PAK000
PETIDIN see DAM700
PETIDION see TLP750

PETIDON see TLP750
PETILEP see TLP750
PETINIMID see ENG500
PETINUTIN see MLP800
PETITGRAIN BIGARADE OIL see OHC150
PETITGRAIN OIL see OHC150
PETITGRAIN OIL SAPONIFIED see OHC150
PETNAMYCETIN see CDP250
PETNIDAN see ENG500
PETRICHLORAL see NCH500
PETRIN see PCR150
PETRISUL see AKO500
PETROGALAR see MQV750
PETROHOL see INJ000
PETROL (DOT) see GBY000
PETROL see PCR250
PETROLATUM, liquid see MQV750
PETROLEUM see PCR250
PETROLEUM ASPHALT see ARO500, PCR500
PETROLEUM BENZIN see NAI500
PETROLEUM BITUMEN see ARO500
PETROLEUM CRUDE see PCR250
PETROLEUM CRUDE OIL (DOT) see PCR250
PETROLEUM-DERIVED NAPHTHA see NAI500
PETROLEUM DISTILLATE see PCS250
PETROLEUM DISTILLATES, CLAY-TREATED HEAVY NAPHTHENIC see
    PCS260
PETROLEUM DISTILLATES, CLAY-TREATED LIGHT NAPHTHENIC see
    PCS270
PETROLEUM DISTILLATES, HYDROTREATED (mild) HEAVY NAPHTHENIC
    see MQV790
PETROLEUM DISTILLATES (NAPHTHA) see NAI500
PETROLEUM DISTILLATES, SOLVENT-DEWAXED HEAVY PARAFFINIC see
    MQV825
PETROLEUM ETHER see PCT250
PETROLEUM GAS, LIQUEFIED see LGM000
PETROLEUM OIL (UN1270) (DOT) see NAI500
PETROLEUM PITCH see ARO500
PETROLEUM ROOFING TAR see ARO500, PCR500
PETROLEUM 60 SOLVENT see PCS750
PETROLEUM 70 SOLVENT see PCT000
PETROLEUM SPIRIT (DOT) see PCT250
PETROLEUM SPIRITS see PCT250
PETROLEUM 50 THINNER see PCT500
PETROL ORANGE Y see PEJ500
PETROL YELLOW C see CMS245
PETROL YELLOW WT see DOT300
PETROSULPHO see IAD000
PETROTHENE see PJS750
PETROTHENE LB 861 see PJS750
PETROTHENE LC 731 see PJS750
PETROTHENE LC 941 see PJS750
PETROTHENE NA 219 see PJS750
PETROTHENE NA 227 see PJS750
PETROTHENE XL 6301 see PJS750
PETUNIDOL see PCU000
PETZINOL see TIO750
PEVARYL see EAE000
PEVIKON D 61 see PKQ059
PEVITON see NCQ900
PEXID see PCH800
PEYRONE'S CHLORIDE see PJD000
PF-1 see DSA800
PF-3 see IRF000
PF-26 see DWK700
PF 38 see FQU875
PF-82 see DWK900
PF 1593 see BON325
PFD see PCG700
PFDA see PCG725
PFEFFERMINZ OEL (GERMAN) see PCB250
PFETFFER'S SUBSTANCE see AMK250
PFH see PNM650
PFIB see OBM000
PFIKLOR see PLA500
PFIZER 1393 see BDE250
PFIZER-E see EDJ500
PFIZERPEN see BFD000
PFIZERPEN A see AIV500
PFIZERPEN VK see PDT750
PFOA see PCH050
PFT see CML835
PG 12 see PML000
PG-501 see MBV100
PG see PML250

PGA2 see MCA025
PGA² see MCA025
5,6-cis-PGA² see MCA025
(155)-PGA² see MCA025
PGA see AHC000
PhGABA see PEE500
PGABA see PEE500
PGDN see PNL000
PGE-1 see POC350
PGE2 see DVJ200
PGE see PFH000
PGF2-α see POC500
PGF2-α METHYL ESTER see DVJ100
PGF2 METHYL ESTER see DVJ100
PGF2-α THAM see POC750
PGF2-α TRIS SALT see POC750
PGF2-α TROMETHAMINE see POC750
PH 60-40 see CJV250
PHACETUR see PEC250
PHALDRONE see CDO000
PHALLOIDIN see PCU350
PHALLOIDINE see PCU350
PHALTAN see TIT250
PHANAMIPHOS see FAK000
PHANANTIN see DKQ000
PHANODORM see TDA500
PHANODORN see TDA500
PHANQUINONE see PCY300
PHANQUINONUM see PCY300
PHANQUONE see PCY300
PHARGAN see DQA400
PHARLON see EDS100
PHARMACID GREEN S see ADF000
PHARMACINE YELLOW R see CMM759
PHARMAGEL A see PCU360
PHARMAGEL AdB see PCU360
PHARMAGEL B see PCU360
PHARMAGLO RED G see CMM325
PHARMANIL RED RB see NAO600
PHARMANIL SCARLET Y see CMM320
PHARMANTHRENE GOLDEN YELLOW see DCZ000
PHARMASORB-COLLOIDAL see PAE750
PHARMATEX YELLOW G see CMM759
PHARMAZOID RED KB see CLK225
PHARMEDRINE see AOB250
PHAROS 100.1 see SMR000
PHARYCIDIN CONCENTRATE see BEL900
PHASEOLUNATIN see GFC100
PHASOLON see BDJ250
PH BC see BQJ500
PHBN see NLN000
PHC see PMY300
M-PHDM see BKL750
PHEASANT'S EYE see PCU375
PHEBUZIN see BRF500
PHELIPAEA CALOTROPIDIS Walp., extract see CMS248
α-PHELLANDRENE (FCC) see MCC000
PHELLOBERIN A see PCU390
PHEM see PCU400
PHEMERIDE see BEN000
PHE-MER-NITE see MCU750
PHEMERNITE see MDH500
PHEMEROL CHLORIDE see BEN000
PHEMEROL CHLORIDE MONOHYDRATE see BBU750
PHEMETONE see ENB500
PHEMITHYN see BEN000
PHEMITON see ENB500
PHEMITONE see ENB500
PHENACAINE see BJO500
PHENACALUM see PEC250
PHENACEMIDE see PEC250
PHENACEREUM see PEC250
PHENACETALDEHYDE DIMETHYL ACETAL see PDX000
p-PHENACETIN see ABG750
PHENACETUR see PEC250
PHENACETYLCARBAMIDE see PEC250
PHENACETYLUREA see PEC250
PHENACHLOR see TIW000
PHENACID see PCU425
PHENACIDE see CDV100
PHENACTYL see CKP250
PHENACYLAMINE see AHR250
PHENACYL-6-AMINOPENICILLINATE see PCU500
PHENACYL CHLORIDE see CEA750
PHENACYLIDENE CHLORIDE see DEN200

PHENACYLPIVALATE see PCV350
PHENADONE see MDO750
PHENADONE HYDROCHLORIDE see MDP750
PHENADOR-X see BGE000
PHENAEMAL see EOK000
PHENAGLYCODOL see CKE750
PHENALCO see MCU750
PHENALENO(1,9-gh)QUINOLINE see PCV500
PHENALGENE see AAQ500
PHENALGIN see AAQ500
PHENALLYMAL see AGQ875
PHENALLYMALUM see AGQ875
PHENALZINE see PFC750
PHENALZINE DIHYDROGEN SULFATE see PFC750
PHENALZINE HYDROGEN SULPHATE see PFC750
PHENAMACIDE HYDROCHLORIDE see PCV750
PHENAMIDE see FAJ150, PCV775
PHENAMINE see AOB250
PHENAMINE BLACK BCN-CF see AQP000
PHENAMINE BLACK CL see AQP000
PHENAMINE BLACK E 200 see AQP000
PHENAMINE BLACK E see AQP000
PHENAMINE BLUE BB see CMO000
PHENAMINE BLUE RW see CMO600
PHENAMINE BROWN D 3G see CMO810
PHENAMINE BROWN 3G see CMO825
PHENAMINE DARK GREEN B see CMO830
PHENAMINE FAST BROWN T see CMO820
PHENAMINE FAST BROWN TWC see CMO820
PHENAMINE FAST RED F see CMO870
PHENAMINE GREEN BG see CMO840
PHENAMINE GREEN C see CMO840
PHENAMINE GREEN G see CMO840
PHENAMINE SCARLET 3B see CMO875
PHENAMINE SKY BLUE A see CMO500
PHENAMINE VISCOSE BLACK RR see CMN240
PHENAMIZOLE HYDROCHLORIDE see DCA600
PHENANTHRA-ACENAPHTHENE see PCW000
9,10-PHENANTHRAQUINONE see PCX250
PHENANTHREN (GERMAN) see PCW250
PHENANTHRENE see PCW250
PHENANTHRENE-1,2-DIHYDRODIOL see DMA000
PHENANTHRENE-3,4-DIHYDRODIOL see PCW500
9,10-PHENANTHRENEDIONE see PCX250
PHENANTHRENE-9,10-EPOXIDE see PCX000
9,10-PHENANTHRENE OXIDE see PCX000
9,10-PHENANTHRENEQUINONE see PCX250
PHENANTHRENEQUINONE see PCX250
PHENANTHRENETETRAHYDRO-3,4-EPOXIDE see ECR500
4,7-PHENANTHROLENE-5,6-QUINONE see PCY300
1,10-o-PHENANTHROLINE see PCY250
1,10-PHENANTHROLINE see PCY250
β-PHENANTHROLINE see PCY250
o-PHENANTHROLINE see PCY250
4,7-PHENANTHROLINE-5,6-DIONE see PCY300
PHENANTHRO(2,1-d)THIAZOLE see PCY400
2-PHENANTHRYLACETAMIDE see AAM250
9-PHENANTHRYLACETAMIDE see PCY750
N-2-PHENANTHRYLACETAMIDE see AAM250
N-(2-PHENANTHRYL)ACETAMIDE see AAM250
N-3-PHENANTHRYLACETAMIDE see PCY500
N-9-PHENANTHRYLACETAMIDE see PCY750
2-PHENANTHRYLACETHYDROXAMIC ACID see PCZ000
N-(2-PHENANTHRYL)ACETOHYDROXAMIC ACETATE see ABK250
N-2-PHENANTHRYLACETOHYDROXAMIC ACID see PCZ000
2-PHENANTHRYLAMINE see PDA250
3-PHENANTHRYLAMINE see PDA500
9-PHENANTHRYLAMINE see PDA750
PHENANTOIN see MKB250
PHENANTRIN see PCW250
PHENAROL see CKF500
PHENARONE see PEC250
PHENARSAZINE CHLORIDE see PDB000
PHENARSAZINE OXIDE see PDB250
10-PHENARSAZINETHIOL, S-ESTER with O,O-DIISOOCTYLPHOSPHORODITH-
   IOATE see PDB300
S-(10-PHENARSAZINYL)-O,O-DIISOOCTYLPHOSPHORODITHIOATE see PDB300
PHENARSEN see OOK100
PHENARSENAMINE see SAP500
PHENASAL see DFV400
PHENATHYL see CKP250
PHENATOINE see DKQ000
PHENATOX see CDV100
PHENAZEPAM see PDB350
PHENAZIN see PDB750

PHENAZINE see DKE800, PDB500
PHENAZINE-5N-OXIDE see PDB750
PHENAZINE-9-OXIDE see PDB750
PHENAZINE-N-OXIDE see PDB750
PHENAZINIUM, 3,7-DIAMINO-2,8-DIMETHYL-5-PHENYL-, CHLORIDE see GJI400
1-PHENAZINOL, 6-METHOXY-, 5,10-DIOXIDE see HLT100
PHENAZIN-5-OXIDE see PDB750
PHENAZIN OXIDE see PDB750
PHENAZO see PDC250
PHENAZO BLACK BH see CMN800
PHENAZO BLACK D see CMN230
PHENAZOCINE HYDROBROMIDE see PMD325
PHENAZODINE see PDC250, PEK250
PHENAZOLINE see PDC000
PHENAZONE (pharmaceutical) see AQN000
PHENAZOPYRIDINE see PEK250
PHENAZOPYRIDINE HYDROCHLORIDE see PDC250
PHENAZOPYRIDINIUM CHLORIDE see PDC250
PHENBENZAMINE see BEM500
PHENBENZAMINE HYDROCHLORIDE see PEN000
PHENBUTAZOL see BRF500
PHENCAPTON see PDC750
PHENCARBAMID see PDC850
PHENCARBAMIDE see PDC850
PHENCARBAMIDE HYDROCHLORIDE see PDC875
PHENCARBAMIDE HYDROCHLORIDE (GERMAN) see PDC875
PHENCAROL see DWM400
PHENCEN see PMI750
PHENCYCLIDINE see PDC890
PHENCYCLIDINE HYDROCHLORIDE see AOO500
PHENDAL see DRR400
PHENDIMETRAZINE BITARTRATE see DKE800
PHENDIMETRAZINE HYDROCHLORIDE see DTN800
PHENDIMETRAZINE TARTRATE see PDD000
PHENDIPHAM see MEG250
PHENE see BBL250
PHENEDRINE see AOB250, BBK000
PHENEENE GERMICIDAL SOLUTION and TINCTURE see AFP250
PHENEGIC see PDP250
PHENELZIN see PFC750
PHENELZINE see PFC500
PHENELZINE ACID SULFATE see PFC750
PHENELZINE BISULPHATE see PFC750
PHENELZINE SULFATE see PFC750
PHENEMALUM see SID000
(v-PHENENYLTRIS(OXYETHYLENE))TRIS(TRIETHYLAMMONIUM IODIDE) see PDD300
PHENERGAN see DQA400
PHENERGAN HYDROCHLORIDE see PMI750
PHENESTERINE see CME250
PHENESTRIN see CME250
PHENETAMINE HYDROCHLORIDE see LFK200
PHENETHAMINE HYDROCHLORIDE see LFK200
PHENETHANOL see PDD750
PHENETHECILLIN POTASSIUM see PDD350
PHENETHECILLIN POTASSIUM SALT see PDD350
PHENETHICILLIN K see PDD350
PHENETHICILLIN K SALT see PDD350
1-PHENETHOXY-1-PROPOXYETHANE see PDD400
β-PHENETHYBIGUANIDE see PDF000
2-PHENETHYL ACETATE see PFB250
β-PHENETHYL ACETATE see PFB250
sec-PHENETHYL ACETATE see SMP600
2-PHENETHYL ALCOHOL see PDD750
α-PHENETHYL ALCOHOL see PDE000
β-PHENETHYL ALCOHOL see PDD750
PHENETHYL ALCOHOL see PDD750
PHENETHYL ALCOHOL, BENZOATE see PFB750
PHENETHYL ALCOHOL, FORMATE see PFC250
PHENETHYL ALCOHOL, α-METHYL- see PGA600
PHENETHYL ALCOHOL, β-METHYL- see HGR600
β-PHENETHYLAMINE see PDE250
β-PHENETHYLAMINE HYDROCHLORIDE see PDE500
PHENETHYLAMINE, β-HYDROXY- see HNF000
PHENETHYLAMINE, α-METHYL-, SULFATE (2:1) see BBK250
β-PHENETHYL-o-AMINOBENZOATE see APJ500
PHENETHYL ANTHRANILATE see APJ500
PHENETHYL BENZOATE see PFB750
1-PHENETHYLBIGUANIDE see PDF000
β-PHENETHYLBIGUANIDE see PDF000
1-PHENETHYLBIGUANIDE HYDROCHLORIDE see PDF250
N'-β-PHENETHYLBIGUANIDE HYDROCHLORIDE see PDF250
PHENETHYLBIGUANIDE HYDROCHLORIDE see PDF250
2-PHENETHYL BUTANOATE see PFB800
β-PHENETHYL N-BUTANOATE see PFB800

PHENETHYL BUTYRATE see PFB800
PHENETHYLCARBAMID (GERMAN) see EFE000
PHENETHYL CHLORACETATE see PDF500
β-PHENETHYL CINNAMATE see BEE250
PHENETHYL CINNAMATE see BEE250
PHENETHYL CYANIDE see HHP100
PHENETHYLDIGUANIDE see PDF000
PHENETHYLENE see SMQ000
PHENETHYLENE OXIDE see EBR000
PHENETHYL ESTER HYDRACRYLIC ACID see PFC100
PHENETHYL ESTER ISOVALERIC ACID see PDF775
N'-β-PHENETHYLFORMAMIDINYLLIMINOUREA see PDF000
PHENETHYL FORMATE see PFC250
PHENETHYLHYDRAZINE see PFC500
PHENETHYLHYDRAZINE SULFATE (1:1) see PFC750
1-(1-PHENETHYL)-IMIDAZOLE-5-CARBOXYLIC ACID, METHYL ESTER, HYDRO-
    CHLORIDE see MQQ750
PHENETHYL ISOBUTYRATE see PDF750
β-PHENETHYL ISOTHIOCYANATE see ISP000
PHENETHYL ISOVALERATE see PDF775
PHENETHYL METHACRYLATE see PFP600
2-PHENETHYL 2-METHYLBUTYRATE see PDF790
PHENETHYL 2-METHYLBUTYRATE see PDF780
PHENETHYLMETHYLETHYLCARBINOL see PFR200
PHENETHYL METHYL KETONE see PDF800
PHENETHYL-8-OXA-1-DIAZA-3,8-SPIRO(4,5)DECANONE-2-HYDROCHLORIDE
    see DAI200
PHENETHYL PHENYLACETATE see PDI000
N-(p-PHENETHYL)PHENYLACETOHYDROXAMIC ACID see PDI250
α-(p-PHENETHYLPHENYL)-1-IMIDAZOLEETHANOL MONOHYDROCHLORIDE
    see DAP880
1-PHENETHYLPIPERIDINE see PDI500
1-PHENETHYL-3-(PIPERIDINOCARBONYL)PIPERIDINE see PDI550
N-(1-PHENETHYL-4-PIPERIDINYL)PROPIONANILIDE DIHYDROGEN CITRATE
    see PDW750
1-PHENETHYL-4-PIPERIDYL-p-AMINOBENZOATE HYDROCHLORIDE see
    PDJ000
1-PHENETHYL-4-PIPERIDYL BENZOATE HYDROCHLORIDE see PDJ250
1-PHENETHYL-3-PIPERIDYL PIPERIDINO KETONE see PDI550
N-(1-PHENETHYL-4-PIPERIDYL)PROPIONANILIDE CITRATE see PDW750
N-(1-PHENETHYL-4-PIPERIDYL)PROPIONANILIDE DIHYDROGEN CITRATE see
    PDW750
PHENETHYL PROPIONATE see PDK000
1-PHENETHYL-4-N-PROPIONYLANILINOPIPERIDINE see PDW500
N-PHENETHYL-4-(N-PROPIONYLANILINO)PIPERIDINE see PDW500
PHENETHYL SALICYLATE see PDK200
1-PHENETHYLSEMICARBAZIDE see PDK300
2-PHENETHYL-3-THIOSEMICARBAZIDE see PDK500
PHENETHYL TIGLATE see PFD250
PHENETHYLUREA see PDK750
PHENETICILLIN POTASSIUM see PDD350
p-PHENETIDIN see PDK890
p-PHENETIDINANTIMONYLTARTRAT (GERMAN) see PDL500
2-PHENETIDINE see PDK819
m-PHENETIDINE see PDK800
o-PHENETIDINE see PDK819
PHENETIDINE see PDK890
m-PHENETIDINE ANTIMONYL TARTRATE see PDL000
o-PHENETIDINE ANTIMONYL TARTRATE see PDK900
p-PHENETIDINE ANTIMONYL TARTRATE see PDL500
p-PHENETIDINE HYDROCHLORIDE see PDL750
PHENETIDINE HYDROCHLORIDE see PDL750
p-PHENETOLCARBAMID (GERMAN) see EFE000
p-PHENETOLCARBAMIDE see EFE000
PHENETOLE see PDM000
p-PHENETOLECARBAMIDE see EFE000
PHENETURIDE see PFB350
p-PHENETYLUREA see EFE000
PHENFLUORAMINE HYDROCHLORIDE see PDM250
PHENFORMINE see PDF000
PHENGLYKODOL see CKE750
PHENHYDREN see PFJ750
PHENIBUT see PEE500
PHENIBUT HYDROCHLORIDE see GAD000
PHENIC ACID see PDN750
PHENICARB see PEC250
PHENICOL see EGQ000
PHENIDONE see PDM500
PHENIDYLATE see MNQ000
PHENIGAM see PEE500
PHENIGAMA see PEE500
PHENIGAMA HYDROCHLORIDE see GAD000
PHENIGAM HYDROCHLORIDE see GAD000
PHENINDAMINE HYDROCHLORIDE see TEQ700
PHENINDAMINE HYDROGEN TARTRATE see PDD000

PHENINDIONE see PFJ750
PHENIODOL see PDM750
PHENIPRAZINE see PDN000
PHENIPRAZINE HYDROCHLORIDE see PDN250
PHENIRAMINE MALEATE see TMK000
PHENISATIN see ACD500
PHENISOBROMOLATE see IOS000
(±)-PHENISOPROPYLAMINE SULFATE see AOB250
PHENISTAN see SPC500
PHENITOL see MCU750
PHENITROTHION see DSQ000
PHENIZIDOLE see PEL250
PHENIZINE see PDN000
PHENLINE see PFC750
PHENMAD see ABU500
PHENMEC see PFS350
PHENMEDIPHAM see MEG250
PHENMEDIPHAME see MEG250
PHENMERZYL NITRATE see MCU750, MDH500
PHENMETHYL TRIMETHYLAMMONIUM IODIDE see BFM750
PHENMETRAZIN see PMA750
PHENMETRAZINE see PMA750
PHENMETRAZINE HYDROCHLORIDE see MNV750
PHENOBAL see EOK000
PHENOBAL SODIUM see SID000
PHENOBARBITAL see EOK000
PHENOBARBITAL ELIXIR see SID000
PHENOBARBITAL Na see SID000
PHENOBARBITAL SODIUM see SID000
PHENOBARBITAL SODIUM SALT see SID000
PHENOBARBITOL and DIPHENYLHDANTOIN see DWD000
PHENOBARBITONE see EOK000
PHENOBARBITONE and PHENOBARBITONE see DWD000
PHENOBARBITONE SODIUM see SID000
PHENOBARBITONE SODIUM SALT see SID000
PHENOBARBITURIC ACID see EOK000
PHENO BLACK EP see AQP000
PHENO BLACK SGN see AQP000
PHENO BLUE 2B see CMO000
PHENOBOLIN see DYF450
PHENO BRIGHT GREEN see CMO840
PHENOCAINE see BQH250
PHENOCHLOR see PJL750
PHENOCLOR see PJL750
PHENOCLOR DP6 see PJN250
PHENODIANISYL see PDN500
PHENODIANISYL HYDROCHLORIDE see PDN500
PHENODIOXIN see DDA800
PHENODODECINIUM BROMIDE see DXX000
PHENODYNE see PFC750
PHENO FAST RED F see CMO870
PHENO FAST SCARLET 4B see CMO875
PHENO FAST SCARLET 9B see CMO885
PHENOFORMINE HYDROCHLORIDE see PDF250
PHENOHEP see HCI000
PHENOL, molten (DOT) see PDN750
PHENOL see PDN750
PHENOL ACETATE see PDY750
PHENOL ALCOHOL see PDN750
PHENOL, 4-ALLYL-2-METHOXY-, FORMATE (ester) see EQS100
PHENOL, 2-AMINO-4-ARSENOSO- see OOK100
PHENOL, 2-AMINO-4-ARSENOSO-, SODIUM SALT see ARJ900
PHENOL, 4-AMINO-3-METHYL- see AKZ000
PHENOL, p-ARSENOSO- see HNG800
PHENOL, 4-ARSENOSO-2-NITRO- see NHE600
PHENOL-p-ARSONIC ACID see PDO250
PHENOL, 4,4'-(3H-2,1-BENZOXATHIOL-3-YLIDENE)BIS-, S,S-DIOXIDE (9CI) see PDO800
PHENOL,4,4'-(3H-2,1-BENZOXATHIOL-3-YLIDENE)BIS(5-METHYL-2-(1-METHYLE-THYL))-, S,S-DIOXIDE see TFX850
PHENOL, 4,4'-(3H-2,1-BENZOXATHIOL-3-YLIDENE)DI-, S,S-DIOXIDE see PDO800
PHENOL, p-(BENZYLAMINO)- see BDY750
PHENOL, 2,4-BIS(ACETOXYMERCURI)- see HNK575
PHENOL, 2,4-BIS(1,1-DIMETHYLETHYL)-6-(1-PHENYLETHYL)- see BJK650
PHENOL, 2,4-BIS(1,1-DIMETHYLETHYL)-6-(PHENYLMETHYL)- see DEG150
PHENOL, 2,6-BIS(1-METHYLETHYL)-(9CI) see DNR800
PHENOL, BROMO- see BNU800
PHENOL, 4-BROMO-2,5-DICHLORO- see LEN050
PHENOL, o-(tert-BUTYL)- see BSE460
PHENOL, 2-sec-BUTYL-4,6-DINITRO-, ACETATE (ESTER) (8CI) see ACE500
PHENOL, 6-t-BUTYL-3-(2-IMIDAZOLIN-2-YLMETHYL)-2,4-DIMETHYL- see ORA100
PHENOL, 4-tert-BUTYL-2-METHYL- see BRU800
PHENOL, 2-tert-BUTYL-5-METHYL- see BQV600

PHENOL, 2-tert-BUTYL-6-METHYL- see BRU790
PHENOL, 4-(3-CARBAZOLYLAMINO)- see CBN100
PHENOLCARBINOL see BDX500
PHENOL, 4-CHLORO- see CJK750
PHENOL, 4-(((2-CHLOROETHYL)AMINO)METHYL)-4-NITRO- see CGQ280
PHENOL, o-(CHLOROMERCURI)- see CHW675
PHENOL, 2-CHLORO-5-METHYL- see CFE500
PHENOL, 4-CHLORO-5-METHYL-2-(1-METHYLETHYL)-(9CI) see CLJ800
PHENOL, 4-CHLORO-2-(PHENYLMETHYL)-, SODIUM SALT see SFB200
PHENOL, 2,4-DIBROMO- see DDR150
PHENOL, 2,6-DIBROMO-4-NITRO- see DDQ500
PHENOL, 2,6-DI-tert-BUTYL- see DEG100
PHENOL, 2,5-DICHLORO- see DFX850
PHENOL, 3,4-DICHLORO- see DFY425
PHENOL, 3,5-DICHLORO- see DFY450
PHENOL, 2,6-DIETHYL- see DJU700
PHENOL, 4,4'-(1,2-DIETHYLIDENE-1,2-ETHANEDIYL)BIS-, (E,E)-(9CI) see DHB550
PHENOL, 2,3-DIMETHYL- see XKJ000
PHENOL, DIMETHYL- see XKA000
PHENOL, 4-((DIMETHYLAMINO)METHYL)-2,6-BIS(1,1-DIMETHYLETHYL)-(9CI) see DEA100, FAB000
PHENOL, p-(α-α-DIMETHYLBENZYL)- see COF400
PHENOL, 3-(1,1-DIMETHYLETHYL)-, METHYLCARBAMATE (9CI) see BSG300
PHENOL, p-(2,4-DINITROANILINO)- see DUW500
PHENOL, 2,6-DINITRO-4-ISOPROPYL- see IOX000
PHENOL, 2,4-DI-tert-PENTYL- see DCI000
PHENOLE (GERMAN) see PDN750
PHENOL, 4-ETHENYL-, ACETATE see ABW550
PHENOL, 4-ETHENYL- (9CI) see VQA200
PHENOL, 4-ETHENYL-2-METHOXY-(9CI) see VPF100
PHENOL, 4-ETHOXY-(9CI) see EFA100
PHENOL, o-ETHYL- see PGR250
PHENOL, p-ETHYL- see EOE100
PHENOL, o-FLUORO- see FKT100
PHENOL GLUCOSIDE see PDO300
PHENOL-GLYCIDAETHER (GERMAN) see PFH000
PHENOL GLYCIDYL ETHER (MAK) see PFH000
PHENOL, o-HEPTYL- see HBP350
PHENOL, 2-HEPTYL-(9CI) see HBP350
PHENOL, p-(HEPTYLOXY)- see HBP285
PHENOL, o-HEXYL- see HFU600
PHENOL, 2-HEXYL-(9CI) see HFU600
PHENOL, 4,4'-IMINOBIS-(9CI) see IBJ100
PHENOL, 4,4'-IMINODI- see IBJ100
PHENOL, 3-IODO- see IEV010
PHENOL, m-IODO- see IEV010
PHENOL, m-ISOPROPYL- see IQX090
PHENOL, o-ISOPROPYL- see IQX100
PHENOL, ISOPROPYLATED, PHOSPHATE (3:1) see DYF500
PHENOL, O-ISOPROPYL-, METHYLCARBAMATE see MIA250
PHENOL, 2-METHOXY-4-METHYL- see MEK250
PHENOL, 2-METHOXY-4-(OXIRANYLMETHYL)- see EBT500
PHENOL, 2-METHOXY-3,4,5,6-TETRACHLORO- see TBQ290
PHENOL, 2-METHOXY-4-VINYL- see VPF100
PHENOL, p-(α-METHYLBENZYL)- see PFD400
PHENOL, 4-METHYL-2,6-BIS(1-PHENYLETHYL)-(9CI) see MHR050
PHENOL, 2-METHYLDINITRO-, SODIUM SALT (9CI) see SGP550
PHENOL, 2,2'-METHYLENEBIS(6-(1,1-DIMETHYLETHYL))-4-ETHYL-(9CI) see MJN250
PHENOL, 3-(1-METHYLETHYL)- see IQX090
PHENOL, 2-(1-METHYLETHYL)-(9CI) see IQX100
PHENOL, 4,4'-(1-METHYLETHYLIDENE)BIS(2-METHYL-(9CI) see IPK000
PHENOL, 2-(1-METHYLETHYL)-, METHYLCARBAMATE (9CI) see MIA250
PHENOL, 4-(1-METHYL-1-PHENETHYL)-(9CI) see COF400
PHENOL, 2-METHYL-4-((4-(PHENYLAZO)PHENYL)AZO)- see CMP090
PHENOL, 2-(1-METHYLPROPYL)-4,6-DINITRO-, ACETATE (ESTER) (9CI) see ACE500
PHENOL, 2-(1-METHYLPROPYL)-, METHYLCARBAMATE see MOV000
PHENOL, 3-METHYL-, SODIUM SALT (9CI) see SJP000
PHENOL, 4-METHYL-, SODIUM SALT (9CI) see SIM100
PHENOL, p-(METHYLTHIO)- see MPV300
PHENOL, p-NITRO-, HYDROGEN PHOSPHATE see BLA600
PHENOL, p-NONYL- see NNC510
PHENOL, o-PENTYL- see AOM325
PHENOL, 2-PENTYL-(9CI) see AOM325
PHENOL, 4-(1-PHENYLETHYL)- see PFD400
PHENOL, 4-(PHENYLMETHYL)-, CARBAMATE see PFR325
PHENOL, o-PHENYL-, SODIUM deriv. see BGJ750
PHENOLPHTHALEIN, 3',3''-DIMETHYL- see CNX400
PHENOLPHTHALEIN see PDO750
PHENOLPHTHALEIN, 4,5,6,7-TETRABROMO-3',3''-DISULFO-, DISODIUM SALT see HAQ600
PHENOL, 4,4'-(2-PYRIDINYLMETHYLENE)BIS-, DIACETATE (ESTER) see PPN100

16-PHENOXY-ω-17,18,19,20-TETRANOR PROSTAGLANDIN E2 METHYLSULFO-
NYLAMIDE see SOU650
PHENOXYTOL see PER000
3-PHENOXYTOLUENE see MNV770
m-PHENOXYTOLUENE see MNV770
4-(3-(α-PHENOXY-p-TOLYL)PROPYL)-MORPHOLINE see PDU250
PHENOXYTRIETHYLSTANNANE see TJV250
PHENOZIN see ZIJ300
PHENSEDYL see DQA400
PHENTALAMINE see PDW400
PHENTANYL see PDW500
PHENTANYL CITRATE see PDW750
PHENTERMINE see DTJ400
PHENTHIAZINE see PDP250
PHENTHIURAM see FAQ930
PHENTHOATE see DRR400
PHENTIN ACETATE see ABX250
PHENTINOACETATE see ABX250
PHENTOLAMINE see PDW400
PHENTOLAMINE MESILATE see PDW950
PHENTOLAMINE MESYLATE see PDW950
PHENTOLAMINE METHANESULFONATE see PDW950
PHENTOLAMINE METHANESULPHONATE see PDW950
PHENTYRIN see BHY500
PHENURIDE see PFB350
PHENURON see PEC250
PHENUTAL see PEC250
PHENVALERATE see FAR100
PHENYBUT HYDROCHLORIDE see GAD000
PHENYCHOLON see EGQ000
PHENYGAM HYDROCHLORIDE see GAD000
PHENYLACETALDEHYDE (FCC) see BBL500
PHENYLACETALDEHYDE 2,4-DIHYDROXY-2-METHYLPENTANE ACETAL see
PFR400
PHENYLACETALDEHYDE DIMETHYL ACETAL see PDX000
PHENYLACETALDEHYDE ETHYLENEGLYCOL ACETAL see BEN250
PHENYLACETALDEHYDE GLYCERYL ACETAL see PDX250
2-PHENYLACETAMIDE see PDX750
α-PHENYLACETAMIDE see PDX750
N-PHENYLACETAMIDE see AAQ500, PDX500
PHENYLACETAMIDOPENICILLANIC ACID see BDY669
2'-PHENYLACETANILIDE see PDY000
3'-PHENYLACETANILIDE see PDY250
4'-PHENYLACETANILIDE see PDY500
p-PHENYLACETANILIDE see PDY500
PHENYL ACETATE see PDY750
PHENYLACETATE SODIUM SALT see SFA200
N'-PHENYLACETHYDRAZIDE see ACX750
ω-PHENYLACETIC ACID see PDY850
PHENYLACETIC ACID see PDY850
PHENYLACETIC ACID ALLYL ESTER see PMS500
PHENYLACETIC ACID AMIDE see PDX750
PHENYLACETIC ACID 2-BENZYLHYDRAZIDE see PDY870
PHENYL ACETIC ACID DIETHYLAMINOETHOXYETHANOL ESTER CITRATE
see BOR350
PHENYLACETIC ACID, ETHYL ESTER see EOH000
PHENYLACETIC ACID, p-METHOXYBENZYL ESTER see APE000
PHENYLACETIC ACID, METHYL ESTER see MHA500
PHENYLACETIC ACID, PHENETHYL ESTER see PDI000
PHENYLACETIC ACID SODIUM SALT see SFA200
PHENYLACETIC ALDEHYDE see BBL500
N-PHENYLACETOACETAMIDE see AAY000
2-PHENYLACETOACETONITRILE see PEA500
α-PHENYLACETOACETONITRILE see PEA500
PHENYL ACETO-ACETONITRILE see PEA500
α-PHENYLACETONE see MHO100
PHENYLACETONE see MHO100
2-PHENYLACETONITRILE see PEA750
PHENYLACETONITRILE, liquid (DOT) see PEA750
PHENYLACETONITRILE see PEA750
2-PHENYLACETOPHENONE see PEB000
4'-PHENYLACETOPHENONE see MHP500
p-PHENYLACETOPHENONE see MHP500
4'-PHENYL-o-ACETOTOLUIDE see PEB250
(PHENYLACETOXY)TRIETHYL PLUMBANE see TJT250
PHENYL-β-ACETYLAMINE see PDX750
PHENYLACETYLENE see PEB750
3-(1'-PHENYL-2'-ACETYLETHYL)-4-HYDROXYCOUMARIN see WAT200
3-(α-PHENYL-β-ACETYLETHYL)-4-HYDROXYCOUMARIN see WAT200
(PHENYL-1 ACETYL-2 ETHYL)-3-HYDROXY-4 COUMARINE (FRENCH) see
WAT200
PHENYLACETYLGLYCINE DIMETHYLAMIDE see PEB775
PHENYL ACETYL NITRILE see PEA750
α-PHENYLACETYLUREA see PEC250
(PHENYLACETYL)UREA see PEC250

PHENYLACETYLUREE (FRENCH) see PEC250
3-PHENYLACROLEIN see CMP969
PHENYLACROLEIN see CMP969
3-PHENYLACRYLAMIDE see CMP970
PHENYLACRYLAMIDE see CMP970
3-PHENYLACRYLIC ACID see CMP975
PHENYLACRYLIC ACID see CMP975
tert-β-PHENYLACRYLIC ACID see CMP975
trans-3-PHENYLACRYLIC ACID see CMP980
3-PHENYLACRYLOPHENONE see CDH000
β-PHENYLACRYLOPHENONE see CDH000
β-PHENYLAETHYLAMIN (GERMAN) see PDE250
1-PHENYL-2-AETHYLAMINO-PROPAN (GERMAN) see EGI500
PHENYLAETHYLBERNSTEINSAEURE-AETHYL-DIAETHYL-AMINOAETHYL-DI-ES-
TER (GERMAN) see SBC700
1-PHENYLAETHYLBIGUANID HYDROCHLORID (GERMAN) see PDF250
PHENYLAETHYLCARBINOL (GERMAN) see EGQ000
5,5-PHENYL-AETHYL-3-(β-DIAETHYLAMINO-AETHYL)-2,4,6-TRIOXO-HEXAHY-
DROPYRIMIDIN-HCl (GERMAN) see HEL500
PHENYLAETHYL-HYDRAZIN see PFC750
PHENYLAETHYLMALONSAEURE-AETHYL-DIAETHYLAMINOAETHYL-DI-ESTER
(GERMAN) see PCU400
PHENYLAETHYLMALONSAEURE-AETHYLESTER-DIAETHYLAMINOAETHYL-
AMID (GERMAN) see FAJ150
PHENYLAETHYLSENFOEL (GERMAN) see ISP000
3-PHENYLALANINE see PEC750
PHENYL-α-ALANINE see PEC750
β-PHENYLALANINE see PEC750
dl-PHENYLALANINE (FCC) see PEC500
(S)-PHENYLALANINE see PEC750
d-β-PHENYLALANINE see PEC500
d-PHENYLALANINE see PEC500
β-PHENYL-α-ALANINE, l- see PEC750
l-β-PHENYLALANINE see PEC750
l-PHENYLALANINE see PEC750
PHENYLALANINE see PEC750
dl-PHENYLALANINE MUSTARD see BHT750
d-PHENYLALANINE MUSTARD see BHU750, SAX200
l-PHENYLALANINE MUSTARD see PED750
o-PHENYLALANINE MUSTARD see BHT250
(±)-o-PHENYLALANINE MUSTARD see BHT250
PHENYLALANINE MUSTARD see BHU750
dl-PHENYLALANINE MUSTARD HYDROCHLORIDE see BHV000
l-PHENYLALANINE MUSTARD HYDROCHLORIDE see BHV250
PHENYLALANINE NITROGEN MUSTARD see PED750
l-PHENYLALANINE, N-((PHENYLMETHOXY)CARBONYL)-, ETHENYL ESTER
see CBR235
PHENYLALANIN-LOST (GERMAN) see BHT750
(S)-d-PHENYLALANYL-N-(4-((AMINOIMINOMETHYL)AMINO)-1-FORMYLBUTYL)-l-
PROLINAMIDE SULFATE (1:1) see PEE100
d-PHENYLALANYL-l-PROLYL-l-ARGININE ALDEHYDE SULFATE (1:1) see
PEE100
γ-PHENYLALLYL ACETATE see CMQ730
3-PHENYLALLYL ALCOHOL see CMQ740
γ-PHENYLALLYL ALCOHOL see CMQ740
5-PHENYL-5-ALLYLBARBITURIC ACID see AGQ875
PHENYLALLYL CINNAMATE see CMQ850
3-PHENYLALLYL ISOVALERATE see PEE200
PHENYLAMINE see AOQ000
PHENYLAMINE HYDROCHLORIDE see BBL000
PHENYLAMINOACETIC ACID ISOAMYL ESTER HYDROCHLORIDE see
PCV750
1-PHENYL-2-AMINO-AETHAN (GERMAN) see PDE250
N-PHENYL-p-AMINOANILINE see PFU500
2-(PHENYLAMINO)BENZOIC ACID see PEG500
4-(PHENYLAMINO)BUTANE see BQH850
PHENYL-α-AMINO-n-BUTYRAMIDE-p-ARSONIC ACID see CBK750
β-PHENYL-γ-AMINOBUTYRATE see PEE500
β-PHENYL-γ-AMINOBUTYRIC ACID see PEE500
N-PHENYLAMINOCARBONYL)AZIRIDINE see PEH250
17-β-PHENYLAMINOCARBONYLOXYOESTRA-1,3,5(10)-TRIENE-3-METHYL
ETHER see PEE600
1-PHENYL-4-AMINO-5-CHLOROPYRIDAZON-(6) (GERMAN) see PEE750
1-PHENYL-4-AMINO-5-CHLORO-6-PYRIDAZONE see PEE750
1-PHENYL-4-AMINO-5-CHLOROPYRIDAZONE-6 see PEE750
1-PHENYL-4-AMINO-5-CHLORPYRIDAZ-6-ONE see PEE750
trans-2-PHENYL-1-AMINOCYCLOPROPANE see PET750
2,2'-(PHENYLAMINO)DIETHANOL see BKD500
4-PHENYLAMINODIPHENYLAMINE see BLE500
p-PHENYLAMINODIPHENYLAMINE see BLE500
1-PHENYL-2-AMINOETHANE see PDE250
1-PHENYL-2-AMINOETHANE HYDROCHLORIDE see PDE500
2-(PHENYLAMINO)ETHANOL see AOR750
1-PHENYL-2-AMINO-1-ETHANOL, HYDROCHLORIDE see PFA750
α-PHENYL-β-AMINOETHANOL HYDROCHLORIDE see PFA750
4-(PHENYLAMINO)-PHENOL see AOT000

N-PHENYL-p-AMINOPHENOL see AOT000
PHENYL-p-AMINOPHENOL see AOT000
2-PHENYL-2-(p-AMINOPHENYL)PROPIONAMIDE see ALX500
1-PHENYL-2-AMINO-PROPAN (GERMAN) see AOA250
d-1-PHENYL-2-AMINOPROPAN (GERMAN) see AOA500
1-PHENYL-2-AMINOPROPANE see AOA250
d-1-PHENYL-2-AMINOPROPANE see AOA500
dl-1-PHENYL-2-AMINOPROPANE see BBK000
1-PHENYL-1-AMINO-PROPANE, HYDROCHLORIDE see PEF750
PHENYL-1-AMINO-1-PROPANE HYDROCHLORIDE see PEF750
PHENYL-2-AMINO-1-PROPANE HYDROCHLORIDE see PEF500
1-PHENYL-2-AMINOPROPANE MONOPHOSPHATE see AOB500
1-PHENYL-2-AMINOPROPANE SULFATE see BBK250
d-1-PHENYL-2-AMINOPROPANE SULFATE see BBK500
l-1-PHENYL-2-AMINOPROPANE SULFATE see BBK750
dl-1-PHENYL-2-AMINOPROPANOL-1 see NNM500
dl-1-PHENYL-2-AMINO-1-PROPANOL MONOHYDROCHLORIDE see PMJ500
2-PHENYLAMINOPROPIONITRILE see AOT100
3-PHENYL-5-AMINO-1,2,4-TRIAZOLYL-(1)-(N,N'-TETRAMETHYL) DIAMIDOPHOS-
    PHONATE see AIX000
2-PHENYLANILINE see BGE250
N-PHENYLANILINE see DVX800
o-PHENYLANILINE see BGE250
p-PHENYLANILINE see AJS100
2-PHENYLANISOLE see PEG000
o-PHENYLANISOLE see PEG000
p-PHENYLANISOLE see PEG250
N-PHENYLANTHRANILIC ACID see PEG500
PHENYLANTHRANILIC ACID see PEG500
PHENYL ARSENIC ACID see BBL750
PHENYLARSENOXIDE see PEG750
PHENYLARSINEDICHLORIDE see DGB600
PHENYL ARSINE OXIDE see PEG750
PHENYLARSONIC ACID see BBL750
PHENYLARSONOUS DIBROMIDE see DDR200
PHENYLARSONOUS DICHLORIDE see DGB600
β-PHENYLATHYLAMINHYDROCHLORID (GERMAN) see PDE500
PHENYL AZIDE see PEH000
N-PHENYL-1-AZIRIDINECARBOXAMIDE see PEH250
α-PHENYL-1-AZIRIDINEETHANOL see PEH500
4'-PHENYLAZOACETANILIDE see PEH750
p-PHENYLAZOACETANILIDE see PEH750
4-(PHENYLAZO)ANILINE see PEI000
N-(PHENYLAZO)ANILINE see DWO800
p-(PHENYLAZO)ANILINE see PEI000
p-(PHENYLAZO)ANILINE HYDROCHLORIDE see PEI250
4-(PHENYLAZO)-o-ANISIDINE see MFF500
1-PHENYLAZO-2-ANTHROL see PEI750
4-(PHENYLAZO)BENZENAMINE see PEI000
4-(PHENYLAZO)-1,3-BENZENEDIAMINE MONOHYDROCHLORIDE see PEK000
3-PHENYLAZO-2,6-DIAMINOPYRIDINE HYDROCHLORIDE see PDC250
β-PHENYLAZO-α,α'-DIAMINOPYRIDINE HYDROCHLORIDE see PDC250
PHENYLAZODIAMINOPYRIDINE HYDROCHLORIDE see PDC250
PHENYLAZO-α,α'-DIAMINOPYRIDINE MONOHYDROCHLORIDE see PDC250
4-(PHENYLAZO)DIPHENYLAMINE see PEI800
N-PHENYLAZO-N-METHYLTAURINE SODIUM SALT see PEJ250
1-(PHENYLAZO)-2-NAPHTHALENAMINE see FAG130
1-(PHENYLAZO)-2-NAPHTHALENOL see PEJ500
1-(PHENYLAZO)-2-NAPHTHOL see PEJ500
1-PHENYLAZO-β-NAPHTHOL see PEJ500
4-PHENYLAZO-1-NAPHTHYLAMINE see PEJ600
1-(PHENYLAZO)-2-NAPHTHYLAMINE see FAG130
PHENYLAZO α-NAPHTHYLAMINE see PEJ600
4-PHENYLAZOPHENOL see HJF000
p-PHENYLAZOPHENOL see HJF000
p-PHENYLAZOPHENYLAMINE see PEI000
(PHENYLAZO-4-PHENYLAZO)-1-ETHYLAMINO-2-NAPHTHALENE see EOJ500
1-(4-PHENYLAZO-PHENYLAZO)-2-ETHYLAMINONAPHTHALENE see EOJ500
1-((4-(PHENYLAZO)PHENYL)AZO)-2-NAPHTHALENOL see OHA000
1-((p-PHENYLAZO)PHENYL)AZO-2-NAPHTHOL see OHA000
4-(PHENYLAZO)-m-PHENYLENEDIAMINE see DBP999
4-PHENYLAZO-m-PHENYLENEDIAMINE see PEK000
4-(PHENYLAZO)-m-PHENYLENEDIAMINE MONOHYDROCHLORIDE see
    PEK000
3-(PHENYLAZO)-2,6-PYRIDINEDIAMINE see PEK250
3-(PHENYLAZO)-2,6-PYRIDINEDIAMINE, HYDROCHLORIDE see PDC250
PHENYLAZOPYRIDINE HYDROCHLORIDE see PDC250
4-(PHENYLAZO)RESORCINOL see CMP600
PHENYLAZO TABLET see PEK250
PHENYLAZO TABLETS see PDC250
(PHENYLAZO)THIOFORMIC ACID, 2-PHENYLHYDRAZIDE see DWN200
N-PHENYLBENEZENAMINE see DVX800
N-PHENYLBENZAMIDRAZONE HYDROCHLORIDE see PEK675
4'-PHENYLBENZANILIDE see BGF000
5-PHENYL-1:2-BENZANTHRACENE see PEK750

PHENYLBENZENE see BGE000
α-PHENYLBENZENEACETIC ACID see DVW800
N-PHENYL-1,4-BENZENEDIAMINE see PFU500
PHENYL-1,2,4-BENZENETRIOL see PEL000
PHENYL BENZHYDRYL KETONE see TMS100
2-PHENYL-1H-BENZIMIDAZOLE see PEL250
2-PHENYLBENZIMIDAZOLE see PEL250
PHENYL BENZOATE see PEL500
4-PHENYLBENZOIC ACID see PEL600
p-PHENYLBENZOIC ACID see PEL600
2-PHENYL-4H-1-BENZOPYRAN-4-ONE see PER700
PHENYL-1,4-BENZOQUINONE see PEL750
2-PHENYLBENZOQUINONE see PEL750
o-PHENYLBENZOQUINONE see PEL750
PHENYL-p-BENZOQUINONE see PEL750
PHENYLBENZOQUINONE see PEL750
2-PHENYL-5-BENZOTHIAZOLEACETIC ACID see PEM000
1-PHENYL-2-BENZOYLETHYLENE see CDH000
β,p-PHENYLBENZOYLPROPIONIC ACID see BGL250
2-PHENYL-BENZYL-AMINO-METHYLIMIDAZOLIN (GERMAN) see PDC000
2-(N-PHENYL-N-BENZYLAMINOMETHYL)IMIDAZOLINE see PDC000
8-(p-PHENYLBENZYL)ATROPINIUM BROMIDE see PEM750
α-PHENYLBENZYLCYANIDE see DVX200
p-(α-PHENYLBENZYLIDENE)-1,1-DIMETHYLPIPERIDINIUM METHYL SULFATE
    see DAP800
PHENYL BENZYL KETONE see PEB000
N-PHENYL-N-BENZYL-N',N'-DIMETHYLETHYLENEDIAMINE HYDROCHLORIDE
    see PEN000
PHENYLBIGUANIDE HYDROCHLORIDE see PEO000
1-PHENYLBIGUANIDE MONOHYDROCHLORIDE see PEO000
N'-PHENYLBIGUANIDE MONOHYDROCHLORIDE see PEO000
4-PHENYLBIPHENYL see TBC750
PHENYLBIS(2-CHLOROETHYLAMINE) see AOQ875
PHENYL BORATE see TMR500
PHENYLBORIC ACID see BBM000
PHENYL BROMIDE see PEO750
PHENYLBUTAMATE see HNJ000
1-PHENYLBUTANE see BQI750
2-PHENYLBUTANE see BQJ000
2-PHENYL-1,2-BUTANEDIOL 1-CARBAMATE see HNJ000
4-PHENYLBUTAN-2-ONE see PDF800
4-PHENYL-2-BUTANONE see PDF800
PHENYLBUTAZON (GERMAN) see BRF500
PHENYLBUTAZONE see BRF500
PHENYLBUTAZONE CALCIUM SALT see PEO750
PHENYLBUTAZONE SODIUM see BOV750
trans-4-PHENYL-3-BUTENE-2-ONE see BAY275
4-PHENYL-3-BUTEN-2-ONE see SMS500
3-PHENYL-BUTIN-1-OL-(3) see EQN230
2-PHENYL-tert-BUTYLAMINE see DTJ400
PHENYLBUTYLCARBINOL see BQJ500
PHENYL(BUTYRATE)MERCURY see BSV750
2-PHENYLBUTYRIC ACID see PEP250
α-PHENYL BUTYRIC ACID see PEP250
2-PHENYLBUTYRIC ACID 2-(2-DIETHYLAMINO)ETHOXY)ETHYL ESTER CI-
    TRATE see BOR350
PHENYLBUTYRIC ACID NITROGEN MUSTARD see CDO500
2-PHENYLBUTYRYLUREA see PFB350
N-(α-PHENYLBUTYRYL)UREA see PFB350
N-PHENYLCARBAMATE D'ISOPROPYLE (FRENCH) see CBM000
N-PHENYLCARBAMIC ACID-3-(DIMETHYLAMINO)PHENYL ESTER HYDRO-
    CHLORIDE see PEQ000
PHENYLCARBAMIC ACID-1-METHYLETHYL ESTER see CBM000
N-PHENYLCARBAMIC ACID-2-(METHYL(1-PHENOXY-2-PROPYL)AMINO) ETHYL
    ESTER HYDROCHLORIDE see PEQ250
N-PHENYLCARBAMIC ACID-3-(TRIMETHYLAMMONIO)PHENYL ESTER, ME-
    THYLSULFATE see HNS500
PHENYLCARBAMIC ESTER OF 3-OXYPHENYLDIMETHYLAMINE HYDROCHLO-
    RIDE see PEQ000
PHENYLCARBAMIC ESTER of 3-OXYPHENYLTRIMETHYLAMMONIUM ME-
    THYLSULFATE see HNS500
PHENYLCARBAMIDE see PGP250
PHENYLCARBAMODITHIOIC ACID METHYL ESTER (9CI) see MJK500
1-PHENYLCARBAMOYLAZIRIDINE see PEH250
PHENYL-N-CARBAMOYLAZIRIDINE see PEH250
2-PHENYL-CARBAMOYLOXY-N-AETHYL-PROPIONAMID (GERMAN) see
    CBL500
(PHENYLCARBAMOYLOXY)-2-N-ETHYLPROPIONAMIDE see CBL500
(3-(N-PHENYLCARBAMOYLOXY)PHENYL)TRIMETHYLAMMONIUM METHYLSUL-
    FATE see HNS500
o-(N-PHENYLCARBAMOYL)-PROPANONOXIM (GERMAN) see PEQ500
o-(N-PHENYLCARBAMOYL)PROPANONOXIME see PEQ500
PHENYLCARBIMIDE see PFK250
PHENYLCARBINOL see BDX500
PHENYL CARBITOL see PEQ750

m-PHENYLENEBIS(1-METHYLETHYLENE)BIS(DIMETHYLETHYLAMMONIUM BROMIDE) see PEX325
p-PHENYLENEBIS(1-METHYLETHYLENE)BIS(DIMETHYLETHYLAMMONIUM) BROMIDE see PEX330
m-PHENYLENEBIS(1-METHYLETHYLENE)BIS(TRIMETHYLAMMONIUM BROMIDE) see PEX350
p-PHENYLENEBIS(1-METHYLETHYLENE)BIS(TRIMETHYLAMMONIUM BROMIDE) see PEX355
(p-PHENYLENEBIS(1-METHYLETHYLENE)BIS(TRIMETHYLAMMONIUM) DIBROMIDE see PEX355
(m-PHENYLENEBIS(1-METHYLETHYLENE)BIS(TRIMETHYLAMMONIUM) DIBROMIDE see PEX350
1,1'-(p-PHENYLENEBIS(OXYETHYLENE))DIHYDRAZINE DIHYDROCHLORIDE see HGL600
2,2'-(1,3-PHENYLENEBIS(OXYMETHYLENE))BISOXIRANE see REF000
1,1'-(m-PHENYLENE)BIS-1H-PYROLE-2,5-DIONE (9CI) see BKL750
4,4'-o-PHENYLENEBIS(3-THIOALLOPHANIC ACID)DIMETHYL ESTER see PEX500
PHENYLENE-p,p'-BIS(2-TRIMETHYLAMMONIUMPROPYL) DIBROMIDE see PEX355
PHENYLENE-m,m'-BIS(2-TRIMETHYLAMMONIUMPROPYL) DIBROMIDE see PEX350
1,1'-(p-PHENYLENEBIS(VINYLENE-p-PHENYLENE))BIS(PYRIDINIUM) DI-p-TOLUENESULFONATE see PEX750
m-PHENYLENEDIACETATE see REA100
1,3-PHENYLENEDIAMINE see PEY000
1,4-PHENYLENEDIAMINE see PEY500
1,2-PHENYLENEDIAMINE (DOT) see PEY250
m-PHENYLENEDIAMINE (DOT) see PEY000
m-PHENYLENEDIAMINE see PEY000
o-PHENYLENEDIAMINE see PEY250
p-PHENYLENEDIAMINE see PEY500
p-PHENYLENEDIAMINE, N,N-DIETHYL-, SULFATE (1:1) see DJV250
1,3-PHENYLENEDIAMINE DIHYDROCHLORIDE see PEY750
1,4-PHENYLENEDIAMINE DIHYDROCHLORIDE see PEY650
o-PHENYLENEDIAMINE DIHYDROCHLORIDE see PEY600
p-PHENYLENEDIAMINE DIHYDROCHLORIDE see PEY650
p-PHENYLENEDIAMINE, N-(1,3-DIMETHYLBUTYL)-N'-PHENYL- see DQV250
m-PHENYLENEDIAMINE HYDROCHLORIDE see PEY750
p-PHENYLENEDIAMINE HYDROCHLORIDE see PEY650
PHENYLENEDIAMINE, META, solid (DOT) see PEY000
p-PHENYLENEDIAMINE, 2-METHOXY-, SULFATE see MEB820
m-PHENYLENEDIAMINE, 4-METHYL-6-(PHENYLAZO)- see CMM760
m-PHENYLENEDIAMINE, 4-NITRO- see NIM550
PHENYLENEDIAMINE, PARA, solid (DOT) see PEY500
1,3-PHENYLENEDIAMINE-4-SULFONIC ACID see PFA250
m-PHENYLENEDIAMINE-4-SULFONIC ACID see PFA250
m-PHENYLENEDIAMINESULFONIC ACID see PFA250
p-PHENYLENEDIAMINESULFONIC ACID see PEY800
1,3-PHENYLENE-DI-4-AMINOPHENYL ETHER see REF070
p-PHENYLENE DIAZIDE see DCL125
1,1'-(1,3-PHENYLENEDICARBONYL)BIS(2-METHYLAZIRIDINE) see BLG400
p-PHENYLENEDICARBONYL DICHLORIDE see TAV250
PHENYLENE-m,m'-DI-(2-DIMETHYLBENZYLAMMONIUMPROPYL) DICHLORIDE see PEX300
PHENYLENE-m,m'-DI(2-DIMETHYLETHYLAMMONIUMPROPYL)DIBROMIDE see PEX325
PHENYLENE-p,p'-DI(2-DIMETHYLETHYLAMMONIUMPROPYL) DIBROMIDE see PEX330
m-PHENYLENE DIISOCYANATE see BBP000
(PHENYLENEDIISOPROPYLIDENE)BIS(tert-BUTYLPEROXIDE) see BHL100
PHENYLENE-1,4-DIISOTHIOCYANATE see PFA500
1,4-PHENYLENEDIISOTHIOCYANIC ACID see PFA500
N,N'-(m-PHENYLENEDIMALEIMIDE) see BKL750
N,N'-(p-PHENYLENEDIMETHYLENE)BIS(2,2-DICHLORO-N-ETHYLACETAMIDE) see PFA600
m-PHENYLENEDIMETHYLENE ISOCYANATE see XIJ000
o-PHENYLENEDIOL see CCP850
4,4'-(m-PHENYLENEDIOXY)DIANILINE see REF070
m-PHENYLENE ISOCYANATE see BBP000
1,10-(1,2-PHENYLENE)PYRENE see IBZ000
1,10-(o-PHENYLENE)PYRENE see IBZ000
2,3-o-PHENYLENEPYRENE see IBZ000
2,3-PHENYLENEPYRENE see IBZ000
PHENYLENE THIOCYANATE see PFA500
o-PHENYLENETHIOUREA see BCC500
PHENYLEPHRINE see NCL500
(−)-PHENYLEPHRINE see NCL500
d-(−)-PHENYLEPHRINE HYDROCHLORIDE see SPC500
R(−)-PHENYLEPHRINE see NCL500
PHENYLEPHRINE HYDROCHLORIDE see SPC500
1-PHENYL-1,2-EPOXYETHANE see EBR000
PHENYL-2,3-EPOXYPROPYL ETHER see PFH000
PHENYLESSIGSAEURE NATRIUM-SALZ see SFA200
PHENYLETHANAL see BBL500

PHENYLETHANE see EGP500
PHENYLETHANEDIOL see SMQ100
PHENYLETHANOIC ACID BUTYL ESTER see BQJ350
1-PHENYLETHANOL see PDE000
2-PHENYLETHANOL see PDD750
β-PHENYLETHANOL see PDD750
N-PHENYLETHANOLAMINE see AOR750
PHENYL ETHANOLAMINE see AOR750
PHENYLETHANOL AMINE HYDROCHLORIDE see PFA750
1-PHENYLETHANONE see ABH000
PHENYLETHENE see SMQ000
4-(2-PHENYLETHENYL)BENZENAMINE,(E) see AMO000
4-(2-PHENYLETHENYL)BENZENAMINE see SLQ900
N-(4-(2-PHENYLETHENYL)PHENYL)ACETAMIDE see SMR500
1-((2-PHENYLETHENYL)PHENYL)ETHANONE see MPL000
2-(4-(2-PHENYLETHENYL)-3-SULFOPHENYL)-2H-NAPHTHO(1,2-d)TRIAZOLE SODIUM SALT see TGE155
PHENYL ETHER see PFA850
PHENYL ETHER-BIPHENYL MIXTURE see PFA860
1-PHENYL ETHER-2,3-DIACETATE GLYCEROL see PFF300
PHENYL ETHER, DICHLORO see DFE800
PHENYL ETHER, HEXACHLORO derivative (8CI) see CDV175
PHENYL ETHER MONO-CHLORO see MRG000
PHENYL ETHER PENTACHLORO see PAW250
PHENYL ETHER TETRACHLORO see TBP250
cis-2-(1-PHENYLETHOXY)CARBONYL-1-METHYLVINYL DIMETHYLPHOSPHATE see COD000
1-(2-PHENYL-2-ETHOXYETHYL)-4-(2-BENZYLOXYPROPYL)PIPERAZINE see PFB000
1-(2-PHENYL-2-ETHOXY)ETHYL-4-(2-BENZYLOXY)PROPYLPIPERAZINE DIHYDROCHLORIDE see ECU550
1-PHENYLETHYL ACETATE see SMP600
2-PHENYLETHYL ACETATE see PFB250
α-PHENYL ETHYL ACETATE see MNT075
α-PHENYLETHYL ACETATE see SMP600
β-PHENYLETHYL ACETATE see PFB250
sec-PHENYLETHYL ACETATE see SMP600
PHENYLETHYLACETYLUREA see PFB350
PHENYLETHYLACETYLUREE (FRENCH) see PFB350
2-PHENYLETHYL ALCOHOL see PDD750
β-PHENYLETHYL ALCOHOL see PDD750
PHENYLETHYL ALCOHOL see PDD750
1-PHENYLETHYLAMINE see ALW250
2-PHENYLETHYLAMINE see PDE250
α-PHENYLETHYLAMINE see ALW250
ω-PHENYLETHYLAMINE see PDE250
PHENYLETHYLAMINE see PDE250
2-PHENYLETHYLAMINE HYDROCHLORIDE see PDE500
β-PHENYLETHYLAMINE HYDROCHLORIDE see PDE500
2-PHENYLETHYL-o-AMINOBENZOATE see APJ500
2-PHENYLETHYLAMINOETHANOL see PFB500
1-PHENYL-2-ETHYLAMINOPROPANE see EGI500
α-PHENYL-β-ETHYLAMINOPROPANE see EGI500
2-PHENYLETHYL ANTHRANILATE see APJ500
PHENYLETHYL ANTHRANILATE see APJ500
PHENYLETHYLBARBITURATE see EOK000
5-PHENYL-5-ETHYLBARBITURIC ACID see EOK000
PHENYL-ETHYL-BARBITURIC ACID see EOK000
PHENYLETHYLBARBITURIC ACID, SODIUM SALT see SID000
2-PHENYLETHYL BENZOATE see PFB750
β-PHENYLETHYL BENZOATE see PFB750
PHENYLETHYL BENZOATE see PFB750
2-PHENYLETHYL BUTYRATE see PFB800
PHENYLETHYL BUTYRATE see PFB800
PHENYLETHYL CARBAMATE see CBL750
N-PHENYL-1-(ETHYLCARBAMOYL-1)-ETHYLCARBAMATE, D ISOMER see CBL500
β-PHENYLETHYL CINNAMATE see BEE250
PHENYLETHYL CINNAMATE see BEE250
2-PHENYLETHYL CYANIDE see HHP100
3-PHENYL-3-ETHYL-2,6-DIKETOPIPERIDINE see DYC800
PHENYLETHYL DIMETHYL CARBINOL see BEC250
3-PHENYL-3-ETHYL-2,6-DIOXOPIPERIDINE see DYC800
PHENYLETHYLENE see SMQ000
1-PHENYLETHYLENE GLYCOL see SMQ100
PHENYLETHYLENE GLYCOL see SMQ100
PHENYLETHYLENE OXIDE see EBR000
β-PHENYLETHYL ESTER HYDRACRYLIC ACID see PFC100
PHENYL ETHYL ETHER see PDM000
2-PHENYLETHYL FORMATE see PFC250
β-PHENYLETHYL FORMATE see PFC250
2-PHENYL-2-ETHYLGLUTARIC ACID IMIDE see DYC800
α-PHENYL-α-ETHYLGLUTARIC ACID IMIDE see DYC800
α-PHENYL-α-ETHYLGLUTARIMIDE see DYC800
5-PHENYL-5-ETHYL-HEXAHYDROPYRIMIDINE-4,6-DIONE see DBB200

2-(1-PHENYLETHYL)HYDRAZIDE BENZOIC ACID see NCQ100
2-PHENYLETHYLHYDRAZINE see PFC500
β-PHENYLETHYLHYDRAZINE see PFC500
β-PHENYLETHYLHYDRAZINE DIHYDROGEN SULFATE see PFC750
2-PHENYLETHYLHYDRAZINE DIHYDROGEN SULPHATE see PFC750
β-PHENYLETHYLHYDRAZINE HYDROGEN SULPHATE see PFC750
α-PHENYLETHYLHYDRAZINE OXALATE see MBV750
β-PHENYLETHYLHYDRAZINE SULFATE see PFC750
PHENYLETHYLHYDRAZINE SULPHATE see PFC750
2-PHENYLETHYL HYDROPEROXIDE see PFC775
4-(PHENYLETHYLIDENE)PHENOL see PFD400
1-(1-PHENYLETHYL)-1H-IMIDAZOLE-5-CARBOXYLIC ACID ETHYL ESTER see HOU100
PHENYLETHYL ISOAMYL ETHER see IHV050
2-PHENYLETHYL ISOBUTYRATE see PDF750
β-PHENYLETHYL ISOBUTYRATE see PDF750
PHENYLETHYL ISOBUTYRATE see PDF750
2-PHENYLETHYL ISOTHIOCYANATE see ISP000
β-PHENYLETHYL ISOTHIOCYANATE see ISP000
PHENYLETHYL ISOTHIOCYANATE see ISP000
β-PHENYLETHYL ISOVALERATE see PDF775
PHENYLETHYL ISOVALERATE see PDF775
PHENYL ETHYL KETONE see EOL500
PHENYLETHYLMALONYLUREA see EOK000
2-PHENYLETHYL METHACRYLATE see PFP600
β-PHENYLETHYL METHACRYLATE see PFP600
5-PHENYL-5-ETHYL-3-METHYLBARBITURIC ACID see ENB500
β-PHENYLETHYL α-METHYLBUTANOATE see PDF780
PHENYLETHYL-α-METHYLBUTENOATE see PFD250
PHENYLETHYL 2-METHYLBUTYRATE see PDF780
2-PHENYLETHYL-3-METHYLBUTYRATE see PDF775
β-PHENYLETHYL METHYL ETHER see PFD325
PHENYLETHYL METHYL ETHER see PFD325
PHENYLETHYL METHYL ETHYL CARBINOL see PFR200
PHENYLETHYL METHYLCARBINYL ACETATE see PFR300
PHENYLETHYLMETHYLHYDANTOIN see MKB250
β-PHENYLETHYL METHYL KETONE see PDF800
PHENYLETHYL 2-METHYL-2-PROPENOATE see PFP600
2-PHENYLETHYL-2-METHYLPROPIONATE see PDF750
PHENYLETHYL MUSTARD OIL see ISP000
8-(2-PHENYLETHYL)-1-OXA-3,8-DIAZASPIRO(4.5)DECAN-2-ONE HYDROCHLO-RIDE see DAI200
4-(1-PHENYLETHYL)PHENOL see PFD400
2-PHENYLETHYL PHENYLACETATE see PDI000
β-PHENYLETHYL PHENYLACETATE see PDI000
N-(β-(4-(β-PHENYLETHYL)PHENYL)-β-HYDROXYETHYL)IMIDAZOLE HYDRO-CHLORIDE see DAP880
β-2-PHENYLETHYLPIPERIDINOETHYL BENZOATE HYDROCHLORIDE see PFD500
γ-2-PHENYLETHYLPIPERIDINOPROPYL BENZOATE HYDROCHLORIDE see PFD750
1-PHENYLETHYL PROPIONATE see PFR000
2-PHENYLETHYL PROPIONATE see PDK000
PHENYLETHYL TIGLATE see PFD250
2-PHENYLETHYL-α-TOLUATE see PDI000
(2-PHENYLETHYL)TRICHLOROSILANE see PFE500
PHENYLETHYNYLCARBINOL CARBAMATE see PGE000
PHENYLETTEN see EOK000
N-PHENYL-2-FLUORENAMINE see PFE900
N-PHENYL-9H-FLUORENAMINE see PFE900
N-PHENYL-N-9H-FLUOREN-2-YLHYDROXYLAMINE see PFF000
N-PHENYL-2-FLUORENYLHYDROXYLAMINE see PFF000
PHENYL FLUORIDE see FGA000
PHENYLFLUOROFORM see BDH500
N-PHENYLFORMAMIDE see FNJ000
PHENYL FORMAMIDE see FNJ000
PHENYLFORMIC ACID see BCL750
PHENYLGAM see PEE500
PHENYLGAMMA see PEE500
PHENYLGAMMA HYDROCHLORIDE see GAD000
PHENYL β-D-GLUCOPYRANOSIDE see PDO300
PHENYL β-D-GLUCOSIDE see PDO300
PHENYLGLYCERYL ETHER DIACETATE see PFF300
PHENYL GLYCIDYL ETHER see PFH000
dl-2-PHENYLGLYCINEDECYLESTERHYDROCHLORIDE see PFF500
dl-2-PHENYLGLYCINEHEPTYLESTERHYDROCHLORIDE see PFF750
dl-2-PHENYLGLYCINEHEXYLESTERHYDROCHLORIDE see PFG000
dl-2-PHENYLGLYCINENONYLESTERHYDROCHLORIDE see PFG250
dl-2-PHENYLGLYCINEOCTYLESTERHYDROCHLORIDE see PFG500
dl-2-PHENYLGLYCINEPENTYLESTERHYDROCHLORIDE see PFG750
dl-2-PHENYLGLYCINISOAMYLESTERHYDROCHLORID (GERMAN) see PCV750
PHENYL GLYCOL see SMQ100
o-PHENYLGLYCOLIC ACID see PDR100
PHENYLGLYCOLIC ACID see MAP000
PHENYLGLYCOLONITRILE see MAP250

PHENYLGLYOXAL see PFH250
PHENYLGLYOXYLONITRILE OXIME-O,O-DIETHYL PHOSPHOROTHIOATE see BAT750
PHENYLGOLD see PFH275
2-PHENYLHYDRACRYLIC ACID-3-α-TROPANYL ESTER see ARR000
PHENYL HYDRATE see PDN750
2-PHENYLHYDRAZIDEBENZENECARBOXIMIDIC ACID MONOHYDROCHLORIDE see PEK675
2-PHENYLHYDRAZIDE, CARBAMIC ACID see CBL000
PHENYLHYDRAZIN (GERMAN) see PFI000
PHENYLHYDRAZINE see PFI000
2-PHENYLHYDRAZINECARBOTHIOAMIDE see PGM750
1-PHENYLHYDRAZINE CARBOXAMIDE see CBL000
2-PHENYLHYDRAZINECARBOXAMIDE see CBL000
PHENYLHYDRAZINE HYDROCHLORIDE see PFI250
PHENYLHYDRAZINE MONOHYDROCHLORIDE see PFI250
PHENYLHYDRAZINE-p-SULFONIC ACID see PFI500
PHENYLHYDRAZIN HYDROCHLORID (GERMAN) see PFI250
PHENYLHYDRAZINIUM CHLORIDE see PFI250
1-PHENYL-2-HYDRAZINOPROPANE see PDN000
PHENYL HYDRIDE see BBL250
2-PHENYLHYDROQUINONE see BGG250
o-PHENYLHYDROQUINONE see BGG250
PHENYLHYDROQUINONE see BGG250
PHENYLHYDROXAMIC ACID see BCL500
PHENYL HYDROXIDE see PDN750
PHENYLHYDROXYACETIC ACID see MAP000
threo-1-PHENYL-1-HYDROXY-2-AMINOPROPANE see NNM510
o-(PHENYLHYDROXYARSINO)BENZOIC ACID see PFI600
1-PHENYL-2-(β-HYDROXYETHYL)AMINOPROPANE see PFI750
2-PHENYL-2-HYDROXYETHYL CARBAMATE see PFJ000
2-PHENYL-2-HYDROXYETHYL, m-CHLOROPHENYL ARSINIC ACID see CKA575
β-PHENYLHYDROXYLAMINE see PFJ250
N-PHENYLHYDROXYLAMINE see PFJ250
PHENYLHYDROXYLAMINIUM CHLORIDE see PFJ275
PHENYL HYDROXYMERCURY see PFN100
1-PHENYL-1-HYDROXYPENTANE see BQJ500
1-PHENYL-2-(p-HYDROXYPHENYL)-3,5-DIOXO-4-BUTYLPYRAZOLIDINE see HNI500
PHENYLIC ACID see PDN750
PHENYLIC ALCOHOL see PDN750
PHENYL-IDIUM 200 see PDC250
PHENYL-IDIUM see PDC250
2-PHENYL-2-IMIDAZOLINE see PFJ300
PHENYLIMIDOCARBONYL CHLORIDE see PFJ400
N-PHENYL-IMIDODICARBONIMIDIC DIAMIDE MONOHYDROCHLORIDE see PEO000
N-PHENYLIMIDOPHOSGENE see PFJ400
N-PHENYLIMINOCARBONYL DICHLORIDE see PFJ400
PHENYLIMINOCARBONYL DICHLORIDE see PFJ400
2,2'-(PHENYLIMINO)DIETHANOL see BKD500
5-PHENYL-2-IMINO-4-OXAZOLIDINONE see IBM000
5-PHENYL-2-IMINO-4-OXOOXAZOLIDINE see IBM000
2-PHENYLINDAN-1,3-DIONE see PFJ750
2-PHENYL-1,3-INDANDIONE see PFJ750
PHENYLINDIONE see PFJ750
PHENYL IODIDE see IEC500
PHENYLIODINE(III) CHROMATE see PFJ775
PHENYLIODINE(III) NITRATE see PFJ780
9-PHENYL-9-IODOFLUORENE see PFK000
PHENYL ISOCYANATE see PFK250
PHENYL ISOHYDANTOIN see IBM000
PHENYLISONITRILE DICHLORIDE see PFJ400
2-PHENYLISOPROPANOL see DTN100
β-PHENYLISOPROPYLAMIN (GERMAN) see AOA250
β-PHENYLISOPROPYLAMINE see AOA250
(PHENYLISOPROPYL)AMINE see AOA250
dl-β-PHENYLISOPROPYLAMINE HYDROCHLORIDE see AOA750
β-PHENYL ISOPROPYLAMINE SULFATE see AOB250
d-β-PHENYLISOPROPYLAMINE SULFATE see BBK500
7-(PHENYL-ISOPROPYL-AMINO-AETHYL)-THEOPHYLLIN-HYDROCHLORID (GER-MAN) see CBF825
N-PHENYL ISOPROPYL CARBAMATE see CBM000
β-PHENYLISOPROPYLHYDRAZINE see PDN000
PHENYLISOPROPYLHYDRAZINE see PDN000
PHENYLISOPROPYLHYDRAZINE HYDROCHLORIDE see PDN250
(—)-PHENYLISOPROPYLMETHYLPROPYNYLAMINE see DAZ125
(+)-PHENYLISOPROPYLMETHYLPROPYNYLAMINE HYDROCHLORIDE see DAZ120
(±)-PHENYLISOPROPYLMETHYLPROPYNYLAMINE HYDROCHLORIDE see DAZ118
3-(β-PHENYLISOPROPYL)-SIDNONIMINE HYDROCHLORIDE see SPA000
PHENYL ISOTHIOCYANATE see ISQ000

4-PHENYL-NITROBENZENE see NFQ000
p-PHENYL-NITROBENZENE see NFQ000
N-PHENYL-p-NITROSOANILINE see NKB500
PHENYL-NITROSO-HANRSTOFF (GERMAN) see NLG500
4-PHENYLNITROSOPIPERIDINE see PFT600
N-PHENYL-N-NITROSOUREA see NLG500
N-PHENYL-N'-1,2,3-THIADIAZOL-5-YL-UREA see TEX600
1-PHENYLOXIRANE see EBR000
2-PHENYLOXIRANE see EBR000
PHENYLOXIRANE see EBR000
1-PHENYL-4-OXO-8-(4,4-BIS(4-FLUOROPHENYL)BUTYL)-1,3,8-TRIAZASPI-
  RO(4,5)DECANE see PFU250
1-PHENYL-3-OXOPYRAZOLIDINE see PDM500
1-PHENYL-2-β-OXY-AETHYL-AMINO-PROPAN (GERMAN) see PFI750
α-PHENYL-β-OXYETHYLAMINOPROPANE see PFI750
β-PHENYL-γ-OXYPROPIONSAEURE-TROPYL-ESTER (GERMAN) see ARR000
S-PHENYL PARATHION see DJS800
1-PHENYLPENTANOL see BQJ500
PHENYLPENTANOL see BQJ500
PHENYL PERCHLORYL see HCC500
2-PHENYLPHENOL see BGJ250
4-PHENYLPHENOL see BGJ500
o-PHENYLPHENOL see BGJ250
p-PHENYLPHENOL see BGJ500
2-PHENYLPHENOL SODIUM SALT see BGJ750
o-PHENYLPHENOL, SODIUM SALT see BGJ750
α-PHENYLPHENYLACETONITRILE see DVX200
N-PHENYL-p-PHENYLENEDIAMINE see PFU500
N-PHENYL-N-(1-(2-PHENYLETHYL)-4-PIPERIDINYL)-PROPANAMIDE (9CI) see
  PDW500
N-PHENYL-N-(1-(PHENYLIMINO)ETHYL)-N'-2,5-DICHLOROPHENYLUREA see
  PFV000
4-PHENYL-1-(2-(PHENYLMETHXY)ETHYL)-4-PIPERIDINECARBOXYLIC ACID
  ETHYL ESTER see BBU625
(s)-3-PHENYL-1'-(PHENYLMETHYL)-(3,4'-BIPIPERIDINE)-2,6-DIONE HYDRO-
  CHLORIDE see DBE000
PHENYLPHOSPHINE see PFV250
PHENYLPHOSPHINE DICHLORIDE see DGE400
PHENYLPHOSPHONIC ACID see PFV500
PHENYLPHOSPHONIC ACID DIOCTYL ESTER see PFV750
PHENYLPHOSPHONIC AZIDE CHLORIDE see PFV775
PHENYLPHOSPHONOCHLORIDOTHIOIC ACID O-ETHYL ESTER see EOM100
PHENYLPHOSPHONOTHIOIC ACID O-(4-BROMO-2,5-BROMO-2,5-DICHLORO-
  PHENYL) O-METHYL ESTER see LEN000
PHENYLPHOSPHONOTHIOIC ACID, O-(2,4-DICHLOROPHENYL), O-ETHYL ES-
  TER see SCC000
PHENYLPHOSPHONOTHIOIC ACID-o-ETHYL ESTER-o-ESTER with p-HYDROX-
  YBENZONITRILE see CON300
PHENYLPHOSPHONOUS ACID DICHLORIDE see DGE400
PHENYLPHOSPHONOUS DICHLORIDE see DGE400
PHENYL PHOSPHONYL DICHLORIDE see PFW100
PHENYL PHOSPHORUS DICHLORIDE (DOT) see DGE400
PHENYL PHOSPHORUS DICHLORIDE see DGE400
PHENYL PHOSPHORUS THIODICHLORIDE see PFW200, PFW210
N-PHENYLPHTHALIMIDINE see PFW750
1-PHENYLPIPERAZINE see PFX000
N-PHENYLPIPERAZINE see PFX000
2-(3-(N-PHENYLPIPERAZINO)-PROPYL)-3-METHYLCHROMONE see PFX250
2-(3-(N-PHENYLPIPERAZINO)-PROPYL)-3-METHYL-7-FLUOROCHROMONE see
  PFX500
4-PHENYLPIPERAZINYL 3,4,5-TRIMETHOXYPHENYL KETONE see PFX600
α-PHENYL-2-PIPERIDINEACETIC ACID METHYL ESTER see MNQ000
4-PHENYL-4-PIPERIDINECARBOXALDEHYDE see PFY100
4-PHENYL-1-PIPERIDINECARBOXAMIDE see PFY105
α-PHENYL-2-PIPERIDINEMETHANOL ACETATE HYDROCHLORIDE see LFO000
1-PHENYL-3-(β-PIPERIDINO-AETHYL)-5-(α'-METHYL-α-)-FURYL-PYRAZOLIN-HCL
  (GERMAN) see MKH750
PHENYL (1-PIPERIDINOCYCLOHEXYL) KETONE see PFY200
α-PHENYL-α-(2-PIPERIDINOETHYL)-β-ETHYLBUTYRIC ACID NITRILE HYDRO-
  CHLORIDE see EQZ000
α-PHENYL-α-(2-PIPERIDINOETHYL)-β-ETHYLBUTYRONITRILE HYDROCHLORIDE
  see EQZ000
2-PHENYL-6-PIPERIDINOHEXYNOPHENONE see PFY750
1-PHENYL-3-PIPERIDINOPROPAN-1-ONE HYDROCHLORIDE see PIV500
1-PHENYL-1-(2-PIPERIDYL)-1-ACETOXYMETHANE HYDROCHLORIDE see
  LFO000
PHENYL-(2-PIPERIDYL)METHYL ACETATE HYDROCHLORIDE see LFO000
1-PHENYL-4-(4-PIPERONYL-1-PIPERAZINYLCARBONYL)-2-PYRROLIDINONE HY-
  DROCHLORIDE see PGA250
3-PHENYL-1-PROPANAL see HHP000
2-PHENYLPROPANAL see COF000
3-PHENYLPROPANAL see HHP000
1-PHENYLPROPANE see IKG000
2-PHENYLPROPANE see COE750
1-PHENYL-1,2-PROPANEDIONE see PGA500

3-PHENYLPROPANENITRILE see HHP100
1-PHENYL-1-PROPANOL see EGQ000
2-PHENYLPROPAN-1-OL see HGR600
1-PHENYLPROPANOL see EGQ000
1-PHENYL-2-PROPANOL see PGA600
3-PHENYLPROPANOL see HHP050
3-PHENYL-1-PROPANOL (FCC) see HHP050
γ-PHENYLPROPANOL see HHP050
3-PHENYL-1-PROPANOL ACETATE see HHP500
1-PHENYL-2-PROPANOL ACETATE see ABU800
PHENYLPROPANOLAMINE see NNM000
PHENYLPROPANOLAMINE HYDROCHLORIDE see NNN000, PMJ500
3-PHENYL-1-PROPANOL CARBAMATE see PGA750
1-PHENYL-1-PROPANONE see EOL500
1-PHENYL-1-PROPANONE see MHO100
3-PHENYL-2-PROPENAL see CMP969
3-PHENYLPROPENAL see CMP969
3-PHENYL-2-PROPENAMIDE see CMP970
3-PHENYLPROPENAMIDE see CMP970
2-PHENYLPROPENE see MPK250
β-PHENYLPROPENE see MPK250
3-PHENYL-2-PROPENOIC ACID see CMP975
(E)-3-PHENYL-2-PROPENOIC ACID see CMP980
3-PHENYLPROPENOIC ACID see CMP975
3-PHENYL-2-PROPENOIC ACID-1,5,DIMETHYL-1-VINYL-4-HEXEN-1-YL ESTER
  see LGA000
3-PHENYL-2-PROPENOIC ACID-1-ETHENYL-1,5-DIMETHYL-4-HEXENYL ESTER
  see LGA000
3-PHENYL-2-PROPENOIC ACID METHYL ESTER (9CI) see MIO500
3-PHENYL-2-PROPENOIC ACID, 2-METHYLPROPYL ESTER see IIQ000
3-PHENYL-2-PROPENOIC ACID PHENYLMETHYL ESTER (9CI) see BEG750
3-PHENYL-2-PROPENOIC ACID 3-PHENYLPROPYL ESTER see PGB800
3-PHENYL-2-PROPEN-1-OL see CMQ740
3-PHENYL-2-PROPEN-1-YL ACETATE see CMQ730
3-PHENYL-2-PROPEN-1-YL ANTHRANILATE see API750
3-PHENYL-2-PROPENYLANTHRANILATE see API750
PHENYLPROPENYL n-BUTYRATE see CMQ800
3-PHENYL-2-PROPEN-1-YL CINNAMATE see CMQ850
3-PHENYL-2-PROPEN-1-YL FORMATE see CMR500
3-PHENYL-2-PROPENYL 3-PHENYL-2-PROPENOATE see CMQ850
3-PHENYL-2-PROPEN-1-YL PROPIONATE see CMR850
3-PHENYL-2-PROPENYL PROPIONATE see CMR850
5-PHENYL-5-(2-PROPENYL)-2,4,6(1H,3H,5H)-PYRIMIDINETRIONE (9CI) see
  AGQ875
α-PHENYLPROPIONALDEHYDE see COF000
β-PHENYLPROPIONALDEHYDE see HHP000
2-PHENYLPROPIONALDEHYDE (FCC) see COF000
3-PHENYLPROPIONALDEHYDE (FCC) see HHP000
2-PHENYLPROPIONALDEHYDE DIMETHYL ACETAL see HII600
3-PHENYLPROPIONITRILE see HHP100
β-PHENYLPROPIONITRILE see HHP100
PHENYLPROPIONITRILE see HHP100
17-β-PHENYLPROPIONYLOXY-4-ESTREN-3-ONE see DYF450
PHENYL(PROPIONYLOXY)MERCURY see PFO000
4-PHENYLPROPIOPHENONE see BGM100
3-PHENYL-1-PROPYL ACETATE see HHP500
2-PHENYLPROPYL ACETATE see PGB750
3-PHENYLPROPYL ACETATE (FCC) see HHP500
PHENYLPROPYL ACETATE see HHP500
1-PHENYLPROPYL ALCOHOL see EGQ000
2-PHENYLPROPYL ALCOHOL see HGR600
3-PHENYLPROPYL ALCOHOL see HHP050
β-PHENYLPROPYL ALCOHOL see HGR600
γ-PHENYLPROPYL ALCOHOL see HHP050
PHENYLPROPYL ALCOHOL see HHP050
3-PHENYLPROPYL ALDEHYDE see HHP000
α-PHENYL-α-PROPYLBENZENEACETIC ACID-2-(DIETHYLAMINO)ETHYL ESTER
  HYDROCHLORIDE see PBM500
γ-PHENYLPROPYLCARBAMAT (GERMAN) see PGA750
γ-PHENYLPROPYL CARBAMATE see PGA750
3-PHENYLPROPYL CINNAMATE see PGB800
PHENYLPROPYL CINNAMATE see PGB800
2-PHENYLPROPYLENE see MPK250
β-PHENYLPROPYLENE see MPK250
3-PHENYL-1-PROPYL FORMATE see HHQ000
PHENYLPROPYL FORMATE see HHQ000
α-(β-PHENYL-PROPYL)-β-HYDROXY-Δ^{α,β}-BUTENOLID (GERMAN) see PGC750
3-(3-PHENYLPROPYL)-4-HYDROXY-2(5H)FURANONE see PGC750
3-PHENYLPROPYL ISOBUTYRATE see HHQ500
3-PHENYLPROPYL PROPIONATE see HHQ550
β-PHENYLPROPYL PROPIONATE see HHQ550
PHENYLPROPYL PROPIONATE see HHQ550
1-PHENYL-2-PROPYNYL CARBAMATE see PGE000
PHENYLPSEUDOHYDANTOIN see IBM000
6-PHENYL-2,4,7-PTERIDINETRIAMINE see UVJ450

PHILBLACK see CBT750
PHILBLACK N 550 see CBT750
PHILBLACK N 765 see CBT750
PHILBLACK O see CBT750
PHILIPS 2133 see CPW250
PHILIPS 2605 see BIK750
PHILIPS-DUPHAR PH 60-40 see CJV250
PHILIPS-DUPHAR V-101 see CKL750
PHILLIPS 1863 see AMJ250
PHILLIPS 1908 see AMA000
PHILLIPS R-11 see BHJ500
PHILODENDRON (various species) see PGQ285
PHILODORM see TDA500
PHILOPON see DBA800, MDT600
PHILOSOPHER'S WOOL see ZKA000
PHILOSTIGMIN BROMIDE see POD000
PHILOSTIGMIN METHYL SULFATE see DQY909
PHISODANV see HCL000
PHISOHEX see HCL000
PHIX see ABU500
PHIXIA see CMS850
PHLEOCIDIN see PGQ350
PHLEOMYCIN see PGQ500
PHLEOMYCIN COMPLEX see PGQ750
PHLEOMYCIN D2 see BLY760
PHLOROGLUCIN see PGR000
PHLOROGLUCINOL see PGR000
PHLOROGLUCINOL TRIMETHYL ETHER see TKY250
PHLOROL see PGR250
PHLORONE see XQJ000
PHLOROPROPIONONE see TKP100
PHLOROPROPIOPHENONE see TKP100
PHLOXIN see CMM000
PHLOXIN B see ADG250
PHLOXINE see CMM000
PHLOXINE B see ADG250
PHLOXINE 2G see CMM300
PHLOXINE G see CMM300
PHLOXINE K see CMM000
PHLOXINE P see ADG250
PHLOXINE RED 20-7600 see BNK700
PHLOXINE TONER B see BNH500
PHLOX RED TONER X-1354 see BNH500
PHOB see EOK000
PHOBEX see BCA000
PHOLATE see AQO000
PHOLCODIN see TCY750
PHOLCODINE see TCY750
PHOLEDRINE see FMS875
PHOLETONE see FMS875
PHOMIN see CQM125
PHOMOPSIN see PGR775
PHONURIT see AAI250
PHORADENDRON RUBRUM see MQW525
PHORADENDRON SEROTINUM see MQW525
PHORADENDRON TOMENTOSUM see MQW525
PHORAT (GERMAN) see PGS000
PHORATE see PGS000
PHORATE-10G see PGS000
PHORBOL see PGS250
PHORBOL ACETATE, CAPRATE see ACZ000
PHORBOL ACETATE, LAURATE see PGS500
PHORBOL ACETATE LAURATE see PGU250
PHORBOL ACETATE, MYRISTATE see PGV000
PHORBOL-12-o-BUTYROYL-13-DODECANOATE see BSX750
PHORBOL CAPRATE, (+)-(S)-2-METHYLBUTYRATE see PGU500
PHORBOL CAPRATE, TIGLATE see PGU750
PHORBOL-12,13-DIACETATE see PGS750
PHORBOL-12,13-DIBENZOATE see PGT000
PHORBOL-12,13-DIDECANOATE see PGT250
PHORBOL-12,13-DIHEXA(Δ-2,4)-DIENOATE see PGT500
PHORBOL-12,13-DIHEXANOATE see PGT750
PHORBOL LAURATE, (+)-S-2-METHYLBUTYRATE see PGU000
PHORBOL MONOACETATE MONOLAURATE see PGS500, PGU250
PHORBOL MONOACETATE MONOMYRISTATE see PGV000
PHORBOL MONODECANOATE (S)-(+)-MONO(2-METHYLBUTYRATE) see PGU500
(E)-PHORBOL MONODECANOATE MONO(2-METHYLCROTONATE) see PGU750
PHORBOL MONOLAURATE MONO(S)-(+)-2-METHYLBUTYRATE see PGU000
PHORBOL MYRISTATE ACETATE see PGV000
PHORBOL-9-MYRISTATE-9a-ACETATE-3-ALDEHYDE see PGV250
PHORBOLOL ACETATE MYRISTATE see PGV500
PHORBOLOL MYRISTATE ACETATE see PGV500
PHORBOL-12-o-TIGLYL-13-BUTYRATE see PGV750

PHORBOL-12-o-TIGLYL-13-DODECANOATE see PGW000
PHORBYOL see CCP250
PHORDENE see DFY800
PHORON (GERMAN) see PGW250
PHORONE see PGW250
PHORTISOLONE see PLY275
PHORTOX see TAA100
PHOSADEN see AOA125
PHOSALON see BDJ250
PHOSALONE see BDJ250
PHOSALON mixed with MILBEX (1:2) see MQT750
PHOSAN-PLUS see PGW500
PHOSAZETIM see BIM000
PHOSCHLOR R50 see TIQ250
PHOSCOLIC ACID see PGW600
PHOSDRIN (OSHA) see MQR750
PHOSFENE see MQR750
PHOSFLEX 179-C see TMO600
PHOSFLEX T-BEP see BPK250
PHOS-FLUR see SHF500
PHOSFOLAN see DXN600, PGW750
PHOSFON D see THY500
PHOSGEN (GERMAN) see PGX000
PHOSGENE see PGX000
PHOSGENE, THIO- see TFN500
PHOSHOROTHIOIC ACID-S-(2-(ETHYLAMINO)-2-OXOETHYL)-O,O-DIMETHYL ESTER see DNX600
PHOSKIL see PAK000
PHOSMET see PHX250
PHOSPHACOL see NIM500
PHOSPHADEN see AOA125
PHOSPHALUGEL see PHB500
PHOSPHAM see PGX250
PHOSPHAMID see DSP400
PHOSPHAMIDON see FAB400
(E)-PHOSPHAMIDON see PGX300
trans-PHOSPHAMIDON see PGX300
PHOSPHATE 100 see EAS000
PHOSPHATE de O,O-DIETHYLE et de O-2-CHLORO-1-(2,4-DICHLOROPHE-NYL) VINYLE (FRENCH) see CDS750
PHOSPHATE de DIETHYLE et de 3-METHYL-5-PYRAZOLYLE (FRENCH) see MOX250
PHOSPHATE de O,O-DIMETHLE et de O-(1,2-DIBROMO-2,2-DICHLORETHYLE) (FRENCH) see NAG400
PHOSPHATE de DIMETHYLE et de (2-CHLORO-2-DIETHYLCARBAMOYL-1-METHYL-VINYLE) see FAB400
PHOSPHATE de DIMETHYLE et de 2,2-DICHLOROVINYLE (FRENCH) see DGP900
PHOSPHATE de DIMETHYLE et de 2-DIMETHYLCARBAMOYL-1-METHYL VI-NYLE (FRENCH) see DGQ875
PHOSPHATE de DIMETHYLE et de 2-METHOXYCARBONYL-1 METHYLVI-NYLE (FRENCH) see MQR750
PHOSPHATE de DIMETHYLE et de 2-METHYLCARBAMOYL 1-METHYL VI-NYLE see MRH209
PHOSPHATES see PGX500
PHOSPHATE, SODIUM HEXAMETA see SHM500
PHOSPHATE de TRICRESYLE (FRENCH) see TNP500
PHOSPHEMOL see PAK000
PHOSPHENE (FRENCH) see MQR750
PHOSPHENOL see PAK000
PHOSPHENTASIDE see AOA125
PHOSPHESTROL see DKA200
PHOSPHIDES see PGX750
PHOSPHINE see PGY000
PHOSPHINE OXIDE, BIS(1-AZIRIDINYL)METHYLAMINO- see MGE100
PHOSPHINE OXIDE, TRIS(p-DIMETHYLAMINOPHENYL)-, compounded with STANNIC CHLORIDE (2:1) see BLT775
PHOSPHINE SELENIDE, TRIMETHYL- see TMD400
PHOSPHINE SELENIDE, TRIPIPERIDINO- see TMX350
PHOSPHINE SULFIDE, TRIBUTYL- see TIA450
PHOSPHINE, TRIBUTYL- see TIA300
PHOSPHINE, TRIBUTYL-, SULFIDE see TIA450
PHOSPHINE, TRIS(DIPROPYLENE GLYCOL)- see OQO000
PHOSPHINIC ACID see PGY250
PHOSPHINIC AMIDE, p,p-BIS(1-AZIRIDINYL)-N-PROPYL- see BGY500
PHOSPHINIC AMIDE, P,P-BIS(1-AZIRIDINYL)-N-ETHYL- see BGX775
PHOSPHINIC AMIDE, P,P-BIS(1-AZIRIDINYL)-N-METHYL- see MGE100
2,2'-PHOSPHINICOBIS(2-HYDROXY-PROPANOIC) ACID (9CI) see PGW600
2,2'-PHOSPHINICODILACTIC ACID see PGW600
PHOSPHINOETHANE see EON000
1,1',1''-PHOSPHINOTHIOYLIDYNETRISAZIRIDINE see TFQ750
1,1',1''-PHOSPHINYLIDYNETRISAZIRIDINE see TND250
1,1',1''-PHOSPHINYLIDYNETRIS(2-METHYL)AZIRIDINE see TNK250
o-PHOSPHOETHANOLAMINE see EED000
2-PHOSPHOLENE, 3-METHYL-1-PHENYL-, 1-OXIDE see MNV800

PHOSPHORODITHIOIC ACID S-(((1,1-DIMETHYLETHYL)THIO)METHYL)-O,O-DIE-THYL ESTER see BSO000

PHOSPHORODITHIOIC ACID, O,O-DIMETHYL S-(MORPHOLINOCARBONYLME-THYL) ESTER see MRU250

PHOSPHORODITHIOIC ACID, O,O-DIMETHYL-S-(2-ETHYLTHIO)ETHYL ESTER see PHI500

PHOSPHORODITHIOIC ACID-O,O-DIMETHYL-S-(2-(METHYLAMINO)-2-OXOE-THYL) ESTER see DSP400

PHOSPHORODITHIOIC ACID-O,O-DIMETHYL-S-(2-((1-METHYLE-THYL)THIO)ETHYL) ESTER see DSK800

PHOSPHORODITHIOIC ACID-S,S'-1,4-DIOXANE-2,3-DIYL O,O,O',O'-TETRAE-THYL ESTER see DVQ709

PHOSPHORODITHIOIC ACID-S-(2-CHLORO-1-(1,3-DIHYDRO-1,3-DIOXO-2H-ISOIN-DOL-2-YL))ETHYL,O,O-DIETHYL ESTER see DBI099

PHOSPHORODITHIOIC ACID-S-((2-CHLORO-1-PHTHALIMIDOETHYL)-O,)-DIE-THYL ESTER see DBI099

PHOSPHORODITHIOIC ACID, ZINC SALT see ZGS000

PHOSPHOROFLUORIDIC ACID see PHJ250

PHOSPHOROFLUORIDIC ACID, DIETHYL ESTER see DJJ400

PHOSPHOROFLUORIDIC ACID, DIISOPROPYL ESTER see IRF000

PHOSPHOROFLUORIDIC ACID, DIMETHYL ESTER see DSA800

PHOSPHOROSELENOIC ACID, Se-(2-(DIETHYLAMINO)ETHYL) O,O-DIETHYL ESTER see DJA325

PHOSPHOROTHIOIC ACID see AMX825

PHOSPHOROTHIOIC ACID, O-(4-(AMINOSULFONYL)PHENYL) O,O-DIMETHYL ESTER (9CI) see CQL250

PHOSPHOROTHIOIC ACID, O-(2-CHLORO-1-ISOPROPYLIMIDAZOL-4-YL) O,O-DIETHYL ESTER see PHK000

PHOSPHOROTHIOIC ACID o-(4-CYANOPHENYL)-9,9-DIMETHYL ESTER see COQ399

PHOSPHOROTHIOIC ACID-o-(4-CYANOPHENYL)-o-ETHYL PHENYL ESTER see CON300

PHOSPHOROTHIOIC ACID, CYCLIC O,O-(METHYLENE-O-PHENYLENE) O-METHYL ESTER see MEC250

PHOSPHOROTHIOIC ACID-O,O-DIETHYL ESTER, -o-NAPHTHALIMIDO DERIVA-TIVE see NAQ500

PHOSPHOROTHIOIC ACID-O,O-DIETHYL ESTER-S-ESTER with ETHYL MER-CAPTOACETATE see DIW600

PHOSPHOROTHIOIC ACID-O,O-DIETHYL-O-(2-(ETHYLTHIO)ETHYL) ESTER, mixed with O,O-DIETHYL S-(2-(ETHYLTHIO)ETHYL) ESTER (7:3) see DAO600

PHOSPHOROTHIOIC ACID, O,O-DIETHYL O-(p-METHYLTHIO)PHENYL ESTER see PHK750

PHOSPHOROTHIOIC ACID-O,O-DIETHYL-o-NAPHTHYLAMIDOESTER see NAQ500

PHOSPHOROTHIOIC ACID, O,O-DIETHYL-O-(4-NITROPHENYL) ESTER see PAK000

PHOSPHOROTHIOIC ACID-O,O-DIETHYL, O-(p-NITROPHENYL)ESTER, mixed with compressed gas see PAK230

PHOSPHOROTHIOIC ACID-O,O-DIETHYL-O-2-PYRAZINYL ESTER see EPC500

PHOSPHOROTHIOIC ACID-O,O-DIMETHYL-O-(4-CYANO-2-METHOXYPHENYL) ESTER see DTQ800

PHOSPHOROTHIOIC ACID-O,O-DIMETHYL ESTER-O-ESTER with VANNILLONI-TRILE see DTQ800

PHOSPHOROTHIOIC ACID, O,O-DIMETHYL S-(ETHYLSULFINYL-(2-ISOPRO-PYL)) ESTER see DSK600

PHOSPHOROTHIOIC ACID, O,O-DIMETHYL S-(2-(METHYLAMINO)-2-OXOE-THYL) ESTER see DNX800

PHOSPHOROTHIOIC ACID-O,O-DIMETHYL-O-(3-METHYL-4-METHYLTHIOPHE-NYLE) (FRENCH) see FAQ900

PHOSPHOROTHIOIC ACID, S-ESTER with 2-((3-AMINOPRO-PYL)AMINO)ETHANETHIOL, HYDRATE see AMC750

PHOSPHOROTHIOIC ACID, O-ETHYL O-METHYL O-(4-NITROPHENYL) ESTER (9CI) see ENI175

PHOSPHOROTHIOIC ACID, O-ETHYL S-(p-TOLYL) ESTER see PHM750

PHOSPHOROTHIOIC ACID, O,O,O-TRI(2-CHLOROETHYL) ESTER see PHN500

PHOSPHOROTHIOIC ACID, O,O,O-TRIS(2-CHLOROETHYL) ESTER see PHN500

PHOSPHOROTHIOIC ACID-S-BENZYL-O,O-DIETHYL ESTER see DIU800

PHOSPHOROTHIOIC ACID-S-(((1-CYANO-1-METHYL-ETH-YL)CARBAMOYL)METHYL)-O,O-DIETHYL ESTER see PHK250

PHOSPHOROTHIOIC ACID-S-2,6-DIAMINOHEXYL ESTER, DIHYDRATE see DBW600

PHOSPHOROTHIOIC ACID, O,O'-(SULFONYLDI-p-PHENYLENE) O,O,O',O'-TET-RAMETHYL ESTER see PHN250

PHOSPHOROTHIOIC ACID TRIETHYLENETRIAMIDE see TFQ750

PHOSPHOROTHIOIC TRICHLORIDE see TFO000

PHOSPHOROTHIONIC TRICHLORIDE see TFO000

PHOSPHOROTRITHIOUS ACID, S,S,S-TRIBUTYL ESTER see TIG250

PHOSPHOROUS ACID, BERYLLIUM SALT see BFS000

PHOSPHOROUS ACID, DIETHYL ESTER see DJW400

PHOSPHOROUS ACID DIMETHYL ESTER see DSG600

PHOSPHOROUS ACID, TRIISOPROPYL ESTER see TKT500

PHOSPHOROUS ACID, TRIPHENYL ESTER see TMU250

PHOSPHOROUS ACID, TRIS(2-CHLOROETHYL)ESTER see PHO000

PHOSPHOROUS ACID, TRIS(DIPROPYLENE GLYCOL) ESTER see OQO000

PHOSPHOROUS ACID, TRIS(2-ETHYLHEXYL) ESTER, COMPLEX with MER-CURY(II) CHLORIDE (1:1) see MCY500

PHOSPHOROUS ACID, TRIS(2-ETHYLHEXYL) ESTER, COMPLEX with MER-CURY(II) BROMIDE (1:1) see MCY250

PHOSPHOROUS ACID TRIS(2-FLUOROETHYLESTER) see PHO250

PHOSPHOROUS ACID, TRIS(1-METHYLETHYL) ESTER see TKT500

PHOSPHOROUS OXYBROMIDE see PHU000

PHOSPHOROUS SULFOCHLORIDE see TFO000

PHOSPHOROUS THIOCHLORIDE see TFO000

PHOSPHOROUS TRICHLORIDE SULFIDE see TFO000

PHOSPHOROUS TRIFLUORIDE see PHQ500

PHOSPHOROUS (WHITE) see PHP010

PHOSPHOROUS YELLOW see PHP010

PHOSPHORPENTACHLORID (GERMAN) see PHR500

PHOSPHORSAEURE-BIS-(p-NITRO-PHENYLESTER) see BLA600

PHOSPHORSAEURELOESUNGEN (GERMAN) see PHB250

PHOSPHORTRICHLORID (GERMAN) see PHT275

PHOSPHORUS, amorphous (DOT) see PHO500

PHOSPHORUS (red) see PHO500

PHOSPHORUS (yellow) see PHP010

PHOSPHORUS ACID, TRIMETHYL ESTER see TMD500

PHOSPHORUS AZIDE DIFLUORIDE see PHP250

PHOSPHORUS AZIDE DIFLUORIDE-BORANE see PHP500

PHOSPHORUS BROMIDE (DOT) see PHT250

PHOSPHORUS CHLORIDE see PHT275

PHOSPHORUS COMPOUNDS, INORGANIC see PHQ000

PHOSPHORUS CYANIDE see PHQ250

PHOSPHORUS FLUORIDE see PHQ500

PHOSPHORUS HEPTASULFIDE, free from yellow or white phosphorus (DOT) see PHQ750

PHOSPHORUS HEPTASULFIDE see PHQ750

PHOSPHORUS(V) OXIDE see PHS250

PHOSPHORUS OXYCHLORIDE see PHQ800

PHOSPHORUS OXYTRICHLORIDE see PHQ800

PHOSPHORUS PENTABROMIDE see PHR250

PHOSPHORUS PENTACHLORIDE see PHR500

PHOSPHORUS PENTAFLUORIDE see PHR750

PHOSPHORUS PENTAOXIDE see PHS250

PHOSPHORUS PENTASULFIDE, free from yellow or white phosphorus (DOT) see PHS000

PHOSPHORUS PENTASULFIDE see PHS000

PHOSPHORUS PENTOXIDE see PHS250

PHOSPHORUS PERCHLORIDE see PHR500

PHOSPHORUS PERSULFIDE see PHS000

PHOSPHORUS SESQUISULFIDE, free from yellow or white phosphorous (DOT) see PHS500

PHOSPHORUS SESQUISULFIDE see PHS500

PHOSPHORUS (III) SULFIDE (IV) see PHS500

PHOSPHORUS SULFOBROMIDE see TFN750

PHOSPHORUS TRIAZIDE see PHT000

PHOSPHORUS TRIBROMIDE see PHT250

PHOSPHORUS TRICHLORIDE see PHT275

PHOSPHORUS TRICYANIDE see PHQ250

PHOSPHORUS TRIFLUORIDE see PHQ500

PHOSPHORUS TRIHYDRIDE see PGY000

PHOSPHORUS TRIHYDROXIDE see PGZ899

PHOSPHORUS TRIOXIDE see PHT500

PHOSPHORUS TRISULFIDE, free from yellow or white phosphorus (DOT) see PHT750

PHOSPHORUS TRISULFIDE see PHT750

PHOSPHORUS, white or yellow, dry or under water or in solution (UN 1381) (DOT) see PHP010

PHOSPHORUS WHITE, molten (UN 2447) (DOT) see PHP010

PHOSPHORUS, YELLOW (ACGIH,OSHA) see PHP010

PHOSPHORWASSERSTOFF (GERMAN) see PGY000

PHOSPHORYL BROMIDE see PHU000

PHOSPHORYL CHLORIDE see PHQ800

PHOSPHORYLETHANOLAMINE see EED000

PHOSPHORYL HEXAMETHYLTRIAMIDE see HEK000

O-PHOSPHORYL-4-HYDROXY-N,N-DIMETHYLTRYPTAMINE see PHU500

PHOSPHORYL TRIBROMIDE see PHU000

PHOSPHOSTIGMINE see PAK000

PHOSPHOTEX see TEE500

PHOSPHOTHION see MAK700

PHOSPHOTOX E see EEH600

PHOSPHOTUNGSTIC ACID see PHU750

PHOSPHURE de MAGNESIUM (FRENCH) see MAI000

PHOSPHURES d'ALUMINUM (FRENCH) see AHE750

PHOSPHURE de SODIUM (FRENCH) see SJI500

PHOSPHURE de STRONTIUM (FRENCH) see SML000

PHOSPHURE de ZINC (FRENCH) see ZLS000

PHOSTEX see BIT250

PHOSTOXIN see AHE750

PHOSVEL see LEN000

PHOSVEL PHENOL see LEN050
PHOSVIN see ZLS000
PHOSVIT see DGP900
PHOTOBILINE see TDE750
PHOTODYN see PHV275
PHOTOL see MGJ750
PHOTOMIREX see MRI750
PHOTOPHOR see CAW250
PHOXIME see BAT750
PHOXIN see BAT750
PHOZALON see BDJ250
PHP see ELL500
PHPH see BGE000
PHPS see DYE700
PHRENOLAN see MDU750
PHRILON see NOH000
PHT see TMB750
PHTALALDEHYDES (FRENCH) see PHV500
PHTALOPHOS see HMV000
p-PHTHALALDEHYDE see TAN500
PHTHALALDEHYDE see PHV500
PHTHALALDEHYDIC ACID see FNK010
PHTHALAMIC ACID, N-1-NAPHTHYL-, SODIUM SALT see SIO500
p-PHTHALAMIDE see TAN600
PHTHALAMIDE see BBO500
PHTHALAMIDE, N,N,N′,N′-TETRAETHYL- see GCE600
PHTHALAMODINE see CLY600
PHTHALAMUDINE see CLY600
1,3-PHTHALANDIONE see PHW750
PHTHALAZINE, 1,4-DIHYDRAZINO- see OJD300
1,4-PHTHALAZINEDIONE, 2,3-DIHYDRO-, DIHYDRAZONE (9CI) see OJD300
PHTHALAZINOL see PHV725
PHTHALAZINOL (PHOSPHODIESTERASE INHIBITOR) see PHV725
1(2H)PHTHALAZINONE see PHV750
PHTHALAZINONE see PHV750
1(2H)-PHTHALAZINONE (1,3-DIMETHYL-2-BUTENYLIDENE)HYDRAZONE see DQU400
1(2H)-PHTHALAZINONE HYDRAZONE see HGP495
1(2H)-PHTHALAZINONE HYDRAZONE HYDROCHLORIDE see HGP500
1(2H)-PHTHALAZINONE, HYDRAZONE, MONOHYDROCHLORIDE see HGP500
3-(1-PHTHALAZINYL)CARBAZIC ACID ETHYL ESTER see CBS000
PHTHALAZOL see PHY750
PHTHALAZONE see PHV750
PHTHALETHAMIDE see GCE600
m-PHTHALIC ACID see IMJ000
PHTHALIC ACID see PHW250
PHTHALIC ACID ANHYDRIDE see PHW750
o-PHTHALIC ACID BIS(DIETHYLAMIDE) see GCE600
PHTHALIC ACID BIS(2-METHOXYETHYL) ESTER see DOF400
PHTHALIC ACID, BIS(2-OCTYL) ESTER see BLB750
o-PHTHALIC ACID, DIALLYL ESTER see DBL200
PHTHALIC ACID, DIALLYL ESTER see DBL200
o-PHTHALIC ACID DIAMIDE see BBO500
PHTHALIC ACID, DICAPRYL ESTER see BLB750
PHTHALIC ACID, DICYCLOHEXYL ESTER see DGV700
PHTHALIC ACID, DIDODECYL ESTER see PHW550
PHTHALIC ACID, DIETHYL ESTER see DJX000
PHTHALIC ACID DIHEXYL ESTER see DKP600
PHTHALIC ACID, DIISOPROPYL ESTER see PHW600
PHTHALIC ACID DINITRILE see PHY000
PHTHALIC ACID, DI-2-OCTYL ESTER see BLB750
PHTHALIC ACID DIOCTYL ESTER see DVL700
PHTHALIC ACID, DIPROPYL ESTER see DWV500
PHTHALIC ACID, DITRIDECYL ESTER see DXQ200
PHTHALIC ACID, HEXAHYDRO-, BIS(2,3-EPOXYPROPYL) ESTER see DKM500
PHTHALIC ACID, HEXAHYDRO-, DIGLYCIDYL ESTER see DKM500
PHTHALIC ACID METHYL ESTER see DTR200
PHTHALIC ACID, TETRACHLORO- see TBT100
PHTHALIC ANHYDRIDE see PHW750
PHTHALIC ANHYDRIDE, TETRACHLORO- see TBT150
m-PHTHALIC DICHLORIDE see IMO000
o-PHTHALIC IMIDE see PHX000
PHTHALIDE 3,3,-BIS(p-HYDROXYPHENYL)- see PDO750
PHTHALIDE, 3-BUTYLIDENE- see BRQ100
PHTHALIDE, 3-(3-FURYL)-3a,4,5,6-TETRAHYDRO-3a,7-DIMETHYL- see FOM200
PHTHALIMETTEN see PDO750
PHTHALIMIDE see PHX000
PHTHALIMIDE, N-(CYCLOHEXYLTHIO)- see CPQ700
PHTHALIMIDE, N-HYDROXY- see HNS600
PHTHALIMIDIMIDE see DNE400
PHTHALIMIDO-O,O-DIMETHYL PHOSPHORODITHIOATE see PHX250
2-PHTHALIMIDOGLUTARIMIDE see TEH500
3-PHTHALIMIDOGLUTARIMIDE see TEH500
α-(N-PHTHALIMIDO)GLUTARIMIDE see TEH500
α-PHTHALIMIDOGLUTARIMIDE see TEH500

PHTHALIMIDOMETHYL-O,O-DIMETHYL PHOSPHORODITHIOATE see PHX250
(PHTHALOCYANINATO(2−))MAGNESIUM see MAI250
PHTHALOCYANINE BLUE 01206 see DNE400
PHTHALOCYANINE GREEN 6G see CMS140
m-PHTHALODINITRILE see PHX550
o-PHTHALODINITRILE see PHY000
p-PHTHALODINITRILE see BBP250
PHTHALODINITRILE see PHY000
PHTHALOGEN see DNE400
PHTHALOL see DJX000
PHTHALONITRILE see PHY000
PHTHALOPHOS see PHX250
m-PHTHALOYL CHLORIDE see IMO000
p-PHTHALOYL CHLORIDE see TAV250
PHTHALOYL DIAZIDE see PHY275
p-PHTHALOYL DICHLORIDE see TAV250
N-PHTHALOYLGLUTAMIMIDE see TEH500
PHTHALOYL PEROXIDE see PHY500
PHTHALOYLSULFATHIAZOLE see PHY750
PHTHALSAEUREANHYDRID (GERMAN) see PHW750
PHTHALSAEUREDIAETHYLESTER (GERMAN) see DJX000
PHTHALSAEUREDIMETHYLESTER (GERMAN) see DTR200
PHTHALTAN see TIT250
2-(N⁴-PHTHALYAMINOBENZENESULFONAMIDE)THIAZOLE see PHY750
2-(p-PHTHALYLAMINOBENZENESULFAMIDO)THIAZOLE see PHY750
2-(N⁴-PHTHALYAMINOBENZENESULFONAMIDO)THIAZOLE see PHY750
N-PHTHALYL-dl-ASPARTIMIDE see PIA000
o-PHTHALYLBIS(DIETHYLAMIDE) see GCE600
N-PHTHALYLGLUTAMIC ACID IMIDE see TEH500
N-PHTHALYL-GLUTAMINSAEURE-IMID (GERMAN) see TEH500
α-N-PHTHALYLGLUTARAMIDE see TEH500
PHTHALYLNORSULFAZOLE see PHY750
2-(N⁴-PHTHALYLSULFANILAMIDO)THIAZOLE see PHY750
2-(p-N-PHTHALYLSULFANILYL)AMINOTHIAZOLE see PHY750
PHTHALYLSULFATHIAZOLE see PHY750
PHTHALYLSULFONAZOLE see PHY750
PHTHALYLSULPHATHIAZOLE see PHY750
PHTHIOCOL see HMI000
PHTIVAZID see VEZ925
PHTIVAZIDE see VEZ925
PHYBAN see MRL750
PHYGON see DFT000
PHYGON PASTE see DFT000
PHYGON SEED PROTECTANT see DFT000
PHYGON XL see DFT000
PHYLCARDIN see TEP500
PHYLLEMBLIN see EKM100
PHYLLINDON see TEP500
PHYLLOCHINON (GERMAN) see VTA000
PHYLLOCORMIN N see HLC000
α-PHYLLOQUINONE see VTA000
PHYLLOQUINONE see VTA000
trans-PHYLLOQUINONE see VTA000
PHYMONE see NAK500
PHYOL see PJA250
PHYONE see PJA250
PHYSALIA PHYSALIS TOXIN see PIA375
PHYSALIN-X see PIA400
PHYSALIS (VARIOUS SPECIES) see JBS100
PHYSEPTONE see MDO750
PHYSEPTONE HYDROCHLORIDE see MDP750
PHYSIC NUT see CNR135
PHYSIOMYCINE see MDO250
PHYSOSTIGMINE see PIA500
PHYSOSTIGMINE SALICYLATE (1:1) see PIA750
PHYSOSTIGMINE SO4 see PIB000
PHYSOSTIGMINE SULFATE (2:1) see PIB000
PHYSOSTIGMINE SULFATE see PIB000
PHYSOSTOL see PIA500
PHYSOSTOL SALICYLATE see PIA750
PHYTAR 138 see HKC000
PHYTAR 560 see HKC000, HKC500
PHYTAR 600 see HKC000
PHYTAR see HKC000
PHYTIC ACID see PIB250
PHYTOGERMINE see VSZ450
PHYTOGLYCOGEN see GHK300
PHYTOL see PIB600
trans-PHYTOL see PIB600
PHYTOLACCA AMERICANA see PJJ315
PHYTOLACCA DODECANDRA, extract see EAQ100
PHYTOMELIN see RSU000
PHYTOMENADIONE see VTA000
PHYTOMYCIN see SLY500
PHYTONADIONE see VTA000

PHYTOSOL see EPY000
PIACCAMIDE see DAB875
PIAFOL see MCB050
PIAMID see MCB050
PIANADALIN see BNK000
PIANIZOL see UTU500
PIAPONON see PMH500
PIATHERM see UTU500
PIATHERM D see UTU500
PIAZINE see POL490
PIBECARB see PCV350
PIC-CLOR see CKN500
PICCOLASTIC D-100 see SMQ500
PICCOLASTIC see SMQ500
PICCOLASTIC A 25 see SMQ500
PICCOLASTIC A 5 see SMQ500
PICCOLASTIC A 50 see SMQ500
PICCOLASTIC A 75 see SMQ500
PICCOLASTIC A see SMQ500
PICCOLASTIC C 125 see SMQ500
PICCOLASTIC D 125 see SMQ500
PICCOLASTIC D 150 see SMQ500
PICCOLASTIC D see SMQ500
PICCOLASTIC E 100 see SMQ500
PICCOLASTIC E 200 see SMQ500
PICCOLASTIC E 75 see SMQ500
PICEA OIL see SLG650
PICENADOL see PIB700
PICENE see PIB750
PICEOL see HIO000
PICFUME see CKN500
PICHUCO (MEXICO) see ROU450
PICIS CARBONIS see CMY800
PICLORAM see PIB900
2-PICOLINAMINE see ALB750
2-PICOLINE see MOY000
3-PICOLINE see PIB920
4-PICOLINE see MOY250
α-PICOLINE see MOY000
β-PICOLINE see PIB920
γ-PICOLINE see MOY250
m-PICOLINE see PIB920
o-PICOLINE see MOY000
p-PICOLINE see MOY250
PICOLINE-2-ALDEHYDE THIOSEMICARBAZONE see PIB925
3-PICOLINE, 6-AMINO- see AMA010
α-PICOLINIC ACID see ILC000
PICOLINIC ACID see PIB930
PICOLINIC ACID, 5-BUTYL-, CALCIUM SALT, HYDRATE see FQR100
PICOLINIC ACID, 3,6-DICHLORO- see DGJ100
2-PICOLYIDENEBIS(p-PHENYL SODIUM SULFATE) see SJJ175
α-PICOLYL ALCOHOL see POR800
β-PICOLYL ALCOHOL see NDW510
γ-PICOLYL ALCOHOL see POR810
2-PICOLYLAMINE see ALB750
4-PICOLYLAMINE see ALC250
2-PICOLYL CHLORIDE HYDROCHLORIDE see PIC000
4,4′-(2-PICOLYLIDENE)BIS(PHENYLSULFURIC ACID) DISODIUM SALT see SJJ175
N-(2-PICOLYL)-N-PHENYL-N-(2-PIPERIDINOETHYL)AMINE HYDROCHLORIDE see CNE375
N-(2-PICOLYL)-N-PHENYL-N-(2-PIPERIDINOETHYL)AMINE TRIPALMITATE see PIC100
PICOPERIDAMINE HYDROCHLORIDE see CNE375
PICOPERINE HYDROCHLORIDE see CNE375
PICOPERINE TRIPALMITATE see PIC100
PICOSULFATE SODIUM see SJJ175
PICOSULFOL see SJJ175
PICRACONITINE see PIC250
PICRAGOL see PID200
PICRAMIC ACID see DUP400
PICRAMIC ACID, SODIUM SALT see PIC500
PICRAMIC ACID, ZIRCONIUM SALT (WET) see PIC750
PICRAMIDE (DOT) see PIC800
PICRAMIDE see PIC800
PICRATES see PIC899
PICRATOL see ANS500
PICRIC ACID (ACGIH,OSHA) see PID000
PICRIC ACID see PID000
PICRIC ACID, AMMONIUM SALT see ANS500
PICRIC ACID, LEAD SALT see PID100
PICRIC ACID, wet, with not <10% water (NA 1344) (DOT) see PID000
PICRIC ACID, SILVER(1+) SALT see PID200
PICRIC ACID, dry or wetted with <30% water, by weight (UN 0154) (DOT) see PID000

PICRIDE see CKN500
PICRITE (the explosive) see NHA500
PICRITE, wetted with not <20% water, by weight (UN 1336) (DOT) see NHA500
PICRITE, dry or wetted with <20% water, by weight (UN 0282) (DOT) see NHA500
PICROLONIC ACID see PIE000
PICRONITRIC ACID see PID000
PICROTIN, compounded with PICROTOXININ (1:1) see PIE500
PICROTOL see PID200
PICROTOXIN see PIE500
PICROTOXINE see PIE500
PICROTOXININ see PIE510
PICROTOXININE see PIE510
PICRYL AZIDE see PIE525
PICRYL CHLORIDE (DOT) see TML325
PICRYLMETHYLNITRAMINE see TEG250
PICRYLNITROMETHYLAMINE see TEG250
2-PICRYL-5-NITROTETRAZOLE see PIE550
PICRYL SULFIDE see BLR750
PID see DVV600, PFJ750
PIDIFIX 303 see MCB050
PIECIOCHLOREK FOSFORU (POLISH) see PHR500
PIED PIPER MOUSE SEED see SMN500
PIELIK see DAA800
PIELIK E see SGH500
PIE PLANT see RHZ600
PIERIS FLORIBUNDA see FBP520
PIERIS JAPONICA see FBP520
PIETIL see OOG000
PIFARNINE METHANESULFONATE see PIE750
PIFAZINE METHANESULFONATE see PIE750
PIGEON BERRY see GIW200
PIGEON-BERRY see PJJ315
PIGEON BERRY see ROA300
PIGLET PRO-GEN V see ARA500
PIGMENT ANTHRAQUINONE DEEP BLUE see IBV050
PIGMENT BLACK 7 see CBT750
PIGMENT BLUE 60 see IBV050
PIGMENT BLUE ANTHRAQUINONE see IBV050
PIGMENT BLUE ANTHRAQUINONE V see IBV050
PIGMENT DEEP BLUE ANTHRAQUINONE see IBV050
PIGMENT FAST ORANGE G see CMS145
PIGMENT FAST YELLOW 2GP see CMS212
PIGMENT FAST YELLOW GP see CMS212
PIGMENT GREEN 15 see LCR000
PIGMENT GREEN 38 see CMS140
PIGMENT ORANGE 13 see CMS145
PIGMENT ORANGE ERH see CMS145
PIGMENT ORANGE G see CMS145
PIGMENT ORANGE ZH see CMS145
PIGMENT PONCEAU R see FMU070
PIGMENT RED 23 see NAY000
PIGMENT RED 3 see MMP100
PIGMENT RED 53 see CMS150
PIGMENT RED 57 see CMS155
PIGMENT RED 48:1 see CMS148
PIGMENT RED 64:1 see CMS160
PIGMENT RED BH see NAY000
PIGMENT RED CD see CHP500
PIGMENT RED GG see CMS150
PIGMENT RED RL see MMP100
PIGMENT RUBINE B see CMS155
PIGMENT RUBINE BCL see CMS155
PIGMENT RUBY see MMP100
PIGMENT SCARLET see MMP100
PIGMENT SCARLET B see MMP100
PIGMENT SCARLET N see MMP100
PIGMENT SCARLET R see MMP100
PIGMENT SCARLET TONER RB see CMS160
PIGMENT WHITE 18 see CAT775
PIGMENT YELLOW 13 see CMS208
PIGMENT YELLOW 14 see CMS212
PIGMENT YELLOW 33 see CAP750
PIGMENT YELLOW 2G see CMS212
PIGMENT YELLOW GGP see CMS212
PIGMENT YELLOW GPP see CMS212
PIGMENT YELLOW GT see DEU000
PIGMENT YELLOW MH see CMS208
PIGMEX see AEY000
PIGTAIL PLANT see APM875
PIG-WRACK see CCL250
PIH see PDN000
PIK-OFF see EEA000
PIKRINEZUUR (DUTCH) see PID000

PIKRINSAEURE (GERMAN) see PID000
PIKRYNOWY KWAS (POLISH) see PID000
PILEWORT see FBS100
PILIOPHEN see TDE750
PILLARDRIN see MRH209
PILLARON see DTQ400
PILLARQUAT see PAJ000
PILLARSTIN see MHC750
PILLARXONE see PAJ000
PILLARZO see CFX000
PILLS (INDIA) see SED400
PILOCARPINE see PIF000
PILOCARPINE HYDROCHLORIDE see PIF250
PILOCARPINE MONOHYDROCHLORIDE see PIF250
PILOCARPINE MONONITRATE see PIF500
PILOCARPINE MURIATE see PIF250
PILOCARPINE NITRATE see PIF500
PILOCARPOL see PIF000
PILOCEL see PIF250
PILOFRIN see PIF500
PILOMIOTIN see PIF250
PILORAL see FOS100
PILOT 447 see MKP500
PILOT HD-90 see DXW200
PILOT SF-40 see DXW200
PILOVISC see PIF250
PILPOPHEN see DQA400
PIMACOL-SOL see NAK500
PIMAFUCIN see PIF750
PIMARICIN see PIF750
PIMELIC ACID see PIG000
PIMELIC ACID DINITRILE see HBD000
PIMELIC KETONE see CPC000
PIMELONITRILE see HBD000
4,4′-(PIMELOYLBIS(IMINO-p-PHENYLENEIMINO))BIS(1-ETHYLPYRIDINIUM) DI-
    PERCHLORATE see PIG250
4,4′-(PIMELOYLBIS(IMINO-p-PHENYLENEIMINO))BIS(1-METHYLPYRIDINIUM) DI-
    BROMIDE see PIG500
PIMENTA BERRY OIL see PIG740
PIMENTA LEAF OIL see PIG730, PIG740
PIMENTA OIL see PIG740
PIMENTA RACEMOSA OIL see LBK000
PIMENT BOUC (HAITI) see PCB275
PIMENT (HAITI) see PCB275
PIMENTO OIL see PIG740
PIMEPROFEN see IAB000
PIMETON see BJP250
PIMENTA del BRAZIL (PUERTO RICO) see PCB300
PIMIENTO de AMERICA (CUBA) see PCB300
PIMM see PFN500
PIMOZIDE see PIH000
PIN see EBD700
PINACIDIL see COS500
PINACOL see TDR000
PINACOLIN see DQU000
PINACOLINE see DQU000
PINACOLONE see DQU000
PINACOLOXYMETHYLPHOSPHORYL FLUORIDE see SKS500
PINACOLYL ALCOHOL (6CI) see BRU300
PINACOLYL METHYLFLUOROPHOSPHONATE see SKS500
PINACOLYL METHYLPHOSPHONOFLUORIDATE see SKS500
PINACOLYL METHYLPHOSPHONOFLUORIDE see SKS500
PINACOLYLOXY METHYLPHOSPHORYL FLUORIDE see SKS500
PINAKOLIN (GERMAN) see DQU000
PINAKOLIN see DQU000
PINAKON see HFP875
PINANG see BFW000
PINAN-2-α-OL see PIH050
α-2-PINANOL see PIH050
cis-2-PINANOL see PIH050
2-PINANOL, cis-(8CI) see PIH050
PINANONA (MEXICO, PUERTO RICO) see SLE890
PINAZEPAM see PIH100
PINDIONE see PFJ750
PINDOLOL see VSA000
PINDON (DUTCH) see PIH175
PINDONE see PIH175
PINEAPPLE KETONE see HKC575
PINEAPPLE SHRUB see CCK675
2-PINENE see PIH250
2(10)-PINENE see POH750
α-PINENE (DOT) see PIH250
β-PINENE (FCC) see POH750
PINE NEEDLE OIL see FBV000
PINE NEEDLE OIL, DWARF see PIH400

PINE NEEDLE OIL, SCOTCH see PIH500
2-PINENE-10-METHYL ACETATE see DTC000
2-PINEN-10-OL, ACETATE (6CI,7CI,8CI) see MSC050
(1R,5R)-(+)-2-PINEN-4-ONE see VIP000
PINE OIL see PIH750
PINERORO see CCR875
PINE TAR see PIH775
PINGYANGMYCIN (CHINESE) see BLY500
PINKROOT see PIH800
PINK 2S see CMM768
PINK WEED see PIH800
PINON (PUERTO RICO, DOMINICAN REPUBLIC) see CNR135
PIN-TEGA see PIJ500
PINUS MONTANA OIL see PIH400
PINUS PUMILIO OIL see PIH400
PIO see PII150
PIODEL see PII100
PIOMBO TETRA-ETILE see TCF000
PIOMY see PII150
PIOMYCIN see PII150
PIOPEN see CBO250
PIPADOX see HEP000
PIPAMAZINE see CJL500
PIPAMPERONE see FHG000
PIPAMPERONE DICHLORHYDRATE see PII200
PIPAMPERONE DIHYDROCHLORIDE see PII200
PIPAMPERONE HYDROCHLORIDE see PII200
PIPANEPERONE see FHG000
PIPANOL see BBV000
PIPA de TURCO (CUBA) see GEW800
PIPECOLATE see HDS300
PIPECOLIC ACID see HDS300
α-PIPECOLINIC ACID see HDS300
PIPECOLINIC ACID see HDS300
16β-PIPECOLINIO-2β-PIPERIDINO-5α-ANDROSTAN-3α,17β-DIOL BROMIDE DI-
    ACETATE see VGU075
PIPECURIUM BROMIDE see PII250
PIPECURONIUM BROMIDE see PII250
PIPEDAC see PIZ000
PIPEMID see PIZ000
PIPEMIDIC ACID see PIZ000
PIPEMIDIC ACID TRIHYDRATE see PII350
PIPENZOLATE BROMIDE see PJA000
PIPENZOLATE METHYLBROMIDE see PJA000
PIPERACETAZINE see PII500
PIPERACILLIN SODIUM see SJJ200
PIPERADROL HYDROCHLORIDE see PII750
PIPERAMIC ACID see PIZ000
PIPERAZIDINE see PIJ000
PIPERAZIN (GERMAN) see PIJ000
PIPERAZINE, anhydrous see PIJ000
PIPERAZINE see PIJ000
PIPERAZINE ADIPATE see HEP000
PIPERAZINE, 1-(4-AMINO-6,7-DIMETHOXY-2-QUINAZOLINYL)-4-((TETRAHYDRO-
    2-FURANYL)CARBONYL )-, MONOHYDROCHLORIDE, DIHYDRATE see
    TEF700
PIPERAZINE, 1-BENZOYL-2,6-DIMETHYL-4-NITROSO-(9CI) see NJL850
PIPERAZINE, 1-BENZYL-4-(p-CHLORO-β-PHENYLHYDROCINNAMOYL)- see
    BFE770
PIPERAZINE, 1-BENZYL-4-(3-(p-CHLOROPHENYL)-3-PHENYLPROPIONYL)- see
    BFE770
PIPERAZINE, 1-(p-tert-BUTYLBENZYL)-4-(3-(p-CHLOROPHENYL)-3-PHENYLPRO-
    PIONYL)- see BQK800
PIPERAZINE, 1-(p-tert-BUTYLBENZYL)-4-(3,4,5-TRIMETHOXYBENZOYL)- see
    BQK830
1-PIPERAZINECARBOXYLIC ACID, ETHYL ESTER see EHG050
1-PIPERAZINECARBOXYLIC ACID, 4-METHYL-, 6-(5-CHLORO-2-PYRIDINYL)-6,7-
    DIHYDRO-7-OXO-5H-PYRROLO(3,4-b)PYRAZIN-5-YL ESTER see ZUA450
PIPERAZINE, 1-(8-CHLORO-10,11-DIHYDRODIBENZO(b,f)THIEPIN-10-YL)-4-
    METHYL- see ODY100
PIPERAZINE, 1-(o-CHLOROPHENYL)-4-(3-(p-CHLOROPHENYL)-3-PHENYLPROPI-
    ONYL)- see CKI020
PIPERAZINE, 1-(o-CHLOROPHENYL)-4-NICOTINOYLMETHYL- see KGK120
PIPERAZINE, 1-(3-(p-CHLOROPHENYL)-3-PHENYLPROPIONYL)-4-(2-HYDROXY-
    PROPYL)- see CKI180
PIPERAZINE, 1-(3-(p-CHLOROPHENYL)-3-PHENYLPROPIONYL)-4-(o-METHOXY-
    PHENYL)- see CKI185
PIPERAZINE, 1-(3-(p-CHLOROPHENYL)-3-PHENYLPROPIONYL)-4-(m-METHYL-
    BENZYL)- see CKI030
PIPERAZINE, 1-(3-(p-CHLOROPHENYL)-3-PHENYLPROPIONYL)-4-PHENETHYL-
    see CKI040
PIPERAZINE, 1-(3-(p-CHLOROPHENYL)-3-PHENYLPROPIONYL)-4-(2-PYRIDYL)-
    see CKI050
PIPERAZINE, 1-(3-(p-CHLOROPHENYL)-3-PHENYLPROPIONYL)-4-(2-PYRIMIDYL)-
    see CKI060

PIPERAZINE, 1-(3-(p-CHLOROPHENYL)-3-PHENYLPROPIONYL)-4-(2-THIAZO-
LYL)- see CKI070
PIPERAZINE, 1-(3-(p-CHLOROPHENYL)-3-PHENYLPROPIONYL)-4-(m-TOLYL)-
see CKI080
PIPERAZINE, 1-(3-(p-CHLOROPHENYL)-3-PHENYLPROPIONYL)-4-(o-TOLYL)-
see CKI090
PIPERAZINE, 1-(3-(p-CHLOROPHENYL)-3-PHENYLPROPIONYL)-4-(p-TOLYL)-
see CKI190
PIPERAZINE, 1-(p-CHLOROPHENYL)-4-(3,4,5-TRIMETHOXYBENZOYL)- see
CKJ100
PIPERAZINE, 1-CINNAMYL-4-(DIPHENYLMETHYL)- see CMR100
PIPERAZINE CITRATE (3:2) see PIJ500
PIPERAZINE CITRATE TELRA see PIJ500
PIPERAZINE DIANTIMONY TARTRATE see PIJ600
PIPERAZINE DIHYDROCHLORIDE see PIK000
PIPERAZINE, 1-(3,4-DIMETHOXYBENZOYL)-4-(1,2,3,4-TETRAHYDRO-2-OXO-6-
QUINOLINYL)- see DLF700
PIPERAZINE, 1,4-DIMETHYL- see LIQ500
PIPERAZINEDIONE see PIK075
PIPERAZINEDIONE 593A see PIK075
2,6-PIPERAZINEDIONE-4,4′-PROPYLENE DIOXOPIPERAZINE see PIK250
PIPERAZINE, 1-(DIPHENYLMETHYL)-4-(3-PHENYL-2-PROPENYL)-(9CI) see
CMR100
1,4-PIPERAZINEDIYLBIS(BIS(1-AZIRIDINYL)PHOSPHINE) OXIDE see BJC250
N,N′-(PIPERAZINEDIYLBIS(2,2,2-TRICHLOROETHYLIDENE)) BIS(FORMAMIDE)
see TKL100
1,1′-(1,4-PIPERAZINEDIYLDIETHYLENE)BIS(1-ETHYLPIPERIDINIUM IODIDE) see
PIK375
(1,4-PIPERAZINEDIYLDIETHYLENE)BIS(TRIETHYLAMMONIUM IODIDE) see
PIK400
1-PIPERAZINEETHANOL see HKY500
PIPERAZINE ETHYLCARBOXYLATE see EHG050
PIPERAZINE HYDROCHLORIDE see PIK000
PIPERAZINE, 1-(o-METHOXYPHENYL)-4-NICOTINOYLMETHYL- see KGK130
PIPERAZINE, 1-(p-METHOXYPHENYL)-4-(3,4,5-TRIMETHOXYBENZOYL)- see
MFH770
PIPERAZINE, 1-(o-METHOXYPHENYL)-4-(3,4,5-TRIMETHOXYBENZOYL)- see
MFH760
PIPERAZINE, 1-PHENYL-4-(3,4,5-TRIMETHOXYBENZOYL)- see PFX600
PIPERAZINE, 1-(2-PYRIDYL)-4-(3,4,5-TRIMETHOXYBENZOYL)- see TKY300
PIPERAZINE, 1-(2-PYRIMIDYL)-4-(TRIMETHOXYBENZOYL)- see PPP550
PIPERAZINE and SODIUM NITRITE (4:1) see PIJ250
PIPERAZINE SULTOSILATE see PIK625
PIPERAZINE, 1-(2-THIAZOLYL)-4-(3,4,5-TRIMETHOXYBENZOYL)- see TEX220
PIPERAZINE, 1-(m-TOLYL)-4-(3,4,5-TRIMETHOXYBENZOYL)- see THF300
PIPERAZINE, 1-(p-TOLYL)-4-(3,4,5-TRIMETHOXYBENZOYL)- see THF310
PIPERAZINIUM, 4-(β-CYCLOHEXYL-β-HYDROXYPHENETHYL)-1,1-DIMETHYL-,
METHYL SULFATE see HFG400
2-(1-PIPERAZINYL)ETHANOL see HKY500
2-(1-PIPERAZINYL)-QUINOLINE (Z)-2-BUTENEDIOATE (1:1) (9CI) see QWJ500
2-(1-PIPERAZINYL)-QUINOLINE MALEATE (1:1) see QWJ500
PIPERIDILATE HYDROCHLORIDE see EOY000
PIPERIDIN (GERMAN) see PIL500
PIPERIDINE see PIL500
PIPERIDINE, 1-(m-(BIS(2-CHLOROETHYL)AMINO)BENZOYL)- see BHP150
PIPERIDINE, 1-(3-(BIS(2-CHLOROETHYL)AMINO)-p-TOLUOYL)- see BIA100
PIPERIDINE, 4-tert-BUTYL-1-NITROSO see NJO300
1-PIPERIDINECARBODITHIOIC ACID, compounded with PIPERIDINE see
PIY500
PIPERIDINE-N-CARBONIC ACID AMIDE see PIL525
1-PIPERIDINECARBOXAMIDE see PIL525
2-PIPERIDINECARBOXAMIDE,1-BUTYL-N-(2,6-DIMETHYLPHE-
NYL)MONOHYDROCHLORIDE MONOHYDRATE see BOO000
2-PIPERIDINECARBOXYLIC ACID (9CI) see HDS300
4-PIPERIDINECARBOXYLIC ACID (9CI) see ILG100
PIPERIDINE, 1-(CYCLOHEXYLCARBONYL)-3-METHYL- see CPI350
PIPERIDINE, 2,6-DIMETHYL- see LIQ555
PIPERIDINE, 2,6-DIMETHYL-1-((2-METHYLCYCLOHEXYL)CARBONYL)- see
DSP650
PIPERIDINE, 3,5-DIMETHYL-1-NITRCSO-, (Z)- see DTA690
PIPERIDINE, 3,5-DIMETHYL-1-NITROSO-, (E)- see DTA700
2,6-PIPERIDINEDIONE, 3-(4-AMINOPHENYL)-3-ETHYL- see AKC600
3-PIPERIDINE-1,1-DIPHENYL-PROPANOL-(1) METHANESULPHONATE see
PIL550
1-PIPERIDINEETHANOL see HKY600
2-PIPERIDINEETHANOL see HKY600
N-PIPERIDINEETHANOL see HKY600
2-PIPERIDINEETHANOL-p-AMINOBENZOATE (ester) HYDROCHLORIDE see
AIQ890
2-PIPERIDINEETHANOL-o-AMINOBENZOATE (ester) HYDROCHLORIDE see
AIQ885
2-PIPERIDINEETHANOL-m-AMINOBENZOATE (ester) HYDROCHLORIDE see
AIQ880
1-PIPERIDINEETHANOL BENZILATE HYDROCHLORIDE see PIM000
2-PIPERIDINEETHANOL CARBANILATE (ester) HYDROCHLORIDE see PIU800

PIPERIDINE, 2-ETHYL-1-(3-METHYL-1-OXO-2-BUTENYL)- see EMY100
PIPERIDINE HYDROCHLORIDE see HET000
4-PIPERIDINEMETHANOL, 1-BROMOACETYL-α-α-DIPHENYL- see BMS300
4-PIPERIDINEMETHANOL, 1-CHLOROACETYL-α-α-DIPHENYL- see CEC300
4-PIPERIDINEMETHANOL, 1-(IODOACETYL)-α-α-DIPHENYL- see IDZ200
PIPERIDINE, 3-((4-METHOXYPHENOXY)METHYL)-1-METHYL-4-PHENYL-, HY-
DROCHLORIDE, (3R-trans)- see FAJ200
PIPERIDINE, 4-METHYL-1-((6-METHYL-3-CYCLOHEXEN-1-YL)CARBONYL)- see
MLL660
PIPERIDINE, 3-METHYL-1-((6-METHYL-3-CYCLOHEXEN-1-YL)CARBONYL)- see
MLL655
PIPERIDINE, 2-METHYL-1-((6-METHYL-3-CYCLOHEXEN-1-YL)CARBONYL)- see
MLL650
PIPERIDINE, 4-METHYL-1-((2-METHYLCYCLOHEXYL)CARBONYL)- see MLM700
PIPERIDINE, 1-(1-METHYL-2-(2-(PHENYLMETHYL)PHENOXY)ETHYL)-(9CI) see
MOA600
PIPERIDINE, 1-(1-METHYL-2-((α-PHENYL-o-TOLYL)OXY)ETHYL)- see MOA600
PIPERIDINE, 1-((1-METHYL-3-PIPERIDINYL)CARBONYL)- see MOG650
PIPERIDINE, 1-((1-(2-PHENYLETHYL)-3-PIPERIDINYL)CARBONYL)- see PDI550
PIPERIDINE, 1-(3-PIPERIDYL)CARBONYL- see PIR200
PIPERIDINE, 2-PROPYL-, (S)- see PNT000
PIPERIDINE, 1-(2-PROPYLVALERYL)- see PNX800
PIPERIDINIC ACID see PIM500
PIPERIDINIUM see PIY500
2-PIPERIDINO-p-ACETOPHENETIDIDE HYDROCHLORIDE see PIM750
2-PIPERIDINO-2′,6′-ACETOXYLIDIDE HYDROCHLORIDE see PIN000
4-PIPERIDINOACETYL-3,4-DIHYDRO-2H-1,4-BENZOXAZINE HYDROCHLORIDE
see PIN100
β-PIPERIDINOAETHYL-(3-CHLOR-4-n-BUTOXY-5-METHYLPHE-
NYL)KETONHYDROCHLORID (GERMAN) see BPI625
β-PIPERIDINOAETHYL-(3-CHLOR-4-PROPOXY-5-METHYLPHENYL)-KETONHY-
DROCHLORID (GERMAN) see CIU325
3-(PIPERIDINOCARBONYL)PIPERIDINE see PIR200
1-PIPERIDINOCYCLOHEXANECARBONITRILE see PIN225
PIPERIDINOCYCLOHEXANECARBONITRILE see PIN225
α-(1-PIPERIDINO)-CYCLOHEXYL PHENYL KETONE see PFY200
6-PIPERIDINO-2,4-DIAMINOPYRIMIDINE-3-OXIDE see DCB000
β-PIPERIDINOETHANOL see HKY600
2-PIPERIDINOETHANOL HYDROCHLORIDE see PIN275
2-(1-PIPERIDINO)-ETHYL BENZILATE ETHYLBROMIDE see PIX800
2-(1-PIPERIDINO)ETHYL BENZILATE HYDROCHLORIDE see PIM000
N-(2-PIPERIDINOETHYL)CARBAMIC ACID, 6-CHLORO-o-TOLYL ESTER, HY-
DROCHLORIDE see PIO000
N-(2-PIPERIDINOETHYL)CARBAMIC ACID, MESITYL ESTER, HYDROCHLORIDE
see PIO250
N-(2-(PIPERIDINO)ETHYL)CARBAMIC ACID, 2,6-XYLYL ESTER, HYDROCHLO-
RIDE see PIO500
PIPERIDINOETHYL CHLORIDE, HYDROCHLORIDE see CHG500
PIPERIDINOETHYL-2-HEPTOXYPHENYLCARBAMOATE HYDROCHLORIDE see
PIO750
PIPERIDINOETHYL-3-METHYLFLAVONE-8-CARBOXYLATE HYDROCHLORIDE
see FCB100
2-PIPERIDINOETHYL-3-METHYL-4-OXO-2-PHENYL-4H-1-BENZOPYRAN-8-CAR-
BOXYLATE HYDROCHLORIDE see FCB100
N-(2-PIPERIDINOETHYL)-N-(2-PYRIDYLMETHYL)ANILINE HYDROCHLORIDE see
CNE375
N-(2-PIPERIDINOETHYL)-N-(2-PYRIDYLMETHYL)ANILINE TRIPALMITATE see
PIC100
4-PIPERIDINOL, 1-(3-(2-ACETYLPHENOTHIAZIN-10-YL)PROPYL)- see HNT200
2-PIPERIDINOMETHYL-1,4-BENZODIOXAN HYDROCHLORIDE see BCI500
PIPERIDINOMETHYLCYCLOHEXANE CAMPHOSULFATE see PIQ750
PIPERIDINOMETHYLCYCLOHEXANE CHLORHYDRATE SALT see PIR000
2-PIPERIDINOMETHYL-4-METHYL-1-TETRALONE HYDROCHLORIDE see
PIR100
α-(PIPERIDINOMETHYL)-5-PHENYL-3-ISOXAZOLEMETHANOL CITRATE see
HNT075
1-PIPERIDINO-2-METHYL-3-(p-TOLYL)-3-PROPANONE see TGK200
1-PIPERIDINO-2-METHYL-3-(p-TOLYL)-3-PROPANONE HYDROCHLORIDE see
MRW125
4-PIPERIDINONE, 2,2,6,6-TETRAMETHYL-(9CI) see TDT770
1-PIPERIDINO-3-(p-OCTYLPHENYL)-3-PROPANONE HYDROCHLORIDE see
PIY000
1-PIPERIDINO-3-(4′-OCTYLPHENYL)-PROPAN-3-ON-HYDROCHLORID (GERMAN)
see PIY000
PIPERIDINOOXY, 2,2,6,6-TETRAMETHYL- see TDT800
1-PIPERIDINO-2-PHENYL-AETHAN (GERMAN) see PDI500
3-PIPERIDINO-1-PHENYL-1-BICYCLOHEPTENYL-1-PROPANOL see BGD500
3-PIPERIDINO-1-PHENYL-1-BICYCLO(2.2.1)HEPTEN-(5)-YL-PROPANOL-
(1)(GERMAN) see BGD500
1-PIPERIDINO-2-PHENYLETHANE see PDI500
PIPERIDINO 3-PIPERIDYL KETONE see PIR200
3-PIPERIDINO-1,2-PROPANEDIOL DICARBANILATE see DVO819
3-PIPERIDINO-1,2-PROPANEDIOL DICARBANILATE HYDROCHLORIDE see
DVV500

3-γ-(1-PIPERIDINO)PROPIL-5,5-DIFENILTIOIDANTOINA CLORIDRATO (ITALIAN) see DWK500
4-(3-PIPERIDINOPROPIONAMIDO) SALICYLIC ACID METHYL ESTER, METH-IODIDE see MOK000
β-PIPERIDINOPROPIOPHENONE HYDROCHLORIDE see PIV500
3-PIPERIDINO-4′-PROPOXYPROPIOPHENONE HYDROCHLORIDE see PNB250
γ-PIPERIDINOPROPYL-p-AMINOBENZOATE HYDROCHLORIDE see PIS000
α-(3-PIPERIDINOPROPYL)BENZHYDROL see DWK200
2-(1-PIPERIDINO)-2-(2-THENYL)ETHYLAMINE MALEATE see PIT250
1-Δ-³-PIPERIDINO-3-o-TOLOXYPROPAN-2-OL HYDROCHLORIDE see TGK225
1-(1-PIPERIDINYL)-CYCLOHEXANECARBONITRILE (9CI) see PIN225
2-(1-PIPERIDINYL)ETHANOL see HKY600
2-(1-PIPERIDINYL)ETHANOL HYDROCHLORIDE see PIN275
PIPERIDINYLETHYLMORPHINE see PIT600
1-PIPERIDINYLOXY, 2,2,6,6-TETRAMETHYL-(9CI) see TDT800
3-(1-PIPERIDINYL)-1-(4-PROPOXYPHENYL)-1-PROPANONE HYDROCHLORIDE see PNB250
3-(2-PIPERIDINYL)PYRIDINE see AON875
(S)-3-(2-PIPERIDINYL)PYRIDINE HYDROCHLORIDE see PIT650
6-(1-PIPERIDINYL)-2,4-PYRIMIDINEDIAMINE-3-OXIDE see DCB000
PIPERIDOLATE HYDROCHLORIDE see EOY000
PIPERIDON (GERMAN) see PIU000
2-PIPERIDONE see PIU000
4-PIPERIDONE, 2,2,6,6-TETRAMETHYL- see TDT770
α-(2-PIPERIDYL)BENZHYDROL see DWK400
α-(2-PIPERIDYL)BENZHYDROL HYDROCHLORIDE see PII750
α-(4-PIPERIDYL)BENZHYDROL HYDROCHLORIDE see PIY750
4-(1-PIPERIDYL)CARBONYL-2,3-TETRAMETHYLENEQUINOLINE see PIU100
3-(1-PIPERIDYL)-1-CYCLOHEXYL-1-PHENYL-1-PROPANOL HYDROCHLORIDE see BBV000
β-PIPERIDYLETHANOL see HKY600
β-PIPERIDYLETHANOL HYDROCHLORIDE see PIN275
β-2-PIPERIDYLETHYL-m-AMINOBENZOATE HYDROCHLORIDE see AIQ880
β-2-PIPERIDYLETHYL-o-AMINOBENZOATE HYDROCHLORIDE see AIQ885
β-2-PIPERIDYLETHYL-p-AMINOBENZOATE HYDROCHLORIDE see AIQ890
α-(2-PIPERIDYLETHYL)BENZHYDROL HYDROCHLORIDE see PMC250
β-2-PIPERIDYLETHYLPHENYLURETHANE HYDROCHLORIDE see PIU800
2-(1-PIPERIDYLMETHYL)-1,4-BENZODIOXAN HYDROCHLORIDE see BCI500
3-(1-PIPERIDYL)-1,2-PROPANE DICARBANILATE see DVO819
3-(1-PIPERIDYL)-1,2-PROPANEDIOL DICARBANILATE HYDROCHLORIDE see DVV500
1-(2-PIPERIDYL)-2-PROPANONE see PAO500
β-(1-PIPERIDYL)PROPIOPHENONE HYDROCHLORIDE see PIV500
1-3-(2′-PIPERIDYL)PYRIDINE see AON875
3-(2-PIPERIDYL)-PYRIDINE see AON875
PIPERILATE HYDROCHLORIDE see PIM000
PIPERIN see PIV600
PIPERINE see PIV600
PIPERITONE see MCF250
PIPER LONGUM L., fruit extract see PIV650
PIPEROCAINE see PIV750
PIPEROCAINE HYDROCHLORIDE see IJZ000
PIPEROCAINIUM CHLORIDE see IJZ000
PIPEROCYANOMAZINE see PIW000
PIPEROLINIC ACID see HDS300
PIPERONAL see PIW250
PIPERONALACETONE see MJR250
PIPERONAL BIS(2-(2-BUTOXYETHOXY)ETHYL)ACETAL see PIZ499
PIPERONALDEHYDE see PIW250
PIPERONYL see FHG000
PIPERONYL ACETATE see PIX000
PIPERONYL ACETONE see MJR250
PIPERONYL ALDEHYDE see PIW250
PIPERONYL BUTOXIDE see PIX250
PIPERONYLIDENEACETONE see MJR250
2-(4-PIPERONYL-1-PIPERAZINYL)-9H-PURINE-9-ETHANOL DIHYDROCHLORIDE see PIX750
2-(4-PIPERONYL-1-PIPERAZINYL)PYRIMIDINE see TNR485
PIPERONYL SULFOXIDE see ISA000
1-PIPERONYL-4-(3,7,11-TRIMETHYL-2,6,10-DODECANTRIENYL)-PIPERAZINE METHANESULFONATE see PIE750
PIPEROXANE HYDROCHLORIDE see BCI500
1-PIPEROYLPIPERIDINE see PIV600
PIPERSAL see DAM600
trans-PIPERYLENE see PBA250
PIPE STEM see CMV390
PIPETHANATE ETHYLBROMIDE see PIX800
PIPETHANATE HYDROCHLORIDE see PIM000
PIPIZAN CITRATE SYRUP see PIJ500
PIPOBROMAN see BHJ250
PIPOCTANONE HYDROCHLORIDE see PIY000
PIPOLPHEN see DQA400
PIP-PIP see PIY500
PIPRACIL see SJJ200
α-PIPRADOL see DWK400

γ-PIPRADOL see PIY750
PIPRADOL see DWK400
PIPRADOL HYDROCHLORIDE see PII750
PIPRADROL HYDROCHLORIDE see PII750
PIPRAM see PIZ000
PIPRIL see SJJ200
PIPRINHYDRINATE see PIZ250
PIPROCTANYLIUMBROMID (GERMAN) see AGE750
PIPROTAL see PIZ499
PIPROZOLIN see EOA500
PIPROZOLINE see EOA500
PIPTAL see PJA000
PIRABUTINA see HNI500
PIRACAPS see TBX250
PIRACETAM see NNE400
PIRAFLOGIN see HNI500
PIRAMIDON see DOT000
PIRARREUMOL ''B'' see BRF500
PIRAZETAM see NNE400
PIRAZINON see DJX400
PIRAZOXON (ITALIAN) see MOX250
PIRBUTEROL DIHYDROCHLORIDE see POM800
PIRECIN see POF500
PIREF see DIZ100
PIRENZEPINE HYDROCHLORIDE see GCE500
PIRETANIDE see PDW250
PIRETRINA 1 (PORTUGUESE) see POO050
PIREVAN see PJA120
PIREXYL see MOA600
PIREXYL PHOSPHATE see PJA130
PIRIA'S ACID see ALI000
PIRIBEDIL see TNR485
PIRIBENZIL see TMP750
PIRIBENZIL METHYL SULFATE see CDG250
PIRID see PDC250, PEK250
PIRIDACIL see PDC250
PIRIDANE see CMA100
7-(2-(PIRIDIL)-METILAMMINO-ETIL)-TEOFILLINO NICOTINATO (ITALIAN) see PPN000
PIRIDINA (ITALIAN) see POP250
PIRIDINOL CARBAMATO (SPANISH) see PPH050
PIRIDISIR see PPP500
PIRIDOL see DOT000
PIRIDOLAN see PJA140
PIRIDOLO see AKO500
PIRIDOSAL see DAM600, DAM700
PIRIDROL see DWK400
PIRIDROL HYDROCHLORIDE see PII750
PIRIEX see TAI500
PIRIMAL-M see ALF250
PIRIMECIDAN see TGD000
PIRIMETAMINA (SPANISH) see TGD000
PIRIMICARB see DOX600
PIRIMIFOSETHYL see DIN600
PIRIMIFOS-METHYL see DIN800
PIRIMIPHOS-ETHYL see DIN600
PIRIMOR see DOX600
PIRINITRAMIDE see PJA140
PIRINIXIL see CLW500
PIRISTIN see POO750
PIRITON see CLX300, TAI500
PIRITRAMIDE see PJA140
PIRMAZIN see SNJ000
PIRMENOL HYDROCHLORIDE see PJA170
PIROAN see PCP250
PIROCRID see POE100
PIROD see UNJ800
PIRODAL see PAF250
PIROFOS see SOD100
PIROHEPTINE see PJA190
PIROHEPTINE HYDROCHLORIDE see TMK150
PIROMEN see PJA200
PIROMIDIC ACID see PAF250
PIROMIDINA see DOT000
PIROSOLVINA see EEM000
PIROXICAM see FAJ100
PIRPROFEN see PJA220
PIRROLIDINOMETIL-TETRACICLINA (ITALIAN) see PPY250
PIRROXIL see NNE400
PIRVINIUM PAMOATE see PQC500
PIRYDYNA (POLISH) see POP250
PISCAROL see IAD000
PISCIDEIN see HJO500
PISCIDIC ACID see HJO500
PISCIOL see IAD000

PISS-A-BED (JAMAICA) see CNG825
PISTACIA LENTISCUS ABSOLUTE see MBU777
PITAYINE see QFS000
PITC see ISQ000
PITCH see CMZ100
PITCH APPLE see BAE325
PITCH, COAL TAR see CMZ100
PITMAL see TLP750
PITOCIN see ORU500
PITON S see ORU500
PITREX see TGB475
PITTCHLOR see HOV500
PITTCIDE see HOV500
PITTCLOR see HOV500
PITTSBURGH PX-138 see DVL700
PITUITARY GLAND ADRENO CORTICO-TROPIC HORMONE see AES650
PITUITARY GROWTH HORMONE see PJA250
PIVACIN see PIH175
PIVADORM see BNP750
PIVADORN see BNP750
PIVAL see PIH175
PIVALDION (ITALIAN) see PIH175
PIVALDIONE (FRENCH) see PIH175
PIVALIC ACID see PJA500
PIVALIC ACID CHLORIDE see DTS400
PIVALIC ACID, HEXYL ESTER see HFP700
PIVALIC ACID LACTONE see DTH000
PIVALOLACTONE see DTH000
PIVALOLYL CHLORIDE see DTS400
PIVALONITRILE see PJA750
PIVALOYL AZIDE see PJB000
PIVALOYL CHLORIDE see DTS400
2-PIVALOYL-INDAAN-1,3-DION (DUTCH) see PIH175
2-PIVALOYL-INDAN-1,3-DION (GERMAN) see PIH175
2-PIVALOYL-1,3-INDANDIONE see PIH175
2-PIVALOYLINDANE-1,3-DIONE see PIH175
PIVALOYLMETHYL BROMIDE see PJB100
PIVALOYLOXYMETHYL d-α-AMINOBENZYLPENICILLINATE HYDROCHLORIDE
  see AOD000
PIVALYL CHLORIDE see DTS400
2-PIVALYL-1,3-INDANDIONE see PIH175
PIVALYL VALONE see PIH175
PIVALYN see PIH175
PIVAMPICILLIN HYDROCHLORIDE see AOD000
PIVATIL see AOD000
PIVMECILLINAM HYDROCHLORIDE see MCB550
PIXALBOL see CMY800
PIX CARBONIS see CMY800
PIX LITHANTHRACIS see CMY800
PJ185 see LIH000
PK 10169 see HAQ550
PK-MERZ see TJG250
PKhNB see PAX000
P 4070L see PJS750
PLACIDAL see EQL000, MNM500
PLACIDAS see MNM500
PLACIDEX see MFD500
PLACIDIL see CHG000
PLACIDOL see CJR909
PLACIDOL E see DJX000
PLACIDON see MQU750
PLACIDYL see CHG000
PLAC OUT see CDT250
PLAFIBRIDA (SPANISH) see PJB500
PLAFIBRIDE see PJB500
PLANADALIN see BNK000
PLANIUM see PJS750
PLANOCAINE see AIT250
PLANOCHROME see MCV000
PLANOFIX see NAK500
PLANOMIDE see TEX250
PLANOMYCIN see FBP300
PLANOTOX see DAA800
PLANT DITHIO AEROSOL see SOD100
PLANTDRIN see MRH209
PLANTFUME 103 SMOKE GENERATOR see SOD100
PLANTGARD see DAA800
PLANTIFOG 160M see MAS500
PLANTOMYCIN see SLY500
PLANT PIN see SOB500
PLANT PROTEASE CONCENTRATE see BMO000
PLANT PROTECTION PP511 see DIN800
PLANTULIN see PMN850
PLANTVAX see DLV200
PLANT WAX see DLV200

PLANUM see CFY750
PLAQUENIL see PJB750
PLASDONE see PKQ250
PLASIL see AJH000
PLASKON 201 see PJY500
PLASKON 3369 see MCB050
PLASKON 3381 see MCB050
PLASKON 3382 see MCB050
PLASKON CTFE see KDK000
PLASKON PP 60-002 see PJS750
PLASMA COAGULASE see CMY725
PLASMASTERIL see HLB400
PLASMOCHIN see RHZ000
PLASMOCIDE see RHZ000
PLASMOCOAGULASE see CMY725
PLASMOQUINE see RHZ000
PLASTANOX 2246 see MJO500
PLASTANOX 425 ANTIOXIDANT see MJN250
PLASTANOX LTDP see TFD500
PLASTANOX LTDP ANTIOXIDANT see TFD500
PLASTANOX STDP see DXG700
PLASTANOX STDP ANTIOXIDANT see DXG700
PLASTAZOTE X 1016 see PJS750
PLASTER of PARIS see CAX500
PLASTHALL 503 see BSB000
PLASTIBEST 20 see ARM268
PLASTICIZER BDP see BQX250
PLASTICIZER G-316 see PJC000
PLASTICIZER 4GO see PJC250
PLASTICIZER GPE see PJC500
PLASTICIZER Z-88 see PJC750
PLASTIFIX PC see PJQ050
PLASTOLEIN 9058 see BJQ500
PLASTOLEIN 9214 see FAB920
PLASTOLEIN 9058 DOZ see BJQ500
PLASTOL ORANGE G see CMS145
PLASTOL RUBINE BC see CMS155
PLASTOL YELLOW GG see CMS212
PLASTOL YELLOW GP see CMS212
PLASTOMOLL DOA see AEO000
PLASTOPAL BT see UTU500
PLASTORESIN ORANGE F 3A see CMP600
PLASTORESIN ORANGE F4A see PEJ500
PLASTORESIN RED F see SBC500
PLASTORESIN RED FR see CMS242
PLASTORESIN VIOLET 5BO see AOR500
PLASTRONGA see PJS750
PLASTYLENE MA 2003 see PJS750
PLASTYLENE MA 7007 see PJS750
PLATENOMYCIN B1 see MBY150
PLATENOMYCIN-B3 see LEV025
PLATH-LYSE see MJM500
PLATIBLASTIN see PJD000
PLATIN (GERMAN) see PJD500
cis-PLATIN see PJD000
PLATINATE(2-), HEXACHLORO-, DIHYDROGEN, HEXAHYDRATE see DLO400
PLATINATE(2), NITROTRICHLORO-, DIPOTASSIUM see PLN050
PLATINATE(2), TETRACHLORO-, DIAMMONIUM see ANV800
PLATINATE(2), TETRAKIS(THIOCYANATO)-, DIPOTASSIUM see PLU590
PLATINATE(1), TRICHLOROETHYLENE-, DIPOTASSIUM see PLW200
PLATINATE(2), TRICHLORO(NITRITO-N-), DIPOTASSIUM (SP-4-2)- see PLN050
PLATINEX see PJD000
PLATINIC AMMONIUM CHLORIDE see ANF250
PLATINIC CHLORIDE see CKO750
PLATINIC POTASSIUM CHLORIDE see PLR000
PLATINIC SODIUM CHLORIDE see SJJ500
PLATINOL see PJD000
PLATINOL AH see DVL700
PLATINOL DOP see DVL700
PLATINOUS CHLORIDE see PJE000
cis-PLATINOUS DIAMMINE DICHLORIDE see PJD000
PLATINOUS POTASSIUM CHLORIDE see PJD250
PLATINUM see PJD500
PLATINUM(II), BIS(METHYL SELENIDE)DICHLORO-, cis- see DEU115
PLATINUM (II), BIS(METHYL SELENIDE)SULFATO-, HYDRATE see SNS100
PLATINUM BLACK see PJD500
PLATINUM(IV) CHLORIDE see PJE250
PLATINUM CHLORIDE see PJE000
PLATINUM COMPOUNDS see PJE500
PLATINUM(II) (CYCLOHEXANE-1,2-DIAMMINE)ISOCITRATO-, (Z)- see PGQ275
cis-PLATINUM(II) DIAMMINEDICHLORIDE see PJD000
trans-PLATINUM(II)DIAMMINEDICHLORIDE see DEX000
cis-PLATINUMDIAMMINE TETRACHLORIDE see TBO776
cis-PLATINUM(IV) DIAMMINOTETRACHLORIDE see TBO776
PLATINUM DIARSENIDE see PJE750

PLATINUM, DICHLOROBIS(METHYL SELENIDE)-, cis- see DEU115
PLATINUM, DICHLOROBIS(SELENOBIS(METHANE))-(SP-4-2) see DEU115
PLATINUM(II) DINITRODIAMMINE see DCF800
PLATINUM ETHYLENEDIAMMINE DICHLORIDE see DFJ000
PLATINUM (2), NITROTRICHLORO-, DIPOTASSIUM see PLN050
PLATINUM SPONGE see PJD500
PLATINUM(II) SULFATE see PJF500
PLATINUM SULFATE see PJF500
PLATINUM(II) SULFATE TETRAHYDRATE see PJF750
PLATINUM SULFATE TETRAHYDRATE see PJF750
PLATINUM TETRACHLORIDE see PJE250
PLATIPHILLIN HYDROCHLORIDE see PJG150
PLATIPHYLLIN HYDROCHLORIDE see PJG150
PLAVOLEX see DBD700
PLAXIDOL see CJR909
PLE 1053 see ASO375
PLECYAMIN see VSZ000
PLEGANGIN see VIZ400
PLEGATIL see MQF750
PLEGECYL see ABH500
PLEGICIN see ABH500
PLEGINE see DKE800, PDD000
PLEGOMAZIN see CKP250, CKP500
PLENASTRIL see PAN100
PLENOLIN see DLO875
PLENUR see LGZ000
PLEOCIDE see ABY900
PLESSY'S GREEN (HEMIHEPTAHYDRATE) see CMK300
PLESTROVIS see QFA250
PLETIL see TGD250
PLEXIGLAS see PKB500
PLICTRAN see CQH650
PLIDAN see DCK759
PLIMASINE see MNQ000
PLIOFILM see PJH500
PLIOFLEX see SMR000
PLIOGRIP see PKL500
PLIOLITE see BOP100
PLIOLITE S5 see SMR000
PLIOVAC AO see AAX175
PLIOVIC see PKQ059
PLISULFAN see AIF000
P-LITE 500 see CAT775
PLIVA see BIE500
PLIVAPHEN see ABH500
PLLETIA see PMI750
PLOMB FLUORURE (FRENCH) see LDF000
PLOSTENE see TKH750
PLOYMANNURONIC ACID see AFL000
PLP see PII100
PLTU see MJZ000
PLUCHEA LANCEOLATA (DC.) Cl., extract excluding roots see PJH550
PLUCKER see NAK500
PLUMBAGIN see PJH610
PLUMBAGO see CBT500
PLUMBAGO ZEYLANICA Linn., root extract see PJH615
PLUMBANE, BIS(ACETYLOXY)DIHEXYL- see DKP200
PLUMBOPLUMBIC OXIDE see LDS000
PLUMBOUS ACETATE see LCJ000, LCV000
PLUMBOUS CHLORIDE see LCQ000
PLUMBOUS CHROMATE see LCR000
PLUMBOUS FLUORIDE see LDF000
PLUMBOUS OXIDE see LDN000
PLUMBOUS PHOSPHATE see LDU000
PLUMBOUS SULFIDE see LDZ000
PLURACOL E-200 see PJT000
PLURACOL E-300 see PJT000
PLURACOL E-400 see PJT000
PLURACOL E-600 see PJT000
PLURACOL E-1500 see PJT000
PLURACOL E-4000 see PJT000
PLURACOL E-6000 see PJT000
PLURACOL E see PJT200
PLURACOL P-410 see PJT000
PLURACOL P-710 see PJT000
PLURACOL P-1010 see PJT000
PLURACOL P-2010 see PJT000
PLURACOL P-3010 see PJT000
PLURACOL P-4010 see PJT000
PLURAFAC RA 43 see DXY000
PLURAGARD see MCB000
PLURAGARD C 133 see MCB000
PLUREXID see CDT250
PLURONIC L-81 see PJH630

PLURYL see BEQ625
PLURYLE see BEQ625
PLUSURIL see BEQ625
PLUTONIUM see PJH750
PLUTONIUM BISMUTHIDE see PJH775
PLUTONIUM COMPOUNDS see PJI000
PLUTONIUM(III) HYDRIDE see PJI250
PLYAMINE HD 1129A see UTU500
PLYAMINE M27 see MCB050
PLYAMINE P 364BL see UTU500
PLYCTRAN see CQH650
PLYMOUTH IPP see IQW000
PLYMOUTM IPM see IQN000
PLYSET TD688 see MCB050
PM245 see BEM500
PM 334 see MNZ000
P.M. 346 see EIT100
P.M. 388 see PBH100
PM 396 see MLP800
PM 671 see ENG500
PM 2763 see DTI600
P.M. 434,526 see DBE885
PMA see ABU500, PGV000
PMAC see ABU500
PMACETATE see ABU500
PMAL see ABU500
PMAS see ABU500
PMB see ECU750
PMC see PFM500
PMCG see PJI575
PMCG HYDROCHLORIDE see PJI575
PMDT see PBG500
P.M.F. see PFN000
PMFP see SKS500
PMH see PAP000
PMMA see PKB500
PMO 10 see PFP100
PMP see PHX250, TGK200
PMP SODIUM GLUCONATE see SHK800
PMS 1.5 see SCR400
PMS 300 see SCR400
PMS see MRW000, SCA750
PMS 154A see SCR400
PMS 200A see SCR400
PMSF see TGO300
PMSG see SCA750
PMS No. 1 see PJW750
PMT see MOA725, PFS500
PM-TC see PMI000
PN6 see AQO000
PN see CFC000
PNA see NEO500
PNAP see NEL600
PNB see NFQ000
PNCB see NFS525
PND see COS500
PNEUMOREL see DAI200
PNNG see NLD000
PNOA see NEQ000
PNOT see NMP500
PNS 25 see SCR400
PNSP see SIV600
PNT see NMO550
PNU see NLO500
PO-20 see PMD800
POA see PER250
POCAN BUSH see PJJ315
PO-DIMETHOATE see DNX800
PODOPHYLLIN see PJJ000
PODOPHYLLINIC ACID LACTONE see PJJ225
PODOPHYLLOTOXIN see PJJ225
PODOPHYLLOTOXIN-o-BENXYLIDENE-β-d-GLUCOPYRANOSIDE see BER500
PODOPHYLLOTOXIN-BENZILIDEN-GLUCOSID (GERMAN) see BER500
PODOPHYLLUM see PJJ000
PODOPHYLLUM PELTATUM see MBU800
PODOPHYLLUM RESIN see PJJ000
POE 20 SORBITAN MONOLAURATE see PKG000
POHA (HAWAII) see JBS100
POINSETTIA see EQX000
POINSETTIA PULCHERRIMA GRAH see EQX000
POINT TWO see SHF500
POIS COCHON (HAITI) see YAG000
POIS MANIOC (HAITI) see YAG000
POISON BULB see SLB250
POISON BUSH see BOO700

POISON de COULEUVRE (CANADA) see BAF325
POISON HEMLOCK see PJJ300
POISON PARSLEY see PJJ300
POISON ROOT see PJJ300
POISON SEGO see DAE100
POISON TOBACCO see HAQ100
POISON TREE see BOO700
POIS PUANTE (HAITI) see CNG825
POIS VALLIERE (HAITI) see SBC550
POKE see PJJ315
POKEBERRY see PJJ315
POKEWEED see PJJ315
POLAAX see OAX000
POLACARITOX see CKM000
POLACTINE G YELLOW see CMS230
POLAKTYN YELLOW G see CMS230
POLAMIDON see MDP250
POLAMIDONE see MDO750
POLAN NAVY BLUE E 2R see CMM080
POLAN RED FS see CMO870
POLAN YELLOW E-3R see SGP500
POLARAMIN see PJJ325
POLARAMINE MALEATE see PJJ325
POLARIS see BLG250
POLARONIL (GERMAN) see TAI500
POLARONIL see CLX300
POLAR RED G see CMM320
POLAR RED GBD see NAO600
POLAR RED G SUPRA see CMM320
POLAR RED R see NAO600
POLAR RED RS see CMM330
POLCARB see CAT775
POLCILLIN see PJJ350
POLCOMINAL see EOK000
POLECAT WEED see SDZ450
POLECTRON 430 see VQK595
POLECTRON 450 see VQK595
POLEON see EID000
POLFOSCHLOR see TIQ250
POLICAPRAN see PJY500
POLICAR MZ see DXI400
POLICAR S see DXI400
POLIDIM see DOR800
POLIDOCANOL see DXY000
POLIFEN see TAI250
POLIFLOGIL see HNI500
POLIFUNGIN see PJJ500
POLIGOSTYRENE see SMQ500
POLIK see IFA000
POLIKARBATSIN (RUSSIAN) see MQQ250
POLINALIN see DOT000
POLISEPTIL see TEX250
POLISIN see BKL250
POLITEF see TAI250
POLITEN see PJS750
POLITEN I 020 see PJS750
POLIURON see BEQ625
POLIVAL see TEX000
POLIVINIT see PKQ059
POLKWEED see PJJ315
POLLACID see HCQ500
POLNOKS R see PJQ750
POLOMEL ME 3 see MCB050
POLOMEL MEC 3 see MCB050
POLONIUM see PJJ750
POLONIUM CARBONYL see PJK000
POLOPIRYNA see ADA725
POLOXAL RED 2B see HJF500
POLPRETAN K 2 see MCB050
POLSTIGMINE see DQY909
POLY see SKN000
POLYAC see PJR500
POLYACRYLAMIDE see PJK350
POLYACRYLATE see ADW200
POLY(ACRYLIC ACID) see ADW200
POLYACRYLONITRILE see ADX750
POLYAETHYLEN see PJS750
POLYAETHYLENGLYCOLE 200 (GERMAN) see PJT200
POLYAETHYLENGLYKOLE 1500 (GERMAN) see PJT500
POLYAETHYLENGLYKOLE #1000 (GERMAN) see PJT250
POLYAETHYLENGLYKOLE 300 (GERMAN) see PJT225
POLYAETHYLENGLYKOLE 400 (GERMAN) see PJT230
POLYAETHYLENGLYKOLE 4000 (GERMAN) see PJT750
POLYAETHYLENGLYKOLE 600 (GERMAN) see PJT240
POLYAETHYLENGLYKOLE 6000 (GERMAN) see PJU000

POLYAMID (GERMAN) see NOH000
POLYAMIDE 6 see PJY500
POLYAMINE D see PJL000
POLYAMINE H SPECIAL see PJL100
POLYAMINE T see FMW330
POLY-p-AMINOBENZALDEHYD (CZECH) see DXS000
POLY(ε-AMINOCAPROIC ACID) see PJY500
POLYBENZARSOL see BCJ150
POLYBIS(2-ETHYLBUTYL)SILOXANE see EHC900
POLYBOR 3 see DXF200
POLYBOR see DXF200, SFF000
POLY[BORANE(1)] see PJL325
POLYBOR-CHLORATE see SFS500
POLYBREME see HCV500
POLYBROMINATED BIPHENYL see FBU509, HCA500
POLYBROMINATED BIPHENYL (FF-1) see FBU509
POLYBROMINATED BIPHENYLS see FBU000, PJL335
POLYBROMINATED SALICYLANILIDE see THW750
POLYBROMOETHYLENE see PKQ000
cis-POLY(BUTADIENE) see PJL350
POLY(1,3-BUTADIENE PEROXIDE) see PJL375
POLYBUTADIENE-POLYSTYRENE COPOLYMER see SMR000
POLY(ε-CAPROAMIDE) see PJY500
POLYCAPROAMIDE see PJY500
POLY(ε-CAPROLACTAM) see PJY500
POLYCAPROLACTAM see PJY500
POLYCARBACIN see MQQ250
POLYCARBACINE see MQQ250
POLYCARBAZIN see MQQ250
POLYCARBAZINE see MQQ250
POLY(CARBON MONOFLUORIDE) see PJL600
POLYCAT 8 see DRF709
POLYCHLORCAMPHENE see CDV100
POLYCHLORINATED BIPHENYL (AROCLOR 1016) see PJL800
POLYCHLORINATED BIPHENYL (AROCLOR 1221) see PJM000
POLYCHLORINATED BIPHENYL (AROCLOR 1232) see PJM250
POLYCHLORINATED BIPHENYL (AROCLOR 1242) see PJM500
POLYCHLORINATED BIPHENYL (AROCLOR 1248) see PJM750
POLYCHLORINATED BIPHENYL (AROCLOR 1254) see PJN000
POLYCHLORINATED BIPHENYL (AROCLOR 1262) see PJN500
POLYCHLORINATED BIPHENYL (AROCLOR 1268) see PJN750
POLYCHLORINATED BIPHENYL (AROCLOR 1260) see PJN250
POLYCHLORINATED BIPHENYL (AROCLOR 2565) see PJO000
POLYCHLORINATED BIPHENYL (AROCLOR 4465) see PJO250
POLYCHLORINATED BIPHENYL (KANECHLOR 300) see PJO500
POLYCHLORINATED BIPHENYL (KANECHLOR 400) see PJO750
POLYCHLORINATED BIPHENYL (KANECHLOR 500) see PJP000
POLYCHLORINATED BIPHENYLS see PJL750
POLYCHLORINATED CAMPHENES see CDV100
POLYCHLORINATED DIBENZOFURANS see PJP100
POLYCHLORINATED TERPHENYL see PJP250
POLYCHLORINATED TRIPHENYL (AROCLOR 5442) see PJP750
POLYCHLOROBENZOIC ACID, DIMETHYLAMINE SALTS see PJQ000
POLYCHLOROBIPHENYL see PJL750
POLY(2-CHLORO-1,3-BUTADIENE) see PJQ050
POLY(2-CHLOROBUTADIENE) see PJQ050
POLYCHLOROCAMPHENE see CDV100
POLYCHLORODICYCLOPENTADIENE see BAF250
POLYCHLORODICYCLOPENTADIENE ISOMERS see BAF250
POLY(CHLOROETHYLENE) see PKQ059
POLYCHLOROPINENE see PJQ250
POLYCHLOROPRENE see PJQ050
POLY(CHLOROTRIFLUOROETHENE) see KDK000
POLY(CHLOROTRIFLUOROETHYLENE) see KDK000
POLYCHOM see ZJS300
POLYCHOME see ZJS300
POLYCIDAL see MFN500
POLYCILLIN see AIV500, AOD125
POLYCILLIN-N see SEQ000
POLYCIZER 332 see BSL600
POLYCIZER 532 see OEU000
POLYCIZER 562 see OEU000
POLYCIZER 962-BPA see DXQ200
POLYCIZER DBP see DEH200
POLYCIZER DBS see DEH600
POLYCLAR L see PKQ250
POLYCLENE see DGB000
POLYCO 2410 see SMR000
POLYCO 2611 see CGW300
POLYCO see SJK000
POLYCO 220NS see SMQ500
POLYCOR DARK GREEN S see CMO830
POLYCOR RED GS see CMM325
POLY C POLY I see PJY750
POLYCRON see BNA750

POLYCTYIDYLIC-POLYINOSINIC ACID see PJY750
POLYCYCLIC MUSK see ACL750
POLYCYCLINE see TBX000
POLYCYCLINE HYDROCHLORIDE see TBX250
POLYCYCLOPENTADIENYLTITANIUM DICHLORIDE see PJQ275
POLYDAZOL see PJQ350
POLYDESIS see PKP750
POLYDIBROMOSILANE see PJQ500
POLYDIBROMOSILYLENE see PJQ500
POLY(1,2-DIHYDRO-2,2,4-TRIMETHYLQUINOLINE) see PJQ750
POLY(DIMERCURYIMMONIUM ACETYLIDE) see PJQ775
POLY(DIMERCURYIMMONIUM BROMATE) see PJQ780
POLY(DIMERCURYIMMONIUM HYDOXIDE) see DNW200
POLY((DIMETHYLIMINIO)HEXAMETHYLENE(DIMETHYLIMINO)TRIMETHYLENE
   DIBROMIDE) see IFT300
POLYDIMETHYLSILOXANE see DTR850
POLYDIMETHYL SILOXANE see PJR000
POLYDIMETHYLSILOXANE RUBBER see PJR250
POLY-p-DINITROSOBENZENE see PJR500
POLY-EM 12 see PJS750
POLY-EM 40 see PJS750
POLY-EM 41 see PJS750
POLYESTER TGM 3 see MDN510
POLY(ESTRADIOL PHOSPHATE) see EDS000
POLYESTRADIOL PHOSPHATE see PJR750
POLYETHYLENE see PJS750
POLYETHYLENE AS see PJS750
POLYETHYLENE 600 DIBENZOATE see PKE750
POLYETHYLENE GLYCOL 1500 see PJT500
POLYETHYLENE GLYCOL #1000 see PJT250
POLYETHYLENE GLYCOL 200 see PJT200
POLYETHYLENE GLYCOL 300 see PJT225
POLYETHYLENE GLYCOL 400 see PJT230
POLYETHYLENE GLYCOL 4000 see PJT750
POLYETHYLENE GLYCOL 600 see PJT240
POLYETHYLENE GLYCOL 6000 see PJU000
POLYETHYLENE GLYCOL see PJT000
POLYETHYLENE GLYCOL CETYL ETHER see PJT300
POLYETHYLENE GLYCOL CHLORIDE 210 see PJU250
POLYETHYLENE GLYCOL 220 DIBENZOATE see PKE750
POLYETHYLENE GLYCOL DIBENZOATE see PKE750
POLYETHYLENE GLYCOL 200 DI(2-ETHYLHEXOATE) see FCD512
POLYETHYLENE GLYCOL 300 DISTEARATE see PJU500
POLYETHYLENE GLYCOL 400 (DI) STEARATE see PJU500
POLYETHYLENE GLYCOL 600 (DI) STEARATE see PJU500
POLYETHYLENE GLYCOL DISTEARATE see PJU500
POLYETHYLENE GLYCOL DODECYL ETHER see DXY000
POLYETHYLENE GLYCOL MONOETHER with p-tert-OCTYLPHENYL see
   PKF500
POLYETHYLENE GLYCOL MONOLEYL ETHER see OIG040
POLYETHYLENE GLYCOL MONO(4-OCTYLPHENYL) ETHER see PKF500
POLYETHYLENE GLYCOL MONO(4-tert-OCTYLPHENYL) ETHER see PKF500
POLYETHYLENE GLYCOL MONO(p-tert-OCTYLPHENYL) ETHER see PKF500
POLYETHYLENE GLYCOL MONO(OCTYLPHENYL) ETHER see GHS000
POLYETHYLENE GLYCOL MONOOLEATE see PJY100
POLYETHYLENE GLYCOL MONOSTEARATE see PJV250
POLYETHYLENE GLYCOL MONO(p-(1,1,3,3-TETRAMETHYLBUTYL)PHENYL)
   ETHER see PKF500
POLYETHYLENE GLYCOL 450 NONYL PHENYL ETHER see PKF000
POLYETHYLENE GLYCOL OCTYLPHENOL ETHER see PKF500
POLYETHYLENE GLYCOL 450 OCTYL PHENYL ETHER see PKF500
POLYETHYLENE GLYCOL p-OCTYLPHENYL ETHER see PKF500
POLYETHYLENE GLYCOL p-tert-OCTYLPHENYL ETHER see PKF500
POLYETHYLENE GLYCOL OCTYLPHENYL ETHER see GHS000
POLYETHYLENE GLYCOL OLEATE see PJY100
POLYETHYLENE GLYCOL 1000 OLEYL ETHER see PJW500
POLYETHYLENE GLYCOL PALMITYL ETHER see PJT300
POLYETHYLENE GLYCOLS MONOHEXADECYL ETHER see PJT300
POLYETHYLENEGLYCOLS MONOSTEARATE see PJW750
POLYETHYLENE GLYCOL p-1,1,3,3,-TETRAMETHYLBUTYLPHENYL ETHER
   see PKF500
POLYETHYLENEIMIN (CZECH) see PJX000
POLYETHYLENE IMINE see PJX000
POLY(ETHYLENE OXIDE) see PJT000
POLYETHYLENE OXIDE CETYL ETHER see PJT300
POLY(ETHYLENE OXIDE) DODECYL ETHER see DXY000
POLYETHYLENE OXIDE HEXADECYL ETHER see PJT300
POLYETHYLENE OXIDE MONOOLEATE see PJY100
POLY(ETHYLENE OXIDE)OCTYLPHENYL ETHER see GHS000
POLY(ETHYLENE OXIDE) OLEATE see PJY100
POLYETHYLENE-POLYPROPYLENE GLYCOLS PLURONIC L-81 see PJH630
POLYETHYLENE RESINS see PJS750
POLYETHYLENE TEREPHTHALATE see PKF750
POLYETHYLENE TEREPHTHALATE FILM see PKF750
POLY(ETHYLENE TETRAFLUORIDE) see TAI250

POLYETHYLENE Y-141-A see PJX750
POLYETHYLENIMINE (10,000) see PJX800
POLYETHYLENIMINE (20,000) see PJX825
POLYETHYLENIMINE (35,000) see PJX835
POLYETHYLENIMINE (40,000) see PJX845
POLY(ETHYLIDENE PEROXIDE) see PJX850
POLYFENE see TAI250
POLYFER see IGS000
POLYFIBRON 120 see SFO500
POLYFIX PM 5 see MCB050
POLYFIX PM 107 see MCB050
POLYFLEX see SMQ500
POLYFLON see TAI250
POLYFOAM PLASTIC SPONGE see PKL500
POLYFOAM SPONGE see PKL500
POLYFUNGIN see PJJ500
POLY G 400 see PJT230
POLY-G see PJT200
POLY-GIRON see TEH500
POLYGLUCIN see DBD700
POLYGLYCOL 1000 see PJT250
POLYGLYCOL 4000 see PJT750
POLYGLYCOL 15-200 see PJX900
POLYGLYCOLAMINE H-163 see AMC250
POLYGLYCOLDIAMINE H 221 see EBV100
POLYGLYCOL DISTEARATE see PJU500
POLYGLYCOL E1000 see PJT250
POLYGLYCOL E-4000 see PJT750
POLYGLYCOL E see PJT200
POLYGLYCOL E-4000 USP see PJT750
POLYGLYCOL LAURATE see PJY000
POLYGLYCOL MONOOLEATE see PJY100
POLYGLYCOL OLEATE see PJY250
POLYGON see SKN000
POLYGONUM HYDROPIPER L., dry powdered whole plant see WAT350
POLYGRIPAN see TEH500
POLY-G SERIES see PJT000
POLY(HEXAMETHYLENE DIISOCYANATE) see HEG300
POLY I:C see PJY750
POLY(IMINOCARBONYLPENTAMETHYLENE) see PJY500
POLY(IMINO(1-OXO-1,6-HEXANEDIYL)) see PJY500
POLYINOSINATE:POLYCYTIDYLATE see PJY750
POLYINOSINIC:POLYCYTIDYLIC ACID COPOLYMER see PJY750
POLYMARCIN see MQQ250
POLYMARCINE see MQQ250
POLYMARSIN see MQQ250
POLYMARZIN see MQQ250
POLYMARZINE see MQQ250
POLYMAT see MQQ250
POLYMER 261 see DTS500
POLYMERIC DIALDEHYDE see PKA000
POLYMER 261LV see DTS500
POLYMERS of EPICHLOROHYDRIN and 2,2-BIS(4-HYDROXYPHEN-
   YL)PIPERAZINE see ECM500
POLYMERS, WATER-INSOLUBLE see PKA850
POLYMERS, WATER-SOLUBLE see PKA860
POLY(METHIBIS(HYDROXYMETHYL)UREYLENE)AMER see UTU500
POLY(2-METHOXY-5-(2-(METHYLAMINO)ETHYL)-m-PHENYLENEMETHYLENE)
   see PMB800
POLY-(METHYLBIS(THIOCYANATO)ARSINE) see MJK750
POLYMETHYLENEPOLYPHENYL ISOCYANATE see PKB100
POLYMETHYLMETHACRYLATE see PKB500
POLYMINE D see BEN000
POLYMIST A12 see PJS750
POLYMON BLUE 3R see IBV050
POLYMONE see DGB000
POLY(MONOCHLOROTRIFLUOROETHYLENE) see KDK000
POLYMO ORANGE GR see CMS145
POLYMO RED FGN see MMP100
POLYMO YELLOW GR see CMS208
POLYMUL CS 81 see PJS750
POLYMYCIN see PKB775
POLYMYXIN see PKC000
POLYMYXIN A see PKC250
POLYMYXIN B1 see PKC550
POLYMYXIN B see PKC500
POLYMYXIN B SULFATE see PKC750
POLYMYXIN D1 see PKD050
POLYMYXIN E see PKD250
POLYNOXYLIN see UTU500
POLY(N,N,N',N'-TETRAMETHYL-N-
   TRIMETHYLENEHEXAMETHYLENEDIAMMONIUM DIBROMIDE) see HCV500
POLYOESTRADIOL PHOSPHATE see EDS000
POLYOX see PJT000
POLYOXIETHYLENE (6) ALKYL (13) ETHER see TJJ250

POLYOXIN AL see PKE100
POLYOXIN B see PKE100
POLY(1-(2-OXO-1-PYRROLIDINYL)ETHYLENE) see PKQ250
POLY(1-(2-OXO-1-PYRROLIDINYL)ETHYLENE)IODINE COMPLEX see PKE250
POLY(OXY(DIMETHYLSILYLENE)) see PJR000
POLY(OXY-1,2-ETHANEDIYL), α-((3-β)-CHOLEST-5-EN-3-YL)-ω-HYDROXY-(9CI) see PKE700
POLY(OXY-1,2-ETHANEDIYL), α-α',α'',α'''-(1,2-ETHANEDIYLBIS(NITRILODI-2,1-ETHANEDIYL))TETRAKIS(ω-HYDROXY)- see EIV750
POLY(OXY-1,2-ETHANEDIYL), α-HEXADECYL-ω-HYDROXY- see PJT300
POLY(OXY-1,2-ETHANEDIYL), α-HYDRO-omega-HYDROXY- see PJT000
POLY(OXY-1,2-ETHANEDIYL), α-(1-OXODODECYL)-ω-HYDROXY-(9CI) see PJY000
POLY(OXY-1,2-ETHANEDIYLOXYCARBONYL-1,4-PHENYLENECARBONYL) see PKF750
POLYOXYETHYLENE 1500 see PJT500
POLYOXYETHYLENE (75) see PJT750
POLYOXYETHYLENE ALKYLAMINE MONO-FATTY ACID ESTER see PKE550
POLYOXYETHYLENE-ALKYL CITRIC DIESTER-TRIETHANOLAMINE see PKE600
POLYOXYETHYLENE (7) ALKYL (14) ETHER see TBY750
POLYOXYETHYLENE-sec-ALKYL ETHER CITRIC DIESTER TRIETHANOLAMINE see PKE600
POLY(OXYETHYLENE)CETYL ETHER see PJT300
POLYOXYETHYLENE CHOLESTERYL ETHER see PKE700
POLYOXYETHYLENE DIBENZOATE see PKE750
POLYOXYETHYLENE DODECANOL see LBT000
POLY(OXYETHYLENE)HEXADECYL ETHER see PJT300
POLYOXYETHYLENE LAURIC ALCOHOL see DXY000
POLYOXYETHYLENE LAURYL ETHER see DXY000
POLY(OXYETHYLENE)MONOCETYL ETHER see PJT300
POLYOXYETHYLENE-20 MONOLAURATE see PJY000
POLYOXYETHYLENE MONOLAURATE see PJY000
POLYOXYETHYLENE MONOOCTYLPHENYL ETHER see GHS000
POLYOXYETHYLENE MONO(OCTYLPHENYL) ETHER see PKF500
POLY(OXYETHYLENE) MONOOLEATE see PJY100
POLYOXYETHYLENE-8-MONOSTEARATE see PJV250
POLYOXYETHYLENE MONOSTEARATE see PJW750
POLYOXYETHYLENE NONYLPHENOL see NND500
POLYOXYETHYLENE (9) NONYL PHENYL ETHER see PKF000
POLY(OXYETHYLENE)OCTYLPHENOL ETHER see GHS000
POLYOXYETHYLENE (13) OCTYLPHENYL ETHER see PKF500
POLYOXYETHYLENE (9) OCTYLPHENYL ETHER see PKF500
POLY(OXYETHYLENE)-p-tert-OCTYLPHENYL ETHER see PKF500
POLY(OXYETHYLENE) OLEATE see PJY100
POLY(OXYETHYLENE) OLEIC ACID ESTER see PJY100
POLYOXYETHYLENE (20) OLEYL ETHER see PJW500
POLY(OXYETHYLENE)OXYTEREPHTHALOYL) see PKF750
POLY(OXYETHYLENE)PALMITYL ETHER see PJT300
POLYOXYETHYLENE (20) SORBITAN MONOLAURATE see PKG000, PKL000
POLYOXYETHYLENE SORBITAN MONOOLEATE see PKL100
POLYOXYETHYLENE 20 SORBITAN MONOPALMITATE see PKG500
POLYOXYETHYLENE SORBITAN MONOPALMITATE see PKG500
POLYOXYETHYLENE 20 SORBITAN MONOSTEARATE see PKL030
POLYOXYETHYLENE SORBITAN MONOSTEARATE see PKL030
POLYOXYETHYLENE SORBITAN OLEATE see PKL100
POLYOXYETHYLENE (20) SORBITAN TRIOLEATE see TOE250
POLYOXYETHYLENE(8)STEARATE see PJV250
POLYOXYL 10 OLEYL ETHER see OIG040
POLYOXYMETHYLENE see TMP000
POLYOXYMETHYLENE GLYCOLS see FMV000
POLY(OXYPROPYLENE) BUTYL ETHER see BRP250
POLYOXYPROPYLENE MONOBUTYL ETHER see BRP250
POLYPEPTIDE, BACILLUS THURINGIENSIS subsp. ISRAELENSIS, crystal preparation see BAC135
POLYPEPTIN see CMS225
POLY(PEROXYISOBUTYROLACTONE) see PKH260
POLYPHENOL FRACTION OF BETEL NUT see BFW010
POLY p-PHENYLENE TEREPTHALAMIDE ARAMID FIBER see PKH850
POLYPHLOGIN see PGG000
POLYPHOS see SHM500
POLYPODINE A see HKG500
POLYPRO 1014 see PMP500
POLYPROPENE see PMP500
POLY(2-PROPYL-m-DIOXANE-4,6-DIYLENE) see PKI000
POLYPROPYLENE see PMP500
POLYPROPYLENE GLYCOL 1200 see PKJ250
POLYPROPYLENE GLYCOL 425 see PKI550
POLYPROPYLENE GLYCOL 750 see PKI750
POLYPROPYLENE GLYCOL see PKI500
POLYPROPYLENE GLYCOL METHYL ETHER see MKS250
POLYPROPYLENE GLYCOL 400, MONOBUTYL ETHER see PKK500
POLYPROPYLENE GLYCOL 800, MONOBUTYL ETHER see PKK750
POLYPROPYLENE GLYCOL MONOBUTYL ETHER see BRP250
POLYPROPYLENE GLYCOL MONOMETHYLETHER see MKS250

POLYPROPYLENGLYKOL (CZECH) see PKI500
POLYQUATERNIUM 6 see DTS500
POLYRAM 80 see MQQ250
POLYRAM see MQQ250
POLYRAM COMBI see MQQ250
POLYRAM M see MAS500
POLYRAM ULTRA see TFS350
POLYRAM 80WP see MQQ250
POLYRAM Z see EIR000
POLYSILICONE see PJR250
POLYSILYLENE see PKK775
POLYSION N 22 see PJS750
POLYSIZER 173 see PKP750
POLY(SODIUM p-STYRENESULFONATE) see SJK375
POLY-SOLV see CBR000
POLY-SOLV DB see DJF200
POLY-SOLV DM see DJG000
POLY-SOLV EB see BPJ850
POLY-SOLV EE see EES350
POLY-SOLV EE ACETATE see EES400
POLY-SOLV EM see EJH500
POLY-SOLVE MPM see PNL250
POLY-SOLV TB see TKL750
POLY-SOLV TE see EFL000
POLY-SOLV TM see TJQ750
POLYSORBAN 80 see PKL100
POLYSORBATE 20 see PKL000
POLYSORBATE 40 see PKG500
POLYSORBATE 65 see SKV195
POLYSORBATE 60 see PKL030
POLYSORBATE 80 see PKL100
POLYSORBATE 80, U.S.P. see PKL100
POLYSTICHOCITRIN see FBY000
POLYSTROL D see SMQ500
POLYSTYRENE see SMQ500
POLYSTYRENE-ACRYLONITRILE see ADY500
POLYSTYRENE BW see SMQ500
POLYSTYRENE LATEX see SMQ500
POLYSTYROL see SMQ500
POLYTAC see PMP500
POLYTAR BATH see CMY800
POLYTEF see TAI250
POLYTETRAFLUOROETHENE see TAI250
POLYTETRAFLUOROETHYLENE see TAI250
POLYTEX 973 see ADW200
POLYTHENE see PJS750
POLYTHERM see PKQ059
POLYTHIAZIDE see PKL250
POLYTOX see DGB000
POLY(TRIFLUOROCHLOROETHYLENE) see KDK000
POLY(TRIFLUOROETHYLENE CHLORIDE) see KDK000
POLY(TRIFLUOROMONOCHLOROETHYLENE) see KDK000
POLY(TRIFLUOROVINYL CHLORIDE) see KDK000
POLYURETHANE A see PKL500
POLYURETHANE ESTER FOAM see PKL500
POLYURETHANE ETHER FOAM see PKL500
POLYURETHANE FOAM see PKL500
POLYURETHANE SPONGE see PKL500
POLYURETHANE Y-195 see PKL750
POLYURETHANE Y-217 see PKM000
POLYURETHANE Y-218 see PKM250
POLYURETHANE Y-221 see PKM500
POLYURETHANE Y-222 see PKM750
POLYURETHANE Y-223 see PKN000
POLYURETHANE Y-224 see PKN250
POLYURETHANE Y-225 see PKN500
POLYURETHANE Y-226 see PKN750
POLYURETHANE Y-227 see PKO000
POLYURETHANE Y-238 see CDV625
POLYURETHANE Y-299 see PKO750
POLYURETHANE Y-290 see PKO500
POLYURETHANE Y-302 see PKP000
POLYURETHANE Y-304 see PKP250
POLYVIDONE see PKQ250
POLYVINOL see PKP750
POLYVINYL ACETATE (FCC) see AAX250
POLYVINYL ACETATE CHLORIDE see PKP500
POLYVINYL ALCOHOL 18/11 see PKP800
POLYVINYL ALCOHOL see PKP750
POLYVINYLBROMIDE see PKQ000
POLYVINYLBUTYRAL (CZECH) see PKI000
POLYVINYL BUTYRAL RESINS see PKI000
POLY(n-VINYLBUTYROLACTAM) see PKQ250
POLYVINYLCHLORID (GERMAN) see PKQ059
POLYVINYL CHLORIDE see PKQ059

POLYVINYLCHLORIDE ACETATE see PKP500
POLYVINYL CHLORIDE–POLYVINYL ACETATE see AAX175
POLY(VINYLPYRIDINE 1-OXIDE) see PKQ100
POLY(VINYLPYRIDINE N-OXIDE) see PKQ100
POLY(1-VINYL-2-PYRROLIDINONE) Hueper's polymer No. 1 see PKQ500
POLY(1-VINYL-2-PYRROLIDINONE) Hueper's polymer No. 2 see PKQ750
POLY(1-VINYL-2-PYRROLIDINONE) Hueper's polymer No. 3 see PKR000
POLY(1-VINYL-2-PYRROLIDINONE) Hueper's polymer No. 4 see PKR250
POLY(1-VINYL-2-PYRROLIDINONE) Hueper's polymer No. 5 see PKR500
POLY(1-VINYL-2-PYRROLIDINONE) Hueper's polymer No. 6 see PKR750
POLY(1-VINYL-2-PYRROLIDINONE) Hueper's polymer No. 7 see PKS000
POLY(1-VINYL-2-PYRROLIDINONE) HOMOPOLYMER see PKQ250
POLYVINYLPYRROLIDONE see PKQ250
POLYVINYL SULFATE POTASSIUM SALT see PKS250
POLYVIOL see PKP750
POLYVIOL M 13/140 see PKP750
POLYVIOL MO 5/140 see PKP750
POLYVIOL W 25/140 see PKP750
POLYVIOL W 40/140 see PKP750
POLYWAX 1000 see PJS750
POLY-ZOLE AZDN see ASL750
POMALUS ACID see MAN000
POMARSOL see TFS350
POMARSOL Z FORTE see BJK500
POMASOL see TFS350
POMEX see CBM750
POMME EPINEUSE (FRENCH) see SLV500
POMMIER (FRENCH) see AQP875
PONALAR see XQS000
PONALID see DWE800
PONALIDE see DWE800
PONCEAU BNA see FMU070
PONCEAU INSOLUBLE OLG see OHA000, XRA000
PONCEAU 3R see FAG018
PONCEAU 4R see FMU080
PONCEAU 4R ALUMINUM LAKE see FMU080
PONCEAU R (BIOLOGICAL STAIN) see FMU070
PONCEAU SX see FAG050
PONCEAU XYLIDINE (BIOLOGICAL STAIN) see FMU070
PONCYL see GKE000
PONDERAL see PDM250
PONDERAX see PDM250
PONDIMIN see PDM250
PONDINIL see PKS500
PONDOCIL see AOD000
PONDOCILLIN see AOD000
PONECIL see AIV500
PONOLITH FAST VIOLET 4RN see DFN450
PONOLITH ORANGE Y see CMS145
PONOXYLAN see UTU500
PONSOL BLUE BCS see DFN425
PONSOL BLUE BF see DFN425
PONSOL BLUE BFD see DFN425
PONSOL BLUE BFDP see DFN425
PONSOL BLUE BFN see DFN425
PONSOL BLUE BFND see DFN425
PONSOL BLUE BFP see DFN425
PONSOL BLUE GZ see IBV050
PONSOL BLUE RCL see IBV050
PONSOL BLUE RPC see IBV050
PONSOL BRILLIANT BLUE R see IBV050
PONSOL BROWN RBT see CMU770
PONSOL JADE GREEN SUPRA D see VGP100
PONSOL NAVY BLUE RA see VGP100
PONSOL NAVY BLUE RAD see VGP100
PONSOL OLIVE AR see DUP100
PONSOL OLIVE ARD see DUP100
PONSOL RED 2B see CMU825
PONSOL RED 2BD see CMU825
PONSOL RP see IBV050
PONSTAN see XQS000
PONSTEL see XQS000
PONSTIL see XQS000
PONSTYL see XQS000
PONTACHROME BLUE BLACK BB see CMP880
PONTACHROME YELLOW GS see SIT850
PONTACYL BRILLIANT BLUE see ADE500
PONTACYL BRILLIANT BLUE A see ERG100
PONTACYL BRILLIANT BLUE V see ADE500
PONTACYL CARMINE 2G see CMM300
PONTACYL FAST BLUE R see ADE750
PONTACYL GREEN BL see FAE950
PONTACYL RUBINE R see HJF500
PONTACYL SCARLET RR see FMU080
PONTACYL SKY BLUE 4BX see CMO500

PONTAL see XQS000
PONTALITE see PKB500
PONTAMINE BLACK E see AQP000
PONTAMINE BLACK EBN see AQP000
PONTAMINE BLUE BB see CMO000
PONTAMINE BLUE 3BX see CMO250
PONTAMINE BLUE RW see CMO600
PONTAMINE BOND BLUE B see CMO650
PONTAMINE BROWN BCW see CMO820
PONTAMINE BROWN BT see CMO820
PONTAMINE BROWN D 3GN see CMO810
PONTAMINE BROWN NCR see CMO810
PONTAMINE BROWN N3G see CMO825
PONTAMINE DEEP BLUE BH see CMN800
PONTAMINE DEVELOPER TN see TGL750
PONTAMINE DIAZO BLACK BHSW see CMN800
PONTAMINE FAST BLUE 7GLN see CMO650
PONTAMINE FAST RED 8BLX see CMO885
PONTAMINE FAST RED F see CMO870
PONTAMINE FAST RED FCB see CMO870
PONTAMINE GREEN BXN see CMO840
PONTAMINE GREEN GXN see CMO840
PONTAMINE GREEN S see CMO830
PONTAMINE PURE YELLOW see CMP050
PONTAMINE PURE YELLOW MN see CMP050
PONTAMINE SCARLET 3B see CMO875
PONTAMINE SKY BLUE 5BX see CMO500
PONTAMINE VIOLET N see CMP000
PONTOCAINE see BQA010
PONTOCAINE HYDROCHLORIDE see TBN000
POOA see PJY100
POOGIPHALAM, nut extract see BFW000
POOR MAN'S TREACLE see WBS850
POP see ELL500
POP-DOCK see FOM100
POPO-HAU (HAWAII) see HGP600
POPROLIN see PMP500
POPULAGE see MBU550
PORAMINE MALEATE see PJJ325
PORAPAK B see VQK595
PORCINE-TRYPSIN see PKS600
PORFIROMYCIN see MLY000
PORFIROMYCINE see MLY000
PORK TRYPSIN see PKS600
POROFOR 57 see ASL750
POROFOR BSH see BBS300
POROFOR-BSH-PULVER see BBS300
POROFOR CHKHC-18 see DVF400
POROFOR ChKhZ 9 see BBS300
POROLEN see PJS750
POROPHOR B see DVF400
PORPHYROMYCIN see MLY000
PORTAMYCIN see SLW475
PORTLAND CEMENT see PKS750
PORTLAND CEMENT SILICATE see PKS750
PORTLAND STONE see CAO000
PORTUGUESE MAN-OF-WAR TOXIN see PIA375
POSEDRAN see BEG000
POSEDRINE see BEG000
POSSUM WOOD see SAT875
POSTAFEN see HGC500
POSTERIOR PITUITARY EXTRACT see ORU500
POSTINOR see NNQ500, NNQ520
POST LOCUST see GJU475
PO-SYSTOX see DAP200
POTABLAN see CGL250
POTALIUM see PLG800
POTASAN see PKT000
POTASH see PLA000
POTASH CHLORATE (DOT) see PLA250
POTASORAL see PLG800
POTASSA see PLJ500
POTASSE CAUSTIQUE (FRENCH) see PLJ500
POTASSIO (CHLORATO di) (ITALIAN) see PLA250
POTASSIO (IDROSSIDO di) (ITALIAN) see PLJ500
POTASSIO (PERMANGANATO di) (ITALIAN) see PLP000
POTASSIUM, metal liquid alloy (DOT) see PKT500
POTASSIUM (liquid alloy) see PKT500
POTASSIUM see PKT250
POTASSIUM ACETATE see PKT750
POTASSIUM ACETYLENE-1,2-DIOXIDE see PKT775
POTASSIUM ACETYLIDE see PKU000
POTASSIUM ACID ARSENATE see ARD250
POTASSIUM ACID FLUORIDE see PKU250
POTASSIUM ACID FLUORIDE (solution) see PKU500

POTASSIUM ACID SULFATE see PKX750
POTASSIUM ACID TARTRATE see PKU600
POTASSIUM ALUM see AHF100
POTASSIUM ALUMINUM SULFATE see AHF100
POTASSIUM AMALGAM see PKU750
POTASSIUM AMIDE see PKV000
POTASSIUM AMYLXANTHATE see PKV100
POTASSIUM n-AMYLXANTHOGENATE see PKV100
POTASSIUM AMYLXANTHOGENATE see PKV100
POTASSIUM ANTIMONYL-d,l-TARTRATE see AQG750
POTASSIUM ANTIMONYL-d-TARTRATE see AQG250, AQG500
POTASSIUM ANTIMONYL-l-TARTRATE see AQH000
POTASSIUM ANTIMONYL-meso-TARTRATE see AQH250
POTASSIUM ANTIMONYL TARTRATE see AQG250
POTASSIUM ANTIMONY TARTRATE see AQG250
POTASSIUM ARSENATE see ARD250
POTASSIUM ARSENATE, MONOBASIC see ARD250
POTASSIUM ARSENITE see PKV500
POTASSIUM ARSENITE solution see FOM050
POTASSIUM l-ASPARTATE see PKV600
POTASSIUM ASPARTATE see PKV600
POTASSIUM AZAOROTATE see AFQ750
POTASSIUM AZIDE see PKW000
POTASSIUM AZIDODISULFATE see PKW250
POTASSIUM AZIDOSULFATE see PKW500
POTASSIUM BENZENEHEXOXIDE see PKW550
POTASSIUM BENZOATE see PKW760
POTASSIUM BENZYLPENICILLIN see BFD000
POTASSIUM BENZYLPENICILLINATE see BFD000
POTASSIUM BENZYLPENICILLIN G see BFD000
POTASSIUM BICARBONATE see PKX100
POTASSIUM BICHROMATE see PKX250
POTASSIUM BIFLUORIDE, solid or solution (DOT) see PKU250
POTASSIUM BIFLUORIDE, solution (DOT) see PKU500
POTASSIUM BIFLUORIDE see PKU250
POTASSIUM BIPHOSPHATE see PLQ405
POTASSIUM BIS(2-HYDROXYETHYL)DITHIOCARBAMATE see PKX500
POTASSIUM BIS(PROPYNYL)PALLADATE see PKX639
POTASSIUM BIS(PROPYNYL)PLATINATE see PKX700
POTASSIUM BISULFATE see PKX750
POTASSIUM BISULPHATE see PKX750
POTASSIUM BITARTRATE see PKU600
POTASSIUM BOROFLUORIDE see PKY000
POTASSIUM BOROHYDRATE see PKY250
POTASSIUM BOROHYDRIDE (DOT) see PKY250
POTASSIUM BROMATE see PKY300
POTASSIUM BROMIDE see PKY500
POTASSIUM-tert-BUTOXIDE see PKY750
POTASSIUM-o-BUTYL XANTHATE see PKY850
POTASSIUM BUTYLXANTHATE see PKY850
POTASSIUM BUTYLXANTHOGENATE see PKY850
POTASSIUM CANRENOATE see PKZ000
POTASSIUM CARBONATE (2:1) see PLA000
POTASSIUM CARBONYL (DOT) see CCB500
POTASSIUM CARBONYL see PKW550
POTASSIUM CHLORATE (DOT) see PLA250
POTASSIUM (CHLORATE de) (FRENCH) see PLA250
POTASSIUM CHLORATE see PLA250
POTASSIUM CHLORATE (UN 1485) (DOT) see PLA250
POTASSIUM CHLORATE, solution (UN 2427) (DOT) see PLA250
POTASSIUM CHLORIDE see PLA500
POTASSIUM CHLORIDE OXIDE see PLI250, PLK000
POTASSIUM CHLORITE see PLA525
POTASSIUM 7-CHLORO-2,3-DIHYDRO-2-OXO-5-PHENYL-1H-1,4-BENZODIAZE-
PINE-3-CARBOXYLATE KOH see CDQ250
POTASSIUM CHLOROPALLADATE see PLA750
POTASSIUM CHLOROPLATINATE see PLR000
POTASSIUM CHLOROPLATINITE see PJD250
POTASSIUM CHROMATE(VI) see PLB250
POTASSIUM CHROMIUM ALUM see CMG850
POTASSIUM CITRATE see PLB750
POTASSIUM CITRATE TRI(HYDROGEN PEROXIDATE) see PLB759
POTASSIUM CLAVULANATE see PLB775
POTASSIUM CLAVULANATE mixed with AMOXICILLIN (1:2) see ARS125
POTASSIUM COLUMBATE see PLL250
POTASSIUM COMPOUNDS see PLC100
POTASSIUM COPPER(I) CYANIDE see PLC175
POTASSIUM CUPROCYANIDE see PLC175
POTASSIUM CYANATE see PLC250
POTASSIUM CYANIDE, solution (DOT) see PLC500
POTASSIUM CYANIDE see PLC500
POTASSIUM CYANIDE-POTASSIUM NITRITE see PLC775
POTASSIUM CYANONICKELATE HYDRATE see TBW250
POTASSIUM CYCLOHEXANEHEXONE-1,3,5-TRIOXIMATE see PLC780
POTASSIUM CYCLOPENTADIENIDE see PLC800

POTASSIUM DICHLOROISOCYANURATE see PLD000
POTASSIUM DICHLORO-s-TRIAZINETRIONE see PLD000
POTASSIUM DICHROMATE(VI) see PKX250
POTASSIUM DICHROMATE, ZINC CHROMATE, and ZINC HYDROXIDE (1:3:1)
see ZFJ150
POTASSIUM DICYANOCUPRATE(1-) see PLC175
POTASSIUM DIETHYNYLPALLADATE(2⁻) see PLD100
POTASSIUM DIETHYNYLPLATINATE(2⁻) see PLD150
POTASSIUM DIFLUOROPHOSPHATE see PLD250
POTASSIUM DIHYDROGEN ARSENATE see ARD250
POTASSIUM DIHYDROGEN PHOSPHATE see PLQ405
POTASSIUM-2,5-DINITROCYCLOPENTANONIDE see PLD550
POTASSIUM DINITROMETHANIDE see PLD575
POTASSIUM DINITROOXALATOPLATINATE(2⁻) see PLD600
POTASSIUM aci-1,1-DINITROPROPANE see PLD700
POTASSIUM-1,1-DINITROPROPANIDE see PLD700
POTASSIUM DINITROSOSULFITE see PLD710
POTASSIUM-3,5-DINITRO-2(1-TETRAZENYL)PHENOLATE see PLD730
POTASSIUM DIOXIDE see PLE260
POTASSIUM DIPEROXY ORTHOVANADATE see PLE500
POTASSIUM ETHOXIDE see PLE575
POTASSIUM O-ETHYL DITHIOCARBONATE see PLF000
POTASSIUM ETHYLMERCURIC THIOGLYCOLLATE see PLE750
POTASSIUM ETHYLXANTHATE see PLF000
POTASSIUM ETHYL XANTHOGENATE see PLF000
POTASSIUM FERRICYANATE see PLF250
POTASSIUM FERRICYANIDE see PLF250
POTASSIUM FERROCYANATE see TEC500
POTASSIUM FERROCYANIDE see PLH100, TEC500
POTASSIUM FLUOBORATE see PKY000
POTASSIUM N-FLUOREN-2-YL ACETOHYDROXAMATE see HIS000
POTASSIUM FLUORIDE, solution (DOT) see PLF500
POTASSIUM FLUORIDE see PKU250, PLF500
POTASSIUM FLUORIDE, DIHYDRATE see PLF750
POTASSIUM FLUOROACETATE see PLG000
POTASSIUM FLUOROBORATE see PKY000
POTASSIUM FLUORURE (FRENCH) see PLF500
POTASSIUM FLUOSILICATE see PLH750
POTASSIUM FLUOTANTALATE see PLH000
POTASSIUM FLUOZIRCONATE see PLG500
POTASSIUM FORMATE see PLG750
POTASSIUM d-GLUCONATE see PLG800
POTASSIUM GLUCONATE see PLG800
POTASSIUM GLUTAMATE see MRK500
POTASSIUM GLUTAMINATE see MRK500
POTASSIUM GLYCEROPHOSPHATE see PLG810
POTASSIUM GRAPHITE see PLG825
POTASSIUM HEPTAFLUOROTANTALATE see PLH000
POTASSIUM HEXACHLOROPLATINATE(IV) see PLR000
POTASSIUM HEXACYANOFERRATE(III) see PLF250
POTASSIUM HEXACYANOFERRATE(II) see PLH100, TEC500
POTASSIUM HEXACYANOFERRATE see TEC500
POTASSIUM HEXAFLUOROARSENATE see PLH500
POTASSIUM HEXAFLUOROSILICATE see PLH750
POTASSIUM HEXAFLUOROTITANATE see PLI000
POTASSIUM HEXAFLUORSTANNATE see PLI250
POTASSIUM HEXAHYDRATE ALUMINATE see PLI500
POTASSIUM HEXANITROCOBALTATE(III) see PLI750
POTASSIUM HEXAOXYXENONATE(3⁻) XENON TRIOXIDE see PLJ000
POTASSIUM HYDRATE (DOT) see PLJ500
POTASSIUM HYDRIDE see PLJ250
POTASSIUM HYDROGEN ARSENATE see ARD250
POTASSIUM HYDROGEN DIFLUORIDE see PKU250
POTASSIUM HYDROGEN FLUORIDE, solution (DOT) see PKU500
POTASSIUM HYDROGEN FLUORIDE see PKU250
POTASSIUM HYDROGEN OXALATE see OLE000
POTASSIUM HYDROGEN SULFATE, solid (DOT) see PKX750
POTASSIUM HYDROGEN TARTRATE see PKU600
POTASSIUM HYDROXIDE, dry, solid, flake, bead, or granular (DOT) see
PLJ500
POTASSIUM HYDROXIDE, liquid or solution (DOT) see PLJ500
POTASSIUM HYDROXIDE see PLJ500
POTASSIUM-4-HYDROXYAMINE-5,7-DINITRO-4,5-DI-HYDROBENZO-FURAZA-
NIDE-3-OXIDE see PLJ775
POTASSIUM (HYDROXYDE de) (FRENCH) see PLJ500
POTASSIUM-4-HYDROXY-5,7-DINITRO-4,5-DIHYDROBENZOFURAZANIDE see
PLJ780
POTASSIUM HYDROXYFLUORONIOBATE see PLN500
POTASSIUM 3-(17-β-HYDROXY-3-OXOANDROSTA-4,6-DIEN-17-YL)PROPIONATE
see PKZ000
POTASSIUM 17-HYDROXY-3-OXO-17-α-PREGNA-4,6-DIENE-21-CARBOXYLATE
see PKZ000
POTASSIUM α-HYDROXYPROPIONATE see PLK650
POTASSIUM HYPERCHLORIDE see PLO500
POTASSIUM HYPOBORATE see PLJ790

POTASSIUM HYPOCHLORITE see PLK000
POTASSIUM HYPOPHOSPHITE see PLQ750
POTASSIUM INDOL-3-YL SULFATE see ICD000
POTASSIUM IODATE see PLK250
POTASSIUM IODIDE see PLK500
POTASSIUM IODOHYDRARGYRATE see NCP500
POTASSIUM ISOAMYL XANTHATE see PLK580
POTASSIUM ISOAMYL XANTHOGENATE see PLK580
POTASSIUM-o-ISOBUTYL XANTHATE see PLK600
POTASSIUM ISOBUTYL XANTHATE see PLK600
POTASSIUM ISOBUTYL XANTHOGENATE see PLK600
POTASSIUM ISOCYANATE see PLC250
POTASSIUM ISOPENTYL XANTHATE see PLK580
POTASSIUM ISOPROPYL XANTHANATE see IRG050
POTASSIUM ISOPROPYL XANTHATE see IRG050
POTASSIUM ISOPROPYL XANTHOGENATE see IRG050
POTASSIUM ISOTHIOCYANATE see PLV750
POTASSIUM KURROL'S SALT see PLR125
POTASSIUM LACTATE see PLK650
POTASSIUM MANGANOCYANIDE see TMX600
POTASSIUM MELARSONYL see PLK800
POTASSIUM MERCURIC IODIDE see NCP500
POTASSIUM METAARSENITE see PKV500
POTASSIUM METABISULFITE (DOT, FCC) see PLR250
POTASSIUM METAPHOSPHATE see PLR125
POTASSIUM METAVANADATE see PLK900
POTASSIUM METHANEDIZOATE see PLK825
POTASSIUM METHOXIDE see PLK850
POTASSIUM-4-METHOXY-1-aci-NITRO-3,5-DINITRO-2,5-CYCLOHEXADIENONE
   see PLK860
POTASSIUM METHYLAMIDE see PLL000
POTASSIUM METHYLDIAZENE OXIDE see PLK825
POTASSIUM-4-METHYLFURAZAN-5-CARBOXYLATE-2-OXIDE see PLL100
POTASSIUM METHYLPHENOXYMETHYLPENICILLIN see PDD350
POTASSIUM MONOCHLORIDE see PLA500
POTASSIUM MONOHYDROGEN DIFLUORIDE see PKU250
POTASSIUM MONOPERSULFATE see PLP750
POTASSIUM MONOSULFIDE see PLT250
POTASSIUM NEUTRAL OXALATE see PLN300
POTASSIUM NIOBATE see PLL250
POTASSIUM NITRATE see PLL500
POTASSIUM NITRATE mixed with CHARCOAL and SULFUR (15:3:2) see
   PLL750
POTASSIUM NITRATE mixed (fused) with SODIUM NITRITE (DOT) see
   PLM000
POTASSIUM NITRATE mixed with SODIUM NITRITE see PLM000
POTASSIUM NITRITE (1:1) see PLM500
POTASSIUM NITRITE (DOT) see PLM500
POTASSIUM-4-NITROBENZENEAZOSULFONATE see PLM550
POTASSIUM NITROCOBALTATE(III) see PLI750
POTASSIUM-6-aci-NITRO-2,4-DINITRO-2,4-CYCLOHEXADIENIMINIDE see
   PLM575
POTASSIUM 1-NITROETHANE-1-OXIMATE see MRK250
POTASSIUM-1-NITROETHOXIDE see PLM650
POTASSIUM-4-NITROPHENOXIDE see PLM700
POTASSIUM NITROSODISULFATE see PLM750
POTASSIUM N-NITROSOHYDROXYLAMINE-N-SULFONATE see PLD710
POTASSIUM NITROTRICHLOROPLATINATE see PLN050
POTASSIUM-O-O-BENZOYLMONOPEROXOSULFATE see PKW775
POTASSIUM OCTACYANODICOBALTATE see PLN100
POTASSIUM cis-9-OCTADECENOIC ACID see OHY000
POTASSIUM OCTATITANATE see PLW150
POTASSIUM OLEATE see OHY000
POTASSIUM OXALATE see PLN300
POTASSIUM 3-(3-OXO-17-β-HYDROXY-4,6-ANDROSTADIEN-17-α-
   YL)PROPANOATE see PKZ000
POTASSIUM OXONATE see AFQ750
POTASSIUM OXOTETRAFLUORONIOBATE(V) see PLN500
POTASSIUM OXYMURIATE see PLA250
POTASSIUM PALLADIUM CHLORIDE see PLN750
POTASSIUM PALLADOUS CHLORIDE see PLN750
POTASSIUM PALMITATE see PLO000
POTASSIUM PENICILLIN G see BFD000
POTASSIUM PENICILLIN V SALT see PDT750
POTASSIUM PENTACARBONYL VANADATE(3-) see PLO100
POTASSIUM PENTACYANODIPEROXOCHROMATE(5-) see PLO150
POTASSIUM PENTAPEROXODICHROMATE see PLO250
POTASSIUM PENTAPEROXYDICHROMATE see PLO250
POTASSIUM PENTYL THIARSAPHENYLMELAMINE see PLK800
POTASSIUM PENTYLXANTHATE see PKV100
POTASSIUM PENTYL XANTHOGENATE see PKV100
POTASSIUM PERCHLORATE, solid or solution (DOT) see PLO500
POTASSIUM PERCHLORATE see PLO500
POTASSIUM PERIODATE see PLO750
POTASSIUM (PERMANGANATE de) (FRENCH) see PLP000

POTASSIUM PERMANGANATE see PLP000
POTASSIUM PEROXIDE see PLP250
POTASSIUM PEROXOFERRATE(2-) see PLP500
POTASSIUM PEROXYDISULFATE see DWQ000
POTASSIUM PEROXYDISULPHATE see DWQ000
POTASSIUM PEROXYFERRATE see PLP500
POTASSIUM PEROXYSULFATE see PLP750
POTASSIUM PERRHENATE(1) see PLQ000
POTASSIUM PERRHENATE see PLQ000
POTASSIUM PERSULFATE (DOT) see DWQ000
POTASSIUM PHENETHICILLIN see PDD350
POTASSIUM (1-PHENOXYETHYL)PENICILLIN see PDD350
POTASSIUM-α-PHENOXYETHYL PENICILLIN see PDD350
POTASSIUM PHENOXYMETHYLPENICILLIN see PDT750
POTASSIUM-6-(α-PHENOXYPROPIONAMIDO)PENICILLANATE see PDD350
POTASSIUM PHENYL DINITROMETHANIDE see PLQ275
POTASSIUM PHOSPHATE, DIBASIC see PLQ400
POTASSIUM PHOSPHATE, MONOBASIC see PLQ405
POTASSIUM PHOSPHATE, TRIBASIC see PLQ410
POTASSIUM PHOSPHINATE see PLQ750
POTASSIUM PICRATE see PLQ775
POTASSIUM PLATINIC CHLORIDE see PLR000
POTASSIUM PLATINOCHLORIDE see PJD250
POTASSIUM POLYMETAPHOSPHATE see PLR125
POTASSIUM POLY(VINYL SULFATE) see PKS250
POTASSIUM-O-PROPIONOHYDROXAMATE see PLR175
POTASSIUM PYROPHOSPHATE see PLR200
POTASSIUM PYROSULFITE see PLR250
POTASSIUM RHENATE see PLR500
POTASSIUM RHENATE (KReO4) see PLQ000
POTASSIUM RHENIUM OXIDE (KReO4) see PLQ000
POTASSIUM RHODANATE see PLV750
POTASSIUM RHODANIDE see PLV750
POTASSIUM SALT of BENZYLPENICILLIN see BFD000
POTASSIUM SALT of SORREL see OLE000
POTASSIUM SELENATE see PLR750
POTASSIUM SELENITE (7CI) see SBO100
POTASSIUM SELENOCYANATE see PLS000
POTASSIUM SILICOFLUORIDE (DOT) see PLH750
POTASSIUM SILVER CYANIDE see PLS250
POTASSIUM SODIUM ALLOY see PLS500
POTASSIUM SORBATE see PLS750
POTASSIUM STANNATE TRIHYDRATE see PLS760
POTASSIUM SULFATE (2:1) see PLT000
POTASSIUM SULFIDE (2:1) see PLT250
POTASSIUM SULFIDE, anhydrous (UN 1382) (DOT) see PLT250
POTASSIUM SULFIDE with <30% water of crystallization (UN 1382) (DOT)
   see PLT250
POTASSIUM SULFIDE, hydrated with not <30% water of crystallization (UN
   1847) (DOT) see PLT250
POTASSIUM SULFOCYANATE see PLV750
POTASSIUM SULFURDIIMIDATE see PLT750
POTASSIUM SULFUR DIIMIDE see PLT750
POTASSIUM SUPEROXIDE (DOT) see PLE260
POTASSIUM TARTRATE see PKU600
POTASSIUM TELLURITE see PLU000
POTASSIUM TETRABROMOPLATINATE see PLU250
POTASSIUM TETRACHLOROPALLADATE see PLN750
POTASSIUM TETRACHLOROPLATINATE(II) see PJD250
POTASSIUM TETRACYANOMERCURATE(II) see PLU500
POTASSIUM TETRACYANONICKELATE(II) see NDI000
POTASSIUM TETRACYANONICKELATE see NDI000
POTASSIUM TETRACYANOTITANATE(IV) see PLU550
POTASSIUM TETRAETHYNYL NICKELATE(2-) see PLU575
POTASSIUM TETRAETHYNYL NICKELATE(4-) see PLU580
POTASSIUM TETRAIODOMERCURATE(II) see NCP500
POTASSIUM TETRAKISTHIOCYANATOPLATINATE see PLU590
POTASSIUM-1,1,2,2-TETRANITROETHANDIIDE see PLU600
POTASSIUM TETRAPEROXYCHROMATE see PLU750
POTASSIUM TETRAPEROXYMOLYBDATE see PLV000
POTASSIUM TETRAPEROXYTUNGSTATE see PLV250
POTASSIUM-1-TETRAZOLACETATE see PLV275
POTASSIUM THALLIUM(I)AMIDE AMMONIATE see PLV500
POTASSIUM THIOCYANATE see PLV750
POTASSIUM THIOCYANIDE see PLV750
POTASSIUM TITANIUM OXIDE see PLW150
POTASSIUM-s-TRIAZINE-2,4-DIONE-6-CARBOXYLATE see AFQ750
POTASSIUM TRICHLOROETHYLENEPLATINATE see PLW200
POTASSIUM TRICYANODIPEROXOCHROMATE (3-) see PLW275
POTASSIUM TRINITROMETHANIDE see PLW300
POTASSIUM-2,4,6-TRINITROPHENOXIDE see PLQ775
POTASSIUM TRIPHOSPHATE see PLW400
POTASSIUM TRIPOLYPHOSPHATE see PLW400
POTASSIUM TROCLOSENE see PLD000
POTASSIUM, metal alloys (UN 1420) (DOT) see PKT250

POTASSIUM VANADIUM TRIOXIDE see PLK900
POTASSIUM WARFARIN see WAT209
POTASSIUM XANTHATE see PLF000
POTASSIUM XANTHOGENATE see PLF000
POTASSIUM XANTHOGENATE BUTYL ETHER see PKY850
POTASSIUM ZINC CHROMATE see CMK400, PLW500
POTASSIUM ZINC CHROMATE HYDROXIDE see PLW500
POTASSURIL see PLG800
POTATO ALCOHOL see EFU000
POTATO BLOSSOMS, GLYCOALKALOID EXTRACT see PLW550
POTATO, GREEN PARTS see PLW750
POTAVESCENT see PLA500
POTCRATE see PLA250
POTENTIATED ACID GLUTARALDEHYDE see GFQ000
POTHOS see PLW800
POTIDE see PLK500
POTOMAC RED see CHP500
POUNCE see AHJ750
POVAL 117 see PKP750
POVAL 120 see PKP750
POVAL 203 see PKP750
POVAL 205 see PKP750
POVAL 217 see PKP750
POVAL 1700 see PKP750
POVAL C 17 see PKP750
POVAN see PQC500
POVANYL see PQC500
POVIDONE-IODINE see PKE250
POVIDONE (USP XIX) see PKQ250
POVIMAL ST see SEA500
POWDERED METALS see PLX000
POWDER GREEN see COF500
POWDER and ROOT see RNZ000
POYAMIN see VSZ000
POZZOLITH 400N see BLX000
PP 5 see PCG700
PP 009 see FDA885
PP-12 see HMN000
PP 062 see DOX600
PP148 see PAJ000
PP149 see BRI750
PP175 see ASD000
PP211 see DIN600
PP383 see RLF350
PP511 see DIN800
PP 557 see AHJ750
PP 563 see GJU600
PP 581 see TAC800
PP588 see BRJ000
PP 675 see BRD000
PP 745 see BJK750
PP781 see MLC250
PP 910 see PAJ250
PP-1466 see RSZ375
P 2070P see PJS750
PPA see NNM000
PPBP see BCP650
PPC 3 see MIA250
PPD see PEY500
PPE 2 see PJS750
PPE201 see PKO500
PP FACTOR see NCQ900
PP-FACTOR see NCR000
P.P. FACTOR-PELLAGRA PREVENTIVE FACTOR see NCQ900
P.P.G. 750 see PKI750
P.P.G. 1200 see PKJ250
PPG-14 BUTYL ETHER see BRP250
PPG-16 BUTYL ETHER see BRP250
PPG-33 BUTYL ETHER see BRP250
P. PHYSALIS TOXIN see PIA375
P. PORPHYRIACUS (AUSTRALIA) VENOM see ARV250
PPTC see PNI750
PPZEIDAN see DAD200
PQ see PCY300
Ph-QA 33 see INU200
PQD see DVR200
PR 703-78 see UTU500
PRACARBAMIN see UVA000
PRACARBAMINE see UVA000
PRACTALOL see ECX100
PRACTOLOL see ECX100
PRADUPEN see BDY669
PRAECIRHEUMIN see BRF500
PRAENITRON see TJL250
PRAENITRONA see TJL250

PRAEQUINE see RHZ000
PRAGMAZONE see CKJ000, THK880
PRAGMOLINE see CMF260
PRAIRIE CROCUS see PAM780
PRAIRIE HEN FLOWER see PAM780
PRAIRIE RATTLESNAKE VENOM see PLX100
PRAJMALINE see PNC875
PRAJMALINE BITARTRATE see DNB000
PRAJMALINE HYDROGEN TARTRATE see DNB000
PRAJMALIUM see PNC875
PRAKTOLOLU (POLISH) see ECX100
PRALIDOXIME CHLORIDE see FNZ000
PRALIDOXIME IODIDE see POS750
PRALIDOXIME MESYLATE see PLX250
PRALIDOXIME METHANESULFONATE see PLX250
PRALIDOXIME METHIODIDE see POS750
PRALIDOXIME METHOSULFATE see PLX250
PRALUMIN see TDA500
PRAMIL see SBN000
PRAMINDOLE see DPX200
PRAMITOL see MFL250
PRAMIVERINE HYDROCHLORIDE see SDZ000
PRAMIVERIN HYDROCHLORIDE see SDZ000
PRAMUSCIMOL see AKG250
PRANDIOL see PCP250
PRANONE see GEK500
PRANONE and DIETHYLSTILBESTROL see EEI050
PRANONE and STILBESTROL see EEI050
PRANOPROFEN see PLX400
PRANTAL see DAP800
PRANTAL METHYLSULFATE see DAP800
PRAPAGEN WK see QAT550
PRAPAGEN WKT see QAT550
PRAPAR see MKU750
PRAPARAT 5968 see HGP500
PRAPARAT 9295 see MKU750
PRASEODYMIUM see PLX500
PRASEODYMIUM CHLORIDE see PLX750
PRASEODYMIUM(III) NITRATE (1:3) see PLY250
PRASTERONE see AOO450
PRASTERONE SODIUM SULFATE see DAL040
PRASTERONE SODIUM SULFATE DIHYDRATE see DAL030
PRAXILENE see NAE100
PRAXIS see IDA400
PRAXITEN see CFZ000
PRAYER BEAD see AAD000
PRAYER BEADS see RMK250
PRAZEPAM see DAP700
PRAZEPINE see DLH600
PRAZIL see CKP250
PRAZINIL see CCK780
PRAZIQUANTEL see BGB400
PRAZOSIN see AJP000
PRAZOSIN HYDROCHLORIDE see FPP100
PRD EXPERIMENTAL NEMATOCIDE see DGL200
PREAN see MBW750
PRECEPTIN see PKF500
PRECIPITATED BARIUM SULPHATE see BAP000
PRECIPITATED CALCIUM PHOSPHATE see CAW120
PRECIPITATED CALCIUM SULFATE see CAX750
PRECIPITATED SILICA see SCL000
PRECIPITATED SULFUR see SOD500
PRECIPITE BLANC see MCW000
PRECISION CLEANING AGENT see TJE100
PRECOCENE 2 see AEX850
PRECOCENE II see AEX850
PRECORT see PLZ000
PRECORTANCYL see PMA000
PRECORTISYL see PMA000
PREDALON-S see SCA750
PREDENT see SHF500
PREDIACORTINE see SOV100
PREDICORT see SOV100
PREDNE-DOME see PMA000
PREDNELAN see PMA000
PREDNELAN-N see SOV100
PREDNESOL see PLY275
PREDNICEN-M see PLZ000
PREDNIDOREN see SOV100
PREDNILONGA see PLZ000
PREDNIS see PMA000
PREDNI-SEDIV see TEH500
PREDNISOLONE see PMA000
PREDNISOLONE 21-ACETATE see SOV100
PREDNISOLONE ACETATE see SOV100

PREDNISOLONE DISODIUM PHOSPHATE see PLY275
PREDNISOLONE 21-PHOSPHATE DISODIUM see PLY275
PREDNISOLONE SODIUM HEMISUCCINATE see PMA100
PREDNISOLONE SODIUM PHOSPHATE see PLY275
PREDNISOLONE 21-SODIUM SUCCINATE see PMA100
PREDNISOLONE SODIUM SUCCINATE see PMA100
PREDNISOLONE 21-SUCCINATE SODIUM see PMA100
PREDNISOLONE-17-VALERATE-21-ACETATE see AAF625
PREDNISOLONE VALERATE ACETATE see AAF625
PREDNISON see PLZ000
PREDNISONE see PLZ000
PREDNIZON see PLZ000
PREDONIN see PMA000
PREDONINE see PMA000
PREDONINE INJECTION see SOV100
PREDONINE SOLUBLE see PMA100
PREDSOL see PLY275
PREDSOLAN see PLY275
PREEGLONE see DWX800
PREFAR see DNO800
PREFAXIL see BBW750
PREFEMIN see CHJ750
PREFERID see BOM520
PREFLAN see BSN000
PREFMID see BSN000
PREFORAN see NIX000
PREFRIN see SPC500
PREGARD see CQG250
PREGICIL see AAF750
PREGLANDIN see CDB775
1,4-PREGNADIENE-17-α,21-DIOL-3,11,20-TRIONE see PLZ000
PREGNA-1,4-DIENE-3,20-DIONE, 21-(ACETYLOXY)-11,17-DIHYDROXY-, (11-β)-
   see SOV100
PREGNA-1,4-DIENE-3,20-DIONE, 7-CHLORO-11-HYDROXY-16-METHYL-17,21-
   BIS(1-OXOPROPOXY)-, (7-α-11-β,16-α)- see AFI980
PREGNA-4,6-DIENE-3,20-DIONE, 6-CHLORO-17-HYDROXY-1-α,2-α-METHYLENE-
   , ACETATE see CQJ500
PREGNA-1,4-DIENE-3,20-DIONE,9-FLUORO-11,16,17,21-TETRAHYDROXY-,(11-
   β,16-α) see AQX250
9-β,10-α-PREGNA-4,6-DIENE-3,20-DIONE and 17-α-HYDROXYPREGN-4-ENE-
   3,20-DIONE see PMA250
PREGNA-1,4-DIENE-3,20-DIONE, 11-β,17,21-TRIHYDROXY-, 21-(DIHYDROGEN
   PHOSPHATE), DISODIUM SALT see PLY275
PREGNA-1,4-DIENE-3,20-DIONE, 11-β,17,21-TRIHYDROXY-, 21-(HYDROGEN
   SUCCINATE), MONOSODIUM SALT see PMA100
1,4-PREGNADIENE-3,20-DIONE-11-β,17-α,21-TRIOL see PMA000
1,4-PREGNADIENE-11-β,17-α,21-TRIOL-3,20-DIONE see PMA000
17-α-2,4-PREGNADIEN-20-YNO(2,3-d)ISOXAZOL-17-OL see DAB830
17-α-PREGNA-2,4-DIEN-20-YNO(2,3-d)ISOXAZOL-17-OL see DAB830
PREGNANT MARE SERUM GONADOTROPIN see SCA750
PREGN-4-EN-17α,21-DIOL-3,11,20-TRIONE see CNS800
3,20-PREGNENE-4 see PMH500
17-α-PREGN-4-ENE-21-CARBOXYLIC ACID, 1-HYDROXY-7-α-MERCAPTO-3-
   OXO-α-LACTONE see AFJ500
4-PREGNENE-17α,21-DIOL-3,11,20-17,21-DIHYDROXY- see CNS800
Δ⁴-PREGNENE-17α,21-DIOL-3,11,20-TRIONE see CNS800
4-PREGNENE-17,α,21-DIOL-3,11,20-TRIONE 21-ACETATE see CNS825
4-PREGNENE-3,20-DIONE see PMH500
PREGN-4-ENE-3,20-DIONE see PMH500
PREGNENE-3,20-DIONE see PMH500
Δ⁴-PREGNENE-3,20-DIONE see PMH500
PREGNENEDIONE see PMH500
4-PREGNENE-3,20-DIONE-21-OL ACETATE see DAQ800
4-PREGNENE-11-β,17-α,21-TRIOL 3,20-DIONE see CNS750
PREGNENINOLONE see GEK500
4-PREGNEN-21-OL-3,20-DIONE see DAQ600
17-α-PREGN-4-EN-20-YNO(2,3-d)ISOXAZOL-17-OL see DAB830
17-α-PREGN-4-EN-20-YN-3-ONE, 17-HYDROXY-, and trans-α-α'-DIETHYL-4,4'-
   STILBENEDIOL see EEI050
PREGNIN see GEK500
PRELIS see SBV500
PRELUDIN see PMA750
PRELUDIN HYDROCHLORIDE see MNV750
PREMALIN see CKD500, DGD600
PREMALOX see CBM700
PREMARIN see ECU750, PMB000
PREMAZINE see BJP000
PREMERGE 3 see BRE500
PREMERGE see BRE500
PREMERGE PLUS see BQI000
PREMGARD see BEP500
PREMOBOLAN see PMC700
PREMODRIN see DBA800
PRENAZONE see PEW000
PRENDEROL see PMB250

PRENDIOL see PMB250
PRENEMA see SOV100
PRENIMON see TND000
PRENITRON see TJL250
PRENOL see MHU110
PRENORMINE see TAL475
PRENOXDIAZINE HYDROCHLORIDE see LFJ000
PRENT see AAE100
PRENTOX see PIX250, RNZ000
PRENYL ACETATE see DOQ350
PRENYL ALCOHOL see MHU110
PRENYLAMINE see PEV750
PRENYLAMINE LACTATE see DWL200
PRENYL BENZOATE see MHU150
α-PRENYL-α-(2-DIMETHYLAMINOETHYL)-1-NAPHTHYLACETAMIDE see
   PMB500
4-PRENYL-1,2-DIPHENYL-3,5-PYRAZOLIDINEDIONE see PEW000
PRENYL SALICYLATE see PMB600
PREP see SFV250
PREPALIN see VSK600
PRE-PAR see RLK700
PREPARATION 125 see DFT800
PREPARATION 84 see TCQ260
PREPARATION 48-80 see PMB800
PREPARATION AF see HEI500
PREPARATION C (the Russian Drug) see HEF200
PREPARATION HE 166 see DUS500
PREPARATION S see HEF200
PREPARED CHALK see CAT775
PREP-DEFOLIANT see SFV250
PREQUINE see RHZ000
PREROIDE see MJE760
PRESAL R60 see MCB050
PRESAMINE see DLH630
PRE-SAN see DNO800
PRE-SATE see ARW750
PRESATE HYDROCHLORIDE see ARW750
PRESCRIN see FAQ950
PRESDATE see HMM500
PRESERVAL M see HJL500
PRESERVAL P see HNU500
PRESERV-O-SOTE see CMY825
PRESFERSUL see FBO000
PRESINOL see DNA800, MJE780
PRESOLISIN see DNA800, MJE780
PRESOMEN see ECU750
PRESOXIN see ORU500
PRESPERSION, 75 UREA see USS000
PRESSAL see MCB050
PRESSALOLO see HMM500
PRESSIMEDIN see RDK000
PRESSITAN see FMS875
PRESSOMIN HYDROCHLORIDE see MDW000
PRESSONEX see HNB875, HNC000
PRESSONEX BITARTRATE see HNC000
PRESSOROL see HNC000
PRESTOCICLINA see MCH525
PRESUREN see VJZ000
PRETILACHLOR see PMB850
PRETILACHLORE see PMB850
PRETONINE see HOA575
PREVANGOR see PBC250
PREVENCILINA P see SLJ050
PREVENOL 56 see CKC000
PREVENOL see CKC000
PREVENTAL see MJM500
PREVENTOL 1 see SKK500
PREVENTOL 56 see CKC000
PREVENTOL see CKC000, MJM500
PREVENTOL CMK see CFE250
PREVENTOL GD see MJM500
PREVENTOL GDC see MJM500
PREVENTOL I see TIV750
PREVENTOL O EXTRA see BGJ250
PREVENTOL-ON see BGJ750
PREVENTOL ON & ON EXTRA see BGJ750
PREVEX see PNI250
PREVICUR see EIH500
PREVICUR-N see PNI250
PREVOCEL #12 see PMC100
PREWEED see CKC000
PREZA see HBT500
PREZERVIT see DSB200
PR-F 36 Cl see CID825
PRIADEL see LGZ000

PRIAMIDE see DAB875
PRIARIE SMOKE see PAM780
PRIATIN see SCA750
PRIAZIMIDE see DAB875
PRIDE see FMQ200
PRIDE of CHINA see CDM325
PRIDE of INDIA see CDM325
PRIDINOL see PMC250
PRIDINOL HYDROCHLORIDE see PMC250
PRIDINOL MESILATE see PMC275
PRIGLONE see PAI990
PRILEPSIN see DBB200
PRILOCAINE HYDROCHLORIDE see CMS250
PRILTOX see PAX250
PRIM see PMD550
PRIMACAINE see AJA500
PRIMACAINE HYDROCHLORIDE see AJA500
PRIMACHIN (GERMAN) see AKR250
PRIMACIONE see DBB200
PRIMACLONE see DBB200
PRIMACOL see NAK500
PRIMACONE see DBB200
PRIMAGRAM see MQQ450
PRIMAL see AHI875
PRIMALAN see MCJ400
PRIMAL ASE 60 see ADW200
PRIMAMYCIN see HAH800
PRIMANTRON see SCA750
PRIMAQUINE see PMC300
PRIMAQUINE DIPHOSPHATE see PMC310
PRIMAQUINE PHOSPHATE see PMC310
PRIMARY AMYL ACETATE see AOD725
PRIMARY AMYL ALCOHOL see AOE000
PRIMARY DECYL ALCOHOL see DAI600
PRIMARY ISOBUTYL IODIDE see IIV509
PRIMARY OCTYL ALCOHOL see OEI000
PRIMARY SODIUM PHOSPHATE see SJH100
PRIMATENE MIST see VGP000
PRIMATOL 25E see MFL250
PRIMATOL see ARQ725, EGD000, MFL250
PRIMATOL-M80 see BQB000
PRIMATOL P see PMN850
PRIMATOL Q see BKL250
PRIMATOL S see BJP000
PRIMAZE see ARQ725
PRIMAZIN see SNJ000
PRIMBACTAM see ARX875
PRIMENE JM-T see PMC400
PRIMEXTRA see MQQ450
PRIMICID see DIN600
PRIMIDOLOL HYDROCHLORIDE see PMC600
PRIMIDON see DBB200
PRIMIDONE see DBB200
PRIMIN see DSK200
PRIMINE see PMI750
PRIMOBOLAN see PMC700
PRIMOBOLONE see PMC700
PRIMOCARCIN see PMC750
PRIMOCORT see DAQ800
PRIMOCORTAN see DAQ800
PRIMODOS see EEH520
PRIMOFOL see EDO000
PRIMOGYN B see EDP000
PRIMOGYN BOLEOSUM see EDP000
PRIMOGYN I see EDP000
PRIMOL 335 see MQV750
PRIMOLUT C see GEK500
PRIMOLUT DEPOT see HNT500
PRIMONABOL see PMC700
PRIMOSTAT see GEK510
PRIMOTEC see DIN600
PRIMOTEST see TBF500
PRIMOVLAR see NNL500
PRIMPERAN see AJH000
PRIMROSE YELLOW see ZFJ100
PRIMUN see FDD100
PRINADOL HYDROBROMIDE see PMD325
PRINALGIN see AGN000
PRINCILLIN see AOD125
PRINCIPEN see AIV500
PRINCIPEN/N see SEQ000
PRINICID see DIN600
PRINODOLOL see VSA000
PRINTEL'S see SMQ500
PRINTEX 60 see CBT750

PRINTEX see CBT750
PRIODAX see PDM750
PRIODERM see MAK700
PRIOSET TD756 see MCB050
PRIOSPEN see PDD350
PRISCOL see BBJ750, BBW750
PRISCOLINE see BBW750
PRISCOLINE HYDROCHLORIDE see BBJ750
PRISILIDENE HYDROCHLORIDE see NEB000
PRISILIDINE see NEA500
PRISMANE see PMD350
PRIST see EJH500
PRISTACIN see CCX000
PRISTANE see PMD500
PRISTIMERIN see PMD525
PRISTINAMYCIN see VRF000
PRISTINAMYCIN IA see VRA700
PRIMAKTON see DBB200
PRIVAPROL see CKL325
PRIVENAL see ERE000
PRIVET see PMD550
PRIVINE see NAH500, NAH550
PRIVINE HYDROCHLORIDE see NCW000
PRIVINE NITRATE see NAH550
PRIZOLE HYDROCHLORIDE see NCW000
PRL-3191 see EHY000
PRM-TC see PPY250
PRN see DTO600
PRO-ACTIDIL see TMX775
PROADIFEN HYDROCHLORIDE see PBM500
PROASMA HYDROCHLORIDE see OJY000
PROAZAIMINE see DQA400
PROAZAMINE see DQA400
PROBAN see CQL250
PROBAN 420B see MCB050
PRO-BAN M see TEH500
PRO-BANTHINE see HKR500
PROBARBITAL see ELX000
PROBARBITONE see ELX000
PROBE see BGD250
PROBECID see DWW000
PROBEDRYL see BBV500
PROBEN see DWW000
PROBENAZOLE see PMD800
PROBENECID ACID see DWW000
PROBENECID SODIUM SALT see DWW200
PROBENEMID see DWW000
PROBESE-P see PMA750
PROBESE-P HYDROCHLORIDE see MNV750
PROBILIN see EOA500
PROBON see PMD825
PROBONAL see PMD825
PROCAINAMIDE see AJN500
PROCAINAMIDE HYDROCHLORIDE see PME000
PROCAINAMIDE SULFATE see AJN750
PROCAINE see AIL750
PROCAINE AMIDE see AJN500
PROCAINE AMIDE HYDROCHLORIDE see PME000
PROCAINE AMIDE SULFATE see AJN750
PROCAINE, BASE see AIL750
PROCAINE BENZYLPENICILLINATE see PAQ200
PROCAINE FLUOBORATE see TCH000
PROCAINE HYDROCHLORIDE see AIT250
PROCAINE PENICILLIN G see PAQ200
PROCALM see BCA000
PROCALMIDOL see MQU750
PROCAMIDE see AJN500
PROCAMIDE HYDROCHLORIDE see PME000
PROCAN-SR HYDROCHLORIDE see PME000
PROCAPAN HYDROCHLORIDE see PME000
PROCARBAZIN (GERMAN) see PME500
PROCARBAZINE see PME250
PROCARBAZINE HYDROCHLORIDE see PME500
PROCARDIA see AEC750
PROCARDIN see POB500
PROCARDINE see DJS200
PROCARDYL HYDROCHLORIDE see PME000
PROCASIL see PNX000
PROCATEROL HYDROCHLORIDE see PME600
PROCENE UF 1.5 see PJS750
PROCHLOROPERAZINE see PMF500
PROCHLOROPROAZINE HYDROGEN MALEATE see PMF250
PROCHLORPEMAZINE see PMF500
PROCHLORPERAZINE see PMF500
PROCHLORPERAZINE BIMALEATE see PMF250

PROCHLORPERAZINE DIMALEATE see PMF250
PROCHLORPERAZINE EDISYLATE see PME700
PROCHLORPERAZINE ETHANE DISULFONATE see PME700
PROCHLORPERAZINE HYDROGEN MALEATE see PMF250
PROCHLORPERAZINE MALEATE see PMF250
PROCHLORPERIZINE MALEATE see PMF250
PROCHLORPROMAZINE see PMF500
PROCHOLON see DAL000
PROCIDLIDINA see CPQ250
PROCILAN see POB500
PROCINOLOL HYDROCHLORIDE see PMF525
(±)-PROCINOLOL HYDROCHLORIDE see PMF535
PROCION BLACK H-N see CMS227
PROCION BLUE MX-R see CMS228
PROCION BRILLIANT BLUE MR see CMS228
PROCION BRILLIANT BLUE MX-R see CMS228
PROCION BRILLIANT BLUE RS see CMS228
PROCION BRILLIANT RED 5BS see PMF540
PROCION BRILLIANT RED M 5B see PMF540
PROCION BRILLIANT RED MX 5B see PMF540
PROCION RED MX 5B see PMF540
PROCION YELLOW MX 4R see RCZ000
PROCIT see DQA400
PROCOL OA-20 see PJW500
PRO-COR see HNY500
PROCORMAN see DJS200
PROCTIN see SEH000
PROCTODON see DVV500
PROCURAN see DAF800
PROCUTENE see TIL500
PROCYAZINE see PMF600
PROCYCLIDINE see CPQ250
PROCYKLIDIN see CPQ250
PROCYMIDONE see PMF750
PROCYTOX see CQC500, CQC650
PRODALUMNOL see SEY500
PRODALUMNOL DOUBLE see SEY500
PRODAN see DXE000
PRODARAM see BJK500
PRODECTINE see PPH050
PRODENIALURE B see TBX300
PRODHYBASE ETHYL see EJM500
PRODHYPHORE B see PJY100
PRO-DIABAN see PMG000
PRODIAMINE see DWS200
PRODICTAZIN see DIR000
PRODIERAZINE see DIR000
PRODILIDINE see DTO200
α-PRODINE see NEA500
α-PRODINE HYDROCHLORIDE see NEB000
PRODIXAMON see HKR500
PRODLURE see FAS100
PRO-DORM see QAK000
PRODOX 131 see IQX100
PRODOX 133 see IQZ000
PRODOX 146 see DEG000
PRODOX 156 see DCI000
PRODOX 340 see BST000
PRODOXAN see GEK500
PRODOXOL see OOG000
PRODOX 146A-85X see DEG000
PRODROMINE see TCY750
PRODROXAN see GEK500
PRODUCER GAS see PMG750
PRODUCT 308 see HCP000
PRODUCT 5022 see MPE250
PRODUCT 5190 see AHL500
PRODUCT DDN see LBU200
PRODUCT HDN see CDF450
PRODUCT No. 161 see SIB600
PRO-DUOSTERONE see NNL500
PRODUXAN see GEK500
PROEPTATRIENE (ITALIAN) see PMH600
PROFAM see CBM000
PROFAMINA see BBK000
PROFARMIL see TEH500
PROFAX see PMP500
PROFAX A 60-008 see PJS750
PROFECUNDIN see VSZ450
PROFEMIN see CHJ750
PROFENAMINA (ITALIAN) see DIR000
PROFENAMINUM see DIR000
PROFENID see BDU500
PROFENOFOS see BNA750
PROFENONE see ELL500

PROFERRIN see IHG000
PROFETAMINE see AOB500
PROFETAMINE PHOSPHATE see AOB500
PROFIROMYCIN see MLY000
PROFLAVIN see DBN600
PROFLAVINE see DBN600, PMH100
PROFLAVINE HEMISULPHATE see PMH100
PROFLAVINE HYDROCHLORIDE see PMH250
PROFLAVINE MONOHYDROCHLORIDE see PMH250
PROFLAVINE MONOHYDROCHLORIDE HEMIHYDRATE see DBN200
PROFLURALIN see CQG250
PROFOLIOL see DBN600, EDO000
PROFORMIPHEN see DBN600
PROFUME A see CKN500
PROFUME (OBS.) see MHR200
PROFUNDOL see AFY500, DBN600
PROFURA see DBN600
PROGALLIN A see EKM100
PROGALLIN LA see DXX200
PROGALLIN P see PNM750
PROGARMED see DBN600
PRO-GASTRON see HKR500
PROGEKAN see PMH500
PRO-GEN see DBN600
PRO-GEN SODIUM see ARA500
PROGESIC see DBN600, FAP100
PROGESTAB see GEK500
PROGESTEROL see PMH500
β-PROGESTERONE see PMH500
PROGESTERONE see PMH500
PROGESTERONE CAPROATE see HNT500
PROGESTERONE mixed with ESTRADIOL BENZOATE (14:1 moles) see EDQ000
PROGESTERONE mixed with ESTRA-1,3,5(10)-TRIENE-3,17-β-DIOL-3-BENZO-ATE (14:1 moles) see EDQ000
PROGESTERONE RETARD PHARLON see HNT500
PROGESTERONUM see PMH500
PROGESTIN see PMH500
PROGESTIN P see GEK500
PROGESTOLETS see GEK500
PROGESTONE see PMH500
PROGESTORAL see GEK500
PRO-GIBB see GEM000
PROGLICEM see DCQ700
PROGLUMETACINA (SPANISH) see POF550
PROGLUMETACIN MALEATE see POF550
PROGLUMIDE see BGC625
PROGLUMIDE SODIUM see PMH575
PROGUANIL see CKB250
PROGUANIL HYDROCHLORIDE see CKB500
PROGYNON see EDO000, EDS100
PROGYNON B see EDP000
PROGYNON BENZOATE see EDP000
PROGYNON-DEPOT see EDS100
PROGYNON-DH see EDO000
PROGYNON-DP see EDR000
PROGYNOVA see EDS100
PROHEPTADIENE see EAH500
PROHEPTADIEN MONOHYDROCHLORIDE see EAI000
PROHEPTATRIENE see PMH600
PROHEPTATRIENE HYDROCHLORIDE see DPX800
PROHEPTATRIEN MONOHYDROCHLORIDE see DPX800
PROKARBOL see DUS700
PROKAYVIT see MMD500
PROLAN see BIN500
PROLAN (CSC) see BIN500
PROLATE see PHX250
PROLAX see GGS000
PROLIDON see PMH500
PROLINOMETHYLTETRACYCLINE see PMI000
PROLINTANE HYDROCHLORIDE see PNS000
PROLIXAN see AQN750, ASA000
PROLIXIN see FMP000
PROLIXINE see TJW500
PROLIXIN ENANTHATE see PMI250
PROLONGAL see IGS000
PROLONGINE see DWW000
PROLOPRIM see TKZ000
PROLOXIN see GGS000
PROLUTOL see GEK500
PROLUTON C see GEK500
PROLUTON DEPOT see HNT500
PROMACID see CKP500
PROMACTIL see CKP250
PROMAMIDE see DTT600

PROMANIDE see AOO800
PROMANTINE see PMI750
PROMAPAR see CKP500
PROMAQUID see FMU039
PROMAR see DVV600
PROMARIT see ECU750
PROMASSOL see AHI875
PROMAXON P60 see CAW850
PROMAZIL see CKP250
PROMAZINAMIDE see DQA400
PROMAZINE see DQA600
PROMAZINE HYDROCHLORIDE see PMI500
PROMECARB see CQI500
PROMEDOL see IMT000
PROMERAN see CHX250
PROMETASIN see DQA400
PROMETAZIN see DQA400
PROMETHAZINE N-(2'-DIMETHYLAMINO-2'-METHYLETHYL)PHENOTHIAZINE
    HYDROCHLORIDE see PMI750
PROMETHAZINE HYDROCHLORIDE see PMI750
PROMETHIAZIN (GERMAN) see PMI750
PROMETHIAZINE see DQA400
PROMETHIN see FMS875
PROMETHIUM see PMJ000
PROMETIN see FMS875
PROMETON see MFL250
PROMETONE see MFL250
PROMETREX see BKL250
PROMETRIN see BKL250
PROMETRYN see BKL250
PROMETRYNE (USDA) see BKL250
PROMEZATHINE see DQA400
PROMIBEN see DLH600, DLH630
PROMIDE (parasympatholytic) see BGC625
PROMIDE HYDROCHLORIDE see PME000
PROMIDIONE see GIA000
PROMIN see AOO800
PROMINAL see ENB500
PROMIN SODIUM see AOO800
PROMIT see DBD700
PROMOTESTON see TBF500
PROMOTIL see PNS000
PROMOTIN see AOO800
PROMPT INSULIN ZINC SUSPENSION see LEK000
PROMPTONAL see EOK000
PROMUL 5080 see HKJ000
PROMURIT see DEQ000
PROMURITE see DEQ000
PROMYR see IQN000
PRONAMIDE see DTT600
PRONASE see PMJ100
PRONDOL see DPX200
PRONE see GEK500
PRONESTYL see AJN500
PRONESTYL HYDROCHLORIDE see PME000
PRONETALOL see INS000
PRONETHALOL see INS000, INT000
PRONETHALOL HYDROCHLORIDE see INT000
PRONEURIN see DXO300
PRONTALBIN see SNM500
PRONTAMID see SNQ000
PRONTODIN see DUO400
PRONTOSIL I see SNM500
PROPACHLOR see CHS500
PROPACHLORE see CHS500
PROPACIL see PNX000
PROPADERM see AFJ625
PROPADRINE see NNM000
PROPADRINE HYDROCHLORIDE see NNN000, PMJ500
PROPAFENONE HYDROCHLORIDE see PMJ525
PROPAFENON HYDROCHLORID see PMJ525
PROPAL see CIR500, IQW000
PROPALDEHYDE see PMT750
PROPALDON see QCS000
PROPALGYL see DPE200
PROPALLYLONAL see QCS000
PRO-PAM see PMJ550
PROPAMIDINE DIHYDROCHLORIDE see DBM400
PROPAMINE D see TDQ750
PROPAMOCARB HYDROCHLORIDE see PNI250
PROPANAL see PMT750
PROPANAL, 2-METHYL-2-(METHYLSULFINYL)-, O-((METHYLAMI-
    NO)CARBONYL)OXIME see MLW750
PROPANAL, 2-METHYL-2-(METHYLTHIO)-, OXIME see MLX800
PROPANAL, 3-(METHYLTHIO)-(9CI) see MPV400, TET900

PROPANALOL see ICB000
PROPANAMIDE, N,N-DIETHYL-(9CI) see DJX250
PROPANAMIDE, N-METHYL-(9CI) see MOS900
2-PROPANAMINE see INK000
PROPANAMINE see PND250
1-PROPANAMINE, 3-DIBENZ(b,e)OXEPIN-11(6H)-YLIDENE-N,N-DIMETHYL-, HY-
    DROCHLORIDE see AEG750
1-PROPANAMINE, 3-(10,11-DIHYDRO-5H-DIBENZO(A,D)CYCLOHEPTEN-5-YLI-
    DENE)-N,N-DI-METHYL-N-OXIDE see AMY000
2-PROPANAMINE, HYDROCHLORIDE (9CI) see INL000
1-PROPANAMINE, 2-METHYL- see IIM000
2-PROPANAMINE, N-(1-METHYLETHYL)- see DNM200
2-PROPANAMINE, 2-METHYL-, HYDROCHLORIDE see MGJ800
2-PROPANAMINE, N,N'-METHANETETRAYLBIS-(9CI) see DNO400
1-PROPANAMINIUM, 3-CARBOXY-2-HYDROXY-N,N,N-TRIMETHYL-, CHLORIDE,
    (R)- (9CI) see CCK660
1-PROPANAMINIUM, 3-CARBOXY-2-HYDROXY-N,N,N-TRIMETHYL-, CHLORIDE,
    (±)-(9CI) see CCK655
1-PROPANAMINIUM, 2-HYDROXY-N,N,N-TRIMETHYL-, CHLORIDE (9CI) see
    MIM300
1,3-PROPAN-BIS-(4-HYDROXYIMINOMETHYL-PYRIDINIUM-(1))-DIBROMIDS (GER-
    MAN) see TLQ500
PROPANE see PMJ750
PROPANE, 1,3-BIS(DIAZO)- see DCQ525
1-PROPANECARBOXYLIC ACID see BSW000
PROPANE, 3-CHLORO-1,1,1-TRIFLUORO- see TJY200
1,3-PROPANEDIAL see PMK000
PROPANEDIAL see PMK000
1,3-PROPANEDIALDEHYDE see PMK000
PROPANEDIAL, ION(1-), SODIUM (9CI) see MAN700
1,2-PROPANEDIAMINE see PMK250
1,3-PROPANEDIAMINE see PMK500
1,3-PROPANEDIAMINE, N-(3-AMINOPROPYL)- see AIX250
1,3-PROPANEDIAMINE, N'-BENZ(c)ACRIDIN-7-YL-N-(2-CHLOROETHYL)-N-ETH-
    YL-, DIHYDROCHLORIDE see EHJ000
PROPANE, 1,2-DIBROMO-1,1,2,3,3,3-HEXAFLUORO- see DDO450
1,3-PROPANEDICARBOXYLIC ACID see GFS000
PROPANE DIETHYL SULFONE see ABD500
PROPANE, 1,3-DIIODO- see TLR050
1,3-PROPANEDIMERCAPTAN see PML350
PROPANE-1,3-DIMETHANESULFONATE see TLR250
PROPANEDINITRILE see MAO250
PROPANEDINITRILE, (3-CHLOROPHENYL)HYDRAZONO)-(9CI) see CKA550
PROPANEDINITRILE((2-CHLOROPHENYL)METHYLENE) see CEQ600
PROPANEDINITRILE, 2,2'-(2,5-CYCLOHEXADIENE-1,4-DIYLIDENE)BIS-(9CI) see
    TBW750
PROPANEDINITRILE, (1-ETHOXYETHYLIDENE)- see EET100
PROPANEDIOIC ACID see CCC750
PROPANEDIOIC ACID, ((3,5-BIS(1,1-DIMETHYLETHYL)-4-HYDROXYPHEN-
    YL)METHYL)BUTYL-, BIS(1,2,2,6,6- PENTAMETHYL-4 see PMK800
PROPANEDIOIC ACID, DIETHYL ESTER see EMA500
PROPANEDIOIC ACID DIMETHYL ESTER (9CI) see DSM200
PROPANEDIOIC ACID, DITHALLIUM SALT see TEM399
PROPANE-1,2-DIOL see PML000
1,2-PROPANEDIOL see PML000
1,3-PROPANEDIOL see PML250
PROPANE-1,3-DIOL see PML250
1,2-PROPANEDIOL-1-ACRYLATE see HNT600
1,2-PROPANEDIOL, ALLYL ETHER see PNK000
1,3-PROPANEDIOL, 2,2-BIS((ACETYLOXY)METHYL)-, DIACETATE (9CI) see
    NNR400
1,3-PROPANEDIOL BIS(α-(p-CHLOROPHENOXY)ISOBUTYRATE) see SDY500
1,3-PROPANEDIOL BIS(2-(4-CHLOROPHENOXY)-2-METHYLPROPIONATE) see
    SDY500
1,3-PROPANEDIOL, 2,2-BIS((NITROOXY)METHYL)-, DINITRATE (ester), mixt.
    with 2-METHYL-1,3,5-TRINITROBENZENE see PBT050
1,2-PROPANEDIOL CARBONATE see CBW500
1,2-PROPANEDIOL, 3-CHLORO-, 1-BENZOATE see CKQ500
1,2-PROPANEDIOL, 3-CHLORO-, DIACETATE see CIL900
1,2-PROPANEDIOL CYCLIC CARBONATE see CBW500
1,1-PROPANEDIOL, 2,3-DICHLORO-, DIACETATE see DBF875
1,3-PROPANEDIOL, 2-ETHYL-2-(HYDROXYMETHYL)- see HDF300
1,3-PROPANEDIOL, 2-ETHYL-2-(HYDROXYMETHYL)-, POLYMER with 1,3-BIS(1-
    ISOCYANATO-1-METHYLETHYL) BENZENE see TDX320
1,3-PROPANEDIOL FORMAL see DVP600
1,3-PROPANEDIOL, 2-(HYDROXYMETHYL)-2-NITRO-, TRINITRATE (ester) see
    NHK650
1,3-PROPANEDIOL, 2-(HYDROXYMETHYL)-2-PROPYL-, CYCLIC ESTER with
    ANTIMONIC ACID see PNX500
1,2-PROPANEDIOL, MONOSTEARATE see SLL000
1,2-PROPANEDIOL, 3-(NITROSO-2-PROPENYLAMINO)- see NJY500
1,2-PROPANEDIOL, 3-(OCTADECYLOXY)- see GGA915
1,2-PROPANEDIOL, 3,3'-OXYDI-, TETRANITRATE see TDY100
1,3-PROPANEDIONE see PMK000
PROPANEDIONE see PQC000

1,3-PROPANEDIONE, 2-BROMO-1,3-DIPHENYL- see BNH100
1,2-PROPANEDITHIOL see PML300
1,3-PROPANEDITHIOL see PML350
1,1'-(1,3-PROPANEDIYL)BIS(4-(HYDROXYIMINO)METHYLPYRIDINIUM) DIBROMIDE see TLQ500
1,1'-(1,3-PROPANEDIYL)BIS((4-(HYDROXYIMINO)METHYL)-PYRIDINIUM DICHLORIDE see TLQ750
4,4'-(1,3-PROPANEDIYLBIS(OXY))BIS-BENZENECARBOXIMIDAMIDE,DIHYDROCHLORIDE see DBM400
1,2-PROPANEDIYL CARBONATE see CBW500
PROPANE, 1,2-EPOXY-3-ETHOXY- see EKM200
PROPANE, 1,2-EPOXY-3-((2-ETHYLHEXYL)OXY)- see GGY100
PROPANE, 1,2-EPOXY-3-(p-NITROPHENOXY)- see NIN050
PROPANE, 1-ETHOXY-(9CI) see EPC125
PROPANE, 2-ETHOXY-2-METHYL-(9CI) see EHA550
PROPANE, HEPTAFLUOROIODO- see HAY300
PROPANE, 1-ISOTHIOCYANATO-3-(METHYLSULFINYL)-(9CI) see MPN100
PROPANE, 1-METHOXY-(9CI) see MOU830
PROPANE, 2-METHOXY-2-METHYL- (9CI) see MHV859
PROPANENITRILE see PMV750
PROPANENITRILE, 3-((2-((2-AMINOETHYL)AMINO)ETHYL)AMINO)-(9CI) see LBV000
PROPANENITRILE, 3-ANILINO- see AOT100
PROPANENITRILE, 3-(ETHYLPHENYLAMINO)-(9CI) see EHQ500
PROPANENITRILE, 2-(β-d-GLUCOPYRANOSYLOXY)-2-METHYL- see GFC100
PROPANENITRILE, 3,3'-(METHYLIMINO)BIS- see MHQ750
PROPANENITRILE, 3-(METHYLNITROSOAMINO)- see MMS200
PROPANENITRILE, 3-(PHENYLAMINO)- see AOT100
PROPANENITRILE, TRICHLORO- see TJC800
α-γ-PROPANE OXIDE see OMW000
PROPANE, 2,2'-OXYBIS(1-CHLORO)- see BII250
PROPANE, 1,1'-OXYBIS-(9CI) see PNM000
PROPANE PEROXOIC ACID see PCO100
PROPANEPEROXOIC ACID, 2-METHYL-, 1,1-DIMETHYLETHYL ESTER see BSC600
1-PROPANESULFONANILIDE, 4'-(4-CARBAMOYL-9-ACRIDINYLAMINO)-3'-METHOXY-, HYDROCHLORIDE see MFN000
1-PROPANESULFONIC ACID-3-HYDROXY-γ-SULTONE see PML400
1,3-PROPANE SULTONE (MAK) see PML400
PROPANE SULTONE see PML400
PROPANE-1-THIOL see PML500
2-PROPANETHIOL see IMU000
PROPANETHIOL see PML500
1,2,3-PROPANETRICARBOXYLIC ACID, 2-(ACETYLOXY)-, TRIBUTYL ESTER see THX100
1,2,3-PROPANETRICARBOXYLIC ACID, 2-(ACETYLOXY)-, TRIETHYL ESTER (9CI) see ADD750
1,2,3-PROPANETRICARBOXYLIC ACID, 2-HYDROXY-, TIN(2+) SALT (2:3) (9CI) see TGC285
1,2,3-PROPANETRICARBOXYLIC ACID, 2-HYDROXY-, TRIBUTYL ESTER see THY100
1,2,3-PROPANETRICARBOXYLIC ACID, 2-HYDROXY-, ZINC SALT (2:3) (9CI) see ZFJ250
1,2,3-PROPANETRIOL see GGA000
1,2,3-PROPANETRIOL, DIACETATE (9CI) see GGA100
1,2,3-PROPANETRIOL, 1,3-DINITRATE (9CI) see GGA200
1,2,3-PROPANETRIOL MONOACETATE see GGO000
1,2,3-PROPANETRIOL TRIACETATE see THM500
1,2,3-PROPANETRIOL, TRINITRATE see NGY000
PROPANETRIOL TRINITRATE see NGY000
1,2,3-PROPANETRIYL NITRATE see NGY000
PROPANEX see DGI000
PROPANID see DGI000
PROPANIDE see DGI000
PROPANIDID see PMM000
PROPANIDIDE see PMM000
PROPANIL see DGI000
PROPANIMIDAMIDE, 2,2'-AZOBIS(2-METHYL-), DIHYDROCHLORIDE (9CI) see ASM050
PROPANOIC ACID see PJA500, PMU750
PROPANOIC ACID, 3-BROMO-2-OXO-(9CI) see BOD550
PROPANOIC ACID BUTYLESTER (9CI) see BSJ500
PROPANOIC ACID, CALCIUM SALT (9CI) see CAW400
PROPANOIC ACID, 2-((4'-CHLORO(1,1'-BIPHENYL)-4-YL)OXY)-2-METHYL-, METHYL ESTER see MIO975
PROPANOIC ACID, 2-(4-CHLORO-2-METHYLPHENOXY)-, POTASSIUM SALT (9CI) see CLO200
PROPANOIC ACID, 2-CHLORO-, SODIUM SALT see CKT100
PROPANOIC ACID, 2,2-DIMETHYL-, n-HEXYL ESTER see HFP700
PROPANOIC ACID, 2,2-DIMETHYL-, HEXYL ESTER (9CI) see HFP700
PROPANOIC ACID, 2,2-DIMETHYL-, ISOOCTADECYL ESTER see ISC550
PROPANOIC ACID, ETHENYL ESTER see VQK000
PROPANOIC ACID, HEXYL ESTER see HFV100
PROPANOIC ACID, 2-HYDROXY- see LAG000
PROPANOIC ACID, 2-HYDROXY-, CALCIUM SALT see CAT600

PROPANOIC ACID, 2-HYDROXY-, (S)-(9CI) see LAG010
PROPANOIC ACID, 2-HYDROXY-2-METHYL-, ETHYL ESTER (9CI) see ELH700
PROPANOIC ACID, 2-HYDROXY-, MONOPOTASSIUM SALT (9CI) see PLK650
PROPANOIC ACID, 2-HYDROXY-, TRIANHYDRIDE with ANTIMONIC ACID (H₃SbO₃) (9CI) see AQE250
PROPANOIC ACID, 3-MERCAPTO-(9CI) see MCQ000
PROPANOIC ACID, 2-METHYL-, 3,7-DIMETHYL-2,6-OCTADIENYL ESTER, (Z)- see NCO200
PROPANOIC ACID, METHYL ESTER see MOT000
PROPANOIC ACID, 2-METHYL-, 2,4-HEXADIENYL ESTER see HCS700
PROPANOIC ACID, 2-METHYL-, 3-HEXENYL ESTER, (Z)- see HFE520
PROPANOIC ACID, 2-METHYL-, HEXYL ESTER (9CI) see HFQ550
PROPANOIC ACID, 2-METHYL-, 1-METHYL-1-(4-METHYL-3-CYCLOHEXEN-1-YL)ETHYL ESTER see MCF515
PROPANOIC ACID, 2-METHYL-, 2-PHENOXYETHYL ESTER see PDS900
PROPANOIC ACID, 2-METHYLPROPYL ESTER see PMV250
PROPANOIC ACID, 2-OXO-(9CI) see PQC100
PROPANOIC ACID-2-PHENYLETHYL ESTER see PDK000
PROPANOIC ACID, SODIUM SALT see SJL500
PROPANOIC ACID, 3,3'-THIOBIS-, DIDODECYL ESTER see TFD500
PROPANOIC ACID, 3,3'-THIOBIS-, DIOCTADECYL ESTER (9CI) see DXG700
PROPANOIC ANHYDRIDE see PMV500
PROPANOL-1 see PND000
1-PROPANOL see PND000
PROPAN-2-OL see INJ000
2-PROPANOL see INJ000
i-PROPANOL (GERMAN) see INJ000
n-PROPANOL see PND000
1,3-PROPANOLAMINE see PMM250
3-PROPANOLAMINE see PMM250
PROPANOLAMINE see PMM250
1-PROPANOL, 3-CHLORO- see CKP725
2-PROPANOL, 1-(CYCLOHEXYLAMINO)- see CPG700
2-PROPANOL, 1-(4-CYCLOPROPYLCARBONYLPHENOXY)-3-(1,2-DIHYDRO-2-IMINO-4-METHYLPYRIDINO)- see PMM300
2-PROPANOL, 1-(4-(2-(CYCLOPROPYLMETHOXY)ETHYL)PHENOXY)-3-((1-METHYLETHYL)AMINO)-, HYDROCHLORIDE see KEA350
1-PROPANOL, 2,3-DIBROMO-, HYDROGEN PHOSPHATE, AMMONIUM SALT see MRH214
1-PROPANOL, 2,3-DIBROMO-, HYDROGEN PHOSPHATE, MAGNESIUM SALT see MAD100
1-PROPANOL, 3-(DIETHYLAMINO)-2,2-DIMETHYL-, TROPATE, PHOSPHATE see AOD250
PROPANOL, 1,1-DIPHENYL-3-PIPERIDINO-, METHANESULFONATE see PIL550
PROPANOLE (GERMAN) see PND000
PROPANOLEN (DUTCH) see PND000
PROPANOL, ETHOXY-(9CI) see EJV000
PROPANOLI (ITALIAN) see PND000
PROPANOLIDE see PMT100
1-PROPANOL, 2-METHYL-2-NITRO-, NITRATE see MMP200
2-PROPANOL NITRITE see IQQ000
PROPANOL NITRITE see PNQ750
PROPANOLONE see PQC000
PROPANOL, OXYBIS- see DWS500
2-PROPANOL, 1,1'-(PENTAMETHYLENEDIOXY)BIS(3-CHLORO- see PBH150
1-PROPANOL, 2-PHENOXY- see PNL300
1-PROPANOL, 3-PHENYL-, PROPIONATE see HHQ550
2-PROPANOL, 1-(2-((3,3,5-TRIMETHYLCYCLOHEXYL)OXY)PROPOXY)-(9CI) see TLO600
2-PROPANONE see ABC750
PROPANONE see ABC750
2-PROPANONE, 1-(4-(CHLOROMETHYL)-1,3-DIOXOLAN-2-YL)- see CIL850
1-PROPANONE, 1-(4-CHLOROPHENYL)-2-METHYL- see IOL100
2-PROPANONE, 1,1-DICHLORO- see DGG500
1-PROPANONE, 1-(3,4-DIMETHOXYPHENYL)- see PMX600
2-PROPANONE, 1-(3,4-DIMETHOXYPHENYL)- see VIK300
2-PROPANONE, 1,1-DIPHENYL- see MJH910
2-PROPANONE, HEXACHLORO- see HCL500
1-PROPANONE, 1-(4-(3-HYDROXYPHENYL)-1-METHYL-4-PIPERIDINYL)-(9CI) see KFK000
1-PROPANONE, 1-(4-(m-HYDROXYPHENYL)-1-METHYL-4-PIPERIDYL)- see KFK000
1-PROPANONE, 1-(2-(2-HYDROXY-3-(PROPYLAMINO)PROPOXY)PHENYL)-3-PHENYL-, HYDROCHLORIDE (9CI) see PMJ525
1-PROPANONE, 1-p-MENTH-6-EN-2-YL- see MCF525
2-PROPANONE, 1-METHOXY- see MDW300, MFL100
2-PROPANONE OXIME see ABF000
2-PROPANONE, 1-PHENYL- see MHO100
2-PROPANONE, o-((PHENYLAMINO)CARBONYL)OXIME see PEQ500
1-PROPANONE, 1-(4-PYRIDYL)- see PMX300
1-PROPANONE, 1-PYRROL-2-YL- see PMX320
1-PROPANONE, 1-(2,4,6-TRIHYDROXYPHENYL)- see TKP100
PROPANOSEDYL see RLU000
PROPANOYL CHLORIDE see PMW500
PROPANTAN see PMM000

PROPANTEL see HKR500
PROPANTHELINE BROMIDE see HKR500
PROPAPHEN see CKP500
PROPAPHENIN see CKP250
PROPAPHENIN HYDROCHLORIDE see CKP500
2-PROPARGILOSSI-5-AMINO-N-(n-BUTIL)-BENZAMIDE (ITALIAN) see AJC000
PROPARGITE (DOT) see SOP000
3-PROPARGLOXYPHENYL-N-METHYL-CARBAMATE see PMN250
PROPARGYL ALCOHOL see PMN450
PROPARGYL ALDEHYDE see PMT250
PROPARGYL BROMIDE see PMN500
PROPARGYL CHLORIDE see CKV275
PROPARGYL ETHER see POA500
5-PROPARGYLFURFURYL CHRYSANTHEMATE see POD875
5-PROPARGYL-2-FURYLMETHYL, dl-cis,trans-CHRYSANTHEMATE see POD875
PROPARGYLIC ACID see PMT275
PROPARSAMIDE see SJG000
PROPARTHRIN see PMN700
PROPASA see AMM250
PROPASIN see PMN850
PROPASOL P see PNB750
PROPASOL SOLVENT B see BPS250
PROPASOL SOLVENT M see PNL250
PROPASOL SOLVENT P see PNB750
PROPASTE 6708 see TAV750
PROPASTE T see SON000
PROPATHENE see PMP500
PROPAVAN see IDA500
PROPAX see CFZ000
PROPAXOLIN CITRATE see POF500
PROPAXOLINE CITRATE see POF500
PROPAZINE see PMN850
PROPAZONE see PMO250
PROPEL see LAG000
PROPELLANT 12 see DFA600
PROPELLANT 22 see CFX500
PROPELLANT 114 see FOO509
PROPELLANT C318 see CPS000
2-PROPENAL see ADR000
PROP-2-EN-1-AL see ADR000
PROPENAL (CZECH) see ADR000
PROPENAL DIETHYL ACETAL see DHH800
2-PROPENAL, 3-(2-FURANYL)-2-METHYL- see FPX050
2-PROPENAL, 2-METHYL- see MGA250
2-PROPENAMIDE see ADS250
PROPENAMIDE see ADS250
2-PROPENAMIDE, N-(BUTOXYMETHYL)-(9CI) see BPM660
2-PROPENAMIDE, N-(1,1-DIMETHYLETHYL)-(9CI) see BPW050
2-PROPENAMIDE, HOMOPOLYMER see PJK350
2-PROPENAMIDE, N-(METHOXYMETHYL)- see MEX300
2-PROPENAMIDE, N-METHYL- (9CI) see MGA300
2-PROPENAMIDE, N-(1-METHYLETHYL)-(9CI) see INH000
2-PROPENAMIDE, N-((2-METHYLPROPOXY)METHYL)-(9CI) see IIE100
2-PROPENAMIDE, 3-PHENYL-(9CI) see CMP970
2-PROPEN-1-AMINE see AFW000
2-PROPENAMINE see AFW000
1-PROPENE see PMO500
PROPENE see PMO500
PROPENE ACID see ADS750
trans-1-PROPENE-1,2-DICARBOXYLIC ACID see MDI250
1-PROPENE, DICHLORO- see DGG700, DGI700
1-PROPENE, 3,3-DIETHOXY-(9CI) see DHH800
1-PROPENE-1,3-DIOL, DIACETATE see PMO800
PROPENE-1,3-DIOL DIACETATE see PMO800
2-PROPENE-1,1-DIOL, 2-METHYL-, DIACETATE see AAW250
1-PROPENE HOMOPOLYMER (9CI) see PMP500
1-PROPENE, 3-IODO-(9CI) see AGI250
1-PROPENE, 2-METHOXY- see MFL300
2-PROPENENITRILE see ADX500
PROPENENITRILE see ADX500
2-PROPENENITRILE HOMOPOLYMER (9CI) see ADX750
2-PROPENENITRILE, POLYMER with CHLOROETHENE (9CI) see ADY250
2-PROPENENITRILE POLYMER with ETHENYLBENZENE see ADY500
PROPENE OXIDE see PNL600
PROPENE OZONIDE see PMP250
PROPENE POLYMERS see PMP500
PROPENE TETRAMER see PMP750
2-PROPENE-1-THIOL see AGJ500
1-PROPENE-1,2,3-TRICARBOXYLIC ACID see ADH000
PROPENE, 3,3,3-TRIFLUORO- see TKH015
1-PROPENE, 3,3,3-TRIFLUORO-(9CI) see TKH015
PROPENE, 3,3,3-TRIFLUORO-2-(TRIFLUOROMETHYL)- see HDC450
PROPENOIC ACID see ADS750
2-PROPENOIC ACID BICYCLO(2,2,1)HEPT-5-EN-2-YLMETHYL ESTER see BFY250

2-PROPENOIC ACID, 1,4-BUTANEDIYL ESTER see TDQ100
2-PROPENOIC ACID 2-BUTOXYETHYL ESTER see BPK500
2-PROPENOIC ACID-2-CHLOROETHYL ESTER see ADT000
2-PROPENOIC ACID, 2-CHLORO-, ETHYL ESTER see EHH200
2-PROPENOIC ACID, 3-CHLORO-, SODIUM SALT, (Z)-(9CI) see SFV250
2-PROPENOIC ACID (9CI) see ADS750
2-PROPENOIC ACID-2-CYANOETHYL ESTER see ADT111
2-PROPENOIC ACID, 2-(DIMETHYLAMINO)ETHYL ESTER (9CI) see DPB300
2-PROPENOIC ACID-2,2-DIMETHYL-1,3-PROPANEDIYL ESTER see DUL200
2-PROPENOIC ACID, 1,2-ETHANEDIYLBIS(OXY-2,1-ETHANEDIYL) ESTER (9CI) see TJQ100
2-PROPENOIC ACID-1,2-ETHANEDIYL ESTER see EIP000
2-PROPENOIC ACID-2-ETHOXYETHYL ESTER see ADT500
2-PROPENOIC ACID-2-ETHYLBUTYL ESTER see EGZ000
2-PROPENOIC ACID, ETHYL ESTER (MAK) see EFT000
2-PROPENOIC ACID-2-ETHYLHEXYL ESTER see ADU250
2-PROPENOIC ACID, HEPTYL ESTER see HBL100
2-PROPENOIC ACID, 1,6-HEXANEDIYL ESTER see HEQ100
2-PROPENOIC ACID, HEXEL ESTER see ADV000
2-PROPENOIC ACID HOMOPOLYMER (9CI) see ADW200
2-PROPENOIC ACID, HOMOPOLYMER, ZINC SALT see ADW100
2-PROPENOIC ACID-2-HYDROXYETHYL ESTER (9CI) see ADV250
2-PROPENOIC ACID-2-(HYDROXYMETHYL)-2-(((1-OXO-2-PROPE-NYL)OXY)METHYL)-1,3-PROPANEDIYL ESTER see PBC750
2-PROPENOIC ACID, 3-(2-HYDROXYPHENYL)-, (E)-(9CI) see CNU850
2-PROPENOIC ACID-2-HYDROXYPROPYL ESTER see HNT600
2-PROPENOIC ACID, 3-(1H-IMIDAZOL-4-YL)-(9CI) see UVJ440
2-PROPENOIC ACID ISODECYL ESTER (9CI) see IKL000
2-PROPENOIC ACID, 2-METHOXYETHYL ESTER see MEM250
2-PROPENOIC ACID, 2-METHYL-(9CI) see MDN250
2-PROPENOIC ACID, 2-METHYL-, 2-(DIETHYLAMINO)ETHYL ESTER (9CI) see DIB300
2-PROPENOIC ACID METHYL ESTER see MGA500
PROPENOIC ACID METHYL ESTER see MGA500
2-PROPENOIC ACID, 2-METHYL-, 1,2-ETHANEDIYLBIS(OXY-2,1-ETHANEDIYL) ESTER(9CI) see MDN510
2-PROPENOIC ACID, (1-METHYL-1,2-ETHANEDIYL)BIS(OXY(METHYL-2,1-ETH-ANEDIYL)) ESTER see TMZ100
2-PROPENOIC ACID, 2-METHYL-, 2-(ETHENYLOXY)ETHYL ESTER (9CI) see VQA150
2-PROPENOIC ACID, 2-METHYL-, 2-((3a,4,5,6,7,7a-HEXAHYDRO-4,7-METHA-NO-1H-INDEN-5-YL)OXY) ETHYL ESTER see DGW450
2-PROPENOIC ACID, 2-METHYL-, 2-HYDROXYPROPYL ESTER (9CI) see HNV500
2-PROPENOIC ACID, 2-METHYL-, METHYL ESTER see MLH750
2-PROPENOIC ACID, 2-METHYL-, OXYBIS(2,1-ETHANEDIYLOXY-2,1-ETHANE-DIYL)ESTER see TCE400
2-PROPENOIC ACID, 2-METHYL-, 2-PHENYLETHYL ESTER see PFP600
2-PROPENOIC ACID-1-METHYL-13-PROPANEDIYL ESTER see BRG500
2-PROPENOIC ACID-2-METHYLPROPYL ESTER see IIK000
2-PROPENOIC ACID-2-(METHYLTHIO)ETHYL ESTER see MPT250
2-PROPENOIC ACID, 3-(4-NITROPHENYL)- see NFT440
2-PROPENOIC ACID OXIRANYLMETHYL ESTER see ECH500
2-PROPENOIC ACID, OXYBIS(2,1-ETHANEDIYLOXY-2,1-ETHANEDIYL)ESTER see ADT050
2-PROPENOIC ACID, OXYBIS(METHYL-2,1-ETHANEDIYL) ESTER see DWS650
2-PROPENOIC ACID, OXYDI-2,1-ETHANEDIYL ESTER (9CI) see ADT250
2-PROPENOIC ACID, 2-PHENOXYETHYL ESTER see PER250
2-PROPENOIC ACID, 3-PHENYL-, (E)-(9CI) see CMP980
2-PROPENOIC ACID, 3-PHENYL-, 1-METHYLETHYL ESTER (9CI) see IOO000
2-PROPENOIC ACID, 3-PHENYL-, 3-PHENYL-2-PROPENYL ESTER (9CI) see CMQ850
2-PROPENOIC ACID, 3-PHENYL-, 3-PHENYLPROPYL ESTER see PGB800
2-PROPENOIC ACID, 3-PHENYL-, SODIUM SALT (9CI) see SFX000
2-PROPENOIC ACID, 3a,4,7,7a-TETRAHYDRO-4,7-METHANO-1H-INDENYL ES-TER see DGW400
2-PROPENOIC ACID TRIDECYL ESTER see ADX000
2-PROPENOIC ACID, 3-(3,4,5-TRIMETHOXYPHENYL)-(9CI) see CMQ100
2-PROPEN-1-OL see AFV500
PROPEN-1-OL-3 see AFV500
1-PROPEN-3-OL see AFV500
PROPENOL see AFV500
2-PROPEN-1-ONE see ADR000
2-PROPEN-1-ONE, 3-(2-CHLOROPHENYL)-1-PHENYL- see CLF150
2-PROPEN-1-ONE, 3-(4-CHLOROPHENYL)-1-PHENYL- see CLF100
2-PROPEN-1-ONE, 1-(4-METHOXYPHENYL)-(9CI) see ONW100
2-PROPEN-1-ONE, 3-(4-METHYLPHENYL)-1-PHENYL- see MIF762
2-PROPEN-1-ONE, 3-(4-NITROPHENYL)-1-PHENYL-(9CI) see NFS505
4-p-PROPENONE-OXY-PHENYLARSONIC ACID see PMQ000
2-PROPENOPHENONE see PMQ250
2-PROPENOYL CHLORIDE see ADZ000
1-PROPEN-2-YL ACETATE see MQK750
2-PROPENYLACRYLIC ACID see SKU000
2-PROPENYL ALCOHOL see AFV500
PROPENYL ALCOHOL see AFV500

4-PROPENYLANISOLE see PMQ750
cis-p-PROPENYLANISOLE see PMR000
p-1-PROPENYLANISOLE see PMQ750
p-PROPENYLANISOLE see PMQ750
(E)-p-PROPENYLANISOLE see PMR250
trans-p-PROPENYLANISOLE see PMR250
5-(1-PROPENYL)-1,3-BENZODIOXOLE see IRZ000
5-(2-PROPENYL)-1,3-BENZODIOXOLE see SAD000
4-PROPENYLCATECHOL METHYLENE ETHER see IRZ000
2-PROPENYL CHLORIDE see AGB250
PROPENYL CHLORIDE see PMR750
PROPENYL CINNAMATE see AGC000
1-PROPENYL CYANIDE see COC300
2-PROPENYL DISULPHIDE see AGF300
(8-β)-6-(2-PROPENYL)ERGOLINE-8-ACETAMIDE TARTRATE see PMR800
PROPENYL ETHER see DBK000
PROPENYLGUAETHOL (FCC) see IRY000
4-PROPENYLGUAIACOL see IKQ000
2-PROPENYL HEPTANOATE see AGH250
2-PROPENYL-N-HEXANOATE see AGA500
2-PROPENYL ISOTHIOCYANATE see AGJ250
2-PROPENYL ISOVALERATE see ISV000
4-(2-PROPENYL)-2-METHOXYPHENYL FORMATE see EQS100
2-PROPENYL 3-METHYLBUTANOATE see ISV000
4-PROPENYL-1,2-METHYLENEDIOXYBENZENE see IRZ000
(2-PROPENYLOXY)BENZENE see AGR000
3-(2-PROPENYLOXY)-1,2-BENZISOTHIAZOLE 1,1-DIOXIDE (9CI) see PMD800
((2-PROPENYLOXY)METHYL)OXIRANE see AGH150
2-(1-PROPENYL)PHENOL see PMS250
2-PROPENYLPHENOL see PMS250
o-PROPENYLPHENOL see PMS250
2-PROPENYL PHENYLACETATE see PMS500
p-PROPENYLPHENYL METHYL ETHER see PMQ750
N-2-PROPENYL-2-PROPEN-1-AMINE see DBI600
(2-PROPENYL)THIOUREA see AGT500
PROP-2-ENYL TRIFLUOROMETHANE SULFONATE see PMS775
2-PROPENYL 3,5,5-TRIMETHYLHEXANOATE see AGU400
2-PROPENYLUREA see AGV000
N-2-PROPENYLUREA see AGV000
4-PROPENYL VERATROLE see IKR000
PROPENZOLATE HYDROCHLORIDE see PMS800
PROPERICIAZINE see PIW000
PROPERIDOL see DYF200
PROPETAMPHOS see MKA000
PROPHAM see CBM000, PMS825
PROPHENAL see AGQ875
PROPHENATIN see DEO600
PROPHENPYRIDAMINE HYDROCHLORIDE see PMS900
PROPHENPYRIDAMINE MALEATE see TMK000
PROPHOS see EIN000
PROPICOL see DNR309
PROPIDIUM DIIODIDE see PMT000
PROPIDIUM IODIDE see PMT000
PROPILDAZINA (ITALIAN) see HHD000
PROPILENTIOUREA see MJZ000
PROPILTHIOURACIL see PNX000
6-PROPIL-TIOURACILE (ITALIAN) see PNX000
PROPINAN DI-n-BUTYLCINICITY (CZECH) see DEB400
PROPINE see MFX590
PROPINEB see ZMA000
PROPINEBE see ZMA000
γ-PROPIOBUTYROLACTONE see HBA550
PROPIOCINE see EDH500
PROPIOKAN see TBG000
1,3-PROPIOLACTONE see PMT100
3-PROPIOLACTONE see PMT100
β-PROPIOLACTONE see PMT100
PROPIOLACTONE see PMT100
PROPIOLALDEHYDE see PMT250
PROPIOLIC ACID see PMT275
PROPIOLIC ACID, TRIPHENYLSTANNYL ESTER see TMW600
PROPIOLOYL CHLORIDE see PMT300
PROPIOMAZINE MALEATE see IDA500
PROPIONALDEHYDE see PMT750
PROPIONALDEHYDE, 3-(p-ISOPROPYLBENZYL)-, DIMETHYL ACETAL see COU525
PROPIONALDEHYDE, 2-METHYL-2-(METHYLSULFINYL)-, O-(METHYLCARBA-MOYL)OXIME see MLW750
PROPIONALDEHYDE, 2-METHYL-2-(METHYLTHIO)-, OXIME see MLX800
PROPIONALDEHYDE, 3-(METHYLTHIO)- see MPV400, TET900
PROPIONALDEHYDE, 3,3,3-TRIFLUORO- see TKH020
PROPIONAMIDE see PMU250
PROPIONAMIDE, N,N-DIBUTYL- see DEH300
PROPIONAMIDE, N,N-DIETHYL- see DJX250
PROPIONAMIDE, N-METHYL- see MOS900

PROPIONAMIDINE, 2,2'-AZOBIS(2-METHYL-), DIHYDROCHLORIDE see ASM050
PROPIONAN SODNY see SJL500
PROPIONATE d'ETHYLE (FRENCH) see EPB500
PROPIONATE de METHYLE (FRENCH) see MOT000
PROPIONE see DJN750
PROPIONIC ACID (ACGIH,DOT,OSHA) see PMU750
PROPIONIC ACID see PMU750
PROPIONIC ACID AMIDE see PMU250
PROPIONIC ACID ANHYDRIDE PMV500
PROPIONIC ACID CHLORIDE see PMW500
PROPIONIC ACID, 2-(m-CHLOROPHENOXY)- see CJQ300
PROPIONIC ACID, 2-((4-CHLORO-o-TOLYL)OXY)-, POTASSIUM SALT see CLO200
PROPIONIC ACID, CINNAMYL ESTER see CMR850
PROPIONIC ACID-3,4-DICHLOROANILIDE see DGI000
PROPIONIC ACID-3,7-DIMETHYL-2,6-OCTADIEN-1-YL ESTER see NCP000
PROPIONIC ACID, α-1,3-DIMETHYL-4-PHENYL-4-PIPERIDYL ESTER see NEA500
PROPIONIC ACID, 3,3'-DISELENODI-, SODIUM SALT see DWZ100
PROPIONIC ACID, ETHYL ESTER see EPB500
PROPIONIC ACID GRAIN PRESERVER see PMU750
PROPIONIC ACID, HEXYL ESTER (6CI,7CI,8CI) see HFV100
PROPIONIC ACID, 2-HYDROXY- see LAG000
PROPIONIC ACID, ISOBUTYL ESTER see PMV250
PROPIONIC ACID, 2-(p-ISOBUTYLPHENYL)-, 2-PYRIDYLMETHYL ESTER see IAB000
PROPIONIC ACID, 3-METHOXY-, METHYL ESTER see MFL400
PROPIONIC ACID, 2-METHYL- see IJU000
PROPIONIC ACID, 2-(8-METHYL-10,11-DIHYDRO-11-OXODIBENZ(b,f)OXEPIN-2-YL)- see DLI650
PROPIONIC ACID, 2-METHYLENE- see MDN250
PROPIONIC ACID, NERYL ESTER see NCP000
PROPIONIC ACID, 3,3'-NITROIMINODI- see NHI600
PROPIONIC ACID, 2-PHENOXY- see PDV725
PROPIONIC ACID-2-PHENOXYETHYL ESTER see EJK500
PROPIONIC ACID, PHENYLMERCURY SALT see PFO000
PROPIONIC ACID, PROPYL ESTER see PNU000
PROPIONIC ACID, 3-SELENINO- see SBN510
PROPIONIC ACID, 3,3'-THIODI-, DIOCTADECYL ESTER see DXG700
PROPIONIC ACID, 2-(p-((5-(TRIFLUOROMETHYL)-2-PYRIDYL)OXY)PHENOXY)-, BUTYL ESTER see FDA885
PROPIONIC ALDEHYDE see PMT750
PROPIONIC AMIDE see PMU250
PROPIONIC ANHYDRIDE see PMV500
PROPIONIC CHLORIDE see PMW500
PROPIONIC ETHER see EPB500
PROPIONIC NITRILE see PMV750
PROPIONITRILE, 3-(ALKYLAMINO)- see AFO200
PROPIONITRILE, 3-((2-((2-AMINOETHYL)AMINO)ETHYL)AMINO)-(8CI) see LBV000
PROPIONITRILE, 3-ANILINO- see AOT100
PROPIONITRILE, 3-(N-ETHYLANILINO)-(6CI,7CI,8CI) see EHQ500
PROPIONITRILE, 2-HYDROXY- see LAQ000
PROPIONITRILE, 3-METHOXY- see MFL750
PROPIONITRILE, TRICHLORO- see TJC800
PROPIONITRILE, 3-(TRIETHOXYSILYL)- see CON250
β-PROPIONOLACTONE see PMT100
PROPIONONITRILE see PMV750
PROPIONYLBENZENE see EOL500
4-(N-PROPIONYL)BENZYLIMINO-1-ETHYL-2,2,6,6-TETRAMETHYLPIPERIDINE see EPM000
PROPIONYL CHLORIDE see PMW500
PROPIONYLCHOLINE IODIDE see PMW750
3-PROPIONYL-10-DIMETHYLAMINO-ISOPROPYLPHENOTHIAZINE MALEATE see IDA500
10-PROPIONYL DITHRANOL see PMW760
N-PROPIONYL-N-2-FLUORENYLHYDROXYLAMINE see FEO000
PROPIONYL HYPOBROMITE see PMW770
N-PROPIONYLINDOLE see ICW100
PROPIONYL OXIDE see PMV500
PROPIONYL PEROXIDE (DOT) see DWQ800
p-PROPIONYLPHENOL see ELL500
PROPIONYLPHLOROGLUCINOL see TKP100
N-PROPIONYL-2-(1-PIPERIDINOISOPROPYL)AMINOPYRIDINE FUMARATE see PMX250
PROPIONYLPROMETHAZINE MALEATE see IDA500
4-PROPIONYLPYRIDINE see PMX300
N-PROPIONYL-N-(2-PYRIDYL)-1-PIPERIDINO-2-AMINOPROPANE FUMARATE see PMX250
2-PROPIONYLPYRROLE see PMX320
PROPIOPHENONE see EOL500
PROPIOPHENONE, 2-BROMO- see BOB550
PROPIOPHENONE, 4'-CHLORO-2-METHYL-(6CI,7CI,8CI) see IOL100
PROPIOPHENONE, 3',4'-DIMETHOXY- see PMX600

PROPIOPHENONE, 4'-(DIMETHYLAMINO)-3-(4-PHENYL-1,2,3,6-TETRAHYDRO-1-PYRIDYL)- see DPS700
PROPIOPHENONE, 4'-PHENYL- see BGM100
PROPIOPHENONE, 2',4',6'-TRIHYDROXY- see TKP100
PROPIOPHLOROGLUCINE see TKP100
PROPIOPROMAZINE MALEATE see PMX500
PROPIOVERATRONE see PMX600
PROPIRAM FUMARATE see PMX250
PROPISAMINE see BBK000
PROPITAN see FHG000, PII200
PROPITOCAINE HYDROCHLORIDE see CMS250
PROPIVANE see PGP500
PROP-JOB see DGI000
PROPOFOL see DNR800
PROPOKSURU (POLISH) see PMY300
PROPOLIN see PMP500
PROPON see TIX500
PROPONESIN HYDROCHLORIDE see TGK225
PROPONEX-PLUS see CIR500
PROPOPHANE see PMP500
PROPOQUIN DIHYDROCHLORIDE see PMY000
PROPOX see PNA500
PROPOXUR see PMY300
PROPOXUR NITROSO see PMY310
4-PROPOXYBENZALDEHYDE see PMY500
p-(N-PROPOXY)BENZALDEHYDE see PMY500
p-PROPOXYBENZALDEHYDE see PMY500
2-N-PROPOXYBENZAMIDE see PMY750
o-PROPOXYBENZAMIDE see PMY750
N-(PROPOXY-5-BENZIMIDAZOLYL)-2, CARBAMATE de METHYLE (FRENCH) see OMY700
N-(2-(5-PROPOXYBENZIMIDAZOLYL)) METHYL CARBAMATE see OMY700
p-PROPOXYBENZOIC ACID 2-(1-PYRROLIDINYL)ETHYL ESTER HYDROCHLORIDE see PNA200
PROPOXYCHEL see PNA500
2-PROPOXYETHANOL see PNG750
2-PROPOXYETHANOL ACETATE see PNA225
2-PROPOXYETHYL ACETATE see PNA225
d-PROPOXYPHENE see DAB879
(+)-PROPOXYPHENE see DAB879
PROPOXYPHENE see PNA250
α-d-PROPOXYPHENE HYDROCHLORIDE see PNA500
α-PROPOXYPHENE HYDROCHLORIDE see PNA500
d-PROPOXYPHENE HYDROCHLORIDE see PNA500
(+)-PROPOXYPHENE HYDROCHLORIDE see PNA500
PROPOXYPHENE HYDROCHLORIDE see PNA500
d-PROPOXYPHENE MONOHYDROCHLORIDE see PNA500
PROPOXYPHENE N see DYB400
PROPOXYPHENE-2-NAPHTHALENESULFONATE see DYB400
PROPOXYPHENE NAPSYLATE see DYB400
d-PROPOXYPHENE NAPSYLATE HYDRATE see DAB880
4'-PROPOXY-3-PIPERIDINO PROPIOPHENONE HYDROCHLORIDE see PNB250
4-PROPOXY-β-(1-PIPERIDYL)PROPIOPHENONE HYDROCHLORIDE see PNB250
PROPOXYPIPEROCAINE see PNB250
PROPOXYPIPEROCAINE HYDROCHLORIDE see PNB250
n-PROPOXYPROPANOL (mixed isomers) see PNB750
dl-PROPRANOLOL see PNB790
racemic-PROPRANOLOL see PNB790
PROPRANOLOL see ICB000, ICC000
(±)-PROPRANOLOL see PNB790
dl-PROPRANOLOL HYDROCHLORIDE see PNB800
PROPRANOLOL HYDROCHLORIDE see ICC000
β-PROPRIOLACTONE (OSHA) see PMT100
β-PROPROLACTONE see PMT100
PROPROP see DGI400
PROPYCIL see PNX000
n-PROPYL ACETAL see AAG850
1-PROPYL ACETATE see PNC250
2-PROPYL ACETATE see INE100
n-PROPYL ACETATE see PNC250
PROPYL ACETATE see PNC250
PROPYLACETIC ACID see VAQ000
N-PROPYL-N-(ACETOXYMETHYL)NITROSAMINE see PNR250
PROPYL ACETOXYMETHYLNITROSAMINE see PNR250
β-PROPYL ACROLEIN see PNC500
PROPYLADIPHENIN see PBM500
S-PROPYL-N-AETHYL-N-BUTYL-THIOCARBAMAT (GERMAN) see PNF500
N-PROPYLAJMALINE see PNC875
N-PROPYLAJMALINE BITARTRATE see DNB000
N-PROPYLAJMALINE BROMIDE see PNC925
N-PROPYLAJMALINE HYDROGEN TARTRATE see DNB000
N⁴-PROPYLAJMALINIUM see PNC875
N-PROPYLAJMALINIUM see PNC875
N-PROPYLAJMALINIUM BITARTRATE see DNB000
1-PROPYLAJMALINIUM BROMIDE see PNC925

N-PROPYLAJMALINIUMHYDROGENTARTRAT (GERMAN) see DNB000
N⁴-PROPYLAJMALINIUM HYDROGEN TARTRATE see DNB000
1-PROPYL ALCOHOL see PND000
sec-PROPYL ALCOHOL (DOT) see INJ000
n-PROPYL ALCOHOL see PND000
PROPYL ALCOHOL see PND000
PROPYL ALDEHYDE see PMT750
i-PROPYLALKOHOL (GERMAN) see INJ000
n-PROPYL ALKOHOL (GERMAN) see PND000
2-PROPYLAMINE see INK000
N-PROPYLAMINE see PND250
sec-PROPYLAMINE see INK000
PROPYLAMINE see PND250
PROPYLAMINE, N-tert-BUTYL-1-METHYL-3,3-DIPHENYL-, HYDROCHLORIDE see TBC200
PROPYLAMINE, 1,2-DIMETHYL- see AOE200
PROPYLAMINE, 2,3-EPOXY-N,N-DIETHYL- see GGW800
PROPYLAMINE, 3,3'-(ETHYLENEDIOXY)BIS- see DCB200
PROPYLAMINE, 3,3'-IMINOBIS- see AIX250
α-PROPYLAMINE-2-METHYL-PROPIONANILIDEHYDROCHLORIDE see CMS250
PROPYLAMINE, 3-(TRIETHOXYSILYL)- see TJN000
PROPYLAMINO-BIS(1-AZIRIDINYL)PHOSPHINE OXIDE see BGY500
2-(PROPYLAMINO)-o-PROPIONOTOLUIDIDE HYDROCHLORIDE see CMS250
4-N-PROPYLANILINE see PNE000
4-PROPYLANILINE see PNE000
p-N-PROPYLANILINE see PNE000
p-PROPYLANILINE see PNE000
4-n-PROPYLANISOLE see PNE250
4-PROPYLANISOLE see PNE250
p-n-PROPYL ANISOLE see PNE250
5-n-PROPYL-1,2-BENZANTHRACENE see PNE750
8-PROPYLBENZ(a)ANTHRACENE see PNE750
PROPYL BENZENE (DOT) see IKG000
n-PROPYLBENZENE see IKG000
5-PROPYL-1,3-BENZODIOXOLE see DMD600
5-PROPYLBENZO(c)PHENANTHRENE see PNF000
2-n-PROPYL-3: 4-BENZPHENANTHRENE see PNF000
3-PROPYLBICYCLO(2.2.1)HEPT-5-ENE-2-CARBOXALDEHYDE see CML620
PROPYL BROMIDE see BNX750
PROPYL BUTANOATE see PNF100
S-PROPYL BUTYLETHYLTHIOCARBAMATE see PNF500
N-PROPYL-N-BUTYLNITROSAMINE see PNF750
PROPYL BUTYRATE see PNF100
PROPYL CAPRYLATE see PNR600
N-PROPYL CARBAMATE see PNG250
PROPYL CARBAMATE see PNG250
PROPYLCARBINOL see BPW500
N-PROPYLCARBINYL CHLORIDE see BQQ750
PROPYL CELLOSOLVE see PNG750
N-PROPYL CHLORIDE see CKP750
1-PROPYL-3-(p-CHLOROBENZENESULFONYL)UREA see CKK000
PROPYL CHLOROCARBONATE see PNH000
N-PROPYL-N-(2-CHLOROETHYL)-2,6-DINITRO-4-TRIFLUOROMETHYLANILINE see FDA900
N-PROPYL-N-(2-CHLOROETHYL)-α,α,α-TRIFLUORO-2,6-DINITRO-p-TOLUIDINE see FDA900
n-PROPYL CHLOROFORMATE (DOT) see PNH000
PROPYL CHLOROFORMATE see PNH000
S-PROPYL CHLOROTHIOFORMATE see PNH150
PROPYLCOPPER(I) see PNH500
PROPYL CYANIDE see BSX250
3-PROPYLDIAZIRINE see PNH550
PROPYLDICHLORARSINE see PNH650
PROPYL-p,p'-DICHLOROBENZILATE see PNH750
PROPYL(4-((DIETHYLCARBAMOYL)METHOXY)-3-METHOXYPHENYL)ACETATE see PMM000
4'-N-PROPYL-4-DIMETHYLAMINOAZOBENZENE see DTT400
2-N-PROPYL-3-DIMETHYLAMINO-5,6-METHYLENEDIOXYINDENE HYDROCHLORIDE see DPY600
2-PROPYL-3-DIMETHYLAMINO-5,6-METHYLENEDIOXYINDENE HYDROCHLORIDE see DPY600
PROPYL(3-(DIMETHYLAMINO)PROPYL)CARBAMATE MONOHYDROCHLORIDE see PNI250
5-n-PROPYL-9,10-DIMETHYL-1,2-BENZANTHRACENE see PNI500
1-n-PROPYL-3,7-DIMETHYL XANTHINE see PNW250
S-PROPYL DIPROPYLTHIOCARBAMATE see PNI750
N-PROPYL-DI-N-PROPYLTHIOLCARBAMATE see PNI750
PROPYL-N,N-DIPROPYLTHIOLCARBAMATE see PNI750
PROPYL DISELENIDE see PNI850
PROPYLENE (DOT) see PMO500
PROPYLENE ALDEHYDE see ADR000, COB260
PROPYLENE BISDITHIOCARBAMATE see MLX850
PROPYLENEBIS(DITHIOCARBAMATO)ZINC see ZMA000
1,2-PROPYLENE CARBONATE see CBW500
PROPYLENE CHLORIDE see PNJ400

PROPYLENECHLOROHYDRIN see CKR500
1,3-PROPYLENEDIAMINE see PMK500
PROPYLENE DIAMINE (DOT) see PMK250
PROPYLENEDIAMINE see PMK250
PROPYLENE DIBROMIDE see DDR400, DDR600
α,β-PROPYLENE DICHLORIDE see PNJ400
PROPYLENE DICHLORIDE see DGG800, PNJ400
PROPYLENE DIMETHANESULFONATE see TLR250
4,4'-PROPYLENEDI-2,6-PIPERAZINEDIONE see RCA375
PROPYLENE EPOXIDE see PNL600
1,2-PROPYLENE GLYCOL see PML000
1,3-PROPYLENE GLYCOL see PML250
α-PROPYLENEGLYCOL see PML000
β-PROPYLENE GLYCOL see PML250
PROPYLENE GLYCOL (FCC) see PML000
PROPYLENE GLYCOL ALGINATE see PNJ750
PROPYLENE GLYCOL, ALLYL ETHER see PNK000
PROPYLENE GLYCOL-n-BUTYL ETHER see BPS250
PROPYLENE GLYCOL CYCLIC CARBONATE see CBW500
α-PROPYLENE GLYCOL DIACETATE see PNK750
PROPYLENE GLYCOL DIACETATE see PNK750
PROPYLENE GLYCOL-1,2-DINITRATE see PNL000
1,2-PROPYLENE GLYCOL DINITRATE see PNL000
PROPYLENE GLYCOL DINITRATE see PNL000
PROPYLENE GLYCOL ETHYL ETHER see EFF500
PROPYLENE GLYCOL METHYL ETHER see PNL250
PROPYLENE GLYCOL MONOACETATE see PNL100
PROPYLENE GLYCOL MONOACRYLATE see HNT600
PROPYLENE GLYCOL MONO-n-BUTYL ETHER see BPS500
PROPYLENE GLYCOL-β-MONOETHYL ETHER see EFG000
β-PROPYLENE GLYCOL MONOETHYL ETHER see EFG000
PROPYLENE GLYCOL MONOETHYL ETHER see EJV000
PROPYLENE GLYCOL MONOMETHYL ETHER (ACGIH,OSHA) see PNL250
α-PROPYLENE GLYCOL MONOMETHYL ETHER see PNL250
β-PROPYLENE GLYCOL MONOMETHYL ETHER see MFL000
PROPYLENE GLYCOL MONOMETHYL ETHER see PNL250
PROPYLENE GLYCOL MONOMETHYL ETHER ACETATE see PNL265
PROPYLENE GLYCOL MONOSTEARATE see SLL000
PROPYLENE GLYCOL MONPROPYL ETHER see PNB750
PROPYLENE GLYCOL PHENYL ETHER see PNL300
PROPYLENE GLYCOL USP see PML000
1,2-PROPYLENEIMINE see PNL400
PROPYLENE IMINE see PNL400
PROPYLENE IMINE, INHIBITED (DOT) see PNL400
1,2-PROPYLENE OXIDE see PNL600
1,3-PROPYLENE OXIDE see OMW000
PROPYLENE OXIDE see PNL600
PROPYLENE OXIDE-METHANOL ADDUCT see MKS250
PROPYLENE POLYMER see PMP500
1,3-PROPYLENE SULFATE see TLR750
PROPYLENE TETRAMER see PMP750
PROPYLENE THIOUREA see MJZ000
PROPYLENGLYKOL-MONOMETHYLAETHER see PNL250
PROPYLENTHIOHARNSTOFF see MJZ000
(8-β)-6-PROPYLERGOLINE-8-ACETAMIDE TARTRATE (2:1) see PNL800
(8S)-6-PROPYLERGOLINE-8-CARBAMIC ACID ETHYL ESTER see EPC115
1-((5R,8S,10R)-6-PROPYL-8-ERGOLINYL)-3,3-DIETHYLUREA see DJX300
PROPYLESTER KYSELINY DUSICNE see PNQ500
PROPYLESTER KYSELINY MASELNE see PNF100
PROPYLESTER KYSELINY MRAVENCI see PNM500
PROPYLESTER KYSELINY OCTOVE see PNC250
n-PROPYL ESTER of 3,4,5-TRIHYDROXYBENZOIC ACID see PNM750
PROPYL ETHER see PNM000
β-PROPYL-α-ETHYLACROLEIN see BRI000
N-PROPYL-N-ETHYL-N-(N-BUTYL)THIOCARBAMATE see PNF500
PROPYL N-ETHYL-N-BUTYLTHIOCARBAMATE see PNF500
PROPYLETHYL-N-BUTYLTHIOCARBAMATE see PNF500
PROPYL-ETHYLBUTYLTHIOCARBAMATE see PNF500
N-PROPYL-N-ETHYL-N-(N-BUTYL)THIOLCARBAMATE see PNF500
PROPYL ETHYLBUTYLTHIOLCARBAMATE see PNF500
N,N-PROPYL ETHYL CARBAMATE see EPC050
PROPYL ETHYL ETHER see EPC125
S-(N-PROPYL)-N-ETHYL-N-N-BUTYLTHIOCARBAMATE see PNF500
PROPYL FORMATE (DOT) see PNM500
n-PROPYL FORMATE see PNM500
PROPYLFORMIC ACID see BSW000
N-n-PROPYL-N-FORMYLHYDRAZINE see PNM650
n-PROPYL GALLATE see PNM750
PROPYL GALLATE see PNM750
4-PROPYLGUAIACOL see MFM750
p-n-PROPYLGUAIACOL see MFM750
p-PROPYLGUAIACOL see MFM750
2-PROPYLHEPTANIC ACID see NMX500
2-PROPYLHEPTANSAEURE (GERMAN) see NMX500
PROPYLHEXADRINE see PNN300

PROPYLHEXEDRINE see PNN400
PROPYLHEXEDRINE HYDROCHLORIDE see PNN300
N-PROPYLHYDRAZINE HYDROCHLORIDE see PNO000
PROPYL HYDRIDE see PMJ750
N-PROPYL-N-(HYDROPEROXYMETHYL)NITROSAMINE see HIE570
n-PROPYL p-HYDROXYBENZOATE see HNU500
PROPYL p-HYDROXYBENZOATE see HNU500
PROPYL-p-HYDROXYBENZOATE, SODIUM SALT see PNO250
PROPYL(4-HYDROXYBUTYL)NITROSAMINE see NLN000
PROPYLIC ALCOHOL see PND000
PROPYLIC ALDEHYDE see PMT750
n-PROPYLIDENE BUTYRALDEHYDE see HBI800
PROPYLIDENE CHLORIDE see DGF400
3-PROPYLIDENE-1(3H)-ISOBENZOFURANONE see PNO500
PROPYLIDENE PHTHALIDE see PNO500
i-PROPYL IODIDE see IPS000
n-PROPYL IODIDE see PNO750
1-PROPYL ISOCYANATE see PNP000
m-PROPYL ISOCYANATE see PNP000
PROPYL ISOCYANATE see PNP000
n-PROPYL ISOMER see PNP250
PROPYL ISOMER see PNP250
2-PROPYLISONICOTINYLTHIOAMIDE see PNW750
PROPYL KETONE see DWT600
PROPYL LITHIUM see PNP275
2-PROPYL MERCAPTAN see IMU000
N-PROPYL MERCAPTAN see PML500
PROPYL MERCAPTAN see PML500
6-PROPYLMERCAPTOPURINE see PNW800
PROPYL METHANOATE see PNM500
PROPYLMETHANOL see BPW500
1-PROPYL-3-METHOXY-4-HYDROXYBENZENE see MFM750
PROPYLMETHYLCARBINYLALLYL BARBITURIC ACID SODIUM SALT see SBN000
PROPYLMETHYLCARBINYLETHYL BARBITURIC ACID SODIUM SALT see NBU000
4-PROPYL-1,2-METHYLENEDIOXYBENZENE see DMD600
6-PROPYL-MP see PNW800
N-PROPYL-N'-(p-CHLOROBENZENESULFONYL)UREA see CKK000
N-PROPYL-N'-p-CHLORPHENYLSULFONYLCARBAMIDE see CKK000
n-PROPYL NITRATE see PNQ500
PROPYL NITRATE see PNQ500
PROPYL NITRITE see PNQ750
4-(PROPYLNITROSAMINO)-1-BUTANOL see NLN000
PROPYLNITROSAMINOMETHYL ACETATE see PNR250
1-(PROPYLNITROSAMINO)PROPYL ACETATE see ABT750
N-PROPYLNITROSOHARNSTOFF (GERMAN) see NLO500
1-PROPYL-1-NITROSOUREA see NLO500
N-PROPYLNITROSOUREA see NLO500
N-PROPYL-N-NITROSOURETHANE see PNR500
N-PROPYL-N'-NITRO-N-NITROSOGUANIDINE see NLD000
N-2-PROPYL-N'-PHENYL-p-PHENYLENEDIAMINE see PFL000
PROPYL OCTANOATE see PNR600
19-PROPYLORVINOL see EQO450
PROPYLORVINOL HYDROCHLORIDE see EQO500
PROPYLOWY ALKOHOL (POLISH) see PND000
PROPYLPARABEN (FCC) see HNU500
PROPYLPARASEPT see HNU500
2-PROPYLPENTAMIDE see PNX600
2-PROPYLPENTANOIC ACID see PNR750
2-PROPYLPENTANOYLTROPINIUM METHYLBROMIDE see LJS000
n-PROPYL PERCARBONATE see DWV400
N-(4-PROPYLPHENAZOL-5-YL)-2-ACETOXYBENZAMIDE see PNR800
PROPYL PHENETHYL ACETAL see PDD400
1-(α-PROPYLPHENETHYL)PYRROLIDINE HYDROCHLORIDE see PNS000
o-PROPYLPHENOL see PNS250
p-PROPYLPHENOL see PNS500
PROPYLPHYLLIN see DNC000
2-PROPYLPIPERIDINE see PNT000
β-PROPYLPIPERIDINE see PNT000
1-PROPYL-4-PIPERIDYL-p-AMINOBENZOATE HYDROCHLORIDE see PNT500
1-PROPYL-4-PIPERIDYL BENZOATE HYDROCHLORIDE see PNT750
6-(PROPYLPIPERONYL)-BUTYL CARBITYL ETHER see PIX250
6-PROPYLPIPERONYL BUTYL DIETHYLENE GLYCOL ETHER see PIX250
N-PROPYL-1-PROPANAMINE see DWR000
PROPYL PROPANOATE see PNU000
n-PROPYL PROPIONATE see PNU000
PROPYL PROPIONATE see PNU000
1-PROPYL-6-((p-(p-((1-PROPYLQUINOLINIUM-6-YL)CARBAMOYL)BENZAMIDO)BENZAMIDO)QUINLINIUM), DI-p-TOLUENE-SULFONATE see PNV250
PROPYL-2-PROPYNYLPHENYLPHOSPHONATE see PNV750
O-n-PROPYL O-(2-PROPYNYL) PHENYLPHOSPHONATE see PNV750
2-PROPYL-4-PYRIDINECARBOTHIOAMIDE see PNW750
PROPYL 4-PYRIDYL KETONE see PNV755

n-PROPYLSELENINIC ACID see PNV760
PROPYL SILANE see PNV775
PROPYL SODIUM see PNV800
1-PROPYL THEOBROMINE see PNW250
((PROPYLTHIO)-5-1H-BENZIMIDAZOLYL-2) CARBAMATE de METHYLE
  (FRENCH) see VAD000
(5-(PROPYLTHIO)-1H-BENZIMIDAZOL-2-YL)CARBAMIC ACID METHYL ESTER
  see VAD000
2-PROPYL-4-THIOCARBAMOYLPYRIDINE see PNW750
5-(PROPYLTHIO)-2-CARBOMETHOXYAMINOBENZIMIDAZOLE see VAD000
2-PROPYL-THIOISONICOTINAMIDE see PNW750
6-(PROPYLTHIO)PURINE see PNW800
6-PROPYL-2-THIO-2,4(1H,3H)PYRIMIDINEDIONE see PNX000
PROPYL THIOPYROPHOSPHATE see TED500
PROPYL-THIORIST see PNX000
1-(PROPYLTHIO)-2,3,6-TRIMETHYL-6-HEPTEN-4-YN-3-OL see TME260
4-PROPYL-2-THIOURACIL see PNX000
6-N-PROPYL-2-THIOURACIL see PNX000
6-PROPYL-2-THIOURACIL see PNX000
6-N-PROPYLTHIOURACIL see PNX000
PROPYLTHIOURACIL see PNX000
PROPYL-THYRACIL see PNX000
PROPYLTRICHLOROSILANE (DOT) see PNX250
n-PROPYLTRICHLOROSILANE see PNX250
n-PROPYL-3,4,5-TRIHYDROXYBENZOATE see PNM750
5-PROPYL-4-(2,5,8-TRIOXA-DODECYL)-1,3-BENZODIOXOL (GERMAN) see
  PIX250
4-PROPYL-2,6,7-TRIOXA-1-STIBABICYCLO(2.2.2)OCTANE see PNX500
1-PROPYLUREA see PNX550
N-PROPYLUREA see PNX550
PROPYLUREA see PNX550
1-PROPYLUREA and SODIUM NITRITE see SIT500
n-PROPYLUREA and SODIUM NITRITE see SIT500
PROPYL URETHANE see PNG250
PROPYL-2-VALERAMIDE see PNX600
2-PROPYLVALERAMIDE see PNX600
α-PROPYLVALERAMIDE see PNX600
2-PROPYLVALERIC ACID see PNR750
2-PROPYLVALERIC ACID CALCIUM SALT (2:1) see CAY675
2-PROPYLVALERIC ACID SODIUM SALT see PNX750
1-(2-PROPYLVALERYL)PIPERIDINE see PNX800
2-PROPYNAL (9CI) see PMT250
2-PROPYN-1-AMINE see POA000
PROPYNE (OSHA) see MFX590
1-PROPYNE-3-OL see PMN450
3-PROPYNETHIOL see PNY275
2-PROPYNOIC ACID see PMT275
PROPYNOIC ACID see PMT275
3-PROPYNOL see PMN450
2-PROPYN-1-OL see PMN450
2-PROPYNYL ALCOHOL see PMN450
2-PROPYNYLAMINE see POA000
1-PROPYNYL COPPER(I) see POA100
1-(2-PROPYNYL)CYCLOHEXANOL CARBAMATE see POA250
PROPYNYLCYCLOHEXANOL CARBAMATE see POA250
1-(2-PROPYNYL)CYCLOHEXYL CARBAMATE see POA250
2-PROPYNYL ETHER see POA500
m-(2-PROPYNYLOXY)PHENYL ESTER METHYLCARBAMIC ACID see PMN250
2-(2-PROPYNYLOXY)PHENYL METHYLCARBAMATE see MIB500
3-(2-PROPYNYLOXY)PHENYL-N-METHYLCARBAMATE see PMN250
2-PROPYNYL(2E,4E)-3,7,11-TRIMETHYL-2,4-DODECADIENOATE see POB000
PROP-2-YNYL-3,7,11-TRIMETHYL-2,4-DODECADIENOATE see POB000
2-PROPYNYL VINYL SULFIDE see POB250
PROPYON see PMY300
PROPYPERONE see FLN000
PROPYPHENAZONE see INY000
N-3'-a-PROPYPHENAZONYL-2-ACETOXYBENZAMIDE see PNR800
PROPYTHIOURACIL see PNX000
PROPYZAMIDE see DTT600
PROQUANIL see MQU750
PROQUAZONE see POB300
P. ROQUEFORTI TOXIN see PAQ875
PRORALONE-MOP see XDJ000
PROREX see DQA400, PMI750
.PROSCILLAN see POB500
PROSCILLARIDIN see POB500
PROSCOMIDE see SBH500
PROSEPTINE see SNM500
PROSEPTOL see SNM500
PROSERIN see DQY909
PROSERINE see POD000
PROSERINE BROMIDE see POD000
PROSERINE METHYL SULFATE see DQY909
PROSEROUT see AAE500

PROSERYL see DPE000
PROSEVOR 85 see CBM750
PROSIL 248 see TNJ500
PRO-SONIL see TDA500
PROSPASMIN see PGP500
PROSPASMINE see PGP500
PROSPASMINE HYDROCHLORIDE see PGP500
PROSPIDIN see POC000
PROSPIDINE see POC000
PROSPIDIUM CHLORIDE see POC000
(+)-PROSTAGLANDIN A² see MCA025
PROSTAGLANDIN A2 see MCA025
PROSTAGLANDIN E1 see POC350
PROSTAGLANDIN E2 see DVJ200
(15S)-PROSTAGLANDIN E2 see DVJ200
(−)-PROSTAGLANDIN E2 see DVJ200
PROSTAGLANDIN F2-α see POC500
PROSTAGLANDIN F2-α METHYL ESTER see DVJ100
PROSTAGLANDIN F2-α-THAM see POC750
PROSTAGLANDIN F2-α THAM SALT see POC750
PROSTAGLANDIN F2a TROMETHAMINE see POC750
PROSTALMON F see POC500
PROSTANDIN see POC350
PROSTAPHILIN see MNV250
PROSTAPHILIN A see SLJ050
PROSTAPHLIN-A see SLJ000
PROSTAPHLYN see DSQ800
PROSTARMON F see POC500
PROSTEARIN see SLL000
PROST-13-EN-1-OIC ACID, 11,16-DIHYDROXY-16-METHYL-9-OXO-, METHYL
  ESTER, (11-α-13E)-(±)- see MJE775
PROSTETIN see ELF100
PROSTIGMIN see NCL100
PROSTIGMIN BROMIDE see POD000
PROSTIGMINE see NCL100
PROSTIGMINE BROMIDE see POD000
PROSTIGMINE METHYLSULFATE see DQY909
PROSTIN see CCC100
PROSTIN E2 see DVJ200
PROSTIN F2-α see POC500
PROSTIN VR see POC350
PROSTOGLANDIN E₂-METHYLHESPERIDIN COMPLEX see KHK100
PROSTOSIN see POB500
PROSTRUMYL see MPW500
PROSULTHIAMINE see DXO300
PROSULTIAMINE see DXO300
PROSYKLIDIN see CPQ250
PROSZIN see POB500
PROTABEN P see HNU500
PROTABOL see TFK300
PROTACELL 8 see SEH000
PROTACHEM 630 see PKF000
PROTACHEM GMS see OAV000
PROTACHEM SMP see MRJ800
PROTACHEM SOC see SKV170
PROTACINE see POF550
PROTACTINIUM see POD500
PROTACTYL see DQA600
PROTAGENT see PKQ250
(−)-PROTAGLANDIN E1 see POC350
PROTALBA see POF000
PROTAMINE SULFATE see POD750
PROTAMINE ZINC INSULIN see IDF325
PROTAMINE ZINC INSULIN INJECTION see IDF325
PROTAMINE ZINC INSULIN SUSPENSION see IDF325
PROTANABOL see PAN100
PROTANAL see SEH000
PROTANDREN see AOO475
PROTARS see CAM300
PROTASIN see POB500
PROTASORB L-20 see PKG000
PROTASORB O-20 see PKL100
PROTATEK see SEH000
PROTAZINE see DQA400, PMI750
PROTEASE, BACILLUS SUBTILIS NEUTRAL see BAC020
PROTECT see NAQ000
PROTECTON see IBQ100
PROTECTONA see DKA600
PROTEINASE, ASPERFILLUS ALKALINE see SBI860
PROTEINASE, ASPERGILLUS TERRICOLA NEUTRAL see TBF350
PROTEK Q see QAT520
PROTELINE see PMJ100
PROTENAZA 1 see PMJ100
PROTERGURIDE see DJX300
PROTERNOL see DMV600

PROTESINE DMU see DTG700
PROTEX (POLYMER) see AAX250
PROTHAZIN see DQA400
PROTHAZIN METHOSULFATE see MRW000
PROTHEOBROMINE see HNY500
PROTHEOPHYLLINE see DNC000
PROTHIADEN see DYC875
PROTHIADENE HYDROCHLORIDE see DPY200
cis-PROTHIADENE-S-OXIDE HYDROGEN MALEATE see POD800
PROTHIADEN HYDROCHLORIDE see DPY200
PROTHIADEN SPOFA see DYC875
PROTHIL see MBZ100
PROTHIOCARB see EIH500
PROTHIONAMIDE see PNW750
PROTHIOPHOS see DGC800
PROTHIPENDYL see DYB600
PROTHIPENDYL HYDROCHLORIDE see DYB800
PROTHIUCIL see PNX000
PROTHIURONE see PNX000
PROTHIZINIC ACID see POE100
PROTHOATE see IOT000
PROTHRIN see POD875
PROTHROMADIN see WAT200
PROTHROMBIN see WAT220
PROTHYCIL see PNX000
PROTHYRAN see PNX000
PROTIOAMPHETAMINE see AOA250
PROTION see PNW750
PROTIONAMID see PNW750
PROTIONAMIDE see PNW750
PROTIONIZINA see PNW750
PROTIRELIN see TNX400
PROTIRELIN TARTRATE see POE050
PROTIURAL see PNX000
PROTIVAR see AOO125
PROTIZINIC ACID see POE100
PROTOAT (HUNGARIAN) see IOT000
PROTOBOLIN see DAL300
PROTOCATECHUALDEHYDE DIMETHYL ETHER see VHK000
PROTOCATECHUIC ACID see POE200
PROTOCATECHUIC ACID, 3-METHYL ESTER see VFF000
PROTOCATECHUIC ALDEHYDE DIMETHYL ETHER see VHK000
PROTOCATECHUIC ALDEHYDE ETHYL ETHER see EQF000
PROTOCATECHUIC ALDEHYDE METHYLENE ETHER see PIW250
PROTOCHLORURE D'IODE (FRENCH) see IDS000
PROTOCOL C see DTG000
PROTOMIN see AOO800
PROTOPAM CHLORIDE see FNZ000
PROTOPAM IODIDE see POS750
PROTOPET see MQV750
PROTOPHENICOL see CDP500
PROTOPINE see FOW000
PROTOPORPHYRIN DISODIUM see DXF700
PROTOPORPHYRIN SODIUM see DXF700
PROTOPORPHYRIN SODIUM SALT see DXF700
PROTOPYRIN see EEM000
PROTOTYPE III SOFT see PKQ059
PROTOVERATRIN see POF000
PROTOVERATRINE see POF000
PROTOVERATRINE A see POF000
PROTOVERATRINE B see NCI600
PROTOX TYPE 166 see ZKA000
PROTOXYL see ARA500
PROTRIPTYLINE see DDA600
PROTRIPTYLINE HYDROCHLORIDE see POF250
PROTRYPTYLINE see DDA600
PROVAMYCIN see SLC000
PROVASAN see EQQ100
PROVENTIL see BQF500
PROVEST see POF275
PROVIGAN see DQA400
PROVITAMIN D$_3$ see DAK600
PROVITAMIN D see CMD750
PROVITAR see AOO125
PROWL see DRN200
PROXAGESIC see DAB879, PNA500
PROXAN SODIUM see SIA000
PROXAZOLE CITRATE see POF500
PROX DW see DTG000
PROXEL PL see BCE475
PROXEN see MFA500
PROXIL see POF550
PROXIMPHAM (GERMAN) see PEQ500
PROXIPHYLLINE see HOA000
PROXITANE 4002 see PCL500

PROX M 3R see MCB050
PROXOL see TIQ250
PROX RX see TCJ900
PROXYPHYLLINE see HOA000
PROZIL see CKP250
PROZIME 10 see SBI860
PROZIN see CKP250
PROZINEX see PMN850
PROZOIN see PMU750
PROZORIN see PLY275
PRT see POE100, POF800
PR TOXIN see PAQ875, POF800
PR TOXINE see POF800
PRULET see PDO750
PRUMYCIN see AFI500
PRUNACETIN A see POG000
PRUNOLIDE see CNF250
PRUNUS (VARIOUS SPECIES) see AQP890
PRURALGAN see DNX400
PRURALGIN see DNX400
PRUSSIAN BLUE see IGY000
PRUSSIAN BROWN see IHD000
PRUSSIC ACID see HHS000
PRUSSIC ACID, UNSTABILIZED see HHS000
PRUSSITE see COO000
PRX 1195 see SMQ500
PRYNACHLOR see CDS275
PRYSKYRICE MH see MCB050
PRYSOLINE see DBB200
PRZEDZIORKOFOS (POLISH) see PDC750
PS 1 see AHE250, SMQ500
PS 2 see SMQ500
PS 200 see SMQ500
PS 209 see SMQ500
PS 454H*see SMQ500
PS 802 see DXG700
PS 2383 see TKX250
PS 100 (carbonate) see CAT775
PS see CKN500
PS-B see SMQ500
PSB-C see SMQ500
PSB-S 40 see SMQ500
PSB-S see SMQ500
PSB-S-E see SMQ500
PSC-801 see POB500
PSC CO-OP WEEVIL BAIT see DXE000
PSEUDECHIS AUSTRALIS VENOM see ARV000
PSEUDECHIS PORPHYRIACUS (AUSTRALIA) VENOM see ARV250
PSEUDECHIS PORPHYRIACUS VENOM see ARV250
PSEUDOACETIC ACID see PMU750
PSEUDOACONITINE see POG250
PSEUDOBUTYLBENZENE see BQJ250
PSEUDOBUTYLENE see BOW255
PSEUDO-BUTYLENE see BOW500
PSEUDOCEF see CCS550
PSEUDOCUMENE see TLL750
PSEUDOCUMIDINE see TLG250
PSEUDOCUMIDINE HYDROCHLORIDE see TLG750
PSEUDOCUMOL see TLL750
p-PSEUDOCUMOQUINONE see POG400
PSEUDOCYANURIC ACID see THS000
PSEUDODIGITOXIN see GEU000
l-(+)-PSEUDOEPHEDRINE see POH000
PSEUDOEPHEDRINE see POH000
1-PSEUDOEPHEDRINE HYDROCHLORIDE see POH500
d-PSEUDOEPHEDRINE HYDROCHLORIDE see POH250
l(+)-PSEUDOEPHEDRINE HYDROCHLORIDE see POH250
(—)-PSEUDOEPHEDRINE HYDROCHLORIDE see POH500
PSEUDOGINSENOSIDE D see PAF450
PSEUDOHEXYL ALCOHOL see EGW000
PSEUDOIONONE see POH525
PSEUDO-α-ISOMETHYL IONONE see TMJ100
PSEUDOLARIC ACID A see POH550
PSEUDOLARIC ACID B see POH600
PSEUDOMETHYLIONONE see DRR700
PSEUDOMONAS AERUGINOSA ENDOTOXIN see POH620
PSEUDOMONAS AERUGINOSA EXOTOXIN see POH630
PSEUDOMONAS AERUGINOSA PHOSPHOLIPASE C see POH640
PSEUDOMONAS AERUGINOSA TOXIN see POH650
PSEUDOMONAS sp. No. 14 POLYSACCHARIDE see POH670
PSEUDOMONAS sp. No. 16 POLYSACCHARIDE see POH680
PSEUDOMONAS POLYSACCHARIDE see PJA200
PSEUDOMONAS PSEUDOMALLEI EXOTOXIN see POH660
PSEUDOMONAS SYRINGAE, ICE-NUCLEATION-ACTIVE, Strain 3/a see IAC100

PSEUDOMONIL see CCS550
PSEUDONAJA TEXTILIS VENOM see TEG650
PSEUDONOREPHEDRINE see NNM510
PSEUDOPINEN see POH750
PSEUDOPINENE see POH750
PSEUDOUREA see USS000
PSEUDOUREA, 2-(p-CHLOROBENZYL)-2-THIO-, MONOHYDROCHLORIDE see CEQ800
PSEUDOUREA, 2-METHYL-2-THIO-, SULFATE see MPV790
PSEUDOXANTHINE see XCA000
PSEUDOTHEOPHYLLINE see TEP000
PSEUDOTHIOUREA see ISR000
PSEUDOTHYMINE see MQI750
PSEUDOTROPINE BENZOATE HYDROCHLORIDE see TNS200
PSICAINE-NEU HYDROCHLORIDE see NCJ000
PSICAIN-NEW HYDROCHLORIDE see NCJ000
PSICHIAL see MDQ250
PSICODISTEN see BET000
PSICOPAX see CFC250, CFZ000
PSICOPERIDOL-R see TKK500
PSICOPLEGIL see MNM500
PSICOSAN see LFK000, MDQ250
PSICOSEDINA see MNM500
PSICOSTEN see PDN000
PSICOSTERONE see AOO450
PSICRONIZER see NMV725
PSIDIUM GUAJAVA Linn., extract excluding roots see POH800
PSIDIUM GUAJAVA see GLW000
PSILOCINE see HKE000
PSILOCIN PHOSPHATE ESTER see PHU500
PSILOCIPIN see PHU500
PSILOTSIBIN see PHU500
PSILOTSIN see HKE000
PSIQUIM see CGA000
PSL see LEN000
PSML see PKG000
PSORADERM see MFN275
PSP see PDO800
PSP (INDICATOR) see PDO800
PS 2 (POLYMER) see SMQ500
PS 5 (POLYMER) see SMQ500
PSV-L 1 see SMQ500
PSV-L 2 see SMQ500
PSV-L see SMQ500
PSV-L 1S see SMQ500
PSYCHAMINE A 66 see PMA750
PSYCHAMINE A 66 HYDROCHLORIDE see MNV750
PSYCHEDRINE see BBK000
PSYCHEDRINUM see AOB250
PSYCHEDRYNA see AOB250
PSYCHODRINE see BBK500
PSYCHOLIQUID see TEH500
PSYCHOPERIDOL see TKK500
PSYCHO-SOMA see MAG050
PSYCHOTABLETS see TEH500
PSYCHOZINE see CKP500
PSYCOPERIDOL HYDROCHLORIDE see TKK750
PSYMOD see PII500
PSYTOMIN see PCK500
PT 01 see OGI300
PT-1 see TBX300
PT-2 see FAS100
PT 155 see DAD040
P 2020T see PJS750
P 2050T see PJS750
P 4007T see PJS750
PTAB see TNI500
PTAP see AON000
PTAQUILOSIDE see POI100
PTB see TMV500
PTC see PCC000, PGN250
PTEGLU see FMT000
PTERIDIUM AQUILINUM see BML000
PTERIDIUM AQUILINUM TANNIN see BML250
PTERIS AQUALINA see BML000
PTEROFEN see UVJ450
PTEROPHENE see UVJ450
PTEROYL-I-GLUTAMIC ACID see FMT000
PTEROYLGLUTAMIC ACID see FMT000
PTEROYL-I-MONOGLUTAMIC ACID see FMT000
PTEROYLMONOGLUTAMIC ACID see FMT000
PTFE see TAI250
PTG see EQP000
PTH see PAK250
PTMS see AKQ000

PTMSA see AKQ000
PTO see HOB500
PTS 2 see PJS750
PTT 119 see POI550
PTU see PGN250
PTU (thyreostatic) see PNX000
PTX see PAE875
PU 603 see CJN250
P 4007EU see PJS750
PUA-HOKU (HAWAII) see SLJ650
PUA KALAUNU (HAWAII) see COD675
PUDDING-PIPE TREE see GIW300
PUD (HERBICIDE) see DTP400
PUERARIA PHASEOLOIDES (Roxb.) Benth., extract excluding roots see POI575
PUFFER POISON, HYDROCHLORIDE see POI600
PUKE WEED see CCJ825
PUKIAWE-LEI (HAWAII) see RMK250
PULANOMYCIN see FBP300
PULARIN see HAQ550
PULEGON see POI615
(+)-(R)-PULEGONE see POI615
(R)-(+)-PULEGONE see POI615
d-PULEGONE see MCF500, POI615
PULEGONE see MCF500, POI615
(+)-PULEGONE see POI615
PULICARIA ANGUSTIFOLIA DC., extract see POI650
PULMICORT see BOM520
PULMOCLASE see CBR675
PULSAN see ICA000
PULSOTYL see FMS875
PUNICINE see PAO500
PURADIN see NGG500
PURAGEL see PCU360
PURALIN see TFS350
PURAPURIDINE see POJ000
PURASAN-SC-10 see ABU500
PURATIZED see TNI500
PURATIZED AGRICULTURAL SPRAY see TNI500
PURATIZEDAT AGRICULTURAL SPRAY see TNI500
PURATIZED B-2 see MDD500
PURATIZED N5E see TNI500
PURATRONIC CHROMIUM CHLORIDE see CMJ250
PURATRONIC CHROMIUM TRIOXIDE see CMK000
PURATURF 10 see ABU500
PURATURF see TNI500
PURECAL see CAT775
PURECALO see CAT775
PURE CHRYSOIDINE YBH see PEK000
PURE CHRYSOIDINE YD see PEK000
PURE EOSINE YY see BNH500
PURE LEMON CHROME L3GS see LCR000
PURE ORANGE CHROME M see LCS000
PURE ORANGE II S see CMM220
PURE QUARTZ see SCI500, SCJ500
PUREX see SFT000
PURE ZINC CHROME see ZFJ100
PURGA see PDO750
PURGEN see PDO750
PURGING BUCKTHORN see MBU825
PURGING FISTULA see GIW300
PURGOPHEN see PDO750
PURIFILIN see DNC000
PURIMETHOL see POK000
1H-PURIN-6-AMINE see AEH000
1H-PURIN-6-AMINE, N-(PHENYLMETHYL)-(9CI) see BDX090
1H-PURIN-6-AMINE PHOSPHATE see POJ100
1H-PURIN-6-AMINE, SULFATE see AEH500
PURIN B see SHU500
7H-PURINE see POJ250
9H-PURINE see POJ250
PURINE see POJ250
1H-PURINE-2-AMINE, 6-((1-METHYL-4-NITRO-1H-IMIDAZOL-5-YL)-THIO)- see AKY250
1H-PURINE-2,6-DIAMINE see POJ500
PURINE-2,6-DIOL see XCA000
9H-PURINE-2,6-DIOL see XCA000
2,6(1,3)-PURINEDION see XCA000
PURINE-2,6-(1H,3H)-DIONE see XCA000
1H-PURINE-2,6-DIONE, 3,7-DIHYDRO-1,7-DIMETHYL-(9CI) see PAK300
1H-PURINE-2,6-DIONE, 3,7-DIHYDRO-1,3-DIMETHYL-, SODIUM SALT (9CI) see SKH000
1H-PURINE-2,6-DIONE, 8-ETHOXY-3,7-DIHYDRO-1,3,7-TRIMETHYL-(9CI) see EEO000
PURINE-3-OXIDE see POJ750

9-PURINE RIBONUCLEOSIDE see RJF000
PURINE RIBONUCLEOSIDE see RJF000
PURINE RIBOSIDE see RJF000
3H-PURINE-6-THIOL see POK000
6-PURINETHIOL see POK000
PURINE-6-THIOL see POK000
6H-PURINE-6-THIONE, 2-AMINO-1,7-DIHYDRO- (9CI) see AMH250
6H-PURINE-6-THIONE, 1,7-DIHYDRO-, MONOHYDRATE (9CI) see MCQ100
PURINETHOL see POK000
1H-PURINE-2,6,8(3H)-TRIONE, 7,9-DIHYDRO- (9CI) see UVA400
1H-PURINE-2,6,8(3H)-TRIONE, 7,9-DIHYDRO-, MONOSODIUM SALT (9CI) see
   SKO575
9H-PURIN-6-OL see DMC000
6(1H)-PURINONE see DMC000
PURIN-6(3H)-ONE see DMC000
PURIN-6-THIOL, MONOHYDRAT see MCQ100
6H-PURIN-6-THION, MONOHYDRAT see MCQ100
PURIVEL see MQR225
PUROCYCLINA see TBX000
PUROMYCIN see AEI000
PUROMYCIN CHLOROHYDRATE see POK250
PUROMYCIN HYDROCHLORIDE see POK250
PUROPHYLLIN see HOA000
PUROSIN-TC see POB500
PUROSTROPHAN see OKS000
PUROVERINE see POF000
PURPLE ALLAMANDA see ROU450
PURPLE MINT PLANT EXTRACT see PCI750
PURPLE 4R see FAG050
PURPLE RED see FMU080
PURPUREA B see DAD650
PURPUREA GLYCOSIDE B see DAD650
PURPUREAGLYKOSID B (GERMAN) see DAD650
PURPURID see DKL800
PURPURIN see TKN750
PURPURINE see TKN750
PURPUROCATECHOL see TNV550
PURPUROGALLIN see TDD500
PURSERPINE see RDK000
PURTALC USP see TAB750
PUTCHUK see CNT350
PUTRESCIN see BOS000
PUTRESCINE see BOS000
PUTRESCINE DIHYDROCHLORIDE see POK325
PUTRESCINE HYDROCHLORIDE see POK325
PVA 008 see PKP750
PVA see AAF625, PKP750
PVBR see PKQ000
PVC (MAK) see PKQ059
P.V. CARBACHOL see CBH250
P. V. CARPINE LIQUIFILM see PIF500
PVC CORDO see AAX175
PV FAST ORANGE GRL see CMU820
PVK see PDT750
PV-ORANGE G see CMS145
PVP 1 see PKQ500
PVP 2 see PKQ750
PVP 3 see PKR000
PVP 4 see PKR250
PVP 5 see PKR500
PVP 6 see PKR750
PVP 7 see PKS000
PVP (FCC) see PKQ250
PVP-IODINE see PKE250
PVPO see PKQ100
PVP 8T see PJS750
PVS 4 see PKP750
PVSK see PKS250
P. VULGARIS see PAM780
PWE of H. INFLUENZAE see HAB710
PWP 8 see MCB050
PWP 15 see MCB050
PX 104 see DEH200
PX-138 see DVL600
PX-238 see AEO000
PX 404 see DEH600
PX 438 see BJS250
PX-538 see DVK800
PX-806 see FAB920
PXO see OMY850
PY 100 see PJS750
PY2763 see SMQ500
PYBUTHRIN see PIX250
PYCARIL see NCR040
PYCAZIDE see ILD000

PYDOXAL see PII100
PYDRIN see FAR100
PYDT see DJY000
PYELOKON-FR see AAN000
PYELOSIL see DNG500
PYKARYL see NCR040
PYKNOLEPSINUM see ENG500
PYLAPRON see ICC000
PYLOSTROPIN see MGR500
PYLUMBRIN see DNG400
PYMAFED see DBM800, WAK000
PYNACOLYL METHYLFLUOROPHOSPHONATE see SKS500
PYNAMIN see AFR250
PYNAMIN-FORTE see AFR250
PYNDT see DTV200
PYNOSECT see BEP500
PYOCEFAL see CCS550
PYOKTANIN see AOR500
PYOLUTEORIN see POK400
PYOPEN see CBO250
PYOPENE see CBO250
PYOSTACINE see VRF000
PYOVERM see AOR500
PYQUITON see BGB400
PYRA see WAK000
PYRABENZAMINE see POO750
PYRABITAL see AMK250
PYRACETAM see NNE400
PYRACETON see DNG400
PYRACRYL ORANGE Y see PEK000
PYRACRYMYCIN-1 see COV125
PYRADEX see DBI200
2H-PYRAN, 3,6-DIHYDRO-6-BUTYL-2,4-DIMETHYL- see GMG100
PYRADONE see DOT000
PYRAHEXYL see HGK500
PYRALCID see ALF250
PYRALENE 1476 see PAV600
PYRALENE see PJL750
PYRALIN see CCU250
PYRAMAL see WAK000
PYRA MALEATE see DBM800
PYRAMEM see NNE400
PYRAN, 2-(2-METHYL-1-PROPENYL)-4-METHYLTETRAHYDRO- see RNU000
PYRAMIDON see DOT000
PYRAMIDONE see DOT000
PYRAMINE see PEE750
PYRAMIN RB see PEE750
PYRAMON see AMK250
PYRAN ALDEHYDE see ADR500
PYRANILAMINE MALEATE see DBM800
PYRANINYL see DBM800
PYRANISAMINE see WAK000
PYRANISAMINE MALEATE see DBM800
PYRANOL see PJL750
2H-PYRAN-2-ONE, 6-HEPTYLTETRAHYDRO- see HBP450
2H-PYRAN-2-ONE, 4-HYDROXY-6-METHYL- see THL800
4H-PYRAN-4-ONE, TETRAHYDRO-2,6-DIMETHYL- see TCQ350
2H-PYRAN-2-ONE, TETRAHYDRO-6-PROPYL- see ODE300
4H-PYRANO(3,2-c)QUINOLINE-2-CARBOXYLIC ACID, 5,6-DIHYDRO-7,8-DIMETH-
   YL-4,5-DIOXO-, ISOPENTYL ESTER see IHR200
PYRANTEL PAMOATE see POK575
PYRANTEL TARTRATE see TCW750
PYRANTON see DBF750
PYRATHIAZINE see PAJ750
PYRATHIAZINE HYDROCHLORIDE see POL000
PYRATHYN see DPJ400, TEO250
PYRAZALONE ORANGE NP 215 see CMS145
PYRAZAPON see POL475
PYRAZINAMIDE see POL500
PYRAZINE see POL490
PYRAZINEAMIDE see POL500
PYRAZINECARBOXAMIDE see POL500
PYRAZINE CARBOXYLAMIDE see POL500
PYRAZINE HEXAHYDRIDE see PIJ000
PYRAZINOIC ACID AMIDE see POL500
PYRAZINOL-O-ESTER with O,O-DIETHYL PHOSPHOROTHIOATE see EPC500
2-PYRAZINYLETHANETHIOL see POM000
PYRAZODINE see PDC250
PYRAZOFEN see PDC250, PEK250
PYRAZOLANTHRONE see APK000
PYRAZOLATE see POM275
PYRAZOL BLUE 3B see CMO250
PYRAZOLE see POM500
PYRAZOLEANTHRONE see APK000
PYRAZOLE, 3,5-DIMETHYL- see DTU850

3,5-PYRIDINEDICARBOXYLIC ACID, 2-CYANO-1,4-DIHYDRO-6-METHYL-4-(3-NI-TROPHENYL)-, 3-METHYL-5-(1-METHYLETHYL) ESTER see NDY650
3,5-PYRIDINEDICARBOXYLIC ACID, 1,4-DIHYDRO-2,6-DIMETHYL-4-(3-NITRO-PHENYL)-, ETHYL METHYL ESTER see EMR600
2,5-PYRIDINEDICARBOXYLIC ACID, DIPROPYL ESTER see EAU500
PYRIDINE, 2-(α-(2-(DIMETHYLAMINO)ETHOXY)-α-METHYLBENZYL)-, SUCCI-NATE (1:1) see PGE775
PYRIDINE, 2-((2-(DIMETHYLAMINO)ETHYL)(SELENOPHENE-2-YLME-THYL)AMINO)- see DPI750
PYRIDINE, 2,4-DIMETHYL-(9CI) see LIY990
PYRIDINE, 2,6-DIMETHYL-(9CI) see LJA010
2-PYRIDINEETHANAMINE, N-METHYL-(9CI) see HGE820
4-PYRIDINEETHANOL see POR500
PYRIDINE, 2-ETHENYL-(9CI) see VQK560
PYRIDINE, 4-(p-ETHOXYBENZOYL)- see EEN600
PYRIDINE, 4-HEPTANOYL- see HBF600
PYRIDINE, HEXAHYDRO-N-NITROSO- see NLJ500
PYRIDINE, 4-HEXANOYL- see HEW050, PBX800
PYRIDINE, 2-(p-HYDROXYBENZOYL)- see HJN700
2-PYRIDINEMETHANETHIOL see POR790
2-PYRIDINEMETHANOL see POR800
3-PYRIDINEMETHANOL see NDW510
4-PYRIDINEMETHANOL see POR810
PYRIDINE METHIODIDE see MOZ000
PYRIDINE, 3-(4-METHOXYBENZOYL)- see MED100
PYRIDINE, 4-(4-METHOXYBENZOYL)- see MFH930
2-PYRIDINEMETHYLAMINE see ALB750
PYRIDINE, 2-(2-(METHYLAMINO)ETHYL)- see HGE820
PYRIDINE, 3-METHYL-, 1-OXIDE see MOY550
PYRIDINE, 4-METHYL-, 1-OXIDE see MOY790
PYRIDINE, 3-(1-METHYL-2-PYRROLIDINYL)-, (S)-, SULFATE (2:1) see NDR500
3-PYRIDINENITRILE see NDW515
PYRIDINE, 4-(p-NITROBENZYL)- see NFO750
PYRIDINE, 3-(1-NITROSO-2-PIPERIDINYL)-, (S)-(9CI) see NJK150
PYRIDINE, 3-(1-NITROSO-2-PYRROLIDINYL)-, (S)- see NLD500
PYRIDINE-1-OXIDE see POS000
PYRIDINE N-OXIDE see POS000
PYRIDINE-1-OXIDE-3-AZO-p-DIMETHYLANILINE see POS250
PYRIDINE-1-OXIDE-4-AZO-p-DIMETHYLANILINE see POS500
PYRIDINE, 2-(PHENYLMETHYL)-(9CI) see BFG600
PYRIDINE, 3-(2-PIPERIDINYL)-, MONOHYDROCHLORIDE, (S)- see PIT650
2-PYRIDINEPROPANAMINE, γ-(4-CHLOROPHENYL)-N,N-DIMETHYL-(9CI) see CLX300
PYRIDINE, 4-PROPIONYL- see PMX300
PYRIDINE, 1,2,3,6-TETRAHYDRO-1-((6-METHYL-3-CYCLOHEXEN-1-YL)CARBONYL)- see TCV375
PYRIDINE, 1,2,3,6-TETRAHYDRO-1-((2-METHYLCYCLOHEXYL)CARBONYL)- see TCV400
PYRIDINE, 3-(TETRAHYDRO-1-METHYLPYRROL-2-YL) see NDN000
PYRIDINE-2-THIOCARBINOL see POR790
2-PYRIDINETHIOL see TFP300
2-PYRIDINETHIOL, 1-OXIDE see MCQ700
2-PYRIDINETHIOL-1-OXIDE SODIUM SALT see MCQ750
2-PYRIDINETHIOL-1-OXIDE, ZINC SALT see ZMJ000
2(1H)-PYRIDINETHIONE see TFP300
PYRIDINE, 2,4,6-TRIMETHYL- see TME272
PYRIDINE, 4-VALERYL- see VBA100
PYRIDINE, 2-VINYL- see VQK560
PYRIDINE, 4-VINYL- see VQK590
PYRIDINIUM ALDOXIME METHOCHLORIDE see FNZ000
PYRIDINIUM-2-ALDOXIME-N-METHYLIODIDE see POS750
PYRIDINIUM, 3,3′-(2-AMINOTEREPHTHALOYLBIS(IMINO-p-PHENYLENECARBO-NYLIMINO))BIS(1-PROPYL)-, DI-p-TOLUENE SULFONAT see POT000
PYRIDINIUM, 3,3′-(BICYCLO(2.2.2)OCTANE-1,4-DIYLBIS(CARBONYLIMINO-4,1-PHENYLENECARBONYLIMINO))BIS-1-PROPYL- see POT500
PYRIDINIUM, 1-BUTYL-3-(p-(p-((p-((1-BUTYLPYRIDINIUM-3-YL)CARBAMOYL)PHENYL)CARBAMOYL)CINNAMAMIDO)BENZAMIDO)-, see POT750
PYRIDINIUM, 3-CARBAMOYL-1-β-D-RIBOFURANOSYL-, HYDROXIDE, 5′,5′-ES-TER with ADENOSINE 2′-(DIHYDROGEN PHOSPHATE) 5′-(TRIHYDROGEN PYROPHOSPHATE), inner salt see CNF400
PYRIDINIUM, 3-CARBAMOYL-1-β-D-RIBOFURANOSYL-, HYDROXIDE, 5′-ESTER with ADENOSINE 5′-5′-(TRIHYDROGEN PYROPHOSPHATE), inner salt see CNF390
PYRIDINIUM, 3,3′-(2-CHLOROTEREPHTHALOYLBIS(IMINO-p-PHENYLENECAR-BONYLIMINO))BIS(1-METHYL-, DI-p-TOLUENE SULFONAT see POU250
PYRIDINIUM, 3,3′-(2-CHLOROTEREPHTHALOYLBIS(IMINO-p-PHENYLENECAR-BONYLIMINO))BIS(1-ETHYL-, DI-p-TOLUENE SULFONAT see POU000
PYRIDINIUM, 1-DODECYL-, CHLORIDE see DXY725, LBX050
PYRIDINIUM, 1,1′-ETHYLENEDI-, DIBROMIDE see EIT100
PYRIDINIUM, 1-ETHYL-3- p-(-((1-ETHYLPYRIDINIUM-3-YL)CARBAMYL)PHENYLOCARBAMOYL)CINNAMAMIDO)BENZAMIDO)-, DI-p see POV250

PYRIDINIUM, 1-ETHYL-3-(p-((p-(1-ETHYLPYRIDINIUM-3-YL)PHENYL)CARBAMOYL)CINNAMAMIDO)PHENYL)-DI-p-TOLUENE S see POV500
PYRIDINIUM, 2-FORMYL-1-METHYL-,CHLORIDE, OXIME mixed with 1,1′-TRI-METHYLENEBIS(4-FORMYLPYRIDINIUM), DICHLORIDE, DIOXI see POV750
PYRIDINIUM, 2-FORMYL-4′-METHYL-1,1′-(OXYDIMETHYLENE)DI-, DICHLORIDE, OXIME see MBY500
PYRIDINIUM, 1-HEXADECYL-, BROMIDE see HCP800
PYRIDINIUM, 1-(2-HYDROXYETHYLCARBAMOYLMETHYL)-, CHLORIDE, DODE-CANOATE see LBD100
PYRIDINIUM, 3,3′-(2-METHOXYTEREPHTHALOYLBIS(IMINO-p-PHENYLENECAR-BONYLIMINO))BIS(1-PROPYL)-DI-p-TOLUENE SULFONAT see POW500
PYRIDINIUM, 3,3′-(2-METHOXYTEREPHTHALOYLBIS(IMINO-p-PHENYLENECAR-BONYLIMINO))BIS(1-ETHYL)-, DI-p-TOLUENE SULFONA see POW000
PYRIDINIUM, 3,3′-(2-METHOXYTEREPHTHALOYLBIS(IMINO-p-PHENYLENECAR-BONYLIMINO))BIS(1-METHYL-DI-p-TOLUENE SULFONAT see POW250
PYRIDINIUM, 1-METHYL-3-(p-(p-((p-((1-METHYLPYRIDINIUM-4-YL)AMINO)PHENYL)CARBAMOYL)BENZAMIDO)BENZAMIDO)-, DI- see POX500
PYRIDINIUM, 1-METHYL-3-(p-(p-((p-((1-METHYLPYRIDINIUM-3-YL)AMINO)PHENYL)CARBAMOYL)BENZAMIDO)BENZAMIDO)-, DIIO see POX250
PYRIDINIUM, 1-METHYL-4-(p-(p-((p-((1-METHYLPYRIDINIUM-4-YL)AMINO)PHENYL)CARBAMOYL)CINNAMAMIDO)ANILINO)-, DI- see POX750
PYRIDINIUM, 1-METHYL-3-(p-(p-((((1-METHYLPYRIDINIUM-3-YL)CARBAMOYL)1,4-NAPHTHYL)CARBAMOYL)BENZAMIDO)BENZAMI see POY250
PYRIDINIUM, 1-METHYL-3-(m-(p-((p-((1-METHYLPYRIDINIUM-3-YL)CARBAMOYL)PHENYL)CARBAMOYL)BENZAMIDO)BENZAMIDO)-, see POY750
PYRIDINIUM, 1-METHYL-3-(p-(p-((p-(1-METHYLPYRIDINIUM-3-YL)PHENYL)CARBAMOYL)CINNAMAMIDO)PHENYL)-, DI-p-TOLUE see PPA250
PYRIDINIUM, 1-METHYL-3-(p-(p-((p-(1-METHYLPYRIDINIUM-3-YL)PHENYL)CARBAMOYL)BENZAMIDO)BENZAMIDO)-, DI-p-TOLU see PPA000
PYRIDINIUM, 3,3′-(2-METHYLTEREPHTHALOYLBIS(IMINO-p-PHENYLENECAR-BONYLIMINO))BIS(1-PROPYL)-, DI-p-TOLUENE SULFONA see PPB250
PYRIDINIUM, 3,3′-(2-METHYLTEREPHTHALOYLBIS(IMINO-p-PHENYLENECAR-BONYLIMINO))BIS(1-ETHYL-DI-p-TOLUENE SULFONATE see PPA750
PYRIDINIUM, 3,3′-(2-METHYLTEREPHTHALOYLBIS(IMINO-p-PHENYLENECAR-BONYLIMINO))BIS(1-METHYL)-DI-p-TOLUENE SULFONATE see PPB000
PYRIDINIUM, 3,3′-(1,4-NAPHTHYLENEBIS(CARBANYLIMINO-p-PHENYLENECAR-BONYLIMINO))BIS(1-METHYL)-DI-p-TOLUENE SULFO see PPB500
PYRIDINIUM NITRATE see PPB550
PYRIDINIUM, 1-(2-OXO-2-((2-((1-OXODODECYL)OXY)ETHYL)-AMINO)ETHYL)-, CHLORIDE see LBD100
PYRIDINIUM, 1,1′-(OXYBIS(METHYLENE))BIS(2-(HYDROXYIMINO)METHYL)), DI-CHLORIDE (9CI) see PPC000
PYRIDINIUM, 1,1′-(OXYDIMETHYLENE)BIS(2-FORMYL)-, DICHLORIDE DIOXIME see PPC000
PYRIDINIUM PERCHLORATE see PPC100
PYRIDINIUM, 4,4′-(p-PHENYLENEBIS(ACRYLOYLIMINO-p-PHENYLENEIMI-NO))BIS(1-ETHYL-, DI-p-TOLUENE SULFONATE) see PPC250
PYRIDINIUM, 1-PROPYL-3-(p-(p-((p-((1-PROPYLPYRIDINIUM-3-YL)CARBAMOYL)PHENYL)CARBAMOYL)CINNAMAMIDO)BENZAMIDO)- see PPD000
PYRIDINIUM, 1-PROPYL-3-(p-(p-((p-(1-PROPYLPYRIDINIUM-3-YL)PHENYL)CARBAMOYL)CINNAMAMIDO)PHENYL)-, DI-p-TOLUE see PPD250
PYRIDINIUM, 4,4′-(SUBEROYLBIS(IMINO-p-PHENYLENEIMINO))BIS(1-METHYL-), DIBROMIDE see PPD500
PYRIDINIUM, 3,3′-(TEREPHTHALOYLBIS(IMINO(3-CHLORO-p-PHENYL-ENE)CARBONYLIMINO))BIS(1-METHYL-DI-p-TOLUENE SULF see PPE500
PYRIDINIUM, 3,3′-(TEREPHTHALOYLBIS(IMINO(3-METHOXY-p-PHENYL-ENE)CARBONYLIMINO))BIS(1-PROPYL)-, DI-p-TOLUENE see PPG250
PYRIDINIUM, 3,3′-(TEREPHTHALOYLBIS(IMINO(3-METHOXY-p-PHENYL-ENE)CARBONYLIMINO))BIS(1-ETHYL-, DI-p-TOLUENE SU see PPF250
PYRIDINIUM, 3,3′-(TEREPHTHALOYLBIS(IMINO-p-PHENYLENECARBONYLIMI-NO))BIS(1-BUTYL-, DI-p-TOLUENESULFONATE) see PPH000
2-PYRIDINOL see OOO200
3-PYRIDINOL see PPH025
4-PYRIDINOL see HOB100
PYRIDINOL CARBAMATE see PPH050
2-PYRIDINOL, 3,5,6-TRICHLORO-, O-ESTER with O,O-DIETHYL PHOSPHO-ROTHIOATE see CMA100
2-PYRIDINONE see OOO200
2(1H)-PYRIDINONE (9CI) see OOO200
2(1H)-PYRIDINONE, 1-METHYL-(9CI) see MPA075
4(1H)-PYRIDINONE, 1-METHYL-3-PHENYL-5-(3-(TRIFLUOROMETHYL)PHENYL)- see FMQ200
α-((2-PYRIDINYLAMINO)METHYL)-BENZENEMETHANOL MONOHYDROCHLO-RIDE see PGG355
1-(2-PYRIDINYL)ETHANONE see PPH200

1-PYRIDINYLETHANONE see PPH200
4,4′-(2-PYRIDINYLMETHYLENE)BISPHENOL BIS(HYDROGEN SULFATE) (ESTER) DISODIUM SALT see SJJ175
PYRIDIPCA see PPK500
PYRIDIUM see PDC250, PEK250
PYRIDIVITE see PDC250
2-PYRIDIYLAMINE see PPH300
PYRIDO(3′,2′:5,6)CHRYSENE see BCQ000
PYRIDO(3′,2′:3,4)FLUORANTHENE see FDP000
5H-PYRIDO(4,3-b)INDOL-3-AMINE, 1-METHYL-1, MONOACETATE see ALE750
9H-PYRIDO(3,4-b)INDOLE see NNR300
1H-PYRIDO(4,3-b)INDOLE, 2,3,4,5-TETRAHYDRO-2,8-DIMETHYL-5-(2-(6-METHYL-3-PYRIDYL)ETHYL)- see TCQ260
9H-PYRIDO(3,4-b)INDOLE, 1,2,3,4-TETRAHYDRO-6-METHOXY-1-METHYL- see MLJ100
3-PYRIDOL see PPH025
PYRIDOMYCIN see PPI775
2-PYRIDONE see OOO200
PYRIDONE-2 see OOO200
2(1H)-PYRIDONE see OOO200
α-PYRIDONE see OOO200
2(1H)-PYRIDONE, 1-METHYL- see MPA075
PYRIDO(2′,3′:4)PYRENE see PCV500
PYRIDOSTIGMINE see PPI800
PYRIDOSTIGMINE BROMIDE see MDL600
12H-PYRIDO(2,3-a)THIENO(2,3-i)CARBAZOLE see PPI815
PYRIDOXALDEHYDE see PII100
PYRIDOXALDEHYDE PHOSPHATE see PII100
PYRIDOXAL HYDROCHLORIDE see VSU000
PYRIDOXAL MONOPHOSPHATE see PII100
PYRIDOXAL 5-PHOSPHATE see PII100
PYRIDOXAL-5′-PHOSPHATE see PII100
PYRIDOXAL PHOSPHATE see PII100
PYRIDOXIN-5′-DISULFID DIHYDROCHLORID HYDRAT (GERMAN) see BMB000
PYRIDOXINE see PPK250
PYRIDOXINE HYDROCHLORIDE (FCC) see PPK500
PYRIDOXINE-5-PHOSPHATE see PPJ900
PYRIDOXINE-5′-PHOSPHATE see PPJ900
PYRIDOXINE PHOSPHATE see PPJ900
PYRIDOXINIUM CHLORIDE see PPK500
PYRIDOXINUM HYDROCHLORICUM (HUNGARIAN) see PPK500
PYRIDOXOL see PPK250
PYRIDOXOL-5-(DIHYDROGEN PHOSPHATE) see PPJ900
PYRIDOXOL HYDROCHLORIDE see PPK500
PYRIDOXOL-5′-PHOSPHATE see PPJ900
PYRIDOXOL-5-PHOSPHATE see PPJ900
PYRIDOXYL PHOSPHATE see PII100
3-PYRIDOYL HYDRAZINE see NDU500
N-3-PYRIDOYLTRYPTAMINE see NDW525
PYRIDROL see DWK400, PII750
PYRIDROLE see DWK400
3-PYRIDYLALDEHYDE see NDM100
4-(PYRIDYL-2-AMIDOSULFONYL)-3′-CARBOXY-4′-HYDROXYAZOBENZENE see PPN750
3-PYRIDYLAMINE see AMI250
4-PYRIDYLAMINE see AMI500
α-PYRIDYLAMINE see AMI000
α-(2-PYRIDYLAMINOMETHYL)BENZYL ALCOHOL see PGG350
α-((2-PYRIDYLAMINO)METHYL)-BENZYL ALCOHOL HYDROCHLORIDE see PGG355
N-(2-PYRIDYL)BENZYLAMINE HYDROCHLORIDE see BEA000
1-(3-PYRIDYL)BUTANE see BSJ550
1-(4-PYRIDYL)-1-BUTANONE see PNV755
2-PYRIDYLCARBINOL see POR800
3-PYRIDYLCARBINOL see NDW510
4-PYRIDYLCARBINOL see POR810
3-PYRIDYLCARBONITRILE see NDW515
3-PYRIDYLCARBOXALDEHYDE see NDM100
1-(PYRIDYL-3-)-3,3-DIAETHYL-TRIAZEN (GERMAN) see DJY000
1-(3-PYRIDYL)-3,3-DIETHYLTRIAZENE see DJY000
1-PYRIDYL-3,3-DIETHYLTRIAZENE see DJY000
1-(PYRIDYL-3)-3,3-DIETHYLTRIAZENE see DJY000
m-PYRIDYL-DIETHYL-TRIAZENE see DJY000
1-(PYRIDYL-3)-3,3-DIMETHYL-TRIAZEN (GERMAN) see PPL500
1-(m-PYRIDYL)-3,3-DIMETHYL-TRIAZENE see PPL500
1-(PYRIDYL-3)-3,3-DIMETHYL TRIAZENE see PPL500
2-(4-PYRIDYL)ETHANOL see POR500
2-(γ-PYRIDYL)ETHANOL see POR500
(2-(2-PYRIDYL)ETHYL)METHYLAMINE see HGE820
N-(2-(2-PYRIDYL)ETHYL)-PHTHALIMIDE HYDROCHLORIDE see PPM550
1-(4-PYRIDYL)-1-HEPTANONE see HBF600
1-(4-PYRIDYL)-1-HEXANONE see HEW050, PBX800
N-2-PYRIDYLMANDELAMIDE HYDROCHLORIDE see PPM725
2-PYRIDYLMETHANOL see POR800
3-PYRIDYLMETHANOL see NDW510

4-PYRIDYLMETHANOL see POR810
N-α-PYRIDYL-N-p-METHOXYBENZYL-N′,N′-DIMETHYLETHYLENEDIAMINE PHOSPHATE see NBV000
(2-PYRIDYLMETHYL)AMINE see ALB750
7-(2-((3-PYRIDYLMETHYL)AMINO)ETHYL)THEOPHYLLINE NICOTINATE see PPN000
1-(2-(N-(2-PYRIDYLMETHYL)ANILINO)ETHYL)PIPERIDINE HYDROCHLORIDE see CNE375
1-(2-N-(2-PYRIDYLMETHYL)ANILINO)ETHYL)PIPERIDINE TRIPALMITATE see PIC100
2-PYRIDYLMETHYLCHLORIDE HYDROCHLORIDE see PIC000
4,4′-(2-PYRIDYLMETHYLENE)DIPHENOLBIS(HYDROGEN SULFATE) (ESTER) DISODIUM SALT see SJJ175
4,4′-(2-PYRIDYLMETHYLENE)DIPHENOL DIACETATE see PPN100
2-PYRIDYL METHYL KETONE see PPH200
2-PYRIDYLMETHYL MERCAPTAN see POR790
2-PYRIDYLMETHYL 2-(p-(2-METHYLPROPYL)PHENYL)PROPIONATE see IAB000
1-(3-PYRIDYLMETHYL)-3-(4-NITROPHENYL)UREA see PPP750
N-3-PYRIDYLMETHYL-N′-p-NITROPHENYLUREA see PPP750
N-(2-PYRIDYLMETHYL)-N-PHENYL-N-2-(PIPERIDINOETHYL)AMINE HYDROCHLORIDE see CNE375
N-(2-PYRIDYLMETHYL)-N-PHENYL-N-2-(PIPERIDINOETHYL)AMINE TRIPALMITATE see PIC100
β-PYRIDYL-α-N-METHYLPYRROLIDINE see NDN000
N′-α-PYRIDYL-N′-BENZYL-N,N-DIMETHYL ETHYLENEDIAMINE MONOHYDROCHLORIDE see POO750
5-(4-PYRIDYL)-1,3,4-OXADIAZOL-2-OL see PPN500
2-PYRIDYL-(4)-1,3,4-OXADIAZOLON-(5) (GERMAN) see PPN500
1-(PYRIDYL-3-N-OXID)-3,3-DIMETHYL-TRIAZEN (GERMAN) see DTV200
1-(PYRIDYL-3-N-OXIDE)-3,3-DIMETHYLTRIAZENE see DTV200
1-(4-PYRIDYL)-1-PENTANONE see VBA100
4-(2-PYRIDYL)PIPERAZINYL 3,4,5-TRIMETHOXYPHENYL KETONE see TKY300
2-(3′-PYRIDYL) PIPERIDINE see AON875
2-(3-PYRIDYL)-PIPERIDINE see AON875
(−)-2-(3′-PYRIDYL)PIPERIDINE see AON875
1-(4-PYRIDYL)-1-PROPANONE see PMX300
4-(4-PYRIDYL)PYRIDINE see BGO600
2-(3-PYRIDYL)PYRROLIDINE see NNR500
5-(4-(2-PYRIDYLSULFAMOYL)PHENYLAZO)-2-HYDROXYBENZOIC ACID see PPN750
5-(p-(2-PYRIDYLSULFAMOYL)PHENYLAZO)SALICYLIC ACID see PPN750
N′-2-PYRIDYLSULFANILAMIDE see PPO000
N-2-PYRIDYLSULFANILAMIDE see PPO000
N′-2-PYRIDYLSULFANILAMIDE SODIUM SALT see PPO250
N-(α-PYRIDYL)-N-(α-THENYL)-N′,N′-DIMETHYLETHYLENEDIAMINE see TEO250
N-(α-PYRIDYL)-N-(β-THENYL)-N′,N′-DIMETHYLETHYLENEDIAMINE see DPJ200
N-(2-PYRIDYL)-N-(2-THIENYL)-N,N′-DIMETHYL-ETHYLENEDIAMINE HYDROCHLORIDE see DPJ400
PYRILAMINE see WAK000
PYRILAMINE HYDROCHLORIDE see MCJ250
PYRILAMINE MALEATE see DBM800
PYRILAX see PPN100
PYRILENE see PBJ000
PYRIMAL see PPP500
PYRIMAL M see ALF250
PYRIMETHAMINE see MQR100
2-PYRIMIDINAMINE (9CI) see PPH300
PYRIMIDINE see PPO750
PYRIMIDINE, 2-AMINO- see PPH300
PYRIMIDINE-DEOXYRIBOSE N1-2′-FURANIDYL-5-FLUOROURACIL see FLZ050
2,4-PYRIMIDINEDIOL see UNJ800
2,4-PYRIMIDINEDIONE see UNJ800
2,4(1H,3H)-PYRIMIDINEDIONE (9CI) see UNJ800
4,6(1H,5H)-PYRIMIDINEDIONE, DIHYDRO-2-THIOXO-(9CI) see MCK500
2,4(1H,3H)-PYRIMIDINEDIONE, 5-(HYDROXYMETHYL)-6-METHYL- (9CI) see HMH300
5-PYRIMIDINEMETHANOL, α-(2-CHLOROPHENYL)-α-(4-FLUOROPHENYL)- see CHJ300
PYRIMIDINE PHOSPHATE see DIN800
2,4,5,6(1H,3H)-PYRIMIDINETETRONE see AFT750
2,4,5,6(1H,3H)-PYRIMIDINETETRONE HYDRATE see MDL500
2,4,5,6(1H,3H)-PYRIMIDINETETRONE 5-OXIME see AFU000
2(1H)-PYRIMIDINETHIONE, 4-AMINO- see TFH800
2,4,6(1H,3H,5H)-PYRIMIDINETRIONE, 1-BENZOYL-5-ETHYL-5-PHENYL-(9CI) see BDS300
2,4,6(1H,3H,5H)-PYRIMIDINETRIONE, 5-(2-BROMO-2-PROPENYL)-5-(1-METHYLPROPYL)-(9CI) see BOR000
2,4,6(1H,3H,5H)-PYRIMIDINETRIONE, 5,5-DIETHYL-, MONOSODIUM SALT (9CI) see BAG250
2,4,6(1H,3H,5H)-PYRIMIDINETRIONE, 5-ETHYL-5-(1-METHYLBUTYL)-, CALCIUM SALT (9CI) see CAV000
2,4,6(1H,3H,5H)-PYRIMIDINETRIONE, 5-ETHYL-5-(3-METHYLBUTYL)-(9CI) see AMX750
2,4,6(1H,3H,5H)-PYRIMIDINETRIONE, MONOSODIUM SALT (9CI) see MRK750

2-PYRROLIDINONE, 1-ETHYL-4-(2-(4-MORPHOLINYL)ETHYL)-3,3-DIPHENYL-(9CI) see ENL100
2-PYRROLIDINONE, 1-(2-HYDROXYETHYL)- see HLB380
2-PYRROLIDINONE, 1-ISOPROPYL- see IRG100
2-PYRROLIDINONE, 1-(1-METHYLETHYL)-(9CI) see IRG100
2-PYRROLIDINONE, 1-OCTYL- see OFU100
2-PYRROLIDINONE, 1-TALLOW ALKYL derivs. see TAC200
2-PYRROLIDINONE, 1-VINYL-, POLYMER with STYRENE (8CI) see VQK595
2-PYRROLIDINYL-p-ACETOPHENETIDIDE HYDROCHLORIDE see PPU750
1-(4-(1-PYRROLIDINYL)-2-BUTYNYL)-2-PYRROLIDINONE see OOY000
4-PYRROLIDINYLCARBONYLMETHYLPIPERAZINYL-3,4,5-TRIMETHOXYCINNA-MYL KETONE MALEATE see PPV750
1-(((1-PYRROLIDINYLCARBONYL)METHYL)-4-(3,4,5-TRIMETHOXYCINNA-MOYL)PIPERAZINE MALEATE see VGK000
trans-2-(1-PYRROLIDINYL)CYCLOHEXYL-3-PENTYLOXYCARBANILATE HYDRO-CHLORIDE see PPW000
N-(2-PYRROLIDINYLETHYL)CARBAMIC ACID, 6-CHLORO-o-TOLYL ESTER, HY-DROCHLORIDE see PPW750
N-(2-PYRROLIDINYLETHYL)CARBAMIC ACID, MESITYL ESTER, HYDROCHLO-RIDE see PPX000
N-(2-(PYRROLIDINYL)ETHYL)CARBAMIC ACID, 2,6-XYLYL ESTER HYDRO-CHLORIDE see PPX250
2-(PYRROLIDINYL)ETHYL-2-CHLORO-6-METHYLCARBANILATE HYDROCHLO-RIDE see CIK500
2-(2-(PYRROLIDINYL)ETHYL)-5-NITRO-1H-BENZ(de)ISOQUINOLINE-1,3(2H)-DI-ONE see MAB055
10-(2-(1-PYRROLIDINYL)ETHYL)PHENOTHIAZINE see PAJ750
3-(1-PYRROLIDINYLMETHYL)(1,1'-BIPHENYL)-4-OL see PPT400
N-(1-PYRROLIDINYLMETHYL)-TETRACYCLINE see PPY250
1-PYRROLIDINYLOXY, 3-CARBOXY-2,2,5,5-TETRAMETHYL- see TDV300
3-(2-PYRROLIDINYL)PYRIDINE see NNR500
3-(2-PYRROLIDINYL)-(s)-PYRIDINE see NNR500
trans-2-(3-(1-PYRROLIDINYL)-1-p-TOLYLPROPENYL)PYRIDINE MONOHYDRO-CHLORIDE see TMX775
4-(1-PYRROLIDINYL)-1-(2,4,6-TRIMETHOXYPHENYL)-1-BUTANONE HYDRO-CHLORIDE see BOM600
PYRROLIDON (GERMAN) see PPT500
2-PYRROLIDONE see PPT500
α-PYRROLIDONE see PPT500
PYRROLIDONE see PPT500
2-PYRROLIDONEACETAMIDE see NNE400
PYRROLIDONE CARBOXYLATE de MAGNESIUM HYDRATE (FRENCH) see PAN775
10-(2-(1-PYRROLIDYL)ETHYL)PHENOTHIAZINE see PAJ750
10-(2-(1-PYRROLIDYL)ETHYL)PHENOTHIAZINE HYDROCHLORIDE see POL000
1-PYRROLINE-2-ACRYLAMIDE see COV125
3-PYRROLINE-2,5-DIONE see MAM750
PYRROLNITRIN see CFC000
5H-PYRROLO(2,2-c)(1,4)BENZODIAZEPIN-5-ONE, 2-ETHYLIDENE-1,2,3,10,11,11A-HEXAHYDRO-8-HYDROXY-7,11-DIMETHOXY- see THG500
(E)-1H-PYRROLO(2,1-C)(1,4)BENZODIAZEPINE-2-ACRYLAMIDE, 5,10,11,11a-TETRAHYDRO-9,11-DIHYDROXY-8-METHYL-5-OXO see API000
PYRROLYLENE see BOP500
PYRROL-2-YL KETONE see PPY300
PYRROL-2-YLPHENYL KETONE see PGF900
3-PYRROL-2-YLPYRIDINE see PQB500
PYRROMECAINE HYDROCHLORIDE see PQB750
PYRROMELITIC ACID DIANHYDRIDE see PQB800
PYRRO(b)MONAZOLE see IAL000
2-PYRRYL ETHYL KETONE see PMX320
α-PYRRYL ETHYL KETONE see PMX320
PYRUVALDEHYDE see PQC000
PYRUVIC ACID see PQC100
PYRUVIC ACID, BROMO- see BOD550
PYRUVIC ALDEHYDE see PQC000
PYRVINIUM EMBONATE see PQC500
PYRVINIUM-4,4'-METHYLENEBIS(3-HYDROXYNAPHTHALENE-2-CARBOXYLATE) see PQC500
PYRVINIUM PAMOATE see PQC500
PYSOCOCCINE see SNM500
PY-TETRAHYDROSERPENTINE see AFG750
PYX see PQC525
PYX EXPLOSIVE see PQC525
PZ-13 see PDB750
PZ 177 see CJW500
PZ 222 see CKD250
PZ 1511 see CCK780
PZ see CAT775
PZh2M see IGK800
PZhO see IGK800
PZO see PDB250
PZT 5 see LED100
PZT 8 see LED100
PZT 574 see LED100

PZT see LED100
PZT-C see LED100

Q-137 see DJC000
Q 174 see TNJ500
QA 71 see PNW800
QB-1 see CMX920
QCB see PAV500
QDO see DVR200
Q-D 86P see DXG625
QG 100 see SCK600
QGH see BDD000
QIDAMP see AIV500
QIDMYCIN see EDJ500
QIDPEN G see BFD000
QIDPEN VK see PDT750
QIDTET see TBX250
QIKRON see BIN000
QINGHAOSU (CHINESE) see ARL375
QING HAU SAU (CHINESE) see ARL375
QM 657 see DGW450
QM-1143 see MOF750
QM-1148 see MLR500
QM-1149 see MOG000
QM-6008 see BAV625
QNB see QVA000
QO THFA see TCT000
QPB see PJA000
QR 483 see MCB050
QR 819 see BLX000
QSAH 7 see PKQ059
QUAALUDE see QAK000
QUADRACYCLINE see TBX250
QUADRATIC ACID see DMJ600
QUADROL see QAT000
QUAMAQUIL see PGA750
QUAMONIUM see HCQ500
QUANTALAN see CME400
QUANTRIL see BCL250
QUANTROVANIL see EQF000
QUANTRYL see BCL250
QUARTZ see SCJ500
QUARTZ GLASS see SCK600
QUARTZ SAND see SCK600
QUARZSAND see SCK600
QUATERNARIO CPC see CCX000
QUATERNARIO LPC see DXY725, LBX050
QUATERNARY AMMONIUM COMPOUNDS, ALKYLBENZYLDIMETHYL, CHLO-RIDES see AFP250
QUATERNARY AMMONIUM COMPOUNDS, BENZYL-$C_{12}$-$C_{16}$-ALKYLDIMETHYL, CHLORIDES see QAT520
QUATERNARY AMMONIUM COMPOUNDS, BIS(HYDROGENATED TALLOW ALKYL)DIMETHYL CHLORIDES see QAT550
QUATERNIUM-12 see DGX200
QUATERNIUM 13 see TCB200
QUATERNIUM 15 see CEG550
QUATERNIUM-18 see QAT550
QUATERNIUM-10 see TLW500
QUATERNIUM 40 see DTS500
QUATERNIUM 5 see DXG625
QUATERNOL 1 see DTC600
QUATRACHLOR see BEN000
QUATRESIN see MSB500
QUATREX see TBX250
QUAZO PURO (ITALIAN) see SCJ500
QUEBRACHIN see YBJ000
QUEBRACHINE see YBJ000
QUEBRACHO TANNIN see QBJ000
QUECKSILBER (GERMAN) see MCW250
QUECKSILBER(I)-CHLORID (GERMAN) see MCW000
QUECKSILBER CHLORID (GERMAN) see MCY475
QUECKSILBER CHLORUER (GERMAN) see MCW000
QUECKSILBEROXID (GERMAN) see MCT500, MDF750
QUEENSGATE WHITING see CAT775
QUELAMYCIN see QBS000
QUELETOX see FAQ900
QUELICIN see CMG250, HLC500
QUELICIN CHLORIDE see HLC500
QUELLADA see BBQ500
QUEMICETINA see CDP250
QUEN see CFZ000
QUERCETIN, 3-(6-DEOXY-α-l-MANNOPYRANOSIDE) see QCJ000
QUERCETIN see QCA000

QUERCETIN-3-(6-o-(6-DEOXY-α-l-MANNOPYRANOSYL)-β-d-GLUCOPYRANO-
 SIDE) see RSU000
QUERCETIN DIHYDRATE see QCA175
QUERCETINE see QCA000
QUERCETIN RHAMNOGLUCOSIDE see RSU000
QUERCETIN-3-RHAMNOGLUCOSINE see RSU000
QUERCETIN-3-(6-o-α-l-RHAMNOPYRANOSYL-β-d-GLUCOPYRANOSIDE) see
 RSU000
QUERCETIN-3-l-RHAMNOSIDE see QCJ000
QUERCETIN-3-RUTINOSIDE see RSU000
QUERCETOL see QCA000
QUERCITIN see QCA000
QUERCITRIN see QCJ000
QUERCUS AEGILOPS L. TANNIN see VCK000
QUERCUS FALCATA PAGODAEFOLIA see CDL750
QUERTINE see QCA000
QUERYL see CNR125
QUESTEX 4 see EIV000
QUESTEX 4H see EIX000
QUESTIOMYCIN A see QCJ275
QUESTRAN see CME400
QUESYL see DEL300
QUETIMID see TEH500
QUETINIL see DCK759
QUIACTIN see OLU000
QUIADON see MCH535
QUIATRIL see DCK759
QUICK see CJJ000
QUICKLIME (DOT) see CAU500
QUICKPHOS see AHE750
QUICKSAN see ABU500
QUICKSET EXTRA see MKA500
QUICKSET SUPER see MKA500
QUICK SILVER see MCW250
QUIDE see PII500
QUIESCIN see RDK000
QUIETAL see QCS000
QUIETALUM see QCS000
QUIETIDON see MQU750
QUIETOPLEX see TEH500
QUI-LEA see QFA250
QUILIBREX see CFZ000
QUILONE see LGO100
QUILON S see CMH300
QUILONUM RETARD see LGZ000
QUIMAR see CML850
QUIMOTRASE see CML850
QUINACETOPHENONE see DMG600
QUINACHLOR see CLD000
QUINACRINE see ARQ250
QUINACRINE DIHYDROCHLORIDE see CFU750
QUINACRINE ETHYL M/2 see QCS875
QUINACRINE ETHYL MUSTARD see DFH000
QUINACRINE HYDROCHLORIDE see CFU750
QUINACRINE MUSTARD see QDJ000, QDS000
QUINACRINE MUSTARD DIHYDROCHLORIDE see QDS000
QUINADOME see DNF600
QUINAGAMINE see CLD000
QUINAGLUTE see QDS225
QUINALBARBITAL see SBM500
QUINALBARBITONE see SBM500
QUINALBARBITONE SODIUM see SBN000
QUINALDIC ACID see QEA000
QUINALDINE see QEJ000
QUINALDINIUM, 1-ETHYL-, IODIDE see EPD600
QUINALIZARIN see TDD000
QUINALIZARINE see TDD000
QUINALPHOS see DJY200
QUINALSPAN see SBN000
QUINAMBICIDE see CHR500
QUINAPYRAMINE see AQN625
QUINAPYRAMINE METHYL SULFATE see AQN625
QUINAZINE see QRJ000
(1H,3H)QUINAZOLINE DIONE-2,4 see QEJ800
2,4(1H,3H)-QUINAZOLINEDIONE see QEJ800
QUINAZOLINE-2,4-DIONE see QEJ800
2,4-QUINAZOLINEDIONE see QEJ800
QUINAZOLINEDIONE see QEJ800
4-QUINAZOLINOL, 2-METHYL-(8CI) see MPF500
4(3H)-QUINAZOLINONE see QFA000
4-QUINAZOLINONE see QFA000
4(3H)-QUINAZOLINONE, 2-METHYL- see MPF500
4(1H)-QUINAZOLINONE, 2-METHYL-(7CI,9CI) see MPF500
10-(3-QUINCULIDINYLMETHYL)PHENOTHIAZINE see MCJ400
QUINDOXIN see QSA000

QUINERCYL see CLD000
QUINESTROL see QFA250
QUINEX see PFN250
QUINEXIN see SAN600
QUINGESTANOL ACETATE see QFA275
QUINGFENGMYCIN see ARP000
QUINHYDRONE see QFJ000
QUINICARDINE see QFS000, QHA000
QUINICINE OXALATE see QFJ300
QUINIDATE see QHA000
QUINIDEX see QFS000, QHA000
α-QUINIDINE see CMP910
(+)-QUINIDINE see QFS000
QUINIDINE see QFS000
QUINIDINE BISULFATE see QFS100
QUINIDINE-d-GLUCONATE (salt) see QDS225
QUINIDINE GLUCONATE see QDS225
QUINIDINE MONO-d-GLUCONATE (salt) see QDS225
QUINIDINE MONOSULFATE see QHA000
QUINIDINE POLYGALACTURONATE see GAX000
QUINIDINE SULFATE (1:1) (salt) see QFS100
QUINIDINE SULFATE (2:1) (salt) see QHA000
QUINIDINE SULFATE see QHA000
QUINIDINE SULPHATE see QHA000
QUINIDINE YELLOW KT see CMM510
QUINILON see CLD000
β-QUININE see QFS000
QUININE see QHJ000
(−)-QUININE see QHJ000
QUININE BIMURIATE see QIJ000
QUININE BISULFATE see QMA000
QUININE CHLORIDE see QJS000
QUININE DIHYDROCHLORIDE see QIJ000
(−)-QUININE DIHYDROCHLORIDE see QIJ000
QUININE ETHIODIDE see QIS000
QUININE FORMATE see QIS300
QUININE, FORMATE (SALT) see QIS300
QUININE HYDROBROMIDE see QJJ100
QUININE HYDROCHLORIDE see QJS000
QUININE HYDROGEN SULFATE see QMA000
QUININE MONOHYDROCHLORIDE see QJS000
QUININE MURIATE see QJS000
QUININE SULFATE see QMA000
QUINIPHEN see IEP200
QUINITEX see QHA000
QUINIZARIN see DMH000
QUINIZARINE GREEN BASE see BLK000
QUINODIMETHAN, TETRACYANO- see TBW750
QUINOFEN see PGG000
QUINOFORM see QIS300
8-QUINOL see QPA000
β-QUINOL see HIH000
QUINOL see HIH000
QUINOL DIMETHYL ETHER see DOA400
4-QUINOLINAMINE, N-HYDROXY-N-METHYL-, 1-OXIDE see HLX550
QUINOLINE see QMJ000
3-QUINOLINEAMINE see AMK725
QUINOLINE, 3-AMINO- see AMK725
QUINOLINE-6-AZO-p-DIMETHYLANILINE see DPR000
6-QUINOLINE CARBONYL AZIDE see QMS000
QUINOLINE-2-CARBOXYLIC ACID see QEA000
2-QUINOLINECARBOXYLIC ACID see QEA000
QUINOLINE, 4,7-DICHLORO- see DGJ250
QUINOLINE, 1,2-DIHYDRO-2,2,4-TRIMETHYL-, HOMOPOLYMER see PJQ750
2,4-QUINOLINEDIOL see QNA000
QUINOLINE, 6-DODECYL-1,2-DIHYDRO-2,2,4-TRIMETHYL- see SAU480
2-QUINOLINEETHANOL see QNJ000
QUINOLINE, 2-ETHYL-4-NITRO-, 1-OXIDE see NGA500
QUINOLINE, 2-METHYL-4-NITRO, 1-OXIDE see MMQ250
5-QUINOLINESULFONIC ACID, 8,8′-((HYDROXYSTIBYLENE)BIS(OXY))BIS(7-
 FORMYL)-, DISODIUM SALT see AQE320
2-QUINOLINE THIOACETAMIDE HYDROCHLORIDE see QOJ000
2-QUINOLINETHIOL see QOJ100
2(1H)-QUINOLINETHIONE see QOJ100
QUINOLINE YELLOW see CMM510
QUINOLINE YELLOW A SPIRIT SOLUBLE see CMS245
QUINOLINE YELLOW BASE see CMS245
QUINOLINE YELLOW EXTRA see CMM510
QUINOLINE YELLOW SPIRIT SOLUBLE see CMS245
QUINOLINE YELLOW SS see CMS245
QUINOLINIUM, 6-AMINO-1-METHYL-4-((4-((((1-METHYLPYRIDINIUM-4-
 YL)AMINO)PHENYL)CARBAMOYL)PHENYL)AMINO)-, DIBROMIDE see
 QOJ250
QUINOLINIUM DIBROMIDE see QOJ250

QUINOLINIUM, 1-ETHYL-4-(3-(1-ETHYL-4(1H)-QUINOLINYLIDENE)-1-PROPENYL)-, IODIDE (9CI) see KHU050
2-QUINOLINOL see CCC250
8-QUINOLINOL see QPA000
8-(QUINOLINOLATO)METHYL MERCURY see MLH000
(8-QUINOLINOLATO)TRIBUTYLSTANNANE see TIB000
8-QUINOLINOL, 5-CHLORO- see CLD600
8-QUINOLINOL, 5,7-DICHLORO- see DEL300
8-QUINOLINOL HYDROGEN SULFATE (2:1) see QPS000
8-QUINOLINOLIUM-4',7'-DIBROMO-3'-HYDROXY-2'-NAPHTHOATE see QPJ000
8-QUINOLINOL, MERCURY COMPLEX see MLH000
8-QUINOLINOL SULFATE see QPS000
8-QUINOLINOL SULFATE (2:1) (SALT) see QPS000
QUINOLOR see DEL300
QUINOLOR COMPOUND see BDS000
N-(4-QUINOLYL)ACETOHYDROXAMIC ACID see QQA000
N-(4-QUINOLYL)HYDROXYLAMINE-1'-OXIDE see HIY500
2-(2-QUINOLYL)-1,3-INDANDIONE DISULFONIC ACID DISODIUM SALT see CMM510
N-(8-QUINOLYLMETHYL)-N-METHYL-2-PROPYLAMINE see IPZ500
α-4-QUINOLYL-5-VINYL-2-QUINUCLIDINEMETHANOL see CMP925
QUINOMETHIONATE see ORU000
QUINOMYCIN A see EAD500
QUINOMYCIN C see QQS075
QUINONDO see BLC250
o-QUINONE see BDC250
p-QUINONE see QQS200
QUINONE see QQS200
QUINONE CHLORIMIDE see CHQ500
p-QUINONE DIOXIME see DVR200
QUINONE DIOXIME see DVR200
QUINONE MONOXIME see NLF200
p-QUINONE OXIME see DVR200
QUINONE OXIME see NLF200
QUINONE OXIME BENZOYLHYDRAZONE see BDD000
QUINON I see BGW750
p-QUINONIMINE see BDD500
QUINOPHEN see PGG000
QUINOPHENOL see QPA000
QUINORA see QHA000
QUINOSCAN see CLD000
QUINOSEPTYL see AKO500
QUINOTILINE see KGK400
QUINOXALINE see QRJ000
QUINOXALINE, 2,3-DICHLORO- see DGJ950
QUINOXALINE, 2,3-DIMETHYL- see DTY700
2,3-QUINOXALINEDIOL see QRS000
QUINOXALINE 1,4-DIOXIDE see QSA000
QUINOXALINE-1,4-DI-N-OXIDE see QSA000
QUINOXALINE DI-N-OXIDE see QSA000
QUINOXALINE DIOXIDE see QSA000
2,3-QUINOXALINEDITHIOL see QSJ000
2,3-QUINOXALINETHIOL see QSJ000
2-QUINOXALINOL see QSJ100
2(1H)-QUINOXALINONE see QSJ100
(2,3-QUINOXALINYLDITHIO)DIMETHYLTIN see QSJ800
(2,3-QUINOXALINYLDITHIO)DIPHENYL STANNANE see QTJ000
(2,3-QUINOXALINYLDITHIO)DIPHENYLTIN see QTJ000
(2-QUINOXALINYLMETHYLENE)-HYDRAZINECARBOXYLIC ACID METHYL ESTER-N,N'-DIOXIDE see FOI000
N'-2-QUINOXALINYLSULFANILAMIDE see QTS000
N-(2-QUINOXALINYL)SULFANILAMIDE see QTS000
N'-2-QUINOXALYLSULFANILAMIDE see QTS000
QUINOXYL see IEP200
QUINPYRROLIDINE see QUJ300
QUINSORB 010 see DMI600
QUINTAR see DFT000
QUINTAR 540F see DFT000
QUINTESS-N see NCG000
QUINTOCENE see PAX000
QUINTOMYCIN C see NCF500
QUINTOMYCIN D see DAS600
QUINTOX see DHB400
QUINTOZEN see PAX000
QUINTOZENE see PAX000
QUINTRATE see PBC250
2-QUINUCLIDINEMETHANOL, α-4-QUINOLYL-5-VINYL- see CMP910
3-QUINUCLIDINOL see QUJ990
3-QUINUCLIDINOL ACETATE see AAE250
3-QUINUCLIDINOL ACETATE (ESTER) HYDROCHLORIDE see QUS000
3-QUINUCLIDINOL BENZILATE see QVA000
3-QUINUCLIDINOL, DIPHENYLACETATE (ESTER), HYDROGEN SULFATE (2:1) DIHYDRATE see QVJ000
3-QUINUCLIDINYL BENZILATE see QVA000
QUINUCLIDYL BENZYLATE see QWJ000

QUINUCLIDYL-3-DIPHENYLCARBINOL HYDROCHLORIDE see DWM400
QUINURONIUM SULFATE see PJA120
QUIPAZINE MALEATE see QWJ500
QUIPENYL see RHZ000
QUIRVIL see PKQ059
QUIXALIN see DEL300
QUOLAC EX-UB see SIB600
QUOTANE see DNX400
QUOTANE HYDROCHLORIDE see DNX400
QX 2168 see CGW300
QYSA see PKQ059
QZ 2 see QAK000

R 2 see BCM250
R 3 see SMQ500
R 5 see DDT300
R 8 see MLF250
R 10 see CBY000
R-11 see BHJ500
R 13 see CLR250
R 14 see CBY250, PKL750
R 21 see BGR750
R-22 see CFX500
R 23 see CBY750
R 30 see MJP450
R 31 see CHI900
R 40 see MIF765
5R04 see ARM268
R-47 see TNF500
R48 see BIF250
R50 see DAD200
R-52 see MAW800
R 54 see DDP600
R 79 see DAB875
88-R see SOP500
R 113 see FOO000
R 114 see FOO509
Rh 116 see RGW100
R 123 see TJY500
R 133a see TJY175
R 143a see TJY900
R161 see FIB000
R 219 see CMS160
R-242 see CKI625
R-246 see TND500
R 261 see DBA200
R-326 see EAU500
R 400 see NCF500
R 516 see CMR100
R-528 see DPH000
R-548 see TIH800
R 561 see PER250
R 661 see BTA325
R 694 see MEK700
R 717 see AMY500
R 744 see CBU250
R-874 see OGA000
R 875 see AFJ400
R 951 see CBP325
R968 see ADW200
R 1067 see CPN750
R 1132 see LIB000
R-1303 see TNP250
R-1492 see MQH750
R 1504 see PHX250
R 1513 see EKN000
R 1575 see CMR100
R-1607 see PNI750
R-1608 see EIN500
R 1625 see CLY500
R 1658 see MNN250
R 1892 see FLL000
R-1910 see EID500
R 1929 see FLU000
R 2028 see HAF400
R-2061 see PNF500
R 2063 see EHT500
R 2167 see HAF400
R 2170 see DAP000
R-2498 see TKK500
R 2498 see TKK750
R 2858 see MRU600
R 3345 see FHG000
R 3365 see PJA140

R-3588 see CNR125
R 3612 see SMQ500
R 4263 see PDW500, PDW750
R 4318 see TKG000
R-4461 see DNO800
R-4572 see EKO500
R 4584 see FGU000, FLK100
R 4749 see DYF200
R-4845 see BCE825
R 4929 see BBU800
R-5,158 see DJA400
R 5240 see PDW750
R-5461 see BJD000
R 6238 see PIH000
R6597 see NHQ950
R 6700 see OAN000
R7000 see MFB400
R 7158 see FLJ000
R 7237 see ACQ790
R 7372 see MEW775
R 7542 see NHR100
R 8299 see TDX750
R 9298 see CLS250
R 9985 see MDV500
R 11333 see BNU725
R-12,564 see LFA020
R 14,827 see EAE000
R 14889 see MQS560
R 14950 see FDD080
R 15,175 see ASD000
R 16341 see PAP250
R 16470 see DBE000
R 16659 see HOU100
R 17635 see MHL000
R 17899 see FDA887
R 17934 see OJD100
Ro 1-7977 see DNV800
R 18134 see MQS550
R 18986 see AIH000
R 23979 see FPB875
R 25061 see TEN750
R-25788 see DBJ600
Ro 2-7113 see AGV890
R-28627 see DNT200
R 31520 see CFC100
R 35443 see OMG000
R 41400 see KFK100
Ro 04-7683 see INY100
R 51619 see CMS237
Ro 05-9129 see NHG000
Ro 7-0582 see NHH500
Ro 10-9359 see EMJ500
Ro 215535 see DMJ400
R12 (DOT) see DFA600
R13 (DOT) see CLR250
R14 (DOT) see CBY250
R21 (DOT) see DFL000
R22 (DOT) see CFX500
R114 (DOT) see DGL600
R500 (DOT) see DFB400
R502 (DOT) see FOO510
R503 (DOT) see FOO515
R1132a (DOT) see VPP000
R 8 (fungicide) see MLF250
Rh-110 see RHF150
Ro-5-6901/3 see FMQ000
R 20 (refrigerant) see CHJ500
R 31 (refrigerant) see CHI900
RA 8 see PCP250
13-RA see VSK955
RA 1226 see ALC250
β-RA see VSK950
RABANO (PUERTO RICO) see DHB309
RABBIT-BELLS see RBZ400
RABBIT FLOWER see FOM100
RABON see RAF100
RABOND see RAF100
RACCOON BERRY see MBU800
RACEMETHORPHAN HYDROBROMIDE see RAF300
RACEMIC LACTIC ACID see LAG000
RACEMIC MANDELIC ACID see MAP000
RACEMOMYCIN A see RAG300
RACEMORPHAN see MKR250
RACEMORPHAN HYDROBROMIDE see MDV250
RACEPHEDRINE see EAW100

RACEPHEDRINE HYDROCHLORIDE see EAX500
RACEPHEN see AOB250, AOB500
(6R-(6-α,7-β(Z)))-3-((ACETYLOXY)METHYL)-7-(((2-AMINO-4-THIAZO-
   LYL)(METHOXYIMIO)ACETYL)AMINO)-8-OXO-5-THIA-1-AZABICY-
   CLO(4,2,0)OCT-2-ENE-2-CARBOXYLIC ACID, SODIUM SALT see CCR950
RACRYL see ADW200
RACUMIN see EAT600
RACUSAN see DSP400
RADAPON see DGI400, DGI600
RADAZIN see ARQ725
RAD-E-CATE 25 see HKC000, HKC500
RAD-E-CATE 35 see HKC500
RAD-E-CATE see HKC500
RADDLE see IHD000
RAD-E-CATE 16 see HKC500
RADEDORM see DLY000
RADEPUR see LFK000
RADIANT YELLOW see CMS212
RADIASURF 7125 see SKV000
RADIASURF 7145 see PKG500
RADIASURF 7155 see SKV100
RADIATION, IONIZING see RAQ010
RADIATION see RAQ000
RADICALISIN see RRA000
RADIOCIN see SPD500
RADIOGRAPHOL see SHX000
RADIOL GERMICIDAL SOLUTION see EKN500
RADIOSTOL see VSZ100
RADIOTETRANE see TDE750
RADIUM see RAV000
RADIUM F see PJJ750
RADIZINE see ARQ725
RADOCON see BJP000
RADOKOR see BJP000
RADON see RBA000
RADONIL see SNK000
RADONIN see SNN300
RADONNA see CHJ750
RADOSAN see MEO750
RADOX see CFK000
RADOXONE TL see AMY050
RADSTERIN see VSZ100
RAFEX see DUS700
d-(+)-RAFFINOSE see RBA100
d-RAFFINOSE see RBA100
RAFFINOSE see RBA100
RAFLUOR see SHF500
RAGADAN see HBK700
RAGUAR (GUAM) see TOA275
RAGWORT see RBA400
RAIN LILY see RBA500
RAISIN de COULEUVRE (CANADA) see MRN100
RAK 1 see GJU050
RAKUTO AMARANTH see FAG020
RALABOL see RBF100
RALGRO see RBF100
RALONE see RBF100
RAM-326 see ORE000
RAMETIN see HMV000
RAMIK see DVV600
RAMIZOL see AMY050
RAMOR see TEI000
R/AMP see RKP000
RAMPART see PGS000
RAMROD see CHS500
RAM'S CLAWS see FBS100
RAMUCIDE see CJJ000
RAMYCIN see FQU000
RANAC see CJJ000
RANCOSIL see SCK600
RANDOLECTIL see BSZ000
RANDOX see CFK000
RANEY ALLOY see NCW500
RANEY COPPER see CNI000
RANEY NICKEL see NCW500
RANGOON YELLOW see DEU000
RANIDIL see RBF400
RANITIDINE HYDROCHLORIDE see RBF400
RANITOL see AFG750
RANKOTEX see CIR500
RANTOX T see CFK000
RANTUDIL see AAE625
RANUNCULUS ACRIS see FBS100
RANUNCULUS BULBOSUS see FBS100
RANUNCULUS SCELERATUS see FBS100

RAPACODIN see DKW800
RAPAMYCIN see RBK000
RAPHATOX see DUS700
RAPHETAMINE see BBK000
RAPHETAMINE PHOSPHATE see AOB500
RAPHONE see CIR250
RAPI-CURE DVE-3 see TDY800
RAPID see DOX600
RAPIDOSEPT see DET000
RAPISOL see DJL000
RAPYNOGEN see ICC000
RARE EARTHS see RBP000
RAS-26 see NIJ500
RASCHIT see CFE250
RASEN-ANICON see CFE250
RASIKAL see SFS000
RASORITE 65 see DXG035
RASPBERIN see SAH000
RASPBERRY KETONE see RBU000
RASPBERRY KETONE METHYL ETHER see MFF580
RASPBERRY RED for JELLIES see FAG020
RASTINON see BSQ000
RATAFIN see ABF500
RATAK see TAC800
RATAK PLUS see TAC800
RAT-A-WAY see ABF500, WAT200
RATBANE 1080 see SHG500
RAT-B-GON see WAT200
RAT-O-CIDE RAT BAIT see RCF000
RAT-GARD see WAT200
RATICATE see NNF000
RATIMUS see BMN000
RATINDAN 1 see DVV600
RAT & MICE BAIT see WAT200
RAT-NIP see PHP010
RATO see TMO000
RATOMET see CJJ000
RATON see RBZ000
RATOX see TEL750
RAT'S END see RCF000
RATS-NO-MORE see WAT200
RATSUL SOLUBLE see WAT220
RATTENGIFTKONSERVE see TEM000
RATTEX see CNV000
RATTLE BOX see RBZ400
RATTLESNAKE WEED see FAB100
RATTLEWEED (JAMAICA) see RBZ400
RATTRACK see AQN635
RAUBASINE see AFG750
RAUBASNE HYDROCHLORIDE see AFH000
RAUCAP see RDK000
RAUCUMIN 57 see EAT600
RAUDIFORD see RDK000
RAUDIXIN see RDK000
RAUDIXOID see RDK000
RAUGAL see RDK000
RAUGALLINE see AFH250
RAULEN see RDK000
RAULOYCIN see RDK000
RAULOYDIN see RDK000
RAUMALINA see AFG750
RAUMANON see ILD000
RAUMORINE see RDK000
RAUNERVIL see RDK000
RAUNORINE see RDF000, RDK000
RAUNORMIN 'ORZAN' see RDK000
RAUNOVA see RCA200, RDK000
RAUPASIL see RDK000
RAUPOID see RDK000
RAUPYROL see TLN500
RAURESCINE see TLN500
RAURINE see RDK000
RAUSAN see RDK000
RAU-SED see RDK000
RAUSEDAN see RDK000
RAUSEDIL see RDK000
RAUSEDYL see RDK000
RAUSERPEN-ALK see RDK000
RAUSERPIN see RDK000
RAUSERPIN-ALK see RDK000
RAUSERPINE see RDK000
RAUSERPOL see RDK000
RAUSINGLE see RDK000
RAUTRAX see RCA275
RAUTRIN see RDK000

RAUVILID see RDK000
RAUVLID see RDK000
RAUWASEDIN see RDK000
RAUWILID see RDK000
RAUWILOID see RDK000
RAUWILOID+ see RDK000
RAUWIPUR see RDK000
RAUWOLEAF see RDK000
RAUWOLFIN see AFH250
RAUWOLFINE see AFH250
RAUWOLSCINE HYDROCHLORIDE see YCA000
RAUWOPUR'BYK' see RDK000
RAV 7 see AGD250
RAVEN 30 see CBT750
RAVEN 420 see CBT750
RAVEN 500 see CBT750
RAVEN 8000 see CBT750
RAVEN see CBT750
RAVIAC see CJJ000
RAVINYL see PKQ059
RAVONA see CAV000
RAVONAL see PBT500
RAVYON see CBM750
RAWETIN see HMV000
RAWILID see RDK000
RAW SHALE OIL see COD750
RAYBAR see BAP000
RAY-GLUCIRON see FBK000
RAYON BLACK G see CMN240
RAYON BLACK GSN see CMN240
RAYON BLACK M see CMN240
RAYON FAST BLACK B see CMN240
RAYOPHANE see CCU150
RAYOX see TGG760
RAYWEB Q see CCU150
RAZIOSULFA see AIF000
RAZOL DOCK KILLER see CIR250
RAZOXANE see RCA375
RAZOXIN see PIK250, RCA375
R 40B1 see MHR200
R-242-B see CKI625
R-451-B see ERD000
RB 1489 see PIZ000
1489 RB see PIZ000
RB 1509 see CGV250
R 114B2 see FOO525
R12B1 (DOT) see BNA250
R13B1 (DOT) see TJY100
R142B (DOT) see CFX250
R12B2 (DOT) see DKG850
RB see RJF400
RBA 777 see CLW000
R 74 BASE see LJD500
R-BASE see CBN100
RB-BL see IHC550
1-RBO-12-A see PNA200
RC 146 see CNG250
R-C 318 see CPS000
RC 5629 see DRR800
RC 61-91 see IAG600
RC 61-96 see BEU800
RC 72-01 see RCA435
RC 72-02 see RCA450
R.C. 27-109 see DGQ500
RC 30-109 see DGQ500
RCA WASTE NUMBER P105 see SFA000
RCA WASTE NUMBER U203 see SAD000
RCA WASTE NUMBER U205 see SBR000
RC COMONOMER DBM see DED600
RC COMONOMER DOF see DVK600
RC COMONOMER DOM see BJR000
RC 172DBM see AHE250
RCH 1000 see PJS750
R 30 730 CITRATE SALT see SNH150
RC PLASTICIZER B-17 see BSL600
RC PLASTICIZER DOP see DVL700
RCRA WASTE NUMBER P111 see TCF250
RCRA WASTE NUMBER P112 see TDY250
RCRA WASTE NUMBER P113 see TEL050
RCRA WASTE NUMBER P114 see TEL500
RCRA WASTE NUMBER P115 see TEM000
RCRA WASTE NUMBER P116 see TFQ000
RCRA WASTE NUMBER P118 see PCF300
RCRA WASTE NUMBER P119 see ANY250
RCRA WASTE NUMBER P011 see ARH500

RCRA WASTE NUMBER P101 see PMV750
RCRA WASTE NUMBER P110 see TCF000
RCRA WASTE NUMBER P121 see ZGA000
RCRA WASTE NUMBER P122 see ZLS000
RCRA WASTE NUMBER P123 see CDV100
RCRA WASTE NUMBER P012 see ARI750
RCRA WASTE NUMBER P102 see PMN450
RCRA WASTE NUMBER P120 see VDU000, VDZ000
RCRA WASTE NUMBER P013 see BAK750
RCRA WASTE NUMBER P103 see SBV000
RCRA WASTE NUMBER P014 see PFL850
RCRA WASTE NUMBER P104 see SDP000
RCRA WASTE NUMBER P015 see BFO750
RCRA WASTE NUMBER P016 see BIK000
RCRA WASTE NUMBER P106 see SGA500
RCRA WASTE NUMBER P017 see BNZ000
RCRA WASTE NUMBER P018 see BOL750
RCRA WASTE NUMBER P108 see SMN500
RCRA WASTE NUMBER P109 see SOD100
RCRA WASTE NUMBER P010 see ARB250
RCRA WASTE NUMBER P001 see WAT200
RCRA WASTE NUMBER P021 see CAQ500
RCRA WASTE NUMBER P022 see CBV500
RCRA WASTE NUMBER P023 see CDY500
RCRA WASTE NUMBER P024 see CEH680
RCRA WASTE NUMBER P026 see CKL000
RCRA WASTE NUMBER P027 see CKT250
RCRA WASTE NUMBER P028 see BEE375
RCRA WASTE NUMBER P029 see CNL000
RCRA WASTE NUMBER P002 see ADD250
RCRA WASTE NUMBER P020 see BRE500
RCRA WASTE NUMBER P031 see COO000
RCRA WASTE NUMBER P033 see COO750
RCRA WASTE NUMBER P034 see CPK500
RCRA WASTE NUMBER P036 see DGB600
RCRA WASTE NUMBER P037 see DHB400
RCRA WASTE NUMBER P039 see DXH325
RCRA WASTE NUMBER P003 see ADR000
RCRA WASTE NUMBER P030 see COI500
RCRA WASTE NUMBER P401 see NIM500
RCRA WASTE NUMBER P042 see VGP000
RCRA WASTE NUMBER P043 see IRF000
RCRA WASTE NUMBER P044 see DSP400
RCRA WASTE NUMBER P404 see EPC500
RCRA WASTE NUMBER P045 see DAB400
RCRA WASTE NUMBER P046 see DTJ400
RCRA WASTE NUMBER P047 see DUS700
RCRA WASTE NUMBER P048 see DUZ000
RCRA WASTE NUMBER P049 see DXL800
RCRA WASTE NUMBER P004 see AFK250
RCRA WASTE NUMBER P040 see DJX400
RCRA WASTE NUMBER P051 see EAT500
RCRA WASTE NUMBER P054 see EJM900
RCRA WASTE NUMBER P056 see FEZ000
RCRA WASTE NUMBER P057 see FFF000
RCRA WASTE NUMBER P058 see SHG500
RCRA WASTE NUMBER P059 see HAR000
RCRA WASTE NUMBER P005 see AFV500
RCRA WASTE NUMBER P050 see EAQ750
RCRA WASTE NUMBER P062 see HCY000
RCRA WASTE NUMBER P063 see HHS000
RCRA WASTE NUMBER P064 see MKX250
RCRA WASTE NUMBER P065 see MDC000
RCRA WASTE NUMBER P066 see MDU600
RCRA WASTE NUMBER P067 see PNL400
RCRA WASTE NUMBER P068 see MKN000
RCRA WASTE NUMBER P069 see MLC750
RCRA WASTE NUMBER P006 see AHE750
RCRA WASTE NUMBER P060 see IKO000
RCRA WASTE NUMBER P071 see MNH000
RCRA WASTE NUMBER P072 see AQN635
RCRA WASTE NUMBER P073 see NCZ000
RCRA WASTE NUMBER P074 see NDB500
RCRA WASTE NUMBER P075 see NDN000
RCRA WASTE NUMBER P076 see NEG100
RCRA WASTE NUMBER P077 see NEO500
RCRA WASTE NUMBER P078 see NGR500
RCRA WASTE NUMBER P007 see AKT750
RCRA WASTE NUMBER P070 see CBM500
RCRA WASTE NUMBER P081 see NGY000
RCRA WASTE NUMBER P082 see NKA600
RCRA WASTE NUMBER P084 see NKY000
RCRA WASTE NUMBER P085 see OCM000
RCRA WASTE NUMBER P087 see OKK000
RCRA WASTE NUMBER P088 see DXD000, EAR000

RCRA WASTE NUMBER P089 see PAK000
RCRA WASTE NUMBER P008 see AMI500, POO050
RCRA WASTE NUMBER P092 see ABU500
RCRA WASTE NUMBER P093 see PGN250
RCRA WASTE NUMBER P094 see PGS000
RCRA WASTE NUMBER P095 see PGX000
RCRA WASTE NUMBER P096 see PGY000
RCRA WASTE NUMBER P097 see FAB600
RCRA WASTE NUMBER P098 see PLC500
RCRA WASTE NUMBER P099 see PLS250
RCRA WASTE NUMBER P009 see ANS500
RCRA WASTE NUMBER U111 see NKB700
RCRA WASTE NUMBER U112 see EFR000
RCRA WASTE NUMBER U113 see EFT000
RCRA WASTE NUMBER U115 see EJN500
RCRA WASTE NUMBER U116 see IAQ000
RCRA WASTE NUMBER U117 see EJU000
RCRA WASTE NUMBER U118 see EMF000
RCRA WASTE NUMBER U119 see EMF500
RCRA WASTE NUMBER U011 see AMY050
RCRA WASTE NUMBER U110 see DWR000
RCRA WASTE NUMBER U101 see XKJ500
RCRA WASTE NUMBER U121 see TIP500
RCRA WASTE NUMBER U122 see FMV000
RCRA WASTE NUMBER U123 see FNA000
RCRA WASTE NUMBER U124 see FPK000
RCRA WASTE NUMBER U125 see FPQ875
RCRA WASTE NUMBER U126 see GGW000
RCRA WASTE NUMBER U127 see HCC500
RCRA WASTE NUMBER U128 see HCD250
RCRA WASTE NUMBER U129 see BBQ500
RCRA WASTE NUMBER U102 see DTR200
RCRA WASTE NUMBER U120 see FDF000
RCRA WASTE NUMBER U131 see HCI000
RCRA WASTE NUMBER U132 see HCL000
RCRA WASTE NUMBER U133 see HGS000
RCRA WASTE NUMBER U134 see HHU500
RCRA WASTE NUMBER U135 see HIC500
RCRA WASTE NUMBER U136 see HKC000
RCRA WASTE NUMBER U137 see IBZ000
RCRA WASTE NUMBER U138 see MKW200
RCRA WASTE NUMBER U139 see IGS000
RCRA WASTE NUMBER U103 see DUD100
RCRA WASTE NUMBER U130 see HCE500
RCRA WASTE NUMBER U141 see IRZ000
RCRA WASTE NUMBER U142 see KEA000
RCRA WASTE NUMBER U143 see LBG000
RCRA WASTE NUMBER U144 see LCV000
RCRA WASTE NUMBER U146 see LCH000
RCRA WASTE NUMBER U147 see MAM000
RCRA WASTE NUMBER U148 see DMC600
RCRA WASTE NUMBER U149 see MAO250
RCRA WASTE NUMBER U014 see IBB000
RCRA WASTE NUMBER U140 see IIL000
RCRA WASTE NUMBER U151 see MCW250
RCRA WASTE NUMBER U152 see MGA750
RCRA WASTE NUMBER U153 see MLE650
RCRA WASTE NUMBER U154 see MGB150
RCRA WASTE NUMBER U155 see TEO250
RCRA WASTE NUMBER U156 see MIG000
RCRA WASTE NUMBER U157 see MIJ750
RCRA WASTE NUMBER U158 see MJM200
RCRA WASTE NUMBER U159 see MKA400
RCRA WASTE NUMBER U015 see ASA500
RCRA WASTE NUMBER U105 see DVH000
RCRA WASTE NUMBER U150 see PED750
RCRA WASTE NUMBER U161 see HFG500
RCRA WASTE NUMBER U162 see MLH750
RCRA WASTE NUMBER U163 see MMP000
RCRA WASTE NUMBER U164 see MPW500
RCRA WASTE NUMBER U165 see NAJ500
RCRA WASTE NUMBER U166 see NBA500
RCRA WASTE NUMBER U167 see NBE700
RCRA WASTE NUMBER U168 see NBE500
RCRA WASTE NUMBER U169 see NEX000
RCRA WASTE NUMBER U016 see BAW750
RCRA WASTE NUMBER U106 see DVH400
RCRA WASTE NUMBER U160 see MKA500
RCRA WASTE NUMBER U171 see NIY000
RCRA WASTE NUMBER U172 see BRY500
RCRA WASTE NUMBER U173 see NKM000
RCRA WASTE NUMBER U174 see NJW500
RCRA WASTE NUMBER U176 see ENV000
RCRA WASTE NUMBER U177 see MNA750
RCRA WASTE NUMBER U178 see MMX250

RCRA WASTE NUMBER U179 see NLJ500
RCRA WASTE NUMBER U017 see BAY300
RCRA WASTE NUMBER U107 see DVL600
RCRA WASTE NUMBER U170 see NIF000
RCRA WASTE NUMBER U181 see NMP500
RCRA WASTE NUMBER U182 see PAI250
RCRA WASTE NUMBER U183 see PAV500
RCRA WASTE NUMBER U184 see PAW500
RCRA WASTE NUMBER U185 see PAX000
RCRA WASTE NUMBER U187 see ABG750
RCRA WASTE NUMBER U188 see PDN750
RCRA WASTE NUMBER U189 see PHS000
RCRA WASTE NUMBER U018 see BBC250
RCRA WASTE NUMBER U108 see DVQ000
RCRA WASTE NUMBER U180 see NLP500
RCRA WASTE NUMBER U191 see MOY000
RCRA WASTE NUMBER U192 see DTT600
RCRA WASTE NUMBER U193 see PML400
RCRA WASTE NUMBER U194 see PND250
RCRA WASTE NUMBER U196 see POP250
RCRA WASTE NUMBER U197 see QQS200
RCRA WASTE NUMBER U019 see BBL250
RCRA WASTE NUMBER U109 see HHG000
RCRA WASTE NUMBER U190 see PHW750
RCRA WASTE NUMBER U001 see AAG250
RCRA WASTE NUMBER U010 see AHK500
RCRA WASTE NUMBER U211 see CBY000
RCRA WASTE NUMBER U212 see TBT000
RCRA WASTE NUMBER U213 see TCR750
RCRA WASTE NUMBER U214 see TEI250
RCRA WASTE NUMBER U215 see TEJ000
RCRA WASTE NUMBER U216 see TEJ250
RCRA WASTE NUMBER U217 see TEK750
RCRA WASTE NUMBER U218 see TFA000
RCRA WASTE NUMBER U219 see ISR000
RCRA WASTE NUMBER U021 see BBX000
RCRA WASTE NUMBER U210 see PCF275
RCRA WASTE NUMBER U201 see REA000
RCRA WASTE NUMBER U221 see TGL500, TGL750
RCRA WASTE NUMBER U222 see TGS500
RCRA WASTE NUMBER U223 see TGM740, TGM750
RCRA WASTE NUMBER U225 see BNL000
RCRA WASTE NUMBER U226 see MIH275
RCRA WASTE NUMBER U227 see TIN000
RCRA WASTE NUMBER U228 see TIO750
RCRA WASTE NUMBER U202 see BCE500
RCRA WASTE NUMBER U022 see BCS750
RCRA WASTE NUMBER U220 see TGK750
RCRA WASTE NUMBER U231 see TIW000
RCRA WASTE NUMBER U232 see TAA100
RCRA WASTE NUMBER U233 see TIX500
RCRA WASTE NUMBER U234 see TMK500
RCRA WASTE NUMBER U235 see TNC500
RCRA WASTE NUMBER U236 see CMO250
RCRA WASTE NUMBER U237 see BIA250
RCRA WASTE NUMBER U238 see UVA000
RCRA WASTE NUMBER U239 see XGS000
RCRA WASTE NUMBER U023 see BFL250
RCRA WASTE NUMBER U230 see TIV750
RCRA WASTE NUMBER U242 see PAX250
RCRA WASTE NUMBER U244 see TFS350
RCRA WASTE NUMBER U246 see COO500
RCRA WASTE NUMBER U247 see MEI450
RCRA WASTE NUMBER U024 see BID750
RCRA WASTE NUMBER U240 see DAA800
RCRA WASTE NUMBER U204 see SBO000, SBQ500
RCRA WASTE NUMBER U025 see DFJ050
RCRA WASTE NUMBER U026 see BIF250
RCRA WASTE NUMBER U206 see SMD000
RCRA WASTE NUMBER U027 see BII250
RCRA WASTE NUMBER U207 see TBN750
RCRA WASTE NUMBER U028 see DVL700
RCRA WASTE NUMBER U208 see TBQ000
RCRA WASTE NUMBER U029 see MHR200
RCRA WASTE NUMBER U209 see TBQ100
RCRA WASTE NUMBER U002 see ABC750
RCRA WASTE NUMBER U020 see BBS750
RCRA WASTE NUMBER U200 see RDK000
RCRA WASTE NUMBER U031 see BPW500
RCRA WASTE NUMBER U032 see CAP500
RCRA WASTE NUMBER U033 see CCA500
RCRA WASTE NUMBER U034 see CDN550
RCRA WASTE NUMBER U035 see CDO500
RCRA WASTE NUMBER U036 see CDR750
RCRA WASTE NUMBER U037 see CEJ125

RCRA WASTE NUMBER U038 see DER000
RCRA WASTE NUMBER U039 see CFE250
RCRA WASTE NUMBER U003 see ABE500
RCRA WASTE NUMBER U041 see EAZ500
RCRA WASTE NUMBER U042 see CHI250
RCRA WASTE NUMBER U043 see VNP000
RCRA WASTE NUMBER U044 see CHJ500
RCRA WASTE NUMBER U045 see MIF765
RCRA WASTE NUMBER U046 see CIO250
RCRA WASTE NUMBER U047 see CJA000
RCRA WASTE NUMBER U048 see CJK250
RCRA WASTE NUMBER U049 see CLK235
RCRA WASTE NUMBER U051 see BAT850, CMY825
RCRA WASTE NUMBER U052 see CNW500, CNW750, CNX000, CNX250
RCRA WASTE NUMBER U053 see COB250, COB260, POD000
RCRA WASTE NUMBER U055 see COE750
RCRA WASTE NUMBER U056 see CPB000
RCRA WASTE NUMBER U057 see CPC000
RCRA WASTE NUMBER U058 see CQC650
RCRA WASTE NUMBER U059 see DAC000
RCRA WASTE NUMBER U050 see CML810
RCRA WASTE NUMBER U005 see FDR000
RCRA WASTE NUMBER U061 see DAD200
RCRA WASTE NUMBER U062 see DBI200
RCRA WASTE NUMBER U063 see DCT400
RCRA WASTE NUMBER U064 see BCQ500
RCRA WASTE NUMBER U066 see DDL800
RCRA WASTE NUMBER U067 see EIY500
RCRA WASTE NUMBER U068 see DDP800
RCRA WASTE NUMBER U069 see DEH200
RCRA WASTE NUMBER U006 see ACF750
RCRA WASTE NUMBER U060 see BIM500
RCRA WASTE NUMBER U071 see DEP800
RCRA WASTE NUMBER U072 see DEP800
RCRA WASTE NUMBER U073 see DEQ600
RCRA WASTE NUMBER U074 see DEV000
RCRA WASTE NUMBER U075 see DFA600
RCRA WASTE NUMBER U076 see DFF809
RCRA WASTE NUMBER U077 see EIY600
RCRA WASTE NUMBER U078 see VPK000
RCRA WASTE NUMBER U079 see ACK000
RCRA WASTE NUMBER U007 see ADS250
RCRA WASTE NUMBER U070 see DEP800
RCRA WASTE NUMBER U081 see DFX800
RCRA WASTE NUMBER U082 see DFY000
RCRA WASTE NUMBER U083 see DGG800, PNJ400
RCRA WASTE NUMBER U084 see DGG950
RCRA WASTE NUMBER U085 see BGA750
RCRA WASTE NUMBER U086 see DJL400
RCRA WASTE NUMBER U088 see DJX000
RCRA WASTE NUMBER U089 see DKA600
RCRA WASTE NUMBER U008 see ADS750
RCRA WASTE NUMBER U080 see MJP450
RCRA WASTE NUMBER U091 see DCJ200
RCRA WASTE NUMBER U092 see DOQ800
RCRA WASTE NUMBER U093 see DOT300
RCRA WASTE NUMBER U094 see DQJ200
RCRA WASTE NUMBER U095 see TGJ750
RCRA WASTE NUMBER U096 see IOB000
RCRA WASTE NUMBER U097 see DQY950
RCRA WASTE NUMBER U098 see DSF400
RCRA WASTE NUMBER U099 see DSF600
RCRA WASTE NUMBER U009 see ADX500
RCRA WASTE NUMBER U090 see DMD600
RCTA see RJA500
RD 8 see SCK600
RD 120 see SCK600
RD 174 see TEC600
RD 292 see CKG000
RD 406 see DGB000
R.D. 1403 see EPD500
RD 1572 see DBV400
RD 2195 see CEP000
R.D. 2786 see WAK000
RD 4593 see CIR500
RD-6584 see RDP300
RD7693 see BAV000
RD 11654 see IJG000
R.D. 13621 see IIU000
R.D. 27419 see MJL250
RDGE see REF000
RDX see CPR800
RDX and HMX MIXTURES, desensitized with not <10% phleg matizer by weight (UN 0391) (DOT) see CPR800

RDX and HMX MIXTURES, wetted with not <15% water by weight (UN 0391) (DOT) see CPR800
RDX, wetted with not <15% water by weight (UN 0072) (DOT) see CPR800
RDX, desensitized (UN 0483) (DOT) see CPR800
RE-4355 see NAG400
RE 5030 see BSG300
RE 5454 see MIA000
RE 5655 see MOU750
RE 1-0185 see ELH600
RE 12420 see DOP600
RE-45550 see EPR100
REACID see COH250
REACTHIN see AES650
REACTIVE BLACK 8 see CMS227
REACTIVE BLUE 19 see BMM500
REACTIVE BLUE 4 see CMS228
REACTIVE BRILLIANT RED 5SKH see PMF540
REACTROL see CMV400
READPRET KPN see DTG000
REALGAR see ARF000
REANIMIL see DNV000
REAZID see COH250
REAZIDE see COH250
REBELATE see DSP400
REBEMID see BCM250
REBONEX see CBT750
REBRAMIN see VSZ000
REBUGEN see IIU000
REC 1-0060 see AJE350
REC 1-0185 see ELH600
REC 7-0040 see FCB100
REC 7/0267 see DNV000, DNV200
REC 7/0268 see DRL600
REC 7-0518 see DNN000
REC 7-0591 see AJO750
REC 15-1533 see DAP880
RE 5305 (CALIFORNIA CHEMICAL) see BSG250
RECANESCIN see RDF000
RECHETON see KGK000
RECINNAMINE see TLN500
RECIPIN see RDK000
RECITENSINA see TLN500
RECOFNAN see CMF350
RECOGNAN see CMF350
RECOLIP see ARQ750
RECOLITE FAST RED RBL see MMP100
RECOLITE FAST RED RL see MMP100
RECOLITE FAST RED RYL see MMP100
RECOLITE FAST YELLOW BLF see CMS208
RECOLITE FAST YELLOW BLT see CMS208
RECOLITE FAST YELLOW B 2T see CMS212
RECOLITE ORANGE G see CMS145
RECOLITE RED LAKE C see CHP500
RECOLITE YELLOW GB see DEU000
RECONIN see FOS100
RECONOX see PDP250
RECOP see CNK559
RECORDIL see ELH600
RECREIN see DPA000
RECTALAD-AMINOPHYLLINE see TEP500
RECTHORMONE OESTRADIOL see EDP000
RECTHORMONE TESTOSTERONE see TBG000
RECTODELT see PLZ000
RECTULES see CDO000
RED #14 see HJF500
RED 102 see CMS155
RED 104 see ADG250
RED 203 see CMS150
RED 219 see CMS160
1306 RED see FAG050
1379 RED see CMM300
1424 RED see CMS231
1427 RED see FAG040
1671 RED see FAG040
1695 RED see FMU070
1860 RED see CHP500
11067 RED see CMS160
11070 RED see CMS155
11411 RED see FAG070
11427 RED see CMS231
11445 RED see BNH500
11554 RED see IHD000
11938 RED see CMS150
11959 RED see HJF500
11969 RED see ADG250

12101 RED see FAG050
12418 RED see MAC250
111440 RED see OHA000
REDAMINA see VSZ000
REDAX see DWI000
RED 10B see CMS231
RED B see XRA000
RED 2B ACID see AJJ250
RED 3B ACID see CLO600
RED 4B ACID see AKQ000
RED BALL see CAT775
RED BASE CIBA IX see CLK220, CLK235
RED BASE CIBA V see NEQ000
RED BASE CIBA VI see KDA050
RED BASE 3 GL see KDA050
RED BASE IRGA IX see CLK220, CLK235
RED BASE IRGA V see NEQ000
RED BASE IRGA VI see KDA050
RED BASE NB see NEQ000
RED BASE NTR see CLK220
RED B BASE see NEQ000
RED BEAD VINE see RMK250
RED BEAN see NBR800
REDBIRD FLOWER OR CACTUS see SDZ475
RED CEDARWOOD OIL see CCR000
RED COPPER OXIDE see CNO000
REDDON see TAA100
REDDOX see TAA100
RED DYE No. 2 see FAG020
REDERGIN see DLL400
RED FOR LAKE C TONER see CMS150
RED FUMING NITRIC ACID see NEE500
RED 2G see CMM300
RED 2G BASE see NEO500
RED 3G BASE see KDA050
RED HOTS see NBR800
REDIFAL see SNN300
REDI-FLOW see BAP000
RED INK PLANT see PJJ315
RED IRON ORE see HAO875
RED IRON OXIDE see IHD000
REDISOL see VSZ000
RED KB BASE see CLK225
RED LEAD see LDS000
RED LEAD CHROMATE see LCS000
RED LEAD OXIDE see LDS000
RED LOBELIA see CCJ825
RED MERCURIC IODIDE see MDD000
RED MOROCCO see PCU375
RED No. 1 see FAG050
RED No. 2 see FAG020
RED No. 4 see FAG050
RED No. 5 see XRA000
RED No. 104 see ADG250
RED NO 213 see FAG070
RED NO. 219 see CMS160
RED No. 227 see CMS231
RED OCHRE see IHD000
RED OIL see OHU000, TOD500
RED OXIDE of MERCURY see MCT500
RED PEPPER see PCB275
RED PRECIPITATE see MCT500
RED R see FMU070
RED ROOT see AHJ875
RED 3R SOLUBLE IN GREASE see SBC500
RED SALT CIBA IX see CLK235
RED SALT IRGA IX see CLK235
RED SCARLET see CHP500
RED-SEAL-9 see ZKA000
REDSKIN see AGJ250
RED SPIDER LILY see REK325
RED SQUILL see RCF000
RED TETRAZOLIUM see TMV500
RED TR BASE see CLK220
RED TRS SALT see CLK235
REDUCED-d-PENICILLAMINE see MCR750
REDUCTO see DKE800
REDUCTONE see SHR500
REDUFORM see NNW500
REDUL see GHK200
REDUTON see RLU000
RED WEED see PJJ315
RED ZH see OHA000
REE see TFF100
REED AMINE 400 see DFY800

REED LV 2,4-D see ILO000
REED LV 400 2,4-D see ILO000
REED LV 600 2,4-D see ILO000
REELON see PAF250
REFINED SOLVENT NAPHTHA see PCT250
REFLEXYN see GKK000
REFORMIN see DJS200
REFOSPOREN see RCK000
REFRACTORY CERAMIC FIBERS see RCK725
REFRIGERANT 12 see DFA600
REFRIGERANT 14 see CBY250
REFRIGERANT 22 see CFX500
REFRIGERANT 112 see TBP050
REFRIGERANT 112a see TBP000
REFRIGERANT 113 see FOO000
REFRIGERANT 502 see FOO510
R 14 (REFRIGERANT) see CBY250
REFRIGERANT R 14 see CBY250
REFUGAL see CMW700
REFUSAL see DXH250
REGAL 99 see CBT750
REGAL 300 see CBT750
REGAL 330 see CBT750
REGAL 600 see CBT750
REGAL see CBT750
REGAL 400R see CBT750
REGAL SRF see CBT750
REGARDIN see ARQ750
REGELAN see ARQ750
REGENERATED CELLULOSE see HHK000
REGENON see DIP600
REGENT see CBT750
REGIANIN see WAT000
REGIM 8 see TKQ250
REGIN 8 see TKQ250
REGION see SNP500
REGITIN see PDW400
REGITINE see PDW400
REGITINE MESYLATE see PDW950
REGITINE METHANESULFONATE see PDW950
REGITIN METHANESULPHONATE see PDW950
REGITIPE see PDW400
REGLISSE (GUADELOUPE and HAITI) see RMK250
REGLON see DWX800
REGLONE see DWX800
REGONAL see MDL600
REGONOL see GLU000
REGONON HYDROCHLORIDE see DIP600
REGROTON see RDK000
REGULIN see BFW250, TBJ275
REGULOX see DMC600
REGULOX 50 W see DMC600
REGULOX W see DMC600
REGULTON see MQS100
REGUTOL see DJL000
REHIBIN see FBP100
REHORMIN see DJS200
REICHSTEIN'S F see MIW500
REICHSTEIN'S SUBSTANCE FA see CNS800
REICHSTEIN'S SUBSTANCE M see CNS750
REIN GUARIN see GLU000
REISE-ENGLETTEN see DYE600
REKAWAN see PLA500
RELA see IPU000
RELACT see DLY000
RELAMINAL see DCK759
RELAN BETA see BEQ625
RELANE see DVP400
RELANIUM see DCK759
RELASOM see IPU000
RELAX see GKK000, IPU000
RELAXAN see PDD300
RELAXANT see GGS000
RELAXAR see GGS000
RELAXYL-G see RLU000
RELBAPIRIDINA see PPO000
RELDAN see CMA250
RELEFACT LH-RH see LIU360
RELESTRID see GKK000
RELIBERAN see MDQ250
RELICOR see DHS200
RELICOR HYDROCHLORIDE see DHS200
RELITON YELLOW C see AAQ250
RELITON YELLOW R see DUW500
RELIVERAN see AJH000

RELON P see PJY500
RELUTIN see HNT500
REMADERM YELLOW HPR see MDM775
REMALAN BRILLIANT BLUE R see BMM500
REMANTADIN see AJU625
REMASAN CHLOROBLE M see MAS500
REMAZIN see HNI500
REMAZOL BLACK B see RCU000
REMAZOL BRILLIANT BLUE R see BMM500
REMAZOL YELLOW G see RCZ000
REMEFLIN see DNV000, DNV200
REMESTAN see CFY750
REMESTYP see MGC350
REMICYCLIN see TBX250
REMID see ZVJ000
REMIN see CKE750
REMKO see IGK800
REMOL TRF see BGJ250
REMONOL see RDZ900
REMSED see PMI750
REMTAL see TJL500
REMYLINE Ac see SLJ500
REMZYME PL 600 see GGA800
RENACIT 1 see NAP500
RENAFUR see NDY000
RENAGLADIN see VGP000
RENAL AC see ALT250
RENAL EG see ALT500
RENALEPTINE see VGP000
RENALINA see VGP000
RENAL MD see TGL750
RENAL PF see PEY500
RENAL SLA see DBO400
RENAL SO see CEG625
RENAMYCIN see ACJ250
RENARCOL see ARW250, GGS000
RENARDIN see DMX200
RENARDINE see DMX200
RENBORIN see DCK759
RENESE R see RDK000
RENGASIL see PJA220
RENOFORM see VGP000
RENOGRAFFIN M-76 see AOO875
RENOGRAFIN see AOO875
RENOLBLAU 3B see CMO250
RENOL MOLYBDATE RED RGS see MRC000
RENO M 60 see AOO875
RENO M see AOO875
RENO-M-DIP see AOO875
RENON see CLY600
RENONCULE (CANADA) see FBS100
RENOSTYPRICIN see VGP000
RENOSTYPTIN see VGP000
RENOSULFAN see SNN500
REN O-SAL see HMY000
RENSTAMIN see DBM800
RENTIAPRIL see FAQ950
RENTOVET see TAF675
RENTYLIN see PBU100
RENUMBRAL see HGB200
RENURIX see AOO875
REOMAX see DFP600
REOMOL DOA see AEO000
REOMOL DOP see DVL700
REOMOL D 79P see DVL700
REOMUCIL see CBR675
REORGANIN see RLU000
REOXYL see HFF500
REP see BCM250
REPAIRSIN see PKB500
β-REPARIL see EDL500
REPARIL see EDK875
REPARIL SODIUM SALT see EDM000
REPEL see DKC800
REPELLENT 612 see EKV000
REPELTIN see AFL500
REPHOXITIN see CCS500
REPICIN see BEQ625
REPOC see PJS750
REPOCAL see CAV000
REPOISE see MFD500
REPOISE MALEATE see BSZ000
REPOSO-TMD see TBF750
REPPER 333 see EAU500
REPPER-DET see DKC800

REPRISCAL see SHL500
REPRODAL see AQH500
REPROMIX see MCA000
REPROTEROL HYDROCHLORIDE see DNA600
REPTILASE see RDA350
REPTILASE R see RDA350
REPTILASE S see CMY725
REPUDIN-SPECIAL see DKC800
REPULSON see PAF550
REQUTOL see DJL000
RERANIL see TBO500
76 RES see ADW200
β-RESACETOPHENONE see DMG400
RESACETOPHENONE see DMG400
RESACTIN A see SMC500
RESALTEX see RDK000
RESAMINE FAST ORANGE G see CMS145
RESAMINE FAST YELLOW GGP see CMS212
RESAMINE RED GB see CMS150
RESAMINE RUBINE BC see CMS155
RESAMINE YELLOW GP see CMS212
RESAMIN 155F see UTU500
RESAMIN HW 505 see UTU500
RESAMIN MW 811 see MCB050
RESANTIN see RDA375
RESARIT 4000 see PKB500
RESAZOIN see HNG500
RESAZURIN see HNG500
RESAZURINE see HNG500
RESCALOID see TLN500
RESCAMIN see TLN500
RESCIDAN see TLN500
RESCIN see TLN500
RESCINNAMINE see TLN500
RESCINPAL see TLN500
RESCINSAN see TLN500
RESCITEN see TLN500
RESCUE SQUAD see SHF500
RESEDA BODY see PFR400
RESEDIN see RDK000
RESEDREX see RDK000
RESEDRIL see RDK000
RESE-LAR see RDK000
RESER-AR see RDK000
RESERBAL see RDK000
RESERCAPS see RDK000
RESERCEN see RDK000
RESERCRINE see RDK000
RESERFIA see RDK000
RESERJEN see RDK000
RESERLOR see RDK000
RESERP see RDK000
RESERPAL see RDK000
RESERPAMED see RDK000
RESERPANCA see RDK000
RESERPATE de METHYLE (FRENCH) see MPH300
RESERPENE see RDK000
RESERPEX see RDK000
RESERPIC ACID-METHYL ESTER, ESTER with 4-HYDROXY-3,5-DIMETHOXY-
   BENZOIC ACID ETHYL CARBONATE see RCA200
RESERPIDEFE see RDK000
RESERPIDINE see RDF000
RESERPIL see RDK000
RESERPIN see RDK000
RESERPINA see RDK000
RESERPINE see RDK000
RESERPINENE see TLN500
RESERPINE PHOSPHATE see RDK050
RESERPINUM see RDK000
RESERPKA see RDK000
RESERPOID see RDK000
RESERPUR see RDK000
RESERP 'WANDER' see RDK000
RESERSANA see RDK000
RESERUTIN see RDK000
RESIATRIC see RDK000
RESIBUFOGENIN see BOM650
RESIDINE see RDK000
RESIDUAL(HEAVY) FUEL OIL see FOP050
RESIDUAL OIL SOLVENT EXTRACT see MQV863
RESIDUAL OILS (PETROLEUM), ACID-TREATED (9CI) see MQV872
RESIDUES (PETROLEUM), THERMAL CRACKED see RDK100
RESIDUES (PETROLEUM), VACUUM see RDK200
RESIL see RLU000
RESIMENE 714 see MCB050

RESIMENE 717 see MCB050
RESIMENE 730 see MCB050
RESIMENE 731 see MCB050
RESIMENE 740 see MCB050
RESIMENE 745 see MCB050
RESIMENE 746 see MCB050
RESIMENE 747 see MCB050
RESIMENE 750 see MCB050
RESIMENE 753 see MCB050
RESIMENE 755 see MCB050
RESIMENE 817 see MCB050
RESIMENE 841 see MCB050
RESIMENE 842 see MCB050
RESIMENE RF 4518 see MCB050
RESIMENE RF 5306 see MCB050
RESIMENE RS 466 see MCB050
RESIMENE X 712 see MCB050
RESIMENE X 714 see MCB050
RESIMENE X 720 see MCB050
RESIMENE X 735 see MCB050
RESIMENE X 730 see MCB050
RESIMENE X 745 see MCB050
RESIMENE X 740 see MCB050
RESIMENE X 764 see MCB050
RESIMENE X 975 see UTU500
RESIMENE X 970 see UTU500
RESIMENE X 980 see UTU500
RESIMINE 975 see UTU500
RESIN 516 see MCB050
RESIN (solution) see RDP000
RESIN ACIDS and ROSIN ACIDS, CALCIUM SALTS see CAW500
S-RESIN AER 20 see UTU500
RESINATED INDO BLUE B 85 see DFN425
RESINA X see UTU500
RESINE see RDK000
RESINOID IRIS see OHC130
RESINOID ORRIS see OHC130
RESINOL BROWN RRN see NBG500
RESINOL ORANGE G see CMP600
RESINOL ORANGE R see PEJ500
RESINOL RED 2B see SBC500
RESINOL RED G see CMS242
RESINOL RRN see NBG500
RESINOL YELLOW GR see DOT300
RESINO RED K see CMS148
RESIN SCARLET 2R see XRA000
RESIN SOLUTION, flammable (DOT) see RDP000
RESIN T see CNH125
RESIN TOLU see BAF000
RESIPAL see TLN500
RESIREN RED TB see AKI750
RESIREN VIOLET TR see DBP000
RESIREN YELLOW TG see AAQ250
RESISAN see RDP300
RESIN, SMA 1440-H see SEA500
RESISTAB see RDU000
RESISTAMINE see POO750, TMP750
RESISTOFLEX see PKP750
RESISTOMYCIN see HAL000
RESISTOMYCIN (BAYER) see KAM000, KAV000
RESISTOPHEN see MNV250
RESITAN see VBK000
RESKINNAMIN see TLN500
RESLOOM HP 50 see MCB050
RESLOOM HP see MCB050
RESLOOM M 75 see HDY000
(+)-cis-RESMETHRIN see RDZ875
d-trans-RESMETHRIN see BEP750
RESMETHRIN see BEP500
(+)-trans-RESMETHRIN see BEP750
RESMETRINA (PORTUGUESE) see BEP500
RESMIT see CGA000
RESOACETOPHENONE see DMG400
RESOBANTIN see DJM800, XCJ000
RESOCALM see RDK000
RESOCHIN see CLD000, CLD250
RESOCHIN DIPHOSPHATE see CLD250
RESOFORM ORANGE G see PEJ500
RESOFORM ORANGE R see XRA000
RESOFORM RED G see SBC500
RESOFORM YELLOW GGA see DOT300
RESOIDAN see CCK125
RESOLIN BLUE FBL see CMP070
RESOLIN RED FB see AKI750
RESOLIN RED FBE see AKI750

RESOLIN YELLOW 5R see CMP090
RESOMINE see RDK000
RESOQUINA see CLD000
RESOQUINE see CLD000, CLD250
β-RESORCALDEHYDE see REF100
RESORCIN see REA000
RESORCIN ACETATE see RDZ900
β-RESORCINALDEHYDE see REF100
RESORCINE see REA000
RESORCINE BROWN J see XMA000
RESORCINE BROWN R see XMA000
RESORCIN MONOACETATE see RDZ900
RESORCINOL see REA000
RESORCINOL, 2-AMINO-, HYDROCHLORIDE see AML600
RESORCINOL, 4-BENZYL- see BFI400
RESORCINOL BIS(2,3-EPOXYPROPYL)ETHER see REF000
RESORCINOL, DIACETATE see REA100
RESORCINOL, DIBENZOATE see BHB100
RESORCINOL DIGLYCIDYL ETHER see REF000
RESORCINOL DIMETHYL ETHER see REF025
RESORCINOL, 2,4-DINITROSO- see DVF300
RESORCINOL, 2-METHYL- see MPH400
RESORCINOL METHYL ETHER see REF050
RESORCINOL, MONOACETATE see RDZ900
RESORCINOL, MONOBENZOATE see HNH500
RESORCINOL MONOMETHYL ETHER see REF050
RESORCINOL OXYDIANILINE see REF070
RESORCINOLPHTHALEIN see FEV000
RESORCINOL PHTHALEIN SODIUM see FEW000
RESORCINOL, 2,4,6-TRINITRO-, BARIUM SALT, HYDRATE, (2:1:1) see
  BAO900
RESORCINYL DIGLYCIDYL ETHER see REF000
RESORCITATE see RDZ900
β-RESORCYLALDEHYDE see REF100
α-RESORCYLIC ACID see REF200
β-RESORCYLIC ALDEHYDE see REF100
RESORIN RED FBE see AKI750
RESOTROPIN see HEI500
RESOXOL see SNN500
RESPAIRE see ACH000
RESPENYL see RLU000
RESPERIN see RDK000
RESPERINE see RDK000
RESPIFRAL see DMV600
RESPILENE see MFG250
RESPIRIDE see DAI200
RESPITAL see RDK000
RESPRAMIN see AJD000
RESTAMIN see BBV500
RESTAMINE see BBV500
RESTAS see FMR100
RESTENIL see MQU750
REST-ON see TEO250
RESTORIL see CFY750
RESTOVAR see LJE000
RESTRAN see RDK000
RESTROL see DAL600
RESTROPIN see SBH500
RESTRYL see TEO250
RESULFON see AHO250
RESURRECTION LILY see REK325
RESYDROL WM 501 see MCB050
RESYDROL WM 461E see MCB050
RETACEL see CMF400
RETALON see DAL600
RETAMA see YAK350
RETAMID see AKO500
RETAR-B₁ see FQJ100
RETARCYL see SAI100
RETARD see DMC600
RETARDER AK see PHW750
RETARDER BA see BCL750
RETARDER ESEN see PHW750
RETARDER J see DWI000
RETARDER PD see PHW750
RETARDER W see SAI000
RETARDEX see BCL750
RETARDILLIN see PAQ200
RETARPEN see BFC750
RETASULFIN see AKO500
RETENS see HGP550
RETENSIN see PDD300
RETICULIN see SLX500
RETIN-A see VSK950
9-cis-RETINAL see VSK975

all-E-RETINAL see VSK985
RETINAL, all-trans- see VSK985
all-trans-RETINAL see VSK985
trans-RETINAL see VSK985
RETINAL (9CI) see VSK985
9-cis-RETINALDEHYDE see VSK975
RETINALDEHYDE see VSK985
RETINENE 1 see VSK985
α-RETINENE see VSK985
RETINENE see VSK985
13-cis-RETINOIC ACID see VSK955
β-RETINOIC ACID see VSK950
all-trans-RETINOIC ACID see VSK950
RETINOIC ACID see VSK950
RETINOIC ACID, SODIUM SALT see SJN000
RETINOID ETRETIN see REP400
all-trans RETINOL see VSK600
RETINOL see VSK600
RETINOL ACETATE see VSK900
RETINOL PALMITATE see VSP000
all-trans-RETINYL ACETATE see VSK900
RETINYL ACETATE see VSK900
all-trans-RETINYLIDENE METHYL NITRONE see REZ200
RETINYL PALMITATE see VSP000
RETOZIDE see ILD000
RETRANGOR see EID200
cis-RETRONECIC ACID ESTER of RETRONECINE see RFP000
cis-RETRONECIC ACID ESTER of RETRONECINE-N-OXIDE see RFU000
RETRONECINE HYDROCHLORIDE see RFK000
RETRORSINE see RFP000
RETRORSINE-N-OXIDE see RFU000
RETROVITAMIN A see VSK600
REUBLONIL see BCP650
REUDO see BRF500
REUFENAC see AGN000
REULATT S.S. see GFM000
REUMACHLOR see CLD000
REUMACIDE see IDA000
REUMALON see OPK300
REUMAQUIN see CLD000
REUMARTRIL see OPK300
REUMASYL see BRF500
REUMATOX see MQY400
REUMAZOL see BRF500
REUMOFENE see IDA400
REUMOFIL see SOU550
REUMOX see HNI500
REUPOLAR see BRF500
REVAC see DJL000
REVACRYL A 191 see ADW200
REVELOX-WIRKSTOFF see RCA200
REVENGE see DGI400
REVERIN see PPY250
REVERTINA see VIZ400
REVIDEX see BBK500
REVIENTA CABALLOS (CUBA) see SLJ650
REVONAL see QAK000
REWOMID DLMS see BKE500
REWOPOL HV-9 see PKF000
REWOPOL NLS 30 see SIB600
REWOPOL TLS 40 see SON000
REXALL 413S see PMP500
REXAN see CKF500
REXCEL see CCU150
REXENE 106 see ADY500
REXENE see EJA379, PMP500
REXOCAINE see BQA010
REXOLITE 1422 see SMQ500
REX REGULANS see GGS000
REYCHLER'S ACID see RFU100
REZERPIN see RDK000
REZIFILM see TFS350
REZIPAS see AMM250
RF 10 see CCU250
RF 46-790 see THH350
RFCNU see RFU600
RFNA see NEE500
Rh 6G see RGW000
RG 235 see AMV375
RG 600 see ARM268
R-GENE see AQW000
RGH-1106 see PII250
RGH 2957 see YGA700
RGH-2958 see PEE100
RGH-4405 see EGM100

RGH-5526 see RFU800
RH 123 see RGP600
RH-124 see BPU000
RH 315 see DTT600
RH-787 see PPP750
RH 893 see OFE000
RH-8218 see EPC175
RH see RHF000
RHABDOPHIS TIGRINUS TIGRINUS VENOM see RFU875
3-β-(α-l-RHAMNOPYRANOSIDE)-5,11-α,14-β-TRIHYDROXY-5-β-CARD(20, see RFZ000
RHAMNOSIDE, STROPHANTHIDIN-3, α-l- see CNH780
3-β-RHAMNOSIDO-14-β-HYDROXY-Δ⁴·²⁰·²²-BUFATRIENOLIDE see POB500
RHAMNUS CALIFORNICA see MBU825
RHAMNUS CATHARTICA see MBU825
RHAMNUS FRANGULA see MBU825
RHATHANI see RGA000
RHEMATAN see PGG000
RHENATE (ReO₄¹')), POTASSIUM, (T-4)- see PLQ000
RHENIC ACID, POTASSIUM SALT see PLR500
RHENIUM see RGF000
RHENIUM(VII) SODIUM OXIDE see SJD500
RHENIUM(VII) SULFIDE see RGK000
RHENIUM TRICHLORIDE see RGP000
RHENOCURE CA see DWN800
RHENOSORB C see CAU500
RHENOSORB F see CAU500
RHEODOL SP-P 10 see MRJ800
RHEONINE B see FAG070
RHEOPYRINE see IGI000
RHEOSMIN see RBU000
RHEUMATOL see CCD750
RHEUM EMODIN see MQF250
RHEUMIN TABLETTEN see ADA725
RHEUMON see HKK000
RHEUMON GEL see HKK000
RHEUMOX see AQN750
RHEUM RHABARBARUM see RHZ600
RHINALAR see FDD085
RHINALL see SPC500
RHINANTIN see NCW000
RHINASPRAY see RGP450
RHINATHIOL see CBR675
RHINAZINE see NAH500
RHINE BERRY see MBU825
RHINOCORT see BOM520
RHINOGUTT see RGP450
RHINOPERD see NCW000
RHINOSPRAY see RGP450
RHIZOCTOL see MGQ750
RHIZOPIN see ICN000
RHODACRYST see VSZ000
RHODALLIN see AGT500
RHODALLINE see AGT500
RHODAMINE 116 see RGW100
RHODAMINE 123 see RGP600
RHODAMINE see FAG070
RHODAMINE 590 CHLORIDE see RGW000
RHODAMINE 69DN EXTRA see RGW000
RHODAMINE F4G see RGW000
RHODAMINE F5G see RGW000
RHODAMINE F5G CHLORIDE see RGW000
RHODAMINE F 5GL see RGW000
RHODAMINE 6GB see RGW000
RHODAMINE 6G (BIOLOGICAL STAIN) see RGW000
RHODAMINE 6GBN see RGW000
RHODAMINE 6G CHLORIDE see RGW000
RHODAMINE 6GCP see RGW000
RHODAMINE 4GD see RGW000
RHODAMINE 6GD see RGW000
RHODAMINE 5GDN see RGW000
RHODAMINE 6 GDN see RGW000
RHODAMINE GDN see RGW000
RHODAMINE 6 GDN EXTRA see RGW000
RHODAMINE 6GEX ETHYL ESTER see RGW000
RHODAMINE 6G EXTRA see RGW000
RHODAMINE 6G EXTRA BASE see RGW000
RHODAMINE 4GH see RGW000
RHODAMINE 6GH see RGW000
RHODAMINE 5GL see RGW000
RHODAMINE 6G LAKE see RGW000
RHODAMINE 6GO see RGW000
RHODAMINE 6GX see RGW000
RHODAMINE J see RGW000
RHODAMINE 6JH see RGW000

RHODAMINE 7JH see RGW000
RHODAMINE LAKE RED 6G see RGW000
RHODAMINE 116 PERCHLORATE see RGW100
RHODAMINE S (RUSSIAN) see FAG070
RHODAMINE WT see RGZ100
RHODAMINE Y 20-7425 see RGW000
RHODAMINE 6Zh-DN see RGW000
RHODAMINE 6ZH see RGW000
RHODAMINE ZH see RGW000
RHODAMIN 6G see RGW000
RHODANDINITROBENZOL see DVF800
RHODANIC ACID see RGZ550
RHODANID see ANW750
RHODANIDE see ANW750, PLV750
RHODANIN (CZECH) see RGZ550
RHODANINE see RGZ550
RHODANINE, N-METHYL- see MPH750
RHODANINIC ACID see RGZ550
RHODIA-6200 see DJA400
RHODIA see DAA800
RHODIACHLOR see HAR000
RHODIACID see BJK500
RHODIACIDE see EEH600
RHODIACUIVRE see CNK559
RHODIANEBE see MAS500
RHODIA RP 11974 see BDJ250
RHODIASOL see PAK500
RHODIATOX see PAK000
RHODIATROX see PAK000
RHODIFAX 16 see CPI250
RHODINE see ADA725
RHODINOL (FCC) see DTF400
RHODINOL see CMT250, DTF410
RHODINOL ACETATE see RHA000
RHODINYL ACETATE see RHA000
RHODINYL BUTYRATE see DTF800
RHODIRUBIN A see RHA125
RHODIRUBIN B see RHA150
RHODIRUBIN C see TAH675
RHODIRUBIN E see MAX000
RHODIUM see RHF000
RHODIUM(II) ACETATE see RHF150
RHODIUM(II) BUTYRATE see RHK250
RHODIUM(III) CHLORIDE (1:3) see RHK000
RHODIUM CHLORIDE see RHK000, RHP000
RHODIUM CHLORIDE, TRIHYDRATE see RHP000
RHODIUM, (1,5-CYCLOOCTADIENE)(2,4-PENTANEDIONATO)- see CPR840
RHODIUM, ((1,2,5,6-eta)-1,5-CYCLOOCTADIENE)(2,4-PENTANEDIONATO-O,O')- see CPR840
RHODIUM DIACETATE see RHF150
RHODIUM DIBUTYRATE see RHK250
RHODIUM METAL (OSHA) see RHF000
RHODIUM(II) PROPIONATE see RHK850
RHODIUM TRICHLORIDE see RHK000
RHODIUM TRICHLORIDE TRIHYDRATE see RHP000
RHODOCIDE see EEH600
RHODODENDRON see RHU500
RHODOL see MGJ750
RHODOLNE see SMQ500
RHODOPAS 6000 see AAX175
RHODOPAS M see AAX250
RHODOQUINE see RHZ000
RHODORA (CANADA) see RHU500
RHODOVIOL 4/125 see PKP750
RHODOVIOL 16/200 see PKP750
RHODOVIOL see PKP750
RHODOVIOL 4-125P see PKP750
RHODOVIOL R 16/20 see PKP750
RHODULINE ORANGE see BJF000
RHODULINE ORANGE NO see BAQ250
RHOMBIC see ELC500
RHOMELLOSE see MIF760
RHOMENE see CIR250
RHOMEX see PIJ500
RHONOX see CIR250
RHODOTOXIN see AOO375
RHOPLEX AC-33 (ROHM and HAAS) see EMF000
RHOPLEX B 85 see PKB500
RHOTEX GS see SJK000
RHOTHANE D-3 see BIM500
RHOTHANE see BIM500
RHODOTYPOS SCANDENS see JDA075
RHUBARB see RHZ600
RHUBARBE (CANADA) see RHZ600
RHUS COPALLINA see SCF000

RHYTHMONORM see PMJ525
RHYTMATON see AFH250
RHYUNO OIL see SAD000
RIANIL see CLO750
RIBALL see ZVJ000
RIBAVIRIN see RJA500
RIBBON CACTUS see SDZ475
RIBIPCA see RIK000
RIBITOL see RIF000
RIBO-AZAURACIL see RJA000
RIBO-AZURACIL see RJA000
RIBODERM see RIK000
RIBOFLAVIN see RIK000
RIBOFLAVIN-ADENINE DINUCLEOTIDE see RIF100
RIBOFLAVINE see RIK000
RIBOFLAVINE-ADENINE DINUCLEOTIDE see RIF100
RIBOFLAVINEQUINONE see RIK000
RIBOFLAVINE SULFATE see RIP000
RIBOFLAVIN 5'-(TRIHYDROGEN DIPHOSPHATE), 5'-5'-ESTER with ADENO-
    SINE (9CI) see RIF100
RIBOFURANOSIDE, GUANINE-9, β-D- see GLS000
RIBOFURANOSIDE, 9H-PURINE-6-THIOL-9 see MCQ500
9-β-d-RIBOFURANOSIDOADENINE see AEH750
1-β-RIBOFURANOSYLCYTOSINE see CQM500
9-β-D-RIBOFURANOSYLGUANINE see GLS000
2-β-d-RIBOFURANOSYLMALEIMIDE see RIU000
5-β-d-RIBOFURANOSYL-1,3-OXAZINE-2,4-DIONE see OMK000
5-β-d-RIBOFURANOSYL-2H-1,3-OXAZINE-2,4(3H)-DIONE see OMK000
9-(β-d-RIBOFURANOSYL)-9H-PURINE see RJF000
9-(β-d-RIBOFURANOSYL)PURINE see RJF000
9-β-d-RIBOFURANOSYL-9H-PURINE-6-THIOL see TFJ825
3-β-d-RIBOFURANOSYL-1H-PYRROLE-2,5-DIONE see RIU000
7-β-d-RIBOFURANOSYL-7H-PYRROLO(2,3-D)PYRIMIDINE-4-AMINE see TNY500
7-β-d-RIBOFURANOSYL-7H-PYRROLO(2,3-d)PYRIMIDIN-4-OL see DAE200
2-β-d-RIBOFURANOSYLTHIAZOLE-4-CARBOXAMIDE see RJF500
2-β-d-RIBOFURANOSYL-4-THIAZOLECARBOXAMIDE see RJF500
2-β-d-RIBOFURANOSYL-1,2,4-TRIAZINE-3,5(2H,4H)-DIONE see RJA000
2-β-d-RIBOFURANOSYL-as-TRIAZINE-3,5(2H,4H)-DIONE see RJA000
2-β-d-RIBOFURANOSYL-as-TRIAZINE-3,5(2H,4H)-DIONE 2',3',5'-TRIACETATE
    see THM750
1-β-d-RIBOFURANOSYL-1,2,4-TRIAZOLE-3-CARBOXAMIDE see RJA500
1-β-d-RIBOFURANOSYLURACIL see UVJ000
RIBOMYCINE see RIP000
RIBONOSINE see IDE000
RIBOSTAMIN see RIP000
RIBOSTAMYCIN see XQJ650
RIBOSTAMYCIN SULFATE see RIP000
β-d-RIBOSYL-6-METHYLTHIOPURINE see MPU000
RIBOSYLPURINE see RJF000
RIBOSYLTHIOGUANINE see TFJ500
RIBOSYL-6-THIOPURINE see MCQ500
RIBOTIDE see RJF400
RIBOXAMIDE see RJF500
RICE STARCH see SLJ500
RICHAMIDE 6310 see BKE500
RICHONATE 1850 see DXW200
RICHONIC ACID B see LBU100
RICHONOL C see SIB600
RICHONOL T see SON000
RICID see DIU800
RICID II see BKS750
RICID P see BKS750
RICIFON see TIQ250
RICIN see RJK000
RICIN (HAITI) see CCP000
RICINIC ACID see RJP000
RICINOLEIC ACID see RJP000
RICINOLEIC ACID, BARIUM SALT see RJU000
RICINOLEIC ACID, SODIUM SALT see SJN500
RICINOLEIC ACID TRIESTER with GLYCEROL 12-ETHER with TRIDECAETHY-
    LENE GLYCOL see EAN800
RICINOLIC ACID see RJP000
RICINO (PUERTO RICO) see CCP000
RICIN-TOXIN CON A see CNH625
RICINUS COMMUNIS see CCP000
RICINUS OIL see CCP250
RICIRUS OIL see CCP250
RICKAMICIN see SDY750
RICKAMICIN SULFATE see APY500
RICKETON see CMC750
RICON 100 see SMR000
RICORTEX see CNS825
RICYCLINE see TBX250
RIDAURA see ARS150
RIDAZOLE see MMN750

RIDDELLIIN see RJZ000
RIDDELLINE see RJZ000
RIDLITE MMT see MCB050
RIDOMIL see MDM100
RIDOMIL 2E see MDM100
RIDZOL P see MMN750
RIFA see RKP000
RIFA ACID ORANGE II see CMM220
RIFADINE see RKP000
RIFAGEN see RKP000
RIFALDAZINE see RKP000
RIFALDIN see RKP000
RIFAMATE see RKP000
RIFAMICINE SV see RKZ000
RIFAMIDE see RKA000
RIFAMPICIN see RKP000
RIFAMPICINE (FRENCH) RKP000
RIFAMPICIN M/14 see RKA000
RIFAMPICINUM see RKP000
RIFAMPIN see RKP000
RIFAMYCIN see RKK000, RKZ000
RIFAMYCIN AMP see RKP000
RIFAMYCIN B see RKK000
RIFAMYCIN B N,N-DIETHYLAMIDE see RKA000
RIFAMYCIN DIETHYLAMIDE see RKA000
RIFAMYCIN M14 see RKA000
RIFAMYCIN O see RKP400
RIFAMYCIN S see RKU000
RIFAMYCIN SV see RKZ000
RIFAPRODIN see RKP000
RIFATHYROIN see TNX400
RIFINAH see RKP000
RIFIT see PMB850
RIFLE POWDER see ERF500, PLL750
RIFLOC RETARD see CCK125
RIFOBAC see RKP000
RIFOCIN see RKZ000
RIFOCINA M see RKA000
RIFOCYN see RKZ000
RIFOLDIN see RKP000
RIFOMIDE see RKA000
RIFOMYCIN B see RKK000
RIFOMYCIN B DIETHYLAMIDE see RKA000
RIFOMYCIN O see RKP400
RIFOMYCIN S see RKU000
RIFOMYCIN SV see RKZ000
RIFORAL see RKP000
RIGEDAL see CCK125
RIGENICID see EPQ000
RIGETAMIN see EDC500
RIGEVIDON see NNL500
RIGIDEX 35 see PJS750
RIGIDEX 50 see PJS750
RIGIDEX see PJS750
RIGIDEX TYPE 2 see PJS750
RIGIDIL see BBV500
RIGIDYL see BBV500
RIKAVARIN see AJV500
RIKELATE CALCIUM see CAR780
RIKEMAL O 71D see GGR200
RIKEMAL OL 100 see GGR200
RIKEMAL S 250 see SKV150
RIKEN RESIN MA 31 see MCB050
RIKER 548 see TIH800
RIKER 595 see BSZ000
RIKER 601 see TND000
RIKER 52G see PAF550
RILANSYL see CKF500
RILAQUIL see CKF500
RILASSOL see CKF500
RILAX see CKF500
RILLASOL see CKF500
RIMACTAN see RKP000
RIMACTAZID see RKP000
RIMADYL see CCK800
RIMANTADINE HYDROCHLORIDE see AJU625
RIMAON see EDW500
RIMAZOLIUM METHYL SULFATE see PMD825
RIMICID see ILD000
RIMIDIN see FAK100
RIMITSID see ILD000
RIMSO-50 see DUD800
RINATIOL see CBR675
RINAZIN see NAH550
RINDEX see MDO250

RINEPTIL see HBM490
RINGER'S GLUCOSE SOLUTION see GFG200
RINO-CLENIL see AFJ625
RIOL see FLZ050
RIOMITSIN see HOH500
RIOMYCIN see FMR500
RIP-15830 see DNG200
RIPAZEPAM see POL475
RIPCORD see RLF350
RIPENTHOL see DXD000
RIPERCOL see TDX750
RIPERCOL-L see LFA020
RIPEREOL see TDX750
RIPOSON see EQL000
RIRILIM see BBW500
RIRIPEN see BBW500
RISE see CKA000
RISELECT see DGI000
RISELF see MFD500
RISERPA see RDK000
RISPASULF see PDC850
RISTAT see HMY000
RISTOGEN see TKH750
RITALIN see MNQ000, RLK000
RITALINE see MNQ000
RITALIN HYDROCHLORIDE see RLK000
RITCHER WORKS see MNQ000
RITMENAL see DKQ000
RITMODAN see DNN600
RITMOS see AFH250
RITODRINE HYDROCHLORIDE see RLK700
RITOSEPT see HCL000
RITROSULFAN see LJD500
RITSIFON see TIQ250
RITUSSIN see RLU000
RIVADORM see NBT500
RIVADORN see NBU000
RIVANOL see EDW500
RIVASED see RDK000
RIVASIN see RDK000
RIVINA HUMILIS see ROA300
RIVINOL see EDW500
RIVIVOL see ILE000
RIVOMYCIN see CDP250
RIVOTRIL see CMW000
RIZABEN see RLK800
RJ 5 see DLJ500
R JUTAN see CAT775
RL-50 see MRL500
RMC see MAC750
RMI 9918 see TAI450
RMI 11002 see FDN000
RMI 71782 see DBE835, DKH875
RMI9,384A see DLS600
RMI 10,482A see MHJ500
RMI-14042A see LIA400
RMI 10024DA see BJA500
RMI 10874DA see BJG500
RMI 11002 DA see FDN000
RMI 11513 DA see XBA000
RMI 11567 DA see DDB400
RMI 11877 DA see DDE000
RO 1-5130 see MDL600
RO-1-5155 see NDW510
RO 1-5431 see MDV250, MKR250
RO 1-5470 see RAF300
RO 1-6463 see DNW400
RO 1-6794 see DBE825
RO-1-7700 see AGI000
RO 1-7788 see LFD200
RO-1-9213 see CQH000
RO 1-9569 see TBJ275
RO 2-1160 see CCH250
RO 2-2222 see TKW500
RO 2-2979 see TMB000
RO 2-2980 see HNK000
RO 2-3208 see FDK000
RO 2-3245 see ARX750
RO 2-3248 see AGD500
RO 2-3308 see QVA000, QWJ000
RO 2-3599 see DLP000
RO 2-3742 see AGE000
RO 2-3951 see ARX770
RO 2-4572 see ILE000
RO 2-5803 see BHR750

RO 2-9757 see FMM000
RO 2-9915 see FHI000
RO 2-9945 see TCQ500
RO 4-0403 see TAF675
RO 4-1385 see GFO200
RO 4-1778 see MDV000
RO 4-2130 see SNK000
RO 4-3476 see SNL800
RO-4-3780 see VSK955
RO 4-3816 see DBK400
RO 4-4393 see AIE500
RO 4-4602 see SCA400
RO 4-5360 see DLY000
RO 4-6316 see DNA200
RO 4-6467 see PME250, PME500
RO 4-6861 see CIF250
RO 4-8180 see CMW000
RO 4-9253 see BMN750
RO 5-0360 see DAR400
RO 5-0690 see MDQ250
RO 5-0831 see IKC000
RO 5-1226 see MNP300
RO 5-2180 see CGA500
RO 5-3059 see DLY000
RO 5-3307 see DAI475
RO 5-3350 see BMN750
RO 5-3438 see FDB100
RO 5-4023 see CMW000
RO 5-4200 see FDD100
RO 5-5345 see CFY750
RO 5-5516 see MLD100
RO 5-6789 see CFZ000
RO 5-6901 see DAB800
RO 5-9963 see NHI000
RO 6-4563 see GFY100
RO 7-0207 see OJS000
RO 7-1554 see IGH000
RO 7-5050 see MIA250
RO 10-1670 see REP400
RO 10-6338 see BON325
RO 11-1781 see TGA275
RO 12-0068 see TAL485
RO 1-5431/7 see DYF000
RO 1-5470/5 see DBE200
RO 1-5470/6 see LFD200
RO 21-9738 see DYE415
RO-5-0810/1 see TJE880
RO 5-3307/1 see IKB000
RO-6-4563/8 see GFY100
RO 8-6270/9 see TNX400
RO 1-9569/12 see TBJ275
ROACH SALT see SHF500
ROAD ASPHALT (DOT) see ARO500, ARO750
ROAD ASPHALT see PCR500
ROAD TAR, liquid (DOT) see ARO750
ROAD TAR (DOT) see ARO500
RO-AMPEN see AIV500, AOD125
ROBAMATE see MQU750
ROBANUL see GIC000
ROBARB see AMX750
ROBAVERON see RLK875
ROBAXAN see GKK000
ROBAXIN see GKK000
ROBAXINE see GKK000
ROBAXON see GKK000
ROBENIDINE see RLK890
ROBICILLIN VK see PDT750
ROBIGRAM see ARQ750
ROBIMYCIN see EDH500
ROBINAX see GKK000
ROBINETIN see RLP000
ROBINIA PSEUDOACACIA see GJU475
ROBINUL see GIC000
ROBIOCINA see SMB000
ROBISELLIN see ILD000
ROBITET see TBX000
ROBITUSSIN see RLU000
ROBORAL see PAN100
ROC-101 see RLU550
ROCALTROL see DMJ400
ROCCAL see AFP100, BBA500
ROCHE 5-9000 see API125
ROCHIPEL see DIR000
RO-CILLIN see PDD350
ROCILLIN-VK see PDT750

ROCIPEL see DIR000
ROCK CANDY see SNH000
ROCK OIL see PCR250
ROCK SALT see SFT000
ROCORNAL see DIO200
RO-CYCLINE see TBX250
RODALON see AFP250, BBA500, BEL900
RODANCA see PLV750
RODANIN S-62 (CZECH) see IAQ000
RODATOX 60 see DVF800
RODENTIN see EAT600
RO-DETH see WAT200
RODEX see FFF000
RO-DEX see SMN500
RODILONE see SNY500
RODINAL see ALT250
RODINE see RCF000
RODINOL see CMT250, DTF410
RODINOLONE see AQX250
RODIPAL see DIR000
RODOCID see EEH600
RODOL 42 see NEM500
RODOL YBA see ALO000
RODY see DAB825
ROE 101 see DBB200
ROERIDORM see CHG000
ROFOB 3 see AHH825
ROGERSINE see TKX700
ROGITINE see PDW400
ROGODIAL see DRR400, DSP400
ROGOR see DSP400
ROGUE see DGI000
ROHAGIT SD 15 see ADW200
ROHM & HAAS RH-218 see EPC175
ROHYDRA see BAU750
RO-HYDRAZIDE see CFY000
ROHYPNOL see FDD100
ROIDENIN see IIU000
ROIPNOL see FDD100
ROKACET see PJY100
ROKACET O 7 see PJY100
ROKANOL L see DXY000
RO-KO see RNZ000
ROKON see BDF000
ROL see RAF100
ROLAMID CD see BKE500
ROLAZINE see HGP500
ROLAZOTE see PAJ750, POL000
ROLICTON see AKL625
ROLITETRACYCLINE see PPY250, SPE000
ROLL-FRUCT see CDS125
ROLODIN see RLZ000
ROLODINE see RLZ000
ROLQUAT CDM/BC see QAT520
ROLSERP see RDK000
ROM 203 see IAG700
ROMACRYL see PKB500
ROMANTHRENE BLUE FRS see IBV050
ROMANTRENE BLUE FBC see DFN425
ROMANTRENE BLUE FRS see IBV050
ROMANTRENE BLUE GGSL see IBV050
ROMANTRENE BLUE RSZ see IBV050
ROMANTRENE BRILLIANT BLUE FR see IBV050
ROMANTRENE BRILLIANT BLUE R see IBV050
ROMANTRENE BROWN FBR see CMU770
ROMANTRENE BROWN FGR see CMU770
ROMANTRENE GOLDEN YELLOW see DCZ000
ROMANTRENE GREY K see IBJ000
ROMANTRENE NAVY BLUE FG see CMU500
ROMANTRENE NAVY BLUE FRA see VGP100
ROMANTRENE OLIVE FR see DUP100
ROMAN VITRIOL see CNP250, CNP500
ROMERGAN see DQA400
ROMETIN see CHR500
ROMEZIN see ALF250
ROMHIDROL M 501 see MCB050
ROMILAR see DBE200
ROMILAR HYDROBROMIDE see DBE200
ROMOPAL O see PJT300
ROMOSOL see ART250
ROMOTAL see TCJ075
ROMPARKIN see BBV000
ROMPHENIL see CDP250
ROMPUN see DMW000
ROMULGIN O see PKL100

RONDAR see CFZ000
RONDOMYCIN see MDO250
RO-NEET see EHT500
RONGALIT see FMW000
RONGALITE C see FMW000
RONIACOL see NDW510
RONIDAZOLE see MMN750
RONIDAZOL-PHARMACHIM (Bulgarian) see MMN750
RONILAN see RMA000
RONIN see PPO000
RONIT see EHT500
RONNEL see RMA500
RONONE see RNZ000
RONTON see ENG500
RONTYL see TKG750
ROOTONE see ICP000, NAK000, NAK500
ROPE BARK see LEF100
ROP 500 F see GIA000
ROPION see FJT100
ROPOL see PJS750
ROPOTHENE OB.03-110 see PJS750
ROPTAZOL see NGG500
ROQUESSINE DIHYDROBROMIDE see DOX000
ROQUINE see CLD000
RORASUL see ASB250, PPN750
RORER 148 see QAK000
ROSACETOL see TIT000
ROSA FRANCESA (CUBA) see OHM875
ROSA LAUREL (MEXICO) see RHU500
ROSAMICIN see RMF000
ROSANIL see DGI000
p-ROSANILINE see MAC500, RMK000
ROSANILINE see MAC250
ROSANILINE BASE see MAC500
ROSANILINE CHLORIDE see MAC250
p-ROSANILINE HCL see RMK020
p-ROSANILINE HYDROCHLORIDE see RMK020
ROSANILINE HYDROCHLORIDE see MAC250
ROSANILINIUM CHLORIDE see MAC250
ROSANOMYCIN A see RMK200
ROSARAMICIN see RMF000
ROSARY PEA see RMK250
ROSCOPENIN see PDT750
ROSCOSULF see SNN300
ROSE ABSOLUTE FRENCH see RMP000
ROSE BAY see OHM875
ROSEBAY see RHU500
ROSE BENGAL SODIUM see RMP175
ROSE CRYSTALS see TIT000
ROSE DE GRASSE see RNA000
ROSE DE MAI see RNA000
ROSE ETHER see PER000
ROSE-FLOWERED JATROPHA see CNR135
ROSE GERANIUM OIL ALGERIAN see GDA000
ROSE LAUREL (MEXICO) see OHM875
ROSE de MAI ABSOLUTE see RMP000
ROSEMARIE OIL see RMU000
ROSEMARY OIL see RMU000
ROSEMIDE see CHJ750
ROSEN OEL (GERMAN) see RNA000
ROSENOL see RNA000
ROSENOXIDE see RNU000
ROSENSTHIEL see MAT250
ROSE OIL see PDD750, RNA000
ROSE OIL BULGARIAN see RNA000, RNF000
ROSE OIL MOROCCAN see RNK000
ROSE OTTO see RNA000
ROSE OXIDE see RNU000
ROSE OXIDE LEVO see RNU000
ROSE QUARTZ see SCI500, SCJ500
ROSES see AOB250
ROSETONE see NAK000
ROSMARIN OIL (GERMAN) see RMU000
ROSOXIDE see RNU000
ROSPAN see PNH750
ROSPIN see PNH750
ROSTONE 2150 see MCB050
RO-SULFIRAM see DXH250
ROTATE see DQM600
ROTAX see BDF000
ROT B see XRA000
ROT C see OHA000
ROTEFIVE see RNZ000
ROTEFOUR see RNZ000
ROTENONA (SPANISH) see RNZ000

ROTENONE see RNZ000
ROTERSEPT see BIM250, CDT250
ROTESAR see BOV825
ROTESSENOL see RNZ000
ROT G see OHA000
ROT GG FETTLOESLICH see XRA000
ROTHANE see BIM500
ROTOCIDE see RNZ000
ROTOX see MHR200
ROTTLERIN see KAJ500
ROUGE see IHD000
ROUGE CERASINE see OHA000
ROUGE de COCHENILLE A see FMU080
ROUGE PLANT see ROA300
ROUGH & READY MOUSE MIX see WAT200
ROUGH & READY RAT BAIT & RAT PASTE see RCF000
ROUGOXIN see DKN400
ROUQUALONE see QAK000
ROUTRAX see TKG750
ROVAMICINA see SLC000
ROVRAL see GIA000
ROWACHOL see ROA400
ROWALIND see NDM100
ROWATIN see ROA425
ROWMATE see DET400, DET600
ROXARSONE (USDA) see HMY000
ROXBURY WAXWORK see AHJ875
ROXEL see RDK000
ROXICAM see FAJ100
ROXIMYCIN see HGP550
ROXINOID see RDK000
ROXION U.A. see DSP400
ROXOSUL TABLETS see SNN500
ROXYNOID see RDK000
ROYAL BLUE see DJO000
ROYAL CBTS see CPI250
ROYAL MBTS see BDE750
ROYAL MH-30 see DMC600
ROYAL SLO-GRO see DMC600
ROYAL SPECTRA see CBT750
ROYALTAC see DAI600, ODE000
ROYAL TMTD see TFS350
ROYAL WHITE LIGHT see CAT775
ROZOL see CJJ000
ROZTOZOL see CKM000
RP-093 see CMY725
866 R.P. see CFU750
RP 2145 see MPQ750
RP 2275 see AHO250
2325 RP see DRV850
2339 RP see BEM500
R.P. 2512 see DBL800
2512 R.P. see DBL800
R.P. 2591 see DFX400
RP 2616 see PPP500
RP 2632 see ALF250
2643-RP see ALF250
RP 2786 see WAK000
RP 2929 see TFH500
2987 R.P. see DII200
RP 2990 see TEX250
RP 3203 see DNG400
3277 RP see DQA400
3277 R.P. see PMI750
RP 3356 see DIR000
3359 RP see CKB500
RP 3377 see CLD000
3389 R.P. see DQA400
RP 3554 see MRW000
RP 3602 see GGS000
RP 3697 see PDD300
3718 RP see CGX625
3735 R.P. see SBE500
RP 3799 see DIW000
4182 R.P. see DQA400
RP 4207 see FNF000
RP 4632 see MFK500
4753 R.P. see TGD000
RP 4909 see CLY750
5171 RP see PBM500
5278 R.P. see SMA000
5337 R.P. see SLC000
6140 RP see PMF500
6710 RP see CIO500
6847 R.P. see AFL750

6909 RP see PIW000
7162 RP see DLH200
7204 RP see COS899
RP7293 see VRF000
RP 7522 see AKO500
7843 R.P. see TFM100
RP 7843 see TFM100
RP 7891 see GFM200
RP 8167 see EEH600
RP 8228 see LFO000
8595 R.P. see DSV800
RP 8599 see FMU039
RP 8823 see MMN250
RP 8908 see PIW000
9159 RP see LEO000
RP 9159 see LEO000
9778 R.P. see PNW750
RP-9921 see PAF550
9965 RP see MPM750
RP 9965 see MQR000
RP 10192 see MIJ500
10257 R.P. see TND000
11,561 RP see CBL500
13,057 R.P. see DAC000
RP 13057 see DAC000
13245 R. P. see PMM000
RP 13907 see TKP100
13907 R.P. see TKP100
RP 16,091 see MNQ500
17190RP see POE100
RP 18,429 see AJV500
19583 RP see BDU500
22050 R.P. see ROZ000
RP 22410 see PMG000
RP 26019 see GIA000
27267 R.P. see ZUA450
RPA 2 see NAP500
RPA NO. 2 see NAP500
RPCNU see ROF200
R-874 PHILLIPS see OGA000
RP 13057 HYDROCHLORIDE see DAC200
RP 22,050 HYDROCHLORIDE see ROZ000
2339 R.P. HYDROCHLORIDE see PEN000
4560 RP HYDROCHLORIDE see CKP500
2786 R.P. MALEATE see DBM800
RP 8599 MESYLATE see FMU039
19583RP-Na see KGK100
19583 RP SODIUM see KGK100
RR 15-12-120 see MCB050
RS 141 see CJJ250
RS 1280 see CBF250
R-3422-S see EMC000
RS-3540 see MFA500
RS 44872 see SNH480
RS-79216 see PCG550
RS see CCU250
R-E-S see RDK000
R SALT see ROF300
RS-1401 AT see SPD500
R.S.NITROCELLULOSE see CCU250
R105 SODIUM see RMP175
RTEC (POLISH) see MCW250
R-010-TK see BIX250
RU 2858 see MRU600
RU-4723 see CIR750
RU-11484 see BEP750
RU 15060 see SOX400
RU 15750 see TKG000
RU 22974 see DAF300
RU 24756 see CCR950
RU 43715 see POB300
RUBATONE see BRF500
RUBBER see ROH900
"522" RUBBER ACCELERATOR see PIY500
RUBBER CEMENT see CCW250
RUBBER FAST YELLOW GA see CMS212
RUBBER FAST YELLOW GRA see CMS208
RUBBER HYDROCHLORIDE see PJH500
RUBBER HYDROCHLORIDE POLYMER see PJH500
RUBBER, NATURAL see ROH900
RUBBER SOLVENT see ROU000
RUBBER VINE see ROU450
RUBEANE see DXO200
RUBEANIC ACID see DXO200
RUBENS BROWN see MAT500

RUBERON see EME100
RUBESCENCE RED MT-21 see NAY000
RUBESCENSLYSIN see AHI500
RUBESOL see VSZ000
RUBIAZOL A see SNM500
RUBIDAZON see ROU800
RUBIDAZONE see ROU800
RUBIDAZONE MONOHYDROCHLORIDE see ROZ000
RUBIDIUM see RPA000
RUBIDIUM ACETYLIDE see RPB100
RUBIDIUM CARBONATE see RPB200
RUBIDIUM CHLORIDE see RPF000
RUBIDIUM DICHROMATE see RPK000
RUBIDIUM FLUORIDE see RPP000
RUBIDIUM HYDRIDE see RPU000
RUBIDIUM HYDROXIDE see RPZ000
RUBIDIUM HYDROXIDE SOLUTION (UN 2677) (DOT) see RPZ000
RUBIDIUM HYDROXIDE (UN 2678) (DOT) see RPZ000
RUBIDIUM IODIDE see RQA000
RUBIDIUM NITRIDE see RQF000
RUBIDOMYCIN see DAC000, DAC200
RUBIDOMYCINE see DAC000
RUBIDOMYCIN HYDROCHLORIDE see DAC200
RUBIFLAVIN see RQF350
RUBIGAN see FAK100
RUBIGEN see TBJ275
RUBIGINE see HHU500
RUBIGO see IHD000
RUBIJERVINE see SKR000
RUBINATE 44 see MJP400
RUBINATE M see PKB100
RUBINATE MF 178 see PKB100
RUBINATE MF 182 see PKB100
RUBINATE TDI 80/20 see TGM740, TGM750
RUBINATE TDI see TGM740
RUBINE RED RR 1253 see CMS155
RUBINE TONER B see CMS148
RUBINE TONER BA see CMS148
RUBINE TONER BT see CMS148
RUBITOX see BDJ250
RUBOMYCIN see RQK000
RUBOMYCIN C 1 see DAC000
RUBOMYCIN C see DAC000, DAC200
RUBRAMIN see VSZ000
RUBRATOXIN B see RQP000
RUBRIMENT see NCR040
RUBRINE C see GHA050
RUBRIPCA see VSZ000
RUBROCITOL see VSZ000
RUBRUM SCARLATINUM see SBC500
RUBUS ELLIPTICUS Smith, extract excluding roots see RQU300
RUCAINA see DHK400
RUCAINA HYDROCHLORIDE see DHK600
RUCOFLEX PLASTICIZER DOA see AEO000
RUCON B 20 see PKQ059
RUDBECKIA BICOLOR nutt., extract see RQU650
RUDOTEL see CGA000
RUELENE 25E see COD850
RUELENE see COD850
RUELENE DRENCH see COD850
RUE OIL see OHE000
RUFOCHROMOMYCIN see SMA000
RUFOCROMOMYCINE see SMA000
RUFOL see MPQ750
RUGULOSIN see RRA000
(+)-RUGULOSIN see RRA000
RUKSEAM see DAD200
RULENE see COD850
RUMAPAX see HNI500
RUMENSIN see MRE230
RUMESTROL 1 see DKA600
RUMESTROL 2 see DKA600
RUMETAN see ZLS000
RUN see PDN000
RUNA RH20 see TGG760
RUNCATEX see CIR500
RUOCID see AHO250
RUSSELL'S VIPER VENOM see VQZ635
RUSSIAN COMFREY LEAVES see RRK000
RUSSIAN COMFREY ROOTS see RRP000
RUTA GRAVEOLENS, extract see RRP675
RUTERGAN HYDROCHLORIDE see DPI000
RUTGERS 612 see EKV000
RUTHENIUM see RRU000
RUTHENIUM CHLORIDE see RRZ000

RUTHENIUM CHLORIDE HYDROXIDE see RSA000
RUTHENIUM COMPOUNDS see RSF000
RUTHENIUMHYDROXIDE TRICHLORIDE see RSA000
RUTHENIUM HYDROXYCHLORIDE see RSA000
RUTHENIUM(VIII) OXIDE see RSK000
RUTHENIUM OXIDE see RSF875
RUTHENIUM SALT of TETRAMETHYLPHENANTHRENE see RSP000
RUTHENIUM TETRAOXIDE see RSK000
RUTHENIUM TRICHLORIDE see RRZ000
RUTHENIUM TRICHLORIDE HYDROXIDE see RSA000
RUTILE see TGG760
RUTIN see RSU000
RUTINIC ACID see RSU000
RUTOSIDE see RSU000
RUTRALIN see BQN600
RUVAZONE see RSU450
RV 1 see SIT850
RVK see DXO200
RVM-FG see PAE750
RX 2557 see CAT775
RX 6029M see TAL325
RYANEXEL see RSZ000
RYANIA see RSZ000
RYANIA POWDER see RSZ000
RYANIA SPECIOSA see RSZ000
RYANICIDE see RSZ000
RYANODINE see RSZ000
RYCELAN see OJY100
RYCELON see OJY100
RYEMYCIN-B2 see TAH675
RYLAM see CBM750
RYLUX PRS (CZECH) see BKO750
RYNACROM see CNX825
RYODIPINE see RSZ375
RYOMYCIN see HOH500
RYSER see RDK000
RYTHMODAN see RSZ600
RYTHMOL see BHR750
RYTHRITOL see PBC250
RYTMONORM see PMJ525
RYUTAN (JAPANESE) see RSZ675
RYZELAN see OJY100


S-6 see MIH300
S 13 see BHD250
S 22 see SAK500
S 30 see PJT300
S-33 see CIF250
S 46 see PFB350
S51 see BBV500
S-62 see ARW750
S 77 see DRC600
S 84 see BNF750
S115 see NCQ900
S 140 see DAM700
S 142 see PNB250
S 151 see EJM500, SAA000
S 154 see BPR500
S 173 see SMQ500
S 187 see DPI700
S 190 see KDU100
S 202 see DHK600
S 222 see BKB250
S 260 see MCB050
Sa 267 see DGW600
S 276 see DXH325
S410 see DSK600
S 421 see OAL000
L-S 519 see GCE500
S 596 see BQE000
S 62-2 see ARW750
S 650 see BQH250
S 767 see FAQ800
S-805 see DCS200
S 816 see CNA750
S-847 see CEW500
S 852 see DBL700
S 940 see HMV000
Sch 1000 see IGG000
S-1000 see SXS375
S 1006 see PJY100
S-1046 see XTJ000
S 1065 see MIB750
S 1096 see GGR200

S 1097 see GGR200
S 1132 see PJY100
S 1520 see IBV100
S 1530 see DLV000
S 1544 see PFC750
S 1688 see AMJ500
S 1702 see DBL700
S 1707 see MCB050
S 1708 see MCB050
S 1710 see MCB050
S 1711 see MCB050
S 1752 see FAQ900
S 1942 see BNL250
S 2225 see EGV500
S 2940 see DRR400
S 2957 see CLJ875
S-3151 see AHJ750
S 32-06 see DAB825
S 3308 see DGC100
S-3440 see BFV765
S 4004 see YCJ200
S 4084 see COQ399
S 4087 see CON300
Sch 4831 see BFV750
S 5057 see MCB050
S 5602 see FAR100
S 5660 see DSQ000
Sch 5705 see PCU400
Sch 5706 see FAJ150
Sch 5712 see BRJ125
6059S see LBH200
S-6115 see CQI750
6315-S see FCN100
S 6437 see ALV000
S 6900 see DRR200
S-6,999 see NNF000
S 7131 see PMF750
S 8527 see CMV700
S-9115 see CQI750
S-9700 see DAB880
S 10165 see DGI000
10275-S see EBD500
10364S see MCH600
Sch 14947 see RMF000
S-15126 see DED000
31252-S see HNT075, PCJ325
S 73 4118 see PDW250
710674-S see CMW550
711389-S see DFM875
Sb-57 see AQI500
Sb-71 see AQC000
S6 (pharmaceutical) see MIH300
S 65 (polymer) see PKQ059
SA 79 see PMJ525
SA96 see MCO775
SA 97 see HGI100
S 99 A see SLD800
SA 111 see SNJ000
S 112A see DSQ000
SA 446 see FAQ950
SA 504 see TGB160
SA 546 see OMY850
SA 1500 see AHH825
SA 20.16 see MCB050
SA 42-548 see MBV250
SA see SAI000
SAA see SAN000
SAATBEIZFUNGIZID (GERMAN) see HCC500
SABACIDE see VHZ000
SABADILLA see VHZ000
SABADININE see EBL000
SABANE DUST see VHZ000
SABARI see HGC500
SABILA (CUBA, PUERTO RICO) see AGV875
SABLIER (HAITI) see SAT875
SACABUCHE (PUERTO RICO) see JBS100
SACANDAGA YELLOW X 2476 see CMS212
SACARINA see BCE500
SACCAHARIMIDE see BCE500
SACCHARATED FERRIC OXIDE see IHG000
SACCHARATED IRON see IHG000
SACCHARIN see SJN700
SACCHARINA see BCE500
SACCHARIN ACID see BCE500
SACCHARIN AMMONIUM see ANT500

SACCHARINATE AMMONIUM see ANT500
SACCHARIN CALCIUM see CAM750
SACCHARINE see BCE500
SACCHARINE SOLUBLE see SJN700
SACCHARINNATRIUM see SJN700
SACCHARINOL see BCE500
SACCHARINOSE see BCE500
SACCHARIN SOLUBLE see SJN700
SACCHARIN, SODIUM see SJN700
SACCHARIN, SODIUM SALT see SJN700
SACCHAROIDUM NATRICUM see SJN700
SACCHAROL see BCE500
SACCHAROLACTIC ACID see GAR000
SACCHAROSE see SNH000
SACCHAROSE ACETATE ISOBUTYRATE see SNH050
SACCHAROSONIC ACID see SAA025
SACCHARUM see SNH000
SACCHARUM LACTIN see LAR000
SACERIL see DKQ000, DNU000
SACERNO see MKB250
SACHSISCHBLAU see FAE100
SACROMYCIN see AHL000
SAD-128 see SAB800
SAD see EDA500
SADAMIN see XCS000
SADH see DQD400
SA 97 DIHYDROCHLORIDE see CJR809
SADOFOS see MAK700
SADOPHOS see MAK700
SADOPLON see TFS350
SADOREUM see IDA000
SAEURE DES PHYTINS (GERMAN) see PIB250
SAEURE FLUORIDE (GERMAN) see FEZ000
SAFARITONE YELLOW G see AAQ250
SAFE-N-DRI see AHF500
SAFFAN see AFK500
SAFFLOWER OIL see SAC000
SAFFLOWER OIL (UNHYDROGENATED) (FCC) see SAC000
SAFFRON YELLOW see DUX800
SAFRANIN see GJI400
SAFRANINE see GJI400
SAFRANINE A see GJI400
SAFRANINE B see GJI400
SAFRANINE G see GJI400
SAFRANINE GF see GJI400
SAFRANINE J see GJI400
SAFRANINE O see GJI400
SAFRANINE OK see GJI400
SAFRANINE SUPERFINE G see GJI400
SAFRANINE T see GJI400
SAFRANINE TH see GJI400
SAFRANINE TN see GJI400
SAFRANINE Y see GJI400
SAFRANINE YN see GJI400
SAFRANINE ZH see GJI400
SAFRANIN T see GJI400
SAFRASIN see SAC875
SAFROL see SAD000
SAFROLE see SAD000
SAFROLE MF see SAD000
SAFROTIN see MKA000, SAD100
SAFSAN see DXE000
SAGAMICIN see MQS579
SAGAMICIN (OBS.) see MQS579
SAGATAL see NBU000
SAGE OIL see SAE500, SAE550
SAGE OIL CLARY see OGQ200
SAGE OIL, DALMATIAN TYPE see SAE500
SAGE OIL, SPANISH TYPE see SAE550
SAGO CYCAS see CNH789
SAGRADO see SAF000
SAH 22 see SEM500
SAH see SAG000
SAIB see SNH050
SAICLATE see DNU100
SAIKOSIDE, CRUDE see SAF300
SAIPHOS see ASD000
SAISAN see MLC250
SAKOLYSIN (GERMAN) see BHT750
SAKURAI No. 864 see YCJ000
SALACETIN see ADA725
SALACHLOR see SHJ000
SALAMID see SAH000
SALAMIDE see SAH000
SAL AMMONIA see ANE500

SAL AMMONIAC see ANE500
SALAMMONITE see ANE500
SALAZOLON see AQN250
SALAZOPYRIN see PPN750
SALAZOSULFAPYRIDINE see PPN750
SALBEI OEL (GERMAN) see SAE500, SAE550
SALBUTAMOL see BQF500
SALBUTAMOL HEMISULFATE see SAF400
SALBUTAMOL SULFATE see SAF400
SALCETOGEN see ADA725
SALCOMIN see BLH250
SALCOMINE POWDER see BLH250
SAL ENIXUM see PKX750
SALETIN see ADA725
SALI see SAF500
SALICAL see SAN000
SALICILAMIDE (ITALIAN) see SAH000
SALICIM see SAH000
SALICINE BLUE BLACK AEF see CMP880
SALICOL see CMG000
SALICOYL HYDRAZIDE see HJL100
SALICRESIN FLUID see CHW675
SALICYCLOHYDRAZINE see HJL100
SALICYLADEHYDE see SAG000
SALICYLAL see SAG000
SALICYLALDEHYDE see SAG000
SALICYLALDEHYDE, 5-CHLORO- see CLD800
SALICYLALDEHYDE ETHYLENEDIIMINE COBALT see BLH250
SALICYLALDEHYDE METHYL ETHER see AOT525
SALICYLALDEHYDE OXIME see SAG500
SALICYLALDOXIME see SAG500
SALICYLAMID see SAH000
SALICYLAMIDE see SAH000
SALICYLAMIDE, 5-(1-HYDROXY-2-((1-METHYL-3-PHENYLPRO-
  PYL)AMINO)ETHYL)-, HYDROCHLORIDE see HMM500
SALICYLANILID see SAH500
SALICYLANILIDE see SAH500
SALICYLANILIDE, 4',5-DIBROMO- see BOD600
SALICYLAZOSULFAPYRIDINE see PPN750
SALICYL-DIAETHYL (GERMAN) see BQH250
SALICYLDIETHYLAMIDE see DJY400
SALICYL-DIMETHYL (GERMAN) see BQH500
SALICYLDIMETHYLAMIDE see DUA800
SALICYL HYDRAZIDE see HJL100
SALICYLHYDROXAMIC ACID see SAL500
m-SALICYLIC ACID see HJI100
p-SALICYLIC ACID see SAI500
SALICYLIC ACID see SAI000
SALICYLIC ACID, 4-AMINO-, 2-(DIETHYLAMINO)ETHYL ESTER, HYDROCHLO-
  RIDE see AMM750
SALICYLIC ACID ANILIDE see SAH500
SALICYLIC ACID, BIMOLECULAR ESTER see SAN000
SALICYLIC ACID, 5-CHLORO- see CLD825
SALICYLIC ACID CHOLINE SALT see CMG000
SALICYLIC ACID, 3,6-DICHLORO- see DGK250
SALICYLIC ACID, DIHYDROGEN PHOSPHATE see PHA575
SALICYLIC ACID, 3,5-DINITRO- see HKE600
SALICYLIC ACID-2-ETHYLHEXYL ESTER see ELB000
SALICYLIC ACID, FLUOROACETATE see FFT000
SALICYLIC ACID-3-HEXEN-1-YL ESTER see SAJ000
SALICYLIC ACID, HYDRAZIDE see HJL100
SALICYLIC ACID, ISOBUTYL ESTER see IJN000
SALICYLIC ACID ISOPROPYL ESTER O-ESTER with O-ETHYL ISOPROPYL-
  PHOSPHORAMIDOTHIOATE see IMF300
SALICYLIC ACID, METHYL ESTER see MPI000
SALICYLIC ACID, METHYL ESTER, ESTER with p-GUANIDINOBENZOIC ACID
  see CBT175
SALICYLIC ACID, MONOAMMONIUM SALT see ANT600
SALICYLIC ACID, compounded with MORPHOLINE (1:1) see SAI100
SALICYLIC ACID, 4-NITRO- see NJF100
SALICYLIC ACID OCTYL ESTER see SAJ500
SALICYLIC ACID PENTYL ESTER see SAK000
SALICYLIC ACID-3-PHENYLPROPYL ESTER see SAK500
SALICYLIC ACID with PHYSOSTIGMINE (1:1) see PIA750
SALICYLIC ACID, SODIUM SALT see SJO000
SALICYLIC ACID, p-TOLYL ESTER see THD850
SALICYLIC ALDEHYDE see SAG000
SALICYLIC ETHER see SAL000
SALICYLIC ETHYL ESTER see SAL000
SALICYLIC HYDRAZIDE see HJL100
SALICYLOHYDROXAMIC ACID see SAL500
SALICYLOHYDROXIMIC ACID see SAL500
SALICYLOYLGLYCINE see SAN200
SALICYLOYLOXYTRIBUTYLSTANNANE see SAM000
SALICYLOYLSALICYLIC ACID see SAN000

SALICYL PHOSPHATE see PHA575
o-SALICYLSALICYLIC ACID see SAN000
SALICYLSALICYLIC ACID see SAN000
SALICYLSULFONIC ACID see SOC500
SALICYLURIC ACID see SAN200
SALICYL-VASOGEN see ANT600
SALICYL YELLOW see SIT850
SALIFEBRIN see SAH500
SALIMED see CPR750
SALIMID see CPR750
SALINA see SAN000
SALINE see SFT000
SALINIDE see SAH500
SALINIDOL see SAH500
SALIPHENAZON see AQN250
SALIPRAN see SAN600
SALIPUR see SAH000
SALIPYRAZOLAN see AQN250
SALIPYRINE see AQN250
SALISAN see CLH750
SALISOD see SJO000
SALITHION see MEC250
SALITHION-SUMITOMO see MEC250
SALIX see CHJ750
SALIZELL see SAH000
SAL de MERCK see CNF000
SALMIAC see ANE500
SALMIDOCHOL see DYE700
SALMINE SULFATE (1:1) see SAO200
SALMINE SULFATE see SAO200
SALMONELLA ENTERITIDIS ENDOTOXIN see SAO250
SALMONELLA TYPHI ENDOTOXIN see EAS100
SALMONELLA TYPHOSA ENDOTOXIN see EAS100
SALNIDE see SAH500
SALOL see PGG750
SALOXIUM see SAN000
SALP see SEM300
SALPETERSAEURE (GERMAN) see NED500
SALPETERZUUROPLOSSINGEN (DUTCH) see NED500
SALPIX see AAN000
SALRIN see SAH000
SALSALATE see SAN000
SALSOCAIN see SAO475
SALSONIN see SJO000
SALT see SFT000
SALT BATHS (NITRATE OR NITRITE) see SAO500
SALT CAKE see SJY000
SALT OF TARTAR see PLA250
SALTPETER see PLL500
SALT of SATURN see LCV000
SALUFER see DXE000
SALUNIL see CLH750
SALUPRES see RDK000
SALURAL see BEQ625
SALURES see BEQ625
SALURETIL see CLH750
SALURETIN see CLY600
SALURIC see CLH750
SALURIN see HII500, SIH500
SALUTENSIN see RDK000
SALVACARD see DJS200
SALVACORIN see DJS200
SALVADERA (CUBA) see SAT875
SALVARSAN see SAP500
SALVIS see SEP000
SALVO see DAA800, HKC000
SALVO LIQUID see BCL750
SALVO POWDER see BCL750
SALYMID see SAH000
SALYRGAN see SIH500
SALYRGAN THEOPHYLLINE see MDH750, SAQ000
SALYSAL see SAN000
SALYZORON see BBW500
SALZBURG VITRIOL see CNP500
SAM see DUA800, SAH000
SAMARIUM see SAQ500
SAMARIUMACETAT (GERMAN) see SAR000
SAMARIUM ACETATE see SAR000
SAMARIUM(III) CHLORIDE see SAR500
SAMARIUM CITRATE see SAS000
SAMARIUM NITRAT (GERMAN) see SAT000
SAMARIUM NITRATE see SAT200
SAMARIUM(III) NITRATE, HEXAHYDRATE (1:3:6) see SAT000
SAMARIUM TRINITRATE see SAT200
SAMARON BLUE FBL see CMP070

SAMARON BLUE 5G see DMM400
SAMARON BRILLIANT VIOLET B see DBY700
SAMARON PINK FBL see AKI750
SAMARON PINK RFL see AKO350
SAMARON RED VIOLET F3B see DBX000
SAMARON YELLOW PA3 see AAQ250
SAMARON YELLOW 5RL see CMP090
SAMBUCUS CAERULEA see EAI100
SAMBUCUS CANADENSIS see EAI100
SAMBUCUS MELANOCARPA see EAI100
SAMBUCUS MEXICANA see EAI100
SAMBUCUS RACEMOSA see EAI100
SAMECIN see HGP550
SAMID see SAH000
SAMURON see INR000
SAN 155 see TFH750
SAN 230 see PHI500
SAN 239 see BAC040
SAN 1551 see TFH750
SANASEED see SMN500
SANASPASMINA see CBA375
SANASTHYMYL see AFJ625
SANATRICHOM see MMN250
SANCAP see EPN500
SANCELER CM-PO see CPI250
SANCHINOSIDE E1 see PAF450
SANCLOMYCINE see TBX000
SAN-CYAN see COI250
SANCYCLAN see DNU100
SAND see SCI500, SCJ500
SAND ACID see SCO500
SANDAMYCIN see AEA750
SANDBOX TREE see SAT875
SAND CORN see DAE100
SANDESIN see AEH750
SANDIMMUN see CQH100
SANDIMMUNE see CQH100
SANDIX see LDS000
SANDOCRYL BLUE BRL see BJI250
SANDOCRYL ORANGE B-G see CMM764
SANDOLAN BLUE P-ARL see CMM090
SANDOLAN RED N-RS see CMM330
SANDOLAN TURQUOISE E-AS see ERG100
SANDOLIN see DUS700
SANDOPARIN see HAQ500
SANDOPEL BLACK EX see AQP000
SANDOPEL DARK GREEN B see CMO830
SANDOPTAL see AGI750
SANDORIN GREEN 8GLS see CMS140
SANDORMIN see TEH500
SANDOSCILL see POB500
SANDOZ 6538 see DJY200
SANDOZ 43-715 see POB300
SANDOZ 52139 see MKA000
SANDRIL see RDK000
SANDRIX see DAB750
SANDRON see RDK000
SANEDRINE see EAW000
SANEGYT see CAY875
SAN-EI BRILLIANT SCARLET 3R see FMU080
SAN-EI TARTRAZINE see FAG140
SANEPIL see DKQ000
SANFURAN see NGE500
SANG gamma see BBQ500
SANGIVAMYCIN see SAU000
SANGUICILLIN see AOD000
SANGUIRITRIN see SAU350
SANGUIRITRINE see SAU350
SANGUIRYTHRINE see SAU350
SANDOTHRENE BLUE NG see DFN425
SANDOTHRENE BLUE NGR see DFN425
SANDOTHRENE BLUE NGW see DFN425
SANDOTHRENE BLUE NRSC see IBV050
SANDOTHRENE BLUE NRSN see IBV050
SANDOTHRENE BROWN NBR see CMU770
SANDOTHRENE DARK BLUE NR see VGP100
SANDOTHRENE OLIVE N2R see DUP100
SANDOTHRENE ORANGE R see CMU815
SANDOTHRENE PRINTING YELLOW see DCZ000
SANDOTHRENE RED N 6B see CMU825
SANDOTHRENE VIOLET N 4R see DFN450
SANDOTHRENE VIOLET N 2RB see DFN450
SANDOTHRENE VIOLET 4R see DFN450
SAN 155 I see TFH750

SAN 244 I see DRR200
SAN 6538 I see DJY200
SAN 6913 I see DRR200
SAN 7107 I see DRR200
SAN 52 139 I see MKA000
SANICLOR 30 see PAX000
SANIPIROL see SEP000
SANIPRIOL-4 see AMM250
SANITIZED SPG see ABU500
SANKUMIN see HGI100
SANLOSE SN 20A see SFO500
SANMARTON see FAR100
SANMORIN OT 70 see DJL000
SANOCHOLEN see DAL000
SANOCHRYSINE see GJG000
SANOCID see HCC500
SANOCIDE see HCC500
SANOCRISIN see FBP100
SANODIAZINE see PPP500
SANODIN see CBO500
SANOHIDRAZINA see ILD000
SANOMA see IPU000
SANOQUIN see CLD000, CLD250
SANORIN see NAH500
SANORIN-SPOFA see NCW000
SANPRENE LQX 31 see PKP000
SANQUINON see DFT000
SANREX see ADY500
SANSDOLOR see GGS000
SANSEL ORANGE G see PEJ500
SANSERT see MQP500
SANSPOR see CBF800
SANTAL OIL see OHG000
α-SANTALOL (FCC) see OHG000
SANTALOZONE see DAF150
SANTALYL ACETATE see SAU400
SANTAR see MCT500
SANTAVY'S SUBSTANCE F see MIW500
SANTHEOSE see TEO500
SANTICIZER 160 see BEC500
SANTICIZER 711 see DXQ400
SANTICIZER M-17 see MOD000
SANTICIZER 141 (MONSANTO) see DWB800
SANTICIZIER B-16 see BQP750
SANTOBANE see DAD200
SANTOBRITE see PAX250, SJA000
SANTOCEL see SCH000
SANTOCHLOR see DEP800
SANTOCURE see CPI250
SANTOCURE MOR see BDG000
SANTOCURE VULCANIZATION ACCELERATOR see CPI250
SANTOFLEX 13 see DQV250
SANTOFLEX 17 see BJT500
SANTOFLEX 36 see PFL000
SANTOFLEX 77 see BJL000
SANTOFLEX 134 see SAU475
SANTOFLEX A see SAV000
SANTOFLEX AW see SAV000
SANTOFLEX DD see SAU480
SANTOFLEX IC see PEY500
SANTOGARD PVI see CPQ700
SANTOMERSE 3 see DXW200
α-SANTONIN see SAU500
l-α-SANTONIN see SAU500
SANTONIN see SAU500
SANTONINIC ANHYDRIDE see SAU500
SANTONOX see TFC600
SANTOPHEN 20 see PAX250
SANTOPHEN see CJU250, PAX250
SANTOPHEN I GERMICIDE see CJU250
SANTOQUIN see SAV000
SANTOQUINE see SAV000
SANTOQUIN EMULSION see TMJ000
SANTOQUIN MIXTURE 6 see TMJ000
SANTOSOL 150 see DQL820
SANTOTHERM see PJL750
SANTOTHERM FR see PJL750
SANTOUAR A see DCH400
SANTOVAR A see DCH400
SANTOWAX see TBC750
SANTOWAX M see TBC620
SANTOWHITE POWDER see BRP750
SANTOX see EBD700
SANULCIN see DAB875
SANVEX see BHL750

SANWAPHYLLIN see HOA000
SANWAX 161P see PJS750
SANYO BENZIDINE ORANGE see CMS145
SANYO BENZIDINE YELLOW see CMS212
SANYO BENZIDINE YELLOW-B see DEU000
SANYO CARMINE L2B see DMG800
SANYO FAST BLUE SALT B see DCJ200
SANYO FAST ORANGE SALT RD see CEG800
SANYO FAST RED 10B see NAY000
SANYO FAST RED 2B see CMS148
SANYO FAST RED B BASE see NEQ000
SANYO FAST RED 2BE see CMS148
SANYO FAST RED SALT TR see CLK235
SANYO FAST RED TR BASE see CLK220
SANYO FAST SCARLET GG BASE see DEO295
SANYO GUM ORANGE A see CMM220
SANYO LAKE RED C see CHP500
SANYO PERMANENT ORANGE D 213 see CMU820
SANYO PERMANENT ORANGE D 616 see CMU820
SANYO SCARLET PURE see MMP100
SANYO SCARLET PURE NO. 1000 see MMP100
SANYO THRENE BLUE IRN see IBV050
SAOLAN see DSK200
SAPECRON see CDS750
SAPEP see AMD000
SAPHICOL see ASD000
SAPHIZON see ASD000
SAPHIZON-DP see ASD000
SAPHOS see ASD000
SAPHRAZINE see SAC875
SAPILENT see DLH200
SAPONA RED LAKE RL-6280 see NAY000
SAPONIN see SAV500
SAPONIN-GYPSOPHILA see GMG000
SAPPILAN see CJT750
SAPPIRAN see CJT750
SAPRECON C see DRP600
SAPROL see TKL100
SAPWOOD of LINDEN see SAW300
SARAN 683 see CGW300
SARAN 746 see CGW300
SARAN see SAX000
SARANAC YELLOW X 2838 see CMS208
SARAN RESIN 683 see CGW300
SARCELL TEL see SFO500
SARCLEX see DGD600
SARCOCLORIN see BHT750
SARCOLACTIC ACID see LAG010
dl-SARCOLYSIN see BHT750
l-SARCOLYSIN see PED750
o-dl-SARCOLYSIN see BHT250
p-l-SARCOLYSIN see PED750
dl-SARCOLYSINE see BHT750
d-SARCOLYSINE see SAX200
m-l-SARCOLYSINE see SAX210
dl-SARCOLYSINE HYDROCHLORIDE see BHV000
l-SARCOLYSINE HYDROCHLORIDE see BHV250
SARCOLYSIN HYDROCHLORIDE see BHV000
SARCOMYCIN see SAX500
SARCOSET GM see DTG000
SARGOTT see YAG000
SARIFA, seed extract see SAX275
SARIN see IPX000
SARIN II see IPX000
SARKOKLORIN see BHV000
SARKOMYCIN see SAX500
SARKOMYCIN B*, SODIUM SALT see SJS500
SARKOMYCIN compounded with NICOTINIC HYDRAZIDE see SAY000
SARKOMYCIN, SODIUM SALT see SJS500
SARLACH GN see CMP620
SARLACH R see SBC500
SARODORMIN see DYC800
SAROLEX see DCM750
SAROMET see DCK759
SAROTEN see EAI000
SAROTENE see EAI000
SARPAGAN see RDK000
SARPAGEN see RDK000
SARPAN see AFG750
SARPIFAN HP 1 see AAX175
SARTOMER SR 206 see BKM250
SARTOMER SR 351 see TLX175
SARTOSONA see ALV000
SARYUURIN see DYE700
S.A.S.-500 see PPN750

SAS 538 see CFY500
SAS 643 see DKV400
SASAPIRIN see SAN000
SASAPYRIN see SAN000
SASAPYRINE see SAN000
SASAPYRINUM see SAN000
SA 96 SODIUM SALT see SAY875
SASP see PPN750
SASSAFRAS see SAY900
SASSAFRAS ALBIDUM see SAY900
SASSAFRAS OIL see OHI000
SASTRIDEX see TKH750
SATANOLON see CCR875
SATECID see CHS500
SATIAGEL GS 350 see CCL250, CCL350
SATIAGUM 3 see CCL250
SATIAGUM STANDARD see CCL250
SATILAN PLZ see CNH125
SATILAN RL 2 see CNH125
SATINITE see CAX750
SATIN SPAR see CAX750
SATOL see OBA000
SATOX 20WSC see TIQ250
SATRANIDAZOLE see SAY950
SATURN see SAZ000
SATURN BROWN LBR see CMO750
SATURNO see SAZ000
SATURN RED see LDS000
SATURN RED B see CMO885
SAUCO BLANCO (CUBA) see EAI100
SAUCO (MEXICO, PUERTO RICO) see EAI100
SAUROL see IAD000
SAUTERALGYL see DAM700
SAUTERZID see ILD000
SAVAC see DNG400
SAVEMIX C 100 see MCB050
SAVENTRINE see DMV600, IMR000
SAVIT see CBM750
SAVORY OIL (summer variety) see SBA000
SAX see SAI000
SAXIN see BCE500, SJN700
SAXITOXIN (8CI) see SBA600
SAXITOXIN DIHYDROCHLORIDE see SBA500
SAXITOXIN HYDRATE see SBA600
SAXITOXIN HYDROCHLORIDE see SBA500
SAXOL see MQV750
SAXOSOZINE see SNN500
SA 50 Y see TGB160
SAYFOR see ASD000
SAYFOS see ASD000
SAYPHOS see ASD000
SAYTEX 102E see PAU500
SAYTEX 102 see PAU500
SAYTEX BCL 462 see DDM300
SAZZIO see AFL000
SB-8 see HJZ000
SB-23 see DQY909
SBa 0108E see BAK000
SB 502 see BOO650
SB 5833 see CFY250
S.B.A. see BPW750
SB 475K see SMQ500
SBO see DJL000
SBP-1382 see BEP500
SBP-1390 see BEP750
SBP-1513 see AHJ750
S.B. PENICK 1382 see BEP500
SBS see SMR000
SBTPC see SOU675
SC 9 see SCQ000
SC10 see DCD000
SC-110 see ABU500
SC. 121 see PAW600
SC 201 see SCQ000
SC-1950 see DRK600
SC-2292 see MEK350
SC 2538 see DIR000
SC-2644 see DOB300
SC-2657 see DOA875
SC-2798 see DOB325
SC 2910 see DJM800, XCJ000
SC/3123 see SBA875
SC-3171 see HKR500
SC-3402 see BKI750
S.C. 3497 see AFW500

SC 4641 see NCI525
SC-4642 see EEH550
SC 4722 see ONO000
SC 7031 see DNN600
SC 8016 see CJL500
SC 9387 see CJL500
SC 9420 see AFJ500
SC 9794 see PII500
SC 10295 see MMN250
SC10363 see VTF000
SC 12937 see ARX800
SC 13957 see RSZ600
SC-14266 see PKZ000
SC 15090 see TEO500
SC 15983 see AFJ500
SC 27166 see DWF700
SC 28762 see VRP200
SC-29333 see MJE775
SCABANCA see BCM000
SCALDIP see DVX800
SCANBUTAZONE see BRF500
SCANDICAIN see CBR250, SBB000
SCANDICAINE see SBB000
SCANDICANE see SBB000
SCANDISIL see SNN300
SCANDIUM see SBB500
SCANDIUM(3+) CHLORIDE see SBC000
SCANDIUM CHLORIDE see SBC000
SCAPUREN see SBE500
SCARCLEX see DGD600
SCARLET BASE CIBA I see DEO295
SCARLET BASE CIBA II see NMP500
SCARLET BASE IRGA II see NMP500
SCARLET BASE NSP see NMP500
SCARLET B FAT SOLUBLE see OHA000
SCARLET G BASE see NMP500
SCARLET GN see CMP620
SCARLET GN SPECIALLY PURE see CMP620
SCARLET LOBELIA see CCJ825
SCARLET PIGMENT RN see MMP100
SCARLET R see FMU070
SCARLET RED see SBC500
SCARLET RED, BIEBRICH see SBC500
SCARLET R (MICHAELIS) see SBC500
SCARLET TR BASE see CLK200
SCARLET WISTERIA TREE see SBC550
SCATOLE see MKV750
SC 9 (CARBIDE) see SCQ000
SCE 129 see CCS550
SCE 1365 HYDROCHLORIDE see CCS300
SCENTENAL see OBW200
SCF see TEE225
SCH 286 see SOU650
SCH 412 see DJO800
SCH 6673 see ABG000
SCH 7307 see IPU500
SCH 9384 see AEX000
SCH 9724 see GCS000
SCH 10159 see TIO800
SCH 10304 see CMX770
SCH 10649 see DLV800
SCH 11527 see TEY000
SCH 12650 see SBC800
SCH 13475 see SDY750
SCH 13521 see FMR050
SCH-17726 see GAA100
SCH 20569 see NCP550, SBD000
SCH 22219 see AFI980
SCH see FMR050
SCHAARDINGER α-DEXTRIN see COW925
SCHAEFFER'S-β-NAPHTHOLSULFONIC ACID see HMU500
SCHARLACH B see PEJ500
SCHARLACHROT see SBC500
SCH 5802 B see SBC700
SCH CHLORIDE see HLC500
SCHEELES GREEN see CNN500
SCHEELE'S MINERAL see CNN500
SCHERCEMOL 85 see ISC550
SCHERCOMEL M see MCB050
SCHERING 34615 see CQI500
SCHERING-35830 see CGL250
SCHERING 36056 see DSO200
SCHERING 36103 see FNE500
SCHERING 36268 see CJJ250, CJJ500
SCHERING 38107 see EEO500

SCHERING-38584 see MEG250
SCHERISOLON see PMA000
SCHEROSON see CNS800, CNS825
SCHEROSON F see CNS750
SCH 31846 HYDROCHLORIDE see ICL500
SCHINOPSIS LORENTZII TANNIN see QBJ000
SCHINUS MOLLE see PCB300
SCHINUS MOLLE OIL see SBE000
SCHINUS TEREBINTHIFOLIUS see PCB300
SCHISANDRIN see SBE400
SCHISANDROL B see SBE450
SCHISTOSOMICIDE see SBE500
SCHIWANOX see AMX750
SCHIZANDRIN see SBE400
SCHIZANDROL B see SBE400, SBE450
SCHLAFEN see CAV000
SCHLEIMSAURE see GAR000
SCHOENOCAULON DRUMMONDII see GJU460
SCHOENOCAULON OFFICIANLIS see GJU460
SCHOENOCAULON TEXANUM see GJU460
SCHOENOPLECTUS ARTICULATUS (Linn.) Palla, extract see SBF550
SCHRADAN see OCM000
SCHRADANE (FRENCH) see OCM000
SCH 21420 SULFATE see SBE800
SCHULTENITE see LCK000
SCHULTZ Nr. 826 see ADE500
SCHULTZ Nr. 836 see ADF000
SCHULTZ Nr. 1309 (GERMAN) see FAE100
SCHULTZ Nr. 208 (GERMAN) see HJF500
SCHULTZ NO. 31 see OHA000
SCHULTZ No. 95 see FMU070
SCHULTZ NO. 541 see SBC500
SCHULTZ NO. 737 see FAG140
SCHULTZ No. 770 see FMU059
SCHULTZ No. 826 see CMM062
SCHULTZ No. 918 see CMM510
SCHULTZ No. 1038 see BJI250
SCHULTZ No. 1041 see AJP250
SCHULTZ No. 1228 see IBV050
SCHULTZ-TAB No. 779 (GERMAN) see RMK020
SCHUTTGELB see MQF250
SCH 15719W see HMM500
SCH 18020W see AFJ625
L-SCHWARZ 1 see BMA000
SCHWEFELDIOXYD (GERMAN) see SOH500
SCHWEFELKOHLENSTOFF (GERMAN) see CBV500
SCHWEFEL-LOST see BIH250
SCHWEFELSAEURELOESUNGEN (GERMAN) see SOI500
SCHWEFELWASSERSTOFF (GERMAN) see HIC500
SCHWEFLIGESAEURE (GERMAN) see SOO500
SCHWEINFURTERGRUEN see COF500
SCHWEINFURT GREEN see COF500
SCILLACRIST see POB500
SCILLA "DIDIER" see POB500
SCILLAGLYKOSID A (GERMAN) see GFC000
SCILLAREN A see GFC000
SCILLARENIN-3,6-DEOXY-4-o-β-d-GLUCOPYRANOSYL-α-l-MANNOPYRANOSIDE see GFC000
α-l-SCILLARENIN-3-RHAMNOPYRANOSIDE see POB500
SCILLARENIN-RHAMNOSE (GERMAN) see POB500
SCILLAREN & RHAMNOSE & GLUCOSE (GERMAN) see GFC000
SCILLA (VARIOUS SPECIES) see SLH200
SCILLIROSID see SBF500
(3-β,6-β)SCILLIROSIDE see SBF500
SCILLIROSIDE see SBF500
SCILLIROSIDE GLYCOSIDE see RCF000
SCILLIROSIDIN + GLUCOSE (GERMAN) see SBF500
3-β-SCILLOBIOSIDO-14-β-HYDROXY-Δ-4,20,22-BUFATRIENOLID (GERMAN) see GFC000
SCINNAMINA see TLN500
SCINTILLAR see XHS000
SCIRPUS ARTICULATUS Linn., extract see SBF550
SCLAIR 59 see PJS750
SCLAIR 2911 see PJS750
SCLAIR 19A see PJS750
SCLAIR 96A see PJS750
SCLAIR 59C see PJS750
SCLAIR 79D see PJS750
SCLAIR 11K see PJS750
SCLAIR 19X6 see PJS750
SCLAVENTEROL see NGG500
SC-11800M see SBA625
SCOBRO see SBG500
SCOBRON see SBG500
SCOBUTIL see SBG500

SCOBUTYL see SBG500
SCOKE see PJJ315
SCOLINE see HLC500
SCOLINE CHLORIDE see HLC500
SCON 5300 see PKQ059
SCONATEX see AAX175, VPK000
SCOPAMIN see HOT500
SCOPARON see DRS800
SCOPARONE see DRS800
SCOPINE TROPATE see SBG000
SCOPOLAMINE see SBG000
(−)-SCOPOLAMINE see SBG000
SCOPOLAMINE BROMIDE see HOT500
(−)-SCOPOLAMINE BROMIDE see HOT500
SCOPOLAMINE BROMOBUTYLATE see SBG500
SCOPOLAMINE BUTOBROMIDE see SBG500
SCOPOLAMINE-N-BUTYL BROMIDE see SBG500
SCOPOLAMINE BUTYL BROMIDE see SBG500
SCOPOLAMINE HYDROBROMIDE see HOT500
(−)-SCOPOLAMINE HYDROBROMIDE see HOT500
SCOPOLAMINE METHOBROMIDE see SBH500
SCOPOLAMINE METHYLBROMIDE see SBH500
(−)-SCOPOLAMINE METHYL BROMIDE see SBH500
SCOPOLAMINE compounded with METHYL NITRATE (1:1) see SBH000
SCOPOLAMINE METHYL NITRATE see SBH000
SCOPOLAMINIUM BROMIDE see HOT500
SCOPOLAMMONIUM BROMIDE see HOT500
SCOPOLAN see SBG500
SCOPOS see HOT500
SCOT see HBT500
SCOTCH ATTORNEY see BAE325
SCOTCH PAR see PKF750
SCOTCH PINE NEEDLE OIL see PIH500
SCOTCIL see BFD000
SC 7031 PHOSPHATE see RSZ600
SCREEN see BEG300
SCROBIN see ARQ750
SCTZ see CHD750
SCURENALINE see VGP000
SCUROCAINE see AIL750, AIT250
SCUTL see ABU500
SCW 1 see SCQ000
SCYAN see SIA500
SD 25 see DHY200
40 SD see IMF300
SD 188 see SMQ500
SD-345 see ADR250
SD 354 see SMR000
SD 709 see DRM000
SD-1750 see DGP900
SD 1836 see CLV375
SD 1897 see DDL800
SD 3450 see HCI475
SD 3562 see DGQ875
SD 4,239 see CEQ500
SD 4294 see COD000
SD 4402 see OAN000
SD 4901 see BDX090
SD 5532 see CDR750
SD 7438 see BES250
SD 7727 see DGD400
SD 7961 see DGM600
SD 8447 see RAF100
SD 8530 see TMD000
SD 9098 see DIX600
SD 9129 see MRH209
SD-10576 see SCD500
SD 14114 see BLU000
SD 15418 see BLW750
SD 171-02 see MQQ050
SD 210-32 see PIR000
SD 210-37 see PIQ750
SD 270-31 see OKW000
SD 30,053 see EGS000
SD 417-06 see DAB825
SD 43775 see FAR100
SD ALCOHOL 23-HYDROGEN see EFU000
S DC 200 see SCR400
SDD see DXH300
SDDC see SGM500
SDDS see AOT125
SDEH see DJL400
SD-GP 6000 see SCQ000
SD-GP 8000 see SCQ000
SD 2124-01 HYDROCHLORIDE see PMF525

(±)-SD 2124-01 HYDROCHLORIDE see PMF535
SDIC see SGG500
SDM see SNN300
SDMH see DSF600
SDM No. 5 see MAW250
SDM No. 23 see PBC250
SDMO see AIE750, SNN300
SDP 640 see PJS750
SDmP see SNI500
SDPH see DNU000
SDT 1041 see FNF000
S. DYSENTERIAE TOXIN see SCD750
SE-1520 see IBV100
SE 1702 see DBL700
20SE60 see MCB050
SEA ANEMONE TOXIN II see ARS000
SEA ANEMONE VENOM see SBI800
SEA COAL see CMY760
SEACYL VIOLET R see DBP000
SEA DAFFODIL see BAR325
SEAKEM CARRAGEENIN see CCL250
SEA-LEGS see HGC500
SEA ONION see SLH200
SEAPROSE S see SBI860
SEARLE 703 see DNN600
SEARLE 27166 see DWF700
SEARLEQUIN see DNF600
SEARS HEAVY DUTY LAUNDRY DETERGENT see SBI875
SEA SALT see SFT000
SEA SNAKE VENOM, AIPYSURUS LAEVIS see SBI880
SEA SNAKE VENOM, ASTROTIA STOKESII see SBI890
SEA SNAKE VENOM, HYDROPHIS CYANOCINCTUS see SBI900
SEA SNAKE VENOM, LAPEMIS HARDWICKII see SBI910
SEA SNAKE VENOM, MICROCEPHALOPHIS GRACILIS see SBI929
SEATREM see CCL250
SEAWATER MAGNESIA see MAH500
SEAZINA see SNJ000
SEBACIC ACID see SBJ500
SEBACIC ACID, DIBUTYL ESTER see DEH600
SEBACIC ACID, DIETHYL ESTER see DJY600
SEBACIC ACID, DIHYDRAZIDE see SBK000
SEBACIC ACID, compd. with 1,6-HEXANEDIAMINE (1:1) see HEG130
SEBACIL see BAT750
SEBACONITRILE see SBK500
SEBACOYLDIOXYBIS(TRIBUTYLSTANNANE) see BLL000
SEBAKAN HEXAMETHYLENDIAMINU see HEG130
SEBAQUIN see DNF600
SEBAR see SBN000
SEBATROL see FMR050
SEBIZON see SNP500
SEBUTHYLAZINE see THR600
SECACORNIN see LJL000
SECAGYN see EDC500
SECALONIC ACID D see EDA500
SECBUBARBITAL see BPF000
SECBUBARBITAL SODIUM see BPF250
SECBUTABARBITAL see BPF000
SECBUTOBARBITONE see BPF000
SECHVITAN see PII100
SECLAR see BEG000
SECO 8 see SBN000
SECOBARBITAL see SBM500
R(+)-SECOBARBITAL SODIUM see SBL500
SECOBARBITAL SODIUM see SBN000
SECOBARBITONE see SBM500, SBN000
SECOBARBITONE SODIUM see SBN000
21,22-SECOCAMPTOTHECIN-21-OIC ACID LACTONE see CBB870
(5Z,7E)-9,10-SECOCHESTA-5.7.10(19)-TRIENE-1-α,3-β,25-TRIOL see DMJ400
9,10-SECOCHOLESTA-5,7,10(19)-TRIENE-1-α,3-β-DIOL see HJV000
(1-α,3-β,5Z,7E)-9,10-SECOCHOLESTA-5,7,10(19)-TRIENE-1,3,25-TRIOL see DMJ400
9,10-SECOCHOLESTA-5,7,10(19)-TRIEN-3-β-OL see CMC750
9,10,SECOERGOSTA-5,7,10(19),22-TETRAEN-3-β-OL see VSZ100
(E-β,5E,7E,10-α,22E)-9,10-SECOERGOSTA-5,7,22-TRIEN-3-OL (9CI) see DME300
16,17-SECOESTRA-1,3,5(10),6,8-PENTAEN-17-OIC ACID, 3-METHOXY- see BIT030
16,17-SECO-13-α-ESTRA-1,3,5,6,7,9-PENTAEN-17-OIC ACID, METHYL ESTER see BIT000
SECOMETRIN see LJL000
SECONAL see SBM500
SECONAL SODIUM see SBN000
SECONDARY AMMONIUM ARSENATE see DCG800
SECONDARY AMMONIUM PHOSPHATE see ANR500
SECONDARY ZINC PHOSPHITE see ZLS200

SECONESINZ see GGS000
SECOPAL OP 20 see GHS000
SECOVERINE see SBN300
SECOVERINE HYDROCHLORIDE see SBN300
SECROSTERON see DRT200
SECROVIN see DNX500
SECT HYDROCHLORIDE see EEQ500
SECTRAL see AAE125
SECUPAN see EDC500
SECURINAN-11-ONE (9CI) see SBN350
SECURININ see SBN350
SECURININE see SBN350
(−)-SECURININE see SBN350
SECURITY see ARB750, LCK000
SEDABAMATE see MQU750
SEDAFORM see ABD000
SEDALANDE see HAF400
SEDALIS SEDI-LAB see TEH500
SEDALON see AMK250
SEDAMYL see ACE000
SEDANTOINAL see MKB250
SEDAPRAN see DAP700
SEDAPSIN see CKE750
SEDARAUPIN see RDK000
SEDARAUPINA see RDK000
SEDA-RECIPIN see RDK000
SEDAREX see EHP000
SEDA-SALUREPIN see RDK000
SEDA-TABLINEN see EOK000
SEDAVIC see HAF400
SEDERAUPIN see RDK000
SEDESTRAN see DKA600
SEDETINE see OAV000
SEDETOL see EJM500
SEDEVAL see BAG000
SEDICAT see EOK000
SEDIMIDE see TEH500
SEDIN see TEH500
SEDIPAM see DCK759
SEDIRANIDO see DBH200
SEDISPERIL see TEH500
SEDIZORIN see EOK000
SEDMYNOL see ACE000
SEDNOTIC see AMX750
SEDOFEN see EOK000
SEDOMETIL see DNA800, MJE780
SEDONAL see EOK000
SEDONEURAL see SFG500
SEDOPHEN see EOK000
SEDOR see SKS700
SEDORMID see IQX000
SED OSTANTHRENOVA M see CMU475
SEDOVAL see TEH500
SEDRENA see BBV000
SEDRESAN see MEP250
SEDSERP see RDK000
SED-TEMS see MDL000
SEDTRAN see ACE000
SEDURAL see PDC250, PEK250
SEDUTAIN see SBN000
SEDUXEN see DCK759
SEED of FIDDLENECK see TAG250
SEEDRIN see AFK250
SEEDTOX see ABU500
SEEDVAX see AKR500
SEEKAY WAX see TIT500
SEENOX DS see DXG700
SEFACIN see TEY000
SEFFEIN see CBM750
SEFONA see IEP200
SEFRIL see SBN440
SEFRIL HYDRATE see SBN450
SEGNALE BRONZE RED P see CMS150
SEGNALE LIGHT ORANGE G see CMS145
SEGNALE LIGHT ORANGE PG see CMS145
SEGNALE LIGHT RED 2B see MMP100
SEGNALE LIGHT RED B see MMP100
SEGNALE LIGHT RED BR see MMP100
SEGNALE LIGHT RED C4R see MMP100
SEGNALE LIGHT RED RL see MMP100
SEGNALE LIGHT RUBINE RG see NAY000
SEGNALE LIGHT YELLOW 2GR see DEU000
SEGNALE LIGHT YELLOW GRX see CMS208
SEGNALE RED C see CMS150
SEGNALE RED GS see CMS148

SEGNALE RED LC see CHP500
SEGNALE RED 3R see CMS155
SEGNALE YELLOW 2GR see CMS212
SEGONTIN see PEV750
S. EGRELTRI ATUNUN (TURKISH) see BML000
SEGUREX see BON325
SEGURIL see CHJ750
SEIKAFAST RED 8040 see CMS148
SEIKAFAST YELLOW 2200 see CMS212
SELACRYN see TGA600
SELECRON see BNA750
SELECTIVE see BPG250
SELECTOMYCIN see SLC000
SELECTROL see SBN475
SELEKTIN see BKL250
SELEKTON B 2 see EIX500
SELEN (POLISH) see SBO500
1,2,5-SELENADIAZOLE-3-CARBOXAMIDE, 4-AMINO- see AMN300
1,2,5-SELENADIAZOLE-3-CARBOXAMIDE, 4-AMINO-N-METHYL- see MGL600
SELENIC ACID, liquid (DOT) see SBN500
SELENIC ACID see SBN500
SELENIC ACID, DIPOTASSIUM SALT see PLR750
SELENIC ACID, DISODIUM SALT, DECAHYDRATE see SBN505
SELENIC ACID, compd. with HYDRAZINE see HGW100
SELENIDE, DIALLYL- see DBL300
SELENINIC ACID, PROPYL- see PNV760
β-SELENINOPROPIONIC ACID see SBN510
SELENINYL BIS(DIMETHYLAMIDE) see SBN525
SELENINYL BROMIDE see SBN550
SELENINYL CHLORIDE see SBT500
SELENIOUS ACID see SBO000
SELENIOUS ACID, DIPOTASSIUM SALT (9CI) see SBO100
SELENIOUS ACID, DISODIUM SALT see SJT500
SELENIOUS ACID DISODIUM SALT PENTAHYDRATE see SJT600
SELENIOUS ANHYDRIDE see SBQ500
SELENIUM (colloidal) see SBP000
SELENIUM see SBO500
SELENIUM ALLOY see SBO500
SELENIUM BASE see SBO500
SELENIUM(IV) CHLORIDE (1:4) see SBU000
SELENIUM CHLORIDE see SBS500
SELENIUM CHLORIDE OXIDE see SBT500
SELENIUM COMPOUNDS see SBP500
SELENIUM CYSTINE see DWY800, SBP600
SELENIUM, DICHLOROBIS(2-CHLOROCYCLOHEXYL)- see DEU100
SELENIUM, DICHLOROBIS(2-ETHOXYCYCLOHEXYL)- see DEU125
SELENIUM DIETHYLDITHIOCARBAMATE see DJD400, SBP900
SELENIUM DIMETHYLDITHIOCARBAMATE see SBQ000
SELENIUM(IV) DIOXIDE (1:2) see SBQ500
SELENIUM DIOXIDE see SBO000, SBQ500
SELENIUM DIOXIDE mixed with ARSENIC TRIOXIDE (1:1) see ARJ000
SELENIUM(IV) DISULFIDE (1:2) see SBR000
SELENIUM(IV) DISULFIDE SHAMPOO (2.5%) see SBR500
SELENIUM DISULFIDE (2.5%) SHAMPOO see SBR500
SELENIUM DISULPHIDE (DOT) see SBR000
SELENIUM DUST see SBO500
SELENIUM ELEMENTAL see SBO500
SELENIUM FLUORIDE see SBS000
SELENIUM HEXAFLUORIDE see SBS000
SELENIUM HOMOPOLYMER see SBO500
SELENIUM HYDRIDE see HIC000
SELENIUM METAL POWDER, NON-PYROPHORIC (DOT) see SBO500
SELENIUM MONOCHLORIDE see SBS500
SELENIUM MONONITRIDE see SBT100
SELENIUM MONOSULFIDE see SBT000
SELENIUM NITRIDE (DOT) see SBT100
SELENIUM NITRIDE see SBT100
SELENIUM OXIDE (DOT) see SBT200
SELENIUM OXIDE see SBQ500, SBT200
SELENIUM OXYCHLORIDE see SBT500
SELENIUM SULFIDE see SBR000, SBT000
SELENIUM SULPHIDE see SBT000
SELENIUM TETRACHLORIDE see SBU000
SELENOASPIRINE see SBU100
2,2′-SELENOBIS(BENZOIC ACID) see SBU150
o,o′-SELENOBIS(BENZOIC ACID) see SBU150
SELENOCYANIC ACID, POTASSIUM SALT see PLS000
SELENO-dl-CYSTINE see SBU200
d,l-SELENOCYSTINE see SBU200
SELENOCYSTINE see DWY800, SBP600, SBU200
4,4′-(SELENODI-2,1-ETHANEDIYL)BISMORPHOLINE DIHYDROCHLORIDE see MRU255
SELENODIGLUTATHIONE see SBU500
2-SELENOETHYLGUANIDINE see SBU600
6-SELENOGUANOSINE see SBU700

6-SELENO-GUANOSINE (9CI) see SBU700
SELENOHOMOCYSTINE see SBU710
SELENOMERCAPTOAETHYLGUANIDIN (GERMAN) see SBU600
SELENOMETHIONINE see SBU725
SELENONIUM, TRIMETHYL- see TME500
SELENONIUM, TRIMETHYL-, CHLORIDE see TME600
SELENOPHOS see SBU900
SELENOSULFURIC ACID, 2-AMINOETHYL ESTER see AJS950
SELENO-TOLUIDINE BLUE see SBU950
SELENOUREA see SBV000
SELENSULFID (GERMAN) see SBT000
SELEPHOS see PAK000
SELES BETA see TAL475
SELEXID see MCB550
SELFER see IDE000
SELF ROCK MOSS see CCL250
SELINON see DUS700
SELLACID YELLOW AEN see SGP500
SELLA FAST RED RS see CMM330
SELLAITE see MAF500
SELOKEN see SBV500
SELSUN see SBW000
SELSUN BLUE see SBR000
SEL-TOX SSO2 and SS-20 see DXG000
SEM (cytostatic) see TND500
SEMAP see PAP250
SEMBRINA see DNA800, MJE780
SEMDOXAN see CQC500, CQC650
SE-MEG see SBU600
SEMERON see INR000
SEMESAN see CHO750
SEMICARBAZIDE see HGU000
SEMICARBAZIDE, 2-(o-CHLOROPHENETHYL)-3-THIO- see CJK100
SEMICARBAZIDE HYDROCHLORIDE see SBW500
SEMICARBAZIDE, 2-(p-METHYLPHENETHYL)-3-THIO- see MNP500
SEMICARBAZIDE, 1-PHENETHYL- see PDK300
SEMICARBAZIDE, 2-PHENETHYL-3-THIO- see PDK500
SEMICARBAZIDE, 1-PHENYL-4-(PHENYLIMINO)-3-THIO- see DWN200
SEMICILLIN see AIV500
p-SEMIDINE see PFU500
SEMIKON see DPJ400, TEO250
SEMIKON HYDROCHLORIDE see DPJ400
SEMILLA de CULEBRA (MEXICO) see RMK250
SEMINOLE BEAD see RMK250
SEMINOLE BREAD see CNH789
SEMOPEN see PDD350
SEMOXYDRINE see DBA800
SEMPERVIVUM (JAMAICA) see AGV875
SEMUSTINE see CHD250
SENAQUIN see CLD000
SENARMONTITE see AQF000
SENCEPHALIN see ALV000
SENCOR see MQR275
SENCORAL see MQR275
SENCORER see MQR275
SENCOREX see MQR275
SENDOXAN see CQC500
SENDRAN see PMY300
SENDUXAN see CQC500, CQC650
SENECA OIL see PCR250
SENECIA JACOBAEA see RBA400
SENECIA LONGILOBUS see RBA400
SENECIO CANNABIFOLIUS, leaves and stalks see SBW950
SENECIOIC ACID see MHT500
SENECIO LONGILOBUS see SBX000
SENECIONAN-11,16-DIONE, 13,19-DIDEHYDRO-12,18-DIHYDROXY-(9CI) see
   RJZ000
SENECIO NEMORENSIS FUCHSII, alkaloidal extract see SBX200
SENECIONINE see ARS500
SENECIO VULGARIS see RBA400
SENECIPHYLLIN see SBX500
SENECIPHYLLINE see SBX500
SENECIPHYLLIN HYDROCHLORIDE see SBX525
SENEGAL GUM see AQQ500
SENEGENIN see SBY000
SENEGIN see SBY000
SENF OEL (GERMAN) see AGJ250
SENFOEL (GERMAN) see ISK000
SENIRAMIN see RCA200
SENKIRKIN see DMX200
SENKIRKINE see DMX200
SENTIPHENE see CJU250
SENTONIL see PDW500
SENTRY see HOV500
SENTRY GRAIN PRESERVER see PMU750

SEOMINAL see RDK000
SEOTAL see SBN000
SEPAN see CMR100
SEPAZON see CMY525
SEPERIDOL see CLS250
SEPEROL see CLS250
SEPPIC MMD see CIR250
SEPPUDY see CNT350
SEPRISAN see HCP800
SEPSINOL see FPI000
SEPTACIDIN see SBZ000
SEPTACIL see ALF250
SEPTAMIDE ALBUM see SNM500
SEPTAMYCIN see SCA000
SEPTEAL see CDT250
SEPTENE see CBM750
SEPTICOL see CDP250
SEPTINAL see SNM500
SEPTIPULMON see PPO000
SEPTISOL see HCL000
SEPTOCHOL see DAQ400
SEPTOFEN see HCL000
SEPTOPLEX see SNM500
SEPTOS see HJL500
SEPTOTAN see PFN000
SEPTRA see SNK000, TKX000
SEPTRAN see SNK000, TKX000
SEPTRIM see TKX000
SEPTRIN see TKX000
SEPTURAL see PAF250
SEPUDDY see CNT350
SEPYRON see DNU100
SEQ 100 see EIX000
SEQUAMYCIN see SLC000
SEQUESTRENE 30A see EIV000
SEQUESTRENE AA see EIX000
SEQUESTRENE Na 4 see EIV000
SEQUESTRENE Na3 see TNL250
SEQUESTRENE NaFe IRON CHELATE see EJA379
SEQUESTRENE SODIUM 2 see EIX500
SEQUESTRENE ST see EIV000
SEQUESTRENE TRISODIUM see TNL250
SEQUESTRENE TRISODIUM SALT see TNL250
SEQUESTRIC ACID see EIX000
SEQUESTROL see EIX000
SEQUILAR see NNL500
SEQUION 40Na32 see DJG700
SEQUOSTAT see NNL500
SERAGON see SCA750
SERAGONIN see SCA750
SERAX see CFZ000
(d-SER(BU(t))-LH-RH-(1-9)NONAPEPTIDE-ETHYLAMIDE see LIU420
SEREEN see HOT500
SERENACE see CLY500
SERENACK see DCK759
SERENAL see CFZ000, OMK300
SERENID see CFZ000
SERENID-D see CFZ000
SERENIL see CHG000
SERENIUM see CGA000
SERENSIL see CHG000
SERENTIL see MON750
SEREN VITA see MDQ250
SERENZIN see DCK759
SEREPAX see CFZ000
SERESTA see CFZ000
SERFIN see RDK000
SERFOLIA see RDK000
SERGOSIN see SHX000
SERIAL see MCA500
SERIBAK see MRM000
SERICOSOL-N see DVR909
SERIEL see GJS200
SERIL see MQU750
SERILENE BRILLIANT RED 2BL see AKI750
SERILENE RED 2BL see AKI750
(S)-SERINE see SCA355
(l)-SERINE see SCA355
l-SERINE see SCA355
SERINE see SCA355
l-SERINE DIAZOACETATE (ester) see ASA500
l-SERINE DIAZOACETATE see ASA500
dl-SERINE 2-(2,3,4-TRIHYDROXYBENZYL)HYDRAZINE HYDROCHLORIDE see
   SCA400

dl-SERINE 2-((2,3,4-TRIHYDROXYPHENYL)METHYL)HYDRAZIDE MONOHYDRO-
CHLORIDE (9CI) see SCA400
SERINGINA see RCA200
SERINYL BLUE 2G see TBG700
SERINYL HOSIERY BLUE see MGG250
SERINYL HOSIERY YELLOW GD see AAQ250
SERIPLAS BLUE GREEN BW see DMM400
SERIPLAS RED X3B see DBX000
SERIPLAS YELLOW GD see AAQ250
SERISOL BLILLIANT RED X 3B see DBX000
SERISOL BRILLIANT BLUE BG see MGG250
SERISOL BRILLIANT VIOLET 2R see DBP000
SERISOL FAST BLUE BGLW see CMP060
SERISOL FAST BLUE GREEN B see DMM400
SERISOL FAST BLUE GREEN BW see DMM400
SERISOL FAST CRIMSON BD see CMP080
SERISOL FAST RED 2B see AKE250
SERISOL FAST VIOLET 6B see AKP250
SERISOL FAST VIOLET B see DBY700
SERISOL FAST YELLOW A see DUW500
SERISOL FAST YELLOW GD see AAQ250
SERISOL FAST YELLOW N 5RD see CMP090
SERISOL ORANGE YL see AKP750
SERISTAN BLACK B see AQP000
SERITOX 50 see DGB000
SERMION see NDM000
SERNAS see CLY500
SERNEL see CLY500
SERNEVIN see EPD500
SERNYL see AOO500
SERNYLAN see AOO500
SERNYL HYDROCHLORIDE see AOO500
SERODEN see FNF000
SEROFINEX see ARQ750
SEROGAN see SCA750
SEROLFIA see RDK000
SEROMYCIN see CQH000
SEROTIN CREATININE SULFATE see AJX750
SEROTINEX see ARQ750
SEROTONIN see AJX500
SEROTONIN BENZYL ANALOG see BEM750
SEROTONIN CREATININE SULFATE MONOHYDRATE see AJX750
SEROTROPIN see SCA750
SERP see RDK000
SERP-AFD see RDK000
SERPAGON see RCA200
SERPALAN see RDK000
SERPALOID see RDK000
SERPANEURONA see RDK000
SERPANRAY see RDK000
SERPASIL see RDK000
SERPASIL APRESOLINE see RDK000
SERPASIL APRESOLINE No. 2 see HGP500
SERPASIL-ESIDREX see RDK000
SERPASIL-ESIDREX K see RDK000
SERPASIL-ESIDREX NO. 1 see RDK000
SERPASIL-ESIDREX NO. 2 see RDK000
SERPASIL PREMIX see RDK000
SERPASOL see RDK000
SERPATE see RDK000
SERPATONE see RDK000
SERPAX see CFZ000, RDK000
SERPAZIL see RDK000
SERPAZOL see RDK000
SERPEDIN see RDK000
SERPEN see RDK000
SERPENA see RDK000
SERPENT see YAK350
SERPENTIL see RDK000
SERPENTIN see RDK000
SERPENTINA see RDK000
SERPENTINE see ARM250, ARM268, SCA475
SERPENTINE CHRYSOTILE see ARM268
SERPENTINE (alkaloid) HYDROXIDE, inner salt see SCA475
SERPENTINE HYDROXIDE, inner salt see SCA475
SERPENTINE 'PHARBIL' see RDK000
SERPENTINIC ACID ISOBUTYL ESTER see SCA525
SERPENTINTARTRAT (GERMAN) see SCA550
SERPICON see RDK000
SERPIL see RDK000
SERPILOID see RDK000
SERPILUM see RDK000
SERPINE see RDK000
SERPINE (Pharmaceutical) see RDK000
SERPIPUR see RDK000

SERPIVITE see RDK000
SERPLEX K see RDK000
SERPOGEN see RDK000
SERPOID see RDK000
SERPONE see RDK000
SERPRESAN see RDK000
SERPYRIT see RDK000
SERRAL see DKA600
SERRATIA EXTRACELLULAR PROTEINASE see SCA625
SERRATIA MARCESCENS ENDOTOXIN see SCA600
SERRATIA PISCVATORUM POLYSACCHARIDE see SCA610
SERRATIO PEPTIDASE see SCA625
SERTABS see RDK000
SERTAN see DBB200
(d-SER(TBU)⁶-EA¹⁰)-LHRH see LIU420
(d-SER(TBU)⁶-EA¹⁰)-LUTEINIZING HORMONE-RELEASING HORMONE see
LIU420
d-SER(TBU⁶)-LH-RH-(1-9)-NONAPEPTIDE ETHYLAMIDE see LIU420
SERTENS see RDK000
SERTENSIN see RDK000
SERTINA see RDK000
SERTINON see EPQ000
SERTOFRAN see DSI709
SERUM GONADOTROPHIN see SCA750
SERUM GONADOTROPIC HORMONE see SCA750
SERUM GONADOTROPIN see SCA750
N-SERVE NITROGEN STABILIZER see CLP750
SERVISONE see PLZ000
SES see CNW000
SESAGARD see BKL250
SESAME OIL see SCB000
SESAMOL see MJU000
SESBANIA SESBAN (L.) Merr. var. BICOLOR W. & A., extract excluding
roots see SCB100
SESBANIA (VARIOUS SPECIES) see SBC550
SESDEN see TGB160
SESONE (ACGIH) see CNW000
SESO VEGETAL (CUBA, PUERTO RICO) see ADG400
SESQUIETHYLALUMINUM CHLORIDE see TJP775
SESQUIMUSTARD see SCB500
SESQUIMUSTARD Q see SCB500
SESQUISULFURE de PHOSPHORE (FRENCH) see PHS500
SET see MDI000
SETACYL BLUE BN see MGG250
SETACYL BLUE 6GN see DMM400
SETACYL BLUE GREEN P-BS see DMM400
SETACYL DIAZO NAVY R see DCJ200
SETACYL PINK 3B see AKE250
SETACYL RED 2B see CMP080
SETACYL RED P-3B see DBX000
SETACYL TURQUOISE BLUE 2G see DMM400
SETACYL TURQUOISE BLUE 4G see DMM400
SETACYL TURQUOISE BLUE G see DMM400
SETACYL TURQUOISE BLUE GD see DMM400
SETACYL VIOLET R see DBP000
SETACYL YELLOW G see AAQ250
SETACYL YELLOW 2GN see AAQ250
SETACYL YELLOW P-BS see DUW500
SETACYL YELLOW P-2GL see AAQ250
SETAMINE US 132 see MCB050
SETAMINE US 141 see MCB050
SETAMINE US 138BB70 see MCB050
SETAMINE US 139BB70 see MCB050
SETHYL see MDL000
SETILE VIOLET 3R see DBP000
SETONIL see DCK759
SETRETE see PFO550
SETTIMA see DAP700
SEVACARB see CBT750
SEVAL see CBT750
S-SEVEN see SCC000
SEVENAL see EOK000
SEVEN BARK see HGP600
SEVICAINE see AIT250
SEVIMOL see CBM750
SEVIN 4 see CBM750
SEVIN (OSHA) see CBM750
SEVIN see CBM750
SEVINOL see TJW500
SEVRON ORANGE G see CMM764
SEWIN see CBM750
SEXADIEN see DAL600
SEXADIENO see FBP100
SEXOCRETIN see DKA600
SEXOVAR see FBP100

SEXOVID see FBP100
SEXTONE see CPC000
SEXTONE B see MIQ740
SEXTRA see SCB000
SF 60 see MAK700
SF-337 see AJO500
SF 733 see XQJ650
SF 837 see MBY150
SF 1173 see OCE100
SF 1202 see DAF350
SF-6505 see HLM000
SF6539 see HDC000
S 7481F1 see CQH100
SF 733 ANTIBIOTIC SULFATE see RIP000
SFERICASE see SCC550
S6F HISTYRENE RESIN see SMR000
SFK 70 see UTU500
SG-67 see SCH000
SGA see SCK600
SGD-SCHA 1059 see BGC500
S. GRISEUS PROTEASE see PMJ100
S. GRISEUS PROTEINASE see PMJ100
SH 100 see OLW400
SH 261 see EGQ000
SH 393 see NNQ000
SH 514 see SKM000
SH 567a see PMC700
SH 567 see PMC700
SH 582 see GEK510
SH 714 see CQJ500
SH 717 see GHK200
SH 771 see AMV375
SH 850 see NNQ500
SH 926 see AAI750
SH 6030 see TNJ500
SH 30858 see FAJ150
SaH 43-715 see POB300
SaH 46-790 see THH350
SH 66752 see PNI250
SH 70850 see NNQ500
SH 71121 see NNL500
SH 80582 see GEK510
SHA see SAL500
SH 213AB see IGD100
SHADOCOL see TDE750
SHAKE-SHAKE see RBZ400
SHALE OIL (DOT) see COD750
SHAMMAH (SAUDI ARABIA) see SED400
SHAMROX see CIR250
SHARSTOP 204 see SGM500
SHAWINIGAN ACETYLENE BLACK see CBT750
SHB 286 see SOU650
SHB 261AB see NNL500
SHB 264AB see NNL500
SHED-A-LEAF see SFS000
SHED-A-LEAF ''L'' see SFS000
SHEEP DIP see ARE250
SHEEP LAUREL see MRU359
SHELL 40 see BQZ000
SHELL 300 see SMQ500
SHELL 345 see ADR250
SHELL 4072 see CDS750
SHELL 4402 see OAN000
SHELL 5520 see PMP500
SHELL ATRAZINE HERBICIDE see ARQ725
SHELL CARBON see CBT750
SHELL GOLD see GIS000
SHELL MIBK see HFG500
SHELL OS 1836 see CLV375
SHELLOYNE H see DLJ500
SHELL SD 345 see ADR250
SHELL SD-3450 see HCI475
SHELL SD-3562 see DGQ875
SHELL SD 4,239 see CEQ500
SHELL SD 4294 see COD000
SHELL SD-5532 see CDR750
SHELL SD 7,438 see BES250
SHELL SD 7,727 see DGD400
SHELL SD-8530 see TMD000
SHELL SD-9098 see DIX600
SHELL SD 9129 see MRH209
SHELL SD-10576 see SCD500
SHELL SD-14114 see BLU000
SHELL SILVER see SDI500
SHELLSOL 140 see NMX000

SHELL UNDRAUTTED A see AFV500
SHELL WL 1650 see OAN000
SHENG BAI XIN (CHINESE) see CMV375
SHERSTAT SLN see SAH500
SHIGATOX see AHO250
SHIGA TOXIN see SCD750
SHIGELLA DYSENTERIAE TOXIN see SCD750
SHIGRODIN see BRF500
SHIKIMATE see SCE000
SHIKIMIC ACID see SCE000
SHIKIMOLE see SAD000
SHIKISO ACID ANTHRACENE RED G see CMM325
SHIKISO ACID FAST YELLOW MR see CMM759
SHIKISO ACID RED PG see CMM320
SHIKISO ACID RED RS see NAO600
SHIKISO AMARANTH see FAG020
SHIKISO DIRECT BRILLIANT BLUE RW see CMO600
SHIKISO DIRECT DARK GREEN B see CMO830
SHIKISO DIRECT SCARLET 3B see CMO875
SHIKISO DIRECT SKY BLUE 5B see CMO500
SHIKISO METANIL YELLOW see MDM775
SHIKOMOL see SAD000
SHIMAZAKI PATENT BLUE AFX see ERG100
SHIMMEREX see ABU500
SHINGLE PLANT see SLE890
SHINING SUMAC see SCF000
SHINKOLITE see PKB500
SHIN-NAITO S see TEH500
SHINNIBROL see TEH500
SHINNIPPON FAST RED B BASE see NEQ000
SHINNIPPON FAST RED GG BASE see NEO500
SHINNIPPON FAST RED 3GL BASE see KDA050
SHIONOGI 6059S see LBH200
SHIPRON A see CAT775
SHIRLAN see SAH500
SHIRLAN AG see SAH500
SHIRLAN EXTRA see SAH500
SHM see SLD900
SHMP see SHM500
SHOALLOMER see PMP500
SHOCK-FEROL see VSZ100
SHOLEX 5003 see PJS750
SHOLEX 5100 see PJS750
SHOLEX 6000 see PJS750
SHOLEX 6002 see PJS750
SHOLEX F 171 see PJS750
SHOLEX F 6050C see PJS750
SHOLEX F 6080C see PJS750
SHOLEX 4250HM see PJS750
SHOLEX L 131 see PJS750
SHOLEX S 6008 see PJS750
SHOLEX SUPER see PJS750
SHOLEX XMO 314 see PJS750
SHOWA CHROME YELLOW GG see SIT850
SHOWA FAST RED B BASE see NEQ000
SHOWDOMYCIN see RIU000
SHOXIN see NNF000
cis-SHP see SCF025
trans(+)-SHP see SCF075
trans(−)-SHP see SCF050
SHRUBBY BITTERSWEET see AHJ875
SHRUB VERBENA see LAU600
SHS see HCP900
SI-6711 see DJV600
SI see LCF000
SIACARB see SAZ000
SIALIDASE see NCQ200
SIARKI CHLOREK (POLISH) see SON510
SIARKI DWUTLENEK (POLISH) see SOH500
SIARKOWODOR (POLISH) see HIC500
SIBEPHYLLIN see DNC000
SIBEPHYLLINE see DNC000
SIBIROMYCIN see SCF500
SIBOL see DKA600
SICILIAN CERISE TONER A-7127 see FAG070
SICOCLOR see FAM000
SICO FAST YELLOW D 1355 see CMS208
SICO FAT RED BG NEW see CMS242
SICOL 150 see DVL700
SICOL 160 see BEC500
SICOL 250 see AEO000
SICO LAKE RED CU see CMS150
SICO LAKE RED 2L see CHP500
SICRON see PKQ059
SIDDIQUI see AFH250

SIDNOFEN see SPA000
SIDVAX see AKR500
501 SIEGFRIED see AOO800
SIEGLE ORANGE S see CMS145
SIEGLE RED 1 see MMP100
SIEGLE RED B see MMP100
SIEGLE RED BB see MMP100
SIENNA see IHD000
SIERRA C-400 see TAB750
SIFERRIT see IHG100
SIGACALM see CFZ000
SIGAPRIN see TKX000
SIGETIN see SPA650
SIGMACELL see CCU150
SIGMAFON see MBW750
SIGMAMYCIN see TBX000
SIGMART see NDL800
SIGNOPAM see CFY750
SIGOPHYL see HOA000
SIGURAN see HGC500
SILACYCLOBUTANE see SCF550
SILAK M 10 see SCR400
SILANAMINE, N,N'-METHANE TETRAYLBIS(1,1,1-TRIMETHYL)- see BLQ950
SILANE see SDH575
SILANE A-163 see MQF500
SILANE, (3-AMINOPROPYL)TRIETHOXY- see TJN000
SILANE, γ-AMINOPROPYLTRIETHOXY- see TJN000
SILANE, BENZYLCHLORODIMETHYL- see BEE800
SILANE, CHLOROTRIPHENYL- see TMU800
SILANE, (3-CYANOPROPYL)DIETHOXY(METHYL)- see COR500
SILANE, (3-CYANOPROPYL)TRICHLORO- see COR750
SILANE, (3-CYANOPROPYL)TRIETHOXY- see COS800
SILANE, DICHLOROMETHYL(3,3,3-TRIFLUOROPROPYL)- see DFS700
SILANE, DIHYDROXYDIPHENYL- see DMN450
SILANE, ETHENYLTRIMETHOXY-(9CI) see TLD000
SILANE, ETHYL TRIETHOXY A-15 see EQA000
SILANE, HYDROXYTRIPHENYL- see HOM300
SILANE, OXYBIS((3-AMINOPROPYL)DIMETHYL)- see OPC100
SILANE, PHENYL- see SDX250
SILANE, p-PHENYLENEBIS(DIMETHYL)- see PEW725
SILANE, 1,4-PHENYLENEBIS(DIMETHYL)-(9CI)) see PEW725
SILANE, PHENYLTRIFLUORO- see PGO500
2,2',2'',2'''-SILANETETRAYLTETRAKISETHANOL see TDH100
SILANETRIAMINE, N,N,N',N',N'',N''-HEXAMETHYL- see TNH300
SILANE, TRI-tert-BUTOXYVINYL- see SDX300
SILANE, TRICHLOROALLYL- see AGU250
SILANE, TRICHLORO(3-CYANOPROPYL)- see COR750
SILANE, TRICHLOROCYCLOHEXYL- see CPR250
SILANE, TRICHLOROETHYL- see EPY500
SILANE, TRICHLOROMETHYL- see MQC500
SILANE, TRIETHOXY(3-CYANOPROPYL)- see COS800
SILANE, TRIETHOXYETHYL- see EQA000
SILANE, TRIETHOXYMETHYL- see MQD750
SILANE, TRIETHOXYPHENYL- see PGO000
SILANE, TRIMETHOXY- see TLB750
SILANE, TRIMETHOXYMETHYL- see MQF500
SILANE, TRIMETHYLCHLORO- see TLN250
SILANE, TRIS(1,1-DIMETHYLETHOXY)ETHENYL-(9CI) see SDX300
SILANE, VINYL TRICHLORO 1-150 see TIN750
SILANE, VINYL TRIETHOXY 1-151 see TJN250
SILANE Y-4086 see EBO000
SILANE-Y-4087 see ECH000
SILANTIN see DKQ000
SILASTIC see PJR250
SILATRANE see TMO750
SILBER (GERMAN) see SDI500
SILBERNITRAT see SDS000
SILBESAN see CLD000
SILENE EF see CAW850
SILICA AEROGEL see SCI000
SILICA, AMORPHOS-FUME (ACGIH) see SCH001
SILICA, AMORPHOUS see SCH000
SILICA, AMORPHOUS-DIATOMACEOUS EARTH (UNCALCINED) (ACGIH) see DCJ800
SILICA, AMORPHOUS FUME see SCH001
SILICA, AMORPHOUS FUMED see SCH000
SILICA, AMORPHOUS-FUSED (ACGIH) see SCK600
SILICA, AMORPHOUS HYDRATED see SCI000
SILICA, CRYSTALLINE see SCI500
SILICA, CRYSTALLINE—CRISTOBALITE see SCJ000
SILICA, CRYSTALLINE—QUARTZ see SCJ500
SILICA, CRYSTALLINE-TRIDYMITE (ACGIH, OSHA) see SCK000
SILICA, CRYSTALLINE—TRIDYMITE see SCK000
SILICA FLOUR see SCI500, SCK500
SILICA FLOUR (powdered crystalline silica) see SCJ500

SILICA, FUSED (OSHA) see SCK600
SILICA, FUSED see SCK600
SILICA GEL see SCI000, SCL000
SILICA, GEL and AMORPHOUS—PRECIPITATED see SCL000
SILICANE see SDH575
SILICANE, CHLOROTRIMETHYL- see TLN250
SILICANE, TRICHLOROETHYL- see EPY500
SILICATE D'ETHYLE (FRENCH) see EPF550
SILICATE(2-), HEXAFLUORO-, BARIUM see BAO750
SILICATE(2-), HEXAFLUORO-, BARIUM (1:1) (9CI) see BAO750
SILICATE(2-), HEXAFLUORO-, CADMIUM (8CI,9CI) see CAG500
SILICATE(2-), HEXAFLUORO-, CALCIUM (1:1) (9CI) see CAX250
SILICATE(2-), HEXAFLUORO-, MAGNESIUM (1:1) see MAF600
SILICATES see SCM500
SILICATE SOAPSTONE see SCN000
SILICA, VITREOUS (9CI) see SCK600
SILICA XEROGEL see SCI000
SILICETANE see SCF550
SILICIC ACID see SCI000, SCL000
SILICIC ACID ALUMINUM SALT see AHF500
SILICIC ACID, BERYLLIUM SALT see SCN500
SILICIC ACID, CYCLIC NITRILOTRIETHYLENE ETHYL ESTER see EFJ600
SILICIC ACID, 2-ETHYLBUTYL ESTER see EHC900
SILICIC ACID, METHYL ESTER of ortho- see MPI750
SILICIC ACID, METHYL ESTER see TDJ100
SILICIC ACID TETRAETHYL ESTER see EPF550
SILICIC ACID, ZIRCONIUM(4+) SALT (1:1) see ZSS000
SILICIC ANHYDRIDE see SCH000, SCJ500
SILICI-CHLOROFORME (FRENCH) see TJD500
SILICIO(TETRACLORURO di) see SCQ500
SILICIUMCHLOROFORM (GERMAN) see TJD500
SILICIUMTETRACHLORID (GERMAN) see SCQ500
SILICIUMTETRACHLORIDE (DUTCH) see SCQ500
SILICIUM(TETRACHLORURE de) (FRENCH) see SCQ500
SILICOBROMOFORM see THX000
SILICOCHLOROFORM see TJD500
SILICOETHANE see DXA000
SILICOFLUORIC ACID see SCO500
SILICON see SCP000
SILICON BROMIDE see SCP500
SILICON CARBIDE (ACGIH, OSHA) see SCQ000
SILICON CARBIDE see SCQ000
SILICON CHLORIDE see SCQ500
SILICON CHLORIDE HYDRIDE see DGK300
SILICON DIBROMIDE SULFIDE see SCR100
SILICON DIOXIDE (FCC) see SCH000
SILICON DIOXIDE see SCI500, SCK600
SILICONE 360 see SCR400
SILICONE 1-174 see TLC250
SILICONE A-172 see TNJ500
SILICONE A-186 see EBO000
SILICONE A-187 see ECH000
SILICONE A-189 see TLC000
SILICONE A-1100 see TJN000
SILICONE A-1120 see TLC500
SILICONE DC 200 see SCR400
SILICONE DC 360 see SCR400
SILICONE DC 360 FLUID see SCR400
SILICONE DIOXIDE see SCK600
SILICONE RELEASE L 45 see SCR400
SILICONE RUBBER see PJR250
SILICONES see SDC000, SDF000
SILICONE SF 1173 see OCE100
SILICONE Y-6607 see SDF000
SILICON FLUORIDE see SDF650
SILICON FLUORIDE BARIUM SALT see BAO750
SILICON METHYLATE see TDJ100
SILICON MONOCARBIDE see SCQ000
SILICON ORGANIC LACQUER MODIFICATOR 113-63 see MRA100
SILICON OXIDE see SDH000
SILICON PHENYL TRICHLORIDE see TJA750
SILICON POWDER, amorphous (DOT) see SCP000
SILICON SF 1202 see DAF350
SILICON SODIUM FLUORIDE see DXE000
SILICON TETRAAZIDE see SDH500
SILICON TETRACHLORIDE (DOT) see SCQ500
SILICON TETRAFLUORIDE (DOT) see SDF650
SILICON TETRAHYDRIDE see SDH575
SILICON TRIETHANOLAMIN see SDH670
SILICON ZINC FLUORIDE see ZIA000
SILIKILL see SCH000
SILIKON ANTIFOAM FD 62 see SCR400
SILK see SDI000
SILK FAST GREEN B see CMM200
SILMOS T see CAW850

SILMURIN see RCF000, SBF500
SILOGOMMA FAST YELLOW 2G see CMS212
SILOGOMMA ORANGE G see CMS145
SILOGOMMA RED RLL see MMP100
SILOMAT see CMW500
SILON see NOH000
SILONIST see CMW500
SILOSAN see DIN800
SILOSOL RED RBN see MMP100
SILOSOL RED RN see MMP100
SILOSUPER PINK B see RGW000
SILOTERMO CARMINE G see CMS155
SILOTERMO ORANGE G see CMS145
SILOTERMO YELLOW G see CMS212
SILOTON ORANGE GT see CMS145
SILOTON RED BRLL see MMP100
SILOTON RED RLL see MMP100
SILOTON RUBINE 2B see CMS155
SILOTON RUBINE B see CMS155
SILOTON YELLOW GTX see DEU000
SILOTRAS BROWN TRN see NBG500
SILOTRAS ORANGE TR see PEJ500
SILOTRAS RED T3B see SBC500
SILOTRAS RED TG see CMS242
SILOTRAS RUBINE TSB see CMP080
SILOTRAS SCARLET TB see OHA000
SILOTRAS YELLOW T2G see DOT300
SILOTRAS YELLOW TSG see AAQ250
SILOXANES see SDC000, SDF000
SILOXANES and SILICONES, DI Me see SCR400
SILTEX see SCK600
SILUBIN see BQL000
SILUNDUM see SCQ000
SILVADENE see PPP500, SNI425
SILVAN (CZECH) see MKH000
SILVER (colloidal) see SDI750
SILVER see SDI500
SILVER(1+) ACETATE see SDI800
SILVER(I) ACETATE see SDI800
SILVER ACETATE see SDI800
SILVER ACETYLIDE (dry) (DOT) see SDJ000
SILVER ACETYLIDE see SDJ000
SILVER ACETYLIDE-SILVER NITRATE see SDJ025
SILVER AMIDE see SDJ500
SILVER 5-AMINOTETRAZOLIDE see SDK000
SILVER AMMONIUM COMPOUNDS see SDK500
SILVER AMMONIUM LACTATE see SDL000
SILVER AMMONIUM NITRATE see SDL500
SILVER AMMONIUM SULFATE see SDM000
SILVER ARSENITE see SDM100
SILVER ATOM see SDI500
SILVER AZIDE (dry) (DOT) see SDM500
SILVER AZIDE see SDM500
SILVER 2-AZIDO-4,6-DINITROPHENOXIDE see SDM525
SILVER AZIDODITHIOFORMATE see SDM550
SILVER BENZO-1,2,3-TRIAZOLE-1-OXIDE see SDM575
SILVER BETA-STANNATE see DXA800
SILVER BOROFLUORIDE see SDN000
SILVER BUSH see NBR800
SILVER BUTEN-3-YNIDE see SDN100
SILVER CHAIN see GJU475
SILVER CHLORATE see SDN399
SILVER CHLORITE, dry see SDN500
SILVER CHLOROACETYLIDE see SDN525
SILVER COMPOUNDS see SDO500
SILVER CUP see CDH125
SILVER CYANATE see SDO525
SILVER CYANIDE see SDP000
SILVER 3-CYANO-1-PHENYLTRIAZEN-3-IDE see SDP025
SILVER CYCLOPROPYLACETYLIDE see SDP100
SILVER DIFLUORIDE see SDQ500
SILVER DINITRITODIOXYSULFATE see SDP500
SILVER DINITROACETAMIDE see SDP550
SILVER 3,5-DINITROANTHRANILATE see SDP600
SILVER FIR NEEDLE OIL see AAC250
SILVER FIR OIL see AAC250
SILVER(II) FLUORIDE see SDQ500
SILVER(I) FLUORIDE see SDQ000
SILVER FLUOROBORATE see SDN000
SILVER FULMINATE, dry see SDR000
SILVER HALIDE SOLVENT (HS103) see BGT750
SILVER 1,3,5-HEXATRIENIDE see SDR150
SILVER 3-HYDROXYPROPYNIDE see SDR175
SILVER MALONATE see SDR350
SILVER MATT POWDER see TGB250

SILVER METHANESULFONATE see SDR500
SILVER 3-METHYLISOXAZOLIN-4,5-DIONE-4-OXIMATE see SDR400
SILVER METHYLSULFONATE see SDR500
SILVER MONOACETATE see SDI800
SILVER MONOACETYLIDE see SDR759
SILVER(I) NITRATE (1:1) see SDS000
SILVER(1+) NITRATE see SDS000
SILVER NITRATE (DOT) see SDS000
SILVER NITRIDE see SDS500
SILVER NITRIDOOSMITE see SDT000
SILVER 4-NITROPHENOXIDE see SDT300
SILVER NITROPRUSSIDE see SDT500
SILVER OSMATE see SDT750
SILVER OXALATE, dry see SDU000
SILVER(1+) OXIDE see SDU500
SILVER PERCHLORATE see SDV000
SILVER PERCHLORYL AMIDE see SDV500
SILVER N-PERCHLORYL BENZYLAMIDE see SDV600
SILVER PEROXIDE see SDV700
SILVER PEROXYCHROMATE see SDW000
SILVER PHENOXIDE see SDW100
SILVER PICRATE (dry) (DOT) see PID200
SILVER PICRATE, wetted with not <30% water, by weight (UN1347) (DOT)
  see PID200
SILVER PINE OIL see AAC250
SILVER POTASSIUM CYANIDE see PLS250
SILVER SULFADIAZINE see SNI425
SILVER SULPHADIAZINE see SNI425
SILVER TETRAFLUOROBORATE see SDN000
SILVER TETRAZOLIDE see SDW500
SILVER TRICHLOROMETHANEPHOSPHONATE see SDW600
SILVER TRIFLUORO METHYL ACETYLIDE see SDX000
SILVER TRIFLUOROPROPYNIDE see SDX000
SILVER TRINITROMETHANIDE see SDX200
SILVER W see CAT775
SILVEX (USDA) see TIX500
SILVIC ACID see AAC500
SILVI-RHAP see TIX500
SILVISAR 510 see HKC000
SILVISAR 550 see MRL750
SILVISAR see HKC500
SILYLBENZENE see SDX250
SILYL BROMIDE see BOE750
SILYL PEROXIDE Y-5712 see SDX300
SILYMARIN see SDX625
SILYMARIN HYDROGEN BUTANEDIOATE SODIUM SALT see SDX630
SILYMARIN SODIUM HEMISUCCINATE see SDX630
SIM see SNK000
SIMAN see CGA000
SIMANEX see BJP000
SIMARUBACEAE see GEW700
SIMATIN(E) see ENG500
SIMAZIN see BJP000
SIMAZINE (USDA) see BJP000
SIMAZINE 80W see BJP000
SIMAZOL see AMY050
SIMEON see POB500
SIMESKELLINA see AHK750
SIMETON see BJP250
SIMETONE see BJP250
SIMFIBRATE see SDY500
SIMPADREN see SPD000
SIMPALON see HLV500
SIMPAMINA-D see BBK500
SIMPATEDRIN see BBK000
SIMPATOBLOCK see HEA000
SIMPATOL see HLV500
SIMPLA see SGG500
SIMPLOTAN see TGD250
SIMULOL 330 M see DXY000
SIN-10 see MRN275
SINAFID M-48 see MNH000
SINALAR see SPD500
SINALGICO see NBS500
SINALOST see TNF500
SINAN see GGS000
SINANOMYCIN see NCP875
SINAXAR see PFJ000
SINBAR see BQT750
SINCALINE see CMF800
SINCICLAN see DKB000
SINCODEEN see BOR350
SINCODEX see BOR350
SINCODIN see BOR350
SINCODIX see BOR350

SINCORTEX see DAQ800
SINCOUMAR see ABF750
SINDERESIN see CJN250
SINDESVEL see QAK000
SINDIATIL see BQL000
SINDRENINA see VGP000
SINECOD see BOR350
SINEFLUTTER see GAX000
SINEFRINA TARTRATO (ITALIAN) see SPD000
SINEQUAN see AEG750
SINESALIN see BEQ625
SINESALIN COMPOSITION see RDK000
SINESTROL see DLB400
SINFIBRATE see SDY500
SINFORIL see CKE750
SINGLE SUPERPHOSPHATE see SOV500
SINGOSERP see RCA200
SINIFIBRATE see SDY500
SINITUHO see PAX250
SINKALIN see CMF800
SINKALINE see CMF800
SINKLE BIBLE (JAMAICA) see AGV875
SINKUMAR see ABF750
SINNAMIN see AQN750
SINNOESTER OGC see GGR200
SINNOQUAT BL 80 see BEO000
SINNOQUAT BL 95 see BEO000
SINOACTINOMYCIN see AEA625
SINOFLUROL see FLZ050
SINOGAN see MCI500
SINOMENI CAULIS et RHIZOMA (LATIN) see SDY600
SINOMENINE A BIS(METHYL IODIDE) see TDX835
SINOMENINE A DIMETHIODIDE see TDX835
SINOMENIUM ACUTUM, crude extract see SDY600
SINOMIN see SNK000
SINORATOX see DSP400
SINOX see DUS700, DUU600
SINOX GENERAL see BRE500
SINOX W see BPG250
SINTECORT see PAL600
SINTESPASMIL see NOC000
SINTESTROL see DKA600
SINTHROM see ABF750
SINTHROME see ABF750
SINTOFENE see OPK300
SINTOMICETINA see CDP250
SINTOSIAN see DYF200
SINTOTIAMINA see DXO300
SINTROM see ABF750
SINTROMA see ABF750
SINUFED see POH250
SINURON see DGD600
SINUTAB see ABG750
SIOCARBAZONE see FNF000
SIONIT see SKV200
SIONON see SKV200
SIPCAVIT see PEX500
SIPERIN see RLF350
SIPEX BOS see TAV750
SIPEX OLS see OFU200
SIPEX OP see SIB600
SIPLARIL see FMO129
SIPLAROL see FMO129
SIPOL L8 see OEI000
SIPOL L10 see DAI600
SIPOL L12 see DXV600
SIPOL O see OBA000
SIPOL S see OAX000
SIPOMER DAM see DBK200
SIPOMER DMM see DSL800
SIPONIC L see DXY000
SIPONIC Y-501 see PJW500
SIPON LT see SON000
SIPONOL S see OAX000
SIPON WD see SIB600
SIPPR-113 see SDY675
SIPTOX I see MAK700
SIRAGAN see CLD000
SIRAN N,N-DIETHYL-p-FENYLENDIAMINU see DJV250
SIRAN HYDRAZINU (CZECH) see HGW500
SIRBIOCINA see SMB000
SIRINGAL see RCA200
SIRINGINA see RCA200
SIRINGONE see RCA200
SIRISERPIN see RCA200

SIRIUS RED 4B see CMO885
SIRIUS RED 4BA see CMO885
SIRLEDI see NHH000
SIRLENE see PML000
SIRMATE see DET400, DET600
SIRNIK AMONNY see ANJ750
SIRNIK FOSFORECNY (CZECH) see PHS000
SIRNIK TRIBENZYLCINICTY (CZECH) see BLK750
SIROKAL see PLG800
SIROMYCIN see VGZ000
SIROTOL see RLU000
SIRUP see GFG000
SISEPTIN see SDY750
SISOLLINE see SDY750
SISOMICIN see SDY750
SISOMICIN HYDROCHLORIDE see SDY755
SISOMICIN SULFATE see APY500
SISOMIN see APY500
SISTALGIN see SDZ000
SISTAN see VFU000
SISTOMETRENOL see LJE000
SISTRURUS MILARIUS BARBOURI VENOM see SDZ300
SITFAST see FBS100
SIX HUNDRED SIX see SAP500
SIXTY-THREE SPECIAL E.C. INSECTICIDE see MNH000, PAK000
SK 1 see MCB050, NCW300
SK 65 see DAB879
SK 74 see CMW700
SK 75 see UTU500
SK-100 see TNF500
SK 555 see BHO250
SK-598 see CFA750
SK1133 see TND500
SK 1150 see AJO500
SK-3818 see TND250
SK 6048 see CKV500
SK 6882 see TFQ750
SK-15673 see PED750
SK 18615 see TFJ500
SK-19849 see BIA250
SK 20501 see CQC650
SK 22591 see HOO500
SK 27702 see BIF750
SK 29836 see DBY500
SK 331 A see XCS000
SK-Apap see HIM000
SK-AMITRIPTYLINE see EAI000
SK-AMPICILLIN see AIV500
SKANE M8 see OFE000
SKATOL see MKV750
SKATOLE see MKV750
ω-SKATOLE CARBOXYLIC ACID see ICN000
SK-BISACODYL see PPN100
SK-CHLORAL HYDRATE see CDO000
SK-CHLOROTHIAZIDE see CLH750
SK-DEXAMETHASONE see SOW000
SK-DIGOXIN see DKN400
SK-DIPHENHYDRAMINE see BAU750
SKDN see WBS675
SKEDULE see CBF250
SKEKhG see EAZ500
SKELAXIN see XVS000
SKELLYSOLVE F see PCT250
SKELLYSOLVE G see PCT250
SKELLY-SOLVE-L see ROU000
SKEROLIP see ARQ750
SK-ERYTHROMYCIN see EDJ500
SK-ESTROGENS see ECU750
SKF 51 see ILM000
SKF 250 see DBA175
SKF 385 see PET750
SKF 478 see DWK200
SKF 1045 see MGD500
SK+F 1340 see DRX400
SKF 1498 see DQA400
SKF 1717 see AOO490
SKF 2170 see AOO425
SKF 2538 see DIR000
SKF-2601 see CKP250
SKF 4740 see DAB875
SKF 5019 see TKK250
SKF 5116 see MCI500
SKF 5137 see AFJ400
SKF 5883 see TFM100
SKF 6539 see HDC000

SKF 7988 see VRA700, VRF000
SKF 8542 see UVJ450
SK&F 14287 see DAS000
SKF 20,716 see PIW000
SKF 24260 see FOO875
SKF 28175 see DMF800
SKF 29044 see BQK000
SKF 30310 see OMY700
SK&F 36914 see CLQ500
SK&F 39162 see ARS150
SKF 41588 see CCS250
SKF 60771 see APT250
SKF 62698 see TGA600
SKF 62979 see VAD000
SKF-69,634 see CMX860
SKF 83088 see CCS350
SKF-88373 see EBE100
SKF 91487 see NNL400
SKF 92334 see TAB250
SK&F 92676 see IBQ075
SKF-183A see ALW000
SKF 525A see PBM500
SKF 688A see DDG800
SKF 16805A see BQB250
SKF 70643-A see DPM200
SKF-501 HYDROCHLORIDE see FEE100
SKF 2208K see CDF400
SK&F No. 478-A see DWK200
SKF No. 769-J² see CBG250
SKF 83088 SODIUM see CCS360
SKG see CBT500
SKI 21739 see BHV000
SKI 24464 see MNA750
SKI 27013 see SAU000
SKI 28404 see CHE750
SKINO #1 see BSU500
SKINO #2 see EMU500
SKIODAN see SHX000
SKLEROMEX see ARQ750
SKLEROMEXE see ARQ750
SKLERO-TABLINEN see ARQ750
SKLERO-TABULS see ARQ750
SK-LYGEN see MDQ250
SKhN6 see ADY250
SK-106N see NGY000
SK-NIACIN see NCQ900
SKOLIN see HLC500
SK-PENICILLIN G see BFD000
SK-PENICILLIN VK see PDT750
SK-PHENOBARBITAL see EOK000
SK 1 (PLASTICIZER) see MCB050
SK-PRAMINE see DLH630
SK-PRAMINE HYDROCHLORIDE see DLH630
SK-PREDNISONE see PLZ000
SK-RESERPINE see RDK000
SKS 85 see SMR000
SKT see CBT500
SKT (ADSORBENT) see CBT500
SK-TETRACYCLINE see TBX000, TBX250
SK-TOLBUTAMIDE see BSQ000
SK-TRIAMCINOLONE see AQX250
SKUNK CABBAGE see FAB100, SDZ450
SK 75V see UTU500
SKY BLUE 4B see CMO500
SKY BLUE 5B see CMO500
SKY FLOWER see GIW200
SL 31 see MNO775
SL-90 see RLU000
SL-236 see FDA885
SL 501 see CMW700
SL-512 see CQF125
SL 700 see CAT775
1000SL see HKS780
SL-6057 see CMB125
SL-75212 see KEA350
SLAB OIL (Obs.) see MQV875
SLAKED LIME see CAT225
SL-D.212 see KEA350
SLEEPAN see TEH500
SLEEPING NIGHTSHADE see DAD880
SLEEPWELL see TEO250
S-426-S (LEPETIT) see EAD500
SLIMICIDE see ADR000
SLIPPER FLOWER see SDZ475
SLIPPER PLANT see SDZ475

SLIPRO see TEH500
SLOE see AQP890
SLO-GRO see DHF200, DMC600
SLOMELAM 2 see MCB050
SLO-PHYLLIN see TEP000
SLOSUL see AKO500
SLOVASOL A see PJY100
SLOW-FE see FBN100
SLOW-K see PLA500
SLS see SIB600
SLUG-TOX see TDW500
SM 67 see MCB050
S 75M see SFO500
SM-307 see DLG000
SM 700 see MCB050
SMA 1000 see SEA500
SMA 1440 see SEA500
SMA 2000 see SEA500
SMA 3000 see SEA500
SMA 17352 see SEA500
SMA see SFU500
SMA 1440A see SEA500
SMA 1725A see SEA500
SMA 2000A see SEA500
SMA 2625A see SEA500
SMA 3000A see SEA500
SMA 4000A see SEA500
SMA 1420AL see SEA500
SMAC-A see SEA500
SMA 1440H see SEA500
SMA 1440-H RESIN see SEA500
S. MARCESCENS LIPOPOLYSACCHARIDE see SEA400
SMA (STYRENE POLYMER) see SEA500
SMCA see SFU500
SMD 3500 see SMQ500
SMDC see SIL550, VFU000
SMEDOLIN see AHP125
SMEESANA see AQN635
SMFPD see MCB050
SMIDAN see PHX250
S. MILARIUS BARBOURI VENOM see SDZ300
SMITHKO KALKARB WHITING see CAT775
SMOG see SEB000
SMOKE BROWN G see TKN750
SMOKE CONDENSATE, cigarette see SEC000
SMOKELESS POWDER see SED000
SMOKELESS TOBACCO see SED400
SMOP see AKO500
SMP see AKO500
S MUSTARD see BIH250
SMUT-GO see HCC500
SN 20 see ADY500
SN 46 see CPK500
S.N. 112 see PPP500
S. N. 166 see AOO800
SN 203 see CMW400
SN 390 see CFU750
SN 407 see CCS625
SN 4075 see MEG250
SN-4395 see PQC500
SN 6718 see CLD000
SN 7618 see CLD000
SN 12,837 see CKB500
SN 13,272 see PMC300
SN 35830 see CGL250
SN 36056 see DSO200
SN 36268 see CJJ250
SN 38107 see EEO500
SN-38584 see MEG250
SN-41703 see EIH500
SN 49537 see TEX600
SN see SMA000
SNAKEBERRY see BAF325
SNAKE FLOWER see VQZ675
SNAKEROOT OIL, CANADIAN see SED500
SNAKE VENOM BITIS ARIETANS see BLV075
SNAKE VENOM BITIS GABONICA see BLV080
SNAKE WEED see PJJ300
SNAPPING HAZEL see WCB000
SNATOWHITE CRYSTALS see TFC600
SN 6771 DIHYDROCHLORIDE see BFX125
SNEEZING GAS see CGN000
S.N.G see NGY000
SNIECIOTOX see HCC500
SNIP see DQZ000

SNIP FLY see DQZ000
SNIP FLY BANDS see DQZ000
SNOMELT see CAO750
SNOW ALGIN H see SEH000
SNOWBERRY see SED550
SNOWCAL see CAT775
SNOWCAL 5SW see BKW100
SNOWDROP see SED575
SNOWFLAKE CRYSTALS see SJT750
SNOWFLAKE WHITE see CAT775
SNOWGOOSE see TAB750
SNOW TEX see AHF500, KBB600
SNOW TOP see CAT775
SNOW WHITE see ZKA000
SNP see PAK000
SNUFF see SED400
SO 15 see SKV170
SO see LCF000
SOAz see SED700
SOAMIN see ARA500
SOAP PLANT see DAE100
SOAPSTONE see SCN000
SOAP YELLOW F see FEV000
SOBELIN see CMV675, CMV690
SOBENATE see SFB000
SOBIC ALDEHYDE see SKT500
SOBIODOPA see DNA200
SOBITAL see DJL000
SOBRIL see CFZ000
SOC see CIL775
SOCAL see CAT775
SOCAL E 2 see CAT775
SOCAREX see PJS750
SOCLIDAN see EQQ100
SODA ALUM see AHG500
SODA ASH see SFO000
SODA CHLORATE (DOT) see SFS000
SODA LIME with >4% sodium hydroxide (DOT) see SEE000
SODA LIME see SEE000
SODA LYE see SHS000, SHS500
SODAMIDE see SEN000
SODA MINT see SFC500
SODANIT see SEY500
SODA NITER see SIO900
SODANTON see DKQ000, DNU000
SODA PHOSPHATE see SJH090
SODAR see DXE600
SODESTRIN-H see ECU750
SODIO (CLORATO di) (ITALIAN) see SFS000
SODIO (DICROMATO di) (ITALIAN) see SGI000
SODIO, FLUORACETATO di (ITALIAN) see SHG500
SODIO(IDROSSIDO di) (ITALIAN) see SHS000
3-SODIO-5-(5′-NITRO-2′-FURFURYLIDENAMINO)IMIDAZOLIDIN-2,4-DIONE see SEE250
SODIOPAS see SEP000
SODIO (PERCLORATO DI) (ITALIAN) see PCE750
SODIO(TRICLOROACETATO di) (ITALIAN) see TII500
SODITAL see NBU000
1719 SODIUM see ASO510
SODIUM, metal liquid alloy (DOT) see SEF600
SODIUM (liquid alloy) see SEF600
SODIUM (dispersions) see SEF500
SODIUM see SEE500
SODIUM-3-ACETAMIDO-2,4,6-TRIIODOBENZOATE see AAN000
SODIUM ACETARSONE see SEG000
SODIUM ACETATE, anhydrous (FCC) see SEG500
SODIUM ACETATE see SEG500
SODIUM ACETATE MONOHYDRATE see SEG650
SODIUM ACETAZOLAMIDE see AAS750
SODIUM ACETRIZOATE see AAN000
SODIUM p-ACETYLAMINOPHENYLANTIMONATE see SLP500
SODIUM-3-ACETYLAMINO-2,4,6-TRIIODOBENZOATE see AAN000
SODIUM ACETYLARSANILATE see AQZ900
SODIUM ACETYL ARSANILATE see ARA000
SODIUM ACETYLIDE see SEG700
SODIUM ACID ARSENATE see ARC000
SODIUM ACID ARSENATE, HEPTAHYDRATE see ARC250
SODIUM ACID CARBONATE see SFC500
SODIUM ACID HEPARIN see HAQ550
SODIUM ACID METHANEARSONATE see MRL750
SODIUM ACID PHOSPHATE see SJH100
SODIUM ACID PYROPHOSPHATE (FCC) see DXF800
SODIUM ACID SULFATE see SEG800
SODIUM ACID SULFATE (solid) see SEG800
SODIUM ACID SULFITE see SFE000

SODIUM ACYCLOVIR see AEC725
SODIUM ADENOSINE-5′-MONOPHOSPHATE see AEM750
SODIUM ADENOSINE-5′-TRIPHOSPHATE see AEM250
SODIUM ADENOSINE TRIPHOSPHATE see AEM250
SODIUM AESCINATE see EDM000
SODIUM ALBAMYCIN see NOB000
SODIUM ALBUCID see SNQ000
SODIUM ALGINATE see SEH000
SODIUM ALGINATE SULFATE see SEH450
SODIUM ALIZARINESULFONATE see SEH475
SODIUM ALIZARIN-3-SULFONATE see SEH475
SODIUM ALLYLBENZYL THIOBARBITURATE see AFX500
SODIUM-5-ALLYL-5-(1-(BUTYLTHIO)ETHYL) BARBITURATE see AGA250
SODIUM-5-ALLYL-5-ISOPROPYLBARBITURATE see BOQ750
SODIUM-5-ALLYL-5-(1-METHYLBUTYL)BARBITURATE see SBN000
SODIUM-5-ALLYL-5-(1-METHYLBUTYL)-2-THIOBARBITURATE see SOX500
SODIUM-dl-5-ALLYL-1-METHYL-5-(1-METHYL-2-PENTYNYL)BARBITURATE see MDU500
SODIUM ALUMINATE, solution (UN 1819) (DOT) see AHG000
SODIUM ALUMINATE, solid (UN 2812) (DOT) see AHG000
SODIUM ALUMINOFLUORIDE see SHF000
SODIUM ALUMINOSILICATE see SEM000
SODIUM ALUMINUM FLUORIDE see SHF000
SODIUM ALUMINUM HYDRIDE (DOT) see SEM500
SODIUM ALUMINUM OXIDE see AHG000
SODIUM ALUMINUM PHOSPHATE, ACIDIC see SEM300
SODIUM ALUMINUM PHOSPHATE, BASIC see SEM305
SODIUM ALUMINUM SULFATE see AHG500
SODIUM ALUMINUM TETRAHYDRIDE see SEM500
SODIUM AMAZOLENE see ADE750
SODIUM AMIDE see SEN000
SODIUM AMINARSONATE see ARA500
SODIUM p-AMINOBENZENEARSONATE see ARA500
SODIUM-p-AMINOBENZENESTIBONATE GLUCOSIDE see NCL000
SODIUM-4-AMINOBENZOATE see SEO500
SODIUM p-AMINOBENZOATE see SEO500
SODIUM (2-AMINO-3-BENZOYLPHENYL)ACETATE MONOHYDRATE see AHK625
SODIUM d-(−)-α-AMINOBENZYLPENICILLIN see SEQ000
SODIUM AMINOPHENOL ARSONATE see ARA500
SODIUM p-AMINOPHENYLARSONATE see ARA500
SODIUM p-AMINOSALICYLATE see SEP000
SODIUM AMINOSALICYLATE see SEP000
SODIUM p-AMINOSALICYLIC ACID see SEP000
SODIUM-7-(2-(2-AMINO-4-THIAZOLYL)-2-METHOXYIMINOACETAMIDO) CEPHAL-OSPORANATE see CCR950
SODIUM AMINOTRIACETATE see SEP500
SODIUM AMIDOTRIZOATE see SEN500
SODIUM AMP see AEM750
SODIUM AMPICILLIN see SEQ000
SODIUM AMYLOBARBITONE see AON750
SODIUM-ANALINE ARSONATE see ARA500
SODIUM ANAZOLENE see ADE750
SODIUM ANDROST-5-EN-17-ONE-3-β-YL SULFATE DIHYDRATE see DAL030
SODIUM-2-ANTHRACHINONESULPHONATE see SER000
SODIUM ANTHRAQUINONE-1,5-DISULFONATE see DLJ700
SODIUM ANTHRAQUINONE-1-SULFONATE see DLJ800
SODIUM-9,10-ANTHRAQUINONE-2-SULFONATE see SER000
SODIUM-2-ANTHRAQUINONESULFONATE see SER000
SODIUM-β-ANTHRAQUINONESULFONATE see SER000
SODIUM ANTIMONY BIS(PYROCATECHOL-2,4-DISULFONATE) see AQH500
SODIUM ANTIMONY (III) BIS-PYROCATECHOL-3,5-DISULFONATE HEPTAHY-DRATE see AQH500
SODIUM ANTIMONY(III)-3-CATECHOL THIOSALICYLATE see SEU000
SODIUM ANTIMONY-2,3-meso-DIMERCAPTOSUCCINATE see AQD750
SODIUM ANTIMONY ERYTHRITOL see SER500
SODIUM ANTIMONY(III) GLUCONATE see AQI000
SODIUM ANTIMONY(V) GLUCONATE see AQI250
SODIUM ANTIMONY GLUCONATE see AQI000
SODIUM ANTIMONYL ADONITOL see SES000
SODIUM ANTIMONYL-d-ARABITOL see SES500
SODIUM ANTIMONYL BISCATECHOL see SET000
SODIUM ANTIMONYL tert-BUTYL CATECHOL see SET500
SODIUM ANTIMONYL CATECHOL THIOSALICYLATE see SEU000
SODIUM ANTIMONYL CITRATE see SEU500
SODIUM ANTIMONYL DIMETHYLCYSTEINE TARTRATE see AQH750
SODIUM ANTIMONYL-d-FUNCITOL see SEV000
SODIUM ANTIMONYL GLUCO-GULOHEPTITOL see SEV500
SODIUM ANTIMONYL GLYCEROL see SEW000
SODIUM ANTIMONYL-d-MANNITOL see SEW500
SODIUM ANTIMONYL-2,5-METHYLENE-d-MANNITOL see SEX000
SODIUM ANTIMONYL-2,4-METHYLENE-d-SORBITOL see SEX500
SODIUM ANTIMONYL TARTRATE see AQI750
SODIUM ANTIMONYL XYLITOL see SEY000

SODIUM ANTIMONY TARTRATE see AQI750
SODIUM ANTIMOSAN see AQH500
SODIUM ARSANILATE (DOT) see ARA500
SODIUM p-ARSANILATE see ARA500
SODIUM ARSANILATE see ARA500
SODIUM ARSENATE (DOT) see ARD750, SEY100
SODIUM ARSENATE see ARC000, ARD500, ARD600, SEY100
SODIUM ARSENATE DIBASIC, anhydrous see ARC000
SODIUM ARSENATE, DIBASIC, HEPTAHYDRATE see ARC250
SODIUM ARSENATE HEPTAHYDRATE see ARC250
SODIUM ARSENITE, solid (DOT) see SEY500
SODIUM ARSENITE, liquid (solution) (DOT) see SEY500
SODIUM ARSENITE see SEY200, SEY500
SODIUM ARSENITE, aqueous solutions (UN 1686) (DOT) see SEY200
SODIUM ARSENITE, solid (UN 2027) (DOT) see SEY200
SODIUM ARSONILATE see ARA500
SODIUM l-ASPARTATE see SEZ350, SEZ355
SODIUM ASPARTATE see SEZ350
SODIUM ASPIRIN see ADA750
SODIUM ATP see AEM100, AEM250
SODIUM AUROTHIOMALATE see GJC000
SODIUM AUROTHIOSULPHATE DIHYDRATE see GJG000
SODIUM AZAPROPAZONE see ASA250
SODIUM AZIDE see SFA000
SODIUM-5-AZIDOTETRAZOLIDE see SFA100
SODIUM AZO-α-NAPHTHOLSULFANILATE see FAG010
SODIUM AZO-α-NAPHTHOLSULPHANILATE see FAG010
SODIUM AZOTOMYCIN see ASO510
SODIUM, AZOTURE de (FRENCH) see SFA000
SODIUM, AZOTURO di (ITALIAN) see SFA000
SODIUM BARBITAL see BAG250
SODIUM BARBITONE see BAG250
SODIUM BARBITURATE see MRK750
SODIUM BENZENEACETATE see SFA200
SODIUM BENZENE HEXOIDE see SFA600
SODIUM BENZENESULFONATE see SJH050
SODIUM-1,2 BENZISOTHIAZOLIN-3-ONE-1,1-DIOXIDE see SJN700
SODIUM BENZOATE see SFB000
SODIUM BENZOATE and CAFFEINE see CAK800
SODIUM BENZOIC ACID see SFB000
SODIUM o-BENZOSULFIMIDE see SJN700
SODIUM BENZOSULFONATE see SJH050
SODIUM-2-BENZOSULPHIMIDE see SJN700
SODIUM-o-BENZOSULPHIMIDE see SJN700
SODIUM BENZOSULPHIMIDE see SJN700
SODIUM-2-BENZOTHIAZOLYLSULFIDE see SFB100
SODIUM o-BENZYL-p-CHLOROPHENATE see SFB200
SODIUM o-BENZYL-p-CHLOROPHENOLATE see SFB200
SODIUM BENZYLPENICILLIN see BFD250
SODIUM BENZYLPENICILLINATE see BFD250
SODIUM BENZYLPENICILLIN G see BFD250
SODIUM BERYLLIUM MALEATE see SFB500
SODIUM BERYLLIUM TARTRATE see SFC000
SODIUM BIBORATE see DXG035, SFF000
SODIUM BIBORATE DECAHYDRATE see SFF000
SODIUM BICARBONATE see SFC500
SODIUM BICHROMATE see SGI000
SODIUM BINOTAL see SEQ000
SODIUM (1,1'-BIPHENYL)-2-OLATE see BGJ750
SODIUM 2-BIPHENYLOLATE see BGJ750
SODIUM, (2-BIPHENYLYLOXY)- see BGJ750
SODIUM BIPHOSPHATE anhydrous see SJH100
SODIUM BIPHOSPHATE see SJH100
SODIUM BIS(2-ETHYLHEXYL)PHOSPHATE see TBA750
SODIUM BIS(2-ETHYLHEXYL) SULFOSUCCINATE see DJL000
SODIUM BISMUTHATE see SFD000
SODIUM BISMUTH THIOGLYCOLATE see BKX750
SODIUM BISMUTH THIOGLYCOLLATE see BKX750
SODIUM-3,5-BIS(aci-NITRO)CYCLOHEXENE-4,6-DIIMINIDE see SFD300
SODIUM BISPROPYLACETATE see PNX750
SODIUM BISULFATE, FUSED see SEG800
SODIUM BISULFIDE see SHR000
SODIUM BISULFITE (1:1) see SFE000
SODIUM BISULFITE, solid (DOT) see SFE000
SODIUM BISULFITE, solution (DOT) see SFE000
SODIUM BISULFITE see SFE000
SODIUM BLUE VRS see ADE500
SODIUM BORATE anhydrous see SFE500
SODIUM BORATE see SFE500
SODIUM BORATE DECAHYDRATE see SFF000
SODIUM BOROHYDRIDE see SFF500
SODIUM BROMATE see SFG000
SODIUM BROMEBRATE see SIK000
SODIUM BROMIDE see SFG500
SODIUM BROMOACETYLIDE see SFG600

SODIUM-5-(2-BROMOALLYL)-5-sec-BUTYLBARBITURATE see BOR250
SODIUM BROMOSULFALEIN see HAQ600
SODIUM BROMOSULFOPHTHALEIN see HAQ600
SODIUM BROMSULPHALEIN see HAQ600
SODIUM BROMSULPHTHALEIN see HAQ600
SODIUM BUTABARBITAL see BPF250
SODIUM BUTANOATE see SFN600
SODIUM BUTAZOLIDINE see BOV750
SODIUM-5-sec-BUTYL-5-ETHYLBARBITURATE see BPF250
SODIUM BUTYLMERCURIC THIOGLYCOLLATE see SFJ500
SODIUM-5-(1-(BUTYLTHIO)ETHYL)-5-ETHYLBARBITURATE see SFJ875
SODIUM-3-BUTYRAMIDO-α-ETHYL-2,4,6-TRIIODOCINNAMATE see EQC000
SODIUM-3-BUTYRAMIDO-α-ETHYL-2,4,6-TRIIODOHYDROCINNAMATE see
  SKO500
SODIUM n-BUTYRATE see SFN600
SODIUM BUTYRATE see SFN600
SODIUM CACODYLATE (DOT) see HKC500
SODIUM CALCIUM ALUMINOSILICATE, HYDRATED see SFN700
SODIUM CALCIUM EDETATE see CAR780
SODIUM CAPRATE see SGB600
SODIUM CAPRINATE see SGB600
SODIUM CAPRYLATE see SIX600
SODIUM CAPRYL SULFATE see OFU200
SODIUM CARBENICILLIN see CBO250
SODIUM CARBOLATE see SJF000
SODIUM CARBONATE (2:1) see SFO000
SODIUM CARBONATE see SJB400
SODIUM CARBOXYMETHYL CELLULOSE see SFO500
SODIUM CARRAGEENAN see SFP000
SODIUM CARRAGEENATE see SFP000
SODIUM CARRAGHEENATE see SFP000
SODIUM CARRIOMYCIN see SFP500
SODIUM CASEINATE see SFQ000
SODIUM CEFACETRIL see SGB500
SODIUM CEFAMANDOLE see CCR925
SODIUM CEFAPIRIN see HMK000
SODIUM CEFAZOLIN see CCS250
SODIUM CEFUROXIME see SFQ300
SODIUM CELLULOSE GLYCOLATE see SFO500
SODIUM CEPHACETRILE see SGB500
SODIUM CEPHALOTHIN see SFQ500
SODIUM CEPHALOTIN see SFQ500
SODIUM CEPHAPIRIN see HMK000
SODIUM CEPHAZOLIN see CCS250
SODIUM CETYL SULFATE see HCP900
SODIUM CEZ see CCS250
SODIUM CHAULMOOGRATE see SFR000
SODIUM CHENODEOXYCHOLATE see CDL375
SODIUM CHENODESOXYCHOLATE see CDL375
SODIUM CHLORAMBUCIL see CDO625
SODIUM CHLORAMINE T see CDP000
SODIUM CHLORAMPHENICOL SUCCINATE see CDP500
SODIUM CHLORATE, aqueous solution (DOT) see SFS000
SODIUM (CHLORATE de) (FRENCH) see SFS000
SODIUM CHLORATE see SFS000
SODIUM CHLORATE BORATE see SFS500
SODIUM CHLORIDE see SFT000
SODIUM CHLORIDE OXIDE see SHU500
SODIUM CHLORITE see SFT500
SODIUM CHLORITE (UN 1496) (DOT) see SFT500
SODIUM CHLORITE, solution with >5% available chlorine (UN 1908) (DOT)
  see SFT500
SODIUM CHLOROACETATE see SFU500
SODIUM-4-CHLOROACETOPHENONE OXIMATE see SFU600
SODIUM CHLOROACETYLIDE see SFV000
SODIUM cis-3-CHLOROACRYLATE see SFV250
SODIUM cis-β-CHLOROACRYLATE see SFV250
SODIUM CHLOROAURATE see GIZ100
SODIUM N-CHLOROBENZENESULFONAMIDE see SFV275
SODIUM-5-(4-CHLOROBENZOYL)-1,4-DIMETHYL-1H-PYRROLE-2-ACETATE DI-
  HYDRATE see ZUA300
SODIUM CHLOROETHYNIDE see SFV000
SODIUM-4-CHLORO-2-METHYL PHENOXIDE see SFV300
SODIUM (4-CHLORO-2-METHYLPHENOXY)ACETATE see SIL500
SODIUM-2-CHLORO-6-PHENYL PHENATE see SFV500
SODIUM CHLOROPLATINATE see SJJ500
SODIUM (Z)-3-CHLORO-2-PROPENOATE see SFV250
SODIUM-2-CHLOROPROPIONATE see CKT100
SODIUM N-CHLORO-4-TOLUENE SULFONAMIDE see SFV550
SODIUM CHOLATE see SFW000
SODIUM CHOLIC ACID see SFW000
SODIUM CHONDROITIN POLYSULFATE see SFW300
SODIUM CHONDROITIN SULFATE see SFW300
SODIUM CHROMATE (DOT) see DXC200
SODIUM CHROMATE (VI) see DXC200

SODIUM CHROMATE see SGI000
SODIUM CHROMATE DECAHYDRATE see SFW500
SODIUM CINCHOPHEN see SJH000
SODIUM CINNAMATE see SFX000
SODIUM CITRATE (FCC) see DXC400
SODIUM CITRATE, anhydrous see TNL000
SODIUM CITRATE see MRL000
SODIUM CITRATE, POTASSIUM CITRATE, CITRIC ACID (2:2:1) see SFX725
SODIUM CLOXACILLIN see SLJ050
SODIUM CLOXACILLIN MONOHYDRATE see SLJ000
SODIUM CMC see SFO500
SODIUM CM-CELLULOSE see SFO500
SODIUM COBALTINITRITE see SFX750
SODIUM COCOMETHYLAMINOETHYL-2-SULFONATE see SFY000
SODIUM COCO METHYL TAURIDE see SFY000
SODIUM COLISTIMETHATE see SFY500
SODIUM COLISTINEMETHANESULFONATE see SFY500
SODIUM COLISTIN METHANESULFONATE see SFY500
SODIUM COMPOUNDS see SFZ000
SODIUM COUMADIN see WAT220
SODIUM m-CRESOLATE see SJP000
SODIUM p-CRESOLATE see SIM100
SODIUM m-CRESOXIDE see SJP000
SODIUM p-CRESOXIDE see SIM100
SODIUM CROMOGLYCATE see CNX825
SODIUM CROMOLYN see CNX825
SODIUM CUMENEAZO-β-NAPHTHOL DISULPHONATE see FAG018
SODIUM CUPROCYANIDE (DOT) see SFZ100
SODIUM CUPROCYANIDE, solid (UN 2316) (DOT) see SFZ100
SODIUM CUPROCYANIDE, solution (UN 2317) (DOT) see SFZ100
SODIUM CYANIDE see SGA500
SODIUM-7-(2-CYANOACETAMIDO)CEPHALOSPORANIC ACID see SGB500
SODIUM CYANURATE see SGB550
SODIUM CYCLAMATE see SGC000
SODIUM CYCLOHEXANESULFAMATE see SGC000
SODIUM CYCLOHEXANESULPHAMATE see SGC000
SODIUM-5-(1-CYCLOHEXEN-1-YL)-1,5-DIMETHYLBARBITURATE see ERE000
SODIUM CYCLOHEXYL AMIDOSULPHATE see SGC000
SODIUM CYCLOHEXYL SULFAMATE see SGC000
SODIUM CYCLOHEXYL SULFAMIDATE see SGC000
SODIUM CYCLOHEXYL SULPHAMATE see SGC000
SODIUM-2,4-D see SGH500
SODIUM DALAPON see DGI600
SODIUM DBDT see SGF500
SODIUM-n-DECANOATE see SGB600
SODIUM DECANOATE see SGB600
SODIUM DECANOIC ACID see SGB600
SODIUM DECYLBENZENESULFONAMIDE see DAJ000
SODIUM DECYLBENZENESULFONATE see DAJ000
SODIUM DECYL SULFATE see SOK000
SODIUM DEDT see SGJ000
SODIUM DEHYDROACETATE (FCC) see SGD000
SODIUM DEHYDROACETIC ACID see SGD000
SODIUM DEHYDROCHOLATE see SGD500
SODIUM DEHYDROEPIANDROSTERONE SULFATE see DAL040
SODIUM DELVINAL see VKP000
SODIUM DEOXYCHOLATE see SGE000
SODIUM DEOXYCHOLIC ACID see SGE000
SODIUM DESOXYCHOLATE see SGE000
SODIUM DEXAMETHASONE PHOSPHATE see DAE525
SODIUM DEXTROTHYROXINE see SKJ300
SODIUM-3,5-DIACETAMIDO-2,4,6-TRIIODOBENZOATE see SEN500
SODIUM DIACETYLDIAMINETRIIODOBENZOATE see SEN500
SODIUM DIARSENATE see SEY100
SODIUM DIATRIZOATE see SEN500
SODIUM-1,2:5,6-DIBENZANTHRACENE-9,10-endo-α,β-SUCCINATE see SGF000
SODIUM DIBUTYLDITHIOCARBAMATE see SGF500
SODIUM DIBUTYLNAPHTHALENE SULFATE see NBS700
SODIUM DIBUTYLNAPHTHALENESULFONATE see NBS700
SODIUM-2,6-DI-tert-BUTYLNAPHTHALENESULFONATE see DEE600
SODIUM DIBUTYLNAPHTHYLSULFONATE see NBS700
SODIUM DICHLORISOCYANURATE see SGG500
SODIUM DICHLOROACETATE see SGG500
SODIUM (o-(2,6-DICHLOROANILINO)PHENYL)ACETATE see DEO600
SODIUM DICHLOROCYANURATE see SGG500
SODIUM 2,6-DICHLOROINDOPHENOL see SGG650
SODIUM 2,6-DICHLOROINDOPHENOLATE see SGG650
SODIUM DICHLOROISOCYANURATE see SGG500
SODIUM-2-((2,6-DICHLORO-3-METHYLPHENYL)AMINO)BENZOATE see SIF425
SODIUM-2,4-DICHLOROPHENOXYACETATE see SGH500
SODIUM-2-(2,4-DICHLOROPHENOXY)ETHYL SULFATE see CNW000
SODIUM-2,4-DICHLOROPHENOXYETHYL SULPHATE see CNW000
SODIUM (o-((2,6-DICHLOROPHENYL)AMINO)PHENYL)ACETATE see DEO600
SODIUM-2,4-DICHLOROPHENYL CELLOSOLVE SULFATE see CNW000
SODIUM-2,2-DICHLOROPROPIONATE see DGI600

SODIUM-α,α-DICHLOROPROPIONATE see DGI600
SODIUM-1,3-DICHLORO-1,3,5-TRIAZINE-2,4-DIONE-6-OXIDE see SGG500
1-SODIUM-3,5-DICHLORO-1,3,5-TRIAZINE-2,4,6-TRIONE see SGG500
1-SODIUM-3,5-DICHLORO-s-TRIAZINE-2,4,6-TRIONE see SGG500
SODIUM DICHLORO-s-TRIAZINETRIONE, dry, containing more than 39% available chlorine (DOT) see SGG500
SODIUM DICHROMATE de (FRENCH) see SGI000
SODIUM DICHROMATE(VI) see SGI000
SODIUM DICHROMATE see SGI000
SODIUM DICHROMATE DIHYDRATE see SGI500
SODIUM DICLOXACILLIN see DGE200
SODIUM DICLOXACILLIN MONOHYDRATE see DGE200
SODIUM-5,5-DIETHYLBARBITURATE see BAG250
SODIUM DIETHYLBARBITURATE see BAG250
SODIUM N,N-DIETHYLDITHIOCARBAMATE see SGJ000
SODIUM DIETHYLDITHIOCARBAMATE see SGJ000
SODIUM DIETHYLDITHIOCARBAMATE TRIHYDRATE see SGJ500
SODIUM DI(2-ETHYLHEXYL)PHOSPHATE see TBA750
SODIUM DI-(2-ETHYLHEXYL) SULFOSUCCINATE see DJL000
SODIUM DIFORMYLNITROMETHANIDE HYDRATE see SGK600
SODIUM DIHEXYL SULFOSUCCINATE see DKP800
SODIUM DIHYDROBIS(2-METHOXYETHOXY)ALUMINATE see SGK800
SODIUM DIHYDROGEN ARSENATE see ARD600
SODIUM DIHYDROGEN CITRATE see MRL000
SODIUM DIHYDROGEN ORTHOARSENATE see ARD600
SODIUM DIHYDROGEN PHOSPHATE (1:2:1) see SJH100
SODIUM DIHYDROGENPHOSPHIDE see SGM000
SODIUM 2,5-DIHYDROXYBENZOATE see GCU050
SODIUM-m-DIISOPROPYLBENZOL (Na-m) DIHYDROPEROXIDE (RUSSIAN) see DNN840
SODIUM-p-DIISOPROPYLBENZOL (Na-p) DIHYDROPEROXIDE (RUSSIAN) see DNN850
SODIUM-2,3-DIMERCAPTOPROPANE-1-SULFONATE see DNU860
SODIUM-4,4-DIMETHOXY-1-aci-NITRO-3,5-DINITRO-2,5-CYCLOHEXADIENE see SGM100
SODIUM-4-(DIMETHYLAMINO)BENZENEDIAZOSULFONATE see DOU600
SODIUM-p-(DIMETHYLAMINO)BENZENEDIAZOSULFONATE see DOU600
SODIUM-4-(DIMETHYLAMINO)BENZENEDIAZOSULPHONATE see DOU600
SODIUM-p-(DIMETHYLAMINO)BENZENEDIAZOSULPHONATE see DOU600
SODIUM-3-(DIMETHYLAMINOMETHYLENEAMINO)-2,4,6-TRIIODOHYDROCINNAMATE see SKM000
SODIUM-(4-(DIMETHYLAMINO)PHENYL)DIAZENESULFONATE see DOU600
SODIUM DIMETHYLARSINATE see HKC500
SODIUM DIMETHYLARSINIC ACID TRIHYDRATE see HKC550
SODIUM DIMETHYLARSONATE see HKC500
SODIUM N,N-DIMETHYLDITHIOCARBAMATE see SGM500
SODIUM-4-(2,4-DINITROANILINO)DIPHENYLAMINE-2-SULFONATE see SGP500
SODIUM DINITRO-o-CRESOLATE, wetted with not <15% water, by weight (UN 1348) (DOT) see SGP550
SODIUM DINITRO-o-CRESOLATE, dry or wetted with <15% water, by weight (UN 0234) (DOT) see SGP550
SODIUM-4,6-DINITRO-o-CRESOXIDE see DUU600
SODIUM DINITRO-o-CRESYLATE see SGP550
SODIUM DINITROMETHANIDE see SGP600
SODIUM-5-DINITROMETHYLTETRAZOLIDE see SGQ000
SODIUM-2,4-DINITROPHENOL see DVA800
SODIUM-2,4-DINITROPHENOLATE see DVA800
SODIUM-2,4-DINITROPHENOXIDE see SGQ100
SODIUM DIOCTYL SULFOSUCCINATE see DJL000
SODIUM DIOCTYL SULPHOSUCCINATE see DJL000
SODIUM DIOXIDE see SJC500
SODIUM-1,1-DIOXOPENICILLANATE see PAP600
SODIUM DIPHENYL-4,4'-BIS-AZO-2''-8''-AMINO-1''-NAPHTHOL-3'',6 '' DISULPHONATE see CMO000
SODIUM DIPHENYLDIAZO-BIS(α-NAPHTHYLAMINESULFONATE) see SGQ500
SODIUM DIPHENYLHYDANTOIN see DNU000
SODIUM-5,5-DIPHENYLHYDANTOINATE see DNU000
SODIUM DIPHENYL HYDANTOINATE see DNU000
SODIUM-5,5-DIPHENYL-2,4-IMIDAZOLIDINEDIONE see DNU000
SODIUM-α,α-DIPROPYLACETATE see PNX750
SODIUM DIPROPYLACETATE see PNX750
SODIUM DISULFIDE see SGR500
SODIUM DISULFITE see SII000
SODIUM-2,3-DITHIOLPROPANESULFONATE see DNU860
SODIUM DITHIONITE (DOT) see SHR500
SODIUM DITOLYLDIAZOBIS-8-AMINO-1-NAPHTHOL-3,6-DISULFONATE see CMO250
SODIUM DITOLYLDIAZOBIS-8-AMINO-1-NAPHTHOL-3,6-DISULPHONATE see CMO250
SODIUM DNP see DVA800
SODIUM DODECANOATE see LBN000
SODIUM DODECYLBENZENESULFONATE (DOT) see DXW200
SODIUM DODECYLBENZENESULFONATE, dry see DXW200
SODIUM DODECYL SULFATE see SIB600
SODIUM EDETATE see EIV000

SODIUM EDTA see EIV000
SODIUM EOSINATE see BNH500
SODIUM EQUILIN 3-MONOSULFATE see ECW520
SODIUM EQUILIN SULFATE see ECW520
SODIUM ESTRONE-3-SULFATE see EDV600
SODIUM ESTRONE SULFATE see EDV600
SODIUM ETASULFATE see TAV750
SODIUM ETHAMINAL see NBU000
SODIUM ETHANEPEROXOATE see SJB000
SODIUM ETHASULFATE see TAV750
SODIUM ETHIDRONATE see DXD400
SODIUM ETHOXIDE see SGR800
SODIUM ETHOXYACETYLIDE see SGS000
SODIUM-6-(2-ETHOXY-1-NAPHTHAMIDO)PENICILLANATE see SGS500
SODIUM ETHYDRONATE see DXD400
SODIUM ETHYLBARBITAL see BAG250
SODIUM-5-ETHYL-5-sec-BUTYLBARBITURATE see BPF250
SODIUM ETHYL-N-BUTYL BARBITURATE see BPF750
SODIUM ETHYLENEDIAMINETETRAACETATE see EIV000
SODIUM ETHYLENEDIAMINETETRAACETIC ACID see EIV000
SODIUM(2-ETHYLHEXYL)ALCOHOL SULFATE see TAV750
SODIUM-5-ETHYL-5-HEXYLBARBITURATE see EKT500
SODIUM-2-ETHYLHEXYL SULFATE see TAV750
SODIUM-2-ETHYLHEXYLSULFOSUCCINATE see DJL000
SODIUM ETHYLISOAMYLBARBITURATE see AON750
SODIUM ETHYLMERCURIC THIOSALICYLATE see MDI000
SODIUM-o-(ETHYLMERCURITHIO)BENZOATE see MDI000
SODIUM ETHYLMERCURITHIOSALICYLATE see MDI000
SODIUM-5-ETHYL-5-(1-METHYL-1-BUTENYL) BARBITURATE see VKP000
SODIUM-5-ETHYL-5-(1-METHYLBUTYL)BARBITURATE see NBU000
SODIUM-5-ETHYL-5-(1-METHYLBUTYL)-2-THIOBARBITURATE see PBT500
SODIUM-5-ETHYL-5-(1-METHYLPROPYL)BARBITURATE see BPF250
SODIUM-7-ETHYL-2-METHYL-4-UNDECANOL SULFATE see EMT500
SODIUM-7-ETHYL-2-METHYLUNDECYL-4-SULFATE see EMT500
SODIUM-5-ETHYL-5-PHENYLBARBITURATE see SID000
SODIUM ETHYLXANTHATE see SHE500
SODIUM ETHYLXANTHOGENATE see SHE500
SODIUM ETHYNIDE see SEG700
SODIUM ETIDRONATE see DXD400
SODIUM EVIPAL see ERE000
SODIUM EVIPAN see ERE000
SODIUM FEREDETATE see EJA379
SODIUM FERRIC EDTA see EJA379
SODIUM FERRIC PYROPHOSPHATE see SHE700
SODIUM FLUCLOXACILLIN see FDA100
SODIUM FLUOACETATE see SHG500
SODIUM FLUOACETIC ACID see SHG500
SODIUM FLUOALUMINATE see SHF000
SODIUM FLUORACETATE de (FRENCH) see SHG500
SODIUM FLUORESCEIN see FEW000
SODIUM FLUORESCEINATE see FEW000
SODIUM FLUORIDE, solid and solution (DOT) see SHF500
SODIUM FLUORIDE see SHF500
SODIUM FLUOROACETATE see SHG500
SODIUM γ-FLUORO-β-HYDROXYBUTYRATE see SHI000
SODIUM FLUOROPHOSPHATE (Na₂PO₃F) see DXD600
SODIUM FLUOROSILICATE see DXE000
SODIUM FLUORURE (FRENCH) see SHF500
SODIUM FLUOSILICATE see DXE000
SODIUM FORMALDEHYDE BISULFITE see SHI500
SODIUM FORMALDEHYDE SULFOXYLATE see FMW000
SODIUM FORMALDEHYDE SULFOXYLATE of 3-AMINO-4-HYDROXYPHENY-
   LARSONIC ACID see SHI625
SODIUM FORMATE see SHJ000
SODIUM FOSFOMYCIN see DXF600
SODIUM FOSFOMYCIN HYDRATE see FOL200
SODIUM FULMINATE see SHJ500
SODIUM FUMARATE see DXD800
SODIUM FUSIDATE see SHK000
SODIUM FUSIDIN see SHK000
SODIUM-GENT see GCU050
SODIUM GENTISATE see GCU050
SODIUM GERMANIDE see SHK500
SODIUM d-GLUCONATE see SHK800
SODIUM GLUCONATE see SHK800
SODIUM GLUCOSULFONE see AOO800
l(+) SODIUM GLUTAMATE see MRL500
SODIUM l-GLUTAMATE see MRL500
SODIUM l-GLUTAMATE see MRL500
SODIUM GMP see GLS800
SODIUM GOLD CHLORIDE see GIZ100
SODIUM GUANOSINE-5'-MONOPHOSPHATE see GLS800
SODIUM-5'-GUANYLATE see GLS800
SODIUM GUANYLATE see GLS800
SODIUM HEPARIN see HAQ550

SODIUM HEPARINATE see HAQ550
SODIUM HEXABARBITAL see ERE000
SODIUM HEXACYCLONATE see SHL500
SODIUM HEXADECANOATE see SIZ025
SODIUM HEXADECYL SULFATE see HCP900
SODIUM HEXAFLUOROALUMINATE see SHF000
SODIUM HEXAFLUOROARSENATE see SHM000
SODIUM HEXAFLUOROSILICATE see DXE000
SODIUM HEXAFLUOSILICATE see DXE000
SODIUM HEXAMETAPHOSPHATE see SHM500, SII500
SODIUM HEXANITROCOBALTATE see SFX750
SODIUM HEXAVANADATE see SHN000
SODIUM HEXESTROL DIPHOSPHATE see SHN150
SODIUM HEXETHAL see EKT500
SODIUM HEXOBARBITONE see ERE000
SODIUM-N-HEXYLETHYL BARBITURATE see EKT500
SODIUM HEXYLETHYL THIOBARBITURATE see SHN275
SODIUM HIPPURATE see SHN500
SODIUM HUMATE see HGM100
SODIUM HYALURONATE see HGN600
SODIUM HYDRATE (DOT) see SHS000
SODIUM HYDRATE, solution see SHS500
SODIUM HYDRAZIDE see SHO000
SODIUM HYDRIDE see SHO500
SODIUM HYDROCORTISONE-21-SUCCINATE see HHR000
SODIUM HYDROCORTISONE SUCCINATE see HHR000
SODIUM HYDROFLUORIDE see SHF500
SODIUM HYDROGEN CARBONATE see SFC500
SODIUM HYDROGEN S-((N-CYCLOOCTYLMETHYLAMIDINO)METHYL) PHOS-
   PHOROTHIOATE HYDRATE (4:4:5) see SHQ000
SODIUM HYDROGEN DIFLUORIDE see SHQ500
SODIUM HYDROGEN FLUORIDE, solution (DOT) see SHQ500
SODIUM HYDROGEN FLUORIDE see SHQ500
SODIUM HYDROGEN PHOSPHATE see SJH090
SODIUM HYDROGEN-S-(2-AMINOETHYL)PHOSPHOROTHIOATE see AKB500
SODIUM HYDROGEN-S-(2-AMINOETHYL)PHOSPHOROTHIOIC ACID see
   AKB500
SODIUM HYDROGEN SULFATE, solid (UN 1821) (DOT) see SEG800
SODIUM HYDROGEN SULFATE, solution (UN 2837) (DOT) see SEG800
SODIUM HYDROGEN SULFIDE see SHR000
SODIUM HYDROGEN SULFITE, solid (DOT) see SFE000
SODIUM HYDROGEN SULFITE, solution (DOT) see SFE000
SODIUM HYDROGEN SULFITE see SFE000
SODIUM HYDROGEN TRILACTATOZIRCONYLATE see ZTA000
SODIUM HYDROSULFIDE see SHR000
SODIUM HYDROSULFIDE, solution (NA 2922) (DOT) see SHR000
SODIUM HYDROSULFIDE, with <25% water of crystallization (UN 2318)
   (DOT) see SHR000
SODIUM HYDROSULFIDE, with not <25% water of crystallization (UN 2949)
   (DOT) see SHR000
SODIUM HYDROSULFITE (DOT) see SHR500
SODIUM HYDROSULPHITE see SHR500
SODIUM HYDROXIDE, bead (DOT) see SHS000
SODIUM HYDROXIDE, dry (DOT) see SHS000
SODIUM HYDROXIDE, flake (DOT) see SHS000
SODIUM HYDROXIDE, granular (DOT) see SHS000
SODIUM HYDROXIDE, solid (DOT) see SHS000
SODIUM HYDROXIDE, solution (FCC) see SHS500
SODIUM HYDROXIDE (liquid) see SHS500
SODIUM HYDROXIDE see SHS000
SODIUM α-HYDROXYACETATE see SHT000
SODIUM HYDROXYACETATE see SHT000
SODIUM-o-HYDROXYBENZOATE see SJO000
SODIUM-4-HYDROXYBUTYRATE see HJS500
SODIUM-γ-HYDROXYBUTYRATE see HJS500
SODIUM(HYDROXYDE de) (FRENCH) see SHS000
SODIUM 2-HYDROXYDIPHENYL see BGJ750
SODIUM-1-HYDROXYETHANESULFONATE see AAH500
SODIUM-2-HYDROXYETHOXIDE see SHT100
SODIUM o-HYDROXYMERCURIBENZOATE see SHT500
SODIUM p-HYDROXYMERCURIBENZOATE see SHU000
SODIUM o-((3-(HYDROXYMERCURI)-2-METHOXYPRO-
   PYL)CARBAMOYL)PHENOXY ACETATE see SIH500
SODIUM-o-(3-HYDROXYMERCURI-2-METHOXYPRO-
   PYL)CARBAMYL)PHENOXYACETATE and THEOPHYLLINE see MDH750
SODIUM-3-HYDROXYMERCURIO-2,6-DINITRO-4-aci-NITRO-2,5-CYCLOHEXA-
   DIENONIDE see SHU175
SODIUM-2-HYDROXYMERCURIO-4-aci-NITRO-2,5-CYCLOHEXADIENONIDE see
   SHU250
SODIUM-2-HYDROXYMERCURIO-6-NITRO-4-aci-NITRO-2,5-CYCLOHEXADIENON-
   IDE see SHU275
SODIUM HYDROXYMETHANESULFINATE see FMW000
SODIUM-1-(HYDROXYMETHYL)CYCLOHEXANEACETATE see SHL500
SODIUM-3-β-HYDROXY-11-OXO-12-OLEANEN-30-OATE SODIUM SUCCINATE
   see CBO500

SODIUM-5(5′-HYDROXYTETRAZOL-3′-YLAZO)TETRAZOLIDE see HOF500
SODIUM-5(5′-HYDROXYTETRAZOL-3′-YLAZO)TETRAZOLIDE see SHU300
SODIUM HYPOBORATE see SHU475
SODIUM HYPOCHLORITE see SHU500
SODIUM HYPOCHLORITE PENTAHYDRATE see SHU525
SODIUM HYPOPHOSPHITE see SHV000
SODIUM HYPOSULFITE see SKI000, SKI500
SODIUM (E)-3-(p-(1H-IMIDAZOL-1-YLMETHYL)PHENYL)-2-PROPENOATE see SHV100
SODIUM-5,5′-INDIGOTIDISULFONATE see FAE100
SODIUM-5′-INOSINATE see DXE500
SODIUM INOSINATE see DXE500
SODIUM IODATE see SHV500
SODIUM IODIDE see SHW000
SODIUM IODINE see SHW000
SODIUM-o-IODOBENZOATE see IEE050
SODIUM-p-IODOBENZOATE see IEE100
SODIUM-2-IODOHIPPURATE see IEE050
SODIUM o-IODOHIPPURATE see HGB200
SODIUM IODOHIPPURATE see HGB200
SODIUM IODOMETHANESULFONATE see SHX000
SODIUM-5-IODO-2-THIOURACIL see SHX500
SODIUM IOPODATE see SKM000
SODIUM IPODATE see SKM000
SODIUM IRON EDTA see EJA379
SODIUM IRON PYROPHOSPHATE see SHE700
SODIUM ISOAMYLETHYL BARBITURATE see AON750
SODIUM-5-ISOBUTYL-5-(METHYLTHIOMETHYL)BARBITURATE see SHY000
SODIUM ISOCYANATE see COI250
SODIUM ISOCYANURATE see SGB550
SODIUM ISONICOTINYL HYDRAZINE METHANSULFONATE see SIJ000
SODIUM ISOPROPOXIDE see SHY300
SODIUM O-ISOPROPYL DITHIOCARBONATE see SIA000
SODIUM ISOPROPYLETHYL THIOBARBITURATE see SHY500
SODIUM O-ISOPROPYL XANTHATE see SIA000
SODIUM ISOPROPYLXANTHATE see SIA000
SODIUM ISOPROPYLXANTHOGENATE see SIA000
SODIUM ISOTHIOCYANATE see SIA500
SODIUM-α-KETOBUTYRATE see SIY600
SODIUM KETOPROFEN see KGK100
SODIUM LACTATE see LAM000
SODIUM LAURATE see LBN000
SODIUM LAURYLBENZENESULFONATE see DXW200
SODIUM LAURYL ETHER SULFATE see SIB500
SODIUM LAURYL ETHOXYSULPHATE see DYA000
SODIUM-N-LAURYL SARCOSINE see DXZ000
SODIUM LAURYL SULFATE see SIB600
SODIUM LAURYL SULFOACETATE see SIB700
SODIUM LAURYL TRIOXYETHYLENE SULFATE see SIC000
SODIUM LEVOTHYROXINE see LFG000
SODIUM LINOLEATE HYDROPEROXIDE see SIC250
SODIUM LINOLEIC ACID HYDROPEROXIDE see SIC250
SODIUM LITHOCHOLATE see SIC500
SODIUM LUMINAL see SID000
SODIUM MALEATE see SIE000
SODIUM MALONATE see SIE500
SODIUM MALONDIALDEHYDE see MAN700
SODIUM MALONYLUREA see BAG250
SODIUM MANNITOL ANTIMONATE see SIF000
SODIUM MCPA see SIL500
SODIUM MECLOFENAMATE see SIF425
SODIUM MECLOPHENATE see SIF425
SODIUM MERALEIN see SIF500
SODIUM MERALLURIDE see SIG000
SODIUM MERCAPTAN see SHR000
SODIUM MERCAPTIDE see SHR000
SODIUM MERCAPTOACETATE see SKH500
SODIUM-2-MERCAPTOBENZOTHIAZOLE see SIG500
SODIUM-2-MERCAPTOETHANESULFONATE see MDK875
SODIUM MERCAPTOMERIN see TFK270
SODIUM MERSALYL see SIH500
SODIUM MERTHIOLATE see MDI000
SODIUM METAARSENATE see ARD500, ARD750
SODIUM METAARSENITE see SEY500
SODIUM METABISULFITE see SII000
SODIUM METABISULPHITE see SII000
SODIUM METABORATE see SII100
SODIUM METAL (DOT) see SEE500
SODIUM, METAL DISPERSION IN ORGANIC SOLVENT see SEF500
SODIUM METAPERIODATE see SJB500
SODIUM METAPHOSPHATE see SII500, TKP750
SODIUM METASILICATE, anhydrous see SJU000
SODIUM METASILICATE see SJU000
SODIUM METAVANADATE see SKP000
SODIUM METHANALSULFOXYLATE see FMW000

SODIUM METHANEARSONATE see DXE600, MRL750
SODIUM METHANESULFONATE see SIJ000
4-SODIUM METHANESULFONATE METHYLAMINE-ANTIPYRINE see AMK500
SODIUM METHARSONATE see DXE600
SODIUM METHIODAL see SHX000
SODIUM METHOHEXITAL see MDU500
SODIUM METHOHEXITONE see MDU500
SODIUM METHOTREXATE see MDV600
SODIUM METHOXIDE see SIK450
SODIUM METHOXYACETYLIDE see SIJ600
SODIUM β-4-METHOXYBENZOYL-β-BROMOACRYLATE see SIK000
SODIUM dl-1-METHYL-5-ALLYL-5-(1-METHYL-2-PENTYNYL)BARBITURATE see MDU500
SODIUM METHYLAMINOANTIPYRINE METHANESULFONATE see AMK500
SODIUM-4-METHYLAMINO-1,5-DIMETHYL-2-PHENYL-3-PYRAZOLONE 4-METH-ANESULFONATE see AMK500
SODIUM METHYLARSONATE see DXE600
SODIUM METHYLATE see SIK450
SODIUM METHYLATE SOLUTIONS in alcohol (UN 1289) (DOT) see SIK450
SODIUM METHYLATE (UN 1431) (DOT) see SIK450
SODIUM-4-METHYLBENZENESULFINATE see SKJ500
SODIUM-p-METHYLBENZENESULFONATE see SKK000
SODIUM (2-METHYL-4-CHLOROPHENOXY)ACETATE see SIL500
SODIUM-N-METHYL CYCLOHEXENYLMETHYBARBITURATE see ERE000
SODIUM N-METHYLTHIOCARBAMATE see VFU000
SODIUM METHYLDITHIOCARBAMATE see VFU000
SODIUM N-METHYLDITHIOCARBAMATE DIHYDRATE see SIL550
SODIUM-2-METHYL-7-ETHYLUNDECANOL-4-SULFATE see EMT500
SODIUM-2-METHYL-7-ETHYLUNDECYL SULFATE-4 see EMT500
SODIUM METHYLHEXABITAL see ERE000
SODIUM METHYLHEXABITOL see ERE000
SODIUM-3-METHYLISOXAZOLIN-4,5-DIONE-4-OXIMATE see SIL600
SODIUM METHYLMERCURIC THIOGLYCOLLATE see SIM000
SODIUM-2-(N-METHYLOLEAMIDO)ETHANE-1-SULFONATE see SIY000
SODIUM N-METHYL-N-OLEOYLTAURATE see SIY000
SODIUM METHYL OLEOYL TAURATE see SIY000
SODIUM 4-METHYLPHENOLATE see SIM100
SODIUM m-METHYLPHENOLATE see SJP000
SODIUM p-METHYLPHENOLATE see SIM100
SODIUM 3-METHYLPHENOXIDE see SJP000
SODIUM 4-METHYLPHENOXIDE see SIM100
SODIUM 1-METHYL-5-SEMICARBAZONO-6-OXO-2,3,5,6-TETRAHYDROINDOLE-3-SULFONATE TRIHYDRATE see AER666
SODIUM-1-METHYL-5-p-TOLUOYLPYRROLE-2-ACETATE DIHYDRATE see SKJ350
SODIUM MOLYBDATE(VI) see DXE800
SODIUM MOLYBDATE see DXE800
SODIUM MOLYBDATE DIHYDRATE see DXE875
SODIUM MONENSIN see MRE230
SODIUM MONOCHLORACETATE see SFU500
SODIUM MONODODECYL SULFATE see SIB600
SODIUM MONOFLUORIDE see SHF500
SODIUM MONOFLUOROACETATE see SHG500
SODIUM MONOHEXADECYL SULFATE see HCP900
SODIUM MONOHYDROGEN ARSENATE see ARD500
SODIUM MONOHYDROGEN PHOSPHATE (2:1:1) see SJH090
SODIUM MONOIODIDE see SHW000
SODIUM MONOIODOACETATE see SIN000
SODIUM MONOIODOMETHANESULFONATE see SHX000
SODIUM MONOOCTADECYL SULFATE see OBG100
SODIUM MONOSTEARYL SULFATE see OBG100
SODIUM MONOSULFIDE see SJY500
SODIUM MONOXIDE, solid (DOT) see SIN500
SODIUM MONOXIDE see SIN500
SODIUM MORPHOLINECARBODITHIOATE see SIN650
SODIUM MORPHOLINEDITHIOCARBAMATE see SIN650
SODIUM MORPHOLINE and NITRITE (1:1) see SIN675
SODIUM MORPHOLINODITHIOCARBAMATE see SIN650
SODIUM MYCOPHENOLATE see SIN850
SODIUM MYRISTATE see SIN900
SODIUM MYRISTYL SULFATE see SIO000
SODIUM,(N')-ACETYLSULFANILAMIDO- see SNQ000
SODIUM NAFCILLIN see SGS500
SODIUM-1,2-NAPHTHOQUINONE-4-SULFONATE see DLK000
SODIUM-β-NAPHTHOQUINONE-4-SULFONATE see DLK000
SODIUM N-1-NAPHTHYLPHTHALAMATE see SIO500
SODIUM N-1-NAPHTHYLPHTHALAMIC ACID see SIO500
SODIUM NEMBUTAL see NBU000
SODIUM-22 NEOPRENE ACCELERATOR see IAQ000
SODIUM NICOTINATE see NDW500
SODIUM NIGERICIN see SIO788
SODIUM NITRATE (1:1) see SIO900
SODIUM NITRATE (DOT) see SIO900
SODIUM NITRIDE see SIP000
SODIUM NITRILOACETATE see SEP500

SODIUM NITRILOTRIACETATE see SEP500, SIP500
SODIUM NITRITE see SIQ500
SODIUM NITRITE mixed with AMINOPYRINE (1:1) see DOT200
SODIUM NITRITE and BENLATE see BAV500
SODIUM NITRITE mixed with BIS(DIETHYLTHIOCARBAMOYL)DISULFIDE see SIS200
SODIUM NITRITE mixed with 1-(p-BROMOPHENYL)-3-METHYLUREA see SIQ675
SODIUM NITRITE and CARBENDAZIM (1:5) see CBN375
SODIUM NITRITE and CARBENDAZIME (1:1) see SIQ700
SODIUM NITRITE mixed with CHLORDIAZEPOXIDE (1:1) see SIS000
SODIUM NITRITE and l-CITRULLINE (1:2) see SIS100
SODIUM NITRITE mixed with 4-(DIMETHYLAMINO)ANTIPYRINE (1:1) see DOT200
SODIUM NITRITE mixed with DIMETHYLDODECYLAMINE (8:7) see SIS150
SODIUM NITRITE and DIMETHYLUREA see DUM400
SODIUM NITRITE mixed with DISULFIRAM see SIS200
SODIUM NITRITE with DODECYL GUANIDINE ACETATE (5:3) see DXX600
SODIUM NITRITE mixed with ETHAMBUTOL (1:1) see SIS500
SODIUM NITRITE mixed with ETHYLENETHIOUREA see IAR000
SODIUM NITRITE and ETHYLUREA (1:2) see EQE000
SODIUM NITRITE mixed with HEPTAMETHYLENEIMINE see SIT000
SODIUM NITRITE and ISOPROPYLUREA see SIS675
SODIUM NITRITE mixed with METHAPYRILENE (2:1) see MDT000
SODIUM NITRITE mixed with N-METHYLADENOSINE (4:1) see SIS650
SODIUM NITRITE and METHYLANILINE (1:1.2) see MGO000
SODIUM NITRITE mixed with N-METHYLANILINE (35:1) see MGO250
SODIUM NITRITE and METHYL-2-BENZIMIDAZOLE CARBAMATE see CBN375
SODIUM NITRITE mixed with N-METHYLBENZYLAMINE (1:1) see MHN000
SODIUM NITRITE mixed with METHYLBENZYLAMINE (3:2) see MHN250
SODIUM NITRITE mixed with 1-METHYL-3-(p-BROMOPHENYL)UREA see SIQ675
SODIUM NITRITE and 1-(METHYLETHYL)UREA see SIS675
SODIUM NITRITE mixed with METHYLGUANIDINE (1:1) see MKJ000
SODIUM NITRITE mixed with 1-METHYL-3-NITROGUANIDINE (1:1) see MML500
SODIUM NITRITE and 2-METHYL-N-NITROSO-BENZIMIDAZOLE CARBAMATE (1:1) see SIQ700
SODIUM NITRITE and 1-METHYL-1-NITROSO-3-PHENYLUREA see SIS700
SODIUM NITRITE and METHYL UREA see MQJ250
SODIUM NITRITE mixed with OCTAHYDROAZOCINE HYDROCHLORIDE (1:1) see SIT000
SODIUM NITRITE and PIPERAZINE (1:4) see PIJ250
SODIUM NITRITE and 1-PROPYLUREA see SIT500
SODIUM NITRITE and n-PROPYLUREA see SIT500
SODIUM NITRITE, mixed with SODIUM NITRATE and POTASSIUM NITRATE see SIT750
SODIUM NITRITE and TRIFORINE see SIT800
SODIUM m-NITROBENZENEAZOSALICYLATE see SIT850
SODIUM p-NITROBENZENEAZOSALICYLATE see SIU000
SODIUM NITROCOBALTATE (III) see SFX750
SODIUM NITROFERRICYANIDE see SIU500
SODIUM NITROMALONALDEHYDE see SIV000
SODIUM aci-NITROMETHANE see SIV500
SODIUM aci-NITROMETHANIDE see SIV500
SODIUM-4-NITROPHENOXIDE see SIV600
SODIUM NITROPRUSSATE see SIU500
SODIUM NITROPRUSSIATE see SIW000
SODIUM NITROPRUSSIDE see SIU500
SODIUM NITROPRUSSIDE DIHYDRATE see SIW500
SODIUM NITROSOPENTACYANOFERRATE (3) see SIW000
SODIUM-4-NITROSOPHENOXIDE see SIW550
SODIUM NITROSYLPENTACYANOFERRATE(III) see SIU500
SODIUM NITROSYLPENTACYANOFERRATE see SIU500
SODIUM NITROSYLPENTACYANOFERRATE(III) DIHYDRATE see SIW500
SODIUM-5-NITROTETRAZOLIDE see SIW600
SODIUM-2-NITROTHIOPHENOXIDE see SIW625
SODIUM NITROXYLATE see SIX000
SODIUM NONYL SULFATE see SIX500
SODIUM NORAMIDOPYRINE METHANESULFONATE see AMK500
SODIUM NORSULFAZOLE see TEX500
SODIUM NOVOBIOCIN see NOB000
SODIUM NPA see SIO500
SODIUM NTA see SEP500
SODIUM OCTADECANOATE see SJV500
SODIUM-OCTADECYL SULFATE see OBG100
SODIUM OCTAHYDROTRIBORATE see SIX550
SODIUM n-OCTANOATE see SIX600
SODIUM OCTANOATE see SIX600
SODIUM OCTYL SULFATE see OFU200
SODIUM OCTYL SULPHATE see OFU200
SODIUM OLEATE see OIA000
SODIUM N-OLEOYL-N-METHYLATAURINE see SIY000
SODIUM N-OLEOYL-N-METHYLTAURATE see SIY000
SODIUM OLEYLMETHYLTAURIDE see SIY000

SODIUM OMADINE see HOC000
SODIUM ORBENIN see SLJ050
SODIU-4-MORPHOLINECARBODITHIOATE see SIN650
SODIUM ORTHOARSENATE see ARD750
SODIUM ORTHOARSENITE see ARJ500
SODIUM ORTHO PHENYLPHENATE see BGJ750
SODIUM ORTHOVANADATE see SIY250
SODIUM OXACILLIN see MNV250
SODIUM OXALATE see SIY500
SODIUM OXIDE see SIN500
SODIUM OXIDE (Na2-O2) see SJC500
SODIUM 2-OXOBUTANOATE see SIY600
SODIUM 2-OXOBUTYRATE see SIY600
SODIUM OXYBATE see HJS500
SODIUM OXYCHLORIDE see SHU500
SODIUM OXYDIACETATE see SIZ000
SODIUM P-50 see SEQ000
SODIUM PALMITATE see SIZ025
SODIUM PALMITYL SULFATE see HCP900
SODIUM PANTOTHENATE see SIZ050
SODIUM PARATOLUENE SULPHONATE see SKK000
SODIUM PATENT BLUE V see ADE500
SODIUM PCP see SJA000
SODIUM PENICILLANATE 1,1-DIOXIDE see PAP600
SODIUM PENICILLIN see BFD250
SODIUM PENICILLIN G see BFD250
SODIUM PENICILLIN II see BFD250
SODIUM-PENT see NBU000
SODIUM PENTABARBITAL see NBU000
SODIUM PENTABARBITONE see NBU000
SODIUM PENTACARBONYL RHENATE see SIZ100
SODIUM PENTACHLOROPHENATE (DOT) see SJA000
SODIUM PENTACHLOROPHENATE see SJA000
SODIUM PENTACHLOROPHENOL see SJA000
SODIUM PENTACHLOROPHENOLATE see SJA000
SODIUM PENTACHLOROPHENOXIDE see SJA000
SODIUM PENTADECANECARBOXYLATE see SIZ025
SODIUM PENTAFLUOROSTANNITE see SJA500
SODIUM-β,β-PENTAMETHYLENE-γ-HYDROXYBUTYRATE see SHL500
SODIUM PENTHIOBARBITAL see PBT500
SODIUM (R)(+)-PENTOBARBITAL see PBS500
SODIUM PENTOBARBITAL see NBU000
SODIUM PENTOBARBITONE see NBU000
SODIUM PENTOBARBITURATE see NBU000
SODIUM PENTOTHAL see PBT500
SODIUM PENTOTHIOBARBITAL see PBT500
SODIUM PERACETATE see SJB000
SODIUM PERBORATE see SJD000
SODIUM PERBORATE TETRAHYDRATE see SJB350
SODIUM PERCARBONATE see SJB400
SODIUM PERCHLORATE (DOT) see PCE750
SODIUM PERCHLORATE see PCE750
SODIUM PERCHLORATE MONOHYDRATE see SJB450
SODIUM PERIODATE see SJB500
SODIUM PERMANGANATE see SJC000
SODIUM PEROXIDE see SJC500
SODIUM PEROXOBORATE see SJD000
SODIUM PEROXYACETATE see SJB000
SODIUM PEROXYBORATE see SJD000
SODIUM PEROXYDICARBONATE see SJB400
SODIUM PEROXYDISULFATE see SJE000
SODIUM PERRHENATE see SJD500
SODIUM PERSULFATE see SJE000
SODIUM PHENATE see SJF000
SODIUM PHENOBARBITAL see SID000
SODIUM PHENOBARBITONE see SID000
SODIUM PHENOLATE, solid (DOT) see SJF000
SODIUM PHENOL TETRABROMOPHTHALEIN see HAQ600
SODIUM PHENOXIDE see SJF000
SODIUM-α-PHENOXYCARBONYLBENZYLPENICILLIN see CBO000
SODIUM PHENYLACETATE see SFA200
SODIUM PHENYLACETYLIDE see SJF500
SODIUM PHENYL-β-AMINOPROPIONAMIDE-p-ARSONATE see SJG000
SODIUM PHENYLBUTAZONE see BOV750
SODIUM-2-PHENYLCINCHONINATE see SJH000
SODIUM-1-PHENYL-2,3-DIMETHYL-4-METHYLAMINOPYRAZOLON-N-METHANE-SULFONATE see AMK500
SODIUM-1-PHENYL-2,3-DIMETHYL-5-PYRAZOLONE-4-METHYLAMINO METH-ANESULFONATE see AMK500
SODIUM PHENYLDIMETHYLPYRAZOLONMETHYLAMINOMETHANE SULFO-NATE see AMK500
SODIUM PHENYLETHYLBARBITURATE see SID000
SODIUM PHENYLETHYLMALONYLUREA see SID000
SODIUM-N-PHENYLGLYCINAMIDE-p-ARSONATE see CBJ750
SODIUM 2-PHENYLPHENATE see BGJ750

SODIUM o-PHENYLPHENATE see BGJ750
SODIUM o-PHENYLPHENOL see BGJ750
SODIUM o-PHENYLPHENOLATE see BGJ750
SODIUM o-PHENYLPHENOXIDE see BGJ750
SODIUM 3-PHENYL-2-PROPENOATE see SFX000
SODIUM PHENYLSULFONATE see SJH050
SODIUM PHOSPHATE, anhydrous see SJH200
SODIUM PHOSPHATE see HEY500, SJH200, TKP750
SODIUM PHOSPHATE, DIBASIC see SJH090
SODIUM PHOSPHATE, MONOBASIC see SJH100
SODIUM PHOSPHATE, TRIBASIC see SJH200
SODIUM PHOSPHIDE see SJI500
SODIUM PHOSPHINATE see SHV000
SODIUM PHOSPHOROFLUORIDATE see DXD600
SODIUM PHOSPHOROFLURIDATE see DXD600
SODIUM PHOSPHOROTHIOATE see TNM750
SODIUM PHOSPHOTUNGSTATE see SJJ000
SODIUM PICOSULFATE see SJJ175
SODIUM PICRAMATE, wetted with not <20% water, by weight (UN 1349) (DOT) see PIC500
SODIUM PICRAMATE, dry or wetted with <20% water, by weight (UN 0235) (DOT) see PIC500
SODIUM PICRATE see SJJ190
SODIUM PIPERACILLIN see SJJ200
SODIUM PLATINIC CHLORIDE see SJJ500
SODIUM POLYACRYLATE see SJK000
SODIUM POLYALUMINATE see AHG000
SODIUM POLYANHYDROMANNURONIC ACID SULFATE see SEH450
SODIUM POLYMANNURONATE see SEH000
SODIUM POLYOXYETHYLENE ALKYL ETHER SULFATE see SJK200
SODIUM POLYPHOSPHATES, GLASSY see SII500
SODIUM POLYSTYRENE SULFONATE see SJK375
SODIUM POTASSIUM ALLOY, liquid and solid (DOT) see PLS500
SODIUM POTASSIUM BISMUTH TARTRATE (SOLUBLE) see BKX250
SODIUM PRASTERONE SULFATE see SJK400
SODIUM PRASTERONE SULFATE DIHYDRATE see SJK410
SODIUM PREDNISOLONE PHOSPHATE see PLY275
SODIUM p-2-PROPANOL-OXY-PHENYLARSONATE see SJK475
SODIUM PROPIONATE see SJL500
SODIUM-2-PROPYLPENTANOATE see PNX750
SODIUM-2-PROPYLVALERATE see PNX750
SODIUM PYRIDINETHIONE see HOC000
SODIUM PYRITHIONE see MCQ750
SODIUM PYRITHIONE SQ 3277 see HOC000
SODIUM PYROARSENATE see SEY100
SODIUM PYROBORATE see SFF000
SODIUM PYROBORATE DECAHYDRATE see SFF000
SODIUM PYROPHOSPHATE (FCC) see TEE500
SODIUM PYROPHOSPHATE see DXF800
SODIUM PYROSULFATE see SEG800
SODIUM PYROSULFITE see SII000
SODIUM PYROVANADATE see SJM500
SODIUM QUINALBARBITONE see SBN000
SODIUM RETINOATE see SJN000
SODIUM RHODANATE see SIA500
SODIUM RHODANIDE see SIA500
SODIUM RICINOLEATE see SJN500
SODIUM RIFOMYCIN SV see SJN650
SODIUM SACCHARIDE see SJN700
SODIUM SACCHARIN see SJN700
SODIUM SACCHARINATE see SJN700
SODIUM SACCHARINE see SJN700
SODIUM SALICYLATE see SJO000
SODIUM SALICYL-(γ-HYDROXYMERCURI-β-METHOXYPROPYL)AMIDE-o-ACETATE see SIH500
SODIUM SALICYLIC ACID see SJO000
SODIUM SALT of CACODYLIC ACID see HKC500
SODIUM SALT of CARBOXYMETHYLCELLULOSE see SFO500
SODIUM SALT-m-CRESOL see SJP000
SODIUM SALT of DICHLORO-s-TRIAZINETRIONE see SGG500
SODIUM SALT of N,N-DIETHYLDITHIOCARBAMIC ACID see SGJ000
SODIUM SALT of 3-(3-DIMETHYLAMINOMETHYLENEAMINO-2,4,6-TRIIODOPHENYL) PROPIONIC ACID see SKM000
SODIUM SALT of 4,6-DINITRO-o-CRESOL see DUU600
SODIUM SALT of ETHYLENEDIAMINETETRAACETIC ACID see EIV000
SODIUM SALT of HEXADECANOIC ACID see SIZ025
SODIUM SALT of HYDROXY-o-CARBOXY-PHENYL-FLUORONE see FEW000
SODIUM SALT of HYDROXYMERCURIACETYLAMI-NO BENZOIC ACID see SJP500
SODIUM SALT of HYDROXYMERCURIPROPANOL-PHENYLACETIC ACID see SJQ000
SODIUM SALT of HYDROXYMERCURIPROPANOL PHENYL MALONIC ACID see SJQ500
SODIUM SALT of ISONICOTINIC ACID see ILF000
SODIUM SALT of MERCURI-BIS SALICYLIC ACID see SJR000

SODIUM SALT of MERCURY DITHIOCARBAMATE PIPERAZINE ACETIC ACID see SJR500
SODIUM SALT of β-4-METHOXYBENZOYL-β-BROMOACRYLIC ACID see SIK000
SODIUM SALT of 4-MORPHOLINECARBODITHIOIC ACID see SIN650
SODIUM SALT of PHENYLBUTAZONE see BOV750
SODIUM SALT of SULFONATED NAPHTHALENEFORMALDEHYDE CONDENSATE see BLX000
SODIUM SALT of 1,1,3,3-TETRACYANOPROPENE see TAI050
SODIUM SALT of 2,4,5-TRICHLOROPHENOL see SKK500
SODIUM SARKOMYCIN see SJS500
SODIUM SECOBARBITAL see SBN000
SODIUM SECONAL see SBN000
SODIUM SELENATE see DXG000
SODIUM SELENIDE see SJT000
SODIUM SELENITE see SJT500
SODIUM SELENITE PENTAHYDRATE see SJT600
SODIUM SESQUICARBONATE see SJT750
SODIUM SILICATE see SJU000
SODIUM SILICIDE see SJU500
SODIUM SILICOALUMINATE see SEM000
SODIUM SILICOFLUORIDE (DOT) see DXE000
SODIUM SORBATE see SJV000
SODIUM SOTRADECOL see EMT500, SIO000
SODIUM STANNOUS TARTRATE see TGF000
SODIUM STEARATE see SJV500
SODIUM STEARYL SULFATE see OBG100
SODIUM STIBANILATE GLUCOSIDE see NCL000
SODIUM STIBINIVANADATE see SJW000
SODIUM STIBOGLUCONATE see AQH800, AQI250
SODIUM-2-(4-STYRYL-3-SULFOPHENYL)-2H-NAPHTHO-(1,2-d)-TRIAZOLE see TGE155
SODIUM SUCARYL see SGC000
SODIUM SUCCINATE see SJW100
SODIUM SULBACTAM see PAP600
SODIUM SULFACETAMIDE see SNQ000
SODIUM SULFACYL see SNQ000
SODIUM SULFADIAZINE see MRM250
SODIUM SULFAETHYLTHIADIAZOLE see SJW300
SODIUM SULFAMERAZINE see SJW475
SODIUM SULFAMETAZINE see SJW500
SODIUM SULFAMETHAZINE see SJW500
SODIUM SULFAMETHIAZINE see SJW500
SODIUM SULFAMEZATHINE see SJW500
SODIUM-2-SULFANILAMIDOTHIAZOLE see TEX500
SODIUM N-SULFANILYLACETAMIDE see SNQ000
SODIUM SULFANILYLACETAMIDE see SNQ000
SODIUM SULFAPYMONOHYDRATE see PPO250
SODIUM SULFAPYRIDINE see PPO250
SODIUM SULFAPYRIMIDINE see MRM250
SODIUM SULFATE (2:1) see SJY000
SODIUM SULFATE anhydrous see SJY000
SODIUM SULFATHIAZOLE see TEX500
SODIUM SULFHYDRATE see SHR000
SODIUM SULFIDE, anhydrous (DOT) see SJY500
SODIUM SULFIDE with <30% water of crystallization (DOT) see SJY500
SODIUM SULFIDE see SJY500
SODIUM SULFITE (2:1) see SJZ000
SODIUM SULFITE, anhydrous see SJZ000
SODIUM SULFOBROMOPHTHALEIN see HAQ600
SODIUM SULFOCYANATE see SIA500
SODIUM SULFOCYANIDE see SIA500
SODIUM SULFODI-(2-ETHYLHEXYL)SULFOSUCCINATE see DJL000
SODIUM SULFOXYLATE see SHR000
SODIUM SULFOXYLATE FORMALDEHYDE see FMW000
SODIUM SULHYDRATE see SFE000
SODIUM SULPHAMERAZINE see SJW475
SODIUM SULPHAPYRIDINE see PPO250
SODIUM SULPHATE see SJY000
SODIUM SULPHIDE see SJY500
SODIUM SULPHITE see SJZ000
SODIUM SULPHOBROMOPHTHALEIN see HAQ600
SODIUM SUPEROXIDE (DOT) see SJZ100
SODIUM SUPEROXIDE see SJZ100
SODIUM SYNTARPEN see SLJ050
SODIUM d-T4 see SKJ300
SODIUM TARTRATE (FCC) see BLC000
SODIUM l-(+)-TARTRATE see BLC000
SODIUM TARTRATE see SKB000
SODIUM TAUROCHOLATE see SKB500
SODIUM TCA INHIBITED see TII500
SODIUM TCA SOLUTION see TII250
SODIUM TELLURATE(IV) see SKC500
SODIUM TELLURATE see SKC000
SODIUM TELLURATE, DIHYDRATE see SKC100

SODIUM TELLURATE VI see SKC000
SODIUM TELLURITE see SKC500
SODIUM TETRABORATE see DXG035, SFF000
SODIUM TETRABORATE DECAHYDRATE see SFF000
SODIUM TETRABORATE (Na₂B₄O₇) see DXG035
SODIUM TETRACHLOROAURATE(1-) see GIZ100
SODIUM TETRACHLOROAURATE(3+) see GIZ100
SODIUM TETRACHLOROAURATE see GIZ100
SODIUM TETRACYANATOPALLADATE(II) see SKD600
SODIUM TETRADECANOATE see SIN900
SODIUM TETRADECYL SULFATE see SIO000
SODIUM TETRAFLUOROBORATE see SKE000
SODIUM TETRAHYDROALUMINATE(1—) see SEM500
SODIUM TETRAHYDROBORATE(1-) see SFF500
SODIUM TETRAHYDROGALLATE see SKE500
SODIUM TETRAPEROXYCHROMATE see SKF000
SODIUM TETRAPEROXYMOLYBDATE see SKF500
SODIUM TETRAPHOSPHATE see HEY500
SODIUM TETRAPOLYPHOSPHATE see HEY500, SII500
SODIUM TETRAVANADATE see SKG500
SODIUM THEOPHYLLINE see SKH000
SODIUM THIAMYLAL see SOX500
SODIUM THIOCYANATE see SIA500
SODIUM THIOCYANIDE see SIA500
SODIUM THIOGLYCOLATE see SKH500
SODIUM THIOGLYCOLLATE see SKH500
SODIUM THIOPENTAL see PBT500
SODIUM THIOPENTOBARBITAL see PBT500
SODIUM THIOPENTONE see PBT500
SODIUM THIOPHOSPHATE see TNM750
SODIUM THIOSULFATE, anhydrous see SKI000
SODIUM THIOSULFATE see SKI000
SODIUM THIOSULFATE, PENTAHYDRATE see SKI500
SODIUM THYROXIN see LFG050
SODIUM THYROXINATE see LFG050
SODIUM d-THYROXINE see SKJ300
SODIUM l-THYROXINE see LFG050
SODIUM THYROXINE see LFG050
SODIUM TOLMETIN see SKJ340
SODIUM TOLMETIN DIHYDRATE see SKJ350
SODIUM-4-TOLUENESULFINATE see SKJ500
SODIUM-p-TOLUENESULFINATE see SKJ500
SODIUM-p-TOLUENE SULFINIC ACID see SKJ500
SODIUM-p-TOLUENESULFONATE see SKK000
SODIUM p-TOLUENESULFONYLCHLORAMIDE see CDP000
SODIUM-p-TOLYL SULFONATE see SKK000
SODIUM TOSYLATE see SKK000
SODIUM TOSYLCHLORAMIDE see CDP000
SODIUM TRIAZIDOAURATE see SKK100
SODIUM (TRICHLORACETATE de) (FRENCH) see TII500
SODIUM TRICHLOROACETATE see TII500
SODIUM-2,4,5-TRICHLOROPHENATE see SKK500
SODIUM-2,4,5-TRICHLOROPHENOXYETHYL SULFATE see SKL000
SODIUM TRIFLUOROSTANNITE see SKL500
SODIUM TRIIODOHYDROCINNAMATE see SKM000
SODIUM-3,7,12-TRIKETOCHOLANATE see SGD500
SODIUM TRIMETAPHOSPHATE see SKM500, TKP750
SODIUM-2,4,6-TRINITROPHENOXIDE see SJJ190
SODIUM-3,7,12-TRIOXO-5-β-CHOLANATE see SGD500
SODIUM TRIPHOSPHATE see SKN000
SODIUM TRIPOLYPHOSPHATE see SKN000
SODIUM TUNGSTATE see SKN500
SODIUM TUNGSTATE, DIHYDRATE see SKO000
SODIUM TUNGSTOPHOSPHATE see SJJ000
SODIUM TYROPANOATE see SKO500
SODIUM URATE see SKO575
SODIUM VALPROATE see PNX750
SODIUM VANADATE see SIY250, SKP000
SODIUM VANADITE see SKQ000
SODIUM VANADIUM OXIDE see SIY250
SODIUM VERONAL see BAG250
SODIUM VERSENATE see EIX500
SODIUM VINBARBITAL see VKP000
SODIUM WARFARIN see WAT220
SODIUM WOLFRAMATE see SKN500
SODIUM XANTHATE see SHE500
SODIUM XANTHOGENATE see SHE500
SODIUM ZIRCONIUM OXIDE SULFATE see ZTS100
SODIUM ZIRCONYL SULPHATE see ZTS100
SODIUM ZOMEPIRAC see ZUA300
SODIURETIC see BEQ625
SODIZOLE see SNN500
SODNA SUL KYSELINY cis-β-4-METHOXYBENZOYL-β-BROMAKRYLOVE see CQK600
SODUXIN see SJW100

SO-FLO see SHF500
SOFRAMYCIN see NCF000
SOFRIL see SOD500
SOFTENIL see TEH500
SOFTENON see TEH500
SOFTIL see DJL000
SOFTON 1000 see CAT775
SOFTRAN see BOM250
SOG-4 see CJN250
SODOTHIOL see SKI500
SOHNHOFEN STONE see CAO000
SOILBROM-40 see EIY500
SOILBROM-85 see EIY500
SOILFUME see EIY500
SOIL FUNGICIDE 1823 see CJA100
SOILSIN see EME100
SOIL STABILIZER 661 see SMR000
SOK see CBM750
SOL see PKB500
SOLABAR see BAO300
SOLACEN see MOV500
SOLACIL see POF500
SOLACIN see MOV500
SOLACTHYL see AES650
SOLACTOL see LAJ000
SOLADREN see VGP000
SOLAESTHIN see MJP450
SOLAMINE see BEN000
SOLAN see SKQ400
SOLANCARPIDINE see POJ000
SOLANDRA (VARIOUS SPECIES) see CDH125
(22r,25s)-5-α-SOLANIDANINE-3-β-OL see SKQ510
(22r,25s)-5-α-SOLANIDAN-3-β-OL see SKQ510
(22s,25r)-5-α-SOLANIDAN-3-β-OL see SKQ500
SOLANID-5-ENE-3-β, 12-α-DIOL see SKR000
(22s,25r)-SOLANID-5-EN-3-β-OL see SKR500
SOLANIDINE-S see POJ000
α-SOLANIN see SKS000
SOLANIN see SKR875
α-SOLANINE see SKS000
SOLANINE see SKS000
SOLANINE HYDROCHLORIDE see SKS100
SOLANIN HYDROCHLORIDE see SKS100
SOLANTAL see CJH750
SOLANTHRENE BLUE B see DFN425
SOLANTHRENE BLUE F-SBA see DFN425
SOLANTHRENE BLUE RS see IBV050
SOLANTHRENE BLUE RSN see IBV050
SOLANTHRENE BLUE SB see DFN425
SOLANTHRENE BRILLIANT ORANGE JR see CMU820
SOLANTHRENE BRILLIANT VIOLET F 2R see DFN450
SOLANTHRENE BRILLIANT YELLOW see DCZ000
SOLANTHRENE BROWN BR see CMU770
SOLANTHRENE BROWN JR see CMU770
SOLANTHRENE GREY BL see CMU320
SOLANTHRENE OLIVE R see DUP100
SOLANTHRENE PRINTING BLACK BL see CMU320
SOLANTHRENE PRINTING BLACK BLN see CMU320
SOLANTHRENE PRINTING BLACK TL see CMU320
SOLANTHRENE PRINTING BLACK TLN see CMU320
SOLANTHRENE R FOR SUGAR see IBV050
SOLANTIN see DKQ000
SOLANTINE BLUE 10GL see CMO650
SOLANTINE RED 8BL see CMO885
SOLANTOIN see DNU000
SOLANTYL see DNU000
SOLANUM SODOMEUM, extract see AQP800
SOLANUM TUBEROSUM L see PLW750
SOLAPRET see MCB050
SOLAPRET MH see MCB050
SOLAR 40 see DXW200
SOLAR BLACK G see CMN240
SOLAR BROWN PL see CMO750
SOLAR FAST RED 3G see CMM300
SOLAR ORANGE see CMM220
SOLAR PURE BLUE VX see CMM062
SOLAR RED B see CMO885
SOLAR RUBINE see HJF500
SOLARSON see CEI250
SOLAR VIOLET 5BN see FAG120
SOLAR WINTER BAN see PML000
SOLASKIL see LFA020
SOLASOD-5-EN-3-β-OL see POJ000
SOLASODINE see POJ000

SOLASON see CEI250
SOLAXIN see CDQ750
SOLBAR see BAO300, BAP000
SOLBASE see PJT200
SOLBROL A see HJL000
SOLBROL B see BSC000
SOLBROL M see HJL500
SOLBUTAMOL see BQF500
SOLCAIN see DHK400
SOLCOSERYL see SKS299
SOLDACTONE see PKZ000
SOLDEP see TIQ250
SOLDESAM see DAE525
SOLDIER'S BUTTONS see MBU550
SOLDIER'S CAP see MRE275
SOLESTRIL see POB500
SOLESTRO see EDP000
SOLEX see CAW850
2-SOLFAMONYL-4,4'-DIAMINOPHENYLSULFONE see AOT125
SOLFARIN see WAT200
SOLFO BLACK B see DUZ000
SOLFO BLACK BB see DUZ000
SOLFO BLACK 2B SUPRA see DUZ000
SOLFO BLACK G see DUZ000
SOLFO BLACK SB see DUZ000
SOLFOCRISOL see GJG000
SOLFONE see AOO800
SOLFO SERPINE see RDK000
SOLFURO di CARBONIO (ITALIAN) see CBV500
SOLGANAL see ART250
SOLGANAL B see ART250
SOLGOL see CNR675
SOLID GREEN CRYSTALS O see AFG500
SOLID GREEN FCF see FAG000
SOLIMIDIN see ZUA200
SOLINDENE ORANGE R see CMU815
SOLIPHYLLINE see CMF500
SOLIUS RED 4B see CMO885
SOLIWAX see DJL000
SOLKA-FIL see CCU150
SOLKA-FLOC see CCU150
SOLKA-FLOC BW 100 see CCU150
SOLKA-FLOC BW 200 see CCU150
SOLKA-FLOC BW 20 see CCU150
SOLKA-FLOC BW 2030 see CCU150
SOLKA-FLOC BW see CCU150
SOLKETAL see DVR600
SOLLICULIN see EDV000
SOLMETHINE see MJP450
SOLNAPYRIN-A see AIB300
SOLNET see PMB850
SOLOCALM see FAJ100
SOLOCHROMATE BLACK 6BN see CMP880
SOLOCHROME BLACK 6BN see CMP880
SOLOCHROME BLUE FB see HJF500
SOLOCHROME YELLOW WN see SIT850
SOLOMON ISLAND IVY see PLW800
SOLOMON'S LILY see ITD050
SOLOSIN see TEP000
SOLOZONE see SJC500
SOL PHENOBARBITAL see SID000
SOL PHENOBARBITONE see SID000
SOLPRENE 300 see SMR000
SOLPRINA see CLD000
SOLPYRON see ADA725
SOL SODOWA KWASU LAURYLOBENZENOSULFONOWEGO (POLISH) see DXW200
SOLSOL NEEDLES see SIB600
SOL. SULFUR BLUE 10 see CMM080
SOLTRIOL see DMJ400
SOLUBACTER see TIL500
SOLUBLE BARBITAL see BAG250
SOLUBLE FLUORESCEIN see FEW000
SOLUBLE GLUSIDE see SJN700
SOLUBLE GUN COTTON see CCU250
SOLUBLE INDIGO see FAE100
SOLUBLE IODOPHTHALEIN see TDE750
SOLUBLE PENTOBARBITAL see NBU000
SOLUBLE PHENOBARBITAL see SID000
SOLUBLE PHENOBARBITONE see SID000
SOLUBLE PHENYTOIN see DNU000
SOLUBLE SACCHARIN see SJN700
SOLUBLE SULFACETAMIDE see SNQ000
SOLUBLE SULFACYL see SNQ000
SOLUBLE SULFADIAZINE see MRM250

SOLUBLE SULFAMERAZINE see SJW475
SOLUBLE SULFAPYRIDINE see PPO250
SOLUBLE SULFATHIAZOLE see TEX500
SOLUBLE TARTRO-BISMUTHATE see BKX250
SOLUBLE THIOPENTONE see PBT500
SOLUBLE VAN DYKE BROWN see MAT500
SOLUBOND 0-869 see RDP000
SOLUBOND 3520 see RDP000
SOLUCORT see PLY275
SOLU-CORTEF see HHR000
SOLUDAGENAN see PPO250
SOLU-DECADRON see DAE525
SOLU-DECORTIN see PMA100
SOLUFILIN see DNC000
SOLUFILINA see CNR125, DIH600
SOLUFYLLIN see DNC000
SOLUGLACIT see NIM500
SOLU-GLYC see HHR000
SOLULAN C-24 see PKE700
SOLUMEDINE see SJW475
SOLU-MEDROL see USJ100
SOLU-MEDRONE see USJ100
SOLUPHYLINE see CNR125
SOLUPHYLLINE see HLC000
SOLUPYRIDINE see DXF400
SOLUROL see SOX875
SOLUSEDIV 2% see HAF400
SOLUSOL-75% see DJL000
SOLUSOL-100% see DJL000
SOLUSTIBOSAN see AQH800, AQI250
SOLUSTIN see AQH800, AQI250
SOLUSURMIN see AQH800, AQI250
SOLUTION CONCENTREE T271 see AMY050
SOLUTION GLYCERYL TRINITRATE see NGY000
SOLUTRAST see IFY000
SOLVANOL see DJX000
SOLVANOM see DTR200
SOLVANT YELLOW 33 see CMS245
SOLVAR see PKP750
SOLVARONE see DTR200
SOLVAT 14 see BFL000
SOLVENT 111 see MIH275
SOLVENT BLUE 78 see BKP500
SOLVENT BLUE 93 see BKP500
SOLVENT-DEWAXED HEAVY NAPHTHENIC DISTILLATE see MQV820
SOLVENT-DEWAXED HEAVY PARAFFINIC DISTILLATE see MQV825
SOLVENT-DEWAXED LIGHT NAPHTHENIC DISTILLATE see MQV835
SOLVENT-DEWAXED LIGHT PARAFFINIC DISTILLATE see MQV840
SOLVENT ETHER see EJU000
SOLVENT ORANGE 15 see BJF000
SOLVENT ORANGE 1 see CMP600
SOLVENT RED 19 see EOJ500
SOLVENT RED 1 see CMS242
SOLVENT REFINED COAL-II, HEAVY DISTILLATE see SLH300
SOLVENT REFINED COAL MATERIALS, HEAVY DISTILLATE see SLH300
SOLVENT-REFINED (mild) HEAVY NAPHTHENIC DISTILLATE see MQV845
SOLVENT-REFINED (mild) HEAVY PARAFFINIC DISTILLATE see MQV850
SOLVENT-REFINED (mild) LIGHT NAPHTHENIC DISTILLATE see MQV852
SOLVENT-REFINED (mild) LIGHT PARAFFINIC DISTILLATE see MQV855
SOLVENT VIOLET 26 see DBX000
SOLVENT YELLOW 14 see PEJ500
SOLVENT YELLOW 1 see PEI000
SOLVIC see PKQ059
SOLVIC 523KC see AAX175
SOLVIREX see DXH325
SOLVOSOL see CBR000
SOLWAY BLUE RB see CMM100
SOLWAY SKY BLUE B see CMM090
SOLWAY SKY BLUE BA see CMM090
SOLYACORD see DJS200
SOLYUSURMIN see AQH800, AQI250
SOMA see IPU000
SOMACTON see PJA250
SOMADRIL see IPU000
SOMALGIT see IPU000
SOMALIA ORANGE A2R see XRA000
SOMALIA ORANGE I see PEJ500
SOMALIA ORANGE 2R see XRA000
SOMALIA RED III see OHA000
SOMALIA RED IV see SBC500
SOMALIA RED PG see CMS242
SOMALIA YELLOW A see DOT300
SOMALIA YELLOW R see AIC250
SOMAN see SKS500
SOMANIL see IPU000

SOMAR see DXE600
SOMATOSTATIN 14 see SKS600
SOMATOSTATIN see SKS600
SOMATOTROPIC HORMONE see PJA250
SOMATOTROPIN see PJA250
SOMAZINA see CMF350
SOMBEROL see QAK000
SOMBREVIN see PMM000
SOMBUCAPS see ERD500
SOMBULEX see ERD500
SOMBUTOL see EOK000
SOMELIN see HAG800
SOMILAN see ENE500
SOMINAT see SKS700
SOMIO see GFA000
SOMIPRONT see DUD800
SOMLAN see DAB800
SOMNAFAC see MDT250, QAK000
SOMNAL see AMX750
SOMNALERT see ERD500
SOMNASED see DLY000
SOMNESIN see EQL000
SOMNEVRIN see CHD750
SOMNIBEL see DLY000
SOMNICAPS see DPJ400
SOMNI SED see CDO000
SOMNITE see DLY000
SOMNOLETTEN see EOK000
SOMNOMED see QAK000
SOMNOPENTYL see NBU000
SOMNOS see CDO000
SOMNOSAN see EOK000
SOMNUROL see BNP750
SOMONAL see EOK000
SOMONIL see DSO000
SOMOPHYLLIN see TEP500
SOMOPHYLLIN O see TEP500
SOMSANIT see HJS500
SONACIDE see GFQ000
SONACON see DCK759
SONAFORM see TDA500
SONAL see QAK000
SONALAN see ENE500
SONALEN see ENE500
SONAPAX see MOO250
SONAR see FMQ200
SONATE see ARA500
SONAZINE see CKP500
SONBUTAL see BOR000
SONEBON see DLY000
SONERILE see BPF500
SONERYL see BPF500
SONG-SAM, ROOT EXTRACT see GEQ425
SONISTAN see NBU000
SONNOLIN see DLY000
SONTEC see CDO000
SONTOBARBITAL NABITONE see NBU000
SOOT see SKS750
SOPANOX see TGX550
SOPAQUIN see CLD000
SOPENTAL see NBU000
SOPHIA see MDL750
SOPHIAMIN see MDQ250
SOPHOCARPINE HYDROBROMIDE see SKS800
SOPHORA SECUNDIFLORA see NBR800
SOPHORA TOMENTOSA see NBR800
SOPHORETIN see QCA000
SOPHORIN see RSU000
SOPHORINE see CQL500
SOPP see BGJ750
SOPRABEL see LCK000, LCK100
SOPRACOL 781 see MLC250
SOPRACOL see MLC250
SOPRANEBE see MAS500
SOPRATHION see EEH600, PAK000
SOPRINAL see BAG250
SOPRINTIN see ABH500
SOPROCIDE see BBQ750
SOPRONTIN see AAF750, ABH500
SOPROTIN see ABH500
SORBALDEHYDE see SKT500
SORBANGIL see CCK125
SORBA-SPRAY Mn see MAU250
SORBESTER P 17 see SKV100
SORBIC ACID see SKU000

SORBIC ACID, COPPER SALT see CNP000
SORBIC ACID, POTASSIUM SALT see PLS750
SORBIC ACID, SODIUM SALT see SJV000
SORBIC ALCOHOL see HCS500
SORBIC OIL see PAJ500
SORBICOLAN see SKV200
SORBID see CCK125
SORBIDE NITRATE see CCK125
SORBIDILAT see CCK125
SORBIDINITRATE see CCK125
SORBIMACROGOL LAURATE 300 see PKG000
SORBIMACROGOL OLEATE see PKL100
SORBIMACROGOL TRIOLEATE 300 see TOE250
SORBIMACROGOL TRISTEARATE 300 see SKV195
SORBINIC ALCOHOL see HCS500
SORBISLO see CCK125
SORBISTAT see SKU000
SORBISTAT-K see PLS750
SORBISTAT-POTASSIUM see PLS750
SORBITAL O 20 see PKL100
SORBITAN, MONODODECANOTE, POLY(OXY-1,2-ETHANEDIYL) DERIVATIVES
   see PKG000
SORBITAN, MONOHEXADECANOATE, POLY(OXY-1,2-ETHANEDIYL) DERIVS.
   see PKG500
SORBITAN, MONOOCTADECANOATE, POLY(OXY-1,2-ETHANEDIYL) DERIVA-
   TIVES see PKL030
SORBITAN C see SKV150
SORBITAN MONODODECANOATE see SKV000
SORBITAN MONOLAURATE see SKV000
SORBITAN MONOOCTADECANOATE see SKV150
SORBITAN MONOOLEATE see SKV100
SORBITAN MONOOLEIC ACID ESTER see SKV100
SORBITAN MONOSTEARATE see SKV150
SORBITAN O see SKV100
SORBITAN OLEATE see SKV100
SORBITAN PALMITATE see MRJ800
SORBITAN STEARATE see SKV150
SORBITAN, (Z)-9-OCTADECENOATE (2:3) see SKV170
SORBITAN, SESQUIOLEATE see SKV170
SORBITAN, TRIOCTADECANOATE, POLY(OXY-1,2-ETHANEDIYL) derivs. (9CI)
   see SKV195
SORBITAN, TRISTEARATE, POLYOXYETHYLENE derivatives. see SKV195
SORBITE see SKV200
d-SORBITOL see SKV200
SORBITOL see SKV200
SORBITRATE see CCK125
SORBO see SKV200
SORBO-CALCIAN see CAL750
SORBO-CALCION see CAL750
SORBOL see SKV200
SORBONIT see CCK125
SORBON S 60 see SKV150
SORBOSTYL see SKV200
SORBYL ACETATE see AAU750
SORBYL ALCOHOL see HCS500
SORCI see FPD100
SORDENAC see CLX250
SORDINOL see CLX250
SOREFLON 604 see TAI250
SORETHYTAN (20) MONOOLEATE see PKL100
SORGEN 30 see SKV170
SORGEN 40 see SKV100
SORGEN 50 see SKV150
SORGEN 70 see MRJ800
SORGHUM GUM see SLJ500
SORGOA see TGB475
SORGOPRIM see BQB000
SORIDERMAL see MNQ500
SORIPAL see MNQ500
SORLATE see PKL100
SORMETAL see CAR780
SORQUAD see CCK125
SORQUAT see CCK125
SORQUETAN see TGD250
SORREL SALT see OLE000
SORROSIE (HAITI) see FPD100
SORSAKA see SKV500
l-SORSAKA, LEAF and STEM EXTRACT see SKV500
SORVILANDE see SKV200
SOS see OFU200
SOSIGON see DOQ400
SOSPITAN see PPH050
SOTACOR see CCK250
SOTALEX see CCK250
SOTALOL see CCK250

SOTALOL HYDROCHLORIDE see CCK250
SOTERENOL see SKW000
SOTERENOL HYDROCHLORIDE see SKW500
SOTIPOX see TIQ250
SOTRADECOL see EMT500
SOTYL see NBU000
SOUCI d'EAU (CANADA) see MBU550
SOUDAN I see PEJ500
SOUDAN II see XRA000
SOUDAN III see OHA000
SOUFRAMINE see PDP250
SOUP see NGY000
SOUTHERN BAYBERRY see WBA000
SOUTHERN BENTONITE see BAV750
SOUTHERN COPPERHEAD VENOM see SKW775
SOVCAIN see NOF500
SOVCAINE see DDT200
SOVCAINE HYDROCHLORIDE see NOF500
SOVERIN see QAK000
SOVIET TECHNICAL HERBICIDE 2M-4C see CIR250
SOVIOL see AAX250
SOVKAIN see NOF500
SOVOCAINE HYDROCHLORIDE see NOF500
SOVOL see PJL750
SOVPRENE see PJQ050
SOWBUG & CUTWORM BAIT see COF500
SOXINAL PZ see BJK500
SOXINOL CZ see CPI250
SOXINOL PZ see BJK500
SOXISOL see SNN500
SOXYSYMPAMINE see MDT600
SOYEX see NIX000
75 SP see DOP600
SP 104 see TCM250
S.P. 147 see LEU000
SP 489 see CGW300
S 640P see APT250
SP 725 see PEF500
SP see AIF000
SPA see DWA600
SPACTIN see BTA325
SPAMOL see CPQ250
SPAN 20 see SKV000
SPAN 40 see MRJ800
SPAN 55 see SKV150
SPAN 60 see SKV150
SPAN 80 see SKV100
SPAN 83 see SKV170
SPANBOLET see SNJ000
SPANISH BROOM see GCM000
SPANISH CARNATION see CAK325
SPANISH FLY see CBE250
SPANISH MARJORAM OIL see MBU500
SPANISH THYME OIL see TFX750
SPANISH WHITE see BKW100
SPANON see CJJ250
SPANONE see CJJ250
SPANTOL see PGA750
SPANTRAN see MQU750
SPARIC see BRE500
SPARICON see SBG500
SPARINE see DQA600
SPARINE HYDROCHLORIDE see PMI500
SPARSOMYCIN see SKX000
SPARTAKON see TDX750
SPARTASE see MAI600
1-SPARTEINE see SKX500
d-SPARTEINE see PAB250
SPARTEINE see SKX500
(−)-SPARTEINE see SKX500
SPARTEINE, d-ISOMER see PAB250
SPARTEINE SULFATE see SKX750
SPARTOCIN see SKX750
SPARTOLOXYN see GGS000
SPARTOSE OM-22 see CCU150
SPASMAMIDE see FAJ150
SPASMEDAL see DAM700
SPASMENTRAL see BBU800
SPASMEX see KEA300
SPASMEXAN HYDROCHLORIDE see LFK200
SPASMIONE see DNU100
SPASMO 3 see KEA300
SPASMOCAN see NOC000
SPASMOCYCLON see DNU100
SPASMOCYCLONE see DNU100

SPASMODOLIN see DAM700
SPASMOLYSIN see HOA000
SPASMOLYTIN see DHX800
SPASMOLYTON see THK000
SPASMO-NIT see PAH250
SPASMOPHEN see ORQ000
SPASURET see FCB100
SPATHE FLOWER see SKX775
SPATHIPHYLLUM (VARIOUS SPECIES) see SKX775
SPATONIN see DIW000
SPAVIT see VSZ450
SP 60 (CHLOROCARBON) PKQ059
(SP-4-2)-trans(+)-DICHLORO(1,2-CYCLOHEXANEDIAMINE-N,N')- (9CI) see DAD050
SPD PHOSPHATE see AMD500
SPEARMINT OIL see SKY000
SPEARWORT see FBS100
SPECIAL BLACK 1V & V see CBT750
SPECIAL BLUE X 2137 see EOJ500
SPECIAL M see AHA275
SPECIAL ORANGE GR see CMM220
SPECIAL ORANGE H see CMM220
SPECIAL SCHWARZ see CBT750
SPECIAL TERMITE FLUID see DEP600
SPECIFEN see EID000
SPECILLINE G see BDY669
SPECTAZOLE see EAE000
SPECTINOMYCIN, DIHYDROCHLORIDE, PENTAHYDRATE see SKY500
SPECTINOMYCIN DIHYDROCHLORIDE see SLI325
SPECTINOMYCIN HYDROCHLORIDE see SLI325
SPECTOGARD see SKY500
SPECTRACIDE see DCM750
SPECTRAR see INJ000
SPECTRA-SORB UV 531 see HND100
SPECTRA-SORB UV 9 see MES000
SPECTROBID see BAB250
SPECTROLENE BLUE B see DCJ200
SPECTROLENE RED KB see CLK225
SPECTROLENE SCARLET 2G see DEO295
SPECTROSIL see SCK600
SPECULAR IRON see IHD000
SPEED see DBA800
"SPEED" see MDQ500
SPENCER 401 see PJY500
SPENCER S-6538 see DJY200
SPENCER S-6900 see DRR200
SPENKEL see PKL500
SPENLITE see PKL500
S(−)-PENTOBARBITAL SODIUM see PBS750
SPENT OXIDE see SKZ000
SPENT SULFURIC ACID (DOT) see SOI500
SPERGON I see TBO500
SPERGON TECHNICAL see TBO500
SPERLOX-S see SOD500
SPERLOX-Z see EIR000
SPERM, HUMAN see HGM000
SPERMIDINE see SLA000
SPERMIDINE PHOSPHATE see AMD500
SPERMINE see DCC400
SPERMINE, PURISS see DCC400
SPERMINE TETRAHYDROCHLORIDE see GEK000
SPERSADOX see DAE525
SPERSUL see SOD500
SPERSUL THIOVIT see SOD500
SP 60 ESTER see AAX250
SPG 827 see BER500
SP G see BER500
SPH see HGN600
SPHEROIDINE see FOQ000
SPHEROMYCIN see SMB000
SPHERON 6 see CBT750
SPHERON see CBT750
SPHYGMOGENIN see VGP000
SPICEBUSH see CCK675
SPICLOMAZINE HYDROCHLORIDE see SLB000
SPICY JATROPHA see CNR135
SPIDER LILY see BAR325, REK325, SLB250
SPIGELIA ANTHELMIA see PIH800
SPIGELIA MARILANDICA see PIH800
SPIKE see BSN000
SPIKE LAVENDER OIL see SLB500
SPILAN see BEL900
SPINACEN see SLG800
SPINACENE see SLG800
SPINOCAINE see AIL750

SPIPERONE see SLE500
SPIRAMIN see AJV500
SPIRAMYCIN, HEXANEDIOATE see SLC300
SPIRAMYCIN see SLC000
SPIRAMYCIN ADIPATE see SLC300
SPIRAMYCINS see SLC000
SPIRESIS see AFJ500
SPIRIDON see AFJ500
SPIRIT of HARTSHORN see AMY500
SPIRIT of NITER (sweet) see SLD500
SPIRIT OF GLONOIN see NGY000
SPIRIT OF GLYCERYL TRINITRATE see NGY000
SPIRIT OF TRINITROGLYCERIN see NGY000
SPIRIT ORANGE see PEJ500
SPIRITS of SALT see HHL000
SPIRITS of TURPENTINE see TOD750
SPIRITS of WINE see EFU000
SPIRIT of TURPENTINE see TOD750
SPIRIT YELLOW I see PEJ500
SPIRO-32 see SLD800
SPIROBROMIN see SLD600
SPIROCID see ABX500
SPIROCTANIE see AFJ500
SPIRO(CYCLOHEXANE-1,5'-HYDANTOIN) see DVO600
SPIRO(17H-CYCLOPENTA(a)PHENANTHRENE-17,2'(5'H)FURAN), PREGN-4-
ENE-21-CARBOXYLIC ACID DERIV. see AFJ500
SPIRO(17H-CYCLOPENTA(a)PHENAUTHRENE-17,2'-(3'H)-FURAN) see AFJ500
SPIROFULVIN see GKE000
SPIROGERMANIUM DIHYDROCHLORIDE see SLD800
SPIROGERMANIUM HYDROCHLORIDE see SLD800
SPIROHYDANTOIN MUSTARD see SLD900
SPIRO(ISOBENZOFURAN)-1(3H),9'-(9H)XANTHENE-3-ONE, 3',6'-DIHYDROXY-DI-
SODIUM SALT see FEW000
SPIROLACTONE see AFJ500
SPIROLAKTON see AFJ500
SPIROLANG see AFJ500
SPIROMUSTINE see SLD900
SPIRO(NAPHTHALENE-2(1H),2'-OXIRANE)-3'-CARBOXALDEHYDE, 3,5,6,7,8,8a-
HEXAHYDRO-7-ACETOXY-5,6-EPOXY-3',8,8a-TRIMETHYL-3-OXO- see
POF800
SPIRONE see AFJ500
SPIRONOLACTONE see AFJ500
SPIRONOLACTONE A see AFJ500
SPIROPENT see VHA350
SPIROPERIDOL see SLE500
SPIROPITAN see SLE500
SPIROPLATIN see SLE875
SPIROZID see ABX500
SPIRT see EFU000
SPIZOFURONE see SLE880
SPLIT LEAF PHILODENDRON see SLE890
SPLOTIN see EPD500
SPMF 4 see MCB050
SPMF 6 see MCB050
SPMF 7 see MCB050
SPOFADAZINE see AKO500
SPOFADRIZINE see PPP500
SPOLAPRET OS see MND550
SPONGIOFORT see PCU360
SPONGOADENOSINE see AEH100
SPONGOCYTIDINE see AQQ750
SPONGOCYTIDINE HYDROCHLORIDE see AQR000
SPONTOX see TAA100
SPOONWOOD IVY see MRU359
SPORAMIN see SBG500
SPORARICIN A see SLF500
SPORILINE see TGB475
SPOR-KIL see ABU500
SPOROSTATIN see GKE000
SPOTRETE see TFS350
SPOTTED ALDER see WCB000
SPOTTED COWBANE see WAT325
SPOTTED HEMLOCK see PJJ300
SPOTTED PARSLEY see PJJ300
SPOTTON see FAQ900
SPP see AIF000
SPRACAL see ARB750
SPRAY-DERMIS see NGE500
SPRAY-FORAL see NGE500
SPRAY-HORMITE see SGH500
SPRAYSET MEKP see MKA500
SPRAY-TROL BRANCH RODEN-TROL see WAT200
SPRENGEL EXPLOSIVES see SLG500
SPRIDINE HYDROCHLORIDE see SLG600
SPRIDINE TRIHYDROCHLORIDE see SLG600

SPRING-BAK see DXD200
SPRITZ-HORMIN/2,4-D see DAA800
SPRITZ-HORMIT see SGH500
SPROUT NIP see CKC000
SPROUT-NIP EC see CKC000
SPROUT-OFF see DAI800
SPROUT/OFF see DMC600
SPROUT-OFF see ODE000
SPROUT-STOP see DMC600
SPRUCE OIL see SLG650
SPS 600 see SMQ500
SPT 50 CPS see EHG100
SPUD-NIC see CKC000
SPUD-NIE see CKC000
SPULMAKO-LAX see PDO750
SPURGE see BRE500
SPURGE LAUREL see LAR500
SPURGE OLIVE see LAR500
SPUTOLYSIN see DXO800
SQ 1089 see HOO500
SQ 1156 see GGS000
SQ 1489 see TFS350
SQ 2113 see HOB500
SQ 2303 see CBK500
SQ 2321 see FNF000
SQ 3277 see HOC000
SQ 4609 see BDX090
SQ 8388 see AQO000
SQ 9453 see DUD800
SQ 10269 see CBQ625
SQ 10,643 see CMP900
SQ 11725 see CNR675
SQ 13396 see IFY000
SQ 14055 see TET800
SQ 14,225 see MCO750
SQ 15,659 see PPY250
SQ-15761 see SKM000
SQ 16,114 see PMI250
SQ 16360 see SHK000
SQ 16423 see MNV250
SQ 16496 see PMC700
SQ 16603 see FQU000
SQ 16819 see HKQ025
SQ-18,566 see HAF300
SQ 21065 see POJ500
SQ 21977 see MPU000
SQ-21983 see IGA000
SQ 22947 see DAR000, TET800
SQ 26,776 see ARX875
SQUALANE see SLG700
SQUALEN see SLG800
SQUALENE see SLG800
(E,E,E,E)-SQUALENE see SLG800
trans-SQUALENE see SLG800
SQUALIDIN see IDG000
SQUALIDINE see IDG000
SQUARIC ACID see DMJ600
SQUAW ROOT see BMA150
SQUIBB 4918 see FMP000
SQUIBB see HNT500
SQUILL see RCF000, SBF500, SLH200
SQUIRREL FOOD see DAE100
SR 73 see DFV600
SR-202 see CMU875
SR 206 see BKM250
SR 209 see TCE400
SR 213 see TDQ100
SR 247 see DUL200
SR 339 see PER250
SR 351 see TLX175
SR406 see CBG000
S 1096R see GGR200
SR 1354 see NHH500
SR 720-22 see ZAK300
SRA 5172 see DTQ400
SRA 7312 see DJY200
SRA 7847 see EIM000
SRA 12869 see IMF300
SRA GOLDEN YELLOW VIII see DUW500
SRC-7 see DWQ850
SRC-15 see BHJ625
SRC-16 see BLD325
SRC-226 see DSH800
SRC-II HEAVY DISTILLATE see CMY75-
SRC-II, HEAVY DISTILLATE see SLH300

SR E2 see SMF000
SRG 95213 see DCQ700
SRI 702 see BFL125
SRI 753 see BRS500
SRI 759 see TFJ500
SRI 859 see MNA750
SRI 1354 see NHH500
SRI 1666 see DRN400
SRI 1720 see BIF750
SRI 1869 see NKL000
SRI 2200 see CGV250
SRI 2489 see IAN000
SRI 2619 see CPL750
SRI 2638 see CHE500
SRI 2656 see CFH750
SRIF see SKS600
SRM 705 see SMQ500
SRM 706 see SMQ500
SRM 1475 see PJS750
SRM 1476 see PJS750
SS-094 see SKS299
SS 578 see DNF600
SS 2074 see ORU000
SS 30 (carbonate) see CAT775
SS 50 (carbonate) see CAT775
SS see CCU250
SSB, see TES800
SSB 100 see CAT775
S.T. 37 see HFV500
ST 90 see SMQ500
ST 155 see CBF250
ST-155 see CMX760
ST 1085 see MQT530
ST-1191 see EQO000
1-ST-2121 see SLI200
ST 720 (FRENCH) see CGB000
ST A 307 see TFK300
STABICILLIN see PDT500
STABILAN see CMF400
STABILENE see BKA000, BRP250
STABILENE FLY REPELLENT see BRP250
STABILISATOR C see DWN800
STABILISATOR SCD see OAT000
STABILISATOR VH see PGP250
STABILIZATOR AR see PFT500
STABILIZER D-22 see DDV600
STABILIZER DLT see TFD500
STABILIZER SCD see OAT000
STABILIZER VH see PGP250
STABILLIN VK SYRUP 125 see PDT750
STABILLIN VK SYRUP 62.5 see PDT750
STABINEX NW 7PS see CAW850
STABINOL see CKK000
STABLE RED KB BASE see CLK225
STAC see TLW500
STADADORM see AMX750
STADALAX see PPN100
STADERM see IAB000
STADOL see BPG325
STAFAC see VRA700, VRF000
STAFAST see NAK500
STA-FAST see TIX500
STAFF TREE see AHJ875
STAFLEN E 650 see PJS750
STAFLEX 500 see OEU000
STAFLEX DBM see DED600
STAFLEX DBP see DEH200
STAFLEX DBS see DEH600
STAFLEX DOP see DVL700
STAFLEX DOX see BJQ500
STAFLEX TOTM see TJR600
STA-FRESH 615 see SGM500
STAGNO (TETRACLORURO di) (ITALIAN) see TGC250
STALFEX DTDP see DXQ200
STALFLEX DOS see BJS250
STALLIMYCIN see SLI300
STALLIMYCIN HYDROCHLORIDE see DXG600
STAM M-4 see DGI000
STAM see DGI000
STAM F 34 see DGI000
STAMINE see WAK000
STAM LV 10 see DGI000
STAMPEDE see DGI000
STAMPEDE 3E see DGI000
STAMPEN see DGE200

STAM SUPERNOX see DGI000
STAMYLAN 900 see PJS750
STAMYLAN 1000 see PJS750
STAMYLAN 1700 see PJS750
STAMYLAN 8200 see PJS750
STAMYLAN 8400 see PJS750
STANCLERET 157 see DEJ100
STANDACOL CARMOISINE see HJF500
STANDACOL SUNSET YELLOW FCF see FAG150
STANDAK see AFK000
STANDAMIDD LD see BKE500
STANDAMUL 1616 see HCP700
STANDAMUL 7061 see IKC050
STANDAMUL DIPA see DNL800
STANDAMUL O20 see PJW500
STANDAPOL 112 CONC see SIB600
STANDAPOL TLS 40 see SON000
STANDARD LEAD ARSENATE see LCK000
STANDOPAL see MCB050
STANEPHRIN see SPC500
STANGEN see WAK000
STANGEN MALEATE see DBM800
STAN-GUARD 156 see BKO250
STANILO see SKY500, SLI325
STAN-MAG MAGNESIUM CARBONATE see MAC650
STANNANE, ((p-ACETAMIDOBENZOYL)OXY)TRIPHENYL- see TMV800
STANNANE, ACETOXYTRICYCLOHEXYL- see ABW600
STANNANE, ACETOXYTRIOCTYL- see ABX150
STANNANE, ACETOXYTRIPENTYL- see ABX175
STANNANE, (ACETYLENECARBONYLOXY)TRIPHENYL- see TMW600
STANNANE, (ACETYLOXY)TRIPROPYL-(9CI) see TNB000
STANNANE, BROMOTRIETHYL-, compounded with 2-PIPECOLINE (1:1) see TJU850
STANNANE, BUTYLDIETHYLIODO- see BRA550
STANNANE, (BUTYLTHIO)TRIOCTYL- see BSO200
STANNANE, (BUTYLTHIO)TRIPROPYL- see TMY850
STANNANE, CYANATOTRIMETHYL- see TMI100
STANNANE, (CYANOACETOXY)TRIPHENYL- see TMV825
STANNANE, CYCLOHEXYLHYDROXYOXO- see MRH212
STANNANE, DICHLORODIMETHYL- see DUG825
STANNANE, DICHLORODIOCTYL- see DVN300
STANNANE, DICYCLOHEXYLOXO- see DGV900
STANNANE, DIMETHYL(2,3-QUINOXALINYLDITHIO)- see QSJ800
STANNANE, ((DIMETHYLTHIOCARBAMOYL)THIO)TRIBUTYL- see TID150
STANNANE, DIOCTYLDICHLORO- see DVN300
STANNANE, (ETHYLTHIO)TRIOCTYL- see EPR200
STANNANE, ETHYNYLENEBIS(CARBONYLOXY)BIS(TRIPHENYL- see BLS900
STANNANE, FLUOROTRIPHENYL- see TMV850
STANNANE, FLUOROTRIS(p-CHLOROPHENYL)- see TNG050
STANNANE, (ISOCYANATO)TRIBUTYL- see THY750
STANNANE, (ISOPROPYLSUCCINYLOXY)TRIBUTYL- see TIE600
STANNANE, (p-NITROPHENOXY)TRIBUTYL- see NII200
STANNANE, OXYBIS(DIBUTYL(2,4,5-TRICHLOROPHENOXY)- see OPE100
STANNANE, TRIBUTYL((4-CHLOROBUTYRYL)OXO)- see TID000
STANNANE, TRIBUTYLISOCYANATO- see THY750
STANNANE, TRIBUTYLSALICYLOYLOXY see SAM000
STANNANE, TRICHLOROOCTYL- see OGG000
STANNANE, TRIISOPROPYL(UNDECANOYLOXY)- see TKT850
STANNIC BROMIDE see TGB750
STANNIC CHLORIDE, anhydrous (DOT) see TGC250
STANNIC CHLORIDE, pentahydrate (DOT) see TGC282
STANNIC CHLORIDE PENTAHYDRATE see TGC282
STANNIC CITRIC ACID see SLI335
STANNIC DICHLORIDE DIIODIDE see TGC280
STANNIC IODIDE see TGD750
STANNIC PHOSPHIDE (DOT) see TGE500
STANNOCHLOR see TGC275
STANN OMF see DVK200
STANNOPLUS see DAE600
STANNORAM see DAE600
STANNOUS CHLORIDE (FCC) see TGC000
STANNOUS CHLORIDE DIHYDRATE see TGC275
STANNOUS CITRIC ACID see TGC285
STANNOUS DIBUTYLDITRIFLUOROACETATE see BLN750
STANNOUS FLUORIDE see TGD100
STANNOUS IODIDE see TGD500
STANNOUS OLEATE see OAZ000
STANNOUS OXALATE see TGE250
STANNOUS POTASSIUM TARTRATE see TGE750
STANNOUS SULFATE see TGF010
STANNOXYL see TGE300
STANNPLOUS see DAE600
STANN RC 40F see DEJ100
STANOMYCETIN see CDP250
STANOZOLOL see AOO400

STANSIN see SNN500
ST. ANTHONY'S TURNIP see FBS100
STANWHITE 500 see CAT775
STANZAMINE see POO750
STAPENOR see DSQ800, MNV250
STAPHCILLIN V see MNV250
STAPHOBRISTOL-250 see SLJ000
STAPHYBIOTIC see SLJ000
STAPHYBIOTIC I see SLJ050
STAPHYLEX see CHJ000
STAPHYLOCOAGULASE see CMY725
STAPHYLOCOCCAL PHAGE LYSATE see SLJ100
STAPHYLOMYCIN see VRA700, VRF000
STAPHYLOMYCINE see VRA700
STAPYOCINE see VRF000
STAR see GGA000
STARAMIC 747 see SLJ500
STAR ANISE OIL see AOU250
STAR CACTUS (HAWAII) see AGV875
α-STARCH see SLJ500
STARCH (OSHA) see SLJ500
STARCH see SLJ500
STARCH, CORN see SLJ500
STARCH DUST see SLJ500
STARCH GUM see DBD800
STARCH HYDROXYETHYL ETHER see HLB400
STAR DUST see CNE750
STARFOL BS-100 see BSL600
STARFOL GMS 450 see OAV000
STARFOL GMS 600 see OAV000
STARFOL GMS 900 see OAV000
STARFOL IPM see IQN000
STARFOL IPP see IQW000
STAR HYACINTH see SLH200
STARIFEN see EOK000
STARLEX L see CAW850
STARLICIDE see CLK230
STAR-OF-BETHLEHEM see SLJ650
STARSOL No. 1 see AQQ500
STARVE-ACRE see FBS100
STA-RX 1500 see SLJ500
ST52-ASTA see DKA200
STATEX see CBT750
STATEX N 550 see CBT750
STATHION see PAK000
STATOBEX see DKE800
STATOMIN see WAK000
STATOMIN MALEATE see DBM800
STAUFFER N-2404 see CJD650
STAUFFER N-3049 see EPY000
STAUFFER N-3051 see BSG000
STAUFFER N-4328 see EOO000
STAUFFER N-4548 see MOB250
STAUFFER ASP-51 see TED500
STAUFFER B-10094 see CON300
STAUFFER B-11163 see DTQ800
STAUFFER CAPTAN see CBG000
STAUFFER MV242 see MLG250
STAUFFER MV-119A see DLK200
STAUFFER N 521 see DSB200
STAUFFER N 2790 see FMU045
STAUFFER R-1,303 see TNP250
STAUFFER R-1492 see MQH750
STAUFFER R 1504 see PHX250
STAUFFER R 1608 see EIN500
STAUFFER R-1910 see EID500
STAUFFER R-2061 see PNF500
STAUFFER R-3413 see DTV400
STAUFFER R-4,572 see EKO500
STAUFFER R-25788 see DBJ600
STAUFFER R-3442-S see EMC000
STAURODERM see FMQ000
STAVELAN CINATY (CZECH) see TGE250
STAVELAN SODNY see SIY500
STAVINOR 30 see CAX350
STAVINOR 40 see BAO825
STAVINOR see LHQ100
STAVINOR 1300SN see DEJ100
STAVINOR SN 1300 see DEJ100
STAY-FLO see SHF500
STAYPRO WS 7 see HBO650
STAZEPIN see DCV200
STB see SKE000
ST. BENNET'S HERB see PJJ300
STCA see TII500

ST 1396 CLORHIDRATO (SPANISH) see SBN475
ST-1512 DIHYDROCHLORIDE see HFG600
STEADFAST see CCP000
STEAPSIN see GGA800
STEARALKONIUM CHLORIDE see DTC600
STEARAMIDE see OAR000
STEARAMINE see OBC000
STEARATE CHROMIC CHLORIDE see CMH300
STEARATO CHROMIC CHLORIDE see CMH300
STEARATO-CHROMIC CHLORIDE COMPLEX see CMH300
STEARATOCHROMIUM CHLORIDE see CMH300
STEAREX BEADS see SLK000
STEARIC ACID see SLK000
STEARIC ACID, ALUMINIUM SALT see AHA250
STEARIC ACID, ALUMINUM SALT see AHH825
STEARIC ACID, AMMONIUM SALT see ANU200
STEARIC ACID, BARIUM CADMIUM SALT (4:1:1) see BAI800
STEARIC ACID, BARIUM SALT see BAO825
STEARIC ACID, CADMIUM SALT see OAT000
STEARIC ACID with CYCLOHEXYLAMINE (1:1) see CPH500
STEARIC ACID-2,3-EPOXYPROPYL ESTER see SLK500
STEARIC ACID, ETHYL ESTER see EPF700
STEARIC ACID, 2-ETHYLHEXYL ESTER see OFU300
STEARIC ACID, ISOBUTYL ESTER see IJN100
STEARIC ACID, ISOHEXADECYL ESTER see IKC050
STEARIC ACID, ISOPROPYL ESTER see IRL100
STEARIC ACID, LEAD SALT see LDX000
STEARIC ACID, LITHIUM SALT see LHQ100
STEARIC ACID, MONOESTER with ETHYLENE GLYCOL see EJM500
STEARIC ACID, MONOESTER with GLYCEROL see OAV000
STEARIC ACID, MONOESTER with 1,2-PROPANEDIOL see SLL000
STEARIC ACID, SODIUM SALT see SJV500
STEARIC ACID, TETRADECYL ESTER see TCB100
STEARIC ACID, ZINC SALT see ZMS000
STEARIC MONOGLYCERIDE see OAV000
STEARIX ORANGE see PEJ500
STEARIX RED 4B see SBC500
STEARIX RED 4S see SBC500
STEARIX SCARLET see OHA000
STEAROL see OAX000
γ-STEAROLACTONE see SLL400
STEARONITRILE see SLL500
STEAROPHANIC ACID see SLK000
1-STEAROYLAZIRIDINE see SLM000
STEAROYL ETHYLENEIMINE see SLM000
STEAR YELLOW JB see DOT300
STEARYL ALCOHOL see OAX000
STEARYL ALCOHOL EO (10) see SLM500
STEARYL ALCOHOL EO (20) see SLN000
STEARYL ALCOHOL condensed with 10 moles ETHYLENE OXIDE see SLM500
STEARYL ALCOHOL condensed with 20 moles ETHYLENE OXIDE see SLN000
n-STEARYLAMINE see OBC000
STEARYLAMINE see OBC000
STEARYLDIMETHYLAMINE see DTC400
STEARYLDIMETHYLBENZYLAMMONIUM CHLORIDE see DTC600
STEARYLTRIMETHYLAMMONIUM CHLORIDE see TLW500
STEAWHITE see TAB750
STEBAC see DTC600
STECKAPFUL (GERMAN) see SLV500
STECKER ASC-4 see THW750
STECLIN see TBX000
STECLIN HYDROCHLORIDE see TBX250
STEDIRIL see NNL500
STEDIRIL D see NNL500
STEINAMID DL 203 S see BKE500
STEINAPOL TLS 40 see SON000
STEINBUHL YELLOW see BAK250, CAP750
STELADONE see CDS750
STELAZINE see TKE500, TKK250
STELAZINE DIHYDROCHLORIDE see TKK250
STELLAMINE see DJS200
STELLARID see POB500
STELLAZINE see TKE500
STELLITE 8 see CNA750
STELLITE 21 see CNA750
STELLITE 23 see CNA750
STELLITE 25 see CNA750
STELLITE 27 see CNA750
STELLITE 30 see CNA750
STELLITE 31 see CNA750
STELLITE 36 see CNA750
STELLITE see VSK000
STELLITE 8A see CNA750

STELLITE C see CNA750
STELLON PINK see PKB500
STEMAX see PAL600
STEMETIL see PMF500
STEMETIL DIMALEATE see PMF250
STEMONE see EMP550
STEMPOR see MHC750
STENANDIOL see AOO410
STENDOMYCIN A see SLN509
STENDOMYCIN SALICLATE see SLO000
STENEDIOL see AOO475
STENIBELL see AOO475
STENOLON see DAL300, MPN500
STENOLONE see DAL300
STENOSINE see DXE600
STENOSTERONE see AOO475
STENOVASAN see TEP500
STENTAL EXTENTABS see EOK000
STEPAN D-50 see IQN000
STEPAN D-70 see IQW000
STEPAN C40 see MLC800
STEPANOL WAQ see SIB600
STEPANOL WAT see SON000
STEPHANIA HERNANDIFOLIA Walp., extract see SLO100
STERAFFINE see OAX000
STERAL see HCL000
STERAMIDE see SNP500
STERANDRYL see TBG000
STERANE see PMA000
STERAQ see DAQ800
STERASKIN see HCL000
STERAZINE see PPP500
STERCORIN see DKW000
STERCULIA FOETIDA OIL see SLO450
STERCULIA GUM see KBK000, SLO500
STERICOL see XKA000
STERIDO see BIM250
STERIGMATOCYSTIN see SLP000
STERILE TETRACAINE HYDROCHLORIDE see TBN000
STERILIZING GAS ETHYLENE OXIDE 100% see EJN500
STERINOL see BEO000
STERINOLU (POLISH) see BEO000
STERISEAL LIQUID #40 see SGM500
STERLING see CBT750, SFT000
STERLING N 765 see CBT750
STERLING NS see CBT750
STERLING SO 1 see CBT750
STERLING WAQ-COSMETIC see SIB600
STERLING WAT see SON000
STERNITE 30 see SMQ500
STERNITE ST 30VL see SMQ500
STEROGENOL see HCP800
STEROGYL see VSZ100
STEROIDS see SLP199
STEROLAMIDE see TKP500
STEROLONE see PMA000
STERONYL see MPN500
STEROX DJ SULFACTANT see TAX500
STESIL see CKE750
STESOLID see DCK759
STEVIA REBAUDIANA Bertoni, extract see SLP350
ST 1396 HYDROCHLORID (GERMAN) see SBN475
ST 1396 HYDROCHLORIDE see SBN475
STIBACETIN see SLP500
STIBAMINE GLUCOSIDE see NCL000
STIBANATE see AQH800, AQI250
STIBANILIC ACID see SLP600
STIBANOSE see AQH800, AQI250
STIBATIN see AQH800, AQI250
STIBENYL see SLP500
STIBIC ANHYDRIDE see AQF750
2,2′,2″-(STIBILIDYNETRIS(THIO)TRIS-BUTANEDIOIC ACID HEXALITHIUM SALT
   see LGU000
STIBILIUM see DKA600
STIBINE see SLQ000
STIBINE OXIDE, TRIPHENYL- see TMQ550
STIBINE SULFIDE, TRIPHENYL- see TMQ600
STIBINE, TRICHLORO- see AQC500
STIBINE, TRIFLUORO-(9CI) see AQE000
STIBINE, TRIS(DODECYLTHIO)- see TNH850
STIBINOL see AQH800, AQI250
STIBIUM see AQB750
STIBNAL see AQI750
STIBOCAPTATE see AQD750
((2-STIBONOPHENYL)THIO)ACETIC ACID see CCH250

((2-STIBONOPHENYL)THIO)ACETIC ACID DIETHANOLAMINE SALT see
   SLQ500
STIBOPHEN see AQH500
STIBOSAMIN see SLP600
STIBSOL see SEU000
STIBUNAL see AQI750
STICKMONOXYD (GERMAN) see NEG100
STICKSTOFFDIOXID (GERMAN) see NGR500
STICKSTOFFLOST see BIE500
STICKSTOFFWASSERSTOFFSAEURE (GERMAN) see HHG500
STIGMANOL BROMIDE see POD000
STIGMANOL METHYL SULFATE see DQY909
5-α-STIGMASTANE-3-β,5,6-β-TRIOL 3-MONOBENZOATE see SLQ625
STIGMATELLIN see SLQ650
STIGMOSAN BROMIDE see POD000
STIGMOSAN METHYL SULFATE see DQY909
STIK see NAK500
STIKSTOFDIOXYDE (DUTCH) see NGR500
STIL see DKA600
STILALGIN see GGS000
STILBAMIDINE DIHYDROCHLORIDE see SLS500
STILBEN (GERMAN) see SLR000
4-N-STILBENAMINE see SLQ900
4-STILBENAMINE see SLQ900
trans-4-N-STILBENAMINE see AMO000
STILBENE 3 see TGE150
STILBENE see SLR000
STILBENE, (E)- see SLR100
(E)-STILBENE see SLR100
trans-4-STILBENE see AMO000
trans-STILBENE see SLR100
α-STILBENECARBONITRILE see DVX600
4,4′-STILBENEDIAMINE see SLR500
4,4′-STILBENEDICARBOXAMIDINE see SLS000
4,4′-STILBENEDICARBOXAMIDINE, DIHYDROCHLORIDE see SLS500
STILBENE, α-α′-DIETHYL-4,4′-DIMETHYL-, (E)- see DRK500
2,2′-STILBENEDISULFONIC ACID, 4,4′-BIS((4-ANILINO-6-((2-HYDROXYETH-
   YL)METHYLAMINO)-s-TRIAZIN- 2-YL)AMINO)-, DISODIUM SALT see TGE100
2,2′-STILBENEDISULFONIC ACID, 4,4′-BIS((4-ANILINO-6-MORPHOLINO-s-TRIAZ-
   IN-2-YL)AMINO)-, DISODIUM SALT see CMP200
2,2′-STILBENEDISULFONIC ACID, 4,4′-BIS((4,6-DIANILINO-s-TRIAZIN-2-
   YL)AMINO)-, DISODIUM SALT see DXB450
2,2′-STILBENEDISULFONIC ACID, 4,4′-BIS(4-PHENYL-2H-1,2,3-TRIAZOL-2-YL)-,
   DIPOTASSIUM SALT see BLG100
2,2′-STILBENEDISULFONIC ACID, 4,4′-BIS(4-PHENYL-1,2,3-TRIAZOL-2-YL), DI-
   POTASSIUM SALT see BLG100
2,2′-STILBENEDISULFONIC ACID, 4,4′-DIAMINO- see FCA100
2,2′-STILBENEDISULFONIC ACID, 4,4′-DIAMINO-, DISODIUM SALT see
   FCA200
2-STILBENESULFONIC ACID, 4-(7-SULFO-2H-NAPHTHO(1,2-d)TRIAZOL-2-YL)-,
   DISODIUM SALT see DXG025
4-STILBENYL-N,N-DIETHYLAMINE see DKA000
STILBENYL-N,N-DIMETHYLAMINE see DUB800
STILBESTROL see DKA600
STILBESTROL DIETHYL DIPROPIONATE see DKB000
STILBESTROL DIMETHYL ETHER see DJB200
STILBESTROL DIPHOSPHATE see DKA200
STILBESTROL DIPROPIONATE see DKB000
STILBESTROL and PRANONE see EEI050
STILBESTROL PROPIONATE see DKB000
STILBESTRONATE see DKB000
STILBESTRONE see DKA600
STILBETIN see DKA600
STILBIOCINA see SMB000
STILBOEFRAL see DKA600
STILBOESTROFORM see DKA600
STILBOESTROL see DKA600
STILBOESTROL DIPROPIONATE see DKB000
STILBOFAX see DKB000
STILBOFOLLIN see DKA600
STILBOL see DKA600
STILBRON see SBG500
STILKAP see DKA600
STILNY see CGA500
STILON see PJY500
STILPHOSTROL see DKA200
STIL-ROL see DKA600
STILRONATE see DKB000
STIMAMIZOL HYDROCHLORIDE see LFA020
STIMATONE see FMS875
STIMINOL see DJS200
STIMULAN see AOB250
STIMULEX see DBA800, DBB000
STIMULEXIN see SLU000
STIMULIN see DJS200

STIMULINA see GFO050
STINERVAL see PFC500, PFC750
STINK DAMP see HIC500
STINKING NIGHTSHADE see HAQ100
STINKING WEED see CNG825
STINKING WILLIE see RBA400
STIPEND see CMA100
STIPINE see SEH000
STIPOLAC see TDE750
STIPTANON see EDU500
STIRAMATO see PFJ000
STIROFOS see RAF100
STIROLO (ITALIAN) see SMQ000
STIROPHOS see RAF100
ST. JOHN'S BREAD see LIA000
STOCKADE see RLF350
STODDARD SOLVENT see SLU500
STOIKON see BCA000
STOMACAIN see DTL200
STOMP see DRN200
STONE RED see IHD000
STOPAETHYL see DXH250
STOP-DROP see NAK500
STOPETHYL see DXH250
STOPETYL see DXH250
STOPGERME-S see CKC000
STOP-MOLD see SFV500
STOPMOLD B see BGJ750
STOP-SCALD see SAV000
STOPSPOT see PFM500
STOPTON ALBUM see SNM500
STOVAINE see AOM000
STOVARSAL see ABX500
STOVARSOL see ABX500
STOVARSOLAN see ABX500
STOXIL see DAS000
STPHCILLIN A BANYU see DGE200
STPP see SKN000
STR see SMD000
STRABOLENE see DYF450
STRAIGHT-CHAIN ALKYL BENZENE SULFONATE see LGF825
STRAIGHT-RUN KEROSENE see KEK000
STRAIGHT RUN MIDDLE DISTILLATE see GBW000
STRAMONA (ITALIAN) see SLV500
STRAMONIUM see SLV500
STRATHION see PAK000
STRAWBERRY ALDEHYDE see ENC000
STRAWBERRY BUSH see CCK675
STRAWBERRY RED A GEIGY see FMU080
STRAWBERRY TOMATO see JBS100
STRAW OIL see LIK000
STRAZINE see ARQ725
STREL see DGI000
STREPAMIDE see SNM500
STREPCEN see SLW500
STREPCIN see SLY500
STREP-GRAN see SLY500
STREPSULFAT see SLY500
STREPTAGOL see SNM500
d-STREPTAMINE, O-3-AMINO-3-DEOXY-α-d-GLUCOPYRANOSYL-(1-6)-O-(2,6-DI-AMINO-2,3,4,6-TETRADEOXY-α-d-erythro-HEXOPYRANOSYL-(1-4))-N'-(4-AMI-NO-2-HYDROXY-1-OXOBUTYL)-2-DEOXY-, (S)-, SULFATE (salt) see HAI600
STREPTASE see SLW450
STREPTOCLASE see SNM500
STREPTOCOCCAL FIBRINOLYSIN see SLW450
STREPTOGRAMIN see VRF000
STREPTOGRAMIN B see VRA700
STREPTOKINASE see SLW450
STREPTOKINASE (ENZYME-ACTIVATING) see SLW450
STREPTOL see SNM500
STREPTOLYDIGIN see SLW475
STREPTOMICINA (ITALIAN) see SLW500
STREPTOMYCES GRISEUS PROTEASE see PMJ100
STREPTOMYCES GRISEUS PROTEINASE see PMJ100
STREPTOMYCES PEUCETIUS see DAC000
STREPTOMYCIN see SLW500
STREPTOMYCIN A see SLW500
STREPTOMYCIN C see SLX500
STREPTOMYCIN compounded with CALCIUM CHLORIDE (1:1) see SLY000
STREPTOMYCIN CALCIUM CHLORIDE see SLY000
STREPTOMYCIN and DIHYDROSTREPTOMYCIN see SLY200
STREPTOMYCINE see SLW500
STREPTOMYCIN SESQUISULFATE see SLY500
STREPTOMYCIN SULFATE see SLY500
STREPTOMYCIN SULPHATE see SLZ000

STREPTOMYCIN SULPHATE B.P. see SLY500
STREPTOMYCINUM see SLW500
STREPTOMYCIN, SULFATE (1:3) SALT see SLZ000
STREPTOMYZIN (GERMAN) see SLW500
STREPTONIGRAN see SMA000
STREPTONIGRIN see SMA000
STREPTONIGRIN METHYL ESTER see SMA500
STREPTONIVICIN see SMB000
STREPTOREX see SLY500
STREPTOSIL see SNM500
STREPTOSILTHIAZOLE see TEX250
STREPTOTHRICIN F see RAG300
STREPTOTHRICIN VI see RAG300
STREPTOVARICIN C see SMB850
STREPTOVIRUDIN see SMC375
STREPTOVITACIN A see SMC500
STREPTOVITACIN E 73 see ABN000
STREPTOZOCIN see SMD000
STREPTOZONE see SNM500
STREPTOZOTICIN see SMD000
STREPTROCIDE see SNM500
STREPVET see SLY500
STRESNIL see FLU000
STRESSON see BON400
STREUNEX see BBQ500
STRIADYNE see ARQ500
STRICNINA (ITALIAN) see SMN500
STRICYLON see NCW000
STRIPED ALDER see WCB000
STROBANE see MIH275, PJQ250, TBC500
STROBANE-T-90 see CDV100
STRONCYLATE see SML500
STRONTIUM see SMD500
STRONTIUM ACETATE see SME000
STRONTIUM ACETYLIDE see SME100
STRONTIUM ARSENITE, solid (DOT) see SME500
STRONTIUM ARSENITE see SME500
STRONTIUM BROMIDE see SMF000
STRONTIUM CHLORATE, solid or solution (DOT) see SMF500
STRONTIUM CHLORATE see SMF500
STRONTIUM CHLORIDE see SMG500
STRONTIUM CHLORIDE, HEXAHYDRATE see SMH525
STRONTIUM CHROMATE (1:1) see SMH000
STRONTIUM CHROMATE 12170 see SMH000
STRONTIUM CHROMATE (VI) see SMH000
STRONTIUM COMPOUNDS see SMH500
STRONTIUM DICHLORIDE HEXAHYDRATE see SMH525
STRONTIUM FLUOBORATE see SMI000
STRONTIUM FLUORIDE see SMI500
STRONTIUM FLUOSILICATE see SMJ000
STRONTIUM IODIDE see SMJ500
STRONTIUM MONOSULFIDE see SMM000
STRONTIUM(II) NITRATE (1:2) see SMK000
STRONTIUM NITRATE (DOT) see SMK000
STRONTIUM PEROXIDE see SMK500
STRONTIUM PHOSPHIDE see SML000
STRONTIUM SALICYLATE see SML500
STRONTIUM SULFIDE see SMM000
STRONTIUM SULPHIDE see SMM000
STRONTIUM YELLOW see SMH000
STROPHANTHIDIN see SMM500
STROPHANTHIDIN-d-CYMAROSID (GERMAN) see CQH750
STROPHANTHIDIN-3-(6-DEOXY-α-l-MANNOPYRANOSIDE) see CNH780
STROPHANTHIDIN-β-d-DIGITOXOSID (GERMAN) see HAN800
k-STROPHANTHIDINE see SMM500
STROPHANTHIDINE see SMM500
STROPHANTHIDIN-GLUCOCYMAROSID (GERMAN) see SMN002
STROPHANTHIDIN-α-l-RHAMNOSIDE see CNH780
α-l-STROPHANTHIDOL-3,6-DEOXY-MANNOPYRANOSIDE see CNH785
β-k-STROPHANTHIN see SMN002
k-STROPHANTHIN see SMN000
STROPHANTHIN see SMN000
STROPHANTHIN G see OKS000
STROPHANTHIN K (crystalline) see SMN002
STROPHANTHIN K see SMN002
β-STROPHANTHOBIOSIDE, STROPHANTHIDIN-3 see SMN002
STROPHANTHUS GRATUS Franch., leaf and stem bark extract see SMN100
STROPHOPERM see OKS000
STROSPESID (GERMAN) see SMN275
STROSPESIDE see SMN275
STROSPESIDE-16-FORMATE see VIZ200
STROSPESIDE TRIFORMATE see TKL175
STRUMACIL see MPW500
STRUMAZOLE see MCO500
STRYCHININE SULFATE see SMP000

STRYCHNIDIN-10-ONE see SMN500
STRYCHNIDIN-10-ONE, 2,3-DIMETHOXY-(9CI) see BOL750
STRYCHNIDIN-10-ONE, SULFATE (2:1) see SMP000
STRYCHNIN (GERMAN) see SMN500
STRYCHNINE, solid and liquid (DOT) see SMN500
STRYCHNINE see NOG800, SMN500
STRYCHNINE, 2,3-DIMETHOXY- see BOL750
STRYCHNINE MONONITRATE see SMO000
STRYCHNINE NITRATE see SMO000
STRYCHNINE SALT (solid) see SMO500
STRYCHNINE SALTS (DOT) see SMO500
STRYCHNINE SULFATE (2:1) see SMP000
STRYCHNOS see SMN500
STRYCHNOS NUX-VOMICA see NOG800
STRYCIN see SLZ000
STRYON 686 see SMQ500
STRYPHNON see MGC350
STRYPHNONE see MGC350
STRYPTIRENAL see VGP000
STRZ see SMD000
STS 153 see PEE600
STS 557 see SMP400
ST 5066/S (GERMAN) see AMU750
STS see EMT500, SIO000
ST-1512 SULFATE see HFG650
STUDAFLUOR see SHF500
STUGERON see CMR100
ST 30UL see SMQ500
STUNTMAN see DMC600
STUPENONE see DLX400
STURCAL D see CAT775
STUTGERON see CMR100
STUTGIN see CMR100
STX DIHYDROCHLORIDE see SBA500
STYLOMYCIN see AEI000, POK250
STYPHNIC ACID see SMP500
STYPHNONE see MGC350
STYPNON see MGC350
STYPTIC WEED see CNG825
STYRACIN see CMQ850
STYRAFOIL see SMQ500
STYRAGEL see SMQ500
STYRALLYL ACETATE see SMP600
STYRALLYL ALCOHOL see PDE000
STYRALLYL PROPIONATE see PFR000
STYRALYL ALCOHOL see PDE000
STYRAMATE see PFJ000
STYREEN (DUTCH) see SMQ000
STYREN (CZECH) see SMQ000
STYREN-ACRYLONITRILEPOLYMER see ADY500
STYRENE see SMQ000
STYRENE-ACRYLONITRILE COPOLYMER see ADY500
STYRENE-1,3-BUTADIENE COPOLYMER see SMR000
STYRENE-BUTADIENE COPOLYMER see SMR000
STYRENE-BUTADIENE POLYMER see SMR000
STYRENE, CHLORO- see CLE600
STYRENE EPOXIDE see EBR000
STYRENE GLYCOL see SMQ100
STYRENE and MALEIC ANHYDRIDE, alternate co-polymer see SEA500
STYRENE-MALEIC ANHYDRIDE COPOLYMER see SEA500
STYRENE-MALEIC ANHYDRIDE POLYMER see SEA500
STYRENE-MALEIC ANHYDRIDE RESIN see SEA500
STYRENE, p-METHYL- see VQK700
STYRENE MONOMER (ACGIH) see SMQ000
STYRENE MONOMER, inhibited (DOT) see SMQ000
STYRENE, β-NITRO- see NMC100
STYRENE-7,8-OXIDE see EBR000
STYRENE OXIDE see EBR000
STYRENE POLYMER see SMQ500
STYRENE POLYMER with 1,3-BUTADIENE see SMR000
STYRENE POLYMERS see SMQ500
STYRENE, α-β,β-TRIFLUORO- see TKH310
STYRENE-VINYLPYRROLIDINONE POLYMER see VQK595
STYRENE-VINYLPYRROLIDONE COPOLYMER see VQK595
STYRENE-N-VINYLPYRROLIDONE POLYMER see VQK595
STYREX C see SMQ500
STYROCELL PM see SMQ500
STYROFAN 2D see SMQ500
STYROFLEX see SMQ500
STYROFOAM see SMQ500
STYROL (GERMAN) see SMQ000
STYROLE see SMQ000
STYROLENE see SMQ000
STYROLFENOL see PFD400
STYROLUX see SMQ500

STYROLYL ACETATE see SMP600
STYROLYL ALCOHOL see SMQ100
STYROLYL PROPIONATE see PFR000
STYROMAL 5 see SEA500
STYROMAL 30 see SEA500
STYROMAL see SEA500
STYRON 475 see SMQ500
STYRON 492 see SMQ500
STYRON 666 see SMQ500
STYRON 678 see SMQ500
STYRON 679 see SMQ500
STYRON 683 see SMQ500
STYRON 685 see SMQ500
STYRON 690 see SMQ500
STYRON 69021 see SMQ500
STYRON see SMQ000, SMQ500
STYRON 440A see SMQ500
STYRON 470A see SMQ500
STYRON 475D see SMQ500
STYRONE see CMQ740
STYRON GP see SMQ500
STYRON 666K27 see SMQ500
STYRON PS 3 see SMQ500
STYRON T 679 see SMQ500
STYRON 666U see SMQ500
STYRON 666V see SMQ500
STYROPIAN see SMQ500
STYROPIAN FH 105 see SMQ500
STYROPOL HT 500 see SMQ500
STYROPOL IBE see SMQ500
STYROPOL JQ 300 see SMQ500
STYROPOL KA see SMQ500
STYROPOR see SMQ000, SMQ500
STYRYL 430 see AMO250
trans-4'-STYRYLACETANILIDE see SMR500
ar-STYRYLACETOPHENONE see MPL000
p-STYRYLANILINE see SLQ900
5-STYRYL-3,4-BENZOPYRENE see SMS000
6-STYRYL-BENZO(a)PYRENE see SMS000
STYRYL CARBINOL see CMQ740
STYRYL METHYL KETONE see SMS500
STYRYL OXIDE see EBR000
N-(p-STYRYLPHENYL)ACETOHYDROXAMIC ACETATE see ABW500
N-(p-STYRYLPHENYL)ACETOHYDROXAMIC ACID see SMT000
trans-N-(p-STYRYLPHENYL)ACETOHYDROXAMIC ACID see SMT500
N-(p-STYRYLPHENYL)ACETOHYDROXAMIC ACID ACETATE see ABW500
trans-N-(p-STYRYLPHENYL)ACETOHYDROXAMIC ACID, COPPER(2+) COM-
   PLEX see SMU000
(E)-N-(p-STYRYLPHENYL)HYDROXYLAMINE see HJA000
trans-N-(4-STYRYLPHENYL)HYDROXYLAMINE see HJA000
trans-N-(p-STYRYLPHENYL)HYDROXYLAMINE see HJA000
STZ see SMD000
SU 2000 see CBT500
SU 3088 see CDY000
SU 3118 see RCA200
SU-4885 see MCJ370
SU 5864 see GKQ000
SU-5864 see GKS000
SU 5879 see CFY000
SU 8341 see CPR750
SU 9064 see MDW100
SU-9064 see MQR200
SU-10568 see DRC600
SU-13320 see TCN250
SU-13437 see MCB500
SU 21524 see PJA220
SUANOVIL see SLC300
SUAVITIL see BCA000
SUBACETATE LEAD see LCH000
SUBAMYCIN see TBX250
SUBARI see HGC500
SUBCHLORIDE of MERCURY see MCW000
SUBDUE see MDM100
SUBDUE 2E see MDM100
SUBDUE 5SP see MDM100
SUBERANE see COX500
SUBERON see SMV000
SUBERONE see SMV000
SUBERONE OXIME see SMV500
SUBERONISOXIM (GERMAN) see SMV500
SUBERONITRILE see OCW050
4,4'-(SUBEROYLBIS(IMINO-p-PHENYLENEIMINO))BIS(1-ETHYLPYRIDINIUM) DI-
   BROMIDE see SMW000
4,4'-(SUBEROYLBIS(IMINO-p-PHENYLENEIMINO))BIS(1-METHYLPYRIDINI-
   UM)DIBROMIDE see PPD500

SUBICARD see PBC250
SUBITEX see BRE500
SUBITOL see IAD000
SUBLIMAT (CZECH) see MCY475
SUBLIMAZE see PDW750
SUBLIMAZE CITRATE see PDW750
SUBLIMED SULFUR see SOD500
SUBLINGULA see HAQ500
SUBSTANCE DS see AJX500
SUBSTANCE F see MIW500
SUBSTANCE H 20 see MFG600
SUBSTANCE H 36 see MEJ775
SUBSTANCE II see AFG750
SUBSTANZ DS see AJX500
SUBSTANZ NR. 2135 see CKE250
SUBSTANZ NR. 1602 (GERMAN) see CJR959
SUBSTANZ NR. 1766 (GERMAN) see CIS000
SUBSTANZ NR. 1925 (GERMAN) see CKE000
SUBSTANZ NR. 1934 (GERMAN) see DSQ600
SUBTIGEN see SMW400
SUBTILIN see SMW500
SUBTILISIN CARLSBURG see BAC000
SUBTILISIN (9CI, ACGIH) see BAC000
SUBTILISIN NOVO see BAC000
SUBTILISINS (ACGIH) see BAB750
SUBTILISINS BPN see BAB750
SUBTILOPEPTIDASE A see BAC000
SUBTILOPEPTIDASE B see BAC000
SUBTILOPEPTIDASE BPN' see BAC000
SUBTILOPEPTIDASE C see BAC000
SUBTOSAN see PKQ250
SUCARYL see CPQ625
SUCARYL ACID see CPQ625
SUCARYL CALCIUM see CAR000
SUCARYL SODIUM see SGC000
SUCCARIL see SGC000, SJN700
SUCCICURAN see HLC500
SUCCIMAL see ENG500
SUCCIMER see DNV800
SUCCIMITIN see ENG500
SUCCINALDEHYDE DISODIUM BISULFITE see SMX000
SUCCINATE SODIQUE de 21-HYDROXYPREGNANDIONE (FRENCH) see VJZ000
SUCCINATO ACIDO DI 1-p-CLOROFENILPENTILE (ITALIAN) see CKI000
SUCCINATO de CLORANFENICOL (SPANISH) see CDP725
SUCCINBROMIMIDE see BOF500
SUCCINCHLORIMIDE see SND500
SUCCINIBROMIMIDE see BOF500
SUCCINIC ACID see SMY000
SUCCINIC ACID ANHYDRIDE see SNC000
SUCCINIC ACID BIS(β-DIMETHYLAMINOETHYL) ESTER BISMETHIODIDE see BJI000
SUCCINIC ACID BIS(β-DIMETHYLAMINOETHYL) ESTER, DIHYDROCHLORIDE see HLC500
SUCCINIC ACID BIS(β-DIMETHYLAMINOETHYL)ESTER DIMETHOCHLORIDE see HLC500
SUCCINIC ACID, BIS(2-(HEXYLOXY)ETHYL) ESTER see SMZ000
SUCCINIC ACID-α-BUTYL-p-CHLOROBENZYL ESTER see CKI000
SUCCINIC ACID, CADMIUM SALT (1:1) see CAI750
SUCCINIC ACID DI-n-BUTYL ESTER see SNA500
SUCCINIC ACID, DIBUTYL ESTER see SNA500
SUCCINIC ACID DICHLORIDE see SNG000
SUCCINIC ACID DIESTER with CHOLINE see CMG250
SUCCINIC ACID DIESTER with CHOLINE CHLORIDE see HLC500
SUCCINIC ACID, DIESTER with CHOLINE IODIDE see BJI000
SUCCINIC ACID, DIETHYL ESTER see SNB000
SUCCINIC ACID, DI-2-HEXYLOXYETHYL ESTER see SMZ000
SUCCINIC ACID, 2,3-DIHYDROXY- see TAF750
SUCCINIC ACID, 2,3-DIMERCAPTO- see DNV610
SUCCINIC ACID, DIMETHYL ESTER see SNB100
SUCCINIC ACID-2,2-DIMETHYLHYDRAZIDE see DQD400
SUCCINIC ACID DINITRILE see SNE000
SUCCINIC ACID DIPROPYL ESTER see DWV800
SUCCINIC ACID, DISODIUM SALT see SJW100
SUCCINIC ACID, (4-ETHOXY-1-NAPTHYLCARBONYLMETHYL) ESTER see SNB500
SUCCINIC ACID, HYDROXY- see MAN000
SUCCINIC ACID, O-ISOPROPYL-O'-TRIBUTYLSTANNYL ESTER see TIE600
SUCCINIC ACID, MERCAPTO-, DIETHYL ESTER, S-ESTER with O,O-DIMETHYLPHOSPHOROTHIOATE see OPK250
SUCCINIC ACID MONOESTER with 7-CHLORO-1,3-DIHYDRO-3-HYDROXY-5-PHENYL-2H-1,4-BENZODIAZEPIN-2-ONE see CFY500
SUCCINIC ACID-α-MONOESTER with d-threo-(−)-2,2-DICHLORO-N-(β-HYDROXY-α-(HYDROXYMETHYL)-p-NITROPHENETHYL)ACETAMIDE see CDP725

SUCCINIC ACID MONOESTER with N-(2-ETHYLHEXYL)-3-HYDROXYBUTYRAMIDE see BPF825
SUCCINIC ACID PEROXIDE (DOT) see SNC500
SUCCINIC ACID, SULFO-, 1,4-DIHEXYL ESTER, SODIUM SALT (8CI) see DKP800
SUCCINIC ACID, SULFO-, 1,4-DIOCTYL ESTER, SODIUM SALT see SOD300
SUCCINIC ANHYDRIDE see SNC000
SUCCINIC ANHYDRIDE, (TETRAPROPENYL)- see TEC600
SUCCINIC CHLORIDE see SNG000
SUCCINIC-1,1-DIMETHYL HYDRAZIDE see DQD400
SUCCINIC DINITRILE see SNE000
SUCCINIC IMIDE see SND000
SUCCINIC PEROXIDE see SNC500
SUCCINIMIDE see SND000
SUCCINIMIDE, N,2-DIMETHYL-2-PHENYL- see MLP800
SUCCINOCHLORIMIDE see SND500
SUCCINOCHOLINE see CMG250
SUCCINODINITRILE see SNE000
SUCCINONITRILE see SNE000
4′-SUCCINOYLAMINO-2,3′-DIMETHYLAZOBENZENE see SNF500
SUCCINOYL CHLORIDE see SNG000
SUCCINOYLCHOLINE see CMG250
SUCCINOYLCHOLINE CHLORIDE see HLC500
SUCCINOYL DIAZIDE see SNF000
4′-SUCCINYLAMINO-2,3′-DIMETHYLAZOBENZOL see SNF500
SUCCINYL-ASTA see HLC500
SUCCINYLBISCHOLINE see CMG250
SUCCINYL BISCHOLINE CHLORIDE see HLC500
SUCCINYLBISCHOLINE DICHLORIDE see HLC500
SUCCINYL CHLORIDE see SNG000
SUCCINYLCHOLINE CHLORIDE see HLC500
SUCCINYLCHOLINE DICHLORIDE see HLC500
SUCCINYLCHOLINE HYDROCHLORIDE see HLC500
SUCCINYL DICHLORIDE see SNG000
SUCCINYLDICHOLINE see CMG250
SUCCINYLDICHOLINE CHLORIDE see HLC500
o,o-SUCCINYLDICHOLINE IODIDE see BJI000
SUCCINYLDICHOLINE IODIDE see BJI000
SUCCINYLFORTE see HLC500
SUCCINYLNITRILE see SNG500
SUCCINYL OXIDE see SNC000
SUCCINYL PEROXIDE see SNC500
SUCCITIMAL see MNZ000
SUCHAR 681 see CBT500
SUCKER PLUCKER see ODE000
SUCKER-STUFF see DMC600
SUCOSTRIN see HLC500
SUCOSTRIN CHLORIDE see HLC500
SUCRA see SJN700
SUCRAPHEN see SPC500
SUCRE EDULCOR see BCE500
SUCRETS see HFV500
SUCRETTE see BCE500
SUCROFER see IHG000
SUCROL see EFE000
SUCROSA see SGC000
SUCROSE see SNH000
SUCROSE ACETATE ISOBUTYRATE see SNH050
SUCROSE ACETOISOBUTYRATE see SNH050
SUCROSE, DIACETATE HEXAISOBUTYRATE see SNH050
SUCROSE, OCTAACETATE see OAF100
SUCROSE OCTANITRATE (dry) (DOT) see SNH125
SUCROSE, OCTANITRATE see SNH125
SUCROSE, POLYALLYL ETHER, POLYMER with ACRYLIC ACID see ADW300
SUDAFED see POH000, POH250
SUDAN AX see XRA000
SUDAN BROWN RR see NBG500
SUDAN BROWN YR see NBG500
SUDAN G see CMP600, OHA000
SUDAN G III see OHA000
SUDAN GREEN 4B see BLK000
SUDAN III see OHA000
SUDAN III (G) see OHA000
SUDAN IV see SBC500
SUDAN ORANGE see XRA000
SUDAN ORANGE G see CMP600
SUDAN ORANGE R see PEJ500
SUDAN ORANGE RPA see XRA000
SUDAN ORANGE RRA see XRA000
SUDAN P see SBC500
SUDAN P III see OHA000
SUDAN R see CMS242
SUDAN RED 290 see CMS242
SUDAN RED see XRA000

SUDAN RED 7B see EOJ500
SUDAN RED 4BA see SBC500
SUDAN RED BB see SBC500
SUDAN RED BBA see SBC500
SUDAN RED G (6CI) see CMS242
SUDAN RED III see OHA000
SUDAN RED IV see SBC500
SUDANROT 7B see EOJ500
SUDAN SCARLET 6G see XRA000
SUDAN X see XRA000
SUDAN YELLOW see DOT300
SUDAN YELLOW AR see CMP600
SUDAN YELLOW R see PEI000
SUDAN YELLOW RRA see AIC250
SUDINE see SNN300
SU 4885 DITARTRATE see MJJ000
SUESSETTE see SGC000
SUESSTOFF see EFE000
SUESTAMIN see SGC000
SUFENTA see SNH150
SUFENTANIL CITRATE see SNH150
SUFENTANYL see SNH150
SUFENTANYL CITRATE see SNH150
SUFFIX 25 see EGS000
SUFFIX see EGS000
SUFRALEM see AOO490
SUGAI CHRYSOIDINE see PEK000
SUGAI CONGO RED see SGQ500
SUGAI FAST SCARLET G BASE see NMP500
SUGAI TARTRAZINE see FAG140
SUGANYL see AHO250
SUGAR see SNH000
SUGAR-BOWLS see CMV390
SUGARIN see SGC000
SUGAR of LEAD see LCV000
SUGARON see SGC000
SUGRACILLIN see BFD000
SU 8629 HYDROCHLORIDE see AKL000
SU 8842 HYDROCHLORIDE see MQR200
SUICALM see FLU000
SUISYNCHRON see MLJ500
SUKHTEH see SNH450
SULADYNE see PDC250
SULAMYD see SNP500
SULAMYD SODIUM see SNQ000
SUL ANILINOVA (CZECH) see BBL000
SULBENICILLIN see SNV000
SULBENICILLIN DISODIUM see SNV000
SULCEPHALOSPORIN see CCS550
SULCOLON see PPN750
SULCONAZOLE NITRATE see SNH480
SULDIXINE see SNN300
SULEMA (RUSSIAN) see MCY475
SULF-10 see SNP500
SULFAAFENAZOLO (ITALIAN) see AIF000
SULFABENZAMIDE see SNH800
SULFABENZID see SNH800
SULFABENZIDE see SNH800
SULFABENZOYLAMIDE see SNH800
SULFABENZPYRAZINE see QTS000
SULFABID see AIF000
SULFABUTIN see AIE750
4-SULFACARBAMIDE see SNQ550
SULFACARBAMIDE see SNQ550
SULFACET see SNP500
SULFACETAMIDE see SNP500
SULFACETAMIDE, SODIUM see SNQ000
SULFACETAMIDE, SODIUM SALT see SNQ000
SULFACETIMIDE see SNP500
SULFACID BRILLIANT BLUE 6J see ADE500
SULFACID BRILLIANT GREEN 1B see FAE950
SULFACID LIGHT YELLOW 5RL see SGP500
SULFACOMBIN see SNI000
SULFACTIN see BAD750
SULFACYL see SNP500
SULFACYL SODIUM see SNQ000
SULFACYL SODIUM SALT see SNQ000
SULFACYL SOL see SNQ000
SULFACYL SOLUBLE see SNQ000
SULFADENE see BDF000
SULFADIAMINE see SNY500
SULFADIAZINE see PPP500
SULFADIAZINE SILVER see SNI425
SULFADIAZINE SILVER SALT see SNI425
SULFADIMERAZINE see SNJ000

SULFADIMETHOXIN see SNN300
SULFADIMETHOXINE see SNN300
SULFADIMETHOXYDIAZINE see SNN300
SULFADIMETHOXYPYRIMIDINE see SNI500
SULFADIMETHYLDIAZINE see SNJ000
SULFADIMETHYLISOXAZOLE see SNN500
SULFADIMETHYLOXAZOLE see AIE750
4-SULFA-2,6-DIMETHYLPYRIMIDINE see SNJ350
SULFADIMETHYLPYRIMIDINE see SNJ000
SULFADIMETINE see SNJ000, SNJ350
SULFADIMETOSSINA (ITALIAN) see SNN300
SULFADIMETOXIN see SNN300
SULFADIMEZINE see SNJ000
SULFADIMIDINE see SNJ000
SULFADIMIDINE SODIUM see SJW500
SULFADINE see SNJ000
SULFADOXINE see AIE500
SULFADSIMESINE see SNJ000
SULFAETHYLTHIADIAZOLE SODIUM see SJW300
SULFAFURAZOL see SNN500
SULFAGAN see SNN500
SULFAGUANIDINE see AHO250
SULFAGUINE see AHO250
SULFA-ISODIMERAZINE see SNJ000
SULFAISODIMERAZINE see SNJ350
SULFAISODIMIDINE see SNJ000, SNJ350
SULFALENE see MFN500
SULFALENE N-METHYLGLUCAMINE SALT see SNJ400
SULFALEX see AKO500
SULFALLATE see CDO250
SULFALONE see SNW500
SULFAMATE see ANU650
SULFAMERADINE see ALF250
SULFAMERAZIN see ALF250
SULFAMERAZINE SODIUM see SJW475
SULFAMETER SODIUM see MFO000
SULFAMETHALAZOLE see SNK000
SULFAMETHAZINE SODIUM see SJW500
SULFAMETHIAZINE see SNJ000
SULFAMETHIN see SNJ000, SNJ350
SULFAMETHIZOL see MPQ750
SULFAMETHIZOLE see MPQ750
SULFAMETHOMIDINE see MDU300
SULFAMETHOPYRAZINE see MFN500
SULFAMETHOXAZOL see SNK000
SULFAMETHOXAZOLE see SNK000
SULFAMETHOXAZOL-TRIMETHOPRIM see TKX000
SULFAMETHOXYDIAZINE SODIUM see MFO000
SULFAMETHOXYPYRAZINE see MFN500
3-SULFA-6-METHOXYPYRIDAZINE see AKO500
SULFA-5-METHOXYPYRIMIDINE SODIUM SALT see MFO000
SULFAMETHYLDIAZINE see ALF250
SULFAMETHYLISOXAZOLE see SNK000
SULFAMETHYLIZOLE see MPQ750
SULFAMETHYLTHIADIAZOLE see MPQ750
SULFAMETOMIDINE see MDU300
SULFAMETOPYRAZINE see MFN500
SULFAMETORINE SODIUM see MFO000
SULFAMETOSSIPIRIDAZINA (ITALIAN) see MFN500
SULFAMETOXIPIRIDAZINE see AKO500
SULFAMETOXYPYRIDAZIN (GERMAN) see MFN500
SULFAMEZATHINE see SNJ000
SULFAMIC ACID see SNK500
SULFAMIC ACID, DIMETHYL-, 5-BUTYL-2-(ETHYLAMINO)-6-METHYL-4-PYRIMI-
    DINYL ESTER see BRJ000
SULFAMIC ACID, MONOAMMONIUM SALT see ANU650
SULFAMIDIC ACID see SNK500
*p*-SULFAMIDOANILINE see SNM500
SULFAMIDYL see SNM500
SULFAMINSAEURE (GERMAN) see ANU650
SULFAMONOMETHOXIN see SNL800
SULFAMONOMETHOXINE see SNL800
SULFAMONOMETOXINE N-METHYLGLUCAMINE SALT see SNL830
SULFAMOXOLE see AIE750
SULFAMOXOLE and TRIMETHOPRIM see SNL850
SULFAMOXOLE-TRIMETHOPRIM mixture see SNL850
SULFAMOXOLUM see AIE750
4-SULFAMOYLBENZOIC ACID see SNL840
*p*-SULFAMOYLBENZOIC ACID see SNL840
5′-SULFAMOYL-2-CHLOROADENOSINE see CEF125
3-SULFAMOYLMETHYL-1,2-BENZISOXAZOLE see BCE750
N-(5-SULFAMOYL-1,3,4-THIADIAZOL-2-YL)ACETAMIDE see AAI250
SULFAMUL see TEX250
N-SULFAMYLBENZAMIDE see SNH800
*p*-SULFAMYLBENZOIC ACID see SNL840

SULFAMYLON ACETATE see MAC000
SULFAN see SOR500
SULFANA see SNM500
SULFANALONE see SNM500
SULFAN BLUE see ADE500
SULFANIL see SNM500
SULFANILACETAMIDE see SNP500
SULFANILAMIDE see SNM500
3-SULFANILAMIDE-6-METHOXYPYRIDAZINE see AKO500
SULFANILAMIDE, N¹-(5-ETHYL-1,3,4-THIADIAZOL-2-YL)-, MONOSODIUM SALT
   see SJW300
SULFANILAMIDE, N¹-2-PYRIDYL-, MONOSODIUM SALT see PPO250
SULFANILAMIDE, N¹-2-PYRIDYL-, N¹-SODIUM deriv. see PPO250
4-SULFANILAMIDO-3,6-DIMETHOXYPYRIDAZINE see DON700
6-SULFANILAMIDO-2,4-DIMETHOXYPYRIMIDINE see SNN300
4-SULFANILAMIDO-5,6-DIMETHOXYPYRIMIDINE see AIE500
5-SULFANILAMIDO-3,4-DIMETHYL-ISOXAZOLE see SNN500
6-SULFANILAMIDO-2,4-DIMETHYLPYRIMIDINE see SNJ350
4-SULFANILAMIDO-2,6-DIMETHYLPYRIMIDINE see SNJ350
2-SULFANILAMIDO-4,6-DIMETHYLPYRIMIDINE see SNJ000
2-SULFANILAMIDO-3-METHOXYPYRAZINE see MFN500
6-SULFANILAMIDO-3-METHOXYPYRIDAZINE see AKO500
3-SULFANILAMIDO-6-METHOXYPYRIDAZINE see AKO500
2-SULFANILAMIDO-5-METHOXYPYRIMIDINE SODIUM SALT see MFO000
3-SULFANILAMIDO-5-METHYLISOXAZOLE see SNK000
2-SULFANILAMIDO-5-METHYL-1,3,4-THIADIAZOLE see MPQ750
5-SULFANILAMIDO-1-PHENYLPYRAZOLE see AIF000
SULFANILAMIDOPYRIMIDINE see PPP500
2-SULFANILAMIDOPYRIMIDINE SODIUM SALT see MRM250
2-SULFANILAMIDOQUINOXALINE see QTS000
2-SULFANILAMIDOTHIAZOLE see TEX250
2-SULFANILAMIDOTHIAZOLE SODIUM SALT see TEX500
SULFANILCARBAMID see SNQ550
N-SULFANILCARBAMIDE see SNQ550
SULFANILGUANIDINE see AHO250
m-SULFANILIC ACID see SNO000
o-SULFANILIC ACID see SNO100
SULFANILIC ACID see SNN600
SULFANILSAEURE see SNN600
N-SULFANILYLACETAMIDE SODIUM see SNQ000
N-SULFANILYLACETAMIDE, SODIUM SALT see SNQ000
2-SULFANILYL AMINOPYRIDINE see PPO000
2-SULFANILYLAMINOPYRIMIDINE see PPP500
2-(SULFANILYLAMINO)THIAZOLE see TEX250
N-SULFANILYLBENZAMIDE see SNH800
SULFANILYLCARBAMIC ACID, METHYL ESTER see SNQ500
SULFANILYLGUANIDINE see AHO250
N¹-SULFANILYL-N²-BUTYLCARBAMIDE see BSM000
N¹-SULFANILYL-N²-BUTYLUREA see BSM000
N-SULFANILYL-N'BUTYLUREE (FRENCH) see BSM000
SULFANILYLUREA see SNQ550
SULFANO see AIE750
SULFANOL NP 1 see SNQ700
SULFANTHRENE ORANGE R see CMU815
SULFANTHRENE ORANGE RS see CMU815
N-SULFANYLACETAMIDE see SNP500
SULFANYLHARNSTOFF see SNQ550
SULFANYLUREE see SNQ550
SULFAPHENAZOLE see AIF000
SULFAPHENAZON see AIF000
SULFAPHENYLPIPAZOL see AIF000
SULFAPHENYLPYRAZOLE see AIF000
SULFAPOL see DXW200
SULFAPOLU (POLISH) see DXW200
SULFAPYRAZINEMETHOXINE see MFN500
SULFAPYRAZINEMETHOXYNE see MFN500
SULFAPYRIDAZINE see AKO500
2-SULFAPYRIDINE see PPO000
SULFAPYRIDINE see PPO000
SULFAPYRIDINE NEUTRAL SOLUBLE see DXF400
SULFAPYRIDINE SODIUM see PPO250
SULFAPYRIDINE, SODIUM SALT see PPO250
SULFAPYRIMIDIN (GERMAN) see PPP500
2-SULFAPYRIMIDINE see PPP500
SULFAQUINOXALINE see QTS000
SULFARLEM see AOO490
SULFARSENOBENZENE see SNR000
SULFARSENOL see SNR000
SULFARSPHENAMINE see SNR000
SULFARSPHENAMINE BISMUTH see BKV250
SULFASALAZINE see PPN750
SULFASAN see BKU500
SULFASAN R POWDER see BKU500
SULFASOL see SNN300
SULFASOMIDINE see SNJ350

SULFASOXAZOLE see SNN500
SULFASTOP see SNN300
SULFATE d'ATROPINE (FRENCH) see ARR500
SULFATE de CUIVRE (FRENCH) see CNP250
SULFATED AMYLOPECTIN see AOM150
SULFATED CASTOR OIL see TOD500
SULFATE DIMETHYLIQUE (FRENCH) see DUD100
SULFATE ESTER of N-HYDROXY-N-2-FLUORENYL ACETAMIDE see FDV000
SULFATE MERCURIQUE (FRENCH) see MDG500
SULFATE de METHYLE (FRENCH) see DUD100
SULFATE de NICOTINE (FRENCH) see NDR500
SULFATEP see SOD100
SULFATE de PLOMB (FRENCH) see LDY000
SULFATES see SNS000
SULFATE de ZINC (FRENCH) see ZNA000
SULFATHALIDINE see PHY750
SULFATHIAZOL see TEX250
SULFATHIAZOLE (USDA) see TEX250
2-SULFATHIAZOLE SODIUM see TEX500
SULFATOBIS(DIMETHYLSELENIDE)PLATINUM(II) HYDRATE see SNS100
cis-SULFATO-1,2-DIAMINOCYCLOHEXANEPLATINUM(II) see SCF025
trans(+)-SULFATO-1,2-DIAMINOCYCLOHEXANEPLATINUM(II) see SCF075
trans(−)-SULFATO-1,2-DIAMINOCYCLOHEXANEPLATINUM(II) see SCF050
SULFAUREA see SNQ550
SULFAVIGOR see AIE750
SULFAZOLE see SNN500
SULFDURAZIN see AKO500
SULFENAMIDE M see BDG000
SULFENAMIDE TS see CPI250
SULFENAX see CPI250
SULFENAX CB 30 see CPI250
SULFENAX CB see CPI250
SULFENAX CB/K see CPI250
SULFENAX MOB (CZECH) see BDF750
SULFENAZIN see TFM100
SULFENONE see CKI625
SULFENTANIL see SNH150
SULFERROUS see FBN100, FBO000
SULFIDAL see SOD500
SULFIDES see SNT000
SULFIDINE see PPO000
SULFIMEL DOS see DJL000
SULFINPYRAZINE see DWM000
SULFINYLBIS(METHANE) see DUD800
SULFINYL BROMIDE see SNT200
SULFINYL CHLORIDE see TFL000
SULFINYL CYANAMIDE see SNT300
SULFIDOTRICHLORID FOSFORECNY see TFO000
SULFISIN see SNN500
SULFISOMEZOLE see SNK000
SULFISOMIDIN see SNJ000, SNJ350
SULFISOMIDINE see SNJ000, SNJ350
SULFISOXAZOLE see SNN500
SULFITE CELLULOSE see CCU150
SULFITE LYE see HIC600
SULFITES see SNT500
SULFIZOLE see SNN500
SULFMETHOXIPIRIDAZINE see AKO500
SULFMIDIL see AIE750
SULFOACETIC ACID see SNU000
SULFOACETIC ACID 1-DODECYL ESTER, SODIUM SALT see SIB700
SULFOACETIC ACID DODECYL ESTER S-SODIUM SALT see SIB700
2-SULFOANTHRAQUINONE SODIUM SALT see SER000
5-SULFOBENZEN-1,3-DICARBOXYLIC ACID see SNU500
SULFOBENZIDE see PGI750
3-SULFOBENZIDINE see BBY250
o-SULFOBENZIMIDE see BCE500
o-SULFOBENZOIC ACID IMIDE see BCE500
SULFOBENZYLPENICILLIN see SNV000
α-SULFOBENZYLPENICILLIN DISODIUM see SNV000
SULFOBROMOPHTHALEIN see HAQ600
SULFOBROMOPHTHALEIN SODIUM see HAQ600
SULFOBROMPHTHALEIN see HAQ600
SULFOCARBANILIDE see DWN800
SULFOCARBOLIC ACID see HJH500
SULFOCIDINE see SNM500
SULFOCILLIN see SNV000
SULFODIAMINE see SNY500
SULFODIMESIN see SNJ000
SULFODIMEZINE see SNJ000
SULFODOR (CZECH) see EPH000
SULFOETHANOIC ACID see SNU000
2-SULFOETHYLAMINE see TAG750
SULFOGAL see AOO490
SULFOGENOL see IAD000

SULFO GREEN J see FAF000
SULFOGUANIDINE see AHO250
SULFOGUENIL see AHO250
SULFOLAN see SNW500
SULFOLANE see SNW500
3-SULFOLENE see DMF000
SULFOL-3-ENE see DMF000
β-SULFOLENE see DMF000
4-SULFOMETANILIC ACID see AIE000
SULFOMETURON METHYL see SNW550
SULFONA see SOA500
SULFONAL see ABD500
SULFONAL NP 1 see SNQ700
SULFONAMIDE see SNM500
4-SULFONAMIDE-4'-DIMETHYLAMINOAZOBENZENE see SNW800
SULFONAMIDE P see SNM500
p-SULFONAMIDOBENZOIC ACID see SNL840
4-SULFONAMIDO-3'-METHYL-4'-AMINOAZOBENZENE see SNX000
2-SULFONAMIDOTHIAZOLE see TEX250
SULFONA P see AOO800
SULFO-3-NAPHTHALENEFURANE see SNX250
2-(4-SULFO-1-NAPHTHYLAZO)-1-NAPHTHOL-4-SULFONIC ACID, DISODIUM
    SALT see HJF500
SULFONATED CASTOR OIL see TOD500
SULFONATES see SNY000
o-SULFONBENZOIC ACID IMIDE SODIUM SALT see SJN700
p-SULFONDICHLORAMIDOBENZOIC ACID see HAF000
SULFONE, BIS(4-FLUORO-3-NITROPHENYL) see DKH250
SULFONE-2,4,4',5-TETRACHLORODIPHENYL see CKM000
SULFONETHYLMETHANE see BJT750
SULFONIC ACID, MONOCHLORIDE see CLG500
SULFONIMIDE see CBF800
SULFONINE ACID BLUE R see ADE750
SULFONINE RED G see CMM320
SULFONINE RED GN see CMM320
SULFONINE RED GS see CMM320
SULFONINE RED RS see NAO600
SULFONINE RED SG see CMM320
SULFONINE YELLOW CSR see CMM759
SULFONIUM, TRIMETHYL-, IODIDE, OXIDE see TMG800
SULFONMETHANE see ABD500
N-SULFONOXY-AAF see FDV000
N-SULFONOXY-N-ACETYL-2-AMINOFLUORENE see FDV000
SULFONPHTHAL see PDO800
4',4'''-SULFONYLBIS(ACETANILIDE) see SNY500
4,4'-SULFONYLBISACETANILIDE see SNY500
p,p'-SULFONYLBISACETANILIDE see SNY500
1,1'-SULFONYLBIS(4-AMINOBENZENE) see SOA500
3,3'-SULFONYLBIS(ANILINE) see SOA000
4,4'-SULFONYLBISANILINE see SOA500
4,4'-SULFONYLBISBENZAMINE see SOA500
p,p-SULFONYLBISBENZAMINE see SOA500
p,p-SULFONYLBISBENZENAMINE see SOA500
1,1'-SULFONYLBISBENZENE see PGI750
1,1'-(SULFONYLBIS(4,1-PHENYLENEIMINO))BIS(1-DEOXY-1-SULFO-d-GLUCITOL)
    DISODIUM SALT see AOO800
4,4'-(SULFONYLBIS(4,1-PHENYLENEOXY))BISBENZENAMINE see SNZ000
4,4'-SULFONYLBIS(4-PHENYLENEOXY)DIANILINE see SNZ000
SULFONYL CHLORIDE see SOT000
SULFONYL CHLORIDE FLUORIDE see SOT500
3,3'-SULFONYLDIANILINE see SOA000
4,4'-SULFONYLDIANILINE see SOA500
p,p'-SULFONYLDIANILINE see SOA500
p,p'-SULFONYLDIANILINE-N,N'-DI-d-GLUCOSE SODIUM BISULFITE see
    AOO800
p,p'-SULFONYLDIANILINE N,N'-DIGLUCOSIDE DISODIUM DISULFONALTE see
    AOO800
4,4'-SULFONYLDIPHENOL see SOB000
O,O'-(SULFONYLDI-1,4-PHENYLENE) O,O,O',O'-TETRAMETHYL DIPHOSPHO-
    ROTHIOATE see PHN250
3-(SULFONYL)-o-((METHYLAMINO)CARBONYL)OXIME-2-BUTANONE see
    SOB500
N-(SULFONYL-p-METHYLBENZENE)-N'-N-BUTYLUREA see BSQ000
(3-β)-3-(SULFOOXY)-ANDROST-5-EN-17-ONE SODIUM SALT (9CI) see SJK400
SULFOPARABLUE see SOB600
SULFOPERAMIDIC ACID see HLN100
1-p-SULFOPHENYLAZO-2-HYDROXYNAPHTHALENE-6-SULFONATE, DISODIUM
    SALT see FAG150
4-SULFOPHENYLAZO-1-NAPHTHOL MONOSODIUM SALT see FAG010
1-p-SULFOPHENYLAZO-2-NAPHTHOL-6-SULFONIC ACID, DISODIUM SALT see
    FAG150
1-(p-SULFOPHENYL)-3-PHENYL-PYRAZOL (GERMAN) see PGE500
SULFOPLAN see SNN300
SULFOPON WA 1 see SIB600
SULFOQUANIDINE see AHO250

SULFORON see SOD500
SULFORTHOMIDINE see AIE500
5-SULFOSALICYLIC ACID see SOC500
SULFOSALICYLIC ACID see SOC500
SULFOSFAMIDE see SOD000
SULFOSUCCINIC ACID 1,4-DIOCTYL ESTER SODIUM SALT see SOD300
SULFOTEP see SOD100
SULFOTEPP see SOD100
SULFOTEX WALA see SIB600
SULFOTHIORINE see SKI500
5-SULFO-TOBIAS ACID see ALH000
SULFOTRIM see TKX000
SULFOTRIMIN see TKX000
SULFOTRINAPHTHYLENOFURAN, SODIUM SALT see SOD200
SULFOX-CIDE see ISA000
SULFOXIDE see ISA000
SULFOXONIUM, TRIMETHYL-, IODIDE see TMG800
SULFOXYL see BDS000, ISA000
2-(6-SULFO-2,4-XYLYLAZO)-1-NAPHTHOL-4-SULFONIC ACID, DISODIUM SALT
    see FAG050
SULFOXYPHENYLPYRAZOLIDINE see DWM000
SULFOZONA see AKO500
SULFRALEM see AOO490
SULFRAMIN 85 see DXW200
SULFRAMIN ACID 1298 see LBU100
SULFRAMIN 40 FLAKES see DXW200
SULFRAMIN 40 GRANULAR see DXW200
SULFRAMIN 1238 SLURRY see DXW200
SULFSTAT see MPQ750
SULFTECH see SJZ000
SULFUNE see AIE750
SULFUNO see AIE750
SULFUR see SOD500
SULFUR BROMIDE see SOE000
SULFUR CHLORIDE see SOG500, SON510
SULFUR CHLORIDE (DI) (DOT) see SON510
SULFUR CHLORIDE FLUORINE see SOF000
SULFUR CHLORIDE (MONO) see SOG500
SULFUR CHLORIDE OXIDE see TFL000
SULFUR CHLORIDE PENTAFLUORIDE see SOF000
SULFUR CHLOROPENTAFLUORIDE see SOF000
SULFUR COMPOUNDS see SOF500
SULFUR DECAFLUORIDE see SOQ450
SULFUR DICHLORIDE see SOG500
SULFUR DIFLUORIDE MONOXIDE see TFL250
SULFUR DIFLUORIDE OXIDE see TFL250
SULFUR DINITRIDE see SOH000
SULFUR DIOXIDE see SOH500
SULFUR DIOXIDE, solution see SOO500
SULFURE de 4-CHLOROBENZYLE et de 4-CHLOROPHENYLE (FRENCH) see
    CEP000
SULFURE de METHYLE (FRENCH) see TFP000
SULFURETED HYDROGEN see HIC500
SULFUR FLOWER (DOT) see SOD500
SULFUR FLUORIDE see SOI000
SULFUR FLUORIDE OXIDE see BLD000
SULFUR HALF-MUSTARD see CHC000
SULFUR HEXAFLUORIDE see SOI000
SULFUR HYDRIDE see HIC500
SULFURIC ACID, fuming <30% free sulfur trioxide (DOT) see SOI520
SULFURIC ACID, aromatic see SOI510
SULFURIC ACID, fuming see SOI520
SULFURIC ACID (mist) see SOI530
SULFURIC ACID see SOI500
SULFURIC ACID, fuming > or =30% free sulfur trioxide (DOT) see SOI520
SULFURIC ACID, sec-ALKYL ESTER, SODIUM SALT see SOJ000
SULFURIC ACID, ALUMINUM SALT (3:2) see AHG750
SULFURIC ACID, AMMONIUM NICKEL(2+) SALT (2:2:1) see NCY050
SULFURIC ACID, BARIUM SALT (1:1) see BAP000
SULFURIC ACID, BERYLLIUM SALT (1:1) see BFU250
SULFURIC ACID, BERYLLIUM SALT (1:1), TETRAHYDRATE see BFU500
SULFURIC ACID, BIS(METHYLMERCURY) SALT see BKS810
SULFURIC ACID, CADMIUM(2+) SALT see CAJ000
SULFURIC ACID, CADMIUM SALT, HYDRATE see CAJ250
SULFURIC ACID, CADMIUM SALT, TETRAHYDRATE see CAJ500
SULFURIC ACID, CALCIUM(2+) SALT, DIHYDRATE see CAX750
SULFURIC ACID, CHROMIUM(3+)POTASSIUM SALT(2:1:1), DODECAHYD-
    RATE see CMG850
SULFURIC ACID, CHROMIUM(3+) SALT (3:2) see CMK415
SULFURIC ACID, CHROMIUM SALT, BASIC see NBW000
SULFURIC ACID, CHROMIUM(3+) SALT (3:2), PENTADECAHYDRATE see
    CMK425
SULFURIC ACID, COBALT(2+) SALT (1:1) see CNE125
SULFURIC ACID, COPPER(2+) SALT (1:1) see CNP250
SULFURIC ACID, COPPER(2+) SALT, PENTAHYDRATE see CNP500

SULFURIC ACID, CYCLIC ETHYLENE ESTER see EJP000
SULFURIC ACID, DECYL ESTER, SODIUM SALT see SOK000
SULFURIC ACID, DIAMMONIUM SALT see ANU750
SULFURIC ACID, DICESIUM SALT see CDE500
SULFURIC ACID, DILITHIUM SALT see LHR000
SULFURIC ACID, DIMETHYL ESTER see DUD100
SULFURIC ACID, DIPOTASSIUM SALT see PLT000
SULFURIC ACID, DISODIUM SALT see SJY000
SULFURIC ACID, DITHALLIUM(1+) SALT (8CI, 9CI) see TEM000
SULFURIC ACID, DODECYL ESTER, TRIETHANOLAMINE SALT see SON000
SULFURIC ACID, GALLIUM SALT (3:2) see GBS100
SULFURIC ACID, INDIUM SALT see ICJ000
SULFURIC ACID, IRON(2⁺) SALT (1:1) see FBN100
SULFURIC ACID LANTHANUM(3+) SALT (3:2) see LBB000
SULFURIC ACID, LAURYL ESTER, AMMONIUM SALT see SOM500
SULFURIC ACID, LEAD(2+) SALT (1:1) see LDY000
SULFURIC ACID, LITHIUM SALT (1:2) see LHR000
SULFURIC ACID, MAGNESIUM SALT (1:1) HEPTAHYDRATE see MAJ500
SULFURIC ACID, MANGANESE(2+) SALT see MAU250
SULFURIC ACID, MANGANESE (2+) SALT (1:1) TETRAHYDRATE see MAU750
SULFURIC ACID, MERCURY(2+) SALT (1:1) see MDG500
SULFURIC ACID MIXTURE with SULFUR TRIOXIDE see SOI520
SULFURIC ACID, MONOAMMONIUM SALT see ANJ500
SULFURIC ACID, MONOANHYDRIDE with NITROUS ACID see NMJ000
SULFURIC ACID, MONODECYL ESTER, SODIUM SALT see SOK000
SULFURIC ACID, MONODODECYL ESTER, AMMONIUM SALT see SOM500
SULFURIC ACID, MONODODECYL ESTER, compounded with 2,2′,2″-NITRI-LOTRIETHANOL (1:1) see SON000
SULFURIC ACID, MONODODECYL ESTER, compounded with 2,2′,2″-NITRI-LOTRIS(ETHANOL) see SON000
SULFURIC ACID, MONODODECYL ESTER, SODIUM SALT see SIB600
SULFURIC ACID, MONO(2-(2-(2-(DODECYLOXY)ETHOXY)ETHOXY)ETHYL) ES-TER, SODIUM SALT see SIC000
SULFURIC ACID, MONOETHENYL ESTER, HOMOPOLYMER, POTASSIUM SALT see PKS250
SULFURIC ACID, MONO(2-ETHYLHEXYL)ESTER see ELB400
SULFURIC ACID, MONO(2-ETHYLHEXYL)ESTER, SODIUM SALT (8CI) see TAV750
SULFURIC ACID, MONONONYL ESTER, SODIUM SALT see SIX500
SULFURIC ACID, MONOOCTADECYL ESTER, SODIUM SALT see OBG100
SULFURIC ACID, MONOOCTYL ESTER, SODIUM SALT see OFU200
SULFURIC ACID, MONOPOTASSIUM SALT see PKX750
SULFURIC ACID, MONOSODIUM SALT see SEG800
SULFURIC ACID, MONOTETRADECYL ESTER, SODIUM SALT see SIO000
SULFURIC ACID, MYRISTYL ESTER, SODIUM SALT see SIO000
SULFURIC ACID, NICKEL(2+) SALT (1:1) see NDK500
SULFURIC ACID, NICKEL(2+)SALT see NDK500
SULFURIC ACID, NICKEL(2+) SALT, HEXAHYDRATE see NDL000
SULFURIC ACID, THALLIUM(1+) SALT (1:2) see TEM000
SULFURIC ACID, THALLIUM(2+) SALT see TEM250
SULFURIC ACID, THALLIUM(3+) SALT see TEM100
SULFURIC ACID, THALLIUM SALT see TEL750
SULFURIC ACID, TIN(2+) SALT (1:1) see TGF010
SULFURIC ACID, TITANIUM(4+) SALT (2:1) see TGH250
SULFURIC ACID, TITANIUM(4+) SALT see TGF220
SULFURIC ACID, VANADIUM SALT see VEA100
SULFURIC ACID, ZINC SALT (1:1) see ZNA000
SULFURIC ACID, ZINC SALT (1:1), HEPTAHYDRATE see ZNJ000
SULFURIC ACID, ZIRCONIUM(4+) SALT (2:1) see ZTJ000
SULFURIC ANHYDRIDE see SOR500
SULFURIC CHLOROHYDRIN see CLG500
SULFURIC OXIDE see SOR500
SULFURIC OXYCHLORIDE see SOT000
SULFURIC OXYFLUORIDE see SOU500
SULFURINE see MPQ750
SULFUR MONOBROMIDE see SOE000
SULFUR MONOCHLORIDE see SON510
SULFUR MUSTARD see BIH250
SULFUR MUSTARD GAS see BIH250
SULFURMYCIN A see SON520
SULFURMYCIN B see SON525
SULFUR NITRIDE see SOO000
SULFUROUS ACID see SOO500
SULFUROUS ACID ANHYDRIDE see SOH500
SULFUROUS ACID, 2-(p-tert-BUTYLPHENOXY)CYCLOHEXYL-2-PROPYNYL ES-TER see SOP000
SULFUROUS ACID, 2-(p-tert-BUTYLPHENOXY)-1-METHYLETHYL-2-CHLOROE-THYL ESTER see SOP500
SULFUROUS ACID, CALCIUM SALT (2:1) (8CI,9CI) see CAN000
SULFUROUS ACID, DIBENZYL ESTER see BFK000
SULFUROUS ACID, cyclic ester with 1,4,5,6,7,7-HEXACHLORO-5-NORBOR-NENE-2,3-DIMETHANOL see EAQ750
SULFUROUS ACID, MONOAMMONIUM SALT see ANB600
SULFUROUS ACID, MONOSODIUM SALT see SFE000

SULFUROUS ACID, SODIUM SALT (1:2) see SJZ000
SULFUROUS ANHYDRIDE see SOH500
SULFUROUS DICHLORIDE see TFL000
SULFUROUS OXIDE see SOH500
SULFUROUS OXYCHLORIDE see TFL000
SULFUROUS OXYFLUORIDE see TFL250
SULFUR OXIDE see SOH500
SULFUR OXIDE (N-FLUOROSULPHONYL)IMIDE see SOQ000
SULFUR PENTAFLUORIDE see SOQ450
SULFUR PHOSPHIDE see PHS000
SULFUR SELENIDE see SBT000
SULFUR SUBCHLORIDE see SON510
SULFUR TETRACHLORIDE see SOQ250
SULFUR TETRAFLUORIDE see SOR000
SULFUR THIOCYANATE see SOI200
SULFUR TRIOXIDE see SOR500
SULFUR TRIOXIDE, uninhibited (NA 1829) (DOT) see SOR500
SULFUR TRIOXIDE, inhibited (UN 1829) (DOT) see SOR500
SULFURYL AZIDE CHLORIDE see SOS500
SULFURYL CHLORIDE see SOT000
SULFURYL CHLORIDE FLUORIDE see SOT500
SULFURYL CHLOROFLUORIDE see SOT500
SULFURYL DIAZIDE see SOU000
SULFURYL FLUORIDE see SOU500
SULFURYL FLUOROCHLORIDE see SOT500
SULGIN see AHO250
SULINDAC see SOU550
SULINOL see SOU550
SULISOBENZONE see HLR700
SULKOL see SOD500
SULMET see SJW500, SNJ000
SULOCTIDIL see SOU600
SULOCTIDYL see SOU600
SULOCTON see SOU600
SULODYNE see PDC250
SULOUREA see ISR000
SULPELIN see SNV000
SULPHABUTIN see BOT250
SULPHACETAMIDE see SNP500
SULPHACETAMIDE SODIUM see SNQ000
SULPHADIAZINE see PPP500
SULPHADIMETHOXINE see SNN300
SULPHADIMETHYLISOXAZOLE see SNN500
SULPHADIMETHYLPYRIMIDINE see SNJ000
SULPHADIMIDINE see SNJ000
SULPHADIONE see SOA500
SULPHAFURAZ see SNN500
SULPHAGUANIDINE see AHO250
SULPHAMERAZINE see ALF250
SULPHAMETHALAZOLE see SNK000
SULPHAMETHIZOLE see MPQ750
SULPHAMETHOXAZOL see SNK000
SULPHAMETHOXAZOLE see SNK000
SULPHAMETHOXYPYRIDAZINE see AKO500
SULPHAMETHYLISOXAZOLE see SNK000
SULPHAMIC ACID (DOT) see SNK500
SULPHAMOPRINE see SNI500
SULPHAN BLUE see ADE500
SULPHANILAMIDE see SNM500
5-SULPHANILAMIDO-3,4-DIMETHYL-ISOXAZOLE see SNN500
3-SULPHANILAMIDO-5-METHYLISOXAZOLE see SNK000
SULPHANILIC ACID see SNN600
SULPHASALAZINE see PPN750
SULPHASIL see SNP500
SULPHASOMIDINE see SNJ350
SULPHATHIAZOLE see TEX250
SULPHAUREA see SNQ550
SULPHEIMIDE see CBF800
SULPHENAZOLE see AIF000
SULPHENONE see CKI625
SULPHENTAL see PDO800
SULPHISOMEZOLE see SNK000
SULPHISOXAZOL see SNN500
2-SULPHOBENZOIC IMIDE see BCE500
SULPHOBENZOIC IMIDE CALCIUM SALT see CAM750
SULPHOBENZOIC IMIDE, SODIUM SALT see SJN700
SULPHOBROMOPHTHALEIN see HAQ600
SULPHOBROMOPHTHALEIN SODIUM see HAQ600
SULPHOCARBONIC ANHYDRIDE see CBV500
SULPHOFURAZOLE see SNN500
SULPHO GREEN 2B see FAE950
SULPHOLANE see SNW500
SULPHON ACID BLUE R see ADE750
SULPHON ACID BLUE RA see ADE750
SULPHONAL NP 1 see SNQ700

1-(4-SULPHO-1-NAPHTHYLAZO)-2-NAPHTHOL-3,6-DISULPHONIC ACID, TRISO-
   DIUM SALT see FAG020
SULPHON-MERE see SOA500
SULPHONOL FAST RED R see CMM330
SULPHONOL RED PG see CMM320
SULPHONOL RED R see CMM330
SULPHONTHAL see PDO800
SULPHON YELLOW RS-CF see CMM759
1,1'-SULPHONYLBIS(4-AMINOBENZENE) see SOA500
4,4'-SULPHONYLBISBENZAMINE see SOA500
p,p-SULPHONYLBISBENZAMINE see SOA500
4,4'-SULPHONYLBISBENZENAMINE see SOA500
p,p-SULPHONYLBISBENZENAMINE see SOA500
p,p-SULPHONYLDIANILINE see SOA500
SULPHONYLDIANILINE see SOA500
1-p-SULPHOPHENYLAZO-2-NAPHTHOL-6-SULPHONIC ACID, DISODIUM SALT
   see FAG150
SULPHORMETHOXINE see AIE500
SULPHOS see PAK000
SULPHOXALINE see SNW500
SULPHOXIDE see ISA000
SULPHUR, molten (DOT) see SOD500
SULPHUR, lump or power (DOT) see SOD500
SULPHUR (DOT) see SOD500
SULPHUR DIOXIDE, LIQUEFIED (DOT) see SOH500
SULPHURIC ACID see SOI500
SULPHURIC ACID, CADMIUM SALT (1:1) see CAJ000
SULPHUR MUSTARD GAS see BIH250
SULPIRID see EPD500
SULPIRIDE see EPD500
SULPRIM see TKX000
SULPROFOS see SOU625
SULPROSTONE see SOU650
SULPYRID see EPD500
SULSOL see SOD500
SULTAMICILLIN TOSILATE see SOU675
SULTANOL see BQF500
SULTIRENE see AKO500
SULTOPRIDE see EPD100
SULTOPRIDE HYDROCHLORIDE see SOU725
SULTOSILATO de PIPERACINA (SPANISH) see PIK625
SULTOSILIC ACID, PIPERAZINE SALT see PIK625
SULTROPRIDE CHLORHYDRATE see SOU725
SULXIN see SNN300
SULZOL see TEX250
SUM 3170 see DCS200
SUMAC TANNIN see SOU750
SUMADIL see DVW700
SUMAPEN VK see PDT750
SUMATRA CAMPHOR see BMD000
SUMATRA YELLOW X 1940 see CMS212
SUMEDINE see ALF250
SUMETROLIM see TKX000
SUMIACRYL ORANGE G see CMM764
SUMICIDIN see FAR100
SUMICURE M see MJQ000
SUMIDUR 44V10 see PKB100
SUMIDUR 44V20 see PKB100
SUMIDUR 44VM see PKB100
SUMIFLOC CL8 see MCB050
SUMIFLY see FAR100
SUMIKANOL 508 see MCB050
SUMIKAPRINT YELLOW GFN see CMS212
SUMIKARON BLUE E-BL see CMP070
SUMIKARON BLUE E-FBL see CMP070
SUMIKARON BLUE R see CMP070
SUMIKARON RED E-FBL see AKI750
SUMIKATHENE see PJS750
SUMIKATHENE F 101-1 see PJS750
SUMIKATHENE F 210-3 see PJS750
SUMIKATHENE F 702 see PJS750
SUMIKATHENE G 201 see PJS750
SUMIKATHENE G 202 see PJS750
SUMIKATHENE G 701 see PJS750
SUMIKATHENE G 801 see PJS750
SUMIKATHENE G 806 see PJS750
SUMIKATHENE HARD 2052 see PJS750
SUMILEX see PMF750
SUMILIGHT BLACK G see CMN240
SUMILIGHT RED 4B see CMO885
SUMILIT BBM see BRP750
SUMILIT EXA 13 see PKQ059
SUMILIT PCX see AAX175
SUMILIZER TPS see DXG700
SUMIMAL 100 see MCB050

SUMIMAL 100C see MCB050
SUMIMAL M 22 see MCB050
SUMIMAL M 55 see MCB050
SUMIMAL M 70 see MCB050
SUMIMAL M 668 see MCB050
SUMIMAL M see MCB050
SUMIMAL M 65B see MCB050
SUMIMAL M 100C see MCB050
SUMIMAL M 504C see MCB050
SUMIMAL M 100D see MCB050
SUMIMAL M 40S see MCB050
SUMIMAL M 30W see MCB050
SUMIMAL M 40W see MCB050
SUMIMAL M 50W see MCB050
SUMIMAL M 62W see MCB050
SUMIMAL 40S see MCB050
SUMINE 2015 see DQP800
SUMINOL BRILLIANT SCARLET DH see CMM325
SUMINOL FAST BLUE PR see CMM100
SUMINOL FAST SKY BLUE B see CMM090
SUMINOL LEVELLING SKY BLUE R see CMM080
SUMINOL MILLING RED GRS see NAO600
SUMINOL MILLING RED RS see CMM330
SUMINOL MILLING YELLOW MR see CMM759
SUMINOL RED PG see CMM320
SUMINOL RED RS see NAO600
SUMIOXON see PHD750
SUMIOXONE see PHD750
SUMIPLEX LG see PKB500
SUMIPOWER see FAR100
SUMIREZ 607 see MCB050
SUMIREZ 613 see MCB050
SUMIREZ 614 see UTU500
SUMIREZ 615 see MCB050
SUMIREZ M613 see MCB050
SUMIREZ RESIN 613 see MCB050
SUMISCLEX see PMF750
SUMISET D see CNH125
SUMISORB 130 see HND100
SUMITAL see AMX750
SUMITEKKUSU REJIN 810 see UTU500
SUMITEX 260 see UTU500
SUMITEX 810 see UTU500
SUMITEX m³ see MCB050
SUMITEX FSK see DTG000
SUMITEX H 10 see PKP750
SUMITEX M6 see MCB050
SUMITEX M10 see MCB050
SUMITEX MC see MCB050
SUMITEX MK see MCB050
SUMITEX MW see MCB050
SUMITEX NF 113 see UTU500
SUMITEX NS see DTG000
SUMITEX RESIN 810 see UTU500
SUMITEX RESIN MC see MCB050
SUMITHIAN see DSQ000
SUMITHION S-ISOMER see MKC250
SUMITOL see BQC250
SUMITOL 80W see BQC250
SUMITOMO FAST SCARLET G see CMM325
SUMITOMO LIGHT GREEN SF YELLOWISH see FAF000
SUMITOMO PATENT PURE BLUE VX see ADE500
SUMITOMO PX 11 see PKQ059
SUMITOMO S 4084 see COQ399
SUMITOMO WOOL GREEN S see ADF000
SUMITOX see MAK700
SUMMETRIN see PAG500
SUMPOCAINE see SPB800
SUNAPTIC ACID B see NAR000
SUNAPTIC ACID C see NAR000
SUNBRELLA see AIH600
SUNCHOLIN see CMF350
SUNCHROMINE BLUE BLACK B see CMP880
SUNCHROMINE YELLOW GG see SIT850
SUNCIDE, nitrosated see PMY310
SUNCIDE see PMY300
SUNFRAL see FLZ050
SUNITOMO S 4084 see COQ399
SUNLIGHT 700 see CAT775
SUNLIGHT see SOV000
SUN ORANGE A GEIGY see FAG150
SUNRABIN see EAU075
SUNSET YELLOW see FAG150
SUNSET YELLOW BSS see FAG150
SUNSET YELLOW FCF see FAG150

SUNSET YELLOW FCF SUPRA see FAG150
SUNSET YELLOW FU see FAG150
SUNSET YELLOW FU SUPRA see FAG150
SUNSET YELLOW LAKE see FAG150
SUNSOFT O 30B see GGR200
SUNTOP M 300 see MCB050
SUNTOP M 420 see MCB050
SUNTOP M700 see MCB050
SUNTOP M701 see MCB050
SUNWAX 151 see PJS750
SUN YELLOW see FAG150
SUN YELLOW A-CE see FAG150
SUN YELLOW A-FDC see FAG150
SUN YELLOW EXTRA CONC. A EXPORT see FAG150
SUN YELLOW EXTRA PURE A see FAG150
SUN YELLOW FCF see FAG150
SUPACAL see CNG835
SUPARI, nut extract see BFW000
SUPARI (INDIA) see AQT650
SUPEOL see GGR200
SUPER 1500 see CAT775
SUPERACRYL AE see PKB500
SUPER AMIDE L-9A see BKE500
SUPERANABOLON see DYF450
SUPERAN BLUE AR see CMM070
SUPERBA see CBT750
SUPER-BECKAMINE see MCB050
SUPER-BECKAMINE G 821 see MCB050
SUPER-BECKAMINE J 1600 see MCB050
SUPER-BECKAMINE J 820 see MCB050
SUPER-BECKAMINE J 840 see MCB050
SUPER-BECKAMINE L 117 see MCB050
SUPER-BECKAMINE L 101 see MCB050
SUPER-BECKAMINE L 121 see MCB050
SUPER-BECKAMINE L 105 see MCB050
SUPER BECKOSOL ODL 131-60 see MCB050
SUPERBRESILINE see LFT800
SUPER-CAID see BMN000
SUPER-CARBOVAR see CBT750
SUPERCEL 3000 see USS000
SUPERCHROME BLUE BC see CMP880
SUPERCHROME BLUE BG see CMP880
SUPERCICLIN see PPY250
SUPERCOAT see CAT775
SUPER COBALT see CNA250
SUPERCOL see LIA000
SUPERCOL G.F. see GLU000
SUPERCOL U POWDER see GLU000
SUPERCORTIL see PLZ000
SUPERCORTYL see SOV100
SUPER COSAN see SOD500
SUPER CRAB-E-RAD-CALAR see CAM000
SUPER DAL-E-RAD see CAM000
SUPER DAL-E-RAD-CALAR see CAM000
SUPER-DENT see SHF500
SUPER-DE-SPROUT see DMC600
SUPER D WEEDONE see DAA800, TAA100
SUPER DYLAN see PJS750
SUPERELGETOL see DUT800
SUPERFLAKE ANHYDROUS see CAO750
SUPERFLOC see PKF750
SUPERFUSTEL see FBW000
SUPERFUSTEL K see FBW000
SUPERGAN see RDK000
SUPER GLUE see MIQ075
SUPER HARTOLAN see CMD750
SUPERIAN YELLOW R see SGP500
SUPERINONE see TDN750
SUPERIOR OIL see MQV855
SUPERIUONE (FRENCH) see TDN750
SUPERLYSOFORM see FMV000
SUPERMAN MANEB F see MAS500
SUPERMIKROKALCIT see CAD800
SUPERMITE see CAT775
SUPER MOSSTOX see MJM500
SUPER MULTIFEX see CAT775
SUPERNOX see DGI000
SUPERNYLITE BLUE BR see CMM090
SUPEROL see GGA000
SUPEROL RED C RT-265 see CHP500
SUPERORMONE CONCENTRE see DAA800
SUPEROX see BDS000
SUPEROXOL see HIB000
SUPER-PFLEX see CAT775
SUPERPHOSPHATE see SOV500

SUPERPREDNOL see SOW000
SUPER PRODAN see DXE000
SUPER RODIATOX see PAK000
SUPER-ROZOL see BMN000
SUPER 3S see CAT775
SUPERSEPTIL see SNJ000
SUPERSORBON IV see CBT500
SUPERSORBON S 1 see CBT500
SUPER-SPECTRA see CBT750
SUPER SPROUT STOP see DMC600
SUPER SSS see CAT775
SUPER SUCKER-STUFF see DMC600
SUPER SUCKER-STUFF HC see DMC600
SUPERTAH see CMY800
SUPER-TREFLAN see DUV600
SUPER VMP see NAI500
SUPICAINE AMIDE SULFATE see AJN750
SUPICANE AMIDE HYDROCHLORIDE see PME000
SUPININ see SOW500
SUPININE see SOW500
SUPLEXEDIL see BPM750
SUPONA see CDS750
SUPONE see CDS750
SUP'OPERATS see BMN000
SUPOTRAN see CKF500
SUPRA see IHD000
SUPRACAPSULIN see VGP000
SUPRACET BLUE GREEN B see DMM400
SUPRACET BRILLIANT BLUE BG see MGG250
SUPRACET BRILLIANT BLUE 2GN see TBG700
SUPRACET BRILLIANT RED 2B see AKE250
SUPRACET BRILLIANT VIOLET 3R see DBP000
SUPRACET DIAZO BLACK A see DPO200
SUPRACET FAST CRIMSON B see CMP080
SUPRACET FAST GREEN BLUE B see DMM400
SUPRACET FAST PINK 3B see DBX000
SUPRACET FAST PINK 2R see AKO350
SUPRACET FAST VIOLET B see DBY700
SUPRACET FAST YELLOW G see AAQ250
SUPRACET FAST YELLOW 2R see DUW500
SUPRACET FAST YELLOW 4R see CMP090
SUPRACET ORANGE R see AKP750
SUPRACET VIOLET 2B see AKP250
SUPRACET YELLOW RR see DUW500
SUPRACHOL see SGD500
SUPRACID see SNQ000
SUPRADIN see VGP000
SUPRAMIKE see BAP000
SUPRAMYCIN see TBX250
SUPRANEPHRANE see VGP000
SUPRANEPHRINE see VGP000
SUPRANEPHRIN SOLUTION see AES500
SUPRANOL see VGP000
SUPRANOL FAST RED 3G see CMM330
SUPRANOL FAST RED GG see CMM330
SUPRANOL FAST RED RX see NAO600
SUPRANOL FAST SCARLET GN see CMM320
SUPRANOL RED PBX-CF see CMM330
SUPRANOL RED PG-CF see CMM320
SUPRANOL RED R see CMM330
SUPRANOL SCARLET BN see CMM320
SUPRANOL SCARLET GS see CMM325
SUPRANOL YELLOW R see CMM759
SUPRARENIN see AES000, VGP000
SUPRARENIN HYDROCHLORIDE see AES500
SUPRASEC 1042 see PKB100
SUPRASEC see LIH000
SUPRASEC DC see PKB100
SUPRASIL see SCK600
SUPRASIL W see SCK600
SUPRASTIN see CKV625
SUPRATHEN see PJS750
SUPRATHEN C 100 see PJS750
SUPRATONIN see MQS100
SUPRAZO RED 4B see CMO885
SUPREFACT see LIU420
SUPREL see VGP000
SUPREMAL see DYE600
SUPREME DENSE see TAB750
SUPREXCEL RED 8BL see CMO885
SUP'R FLO see DXQ500
SUP 'R FLO see MAS500
SUP'R FLO FERBAM FLOWABLE see FAS000
SUPRIFEN see HKH500
SUPRIFENE see HKH500

SUPRIFEN PSB HYDROCHLORIDE see DNU200
SUPRILENT see VGA300
SUPRIMAL see HGC500
SUPRIN see TKX000
SUPRISTOL see SNL850
SUPROFEN see TEN750
SUPROTAN see CKF500
SURAMETHINIUM see HLC500
SURAMIN see BAT000
SURAMINE see BAT000
SURAUTO see IDW000
SURCHLOR see SHU500
SURCOPUR see DGI000
SUREAU (CANADA, HAITI) see EAI100
SURECIDE see CON300
SUREM see DLY000, PEU000
SURENINE see VGP000
SURE-SET see CJN000
SURESTRINE see BIT000
SURESTRYL see BIT000, MRU600
SURFACTANT NF see BLX000
SURFACTANT WK see DXY000
SURFEX MM see CAT775
SURFIL S see CAT775
SURFLAN see OJY100
SURGAM see SOX400
SURGEX see BET000
SURGICAL SIMPLEX see PKB500
SURGI-CEN see HCL000
SURHEME see PEU000
SURIKA see TKH750
SURINAM GREENHEART WOOD see HLY500
SURIRENE see AKO500
SURITAL see AGL375, SOX500
SURITAL SODIUM (derivative) see SOX500
SURITAL SODIUM see SOX500
SURITAL SODIUM SALT see SOX500
SURMONTIL see DLH200
SURMONTIL MALEATE see SOX550
SUROFENE see HCL000
SURPLIX see DLH600, DLH630
SURPRACIDE see DSO000
SURPUR see DGI000
SURRECTAN see AJY250
SURSUM see CJN250
SURSUMID see EPD500
SU SEGURO CARPIDOR see DUV600
SUSPEN see PDT750
SUSPENDOL see ZVJ000
SUSPENSO see CAT775
SUSPHRINE see VGP000
SUSTANE see BFW750, BQI000, BRM500
SUSTANE 1-F see BQI000
SUSTANONE see TBF500
SUSVIN see MRH209
SUTAN see EID500
SUTICIDE see HCQ500
SUTOPROFEN see TEN750
SUVREN see BRS000
SUXAMETHONIUM CHLORIDE see HLC500
SUXAMETHONIUM see CMG250
SUXAMETHONIUM CHLORIDE see HLC500
SUXAMETHONIUM DICHLORIDE see HLC500
SUXAMETHONIUM IODIDE see BJI000
SUXCERT see HLC500
SUXEMETHONIUM see CMG250
SUXETHONIUM CHLORIDE see HLC500
SUXIBUZONE see SOX875
SUXIL see SNE000
SUXILEP see ENG500
SUXIMAL see ENG500
SUXIMER see DNV610
SUXIN see ENG500
SUXINUTIN see ENG500
SUXINYL see HLC500
SUY-B 2 see IGK800
SUZORITE MICA see MQS250
SUZU see ABX250
SUZU H see HON000
SV-052 see SMW400
SV 1017 see DKP800
SV-1522 see ABH500
SVC see ABX500
SVITPREN see PJQ050
SVKh 1 see CGW300

SVKh 40 see CGW300
SVO 9 see PKL100
SW 400 see CAW850
SW-751 see POM275
SWALLOW WORT see CCS650
SWAMP HELLEBORE see FAB100
SWAMP WOOD see LEF100
SWANOL AM 301 see LBU200
SWAT see SOY000
SWEBATE see TAL250
SWEDISH GREEN see CNN500, COF500
SWEENEY'S ANT-GO see ARD750
SWEEP see PAJ000, TBQ750
SWEETA see SJN700
SWEET BELLS see DYA875
SWEET BETTIE see CCK675
SWEET BIRCH OIL see MPI000, SOY100
SWEET DIPEPTIDE see ARN825
SWEETENED NAPHTHA (PETROLEUM) see NAQ570
SWEET GUM see SOY500
SWEET MYRTLE see WBA000
SWEET ORANGE OIL see OGY000
SWEET PEA SEEDS see SOZ000
SWEET POTATO PLANT see CCO680
SWEET SHRUB see CCK675
SWEP see DEV600
SWIETENIA MAHAGONI see MAK300
SWINE PROSTATE EXTRACT see RLK875
SWISS BLUE see BJI250
SWISS CHEESE PLANT see SLE890
SWITCH IVY see DYA875
SWP (ANTIOXIDANT) see BRP750
SY-83 see LAG000
SYBIROMYCIN see SCF500
SYCOTROL see PIM000
SYD 230 see CGB250
SYDNONE IMINE, 3-(1-METHYL-2-PHENYLETHYL)-, MONOHYDROCHLORIDE see SPA100
SYDNOPHENE see SPA000
SYDNOPHEN HYDROCHLORIDE see SPA000
SYEP see SPA500
SYGETHIN see SPA650
SYGETIN see SPA650
SYKOSE see BCE500, SJN700
SYLACAUGA 88B see CAT775
SYLANTOIC see DKQ000, DNU000
SYLGARD 184 CURING AGENT see SPB000
SYLLIT see DXX400
SYLODEX see ARM268
SYMBIO see SNN300
SYMCLOSEN see TIQ750
SYMCLOSENE see TIQ750
SYMETRA see DKE800
SYMETRYCZNA DWUMETYLOHYDRAZYNA (POLISH) see DSF600
SYMMETREL see AED250, TJG250
SYMMETRIC DIMETHYLUREA see DUM200
SYMPAMINA-D see BBK500
SYMPAMINE see BBK000
SYMPATEDRINE see BBK000
SYMPATEKTOMAN see TCC000
SYMPATHIN E see NNO500
SYMPATHIN I see VGP000
m-SYMPATHOL see NCL500, SPC500
SYMPATHOL see HLV500, SPD000
SYMPATHOLYTIN see DCR200, DCT050
m-SYMPATOL see NCL500, SPC500
SYMPATOL see SPD000
SYMPATOL TARTRATE see SPD000
SYMPHORICARPOS (various species) see SED550
SYMPHYTINE see SPB500
SYMPHYTUM OFFICINALE L see RRK000, RRP000
SYMPLOCARPUS FOETIDUS see SDZ450
SYMPOCAINE HYDROCHLORIDE see SPB800
SYMPROPAMIN see FMS875
SYMPTOM 2 see POH250
SYMULER BRILLIANT SCARLET G see CMS160
SYMULER EOSIN TONER see BNH500
SYMULER FAST BLUE 6011 see IBV050
SYMULER FAST ORANGE GRD see CMU820
SYMULER FAST PYRAZOLONE ORANGE G see CMS145
SYMULER FAST SCARLET 4R see MMP100
SYMULER FAST VIOLET R see DFN450
SYMULER FAST YELLOW 4090G see CMS212
SYMULER FAST YELLOW 5GF see CMS212
SYMULER FAST YELLOW GF see DEU000

SYMULER FAST YELLOW GRF see CMS208
SYMULER FAST YELLOW GRTF see CMS208
SYMULER LAKE RED C see CHP500
SYMULER ORANGE LAKE 43 see CMM220
SYMULER RED 3023 see CMS148
SYMULER RED NRY see CMS148
SYMULEX MAGENTA F see FAG070
SYMULEX PINK F see FAG070
SYMULON ACID BRILLIANT SCARLET 3R see FMU080
SYMULON ACID FAST YELLOW MR see CMM759
SYMULON ACID ORANGE II see CMM220
SYMULON ACID RED PG see CMM320
SYMULON DIRECT BLACK BH see CMN800
SYMULON METANIL YELLOW see MDM775
SYMULON RED B BASE see NEQ000
SYMULON SCARLET G BASE see NMP500
SYMULON SCARLET 2G SALT see DEO295
SYNADRIN see PEV750
SYNALAR see SPD500
SYNALGOS see DQA400
SYNAMOL see SPD500
SYNANCEJA HORRIDA Linn. VENOM see SPB875
SYNANDONE see SPD500
SYNANDRETS see MPN500
SYNANDROL see TBG000
SYNANDROL F see TBF500
SYNANDRONE see SPD500
SYNANDROTABS see MPN500
SYNAPAUSE see EDU500
SYNAPEN see PDD350
SYNASAL see SPC500
SYNASTERON see PAN100
SYNATE see SBN000
SYNBETAN P see MEG250
SYNCAINE see AIT250
SYNCAL see BCE500
SYNCELOSE see MIF760
SYNCILLIN see PDD350
SYNCL see ALV000
SYNCORDAN see FMS875
SYNCORT see DAQ800
SYNCORTA see DAQ800
SYNCORTYL see DAQ800
SYNCOUMAR see ABF750
SYNCUMAR see ABF750
SYNCURARINE see PDD300
SYNCURINE see DAF600
SYNDIOL see EDO000
SYNDIOTACTIC POLYPROPYLENE see PMP500
SYNDROX see MDQ500, MDT600
SYNELAUDINE see DAM700
SYNEPHRIN see HLV500
dl-p-SYNEPHRINE see SPD000
dl-SYNEPHRINE see SPD000
m-SYNEPHRINE see NCL500, SPC500
p-SYNEPHRINE see HLV500
m-SYNEPHRINE HYDROCHLORIDE see SPC500
SYNEPHRINE TARTRATE see SPD000
(+−)-SYNEPHRINE TARTRATE see SPD000
SYNERGID R see DWW000
SYNERGIST 264 see OES000
SYNERONE see TBG000
SYNERPENIN see PDD350
SYNESTRIN see DKA600, DKB000
SYNESTROL see DAL600, DLB400
SYNETHENATE see SPC500
SYNFAT 1006 see CAU000
SYNFEROL AH EXTRA see SPD100
SYNFLORAN see DUV600
SYNGACILLIN see AJJ875
SYNGESTERONE see PMH500
SYNGESTROTABS see GEK500
SYNGUM D 46D see GLU000
SYNGYNON see HNT500
SYNHEXYL see HGK500
SYNISTAMIN see TAI500
SYNKAMIN see AKX500
SYNKAMIN BASE see AKX500
SYNKAY see MMD500
SYNKLOR see CDR750
SYNMIOL see DAS000
SYNOESTRON see DKB000
SYNOPEN see CKV625
SYNOPEN R see CKV625
SYNOTODECIN see PPY250

SYNOTOL L-60 see BKE500
SYNOVEX S see PMH500
SYNOX 5LT see MJO500
SYNOX TBC see BSK000
SYNPEN see CKV625
SYNPENIN see AIV500
SYNPERONIC OP see GHS000
SYNPITAN see ORU500
SYNPOL 1500 see SMR000
SYNPOR see CCU250
SYNPREN-FISH see PIX250
SYNPRO STEARATE see CAX350
SYNSAC see SPD500
SYNSTIGMIN BROMIDE see POD000
SYNSTIGMINE see DER600
SYNTAR see CMY800
SYNTARIS see FDD085
SYNTARPEN see DGE200
SYNTASE 62 see MES000
SYNTASE 100 see DMI600
SYNTEDRIL see BBV500
SYNTEFIX see CNH125
SYNTEN YELLOW 2G see AAQ250
SYNTEN YELLOW P 2R see DUW500
SYNTES 12A see EIV000
SYNTESTRIN see DKB000
SYNTESTRINE see DKB000
SYNTETREX see PPY250
SYNTETRIN see PPY250
SYNTETRIN NITRATE see SPE000
SYNTEXAN see DUD800
SYNTHAMIDE 5 see SPE500
SYNTHARSOL see ACN250
SYNTHECILLIN see PDD350
SYNTHECILLINE see PDD350
SYNTHEMUL 90-588 see ADW200
SYNTHENATE see HLV500
SYNTHETIC 3956 see CDV100
SYNTHETIC EUGENOL see EQR500
SYNTHETIC GLYCERIN see GGA000
SYNTHETIC IRON OXIDE see IHD000
SYNTHETIC LH-RH see LIU360
SYNTHETIC MUSTARD OIL see AGJ250
SYNTHETIC OXYTOCIN see ORU500
SYNTHETIC PYRETHRINS see AFR250
SYNTHETIC TRF see TNX400
SYNTHETIC TRH see TNX400
SYNTHETIC TSH-RELEASING FACTOR see TNX400
SYNTHETIC TSH-RELEASING HORMONE see TNX400
SYNTHETIC WINTERGREEN OIL see MPI000
SYNTHILA see DJB200
SYNTHOESTRIN see DKA600
SYNTHOFOLIN see DKA600
SYNTHOMYCINE see CDP250
SYNTHOPHYLLINE see DNC000
SYNTHOSTIGMINE BROMIDE see POD000
SYNTHOSTIGMINE METHYL SULFATE see DQY909
SYNTHOVO see DLB400
SYNTHRIN see BEP500
SYNTHROID see LFG050
SYNTHROID SODIUM see LFG050
SYNTOCIN see ORU500
SYNTOCINON see ORU500
SYNTOCINONE see ORU500
SYNTODRIL see BBV500
SYNTOFOLIN see DKA600
SYNTOLUTAN see PMH500
SYNTOMETRINE see LJL000
SYNTON YELLOW 2G see AAQ250
SYNTOPHEROL see VSZ450
SYNTOPHEROL ACETATE see TGJ055
SYNTOSTIGMIN see DER600
SYNTOSTIGMIN (tablet) see POD000
SYNTOSTIGMIN BROMIDE see POD000
SYNTOSTIGMINE BROMIDE see POD000
SYNTROGENE see DLB400
SYNTROM see ABF750
SYNTRON B see EIV000
SYNTROPAN see AOD250
SYN-U-TEX 4113E see MCB050
SYPHOS see ASD000
SYRAP see CMG000
SYRAPRIM see TKZ000
SYRIAN BEAD TREE see CDM325
SYRINGALDEHYDE see DOF600

SYRINGEALDEHYDE see DOF600
SYRINGIC ACID see SPE700
SYRINGIC ACID ETHYL CARBONATE ESTER with METHYL RESERPATE see
  RCA200
SYRINGIC ALDEHYDE see DOF600
SYRINGOL see DOJ200
SYRINGOPINE see RCA200
SYRINGYLALDEHYDE see DOF600
SYROSINGOPIN see RCA200
SYROSINGOPINE see RCA200
SYRUP of IPECAC, U.S.P. see IGF000
SYS 67ME see SIL500
SYS 67MPROP see CLO200
SYSTAM see OCM000
SYSTANATE MR see PKB100
SYSTANAT MR see PKB100
SYSTEMOX see DAO600
SYSTODIN see QHA000
SYSTOGENE see TOG250
SYSTOPHOS see OCM000
SYSTOX see DAO600
SYSTOX SULFONE see SPF000
SYSTRAL see CIS000
SYTAM see OCM000
SYTASOL see CBW000
SYTOBEX see VSZ000
SYTON FAST ORANGE G see CMS145
SYTON FAST SCARLET RB see MMP100
SYTON FAST SCARLET RD see MMP100
SYTON FAST SCARLET RN see MMP100
SYTRON see EJA379
SYZYGIUM JAMBOS (Linn.) Alston, extract excluding roots see SPF200
SZ 6300 see TLD000
SZESCIOMETYLENODWUIZOCYJANIAN see DNJ800
SZKLARNIAK see DGP900


T-2 see TIK500
T3 see LGK050
T₃ see LGK050
l-T4 see TFZ275
T4 see CPR800
T10 see DSF300
Ts 14 see PJT300
Ts 20 see PJT300
Ts 30 see PJT300
Ts 35 see PJT300
T 40 see TGF250
Ts 40 see PJT300
Ts 55 see PJT300
Ts 62 see PJT300
T 72 see PNX000
T 100 see EHG100, TGM740
T 101 see UTU500
T-113 see BPG000
T-125 see TBX000
T-144 see IPX000
2,4,5-T see TAA100
864T see IBQ100
T-1035 see DSA800
T-1036 see DJJ400
T-1088 see HNP000
T-1123 see TGH665
T-1125 see HNO000
T-1152 see HNO500
Th 1165a see FAQ100
T 1220 see SJJ200
T 1384 see NCP875
T-1551 see CCS369
T-1690 see HNN000
T-1703 see IRF000
T-1768 see DPL900
T-1770 see MIA775
T 1824 see BGT250
T-1835 see DEC200
T-1843 see MIC250
T-1982 see TAA400
T-2002 see BJE750
T-2104 see EIF000
T-2106 see IPX000
T-2588 see TAA420
79T61 see SBE500
T-42082 see SFP500
T 130-2500 see CAT775
α-T see MIH275

β-T see TIN000
TDOT see TCQ275
T4 (hormone) see TFZ275
Ts 35 (polymer) see PJT300
TA 1 see MRN675
TA 12 see TAN750
Ts A 16 see PJT300
TA 064 see DAP850
TA see THS825
TAA see TFA000
TA-AZUR see THM750
TABAC (FRENCH) see TGI100
TABAC du DIABLE (CANADA) see SDZ450
TABACHIN (MEXICO) see CAK325
TABACO (SPANISH) see TGI100
TABALGIN see HIM000
TABASCO HOT PEPPER SAUCE see TAA875
TABASCO PEPPER see PCB275
TABATREX see SNA500
TABILIN see BFD000
TABLE SALT see SFT000
TABLOID see AMH250
TABOON A see EIF000
TABUN see EIF000
TABUTREX see SNA500
TAC-28 see VIK150
TAC 121 see TGG250
TAC 131 see TGG250
TACARYL see MDT500, MPE250
TACAZYL see MPE250
TACE see CLO750
TACE-FN see CLO750
TACHIGAREN see HLM000
TACHIONIN see HII500
TACHMALIN see AFH250
TACITIN see BCH750
TACKLE see CLS075
TACOSAL see DKQ000, DNU000
TACP see TNC725
TACRINE see TCJ075
TACRYL see MPE250
TACTARAN see TAF675
TACUMIL see TKX000
TAD see TEP500
TAENIATOL see MJM500
TAFASAN see MEP250
TAFASAN 6W see MEP250
TAFAZINE see BJP000
TAFIL see XAJ000
TAG-39 see ECU750
TAG 331 see ABU500
TAG see ABU500
TAGAMET see TAB250
TAGAT see SEH000
TAGATHEN see CHY250
TAGETES OIL see TAB275
TAG FUNGICIDE see ABU500
TAG HL 331 see ABU500
TAGN see THN800
TAHMABON see DTQ400
dl-TAI 284 see CFH825
(±)-TAI 284 see CFH825
TAI-284 see CFH825
TAI 284 see CMV500
TAIC see THS100
TAIFEN see ZMA000
TAIGUIC ACID see HLY500
TAIGU WOOD see HLY500
TAIL FLOWER see APM875
TAJMALIN see AFH250
TAK see MAK700
TAKACIDIN see TAB300
TAKACILLIN see LEJ500
TAKAMINA see VGP000
TAKAMINE HT (BACILLUS SUBTILIS) see TAB400
TAKANARUMIN see ZVJ000
TAKAOKA ACID RED RS see NAO600
TAKAOKA AMARANTH see FAG020
TAKAOKA BRILLIANT SCARLET 3R see FMU080
TAKAOKA METANIL YELLOW see MDM775
TAKAOKA RHODAMINE B see FAG070
TAKATHENE see PJS750
TAKATHENE P 12 see PJS750
TAKATHENE P 3 see PJS750
TAKEDO 1969-4-9 see GGA800

TAKENATE 500 see XIJ000
TAKENATE see XIJ000
TAKENATE 300C see PKB100
TAKESULIN see CCS550
TAKILON see PKQ059
TAKTIC see MJL250
TAKYCOR see AFH250
TALADREN see DFP600
TALAMO see AMX750
TALAN see CBW000
TALARGAN see TEH500
TALATROL see TEM500
TALBOT see LCK000, LCK100
TALBUTAL see AFY500
TALC, containing asbestos fibers see TAB775
TALC see TAB750
TALCORD see AHJ750
TALCUM see TAB750
TALIBLASTIN see TEH250
TALIBLASTINE see TEH250
TALIMOL see TEH500
TALISOMYCIN see TAC500
TALISOMYCIN B see TAC750
TALISOMYCIN S$^{10b}$ see TAB785
TALL OIL see TAC000
TALLOL see TAC000
TALLOW BENZYL DIMETHYLAMMONIUM CHLORIDE see DTC600
N-TALLOWPYRROLIDINONE see TAC200
TALLYSOMYCIN A see TAC500
TALLYSOMYCIN B see TAC750
TALLYSOMYCIN S$^{10b}$ see TAB785
TALMON see MAO350
TALODEX see FAQ900
TALOFLOC see DXG625
TALON see TAC800
TALON RODENTICIDE see TAC800
TALPHENO see EOK000
TALUCARD see POB500
TALUSIN see POB500
TALWAN see DOQ400
TALWIN see DOQ400
TAMA PEARL TP 121 see CAT775
TAMARIZ see SAZ000
TAMARON see DTQ400
TAMAS see BBW500
TAMBALISA (CUBA) see NBR800
TAMCHA see AJV500
TAMETIN see TAB250
TAMEX see BQN600
TAMILAN see PAP000
TAMOFEN see TAD175
TAMOL L see BLX000
TAMOL SN see BLX000
TAMOXASTA see TAD175
TAMOXIFEN see NOA600
TAMOXIFEN CITRATE see TAD175
TA-33MP see TAN750
TAMPOVAGAN STILBOESTROL see DKA600
TAMPULES see CDP000
TAMRAGHOL see CNK559
TANAFOL see CKF500
TANAK m$^3$ see MCB050
TANAKAN see CLD000, CLD250
TANAK MRX see MCB050
TANAMICIN see HGP550
TANAN see TDT800
TANANE see TDT800
TANASUL see MDU300
TANCAL 100 see CAT775
TANDACOTE see HNI500
TANDALGESIC see HNI500
TANDEARIL see HNI500
TANDEM see TJK100
TANDERAL see HNI500
TANDEX see DUM800
TANDIX see IBV100
TANGANTANGAN OIL see CCP250
TANGARINE LAKE X-917 see CMM220
TANGELO OIL see TAD250
TANGERINE OIL see TAD500
TANGERINE OIL, COLDPRESSED (FCC) see TAD500
TANGERINE OIL, EXPRESSED (FCC) see TAD500
TANICAINE see BJO500
TANIDIL see DQA400
TANNEX see IDA000

TANNIC ACID see TAD750
TANNIN see TAD750
TANNIN from ACORN see ADI625
TANNIN from BETEL NUT see BFW050
TANNIN from BRACKEN FERN see BML250
TANNIN from CHERRY BARK OAK see CDL750
TANNIN from CHESTNUT see CDM250
TANNIN-FREE FRACTION of BRACKEN FERN see TAE250
TANNIN from LIMONIUM NASHII see MBU750
TANNIN from MARSH ROSEMARY see MBU750
TANNIN from MIMOSA see MQV250
TANNIN from MYRABOLAM see MRZ100
TANNIN from MYROBALANS see MSB750
TANNIN from MYRTAN see MSC000
TANNIN from PERSIMMON see PCP500
TANNIN from QUEBRACHO see QBJ000
TANNIN from SUMAC see SOU750
TANNIN from SWEET GUM see SOY500
TANNIN from VALONEA see VCK000
TANNIN from WAX MYRTLE see WBA000
TANONE see DRR400
TANRUTIN see RSU000
TANSTON see XQS000
TANSY OIL see TAE500
TANSY RAGWORT see RBA400
TANTALIC ACID ANHYDRIDE see TAF500
TANTALUM-181 see TAE750
TANTALUM see TAE750
TANTALUM CHLORIDE see TAF000
TANTALUM FLUORIDE see TAF250
TANTALUM(V) OXIDE see TAF500
TANTALUM OXIDE see TAF500
TANTALUM PENTACHLORIDE see TAF000
TANTALUM PENTAFLUORIDE see TAF250
TANTALUM PENTAOXIDE see TAF500
TANTALUM PENTOXIDE see TAF500
TANTALUM POTASSIUM FLUORIDE see PLH000
TANTAN (PUERTO RICO) see LED500
TANTARONE see MJE760
TANTUM see BBW500
TAOMYCIN see HOH500
TAOMYXIN see HOH500
TAORYL see CBG250
TAP 85 see BBQ500
TAP see MPN000
TAPAR see HIM000
TAPAZOLE see MCO500
T.A.P.E. see NNR400
TAPHAZINE see BJP000
TAPIOCA see CCO680, DBD800
TAPIOCA STARCH see SLJ500
TAPIOCA STARCH HYDROXYETHYL ETHER see HLB400
TAPON see SLJ500
TAP 9VP see DGP900
TAR see CMY800
TAR, from tobacco see CMP800
TARACTAN see TAF675
TARAPACAITE see PLB250
TARAPON K 12 see SIB600
TARARACO see AHI635
TARARACO BLANCO (CUBA) see BAR325
TARARACO DOBLE (CUBA) see AHI635
TARASAN see TAF675
TAR CAMPHOR see NAJ500
TAR, COAL see CMY800
TARDAMID see AIE750
TARDAMIDE see AIE750
TARDEX 100 see PAU500
TARDIGAL see DKL800
TARDOCILLIN see BFC750
TARFLEN see TAI250
TARGET MSMA see MRL750
TARICHATOXIN see FOQ000
TARIMYL see SOA500
TARIVID see OGI300
TARLON XB see PJY500
TARNAMID T see PJY500
TARO see EAI600
TAROCTYL see CKP500
TARODYL see GIC000
TARODYN see GIC000
TAR OIL see CMY825
TARO VINE see PLW800
TARPAN see CMW250
TAR, PINE see PIH775

TARRAGON see AFW750
TARRAGON OIL see TAF700
TARTAGO (PUERTO RICO) see CNR135
TARTAN see PHK250
TARTAR see PKU600
TARTAR CREAM see PKU600
TARTAR EMETIC see AQG250
l-(+)-TARTARIC ACID see TAF750
TARTARIC ACID see TAF750
l-TARTARIC ACID, AMMONIUM SALT see DCH000
dl-TARTARIC ACID, ANTIMONY POTASSIUM SALT see AQG750
l-TARTARIC ACID, ANTIMONY POTASSIUM SALT see AQH000
TARTARIC ACID, DIAMMONIUM SALT see DCH000
TARTARIC ACID, DIBENZOATE see DDE300
meso-TARTARIC ACID, ION(2−) see TAF775
TARTARIC ACID, MONOSODIUM SALT see SKB000
TARTARIZED ANTIMONY see AQG250
TARTAR YELLOW FS see FAG140
TARTAR YELLOW N see FAG140
TARTAR YELLOW PF see FAG140
TARTAR YELLOW S see FAG140
TARTRAN YELLOW see FAG140
TARTRAPHENINE see FAG140
meso-TARTRATE see TAF775
TARTRATE ANTIMONIO-POTASSIQUE (FRENCH) see AQG250
TARTRATED ANTIMONY see AQG250
TARTRATE de NICOTINE (FRENCH) see NDS500
TARTRAZINE see FAG140
TARTRAZINE A EXPO T see FAG140
TARTRAZINE B see FAG140
TARTRAZINE B.P.C. see FAG140
TARTRAZINE EXTRA PURE A see FAG140
TARTRAZINE FD & C YELLOW #5 see FAG140
TARTRAZINE FQ see FAG140
TARTRAZINE G see FAG140
TARTRAZINE LAKE see FAG140
TARTRAZINE LAKE YELLOW N see FAG140
TARTRAZINE M see FAG140
TARTRAZINE MCGL see FAG140
TARTRAZINE N see FAG140
TARTRAZINE NS see FAG140
TARTRAZINE O see FAG140
TARTRAZINE T see FAG140
TARTRAZINE XX see FAG140
TARTRAZINE XXX see FAG140
TARTRAZINE YELLOW see FAG140
TARTRAZOL BPC see FAG140
TARTRAZOL YELLOW see FAG140
TARTRINE YELLOW O see FAG140
TARWEED see TAG250
TARZOL see DGA200
TASK see DGP900
TASK TABS see DGP900
TASMIN see AOO800
TAT-1 see DIK000
TAT CHLOR 4 see CDR750
TATD see DXH250, GEK200
TATERPEX see CKC000
TATHIONE see GFW000
TAT-3 HYDROCHLORIDE see CNE375
TATTOO see DQM600
TAT-3 TRIPALMITATE see PIC100
TATURIL see UVJ450
TAUPHON see TAG750
TAURE(o)DON see GJC000
TAURINE see TAG750
TAURINOPHENETIDINE HYDROCHLORIDE see TAG875
TAUROCHOLATE see TAH250
TAUROCHOLIC ACID see TAH250
TAUROMYCETIN see TAH500
TAUROMYCETIN-III see TAH650
TAUROMYCETIN-IV see TAH675
TAUTUBA (PUERTO RICO) see CNR135
TAVEGIL see FOS100
TAVEGYL see FOS100
TA-VERM see PIJ500
TAVOR see CFC250
TAVROMYCETIN III see TAH650
TAVROMYCETIN-IV see TAH675
TAXIFOLIN see DMD000
TAXIFOLIOL see DMD000
TAXILAN see PCK500
TAXIN (GERMAN) see TAH750
TAXINE see TAH750
TAXOL see TAH775

TAXUS BACCATA LINN., LEAF EXTRACT see KCA100
TAXUS (VARIOUS SPECIES) see YAK500
TAYO BAMBOU (HAITI) see EAI600
TAYSSATO see MEP250
TAZEPAM see CFZ000
TAZONE see BRF500
TB 2 see TFD000
TB see CMO250
2,3,6-TBA (herbicide) see TIK500
TBB see THX500
TBBA see BQK500
TB 1 (BAYER) see FNF000
TBDZ see TEX000
TBE see ACK250
TBEP see BPK250
TBF-43 see TKP850
TBHP-70 see BRM250
TBHQ (FCC) see BRM500
TBOT see BLL750
TBP see TFD250, TIA250
TBPMC see BSG300
TBS 95 see THW750
TBS see THW750
TBT see BSP500
TBTO see BLL750
2,4,5-TC see TIX500
TC 3-30 see SMQ500
T 1 (Catalyst) see DBF800
T-4CA see TEV000
TCA see TII250, TII500
TCAB see TBN500
TCAOB see TBN550
TCA-PE see AMU000
TCA-PR see AMU500
T-250 CAPSULES see TBX250
TCA SODIUM see TII500
TCAT see TIJ500
2,3,6-TCB see TIK500
TCB see TBO700
2,3,6-TCBA see TIK500
TCBC see TIL250
TCBN see TBS000
TCC see TIL500, TIQ000
TCC mixed with TFC (2:1) see TIL526
TCDBD see TAI000
2,3,7,8-TCDD see TAI000
TCDD see TAI000
1,1,1-TCE see MIH275
TCE see TBQ100
TCEO see ECT600
TCH see TFE250
TC HYDROCHLORIDE see TBX250
TCIN see TBQ750
TCM see CHJ500, TNE775
TCMTB see BOO635
TCNA see TBR250
TCNB see TBR750
TCNP SODIUM SALT see TAI050
TCNQ see TBW750
TCP see TBT000
TCPA see TIY500
TCPE see TIX000
m-TCPN see TBQ750
o-TCPN see TBT200
TCPO see TJB750
TCPP see FQU875
2,4,5-TCPPA see TIX500
TCSA see TBV000
TCT see TFZ000
TCTH see CQH650
TCTNB see TJE200
TCTP see TBV750
TCV-3B see EGM100
TD-183 see TBV750
TD-758 see TBC200
TD 1771 see PEX500
TD-5032 see HEE500
TDA see TGL750
TDBP (CZECH) see TNC500
TDCPP see FQU875, TNG750
TDE (DOT) see BIM500
o,p-TDE see CDN000
o,p'-TDE see CDN000
p,p'-TDE see BIM500
TDE see TJQ333

T-DET see DXY600
2,4-TDI see TGM750
2,6-TDI see TGM800
TDI-80 see TGM740, TGM750
TDI 80-20 see TGM740
TDI (OSHA) see TGM750
TDI see TGM740
T-82 DIFUMARATE see MDU750
TDPA see BHM000
2,5-TDS see TGM400
TDS see CHK825, TES800
TDS (NEUROTROPE) see TES800
TE 114 see HNM000
TE see TBF750
TEA see TCB725, TCC000, TJN750
TEAB see TCC000
TEABERRY OIL see MPI000
TEAC see TCC250
TEA CHLORIDE see TCC250
TEAEI see TAI100
TEA LAURYL SULFATE see SON000
T.E.A.S. see SDH670
TEA TREE OIL see TAI150
TEB see TDG500
TEBAC see BFL300
TEBALON see FNF000
TEBE see HNY500
TEBECID see ILD000
TEBECURE see FNF000
TEBEFORM see PNW750
TEBEMAR see FNF000
TEBERUS see EPQ000
TEBESONE I see FNF000
TEBETHIONE see FNF000
TEBEXIN see ILD000
TEBEZON see FNF000
TEBLOC see LIH000
TEBRAZID see POL500
TEBULAN see BSN000
TEBUTHIURON see BSN000
TEC see TJP750
TECACIN see HGP550
TECH DDT see DAD200
TECHNETIUM TC 99M SULFUR COLLOID see SOD500
TECHNICAL BHC see BBQ750
90 TECHNICAL GLYCERINE see GGA000
TECHNICAL HCH see BBQ750
TECHNOPOR see PKQ059
TECH PET F see MQV750
TECNAZEN (GERMAN) see TBR750
TECNAZENE see TBR750
TECODIN see DLX400
TECODINE see DLX400
TECOFLEX HR see PKN000
TECOMIN see HLY500
TECPOL see ADW200
TECQUINOL see HIH000
TECRAMINE see TKH750
TECSOL see EFU000
TECTILON ORANGE 3GT see SGP500
TECTO see TEX000
TECZA see TJR000
TEDIMON 31 see PKB100
TEDION see CKM000
TEDION V-18 see CKM000
TEDMA see MDN510
TEDP (OSHA) see SOD100
TEDP see SOD100
TEDTP see SOD100
TEEBACONIN see ILD000
TEF see TND250
TEFAMIN see TEP000, TEP500
TEFASERPINA see RDK000
TEFILAN see DNC000
TEFILIN see TBX250
TEFLEX see KDK000
TEFLON see TAI250
TEFLON (various) see TAI250
TEFSIEL C see FLZ050
TEG see TJQ000
TEGAFUR see FLZ050
TEGAFUR mixture with URACIL (1:4) see UNJ810
TEGDH see TCE375
TEGDN see TJQ500
TEGENCIA HYDROCHLORIDE see DAI200

TEGESTER see IQN000
TEGESTER BUTYL STEARATE see BSL600
TEGESTER ISOPALM see IQW000
TEGIN 503 see OAV000
TEGIN 515 see OAV000
TEGIN see OAV000
TEGIN P see SLL000
TEGO 51 see DYA850
TEGOLAN see CMD750
TEGON see TJQ500
TEGO-OLEIC 130 see OHU000
TEGOPEN see AOC500, SLJ000, SLJ050
TEGOSEPT B see BSC000
TEGOSEPT E see HJL000
TEGOSEPT M see HJL500
TEGOSEPT P see HNU500
TEGO-STEARATE see EJM500
TEGOSTEARIC 254 see SLK000
TEGRETAL see DCV200
TEGRETOL see DCV200
TEIB see TND000
TEICHOMYCIN A2 see TAI400
TEKKAM see NAK500
TEKODIN see DLX400
TEKRESOL see CNW500
TEKWAISA see MNH000
TEL see TCF000
TELAGAN see TEH500
TELARGAN see TEH500
TELARGEAN see TEH500
TELCOTENE see PJS750
TELDANE see TAI450
TELDRIN see CLD250, TAI500
TELEBRIX 38 see TAI600
TELEBRIX 300 see MQR300
TELECOTHENE see PJS750
TELEFOS see IOT000
TELEMID see MDU300
TELEMIN see PPN100
TELEMIN-SOFT see TAI725
TELEPAQUE see IFY100
TELEPATHINE see HAI500
TELEPRIN see TKX000
TELESMIN see DCV200
TELETRAST see IFY100
TELGIN-G see FOS100
TELIDAL see HNI500
TELINE see TBX250
TELIPEX see TBG000
TELLOY see TAJ000
TELLUR (POLISH) see TAJ000
TELLURANE-1,1-DIOXIDE see TAI735
TELLURATE see TAI750
TELLURIC(VI) ACID see TAI750
TELLURIC ACID see TAI750
TELLURIC ACID, DISODIUM SALT, PENTAHYDRATE see TAI800
TELLURIC CHLORIDE see TAJ250
TELLURIUM (dust or fume) see TAJ010
TELLURIUM see TAJ000
TELLURIUM CHLORIDE see TAJ250
TELLURIUM COMPOUNDS see TAJ500
TELLURIUM DIETHYLDITHIOCARBAMATE see EPJ000
TELLURIUM DIOXIDE see TAJ750
TELLURIUM HEXAFLUORIDE see TAK250
TELLURIUM HYDROXIDE see TAI750
TELLURIUM NITRIDE see TAK500
TELLURIUM OXIDE see TAJ750
TELLURIUM TETRABROMIDE see TAK600
TELLURIUM TETRACHLORIDE see TAJ250
1,1'-TELLUROBISETHANE see DKB150
TELLUROUS ACID, DISODIUM SALT see SKC500
TELMICID see DJT800
TELMID see DJT800
TELMIDE see DJT800
TELMIN see MHL000
TELOCIDIN B see TAK750
TELODRIN see OAN000
TELODRON see CLD250
TELOMYCIN see TAK800
TELON see BCP650
TELON BLUE BL see CMM090
TELON BLUE RRL see CMM080
TELONE see DGG000, DGG950
TELONE II SOIL FUMIGANT see DGG950
TELON FAST BLACK E see AQP000

TELON FAST BLACK PE see CMN230
TELON FAST RED GG see CMM330
TELON FAST SCARLET N see CMM320
TELOTREX see TBX250
TELTOZAN see CAL750
TELVAR see CJX750, DXQ500
TELVAR DIURON WEED KILLER see DXQ500
TELVAR MONURON WEEDKILLER see CJX750
TEMARIL see TAL000
TEMASEPT see BOD600, THW750
TEMASEPT II see THW750
TEMAZEPAM see CFY750
TEMECHINE see TAL275
TEMED see TDQ750
TEMEFOS see TAL250
TEMENTIL see PMF500
TEMEPHOS see TAL250
TEMEQUINE see TAL275
TEMESTA see CFC250
TEMETEX see DKF130
TEMGESIC see TAL325
TEM-HISTINE see DPJ400
TEMIC see CBM500
TEMIK see CBM500
TEMIK G10 see CBM500
TEMIK OXIME see MLX800
TEMIK SULFOXIDE see MLW750
TEMLO see HIM000
TEMOPHOS see TAL250
TEMPANAL see HIM000
TEMPARIN see BJZ000
TEMPLIN OIL see AAC250
TEMPO see TDT800
TEMPODEX see BBK500
TEMPONITRIN see NGY000
TEMPO-RESERPINA see RDK000
TEMPOSERPINE see RDK000
TEMPRA see HIM000
TEMUR see TDX250
TEMUS see BMN000
TEN see TJO000
TENAC see DGP900
TENALIN see TEO250
TENAMENE 2 see DEG200
TENAMENE 31 see BJT500
TENAMENE see BJL000
TENAMINE see TLN500
TENAPLAS see PJS750
TENATHAN see BFW250
TENCILAN see CDQ250
TENDEARIL see HNI500
TENDIMETHALIN see DRN200
TENDOR see IKB000
TENDOSCEN-COMPR. see RDK000
TENDUST see NDN000
TENEBRIMYCIN see NBR500, TAL350
TENEMYCIN see NBR500, TAL350
TENESDOL see CCR875
TENFIDIL see DPJ200
TENIATHANE see MJM500
TENIATOL see MJM500
TENICID see PDM750
TENIPOSIDE see EQP000
TENITE 423 see PMP500
TENITE 800 see PJS750
TENITE 1811 see PJS750
TENITE 2910 see PJS750
TENITE 2918 see PJS750
TENITE 3300 see PJS750
TENITE 3340 see PJS750
TENNECETIN see PIF750
TENNECO 1742 see PKQ059
TENN-PLAS see BCL750
TENNUS 0565 see AAX175
TENORAN see CJQ000
TENORMIN see TAL475
TENOSIN-WIRKSTOFF see TOG250
TENOX BHA see BQI000
TENOX BHT see BFW750
TENOX HQ see HIH000
TENOXICAM see TAL485
TENOX PG see PNM750
TENOX P GRAIN PRESERVATIVE see PMU750
TENOX TBHQ see BRM500
TENSANYL see RDK000

TENSERLIX see RDK000
TENSERPINE 'ASSIA' see RDK000
TENSERPINIE see RDK000
TENSIBAR see DIH000
TENSICOR see DNM400
TENSIFEN see HNJ000
TENSILON see EAE600, TAL490
TENSILON BROMIDE see TAL490
TENSILON CHLORIDE see EAE600
TENSINASE D see DWF200
TENSINYL see MDQ250
TENSIONAL see RDK000
TENSIONORME see RDK000
TENSIVAL see TEH500
TENSOL 7 see PKB500
TENSOPAM see DCK759
TENSYL see AFG750
TENTON see MFK500
TENTONE see MFK500
TENTONE MALEATE see MFK750
TENTRATE-20 see PBC250
TENUATE see DIP600
TENUATE HYDROCHLORIDE see DIP600
l-TENUAZONIC ACID see VTA750
TENUAZONIC ACID see VTA750
TENULIN see TAL550
TENURID see DXH250
TENUTEX see DXH250
TEOBROMIN see TEO500
TEODRAMIN see DYE600
TEOF see TJL600
TEOFILCOLINA see CMF500
2-(7'-TEOFILLINMETIL)-1,3-DIOSSOLANO (ITALIAN) see TEQ175
TEOFYLLAMIN see TEP000
TEOHARN see MCB000
TEOKOLIN see CMF500
TEONANACATL see PHU500
TEONICON see PPN000
TEOPROLOL see TAL560
TEOS see EPF550
TEP see TJT750
TEPA see TND250
TEPANIL see DIP600
TEPERINE see DLH630
TEPIDONE see SGF500
TEPIDONE RUBBER ACCELERATOR see SGF500
TEPILTA see DTL200
TEPOGEN see SLJ000
TEPP (ACGIH) see TCF250
TEPSERPINE see RDK000
TEQUINOL see HIH000
TERABOL see MHR200
TERALEN see AFL500
TERALIN see DPJ400
TERALLETHRIN see TAL575
TERALUTIL see HNT250
TERAMYCIN HYDROCHLORIDE see HOI000
TERANOL see ECX100
TERAPINYL see FMS875
TERAPRINT see AKI750
TERASIL BLUE GREEN CB see DMM400
TERASIL BRILLIANT BLUE 3RL see CMP070
TERASIL BRILLIANT PINK 4BN see DBX000
TERASIL BRILLIANT VIOLET 3B see DBY700
TERASIL TURQUOISE BLUE G see DMM400
TERASIL YELLOW GBA EXTRA see AAQ250
TERASIL YELLOW 2GC see AAQ250
TERAZOSIN see FPP100
TERBACIL see BQT750
TERBAM see BSG300
TERBENOL see NOA000
TERBENOL HYDROCHLORIDE see NOA500
TERBENZENE see TBD000
TERBIUM see TAL750
TERBIUM CHLORIDE see TAM000
TERBIUM CITRATE see TAM500
TERBIUM OXIDE see TAN000
TERBOLAN see RDK000
(−)-TERBUCLOMINE see PAP230
TERBUFOS see BSO000
TERBUMETON see BQC500
TERBUTALIN see TAN100
TERBUTALINE see TAN100
TERBUTALINE SULFATE see TAN250
TERBUTALINE SULPHATE see TAN250

TERBUTHYLAZINE see BQB000
TERCININ see CMV000
TERCYL see CBM750
TEREBENTHINE see TOD750
TEREFTALODINITRIL (CZECH) see BBP250
TEREPHTAHLIC ACID-ETHYLENE GLYCOL POLYESTER see PKF750
TEREPHTALDEHYDE see TAN500
TEREPHTALDEHYDES (FRENCH) see TAN500
TEREPHTHALALDEHYDE see TAN500
TEREPHTHALALDEHYDONITRILE see COK250
TEREPHTHALAMIDE see TAN600
TEREPHTHALAMIDIC ACID see TAN600
TEREPHTHALDIAMIDE see TAN600
TEREPHTHALIC ACID see TAN750
TEREPHTHALIC ACID, BIS(2-ETHYLHEXYL)ESTER see BJS500
TEREPHTHALIC ACID CHLORIDE see TAV250
TEREPHTHALIC ACID DIAMIDE see TAN600
TEREPHTHALIC ACID, DIBUTYL ESTER see DEH700
TEREPHTHALIC ACID DICHLORIDE see TAV250
TEREPHTHALIC ACID, DIETHYL ESTER see DKB160
TEREPHTHALIC ACID ISOPROPYLAMIDE see IRN000
TEREPHTHALIC ACID METHYL ESTER see DUE000
TEREPHTHALIC ALDEHYDE see TAN500
TEREPHTHALIC DIAMIDE see TAN600
TEREPHTHALIC DICHLORIDE see TAV250
TEREPHTHALONITRILE see BBP250
5,5'-(TEREPHTHALOYLBIS(IMINO-p-PHENYLENE)BIS(2,4-DIAMINO-1-ETHYLPYRI-MIDINIUM)-DI-p-TOLUENESULFONATE see TAO750
5,5'-(TEREPHTHALOYLBIS(IMINO-p-PHENYLENE)BIS(2,4-DIAMINO-1-METHYLPY-RIMIDINIUM)-DI-p-TOLUENESULFONATE see TAP000
5,5'-(TEREPHTHALOYLBIS(IMINO-p-PHENYLENE)BIS(2,4-DIAMINO-1-PROPYLPY-RIMIDINIUM)-DI-p-TOLUENESULFONATE see TAP250
3,3'-(TEREPHTHALOYLBIS(IMINO-p-PHENYLENE)BIS(1-PROPYLPYRIDINIUM)-DI-p-TOLUENESULFONATE see TAR000
4,4'-(TEREPHTHALOYLBIS(IMINO-p-PHENYLENE)BIS(1-PROPYLPYRIDINIUM)-DI-p-TOLUENESULFONATE see TAR250
TEREPHTHALOYL DICHLORIDE see TAV250
3,3'-(TEREPHTHALOYLDIIMINOBIS(p-PHENYLENE)BIS(1-PROPYLPYRIDINIUM)-DI-p-TOLUENESULFONATE see TAR000
TERETON see MFW100
TERFAN see PKF750
TERFENADINE see TAI450
TERFLUZINE see TKE500, TKK250
TERFLUZINE DIHYDROCHLORIDE see TKK250
TERGAL see PKF750
TERGEMIST see TAV750
TERGIMIST see TAV750
TERGITOL 4 see SIO000
TERGITOL 08 see TAV750
TERGITOL see EMT500
TERGITOL ANIONIC 4 see EMT500
TERGITOL ANIONIC 7 see HCP900
TERGITOL ANIONIC 08 see TAV750
TERGITOL ANIONIC P-28 see TBA750
TERGITOL MIN-FOAM 1X see TBA800
TERGITOL NONIONIC XD see GHY000
TERGITOL NP-14 see TAW250
TERGITOL NP-27 see TAW500
TERGITOL NP-35 (nonionic) see TAX000
TERGITOL NP-40 (nonionic) see TAX250
TERGITOL NPX see NND500
TERGITOL 12-P-9 see TAX500
TERGITOL PENETRANT 4 see EMT500
TERGITOL PENETRANT 7 see TBB750
TERGITOL 15-S-5 see TAY250
TERGITOL 15-S-20 see TBA500
TERGITOL 15-S-12 (nonionic) see TAZ100
TERGITOL 15-S-9 (nonionic) see TAZ000
TERGITOL TMN-10 see TBB775
TERGITOL TP-9 (NONIONIC) see PKF000
TERGITOL XD (nonionic) see GHY000
TERGURID see DLR100
TERGURIDE see DLR100
TERGURIDE HYDROGEN MALEATE see DLR150
TERIAM see UVJ450
TERIDAX see IFZ800
TERIDIN see UVJ450
TERIDOX see DSM500
TERIMON see TAD175
TERININ see CMV000
TERIT see BAR800
TERMIL see TBQ750
TERMINALIA CHEBULA RETZ TANNING see MSB750
TERMITKIL see DEP600
TERM-I-TROL see PAX250

TERMOFLEKS A see BKO600
TERMOSOLIDO RED LCG see CHP500
TERODILINE CHLORIDE see TBC200
TERODILINE HYDROCHLORIDE see TBC200
TEROFENAMATE see DGN000
TEROM see PKF750
TEROXIRONE see TBC450
$\Delta^{6,8}$-(9)-TERPADIENONE-2 see MCD250
TERPENE POLYCHLORINATES see TBC500
TERPENTINOEL (GERMAN) see PIH750
TERPENTIN OEL (GERMAN) see TOD750
1,3-TERPHENYL see TBC620
m-TERPHENYL see TBC620
o-TERPHENYL see TBC640
p-TERPHENYL see TBC750
TERPHENYLS see TBD000
p-TERPHENYL-4-YLACETAMIDE see TBD250
α-TERPINENE (FCC) see MLA250
γ-TERPINENE (FCC) see MCB750
TERPINENOL-4 see TBD825
4-TERPINEOL see TBD825
α-TERPINEOL see TBD750
β-TERPINEOL see TBD775
α-TERPINEOL (FCC) see TBD500
TERPINEOL see TBD500
α-TERPINEOL ACETATE see TBE250
TERPINEOLS see TBD500
TERPINEOL SCHLECHTHIN see TBD750
TERPINOLENE see TBE000
TERPINYL ACETATE see TBE250
TERPINYL FORMATE see TBE500
TERPINYL PROPIONATE see TBE600
TERPINYL THIOCYANOACETATE see IHZ000
Δ-1,8-TERPODIENE see MCC250
TERRA ALBA see CAX750
TERRACHLOR see PAX000
TERRACHLOR-SUPER X see EFK000
TERRACOAT see EFK000
TERRACUR P see FAQ800
TERRAFLO see EFK000
TERRAFUN see PAX000
TERRAFUNGINE see HOH500
TERRAKLENE see PAJ000
TERRAMITSIN see HOH500
TERRAMYCIN see HOH500
TERRAMYCIN SODIUM see TBF000
TERRA-SYSTAM see BJE750
TERRA-SYTAM see BJE750
TERRASYTUM see BJE750
TERRAZOLE see EFK000
TERREIC ACID see TBF325
TERR-O-GAS 100 see MHR200
TERRILITIN see TBF350
TERRILYTIN see TBF350
TERSAN 1991 see BAV575
TERSAN see TFS350
TERSAN-LSR see MAS500
TERSAN-SP see CJA100
TERSASEPTIC see HCL000
TERSAVIN see PAQ120
TERSETILE BLUE RBL see CMP070
TERSETILE RUBINE FL see AKI750
TERSETILE YELLOW 5R see CMP090
TERSETILE YELLOW 5RL see CMP090
TERTRACID BRILLIANT LIGHT BLUE R see CMM080
TERTRACID CARMINE BLUE A see ERG100
TERTRACID FAST BLUE SR see ADE750
TERTRACID FAST BLUE TR see CMM100
TERTRACID LIGHT YELLOW 2R see SGP500
TERTRACID MILLING RED AGE see CMM325
TERTRACID MILLING RED G see CMM320
TERTRACID MILLING YELLOW R see CMM759
TERTRACID ORANGE I see FAG010
TERTRACID ORANGE II see CMM220
TERTRACID PONCEAU 2R see FMU070
TERTRACID RED A see FAG020
TERTRACID RED CA see HJF500
TERTRACID YELLOW M see MDM775
TERTRAL D see PEY500
TERTRAL EG see ALT500
TERTRAL ERN see NAW500
TERTRAL G see TGL750
TERTRAL P BASE see ALT250
TERTRANESE YELLOW N-2GL see AAQ250

2,4,5,7-TETRABROMO-3,6-FLUORANDIOL see BMO250, BNH500
2',4',5',7'-TETRABROMOFLUORESCEIN see BMO250
TETRABROMOFLUORESCEIN see BMO250, BNH500
2',4',5',7'-TETRABROMOFLUORESCEIN DISODIUM SALT see BNH500
TETRABROMOFLUORESCEIN S see BNH500
TETRABROMOFLUORESCEIN SOLUBLE see BNH500
2-(2,4,5,7-TETRABROMO-6-HYDROXY-3-OXO-3H-XANTHENE-9-YL)BENZOIC ACID, DISODIUM SALT see BNH500
TETRABROMOMETHANE see CBX750
TETRABROMOPHENOLSULFOPHTHALEIN see HAQ600
TETRABROMO-PLATINUM(2), DIPOTASSIUM see PLU250
TETRABROMOSILANE see SCP500
TETRABROMOSILICANE see SCP500
TETRABROMOSULFOPHTHALEIN see HAQ600
TETRABROMO-p-XYLEN (CZECH) see TBK250
α,α,α',α'-TETRABROMO-m-XYLENE see TBK000
TETRABROMO-p-XYLENE see TBK250
TETRABROMSULFTHALEIN see HAQ600
1,1,2,2-TETRABROOMETHAAN (DUTCH) see ACK250
TETRA-N-BUTYLAMMONIUM BROMIDE see TBK500
TETRABUTYLAMMONIUM BROMIDE see TBK500
TETRA-n-BUTYLAMMONIUM HYDROXIDE see TBK750
TETRABUTYLAMMONIUM HYDROXIDE see TBK750
TETRABUTYLAMMONIUM IODIDE see TBL000
TETRA-N-BUTYLAMMONIUMJODID (CZECH) see TBL000
TETRA-n-BUTYLAMMONIUM NITRATE see TBL250
TETRABUTYLAMMONIUM NITRATE see TBL250
TETRA-n-BUTYLCIN (CZECH) see TBM250
TETRABUTYL DICHLOROSTANNOXANE see TBL500
TETRA-N-BUTYLPHOSPHONIUM BROMIDE see TBL750
TETRA-N-BUTYLPHOSPHONIUM CHLORIDE see TBM000
TETRABUTYLSTANNANE see TBM250
TETRABUTYLTHIURAM DISULPHIDE see TBM750
TETRABUTYLTIN see TBM250
1,1,3,3-TETRABUTYLUREA see TBM850
TETRABUTYLUREA see TBM850
TETRACAINE see BQA010
TETRACAINE HYDROCHLORIDE see TBN000
TETRACAP see PCF275
TETRACAPS see TBX250
TETRACARBON MONOFLUORIDE see TBN100
TETRACARBONYLMOLYBDENUM DICHLORIDE see TBN150
TETRACARBONYL NICKEL see NCZ000
TETRACARBONYL(TRIFLUOROMETHYLTHIO)MANGANESE dimer see TBN200
1,2,4,5-TETRACARBOXYBENZENE see PPQ630
TETRACEMATE DISODIUM see EIX500
TETRACEMIN see EIV000
TETRACENE see NAI000, TEF500
TETRACENE EXPLOSIVE see TEF500
2,4,4',5-TETRACHLOOR-DIFENYL-SULFON (DUTCH) see CKM000
1,1,2,2-TETRACHLOORETHAAN (DUTCH) see TBQ100
TETRACHLOORETHEEN (DUTCH) see PCF275
TETRACHLOORKOOLSTOF (DUTCH) see CBY000
TETRACHLOORMETAAN see CBY000
1,1,2,2-TETRACHLORAETHAN (GERMAN) see TBQ100
TETRACHLORAETHEN (GERMAN) see PCF275
N-(1,1,2,2-TETRACHLORAETHYLTHIO)CYCLOHEX-4-EN-1,4-DIACARBOXIMID (GERMAN) see CBF800
N-(1,1,2,2-TETRACHLORAETHYLTHIO)TETRAHYDROPHTHALAMID (GERMAN) see CBF800
TETRACHLORDIAN (CZECH) see TBO750
3,4,6,R'-TETRACHLOR-DIPHENYLSULFID (GERMAN) see CKL750
2,4,4',5-TETRACHLOR-DIPHENYL-SULFON (GERMAN) see CKM000
1,1,2,2-TETRACHLORETHANE (FRENCH) see TBQ100
TETRACHLORETHANE see TBQ100
TETRACHLORKOHLENSTOFF, (GERMAN) see CBY000
TETRACHLORMETHAN (GERMAN) see CBY000
2,3,5,6-TETRACHLOR-3-NITROBENZOL (GERMAN) see TBR750
1,1,3,3-TETRACHLOROACETONE see TBN300
TETRACHLOROACETONE see TBN250
N,N,N',N'-TETRACHLOROADIPAMIDE see TBN400
TETRACHLOROAURIC(3+) ACID, SODIUM SALT see GIZ100
3,3',4,4'-TETRACHLOROAZOBENZENE see TBN500
3,4,3',4'-TETRACHLOROAZOBENZENE see TBN500
3,3',4,4'-TETRACHLOROAZOXYBENZENE see TBN550
3,4,3',4'-TETRACHLOROAZOXYBENZENE see TBN550
1,2,3,4-TETRACHLOROBENZENE see TBN740
1,2,3,5-TETRACHLOROBENZENE see TBN745
1,2,4,5-TETRACHLOROBENZENE see TBN750
3,4,5,6-TETRACHLORO-1,2-BENZENEDICARBOXYLIC ACID see TBT100
2,3,5,6-TETRACHLORO-1,4-BENZENEDICARBOXYLIC ACID, DIMETHYL ESTER see TBV250
3,4,5,6-TETRACHLORO-1,2-BENZENEDIOL see TBU500
2,2',5,5'-TETRACHLOROBENZIDINE see TBO000

3,3',6,6'-TETRACHLOROBENZIDINE see TBO000
TETRACHLOROBENZIDINE see TBO000
2,3,5,6-TETRACHLORO-1,4-BENZOQUINONE see TBO500
TETRACHLORO-1,4-BENZOQUINONE see TBO500
2,3,5,6-TETRACHLORO-p-BENZOQUINONE see TBO500
TETRACHLORO-p-BENZOQUINONE see TBO500
TETRACHLOROBENZOQUINONE see TBO500
2,2',6,6'-TETRACHLOROBIPHENYL see TBO600
2,6,2',6'-TETRACHLOROBIPHENYL see TBO600
3,3',4,4'-TETRACHLOROBIPHENYL see TBO700
3,4,3',4'-TETRACHLOROBIPHENYL see TBO700
2,2',5,5'-TETRACHLORO-(1,1'-BIPHENYL)-4,4'-DIAMINE, (9CI) see TBO000
2,2',6,6'-TETRACHLOROBISPHENOL A see TBO750
1,1,2,3-TETRACHLORO-1,3-BUTADIENE see TBO760
1,1,4,4-TETRACHLOROBUTATRIENE see TBO765
TETRACHLOROCARBON see CBY000
9-(3',4',5',6'-TETRACHLORO-o-CARBOXYPHENYL)-6-HYDROXY-2,4,5,7-TETRAIODO-3-ISOXANTHONE•2Na see RMP175
TETRACHLOROCATECHOL see TBU500
2,4,5,6-TETRACHLORO-3-CYANOBENZONITRILE see TBQ750
2,3,4,5-TETRACHLORO-2-CYCLOBUTEN-1-ONE see PCF250
2,3,5,6-TETRACHLORO-2,5-CYCLOHEXADIENE-1,4-DIONE see TBO500
2,2',5,5'-TETRACHLORO-4,4'-DIAMINODIPHENYL see TBO000
cis-TETRACHLORODIAMMINE PLATINUM(IV) see TBO776
trans-TETRACHLORODIAMMINE PLATINUM(IV) see TBO777
TETRACHLORODIAZOCYCLOPENTADIENE see TBO778
2,3,7,8-TETRACHLORODIBENZO(b,e)(1,4)DIOXAN see TAI000
2,3,7,8-TETRACHLORODIBENZO-1,4-DIOXIN see TAI000
1,2,3,8-TETRACHLORODIBENZO-p-DIOXIN see CDV125
2,3,6,7-TETRACHLORODIBENZO-p-DIOXIN see TAI000
2,3,7,8-TETRACHLORODIBENZO-p-DIOXIN see TAI000
2,3,7,8-TETRACHLORODIBENZOFURAN see TBO780
1,1,2,2-TETRACHLORO-1,2-DIFLUOROETHANE see TBP050
1,1,1,2-TETRACHLORO-2,2-DIFLUOROETHANE see TBP000
3,3,4,5-TETRACHLORO-3,6-DIHYDRO-1,2-DIOXIN see TBP075
4,5,6,7-TETRACHLORO-2-(2-DIMETHYLAMINOETHYL)-ISOINDOLINE DIMETHOCHLORIDE see CDY000
TETRACHLORODIPHENYLETHANE see BIM500
TETRACHLORODIPHENYL OXIDE see TBP250
2,4,4',5-TETRACHLORODIPHENYL SULFIDE see CKL750
2,4,5,4'-TETRACHLORODIPHENYL SULFIDE see CKL750
2,4,5,4'-TETRACHLORODIPHENYL SULFONE see CKM000
2,4,5,4'-TETRACHLORODIPHENYLSULPHONE see CKM000
TETRACHLORODIPHOSPHANE see TBP500
TETRACHLOROEPOXYETHANE see TBQ275
1,1,1,2-TETRACHLOROETHANE see TBQ000
1,1,2,2-TETRACHLOROETHANE see TBQ100
TETRACHLOROETHANE see TBP750
sym-TETRACHLOROETHANE see TBQ100
TETRACHLOROETHENE see PCF275
1,1,2,2-TETRACHLOROETHYLENE see PCF275
TETRACHLOROETHYLENE (DOT) see PCF275
TETRACHLOROETHYLENE CARBONATE see TBQ255
TETRACHLOROETHYLENE OXIDE see TBQ275
N-1,1,2,2-TETRACHLOROETHYLMERCAPTO-4-CYCLOHEXENE-1,2-CARBOXIMIDE see CBF800
N-((1,1,2,2-TETRACHLOROETHYL)SULFENYL)-cis-4-CYCLOHEXENE-1,2-DICARBOXIMIDE see CBF800
N-(1,1,2,2-TETRACHLOROETHYLTHIO)-4-CYCLOHEXENE-1,2-DICARBOXIMIDE see CBF800
TETRACHLOROGUAIACOL see TBQ290
2,3,4,5-TETRACHLOROHEXATRIENE see TBQ300
TETRACHLOROHYDROQUINONE see TBQ500
4,5,6,7-TETRACHLORO-1,3-ISOBENZOFURANDIONE see TBT150
TETRACHLOROISOPHTHALONITRILE see TBQ750
TETRACHLOROMETHANE see CBY000
1,2,4,5-TETRACHLORO-3-METHOXY-6-NITROBENZENE (9CI) see TBR250
TETRACHLORONAPHTHALENE see TBR000
2,3,5,6-TETRACHLORO-4-NITROANISOLE see TBR250
TETRACHLORONITROANISOLE see TBR250
2,3,4,5-TETRACHLORONITROBENZENE see TBS000
2,3,4,6-TETRACHLORONITROBENZENE see TBR500
2,3,5,6-TETRACHLORONITROBENZENE see TBR750
1,2,4,5-TETRACHLORO-3-NITROBENZENE see TBR750
1,2,3,5-TETRACHLORO-4-NITROBENZENE see TBR500
1,2,3,4-TETRACHLORO-5-NITROBENZENE see TBS000
TETRACHLOROPALLADATE(2) DIPOTASSIUM see PLN750
2,3,4,5-TETRACHLOROPHENOL see TBS500
2,3,4,6-TETRACHLOROPHENOL see TBT000
2,3,5,6-TETRACHLOROPHENOL see TBS750
2,4,5,6-TETRACHLOROPHENOL see TBT000
TETRACHLOROPHENOL see TBS250
TETRACHLOROPHENYL ETHER see TBP250
TETRACHLOROPHTHALIC ACID see TBT100
TETRACHLOROPHTHALIC ANHYDRIDE see TBT150

o-TETRACHLOROPHTHALODINITRILE see TBT200
3,4,5,6-TETRACHLOROPHTHALONITRILE see TBT200
m-TETRACHLOROPHTHALONITRILE see TBQ750
1,1,1,3-TETRACHLOROPROPANE see TBT250
1,1,3,3-TETRACHLOROPROPANONE see TBN300
1,1,3,3-TETRACHLORO-2-PROPANONE see TBN300
1,1,2,3-TETRACHLOROPROPENE see TBT500
2,3,4,5-TETRACHLOROPYRIDINE see TBT750
2,3,5,6-TETRACHLOROPYRIDINE see TBU000
2,4,5,6-TETRACHLOROPYRIMIDINE see TBU250
TETRACHLOROPYRIMIDINE see TBU250
TETRACHLOROPYROCATECHOL see TBU500
TETRACHLOROPYROCATECHOL HYDRATE see TBU750
TETRACHLORO-p-QUINONE see TBO500
TETRACHLOROQUINONE see TBO500
5,6,7,8-TETRACHLOROQUINOXALINE see CMA500
3,3',4',5-TETRACHLOROSALICYLANILIDE see TBV000
3,5,3',4'-TETRACHLOROSALICYLANILIDE see TBV000
TETRACHLOROSILANE see SCQ500
TETRACHLOROSTANNANE PENTAHYDRATE see TGC282
TETRACHLOROTELLURIUM see TAJ250
TETRACHLOROTEREPHTHALIC ACID DIMETHYL ESTER see TBV250
4,5,6,7-TETRACHLORO-2',4',5',7'-TETRAIODOFLUORESCEIN DISODIUM SALT
    see RMP175
TETRACHLOROTHIOFENE see TBV750
2,3,4,5-TETRACHLOROTHIOPHENE see TBV750
TETRACHLOROTHIOPHENE see TBV750
TETRACHLOROTHORIUM see TFT000
α-α-α-4-TETRACHLOROTOLUENE see TIR900
p,α-α-α-TETRACHLOROTOLUENE see TIR900
4,5,6,7-TETRACHLORO-2-(TRIFLUOROMETHYL)BENZIMIDAZOLE see TBW000
TETRACHLOROTRIFLUOROMETHYLPHOSPHORANE see TBW025
5-endo,6-exo-2,2,5,6-TETRACHLORO-1,7,7-TRIS(CHLOROMETHYL)-BICY-
    CLO(2.2.1) HEPTANE see THH575
2,3,5,6-TETRACHLORPHTHALSAURE-DIMETHYLESTER (GERMAN) see
    TBV250
TETRACHLORPYROKATECHIN (CZECH) see TBU750
TETRACHLORURE d'ACETYLENE (FRENCH) see TBQ100
TETRACHLORURE de CARBONE (FRENCH) see CBY000
TETRACHLORURE de SILICIUM (FRENCH) see SCQ500
TETRACHLORURE de TITANE (FRENCH) see TGH350
TETRACHLORVINPHOS see TBW100
TETRACICLINA CLORIDRATO (ITALIAN) see TBX250
TETRACICLINA-l-METILENLISINA (ITALIAN) see MRV250
TETRACID see MPQ750
TETRACID CARMINE BLUE V see ADE500
TETRACID MILLING RED B see CMM330
TETRACID MILLING RED G see CMM330
2,4,4',5-TETRACLORO-DIFENIL-SOLFONE (ITALIAN) see CKM000
1,1,2,2-TETRACLOROETANO (ITALIAN) see TBQ100
TETRACLOROETENE (ITALIAN) see PCF275
TETRACLOROMETANO (ITALIAN) see CBY000
TETRACLORURO di CARBONIO (ITALIAN) see CBY000
TETRACOMPREN see TBX250
2,6,10,14,18,22-TETRACOSAHEXENE, 2,6,10,15,19,23-HEXAMETHYL-, (all-E)-
    see SLG800
TETRACOSAHYDRO DI-
    BENZ(b,n)(1,4,7,10,13,16,19,22)OCTAOXACYCLOTETRACOSIN see DGT300
TETRACOSANE, 2,6,10,15,19,23-HEXAMETHYL- see SLG700
(1,1,3,3-TETRACYANOALLYL) SODIUM see TAI050
1,1,2,2-TETRACYANOETHENE see EEE500
TETRACYANOETHENE see EEE500
1,1,2,2-TETRACYANOETHYLENE see EEE500
TETRACYANOETHYLENE see EEE500
TETRACYANONICKELATE(2−) DIPOTASSIUM, HYDRATE see TBW250
TETRACYANOOCTAETHYLTETRAGOLD see TBW500
N,N,N',N'-TETRACYANOMETHYLAETHYLENEDIAMIN (GERMAN) see EJB000
TETRACYANOQUINODIMETHAN see TBW750
TETRACYANOQUINODIMETHAN(e) see TBW750
7,7',8,8'-TETRACYANOQUINODIMETHANE see TBW750
7,7,8,8-TETRACYANOQUINODIMETHANE see TBW750
TETRACYANO-p-QUINODIMETHANE see TBW750
TETRACYCLINE see TBX000
TETRACYCLINE CHLORIDE see TBX250
TETRACYCLINE, 7-CHLORO-6-DEMETHYL- see MIJ500
TETRACYCLINE HYDROCHLORIDE see TBX250
TETRACYCLINE, 5-HYDROXY- see HOH500
TETRACYCLINE I see TBX000
TETRACYCLINE-l-METHYLENE LYSINE see MRV250
TETRACYDIN see ABG750
TETRACYN see TBX000
TETRA-D see TBX250
9Z,12E-TETRADECADIENYL ACETATE see TBX300
9,11-TETRADECADIEN-1-YL, ACETATE, (E,Z)- see FAS100

9,12-TETRADECADIEN-1-YL, ACETATE, (E,Z)- see TBX300
1-TETRADECANAL see TBX500
TETRADECANAL see TBX500
1-TETRADECANAMINIUM, N,N,N-TRIMETHYL-, BROMIDE (9CI) see TCB200
TETRADECANE see TBX750
TETRADECANOIC ACID see MSA250
TETRADECANOIC ACID, BUTYL ESTER see MSA300
TETRADECANOIC ACID, ISOPROPYL see IQN000
TETRADECANOIC ACID, 1-METHYLETHYL ESTER see IQN000
TETRADECANOIC ACID, SODIUM SALT (9CI) see SIN900
N-TETRADECANOL-1 see TBY250
1-TETRADECANOL see TBY250
TETRADECANOL, mixed isomers see TBY500
TETRADECANOL condensed with 7 moles ETHYLENE OXIDE see TBY750
2-TETRADECANOL, 1-(2-HYDROXYETHYLAMINO)-, and 1-(2-HYDROXYETHYL)-
    2-DODECANOL (1:1) see HKS400
TETRADECANOYLETHYLENEIMINE see MSB000
12-TETRADECANOYLPHORBOL-13-ACETATE see PGV000
N-TETRADECANOYL-N-TETRADECANOYLOXY-2-AMINOFLUORENE see
    MSB250
12-o-TETRADECA-2-cis-4-trans-6,8-TETRAENOYLPHORBOL-13-ACETATE see
    TCA250
4,8,12-TETRADECATRIENOIC ACID, 5,9,13-TRIMETHYL-, 3,7-DIMETHYL-2,6-
    OCTADIENYLESTER, (E,E,E)- see GDG200
TETRADECIN see TBX000
n-TETRADECOIC ACID see MSA250
N-TETRADECYL ALCOHOL see TBY250
TETRADECYL ALCOHOL see TBY250, TBY500
1-TETRADECYL ALDEHYDE see TBX500
TETRADECYL DIMETHYL BENZYLAMMONIUM CHLORIDE see TCA500
TETRADECYLHEPTAETHOYLATE see TCA750
TETRADECYL OCTADECANOATE see TCB100
TETRADECYL PHOSPHONIC ACID see TCB000
TETRADECYL SODIUM SULFATE see SIO000
TETRADECYL STEARATE see TCB100
TETRADECYL SULFATE, SODIUM SALT see SIO000
TETRADECYLTRIMETHYLAMMONIUM BROMIDE see TCB200
1,2,6,7-TETRADEHYDRO-14,17-DIHYDRO-3-METHOXY-16(15H)-OXAERYTHRI-
    NAN-15-ONE, HYDROCHLORIDE see EDH000
7,8,13,13A-TETRADEHYDRO-9,10-DIMETHOXY-2,3-(METHYLENEDIOX-
    Y)BERBINIUM SULFATE TRIHYDRATE see BFN750
TETRADEHYDRODOISYNOLIC ACID METHYL ETHER see BIT000
6,7,8,14-TETRADEHYDRO-4,5-α-EPOXY-3,6-DIMETHOXY-17-METHYLMORPHI-
    NAN see TEN000
6,7,8,14-TETRADEHYDRO-4,5-α-EPOXY-3,6-DIMETHOXY-17-METHYLMORPHI-
    NAN HYDROCHLORIDE see TEN100
4,5,6,6a-TETRADEHYDRO-9-HYDROXY-1,2,10-TRIMETHOXYNORAPORPHIN-7-
    ONE see ARQ325
(3-β)-1,2,6,7-TETRADEHYDRO-3-METHOXYERYTHRINAN-16-OL HYDROCHLO-
    RIDE see EDE500
(3-β)-1,2,6,7-TETRADEHYDRO-3-METHOXY-15,16-(METHYLENE-
    BIS(OXY))ERYTHRINAN, HYDROBROMIDE see EDF000
7,8,13,13a-TETRADEHYDRO-2,3,9,10-TETRAMETHOXYBERBINIUM HYDROXIDE
    see PAE100
12-o-TETRADEKANOYLPHORBOL-13-ACETAT (GERMAN) see PGV000
2,3,4,6-TETRADEOXY-4-((5,6,11,12,13,14,15,16,16A,17,17A,18,21,22-TETRADE-
    CAHYDRO-2,12,14,16-TETRAHYDROXY-10-METHOXY-3,6,11,13,15,20-HEX-
    AMETHYL-5,18,21,26-TETRAOXO-4,6-EPOXY-1,23-METHANOBAN-
    ZO(d)CYCLOP ROP(n)(1,9)OXAAZACYCLOTETRACOSIN-25(10H)-
    YLIDENE)AMINO)-l-erythro-HEXOPYRANOSE, 12-ACETATE see THD300
TETRADICHLONE see CKM000
TETRADIFON see CKM000
TETRADIN see DXH250
TETRADINE see DXH250
TETRADIOXIN see TAI000
TETRADIPHON see CKM000
TETRADONIUM BROMIDE see TCB200
TETRADOX see HGP550
TETRAETHANOL AMMONIUM HYDROXIDE see TCB500
1,1,3,3-TETRAETHOXYPROPANE see MAN750
TETRAETHOXY PROPANE see MAN750
TETRAETHOXYSILANE see EPF550
TETRAETHYLAMMONIUM see TCB725
TETRAETHYLAMMONIUM BOROHYDRIDE see TCB750
TETRAETHYL AMMONIUM BROMIDE see TCC000
TETRAETHYLAMMONIUM CHLORIDE see TCC250
TETRAETHYLAMMONIUM HYDROXIDE see TCC500
TETRAETHYLAMMONIUM IODIDE see TCC750
TETRAETHYLAMMONIUM PERCHLORATE (dry) (DOT) see TCD002
TETRAETHYLAMMONIUM PERCHLORATE see TCD000
N,N,N',N'-TETRAETHYL-1,2-BENZENEDICARBOXAMIDE see GCE600
TETRA(2-ETHYLBUTOXY) SILANE see TCD250
N,N,N',N'-TETRAETHYL-2-BUTYNYLENEDIAMINE see BJA250
TETRA(2-ETHYLBUTYL) ORTHOSILICATE see TCD250

1,2,3,4-TETRAHYDRODIBENZ(a,h)ANTHRACENE see TCN750
1,2,3,4-TETRAHYDRODIBENZ(a,j)ANTHRACENE see TCO000
exo-TETRAHYDRODI(CYCLOPENTADIENE) see TLR675
1-(TETRAHYDRODICYCLOPENTADIENYL)-3,3-DIMETHYLUREA see HDP500
7,8,9,10-TETRAHYDRO-N,N-DIETHYL-6H-CYCLOHEPTA(b)QUINOLINE-11-CARBOXAMIDE, see TCO100
(+)-Z-7,8,9,10-TETRAHYDRO-7-α,8-β-DIHDYROXY-9-α,10-α-EPOXYBENZO(a)PYRENE see DMP600
(−)-Z-7,8,9,10-TETRAHYDRO-7-α,8-β-DIHDYROXY-9-α,10-α-EPOXYBENZO(a)PYRENE see DMP800
(E)-8,9,10,11-TETRAHYDRO-8-β,9-α-DIHYDROXY-10-α,11-α-BENZ(a)ANTHRACENE see DMO600
(E)-1,2,3,4-TETRAHYDRO-3-α,4-β-DIHYDROXY-1-α,2-α-EPOXYBENZ(a)ANTHRACENE see DLE000
(E)-(±)-8,9,10,11-TETRAHYDRO-8-β,9-α-DIHYDROXY-10-β,11-β-EPOXYBENZ(a)ANTHRACENE see DMP200
8,9,10,11-TETRAHYDRO-10,11-DIHYDROXY-8-α,9-α-EPOXYBENZ(a)ANTHRACENE see BAD000
(±)-7,8,9,10-TETRAHYDRO-7-α,8-β-DIHYDROXY-9-α,10-α-EPOXYBENZO(a)PYRENE see DMR000
(−)-7,8,9,10-TETRAHYDRO-7-β,8-α-DIHYDROXY-9-α,10-α-EPOXY-BENZO(a)PYRENE see DVO175
(+)-E-7,8,9,10-TETRAHYDRO-7-α,8-β-DIHYDROXY-9-β,19-β-EPOXY-BENZO(a)PYRENE see BMK620
(+)-cis-7,8,9,10-TETRAHYDRO-7-β,8-α-DIHYDROXY-9-β,10-β-EPOXYBENZO(a)PYRENE see DMP600
(−)-Z-7,8,9,10-TETRAHYDRO-7-β,8-α-DIHYDROXY-9-β,10-β-EPOXYBENZO(a)PYRENE see DMP800
1,2,3,11a-TETRAHYDRO-3,8-DIHYDROXY-7-METHOXY-5H-PYRROLO(2,1-c)(1,4)BENZODIAZEPIN-5-ONE-(3S-cis) see TCP000
TETRAHYDRO-4,6-DIIMINO-s-TRIAZINE-2(1H)-THIONE see DCF000
(s)-5,8,13,13a-TETRAHYDRO-9,10-DIMETHOXY-6H-BENZO(g)-1,3-BENZODIOXOLO(5,6-a)QUINOLIZINE see TCJ800
(R)-5,6,6a,7-TETRAHYDRO-1,2-DIMETHOXY-6-METHYL-4H-DIBENZO(de,g)QUINOLINE see NOE500
(S)-5,6,6a,7-TETRAHYDRO-1,10-DIMETHOXY-6-METHYL-4H-DIBENZO(de,g)QUINOLINE-2,9-DIOL see DNZ100
1,2,3,4-TETRAHYDRO-7,12-DIMETHYLBENZ(a)ANTHRACENE see TCP600
8,9,10,11-TETRAHYDRO-7,12-DIMETHYLBENZ(a) ANTHRACENE see DUF000
1,2,3,6-TETRAHYDRO-1,3-DIMETHYL-2,6-DIOXO-7H-PURINE-7-ACETIC ACID compounded with N-((4-METHOXYPHENYL)METHYL)-N′,N′-DIMETHYL-N-2-PYRIDINYL-1,2-ETHANEDIAMINE (1:1) see MCJ300
TETRAHYDRODIMETHYLFURAN see TCQ250
TETRAHYDRODIMETHYL FURANE see TCQ250
2,3,4,5-TETRAHYDRO-2,8-DIMETHYL-5-(2-(6-METHYL-3-PYRIDYL)ETHYL)-1H-PYRID O(4,3-b)INDOLE see TCQ260
TETRAHYDRO-3,5-DIMETHYL-4H,1,3,5-OXADIAZINE-4-THIONE see TCQ275
TETRAHYDRO-2,6-DIMETHYL-4H-PYRAN-4-ONE see TCQ350
TETRAHYDRO-2H-3,5-DIMETHYL-1,3,5-THIADIAZINE-2-THIONE see DSB200
TETRAHYDRO-3,5-DIMETHYL-2H-1,3,5-THIADIAZINE-2-THIONE see DSB200
TETRAHYDRO-2,4-DIMETHYLTHIOPHENE-1,1-DIOXIDE see DUD400
TETRAHYDRO-1,4-DIOXIN see DVQ000
TETRAHYDRO-p-DIOXIN see DVQ000
1,2,3,4-TETRAHYDRO-2,4-DIOXO-5-FLUORO-N-HEXYL-1-PYRIMIDINECARBOXAMIDE see CCK630
1,2,3,6-TETRAHYDRO-2,6-DIOXO-5-FLUORO-4-PYRIMIDINECARBOXYLIC ACID see TCQ500
TETRAHYDRO-2,5-DIOXOFURAN see SNC000
1,2,3,6-TETRAHYDRO-3,6-DIOXOPYRIDAZINE see DMC600
1,4,5,6-TETRAHYDRO-4,6-DIOXO-s-TRIAZINE-2-CARBOXYLIC ACID, POTASSIUM SALT see AFQ750
5′,6′,7′,8′-TETRAHYDRODISPIRO(CYCLOHEXANE-1,2′(3′H)-QUINAZOLINE-4′,1′′(4 a′H)-CYCLOHEXANE) see TCQ600
TETRAHYDRO-2,5-DIVINYL-2H-PYRAN see DXR400
TETRAHYDRO-2,5-DIVINYLPYRAN see DXR400
1,2,3,4-TETRAHYDRO-DMBA see TCP600
9,10,11,12-TETRAHYDRO-9,10-EPOXY-BENZO(e)PYRENE see ECQ150
7,8,9,10-TETRAHYDRO-9,10-EPOXY-BENZO(a)PYRENE see ECQ100
(±)-1,2,3,4-TETRAHYDRO-3,α,4,α-EPOXY-1,β,2,α-CHRYSENEDIOL see DMS200
(±)-1,2,3,4-TETRAHYDRO-3,β,4,β-EPOXY-1-β,2,α-CHRYSENEDIOL see DMS400
TETRAHYDRO-3,4-EPOXYTHIOPHENE-1,1-DIOXIDE see ECP000
1,2,3,4-TETRAHYDRO-6-ETHYL-1,1,4,4-TETRAMETHYLNAPHTHALENE see TCR250
5,6,7,8-TETRAHYDRO-3-ETHYL-5,5,8,8-TETRAMETHYL-2-NAPHTHALENECARBOXALDEHYDE see FNK200
5,6,7,8-TETRAHYDROFOLIC ACID see TCR400
TETRAHYDROFOLIC ACID see TCR400
TETRAHYDROFTALANHYDRID (CZECH) see TDB000
TETRAHYDROFURAAN (DUTCH) see TCR750
TETRAHYDROFURAN see TCR750
TETRAHYDRO-2-FURANCARBINOL see TCT000
TETRAHYDRO-2-FURANMETHANOL see TCT000
TETRAHYDROFURANNE (FRENCH) see TCR750
TETRAHYDRO-3-FURANOL see HOI200
TETRAHYDRO-2-FURANONE see BOV000

TETRAHYDRO-2-FURANPROPANOL see TCS000
1-(TETRAHYDROFURAN-2-YL)-5-FLUOROURACIL see FLZ050
TETRAHYDROFURFURYL ALCOHOL see TCT000
TETRAHYDROFURFURYLAMINE see TCS000
TETRAHYDROFURFURYL BROMIDE see TCS550
α-TETRAHYDROFURFURYL-1-NAPHTHALENEPROPIONIC ACID 2-(DIETHYLAMINO)ETHYL ESTER see NAE000
TETRAHYDROFURYLALKOHOL (CZECH) see TCT000
N′-(2-TETRAHYDROFURYL)-5-FLUOROURACIL see FLZ050
1-(2-TETRAHYDROFURYL)-5-FLUOROURACIL mixture with URACIL (1:4) see UNJ810
2-TETRAHYDROFURYL HYDROPEROXIDE see TCS750
TETRAHYDRO-2-FURYLMETHANOL see TCT000
TETRAHYDROGERANIOL see DTE600
TETRAHYDROGERANYL ACETATE see DTE800
TETRAHYDROGERANYL BUTYRATE see DTF000
1′,3′-α,4′,7′-α-TETRAHYDRO-7′-α- β-3-(GLUCOPYRANOSYLOXY)-4′-HYDROXY-2′,4′,6′-TRIMETHYL-SPIRO(CYCLOPROPANE-1,5′-(5H)INDEN)-3′,(2′H)-ONE see POI100
TETRAHYDROHARMINE see TCT250
2,3-cis-1,2,3,4-TETRAHYDRO-5-((2-HYDROXY-3-tert-BUTYLAMINO)PROPOXY)-2,3-NAPHTHALENEDIOL see CNR675
(E)-5,10,11,11a-TETRAHYDRO-9-HYDROXY-11-METHOXY-8-METHYL-5-OXO-1H-PYRROLO(2,1-c)(1,4)BENZODIAZEPINE-2-ACRYLAMIDE MONOHYDRATE see API125
TETRAHYDRO-7-α-(1-HYDROXY-1-METHYLBUTYL)-6,14-endo-ETHENOORIPAVINE see EQO450
1-(TETRAHYDRO-5-HYDROXY-4-METHYL-3-FURANYL)-ETHANONE (9CI) see BMK290
1,2,3,4-TETRAHYDRO-5-(HYDROXYMETHYL)-2-(ISOPROPYLAMINO)-1,6-NAPHTHALENEDIOL SESQUIHYDRATE (6-E), HYDROCHLORIDE see TCU000
TETRAHYDRO-7-α-(2-HYDROXY-2-PENTYL)-6,14-endo-ETHENOORIPAVINE see EQO450
3a,4,7,7a-TETRAHYDRO-1H-INDENE see TCU250
3a,4,7,7a-TETRAHYDROINDENE see TCU250
7,7,8,9-TETRAHYDROINDENE see TCU250
TETRAHYDROINDENE (RUSSIAN) see TCU250
TETRAHYDRO-3-IODOTHIOPHENE-1,1-DIOXIDE see IFG000
3a,4,7,7a-TETRAHYDRO-1,3-ISOBENZOFURANDIONE see TDB000
1,2,3,4-TETRAHYDRO-2-((ISOPROPYLAMINO)METHYL)-7-NITRO-6-QUINOLINEMETHANOL see OLT000
1,2,3,4-TETRAHYDROISOQUINOLINE see TCU500
TETRAHYDROISOQUINOLINE see TCU500
TETRAHYDRO-1,4-ISOXAZINE see MRP750
TETRAHYDRO-p-ISOXAZINE see MRP750
TETRAHYDROLINALOOL see LFY510, TCU600
TETRAHYDRO-4,7-METHANOINDAN-5(4H)-ONE see OPC000
3a,4,7,7a-TETRAHYDRO-4,7-METHANOINDENE see DGW000
TETRAHYDRO-4,7-METHANOINDENOL see HKB200
1,4,4a,8a-TETRAHYDRO-1,4-METHANONAPHTHALENE-5,8-DIONE see TCU650
1,2,3,4-TETRAHYDRO-1,4-METHANONAPHTHALEN-6-YL N-METHYL-N-(m-TOLYL)CARBAMOTHIOATE see MQA000
o-(1,2,3,4-TETRAHYDRO-1,4-METHANONAPHTHALEN-6-YL)-m,N-DIMETHYL-THIO-CARBANILATE see MQA000
o-(5,6,7,8-TETRAHYDRO-5,8-METHANO-2-NAPHTHYL)-N-METHYL-N-(m-METHYLPHENYL)THIOCARBAMATE see MQA000
5,6,7,8-TETRAHYDRO-4-METHOXY-6-METHYL-1,2-DIOXOLO-(4,5-g)ISOQUINOLIN-5-OL (9CI) see CNT625
1,2,3,4-TETRAHYDRO-7-METHOXY-1-METHYL-9H-PYRIDO(3,4-b)INDOLE HYDROCHLORIDE see TCT250
1,2,3,4-TETRAHYDRO-6-METHOXYQUINOLINE see TCU700
4,6,7,14-TETRAHYDRO-5-METHYL-BIS(1,3)BENZODIOXOLO(4,5-c: 5′,6′-g)AZECIN-13(5H)-ONE see FOW000
1,2,3,4-TETRAHYDRO-11-METHYLCHRYSEN-1-ONE see MNG250
1,2,3,4-TETRAHYDRO-1-((6-METHYL-2-CYCLOHEXEN-1-YL)CARBONYL)PYRIDINE see TCV375
1,2,3,6-TETRAHYDRO-1-((2-METHYLCYCLOHEXYL)CARBONYL)PYRIDINE see TCV400
5,6,7,8-TETRAHYDRO-6-METHYL-1,3-DIOXOLO-(4,5-g)ISOQUINOLIN-5-OL see HGR000
1,2,3,4-TETRAHYDRO-2-(((1-METHYL-ETHYL)AMINO)METHYL)-7-NITRO-6-QUINOLINEMETHANOL see OLT000
4a,5,7a,8-TETRAHYDRO-12-METHYL-9H-9,9c-IMINOETHANOPHENANTHRO(4,5-bcd)FURAN-3,5-DIOL see MRO500
3a,4,7,7a-TETRAHYDRO-3a-METHYL-1,3-ISOBENZOFURANDIONE see EBW000
TETRAHYDRO-4-METHYL-2-(2-METHYLPROPEN-1-YL)PYRAN see RNU000
1,2,5,6-TETRAHYDRO-1-METHYLNICOTINIC ACID, METHYL ESTER see AQT750
1,2,5,6-TETRAHYDRO-1-METHYLNICOTINIC ACID, METHYL ESTER, HYDROBROMIDE see AQU000
TETRAHYDRO-3-METHYL-4-((5-NITROFURFURYLIDENE)AMINO)-2H-1,4-THIAZINE-1,1-DIOXIDE see NGG000
TETRAHYDRO-N-METHYL-N-NITROSO-3-THIOPHENAMINE 1,1-DIOXIDE see NKQ500

3,4,5,6,7-TETRAHYDRO-5-METHYL-1-PHENYL-1H-2,5-BENZOXAZOCINE see FAL000

2,3,4,9-TETRAHYDRO-2-METHYL-9-PHENYL-1H-INDENO(2,1-c)PYRIDINE HYDROCHLORIDE see TEQ700

1,2,3,4-TETRAHYDRO-2-METHYL-4-PHENYL-8-ISOQUINOLINAMINE (9CI) see NMV700

2,3,4,5-TETRAHYDRO-2-METHYL-5-(PHENYLMETHYL)-1H-PYRIDO(4,3-b)INDOLE see MBW100

TETRAHYDRO-2-METHYL-6-(TETRAHYDRO-2,5-DIOXO-3-FURYL)PYRAN-3,4-DICARBOXYLIC ANHYDRIDE POLYMER see TCW500

(E)-4,5,6-TETRAHYDRO-1-METHYL-2-(2-(2-THIENYL)ETHENYL)PYRIMIDINE see TCW750

(E)-1,4,5,6-TETRAHYDRO-1-METHYL-2-(2-(2-THIENYL)VINYL)PYRIMIDINE TARTRATE (1:1) see TCW750

5,6,7,8-TETRAHYDRO-1-NAFTYL-N-METHYLKARBAMAT (CZECH) see TCY275

TETRAHYDRO-α-(1-NAPHTALENYLMETHYL)-2-FURANPROPANOIC ACID 2-(DIETHYLAMINO)ETHYL ESTER see NAE000

1,2,3,4-TETRAHYDRONAPHTHALENE see TCX500

TETRAHYDRONAPHTHALENE see TCX500

1,2,3,4-TETRAHYDRO-1-NAPHTHOL see TDI350

1,2,3,4-TETRAHYDRO-2-NAPHTHOL see TCX750

5,6,7,8-TETRAHYDRO-1-NAPHTHOL see TCY000

1,2,3,4-TETRAHYDRO-α-NAPHTHOL see TDI350

5,6,7,8-TETRAHYDRO-β-NAPHTHOL see TCY000

N-d,l-ac-TETRAHYDRO-β-NAPHTHYL-N-ALLYL-β-ALANINDIAETHYLAMID-HYDROCHLORID (GERMAN) see AGS375

1,2,3,4-TETRAHYDRO-2-NAPHTHYLAMINE see TCY250

β-1,2,3,4-TETRAHYDRONAPHTHYLAMINE see TCY250

β-TETRAHYDRONAPHTHYLAMINE see TCY250

TETRAHYDRO-β-NAPHTHYLAMINE see TCY250

2-(1,2,3,4-TETRAHYDRO-1-NAPHTHYLAMINO)-2-IMIDAZOLINE HYDROCHLORIDE see TCY260

TETRAHYDRONAPHTHYLAMINOIMIDAZOLINE HYDROCHLORIDE see THJ825

2'-(5,6,7,8-TETRAHYDRO-1-NAPHTHYLAMINO)IMIDAZOLINE HYDROCHLORIDE see THJ825

2-((5,6,7,8-TETRAHYDRO-1-NAPHTHYL)AMINO)-2-IMIDAZOLINE HYDROCHLORIDE HYDRATE see RGP450

3-(1,2,3,4-TETRAHYDRO-1-NAPHTHYL)-4-HYDROXYCUMARIN (GERMAN) see EAT600

2-(1,2,3,4-TETRAHYDRO-1-NAPHTHYL)-2-IMIDAZOLINE HYDROCHLORIDE see VRZ000

2-(1,2,3,4-TETRAHYDRO-1-NAPHTHYL)-2-IMIDAZOLINE MONOHYDROCHLORIDE see VRZ000

5,6,7,8-TETRAHYDRO-1-NAPHTHYL METHYLCARBAMATE see TCY275

TETRAHYDRO-α-(1-NAPHTHYLMETHYL)-2-FURANPROPANOIC ACID 2-(DIETHYLAMINO)ETHYL ESTER see NAE000

3-(1,2,3,4-TETRAHYDRO-1-NAPHTYL)-4-HYDROXYCOUMARINE (FRENCH) see EAT600

TETRAHYDRONERAL see TCY300

dl-TETRAHYDRONICOTYRINE see NDN000

TETRAHYDRO-1-((5-NITROFURFURYLIDENE)AMINO)-2(1H)-PYRIMIDINONE see FPO100

TETRAHYDRO-1-(5-NITROFURFURYLIDENEAMINO)-2-PYRIMIDONE see FPO100

TETRAHYDRO-2-NITROSO-2H-1,2-OXAZINE see NLT500

1,2,5,6-TETRAHYDRO-1-NITROSO-3-PYRIDINECARBOXYLIC ACID METHYL ESTER see NKH500

TETRAHYDRO-N-NITROSOPYRROLE see NLP500

TETRAHYDRONORHARMAN see TCY500

1,2,3,4-TETRAHYDRO-2-OCTANOYLISOQUINOLINE see ODO000

TETRAHYDRO-1,4-OXAZINE see MRP750

TETRAHYDRO-2H-1,4-OXAZINE see MRP750

TETRAHYDRO-1,4-OXAZINYLMETHYLCODEINE see TCY750

1,2,3,4-TETRAHYDRO-2-(1-OXOOCTYL)ISOQUINOLINE see ODO000

N-(TETRAHYDRO-2-OXO-3-THIENYL)ACETAMIDE see TCZ000

TETRAHYDROPHENOBARBITAL see TDA500

3,4,5,6-TETRAHYDRO-6-PHENOXYMETHYL-2H-1,3-OXAZINE-2-THIONE see PDU750

TETRAHYDRO-6-(PHENOXYMETHYL)-2H-1,3-OXAZINE-2-THIONE see PDU750

l-(−)-2,3,5,6-TETRAHYDRO-6-PHENYL-IMIDAZO(2,1-B)THIAZOLE HYDROCHLORIDE see LFA020

2,3,5,6-TETRAHYDRO-6-PHENYLIMIDAZO(2,1-b)THIAZOLE see LFA000

(−)-2,3,5,6-TETRAHYDRO-6-PHENYLIMIDAZO(2,1-b)THIAZOLE HYDROCHLORIDE see LFA020

(±)-2,3,5,6-TETRAHYDRO-6-PHENYLIMIDAZO(2,1-b)THIAZOLE MONOHYDROCHLORIDE see TDX750

7,8,9,10-TETRAHYDRO-11-(4-PHENYL-1-PIPERAZINYL)-6H-CYCLOHEPTA(b)QUINOLINE DIHYDROCHLORIDE see CCW800

TETRAHYDROPHTHALIC ACID ANHYDRIDE see TDB000

TETRAHYDROPHTHALIC ACID IMIDE see TDB100

1,2,3,6-TETRAHYDRO PHTHALIC ANHYDRIDE see TDB000

TETRAHYDROPHTHALIC ANHYDRIDE see TDB000

Δ⁴-TETRAHYDROPHTHALIC ANHYDRIDE see TDB000

1,2,3,6-TETRAHYDROPHTHALIMIDE see TDB100

TETRAHYDROPHTHALIMIDE see TDB100

Δ⁴-TETRAHYDROPHTHALIMIDE see TDB100

α-(1,2,3,6-TETRAHYDROPHTHALIMIDO)GLUTARIMIDE see TDB200

TETRAHYDRO-6-PROPYL-2H-PYRAN-2-ONE see OCE000

2,3,4,5-TETRAHYDRO-6-PROPYL-PYRIDINE see CNH730

TETRAHYDROPTEROYLGLUTAMIC ACID see TCR400

TETRAHYDROPYRAN-2-METHANOL see MDS500

4'-O-TETRAHYDROPYRANYLADRIAMYCIN HYDROCHLORIDE see TDB600

(2''-R)-4'-O-TETRAHYDROPYRANYLADRIAMYCIN MONOHYDROCHLORIDE see TDB600

TETRAHYDROPYRANYL-2-METHANOL see MDS500

1-(1,2,3,6-TETRAHYDROPYRIDINO)-3-o-TOLYLOXYPROPAN-2-OL HYDROCHLORIDE see TGK225

1,2,3,4-TETRAHYDRO-9H-PYRIDO(3,4-b)INDOLE, HYDROCHLORIDE see TCY500

TETRAHYDRO-2(1H)-PYRIMIDINETHIONE see TLS000

TETRAHYDROPYRROLE see PPS500

1,2,3,9-TETRAHYDROPYRROLO(2,1-B)QUINAZOLIN-3-OL see VGA000

TETRAHYDROSERPENTINE see AFG750

TETRAHYDROSPIRAMYCIN see TDB750

1,2,3,4-TETRAHYDROSTYRENE see CPD750

2,4,4a,10b-TETRAHYDRO-3,4,8,10-TETRAHYDROXY-2-(HYDROXYMETHYL)-9-METHOXY-PYRANO(3,2-o)(2)BENZOPYRAN-6(2H)-ONE HYDRATE see BFO100

(S)-5,8,13,13a-TETRAHYDRO-2,3,9,10-TETRAMETHOXY-6H-DIZENZO(a,g)QUINOLIZINE HYDROCHLORIDE see GEO600

(s)-5,6,6a,7-TETRAHYDRO-1,2,9,10-TETRAMETHOXY-6-METHYL-4H-DIBENZO(de,g)QUINOLINE see TDI475

(s)-5,6,6a,7-TETRAHYDRO-1,2,9,10-TETRAMETHOXY-6-METHYL-4H-DIBENZO(de,g)QUINOLINE (9CI) see TDI475

N-(5,6,7,9-TETRAHYDRO-1,2,3,10-TETRAMETHOXY-9-OXOBENZO(α)HEPTALEN-7-YL)-ACETAMIDE see CNG830

6,7,8,9-TETRAHYDRO-5H-TETRAZOLOAZEPINE see PBI500

o-((4,5,6,7-TETRAHYDROTHIENO(3,2-c)PYRIDIN-5-YL)METHYL)BENZONITRILE METHANESULFONATE see TDC725

TETRAHYDROTHIOFEN see TDC730

TETRAHYDROTHIOFEN-1,1-DIOXID see SNW500

TETRAHYDROTHIOPHENE see TDC730

2,3,4,5-TETRAHYDROTHIOPHENE-1,1-DIOXIDE see SNW500

TETRAHYDROTHIOPHENE 1,1-DIOXIDE see SNW500

TETRAHYDROTHIOPHENE DIOXIDE see SNW500

TETRAHYDRO-THIOPHENE-1-OXIDE see TDQ500

TETRAHYDRO-2-THIOPHENONE see TDC800

1,2,3,7-TETRAHYDRO-2-THIOXO-6H-PURIN-6-ONE see TFL500

3a,4,7,7a-TETRAHYDRO-N-(TRICHLOROMETHANESULPHENYL)PHTHALIMIDE see CBG000

3a,4,7,7a-TETRAHYDRO-2-((TRICHLOROMETHYL)THIO)-1H-ISOINDOLE-1,3(2H)-DIONE see CBG000

1,2,3,6-TETRAHYDRO-N-(TRICHLOROMETHYLTHIO)PHTHALIMIDE see CBG000

TETRAHYDRO-2-(α,α,α-TRIFLUORO-m-TOLYL)-1,4-OXAZINE see TKJ500

(+)-1,2,3,4-TETRAHYDRO-1-(3,4,5-TRIMETHOXYBENZYL)-6,7-ISOQUINOLINEDIOL HYDROCHLORIDE see TMJ800

(±)-1,2,3,4-TETRAHYDRO-1-(3,4,5-TRIMETHOXYBENZYL)-6,7-ISOQUINOLINEDIOL HYDROCHLORIDE see TKX125

(s)-5,6,6a,7-TETRAHYDRO-1,2,10-TRIMETHOXY-6-METHYL-4H-DIBENZO(de,g)QUINOLIN-9-OL see TKX700

3a,5,5a,9b-TETRAHYDRO-3,5a,9-TRIMETHYL-NAPHTHO(1,2-b)PURAN-2,8(3H,4H)-DIONE see SAU500

TETRAHYDROXYADIPIC ACID see GAR000

1,4,9,10-TETRAHYDROXYANTHRACENE see LEX200

1,2,5,8-TETRAHYDROXY-9,10-ANTHRACENEDIONE see TDD000

1,4,5,8-TETRAHYDROXY-9,10-ANTHRACENEDIONE see TDD250

1,4,5,8-TETRAHYDROXYANTHRACHINON (CZECH) see TDD250

1,2,5,8-TETRAHYDROXYANTHRAQUINONE see TDD000

1,4,5,6-TETRAHYDROXYANTHRAQUINONE see TDD000

1,4,5,8-TETRAHYDROXYANTHRAQUINONE see TDD250

2,3,4,6-TETRAHYDROXY-5H-BENZOCYCLOHEPTENE-5-ONE see TDD500

2,2',4,4'-TETRAHYDROXY BENZOPHENONE see BCS325

2,4,2',4'-TETRAHYDROXYBENZOPHENONE see BCS325

TETRA-(2-HYDROXYETHYL)SILANE see TDH100

3',4',5,7-TETRAHYDROXYFLAVAN-3-OL see QCA000

3,3',4',7-TETRAHYDROXYFLAVONE see FBW000

11-β,16-α,17-α,21-TETRAHYDROXY-9-α-FLUORO-1,4-PREGNADIENE-3,20-DIONE see AQX250

TETRAHYDROXYMETHYLMETHANE see PBB750

Δ²⁰·²²-3-β,12-β,14,21-TETRAHYDROXYNORCHOLENIC ACID LACTONE see DKN300

3-α,4-β,7-α,15-TETRAHYDROXYSCIRP-9-EN-8-ONE see NMV000

7-β,8-α-9-β,10-β-TETRAHYDROXY-7,8,9,10-TETRAHYDROBENZO(a)PYRENE see TDD750

dl-TETRAHYDROZOLINE HYDROCHLORIDE see VRZ000

TETRAHYDROZOLINE HYDROCHLORIDE see VRZ000

TETRAIDROFURANO (ITALIAN) see TCR750

TETRAIODE see TDE750

TETRAIODO-α,α'-DIETHYL-4,4'-STILBENEDIOL see TDE000

TETRAIODOETHYLENE see TDE250

2',4',5',7'-TETRAIODOFLUORESCEIN, DISODIUM SALT see FAG040

TETRAIODOFLUORESCEIN SODIUM SALT see FAG040
TETRAIODOMERCURATE(2+), DIPOTASSIUM see NCP500
TETRAIODOMETHANE see CBY500
3′,3″,5′,5″-TETRAIODOPHENOLPHTHALEIN see IEU100
TETRAIODOPHENOLPHTHALEIN see IEU100
TETRAIODOPHENOLPHTHALEIN SODIUM see TDE750
3:5:3′:5′-TETRAIODO-d-THYRONINE SODIUM see SKJ300
3,3′,5,5′-TETRAIODO-l-THYRONINE, SODIUM SALT see LFG050
TETRAIODOTHALEIN SODIUM see TDE750
TETRAIODOTHYRONINE see TFZ275
TETRAIRON TRIS(HEXACYANOFERRATE) see IGY000
TETRAISOAMYLSTANNANE see TDE765
TETRAISOBUTYLSTANNANE see TDE775
TETRAISOBUTYLTIN see TDE775
TETRAISOPENTYLAMMONIUM IODIDE see TDF000
TETRAISOPENTYLSTANNANE see TDE765
TETRAISOPROPOXIDE TITANIUM see IRN200
TETRAISOPROPOXYTITANIUM see IRN200
TETRAISOPROPYL DITHIONOPYROPHOSPHATE see TDF250
TETRAISOPROPYL GERMANE see TDF500
TETRAISOPROPYL ORTHOTITANATE see IRN200
TETRAISOPROPYL PYROPHOSPHATE see TDF750
TETRAISOPROPYLSTANNANE see TDG000
TETRAISOPROPYL THIOPYROPHOSPHATE see TDF250
TETRAISOPROPYLTIN see TDG000
TETRAISOPROPYL TITANATE see IRN200
TETRAJODPHENOLPHTHALEIN (GERMAN) see IEU100
1,1,2,2-TETRAKIS(ALLYLOXY)ETHANE see TDG250
2,3,5,6-TETRAKIS(1-AZIRIDINYL)-1,4-BENZOQUINONE see TDG500
N,N,N′N′-TETRAKIS(2-CHLOROETHYL)ETHYLENE DIAMINE DIHYDROCHLO-
   RIDE see TDG750
TETRAKIS(CHLOROETHYNYL)SILANE see TDG775
TETRAKIS-(N,N-DICHLOROAMINOMETHYL)METHANE see TDH000
TETRAKIS(DIETHYLCARBAMODITHIOATO-S,S′)SELENIUM see SBP900
TETRAKIS(DIETHYLCARBAMODITHIOATO-S,S′)TELLURIUM see EPJ000
TETRAKIS(DIETHYLDITHIOCARBAMATO)SELENIUM see SBP900
TETRAKIS(DIETHYLDITHIOCARBAMATO)TELLURIUM see EPJ000
TETRAKISDIMETHYLAMINOPHOSPHONOUS ANHYDRIDE see OCM000
TETRAKIS(DIMETHYLCARBAMODITHIOATO-S,S′)SELENIUM see SBQ000
TETRAKIS-(2-HYDROXYETHYL)SILANE see TDH100
TETRAKIS(HYDROXYMETHYL)METHANE see PBB750
TETRAKIS(HYDROXYMETHYL)PHOSPHONIUM ACETATE see TDH250
TETRAKIS(HYDROXYMETHYL)PHOSPHONIUM ACETATE mixed with TETRAK-
   IS(HYDROXYMETHYL)PHOSPHONIUM DIHYDROGEN PHOSPHATE (76:24)
   see TDH500
TETRAKIS(HYDROXYMETHYL)PHOSPHONIUM CHLORIDE see TDH750
TETRAKIS(HYDROXYMETHYL)PHOSPHONIUM NITRATE see TDH775
TETRAKIS(HYDROXYMETHYL)PHOSPHONIUM SULFATE see TDI000
N,N,N′,N′-TETRAKIS(2-HYDROXY-PROPYL)ETHYLENEDIAMINE see QAT000
TETRAKIS(ISOPROPOXY)TITANIUM see IRN200
1,3,4,6-TETRAKIS(2-METHYLTETRAZOL-5-YL)-HEXAAZA-1,5-DIENE see TDI100
2,2,6,6-TETRAKIS(NICOTINOYLOXYMETHYL)CYCLOHEXANOL see CME675
TETRAKIS(p-PHENOXYPHENYL)STANNANE see TDI250
TETRAKIS(p-PHENOXYPHENYL)TIN see TDI250
α,α,α′,α′-TETRAKIS(PHENYLMETHYL)-2,6-PIPERIDINEDIMETHANOL see BIV000
TETRAKIS(μ-(PROPANOATO-O: O′))DI-RHODIUM (Rh-Rh) see RHK850
TETRAKIS(μ-(PROPIONATO))DI-RHODIUM (Rh-Rh) see RHK850
TETRAKIS(THIOUREA)MANGANESE(II) PERCHLORATE see TDI300
TETRAKYANETHYLEN see EEE500
TETRALENO see PCF275
TETRALEX see PCF275
TETRALIN see TCX500
TETRALINA (POLISH) see TCX500
TETRALINE see TCX500
TETRALISAL see MRV250
TETRALIT see TEG250
TETRALITE see TEG250
TETRALLOBARBITAL see AGI750
α-TETRALOL see TDI350
β-TETRALOL see TCX750
ac-β-TETRALOL see TCY000
TETRALOL see TCY000
1-TETRALONE see DLX200
α-TETRALONE see DLX200
3-(α-TETRAL)-4-OXYCOUMARIN see EAT600
TETRALUTION see TBX250
3-(α-TETRALYL)-4-HYDROXYCOUMARIN see EAT600
3-(d-TETRALYL)-4-HYDROXYCOUMARIN see EAT600
TETRALYSAL see MRV250
TETRAM see AMX825, DJA400
TETRAMEEN see TDQ750
1,2,5,6-TETRAMESYL-d-MANNITOL see MAW800
1,2,5,6-TETRAMETHANESULFONATE-d-MANNITOL (9CI) see MAW800
1,2,5,6-TETRAMETHANESULFONYL-d-MANNITOL see MAW800
1,2,9,10-TETRAMETHOXY-6a-α-APORPHINE see TDI475

1,2,9,10-TETRAMETHOXY-6a-α-APORPHINE HYDROCHLORIDE see TDI500
dl-1,2,9,10-TETRAMETHOXY-6a-α-APORPHINE PHOSPHATE see TDI600
6,7,3′,4′-TETRAMETHOXY-1-BENZYLISOQUINOLINE HYDROCHLORIDE see
   PAH250
2,3,9,10-TETRAMETHOXY-13a,α-BERBINE HYDROCHLORIDE see GEO600
dl-2,3,9,10-TETRAMETHOXYBERBIN-1-OL see TDI750
3,6,11,14-TETRAMETHOXYDIBENZO(g,p)CHRYSENE see TDJ000
(1-β)-6,6′,7,12-TETRAMETHOXY-2,2′-DIMETHYL-BERBAMAN (9CI) see TDX830
6′,7′,10,11-TETRAMETHOXYEMETAN see EAL500
6′,7′,10,11-TETRAMETHOXYEMETAN DIHYDROCHLORIDE TETRAHYDRATE
   see EAN500
2,3,10,11-TETRAMETHOXY-8-METHYL-DIBENZO(a,g)QUINOLIZINIUM SALT with
   SULFOACETIC ACID (1:1) see CNR250
1,2,9,10-TETRAMETHOXY-6a-α-NORAPORPHINE see NNR200
TETRAMETHOXYSILANE see MPI750
TETRAMETHOXYSILANE POLYMER see TDJ100
(1-β)-6,6′,7,12-TETRAMETHOXY-2,2,2′,2′-TETRAMETHYL-BERBAMANIUM DIIOD-
   IDE (9CI) see TDX835
6,6′,7′,12′-TETRAMETHOXY-2,2,2′,2′-TETRAMETHYLTUBOCURARANIUM DIIOD-
   IDE see DUM000
N,N,N′-TETRAMETHYL-3,6-ACRIDINEDIAMINE see BJF000
N,N,N′,N′-TETRAMETHYL-3,6-ACRIDINEDIAMINE MONOHYDROCHLORIDE (9CI)
   see BAQ250
((R-R*,R*))-N,N,9,9-TETRAMETHYL-10(9H)-ACRIDINEPROPANAMINE-2,3-DIHY-
   DROXYBUTANEDIOATE (1:1) see DRM000
3,2′,4′,6′-TETRAMETHYLAMINODIPHENYL see TDJ250
TETRAMETHYLAMMONIUM AMIDE see TDJ300
TETRAMETHYLAMMONIUM AZIDOCYANATOIODATE(I) see TDJ325
TETRAMETHYLAMMONIUM AZIDOCYANOIODATE(I) see TDJ350
TETRAMETHYLAMMONIUM AZIDOSELENOCYANATOIODATE(I) see TDJ375
TETRAMETHYLAMMONIUM BOROHYDRATE see TDJ500
TETRAMETHYLAMMONIUM BROMIDE see TDJ750
TETRAMETHYLAMMONIUM CHLORIDE see TDK000
TETRAMETHYLAMMONIUM CHLORITE see TDK250
TETRAMETHYLAMMONIUM DIAZIDOIODATE(I) see TDK300
TETRAMETHYLAMMONIUM HYDROXIDE, liquid (DOT) see TDK500
TETRAMETHYLAMMONIUM HYDROXIDE see TDK500
TETRAMETHYLAMMONIUM IODIDE see TDK750
2,3,9,10-TETRAMETHYLANTHRACENE see TDK885
N,N,10,10-TETRAMETHYL-Δ(9(10),γ)-ANTHRACENEPROPYLAMINE see AEG875
N,N,10,10-TETRAMETHYL-Δ(9(10H),γ)-ANTHRACENEPROPYLAMINE HYDROCHLO-
   RIDE see TDL000
TETRAMETHYLARSONIUM IODIDE see TDL250
1,3,4,10-TETRAMETHYL-7,8-BENZACRIDINE (FRENCH) see TDL500
7,8,9,11-TETRAMETHYLBENZ(c)ACRIDINE see TDL500
5,6,9,10-TETRAMETHYL-1,2-BENZANTHRACENE see TDL750
6,7,9,10-TETRAMETHYL-1,2-BENZANTHRACENE see TDM000
7,8,9,12-TETRAMETHYLBENZ(a)ANTHRACENE see TDL750
7,9,10,12-TETRAMETHYLBENZ(a)ANTHRACENE see TDM000
1,2,3,4-TETRAMETHYLBENZENE see TDM250
1,2,3,5-TETRAMETHYLBENZENE see TDM500
1,2,4,5-TETRAMETHYLBENZENE see TDM750
3,3′,5,5′-TETRAMETHYLBENZIDINE see TDM800
3,5,3′,5′-TETRAMETHYLBENZIDINE see TDM800
3,2′,4′,6′-TETRAMETHYLBIPHENYLAMINE see TDJ250
1,3,5,8-TETRAMETHYL-2,4-BIS(α-HYDROXYETHYL)PROPHINE-6,7-DIPROPIONIC
   ACID see PHV275
TETRAMETHYL BUTANEDIAMINE see TDN000
1,1,3,3-TETRAMETHYLBUTYLAMINE see TDN250
N,N,N′,N′-TETRAMETHYL-1,3-BUTANEDIAMINE see TDN000
p-(1′,1′,3′,3′-TETRAMETHYLBUTYL)FENOL see TDN500
p-(1,1,3,3-TETRAMETHYLBUTYL)PHENOL see TDN500
p-(1,1,3,3-TETRAMETHYLBUTYL)PHENOL, polymer with ETHYLENE OXIDE
   and FORMALDEHYDE see TDN750
4-(1,1,3,3-TETRAMETHYLBUTYL)PHENOL, polymer with FORMALDEHYDE and
   OXIRANE see TDN750
α-((1,1,3,3-TETRAMETHYLBUTYL)PHENYL)-ω-HYDROXY-POLY(OXY-1,2-ETHAN-
   EDIYL) see GHS000
(2-(2,2,3,3-TETRAMETHYLBUTYLTHIO)ACETOXY)TRIBUTYLSTANNANE see
   TDO000
TETRAMETHYLCYCLOBUTA-1,3-DIONE see TDO250
2,2,4,4-TETRAMETHYL-1,3-CYCLOBUTANEDIONE see TDO250
TETRAMETHYL-1,3-CYCLOBUTANEDIONE see TDO250
α-2,2,6-TETRAMETHYLCYCLOHEXENEBUTANAL see TDO255
4-(2,5,6,6-TETRAMETHYL-2-CYCLO-HEXEN-1-YL)-3-BUTEN-2-ONE see IGW500
2,2,9,9-TETRAMETHYL-1,10-DECANEDIOL see TDO260
TETRAMETHYLDIALUMINUM DIHYDRIDE see TDO500
TETRAMETHYLDIAMIDOPHOSPHORIC FLUORIDE see BJE750
TETRAMETHYLDIAMINOBENZHYDROL see TDO750
TETRAMETHYLDIAMINOBENZOPHENONE see MQS500
TETRAMETHYLDIAMINODIPHENYLACETIMINE see IBB000
4,4′-TETRAMETHYLDIAMINODIPHENYLMETHANE see MJN000
p,p-TETRAMETHYLDIAMINODIPHENYLMETHANE see MJN000
TETRAMETHYLDIAMINODIPHENYLMETHANE see MJN000
TETRAMETHYL DIAPARA-AMIDO-TRIPHENYL CARBINOL see AFG500

TETRAMETHYLDIARSANE see TDP250
TETRAMETHYLDIBORANE see TDP275
TETRAMETHYLDIGALLANE see TDP300
TETRAMETHYLDIGOLD DIAZIDE see TDP500
TETRAMETHYL-1,2-DIOXETANE see TDP325
N,N,N,2-TETRAMETHYL-1,3-DIOXOLANE-4-METHANAMINIUM IODIDE (9CI) see MJH250
TETRAMETHYLDIPHOSPHANE see TDP525
TETRAMETHYLDIPHOSPHINE see TDP525
1,1,3,3-TETRAMETHYLDISILOXANE see TDP775
sym-TETRAMETHYLDISILOXANE see TDP775
TETRAMETHYLDISTIBINE see TDQ000
TETRAMETHYLDIURANE SULPHITE see TFS350
N,N,N′,N′-TETRAMETHYLDEUTEROFORMAMIDINIUM PERCHLORATE see TDO275
N,N,N′,N′-TETRAMETHYL-p,p′-DIAMIDODIPHOSPHORIC ACID DIETHYL ESTER see DIV200
N,N,N′,N′-TETRAMETHYL-DIAMIDO-FOSFORZUUR-FLUORIDE (DUTCH) see BJE750
N,N,N′,N′-TETRAMETHYL-DIAMIDO-PHOSPHORSAEURE-FLUORID (GERMAN) see BJE750
N,N′-(N,N′-TETRAMETHYL)-1-DIAMINODIPHENYLNAPHTHYLAMINOMETHANE HYDROCHLORIDE see VKA600
N,N,N′,N′-TETRAMETHYL-1,2-DIAMINOETHANE see TDQ750
N,N,N′,N′-TETRAMETHYLDIAMINOMETHAN (GERMAN) see TDR750
N,N,N′,N′-TETRAMETHYL-DIPROPYLENETRIAMINE see TDP750
1,4-TETRAMETHYLEN-BIS-(N-AETHYLPIPERIDINIUM)-DIBROMID (GERMAN) see TDQ255
1,4-TETRAMETHYLEN-BIS-(N-AETHYLPYRROLIDINIUM)-DIBROMID (GERMAN) see TDQ260
1,4-TETRAMETHYLEN-BIS-(N-METHYLPIPERIDINIUM)-DIBROMID (GERMAN) see TDQ263
1,4-TETRAMETHYLEN-BIS-(N-METHYLPYRROLIDINIUM)-DIBROMID (GERMAN) see TDQ265
1,4-TETRAMETHYLEN-BIS-(PYRIDINIUM)-DIBROMID (GERMAN) see TDQ325
1,1′-TETRAMETHYLENBIS-PYRIDINIUM DIBROMIDE see TDQ325
TETRAMETHYLENE see COW000
TETRAMETHYLENE ACRYLATE see TDQ100
N,N′-TETRAMETHYLENEBIS(1-AZIRIDINECARBOXAMIDE) see TDQ225
5,5′-(TETRAMETHYLENEBIS(CARBONYLIMINO))BIS(N-METHYL-2,4,6-TRIIODOISOPHTHALAMIC ACID) see TDQ230
1,1′-TETRAMETHYLENEBIS(3-(2-CHLOROETHYL)-3-NITROSOUREA) see TDQ250
1,1′-TETRAMETHYLENEBIS(1-ETHYLPIPERIDINIUM)DIBROMIDE see TDQ255
1,1′-TETRAMETHYLENEBIS(1-ETHYL-PYRROLIDINIUM) DIBROMIDE see TDQ260
TETRAMETHYLENE BIS(METHANESULFONATE) see BOT250
1,1′-TETRAMETHYLENEBIS(1-METHYLPIPERIDINIUM)DIBROMIDE see TDQ263
1,1′-TETRAMETHYLENEBIS(1-METHYL-PYRROLIDINIUM) DIBROMIDE see TDQ265
1,1′-TETRAMETHYLENEBIS(PYRIDINIUM BROMIDE) see TDQ325
TETRAMETHYLENE CHLOROHYDRIN see CEU500
TETRAMETHYLENE CYANIDE see AER250
TETRAMETHYLENE DIACRYLATE see TDQ100
1,4-TETRAMETHYLENEDIAMINE see BOS000
TETRAMETHYLENEDIAMINE see BOS000
TETRAMETHYLENEDIAMINE, N,N-DIETHYL-4-METHYL- see DJQ850
TETRAMETHYLENE DIIODIDE see TDQ400
TETRAMETHYLENE DIMETHANE SULFONATE see BOT250
TETRAMETHYLENEDISULFOTETRAMINE see TDX500
1,4-TETRAMETHYLENE GLYCOL see BOS750
TETRAMETHYLENE GLYCOL DIACRYLATE see TDQ100
TETRAMETHYLENE IODIDE see TDQ400
TETRAMETHYLENE OXIDE see TCR750
TETRAMETHYLENEOXIRANE see CPD000
6,7-TETRAMETHYLENE-5-PHENYL-1,2-DIHYDRO-3H-THIENO(2,3-e)(1,4)DIAZEPIN-2-ONE see BAV625
TETRAMETHYLENESULFIDE see TDC730
TETRAMETHYLENE SULFONE see SNW500
TETRAMETHYLENE SULPHOXIDE see TDQ500
TETRAMETHYLENETETRANITRAMINE see CQH250
TETRAMETHYLENETHIURAM DISULPHIDE see TFS350
TETRAMETHYLENIMINE see PPS500
3,3,5,5-TETRAMETHYL-4-ETHOXYVINYLCYCLOHEXANONE see EES370
TETRAMETHYL ETHYLENE DIAMINE see TDQ750
N,N,N′,N′-TETRAMETHYLETHYLENEDIAMINE see TDQ750
TETRAMETHYLETHYLENE GLYCOL see TDR000
TETRAMETHYLETHYLFORMYLTETRALIN see FNK200
E,E-3,7,11,15-TETRAMETHYL-1,6,10,14-HEXADECATETRAEN-3-OL see GDG300
3,7,11,15-TETRAMETHYL-2-HEXADECEN-1-OL see PIB600
N,N,N′,N′-TETRAMETHYLHEXAMETHYLENE DIAMINE see TDR100
N,N,N′,N′-TETRAMETHYLHEXAMETHYLENEDIAMINE-1,3-DIBROMOPROPANE copolymer see HCV500
N,N,N′,N′-TETRAMETHYL-1,6-HEXANEDIAMINE see TDR100

TETRAMETHYLHYDRAZINE HYDROCHLORIDE see TDR250
TETRAMETHYL LEAD see TDR500
N,N,N′,N′-TETRAMETHYLMETHANEDIAMINE see TDR750
TETRAMETHYL METHYLENE DIAMINE (DOT) see TDR750
2,2,6,6-TETRAMETHYLNITROSOPIPERIDINE see TDS000
N,2,3,3-TETRAMETHYL-2-NORBORNAMINE see VIZ400
N,N,2,3-TETRAMETHYL-2-NORBORNANAMINE see TDS250
N,2,3,3-TETRAMETHYL-2-NORBORNANAMINE HYDROCHLORIDE see MQR500
N,2,3,3-TETRAMETHYL-2-NORCAMPHANAMINE see VIZ400
1,(5,5,7,7-TETRAMETHYL-2-OCTANYL)-2-METHYL-5-ETHYLPYRIDINIUM CHLORIDE see TDS500
p-1′,1′,4′,4′-TETRAMETHYLOKTYLBENZENSULFONAN SODNY (CZECH) see DXW200
TETRAMETHYLOLMETHANE see PBB750
2,6,10,14-TETRAMETHYLPENTADECANE see PMD500
1:2:3:4-TETRAMETHYLPHENANTHRENE see TDS750
N,N,N′,N′-TETRAMETHYL-o-PHENYLENEDIAMINE see TDT000
N,N,N′,N′-TETRAMETHYL-p-PHENYLENEDIAMINE see BJF500
N,N,N′,N′-TETRAMETHYL-p-PHENYLENEDIAMINE DIHYDROCHLORIDE see TDT250
N,N,N-α-TETRAMETHYL-10H-PHENOTHIAZINE-10-ETHANAMINIUM METHYL SULFATE see MRW000
N,N,N′,N′-TETRAMETHYL-1,3-PROPANEDIAMINE see TDU500
2-(2,3,5,6-TETRAMETHYLPHENOXY)PROPIONIC ACID see BHM000
TETRAMETHYL-p-PHENYLENEDIAMINE see BJF500
(2,3,5,6-TETRAMETHYLPHENYL)MERCURY ACETATE see AAS500
TETRAMETHYLPHOSPHONIUM IODIDE see TDT500
N,N,N,N-TETRAMETHYLPHOSPHORODIAMIDIC FLUORIDE see BJE750
TETRAMETHYLPHOSPHORODIAMIDIC FLUORIDE see BJE750
2,2,6,6-TETRAMETHYLPIPERIDINE see TDT750
TETRAMETHYLPIPERIDINE NITROXIDE see TDT800
2,2,6,6-TETRAMETHYLPIPERIDINE-1-OXYL see TDT800
2,2,6,6-TETRAMETHYLPIPERIDINONE see TDT770
2,2,6,6-TETRAMETHYL-4-PIPERIDINONE see TDT770
2,2,6,6-TETRAMETHYLPIPERIDINOOXY see TDT800
TETRAMETHYLPIPERIDINOOXY see TDT800
2,2,6,6-TETRAMETHYLPIPERIDINOOXYL see TDT800
2,2,6,6-TETRAMETHYL-1-PIPERIDINYLOXY see TDT800
2,2,6,6-TETRAMETHYLPIPERIDINYLOXY see TDT800
2,2,6,6-TETRAMETHYLPIPERIDONE see TDT770
2,2,6,6-TETRAMETHYL-4-PIPERIDONE see TDT770
2,2,6,6-TETRAMETHYL-4-PIPERIDONE OXIME see TDU000
2,2,6,6-TETRAMETHYLPIPERIDOXYL see TDT800
TETRAMETHYLPLATINUM see TDU250
TETRAMETHYLPLUMBANE see TDR500
N,N,N′,N′-TETRAMETHYL PROPENE-1,3-DIAMINE see TDU750
N,N,N′,N′-TETRAMETHYL-1,3-PROPENYLDIAMINE see TDU750
2,3,5,6-TETRAMETHYL PYRAZINE (FCC) see TDV725
TETRAMETHYLPYRAZINE see TDV725
N,N,2,3-TETRAMETHYL-4-(4′-(PYRIDYL-1′-OXIDE)AZO)ANILINE see DQF200
N,N,2,5-TETRAMETHYL-4-(4′-(PYRIDYL-1′-OXIDE)AZO)ANILINE see DQF400
N,N,2,6-TETRAMETHYL-4-(4′-(PYRIDYL-1′-OXIDE)AZO)ANILINE see DQF600
TETRAMETHYLPYROPHOSPHATE see TDV000
2,2,5,5-TETRAMETHYLPYRROLIDINE see TDV250
2,2,5,5-TETRAMETHYL-1-PYRROLIDINYLOXY-3-CARBOXYLIC ACID see TDV300
2,2,6,6-TETRAMETHYLQUINUCLIDINE HYDROBROMIDE see TAL275
TETRAMETHYLSILANE see TDV500
TETRAMETHYL SILICATE see MPI750
TETRAMETHYLSILIKAT see MPI750
TETRAMETHYLSTANNANE see TDV750
TETRAMETHYLSTIBONIUM IODIDE see TDW000
TETRAMETHYLSUCCINONITRILE see TDW250
TETRA-o-METHYL-SULPHONYL-d-MANNITOL see MAW800
1,1,4,4-TETRAMETHYL-1,2,3,4-TETRAHYDRO-6-ETHYLNAPHTHALENE see TCR250
2,2,5,5-TETRAMETHYLTETRAHYDRO-3-HYDROXY-3-FURANYL METHYL KETONE see HOI400
3,3,6,6-TETRAMETHYL-1,2,4,5-TETRAOXANE see TDW275
TETRAMETHYL-2-TETRAZENE see TDW300
2,4,6,8-TETRAMETHYL-1,3,5,7-TETROXOCANE see TDW500
TETRAMETHYLTHIOCARBAMOYLDISULPHIDE see TFS350
O,O,O′O′-TETRAMETHYL-O,O′-THIODI-p-PHENYLENE PHOSPHOROTHIOATE see TAL250
TETRAMETHYL-O,O′-THIODI-p-PHENYLENE PHOSPHOROTHIOATE see TAL250
TETRAMETHYLTHIONINE CHLORIDE see BJI250
TETRAMETHYLTHIORAMDISULFIDE (DUTCH) see TFS350
1,1,3,3-TETRAMETHYLTHIOUREA see TDX000
TETRAMETHYLTHIOUREA see TDX000
TETRAMETHYL-THIRAM DISULFID (GERMAN) see TFS350
TETRAMETHYLTHIURAM BISULFIDE see TFS350
TETRAMETHYLTHIURAM DISULFIDE see TFS350
TETRAMETHYL THIURAM DISULFIDE see TFS350

TETRAMETHYLTHIURAM DISULFIDE mixed with FERRIC NITROSODIMETHYLDITHIOCARBAMATE see FAZ000
N,N-TETRAMETHYLTHIURAM DISULPHIDE see TFS350
N,N,N',N'-TETRAMETHYLTHIURAM DISULPHIDE see TFS350
TETRAMETHYLTHIURAMMONIUM SULFIDE see BJL600
TETRAMETHYLTHIURAM MONOSULFIDE see BJL600
TETRAMETHYLTHIURAM SULFIDE see BJL600
TETRAMETHYL THIURANE DISULFIDE see TFS350
TETRAMETHYL TIN see TDV750
(−)-2,2,5,5-TETRAMETHYL-α-((o-TOLYLOXY)METHYL)-1-PYRROLIDINEETHANOL HYDROCHLORIDE see TGJ625
TETRAMETHYLTRITHIO CARBAMIC ANHYDRIDE see BJL600
N,N',o,o-TETRAMETHYL-(+)-TUBOCURINE see DUL800
1,1,3,3-TETRAMETHYLUREA see TDX250
TETRAMETHYLUREA see TDX250
TETRAMETHYLUREE (FRENCH) see TDX250
m-TETRAMETHYLXYLENE DIISOCYANATE see TDX300
m-TETRAMETHYLXYLENE DIISOCYANATE/TRIMETHYLOL PROPANE ADDUCT see TDX320
N,N,N',N'-TETRAMETIL-FOSFORODIAMMIDO-FLUORURO (ITALIAN) see BJE750
TETRAMIN see VMA000
TETRAMINE see HOI000, TDX500
TETRAMINE FAST BROWN BRS see CMO750
TETRAMINE FAST RED 8B see CMO885
TETRAMINE PLATINUM(II) CHLORIDE see ANV800
1,4,5,8-TETRAMINOANTHRAQUINONE see TBG700
1-TETRAMISOLE HYDROCHLORIDE see LFA020
dl-TETRAMISOLE HYDROCHLORIDE see TDX750
TETRAMISOLE HYDROCHLORIDE see TDX750
TETRAMISOL HYDROCHLORIDE see TDX750
TETRAM MONOOXALATE see AMX825
TETRAMON see TCB725
TETRAMON J see TCC750
TETRAN see HOH500
TETRANAP see TCX500
TETRANATRIUMPYROPHOSPHAT (GERMAN) see TEE500
d-TETRANDRINE see TDX830
(+)-TETRANDRINE see TDX830
TETRANDRINE see TDX830
TETRANDRINE DIMETHIODIDE see TDX835
TETRANGOMYCIN see TDX840
TETRAN HYDROCHLORIDE see HOI000
TETRANICOTINIC ACID-2-HYDROXYCYCLOHEXA-1,1,3,3-TETRAMETHYL ESTER see CME675
TETRANICOTYLFRUCTOFURANOSE see TDX860
TETRANICOTYLFRUCTOSE see TDX860
TETRANITRANILINE (FRENCH) see TDY000
2,3,4,6-TETRANITROANILINE see TDY000
N,2,3,5-TETRANITROANILINE see TDY050
N,2,4,6-TETRANITROANILINE see TDY075
TETRANITROANILINE see TDY000
1,3,6,8-TETRANITROCARBAZOLE see TDY120
2,4,5,7-TETRANITROCHRYSAZIN see CML600
TETRANITRO DIGLYCERIN see TDY100
1,3,6,8-TETRANITROKARBAZOL see TDY120
TETRANITROMETHANE see TDY250
N,2,4,5-TETRANITRO-N-METHYLANILINE see TEG250
1,3,6,8-TETRANITRO NAPHTHALENE see TDY500
TETRANITROPENTAERYTHRITE see PBC250
2,3,4,6-TETRANITROPHENOL see TDY600
trans-1,4,5,8-TETRANITRO-1,4,5,8-TETRAAZADECAHYDRONAPHTHALENE see TDY775
TETRAN PTFE see TAI250
TETRA OLIVE N2G see APG500
2,5,7,8-TETRAOXA[4.2.0]BICYCLOOCTANE see DVQ759
1,4,7,10-TETRAOXACYCLODODECANE see COD475
2,4,8,10-TETRAOXA-3,9-DIPHOSPHASPIRO(5.5)UNDECANE, 3,9-BIS(2,4-BIS(1,1-DIMETHYLETHYL)PHENOXY)- see COV800
2,5,8,11-TETRAOXADODECANE see TKL875
2,2'-(2,5,8,11-TETRAOXA-1,12-DODECANE DIYL)BISOXIRANE see TJQ333
2,2'-(2,5,8,11-TETRAOXA-1,2-DODECANEDIYL)BISOXIRANE see TJQ333
2,4,10,12-TETRAOXA-6,16,17,18-TETRAAZA-3,11-DISTIBATRICYCLO(11.3.1.1^{5,9})OCTADECA-1(17),5,7,9(18),13,15-HEXAENE, 3,11-DIHYDROXY- see AQE305
2,4,10,12-TETRAOXA-6,16,17,18-TETRAAZA-3,11-DISTIBATRICYCLO(11.3.1.1^{5,9})OCTADECA-1(17),5,7,9(18),13,15-HEXAENE-8,14-DIMETHANOL, 3,11-DIHYDROXY- see AQE300
3,6,9,12-TETRAOXATETRADECA-1,13-DIENE see TDY800
2,5,8,10-TETRAOXATRIDEC-12-ENOIC ACID, 9-OXO-, 2-PROPENYL ESTER (9CI) see AGD250
(±)-(3,5,3',5'-TETRAOXO)-1,2-DIPIPERAZINOPROPANE see PIK250
2,4,5,6-TETRAOXOHEXAHYDROPYRIMIDINE see AFT750
2,4,5,6-TETRAOXOHEXAHYDROPYRIMIDINE HYDRATE see MDL500
TETRA-N-PENTYLAMMONIUM BROMIDE see TEA250

TETRAPENTYLAMMONIUM BROMIDE see TEA250
TETRAPENTYLSTANNANE see TBI600
TETRAPHAN see TEF775
TETRAPHENE see BBC250
TETRAPHENYLARSENIUM CHLORIDE see TEA300
TETRAPHENYLARSONIUM CHLORIDE see TEA300
1,1,4,4-TETRAPHENYLBUTADIENE see TEA400
TETRAPHENYLBUTADIENE see TEA400
1,1,4,4-TETRAPHENYL-1,4-BUTANEDIOL see TEA500
1,1,2,2-TETRAPHENYL-1,2-ETHANEDIOL see TEA600
TETRAPHENYL-1,2-ETHANEDIOL see TEA600
1,1,2,2-TETRAPHENYLETHYLENE GLYCOL see TEA600
TETRAPHENYLETHYLENE GLYCOL see TEA600
TETRAPHENYL METHANE see TEA750
TETRAPHOSPHATE HEXAETHYLIQUE (FRENCH) see HCY000
TETRAPHOSPHOR (GERMAN) see PHP010
TETRAPHOSPHORUS IODIDE see TEB750
TETRAPHOSPHORUS TRISELENIDE see TEC000
TETRAPHOSPHORUS TRISULFIDE see PHS500
TETRA PINK B see DNT300
TETRAPOM see TFS350
TETRAPOTASSIUM ETIDRONATE see TEC250
TETRAPOTASSIUM FERROCYANIDE see TEC500
TETRAPOTASSIUM HEXACYANOFERRATE(4-) see TEC500
TETRAPOTASSIUM HEXACYANOFERRATE(II) see TEC500
TETRAPOTASSIUM HEXACYANOFERRATE see TEC500
TETRAPOTASSIUM PYROPHOSPHATE see PLR200
TETRAPROPENYLSUCCINIC ANHYDRIDE see TEC600
TETRAPROPYLAMMONIUM BROMIDE see TEC750
TETRA-N-PROPYLAMMONIUM HYDROXIDE see TED000
TETRAPROPYLAMMONIUM HYDROXIDE see TED000
TETRA-N-PROPYLAMMONIUM IODIDE see TED250
TETRAPROPYLAMMONIUM IODIDE see TED250
TETRAPROPYLAMMONIUM OXIDE see TED000
TETRA-n-PROPYL DITHIONOPYROPHOSPHATE see TED500
TETRA-n-PROPYL DITHIOPYROPHOSPHATE see TED500
O,O,O,O-TETRAPROPYL DITHIOPYROPHOSPHATE see TED500
TETRAPROPYLENE see PMP750
TETRAPROPYLGERMANE see TED650
TETRAPROPYLGERMANIUM see TED650
TETRAPROPYL LEAD see TED750
TETRASELENIUM TETRANITRIDE see TEE000
TETRASEPTAN see BEL900
TETRASILVER DIIMIDODIOXOSULFATE see SDP500
TETRASILVER DIIMIDOTRIPHOSPHATE see TEE100
TETRASILVER ORTHODIAMIDOPHOSPHATE see TEE125
TETRASIPTON see TFS350
TETRASODIUM BIS(CITRATE(3-)FERRATE(4-)) see TEE225
TETRASODIUM BISCITRATO FERRATE see TEE225
TETRASODIUM DIARSENATE see SEY100
TETRASODIUM DIETHYLSTILBESTROL PHOSPHATE see TEE300
TETRASODIUM DIPHOSPHATE see TEE500
TETRASODIUM EDTA see EIV000
TETRASODIUM ETHYLENEDIAMINETETRAACETATE see EIV000
TETRASODIUM ETHYLENEDIAMINETETRACETATE see EIV000
TETRASODIUM (ETHYLENEDINITRILO)TETRAACETATE see EIV000
TETRASODIUM ETIDRONATE see TEE250
TETRASODIUM FOSFESTROL see TEE300
TETRASODIUM PYROPHOSPHATE see TEE500
TETRASODIUM PYROPHOSPHATE, ANHYDROUS see TEE500
TETRASODIUM SALT of EDTA see EIV000
TETRASODIUM SALT of ETHYLENEDIAMINETETRACETIC ACID see EIV000
TETRASOL see CBY000
TETRASTIGMINE see TCF250
TETRASUL see CKL750
TETRASULE see PBC250
TETRASULFIDE, BIS(ETHOXYTHIOCARBONYL) see BJU250
TETRASULFUR DINITRIDE see TEE750
TETRASULFUR TETRANITRIDE see SOO000
TETRA SYTAM see BJE750
TETRATELLURIUM TETRANITRIDE see TNO250
TETRATHIURAM DISULFIDE see TFS350
TETRATHIURAM DISULPHIDE see TFS350
TETRAVEC see PCF275
TETRAVERINE see TBX000
TETRAVINYLLEAD see TEF250
TETRAVOS see DGP900
TETRA-WEDEL see TBX250
7,8,9,10-TETRAZABICYCLO(5.3.0)-8,10-DECADIENE see PBI500
1,2,3,3A-TETRAZACYCLOHEPTA-8A,2-CYCLOPENTADIENE see PBI500
TETRAZENE see TEF500
1-TETRAZENE, 4-AMIDINO-1-(NITRSOAMINOAMIDINO)-(8CI) see TEF500
TETRAZEPAM see CFG750
TETRAZIDO-1,4-BENZOQUINONE see TBJ250
1,2,4,5-TETRAZINE see TEF600

TETRAZINE (dry) (DOT) see TEF600
TETRAZOBENZENE-β-NAPHTHOL see OHA000
TETRAZO DEEP BLACK G see AQP000
TETRAZO DEEP BLACK GC EXTRA see CMN240
TETRAZOL see TEF650
1H-TETRAZOL-5-AMINE see AMR000
TETRAZOLE-5-DIAZONIUM CHLORIDE see TEF675
TETRAZOSIN HYDROCHLORIDE DIHYDRATE see TEF700
TETRIL see TEG250
TETRINE see EIV000
TETRINE ACID see EIX000
TETROCARCIN A see TEF725
TETRODIRECT BLACK EFD see AQP000
TETRODIRECT BLACK V see CMN240
TETRODONTOXIN see FOQ000
TETRODOXIN see FOQ000
TETROFAN see TEF775
TETROGUER see PCF275
TETROLE see FPK000
TETRON-100 see TCF250
TETRON see TCF250
TETRONE A see TEF750
TETRODOTOXIN see FOQ000
TETROPHAN see TEF775
TETROPHENE GREEN M see AFG500
TETROPHINE see TEF775
TETROPIGMENT FAST YELLOW BG see CMS212
TETROPIGMENT FAST YELLOW VG see CMS212
TETROPIL see PCF275
TETROSAN see AFP750
TETROSIN OE see BGJ250
TETROSOL see TBX250
2,4,6-TETRYL see TEG250
TETRYL see TEG250
TETRYLAMMONIUM see TCB725
TETRYLAMMONIUM BROMIDE see TCC000
TETRYL FORMATE see IIR000
TETRZOLIUM CHLORIDE see TMV500
TETTERWORT see CCS650
TETURAM see DXH250
TETURAMIN see DXH250
TEVCOCIN see CDP250
TEVCODYNE see BRF500
TEXACO LEAD APPRECIATOR see BPV100
TEXANOL see TEG500
TEXAN RED TONER D see CHP500
TEXAPON T-42 see SON000
TEXAPON ZHC see SIB600
TEXAS see CBT750
TEXAS MOUNTAIN LAUREL see NBR800
TEXAS SARSAPARILLA see MRN100
TEXAS UMBRELLA TREE see CDM325
TEXCRYL see ADW200
TEXIN 192A see PKO500
TEXIN 445D see PKM250
TEXTILE see SFT500
TEXTILE RED WD-263 see NAY000
TEXTILON (OBS.) see TEG650
TEXTILOTOXIN see TEG650
TEXTONE see SFT500
TEX WET 1001 see DJL000
TF 1169 see FDA885
TF see GJS200, TJE100
2,3,4-TFA see TJX900
TFA see TJX900
TFC mixed with TCC (1:2) see TIL526
TFDU see TKH325
TFE see TKA350
TFF see TNT500
T. FLAVOVIRIDIS VENOM see TKW100
T-FLUORIDE see SHF500
TFPHM see TCI500
TFT THILO see TKH325
ThG see AMH250
TG see AMH250
TGA 2 see ADT250
TGA-EXTRACT see PLW550
T-GAS see EJN500
TGD 5161 see SMQ500
β-TGDR see TFJ250
T-GELB BZW, GRUN 1 see QCA000
TGM 3 see MDN510
TGM 4 see TCE400
TGM 3PC see MDN510
TGM 3S see MDN510

TGR see TFJ500
TGS see TFJ500
TH/100 see TEP500
TH-152 see MDM800
TH-218 see EPC175
TH 564 see DTU850
1064 TH see DUO400
TH 1314 see EPQ000
1321 TH see PNW750
TH-1395 see GFM200
2180 TH see PMM000
TH 346-1 see DRR400
TH 6040 see CJV250
THA see TCJ075
THACAPZOL see MCO500
THAIMYCIN B see TEG750
THAIMYCIN C see TEH000
THALAMONAL see DYF200
THALIBLASTINE see TEH250
THALICARPIN see TEH250
(+)-THALICARPINE see TEH250
THALICARPINE see TEH250
THALIDOMIDE see TEH500
(+)-THALIDOMIDE see TEH510
R-(+)-THALIDOMIDE see TEH510
(±)-THALIDOMIDE see TEH520
s-(−)-THALIDOMIDE see TEH520
(−)-THALIDOMIDE see TEH520
THALIN see TEH500
THALINETTE see TEH500
THALLIC OXIDE see TEL050
THALLIC SULFATE see TEM100
THALLINE see TCU700
THALLIUM see TEI000
THALLIUM(1+) ACETATE see TEI250
THALLIUM(I) ACETATE see TEI250
THALLIUM ACETATE see TEI250
THALLIUM(I) AZIDE see TEI500
THALLIUM(I) AZIDODITHIOCARBONATE see TEI600
THALLIUM BROMATE see TEI625
THALLIUM BROMIDE see TEI750
THALLIUM(I) CARBONATE (2:1) see TEJ000
THALLIUM CHLORATE see TEJ100
THALLIUM(1+) CHLORIDE see TEJ250
THALLIUM CHLORIDE see TEJ250
THALLIUM COMPOUNDS see TEJ500
THALLIUM(I) DITHIOCARBONAZIDATE see TEI600
THALLIUM(I) FLUOBORATE see TEJ750
THALLIUM(I) FLUORIDE see TEK000
THALLIUM(I) FLUOSILICATE see TEK250
THALLIUM FULMINATE see TEK300
THALLIUM(1+) IODIDE see TEK500
THALLIUM(I) IODIDE see TEK500
THALLIUM IODIDE see TEK500
THALLIUM(I) IODOACETYLIDE see TEK525
THALLIUM MALONATE see TEM399
THALLIUM MONOACETATE see TEI250
THALLIUM MONOCHLORIDE see TEJ250
THALLIUM MONOFLUORIDE see TEK000
THALLIUM MONOIODIDE see TEK500
THALLIUM MONONITRATE see TEK750
THALLIUM MONOSELENIDE see TEL500
THALLIUM NITRATE see TEK750
THALLIUM(I) NITRIDE see TEL000
THALLIUM(3+) OXIDE see TEL050
THALLIUM(III) OXIDE see TEL050
THALLIUM OXIDE see TEL040, TEL050
THALLIUM PEROXIDE see TEL050
THALLIUM(I) PEROXODIBORATE see TEL100
THALLIUM aci-PHENYLNITROMETHANIDE see TEL150
THALLIUM SELENIDE see TEL500
THALLIUM SESQUIOXIDE see TEL050
THALLIUM(II) SULFATE (1:1) see TEM250
THALLIUM(I) SULFATE (2:1) see TEM000
THALLIUM SULFATE, solid (DOT) see TEL750
THALLIUM(III) SULFATE see TEM100
THALLIUM SULFATE see TEL750
THALLOUS ACETATE see TEI250
THALLOUS CARBONATE see TEJ000
THALLOUS CHLORIDE see TEJ250
THALLOUS FLUORIDE see TEK000
THALLOUS IODIDE see TEK500
THALLOUS MALONATE see TEM399
THALLOUS NITRATE see TEK750
THALLOUS OXIDE see TEL040

5-THIA-1-AZABICYCLO(4.2.0)OCT-2-ENE-2-CARBOXYLIC ACID, 7-(((2-AMINO-4-THIAZOLYL) (METHOXYIMINO)ACETYL)AMINO)-3-((5-METHYL-2H-TETRAZOL-2-YL)METHYL)-8-OXO-, (2,2-DIMETHYL-1- OXOPROPOXY)METHYL ESTER, (6R-(6-α-7-β(Z))- see TAA420

5-THIA-1-AZABICYCLO(4.2.0)OCT-2-ENE-2-CARBOXYLIC ACID, 7-(((2-AMINO-4-THIAZOLYL) (METHOXYIMINO)ACETYL)AMINO)-8-OXO-3-((1,2,3-THIADIAZOL-5-YLTHIO)METHYL)-, SODIUM SALT, (6R-(6-α-7-β(Z)))- see CCS635

5-THIA-1-AZABICYCLO(4.2.0)OCT-2-ENE-2-CARBOXYLIC ACID, 3-(HYDROXY-METHYL)-8-OXO-7-(2-(2-IHIENYL)ACETAMIDO)-, ACETATE, MONOSODIUM SALT see SFQ500

1-THIA-3-AZAINDENE see BDE500

THIABEN see TEX000

THIABENDAZOLE (USDA) see TEX000

THIABENDAZOLE HYDROCHLORIDE see TER500

THIABENZOLE see TEX000

3-THIABUTAN-2-ONE, O-(METHYLCARBAMOYL)OXIME see MDU600

THIACETAMIDE see TFA000

THIACETARSAMIDE see TFA350

THIACETAZONE see FNF000

THIACETIC ACID see TFA500

THIACOCCINE see TEX250

THIACTIN see TFQ275

THIACYCLOPENTADIENE see TFM250

THIACYCLOPENTANE see TDC730

THIACYCLOPENTANE DIOXIDE see SNW500

THIACYCLOPENTAN-2-ONE see TDC800

THIACYCLOPENTANONE-2 see TDC800

THIACYCLOPROPANE see EJP500

1,3,4-THIADIAZOL-2-AMINE see AMR250

1,3,4-THIADIAZOLE-2-ACETAMIDO see TES000

1,3,4-THIADIAZOLE-2,5-DITHIOL see TES250

(N-1,2,3-THIADIAZOLYL-5)-N'-PHENYLUREA see TEX600

THIADIPONE see BAV625

9-THIAFLUORENE see TES300

4-THIAHEPTANEDIOIC ACID see BHM000

THIALBARBITAL see TES500

THIALBARBITONE see TES500

THIALPENTON see TES500

THIAMAZOLE see MCO500

THIAMBUIENE HYDROCHLORIDE see DJP500

THIAMETON see PHI500

THIAMIDIN F see TES800

THIAMIN see TES750

THIAMIN DISULFIDE see TES800

THIAMINDISULFID-MONOOROTAT (GERMAN) see TET250

THIAMINE CHLORIDE see TES750

THIAMINE CHLORIDE HYDROCHLORIDE see TET300

THIAMINE DICHLORIDE see TET300

THIAMINE DISULFIDE see TES800

THIAMINE DISULFIDE, OROTATE see TET250

THIAMINE HYDROCHLORIDE see TET300

THIAMINE MONOCHLORIDE see TES750

THIAMINE MONONITRATE see TET500

THIAMINE NITRATE see TET500

THIAMINE PROPYL DISULFIDE see DXO300

THIAMINE PROPYL DISULFIDE HYDROCHLORIDE see DXO400

THIAMINEPYROPHOSPHATE see TET750

THIAMINEPYROPHOSPHATECHLORIDE see TET750

THIAMINEPYROPHOSPHORICESTER see TET750

THIAMINE TETRAHYDROFURFURYL DISULFIDE see FQJ100

THIAMINE, TRIHYDROGEN PYROPHOSPHATE (ESTER) see TET750

THIAMIN HYDROCHLORIDE see TET300

THIAMINIUM CHLORIDE HYDROCHLORIDE see TET300

THIAMIN PROPYL DISULFIDE see DXO300

THIAMINPYROPHOSPHATE see TET750

THIAMIPRINE see AKY250

THIAMIZIDE see CIP500

THIAMPHENICOL see MPN000

THIAMPHENICOL AMINOACETATE HYDROCHLORIDE see TET780

THIAMPHENICOL GLYCINATE HYDROCHLORIDE see UVA150

THIAMUTILIN see TET800

THIAMYLAL see AGL375

THIAMYLAL SODIUM see SOX500

THIANIDE see EPQ000

THIANONANE-5 see BSM125

5-THIANONANE see BSM125

THIANTAN see DHF600, DII200

THIA-4-PENTANAL (DOT) see MPV400, TET900

3-THIAPENTANE see EPH000

THIAPHENE see TFM250

2-THIAPROPANE see TFP000

THIARETIC see CFY000

THIASIN see SNN500

THIATE H see DKC400

THIATE U see DEI000

THIATON see TGF075

3,7-THIAXANTHENEDIAMINE-5,5-DIOXIDE see TEU000

THIAZAMIDE see TEX250

THIAZESIM HYDROCHLORIDE see TEU250

THIAZESIUM HYDROCHLORIDE see TEU250

THIAZIDE see CLH750

THIAZINAMIUM METHYL SULFATE see MRW000

THIAZIPIDICO see BEQ625

2-THIAZOLAMINE see AMS250

2-THIAZOLAMINE, 4,5-DIHYDRO-(9CI) see TEV600

THIAZOLE see TEU300

THIAZOLE, 2-ACETAMIDO- see TEW100

5-THIAZOLECARBOXYLIC ACID, 2-AMINO-4-(TRIFLUOROMETHYL)-, ETHYL ESTER see AMU550

5-THIAZOLECARBOXYLIC ACID, 2-CHLORO-4-(TRIFLUOROMETHYL)-, PHE-NYLMETHYL ESTER see BEG300

THIAZOLE, 4,5-DIHYDRO-2-METHYL- see DLV900

2,4(3H,5H)-THIAZOLEDIONE see TEV500

THIAZOLE, 4-METHYL-5-VINYL- see MQM800

THIAZOLE YELLOW see CMP050

THIAZOLE YELLOW G see CMP050

THIAZOLIDINE-4-CARBOXYLIC ACID see TEV000

4-THIAZOLIDINECARBOXYLIC ACID see TEV000

THIAZOLIDINECARBOXYLIC ACID see TEV000

4-THIAZOLIDINECARBOXYLIC ACID, 2-(2-HYDROXYPHENYL)-3-(3-MERCAPTO-1-OXOPROPYL)-, (2R-cis)- see FAQ950

THIAZOLIDINEDIONE-2,4 see TEV500

2,4-THIAZOLIDINEDIONE see TEV500

2-THIAZOLIDINIMINE see TEV600

1-(2-THIAZOLIDINYL)1,2,3,4,5-PENTANEPENTOL see TEW000

4-THIAZOLIDONE-2-CAPROIC ACID see CCI500

epsilon-(2-(4-THIAZOLIDONE))HEXANOIC ACID see CCI500

2-THIAZOLINE, 2-AMINO- see TEV600

2-THIAZOLINE, 2-METHYL- see DLV900

THIAZOLIUM, 5-(2-HYDROXYETHYL)-3-((4-HYDROXY-2-METHYL-5-PYRIMIDIN-YL)METHYL)-4-METHYL- see ORS200

THIAZOLIUM, 3-METHYL-2-((1-METHYL-2-PHENYL-1H-INDOL-3-YL)AZO)-, CHLO-RIDE see CMM770

THIAZOL YELLOW see CMP050

THIAZOL YELLOW G see CMP050

THIAZOL YELLOW GGM see CMP050

THIAZOL YELLOW R see CMP050

THIAZOL YELLOW Z see CMP050

N-2-THIAZOLYLACETAMIDE see TEW100

2-THIAZOLYLAMINE see AMS250

2-(4-THIAZOLYL)-1H-BENZIMIDAZOLE see TEX000

2-(4'-THIAZOLYL)BENZIMIDAZOLE see TEX000

2-(4-THIAZOLYL)BENZIMIDAZOLE see TEX000

2-(THIAZOL-4-YL)BENZIMIDAZOLE see TEX000

2-(4-THIAZOLYL)-5-BENZIMIDAZOLECARBAMIC ACID METHYL ESTER see TEX200

2-(4-THIAZOLYL)-BENZIMIDAZOLE, HYDROCHLORIDE see TER500

4-(2-THIAZOLYL)PIPERAZINYL 3,4,5-TRIMETHOXYPHENYL KETONE see TEX220

4'-(2-THIAZOLYLSULFAMOYL)PHTHALANILIC ACID see PHY750

4'-(2-THIAZOLYLSULFAMYL)PHTHALANILIC ACID see PHY750

N'-2-THIAZOLYLSULFANIDAMIDE SODIUM SALT see TEX500

(N'-2-THIAZOLYLSULFANIDAMIDO)SODIUM see TEX500

N'-2-THIAZOLYLSULFANILAMIDE see TEX250

1-(2-THIAZOLYL)-4-(3,4,5-TRIMETHOXYBENZOYL)PIPERAZINE see TEX220

N'-2-THIAZOLYSULFANILAMIDE SODIUM SALT see TEX500

THIAZON see DSB200

THIAZONE see DSB200

2-THIAZYLAMINE see AMS250

THIBENZOLE see TEX000

THIBETINE see TJL250

THIBONE see FNF000

2,3,3-THICHLORO-2-PROPENE-1-THIOL, DIISOPROPYLCARBAMATE see DNS600

THIDIAZURON see TEX600

THIDICUR see MPQ750

1H-THIENO(3,4-d)IMIDAZOLE-4-PENTANOIC ACID, HEXAHYDRO-2-OXO-, (3aS-(3a-α-4-β, 6a-α))- see VSU100

5-(2-THIENOYL)-2-BENZIMIDAZOLECARBAMIC ACID METHYL ESTER see OJD100

N-(5-(2-THIENOYL)-2-BENZIMIDAZOLYL)CARBAMIC ACID METHYL ESTER see OJD100

7-(2-THIENYLACETAMIDO)CEPHALOSPORANIC ACID see CCX250

7-((2-THIENYL)ACETAMIDO)-3-(1-PYRIDYLMETHYL)CEPHALOSPORANIC ACID see TEY000

7-(α-(2-THIENYL)ACETAMIDO)-3-(1-PYRIDYLMETHYL)-3-CEPHEM-4-CARBOXYLIC ACID BETAINE see TEY000

2-THIENYLALANINE see TEY250

β-2-THIENYLALANINE see TEY250

dl-β-2-THIENYLALANINE see TEY250

2-THIENYLALDEHYDE see TFM500
5-(2-THIENYLCARBONYL)-2-BENZIMIDAZOLECARBAMIC ACID METHYL ESTER
  see OJD100
(5-(2-THIENYLCARBONYL)-1H-BENZIMIDAZOL-2-YL)-CARBAMIC ACID METHYL
  ESTER see OJD100
2-THIENYLCARBOXALDEHYDE see TFM500
THIENYLIC ACID see TGA600
4-(2-THIENYLKETO)-2,3-DICHLOROPHENOXYACETIC ACID see TGA600
2-(2-THIENYL)MORPHOLINE HYDROCHLORIDE see TEY600
1-α-THIENYL-1-PHENYL-3-N-METHYLMORPHOLINIUM-1-PROPANOL IODIDE see
  HNM000
THIEPAN-2-ONE see TEY750
2-THIEPANONE see TEY750
THIERGAN see DQA400
THIETHYLPERAZINE DIMALEATE see TEZ000
THIETHYLPERAZINE MALEATE see TEZ000
THIFOR see EAQ750
THIIRANE see EJP500
THILANE see TDC730
THILAVEN see IAD000
THILLATE see TFS350
THILOPEMAL see ENG500
THILOPHENYL see DKQ000
THILOPHENYT see DNU000
THIMBLES see FOM100
THIMBLEWEED see PAM780
THIMECIL see MPW500
THIMER see TFS350
THIMEROSALATE see MDI000
THIMEROSOL see MDI000
THIMET see PGS000
THIMUL see EAQ750
50 THINNER see PCT500
THIOACETAMIDE see TFA000
THIOACETANILIDE see TFA250
THIOACETARSAMIDE see TFA350
THIOACETAZONE see FNF000
THIOACETIC ACID see TFA500
THIOALKOFEN BM 4 see TFC600
THIOALKOFEN BM see TFD000
THIOALLATE see CDO250
THIOALLYL ETHER see AGS250
THIOAMI see DXN709
THIOAMIDE see EPQ000
4,4'-THIOANILINE see TFI000
THIOANILINE see TFI000
p-THIOANISIDINE see AMS675
THIOANISOLE see TFC250
THIOANTIMONIC(III) ACID, TRIESTER with MERCAPTO SUCCINIC ACID DILI-
  THIUM SALT, NONAHYDRATE see AQE500
THIOARSENITE see TFA350
(THIOARSENOSO)METHANE see MGQ750
THIOARSMINE see SNR000
THIOAURIN see TFC500
2-THIOBARBITURIC ACID see MCK500
THIOBARBITURIC ACID see MCK500
THIOBEL see BHL750
THIOBENCARB see SAZ000
THIOBENZAMIDE see BBM250
THIOBENZOIC ACID see TFC550
THIOBENZYL ALCOHOL see TGO750
2-THIO-2-BENZYL-PSEUDOUREA HYDROCHLORIDE see BEU500
4,4'-THIOBIS(ANILINE) see TFI000
4,4'-THIOBISBENZENAMINE see TFI000
1,1'-THIOBIS(BENZENE) see PGI500
4,4'-THIOBIS(6-tert-BUTYL-m-CRESOL) see TFC600
4,4'-THIOBIS(6-tert-BUTYL-o-CRESOL) see TFD000
4,4'-THIOBIS(6-tert-BUTYL-3-METHYLPHENOL) see TFC600
4,4'-THIOBIS(2-tert-BUTYL-5-METHYLPHENOL) see TFC600
1,1'-THIOBIS(2-CHLOROETHANE) see BIH250
2,2'-THIOBIS(4,6-DICHLOROPHENOL) see TFD250
4,4'-THIOBIS(2-(1,1-DIMETHYLETHYL))-6-METHYLPHENOL see TFD000
1,1'-THIOBIS(N,N-DIMETHYLTHIO)FORMAMIDE see BJL600
THIOBIS(DODECYL PROPIONATE) see TFD500
1,1'-THIOBISETHANE see EPH000
4,4'-THIOBIS(3-METHYL-6-tert-BUTYLPHENOL) see TFC600
1,1'-THIOBIS(2-METHYL-4-HYDROXY-5-tert-BUTYLBENZENE) see TFC600
THIOBISMOL see BKX750, BKY500
3,3-THIOBIS(1-PROPENE) see AGS250
THIOBIS(TRIBENZYL-TIN) (8CI) see BLK750
THIOBORIC ACID, ESTER with 2,2'-((DIBUTYLSTANNY-
  LENE)DIOXY)DIETHANETHIOL (2:3) see TNG250
4-THIOBUTYROLACTONE see TDC800
γ-THIOBUTYROLACTONE see TDC800
epsilon-THIOCAPROLACTONE see TEY750

THIOCARB see SGJ000
THIOCARBAMATE see ISR000
THIOCARBAMIC ACID-S,S-(2-(DIMETHYLAMINO)TRIMETHYLENE)ESTER HY-
  DROCHLORIDE see BHL750
THIOCARBAMIDE see ISR000
THIOCARBAMISIN see TFD750
THIOCARBAMIZINE see TFD750
THIOCARBAMYLHYDRAZINE see TFQ000
THIOCARBANIL see ISQ000
THIOCARBANILIDE see DWN800
THIOCARBARSONE see CBI250
THIOCARBAZIDE see TFE250
THIOCARBAZIL see FNF000
THIOCARBOHYDRAZIDE see TFE250
THIOCARBONIC DICHLORIDE see TFN500
THIOCARBONIC DIHYDRAZIDE see TFE250
THIOCARBONOHYDRAZIDE see TFE250
THIOCARBONYL AZIDE THIOCYANATE see TFE275
THIOCARBONYL CHLORIDE see TFN500
THIOCARBONYL DICHLORIDE see TFN500
THIOCARBOSTYRIL see QOJ100
THIOCHLORID FOSFORECNY see TFO000
THIOCHROMAN-4-ONE, OXIME see TEJ000
THIOCHRYSINE see GJG000
10-THIOCOLCHICOSIDE see TFE325
THIOCOLCHICOSIDE see TFE325
THIO-COLCIRAN see DBA200
3-THIOCRESOL see TGO800
4-THIOCRESOL see TGP250
m-THIOCRESOL see TGO800
o-THIOCRESOL see TGP000
p-THIOCRESOL see TGP250
THIOCRON see AHO750
THIOCTACID see DXN800
THIOCTAMID see DXN709
THIOCTAMIDE see DXN709
6,8-THIOCTIC ACID see DXN800
6-THIOCTIC ACID see DXN800
THIOCTIC ACID see DXN800
THIOCTIC ACID AMIDE see DXN709
THIOCTIDASE see DXN800
THIOCTSAN see DXN800
THIOCYAN see CAY250
THIOCYANATES see TFE500
THIOCYANATE SODIUM see SIA500
THIOCYANATOACETIC ACID CYCLOHEXYL ESTER see TFF000
THIOCYANATOACETIC ACID ETHYL ESTER see TFF100
THIOCYANATOACETIC ACID ISOBORNYL ESTER see IHZ000
1-THIOCYANATODODECANE see DYA200
THIOCYANATOETHANE see EPP000
α-THIOCYANATOTOLUENE see BFL000
THIOCYANIC ACID, ALLYL ESTER see AGS750
THIOCYANIC ACID, AMYL ESTER see AON500
THIOCYANIC ACID, 2-(BENZOTHIAZOLYLTHIO)METHYL ESTER see BOO635
THIOCYANIC ACID-p-DIMETHYLAMINOPHENYL ESTER see TFH500
THIOCYANIC ACID, DODECYL ESTER see DYA200
THIOCYANIC ACID, ETHYL ESTER see EPP000
THIOCYANIC ACID compounded with GUANIDINE (1:1) see TFF250
THIOCYANIC ACID, 2-HYDROXYETHYL ESTER, LAURATE see LBO000
THIOCYANIC ACID, LITHIUM SALT (1:1) see TFF500
THIOCYANIC ACID, MERCURY(2+) SALT see MCU250
THIOCYANIC ACID, 4-METHOXY-2-NITROPHENYL ESTER see MFB500
THIOCYANIC ACID, TRIMETHYLSTANNYL ESTER see TMI750
THIOCYANIC ACID, TRIPHENYLSTANNYL ESTER see TMV750
1-THIOCYANOBUTANE see BSN500
4-THIOCYANO-N,N-DIMETHYLANILINE see TFH500
p-THIOCYANODIMETHYLANILINE see TFH500
THIOCYANO-ESSIGSAEURE-AETHYL-ESTER (GERMAN) see TFF100
2-THIOCYANOETHYL COCONATE see LBO000
2-THIOCYANOETHYL DODECANOATE see LBO000
2-THIOCYANOETHYL LAURATE see LBO000
β-THIOCYANOETHYL LAURATE see LBO000
THIOCYANOGEN see TFH600
2-(THIOCYANOMETHYLTHIO)BENZOTHIAZOLE, 60% see BOO635
THIOCYCLAM (ETHANEDIOATE 1:1) see TFH750
THIOCYCLAM HYDROGEN OXALATE see TFH750
THIOCYCLOPENTANE-1,1-DIOXIDE see SNW500
THIOCYMETIN see MPN000
THIOCYNAMINE see AGT500
2-THIOCYTOSINE see TFH800
THIOCYTOSINE see TFH800
THIODAN see EAQ750
THIODELONE see MCH600
THIODEMETON see DAO500, DXH325
THIODEMETRON see DXH325

6-THIODEOXYGUANOSINE see TFJ500
THIODERON see MCH600
4,4'-THIODIANILINE see TFI000
p,p-THIODIANILINE see TFI000
2,2'-THIODIETHANOL see TFI500
THIODIETHYLENE GLYCOL see TFI500
THIODIFENYLAMINE (DUTCH) see PDP250
β-THIODIGLYCOL see TFI500
THIODIGLYCOL see TFI500
2,2'-THIODIGLYCOLIC ACID see MCM750
β,β'-THIODIGLYCOLIC ACID see MCM750
THIODIGLYCOLIC ACID see MCM750
THIODIGLYCOLLIC ACID see MCM750
2-THIO-3,5-DIMETHYLTETRAHYDRO-1,3,5-THIADIAZINE see DSB200
4,4'-THIODIPHENOL see TFJ000
THIODIPHENYLAMIN (GERMAN) see PDP250
THIODIPHENYLAMINE see PDP250
O,O'-(THIODI-4,1-PHENYLENE)BIS(O,O-DIMETHYL PHOSPHOROTHIOATE) see TAL250
THIODI-p-PHENYLENEDIAMINE see TFI000
O,O'-(THIODI-p-PHENYLENE)-O,O,O',O'-TETRAMETHYL BIS(PHOSPHOROTHIOATE) see TAL250
3,3'-THIODIPROPIONIC ACID see BHM000
β,β'-THIODIPROPIONIC ACID see BHM000
THIODIPROPIONIC ACID see BHM000
β,β'-THIODIPROPIONITRILE see DGS600
THIODOW see EIR000
THIODRIL see CBR675
THIODROL see EBD500
2-THIOETHANOL see MCN250
THIOETHANOL see EMB100
THIOETHANOLAMINE see AJT250
THIOETHYL ALCOHOL see EMB100
(THIOETHYLENE)BIS(BIS(2-HYDROXYETHYL)SULFONIUM) DICHLORIDE see BHD000
THIOETHYL ETHER see EPH000
THIOFACO M-50 see EEC600
THIOFACO T-35 see TKP500
THIOFAN see TDC730
THIOFANATE see DJV000
THIOFANOX see DAB400
THIOFENOL see PFL850
THIOFIDE see BDE750
THIOFOR see EAQ750
THIOFORM (CZECH) see TLS500
THIOFORMIC ACID, PHENYLAZO-, PHENYLHYDRAZIDE see DWN200
THIOFOSGEN (CZECH) see TFN500
THIOFOZIL see TFQ750
THIOFURAM see TFM250
THIOFURAN see TFM250
THIOFURFURAN see TFM250
(1-THIO-d-GLUCOPYRANOSATO)GOLD see ART250
5-THIO-α-d-GLUCOPYRANOSE see GFK000
1-THIO-GLUCOPYRANOSE, MONOGOLD(1+) SALT see ART250
5-THIO-d-GLUCOSE see GFK000
THIOGLUCOSE d'OR (FRENCH) see ART250
THIOGLYCERIN see MRM750
1-THIOGLYCEROL see MRM750
α-THIOGLYCEROL see MRM750
THIOGLYCOL (DOT) see MCN250
THIOGLYCOLANILIDE see MCK000
THIOGLYCOLATESODIUM see SKH500
2-THIOGLYCOLIC ACID see TFJ100
THIOGLYCOLIC ACID see TFJ100
THIOGLYCOLIC ACID ANILIDE see MCK000
THIOGLYCOLIC ACID ETHYL ESTER see EMB200
THIOGLYCOLIC ACID-2-ETHYLHEXYL ESTER see EKW300
THIOGLYCOLIC ACID METHYL ESTER see MLE750
THIOGLYCOLIC ACID, SODIUM SALT see SKH500
THIOGLYCOLLIC ACID see TFJ100
THIOGLYCOLLIC ACID, AMMONIUM SALT see ANM500
THIOGLYKOLSAEURE-AETHYLESTER (GERMAN) see EMB200
THIOGLYKOLSAEURE-2-AETHYLHEXYL ESTER (GERMAN) see EKW300
THIOGLYKOLSAEURE-METHYLESTER (GERMAN) see MLE750
THIOGUAIACOL see MFQ300
6-THIOGUANINE see AMH250
THIOGUANINE see AMH250
β-THIOGUANINE DEOXYRIBOSIDE see TFJ250
6-THIOGUANINE RIBONUCLEOSIDE see TFJ500
THIOGUANINE RIBOSIDE see TFJ500
6-THIOGUANOSINE see TFJ500
THIOGUANOSINE see TFJ500
THIOHEXAM see CPI250
THIOHEXITAL see AGL875
2-THIOHYDANTOIN see TFJ750

2-THIO-4-HYDRAZINOURACIL see HHF500
2-THIO-6-HYDROXY-8-AZAPURINE see HOJ100
THIOHYPOXANTHINE see POK000
THIOINDIGO see DNT300
THIOINDIGO BLACK see CMU320
THIOINDIGO ORANGE KKh see CMU815
THIOINDIGO RED B see DNT300
THIOINDIGO RED S see DNT300
THIOINOSINE see TFJ825
THIOISONICOTINAMIDE see TFK000
THIOKARBONYLCHLORID (CZECH) see TFN500
2-THIO-4-KETOIHIAZOLIDINE see RGZ550
THIOKOL NVT see ROH900
THIOKTSAFURE (GERMAN) see DXN800
THIOLA see MCI375
THIOLACETIC ACID see TFA500
2-THIOLACTIC ACID see TFK250
THIOLANE see TDC730
THIOLANE-1,1-DIOXIDE see SNW500
THIOLAN-2-ONE see TDC800
THIOLDEMETON see DAP200
2-THIOL-DIHYDROGLYOXALINE see IAQ000
THIOLE see TFM250
THIOLIN see IAD000
THIOLITE see CAX500
THIOLMECAPTOPHOS see DAO500
THIOL SYSTOX see DAP200
THIOLUTIN see ABI250
THIOLUX see SOD500
2-THIO-MALIC ACID see MCR000
THIOMALIC ACID see MCR000
THIOMEBUMAL see PBT250
THIOMEBUMAL SODIUM see PBT500
THIOMECIL see MPW500
THIOMERIN SODIUM see TFK270
THIOMERSALATE see MDI000
THIOMESTERONE see TFK300
THIOMESTRONE see TFK300
THIOMETAN see DSK600
THIOMETHANOL see MLE650
p-THIOMETHOXYANILINE see AMS675
2-THIO-6-METHYL-1,3-PYRIMIDIN-4-ONE see MPW500
6-THIO-4-METHYLURACIL see MPW500
THIOMETON see PHI500
THIOMICID see FNF000
THIOMIDIL see MPW500
THIOMONOGLYCOL see MCN250
THIOMUL see EAQ750
THIOMYLAL SODIUM see SOX500
2-THIONAPHTHOL see NAP500
THIO-β-NAPHTHOL see NAP500
β-THIONAPHTHOL see NAP500
THIONAPHTHOL see NAP500
THIONAZIN see EPC500
THIONEMBUTAL see PBT500
THIONEX see BJL600, EAQ750
THIONEX RUBBER ACCELERATOR see BJL600
THIONICID see FNF000
THIONIDEN see EPQ000
THIONIN see AKK750
THIONINE see AKK750
THIONOACETIC ACID see TFA500
THIONOBENZENEPHOSPHONIC ACID ETHYL-p-NITROPHENYL ESTER see EBD700
THIONODEMETON SULFONE see SPF000
THIONOSINE see MCQ500
5-THIONO-v-TRIAZOLO(4,5-d)PYRIMIDIN-7(4H,6H)-ONE see HOJ100
THIONTAN see DHF600
THIONYLAN see TEO250
THIONYL BROMIDE see SNT200
THIONYL CHLORIDE see TFL000
THIONYL DICHLORIDE see TFL000
THIONYL DIFLUORIDE see TFL250
THIONYL FLUORIDE see TFL250
THIOOCTANOIC ACID see DXN800
THIOORATIN see TET250
2-THIO-4-OXO-6-METHYL-1,3-PYRIMIDINE see MPW500
2-THIO-4-OXO-6-PROPYL-1,3-PYRIMIDINE see PNX000
2-THIO-6-OXYPURINE see TFL500
2-THIO-6-OXYPYRIMIDINE see TFR250
THIOPARAMIZONE see FNF000
THIOPENTAL see PBT250
THIOPENTAL SODIUM see PBT500
THIOPENTAL SODIUM SALT see PBT500
THIOPENTEX see CPK000

THIOTHIXENE see NBP500
THIOTHIXENE DIHYDROCHLORIDE see TFQ600
THIOTHIXINE see NBP500
THIOTHYMIN see MPW500
2-THIOTHYMINE see TFQ650
THIOTHYMINE see TFQ650
THIOTHYRON see MPW500
6,8-THIOTIC ACID see DXN800
6-THIOTIC ACID see DXN800
3-THIOTOLENE see MPV250
THIOTOMIN see DXN709
THIOTOX see TFS350
5-THIO-1H-υ-TRIAZOLO(4,5-d)PYRIMIDINE-5,7(4H,6H)-DIONE see MLY000
THIOTRIETHYLENEPHOSPHORAMIDE see TFQ750
THIOTRITHIAZYL NITRATE see TFR000
2-THIOURACIL see TFR250
6-THIOURACIL see TFR250
THIOURACIL see TFR250
2-THIOUREA see ISR000
THIOUREA (DOT) see ISR000
THIOUREA, N,N'-BIS(1-METHYLETHYL)-(9CI) see DNS800
THIOVANIC ACID see TFJ100
THIOVANOL see MRM750
THIOVIT see SOD500
THIOXAMYL see DSP600
p-THIOXANE see OLY000
3-(THIOXANTHEN-9-YL)METHYL-1-PIPECOLINE, HYDROCHLORIDE see THL500
THIOXIDIL see GGS000
THIOXIDRENE see TCZ000
6-THIOXOPURINE see POK000
2-THIOXO-4-THIAZOLIDINONE see RGZ550
4-(THIOXO(3,4,5-TRIMETHOXYPHENYL)METHYL)MORPHOLINE see TNP275
THIOZAMIDE see TEX250
THIOZIN see IAD000, ZJS300
2-THIOZOLIDINETHIONE see TFS250
THIPENTAL SODIUM see PBT500
THIPHEN see DHY400
THIPHENAMIL HYDROCHLORIDE see DHY400
THIRAM see TFS350
THIRAMAD see TFS350
THIRAME (FRENCH) see TFS350
THIRASAN see TFS350
THIRERANIDE see DXH250
THISTROL see CLN750, CLO000
THIULIX see TFS350
THIURAD see TFS350
THIURAGYL see PNX000
THIURAM see TFS350
THIURAM DISULFIDE see TFS500
THIURAM E see DXH250
THIURAMIN see TFS350
THIURAMYL see TFS350
THIURANIDE see DXH250
THIURETIC see CFY000
THIURYL see MPW500
THIXOKON see AAN000
THIZONE see FNF000
THLARETIC see CFY000
THN see TCY250
THOMAPYRIN see ARP250
THOMAS BALSAM see BAF000
THOMBRAN see CKJ000, THK880
THOMIZINE see THG600
THOMPSON-HAYWARD TH6040 see CJV250
THOMPSON'S WOOD FIX see PAX250
THONZYLAMINE HYDROCHLORIDE see RDU000
THONZYLAMINIUM CHLORIDE see RDU000
THORAZINE see CKP250, CKP500
THORAZINE HYDROCHLORIDE see CKP500
THORIA see TFT750
THORIDAZINE HYDROCHLORIDE see MOO500
THORIUM-232 see TFS750
THORIUM see TFS750
THORIUM CHLORIDE see TFT000
THORIUM DICARBIDE see TFT100
THORIUM DIOXIDE see TFT750
THORIUM HYDRIDE see TFT250
THORIUM METAL, pyrophoric (DOT) see TFS750
THORIUM (4+) NITRATE see TFT500
THORIUM(IV) NITRATE see TFT500
THORIUM OXIDE see TFT750
THORIUM OXIDE SULFIDE see TFU000
THORIUM TETRACHLORIDE see TFT000
THORIUM TETRANITRATE see TFT500
THORN APPLE see SLV500

THOROTRAST see TFT750
THORTRAST see TFT750
THPA see TDB000
THPC see TDH750
THPS see TDI000
THQ see MKO250
THREADLEAF GROUNDSEL see RBA400
THREAMINE see AMA500
THREARIC ACID see TAF750
THREE ELEPHANT see BMC000
l-THREITOL-1,4-BISMETHANESULFONATE see TFU500
THRENE BRILLIANT ORANGE GR see CMU820
d-THREO-α-BENZYL-N-ETHYLTETRAHYDROFURFURYLAMINE see ZVA000
l-THREONINE see TFU750
THREONINE see TFU750
THRETHYLENE see TIO750
THROATWORT see FOM100
THROMBASE see TFU800
THROMBIN see TFU800
THROMBIN-C see TFU800
THROMBIN COAGULASE see CMY725
THROMBOCID see SEH450
THROMBOCYTIN see AJX500
THROMBOFORT see TFU800
THROMBOLIQUINE see HAQ500
THROMBOTONIN see AJX500
THS-101 see DBC500
THS-201 see HAG325
THS-839 see DAP812
THU see ISR000
(1S,4R,5R)-(−)-3-THUJANONE see TFW000
THUJA OIL see CCQ500
β-THUJAPLICIN see IRR000
γ-THUJAPLICIN see TFV750
β-THUJAPLICINE see IRR000
γ-THUJAPLICINE see TFV750
THUJON see TFW000
α-THUJONE see TFW000
l-THUJONE see TFW000
THUJONE see TFW000
(−)-THUJONE see TFW000
THULIUM see TFW250
THULIUM CHLORIDE see TFW500
THULIUM EDETATE see TFX000
THULIUM(III) NITRATE, HEXAHYDRATE (1:3:6) see TFX250
THULOL see EDU500
THURICIDE see BAC040
THURINGIENSIN see BAC125
THURINGIENSIN A see BAC125
THURINGIN see BAC040
THURINTOX see BAC125
THX see TFZ275
THYALONE see BAG250
THYCAPSOL see MCO500
THYLATE see TFS350
THYLFAR M-50 see MNH000
THYLOGEN see WAK000
THYLOGEN MALEATE see DBM800
THYLOQUINONE see MMD500
THYME CAMPHOR see TFX810
THYME OIL see TFX500
THYME OIL RED see TFX750
THYMEOL see AEG875
THYMIAN OEL (GERMAN) see TFX500
THYMIC ACID see TFX810
THYMIDIN see TFX790
THYMIDINE see TFX790
THYMINE see TFX800
THYMINE-2-DEOXYRIBOSIDE see TFX790
THYMINEDEOXYRIBOSIDE see TFX790
THYMINE, 2-THIO- see TFQ650
THYMIN (PURINE BASE) see TFX800
THYM OIL see TFX500
m-THYMOL see TFX810
o-THYMOL see CCM000
THYMOL see TFX810
THYMOL, 6,6'-(3H-2,1-BENZOXATHIOL-3-YLIDENE)DI-, S,S-DIOXIDE see TFX850
THYMOL BLUE see TFX850
THYMOL, 6-CHLORO- see CLJ800
THYMOL METHYL ETHER see MPW650
THYMOLSULFONEPHALEIN see TFX850
THYMOLSULFONEPHTHALEIN see TFX850
THYMOLSULFOPHTHALEIN see TFX850
THYMOLSULPHONPHTHALEIN see TFX850

THYMOQUINONE see IQF000
THYMOSULFONPHTHALEIN see TFX850
THYMOXAMINE HYDROCHLORIDE see TFY000
THYMUS KOTSCHYANUS, OIL EXTRACT see TFY100
THYMYL METHYL ETHER see MPW650
THYNESTRON see EDV000
THYNON see DLK200
THYRADIN see TFZ100
THYREOIDEUM see TFZ275
THYREONORM see MPW500
THYREOSTAT see MPW500
THYREOSTAT II see PNX000
THYRIL see MPW500
THYROCALCITONIN see TFZ000
THYROID see TFZ100
THYROID RELEASING HORMONE see TNX400
THYROLIBERIN see TNX400
THYROTROPIC-RELEASING FACTOR see TNX400
THYROTROPIC RELEASING HORMONE see TNX400
THYROTROPIN-RELEASING FACTOR see TNX400
THYROTROPIN-RELEASING HORMONE see TNX400
THYROXEVAN see LFG050
l-THYROXIN see TFZ275
THYROXIN see TFZ275
l-THYROXINE see TFZ275
THYROXINE see TFZ275, TFZ300
(−)-THYROXINE see TFZ275
l-THYROXINE MONOSODIUM SALT see LFG050
d-THYROXINE SODIUM see SKJ300
d-THYROXINE SODIUM SALT see SKJ300
l-THYROXINE SODIUM SALT see LFG050
THYROXINE SODIUM SALT see LFG050
TI-8 see TCA250
TI-1258 see BHL750
TIACETAZON see FNF000
TIN, ACETOXYTRIPROPYL-(7CI) see TNB000
TIADIPONE see BAV625
TIAMIPRINE see AKY250
TIAMIZID see CIP500
TIAMIZIDE see CIP500
TIAMULIN see TET800
TIAMULINA (ITALIAN) see TET800
TIAMUTIN see DAR000
TIAPAMIL see TGA275
TIAPRIDE HYDROCHLORIDE see TGA375
TIAPROFENIC ACID see SOX400
TIARAMIDE HYDROCHLORIDE see CJH750
TIAZOFURIN see RJF500
TIAZON see DSB200
TIB see TKQ250
2,3,5-TIBA see TKQ250
TIBA see TKQ250
TIBAMATO see MOV500
TIBAZIDE see ILD000
TIBENZATE see BFK750
TIBERAL see OJS000
TIBEXIN see LFJ000
TIBEY (PUERTO RICO) see SLJ650
TIBICUR see FNF000
TIBINIDE see ILD000
TIBIONE see FNF000
TIBIVIS see ILD000
TIBIZAN see FNF000
TIBRIC ACID see CGJ250
TIBUTOL see TGA500
TIC see TKJ250
TICARCILLIN SODIUM see TGA520
TICKLE WEED see FAB100
TICLID see TGA525
TICLOBRAN see ARQ750
TICLODIX see TGA525
TICLODONE see TGA525
TICLOPIDINE HYDROCHLORIDE see TGA525
TIC MUSTARD see IAN000
TICOLIN see DXN709
TIN, COMPLEX with PYROCATECHOL see TGC300
TICREX see TGA600
TICRYNAFEN see TGA600
TIDEMOL see BQL000
TIN, DIETHYL-, DIIODIDE see DJB000
TIN, DIMETHYL-, DICHLORIDE see DUG825
TIN, DIOCTYL-, DICHLORIDE see DVN300
TIEMONIUM IODIDE see HNM000
TIEMOZYL see HNM000
TIEMPE see TKZ000

TIENILIC ACID see TGA600
2-(2-TIENYL)MORFOLINA CLORIDRATE (ITALIAN) see TEY600
TIEZENE see EIR000
TIFEMOXONE see PDU750
TIFEN see DHY400
TIFERRON see DXH300
TIN, FLUOROTRIPHENYL- see TMV850
TIFOMYCINE see CDP250
TIGAN see TKW750
TIGAN HYDROCHLORIDE see TKW750
TIGASON see EMJ500
TIGLIC ACID see TGA700
TIGLIC ACID, ETHYL ESTER (6CI,7CI) see TGA800
TIGLIC ACID, GERANIOL ESTER see GDO000
TIGLIC ACID, METHYL ESTER (6CI,7CI) see MPW700
TIGLINIC ACID see TGA700
12-o-TIGLYL-PHORBOL-13-BUTYRATE see PGV750
12-o-TIGLYL-PHORBOL-13-DODECANOATE see PGW000
7-TIGLYLRETRONECINE VIRIDIFLORATE see SPB500
7-TIGLYL-9-VIRIDIFLORYLRETRONECINE see SPB500
TIGUVON see FAQ900
TIKLID see TGA525
TIKOFURAN see NGG500
TILCAREX see PAX000
TILCOTIL see TAL485
TILDIEM see DNU600
TILDIN see BNK000
TILIDATE see EIH000
TILIDINE HYDROCHLORIDE see EIH000
TIL4K8 see TOC250
TILLAM (RUSSIAN) see PNF500
TILLAM-6-E see PNF500
TILLANTOX see BDD000
TILLRAM see DXH250
TILMAPOR see CCS550
TILORONE HYDROCHLORIDE see TGB000
TIMACOR see TGB185
TIMAZIN see FMM000
TIMBER RATTLESNAKE VENOM see TGB150
TIMELIT see IFZ900
TIMEPIDIUM BROMIDE see TGB160
TIMET see PGS000
TIMIPERONE see TGB175
TIMOLET see DLH600, DLH630
l-TIMOLOL MALEATE see TGB185
TIMOLOL MALEATE see TGB185
TIMONIL see DCV200
TIMONOX see AQF000
TIMOPED see TGB475
TIMOPTOL see TGB185
TIMOSIN see MDQ250
TIN (α) see TGB250
TIN see TGB250
TINACTIN see TGB475
TINADERM see TGB475
TINA GREY BL see CMU320
TINAJA (PUERTO RICO) see CNR135
TINA ORANGE R see CMU815
TINA PINK B see DNT300
TINA PRINTING BLACK BL see CMU320
TIN AZIDE TRICHLORIDE see TGB500
TIN BIFLUORIDE see TGD100
TIN(IV) BROMIDE (1:4) see TGB750
TIN(II) CHLORIDE (1:2) see TGC000
TIN(IV) CHLORIDE (1:4) see TGC250
TIN CHLORIDE, fuming (DOT) see TGC250
TIN(II) CHLORIDE DIHYDRATE (1:2:2) see TGC275
TIN CHLORIDE IODIDE see TGC280
TIN(IV) CHLORIDE, PENTAHYDRATE (1:4:5) see TGC282
TIN(2+) CITRATE see TGC285
TIN CITRATE (7CI) see TGC285
TIN COMPOUNDS see TGC500
TINDAL see ABG000
TIN DIBUTYL DILAURATE see DDV600
TIN DIBUTYLDITRIFLUOROACETATE see BLN750
TIN DIBUTYL MERCAPTIDE see DEI200
TIN DICHLORIDE see TGC000
TIN DIFLUORIDE see TGD100
TINDURIN see TGD000
TINESTAN see ABX250
TINESTAN 60 WP see ABX250
TIN FLAKE see TGB250
TIN FLUORIDE see TGD100
TINGENIN A see TGD125
TINGENON see TGD125

TINGENONE see TGD125
TINIC see NCQ900
TINIDAZOL see TGD250
TINIDAZOLE see TGD250
TIN(IV) IODIDE (1:4) see TGD750
TIN(II) IODIDE see TGD500
TINMATE see CLU000
TIN(II) NITRATE OXIDE see TGE000
TINOFEDRINE HYDROCHLORIDE see DXI800
TINOLITE see CBT750
TINON BLUE GF see DFN425
TINON BLUE GL see DFN425
TINON BLUE RS see IBV050
TINON BLUE RSN see IBV050
TINON BRILLIANT ORANGE GR see CMU820
TINON BROWN BR see CMU770
TINON GOLDEN YELLOW see DCZ000
TINON NAVY BLUE RA see VGP100
TINON OLIVE 2R see DUP100
TINON RED 6B see CMU825
TINON VIOLET B 4RP see DFN450
TINON VIOLET 4R see DFN450
TINON VIOLET 2RB see DFN450
TINOPAL AMS see CMP200
TINOPAL BHS see FCA100
TINOPAL 5BM see TGE100
TINOPAL CBS see TGE150
TINOPAL CBS-X see TGE150
TINOPAL EMS see CMP200
TINOPAL RBS 200 see TGE155
TINOPAL RBS see TGE155
TINORIDINE HYDROCHLORIDE see TGE165
TINOSTAT see DDV600
TINOX see MIW250
TIN(2+) OXALATE see TGE250
TIN(II) OXALATE see TGE250
TIN OXALATE see TGE250
TIN OXIDE see TGE300
TIN P see HML500
TIN PERBROMIDE see TGB750
TIN PERCHLORIDE (DOT) see TGC250
TIN(IV) PHOSPHIDE see TGE500
TIN POTASSIUM TARTRATE see TGE750
TIN POWDER see TGB250
TIN PROTOCHLORIDE see TGC000
TIN-SAN see TIC500
TINSET see OMG000
TIN SODIUM TARTRATE see TGF000
TIN(2+) SULFATE see TGF010
TIN(II) SULFATE see TGF010
TIN TETRABROMIDE see TGB750
TIN TETRACHLORIDE, anhydrous (DOT) see TGC250
TINTETRACHLORIDE (DUTCH) see TGC250
TIN TETRAIODIDE see TGD750
TINTORANE see WAT220
TIN TRIPHENYL ACETATE see ABX250
TINUVIN 144 see PMK800
TINUVIN 1130 see TGF025
TINUVIN P see HML500
TIOACETAZON see FNF000
TIOALKOFEN BM see TFD000
TIOBENZAMIDE (ITALIAN) see BBM250
TIOCARONE see FNF000
TIOCONAZOLE see TGF050
TIOCTACID see DXN800
TIOCTAN see DXN709
TIOCTIDASI see DXN800
TIOCTIDASI ACETATE REPLACING FACTOR see DXN800
TIN, OCTYL-, TRICHLORIDE see OGG000
TIN, OCTYL-, TRIS(ISOOCTYLTHIO GLYCOLLATE) see OGI000
TIODIFENILAMINA (ITALIAN) see PDP250
TIOFANATE ETILE (ITALIAN) see DJV000
TIOFINE see TGG760
TIOFOS see PAK000
TIOFOSFAMID see TFQ750
TIOFOZIL see TFQ750
TIOGUANIN see AMH250
TIOGUANINE see AMH250
TIOINOSINE see MCQ500
TIOMERACIL see MPW500
TIOMESTERONE see TFK300
TIONAL see BJT750
TIOPENTALE (ITALIAN) see PBT250
TIOPENTAL SODIUM see PBT500
TIOPRONIN see MCI375

TIOPROPEN see AOO490
TIORALE M see MPW500
TIORIDAZIN see MOO500
TIOTIAMINE see DXO300
TIOTIRON see MPW500
TIOTIXENE see NBP500
TIOTRIFAR see AOO490
TIOURACYL (POLISH) see TFR250
TIOVEL see EAQ750
TIOVIT-B₁ see DXO300
TIOXANONA see TLP750
TIOXIDE see TGG760
TIOXIDE A-HR see OBU100
TIOXIN see ZJS300
TIPEDINE see BLV000
TIPEPIDINE see BLV000
TIPHEN see DHY400
TIPIDI see DXO300
TIP-OFF see NAK500
TIPPON see TAA100
TIPPS see TDE750
TIQUIZIUM BROMIDE see TGF075
TIRAMPA see TFS350
TIRANTIL see SBG500
TIRIAN HYDROCHLORIDE see CKB500
TIROIDINA see TFZ100
TIRON see DXH300
TIROPRAMIDA (SPANISH) see TGF175
TIROPRAMIDE see TGF175
TIRPATE see DRR000
TISERCIN see MCI500
TISIN see ILD000
TISKAN (CZECH) see NFC500
TISOMYCIN see CQH000
2,4,5-T ISOOCTYL ESTER see TGF200
TISPERSE MB-58 see BHA750
TISPERSE MB-2X see NBL000
TISULAC see LAG010
TITAANTETRACHLORID (DUTCH) see TGH350
TITANATE see TGF250
TITANDIOXID (SWEDEN) see TGG760
TITANE (TETRACHLORURE de) (FRENCH) see TGH350
TITANIC ACID, LEAD SALT see LED000
TITANIC SULFATE see TGF220
TITANIC SULPHATE see TGF220
TITANIO TETRACHLORURO di (ITALIAN) see TGH350
TITANIUM (wet powder) see TGF500
TITANIUM see TGF250
TITANIUM 50A see TGF250
TITANIUM ACETONYL ACETONATE see BGQ750
TITANIUM ALLOY see TGF250
TITANIUM AZIDE TRICHLORIDE see TGF750
TITANIUM compounded with BERYLLIUM (1:12) see BFR000
TITANIUM CARBIDE see TGG000
TITANIUM(III) CHLORIDE see TGG250
TITANIUM CHLORIDE see TGG250, TGH350
TITANIUM COMPOUNDS see TGG500
TITANIUM DIAZIDE DIBROMIDE see TGG600
TITANIUM DIBROMIDE see TGG625
TITANIUM DICHLORIDE see TGG750
TITANIUM DIOXIDE see TGG760
TITANIUM DISULFATE see TGH250
TITANIUM FERROCENE see DGW200
TITANIUM(4+) ISOPROPOXIDE see IRN200
TITANIUM ISOPROPYLATE see IRN200
TITANIUM METAL POWDER, WET (DOT) see TGF500
TITANIUM(III) METHOXIDE see TGF775
TITANIUM NICKEL OXIDE see NDL500
TITANIUM OXIDE see TGG760
TITANIUM OXIDE BIS(ACETYLACETONATE) see BGQ750
TITANIUM, OXOBIS(2,4-PENTANEDIONATO-O,O') see BGQ750
TITANIUM POTASSIUM FLUORIDE see PLI000
TITANIUM SULFATE see TGH250
TITANIUM SULFATE SOLUTION (DOT) see TGH250
TITANIUM TETRACHLORIDE see TGH350
TITANIUM TETRAISOPROPOXIDE see IRN200
TITANIUM TETRAISOPROPYLATE see IRN200
TITANIUM TETRA-n-PROPOXIDE see IRN200
TITANIUM TRICHLORIDE MIXTURES, pyrophoric (UN 2441) (DOT) see TGG250
TITANIUM TRICHLORIDE MIXTURES (UN 2869) (DOT) see TGG250
TITANIUM TRICHLORIDE, pyrophoric (UN 2441) (DOT) see TGG250
TITANOCENE see DGW200, TGH500
TITANOCENE, DICHLORIDE see DGW200
TITANOUS CHLORIDE see TGG250

TITANTETRACHLORID (GERMAN) see TGH350
TITAN YELLOW see CMP050
TITAN YELLOW DYE see CMP050
TITAN YELLOW G see CMP050
TITANYL BIS(ACETYLACETONATE) see BGQ750
TIN, TRIBUTYL-, ISOTHIOCYANATE see THY750
TITRIPLEX see EIX000
TITRIPLEX I see AMT500
TITRIPLEX III see EIX500
TIURAM (POLISH) see TFS350
TIURAM see DXH250
TIURAMYL see TFS350
TIUROLAN see BSN000
TIVIST see FOS100
TIXANTONE see SBE500
TIXOTON see BAV750
TIZANIDINE HYDROCHLORIDE see TGH600
TJB see DWI000
TK 1000 see PKQ059
TK-5477 see MJL750
TK 10352 see BOS100
TK 10408 see BRK750
TK 12627 see BJK550
Ts36Khr see ZFJ130
TKB see NKB500
TL 59 see PER600
TL 69 see DGB600
TL 70 see SOQ450
TL 75 see PHQ500
TL 77 see EEC000
TL 78 see DNJ800
TL 80 see DDL400
TL 85 see BJF500
TL 86 see SCB500
TL 108 see OLK000
TL 145 see TNF250
TL 146 see BIE250
TL 154 see CHF500
TL 158 see DKB119
TL 161 see BHO250
TL 186 see MIH250
TL 189 see FOY000
TL 190 see MPI750
TL 207 see CGU199
TL 212 see SOT500
TL 214 see DFH200
TL 229 see HBN600
TL 231 see HFP600
TL 262 see TFO000
TL 263 see TFO500
TL 266 see NLJ500
TL 294 see DFP200
TL 295 see PNH650
TL-301 see IOF300
TL 311 see DSA800
TL 314 see ADX500
TL 329 see BID250
TL 331 see TFO250
TL 336 see BOB550
TL 337 see EJM900
TL 340 see KDK000
TL 345 see DJJ400
TL 350 see BOX500
TL 365 see TNC250
TL 367 see FBS000
TL 373 see EOQ000
TL 379 see DQS600
TL 385 see TEJ750
TL 389 see DQY950
TL 399 see TIY800
TL 401 see CHD875
TL 423 see EHK500
TL 429 see MRI000
TL 457 see CBH250
TL 465 see EIC000
TL 466 see IRF000
TL 472 see DEP400
TL 476 see AOQ875
TL 478 see BRZ000
TL 513 see DEV300
TL 524 see BQL500
TL 525 see BIE750
TL 568 see BQL750
TL-599 see HJZ000
TL 671 see CGY825

TL 741 see FIE000
TL 751 see FIH100
TL 783 see MEN775
TL 790 see FIS000
TL 792 see BJE750
TL 797 see DXR200
TL 821 see FIH000
TL 822 see COO500
TL 833 see PHO250
TL 855 see FIM000
TL 869 see SHG500
TL 898 see MCY475
TL 944 see BEU500
TL 965 see BIA750
TL 1002 see MHQ500
TL 1026 see CAG000
TL 1055 see FPX000
TL 1070 see CAG500
TL 1091 see NDC000
TL-1097 see HNP000
TL 1139 see SMI000
TL 1149 see BID250
TL 1163 see TLN250
TL 1178 see HNO500
TL 1182 see CAI000
TL 1183 see MKE250
TL-1185 see TGH650
TL-1186 see TGH655
TL-1188 see TGH660
TL 1217 see MID250
TL-1226 see HNN000
TL-1238 see HNK500
TL 1266 see DEC200
TL-1299 see TGH665
TL 1312 see FFH000
TL-1317 see HNK550
TL 1333 see MKE750
TL-1345 see HJY500
TL-1350 see BKR000
TL-1380 see PIA750
TL-1394 see DQY909
TL-1422 see TGH670
TL 1428 see MFW750
TL-1434 see TGH675
TL-1435 see TGH680
TL-1448 see TGH685
TL 1450 see MKX250
TL 1473 see DGJ250
TL 1483 see EEV200
TL 1503 see DJN400
TL-1504 see MID900
TL 1505 see ABO000
TL 1578 see EIF000
TL 1618 see IPX000
TLA see BPV100
TLCK see THH500
TLD 100 see LHF000
TL 301 HYDROCHLORIDE see IPG000
TL 329 HYDROCHLORIDE see EGU000
TL 513 HYDROCHLORIDE see BIB250
TL4N see DJD600
TLP-607 see PAP250
TM 3 see TGH690
β-TM10 see TLQ000
TM 10 see XSS900
TM 30 see CBT750
TM 723 see AAC875
TM 906 see TKU675
TM-4049 see MAK700
TM 12008 see DJN600
TM 1 (filler) see CAT775
TM see TDK500
TMA see TKU700, TKV000, TLD500
TMAB see BMB250, TDJ500
TMAC see TDQ225
TMAN see TKV000
TMAT see TNK000
TMB see TDM800, TLM050
7,8,12-TMBA see TLK750
TMBAC see BFM250
TMB-4 DIBROMIDE see TLQ500
TMB-4 DICHLORIDE see TLQ750
TMCA see DBA175, TLA250, TLN750
TMCA METHYL ESTER d-TARTRATE, HYDRATE see DBA175
TMCA METHYL ETHER see TLO000

TMDE 6500 see SMQ500
TMEDA see TDQ750
m-TMI see IKG800
TriML see TLU175
TML see TDR500
TMM see MAP300
TMP see HDF300, TMD250
TMP (ALCOHOL) see HDF300
TMPD see BJF500, TLY750
TMPH see HNN500
TMPTA see TLX175
TMS 480 see TKX000
TMSM see MAW800
TMSN see TDW250
TMTD see TFS350
TMTDS see TFS350
TMTM see BJL600
TMTMS see BJL600
TMTU see TDX000
TMU see TDX250
TMV-4 see TLQ500
m-TMXDI see TDX300
m-TMXDI/TMP ADDUCT see TDX320
T-2 MYCOTOXIN see FQS000
TN 762 see TEN750
TNA see TDY000
TNB see TMK500
TNCS 53 see CNP250
TNG see NGY000
TN 3J see DEJ100
TNM see TDY250
TNO 6 see SLE875
α-TNT see TMN490
TNT (OSHA) see TMN490
TNT see TMN490
TNT-TOLITE (FRENCH) see TMN490
TNT, dry or wetted with <30% water, by weight (UN 0209) (DOT) see TMN490
TO 125 see SAN600
TOA BLA-S see BLX500
TOABOND 40H see AAX250
TOBACCO see TGH725
TOBACCO LEAF ABSOLUTE see TGH750
TOBACCO LEAF, NICOTIANA GLAUCA see TGI000
TOBACCO LEAF, NICOTIANA TABACUM see TGH725
TOBACCO PLANT see TGI100
TOBACCO REFINED TAR see CMP800
TOBACCO SMOKE CONDENSATE see SEC000
TOBACCO SUCKER CONTROL AGENT 148 see DAI800, ODE000
TOBACCO SUCKER CONTROL AGENT 504 see ODE000
TOBACCO TAR see CMP800, SEC000
TOBACCO WOOD see WCB000
TOBIAS ACID see ALH750
TOBRA see TGI500
TOBRADISTIN see TGI250
TOBRAMYCIN see TGI250
TOBRAMYCIN SULPHATE see TGI500
TOBREX see TGI250
TOCAINIDE see TGI699
TOCE see DJT400
TOCEN see DJT400
TOCHERGAMINE see TGI725
TOCHLORINE see CDP000
TOCOFEROL-2-(p-CHLOROPHENOXY)-2-METHYLPROPIONATE see TGJ000
TOCOFIBRATE see TGJ000
TOCOFIBRATO (SPANISH) see TGJ000
(2R,4'R,8'R)-α-TOCOPHEROL see VSZ450
α-TOCOPHEROL see VSZ450
d-α-TOCOPHEROL (FCC) see VSZ450
dl-α-TOCOPHEROL (FCC) see VSZ450
(R,R,R)-α-TOCOPHEROL see VSZ450
(+-)-α-TOCOPHEROL ACETATE see TGJ055
d,l-α-TOCOPHEROL ACETATE see TGJ055
dl-α-TOCOPHERYL ACETATE see TGJ055
TOCOSINE see TOG250
TOCP see TMO600
T-OCTYL MERCAPTAN see MKJ250, OFE030
TODALGIL see BRF500
TODRALAZINA (ITALIAN) see CBS000
TODRALAZINE see CBS000
TODRALAZINE HYDROCHLORIDE see TGJ150
TODRALAZINE HYDROCHLORIDE HYDRATE see TGJ150
TOE see TIH800
TOF see TNI250
TOFENACINE HYDROCHLORIDE see TGJ250

TOFENACIN HYDROCHLORIDE see TGJ250
TOFISOPAM see GJS200
TOFK see TMO600
TOFRANIL see DLH600, DLH630
TOFRANILE see DLH630
TOFURON see FQN000
TOGAL see TGJ350
TOGAMYCIN see SKY500
N-TOIN see NGE000
TOIN UNICELLES see DKQ000
TOK-2 see DFT800
TOK see DFT800
TOKAMINA see VGP000
TOKAWHISKER see SCQ000
TOK E-25 see DFT800
TOK E 40 see DFT800
TOK E see DFT800
TOKIOCILLIN see AIV500
TOKKORN see DFT800
TOKOKIN see EDV000
TOKOPHARM see VSZ450
TOKSOBIDIN see BGS250
TOKUTHION see DGC800
TOK WP-50 see DFT800
TOKYO ANILINE ASTRAPHLOXINE FF see CMM765
TOKYO ANILINE BRILLIANT GREEN see BAY750
TOLADRYL see TGJ475
2,4-TOLAMINE see TGL750
TOLAMINE see CDP000
TOLANSIN see GGS000
TOLAPIN see PQC500
TOLAVAD see BBJ750
TOLAZAMIDE see TGJ500
TOLAZOLINE see BBW750
TOLAZOLINE CHLORIDE see BBJ750
TOLAZOLINE HYDROCHLORIDE see BBJ750
TOLAZUL see AJP250
TOLBAN see CQG250
TOLBUSAL see BSQ000
TOLBUTAMID see BSQ000
TOLBUTAMIDE see BSQ000
TOLCAINE HYCROCHLORIDE see TGJ625
TOLCASONE see HII500
TOLCICLATE see MQA000
TOLCIL see GGS000
TOLCYCLAMIDE see CPR000
TOLECTIN see SKJ340
TOLERON see FBJ100
TOLESTAN see CMY525
TOLFENAMIC ACID see CLK325
TOLFERAIN see FBJ100
TOLHEXAMIDE see CPR000
2-TOLIDIN (GERMAN) see TGJ750
o-TOLIDIN see TGJ750
2-TOLIDINA (ITALIAN) see TGJ750
2-TOLIDINE see TGJ750
3,3'-TOLIDINE see TGJ750
o,o'-TOLIDINE see TGJ750
o-TOLIDINE see TGJ750
TOLIDINE see TGJ750
o-TOLIDINE DIHYDROCHLORIDE see DQM000
TOLIFER see FBJ100
TOLIMAN see CMR100
TOLINASE see TGJ500
TOLIT see TMN490
TOLITE see TMN490
TOLL see MNH000
TOLMETIN see TGJ850
TOLMETINE see TGJ850
TOLMETIN SODIUM see SKJ340
TOLMETIN SODIUM DIHYDRATE see SKJ350
TOLNAFTATE see TGB475
TOLNAPHTHATE see TGB475
TOLNIDAMIDE see TGJ875
TOLOFREN see GGS000
TOLOMOL see AIV500
TOLONIDINE NITRATE see TGJ885
TOLONIUM CHLORIDE see AJP250
TOLOPELON see TGB175
(−)-1-(o-TOLOSSI-3-(2,2,5,5-TETRAMETIL-PIRROLIDIN-1-IL))-PROPAN-2-OLO CLORIDRATO (ITALIAN) see TGJ625
3-o-TOLOXY-2-HYDROXYPROPYL-1-CARBAMATE see CBK500
3-o-TOLOXY-1,2-PROPANEDIOL see GGS000
3-o-TOLOXY-1,2-PROPANEDIOL-1-CARBAMIC ACID ESTER see CBK500
TOLPAL see BBJ750

TOLPERISONE see TGK200
TOLPERISONE HYDROCHLORIDE see MRW125
TOLPRONINE HYDROCHLORIDE see TGK225
TOLSANIL see TGB475
TOLSERAM see CBK500
TOLSEROL see GGS000
TOLSERON see RLU000
TOLU see BAF000
α-TOLUALDEHYDE see BBL500
TOLUALDEHYDE see TGK250
TOLUALDEHYDE GLYCERYL ACETAL see TGK500
α-TOLUAMIDE see PDX750
TOLU BALSAM GUM see BAF000
TOLU BALSAM TINCTURE see BAF000
p-TOLUBENZYL ACETATE see XQJ700
TOLUBUTEROL HYDROCHLORIDE see BQE250
TOLUEEN (DUTCH) see TGK750
TOLUEEN-DIISOCYANAAT see TGM750
TOLUEN (CZECH) see TGK750
TOLUEN-DISOCIANATO see TGM750
TOLUENE see TGK750
TOLUENE, p-ARSENOSO- see TGV100
TOLUENE-2-AZONAPHTHOL-2 see TGW000
o-TOLUENE-1-AZO-2-NAPHTHYLAMINE see FAG135
o-TOLUENEAZO-o-TOLUENEAZO-β-NAPHTHOL see SBC500
o-TOLUENEAZO-o-TOLUENE-β-NAPHTHOL see SBC500
o-TOLUENEAZO-o-TOLUIDINE see AIC250
N,N'-2,4-TOLUENEBISMALEIMIDE see TGY770
TOLUENE-2,4-BISMALEIMIDE see TGY770
TOLUENE, m-BROMO- see BOG300
TOLUENE, o-BROMO- see BOG260
TOLUENE, p-BROMO- see BOG255
TOLUENECARBOXALDEHYDE see TGK250
TOLUENE, ar-CHLORO- see CLK130
TOLUENE, 4-CHLORO-3,5-DINITRO-α-α-α-TRIFLUORO- see CGM225
TOLUENE, 2-CHLORO-4-NITRO- see CJG800
TOLUENE, 2-CHLORO-6-NITRO- see CJG825
TOLUENE-2,3-DIAMINE see TGY800
TOLUENE-2,4-DIAMINE see TGL750
2,4-TOLUENEDIAMINE see TGL750
TOLUENE-2,5-DIAMINE see TGM000
TOLUENE-2,6-DIAMINE see TGM100
TOLUENE-3,4-DIAMINE see TGM250
2,4-TOLUENEDIAMINE (DOT) see TGL750
m-TOLUENEDIAMINE see TGL750
p-TOLUENEDIAMINE see TGM000
TOLUENEDIAMINE see TGL500
p-TOLUENEDIAMINE DIHYDROCHLORIDE see DCE200
TOLUENE-2,4-DIAMINE, 5-(PHENYLAZO)-(6CI) see CMM760
2,5-TOLUENEDIAMINE SULFATE see DCE600
p-TOLUENEDIAMINE SULFATE see DCE600, TGM400
TOLUENE-2,5-DIAMINE, SULFATE (1:1) (8CI) see DCE600
TOLUENE-2,5-DIAMINE SULPHATE see DCE600
p-TOLUENEDIAMINE SULPHATE see DCE600
TOLUENE-3,5-DIAMINE, α-α-α-TRIFLUORO-(8CI) see TKB775
2-TOLUENEDIAZONIUM BROMIDE see TGM425
p-TOLUENEDIAZONIUM FLUOROBORATE see TGM450
2-TOLUENEDIAZONIUM PERCHLORATE see TGM525
o-TOLUENEDIAZONIUM PERCHLORATE see TGM525
2-TOLUENEDIAZONIUM SALTS see TGM550
3-TOLUENEDIAZONIUM SALTS see TGM575
4-TOLUENEDIAZONIUM SALTS see TGM580
p-TOLUENEDIAZONIUM, TETRAFLUOROBORATE (1⁻) see TGM450
p-TOLUENEDIAZONIUM TETRAFLUOROBORATE (7CI) see TGM450
4-TOLUENEDIAZONIUM TRIIODIDE see TGM590
TOLUENE, 2,4-DICHLORO- see DGM700
TOLUENE, α-α-DICHLORO- see BAY300
TOLUENE, p,α-DICHLORO- see CFB600
TOLUENE, α-α-DICHLORO-α-FLUORO- see DFL100
TOLUENE-1,3-DIISOCYANATE see TGM740
TOLUENE-2,4-DIISOCYANATE see TGM750
2,4-TOLUENEDIISOCYANATE see TGM750
2,6-TOLUENE DIISOCYANATE see TGM800
TOLUENE-2,6-DIISOCYANATE see TGM800
TOLUENE DIISOCYANATE see TGM740, TGM750
TOLUENE, ar,ar-DINITRO- see DVG600
2,3-TOLUENEDIOL see DNE000
2,5-TOLUENEDIOL see MKO250
TOLUENE-3,4-DITHIOL see TGN000
TOLUENE-α,α-DITHIOL BIS(O,O-DIMETHYL PHOSPHORODITHIOATE) see BES250
TOLUENE, o-FLUORO- see FLZ100
TOLUENE HEXAHYDRIDE see MIQ740
m-TOLUENENITRILE see TGT250
p-TOLUENENITRILE see TGT750

4-TOLUENESULFANAMIDE see TGN500
p-TOLUENESULFANAMIDE see TGN500
p-TOLUENESULFANILIDE see TGN600
4-TOLUENESULFINIC ACID SODIUM SALT see SKK000
4-TOLUENESULFINYL AZIDE see TGN100
TOLUENE-2-SULFONAMIDE see TGN250
TOLUENE-4-SULFONAMIDE see TGN500
o-TOLUENESULFONAMIDE see TGN250
p-TOLUENESULFONAMIDE see TGN500
p-TOLUENESULFONAMIDE, N-(α-(CHLOROACETYL)PHENETHYL)-, (–)- see THH450
p-TOLUENESULFONAMIDE, N-(3-CHLORO-1-sec-BUTYLACETONYL)- see THH470
p-TOLUENESULFONAMIDE, N-(3-CHLORO-1-ISOPROPYLACETONYL)- see THH555
p-TOLUENESULFONAMIDE, N-(3-CHLORO-1-METHYLACETONYL)-, L- see THH360
p-TOLUENESULFONAMIDE, N-(4-CHLORO-3-OXOBUTYL)- see THH370
p-TOLUENESULFONAMIDE, N-(1-(DIAZOACETYL)-2-METHYLBUTYL)- see THH480
p-TOLUENESULFONAMIDE, N-(3-DIAZO-1-ISOBUTYLACETONYL)- see THH490
p-TOLUENESULFONAMIDE, N-(3-DIAZO-1-METHYLACETONYL)- see THH380
p-TOLUENESULFONAMIDE, N-(3-DIAZO-3-OXOBUTYL)- see THH390
p-TOLUENESULFONAMIDE, N-(7-DIAZO-6-OXOHEPTYL)- see THH460
p-TOLUENESULFONANILIDE see TGN600
4-TOLUENESULFONIC ACID see TGO000
p-TOLUENESULFONIC ACID see TGO000
TOLUENESULFONIC ACID see TGO000
m-TOLUENESULFONIC ACID, 6-((4-AMINO-3-BROMO-1-ANTHRAQUINO-NYL)AMINO)-, MONOSODIUM SALT see CMM090
p-TOLUENE SULFONIC ACID CHLORIDE see TGO250
p-TOLUENESULFONIC ACID HYDRAZIDE see THE250
o-TOLUENESULFONYLAMIDE see TGN000
p-TOLUENESULFONYLAMIDE see TGN500
p-TOLUENESULFONYLANILIDE see TGN600
1-p-TOLUENESULFONYL-3-BUTYLUREA see BSQ000
p-TOLUENESULFONYL CHLORIDE see TGO250
TOLUENE-4-SULFONYL FLUORIDE see TGO500
α-TOLUENESULFONYL FLUORIDE see TGO300
p-TOLUENESULFONYL FLUORIDE see TGO500
4-TOLUENESULFONYL HYDRAZIDE see TGO550
o-4-TOLUENE SULFONYL HYDROXYLAMINE see TGO575
TOLUENE-p-SULFONYLMETHYLNITROSAMIDE see THE500
N-(p-TOLUENESULFONYL)-N'-HEXAMETHYLENIMINOUREA see TGJ500
TOLUENE-p-SULPHONAMIDE see TGN500
p-TOLUENESULPHONIC ACID see TGO000
TOLUENE-p-SULPHONYL FLUORIDE see TGO500
TOLUENE, α-α-α-p-TETRACHLORO- see TIR900
2-TOLUENETHIOL see TGP000
4-TOLUENETHIOL see TGP250
α-TOLUENETHIOL see TGO750
m-TOLUENETHIOL see TGO800
o-TOLUENETHIOL see TGP000
p-TOLUENETHIOL see TGP250
TOLUENE TRICHLORIDE see BFL250
TOLUENE, 2,3,6-TRICHLORO- see TJD600
TOLUENE, α-α',p-TRICHLORO- see TJD650
TOLUENE, VINYL-(mixed isomers) see VQK650
3-TOLUENKARBONITRIL see TGT250
o-TOLUENO-AZO-β-NAPHTHOL see TGW000
α-TOLUENOL see BDX500
ar-TOLUENOL see CNW500
o-TOLUENSULFAMID (CZECH) see TGN000
p-TOLUHYDROQUINOL see MKO250
α-TOLUIC ACID see PDY850
m-TOLUIC ACID see TGP750
o-TOLUIC ACID see TGQ000
p-TOLUIC ACID see TGQ250
m-TOLUIC ACID DIETHYLAMIDE see DKC800
α-TOLUIC ACID, ETHYL ESTER see EOH000
α-TOLUIC ACID, α-HYDROXY- see MAP000
α-TOLUIC ALDEHYDE see BBL500
o-TOLUIC NITRILE see TGT500
p-TOLUIC NITRILE see TGT750
TOLUIDENE BLUE O CHLORIDE see AJP250
m-TOLUIDIN (CZECH) see TGQ500
o-TOLUIDIN (CZECH) see TGQ750
p-TOLUIDIN (CZECH) see TGR000
2-TOLUIDINE see TGQ750
3-TOLUIDINE see TGQ500
4-TOLUIDINE see TGR000
m-TOLUIDINE see TGQ500
o-TOLUIDINE see TGQ750
p-TOLUIDINE see TGR000

o-TOLUIDINE, 4-((p-AMINOPHENYL)(4-IMINO-2,5-CYCLOHEXADIEN-1-YLI-DENE)METHYL)- see MAC500
m-TOLUIDINE ANTIMONYL TARTRATE see TGR250
o-TOLUIDINE ANTIMONYL TARTRATE see TGR500
p-TOLUIDINE ANTIMONYL TARTRATE see TGR750
TOLUIDINE BLUE see AJP250, TGS000
TOLUIDINE BLUE O see AJP250
o-TOLUIDINE, 6-CHLORO- see CLK227
m-TOLUIDINE, 6-CHLORO-α-α-α-TRIFLUORO- see CEG800
m-TOLUIDINE, N,N-DIMETHYL- see TLG100
p-TOLUIDINE, N,N-DIMETHYL- see TLG150
o-TOLUIDINE, 6-ETHYL- see MJY000
2-TOLUIDINE HYDROCHLORIDE see TGS500
m-TOLUIDINE HYDROCHLORIDE see TGS250
o-TOLUIDINE HYDROCHLORIDE see TGS500
p-TOLUIDINE HYDROCHLORIDE see TGS750
o-TOLUIDINE, 5-NITRO-, HYDROCHLORIDE see NMP600
p-TOLUIDINE, α-(PHENYLTHIO)- see PGN100
TOLUIDINE RED 10451 see MMP100
TOLUIDINE RED see MMP100
TOLUIDINE RED 3B see MMP100
TOLUIDINE RED BFB see MMP100
TOLUIDINE RED BFGG see MMP100
TOLUIDINE RED D 28-3930 see MMP100
TOLUIDINE RED LIGHT see MMP100
TOLUIDINE RED M 20-3785 see MMP100
TOLUIDINE RED 4R see MMP100
TOLUIDINE RED R see MMP100
TOLUIDINE RED RT-115 see MMP100
TOLUIDINE RED TONER see MMP100
TOLUIDINE RED XL 20-3050 see MMP100
p-TOLUIDINE-m-SULFONIC ACID see AKQ000
TOLUIDINE TONER see MMP100
TOLUIDINE TONER DARK 5040 see MMP100
TOLUIDINE TONER HR X-2741 see MMP100
TOLUIDINE TONER KEEP HR X-2742 see MMP100
TOLUIDINE TONER L 20-3300 see MMP100
TOLUIDINE TONER RT-252 see MMP100
TOLUIDINE TONER 4R X-2700 see MMP100
p-TOLUIDINIUM CHLORIDE see TGS750
2-o-TOLUIDINOETHANOL see TGT000
o-TOLUIDYNA (POLISH) see TGQ750
TOLUILENODWUIZOCYJANIAN see TGM750
α-TOLUIMIDIC ACID see PDX750
TOLUINA see BSQ000
TOLULOX see GGS000
TOLUMETIN SODIUM DIHYDRATE see SKJ350
TOLUMID see BSQ000
p-TOLUNITRIL (CZECH) see TGT750
3-TOLUNITRILE see TGT250
4-TOLUNITRILE see TGT750
α-TOLUNITRILE see PEA750
m-TOLUNITRILE see TGT250
o-TOLUNITRILE see TGT500
p-TOLUNITRILE see TGT750
TOLUOL (DOT) see TGK750
m-TOLUOL see CNW750
o-TOLUOL see CNX000
p-TOLUOL see CNX250
o-TOLUOL-AZO-o-TOLUIDIN (GERMAN) see AIC250
TOLUOLO (ITALIAN) see TGK750
p-TOLUOLSULFONSAEUREAETHYL ESTER (GERMAN) see EPW500
p-TOLUOLSULFONSAEURE METHYL ESTER (GERMAN) see MLL250
N-TOLUOLSULPHONYL HYDRAZINE see THE250
TOLUON see BMA625
m-TOLUONITRILE see TGT250
o-TOLUONITRILE see TGT500
p-TOLUONITRILE see TGT750
p-TOLUOYL CHLORIDE see TGU000
TOLUQUINOL see MKO250
p-TOLUQUINOLINE see MPF800
1,4-TOLUQUINONE see MHI250
p-TOLUQUINONE see MHI250
TOLU RESIN see BAF000
TOLUSAFRANINE see GJI400
TOLU-SOL see TGK750
TOLUVAN see BSQ000
m-TOLUYLENDIAMIN (CZECH) see TGL750
p-TOLUYLENDIAMINE see TGM000
TOLUYLENE BLUE (BASIC DYE) see DCE800
TOLUYLENE BLUE MONOHYDRATE see TGU500
2,3-TOLUYLENEDIAMINE see TGY800
2,4-TOLUYLENEDIAMINE see TGL750
TOLUYLENE-2,5-DIAMINE see TGM000
2,6-TOLUYLENEDIAMINE see TGM100

3,4-TOLUYLENEDIAMINE see TGM250
2,4-TOLUYLENEDIAMINE (DOT) see TGL750
m-TOLUYLENEDIAMINE see TGL750
TOLUYLENE-2,5-DIAMINE SULPHATE see DCE600
p-TOLUYLENEDIAMINE SULPHATE see DCE600
TOLUYLENE-2,4-DIISOCYANATE see TGM750
TOLUYLENE RED see AJQ250
m-TOLUYLIC ACID see TGP750
o-TOLUYLIC ACID see TGQ000
p-TOLUYLIC ACID see TGQ250
TOLVIN see BMA625
TOLVON see BMA625
TOLYCAINE HYDROCHLORIDE see BMA125
p-TOLYCARBAMIDE see THG000
TOLYDRIN see GGS000
m-TOLYENEDIAMINE see TGL750
N-m-TOLYLACETAMIDE see ABI750
m-TOLYLACETAMIDE see ABI750
p-TOLYL ACETATE see MNR250
p-TOLYL ALCOHOL see CNX250
TOLYL ALDEHYDE see TGK250
α-TOLYL ALDEHYDE DIMETHYL ACETAL see PDX000
TOLYLALDEHYDE GLYCERYL ACETAL see TGK500
m-TOLYLAMINE see TGQ500
o-TOLYLAMINE see TGQ750
p-TOLYLAMINE see TGR000
TOLYLAMINE see TGR000
o-TOLYLAMINE HYDROCHLORIDE see TGS500
N-(p-TOLYL)ANTHRANILIC ACID see TGV000
4-TOLYLARSENOUS ACID see TGV100
N-(p-TOLYL)-1-AZIRIDINECARBOXAMIDE see TGV250
m-TOLYLAZOACETANILIDE see TGV500
5-(o-TOLYLAZO)-2-AMINOTOLUENE see AIC250
p-(p-TOLYLAZO)-ANILINE see TGV750
p-(o-TOLYLAZO)-o-CRESOL see HJG000
4-o-TOLYLAZO-o-DIACETOTOLUIDE see ACR300
4'-(o-TOLYLAZO)-o-DIACETOTOLUIDIDE see ACR300
1-(o-TOLYLAZO)-2-NAPHTHOL see TGW000
1-(o-TOLYLAZO)-β-NAPHTHOL see TGW000
1-(o-TOLYLAZO)-2-NAPHTHYLAMINE see FAG135
4-(p-TOLYLAZO)-m-TOLUIDINE see AIC500
4-(o-TOLYLAZO)-o-TOLUIDINE see AIC250
4-(p-TOLYLAZO)-o-TOLUIDINE see TGW750
2-(o-TOLYLAZO)-p-TOLUIDINE see TGW500
1-((4-TOLYLAZO)TOLYLAZO)-2-NAPHTHOL see TGX000
1-(4-o-TOLYLAZO-o-TOLYLAZO)-2-NAPHTHOL see SBC500
o-TOLYLAZO-o-TOLYLAZO-2-NAPHTHOL see SBC500
o-TOLYLAZO-o-TOLYLAZO-β-NAPHTHOL see SBC500
p-TOLYL BENZOATE see TGX100
3-(α(o-TOLYL)BENZYLOXY)TROPANE HYDROBROMIDE see TGX500
1-o-TOLYLBIGUANIDE see TGX550
o-TOLYLBIGUANIDE see TGX550
2-TOLYL BROMIDE see BOG260
m-TOLYL BROMIDE see BOG300
o-TOLYL BROMIDE see BOG260
m-TOLYLCARBAMIDE see THF750
p-TOLYL-N-CARBAMOYLAZIRIDINE see TGV250
4-TOLYLCARBINOL see MHB250
p-TOLYLCARBINOL see MHB250
o-TOLYL CHLORIDE see CLK100
p-TOLYL CHLORIDE see TGY075
TOLYL CHLORIDE see BEE375
2-TOLYLCOPPER see TGY250
m-TOLYLDIETHANOLAMINE see DHF400
o-TOLYLDIETHANOLAMINE see DMT800
o-TOLYLDIGUANIDE see TGX550
TOLYL DIPHENYL PHOSPHATE see TGY750
2,4-TOLYLENEBIS(MALEIMIDE) see TGY770
2,3-TOLYLENEDIAMINE see TGY800
2,4-TOLYLENEDIAMINE see TGL750
2,6-TOLYLENEDIAMINE see TGM100
3,4-TOLYLENEDIAMINE see TGM250
4-m-TOLYLENEDIAMINE see TGL750
m-TOLYLENEDIAMINE see TGL750
p,m-TOLYLENEDIAMINE see TGM000
p-TOLYLENEDIAMINE SULPHATE see DCE600
2,4-TOLYLENEDIISOCYANATE see TGM750
TOLYLENE-2,4-DIISOCYANATE see TGM750
TOLYLENE-2,6-DIISOCYANATE see TGM800
m-TOLYLENE DIISOCYANATE see TGM750, TGM800
TOLYLENE DIISOCYANATE see TGM740
TOLYLENE ISOCYANATE see TGM740
m-TOLYLESTER KYSELINY METHYLKARBAMINOVE see MIB750
p-TOLYL ETHANOATE see MNR250
1-(p-TOLYL)ETHANOL see TGZ000

o-TOLYL ETHANOLAMINE see TGT000
1-p-TOLYLETHENE see VQK700
1-o-TOLYLGLYCEROL ETHER see GGS000
α-(o-TOLYL)GLYCERYL ETHER see GGS000
TOLYL GLYCIDYL ETHER see TGZ100
TOLYLHYDROQUINONE see MKO250
N-(p-TOLYL)-4-HYDROXY-1-ANTHRAQUINONYLAMINE see HOK000
N-(o-TOLYL)HYDROXYLAMINE see THA000
o-TOLYLHYDROXYLAMINE see THA000
2-(N'-p-TOLYL-N'-m-HYDROXYPHENYLAMINOMETHYL)-2-IMIDAZOLINE see
    PDW400
2,2'-(m-TOLYLIMINO)DIETHANOL see DHF400
p-TOLYL ISOBUTYRATE see THA250
β-p-TOLYL-ISOPROPYLAMINE SULFATE see AQQ000
α-TOLYL MERCAPTAN see TGO750
m-TOLYLMERCAPTAN see TGO800
o-TOLYL MERCAPTAN see TGP000
p-TOLYL MERCAPTAN see TGP250
3-TOLYL-N-METHYLCARBAMATE see MIB750
m-TOLYL N-METHYLCARBAMATE see MIB750
p-TOLYLMETHYLCARBINOL (GERMAN) see TGZ000
p-TOLYLMETHYL CHLORIDE see MHN300
p-TOLYL METHYL ETHER see MGP000
N-p-TOLYL-N-METHYLMERCURIAMID KYSELINY 2-METHYL-5-ISOPROPYLBEN-
    ZENSULFONOVE (CZECH) see MLB000
1-(p-TOLYL)-3-METHYLPYRAZOLONE-5 see THA300
TOLYLMYCIN Y see THD300
o-TOLYLNITRILE see TGT500
p-TOLYLNITRILE see TGT750
p-TOLYL OCTANOATE see THB000
2-((4-(p-TOLYLOXY)BUTYL)AMINO)ETHANETHIOL HYDROGEN SULFATE (ES-
    TER) see THB250
S-2-((4-(p-TOLYLOXY)BUTYL)AMINO)ETHYL THIOSULFATE see THB250
(2-o-TOLYLOXYETHYL)HYDRAZINE HYDROCHLORIDE see THC250
(2-p-TOLYLOXYETHYL)HYDRAZINE HYDROCHLORIDE see THC500
3-o-TOLYLOXY-2-HYDROXYPROPYL-1-CARBAMATE see CBK500
3-(o-TOLYLOXY)PROPANE-1,2-DIOL see GGS000
o-TOLYL PHOSPHATE see TMO600
7-(4-(m-TOLYL)-1-PIPERAZINYL)-4-NITRO-BENZOFURAZAN-1-OXIDE see
    THD275
4-(m-TOLYL)PIPERAZINYL 3,4,5-TRIMETHOXYPHENYL KETONE see THF300
4-(p-TOLYL)PIPERAZINYL 3,4,5-TRIMETHOXYPHENYL KETONE see THF310
2-(p-TOLYL)PROPIONIC ALDEHYDE see THD750
p-TOLYL SALICYLATE see THD850
(((p-TOLYL)SULFAMOYL)IMINO)BIS(METHYLMERCURY) see MLH100
p-TOLYLSULFONAMIDE see TGN500
TOLYLSULFONAMIDE see TGN500
p-TOLYLSULFONIC ACID see TGO000
3-(p-TOLYL-4-SULFONYL)-1-BUTYLUREA see BSQ000
TOLYLSULFONYLBUTYLUREA see BSQ000
1-(p-TOLYLSULFONYL)-3-CYCLOHEXYLUREA see CPR000
4-(p-TOLYLSULFONYL)-1,1-HEXAMETHYLENESEMICARBAZIDE see TGJ500
p-TOLYLSULFONYLHYDRAZINE see THE250
p-TOLYLSULFONYL-METHYL-NITROSAMID (GERMAN) see THE500
p-TOLYLSULFONYLMETHYLNITROSAMIDE see THE500
p-TOLYLSULFONYLMETHYLNITROSAMINE see THE500
N-(p-TOLYLSULFONYL)-N'-BUTYLCARBAMIDE see BSQ000
p-TOLYLTHIOL see TGP250
1-o-TOLYL-2-THIOUREA see THF250
o-TOLYL THIOUREA see THF250
N-(4'-o-TOLYL)-o-TOLYLAZOSUCCINAMIC ACID see SNF500
1-(m-TOLYL)-4-(3,4,5-TRIMETHOXYBENZOYL)PIPERAZINE see THF300
1-(p-TOLYL)-4-(3,4,5-TRIMETHOXYBENZOYL)PIPERAZINE see THF310
3-TOLYLUREA see THF750
m-TOLYLUREA see THF750
p-TOLYLUREA see THG000
p-TOLYMERCURY IODIDE see IFL000
TOLYN see RLU000
m-TOLYNITRILE see TGT250
TOLYNOL see GGS000
TOLYSPAZ see GGS000
p-TOLYUREA see THG000
TOMANOL-WIRKSTOFF see INM000
TOMATES (MEXICO) see JBS100
TOMATHREL see CDS125
TOMATIDINE GLYCOSIDE see THG250
TOMATIN see THG250
α-TOMATINE see THG250
TOMATINE see THG250
TOMATO FIX CONCENTRATE see CJN000
TOMATO HOLD see CJN000
TOMATOTONE see CJN000
TOMAYMYCIN see THG500
TOMBRAN see CKJ000, THK880
TOMIL see DIR000

TOMIZINE see THG600
TOMOBIL see TGZ000
TOMOFAN see CCU150
TOMOXIPROL see THG700
TOMOXIPROLE see THG700
TONARIL see TMP750
TONARSEN see DXE600
TONASO see CAT775
TONCARINE see MIP750
TONCO-70 see OBG000
TONEDRIN see DBA800
TONEDRON see MDT600
TONER LAKE RED C see CHP500
TONEXOL see TBN000
TONEY RED see OHA000
TONING ORANGE see CMS150
TONITE see CDN200
TONKA ABSOLUTE see THG750
TONKA BEAN CAMPHOR see CNV000
TONKALIDE see HDY600
TONOCARD see DJS200, TGI699
TONOCHOLIN B see CMF260
TONOCOR see DJS200
TONOFTAL see TGB475
TONOGEN see VGP000
TONOLYSIN see OJD300
TONOLYT ISOPROPYL MEPROBAMATE see IPU000
TONOX see MJQ000
TO NTU see BMR750
TONY RED see OHA000
TOOSENDANIN see CML825
TOOT POISON see TOE175
TOPANE see BGJ750
TOPANEL see COB260
TOPANOL see BFW750
TOPAZ see THH000
TOPAZONE see NGG500
TOPAZ TONER R 5 see CMS160
TOPCAINE see EFX000
TOPCIDE see ANB000
TOP COP TRI BASIC see CNM600
TOPE-TOPES (DOMINICAN REPUBLIC) see JBS100
TOPEX see BDS000
TOP FLAKE see SFT000
TOP FORM WORMER see TEX000
TOPHOL see PGG000
TOPICAIN see DTL200
TOPICHLOR 20 see CDR750
TOPICLOR 20 see CDR750
TOPICLOR see CDR750
TOPICON see HAG325
TOPICYCLINE see TBX250
TOPITOX see CJJ000
TOPITRACIN see BAC250
TOPOKAIN see AIT250
TOPOREX 500 see SMQ500
TOPOREX 830 see SMQ500
TOPOREX 550-02 see SMQ500
TOPOREX 850-51 see SMQ500
TOPOREX 855-51 see SMQ500
TOPSIN see DJV000
TOPSIN WP METHYL see PEX500
TOPSYM see DUD800
TOPUSYN see INR000
TOPZOL see RCF000
TORAK see DBI099
TORATE see BPG325
TORAZINA see CKP250, CKP500
TORBIN see EIN500
TORBUTROL see BPG325
TORCH see XPJ000
TORCH BRAND see CBT750
TORDON see PIB900
TORDON 10K see PIB900
TORDON 22K see PIB900
TORDON 101 MIXTURE see PIB900
TORECAN see TEZ000
TORECAN BIMALEATE see TEZ000
TORECAN DIMALEATE see TEZ000
TORECAN MALEATE see TEZ000
TORELLE see PHE250
TORENTAL see PBU100
TORESTEN see TEZ000
TORINAL see QAK000
TORMONA see BSQ750, TAA100

TORMOSYL see THH350
TOROMYCIN see GEO200
TORQUE see BLU000
TORSITE see BGJ250
TORULOX see GGS000
TORYN see CBG250
TOSIC ACID see TGO000
5-TOSILOXI 2-HYDROXIBENCENO SULFONATO de PIPERACINA (SPANISH) see PIK625
TOSTRIN see TBG000
TOSYL see CDO000
N-TOSYL-β-ALANINE CHLOROMETHYL KETONE see THH370
N-TOSYL-I-ALANINE CHLOROMETHYL KETONE see THH360
N-TOSYL-β-ALANINE DIAZOMETHYL KETONE see THH390
N-TOSYL-I-ALANINE DIAZOMETHYL KETONE see THH380
p-TOSYLAMIDE see TGN500
TOSYLAMIDE see TGN500
I-1-TOSYLAMIDO-2-PHENYLETHYL CHLOROMETHYL KETONE see THH450
6-(N-TOSYL)AMINOCAPROIC ACID DIAZOMETHYL KETONE see THH460
N-TOSYLANILINE see TGN600
TOSYLCHLORAMIDE SODIUM see CDP000
TOSYL CHLORIDE see TGO250
TOSYL FLUORIDE see TGO500
N-TOSYL-d,I-ISOLEUCINE CHLOROMETHYL KETONE see THH470
N-TOSYL-d,I-ISOLEUCINE DIAZOMETHYL KETONE see THH480
N-TOSYL-I-LEUCINE DIAZOMETHYL KETONE see THH490
TOSYL-L-LYSINE CHLOROMETHYL KETONE see THH500
α-TOSYL-L-LYSYLCHLOROMETHYL KETONE see THH500
TOSYLLYSINE CHLOROMETHYL KETONE see THH500
N-α-TOSYL-I-LYSYL-CHLOROMETHYLKETONE see THH500
TOSYLLYSYL CHLOROMETHYL KETONE see THH500
N-TOSYL-I-PHENYLALANINE CHLOROMETHYL KETONE see THH550
TOSYL-I-PHENYLALANYLCHLOROMETHYL KETONE see THH550
N-TOSYL-I-VALINE CHLOROMETHYL KETONE see THH555
TOTACEF see CCS250
TOTACILLIN see AIV500
TOTACOL see PAJ000
TOTAL see PAJ000
TOTALCICLINA see AIV500
TOTAPEN see AIV500
TOTAZINA see PKD250
TOTM see TJR600
TOTOCAINE HYDROCHLORIDE see AIT750
TOTOMYCIN see TBX250
TOTP see TMO600
TOTRIL see DNG200, HKB500
TOUKALIDE see HDY600
2,5-TOULENEDIAMINE SULFATE see TGM400
TOWK see RGP450
TOX 47 see PAK000
TOXADUST see CDV100
TOXAFEEN (DUTCH) see CDV100
TOXAKIL see CDV100
TOXALBUMIN see AAD000
TOXAN see CBM750
TOXAPHEN (GERMAN) see CDV100
TOXAPHENE see CDV100
TOXAPHENE TOXICANT A see THH560
TOXAPHENE TOXICANT B see THH575
TOXB-DIF see CMY090
TOXER TOTAL see PAJ000
TOXIN, ADENIA DIGITATA (MODECCA DIGITATA), MODECCIN see MRA000
TOXIN, ANABAENA FLOS-AQUAE NRC-44-1 see AON825
TOXIN, BACTERIUM CLOSTRIDIUM TETANI, TETANUS see TBG100
TOXIN, BACTERIUM CORYNE-BACTERIUM DIPHTHERIAE, DIPTHERIA see DWP300
TOXICHLOR see CDR750
TOXIN, CLOSTRIDIUM BOTULINUM see CMY030
TOXIN, CLOSTRIDIUM DIFFICILE see CMY070
α-TOXIN, CLOSTRIDIUM NOVYI see CMY130
TOXIN, CLOSTRIDIUM OEDEMATIENS, TYPE A see CMY150
β-TOXIN, CLOSTRIDIUM PERFRINGENS see CMY190
TOXIN, CLOSTRIDIUM SORDELLII see CMY240
TOXIFERINE DICHLORIDE see THI000
TOXIFREN see FBY000
TOXIN, JELLYFISH, CHIRONEX FLECKERI see CDM700
TOXIN, JELLYFISH, SCYPHOZOAN, CYANEA CAPILLATA see COI125
TOXILIC ACID see MAK900
TOXILIC ANHYDRIDE see MAM000
TOXIN B, CLOSTRIDIUM DIFFICILE see CMY090
TOXIN, NEMATOCYST, PHYSALIA PHYSALIS see PIA375
TOXINE TETANIQUE see TBG100
TOXIN HT 2 see THI250
TOXIN (PENICILLIUM ROQUEFORTII) see POF800
TOXIN PR see POF800

TOXIN T2 see FQS000
TOXIN T-2 TRIOL see DAD600
TOXIN, PENICILLIUM ROQUEFORTI see PAQ875
TOXIN, SHIGELLA DYSENTERIAE see SCD750
TOX-α-NOV see CMY130
TOXOBIDIN see BGS250
TOXOFACTOR see THI425
TOXOGONIN see BGS250
TOXOGONIN DICHLORIDE see BGS250
TOXOGONINE see BGS250
TOXON 63 see CDV100
TOXYLON POMIFERUM see MRN500
TOXYNON see SJP500
TOXYPHEN see CDV100
2,2'-(o-TOYLYIMINO)DIETHANOL see DMT800
TOYO ACID PHLOXINE see CMM000
TOYO AMARANTH see FAG020
TOYOCAMYCIN NUCLEOSIDE see VGZ000
TOYO EOSINE G see BNH500
TOYOFINE A see CAW850
TOYOFINE TF-X see CAT775
TOYOMYCIN see CMK650
TOYO OIL ORANGE see PEJ500
TOYO OIL RED BB see SBC500
TOYO OIL YELLOW G see DOT300
TOYO ORIENTAL OIL BLUE G see BLK000
TP-21 see MOO250, MOO500
TP-95 see DDT500
2,4,5-TP see TIX500
TP540 see DTQ800
TP 121 (filler) see CAT775
TP see TBG000
TPA see PGV000
TPA-3-β-OL see PGV500
TP 90B see BHK750
TPB see TEA400
TsPB see HCP800
TPBO see BAY275
TPD see DXO300
2,4,5-T PGBEE see THJ100
TPG HYDROCHLORIDE see UVA150
TPIA see MCB500
β-TPN see CNF400
TPN (pesticide) see TBQ750
TPN see CNF400
TPN (NUCLEOTIDE) see CNF400
T 45 (POLYGLYCOL) see GHS000
T 100 (POLYSACCHARIDE) see EHG100
TPP see TMT750
2,4,5-T PROPYLENE GLYCOL BUTYL ETHER ESTER see THJ100
TPS23 see MON750
TPTA see ABX250
TPTC see CLU000
TPTH see HON000
TPTZ see TMV500
TPU 10M see PKM250
TPU 2T see PKO500
TPZA see ABX250
TR 201 see SMR000
TRABEST see FOS100
TRACHOSEPT see DBX400
TRADENAL see POB500
TRAFARBIOT see AOD125
TRAGACANTH see THJ250
TRAGACANTH GUM see THJ250
TRAGAYA see SEH000
TRA-KILL TRACHEAL MITE KILLER see MCF750
TRAKIPEAL see MDQ250
TRALGON see HIM000
TRAMACIN see AQX500
TRAMADOL (2) see THJ755
TRAMADOL see THJ500
TRAMADOL HYDROCHLORIDE see THJ750
TRAMAL see THJ500, THJ750
TRAMAZOLINE HYDROCHLORIDE see THJ825
TRAMAZOLINE HYDROCHLORIDE MONOHYDRATE see RGP450
TRAMETAN see TFS350
TRAMISOL see LFA020
TRAMISOLE see LFA020
TRANCALGYL see EEM000
TRANCOLON see CBF000
TRANCOPAL see CKF500
TRANCYLPROMINE SULFATE see PET500
TRANDATE see HMM500

TRANEX see AJV500
TRANEXAMIC ACID see AJV500
TRANHEXAMIC ACID see AJV500
TRANID see CFF250
TRANILAST see RLK800
TRANILCYPROMINE see PET750
TRANIMUL see DCK759
TRANITE D-LAY see PBC250
TRANK see AOO500
TRAN-Q see CJR909
TRAN-Q DIHYDROCHLORIDE see HOR470
TRANQUIL see PGA750
TRANQUILAN see MQU750, TAF675
TRANQUILAX see CGA000
TRANQUILLIN see BCA000
TRANQUINIL see RDF000
TRANQUIRIT see DCK759
TRANQUIZINE DIHYDROCHLORIDE see HOR470
TRANQUO-BUSCOPAN-WIRKSTOFF see CFZ000
TRANS-AID see ANW750
TRANSALLYL CR 39 see AGD250
TRANSAMINE see DAA800, PET750, TAA100
TRANSAMINE SULFATE see PET500
TRANSAMLON see AJV500
TRANSANNON see ECU750
TRANSBRONCHIN see CBR675
TRANSCYCLINE see PPY250
TRANSENE see CDQ250
TRANSENTINE see DHX800, THK000
TRANSERGAN see THK600
TRANSERPIN see RDK000
TRANSETILE BLUE P-FER see MGG250
TRANSETILE RUBINE P-FL see AKI750
TRANSETILE VIOLET P 3R see DBP000
TRANSILIUM see CDQ250
TRANSIT see CHJ750
TRANSPARENT BRONZE SCARLET see CHP500
TRANSPLANTONE see NAK500
TRANSPOISE see MFD500
TRANSVAALIN see GFC000
TRANS-VERT see MRL750
TRANXENE see CDQ250
TRANXILEN see CDQ250
TRANXILENE see CDQ250
TRANXILIUM see CDQ250
TRANYLCYPRAMINE see PET750
TRANYLCYPRAMINE SULFATE see PET500
TRANYLCYPROMINE see PET750
TRANYLCYPROMINE SULFATE see PET500
TRANYLCYPROMINE SULPHATE see PET500
TRANZER see EHP000
TRANZETIL see DHX800
TRANZINE see CKP500
TRAPANAL see PBT500
TRAPANAL SODIUM see PBT500
TRAPEX-40 see ISE000
TRAPEX see ISE000, VFU000
TRAPEXIDE see ISE000
TRAPIDIL see DIO200
TRAPYMIN see DIO200
TRAQUILAN see TAF675
TRAQUIZINE see CJR909
TRASAN see CIR250
TRASENTIN see DHX800
TRASENTINE see DHX800
TRASICOR see THK750
TRASULPHANE see IAD000
TRASYLOL see PAF550
TRATUL see TAB250
TRAUMANASE see BMO000
TRAUMASEPT see PKE250
TRAUSABUM see AEG875
TRAUSABUN see AEG875, TDL000
TRAVAD see BAP000
TRAVELERS JOY see CMV390
TRAVELIN see DYE600
TRAVELMIN see DYE600
TRAVELON see HGC500
TRAVENON see SKJ300
TRAVEX see SFS000
TRAWOTOX see CDO000
TRAXANOX SODIUM PENTAHYDRATE see THK850
TRAZENTYNA (POLISH) see THK000
TRAZENTYNA see DHX800
TRAZININ see PAF550

TRAZODON see THK875
TRAZODONE see THK875
TRAZODONE HYDROCHLORIDE see CKJ000, THK880
TRAZOLINONE see EQO000
TRECALMO see CKA000
TRECATOR see EPQ000
TREDIONE see TLP750
TREDUM see CKG000
TREESAIL see GJU475
TREFANOCIDE see DUV600
TREFICON see DUV600
TREFLAM see DUV600
TREFLAN see DUV600
TREFLANOCIDE ELANCOLAN see DUV600
TRE-HOLD see NAK500
TRELMAR see MQU750
TREMARIL HYDROCHLORIDE see THL500
TREMARIL WONDER see THL500
TREMBLEX see DBE000
TREMERAD see CGB250
TREMIN see BBV000
TREMOLITE ASBESTOS see ARM280
TREMORINE see DWX600
TREMORTIN A see PAR250
TRENAMINE D-201 see OHY000
TRENAMINE D-200 see OHY000
TRENIMON see TND000
TRENTADIL see THL750
TRENTADIL HYDROCHLORIDE see THL750
TRENTAL see PBU100
TREOMICETINA see CDP250
TREOSULFAN see TFU500
TREPENOL WA see SIB600
TREPIBUTONE see CNG835
TREPIDAN see DAP700
TREPIDONE see MFD500
TREPOL (FRENCH) see BKX250
TRESANIL see TNP275
TRESCATYL see EPQ000
TRESCAZIDE see EPQ000
TRESITOPE see LGK050
TRESOCHIN see CLD000
TRESORTIL see GKK000
TRESPAPHAN see PMP500
TREST see THL500
TRESTEN see TEZ000
TRESULFAN see TFU500
TRETAMINE see TND500
TRETINOIN see VSK950
TRET-O-LITE WF 88 see QAT520
TRET-O-LITE WF 828 see QAT520
TRET-O-LITE XC 511 see AFP100
TREVINTIX see PNW750
TRF see TNX400
TRH-T see POE050
TRI-4 see DUV600
TRIABARB see EOK000
TRI-ABRODIL see AAN000
TRIACETALDEHYDE (FRENCH) see PAI250
TRIACETATE d'HYDROXYHYDROQUINONE see HLF600
TRIACETIC ACID LACTONE see THL800
TRIACETIN (FCC) see THM500
TRIACETINASE see GGA800
TRIACETONAMIN see TDT770
TRIACETONAMINE see TDT770
TRIACETONE AMINE see TDT770
TRIACETONITRILE TUNGSTEN TRICARBONYL see THM250
1,8,9-TRIACETOXYANTHRACENE see APH500
1,2,4-TRIACETOXYBENZENE see HLF600
4-β,8-α-15-TRIACETOXY-3-α-HYDROXY-12,13-EPOXYTRICHOTHEC-9-ENE see ACS500
2′,3′,5′-TRIACETYL-6-AZAURIDINE see THM750
2′,3′,5′-TRI-o-ACETYL-6-AZAURIDINE see THM750
TRIACETYL-6-AZAURIDINE see THM750
TRIACETYLDIPHENOLISATIN see ACD500
TRIACETYL GLYCERIN see THM500
2-(2′,3′,5′-TRIACETYL-β-d-RIBOFURANOSYL)-as-TRIAZINE-3,5-(2H,4H)-DIONE see THM750
TRIACRYLFORMAL see THM900
1,3,5-TRIACRYLOYLHEXAHYDROTRIAZINE see THM900
TRIACRYLOYLHEXAHYDROTRIAZINE see THM900
TRI(N-ACRYLOYL)HEXAHYDROTRIAZINE see THM900
TRIACRYLOYLHEXAHYDRO-s-TRIAZINE see THM900
TRIACRYLOYLPERHYDROTRIAZINE see THM900
TRIACYLGLYCEROL HYDROLASE see GGA800

TRIACYLGLYCEROL LIPASE see GGA800
TRIAD see TIO750
TRIADENYL see ARQ500
TRIADIMEFON see CJO250
TRIADIMENOL see MEP250
TRIAETHANOLAMIN-NG see TKP500
TRIAETHYLAMIN (GERMAN) see TJO000
TRIAETHYLENMELAMIN (GERMAN) see TND500
TRIAETHYLENPHOSPHORSAEUREAMID (GERMAN) see TND250
TRIAETHYLZINNACETAT (GERMAN) see ABW750
TRIAETHYLZINNSULFAT (GERMAN) see BLN500
TRIALLAT (GERMAN) see DNS600
TRIALLATE see DNS600
TRIALLYLAMINE see THN000
TRIALLYL BORATE see THN250
TRIALLYL CYANURATE see THN500
1,3,5-TRIALLYLISOCYANURATE see THS100
TRIALLYL ISOCYANURATE see THS100
1,3,5-TRIALLYLISOCYANURIC ACID see THS100
TRIALLYL PHOSPHATE see THN750
TRIAMCINCOLONE ACETONIDE see AQX500
TRIAMCINOLONE see AQX250
TRIAMCINOLONE-16,17-ACETONIDE see AQX500
TRIAMCINOLONE ACETONIDE see AQX500
TRIAMELIN see TND500
TRIAMIFOS (GERMAN, DUTCH, ITALIAN) see AIX000
TRIAMINE BLACK D see CMN230
1,3,5-TRIAMINOBENZENE see THN775
s-TRIAMINOBENZENE see THN775
sym-TRIAMINOBENZENE see THN775
2,4,7-TRIAMINO-6-FENILPTERIDINA (ITALIAN) see UVJ450
TRIAMINOGUANIDINE NITRATE see THN800
TRIAMINOGUANIDINIUM PERCHLORATE see THO250
2,4,6-TRIAMINOPHENOL TRIHYDROCHLORIDE see THO500
TRIAMINOPHENYL PHOSPHATE see THO550
2,4,7-TRIAMINO-6-PHENYLPTERIDINE see UVJ450
2,4,6-TRIAMINO-1,3,5-TRIAZINE see MCB000
2,4,6-TRIAMINO-s-TRIAZINE see MCB000
2,4,6-TRIAMINO-s-TRIAZINE compounded with s-TRIAZINE-TRIOL see THO750
1,3,5-TRIAMINOTRINITROBENZENE see THO775
4,4',4''-TRIAMINOTRIPHENYLMETHANE see THP000
p,p',p''-TRIAMINOTRIPHENYLMETHANE see THP000
TRIAMINOTRIPHENYLMETHANE see THP000
4,4'4''-TRIAMINOTRIPHENYLMETHAN-HYDROCHLORID (GERMAN) see RMK020
TRIAMIPHOS see AIX000
TRIAMMINEDIPEROXOCHROMIUM(IV) see THP250
TRIAMMINE GOLDTRIHYDROXIDE see THP500
TRIAMMINENITRATOPLANTINUM(II) NITRATE see THP750
cis-TRIAMMINETRICHLORORUTHENIUM(III) see THQ000
TRIAMMINETRINITROCOBALT(III) see THQ250
TRIAMMONIUM ALUMINUM HEXAFLUORIDE see THQ500
TRIAMMONIUM AURINTRICARBOXYLATE see AGW750
TRIAMMONIUM HEXAFLUOROALUMINATE see THQ500
TRIAMMONIUM TRIS-(ETHANEDIOATO(2-)-O,O')FERRATE(3-1) see ANG925
TRIAMPHOS see AIX000
TRIAMPUR see UVJ450
TRIAMTEREN see UVJ450
TRIAMTERENE see UVJ450
TRIAMTERIL see UVJ450
TRIAMTERIL COMPLEX see UVJ450
TRI-n-AMYL BORATE see TMQ000
TRIANATE see TJL250
TRIANGLE see CBT750, CNP500
TRI-p-ANISYLCHLOROETHYLENE see CLO750
TRIANOL DIRECT BLUE 3B see CMO250
TRIANON see PPO000
1,4-TRIANTHRIMID (CZECH) see APL750
1,5-TRIANTHRIMID (CZECH) see APM000
TRIANTINE FAST RED 4BN see CMO885
TRIANTINE LIGHT BROWN 3RN see FMU059
TRIANTINE LIGHT RED 4BN see CMO885
TRIANTOIN see MKB250
TRIASOL see TIO750
TRIASYN see DEV800
TRIATIX see MJL250
TRIATOMIC OXYGEN see ORW000
TRIATOX see MJL250
3,5,7-TRIAZA-1-AZONIAADAMANTANE, 1-(3-CHLOROALLYL)-, CHLORIDE see CEG550
1,2,3-TRIAZAINDENE see BDH250
3,5,7-TRIAZAINDOLE see POJ250
3,6,9-TRIAZATETRACYCLO[6.1.0.0$^{2,4}$.O$^{5,7}$] NONANE see THQ600
1-TRIAZENE see THQ750
TRIAZENE see THQ750

1-TRIAZENE, 1,3-DIETHYL-(9CI) see DKD200
p,p'-TRIAZENYLENEDIBENZENESULFONAMIDE see THQ900
TRIAZICHON (GERMAN) see TND000
TRIAZIDOBORANE see BMG325
TRIAZIDOMETHYLIUM HEXACHLOROANTIMONATE see THR100
2,4,6-TRIAZIDO-1,3,5-TRIAZINE see THR250
TRIAZIN see DEV800
1,2,4-TRIAZIN-3-AMINE, 6-(2-(5-NITRO-2-FURANYL)ETHENYL)-(9CI) see FPF000
TRIAZINATE see THR500
1,3,5-TRIAZINE see THR525
TRIAZINE (pesticide) see DEV800
sym-TRIAZINE see THR525
TRIAZINE A 1294 see ARQ725
1,3,5-TRIAZINE-2-AMINE, 4-METHOXY-6-METHYL-(9CI) see MGH800
s-TRIAZINE, 2-AMINO-4-METHOXY-6-METHYL- see MGH800
s-TRIAZINE, 2-(sec-BUTYLAMINO)-4-CHLORO-6-(ETHYLAMINO)- see THR600
1,3,5-TRIAZINE (9CI) see THR525
1,3,5-TRIAZINE-2,4-DIAMINE, 6-CHLORO-N-ETHYL-N'-(1-METHYLPROPYL)-(9CI) see THR600
1,3,5-TRIAZINE-2,4-DIAMINE, 6-CHLORO-N,N,N'-TRIETHYL- see TJL500
s-TRIAZINE, 1,2-DIHYDRO-1-(p-CHLOROPHENYL)-4,6-DIAMINO-2,2-DIMETHYL- see COX400
s-TRIAZINE-3,5(2H,4H)-DIONE see THR750
s-TRIAZINE, HEXAHYDRO-1,3,5-TRIACRYLOYL- see THM900
1,3,5-TRIAZINE, HEXAHYDRO-1,3,5-TRIMETHYL- see HDW100
s-TRIAZINE, HEXAHYDRO-1,3,5-TRIMETHYL- see HDW100
1,3,5-TRIAZINE, HEXAHYDRO-1,3,5-TRINITRO-(9CI) see CPR800
1,3,5-TRIAZINE, HEXAHYDRO-1,3,5-TRIS(1-OXO-2-PROPENYL)-(9CI) see THM900
1,3,5-TRIAZINE-2,4,6-TRIAMINE see MCB000
1,3,5-TRIAZINE-2,4,6-TRIAMINE, polymer with FORMALDEHYDE (9CI) see MCB050
s-TRIAZINE, 2,4,6-TRIAMINO- see MCB000
s-TRIAZINE TRICHLORIDE see TJD750
s-TRIAZINE-1,3,5(2H,4H,6H)-TRIETHANOL see THR820
1,3,5-TRIAZINE-1,3,5(2H,4H,6H)-TRIETHANOL (9CI) see THR820
1,3,5-TRIAZINE-2,4,6-TRIMERCAPTAN see THS250
s-TRIAZINE-2,4,6-TRIOL see THS000
s-2,4,6-TRIAZINETRIOL see THS000
sym-TRIAZINETRIOL see THS000
s-TRIAZINE-2,4,6-TRIOL, TRI(2,3-EPOXYPROPYL) ESTER see TKL250
s-TRIAZINE-2,4,6(1H,3H,5H)-TRIONE see THS000
s-TRIAZINE-2,4,6(1H,3H,5H)-TRIONE, DICHLORO-, POTASSIUM DERIV see PLD000
1,3,5-TRIAZINE-2,4,6(1H,3H,5H)-TRIONE, 1,3-DICHLORO-, POTASSIUM SALT see PLD000
s-TRIAZINE-2,4,6(1H,3H,5H)-TRIONE, MONOSODIUM SALT see SGB550
1,3,5-TRIAZINE-2,4,6(1H,3H,5H)-TRIONE, MONOSODIUM SALT (9CI) see SGB550
s-TRIAZINE-2,4,6(1H,3H,5H)-TRIONE, 1,3,5-TRIALLYL- see THS100
s-TRIAZINE-2,4,6(1H,3H,5H)-TRIONE, TRIALLYL- see THS100
1,3,5-TRIAZINE-2,4,6(1H,3H,5H)-TRIONE, 1,3,5-TRI-2-PROPENYL-(9CI) see THS100
2,4,6-TRIAZINETRITHIOL see THS250
s-TRIAZINE-2,4,6-TRITHIOL see THS250
1,3,5-TRIAZINE-2,4,6(1H,3H,5H)-TRITHIONE see THS250
(1,3,5-TRIAZINE-2,4,6-TRIYLTRINITRILO)HEXAKIS METHANOL see HDY000
(s-TRIAZINE-2,4,6-TRIYLTRINITRILO)HEXAMETHANOL see HDY000
1,1',1''-s-TRIAZINE-2,4,6-TRIYLTRISAZIRIDINE see TND500
1,3,5-TRIAZIN-2(1H)-ONE, TETRAHYDRO-1,3-BIS(HYDROXYMETHYL)-5-ETHYL- see TCJ900
s-TRIAZIN-2,4,6-TRIYLTRIAMINOMETHANESULFONIC ACID TRISODIUM SALT see THS500
(s-TRIAZIN-2,4,6-TRIYLTRIAMINO)TRISMETHANESULFONIC ACID TRISODIUM SALT see THS500
TRIAZIQUINONE see TND000
TRIAZIQUONE see TND000
TRIAZIRIDINOPHOSPHINE OXIDE see TND250
2,3,5-TRI-(1-AZIRIDINYL)-p-BENZOQUINONE see TND000
TRI(-1-AZIRIDINYL)PHOSPHINE OXIDE see TND250
TRI(AZIRIDINYL)PHOSPHINE OXIDE see TND250
TRIAZIRIDINYLPHOSPHINE SULFIDE see TFQ750
TRIAZIRIDINYL TRIAZINE see TND500
TRIAZOFOSZ (HUNGARIAN) see THT750
TRIAZOIC ACID see HHG500
TRIAZOLAM see THS800
1H-1,2,4-TRIAZOL-3-AMINE see AMY050
TRIAZOLAMINE see AMY050
TRIAZOL BROWN B see CMO820
1,2,3-TRIAZOLE see THS850
s-TRIAZOLE see THS825
1H-1,2,4-TRIAZOLE (9CI) see THS825
1H-1,2,4-TRIAZOLE, 3,5-DIPHENYL- see DWO950
s-TRIAZOLE, 3,5-DIPHENYL- see DWO950

1H-1,2,4-TRIAZOLE-1-ETHANOL, β-((2,4-DICHLOROPHENYL)METHYLENE)-α-(1,1-DIMETHYLETHYL)-, (E)- see DGC100
1H-1,2,4-TRIAZOLE-3-THIOL see THT000
TRIAZOL FAST RED 8B see CMO880
TRIAZOL FAST SCARLET 3B see CMO875
TRIAZOLOGUANINE see AJO500
1,2,4-TRIAZOL-5-ONE, 3-NITRO- see NMP620
1,2,4-TRIAZOLO[4,3-a]PYRIDINE-SILVER NITRATE see THT250
s-TRIAZOLO(4,3-a)PYRIDIN-3(2H)-ONE, 2-(3-(4-(m-CHLOROPHENYL)-1-PIPERAZINYL)PROPYL)-, MONOHYDROCHLORIDE see THK880
1H-v-TRIAZOLO(4,5-d)PYRIMIDIN-7-AMINE see AIB340
v-TRIAZOLO(4,5-d)PYRIMIDINE-5,7-DIOL see THT350
1H-v-TRIAZOLO(4,5-d)PYRIMIDINE-5,7(4H,6H)-DIONE see THT350
7H-1,2,3-TRIAZOLO(4,5-d)PYRIMIDIN-7-ONE, 1,4,5,6-TETRAHYDRO-5-THIOXO-(9CI) see HOJ100
3H-v-TRIAZOLO(4,5-d)PYRIMIDIN-7(4H,6H)-ONE, 5-THIONO- see HOJ100
(1H-1,2,4-TRIAZOLYL)TRICYCLOHEXYLSTANNANE see THT500
1H-1,2,4-(TRIAZOL-1-YL)TRIPHENYLSTANNANE see TMX000
(1H-1,2,4-TRIAZOLYL-1-YL)TRICYCLOHEXYLSTANNANE see THT500
TRIAZOPHOS see THT750
TRIAZOTION (RUSSIAN) see EKN000
TRIAZURE see THM750
TRIB see SNK000
TRIBAC see TIK500
TRI-BAN see PIH175
TRIBASIC ALUMINUM STEARATE see AHH825
TRIBASIC COPPER SULFATE see CNM600
TRIBASIC SODIUM PHOSPHATE see SJH200
(1,2,4,5,7,8)TRIBENZOPYRENE see DDC200
1,2:4,5:8,9-TRIBENZOPYRENE see DDC200
(1,2,4,5,8,9)TRIBENZOPYRENE see DDC200
TRIBENZO(a,e,i)PYRENE see DDC200
TRIBENZYLARSINE see THU000
TRIBENZYLCHLOROSTANNANE see CLP000
TRIBENZYLSULFONIUM IODIDE MERCURIC IODIDE see THU250
TRIBENZYLSULFONIUM IODIDE, compounded with MERCURY IDODIDE (1:1) see THU250
TRIBENZYLTIN CHLORIDE see CLP000
TRIBENZYLTIN FORMATE see FOE000
TRIBONATE see DVC200
TRIBORON PENTAFLUORIDE see THU275
TRIBROMAMINE HEXAAMMONIATE see NGV500
TRIBROMETHANOL see ARW250, THV000
TRIBROMMETHAAN (DUTCH) see BNL000
TRIBROMMETHAN (GERMAN) see BNL000
TRIBROMOALDEHYDE HYDRATE see THU500
TRIBROMOALUMINUM see AGX750
2,4,6-TRIBROMOANILINE see THU750
sym-TRIBROMOANILINE see THU750
TRIBROMOARSINE see ARF250
2,2,2-TRIBROMOETHANOL see ARW250
TRIBROMOETHANOL see THV000
TRIBROMOETHENE see THV100
2,2,2-TRIBROMOETHYL ALCOHOL see ARW250
TRIBROMOETHYL ALCOHOL see ARW250, THV000
1,1,2-TRIBROMOETHYLENE see THV100
TRIBROMOETHYLENE see THV100
1,3,7-TRIBROMOFLUOREN-2-AMINE see THV250
1,3,7-TRIBROMO-2-FLUORENAMINE see THV250
2,4,5-TRIBROMOIMIDAZOLE see THV450
2,4,5-TRIBROMOIMIDAZOLE CADMIUM SALT (2:1) see THV500
TRIBROMOMETAN (ITALIAN) see BNL000
TRIBROMOMETHANE see BNL000
TRIBROMONITROMETHANE see NMQ000
2,4,6-TRIBROMOPHENOL see THV750
TRIBROMOPHENOL see THV750
TRIBROMOPHOSPHINE see PHT250
1,2,3-TRIBROMOPROPANE see GGG000
sym-TRIBROMOPROPANE see GGG000
3,4′,5-TRIBROMOSALICYLANILIDE see THW750
TRIBROMOSILANE see THX000
TRIBROMOSTIBINE see AQK000
TRIBROMOTRIMETHYLDIALUMINUM see MGC225
TRIBROMSALAN see THW750
TRIBUFON see BPG000
TRIBURON see TJE880
TRIBURON CHLORIDE see TJE880
TRIBUTILFOSFATO (ITALIAN) see TIA250
TRIBUTON see DAA800, TAA100
TRI-n-BUTOXYBORANE see THX750
TRIBUTOXYBORANE see THX750
TRI(2-BUTOXYETHANOL PHOSPHATE) see BPK250
TRI(2-BUTOXYETHYL) PHOSPHATE see BPK250
TRIBUTOXYETHYL PHOSPHATE see BPK250
TRIBUTYL O-ACETYLCITRATE see THX100

TRIBUTYL ACETYLCITRATE see THX100
TRIBUTYL 2-(ACETYLOXY)-1,2,3-PROPANETRICARBOXYLATE see THX100
TRI-n-BUTYLAMINE see THX250
TRIBUTYLAMINE see THX250
TRI-n-BUTYL BORANE see THX500
TRI-n-BUTYL BORATE see THX750
TRI-sec-BUTYL BORATE see THY000
TRIBUTYL BORATE see THX750
TRIBUTYLBORINE see THX500
N,N,N-TRIBUTYL-1-BUTANAMINIUM HYDROXIDE see TBK750
N-N-N-TRIBUTYL-1-BUTANAMINIUM NITRATE see TBL250
TRIBUTYL CELLOSOLVE PHOSPHATE see BPK250
TRIBUTYLCHLOROGERMANE see CLP250
TRIBUTYLCHLOROSTANNANE see CLP500
TRI-n-BUTYL CITRATE see THY100
TRIBUTYL CITRATE see THY100
TRIBUTYL CITRATE ACETATE see THX100
TRIBUTYL(2,4-DICHLOROBENZYL)PHOSPHONIUM CHLORIDE see THY500
TRI-n-BUTYL-2,4-DICHLOROPHENOXYTIN see DGB400
TRIBUTYLE (PHOSPHATE de) (FRENCH) see TIA250
TRIBUTYL((2-ETHYLHEXANOYL)OXY)STANNANE see TID250
TRIBUTYL((2-ETHYL-1-OXOHEXYL)OXY)STANNANE see TID250
TRIBUTYLFOSFAAT (DUTCH) see TIA250
TRIBUTYLFOSFIN see TIA300
TRIBUTYLFOSFINSULFID see TIA450
TRIBUTYL(GLYCOLOYLOXY)STANNANE see GHQ000
TRIBUTYL(GLYCOLOYLOXY)TIN see GHQ000
TRIBUTYL 2-HYDROXY-1,2,3-PROPANETRICARBOXYLATE see THY100
TRIBUTYLHYDROXYSTANNANE see TID500
TRIBUTYLISOCYANATOSTANNANE see THY750
TRIBUTYLLEAD ACETATE see THY850
TRIBUTYL(METHACRYLOXY)STANNANE see THZ000
TRIBUTYL(METHACRYLOYLOXY)STANNANE see THZ000
TRIBUTYL(METHACRYLOYLOXY)-STANNANE POLYMER with METHYL METHACRYLATE (8CI) see OIY000
TRIBUTYL((2-METHYL-1-OXO-2-PROPENYL)OXY)STANNANE see THZ000
TRIBUTYL(NEODECANOYLOXY)STANNANE see TIF250
TRIBUTYL(OLEOYLOXY)STANNANE see TIA000
TRIBUTYL(OLEOYLOXY)TIN see TIA000
TRIBUTYL((1-OXODODECYL)OXY)STANNANE (9CI) see TIE750
(TRIBUTYL)PEROXIDE see BSC750
TRIBUTYL-o-PHENYLPHENOXYTIN see BGK000
TRIBUTYLPHOSPHAT (GERMAN) see TIA250
TRI-n-BUTYL PHOSPHATE see TIA250
TRIBUTYL PHOSPHATE see TIA250
TRI-n-BUTYLPHOSPHINE see TIA300
TRIBUTYLPHOSPHINE see TIA300
TRIBUTYL-PHOSPHINE compounded with NICKELCHLORIDE (2:1) see BLS250
TRIBUTYLPHOSPHINE SULFIDE see TIA450
TRIBUTYL PHOSPHITE see TIA750
S,S,S-TRIBUTYL PHOSPHOROTRITHIOATE see BSH250
S,S,S-TRIBUTYL PHOSPHOROTRITHIOITE see TIG250
TRIBUTYL PHOSPHOROTRITHIOITE see TIG250
TRI-n-BUTYLPLUMBYL ACETATE see THY850
TRIBUTYL(8-QUINOLINOLATO)TIN see TIB000
TRIBUTYLSTANNANECARBONITRILE see TIB250
TRIBUTYLSTANNANE FLUORIDE see FME000
TRI-n-BUTYLSTANNANE HYDRIDE see TIB500
TRI-n-BUTYL-STANNANE OXIDE see BLL750
TRIBUTYLSTANNIC HYDRIDE see TIB500
TRIBUTYLSTANNYL ISOCYANATE see THY750
TRIBUTYLSTANNYL METHACRYLATE see THZ000
TRIBUTYLTIN-p-ACETAMIDOBENZOATE see TIB750
TRIBUTYLTIN ACETATE see TIC000
TRIBUTYLTIN BENZOATE see BDR750
TRI-n-BUTYLTIN BROMIDE see TIC250
TRI-n-BUTYLTIN CHLORIDE see CLP500
TRIBUTYLTIN CHLORIDE COMPLEX see TIC500
TRIBUTYLTIN CHLOROACETATE see TIC750
TRIBUTYLTIN-γ-CHLOROBUTYRATE see TID000
TRIBUTYLTIN CYANATE see COI000
TRI-n-BUTYLTIN CYANIDE see TIB250
TRIBUTYLTIN CYCLOHEXANECARBOXYLATE see TID100
TRIBUTYLTIN-S,S′-DIBUTYLDITHIOCARBAMATE see DEB600
TRIBUTYLTIN DIMETHYLDITHIOCARBAMATE see TID150
TRIBUTYLTIN DODECANOATE see TIE750
TRIBUTYLTIN-2-ETHYLHEXANOATE see TID250
TRIBUTYLTIN FLUORIDE see FME000
TRI-n-BUTYLTIN HYDRIDE see TIB500
TRIBUTYLTIN HYDRIDE see TIB500
TRIBUTYLTIN HYDROXIDE see TID500
TRI-N-BUTYL TIN IODIDE see IFM000
TRIBUTYLTIN IODOACETATE see TID750
TRIBUTYLTIN-o-IODOBENZOATE see TIE000

TRIBUTYLTIN-p-IODOBENZOATE see TIE250
TRIBUTYLTIN-β-IODOPROPIONATE see TIE500
TRI-n-BUTYLTIN ISOCYANATE see THY750
TRIBUTYLTIN ISOCYANATE see THY750
TRIBUTYLTIN ISOOCTYLTHIOACETATE see TDO000
TRIBUTYLTIN ISOPROPYLSUCCINATE see TIE600
TRIBUTYLTIN ISOTHIOCYANATE see THY750
TRIBUTYLTIN LAURATE see TIE750
TRIBUTYL TIN LINOLEATE /CS LGJ000
TRIBUTYL TIN METHACRYLATE see THZ000
TRI-n-BUTYLTIN METHANESULFONATE see TIF000
TRIBUTYLTIN MONOLAURATE see TIE750
TRIBUTYLTIN NEODECANOATE see TIF250
TRIBUTYLTIN NONANOATE see TIF500
TRI-N-BUTYLTIN OLEATE see TIA000
TRIBUTYLTIN OXIDE see BLL750
TRIBUTYLTIN-o-PHENYLPHENOXIDE see BGK000
TRI-N-BUTYLTIN SALICYLATE see SAM000
TRIBUTYLTIN SALICYLATE see SAM000
TRIBUTYLTIN SULFATE see TIF600
TRIBUTYLTIN SULFIDE see HCA700
TRIBUTYLTIN-α-(2,4,5-TRICHLOROPHENOXY)PROPIONATE see TIF750
TRI-n-BUTYLTIN UNDECYLATE see TIG500
TRIBUTYLTIN UNDECYLENATE see TIG500
TRIBUTYL(2,4,5-TRICHLOROPHENOXY)STANNANE see TIG000
TRIBUTYL(2,4,5-TRICHLOROPHENOXY)TIN see TIG000
S,S,S-TRIBUTYL TRITHIOPHOSPHATE see BSH250
S,S,S-TRIBUTYL TRITHIOPHOSPHITE see TIG250
TRIBUTYL(UNDECANOYLOXY)STANNANE see TIG500
TRI-n-BUTYL-ZINN-ACETAT (GERMAN) see TIC000
TRI-N-BUTYL-ZINN BENZOATE (GERMAN) see BDR750
TRI-n-BUTYLZINN-CHLORID (GERMAN) see CLP500
TRI-n-BUTYLZINN-LAURAT (GERMAN) see TIE750
TRI-N-BUTYL-ZINN OLEAT (GERMAN) see TIA000
TRI-N-BUTYL-ZINN SALICYLAT (GERMAN) see SAM000
TRI-n-BUTYL-ZINN UNDECYLAT (GERMAN) see TIG500
TRIBUTYRASE see GGA800
TRIBUTYRIN see TIG750
TRIBUTYRINASE see GGA800
TRIBUTYRIN ESTERASE see GGA800
TRIBUTYROIN see TIG750
TRICADMIUM DINITRIDE see TIH000
TRICAINE see EFX500
TRICAINE METHANE SULFONATE see EFX500
TRICALCIUMARSENAT (GERMAN) see ARB750
TRICALCIUM ARSENATE see ARB750
TRICALCIUM DINITRIDE see TIH250
TRICALCIUM PHOSPHATE see CAW120
TRICALCIUM SILICATE see TIH600
TRICAMBA see TIK000
TRICAPROIN see GGK000
TRICAPRONIN see GGK000
TRICAPROYLGLYCEROL see GGK000
TRICAPRYLIC GLYCERIDE see TMO000
TRICAPRYLIN see TMO000
TRICAPRYLMETHYLAMMONIUM CHLORIDE see MQH000
TRICAPRYLYLAMINE see DVL000
TRICAPRYLYLMETHYLAMMONIUM CHLORIDE see MQH000
TRICARBALLYLIC ACID, β-ACETOXYTRIBUTYL ESTER see ADD750
TRICARBAMIX Z see BJK500
TRICARBONYL(METHYLCYCLOPENTADIENYL)MANGANESE see MAV750
1,2,4-TRICARBOXYBENZENE see TKU700
TRICARNAM see CBM750
TRICESIUM NITRIDE see TIH750
TRICESIUM TRICHLORIDE see CDD000
TRICESIUM TRIFLUORIDE see CDD500
TRICESIUM TRIIODIDE see CDE000
TRICETAMIDE see TIH800
TRICETO 3-7-12 CHOLANATE de Na (FRENCH) see SGD500
TRICHAZOL see MMN250
TRICHLOORAZIJNZUUR (DUTCH) see TII250
1,1,1-TRICHLOOR-2,2-BIS(4-CHLOOR FENYL)-ETHAAN (DUTCH) see DAD200
2,2,2-TRICHLOOR-1,1-BIS(4-CHLOOR FENYL)-ETHANOL (DUTCH) see BIO750
1,1,1-TRICHLOORETHAAN (DUTCH) see MIH275
TRICHLOORETHEEN (DUTCH) see TIO750
TRICHLOORETHYLEEN (DUTCH) see TIO750
(2,4,5-TRICHLOOR-FENOXY)-AZIJNZUUR (DUTCH) see TAA100
2-(2,4,5-TRICHLOOR-FENOXY)-PROPIONZUUR (DUTCH) see TIX500
O-(2,4,5-TRICHLOOR-FENYL)-O,O-DIMETHYL-MONOTHIOFOSFAAT (DUTCH)
  see RMA500
TRICHLOORFON (DUTCH) see TIQ250
TRICHLOORMETHAAN (DUTCH) see CHJ500
TRICHLOORMETHYLBENZEEN (DUTCH) see BFL250
TRICHLOORNITROMETHAAN (DUTCH) see CKN500
TRICHLOORSILAAN (DUTCH) see TJD500

TRICHLORACETALDEHYD-HYDRAT (GERMAN) see CDO000
TRICHLOR-ACETONITRIL (GERMAN) see TII750
TRICHLORACRYLYL CHLORIDE see TIJ500
TRICHLORAD see ABY900
1,1,1-TRICHLORAETHAN (GERMAN) see MIH275
TRICHLORAETHEN (GERMAN) see TIO750
TRICHLORAETHYLEN, (GERMAN) see TIO750
2,3,3-TRICHLORALLYL-N,N-(DIISOPROPYL)-THIOCARBAMAT (GERMAN) see
  DNS600
TRICHLORAMINE see NGQ500
TRICHLORAN see TIO750
2,3,6-TRICHLORBENZOESAEURE (GERMAN) see TIK500
1,1,1-TRICHLOR-2,2-BIS(4-CHLOR-PHENYL)-AETHAN (GERMAN) see DAD200
2,2,2-TRICHLOR-1,1-BIS(4-CHLOR-PHENYL)-AETHANOL (GERMAN) see
  BIO750
1,1,1-TRICHLOR-2,2-BIS(4-CHLORPHENYL)-AETHANOL (GERMAN) see BIO750
1,1,1-TRICHLOR-2,2-BIS(4-METHOXY-PHENYL)-AETHAN (GERMAN) see
  MEI450
TRICHLORESSIGSAEURE (GERMAN) see TII250
TRICHLORESSIGSAURES NATRIUM (GERMAN) see TII500
1,1,2-TRICHLORETHANE see TIN000
TRICHLORETHANOL see TIN500
TRICHLORETHENE (FRENCH) see TIO750
TRICHLORETHYLENE, (FRENCH) see TIO750
TRICHLORETHYL PHOSPHATE see CGO500
2,4,6-TRICHLORFENOL (CZECH) see TIW000
TRICHLORFENSON (OBS.) see CJT500
2,4,6-TRICHLORFENYLESTER KYSELINY OCTOVE (CZECH) see TIY250
TRICHLORFON (USDA) see TIQ250
TRICHLORINATED ISOCYANURIC ACID see TIQ750
TRICHLORINE NITRIDE see NGQ500
TRICHLOR-3-KYANPROPYLSILAN see COR750
TRICHLORMETAFOS-3 see TIR250
TRICHLORMETAZID see HII500
TRICHLORMETHAN (CZECH) see CHJ500
TRICHLORMETHIAZIDE see HII500
TRICHLORMETHINE see TNF250, TNF500
TRICHLORMETHINIUM CHLORIDE see TNF500
TRICHLORMETHYLBENZOL (GERMAN) see BFL250
TRICHLORMETHYLESTER KYSELINY CHLORMRAVENCI see TIR920
TRICHLORMETHYLESTER KYSELINY DICHLORMETHANTHIOSULFONOVE
  (CZECH) see DFS600
TRICHLOR-METHYLSILAN see MQC500
N-(TRICHLOR-METHYLTHIO)-PHTHALAMID (GERMAN) see TIT250
N-(TRICHLOR-METHYLTHIO)-PHTHALIMID (GERMAN) see CBG000
TRICHLORNITROMETHAN (GERMAN) see CKN500
2,2,2-TRICHLOROACETALDEHYDE see CDN550
TRICHLOROACETALDEHYDE see CDN550
TRICHLOROACETALDEHYDE HYDRATE see CDO000
TRICHLOROACETALDEHYDE MONOETHYLACETAL see TIO000
TRICHLOROACETALDEHYDE MONOHYDRATE see CDO000
TRICHLOROACETALDEHYDE OXIME see TIH825
2,2,2-TRICHLOROACETAMIDE see TII000
α-α-α-TRICHLOROACETAMIDE see TII000
TRICHLOROACETAMIDE see TII000
TRICHLOROACETIC ACID see TII250
TRICHLOROACETIC ACID CHLORIDE see TIJ150
TRICHLOROACETIC ACID compounded with N'-(4-CHLOROPHENYL)-N,N-DI-
  METHYLUREA (1:1) see CJY000
TRICHLOROACETIC ACID SODIUM SALT see TII500
TRICHLOROACETIC ACID-2-(2,4,5-TRICHLOROPHENOXY)ETHYL ESTER see
  TIX250
TRICHLOROACETIC ACID TRIPROPYLSTANNYL ESTER see TNC000
TRICHLOROACETIC ACID (UN 1839) (DOT) see TII250
TRICHLOROACETIC ACID, solution (UN 2564) (DOT) see TII250
TRICHLOROACETOCHLORIDE see TIJ150
1,1,3-TRICHLOROACETONE see TII550
α,α',α'-TRICHLOROACETONE see TII550
TRICHLOROACETONITRILE see TII750
(TRICHLOROACETOXY)TRIPROPYLSTANNANE see TNC000
10-TRICHLOROACETYL-1,2-BENZANTHRACENE see TIJ000
TRICHLOROACETYL CHLORIDE see TIJ150
TRICHLOROACETYL FLUORIDE see TIJ175
2,3,3-TRICHLOROACROLEIN see TIJ250
2,3,3-TRICHLOROACRYLOYL CHLORIDE see TIJ500
TRICHLOROACRYLOYL CHLORIDE see TIJ500
TRICHLOROACRYLYL CHLORIDE see TIJ500
2,3,3-TRICHLOROALLYL DIISOPROPYLTHIOCARBAMATE see DNS600
S-2,3,3-TRICHLOROALLYL-N,N-DIISOPROPYLTHIOCARBAMATE see DNS600
TRICHLOROALLYLSILANE see AGU250
TRICHLOROALUMINUM see AGY750
3,5,6-TRICHLORO-4-AMINOPICOLINIC ACID see PIB900
2,4,6-TRICHLOROANILINE see TIJ750
sym-TRICHLOROANILINE see TIJ750
3,5,6-TRICHLORO-o-ANISIC ACID see TIK000

TRICHLOROARSINE see ARF500
2,4,6-TRICHLOROBENZENAMINE see TIJ750
1,2,3-TRICHLOROBENZENE see TIK100
1,2,4-TRICHLOROBENZENE see TIK250
1,2,5-TRICHLOROBENZENE see TIK250
1,2,6-TRICHLOROBENZENE see TIK100
1,3,4-TRICHLOROBENZENE see TIK250
1,3,5-TRICHLOROBENZENE see TIK300
1,2,4-TRICHLOROBENZENE (ACGIH,OSHA) see TIK250
s-TRICHLOROBENZENE see TIK300
sym-TRICHLOROBENZENE see TIK300
unsym-TRICHLOROBENZENE see TIK250
vic-TRICHLOROBENZENE see TIK100
2,3,6-TRICHLOROBENZENEACETIC ACID see TIY500
2,4,5-TRICHLOROBENZENEDIAZO p-CHLOROPHENYL SULFIDE see CDS500
2,3,6-TRICHLOROBENZOIC ACID see TIK500
TRICHLOROBENZOIC ACID see TIK500
2,3,6-TRICHLOROBENZOIC ACID, DIMETHYLAMINE SALT see DOR800
1,2,4-TRICHLOROBENZOL see TIK250
N,2,6-TRICHLORO-p-BENZOQUINONE IMINE see CHR000
TRICHLOROBENZYL CHLORIDE see TIL250
1,1,1-TRICHLORO-2,2-BIS(p-ANISYL)ETHANE see MEI450
TRICHLOROBIS (4-CHLOROPHENYL) ETHANE see DAD200
2,2,2-TRICHLORO-1,1-BIS(4-CHLOROPHENYL)-ETHANOL (FRENCH) see BIO750
2,2,2-TRICHLORO-1,1-BIS(4-CLORO-FENIL)-ETANOLO (ITALIAN) see BIO750
1,1,1-TRICHLORO-2,2-BIS(p-HYDROXYPHENYL)ETHANE see BKI500
1,1,1-TRICHLORO-2,2-BIS(p-METHOXYPHENOL)ETHANOL see MEI450
1,1,1-TRICHLORO-2,2-BIS(p-METHOXYPHENYL)ETHANE see MEI450
1,1,2-TRICHLOROBUTADIENE see TIL350
1,1,2-TRICHLORO-1,3-BUTADIENE see TIL350
2,3,4-TRICHLOROBUTENE-1 see TIL360
β,β,β-TRICHLORO-tert-BUTYL ALCOHOL see ABD000
tert-TRICHLOROBUTYL ALCOHOL see ABD000
TRICHLORO-tert-BUTYL ALCOHOL see ABD000
TRICHLOROBUTYLENE OXIDE see ECT500
TRICHLOROBUTYLSILANE see BSR000
3,4,4'-TRICHLOROCARBANILIDE see TIL500
3,4,4'-TRICHLOROCARBANILIDE mixed with 3-TRIFLUOROMETHYL-4,4'-DI-CHLOROCARBANILIDE (2:1) see TIL526
2,4,5-TRICHLORO-α-(CHLOROMETHYLENE)BENZYL PHOSPHATE see TBW100
2,4,5-TRICHLORO-α-(CHLOROMETHYLENE)BENZYL PHOSPHATE ESTER see RAF100
TRICHLORO(CHLOROMETHYL)SILANE (9CI) see CIY325
1,1,1-TRICHLORO-2-(o-CHLOROPHENYL)-2-(p-CHLOROPHENYL)ETHANE see BIO625
1,2,4-TRICHLORO-5-((4-CHLOROPHENYL)SULFONYL)-BENZENE see CKM000
1,2,4-TRICHLORO-5-((4-CHLOROPHENYL)THIO)-BENZENE see CKL750
TRICHLOROCHROMIUM see CMJ250
TRICHLOROCTAN SODNY (CZECH) see TII500
TRICHLOROCYANIDINE see TJD750
TRICHLOROCYANURIC ACID see TIQ750
TRICHLORO-3-CYCLOHEXENYLSILANE see CPE500
1,2,4-TRICHLORO DIBENZO-p-DIOXIN see CDV125
N-(2,2,2-TRICHLORO-1-(3,4-DICHLOROANILINO))ETHYLFORMAMIDE see CDP750
2,3,5-TRICHLORO-N-(3,5-DICHLORO-2-HYDROXYPHENYL)-6-HYDROXYBEN-ZAMIDE see DMZ000
4,5,6-TRICHLORO-2-(2,4-DICHLOROPHENOXY)PHENOL see TIL750
1,1,1-TRICHLORO-2,2-DI(4-CHLOROPHENYL)-ETHANE see DAD200
2,2,2-TRICHLORO-1,1-DI-(4-CHLOROPHENYL)ETHANOL see BIO750
TRICHLORO(DICHLOROPHENYL)SILANE see DGF200
1,1,2-TRICHLORO-2,2-DIFLUOROETHANE see TIM000
1,1,1-TRICHLORO-2,2-DI(4-METHOXYPHENYL)ETHANE see MEI450
1,2,3-TRICHLORO-4,6-DINITROBENZENE see DVI600
TRICHLORO DIPHENYL ETHER see CDV175
TRICHLORO DIPHENYL OXIDE see CDV175
3,4,4'-TRICHLORODIPHENYLUREA see TIL500
TRICHLORODODECYLSILANE see DYA800
4,4,4-TRICHLORO-1,2-EPOXYBUTANE see ECT500
1,1,2-TRICHLOROEPOXYETHANE see ECT600
1,1,1-TRICHLORO-2-3-EPOXYPROPANE see TJC250
TRICHLORO ESTERTIN see TIM500
TRICHLOROETHANAL see CDN550
1,1,1-TRICHLOROETHANE see MIH275
1,1,2-TRICHLOROETHANE see TIN000
1,2,2-TRICHLOROETHANE see TIN000
α-TRICHLOROETHANE see MIH275
β-TRICHLOROETHANE see TIN000
TRICHLORO-1,1,1-ETHANE (FRENCH) see MIH275
2,2,2-TRICHLORO-1,1-ETHANEDIOL see CDO000
TRICHLOROETHANOIC ACID see TII250
2,2,2-TRICHLOROETHANOL see TIN500
TRICHLOROETHANOL see TIN500

2,2,2-TRICHLOROETHANOL CARBAMATE see TIO500
TRICHLOROETHENE see TIO750
TRICHLOROETHENYLSILANE see TIN750
2,2,2-TRICHLORO-1-ETHOXYETHANOL see TIO000
2,2,2-TRICHLOROETHYL ALCOHOL see TIN500
TRICHLOROETHYL ALCOHOL see TIN500
TRI-(2-CHLOROETHYL)AMINE see TNF250
TRI-(2-CHLOROETHYL)AMINE HYDROCHLORIDE see TNF500
TRI(β-CHLOROETHYL)AMINE HYDROCHLORIDE see TNF500
TRICHLOROETHYL CARBAMATE see TIO500
1,2,2-TRICHLOROETHYLENE see TIO750
TRICHLOROETHYLENE see TIO750
TRICHLOROETHYLENE EPOXIDE see ECT600
TRICHLOROETHYLENE OXIDE see ECT600
1,1'-(2,2,2-TRICHLOROETHYLIDENE)BIS(4-CHLOROBENZENE) see DAD200
1,1'-(2,2,2-TRICHLOROETHYLIDENE)BIS(4-METHOXYBENZENE) see MEI450
1,2-o-(2,2,2-TRICHLOROETHYLIDENE)-α-d-GLUCOFURANOSE see GFA000
α-TRICHLOROETHYLIDENE GLYCEROL see TIP000
β-TRICHLOROETHYLIDENE GLYCEROL see TIP250
2,2,2-TRICHLOROETHYL PHOSPHATE see TIO800
TRI(2-CHLOROETHYL)PHOSPHATE see CGO500
TRI-β-CHLOROETHYL PHOSPHATE see CGO500
TRICHLOROETHYL PHOSPHATE see TIO800
TRICHLOROETHYLSILANE see EPY500
TRICHLOROETHYLSILICANE see EPY500
TRICHLOROETHYLSTANNANE see EPS000
TRICHLOROETHYLTIN see EPS000
2,3,5-TRICHLOROFENOLAT ZINECNATY (CZECH) see BLN000
TRICHLOROFLUOROMETHANE see TIP500
TRICHLOROFORM see CHJ500
TRICHLOROHEXADECYLSILANE see HCQ000
TRICHLOROHYDRIN see TJB600
4',4'',5-TRICHLORO-2-HYDROXY-3-BIPHENYLCARBOXANILIDE see TIP750
2,4,4'-TRICHLORO-2'-HYDROXYDIPHENYL ETHER see TIO000
(2,2,2-TRICHLORO-1-HYDROXYETHYL) DIMETHYLPHOSPHONATE see TIQ250
2,2,2-TRICHLORO-1-HYDROXYETHYL-PHOSPHONATE, DIMETHYL ESTER see TIQ250
(2,2,2-TRICHLORO-1-HYDROXYETHYL)PHOSPHONIC ACID DIMETHYL ESTER see TIQ250
TRICHLOROISOCYANIC ACID see TIQ750
1,3,5-TRICHLOROISOCYANURIC ACID see TIQ750
N,N',N''-TRICHLOROISOCYANURIC ACID see TIQ750
TRICHLOROISOCYANURIC ACID see TIQ750
TRICHLOROISOPROPANOL see IMQ000
1,1,1-TRICHLOROISOPROPYL ALCOHOL see IMQ000
TRICHLOROMELAMINE see TNE775
TRICHLOROMETAFOS see RMA500
TRICHLOROMETAPHOS-3 see TIR250
TRICHLOROMETHANE see CHJ500
TRICHLOROMETHANE SULFENYL CHLORIDE see PCF300
TRICHLORO-3-METHAPHOS see TIR250
TRICHLOROMETHIADIAZIDE see HII500
TRICHLOROMETHIAZIDE see HII500
3,5,6-TRICHLORO-2-METHOXYBENZOIC ACID see TIK000
2,3,5-TRICHLORO-6-METHOXYBENZOIC ACID see TIK000
TRICHLOROMETHYL ALLYL PERTHIOXANTHATE see TIR750
1-(TRICHLOROMETHYL)BENZENE see BFL250
TRICHLOROMETHYLBENZENE see BFL250
α-(TRICHLOROMETHYL)BENZENEMETHANOL see TIR800
α-(TRICHLOROMETHYL)BENZENEMETHANOL, ACETATE (9CI) see TIT000
α-(TRICHLOROMETHYL)BENZYL ALCOHOL see TIR800
α-(TRICHLOROMETHYL)BENZYL ALCOHOL, ACETATE (9CI) see TIT000
p-TRICHLOROMETHYLCHLOROBENZENE see TIR900
TRICHLOROMETHYL CHLOROFORMATE see TIR920
TRICHLOROMETHYL CYANIDE see TII750
α-2-(TRICHLOROMETHYL)-1,3-DIOXOLANE-4-METHANOL see TIP000
β-2-(TRICHLOROMETHYL)-1,3-DIOXOLANE-4-METHANOL see TIP250
3-(TRICHLOROMETHYL)-5-ETHOXY-1,2,4-THIADIAZOLE see EFK000
N-TRICHLOROMETHYLMERCAPTO-4-CYCLOHEXENE-1,2-DICARBOXIMIDE see CBG000
N-(TRICHLOROMETHYLMERCAPTO)PHTHALIMIDE see TIT250
N-(TRICHLOROMETHYLMERCAPTO)-Δ⁴-TETRAHYDROPHTHALIMIDE see CBG000
TRICHLOROMETHYL METHYL PERTHIOXANTHATE see TIS500
TRICHLOROMETHYLNITRILE see TII750
(TRICHLOROMETHYL)OXIRANE see TJC250
TRICHLOROMETHYL PERCHLORATE see TIS750
TRICHLOROMETHYLPHENYL CARBINOL see TIR800
TRICHLOROMETHYLPHENYLCARBINYL ACETATE see TIT000
1,1,1-TRICHLORO-2-METHYL-2-PROPANOL see ABD000
TRICHLOROMETHYLSTANNANE see MQC750
TRICHLOROMETHYLSULFENYL CHLORIDE see PCF300
TRICHLOROMETHYLSULPHENYL CHLORIDE see PCF300
N-TRICHLOROMETHYLTHIO-3A,4,7,7A-TETRAHYDROPHTHALIMIDE see CBG000

N-TRICHLOROMETHYLTHIO-cis-Δ⁴-CYCLOHEXENE-1,2-DICARBOXIMIDE see CBG000
N-TRICHLOROMETHYLTHIOCYCLOHEX-4-ENE-1,2-DICARBOXIMIDE see CBG000
N-((TRICHLOROMETHYL)THIO)-4-CYCLOHEXENE-1,2-DICARBOXIMIDE see CBG000
2-((TRICHLOROMETHYL)THIO)-1H-ISOINDOLE-1,3(2H)-DIONE see TIT250
N-(TRICHLOROMETHYLTHIO)PHTHALIMIDE see TIT250
TRICHLOROMETHYLTHIO-1,2,5,6-TETRAHYDROPHTHALAMIDE see CBG000
N-((TRICHLOROMETHYL)THIO)TETRAHYDROPHTHALIMIDE see CBG000
TRICHLOROMETHYLTIN see MQC750
5-TRICHLOROMETHYL-1-TRIMETHYLSILYLTETRAZOLE see TIT275
TRICHLOROMONOFLUOROMETHANE see TIP500
TRICHLOROMONOSILANE see TJD500
TRICHLORONAPHTHALENE see TIT500
TRICHLORONAT see EPY000
2,4,5-TRICHLORONITROBENZENE see TIT750
1,2,4-TRICHLORO-5-NITROBENZENE see TIT750
2′,4′,6′-TRICHLORO-4-NITROBIPHENYL ETHER see NIW500
2,4,6-TRICHLORO-4′-NITRODIPHENYL ETHER see NIW500
TRICHLORONITROMETHANE see CKN500
1,3,5-TRICHLORO-2-(4-NITROPHENOXY)BENZENE see NIW500
TRICHLOROOCTADECYLSILANE see OBI000
TRICHLORO-OXIRANE see ECT600
TRICHLOROOXOVANADIUM see VDP000
TRICHLOROPEROXYACETIC ACID see TIV275
TRICHLOROPHENE see HCL000
2,3,6-TRICHLOROPHENOL see TIV500
2,4,5-TRICHLOROPHENOL see TIV750
2,4,6-TRICHLOROPHENOL see TIW000
2,4,6-TRICHLOROPHENOL ACETATE see TIY250
2,4,5-TRICHLOROPHENOL, O-ESTER with O,O-DIMETHYL PHOSPHOROTH-IOATE see RMA500
2,4,5-TRICHLOROPHENOL-O-ESTER with O-ETHYL ETHYLPHOSPHONOTH-IOATE see EPY000
2,4,5-TRICHLOROPHENOL, SODIUM SALT see SKK500
2,4,5-TRICHLOROPHENOXYACETIC ACID see TAA100
(2,4,5-TRICHLOROPHENOXY)ACETIC ACID-2-BUTOXYETHYL ESTER see TAH900
2,4,5-TRICHLOROPHENOXYACETIC ACID, BUTYL ESTER see BSQ750
2,4,5-TRICHLOROPHENOXY, ACETIC ACID, ISOOCTYL ESTER see TGF200
(2,4,5,-TRICHLOROPHENOXY)ACETIC ACID, ISOPROPYL ESTER see TGF210
(2,4,5,-TRICHLOROPHENOXY)ACETIC ACID-1-METHYL ESTER (9CI) see TGF210
2,4,5-TRICHLOROPHENOXYACETIC ACID, PROPYLENE GLYCOL BUTYL ETHER ESTERS see THJ100
4-(2,4,5-TRICHLOROPHENOXY)BUTYRIC ACID see TIW750
2-(2,4,5-TRICHLOROPHENOXY)ETHANOL see TIX000
2-(2,4,5-TRICHLOROPHENOXY)ETHYL-2,2-DICHLOROPROPIONATE see PBK000
2,4,5-TRICHLOROPHENOXYETHYL-α,α-DICHLOROPROPIONATE see PBK000
2,4,5-TRICHLOROPHENOXYETHYL-α,α,α-TRICHLOROACETATE see TIX250
2-(2,4,5-TRICHLOROPHENOXY)PROPIONIC ACID see TIX500
α-(2,4,5-TRICHLOROPHENOXY)PROPIONIC ACID see TIX500
2,4,5-TRICHLOROPHENOXY-α-PROPIONIC ACID see TIX500
2-(2,4,5-TRICHLOROPHENOXY)PROPIONIC ACID PROPYLENE GLYCOL BU-TYL ETHER ESTER see TIX750
2-(2,4,5-TRICHLOROPHENOXY)PROPIONIC ACID TRIBUTYLSTANNYL ESTER see TIF750
(2,4,5-TRICHLOROPHENOXY)SODIUM see SKK500
2,4,6-TRICHLOROPHENYL ACETATE see TIY250
2,3,6-TRICHLOROPHENYLACETIC ACID see TIY500
2,4,5-TRICHLOROPHENYLAZO-4′-CHLOROPHENYL-SULFIDE see CDS500
TRI-o-CHLOROPHENYL BORATE see TIY750
2,4,6-TRICHLORO-PHENYL CHLOROFORMATE see TIY800
2,4,6-TRICHLORO-PHENYLDIMETHYLTRIAZENE see TJA000
1-(2,4,6-TRICHLOROPHENYL)-3,3-DIMETHYLTRIAZENE see TJA000
2,4,5-TRICHLOROPHENYL IODOPROPARGYL ETHER see IFA000
2,4,5-TRICHLOROPHENYL-γ-IODOPROPARGYL ETHER see IFA000
TRICHLOROPHENYLMETHANE see BFL250
1-(2,4,6-TRICHLOROPHENYL)-3-p-NITROANILINO-2-PYRAZOLIN-5-ONE see TJA500
2,4,6-TRICHLOROPHENYL-4-NITROPHENYL ETHER see NIW500
TRICHLOROPHENYLSILANE see TJA750
TRICHLOROPHON see TIQ250
TRICHLOROPHOSPHINE SULFIDE see TFO000
2,4,6-TRICHLORO-PMDT see TJA000
1,1,1-TRICHLOROPROPANE see TJB000
1,1,2-TRICHLOROPROPANE see TJB250
1,2,2-TRICHLOROPROPANE see TJB500
1,2,3-TRICHLOROPROPANE see TJB600
1,2,3-TRICHLOROPROPANE-2,3-OXIDE see TJB750
TRICHLOROPROPANE OXIDE see TJC250
1,1,1-TRICHLORO-2-PROPANOL see IMQ000
1,1,3-TRICHLORO-2-PROPANONE see TII550

2,3,3-TRICHLORO-2-PROPENAL see TIJ250
2,3,3-TRICHLOROPROPENAL see TIJ250
1,2,3-TRICHLOROPROPENE see TJC000
1,1,1-TRICHLOROPROPENE OXIDE see TJC250
1,1,1-TRICHLOROPROPENE-2,3-OXIDE see TJC250
3,3,3-TRICHLOROPROPENE OXIDE see TJC250
TRICHLOROPROPENE OXIDE see TJC250
2,3,3-TRICHLORO-2-PROPEN-1-OL see TJC500
2,3,4-TRICHLORO-2-PROPENOYL CHLORIDE see TIJ500
2,2,3-TRICHLOROPROPIONALDEHYDE see TJC750
TRICHLOROPROPIONITRILE see TJC800
TRICHLOROPROPYLENE see TJB750
1,1,1-TRICHLOROPROPYLENE OXIDE see TJC250
3,3,3-TRICHLOROPROPYLENE OXIDE see TJC250
TRICHLOROPROPYLSILANE see PNX250
3,5,6-TRICHLORO-2-PYRIDYLOXYACETIC ACID see TJE890
TRICHLOROSILANE see TJD500
TRICHLOROSTIBINE see AQC500
2,2,2-TRICHLORO-1-(2-THIAZOLYLAMINO)ETHANOL see CMB675
TRICHLOROTITANIUM see TGG250
2,3,6-TRICHLOROTOLUENE see TJD600
α,α,α-TRICHLOROTOLUENE see BFL250
α-α-p-TRICHLOROTOLUENE see TJD650
ω,ω,ω-TRICHLOROTOLUENE see BFL250
N,N′,N″-TRICHLORO-2,4,6-TRIAMINE-1,3,5-TRIAZINE see TNE775
2,4,6-TRICHLORO-1,3,5-TRIAZINE see TJD750
1,3,5-TRICHLOROTRIAZINE see TJD750
2,4,6-TRICHLOROTRIAZINE see TJD750
2,4,6-TRICHLORO-s-TRIAZINE see TJD750
TRICHLORO-s-TRIAZINE see TJD750
sym-TRICHLOROTRIAZINE see TJD750
1,3,5-TRICHLORO-1,3,5-TRIAZINETRIONE see TIQ750
TRICHLORO-s-TRIAZINE-2,4,6(1H,3H,5H)-TRIONE see TIQ750
TRICHLORO-s-TRIAZINETRIONE see TIQ750
2,2′,2″-TRICHLOROTRIETHYLAMINE see TNF250
2,2′,2″-TRICHLOROTRIETHYLAMINE HYDROCHLORIDE see TNF500
TRICHLOROTRIETHYLDIALUMINIUM see TJP775
TRICHLOROTRIETHYLDIALUMINUM see TJP775
1,3,5-TRICHLORO-2,4,6-TRIFLUOROBORAZINE see TJE050
1,1,1-TRICHLORO-2,2,2-TRIFLUOROETHANE see TJE100
1,1,2-TRICHLORO-1,2,2-TRIFLUOROETHANE (OSHA, ACGIH, MAK) see FOO000
TRICHLOROTRIFLUOROETHANE see FOO000, TJE100
TRICHLOROTRIMETHYLDIALUMINUM see MGC230
TRICHLORO-1,3,5-TRINITROBENZENE see TJE200
1,3,5-TRICHLORO-2,4,6-TRINITROBENZENE see TJE200
sym-TRICHLOROTRINITROBENZENE see TJE200
1,3,5-TRICHLORO-2,4,6-TRIOXOHEXAHYDRO-s-TRIAZINE see TIQ750
TRICHLORO(VINYL)SILANE see TIN750
TRICHLOROVINYL SILICANE see TIN750
TRICHLORPHENE see TIQ250
(2,4,5-TRICHLOR-PHENOXY)-ESSIGSAEURE (GERMAN) see TAA100
2-(2,4,5-TRICHLOR-PHENOXY)-PROPIONSAEURE (GERMAN) see TIX500
O-(2,4,5-TRICHLOR-PHENYL)-O,O-DIMETHYL-MONOTHIOPHOSPHAT (GERMAN) see RMA500
2,3,6-TRICHLORPHENYLESSIGSAEURE (GERMAN) see TIY500
TRICHLORPHON see TIQ250
TRICHLORPHON FN see TIQ250
TRICHLORSILAN (GERMAN) see TJD500
TRICHLOR-TRIAETHYLAMIN-HYDROCHLORID (GERMAN) see TNF500
TRICHLORURE d′ANTIMOINE see AQC500
TRICHLORURE d′ARSENIC (FRENCH) see ARF500
sym-TRICHLOTRIAZIN (CZECH) see TJD750
TRICHLOURACETONITRIL (DUTCH) see TII750
TRICHOCHROMOGENIC FACTOR see AIH600
TRICHOCID see ABY900
TRICHOCIDE see MMN250
TRICHOFURON see NGG500
TRICHOMAN see ABY900
TRICHOMOL see MMN250
TRICHOMONACID "PHARMACHIM" see MMN250
TRICHOPOL see MMN250
TRICHORAD see ABY900
TRICHORAL see ABY900
TRICHOSANTHIN see TJF350
TRICHOTHEC-9-ENE, 12,13-EPOXY-4-β,8,8-α-15-TRIACETOXY-3-α-HYDROXY- see ACS500
TRICHOTHECIN see TJE750
TRICICLIDINA see CPQ250
TRICILOID see CPQ250
TRICIONE see TLP750
TRICIRIBINE see TJE870
TRICIRIBINE PHOSPHATE HYDRATE see TJE875
TRICLABENDAZOLE see CFL200
TRI-CLENE see TIO750

TRICLOBISONIUM CHLORIDE see TJE880
TRICLOCARBAN see TIL500
TRICLOFOS see TIO800
TRICLOPYR see TJE890
TRI-CLOR see CKN500
TRICLORDIURIDE see HII500
TRICLORETENE (ITALIAN) see TIO750
TRICLORMETIAZIDE (ITALIAN) see HII500
1,1,1-TRICLORO-2,2-BIS(4-CLORO-FENIL)-ETANO (ITALIAN) see DAD200
1,1,1-TRICLOROETANO (ITALIAN) see MIH275
TRICLOROETILENE (ITALIAN) see TIO750
O-(2,4,5-TRICLORO-FENIL)-O,O-DIMETIL-MONOTIOFOSFATO (ITALIAN) see RMA500
TRICLOROMETANO (ITALIAN) see CHJ500
TRICLOROMETILBENZENE (ITALIAN) see BFL250
TRICLORO-NITRO-METANO (ITALIAN) see CKN500
TRICLOROSILANO (ITALIAN) see TJD500
TRICLOROTOLUENE (ITALIAN) see BFL250
TRICLOS see TIO800
TRICLOSAN see TIQ000
TRICLOSE see TJF000
TRICOBALT TETRAOXIDE see CND020
TRICOBALT TETROXIDE see CND020
TRICOFURON see NGG500
TRICOGEN see ABY900
TRICOLAM see TGD250
TRICOLAVAL see ABY900
TRICOLOID see CPQ250
TRICOLOID CHLORIDE see EAI875
TRICOM see MMN250
TRICON BW see EIX000
TRICOP 50 see CNK559
TRICORAL see ABY900
TRI-CORNOX SPECIAL see BAV000
TRICORYL see TJL250
12-TRICOSANONE see TJF250
TRICOSANTHIN see TJF350
9-TRICOSENE, (Z)- see TJF400
(Z)-9-TRICOSENE see TJF400
9-TRICOSENE see TJF400
cis-9-TRICOSENE see TJF400
TRICOSTERIL see ABY900
TRICOWAS B see MMN250
2,4,6-TRICI-PDMT see TJA000
TRICRESILFOSFATI (ITALIAN) see TNP500
TRICRESOL see CNW500
TRI-o-CRESYL BORATE see TJF500
TRICRESYLFOSFATEN (DUTCH) see TNP500
TRICRESYLPHOSPHATE, with more than 3% ortho isomer (DOT) see TNP500
TRI-o-CRESYL PHOSPHATE see TMO600
TRICRESYL PHOSPHATE see TMO600, TNP500
TRI-o-CRESYL PHOSPHITE see TJF750
TRI-p-CRESYL PHOSPHITE see TJG000
TRICTAL see TAF675
TRICURAN see PDD300
TRICYANIC ACID see THS000
TRICYANOGEN CHLORIDE see TJD750
TRICYCLAMOL see CPQ250
TRICYCLAMOL CHLORIDE see EAI875
TRICYCLAMOL METHOCHLORIDE see EAI875
TRICYCLAMOL SULFATE see TJG225
TRICYCLAZOLE see MQC000
TRICYCLAZONE see MQC000
TRICYCLIC ANTIDEPRESSANTS see TJG239
TRICYCLO(3.3.1.1$^{3,7}$)DECAN-1-AMINE see TJG250
TRICYCLO(3.3.1.1.$^{(3,7)}$)DECAN-1-AMINE, HYDROCHLORIDE (9CI) see AED250
exo-TRICYCLO(5.2.1.0$^{2,6}$)DECANE see TLR675
TRICYCLODECANE(5.2.1.0$^{2,6}$)-3,10-DIISOCYANATE see TJG500
TRICYCLODECEN-4-YL-8-ACETATE see DLY400
TRICYCLODECENYL PROPIONATE see TJG600
4-TRICYCLODECYLIDENE BUTANAL see OBW100
TRICYCLO(6.3.1.0$^{2,5}$))DODECAN-1-OL, 4,4,8-TRIMETHYL-, ACETATE, (1R-(1-α-2-α-5-β,8-β))- see CCN050
TRI(2-CYCLOHEXYLCYCLOHEXYL)BORATE see TJG750
TRICYCLOHEXYLHYDROXYSTANNANE see CQH650
TRICYCLOHEXYLHYDROXYTIN see CQH650
1-(TRICYCLOHEXYLSTANNYL)-1H-1,2,4-TRIAZOLE see THT500
TRICYCLOHEXYLTIN HYDROXIDE see CQH650
TRICYCLOHEXYLZINNHYDROXID (GERMAN) see CQH650
TRICYCLOQUINAZOLINE see TJH250
n-TRIDECANE see TJH500
TRIDECANE see TJH500
TRIDECANE, 1-BROMO- see BOI500
1-TRIDECANECARBOXYLIC ACID see MSA250

1-TRIDECANECARBOXYLIC ACID, ISOPROPYL ESTER see IQN000
TRIDECANEDIOIC ACID, CYCLIC ETHYLENE ESTER see EJQ500
TRIDECANENITRILE see TJH750
TRIDECANITRILE (mixed isomers) see TJI000
TRIDECANOIC ACID see TJI250
TRIDECANOIC ACID-2,3-EPOXYPROPYL ESTER see TJI500
1-TRIDECANOL see TJI750
n-TRIDECANOL see TJI750
TRIDECANOL see TJI750
TRIDECANOL condensed with 6 moles ETHYLENE OXIDE see TJJ250
1-TRIDECANOL PHTHALATE see DXQ200
7-TRIDECANONE see TJJ300
2-TRIDECENAL see TJJ400
n-TRIDECOIC ACID see TJI250
TRIDECYL ACRYLATE see ADX000
n-TRIDECYL ALCOHOL see TJI750
TRIDECYL ALCOHOL see TJI750
N-TRIDECYL-2,6-DIMETHYLMORPHOLIN (GERMAN) see DUJ400
4-TRIDECYL-2,6-DIMETHYLMORPHOLINE see DUJ400
N-TRIDECYL-2,6-DIMETHYLMORPHOLINE see DUJ400
TRIDECYLIC ACID see TJI250
TRIDEMORF see TJJ500
TRIDEMORPH see TJJ500
TRIDESTRIN see EDU500
TRIDEUTEROMETHYL ACETOXYMETHYLNITROSAMINE see MMS000
TRIDEX see DUN600
TRIDEZIBARBITUR see EOK000
1,2,3-TRI(β-DIETHYLAMINOETHOXY)BENZENE TRIETHIODIDE see PDD300
TRI(β-DIETHYLAMINOETHOXY)-1,2,3-BENZENE TRI-IODOETHYLATE see PDD300
TRI-DIGITOXOSIDE (GERMAN) see DKL800
TRI(DIISOBUTYLCARBINYL) BORATE see TJJ750
TRIDILONA see TLP750
2,4,6-TRI(DIMETHYLAMINOMETHYL)PHENOL see TNH000
TRI(p-DIMETHYLAMINOPHENYL)METHANOL see TJK000
TRI(DIMETHYLAMINO)PHOSPHINE OXIDE see HEK000
TRIDIMITE (FRENCH) see SCK000
TRIDIONE see TLP750
TRIDIPAM see TFS350
TRIDIPHANE see TJK100
TRI-n-DODECYL BORATE see TJK250
TRI-(3-DODECYL-1-METHYL-2-PHENYLBENZIMIDAZOLIUM) FERRICYANIDE see TNH750
TRIDONE see TLP750
TRIDYMITE 118 see SCK000
α-TRIDYMITE see SCK000
TRIDYMITE see SCI500, SCK000
TRIELINA (ITALIAN) see TIO750
TRIEN see TJR000
TRI-ENDOTHAL see DXD000, EAR000
TRIENTINE see TJR000
1,2,4,5,9,10-TRIEPOXYDECANE see TJK500
TRI-ERVONUM see MCA500
TRIESIFENIDILE see BBV000
TRIESTE FLOWERS see POO250
TRIETAZINE see TJL500
TRI-ETHANE see MIH275
N,N′,N″-TRI-1,2-ETHANEDIYL PHOSPHORIC TRIMIDE see TND250
N,N′,N″-TRI-1,2-ETHANEDIYLPHOSPHOROTHIOIC TRIAMIDE see TFQ750
N,N′,N″-TRI-1,2-ETHANEDIYLTHIOPHOSPHORAMIDE see TFQ750
TRIETHANOLAMIN see TKP500
TRIETHANOLAMINE (ACGIH) see TKP500
TRIETHANOLAMINE BORATE see TJK750
TRIETHANOLAMINE DODECYLBENZENE SULFONATE see TJK800
TRIETHANOLAMINE DODECYL SULFATE see SON000
TRIETHANOLAMINE LAURYL SULFATE see SON000
TRIETHANOLAMINE SILICIEE (FRENCH) see SDH670
TRIETHANOLAMINE TRINITRATE BIPHOSPHATE see TJL250
TRIETHANOLAMINE TRINITRATE DIPHOSPHATE see TJL250
TRIETHANOMELAMINE see TND500
TRIETHAZINE see TJL500
TRIETHOXONIUM FLUOROBORATE see TJL600
TRIETHOXY(3-AMINOPROPYL)SILANE see TJN000
3-(2,4,5-TRIETHOXYBENZOYL)PROPIONIC ACID see CNG835
TRIETHOXYDIALUMINUM TRIBROMIDE see TJL775
TRIETHOXY-ETHYLSILANE see EQA000
TRIETHOXYFENYLSILAN see PGO000
1,1,3-TRIETHOXYHEXANE see TJM000
TRIETHOXY-2-KYANETHYLSILAN see CON250
TRIETHOXY-3-KYANPROPYLSILAN see COS800
TRIETHOXYMETHANE see ENY500
TRIETHOXYMETHYLSILANE see MQD750
2,4,5-TRIETHOXY-γ-OXOBENZENEBUTANOIC ACID see CNG835
TRIETHOXYPHENYLSILANE see PGO000
1,3,3-TRIETHOXYPROPANE see TJM250

TRIFLUMEN see HII500
TRIFLUMETHAZINE see TJW500
TRIFLUOMETHYLTHIAZIDE see TKG750
TRIFLUOPERAZINA (ITALIAN) see TKE500
TRIFLUOPERAZINE see TKE500
TRIFLUOPERAZINE DIMALEATE see TJW600
TRIFLUOPERAZINE HYDROCHLORIDE see TKK250
TRIFLUORACETIC ACID see TKA250
TRIFLUORALIN (USDA) see DUV600
3-(5-TRIFLUORMETHYLPHENYL)-,1-DIMETHYLHARNSTOFF (GERMAN) see DUK800
TRIFLUOROACETALDEHYDE HYDRATE see TJZ000
2-(2,2,2-TRIFLUOROACETAMIDO)-4-(5-NITRO-2-FURYL)THIAZOLE see NGN500
TRIFLUOROACETIC ACID (DOT) see TKA250
TRIFLUOROACETIC ACID ANHYDRIDE see TJX000
TRIFLUOROACETIC ACID TRIETHYLSTANNYL ESTER see TJX250
TRIFLUOROACETIC ANHYDRIDE see TJX000
N-TRIFLUOROACETYLADRIAMYCIN-14-VALERATE see TJX350
TRIFLUOROACETYLADRIAMYCIN-14-VALERATE see TJX350
2-TRIFLUOROACETYLAMINOFLUORENE see FER000
2-TRIFLUOROACETYLAMINOFLUOREN-9-ONE see TKH000
TRIFLUOROACETYL ANHYDRIDE see TJX000
TRIFLUOROACETYL AZIDE see TJX375
TRIFLUOROACETYL CHLORIDE see TJX500
2-TRIFLUOROACETYL-1,3,4-DIOXAZALONE see TJX600
TRIFLUOROACETYL HYPOCHLORITE see TJX650
TRIFLUOROACETYL HYPOFLUORITE see TJX750
TRIFLUOROACETYLIMINOIODOBENZENE see TJX775
TRIFLUOROACETYL NITRITE see TJX780
O-TRIFLUOROACETYL-S-FLUOROFORMYL THIOPEROXIDE see TJX625
TRIFLUOROACETYL TRIFLUOROMETHANE SULFONATE see TJX800
TRIFLUOROACRYLOYL FLUORIDE see TJX825
TRIFLUOROAMINE OXIDE see NGS500
2,3,4-TRIFLUOROANILINE see TJX900
TRIFLUOROANTIMONY see AQE000
TRIFLUOROARSINE see ARI250
1,1,2-TRIFLUORO-1-BROMO-2-CHLOROETHANE see TJY000
1,1,1-TRIFLUORO-2-BROMO-2-CHLOROETHANE see HAG500
TRIFLUOROBROMOETHYLENE see BOJ000
TRIFLUOROBROMOMETHANE see TJY100
2,2,2-TRIFLUORO-1-CHLORO-1-BROMOETHANE see HAG500
1,1,1-TRIFLUORO-2-CHLORO-2-BROMOETHANE see HAG500
1,1,1-TRIFLUORO-2-CHLOROETHANE see TJY175
2,2,2-TRIFLUOROCHLOROETHANE see TJY175
1,1,2-TRIFLUORO-2-CHLOROETHYLENE see CLQ750
TRIFLUOROCHLOROETHYLENE (DOT) see CLQ750
TRIFLUOROCHLOROETHYLENE POLYMER see KDK000
TRIFLUOROCHLOROMETHANE (DOT) see CLR250
1,1,1-TRIFLUORO-3-CHLOROPROPANE see TJY200
α,α,α-TRIFLUORO-4-CHLOROTOLUENE see CEM825
α,α,α-TRIFLUORO-m-CRESOL see TKE750
5-TRIFLUORO-2'-DEOXYTHYMIDINE see TKH325
2,2,2-TRIFLUORODIAZOETHANE see TJY275
1,1,2-TRIFLUORO-1,2-DICHLOROETHANE see TJY750
1,1,1-TRIFLUORO-2,2-DICHLOROETHANE see TJY500
2',4',6'-TRIFLUORO-4-DIMETHYLAMINOAZOBENZENE see DUK200
α,α,α-TRIFLUORO-2,6-DINITRO-N,N-DIPROPYL-p-TOLUIDINE see DUV600
1,1,1-TRIFLUOROETHANE see TJY900
2,2,2-TRIFLUORO-1,1-ETHANEDIOL see TJZ000
TRIFLUOROETHANOIC ACID see TKA250
2,2,2-TRIFLUOROETHANOL see TKA350
(TRIFLUOROETHENYL)BENZENE see TKH310
(2,2,2-TRIFLUOROETHOXY)ETHENE see TKB250
2,2,2-TRIFLUOROETHYLAMINE see TKA500
TRIFLUOROETHYLAMINE HYDROCHLORIDE see TKA750
1,1,1-TRIFLUOROETHYL CHLORIDE see TJY175
2,2,2-TRIFLUOROETHYL VINYL ETHER see TKB250
2,2,2-TRIFLUORO-N-(FLUOREN-2-YL)ACETAMIDE see FER000
1,1,1-TRIFLUOROFORM see TJY900
N,N,N'-TRIFLUOROHEXANAMIDINE see TKB275
(3,3,3-TRIFLUORO-2-HYDROXY-2-(TRIFLUOROMETHYL))PROPYL BENZYL KETONE see TKB285
TRIFLUOROMETHANE see CBY750
TRIFLUOROMETHANESULFENYL CHLORIDE see TKB300
TRIFLUOROMETHANE SULFONIC ACID see TKB310
TRIFLUOROMETHANE (UN 1984) (DOT) see CBY750
TRIFLUOROMETHANE, refrigerated, liquid (UN 3136) (DOT) see CBY750
1,3,3-TRIFLUORO-2-METHOXYCYCLOPROPENE see TKB325
4-TRIFLUOROMETHYL ALLYLOXY-2 N-(β-DIETHYLAMINO-ETHYL)BENZAMIDE (FRENCH) see FDA875
3-(TRIFLUOROMETHYL)ANILINE see AID500
m-(TRIFLUOROMETHYL)ANILINE see AID500
p-TRIFLUOROMETHYLANILINE see TKB750
2-(3-(TRIFLUOROMETHYL)ANILINO)NICOTINIC ACID see NDX500
3-(TRIFLUOROMETHYL)BENZENAMINE see AID500

(TRIFLUOROMETHYL)BENZENE see BDH500
3-(TRIFLUOROMETHYL)BENZENEACETONITRILE see TKF000
5-(TRIFLUOROMETHYL)-1,3-BENZENEDIAMINE see TKB775
4-TRIFLUOROMETHYL-6H-BENZO(e)(1)BENZOTHIOPYRANO(4,3-b)INDOLE see TKC000
m-TRIFLUOROMETHYL BENZOIC ACID, THALLIUM SALT see TKH500
6-(TRIFLUOROMETHYL)-1,2,4-BENZO-THIADIAZINE-7-SULFONAMIDE-1,1-DIOX-IDE see TKG750
6-(TRIFLUOROMETHYL)-1,4,2-BENZOTHIADIAZINE-7-SULFONAMIDO-1,1-DIOX-IDE see TKG750
6-TRIFLUOROMETHYL-3-BENZYL-7-SULFAMYL-3,4-DIHYDRO-1,2,4-BENZOTHIA-DIAZINE-1,1-DIOXIDE see BEQ625
3-(TRIFLUOROMETHYL)BROMOBENZENE see BOJ500
m-(TRIFLUOROMETHYL)BROMOBENZENE see BOJ500
o-(TRIFLUOROMETHYL)BROMOBENZENE see BOJ750
TRIFLUOROMETHYL CHLORIDE see CLR250
p-(TRIFLUOROMETHYL)CHLOROBENZENE see CEM825
6-TRIFLUOROMETHYLCYCLOPHOSPHAMIDE see TKD000
5-(TRIFLUOROMETHYL)-2'-DEOXYURIDINE see TKH325
5-(TRIFLUOROMETHYL)-2-DEOXYURIDINE see TKH325
5-(TRIFLUOROMETHYL)DEOXYURIDINE see TKH325
TRIFLUOROMETHYLDEOXYURIDINE see TKH325
8-TRIFLUOROMETHYL-7-DESCHLOROGLAFENINE see TKG000
3-TRIFLUOROMETHYL-4,4'-DICHLOROCARBANILIDE mixed with 3,4,4'-TRI-CHLOROCARBANILIDE (1:2) see TIL526
3'-TRIFLUOROMETHYL-4-DIMETHYLAMINOAZOBENZENE see TKD325
4-(TRIFLUOROMETHYL)-2,6-DINITRO-N,N-DIPROPYLANILINE see DUV600
3-TRIFLUOROMETHYLDIPHENYLAMINE-2-CARBOXYLIC ACID see TKH750
TRIFLUOROMETHYLETHYLENE see TKH015
3-(TRIFLUOROMETHYL)-N-ETHYL-α-METHYL PHENETHYL AMINE see ENJ000
TRIFLUOROMETHYL-3-FLUOROCARBONYL HEXA- FLUORO-PEROXYBUTY-RATE see TKD350
2-TRIFLUOROMETHYL-9-(3-(4-(β-HYDROXYETHYL)-1-PIPERAZI-NYL)PROPYLIDENE)THIOXANTHENE see FMO129
TRIFLUOROMETHYL HYPOFLUORITE see TKD375
TRIFLUOROMETHYL-10-(3'-(1-METHYL-4-PIPERAZI-NYL)PROPYL)PHENOTHIAZINE see TKE500
3-TRIFLUOROMETHYLNITROBENZENE see NFJ500
m-(TRIFLUOROMETHYL)NITROBENZENE see NFJ500
α,α,α-TRIFLUORO-2-METHYL-4'-NITRO-m-PROPIONOTOLUIDIDE see FMR050
TRIFLUOROMETHYLPERAZINE see TKE500
TRIFLUOROMETHYL PEROXONITRATE see TKE525
TRIFLUOROMETHYL PEROXYACETATE see TKE550
3-(TRIFLUOROMETHYL)PHENOL see TKE750
2-(TRIFLUOROMETHYL)PHENOTHIAZINE see TKE775
4-(3-(2-(TRIFLUOROMETHYL)PHENOTHIAZIN-10-YL)PROPYL)-1-PIPERAZINEETH-ANOL, DIHYDROCHLORIDE see FMP000
4-(3-(2-TRIFLUOROMETHYL-10-PHENOTHIAZINYL)PROPYL)-1-PIPERAZINEETHA-NOL ENANTHATE see PMI250
2-((4-(3-(2-(TRIFLUOROMETHYL)PHENOTHIAZIN-10-YL)PROPYL)-1-PIPERAZI-NYL))ETHYL HEPTANOATE see PMI250
4-(3-(2-TRIFLUOROMETHYL-10-PHENOTHIAZYL)-PROPYL)-1-PIPERAZINEETHA-NOL see TJW500
m-TRIFLUOROMETHYLPHENYLACETONITRILE see TKF000
N-(m-TRIFLUOROMETHYLPHENYL)-2-AMINOBENZOIC ACID see TKH750
2-(3-(TRIFLUOROMETHYL)-PHENYL)AMINONICOTINIC ACID see NDX500
N-(3-TRIFLUOROMETHYLPHENYL) ANTHRANILIC ACID see TKH750
3-(TRIFLUOROMETHYL)PHENYL BROMIDE see BOJ500
m-(TRIFLUOROMETHYL PHENYL BROMIDE see BOJ500
p-TRIFLUOROMETHYLPHENYL CHLORIDE see CEM825
4-(2-TRIFLUOROMETHYLPHENYL)-3,5-DICARBETHOXY-2,6-DIMETHYL-1,4-DI-HYDROPYRIDINE see FOO875
3-(m-TRIFLUOROMETHYLPHENYL)-1,1-DIMETHYLUREA see DUK800
1-(3-TRIFLUOROMETHYLPHENYL)-2-ETHYLAMINOPROPANE HYDROCHLORIDE see PDM250
1-(m-TRIFLUOROMETHYLPHENYL)-3-(2-HYDROXYETHYL)-QUINAZOLINE-2,4(1H,3H)-DIONE see TKF250
1-(m-TRIFLUOROMETHYLPHENYL)-3-(2'-HYDROXYETHYL)QUINAZOLINE-2,4-DI-ONE see TKF250
2-((3-(TRIFLUOROMETHYL)PHENYL)-4-ISOPROPYL-TETRAHYDRO-1,4-OXAZINE see IRP000
2-TRIFLUOROMETHYLPHENYL MAGNESIUM BROMIDE see TKF525
3-TRIFLUOROMETHYLPHENYL MAGNESIUM BROMIDE see TKF530
4-TRIFLUOROMETHYLPHENYL MAGNESIUM BROMIDE see TKF535
N-(m-TRIFLUOROMETHYLPHENYL)-N',N'-DIMETHYLUREA see DUK800
1-(m-TRIFLUOROMETHYLPHENYL)-N-NITROSOANTHRANILIC ACID see TKF699
N-(3-TRIFLUOROMETHYLPHENYL)-N'-N'-DIMETHYLUREA see DUK800
1,1,1-TRIFLUORO-N-(2-METHYL-4-(PHENYLSULFO-NYL)PHENYL)METHANESULFONAMIDE see TKF750
TRIFLUOROMETHYL PHOSPHINE see TKF775
2-(8'-TRIFLUOROMETHYL-4'-QUINOLYLAMINO)BENZOIC ACID, 2,3-DIHY-DROXY PROPYL ESTER see TKG000
6-TRIFLUOROMETHYL-7-SULFAMOYL-4H-1,4,2-BENZOTHIADIAZINE-1,1-DIOXIDE see TKG750

6-TRIFLUOROMETHYL-7-SULFAMYL-1,2,4-BENZOTHIADIAZINE-1,1-DIOXIDE see TKG750

N-(TRIFLUOROMETHYLSULFINYL)TRIFLUOROMETHYL IMIDOSULFINYL AZIDE see TKG275

TRIFLUOROMETHYLSULFONYL AZIDE see TKG525

TRIFLUOROMETHYLTHIAZIDE see TKG750

4-(3-(2-(TRIFLUOROMETHYL)THIOXANTHEN-9-YLIDENE)PROPYL)-1-PIPERAZI-NEETHANOL DIHYDROCHLORIDE see FMO150

4-(3-(2-(TRIFLUOROMETHYL)-9H-THIOXANTHEN-9-YLIDENE)PROPYL)-1-PIPERA-ZINEETHANOL see FMO129

4-(3-(2-(TRIFLUOROMETHYL)THIOXANTHEN-9-YLIDENE)PROPYL)-1-PIPERAZI-NEETHANOL see FMO129

4-(3-(2-TRIFLUOROMETHYLTHIOXANTH-9-YLIDENE)PROPYL)-1-PIPERAZINEETH-ANOL DIHYDROCHLORIDE see FMO150

TRIFLUOROMETHYL TRIFLUOROVINYL ETHER see TKG800

1-TRIFLUOROMETHYL-1,2,2,3,3,4,4,5,5,6,6-UNDECAFLUORO CYCLOHEXANE see PCH290

TRIFLUOROMONOBROMOMETHANE see TJY100

TRIFLUOROMONOCHLOROCARBON see CLR250

TRIFLUOROMONOCHLOROETHYLENE see CLQ750

2,2,2-TRIFLUORO-N-(4-(5-NITRO-2-FURYL)-2-THIAZOLYL)ACETAMIDE see NGN500

α,α,α-TRIFLUORO-m-NITROTOLUENE see NFJ500

2,2,2-TRIFLUORO-N-(9-OXOFLUOREN-2-YL)ACETAMIDE see TKH000

TRIFLUOROPERAZINE DIHYDROCHLORIDE see TKK250

TRIFLUOROPHOSPHINE see PHQ500

1,1,1-TRIFLUOROPROPENE see TKH015

3,3,3-TRIFLUORO-1-PROPENE see TKH015

3,3,3-TRIFLUOROPROPENE see TKH015

3,3,3-TRIFLUOROPROPIONALDEHYDE see TKH020

TRIFLUOROPROPIONALDEHYDE see TKH020

N,N,N′-TRIFLUOROPROPIONAMIDINE see TKH025

3,3,3-TRIFLUOROPROPYLENE see TKH015

3,3,3-TRIFLUOROPROPYNE see TKH030

17-(3,3,3-TRIFLUORO-1-PROPYNYL)ESTRA-1,3,5(10)-TRIEN-3,17-β-DIOL see TKH050

TRIFLUOROPYRAZIN DIHYDROCHLORIDE see TKK250

TRIFLUORO SELENIUM HEXAFLUORO ARSENATE see TKH250

TRIFLUOROSTANNITE HEXADECYLAMINE see TKH300

TRIFLUOROSTANNITE OF HEXADECYLAMINE see TKH300

TRIFLUOROSTIBINE see AQE000

α-β,β-TRIFLUOROSTYRENE see TKH310

α,α,α-TRIFLUOROTHYMIDINE see TKH325

TRIFLUOROTHYMIDINE see TKH325

α,α,α-TRIFLUOROTOLUENE see BDH500

ω-TRIFLUOROTOLUENE see BDH500

α-α-α-TRIFLUOROTOLUENE-3,5-DIAMINE see TKB775

α,α,α-TRIFLUORO-m-TOLUIC ACID THALLIUM(I) SALT see TKH500

α,α,α-TRIFLUORO-p-TOLUIDINE see TKB750

N-(α,α,α-TRIFLUORO-m-TOLYL)ANTHRANILIC ACID see TKH750

N-(α,α,α-TRIFLUORO-m-TOLYL)ANTHRANILIC ACID BUTYL ESTER see BRJ325

4-(2-(α-(α,α,α-TRIFLUORO-m-TOLYL)BENZYLOXY)ETHYL)MORPHOLINE FUMA-RATE see TKI000

3-(α-(α,α,α-TRIFLUORO-p-TOLYL)BENZYLOXY)TROPANE FUMARATE see TKI750

(α,α,α-TRIFLUORO-m-TOLYL) ISOCYANATE see TKJ250

2-(α,α,α-TRIFLUORO-m-TOLYL)MORPHOLINE see TKJ500

5-(((α,α,α-TRIFLUORO-m-TOLYL)OXY)METHYL)-2-OXAZOLIDINETHIONE see TKJ750

5-(α,α,α-TRIFLUORO-m-TOLYOXYMETHYL)-2-OXAZOLIDINETHIONE see TKJ750

2,4,6-TRIFLUORO-s-TRIAZINE see TKK000

3,3,3-TRIFLUORO-2-(TRIFLUOROMETHYL)PROPENE see HDC450

1,3,5-TRIFLUOROTRINITROBENZENE see TKK050

TRIFLUOROVINYLBROMIDE see BOJ000

TRIFLUOROVINYL CHLORIDE see CLQ750

TRIFLUORURE de CHLORE (FRENCH) see CDX750

TRIFLUPERAZINE see TKE500

TRIFLUPERAZINE DIHYDROCHLORIDE see TKK250

TRIFLUPERIDOL see TKK500

TRIFLUPERIDOL HYDROCHLORIDE see TKK750

TRIFLUPERIDOLO (ITALIAN) see TKK500

TRIFLUPROMAZINE see TKL000

TRIFLURALIN see DUV600

TRIFLURALINA 600 see DUV600

TRIFLURALINE see DUV600

TRIFLURIDINE see TKH325

TRIFOCIDE see DUS700

TRIFOLEX see CLN750

TRIFORINE see TKL100

TRIFORINE and SODIUM NITRITE see SIT800

TRIFORMOL see PAI000

TRIFORMYL-STROSPESIDE see TKL175

TRIFOSFAMIDE see TNT500

TRIFRINA see DUS700

TRIFTAZIN see TKK250

TRIFUNGOL see FAS000

TRIFUREX see DUV600

TRIGEN see TJQ000

TRIGLYCERIDE HYDROLASE see GGA800

TRIGLYCERIDE LIPASE see GGA800

TRIGLYCIDYL CYANURATE see TKL250

α-TRIGLYCIDYL ISOCYANURATE see TBC450

TRIGLYCINE see AMT500

TRIGLYCOL see TJQ000

TRIGLYCOL, DIACETATE see EJB500

TRIGLYCOL DICAPROATE see TJQ250

TRIGLYCOL DICHLORIDE see TKL500

TRIGLYCOL DIHEXOATE see TJQ250

TRIGLYCOLLAMIC ACID see AMT500

TRIGLYCOL MONOBUTYL ETHER see TKL750

TRIGLYCOL MONOETHYL ETHER see EFL000

TRIGLYCOL MONOMETHYL ETHER see TJQ750

TRIGLYME see TKL875

TRIGONOX 40 see PBL600

TRIGONOX 25/75 see BSD250

TRIGONOX 101-101/45 see DRJ800

TRIGONOX A-75 (CZECH) see BRM250

TRIGONOX 25-C75 see BSD250

TRIGONOX C see BSC500

TRIGONOX F-C50 see BSC250

TRIGONOX HM 80 see IJB100

TRIGONYL see TKX000

TRIGOSAN see ABU500

TRIHERBICIDE CIPC see CKC000

TRIHERBIDE see CBM000

TRIHERBIDE-IPC see CBM000

TRIHEXANOIN see GGK000

TRIHEXANOYLGLYCEROL see GGK000

TRI-n-HEXYL BORATE see TKM000

TRIHEXYL BORATE see TKM000

TRIHEXYLENE GLYCOL BIBORATE see TKM250

TRIHEXYLPHENIDYL see PAL500

TRIHEXYLPHENIDYL HYDROCHLORIDE see BBV000

TRI-n-HEXYLPHOSPHINE OXIDE see TKM500

TRIHEXYLTIN ACETATE see ABX000

TRI-N-HEXYLZINNACETAT (GERMAN) see ABX000

TRIHISTAN see CDR000

TRIHYDRATED ALUMINA see AHC000

TRIHYDRAZINECOBALT(II) NITRATE see TKM750

TRIHYDRAZINENICKEL(II) NITRATE see TKN000

TRIHYDROGENPYROPHOSPHATE(ESTER)THIAMINE see TET750

2,3,4-TRIHYDROXYACETOPHENONE see TKN250

3,7,15-TRIHYDROXY-4-ACETOXY-8-OXO-12,13-EPOXY-Δ⁹-TRICHOTHECENE see FQR000

1,8,9-TRIHYDROXYANTHRACENE see APH250

1,2,3-TRIHYDROXY-9,10-ANTHRACENEDIONE see TKN500

1,2,4-TRIHYDROXY-9,10-ANTHRACENEDIONE see TKN750

1,2,4-TRIHYDROXYANTHRACHINON (CZECH) see TKN750

1,2,3-TRIHYDROXYANTHRAQUINONE see TKN500

1,2,4-TRIHYDROXYANTHRAQUINONE see TKN750

2,3,4-TRIHYDROXYBENZALDEHYDE see TKN800

1,2,3-TRIHYDROXYBENZEN (CZECH) see PPQ500

1,2,3-TRIHYDROXYBENZENE see PPQ500

1,2,4-TRIHYDROXYBENZENE see BBU250

1,3,5-TRIHYDROXYBENZENE see PGR000

s-TRIHYDROXYBENZENE see PGR000

sym-TRIHYDROXYBENZENE see PGR000

3,4,5-TRIHYDROXYBENZENE-1-PROPYLCARBOXYLATE see PNM750

2,4,6-TRIHYDROXYBENZOIC ACID see TKO000

3,4,5-TRIHYDROXYBENZOIC ACID see GBE000

3,4,5-TRIHYDROXYBENZOIC ACID, n-PROPYL ESTER see PNM750

2-(2,3,4-TRIHYDROXYBENZYL)HYDRAZIDE SERINE MONOHYDROCHLORIDE, dl- see SCA400

3-β,14,16-β-TRIHYDROXY-5-β-BUFA-20,22-DIENOLIDE see BOM655

3-β,14,16-β-TRIHYDROXY-5-β-BUFA-20,22-DIENOLIDE-16-ACETATE see BON000

2′,4′,5′-TRIHYDROXYBUTYROPHENONE see TKO250

2,4,5-TRIHYDROXYBUTYROPHENONE see TKO250

3-β,12-β,14-TRIHYDROXY-CARD-20(22)-ENOLIDE see DKN300

3-β,12-β,14-TRIHYDROXY-5-β-CARD-20(22)-ENOLIDE-3,12-DIFORMATE see FNK075

3β,12β,14β-TRIHYDROXY-5-β-CARD-20(22)-ENOLIDE-3-(4″)-o-(METHYL-TRIDIGI-TOXOSIDE) see MJD300

3-β,12-β,14-β-TRIHYDROXY-5-β-CARD-20(22)-ENOLIDE-3-(R‴-O-METHYLTRIDI-GITOXOSIDE), ACETONE (2:1) see MJD500

TRIHYDROXY 3-7-12 CHOLANATE de Na see SFW000

3-α,7-α,12-α-TRIHYDROXY-5-β-CHOLAN-24-OIC ACID see CME750

3,7,12-TRIHYDROXY-CHOLAN-24-OIC ACID (3-α,5-β,7-α,12-α) see CME750

TRILOSTANE see EBY600
TRILUDAN see TAI450
TRIM see DUV600
TRIMAGNESIUM PHOSPHATE see MAH780
TRIMANGANESE TETRAOXIDE see MAU800
TRIMANGANESE TETROXIDE see MAU800
TRIMANGOL 80 see MAS500
TRIMANGOL see MAS500
TRIMANYL see TKZ000
TRIMAR see TIO750
TRIMATON see SIL550, VFU000
TRI-ME see MQH750
TRIMEBUTINE see TKU650
TRIMEBUTINE MALEATE see TKU675
TRIMECAINE see DHL800
TRIMECAINE HYDROCHLORIDE see DHL800
TRIMEDAL see TLP750
TRIMEDLURE see BQT600
TRIMEDONE see TLP750
TRIMEDOXIME DICHLORIDE see TLQ750
TRIMEGLAMIDE see TIH800
TRIMEKAIN HYDROCHLORIDE see DHL800
TRIMEKS see MMN250
TRIMELARSEN see PLK800
TRIMELLIC ACID-1,2-ANHYDRIDE see TKV000
TRIMELLIC ACID ANHYDRIDE see TKV000
TRIMELLITIC ACID see TKU700
TRIMELLITIC ACID CYCLIC-1,2-ANHYDRIDE see TKV000
TRIMELLITIC ANHYDRIDE see TKV000
TRIMEPERDINE see IMT000
TRIMEPRAMINE HYDROCHLORIDE see TAL000
TRIMEPRAZINE see AFL500
TRIMEPRIMINA (ITALIAN) see DLH200
TRIMEPRIMINE MALEATE see SOX550
TRIMEPRIMINE MONOMALEATE see SOX550
TRIMEPROPIMINE see DLH200
TRIMEPROPIMINE HYDROCHLORIDE see TAL000
TRIMEPROPRIMINE HYDROCHLORIDE see TAL000
2,4,6-TRIMERCAPTO-S-TRIAZINE see THS250
1,3,5-TRIMERCAPTOTRIAZINE see THS250
TRIMERCURY DINITRIDE see TKW000
TRIMERESURUS FLAVOVIRIDIS VENOM see TKW100
TRIMESULF see TKX000
TRIMETADIONE see TLP750
TRIMETAPHAN CAMPHORSULFONATE see TKW500
TRIMETAPHAN CAMSILATE see TKW500
TRIMETAPHOSPHATE SODIUM see SKM500
TRIMETAZIDINE DIHYDROCHLORIDE see YCJ200
TRIMETAZIDINE HYDROCHLORIDE see YCJ200
TRIMETHADIONE see TLP750
TRIMETHAPHAN-10-CAMPHORSULFONATE see TKW500
TRIMETHAPHAN CAMPHORSULFONATE see TKW500
TRIMETHAPHAN CAMSYLATE see TKW500
TRIMETHIDINIUM BIMETHOSULFATE see TLG000
TRIMETHIDINIUM METHOSULFATE see TLG000
TRIMETHIN see TLP750
TRIMETHIOPHANE see TKW500
TRIMETHOATE see IOT000
TRIMETHOBENZAMIDE HYDROCHLORIDE see TKW750
TRIMETHOPRIM see TKZ000
TRIMETHOPRIMSULFA see TKX000
TRIMETHOPRIM and SULFAMOXOLE see SNL850
TRIMETHOPRIM-SULFAMOXOLE mixture see SNL850
TRIMETHOPRIM and SULPHAMETHOXAZOLE see TKX000
TRIMETHOPRIOM see TKZ000
TRIMETHOQUINOL see IDD100
(±)-TRIMETHOQUINOL see TKX125
(−)-TRIMETHOQUINOL see IDD100
TRIMETHOXAZINE see TKX250
3,4,5-TRIMETHOXPHENETHYLAMINE SULFATE see MDJ000
3,4,5-TRIMETHOXYAMPHETAMINE see MLH250
3,4,5-TRIMETHOXYAMPHETAMINE HYDROCHLORIDE see TKX500
1,2,10-TRIMETHOXY-6a-α-APORPHIN-9-OL see TKX700
3,4,5-TRIMETHOXYBENZAMIDE see TKY000
epsilon-(3,4,5-TRIMETHOXYBENZAMIDO)CAPROIC ACID SODIUM SALT see
    CBF625
epsilon-(3,4,5-TRIMETHOXYBENZAMIDO)CAPRONSAEURE NATRIUM (GER-
    MAN) see CBF625
5-(3,4,5-TRIMETHOXYBENZAMIDO)-2-METHYL-trans-DECAHYDROISOQUINO-
    LINE see DAE700
trans-9,10-t-5-H-5-(3,4,5-TRIMETHOXYBENZAMIDO)-2-METHYL DECAHYDROISO-
    QUINOLINE see DAE700
cis-5,8,10-H-5-(3,4,5-TRIMETHOXYBENZAMIDO)-2-METHYL DECAHYDROISO-
    QUINOLINE see DAE695
1,3,5-TRIMETHOXYBENZENE see TKY250

3,4,5-TRIMETHOXYBENZENEETHANAMINE see MDI500
3,4,5-TRIMETHOXY-BENZOIC ACID 2-(4-(3-(2-CHLOROPHENOTHIAZIN-10-
    YL)PROPYL)-1-PIPERAZINYL)ETHYL ESTER, DIFUMARATE see MDU750
3,4,5-TRIMETHOXYBENZOIC ACID 3-((3,3-DIPHENYLPROPYL)AMINO)PROPYL
    ESTER HYDROCHLORIDE see DWK700
3,4,5-TRIMETHOXYBENZOIC ACID, METHYL ESTER see MQF300
6-((3,4,5-TRIMETHOXYBENZOYL)AMINO)HEXANOIC ACID SODIUM SALT see
    CBF625
N-(3,4,5-TRIMETHOXYBENZOYL)GLYCINE DIETHYLAMIDE see TIH800
3,4,5-TRIMETHOXYBENZOYL METHYL RESERPATE see RDK000
4-(3,4,5-TRIMETHOXYBENZOYL)MORPHOLINE see TKX250
N-(3,4,5-TRIMETHOXYBENZOYL)MORPHOLINE see TKX250
1-(3,4,5-TRIMETHOXYBENZOYL)-4-(2-PYRIDYL)PIPERAZINE see TKY300
N-(3,4,5-TRIMETHOXYBENZOYL)TETRAHYDRO-1,4-OXAZINE see TKX250
3,4,5-TRIMETHOXYBENZOYL-N-TETRAHYDROXAZINE see TKX250
3,4,5-TRIMETHOXY-N-BENZOYLTETRAHYDROXAZINE see TKX250
5-(3,4,5-TRIMETHOXYBENZYL)-2,4-DIAMINOPYRIMIDINE see TKZ000
dl-1-(3,4,5-TRIMETHOXYBENZYL)-6,7-DIHYDROXY-1,2,3,4-TETRAHYDROISOQUI-
    NOLINE HYDROCHLORIDE see TKX125
1-1-(3,4,5-TRIMETHOXYBENZYL)-6,7-DIHYDROXY-1,2,3,4-TETRAHYDROISOQUI-
    NOLINE HYDROCHLORIDE see IDD100
1-(2,3,4-TRIMETHOXYBENZYL)PIPERAZINE DIHYDROCHLORIDE see YCJ200
TRIMETHOXYBORINE see TLN000
TRIMETHOXYBOROXIN see TKZ100
TRIMETHOXYBOROXINE see TKZ100
1,1,3-TRIMETHOXYBUTANE see TLA000
3,4,5-TRIMETHOXYCINNAMALDEHYDE see TLA250
3,4,5-TRIMETHOXYCINNAMIC ACID see CMQ100
4-(3′,4′,5′-TRIMETHOXYCINNAMOYL)-1-(ETHOXYCARBONYLEME-
    THYL)PIPERAZINE HYDROCHLORIDE see TLA500
(3′,4′,5′-TRIMETHOXYCINNAMOYL)-1-(N-ISOPROPYL AMINO CARBONYL
    METHYL)-4 PIPERAZINE MALEATE see IRQ000
TRIMETHOXYCINNAMOYL METHYL RESERPATE see TLN500
4-(3,4,5-TRIMETHOXYCINNAMOYL)-1-PIPERAZINEACETIC ACID ETHYL ESTER
    MALEATE see TLA525
4-(3,4,5-TRIMETHOXYCINNAMOYL)-1-PIPERAZINEACETIC ACID ETHYL ESTER
    HYDROCHLORIDE see TLA500
((TRIMETHOXY-3′,4′,5′ CINNAMOYL)-4 PIPERAZINYL)-2 ACETATE D'ETHYLE
    (MALEATE) (FRENCH) see TLA525
1-(4-((3′,4′,5′-TRIMETHOXYCINNAMOYL)-1-PIPERAZINYL)ACETYL)PYRROLIDINE
    MALEATE see VGK000
4-(3,4,5-TRIMETHOXYCINNAMOYL)-1-(1-PYRROLIDI-
    NYL)CARBONYLMETHYLPIPERAZINE MALEATE see VGK000
TRI-(2-METHOXYETHANOL)PHOSPHATE see TLA600
TRIMETHOXYFOSFIN see TMD500
TRIMETHOXYMETHANE see TLX600
3,4,5-TRIMETHOXY-α-METHYL-β-PHENYLETHYLAMINE HYDROCHLORIDE see
    TKX500
TRIMETHOXYMETHYLSILANE see MQF500
1,2,10-TRIMETHOXY-6a-α-NORAPORPHIN-6-OL see LBO200
3,4,5-TRIMETHOXYPHENETHYLAMINE see MDI500
3,4,5-TRIMETHOXYPHENETHYLAMINE HYDROCHLORIDE see MDI750
3,4,5-TRIMETHOXYPHENYLACRYLIC ACID see CMQ100
1-(3,4,5-TRIMETHOXYPHENYL)-2-AMINOPROPANE see TKX500
3,4,5-TRIMETHOXYPHENYL-β-AMINOPROPANE see MLH250
TRIMETHOXYPHENYL-β-AMINOPROPANE see MLH250
3,4,5-TRIMETHOXY-β-PHENYLETHYLAMINE HYDROCHLORIDE see MDI750
1-((2,3,4-TRIMETHOXYPHENYL)METHYL)PIPERAZINE DIHYDROCHLORIDE (9CI)
    see YCJ200
5-((3,4,5-TRIMETHOXYPHENYL)-METHYL)-2,4-PYRIMIDINEDIAMINE see TKZ000
3-(3,4,5-TRIMETHOXYPHENYL)-2-PROPENAL see TLA250
3-(3,4,5-TRIMETHOXYPHENYL)-2-PROPENOIC ACID see CMQ100
TRIMETHOXYPHOSPHINE see TMD500
TRIMETHOXYSILANE (DOT) see TLB750
TRIMETHOXY SILANE see TLB750
2-(TRIMETHOXYSILYL)ETHANETHIOL see MCO250
(TRIMETHOXYSILYL)ETHENE see TLD000
TRIMETHOXYSILYLPROPANETHIOL see TLC000
3-(TRIMETHOXYSILYL)-1-PROPANOL METHACRYLATE see TLC250
TRIMETHOXYSILYL-3-PROPYLESTER KYSELINY METHAKRYLOVE (CZECH)
    see TLC250
3-(TRIMETHOXYSILYL)PROPYL ESTER METHACRYLIC ACID see TLC250
N-(3-TRIMETHOXYSILYLPROPYL)-ETHYLENEDIAMINE see TLC500
4-(3,4,5-TRIMETHOXYTHIOBENZOYL)MORPHOLINE see TNP275
4-(3,4,5-TRIMETHOXYTHIOBENZOYL)TETRAHYDRO-1,4-OXAZINE see TNP275
3,4,5-TRIMETHOXY-α-VINYLBENZYL ALCOHOL ACETATE see TLC850
TRIMETHOXYVINYLSILANE see TLD000
2′,4′,6′-TRIMETHYLACETANILIDE see TLD250
2,4,6-TRIMETHYLACETANILIDE see TLD250
TRIMETHYLACETHYDRAZIDE AMMONIUM CHLORIDE see GEQ500
TRIMETHYLACETIC ACID see PJA500
TRIMETHYLACETONITRILE see PJA750
TRIMETHYL-β-ACETOXYPROPYLAMMONIUM CHLORIDE see ACR000
TRIMETHYLACETYL AZIDE see PJB000
TRIMETHYL ACETYL CHLORIDE (DOT) see DTS400

3,5,5-TRIMETHYL-2-CYCLOHEXEN-1-ON (GERMAN, DUTCH) see IMF400
4-(2,6,6-TRIMETHYL-2-CYCLOHEXEN-1-YL)-2-BUTANONE see DLP800
4-(2,6,6-TRIMETHYL-1-CYCLOHEXEN-1-YL)-3-BUTEN-2-OL see IFT500
4-(2,6,6-TRIMETHYL-2-CYCLOHEXEN-1-YL)-3-BUTEN-2-OL see IFT400
4-(2,6,6-TRIMETHYL-2-CYCLOHEXEN-1-YL)-3-BUTEN-2-OL ACETATE see IFX300
4-(2,6,6-TRIMETHYL-1-CYCLOHEXEN-1-YL)-3-BUTEN-2-ONE see IFX000
4-(2,6,6-TRIMETHYL-2-CYCLOHEXEN-1-YL)-3-BUTEN-2-ONE see IFW000
4-(2,4,6-TRIMETHYL-3-CYCLOHEXEN-1-YL)-3-BUTEN-2-ONE see IGJ600
1-(2,6,6-TRIMETHYL-2-CYCLOHEXEN-1-YL)-1,6-HEPTADIEN-3-ONE see AGI500
1-(2,6,6-TRIMETHYL-2-CYCLOHEXEN-1-YL)-3-PENTANONE see TLO530
3,5,5-TRIMETHYLCYCLOHEXYL AMYGDALATE see DNU100
3,3,5-TRIMETHYLCYCLOHEXYL DIPROPYLENE GLYCOL see TLO600
3,3,5-TRIMETHYLCYCLOHEXYL MANDELATE see DNU100
11,12-17-TRIMETHYLCYCLOPENTA(a)PHENANTHRENE see TLP000
3,8,13-TRIMETHYLCYCLOQUINAZOLINE see TLP250
3,7,9-TRIMETHYL-1,6-DECADIEN-3-OL see IIW100
TRIMETHYLDIBORANE see TLP275
2,2,4-TRIMETHYL-1,2-DIHYDROCHINOLIN see TLP500
11,12,17-TRIMETHYL-16,17-DIHYDRO-15H-CYCLOPENTA(a)PHENANTHRENE see DMF600
2,2,4-TRIMETHYL-1,2-DIHYDROQUINOLINE see TLP500
TRIMETHYL-1,2-DIHYDROQUINOLINE see TLP500
2,2,4-TRIMETHYL-1,2-DIHYDROQUINOLINE POLYMER see PJQ750
TRIMETHYLDIHYDROQUINOLINE POLYMER see PJQ750
3,3,5-TRIMETHYL-2,4-DIKETOOXAZOLIDINE see TLP750
TRIMETHYL(2-(2,6-DIMETHYLPHENOXY)PROPYL)AMMONIUM CHLORIDE MONOHYDRATE see TLQ000
N,N,N-TRIMETHYL-1,3-DIOXOLANE-4-METHANAMINIUM IODIDE see FMX000
1,3,7-TRIMETHYL-2,6-DIOXOPURINE see CAK500
1-(1,3,7-TRIMETHYL-2,6-DIOXOPURIN-8-YL)-4-(2-PHENYL-1-METHYL)ETHYL-4-METHYL-ETHYLENEDIAMINE, HYDROCHLORIDE see FAM000
3,7,11-TRIMETHYL-2,4-DODECADIENOIC ACID 2-PROPYNYL ESTER see POB000
3,7,11-TRIMETHYL-2,6,10-DODECATRIEN-1-OL see FAB800
3,7,11-TRIMETHYL-1,6,10-DODECATRIEN-3-YL ACETATE see NCN800
TRIMETHYLEENTRINITRAMINE (DUTCH) see CPR800
1,3-TRIMETHYLEN-BIS-(4-HYDROXIMINOFORMYLPYRIDINIUM)-DIBROMID (GERMAN) see TLQ500
N,N-TRIMETHYLEN-BIS-(PYRIDINIUM-4-ALDOXIM)-DIBROMID (GERMAN) see TLQ500
TRIMETHYLENE see CQD750
5:10-TRIMETHYLENE-1:2-BENZANTHRACENE see DKT400
1,1'-TRIMETHYLENEBIS(4-FORMYLPYRIDINIUM BROMIDE)DIOXIME see TLQ500
1,1'-TRIMETHYLENEBIS(4-FORMYLPYRIDINIUM CHLORIDE) DIOXIME see TLQ750
1,1'-TRIMETHYLENEBIS(4-FORMYLPYRIDINIUM) DIOXIME DICHLORIDE see TLQ750
1,1'-TRIMETHYLENEBIS(4-(HYDROXYIMINOMETHYL)PYRIDINIUM CHLORIDE) see TLQ750
N,N-TRIMETHYLENE BIS(PYRIDINIUM-4-ALDOXIME)DICHLORIDE see TLQ750
TRIMETHYLENE BROMIDE see TLR000
TRIMETHYLENE BROMIDE CHLORIDE see BNA825
TRIMETHYLENE CHLOROBROMIDE see BNA825
TRIMETHYLENE CHLOROHYDRIN see CKP725
1:12-TRIMETHYLENECHRYSENE see DKT400
TRIMETHYLENEDIAMINE see PMK500
TRIMETHYLENE DIBROMIDE see TLR000
TRIMETHYLENE DICHLORIDE see DGF800
TRIMETHYLENE DIIODIDE see TLR050
TRIMETHYLENE DIMERCAPTAN see PML350
TRIMETHYLENEDIMETHANESULFONATE see TLR250
TRIMETHYLENE DIMETHANESULPHONATE see TLR250
1,3-TRIMETHYLENEDINITRILE see TLR500
4,4'-(TRIMETHYLENEDIOXY)DIBENZAMIDINE DIHYDROCHLORIDE see DBM400
TRIMETHYLENEDITHIOGLYCOL see PML350
TRIMETHYLENEDITHIOL see PML350
TRIMETHYLENEETHYLENEDIAMINE see HGI900
TRIMETHYLENEFUROXAN see CPW325
TRIMETHYLENE GLYCOL see PML250
exo-5,6-TRIMETHYLENENORBORNANE see TLR675
exo-TRIMETHYLENENORBORNANE see TLR675
TRIMETHYLENE OXIDE see OMW000
TRIMETHYLENE SULFATE see TLR750
N,N'-TRIMETHYLENETHIOUREA see TLS000
TRIMETHYLENETRINITRAMINE see CPR800
sym-TRIMETHYLENETRINITRAMINE see CPR800
TRIMETHYLENE TRISULFIDE see TLS500
TRIMETHYLENOXID (GERMAN) see OMW000
TRIMETHYLENTRISULFID (CZECH) see TLS500
TRIMETHYLESTER KYSELINY BORITE see TLN000
N,N,N-TRIMETHYLETHANAMINIUM IODIDE see EQC600
N,N,N-TRIMETHYLETHENAMINIUM HYDROXIDE see VQR300
2,2,4-TRIMETHYL-6-ETHOXY-1,2-DIHYDROQUINOLINE see SAV000

TRIMETHYLETHYLENE OXIDE see TLY175
TRIMETHYLFOSFIT see TMD500
N,N,N-TRIMETHYL-2-FURANMETHANAMINIUM IODIDE see FPY000
TRIMETHYLFURFURYLAMMONIUM IODIDE see FPY000
TRIMETHYLGALLIUM see TLT000
TRIMETHYLGERMYL PHOSPHINE see TLT100
TRIMETHYLGLYCIDYLAMMONIUM CHLORIDE see GGY200
TRIMETHYLGLYCINE see GHA050
TRIMETHYLGLYCOCOLL see GHA050
TRIMETHYL GLYCOL see PML000
TRIMETHYLHARNSTOFF (GERMAN) see TMJ250
N,N,N-TRIMETHYL-1-HEXADECANAMINIUM BROMIDE see HCQ500
TRIMETHYLHEXADECYLAMMONIUM BROMIDE see HCQ500
1,3,5-TRIMETHYLHEXAHYDRO-1,3,5-TRIAZINE see HDW100
1,3,5-TRIMETHYLHEXAHYDRO-sym-TRIAZINE see HDW100
3,5,5-TRIMETHYLHEXANAL see ILJ100
3,5,5-TRIMETHYLHEXANOIC ACID ALLYL ESTER see AGU400
N-1,5-TRIMETHYL-4-HEXENYLAMINE see ILK000
3,5,5-TRIMETHYLHEXYL ACETATE see TLT500
3,5,5-TRIMETHYLHEXYL ACETIC ACID see TLT500
TRIMETHYLHYDRAZINE HYDROCHLORIDE see TLT750
TRIMETHYLHYDROXYLAMMONIUM PERCHLORATE see TLE500
TRIMETHYL INDIUM see TLT775
TRIMETHYLIODOMETHANE see TLU000
TRI(METHYLISOBUTYLCARBINYL) BORATE see BMC750
TRIMETHYL-ε-LACTONE (mixed isomers) see TLY000
TRIMETHYL LEAD CHLORIDE see TLU175
TRIMETHYL((2-METHYL-1,3-DIOXOLAN-4-YL)METHYL)AMMONIUM IODIDE (8CI) see MJH250
1,3,3-TRIMETHYL-2-METHYLENEINDOLINE see TLU200
4,11,11-TRIMETHYL-8-METHYLENE-5-OXATRICYCLO(8.2.0.0(4,6))DODECANE see CCN100
TRIMETHYL(1-METHYL-2-(10-PHENOTHIAZINYL)ETHYL)AMMONIUM METHYL SULFATE see MRW000
TRIMETHYL (1-METHYL-2-PHENOTHIAZIN-10-YLETHYL)AMMONIUM METHYL SULFATE see MRW000
N,N,N'-TRIMETHYL-N'-(5-METHYL-3-PHENYL-1H-INDOL-1-YL)-1,2-ETHANEDIA-MINE HYDROCHLORIDE see MNU750
N,N,N'-TRIMETHYL-N'-(3-PHENYL-1H-INDOL-1-YL)-1,2-ETHANEDIAMINE MONO-HYDROCHLORIDE see BGC500
N,N,2-TRIMETHYL-4-(4'-(3'-METHYLPYRIDYL-1'-OXIDE)AZO)ANILINE see DQE000
N,N,3-TRIMETHYL-4-(4'-(3'-METHYLPYRIDYL-1'-OXIDE)AZO)ANILINE see DQE200
$N^2,N^4,N^6$-TRIMETHYLMELAMINE see TNJ825
TRIMETHYLMYRISTYLAMMONIUM BROMIDE see TCB200
1,3,3-TRIMETHYL-6'-NITROINDOLINE-2-SPIRO-2'-BENZOPYRAN see TLU500
TRIMETHYLNITROSOHARNSTOFF (GERMAN) see TLU750
N,2,2-TRIMETHYL-N-NITROSO-1-PROPYLAMINE see NKU570
1,1,3-TRIMETHYL-3-NITROSOUREA see TLU750
N-TRIMETHYL-N-NITROSOUREA see TLU750
2,6,8-TRIMETHYLNONANOL-4 see TLV000
2,6,8-TRIMETHYL-4-NONANOL see TLV000
TRIMETHYL NONANONE see TLV250
2,6,8-TRIMETHYLNONYL VINYL ETHER see VQU000
(—)-endo-1,3,3-TRIMETHYL-2-NORBORNANOL see TLW000
1,3,3-TRIMETHYL-2-NORBORNANOL ACETATE see FAO000
1,3,3-TRIMETHYL-2-NORBORNANONE see TLW250
d-1,3,3-TRIMETHYL-2-NORBORNANONE see FAM300
1,3,3-TRIMETHYL-2-NORBORNANYL ACETATE see FAO000
1,3,3-TRIMETHYL-2-NORCAMPHANONE see TLW250
d-1,3,3-TRIMETHYL-2-NORCAMPHANONE see FAM300
1,7,7-TRIMETHYLNORCAMPHOR see CBA750
3,7,7-TRIMETHYL-3-NORCARENE see CCK500
4,7,7-TRIMETHYL-3-NORCARENE see CCK500
TRIMETHYLOCTADECYLAMMONIUM CHLORIDE see TLW500
TRIMETHYLOLAMINOMETHANE see TEM500
TRIMETHYLOLMELAMINE see TLW750
TRIMETHYLOLNITROMETHANE see HMJ500
TRIMETHYLOL NITROMETHANE TRINITRATE (DOT) see NHK650
1,1,1-TRIMETHYLOLPROPANE see HDF300
TRIMETHYLOLPROPANE see HDF300
TRIMETHYLOLPROPANE DIALLYL ETHER see TLX000
TRIMETHYLOLPROPANE ETHOXYTRIACRYLATE see TLX100
TRIMETHYLOLPROPANE TRIACRYLATE see TLX175
TRIMETHYLOLPROPANE TRIMETHACRYLATE see TLX250
TRIMETHYLOLPROPANE TRIMETHANCRYLATE see TLX250
TRIMETHYLOPROPANE MONOALLYL ETHER see TLX110
TRIMETHYL ORTHOFORMATE see TLX600
1,3,3-TRIMETHYL-2-OXABICYCLO(2.2.2)OCTANE see CAL000
TRIMETHYLOXACYCLOPROPANE see TLY175
3,5,5-TRIMETHYL-2,4-OXAZOLIDINEDIONE see TLP750
TRIMETHYL-2-OXEPANONE (mixed isomers) see TLY000
2,2,3-TRIMETHYLOXIRANE see TLY175
TRIMETHYLOXIRANE see TLY175

TRIMOSULFA see TKX000
TRIMPEX see TKZ000
TRIMSTAT see DKE800
TRIMTABS see DKE800
TRIMUSTINE see TNF500
TRIMUSTINE HYDROCHLORIDE see TNF500
TRIMYSTEN see MRX500
TRINAGLE see CNP250
TRINALGON see NGY000
TRINATRIUMPHOSPHAT (GERMAN) see SJH200
TRINEX see TIQ250
TRINITRIN see NGY000
TRINITROACETONITRILE see TMK250
TRINITROANILINE (DOT) see PIC800
2,4,6-TRINITROANISOLE see TMK300
TRINITROBENZEEN see TMK500
1,3,5-TRINITROBENZENE see TMK500
TRINITROBENZENE see TMK500
2,3,5-TRINITROBENZENEDIAZONIUM-4-OXIDE see TMK775
2,4,6-TRINITROBENZENE-1,3-DIOL see SMP500
2,4,6-TRINITRO-1,3-BENZENEDIOL see SMP500
2,4,6-TRINITROBENZENE-1,3,5-TRIOL see TMM775
TRINITROBENZENE, wetted with not <30% water, by weight (UN 1354)
 (DOT) see TMK500
TRINITROBENZENE, dry or wetted with <30% water, by weight (UN 0214)
 (DOT) see TMK500
TRINITROBENZOIC ACID (dry) see TML000
TRINITROBENZOIC ACID, wetted with not <30% water, by weight (UN 1355)
 (DOT) see TML000
TRINITROBENZOIC ACID, dry or wetted with <30% water, by weight (UN
 0215) (DOT) see TML000
TRINITROBENZOL (GERMAN) see TMK500
TRINITROCHLOROBENZENE see TML325
2,4,6-TRINITRO-m-CRESOL see TML500
TRINITRO-m-CRESOL see TML500
TRINITRO-m-CRESOLIC ACID see TML500
TRINITROCYCLOTRIMETHYLENE TRIAMINE see CPR800
2,4,6-TRINITRO-1,3-DIMETHYL-5-tert-BUTYLBENZENE see TML750
2,4,6-TRINITRO-3,5-DIMETHYL-tert-BUTYLBENZENE see TML750
2,2,2-TRINITROETHANOL see TMM000
TRINITROETHANOL (DOT) see TMM000
2,4,6-TRINITROFENOL (DUTCH) see PID000
2,4,6-TRINITROFENOLO (ITALIAN) see PID000
2,4,7-TRINITROFLUOREN-9-ONE see TMM250
2,4,7-TRINITRO-9-FLUORENONE see TMM250
2,4,7-TRINITROFLUORENONE (MAK) see TMM250
TRINITROGLYCERIN see NGY000
TRINITROGLYCEROL see NGY000
TRINITROL see NGY000
TRINITROMETACRESOL see TML500
TRINITROMETHANE see TMM500
1,3,5-TRINITRONAPHTHALENE see TMM600
1,3,5-TRINITROPHENOL see PID000
2,4,6-TRINITROPHENOL AMMONIUM SALT see ANS500
2,4,6-TRINITROPHENOL COMPOUND with 1,2-DIHYDRO-3-METHYL-
 BENZ(j)ACEANTHRYLENE see MIL750
TRINITROPHENOL, wetted with not <30% water, by weight (UN 1344) (DOT)
 see PID000
TRINITROPHENOL (UN 0154) (DOT) see PID000
2,4,6-TRINITROPHENYL (OSHA) see PID000
2,4,6-TRINITROPHENYL AZIDE see PIE525
2,4,6-TRINITROPHENYLMETHYLNITRAMINE see TEG250
2,4,6-TRINITROPHENYL-N-METHYLNITRAMINE see TEG250
TRINITROPHENYLMETHYLNITRAMINE see TEG250
2,4,6-TRINITROPHENYL NITRAMINE (DOT) see TDY075
TRINITROPHLOROGLUCINOL see TMM775
1,3,6-TRINITROPYRENE see TMN000
TRINITROPYRENE see TMN000
2,4,6-TRINITRORESORCINOL see SMP500
TRINITRORESORCINOL (DOT) see SMP500
TRINITRORESORCINOL, wetted with less than 20% water (DOT) see SMP500
TRINITRORESORCINOL, DRY (DOT) see SMP500
1,3,5-TRINITROSO-1,3,5-TRIAZACYCLOHEXANE see HDV500
TRINITROSOTRIMETHYLENETRIAMINE see HDV500
TRINITROSOTRIMETHYLENTRIAMIN (GERMAN) see HDV500
2,4,6-TRINITROTOLUEEN (DUTCH) see TMN490
2,4,5-TRINITROTOLUENE see TMN400
2,4,6-TRINITROTOLUENE see TMN490
2,4,6-TRINITROTOLUENE (ACGIH,OSHA) see TMN490
TRINITROTOLUENE see TMN490
s-TRINITROTOLUENE see TMN490
sym-TRINITROTOLUENE see TMN490
TRINITROTOLUENE, wetted with not <30% water, by weight (UN 1356)
 (DOT) see TMN490
TRINITROTOLUENE (UN 0209) (DOT) see TMN490

2,4,6-TRINITROTOLUOL (GERMAN) see TMN490
s-TRINITROTOLUOL see TMN490
sym-TRINITROTOLUOL see TMN490
1,3,5-TRINITRO-1,3,5-TRIAZACYCLOHEXANE see CPR800
TRINITROTRIETHANOLAMINE see TJL250
TRINOXOL see DAA800, TAA100, TAH900
TRIOCTADECYL BORATE see TMN750
TRIOCTANOIN see TMO000
TRIOCTANOYLGLYCEROL see TMO000
TRI-n-OCTYLAMINE see DVL000
TRI(2-OCTYL)BORATE see TMO500
TRI-n-OCTYL BORATE see TMO250
TRIOCTYL(BUTYLTHIO)STANNANE see BSO200
TRIOCTYL(ETHYLTHIO)STANNANE see EPR200
TRIOCTYLMETHYLAMMONIUM CHLORIDE see MQH000
TRIOCTYL PHOSPHATE see TNI250
TRIOCTYL PHOSPHITE see TNI300
TRIODURIN see EDU500
TRIOLEIN HYDROLASE see GGA800
TRIOLEYL BORATE see BMC500
TRIOMBRIN see SEN500
TRIONAL see BJT750
TRIOPAC 200 see AAN000
TRIOPAS see AAN000
TRIORTHOCRESYL PHOSPHATE see TMO600
TRIOSSIMETELENE (ITALIAN) see TMP000
TRIOTHYRONE see LGK050
TRIOVEX see EDU500
2,6,7-TRIOXA-1-ARSABICYCLO(2.2.2)OCTANE, 4-ETHYL- see EQD100
2,6,7-TRIOXA-1-ARSABICYCLO(2.2.2)OCTANE, 4-ISOPROPYL- see IRQ100
2,6,7-TRIOXA-1-ARSABICYCLO(2.2.2)OCTANE, 4-METHYL- see MQH100
2,8,9-TRIOXA-5-AZA-1-SILABICYCLO(3.3.3)UNDECANE see TMO750
2,8,9-TRIOXA-5-AZA-1-SILABICYCLO(3.3.3)UNDECANE, 1-ETHENYL- see
 VQK600
2,8,9-TRIOXA-5-AZA-1-SILABICYCLO(3.3.3)UNDECANE, 1-ETHOXY- see
 EFJ600
2,8,9-TRIOXA-5-AZA-1-SILABICYCLO(3.3.3)UNDECANE, 1-METHYL- see MPI650
2,8,9-TRIOXA-5-AZA-1-SILABICYCLO(3.3.3)UNDECANE, 1-VINYL- see VQK600
TRIOXACARCIN C see TMO775
3,6,9-TRIOXADECYLESTER KYSELINY OCTOVE see AAV500
1,3,5-TRIOXANE see TMP000
TRIOXANE see TMP000
s-TRIOXANE see TMP000
sym-TRIOXANE see TMP000
1,3,5-TRIOXANE, 2,4,6-TRIPROPYL- see TNC100
s-TRIOXANE, 2,4,6-TRIPROPYL- see TNC100
TRIOXANONA see PEC250
5,8,11-TRIOXAPENTADECANE see DDW200
3,6,9-TRIOXAUNDECANE see DIW800
3,3'-(3,6,9-TRIOXAUNDECANEDIOYLDIAMINO)BIS(2,4,6-TRIIODOBENZOIC ACID)
 see IGD100
TRIOXAZIN see TKX250
TRIOXIDE(S) see TGG760
TRIOXIFENE MESYLATE see TMP175
3,7,12-TRIOXOCHOLANIC ACID see DAL000
3,7,12-TRIOXO-5-β-CHOLAN-24-OIC ACID see DAL000
3,7,12-TRIOXO-5-β-CHOLAN-24-OIC ACID MONOSODIUM SALT see SGD500
1,2,4-TRIOXOLANE see EJO500
TRIOXOLANES see ORY499
TRIOXON see TAA100
TRIOXONE see BSQ750, TAA100
2,6,8-TRIOXOPURINE see UVA400
(1,5-(2,4,6-TRIOXO-(1H,3H,5H)-PYRIMIDYLENE))BIS((2-METHOXYPRO-
 PYL)HYDROXYMERCURY SODIUM SALT see BKG500
(2,4,6-TRIOXO)-s-TRIAZINETRIYLTRIS(TRIBUTYLSTANNANE) see TMP250
3-β,12,14-TRIOXY-CARDEN-(20:22)-OLID (GERMAN) see DKN300
3-β,12,14-TRIOXY-DIGEN-(20:22)-OLID (GERMAN) see DKN300
TRIOXYMETHYLEEN (DUTCH) see TMP000
TRIOXYMETHYLEN (GERMAN) see TMP000
TRIOXYMETHYLENE see PAI000, TMP000
3-β,5,14-TRIOXY-19-OXO-CARDEN-(20:22)-OLID (GERMAN) see SMM500
3-β,5,14-TRIOXY-19-OXO-DIGEN-(20:22)-OLID (GERMAN) see SMM500
2,6,8-TRIOXYPURINE see UVA400
TRIOZANONA see TLP750
TRIPAN BLUE see CMO250
TRIPARANOL see TMP500
TRI-PCNB see PAX500
TRI-PE see DUN600
TRIPELENAMINE see TMP750
TRIPELENNAMINA (ITALIAN) see TMP750
TRIPELENNAMINE see TMP750
TRIPELENNAMINE HYDROCHLORIDE see POO750
TRIPELENNAMINE MONOHYDROCHLORIDE see POO750
TRI-PENAR see DRS000
TRI-n-PENTYL BORATE see TMQ000

TRI-N-PENTYLTIN BROMIDE see BOK250
TRI(PERFLUOROBUTYL)AMINE see HAS000
TRIPERIDOL see TKK500, TKK750
TRIPERIDOL HYDROCHLORIDE see TKK750
TRIPERVAN see VLF000
TRIPHACYCLIN see TBX250
TRIPHEDINON see BBV000
TRIPHENIDYL see BBV000, PAL500
m-TRIPHENYL see TBC620
p-TRIPHENYL see TBC750
TRIPHENYL see TBD000
TRIPHENYLACETO STANNANE see ABX250
2,3,3-TRIPHENYLACRYLONITRILE see TMQ250
α,β,β-TRIPHENYLACRYLONITRILE see TMQ250
TRIPHENYLACRYLONITRILE see TMQ250
TRIPHENYLAMINE see TMQ500
TRIPHENYLANTIMONY see TMV250
TRIPHENYLANTIMONY DICHLORIDE see DGO800
TRIPHENYLANTIMONY OXIDE see TMQ550
TRIPHENYL ANTIMONY SULFIDE see TMQ600
1,3,5-TRIPHENYLBENZENE see TMR000
TRIPHENYLBENZENE see TMR000
sym-TRIPHENYLBENZENE see TMR000
TRIPHENYLBISMUTH see TMR250
TRIPHENYLBISMUTHINE see TMR250
TRIPHENYL BORATE see TMR500
TRIPHENYLCHLOROSTANNANE see CLU000
TRIPHENYLCHLOROTIN see CLU000
TRIPHENYLCYANOETHYLENE see TMQ250
TRIPHENYLCYCLOHEXYL BORATE see TMR750
TRIPHENYLENE see TMS000
1,2,2-TRIPHENYLETHANONE see TMS100
1,1,2-TRIPHENYLETHYLENE see TMS250
TRIPHENYLETHYLENE see TMS250
N,N',N''-TRIPHENYLGUANIDINE see TMS500
TRIPHENYLGUANIDINE see TMS500
2,4,5-TRIPHENYLIMIDAZOLE see TMS750
TRIPHENYLLEAD ACETATE see TMT000
TRIPHENYL LEAD(1+) HEXAFLUOROSILICATE see TMT250
TRIPHENYLMETHANETHIOL see TMT500
S-TRIPHENYLMETHYL-l-CYSTEINE see TNR475
TRIPHENYL PHOSPHATE see TMT750
TRIPHENYLPHOSPHINE see TMU000
TRIPHENYL PHOSPHITE see TMU250
TRIPHENYL-2-PROPENYL-STANNANE (9CI) see AGU500
1,3,4-TRIPHENYLPYRAZOLE-5-ACETIC ACID SODIUM SALT see TMU750
TRIPHENYLSILYL CHLORIDE see TMU800
TRIPHENYLSTANNANE see TMV775
TRIPHENYLSTANNANE SULFATE (2:1) see BLT000
TRIPHENYLSTANNYL BENZOATE see TMV000
TRIPHENYLSTANNYL HYDRIDE see TMV775
TRIPHENYL STIBINE see TMV250
2,3,5-TRIPHENYL-2H-TETRAZOLIUM CHLORIDE see TMV500
TRIPHENYLTHIOCYANATOSTANNANE see TMV750
TRIPHENYLTIN see TMV775
TRIPHENYLTIN p-ACETAMIDOBENZOATE see TMV800
TRIPHENYLTIN ACETATE see ABX250
TRIPHENYLTIN BENZOATE see TMV000
TRIPHENYLTIN CHLORIDE see CLU000
TRIPHENYLTIN CYANOACETATE see TMV825
TRIPHENYLTIN FLUORIDE see TMV850
TRIPHENYLTIN HYDRIDE see TMV775
TRIPHENYLTIN HYDROPEROXIDE see TMW000
TRIPHENYLTIN HYDROXIDE (USDA) see HON000
TRIPHENYLTIN IODIDE see IFO000
TRIPHENYLTIN LEVULINATE see TMW250
TRIPHENYLTIN METHANESULFONATE see TMW500
TRIPHENYLTIN OXIDE see HON000
TRIPHENYLTIN PROPIOLATE see TMW600
TRIPHENYL TIN THIOCYANATE see TMV750
TRIPHENYL-1H-1,2,4-TRIAZOL-1-YL TIN see TMX000
TRIPHENYL-ZINNACETAT (GERMAN) see ABX250
TRIPHENYL-ZINNHYDROXID (GERMAN) see HON000
TRIPHOSADEN see ARQ500
TRIPHOSPHADEN see ARQ500
TRIPHOSPHOPYRIDINE NUCLEOTIDE see CNF400
TRIPHOSPHORIC ACID ADENOSINE ESTER see ARQ500
TRIPHOSPHORIC ACID, SODIUM SALT see SKN000
TRIPHTHAZINE see TKE500, TKK500
TRIPHTHAZINE DIHYDROCHLORIDE see TKK250
TRIPIPERAZINE DICITRATE see PIJ500
TRIPIPERIDINOPHOSPHINE OXIDE see TMX250
TRIPIPERIDINOPHOSPHINE SELENIDE see TMX350
TRIPLA-ETILO see DBX400
TRIPLEX III see EIX500

TRI-PLUS see TIO750
TRIPOLI see SCI500, TMX500
TRIPOLY see SKN000
TRIPOLYPHOSPHATE see SKN000
TRIPOMOL see TFS350
TRIPOTASSIUM CITRATE MONOHYDRATE see PLB750
TRIPOTASSIUM HEXACYANOFERRATE see PLF250
TRIPOTASSIUM HEXACYANOMANGANATE(3-) see TMX600
TRIPOTASSIUM NITRILOTRIACETATE see TMX750
TRIPOTASSIUM PHOSPHATE see PLQ410
TRIPOTASSIUM TRICHLORIDE see PLA500
TRIPROLIDINE HYDROCHLORIDE see TMX775
TRIPROPARGYL CYANURATE see THN500
TRIPROPIONIN see TMY000
TRIPROPIONINE see TMY000
TRIPROPYLALUMINUM (DOT) see TMY100
TRIPROPYL ALUMINUM see AHH750
TRIPROPYLALUMINUM see TMY100
TRIPROPYLAMINE (DOT) see TMY250
TRI-N-PROPYLAMINE see TMY250
TRI-n-PROPYL BORATE see TMY750
TRIPROPYL(BUTYLTHIO)STANNANE see TMY850
TRIPROPYLENE GLYCOL see TMZ000
TRIPROPYLENEGLYCOL DIACRYLATE see TMZ100
TRIPROPYLENE GLYCOL, METHYL ETHER see TNA000
TRIPROPYL INDIUM see TNA250
TRIPROPYL LEAD see TNA500
TRI-n-PROPYL LEAD CHLORIDE see TNA750
TRIPROPYL LEAD CHLORIDE see TNA750
TRIPROPYL PLUMBANE see TNA500
N,N,N-TRIPROPYL-1-PROPANAMINIUM HYDROXIDE see TED000
TRIPROPYLSTANNIUM ACETATE see TNB000
TRIPROPYLTIN ACETATE see TNB000
TRI-N-PROPYLTIN BROMIDE see BOK750
TRI-n-PROPYLTIN CHLORIDE see CLU250
TRIPROPYLTIN CHLORIDE see CLU250
TRIPROPYLTIN IODIDE see TNB250
TRIPROPYLTIN IODOACETATE see TNB500
TRIPROPYLTIN-o-IODOBENZOATE see IEF000
TRIPROPYLTIN ISOTHIOCYANATE see TNB750
TRIPROPYLTIN TRICHLOROACETATE see TNC000
2,4,6-TRIPROPYL-s-TRIOXANE see TNC100
TRI-n-PROPYLZINNACETAT see TNB000
2,4,6-TRIPROP-2-YNYLOXY-s-TRIAZINE see THN500
TRIPTIDE see GFW000
TRIPTIL see DDA600
TRIPTIL HYDROCHLORIDE see POF250
TRIPTOLID see TNC175
TRIPTOLIDE see TNC175
TRIPTONE see HOT500
(TRI-2-PYRIDYL)STIBINE see TNC250
TRIQUILAR see NNL500
TRIQUINOL see IDD100
TRIRODAZEEN see DVF800
TRI-RODAZENE see DVF800
TRIS (flame retardant) see TNC500
TRIS see TEM500, TNC500
TRIS(ACETONITRILE)TRICARBONYLTUNGSTEN see THM250
TRIS(ACETYLACETONATO)ALUMINUM see TNN000
TRIS(ACETYLACETONATO)CHROMIUM(III) see TNN250
TRIS(ACETYLACETONATO)CHROMIUM see TNN250
TRIS(ACETYLACETONE)ALUMINUM see TNN000
TRIS(ACETYLACETONYL)ALUMINUM see TNN000
TRIS(N-ACRYLOYL)HEXAHYDROTRIAZINE see THM900
TRIS(ACRYLOYL)HEXAHYDRO-s-TRIAZINE see THM900
TRISAETHYLENIMINOBENZOCHINON (GERMAN) see TND000
2,4,6-TRIS(ALLYLOXY)TRIAZINE see THN500
TRISAMINE see TEM500
TRIS-AMINO see TEM500
TRIS-4-AMINOFENYLMETHAN (CZECH) see THP000
TRISAMINOL see TEM500
TRIS(p-AMINOPHENYL)CARBONIUM PAMOATE see TNC725, TNC750
TRIS(p-AMINOPHENYL)CARBONIUM SALT with 4,4-METHYLENEBIS(3-HY-DROXY-2-NAPHTHOIC ACID) (2:1) see TNC725
TRIS-(p-AMINOPHENYL)CARBONIUM SALT with 4,4-METHYLENEBIS(3-HY-DROXY-2-NAPHTHOIC ACID) (2:1) see TNC750
TRIS(p-AMINOPHENYL)METHYLIUM SALT with 4,4'-METHYLENEBIS(3-HY-DROXY-2-NAPHTHOIC ACID) (2:1) see TNC725, TNC750
TRIS(4-AMINOPHENYL) PHOSPHATE see THO550
TRISATIN see ACD500
TRIS(2-AZIDOETHYL)AMINE see TNC800
1,1,1-TRIS(AZIDOMETHYL)ETHANE see TNC825
TRIS(1-AZIRIDINO)PHOSPHINE OXIDE see TND250
2,3,5-TRIS(AZIRIDINO)-1,4-BENZOQUINONE see TND000
2,3,5-TRIS(1-AZIRIDINO)-p-BENZOQUINONE see TND000

TRIS(OCTAMETHYLPYROPHOSPHORAMIDE)NICKEL(2+), DIPERCHLORATE see PCE250
TRISODIUM-4'-ANILINO-8-HYDROXY-1,1'-AZONAPHTHALENE-3,6,5'-TRISULFO-NATE see ADE750
TRISODIUM ARSENATE, HEPTAHYDRATE see ARE000
TRISODIUM ARSENITE see SEY200
TRISODIUM 3-CARBOXY-5-HYDROXY-1-p-SULFOPHENYL-4-p-SULFOPHENYLA-ZOPYRAZOLE see FAG140
TRISODIUM CITRATE see TNL000
TRISODIUM EDETATE see TNL250
TRISODIUM EDTA see TNL250
TRISODIUM ETHYLENEDIAMINETETRAACETATE see TNL250
TRISODIUM ETHYLENEDIAMINETETRAACETATE TRIHYDRATE see TNL500
TRISODIUM ETIDRONATE see TNL750
TRISODIUM HEXANITRITOCOBALTATE(3-) see SFX750
TRISODIUM HEXANITRITOCOBALTATE see SFX750
TRISODIUM HEXANITROCOBALTATE(3-) see SFX750
TRISODIUM HEXANITROCOBALTATE see SFX750
TRISODIUM HYDROGEN ETHYLENEDIAMINETETRAACETATE see TNL250
TRISODIUM HYDROGEN (ETHYLENEDINITRILO)TETRAACETATE see TNL250
TRISODIUM-1-HYDROXY-3,6,8-PYRENETRISULFONATE see TNM000
TRISODIUM MONOTHIOPHOSPHATE see TNM750
TRISODIUM NITRILOTRIACETATE see SIP500
TRISODIUM NITRILOTRIACETATE MONOHYDRATE see NEI000
TRISODIUM NITRILOTRIACETIC ACID see SIP500
TRISODIUM ORTHOPHOSPHATE see SJH200
TRISODIUM ORTHOVANADATE see SIY250
TRISODIUM PHOSPHATE see SJH200
TRISODIUM PHOSPHOROTHIOATE see TNM750
TRISODIUM SALT of 3-CARBOXY-5-HYDROXY-1-SULFOPHENYLAZOPYRA-ZOLE see FAG140
TRISODIUM SALT of 1-(4-SULFO-1-NAPHTHYLAZO)-2-NAPHTHOL-3,6-DISUL-FONIC ACID see FAG020
TRISODIUM THIOPHOSPHATE see TNM750
TRISODIUM TRIFLUORIDE see SHF500
TRISODIUM TRIMETAPHOSPHATE see TKP750
TRISODIUM VERSENATE see TNL250
TRISODIUM ZINC DTPA see TNM850
TRISOMNIN see SBM500, SBN000
(α)-1,3,5-TRIS(OXIRANYLMETHYL)-1,3,5-TRIAZINE-2,4,6-(1H,3H,5H)-TRIONE see TBC450
TRI-SPAN see UVJ450
TRIS(2,4-PENTANEDIONATO)ALUMINUM see TNN000
TRIS(2,4-PENTANEDIONATO)CHROMIUM see TNN250
TRIS(2,4-PENTANEDIONATO)CHROMIUM(3+) see TNN250
TRIS(2,4-PENTANEDIONATO)INDIUM see ICG000
TRIS(2,4-PENTANEDIONATO)IRON see IGL000
TRIS(2,4-PENTANEDIONE)ALUMINUM see TNN000
TRIS(1-PHENYL-1,3-BUTANEDIONATO)CHROMIUM see TNN500
TRIS(1-PHENYL-1,3-BUTANEDIONATO)CHROMIUM(3+) see TNN500
TRIS(1-PHENYL-1,3-BUTANEDIONATO-O,O')CHROMIUM see TNN500
TRIS(1-PHENYL-1,3-BUTANEDIONO)CHROMIUM(III) see TNN500
TRISPHOSPHAMIDE see TNT500
TRIS(PHOSPHONOMETHYL)AMINE see NEI100
TRISPUFFER see TEM500
TRIS-STERIL see TEM500
TRISTAR see DUV600
TRISTEARYL BORATE see TMN750
TRIS(TOLYLOXY)PHOSPHINE OXIDE see TNP500
TRIS(o-TOLYL)-PHOSPHATE see TMO600
TRIS-(p-TOLYL)PHOSPHINE see TNN750
1,3,5-TRIS(TRIBUTYLTIN)-s-TRIAZINE-2,4,6-TRIONE see TMP250
1,2,3-TRIS(2-TRIETHYLAMMONIUM ETHOXY)BENZENE TRIIODIDE see PDD300
TRIS(TRIFLUOROMETHYL)NITROSOMETHANE see PCH300
TRIS(TRIFLUOROMETHYL)PHOSPHINE see TNN775
TRIS(N-(α,α,α-TRIFLUORO-m-TOLYL)ANTHRANILATO)ALUMINUM see AHA875
TRIS(3,3,5-TRIMETHYLCYCLOHEXYL)ARSINE see TNO000
TRIS-(TRIMETHYLSILYL)FOSFAT (CZECH) see PHE750
TRISULFIDE, BIS(ETHOXYTHIOCARBONYL)- see DKE400
TRISULFON BLUE RW see CMO600
TRISULFON CONGO RED see SGQ500
TRISULFON VIOLET N see CMP000
TRISULFURATED PHOSPHORUS see PHS500
TRISULPHONE BROWN B see CMO820
TRISUSTAN see TJL250
TRITELLURIUM TETRANITRIDE see TNO250
TRITEREN see UVJ450
TRITERPENE SAPONINS mixture from AESCULUS HIPPOCASTONUM see TNO275
TRITHENE see CLQ750
TRITHEON see ABY900
1,3,5-TRITHIACYCLOHEXANE see TLS500
sym-TRITHIAN (CZECH) see TLS500
1,3,5-TRITHIANE see TLS500
3,4,5-TRITHIATRICYCLO(5.2.1.0²·⁶)DECANE see TNO300

TRITHIO see AOO490
TRITHIOACETONE see HEL000
TRITHIOANETHOLE see AOO490
TRITHIOBIS(TRICHLOROMETHANE) see BLM750
TRITHIOCARBONIC ACID, CYCLIC ETHYLENE ESTER see EJQ100
TRITHIOCYANURIC ACID see THS250
TRITHIOFORMALDEHYDE see TLS500
TRITHIO-(p-METHOXYPHENYL)PROPENE see AOO490
TRITHION see TNP250
TRITHION MITICIDE see TNP250
TRITHIOZINE see TNP275
TRITICOL see MHC750
TRITIOZINA (ITALIAN) see TNP275
TRITIOZINE see TNP275
TRITISAN see PAX000
TRITOFTOROL see EIR000
TRITOL see TMN490
TRI-2-TOLYL PHOSPHATE see TMO600
TRI-o-TOLYL PHOSPHATE see TMO600
TRITOLYL PHOSPHATE see TNP500
TRITON N-100 see NND500
TRITON A-20 see TDN750
TRITON GR-5 see DJL000
TRITON K-60 see AFP250
TRITON WR 1339 see TDN750
TRITON X 15 see GHS000
TRITON X 35 see PKF500
TRITON X-40 see DTC600
TRITON X 45 see PKF500
TRITON X100 see OFQ000
TRITON X 100 see PKF500
TRITON X 102 see PKF500
TRITON X 165 see PKF500
TRITON X 305 see PKF500
TRITON X 405 see PKF500
TRITON X 705 see PKF500
TRITOX see TII750
TRITROL see CLN750
TRITTICO see CKJ000, THK880
TRITYL BROMIDE see TNP600
TRITYL CHLORIDE see CLT500
S-TRITYLCYSTEINE see TNR475
S-TRITYL-I-CYSTEINE see TNR475
3-(TRITYLTHIO)-I-ALANINE see TNR475
3-TRITYLTHIO-I-ALANINE see TNR475
TRITYLTHIOL see TMT500
TRIUMBREN see AAN000
TRIUMPH see PHK000
TRIUROL see AAN000
TRIUROPAN see AAN000
TRIVALENT SODIUM ANTIMONYL GLUCONATE see AQI000
TRIVASTAL see TNR485
TRIVASTAN see TNR485
TRIVAZOL see MMN250
TRI-VC 13 see DFK600
TRIVINYLANTIMONY see TNR490
TRIVINYLBISMUTH see TNR500
TRIVINYLTIN CHLORIDE see CLU500
TRIVITAN see CMC750
TRIZILIN see DFT800
TRIZIMAN see DXI400
TRIZIMAN D see DXI400
TR5379M see COW675
TROBICIN see SKY500
TROCHIN see CLD000
TROCINAT see DHY400
TROCINATE see DHY400
TROCINATE HYDROCHLORIDE see DHY400
TROCLOSENE see DGN200
TROCLOSENE POTASSIUM see PLD000
TROCOSONE see ECU750
TRODAX see HLJ500
TROFORMONE see AOO475
TROFOSFAMID see TNT500
TROFURIT see CHJ750
TROGAMID T see NOH000
TROGUM see SLJ500
TROJACETONOAMINY see TDT770
TROJCHLOREK FOSFORU (POLISH) see PHT275
TROJCHLOROBENZEN (POLISH) see TIK250
TROJCHLOROBENZEN see TIK250
TROJCHLOROETAN(1,1,2) (POLISH) see TIN000
TROJCHLOROWODOREK 4-ACETYLO-4-(3-CHLOROFENYLO)-1-(3-(4-METYLO-PIPERAZYNO)-PROPYLO)-PIPERYDYNY see ACG125
TROJKREZYLU FOSFORAN (POLISH) see TMO600

TROJNITROTOLUEN (POLISH) see TMN490
TROLAMINE see TKP500
TROLEN see RMA500
TROLENE see RMA500
TROLITUL see SMQ500
TROLMINE see TJL250
TROLNITRATE PHOSPHATE see TJL250
TROLOVOL see MCR750
TROLOX see TNR625
TROLOX C see TNR625
TROMASEDAN see BEA825
TROMASIN see PAG500
TROMBARIN see BKA000
TROMBAVAR see SIO000
TROMBIL see BKA000
TROMBOLYSAN see BKA000
TROMBOSAN see BJZ000
TROMBOVAR see EMT500, SIO000
TROMEDONE see TLP750
TROMETAMOL see TEM500
TROMETE see SJH200
TROMETHAMINE see TEM500
TROMETHAMINE PROSTAGLANDIN F2-α see POC750
TROMETHANE see TEM500
TROMETHANMIN see TEM500
TROMEXAN see BKA000
TROMEXAN ETHYL ACETATE see BKA000
TRONA see BMG400, SFO000, SJY000
TRONAMANG see MAP750
TRONOX see TGG760
TROPACAINE HYDROCHLORIDE see TNS200
TROPACOCAINE HYDROCHLORIDE see TNS200
TROPAEOLIN 1 see FAG010
TROPAEOLIN see MND600
TROPAEOLIN D see DOU600
TROPAEOLIN G see MDM775
TROPAEOLIN OOO 2 see CMM220
TROPAEOLIN OOO see CMM220
TROPAKOKAIN HYDROCHLORID (GERMAN) see TNS200
2-β-TROPANECARBOXYLIC ACID, 3-β-HYDROXY-, METHYL ESTER, BENZO-ATE (ESTER) see CNE750
1-α-H,5-α-H-TROPAN-3-α-OL, ATROPATE (ESTER) see AQO250
1-α-H,5-α-H-TROPAN-3-α-OL, MANDELATE (ester), HYDROBROMIDE see HGH150
1-α-H,5-α-H-TROPAN-3-α-OL (±)-TROPATE (ESTER) see ARR000
1-α-H,5-α-H-TROPAN-3-α-OL (±)-TROPATE (ESTER), SULFATE (2:1) SALT see ARR500
1-α-H,5-α-H-TROPAN-3-α-OL, (–)-TROPATE (ester), SULFATE (2:1) see HOT600
3-TROPANYLBENZOATE-2-CARBOXYLIC ACID METHYL ESTER see CNE750
dl-TROPANYL-2-HYDROXY-1-PHENYLPROPIONATE see ARR000
dl-TROPANYL-2-HYDROXY-1-PHENYLPROPIONATE SULFATE see ARR500
3-α-TROPANYL (–)-2-METHYL-2-PHENYLHYDRACRYLATE HYDROCHLORIDE see LFC000
dl-TROPASAEUREESTER DES 3-DIAETHYLAMINO-2,2-DIMETHYL-1-PROPANOL PHOSPHAT (GERMAN) see AOD250
TROPASAEUREESTER DES 3-TRIAETHYLAMMONIUM-2,2-DIMETHYL-1-PROPA-NOLBROMID (GERMAN) see DSI200
(–+)-TROPATE-3-α-HYDROXY-8-OCTYL-1-α-H,5-α-H-TROPANIUM see OEM000
TROPAX see OPK000
TROPEOLIN see BFL000
TROPHICARD see MAI600
TROPHICARDYL see IDE000
TROPHOSPHAMID see TNT500
TROPHOSPHAMIDE see TNT500
TROPIC ACID, ESTER with SCOPINE see SBG000
TROPIC ACID, ESTER with TROPINE see ARR000
(–)-TROPIC ACID ESTER with TROPINE see HOU000
TROPIC ACID, 9-METHYL-3-OXA-9-AZATRICYCLO(3.3.1.0²ʼ⁴)NON-7-YL ESTER see SBG000
TROPIC ACID, 3-α-TROPANYL ESTER see AQO250
TROPIC ACID-3-α-TROPANYL ESTER see ARR000
TROPIDECHIS CARINATUS VENOM see ARV375
TROPILIDENE see COY000
TROPILIDIN see COY000
TROPIN see MGR250
TROPINE, ATROPATE (ESTER) see AQO250
TROPINE BENZOHYDRYL ETHER METHANESULFONATE see TNU000
TROPINE-4-CHLOROBENZHYDRYL ETHER HYDROCHLORIDE see CMB125
TROPINE (–)-α-METHYLTROPATE HYDROCHLORIDE see LFC000
TROPINE TROPATE see ARR000
TROPINIUM METHOBROMIDE MANDELATE see MDL000
TROPINTRAN see ARR500
TROPISTON see PGP500
TROPITAL see PIZ499

TROPOCOCAIN HYDROCHLORIDE see TNS200
TROPOLONE see TNV550
TROPOTASIN see TFU800
TROPOTOX see CLN750, CLO000
TROPYLIUM PERCHLORATE see TNV575
dl-TROPYLTROPATE see ARR000
(±)-TROPYL TROPATE see ARR000
TROSINONE see GEK500
TROSPIUM CHLORIDE see KEA300
TROTOX see CLN750
TROTYL see TMN490
TROTYL OIL see TMN490
TROVIDUR see PKQ059, VNP000
TROVIDUR PE see PJS750
TROVITHERN HTL see PKQ059
TROXIDONE see TLP750
TROXILAN see AFJ400
TROXONIUM TOSILATE see TNV625
TROXONIUM TOSYLATE see TNV625
TROYKYD ANTI-SKIN B see EMU500
TROYKYD ANTI-SKIN BTO see BSU500
TROYSAN 142 see DSB200
TROYSAN ANTI-MILDEW O see TIT250
TROYSAN COPPER 8% see NAS000
TRP-P-1 see TNX275
TRP-P-2 see ALD500
TRP-P-1 (ACETATE) see AJR500
TRP-P-2(ACETATE) see ALE750
TRU see PQC500
TRUBAN see EFK000
TRUCIDOR see MJG500
TRUE AMMONIUM SULFIDE see ANJ750
TRUFLEX DOA see AEO000
TRUFLEX DOP see DVL700
TRUFLEX DOX see BJQ500
TRUFLEX DTDP see DXQ200
TRUMPET PLANT see CDH125
TRUSONO see EQL000, MNM500
TRUXAL see TAF675
TRUXALETTEN see TAF675
TRUXIL see TAF675
TRYBEN see TIK500
TRYCITE 1000 see SMQ500
TRYCOL HCS see PJW500
TRYCOL LAL SERIES see DXY000
TRYCOL NP-1 see NND500
TRYDET OS SERIES see PJY100
TRYDET SA SERIES see PJV250
TRYOPANOATE SODIUM see SKO500
TRYPAFLAVIN see XAK000
TRYPAFLAVINE see DBX400
TRYPANBLAU (GERMAN) see CMO250
TRYPAN BLUE see CMO250
TRYPAN BLUE SODIUM SALT see CMO250
TRYPARSAMIDE see CBJ750
TRYPOXYL see ARA500
TRYPSIN see TNW000
TRYPSIN INHIBITOR see PAF550
TRYPSIN INHIBITOR, PANCREATIC BASIC see PAF550
TRYPSIN-KALLIKREIN INHIBITOR (KUNITZ) see PAF550
TRYPTAMIDE see NDW525
TRYPTAMINE see AJX000
TRYPTAMINE HYDROCHLORIDE see AJX250
TRYPTAR see TNW000
TRYPTAZINE DIHYDROCHLORIDE see TKK250
TRYPTIZOL see EAH500, EAI000
TRYPTIZOL HYDROCHLORIDE see EAI000
dl-TRYPTOPHAN, pyrolyzate 1 see TNX275
l-TRYPTOPHAN, pyrolyzate see TNW950
l-TRYPTOPHAN, 5-HYDROXY-, (9CI) see HOA600
l-TRYPTOPHAN (FCC) see TNX000
dl-TRYPTOPHAN see TNW500
d-TRYPTOPHAN see TNW250
l-TRYPTOPHAN see TNX000
(–)-TRYPTOPHAN see TNX000
d-TRYPTOPHAN see TNW250
TRYPTOPHANE see TNX000
TRYPTOPHAN P1 see TNX275
TRYPTOPHAN P2 see ALD500
9-l-TRYPTOPHAN-TRYCODINE A HYDROCHLORIDE (9CI) see TOF825
TRYPURE see TNW000
TRYSBEN 200 see DOR800, TIK500
TS 16 see AGD250
TS 160 see TNF250
TS-160 see TNF500

TS 219 see NIM500
TS 1801 see AHA875
TS-7236 see FDA885
TSAA 291 see ELF100
TSAA-328 see ELF110
TSA-HP see TGO000
TSA-MH see TGO000
TSAPOLAK 964 see CCU250
TSC see TFQ000
TSD see MCR250
TSELATOX see TNX375
TSERENOX see BDD000
T-SERP see RDK000
TSH-RELEASING FACTOR see TNX400
TSH-RELEASING HORMONE see TNX400
TSH-RF see TNX400
TSIAZID see COH250
TSIDIAL see DRR400
TSIKLAMID see ABB000
TSIKLODOL see BBV000
TSIKLOMETIAZID see CPR750
TSIKLOMITSIN see TBX000
TSIM see TMF250
TSIMAT see BJK500
TSINEB (RUSSIAN) see EIR000
TSIPROMAT (RUSSIAN) see ZMA000
TSIRAM (RUSSIAN) see BJK500
TSITREX see DXX400
TSIZP 34 see ISR000
TSL 8123 see MQD750
TSL 8370 see TNJ500
TSLT see TNX375
d-T4 SODIUM see SKJ300
TSP see SJH200
TSPA see TFQ750
TSPP see TEE500
TST see EIV000
TSTS 19 see LED100
TSTS 21 see LED100
TSTS 22 see LED100
TSTS 23 see LED100
T-STUFF see HIB000
TSUDOHMIN see DEO600
TSUGA OIL see SLG650
TSUMACIDE see MIB750
TSUMAUNKA see MIB750
TSUSHIMYCIN see TNX650
TT see TMV500
TTC see TMV500
TTD see DXH250, TFS350
TTFB see TBW000
TTFD see FQJ100
T²-TRICHOTHECENE see FQS000
TTS see DXH250
TTT see HDV500
TTX see FOQ000
2-TU see TFR250
TU see TFR250
TUADS see TFS350
TUAMINE see HBM490
TUAMINE SULFATE see AKD600
TUAMINOHEPTANE see HBM490
TUAMINOHEPTANE SULFATE see AKD600
TUAREG see TLN500
TUASOL 100 see THW750
TUATUA (PUERTO RICO) see CNR135
TUAZOLE see MDT250, QAK000
TUAZOLONE see QAK000
TUBADIL see TOA000
TUBARINE see TOA000
TUBA ROOT see DBA000
TUBATOXIN see RNZ000
TUBAZIDE see ILD000
TUBERACTINOMYCIN B see VQZ000
TUBERACTINOMYCIN-N SULFATE see VQZ100
TUBERACTIN SULFATE see VQZ100
TUBERCID see ILD000
TUBERCIDIN see TNY500
TUBEREX see PNW750
TUBERGAL see AGQ875
TUBERIT see CBM000
TUBERITE see CBM000
TUBERMIN see EPQ000
TUBEROID see EPQ000
TUBEROSON see EPQ000

TUBERSAN see SEP000
TUBEX see AES650
TUBICON see ILD000
TUBIGAL see FNF000
TUBIN see FNF000
TUBOCIN see RKP000
TUBOCURARIN see TNY750
d-TUBOCURARINE see TNY750
(+)-TUBOCURARINE see TNY750
TUBOCURARINE see TNY750
d-TUBOCURARINE CHLORIDE see TOA000
TUBOCURARINE CHLORIDE see TOA000
(+)-TUBOCURARINE CHLORIDE see TOA000
TUBOCURARINE, CHLORIDE, HYDROCHLORIDE, (+)- (8CI) see TOA000
d-TUBOCURARINE CHLORIDE PENTAHYDRATE see TNZ000
d-TUBOCURARINE DICHLORIDE see TOA000
(+)-TUBOCURARINE DICHLORIDE PENTAHYDRATE see TNZ000
TUBOCURARINE DIMETHYL ETHER IODIDE see DUM000
d-TUBOCURARINE HYDROCHLORIDE see TOA000
(+)-TUBOCURARINE HYDROCHLORIDE see TOA000
TUBOCURARINE HYDROCHLORIDE see TOA000
d-TUBOCURARINE IODIDE DIMETHYL ETHER see DUM000
TUBOPHAN see DWW000
TUBOTHANE see MAS500
TUBOTIN see ABX250, HON000
TU CILLIN see BFD000
TUEX see TFS350
TUFF-LITE see PMP500
TUFT ROOT see DHB309
TUGON see TIQ250
TUGON FLIEGENKUGEL see PMY300
TUGON FLY BAIT see TIQ250
TUGON STABLE SPRAY see TIQ250
TULABASE FAST GARNET GB see AIC250
TULABASE FAST GARNET GBC see AIC250
TULABASE FAST RED TR see CLK220
TULADISPERSE FAST YELLOW 2G see AAQ250
TULA (PUERTO RICO) see COD675
TULASTERON FAST YELLOW 5R-B see CMP090
TULISAN see TFS350
TULLIDORA see BOM125
TULUYLENDIISOCYANAT see TGM750
TULYL see RLU000
TUMBLEAF see SFS000
TUMENOL see IAD000
TUMESCAL OPE see BGJ250
TUMEX see QPA000
TUNG NUT see TOA275
TUNG NUT MEALS see TOA500
TUNG NUT OIL see TOA510
TUNG OIL TREE see TOA275
TUNGSTEN see TOA750
TUNGSTEN AZIDE PENTABROMIDE see TOB000
TUNGSTEN AZIDE PENTACHLORIDE see TOB250
TUNGSTEN BLUE see TOC750
TUNGSTEN CARBIDE see TOB500
TUNGSTEN CARBIDE, mixed with COBALT (85%:15%) see TOB750
TUNGSTEN CARBIDE, mixed with COBALT (92%:8%) see TOC000
TUNGSTEN CARBIDE, mixed with COBALT and TITANIUM (78%:14%:8%) see TOC250
TUNGSTEN COMPOUNDS see TOC500
TUNGSTEN FLUORIDE see TOC550
TUNGSTEN HEXAFLUORIDE see TOC550
TUNGSTEN-NICKEL CATALYST DUST see TOC600
TUNGSTEN OXIDE see TOC750
TUNGSTEN TRIOXIDE see TOC750
TUNGSTIC ACID see TOD000
TUNGSTIC ACID, SODIUM SALT, DIHYDRATE see SKO000
TUNGSTIC ANHYDRIDE see TOC750
TUNGSTIC OXIDE see TOC750
TUNGSTOPHOSPHORIC ACID (8CI) see PHU750
TUNGSTOPHOSPHORIC ACID, SODIUM SALT see SJJ000
TUNIC see BGD250
TUNICIN see CCU150
TUPHETAMINE see BBK500
TUR see CMF400
TURBINAIRE see DAE525
TURBSVIL see BQT750
TURBULETHYLAZIN (GERMAN) see BQB000
TURCAM see DQM600
TURF-CAL see ARB750
3336 TURF FUNGICIDE see DJV000
TURGEX see HCL000
TURIMYCIN A3 see LEV025
TURIMYCIN A5 see JDS200

TURIMYCIN P₃ see MBY150
TURINGIN-1 see BAC125
TURISYNCHRON see MLJ500
TURKEY RED see DMG800
TURKEY-RED OIL see TOD500
TURMERIC rhizome extract see COF850
TURMERIC see TOD625
TURMERIC OIL see COG000
TURMERIC OLEORESIN see COG000
TURPENTINE, steam distilled see TOD750
TURPENTINE see TOD750
TURPENTINE OIL see TOD750
TURPENTINE OIL, RECTIFIER see TOD750
TURPENTINE STEAM DISTILLED see TOD750
TURPENTINE SUBSTITUTE (UN 1300) (DOT) see TOD750
TURPENTINE (UN 1299) (DOT) see TOD750
TURPETH MINERAL see MDG000
TURPINAL SL see HKS780
TUS-1 see XAJ000
TUSILAN see DBE200
TUSSADE see DBE200
TUSSAL see MDP750
TUSSAPAP see HIM000
TUSSAPHED see POH250
TUSSCAPINE see NOA000
TUSSCAPINE HYDROCHLORIDE see NOA500
TUTANE see BPY000
TUTIN see TOE175
TUTINE see TOE175
TUTOCAINE see TOE150
TUTOCAINE HYDROCHLORIDE see AIT750
TUTOFUSIN TRIS see TEM500
TUTTOMYCIN see NCD550
TUTU see TOE175
TUZET see USJ075
TV 485 see HKK000
825TV see SMQ500
TV 1322 see AAE625
TVOPA (DOT) see TOE200
TVOPA see TOE200
825TV-PS see SMQ500
TVS 8105 see DVK200
TVS-MA 300 see DEJ100
TVS-N 2000E see DEJ100
TVX 485 see HKK000
T-WD602 see TJE100
TWEEN 20 see PKG000
TWEEN 40 see PKG500
TWEEN 60 see PKL030
TWEEN 65 see SKV195
TWEEN 80 see PKL100
TWEEN 85 see TOE250
TWEENASE see GGA800
TWEEN ESTERASE see GGA800
TWEEN HYDROLASE see GGA800
TWINKLING STAR see AQF000
TWIN LIGHT RAT AWAY see WAT200
TX 100 see PKF500
TYBAMATE see MOV500
TYBATRAN see MOV500
TYBON N 1765A see MCB050
TYCLAROSOL see EIV000
TYDEX see BBK500
TYGON see AAX175
TYLAN see TOE600
TYLANDRIL see RDK000
TYLENOL see HIM000
TYLON see TOE600
TYLOROL LT 50 see SON000
TYLOSE 444 see MIF760
TYLOSE 666 see SFO500
TYLOSE A4S see MIF760
TYLOSE H 20 see HKQ100
TYLOSE H 300 see HKQ100
TYLOSE H SERIES see HKQ100
TYLOSE MB see HKQ100
TYLOSE MF see MIF760
TYLOSE MH20 see MIF760
TYLOSE MH50 see MIF760
TYLOSE MH300 see MIF760
TYLOSE MH1000 see MIF760
TYLOSE MH2000 see MIF760
TYLOSE MH4000 see MIF760
TYLOSE MH see HKQ100, MIF760
TYLOSE MHB see HKQ100

TYLOSE MHB-Y see HKQ100
TYLOSE MHB-YP see HKQ100
TYLOSE MH-K see HKQ100
TYLOSE MH300P see MIF760
TYLOSE MH-XP see HKQ100
TYLOSE P see HKQ100
TYLOSE PS-X see HKQ100
TYLOSE P-X see HKQ100
TYLOSE P-Z SERIES see HKQ100
TYLOSE SAP see MIF760
TYLOSE SL 100 see MIF760
TYLOSE SL 400 see MIF760
TYLOSE SL 600 see MIF760
TYLOSE SL see MIF760
TYLOSE TWA see MIF760
TYLOSIN see TOE600
TYLOSIN HYDROCHLORIDE see TOE750
TYLOSTERONE see DKA600
TYLOXAPOL see TDN750
TYLOXYPAL see TDN750
TYMELYT see IFZ900
TYOX A see BHM000
TYOX B see TFD500
TYPOGEN BROWN N see NBG500
TYPOGEN CARMINE see EOJ500
p-TYRAMINE see TOG250
TYRAMINE see TOG250
TYRAMINE, 3-DIAZO-, HYDROCHLORIDE see DCQ575
TYRAMINE MONOCHLORIDE see TOF750
TYRANTON see DBF750
TYRIAN BLUE I-RSN see IBV050
TYRIAN BRILLIANT BLUE I-R see IBV050
TYRIAN BROWN I-BR see CMU770
TYRIAN OLIVE I-R see DUP100
TYRIAN ORANGE A-RF see CMU815
TYRIAN RED A-5B see DNT300
TYRIL see ADY500
TYRIMIDE see DAB875
TYRIN see PJS750
TYRION YELLOW see DCZ000
TYROCIDIN B see TOF825
TYROCIDINE A, HYDROCHLORIDE see GJQ100
TYROCIDINE B, HYDROCHLORIDE see TOF825
TYROSAMINE see TOG250
dl-m-TYROSINE see TOG275
m-TYROSINE, dl- see TOG275
l-p-TYROSINE see TOG300
l-TYROSINE see TOG300
p-TYROSINE see TOG300
TYROSINE see TOG300
l-TYROSINE, O-(4-HYDROXY-3-IODOPHENYL)-3,5-DIIODO- (9CI) see LGK050
TYROTHRICIN see TOG500
TYVEK see PJS750
TYVID see ILD000
TYZANOL HYDROCHLORIDE see VRZ000
TYZINE see VRZ000
TYZINE HYDROCHLORIDE see VRZ000
TYZOR TPt see IRN200
TZT see THR500
TZU-0460 see HNT100


U 02 see CBT500
U-14 see POC750
U-0045 see BKP200
U 46 see CIR500, DAA800
U46 see DGB000
U 46 see TAA100
U-0172 see DPO100
U-197 see MQR100
U-0290 see UAG000
475U see SMQ500
U625 see SMQ500
666U see SMQ500
U 963 see UTU500
U-1149 see FOU000
U-1247 see BOQ500
U 1363 see DVV600
U-1434 see EEI000
U-1804 see UAG025
U-2069 see RDP300
U 2134 see MGC350
U-2363 see DTS625, UAG050
U-2397 see UAG075
U-3818 see CKK000

U-3886 see SFA000
U-4224 see DSB000
U 4513 see DRP800
U-4527 see CPE750
U-4748 see POK000
U-4761 see AHL000
U-4783 see ACJ500
U-4858 see TNW000
U 4905 see HHR000
U-5043 see DAA800
U-5227 see AQN635
U 5446 see PBM500
U-5897 see CDT750
U-5954 see DOO800
U 5963 see FHH100
U-5965 see TBX250
U 6020 see PLZ000
U 6040 see AOO275
U-6062 see CDP250
U-6233 see BDH250
U 6245 see DTU850
U 6324 see PGN250
U-6421 see NGE500
U-6591 see NOB000
U-6658 see AOB875
U-6780 see CMB000
U 6987 see BSM000
U-7118 see FBP300
U-7257 see CDK250
U 7524 see ALQ650
U-7726 see KFA100
U-7743 see CHW675
U 8210 see DAZ117
U-8344 see BIA250
U 8771 see EHP000
U-8774 see ALL250
U-8953 see FMM000
U 9088 see USJ100
U-9361 see SMC500
U-9889 see SMD000
U-10149 see LGD000
U-10,858 see DCB000
U 10997 see MQS225
U 11100 see NAD750
U-11,634 see TKJ750
U-12062 see DVJ200
U 12241 see CMS232
U 12927 see CGI500
U-14583 see POC500
U-14743 see MLY000
U 15030 see DKC400
U-15167 see NMV500
U-15,646 see PBH150
U-15800 see PAB500
U-17004 see CGI500, DET600
U 17556 see MIA500
U-17835 see TGJ500
U 18496 see ARY000
U-19183 see SKX000
U 19571 see AGT500
U-19,646 see CJQ250
U-19,920 see AQQ750
U 21221 see DSK950
U-21,251 see CMV675
U-24544 see PPI775
U-24973 see AEG875
U 25,354 see DGG400
U 26452 see CEH700
U 27,574 see CKQ500
U-28,508 see CMV690
U-29135 see GCI000
U-29,409 see DLX300
U 31889 see XAJ000
U-32.104 see MHC750
U-33,030 see THS800
U-36059 see MJL250
U 48160 see QQS075
U-52047 see MCB600
U-54461 see AIY850
U 75630 see IAB000
U 631963 see CCS525
U 1 (polymer) see PKQ059
UA 1 see SCQ000
UA 2 see SCQ000
UA 3 see SCQ000

UA 4 see SCQ000
U 11100A see NAD750
U-11555A see MFG260
U-19920 A see AQQ750
U-22,304A see LFG100
U-22394A see HDQ500
U-22,559A see DVP400
U-24973A see TDL000
UBATOL U 2001 see SMQ500
UBIDECARENONE see UAH000
UBIQUINONE 10 see UAH000
UBIQUINONE 50 see UAH000
UBRETID see DXG800
UBRITIL see DXG800
UC 7207 see OCE100
UC 7744 see CBM750
UC 8305 see CGM400
UC 8,454 see TCY275
UC 9880 see CQI500
UC 10854 see COF250
UC 19786 see CBW000
UC 20047 see CFF250
UC 20,299 see SFV250
UC-21149 see CBM500
UC-21865 see AFK000
UC 22,463 see DET400, DET600
UC-25074 see FNE500
UC 26089 see CFF250
UC 20,047A see CFF250
UCANE ALKYLATE 12 see UAK000
UCAR 17 see EJC500
UCAR 130 see AAX250
UCAR AMYL PHENOL 4T see AON000
UCAR BUTYLENE OXIDE 12 see UAK100
UCAR BUTYLPHENOL 4-T see BSE500
UCAR SOLVENT 2LM see DWT200
UCAR SOLVENT LM (OBS.) see PNL250
UCAR TRIOL HG-170 see UBA000
UCB 170 see HGC500
UCB 492 see CJR909
UCB 1402 see BBV750
U.C.B. 1474 see CDQ500
UCB 2493 see CJL409
UCB 3412 see MKQ000
UCB 4208 see COL250
UCB 4268 see HDS200
UCB 4445 see BOM250
U.CB 4492 see CJR909
UCB 6215 see NNE400
UCB 6249 see DIF600
UCB 1545 HYDROCHLORIDE see LFK200
UCC 974 see DSB200
UCC 6863 see SMQ500
UCET see CBT750
UCET TEXTILE FINISH 11-74 (OBS.) see VOA000
UC LIQUID G see SCR400
U-COMPOUND see UVA000
UCON 12 see DFA600
UCON 22 see CFX500
UCON 112 see TBP050
UCON 113 see FOO000
UCON 114 see FOO509
UCON FLUID AP-1 see UBJ000
UCON FLUOROCARBON 113 see FOO000
UCON FLUOROCARBON 122 see TIM000
UCON 113/HALOCARBON 113 see FOO000
UCON 12/HALOCARBON 12 see DFA600
UCON 22/HALOCARBON 22 see CFX500
UCON 500/HALOCARBON 500 see DFB400
UCON 50-HB-55 see UBS000
UCON 50-HB-100 see UCA000
UCON 50-HB-260 see UCJ000
UCON 50-HB-400 see UDA000
UCON 50-HB-660 see UDJ000
UCON 50-HB-2000 see UDS000
UCON 50-HB-3520 see UEA000
UCON 50-HB-5100 see UEJ000
UCON 50-HB-280-X see UFA000
UCON LB-250 see BRP250
UCON LB 1145 see BRP250
UCON LB-1715 see MKS250
UCON LB 1800X see BRP250
UCON LO-500 see UHA000
UCON LUBRICANT DLB-140-E see UIJ000
UCON LUBRICANT DLB-200-E see UIS000

UCON LUBRICANT DLB-62-E see UIA000
UCON REFRIGERANT 11 see TIP500
UDANTOL HYDROCHLORIDE see AEG625
UDCA see DMJ200
UDICIL see AQQ750
UDMH (DOT) see DSF400
UDOLAC see SOA500
U 46DP see DAA800
U46 DP-FLUID see DGB000
U 32921E see CCC100
UF 1 see DMI600
UF 4 see HND100
UF 15 see SCQ000
UF 33 see UTU500
UF 240 see UTU500
U 46 M-FLUID see CIR250
UFORMITE 700 see UTU500
UFORMITE F 240N see UTU500
UFORMITE MM 46 see MCB050
UFORMITE MM 47 see MCB050
UFORMITE MM 83 see MCB050
UFORMITE QR 336 see MCB050
UFT see UNJ810
UG 767 see PAR600
UGM 3 see MCB050
UGUROL see AJV500
UJOVIRIDIN see CCK000
UK-738 see DWE800
UK 4261 see OLT000
UK 4271 see OLT000
UKARB see CBT750
UKOPEN see AOD125
UKS 72 see UTU500
UKS 73 see UTU500
U 46 KV-ESTER see CIR500
U 46 KV-FLUID see CIR500
U46KW see BSQ750
ULACORT see PMA000
ULATKAMBAL ROOT EXTRACT see UIS300
ULCEDINE see TAB250
ULCERFEN see TEH500
ULCIMET see TAB250
ULCINE see DJM800, XCJ000
ULCOLAX see PPN100
ULCOLIND see CLY500
ULCOMET see TAB250
ULCUDEXTER see DJM800, XCJ000
ULCUS-TABLINEN see CBO500
ULCUTIN see BGC625
ULEXINE see CQL500
ULFARET see CCS550
ULIOLIND see CLY500
ULMENIDE see CDL325
ULO see CMW700
ULOID 22 see UTU500
ULOID 100 see UTU500
ULOID 230 see MCB050
ULOID 301 see UTU500
ULOID 344 see MCB050
ULOID U 755 see MCB050
ULOID UL213-2 see MCB050
ULONE see CMW700
UL 52R see UTU500
ULSTRON see PMP500
ULTANDREN see AOO275
ULTANDRENE see AOO275
ULTRABIL see BGB315
ULTRABION see AIV500
ULTRA BRILLIANT BLUE P see DSY600
ULTRABRON see AIV500
ULTRACIDE see DSO000
ULTRACILLIN see AJJ875
ULTRACORTEN see PLZ000
ULTRACORTENE-H see PMA000
ULTRADINE see PKE250
ULTRAFUR see FPI000
ULTRALENTE INSULIN see LEK000
ULTRA LENTE ISZILIN see LEK000
ULTRAMARINE BLUE see UJA200
ULTRAMARINE GREEN see CMJ900
ULTRAMARINE YELLOW see BAK250
ULTRAMID BMK see PJY500
ULTRAN see CKE750
ULTRANOX 624 see COV800
ULTRANOX 626 see COV800

ULTRAPAL CHLORIDE see HLC500
ULTRA-PFLEX see CAT775
ULTRASUL see MPQ750
ULTRA SULFATE SL-1 see SIB600
ULTRAWET K see DXW200
ULTRON see PKQ059
ULUP see FMM000
ULV see DAO600
ULVAIR see MRH209
UM-792 see CQF099
UMALUR see UTU500
UMBELLATINE SULFATE TRIHYDRATE see BFN750
UMBETHION see CNU750
UMBRADIL see DNG400
UMBRATHOR see TFT750
UMBRELLA LEAF see MBU800
UMBRIUM see DCK759
U46 MCPB see CLN750
U 27 METHANESULFONATE see PIE750
UM-G see UTU500
UML 491 see MLD250
UNADS see BJL600
UNAMIDE J-56 see BKE500
UNAMYCIN-B see VGZ000
1,2,3,4,5,5,6,7,9,10,10-UNDECACHLOROPENTACYCLO(5.3.0.O$^{2,6}$.O$^{3,9}$.O$^{4,8}$)DEC-
    ANE see MRI750
5,9-UNDECADIEN-2-ONE, 6,10-DIMETHYL- see GDE400
γ-UNDECALACTONE see HBN200, UJA800
Δ-UNDECALACTONE see UKJ000
1-UNDECANAL see UJJ000
n-UNDECANAL see UJJ000
UNDECANAL see UJJ000
UNDECANALDEHYDE see UJJ000
n-UNDECANE see UJS000
UNDECANE see UJS000
1-UNDECANECARBOXYLIC ACID see LBL000
1,1'-UNDECANEDICARBOXYLIC ACID ESTER with ETHYLENE GLYCOL see
    EJQ500
UNDECANE, 1,1-DIMETHOXY-2-METHYL- see MNB600
UNDECANOIC ACID see UKA000
UNDECANOIC ACID, 4-HYDROXY-, γ-LACTONE see HBN200
UNDECANOIC ACID, TRIBUTYLSTANNYL ESTER see TIG500
UNDECANOIC ACID, TRIISOPROPYLSTANNYL ESTER see TKT850
n-UNDECANOL see UNA000
γ-UNDECANOLACTONE see HBN200
1,4-UNDECANOLIDE see HBN200
UNDECANOLIDE-1,5 see UKJ000
4-UNDECANOLIDE see HBN200
γ-UNDECANOLIDE see HBN200
2-UNDECANONE see UKS000
6-UNDECANONE see ULA000
UNDECAN-6-ONE see ULA000
1,3,5-UNDECATRIENE see ULA100
3,5,9-UNDECATRIEN-2-ONE, 6,10-DIMETHYL- see POH525
3,5,9-UNDECATRIEN-2-ONE, 3,6,10-TRIMETHYL- see TMJ100
10-UNDECENAL see ULJ000
1-UNDECEN-10-AL see ULJ000
10-UNDECENAL DIGERANYL ACETAL see UNA100
9-UNDECENAL, 2,6,10-TRIMETHYL- see TMJ150
1-UNDECENE, 11,11-BIS((3,7-DIMETHYL-2,6-OCTADIENYL)OXY)- see UNA100
10-UNDECENOIC ACID see ULS000
10-UNDECENOIC ACID, BUTYL ESTER see BSS100
10-UNDECENOIC ACID, ETHYL ESTER see EQD200
9-UNDECENOIC ACID, METHYL ESTER see ULS400
10-UNDECENOL see UMA000
1-UNDECEN-11-OL see UMA000
10-UNDECENOYL CHLORIDE see UMJ000
10-UNDECENYL ACETATE see UMS000
UNDECENYL ACETATE see UMS000
ω-UNDECENYL ALCOHOL see UMA000
n-UNDECOIC ACID see UKA000
UNDECYL ALCOHOL see UNA000
n-UNDECYL ALDEHYDE see UJJ000
UNDECYL ALDEHYDE see UJJ000
UNDECYLENALDEHYDE see ULJ000
10-UNDECYLENEALDEHYDE see ULJ000
UNDECYL-10-ENIC ACID see ULS000
10-UNDECYLENIC ACID see ULS000
9-UNDECYLENIC ACID see ULS000
UNDECYLENIC ACID see ULS000
ω-UNDECYLENIC ACID CHLORIDE see UMJ000
UNDECYLENIC ALCOHOL see UMA000
UNDECYLENIC ALDEHYDE see ULJ000
UNDECYLENIC ALDEHYDE DIGERANYL ACETAL see UNA100
10-UNDECYLENOYL CHLORIDE see UMJ000

UNDECYLIC ACID see UKA000
UNDECYLIC ALDEHYDE see UJJ000
γ-UNDEKALAKTON see HBN200
UNDEN see EDV000, PMY300
UNFINISHED LUBRICATING OIL see COD750
UNGARISCHES GELBHOLZ see FBW000
UNGEREMINE see LJB800
UNIBARYT see BAP000
UNIBOLDINA see DNZ100
UNIBUR 70 see CAT775
UNICA F 730 see MCB050
UNICA 380K see MCB050
UNICA RESIN 380K see MCB050
UNICELLES see BPF000
UNICEL-ND see DVF400
UNICEL NDX see DVF400
UNICHEM see PKQ059
UNICIN see TBX250
UNICOCYDE see ILD000
UNICROP CIPC see CKC000
UNICROP DNBP see BRE500
UNICROP MANEB see MAS500
UNIDERM WGO see WBJ700
UNIDIGIN see DKL800
UNIDOCAN see DAQ800
UNIDRON see DXQ500
UNIFLEX BYO see BSB000
UNIFLEX BYS see BSL600
UNIFLEX DOS see BJS250
UNIFOS DYOB S see PJS750
UNIFOS EFD 0118 see PJS750
UNIFUME see EIY500
UNIFUR see FPI000
UNI-GUAR see GLU000
UNILAX see ACD500
UNILORD see RDK000
UNIMATE GMS see OAV000
UNIMATE IPM see IQN000
UNIMATE IPP see IQW000
UNIMOLL 66 see DGV700
UNIMOLL BB see BEC500
UNIMYCETIN see CDP250
UNIMYCIN see TBX250
UNION BLACK EM see AQP000
UNION CARBIDE 1-174 see TLC250
UNION CARBIDE 1-189 see TLC000
UNION CARBIDE 7207 see OCE100
UNION CARBIDE 7,744 see CBM750
UNION CARBIDE 19786 see CBW000
UNION CARBIDE 20299 see SFV250
UNION CARBIDE A-151 see TJN250
UNION CARBIDE A-15 see EQA000
UNION CARBIDE A-150 see TIN750
UNION CARBIDE A-162 see MQD750
UNION CARBIDE A-163 see MQF500
UNION CARBIDE A-186 see EBO000
UNION CARBIDE A-187 see ECH000
UNION CARBIDE LIQUID G see SCR400
UNION CARBIDE 7158 SILICONE FLUID see DAF350
UNION CARBIDE UC-10,854 see COF250
UNION CARBIDE UC 20047 see CFF250
UNION CARBIDE UC-8305 see CGM400
UNION CARBIDE UC-8454 see TCY275
UNION CARBIDE UC-9880 see CQI500
UNION DARK GREEN B see CMO830
UNION FAST NAVY BLUE DS see CMN800
UNION FAST RED 3B see CMO870
UNION FAST SCARLET 3B see CMO875
UNIPAQUE see AOO875
UNIPEN see SGS500
UNIPHAT A20 see MHY800
UNIPHAT A30 see MHY650
UNIPHAT A40 see MLC800
UNIPINE see PIH750
UNIPON see DGI400, DGI600
UNIPROFEN see OJI750
UNIROYAL see DFT000
UNIROYAL D014 see SOP000
UNIROYAL F849 see AKR500
UNISEDIL see DCK759
UNISEPT see CDT250
UNISOL 4-O see PJY100
UNISOL RH see SFO500
UNISOM see PGE775
UNISOMNIA see DLY000

UNISPIRAN see GCE600
UNISTRADIOL see EDP000
UNISULF see SNN500
UNITANE O-110 see TGG760
UNITED see CBT750
UNITED CHEMICAL DEFOLIANT No. 1 see SFS000
UNITENE see MCC250
UNITENSEN see CKP500, RDK000
UNITERTRACID GREEN BS see ADF000
UNITERTRACID LIGHT BLUE AB see CMM070
UNITERTRACID LIGHT YELLOW RR see SGP500
UNITERTRACID RED 2G see CMM300
UNITERTRACID YELLOW TE see FAG140
UNITESTON see TBG000
UNITHIOL see DNU860
UNITIOL see DNU860
UNITOL see DNU860
UNITOX see CDS750
UNIVERM see CBY000
UNIVOL U 316S see MSA250
UNJECOL 50 see OBA000
UNJECOL 70 see OBA000
UNJECOL 90 see OBA000
UNJECOL 110 see OBA000
UNLEADED GASOLINE see GCE100
UNLEADED MOTOR GASOLINE see GCE100
UNON P see TAI250
UNOSPASTON see DGW600
UNOX 201 see ECB000
UNOX 207 see BGA250
UNOXAT EPOXIDE 269 see LFV000
UNOX EPOXIDE 201 see ECB000
UNOX EPOXIDE 206 see VOA000
UNOX EPOXIDE 207 see BGA250
UNOX 207X see BGA250
UNYSH A see FAQ930
UOP 88 see BJT500
UP 1 see SMQ500
UP 2 see SMQ500
UP 27 see SMQ500
UP 83 see NDX500
UP 925 see CGW300
UPALET see CNT350
UP 1E see SMR000
UPIOL see BNP750
UPJOHN U-12,927 see CGI500
UPJOHN U-18120 see MDX250
UPJOHN U-22023 see MLX000
UPJOHN U-32714 see CON300
UPJOHN U-36059 see MJL250
UPM703 see SMQ500
UPM see SMQ500
UPM508L see SMQ500
200U/P-RVM see PAE750
UR 606 see CEH700
UR 1522 see PHA575
URACIL see UNJ800
URACIL, 6-AMINO-2-THIO- see AMS750
6-URACILCARBOXYLIC ACID see OJV500
URACIL mixture with FT (4:1) see UNJ810
URACIL, 5-(HYDROXYMETHYL)-6-METHYL- see HMH300
URACILLOST see BIA250
URACILMOSTAZA see BIA250
URACIL MUSTARD see BIA250
URACIL RIBOSIDE see UVJ000
URACIL mixture with TEGAFUR (4:1) see UNJ810
URACIL mixture with 1-(2-TETRAHYDROFURYL)-5-FLUOROURACIL (4:1) see
   UNJ810
URACTONE see AFJ500
URACTYL see SNQ550
URADAL see BNK000
URAGAN see BMM650, SIH500
URAGON see BMM650
URALENIC ACID see GIE000
URALGIN see EID000
URALITE see UTU500
URALITE (POLYMER) see UTU500
URALYT-U see SFX725
URAMID see SNQ550
URAMINE T 80 see HLU500
URAMINE T101 see UTU500
URAMINE T105 see UTU500
URAMINE TSL 58 see UTU500
U-RAMIN P 6100 see MCB050
U-RAMIN P 6300 see MCB050

U-RAMIN T 33 see MCB050
U-RAMIN T 34 see MCB050
URAMITE see UTU500
URAMUSTIN see BIA250
URAMUSTINE see BIA250
URAMYCIN B see VGZ000
URANIN see FEW000
URANINE A EXTRA see FEW000
URANINE USP XII see FEW000
URANINE YELLOW see FEW000
URANIUM see UNS000
URANIUM ACETATE see UPS000
URANIUM AZIDE PENTACHLORIDE see UOA000
URANIUM, BIS(ACETATO)DIOXO-, DIHYDRATE see UQT700
URANIUM, BIS(ACETO-O)DIOXO-, DIHYDRATE (9CI) see UQT700
URANIUM, BIS(NITRATO-O,O')DIOXO-, (OC-6-11)- see URA100
URANIUM CARBIDE see UOB100
URANIUM(IV) CHLORIDE see UQJ000
URANIUM DICARBIDE see UOC200
URANIUM FLUORIDE (fissile) see UOJ000
URANIUM FLUORIDE OXIDE see UQA000
URANIUM HEXAFLUORIDE, fissile excepted or non-fissile (UN 2978) (DOT)
   see UOJ000
URANIUM HEXAFLUORIDE, fissile (containing >1% U-235) (UN 2977) (DOT)
   see UOJ000
URANIUM(III) HYDRIDE see UPA000
URANIUM METAL, pyrophoric (DOT) see UNS000
URANIUM(IV) OXIDE see UPJ000
URANIUM OXYACETATE see UPS000
URANIUM OXYFLUORIDE see UQA000
URANIUM TETRACHLORIDE see UQJ000
URANIUM(III) TETRAHYDROBORATE see UQS000
URANIUM(IV) TETRAHYDROBORATE see UQT300
URANYL ACETATE see UPS000
URANYL ACETATE DIHYDRATE see UQT700
URANYL CHLORIDE see URA000
URANYL FLUORIDE see UQA000
URANYL NITRATE see URA100
URANYL NITRATE (solid) see URA200
URANYL NITRATE HEXAHYDRATE, solution (DOT) see URS000
URANYL NITRATE HEXAHYDRATE see URS000
URAPIDIL see USJ000
URAPRINT 62-126 see CAW500
URARI see COF750
URAZIUM see PDC250
URBACID see USJ075
URBACIDE see USJ075
URBANYL see CIR750
URBASON see MOR500
URBASON CRYSTAL SUSPENSION see DAZ117
URBASONE see MOR500
URBASON SOLUBLE see USJ100
URBASULF see MGQ750
URBAZID see USJ075
URBIL see MQU750
URBOL see ZVJ000
UREA see USS000
UREA, 1-AMIDINO-3-(p-NITROPHENYL)-, MONOHYDROCHLORIDE see NIJ400
UREA ANTIMONYL TARTRATE see UTA000
UREA, N-(1,3-BIS(HYDROXYMETHYL)-2,5-DIOXO-4-IMIDAZOLIDINYL)-N,N'-
   BIS(HYDROXYMETH YL)- see IAS100
UREA, N-(4-BROMOPHENYL)-N'-METHYL-(9CI) see MHS375
UREA, sec-BUTYL- see BSS300
UREA, tert-BUTYL- see BSS310
UREA, N-(4-CHLOROPHENYL)-N'-(3,4-DICHLOROPHENYL)-(9CI) see TIL500
UREA, 1-((p-CHLOROPHENYL)SULFONYL)-3-CYCLOHEXYL- see CDR550
UREA, 1-((p-CHLOROPHENYL)SULFONYL)-3-ISOPROPYL- see CDY100
UREA, N-(3,4-DICHLOROPHENYL)-N'-HYDROXY- see DGD085
UREA, 1,1-DIETHYL- see DKD650
UREA, N,N-DIETHYL-(9CI) see DKD650
UREA, N,N-DIETHYL-N'-((8-α)-6-METHYLERGOLIN-8-YL)- see DLR100
UREA, N,N-DIETHYL-N'-((8-α)-6-METHYLERGOLIN-8-YL)-, (Z)-2-BUTENEDIOATE
   (1:1) see DLR150
UREA, N,N-DIETHYL-N'-((8-α)-6-PROPYLERGOLIN-8-YL)- see DJX300
UREA, N,N-DIETHYL-N'-((8-α)-6-PROPYLERGOLIN-8-YL)-, (Z)-2-BUTENEDIOATE
   (1:1) see DJX350
UREA, 1,1-DIMETHYL- see DUM150
UREA, (1,1-DIMETHYLETHYL)-(9CI) see BSS310
UREA DIOXIDE see HIB500
UREA, 1,1'-ETHYLENEDI- see EJC100
UREA-FORMALDEHYDE ADDUCT see UTU500
UREA-FORMALDEHYDE CONDENSATE see UTU500
UREA-FORMALDEHYDE COPOLYMER see UTU500
UREA-FORMALDEHYDE OLIGOMER see UTU500
UREA-FORMALDEHYDE POLYMER see UTU500

UREA-FORMALDEHYDE PRECONDENSATE see UTU500
UREA-FORMALDEHYDE PREPOLYMER see UTU500
UREA-FORMALDEHYDE RESIN see UTU500
UREA HYDROGEN PEROXIDE (DOT) see HIB500
UREA HYDROGEN PEROXIDE SALT see HIB500
UREA HYDROPEROXIDE see HIB500
UREA, ISOPROPYL- see IRR100
UREA, N-(2-METHOXYETHYL)-N-NITROSO- see NKO900
UREA, (1-METHYLETHYL)-(9CI) see IRR100
UREA, 1-METHYL-2-THIO- see MPW600
UREA, MONONITRATE (8CI,9CI) see UTJ000
UREA, N,N'-BIS(4-NITROPHENYL)-, compd. with 4,6-DIMETHYL-2(1H)-PYRIMIDI-
   NONE (1:1) (9CI) see NCW100
UREA, N,N'-DIMETHYL-N,N'-DIPHENYL-(9CI) see DRB200
UREA, N,N''-1,2-ETHANEDIYLBIS-(9CI) see EJC100
UREA NITRATE see UTJ000
UREA NITRATE (wet) see UTJ000
UREA NITRATE, wetted with not <20% water, by weight (UN 1357) (DOT)
   see UTJ000
UREA NITRATE, dry or wetted with <20% water, by weight (UN 0220)
   (DOT) see UTJ000
UREAPAP W see UTU500
UREA PERCHLORATE see UTU400
UREA PEROXIDE (DOT) see HIB500
UREAPHIL see USS000
UREA, POLYMER with FORMALDEHYDE see UTU500
UREA, 2-PROPENYL-(9CI) see AGV000
UREA, PROPYL- see PNX550
UREASE see UTU550
UREA, SULFANILYL- see SNQ550
UREA, SULFATE (1:1) see UTU600
UREA, 1,1,3,3-TETRABUTYL- see TBM850
UREA, TETRABUTYL- see TBM850
URECHITES LUTEA see YAK300
URECHOLINE see HOA500
URECHOLINE CHLORIDE see HOA500
URECOLI S see UTU500
URECOLL K see UTU500
URECOLL KL see UTU500
UREGIT see DFP600
p-UREIDOBENZENEARSONIC ACID see CBJ000
(p-UREIDOBENZENEARSYLENEDITHIO)DI-o-BENZOIC ACID see TFD750
4-UREIDO-1-PHENYLARSONIC ACID see CBJ000
(p-UREIDOPHENYLARSYLENEDITHIO)DIACETIC ACID see CBI250
(p-UREIDOPHENYLARSYLENEDITHIO)DI-o-BENZOIC ACID see TFD750
URELIT C see UTU500
URELIT HM see UTU500
URELIT R see UTU500
URENIL see SNQ550
UREOL P see DTG700
UREOPHIL see USS000
UREPRET see UTU500
URESE see BDE250
URETAN ETYLOWY (POLISH) see UVA000
URETHAN see UVA000
URETHANE see UVA000
URETHANE POLYMERS see PKL500
URETHYLANE see MHZ000
UREVERT see USS000
UREX see CHJ750
6,6'-UREYLENEBIS(1,1'-DIMETHYLQUINOLINIUM) SULFATE see PJA120
6,6'-UREYLENEBIS(1-METHYLQUINOLINIUM)BIS(METHOSULFATE) see PJA120
URFAMICIN HYDROCHLORIDE see UVA150
URGENEA MARITIMA see RCF000
URGILAN see POB500
URI see SPC500
URIBEN see EID000
URIC ACID see UVA400
URIC ACID, MONOSODIUM SALT see SKO575
URICEMIL see ZVJ000
URICOSID see DWW000
URICOVAC see DDP200
URIDINAL see PDC250, PEK250
β-URIDINE see UVJ000
URIDINE see UVJ000
URIDION see UVJ400
URINARY INDICAN see ICD000
URINEX see CLH750
URIODONE see DNG400
URIPLEX see PDC250
URISOXIN see SNN500
URISPAS see FCB100
URITAS see ZVJ000
URITONE see HEI500
URITRATE see OOG000

URITRISIN see SNN500
URIZEPT see NGE000
URLEA see BEQ625
URNER'S LIQUID see DEL000
URO-ALVAR see OOG000
UROANTHELONE see UVJ475
UROBENYL see ZVJ000
UROBIOTIC-250 see PDC250
UROCALUM see UVJ425
UROCANIC ACID see UVJ440
UROCANINIC ACID see UVJ440
UROCAUDAL see UVJ450
UROCHECK see TMV500
UROCONTRAST see HGB200
UROCYDAL see MPQ750
URODIATON see MPQ750
URODIAZIN see CFY000
URODINE see PDC250, PEK250
URODIXIN see EID000
UROENTERONE see UVJ475
UROFEEN see PDC250
UROFIX see UTU500
UROGAN see SNN500
UROGASTRON see UVJ475
UROGASTRONE see UVJ475
UROGRAFIN ACID see DCK000
UROGRANOIC ACID see DCK000
UROKINASE see UVS500
UROKINASE (ENZYME-ACTIVATING) see UVS500
UROKON SODIUM see AAN000
UROLUCOSIL see MPQ750
UROMALINE see MAP000
UROMAN see EID000
UROMIDE see PDC250
UROMIDIN see PIZ000
UROMIRO 380 see MCA775
UROMIRO see AAI750
UROMIRON see AAI750
UROMITEXAN see MDK875
UROMUCAESTHIN see BQA010
UROMYCINE see GCO000
URONAL see BAG000
URONEG see EID000
URONIUM PERCHLORATE see UTU400
UROPHENYL see PDC250
UROPYRIDIN see PDC250
UROPYRINE see PDC250
UROSCREEN see TMV500
UROSEMIDE see CHJ750
URO-SEPTRA see TKX000
UROSIN see ZVJ000
UROSULFAN see SNQ550
UROSULFANE see SNQ550
UROSULFON see SNP500
UROSULFONE see SNP500
UROTRAST see DCK000
UROTRATE see OOG000
UROTROPIN see HEI500
UROTROPINE see HEI500
UROVISON see SEN500
UROVIST see AOO875
UROX 379 see CJY000
UROX B WATER SOLUBLE CONCENTRATE WEED KILLER see BMM650
UROX HX GRANULAR WEED KILLER see BMM650
UROXIN see AFU250
UROXOL see OOG000
URSACOL see DMJ200
URSO see DMJ200
URSOCHOL see DMJ200
URSODEOXYCHOL see DMJ200
URSODEOXYCHOLIC ACID see DMJ200
URSODESOXYCHOLIC ACID see DMJ200
URSOFALK see DMJ200
URSOFERRAN see IGS000
URSOL BROWN O see CEG600
URSOL BROWN RR see ALL750
URSOL D see PEY500
URSOL EG see ALT500
URSOL ERN see NAW500
URSOL OLIVE 6G see CFK125
URSOL P see ALT250
URSOL P BASE see ALT250
URSOL SLA see DBO400
URSOLVAN see DMJ200
URSOL YELLOW BROWN A see ALO000

URTOSAL see SAH000
US 2 see CNH125
USACERT BLUE No. 1 see FAE000
USACERT BLUE No. 2 see FAE100
USACERT FD & C RED No. 4 see FAG050
USACERT FD & C YELLOW NO. 6 see FAG150
USACERT RED No. 1 see FAG018
USACERT RED No. 3 see FAG040
USACERT YELLOW NO. 5 see FAG140
USACERT YELLOW NO. 6 see FAG150
USAF D-1 see IDW000
USAF D-3 see CHU500
USAF D-4 see POD750
USAF D-5 see BSO500
USAF D-9 see CBM000
USAF M-2 see MCR000
USAF M-4 see ISQ000
USAF M-5 see ALI000
USAF M-6 see FOF000
USAF M-7 see CEI500
USAF A-233 see AIR250
USAF A-3701 see TEV000
USAF A-4600 see MAO250
USAF A-6598 see DXP200
USAF A-8354 see TES250
USAF A-8564 see BIQ500
USAF A-8565 see HIM500
USAF A-8798 see CKT250
USAF A-9230 see EEW000
USAF A-9442 see SNE000
USAF A-9789 see DVX600
USAF A-11074 see DXN400
USAF A-14980 see MIV300
USAF A -15972 see DGX000
USAF A-19120 see BEX500
USAF AB-315 see DXJ800
USAF AM-1 see DJI400
USAF AM-2 see DNP700
USAF AM-3 see EMU500
USAF AM-4 see MKW750
USAF AM-5 see AAH250
USAF AM-6 see BSU500
USAF AM-7 see MKW500
USAF AM-8 see IJT000
USAF AN-7 see MFC700
USAF AN-8 see TDK000
USAF AN-9 see DOA400
USAF AN-11 see CAM000
USAF B-7 see ADC750
USAF B-15 see TFC600
USAF B-17 see BKU500
USAF B-19 see DWC600
USAF B-21 see DCH400
USAF B-22 see TFD250
USAF B-24 see SAV000
USAF B-30 see TFS350
USAF B-31 see TEF750
USAF B-32 see BJL600
USAF B-33 see BDE750, DXH250
USAF B-35 see SGF500
USAF B-40 see MRM750
USAF B-44 see BLJ250, DXL800
USAF B-45 see DCF000
USAF B-51 see PAY500
USAF B-58 see FPM000
USAF B-59 see TGN000
USAF B-100 see DJY800
USAF B-121 see MAL250
USAF BE-25 see TFJ750
USAF BE-0405 see ADC750
USAF BO-1 see TKM250
USAF BO-2 see BBM000
USAF BV-8 see CJF500
USAF C-1 see TEV000
USAF CB-2 see ILD000
USAF CB-7 see BAC250
USAF CB-10 see AEH750
USAF CB-11 see GLS000
USAF CB-13 see FMT000
USAF CB-17 see XCA000
USAF CB-18 see AEH000
USAF CB-19 see NCG000
USAF CB-20 see TET300
USAF CB-21 see TFA000
USAF CB-22 see NBE500

USAF CB-25 see HOJ100
USAF CB-26 see THT350
USAF CB-27 see RDK000
USAF CB-29 see ICW000
USAF CB-30 see THR750
USAF CB-34 see CQJ750
USAF CB-35 see TFJ100
USAF CB-36 see MCM750
USAF CB-37 see MRM750
USAF CB-96 see HOO100
USAF CF-2 see QCJ000
USAF CF-3 see HBU000
USAF CF-5 see RSU000
USAF CS-1 see DPM400
USAF CS-4 see AKS750
USAF CS-6 see NNM000
USAF CY-2 see CAQ250
USAF CY-4 see NAP500
USAF CY-5 see BDE750
USAF CY-6 see MJN250
USAF CY-7 see BDG000
USAF CY-9 see MES000
USAF CY-10 see NBA500
USAF CY-14 see DCF000
USAF CZ-1 see LFH000
USAF DO-1 see CEB250
USAF DO-4 see DES000
USAF DO-5 see HHW000
USAF DO-6 see BMY500
USAF DO-10 see GHA100
USAF DO-11 see ELB000
USAF DO-12 see HHR500
USAF DO-14 see HKI500
USAF DO-17 see PDQ750
USAF DO-19 see CPG700
USAF DO-20 see BQW000
USAF DO-21 see EKR500
USAF DO-22 see HKY500
USAF DO-23 see AGR000
USAF DO-28 see DMI600
USAF DO-29 see CDY850
USAF DO-30 see BGG500
USAF DO-32 see TCC000
USAF DO-36 see DVF200
USAF DO-37 see HKM500
USAF DO-38 see DVR000
USAF DO-40 see BMR100
USAF DO-41 see TNC500
USAF DO-42 see DDS000
USAF DO-43 see THU750
USAF DO-44 see EMU500
USAF DO-45 see AAG000
USAF DO-46 see AKB000
USAF DO-47 see AGO000
USAF DO-49 see HDH200
USAF DO-50 see MGG000
USAF DO-51 see BOB500
USAF DO-52 see MKU250
USAF DO-53 see BHB300
USAF DO-54 see MHF750
USAF DO-55 see IBH000
USAF DO-59 see PGH000
USAF DO-61 see BHJ000
USAF DO-62 see TBQ500
USAF DO-63 see DWY200
USAF DO-65 see HCE000
USAF DO-68 see DGK200
USAF E-1 see PEW250
USAF E-2 see MCM750
USAF E-4 see DHO400
USAF EA-1 see NGG500
USAF EA-2 see NGE000
USAF EA-3 see FPE100
USAF EA-4 see NGE500
USAF EA-5 see NGC000
USAF EA-14 see NGC400
USAF EE-3 see MCN750
USAF EK-3 see AAQ500
USAF EK-206 see PEY750
USAF EK-218 see QMJ000
USAF EK-245 see DWN800
USAF EK-338 see DOT300
USAF EK-356 see HIH000
USAF EK-394 see PEY500
USAF EK-442 see BBL000

USAF EK-488 see ABE500
USAF EK-496 see ABH000
USAF EK-497 see ISR000
USAF EK-510 see TGP250
USAF EK-534 see CBM250
USAF EK-572 see IEE000
USAF EK-600 see CBN000
USAF EK-631 see AHR240
USAF EK-660 see MCK500
USAF EK-678 see PEY600
USAF EK-695 see CCC750
USAF EK-704 see ASL250
USAF EK-705 see TFF250
USAF EK-743 see DVY000
USAF EK-749 see GKY000
USAF EK-794 see QPA000
USAF EK-906 see EMU500
USAF EK-982 see MHK500
USAF EK-1047 see DJC400
USAF EK-1235 see EOL500
USAF EK-1239 see AAY000
USAF EK-1270 see DWC600
USAF EK-1275 see TFQ000
USAF EK-1375 see PEI000
USAF EK-1509 see TGO750
USAF EK-1569 see PGN250
USAF EK-1597 see EEC600
USAF EK-1651 see DXP600
USAF EK-1719 see TFA000
USAF EK-1803 see DKC400
USAF EK-1853 see MHI750
USAF EK-1860 see TFM250
USAF EK-1902 see TFA250
USAF EK-1995 see COH500
USAF EK-2070 see EMB200
USAF EK-2089 see TFS350
USAF EK-2122 see HBD500
USAF EK-2124 see BEU500
USAF EK-2138 see DEI000
USAF EK-2219 see BGJ250
USAF EK-2596 see SGJ000
USAF EK-2635 see DJD000
USAF EK-2676 see TGP000
USAF EK-2680 see TGO800
USAF EK-2784 see CEM500
USAF EK-3092 see DWN200
USAF EK-3110 see DWN400
USAF EK-3302 see ELL500
USAF EK-3941 see AIS500
USAF EK-3967 see HHB500
USAF EK-4037 see AIG000
USAF EK-4196 see MCN250
USAF EK-4376 see AIF500
USAF EK-4394 see DXO200
USAF EK-4628 see HES000
USAF EK-4733 see IAL000
USAF EK-4812 see BDE500
USAF EK-4890 see ADD250
USAF EK-5017 see BDI500
USAF EK-5185 see MMD500
USAF EK-5199 see SKH500
USAF EK-5296 see CKT500
USAF EK-5426 see PGN000
USAF EK-5429 see BDJ000
USAF EK-5432 see BDE750
USAF EK-5496 see TEV500
USAF EK-6232 see QRS000
USAF EK-6279 see SIN000
USAF EK-6454 see MPW500
USAF EK-6540 see BCC500
USAF EK-6561 see ALQ000
USAF EK-6583 see MCK000
USAF EK-6754 see HNI000
USAF EK-6775 see DWI200
USAF EK-7087 see DWO000
USAF EK-7094 see QRJ000
USAF EK-7119 see MLE750
USAF EK-7317 see QSJ000
USAF EK-7372 see TFE250
USAF EK-8413 see AJT500
USAF EK see TKO250
USAF EK-P-433 see ANW750
USAF EK-P-583 see FOU000
USAF EK-P-737 see TFA500
USAF EK-P-4382 see CIH100

USAF EK-P-5430 see ACQ250
USAF EK-P-5501 see AMS250
USAF EK-P-5976 see AQN635
USAF EK-P-6255 see BJL600
USAF EK-P-6281 see BLJ250, DXL800
USAF EK-P-6297 see MCR000
USAF EK-T-434 see SIA500
USAF EK-T-2805 see MCK750
USAF EK-T-6645 see BKU500
USAF EL-23 see PAP550
USAF EL-30 see MCO500
USAF EL-42 see EEH000
USAF EL-44 see CKE750
USAF EL-45 see PGN000
USAF EL-52 see MFC500
USAF EL-54 see BEQ500
USAF EL-57 see IAO000
USAF EL-62 see IAQ000
USAF EL-76 see PDE500
USAF EL-78 see DQQ000
USAF EL-82 see BDY000
USAF EL-101 see BQP250
USAF EL-108 see MNM500
USAF FA-4 see TES250
USAF FA-5 see BCP500
USAF FO-1 see BQP250
USAF GE-1 see MNV750
USAF GE-12 see PCP250
USAF GE-13 see DWM000
USAF GE-14 see HNI500
USAF GY-1 see DXP400
USAF GY-2 see BLE500
USAF GY-3 see BDF000
USAF GY-5 see BIX000
USAF GY-7 see BHA750
USAF H-1 see QSA000
USAF HA-2 see RGZ550
USAF HA-5 see DGS600
USAF HC-1 see SBJ500
USAF IN-399 see MKW000
USAF KE-3 see MEQ000
USAF KE-5 see IQW000
USAF KE-7 see OAV000
USAF KE-8 see HKJ000
USAF KE-11 see EJM500
USAF KE-13 see SLL000
USAF KE-20 see HIN500
USAF KF-1 see COK250
USAF KF-2 see MCK750
USAF KF-3 see PES750
USAF KF-5 see CDN500
USAF KF-10 see EEW000
USAF KF-11 see CEQ600
USAF KF-13 see DVX200
USAF KF-14 see COJ250
USAF KF-15 see HIO000
USAF KF-17 see COJ500
USAF KF-18 see COH250
USAF KF-21 see PEA750
USAF KF-22 see MIQ000
USAF KF-25 see EHP500
USAF KF-26 see MAN750
USAF LO-3 see MCR250
USAF MA-1 see NEL500
USAF MA-2 see BDH000
USAF MA-3 see CEK000
USAF MA-4 see AID500
USAF MA-5 see NFJ500
USAF MA-9 see BLA800
USAF MA-10 see CDQ750
USAF MA-12 see AJF500
USAF MA-13 see CEG800
USAF MA-16 see BDH500
USAF MA-17 see MLX750
USAF ME-1 see BAD750
USAF MK-1 see DDF700
USAF MK-5 see BKD800
USAF MK-6 see DXO200
USAF MK-43 see BJJ000
USAF MO-2 see ANM500
USAF NB-1 see CEG750
USAF ND-09 see PHY000
USAF ND-54 see DMI600
USAF ND-59 see DMH400
USAF ND-60 see SOB600

USAF P-2 see BJK500
USAF P-5 see TFS350
USAF P-7 see DXQ500
USAF P-8 see CJX750
USAF P-220 see QQS200
USAF PD-20 see DOT800
USAF PD-25 see AKM000
USAF PD-57 see TEV600
USAF PE-1 see PEA500
USAF Q-1 see FPW000
USAF Q-2 see TCS500
USAF RH-1 see MDN500
USAF RH-3 see DPG600
USAF RH-4 see MCD750
USAF RH-6 see OFK000
USAF RH-7 see HGP000
USAF RH-8 see MLC750
USAF SC-2 see GKE000
USAF SN-9 see TEX250
USAF SN-31 see DVB850
USAF SO-1 see EGL500
USAF SO-2 see EPK000
USAF ST-40 see MGA750
USAF SZ-1 see BNM250
USAF SZ-3 see MOO500
USAF SZ-B see MOO500
USAF T-2 see MPH750
USAF TH-3 see THS250
USAF TH-9 see DXM600
USAF UCTL-7 see AES639
USAF UCTL-8 see MJV750
USAF UCTL-958 see BHY750
USAF UCTL-974 see BII500
USAF UCTL-1766 see MCL500
USAF UCTL-1791 see DOA400
USAF UCTL-1856 see AMC000
USAF VI-6 see PDF250
USAF WI-1 see MRN000
USAF WI-3 see MLY500
USAF XF-21 see BCC500
USAF XR-10 see DBL800
USAF XR-19 see PFL850
USAF XR-20 see PPQ630
USAF XR-22 see AMY050
USAF XR-27 see AIS500
USAF XR-29 see BDF000
USAF XR-30 see AIS550
USAF XR-31 see AJY250
USAF XR-32 see AJY500
USAF XR-35 see MCK750
USAF XR-41 see CJX750
USAF XR-42 see DXQ500
USALAKE FD & C YELLOW NO. 6 LAKE see FAG150
USAN-DIAZIQUONE see ASK875
USB-3153 see DWS200
USB-3584 see CNE500
U.S. BLENDED LIGHT TOBACCO CIGARETTE REFINED TAR see CMP800
USNEIN see UWJ000
USNIACIN see UWJ000
d-USNIC ACID see UWJ100
USNIC ACID see UWJ000
(+)-USNIC ACID see UWJ100
USNIC ACID, (R)-(8CI) see UWJ100
d-USNINIC ACID see UWJ100
USNINIC ACID see UWJ000
(+)-USNINIC ACID see UWJ100
USNINSAEURE (GERMAN) see UWJ000
U.S.P. MENTHOL see MCG250
USP METHYLCELLULOSE see MIF760
USP SODIUM CHLORIDE see SFT000
USPULUM see CHW675
USP XIII STEARYL ALCOHOL see OAX000
USR 604 see DFT000
U.S. RUBBER 604 see DFT000
U.S. RUBBER D-014 see SOP000
UST see UTU500
USTINEX see CIR250
U 46T see TGF200
UTEDRIN see ORU500
UTERACON see ORU500
UTERAMINE see TOG250
UTIBID see OOG000
UTICILLIN see CBO000
UTICILLIN VK see PDT750
UTIL see EQL000

UTOPAR see RLK700
UTOSTAN see PDC250
UTRASUL see MPQ750
UV 1 see HND100
UV 531 see HND100
UVA 1 see HND100
UV ABSORBER-1 see HML500
UV ABSORBER-2 see DGX000
UVALERAL see BNP750
U-VAN 28 see MCB050
U-VAN 62 see MCB050
U-VAN 102 see MCB050
U-VAN 120 see MCB050
U-VAN 122 see MCB050
U-VAN 128 see MCB050
U-VAN 220 see MCB050
U-VAN 221 see MCB050
U-VAN 225 see MCB050
U-VAN 2020 see MCB050
U-VAN 20HS see MCB050
U-VAN 28N see MCB050
U-VAN 20N60 see MCB050
U-VAN 21R see MCB050
U-VAN 22R see MCB050
U-VAN 60R see MCB050
U-VAN 20S see MCB050
U-VAN 20SA see MCB050
U-VAN 20SB see MCB050
U-VAN 20SE see MCB050
U-VAN 28SE see MCB050
U-VAN 20SE50 see MCB050
U-VAN 20SE60 see MCB050
U-VAN 21HV see MCB050
UV CHEK AM 104 see BIW750
UVINOL D-50 see BCS325
UVINUL 400 see DMI600
UVINUL 408 see HND100
UVINUL D-50 see BCS325
UVINUL M 40 see MES000
UV 20SR see MCB050
UZONE see BRF500

V 7 see TKX250
V-18 see CKM000
V 50 see ASM050
V 252 see CGD000
V 255 see CKN750
V 315 see CIT750
V 316 see CFU500
V 317 see DIK800
V331 see DIM400
V 340 see DIS000
V 343 see DHJ400
V 346 see DIK600
V 374 see DTR800
V 375 see CGI750
V 377 see DNM600
V 4917 see TLD000
V 17004 see XTJ000
VA 0112 see AAX250
VA see VFF000
VABEN see CFZ000
VABROCID see NGE500
VAC see VLU250
VACATE see CIR250
VACOR see PPP750
VACUUM RESIDUUM see MQV755
VACUUM RESIDUUM (PETROLEUM) see RDK200
VADEBEX see NOA000
VADEBEX HYDROCHLORIDE see NOA500
VADILEX see IAG625
VADOSILAN see VGA300
VADROCID see NGE500
VAFLOL see VSK600
VAGAMIN see DJM800, XCJ000
VAGANTIN see DJM800, XCJ000
VAGD see AAX175
VAGESTROL see DKA600
VAGILEN see MMN250
VAGILIA see SNN500
VAGIMID see MMN250
VAGISEPT see ABX500
VAGOFLOR see ABX500
VAGOPHEMANIL see DAP800

VAGOPHEMANIL METHYL SULFATE see DAP800
VAGOSIN see CPQ250
VAGOSTIGMIN see DER600
VAGOSTIGMINE see NCL100
VAGOSTIGMINE BROMIDE see POD000
VAGOSTIGMINE METHYL SULFATE see DQY909
VAL 13081 see VCK100
VALACIDIN see SMA000
VALADOL see HIM000
VALAMINA see EEH000
VALAMINE see IIM000
VALAMINETTEN see EEH000
VALAN see MRW000
VALANAS see MFD500
VALBAZEN see VAD000
VALBIL see BPP250
VALCOR see CPP000
VALDRENE see BAU750
VAL-DROP see SFS000
VALECOR see PEV750
VALENTINITE see AQF000
VALEO see DCK759
VALERAL see VAG000
n-VALERALDEHYDE see VAG000
VALERAMIDE, 2-(2-ACETAMIDO-4-METHYLVALERAMIDO)-N-(1-FORMYL-4-
    GUANIDINOBUTYL)-4-METHYL-(S)- see AAL300
VALERAMIDE-OM see DOY400
VALERAN DI-n-BUTYLCINICITY (CZECH) see DEA600
VALERIANIC ACID see VAQ000
VALERIANIC ALDEHYDE see VAG000
n-VALERIC ACID see VAQ000
VALERIC ACID see VAQ000
VALERIC ACID ALDEHYDE see VAG000
VALERIC ACID, 2-AMINO-3-METHYL- see IKX000
VALERIC ACID, 2-AMINO-4-METHYL- see LES000
VALERIC ACID, 4,4'-AZOBIS(4-CYANO)- see ASL500
VALERIC ACID, 3-HEXENYL ESTER, (Z)- see HFE800
VALERIC ACID, METHYL ESTER see VAQ100
VALERIC ALDEHYDE see VAG000
4-VALEROLACTONE see VAV000
γ-VALEROLACTONE (FCC) see VAV000
VALERON see PJS750
VALERONE see DNI800
n-VALERONITRILE see VAV300
VALERONITRILE see VAV300
Δ-VALEROSULTONE see BOU250
VALERYLALDEHYDE see VAG000
VALERYL CHLORIDE see VBA000
4-VALERYLPYRIDINE see VBA100
VALETAN see DEO600
VALETHAMATE see VBK000
VALETHAMATE BROMIDE see VBK000
VALEXONE see BAT750
VALFLON see TAI250
VALFOR see AHF500
VALGESIC see HIM000
VALGIS see TEH500
VALGRAINE see TEH500
VALI FAST RED 1308 see RGW000
l-VALINE (FCC) see VBP000
d-VALINE see VBU000
VALINE see VBP000
VALINE ALDEHDYE see IJS000
VALINOMYCIN see VBZ000
VALIOIL see HNI500
VALISONE see VCA000
VALITRAN see DCK759
VALIUM see DCK759
VALKACIT CA see DWN800
VALLADAN see BCA000
VALLENE see MBW750
VALLERGINE see DQA400
VALMAGEN see TAB250
VALMETHAMIDE see ENJ500
VALMID see EEH000
VALMIDATE see EEH000
VALOID see EAN600
VALONEA TANNIN see VCK000
VALONTIN see CAY675
VALOPRIDE see VCK100
VALORON HYDROCHLORIDE see EIH000
VALPIN see LJS000
VALPROATE see PNR750
VALPROATE SODIUM see PNX750
VALPROIC ACID see PNR750

VALPROIC ACID CALCIUM SALT see CAY675
VALPROIC ACID HEMI-CALCIUM SALT see CAY675
VALPROIC ACID SODIUM SALT see PNX750
VALPROMIDE see PNX600
VALSPEX 155-53 see PJS750
VALSYN see FPI000
VALYL GRAMICIDIN A see GJO025
VALZIN see EFE000
VAMIDOATE see MJG500
VAMIDOTHION see MJG500
V. AMMODYTES VENOM see VQZ425
VANADIC ACID, AMMONIUM SALT see ANY250
VANADIC ACID, MONOSODIUM SALT see SKP000
VANADIC ACID, POTASSIUM SALT see PLK900
VANADIC(II) ACID, TRISODIUM SALT see SIY250
VANADIC ANHYDRIDE see VDU000
VANADIC OXIDE see VEA000
VANADIO, PENTOSSIDO di (ITALIAN) see VDU000
VANADIOUS(4+) ACID, DISODIUM SALT see SKQ000
VANADIUM (OSHA) see VDZ000
VANADIUM see VCP000
VANADIUM AZIDE TETRACHLORIDE see VCU000
VANADIUM compounded with BERYLLIUM (1:12) see BFR250
VANADIUM BROMIDE see VEK000
VANADIUM(III) CHLORIDE see VEP000
VANADIUM CHLORIDE see VEF000
VANADIUM COMPOUNDS see VCZ000
VANADIUM DICHLORIDE see VDA000
VANADIUM DUST and FUME (ACGIH) see VDU000, VDZ000
VANADIUM ORE see VDF000
VANADIUM(V) OXIDE see VDU000
VANADIUM OXIDE see VEA000
VANADIUM OXIDE TRIISOBUTOXIDE see VDK000
VANADIUM OXYTRICHLORIDE see VDP000
VANADIUM PENTAOXIDE see VDU000
VANADIUMPENTOXID (GERMAN) see VDU000
VANADIUM PENTOXIDE, non-fused form (DOT) see VDU000
VANADIUM PENTOXIDE, nonfused form (DOT) see VDZ000
VANADIUM PENTOXIDE (dust) see VDU000
VANADIUM PENTOXIDE (fume) see VDZ000
VANADIUMPENTOXYDE (DUTCH) see VDU000
VANADIUM, PENTOXYDE de (FRENCH) see VDU000
VANADIUM SESQUIOXIDE see VEA000
VANADIUM SULFATE see VEA100
VANADIUM SULPHATE see VEA100
VANADIUM TETRACHLORIDE see VEF000
VANADIUM TRIBROMIDE see VEK000
VANADIUM TRIBROMIDE OXIDE see VEK100
VANADIUM TRICHLORIDE see VEP000
VANADIUM TRICHLORIDE OXIDE see VDP000
VANADIUM TRIOXIDE see VEA000
VANADYL AZIDE DICHLORIDE see VEU000
VANADYL SULFATE see VEZ000
VANADYL SULFATE PENTAHYDRATE see VEZ100
VANADYL TRICHLORIDE see VDP000
VANANOTE see MDW750
VANAY see THM500
VANCENASE see AFJ625
VANCERIL see AFJ625
VANCIDA TM-95 see TFS350
VANCIDE 89 see CBG000
VANCIDE see VEZ925
VANCIDE BL see TFD250
VANCIDE FE95 see FAS000
VANCIDE KS see HON000
VANCIDE MANEB 80 see MAS500
VANCIDE MZ-96 see BJK500
VANCIDE P see ZMJ000
VANCIDE PA see BLG500
VANCIDE PA DISPERSION see BLG500
VANCIDE PB see DVI600
VANCIDE TH see HDW000
VANCIDE TM see TFS350
VANCOCIN see VFA000
VANCOCINE HYDROCHLORIDE see VFA050
VANCOCIN HYDROCHLORIDE see VFA050
VANCOMYCIN see VFA000
VANDEX see SBO500
VAN DYK 264 see OES000
VAN DYKE BROWN see MAT500
VANGARD see MEG250
VANGARD K see CBG000
VANGUARD GF see FAZ000
VANGUARD N see BIW750
VANICID see VEZ925

VANICIDE see CBG000
VANILLA see VFK000
VANILLABERON see VEZ925
VANILLAL see EQF000
VANILLALDEHYDE see VFK000
VANILLA PLANT see DAJ800
VANILLA TINCTURE see VFA200
o-VANILLIC ACID see HJB500
p-VANILLIC ACID see VFF000
VANILLIC ACID see VFF000
VANILLIC ACID DIETHYLAMIDE see DKE200
VANILLIC ACID-N,N-DIETHYLAMIDE see DKE200
VANILLIC ALDEHYDE see VFK000
o-VANILLIN see VFP000
p-VANILLIN see VFK000
VANILLIN see VFK000
VANILLIN METHYL ETHER see VHK000
VANILLINSAEURE-DIAETHYLAMID (GERMAN) see DKE200
VANILLYL ACETONE see VFP100
VANILOL see VFP200
VANIROM see EQF000
VANITROPE see IRY000
VANIZIDE see VEZ925
VANLUBE PCX see BFW750
VANNAX NS see BQK750
VANOBID see LFF000
VANOXIDE see BDS000
VANQUIN see PQC500
VANSIL see OLT000
VANSIL W 10 see WCJ000
VANSIL W 20 see WCJ000
VANSIL W 30 see WCJ000
VANTYL see PGG000
VANZOATE see BCM000
VAPAM see SIL550, VFU000
VAPIN see LJS000
VAPO-N-ISO see DMV600
VAPO-ISO see IMR000
VAPONA see DGP900
VAPONEFRIN see VGP000
VAPONITE see DGP900
VAPOPHOS see PAK000
VAPOROLE see IMB000
VAPOROOTER see SIL550
VAPORPAC see ILM000
VAPOTONE see TCF250
VARACILLIN see LEJ500
VARAMID ML 1 see BKE500
VARDHAK see NAK500
VARFINE see WAT220
VARIAMINE BLUE SALT RT see PFU500
VARICOL see EMT500
VARIEGATED PHILODENDRON see PLW800
VARIOFORM I see ANN000
VARIOFORM II see USS000
VARISOFT 100 see DXG625
VARISOFT SDC see DTC600
VARITOL see FMS875
VARITON see DAP800
VARITOX see TII250, TII500
VARNISH MARKER'S NAPHTHA see PCT250
VARNOLINE see SLU500
VAROX see DRJ800
VAROX DCP-R see DGR600
VAROX DCP-T see DGR600
VASAL see PAH250
VASALGIN see ELX000
VASAPRIL see PPH050
VASAZOL see DJS200
VASCARDIN see CCK125
VASCOPIL see HNY500
VASCUALS see VSZ450
VASCULAT see BOV825
VASCULIT see BOV825
VASCULOPATINA see DNM400
VASCUNICOL see BOV825
VASE FLOWER see CMV390
VASE VINE see CMV390
VASICIN (GERMAN) see VGA000
VASICINE see VGA000
VASICINONE HYDROCHLORIDE see VGA025
VASIMID see BBW750
VASIODONE see DNG400
VASITOL see PBC250
VASKULAT see BOV825

VASOBRIX 32 see VGA100
VASOCIL see PPH050
VASOCON see NCW000
VASOCONSTRICTINE see VGP000
VASOCONSTRICTOR see VGP000
VASODIATOL see PBC250
VASODIL see BBW750
VASODILAN see VGA300, VGF000
VASODILATAN see BBW750
VASODILATATEUR 2249F see FMX000
VASODILIAN see VGF000
VASODISTAL see VGK000
VASODRINE see VGP000
VASOFILINA see TEP500
VASOGLYN see NGY000
VASOKELLINA see AHK750
VASOLAN see IRV000, VHA450
VASOMED see TJL250
VASOPERIF see CBH250
VASOPLEX see VGA300
VASORBATE see CCK125
VASORELAX see POD750
VASOROME see AOO125
VASOSPAN see PAH250
VASOTON see VGP000
VASOTONIN see VGP000
VASOTRAN see VGA300
VASOTRATE see CCK125
VASO-80 UNICELLES see PBC250
VASOVERIN see PPH050
VASOXINE see MDW000
VASOXINE HYDROCHLORIDE see MDW000
VASOXYL HYDROCHLORIDE see MDW000
VASTAREL see YCJ200
VASTCILLIN see AJJ875
VAT BLACK 1 see CMU320
VAT BLUE 4 see IBV050
VAT BLUE 6 see DFN425
VAT BLUE 18 see VGP100
VAT BLUE KD see DFN425
VAT BLUE O see IBV050
VAT BLUE OD see IBV050
VAT BRIGHT VIOLET K see DFN450
VAT BRILLIANT ORANGE see CMU820
VAT BRILLIANT VIOLET K see DFN450
VAT BRILLIANT VIOLET KD see DFN450
VAT BRILLIANT VIOLET KP see DFN450
VATERITE see CAO000
VAT FAST BLUE BCS see DFN425
VAT FAST BLUE R see IBV050
VAT GOLDEN YELLOW see DCZ000
VAT GRAY S see CMU475
VAT GREEN 3 see CMU810
VAT GREEN B see DFN425
VAT GREY S see CMU475
VAT ORANGE R see CMU815
VAT ORANGE RF see CMU815
VAT PRINTING ORANGE R see CMU815
VATRACIN see AJJ875
VATRAN see DCK759
VAT RED 13 see CMU825
VAT RED 5B see DNT300
VATROLITE see SHR500
VAT SCARLET 2Zh see CMU820
VAT SKY BLUE K see DFN425
VAT SKY BLUE KD see DFN425
VAT SKY BLUE KP 2F see DFN425
VATSOL OT see DJL000
VAT YELLOW 2 see VGP200
VAZADRINE see ILD000
VAZO 64 see ASL750
VAZOFIRIN see PBU100
V-BRITE see SHR500
V-C 3-670 see DWU200
VC see VNP000
V-C CHEMICAL V-C 9-104 see EIN000
V-CIL see PDT500
V-CIL-K see PDT750
V-CILLIN see PDT500
V-CILLIN K see MNV250, PDT750
VCM see VNP000
VCN see ADX500
VC13 NEMACIDE see DFK600
VCR see LEY000
VCR SULFATE see LEZ000

VCS 438 see BGD250
VDC see VPK000
VDF see VPP000
VDM see VFU000
VD 1827 p-TOLUENESULFONATE see SOU675
VEATCHINE HYDROCHLORIDE see VGU000
VEBECILLIN see PDT500
VECTAL see ARQ725
VECTAL SC see ARQ725
VECTREN see CIP500
VECURONIUM BROMIDE see VGU075
VEDERON see ILD000
VEDITA 250 see THR525
VEDRIL see PKB500
VEE GEE GELATIN see PCU360
VEETIDS see PDT750
VEGABEN see AJM000
VEGADEX see CDO250
VEGADEX SUPER see CDO250
VEGANTINE see DHX800, THK000
VEGASERPINA see RCA200
VEGENTINE FAST BROWN B see CMO820
VEGETABLE GUM see DBD800
VEGETABLE-HUMMING-BIRD see SBC550
VEGETABLE OIL 1400 see CBF710
VEGETABLE OIL see VGU200
VEGETABLE OIL MIST (OSHA) see VGU200
VEGETABLE PEPSIN see PAG500
VEGETOX see BHL750
VEGFRU see PGS000
VEGFRU FOSMITE see EEH600
VEGFRU MALATOX see MAK700
VEGOLYSEN see HEA000
VEGOLYSIN see HEA000
VEHAM-SANDOZ see EQP000
VEHEM see EQP000
VEJIGA de PERRO (CUBA) see JBS100
VEL 3973 see DUK000
VEL 4283 see MKA000
VEL 4284 see DRR200
VELACICLINE see PPY250
VELACYCLINE see PPY250
VELARDON see PAG500
VELBAN see VLA000
VELBE see VLA000
VELDOPA see DNA200
VELFLON see TAI250
VELLORITA (CUBA) see CNX800
VELMOL see DJL000
VELON see CGW300
VELO de NOVIA (MEXICO) see GIW200
VELOSEF see SBN440
VELOSEF HYDRATE see SBN450
VELPAR see HFA300
VELPAR WEED KILLER see HFA300
VELSICOL 104 see HAR000
VELSICOL 506 see LEN000
VELSICOL 1068 see CDR750
VELSICOL see TIK000
VELSICOL COMPOUND C see TIK000
VELSICOL COMPOUND "R" see MEL500
VELSICOL 58-CS-11 see MEL500
VELSICOL 53-CS-17 see EBW500
VELSICOL FCS-303 see BND750
VELSICOL VCS 506 see LEN000
VELUSTRAL KPA see PJS750
VELVETEX see CBT750
VENA see BBV500
VENACIL see VGU700
VENCIPON see EAW000
VENDESINE SULFATE see VGU750
VENDEX see BLU000
VENETIAN RED see IHD000
VENETLIN see BQF500
VENGICIDE see VGZ000
VENOM, AUSTRALIAN ELAPIDAE SNAKE, ACANTHOPHIS ANTARCTICUS see ARU875
VENOM, AUSTRALIAN ELAPIDAE SNAKE, AUSTRELAPS SUPERBA see ARU750
VENOM, AUSTRALIAN ELAPIDAE SNAKE, OXYURANUS SCUTELLATUS see ARV500
VENOM, AUSTRALIAN ELAPIDAE SNAKE, PSEUDECHIS PORPHYRIACUS see ARV250
VENOM, AUSTRALIAN SNAKE, AUSTRELAPS SUPERBA see ARV625
VENOM, AUSTRALIAN SNAKE, NOTECHIS SCUTATUS see ARV550

VENOM, AUSTRALIAN SNAKE, OPHIOPHAGUS HANNAH see ARV125
VENOM, AUSTRALIAN SNAKE, PSEUDECHIS see ARV000
VENOM, AUSTRALIAN SNAKE, PSEUDONAJA TEXTILIS see TEG650
VENOM, AUSTRALIAN SNAKE, TROPIDECHIS CARINATUS see ARV375
VENOM, COSTA RICAN SNAKE, BOTHROPS ASPER see BMI000
VENOM, COSTA RICAN SNAKE, BOTHROPS ATROX see BMI125
VENOM, COSTA RICAN SNAKE, BOTHROPS GODMANI see BMI500
VENOM, COSTA RICAN SNAKE, BOTHROPS LATERALIS see BMI750
VENOM, COSTA RICAN SNAKE, BOTHROPS NASUTUS see BMJ000
VENOM, COSTA RICAN SNAKE, BOTHROPS NIGROVIRIDIS NIGROVIRIDIS
see BMJ250
VENOM, COSTA RICAN SNAKE, BOTHROPS NUMMIFER MEXICANUS see
BMJ500
VENOM, COSTA RICAN SNAKE, BOTHROPS OPHRYOMEGA see BMJ750
VENOM, COSTA RICAN SNAKE, BOTHROPS PICADOI see BMK000
VENOM, COSTA RICAN SNAKE, BOTHROPS SCHLEGLII see BMK250
VENOM, COSTA RICAN SNAKE, MICRURUS ALLENI YATESI see MQS750
VENOM, COSTA RICAN SNAKE, MICRURUS MIPARTITUS HERTWIGI see
MQT250
VENOM, COSTA RICAN SNAKE, MICRURUS NIGROCINCTUS see MQT500
VENOM, EGYPTIAN COBRA, NAJA HAJE ANNULIFERA see NAE512
VENOM, FORMOSAN BANDED KRAIT, BUNGARUS MULTICINCTUS see
BON370
VENOM, INDIAN COBRA, NAJA NAJA NAJA see ICC700
VENOM, LIZARD, HELODERMA SUSPECTUM see HAN625
VENOM, MIDGET FADED RATTLESNAKE, CROTALUS VIRIDIS CONCOLOR
see CNY750
VENOM, MIDGET FADED RATTLESNAKE, CROTALUS VIRIDIS HELLERI see
COA000
VENOM, MIDGET FADED RATTLESNAKE, CROTALUS VIRIDUS CERBERUS
see CNY390
VENOM, ORIENTAL HORNET, VESPA ORIENTALIS see VJP900
VENOM, SCORPION, ANDROCTONUS AMOREUXI see AOO250
VENOM, SCORPION, ANDROCTONUS AUSTRALIS HECTOR see AOO265
VENOM, SCORPION, CENTRUROIDES SUFFUSUS SUFFUSUS see CCW925
VENOM, SEA ANEMONE, AIPTASIA PALLIDA see SBI800
VENOM, SEA SNAKE, AIPYSURUS LAEVIS see SBI880
VENOM, SEA SNAKE, AIPYSURUS LAEVIS (AUSTRALIA) see AFG000
VENOM, SEA SNAKE, ASTROTIA STOKESII see SBI890
VENOM, SEA SNAKE, ENHYDRINA SCHISTOSA see EAU000
VENOM, SEA SNAKE, HYDROPHIS CYANOCINCTUS see SBI900
VENOM, SEA SNAKE, HYDROPHIS ELEGANS see HIG000
VENOM, SEA SNAKE, LAPEMIS HARDWICKII see SBI910
VENOM, SEA SNAKE, LATICAUDA SEMIFASCIATA see LBI000
VENOM, SEA SNAKE, MICROCEPHALOPHIS GRACILIS see SBI929
VENOM, SEA SNAKE, NAJA NAJA ATRA see NAF000
VENOM, SNAKE, AGKISTRODON ACUTUS see HGM600
VENOM, SNAKE, AGKISTRODON CONTORTRIX see AEY125
VENOM, SNAKE, AGKISTRODON CONTORTRIX CONTORTRIX see SKW775
VENOM, SNAKE, AGKISTRODON CONTORTRIX MOKASEN see NNX700
VENOM, SNAKE, AGKISTRODON PISCIVORUS see AEY130
VENOM, SNAKE, AGKISTRODON PISCIVORUS PISCIVORUS see EAB200
VENOM, SNAKE, AGKISTRODON RHODOSTOMA see AEY135
VENOM, SNAKE, ANCISTRODON PISCIVORUS see AOO135
VENOM, SNAKE, BOTHROPS COLOMBIENSIS see BMI250
VENOM, SNAKE, BUNGARUS CAERULEUS see BON365
VENOM, SNAKE, BUNGARUS FASCIATUS see BON367
VENOM, SNAKE, CERASTES CERASTES see CCX620
VENOM, SNAKE, CROTALUS ADAMANTEUS see EAB225
VENOM, SNAKE, CROTALUS ATROX see WBJ600
VENOM, SNAKE, CROTALUS CERASTES see CNX830
VENOM, SNAKE, CROTALUS DURISSUS TERRIFICUS see CNY000
VENOM, SNAKE, CROTALUS HORRIDUS see CNY300
VENOM, SNAKE, CROTALUS HORRIDUS HORRIDUS see TGB150
VENOM, SNAKE, CROTALUS RUBER RUBER see CNY325
VENOM, SNAKE, CROTALUS SCUTULATUS SCUTULATUS see CNY350,
CNY375
VENOM, SNAKE, CROTALUS VIRIDIS VIRIDIS see PLX100
VENOM, SNAKE, DENDROASPIS ANGUSTICEPS see DAP815
VENOM, SNAKE, DENDROASPIS JAMESONI see DAP820
VENOM, SNAKE, DENDROASPIS VIRIDIS see GJU500
VENOM, SNAKE, DISPHOLIDUS TYPHUS see DXG150
VENOM, SNAKE, ECHIS CARINATUS see EAD600
VENOM, SNAKE, ECHIS COLORATUS see EAD650
VENOM, SNAKE, HEMACHATUS HAEMCHATES see HAO500
VENOM, SNAKE, NAJA FLAVA see NAE510
VENOM, SNAKE, NAJA HAJE see NAE515
VENOM, SNAKE, NAJA MELANOLEUCA see NAE875
VENOM, SNAKE, NAJA NAJA KAOUTHIA see NAF200
VENOM, SNAKE, NAJA NIGRICOLLIS see NAG000
VENOM, SNAKE, NICRURUS FULVIUS see MQT100
VENOM, SNAKE, RHABDOPHIS TIGRINUS TIGRINUS see RFU875
VENOM, SNAKE, SISTRURUS MILARIUS BARBOURI see SDZ300
VENOM, SNAKE, TRIMERESURUS FLAVOVIRIDIS see TKW100
VENOM, SNAKE, VIPERA AMMODYTES see VQZ425

VENOM, SNAKE, VIPERA ASPIS see VQZ475
VENOM, SNAKE, VIPERA BORNMULLERI see VQZ500
VENOM, SNAKE, VIPERA LATIFII see VQZ525
VENOM, SNAKE, VIPERA LEBETINA see VQZ550
VENOM, SNAKE, VIPERA PALESTINAE see VQZ575
VENOM, SNAKE, VIPERA RUSSELLII see VQZ635
VENOM, SNAKE, VIPERA RUSSELLII FORMOSENSIS see VQZ625
VENOM, SNAKE, VIPERA XANTHINA PALAESTINAE see VQZ650
VENOM, SNAKE, WALTERINNESIA AEGYPTIA see WAT100
VENOM, SOUTH AFRICAN CAPE COBRA, NAJA NIVEA see NAG200
VENOM, STONEFISH, SYNANCEJA HORRIDA Linn. see SPB875
VENOM, THAILAND COBRA, NAJA NAJA SIAMENSIS see NAF250
VENOPEX see TAB250
VENOPIRIN see VHA275
VENTILAT see ONI000
VENTIN SUMACH see FBW000
VENTIPULMIN see VHA350
VENTOLIN see BQF500
VENTOX see ADX500
VENTRAMINE see DJS200
VENTUROL see DXX400
VENZONATE see BCM000
VEON 245 see TAA100
VEPEN see PDT750
VEPESID see EAV500
VER A see MRV500
VERACILLIN see DGE200
VERACTIL see MCI500
VERAMID see AMK250
VERAMIN ED 4 see EIV750
VERAMIN ED 40 see EIV750
VERAMON see AMK250
VERANTEROL see PPH050
VERANTIN see TKN750
VERAPAMIL see IRV000
VERAPAMIL HYDROCHLORIDE see VHA450
VERAPRET DH see DTG000
VERATENSINE see VHF000
VERATRALDEHYDE see VHK000
VERATRAMINE see VHP500
VERATRIC ACID see VHP600
VERATRIC ALDEHYDE see VHK000
VERATRIDINE see VHU000, VHZ000
VERATRIN (GERMAN) see VHZ000
VERATRIN see NCI600
VERATRINE (amorphous) see VHU000
VERATRINE (crystallized) see CDG000
VERATRINE see CDG000, NCI600, VHZ000
VERATROL see DOA200
VERATROLE see DOA200
VERATROLE METHYL ETHER see AGE250
N-(o-VERATROYL)GLYCINOHYDROXAMIC ACID see VIA875
3-VERATROYLVERACEVINE see VHU000
VERATRUM ALBUM see FAB100
VERATRUM CALIFORNICUM see FAB100
VERATRUM VIRIDE see FAB100, VIZ000
VERATRUM VIRIDE ALKALOIDS EXTRACT see VIZ000
VERATRYL ALDEHYDE see VHK000
VERATRYLAMINE see VIK050
VERATRYL CHLORID (GERMAN) see BKM750
VERATRYL CHLORIDE see BKM750
VERATRYL CYANIDE see VIK100
VERATRYLHYDRAZINE see VIK150
1-VERATRYLPIPERAZINE DIHYDROCHLORIDE see VIK200
VERATRYL-2-PROPANONE see VIK300
VERAX see BBW500
VERAZINC see ZNA000
VERBENA ABSOLUTE see VIK500
VERBENA HYBRIDA Cornol. & Rpl., extract see VIK400
VERBENA OIL see VIK500
d-VERBENONE see VIP000
VERBINDUNG S 557 HCl see AMM750
VERCIDON see DJT800
VERCYTE see BHJ250
VERDANTIOL see LFT100
VERDICAN see DGP900
VERDIPOR see DGP900
VERDONE see CIR250
VERDORACINE see VIP100
VERDOX see BQW490
VERDYL ACETATE see DLY400
VEREDEN see MNM500
VERESENE DISODIUM SALT see EIX500
VERGEMASTER see DAA800
VERGFRU FORATOX see PGS000

VERILOID see RDK000, VIZ000
VERINA see DNU200
3427 VERI PUR PINK see ADG250
VERITAIN see FMS875
VERITOL see FMS875
VERMAGO see PIJ500
VERMICID see AOR500
VERMICIDE BAYER 2349 see TIQ250
VERMICIDIN see MHL000
VERMICOMPREN (TABL.) see HEP000
VERMICULIN see VIZ100
VERMICULINE see VIZ100
VERMILASS see HEP000
VERMINUM see BQK000
VERMIRAX see MHL000
VERMITIN see DFV400, PDP250
VERMIZYM see PAG500
VERMOESTRICID see CBY000
VERMOX see MHL000
VERNALDEHYDE see VIZ150
VERNAM see PNI750
VERNAMYCIN see VRF000
VERNAMYCIN BA see VRA700
VERNINE see GLS000
VERNOLATE see PNI750
VERODIGEN see GES000
VERODOXIN see VIZ200
VEROL see MGJ750
VEROLETTIN see BAG000
VERONAL see BAG000
VERONAL SODIUM see BAG250
VERONA PAPER RED see CMM765
VERONA YELLOW X-1791 see DEU000
VERON P 130/1 see PKQ059
VEROPHEN see DQA600
VEROS 030 see VIZ250
VEROSPIRON see AFJ500
VEROSPIRONE see AFJ500
VERPANYL see MHL000
VERROL see VSZ450
VERRUCARIN A see MRV500
VERSAL BLUE GGSL see IBV050
VERSAL FAST YELLOW PG see CMS212
VERSALIDE see ACL750
VERSAL SCARLET PRNL see MMP100
VERSAL SCARLET RNL see MMP100
VERSAMINE see VIZ400
VERSAR DSMA LQ see DXE600
VERSATIC 9-11 see VIZ500
VERSATIC ACID 911 see VIZ500
VERSATIC 9-11 ACID see VIZ500
VERSENE 9 see TNL250
VERSENE 100 see EIV000
VERSENE ACID see EIX000
VERSENE CA see CAR780, CAR800
VERSENE NTA ACID see AMT500
VERSENE POWDER see EIV000
VERSENE SODIUM 2 see EIX500
VERSENOL 120 see HKS000
VERSENOL see HKS000
VERSICOL E 7 see ADW200
VERSICOL E9 see ADW200
VERSICOL E15 see ADW200
VERSICOLORIN A see TKO500
VERSICOLORIN B see VJP800
VERSICOL S 25 see ADW200
VERSIDYNE see MDV000
VERSNELLER NL 63/10 see DQF800
VERSOMNAL see EOK000
VERSOTRANE see PFN000
VERSTRAN see DAP700
VERSULIN see CDH250
VERSUS see BAV325
VERTAC 90% see CDV100
VERTAC see BQI000, DGI000
VERTAC DINITRO WEED KILLER see BRE500
VERTAC GENERAL WEED KILLER see BRE500
VERT ACIDE BRILLANT BS see ADF000
VERTAC METHYL PARATHION TECHNISCH 80% see MNH000
VERTAC SELECTIVE WEED KILLER see BRE500
VERTAC TOXAPHENE 90 see CDV100
VERTAVIS see VIZ000
VERTENEX see BQW500
VERTHION see DSQ000
VERTIGON see PMF250

VERTISAL see MMN250
VERTOFIX COEUR see ACF250
VERTOLAN see SNJ000
VERTON 2D see DAA800
VERTON D see DAA800
VERTON 2T see TAA100
VERTRON 2D see DAA800
VESADIN see SJW500
VESAKONTUHO MCPA see CIR250
VESALIUM see CLY500
VESAMIN see AAN000
VESPA ORIENTALIS VENOM see VJP900
VESPARAZ-WIRKSTOFF see CJR909
VESPERAL see BAG000
VESPRIN see TKL000
VESTIN see PDC250
VESTINOL AH see DVL700
VESTINOL OA see AEO000
VESTOLEN see PJS750
VESTOLEN A 616 see PJS750
VESTOLEN A 6016 see PJS750
VESTOLEN P 5232G see SMQ500
VESTOLIT B 7021 see PKQ059
VESTROL see TIO750
VESTYRON 512 see SMQ500
VESTYRON 114-12 see SMQ500
VESTYRON see SMQ500
VESTYRON HI see SMR000
VESTYRON MB see SMQ500
VESTYRON N see SMQ500
VETACALM see TAF675
VETAFLAVIN see DBX400
VETALAR see CKD750
VETALOG see AQX500
VETA-MERAZINE see ALF250
VETAMOX see AAI250
VETARCILLIN see BFC750
VETARSENOBILLON see NCJ500
VETBUTAL see NBU000
VETERINARY NITROFURAZONE see NGE500
VETICILLIN see BFD250
VETICOL see CDP250
VETIDREX see CFY000
VETIOL see MAK700
VETIVER ACETATE see AAW750
VETIVEROL ACETATE see AAW750
VETIVERT ACETATE see AAW750
VETIVERT OIL see VJU000
VETIVERYL ACETATE see AAW750
VETKALM see DYF200
VETOL see MAO350
VETOX see CBM750
VETQUAMYCIN-324 see TBX250
VETRANQUIL see ABH500
VETRAZIN see VIK150
VETRAZINE see VIK150
VETREN see HAQ500
VETSIN see MRL500
VETSPEN see PAQ200
VETSTREP see SLY500
VEVETONE see CAT775
VFR 3801 see PKF750
VH see PGP250
VIADRIL see VJZ000
VIALIDON see XQS000
VI-ALPHA see VSK600
VIAMIN MF 514 see MCB050
VIAMIN MF 754 see MCB050
VIANIN see AOR500
VIANSIN see MDQ250
VIARESPAN HYDROCHLORIDE see DAI200
VIAROX see AFJ625
VIBALT see VSZ000
VIBATEX S see PKP750
VIBAZINE see BOM250, HGC500
VIBISONE see VSZ000
VIBRADOX see HGP550
VIBRAMYCIN see DYE425
VIBRAMYCIN HYCLATE see HGP550
VIBRA-TABS see HGP550
VIBRIO CHOLERAE ENDOTOXIN see VJZ100
VI-CAD see CAE250
VICALIN see BKW100
VICCILLIN see AIV500
VICCILLIN S see AIV500

VICILAN see VKF000
VICILLIN see AIV500
VICIN see BFC750
VICKNITE see PLL500
VICRON 31-6 see CAT775
VICRON see CAT775
VICTAN see EKF600
VICTORIA BLUE R see VKA600
VICTORIA BLUE RS see VKA600
VICTORIA GREEN see AFG500
VICTORIA LAKE BLUE R see VKA600
VICTORIA ORANGE see DUT600
VICTORIA RUBINE O see FAG020
VICTORIA SCARLET 3R see FMU080
VICTORIA YELLOW see DUT600
VICTOR TSPP see TEE500
VIDANGA DRIED BERRY EXTRACT see VKA650
VIDARABIN see AQQ900
VIDARABINE see AEH100, AQQ900
VIDDEN D see DGG000, DGG950
VIDEOBIL see IGA000
VIDEOPHEL see TDE750
VIDINE see CMF800
VIDLON see PJY500
VIDON 638 see DAA800
VIDOPEN see AOD125
VIDR-2GD see VKA675
VIENNA GREEN see COF500
VIENNA WHITE see CAT775
VIGANTOL see VSZ100
VIGORSAN see CMC750
VIGOT 15 see CAT775
VIKh 65 see CGW300
VIKANE see SOU500
VIKANE FUMIGANT see SOU500
VIKROL RQ see AFP250
VILESCON see PNS000
VILLESCON see PNS000
VILLIAUMITE see SHF000, SHF500
VILOXAZIN see VKA875
VILOXAZINE see VKA875
VILOXAZINE HYDROCHLORIDE see VKF000
VINAC B 7 see AAX250
VINACETIN A see VQZ000
VINACOL MH see PKP750
VINADINE see OMY850
VINALAK see PKP750
VINAMAR see EQF500
VINAMUL N 710 see SMQ500
VINAMUL N 7700 see SMQ500
VINAROL see PKP750
VINAROL DT see PKP750
VINAROLE see PKP750
VINAROL ST see PKP750
VINAVILOL 2-98 see PKP750
VINBARBITAL SODIUM see VKP000
VINBLASTIN see VKZ000
VINBLASTINE see VKZ000
VINBLASTINE SULFATE see VLA000
VINCADAR see VLF000
VINCAFOLINA see VLF000
VINCAFOR see VLF000
VINCAGIL see VLF000
VINCAIN see AFG750
VINCALEUCOBLASTIN see VKZ000
VINCALEUKOBLASTINE see VKZ000
VINCALEUKOBLASTINE SULFATE see VLA000
VINCALEUKOBLASTINE SULFATE (1:1) (SALT) see VLA000
VINCAMIDOL see VLF000
VINCAMIN COMPOSITUM see ARQ750
(+)-VINCAMINE see VLF000
VINCAMINE see VLF000
VINCA MINOR L., TOTAL ALKALOIDS see VLF300
VINCAPAN see VLF000
VINCAPRONT see VLF000
VINCASAUNIER see VLF000
VINCEINE see AFG750
VINCES see AKO500
VINCIMAX see VLF000
VINCLOZOLIN (GERMAN) see RMA000
VINCOBLASTINE see VKZ000
VINCRISTINE see LEY000
VINCRISTINE SULFATE ONCORIN see LEZ000
VINCRISTINSULFAT (GERMAN) see LEZ000
VINCRISUL see LEZ000

VINCRYSTINE see LEY000
VINCUBINA see TDT770
VINCUBINE see TDT770
VINDESINA SULFATO (SPANISH) see VGU750
VINEGAR ACID see AAT250
VINEGAR NAPHTHA see EFR000
VINEGAR SALTS see CAL750
VINESTHENE see VOP000
VINESTHESIN see VOP000
VINETHEN see VOP000
VINETHENE see VOP000
VINETHER see VOP000
VINFOS see RAF100
VINICIZER 85 DVL600
VINICIZER 80 see DVL700
VI-NICOTYL see NCR000
VI-NICTYL see NCR000
VINIDEN 60 see CGW300
VINIDYL see VOP000
VINIKA KR 600 see PKQ059
VINIKULON see PKQ059
VINILE (ACETATO di) (ITALIAN) see VLU250
VINILE (BROMURO di) (ITALIAN) see VMP000
VINILE (CLORURO di) (ITALIAN) see VNP000
VINIPLAST see PKQ059
VINIPLEN P 73 see PKQ059
VINISIL see PKQ250
VINKEIL 100 see EIX000
VINKRISTIN see LEY000
VINNAROL see PKP750
VINNOL E 75 see PKQ059
VINNOL H 10/60 see AAX175
VINODREL RETARD see VLF000
VINOFLEX see PKQ059
VINOFLEX MO 400* see IJQ000
VINOL 125 see PKP750
VINOL 205 see PKP750
VINOL 351 see PKP750
VINOL 523 see PKP750
VINOL see PKP750
VINOL UNISIZE see PKP750
VINOTHIAM see TES750
VINPOCETINE see EGM100
VINSTOP see SGM500
l-VINTHIONINE see VLU210
VINTHIONINE see VLU200
VINYDAN see VOP000
VINYLACETAAT (DUTCH) see VLU250
VINYLACETAT (GERMAN) see VLU250
VINYL ACETATE, inhibited (DOT) see VLU250
VINYL ACETATE see VLU250
VINYL ACETATE HOMOPOLYMER see AAX250
VINYL ACETATE H.Q. see VLU250
VINYL ACETATE OZONIDE see VLU310
VINYL ACETATE POLYMER see AAX250
VINYL ACETATE RESIN see AAX250
VINYL ACETATE–VINYL CHLORIDE COPOLYMER see AAX175
VINYL ACETATE–VINYL CHLORIDE POLYMER see AAX175
VINYLACETONITRILE see BOX500
VINYL ACETYLENE see BPE109
α-VINYL AE see VMA000
VINYL ALCOHOL POLYMER see PKP750
VINYL AMIDE see ADS250
VINYL A MONOMER see VLU250
VINYL AZIDE see VLY300
1-VINYL AZIRIDINE see VLZ000
α-VINYL-1-AZIRIDINEETHANOL see VMA000
7-VINYLBENZ(a)ANTHRACENE see VMF000
VINYLBENZEN (CZECH) see SMQ000
VINYLBENZENE see SMQ000
VINYLBENZENE POLYMER see SMQ500
VINYL BENZOATE see VMK000
VINYLBENZOL see SMQ000
VINYL(β-BIS(β-CHLOROETHYL)AMINO)ETHYL SULFONE see BHP500
VINYLBROMID (GERMAN) see VMP000
VINYL BROMIDE, inhibited (DOT) see VMP000
VINYL BROMIDE see VMP000
VINYL 2-BUTENOATE see VNU000
VINYL-2-(BUTOXYETHYL) ETHER see VMU000
VINYL-n-BUTYL ETHER see VMZ000
VINYL BUTYL ETHER see VMZ000
VINYL 2-(BUTYLMERCAPTOETHYL) ETHER see VNA000
VINYL BUTYRATE see VNF000
VINYL BUTYRATE, INHIBITED (DOT) see VNF000
VINYLBUTYROLACTAM see EEG000

N-VINYLBUTYROLACTAM POLYMER see PKQ250
VINYL CARBAMATE see VNK000
9-VINYLCARBAZOLE see VNK100
N-VINYLCARBAZOLE see VNK100
VINYLCARBAZOLE see VNK100
VINYLCARBINOL see AFV500
VINYL CARBINYL BUTYRATE see AFY250
VINYL CARBINYL CINNAMATE see AGC000
VINYLCHLON 4000LL see PKQ059
VINYLCHLORID (GERMAN) see VNP000
VINYL CHLORIDE see VNP000
VINYL CHLORIDE ACETATE COPOLYMER see PKP500
VINYL CHLORIDE COPOLYMER with VINYLIDENE CHLORIDE see CGW300
VINYL CHLORIDE-1,1-DICHLOROETHYLENE COPOLYMER see CGW300
VINYL CHLORIDE HOMOPOLYMER see PKQ059
VINYL CHLORIDE MONOMER see VNP000
VINYL CHLORIDE POLYMER see PKQ059
VINYL CHLORIDE VINYL ACETATE COPOLYMER see PKP500
VINYL CHLORIDE–VINYL ACETATE POLYMER see AAX175
VINYL CHLORIDE-VINYLIDENE CHLORIDE COPOLYMER see CGW300
VINYL CHLORIDE-VINYLIDENE CHLORIDE POLYMER see CGW300
VINYL-2-CHLOROETHYL ETHER see CHI250
VINYL-β-CHLOROETHYL ETHER see CHI250
VINYL C MONOMER see VNP000
VINYL CROTONATE see VNU000
VINYL CYANIDE see ADX500
VINYLCYCLOHEXANE MONOXIDE see VNZ000
4-VINYL-1-CYCLOHEXENE see CPD750
4-VINYLCYCLOHEXENE-1 see CPD750
1-VINYLCYCLOHEXENE see VNZ990
1-VINYLCYCLOHEXENE-3 see CPD750
1-VINYLCYCLOHEX-3-ENE see CPD750
4-VINYLCYCLOHEXENE see CPD750
4-VINYL-1,2-CYCLOHEXENE DIEPOXIDE see VOA000
4-VINYL-1-CYCLOHEXENE DIEPOXIDE see VOA000
4-VINYLCYCLOHEXENE DIEPOXIDE see VOA000
VINYL CYCLOHEXENE DIEPOXIDE see VOA000
1-VINYL-3-CYCLOHEXENE DIOXIDE see VOA000
4-VINYLCYCLOHEXENE DIOXIDE see VOA000
4-VINYL-1-CYCLOHEXENE DIOXIDE (MAK) see VOA000
VINYL CYCLOHEXENE DIOXIDE see VOA000
4-VINYLCYCLOHEXENE-1,2-EPOXIDE see VNZ000
4-VINYLCYCLOHEXENE MONOXIDE see VNZ000
VINYLCYCLOHEXENE MONOXIDE see VNZ000
5-VINYL-DEOXYURIDINE see VOA550
VINYLDIAZOMETHANE see DCQ550
VINYL DIHYDROGEN PHOSPHATE see VQA400
VINYL-2-(N,N-DIMETHYLAMINO)ETHYL ETHER see VOF000
VINYLE (ACETATE de) (FRENCH) see VLU250
VINYLE (BROMURE de) (FRENCH) see VMP000
VINYLE (CHLORURE de) (FRENCH) see VNP000
VINYLENE CARBONATE see VOK000
4,4'-VINYLENEDIANILINE see SLR500
4,4'-VINYLENEDIBENZAMIDINE, DIHYDROCHLORIDE see SLS500
N,N'-VINYLENEFORMAMIDINE see IAL000
1-VINYL-3,4-EPOXYCYCLOHEXANE see VNZ000
VINYLESTER KYSELINY MASELNE see VNF000
VINYLESTER KYSELINY OCTOVE see VLU250
VINYL ETHANOATE see VLU250
VINYL ETHER see VOP000
VINYLETHYLENE see BOP500
N-VINYLETHYLENEIMINE see VLZ000
VINYL ETHYL ETHER, inhibited (DOT) see EQF500
VINYL ETHYL ETHER see EQF500
VINYL-2-ETHYLHEXOATE see VOU000
VINYL-2-ETHYLHEXYL ETHER see ELB500
VINYLETHYLNITROSAMIN (GERMAN) see NKF000
VINYLETHYLNITROSAMINE see NKF000
VINYL FLUORIDE, inhibited (DOT) see VPA000
VINYL FLUORIDE see VPA000
VINYL FORMATE see VPF000
VINYLFORMIC ACID see ADS750
VINYLGLYCOLONITRILE see HJQ000
4-VINYLGUAIACOL see VPF100
p-VINYLGUAIACOL see VPF100
S-VINYL-dl-HOMOCYSTEINE see VLU200
S-VINYL-l-HOMOCYSTEINE see VLU210
VINYLIDENE CHLORIDE (II) see VPK000
VINYLIDENE CHLORIDE see VPK000
VINYLIDENE CHLORIDE-VINYL CHLORIDE POLYMER see CGW300
VINYLIDENE DICHLORIDE see VPK000
VINYLIDENE DIFLUORIDE see VPP000
VINYLIDENE FLUORIDE see VPP000
VINYLIDINE CHLORIDE see VPK000
VINYL ISOBUTYL ETHER, inhibited (DOT) see IJQ000

VINYL ISOBUTYL ETHER (DOT) see IJQ000
VINYLITE VYDR 21 see AAX175
N-VINYLKARBAZOL see VNK100
VINYLKYANID see ADX500
VINYLLITHIUM see VPU000
VINYL-2-METHOXYETHYL ETHER see VPZ000
VINYL METHYLADIPATE see MQL000
VINYL METHYL ETHER (DOT) see MQL750
VINYL METHYL KETONE see BOY500
5-VINYL-2-NORBORNENE see VQA000
2-VINYLNORBORNENE see VQA000
5-VINYLNORBORNENE see VQA000
VINYLNORBORNENE see VQA000
17-α-VINYL-19-NORTESTOSTERONE see NCI525
17-VINYL-19-NORTESTOSTERONE see NCI525
VINYLOFOS see DGP900
VINYLON FILM 2000 see PKP750
VINYLOPHOS see DGP900
3-VINYL-7-OXABICYCLO(4.1.0)HEPTANE see VNZ000
R-5-VINYL-2-OXAZOLIDINETHIONE see VQA100
VINYLOXIRANE see EBJ500
2-(VINYLOXY)ETHANOL see EJL500
VINYLOXYETHANOL see EJL500
2-(VINYLOXY)ETHYL METHACRYLATE see VQA150
VINYLPHATE see CDS750
4-VINYLPHENOL see VQA200
p-VINYLPHENOL see VQA200
p-VINYLPHENOL ACETATE see ABW550
4-VINYLPHENYL ACETATE see ABW550
VINYLPHOSPHATE see RAF100
VINYL PHOSPHATE see VQA400
VINYLPHOSPHONIC ACID BIS(2-CHLOROETHYL) ESTER see VQF000
5-VINYL-2-PICOLINE see MQM500
α-VINYLPIPERONYL ALCOHOL see BCJ000
VINYL PRODUCTS R 10688 see AAX250
VINYL PRODUCTS R 3612 see SMQ500
VINYL PROPIONATE see VQK000
1-VINYLPYRENE see EEF000
3-VINYLPYRENE see EEF000
4-VINYLPYRENE see EEF500
2-VINYLPYRIDINE see VQK560
4-VINYLPYRIDINE see VQK590
α-VINYLPYRIDINE see VQK560
VINYL PYRIDINE see VQK550
1-VINYL-2-PYRROLIDINONE see EEG000
N-VINYL-2-PYRROLIDINONE see EEG000
N-VINYLPYRROLIDINONE see EEG000
1-VINYL-2-PYRROLIDINONE POLYMER, compounded with IODINE see PKE250
1-VINYL-2-PYRROLIDINONE POLYMER with STYRENE see VQK595
VINYLPYRROLIDINONE-STYRENE POLYMER see VQK595
1-VINYL-2-PYRROLIDONE see EEG000
N-VINYL-2-PYRROLIDONE see EEG000
N-VINYLPYRROLIDONE see EEG000
VINYLPYRROLIDONE see EEG000
N-VINYLPYRROLIDONE POLYMER see PKQ250
α-(5-VINYL-2-QUINOLYL)-2-QUINUCLIDINEMETHANOL see CMP925
VINYLSILATRAN see VQK600
1-VINYLSILATRANE see VQK600
VINYLSILATRANE see VQK600
VINYLSILICON TRICHLORIDE see TIN750
m-VINYLSTYRENE see DXQ745
VINYLSTYRENE see DXQ740
VINYL SULFONE see DXR200
4-VINYLTOLUENE see VQK700
VINYL TOLUENE, inhibited mixed isomers (DOT) see VQK650
3-and 4-VINYL TOLUENE (mixed isomers) see VQK650
p-VINYLTOLUENE see VQK700
VINYL TOLUENE see VQK650
VINYL TRICHLORIDE see TIN000
VINYL TRICHLOROSILANE (DOT) see TIN750
VINYL TRICHLOROSILANE, INHIBITED (DOT) see TIN750
VINYLTRIETHOXYSILANE see TJN250
VINYL TRIMETHOXY SILANE see TLD000
VINYLTRIMETHYLAMMONIUM HYDROXIDE see VQR300
VINYL-2,6,8-TRIMETHYLNONYL ETHER see VQU000
1-VINYL-2,8,9-TRIOXA-5-AZA-1-SILABICYCLO(3.3.3)UNDECANE see VQK600
VINYLTRIS(2-METHOXYETHOXY)SILANE see TNJ500
VINYLTRIS(β-METHOXYETHOXY)SILANE see TNJ500
VINYLTRIS(METHOXYETHOXY)SILANE see TNJ500
VINYON N see ADY250
VINYZENE bp 5 see OMY850
VINYZENE bp 5-2 see OMY850
VINYZENE (pesticide) see OMY850
VINYZENE see OMY850

VINYZENE SB 1 see OMY850
VIOACTANE see VQZ000
VIOCID see AOR500
VIOCIN see VQZ000
VIOFORM see CHR500
VIOFORM N.N.R. see CHR500
VIOFURAGYN see NGG500
VIOLACETIN see VQU450
VIOLANTHRONE see DCU800
VIOLAQUERCITRIN see RSU000
VIOLET 3 see XGS000
11092 VIOLET see HOK000
12416 VIOLET see AOR500
VIOLET 6BN see AOR500
VIOLET BNP see VQU500
VIOLET 5BO see AOR500
VIOLET CP see AOR500
VIOLET DISPERZNI 4 see AKP250
VIOLET GENCIANOVA see AOR500
VIOLET KRYSTALOVA see AOR500
VIOLET KYPOVA 1 see DFN450
VIOLET LEAF ALCOHOL see NMV780
VIOLET LEAF ALDEHYDE see NMV760
VIOLET PIGMENT 31 see DFN450
VIOLET POWDER H 2503 see MQN025
VIOLET ROZPOUSTEDLOVA 12 see AKP250
VIOLET ROZPOUSTEDLOVA 26 see DBX000
VIOLET 2S see DBY700
VIOLET XXIII see AOR500
VIOLET ZASADITA 14 see MAC250
VIOLET ZASADITA 3 see AOR500
VIOLOGEN, METHYL- see PAJ000
VIOLURIC ACID see AFU000
VIOMICINAE (ITALIAN) see VQZ000
VIOMYCIN, 1-(threo-4-HYDROXY-L-3,6-DIAMINOHEXANOIC ACID)-6-(L-2-(2-AMI-
    NO-1,4,5,6-TETRAHYDRO- 4-PYRIMIDINYL)GLYCINE)-, (R)-, SESQUISULFATE
    see VQZ100
VIOMYCIN see VQZ000
VIOMYCIN SULFATE SALT (2:3) see VQZ100
VIOPSICOL see LFK000, MDQ250
VIO-SERPINE see RDK000
VIOSTEROL see VSZ100
VIOXAN see CBM750
VIOZENE see RMA500
VI-PAR see CIR500
VIPERA AMMODYTES VENOM see VQZ425
VIPERA BERUS VENOM see VQZ475
VIPERA BORNMULLERI VENOM see VQZ500
VIPERA LATIFII VENOM see VQZ525
VIPERA LEBETINA VENOM see VQZ550
VIPERA PALESTINAE VENOM see VQZ575
VIPERA RUSSELLII FORMOSENSIS VENOM see VQZ625
VIPERA RUSSELLII VENOM see VQZ635
VIPERA XANTHINA PALAESTINAE VENOM see VQZ650
VIPERINE (CANADA) see VQZ675
VIPER'S BUGLOSS see VQZ675
VI-PEX see CIR500, CLO200
VIPICIL see AJJ875
VIPRYNIUM EMBONATE see PQC500
VIRA-A see AEH100
VIRACTIN see VRA000
VIRAMID see RJA500
VIRAZOLE see RJA500
VIRCHEM see SHR500
VIRGIMYCIN see VRA700, VRF000
VIRGINIA-CAROLINA 3-670 see DWU200
VIRGINIA CAROLINA VC 9-104 see EIN000
VIRGINIAMYCIN see VRF000
VIRGIN'S BOWER see CMV390
VIRICUIVRE see CNK559
VIRIDICATUMTOXIN see VRP000
VIRIDINE see PDX000
VIRIDITOXIN see VRP200
VIROFRAL see AED250
VIROPHTA see TKH325
VIROPTIC see TKH325
VIRORMONE see TBF500
VIROSIN see AQM250
VIROSTERONE see TBF500
VIRSET 656-4 see MCB000
VIRTEX CC see SHR500
VIRTEX D see SHR500
VIRTEX L see SHR500
VIRTEX RD see SHR500
VIRUBRA see VSZ000

VIRUSTOMYCIN A see VRP775
VIRUZONA see MKW250
VISADRON see NCL500, SPC500
VISCALEX HV 30 see ADW200
VISCARIN 402 see CAO250
VISCARIN see CCL250, SFP000
VISCERALGIN see HNM000
VISCERALGINE see HNM000
VISCLAIR see MBX800
VISCO BLACK N see CMN240
VISCOL see MIF760
VISCOLEO OIL see VGU200
VISCOL 350P see PMP500
VISCON 103 see ADW200
VISCONTRAN L52 see MIF760
VISCOSE BLACK G see CMN240
VISCOSE BLACK GNA see CMN240
VISCOSE BLACK J see CMN240
VISCOSE BLACK N see CMN240
VISCOSE BLACK NG see CMN240
VISCOSOL see MIF760
VISCOTOXIN see VRU000
VISCUM ALBUM see ERA100
VISET see FBW000
VISINE see VRZ000
VISINE HYDROCHLORIDE see VRZ000
VISIREN see OGI300
VISKEN see VSA000
VISKING CELLOPHANE see CCT250
VISKO-RHAP see DAA800, DGB000
VISKO-RHAP DRIFT HERBICIDES see DAA800
VISKO RHAP LOW VOLATILE ESTER see TAA100
VISKO-RHAP LOW VOLATILE 4L see DAA800
VISNAGALIN see AHK750
VISOTRAST see AAN000
VISTABAMATE see MQU750
VISTAMYCIN see RIP000, XQJ650
VISTAR see DUK000
VISTAR HERBICIDE see DUK000
VISTARIL HYDROCHLORIDE see VSF000
VISUBUTINA see HNI500
VITABLEND see PAQ200
VITACAMPHER see VSF400
VITACIN see ARN000
VITAHEXIN P see PII100
VITALLIUM see CNA750, VSK000
VITALOID see VQR300
VITAMIN A1 see VSK600
VITAMIN A see VSK600
VITAMIN A ACETATE see VSK900
trans-VITAMIN A ACETATE see VSK900
13-cis-VITAMIN A ACID see VSK955
VITAMIN A ACID see VSK950
VITAMIN A1 ALCOHOL see VSK600
all-trans-VITAMIN A ALCOHOL see VSK600
VITAMIN A ALCOHOL ACETATE see VSK900
VITAMIN A1 ALDEHYDE see VSK985
9-cis-VITAMIN A ALDEHYDE see VSK975
VITAMIN A ALDEHYDE see VSK985
trans-VITAMIN A ALDEHYDE see VSK985
VITAMIN AB see HAQ500
VITAMIN A PALMITATE see VSP000
VITAMIN B1 see TES750
VITAMIN B$^1$ see TET300
VITAMIN B2 see RIK000
VITAMIN B3 see NCR000
VITAMIN B4 see AEH000
VITAMIN B-5 see CAU750
VITAMIN B6 see PPK250
VITAMIN B$_7$ see VSU100
VITAMIN B12 (FCC) see VSZ000
VITAMIN Bc see FMT000
VITAMIN B$_{12}$ COMPLEX see VSZ000
VITAMIN B$_1$ DISULFIDE see TES800
VITAMIN B6-HYDROCHLORIDE see PPK500
VITAMIN B$_5$ HYDROCHLORIDE see VSU000
VITAMIN B HYDROCHLORIDE see TET300
VITAMIN B$_{12}$ METHYL see VSZ050
VITAMIN B1 MONONITRATE see TET500
VITAMIN B1 NITRATE see TET500
VITAMIN B$_1$ PROPYL DISULFIDE see DXO300
VITAMIN BX see AIH600
VITAMIN C see ARN000
VITAMIN D2 see VSZ100
VITAMIN D3 see CMC750

VITAMIN D³ see HJV000
VITAMIN D see VSZ095
VITAMIN E see VSZ450
VITAMIN G see RIK000
VITAMIN H see AIH600, VSU100
VITAMIN K1 see VTA000
VITAMIN K3 see MMD500
VITAMIN K5 see AKX500
VITAMIN K2₂₀ see VTA650
VITAMIN K see VSZ500
VITAMIN K3-NATRIUM-BISULFIT TRIHYDRAT (GERMAN) see MMD750
VITAMIN K2(O) see MMD500
VITAMIN L see API500
VITAMIN M see FMT000
VITAMIN MK 4 see VTA650
VITAMIN P see RSU000
VITAMIN PP see NCR000
VITAMISIN see ARN000
VITANEURON see TES750
VITAPLEX E see VSZ450
VITAPLEX N see NCQ900
VITARUBIN see VSZ000
VITA-RUBRA see VSZ000
VITASCORBOL see ARN000
VITASTAIN see TMV500
VITAVAX see CCC500
VITAVEL-A see VSK600
VITAVEL-D see VSZ100
VITAVEX see DLV200
VITAYONON see VSZ450
VITAZECHS see PII100
VITEOLIN see VSZ450
VITESTROL see DLB400
VITICOSTERONE see HKG500
VITIGRAN see CNK559
VITIGRAN BLUE see CNK559
VITINC DAN-DEE-3 see CMC750
VITON see BBQ500
VITPEX see VSK600
VITRAL see VSZ000
VITRAN see TIO750
VITREOSIL IR see SCK600
VITREOUS QUARTZ see SCK600
VITREOUS SILICA see SCK600
VITREX see PAK000
VITRIFIED SILICA see SCK600
VITRIOL BROWN OIL see SOI500
VITRIOL, OIL OF (DOT) see SOI500
VITRIOL RED see IHD000
VITRUM AB see HAQ500
VITUF see PKF750
VI-TWEL see VSZ000
VIVACTIL see DDA600, POF250
VIVAL see DCK759
VIVALAN see VKF000
VIVICIL see FMR500
VIVOTOXIN see VTA750
VK-738 see DWE800
VKhVD 40 see CGW300
VLB see VKZ000
VLB MONOSULFATE see VLA000
VLVF see AAX175
VM-26 see EQP000
VMCC see AAX175
VMI 10-3 see AMI500
VML 2 see MCB050
VM & P NAPHTHA (ACGIH,OSHA) see PCT250
VM and P NAPHTHA see PCT250
VM & P NAPHTHA see PCT250
VOFATOX see MNH000
VOGALENE see MQR000
VOGAN see VSK600
VOGAN-NEU see VSK600
VOGEL'S IRON RED see IHD000
VOLAMIN see EEH000
VOLATILE OIL of MUSTARD see AGJ250
VOLATON see BAT750
VOLCLAY see BAV750
VOLCLAY BENTONITE BC see BAV750
VOLDYS see MCA500
VOLFARTOL see TIQ250
VOLFAZOL see COD000
VOLID see TAC800
VOLIDAN see MCA500, VTF000
VOLPO 20 see PJW500

VOLTALEF 202 see KDK000
VOLTALEF see KDK000
VOLTALEF IS see KDK000
VOLTALEF 300LD see KDK000
VOLTALEF 300UF see KDK000
VOLTAREN see DEO600
VOLTAROL see DEO600
VOLUNTAL see TIO500
VOMEX A see DYE600
VOMISSELS see HGC500
VOMITOXIN see VTF500
VONAMYCIN POWDER V see NCE000
VONDACEL BLACK D see CMN230
VONDACEL BLACK N see AQP000
VONDACEL BLACK VG see CMN240
VONDACEL BLACK VN see CMN240
VONDACEL BLUE 2B see CMO000
VONDACEL BLUE HH see CMO500
VONDACEL BROWN S see CMO820
VONDACEL BROWN SP see CMO820
VONDACEL DARK BLUE BH see CMN800
VONDACEL GREEN B see CMO840
VONDACEL GREEN DB see CMO830
VONDACEL RED CL see SGQ500
VONDACEL RED FN see CMO870
VONDACID FAST BLUE BR see CMM090
VONDACID FAST YELLOW AE see SGP500
VONDACID GREEN L see FAE950
VONDACID GREEN S see ADF000
VONDACID LIGHT RED NG see CMM300
VONDACID LIGHT YELLOW AE see SGP500
VONDACID METANIL YELLOW G see MDM775
VONDACID ORANGE II see CMM220
VONDACID TARTRAZINE see FAG140
VONDALHYDE see DMC600
VONDAMOL BRILLIANT RED G see CMM325
VONDAMOL FAST BLUE R see ADE750
VONDAMOL FAST RED G see CMM320
VONDAMOL FAST RED RS see CMM330
VONDCAPTAN see CBG000
VONDODINE see DXX400
VONDRAX see DMC600
VONDURON see DXQ500
VONEDRINE see DBA800
VONTERYL VIOLET 2B see DBY700
VONTERYL YELLOW G see AAQ250
VONTERYL YELLOW 3R see CMP090
VONTERYL YELLOW R see AAQ250
VONTIL see TFM100
VONTROL see DWK200
VOPCOLENE 27 see OHU000
VORANIL see DRC600
VOREN see DBC510
V. ORIENTALIS VENOM see VJP900
VORLEX see ISE000, MLC000
VORONITE see FQK000
VOROX see AMY050
VOROX AA see AMY050
VOROX AS see AMY050
VORTEX see ISE000
VOTEXIT see TIQ250
VP 1940 see ABX250
VP 16213 see EAV500
VPK 402 see DTS500
VPM see SIL550, VFU000
V-PYROL see EEG000
VROKON see AAN000
VS 7158 see DAF350
VS 7207 see OCE100
V-SERP see RDK000
VT 1 see TGF250
VTI 1 see AOT000
VTS-M see TLD000
VUAGT-I-4 see TFS350
VUCINE see EDW500
VUFB 3511 see PCI250
VUFB 6638 see DLR150
VULCACEL B-40 see DVF400
VULCACEL BN see DVF400
VULCACID D see DWC600
VULCACURE see BIX000, BJC000, SGF500
VULCAFIX FAST BLUE SD see IBV050
VULCAFIX ORANGE J see CMS145
VULCAFIX ORANGE JV see CMS145
VULCAFIX SCARLET R see CHP500

VULCAFOR BSM see BDG000
VULCAFOR CBS see CPI250
VULCAFOR FAST BLUE 3R see IBV050
VULCAFOR FAST ORANGE G see CMS145
VULCAFOR FAST ORANGE GA see CMS145
VULCAFOR FAST YELLOW 2G see CMS212
VULCAFOR FAST YELLOW GTA see DEU000
VULCAFOR HBS see CPI250
VULCAFOR SCARLET A see MMP100
VULCAFOR TMTD see TFS350
VULCALENT A see DWI000
VULCAMEL TBN see NAP500
VULCAN see CBT750
VULCAN FAST ORANGE G see CMS145
VULCAN FAST ORANGE GA see CMS145
VULCAN FAST ORANGE GN see CMS145
VULCAN FAST YELLOW G see CMS212
VULCAN FAST YELLOW GR see CMS208
VULCAN FAST YELLOW GRA see CMS208
VULCAN FAST YELLOW GRN see CMS208
VULCANOSINE FAST BLUE GG see IBV050
VULCAN RED LC see CHP500
VULCAN RED LCN see CMS150
VULCATARD see DWI000
VULCOL BLUE BZ see ERG100
VULCOL FAST BLUE S see IBV050
VULCOL FAST ORANGE G see CMS145
VULCOL FAST RED C see CMS150
VULCOL FAST RED L see CHP500
VUL-CUP see BHL100
VUL-CUP 40KE see BHL100
VUL-CUP R see BHL100
VULCUREN 2 see BDF750
VULKACIT 1000 see TGX550
VULKACIT 4010 (CZECH) see PET000
VULKACIT C see CPI250
VULKACIT CZ see CPI250
VULKACIT CZ/C see CPI250
VULKACIT CZ/K see CPI250
VULKACIT D/C see DWC600
VULKACIT DM see BDE750
VULKACIT DM/MGC see BDE750
VULKACIT DOTG/C see DXP200
VULKACIT LDA see BJC000
VULKACIT LDB/C see BIX000
VULKACIT MERCAPTO see BDF000
VULKACIT MTIC see TFS350
VULKACIT NPV/C2 see IAQ000
VULKACIT P EXTRA N see ZHA000
VULKACIT THIURAM see TFS350
VULKACIT THIURAM/C see TFS350
VULKACIT THIURAM MS/C see BJL600
VULKACIT ZM see BHA750
VULKALENT A (CZECH) see DWI000
VULKANOX 4020 see DQV250, PEY500
VULKANOX BKF see MJO500
VULKANOX HS/LG see TLP500
VULKANOX HS/POWDER see TLP500
VULKANOX PAN see PFT250
VULKASIL see SCH000
VULKAZIT see DWC600
VULKLOR see TBO500
VULNOC AB see ANB100
VULNOPOL NM see SGM500
VULTROL see DWI000
VULVAN see TBG000
VUMON see EQP000
VU 51-3N see MCB050
VU 59-3N see MCB050
VU 5711N see MCB050
VX see EIG000
VYAC see VLU250
VYDATE see DSP600
VYDATE L INSECTICIDE/NEMATICIDE see DSP600
VYDATE L OXAMYL INSECTICIDE/NEMATOCIDE see DSP600
VYDYNE see NOH000
VYGEN 85 see PKQ059
VYNAMON BLUE 3R see IBV050
VYNAMON CLARET Y see CMS155
VYNAMON GREEN 6Y see CMS140
VYNAMON ORANGE CR see LCS000
VYNAMON ORANGE G see CMS145
VYNAMON SCARLET BY see MRC000
VYNAMON YELLOW 2G see CMS212
VYNAMON YELLOW 2GE see CMS212

VYNAMON YELLOW GRE see CMS208
VYNAMON YELLOW GRES see CMS208
VYNW see AAX175

W 37 see BRA625
W 45 see ALC250
W 70 see UTU500
W 101 see PMP500
W 483 see DIR000
W 491 see CCP500
W 524 see TKL100
W 713 see MOV500
W-801 see CMY650
W 1544 see PFC500
W 1655 see PDC250, PEK250
W 1929 see SFY500
2317-W see DVF800
W-2395 see CIL500
W 2426 see ARW750
W-2429 see DLQ400
W 2451 see MJE250
W 3566 see QFA250
W 3699 see EOA500
W 4565 see OOG000
W 5219 see BGC625
W5769 see CFC750
W 6309 see DKJ300
W 7320 see AGN000
W 7618 see CLD000
W 2900 A see MNG000
W 5759A see EIH000
WACHOLDERBEER OEL (GERMAN) see JEA000
WACKER S 14/10 see BJE750
W. AEGYPTIA VENOM see WAT100
WA 335 HYDROCHLORIDE see WAJ000
WAIT'S GREEN MOUNTAIN ANTIHISTAMINE see WAK000
WALKO-NESIN see GGS000
WALNUT EXTRACT see WAT000
WALNUT STAIN see MAT500
WALSRODER MC 20000S see MIF760
WALTERINNESIA AEGYPTIA VENOM see WAT100
WAMPOCAP see NCQ900
WANADU PIECIOTLENEK (POLISH) see VDU000
WANDAMIN HM see MJQ260
WANSAR see CCR875
WAPNIOWY TLENEK (POLISH) see CAU500
WARAN see WAT220
WARBEX see FAB600
WARBEXOL see FAB600
WARCO F 71 see CNH125
WARCOUMIN see WAT220
WARDAMATE see MQU750
WARDUZIDE see CLH750
WARECURE C see IAQ000
WAREFLEX see DDT500
WARFARIN see WAT200
WARFARINE (FRENCH) see WAT200
WARFARIN K see WAT209
WARFARIN POTASSIUM see WAT209
WARFARIN SODIUM see WAT220
WARFILONE see WAT220
WARKEELATE ACID see EIX000
WARKEELATE PS-43 see EIV000
WASH OIL see CMY825
WASSERINA see CQH000
WASSERSTOFFPEROXID (GERMAN) see HIB000
WATAPANA SHIMARON see AAD750
WATCHUNG RED Y see CMS148
WATER see WAT259
WATER ARUM see WAT300
WATER BLACK 100 see CMN240
WATER BLACK 200L see CMN240
WATER BLACK P 200 see CMN240
WATERCARB see CBT500
WATER CROWFOOT see FBS100
WATER DRAGON see WAT300
WATER DROPWORT see WAT315
WATER GLASS see SJU000
WATER GOGGLES see MBU550
WATER GREEN SX see ADF000
WATER HEMLOCK see WAT325
WATER LILY see DAE100
WATER-PEPPER HERB see WAT350
WATER QUENCH PYROLYSIS FUEL OIL see FOP100

WATERSOL S 683 see MCB050
WATERSOL S 685 see MCB050
WATERSOL S 695 see MCB050
WATERSTOFPEROXYDE (DUTCH) see HIB000
WATERWEED see PIH800
WATTLE GUM see AQQ500
WAXAKOL ORANGE GL see PEJ500
WAXAKOL RED BL see SBC500
WAXAKOL VERMILION L see XRA000
WAXAKOL YELLOW NL see AIC250
WAXBERRY see SED550
WAX LE see PJS750
WAX MYRTLE see WBA000
WAXOLINE GREEN see BLK000
WAXOLINE ORANGE A see BJF000
WAXOLINE PURPLE A see HOK000
WAXOLINE RED A see MAC500
WAXOLINE RED O see SBC500
WAXOLINE RED OM see SBC500
WAXOLINE RED OS see SBC500
WAXOLINE YELLOW AD see DOT300
WAXOLINE YELLOW I see PEJ500
WAXOLINE YELLOW O see IBB000
WAXOLINE YELLOW T see CMS245
WAXSOL see DJL000
WAYNECOMYCIN see LGC200
WAYNE RED X-2486 see CHP500
WAYPLEX 55S see DJG700
W3207B see MOM750
WBA 8119 see TAC800
WD 67/2 see WBA600
WE 941 see LEJ600
WEATHERBEE MUSTARD see BHB750
WEATHER PLANT see RMK250
WE 941-BS see LEJ600
WEC 50 see TIQ250
WECKAMINE see BBK000
WECOLINE 1295 see LBL000
WECOLINE OO see OHU000
WEDDING BELLS see DJO000
WEECON see COI250
WEED 108 see MRL750
WEED-AG-BAR see DAA800
WEEDANOL CYANOL see PLC250
WEEDAR-64 see DAA800
WEEDAR see TAA100
WEEDAR ADS see AMY050
WEEDAR AT see AMY050
WEEDAR MCPA CONCENTRATE see CIR250
WEEDAZIN see AMY050
WEEDAZIN ARGINIT see AMY050
WEEDAZOL see AMY050
WEEDAZOL GP2 see AMY050
WEEDAZOL SUPER see AMY050
WEEDAZOL T see AMY050
WEEDAZOL TL see AMY050, ANW750
WEEDBEADS see SJA000
WEED-B-GON see DAA800, TIX500
WEED BROOM see DXE600
WEED DRENCH see AFV500
WEEDEX A see ARQ725
WEEDEX GRANULAT see AMY050
WEEDEZ WONDER BAR see DAA800
WEED-HOE see DXE600, MRL750
WEEDMASTER GRASS KILLER see TII500
WEEDOCLOR see AMY050
WEEDOL see PAJ000
WEEDONE 128 see IOY000
WEEDONE 170 see DGB000
WEEDONE see PAX250, TAA100
WEEDONE CRAB GRASS KILLER see PLC250
WEEDONE DP see DGB000
WEEDONE LV4 see DAA800
WEEDONE MCPA ESTER see CIR250
WEED-E-RAD see DXE600, MRL750
WEED-RHAP see CIR250
WEED TOX see DAA800
WEEDTRINE-D see DWX800
WEEDTRINE-II see ILO000
WEEDTROL see DAA800
WEEVILTOX see CBV500
WEG 147 see CJP500
WEG 148 see MNR100
WEGANTYNA (POLISH) see THK000
WEGANTYNA see DHX800

WEGLA DWUSIARCZEK (POLISH) see CBV500
WEGLA TLENEK (POLISH) see CBW750
WEHYDRYL see BAU750
WEIFACODINE see TCY750
WEISS PHOSPHOR (GERMAN) see PHP010
WEISSPIESSGLANZ see AQF000
WELDING FUMES see WBJ000
WELFURIN see NGE000
WELLBATRIN see WBJ500
WELLCOME PREPARATION 47-83 see EAN600
WELLCOME-248U see AEC700
WELLCOME U3B see AMH250
WELLCOPRIM see TKZ000
WELLDORM see SKS700
WELVIC G 2/5 see PKQ059
WEMCOL see IOF200
WEPSIN see AIX000
WEPSYN 155 see AIX000
WEPSYN see AIX000
WESCOZONE see BRF500
WESPURIL see MJM500
WESTERN DIAMONDBACK RATTLESNAKE VENOM see WBJ600
WEST INDIAN BAY OIL see LBK000
WEST INDIAN LEMONGRASS OIL see LEH000
WEST INDIAN LILAC see CDM325
WEST INDIAN PINKROOT see PIH800
WEST INDIAN SANDALWOOD OIL see WBJ650
WESTOCAINE see AIT250
WESTON 626 see COV800
WESTON MDW 626 see COV800
WESTRON see TBQ100
WESTROSOL see TIO750
WETAID SR see DJL000
WET-TONE B see MSB500
WEX 1242 see PMP500
WF-104 see OKK500
WFNA see NEF000
W-GUM see SLJ500
WH 5668 see PMM000
WH 7286 see DMW000
WH 7508 see PIW000
WHATMAN CC-31 see CCU150
WHEAT GERM OIL see WBJ700
WHEAT HUSK OIL see WBJ700
WHEY FACTOR see OJV500
WHICA BA see CAT775
WHISKEY see WBS000
WHITCARB W see CAT775
1700 WHITE see TGG760
WHITE ACID (DOT) see ANG250
WHITE ANTHURIUM see SKX775
WHITE ARSENIC see ARI750
WHITE ASBESTOS, (chrysotile, actinolite, anthophyllite, tremolite) (DOT) see ARM268
WHITE ASBESTOS see ARM268
WHITE CAMPHOR OIL see CBB500
WHITE CAUSTIC see SHS000
WHITE CAUSTIC, solution see SHS500
WHITE CEDAR see CDM325
WHITE CEDAR OIL see CCQ500
WHITE COPPERAS see ZNA000
WHITE CRYSTAL see SFT000
WHITE FUMING NITRIC ACID see NEF000
WHITE HELLEBORE see FAB100
WHITE HONEY FLOWER see GJU475
WHITE KERRIA see JDA075
WHITE LEAD see LCP000
WHITE LOCUST see GJU475
WHITE MERCURY PRECIPITATED see MCW500
WHITE MINERAL OIL see MQV750, MQV875
WHITE OIL of CAMPHOR see CBB500
WHITE OSTER see DYA875
WHITE PHOSPHORUS see PHP010
WHITE POPINAC see LED500
WHITE-POWDER see CAT775
WHITE PRECIPITATE see MCW500
WHITE SEAL-7 see ZKA000
WHITESET see MCB050
WHITE SPIRIT see WBS675
WHITE SPIRIT, DILUTINE 5 see WBS700
WHITE SPIRITS see SLU500
WHITE STAR see AQF000
WHITE STREPTOCIDE see SNM500
WHITE STUFF see HBT500
WHITE TAR see NAJ500

WHITE VITRIOL see ZNA000, ZNJ000
WHITING see CAT775
WHITON 450 see CAT775
WHORTLEBERRY RED see FAG020
W-53 HYDROCHLORIDE see DPJ400
WHYO TREE see GJU475
WICKENOL 101 see IQN000
WICKENOL 111 see IQW000
WICKENOL 116 see DNL800
WICKENOL 122 see BSL600
WICKENOL 127 see IRL100
WICKENOL 131 see IPS450
WICKENOL 141 see MSA300
WICKENOL 155 see OFE100
WICKENOL 156 see OFU300
WICKENOL 158 see AEO000
WICKENOL 324 see AHA000
WICKERBY BUSH see LEF100
WICKUP see LEF100
WICKY see MRU359
WICOPY see LEF100
WICRON see HGI100
WIDLON see PJY500
WIE OBEN see DJY200
WIJS' CHLORIDE see IDS000
WILD ALLAMANDA (FLORIDA) see YAK300
WILD BALSAM APPLE see FPD100
WILD CALLA see WAT300
WILD COFFEE see CNG825
WILD CROCUS see PAM780
WILD GARLIC see WBS850
WILD GINGER OIL see SED500
WILD JALAP see MBU800
WILD LEMON see MBU800
WILD LICORICE see RMK250
WILD MAMEE see BAE325
WILD NIGHTSHADE see YAK300
WILD ONION see DAE100, WBS850
WILD TAMARIND (HAWAII, PUERTO RICO) see LED500
WILD TOBACCO see CCJ825
WILD UNCTION (BAHAMAS) see YAK300
WILD YAM BEAN see YAG000
WILELAIKI (HAWAII) see PCB300
WILKINITE see BAV750
WILLBUTAMIDE see BSQ000
WILLESTROL see DKB000
WILLNESTROL see DAL600
WILLOSETTEN see SJN700
WILPO see DTJ400
WILT PRUF see PKQ059
WILTZ-65 see NAS000, NBW500
WIN 244 see CLD000
WIN 357 see BEL900
WIN 833 see HLK000
WIN 1258 see PJB750
WIN 1539 see KFK000
WIN 1593 see CPM750
WIN 1701 see DOS800
WIN 2022 see AMM750
WIN 2661 see WBS855
WIN 2663 see WBS860
WIN-2848 see DPJ200, TEO000
WIN 3074 see PPM550
WIN 3706 see SPB800
WIN 4369 see PBS000
WIN 4981 see BJA825
WIN 5095 see AFY500
WIN 5162 see DMV600
WIN 5501 see AKT500
WIN 5512 see AKT250
WIN 5606 see BCA000
WIN 6703 see MGJ600
WIN 8077 see MSC100
WIN 11450 see SAN600
WIN-1161-3 see DWC100
WIN 12267 see DEM000
WIN 13,099 see PFA600
WIN 14833 see AOO400
WIN 16568 see DAP812
WIN 17,416 see HEF300
WIN 17757 see DAB830
WIN 18,320 see EID000
WIN 18,441 see BIX250
WIN 18,446 see BIX250
WIN 20228 see DOQ400

WIN 20740 see COV500
WIN 22005 see UVS500
WIN 24450 see EBY600
WIN 24933 see LIM000
WIN 32729 see EBH400
WIN 32784 see BLV125
WIN 39103 see MQR300
WIN 40680 see AOD375
WIN-5063-2 see MPN000
WIN 8308-3 see SEN500
WIN 8851-2 see SKO500
WIN 18501-2 see ECW600
WINACET D see AAX250
WINCORAM see AOD375
WINDFLOWER see PAM780
WINDOWLEAF see SLE890
WINDOW PLANT see SLE890
WINE see WCA000
WINE ETHER see ENW000
WINE PALM see FBW100
WINE PLANT see RHZ600
WING STOP B see SGM500
WIN-2299 HYDROCHLORIDE see DHW400
WIN 2848 HYDROCHLORIDE SALT see DPJ400
WINIDEN 60 see CGW300
WINIDUR see PKQ059
WIN-KINASE see UVS500
WINNOFIL S see CAT775
WINOBANIN see DAB830
WINSTROL see AOO400
WINSTROL V see AOO400
WINTER BLOOM see WCB000
WINTER CHERRY see JBS100
WINTER FERN see PJJ300
WINTERGREEN OIL (FCC) see MPI000
WINTERGREEN OIL, SYNTHETIC see MPI000
WINTERMIN see CKP250
WINTERSTEINER'S COMPOUND F see CNS800
WINTERSWEET see BOO700
WINTERWASH see DUS700
WINTOMYLON see EID000
WINYLU CHLOREK (POLISH) see VNP000
WINZER SOLUTION see BEL900
WIRKSTOFF 37289 see EPY000
WIRNESIN see POB500
WISTARIA see WCA450
WISTERIA see WCA450
WISTERIA FLORIBUNDA see WCA450
WISTERIA SINENSIS see WCA450
WITAMINA PP see NCR000
WITAMOL 320 see AEO000
WITCARB 940 see CBT500
WITCARB see CAT775
WITCARB P see CAT775
WITCARB REGULAR see CAT775
WITCHES' THIMBLES see FOM100
WITCH HAZEL see WCB000
WITCIZER 200 see BSL600
WITCIZER 201 see BSL600
WITCIZER 300 see DEH200
WITCIZER 312 see DVL700
WITCO 31 see PJY100
WITCO see CBT750
WITCOBLAK NO. 100 see CBT750
WITCO G 339S see CAX350
WITCONATE 5725 see TJK800
WITCONATE S-1280 see TJK800
WITCONATE see AFO250
WITCONATE 60L see TJK800
WITCONATE 79S see TJK800
WITCONATE 60T see TJK800
WITCONATE TAB see TJK800
WITCONOL MS see OAV000
WITCONOL MST see OAV000
WITCO 1298 SULFONIC ACID see LBU100
WITEPSOL E-75 see WCB100
WITTOX C see NAS000
WJG 11 see PJS750
WL 19 see PDK500
WL 20 see NNL500, PDK300
WL 23 see MNP400
WL 28 see CJK000
WL 29 see MFG200
WL 31 see MNU100
WL 32 see MNU150

WL 33 see NNL500
WL 34 see CJK100
WL 39 see MNP500
WL 1650 see OAN000
WL-5792 see DGM600
WL 17,731 see EGS000
WL 18236 see MDU600
WL 19805 see BLW750
WL 417-06 see DAB825
WL 43467 see RLF350
WL 43479 see AHJ750
WL 43775 see FAR100
WM 100 see MCB050
W-2946M see DNA600
WMO see ELB000
WN 12 see ISE000
WNF 15 see PJS750
WNM see DTG000
WNYESTRON see EDV000
WOCHEM No. 320 see OHU000
WODE WHISTLE see PJJ300
WOFACAIN A see BQH250
WOFATOS see MNH000
WOFATOX see MNH000
WOFAVERDIN see CCK000
WOFOTOX see MNH000
WOJTAB see PLZ000
WOLFEN see CNL500
WOLFRAM see TOA750
WOLFRAMITE see TOC750
WOLFSBANE see AQY500, MRE275
WOLLASTOKUP see WCJ000
WOLLASTONITE see WCJ000
WONDER BULB see CNX800
WONDER TREE see CCP000
WONUK see ARQ725
WOOD ALCOHOL (DOT) see MGB150
WOOD ETHER see MJW500
WOOD LAUREL see MRU359
WOOD NAPHTHA see MGB150
WOOD OIL see GME000
WOOD SPIRIT see MGB150
WOOD VINE see YAK100
WOOL BLUE RL see ADE750
WOOL BORDEAUX 6RK see FAG020
WOOL BRILLIANT GREEN SF see FAF000
WOOL FAST BLUE R see ADE750
WOOL GREEN 5 see ADF000
WOOL GREEN B see ADF000
WOOL GREEN BS see ADF000
WOOL GREEN BSNA see ADF000
WOOL GREEN MS see ADF000
WOOL GREEN S see ADF000
WOOL GREEN S (BIOLOGICAL STAIN) see ADF000
WOOL GREEN SG see ADF000
WOOLLEY'S ANTISEROTONIN see BEM750
WOOL MORDANT see CMK425
WOOL ORANGE A see CMM220
WOOL RED see FAG020
WOOL VIOLET see FAG120
WOOL YELLOW see FAG140
WOOL YELLOW G see CMM758
WOORALI see COF750
WOORARI see COF750
WORM-AGEN see HFV500
WORM-CHEK see LFA020
WORM GRASS see PIH800
WORM GUARD see BQK000
WORMWOOD see SMY000
WORMWOOD ACID see SMY000
WORMWOOD PLANT see SAF000
WORT-WEED see CCS650
WOTEXIT see TIQ250
WOURARA see COF750
WP 155 see AIX000
W-PYRIDINYLMETHANOL see POR800
WR 141 see GJS000
WR 448 see SOA500
WR 638 see AKB500
WR 2721 see AMD000
WR 2978 see TGD000
WR 3121 see THB250
WR 5473 see COX400
WR 09792 see FMH000
WR 9838 see KFA100

WR 10019 see SAH500
WR 14,997 see AJK250
WR 81844 see WCJ750
WR-171,669 see HAF500
W 45 RASCHIG see ALC250
WRIGHTINE DIHYDROBROMIDE see DOX000
WS 24 see ADW200
WS 102 see NDU500
WS 801 see ADW200
WS 2434 see MJH900
W-13 STABILIZER see SLJ500
W-T SASP ORAL see PPN750
WURM-THIONAL see PDP250
WURSTER'S BLUE see BJF500
WURSTER'S REAGENT see BJF500
WUWEIZI ALCOHOL A see SBE400
WUWEIZI ALCOHOL B see SBE450
WUWEIZICHUN A see SBE400
WUWEIZICHUN B see SBE450
WV 365 see TFY000
WV 761 see BMW000
WVG 23 see PJS750
W VII/117 see FQK000
WY 509 see DQA400
WY 554 see DAM700
WY 806 see DTL200
WY 1094 see DQA600
WY-1172 see ABH500
WY-1395 see TLG000
WY 1485 see CHD750
WY 2149 see CMB125
WY 2917 see CFY750
WY-3263 see DPX200
WY 3277 see SGS500
WY-3478 see HJS500
WY-3498 see CFZ000
WY 3707 see NNQ500
WY 3917 see CFY750
WY 4036 see CFC250
WY 4508 see AJJ875
WY-5090 see HGI700
WY-5103 see AIV500
WY 5244 see CKA500
WY 5256 see AKN750
WY-8678 see GKO750
WY-14,643 see CLW250
WY-21743 see OLW600
WY-42956 see BCP250
WY 8678 ACETATE see GKO750
WYACORT see MOR500
WYCILLINA see BFC750
WY-E 104 see NNL500
WYEX see CBT750
WY-4355 mixed with MESTRANOL (20:1) see CHI750
WYPAX see CFC250
WYSEALS see MQU750
WYVITAL see AJJ875
WZ 884 see PMJ525


X 34 see IBC000
X 79 see POJ500
X 119 see DTP000
X 134 see TES000
X 146 see TFQ275
X 149 see BOT250
X 201 see AJK250
X 250 see PAE750
X-340 see HAL000
X 600 see SMQ500
X 3387 see MCB050
XA 2 see CFA750
X-AB see PKQ059
X-ALL Liquid see AMY050
XAMAMINA see DYE600
XAMOTEROLFUMARATE see XAH000
XAN see XCA000
XANAX see XAJ000
XANTELINE see DJM800, XCJ000
XANTHACRIDINE see XAK000
XANTHACRIDINUM see DBX400
XANTHAHYDROGEN see IBL000
XANTHANOL see XBJ000
XANTHAURINE see QCA000
9H-XANTHENE see XAT000

XANTHENE see XAT000
XANTHENE-9-CARBOXYLIC ACID ESTER with DIETHYL(2-HYDROXYETHYL)
  METHYLAMMONIUM BROMIDE see DJM800
XANTHENE-9-CARBOXYLIC ACID, ESTER with DIETHYL(2-HYDROXYETH-
  YL)METHYLAMMONIUM BROMIDE see XCJ000
XANTHENE-9-CARBOXYLIC ACID, ESTER with (2-HYDROXYETH-
  YL)DIISOPROPYLMETHYLAMMONIUM BROMIDE see HKR500
1,1′-(9H-XANTHENE-2,7-DIYL)BIS(2-(DIMETHYLAMINO))ETHANONE DIHYDRO-
  CHLORIDE SESQUIHYDRATE see XBA000
XANTHEN-9-OL see XBJ000
9-XANTHENONE see XBS000
XANTHIC ACID, COPPER(II) SALT see CBW200
XANTHIC ACID, ISOPROPYL-, POTASSIUM SALT see IRG050
XANTHIC ACID, ISOPROPYL-, SODIUM SALT see SIA000
XANTHIC ACID, PENTYL-, POTASSIUM SALT see PKV100
XANTHIC OXIDE see XCA000
XANTHINE see XCA000
XANTHINE BROMIDE see XCJ000
XANTHINE-3-N-OXIDE see HOP000
XANTHINE-7-N-OXIDE see HOP259
XANTHINE-x-N-OXIDE see HOP000
XANTHINOL NIACINATE see XCS000
XANTHINOL NICOTINATE see XCS000
XANTHIUM CANADENSE, KERNAL EXTRACT see XCS400
XANTHOCILLIN see DVU000
XANTHOCILLIN Y 1 see XCS680
XANTHOCILLIN Y 2 see XCS700
XANTHOPUCCINE see TCJ800
XANTHOSOMA (various species) see XCS800
XANTHOTOXIN see XDJ000
XANTHURENIC ACID see DNC200
XANTHURENIC ACID-8-METHYL ETHER see HLT500
XANTHYDROL see XBJ000
XANTHYLIUM, 9-(2-CARBOXYPHENYL)-3,6-BIS(METHYLAMINO)-, PERCHLO-
  RATE see RGW100
XANTHYLIUM, 3,6-DIAMINO-9-(2-(METHOXYCARBONYL)PHENYL)-, CHLORIDE
  see RGP600
XANTHYLIUM, 9-(2-(ETHOXYCARBONYL)PHENYL)-3,6-BIS(ETHYLAMINO)-2,7-DI-
  METHYL-, CHLORIDE see RGW000
XANTURAT see ZVJ000
XANTYRID see DVU000
XARIL see ABG750
X465A SODIUM SALT see CDK500
XAVIN see XCS000
XAXA see ADA725
XB 2615 see TBC450
XB 2793 see DKM130
1-X-5-BAA (RUSSIAN) see BDK750
M-XDI see XIJ000
XE 340 see CBT500
XENENE see BGE000
o-XENOL see BGJ250
XENON see XDS000
XENON(II) FLUORIDE METHANESULFONATE see XEA000
XENON(II) FLUORIDE TRIFLUOROACETATE see XEJ000
XENON(II) FLUORIDE TRIFLUOROMETHANESULFONATE see XEJ100
XENON HEXAFLUORIDE see XEJ300
XENON(II) PERCHLORATE see XES000
XENON TETRAFLUORIDE see XFA000
XENON(II) TRIFLUOROACETATE see XFJ000
XENON TRIOXIDE see XFS000
XENON (UN 0236) (DOT) see XDS000
XENON, refrigerated liquid (cryogenic liquids) (UN 2591) (DOT) see XDS000
XENYLAMIN (CZECH) see AJS100
XENYLAMINE see AJS100
1-XENYL-3,3-DIMETHYLTRIAZIN (CZECH) see BGL500
XENYTROPIUM BROMIDE see PEM750
XERAC see BDS000
XERAL see GGS000
XERENE see MFD500
XEROSIN see XFS600
XF-408 see XGA000
XF-13-563 see SCR400
XF 4175L see CBT500
m-XHQ see DSG700
XIBENOL HYDROCHLORIDE see XGA500
XILENOLI (ITALIAN) see XKA000
XILIDINE (ITALIAN) see XMA000
XILOBAM see XGA725
XILOCAINA (ITALIAN) see DHK400
XILOLI (ITALIAN) see XGS000
XIPAMID see CLF325
XIPAMIDE see CLF325
XITIX see ARN000
X 242K see MCB050

XK-62-2 see MQS579
XK-70-1 see FOK000
XL 7 see TFD250
XL-50 see PDP250
XL-90 see RLU000
XL 1246 see PJS750
XL 335-1 see PJS750
XL ALL INSECTICIDE see NDN000
XM 1116 see MCB050
XM 1130 see MCB050
3,5-XMC see DTN200
XMNA see NAC000
XN see XCS000
XNM 68 see PJS750
XO 440 see PJS750
XOE 2982 (RUSSIAN) see HBK700
XOLAMIN see FOS100
XORU-OX see CJY000
XPA see ADW200
X-539-R see PJY100
XRM 3972 see DGJ100
XS-89 see SMM500
XU 292A see BKL800
XUMBRADIL see DNG400
XX 212 see DAD050
XYCAINE HYDROCHLORIDE see DHK600
XYCIANE see DHK400
XYDE see BGC625
XYDURIL see ARQ750
XYLAMIDE (gastroprotective agent) see BGC625
XYLAMIDE see BGC625
XYLAZINE (USDA) see DMW000
1,2-XYLENE see XHJ000
1,3-XYLENE see XHA000
1,4-XYLENE see XHS000
m-XYLENE see XHA000
o-XYLENE see XHJ000
p-XYLENE see XHS000
XYLENE see XGS000
XYLENE ACID MILLING GREEN BL see CMM200
XYLENE BLACK F see BMA000
XYLENE BLUE AS see ERG100
XYLENE BLUE VS see ADE500
XYLENE BLUE VSG see FMU059
m-XYLENE, 5-tert-BUTYL-2,4,6-TRINITRO- see TML750
m-XYLENE-α,α′-DIAMINE see XHS800
m-XYLENE DIISOCYANATE see XIJ000
XYLENE FAST YELLOW ES see SGP500
XYLENE FAST YELLOW GT see FAG140
XYLENE MILLING RED G see CMM325
XYLENE MILLING YELLOW SH see CMM759
XYLENE MUSK see TML750
XYLENEN (DUTCH) see XGS000
p-XYLENE, 2-NITRO- see NMS520
2,4-XYLENESULFONIC ACID see XJJ000
m-XYLENE-4-SULFONIC ACID see XJJ000
m-XYLENESULFONIC ACID see XJJ000
XYLENESULFONIC ACID see XJA000
2,3-XYLENOL see XKJ000
2,4-XYLENOL see XKJ500
2,5-XYLENOL see XKS000
2,6-XYLENOL see XLA000
3,4-XYLENOL see XLJ000
3,5-XYLENOL see XLS000
1,2,5-XYLENOL see XKS000
1,3,4-XYLENOL see XLJ000
1,3,5-XYLENOL see XLS000
m-XYLENOL see XKJ500
o-XYLENOL see XKJ000
p-XYLENOL see XKS000
XYLENOL see XKA000
2,6-XYLENOL, 4-BROMO- see BOL303
XYLENOLEN (DUTCH) see XKA000
3,5-XYLENOL METHYLCARBAMATE see DTN200
XYLENOLS (DOT) see XKA000
3,5-XYLENOL-N-METHYLCARBAMATE see DTN200
XYLESTESIN see DHK400
XYLESTESIN HYDROCHLORIDE see DHK600
2,4-XYLIDENE (MAK) see XMS000
m-XYLIDINE DIISOCYANATE see XIJ000
2,6-XYLIDIDE of 2-PYRIDONE-3-CARBOXYLIC ACID see XLS300
2,6-XYLIDID der 2-PYRIDON-3-CARBOXYLSAURE see XLS300
2,3-XYLIDINE see XMJ000
2,4-XYLIDINE see XMS000
2,5-XYLIDINE see XNA000

2,6-XYLIDINE see XNJ000
3,4-XYLIDINE see XNS000
3,5-XYLIDINE see XOA000
p-XYLIDINE (DOT) see XNA000
m-4-XYLIDINE see XMS000
m-XYLIDINE see XMS000
o-XYLIDINE see XMJ000, XNJ000
XYLIDINE see XMA000
2,4-XYLIDINE HYDROCHLORIDE see XOJ000
2,5-XYLIDINE HYDROCHLORIDE see XOS000
m-XYLIDINE HYDROCHLORIDE see XOJ000
p-XYLIDINE HYDROCHLORIDE see XOS000
XYLIDINEN (DUTCH) see XMA000
XYLIDINE PONCEAU see FMU070
XYLITE see XPJ000
XYLITE (SUGAR) see XPJ000
XYLITOL see XPJ000
XYLITON see XPJ000
l-XYLOASCORBIC ACID see ARN000
XYLOCAIN see DHK400
XYLOCAINE HYDROCHLORIDE see DHK600
XYLOCARD see DHK600
XYLOCHOLINE see XSS900
XYLOCHOLINE BROMIDE see XSS900
XYLOCITIN see DHK400
XYLOCITIN HYDROCHLORIDE see DHK600
2,6-XYLOHYDROQUINONE see DSG700
m-XYLOHYDROQUINONE see DSG700
XYLOIDIN see CCU250
m-XYLOL (DOT) see XHA000
p-XYLOL (DOT) see XHS000
XYLOL (DOT) see XGS000
o-XYLOL see XHJ000
XYLOLE (GERMAN) see XGS000
5-(3,5-XYLOLOXYMETHYL)OXAZOLIDIN-2-ONE see XVS000
XYLOMETAZOLINE HYDROCHLORIDE see OKO500
XYLONEURAL see DHK600
2,6-XYLOQUINOL see DSG700
p-XYLOQUINONE see XQJ000
XYLOSE, D- see XQJ300
XYLOSTATIN see XQJ650
XYLOTOX see DHK400
XYLOTOX HYDROCHLORIDE see DHK600
p-XYLYL ACETATE see XQJ700
2,3-XYLYLAMINE see XMJ000
2,6-XYLYLAMINE see XNJ000
3,4-XYLYLAMINE see XNS000
3,5-XYLYLAMINE see XOA000
N-(2,3-XYLYL)-2-AMINOBENZOIC ACID see XQS000
N-(2,3-XYLYL)ANTHRANILIC ACID see XQS000
1-(2,4-XYLYLAZO)-2-NAPHTHOL see XRA000
1-(2,5-XYLYLAZO)-2-NAPHTHOL see FAG080
1-(o-XYLYLAZO)-2-NAPHTHOL see XRA000
1-XYLYLAZO-2-NAPHTHOL see FAG080, XRA000
1-XYLYLAZO-2-NAPHTHOL-3,6-DISULFONIC ACID, DISODIUM SALT see
    FMU070
1-(2,4-XYLYLAZO)-2-NAPHTHOL-3,6-DISULPHONIC ACID, DISODIUM SALT see
    FMU070
1-XYLYLAZO-2-NAPHTHOL-3,6-DISULPHONIC ACID, DISODIUM SALT see
    FMU070
XYLYL BROMIDE see XRS000
(((2,6-XYLYLCARBAMOYL)METHYL)IMINO)DI-ACETIC ACID see LFO300
p-XYLYL CHLORIDE see MHN300
m-XYLYLENDIAMIN (CZECH) see XHS800
p-XYLYLENDIAMINE (CZECH) see PEX250
XYLYLENDIISOKYANAT (CZECH) see XIJ000
m-XYLYLENDIISOKYANAT see XIJ000
1,1'-(p-XYLYLENE)BIS(3-(1-AZIRIDINYL)UREA) see XSJ000
p-XYLYLENEBIS(TRIPHENYLPHOSPHONIUM CHLORIDE) see XSS000
XYLYLENE CHLORIDE see XSS250
m-XYLYLENE DICHLORIDE see DGP200
o-XYLYLENE DICHLORIDE see DGP400
p-XYLYLENE DICHLORIDE see DGP600
XYLYLENE DICHLORIDE see XSS250
m-XYLYLENE DIISOCYANATE see XIJ000
p-XYLYLENE ISOCYANATE see XSS260
s,6-XYLYL ESTER of 1-PIPERIDINEACETIC ACID HYDROCHLORIDE see
    XSS375
2,6-XYLYL ETHER BROMIDE see XSS900
3,4-XYLYL METHYLCARBAMATE see XTJ000
3,5-XYLYL-N-METHYLCARBAMATE see DTN200
3,4-XYLYL-N-METHYLCARBAMATE, nitrosated see XTS000
p-XYLYLMETHYLCARBINAMINE SULFATE see AQQ000
(2-(3,5-XYLYLOXY)ETHYL)HYDRAZINE HYDROCHLORIDE see XVA000
(2-(3,4-XYLYLOXY)ETHYL)HYDRAZINE MALEATE see XVJ000

5-((3,5-XYLYLOXY)METHYL)-2-OXAZOLIDINONE see XVS000
2-(2,6-XYLYLOXY)TRIETHYLAMINE HYDROCHLORIDE see XWS000
XYLZIN see DMW000
XYMOSTANOL see DKW000


Y 2 see CBM000
Y 3 see CKC000
Y-4 see FAG140
Y 40 see SCK600
Y 195 see PKL750
Y 218 see PKM250
Y 221 see PKM500
Y-223 see PKN000
Y-238 see CDV625
Y 299 see PKO750
Y 302 see PKP000
Y-3642-HCl see AJU000, TGE165
Y-4153 see DCV400
Y 4302 see TLD000
Y 5712 see SDX300
Y 6047 see CKA000
Y-7131 see EQN600
Y-8004 see PLX400
Y-9000 see ACU125
Y-9213 see IAY000
YADALAN see CLH750
YAGEINE see HAI500
YAJEINE see HAI500
YALAN see EKO500
YALTOX see CBS275
YAMAFUL see CCK630
YAMAMOTO METHYLENE BLUE B see BJI250
YAM BEAN see YAG000
YANOCK see FFF000
YARMOR see PIH750
YARMOR PINE OIL see PIH750
YASOKNOCK see SHG500
YATREN see IEP200
YATROCIN see NGE500
YATROZIDE see ILE000
YAUPON see HGF100
YAUTIA MALANGA (PUERTO RICO) see EAI600
YAUTIA (PUERTO RICO) see XCS800
YAZUMYCIN A see RAG300
dl-YB2 see ICA000
YuB-25 see NBO525
YC 93 see PCG550
YE 5626 see SCQ000
YEAST (active) see YAK000
YEAST, EXT. see YAK050
YEAST EXTRACT see YAK050
YEDRA (CUBA) see AFK950
YEH-YAN-KU see ARW000
YELLON see IEP200
YELLOW 204 see CMS245
1310 YELLOW see FAG140
1351 YELLOW see FAG150
1409 YELLOW see FAG140
1504 YELLOW see CMP600
1899 YELLOW see FAG150
11363 YELLOW see MDM775
11712 YELLOW see FEV000
11824 YELLOW see FEW000
12417 YELLOW see FEW000
YELLOW see DCZ000
YELLOW AAMX see CMS208
YELLOW AB see FAG130
YELLOW ACID see CMM758
YELLOW ALLAMANDA see AFQ625
YELLOW CROSS LIQUID see BIH250
YELLOW CUPROCIDE see CNO000
YELLOW FALSE JESSAMINE see YAK100
YELLOW FAST DYE 4K see CMP090
YELLOW FERRIC OXIDE see IHD000
YELLOW GINSENG see BMA150
YELLOW GOWAN see FBS100
YELLOW G SOLUBLE in GREASE see DOT300
YELLOW HENBANE see JBS100
YELLOW JESSAMINE see YAK100
YELLOW LAKE 69 see FAG140
YELLOW LEAD OCHER see LDN000
YELLOW LOCUST see GJU475
YELLOW MERCURIC OXIDE see MCT500
YELLOW MERCURY IODIDE see MDC750

YELLOW M SOLUBLE IN GREASE see CMP600
YELLOW NIGHTSHADE see YAK300
YELLOW No. 2 see FAG130
YELLOW NO. 5 see FAG140
YELLOW NO. 6 see FAG150
YELLOW NO. 5 FDC see FAG140
YELLOW OB see FAG135
YELLOW OLEANDER see YAK350
YELLOW ORANGE S see FAG150
YELLOW ORANGE SPECIALLY PURE 85 see FAG150
YELLOW ORANGE S SPECIALLY PURE see FAG150
YELLOW OXIDE of IRON see IHD000
YELLOW OXIDE of MERCURY see MCT500
YELLOW PARILLA (CANADA) see MRN100
YELLOW PHOSPHORUS see PHP010
1903 YELLOW PINK see BNH500
YELLOW POLAKTIN G sce CMS230
YELLOW PRECIPITATE see MCT500
YELLOW PYOCTANINE see IBB000
YELLOW RELITON G see AAQ250
YELLOW SAGE see LAU600
YELLOW SARSAPARILLA see MRN100
YELLOW SF FOR FOOD see FAG150
YELLOW STABLE 4K see CMP090
YELLOW SUN see FAG150
YELLOW SY FOR FOOD see FAG150
YELLOW TONER YB5 see CMS208
YELLOW ULTRAMARINE see CAP500
YELLOW Z see AAQ250
YERBA HEDIONDA (CUBA) see CNG825
YERBA de PORDIOSEROS (CUBA) see CMV390
YERMONIL see EEH575
YESDOL see CCR875
YESPAZINE see TJW500
YESQUIN (HAITI) see FPD100
YETRIUM see AFJ400
YEW see YAK500
Y-GUTTIFERIN see GME300
Y-3642 HYDROCHLORIDE see TGE165
YL-704 A3 see JDS200
YLANG YLANG OIL see YAT000
YL 704 B₁ see MBY150
YL 704 B3 see LEV025
YM-08054 see IBW500
YM-08310 see AMD000
YM-09229 see FAP100
YM 09330 see CCS371, CCS373
YM-09538 see AOA075
YM-11170 see FAB500
YM-08054-1 see IBW500
YOCLO see ARQ750
YODOCHROME METANIL YELLOW see MDM775
YODOMIN see TEH500
YODOXIN see DNF600
YOHIMBAN-16-CARBOXYLIC ACID derivative of BENZ(g)INDOLO(2,3-a)QUINOLIZINE see RDK000
3-β,20-α-YOHIMBAN-16-β-CARBOXYLIC ACID, 18-β-HYDROXY-11,17-α-DIMETHOXY-METHYL ESTER, 3,4,5-TRIMETHOXYBENZOATE (ESTER), TARTRATE see SCA550
3-β,20-α-YOHIMBAN-16-β-CARBOXYLIC ACID, 18-β-HYDROXY-10,17-α-DIMETHOXY-, METHYL ESTER, 3,4,5-TRIMETHOXYBENZOATE (ester) see MEK700
YOHIMBIC ACID METHYL ESTER see YBJ000
YOHIMBINE see YBJ000
Δ-YOHIMBINE see AFG750
γ-YOHIMBINE HYDROCHLORIDE see AFH000
YOHIMBINE HYDROCHLORIDE see YBS000
YOHIMBINE MONOHYDROCHLORIDE see YBS000
α-YOHIMBIN HYDROCHLORIDE see YCA000
YOMESAN see DFV400
YORK WHITE see CAT775
YOSHI 864 see YCJ000
YOSHINOX 425 see MJN250
YOSHINOX DSTDP see DXG700
YOSHINOX S see TFC600
YOSIMILON see YCJ200
YOUNG FUSTIC see FBW000
YOUNG FUSTIC CRYSTALS see FBW000
YPERITE see BIH250
YPERITE SULFONE see BIH500
Y-5712 SILYL PEROXIDE see SDX300
YTTERBIUM see YDA000
YTTERBIUM CHLORIDE see YDJ000
YTTERBIUM NITRATE see YDS800
YTTERBIUM(III) NITRATE, HEXAHYDRATE (1:3:6) see YEA000

YTTERBIUM TRICHLORIDE see YDJ000
YTTRIA see YGA000
YTTRIUM-89 see YEJ000
YTTRIUM see YEJ000
YTTRIUM CHLORIDE see YES000
YTTRIUM CITRATE see YFA000
YTTRIUM EDETATE complex see YFA100
YTTRIUMNITRAT (GERMAN) see YFS000
YTTRIUM(III) NITRATE (1:3) see YFJ000
YTTRIUM(III) NITRATE HEXAHYDRATE (1:3:6) see YFS000
YTTRIUM OXIDE see YGA000
YTTRIUM TRICHLORIDE see YES000
YU 7802 see BDN125
YUBAN 10S see UTU500
YUBAN 10HV see UTU500
YUCA see CCO680
YUCA BRAVA see CCO680
YUDOCHROME YELLOW GGN see SIT850
YUGILLA (CUBA) see CNH789
YUGOVINYL see PKQ059
YUKALON EH 30 see PJS750
YUKALON HE 60 see PJS750
YUKALON K 3212 see PJS750
YUKALON LK 30 see PJS750
YUKALON MS 30 see PJS750
YUKALON PS 30 see PJS750
YUKALON YK 30 see PJS750
YULAN see EKO500
YUROTIN A see ERC600
YUTAC see YGA700
YUTOPAR see RLK700


Z-4 see DOX100
Z 6 see PKV100
Z 11 see SIA000
Z-78 see EIR000
Z 88 see PJC750
Z 102 see BEQ500
Z-134 see BRE255
Z-905 see PIH100
Z 4828 see TNT500
Z 4942 see IMH000
Z 6030 see TNJ500
Z 7557 see ARP625
ZABILA (MEXICO, DOMINICAN REPUBLIC) see AGV875
ZACLONDISCOIDS see HHS000
ZACTANE CITRATE see MIE600
ZACTIRIN COMPOUND see ABG750
ZACTRAN see DOS000
ZADINE see DLV800
ZADITEN see KGK200
ZADONAL see EOK000
ZAFFRE see CND125
ZAGREB see BIE500
ZAHARINA see BCE500
ZAHLREICHE BEZEICHNUNGEN (GERMAN) see DUS700
ZAMBESI BLACK D see CMN230
ZAMBESI BLACK DA-CF see CMN230
ZAMBESI DARK BLUE BH see CMN800
ZAMBESIL see CLY600
ZAMI 905 see PIH100
dl-ZAMI 1305 see BQF750
ZAMI 1305 see BQF750
ZAMIA DEBILIS see ZAK000
ZAMIA PUMILA see CNH789
ZAMINE see BBK500
ZANIL see DMZ000
ZANILOX see DMZ000
ZANOSAR see SMD000
ZANTAC see RBF400
ZANTEDESCHIA AETHIOPICA see CAY800
ZANTE FUSTIC see FBW000
ZAPRAWA NASIENNA PLYNNA see MLF250
ZARAONDAN see ENG500
ZARCILLA (PUERTO RICO) see LED500
ZARDA (INDIA) see SED400
ZARDANCHU see RQU300
ZARDEX see HCP500
ZARODAN see ENG500
ZARONDAN-SAFT see ENG500
ZARONTIN see ENG500
ZAROXOLYN see ZAK300
ZARTALIN see ENG500
ZASSOL see COI250

ZAVILA (PUERTO RICO) see AGV875
Z-9-DDA see GJU050
ZEAPUR see BJP000
ZEARALANOL see RBF100
(10s)-ZEARALENONE see ZAT000
ZEARALENONE see ZAT000
(s)-ZEARALENONE see ZAT000
trans-ZEARALENONE see ZAT000
(−)-ZEARALENONE see ZAT000
ZEARANOL see RBF100
ZEAZIN see ARQ725
ZEAZINE see ARQ725
ZEBENIDE see EIR000
ZEBTOX see EIR000
ZECTANE see DOS000
ZECTRAN see DOS000
ZEFRAN, combustion products see ADX750
ZEIDANE see DAD200
ZELAN see CIR250
ZELAZA TLENKI (POLISH) see IHF000
ZELEN ALIZARÍNOVA BRILANTNI G-EXTRA (CZECH) see APL500
ZELEN KYPOVA 3 see CMU810
ZELEN KYSELA 3 see FAE950
ZELEN KYSELA 40 see CMM200
ZELEN KYSELA 50 see ADF000
ZELEN KYSELA BS see ADF000
ZELEN MIDLONOVA BLS (CZECH) see APM250
ZELEN MORIDLOVA 4 see NLB000
ZELEN OLIVOVA OSTANTHRENOVA B see CMU810
ZELEN OSTANTHRENOVA BRILANTNI FFB (CZECH) see JAT000
ZELEN POTRAVINARSKA 1 see FAE950
ZELEN POTRAVINARSKA 4 see ADF000
ZELIO see TEL750
ZENADRID (VETERINARY) see PLZ000
ZENALOSYN see PAN100
ZENITE see BHA750
ZENITE SPECIAL see BHA750
ZENTAL see VAD000
ZENTINIC see PAG200
ZENTRONAL see DKQ000
ZENTROPIL see DKQ000, DNU000
ZEOGEL see PAE750
ZEOTIN see MBX800
ZEPC see EOK550
ZEPELIN see PEW000
ZEPHIRAN CHLORIDE see AFP250, BBA500
ZEPHIROL see BBA500
ZEPHROL see EAW000
ZEPHYRANTHES ATAMASCO see RBA500
ZEPHYR LILY see RBA500
ZERANOL (USDA) see RBF100
ZERDANE see DAD200
ZERLATE see BJK500
ZERTELL see CMA250
ZESET T see VLU250
ZEST see MRL500
ZESTE see ECU750
ZET see ZBA000
ZETAR see CMY800
ZETAR EMULSION see ZBA000
ZETAX see BHA750
ZETIFEX ZN see TNC500
ZETTYN CHLORIDE see BEL900
ZEXTRAN see DOS000
ZF 36 see PJS750
Z 6 (Flotation agent) see PKV100
ZG 301 see CAT775
Z-GLY see CBR125
ZHENGGUANGMYCIN A2 (CHINESE) see BLY250
ZHENGGUANGMYCIN A5 (CHINESE) see BLY500
Z-3-HEXENYL VALERATE see HFE800
ZIARNIK see ABU500
ZIAVETINE see BQL000
ZIDAN see EIR000
ZIDE see CFY000
ZIGADENUS (VARIOUS SPECIES) see DAE100
ZIKHOM see ZJS300
ZILDASAC see BAV325
ZIMALLOY see CNA750
ZIMANAT see DXI400
ZIMANEB see DXI400
ZIMATE see BJK500, EIR000
ZIMATE METHYL see BJK500
ZIMCO see VFK000
cis-ZIMELIDINE see ZBA500

ZIMELIDINE see ZBA500
(Z)-ZIMELIDINE see ZBA500
ZIMELIDINE DIHYDROCHLORIDE see ZBA525
ZIMELIDINE HYDROCHLORIDE see ZBA525
ZIMMAN-DITHANE see DXI400
ZIMTALDEHYDE see CMP969
ZIMTSAEURE (GERMAN) see CMP975
ZINACEF see CCS600
ZINADON see ILD000
ZINC see ZBJ000
ZINC ACETATE see ZBS000
ZINC ACETATE, DIHYDRATE see ZCA000
ZINC ACETOACETONATE see ZCJ000
ZINC ALLYL DITHIO CARBAMATE see ZCS000
ZINC AMMONIUM NITRITE see ZDA000
ZINC ARSENATE see ZDJ000
ZINC ARSENATE, BASIC see ZDJ000
ZINC ARSENATE FLUORIDE see ZDJ100
ZINC ARSENITE, solid (DOT) see ZDS000
ZINC-m-ARSENITE see ZDS000
ZINC ASHES (UN 1435) (DOT) see ZBJ000
ZINC BENZIMIDAZOLE-2-THIOLATE see ZIS000
ZINC-2-BENZOTHIAZOLETHIOLATE see BHA750
ZINC BENZOTHIAZOLYL MERCAPTIDE see BHA750
ZINC BENZOTHIAZOL-2-YLTHIOLATE see BHA750
ZINC BENZOTHIAZYL-2-MERCAPTIDE see BHA750
ZINC BERYLLIUM SILICATE see BFV250
ZINC-BIBUTYLDITHIOCARBAMATE see BIX000
ZINC BICHROMATE see ZFA102
ZINC BIS(1H-BENZIMIDAZOLE-2-THIOLATE) see ZIS000
ZINC BIS(DIMETHYLDITHIOCARBAMATE) see BJK500
ZINC BIS(DIMETHYLDITHIOCARBAMATE)CYCLOHEXYLAMINE COMPLEX see
  ZEA000
ZINC BIS(DIMETHYLDITHIOCARBAMOYL)DISULPHIDE see BJK500
ZINC,((BIS(d-GLUCONATO-O¹),O²))- (9CI) see ZIA750
ZINC BUTTER see ZFA000
ZINC CAPRYLATE see ZEJ000
ZINC CARBONATE (1:1) see ZEJ050
ZINC CHLORATE see ZES000
ZINC CHLORIDE (ACGIH,OSHA) see ZFA000
ZINC CHLORIDE see ZFA000
ZINC CHLORIDE, solution (UN 1840) (DOT) see ZFA000
ZINC CHLORIDE, anhydrous (UN 2331) (DOT) see ZFA000
ZINC (CHLORURE de) (FRENCH) see ZFA000
ZINC CHROMATE see ZFA100, ZFA102, ZFJ100
ZINC CHROMATE(VI) HYDROXIDE see CMK500, ZFJ100
ZINC CHROMATE HYDROXIDE see CMK500
ZINC CHROMATE, POTASSIUM DICHROMATE, and ZINC HYDROXIDE (3:1:1)
  see ZFJ150
ZINC CHROMATE with ZINC HYDROXIDE and CHROMIUM OXIDE (9:1) see
  ZFJ120
ZINC CHROME see PLW500
ZINC CHROME YELLOW see ZFJ100
ZINC CHROMITE see ZFA100, ZFJ130
ZINC CHROMIUM OXIDE see ZFA100, ZFA102, ZFJ100
ZINC CITRATE see ZFJ250
ZINC COMPOUNDS see ZFS000
ZINC-COPPER CHROMATE COMPLEX see CNQ750
ZINC CYANIDE see ZGA000
ZINC DIACETATE see ZBS000
ZINC DIACETATE, DIHYDRATE see ZCA000
ZINC-N,N-DIBUTYLDITHIOCARBAMATE see BIX000
ZINC-DIBUTYLDITHIOCARBAMATE see BIX000
ZINC DICHLORIDE see ZFA000
ZINC DICHROMATE (VI) see ZFA102
ZINC DICHROMATE see ZFA102
ZINC DICYANIDE see ZGA000
ZINC-N,N-DIETHYLDITHIOCARBAMATE see BJC000
ZINC DIETHYLDITHIOCARBAMATE see BJC000
ZINC DIHYDRAZIDE see ZGJ000
ZINC N,N-DIMETHYLDITHIOCARBAMATE see BJK500
ZINC DIMETHYLDITHIOCARBAMATE see BJK500
ZINC, DIMETHYLDITHIOCARBAMATE CYCLOHEXYLAMINE COMPLEX see
  ZEA000
ZINC DISTEARATE see ZMS000
ZINC DITHIONITE (DOT) see ZIJ100
ZINC DITHIONITE see ZGJ100, ZIJ100
ZINC DITHIOPHOSPHATE see ZGS000
ZINC DUST (DOT) see ZBJ000
ZINC DUST see ZBJ000
ZINC(II) EDTA COMPLEX see ZGS100
ZINC ETHIDE see DKE600
ZINC ETHOXIDE see ZGW100
ZINC ETHYL (DOT) see DKE600
ZINC ETHYLENE-1,2-BISDITHIOCARBAMATE see EIR000

ZINC ETHYLENEBISDITHIOCARBAMATE see EIR000
ZINC ETHYLPHENYLDITHIOCARBAMATE see ZHA000
ZINC ETHYLPHENYLTHIOCARBAMATE see ZHA000
ZINC-N-FLUOREN-2-YLACETOHYDROXAMATE see ZHJ000
ZINC FLUORIDE see ZHS000
ZINC FLUOROSILICATE (DOT) see ZIA000
ZINC FLUOROSILICATE see ZIA000
ZINC FLUORURE (FRENCH) see ZHS000
ZINC FLUOSILICATE see ZIA000
ZINC GLUCONATE see ZIA750
ZINC HEXAFLUOROSILICATE see ZIA000
ZINC HYDRIDE see ZIJ000
ZINC HYDROSULFITE (DOT) see ZGJ100
ZINC HYDROSULFITE see ZGJ100, ZIJ100
ZINC p-HYDROXYBENZENESULFONATE see ZIJ300
ZINC HYDROXYCHROMATE see CMK500, ZFJ100
ZINC INSULIN see LEK000
ZINCITE see ZKA000
ZINC MANGANESE BERYLLIUM SILICATE see BFS750
ZINCMATE see BJK500
ZINC MERCAPTOBENZIMIDAZOLATE see ZIS000
ZINC MERCAPTOBENZIMIDAZOLE see ZIS000
ZINC MERCAPTOBENZOTHIAZOLATE see BHA750
ZINC-2-MERCAPTOBENZOTHIAZOLE see BHA750
ZINC MERCAPTOBENZOTHIAZOLE SALT see BHA750
ZINC MERCURY CHROMATE COMPLEX see ZJA000
ZINC METAARSENITE see ZDS000
ZINC METHARSENITE see ZDS000
ZINC METIRAM see MQQ250
ZINC MURIATE, solution (DOT) see ZFA000
ZINC NAPHTHENATE see NAT000
ZINC NITRATE see ZJJ000
ZINC(II) NITRATE, HEXAHYDRATE (1:2:6) see NEF500
ZINC NITRILOTRIMETHYLPHOSPHONIC ACID TRISODIUM TETRAHYDRATE
    see ZJJ200
ZINC NITROSYLPENTACYANOFERRATE see ZJJ400
ZINC (N,N'-PROPYLENE-1,2-BIS(DITHIOCARBAMATE)) see ZMA000
ZINCO (CLORURO di) (ITALIAN) see ZFA000
ZINC OCTADECANOATE see ZMS000
ZINCO (FOSFURO di) (ITALIAN) see ZLS000
ZINCOID see ZKA000
ZINC OLEATE (1:2) see ZJS000
ZINC OMADINE see ZMJ000
ZINCOP see ZJS300
ZINC OXIDE (ointment) see ZKJ000
ZINC OXIDE see ZKA000
ZINC OXIDE FUME (MAK) see ZKA000
ZINC PANTOTHENATE see ZKS000
ZINC PENTACYANONITROSYLFERRATE(2−) see ZJJ400
ZINC 2,4-PENTANEDIONATE see ZCJ000
ZINC PERCHLORATE HEXAHYDRATE see ZKS100
ZINC PERMANGANATE see ZLA000
ZINC PEROXIDE see ZLJ000
ZINC p-PHENOL SULFONATE see ZIJ300
ZINC PHENOLSULFONATE see ZIJ300
ZINC PHOSPHIDE see ZLS000
ZINC PHOSPHITE see ZLS200
ZINC (PHOSPHURE de) (FRENCH) see ZLS000
ZINC POLYACRYLATE see ADW100
ZINCPOLYANEMINE see ZMJ000
ZINC POLYCARBOXYLATE see ADW100
ZINC POTASSIUM CHROMATE see CMK400
ZINC POWDER (DOT) see ZBJ000
ZINC POWDER see ZBJ000
ZINC PROTAMINE INSULIN see IDF325
ZINC PT see ZMJ000
ZINC PYRIDINE-2-THIOL-1-OXIDE see ZMJ000
ZINC PYRIDINETHIONE see ZMJ000
ZINC PYRION see ZMJ000
ZINC PYRITHIONE see ZMJ000
ZINC STEARATE see ZMS000
ZINC SULFATE see ZNA000, ZNJ000
ZINC SULFATE HEPTAHYDRATE (1:1:7) see ZNJ000
ZINC SULFATE (1:1) HEPTAHYDRATE see ZNJ000
ZINC SULFOCARBOLATE see ZIJ300
ZINC SULFOPHENATE see ZIJ300
ZINC SULPHATE see ZNA000
ZINC SUPEROXIDE see ZLJ000
ZINC TETRAOXYCHROMATE 76A see ZFJ100
ZINC-TOX see ZLS000
ZINC TRIFLUOROSTANNITE, HEPTAHYDRATE see ZNS000
ZINC TRISODIUM DIETHYLENETRIAMINEPENTAACETATE see TNM850
ZINC TRISODIUM DTPA see TNM850
ZINC UVERSOL see NAT000
ZINC VITRIOL see ZNA000, ZNJ000

ZINC WHITE see ZKA000
ZINC YELLOW see CMK500, PLW500, ZFJ100, ZFJ120
ZINEB see EIR000
ZINEB-COPPER OXYCHLORIDE mixture see ZJS300
ZINEB-ETHYLENE THIURAM DISULFIDE ADDUCT see MQQ250
ZINGERONE see VFP100
ZINGIBERONE see VFP100
ZINGIBER ROSEUM (Roxb.) Rosc., extract see ZNS200
ZINK-BIS(N,N-DIMETHYL-DITHIOCARBAMAAT) (DUTCH) see BJK500
ZINK-BIS(N,N-DIMETHYL-DITHIOCARBAMAT) (GERMAN) see BJK500
ZINKCARBAMATE see BJK500
ZINKCHLORID (GERMAN) see ZFA000
ZINKCHLORIDE (DUTCH) see ZFA000
ZINK-(N,N-DIMETHYL-DITHIOCARBAMAT) (GERMAN) see BJK500
ZINKFOSFIDE (DUTCH) see ZLS000
ZINK-(N,N'-AETHYLEN-BIS(DITHIOCARBAMAT)) (GERMAN) see EIR000
ZINK-(N,N'-PROPYLEN-1,2-BIS(DITHIOCARBAMAT)) (GERMAN) see ZMA000
ZINKOSITE see ZNA000
ZINKPHOSPHID (GERMAN) see ZLS000
ZINN (GERMAN) see TGB250
ZINNTETRACHLORID (GERMAN) see TGC250
ZINOCHLOR see DEV800
ZINOPHOS see EPC500
ZINOSAN see EIR000
ZINOSTATIN see NBV500
ZINPOL see ADW200, PJS750
ZIPAN see DCK759
ZIPEPROL see MFG250
ZIPEPROL DIHYDROCHLORIDE see MFG250
ZIPROMAT see ZMA000
ZIRAM see BJK500
ZIRAM CYCLOHEXYLAMINE COMPLEX see ZEA000
ZIRAMVIS see BJK500
ZIRASAN see BJK500
ZIRBERK see BJK500
ZIRCAT see ZOA000
ZIRCON see ZSS000
ZIRCONATE, BARIUM (1:1) see BAP750
ZIRCONATE(2-), OXODISULFATO-, DISODIUM (8CI) see ZTS100
ZIRCONIUM (ACGIH,OSHA) see ZOA000
ZIRCONIUM see ZOA000
ZIRCONIUM(IV) CHLORIDE (1:4) see ZPA000
ZIRCONIUM(II) CHLORIDE see ZQJ000
ZIRCONIUM CHLORIDE see ZPA000
ZIRCONIUM CHLORIDE HYDROXIDE see ZPJ000
ZIRCONIUM CHLORIDE OXIDE OCTAHYDRATE see ZPS000
ZIRCONIUM CHLOROHYDRATE see ZPJ000
ZIRCONIUM COMPOUNDS see ZQA000
ZIRCONIUM DIBROMIDE see ZQB100
ZIRCONIUM DICARBIDE see ZQC200
ZIRCONIUM DICHLORIDE see ZQJ000
ZIRCONIUM, DICHLORO-DI-pi-CYCLOPENTADIENYL- see ZTK400
ZIRCONIUM FLUORIDE see ZQS000
ZIRCONIUM GLUCONATE see ZQS100
ZIRCONIUM HYDRIDE see ZRA000
ZIRCONIUM HYDROXYCHLORIDE see ZPJ000
ZIRCONIUM(III) LACTATE (1:3) see ZRJ000
ZIRCONIUM(IV) LACTATE see ZRS000
ZIRCONIUM METAL, dry, chemically produced, finer than 20 mesh particle
    size (UN 2008) see ZOA000
ZIRCONIUM NITRATE see ZSA000
ZIRCONIUM OXYCHLORIDE see ZSJ000
ZIRCONIUM PICRAMATE, wetted with not <20% water, by weight (UN 1517)
    (DOT) see PIC750
ZIRCONIUM PICRAMATE, dry or wetted with <20% water, by weight (UN
    0236) (DOT) see PIC750
ZIRCONIUM POTASSIUM FLUORIDE see PLG500
ZIRCONIUM POWDER, wetted with not <25% water (UN 1358) (DOT) see
    ZOA000
ZIRCONIUM POWDER, dry (UN 2008) (DOT) see ZOA000
ZIRCONIUM SCRAP (UN 1932) (DOT) see ZOA000
ZIRCONIUM(IV) SILICATE (1:1) see ZSS000
ZIRCONIUM SODIUM LACTATE see ZTA000
ZIRCONIUM(IV) SULFATE (1:2) see ZTJ000
ZIRCONIUM TETRACHLORIDE (DOT) see ZPA000
ZIRCONIUM TETRACHLORIDE, solid (DOT) see ZPA000
ZIRCONIUM TETRAFLUORIDE see ZQS000
ZIRCONIUM(IV) TETRAHYDROBORAT see ZTK300
ZIRCONIUM, dry, coiled wire, finished metal sheets, strip (UN 2858) (DOT)
    see ZOA000
ZIRCONIUM, dry, finished sheets, strip or coiled wire (UN 2009) (DOT) see
    ZOA000
ZIRCONOCENE, DICHLORIDE see ZTK400
ZIRCONYL ACETATE see ZTS000
ZIRCONYL CHLORIDE see ZSJ000

ZIRCONYL CHLORIDE OCTAHYDRATE see ZPS000
ZIRCONYL HYDROXYCHLORIDE see ZPJ000
ZIRCONYL NITRATE see BLA000
ZIRCONYL SODIUM SULPHATE see ZTS100
ZIRCONYL SULFATE see ZTJ000
ZIREX 90 see BJK500
ZIRIDE see BJK500
ZIRPON see MQU750
ZIRTHANE see BJK500
ZITEX H 662-124 see TAI250
ZITHIOL see MAK700
ZITOSTOP see MAW800
ZITOX see BJK500
ZITOXIL see MFG250
ZITRONELL OEL (GERMAN) see CMT000
ZITRONEN OEL (GERMAN) see LEI000
ZK 31224 see DLR100
ZK 57671 see SOU650
ZLUT CHINOLONOVA see CMM510
ZLUT DISPERZNI 3 see AAQ250
ZLUT KYPOVA 2 see VGP200
ZLUT KYSELA 23 see FAG140
ZLUT KYSELA 3 see CMM510
ZLUT KYSELA 9 see CMM758
ZLUT MARCIOVA see DUX800
ZLUT MASELNA (CZECH) see DOT300
ZLUT MASELNA OB see FAG135
ZLUT NAFTOLOVA see DUX800
ZLUT OSTANTHRENOVA GC see VGP200
ZLUT PIGMENT 100 see FAG140
ZLUT POTRAVINARSKA 13 see CMM510
ZLUT POTRAVINARSKA 2 see CMM758
ZLUT POTRAVINARSKA 3 see FAG150
ZLUT POTRAVINARSKA 4 see FAG140
ZLUT PRIRODNI 11 see MRN500
ZLUT PRIRODNI 8 see MRN500
ZLUT ROZPOUSTEDLOVA 6 see FAG135
ZLUT ROZPOUSTEDLOVA 77 see AAQ250
ZMA see ZDS000
ZnMB see BHA750
ZMBT see BHA750
ZN 6 see SHK000
ZN 6-NA see SHK000
ZN-0312 T 1/4″ see ZFA100
ZOALENE see DUP300
ZOAMIX see DUP300
ZOAPATLE, crude leaf extract see ZTS600
ZOAPATLE, semi-purified leaf extract see ZTS625
ZOAQUIN see DNF600
ZOBA BLACK D see PEY500
ZOBA BROWN P BASE see ALT250
ZOBA BROWN RR see ALL750
ZOBA EG see ALT500
ZOBA ERN see NAW500
ZOBA 3GA see ALT000
ZOBA GKE see TGL750
ZOBA 4R see DUP400
ZOBAR see PJQ000, TIK500
ZOBA SLE see DBO400
ZOGEN DEVELOPER H see TGL750
ZOLAMINE HYDROCHLORIDE see ZUA000
ZOLAPHEN see BRF500
ZOLCIENI POLACTYNOWEJ G (POLISH) see CMS230
ZOLICEF see CCS250
ZOLIDINUM see BRF500
ZOLIMIDIN see ZUA200

ZOLIMIDINE see ZUA200
ZOLIRIDINE see ZUA200
ZOLON see BDJ250
ZOLONE see BDJ250
ZOLONE PM see BDJ250
ZOMAX see ZUA300
ZOMEPIRAC see ZUA300
ZOMEPIRAC SODIUM see ZUA300
ZOMEPIRAC SODIUM SALT see ZUA300
ZONAZIDE see ILD000
ZONDEL see NNT100
ZONGNON (HAITI) see WBS850
ZONIFUR see NGB700
ZOOCOUMARIN (RUSSIAN) see WAT200
ZOOFURIN see NGE000
ZOOLON see BDJ250
ZOPAQUE see TGG760
ZOPICLONE see ZUA450
ZORANE see BRF500, XVS000
ZORIFLAVIN see DBX400
ZORUBICIN see ROU800
ZORUBICIN HYDROCHLORIDE see ROZ000
ZOTEPINE see ZUJ000
ZOTHELONE see PJA120
ZOTIL see ABF750
ZOTOX see ARB250, ARH500
ZOTOX CRAB GRASS KILLER see ARB250
ZOVIRAX see AEC700
ZOVIRAX SODIUM see AEC725
ZOXAMIN see AJF500
ZOXAZOLAMINE see AJF500
ZOXINE see AJF500
3ZhP see IGK800
ZP see ZLS000
Z 3 (PESTICIDE) see PLF000
ZR 512 see EQD000
ZR 515 see KAJ000
ZR-777 see POB000
ZR-856 see HCP500
Z10-TR see CFZ000
ZUMARIL see AGN000
ZUTRACIN see BAC250
ZWAVELWATERSTOF (DUTCH) see HIC500
ZWAVELZUUROPLOSSINGEN (DUTCH) see SOI500
ZWITSALAX see DMH400
ZY 15021 see BDJ600
ZYBAN see PEX500
ZYGADENUS (VARIOUS SPECIES) see DAE100
ZYGOMYCIN A1 see NCF500
ZYGOSPORIN A see ZUS000
N-(ZYKLOHEXYL)-SYDNONIMIN HYDROCHLORID (GERMAN) see CPQ650
ZYKLOPHOSPHAMID (GERMAN) see CQC650
ZYLOFURAMINE see ZVA000
ZYLOPRIM see ZVJ000
ZYLORIC see ZVJ000
ZYMOFREN see PAF550
ZYNOPLEX see TAD175
ZYTEL 211 see PJY500
ZYTOX see MHR200
ZYTRON see DGD800
ZZL-0810 see ZVS000
ZZL-0814 see EBU500
ZZL-0816 see EBV000
ZZL-0822 see EBV100
ZZL-0854 see EBV500
ZZLA-0334 see EBW000

# SECTION 4
# REFERENCES

This section lists the general toxicological bibliography of nearly 2200 keyed references to an alpha numeric system.

**AAAHAN** Australian Journal of Experimental Agriculture and Animal Husbandry. (Commonwealth Scientific and Industrial Research Organization, POB 89, E. Melbourne, Victoria 3002, Australia) V.1- 1961-

**AAATAP** Annali dell'Accademia di Agricoltura di Torino (Academia de Agricoltura, Via Andrea Doria 10, Turin, Italy) V.119- 1976/77-

**AABIAV** Annals of Applied Biology. (Biochemical Society, P.O. Box 32, Commerce Way, Whitehall Industrial Estate, Colchester CO2 8HP, Essex, England) V.1- 1914-

**AACHAX** Antimicrobial Agents and Chemotherapy. (Ann Arbor, MI) 1961-70. For publisher information, see AMACCQ

**AACRAT** Anesthesia and Analgesia; Current Research. (International Anesthesia Research Society, 3645 Warrensville Center Rd., Cleveland, OII 44122) V.36- 1957-

**AAGAAW** Antimicrobial Agents Annual. (New York, NY) 1960-60. For publisher information, see AMACCQ

**AAGEAA** Arkhiv Anatomii, Gistologii i Embriologii. Archives of Anatomy, Histology and Embryology. (v/o Mezhdunarodnaya Kniga, Kuznetskii Most 18, Moscow G-200, USSR) V.1- 1916-

**AAHEAF** Archives d'Anatomie, d'Histologie et d'Embryologie. (Editions Alsatia, 10 rue Bartholdi, 68001 Colmar, France) V.1- 1922-

**AAJRDX** AJR, American Journal of Roentgenology. (Charles C. Thomas Publisher, 301-327 E. Lawrence Ave., Springfield, IL 62717) V.126- 1976-

**AAMLAR** Atti della Accademia Medica Lombarda. (The Academy, c/o Prof. W. Montorsi, Festa del Perdono 3, Milan, Italy) V.20-29, 1931-40; V.15- 1960-

**AAMMAU** Archives d'Anatomie Microscopique et de Morphologie Experimentale. (Masson Publishing USA, Inc. 14 E. 60th St., New York, NY 10022) V.36- 1947-

**AANEAB** Acta Anaesthesiologica Scandinavica. (Munksgaard, 35 Noerre Soegade, DK-1370, Copenhagen K, Denmark) V.1- 1957-

**AANFAE** Annales de l'Anesthesiologie Francaise. (Doin Editeurs, 8 Place de l'Odeon, F-75006 Paris, France) V.1- 1960-

**AANLAW** Atti della Accademia Nazionale dei Lincei, Rendiconti della Classe di Scienze Fisiche, Matematiche e Naturali. (Academia Nazionale dei Lincei, Ufficio Pubblicazioni, Via della Lungara, 10, I-00165 Rome, Italy) V.1- 1946-

**AAOPAF** AMA Archives of Ophthalmology. (Chicago, IL) V.44, No. 4-63, 1950-60. For publisher information, see AROPAW

**AAREAV** Anesthesie, Analgesie, Reanimation. (Masson et Cie, Editeurs, 120 Blvd. Saint-Germain, P-75280, Paris 06, France) V.14- 1957-

**ABAHAU** Acta Biologica Academiae Scientiarum Hungaricae. (Kultura, P.O. Box 149, H-1389 Budapest, Hungary) V.1- 1950-

**ABANAE** Antibiotics Annual. (New York, NY) 1953-60. For publisher information, see AMACCQ

**ABBIA4** Archives of Biochemistry and Biophysics. (Academic Press, 111 5th Ave., New York, NY 10003) V.31- 1951-

**ABCHA6** Agricultural and Biological Chemistry. (Maruzen Co. Ltd., P.O.Box 5050 Tokyo International, Tokyo 100-31, Japan) V.25- 1961-

**ABEMAV** Annals of Biochemistry and Experimental Medicine. (Calcutta, India) V.1-23, 1941-63. For publisher information, see IJEBA6

**ABHUE6** Acta Biologica Hungarica. (Kultura, POB 149, H-1389 Budapest, Hungary) V.34- 1983-

**ABHYAE** Abstracts on Hygiene. (Bureau of Hygiene and Tropical Diseases, Keppel St., London WC1E 7HT, England) V.1- 1926-

**ABILAE** Archives de Biologie. (Vaillant-Carmanne, SA rue Fond Saint-Servais, 14 B.P.22, B-4000 Liege, Belgium) V.1- 1880-

**ABMGAJ** Acta Biologica et Medica Germanica. (Berlin, Germany) V.1-41, 1958-82. For publisher information, see BBIADT

**ABMHAM** Archives Belges de Medicine Social, Hygiene, Medicine du Trevail et Medicine Legale. (Publications Acta Medica Belgica, rue des Champs-Elysees 43, 1050 Brussels 5, Belgium) V.4- 1946-

**ABMPAC** Advances in Biological and Medical Physics. (Academic Press, 111 5th Ave., New York, NY 10003) V.1- 1948-

**ABPAAG** Acta Biologiae Experimentalis (Warsaw). (Warsaw, Poland) V.1-13, 1928-39; V.14-29, 1947-69

**ACATA5** Acta Anatomica. (S. Karger AG, Arnold-Boecklin-St 25, CH-4000 Basel 11, Switzerland) V.1- 1945-

**ACCBAT** Acta Clinica Belgica. (Association des Societes Scientifiques Medicales Belges, rue des Champs-Elysees, 43, 1050 Brussels 5, Belgium) V.1- 1946-

**ACEDAB** Acta Endocrinologica, Supplementum. (Periodica, Skolegade 12E, 2500 Copenhagen-Valby, Denmark) No. 1- 1948-

**ACENA7** Acta Endocrinologica. (Periodica, Skolegade 12E, 2500 Copenhagen-Valby, Denmark) V.1- 1948-

**ACGHD2** Annals of the American Conference of Governmental Industrial Hygienists. (American Conference of Governmental Industrial Hygienists, Inc., 6500 Glenway Ave., Bldg. D-5, Cincinnati, OH, 54211) V.1- 1981-

**ACHAAH** Acta Haematologica. (S. Karger AG, Arnold-Boecklin-St 25, CH-4000 Basel 11, Switzerland) V.1- 1948-

**ACHCBO** Acta Histochemica et Cytochemica. (Japan Society of Histochemistry and Cytochemistry, Dept. Anatomy, Faculty of Medicine, Kyoto University, Konoe-cho, Yoshida, Sakyo-ku, Kyoto, 606, Japan) V.1- 1968-

**ACHTA6** Antibiotica et Chemotherapia. (Basel, Switzerland) V.1-16, 1954-70. For publisher information, see ANBCB3

**ACIEAY** Angewandte Chemie, International Edition in English. (Verlag Chemie GmbH, Postfach 1260/1280, D6940, Weinheim, Germany) V.1- 1962-

**ACLRBL** Annals of Clinical Research. (Finnish Medical Society, Duodecim Runeberginkatu 47A, SF-00260 Helsinki 26, Finland) V.1- 1969-

**ACLSCP** Annals of Clinical Laboratory Science. (1833 Delancey Pl., Philadelphia, PA 19103) V.1- 1971-

**ACNSAX** Activitas Nervosa Superior. (Avicenum, Malostranske nam. 28, Prague 1, Czechoslovakia) V.1- 1959-

**ACPAAN** Acta Paediatrica. (Stockholm, Sweden) V.1-53, 1921-64. For publisher information, see APSVAM

**ACPADQ** Acta Pathologica, Microbiologica et Immunologica Scandinavica, Section A: Pathology. (Munksgaard, 35 Noerre Soegade, DK-1370, Copenhagen K, Denmark) V.90- 1982-

**ACPMAP** Actualites Pharmacologiques. (Paris, France) No. 1-33, 1949-81. For publisher information, see JNPHAG

**ACRAAX** Acta Radiologica. (Stockholm, Sweden) V.1-58, 1921-62. For publisher information, see ACRDA8

**ACRDA8** Acta Radiologica: Diagnosis. (Box 7449, S-10391 Stockholm, Sweden) NS.V.1- 1963-

**ACRSAJ** Advances in Cancer Research. (Academic Press, 111 5th Ave., New York, NY 10003) V.1- 1953-

**ACRSDM** Alcoholism: Clinical and Experimental Research. (Grune and Stratton, Inc., 111 5th Ave., New York, NY 10003) V.1- 1977-

**ACTRAQ** Acta Tropica. (Schwabe and Co., Steintorstr., 13, CH-4010 Basel, Switzerland) V.1 1944-

**ACTTDZ** Acta Therapeutica. (Pl. de Bastogne 8, B13, 1080 Brussels, Belgium) V.1- 1975-

**ACVIA9** Acta Vitaminologica. (Milan, Italy) V.1-20, 1947-66. For publisher information, see AVEZA6

**ACVTA8** Acta Veterinaria. (Jugoslovenska Knjiga, Terazije 27/II, Belgrade, Yugoslavia) V.1- 1951-

**ACYTAN** Acta Cytologica. (Science Printers and Pub., Inc., 2 Jacklynn Ct., St. Louis, MO 63132) V.1- 1957-

**ADBBBW** Advances in Behavioral Biology. (Plenum Publishing Corp., 233 Spring Spring St., New York, NY 10013) V.1- 1971-

**ADCHAK** Archives of Disease in Childhood. (British Medical Journal, 1172 Commonwealth Avenue, Boston, MA 02134) V.1- 1926-

**ADCSAJ** Advances in Chemistry Series. (American Chemical Society, 1155 16th St., N.W., Washington, DC 20036) No. 1- 1950-

**ADIRDF** Advances in Inflammation Research. (Raven Press, 1140 Avenue of the Americas, New York, NY 10036) V.1- 1979-

**ADMFAU** Archiv fuer Dermatologische Forschung. (Secaucus, NJ) V.240-252, No. 4, 1971-75. For publisher information, see ADREDL

**ADRCAC** Annali Italiani di Dermatologia Clinica e Sperimentale. (Clinica Dermosifilopatica Universita degli Studi, Policlinica Monteluce, 06100 Perugia, Italy) V.16- 1962-

**ADREDL** Archives of Dermatological Research. (Springer-Verlag New York, Inc., Service Center, 44 Hartz Way, Secaucus, NJ 07094) V.253- 1975-

**ADRPBI** Advances in Reproductive Physiology. (Academic Press, 111 Fifth Ave., New York, NY 10003) V.1-6, 1966-73, Discontinued

**ADSYAF** Archives of Dermatology and Syphilology. (Chicago, IL) V.1-62, 1920-60. For publisher information, see ARDEAC

**ADTEAS** Advances in Teratology. (Academic Press, 111 5th Ave., New York, NY 10003) V.1-5, 1966-72, Discontinued

**ADVEA4** Acta Dermato-Venereologica. (Almqvist and Wiksell International, Box 62, S-101 20 Stockholm, Sweden) V.1- 1920-

**ADVED7** Annales de Dermatologie et de Venereologie. (Masson Publishing USA Inc., 133 E. 58th St., New York, NY 10022) V.104- 1977-

**ADVPA3** Advances in Pharmacology. (New York, NY) V.1-6, 1962-68. For publisher information, see AVPCAQ

**ADVPB4** Advances in Planned Parenthood. (Excerpta Medica, Inc., POB 3085 Princeton, NJ 08540) V.1- 1967-

**ADWMAX** Abhandlungen der Deutschen Akademie der Wissenschaften zu Berlin, Klasse fuer Medizin. (Akademie-Verlag GmbH, Leipziger Str. 3-4, DDR-108 Berlin, Germany) 1950-

**AECTCV** Archives of Environmental Contamination and Toxicology. (Springer-Verlag New York, Inc., Service Center, 44 Hartz Way, Secaucus, NJ 07094) V.1- 1973-

**AEEDDS** Adverse Effects of Environmental Chemicals and Psychotropic Drugs. (Elsevier North Holland, Inc., 52 Vanderbilt Ave., New York, NY 10017) V.1- 1973-

**AEEXAH** Acta Embryologiae Experimentalis. (Via Archirafi 18, 90123 Palermo, Italy) No. 1- 1969-

**AEFTAA** Acta Europaea Fertilitatis. (Piccin Medical Books, Via Brunacci, 12, 35100 Padua, Italy) V.1- 1969-

**AEHA\*\*** U.S. Army, Environmental Hygiene Agency Reports. (Edgewood Arsenal, MD 21010)

**AEHLAU** Archives of Environmental Health. (Heldreff Publications, 4000 Albemarle St., N.W., Washington, DC 20016) V.1- 1960-

**AEMBAP** Advances in Experimental Medicine and Biology. (Plenum Publishing Corp., 233 Spring St., New York, NY 10013) V.1- 1967-

**AEMED3** Annals of Emergency Medicine. (American College of Emergency Physicians, 1125 Executive Circle, Irving, TX 75038)

**AEMIDF** Applied and Environmental Microbiology. (American Society for Microbiology, 1913 I St., N.W., Washington, DC 20006) V.31- 1976-

**AEPPAE** Naunyn-Schmiedeberg's Archiv fuer Experimentelle Pathologie und Pharmakologie. (Berlin, Germany) V.110-253, 1925-66. For publisher information, see NSAPCC

**AESAAI** Annals of the Entomological Society of America. (Entomological Society of America, 4603 Calvert Road, College Park, MD 20740) V.1- 1908-

**AETODY** Advances in Modern Environmental Toxicology. (Senate Press, Inc., P.O. Box 252, Princeton Junction, NJ 08550) V.1- 1980-

**AEXPBL** Archiv fuer Experimentelle Pathologie und Pharmakologie. (Leipzig, Germany) V.1-109, 1873-1925. For publisher information, see NSAPCC

**AEZRA2** Advances in Enzyme Regulation. (Pergamon Press, Headington Hill Hall, Oxford OX3 OBW, England) V.1- 1963-

**AFCPDR** Annual Symposium on Fundamental Cancer Research, Proceedings. (University of Texas System Cancer Center, M.D. Anderson Hospital and Tumor Institute, Houston, TX 77030) V.30- 1978-

**AFDOAQ** Association of Food and Drug Officials of the United States, Quarterly Bulletin. (Editorial Committee of the Association, P.O. Box 20306, Denver, CO 80220) V.1- 1937-

**AFECAT** Annales des Falsifications et de l'Expertise Chimique. (Paris, France) V.53, No. 613-No. 779, 1960-79

**AFPCAG** Acta Facultatis Pharmaceuticae, Universitatis Comenianae. (Ustredna Kniznica Farmaceuticka Fakulta Univerzity Komenskeho, Kolinciakova 8, 886 34 Bratislava, Czechoslovakia) V.14- 1967-

**AFPEAM** Archives Francaises de Pediatrie. (Doin Editeurs, 8 Place de l'Odeon, F-75006 Paris, France) V.1- 1942-

**AFPYAE** American Family Physician. (American Academy of Family Physicians, 1740 W. 92nd Street, Kansas City, MO 64114) V.2-17, 1962-1969; New series: V.2- 1970-

**AFREAW** Advances in Food Research. (Academic Press, 111 5th Ave., New York, NY 10003) V.1- 1948-

**AFSPA2** Archivo di Farmacologia Sperimentale e Scienze Affini. (Rome, Italy) V.1-82, 1902-1954. Discontinued

**AFTOD7** Archivos de Farmacologia y Toxicologia. (Universidad Complutense, Facultad de Medicina, Departamento Coordinado de Farmacologia, Ciudad Universitaria, Madrid 3, Spain) V.1- 1975-

**AGACBH** Agents and Actions, A Swiss Journal of Pharmacology. (Birkhaeuser Verlag, P.O. Box 34, Elisabethenst 19, CH-4010, Basel, Switzerland) V.1- 1969/70-

**AGDJAI** Journal of the Academy of General Dentistry. (Academy of General Dentistry, 211 E. Chicago Ave., Suite 1200, Chicago, IL 60611) V.1- 1952(?)-

**AGGHAR** Archiv fuer Gewerbepathologie und Gewerbehygiene. (Berlin, Germany) V.1-18, 1930-61. For publisher information, see IAEHDW

**AGMGAK** Acta Geneticae Medicae et Gemellologiae. (Cappelli Editore, Via Marsili 9, I-40124 Bologna, Italy) V.1- 1952-

**AGPSA3** AMA Archives of General Psychiatry. (Chicago, IL) V.1-2, 1959-60. For publisher information, see ARGPAQ

**AGSOA6** Agressologie. Revue Internationale de Physio-Biologie et de Pharmacologie Appliquees aux Effets de l'Agression. (Masson et Cie, Editeurs, 120 Blvd. Saint-Germain, P-75280, Paris 06, France) V.1- 1960-

**AGTQAH** Annales de Genetique. (Expansion Scientifique Francaise, 5 rue Saint- Benoit, F-75278, Paris 06, France) V.1- 1958-

**AHBAAM** Archiv fuer Hygiene und Bakteriologie. (Munich, Germany) V.101-154, 1929-71. For publisher information, see ZHPMAT

**AHEMA5** Anatomia, Histologia, Embryologia. (Verlag Paul Parey, Linderstr. 44-47, D-1000 Berlin 61, Germany) V.2- 1973-

**AHJOA2** American Heart Journal. (C.V. Mosby Co., 11830 Westline Industrial Dr., St. Louis, MO 63141) V.1- 1925-

**AHRTAN** Arhiv za Higijenu Rada i Toksikologiju. Archives of Industrial Hygiene and Toxicology. (Jugoslovenska Knjiga, P.O. Box 36, Terazije 27, 11001, Belgrade, Yugoslavia) V.7- 1956-

**AHYGAJ** Archiv fuer Hygiene. (Munich, Germany) V.1-100, 1883-1928. For publisher information, see ZHPMAT

**AICCA6** Acta Unio Internationalis Contra Cancrum. (Louvain, Belgium) V.1-20, 1936-64. For publisher information, see IJCNAW

**AIDZAC** Aichi Ika Daigaku Igakkai Zasshi. Journal of the Aichi Medical Univ. Assoc. (Aichi Ika Daigaku, Yazako, Nagakute-machi, Aichi-gun, Aichi- Ken 480-11, Japan) V.1- 1973-

**AIHAAP** American Industrial Hygiene Association Journal. (AIHA, 475 Wolf Ledges Pkwy., Akron, OH 44311) V.19- 1958-

**AIHAM\*** Annual Meeting of American Industrial Hygiene Association. For publisher information, see AIHAAP

**AIHOAX** Archives of Industrial Hygiene and Occupational Medicine. (Chicago, IL) V.1-2, No. 3, 1950. For publisher information, see AEHLAU

**AIHQA5** American Industrial Hygiene Association Quarterly. (Baltimore, MD) V.7-18, 1946-57. For publisher information, see AIHAAP

**AIMDAP** Archives of Internal Medicine. (American Medical Association, 535 N. Dearborn St., Chicago, IL 60610) V.1- 1908-

**AIMEAS** Annals of Internal Medicine. (American College of Physicians, 4200 Pine St., Philadelphia, PA 19104) V.1- 1927-

**AIMJA9** Ain Shams Medical Journal (Ain Shams Clinical and Scientific Society, c/o Ain Shams University Hospital, Abbassia, Cairo, Egypt) V.1- 1950-

**AIPAAV** Annales de l'Institut Pasteur. (Paris, France) V.1-123, 1887-1972. For publisher information, see ANMBCM

**AIPSAH** Archives de l'Institut Pasteur d'Algerie. (Bibliotheque, Institut Pasteur d'Algerie, rue du Docteur-Laveran, Algiers, Algeria) V.1- 1923-

**AIPTAK** Archives Internationales de Pharmacodynamie et de Therapie. (Editeurs, Institut Heymans de Pharmacologie, De Pintelaan 135, B-9000 Ghent, Belgium) V.4- 1898-

**AIPUAN** Archivio Italiano di Patologia e Clinica dei Tumori. (Istituto di Farmacologia della Universita, Via Vanvitelli 32, 20129 Milan, Italy) V.1-15, 1957-72, Discontinued

**AISFAR** Archivio Italiano di Scienze Farmacologiche. (Modena, Italy) 1932-65. Discontinued

**AISMAE** Archivio Italiano di Scienze Mediche Tropical e di Parassitologia. Italian Archives of the Science of Tropical Medicine and Parasitology. (Rome, Italy) V.1-54, 1920-73. Suspended

**AISSAW** Annali dell'Istituto Superiore di Sanita. (Istituto Poligrafico dello Stato, Libreria dello Stato, Piazza Verdi, 10 Rome, Italy) V.1- 1965-

**AITDAQ** Archiwum Immunologii i Terapii Doswiadczalnej. (Warsaw, Poland) V.1-9, 1953-61. For publisher information, see AITEAT

**AITEAT** Archivum Immunologiae et Therapiae Experimentalis. (Ars Polona-RUCH, P.O. Box 1001, P-00 068 Warsaw, 1, Poland) V.10- 1962-

**AIVMBU** Archivos de Investigacion Medica. (POB 12,631, Col. Narvarte, Deleg. Benito Juarez, 03020 Mexico City, DF, Mexico, V.1- 1970-

**AJANA2** American Journal of Anatomy. (Alan R. Liss Inc., 150 5th Ave., New York, NY 10011) V.1- 1901-

**AJBSAM** Australian Journal of Biological Sciences. (Commonwealth Scientific and Industrial Research Organization, POB 89, E. Melbourne, Victoria, Australia) V.6- 1953-

**AJCAA7** American Journal of Cancer. (New York, NY) V.15-40, 1931-40. For publisher information, see CNREA8

**AJCDAG** American Journal of Cardiology. (Technical Publishing Co., 666 5th Ave., New York, NY 10103) V.1- 1958-

**AJCNAC** American Journal of Clinical Nutrition. (American Society for Clinical Nutrition, Inc., 9650 Rockville Pike, Bethesda, MD 20814) V.2- 1954-

**AJCPAI** American Journal of Clinical Pathology. (J.B. Lippincott Co., Keystone Industrial Park, Scanton, PA 18512) V.1- 1931-

**AJDCAI** American Journal of Diseases of Children. (American Medical Association, 535 N. Dearborn St., Chicago, IL 60610) V.1-80(3), 1911-50; V.100- 1960-

**AJDDAL** American Journal of Digestive Diseases. (Plenum Publishing Corp., 233 Spring St., New York, NY 10013) V.5-22, 1938-55, V.1-23, 1956-78

**AJDEBP** Australasian Journal of Dermatology. (Australasian College of Dermatologists, 271 Bridge Rd., Glebe, NSW 2037, Australia) V.9- 1967-

**AJEBAK** Australian Journal of Experimental Biology and Medical Science. (Univ. of Adelaide Registrar, Adelaide, S.A. 5000, Australia) V.1- 1924-

**AJEMEN** American Journal of Emergency Medicine. (WB Saunders, Philadelphia, PA) V.1- 1983-

**AJEPAS** American Journal of Epidemiology. (Johns Hopkins Univ., 550 N. Broadway, Suite 201, Baltimore, MD 21205) V.81- 1965-

**AJGAAR** American Journal of Gastroenterology. (American College of Gastroenterology, Inc., 299 Broadway, New York, NY 10007) V.21- 1954-

**AJHEAA** American Journal of Public Health. (American Public Health Association, Inc., 1015 15th St., N.W., Washington, DC 20005) V.2-17, 1912-27; V.61- 1971-

**AJHPA9** American Journal of Hospital Pharmacy. (American Society of Hospital Pharmacists, 4630 Montgomery Ave., Washington, DC 20014) V.15- 1958-

**AJHYA2** American Journal of Hygiene. (Baltimore, MD) V.1-80, 1921-64. For publisher information, see AJEPAS

**AJIMD8** American Journal of Industrial Medicine. (Alan R. Liss, Inc., 150 Fifth Ave., New York, NY 10011) V.1- 1980-

**AJINO\*** Ajinomoto Co., Inc. (9 W. 57th St., Suite 4625, New York, NY 10019)

**AJKDDP** American Journal of Kidney Diseases. (Grune & Stratton, Inc., Journal Subscription Department, POB 6280, Duluth, MN 55806) V.1- 1981-

**AJMEAZ** American Journal of Medicine. (Yorke Medical Group, 666 5th Ave., New York, NY 10103) V.1- 1946-

**AJMSA9** American Journal of the Medical Sciences. (Charles B. Slack, Inc., 6900 Grove Rd., Thorofare, NJ 08086) V.1- 1841-

**AJNED9** American Journal of Nephrology. (S. Karger Pub., Inc., 79 Fifth Ave., New York, NY 10003) V.1- 1981-

**AJOGAH** American Journal of Obstetrics and Gynecology. (C.V. Mosby Co., 11830 Westline Industrial Dr., St. Louis, MO 63141) V.1- 1920-

**AJOMAZ** Alabama Journal of Medical Sciences. (University of Alabama Medical Center, Univ. Station, Birmingham, AL 35294) V.1- 1964-

**AJOPAA** American Journal of Ophthalmology. (Ophthalmic Publishing Co., 435 N. Michigan Ave., Chicago, IL 60611) V.1- 1918-

**AJPAA4** American Journal of Pathology. (Lippincott/Harper, Journal Fulfillment Dept., 2350 Virginia Ave., Hagerstown, MD 21740) V.1- 1925-

**AJPEAG** American Journal of Public Health and the Nation's Health. (New York, NY) V.18-60, 1928-60. For publisher information, see AJHEAA

**AJPHAP** American Journal of Physiology. (American Physiological Society, 9650 Rockville Pike, Bethesda, MD 20814) V.1- 1898-

**AJPRAL** American Journal of Pharmacy and the Sciences Supporting Public Health. (Philadelphia College of Pharmacy and Science, 43rd St., Woodland Ave. and Kingsessing Mall, Philadelphia, PA 19104) V.109-150, 1937-78

**AJPSAO** American Journal of Psychiatry. (American Psychiatric Association, Publications Services Div., 1700 18th St., N.W., Washington, DC 20009) V.78- 1921-

**AJRRAV** American Journal of Roentgenology, Radium Therapy and Nuclear Medicine. (Springfield, IL) V.67-125, 1952-75. For publisher information, see AAJRDX

**AJSGA3** American Journal of Syphilis, Gonorrhea and Venereal Diseases. (St. Louis, MO) V.20-38, 1936-54. Discontinued

**AJSNAO** American Journal of Syphilis and Neurology. (St. Louis, MO) V.1-19, 1917-35. For publisher information, see AJSGA3

**AJTHAB** American Journal of Tropical Medicine and Hygiene. (Allen Press, 1041 New Hampshire St., Lawrence, KS 66044) V.1- 1952-

**AJTMAQ** American Journal of Tropical Medicine. (Baltimore, MD) V.1-31, 1921-51. For publisher information, see AJTHAB

**AJVRAH** American Journal of Veterinary Research. (American Veterinary Medical Association, 930 N. Meacham Road, Schaumburg, IL 60196) V.1- 1940-

**AKBNAE** Arkhiv Biologicheskikh Nauk. Archives of Biological Sciences. (Moscow, USSR) V.1-64, 1892-1941. Discontinued

**AKEDAX** Archiv fuer Klinische und Experimentelle Dermatologie. (Berlin, Germany) V.201-239, 1955-71. For publisher information, see ADMFAU

**AKGIAO** Akushcherstvo i Ginekologiya. (v/o Mezhdunarodnaya Kniga, Kuznetskii Most 18, Moscow G-200, USSR) No. 1- 1936-

**ALACBI** Aldrichimica Acta. (Aldrich Chemical Co., Inc., 940 W. St. Paul Ave., Milwaukee, WI 53233) V.1- 1968-

**ALBRW\*** Albright and Wilson Inc., P.O. Box 26229, Richmond, VA 23260-6229

**ALEPA8** Acta Leprol. (Publisher unknown) V.1- 1920(?)-

**ALLVAR** Alimentation et la Vie. (Societe Scientifique d'Hygiene Alimentaire, 16, rue de l'Estrapade, 75005 Paris, France) V.39- 1951-

**AMACCQ** Antimicrobial Agents and Chemotherapy. (American Society for Microbiology, 1913 I St., N.W., Washington, DC 20006) V.1- 1972-

**AMASA4** Acta Medica Academiae Scientiarum Hungaricae. (Kultura, POB 149 H-1389, Budapest, Hungary) V.1- 1950-

**AMBNAS** Acta Medica et Biologica. (Niigata University School of Medicine, 1 Asahi-machi-dori, Niigata 951, Japan) V.1- 1953-

**AMBOCX** Ambio. A Journal of the Human Environment, Research and Management. (Pergamon Press, Inc., Maxwell House, Fairview Park, Elmsford, NY 10523) V.1- 1972-

**AMBPBZ** Acta Pathologica et Microbiologica Scandinavica, Section A: Pathology. (Copenhagen K, Denmark) V.78-89, 1970-81. For publisher information, see ACPADQ

**AMBUCH** Acta Medica Bulgarica. (Durzhavno Izdatelstvo Meditsina i Fizkultura, Pl. Slaveikov 11, Sofia, Bulgaria) V.1- 1973-

**AMCTAH** Antibiotic Medicine and Clinical Therapy. (New York, NY) V.3-8, 1956-61. For publisher information, see CLMEA3

**AMCYC\*** Toxicological Information on Cyanamid Insecticides. (American Cyanamid Co., Agricultural Division, Princeton, NJ 08540)

**AMDCA5** AMA Journal of Diseases of Children. (Chicago, IL) V.91-99, 1956-60. For publisher information, see AJDCAI

**AMEBA7** Annales Medicinae Experimentalis et Biologiae Fenniae. (Mikonkatu 8, Helsinki, Finland) V.25- 1947-

**AMICCW** Archives of Microbiology. (Springer-Verlag New York, Inc., Service Center, 44 Hartz Way, Secaucus, NJ 07094) V.95- 1974-

**AMIHAB** AMA Archives of Industrial Health. (Chicago, IL) V.11-21, 1955-60. For publisher information, see AEHLAU

**AMIHBC** AMA Archives of Industrial Hygiene and Occupational Medicine. (Chicago, IL) V.2-10, 1950-54. For publisher information, see AEHLAU

**AMILAN** Annales de Medecine Legale. Criminologie, Police Scientifique et Toxicologie. (Paris, France) V.1, 1920-51; V.39-47, 1958-67. For publisher information, see MLDCAS

**AMIUAG** Acta Medica Iugoslavica. (Akademija Zbora Lijecnika Hrvatske, Subiceva 9/1, Zagreb, Yugoslavia) V.1- 1947-

**AMJPA6** American Journal of Pharmacy (1835-1936). (Philadelphia, PA) V.1-108, 1835-1936. For publisher information, see AJPRAL

**AMLTAS** Annales de Medecine Legale et de Criminologie. (Paris, France) V. 31-8, 1951-8. For publisher information, see MLDCAS

**AMNIB6** Acta Manilana, Series A: Natural and Applied Sciences. (University of Santo Tomas Research Center, Manila, D-403, Philippines) V.4- 1968-

**AMNTA4** American Naturalist. (University of Chicago Press, 5801 S. Ellis Ave., Chicago, IL 60637) V.1- 1867-

**AMOKAG** Acta Medicia Okayama. (Okayama University Medical School, 2-5-1 Shikata-cho, Okayama 700, Japan) V.8- 1952-

**AMONDS** Applied Methods in Oncology. (Elsevier North Holland, Inc., 52 Vanderbilt Ave., New York, NY 10017) V.1- 1978-

**AMPLAO** AMA Archives of Pathology. (American Medical Association, 535 N. Dearborn St., Chicago, IL 60610) V.50,No. 4-V.69, 1950-60

**AMPMAR** Archives des Maladies Professionnelles de Medecine du Travail et de Securite Sociale. (Masson et Cie, Editeurs, 120 Blvd. Saint-Germain, P-75280, Paris 06, France) V.7- 1946-

**AMPYAT** Annales Medico-Psychologiques. (Masson et Cie, Editeurs, 120 Blvd. Saint-Germain, P-75280, Paris 06, France) V.80- 1922-

**AMRL\*\*** Aerospace Medical Research Laboratory Report. (Aerospace Technical Div., Air Force Systems Command, Wright-Patterson Air Force Base, OH 45433)

**AMSHAR** Acta Morphologica Academiae Scientiarum Hungaricae. (Akademiai Kiado, P.O. Box 24, H-1389 Budapest 502, Hungary) V.1- 1951-

**AMSSAQ** Acta Medica Scandinavica, Supplement. (Almqvist and Wiksell, P.O. Box 62, 26 Gamla Brogatan, S-101, 20 Stockholm, Sweden) No. 1- 1921-

**AMSVAZ** Acta Medica Scandinavica. (Almqvist and Wiksell, P.O. Box 62, 26 Gamla Brogatan, S-101, 20 Stockholm, Sweden) V.52- 1919-

**AMTUA3** Acta Medica Turcica. (Dr. Ayhan Okcuoglu, Cocuk Hastalikari Klinig i, c/o Ankara University Tip Facultesi, PK 48, Cebeci, Ankara, Turkey) V.1- 1964-

**AMUK\*\*** Acta Medica University Kyoto. (Kyoto, Japan)

**AMZOAF** American Zoologist. (American Society of Zoologists, Box 2739, California Lutheran College, Thousand Oaks, CA 91360) V.1- 1961-

**ANAEA3** Annals of Allergy. (American College of Allergists, Box 20671, Bloomington, MN 55420) V.1- 1943-

**ANANAU** Anatomischer Anzeiger. (VEB Gustav Fischer Verlag, Postfach 176, DDR-69, Jena, Germany) V.1 1886-

**ANASAB** Anaesthesia. (Blackwell Scientific, Osney Mead, Oxford OX2 OEL, England) V.1- 1946-

**ANATAE** Anaesthesist. (Springer-Verlag, Heidelberger Pl. 3, D-1000 Berlin 33, Germany) V.1- 1952-

**ANBCA2** Analytical Biochemistry. (Academic Press, 111 5th Ave., New York, NY 10003) V.1- 1960-

**ANBCB3** Antibiotics and Chemotherapy. (S. Karger, AG, Arnold-Boecklin-St 25, Postfach CH-4009 Basel, Switzerland) V.17- 1971-

**ANCHAM** Analytical Chemistry. (American Chemical Society, 1155 16th St., N.W., Washington, DC 20036) V.19- 1947-

**ANDRDQ** Andrologia. (Grosse Verlag, Kurfuerstendamm 152, D-1000 Berlin 31, Germany) V.6- 1974-

**ANENAG** Annales d'Endocrinologie. (Masson et Cie, Editeurs, 120 Blvd. Saint-Germain, P-75280, Paris 06, France) V.1- 1939-

**ANESAV** Anesthesiology. (J.B. Lippincott Co., Keystone Industrial Park, Scranton, PA 18512) V.1- 1940-

**ANGIAB** Angiology. (Williams & Wilkins Co., 428 E. Preston St., Baltimore, MD 21202) V.1- 1950-

**ANIFAC** Annales de Chirurgie Infantile. (Paris, France) V.1- 1960(?)-

**ANMBCM** Annales de Microbiologie (Paris). (Masson et Cie, Editeurs, 120 Blvd. Saint-Germain, P-75280, Paris 06, France) V.124- 1973-

**ANOPB5** Annals of Ophthalmology. (American Society of Contemporary Ophthalmology, 211 E. Chicago Ave., Suite 1044, Chicago, IL 60611) V.1- 1969-

**ANPBAZ** Acta Neurologia et Psychiatrica Belgica. (Brussels, Belgium) V.48-69, 1948-69. For publisher information, see ANUBBR

**ANPSAI** Archives of Neurology and Psychiatry. (Chicago, IL) V.1-64, 1919-50. For publisher information, see ARNEAS

**ANPTAL** Acta Neuropathologica. (Springer-Verlag New York, Inc., Service Center, 44 Hartz Way, Secaucus, NJ 07094) V.1- 1961-

**ANREAK** Anatomical Record. (Alan R. Liss, Inc., 150 5th Ave., New York, NY 10011) V.1- 1906/08-

**ANSUA5** Annals of Surgery. (J.B. Lippincott Co., Keystone Industrial Park, Scranton, PA 18512) V.1- 1885-

**ANTBAL** Antibiotiki. (Moscow, USSR) V.1-29, 1956-84. For publisher information, see AMBIEH

**ANTCAO** Antibiotics and Chemotherapy. (Washington, DC) V.1-12, 1951-62. For publisher information, see CLMEA3

**ANTRD4** Anticancer Research. (Anticancer Research, 5 Argyropoulou St., Kato Patissia, Athens 907, Greece) V.1- 1981-

**ANUBBR** Acta Neurologica Belgica. (Association des Societes Scientifiques, Medicales Belges, rue des Champs-Elysees 43, B-1050 Brussels, Belgium) V.1- 1900-

**ANYAA9** Annals of the New York Academy of Sciences. (The Academy, Exec. Director, 2 E. 63rd St., New York, NY 10021) V.1- 1877-

**ANZJA7** Australian and New Zealand Journal of Surgery. (Blackwell Scientific Publications, College of Surgeons' Gardens, 99 Barry St., Carlton 3053, Australia) V.1- 1931-

**ANZJB8** Australian and New Zealand Journal of Medicine. (Modern Medicine of Australia Pty., Ltd., 100 Pacific Highway, North Sydney, 2060, Australia) V.1- 1971-

**AOBIAR** Archives of Oral Biology. (Pergamon Press, Headington Hill Hall, Oxford OX3 OBW, England) V.1- 1959-

**AOGLAR** Acta Obstetrica et Gynaecologica Japonica, English Edition. (Tokyo, Japan) V.16-23, 1969-76. For publisher information, see NISFAY

**AOGMAU** Annali di Ostetricia, Ginecologia, Medicina Perinatale. (Via Commenda, 12, 20122 Milan, Italy) V.93- 1972-

**AOGNAX** Archivio di Ostetricia e Ginecologia. (C.C. Postale N 6/19773, Naples, Italy) V.1- 1937-

**AOGSAE** Acta Obstetricia et Gynecologica Scandinavica. (Kvinnokliniken, Lasarettet, Lund, Sweden) V.1- 1921-

**AOHYA3** Annals of Occupational Hygiene. (Pergamon Press, Headington Hill Hall, Oxford OX3 OBW, England) V.1- 1958-

**AOISDR** Annual Report of Osaka City Institute of Public Health and Environmental Sciences. (Osaka-shiritsu Kankyo Kagaku Kenkyusho, 8-34 Tojo-cho, Tennoji-ku, Osaka 543, Japan) No. 43- 1980-

**AORLCG** Archives of Oto-Rhino-Laryngology. (Springer-Verlag New York, Inc., Service Center, 44 Hartz Way, Secaucus, NJ 07094) V.206- 1973-

**AOUNAZ** Archiv fuer Orthopaedische und Unfall-Chirurgie. (Munich, Germany) V.1-90 1903-77. see AOTSDE

**APACAB** Acta Physiologica Academiae Scientiarum Hungaricae. (Akademiai Kiado, P.O. Box 24, H-1389 Budapest 502, Hungary) V.1- 1950-

**APAVAY** Virchows Archiv fuer Pathologische, Anatomie und Physiologie, und fuer Klinische Medizin. (Berlin, Germany) V.1-343, 1847-1967. For publisher information, see VAAPB7

**APBDAJ** Archiv der Pharmazie und Berichte der Deutschen Pharmazeutischen Gesellschaft. (Weinheim, Germany) V.262-304, 1924-71. For publisher information, see ARPMAS

**APBOAI** Acta Facultatis Pharmaceuticae Bohemoslovenicae. (Bratislava, Czechoslovakia) V.4-13, 1961-67. For publisher information, see AFPCAG

**APCRAW** Advances in Pest Control Research. (New York, NY) V.1-8, 1957-68. Discontinued

**APDCDT** Advances in Tumour Prevention, Detection and Characterization. (Elsevier North Holland, Inc., 52 Vanderbilt Ave., New York, NY 10017) V.1- 1974-

**APEPA2** Naunyn-Schmiedebergs Archiv fuer Pharmakologie und Experimentelle Pathologie. (Berlin, Germany) V.254-263, 1966-69. For publisher information, see NSAPCC

**APFRAD** Annales Pharmaceutiques Francaises. (Masson et Cie, Editeurs, 120 Blvd. Saint-Germain, P-75280, Paris 06, France) V.1- 1943-

**APHGAO** Acta Pharmaceutica Hungarica. (Kultura, POB 149, H-1389 Butapest, Hungary) 1953- Adopts V.24 in 1955

**APHGBP** Acta Pathologica (Belgrade). (Belgrade Univerzitet, Institute de Pathologie, Belgrade, Yugoslavia) V.2/3- 1964-

**APJAAG** Acta Pathologica Japonica. (Nippon Byori Gakkai, 7-3-1, Hongo, Bunkyo-Ku, Tokyo 113, Japan) V.1- 1951-

**APJUA8** Acta Pharmaceutica Jugoslavica. (Jugoslovenska Knjiga, P.O. Box 36, Terazije 27, YU-11001 Belgrade, Yugoslavia) V.1- 1951-

**APLAAQ** Acta Paediatrica Latina. (Editrice Arti Grafiche Emiliane, postale 10110427, 42100 Reggio Emilia, Italy) V.1- 1948-

**APLMAS** Archives of Pathology and Laboratory Medicine. (American Medical Association, 535 N. Dearbon St., Chicago, IL. 60610) V.1-5, No. 2, 1926-28; V.100- 1976-

**APMBAY** Applied Microbiology. (Washington, DC) V.1-30, 1953-75. For publisher information, see AEMIDF

**APMIAL** Acta Pathologica et Microbiologica Scandinavica. (Copenhagen, Denmark) V.1-77, 1924-69. For publisher information, see AMBPBZ

**APMIBM** Acta Pathologica et Microbiologica Scandinavica, Section B: Microbiology and Immunology. (Copenhagen, Denmark) V.78B-82B, 1970-74. For publisher information, see APSCD2

**APMUAN** Acta Pathologica et Microbiologica Scandinavica, Supplementum. (Munksgaard, 35 Noerre Soegade, DK-1370 Copenhagen K, Denmark) No. 1- 1926-

**APPBDI** Acta Physiologica et Pharmacologica Bulgarica. (Izdatelstvo na Bulgarskata Akademiya na Naukite, St. Geo Milev ul 36, Sofia 13, Bulgaria) V.1- 1974-

**APPHAX** Acta Poloniae Pharmaceutica. (Ars Polona-RUCH, P.O. Box 1001, P-00 068 Warsaw, 1, Poland) V.1- 1937-

**APPNAH** Acta Physiologica et Pharmacologica Neerlandica. (Amsterdam, Netherlands) V.1-15, 1950-69. For publisher information, see EJPHAZ

**APPYAG** Annual Review of Phytopathology. (Annual Reviews, Inc., 4139 El Camino Way, Palo Alto, CA 94306) V.1- 1963-

**APRCAS** American Perfumer and Cosmetics. (Oak Park, IL) V.77-86, 1962-71. For publisher information, see CSPEAX

**APSCAX** Acta Physiologica Scandinavica. (Karolinska Institutet, S-10401 Stockholm, Sweden) V.1- 1940-

**APSCD2** Acta Pathologica et Microbiologica Scandinavica, Section C: Immunology. (Munksgaard, 35 Noerre Soegade, DK-1370, Copenhagen CK, Denmark) V.83C- 1975-

**APSVAM** Acta Paediatrica Scandinavica. (Almqvist and Wiksell, P.O. Box 62, 26 Gamla Brogatan, S-101, 20 Stockholm, Sweden) V.54- 1965-

**APSXAS** Acta Pharmaceutica Suecdca. (Apotekarsocieteten, Wallingatan 26, Box 1136, S-111, 81 Stockholm, Sweden) V.1- 1964-

**APTOA6** Acta Pharmacologica et Toxicologica. (Munksgaard, 35 Noerre Soegade, DK-1370, Copenhagen K, Denmark) V.1- 1945-

**APTOD9** Abstracts of Papers, Society of Toxicology. Annual Meetings. (Academic Press, 111 5th Ave., New York, NY 10003)

**APTRDI** Advances in Prostaglandin and Thromboxane Research. (Raven Press 1140 Ave. of the America, New York, NY 10036) V.1- 1976-

**APTSAI** Acta Pharmacologica et Toxicologica, Supplementum. (Munksgaard, 35 Noerre Soegade, DK-1370, Copenhagen K, Denmark) No. 1- 1947-

**APYPAY** Acta Physiologica Polonica. (Panstwowy Zaklad Wydawnictw Lekarskich, ul. Dluga 38-40, P-00 238 Warsaw, Poland) V.1- 1950-

**AQMOAC** Air Quality Monographs. (American Petroleum Institute, 2101 L St., N.W., Washington, DC 20037) No. 69-1- 1969-

**ARANDR** Archives of Andrology. (Elsevier North Holland, Inc., 52 Vanderbilt Ave., New York, NY 10017) V.1- 1978-

**ARCGDG** Archives of Gynecology. (Springer-Verlag New York, Inc., Service Center, 44 Hartz Way, Secaucus, NJ 07094) V.226- 1978-

**ARCVBP** Annales des Recherches Veterinaires. (Institut National de la Recherche Agronomique, Service des Publ., route de Saint-Cyr, 78000 Versailles, France) V.1- 1970-

**ARDEAC** Archives of Dermatology. (American Medical Association, 535 N. Dearborn St., Chicago, IL 60610) V.82- 1960-

**ARDIAO** Annals of the Rheumatic Diseases. (BMA or British Medical Journal, 1172 Commonwealth Ave., Boston, MA. 02134) V.1- 1939-

**ARDSBL** American Review of Respiratory Disease. (American Lung Association, 1740 Broadway, New York, NY 10019) V.80- 1959-

**AREAD8** Anesteziologiya i Reanimatologiya. (v/o Mezhdunarodnaya Kniga, Kuznetskii Most 18, Moscow G-200, USSR) No. 1- 1977-

**ARGEAR** Archiv fuer Geschwulstforschung. (VEB Verlag Volk und Gesundheit Neue Gruenstr. 18, DDR-102 Berlin, Germany) V.1- 1949-

**ARGPAQ** Archives of General Psychiatry. (American Medical Association, 535 N. Dearborn St., Chicago, IL 60610) V.3- 1960-

**ARGYAJ** Archiv fuer Gynaekologie. (Munich, Germany) V.1-225, 1870-1978. For publisher information, see ARCGDG

**ARHEAW** Arthritis and Rheumatism. (Arthritis Foundation, 3400 Peachtree Road, N.E., Atlanta, GA 30326) V.1- 1958-

**ARINAU** Annual Report of the Research Institute of Environmental Medicine, Nagoya University. (Nagoya, Japan) V.1-25, 1951-80

**ARKIAP** Archiv fuer Kinderheilkunde. (Stuttgart, Germany.) V.1-183, 1880-1971

**ARMCAH** Annual Review of Medicine. (Annual Reviews, Inc., 4139 El Camino Way, Palo Alto, CA 94306) V.1- 1950-

**ARMIAZ** Annual Review of Microbiology. (Annual Reviews, Inc., 4139 El Camino Way, Palo Alto, CA 94306) V.1- 1947-

**ARMKA7** Archiv fuer Mikrobiologie. (Springer (Berlin)) V.1-13, 1930-43; V.14-94, 1948-73. For publisher information, see AMICCW

**ARNEAS** Archives of Neurology. (American Medical Association, 535 N. Dearborn St., Chicago, IL 60610) V.3- 1960-

**AROPAW** Archives of Ophthalmology. (American Medical Association., 535 N. Dearborn St., Chicago, IL 60610) V.1-44, No. 3, 1929-50; V.64- 1960-

**AROTAA** Archives of Otolaryngology. (American Medical Association, 535 N. Dearborn St., Chicago, IL 60610) V.1-52, 1925-50; V.72- 1960-

**ARPAAQ** Archives of Pathology. (American Medical Association., 535 N. Dearborn St., Chicago, IL 60610) V.5, No. 3-V.50, No. 3, 1928-50; V.70-99, 1960-75

**ARPMAS** Archiv der Pharmazie. (Verlag Chemie GmbH, Postfach 1260/1280 D-6940 Weinheim, Germany) V.51-261, 1835-1923; V.305- 1972-

**ARPTAF** Arkhiv Patologii. Archives of Pathology. (v/o Mezhdunarodnaya Kniga, Kuznetskii Most 18, Moscow G-200 USSR.) V.1- 1959-

**ARPTDI** Annual Review of Pharmacology and Toxicology. (Annual Reviews, Inc., 4139 El Camino Way, Palo Alto, CA 94306) V.16· 1976·

**ARSIM\*** Agricultural Research Service, USDA Information Memorandum. (Beltsville, MD 20705)

**ARSUAX** Archives of Surgery. (American Medical Association, 535 N. Dearborn St., Chicago, IL 60610) V.1-61, 1920-50; V.81· 1960·

**ARTODN** Archives of Toxicology. (Springer-Verlag, Heidelberger Pl. 3, D-1 Berlin 33, Germany) V.32· 1974·

**ARTUA4** American Review of Tuberculosis. (New York, NY) V.1-70, 1917-54. For publisher information, see ARDSBL

**ARVPAX** Annual Review of Pharmacology. (Palo Alto, CA) V.1-15, 1961-75. For publisher information, see ARPTDI

**ARZFAN** Aerztliche Forschung. (Munich, Germany) V.1-26, 1947-72. Discontinued

**ARZNAD** Arzneimittel-Forschung. Drug Research. (Editio Cantor Verlag, Postfach 1255, W-7960 Aulendorf, Germany) V.1· 1951·

**ASBDD9** Advances in the Study of Birth Defects. (University Park Press, 233 E. Redwood St., Baltimore, MD 21202) V.1· 1979·

**ASBIAL** Archivio di Science Biologiche. (Cappelli Editore, Via Marsili 9, I-40124 Bologna, Italy) V.1· 1919·

**ASBUAN** Archives des Sciences Biologiques. (Leningrad, USSR.) V.1-22, 1892-1922. Discontinued

**ASMUAA** Acta Scholae Medicinalis Universitatis in Kyoto. (Kyoto, Japan) V.27-40, 1947-70. Discontinued

**ASPHAK** Archives des Sciences Physiologique. (Paris, France) V.1-28, 1947-74. Discontinued

**ASTTA8** ASTM Special Technical Publication. (American Society for Testing Materials, 1916 Race St., Philadelphia, PA 19103) No. 1· 1911·

**ASUPAZ** Acta Societatis Medicorum Upsaliensis. (Uppsala, Sweden) V.55-76, 1950-71. For publisher information, see UJMSAP

**ATAREK** AAMI Technology Assessment Report. (Association for the Advancement of Medical Instrumentation, 1901 N. Ft. Myer Dr., Suite 602, Arlington, VA 22209) No. 1-81· 1981·

**ATENBP** Atmospheric Environment. Air Pollution, Industrial Aerodynamics, Micrometerology, Aerosols. (Pergamon Press, Headington Hill Hall, Oxford OX3 0BW, England) V.1· 1967·

**ATHBA3** Acta Radiologica, Therapy, Physics, Biology. (P.O. Box 7449, S-103 91 Stockholm, Sweden) V.1-16, No. 6, 1963-77

**ATHSBL** Atherosclerosis (Shannon, Ireland). (Elsevier/North-Holland Scientific Publishers Ltd., POB 85, Limerick, Ireland) V.11· 1970·

**ATMPA2** Annals of Tropical Medicine and Parasitology. (Academic Press, 24-28 Oval Rd., London NW1 7DX, England) V.1· 1907·

**ATPNAB** Ateneo Parmense, Acta Naturalia. (Ospedale Maggiore, Via Gramsci 14, 43100 Parma, Italy) V.1· 1965·

**ATSUDG** Archives of Toxicology, Supplement. (Springer-Verlag, Heidelberger Pl. 3, D-1000 Berlin 33, Germany) No. 1· 1978·

**ATXKA8** Archiv fuer Toxikologie. (Berlin, Germany) V.15-31, 1954-74. For publisher information, see ARTODN

**AUAAB7** Acta Universitatis Agriculturae, Facultas Agronomica (Brno) (Ustredni Knihovna Vysoke Skoly Zemedelske, Zemedelska 1, 662-65 Brno, Czechoslovakia) V.15· 1967·

**AUCMBJ** Acta Universitatis Carolinae, Medica, Monographia. (Univerzita Karlova, Ovocny Trh.3, CS-116-36 Prague 1, Czechoslovakia) No. 1· 1954·

**AUODDK** Acta Universitatis Ouluensis, Series D: Medica. (Oulu University Library, Box 186, SF-90101 Oulu 10, Finland) No. 1· 1972·

**AUPJB7** Australian Paediatric Journal. (Royal Childrens Hospital, Parkville, Victoria 3052, Australia) V.1· 1965·

**AUVJA2** Australian Veterinary Journal. (Australian Veterinary Association, Executive Director, 134-136 Hampden Road, Artarmon, NSW 2064, Australia) V.2· 1927·

**AVBIB9** Advances in the Biosciences. (Pergamon Press Ltd., Headington Hill Hall, Oxford OX3 0BW, England) V.1· 1969·

**AVBNAN** Arhiv Bioloskih Nauka. (Jugoslovenska Knjigu, P.O. Box 36, Terazije 27, 11001, Belgrade, Yugoslavia)

**AVERAG** American Veterinary Review. (Chicago, IL) V.1-47, 1877-1915. For publisher information, see JAVMA4

**AVEZA6** Acta Vitaminologica et Enzymologica. (Gruppo Lepetit SpA, Via R. Lepetit N. 8, 20124 Milan, Italy) V.21· 1967·

**AVPCAQ** Advances in Pharmacology and Chemotherapy. (Academic Press, 111 5th Ave., New York, NY 10003) V.7· 1969·

**AVSUAR** Acta Dermato-Venereologica, Supplementum. (Almqvist and Wiksell Periodical Co., P.O. Box 62, 26 Gamla Brogatan, S-101 20 Stockholm, Sweden) No. 1· 1929·

**AWLRAO** Australian Wildlife Research. (Commonwealth Scientific and Industrial Research Organization, POB 89, E. Melbourne, Victoria 3002, Australia) V.1· 1974·

**AXVMAW** Archiv fuer Experimentelle Veterinaermedizin. (S. Hirzel Verlag, Postfach 506, DDR-701 Leipzig, Germany) V.6· 1952·

**AZMZA6** Azerbaidzhanskii Meditsinskii Zhurnal. Azerbaidzhan Medical Journal. (v/o Mezhdunarodnaya Kniga, Kuznetskii Most 18, Moscow G-200, USSR.) 1928-41; 1955-

**BAFEAG** Bulletin de l'Association Francaise pour l'Etude du Cancer. (Paris, France) V.1-52, 1908-65. For publisher information, see BUCABS

**BANMAC** Bulletin de l'Academie Nationale de Medecine. (Masson et Cie, Editeurs, 120 Blvd. Saint-Germain, P-75280, Paris 06, France) V.1· 1836·

**BANRDU** Banbury Report. (Cold Spring Harbor Laboratory, POB 100, Cold Spring Harbor, NY 11724) V.1· 1979·

**BAPBAN** Bulletin de l'Academie Polonaise des Sciences, Series des Sciences Biologiques. (Ars Polona-RUCH, P.O. Box 1001 P-00 068 Warsaw 1, Poland) V.5· 1957·

**BATTL\*** Reports produced for the National Institute for Occupational Safety and Health by Battelle Pacific Northwest Laboratories, Richland, WA 99352

**BAXXDU** British UK Patent Application. (U.S. Patent Office, Science Library, 2021 Jefferson Davis Highway, Arlington, VA 22202)

**BBACAQ** Biochimica et Biophysica Acta. (Elsevier Publishing Co., POB 211, Amsterdam C, Netherlands) V.1· 1947·

**BBGED3** Revista Brasileira de Genetica. Brazilian Journal of Genetics. (Dr. F.A. Moura Duarte, Dep. de Genetica, Faculdade de Medicina de Ribeirao Preto, 14.100 Ribeirao Preto, Sao Paulo, Brazil) V.1· 1978·

**BBIADT** Biomedica Biochimica Acta. (Akademie-Verlag GmbH, Postfach 1233, DDR-1086 Berlin, Germany) V.42· 1983·

**BBMS\*\*** "Medicaments du Systeme Nerveux Vegetalif" Bovet, D., and F. Bovet-Nitti, New York, NY, S. Karger, 1948

**BBRCA9** Biochemical and Biophysical Research Communications. (Academic Press Inc., 111 5th Ave., New York, NY 10003) V.1· 1959·

**BCFAAI** Bollettino Chimico Farmaceutico. (Societa Editoriale Farmaceutica, Via Ausonio 12, 20123 Milan, Italy) V.33· 1894·

**BCPCA6** Biochemical Pharmacology. (Pergamon Press Inc., Maxwell House, Fairview Park, Elmsford, NY 10523) V.1· 1974·

**BCPHBM** British Journal of Clinical Pharmacology. (Macmillan Journals, Houndmills Estate, Basingstoke, Hants RG21 2XS, England) V.1· 1974·

**BCSTB5** Biochemical Society Transactions. (Biochemical Society, P.O. Box 32, Commerce Way, Whitehall Rd., Industrial Estate, Colchester CO2 8HP, Essex, England) V.1· 1973·

**BCSYDM** Bristol-Myers Cancer Symposia. (Academic Press, 111 Fifth Ave., New York, NY 10003) V.1· 1979·

**BCTKAG** Bromatologia i Chemia Toksykologiczna. (Ars Polona-RUCH, P.O. Box 1001, P-00 068 Warsaw, 1, Poland) V.4· 1971·

**BCTRD6** Breast Cancer Research and Treatment. (Kluwer Academic Publishers Group, Distribution Center, POB 322, 3300 AH Dordrecht, Netherlands) V.1· 1981·

**BDCGAS** Berichte der Deutschen Chemischen Gesellschaft. (Leipzig/Berlin) V.1-61, 1868-1928. For publisher information, see CHBEAM

**BDHU\*\*** Nachprufung der Toxicitat von Novocain und Tutocain bei Chloralisierten Tieren, August Barke Dissertation. (Pharmakologischen Institut der Tierarztlichen Hochschule zu Hannover, Germany, 1936)

**BDKS\*\*** Studien uber die Pharmakologie des Pinakolins und einiger seiner Derivate mit besonderer Berucksichtigung der Kreislaufwirkung, Lennart Bang Dissertation. (Pharmakologischen Abteilung des Karolinischen Instituts, Stockholm, Sweden, 1934)

**BDVU\*\*** Uber die Pharmakologische Wirkung des 1,3-Dioxy-2-Nicotinsaureamid-Tetrazols. Eine Experimentelle Studie zur Kenntnis der Wirkung neuer Tetrazolderivate, Rudolf Barwanietz Dissertation. (Institut fuer Veterinar-Pharmakolgie der Universitat Berlin, Germany, 1937)

**BEBMAE** Byulleten' Eksperimental'noi Biologii i Meditsiny. Bulletin of Experimental Biology and Medicine. (v/o Mezhdunarodnaya Kniga, Kuznetskii Most 18, Moscow G-200, USSR.) V.1· 1936·

**BECCAN** British Empire Cancer Campaign Annual Report. (Cancer Research Campaign, 2 Carlton House Terrace, London SW1Y 5AR, England) V.1· 1924·

**BECTA6** Bulletin of Environmental Contamination and Toxicology. (Springer-Verlag New York, Inc., Service Center, 44 Hartz Way, Secaucus, NJ 07094) V.1· 1966·

**BEGMA5** Beitraege zur Gerichtlichen Medizin. (Verlag Franz Deuticke, Helferstorferstr 4, A-1010 Vienna, Austria) V.1· 1911·

**BEMTAM** Berliner und Muenchener Tieraerztliche Wochenschrift. (Verlag Paul Parey, Lindenstr. 44-47, D-1000 Berlin 61, Germany) 1938-43; No. 1-1946·

**BENPBG** Behavioral Neuropsychiatry. (Behavioral Neuropsychiatry Medical Publ., Inc., 61 E. 86th St., New York, NY 10028) V.1· 1969·

**BESAAT** Bulletin of the Entomological Society of America. (The Society, 4603 Calvert Rd., College Park, MD 20740) V.1· 1955·

**BEXBAN** Bulletin of Experimental Biology and Medicine. Translation of BEBMAE. (Plenum Publishing Corp., 233 Spring St., New York, NY 10013) V.41· 1956·

**BEXBBO** Biochemistry and Experimental Biology. (Piccin Medical Books, Via Brunacci, 12, 35100 Padua, Italy) V.10· 1971/72·

**BHJUAV** British Heart Journal. (British Medical Journal, 1172 Commonwealth Ave., Boston, MA 02134) V.1· 1939·

**BIALAY** Biochimica Applicata. (Parma, Italy) V.1-19, 1954-72(?). Discontinued

**BIANA6** Bibliotheca Anatomica. (S. Karger AG, Postfach CH-4009, Basel, Switzerland) No. 1· 1961·

**BIATDR** Bulletin of the International Association of Forensic Toxicologists

**BIBIAU** Biotechnology and Bioengineering. (John Wiley & Sons, 605 3rd Ave., New York, NY 10016) V.4· 1962·

**BIBUBX** Biological Bulletin (Woods Hole, MA). (Biological Bulletin, Marine Biological Laboratories, Woods Hole, MA 02543) V.1· 1898·

**BUMMAB** Bulletin of the University of Miami School of Medicine and Jackson Memorial Hospital. (Coral Gables, FL) V.1-20, 1939-66. Discontinued

**BurLW#** Personal Communication from Mr. L.W. Burnette, Material Safety Dept., GAF Corp., 1361 Alps Rd., Wayne, NJ 07470, to Dr. A. Friedman, Tracor Jitco, Inc., November 2, 1978

**BUYRAI** Bulletin of Parenteral Drug Association. (The Association, Western Saving Fund Bldg., Broad and Chestnut Sts., Philadelphia, PA 19107) V.1-1946-

**BVIPA7** Bulletin of the Veterinary Institute in Pulawy. (Instytut Weterynaryjny, Aleja Partyzantow 55, Pulawy, Poland) V.1- 1957-

**BVJOA9** British Veterinary Journal. (Bailliere Tindall, 35 Red Lion Sq., London WCIR 4SG, England) V.105- 1949-

**BWHOA6** Bulletin of the World Health Organization. (WHO, 1211 Geneva 27, Switzerland) V.1- 1947-

**BYYADW** Byoin Yakugaku. Hospital Pharmacology. (Yakuji Nippon Sha, 1-11 Izumi-cho, Kanda, Chiyoda-ku, Tokyo 101, Japan) V.1- 1975-

**BZARAZ** Biologicheskii Zhurnal Armenii. Biological Journal of Armenia. (v/o Mezhdunarodnaya Kniga, Kuznetskii Most 18, Moscow G-200, USSR) V.19- 1966-

**CAANAT** Comptes Rendus de l'Association des Anatomistes. (Paris, France) V.1- 1916(?)-

**CAES\*\*** Reports supported by the Pennwalt Corp. and the Center for Air Environment Studies at the Pennsylvania State University

**CAJPBD** Proceedings of the Congenital Anomalies Research Association of Japan. (Kyoto, Japan) No. 1- 1961-

**CALEDQ** Cancer Letters (Shannon, Ireland). (Elsevier Scientific Pub. Ireland Ltd., POB 85, Limerick, Ireland) V.1- 1975-

**CAMEAS** California Medicine. (San Francisco, CA) V.65-119, 1946-73

**CANCAR** Cancer. (J.B. Lippincott Co., E. Washington Sq., Philadelphia, PA 19105) V.1- 1948-

**CANJAE** Canadian Anaesthetist's Society Journal. (178 St. George St., Toronto 5, Ontario M5R 2M7, Canada) V.1- 1954-

**CAREBK** Caries Research. (S. Karger AG, Postfach, CH-4009, Basel, Switzerland) V.1- 1967-

**CARYAB** Caryologia. (Caryologia, Via Lamarmora 4, 50121 Florence, Italy) V.1- 1948-

**CAXXA4** Canadian Patents. (U.S. Patent Office, Science Library, 2021 Jefferson Davis Highway, Arlington, VA 22202)

**CBCCT\*** "Summary Tables of Biological Tests" National Research Council Chemical-Biological Coordination Center. (National Academy of Science Library, 2101 Constitution Ave., N.W., Washington, DC 20418)

**CBINA8** Chemico-Biological Interactions. (Elsevier Publishing, P.O. Box 211, Amsterdam C, Netherlands) V.1- 1969-

**CBPBB8** Comparative Biochemistry and Physiology, B: Comparative Biochemistry. (Pergamon Press, Headington Hill Hall, Oxford OX3 OBW, England) V.38- 1971-

**CBPCBB** Comparative Biochemistry and Physiology, C: Comparative Pharmacology. (Oxford, England) V.50-73, 1975-82

**CBTIAE** Contributions from Boyce Thompson Institute. (Yonkers, NY) V.1-24, 1925-71. Discontinued

**CBTOE2** Cell Biology and Toxicology. (Princeton Scientific Publishers, Inc., 301 N. Harrison St., CN 5279, Princeton, NJ 08540) V.1- 1984-

**CCCCAK** Collection of Czechoslovak Chemical Communications. (Academic Press, 24-28 Oval Rd., London NW1 7DX, England) V.1- 1929-

**CCECAU** Critical Reviews in Environmental Control. (Chemical Rubber Company, Cleveland, OH) V.1-1970-

**CCHCDE** Zhonghua Jiehe He Huxixi Jibing Zazhi. Chinese Journal of Tuberculosis and Respiratory Diseases. (China International Book Trading Corp., POB 2820, Beijing, People's Republic of China) V.1- 1978-

**CCLCDY** Zhonghua Zhongliu Zazhi. Chinese Journal of Oncology. (Guozi Sudian, Beijing, People's Republic of China) V.1- 1978-

**CCPHDZ** Cancer Chemotherapy and Pharmacology. (Springer-Verlag, Heidelberger Pl. 3, D-1 Berlin 33, Germany) V.1- 1978-

**CCPTAY** Contraception. (Geron-X, Publishers, P.O. Box 1108, Los Altos, CA 94022) V.1- 1970-

**CCROBU** Cancer Chemotherapy Reports, Part 1. (Washington, DC) V.52, No. 6-V.59, 1968-75. For publisher information, see CTRRDO

**CCSUBJ** Cancer Chemotherapy Reports, Part 2. (Washington, DC) V.1-5, 1964-75. For publisher information, see CTRRDO

**CCSUDL** Carcinogenesis-A Comprehensive Survey (Raven Press, 1140 Ave. of the Americas, New York, NY 10036) V.1- 1976-

**CCYPBY** Cancer Chemotherapy Reports, Part 3. (Washington, DC) V.1-6, 1968-75. For publisher information, see CTRRDO

**CDESDK** Contraceptive Delivery Systems. (Kluwer Academic Publishers Group Distribution Centre, POB 322, 3300 AH Dordrecht, Netherlands) V.1- 1980-

**CDGU\*\*** Uber die Pharmakologische Wirkung eines dem Pentamethylentetrazol (Cardiazol) nahestehenden stickstoffhaltigen Kampherab-Kommlings, Hans Hermann Czygan Dissertation. (Hessischen Ludwigs-Universitat zu Giessen, Germany, 1934)

**CDPRD4** Cancer Detection and Prevention. (Marcel Dekker, Inc., POB 11305, Church St. Station, New York, NY 10249) V.1- 1979-

**CECED9** Commission of the European Communities, Report EUR. (Office for Official Publications, POB 1003, Luxembourg 1, Luxembourg) 1967-

**CEDEDE** Clinical and Experimental Dermatology. (Blackwell Scientific Publications Ltd., Osney Mead, Oxford OX2 0EL, England) V.1- 1976-

**CEFYAD** Ceskoslovenska Fysiologie. (Academia, Vodickova 40, Prague 1, Czechoslovakia) V.1- 1952-

**CEHYAN** Ceskoslovenska Hygiena. Czechoslovak Hygiene. (Ve Smeckach 30, Prague 1, Czechoslovakia). V.1- 1956-

**CEKNA5** Chiba-ken Eisei Kenkyusho Nenpo. (666-2 Nitona-cho, Chiva 280, Japan)

**CELLB5** Cell (Cambridge, MA.). (Massachusetts Institute of Technology Press, 28 Carleton St., Cambridge, MA 02142) V.1- 1974-

**CENEAR** Chemical and Engineering News. (American Chemical Society, 1155 16th St., N.W., Washington, DC 20036) V.20- 1942-

**CEXIAL** Clinical and Experimental Immunology. (Blackwell Scientific Publications Ltd., Osney Mead, Oxford OX2 0EL, England) V.1- 1966-

**CFRGBR** Code of Federal Regulations. (U.S. Government Printing Office, Supterintendent of Documents, Washington, DC 20402)

**CGCGBR** Cytogenetics and Cell Genetics. (S. Karger AG, Arnold-Boecklin Str. 25, CH-4011 Basel, Switzerland) V.12- 1973-

**CGCYDF** Cancer Genetics and Cytogenetics. (Elsevier North Holland, Inc., 52 Vanderbilt Ave., New York, NY 10017) V.1- 1979-

**CHABA8** Chemical Abstracts. (Chemical Abstracts Service, Box 3012, Columbus, OH 43210) V.1- 1907-

**ChaEB#** Personal Communication from Dr. E.B. Chappel, Abbott Labs., N. Chicago, IL, to I. Pigman, Tracor Jitco, Inc., January 19, 1978

**CHBEAM** Chemische Berichte. (Verlag Chemie GmbH, Postfach 129/149, Pallelallee 3, D-6940, Weinheim/Bergst, Germany) V.80- 1947-

**CHDDAT** Comptes Rendus Hebdomadaires des Seances de l'Academie des Sciences, Serie D. (Centrale des Revues Dunod-Gauthier-Villars, 24-26 Blvd. de l'Hopital, 75005 Paris, France) V.262- 1966-

**CHETBF** Chest: The Journal of Circulation, Respiration and Related Systems. (American College of Chest Physicians, 911 Busse Hwy., Park Ridge, IL 60068) V.57- 1970-

**CHHTAT** Zhonghua Yixue Zazhi. Chinese Medical Journal. (Guozi Shudian, Beijing, People's Republic of China) V.1- 1915-

**CHIMAD** Chimia. (Sauerlaender AG, 5001 Aarau, Switzerland) V.1- 1947-

**CHINAG** Chemistry and Industry. (Society of Chemical Industry, 14 Belgrave Sq., London SW1X 8PS, England) V.1-21, 1923-43; No. 1- 1944-

**CHIP\*\*** Chemical Hazard Information Profile. Draft Report. (U.S. Environmental Protection Agency, Office of Toxic Substances, 401 M St., S.W., Washington, DC 20460)

**CHMTBL** Chemical Technology. (American Chemical Society, 1155 16th St., N.W., Washington, DC 20036) V.1- 1917-

**CHPUA4** Chemicky Prumysl. Chemical Industry. (ARTIA, Ve Smeckach 30, 111-27 Prague 1, Czechoslovakia) V.1- 1951-

**CHREAY** Chemical Reviews. (American Chemical Society, 1155 16th St., N.W., Washington, DC 20036) V.1- 1924-

**CHROAU** Chromosoma. (Springer-Verlag, Heidelberger Pl. 3, D-1000 Berlin 33, Germany) V.1- 1939-

**CHRTBC** Chromosomes Today. (Elsevier Scientific Publishing Co., P.O. Box 211, Amsterdam, Netherlands) V.1- 1966-

**CHTHBK** Chemotherapy. (S. Karger AG, Arnold-Boecklin-St 25, CH-4000, Basel 11, Switzerland) V.13- 1968-

**CHTPBA** Chimica Therapeutica. (Editions DIMEO, Arcueil, France) V.1-8, 1965-73

**CHWKA9** Chemical Week. (McGraw-Hill, Inc., Distribution Center, Princeton, NJ, Hightstown, NJ 08520) V.68- 1951-

**CHYCDW** Zhonghua Yufangyixue Zazhi. Chinese Journal of Preventive Medicine. (42 Tung Szu Hsi Ta Chieh, Beijing, People's Republic of China) Beginning history not known

**CIGET\*** Ciba-Geigy Toxicology Data/Indexes, 1977. (Ciba-Geigy Corp., Ardsley, NY 10502)

**CIGZAF** Chiba Igakkai Zasshi. Journal of the Chiba Medical Society. (Chiba, Japan) V.1-49, 1923-73. For publisher information, see CIZAAZ

**CIHPDR** Zhongguo Yixue Kexueyuan Xuebao. Journal of the Chinese Academy of Medicine. (China Book Trading Corp., POB 2820, Beijing, People's Republic of China) V.1- 1979-

**CIIT\*\*** Chemical Industry Institute of Toxicology, Docket Reports. (POB 12137, Research Triangle Park, NC 27709)

**CIRUAL** Circulation Research. (American Heart Association, Publishing Director, 7320 Greenville Ave., Dallas, TX 75231) V.1- 1953-

**CISCB7** CIS, Chromosome Information Service. (Maruzen Co. Ltd., POB 5050, Tokyo International, Tokyo 100-31, Japan) No. 1- 1961-

**CIWYAO** Carnegie Institute of Washington, Year Book. (Carnegie Institution of Washington, 1530 P St., N.W., Washington, DC 20005) V.1- 1902-

**CIYPDA** Sichuan Yixueyuan Xuebao. Acta Akademiae Medicinae Sichuan. (Guozi Sudian, Beijing, People's Republic of China) V.1- 1970-

**CIZAAZ** Chiba Igaku Zasshi. Chiba Medical Journal. (Chiba Igakkai, Inohana 1-8-8, Chiba 280, Japan) V.50- 1974-

**CJBBDU** Canadian Journal of Biochemistry and Cell Biology. (National Research Council of Canada, Ottawa, Ontario, KIA OR6 Canada) V.61-1983-

**CJBIAE** Canadian Journal of Biochemistry. (Ontario, Canada) V.42-60, 1964-82. For publisher information, see CJBBDU

**CJBPAZ** Canadian Journal of Biochemistry and Physiology. (Ottawa, Canada) V.32-41, 1954-63. For publisher information, see CJBIAE

**CJCHAG** Canadian Journal of Chemistry. (National Research Council of Canada, Ottawa, Ontario, K1A OR6 Canada) V.29- 1951-

**CJCMAV** Canadian Journal of Comparative Medicine. (360 Bronson Ave., Ottawa, Ontario, K1R 6J3 Canada) V.1-3, 1937-39; V.32- 1968-

**CJMIAZ** Canadian Journal of Microbiology. (National Research Council of Canada, Administration, Ottawa, Ontario K1A OR6 Canada) V.1- 1954-

**CJPEA4** Canadian Journal of Public Health. (Canadian Public Health Association, 1335 Carling Avenue, Suite 210, Ottawa, Ontario K1Z 8N8, Canada) V.20- 1929-

**CJPPA3** Canadian Journal of Physiology and Pharmacology. (National Research Council of Canada, Ottawa, Ontario, K1A OR6 Canada) V.42- 1964-

**CJPSDF** Canadian Journal of Psychiatry. (Keith Health Care Communications, 289 Rutherford Road S., Suite 11, Brampton, Ontario L6W 3R9, Canada)

**CJVRE9** Canadian Journal of Veterinary Research. (339 Booth St. Ottawa, Ontario K1R 7K1, Canada) V.50- 1986-

**CKFRAY** Ceskoslovenska Farmacie. (PNS-Ustredni Expedice Tisku, Jindriska 14, Prague 1, Czechoslovakia) V.1- 1952-

**CLBIAS** Clinical Biochemistry. (Canadian Society of Clinical Chemists, 151 Slater St., Ottawa, Ontario, K1P 5H3 Canada) V.1- 1967-

**CLCEAL** Casopis Lekaru Ceskych. Journal of Czech Physicians. (ARTIA, Ve Smeckach 30, Prague 1, Czechoslovakia) V.1- 1862-

**CLCHAU** Clinical Chemistry. (American Association of Clinical Chemists, 1725 K St., N.W., Washington, DC 20006) V.1- 1955-

**CLDFAT** Cell Differentiation. (Elsevier/North-Holland Scientific Publishers Ltd., POB 85, Limerick, Ireland) V.1- 1972-

**CLDND\*** Compilation of LD50 Values of New Drugs. (J.R. MacDougal, Dept. of National Health and Welfare, Food and Drug Divisions, 35 John St., Ottawa, Ontario, Canada)

**CLMEA3** Clinical Medicine. (Clinical Medicine Publications, 444 Frontage Rd., Northfield, IL 60093) V.69- 1962-

**CLNHBI** Clinical Nephrology. (Dustri-Verlag Dr. Karl Feistle, Postfach 49, D-8024 Munich-Deisenhofen, Germany) V.1- 1973-

**CLONEA** Clinics in Oncology. (W.B. Saunders Co., W. Washington Sq., Philadelphia, PA 19105) V.1- 1982-

**CLPTAT** Clinical Pharmacology and Therapeutics. (C.V. Mosby Co., 11830 Westline Industrial Dr., St. Louis, MO 63141) V.1- 1960-

**CLREAS** Clinical Research. (American Federation for Clinical Research, 6900 Grove Rd., Thorotare, NJ 08086) V.6- 1958-

**CMAJAX** Canadian Medical Association Journal. (CMA House, Box 8650, Ottawa, Ontario, K1G OG8 Canada) V.1- 1911-

**CMBID4** Cellular and Molecular Biology. (Pergamon Press Ltd., Headington Hill Hall, Oxford OX3 0BW, England) V.22- 1977-

**CMEP\*\*** "Clinical Memoranda on Economic Poisons" US Dept. HEW, Public Health Service, Communicable Disease Center, Atlanta, GA, 1956

**CMJOAP** Chinese Medical Journal. (Peking, China) V.46-85, 1932-66. For publisher information, see CHMEBA

**CMJODS** Chinese Medical Journal (Beijing, English Edition). New Series. (Guozi Shudian, Beijing, People's Republic of China) V.1- 1975- (Adopted vol. No. 92 in 1979)

**CMJRAY** Calcutta Medical Journal. (Calcutta Medical Club, CMC House, 91-B Chittaranjan Ave., Calcutta, India) V.1- 1906-

**CMLMDW** Collection de Medecine Legale et de Toxicologie Medicale. (Masson et Cie, Editeurs, 120 Blvd. Saint-Germain, F-75280 Paris 06, France) No. 53- 1970-

**CMMUAO** Chemical Mutagens. Principles and Methods for Their Detection (Plenum Publishing Corp., 233 Spring St., New York, NY 10013) V.1- 1971-

**CMROCX** Current Medical Research and Opinion. (Clayton-Wray Publications, 27 Sloane Sq., London SW1W 8AB, England) V.1- 1973-

**CMSHAF** Chemosphere. (Pergamon Press Inc., Maxwell House, Fairview Park, Elmsford, NY 10523) V.1- 1971-

**CMTRAG** Chemotherapia. (Basel, Switzerland) V.1-12, 1960-67. For publisher information, see CHTHBK

**CNCRA6** Cancer Chemotherapy Reports. (Bethesda, MD) V.1-52, 1959-68. For publisher information, see CCROBU

**CNJGA8** Canadian Journal of Genetics and Cytology. (Genetics Society of Canada, 151 Slater St., Suite 907, Ottawa, Ontario, K1P 5H4 Canada) V.1- 1959-

**CNJMAQ** Canadian Journal of Comparative Medicine and Veterinary Science. (Gardenvale, Quebec, Canada) V.4-32, 1940-68. For publisher information, see CJCMAV

**CNREA8** Cancer Research. (Public Ledger Building, Suit 816, 6th & Chestnut Sts., Philadelphia, PA 19106) V.1- 1941-

**CNRMAW** Canadian Journal of Research, Section E, Medical Sciences. (Ottawa, Canada) V.22-28, 1944-50. For publisher information, see CJBIAE

**CNVJA9** Canadian Veterinary Journal. (Canadian Veterinary Journal, 360 Bronson Ave., Ottawa, Ontario K1R 6J3, Canada) V.1- 1960-

**CODEDG** Contact Dermatitis. Environmental and Occupational Dermatitis. (Munksgaard, 35 Norre Sogade, DK 1370 Copenhagen K, Denmark) V.1- 1975-

**COINAV** Colloques Internationaux du Centre National de la Recherche Scientifique. (Centre National de la Recherche Scientifique, 15, Quai Anatole-France, F-75700 Paris, France) V.1- 1946-

**CONEAT** Confinia Neurologica. (S. Karger AG, Arnold-Boecklin-St 25, CH-4000, Basel 11, Switzerland) V.1- 1938-

**COREAF** Comptes Rendus Hebdomadaires des Seances de l'Academie des Sciences. (Paris, France) V.1-261, 1835-1965. For publisher information, see CHDDAT

**CORTBR** Clinical Orthopaedics and Related Research. (J.B. Lippincott Co., E. Washington Sq., Philadelphia, PA 19105) No. 26- 1963-

**COTODO** Journal of Fire and Flammability/Combustion Toxicology Supplement. (Technomic Publishing Co., 265 Post Rd. W., Westport, CT 06880) V.1-2, 1974-75. For Publisher information, see JCTODH

**COVEAZ** Cornell Veterinarian. (Cornell University, New York State College of Veterinary Medicine, Ithaca, NY 14853) V.1- 1911-

**CPAJAK** Canadian Psychiatric Association Journal. (Suite 103, 225 Lisgar St., Ottawa, Ontario, K2P OC6, Canada) V.1- 1956-

**CPBTAL** Chemical and Pharmaceutical Bulletin. (Pharmaceutical Society of Japan, 12-15-501, Shibuya 2-chome, Shibuya-ku, Tokyo, 150, Japan) V.6- 1958-

**CPCHAO** Clinical Proceedings of the Children's Hospital of the District of Columbia. (Washington, DC) V.1-27, 1971. For publisher information, see CPNMAQ

**CPEDAM** Clinical Pediatrics. (J.B. Lippincott Co., E. Washington Sq., Philadelphia, PA 19105) V.1- 1962-

**CPGPAY** Comparative and General Pharmacology. (New York, NY) V.1-5, 1970-74. For publisher information, see GEPHDP

**CPHADV** Clinical Pharmacy. (American Society of Hospital Pharmacists, 4630 Montgomery Ave., Bethesda, MD 20814) V.1- 1982-

**CPHPA5** Jiepou Xuebao. Journal of Anatomy. (Guozi Shudian, Beijing, People's Republic of China) V.1-9, 1953-66; V.10- 1979-

**CPNMAQ** Clinical Proceedings, Children's Hospital National Medical Center. (2125 13th St., N.W., Washington, DC 20009) V.28- 1972-

**CRAAA7** Current Researches in Anesthesia and Analgesia. (Cleveland, OH) V.1-35, 1922-56. For publisher information, see AACRAT

**CRDLP\*** U.S. Army Chemical Research and Development Laboratory, Special Publication. (Edgewood Arsenal, MD 21010)

**CRDLR\*** U.S. Army Chemical Research and Development Laboratory, Technical Report. (Edgewood Arsenal, MD 21010)

**CRNGDP** Carcinogenesis. (Information Retrieval, 1911 Jefferson Davis Highway, Arlington, VA 22202) V.1- 1980-

**CroHP#** Personal Communication from H.P. Crocker, Ultimate Holding Co.-Reckitt and Colman Ltd., London, England, to R.L. Tatken, NIOSH, Cincinnati, OH, September 2, 1980

**CRSBAW** Comptes Rendus des Seances de la Societe de Biologie et de Ses Filiales. (Masson et Cie, Editeurs, 120 Blvd. Saint-Germain, P-75280, Paris 06, France) V.1- 1849-

**CRSHAG** Circulatory Shock. (Alan R. Liss Inc., 150 5th Ave., New York, NY 10011) V.1- 1974-

**CRSUBM** Cancer Research Supplement (Williams & Wilkins Company, 428 E. Preston St., Baltimore, MD 21202) No. 1-4, 1953-56

**CRTXB2** CRC Critical Reviews in Toxicology. (CRC Press, Inc., 2000 N.W. 24th St., Boca Raton, FL 33431) V.1- 1971-

**CSHCAL** Cold Spring Harbor Conferences on Cell Proliferation. (Cold Spring Harbor Laboratory, POB 100, Cold Spring Harbor, NY 11724) V.1- 1974-

**CSHSAZ** Cold Spring Harbor Symposia on Quantitative Biology. (Cold Spring Harbor Laboratory of Quantitative Biology, Cold Spring Harbor, NY 11724) V.1- 1933-

**CSLNX\*** U.S. Army Armament Research & Development Command, Chemical Systems Laboratory, NIOSH Exchange Chemicals. (Aberdeen Proving Ground, MD 21010)

**CSPEAX** Cosmetics and Perfumery. (Allured Publishing Corp., Box 318, Wheaton, IL 60187) V.88- 1973-

**CTCEA9** Current Therapeutic Research, Clinical and Experimental. (Therapeutic Research Press, P.O. Box 514, Tenafly, NJ 07670) V.1- 1959-

**CTKIAR** Cell and Tissue Kinetics. (Blackwell Scientific Publications Ltd., Osney Mead, Oxford OX2 0EL, England) V.1- 1968-

**CTOIDG** Cosmetics and Toiletries. (Allured Publishing Corp., P.O. Box 318, Wheaton, IL 60187) V.91- 1976-

**CTOXAO** Clinical Toxicology. (New York, NY) V.1-18, 1968-81. For publisher information, see JTCTDW

**CTRRDO** Cancer Treatment Reports. (U.S. Government Printing Office, Superintendent of Documents, Washington, DC 20402) V.60- 1976-

**CTYAD8** Zhongcaoyao. Chinese Herbal Medicine. (Tsa Chih Pien Chi Pu, Hu-nan Yao Kung Yeh Yen Chiu So, Shao-Yang, Hu-nan, People's Republic of China) V.11- 1980-

**CUMIDD** Current Microbiology. (Springer-Verlag New York, Inc., Service Center, 44 Hartz Way, Secaucus, NJ 07094) V.1- 1978-

**CUSCAM** Current Science. (Current Science Association, Mgr., Raman Research Institute, Bangalore 6, India) V.1- 1932-

**CUTIBC** CUTIS; Cutaneous Medicine for the Practitioner. (Technical Pub., 875 Third Ave., New York, NY 10022) 1965-

**CWLTM\*** Chemical Warfare Laboratories Technical Memorandum. (U.S. Army Chemical Center, Edgewood Arsenal, MD 21010)

**CYGEDX** Cytology and Genetics. English Translation of Tsitologiya i Genetika. (Allerton Press, Inc., 150 Fifth Ave., New York, NY 10011) V.8- 1974-

**CYLPDN** Zhongguo Yaoli Xuebao. Acta Pharmacologica Sinica. (Shanghai K'o Hsueh Chi Shu Ch'u Pan She, 450 Shui Chin Erh Lu, Shanghai 200020, People's Republic of China) V.1- 1980-

**CYTBAI** Cytobios. (The Faculty Press, 88 Regent St., Cambridge, England) V.1- 1969-

**CYTOAN** Cytologia. (Maruzen Co. Ltd., P.O. Box 5050, Tokyo International, Tokyo 100-31, Japan) V.1- 1929-

**CYTZAM** Cytobiologie. (Stuttgart 1, Germany) V.1-18, 1969-79. For publisher information, see EJCBDN

**DABBBA** Dissertation Abstracts International, B: The Sciences and Engineering. (University Microfilms, A Xerox Co., 300 N. Zeeb Rd., Ann Arbor, MI 48106) V.30- 1969-

**DABSAQ** Dissertation Abstracts, B: The Sciences and Engineering. (University Microfilms, A Xerox Co., 300 N. Zeeb Rd., Ann Arbor, MI 48106) V.27-29, 1966-69

**DADEDV** Drug and Alcohol Dependence. (Elsevier Sequoia SA, POB 851, CH-1001 Lausanne, Switzerland) V.1- 1975-

**DAKMAJ** Deutsches Archiv fuer Klinische Medizin. (Munich, Germany) V.1- 211, 1865-1965. For publisher information, see EJCIB8

**DANKAS** Doklady Akademii Nauk S.S.S.R. (v/o Mezhdunarodnaya Kniga, Kuznetskii Most 18, Moscow G-200, USSR.) V.1- 1933-

**DANND6** Dopovidi Akademii Nauk Ukrains'koi RSR, Seriya B: Geologichini, Khimichni ta Biologichni Nauki. (v/o Mezhdunarodnaya Kniga, 121200 Moscow, USSR) 1976-

**DAZEA2** Deutsche Apotheker-Zeitung. (Deutscher Apotheker-Verlag, Postfach 40, D-7000 Stuttgart 1, Germany) V.49- 1934-

**DAZRA7** Doklady Akademii Nauk Azerbaidzhnskoi SSR. Proceedings of the Academy of Sciences of the Azerbaidzhan SSR. (v/o Mezhdunarodnaya Kniga, Kuznetskii Most 18, Moscow G-200, USSR) V.1- 1945-

**DBABEF** Doga Bilim Dergisi, Seri A2: Biyoloji. Natural Science Journal, Series A2. (Turkiye Bilimsel ve Teknik Arastirma Kurumu, Ataturk Bul. No. 221, Kavaklidere, Ankara, Turkey) V.8- 1984-

**DBANAD** Doklady Bolgarskoi Akademii Nauk. (Hemus, Blvd. Russki 6, Sofia, Bulgaria) V.1- 1948-

**DBLRAC** Doklady Akademii Nauk BSSR. Proceedings of the Academy of Sciences of the Belorussian SSR. (v/o Mezhdunarodnaya Kniga, Kuznetskii Most 18, Moscow G-200, USSR) V.1- 1957-

**DBTEAD** Diabete. (Le Raincy, France) V.1-22, 1953-1974

**DBTGAJ** Diabetologia. (Springer-Verlag New York, Inc., Service Center, 44 Hartz Way, Secaucus, NJ 07094) V.1- 1965-

**DCTODJ** Drug and Chemical Toxicology. (Marcel Dekker, POB 11305, Church St. Station, New York, NY 10249) V.1- 1977/78-

**DDEVD6** Drug Development and Evaluation. (Gustav Fischer Verlag, Postfach 720143, D-7000 Stuttgart 70, Germany) V.1- 1977-

**DDIPD8** Drug Development and Industrial Pharmacy. (Marcel Dekker, POB 11305, Church St. Station, New York, NY 10249) V.3- 1977-

**DDREDK** Drug Development Research. (Alan R. Liss, Inc., 150 5th Ave., New York, NY 10011) V.1- 1981-

**DDSCDJ** Digestive Diseases and Sciences. (Plenum Publishing Corp., 233 Spring St., New York, NY 10013) V.24- 1979-

**DEBIAO** Developmental Biology. (Academic Press, Inc., 111 5th Ave., New York, NY 10003) V.1- 1959-

**DEBIDR** Developments in Biochemistry. (Elsevier North Holland, Inc., 52 Vanderbilt Ave., New York, NY 10017) V.1- 1978-

**DECRDP** Drugs under Experimental and Clinical Research. (J.R. Prous Pub., Apartado de Correos 1179, Barcelona, Spain) V.1- 1977-

**DEGEA3** Deutsche Gesundheitswesen. (VEB Verlag Volk und Gesundheit, Neue Gruenstr 18, 102 Berlin, Germany) V.1- 1946-

**DEMAEP** Dental Materials. (Munksgaard International Pub., POB 2148, DK-1016 Copenhagen K, Denmark) V.1-1985-

**DEPBA5** Developmental Psychobiology. (John Wiley & Sons Ltd., Baffins Lane, Chichester, Sussex P01 1UD, England) V.1- 1968-

**DERAAC** Dermatologica. (Albert J. Phiebig, Inc., P.O. Box 352, White Plains, NY 10602) V.79- 1939-

**DFSCDX** Developments in Food Science. (Elsevier Science Pub. Co., Inc., 52 Vanderbilt Ave., New York, NY 10017) V.1- 1978-

**DGDFA5** Development, Growth and Differentiation. (Maruzen Co. Ltd., P.O. Box 5050, Tokyo International, Tokyo 100-31, Japan) V.11- 1969-

**DHEFDK** HEW Publication (FDA. United States). (Washington, DC) 19??-1979(?). For publisher information, see HPFSDS

**DIAEAZ** Diabetes. (American Diabetes Association, 600 5th Ave., New York, NY 10020) V.1- 1952-

**DICHAK** Diseases of the Chest. (Chicago, IL) V.1-56, 1935-69. For publisher information, see CHETBF

**DICPBB** Drug Intelligence and Clinical Pharmacy. (Drug Intelligence and Clinical Pharmacy, Inc., University of Cincinnati, Cincinnati, OH 45267) V.3- 1969-

**DICRAG** Diseases of the Colon and Rectum. (Harper and Row, Publishers, 10 E. 53rd St., New York, NY 10022) V.1- 1958-

**DIGEBW** Digestion. (S. Karger AG, Arnold-Boecklin Street 25, CH-4011 Basel, Switzerland) V.1- 1968-

**DIPHAH** Dissertationes Pharmaceuticae. (Warsaw, Poland) V.1-17, 1949-65. For publisher information, see PJPPAA

**DKBSAS** Doklady Biological Sciences (English Translation). (Plenum Publishing Corp., 233 Spring St., New York, NY 10013) V.112- 1957-

**DMBUAE** Danish Medical Bulletin. (Ugeskrift for Laeger, Domus Medica, 2100 Copenhagen, Denmark) V.1- 1954-

**DMWOAX** Deutsche Medizinische Wochenschrift. (Georg Thieme Verlag, Herdweg 63, Postfach 732, 7000 Stuttgart 1, Germany) V.1- 1875-

**DNEUD5** Developments in Neuroscience (Amsterdam). (Elsevier North Holland, Inc., 52 Vanderbilt Ave., New York, NY 10017) V.1- 1977-

**DNSSAW** Diseases of the Nervous System, Supplement. (Irvington, NJ) For publisher information, see DNSYAG

**DNSYAG** Diseases of the Nervous System. (Physicians Postgraduate Press, Box 38293, Memphis, TN 38138) V.1-38, 1940-77

**DOEAAH** Down to Earth. A Review of Agricultural Chemical Progress. (Dow Chemical U.S.A., 1703 S. Saginaw Rd., Midland, MI 48640) V.1- 1945-

**DOESD6** DOE Symposium Series. (NTIS, 5285 Port Royal Rd., Springfield, VA 22161) No. 45- 1978-

**DOVEAA** Defense des Vegetaux. (Federation Nationale des Groupements de Protection des Cultures, 149, rue de Bercy, 75595 Paris Cedex, 12, France)V.1-1947-

**DOWCC\*** Dow Chemical Company Reports. (Dow Chemical U.S.A., Health and Environment Research, Toxicology Research Lab., Midland, MI 48640)

**DPHFAK** Dissertationes Pharmaceuticae et Pharmacologicae. (Warsaw, Poland) V.18-24, 1966-72. For publisher information, see PJPPAA

**DPIRDU** Dangerous Properties of Industrial Materials Report. (Van Nostrand Reinhold Co., Inc., 115 Fifth Avenue, New York, NY 10003) V.1- 1981-

**DPTHDL** Developmental Pharmacology and Therapeutics. (S. Karger AG, Postfach CH-4009 Basel, Switzerland) V.1- 1980-

**DRFUD4** Drugs of the Future. (J.R. Prous, S.A. International Publishers, Apartado de Correos 1641, Barcelona, Spain) V.1- 1975/76-

**DRISAA** Drosophila Information Service. (Cold Spring Harbor Laboratory, POB 100, Cold Spring Harbor, NY 11724) No. 1- 1934-

**DRSAEA** Drug Safety. (Adis International Ltd., Private Bag 65901, Mairangi Bay, Auckland 10, New Zealand) V.5-1990-

**DRSTAT** Drug Standards. (Washington, DC) V.19-28, 1951-60. For publisher information, see JPMSAE

**DRUGAY** Drugs. International Journal of Current Therapeutics and Applied Pharmacology Reviews. (ADIS Press Ltd., 18/F., Tung Sun Commercial Centre, 194-200 Lockhart Road, Wanchai, Hong Kong)

**DTESD7** Developments in Toxicology and Environmental Science. (Elsevier, Scientific Publishing Co., POB 211, 1000 AE Amsterdam, Netherlands) V.1- 1977-

**DTLVS\*** "Documentation of Threshold Limit Values for Substances in Workroom Air." For publisher information, see 85INA8

**DTLWS\*** "Documentation of the Threshold Limit Values for Substances in Workroom Air" Supplements. For publisher information, see 85INA8

**DTTIAF** Deutsche Tieraerztliche Wochenschrift. (Verlag M. und H. Schaper, Postfach 260669, 3 Hanover 26, Germany) V.1- 1893-

**DUPON\*** E.I. Dupont de Nemours and Company, Technical Sheet. (1007 Market St., Wilmington, DE 19898)

**DZZEA7** Deutsche Zahnaerztliche Zeitschrift. (Carl Hanser Verlag, Postfach 860420, D-8000 Munich 86, Germany) V.1- 1946-

**EAGRDS** Experimental Aging Research. (Beech Hill Publishing Co., P.O. Box 136, Southwest Harbor, ME 04679) V.1- 1975-

**EAMJAV** East African Medical Journal (P.O. Box 41632, Nairobi, Kenya) V.9- 1932-

**EAPHA6** Eastern Pharmacist. (Eastern Pharmacist, 507, Ashok Bhawan, 93, Neru Place, New Delhi 110019, India) V.1- 1958-

**EATR\*\*** U.S. Army, Edgewood Arsenal Technical Report. (Aberdeen Proving Ground, MD 21010)

**EbeAG#** Personal Communication to NIOSH from A.G. Ebert, International Glutamate Technical Committee, 85 Walnut St., Watertown, MA 02172

**ECBUDQ** Ecological Bulletins. (Swedish National Science Research Council, Stockholm, Sweden) Number 19- 1975-

**ECEBDI** Experimental Cell Biology. (Phiebig Inc., POB 352, White Plains, NY 10602) V.44- 1976-

**ECJPAE** Endocrinologia Japonica. (Japan Publications Trading Co., 1255 Howard St., San Francisco, CA 94103) V.1- 1954-

**ECNEAZ** Electroencephalography and Clinical Neurophysiology. (Elsevier Scientific Publishing Co., POB 211, 1000 AE Amsterdam, Netherlands) V.1- 1949-

**ECREAL** Experimental Cell Research. (Academic Press, 111 5th. Ave., New York, NY 10003) V.1- 1950-

**EDRCAM** Endocrine Research Communications. (Marcel Dekker, POB 11305, Church St. Station, New York, NY 10249) V.1- 1974-

**EDWU\*\*** Beitrag zur Toxikologie Technischer Weichmachungsmittel, Heinrich Eller Dissertation. (Pharmakologischen Institut der Universitat Wurzburg, Germany, 1937)

**EESADV** Ecotoxicology and Environmental Safety. (Academic Press, 111 5th Ave., New York, NY 10003) V.1- 1977-

**EGESAQ** Egeszsegtudomany. (Kultura, POB 149, H-1389 Budapest, Hungary) V.1- 1957-

**EGJBAY** Egyptian Journal of Botany. (National Information and Documentation Centre, Al-Tahrir St., Awgaf P.O. Dokki, Cairo, Egypt) V.1- 1958-

**EJBCAI** European Journal of Biochemistry. (Springer-Verlag, Heidelberger Pl. 3, D-1 Berlin 33, Germany) V.1- 1967-

**EJBLAB** Egyptian Journal of Bilharziasis. (National Information Documentation Center, Tahrir St., Dokki, Cairo, Egypt) V.1- 1974-

**EJCAAH** European Journal of Cancer. (Pergamon Press, Headington Hill Hall, Oxford OX3 OEW, England) V.1- 1965-

**EJCBDN** European Journal of Cell Biology. (Wissenschaftliche Verlagsgesellschaft mbH, Postfach 40, D-7000, Stuttgart l, Germany) V.19- 1979-

**EJCIB8** European Journal of Clinical Investigation. (Springer-Verlag, Heidelberger Pl. 3, D-1 Berlin 33, Germany) V.1- 1970-

**EJCODS** European Journal of Cancer and Clinical Oncology. (Pergamon Press Ltd., Headington Hill Hall, Oxford OX3 0BW, England) V.17, No. 7- 1981-

**EJCPAS** European Journal of Clinical Pharmacology. (Springer-Verlag, Heidelberger Pl. 3, D-1 Berlin 33, Germany) V.3- 1970-

**EJMBA2** Egyptian Journal of Microbiology. (National Information and Documentation Centre, A1-Tahrir St., Awqaf P.O. Dokki, Cairo, Egypt) V.7- 1972-

**EJMCA5** European Journal of Medicinal Chemistry. Chimie Therapeutique. (Center National de la Recherche Scientifique, 3 rue J.B. Clement, F-92290 Chatenay-Malabry, France) V.9- 1974-

**EJNMD9** European Journal of Nuclear Medicine. (Springer-Verlag New York, Inc., Service Center, 44 Hartz Way, Secaucus, NJ 07094) V.1- 1975-

**EJPEDT** European Journal of Pediatrics. (Springer-Verlag New York, Inc., Service Center, 44 Hart Way, Secaucus, NJ 07094) V.121- 1975-

**EJPHAZ** European Journal of Pharmacology. (North-Holland Publishing, POB 211, 1000AE Amsterdam, Netherlands) V.1- 1967-

**EJRDD2** European Journal of Respiratory Diseases. (Munksgaard, 35 Norre Sogade, DK 1370 Copenhagen K, Denmark) V.61- 1980-

**EJTXAZ** European Journal of Toxicology and Environmental Hygiene. (Paris, France) V.7-9, 1974-76. For publisher information, see TOERD9

**EKFMA7** Eksperimental'naya i Klinicheskaya Farmakoterapiya. (Akademiya Nauk Latviiskoi SSR, Inst., Organicheskogo Sinteza, ul. Aizkraukles 21, Riga, USSR.) No. 1- 1970-

**EKMMA8** Eksperimentalna Meditsina i Morfologiya. (Hemus, Blvd. Russki 6, Sofia, Bulgaria) V.1- 1962-

**EKSODD** Eksperimental'naya Onkologiya. Experimental Oncology. (v/o Mezhdunarodnaya Kniga, 121200 Moscow, USSR) V.1- 1979-

**EMERY\*** Emery Industries, Inc., Data Sheets. (4900 Este Ave., Cincinnati, OH 45232)

**EMMUEG** Environmental and Molecular Mutagenesis. (Alan R. Liss, Inc., 4 E. 11th St., New York, NY 10003) V.10- 1987-

**EMPSAL** Experimental and Molecular Pathology, Supplement. (Academic Press, 111 5th Ave., New York, NY 10003) No. 1- 1963-

**EMSUA8** Experimental Medicine and Surgery. (Brooklyn Medical Press, 600 Lafayette Ave., Brooklyn, NY 11216) V.1- 1943-

**ENDKAC** Endokrinologie. (Leipzig, Germany) V.1-80, 1928-82. For publisher information, see EXCEDS

**ENDOAO** Endocrinology (Baltimore). (Williams & Wilkins Co., 428 E. Preston St., Baltimore, MD 21203) V.1- 1917-

**ENMUDM** Environmental Mutagenesis. (New York, NY) V.1-9, 1979-87. For publisher information, see EMMUEG. V.1- 1979-

**ENPBBC** Environmental Physiology and Biochemistry. (Copenhagen, Denmark) V.2-5, No. 6, 1972-5, Discontinued

**ENVIDV** Environmental Internationl. (Pergamon Press, Ltd., Headington Hill Hall, Oxford OX3 0BW, England) V.1- 1978-

**ENVRAL** Environmental Research. (Academic Press, 111 5th Ave., New York, NY 10003) V.1- 1967-

**ENZYAS** Enzymologica. (Dr. W. Junk bv Publishers, POB 13713, 2501 ES The Hague, Netherlands) V.1-43, 1936-72

**EOGRAL** European Journal of Obstetrics, Gynecology and Reproductive Biology. (Elsevier Science Publishing BV, POB 211,1000 AE Amsterdam, Netherlands) V3- 1973-

**EPASR\*** United States Environmental Protection Agency, Office of Pesticides and Toxic Substances. (U.S. Environmental Protection Agency, 401 M St., S.W., Washington, DC 20460) History Unknown

**EPILAK** Epilepsia. (Elsevier Publishing Co., P.O. Box 211, Amsterdam C, Netherlands) V.1- 1959-

**EPXXDW** European Patent Application. (U.S. Patent Office, Science Library, 2021 Jefferson Davis Highway, Arlington, VA 22202)

**EQSFAP** Environmental Quality and Safety. (Academic Press, 111 5th Ave., New York, NY 10003) V.1- 1972-

**EQSSDX** Environmental Quality and Safety, Supplement. (Academic Press, 111 5th Ave., New York, NY 10003) V.1- 1975-

**ERNFA7** Ernaehrungsforschung. (Akademie-Verlag GmbH, Leipziger St. 3-4, 108 Berlin, Germany) V.1- 1956-

**ESDBAK** Eisei Dobutsu. Sanitary Zoology. (Japan Society of Sanitary Zoology, c/o Institute of Medical Science, Unversity of Tokyo, 4-6- 1 Shiroganedai, Minato-ku, Tokyo 108, Japan) V.1- 1950-

**ESKGA2** Eisei Kagaku. (Nippon Yakugakkai, 2-12-15 Shibuya, Shibuya-Ku, Tokyo 150, Japan) V.1- 1953-

**ESKHA5** Eisei Shikenjo Hokoku. Bulletin of the National Hygiene Sciences. (Kokuritsu Eisei Shikenjo, 18-1 Kamiyoga 1 chome, Setagaya-ku, Tokyo, Japan) V.1- 1886-

**ETATAW** Eesti N.S.V. Teaduste Akadeemia Toimetised, Biologia. (v/o Mezhdunarodnaya Kniga, Kuznetskii Most 18, Moscow G-200, USSR.) V.5- 1956-

**EUURAV** European Urology. (S. Karger Publishers, Inc., 150 Fifth Ave., Suite 1105, New York, NY 10011) V.1- 1975-

**EVETBX** Environmental Entomology. (Entomological Society of America, 4603 Calvert Rd., College Park, MD 20740) V.1- 1972-

**EVHPAZ** EHP, Environmental Health Perspectives. Subseries of DHEW Publications. (U.S. Government Printing Office, Superintendent of Documents, Washington, DC 20402) No. 1- 1972-

**EVPHBI** Environmental Physiology. (Copenhagen, Denmark) V.1 (No. 1-4), 1971, For publisher information, see ENPBBC

**EVSRBT** Environmental Science Research. (Plenum Publishing Corp., 233 Spring St., New York, NY 10013) V.1- 1972-

**EVSSAV** Environmental Space Science. English Translation of Kosmicheskaya Biologiya Meditsina. 1967-70

**EXCEDS** Experimental and Clinical Endocrinology. (Johann Ambrosius Barth Verlag, Postfach 109, DDR-7010 Leipzig, Germany) V.81- 1983-

**EXERA6** Experimental Eye Research. (Academic Press, Inc., 1 E. 1st., Duluth, MN 55802) V.1- 1961-

**EXMDA4** International Congress Series - Excerpta Medica. (Elsevier North Holland, Inc., 52 Vanderbilt Ave., New York, NY 10017) No. 1- 1952-

**EXMPA6** Experimental and Molecular Pathology. (Academic Press, 111 5th Ave., New York, NY 10003) V.1- 1962-

**EXNEAC** Experimental Neurology. (Academic Press, 111 5th Ave., New York, NY 10003) V.1- 1959-

**EXPAAA** Experimental Parasitology. (Academic Press, 111 5th Ave., New York, NY 10003) V.1- 1951-

**EXPADD** Experimental Pathology. (Elsevier Scientific Publishers Ireland Ltd., POB 85, Limerick, Ireland) V.19- 1981-

**EXPEAM** Experientia. (Birkhaeuser Verlag, P.O. Box 34, Elisabethenst 19, CH-4010, Basel, Switzerland) V.1- 1945-

**EXPTAX** Experimentelle Pathologie. (Jena, Germany) V.1-18, 1967-80. For publisher information, see EXPADD

**FAATDF** Fundamental and Applied Toxicology. (Academic Press, Inc., 1 E. First St., Duluth, MN 55802) V.1- 1981-

**FACOEB** Food Additives and Contaminants. (Taylor and Francis Inc., 242 Cherry St., Philadelphia, PA 19106) V.1- 1984-

**FAONAU** Food and Agriculture Organization of United Nations, Report Series. (FAO-United Nations, Room 101, 1776 F Street, NW, Washington, DC 20437)

**FATOAO** Farmakologiya i Toksikologiya (Moscow). (v/o Mezhdunarodnaya Kniga, Kuznetskii Most 18, Moscow G-200, USSR.) V.2- 1939- For English translation, see PHTXA6 and RPTOAN

**FATOBP** Farmakologiya i Toksikologiya (Kiev). Pharmacology and Toxicology. (Kievskii Nauchno-Issledovatel'skii Institut Farmakologii i Toksikologii, Kiev, USSR.) No. 1- 1964-

**FAVUAI** Fiziologicheski Aktivnye Veshchestva. Physiologically Active Substances. (Akademiya Nauk Ukrainskoi S.S.R., Kiev, USSR.) No. 1- 1966-

**FAZMAE** Fortschritte der Arzneimittelforschung. (Birkhauser Verlag, P.O. Box 34, Elisabethenst 19, CH-4010, Basel, Switzerland) V.1- 1959-

**FCHNA9** Food Chemical News. (Food Chemical News, Inc., 1101 Pennsylvania Ave., S.E., Washington, DC 20003) V.1- 1959-

**FCTOD7** Food and Chemical Toxicology. (Pergamon Press, Headington Hill Hall, Oxford OX3 OBW, England) V.20- 1982-

**FCTXAV** Food and Cosmetics Toxicology. (London, UK) V.1-19, 1963-81. For publisher information, see FCTOD7

**FDHU\*\*** Uber die lokalanasthetischen Eigenschaften des salzsauren Salzes des Benzoyl-d-ekgonin-n-propylesters, Gerda Felgner Dissertation. (Pharmakologischen Institut der Martin-Luther-Universitat Halle-Wittenburg, Germany, 1933)

**FDMU\*\*** Zur Pharmakologie des Octins, Adolf Fiegenbaum Dissertation. (Pharmakologischen Institut der Westfalischen Wilhelms-Universitat, Munster, Germany, 1935)

**FDRLI\*** Food and Drug Research Labs., Papers. (Waverly, NY 14892)

**FDWU\*\*** Uber die Wirkung Verschiedener Gifte Auf Vogel, Ludwig Forchheimer Dissertation. (Pharmakologischen Institut der Universitat Wurzburg, Germany, 1931)

**FEBLAL** FEBS Letters. (Elsevier Scientific Publishing Co., POB 211, 1000 AE Amsterdam, Netherlands) V.1- 1968-

**FEPRA7** Federation Proceedings, Federation of American Societies for Experimental Biology. (9650 Rockville Pike, Bethesda, MD 20014) V.1- 1942-

**FESTAS** Fertility and Sterility. (American Fertility Society, 1608 13th Ave. S., Birmingham, AL 35205) V.1- 1950-

**FGIGDO** Fujita Gakuen Igakkaishi. Journal of the Fujita Gakuen Medical Society. (Nagoya Hoken Eisei Daigaku Fujita Gakuen Igakkai, 1-98 Dengakuga, Kubo, Kutsukake-Cho, Toyoake, Aichi-Ken 470-11, Japan) V. 1- 1977-

**FHCYAI** Folia Histochemica et Cytochemica. (Ars Polona-RUCH, P.O. Box 1001, P-00 068 Warsaw, 1, Poland) V.1- 1963-

**FKGCAR** Fortschritte der Kiefer und Gesichts-Chirurgie. (Georg Thieme Verlag, Stuttgart, Germany) V.1- 1955-

**FKIZA4** Fukuoka Igaku Zasshi. (Fukuoka Igakkai, c/o Kyushu Daigaku Igakubu, Tatekasu, Fukuoka-shi, Fukuoka, Japan) V.33- 1940-

**FLABAZ** Fluoride Abstracts. (Kettering Lab., University of Cincinnati, College of Medicine, Dept. of Environmental Health, Cincinnati, OH 45219) No. 1- 1955/58-

**FLCRAP** Fluorine Chemistry Reviews. (Marcel Dekker Inc., 305 E. 45th St., New York, NY 10017) V.1- 1967-

**FLUOA4** Fluoride. (International Society for Fluoride Research, P.O. Box 692, Warren, MI 48090) V.3- 1970-

**FMCHA2** Farm Chemicals Handbook. (Meister Publishing, 37841 Euclid Ave., Willoughy, OH 44094)

**FMDZAR** Fortschritte der Medizin. (Dr. Schwappach und Co., Wessobrunner Str 4, 8035 Gauting vor Muenchen, Germany) V.1- 1883-

**FMLED7** FEMS Microbiology Letters. (Elsevier Scientific Publishing Co., POB 211, Amsterdam, Netherlands) V.1- 1977-

**FMORAO** Folia Morphologica (Prague). (Plenum Publishing Corp. 233 Spring St., New York, NY 10013) V.13- 1965-

**FMTYA2** Farmatsiya (Sofia). (Hemus, Blvd. Russki 6, Sofia, Bulgaria) V.1- 1951-

**FNSCA6** Forensic Science. (Elsevier Sequoia SA, P.O. Box 851, CH 1001, Lausanne 1, Switzerland) V.1- 1972-

**FOBGA8** Folia Biologica (Krakow). (Ars Polona-RUCH, POB 154, Warsaw 1, Poland) V.1- 1953-

**FOBLAN** Folia Biologica (Prague). (Academic Press, 111 Fifth Ave., New York, NY 10003) V.1- 1955-

**FOHEAW** Folia Haematologica (Leipzig). (Buchexport, Postfach 160, Leninstr. 16, DDR-701, Leipzig, Germany) V.69- 1949-

**FOMDAK** Folia Medica. (Via Raffaele de Caesare, 31 Naples, Italy) V.1- 1915-

**FOMIAZ** Folia Microbiologica (Prague). (Academia, Vodickova 40, CS-112 29 Prague 1, (Czechoslavakia) V.4- 1959-

**FOMOAJ** Folia Morphologica (Warsaw). (Panstwowy Zaklad Wydawnictw Lekarskich, ul. Druga 38-40, P-00 238 Warsaw, Poland) V.1- 1929-

**FOREAE** Food Research. (Champaign, IL) V.1-25, 1936-60. For publisher information, see JFDSAZ

**FOTEAO** Food Technology. (Institute of Food Technologists, 221 N. La Salbe St., Chicago, IL 60601) V.1- 1947-

**FPNJAG** Folia-Psychiatrica et Neurologica Japonica. (Folia Publishing Society, Todai YMCA Bldg., 1-20-6 Mukogaoka, Bunkyo-Ku, Tokyo 113, Japan) 1947-

**FRBGAT** Frontiers of Biology. (Elsevier North Holland Inc., 52 Vanderbilt Ave., New York, NY 10017) V.1- 1966-

**FRMBAZ** Farmacia. (Rompresfilatelia, P.O. Box 2001, Calea Grivitei 64-66, Bucharest, Romania) V.1- 1953-

**FRMTAL** Farmatsiya (Moscow). Pharmacy. (v/o Mezhdunarodnaya Kniga, 121200 Moscow, USSR) V.16- 1967-

**FRPPAO** Farmaco, Edizione Pratica. (Casella Postale 114, 27100 Pavia, Italy) V.8- 1953-

**FRPSAX** Farmaco, Edizione Scientifica. (Casella Postale 114, 27100 Pavia, Italy) V.8- 1953-

**FRXXBL** French Demande Patent Document. (Commissioner of Patents and Trademarks, Washington, DC 20231)

**FRZKAP** Farmatsevtichnii Zhurnal (Kiev). (v/o Mezhdunarodnaya Kniga, Kuznetskii Most 18, Moscow G-200, USSR) V.3- 1930-

**FSASAX** Fette, Seifen, Anstrichmittel. (Industrieverlag von Hernhaussen KG, Roedingsmarkt 24, 2 Hamburg 11, Germany) V.55- 1953-

**FSDZD4** Fukuoka Shika Daigaku Gakkai Zasshi. Journal of Fukuoka Dental College. (700 Ta, O-aza, Sawara-ku, Fukuoka, 814-01, Japan) V.1- 1974-

**FSIZAQ** Fukushima Igaku Zasshi. (c/o Fukushima-Kenritsu Ika Daigaku Toshokan, 5175 Sugizuma-cho, Fukushima, Japan) V.1- 1951-

**FSTEAI** Farmaco (Pavia); Scienza e Tecnica. (Pavia, Italy) V.1-7, 1946-52. For publisher information, see FRPSAX

**FZPAAZ** Frankfurter Zeitschrift fuer Pathologie. (Munich, Germany) V.1-77, 1907-67. For publisher information, see VAAZA2

**GAFCC\*** GAF Material Safety Data Sheet. (GAF Chemicals Corporation, 1361 Alps Road, Wayne, NJ 07470)

**GANMAX** Gann Monograph. (Tokyo, Japan) No. 1-10, 1966-71. For publisher information, see GMCRDC

**GANNA2** Gann. Japanese Journal of Cancer Research. (Tokyo, Japan) V.1-75, 1907-84. For publisher information, see JJCREP

**GANRAE** Gan No Rinsho. Cancer Clinics. (Ishiyaku Publishers, Inc., 7-10 Honkomagome 1-chome, Bunkyo-ku, Tokyo, Japan) V.1- 1954-

**GASTAB** Gastroenterology. (Elsevier North Holland, Inc., 52 Vanderbilt Avenue, New York, NY 10017) V.1- 1943-

**GCENA5** General and Comparative Endocrinology. (Academic Press, 111 Fifth Ave., New York, NY 10003) V.1- 1961-

**GCTB\*\*** Givaudan Corporation, Corporate Communications. (100 Delawanna Ave., Clifton, NJ 07014)

**GDIKAN** Gifu Daigaku Igakubu Kiyo. Papers of the Gifu University School of Medicine. (Gifu Daigaku Igakubu, 40 Tsukasa-cho, Gifu, Japan) V.15- 1967-

**GEFRA2** Geburtshilfe und Frauenheilkunde. (Intercontinental Medical Books Corp., 381 Park Ave., S., New York, NY 10016) V.1- 1939-

**GEIRDK** Gendai Iryo. Modern Medical Care. (Gendai Iryosha, Kandabashi Bldg., 1-21 Nishiki-cho, Kanada Chiyoda-ku, Tokyo 101, Japan) V.1- 1969-

**GENEA3** Genetica (The Hague). (Dr. W. Junk bv Publishers, POB 13713, 2501 ES The Hague, Netherlands) V.1- 1919-

**GENRA8** Genetical Research. (Cambridge University Press, P.O. Box 92, Bentley House, 200 Euston Rd., London NW1 2DB, England) V.1- 1960-

**GENTAE** Genetics. (P.O. Drawer U, University Station, Austin, TX 78712) V.1- 1916-

**GEPHDP** General Pharmacology. (Pergamon Press Inc., Maxwell House Fairview Park, Elmsford, NY 10523) V.1- 1970-

**GERNDJ** Gerontology. (S. Karger AG, Postfach CH-4009, Basel, Switzerland) V.22- 1976-

**GESKAC** Genetika i Selektsiya. Genetics and Plant Breeding. (Izdatelstvo na Naukite, ul. Akad. G. Bonchev, 1113 Sofia, Bulgaria) V.1- 1968-

**GETRE8** Gematologiya i Transfuziologiya. Hematology and Transfusion Science. (v/o Mezhdunarodnaya Kniga, Kuznetskii Most 18, Moscow G-200, USSR) V.28- 1983-

**GEXXA8** German (East) Patent Document. (U.S. Patent Office, Science Library, 2021 Jefferson Davis Highway, Arlington, VA 22202)

**GICTAL** Giornale Italiano di Chemioterapia. Italian Journal of Chemotherapy. (Edizioni Minerva Medica, Casella Postale 491, I-10126 Turin, Italy) V.1-5, 1954-58; V.5- 1962-

**GISAAA** Gigiena i Sanitariya. For English translation, see HYSAAV. (V/O Mezhdunarodnaya Kniga, 113095 Moscow, USSR) V.1- 1936-

**GMCRDC** Gann Monograph on Cancer Research. (Japan Scientific Societies Press, Hongo 6-2-10, Bunkyo-ku, Tokyo 113, Japan) No. 11- 1971-

**GMITAB** Gazzetta Medica Italiana. (Minerva Medica, Casella Postale 491, Turin, Italy) V.95- 1936-

**GMJOAZ** Glasgow Medical Journal. (Glasgow, Scotland) 1828-1955. For publisher information, see SMDJAK

**GNAMAP** Gigiena Naselennykh Mest. Hygiene in Populated Places. (Kievski Nauchno-Issledovatel'skii Institut Obshchei i Kommunol'noi Gigieny, Kiev, USSR) V.7- 1967-

**GNKAA5** Genetika. (see SOGEBZ for English Translation) (v/o "Mexhdunarodnaya Kniga" Kuznetskii Most 18, Moscow G-200, USSR) No. 1- 1965-

**GNRIDX** Gendai no Rinsho. (Tokyo, Japan) V.1-10, 1967-76(?)

**GOBIDS** Gynecologic and Obstetric Investigation. (S. Karger AG, Postfach CH-4009, Basel, Switzerland) V.9- 1978-

**GRCSB\*** Goodrich Chemical Co., Service Bulletin (B.F. Goodrich Chemical Group, 6100 Oak Tree Blvd., Cleveland, OH 44131)

**GROWAH** Growth. (Southern Bio-Research Institute, Florida Southern College, Lakeland, FL 33802) V.1- 1937-

**GSLNAG** Gaslini(Genoa). Rivista de Pediatria e di Specialita Pediatriche. (Istituto "Giannina Gaslini", Via Cinque Maggio, 39, 16148 Genoa, Italy) V.1- 1969-?

**GTKRDX** Gan to Kagaku Ryoho. Cancer and Chemotherapy. (Gan to Kagaku Ryohosha, Yaesu Bldg., 1-8-9 Yaesu, Chuo-Ku, Tokyo 103, Japan) V.1- 1974-

**GTPPAF** Gigiena Truda i Professional'naya Patologiya v Estonskoi SSR. Labor Hygiene and Occupational Pathology in the Estonian SSR. (Institut Eksperimental'noi i Klinicheskoi Meditsiny Ministerstva Zdravookhraneniya Estonskoi SSR, Tallinn, USSR) V.8- 1972-

**GTPZAB** Gigiena Truda i Professional'nye Zabolevaniia. Labor Hygiene and Occupational Diseases. (v/o Mezhdunarodnaya Kniga, Kuznetskii Most 18, Moscow G-200, USSR) V.1- 1957-

**GUCHAZ** "Guide to the Chemicals Used in Crop Protection" Information Canada, 171 Slater St., Ottawa, Ontario, Canada

**GUTTAK** Gut. (British Medical Journal, 1172 Commonwealth Ave., Boston, MA 02134) V.1- 1960-

**GWXXAW** German Patent Document. (U.S. Patent Office, Science Library, 2021 Jefferson Davis Highway, Arlington, VA 22202)

**GWXXBX** German Offenlegungsschrift Patent Document. (U.S. Patent Office, Science Library, 2021 Jefferson Davis Highway, Arlington, VA 22202)

**GWZHEW** Gongye Weisheng Yu Zhiyebing. Industrial Health and Occupational Diseases. (China International Book Trading Corp., POB 2820, Beijing, People's Republic of China) V.1- 1973-

**GYNOA3** Gynecologic Oncology. (Academic Press, 111 Fifth Ave., New York, NY 10003) V.1- 1972-

**HAEMAX** Haematologica. (Il Pensiero Scientifico, Via Panama 48, I-00198, Rome, Italy) V.1- 1920-

**HAONDL** Hematological Oncology. (John Wiley & Sons Ltd., Baffins Lane, Chichester, Sussex, PO19 1UD, UK) V.1- 1983-

**HAREA6** Harefuah Medicine. (Israel Medical Association, 39 Shaul Hamlech Blvd., Tel Aviv-Jaffa, Israel) V.1- 1924

**HarPN#** Personal Communication to Henry Lau, Tracor Jitco, from Paul N. Harris, M.D., Eli Lily Co., Greenfield, IN 46140

**HAZL\*\*** Hazelton Laboratories, Reports. (Falls Church, VA)

**HBAMAK** "Abderhalden's Handbuch der Biologischen Arbeitsmethoden." (Leipzig, Germany)

**HBTXAC** "Handbook of Toxicology, Volumes I-V." W.B. Saunders, Philadelphia, PA, 1956-59

**HCACAV** Helvetica Chimica Acta. (Verlag Helvetica Chimica Acta, Postfach, CH-4002 Basel, Switzerland) V.1- 1918-

**HCMYAL** Histochemistry. (Springer-Verlag New York, Inc., Service Center, 44 Hartz Way, Secaucus, NJ 07094) V.38- 1974-

**HDIZAB** Hiroshima Daigaku Zasshi, Medical Journal of the Hiroshima University (Hiroshima Daigaku Igakubu Saikingaku Kyoshitsu, Kasumi-cho, Hiroshima, Japan) V.10- 1962-

**HDKU\*\*** Die Krampfmischung der Lokalanaesthetica, ihre Beeinflussung durch Mineralsalze und Adrenalin, Gerhard Hoppe Dissertation. (Pharmakologischen Institut der Albertus-Universitat zu Konigsberg Pr., Germany, 1933)

**HDSKEK** Hiroshima Daigaku Sogo Kagakubu Kiyo, 4: Kiso, Kankyo Kagaku Kenkyu. Bulletin of the Faculty of Integrated Arts and Sciences, Hiroshima University, Series 4: Fundamental and Environmental Sciences. (1-1-89 Higashisenda-machi, Naka-ku, Hiroshima, 730, Japan) V.6- 1980-

**HDTU\*\*** Pharmakologische Prufung von Analgetika, Gunter Herrlen Dissertation. (Pharmakologischen Institut der Universitat Tubingen, Germany, 1933)

**HDWU\*\*** Beitrage zur Toxikologie des Athylenoxyds und der Glykole, Arnold Hofbauer Dissertation. (Pharmakologischen Institut der Universitat Wurzburg, Germany, 1933)

**HEADAE** Headache. (American Association for the Study of Headache, Rm 537, 621 S. New Ballas Rd., St. Louis, MO 63141) V.1- 1961-

**HEGAD4** Hepatogastroenterology. (Georg Thieme Verlag, Postfach 732, Herdweg 63, D-7000 Stuttgart 1, Germany) V.1- 1980-

**HEPHD2** Handbook of Experimental Pharmacology. (Springer-Verlag, Heidelberger Pl. 3, D-1000 Berlin 33, Germany) V.50- 1978-

**HERBU\*** Hercules Incorporated, Hercules Bulletin. (Hercules Incorporated, Pine and Paper Chemicals Department, Wilmington, DE)

**HEREAY** Hereditas. (J.L. Toernqvist Book Dealers, S-26122 Landskrona, Sweden) V.1- 1947-

**HETOEA** Human & Experimental Toxicology. (Macmillan Press Ltd., Brunel Road, Houndmills, Basingstoke, Hampshire, RG21 2XS, UK) V.9- 1990-

**HGANAO** Haigan. Lung Cancer. (Nippon Haigan Gakkai, c/o Chiba Daigaku Igakubu Haigan Kenkyusho, 1-8-1 Inohona, Chiba 280, Japan)

**HIFUAG** Hifu. Skin. (Nihon Hifuka Gakkai Osaka Chihokai, c/o Osaka Daigaku Hifuka Kyoshitsu, 3-1, Dojima Hamadori, Fukushima-ku, Osaka 553, Japan) V.1- 1959-

**HIKYAJ** Hinyokika Kiyo. (Acta Urologica Japonica). (Kyoto University Hospital, Department of Urology) V.1- 1955-

**HINAAU** Hindustan Antibiotics Bulletin. (Hindustan Antibiotics, Pimpri, India) V.1- 1958-

**HINEL\*** Hine Laboratories Reports. (San Francisco, CA)

**HIRIA6** Hirosaki Igaku. (Hirosaki Daigaku Igakubu, Hirosaki, Japan) V.1- 1950-

**HIUN\*\*** Department of Cancer Research, Research Institute for Nuclear Medicine and Biology, Hiroshima University, Kasumi 1-2-3, Hiroshima 734, Japan

**HKXUDL** Huanjing Kexue Xuebao. Environmental Sciences Journal. (Guoji Shudian, POB 399, Beijing, People's Republic of China) V.1- 1981-

**HLTPAO** Health Physics. (Pergamon Press, Maxwell House, Fairview Park, Elmsford, NY 10523) V.1- 1958-

**HMMRA2** Hormone and Metabolic Research. (Georg Thieme Verlag, Postfach 732, Herdweg 63, D-7000 Stuttgart 1, Germany) V.1- 1969-

**HOEKAN** Hokkaidoritsu Eisei Kenkyushoho. (Hokkaidoritsu Eisei Kenkyusho, Nishi-12-chome, Kita-19-jo, Kita-ku, Sapporo, Japan) No. 1- 1951-

**HOIZAK** Hokkaido Igaku Zasshi. Hokkaido Journal of Medical Science. (Hokkaido Daigaku Igakubu, Nishi-5-chome, Kita-12-jo, Sapporo, Japan) V.1- 1923-

**HOKBAQ** Hoken Butsuri. Journal of the Japan Health Physics Society. (Nihon Hoken Butsuri Kyogikai, Sakamoto Bldg., Kitaaoyama 2-12-4, Minato-ku, Tokyo, Japan) V.1- 1966-

**HPCQA4** Human Pathology. (W. B. Saunders Co., W. Washington Sq., Philadelphia, PA 19105) V.1- 1970-

**HPFSDS** HHS Publication (FDA. United States). (U.S. Government Printing Office, Superintendent of Documents, Washington, DC 20402) 1980-

**HPPAAL** Helvetica Physiologia et Pharmacologica Acta. (Basel, Switzerland) V.1-26, 1943-68. Discontinued

**HRMRA3** Hormone Research. (S. Karger AG, Postfach CH-4009, Basel, Switzerland) V.4- 1978-

**HSZPAZ** Hoppe-Seyler's Zeitschrift fuer Physiologische Chemie. (Walter de Gruyter and Co., Genthiner Street 13, D-1000, Berlin 30, Germany) V.21- 1895/96-

**HUGEDQ** Human Genetics. (Springer-Verlag, Neuenheimer Landst 28-30, D-6900 Heidelberg 1, Germany) V.31- 1976-

**HUMAA7** Humangenetik. (Heidelberg, Germany) V.1-30, 1964-1975. For publisher information, see HUGEDQ

**HunNJ#** Personal Communication from Nancy J. Hunt, E.I. DuPont de Nemours and Co., 1007 Market St., Wilmington, DE 19898, to Henry Lau, Tracor Jitco, Inc., May 10, 1977

**HUPSEC** Human Psychopharmacology. (John Wiley & Sons Ltd., Baffins Lane, Chichester, W. Sussex PO19 1UD, UK) V.1-1986-

**HURC\*\*** Huntingdon Research Center Reports. (Box 527, Brooklandville, MD 21022)

**HUTODJ** Human Toxicology. (Macmillan Press Ltd., Houndmills, Bassingstoke, Hants., RG21 2XS, UK) V.1- 1981-

**HXPHAU** Handbuch der Experimentellen Pharmakologie. (Berlin, Germany) V.1-49, 1935-78. For publisher information, see HEPHD2

**HYDKAK** Hoshi Yakka Daigaku Kiyo. Proceedings of the Hoshi College of Pharmacy. (2-chome, Ebara, Shinagawa-ku, Tokyo, Japan) No. 1- 1951-

**HYDRDA** Hydrometallurgy. (Elsevier Science Publishers B.V., POB 211, 1000 AE Amsterdam, Netherlands) V.1- 1975-

**HYDXET** Huaxi Yike Daxue Xuebao. Journal of West China University of Medical Sciences. (China International Book Trading Corp., POB 2820, Beijing, People's Republic of China) V.17- 1986-

**HYSAAV** Hygiene and Sanitation: English Translation of Gigiena Sanitariya. (Springfield, VA) V.29-36, 1964-71. Discontinued

**IAAAAM** International Archives of Allergy and Applied Immunology. (S. Karger, Postfach CH-4009, Basel, Switzerland) V.1- 1950-

**IAANBS** Internationales Archiv fuer Arbeitsmedizin. (Berlin, Germany) V.26-34, 1970-74. For publisher information, see IAEHDW

**IAEC\*\*** Interagency Collaborative Group on Environmental Carcinogenesis, National Cancer Institute, Memorandum, June 17, 1974

**IAEHDW** International Archives of Occupational and Environmental Health. Springer-Verlag, Heidelberger Pl 3, D-1000 Berlin 33, Germany) V.35- 1975-

**IAPUDO** IARC Publications. (World Health Organization, CH-1211 Geneva 27, Switzerland) No. 27- 1979-

**IAPWAR** International Journal of Air and Water Pollution. (London, England) V.4, No. 1-4, 1961. For publisher information, see ATENBP

**IARC\*\*** IARC Monographs on the Evaluation of Carcinogenic Risk of Chemicals to Man. (World Health Organization, Internation Agency for Research on Cancer, Lyon, France) (Single copies can be ordered from WHO Publications Centre U.S.A., 49 Sheridan Avenue, Albany, NY 12210)

**IARCCD** IARC Scientific Publications. (Geneva Switzerland) V.1-No. 26, 1971-78, For publisher information, see IAPUDO

**ICHAA3** Inorganica Chimica Acta. (Elsevier Sequoia SA, POB 851, CH-1001 Lausanne 1, Switzerland) V.1- 1967-

**ICMED9** Intensive Care Medicine. (Springer-Verlag New York, Inc., Service Center, 44 Hart Way, Secaucus, NJ 07094) V.3- 1977-

**IDZAAW** Idengaku Zasshi. Japanese Journal of Genetics. (Genetics Society of Japan, Nippon Iden Gakkai, Tanida 111, Mishima-shi, Shizuoka 411, Japan) V.1- 1921-

**IECHAD** Industrial and Engineering Chemistry. (Washington, DC) V.15-62, 1923-70. For publisher information, see CHMTBL

**IGAYAY** Igaku No Ayumi. Progress in Medicine. (Ishiyaku Shuppan K.K., 7-10, Honkomagone 1 chome, Bunkyo-ku, Tokyo, Japan) V.1- 1946-

**IGIBA5** Igiena. (Editura Medicala, St 13 Decembrie 14, Bucharest, Romania) V.5- 1956-

**IGKEAO** Igaku Kenkyu. (Daido Gakkan Shuppanbu, Kyushu Daigaku Igakubu, Hoigaku Kyoshitsu, Fukuoka 812, Japan) V.1- 1927-

**IGSBAL** Igaku to Seibutsugaku. Medicine and Biology. (1-11-4 Higashi-Kanda, Chiyoda, Tokyo 101, Japan) V.1- 1942-

**IGSBDO** Izvestiya Akademii Nauk Gruzinskoi SSR, Seriya Biologicheskaya. Proceedings of the Academy of Sciences of the Georgian SSR, Biological Series. (v/o Mezhdunarodnaya Kniga, Kuznetskii Most 18, Moscow G-200, USSR) V.1- 1975-

**IHFCAY** Industrial Hygiene Foundation of America, Chemical and Toxicological Series, Bulletin. (5231 Centre Ave., Pittsburgh, PA 15232) 1947-69

**IIFBA4** Izvestiya na Instituta po Fiziologiya, Bulgarska Akademiya na Naukite. (Izdatelstvo na Bulgarskata Akademiya na Naukite, St. Geo. Milev ul 36, Sofia 13, Bulgaria) V.4- 1960-

**IIZAAX** Iwate Igaku Zasshi. Journal of the Iwate Medical Association. (Iwate Igakkai, c/o Iwate Ika Daigaku, Uchimaru, Morioka, Japan) V.1- 1947-

**IJANDP** International Journal of Andrology. (Scriptor Publisher ApS, 15 Gasvaerksvej, DK-1656 Copenhagen V, Denmark) V.1- 1978-

**IJBBBQ** Indian Journal of Biochemistry and Biophysics. (Council of Scientific and Industrial Research, Publication and Information Director, Hillside Rd., New Delhi 110012, India) V.8- 1971-

**IJCAAR** Indian Journal of Cancer. (Dr. D.J. Jussawalla, Hospital Ave, Parel, Bombay 12, India) V.1- 1963-

**IJCBDX** International Journal of Clinical Pharmacology and Biopharmacy. (Urban and Schwarzenberg, Pettenkoferst 18, D-8000 Munich 15, Germany) V.11- 1975-

**IJCNAW** International Journal of Cancer. (International Union Against Cancer, 3 rue du Conseil-General, 1205 Geneva, Switzerland) V.1- 1966-

**IJCPB5** International Journal of Clinical Pharmacology, Therapy and Toxicology. (Munich, Germany) V.6-10, 1972-74. For publisher information, see IJCBDX

**IJCREE** International Journal of Crude Drug Research. (Swets and Zeitlinger B.V., POB 825, 2160 SZ Lisse, Netherlands) V.20- 1982-

**IJEBA6** Indian Journal of Experimental Biology. V.1- 1963-. For publisher information, see IJBBBQ

**IJEVAW** International Journal of Environmental Studies. (Gordon and Breach Science Publishers Ltd., 41-42 William IV St., London WC2N 4DE, England) V.1- 1970-

**IJIMDS** International Journal of Immunopharmacology. (Pergamon Press Ltd., Headington Hill Hall, Oxford OX3 0BW, England) V.1- 1979-

**IJLEAG** International Journal of Leprosy. (Box 39088, Washington, DC 20016) V.1- 1933-

**IJMDAI** Israel Journal of Medical Sciences. (P.O. Box 2296, Jerusalem, Israel) V.1- 1965-

**IJMRAQ** Indian Journal of Medical Research. (Indian Council of Medical Research, P.O. Box 4508, New Delhi 110016, India) V.1- 1913-

**IJMSAT** Irish Journal of Medical Science. (Royal Academy of Medicine, 6 Kildare St., Dublin, Ireland) V.1- 1968-

**IJNEAQ** International Journal of Neuropharmacology. (Oxford, England/New York, NY) V.1-8, 1962-69. For publisher information, see NEPHBW

**IJNMCI** International Journal of Nuclear Medicine and Biology. (Pergamon Press Inc., Maxwell House, Fairview Park, Elmsford, NY 10523) V.1- 1973-

**IJNUB7** International Journal of Neuroscience. (Gordon and Breach Science Publication S.A., 50 W. 23rd St., New York, NY 10010) V.1- 1970-

**IJOCAP** Indian Journal of Chemistry. (Council of Scientific and Industrial Research, Publication and Information Director, Hillside Rd., New Delhi 110012, India) V.1-13, 1963-75

**IJPAAO** Indian Journal of Pharmacy. (Bombay, India) V.1-40, No. 1, 1939-78. For publisher information, see IJSIDW

**IJPBAR** Indian Journal of Pathology and Bacteriology. (Dr. S.G. Deodhare, Medical College, Miraj Maharashtra, India) V.1- 1958-

**IJPPAZ** Indian Journal of Physiology and Pharmacology. (Indian Journal of Physiology and Pharmacology, All India Institute of Medical Sciences, Department of Physiology, Ansari Nagar, New Delhi 110016, India) V.1- 1957-

**IJRBA3** International Journal of Radiation Biology and Related Studies in Physics, Chemistry and Medicine. (Taylor and Francis Ltd., 4 John Street, London WC1N 2ET, UK) V.1- 1959-

**IJSIDN** Journal of Traditional Chinese Medicine (English Edition). (All-China Association of Traditional Chinese Medicine, Academy of Traditional Chinese Medicine) V.1- 1981-

**IJSIDW** Indian Journal of Pharmaceutical Sciences. (Indian Journal of Pharmaceutical Sciences, Kalina, Santa Cruz (East), Bombay 400 029, India) V.40, No. 2- 1978-

**IJTEDP** International Journal on Tissue Reactions. Cell and Tissue Injuries, Experimental and Clinical Aspects. (Biosciences Ediprint, Rue Winkelried 8, 1211 Geneva 1, Switzerland) V.1- 1979-

**IJVNAP** International Journal for Vitamin and Nutrition Research. (Verlag Hans Huber, Laenggasstr. 76, CH-3000 Bern 9, Switzerland) V.41- 1971-

**IMEMDT** IARC Monographs on the Evaluation of Carcinogenic Risk of Chemicals to Man. (World Health Organization, Internation Agency for Research on Cancer, Lyon, France) (Single copies can be ordered from WHO Publications Centre U.S.A., 49 Sheridan Avenue, Albany, NY 12210)

**IMGAAY** Indian Medical Gazette. (Indian Medical Gazette, Block F, 105-C, New Alipore, Calcutta 700053, India) V.1-90, 1866-1955; new series V.1- 1961-

**IMLCAV** Immunological Communications. (Marcel Dekker, POB 11305, Church St. Station, New York, NY 10249) V.1- 1972-

**IMMUAM** Immunology. (Blackwell Scientific Publications, Osney Mead, Oxford OX2 OEL, England) V.1- 1958-

**IMSCE2** IRCS Medical Science. (Lancaster, UK) V.12-14 1984-86

**IMSUAI** Industrial Medicine and Surgery. (Chicago, IL/Miami, FL) V.18-42, 1949-73. For publisher information, see IOHSA5

**INDRBA** Indian Drugs. (Indian Drugs Manufacturers' Associations c/o Dr. A. Patani, 102B, Poonam Chambers, Dr. A.B. Rd., Worli Bobmay 400 018, India) V.1- 1963-

**INFIBR** Infection and Immunity. (American Society for Microbiology, 1913 I St., N.W., Washington, DC 20006) V.1- 1970-

**INHEAO** Industrial Health. (2051 Kizukisumiyoshi-cho, Nakahara-ku, Kawasaki, Japan) V.1- 1963-

**INJFA3** International Journal of Fertility. (Allen Press, 1041 New Hampshire St., Lawrence, KS 66044) V.1- 1955-

**INJHA9** Indian Journal of Heredity. (Genetic Association of India, Indian Veterinary Research Institute, Izatnagar, India) V.1- 1969-

**INJPD2** Indian Journal of Pharmacology. (Indian Pharmacological Society, Jawaharlal Institute of Postgraduate Medical Education and Research, Pharmacology Dept., Pondicherry 605006, India) V.1- 1968(?)-

**INMEAF** Industrial Medicine. (Chicago, IL) V.1-18, 1932-49. For publisher information, see IOHSA5

**INNDDK** Investigational New Drugs. The Journal of New Anticancer Agents. (Kluwer Boston Inc., 190 Old Derby St., Hingham, MA 02043) V.1- 1983-

**INOPAO** Investigative Ophthalmology. (C.V. Mosby Co., 11830 Westline Industrial Dr., St. Louis, MO 63141) V.1- 1962-

**INPDAR** Indian Pediatrics. (46-B Garcha Rd., Calcutta, India) V.1- 1964-

**INPHB6** International Pharmacopsychiatry. (S. Karger AG, Postfach CH-4009, Basel, Switzerland) V.1- 1968-

**INSSDM** INSERM Symposium. (Elsevier North Holland, Inc., 52 Vanderbilt Ave., New York, NY 10017) No. 1- 1975-

**INTSAO** International Surgery. (International College of Surgeons, 1516 N. Lake Shore Dr., Chicago, IL 60610) V.45- 1966-

**INURAQ** Investigative Urology. (Williams & Wilkins Co., 428 E. Preston St., Baltimore, MD 21202) V.1- 1963-

**INVRAV** Investigative Radiology. (J.B. Lippincott Co., E. Washington Sq., Philadelphia, PA 19105) V.1- 1966-

**IOBPD3** International Journal of Radiation Oncology, Biology and Physics. Pergamon Press Ltd., Headington Hill Hall, Oxford OX3 0BW, England) V.1- 1975-

**IOHSA5** International Journal of Occupational Health and Safety. (Medical Publications, Inc., 3625 Woodhead Dr., Northbrook, IL 60093) V.43- 1974-

**IOVSDA** Investigative Ophthalmology and Visual Science. (C.V. Mosby Co., 11830 Westline Industrial Dr., St. Louis, MO 63146) V.16- 1977-

**IPCLBZ** International Pest Control. (McDonald Publications, 268 High St., Uxbridge UB8 1UA, Middlesex., England) V.5- 1962-

**IPPABX** Iugoslavica Physiologica et Pharmacologica Acta. (Unija Bioloskih Naucnik Drustava Jugoslavije, Postanski fah 127, Belgrade-Zemun, Yugoslavia) V.1- 1965-

**IPRAA8** Indian Practitioner. (231 Dr. Dadabhoy Naoroji Rd., Bombay 1, India) V.1- 1947/48-

**IPSTB3** International Polymer Science and Technology. (Rapra Technology Ltd., Shawbury, Shrewsbury, Shropshire SY4 4NR, UK)

**IRGGAJ** Internationales Archiv fuer Gewerbepathologie und Gewerbehygiene. (Heidelberg, Germany) V.19-25, 1962-69. For publisher information, see IAEHDW

**IRLCDZ** IRCS Medical Science: Library Compendium. (MTP Press Ltd., St. Leonards House, St. Leonards Gate, Lancaster LA1 3BR, England) V.3-11, 1975-83

**IRMEA9** International Record of Medicine. (New York, NY) V.170-174, 1957-61. For publisher information, see CLMEA3

**IRNEAE** International Review of Neurobiology. (Academic Press, 111 5th Ave., New York, NY 10003) V.1- 1959-

**ISBNBN** Izvestiya Sibirskogo Otdeleniya Akademii Nauk S.S.S.R., Seriya Biologicheskikh Nauk. (v/o Mezhdunarodnaya Kniga, Kuznetskii Most 18, Moscow G-200, USSR.) No. 1- 1969-

**ISMJAV** Israel Medical Journal. (Jerusalem, Israel) V.17-23, 1958-64. For publisher information, see IJMDAI

**ISYAM*** "Amphetamines and Related Compounds" E. Costa, ed., Proceedings of the Mario Negri Institute for Pharmacological Research, Milan, Italy, New York, NY, Raven Press, 1970

**ITCSAF** In Vitro. (Tissue Culture Association, 12111 Parklawn Dr., Rockville, MD 20852) V.1- 1965-

**ITMZBJ** Intensivmedizin. (Dr. Dietrich Steinkopff Verlag, Postfach 11-10-08, D-6100 Darmstadt 11, Germany) V.1- 1964-

**ITOBAO** Izvestiya Akademii Nauk Tadzhikskoi SSR, Otdelenie Biologicheskikh Nauk. Proceedings of the Academy of Sciences of the Tadzhik SSR, Department of Biological Sciences. (v/o Mezhdunarodnaya Kniga, Kuznetskii Most 18, Moscow G-200, USSR) No. 1- 1962-

**IUSMDJ** ICN-UCLA Symposia on Molecular and Cellular Biology. (Academic Press, 111 5th Ave., New York, NY 10003) V.1- 1974-

**IVEBBN** Beiheft zur Internationalen Zeitschrift fuer Vitamin-und Ernaehrungsforschung. (Verlag Hans Huber Laenggastr. 76 CH-3000 Bern 9, Switzerland) No. 12- 1972-

**IVEJAC** Indian Veterinary Journal. (Indian Veterinary Journal, Dr. R. Krishnamurti, G.M.V.C., 10 Avenue Rd., Madras 600 034, India) V.1- 1924-

**IVIVE4** In Vivo. (POB 51359, Kiffisia, GR-145 10, Greece) V.1- 1987-

**IYKEDH** Iyakuhin Kenkyu. Study of Medical Supplies. (Nippon Koteisho Kyokai, 12-15, 2-chome, Shibuya, Shibuya-ku, Tokyo 150, Japan) V.1- 1970-

**IZKPAK** Internationale Zeitschrift fuer Klinische Pharmakologie, Therapie und Toxikologie. (Munich, Germany) V.1-5, 1967-72. For publisher information, see IJCBDX

**IZSBAI** Izvestiya Sibirskogo Otdeleniya Akademii Nauk S.S.S.R., Seriya Biologomeditsinskikh Nauk. (Novosibirsk, USSR) 1963-68. For publisher information, see ISBNBN

**IZVIAK** Internationale Zeitschrift fuer Vitaminforschung. International Review of Vitamin Research. (Bern, Switzerland) V.19-40, 1947-70. For publisher information, see IJVNAP

**JAADDB** Journal of the American Academy of Dermatology. (C.V. Mosby Co., 11830 Westline Industrial Dr., St. Louis, MO 63141) 1979-

**JACHDX** Journal of Antimicrobial Chemotherapy. (Academic Press, 24-28 Oval Rd., London NW1 7DX, England) V.1- 1975-

**JACIBY** Journal of Allergy and Clinical Immunology. (C.V. Mosby Co., 11830 Westline Industrial Dr., St. Louis, MO 63141) V.48- 1971-

**JACSAT** Journal of the American Chemical Society. (American Chemical Society Publications, 1155 16th St., N.W., Washington, DC 20036) V.1- 1879-

**JACTDZ** Journal of the American College of Toxicology. (Mary Ann Liebert, Inc., 500 East 85th St., New York, NY 10028) V.1- 1982-

**JADSAY** Journal of the American Dental Association. (American Dental Association, 211 E. Chicago Ave., Chicago, IL 60611) V.9- 1922-

**JAFCAU** Journal of Agricultural and Food Chemistry. (American Chemical Society Publications, 1155 16th St., N.W., Washington, DC 20036) V.1- 1953-

**JAGRAC** Journal of Agricultural Research. (Washington, DC) V.1-78, 1913-49. Discontinued

**JAGSAF** Journal of the American Geriatrics Society. (American Geriatrics Society, Inc., 10 Columbus Circle, New York, NY 10019) V.1- 1953-

**JAINAA** Journal of the Anatomical Society of India. (Dept. of Anatomy, Medical College, Aurangabad MS, India) V.1- 1952-

**JAJAAA** Journal of Antibiotics, Series A. (Tokyo, Japan) V.6-20, 1953-67. For publisher information, see JANTAJ

**JAMAAP** JAMA, Journal of the American Medical Association. (American Medical Association, 535 N. Dearborn St., Chicago, IL 60610) V.1- 1883-

**JAMPA2** Journal of Animal Morphology and Physiology. (Society of Animal Morphologists and Physiologists, Dept. of Zoology, Faculty of Science, M.S. Univ. of Baroda, Baroda 2, India) V.1- 1954-

**JANCA2** Journal of the Association of Official Analytical Chemists. (Association of Official Analytical Chemists, Box 540, Benjamin Franklin Station, Washington, DC 20044) V.49- 1966-

**JANSAG** Journal of Animal Science. (American Society of Animal Science, 309 West Clark Street, Champaign, IL 61820) V.1- 1942-

**JANTAJ** Journal of Antibiotics. (Japan Antibiotics Research Association, 2-20-8 Kamiosaki, Shinagawa-ku, Tokyo, Japan) V.2-5, 1948-52; V.21- 1968-

**JAOCA7** Journal of the American Oil Chemists' Society. (American Oil Chemists' Society, 508 South 6th St., Champaign, IL 61820) V.24-56, 1947-79

**JAPHAR** Journal of Anatomy and Physiology. (London, England) V.1-50, 1916. For publisher information, see JOANAY

**JAPMA8** Journal of the American Pharmaceutical Association, Scientific Edition. (Washington, DC) V.29-49, 1940-60. For publisher information, see JPMSAE

**JAPYAA** Journal of Applied Physiology. (American Physiological Society, 9650 Rockville Pike, Bethesda, MD 20014) V.1- 1948-

**JASIAB** Journal of Agricultural Science. (Cambridge University Press, London or Cambridge University Press, New York) V.1- 1905-

**JATOD3** Journal of Analytical Toxicology. (Preston Publications, P.O. Box 48312, Niles, IL 60648) V.1- 1977-

**JAVMA4** Journal of the American Veterinary Medical Association. (American Veterinary Medical Association, 600 S. Michigan Ave., Chicago, IL 60605) V.48- 1915-

**JBCHA3** Journal of Biological Chemistry. (American Society of Biological Chemists, Inc., 428 E. Preston St., Baltimore, MD 21202) V.1- 1905-

**JBJSA3** Journal of Bone and Joint Surgery. American Volume. (10 Shattuck St., Boston, MA 02115) V.30- 1948-

**JBJSB4** Journal of Bone and Joint Surgery. (Boston, MA) V.20-45, 1922-47. For publisher information, see JBJSA3

**JBMRBG** Journal of Biomedical Materials Research. (John Wiley & Sons, 605 3rd Ave., New York, NY 10016) V.1- 1967-

**JCCCA5** Journal of the Society of Cosmetic Chemists. (Soc. of Cosmetic Chemists, 1995 Broadway, Suite 1701, New York, NY 10023) V.1- 1947-

**JCCPAY** Journal of Cellular and Comparative Physiology. (Philadelphia, PA) V.1-66, 1932-65. For publisher information, see JCLLAX

**JCEMAZ** Journal of Clinical Endocrinology and Metabolism. (Williams & Wilkins Co., 428 East Preston St., Baltimore, MD 21202) V.12- 1952-

**JCENA4** Journal of Clinical Endocrinology. (Springfield, IL) V.1-11, 1941-51. For publisher information, see JCEMAZ

**JCGADC** Journal of Clinical Gastroenterology. (Raven Press, 1185 Avenue of the Americas, New York, NY 10036)

**JCGBDF** Journal of Craniofacial Genetics and Developmental Biology. (Alan R. Liss, Inc., 150 Fifth Ave., New York, NY 10011) V.1- 1980-

**JCGEDO** Journal of Cytology and Genetics. (Society of Cytologists and Geneticists, Treasurer, Professor M.S. Chennaveeraiah, Karnatak University, Department of Botany, Dharwar 580003, India) V.1- 1966-

**JCHAAE** Journal of Chemotherapy and Advanced Therapeutics. (Philadelphia, PA) V.3,1926; V.11(2-15), 1934-39

**JCHODP** Journal of Clinical Hematology and Oncology. (Wadley Institutes of Molecular Medicine, 9000 Harry Hines Blvd., Dallas, TX 75235) V.5, No. 3- 1975-

**JCINAO** Journal of Clinical Investigation. (Rockefeller University Press, 1230 York Avenue, New York, NY 10021) V.1- 1924-

**JCLBA3** Journal of Cell Biology. (Rockefeller University Press, 1230 York Avenue, New York, NY 10021) V.12- 1962-

**JCLLAX** Journal of Cellular Physiology. (Alan R. Liss, Inc., 150 Fifth Avenue, New York, NY 10011) V.67- 1966-

**JCLPDE** Journal of Clinical Psychiatry. (Physician Postgraduate Press, Inc., POB 24008, Memphis TN 38124) V.39- 1978-

**JCNDBK** Journal of Clinical Pharmacology and New Drugs. (Stamford, CT) V.11-13, No. 4, 1971-3, For publisher information, see JCPCBR

**JCPAAK** Journal of Clinical Pathology. (British Medical Association, Tavistock Sq., London WC1H 9JR, England) V.1- 1947-

**JCPCBR** Journal of Clinical Pharmacology. (Hall Associates, P.O. Box 482, Stamford, CN 06904) V.13- 1973-

**JCPCDT** Journal of Cardiovascular Pharmacology. (Raven Press, 1140 Ave. of the Americas, New York, NY 10036) V.1- 1979-

**JCPHB8** Journal of Clinical Pharmacology and Journal of New Drugs. (Albany, NY) V.7-10, 1967-70. For publisher information, see JCPCBR

**JCPPAV** Journal of Comparative and Physiological Psychology. (American Psychological Association, 1200 17th St., N.W., Washington, DC 20036) V.40- 1947-

**JCPRB4** Journal of the Chemical Society, Perkin Transactions 1. (Chemical Society, Publications Sales Office, Burlington House, London W1V 0BN, England) No. 1- 1972-

**JCPTA9** Journal of Comparative Pathology and Therapeutics. (Liverpool, England) V.1-74, 1883-1964. For publisher information, see JCVPAR

**JCPYDR** Journal of Clinical Pyschopharmacology. (Williams & Wilkins Co., 428 E. Preston St., Baltimore, MD 21202) V.1- 1981-

**JCREA8** Journal of Cancer Research. (Baltimore, MD) V.1-14, 1916-30. For publisher information, see CNREA8

**JCROD7** Journal of Cancer Research and Clinical Oncology. (Springer-Verlag, Heidelberger Pl. 3, D-1 Berlin 33, Germany) V.93- 1979-

**JCSOA9** Journal of the Chemical Society. (London, England) 1926-65. For publisher information, see JCPRB4

**JCTODH** Journal of Combustion Toxicology. (Technomic Publishing Co., 265 Post Rd. W., Westport, CT 06880) V.3, No. 1- 1976-

**JCUPBN** Journal of Cutaneous Pathology. (Munksgaard, 35 Noerre Soegade, DK-1370 Copenhagen K, Denmark) V.1- 1974-

**JCVPAR** Journal of Comparative Pathology. (Academic Press Inc., Ltd., 24-28 Oval Rd., London NW1 7DX, England) V.75- 1965-

**JDGRAX** Journal of Drug Research. (Drug Research and Control Center, 6, Abou Hazem St., Pyramids Ave., POB 29, Cairo, Egypt) V.2- 1969-

**JDREAF** Journal of Dental Research. (American Association for Dental Research, 734 15th St., NW, Suite 809, Washington, DC 20005) V.1- 1919-

**JEDIDP** Journal of Environmental Biology. (Academy of Environmental Biology, India, 657/5, Civil Lines (South), Muzaffarnagar, 251001, India) V.1- 1980-

**JEEMAF** Journal of Embryology and Experimental Morphology. (P.O. Box 32, Commerce Way, Colchester CO2 8HP, Essex. UK) V.1- 1953-

**JEENAI** Journal of Economic Entomology. (Entomological Society of America, 4603 Calvert Rd., College Park, MD 21201) V.1- 1908-

**JEMAAJ** Journal of the Egyptian Medical Association. (The Egyptian Medical Association, 42 Sharia Kasr-el-aini, Cairo, Egypt) V.11- 1928-

**JEMEAV** Journal of Experimental Medicine. (Rockefeller University Press, 1230 York Avenue, New York, NY 10021) V.1- 1896-

**JEPOEC** Journal of Environmental Pathology, Toxicology and Oncology. (Chem-Orbital, POB 134, Park Forest, IL 60466) V.5(4)- 1984-

**JEPTDQ** Journal of Environmental Pathology and Toxicology. (Park Forest South, IL) V.1-5, 1977-81

**JESEDU** Journal of Environmental Science and Health, Part A: Environmental Science and Engineering. (Marcel Dekker, POB 11305, Church St. Station, New York, NY 10249) V.A11- 1976-

**JETHDA** Journal of Ethnopharmacology. (Elsevier Sequoia SA, POB 851, CH-1001 Lausanne 1, Switzerland) V.1- 1979-

**JETOAS** Journal Europeen de Toxicologie. (Paris, France) V.1-6, 1968-72. For publisher information, see TOERD9

**JETPEZ** Journal of Engineering for Gas Turbines and Power. (American Society of Mechanical Engineers. Order Dept., POB 2300, Fairfield, NJ 07007) V.106- 1984-

**JEZOAO** Journal of Experimental Zoology (Alan. R. Liss, Inc., 150 Fifth Avenue, New York, NY 10011) V.1- 1904-

**JFALAX** Journal of the Faculty of Agriculture, Tottori University. (Maruzen Co., Ltd., Export Dep., P.O. Box 5050, Tokyo Int., 100-31 Tokyo, Japan) V.1- 1951-

**JFDSAZ** Journal of Food Science. (Institute of Food Technologists, Subscrip. Dept., Suite 2120, 221 N. La Salle St., Chicago, IL 60601) V.26- 1961-

**JFIBA9** Journal of Fish Biology. (Academic Press, 24-28 Oval Rd., London NW1 7DX, England) V.1- 1969-

**JFLCAR** Journal of Fluorine Chemistry. (Elsevier Sequoia SA, P.O. Box 851, CH 1001, Lausanne 1, Switzerland) V.1- 1971-

**JFMAAQ** Journal of the Florida Medical Association. (Florida Medical Association, POB 2411, Jacksonville, FL 32203) V.1- 1914-

**JFSCAS** Journal of Forensic Sciences. (American Society for Testing and Materials, 1916 Race St., Philadelphia, PA 19103) V.1- 1956-

**JGAMA9** Journal of General and Applied Microbiology. (Microbiology Research Foundation, c/o Japan Academic Societies Center Bldg., 4-16, Tokyo 113, Japan) V.1- 1955-

**JGHUAY** Journal de Genetique Humaine. (Editeurs Medecine et Hygiene, 78, ave de la Rosaraie, 1211 Geneva 4, Switzerland) V.1- 1952-

**JGMIAN** Journal of General Microbiology (P.O. Box 32, Commerce Way, Colchester CO2 8HP, UK) V.1- 1947-

**JGOBAC** Journal de Gynecologie Obstetrique et Biologie de la Reproduction. (Masson et Cie, Editeurs, 120 Blvd. Saint-Germain, P-75280, Paris 06, France) V.1- 1972-

**JHEMA2** Journal of Hygiene, Epidemiology, Microbiology and Immunology. (Avicenum, Zdravotnicke Nakladatelstvi, Malostranske namesti 28, Prague 1, Czechoslovakia) V.1- 1957-

**JHHBAI** Bulletin of the Johns Hopkins Hospital. (Baltimore, MD) V.1-119, 1889-1966. For publisher information, see JHMJAX

**JHMJAX** Johns Hopkins Medical Journal. (Journals Dept., Johns Hopkins Univ. Press, Baltimore, MD 21218) V.120-151, 1967-82. Discontinued

**JHTCAO** Journal of Heterocyclic Chemistry. (Hetero Corporation, POB 16000 MH, Tampa, FL 33687) V.1- 1964-

**JICSAH** Journal of the Indian Chemical Society. (Indian Chemical Society, 92, Acharya Prafulla Chandra Rd., Calcutta 700009, India) V.5- 1928-

**JIDEAE** Journal of Investigative Dermatology. (Williams & Wilkins Co., 428 E. Preston St., Baltimore, MD 21202) V.1- 1938-

**JIDHAN** Journal of Industrial Hygiene. (Baltimore, MD/New York, NY) V.1-17, 1919-35. For publisher information, see AEHLAU

**JIDIAQ** Journal of Infectious Diseases. (University of Chicago Press, 5801 S. Ellis Ave., Chicago, IL 60637) V.1- 1904-

**JIDOAA** Jikken Dobutsu. Experimental Animals. (Nippon Jikken Dobutsu Kenkyukai, c/o Tokyo Daigaku Ikngaku Kenkyusho, 6-1, Shiroganedai, Minato-ku, Tokyo 108, Japan) V.1- 1952-

**JIDXA3** Journal of the Indiana State Medical Association. (3935 N. Meridian, Indianapolis, IN 46208) V.1- 1908-

**JIHTAB** Journal of Industrial Hygiene and Toxicology. (Baltimore, MD/New York, NY) V.18-31, 1936-49. For publisher information, see AEHLAU

**JIMAAD** Journal of the Indian Medical Association. (Indian Medical Association, 53, Creek Row, Calcutta 700014, India) V.1- 1931-

**JIMRBV** Journal of International Medical Research. (Secy., Cambridge Medical Publications, Ltd., 435/437 Wellingborough Rd., Northampton NN1 4EZ, England) V.1- 1972-

**JIMSAX** Journal of the Irish Medical Association. (Irish Medical Association., I.M.A. House, 10 Fitzwilliam Pl., Dublin, Ireland) V.28- 1951-

**JINCAO** Journal of Inorganic Nuclear Chemistry. (Pergamon Press Ltd., Headington Hill Hall, Oxford OX3 0BW, England) V.1- 1955-

**JISMAB** Journal of the Iowa State Medical Society. (Iowa Medical Society, 1001 Grand Ave., West Des Moines, IA 50265) V.1- 1911-

**JJANAX** Japanese Journal of Antibiotics. (Japan Antibiotics Research Association, 2-20-8 Kamiosaki, Shinagawa-ku, Tokyo, Japan) V.21- 1968-

**JJATDK** JAT, Journal of Applied Toxicology. (Heyden and Son, Inc., 247 S. 41st St., Philadelphia, PA 19104) V.1- 1981-

**JJCREP** Japanese Journal of Cancer Research (Gann). (Elsevier Science Publishers B.V., POB 211, 1000 AE Amsterdam, Netherlands) V.76- 1985-

**JJEMAG** Japanese Journal of Experimental Medicine. (Editorial Office, The Institute of Medical Science, University of Tokyo, Shirokanedai, Minatoku, Tokyo, Japan) V.7- 1928-

**JJIND8** JNCI, Journal of the National Cancer Institute. (U.S. Government Printing Office, Superintendent of Documents, Washington, DC 20402) V.61- 1978-

**JJMCAQ** Japanese Journal of Medical Science and Biology. (National Institute of Health, 2-chome, Kamiosaki, Shinagawa-ku, Tokyo 141, Japan) V.5- 1952-

**JJMDAT** Japanese Journal of Medicine. (Nankodo Co., Ltd., POB 5272, Tokyo International 100-31, Japan) V.1- 1962-

**JJPAAZ** Japanese Journal of Pharmacology. (Nippon Yakuri Gakkai, c/o Kyoto Daigaku Igakubu Yakurigaku Kyoshitu Sakyo-ku, Kyoto 606, Japan) V.1- 1951-

**JKXXAF** Japanese Kokai Tokkyo Koho Patents. (U.S. Patent Office, Science Library, 2021 Jefferson Davis Highway, Arlington, VA 22202)

**JLCMAK** Journal of Laboratory and Clinical Medicine. (C.V. Mosby Co., 11830 Westline Industrial Dr., St. Louis, MO 63141) V.1- 1915-

**JLSMAW** Journal of the Louisiana State Medical Society. (New Orleans, LA) V.105-134, 1953-82

**JMCMAR** Journal of Medicinal Chemistry. (American Chemical Society Pub., 1155 16th St., N.W., Washington, DC 20036) V.6- 1963-

**JMEJAS** Jikeikai Medical Journal. (Pharmacological Institute, The Jikei University School of Medicine, Minato-Ku, Tokyo, Japan) V.1- 1954-

**JMGZAI** Japan Medical Gazette. (Tokyo, Japan) V.1-18(?), 1964-81. Discontinued

**JMIABM** Journal of Medicine and International Medical Abstracts and Reviews. (Hemangini Publications, P.O. Box 16705, Calcutta, India) V.19-31, 1956-66(?)

**JMICAR** Journal of Microscopy. (Blackwell Scientific Publications, Ltd., Osney Mead, Oxford OX2 OEL, England) V.89- 1969-

**JMMIAV** Journal of Medical Microbiology. (Longman Group Journal Division, 4th Ave., Harlow Essex CM20 2JE, England) V.1- 1968-

**JMOBAK** Journal of Molecular Biology. (Academic Press, 111 Fifth Ave., New York, NY 10003) V.1- 1959-

**JMPCAS** Journal of Medicinal and Pharmaceutical Chemistry. (Washington, DC) V.1-5, 1959-62. For publisher information, see JMCMAR

**JMSCA9** Journal of Mental Science. (London, England) V.4-108, 1857-1962. For publisher information, see BJPYAJ

**JMSHAO** Journal of the Mount Sinai Hospital. (New York, NY) V.1-36, 1934-69. For publisher information, see MSJMAZ

**JMSUAT** Journal of the Madras Agricultural Students' Union. (Madras Agricultural Journal, Tamil Nadu Agricultural University Campus, Coimbatore 641003, India) V.1-16, 1912-28

**JMTHBU** Journal of the Medical Association of Thailand. (Medical Association of Thailand, New Pejburi Rd., Bangkok 10, Thailand) V.1- 1918-

**JMXSAE** Journal of Maxillofacial Surgery. (Georg Thieme Verlag, Postfach 732, Herdweg 63, D-7000 Stuttgart 1, Germany)

**JNCIAM** Journal of the National Cancer Institute. (Washington, DC) V.1-60, No. 6, 1940-78. For publisher information, see JJIND8

**JNCSAI** Journal of Cell Science. (P.O. Box 32, Commerce Way, Colchester CO2 8HP. Essex. UK) V.1- 1966-

**JNDRAK** Journal of New Drugs. (Albany, NY) V.1-6, 1961-66. For publisher information, see JCPCBR

**JNEUAY** Journal of Neuropsychiatry. (Chicago, IL) V.1-5, 1959-64. For publisher information, see BENPBG

**JNMAAE** Journal of the National Medical Association. (Appelton and Lange, 25 Van Zant St., E. Norwalk, CT 06855) V.1- 1909-

**JNMDAN** Journal of Nervous and Mental Disease. (Williams & Wilkins Co., 428 E. Preston St., Baltimore, MD 21202) V.3- 1876-

**JNMDBO** Journal of Medicine. (PJD Publications, POB 966, Westbury, NY 11590) V.1- 1970-

**JNNPAU** Journal of Neurology, Neurosurgery and Psychiatry. (British Medical Association, Tavistock Square, London WC1H 9JR, UK) V.7- 1944-

**JNPHAG** Journal de Pharmacologie. (Masson Publishing USA, Inc. (Journals,) 211 E. 43rd St., Suite 1306, New York, NY 10017) V.1- 1970-

**JNRYA9** Journal of Neurology. (Springer-Verlag New York, Inc., Service Center, 44 Hartz Way, Secaucus, NJ 07094) V.207- 1974-

**JNSVA5** Journal of Nutritional Science and Vitaminology. (Business Center Acad. Soc., Japan) V.19- 1973-

**JOALAS** Journal of Allergy, Including Allergy Abstracts. (St. Louis, MO) V.1-47, 1929-71. For publisher information, see JACIBY

**JOANAY** Journal of Anatomy. (Cambridge University Press, The Pitt Building, Trumpington Street, Cambridge CB2 1RP, UK) V.51- 1916-

**JOAND3** Journal of Andrology. (Lippincott/Harper, Journal Fulfillment Dept., 2350 Virginia Ave., Hagerstown, MD 21740) V.1- 1980-

**JOBAAY** Journal of Bacteriology. (American Society for Microbiology, 1913 I St., N.W., Washington, DC 20006) V.1- 1916-

**JOBIAO** Journal of Biochemistry (Tokyo). (Japanese Biochemical Society, Gakkai Senta Bldg., 2-4-16 Yayoi, Bunkyo-Ku, Tokyo 113, Japan or Japan Publication USA or Maruzen) V.1- 1922-

**JOBSDN** Journal of Biological Sciences. (Indian Academy of Sciences, POB 8005, Bangalore 560080, India) V.1- 1979-

**JOCEAH** Journal of Organic Chemistry. (American Chemical Society Pub., 1155 16th St., N.W., Washington, DC 20036) V.1- 1936-

**JOCMA7** Journal of Occupational Medicine. (American Occupational Medical Association, 150 N. Wacker Dr., Chicago, IL 60606) V.1- 1959-

**JODUA2** Journal of Osaka Dental University. (Osaka Dental Univ., 1-47, Kyobashi Higashi-Ku Osaka 540, Japan) V.1- 1967-

**JOENAK** Journal of Endocrinology. (Biochemical Society Publications, P.O. Box 32, Commerce Way, Whitehall Industrial Estate, Colchester CO2 8HP, Essex, England) V.1- 1939-

**JOETD7** Journal of Ethnopharmacology. (Elsevier Sequoia SA, POB 1304, CH-1001, Lausanne, Switzerland) V.1- 1979-

**JOGBAS** Journal of Obstetrics and Gynaecology of the British Commonwealth. (London, England) V.68-81, 1961-74. For publisher information, see BJOGAS

**JOGNAU** Journal of Genetics. (Asit Kr. Bhattacharyya, M.A., 18/1, Barrackpore Trunk Rd, Belghoria, India) V.1- 1910-

**JOHEA8** Journal of Heredity. (American Genetic Association, 818 18th St. N.W., Washington, DC 20006) V.5- 1914-

**JOHYAY** Journal of Hygiene. (Cambridge University Press, P.O. Box 92, Bentley House, 200 Euston Rd., London NW1 2DB, England) V.1- 1901-

**JOIMA3** Journal of Immunology. (Williams & Wilkins Co., 428 E. Preston St., Baltimore, MD 21202) V.1- 1916-

**JOIMD6** Journal of Immunopharmacology. (Marcel Dekker, POB 11305, Church St. Station, New York, NY 10249) V.1- 1978/79-

**JOMAAL** Journal of Mammalogy. (The American Museum of Natural History, Central Park W at 79th St., New York, NY 10024) V.1- 1919-

**JONRA9** Journal of Neurochemistry. (Pergamon Press Ltd., Headington Hill Hall, Oxford OX3 0BW, England) V.1- 1956-

**JONUAI** Journal of Nutrition. (Journal of Nutrition, Subscription Dept., 9650 Rockville Pike, Bethesda, MD 20014) V.1- 1928-

**JOPDAB** Journal of Pediatrics. (C.V. Mosby Co., 11830 Westline Industrial Dr., St. Louis, MO 63141) V.1- 1932-

**JOPHAN** Journal de Physiologie. (Masson et Cie, Editeurs, 120 Blvd. Saint-Germain, P-75280, Paris 06, France) V.39- 1946-

**JOPHDQ** Journal of Pharmacobio-Dynamics. (Pharmaceutical Society of Japan, 12-15-501, Shibuya 2-chome, Shibuya-Ku, Tokyo 150, Japan) V.1- 1978-

**JOPRAJ** Journal of Periodontology. (American Academy of Periodontology, 211 E. Chicago Ave., Chicago, IL 60611) V.1- 1930-

**JOPSAM** Journal of Psychology. (Journal Press, 2 Commercial St., Provincetown, MA 02657) V.1- 1935/36-

**JORCAI** Journal of Organometallic Chemistry. (Elsevier Sequoia SA, POB 851, CH-1001 Lausanne l, Switzerland) V.1- 1963-

**JOSCDQ** Journal of SCCJ (Society of Cosmetic Chemists of Japan). (Nippon Keshohin Gijutsushakai, c/o Shiseido, 7-5-5 Ginza, Chuoku, Tokyo, 104, Japan) V.10- 1976-

**JOSUA9** Journal of Oral Surgery. (American Dental Association, 211 E. Chicago Ave., Chicago, IL 60611) V.1-16, 1943-58; V.23- 1965-

**JOUOD4** Journal of UOEH (University of Occupational Environmental Health). (University of Occupational and Environmental Health, 1-1 Iseigaoka, Yahata-nishi-ku, Kitakyushu, 807, Japan) V.1- 1979-

**JOURAA** Journal of Urology. (Williams & Wilkins Co., 428 E. Preston St., Baltimore, MD 21202) V.1- 1917-

**JPBAA7** Journal of Pathology and Bacteriology. (London, England) V.1-96, 1892-1968. For publisher information, see JPTLAS

**JPBEAJ** Journal de Pharmacie de Belgique. (Masson et Cie, Editeurs, 120 Blvd. Saint-Germain, P-75280, Paris 06, France) V.1- 1945/46-

**JPCAAC** Journal of the Air Pollution Control Association. (Air Pollution Control Association, 4400 5th Ave., Pittsburgh, PA 15213) V.5- 1955-

**JPCEAO** Journal fuer Praktische Chemie. (Johann Ambrosius Barth Verlag, Postfach 109, DDR-701, Leipzig, Germany) V.1- 1834- (Several new series, but continuous vol. nos. also used)

**JPEMAO** Journal of Perinatal Medicine. (Walter de Gruyter, Inc., 200 Sawmill River Rd., Hawthorne, MA 10532) V.1- 1973

**JPETAB** Journal of Pharmacology and Experimental Therapeutics. (Williams & Wilkins Co., 428 E. Preston St., Baltimore, MD 21202) V.1- 1909/10-

**JPFCD2** Journal of Environmental Science and Health, Part B: Pesticides, Food Contaminants, and Agricultural Wastes. (Marcel Dekker, POB 11305, Church St. Station, New York, NY 10249) V.B11- 1976-

**JPHAA3** Journal of the American Pharmaceutical Association. (American Pharmaceutical Association, 2215 Constitution Ave., N.W., Washington, DC 20037) V.1-28, 1912-39; New Series V.1-17, 1961-77

**JPHYA7** Journal of Physiology. (Cambridge University Press, P.O. Box 92, Bentley House, 200 Euston Rd., London NW1 2DB, England) V.1- 1878-

**JPIFAN** Japan Pesticide Information. (Japan Plant Protection Association, 1-43-11, Komagome, Toshima-ku, Tokyo 170, Japan) No. 1-1969-

**JPMRAB** Japanese Journal of Medical Sciences, Part IV: Pharmacology. (Tokyo, Japan) V.1-16, 1926-43. For publisher information, see JJPAAZ

**JPMSAE** Journal of Pharmaceutical Sciences. (American Pharmaceutical Association, 2215 Constitution Ave., N.W., Washington, DC 20037) V.50- 1961-

**JPPGAR** Journal de Physiologie et de Pathologie Generale. (Paris, France) V.1-38, 1899-1945. For publisher information, see JOPHAN

**JPPMAB** Journal of Pharmacy and Pharmacology. (Pharmaceutical Society of Great Britain, 1 Lambeth High Street, London SEI 5JN, England) V.1- 1949-

**JPROAR** Journal of Protozoology. (Society of Protozoologists, P.O. Box 368, Lawrence, KS 66044) V.1- 1954-

**JPTLAS** Journal of Pathology. (Longman Group Ltd., Subscriptions Journals Department, Fourth Avenue, Harlow, Essex, CM19 5AA, UK) V.97- 1969-

**JRARAX** Journal of Radiation Research. (c/o Tokai Daigaku Igakubu Bunshi Seibutsugaku Kyoshitsu, Tenbodai, Isehara, 259-11, Japan) V.1- 1960-

**JRBED2** Journal of Reproductive Biology and Comparative Endocrinology. (P.G. Institute of Basic Medical Sciences, Dept. of Endocrinology, Taramani, 600 113, India) V.1- 1981-

**JRFSAR** Journal of Reproduction and Fertility, Supplement. (Journal of Reproduction and Fertility Ltd., 22 New Market Rd., Cambridge CB5 8D7, England) No. 1- 1966-

**JRHUA9** Journal of Rheumatology. (Business Office, Suite 420, 920 Yonge St., Toronto, Ontario M4W 3J7 Canada) V.1- 1974-

**JRIHDC** Journal of Research in Indian Medicine Yoga and Homoeopathy. (Banaras Hindu University, Varanasi-5, India) V.11- 1976-

**JRIMAO** Journal of Research in Indian Medicine. (Varanasi, India) V.1-10, 1966-75. For publisher information, see JRIHDC

**JRMSAS** Journal of the Royal Microscopical Society. (London, England) V.47-88, 1927-68. For publisher information, see JMICAR

**JRPFA4** Journal of Reproduction and Fertility. (Journal of Reproduction and Fertility Ltd., 22 New Market Rd., Cambridge CB5 8D7, England) V.1- 1960-

**JRPMAP** Journal of Reproductive Medicine. (Box 56, 5841 Maryland Ave., Chicago, IL 60637) V.3- 1969-

**JRSMD9** Journal of the Royal Society of Medicine. (Academic Press Inc., Ltd., 24-28 Oval Road, London NW1 7DX, England) V.71- 1978-

**JSCCA5** Journal of the Society of Cosmetic Chemists. (Society of Cosmetic Chemists, 50 E. 41st St., New York, NY 10017) V.1- 1947-

**JSFAAE** Journal of the Science of Food and Agriculture. (Society of Chemical Industry, 14 Belgrave Sq., London SW1X 8PS, England) V.1- 1950-

**JSGRA2** Journal of Surgical Research. (Academic Press, 111 Fifth Ave., New York, NY 10003) V.1- 1961-

**JSICAZ** Journal of Scientific and Industrial Research, Section C: Biological Sciences. (New Delhi, India) V.14-21, 1955-62. For publisher information, see IJEBA6

**JSOMBS** Journal of the Society of Occupational Medicine. (John Wright and Sons, 42-44 Triangle W., Bristol BS8 1EX, England)

**JSONAU** Journal of Surgical Oncology. (Alan R. Liss, Inc., 150 5th Ave., New York, NY 10011) V.1- 1969-

**JSONDX** Journal of Soviet Oncology. (New York) V. 1-5, 1980-84

**JSTBBK** Journal of Steroid Biochemistry. (Pergamon Press Ltd., Headington Hill Hall, Oxford OX3 0BW, England) V.1- 1969-

**JSXRAJ** Journal of Sex Research. (Society for the Scientific Study of Sex, Inc., Mt. Royal and Guilford Aves., Baltimore, MD 21202) V.1- 1965-

**JTASAG** Journal of the Tennessee Academy of Science. (Tennessee Academy of Science, Box 153, George Peabody College, Nashville, TN 37203) V.1- 1926-

**JTBIDS** Journal of Thermal Biology. (Pergamon Press Ltd., Headington Hill Hall, Oxford OX3 0BW, England) V.1- 1975-

**JTCEEM** Journal de Toxicologie Clinique et Experimentale. (SPPIF, B. P. 22, F-41353 Vineuil, France) V.5- 1985-

**JTCSAQ** Journal of Thoracic and Cardiovascular Surgery. (C.V. Mosby Co., 11830 Westline Industrial Dr., St. Louis, MO 63141) V.38- 1959-

**JTCTDW** Journal of Toxicology, Clinical Toxicology. (Marcel Dekker, POB 11305, Church St. Station, New York, NY 10249) V.19- 1982-

**JTEHD6** Journal of Toxicology and Environmental Health. (Hemisphere Publ., 1025 Vermont Ave., N.W., Washington, DC 20005) V.1- 1975/76-

**JTMHA9** Journal of Tropical Medicine and Hygiene. (Staples and Staples Ltd., 94-96 Wigmore St., London, England) V.10- 1907-

**JTOTDO** Journal of Toxicology, Cutaneous and Ocular Toxicology. (Marcel Dekker, POB 11305, Church St. Station, New York, NY 10249) V.1- 1982-

**JTSCDR** Journal of Toxicological Sciences. (Editorial Office, Higashi Nippon Gakuen Univ., 7F Fuji Bldg., Kita 3, Nishi 3, Sapporo 060, Japan) V.1- 1976-

**JUIZAG** Juntendo Igaku. Juntendo Medicine. (Juntendo Igakkai, 2-1-1, Hongo, Bunkyo-ku, Tokyo, 113, Japan) V.1- 1955-

**JULRA7** Journal of Ultrastructure Research. (Academic Press, 111 Fifth Ave., New York, NY 10003) V.1- 1975-

**JWMAA9** Journal of Wildlife Management. (Wildlife Society, Suite 611, 7101 Wisconsin Ave., Washington, DC 20014) V.1- 1937-

**JZKEDZ** Jitchuken, Zenrinsho Kenkyuho. Central Institute for Experimental Animals, Preclinical Reports. (The Institute, 1433 Nogawa, Takatsu-Ku, Kawasaki 211, Japan) V.1- 1974/75-

**KAEMAW** Kitasato Archives of Experimental Medicine. (Kitasato Institute, 5-9-1 Shirokane, Minato-ku, Tokyo 108, Japan) V.1- 1917-

**KAIZAN** Kaibogaku Zasshi. Journal of Anatomy. (Nihon Kaibo Gakkai, c/o Tokyo Daigaku Igakubu Kaibogaku Kyoshitsu, 7-3-1, Hongo, Bunkyo-ku, Tokyo, Japan) V.1- 1928-

**KAMJDW** Kawasaki Medical Journal. (Kawasaki Medical School, Kurashiki 701-01, Japan) V.1- 1975-

**KBAMAJ** Kosmicheskaya Biologiya I Aviakosmicheskaya Meditsina Space Biology and Aerospace Medicine. (Mezhdunarodnaya Kniga, Kuznetskii Most 18, Moscow G-200, USSR) V.8- 1974-

**KBMEAL** Kosmicheskaya Biologiya i Meditsina. Space Biology and Medicine. (Moscow, USSR) V.1-7, 1967-73. For publisher information, see KBAMAJ

**KCRZAE** Kauchuk i Rezina. (v/o Mezhdunarodnaya Kniga, Kuznetskii Most 18, Moscow G-200, USSR) V.11- 1937-

**KDIZAA** Kagoshima Daigaku Igaku Zasshi. Medical Journal of Kagoshima University. (Kagoshima Daigaku Igakkai, 1208-1, Usuki-cho, Kagoshima 890, Japan) V.1- 1945-

**KDPU\*\*** Beitrag zur Toxikologischen Wirkung Technischer Losungsmittel, Otto Klimmer Dissertation. (Pharmakologischen Institut der Universitat Wurzburg, Germany, 1937)

**KEKHA7** Koshu Eiseiin Kenkyu Hokoku. Bulletin of the Institute of Public Health. (Kokuritsu Koshu Eiseiin, 4-6-1 Shirokanedai Minato-ku, Tokyo, 108, Japan) V.1- 1951-

**KEKHB8** Kanagawa-ken Eisei Kenkyusho Kenkyu Hokoku. Bulletin of Kanagawa Prefectural Public Health Laboratories. (Kanagawa Prefectural Public Health Laboratories, 52-2, Nakao-cho, Asahi-ku, Yokohama 221, Japan) No. 1- 1971-

**KFIZAO** Kyoto-furitsu Ika Daigaku Zasshi. Journal of the Kyoto Prefectural School of Medicine. (Kyoto-furitsu Ika Daigaku Igakkai, Hirokoji, Kawaramachidori, Kamiyoku, Kyoto, Japan) V.1- 1927-

**KGGZAL** Koku Geka Gakkai Zasshi. Oral Surgery. (Tokyo, Japan)

**KHFKDF** K'o Hsueh Fa Chan Yueh K'an. Progress in Sciences. (Kuo Chia K'o Hsueh Wei Yuan Hui, 2 Canton St., Taipei 107, Taiwan) V.1- 1973-

**KHFZAN** Khimiko-Farmatsevticheskii Zhurnal. Chemical Pharmaceutical Journal. (v/o Mezhdunarodnaya Kniga, Kuznetskii Most 18, Moscow G-200, USSR.) V.1- 1967-

**KHZDAN** Khigiena i Zdraveopazvane. (Hemus, blvd Russki 6, Sofia, Bulgaria) V.9- 1966-

**KIDZAK** Kansai Ika Daigaku Zasshi. Journal of the Kansai Medical School. (Kansai Ika Daigaku Igakkai, 1, Fumizono-cho, Moriguchi 570, Osaka, Japan) V.8- 1956-

**KIHSDM** Guangxi Yixue. Kuanghsi Medicine. (Kuang-hsi I Hsueh K'o Hsueh Ch'ing Pao Yen Chin So, 19 T'ien T'ao Lu, Nan-ning, Kuang-hsi, People's Republic of China) V.?- 1979-

**KIKNAJ** Kokuritsu Idengaku Kenkyusho Nempo. Annual Report of the National Institute of Genetics. (Kokuritsu Idengaku Kenkyusho, 1111 Yata, Mishima, Shizuoka-ken 411, Japan) No. 1- 1949-

**KIZAAL** Kurume Igakkai Zasshi. Journal of the Kurume Medical Association. (Kurume Igakkai, c/o Kurume Daigaku Igakubu, 67, Asahi-machi, Kurume 830, Japan) V.9- 1946-

**KIZSB8** Kyorin Igakukai Zasshi. Journal of the Kyorin Medical Society. (6-20-2 Shinkawa, Mitaka City, Tokyo, Japan) V.1- 1970-

**KJMDA6** Kobe Journal of the Medical Sciences. (Kobe Daigaku Igakubu, Editorial Board, Kobe, Japan) V.1- 1951-

**KJMEA9** Keio Journal of Medicine. (Keio University, School of Medicine, 35 Shinano-machi, Shinjuku-Ku, Tokyo 160, Japan) V.1- 1952-

**KJMSAH** Kyushu Journal of Medical Science. (Fukuoka, Japan) V.6-15, 1955-64. Discontinued

**KLWOAZ** Klinische Wochenschrift. (Springer-Verlag, Heidelberger Pl. 3, D-1 Berlin 33, Germany) V.1- 1922-

**KNZOAU** Kanzo. Liver. (Nippon Kanzo Gakkai, c/o Toyo Bunko, 28-21, 2-chome, Hon Komagome, Bunkyo-ku, Tokyo 113, Japan) V.1- 1960-

**KOBUA3** Kobunshi. High Polymers. (Kobunshi Gakkai, 5-12-8, Ginzai, Chuo-Ku, Tokyo 104, Japan) V.1- 1952-

**KODAK\*** Kodak Company Reports. (343 State St., Rochester, NY 14650)

**KOKABN** Kobunshi Kako. Polymer Applications. (Kobunshi Kankokai, Chiekoin Maruta-machi Kundaru, Kamigyo-ku, Kyoto, 602, Japan) V.13- 1964-

**KONODE** Kongetsu No Noyaku. Agricultural Chemicals Monthly. (Kagoku Kogyo Nipposha, 3-19-16 Shibaura, Minato-ku, Tokyo 108, Japan) V.1- 1953(?)-

**KorCJ#** Personal Communication to NIOSH, from C.J. Korpics, Sherwin-Williams Chemicals, 1310 Expressway Drive, Toledo, OH 43608, August 22, 1974

**KPJBAR** Kalikasan. The Philippine Journal of Biology. (National Publishing Cooperative Inc., 2nd Fl., Santander Bldg., 20 M Hemady St. Cor Aurora Blvd., Quezon City, Philippines) V.1- 1972-

**KRANAW** Krankheitsforschung. (Leipzig, Germany) V.1-9, 1925-32. Discontinued

**KRKRDT** Kriobiologiya i Kriomeditsina. (Izdatel'stvo Naukova Dumka, ul Repina 3, Kiev, USSR) No. 1- 1975-

**KRMJAC** Kurume Medical Journal. (Kurume Igakkai, c/o Kurume Daigaku Igakubu, 67, Asahi-machi, Kurume, Japan) V.1- 1954-

**KSGZA3** Kyushu Shika Gakkai Zasshi. Journal of the Kyushu Dental Society. (c/o Kyushu Shika Daigaku, 2-6-1 Manazuru, Kokurakita-ku, Kitakyushu, Japan) V.1- 1933-35; 1939-40; 1951-

**KSKZAN** Khimiya v Sel'skom Khozyaistve. Chemistry in Agriculture. (v/o Mezhdunarodnaya Kniga, Kuznetskii Most 18, Moscow G-200, USSR.) No. 1- 1963-

**KSRNAM** Kiso to Rinsho. Clinical Report. (Yubunsha Co., Ltd., 1-5, Kanda Suda-Cho, Chiyoda-ku, KS Bldg., Tokyo 101, Japan) V.1- 1960-

**KTHYAC** Kuo Li Tai-Wan Ta Hsueh I Hsueh Yuan Yen Chiu Pao Kao. The memoirs of the College of Medicine of the National Taiwan University. (Kuo Li Tai-Wan Ta Hsueh I Hsueh Yuan, Taipei, Taiwan) 1947-

**KTUNAA** K'at'ollik Taehak Uihakpu Nonmunjip. Journal of Catholic Medical College. (Catholic Medical College, Seoul, South Korea) V.1- 1957-

**KTUWDD** Koryo Taehakkyo Uikwa Taehak Nonmunjip. Collection of Papers of the Medical School, Korea University

**KUMJAX** Kumamoto Medical Journal. (Kumamoto Daigaku Igakubu, Library, Kumamoto, Japan) V.1- 1938-

**KYDKAJ** Kyoritsu Yakka Daigaku Kenkyu Nempo. Annual Report of the Kyoritsu College of Pharmacy. (Kyoritsu Yakka Daigaku, 1-5-30 Shibakoen Minato-ku, Tokyo, Japan) V.1- 1955-

**LacHB#** Personal Communication from Mr. H.B. Lackey, Chemical Products Div., Crown Zellerbach, Camas, WA 98607, to Dr. H.E. Christensen, NIOSH, Rockville, MD 20852, June 9, 1978

**LAINAW** Laboratory Investigation. (Williams & Wilkins Co., 428 E. Preston St., Baltimore, MD 21202) V.1- 1952-

**LAKAA3** Laekartidningen. Medical News. (Swedish Medical Association, POB 5610, 11486 Stockholm 5, Sweden) V.62- 1965-

**LAMEDS** LARC Medical. (19 Bis, Rue d' Inkermann, 59000 Lille, France) V.1- 1981-

**LANCAO** Lancet. (7 Adam St., London WC2N 6AD, England) V.1- 1823-

**LAPPA5** Lavori dell'Istituto di Anatomia e Istologia Patologica della Universita Degli Studi Perugia. (Istituto di Anatomia e Isologia Patologica, Caselle Postale 327, 06100 Perugia, Italy) V.1- 1939-

**LARYA8** Laryngoscope. (222 Pine Lake Rd., Collinsville, IL 62234) V.1- 1896-

**LBANAX** Laboratory Animals. (Biochemical Society Book Depot, POB 32, Commerce Way, Colchester, Essex CO2 8HP, England) V.1- 1967-

**LBASAE** Laboratory Animal Science. (American Association for Laboratory Animal Science, 210 N. Hammes Ave., Suite 205, Joliet, IL 60435) V.21- 1971-

**LDBU\*\*** Langer Dissertation. (Breslow, 1932)

**LDTU\*\*** Narkoseversuche mit Hoheren Alkoholen und Stickstoffderivaten, Fritz Leube Dissertation. (Pharmakologischen Institut der Universitat Tubingen, Germany, 1931)

**LIFSAK** Life Sciences. (Pergamon Press, Maxwell House, Fairview Park, Elmsford, NY 10523) V.1-8, 1962-69; V.14- 1974-

**LilPW#** Personal Communication from P.W. Lilley, Sun Company, Inc., P.O. Box 1135, Marcus Hook, PA 19061, to Richard Lewis, NIOSH, November 1, 1983

**LitL##** Personal Communication from Larry Little, Occidental Chemical Corp., Hooker Chemical Center, 360 Rainbow Blvd. South, Box 728, Niagara Falls, NY 14302, to Doris Sweet, NIOSH, Cincinnati, OH 45226, May 6, 1985

**LITRC\*** Literature Research Co. Translation. (Annandale, VA)

**LMSED6** I.E.R.S. Monograph Series. Laboratories d'Etudes et de Recherches Synthelabo Monograph Series. (Raven Press, 1185 Ave. of Americas, New York, NY 10036) V.1- 1983-

**LONZA#** Personal Communication from LONZA Ltd., CH-4002, Basel, Switzerland, to NIOSH, Cincinnati, OH 45226

**LPDSAP** Lipids. (American Oil Chemists' Society, 508 South Sixth St., Champaign, IL 61820) V.1- 1966-

**LPPTAK** Labo-Pharma-Problemes et Techniques (19, rue Louis-le-Grand, Paris, France) V.12- 1965-

**LROTDW** Laryngo-Rhinologie und Otolaryngologie. (Georg Thieme Verlag, Postfach 732, Herdweg 63, 7000 Stuttgart, Germany) V.54- 1975-

**LSBGAY** Life Sciences, Part 2: Biochemistry, General and Molecular Biology. (New York, NY) V.9-13, 1970-73. For publisher information, see LIFSAK

**LSPPAT** Life Sciences, Part 1: Physiology and Pharmacology. (New York, NY) V.9-13, 1970-73. For publisher information, see LIFSAK

**LYPHAD** Lyon Pharmaceutique. (Edition Paul Chatelain, 63 rue de la Republique, 69 Lyons 2, France) V.1- 1950-

**MACPAJ** Mayo Clinic Proceedings. (Mayo Foundation, Plummer Bldg., Mayo Clinic, Rochester, MN 55901) V.39- 1964-

**MADCAJ** Medical Annals of the District of Columbia. (Washington, DC) V.1- 43, 1932-74. Discontinued

**MAGDA3** Mechanisms of Ageing and Development. (Elsevier Sequoia SA, POB 851, Ch 1001, Lausanne 1, Switzerland) V.1- 1972-

**MAGJAL** Malaysian Agricultural Journal. (Ministry of Agriculture and Fisheries, Business Mgr., Kuala Lumpur, Malaysia) V.45- 1965-

**MahWM#** Personal Communication from W.M. Mahlburg, Hopkins Agricultural Chemical Co., P.O. Box 7532, Madison, WI 53707, to NIOSH, Cincinnati, OH 45226, November 16, 1982

**MAIKD3** Maikotokishin (Tokyo). Mycotoxin. (Maikotokishin Kenkyukai, c/o Tokyo Rika Daigaku Yakugakubu, 12 Funagawara-machi, Ichigaya, Shinjuku-ku, Tokyo 162, Japan) No. 1- 1975-

**MAIZAB** Manshu Igaku Zasshi. Manthuria Medical Journal. (Dairen/Shimmeicho, South Manchuria) V.1-33, 1923-40. Discontinued

**MarJV#** Personal Communication from Josef V. Marhold, VUOS, 539-18, Pardubice, Czechoslovakia, to the Editor of RTECS, Cincinnati, OH, March 29, 1977

**MASODV** Masui to Sosei. Anesthesia and Resuscitation. (Hiroshima Masui Igakkai Hiroshima Daigaku Igakubu Masuigaku Kyoshitsu, 1-2-3 Kasumi, Hiroshima 734, Hiroshima, Japan) V.6- 1970-

**MCBIA7** Microbios. (Faculty Press, 88 Regent St., Cambridge, England) V.1- 1969-

**MccSB#** Personal Communication from Susan B. McCollister, Dow Chemical U.S.A., Midland, MI 48640, to NIOSH, Cincinnati, OH 45226, June 15, 1984

**MCEBD4** Molecular and Cellular Biology. (American Society for Microbiology, 1913 I St., NW, Washington, DC 20006) V.1- 1981-

**MDACAP** Medicamentos de Actualidad. (Medicamentos de Actualidad, Apartado de Correos 540, Barcelona, Spain) V.1- 1965-

**MDCHAG** Medicinal Chemistry: A Series of Monographs. (Academic Press, 111 5th Ave., New York, NY 10003) V.1- 1963-

**MDMIAZ** Medycyna Doswiadczalna i Mikrobiologia. (Ars Polona-RUCH, POB 1001, 1, P-00068 Warsaw, 1, Poland) V.1- 1949- For English Translation, see EXMMAV

**MDREP\*** U.S. Army, Chemical Corps Medical Division Reports. (Army Chemical Center, MD)

**MDSR\*\*** U.S. Army, Chemical Corps Medical Division Special Report. (Army Chemical Center, MD)

**MDZEAK** Medizin und Ernaehrung. (Stuttgart, Germany) V.1-13, 1959-72. Discontinued

**MECHAN** Medicinal Chemistry, A Series of Reviews. (New York, NY) V.1-6, 1951-63. Discontinued

**MEDIAV** Medicine. (Williams & Wilkins, 428 E. Preston St., Baltimore, MD 21202) V.1- 1922-

**MEHYDY** Medical Hypotheses. (Churchill Livingstone Inc., 19 W. 44 St., New York, NY 10036) V.1- 1975-

**MEIEDD** Merck Index. (Merck and Co., Inc., Rahway, NJ 07065) 10th ed. 1983-

**MEKLA7** Medizinische Klinik. (Urban and Schwarzenberg, Pettenkoferst 18, D-8000 Munich 15, Germany) V.1- 1904-

**MELAAD** Medicina del Lavoro. Industrial Medicine. (Via S. Barnaba, 8 Milan, Italy) V.16- 1925-

**MEMOAQ** Medizinische Monatsschrift. (Stuttgart, Germany) V.1-31, 1947-77. Discontinued

**MEPAAX** Medycyna Pracy. Industrial Medicine. (ARs-Polona-RUSH, POB 1001, 00-068 Warsaw 1, Poland) V.1- 1950-

**MEPHDN** Medical Pharmacy. (Daiichi Seiyaku K.K., 3-14-10 Nihonbashi, Chuo-ku, Tokyo 103, Japan) V.1- 1966-

**METAAJ** Metabolism, Clinical and Experimental. (Grune and Stratton, Inc., 111 Fifth Ave., New York, NY 10003) V.1- 1952-

**METRA2** Medecine Tropicale (Marseille). (Parc du Pharo, 13007 Marseille, France) V.1- 1941-

**MEWEAC** Medizinische Welt. (F.K. Schattauer Verlag, Postfach 2945, D-7000 Stuttgart 1, Germany) V.1-18, 1927-1944; V.1-1950-

**MEXPAG** Medicina Experimentalis. (Basel, Switzerland) V.1-11, 1959-64; V.18-19, 1968-69. For publisher information, see JNMDBO

**MFEPDX** Methods and Findings in Experimental and Clinical Pharmacology. (Methods and Findings, Sub. Dept., Apartado Correos 1179, Barcelona, Spain) V.1- 1979-

**MFLRA3** Mededelingen van de Faculteit Landbouwwetenschappen, Rijksuniversiteit Gent. Communications of the Faculty of Agricultural Sciences, State University of Ghent. (Bibliotheek, Faculteit Labdbouwwettenschappen, Rijksuniversiteit Gent, Coupure 533, B-9000, Ghent, Belgium) V.35- 1970-

**MGBUA3** Microbial Genetics Bulletin. (Ohio State University, College of Biological Science, Dept. of Microbiology, 484 W. 12th Ave., Columbus, OH 43210) No. 1- 1950-

**MGGEAE** Molecular and General Genetics. (Springer-Verlag, Heidelberger Pl. 3, D-1 Berlin 33, Germany) V.99- 1967-

**MGLHAE** Mitteilungen Aus Dem Gebiete Der Lebensmitteluntersuchung und Hygiene. (Eidgenoessiche Drucksachen und Materialzentrale, 3000 Bern, Switzerland) V.1- 1910-

**MGONAD** Magyar Onkologia. Hungarian Onkology. (Kultura, P.O. Box 149, H-1389 Budapest, Hungary) V.1- 1957-

**MIBLAO** Microbiology (Moscow). (Plenum Publishing Corp., 233 Spring St., New York, NY 10013) V.26- 1957-

**MIDAD4** NIDA Research Monograph. (National Institute on Drug Abuse, Division of Research, 5600 Fishers Lane, Rockville, MD 20857) No. 1- 1975-

**MIFAAB** Minerva Farmaceutica. (Turin, Italy) V.1-13, 1952-64. Discontinued

**MIIMDV** Microbiology and Immunology. (Center for Academic Publications Japan, 4-16, Yayoi 2-chome, Bunkyo-ku, Tokyo 113, Japan) V.21- 1977-

**MIKBA5** Mikrobiologiya. (v/o Mezhdunarodnaya Kniga, Kuznetskii Most 18, Moscow G-200, USSR) V.1- 1932-

**MILEDM** Microbios Letters. (Faculty Press, 88 Regent St., Cambridge, England) V.1- 1976-

**MIMEAO** Minerva Medica. (Edizioni Minerva Medica, Casella Postale 491, Turin, Italy) V.1- 1909-

**MIPEA5** Minerva Pediatrica. (Edizioni Minerva Medica, Cassella Postale 491, I-10126 Turin, Italy) V.1- 1949-

**MIVRA6** Microvascular Research. (Academic Press, 111 Fifth Ave., New York, NY 10003) V.1- 1968-

**MJAUAJ** Medical Journal of Australia. (P.O. Box 116, Glebe 2037, NSW 2037, Australia) V.1- 1914-

**MJDHDW** Mukogawa Joshi Daigaku Kiyo, Yakugaku Hen. Bulletin of Mukogawa Women's College, Food Science Series. (Mukogawa Joshi Daigaku, 6-46, Ikebiraki-cho, Nishinomiya 663, Japan) No. 19- 1972-

**MJOUAL** Medical Journal of Osaka University. (The University, 33, Joan-cho, Kita-Ku, Osaka, Japan) V.1- 1949-

**MLDCAS** Medecine Legale et Dommage Corporel. (Paris, France) V.1-7, 1968-74. Discontinued

**MLSR\*\*** U.S. Army, Chemical Corps Medical Laboratories Special Reports. (Army Chemical Center, MD)

**MMAPAP** Mycopathologia et Mycologia Applicata. (The Hague, Netherlands) V.5-54, No. 4, 1950-74, For publisher information, see MYCPAH

**MMDPA6** Materia Medica Polona (English Edition). (Ars Polona-RUCH, P.O. Box 1001, P-00 068 Warsaw, 1, Poland) V.1- 1969-

**MMEDA9** Military Medicine. (Association of Military Surgeons of the United States, Box 104, Kensington, MD 20795) V.116- 1955-

**MMIYAO** Medical Microbiology and Immunology. (Springer-Verlag, Heidelberger Pl. 3, D-1 Berlin 33, Germany) V.157- 1971-

**MMJJAI** Mie Medical Journal. (Mie Medical Society, Mie Prefectual Univ., School of Medicine, Tsu, Japan) V.3- 1952-

**MMWOAU** Muenchener Medizinische Wochenschrift. (Munich, Germany) V.33-115, 1886-1973

**MOLAAF** Monatsschrift fuer Ohrenheilkunde und Laryngo-rhinologie. (Vienna, Austria) V.1-108, 1867-1974. For publisher information, see LROTDW

**MONS\*\*** Monsanto Co. Toxicity Information. (Monsanto Industrial Chemicals Co., Bancroft Bldg., Suite 204, 3411 Silverside Rd., Wilmington, DE 19810)

**MOPMA3** Molecular Pharmacology. (The American Society for Pharmacology and Experimental Therapeutics, 9650 Rockville Pike, Bethesda, MD 20014) V.1- 1965-

**MosJN#** Personal Communication from J.N. Moss, Toxicology Department, Rohm and Haas Co., Spring House, PA 19477, to R. J. Lewis, Sr., NIOSH, Cincinnati, OH 45226, August 15, 1979

**MPHEAE** Medicina et Pharmacologia Experimentalis. (Basel, Switzerland) V.12-17, 1965-67. For publisher information, see PHMGBN

**MPPBAB** Meditsinskaya Parazitologiya i Parazitarnye Bolezni. Medical Parasitology and Parasitic Diseases. (v/o Mezhdunarodnaya Kniga, Kuznetskii Most 18, Moscow G-200, USSR.) V.1- 1932-

**MPPPBK** Modern Problems of Pharmacopsychiatry. (S. Karger AG, Postfach, CH-4009, Basel, Switzerland) V.1- 1968-

**MRBUDF** MARDI Research Bulletin. (MARDI, Secy., Publication Committee, POB 208, Serdang, Selangor, Malaysia) V.1- 1973-

**MRCSAB** Medical Research Council, Special Report Series. (Her Majesty's Stationery Office, P.O. Box 569, London SE1 9NH, England) SRS1- 1915-

**MRLAB3** Mededelingen Rijksfaculteit Landbouwwetenschappen, Gent. Communications of the State University of Agricultural Sciences, Ghent. (Ghent, Belgium) V.31-34, 1966-69. For publisher information, see MFLRA3

**MRLEDH** Mutation Research Letters. (Elsevier/North-Holland Biomedical Press, POB 211, 1000 AE Amsterdam, Netherlands)

**MRLR\*\*** U.S. Army, Chemical Corps Medical Laboratories Research Reports. (Army Chemical Center, Edgewood Arsenal, MD)

**MSCREJ** Medical Science Research. (Elsevier Applied Science Publishing Ltd., Crown House, Linton Rd., Barking, Essex IG11 8JU, UK) V.15 1987-

**MSERDS** Microbiology Series. (Marcel Dekker, Inc., POB 11305, Church St. Station, New York, NY 10249) V.1- 1973-

**MSJMAZ** Mount Sinai Journal of Medicine, New York. (The Annenberg Building, Room 10-35, 5th Ave. and 100th St., New York, NY 10029) V.37- 1970-

**MUREAV** Mutation Research. (Elsevier Science Publications B.V., POB 211, 1000 AE Amsterdam, Netherlands) V.1- 1964-

**MUTAEX** Mutagenesis. (IRL Press Ltd. 1911 Jefferson Davis Highway, Suite 907, Arlington, VA 22202) V.1- 1986-

**MVCRB3** "Fluorescent Whitening Agents, Proceedings of A Symposium." MVC-Report, Miljoevardscentrum, Stockholm Center for Environmental Sciences, No. 2, 1973

**MVMZA8** Monatshefte fuer Veterinaermedizin. (VEB Gustav Fischer Verlag, Postfach 176, Villengang 2, 69 Jena, Germany) V.1- 1946-

**MYCPAH** Mycopathologia. (Dr. W. Junk bv Publishers, POB 13713, 2501 ES The Hague, Netherlands) V.1- 1938-

**MZHUDX** Mikrobiologicheskii Zhurnal (Kiev). Journal of Microbiology. (v/o Mezhdunarodnaya Kniga, Kuznetskii Most 18, Moscow G-200, USSR.) V.40- 1978-

**MZUZA8** Meditsinskii Zhurnal Uzbekistana (v/o Mezhdunarodnaya Kniga, Kuznetskii Most 18, Moscow G-200, USSR.) No. 1- 1957-

**NAGZAC** Nagasaki Igakkai Zasshi. Journal of Nagasaki Medical Association. (Nagasaki Igakkai, c/o Nagasaki Daigaku Igakubu, 12-4 Sakamoto-machi, Nagasaki 852, Japan) V.1- 1923-

**NAHRAR** Nahrung. Chemistry, Biochemistry, Microbiology, Technology, Nutrition. (Akademie-Verlag GmBH, Leipziger St. 3-4, 108 Berlin, Germany) V.1- 1957-

**NAIZAM** Nara Igaku Zasshi. Journal of the Nara Medical Association. (Nara Kenritsu Ika Daigaku, Kashihara, Nara, Japan) V.1- 1950-

**NALSDJ** NATO Advanced Study Institute Series, Series A: Life Sciences. (Plenum Publishing Corp., 233 Spring St., New York, NY 10013) V.53- 1983-

**NAREA4** Nutrition Abstracts and Reviews. (Central Sales Branch, Commonwealth Agricultural Bureaux, Farnham Royal, Slough SL2 3BN, England) V.1- 1931-

**NARHAD** Nucleic Acids Research. (Information Retrieval Inc., 1911 Jefferson Davis Highway, Arlington, VA 22202) V.1- 1974-

**NASDA6** Nagoya Shiritsu Daigaku Igakkai Zasshi. Journal of the Nagoya City University Medical Association. (The University, Nagoya, Japan) V.1- 1950-

**NATUAS** Nature. (Macmillan Journals Ltd., Brunel Rd., Basingstoke RG21 2XS, UK) V.1- 1869-

**NATWAY** Naturwissenschaften. (Springer-Verlag, Heidelberger Platz 3, D-1000 Berlin 33, Germany) V.1- 1913-

**NCDREP** New Cardiovascular Drugs. (Ravencrest, 1185 Avenue of the Americas, New York, NY 10036) 1985-

**NCIAL\*** Progress Report Submitted to the National Cancer Institute by Arthur D. Little, Inc. (15 Acorn Park, Cambridge, MA 02140)

**NCIBR\*** Progress Report for Contract No. NIH-NCI-E-68-1311, Submitted to the National Cancer Institute by Bio-Research Consultants, Inc. (9 Commercial Ave., Cambridge, MA 02141)

**NCICP\*** Progress Report Submitted to the National Cancer Institute by Charles Pfizer and Company

**NCIHL\*** Progress Report Submitted to the National Cancer Institute by Hazelton Laboratories, Inc

**NCIIR\*** Progress Report for Contract No. N01-CP-12338, Submitted to the National Cancer Institute by IIT Research Institute. (Chicago, IL)

**NCILB\*** Progress Report for Contract No. NIH-NCI-E-C-72-3252, Submitted to the National Cancer Institute by Litton Bionetics, Inc. (Bethesda, MD)

**NCIMAV** National Cancer Institute, Monograph. (U.S. Government Printing Office, Superintendent of Documents, Washington, DC 20402) No. 1- 1959-

**NCIMR\*** Progress Report Submitted to the National Cancer Institute by Mason Research Institute. (Worcester, MA)

**NCINS\*** National Cancer Institute Report. (Bethesda, MD 20014)

**NCIRI\*** Progress Report Submitted to the National Cancer Institute by Piason Research Institute

**NCISA\*** Progress Report for Contract No. PH-43-63-1132, Submitted to the National Cancer Institute by Scientific Associates, Inc. (6200 S. Lindberg Blvd., St. Louis, MO 63123)

**NCISP\*** National Cancer Institute Screening Program Data Summary, Developmental Therapeutics Program, Bethesda, MD 20205

**NCISS\*** Progress Report Submitted to the National Cancer Institute by South Shore Analytical and Research Laboratory

**NCITR\*** National Cancer Institute Carcinogenesis Technical Report Series. (Bethesda, MD 20014) No. 0-205. For publisher information, see NTPTR\*

**NCIUS\*** Progress Report for Contract No. PH-43-64-886, Submitted to the National Cancer Institute by the Institute of Chemical Biology, University of San Francisco. (San Francisco, CA 94117)

**NCNSA6** National Academy of Sciences, National Research Council, Chemical-Biological Coordination Center, Review. (Washington, DC)

**NCPBBY** National Clearinghouse for Poison Control Centers, Bulletin. U.S. Department of Health, Education, and Welfare (Washington, DC)

**NDADD8** New Drugs Annual: Cardiovascular Drugs. (New York, NY) V.1-2 1983-84. For publisher information, see NCDREP

**NDKIA2** Kankyo Igaku Kenkyusho Nenpo (Nagoya Daigaku). Annual Report of the Research Institute of Environmental Medicine, Nagoya University. (The University, Furo-cho, Chikosa-Ku, Nagoya, Japan) V.1- 1947/48(Pub. 1949)-

**NDRC\*\*** National Defense Research Committee, Office of Scientific Research and Development, Progress Report

**NEACA9** News Edition, American Chemical Society (Easton, PA) V.18-19, 1940-41. For publisher information, see CENEAR

**NEAGDO** Neurobiology of Aging. (ANKHO International Inc., POB 426, Fayetteville, NY 13066) V.1- 1980-

**NEJMAG** New England Journal of Medicine. (Massachusetts Medical Society, 10 Shattuck St., Boston, MA 02115) V.198- 1928-

**NEOLA4** Neoplasma. (Karger-Libri AG, Scientific Booksellers, Arnold-Boecklin-Strasse 25, CH-4000 Basel 11, Switzerland) V.4- 1957-

**NEPHBW** Neuropharmacology. (Pergamon Press, Headington Hill Hall, Oxford OX3 OBW, England) V.9- 1970-

**NEPSBV** Neuropsychopharmacology, Proceedings of the Meeting of the Collogium Internationale, Neuropsychopharmacologicum. (Excerpta Medica Foundation, P.O. Box 1126, Amsterdam, Netherlands) V.1- 1959-

**NEREDZ** Neurochemical Research. (Plenum Publishing Corp., 233 Spring St., New York, NY 10013) V.1- 1976-

**NETOD7** Neurobehavioral Toxicology. (ANKHO International, Inc., P.O. Box 426, Fayetteville, NY 13066) V.1-2, 1979-80, For publisher information, see NTOTDY

**NEURAI** Neurology. (Modern Medicine Publications, Inc., 757 Third Avenue, New York, NY 10017) V.1- 1951-

**NEZAAQ** Nippon Eiseigaku Zasshi. Japanese Journal of Hygiene. (Nippon Eisei Gakkai, c/o Kyoto Daigaku Igakubu Koshu Eiseigaku Kyoshita, Yoshida Konoe-cho, Sakyo-ku, Kyoto, Japan) V.1- 1946-

**NEZTAF** New Zealand Veterinary Journal. (New Zealand Veterinary Association, Massey Univ., Palmerston North, New Zealand) V.1- 1952-

**NFGZAD** Nippon Funin Gakkai Zasshi. Japanese Journal of Fertility and Sterility. (Nippon Funin Gakkai, 1-1 Sadohara-cho, Ichigaya, Shinjuku-ku, Toyko 162, Japan) V.1- 1956-

**NGCJAK** Nippon Gan Chiryo Gakkai-shi. Journal of Japan Society for Cancer Therapy. (Nihon Gan Chiryo Gakkai, Kyoto, Japan) V.1- 1966-

**NGGKED** Nippon Gan Gakkai Sokai Kiji. Proceedings of the Annual Meeting of the Japanese Cancer Association. (Japanese Cancer Association, Tokyo, Japan) V.1- 1956-

**NGGZAK** Nippon Geka Gakkai Zasshi. Journal of the Japanese Surgical Society. (Nippon Geka Gakkai, 2-3-10 Koraku, Bunkyo-ku, Tokyo 112, Japan) V.8- 1908-

**NHOZAX** Nippon Hoigaku Zasshi. Japanese Journal of Legal Medicine. (Nippon Hoi Gakkai, c/o Tokyo Daigaku Igakubu Hoigaku Kyoshitsu, 7-3-1, Hongo, Bunkyo-ku, Tokyo, Japan) V.1- 1944-

**NIAND5** NIPH Annals. (National Institute of Public Health, Postuttak, Oslo, 1, Norway) V.1- 1978-

**NIBKAW** Nippon Byori Gakkai Kaishi. Journal of the Japanese Pathological Society. (c/o Tokyo Daigaku Igakubu Byorigaku Kyoshitsu, 7-3-1 Hongo, Bunkyo-ku, Tokyo 113, Japan) V.1- 1911-

**NICHAS** Nichidai Igaku Zasshi. (Nihon Daigaku Igakkai, 30, Oyaguchi kami-machi, Itabashi-ku, Tokyo 173, Japan) V.1- 1937-

**NIGHAE** Archiv Fuer Japanische Chirurgie. (Nippon Geka Hokan Henshushitsu, c/o Kyoto Daigaku Igakubu Geka Seikei Geka Kyoshitsu, 54 Kawara- machi, Shogoin, Sakyo-ku, Kyoto 606, Japan) V.1- 1924-

**NIGZAY** Niigata Igakkai Zasshi. Niigata Medical Journal. (Niigata Daigaku, 1 Asahi-Machi, Niigata, Japan) V.60- 1946-

**NIHBAZ** National Institutes of Health, Bulletin. (Bethesda, MD)

**NIIHAO** Nichidoku Iho. Japanese-German Medical Reports. (Nihon Schering Co., Ltd., 6-64, Nishimiyahara, 2-chome, Yodogawa-ku, Osaka 532, Japan) V.1- 1956-

**NIIRDN** "Drugs in Japan. Ethical Drugs, 6th Edition 1982" Edited by Japan Pharmaceutical Information Center. (Yakugyo Jiho Co., Ltd., Tokyo, Japan)

**NIOSH\*** National Institute for Occupational Safety and Health, U.S. Dept. of Health, Education, and Welfare, Reports and Memoranda

**NIPAA4** Nippon Shokakibyo Gakkai Zasshi. Journal of the Japanese Society of Gastroenterology. (Nippon Shokakibyo Gakkai, 4-12 7-Chome, Ginza, Chuo-Ku, Tokyo 104, Japan) V.1- 1899-

**NIPDAD** Nihon Daigaku No-Juigakubu Gakujutsu Kenkyu Hokoku. Research Reports of the College of Agriculture and Veterinary Medicine, Nihon Univ. (The Univ., 34-1 Shimouma, 3-chome, Setagaya-ku, Tokyo 154, Japan) No. 1- 1953-

**NIPOAC** Nippon Shonika Gakkai Zasshi. (Nippon Shonika Gakkai, 1-29-8 Shinjuku, Shinjuku-Ku, Tokyo 160, Japan) V.55- 1951-

**NISFAY** Nippon Sanka Fujinka Gakkai Zasshi. Journal of Japanese Obstetrics and Gynecology. (Nippon Sanka Fujinka Gakkai, c/o Hoken Kaikan Building., 1-1 Sadohara-cho, Ichigaya, Shinjuku-ku, Tokyo 162, Japan) V.1- 1949-

**NIVAAY** Notiziario dell'Istituto Vaccinogeno Antitubercolare. Bulletin of the Institute for Antitubercular Vaccinogens. (l'Instituto, Via Clericetti, 45 Milan, Italy) V.1-11, 1951-61

**NJGKBV** Snake. (Japan Snake Institute, Hon-machi, Yaduzuka, Nitta-gun, Gunma-ken, Japan) V.1- 1969-

**NJMSAG** Nagoya Journal of Medical Science. (Nagoya University School of Medicine, 65 Tsuruma-cho, Showa-ku, Nagoya 466, Japan) V.2- 1927-

**NJUZA9** Japanese Journal of Veterinary Science. (Nippon Jui Gakkai, 1-37-20, Yoyogi, Shibuya-ku, Tokyo 151, Japan) V.1- 1939-

**NKEZA4** Nippon Koshu Eisei Zasshi. Japanese Journal of Public Health. (Nippon Koshu Eisei Gakkai, 1-29-8 Shinjuku, Shinjuku-ku, Tokyo 160, Japan) V.1- 1954-

**NKGZAE** Nippon Ketsueki Gakkai Zasshi. Journal of Japan Haematological Society. (Kyoto Univ. Hospital, Faculty of Medicine, Kyoto, Japan) V.1- 1937-

**NKOGAV** Nippon Kokuka Gakkai Zasshi. Journal of the Japanese Stomatological Society. (7-3-1, Hongo, Bunkyo-Ku, Tokyo, Japan) V.1- 1952-

**NKRZAZ** Chemotherapy (Tokyo). (Nippon Kagaku Ryoho Gakkai, 2-20-8 Kamiosaki, Shinagawa-Ku, Tokyo, 141, Japan) V.1- 1953-

**NMJOAA** Nagoya Medical Journal. (Nagoya City University Medical School, 2-38 Nagarekawa-machi, Naka-ku, Nagoya 460, Japan) V.1- 1953-

**NNAPBA** Naunyn-Schmiedebergs Archiv fuer Pharmakologie. (Berlin, Germany) V.264-271, 1969-71. For publisher information, see NSAPCC

**NNBYA7** Nature: New Biology. (Macmillan Journals Ltd., Houndmills Estate, Basingstoke, Hants RG21 2XS, England) V.229-246, 1971-73

**NNGAAS** Nippon Naika Gakkai Zasshi. Journal of the Japanese Society of Internal Medicine. (Nippon Naika Gakkai, 33-5, 3-Chome, Hongo, Bunkyo-ku, Tokyo 13) V.1- 1913-

**NNGADV** Nippon Noyaku Gakkaishi. (Pesticide Science Society of Japan, 43-11, 1-Chome, Komagome, Toshima-ku, Tokyo 170, Japan) V.1- 1976-

**NNGZAZ** Nippon Naibumpi Gakkai Zasshi. Journal of the Japan Endocrine Society. (Nippon Naibumpi Gakkai, c/o Kyoto-Furitsu Ika Daigaku, Kojinbashi Nishizume-Sagaru, Kamigyo-ku, Kyoto 602, Japan) V.1- 1925-

**NOMDA6** Northwest Medicine. (Seattle, WA) V.1-73, 1903-73. Discontinued

**NPIRI\*** Raw Material Data Handbook, Vol.1: Organic Solvents, 1974. (National Association of Printing Ink Research Institute, Francis McDonald Sinclair Memorial Laboratory, Lehigh Univ., Bethlehem, PA 18015)

**NPMDAD** Nouvelle Presse Medicale. (Masson et Cie, Editeurs, 120 Blvd. Saint-Germain, P-75280, Paris 06, France) V.1- 1972-

**NPRNAY** Nephron. (S. Karger Publishers, Inc., 150 Fifth Ave., Suite 1105, New York, NY 10011) V.1- 1964-

**NRINA3** Nippon Rinsho. Japanese Clinical Medicine. (Nippon Rinsho Sha, 3-1 Dosho-machi, Higashi-ku, Osaka 541, Japan) V.1- 1943-

**NRSCDN** Neuroscience. (Pergamon Press Ltd., Headington Hill Hall, Oxford OX3 OBW, England) V.1- 1976-

**NRTTA8** NRC Technical Translation. (The National Research Council of Canada, Ottawa, Ontario K1A 0R6, Canada) No. 1- 1949-

**NRTXDN** Neurotoxicology. (Pathotox Publishers, Inc., 2405 Bond St., Park Forest South, IL 60464) V.1- 1979-

**NSAPCC** Naunyn-Schmiedeberg's Archives of Pharmacology. (Springer-Verlag, Heidelberger Pl. 3, D-1 Berlin 33, Germany) V.272- 1972-

**NSMZDZ** Nippon Shika Masui Gakkai Zasshi. (Tokyo Ikashika Daigaku Shika Masuigaku Kyoshitsu, 1-5-45 Yushima, Bunkyo-Ku, Tokyo 113, Japan) V.1- 1973-

**NTIMBF** Nauchnye Trudy, Irkutskii Gosudarstvennyi Meditsinskii Institut. Scientific Works, Irkutsk State Medical Institute. (Irkutskii Gosudarstvennyi Meditsinskii Institut, Irkutsk, USSR) No. 80- 1967-

**NTIS\*\*** National Technical Information Service. (Springfield, VA 22161) (Formerly U.S. Clearinghouse for Scientific and Technical Information)

**NTOTDY** Neurobehavioral Toxicology and Teratology. (ANKHO International Inc., P.O. Box 426, Fayetteville, NY 13066) V.3- 1981-

**NTPTB\*** NTP Technical Bulletin. (National Toxicology Program, Landow Bldg. 3A-06, 7910 Woodmont Ave., Bethesda, MD 20205)

**NTPTR\*** National Toxicology Program Technical Report Series. (Research Triangle Park, NC 27709) No. 206-

**NUCADQ** Nutrition and Cancer. (Franklin Institute Press, POB 2266, Philadelphia, PA 19103) V.1- 1978-

**NULSAK** Nucleus (Calcutta). (Dr. A.K. Sharma, c/o Cytogenetics Laboratory, Department of Botany, University of Calcutta, 35 Ballygunge Circular Rd., Calcutta 700 019, India) V.1- 1958-

**NUNDAJ** Neuroendocrinology. (S. Karger AG, Postfach CH-4009 Basel, Switzerland) V.1- 1965/66-

**NUPOBT** Neuropatologia Polska. (Ars-Polona-RUCH, POB 1001, 00-068 Warsaw 1, Poland) V.1- 1963-

**NURIBL** Nutrition Reports International. (Geron-X, Inc., POB 1108, Los Altos, CA 94022) V.1- 1970-

**NWSCAL** New Scientist. (IPC Magazines Ltd., Tower House, Southampton St., London WC2E 9QX, England) V.1- 1956-

**NYKGA7** Noyaku Kagaku. Pesticide Science. (Pesticide Science Society of Japan, 43-11, 1-Chome, Komagome, Toshima-ku, Tokyo 170, Japan) V.1-3, 1973-76

**NYKZAU** Nippon Yakurigaku Zasshi. Japanese Journal of Pharmacology. (Nippon Yakuri Gakkai, 2-4-16, Yayoi, Bunkyo-Ku, Tokyo 113, Japan) V.40- 1944-

**NYSJAM** New York State Journal of Medicine. (Medical Society of the State of New York, 420 Lakeville Rd., Lake Success, NY 11040) V.1- 1901-

**NZMJAX** New Zealand Medical Journal. (Otago Daily Times and Witness Newspapers, P.O. Box 181, Dunedin C1, New Zealand) V.1- 1900-

**OBGNAS** Obstetrics and Gynecology. (Elsevier/North Holland, Inc., 52 Vanderbilt Avenue, New York, NY 10017) V.1- 1953-

**OCHRAI** Occupational Health Review. (Ottawa, Canada) V.4-22, 1953-71. Discontinued

**OCMJAJ** Osaka City Medical Journal. (Osaka City Medical Center, 1-4-54, Asahimachi, Abenoku, Osaka 545, Japan) V.1- 1954-

**OCRAAH** Organometallic Chemistry Reviews, Section A: Subject Reviews. (Lausanne, Switzerland) V.3-8, 1968-72. For publisher information, see JORCAI

**ODFU\*\*** Vergleichende pharmakologische Prufung Zweier neuer Lokalanasthetika: Perkain und Pantokain, Wilhelm Ost Dissertation. (Pharmakologischen Institut der Universitat zu Frankfurt am Main, Germany, 1931)

**OEKSDJ** Osaka-furitsu Koshu Eisei Kenkyusho Kenkyu Hokoku, Shokuhin Eisei Hen. (Osaka-furitsu Koshu Eisei Kenkyusho, 1-3-69 Nakamichi, Higashinari-ku, Osaka 537, Japan) No. 1- 1970-

**OFAJAE** Okajimas Folia Anatomica Japonica. (Japan Publications Trading Co., 175 5th Ave., New York, NY 10010) V.14- 1936-

**OGSUA8** Obstetrical and Gynecological Survey. (Williams & Wilkins, 428 E. Preston St., Baltimore, MD 21202) V.1- 1946-

**OHSLAM** Occupational Health and Safety Letter. (Environews, Inc., 1097 National Press Bldg., Washington, DC 20045) V.1- 1970-

**OIGZDE** Osaka-shi Igakkai Zasshi. Journal of Osaka City Medical Association. (Osaka-shi Igakkai, c/o Osaka-shiritsu Daigaku Igakubu, 1-4-54 Asahi-cho, Abeno-ku, Osaka, 545, Japan) V.24- 1975-

**OIGZSE** Osaka-shi Igakkai Zasshi. Journal of Osaka City Medical Association. (Osaka-shi Igakkai, c/o Osaka-shiritru Daigaku Igakubu, 1-4-54 Asahi-cho, Abeno-ku, Osaka, 545, Japan) V.24- 1975-

**OIZAAV** Osaka Igaku Zasshi. (Osaka, Japan)

**OJVRAZ** Onderstepoort Journal of Veterinary Research. (Div. of Agricultural Inform., Dept. of Agricultural Tech. Ser., Private Bag X-144, Pretoria, S. Africa) V.25- 1951-

**OJVSA4** Onderstepoort Journal of Veterinary Science and Animal Industry. (Pretoria, S. Africa) V.1-24, 1933-50. For publisher information, see OJVRAZ

**OKEHDW** Osaka-furitsu Koshu Eisei Kenkyusho Kenkyu Hokoku, Koshu Eisei Hen. Research Reports of the Osaka Prefectural Institute of Public Health, Public Health Section. (1-3-69 Nakamichi, Higashinari-ku, Osaka, 537, Japan) No. 3- 1966-

**OMCDS\*** Olin Chemicals Data Sheet. (Industrial Development, 745 5th Ave., New York, NY 10022)

**ONCOAR** Oncologia. (Basel, Switzerland) V.1-20, 1948-66. For publisher information, see ONCOBS

**ONCOBS** Oncology. (S. Karger AG, Postfach CH-4009 Basel, Switzerland) V.21- 1967-

**ONCODU** Revista de Chirurgui, Oncologie, Radiologie, ORL, Oftalmologie, Stomatologie, Seria: Oncologia.(Rompresfilatelia, ILEXIM, POB 136-137, Bucharest, Romania) V.13, No. 4- 1974-

**ONGZAC** Ontogenez (Moscow). For English translation, see SJDBA9. (v/o Mezhdunarodnaya Kniga, 113095, Moscow, USSR) V.1- 1970-

**OPHTAD** Ophtalmologica. (S. Karger AG, Postfach CH-4009 Basel, Switzerland) V.96- 1978-

**OSDIAF** Osaka Shiritsu Daigaku Igaku Zasshi. Journal of the Osaka City Medical Center. (Osaka-Shiritsu Daigaku Igakubu, 1-4-54, Asahi-cho, Abeno-Ku, Osaka, Japan) V.4-23, 1955-74

**OSMJAT** Ohio State Medical Journal. (Ohio State Medical Association, 600 S. High St., Suite 500, Columbus, OH 43215) V.1- 1905-

**OSOMAE** Oral Surgery, Oral Medicine and Oral Pathology. (C.V. Mosby Co., 11830 Westline Industrial Dr., St Louis, MO. 63141) V.1- 1948-

**OYYAA2** Oyo Yakuri. Pharmacometrics. (Oyo Yakuri Kenkyukai, Tohoku Daigaku, Kitayobancho, Sendai 980, Japan) V.1- 1967-

**OZSEDS** Ozone: Science & Engineering. (Pergamon Press Inc., Maxwell House, Fairview Park, Elmsford, NY 10523) V.1- 1979-

**PAACA3** Proceedings of the American Association for Cancer Research. (Waverly Press, 428 E. Preston St., Baltimore, MD 21202) V.1- 1954-

**PAASAH** Publication, American Association for the Advancement of Science. (AAAS, 1515 Massachusetts Ave., N.W., Washington, DC 20005) No. 1-94, 1934-73. Discontinued

**PABIAQ** Pathologie et Biologie. (Paris, France) V.1-16, 1953-68. For publisher information, see PTBIAN

**PACHAS** Pure and Applied Chemistry. (Butterworth and Co., Ltd., Borough Green, Sevenoaks, Kent TN15 8PH, England) V.1- 1960-

**PAFEAY** Patologicheskaya Fiziologiya i Eksperimental'naya Terapiya. (v/o Mezhdunarodnaya Kniga, Kuznetskii Most 18, Moscow G-200, USSR) V.1- 1957-

**PAHEAA** Pharmaceutica Acta Helvetiae. (Case Postale 210, CH 1211 Geneva 1, Switzerland) V.1- 1926-

**PAMIAD** Pathologia et Microbiologia. V.23-43, 1960-75. For publisher information, see ECEBDI

**PAPOAC** Patologia Polska. (Ars-Polona-RUSH, POB 1001, 00-068 Warsaw 1, Poland) V.1- 1950-

**PAREAQ** Pharmacological Reviews. (Williams & Wilkins, 428 E. Preston St., Baltimore, MD 21202) V.1- 1949-

**PARPDS** Pathology, Research and Practice. (Gustav Fisher Verlag, Postfach 72 01 43, D-7000 Stuttgart 70, Germany) V.162- 1978-

**PARWAC** Polskie Archiwum Weterynaryjne. Polish Archives of Veterinary Medicine. (Panstwowe Wydawnictwo Naukowe, POB 391, P-00251 Warsaw, Poland) V.1- 1951-

**PATHAB** Pathologica. (Via Alessandro Volta, 8 Casella Postale 894, 16128 Genoa, Italy) V.1- 1908-

**PAVEAC** Pathologia Veterinaria. (Basel, Switzerland) V.1-7, 1964-70. For publisher information, see VTPHAK

**PBBHAU** Pharmacology, Biochemistry and Behavior. (ANKHO International Inc., P.O. Box 426, Fayetteville, NY 13066) V.1- 1973-

**PBFMAV** Problemi na Farmatsiyata. Problems in Pharmacy. (Durzhavno Izdatel'stvo Meditsina i Fizkultura, Pl. Slaveikov II, Sofia, Bulgaria) V.1- 1973-

**PBPHAW** Progress in Biochemical Pharmacology. (S. Karger AG, Postfach CH-4009 Basel, Switzerland) V.1- 1965-

**PBPSDY** "Pharmacological and Biochemical Properties of Drug Substances" Morton E. Goldberg, ed., Washington, DC, American Pharmaceutical Association, V.1- 1977-

**PCBPBS** Pesticide Biochemistry and Physiology. (Academic Press, 111 5th Ave., New York, NY 10003) V.1- 1971-

**PCBRD2** Progress in Clinical and Biological Research. (Allan R. Liss, Inc., 150 5th Ave., New York, NY 10011) V.1- 1975-

**PCCRA4** Proceedings of the Canadian Cancer Research Conference. (Univ. of Toronto Press, Toronto, Ontario M55 1A6, Canada) V.1- 1954-

**PCIPDV** Beijing Yixueyuan Xuebao. Journal of Peking Medical College. (Beijing Yixueyuan, Beijiaohaidianqu, Beijing, People's Republic of China) Beginning history not known

**PCJOAU** Pharmaceutical Chemistry Journal (English Translation). Translation of KHFZAN. (Plenum Publishing Corp., 233 Spring St., New York, NY 10013) No. 1- 1967-

**PCOC\*\*** Pesticide Chemicals Official Compendium, Association of the American Pesticide Control Officials, Inc. (Topeka, KS, 1966)

**PEDIAU** Pediatrics. (P.O. Box 1034, Evanston, IL 60204) V.1- 1948-

**PEMNDP** Pesticide Manual. (The British Crop Protection Council, 20 Bridport Rd., Thornton Heath CR4 7QG, UK) V.1- 1968-

**PENDAV** Polish Endocrinology. English translation of Endokrynologia Polska. (Springfield, VA) V.13-23, 1962-72. Discontinued

**PENNS\*** Pennsalt Chemicals Corporation, Technical Div., New Products. (Philadelphia, PA)

**PEREBL** Pediatric Research. (Williams & Wilkins Co., 428 E. Preston St., Baltimore, MD 21202) V.1- 1967-

**PESTC\*** Pesticide and Toxic Chemical News. (Food Chemical News, Inc., 400 Wyatt Bldg., 777 14th St. N.W., Washington, DC 20005) V.1- 1972-

**PESTD5** Proceedings of the European Society of Toxicology. (Amsterdam, Netherlands) V.16-18, 1975-77. Discontinued

**PetKP#** Personal Communication from Dr. K.P. Petersen, DAK Labs., 59 Lergravsvej, DK-2300, Copenhagen, Demark, to B. Jones, Tracor Jitco, Inc., December 22, 1977

**PEXTAR** Progress in Experimental Tumor Research. (S. Karger AG, Postfach CH-4009 Basel, Switzerland) V.1- 1960-

**PFLABK** Pfluegers Archiv. European Journal of Physiology. (Springer-Verlag, Heidelberger Pl. 3, 1 Berlin 33, Germany) V.302- 1968-

**PGMJAO** Postgraduate Medical Journal. (Blackwell Scientific Publications, Osney Mead, Oxford OX2 OEL, England) V.1- 1925-

**PGPKA8** Problemy Gematologii i Perelivaniia Krovi. Problems of Hematology and Blood Transfusion. (v/o Mezhdunarodnaya Kniga, Kuznetskii Most 18, Moscow G-200, USSR.) V.1- 1956-

**PGTCA4** Pigment Cell. (S. Karger AG, Arnold-Boecklin St., 25 CH-4011 Basel, Switzerland) V.1- 1973-

**PHABDI** Proceedings of the Hungarian Annual Meeting for Biochemistry. (Magyar Kemikusok Egyesulete, Anker Koz 1, 1061 Budapest, Hungary) V.1- 1961-

**PHARAT** Pharmazie. (VEB Verlag Volk und Gesundheit, Neue Gruenstr 18, 102 Berlin, Germany) V.1- 1946-

**PHBHA4** Physiology and Behavior. (Pergamon Press Inc., Maxwell House, Fairview Park, Elmsford, NY 10523) V.1- 1966-

**PHBTH\*** "Pharmacology: Basis of Therapy" Goodman, L.S., ed., 3rd ed., New York, NY, Macmillan, 1967

**PHBUA9** Pharmaceutical Bulletin. (Tokyo, Japan) V.1-5, 1953-57. For publisher information, see CPBTAL

**PHINDQ** Pharmacy International. (Elsevier Science Publications Co., Inc., 52 Vanderbilt Ave., New York, NY 10017) V.1- 1990-

**PHJOAV** Pharmaceutical Journal. (Pharmaceutical Press, 17 Bloomsbury Sq., London WC1A 2NN, England) V.131- 1933-

**PHMCAA** Pharmacologist. (American Society for Pharmacology and Experimental Therapeutics, 9650 Rockville Pike, Bethesda, MD 20014) V.1- 1959-

**PHMGBN** Pharmacology: International Journal of Experimental and Clinical Pharmacology. (S. Karger AG, Postfach CH-4009 Basel, Switzerland) V.1- 1968-

**PHPHA6** Phytiatrie-Phytopharmacie. (Societe Francaise de Phytiatrie et de France) V.1- 1952-. Phytopharmacie, Etoile de Choisy, Route de St-Cyr, 78000 Versailles.

**PHREA7** Physiological Reviews. (9650 Rockville Pike, Bethesda, MD 20014) V.1- 1921-

**PHREEB** Pharmaceutical Research. (Thieme Inc., 381 Park Ave. S, New York, NY 10016) No. 1- 1984-

**PHRPA6** Public Health Reports. (U.S. Government Printing Office, Superintendent of Documents, Washington, DC 20402) V.1- 1878-

**PHTHDT** Pharmacology and Therapeutics. (Pergamon Press Ltd., Headington Hill Hall, Oxford OX3 0BW, England) V.4- 1979-

**PHTXA6** Pharmacology and Toxicology. Translation of FATOAO. (New York, NY) V.20-22, 1957-59. Discontinued

**PHYTAJ** Phytopathology. (Phytopathological Society, 3340 Pilot Knob Rd., St. Paul, MN 55121) V.1- 1911-

**PIAIA9** Proceedings of the Iowa Academy of Science. (Iowa Academy of Science, Univ. of Northern Iowa, Cedar Falls, IA 50613) V.1- 1887/93-

**PIATA8** Proceedings of the Imperial Academy of Tokyo. (Tokyo, Japan) V.1-21, 1912-45. For publisher information, see PJACAW

**PISCAD** Proceedings of the Indian Science Congress. (Indian Science Congress Association, 14, Dr. Biresh Guha St., Calcutta, 700017, India) 1st- 1914-

**PIXXD2** PCT (Patent Cooperation Treaty) International Application. (U.S. Patent and Trademark Office, Foreign Patents, Washington, DC 20231)

**PJABDW** Proceedings of the Japan Academy, Series B: Physical and Biological Sciences. (Maruzen Co., Ltd., P.O. Box 5050, Tokyo International 100-31, Japan) V.53- 1977-

**PJACAW** Proceedings of the Japan Academy. (Tokyo, Japan) V.21-53, 1945-77. For publisher information, see PJABDW

**PJPHEO** Pakistan Journal of Pharmacology. (c/o Univ. of Karachi, Faculty of Pharmacy, Karachi, 32, Pakistan) V.1- 1984-

**PJPPAA** Polish Journal of Pharmacology and Pharmacy. (ARS-Polona-Rush, POB 1001, 00-068 Warsaw 1, Poland) V.25- 1973-

**PLCHB4** Physiological Chemistry and Physics. (U.S. Government Printing Office, Superintendent of Documents, Washington, DC 20402) V.1- 1969-

**PLENBW** Pollution Engineering. (1301 S. Grove Ave., Barrington, IL 60010) V.1- 1969-

**PLIRDW** Progress in Lipid Research. (Pergamon Press Ltd., Headington Hill Hall, Oxford OX3 OBW, England) V.17- 1978-

**PLMEAA** Planta Medica. (Hippokrates-Verlag GmbH, Neckarstr 121, 7 Stuttgart, Germany) V.1- 1953-

**PLMEDD** Prostaglandins, Leukotrienes and Medicine. (Longman Inc., 19 W. 44th St., New York, NY 10036) V.8- 1982-

**PLPSAX** Physiological Psychology. (Psychonomic Society, Inc., 1108 W. 34th Ave., Austin, TX 78705) V.1- 1973-

**PLRCAT** Pharmacological Research Communications. (Academic Press, 111 5th Ave., New York, NY 10003) V.1- 1969-

**PMARAU** Prensa Medica Argentina. (Junin 845, Buenos Aires, Argentina) V.1 1914-

**PMDCAY** Progress in Medical Chemistry. (American Elsevier Publishing Co., 52 Vanderbilt Ave., New York, NY 10017) V.1- 1961-

**PMDFA9** Pakistan Medical Forum. (Karachi, Pakistan) V.1-8, 1966-73. Discontinued

**PMJMAQ** Proceedings of the Annual Meeting of the New Jersey Mosquito Extermination Association. (New Brunswick, NJ) V.1-61, 1914-74

**PMRSDJ** Progress in Mutation Research. (Elsevier North Holland, Inc., 52 Vanderbilt Ave., New York, NY 10017) V.1- 1981-

**PMSBA4** Progress in Molecular and Subcellular Biology. (Springer-Verlag, Heidelberger Pl. 3, D-1000 Berlin 33, Germany) V.1- 1969-

**PNASA6** Proceedings of the National Academy of Sciences of the United States of America. (The Academy, Printing and Publishing Office, 2101 Constitution Ave., Washington, DC 20418) V.1- 1915-

**PNCCA2** Proceedings, National Cancer Conference. (Philadelphia, PA) V.1-7, 1949-72, For publisher information, see CANCAR

**POASAD** Proceedings of the Oklahoma Academy of Science. (Oklahoma Academy of Science, c/o James F. Lowell, Executive Secretary-Treasurer, Southwestern Oklahoma State University, Weatherford, OK 73096) V.1- 1910/1920-

**POMDAS** Postgraduate Medicine. (McGraw-Hill, Inc., Distribution Center, Princeton Rd., Hightstown, NJ 08520) V.1- 1947-

**POMJAC** Polish Medical Journal. (Warsaw, Poland) V.1-11, 1962-72. Discontinued

**POSCAL** Poultry Science. (Poultry Science Association, Texas A and M University, College Station, TX 77843) V.1- 1921-

**PPASAK** Proceedings of the Pennsylvania Academy of Science. (Pennsylvania Academy of Science, c/o Stanley Zagorski, Tr., Gannon College, Perry Sq., Erie, PA 16501) V.1- 1924/1926-

**PPHAD4** Pediatric Pharmacology. (Alan R. Liss, Inc., 150 5th Ave., New York, NY 10011) V.1- 1980-

**PPRPAS** Produits and Problemes Pharmaceutiques. (Paris, France) V.17-28, 1962-73. Suspended

**PPTCBY** Proceedings of the International Symposium of the Princess Takamatsu Cancer Research Fund. (Japan Scientific Societies Press, 2-10, Hongo 6-chome, Bunkyo-ku, Tokyo 113, Japan) (1st)- 1970(Pub. 1971)-

**PRACAK** Practitioner. (5 Bentinck St., London, England) V.1- 1868-

**PRBIDC** Progress in Reproductive Biology. (S. Karger AG, Arnold-Boecklin Str. 25, CH-4011, Basel, Switzerland) V.1- 1976-

**PREBA3** Proceedings of the Royal Society of Edinburgh, Section B. (Royal Society of Edinburgh, 22 George St., Edinburgh, Scotland) V.61- 1943-

**PREPAB** Przeglad Epidemiologiczny. (Panstwowy Zaklad Wydawnictw Lekarskich, ul. Dluga 38-40, 00-238 Warsaw, Poland) V.1-1947-

**PRGLBA** Prostaglandins. (Geron-X, Inc., P.O. Box 1108, Los Altos, CA 94022) V.1- 1972-

**PRKHDK** Problemi na Khigienata. Problems in Hygiene. (Durzhavno Izdatel'stvo Meditsina i Zizkultura, Pl. Slaveikov 11, Sofia, Bulgaria) V.1- 1975-

**PRLBA4** Proceedings of the Royal Society of London, Series B, Biological Sciences. (The Society, 6 Carlton House Terrace, London SW1Y 5AG, England) V.76- 1905-

**PROEAS** Problemy Endokrinologii. (v/o Mezhdunarodnaya Kniga, Kuznetskii Most 18, Moscow G-200, USSR.) V.1-6, 1936-41; V.13- 1967-

**PROTA\*** "Problemes de Toxicologie Alimentaire," Truhaut, R., Paris, France, L'evolution Pharmaceutique, (1955?)

**PRPHA8** Produits Pharmaceutiques. (Paris, France) V.1-16, 1946-61. For publisher information, see PPRPAS

**PRSMA4** Proceedings of the Royal Society of Medicine. (Grune and Stratton Inc., 111 5th Ave., New York, NY 10003) V.1- 1907-

**PSCBAY** Psychopharmacology Service Center, Bulletin. (Bethesda, MD) V.1-3, 1961-65. For publisher information, see PSYBB9

**PSCHDL** Psychopharmacology (Berlin). (Springer-Verlag New York, Inc., Service Center, 44 Hartz Way, Secaucus, NJ 07094) V.47- 1976-

**PSDAA2** Proceedings of the South Dakota Academy of Science. (Univ. of South Dakota, Vermillion, SD 57069) V.1- 1916-

**PSDTAP** Proceedings of the European Society for the Study of Drug Toxicity. (Princeton, NJ 08540) V.1-15, 1963-74. For publisher information, see PESTD5

**PSEBAA** Proceedings of the Society for Experimental Biology and Medicine. (Academic Press, 111 5th Ave., New York, NY 10003) V.1- 1903/04-

**PSSCBG** Pesticide Science. (Blackwell Scientific Publications Ltd., Osney Mead, Oxford OX2 OEL, England) V.1- 1970-

**PSSID2** Pergamon Series on Environmental Science. (Pergamon Press Ltd., Headington Hill Hall, Oxford OX3 0BW, England) V.1- 1978-

**PSSYDG** Proceedings of the Serono Symposia. (Academic Press Inc. Ltd., 24-28 Oval Rd., London NW1 7DX, England) V.1- 1973-

**PSTDAN** Pesticides. (Colour Publications Pvt. Ltd., 126-A Dhuruwadi, Off Dr. Nariman Rd., Bombay 400-025, India) V.1- 1967-

**PSTGAW** Proceedings of the Scientific Section of the Toilet Goods Association. (The Toilet Goods Association, Inc., 1625 I St., N.W., Washington, DC 20006) No. 1-48, 1944-67. Discontinued

**PSYBB9** Psychopharmacology Bulletin. (U.S. Government Printing Office, Superintendent of Documents, Washington, DC 20402) V.3- 1966-

**PSYPAG** Psychopharmacologia (Berlin). (Berlin, Ger.) V.1-46, 1959-76. For publisher information, see PSCHDL

**PTBIAN** Pathologie-Biologie. (Expansion Scientifique Francaise, 15 rue St. Benoit, Paris 6, France) V.17- 1969-

**PTEUA6** Pathologia Europaea. (Presses Academiques Europeennes, 98, Chaussee de Charleroi, Brussels, Belgium) V.1- 1966-

**PTLGAX** Pathology. (Royal College of Pathologists of Australia, 82 Windmill St., Sydney, NSW 2000, Australia) V.1- 1969-

**PTRMAD** Philosophical Transactions of the Royal Society of London, Series A: Mathematical and Physical Sciences. (Royal Society of London, 6 Carlton House Terrace, London SW1Y 5AG, England) V.178- 1887-

**PUMTAG** Trace Substances in Environmental Health. Proceedings of University of Missouri's Annual Conference on Trace Substances in Environmental Health. (Environmental Trace Substances Research Center, Univ. of Missouri, Columbia, MO 65201) V.1- 1967-

**PUOMA5** Proceedings of the University of Otago Medical School. (Otago Medical School Research Society, Box 913, Dunedin, New Zealand) V.1- 1922-

**PWPSA8** Proceedings of the Western Pharmacology Society. (Univ. of California, Dept. of Pharmacology, Los Angeles, CA 94122) V.1- 1958-

**PYRTAZ** Psychological Reports. (Southern Universities Press, Baton Rouge, LA 70813) V.1- 1952-

**PYTCAS** Phytochemistry. An International Journal of Plant Biochemistry. (Pergamon Press Ltd., Headington Hill Hall, Oxford OX3 OEW, England) V.1- 1961-

**QJDRAZ** Quarterly Journal of Crude Drug Research. (Lisse, Netherlands) V.1-19, 1961-81 For publisher information, see IJCREE

**QJMEA7** Quarterly Journal of Medicine. (Oxford University Press, Press Road, Neasden, London NW 10 0DD, England) V.1- 1932-

**QJPPAL** Quarterly Journal of Pharmacy and Pharmacology. (London, England) V.2-21, 1929-48. For publisher information, see JPPMAB

**QJSAAP** Quarterly Journal of Studies on Alcohol. (New Brunswick, NJ) V.1-28, 1940-67. For publisher information, see QJSOAX

**QJSOAX** Quarterly Journal of Studies on Alcohol, Part A: Originals. (Rutgers State University, New Brunswick, NJ 08903) V.29- 1968-

**QUNUAZ** Quaderni della Nutrizione. (Moore-Cottrell Subscription Agencies Inc., North Cohocton, NY 14868) V.1- 1934-

**RADLAX** Radiology. (Radiological Society of North America, 20th and Northampton Sts., Easton, PA 18042) V.1- 1923-

**RADOA8** Radiobiologiya. (v/o Mezhdunarodnaya Kniga, Kuznetskii Most 18, Moscow G-200, USSR) V.1- 1961-

**RalRL#** Personal Communication to NIOSH from Robert L. Raleigh, M.D., Assistant Director of Health and Safety Lab., Eastman Kodak Company, Rochester, NY

**RAMAAB** Revista de la Asociacion Medica Argentina. (Santa Fe 1171, Buenos Aires, Argentina) V.1- 1915-

**RAREAE** Radiation Research. (Academic Press, 111 Fifth Ave., New York, NY 10003) V.1- 1954-

**RARSAM** Radiation Research, Supplement. (Academic Press, 111 5th Ave., New York, NY 10003) No. 1- 1959-

**RBBIAL** Revista Brasileira de Biologia. (Caixa Postal 1587, ZC-00 Rio de Janeiro, Brazil) V.1- 1941-

**RBPMAZ** Revue Belge de Pathologie et de Medecine Experimentale. (Brussels, Belgium) V.18-31, 1947-65. For publisher information, see PTEUA6

**RCBIAS** Revue Canadienne de Biologie. (Les Presses de l'Universite de Montreal, P.O. Box 6128, 101 Montreal 3, Quebec, Canada) V.1- 1942-

**RCOCB8** Research Communications in Chemical Pathology and Pharmacology. (PJD Publications, P.O. Box 966, Westbury, NY 11590) V.1- 1970-

**RCPBDC** Research Communications in Psychology, Psychiatry and Behavior. (PJD Publications, P.O. Box 966, Westbury, NY 11590) V.1- 1976-

**RCPRAN** Record of Chemical Progress. (Detroit, MI) V.1-32, 1939-71. Discontinued

**RCRVAB** Russian Chemical Reviews. (Chemical Society, Publications Sales Office, Burlington House, London W1V 0BN, England) V.29- 1960-

**RCSADO** Research Communications in Substances Abuse. (PJD Publications Ltd., Box 966, Westbury, NY 11590) V.1- 1980-

**RCTEA4** Rubber Chemistry and Technology. (Div. of Rubber Chemistry, American Chemical Society, University of Akron, Akron, OH 44325) V.1- 1928-

**RDBGAT** Radiobiologia, Radiotherapia. (VEB Verlag Volk und Gesundheit, Neue Gruenstr. 18, DDR-102 Berlin, Germany) V.1- 1960-

**RDCNBM** Reproduction. (Calle Puerto de Bermeo, 11 Madrid 34, Spain) V.1- 1974-

**RDMIDP** Quarterly Reviews on Drug Metabolism and Drug Interactions. (Freund Pub. House, Suite 500, Chestham House, 150 Regent St., London W1R 5FA, UK) V.3- 1980-

**RDWU\*\*** Beitrage zur Pharmakologie des Berylliums, Ursula Richter Dissertation. (Pharmakologischen Institut der Universitat Wurzburg, Germany, 1930)

**REANBJ** Revista Espanola de Anestesiologia y Reanimacion. (Sociedad Espanola de Anestesiologia y Reanimacion, Mallorca, 189 Barcellona 36, Spain) V.1- 1954-

**RECYAR** Revue Roumaine d'Embryologie et de Cytologie, Serie d'Embryologie. (Bucharest) V.1-8, 1964-71

**REDH\*\*** J.D. Riedel-E. de Haen A.-G., Laboratory. (Berlin, Germany)

**REEBB3** Revue Europeenne d'Etudes Cliniques et Biologiques. European Journal of Clinical and Biological Research. (Paris, France) V.15-17, 1970-72. For publisher information, see BIMDB3

**REMBA8** Revista Ecuatoriana de Medicina y Ciencias Biologicas. (Facultad de Ciencias Medicas, Quito, Ecuador) V.1- 1963-

**REMVAY** Revue d'Elevage et de Medecine Veterinaire des Pays Tropicaux. (Editions Vigot Freres, 23 rue de l'Ecole-de-Medecine, 75006 Paris 6, France) V.1- 1947-

**REONBL** Revista Espanola de Oncologia. (Instituto Nacional de Oncologia del Cancer, Ciudad Universitaria, Madrid 3, Spain) V.1- 1952-

**REPMBN** Revue d'Epidemiologie, Medecine Sociale et Sante Publique. (Masson et Cie, Editeurs, 120 Blvd. Saint-Germain, P-75280, Paris 06, France) V.1- 1953-

**REPTED** Reproductive Toxicology. (Pergamon Press Inc., Maxwell House, Fairview Park, Elmsford, NY 10523) V.1- 1987

**RESJAS** Journal of the Reticuloendothelial Society. (1964) (Res: Journal of the Reticuloendothelial Society, New York, NY) V.1(1)-14(6), 1964-73

**RETOAE** Research Today. (Eli Lilly Co., Indianapolis, IN) V.1-16, 1944-60. Discontinued

**REXMAS** Research in Experimental Medicine. (Springer-Verlag, Heidelberger Pl. 3, D-1 Berlin 33, Germany) V.157- 1972-

**RFCTAJ** Rassegna di Fisiopatologia Clinica e Terapeutica. Review of Clinical and Therapeutic Physiopathology. (Rome, Italy) V.9-42, 1937-70. Discontinued

**RFECAC** Revue Francaise d'Etudes Cliniques et Biologiques. (Paris, France) V.1-14, 1956-69. For publisher information, see BIMDB3

**RFGOAO** Revue Francaise de Gynecologie et d'Obstetrique. (Masson Publishing USA, Inc., 14 E. 60th St., New York, NY 10022) V.20- 1920-

**RHPC\*\*** Rohm and Haas Company Petroleum Chemicals. (Philadelphia, PA 19105)

**RIHYAC** Rinsho Hinyokika. Clinical Urology. (Igakushoin Medical Publishers, Inc., 1140 Avenue of the Americas, New York, NY 10036) V.21- 1967-

**RIMAAX** Rivista di Malariologia. (Rome, Italy) V.5-46, 1926-67. Discontinued

**RISSAF** Rendiconti Istituto Superiore di Sanita (Italian Edition). (Rome, Italy) V.4-27, 1941-64

**RJARAV** Rhodesian Journal of Agricultural Research. (Rhodesian Ministry of Agriculture, P.O. Box 8108, Causeway, Salisbury, Rhodesia) V.1- 1963-

**RKGEDW** Rinsho Kyobu Geka. Japanese Annals of Thoracic Surgery

**RMCHAW** Revista Medica de Chile. (Sociedad Medica de Santiago, Esmeralda 678, Casilla 23-d, Santiago, Chile) V.1- 1872-

**RMEMDQ** Revue Roumaine de Morphologie, d'Embryologie et de Physiologie, Serie Morphologie et Embryologie. (ILEXIM, POB 136-137, Bucharest, Romania) V.21- 1975-

**RMISDU** International Congress and Symposium Series-Royal Society of Medicine Services Limited. (Royal Society of Medicine Services Ltd., 7 E. 60th St., New York, NY 10022) No. 1- 1978-

**RMLIAC** Revue Medicale de Liege. (Hopital de Baviere, Universite de Liege, Pl. du XX Aout 7, Liege, Belgium) V.1- 1946-

**RMMJAK** Rocky Mountain Medical Journal. (Denver, CO) V.35-76, 1938-79

**RMNIBN** Revista Medico-Chirurgicala. (Societatea de Medici si Naturalisti, Bulevardul Independentei, Iasi, Romania) V.35- 1924-

**RMSRA6** Revue Medicale de la Suisse Romande. (Societe Medicale de La Suisse Romande, 2 Bellefontaine, 1000 Lausanne, Switzerland) V.1- 1881-

**RMVEAG** Recueil de Medecine Veterinaire. (Editions Vigot Freres, 23 rue de l'Ecole-de-Medecine, Paris 6, France) V.1- 1824-

**ROHM\*\*** Rohm and Haas Company Data Sheets (Philadelphia, PA 19105)

**RPHRA6** Recent Progress in Hormone Research. Proceedings of the Laurentian Hormone Conference. (Academic Press, 111 5th Ave., New York, NY 10003) V.1- 1947-

**RPOBAR** Research Progress in Organic Biological and Medicinal Chemistry. (New York, NY) V.1-3, 1964-72. Discontinued

**RPTOAN** Russian Pharmacology and Toxicology. Translation of FATOAO. (Euromed Publications, 97 Moore Park Rd., London SW6 2DA, England) V.30- 1967-

**RPZHAW** Roczniki Panstwowego Zakladu Higieny. (Ars Polona-RUSH, POB 1001, 00-068 Warsaw 1, Poland) V.1- 1950-

**RRBCAD** Revue Roumaine de Biochimie. (Romprestilatelia, POB 2001, Calea Grivitei 64-66, Bucharest, Romania) V.1- 1964-

**RRCRBU** Recent Results in Cancer Research. (Springer-Verlag New York, Inc., Service Center, 44 Hartz Way, Secaucus, NJ 07094) V.1- 1965-

**RRENAR** Revue Roumaine d'Endocrinologie. (Bucharest, Romania) V.1-11, 1964-74. For publisher information, see RRENDU

**RRENDU** Revue Roumaine de Medecine, Serei Endocrinologie. (ILEXIM, POB 136-137, R-70116 Bucharest, Romania) V.13- 1975-

**RREVAH** Residue Reviews. (Springer-Verlag New York, Inc., Service Center, 44 Hartz Way, Secaucus, NJ 07094) V.1- 1962-

**RSABAC** Revista de la Sociedad Argentina de Biologia. (Associacion Medica Argentina, Santa Fe 1171, Buenos Aires, Argentina) V.1- 1925-

**RSPSA2** Rassegna di Studi Psichiatrici. Review of Psychiatric Studies. (Ospedale Psichiatrico, Calata CapodichiNo. 232, 80141 Naples, Italy) V.1- 1911-

**RSTUDV** Rivista di Scienza e Technologia degli Alimenti e di Nutrizione Umana. Review of Science and Technology of Food and Human Nutrition. (Bologna, Italy) V.5-6, 1975-76. Discontinued

**RTOPDW** Regulatory Toxicology and Pharmacology. (Academic Press Inc., 111 Fifth Ave., New York, NY 10003) V.1- 1981-

**RTPCAT** Rassegna di Terapia e Patologia Clinica. (Rome, Italy) V.1-8, 1929-36. For publisher information, see RFCTAJ

**RVFTBB** Rivista di Farmacologia e Terapia. (Prof. William Ferrari, Via G. Campi 287, 41100 Modena, Italy) V.1- 1970-

**RVTSA9** Research in Veterinary Science. (Blackwell Scientific Publications, Ltd., Osney Mead, Oxford OX2 OEL, England) V.1- 1960-

**RZOVBM** Rivista di Zootecnia e Veterinaria. Zootechnical and Veterinary Review. (Rivista di Zootecnia e Veterinaria, Via Viotti 3/5, 20133 Milan, Italy) No. 1- 1973-

**SAAMDZ** Substance and Alcohol Actions/Misuse. (Pergamon Press Inc., Maxwell House, Fairview Park, Elmsford, NY 10523) V.1- 1980-

**SACAB7** South African Cancer Bulletin. (National Cancer Association of South Africa, 9 Jubillee Rd., Parktown, Johannesburg, S. Africa) V.1- 1957-

**SAIGAK** Saishin Igaku. Modern Medicine. (Saishin Igakusha, Senchuri Bldg., 3-6-1 Hirano-machi, Higashi-ku, Osaka 541, Japan) V.19- 1945-

**SAIGBL** Sangyo Igaku. Japanese Journal of Industrial Health. (Japan Association of Industrial Health, c/o Public Health Building, 78 Shinjuku 1-29-8, Shinjuku-Ku, Tokyo, Japan) V.1- 1959-

**SAKNAH** Soobshcheniia Akademii Nauk Gruzinskoi S.S.R. (v/o Mezhdunarodnaya Kniga, Kuznetskii Most 18, Moscow G-200, USSR.) V.2- 1941-

**SAMJAF** South African Medical Journal. (Medical Association of South Africa, Secy., P.O. Box 643, Cape Town, S. Africa) V.6- 1932-

**SAPHAO** Skandinavisches Archiv fuer Physiologic. (Karolinska Institutet, Editorial Office, Stockholm, Sweden) V.1-83, 1899-1940. Superseded by APSCAX

**SAVEAB** Sammlung von Vergiftungsfaellen. (Berlin, Germany) V.1-13, 1930-1944; V.14, 1952-54. For publisher information, see ARTODN

**SBLEA2** Sbornik Lekarsky. (PNS-Ustredni Expedice Tisku, Jindriska 14, Prague 1, Czechoslovakia) V.1- 1887-

**SCALA9** Scalpel. (Brussels, Belgium) 1920-V.124, 1971. Discontinued

**SCAMAC** Scientific American. (Scientific American, Inc., 415 Madison Ave., New York, NY 10017) V.1- 1859-

**SCCSC8** Soap, Cosmetics, Chemical Specialties. (MacNair-Dorland, 101 W. 31st St., New York, NY 10001) V.47- 1971-

**SCCUR\*** Shell Chemical Company. Unpublished Report. (2401 Crow Canyon Rd., San Romon, CA 94583)

**SchF##** Personal Communication from F.W. Schaller, Inco Ltd., Park 80 West Plaza Two, Saddle Brook, NJ 07662, May 16, 1986

**SchP##** Personal Communication from Dr. P. Schmitz, Bayer AG, 5090 Leverkusen, Bayerwerk, Germany, to NIOSH, April 4, 1986

**SCHSAV** Soap and Chemical Specialties. (New York, NY) V.30-47, 1954-71. For publisher information, see SCCSC8

**SCIEAS** Science. (American Association for the Advancement of Science, 1515 Massachusetts Ave., NW, Washington, DC 20005) V.1- 1895-

**SCJUAD** Science Journal. (London, England) V.1-7, 1965-71. For publisher information, see NWSCAL

**SCMBE9** Spokane County Medical Society Bulletin. (Spokane, WA) 1929-81

**SCMGDN** Somatic Cell and Molecular Genetics. (Plenum Publishing Corp., 233 Spring St., New York, NY 10013) V.10- 1984-

**SCNEBK** Science News. (Science Service, Inc., 1719 N. St., NW, Washington, DC 20036) V.1- 1921-

**SCPHA4** Scientia Pharmaceutica. (Oesterreichische Apotheker-Verlagsgesellschaft MBH, Spitalgasse 31, 1094 Vienna 9, Austria) V.1- 1930-

**SCYYDZ** Shengzhi Yu Biyun. Reproduction and Contraception. (China International Book Trading Corp., POB 2820, Beijing, People's Republic of China) 1980-

**SDMEAL** South Dakota Journal of Medicine. (South Dakota Medical Association, 608 West Ave. N., Sioux Falls, SD 57104) V.18- 1965-

**SDMU\*\*** Zur Pharmakologie einiger Triazoliumverbindungen I, Erhart Schulze Dissertation. (Pharmakologischen Institut der Martin Luther-Universitat Halle-Wittenberg, Germany, 1936)

**SDSTBT** Scientific and Technical Report-Soap and Detergent Association. (Soap and Detergent Association, 485 Madison Ave., New York, NY 10022) No. 1- 1965-

**SDUU\*\*** Ueber die Beeinflussung der Salvarsantoxizitat im Tierversuch, Jurgen Steudemann Dissertation. (University of Hamburg, Germany, 1934)

**SEIJBO** Senten Ijo. Congenital Anomalies. (Nihon Senten Ijo Gakkai, Kyoto 606, Japan) V.1- 1960-

**SEMEAS** Semana Medica. (Sociedad Argentina de Gastroenterologica, Santa Fe 1171, Buenos Aires, Argentina) V.1- 1894-

**SFCRAO** Symposium on Fundamental Cancer Research (Williams & Wilkins Co., 428 E. Preston St., Baltimore, MD 21202) V.1- 1947-

**SFTIAE** Svensk Farmaceutisk Tidskrift. Swedish Journal of Pharmacy. (Apotekarsocieteten, Upplandsgatan 6A, Stockholm C, Sweden) V.1- 1897-

**SGOBA9** Surgery, Gynecology and Obstetrics. (Franklin H. Martin Memorial Foundation, 55 E. Erie St., Chicago, IL 60611) V.1- 1905-

**SHBOAO** Shokubutsu Boeki. (Nippon Shokubutsu Boeki Kyokai, 43-11, Komagome 1-chome, Toshima-Ku, Tokyo, Japan) V.1- 1947-

**SheCW#** Personal Communication from Dr. C.W. Sheu, Genetic Toxicology Branch, FDA, to H. Lau, Tracor Jitco, March 25, 1977

**SHELL\*** Shell Chemical Company, Technical Data Bulletin. (Agricultural Div., Shell Chemical Co., 2401 Crow Canyon Rd., San Ramon, CA 94583)

**SHGKA3** Shika Gakuho. Journal of Dentistry. (Tokyo Shika Daigaku Gakkai, 9-18, 2-chome, Sanzaki-cho, Chiyuoda-ku, Tokyo, Japan) V.1- 1895-

**SHHUE8** Shiyou Huagong. Petrochemical Technology. (Beijing Huagon Yanjiuyan, Beikou, Hepingjie, Beijing, People's Republic of China)

**SHIGAZ** Shigaku. Ondotology. (Nippon Shika Daigaku Shigakkai, 1-9-20 Fujimi, Chiyodaku, Tokyo 102, Japan) V.38- 1949-

**SHKKAN** Shika Kiso Igakkai Zasshi. Journal of the Japanese Association for Basic Dentistry. (Shika Kiso Igakkai, Nihon Daigaku Shigakubu, 1-8 Surugadai, Kanda, Chiyoda-ku, Tokyo 101, Japan) V.1- 1959-

**SHNSAS** Shinryo to Shinyaku. Medical Consultation and New Remedies. (Iji Shuppan Sha, 1-2-8 Shinkawa, Chuo-Ku, Tokyo 104, Japan) V.1- 1964-

**SHYCD4** Shengwu Huaxue Yu Shengwu Wuli Jinzhan. Progress in Biochemistry and Biophysics. (Kexue Chubanshe, 137 Zhaoyangmennei Dajie, Beijing, People's Republic of China) No. 1- 1974(?)-

**SHZAAY** Shoyakugaku Zasshi. (Kyoto Daigaku Yakugakubu, Yoshida Shimoadachi-machi, Sakyo-Ku, Kyoto, Japan) V.6- 1952-

**SIGAAE** Shika Igaku. Odontology. (Osaka Shika Gakkai, 1 Kyobashi, Higashi-Ku, Osaka, Japan) V.1- 1930-

**SIGZAL** Showa Igakkai Zashi. Journal of the Showa Medical Association. (Showa Daigaku, Showa Igakki, 1-5-8, Hatanodai, Shinagawa-ku, Tokyo 142, Japan) V.1- 1939-

**SIHSD8** Shang-hai I Hsueh. (Shanghi, China) V.1- 1978(?)-

**SIIPD4** Shanghai Diyi Yixueyuan Xuebao. Journal of the Shanghai First Medical College. (Beijing, Pep. Rep China) 1956- V.12 1985

**SinJF#** Personal Communication from J.F. Sina, Merck Institute for Therapeutic Research, West Point, PA 19486, to the Editor of RTECS, Cincinnati, OH, on October 26, 1982

**SIZSAR** Sapporo Igaku Zasshi. Sapporo Medical Journal. (Sapporo Igaku Daigaku, Nishi-17-Chome, Minami-1-jo, Chuo-ku Sapporo 060, Japan) V.3- 1952-

**SJDBA9** Soviet Journal Developmental Biology (English Translation). (Plenum Publishing Corp., 233 Spring St., New York, NY 10013) V.1- 1970-

**SJHAAQ** Scandinavian Journal of Haematology. (Munksgaard, 35 Norre Sogade, DK 1370 Copenhagen K, Denmark) V.1- 1964-

**SJHSBD** Scandinavian Journal of Haematology. Supplementum. (Munksgaard, Noerre Soegade, DK-1370 Copenhagen K, Denmark) No. 1- 1964-

**SJRDAH** Scandinavian Journal of Respiratory Diseases. (Munksgaard, 6 Norregade, DK-1165 Copenhagen K, Denmark) V.47- 1966-

**SJRHAT** Scandinavian Journal of Rheumatology. (Almqvist and Wiksell, POB 45150, S-10430 Stockholm, Sweden) V.1- 1972-

**SJUNAS** Scandinavian Journal of Urology and Nephrology. (Almqvist and Wiksell Periodical Co., POB 62, 26 Gamla Brogatan, S-101-20 Stockholm, Sweden) V.1- 1967-

**SKEZAP** Shokuhin Eiseigaku Zasshi. Journal of the Food Hygiene Society of Japan. (Nippon Shokuhin Eisei Gakkai, c/o Kokuritsu Eisei Shikenjo, 18-1, Kamiyoga 1-chome, Setagaya-Ku, Tokyo, Japan) V.1- 1960-

**SKIZAB** Shikoku Igaku Zasshi. Shikoku Medical Journal. (Tokushima Daigaku Igaku, Kuramoto-cho, Tokushima, Japan) V.1- 1950-

**SKKNAJ** Sankyo Kenkyusho Nempo. Annual Report of Sankyo Research Laboratories (Sankyo K.K. Chuo Kenkyusho, 1-2-58 Hiro-machi, Shinagawa-ku, Tokyo 140, Japan) No. 15- 1963-

**SKNEA7** Shionogi Kenkyusho Nempo. Annual Report of Shionogi Research Laboratory. (Shionogi Seiyaku K.K. Kenkyusho, 2-47 Sagisukami, Fukushima, Osaka, Japan) No. 1- 1951-

**SMBUA9** Stanford Medical Bulletin. (Palo Alto, CA) V.1-20, 1942-62. Discontinued

**SMDJAK** Scottish Medical Journal. (Longman Group Journal Division, 4th Ave., Harlow, Essex CM20 2JE, England) V.1- 1956-

**SMEZA5** Sudebno-Meditsinskaya Ekspertiza. Forensic Medical Examination. (v/o Mezhdunarodnaya Kniga, Kuznetskii Most 18, Moscow G-200, USSR.) V.1- 1958-

**SMJOAV** Southern Medical Journal. (Southern Medical Association, 2601 Highland Ave., Birmingham, AL 35205) V.1- 1908-

**SMSJAR** Scottish Medical and Surgical Journal. (Edinburgh, Scotland). For publisher information, see SMDJAK

**SMWOAS** Schweizerische Medizinische Wochenschrift. (Schwabe and Co., Steintorst 13, 4000 Basel 10, Switzerland) V.50- 1920-

**SNSHBT** Senshokutai. Chromosome. (Senshokutai Gakkai, Kaokusai Kirisutokyo Daigaku, Mitaka, Tokyo-to, Japan) No. 1- 1946- Adopted new numbering in 1976

**SOGEBZ** Soviet Genetics. Translation of GNKAA5. (Plenum Publishing Corp., 233 Spring St., New York, NY 10013). V.2- 1966-

**SOMEAU** Sovetskaia Meditsina. (v/o Mezhdunarodnaya Kniga, Kuznetskii Most 18, Moscow G-200, USSR.) V.1- 1937-

**SOVEA7** Southwestern Veterinarian. (College of Veterinary Medicine, Texas A and M University, College Station, TX 77843) V.1- 1948-

**SPCOAH** Soap, Perfumery and Cosmetics. (United Trade Press, Ltd., 33 Bowling Green Lane, London EC1R0DA, England) V.8, No. 7- 1935-

**SPEADM** Special Publication of the Entomological Society of America. (4603 Calvert Rd., College Park, MD 20740)

**SPERA2** Sperimentale. (Universita degli Studi Piazza San Marco, 4, 50121 Florence, Italy) V.50- 1896-

**SpiEW#** Personal Communication to Roger Tatken, NIOSH, from Dr. E.W. Spitz, E.P.A. on February 13, 1980

**SRTCDF** Scientific Report. (Dr. C.R. Krishna Murti, Director, Industrial Toxicology Research Centre, Luchnow-226001, India). History unknown

**SSBSEF** Scientia Sinica, Series B: Chemical, Biological, Agricultural, Medical, and Earth Sciences (English Edition). (Scientific and Technical Books Service Ltd., POB 197, London WC2N 4DE, England) V.25- 1982-

**SSCHAH** Soap and Sanitary Chemicals. (New York, NY) V.14-30, 1938-54. For publisher information, see SCCSC8

**SSCMBX** Scripta Scientifica Medica. (Durzhavno Izdatelstvo Meditsina i Fizkultura, P1. Slaveikov II, Sofia, Bulgaria) V.4(2)- 1965-

**SSEIBV** Sumitomo Sangyo Eisie. Sumitomo Industrial Health. (Sumitomo Byoin Sangyo Eisei Kenkyushitsu, 5-2-2, Nakanoshima, Kita-ku, Osaka, 530, Japan) No. 1- 1965-

**SSINAV** Scientia Sinica. (Peking, People's Republic of China) V.3-24, 1954-81. For publisher information, see SSBSEF

**SSSEAK** Studi Sassaresi, Sezione 2. (Societa Sassarese di Scienze Mediche e Naturali, Viale S, Pietro 12, I-07100 Sassari, Italy) V.42- 1964-

**SSZBAC** Sbornik Vysoke Skoly Zemedelske v Brne, Rada B: Spisy Veterinarni. (Brno, Czechoslovakia) 1953-66. For publisher information, see AUAAB7

**STBIBN** Studia Biophysica. (Akademie-Verlag GmbH, Liepziger Str. 3-4, DDR-108 Berlin, Germany) V.1- 1966-

**STCC\*\*** Stauffer Chemical Company, Product Safety Information Sheets. (Stauffer Chemical Company, Agricultural Chemical Division, Westport, CT 06880)

**STEDAM** Steroids. (Holden-Day Inc., 500 Sansome St., San Francisco, CA 94111) V.1- 1963-

**STEVA8** Science of the Total Environment. (Elsevier Scientific Publishing Co., P.O. Box 211, Amsterdam C, Netherlands) V.1- 1972-

**STOAAT** Stomatologiia (Moscow). (v/o Mezhdunarodnaya Kniga, Kuznetskii Most 18, Moscow G-200, USSR) V.1- 1923-

**StoGD#** Personal Communication to NIOSH from Fr. Gary D. Stoner, Dept. of Community Medicine, School of Medicine, University of California, La Jolla, CA 92037, May 25, 1975

**STRAAA** Strahlentherapie. (Urban and Schwarzenberg, Pattenkoferst 18, D-8000 Munich 15, Germany) V.1- 1912-

**STRHAV** Staub-Reinhaltung der Luft. (VDI-Verlag GmbH, Postfach 1139, D-4000 Duesseldorf 1, Germany) V.26- 1966-

**SUFOAX** Surgical Forum. (American College of Surgeons, 55 E. Erie St., Chicago, IL 60611) V.1- 1950-

**SUNCO*** Sun Company Material Safety Data Sheets. (1608 Walnut St., Philadelphia, PA 19103)

**SURGAZ** Surgery. (C.V. Mosby Co., 11830 Westline Industrial Dr., St. Louis, MO 63141) V.1- 1937-

**SURR**** Szechuan University Research Report, Natural Science. (Szechuan University, Szechuan, China) No. 2/3- 1978-

**SVLKAO** Sbornik Vedeckych Praci Lekarske Fakulty Univerzity Karlovy v Hradci Kralove. Collection of Scientific Works of the Charles University Faculty of Medicine in Hradec Kralove. (Univerzita Karlova, Ovocny Trh. 3, CS-116-36 Prague 1, Czechoslovakia) V.1- 1958-

**SWEHDO** Scandinavian Journal of Work, Environment and Health. (Haartmaninkatu 1, FIN-00290 Helsinki 29, Finland) V.1 1975-

**SWSPBE** Proceedings, Southern Weed Science Society. (Southern Weed Science Society, 309 W. Clark St., Champaign, IL 61820) V.22-1969-

**SYHJAM** Saengyak Hakhoechi. Journal of the Society of Pharmacognosy. (c/o Natural Products Research Institute, Seoul National University, 28 Yunkeon-Dong, Chong-ro-ku, Seoul 110, S. Korea) V.9- 1978-

**SYSWAE** Shiyan Shengwu Xuebao. Journal of Experimental Biology. (Guozi Shudian, China Publications Center, P.O. Box 399, Peking, People's Republic of China) V.1- 1953- (Suspended 1966-77)

**SYXUD2** Shandong Yixueyuan Xuebao. Journal of Shandong Medical College (China International Book Trading Corp., POB 2820, Beijing, People's Republic of China)

**SYXUE3** Shenyang Yaoxueyuan Xuebao. Journal of Shenyang College of Pharmacy. (Shenyang Yaoxueyuan, 7, 2-duan Wenhualu, Shenhequ, Shenyang, People's Republic of China) V.1- 1984-

**SZAPAC** Schweizerische Zietschrift fuer Allgemeine Pathologie und Bakteriologie. (White Plains, NY) V.1-3, 1938-40; V.13-22, 1950-59. For publisher information, see ECEBDI

**SZPMAA** Sozial-und Praeventivmedizin. (Art. Institut Orell Fuessli AG, Dietzingerstrasse 3, 8003 Zurich, Switzerland) V.19- 1974-

**SZTPA5** Schweizerische Zeitschrift fuer Tuberkulose und Pneumonologie. (Basel, Switzerland) V.1- 18, 1944-1961

**TABIA2** Tabulae Biologicae. (The Hague, Netherlands) V.1-22, 1925-63. Discontinued

**TAFSAI** Transactions of the American Fisheries Society. (American Fisheries Society, 5410 Grosvenor Lane, Bethesda, MD 20014) V.1- 1870-

**TAGTBR** Trudy Azerbaidzhanskogo Nauchno-Issledovatel'skogo Instituta Gigieny Truda i Professional'nykh Zabolevanii. Transactions of the Azerbaidzhan Scientific Research Institute of Labor Hygiene and Occupational Diseases. (The Institute, Baku, USSR.) No. 1- 1966-

**TAKHAA** Takeda Kenkyusho Ho. Journal of the Takeda Research Laboratories. (Takeda Yakuhin Kogyo K. K., 4-54 Juso-nishino-cho, Higashi Yodogawa-Ku, Osaka 532, Japan) V.29- 1970-

**TCMUD8** Teratogenesis, Carcinogenesis, and Mutagenesis. (Alan R. Liss, Inc., 150 Fifth Ave., New York, NY 10011) V.1- 1980-

**TCMUE9** Topics in Chemical Mutagenesis. (Plenum Publishing Corp., 233 Spring Street, New York, NY 10013) V.1- 1984-

**TDBU**** Ueber die Wirkung des Braunsteins, Paul Thissen Dissertation. (Pharmakologischen Institut der Tierarztlichen Hochschule zu Berlin, Germany, 1933)

**TDKNAF** Takeda Kenkyusho Nempo. Annual Report of the Takeda Research Laboratories. (Osaka, Japan) V.1-28, 1936-69. For publisher information, see TAKHAA

**TEARAI** Terapevticheskii Archiv. Archives of Therapeutics. v/o Mezhdunarodnaya Kniga, Kuznetskii Most 18, Moscow G-200, USSR) V.1- 19, 1923-41; V.20- 1948-

**TECSDY** Toxicological and Environmental Chemistry. (Gordon and Breach Science Pub. Inc., 1 Park Ave., New York, NY 10016) V.3(3/4)- 1981-

**TELEAY** Tetrahedron Letters. (Pergamon Press Ltd., Headington Hill Rd., Oxford OX3 OBW, England) 1959-

**TFAKA4** Trudy Instituta Fiziologii, Akademiya Nauk Kazakhshoi SSR. Transactions of the Institute of Physiology, Academy of Sciences of the Kazakh. SSR. (The Academy, Alma-Ata, USSR) V.1- 1955-

**TGANAK** Tsitologiya i Genetika. (v/o Mezhdunarodnaya Kniga, Kuznetskii Most 18, Moscow G-200, USSR) V.1- 1967-

**TGMEAJ** Tropical and Geographical Medicine. (DeErven F. Bohn NV, Box 66, Harlem, Netherlands) V.10- 1958-

**TGNCDL** "Handbook of Organic Industrial Solvents" 2nd ed., Chicago, IL, National Association of Mutual Casualty Companies, 1961

**THAGA6** Theoretical and Applied Genetics. (Springer-Verlag New York, Inc., Service Center, 44 Hartz Way, Secaucus, NJ 07094) V.38- 1968-

**THBRAA** Thrombosis Research. (Pergamon Press, Maxwell House, Fairview Park, Elmsford, NY 10523) V.1- 1972-

**THERAP** Therapie. (Doin, Editeurs, 8 Place de l'Odeon, Paris 6, France) V.1- 1946-

**THEWA6** Therapiewoche. (G. Braun GmbH, Postfach 1709, 7500 Karlsruhe 1, Germany) V.1- 1950-

**THGNBO** Theriogenology. (Geron-X, Inc., P.O. Box 1108, Los Altos, CA 94022) V.1- 1974-

**TIDZAH** Tokyo Ika Daigaku Zasshi. Journal of Tokyo Medical College. (Tokyo Ika Daigaku Igakkai, 1-412, Higashi Okubo, Shinjuku-ku, Tokyo, Japan) 1918-

**TIEUA7** Tieraerztliche Umschau. (Terra-Verlag, Neuhauser. Str. 21, 7750 Constance, Germany) V.1- 1946-

**TIHHAH** Tai-wan I Hsueh Hui Tsa Chih. Journal of the Formosan Medical Association. (Tai-wan I Hsueh Hui, 1, Sect.1 Jen-ai Road, Taipei, Taiwan) V.45- 1946-

**TIUSAD** Tin and Its Uses. (Tin Research Institute, 483 W. 6th Ave., Columbus, OH 43201) No. 1- 1939-

**TIVSAI** Trudy, Vsesoyuznyi Nauchno-Issledovatel'skii Institut Veterinarnoi Sanitarii. Transactions, All-Union. Scientific Research Institute of Veterinary Sanitation. (The Institute, Moscow, USSR.) V.11- 1957-

**TIYADG** Tianjin Yiyao. Tianjin Medical Journal. (Guoji Shudian, China Publications Center, POB 399, Peking, China) V.1- 1973-

**TJADAB** Teratology, A Journal of Abnormal Development. (Wistar Institute Press, 3631 Spruce St., Philadelphia, PA 19104) V.1- 1968-

**TJEMAO** Tohoku Journal of Experimental Medicine. (Maruzen Co., Export Dept., P.O. Box 5050, Tokyo Int., 100-31 Tokyo, Japan) V.1- 1920-

**TJIDAH** Tokyo Jikeikai Ika Daigaku Zasshi. Tokyo Jikeikai Medical Journal. (3-25-8, Nishi Shinbashi, Minato-ku, Tokyo, Japan) V.66- 1951-

**TJIZAF** Tokyo Joshi Ika Daigaku Zasshi. Journal of Tokyo Women's Medical College. (Society of Tokyo Women's Medical College, c/o Tokyo Joshi Ika Daigaku Toshokan, 10, Kawada-cho, Shinjuku-ku, Tokyo 162, Japan) V.1- 1931-

**TJSGA8** Tijdschrift Voor Sociale Geneeskunde. (B.V. Uitgeversmaatschappij. Reflex, Mathenesserlaan 310, Rotterdam 3003, Netherlands) V.1- 1923-

**TJXMAH** Tokushima Journal of Experimental Medicine. (Tokushima Diagaku Igakubu, 3, Kuramoto-cho, Tokushima, Japan) V.1- 1954-

**TKIZAM** Tokyo Igaku Zasshi. Tokyo Journal of Medical Sciences. (Tokyo, Japan) V.59-76, 1951-68

**TKORAS** Trudy Kazakhskogo Nauchno-Issledovatel'skogo Instituta Onkologii i Radiologii. (The Institute, Alma-Ata, USSR.) V.1- 1965-

**TMPBAX** Trudy Moskovskogo Obshchestva Ispytatelei Prirody. Transactions of the Moscow Society of Naturalists. (Moskovskoe Obshchestvo Ispytaltelei, Prirody, Moscow, USSR) V.1- 1951-

**TMPRAD** Tropenmedizin und Parasitologie. (Georg Thieme Verlag, Postfach 732, Herdweg 63, 7000 Stuttgart, Germany) V.25- 1974-

**TNEOAO** Transactions of the New England Obstetrical and Gynecological Society. (Publisher unknown)

**TNICS*** Toxicology of New Industrial Chemical Substances. (USSR. Academy of Medical Sciences, Moscow-Meditsina)

**TOANDB** Toxicology Annual. (Marcel Dekker, POB 11305, Church St. Station, New York, NY 10249) 1974/75-

**TobJS#** Personal Communication to the Editor, Toxic Substances List from Dr. J.S. Tobin, Director, Health and Safety, FMC Corporation, New York, NY 14103, November 9, 1973

**TOERD9** Toxicological European Research. (Editions Ouranos, 12 bis, rue Jean-Jaures, 92807 Puteaux, France) V.1- 1978-

**TOFOD5** Tokishikoroji Foramu. Toxicology Forum. (Saiensu Foramu, c/o Kida Bldg., 1-2-13 Yushima, Bunkyo-ku, Tokyo, 113, Japan) V.6- 1983-

**TOIZAG** Toho Igakkai Zasshi. Journal of Medical Society of Toho University. (Toho Daigaku Igakubu Igakkai, 5-21-16, Omori, Otasku, Tokyo, Japan) V.1- 1954-

**TOLED5** Toxicology Letters. (Elsevier Scientific Publishing Co., P.O. Box 211, Amsterdam, Netherlands) V.1- 1977-

**TOPADD** Toxicologic Pathology. (E.I. du Pont de Nemours Co., Inc., Elkton Rd., Newark, DE 19711) V.6(3/4)- 1978-

**TORAAK** El Torax. (Montevideo, Uruguay) V.1- 1952-

**TOSUAH** Transactions of the Ophthalmological Societies of the UK. (Churchill Livingstone Inc., 19 W. 44th St., New York, NY 10036) V.1- 1880-

**TOXIA6** Toxicon. (Pergamon Press, Headington Hill Hall, Oxford OX3 OBW, England) V.1- 1962-

**TOXID9** Toxicologist. (Soc. of Toxicology, Inc., 475 Wolf Ledge Parkway, Akron, OH 44311) V.1- 1981-

**TPHSDY** Trends in Pharmacological Sciences. (Elsevier North Holland, Inc., 52 Vanderbilt Ave., New York, NY 10017) V.1- 1979-

**TPKVAL** Toksikologiya Novykh Promyshlennykh Khimicheskikh Veshchestv. Toxicology of New Industrial Chemical Sciences. (Akademiya Meditsinskikh Nauk S.S.R., Moscow, USSR.) No. 1- 1961-

**TRBMAV** Texas Reports on Biology and Medicine. (Texas Reports, University of Texas Medical Branch, Galveston, TX 77550) V.1- 1943-

**TRENAF** Kenkyu Nenpo-Tokyo-toritsu Eisei Kenkyusho. Annual Report of Tokyo Metropolitan Research Laboratory of Public Health. (24-1, 3 Chome, Hyakunin-cho, Shin-Juku-Ku, Tokyo, Japan) V.1- 1949/50-

**TRIPA7** Tin Research Institute, Publication. (The Institute, Fraser Rd., Greenford UB6 7AQ, Middlesex, England) No. 1- 1934-

**TRPLAU** Transplantation. (Williams & Wilkins, Co., 428 E. Preston St., Baltimore, MD 21202) V.1- 1963-

**TRSTAZ** Transactions of the Royal Society of Tropical Medicine and Hygiene. (The Society, Manson House, 26 Portland Pl., London W1N 4EY, England) V.14- 1920-

**TSCAT\*** Office of Toxic Substances Report. (U.S. Environmental Protection Agency, Office of Toxics Substances, 401 M Street SW, Washington, DC 20460)

**TSITAQ** Tsitologiya. Cytology. (v/o Mezhdunarodnaya Kniga, Kuznetskii Most 18, Moscow G-200, USSR) V.1 1959-

**TSPMA6** Travaux de la Societe de Pharmacie de Montpellier. (Soc. de Pharmacie de Montpellier, Faculte de Pharmacie de Montpellier, Ave. Ch.-Flahault, 34060 Montpellier, France) V.1- 1942-

**TUBEAS** Tubercle. (Williams & Wilkins, 428 E. Preston Street, Baltimore, MD 21202) V.1- 1919-

**TUMOAB** Tumori. (Casa Editrice Ambrosiana, Via G. Frua 6, 20146 Milan, Italy) V.1- 1911-

**TVLAA5** Trudy Vsesoyuznogo Nauchno-Issledovatel'skogo Instituta Lekarstvennykh Rastenii. Transactions of the All-Union Scientific Research Institute of Medicinal Plants. (Vsesoyuznyi Nauchno-Issledovatel'skii Institut Lekarstvennykh Rastenii, Bittsa, USSR.) No. 5- 1936-

**TWHPA3** Tung Wu Hsueh Pao. Journal of Zoology. (Guozi Shudian, China Publications Center, POB 399, Peking, People's Republic of China) V.1- 1949-

**TXAPA9** Toxicology and Applied Pharmacology. (Academic Press, 111 5th Ave., New York, NY 10003) V.1- 1959-

**TXCYAC** Toxicology. (Elsevier Scientific Pub. Ireland, Ltd., POB 85, Limerick, Ireland) V.1- 1973-

**TXMDAX** Texas Medicine. (Texas Medical Association, 1905 N. Lamar Blvd., Austin, TX 78705) V.60- 1964-

**TYKNAQ** Tohoku Yakka Daigaku Kenkyu Nempo. Annual Report of the Tohoku College of Pharmacy. (77, Odawara Minamikotaku, Hara-machi, Sendai, Japan) No. 10- 1963-

**UBZHAZ** Ukrains'kii Biokhimichnii Zhurnal (1946-77). Ukranian Biochemical Journal. (Kiev, USSR) V.18-49, 1946-77. For publisher information, see UBZHD4

**UBZHD4** Ukrainskii Biokhimicheskii Zhurnal. Ukranian Biochemical Journal. v/o Mezhdunarodnaya Kniga, Kuznetskii Most 18, (Moscow G-200, USSR) V.50- 1978-

**UCDS\*\*** Union Carbide Data Sheet. (Industrial Medicine and Toxicology Dept., Union Carbide Corp., 270 Park Ave., New York, NY 10017)

**UCMH\*\*** Union Carbide Manual Hazard Sheet. (Industrial Medicine and Toxicology Dept., Union Carbide Corp., 270 Park Ave., New York, NY 10017)

**UCPHAQ** University of California, Publications in Pharmacology. (Berkeley, CA) V.1-3, 1938-57. Discontinued

**UCREAR** Urologic and Cutaneous Review. (St. Louis, MO) V.17-56, 1913-52. Discontinued

**UCRR\*\*** Union Carbide Research Report. (Union Carbide Corp., Old Ridgebury Rd., Danbury, CT 06817)

**UDHU\*\*** Zur Frage der Entgiftung des Thalliums, Gerhart Urban Dissertation. (Pharmakologischen Institut der Martin Luther-Universitat Halle-Wittenberg, Germany, 1936)

**UGLAAD** Ugeskrift for Laeger. (Almindelige Danske Laegeforening, Kristianiagade 12A, 2100 Copenhagen, Denmark) V.1- 1839-

**UICMAI** UICC Monograph Series. (Springer-Verlag New York, Inc., Service Center, 44 Hartz Way, Secaucus, NJ 07094)

**UJMSAP** Upsala Journal of Medical Sciences. (Almqvist and Wiksell, Periodical Co., POB 62, 26 Gamla Brogatan, S-101, 20 Stockholm, Sweden) V.77- 1972-

**UMJOAJ** Ulster Medical Journal. (Ulster Medical Society, Belfast City Hospital, Belfast BT9 7AD, N. Ireland) V.1- 1932-

**UNMI\*\*** University Microfilms International. (300 N. Zeeb Rd., Ann Arbor, MI 48106)

**UPJOH\*** "Compounds Available for Fundamental Research, Volume II-6, Antibiotics, A Program of Upjohn Company Research Laboratory." (Kalamazoo, MI 49001)

**URGABW** Urologe Ausgabe A. (Springer-Verlag, Heidelberger Pl. 3, D-1000 Berlin 33, Germany) V.9- 1970-

**URLRA5** Urological Research. (Springer-Verlag New York, Inc., Service Center, 44 Hartz Way, Secaucus, NJ 07094) V.1- 1973-

**UROTAQ** Urologia. (Libreria Editrice Canova, Via Panciera, 3b, 31100 Treviso, Italy) V.1- 1934-

**USBCC\*** U.S. Borax and Chemical Company Reports. (New York, NY)

**USXXAM** United States Patent Document. (Commissioner of Patents and Trademarks, Washington, DC 20231)

**UZBZAZ** Uzbekskii Biologicheskii Zhurnal. (v/o Mezhdunarodnaya Kniga, Kuznetskii Most 18, Moscow G-200, USSR) No. 1- 1958-

**VAAPB7** Virchows Archiv, Abteilung A: Pathologische Anatomie. (Berlin, Germany) V.344-361, 1968-73. For publisher information, see VAPHDQ

**VAAZA2** Virchows Archiv, Abteilung B: Zellpathologie, Cell Pathology. (Springer-Verlag, Heidelberger Pl. 3, D-1 Berlin 33, Germany) V.1- 1968-

**VAMNAG** Vestnik Akademii Meditsinskikh Nauk S.S.S.R. Journal of the Academy of Medical Sciences of the USSR. (v/o Mezhdunarodnaya Kniga, Kuznetskii Most 18, Moscow G-200, USSR) V.1- 1946-

**VAPHDQ** Virchows Archiv, Abteilung A: Pathological Anatomy and Histology. (Springer-Verlag, Heidelberger pl.3, D-1 Berlin. 33, Germany) V.362- 1974-

**VCTDC\*** Vanderbilt, R.T., Co., Technical Data Sheet. (230 Park Ave., New York, NY 10017)

**VDGIA2** Verhandlungen der Deutschen Gesellschaft fuer Innere Medizin. (J.F. Bergmann Verlag, Trogerstrasse 56, 8000 Munich 80, Germany) V.33-52, 1921-40; V.54- 1948-

**VDGPAN** Verhandlungen der Deutschen Gesellschaft fuer Pathologie. (Gustav Fischer Verlag, Postfach 53, Wollgrasweg 49, 7000 Stuttgart-Hohenheim, Germany) V.1- 1898-

**VEARA6** Veterinarski Arhiv. (Jugoslovenska Knjiga, P.O. Box 36, Terazije 27, 11001, Belgrade, Yugoslavia) V.1- 1931-

**VELPB\*** Velsicol Chemical Corporation, Product Bulletin. (341 E. Ohio St., Chicago, IL 60611)

**VEMJA8** Veterinary Medical Journal. (Cairo Univ., Faculty of Veterinary Medicine, Giza, Egypt) V.1- 1953-

**VETNAL** Veterinariya. (v/o Mezhdunarodnaya Kniga, Kuznetskii Most 18, Moscow G-200, USSR) V.1- 1924-

**VETRAX** Veterinary Record. (7 Mansfield St., London, W1M 0AT, England) V.1- 1888-

**VHAGAS** Verhandlungen der Anatomischen Gesellschaft. (VEB Gustav Fischer, Postfach 176, DDR-69 Jena, Germany) V.1- 1887-

**VHTODE** Veterinary and Human Toxicology. (American College of Veterinary Toxicologists, Office of the Secretary-Treasurer, Comparative Toxicology Laboratory, KS State University, Manhatten, Kansas 66506) V.19- 1977-

**VIHOAQ** Vitamins and Hormones. (Academic Press, 111 5th Ave., New York, NY 10003) V.1- 1943-

**VINIT\*** Vsesoyuznyi Institut Nauchnoi i Tekhnicheskoi Informatsii (VINITI). All-Union Institute of Scientific and Technical Information. (Moscow, USSR)

**VIRLAX** Virology. (Academic Press, 111 Fifth Ave., New York, NY 10003) V.1- 1955-

**VKMGA7** Voprosy Kommunal'noi Gigieny. Problems of Communal Hygiene. (Kiev, USSR.) V.1-6, 1963-66. For publisher information, see GNAMAP

**VMDNAV** Veterinarno-Meditsinski Nauki. (Hemus, Blvd. Russki 6, Sofia, Bulgaria) V.1- 1964-

**VOONAW** Voprosy Onkologii. Problems of Onkology. (v/o Mezhdunarodnaya Kniga, Kuznetskii Most 18, Moscow G-200, USSR.) V.1- 1955-

**VPITAR** Voprosy Pitaniya. Problems of Nutrition. (v/o Mezhdunarodnaya Kniga, Kuznetskii Most 18, Moscow G-200, USSR.) V.1- 1932-

**VRDEA5** Vrachebnoe Delo. Medical Profession. (v/o Mezhdunarodnaya Kniga, Kuznetskii Most 18, Moscow G-200, USSR.) No. 1- 1918-

**VRRAAT** Vestnik Rentgenologii i Radiologii. Bulletin of Roentgenology and Radiology. (v/o Mezhdunarodnaya Kniga, Kuznetskii Most 18, Moscow G-200, USSR.) V.1- 1920-

**VTPHAK** Veterinary Pathology. (S. Karger AG, Postfach CH-4009 Basel, Switzerland) V.8- 1971-

**VTYMAC** VM/SAC, Veterinary Medicine and Small Animal Clinician. (Veterinary Medicine Publishing Co., 690 S. 4th St., POB 13265, Edwardsville, KS 66113) V.59-1964-

**VVIRAT** Voprosy Virusologii. (v/o Mezhdunarodnaya Kniga, Kuznetskii Most 18, Moscow G-200, USSR) V.1- 1956-

**WADTAA** Wright Air Development Center, Technical Report, U.S. Air Force. (Wright-Patterson Air Force Base, OH 45433)

**WATRAG** Water Research. (Pergamon Press Ltd., Headington Hill Hall, Oxford OX3 0BW, England) V.1- 1967-

**WDMBAM** Wadley Medical Bulletin. (Dallas, TX) V.1-5, 1970-75. For publisher information, see JCHODP

**WDMU\*\*** Vergleichend-pharmakologische Untersuchungen uber die haemolytische Wirkung und die Toxizitat der Lokalanaesthetika, Martin Walther Dissertation. (Pharmakologischen Institut der Martin Luther-Universitat Halle-Wittenburg, Germany, 1936)

**WEHRBJ** Work, Environment, Health. (Helsinki, Finland) V.1-11, 1962-74. For publisher information, see SWEHDO

**WHOTAC** World Health Organization, Technical Report Series. (1211 Geneva, 27, Switzerland) No. 1- 1950-

**WHYHAQ** Wuhan Yixueyuan Xuebao. (China International Book Trading Corp., POB 2820, Beijing, People's Republic of China) No. 1-1957- Suspended 1959-76(?)

**WILEAR** Wiadomosci Lekarskie. (Ars Polona-RUSH, POB 1001, P-00068 Warsaw 1, Poland) V.1- 1948-

**WIMJAD** West Indian Medical Journal. (West Indian Medical Journal, University of the West Indies, Faculty of Medicine, Kingston 7, Jamaica) V.1- 1951-

**WJMDA2** Western Journal of Medicine. (California Medical Association, 731 Market St., San Francisco, CA 94103) V.120- 1974-

**WKMRAH** Wakayama Medical Report. (Wakayama Medical College Library, 9,9-bancho, Wakayama, 640, Japan) V.1- 1953-

**WKWOAO** Wiener Klinische Wochenschrift. (Springer-Verlag, Postfach 367, A-1011 Vienna, Austria) V.1- 1888-

**WMWOA4** Wiener Medizinische Wochenschrift. (Brueder Hollinek, Gallgasse 40A, A-1130 Vienna, Austria) V.1- 1851-

**WolMA#** Personal Communication to Henry Lau, Tracor Jitco, Inc., from Mark A. Wolf, Dow Chemical Co., Midland, MI 48640

**WQCHM\*** "Water Quality Characteristics of Hazardous Materials" W. Hann, and P.A. Jensen, Environmental Engineering Division, Civil Engineering Department, Texas A and M University, vol. 1-4, 1974

**WRABDT** Wilhelm Roux's Archives of Developmental Biology. (Springer-Verlag New York, Inc., Service Center, 44 Hartz Way, Secaucus, NJ 07094) V.177- 1975-

**WRPCA2** World Review of Pest Control. (London, England) V.1-10, 1962-71. Discontinued

**WTMOA3** Wiener Tieraerztliche Monatsschrift. (Ferdinand Berger and Soehne OHG, Wiene Str. 21-23, A-3580 Horn, Austria) V.1- 1914-

**WZERDH** Wissenschaftliche Zeitschrift der Ernst-Moritz-Arndt-Universitaet Greifswald, Medizinische Reihe. (Buchexport, Postfach 160, Leipzig DDR-7010, Germany) V.21- 1972-

**XAWPA2** U.S. Army Chemical Warfare Laboratories. (Army Chemical Center, MD)

**XENOBH** Xenobiotica. (Taylor and Francis Ltd., 4 John St., London WC1N 2ET, England) V.1- 1971-

**XEURAQ** U.S. Atomic Energy Commission, University of Rochester, Research and Development Reports. (Rochester, NY)

**XPHBAO** U.S. Public Health Service, Public Health Bulletin. (Washington, DC)

**XPHCI\*** U.S. Public Health Service, Current Intelligence Bulletin. (National Institute for Occupational Safety and Health, 5600 Fishers Lane, Rockville, MD 20857)

**XPHPAW** U.S. Public Health Service Publication. (U.S. Government Printing Office, Superintendent of Documents, Washington, DC 20402) No. 1-1950-

**YACHDS** Yakuri to Chiryo. Pharmacology and Therapeutics. (Raifu Saiensu Pub. Co., 5-3-3 Yaesu, Chuo-ku, Tokyo 104, Japan) V.1- 1973-

**YAHOA3** Yakhak Hoeji. Journal of the Pharmaceutical Society of Korea. (Seoul Taehakkyo Yakhak Taehak, 28 YonGondong, Chongnogu, Seoul, S. Korea) V.1- 1956(?)-

**YAKUD5** Gekkan Yakuji. Pharmaceuticals Monthly. (Yakugyo Jihosha, Inaoka Bldg., 2-36 Jinbo-cho, Kandu, Chiyoda-ku, Tokyo 101, Japan) V.1-1959-

**YFZADL** Yaowu Fenxi Zazhi. Journal of Pharmaceutical Analysis. (Guoji Shudian, POB 2820, Beijing, People's Republic of China) V.1- 1981-

**YHHPAL** Yaoxue Xuebao. Acta Pharmaceutica Sinica. Pharmaceutical Journal. (China International Book Trading Corp., POB 2820, Beijing, People's Republic of China) V.1- 1953- (Suspended 1966-78)

**YHTPAD** Yaoxue Tongbao. Bulletin of Pharmacology. (China International Book Trading Corp., POB 2820, Beijing, People's Republic of China) V.1-12, 1953-66; V.13- 1978-

**YIGODN** Yiyao Gongye. Pharmaceutical Industry. (c/o Shanghai Yiyao Gongye Yanjiuyuan, 1320 Beijing-Xilu, Shanghai 200040, People's Republic of China). Beginning history not known

**YIKUAO** Yamaguchi Igaku. (Yamaguchi Daigaku Igakkai, Kogushi Ube, Yamaguchi-Ken 755, Japan) V.1- 1953-

**YJBMAU** Yale Journal of Biology and Medicine. (The Yale Journal of Biology and Medicine, Inc., 333 Cedar Street, New Haven, CT 06510) V.1-1928-

**YKGKAM** Yukagaku. Journal of Japan Oil Chemists' Society. (Nippon Yukagaku Kyokai, c/o Yushi Kogyo Kaigan, 3-13-11 Nihonbashi Chuo-Ku, Tokyo, Japan) V.5- 1956-

**YKIGAK** Yokohama Igaku. (Yokohama-Shiritsu Daigaku Igakkai, 2-33 Urafune-cho, Minami-Ku, Yokohama 232, Japan) V.1- 1948-

**YKKZAJ** Yakugaku Zasshi. Journal of Pharmacy. (Nippon Yakugakkai, 12-15-501, Shibuya 2-chome, Shibuya-ku, Tokyo 150, Japan) No. 1- 1881-

**YKRYAH** Yakubutsu Ryoho. Medicinal Treatment. (Iji Nipposha, Kamiya Bldg., 4-11-7 Hatchobori, Chuo-Ku, Tokyo 104, Japan) V.1- 1968-

**YKYUA6** Yakkyoku. Pharmacy. (Nanzando, 4-1-11, Yushima, Bunkyo-ku, Tokyo, Japan) V.1- 1950-

**YMBUA7** Yokohama Medical Bulletin. (Npg. Yokohama Ika Daigaku, Urafune-cho, Minato-ku, Yokohama, Japan) V.1- 1950-

**YOIZA3** Yonago Igaku Zasshi. Journal of the Yonago Medical Association. (c/o Tottori Daigaku Igakubu, Nishi-machi, Yonago, Japan) V.1- 1948-

**YOMJA9** Yonsei Medical Journal. (Yonsei University College of Medicine, P.O. Box 71, Seoul, S. Korea) V.1- 1960-

**YRTMA6** Yonsei Reports on Tropical Medicine. (Yonsei University, Institute of Tropical Medicine, Box 17, Seoul, S. Korea) V.1- 1970-

**ZAARAM** Zentralblatt fuer Arbeitsmedizin und Arbeitsschutz. (Dr. Dietrich Steinkopff Verlag, Saalbaustr 12,6100 Darmstadt, Germany) V.1- 1951-

**ZACCAL** Zacchia. Archivio di Medicina Legale, Sociale e Criminologica. Zacchia. Archives of Forensic, Social and Criminological Medicine. (Zacchia, Viale Regina Elena 336, Rome 00161, Italy) V.1- 1921-

**ZANZA9** Zeitschrift Angewandte Zoologie. (Duncker und Humblot, Postfach 410329, D-1000 Berlin 41, Germany) V.41- 1954-

**ZAPOAK** Zeitschrift fuer Allgemeine Mikrobiologie. Morphologie, Physiologie, Genetik und Oekologie der Mikroorganismen. (Akademie-Verlag GmbH, Liepziger Str. 3-4, DDR-108 Berlin, Germany) V.1- 1960-

**ZAPPAN** Zentralblatt fuer Allgemeine Pathologie und Pathologische Anatomie. (VEB Gustav Fischer Verlag, Postfach 176, Villengang 2, 69 Jena, Germany) V.1- 1890-

**ZBPHA6** Zentralblatt fuer Bakteriologie, Parasitenkunde, Infektionskrankheiten und Hygiene, Abteilung I: Medizinisch-Hygienische Bakteriologie, Virusforschung und Parasitologie, Originale. (Stuttgart, Germany) V.31-151, 1902-45; V.152-216, 1947-71. For publisher information, see ZMMPAO

**ZBPIA9** Zentralblatt fuer Bakteriologie, Parasitenkunde, Infektionskrankheiten und Hygiene, Abteilung II. Naturwissenschattliche: Allgemeine, Landwirtschattliche und Technische Miknobiologie. (VEB Gustav Fischer Verlag, Postfach 176, DDR-69 Jena, Germany) V.107-132, 1952-77

**ZBPMAL** Zentralblatt fuer Pharmazie. (Nurenburg, Germany) V.1-29, 1904-33. Discontinued

**ZDBEA9** Zdravookhranenie Belorussii. Public Health Service of White Russia. (v/o Mezhdunarodnaya Kniga, Kuznetskii Most 18, Moscow G-200, USSR) V.1- 1955-

**ZDKAA8** Zdravookhranenie Kazakhstana. Public Health of Kazakhstan. (v/o Mezhdunarodnaya Kniga, Kuznetskii Most 18, Moscow G-200, USSR.) V.1-1941-

**ZDTUAB** Zdravookhranenie Turkmenistana. Public Health of Turkmenistan. (v/o Mezhdunarodnaya Kniga, 113095 Moscow, USSR) V.1- 1957-

**ZDVEA7** Zdravstveni Vestnik. Medical Journal. (Jugoslovenska Knjiga, POB 36, Terazije 27, YU-11001 Belgrade, Yugoslavia) V.15- 1946-

**ZDVKAP** Zdravookhranenie. Public Health. (v/o Mezhdunarodnaya Kniga, Kuznetskii Most 18, Moscow G-200, USSR) V.1- 1958-

**ZEGYAX** Zentralblatt fuer Gynaekologie. (Johann Ambrosius Barth Verlag, Postfach 109, DDR-7010, Leipzig, Germany) V.1- 1877-

**ZEKBAI** Zeitschrift fuer Krebsforschung. (Berlin, Germany) V.1-75, 1903-71. For publisher information, see JCROD7

**ZEKIA5** Zeitschrift fuer Kinderheilkunde. (Springer-Verlag, Heidelberger Pl. 3, D-1 Berlin 33, Germany) V.1- 1910-

**ZENBAX** Zeitschrift fuer Naturforschung, Ausgabe B, Chemie, Biochemie, Biophysik, Biologie und Verwandten Gebieten. (Verlag der Zeitschrift fuer Naturforschung, Postfach 2645, D-7400, Tuebingen, Germany) V.2-1947-

**ZEPHAR** Zentralblatt fuer Physiologie. (Leipzig, Germany) V.1-34, 1887-1921. For publisher information, see PFLABK

**ZEPRAN** Zeitschrift fuer Praeventivmedizin. (Zurich, Switzerland) V.1-13, 1956-68. For publisher information, see SZPMAA

**ZEPTAT** Zeitschrift fuer Experimentelle Pathologie und Therapie. (Berlin, Germany) V.1-22, 1905-21. For publisher information, see REXMAS

**ZERNAL** Zeitschrift fuer Ernaehrungswissenschaft. (Steinkopff Verlag, Postfach 1008, 6100 Darmstadt, Germany) V.1- 1960-

**ZEVBA5** Zeitschrift fuer Verebungslehre. (Springer-Verlag, Heidelberger Pl. 3, D-1 Berlin 33, Germany) V.89-98, 1958-66

**ZGEMAZ** Zeitschrift fuer die Gesamte Experimentelle Medizin. (Berlin, Germany) V.1-139, 1913-65. For publisher information, see REXMAS

**ZGIMAL** Zeitschrift fuer die Gesamte Innere Medizin und Ihre Grenzgebiete. (VEB Georg Thieme, Hainst 17/19, Postfach 946, 701 Leipzig, Germany) V.1- 1946-

**ZGSHAM** Zeitschrift fuer Gesundheitstechnik und Staedtehygiene. (Berlin, Germany) V.21-27, 1929-35. For publisher information, see ZANZA9

**ZHINAV** Zeitschrift fuer Hygiene und Infektionskrankheiten. (Berlin, Germany) V.11-151, 1892-1965. For publisher information, see MMIYAO

**ZHPMAT** Zentralblatt fuer Bakteriologie, Parasitenkunde, Infektionskran- heiten und Hygiene, Abteilung 1: Originale, Reihe B: Hygiene, Praeventive Medizin. (Gustav Fischer Verlag, Postfach 72-01-43, D-7000 Stuttgart 70, Germany) V.155- 1971-

**ZHYGAM** Zeitschrift fuer die Gesamte Hygiene und Ihre Grenzgebiete. (VEB Georg Thieme, Hainst 17/19, Postfach 946, 701 Leipzig, Germany) V.1- 1955-

**ZIEKBA** Zeitschrift fuer Immunitaetsforschung, Experimentelle und Klinische Immunologie. (Fischer, Gustav, Verlag, Postfach 53, Wollgrasweg 49, 7000 Stuttgart-Hohenheim, Germany) V.141- 1970-

**ZIETA2** Zeitschrift fuer Immunitaetsforschung und Experimentelle Therapie. (Stuttgart, Germany) 1924-V.124, 1962. For publisher information, see ZIEKBA

**ZKKOBW** Zeitschrift fuer Krebsforschung und Klinische Onkologie. (Berlin, Germany) V.76-92, 1971-78. For publisher information, see JCROD7

**ZKMAAX** Zhurnal Eksperimental'noi i Klinicheskoi Meditsiny. (v/o Mezhdunarodnaya Kniga, 121200 Moscow, USSR) V.2- 1962-

**ZKMEAB** Zeitschrift fuer Klin Medicine. (Berlin, Germany) V.1-158, 1879-1965. For publisher information, see EJCIB8

**ZLUFAR** Zeitschrift fuer Lebensmittel-Untersuchung und-Forschung. (Springer-Verlag, Heidelberger Pl. 3, D-1 Berlin 33, Germany) V.86-1943-

**ZMEIAV** Zhurnal Mikrobiologii, Epidemiologii i Immunobiologii. (v/o Mezhdunarodnaya Kniga, Kuznetskii Most 18, Moscow G-200, USSR.) V.14- 1935-

**ZMMPAO** Zentralblatt fuer Bakteriologie, Parasitenkunde, Infektionskrankh eiten und Hygiene, Abteilung 1: Originale, Reihe A: Medizinische Mikrobiologie und Parasitologie. (Gustav Fischer Verlag, Postfach 72-01-43, D-7000 Stuttgart 70, Germany) V.217- 1971-

**ZMWIAJ** Zentralblatt fuer die Medizinischen Wissenschaften. (Berlin, Germany)

**ZNCBDA** Zeitschrift fuer Naturforschung, Section C: Biosciences. (Verlag der Zeitschrift fuer Naturforschung, Postfach 2645, D-7400 Tuebingen, Germany) V.29C- 1974-

**ZNTFA2** Zeitschrift fuer Naturforschung. (Wiesbaden, Germany) V.1, No. 1-12 1946. For publisher information, see ZENBAX

**ZOBIAU** Zhurnal Obshchei Biologii. Journal of General Biology. (v/o Mezhdunarodnaya Kniga, Kuznetskii Most 18, Moscow G-200, USSR.) V.1-1940-

**ZolH##** Personal Communication from H. Zollner, Institut fur Biochemie, Der Universitat, Gras, to H. Lau, Tracor Jitco, Inc., October 23, 1975

**ZPPLBF** Zentralblatt fuer Pharmazie, Pharmakotherapie und Laboratoriums-diagnostik. (VEB Verlag Volk und Gesundheit, Neue Gruenstr 18, 102 Berlin, Germany) V.109- 1970-

**ZSNUAI** Zeitschrift fuer Neurologie. (Berlin) V.198-206, 1970-74

**ZTMPA5** Zeitschrift fuer Tropenmedizin und Parasitologie. (Stuttgart, Germany) V.1-24, 1949-73. For publisher information, see TMPRAD

**ZUBEAQ** Zeitschrift fuer Unfallmedizin und Berufskrankheiten. (Revue de Medecine des Accidents et des Maladies Professionelles. (Verlag Berichthaus, Zwingliplatz 3, Postfach, CH-8022, Zurich, Switzerland) V.1-1907-

**ZVKOA6** Zhurnal Vsesoyuznogo Khimicheskogo Obshchestva Imeni D.I. Mendeleeva. Journal of the D.I. Mendeleeva All-Union Chemical Society. (v/o Mezhdunarodnaya Kiga, Kuznetskii Most 18, Moscow G-200, USSR.) V.5- 1960-

**ZVRAAX** Zentralblatt fuer Veterinaermedizin, Reihe A. (Paul Parey, Linderst 44-47, 1000 Berlin 61, Germany) V.10- 1963-

**ZYDXDM** Zhejiang Yike Daxue Xuebao. Journal of Zhejiang Medical University. (Zhejiang Yike Daxue, Yan'an Lu, Hanzhou, Zhejiang, People's Republic of China) V.1- 1972-

**11FYAN** "Fluorine Chemistry, Volume III. Biological Effects of Organic Fluorides" J.H. Simons, ed., New York, NY, Academic Press, 1963

**13BYAH** "Morphological Precursors of Cancer" Lucio Severi ed., Proceedings of an International Conference, Universita Degli Studi, Perugia, Italy, June 26-30, 1961, Perugia, Italy, Univ. of Perugia, 1962

**13OPAL** "Control of Ovulation" Proceedings of the Conference, Massachusetts Institute of Technology, Edicott House, Pergamon Press, 1961

**13ZGAF** "Gigiena i Toksikologiya Novykh Pestitsidov i Klinika Otravlenii, Doklady Vsesoyuznoi Konferentsii" Komiteta Po Izucheniiu i Reglamentatsii Indokhimikatov Glavnoi Gosudarstvennoi Sanitarnoi Inspektsii USSR., 1962

**14CYAT** "Industrial Hygiene and Toxicology, 2nd rev. ed.," Patty, F.A., ed., New York, NY, Interscience Publishers, 1958-63

**14FHAR** "Venomous and Poisonous Animals and Noxious Plants of the Pacific Region" Keegan, H.L., and W.V. Macfarlane, eds., New York, NY, Pergamon Press Inc., 1963

**14JTAF** "Mycotoxins in Foodstuffs" Proceedings of the Symposium held at the Massachusetts Institute of Technology, Mar. 18-19, 1964, Wogan, G.N., ed., Cambridge, MA, MIT Press, 1965

**14KTAK** "Boron, Metallo-Boron Compounds and Boranes" R.M. Adams, New York, NY, Wiley, 1964

**14XBAV** "International Congress of Chemotherapy" Proceedings of the 3rd Congress, July 22-27, 1963, Stuttgart, Thieme, 1964

**14YA8** "Induction of Mutations and the Mutation Process" Veleminsky, Jiri, ed., Proceedings of a Symposium held in Prague, Czechoslovak, 26-28, 1963, Prague, Czechoslavokia , Publishing House of the Czechoslovak Academy of Science, 1965

**15MBAH** "Proceedings of the International Symposium on Chemotherapy of Cancer" Lugano, Switzerland, April 28-May 1, 1964, Plattner, P.A., ed., New York, NY, American Elsevier Publishing Co., 1964

**15QWAW** "Agents Affecting Fertility, a Symposium" 1964, Austin, C.R., and J.S. Perry, eds., Boston, MA, Little, Brown and Co., 1965

**17QLAD** "Ratsionl'noe Pitanie" (Rational Nutrition) Leshchenko, P.D., ed., USSR, 1967

**17TVAO** "Radiation-Induced Cancer, Proceedings of a Symposium, Athens, 1969," Lanham, MD, Bernan-UNIPUB, 1969

**17XKAB** "International Symposium on Amphetamines and Related Compounds, Proceedings, Mario Negri Institute for Pharmacological Research, Milan, 1969," Costa, E., and S. Garattini, eds., New York, NY, Raven Press, 1970

**19DDA6** "Animal Toxins, a Collection of Papers Presented at the International Symposium on Animal Toxins, 1st, 1966" Russell, F.E., and P.R. Saunders, eds., New York, NY, Pergamon Press Inc., 1967

**19FKA3** "Biochemistry of Some Foodborne Microbial Toxins" Symposium on Microbial Toxins, New York, NY, Cambridge, MA, MIT Press, 1967

**19UQAS** "Khimicheskie Paktory Vneshnei Sredy I Ikh Gigienicheskoe Znachenie, Materialy Nauchnoi Konferentsii" 2nd, Pervyi Moskovskii Meditsinskii Institut, Moscow, USSR, June 22, 1965

**20PKA3** "Novye Dannye Po Toksikologii Redkikh Metalov Ikh Soedinenii" 1967

**20ZJAG** "Toksikologiia Novykh Khimicheskikh Veshchesty, Vnedriaemykh V" Rezinovuiu I Shinnaiu Promyslennost, 1968

**21ACAB** "Imifos" Giller, S.A., ed., Izd, Riga, USSR, 1968

**21NDAB** "Symposium on Carbenoxolone Sodium" Robson, J.M., and F.M. Sullivan, eds., London, UK, Butterworth, 1968

**22XWAN** "Chemical Mutagenesis in Mammals and Man" Vogel, F. and G. Roehrborn, eds., Berlin, Germany, Springer-Verlag, 1970

**23EIAT** "Venomous Animals and Their Venoms" Buecherl, W., and E.E. Buckley, eds., New York, NY, Academic Press, Inc., 3 vols., 1968-71

**23HZAR** "Carcinoma of the Colon and Antecedent Epithelium" W.J. Burdette, Springfield, IL, C.C. Thomas, 1970

**25NJAN** "International Cancer Congress, Abstracts, 10th" Houston, TX, May 22-29, 1970, Cumley, R.W. and J.E. McCay, eds., Chicago, IL, Year Book Medical Publishers, Inc., 1970

**26QZAP** "Advances in Antimicrobial and Antineoplastic Chemotherapy, Progress in Research and Clinical Application" 7th Proceedings of the International Congress of Chemotherapy, Hejzlar, M. et al., eds., Prague, Czechoslovakia, Aug. 23-28, 1971, 3 vols., Prague, Czechoslovakia, University Press, 1972

**26RAAN** "Benzodinzenines" S. Garattini, ed., New York, NY, Raven Press, 1973

**26UYA8** "Psoriasis" Proceedings of the International Symposium, Stanford, CA, July 7-10, 1971, Farber, E.M., and A.J. Cox, eds., Stanford, CA, Stanford University Press, 1971

**26UZAB** "Pesticides Symposia" collection of papers presented at the Sixth and Seventh Inter-American Conferences on Toxicology and Occupational Medicine, Miami, FL, Univ. of Miami, School of Medicine, 1968-70

**27CWAL** "Medical Primatology" Selected Papers from the 3rd Conference on Experimental Medicine and Surgery in Primates, Lyons, France, June 21-23, 1972, Goldsmith, E.I., and J. Moor-Jankowski, eds., White Plains, NY, Phiebig, 1972

**27ZIAQ** "Drug Dosages in Laboratory Animals-A Handbook" C.D. Barnes and L.G. Eltherington, Berkeley, CA, Univ. of California Press, 1965, 1973

**27ZMA4** "Psychopharmacological Agents, Volume 2" Gorden, M., ed., New York, NY, Academic Press, 1967

**27ZQAG** "Psychotropic Drugs and Related Compounds" E. Usdin and D.H. Efron, 2nd ed., Washington, DC, 1972

**27ZTAP** "Clinical Toxicology of Commercial Products-Acute Poisoning" Gleason, et al., 3rd Ed., Baltimore, MD, Williams, and Wilkins, 1969

**27ZWAY** "Heffter's Handbuch der Experimentelle Pharmakologie."

**27ZXA3** "Handbook of Poisoning: Diagnosis and Treatment" Dreisbach, R.H., Los Altos, CA, Lange Medical Publications

**27ZZA9** "Manual of Pharmacology and Its Applications to Therapeutics and Toxicology" Sollman, T., 8th ed., PA, W.B. Saunders, 1957

**28QFAD** "Laboratory Animal in Drug Testing" Symposium of the 5th International Committee on Laboratory Animals, Hanover, Sept. 19-21, 1972, Speigel, A., ed., Stuttgart, Germany, Fischer, 1973

**28ZEAL** "Pesticide Index," Frear, E.H., ed., State College, PA, College Science Pub., 1969

**28ZLA8** "Toxicity of Industrial Metals" Browning, E., London, UK, Butterworths, 1961

**28ZNAE** "Public Health Reports, Supplement." (Baltimore, MD/Washington, DC)

**28ZOAH** "Chemicals in War" Prentiss, A.M., 1937

**28ZPAK** "Sbornik Vysledku Toxixologickeho Vysetreni Latek A Pripravku" Marhold, J.V., Institut Pro Vychovu Vedoucicn Chemickeho Prumyclu Praha, Czechoslovakia, 1972

**28ZRAQ** "Toxicology and Biochemistry of Aromatic Hydrocarbons" Gerarde, H., New York, NY, Elsevier, 1960

**28ZSAT** "Drugs in Research" DeHaen, P., New York, NY, Paul DeHaen, 1964

**29QHAQ** "Principles of Medicinal Chemistry" Foye, W.O., eds., Philadelphia, PA, Lea and Febiger, 1974

**29QKAZ** "Toxins of Animal and Plant Origin" de Vries, A., and E. Kochva, eds., New York, NY, Gordon and Breach Science Publishers, 1971

**29ZSA2** "Trifluoperazine: Further Clinical and Laboratory Studies" Moyer, J.E., ed., Philadelphia, PA, Lea and Febiger, 1959

**29ZUA8** "Toxicity and Metabolism of Industrial Solvents" Browning, E., New York, NY, Elsevier, 1965

**29ZVAB** "Handbook of Analytical Toxicology," Sunshine, I., ed., Cleveland, OH, Chemical Rubber Co., 1969

**29ZWAE** "Practical Toxicology of Plastics" Lefaux, R., Cleveland, OH, Chemical Rubber Company, 1968

**30FDAB** "Neuropoisons: Their Pathophysiological Actions, 1971-1974" Simpson, L.L., and D.R. Curtis, eds., New York, NY, Plenum Pub. Corp., 1974

**30ZDA9** "Chemistry of Pesticides." Melnikov, N.N., New York, NY, Springer-Verlag, 1971

**30ZFAF** "Toxic Aliphatic Fluorine Compounds" Pattison, F.L.M., London, UK, Elsevier Publishing, 1959

**30ZIAO** "Vanadium Toxicology and Biological Significance" Faulkner, T.G., Hudson, NY, Elsevier Publishing Company, 1964

**31BYAP** "Experimental Lung Cancer: Carcinogenesis Bioassays, International Symposium, 1974" New York, NY, Springer, 1974

**31TFAO** "Progress in Chemotherapy" Proceedings of the 8th International Congress on Chemotherapy, Athens, Greece, 1973, Daikos, G.K. ed., Athens, Greece. Hill. Society of Chemotherapy, 1974

**31ZOAD** "Pesticide Manual" Worcesteshire, England, British Crop Protection Council, 1968

**31ZPAG** "Synthetic Analgesics" Oxford, UK, Pergamon Press, 1966

**31ZTAS** "Alcohols: Their Chemistry, Properties and Manufacture" J.A. Monick, New York, Reinhold Book, 1968

**31ZUAV** "Investigations in the Field of Organogermanium Chemistry" Rijkens, F., and G.J.M. Van Der Kerk, Utrecht, Netherlands, Schotanus & Jens Utrecht N.V., 1964

**32OAAP** "Pesticide and the Environment: a Continuing Controversy" Papers Presented and Papers Reviewed at the 8th Inter-American Conference on Toxicology and Occupational Medicine, Miami, FL, July 1973, Deichmann, W.B., ed., New York, NY, Stratton, 1973

**32XPAD** "Teratology" C.L. Berry, ed., New York, NY, Springer-Verlag, 1975

**32YWA5** "Radiation Research, Biomedical, Chemical, and Physical Perspectives, Proceedings of the International Congress of Radiation Research, 5th, Seattle, WA, July 14-20, 1974 (Pub 1975)" Nygaard, O.F., et al., ed., New York, NY, Academic, 1975

**32ZCAI** "Ganglion-Blocking and Ganglion-Stimulation Agents" Kharkevich, D.A, Moscow, USSR, First Moscow Medical Institute, 1967

**32ZDAL** "Aldrin Dieldrin Endrin and Telodrin: An Epidemiological and Toxicological Study of Long-Term Occupational Exposure" Jager, K.W., New York, NY, Elsevier, 1970

**32ZWAA** "Handbook of Poisoning: Diagnosis and Treatment" Dreishach, R.H, 8th ed., Los Altos, CA Lange Medical Publications, 1974

**32ZXAD** "Transactions of the 36th Annual Meeting of the American Conference of Governmental Industrial Hygienists" Miami Beach, FL., May 12-17, 1974

**33IUAS** "Mycotoxins" Purchase, I.F.H., ed., Amsterdam, Netherlands, Elsevier, 1974

**33NFA8** "Animal Models in Dermatology" Maibach, H., ed., New York, NY, Churchill Livingston, 1975

**34LXAP** "Insecticide Biochemistry and Physiology" Wilkinson, C.F., ed., New York, NY, Plenum, 1976

**34TRAD** "Food-Drugs from the Sea" Proceedings of the Conference, 4th, University of Puerto Rico, Mayaguez, PR, November 17-21, 1974, Webber, H.H., and G.D. Ruggieri, eds., Washington, DC, Marine Technology Society, 1976

**34ZHAD** "Symposium on Mycotoxins in Human Health" Purchase, I.F.H., London, UK, Macmillan, 1971

**34ZIAG** "Toxicology of Drugs and Chemicals," Deichmann, W.B., New York, NY, Academic Press, Inc., 1969

**34ZRA9** "Proceedings Quadrennial Conference on Cancer" Severi, L., ed., Perugia, Italy, University of Perugia, 1966

**35FUAR** "Animal, Plant, and Microbial Toxins" Ohsaka, Akira, et al., eds., New York, NY, Plenum Press, 1976

**35WYAM** "In Vitro Metabolic Activation in Mutagenic Testing" Proceedings of the Symposium on the Role of Metabolic Activation in Producing Mutagenic and Carcinogenic Environmental Chemicals, Research Triangle Park, NC, Feb. 9-11, 1976, De Serres, F.J. et al., eds., New York, NY, Elsevier North Holland, 1976

**36PYAS** "Pharmacology of Steroid Contraceptive Drugs" Garattini, S. and H.W. Berendes, eds., New York, NY, Raven Press, 1977. (Monograph Mario Negri Inst. Pharm. Res.)

**36SBA8** "Antifungal Compounds" Siegel, M.R. and H.D. Sisler, eds., 2 vols., New York, NY, Marcel Dekker, 1977

**36THAV** "Management of the Poisoned Patient" Rumack, B.H., and A.R. Temple, eds., Princeton, NJ, Science Press, 1977

**36YFAG** "Biological Reactive Intermediates, Formation, Toxicity and Inactivation" Proceedings of the International Conference on Active Intermediates, Formation, Toxicity and Inactivation, University of Turku, Turku, Finland, July 26-27, 1975. New York, NY, Plenum, 1977

**37ASAA** "Kirk-Othmer Encyclopedia of Chemical Technology, 3rd Edition" Grayson, M. and D. Eckroth, eds., New York, NY, Wiley, 1978

**37KXA7** "Water Chlorination: Environmental Impact and Health Effects" Proceedings of the Conferences on the Environment, Ann Arbor, MI, Ann Arbor Science Pub. Inc., 1975-

**37QLAZ** "Current Approaches in Toxicology" Ballantyne, B., ed., Bristol, UK, John Wright & Sons, Ltd., 1977

**37XLA2** "Current Chemotherapy" Proceedings of the 10th International Congress of Chemotherapy, Zurich, Switzerland, Sept. 18-23, 1977, Siegenthaler,W. and R. Luethy, eds., 2 vols., Washington, DC, American Society for Microbiology, 1978

**38KLAC** "Structure-Activity Relationship among the Semisynthetic Antibiotics." Perlman, D., ed., New York, NY, Academic Press, 1977

**38MKAJ** "Patty's Industrial Hygiene and Toxicology" 3rd rev. ed., Clayton, G.D., and F.E. Clayton, eds., New York, NY, John Wiley & Sons, Inc., 1978-82. Vol. 3 originally pub. in 1979; pub. as 2nd rev. ed. in 1985

**40QBA3** "Proceedings of the International Congress on Toxicology, Toxicology as a Predictive Science, 1st, Toronto, 1977" Plaa, G.L., and W.A. Duncan, eds., New York, NY, Academic Press, Inc., 1978

**40RMA7** "Health and Sugar Substitutes" Proceedings of the ERGOB Conference on Sugar Substitutes, Geneva, Switzerland, 1978, Guggenheim, B., ed., Basel, Switzerland, S. Karger AG, 1979

**40WDA5** "Bleomycin: Current Status and New Development" Papers Presented in a Symposium, Oakland, CA, 1977, Carter, S.K., et al. eds., New York, NY, Academic Press, 1978

**40YJAX** "Evaluation of Embryotoxicity, Mutagenicity and Carcinogenicity Risks In New Drugs" Proceedings of the Symposium on Toxicological Testing for Safety of New Drugs, 3rd, Prague, Czechoslovakia, Apr. 6-8, 1976, Benesova, O., et al., eds., Prague, Czechoslovakia, Univerzita Karlova, 1979

**41CIAR** "Effects of Poisonous Plants on Livestock" Proceedings of a Joint U.S.-Australian Symposium on Poisonous Plants, Logan, UT, 1978, Keeler, R.F., et al. eds., New York, NY, Academic Press, 1978

**41HTAH** "Aktual'nye Problemy Gigieny Truda. Current Problems of Labor Hygiene" Tarasenko, N.Y., ed., Moscow, USSR, Pervyi Moskovskii Meditsinskii Inst., 1978

**41KEAL** "Toxicology, Biochemistry and Pathology of Mycotoxins", Uraguchi, K. and M. Yamazaki, eds., New York, NY, Wiley, 1978

**43FLAV** "Burger's Medicinal Chemistry" Wolff, M.E., ed., 4th ed., New York, NY, John Wiley & Sons, 1981

**43GRAK** "Dusts and Disease" Proceedings of the Conference on Occupational Exposures to Fibrous and Particulate Dust and Their Extension into the Environment, 1977, Lemen, R., and J.M. Dement, eds., Park Forest South, IL, Pathotox Publishers, 1979

**43MKAT** "Current Chemotherapy and Infectious Diseases" Proceedings of the 11th International Congress of Chemotherapy and the 19th Interscience Conference on Antimicrobial Agents and Chemotherapy, Boston, MA, Oct. 1-5, 1979, American Society for Microbiology, Washington, DC, 1980

**43XWAI** "Brain Tumors and Chemical Injuries to the Central Nervous System" Proceedings of the International Neuropathological Symposium, Warsaw, Poland, Sep. 23-26, 1976, Mossakowski, M.J., ed., Warsaw, Poland, Panstwowe Wydawnictwo Naukowe, 1978

**43ZRAD** "Hemoperfusion: Kidney and Liver Support and Detoxification" Proceedings of the International Symposium, Haifa, Israel, 1979, Sideman, S., and T.M.S. Chang, eds., Washington, DC, Hemisphere Pub. Corp., 1980

**44LQAF** "International Sulprostone Symposium, (Papers)" Vienna, Austria, 1978, Friebel, K., et al., eds., Berlin, Germany, Schering AG, 1979

**45ICAX** "Nickel Toxicology" 2nd International Conference on Nickel Toxicology. Swansea, UK, Sept. 3-5, 1980. London, UK, Academic Press, 1980

**45JVAR** "Research Frontiers in Fertility Regulation" Proceedings of an International Workshop, 1980, Zatuchni, G.I., et al., eds., Hagerstown, MD, Harper and Row Medical Dep., 1980

**45OHAA** "Short-Term Test Systems for Detecting Carcinogens" Proceedings of the Symposium, 1978, Norpoth, K.H., and R.G. Garner, eds., Berlin, Germany, Springer-Verlag, 1980

**45OJAG** "Radiation Sensitizers: Their Use in the Clinical Management of Cancer" Proceedings of the Conference, Key Biscayne, FL., Oct. 3-6, 1979, Brady, L.W., ed., New York, NY, Masson Publishing USA, Inc., 1980

**45REAG** "Teratology of the Limbs" Symposium on Prenatal Development, 4th, 1980, Merker, H.J., et al., eds., Berlin, Germany. Walter de Gruyter and Co., 1980

**46GFA5** "Safety Problems Related to Chloramphenicol and Thiamphenicol Therapy" Najean, Y., et al., eds., New York, NY, Raven Press, 1981

**46IZA7** "Retinoids, Advances in Basic Research and Therapy" Proceedings of the International Dermatology Symposium, Berlin, Germany, 1980, Orfanos, C.E., et al., eds., New York, NY, Springer-Verlag, 1981

**46OJAN** "Chemical Analysis and Biological Fate: Polynuclear Aromatic Hydrocarbons" 5th International Symposium, Battelle's Labs., Oct. 1980. Columbus, OH, Battelle Press, 1981

**47JMAE** "Cytogenetic Assays of Environmental Mutagens" Hsu, T.C., ed., Totowa, NJ, Allanheld, Osmun and Co., 1982

**47YKAF** "Sporulation and Germination" Proceedings of the Eighth International Spore Conference, 1980, Washington, DC, American Society for Microbiology, 1981

**48ECAY** "Spurenelement-Symposium: Nickel, 3rd" Jena, Germany, Anke, M., et al., eds., Jena, Germany, Friedrich-Schiller-Universitaet, 1980

**48HGAR** "Current Chemotherapy and Immunotherapy" Proceedings of the International Congress of Chemotherapy, 12th, 1981, Periti, P., and G. Gialdroni-Grassi, eds., Washington, DC, American Society for Microbiology, 1982

**48NTAS** "Advances in Genetics, Development, and Evolution of Drosophila" Proceedings of the European Drosophila Research Conference, 7th, 1981, Lakovaara, S., ed., New York, NY, Plenum Publishing Corp., 1982

**48RKAL** "Industrial and Environmental Xenobiotics: Metabolism and Pharmacokinetics of Organic Chemicals and Metals, Proceedings of an International Conference, Prague," 1980, Gut, I., et al., Berlin, Fed. Rep, Ger., Springer-Verlag, 1981

**48THAM** "Proceedings of the International Symposium on Cyclodextrins" 1st, 1981, Szejtli, J., ed., Hingham, MA, Kluwer Boston Inc., 1982

**49RQAC** "Chemistry and Biology of Hydroxamic Acids" Proceedings of the International Symposium, 1st, Dayton, OH, 1981, Kehl, H., ed., Basel, Switzerland, S. Karger AG, 1982

**49VWAG** "Female Transcervical Sterilization" Proceedings of an International Workshop on Non-Surgical Methods for Female Tubal Occlusion, Chicago, IL, June 22-24, 1982, Zatuchni, G.I., et al., eds. New York, NY, Harper and Row, 1983

**49WEAZ** "Pesticide Chemistry: Human Welfare and the Environment" Proceedings of the International Congress of Pesticide Chemistry, 5th, Kyoto, Japan, Aug. 29-Sept 4, 1982, Miyamoto, J. and P.C. Kearney, eds., Oxford, UK, Pergamon Press Ltd., 1983

**50EXAK** "Formaldehyde Toxicity" Conference, 1980, Gibson, J.E., ed., Washington, DC, Hemisphere Publishing Corp., 1983

**50EYAN** "Structure-Activity Correlation as a Predictive Tool in Toxicology" Papers from a Symposium, 1981, Golberg, L., ed., Washington, DC, Hemisphere Publishing Corp., 1983

**50NNAZ** "Polynuclear Aromatic Hydrocarbons: Mechanisms, Methods and Metabolism, Papers of the 8th International Symposium, Columbus, OH, 1983," Cooke, M., and A.J. Dennis, eds., Columbus, OH, Battelle Press, 1985

**50WKA3** "Reproduction and Developmental Toxicity of Metals" Proceedings of a Joint Meeting, Rochester NY, 1982, Clarkson, T.W., et al., eds., New York, NY, Plenum Publishing Corp., 1983

**51UDAB** "Heavy Metals in the Environment, International Conference, 4th, 1983" Heidelberg, S., 2 vols., Edinburgh, UK, CEP Consultants Ltd., 1983

**51ZKAW** "The Retinoids, Vol. 2" Sporn, M.B., et al., eds., New York, Academic Press, Inc., 1984

**52EDA6** "Nitrendipine Present Int Workshop Nitrendipine 1st", New York, NY, 1982, Scriabine, A., et al., eds., Baltimore, MD, Urban & Schwarzenberg, 1984

**52GUAX** "Anorganische Stoffe in der Toxikologie und Kriminalistik, Symposium, Mosbach, Germany, 1983" Heppenheim, Germany, Verlag Dr. Dieter Helm, 1983

**52MLA2** "Toxicology of Petroleum Hydrocarbons, Proceedings of the Symposium, 1st, 1982" MacFarland, H.N., et al., eds., Washington, DC, American Petroleum Institute, 1983

**52OLAC** "Veterinarnaya Entomologiya i Akarologiya" Veterinary Entomology and Acarology, Nepoklonov, A.A., ed., Moscow, USSR, Izdatel'stovo Kolos, 1983

**53HJAC** "Mouse Liver Neoplasia: Current Perspectives" Popp, J.A., ed., New York, NY, Hemisphere Pub. Corp., 1984

**54DEAI** "Occupational Health in the Chemical Industry, Proceedings of the International Congress, 11th, Alberta, Canada, 1983" Oxford, R.R., et al., eds., Calgary, Alberta, Canada, Univ. of Calgary, 1984

**55CXA9** "Trichothecenes and Other Mycotoxins, Proceedings of the International Mycotoxin Symposium, 1984" Lacey, J., ed., Chichester, UK, John Wiley & Sons Ltd., 1985

**55DXAE** "Progress in Nickel Toxicology, Proceedings of the International Conference on Nickel Metabolism and Toxicology, 3rd, 1984" Brown, S.S., and F.W. Sunderman, Jr., eds., Oxford, UK, Blackwell Scientific Pub. Ltd., 1985

**85AEA9** "Environmental Hazards of Metals" I.T. Brakhnova, trans. by Slep, J.H., New York, NY, Consultants Bureau, 1975

**85AGAF** "Effects and Dose-response Relationships of Toxic Metals" Nordberg, G.F., ed., Proceedings from an International Meeting Organized by the Subcommittee on the Toxicology of Metals of the Permanent Commission and International Association on Occupational Health, Tokyo, Japan, November 18-23, 1974, New York, NY, Elsevier Scientific, 1976

**85AIAL** "Toxicity of Pure Foods" Boyd, E.M., Cleveland, OH, CRC Press, 1973

**85ALAU** "Carbamate Insecticides: Chemistry, Biochemistry, and Toxicology" Kuhr, R.J. and H.W. Dorough, Cleveland, OH, CRC Press, 1976

**85ARAE** "Agricultural Chemicals, Books I, II, III, and IV" W.T. Thomson, Fresno, CA, Thomson Publications, 1976/77 revision

**85AREA** "Agricultural Chemicals," Thomson, W.T., 4 vols., Fresno, CA, Thomson Publications, 1976/77 revision

**85CVA2** "Oncology 1970" Proceedings of the Tenth International Cancer Congress, Chicago, IL, Year Book Medical Publishers, 1971

**85CYAB** "Chemistry of Industrial Toxicology" H.B. Elkins, 2nd Ed., New York, John Wiley & Sons, 1959

**85DAAC** "Bladder Cancer, A Symposium" Lampe K.F. et al., eds., Fifth Inter-American Conference on Toxicology and Occupational Medicine, University of Miami, School of Medicine, Coral Gables, FL, Aesculapius Pub., 1966

**85DCAI** "Poisoning; Toxicology, Symptoms, Treatments" Arena, J.M. 2nd ed., Springfield, IL, C. C. Thomas, 1970

**85DGAU** "Catalog of Teratogenic Agents" Shepard, T.H., 2nd Ed., Baltimore, MD, Johns Hopkins University Press, 1976

**85DHAX** "Medical and Biologic Effects of Environmental Pollutants Series" Washington, DC, National Academy of Sciences, 1972-77

**85DJA5** "Malformations Congenitales des Mammiferes" Tuchmann-Duplessis, H., Paris, France, Masson et Cie, 1971

**85DKA8** "Cutaneous Toxicity" Drill, V.A. and P. Lazar, eds., New York, NY, Academic Press, 1977

**85DLAB** "Studies on Chemical Carcinogenesis by Diels-Alder Adducts of Carcinogenic Aromatic Hydrocarbons" Earhart, Jr., R.H., Ann Arbor, MI, Xerox Univ. Microfilms, 1975

**85DPAN** "Wirksubstanzen der Pflanzenschutz und Schadlingsbekampfungsmittel" Werner Perkow, Berlin, Germany, Verlag Paul Parey, 1971-1976

**85DUA4** "Chemical Tumour Problems" Nakahara, W., ed., Tokyo, Japan, Japanese Society for the Promotion of Science, 1970

**85DZAJ** "Principles of Insect Chemosterilization" Labrecque, G.C. and C.N. Smith, eds., New York, NY, Appleton-Century-Crofts, 1968

**85ECAN** "Metal Toxicity in Mammals, Vol. 2: Chemical Toxicity of Metals and Metalloids" Venugopal, B. and T.D. Luckey, New York, NY, Plenum Press, 1978

**85EGD4** "Toxins: Animal, Plant and Microbial. Proceedings of the International Symposium" Rosenberg, P., ed., New York, NY, Pergamon Press, 1978

**85ELDJ** "Die Herzwirksamen Glykoside" Baumgarten, G. and W. Forster, Leipzig, Germany, VEB Georg Thieme, 1963

**85ERAY** "Antibiotics: Origin, Nature, and Properties" Korzyoski, T., Z. Kowszyk-Gindifer, and W. Kurylowicz, eds., Washington DC, American Society for Microbiology, 1978

**85ESA3** "Merck Index; an Encyclopedia of Chemicals, Drugs, and Biologicals," 11th ed., Rahway, NJ 07065, Merck & Co., Inc. 1989

**85FZAT** "Index of Antibiotics from Actinomycetes" Umezawa, H., et al., eds., Tokyo, Japan, University of Tokyo Press, 1967

**85GDA2** "CRC Handbook of Antibiotic Compounds, Volumes 1-9" Berdy, Janos. Boca Raton, FL, CRC Press, 1980

**85GLAQ** "A Survey of Antimalarial Drugs, 1941-1945" Wiselogle, F.Y., 2 vols., Ann Arbor, MI, J.W. Edwards, 1946

**85GMAT** "Toxicometric Parameters of Industrial Toxic Chemicals Under Single Exposure" Izmerov, N.F., et al., Moscow, USSR, Centre of International Projects, GKNT, 1982

**85GNAW** "Pharmacodynamie de l'Acide Dipropylacetique (ou Propyl-2-Pentanoique) et de ses Amides" Carraz, G., Imprimerie Eymond, 1968

**85GRAA** "The Control of Fertility" Pincus, G., New York, NY, Academic Press, 1965

**85GUAJ** "Antifertility Compounds in the Male and Female" Jackson, H., Springfield, IL, Charles C. Thomas, 1966

**85GYAZ** "Pflanzenschutz-und Schaedlingsbekaempfungsmittel: Abriss einer Toxikologie und Therapie von Vergiftungen" 2nd ed., Klimmer, O.R., Hattingen, Germany, Hundt-Verlag, 1971

**85HBAZ** "Environmental Variables in Oral Disease;" A Symposium Presented at the Montreal Meeting of the American Association for the Advancement of Science, 1964, Kreshover, S.J., et al., eds., Washington, DC, American Association for the Advancement of Science, 1966

**85INA8** "Documentation of the Threshold Limit Values and Biological Exposure Indices" 5th ed., Cincinnati, OH, American Conference of Governmental Industrial Hygienists, Inc., 1986

**85IPAE** "Modern Pharmaceuticals of Japan, V" Tokyo, Japan, Pharmaceutical, Medical and Dental Supply Exporters' Association, 1975

**85IVAW** "Possible Long-Term Health Effects of Short-Term Exposure to Chemical Agents" National Research Council, 3 vols., Washington, DC, National Academy Press, 1982-85

**85IXA4** "Structure et Activite Pharmacodyanmique des Medicaments du Systeme Nerveux Vegetatif" Bovet, D., and F. Bovet-Nitti, New York, NY, S. Karger, 1948

**85JCAE** "Prehled Prumyslove Toxikologie; Organicke Latky" Marhold, J., Prague, Czechoslovakia, Avicenum, 1986

**85JDAH** "Organophosphorus Pesticides: Organic and Biological Chemistry" Eto, M., Cleveland, OH, CRC Press, Inc., 1974

**85JFAN** "Agrochemicals Handbook," with updates, Hartley, D., and H. Kidd, eds., Nottingham, UK, Royal Soc. of Chemistry, 1983-86

# A Guide to Using This Book

**Entry Code:** – Entries are indexed in order by this alphanumeric code.
*See Introduction: paragraph I, p. xiii.*

**Entry Name**: – A complete entry name and synonym cross-index is located in Section 3.
*See Introduction: paragraph 2, p. xiii.*

**DOT**: – The four-digit hazard code assigned by the U.S. DOT. A complete DOT cross-index is located in Section 1.
*See Introduction: paragraph 5, p. xiii.*

**mf**: – The molecular formula
**mw**: – The molecular weight
*See Introduction: paragraphs 6 and 7, p. xiv.*

**PROP**: – Physical properties, including solubility and flammability data. May contain a definition of the entry.
*See Introduction: paragraph 9, p. xiv.*

**SYNS**: – Synonyms for the entry. A complete synonym cross-index is located in Section 3.
*See Introduction: paragraph 10, p. xiv.*

**Toxicity Data**: – Data for skin and eye irritation, mutation, teratogenic, reproductive, carcinogenic, human, and acute lethal effects.
*See Introduction: paragraphs 11 - 15 pp. xiv-xxii.*

**Standards and Recommendations**: – OSHA PEL, ACGIH TLV, DFG MAK, NIOSH REL workplace air levels, and U.S. DOT Classification and labels are listed here.
*See Introduction: paragraph 18, p. xxv.*